U0480671

刘人怀院士文集

第一卷

刘人怀◎著

科学出版社
北京

内 容 简 介

本文集由著者在力学和管理科学领域 60 多年中所发表的文章汇编而成，分为上、下篇。上篇是力学部分，结合工程需要，对单层板壳、波纹板壳、双层板壳、夹层板壳、复合材料层合板壳和网格扁壳等六类板壳的非线性弯曲、稳定和振动问题，以及厚、薄板壳弯曲问题进行理论探索。下篇是管理科学部分，联合实际，对工程管理、公共管理、工商管理、科技管理和教育管理的问题进行了研究。

本文集可作为高校力学和管理科学专业，以及相关专业老师、研究生、本科生的参考书，适合政府、高等学校、科研院所、科技社团、企业等领域的领导、管理人员和科研人员参考阅读，又可作为航天、航空、航海、机械、建筑和交通等工程设计师与工程师的设计制造指导书。

GS京（2025）0637号

图书在版编目（CIP）数据

刘人怀院士文集 / 刘人怀著. -- 北京 ：科学出版社，2025. 4. -- ISBN 978-7-03-080897-4

Ⅰ. O3-53；C93-53

中国国家版本馆 CIP 数据核字第 2024JH1723 号

责任编辑：陈会迎/ 责任校对：姜丽策

责任印制：张　伟/ 封面设计：有道设计

科学出版社出版

北京东黄城根北街16号

邮政编码：100717

http://www.sciencep.com

北京中科印刷有限公司印刷

科学出版社发行　各地新华书店经销

*

2025 年 4 月第　一　版　开本：787×1092 1/16

2025 年 4 月第一次印刷　印张：153 3/4

字数：3 646 000

定价：698.00 元（全五卷）

（如有印装质量问题，我社负责调换）

刘人怀　中国工程院院士

前　言

力学属于自然科学，其应用范围十分广泛，又属于技术学科，国民经济的各个产业，哪里有技术难题，哪里就有力学难题。力学又是一门历史悠久的学科，人类文明有多久，力学就有多久，并随着人类的文明进步而不断发展。

管理科学属于社会科学，是一门综合性交叉科学，涉及人类社会各个领域，对人类的文明发展至关重要。从人类诞生以来，管理作为实践活动就一直存在。但作为一门管理学科，则从 20 世纪初才开始出现。在我国，直到 1978 年改革开放后，管理科学专业才开始在高校设立，才受到从上到下的高度重视。

我有幸与这两个学科有缘。从 1960 年 9 月起，在兰州大学，我从数学专业转到力学专业读书，加入到力学行列，直至今日，始终未变，已经 62 年了。一直从事板壳力学的教学和研究工作，培养本科生、硕士生和博士生。担任过教育部高等学校力学教学指导委员会主任委员、中国工程院机械与运载工程学部常委、中国振动工程学会理事长、中国力学学会副理事长、中国复合材料学会副理事长等职务。从 1978 年 9 月起，我被群众推选为中国科技大学近代力学系飞行器结构力学教研室副主任，开始从事基层管理工作。不久，又任安徽省副省长顾问，接着又任上海工业大学（现上海大学）副校长，创办上海工业大学经济管理学院。然后，又任暨南大学校长，先后创办暨南大学深圳旅游学院和暨南大学管理学院。担任过教育部科技委员会管理科学部主任、中国工程院工程管理学部副主任和首席咨询专家、上海高校管理学院院长联谊会首任会长以及广东省人民政府参事等职务。40 多年来，一直从事管理科学的教学和研究工作，培养本科生、硕士生和博士生。

在力学领域，结合工程需要，对单层板壳、波纹板壳、双层扁壳、夹层板壳、复合材料层合板壳和网格扁壳等六类板壳的非线性弯曲、稳定和振动问题以及厚、薄板壳弯曲问题进行理论探索。

在管理科学领域，联系实际，对工程管理、公共管理、工商管理、科技管理和教育管理的问题进行了研究。

现将这些工作汇编于此，以供借鉴。借此机会，衷心感谢在上述工作中与我合作的老师、同事、朋友、助手和学生。

书中难免有缺点和不足，恳请专家和读者指正。

本文集在编辑过程中得到中国工程院的大力支持，在此谨表谢忱。

刘人怀
2022 年 10 月 5 日

总 目 录

前言 …… i

上篇　板壳力学理论与工程应用

第一章　概述 …… 3

板壳分析与应用 …… 3

板壳非线性力学研究的最新进展 …… 14

板壳非线性力学 …… 29

The effect of the Göttingen school on investigation of flexible plates and shells in China …… 32

多姿多彩的板壳结构 …… 34

弹性元件国内外理论发展概况 …… 36

第二章　单层板壳非线性力学 …… 52

在内边缘均布力矩作用下中心开孔圆底扁球壳的非线性稳定问题 …… 52

在边缘载荷作用下中心开孔圆底扁薄球壳的轴对称稳定性 …… 56

集中载荷作用下开顶扁球壳的非线性稳定问题 …… 66

均布载荷作用下开顶扁球壳的非线性稳定问题 …… 77

On the nonlinear stability of a truncated shallow spherical shell under axisymmetrically distributed load …… 91

均布载荷作用下具有硬中心的开顶扁球壳的非线性屈曲 …… 106

中心开孔扁球壳在均布载荷作用下的非线性屈曲 …… 112

A new approach to nonlinear vibration of orthotropic thin circular plates …… 120

正交异性扁薄球壳的非线性轴对称振动 …… 128

第三章　波纹板壳非线性力学 …… 144

波纹圆板的特征关系式 …… 144

具有光滑中心的波纹圆板的特征关系式 …… 151

波纹环形板的非线性弯曲 …… 163

Large deflection of corrugated circular plate with a plane central region under the action of concentrated loads at the center …… 171

Large deflection of corrugated circular plate with plane boundary region …… 185

在复合载荷作用下波纹环形板的非线性弯曲 …… 202

On large deflection of corrugated annular plates under uniform pressure …… 212

复合载荷下波纹圆板的非线性分析 …… 223

On the non-linear bending and vibration of corrugated circular plates …… 238

U 型波纹管非线性变形的刚度和应力分析理论 …… 252
Non-linear bending of a corrugated annular plate with a plane boundary region and a non-deformable rigid body at the center under compound load …… 266
均布载荷作用下波纹环形板的非线性弯曲 …… 281
波纹管的制造与理论研究概况 …… 291
中心受载下具有平面边缘区域的固支波纹环形板的非线性分析 …… 298
Nonlinear bending of corrugated annular plate with large boundary corrugation …… 310
Nonlinear deformation analysis of a U-shaped bellows with varying thickness …… 321
波纹扁球壳的大挠度方程 …… 333
Nonlinear stability of corrugated shallow spherical shell …… 339
第四章　双层扁壳非线性力学 …… 353
双层金属截头扁锥壳的热稳定性 …… 353
双层金属中心开孔扁球壳的非线性热稳定问题 …… 364
Non-linear thermal stability of bimetallic shallow shells of revolution …… 378
第五章　夹层板壳非线性力学 …… 399
在边缘力矩作用下夹层圆板的非线性轴对称弯曲问题 …… 399
夹层圆板的非线性弯曲 …… 410
夹层圆板大挠度问题的精确解 …… 428
Nonlinear bending of circular sandwich plates under the action of axisymmetric uniformly distributed line loads …… 441
夹层圆板大挠度问题的进一步研究 …… 464
夹层矩形板的非线性振动 …… 472
简支夹层矩形板的非线性弯曲 …… 485
On the non-linear buckling of circular shallow spherical sandwich shells under the action of uniform edge moments …… 500
Nonlinear vibration of shallow conical sandwich shells …… 512
复合材料面层夹层板中转动一致有效理论 …… 525
夹层环形板的非线性弯曲 …… 531
Nonlinear theory of sandwich shells part Ⅰ: Exact kinematics of moderately thick shells …… 541
Nonlinear theory of sandwich shells part Ⅱ: Approximate theories …… 558
On nonlinear stability of shallow conical sandwich shells …… 575
考虑横向剪应力连续的复合材料面层夹层壳非线性一致有效理论 …… 589
Large deflection of annular sandwich plates …… 599
第六章　复合材料层合板壳非线性力学 …… 611
四边简支对称正交层合矩形板的非线性弯曲问题 …… 611

对称圆柱正交异性层合扁球壳的非线性稳定问题 …… 618
考虑横向剪切的对称层合圆柱正交异性扁球壳的非线性稳定问题 …… 626
A simple theory for non-linear bending of laminated composite rectangular plates including higher-order effects …… 636
考虑横向剪切的对称层合圆柱正交异性中心开孔扁球壳的非线性屈曲 …… 647
Non-linear buckling of symmetrically laminated, cylindrically orthotropic, shallow, conical shells considering shear …… 660
On non-linear buckling of symmetrically laminated, cylindrically orthotropic, truncated, shallow, spherical shells under uniform pressure including shear effects …… 672
复合材料层合扁锥壳的非线性稳定问题 …… 687
复合材料层合扁球壳的非线性强迫振动 …… 693
Large deflection bending of symmetrically laminated rectilinearly orthotropic elliptical plates including transverse shear …… 699
Nonlinear dynamic buckling of symmetrically laminated cylindrically orthotropic shallow spherical shells …… 715
Nonlinear stability of symmetrically laminated cylindrically orthotropic truncated shallow conical shells including transverse shear …… 725
复合材料层合板壳非线性力学的研究进展 …… 741
第七章 网格扁壳非线性力学 …… 769
网格扁壳的非线性弯曲理论 …… 769
网格扁壳的非线性自由振动分析 …… 776
Non-linear buckling of squarely-latticed shallow spherical shells …… 781
双层正交正放网格扁壳结构的非线性弯曲理论 …… 802
矩形底双层网格扁壳的非线性弯曲 …… 807
矩形底双层网格扁壳的非线性屈曲 …… 812
第八章 厚、薄板壳弯曲分析 …… 818
高压聚乙烯反应器厚壁筒体径向开孔的应力计算 …… 818
尿素合成塔底部球形封头开孔的应力计算 …… 831
铂重整装置反应器椭球封头中心开孔接管的强度问题 …… 842
500 万吨/年常减压装置减压塔下端部分壳体的应力分析 …… 854
在轴向压力与均匀外压力共同作用下薄壁截头圆锥形壳的稳定性 …… 872
加氢反应器顶部厚壁壳体的应力分析 …… 879
双锥密封中的内力分析 …… 897
厚壁圆柱壳轴对称变形近似理论的应力公式 …… 900
双层套箍式厚壁压力容器环沟部位的应力状态 …… 902
厚壁球壳的弯曲理论及其在高压容器上的应用 …… 916
厚管板的设计 …… 927
固定式厚管板的弯曲问题 …… 945

层合圆薄板的轴对称弯曲问题 …… 958
焦炭塔鼓胀与开裂变形机理及疲劳断裂寿命预测的研究进展 …… 964
膜盒基体的理论与设计 …… 974

下篇　管理科学理论与应用

第九章　管理学科 …… 985
谈谈创建现代管理科学中国学派的若干问题 …… 985
再谈创建现代管理科学中国学派的若干问题 …… 995
三谈创建现代管理科学中国学派的若干问题：四条定义与三点建议 …… 1004
大平台、聚义厅及其他——四谈创建现代管理科学中国学派的若干问题 …… 1013
《学科目录》第 12 学科门类与管理科学话语体系——五谈创建现代管理科学中国学派的若干问题 …… 1021
当前管理科学研究中的若干问题——几个疑点的澄清和两种研究方法的评析 …… 1030
中国管理科学的现状和走向 …… 1040
传统文化基因与中国本土管理研究的对接：现有研究策略与未来探索思路 …… 1041
Research on Chinese school of modern GUANLI science …… 1056
第十章　工程管理 …… 1068
上海浦东新区建设工程 …… 1068
中国制造业的生存哲学 …… 1097
东水西调工程 …… 1100
一、关于改善我国北方水资源缺乏的建议 …… 1100
二、关于实施“东水西调”工程的建议 …… 1100
关于“发展中国家的工业化道路”论坛的讨论 …… 1133
绿色制造与学科会聚 …… 1140
工程管理是管理对国民经济的深度介入 …… 1144
城市矿产工程 …… 1147
一、关于治理垃圾的建议 …… 1147
二、推进“垃圾分类，从我做起”科普宣传活动 …… 1148
三、开展“垃圾分类，从我做起”科普资源包研发及宣讲活动 …… 1150
四、开展“垃圾分类”宣讲与培训服务 …… 1152
五、关于推进“餐厨废弃物变废为宝”的建议 …… 1155
六、开展美丽广东科普教育示范户工作 …… 1157
七、早日实现“美丽广东梦” …… 1158
八、促进城市矿产资源化利用，共建美丽中国 …… 1159
九、绿色再制造的探索 …… 1159
十、珠三角城乡生活垃圾统筹治理战略研究 …… 1168
十一、城市餐厨垃圾回收逆向物流系统构建的研究 …… 1184
十二、让人民过上好日子 …… 1192

十三、关于推进餐厨垃圾治理装备高技术制造业发展的建议 …… 1193
大型工程项目管理的中国特色及与美苏的比较 …… 1195
建设低碳社会关键在制造创新 …… 1207
工程管理信息化的内涵与外延探讨 …… 1209
转变经济发展方式关键在制造业创新 …… 1215
公共安全工程 …… 1217
一、公共安全相关工程科技研究的重要性 …… 1217
二、信息安全相关工程科技发展战略 …… 1219
三、生物安全相关工程科技发展战略 …… 1224
四、食品安全相关工程科技发展战略 …… 1232
五、土木工程安全相关工程科技发展战略 …… 1244
六、生产安全相关工程科技发展战略 …… 1256
七、社会安全相关工程科技发展战略 …… 1268
八、生态安全相关工程科技发展战略 …… 1279
工程管理信息化架构研究 …… 1292
大力促进科技与经济融合 …… 1300
勇担重任　敢攀高峰 …… 1301
有效推动轨道交通结构健康监测和整治修复技术的发展提升 …… 1303
第十一章　公共管理 …… 1304
征求意见　完善报告 …… 1304
领导用人标准 …… 1306
有效利用外资　扩大对外开放 …… 1309
如何防止公共关系庸俗化 …… 1314
领导科学与领导艺术 …… 1315
城市政府工作目标管理与治理整顿、深化改革 …… 1329
实施城市管理系统工程建设　开创广州可持续发展新格局 …… 1332
建议重视我省仪器仪表工业的发展，以迎接21世纪挑战 …… 1338
广东省发展高新技术的若干意见和建议 …… 1340
珠海前山镇（街）转型与社区建设研究 …… 1345
广东的治安状况与投资环境 …… 1368
教育彩票作用惊人 …… 1370
建设广州石牌大学城 …… 1372
办好院士之家 …… 1374
坦诚建言 …… 1376
光大华侨文化　建设文化大省 …… 1377
传承岭南文化　服务文化大省 …… 1380
建立粤港澳综合协调机构 …… 1381
泛珠三角：推进科技、教育和文化的区域合作 …… 1382

促进广东省职业教育发展 …… 1385
关于尽快制定国家统一法的建议 …… 1392
激励民办专科学校升为本科学校 …… 1393
让象牙塔成为顶梁柱 …… 1394
大力发展我省高中阶段教育 …… 1397
关于允许市民在节假日有条件燃放烟花的建议 …… 1404
在推进和谐社会建设中切实解决“农民工”身份问题 …… 1405
关于改善财政宏观调控深化分税制财政体制改革的调研报告 …… 1407
岁月留声 …… 1418
黄石应该在建设特色城市中凸出优势 …… 1420
关于将清明节设为国家法定假日的建议 …… 1421
爱心和匠心 …… 1423
发展“乡村旅游”促进广东新农村建设 …… 1426
关于城市基础设施建设投融资体制改革研究 …… 1428
关于将香港、澳门特别行政区的所有统计数据纳入全国性统计数据的建议 …… 1437
大规模引进和培训人才为广东产业结构优化升级服务 …… 1438
大力推动“政产学研金”合作创新为广东省经济社会发展做贡献 …… 1447
积极推进知识产权事业发展 …… 1449
关于完善我省应对台风灾害预防措施的建议 …… 1455
坚持重点　保持特色 …… 1456
中国的过去和现在 …… 1457
系统工程与领导科学 …… 1473
关于实行九年一贯制办校的建议 …… 1474
2012 年中国工程院院士广州咨询活动中心工作总结 …… 1476
百年追梦　科技兴国 …… 1481
充分发挥桥头堡作用，推进孟中印缅经济走廊建设 …… 1488
建设低碳社会　托起美丽中国梦 …… 1490
研究传染病突发事件的危机管理十分重要 …… 1492
以智能制造促进产业转型升级 …… 1493
网络强国战略与实践 …… 1502
两化深度融合与质量管理 …… 1503
推动广东非开挖产业快速发展 …… 1512
谈谈标准化工作对当前的重要意义 …… 1517
为推动仪器仪表产业发展做出新贡献 …… 1522
加强数字产业化和产业数字化双轮驱动 …… 1526
第十二章　工商管理 …… 1529
上海旅游交通的症结与对策研究 …… 1529
上海华亭集团旅游宾馆摆脱当前困境的对策研究 …… 1535

人才开发是搞好企业管理的关键 …… 1544
上海——旅游业的春光与希望 …… 1545
旅游工程理论及其在浦东开发中的应用 …… 1548
上海旅游交通中的若干问题 …… 1552
旅游工程学的提出 …… 1561
开启思想的眼睛 …… 1564
质量是企业生存的根本 …… 1566
对某跨国公司绩效考评系统的评价 …… 1571
公司治理：理论演进与实践发展的分析框架 …… 1575
以价值工程方法全面提升荔湾商旅核心竞争力 …… 1581
绿色化是中国制造业的必由之路 …… 1633
旅游教材为旅游教学之本 …… 1638
关系质量研究述评 …… 1640
消费者行为研究是营销理论的基石 …… 1648
从 CSSCI 旅游研究文献看旅游学学科发展 …… 1649
旅游中间产品转移价格的确定 …… 1657
我国旅游价值链管理探讨 …… 1665
旅游业零负团费的运行机制及危害性探析 …… 1669
文化生产力：管理的视角 …… 1673
旅游交通与航空运输规划 …… 1680
办好旅游教育至关重要 …… 1682
旅游标志景区研究有意义 …… 1683
Preface …… 1685
产业工人的中国梦：从低技能劳工到专业技术工人的人资转型升级战略 …… 1687
企业战略管理概要 …… 1698
“技工荒”困扰中国高尔夫球具代工产业 …… 1715
系统论视角下的旅游发展与旅游研究——中国工程院工程管理学部刘人怀院士访谈 …… 1720
旅游城市品牌的塑造 …… 1725
保护遗产至关重要 …… 1726
城市遗产旅游景观的研究 …… 1727
应用旅游工程理论探讨阳江旅游发展规划 …… 1728
创业拼凑对创业学习的影响研究——基于创业导向的调节作用 …… 1733
澳门会展业的经济效应与发展策略研究 …… 1747
创业拼凑、创业学习与新企业突破性创新的关系研究 …… 1755
数字经济 创新未来 …… 1768
互补性资产对双元创新的影响及平台开放度的调节作用 …… 1770
系统论视角下旅游学科提升发展的思考 …… 1782

Guest editorial …… 1785
发挥社会创业的重要作用 …… 1789
第十三章　科技管理 …… 1790
谈谈科研中的几个问题 …… 1790
青年人的奋斗方向 …… 1796
“科学技术是第一生产力”的趋势与我国高新技术发展的战略 …… 1798
人才是振兴国家的关键 …… 1804
传播科学思想　提高国民素质 …… 1805
大力重视非开挖工程技术 …… 1806
认真评选科技奖 …… 1807
与时俱进　开拓创新 …… 1808
以科技创新加快推进全面建设小康社会步伐 …… 1809
加强基础研究　实现科技强省 …… 1815
成立广东省仪器仪表学会是紧迫的历史使命 …… 1819
培养青少年的创新精神 …… 1822
大力推动科技事业的发展 …… 1824
迎接新挑战 …… 1826
同心同德　开拓前进 …… 1827
创新路上的感想 …… 1829
为提高全民族科学素质做出新贡献 …… 1844
稳步推进科普志愿服务事业 …… 1847
努力开创仪表元件分会工作的新局面 …… 1851
关注世界科技创新态势 …… 1861
推进振动工程学科发展 …… 1874
积极投身科技创新活动 …… 1876
认真做好“2011 计划”重大战略部署工作 …… 1877
组建“政产学研金”合作平台　推动协同创新迈上新台阶 …… 1878
开拓创新　做好科普工作 …… 1881
复合材料创新成果丰硕 …… 1884
肩负使命与责任　继续向前 …… 1886
去除浮躁现象　建立公正科技评价机制 …… 1889
切实做好科技评价工作 …… 1891
专家学者不妨多点科普和传播意识 …… 1892
推动力学学科发展 …… 1896
汇聚人才为广东发展效力 …… 1897
扎实开展科普志愿服务工作，为全面提高我省公民科学素质而努力 …… 1899
应该高度关注青少年的视觉健康问题 …… 1904
为中华民族伟大复兴做出更多更大的贡献 …… 1905

全省科技工作者应履行科普社会责任 …… 1907
坚守初心　坚持创新 …… 1909
追求真理　奉献国家 …… 1911
让未来祖国的科技天地群英荟萃 …… 1912
众人拾柴火焰高 …… 1913
努力推动压力容器技术进步 …… 1914
大力加强高校科技创新 …… 1917
创新争先　自立自强 …… 1918
第十四章　教育管理 …… 1921
欢迎新同学们 …… 1921
谈谈课堂教学中的几个问题 …… 1923
谈谈大学的学习生活 …… 1924
前进中的经济管理学院 …… 1927
情牵母校 …… 1930
（日本）新世纪中文电视学校校长致辞 …… 1931
当代科技发展与大学理念和人才培养 …… 1932
加快我国高等教育进入世界先进行列 …… 1940
关于高考的一点浅见 …… 2019
认真做好高校力学教学指导工作 …… 2021
我国力学专业教育现状与思考 …… 2022
浅谈高等学校科学管理“三”字经 …… 2027
我的语文观 …… 2040
爱低碳生活　创绿色校园 …… 2041
深化力学课程改革 …… 2043
从学生会主席到大学校长之路 …… 2045
试答“钱学森之问” …… 2048
认真开展高等学校教育教学研究 …… 2055
加强重大工程结构安全领域国际合作 …… 2056
知识交融　创新成才 …… 2058
从教感想 …… 2065
珍惜时间　勤奋成才 …… 2068
以我个人经历及家庭经历浅谈家庭教育 …… 2072
教化育人　师法天地 …… 2075
感谢党和国家的培养和教育 …… 2078
澳门民办高校发展现状与前景 …… 2081
八十有感 …… 2087
新都二中——我的母校 …… 2091
推动暨南大学“重大工程灾害与控制”教育部重点实验室跨越式发展 …… 2092

高等教育管理实例：暨南大学“侨校＋名校”之路（1995—2006） …… 2101
1. 暨南大学校长任职仪式讲话 …… 2101
2. 加强基础　从严治校　培养高素质人才 …… 2104
3. 根据侨校特点改进教学工作　采取有效措施提高教学质量 …… 2107
4. 提高认识　办好成人教育 …… 2116
5.《红杏枝头春意闹——暨南大学成人高等教育毕业生业绩选》序 …… 2121
6. 暨南大学“211 工程”部门预审汇报 …… 2122
7. 在暨南大学“211 工程”部门预审总结会上的讲话 …… 2136
8. 庆贺九十华诞　创建一流大学 …… 2137
9. 开拓创新　共建附属医院 …… 2138
10. 办出特色　办出水平 …… 2139
11. 总结经验　深化改革　加快科技发展 …… 2147
12. 积极主动地为侨务工作服务 …… 2150
13. 用现代化管理促进高等教育事业的发展 …… 2156
14. 深化教育改革　提高教学质量 …… 2158
15. 暨南大学兴办高等华侨教育的历史回顾与展望 …… 2160
16. 华文学院越办越好 …… 2164
17. 在高校党的建设中贯彻落实邓小平“从严治党”的思想 …… 2165
18. 脚踏实地　循序渐进 …… 2169
19. 举行全国 100 所“211 工程”学校赠书仪式 …… 2175
20. 转变观念　量化考核　优劳优酬 …… 2176
21. 同质同水平异地办学 …… 2181
22. 暨南大学国际化之路 …… 2183
23. 面向新世纪的创新教育 …… 2186
24. 坚持社会主义办学方向　办好华侨高等教育　为海内外培养高素质人才 …… 2191
25. 辉煌与梦想 …… 2203
26. 狠抓办学质量　走“侨校＋名校”之路 …… 2204
27. 暨南大学的特点和优势 …… 2214
28. 弘扬中华民族文化 …… 2216
29. 珍惜“暨南人”光荣称号 …… 2219
30. 饮水思源 …… 2221
31. 为祖国侨务事业和华侨高等教育做出新的贡献 …… 2223
32. 建设国际化、现代化、综合化的高水平社会主义华侨大学 …… 2227
33. 深深感谢校友深情 …… 2229
34. 办好研究生教育至关重要 …… 2231
35. 进一步提高干部人事档案工作的管理水平 …… 2233
36. 胸怀世界　放眼未来 …… 2237

37. 持之以恒　依法治校 …… 2240
38. 全球化进程与华侨高等学府的重要使命 …… 2241
39. 努力完成高校扩招任务 …… 2247
40. 答谢珠海人民 …… 2254
41. 坚决反对和防止腐败是学校重大的政治任务 …… 2255
42. 寄语中层干部 …… 2257
43. 暨南大学的创新发展之路 …… 2266
44. 弘扬中华文化　发展华文教育　传播华夏文明　促进文化交流 …… 2274
45. 为迎接暨南大学百年庆典增添新的光彩 …… 2276
46. 发挥优势　深化改革　保证重点　改善条件　提高质量 …… 2277
47. “侨校＋名校”的发展定位 …… 2283
48. 教学是学校的生命 …… 2290
49. 大力发展具有侨校特色的研究生教育 …… 2298
50. 贺暨南大学澳门校友会会长就职 …… 2306
51. 忍耐是一个人成功的秘诀——与暨南学子谈成才 …… 2307
52. 标准学分制的研究与实践 …… 2313
53. 积极服务海外华侨华人社会 …… 2318
54. 暨南大学百年校庆公告（第一号） …… 2320
55. 面向海外　面向港澳台　为祖国统一大业服务 …… 2321
56. 坚持大力发展研究生教育 …… 2331
57. 广纳贤才　全球招聘院长 …… 2333
58. 学会普通话　走遍天下都不怕 …… 2336
59. 亚洲　青春　竞技 …… 2337
60. 贺暨南大学香港社会学同学会成立 …… 2338
61. 立足侨校　服务学生　全面推进我校学生德育工作 …… 2340
62. 图书馆是大学的心脏 …… 2343
63. 生逢其时的学院 …… 2344
64. 述职报告 …… 2345
65. 衷心祝愿暨南大学的明天更美好 …… 2361

附录一　刘人怀大事年表 …… 2364
附录二　刘人怀主要论著目录 …… 2384

目　　录

上篇　板壳力学理论与工程应用

第一章　概述 …… 3

板壳分析与应用 …… 3

板壳非线性力学研究的最新进展 …… 14

板壳非线性力学 …… 29

The effect of the Göttingen school on investigation of flexible plates and shells in China …… 32

多姿多彩的板壳结构 …… 34

弹性元件国内外理论发展概况 …… 36

第二章　单层板壳非线性力学 …… 52

在内边缘均布力矩作用下中心开孔圆底扁球壳的非线性稳定问题 …… 52

在边缘载荷作用下中心开孔圆底扁薄球壳的轴对称稳定性 …… 56

集中载荷作用下开顶扁球壳的非线性稳定问题 …… 66

均布载荷作用下开顶扁球壳的非线性稳定问题 …… 77

On the nonlinear stability of a truncated shallow spherical shell under axisymmetrically distributed load …… 91

均布载荷作用下具有硬中心的开顶扁球壳的非线性屈曲 …… 106

中心开孔扁球壳在均布载荷作用下的非线性屈曲 …… 112

A new approach to nonlinear vibration of orthotropic thin circular plates …… 120

正交异性扁薄球壳的非线性轴对称振动 …… 128

第三章　波纹板壳非线性力学 …… 144

波纹圆板的特征关系式 …… 144

具有光滑中心的波纹圆板的特征关系式 …… 151

波纹环形板的非线性弯曲 …… 163

Large deflection of corrugated circular plate with a plane central region under the action of concentrated loads at the center …… 171

Large deflection of corrugated circular plate with plane boundary region …… 185

在复合载荷作用下波纹环形板的非线性弯曲 …… 202

On large deflection of corrugated annular plates under uniform pressure …… 212

复合载荷下波纹圆板的非线性分析 …… 223

On the non-linear bending and vibration of corrugated circular plates …… 238

U 型波纹管非线性变形的刚度和应力分析理论 …… 252

Non-linear bending of a corrugated annular plate with a plane boundary region and a non-deformable rigid body at the center under compound load …… 266
均布载荷作用下波纹环形板的非线性弯曲 …… 281
波纹管的制造与理论研究概况 …… 291
中心受载下具有平面边缘区域的固支波纹环形板的非线性分析 …… 298
Nonlinear bending of corrugated annular plate with large boundary corrugation …… 310
Nonlinear deformation analysis of a U-shaped bellows with varying thickness …… 321
波纹扁球壳的大挠度方程 …… 333
Nonlinear stability of corrugated shallow spherical shell …… 339
第四章　双层扁壳非线性力学 …… 353
双层金属截头扁锥壳的热稳定性 …… 353
双层金属中心开孔扁球壳的非线性热稳定问题 …… 364
Non-linear thermal stability of bimetallic shallow shells of revolution …… 378

上篇　板壳力学理论与工程应用

第一章　概　　述

板壳分析与应用①

一、引言

板壳结构分析是现代固体力学中特别引人注目的一个分支，近几十年，随着科学技术的突飞猛进，其发展异常迅速。这门学科几乎与一切工程设计都有关联，对航天、航空、航海、机械、石化、建筑、水利、动力、仪表、交通等工程设计，尤其具有指导意义。

板壳是平板和壳体的总称，是最常见的物体形式。其外形特点是厚度比其余两个方向尺寸在数量级上小得多。平分物体厚度的分界面称为中面。若中面是平面，则称此物体为平板；若中面是曲面，则称此物体为壳体。

由于厚度小、质量轻、耗材少、性能好，板壳成为具有优良特性的结构元件，不仅广泛应用于各种工程结构作为最基本和最主要的构件，而且在自然界和日常生活中也常常碰见，它们与每个人的生活休戚相关，与人类的生存紧密相连。

板壳结构分析包括板壳静力学和板壳动力学两大部分。

板壳静力学是研究板壳在静荷载作用下所产生的应力和变形，亦即通常所说的刚度、强度和稳定问题。通过分析计算，使板壳设计得既美观大方又安全经济。

板壳动力学是研究板壳在动荷载作用下的结构反应。其中一个重要问题是板壳的振动问题。

按照厚度的大小，板壳可分为薄板壳和厚板壳两大类，而大多数板壳属于薄板壳范畴。

按照隶属的理论范畴，当板壳弯曲变形时，其挠度相对于厚度是小量，所建立的微分方程属线性性质，则纳入板壳线性理论范畴；反之，若挠度不是小量，所建立的微分方程属非线性的，则纳入板壳非线性理论范畴。

板结构分析随着工业的发展起源于 18 世纪，Euler 最先探索板的弯曲问题[1]。但是，直到 1850 年，Kirchhoff 才给出第一个完善的板的弯曲理论[2]。接着 Aron 做了薄板壳的分析工作[3]。此后，特别是在 20 世纪，工业的飞跃发展，极大地推动了板壳结构分析的发展和应用。现今，经典的薄板壳线性理论已较成熟，并在各种工程设计中起着指导作用。然而，在薄板壳非线性领域（以下简称为板壳非线性问题）和厚板壳线性领域，还有许多问题尚未解决。从 1962 年开始，在前人的基础上，作者结合工程

---

① 本文原载《中国工程科学》，2000，2（11）：60-67.

需要，在这些领域进行了一些探索。本文即是这些工作的小结。

二、板壳非线性理论与工程应用

1. 概述

从本质上讲，板壳理论作为精确理论而言，应该是非线性的。板壳非线性理论的奠基者是 20 世纪杰出的科学家 von Kármán，1910 年，他最先给出板的大挠度理论的微分方程[4]。由于工程上未提出应用精确理论的迫切要求，加之数学问题求解的巨大困难，所以此后的发展是缓慢的。直到 20 世纪 60 年代，随着工业的发展，工程上大量提出了非线性现象与问题，它比线性情形更复杂，描述的现象更丰富，更具有挑战性，板壳非线性理论的研究蓬勃兴起，直到今天，它仍然作为固体力学研究的一个最活跃的领域而备受人们关注，并推动非线性科学的发展。目前研究的中心课题是板壳的几何非线性弯曲、稳定和振动问题。

2. 修正幂级数法和修正迭代法

几何非线性关系的引入，导致板壳弯曲理论的微分方程是非线性的，这在数学研究上存在极大的困难，因而在 von Kármán 提出板的大挠度方程后，研究进展十分缓慢，寻求其解法便成为解决问题的关键。人们将解法分为两类：解析法和数值法。数值法常见的有有限元法、边界元法和有限差分法等；解析法有精确解法和近似解法。常见的精确解法为幂级数法和三角级数法等；常见的近似解法有摄动法、奇异摄动法、逐次逼近法、Ritz 法、Галёркин 法等。作者的工作限于解析法方面。

寻求精确解一直是人们的希望，因为它不仅使问题得以圆满解决，为工程设计提供最可靠的依据，而且为近似解提供一个检验标准。1934 年，Way 最先使用幂级数方法求解了板壳非线性力学的经典问题：圆板大挠度问题，给出其精确解的计算结果[5]。1989 年，作者发展了 Way 的方法，提出修正幂级数法，求解了计及表层抗弯刚度的夹层圆板的大挠度问题[6]。该问题十分复杂，控制方程的最高阶导数前有一个小参数，属于边界层型，微分方程组为

$$\varepsilon L^2\Phi-\left(1+\frac{\varepsilon}{\lambda}\right)L\Phi-\lambda L(S_r\Phi)+S_r\Phi+P\rho=0$$

$$L(\rho S_r)+\frac{\Phi^2}{\rho}=0$$

这里 ε 是一个远小于 1 的小参数，Φ 是挠度的一阶导数，S_r 是径向应力，P 是均布荷载，λ 是参数，ρ 是径向坐标，L 是二阶微分算子。

此微分方程组不仅变系数和阶数高，而且为非线性，用奇异摄动理论求解将是十分困难的。倘若用 Way 的传统的幂级数法，则因最高阶导数前有一小参数而无法求解。为此，作者在综合了传统的幂级数法和奇异摄动法的特点后，提出了修正幂级数法，从而解决了这个难题。这一方法的关键技术在于对未知量 Φ 和 S_r 的幂级数系数 a_i 和 b_i 的迭代方式和程序进行了改变，依照奇异摄动法的思想，提出了如下的迭代公式

$$a_0^{(n+1)}=-\sum_{i=1}^{\infty}a_{2i}^{(n)},$$

$$b_1^{(n+1)} = \sum_{i=1}^{\infty}[4\varepsilon i(i+1)-1]b_{2i+1}^{(n)} - \lambda P,$$

$$b_3^{(n+1)} = \frac{1}{8(1+\varepsilon/\lambda+a_0\lambda)}\{192\varepsilon b_5^{(n)} + \lambda[b_1^{(n+1)}]^3 + a_0^{(n+1)}b_1^{(n+1)} + P\},$$

$$a_{2i}^{(n+1)} = -\frac{\beta_{2i-1}^{(n+1)}}{4i(i+1)},(i=0,1,2,\cdots)$$

$$b_{2i-1}^{(n+1)} = \frac{1}{4i(i-1)(1+\varepsilon/\lambda+a_0\lambda)}[16\varepsilon i^2(i^2-1)b_{2i+1}^{(n)}$$

$$+\alpha_{2i-3}^{(n+1)} - 4\lambda i(i-1)\sum_{k=0}^{i-2}a_{2(i-k-1)}^{(n+1)}b_{2k+1}^{(n+1)}],(i=3,4,5,\cdots)$$

$$\alpha_{2i+1}^{(n+1)} = \sum_{k=0}^{i}a_{2(i-k)}^{(n+1)}b_{2k+1}^{(n+1)},(i=0,1,2,\cdots)$$

$$\beta_{2i+1}^{(n+1)} = \sum_{k=0}^{i}b_{2(i-k)+1}^{(n+1)}b_{2k+1}^{(n+1)},(i=0,1,2,\cdots)$$

$$(n=0,1,2,\cdots)$$

在近似求解方面，代表性的工作为 Vincent 和钱伟长在求解圆板大挠度问题中分别提出的以荷载和中心挠度为摄动参数的摄动法[7,8]。然而若使用此法去处理壳体非线性问题中的经典问题，即扁球壳的非线性稳定问题，则异常困难。1965 年，叶开沅和作者创立了修正迭代法，克服了求解的困难[9-11]。这一方法结合了钱伟长摄动法和逐次逼近法的优点，不仅程序简单、计算量小，而且收敛快、所获解析解的精度高。为简单起见，现以著名的 Kármán 方程，即矩形板的非线性微分方程组为例，介绍这一方法。该方程组为

$$\frac{D}{h}\nabla^2\nabla^2 W = H(W,\Phi) + \frac{q}{h} \tag{1a}$$

$$\frac{1}{E}\nabla^2\nabla^2\Phi = -\frac{1}{2}H(W,W) \tag{1b}$$

式中，D 是抗弯刚度，h 是厚度，E 是弹性模量，W 是挠度，Φ 是应力函数，q 是横向荷载，∇^2是拉普拉斯算子。把算子 H 应用到函数 W 和 Φ，有

$$H(W,\Phi) = \frac{\partial^2 W}{\partial x^2}\frac{\partial^2 \Phi}{\partial y^2} + \frac{\partial^2 W}{\partial y^2}\frac{\partial^2 \Phi}{\partial x^2} - 2\frac{\partial^2 W}{\partial x\partial y}\frac{\partial^2 \Phi}{\partial x\partial y} \tag{2}$$

这里 x 和 y 是直角坐标。

对于第一次近似，首先略去方程（1a）右端的非线性项 H（W，Φ），便有如下的线性微分方程

$$\frac{D}{h}\nabla^2\nabla^2 W_1 = \frac{q}{h} \tag{3}$$

解此方程，并使用相应的边界条件，可得解

$$W_1 = f_1(x,y,q) \tag{4}$$

以矩形板的中心挠度 W_0 作为迭代参数，解（4）便可转换为

$$W_1 = g_1(x,y,W_0) \tag{5}$$

将此解代入方程（1b）的右端，则得第一次近似中关于 Φ 的线性微分方程

$$\frac{1}{E}\nabla^2\nabla^2\Phi_1=-\frac{1}{2}H(W_1,W_1) \tag{6}$$

解此方程后，再将解 Φ_1 和 W_1 形式地代入方程（1a）的右端，于是又可得二次近似中关于 W 的线性微分方程。重复上述步骤，便可得更高近似的解。经过证明，这一方法是收敛的。

应用修正迭代法，作者和其他学者成功地解决了一系列板壳非线性理论中的难题。

3. 波纹板壳

精密仪器仪表工业的发展程度是一个国家科学技术发展水平的重要标志。波纹膜片和波纹管称为弹性敏感元件，是多种传感器和精密仪表的核心，应用十分广泛。当今，传感器技术已成为信息时代三大技术支柱之一。因此，作为传感器敏感器件的弹性敏感元件，更有用武之地。最近，波纹管又作为膨胀节，在工业管道中起着特殊作用。

波纹膜片，又称为波纹圆板，是一种采用铍青铜、不锈钢等高弹性金属材料制造的压有同心折皱的圆板。严格地讲，它是一种特殊形状的壳体。其折皱主要为正弦形、圆形、梯形和锯齿形。在荷载相当小时，它就能给出明显而容易被测量的挠度。这一优良特性被用来作为敏感元件。它相当于人的眼睛、耳朵等感觉器官，能灵敏地感受力、流量、速度等环境的变化，发出信号，由控制系统进行控制。由于结构形状复杂、种类繁多，且参数多，如半径、厚度、波宽、波深和波数等，加之又归结为非线性微分方程组求解，所以研究极为困难。1941 年，Панов 第一次讨论了浅正弦波纹膜片的非线性弯曲问题[12]。但是，直到 20 世纪 70 年代，国际上仍只有少数国家（苏联、美国和日本）的学者从事这一领域中最简单问题的研究，仅涉及全波纹膜片和具有刚性中心的环形波纹膜片两种，所给出的解的精确度不够，远不能满足仪表和传感器设计的要求。因此，工程设计时只好用实验研究和经验来弥补理论的严重不足。即使进行实验，难度也极大，常常达不到目的。1964 年，在我国研制飞机测高计时遇到难题：要研究分析此仪表心脏元件锯齿形波纹膜片的非线性特征。经过作者四年的努力，应用圆柱正交各向异性圆板大挠度理论和修正迭代法，首次成功地研究了波纹膜片问题，获得了既简单又精确的优于苏联著名科学家 Феодосьев 的公式[13]。以一个无光滑中心的正弦截面波纹圆板为例，其半径为 76mm，厚度 h 为 0.33mm，作者的特征曲线的理论值与实验值非常一致。在中心挠度 $w_0=8.79h$ 的情况下，作者的理论值与实验值误差仅为 0.90%，而 Феодосьев 公式结果与实验值误差高达 39.0%。

这一工作在 1978 年复刊的《力学学报》第 1 期上发表[14]。此后，作者又继续探索，系统地在国内外学术刊物上发表了十几篇论文，如文献［15］和［16］等，共讨论了全波纹圆板、具有光滑中心的波纹圆板、具有光滑中心和平面边缘的波纹圆板、具有光滑中心和圆柱壳边缘的波纹圆板、具有刚性中心的波纹环形板、具有刚性中心和平面边缘的波纹环形板以及具有刚性中心和大边缘波纹的波纹环形板等 7 种波纹膜片的弯曲和振动问题。这些工作大部分是创新性研究，所得到的公式和程序，已被国内主要设计部门和工厂采用。

波纹管是一种圆柱形的薄壁金属软管，沿着圆周做成有波纹的折皱，这些折皱的形状主要为 U 型、V 型和 Ω 型。常用的是 U 型波纹管。由于形状复杂，研究十分困难。理论研究已有50多年的历史，大多是讨论线性变形问题，理论结果与实验很不一致。为此，Андреева首先用差分法研究了U型波纹管的非线性变形问题[17]，接着钱伟长、Axelrad等又使用摄动法进行研究，理论值与实验值的误差有所缩小[18,19]，但仍不能满足工程设计的要求。1986年，作者承担国家“七五”科技攻关任务，将U型波纹管视为圆环壳和变厚度截头扁锥壳的组合，采用整体非线性分析方法进行处理，所获结果精确度已能满足工程设计要求，其公式和程序已供国内主要设计部门和工厂使用[20]。

4. 单层板壳

在近代工程结构中，构件的稳定性设计至关重要。1744年，著名法国科学家Euler第一次提出并解决了压杆横向屈曲的弹性稳定问题[21]，开创了弹性构件稳定性研究，作出了划时代贡献。此后，在长达200年的时间里，人们陆续研究了杆、板、壳以及组合结构的稳定问题。由于生产力水平以及数学的限制，发展是不快的。此间对杆、板研究较深，对壳体则研究得很不深透，使用线性理论研究壳体稳定问题，其结果无法与实验一致。1939年，von Kármán和钱学森第一次用非线性理论讨论了均布外压下的固定边扁球壳的稳定问题[22]，为迷雾中的壳体稳定性研究指明了正确方向。但是，由于解法问题，此后的研究进展缓慢。扁球壳的非线性稳定是固体力学的经典问题，引起国际众多学者争相研究。1965年，叶开沅和作者提出了修正迭代法[9,10]，成功地研究了扁球壳的非线性稳定问题，其结果与钱伟长等的实验值吻合[23]。同时，作者又研究了尚无人研究过的难度更大的开顶扁球壳[11]。此后，又继续处理了这一领域里几个有意义的问题，如文献［24］等，这一类问题的研究对于作为精密仪表自控元件的跳跃膜片和大型屋盖等的设计有指导意义。此外，还讨论了作为仪表弹性元件的圆板的非线性振动问题[25]。

5. 双层金属旋转扁壳

在弹性敏感元件中，有一类采用双层金属旋转扁壳做成的热敏弹性元件，用于自动控制仪表。这类元件具有这样一种特性：在均匀加热情况下，由于两层金属的热膨胀系数不同，将产生弯曲，以致发生跳跃现象，自动控制仪表便应用这一特性作为控制信号。此问题最早由Панов研究过[26]，结果不尽如人意。作者应用Королев和Radkowski的选择多层壳体坐标参考面的方法，将双层旋转扁壳的基本方程表示成单层壳体理论的形式[27-29]，从而建立了在均匀温度作用下双层金属旋转扁壳的大挠度弯曲理论，并应用修正迭代法得到了临界温度的解析解。解的精确度高，公式远较前人结果简单，易于工程设计时应用。

6. 网格扁壳

网格扁壳是由较短构件按实体壳的形式布置成的一种空间构架，包括单层和双层两种形式。它具有空间刚度大、受力合理等力学特点，成本低、轻质省材、易于包装、易于运输、易于安装等经济特点，以及造型新颖、美观等结构特点，近年来在体育馆、展览馆、电影院等的屋盖和大型储油罐顶盖以及空间站等结构中得到广泛应用。但因结构复杂，其非线性分析始终无人研究，因而不能满足现代工业的要求。作者应用等效原理，建立了单层和双层网格扁壳非线性弯曲理论，并获得了一些弯曲、稳定和振动问题的解析解[30,31]。特别是1985年为我国大型10 000m^{3}储油罐的新型顶盖稳定设计提供了公式和数据，节省材料 20%，节省施工费用，有利于安全使用。

7. 夹层板壳

夹层板壳由三层材料组合而成，上、下两块表面层很薄，强度很高，中间是一块软而轻的厚夹心。这是一种新型的结构元件，具有质量轻、强度高、刚度大等许多突出的优点，在航天、航空、包装等工程中起了重大作用。这方面的理论与应用研究文献十分丰富，杜庆华提出了有名的考虑剪切影响的夹层板理论[32]。然而，由于夹层结构复杂和在非线性数学上的困难，国际上基本处于线性分析研究阶段，非线性分析的论文如凤毛麟角，远不能满足科学技术迅猛发展的需要。Reissner 于 1948 年首次给出具有软夹心的夹层板的大挠度方程[33]。由于分析困难，研究进展十分缓慢。作者最先建立了比 Reissner 理论更一般性的具有软夹心的夹层圆板非线性弯曲理论[6,34]，随后又分析了表层抗弯刚度的影响，给出有意义的结论：对于一般工程，应用忽略表层抗弯刚度的夹层板理论已足够精确。接着，作者系统地研究了夹层圆形板、夹层环形板、夹层矩形板、夹层扁球壳、夹层扁锥壳和夹层圆柱壳的非线性弯曲、稳定和振动问题[35-39]，并按照“能量误差一致原则”，采用张量分析工具，建立了一致有效的夹层壳大挠度理论，同时还研究了表层是层合复合材料的更复杂情况。

8. 复合材料层合板壳

复合材料是一类新型材料，其强度高、刚度大、质量轻，具有抗疲劳、减振、耐高温、可设计等优点，近 30 年来在航空、航天、造船、能源、交通、建筑、机械、生物医学和体育等部门已有广泛的应用。可以预言，21 世纪将进入复合材料的时代。随着复合材料的开发和应用，复合材料层合板壳的力学分析提上了日程，已有大量文献。这些研究大多属于线性理论，少量非线性分析也仅限于圆柱壳等。作者率先涉及特别复杂的复合材料层合旋转壳体领域，建立了计及横向剪切影响的对称层合扁球壳和扁锥壳的非线性稳定和振动理论，求解了几个有意义的问题，得出了重要的结论：若使用不计及横向剪切影响的经典理论进行工程设计，那将是保守的[40-42]。同时，在计及横向剪切影响下，进一步系统地研究了复合材料层合圆板、椭圆板以及矩形板的非线性弯曲问题[43,44]。这些结果可直接用于工程设计。此外，作者还提出了一个计及高阶影响的层合矩形板的非线性弯曲理论[45]。

三、厚板壳弯曲理论与工程应用

厚板壳在化工、石油、动力、水利、建筑等工程中应用广泛，但厚板壳理论研究甚为困难。作者结合我国工程实际，提出了简单而实用的厚板壳弯曲理论，解决了工程中的实际问题。

1. 厚圆板与高压热交换器厚管板

在不同温度的流体间传递热能的热交换器（简称换热器）在工厂中大量使用，是许多行业的通用设备，它占到石油、化工工厂设备总质量的 40%。随着工业的发展，对能源利用、开发和节约的要求不断提高，对换热器的要求也越来越高。在换热器中，最常见的一类是列管式换热器，而管板是它的关键部件。管板是一个满布孔洞的圆板，其强度设计十分重要。迄今，基于薄板理论方面的研究成果，对中、低压换热器的管板设计具有实际意义。这中间，以乌班诺夫斯基的工作最为出色[46]。20世纪60年代以来，工业部门开始采用高压换热器。这种换热器的管板厚度很大，已超出薄板理论的研究范围。但是，却没有一个能供工程设计应用的公式，因而工程上只得沿用薄管板公式进行粗略设计。显然，这是不合理的。

1970年，我国试制第一台大型换热分离氨组合设备，即四合一氨合成塔（长度28.467m）。由于合成塔中的水冷却器管板尺寸很大（厚度230mm，直径700mm），处于200℃高温和32MPa高压工作状态，若按国际王牌产品标准尺寸进行设计，则我国没有一家工厂具备此大型构件的制造能力；若用薄管板公式进行设计，则结果偏于保守，更无法进行试制。作者采用Reissner厚板理论，足够精确地建立了高压固定式换热器厚管板的弯曲理论，为工程设计提供了可靠的设计公式[47]。计算结果表明，厚管板厚度可进行减薄设计，定为190mm，既节省材料，又使产品在我国试制成功。

2. 厚圆柱壳和薄椭球壳与铂重整装置

铂重整装置是石油炼制工业的重要装置，用于生产航空煤油，属高温（设计温度300℃）中压（设计压力8MPa）容器。1969年，我国按照国际王牌产品水平进行第一次产品试制（长10.729m，直径1.910m），共生产6台。由于设计不当，其中椭球封头（长半轴930mm，短半轴480mm，厚60mm）厚壁接管（直径423mm，厚90mm）根部强度不够，造成产品水压试验压力仅达到6MPa，致使产品报废。作者建立了厚壁圆柱壳弯曲理论，并采用薄椭球壳弯曲理论，给出了椭球封头中心孔边缘的应力计算公式，提出了在中心孔边焊接32mm厚度护强板的补救措施，仅少许花费，便复活了6台产品[48]。

3. 厚球壳与尿素合成塔

大型尿素合成塔是生产优质化肥尿素的关键设备，是我国20世纪六七十年代农业的急需设备之一。1970年，瞄准国际王牌产品，试制我国第一台产品（长28.295m，直径1.594m，工作压力22MPa），在底部球形封头同心圆上，等距离开了三个孔洞，此处成为新产品的关键部位，需要精心设计。在原设计制造方案中，是在厚度190mm的球壳开孔洞（直径145mm）的内缘，焊接一块厚达80mm的补强圈。这样做，焊接工作量大，操作条件恶劣，封头易产生裂纹，产品质量得不到保证。为此，作者建立了一个较适用的厚球壳弯曲理论，给出了孔边缘应力计算公式[49]。最后按照作者的建议，取消了边缘的补强圈，使产品得以试制成功。

4. 厚球壳和厚圆柱壳与加氢反应器

加氢反应器是炼油工业用于裂解的关键设备。1974年，试制我国第一台直径最大的加氢反应器，长度26.13m，直径2.1m，设计压力21MPa。除了反应器大外，还要在这台反应器上采用两项新技术：a. 在封头上开大孔，孔径为800mm，突破了高压容器传统制造工艺上开大孔的禁忌；b. 高压容器壁厚大，而我国又无能力制造厚合金钢板材，因此，想用两块较薄的合金钢板材，采用热套技术来制造，亦即反应器筒体采用双层套箍式工艺制造。这两项新技术都没有现成理论可供使用。作者创立了适用的厚壁圆柱壳弯曲理论[50,51]，与以前提出的厚壁球壳理论一起，为加氢反应器正确设计提供了可靠的公式和数据，使产品试制成功。

5. 厚圆柱壳与高压聚乙烯反应器

1970 年，在研制我国最高压力容器——高压聚乙烯反应器（长度 3.505m，设计压力 230MPa）中，作者接受了研究反应器筒体关键部位径向开孔（直径 51mm）应力集中问题的任务。难度是在厚壁圆柱壳上开径向孔，尚无理论分析文献。作者应用自己提出的厚壁圆柱壳理论，采用复变函数方法，给出了问题的解析解①。应力集中的数据与随后实验所给数据十分吻合，为反应器研制提供了可靠的设计依据。

6. 变厚度厚锥壳与铁路高桥墩

1975 年我国进行最高的铁路桥梁的设计，桥墩属变厚度厚壁截头锥壳，高 69.4m，底部直径 696cm，顶部直径 240cm。由于无设计理论，作者提出了工程适用的变厚度的厚壁锥壳理论，给出了这一高桥墩的设计公式，为大桥的设计提供了可靠的依据②。

7. 厚矩形板与新型油井钻头

对一口油井而言，目前的采油技术水平只能采出石油储量的三分之一左右。因此，提高采油技术水平是当务之急。从 1998 年开始，通过深入工程实际，对由厚矩形板构成的油井大直径扩孔钻头进行力学分析，为采用高压水射流超短半径水平井钻井新技术提供了可靠的理论分析③。新技术的成功，有望使油井产油量大幅度提高。

四、薄板壳弯曲理论与工程应用

1974 年，在我国最大的 5Mt/a 炼油厂的设计中，关键设备之一是常减压蒸馏装置的高达 34m 的减压塔，它是变截面的塔器，自上而下的直径为 6.4m/10m/6.4m/2m。塔的壁厚为 26mm，操作质量高达 479.831t。以往的减压塔裙座均与下封头搭接。此次设计，由于考虑到基础安装、管线连接和节约材料等，拟将裙座支承于该塔下端 ϕ6.4m 处，但如此大型的塔器，在负压操作情况下，这种支承方式引起的应力分析相当复杂，塔器是否处于稳定状态，均是设计难题。鉴于现有设计规范与通常的工程方法已不适于该塔的强度和稳定设计，作者应用组合壳体理论给出塔的应力分析公式[52,53]，对塔的一些强度不足的薄弱处提出了改进建议。同时，给出了稳定设计的临界荷载公式，使塔的稳定设计有了可靠的依据。

此外，还提出了一种新的复合材料层合板理论[54]。这一理论可满足层间位移和横向剪应力连续条件以及上下表面横向剪应力协调条件，其控制方程只有 5 个未知量。数值计算表明，这一理论具有很高的精确度。

五、结束语

21 世纪即将到来，世界正面临一场新的技术革命。现代工业、现代国防和现代科技的更大发展，将对板壳结构分析提出更多和更高的要求[55,56]。显然，目前的理论与方法不能满足这些要求。因此，紧密结合工程需要，推动我国板壳结构分析与应用事

---

① 刘人怀．一机部和化工部高压聚乙烯反应器试制组研究报告，1970.

② 刘人怀．铁道部第一设计院白水河大桥设计组研究报告，1975.

③ 刘人怀，王璠，肖潭．辽河油田勘探局钻井研究所研究报告，1998.

业继续向前发展是一项重要任务。

参考文献

[1] Euler L. De motu vibvatorio tympanorum. Novi commentari Acad. Petropolit，1766，(10)：243-260.

[2] Kirchhoff G. Über das Gleichgewicht und die Bewegung einer elastischen Seheibe. Journal für die reine und angewandte Math.，1850，40：51-58.

[3] Aron H. Das Gleichgewicht und die Bewegung einer unendlich dünnen beliebig gekrümmten elastischen Schale. Journal für die reine und angewandte Math.，1874，78：136-174.

[4] von Kármán T. Festigkeitsprobleme im Maschinenbau. Encyd. Math. Wiss.，1910，4 (4)：348-352.

[5] Way S. Bending of circular plates with large deflection. Trans. ASME，1934，56 (7)：627-633.

[6] Liu Renhuai and Zhu Gaoqiu. Further study on large deflection of circular sandwich plates. Applied Mathematics and Mechanics：English Edition，1989，10 (12)：1099-1106.

[7] Vincent J J. The bending of a thin circular plate. Phil. Mag.，1931，(12)：185-196.

[8] Chien Weizang. Large deflection of a circular clamped plate under uniform pressure. Chinese J. Phys.，1947，7 (2)：102-107.

[9] 叶开沅，刘人怀，平庆元，等. 在对称线布载荷作用下的圆底扁薄球壳的非线性稳定问题. 科学通报，1965，(2)：142-145.

[10] 叶开沅，刘人怀，张传智，等. 圆底扁薄球壳在边缘力矩作用下的非线性稳定问题. 科学通报，1965，(2)：145-147.

[11] 刘人怀. 在内边缘均布力矩作用下中心开孔圆底扁薄球壳的非线性稳定问题. 科学通报，1965，(3)：253-255.

[12] Панов Д Ю. О больших прогибах круглых мембран со слабым гофром. Прикл. Матем. Мех.，1941，5 (2)：303-318.

[13] Феодосьев В И. Упругие Элементы Точного Приборостроения. Москва：Оборонгиз，1949.

[14] 刘人怀. 波纹圆板的特征关系式. 力学学报，1978，(1)：47-52.

[15] Liu Renhuai. Large deflection of corrugated circular plate with a plane central region under the action of concentrated loads at the center. International Journal of Non-Linear Mechanics，1984，19 (5)：409-419.

[16] Liu Renhuai，Li Dong. On the non-linear bending and vibration of corrugated circular plates. Int. J. Non-Linear Mech.，1989，24 (3)：165-176.

[17] Андреева Л Е. Сильфоны (Расчёт и Проектирование). Москва：Машиностроение，1975.

[18] 钱伟长，吴明德. U 型波纹管的非线性特性摄动法计算. 应用数学和力学，1983，4 (5)：649-665.

[19] Axelrad E L. Theory of Flexible Shells. Amsterdem：North-Holland，1987.

[20] Liu Renhuai and Wang Zhiwei. Nonlinear deformation analysis of a U-shaped bellows with varying thickness. Archive of Applied Mechanics，2000，70：366-376.

[21] Euler L. Methodus inveniendi lineas curvas maximi minimive proprietate gandentes. Lausanne and Geneve，1744.

[22] von Kármán T, Tisen Huseshen. The buckling of spherical shells by external pressure. J. Aeron. Sei., 1939, (7): 43-50.

[23] Chien Weizang, Hu Heichang. On the snapping of a thin spherical cap. Int. Cong. Appl. Mech. Brussels, 1956.

[24] Liu Renhuai and Zhu Gaoqiu. Further study on large deflection of circular sandwich plates. Applied Mathematics and Mechanics: English Edition, 1989, 10 (12): 1099-1106.

[25] Liu Renhuai and Li Dong. A new approach to nonlinear vibration of orthotropic thin circular plates. Proceedings of the International Conference on Vibration Engineering, Shenyang: Northeastern University Press, 1998: 248-252.

[26] Панов Д Ю. Об устойчивости биметаллической оболчки при нагреве. Прикл. Матем. Мех., 1947, 11 (6): 603-610.

[27] Liu Renhuai. Non-linear thermal stability of bimetallic shallow shells of revolution. International Journal of Non-Linear Mechanics, 1983, 18 (5): 409-429.

[28] Королев В И. Тонкие двухслойные пластины и оболочки. Инж. Сборник, 1955, 22: 98-110.

[29] Radkowski P P. Buckling of thin single and multi-layer conical and cylindrical shells with rotationally symmetric stress. Proc. of 3rd U. S. National Congress of Appl. Mech., 1958: 443-450.

[30] Liu Renhuai, Li Dong, Nie Guohua, et al. Non-linear buckling of squarely-latticed shallow spherical shells. Int. J. Non-Linear Mech., 1991, 26 (5): 547-565.

[31] 刘人怀，肖潭．双层正交正放网格扁壳结构的非线性弯曲理论//中国力学学会．现代力学与科技进步，第3卷．北京：清华大学出版社，1997：1212-1215.

[32] Du Qinghua. General equations of sandwich plates under transverse loads and edgewise shears and compressions. Sciencia Sinica, 1955, 4 (4): 71-88.

[33] Reissner E. Finite deflection of sandwich plates. J. Aeron. Sci., 1948, 15 (7): 435-440; 1950, 17 (2): 125.

[34] Liu Renhuai. Nonlinear bending of circular sandwich plates. Applied Mathematics and Mechanics: English Edition, 1981, 2 (2): 189-208.

[35] 刘人怀，吴建成．夹层矩形板的非线性振动．中国科学，A辑，1991，(10)：1075-1086.

[36] Liu Renhuai and Cheng Zhenqiang. On the non-linear buckling of circular shallow spherical sandwich shells under the action of uniform edge moments. International Journal of Non-Linear Mechanics, 1995, 30 (1): 33-43.

[37] Liu Renhuai and Zhu Jinfu. Nonlinear theory of sandwich shells, Part Ⅰ-Exact kinematics of moderately thick shells. Applied Mechanics and Engineering, 1997, 2 (2): 213-240.

[38] Liu Renhuai and Zhu Jinfu. Nonlinear theory of sandwich shells, Part Ⅱ-Approximate theories. Applied Mechanics and Engineering, 1997, 2 (2): 241-269.

[39] 刘人怀，王志伟．复合材料面层夹层板中转动一致有效理论．上海力学，1996，17 (3)：222-228.

[40] 刘人怀．考虑横向剪切的对称层合圆柱正交异性扁球壳的非线性稳定问题．中国科学，A辑，1991，(7)：742-751.

[41] Liu Renhuai. Non-linear buckling of symmetrically laminated, cylindrically orthotropic, shallow, conical shells considering shear. International Journal of Non-Linear Mechanics, 1996, 31 (1): 89-99.

[42] Liu Renhuai and Wang Fan. Nonlinear dynamic buckling of symmetrically laminated cylindrically orthotropic shallow spherical shells. Archive of Applied Mechanics, 1998, 68 (6): 375-384.

[43] Liu Renhuai and He Linghui. Nonlinear bending of simply supported symmetric laminated cross-ply rectangular plates. Applied Mathematics and Mechanics: English Edition, 1990, 11 (9): 801-807.

[44] Liu Renhuai, Xu Jiachu and Zhai Shangzhong. Large-deflection bending of symmetrically laminated rectilinearly orthotropic elliptical plates including transverse shear. Archive of Applied Mechanics, 1997, 67 (7): 507-520.

[45] Liu Renhuai and He Linghui. A simple theory for non-linear bending of laminated composite rectangular plates including higher-order effects. International Journal of Non-Linear Mechanics, 1991, 26 (5): 537-545.

[46] W·乌班诺夫斯基. 热交换器管板的设计. 力学学报, 1960, 4 (2): 94-111.

[47] 刘人怀. 固定式厚管板的弯曲问题. 力学学报, 1982, (2): 166-179.

[48] 刘人怀, 陈山林. 铂重整装置反应器椭球封头中心开孔接管的强度问题. 科技专刊 (兰州大学), 1973, (1): 14-28.

[49] 刘人怀. 厚壁球壳的弯曲理论及其在高压容器上的应用. 化工炼油机械通讯, 1980, (3): 1-10.

[50] 刘人怀. 加氢反应器顶部厚壁壳体的应力分析. 化工炼油机械通讯, 1975, (6): 40-59.

[51] 刘人怀. 双层套箍式厚壁压力容器环沟部位的应力状态. 兰州大学学报, 1977, (4): 9-25.

[52] 刘人怀, 王凯. 500 万吨/年常减压装置减压塔下端部分壳体的应力分析. 压力容器, 1975, (2): 1-16.

[53] 刘人怀. 在轴向压力与均匀外压力共同作用下薄壁截头圆锥形壳的稳定性. 兰州大学学报 (自然科学版), 1975, (2): 16-25.

[54] 何陵辉, 刘人怀. 一种考虑层间位移和横向剪应力连续条件的层合板理论. 固体力学学报, 1994, 15 (4): 319-326.

[55] 刘人怀. 板壳力学. 北京: 机械工业出版社, 1990.

[56] Liu Renhuai. Study on Nonlinear Mechanics of Plates and Shells. New York: Science Press and Jinan University Press, 1998.

板壳非线性力学研究的最新进展①

一、引言

由于近代科学技术的飞跃发展，板壳非线性力学已成为当前固体力学研究的一个最活跃的领域，倍受人们关注。目前研究的中心课题是板壳的几何非线性问题。

板壳非线性力学的奠基者是 von Kármán。他于 1910 年提出了著名的薄板大挠度方程[1]。在 1939 年，和钱学森[2]一起用非线性理论讨论了扁球壳的屈曲问题。这以后，板壳非线性理论得到不断的发展。

20 世纪 60 年代以来，由于航空、航天等工程中结构元件设计的需要，高强、轻质的复合材料获得广泛的应用。复合材料板壳的非线性问题的分析由此成为研究者关心的一个热点。本文将以较多的篇幅评述这一领域的发展。

板壳方面的文献可以说是浩如烟海。据统计，近年来仅壳体的文章就以每年数百篇甚至上千篇的速度递增，内容涉及静态和动态的各个方面。要对其作全面的评述是困难的。本文就近十几年有关板壳非线性问题的研究进展作一个简要的介绍。

二、板的非线性理论与应用

1. 弯曲问题

近年来，人们对复合材料板的研究兴趣越来越大。进行分析时运用较广的是 von Kármán 非线性理论，所使用的方法有摄动法、广义 Fourier 级数法、伽辽金法和动态松弛法（DR）等。Lekhnitskii[3]首先给出了正交各向异性板的 von Kármán 型大挠度方程。Chia[4]考虑了正交各向异性矩形板受均布载荷作用下的大挠度问题，并用钱伟长的摄动法进行求解。其后，Prabhakara 和 Chia[5]进一步研究了任意横向载荷作用下正交各向异性板的大挠度问题。文献［6］也同时考虑了受边界压力和横向载荷联合作用的这种板的弯曲，分析中利用双重 Fourier 级数将控制方程转化为非线性代数方程组，最后采用 Newton-Raphson 法求得级数系数。对正交各向异性三角板、椭圆板和圆板的研究参见文献［7］～［9］。

近年，Jiang 和 Chia[10]考虑了两边自由另两边具有非均匀转动弯曲的正交各向异性矩形板的非线性弯曲，Chia[11,12]研究了具有边界转动的弹性约束和固定边界条件的非对称角交层合矩形板的大挠度问题，采用的都是双重 Fourier 级数法。

Gandhi 等[13]通过正交配点法求解了正交各向异性圆板的非线性轴对称弯曲问题。计算的结果与各向同性的情况相吻合。

此外，Turvey[14,15]用动态松弛法分析了正交细长层合板和复合材料圆板的非线性弯曲，考虑了伸长和弯曲的耦合效应。

---

① 本文原载《上海力学》，1989，10（3）：19-32. 作者：刘人怀，聂国华.

Gorji[16]在分析静载作用下的对称复合材料板时，将横向位移的非线性项当作一组附加的横向载荷作用在物体上，因而把原 Kármán 型板归结为一个等效的小位移型板来求解。文中计及了横向剪切的影响。Reddy 和 Chao[17]，Huang 和 Dong[18]以及 Gordaninejad[19]还对由双模量材料组成的复合材料板的非线性问题进行了分析。采用的是有限元法、DR 法等。Mottram[20]也用有限元法计算了多层矩形板的大挠度问题。文献［21］用五次 B-样条函数作为试函数利用加权残数法考虑了各向异性层合板的非线性问题。文献［22］用加权残数法探讨了对称复合材料矩形层合板的非线性弯曲。刘人怀和何陵辉[23]用 Fourier 级数法对四边铰支对称正交层合矩形板在任意横向载荷作用下的非线性弯曲问题给予了分析。

众所周知，大挠度复合材料板的 von Kármán 方程是非线性和耦合的，这给求解带来很大困难。为此，Berger[24]在分析各向同性板的非线性静态特性时提出了一种新方法。该方法基于这样一个假设，即薄膜应变的二阶不变量对板的应变能的贡献可忽略而不影响结果的精度。据此假设，从 Lagrange 变分方程导出的 Euler-Lagrange 方程较 Kármán 方程更为简单，并且是解耦的。Berger 本人和其他学者的工作，使这种方法获得了广泛的应用。Iwinski 和 Nowinski[25]首先运用这一假设分析了正交各向异性矩形板的大挠度问题。正交各向异性圆板的工作见文献［9］。然而，由于难以找到这种方法的理性力学基础，文献［26］～［29］对其实用性提出疑问。Nowinski 和 Ohnabe[26]指出，对于具有面内运动约束的边界，Berger 方法是适用的，但对自由可动边界，方法的精度是有疑问的，甚至将导致荒谬的结果。值得指出的是，因为 Berger 方法在处理许多问题时结果令人满意，因此仍不失为一种有效的方法。我们将在以后介绍其在动力分析（振动）方面的应用。

国内学者除上述有关复合材料板的研究[18,21-23]外，对一般板（各向同性）的大挠度分析作了许多有益的工作。早期的工作有钱伟长[30,31]、胡海昌[32]及钱伟长和叶开沅[33]对圆板和矩形板的大挠度分析，采用的是摄动法。叶开沅和刘人怀[34-36]在处理圆底扁球壳的非线性稳定问题时提出了修正迭代法。这种方法吸收了钱伟长的摄动法以及逐次逼近法的优点，计算程序简单明确，计算量小且结果精确，是解决板壳非线性问题的一种比较有效的方法。

近年，叶开沅、郑晓静等[37,38]在研究集中载荷作用下圆板的大挠度问题时，通过先将无量纲的 Kármán 方程用格林函数法化为积分方程，引入解空间，选取迭代格式后，最后分析得到该问题的精确解，并给出了解的收敛性证明。对受集中和均布载荷联合作用的圆板，郑晓静和周又和[39]找到了大挠度问题的精确解。他们还获得了弹性地基圆板受集中载荷作用的大挠度问题的精确解[40]。叶开沅等[41,42]还用修正迭代法和摄动法分别考虑了变厚度圆板和矩形板的大挠度问题。类似的问题还有文献［43］和［44］。对摄动参数的选取和摄动法、修正迭代法收敛性分析参见文献［45］～［50］。此外，李增福和冯正农[51]用边界元法来处理薄板大挠度弯曲问题，对正三角形、半圆形和半椭圆形板也给予了数值计算。朱正佑和程昌钧[52]在分析开孔薄板大挠度问题时，求得使 Kármán 方程中的应力函数成为单值的一个充要条件，并证明了位移单值性要求等价某些约束方程。

除了单层板外，夹层板的研究也取得了很大进展。夹层板是由一块厚夹心和两块薄板组成的。由于这种板具有较高的刚度和较轻的质量，因而成为航空、宇航和船舶工业中的重要结构元件。Reissner[53]首先建立了具有软夹心和极薄表板的夹层矩形板的大挠度理论。Kan 和 Huang[54]对夹层矩形板的大挠度进行了求解。

中国科学院力学研究所固体力学研究室板壳组[55]对夹层圆板的非线性弯曲问题用摄动法进行了分析。刘人怀[56,57]导出了均布载荷和边缘力矩作用下夹层圆板的非线性弯曲方程。随后，刘人怀和施云方[58]应用幂级数方法得到了承受均布载荷的夹层圆板的精确解。最近，刘人怀[59]用修正迭代法和 Heaviside 函数分析了受线布载荷的夹层圆板的非线性弯曲问题。均布载荷作用下的夹层环板和夹层矩形板的大挠度分析见文献［60］和［61］。上述工作都是将夹层板的表板视为薄膜，即忽略了表板的抗弯刚度。计及这一因素的影响，刘人怀和朱高秋[62]提出一种修正幂级数法，用以对夹层圆板作进一步的研究。

另外，Rajagopal 等[63]对多层的夹层板进行了分析，使用的是二次矩形有限元法。

2. 稳定性问题

自从 Euler[64]第一次提出并解决压杆横向屈曲的弹性稳定问题以来，在长达 200 年的时间里，人们陆续对杆、板、壳及其组合结构的稳定问题进行了研究，给出了各种载荷作用下的临界载荷，并提出了各种稳定理论。

对复合材料层合板的屈曲分析，依据铺层顺序和取向有三种相应的几何非线性理论：①解耦的正交各向异性板理论；②解耦的各向异性板理论；③耦合理论（包括弯曲和伸长的耦合）。对稳定问题进行分析较早的是 Whitney 和 Leissa[65]。他们得到了具有简支边界的反对称正交和角交矩形板的封闭形式的精确解。同年，Chamis[66]用伽辽金法分析了正交各向异性矩形层合板的屈曲问题。这里，板的边界不平行于正交各向异性的轴向。Harris[67]对受双轴向载荷具有简支边界的复合材料板的屈曲进行了研究。文中计及了弯曲和伸长的耦合效应。对正交各向异性环板的屈曲和后屈曲行为的讨论见文献［68］和［69］。

近年，Bhattacharya[70]分析了正交层合板的后屈曲，板具有弹性约束边界和初始曲率。求解所用的是布勒诺夫—伽辽金方法。Stein[71]在分析正交各向异性板时找到了确定后屈曲的两个参数。Singh 等[72]还用有限元法研究了厚角交矩形板的非线性稳定性。详细讨论了纤维取向、材料特性、分层情况和边界条件对屈曲载荷的影响。

Hui[73]对计及弯曲和伸长耦合的层合板相对缺陷的敏感性进行了探讨。屈曲和初始后屈曲问题的分析基于 Koiter 稳定理论。关于 Koiter 稳定理论的详细论述可参阅文献［74］。

Shilkrut[75]用小振动方法对正交各向异性圆板的非线性稳定问题给予了分析。有关环板的后屈曲问题的处理可以从 Reddy 和 Alwar[76]以及 Dumir[77]的工作中了解到。

3. 动态分析

几何非线性弹性板的动力问题的理论和应用近 30 年来得到了不断发展。Herrmann[78]分析高幅对弹性板的弯曲运动影响时获得了动力 Kármán 方程。随之，Medwadowski[79]将这一动力方程推广到计及横向剪切和转动惯量影响的正文各向异性板的分析之中。Rostovtsev[80]后来也做了这方面的工作。Nowinski[81]得到了正交各向异性圆板的 Kármán 非线性方程。对各向异性板的情况，Ebcioglu[82]也作了研究。另一方面，Nash 和 Modeer[83]、Wah[84]以及 Wu 和 Vinson[85,86]根据 Berger 假设分析了板的非线性动力问题。

近年来的研究工作有，Sathyamoorthy 和 Chia[87]利用伽辽金法和 Runge-Kutta 数值方法求出了简支和夹支条件下各向异性矩形板非线性振动问题的解。文章考虑了横向剪切和转动惯量的影响。Niyogi 和 Meyers[88]研究了正交各向异性矩形板的摄动解。摄动参数是选用板的中心挠度。文献［89］采用伽辽金法和摄动法分析了矩形层合板的非线性振动。对层合板的振动问题，Reddy 和 Chao[90]用有限元法也进行了分析。Chia[91]还对具有非均匀边界约束的各向异性矩形板的非线性振动用伽辽金法作了研究，给出了高弹复合材料对于各种取向角、纵横尺寸比和边界条件的幅频响应的数值结果。Chia[92]还就两邻边夹支、另两邻边简支的非对称边界条件分析了正交各向异性矩形板的非线性振动。具有集中质量的夹支边界条件的正交各向异性方板，Banerjee[93]首次作了探讨。层合矩形板的随机响应分析见文献［94］和［95］。

对于斜交板的非线性振动问题，Chia 和 Sathyamoorthy[96]用 Hamilton 原理导出了 Kármán 方程。同年，运用与文献［87］相同的方法，他们进一步考虑了横向剪切和转动惯量的影响[97,98]。计算结果表明，横向剪切变形及各种复合材料几何特性和斜交角对高幅振动具有重要影响。Sathyamoorthy[99]以及 Sivakumaran 和 Chia[100,101]也做了类似的工作。

对圆板、椭圆板和环板的分析，近年来也受到重视。1980 年，Nath 和Alwar[102]考虑了正交各向异性圆板承受步函数、正弦函数和 N 型脉冲三种形式的载荷作用而引起的动力响应。取 Chebyshev 级数展开的前五项，就得到了问题的较精确结果，这是运用幂级数和三角级数展开所不及的。同年，Rao 和 Raju[103]用有限元法分析了具有转动约束的正交各向异性圆板的非线性振动。次年，Sathyamoorthy[104]利用 Berger 假设分析了横向剪切和转动惯量对振动的影响，计算较用 Kármán 理论更简单且结果更好。基于 Kármán 和 Berger 两种理论，他还对正交各向异性椭圆板的非线性振动进行了分析[105,106]。文献［107］还用 Berger 理论研究了正交各向异性厚圆板的问题。通过计算得知，当半径和厚度之比大于 10 时，横向剪切和转动惯量的影响很小。用 Kármán 理论，Biswas 和 Kapoor[108]. Ruei 等[109]分别考虑了正交各向异性圆板的非线性自由振动和受迫振动。同时，Dumir[110]等对正交各向异性薄环板作了瞬态分析，并考虑了具有刚性集中质量条件下的瞬态响应和振动[111,112]。Dumir 等[113]还就置于 winkler，Pasternak 和非线性 winkler 三种地基上的圆板振动进行了研究。具有弹性约束的薄圆板的瞬态响应和振动参见文献［114］和［115］。此外，Sathyamoorthy 等[116,117]在分析正交各向异性椭圆板的单型解的基础上分析了多型解。两种解符合得很好。圆板的多型解分析见文献［118］。

综观各种形状的板的几何非线性静态和动态分析，在用 Kármán 或 Berger 理论时，一般都要假定有关横向位移的一个合适的形状函数，然后运用上述各种方法使其近似满足控制方程。对于许多非线性问题来说，不仅难以得到封闭解，且近似的精度也难以估计。对此，Sathyamoorthy[119]提出了一种自生函数的新方法来分析复合材料板的非线性动力问题。该方法的特点是，零阶自生函数是四阶多项式的形式，一阶函数的多项式为八阶，二阶函数的多项式为十二阶，依此类推。自生函数的阶数和多项式形式依板的边界情况而定。高阶自生函数适合于更复杂的边界条件。这种方法在有关梁的动力问题中已经成功地得到应用[120]。不过，就目前的进展情况看，该方法并未引起重视。

三、壳体的非线性理论与应用

1. 弯曲问题

在精密仪器的弹性元件中，有一种特殊的壳体——波纹圆板，占有极其突出的地位，在仪表结构中起核心作用。因为它在作用力相当小的情况下，就能给出明显而容易被测量的挠度，因而常被作为敏感元件。

波纹圆板的非线性弯曲问题的研究近年来取得了很大进展。刘人怀[121]讨论了在均布载荷作用下的全波纹圆板的非线性弯曲问题，得到了与实验结果相吻合的特征曲线。对具有刚性中心的波纹环形板，刘人怀[122-124]研究了集中、均布以及集中和均布联合作用的三种载荷影响下该类波纹板的非线性弯曲问题。对具有光滑中心的波纹圆板的大挠度问题，就上述三种形式的载荷，刘人怀[125-128]也分别得到了精确的特征关系式。具有光滑中心和平面边缘的波纹圆板承受均布和集中载荷作用的非线性弯曲分析可参见文献［129］和［130］。

上述一系列问题的研究，采用的是正交各向异性圆板大挠度理论，运用的是修正迭代法。

此外，最近（1987年），Stumpf 和 Makowski[131]提出了由强非线性材料组成的大应变、大伸长变形的非线性壳理论。文中采用了修正的 Kirchhoff 假设，计及了厚度变化和壳中面位置的偏移。早其两年，Taber[132]也使用了 Kirchhoff 修正假定，考虑厚度改变但忽略横向剪切变形，研究了轴对称大应变弹性旋转壳的一组几何和本构关系。壳材料大变形问题目前已开始引起人们的兴趣，有关文献可参阅文献［133］～［137］。

对球壳的非线性弯曲问题，Cagan 和 Taber[138]用 Newton-Raphson 方法进行了分析。文献［139］和［140］还利用有限元法考虑了轴对称和局部缺陷对薄球壳应力的影响。此外，黄黔[141]用摄动法分析了轴对称壳的非线性问题。Szwabowicz[142]对几阿非线性薄壳作了变分分析。

具有矩形开口的圆柱壳的非线性变形问题，Fridman[143]用渐近法进行了研究。Kryukov[144]考虑了正交各向异性非圆的柱壳，使用 Vlabor-Kantorovich 方法对其非线性变形进行了分析。

另外，文献［145］讨论了受面内载荷作用的简支层合曲板的非线性问题。结果表明，具有非对称正交和角交层合曲板分别在受压和受剪时都会有横向弯曲产生。

2. 稳定性分析

在壳体的稳定理论中，最受人注意的经典问题有两个，即均布外压作用下的扁球壳和轴压作用下的圆柱壳的稳定问题。文献［2］的工作开拓了用非线性理论研究壳体稳定问题的正确途径。

近年来，国内学者在扁薄壳非线性稳定问题方面做了不少工作。1980 年，叶开沅和顾淑贤[146]在分析均布载荷作用下圆底薄球壳的非线性稳定性时，提出了解析-电算法，使得计算精度大为提高。叶志明[147]对变厚度圆底扁薄球壳的非线性稳定问题进行了分析。顾淑贤[148]用三次 B 样条函数和逐步迭代法求解了受线布荷载作用的扁球壳的稳定问题。此外，叶开沅和宋卫平[149]还讨论了均布压力作用下扁圆锥壳的非线性稳定问题。对中心开孔扁球壳的非线性稳定问题，刘人怀[36,150]率先研究。最近，刘人怀和他的学生对具有刚性中心的开顶扁球壳的非线性稳定问题进行了系统的研究。使用修正迭代法，获得了临界载荷的计算公式。集中载荷、均布载荷、轴对称分布的载荷以及集中和均布联合作用的载荷的四种承载方式，文献 [151]～[155] 分别给予了详细的分析。

以上涉及的都是单层扁壳。对双层扁壳的基本的热稳定性，刘人怀也进行了研究。他利用 Radkowski 等的选择多层壳体坐标参考面的方法，将双层旋转扁壳的基本方程表示成单层壳体理论的形式，从而建立了在均匀温度作用下该类扁壳的大挠度弯曲理论。采用这一理论，他运用修正迭代法分析了双层金属扁球壳的非线性热稳定问题[156]。利用同样的方法，还分析了双层金属截头扁锥壳的热稳定性[157]。有关这类问题的论述还可参阅文献 [158]。

目前，国外学者对各类壳的非线性稳定问题也进行了研究。Cagan 和 Taber[138]用 Newton-Raphson 方法分析了环压作用下的球壳的非线性屈曲问题。Subbiah[159]采用有限元法对含几何缺陷的加筋旋转壳的非线性稳定问题进行了研究。具体考虑了加筋圆柱壳。

对复合材料壳体，In 和 Chia[160]考虑了非对称正交圆柱壳的后屈曲。使用的是多振型方法。Zhang 和 Matthews[161]对复合材料圆柱形曲板的后屈曲进行了分析。求解采用了 von Kármán 型控制方程，运用基于修正牛顿法的一种迭代过程进行计算，并考虑了初始几何缺陷的影响。文献 [162] 也考虑了这种曲板的非线性屈曲。刘人怀[163]在对圆柱形正交对称层合扁球壳的非线性稳定问题进行分析时，用修正迭代法首次得到了这一领域的一个解析解。

3. 振动问题

壳体的非线性振动问题非常复杂，近年来的文章不是很多。下面简单介绍一下这方面的目前研究进展。

建立在基于 von Kármán 动力方程的模态关系基础之上，Yasuda 和 Kushidz[164]研究了扁球壳在谐和激励下的非线性受迫振动。Mukherjee 和 Chakraborty[165]运用有限变形弹性理论得到了薄球壳高幅自由和受迫振动的精确解。Sinharay 和 Banerjee[166]对变厚度的扁薄球壳和圆柱壳的高幅自由振动给予了分析。无厚度变化的同一问题可参见文献 [167]。此外，Chrzeszczyk[168]还探讨了非均匀弹性薄壳振动的非线性方程的一般解。文中计及了初挠度、初应力和温度的影响。

刘人怀和李东[169]使用伽辽金法研究了波纹圆板的非线性自由振动问题，获得了固有周期和振幅的关系式，并通过数值计算讨论了光滑中心以及波纹深度对波纹板非线性振动周期的影响。

对复合材料层合壳体的非线性振动，Wu 和 Yang[170] 用高阶曲壳有限元进行了分析。Bhattacharya[171] 针对 Pagternak 型弹性地基上的圆柱形曲板的非线性振动，应用伽辽金法和摄动法进行了求解，并考虑了初始缺陷的影响。同时，Hui[172] 应用 Donnell 方程并采用伽辽金法分析了初始缺陷对各向同性圆柱曲板非线性振动的影响。文献 [173] 还考虑了 Winkler-Pasternak 型地基上层合曲板的问题。采用动力 Marguerre 型控制方程，文献 [174] 分析了具有转动的弹性约束的层合曲板的高幅振动。对具有与文献 [92] 所涉及的相同边界条件的层合曲板，可参阅文献 [175] 的分析。另外，Nath 和 Jain[176,177] 还讨论了 Winkler 和 Pasternak 型弹性地基对正交各向异性扁球壳非线性动态响应的影响。

四、结束语

20 世纪 60 年代以来，复合材料的广泛应用，促进了其常用结构——复合材料板壳力学的发展。人们从研究较简单的正交各向异性单层复合材料到多层、夹层和加筋板壳；从分析经典的边界条件（简支、夹支和自由）到复杂的弹性边界约束（转动约束、弹性地基等）；从忽略到计及横向剪切变形、转动惯量、几何初始缺陷及开口的影响，把复合材料板壳力学的分析研究推向高潮。随着复合材料的进一步发展和应用，将需要更精确的力学理论和方法，用来正确反映该类材料的力学性能。

另外，人们对一般（各向同性）板壳非线性问题的分析也在不断深入。尤其是具有非对称边界条件的单层、双层和多层板壳在复杂载荷作用下的非线性问题将是今后的一个热门课题。

到目前为止，板壳的非线性问题的研究主要基于小应变条件。对大应变材料的分析，目前还处于初始阶段。Taber[132] 和 Stumpf[131] 最近已提出了这方面的理论。这一领域里的问题，如橡胶和一些软生物组织（心脏、红血细胞等）所发生的大应变变形，相信会得到国内外研究者的重视。

参 考 文 献

[1] Th. von Kármán. The festigkeit problem in maschineuban. Enzyklopadie der Mathematischen Wissenschaften，1910，4（4）：349.

[2] Th. von Kármán and Tsien Huse-Shen. The buckling of spherical shells by external pressure. J. Aero. Sci.，1939，7：43.

[3] Lekhnitskii S G. Theory of bending of anisotropic plates. Anisotropic plates，New York：Gordon and Breach Science Publishers，1968，273.

[4] Chia C Y. Large deflection of rectangular orthotropic plates. J. Eng. Mech. Dir. Proc. ASCE93，EM5，1972，1285.

[5] Prabhakara M K and Chia C Y. Large deflection of arbitrarity loaded plates. J. Eng. Mech. Div. Pro. ASCE100，EM6，1974，1282.

[6] Chia C Y and Prabhakara M E. Nonlinear analysis of orthotropic plates. J. Mech. Eng. Sci.，1975，17：133.

[7] Datta，S. Large deflection of a triangular orthotropic plate on elastic foundation. Def. Sci. J.，

1975, 25: 115.

[8] Prabhakara, M K and Chia, C Y. Bending of elliptical orthotropic plates with large deflection. Acta Mech., 1975, 21: 29.

[9] Banerjee, B. Note on the large deflection of an orthotropic circular plate with clamped edge under symmetrical load. J. Indian Inst. Sci., 1976, 58: 175.

[10] Jiang, C and Chia, C Y. Nonlinear bending of rectangular orthotropic plate with two free edges and the other edges having nonuniform rotational flexibility. Int. J. Solids Struct., 1984, 20: 1009.

[11] Chia, C Y. Nonlinear bending of unsymmetric angle-ply plates with edges elastically restrained against rotation. Acta Mech., 1984, 53: 201.

[12] Chia, C Y. Large deflection of unsymmetric laminates with mixed boundary conditions. Int. J. Non-linear Mech., 1985, 20: 273.

[13] Gandhi, M L, Dumir, P C. and Nath, Y. Nonlinear axisymmetric static analysis of orthotropic thin circular plates with elastically restrained edge. Comput. Struct., 1985, 20: 841.

[14] Turvey, G J. Large deflection cylindrical bending analysis of crossply laminated strips. J. Mech. Eng. Sci., 1981, 23: 21.

[15] Turvey, G J. Axisymmetric elastic large deflection analysis of composite circular plates. Fibre Sci. Technol., 1982, 16: 191.

[16] Gorji, M. On large deflection of symmetric composite plates under static loading. Proc. Inst. Mech. Engrs. 200, No. Cl, 1986, 13.

[17] Reddy, J N. and Chao, W C. Nonlinear bending of bimodular material plate. Int. J. Solids struct., 1983, 19: 229.

[18] Huang, X. and Dong, W. Effects of transverse shear on the nonlinear bending of rectangular plates laminated of bimoclular composite materials. Proc. Int: Symp. Compos. Mat. Struct., Beijing: Technomic Publish Co., pa., 1986, 239.

[19] Gordaninejad, F. Large deflection of anisotropic bimodular composite material plates. Proc. 4th Int. Conf. Compos. Struct., London: Elsevier Applied Science, 1987, Vol. 1, 1.152.

[20] Mottram, J T. A simple non-linear analysis of multi-layered rectangular plates. Comput. Struct., 1987, 26: 597.

[21] Xu, G and Shen, D. Geometrical nonlinear analysis of laminated anisotropic plates by weighted-residual method. Proc. Int. Symp. Compos. Mat. Struct, Beijing: Technomic Publishing Co. 1986, 416.

[22] Zhou, C Q. Nonlinear bendings of symmetrically layered anisotropic rectangular plates. Appl. Math. Mech., 1988, 9 (3): 295.

[23] 刘人怀，何陵辉. 四边铰支对称正交铺设层合矩形板的非线性弯曲. 第六届华东固体力学学术讨论会论文，镇江，1989.

[24] Berger, H M. A new approach to the analysis of large deflection of plates. J. Appl. Mech., 1955, 22: 464.

[25] Iwinski, T and Nowinski, J. Problems of large deflection of orthotropic plates. Bulletin de l' Academic polonaise des Sciences, 1957, 5: 335.

[26] Nowinski, J L and Ohnabe, N. On certain inconsistencies in Berger equations for the large deflections of elastic plates. Inter. J. Mech. Sci., 1972, 14: 165.

[27] Vendhan, C P. A study of the Berger eguations applied to the nonlinear viration of elastic plates. Inter. J. Mech. Sci.，1975，17：451.

[28] Prathap，G and Varadan，T K. On the non-linear vibration of rectangular plates. J. Sound and Vibration，1978，56：521.

[29] Prathap，G. On the Berger approximation：A Critical re-examination. J. Sound and Vibration，1979，66：149.

[30] Chien Wei-Zang. Large deflection of a circular clamped plate under uniform pressure. Acta Physica Sinnica，1947，7（2）：102.

[31] Chien Wei-Zang. Asymptotic behavior of a thin clamped circular plate under uniform normal pressure at very large deflection. Tsinghua University Science Report，1948，5（1）：1.

[32] 胡海昌. 在均布及中心集中载荷下圆板的大挠度问题. 物理学报，1954，10（4）：383.

[33] Chien Wei-Zang and Yeh Kai-Yuan. On the large deflection of rectangular plate. Proceedings of Ninth inter. congress on Appl. Mech.，Brussels，Belgium，1957，6：403.

[34] 叶开沅，刘人怀，平庆元，等. 在对称线布载荷下的圆底扁薄球壳的非线性稳定问题. 科学通报，1965，(2)：142；兰州大学学报，1965，(2)：10.

[35] 叶开沅，刘人怀，张传智，等. 圆底扁薄球壳在边缘力矩作用下的非线性稳定问题. 科学通报，l965，(2)：145.

[36] 刘人怀. 在内边缘均布力矩作用下中心开孔圆底扁球壳的非线性稳定问题. 科学通报，1965，(3)：253.

[37] 叶开沅，郑晓静，周又和. An analytical formula of exact solution to Kármán's equations of circular plate under a concentrated load. Proc. ICNM（Shanghai），Beijing：Science Press，1985，386.

[38] 叶开沅，郑晓静，王新志. On some properties and calculation of the exact solution to Kármán's eguations of circular plate under a concentrated load ibid，1985，379.

[39] 郑晓静，周又和. 复合载荷下圆薄板大挠度问题的精确解. 中国科学，A辑，1986，（10）：1045.

[40] 郑晓静，周又和. 集中载荷下弹性地基圆板大挠度问题的精确解. 力学学报，1988，（2）：161-172.

[41] 叶开沅，王新志. 用修正迭代法解变厚度圆板的大挠度问题. 国际非线性力学会议讨论文集，上海，1985，398.

[42] 叶开沅，房居贤，王云. 用摄动法求解四边固定夹紧受均布载荷的矩形板的大挠度问题. 《应用数学进展》，荷兰 Nartinus Nijhoff Publishers 出版，1987，245.

[43] 叶开沅，王新志，王林详. 锥形圆薄板的非线性问题. 甘肃工业大学学报，1983，(2)：1.

[44] 叶开沅，叶志明. 对“变厚度圆薄板在均布载荷下大挠度问题”解法的讨论. 应用数学和力学，1985，6（3）：285.

[45] 陈山林，光积昌. 圆薄板大挠度问题的摄动参数. 应用数学和力学，1981，2（1）：131.

[46] 周焕文. 圆板非线性理论的一种摄动解. 应用数学和力学，1981，2（5）：475.

[47] 陈山林. 圆板大挠度的钱伟长解及其渐近特性. 应用数学和力学，1982，3（4）：513.

[48] 叶开沅，周又和. 关于钱氏摄动法的高阶解的计算机求解和收敛性的研究. 应用数学和力学，1986，7（4）：285.

[49] 周又和，郑晓静. Vincent 解的收敛上界. 力学学报，1988，(5)：473.

[50] 周又和. 关于修正迭代解与钱氏摄动解的关系. 应用数学和力学，1989，(1)：59.

[51] 李增福，冯正农. 边界元法在薄板大挠度弯曲问题中的应用. 工程力学，1987，4 (3)：1.
[52] 朱正佑，程昌钧. 关于开孔薄板大挠度问题的一般数学理论. 力学学报，1986，18 (2)：123.
[53] Reissner，E. Finite deflections of sandwich plates. J. Aeron. Sci.，1948，15 (7)：435；1950，17 (2)：125.
[54] Kan，H and Huang，J. Large deflection of rectangular sandwich plates. AIAA J.，1967，5 (9)：1706.
[55] 中国科学院北京力学研究所固体力学研究室板壳组. 夹层板壳的弯曲、稳定和振动. 北京：科学出版社，1977，42.
[56] 刘人怀. 在边缘力矩作用下夹层圆板的非线性轴对称弯曲问题. 中国科学技术大学学报，1980，10 (2)：56.
[57] 刘人怀. 夹层圆板的非线性弯曲. 应用数学和力学，1981，2 (2)：173.
[58] 刘人怀，施云方. 夹层圆板大挠度问题的精确解. 应用数学和力学，1982，3 (1)：11.
[59] Liu Ren-huai. Nonlinear bending of circular sandwich plates under the action of axisymmetric uniformly distributed line loads. Progress in Applied Mechanics，Dordrecht：Martinus Nijhoff Publishers，1987，293.
[60] 刘人怀，朱金福，张小果. 夹层环形板的非线性弯曲. 暨南大学学报，1997，18 (1)：1.
[61] 刘人怀，成振强. 简支夹层矩形板的非线性弯曲. 应用数学和力学，1993，14 (3)：203.
[62] 刘人怀，朱高秋. 夹层圆板大挠度问题的进一步研究. 应用数学和力学，1989，10 (12)：1041.
[63] Rajagopal，S V，Singh G，Y V K. Sadasiva，Rao. Large deflection and nonlinear vibration of multilayered sandwich plates. AIAA J.，1987，25 (1)：130.
[64] Euler，L. Methodus inveniendi Lineas curvas maximi minimive proprietate gaudentes. Lausanne and Geneva，1744.
[65] Whitney，J M and Leissa，A W. Analysis of heterogeneous anisotropic plates. Trans. ASME，J. App. Mech.，1969，36：261.
[66] Chamis. C C. Buckling of anisotropic composite plates. Proc. ASCE，J. Struc. Div.，1969，95：2119.
[67] Harris，G. Z. The buckling and post-buckling behaviour of composite plates under biaxial loading. Int. J. Mech. Sci.，1975，17：187.
[68] Vthgenannt，E B and Brand，R S. Postbuckling of orthotropic annular plates. ASME J. Appl. Mech.，1973，40：559.
[69] Vthgenannt，E B and Brand，R S. Buckling of orthotropic annular plates. AIAA J.，1970，8 (11)：2102.
[70] Bhattacharya，A P. Note on the postbuckling analysis of cross-ply laminated plates with elastically restrained edges and initial curvatures. J. Struct. Mech.，1982-83，10：359.
[71] Stein，M. Postbuckling of orthotropic composite plates loaded in compression. AIAA J.，1983，21：1729.
[72] Singh，G，Y V K. Sadasiva，Rao. Stability of thick angle-ply composite plates. Comput. Struct. 1988，29 (2)：317.
[73] Hui，D. Imperfection sensitivity of axially compressed laminated flat plates due to bending-stretching coupling. Int. J. Solid Struct. 1986，22：13.
[74] 黄宝宗，任文敏. Koiter 稳定理论及其应用. 力学进展，1987，17 (1)：30.

[75] Shilkrut, D. Stability and vibration of geometrically nonlinear cylindrically orthotropic circular plates. ASME J. App. Mech., 1984, 51: 354.

[76] Reddy, B S and Alwar, R S. Postbuckling analysis of orthotropic annular plates. Acta Mech., 1981, 39: 289.

[77] Dumir, P C. Axisymmetric postbuckling of orthotropic tapered thick annular plates. ASME J. Appl. Mech., 1985, 52: 725.

[78] Herrmann, G. Influence of large amplitudes on flexural motions of elastic plates. National Advisory Committee for Aeronautics, Technical Note 3578, Washington, D. C., 1955.

[79] Medwadowski, S J. A refined theory of elastic, orthotropic plates. ASME J. Appl. Mech., 1958, 25: 437.

[80] Rostovtsev, S G. Anisotropic plates. 2nd Edition, Gordon and Breach Science Publishers, 1968.

[81] Nowinski, J L. Nonlinear vibrations of elastic circular plates exhibiting rectilinear orthotropy. Z. Angew. Math. Phys., 1963, 14: 113.

[82] Ebcioglu, I K. A large deflection theory of anisotropic plates: Ingenieur-Archiv., 1964, 33: 396.

[83] Nash, W A and Modeer, J R. Certain approximate analysis of the nonlinear behavior of plates and shells. Proc. Symp. on the Theory of Thin Elastic Shells, 1950, 331.

[84] Wah, T. The normal, modes of vibration of certain nonlinear continuous systems. J. Appl. Mech., 1964, 31: 139.

[85] Wu, C I and Vinson, J R. Influences of large amplitudes, transverse shear deformation, and rotatory inertia on lateral vibrations of transversely isotropic plates. ASME J. Appl. Mech., 1969, 36: 254.

[86] Wu. C I and Vinson, J R. On the nonlinear oscillations of plates composed of composite materials. J. Composite Materials, 1969, 3: 548.

[87] Sathyamoorthy, M and Chia, C Y. Nonlinear vibration of anisotropic rectangular plates including shear and rotatory inertia. Fibre Sci. Technol., 1980, 13: 337.

[88] Niyogi, A K and Meyers, B L. Perturbation solution of nonlinear vibration of rectangular orthotropic plates. Inter. J. Nonlinear Mech., 1981, 16: 401.

[89] Chia, C Y. Large amplitude vibration of laminated rectangular plates. Fibre Sci. Technol., 1982, 17: 123.

[90] Reddy, J N and Chao, W C. Large-deflection and large-amplitude free vibrations of laminated composite-material plates. Comput. Struct., 1981, 13: 341.

[91] Chia, C Y. Non-linear vibration of anisotropic rectangular plates with non-uniform edge constraints. J. Sound. Vib., 1985, 101: 539.

[92] Chia, C Y. Nonlinear vibration and postbuckling of orthotropic rectangular plate with two adjacent edges clamped and the other edges simply supported. Proc. Int. Symp. Compos. Mat. Struct., Beijing: Technomic Publishing Co., Pa., 1986, 174.

[93] Banerjee, B. Large amplitude vibrations of a clamped orthotropic square plate carrying a concentrated mass. J. Sound Vib., 1982, 82: 329.

[94] Mei, C and Wentz, K R. Nonlinear response of angle-ply laminated plates to random loads. Proc. lst Int. Conf. Compos. Struct., Applied Science Publishers, 1981, 656.

[95] Mei, C and Wentz, K R. Large amplitude random response of angle-ply laminated composite plates. AIAA J. 1982, 20: 1450.

[96] Chia, C Y and Sathyamoorthy, M. Nonlinear vibration of anisotropic skew plates. Fibre Sci. Technol., 1980, 13: 81.

[97] Sathyamoorthy, M and Chia, C Y. Effects of transverse shear and rotatory inertia on large amplitude vibration of anisotropic skew plates. I-theory. ASME J. App. Mech., 1980, 47: 128.

[98] Sathyamoorthy, M and Chia, C Y. Effects of transverse shear and rotatory inertia on large amplitude vibration of anisotropic skew plates. II-numerical results. ASME. J. App. Mech., 1980, 47: 133.

[99] Sathyamoorthy, M. Large amplitude vibration of skew orthotropic plates. ASME J. App. Mech., 1980, 47: 675.

[100] Sivakumaran, K S and Chia, C Y. Nonlinear vibration of generally laminated anisotropic thick plates. Ing. Arch., 1984, 54: 220.

[101] Sivakumaran, K S and Chia, C Y. Large-amplitude oscillations of unsymmetrically laminated anisotropic rectangular plates including shear, rotatory, and transverse normal stress. ASME J. App. Mech., 1985, 52: 536.

[102] Nath, Y and Alwar, R S. Nonlinear dynamic analysis of orthotropic circular plates. Int. J. Solids Struct., 1980, 16: 433.

[103] Rao, G V and Raju, K K. Large amplitude axisymmetric vibrations of orthotropic circular plates elastically restrained against rotation. J. Sound Vib., 1980, 69: 175.

[104] Sathyamoorthy, M. Transverse shear and rotatory inertia effects on nonlinear vibration of orthotropic circular plates. Comput. Struct., 1981, 14: 129.

[105] Sathyamoorthy, M. Effects of large amplitude, transverse shear and rotatory inertia on vibration of orthotropic elliptical plates. Int. J. Nonlinear Mech., 1981, 16: 327.

[106] Sathyamoorthy, M. Nonlinear vibration of orthotropic elliptical plates with attention to shear and rotatory inertia Fibre Sci. Technol., 1981, 15: 79.

[107] Sathyamoorthy, M and Chia, C Y. Nonlinear flexural vibration of moderately thick orthotropic circular plates. Ing. Arch., 1982, 52: 237.

[108] Biswas, P and Kapoor, P. Nonlinear free vibration of orthotropic circular plates at elevated temperature. Indian Inst. Sci. J. 1984, 65: 87.

[109] Ruei, K H, Jiang, C and Chia, C Y. Dynamic and static nonlinear analysis of cylindrically orthotropic circular plates with nonuniform edge constraints. J. Appl. Math. Phys., (ZAMP), 1984, 35: 387.

[110] Dumir, P C, Gandhi, M L and Nath, Y. Nonlinear axisymmetric transient analysis of orthotropic thin annular plates. Fibre Sci. Technol., 1984, 21: 23.

[111] Dumir, P C, Nath, Y and Gandhi, M L. Nonlinear axisymmetric transient analysis of orthotropic with annular plates with a rigid central mass. J. Sound Vib., 1984, 97: 387.

[112] Dumir, P C, Kumar, C R and Gandhi, M L. Nonlinear axisymmetric vibration of orthotropic thin annular plates with a rigid central mass. J. Acoust. Sot. Amer. 1985, 77: 2184.

[113] Dumir, P C, Kumar, C R and Gandhi, M L. Nonlinear axisymmetric vibration of orthotropic thin circular plates on elastic foundations. J. Sound Vib., 1985, 103: 273.

[114] Nath, Y, Dumir, P C and Gandhi, M L. Nonlinear axisymmetric transient analysis of an or-

thotropic thin circular plate with an elastically restrained edge. J. Sound Vib., 1984, 95: 85.

[115] Dumir, P C, Kumar, C R and Gandhi, M L. Nonlinear axisymmetric vibration of orthotropic thin circular plates with elastically restrained edges. Comput. Struct., 1986, 22: 677.

[116] Sathyamoorthy, M and Chia, C Y. Large amplitude vibration of orthotropic elliptical plates. Acta Mech., 1982, 37: 247.

[117] Sathyamoorthy, M. Nonlinear dynamic analysis of orthotropic elliptical plates. Fibre Sci. Technol., 1984, 20: 135.

[118] Sathyamoorthy, M. Multiple-mode large amplitude vibration of orthotropic circular plates. Int. J. Nonlinear Mech., 1984, 19: 341.

[119] Sathyamoorthy, M. A new approach to nonlinear dynamic analysis of composite plates. Proc. 2nd Int. Cont. Compos. Struct, London: Applied Science Publishers, 1983, 128.

[120] Sathyamoorthy, M. Large amplitude vibrations of moderately thick beams. Proceedings of the International Modal Analysis Conference, Orlando, Florida, 1982, 136.

[121] 刘人怀，波纹圆板的特征关系式. 力学学报，1978，(1)：47.

[122] 刘人怀，波纹环形板的非线性弯曲. 中国科学，A辑，1984，(3)：147.

[123] Liu Ren-huai. On large deflection of corrugated annular plates under uniform pressure. The Advances of Applied Mathematics and Mechanics in China, Beijing: China Academic Publishers, 1987, 138.

[124] 刘人怀. 在复合载荷作用下波纹环形板的非线性弯曲. 中国科学，A辑，1985，(6)：537.

[125] Liu Ren-huai. Large deflection of corrngated circular plate under the action of concentrated loads at the center. International J. Non-Linear Mechanics, 1984, 19 (5): 409.

[126] 刘人怀. 具有光滑中心的波纹圆板的特征关系式. 中国科学技术大学学报，1979，9 (2)：75.

[127] Liu Ren-huai. Nonlinear bending of corrugated circular plate under the combined action of uniformly distributed load and concentrated load at the center. Proceedings of the International Conference on Nonlinear Mechanics. Shanghai, 1985, 271.

[128] 刘人怀. 复合载荷下波纹圆板的非线性分析. 应用数学和力学，1988，9 (8)：661.

[129] Liu Ren-huai. Large deflection of corrugated circular plate with plane boundary region. Solid Mechanics Archives, 1984, 9 (4): 383.

[130] 刘人怀，宗赴传. 集中载荷作用下具有平面边缘的波纹圆板的非线性弯曲问题. 首届全国敏感元件与传感器学术会议，北京，1989.

[131] Stumpf, H and Makowski, J. Large strain deformations of shells. Acta Mech., 1987, 65 (1): 153.

[132] Taber, L A. On approximate large strain relations for a shell of revolution. Int. J. Nonlinear Mechanics, 1985, 20 (1): 27.

[133] Biricikoglu, V and Kalnins, A. Large elastic deformations of shells with the inclusion of transverse normal strain. Int. J. Solid-Struct., 1971, 7 (5): 431.

[134] Libai, A and Simmonds, J G. Large-strain constitutive laws for the cylindrical deformation of shells. Int. J. Nonlinear Mech., 1981, 16: 91.

[135] Simmonds, J G. The strain energy density of rubber-like shells. Int. J. Solids Struct., 1985, 21 (1): 67-77.

[136] Libai, A and Simmonds, J G. Highly non-linear cylindrical deformations of rings and shells. Int. J. Nonlinear Mech. 1983, 18 (3): 181.

[137] Taber, L A. Large elastic deformation of shear deformable shells of revolution: Theory and Analysis. J. Appl. Mech., 1987, 54 (3): 578.

[138] Cagan, J and Taber, L A. Large deflection stability of spherical shells with ring loads. J. Appl. Mech., 1986, 53 (4): 897.

[139] Godoy, L A and Flores, F G. Stresses in thin spherical shells with imperfections, Part 1: Influence of axisymmetric imperfections. Thin-Walled Struct., 1987, 5 (1): 5.

[140] Godoy, L A. and Flores, F G. Stresses in thin spherical shells with imperfections, Part 2: Influence of local imperfections. Thin-Walled Struct., 1987, 5 (2): 145.

[141] Hwang C. Perturbation initial parameter method for solving the geometrical nonlinear problem of axisymmetrical shells. App. Math. Mech., 1986, 7 (6): 573.

[142] Szwabowicz, M L. Variational formulation in the geometrically nonlinear thin elastic shell theory. Int. J. Solids Struct., 1986, 22 (11): 1161.

[143] Fridman, A D. Asymptotic method of investigating the deformation of cylindrical shells with holes. Sov. Appl. Mech., 1987, 22 (12): 1181.

[144] Kryukov, N N. Nonlinear deformation of noncircular orthotropic cylindrical shells of variable rigidity. Sov. Appl. Mech., 1986, 22 (5): 435.

[145] Zhang, Y and Matthews, F L. Large deflection behaviour of simple supported laminated panels under in-plane loading. ASME J. Appl. Mech., 1985, 52 (3): 553.

[146] 叶开沅，顾淑贤．均布载荷作用下圆底薄球壳的非线性稳定性．中国力学学会 1980 年全国计算力学会议论文集，北京：北京大学出版社，1981，280.

[147] 叶志明．变厚度圆底扁薄球壳的非线性稳定问题．力学学报，1984，16 (6)：634.

[148] 顾淑贤．扁球壳在对称线布荷载作用下的稳定性问题．固体力学学报，1988，9 (1)：39.

[149] 叶开沅，宋卫平．圆锥扁壳在均布压力作用下的非线性稳定性问题．兰州大学学报（自然科学版），1983，19：134.

[150] 刘人怀．在边缘载荷作用下中心开孔圆底扁薄球壳的轴对称稳定性．力学学报，1977，(3)：206.

[151] 刘人怀，成振强．集中载荷作用下开顶扁球壳的非线性稳定问题．应用数学和力学，1988，9 (2)：95.

[152] 刘人怀，李东．均布载荷作用下开顶扁球壳的非线性稳定问题．应用数学和力学，1988，9 (3)：205.

[153] 刘人怀，张小果．具有硬中心的开顶扁球壳在均布载荷作用下的非线性稳定问题．第五届华东固体力学学术讨论会，无锡，1988 年 9 月.

[154] Liu Ren-huai, He Ling-hui. Nonlinear stability of truncated shallow spherical shells under axisymmetric distributed loads. Solid Mechanics Archives, 1989, 14 (2): 81.

[155] 刘人怀，宗赴传．复合载荷作用下开顶扁球壳的非线性稳定问题．第五届华东固体力学学术讨论会，无锡，1988 年 9 月.

[156] 刘人怀．双层金属中心开孔扁球壳的非线性热稳定问题．中国科学技术大学学报，1981，11 (1)：84.

[157] 刘人怀．双层金属截头扁锥壳的热稳定性．力学学报，特刊，1981，72.

[158] Liu Ren-huai. Nonlinear thermal stability of bimetallic shallow shells of revolution. International J. Nonlinear Mechanics, 1983, 18 (5): 409.

[159] Subbiah, J. Nonlinear analysis of geometrically imperfect stiffened shells of revolution. J. Ship Res., 1988, 32 (1): 29.

[160] In, V P and Chia, C Y. Nonlinear vibration and postbuckling of unsymmetric cross-ply circular cylindrical shells. Int. J. Solids Struct., 1988, 24 (2): 195.

[161] Zhang, Y and Matthews, F L. Postbuckling brhariour of cylindrical curred panels of generally layered composite materials with small initial imperfections of geometry. Proc. 2nd Int. Lonf. Compos. Struct., London: Appl. Sci. Pubishers, 1983, 428.

[162] Wang, Z and You, S. The nonlinear elastic buckling problem of laminated cylindrical panels of composite materials under the action of axial load. Proc. Int. Symp. Compos. Mat. Struct., Beijing: Technomic Publishing Co., Pa, 1986, 393.

[163] Liu Ren-huai. Nonlinear stability of symmetrically laminated cylindrically orthotropic shallow spherical shells including transverse shear. Science in China, Series A, 1992, 35 (6): 734.

[164] Yasuda, K and Kushidz, G. Nonlinear forced oscillations of a shallow spherical shell. Bull. JSME., 1985, 27 (232): 2233.

[165] Mukherjee, K and Chakraborty, S K. Exact solution for large amplitude free and forced oscillation of a thin spherical shell. J. Sound Vib., 1985, 100 (3): 339.

[166] Sinharay, G C, and Banerjee, B. Large amplitude free vibrations of shells of variable thickness: A new approach. AIAA J., 1986, 24 (6): 998.

[167] Sinharay, G C, and Banerjee, B. Large amplitude free vibrations of shallow and cylindrical shell: A new approach. Int. J. Nonlinear Mechanics, 1985, 20 (2): 69.

[168] Chrzeszczyk, A. Generalized solutions of dynamical equations in nonlinear theory of thin elastic shells. Arch. Mech., 1983, 35 (5-6): 555.

[169] Liu Ren-huai and Li Dong. On the nonlinear bending and vibration of corrugated circular plates. Int. J. Nonlinear Mechanics, 1989, 24 (3): 165.

[170] Wu, Y C and Yang, T Y. Free and forced nonlinear dynamics of shell structures. J. Compopite Materials, 1987, 21 (10): 898.

[171] Bhattacharya, A P. Large amplitude vibrations of imperfect cross-ply laminated cylindrical shell panels with elastically restrained edges and resting on elastic foundation. Fibre Sci. Technol., 1984, 21: 205.

[172] Hui, D. Influence of geometric imperfections and in-plane constraints on nonlinear vibrations of simple supported cylindrical panel. J. App. Mech., 1984, 51: 383.

[173] Chia, C Y. Nonlinear vibration and postbuckling of unsymmetrically laminated imperfect shallow cylindrical panels with mixed boundary conditions resting on elastic foundation. Int. J. Eng. Sci., 1987, 25: 427.

[174] Chia, C Y. Large amplitude vibration of initially imperfect unsymmetrically laminated shallow cylindrical panels with edges elastically restrained against rotation. Proc. Int. Conf. Vib. Prob. Eng., Xian, China, 1986, 260.

[175] Chia, C Y. Nonlinear free vibration and post-buckling of symmetrically laminated orthotropic imperfect shallow cylindrical panels with two adjacent edges simply supported and the other edges clamped. Int. J. Solids Struct., 1987, 23 (8): 1123.

[176] Nath, Y and Jain, R K. Influence of foundation mass on nonlinear damped respone of orthotropic shallow spherical shells. Int. J. Mech. Sci., 1985, 27 (7-8): 471.

[177] Nath, Y and Jain, R K. Non-linear studies of orthotropic shallow spherical shells on elastic foundation. Int. J. Nonlinear Mechanics, 1986, 21 (6): 447.

板壳非线性力学①

由于近代科学技术的飞跃发展，板壳非线性力学已成为当前固体力学研究的一个最活跃的领域，倍受人们关注。目前研究的中心课题是板壳的几何非线性问题。

板壳非线性力学的奠基者是 von Kármán。他于 1910 年提出著名的薄板大挠度方程，于 1939 年，又和钱学森一起用非线性理论讨论了扁球壳的屈曲问题。

板壳非线性力学问题的基本方程是非线性的，在数学研究上存在巨大的困难。因此，如何求解便成为解决这类问题的关键。求解方法可分为两类，即解析求解和数值求解。在数值求解方面，常见的有有限元法、边界元法、有限差分法和加权残数法等。在解析求解方面，有精确解法和近似解法。

求精确解一直是人们的期望，因为它不仅使问题得以圆满解决，为工程设计提供最可靠的依据，而且为近似解法提供一个检验标准。早期 Way 和 Levy 分别用幂级数方法和三角级数法求解了圆板和简支矩形板的大挠度问题。最近，刘人怀和施云方用幂级数方法研究了更为困难的夹层圆板的大挠度问题[1]。郑晓静系统地讨论了圆薄板和弹性地基上的圆薄板问题[2]。

在近似求解方面，最受人注意的经典问题有四个，即圆板和矩形板的大挠度问题，扁球壳和圆柱壳和非线性稳定问题。早期的代表性工作为：Vincent 和钱伟长分别用以载荷和中心挠度为摄动参数的摄动法讨论圆板大挠度问题，钱伟长用奇异摄动法处理圆板大挠度的边界层问题。钱伟长的方法为推动板壳非线性力学的发展做出了杰出贡献。钱伟长用变分法，Kaplan 和 Fung 用摄动法讨论了扁球壳的非线性稳定问题。Donnell 和钱学森研究了轴向压缩下的圆柱壳非线性稳定问题。1965 年叶开沅和刘人怀提出更为优越的修正迭代法讨论了扁球壳的非线性稳定问题[3-5]。此后，刘人怀等用此方法处理了一系列板壳非线性问题，包括中心开孔扁球壳[6-10]、双层金属旋转扁壳[11-13]、夹层板壳[14-17]、复合材料层合扁球壳[18]以及波纹圆板[19-29]。叶开沅[30]于 20 世纪 50 年代以来用摄动法系统地研究了圆板和矩形板大挠度问题。Chia 于近十几年采用级数方法系统地处理了复合材料层合圆板和圆柱壳的非线性弯曲、稳定和振动问题[31-33]。此外，对于椭圆板、扇形板、变厚度板以及扁锥壳、圆环壳等已有许多研究。

参考文献

[1] 刘人怀，施云方. 夹层圆板大挠度问题的精确解. 应用数学和力学，1982，3 (1)：11-23.

[2] 郑晓静，任意载荷下轴对称 Kármán 方程的精确解及其近似解析求解法的研究. 博士学位论文，兰州大学，1987.

[3] 叶开沅，刘人怀，严庆元，等. 在对称线布载荷作用下的圆底扁薄球壳的非线性稳定问题. 科学通报，1965，(2)：142.

① 本文原载《自然科学年鉴（1989)》(纪念创刊十周年专辑)，上海：上海翻译出版公司，1990，371.

[4] 叶开沅，刘人怀，张传智，等. 圆底扁薄球壳在边缘力矩作用下的非线性稳定问题. 科学通报，1965，(2)：145.

[5] 刘人怀. 在内边缘均布力矩作用下中心开孔圆底扁球壳的非线性稳定问题. 科学通报，1965，(3)：253-255.

[6] 刘人怀. 在边缘载荷作用下中心开孔圆底扁薄球壳的轴对称稳定性. 力学学报，1977，(3)：206-212.

[7] 刘人怀，成振强. 集中载荷作用下开顶扁球壳的非线性稳定问题. 应用数学和力学，1988，9 (2)：95-106.

[8] 刘人怀，李东. 均布载荷作用下开顶扁球壳的非线性稳定问题. 应用数学和力学，1988，9 (3)：205-217.

[9] 刘人怀. 扁球壳非线性稳定问题研究. 近代数学和力学讨论会文集，上海 (1987).

[10] Liu R. H.，He L. H. On the nonlinear stability of a truncated shallow spherical shell under axisymmetrically distributed load. Solid Mechanics Archives，1989，14 (2)：81-102.

[11] 刘人怀. 双层金属中心开孔扁球壳的非线性热稳定问题. 中国科学技术大学学报，1981，11 (1)：84-99.

[12] 刘人怀. 双层金属截头扁锥壳的热稳定性. 力学学报，1981，特刊：172-180.

[13] Liu Renhuai. Non-linear thermal stability of bimetallic shallow shells of revolution. International Journal of Non-Linear Mechanics，1983，18 (5)：409-429.

[14] 刘人怀. 在边缘力矩作用下夹层圆板的非线性轴对称弯曲问题. 中国科学技术大学学报，1980，10 (2)：56-67.

[15] 刘人怀. 夹层圆板的非线性弯曲. 应用数学和力学，1981，2 (2)：173-190.

[16] Liu Renhuai. Nonlinear bending of circular sandwich plates under the action of axisymmetric uniformly distributed line loads. Progress in Applied Mechanics，Dordrecht：Martinus Nijihoff Publishers，1987，293-321.

[17] 刘人怀，朱金福. 夹层壳非线性理论. 北京：机械工业出版社，1993.

[18] Liu R. H. Nonlinear stability of symmetrically laminated cylindrically orthotropic shallow spherical shells including transverse shear，Science in China，Series A，1992，35 (6)：734-746.

[19] 刘人怀. 波纹圆板的特征关系式. 力学学报，1978，(1)：47-52.

[20] 刘人怀. 具有光滑中心的波纹圆板的特征关系式. 中国科学技术大学学报，1979，9 (2)：75-86.

[21] 刘人怀. 波纹环形板的非线性弯曲. 中国科学，A 辑，1984，(3)：247-253.

[22] Liu Renhuai. Large deflection of corrugated circular plate with a plane central region under the action of concentrated loads at the center. International Journal of Non-Linear Mechanics，1984，19 (5)：409-419.

[23] Liu Renhuai. Large deflection of corrugated circular plate with plane boundary region. Solid Mechanics Archives，1984，9 (4)：383-406.

[24] Liu Renhuai. Nonlinear bending of corrugated circular plate under the combined action of uniformly distributed load and concentrated load at the center. Proceedings of the International Conference on Nonlinear Mechanics. Beijing：Science Press，1985，271-278.

[25] 刘人怀. 在复合载荷作用下波纹环形板的非线性弯曲. 中国科学，A 辑，1985，(6)：537-545.

[26] Liu Renhuai. On large deflection of corrugated annular plates under uniform pressure. The Ad-

vances of Applied Mathematics and Mechanics in China，1987，1：138-152.

[27] Liu Renhuai. The study on nonlinear bending problems of corrugated circular plates. The Symposium of Alexander von Humboldt Foundation，Shanghai，1987-10-17.

[28] 刘人怀. 复合载荷下波纹圆板的非线性分析. 应用数学和力学，1988，9 (8)：661-674.

[29] Liu Renhuai and Li Dong. On the non-linear bending and vibration of corrugated circular plates. International Journal of Non-Linear Mechanics，1989，24 (3)：165-176.

[30] 叶开沅. 薄板大挠度问题. 第六届全国弹性元件学术交流会，厦门，1981.

[31] Chia C. Y. Nonlinear Analysis of Plates，Mc Graw Hill，New York，1980.

[32] Chia C. Y. Nonlinear vibration and postbuckling of unsymmetrically laminated imperfect shallow cylindrical panels. Intern. J. of Engng. Sci.，1987，(25)：427-441.

[33] Chia C. Y. Nonlinear analysis of doubly curved symmetrically laminated shallow shells with rectangular platform. Ingenieur-Archiv，1988，(58)：252-264.

The effect of the Göttingen school on investigation of flexible plates and shells in China①

中 文 摘 要

本文讨论了德国哥廷根学派对中国板壳力学研究的影响。

It is common knowledge that a structural element by the name of thin plates and shells is applied widely in various industries such as aeronautical, astronautical, naval, chemical, civil, and mechanical engineering. Therefore, a considerable amount of research has been completed. However, as the theoretical analysis is very difficult, investigators have discussed only the problem of the linear theory for thin plates and shells. Obviously, this cannot satisfy the needs of the rapid development of modern technology.

It was a successor of the world-famous Göttingen school, a great American scientist in the twentieth century, Prof. Th. von Kármán[1] (1881—1963) who first suggested the problem of large deflection of thin plates in 1910, and established a new important region by the name of nonlinear mechanics of plates and shells. His teacher Professor L. Prandtl (1875—1953) is a great German scientist from Göttingen University, and the Father of air dynamics. In 1939, the importance of the nonlinear features in the thin shell buckling problem was first pointed out in a most spectacular manner by von Kármán and his Chinese student Tsien Hsue-shen[2]. Next they[3] studied another difficult problem for the buckling of thin cylindrical shells. Dr. Tsien is an honour to my motherland, and goes by the name of the Father of the Chinese rocket. He was concurrently director of the Department of Modern Mechanics, University of Science and Technology of China from 1958 to 1983. At that time, I worked in the department.

On account of the requirements of modern industries, this kind of problem is noticed greatly. The difficulty of nonlinear mathematics is so great that progress has been slow after von Kármán's attempts. However, nonlinear mechanics of plates and shells become a key region of solid mechanics. Thus, these problems have been attracting the attention of numerous investigators in the world. As we know, Chinese scientists make greater contributions to this cause.

Dr. Chien Wei-zang is a student of Prof. J. L. Synge who is the director of the Department of Applied Mathematics of University of Toronto, and a successor of the Göttingen school too. In 1942, Dr. Chien became an assistant of Prof. von Kármán.

---

① Reprinted from The Symposium on International Scientific Cooperation for Developing Countries, Bonn, Germany, 1997-04-15.

After that time, he and his students, such as Prof. Yeh Kai-yuan completed many works about the nonlinear problem of plates and shells. Prof. Yeh is my teacher. From 1986 to 1991, I was Dr. Chien's assistant. In 1981, I was supported by the Humboldt Foundation, and engaged in research work at Prof. W. Zerna's Institute of Ruhr University.

Obviously, on the one hand, the Göttingen school has a great pushing effect on the investigation of flexible plates and shells in China. And on the other hand, the Chinese investigation brings about a great advance in industrial production of the world.

References

[1] Th. von Kármán, Festigkeitsprobleme im Mashinenbau, Enzyklopadit der Mathematischen Wissenschaften, 1910, 4: 348-351.

[2] Th. von Kármán and Tsien, Hsue-shen. The buckling of spherical shell by external pressure. J. Aeron. Sci., 1939, 7: 43-50.

[3] Th. von Kármán and Tsien, Hsue-shen. The buckling of thin cylindrical shells under axial compression. J. Aeron. Sci. 1941, 8 (8): 303-312.

多姿多彩的板壳结构[①]

尽管世界上的物体千变万化，具有各种各样的形式，但是，最常见的物体大都是平板和壳体，其外形特点是厚度比其余两个方向尺寸在数量级上小得多。平分物体厚度的分界面称为中面。若中面是平面，则称此物体为平板；若中面是曲面，则称此物体为壳体。

由于厚度小、重量轻、耗材料少和性能好，板壳成为具有优良特性的结构元件，不仅广泛应用于各种工程结构作为最基本和最主要的构件，而且在自然界和日常生活中也常常碰见，它们与每个人自己和每个人的生活休戚相关，与人类的生存紧密相连。

就以人类居住的地球为例，它是一个略微有点扁的圆球，平均半径为 6371km，由三个圆层，即地壳、地幔和地核组成。地壳主要由岩石构成，平均厚度为 33km。地幔是介于地壳和地核之间的圆层，厚度约为 2865km，主要是一种密度很大和温度很高的半黏性流动物质。显然，刚性的地壳与整个地球相比，只不过是很薄的一层表皮，漂浮在致密和高温的半黏性流动的地幔之上，是一个天然的类似于圆球的壳体。在地球表面上，有绵延的高山、交错的深谷、星罗棋布的岛屿。这种复杂奇特的地质形态，正是地壳这个巨大壳体在外力作用下发生变形的结果。

在自然界里，鸟卵、蚌壳、植物种子、花瓣等都呈壳形，就连我们人类自己身体的骨骼、眼角膜和细胞壁等也呈壳形，这是经过长期自然选择形成的，是以最少的材料造成的坚强结构。

人们通过与自然界的斗争，逐渐认识到板壳结构的优越性，因而在工程中，确切地说，几乎在世界上一切工程部门都得到了应用，其重要性十分突出。

千百年来，人们渴望着去宇宙飞行。我国就有嫦娥奔月的美丽故事，驰名中外。为实现这一理想，我们的祖先早在 10 世纪宋朝时代，便发明了人类第一支火箭，进行了通向空间的第一次尝试。今天，科学技术的巨手已经填平了理想与现实的鸿沟。1970 年，我国第一颗重达 173kg 的人造地球卫星飞上了天空。美国则于 1969 年用阿波罗 11 号飞船将两名宇航员送上了距地球 38 万 km 的月球，在月球上第一次留下人类的足迹。最近，举世瞩目的我国神舟三号无人飞船发射成功，为我国宇航员未来几年内进入宇宙创造了条件。同时，在航空工程中，飞机的出现已有 100 年历史，目前已有起飞重量超过 300 吨、尺寸达大半个足球场的民航飞机。在进行这些飞行时，使用了火箭、卫星、宇宙飞船和飞机，它们的外壳是壳体；一些部件，如尾翼、太阳能翼板、机舱地板等都是平板。

板壳在建筑工程中，更具有特殊的意义。门、楼板、墙壁、地基是平板。一座宽敞、无障碍的大跨度建筑物，如会议厅、展览馆、体育馆、工厂车间、剧场、车库、仓库、飞机库等，则以壳体屋盖为最好。

---

① 本文原载《高校招生》，2002，(7)：1.

在旧社会，我国自己没有设计过大型壳体建筑，直到中华人民共和国成立以后，壳体结构才开始采用，并得到迅速发展。许多工厂车间和民用住宅采用圆柱壳形屋盖，较大的跨度达 33m。使用双曲扁壳屋盖情况也较多，如北京火车站大厅和北京网球馆，其中网球馆地面尺寸为 42m×42m。

在国外，最大的圆球形屋盖大约要算是美国的宾夕法尼亚州的大会堂，其平面直径达到 126.6m，而最有名的美观的壳体屋盖则要数澳大利亚悉尼歌剧院。

除了上述领域，在航海、石油化工、动力、交通、仪器仪表、军事等工程中，以至人们的体育运动（如大家喜爱的足球）和日常生活中，我们还可以举出无数的板壳例子。总之，作为构件，板壳具有极大的优越性和重要性，我们处处需要，处处可见。它们在人类征服自然并提高自己衣食住行生活质量的进程中，立下了特殊的功勋。对于板壳而言，由于几何形状和构成材料的不同，其结构的承载能力也就不同，这就需要用板壳结构分析知识去进行分析和设计。特别是在人类当代几大难题，如宇宙的变化、地球的起源和生命的奥秘的探索中，都离不开板壳结构分析知识。现有的板壳理论还很不完善，需要发展。这是一个大有可为的领域，值得青年人参加到这一领域来大显身手。

弹性元件国内外理论发展概况①

一、引言

在仪器仪表中，利用材料的弹性特性实现各种功能的元件称为弹性元件。常用的有弹簧、平膜片、跳跃膜片、热敏双金属膜片、波纹膜片、波纹管等。

弹性元件能够完成测量、变换、隔离、密封、补偿、贮能、连接等各种不同的功能，其质量直接影响到仪器仪表的精度与可靠性。弹性元件的结构简单，价格低廉。因此，弹性元件在仪器仪表中的应用十分广泛，是仪器仪表的主要基础元件之一。

弹性元件能将一些难以直接测量的物理量（如压力、流量、温度等）转换成便于测量的长度、角度等参量。在变形不大的情况下，各种弹性元件的弹性特性基本符合胡克定律，其载荷与位移之间具有一定的函数关系，利用这种特性就可以测量力、压力、压差和力矩等参量。同时，弹性元件还能很方便地将很多物理量（如流量、液位、温度、电流、压力等）转换为力、压力和力矩等参量来进行测量。文中重点综述平膜片、跳跃膜片、热敏双金属跳跃膜片、波纹膜片及波纹管的非线性弯曲、振动和稳定性研究情况。

二、平膜片

平膜片是将两种压力不等的流体隔开而具有挠性的圆形薄板或薄膜，通常分为金属和非金属两类，它是一种简单可靠的压力测量元件。

许多时候，由于膜片变形很小，使用薄圆板小挠度理论就可以了。这个理论的解最早是 Poisson[1]给出的，详细的解答可参见文献［2］和［3］。

薄圆板的大挠度理论来源于 von Kármán[4]，在此后 10 年里，人们用各种方法研究薄圆板。

Hencky[5]首先研究了受均布压力的圆形薄板。接着 Vincent[6]用载荷摄动法讨论了受均布压力的圆板大挠度问题。Timoshenko 等[2]用变分法，Way[7]用幂级数方法，Panov[8]用伽辽金法等分别求解了同一个问题。直到钱伟长[9]采用中心挠度为参数的摄动法，才得到了最满意的解答。此后，钱伟长又和叶开沅[10-16]获得了圆板和环形板一系列问题的解。后来，李东和刘人怀[17]又研究了薄圆板的非线性振动问题，王永亮等[18]讨论了弹性支持环形板的非线性弯曲问题。

三、跳跃膜片

1. 跳跃膜片的性质及其应用

用高弹性极限材料做成微小倾度的薄壁球形圆顶，在仪器制造中通常称为跳跃膜

① 本文原载《仪表技术与传感器》，2011，(9)：1-8，29. 作者：刘人怀，袁鸿.

片。这种零件在某种力的作用下，尤其是沿凸面分布的压力，在一定条件下会失掉稳定性。当薄膜的壁足够薄而又是用高弹性极限材料做成时，在超过临界力以后仍能使膜片保有弹性。类似于这种并不引起零件破坏的稳定性丧失，称为跳跃。

2. 跳跃膜片几何非线性理论发展概述

von Kármán 和钱学森[19]首先指出扁球壳在外压下的屈曲问题是一个非线性问题，这篇论文被誉为 20 世纪板壳非线性力学的开创性论文。接着，钱伟长使用能量法处理了一个更困难的问题，即在对称线布载荷作用下扁球壳的跳跃问题，获得了解析公式以及与实验吻合的半经验公式，这一成果于 1956 年才公开发表[20]。

Feodosev[21]使用伽辽金法研究了均布压力下的扁球壳屈曲问题，Simons[22]和 Reiss 等[23]使用幂级数法研究了这一问题。Kaplan 和冯元桢[24]以及 Krivoshchier 和 Mushtary[25]再用摄动法研究了这一问题。但是，上述研究都无法满足工程需求，工程上需要一个计算程序简单而精确度比较高的解法。

1965 年，叶开沅和刘人怀等[26-28]结合了钱伟长以中心挠度为摄动参数的摄动法和逐次逼近法的优点，提出了修正迭代法。这个方法迭代程序明确、简单、精确度高。用这个方法分别求解了在线布载荷作用下和边缘力矩作用下的扁球壳非线性稳定问题，给出了更精确的解析解。同时，刘人怀[28]又在国际上首次研究了难度更大的开顶扁球壳的非线性稳定问题。随后，刘人怀等[29-36]系统地求解了开顶扁球壳的非线性稳定问题。叶开沅和顾淑贤[37]再次精确地求解了均布载荷作用下扁球壳的非线性稳定问题。

四、热敏双金属跳跃膜片

用双层金属扁壳做成的热敏弹性元件，被用于自动控制仪表中。这类元件具有这样的特性：在均匀加热的情况下，由于两层金属的热膨胀系数不同，将产生弯曲，以至于发生跳跃现象。自动控制仪表便应用此跳跃现象作为控制信号。

Panov[38]首先用伽辽金法讨论了双金属扁球壳的非线性热稳定问题，接着 Wittrick 等[39]、Grigolyuk[40]和罗祖道等[41]又用级数法等方法再次讨论了同一问题，结果均不理想。为此，刘人怀[42-44]建立了双层金属旋转扁球壳在均匀温度作用下的非线性热稳定性的一般理论，使用修正迭代法不仅求解了前人已研究过的封顶和开顶双金属扁球壳的热稳定问题，获得了令人满意的解析解，而且首次获得了封顶和截头双金属扁锥壳的热稳定问题的解析解。

五、波纹膜片

波纹膜片属于波纹壳体，由于其具有灵敏度高、线性度好等优点，在传感器中占有重要地位。波纹膜片是广泛使用的一种弹性敏感元件，它既用于测量高于大气压的压力计中，也用于测量低于大气压的真空计中。

1. 波纹膜片的构造

波纹膜片表面的波纹是多种多样的，与具体情况下所必须得到的特征（力与挠度之间的关系）的形式有关。最常见的波纹膜片分三种：平面型波纹膜片（图 1）；带大边缘波纹的波纹膜片（图 2）；壳型波纹膜片（图 3）。

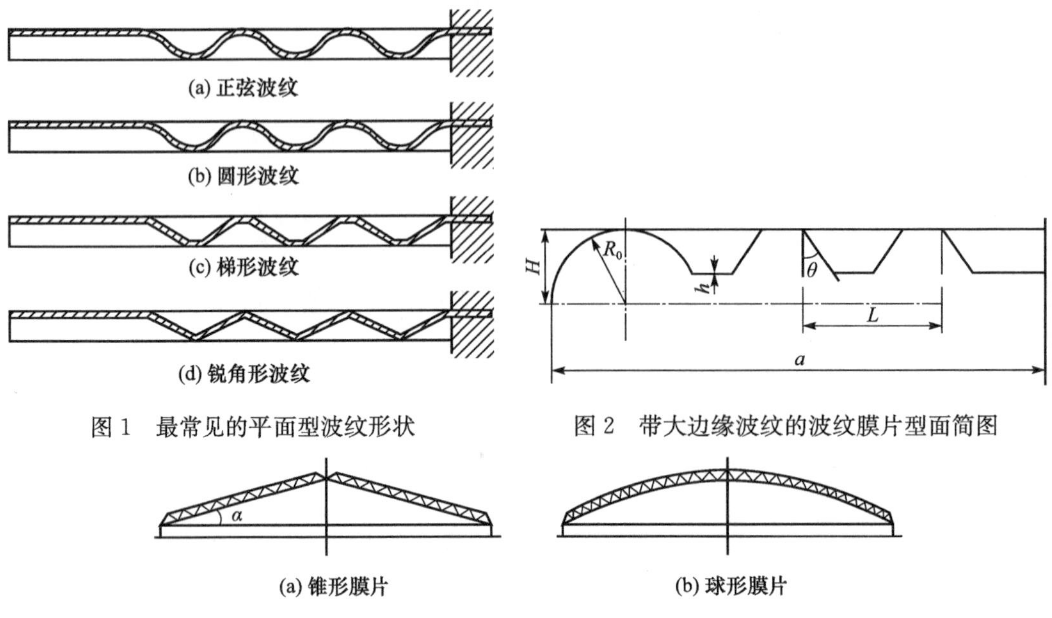

图 1　最常见的平面型波纹形状

图 2　带大边缘波纹的波纹膜片型面简图

图 3　壳型波纹膜片

波纹膜片参数很多，又相互制约，设计很复杂。波纹膜片的参数主要指两方面：膜片参数；型面参数。膜片参数主要有膜片材料、膜片厚度和膜片工作直径等。型面参数是指膜片波纹形状有关的参数，主要有波纹深度、波纹形状、边缘波纹、型面锥度、波纹倾角、波距等。

波纹膜片型面简图如图 4 所示。

对于膜片中心挠度，以数学形式表达如下：

$$W = f(p, E, \nu, h, a, H, \alpha, \theta, L, R_0, \cdots)$$

式中，p 为工作压力；E 为材料的弹性模量；ν 为泊松系数；h 为膜片厚度；a 为膜片工作半径；H 为波纹深度；α 为型面锥度；θ 为波纹倾角；L 为波距；R_0 为边缘波纹半径。

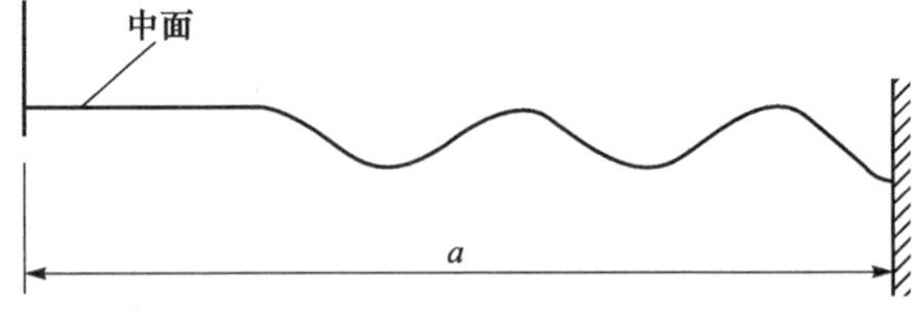

图 4　波纹膜片型面简图

2. 研究现状

波纹膜片以其独特的优点在各领域里广泛使用，尤其是作为低压敏感元件被广泛应用于精密仪器仪表和传感器中，自从 Hersey[45] 和 Griffith[46] 各自展开了对波纹圆板的实验和理论研究以来，世界各国的科学家在这方面做了大量的工作。这种弹性元件的理论研究十分困难，造成这种状况的主要原因是波纹膜片本身形状复杂，参数很多，特别是大挠度非线性微分方程组在数学上求解极其困难。以往的理论计算可以分为两种：采用壳体（或者扁壳）的大挠度（或者小挠度）方程来求解；将波纹壳（或者波纹圆板）看成是结构上的正交各向异性壳（或者正交各向异性圆平板），采用正交各向异性壳（或者正交各向异性圆平板）的大挠度（或者小挠度）方程来进行研究。

1）正交各向异性板壳理论

Haringx[47-50]首先提出将波纹圆板问题转化成等效的正交各向异性板问题的思想，为波纹圆板的理论分析提供了一种新的手段。文献［51］根据波纹圆板在径向和周向有不同的刚度，以各向异性圆板代替波纹圆板，该板在径向和周向的刚度对应于波纹圆板的刚度，而厚度、半径及材料是相同的，采用各向异性圆板大挠度方程，用Galerkin 方法获得了适应各种波纹圆板的特征关系式，Akasaka[52]也建立了各向异性圆板与波纹圆板弹性常数的等价关系，对于小挠度问题，求得了精确解，对于大挠度问题，采用了能量法，求得了近似解。

Akasaka 和 Takagishi[53]讨论了波纹膜片的线性振动，求出了轴对称振型和反对称振型的固有频率，并且用能量法分析了初始拉应力对固有频率的影响。Rubin[54]采用圆柱正交异性线性板理论探讨了中心带有轴的波纹膜片的优化设计，假定轴经历了小的轴向位移或者小的角位移，那么，最大切向应力的位置将依赖于均匀压力。

由于上述解以及采用壳体理论所获得的解的精度不够，造成设计制造时只能靠理论和经验。为此，刘人怀[55]从 1964 年开始研究这一问题，并于 1968 年用修正迭代法获得了十分精确的、可供设计直接应用的解析解。随后，刘人怀等[56-67]求得了各种波纹圆板（全波纹圆板、具有光滑中心的波纹圆板、具有光滑中心和平面边缘的波纹圆板、具有刚性中心的波纹圆板）在各种载荷（均匀载荷、集中载荷、复合载荷）下的特征关系式，所得结果与实验值十分符合，对波纹圆板的设计具有指导作用。最近，刘人怀[68]采用厚板理论分析了膜盒基体，给出了解析解。

卢耘耘等[69-70]以正交异性板理论为基础，提出了一种波纹圆板非线性弯曲的 Chebyshev 级数解法，推导出具有光滑中心的波纹圆板在任意轴对称荷载作用下的弹性特征方程。王新志等[71]用最小作用量原理推导出波纹圆板的变分方程，选取波纹圆板中心最大振幅为摄动参数，采用摄动变分法，一次近似求得了波纹圆板线性振动时的固有频率，继而求得了波纹板的非线性固有频率，但只以圆平板作为例子给出了数值解。刘人怀和李东[72]使用 Galerkin 方法得到了波纹圆板的特征关系及非线性自由振动的固有频率。李东和刘人怀[73]使用修正迭代法成功地解决了全波纹圆板的非线性自由振动问题，得到了幅频关系式。Шалашилин[74]将正交各向异性板的思想推广到波纹壳中，探讨了将波纹壳等效为正交异性壳体问题，求出了等价的弹性常数。刘人怀和王璠[75-79]从正交异性壳理论出发，研究了波纹扁球壳的非线性稳定、非线性动态屈曲和非线性强迫振动问题。

2）壳体理论

众多学者直接从壳体理论出发，讨论波纹板壳的特性。Панов[80]第一个讨论了正弦波纹膜片的大挠度问题。Феодосьев[81-82]使用扁壳理论，采用 Galerkin 方法讨论了浅正弦波纹圆板的大挠度问题。Grover 和 Bell[83]对图 4 所示波纹膜片进行了实验和理论分析，假定膜片是由圆板、圆环壳和锥壳组成的，圆板部分使用 Way[84]的非线性解，圆环壳部分使用 Stange[85]的线性解，锥壳部分采用 Timoshenko[86]导出的线性解。Wildhack[87]和 Dressler[88]讨论了 Grover-Bell 膜片，采用龙格-库塔方法，直接积分求解了线性壳体问题，这种计算适用于任意形状的膜片，但是，由于计算困难而没有考虑到非线性因素。Dressler[88]对于另外两种厚度的 Grover-Bell 膜片，分析了膜片的应力和位移，定量地表明了厚度对膜片中弯曲和拉伸的影响。Аксельрад[89-93]推导了旋转壳的非线性微分方程，借助 Galerkin 方法，分析了深正弦波纹膜片的线性和非线性弯曲，但没有将特征关系式的曲线用图表示出来。Hamada 和 Seguchi[94-95]使用 Reissner 方程，采用有限差分法，研究了圆板和扁锥壳的非线性弯曲。Hamada[96]将这一方法应用到波纹膜片上，求解了 Grover-Bell 膜片及 NBS 膜片的特征曲线，获得了膜片各点的应力值。Bihari 和 Elbert[97]考虑了带边缘波纹膜片的弯曲，将问题化为 6 个一阶非线性常微分方程，直接从微分方程求解得到了波纹膜片的挠度和径向位移，从数值分析观点考察，解决的是一个多点问题。陈山林[98]用修正迭代法求解了波纹圆板壳体的大挠度方程，得到了具有光滑中心的浅正弦波纹圆板在均匀载荷下的弹性特征，为了得到一个较好的初始近似，引入了初始近似修正系数。Andryewa[99]应用差分方法研究了具有型面锥度的波纹膜片的非线性弯曲问题。周哲伟和王蜀[100]讨论了圆弧形波纹膜片的特征曲线，在圆弧部分应用钱伟长等[101]关于圆环壳的一般线性解，并采用中心圆板部分的最大相对挠度为摄动系数，用摄动法和矩阵联乘法联合求解，获得了较为满意的结果。宋卫平和叶开沅[102]采用旋转扁壳大挠度理论研究波纹圆板在中心力作用下的变形、应力和稳定性，利用牛顿-样条函数法求解了浅正弦波纹圆板的非线性壳体方程，获得了屈曲前和屈曲后的解。

刘人怀和袁鸿[103]采用格林函数方法，将简化的 Reissner 方程化为积分方程，成功地求解了带边缘大波纹膜片的非线性弯曲问题。随后，袁鸿和刘人怀等[104-114]又继续用此方法研究了带大波纹膜片、波纹壳、波纹膜片的非线性弯曲和振动问题，并与其他理论和实验结果进行比较。最近，袁鸿等[115,116]采用摄动法和幂级数方法，得到了具有光滑中心的浅正弦波纹圆板以及具有光滑中心的锯齿形和梯形波纹壳具有中心挠度二次项的弹性特征。Wang 等[117]研究了温度变化下正弦波纹膜片的非线性自由振动。

另外，Chernopiskii[118]、Гузь 等[119]研究了厚壁波纹壳体，以波纹幅度作为小参数，使用摄动法对三维弹性体进行求解。Liew 等[120]采用无网格伽辽金方法和正交异性板理论，研究了矩形波纹板的非线性问题。Gulgazaryan 等[121]研究了波纹圆柱壳的非线性问题。有关波纹膜片的设计计算可以参考 Вольмир[122]、Пономарев 和 Андреева[123]、Andryewa[99]、刘广玉和庄肇康[124]、翁善臣[125]和刘人怀[126,127]的著作。

3）理论比较

用正交各向异性壳（板）处理波纹壳（圆板）问题有效简单，能同时讨论各种形状的波纹壳（圆板），便于工程应用。然而，简化本身也带来一些不足：对于波纹圆板，它得到的特征缺少中心挠度的偶次项，因而不能反映载荷反向时波纹壳（圆板）不同的刚度特征，而当波纹数较少及挠度较大时，这种差别可以是显著的；它的解答不能用于研究波纹壳（圆板）的应力分布和局部失稳现象；它只能解决分段均匀的波纹壳（圆板）问题，即每个波的波纹深度、波纹形状必须一致，而不能解决带有边缘波纹、波纹深浅不同的波纹壳问题。使用薄壳非线性理论进行分析能解决以上问题，但由于壳体大挠度方程本身的非线性和复杂性，增加了求解难度。钱伟长等[128]对采用环壳理论与正交异形板理论计算三圆弧波纹膜片进行了比较。

六、波纹管

1. 波纹管的构造

波纹管是一种环形波纹的圆柱型薄壁弹性壳体，具有刚度低、疲劳性能好、补偿能力强等特点（图 5）。

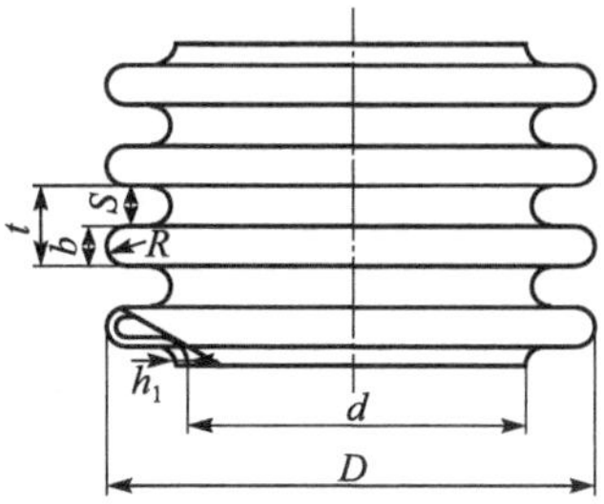

图 5 中，R 为波纹圆弧半径；D 为外径；d 为内径；t 为波距；h_1 为壁厚；b 为波纹厚度；S 为波间距。

图 5　波纹管

几种常见波纹管中，U 形波纹管应用最广泛，另外还有 C 形、Ω 形、S 形、锯齿形、正弦波形和矩形等（图 6）。每种波纹管都有其力学特点。

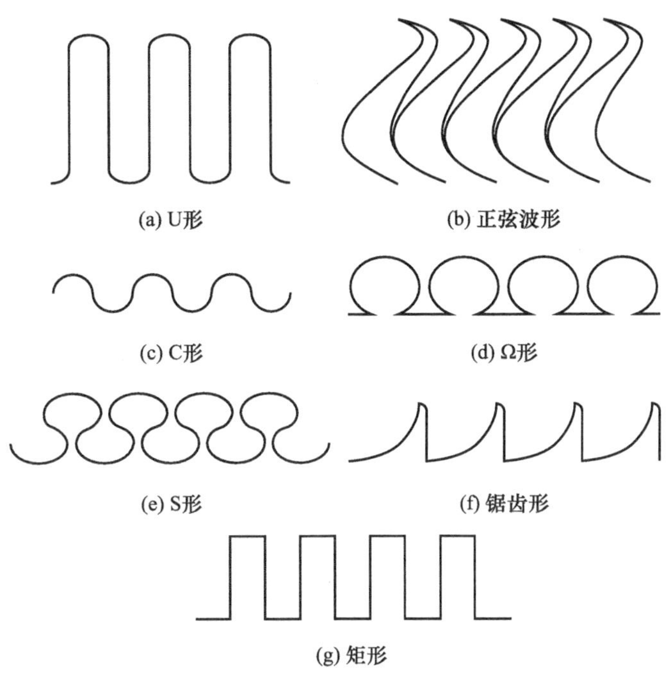

图 6　几种常见波纹管子午线

一般来说，波纹管的各几何参数是根据设计要求和工艺条件确定的，影响波纹管性能的参数主要来源于以下几个方面[129]：内径和外径的比值；波距和外径的比值；波纹厚度、波间宽与外径的比值；1 个波纹的允许位移与外径的比值。波纹管在仪器仪表中应用广泛，主要用途是压力测量仪表的测量元件，将压力转换成位移或力。此类波纹管管壁较薄，灵敏度较高，测量范围为数十帕至数十兆帕。另外，波纹管也可以用作密封隔离元件，将两种介质分开或防止有害流体进入设备的测量部分；它还可以用作补偿元件，利用其体积的可变性补偿仪器的温度误差；也用作为两个零件的弹性联接接头等。波纹管在各种技术领域里使用广泛，尤其是作为弹性敏感元件和各类管道的联结元件被用于矿山、石油、化工、冶金、电力、热力、航海、航天等工程设备中。

2. 研究现状

目前，大多数学者都利用轴对称单元来分析波纹管的稳定性，用解析法、数值分析、工程计算和实验应力等方法来探索波纹管的应力分布[130]。解析法是单层波纹管计算的基本方法，也是波纹管应力分析的力学基础。解析法主要包括能量法和渐进积分法。能量法是采用傅里叶级数表示各变量（即子午线弯曲变形），求出势能或余能的极小值来截取级数项，并求出余项级数；渐进积分法用渐进积分求解一般旋转壳的二阶微分方程。工程近似法是研究单层波纹管的另一种重要方法，它多采用直梁或曲梁模型对波纹管进行简化处理，应用材料力学的方法给出一些简单的设计公式和图表，以供工程使用。有限元法、有限差分法、边界元法和加权余量法等数值法也广泛运用到波纹管的力学研究中。对于多层波纹管，一般采用轴对称的简化模型，以减少计算难度；而实际波纹管各个方向的受力大多是非对称的，采取轴对称模型简化可带来几何形状的离散误差。

1）U 形波纹管力学分析

U 形波纹管是一种典型的弹性元件，在仪表工业中有着广泛的应用，其力学性能的研究已有 60 多年的历史。最早的分析是将其视为梁进行研究[130-132]。接着，把 U 形波纹管简化为组合元件，即由圆环管和圆形板组合，Turner[133]、Hamada 和 Takezono[134]、Anderson[135]和 Clark[136]等研究了线性变形，其结果与实验值相差甚大。

钱伟长和郑思梁[137-138]给出了周对称圆环壳的一般解，为用解析方法分析圆环壳进行了开创性的工作。此后，许多中国学者应用这一结果对 U 形波纹壳进行了研究。

钱伟长和吴明德[139]将 U 形波纹管的半个波分成内环壳、外环壳和环板 3 个部分，对环壳用已得的线性解，对环板用摄动法求大挠度解，由此得到 U 形波纹管解，较已有长度解有很大改善，但与实验值相差仍较大。接着，徐志翘等[140]在此基础上，分析了变厚度对刚度的影响。

刘人怀[141]将波纹管的半个波分成等厚度内环壳、外环壳和变厚度的截头扁锥壳，真实地反映 U 形波纹管的截面形式，并采用摄动法，进行了整体非线性分析计算，获得了与实验吻合的结果。接着，刘人怀和王志伟[142-143]再按壳体的几何非线性理论，应用积分方程迭代法和梯度法对 U 形波纹管进行了新的研究，获得了与实验值完全吻合的理论结果，此结果可直接供波纹管的设计使用。

此外，朱卫平[144]和于长波等[145]用有限元法分别研究波纹管的屈曲和疲劳问题。

2）Ω 形波纹管力学分析

Ω 形波纹管也称作 Ω 形膨胀节，是由圆环截面的波纹壳与附在开口波谷处直边段上的加强环所组成的。由几何学可知，由圆截面所产生的回转表面称为圆环或一般称为 Ω 形。Ω 形波纹管比 U 形波纹管承受内压能力高，其安全性和经济性显著。Ω 形膨胀节波纹壳本身为圆环形壳体，在内压作用下会产生环向与径向薄膜力，沿截面均匀分布[146]。

黎廷新和李添祥[147]指出，Ω形膨胀节比常用U形膨胀节承受内压能力高，在较高内压场合使用，其安全性和经济性显著。1993年，Ω形膨胀节首次列入了美国膨胀节制造商协会（EJMA）标准中。然而，对于Ω形膨胀节由内压载荷和位移载荷产生的应力，人们的研究和EJMA标准的计算仅限于理想圆环的Ω形膨胀节。现实中，多波纹整体无模液压成型的Ω形膨胀节的波纹圆环截面，制造后通常成为近似椭圆形，有一定的椭圆度，椭圆度影响着载荷应力的大小和分布，特别是较大地影响内压应力的大小和分布，从而影响着膨胀节的强度与疲劳寿命。有关圆环椭圆度对应力的影响，可以得出两个结论：由于多波整体无模液压制成的Ω形膨胀节圆环的截面形状与椭圆形状较为近似，因此以椭圆方程表示的圆环形状来计算Ω形膨胀节圆环的应力，其结果与实际Ω形膨胀节的实测应力结果较符合；Ω形膨胀节圆环椭圆度的大小，影响着Ω形膨胀节内压应力的大小和位移应力的大小，特别是对内压应力大小影响较大。为保障Ω形膨胀节的强度和疲劳寿命，制造中应大力降低Ω形膨胀节圆环椭圆度，如控制在15%范围以内或更小。

Clark[148]应用渐进积分法，得到轴对称载荷作用下环形壳体的弯曲微分方程的解，分析了Ω形波纹管承受轴向载荷和侧向压力时的应力。吴培媛和黎廷新[149]采用了黄黔的摄动初参数法，分析了Ω形波纹管圆环开口量对各向应力最大值和单波轴向刚度值的影响。易南伟等[150]论述了椭圆截面Ω形膨胀节内压应力和位移应力的弹塑性有限元计算，并对其进行了试验验证和分析。苏旭[151]通过试验测定Ω形波纹管的轴向载荷与位移量的关系曲线，得出波纹管的实测初始刚度值，验证Ω形波纹管理论计算刚度与实测刚度的一致性。

3）S形波纹管力学分析

S形波纹管子午线由圆弧段周期连结而成，连结点中面线是光滑相连的。陈山林等[152-154]用初参数积分方程法求解了S形波纹管问题，计算并研究了S形波纹管的轴对称应力和位移，载荷考虑内压和轴力。陈民[155]将圆环壳弯曲问题一般解应用于处理S形波纹管纯弯曲应力和位移的计算，得到较精确的应力和位移结果，为求解S形波纹管一般弯曲问题奠定了基础。计算分析表明：S形波纹管最大径向应力总是发生在波谷附近区域，且环壳截面半径及初相角对位移的影响很大。S形波纹管最佳柔韧性状态与最不利应力状态相对应，但只要取适当的初相角，便可获得最佳柔韧性。陈山林和黄骏[152]通过假定变形的对称性，给出了S形波纹管问题的一种初参数积分方程数值解法，这对大多数实际的波纹管问题是适用的。

4）C形波纹管力学分析

陈民[156]引用圆环壳弯曲方程一般解，对C形波纹管在端部弯矩作用下变形进行了系统分析计算，提出了实际工程中刚度设计公式，并对精确分析与简化分析弯曲刚度的差异进行了研究，得出了简化分析刚度设计公式的修正式，从而可方便地将弯曲刚度设计转化为轴对称刚度设计。朱卫平[157]讨论了C形波纹管在子午面内整体弯曲的应力分布和端面角位移，从柔性圆环壳在子午面内整体弯曲的复变量方程出发[158]，化边值问题为初值问题，用S. Gill法得到了该波纹管的数值解，并指出了解答的适用范围。

王永岗等[159]基于B. B. 诺沃日洛夫薄壳理论，用有限元法研究了C形波纹管非轴对称自由振动的广义特征值问题，通过选取两结点非协调曲边旋转壳单元作为离散单元，解决了C形波纹管因子午线曲率有突变而在求解上造成的困难，将所有相关变量沿其环向进行了傅里叶展开，得出了任意子午线形状的波纹管在任意环向谐波时的特征值，并对两种端部条件下的C形波纹管进行了特征值分析。朱卫平和黄黔[160]计算了C形波纹管的角向刚度、横向刚度和相应的应力分布，将波纹管的凸面和凹面分开处理，分别应用中细柔性圆环壳整体弯曲的一般解[26]，使连接点满足内力和变形连续性条件，并将所得结果与相应的数值积分解、其他理论解及实验进行了比较。结果表明，中细柔性圆环壳的方程和一般解准确可信。

5）矩形波纹管力学分析

矩形波纹管由圆柱壳和圆环板组成。曹红芍等[161]用EJ-MA公式对矩形波纹管进行了应力计算，并用有限元法和试验法进行了应力分析研究，指出矩形波纹管最大应力位于接角焊缝处。

七、结束语

精密仪器仪表弹性元件理论的研究已取得长足进展，但仍需要继续努力，解决工程设计制造中的一些难题。

参考文献

[1] Poisson S D. Mémoire sur l'equilibre et le movememt des corps élastiquse. paris，Mem. de l'Acad，1829.

[2] 铁摩辛柯S，沃诺斯基S. 板壳理论. 北京：科学出版社，1977.

[3] 刘人怀. 板壳力学. 北京：机械工业出版社，1990.

[4] von Kármán T. Festigkeits problem in Mashinenbau. Enzyklopadie der mathematischen Wissenschaften，1910，4（27）：348-351.

[5] Hencky H. Über den Spannungszustand in Kreisrunden Platten mit Versehwindender Biegungstreifikeit. Zeit. f. Math. n. Phyzik，1915，63：311-317.

[6] Vincent J J. The bending of a thin circular plate. Phil.，1931，12（75）：185-196.

[7] Way S. Bending of circular plate with large deflection. ASME Transactions，Applied Mech.，1934，56：627-636.

[8] Panov D Y. Nonlinear analysis on a problem of elastic theory by Galerkin's method. Prikl. Mat. Mekh.，1939，3（2）：139.

[9] Chien W Z. Large deflection of a circular clamped plate under uniform pressure. The Chinese Journal of Physics，1947，7（2）：102-113.

[10] Chien W Z. Asymptotic behavior of a thin clamped circular plate under uniform normal pressure at very large deflection. The Science Report of National Tsinghua University，1948，5（1）：1-24.

[11] 钱伟长. 弹性圆薄板大挠度问题——轴对称圆薄板在大挠度情形下的一般理论. 北京：中国科学院，1954：1-22.

[12] 钱伟长. 弹性圆薄板大挠度问题——圆薄板大挠度问题的摄动法. 北京：中国科学院，1954：37-55.

[13] 钱伟长，叶开沅. 圆薄板大挠度问题. 中国物理学报，1945，(3)：209-238.

[14] 钱伟长，叶开沅. 圆薄板大挠度问题的设计资料. 中国机械工程学报，1955，(1)：15-35.

[15] Chien W Z. Problem of large deflection of circular plate. Archiwum Mechaniki Stosowanej，Warszawa，1956，8 (1)：1-12.

[16] 叶开沅. 边缘载荷下环形薄板大挠度问题. 物理学报，1953，(2)：110-129.

[17] Li D，Liu R H. Nonlinear Vibration of Thin Circular Plates，Advances in Applied Mathematics and Mechanics in China. Oxford：International Academic Publishers and Pergamon Press，1991，15-23.

[18] 王永亮，刘人怀，王鑫伟. 弹性支承环形板的非线性弯曲. 固体力学学报，1998，19 (增刊)：107-110.

[19] von Kármán，Tsien H S. The buckling of spherical shells by external pressure. J. of Aeronautical Sciences，1939，7 (2)：43-50.

[20] Chien W Z，Hu H C. On the snapping of a thin spherical cap. 9th International Congress of Applied Mechanics，Brussels，1956.

[21] Feodosev V I. Calculation of thin clicking membranes. Prikl. Mat. Mekh.，1946，10 (2)：295-300.

[22] Simons R M. A power series solution of the nonlinear equations for axi-symmetrical bending of shallow spherical shells. J. Math. Phys.，1956，35 (2)：164.

[23] Reiss E L，Greenberg H J，Keller H B. Nonlinear deflections of shallow spherical shells. J. Aero. Sci.，1957，24 (7)：533-543.

[24] Kaplan A，Feng Y Z. A nonlinear theory of bending and bucking of thin elastic shallow spherical shells. NACA TN 3212，1954.

[25] Krivoshchier N I，Mushtary H M. On the bending of shallow spherical segment under the action of the external normal pressure. Izv，Branch of Kazan，AN SSSR，Ser. Fiz-matem. i Tekhn.，1957，12：69-84.

[26] 叶开沅，刘人怀，平庆元，等. 在对称线布载荷作用下的圆底扁薄壳的非线性稳定性. 科学通报，1965，(2)：142-144.

[27] 叶开沅，刘人怀，张传智，等. 圆底扁薄球壳在边缘力矩作用下的非线性稳定问题. 科学通报，1965，(2)：145-147.

[28] 刘人怀. 在内边缘均布力矩作用下中心开孔圆底扁球壳的非线性稳定问题. 科学通报，1965，(3)：253-255.

[29] 刘人怀. 在边缘载荷作用下中心开孔圆底扁薄球壳的轴对称稳定性. 力学学报，1977，(3)：206-212.

[30] Liu R H，Cheng Z Q. On the nonlinear stability of a truncated shallow spherical shell under a concentrated load. Applied Mathematics and Mechanics，1988，9 (2)：101-112.

[31] Liu R H，Li D. On the nonlinear stability of a truncated shallow spherical shell under a uniformly distributed load. Applied Mathematics and Mechanics，1988，9 (3)：227-240.

[32] Liu R H，He L H. On the nonlinear stability of a truncated shallow spherical shell under axisymmetrically distributed load. Solid Mechanics Archives，1989，14 (2)：81-102.

[33] 刘人怀，宗赴传. 复合载荷作用下开顶扁球壳的非线性稳定问题. 华东第五届固体力学学术会议，无锡，1988.

[34] 刘人怀. 新型跳跃薄膜的研究. 仪表技术与传感器，1991，(3)：10-11.

[35] 刘人怀，张小果. 均布载荷作用下具有硬中心的开顶扁球壳的非线性屈曲. 工程力学学报，1995，3（增刊）：1839-1844.

[36] 刘人怀，梅魁银. 中心开孔扁球壳在均布载荷作用下的非线性屈曲. 暨南大学学报（自然科学版），1996，(5)：1-7.

[37] 叶开沅，顾淑贤. 均布载荷作用下圆底扁薄壳的非线性稳定性. 北京：北京大学出版社，1981：280-287.

[38] Panov D Y. On the stability of a bimetallic membrane on heating. Prikl. Mat. Mekh.，1947，11（6）：603-610.

[39] Wittrick W H，Myers D M，Blunden W R. Stability of a bimetallic dick. Q. J. Mech. Appl. Math.，1953，6（1）：15-31.

[40] Grigolyuk E I. Thin bimetallic shells and plates. Engineer's Collection，1953.

[41] 罗祖道，聂德耀，刘汉东，等. 双层金属球面扁壳的热稳定性. 力学学报，1996，(1)：1-13.

[42] 刘人怀. 双层金属中心开孔扁球壳的非线性热稳定问题. 中国科学技术大学学报，1981，(1)：84-99.

[43] 刘人怀. 双层金属截头扁锥壳的热稳定性. 力学学报（特刊），1981：172-180.

[44] Liu R H. Non-linear thermal stability of bimetallic shallow shells of revolution. International Journal of Non-linear Mechanics，1983，18（5）：409-429.

[45] Hersey M D. Diaphragms for aeronautic instruments. NACA TR，1925：165.

[46] Griffith A A. The theory of pressure capsules. Great Britain Aeronautical Research Council，R&M，1928：1136.

[47] Haringx J A. The rigidity of corrugated diaphragms. Applied Scientific Research，Series A，1951，2：299-325.

[48] Haringx J A. Stresses in corrugated diaphragms. Anniversary Volume on Applied Mechanics dedicated to C. B. Biezeno. H. stam，Haarlem，The Netherlands，1953：199-213.

[49] Haringx J A. Nonlinearity of corrugated diaphragms. Applied Scientific Research，Series A，1956，6：45-52.

[50] Haringx J A. Design of corrugated diaphragms. Trans. of the ASME，1957，79（1）：55-61.

[51] Анлреева Д Е. Расчет гофрированных мембран как анизотропных пластинок. Инженернлй Собрник，1955，21：128-141.

[52] Akasaka T. On the elastic properties of the corrugated diaphragm. J. of the Japan Society of Aeronautical Engineering，1955，3：279-288.

[53] Akasaka T，Takagishi T. Vibration of corrugated diaphragm. Bulletin of JSME，1958，1（3）：215-221.

[54] Rubin C. Optimum design for a corrugated diaphragm clamped to a shaft undergoing axial or angular displacement. J. of Engineering Materials and Technology，1975，97（4）：363-366.

[55] 刘人怀. 波纹圆板的特征关系式. 力学学报，1978，(1)：47-52.

[56] 刘人怀. 具有光滑中心的波纹圆板的特征关系式. 中国科学技术大学学报，1979，(2)：75-86.

[57] 刘人怀. 波纹环形板的非线性弯曲. 中国科学 A 辑，1984，(3)：247-253.

[58] Liu R H. Large deflection of corrugated circular plate with a plane central region under the action of concentrated loads at the center. Int. J. Non-Linear Mechanics，1984，19（5）：409-419.

[59] Liu R H. Large deflection of corrugated circular plate with plane boundary region. Solid Mechanics Archives，1984，9（4）：383-406.

[60] 刘人怀. 在复合载荷作用下波纹环形板的非线性弯曲. 中国科学 A 辑，1984，(6)：537-545.

[61] Liu R H. On large deflection of corrugated annular plates under uniform pressure. The Advances of Applied Mathematics and Mechanics in China，1987：139-152.

[62] 刘人怀. 复合载荷作用下波纹圆板的非线性分析. 应用数学和力学，1988，(8)：661-674.

[63] 刘人怀，宗赴传. 集中载荷作用下具有平面边缘的波纹圆板的非线性弯曲问题. 首届全国敏感元件与传感器学术会议，上海，1989：547-550.

[64] 刘人怀. 波纹膜片理论的研究. 仪表技术与传感器，1991，(5)：9-11.

[65] Liu R H，Zou R P. Non-linear bending of a corrugated annular plate with a plane boundary region and a non-deformable rigid body at the center under compound load. Int. J. Non-linear Mechanics，1993，28 (3)：353-364.

[66] 刘人怀，翟赏中. 均布载荷作用下波纹环形板的非线性弯曲. 华南理工大学学报（自然科学版），1994，22（增刊）：1-10.

[67] 刘人怀，徐加初. 中心受载下具有平面边缘区域的固支波纹环形板的非线性分析. 暨南大学学报（自然科学版），1995，(1)：1-13.

[68] 刘人怀. 膜盒基体的理论与设计. 澳门科技大学学报，2009，(1)：111-116.

[69] 卢耘耘，王秀喜. 复合载荷作用下波纹圆板和环形板的大挠度问题. 中国科学技术大学学报，1989，(2)：200-210.

[70] 卢耘耘，王秀喜，黄茂光. 任意载荷下波纹圆板大挠度弹性特征的级数解法. 应用数学和力学，1988，(12)：1097-1108.

[71] 王新志，王林祥，胡小方. 波纹圆薄板的非线性振动. 应用数学和力学，1987，(3)：237-245.

[72] Liu R H，Li D. On the non-linear bending and vibrating of corrugated circular plates. Int. J. Non-Linear Mechanics，1989，24 (3)：165-176.

[73] 李东，刘人怀. 修正迭代法在波纹圆板非线性振动问题中的应用. 应用数学和力学，1990 (1)：13-21.

[74] Шалашилин В И. К расчету оболчек выполненных нз гофрированного материала. В кн.，Проблемы устойчвостн в строительной механике. Москва：Стройиздат，1965：339-346.

[75] 刘人怀，王璠. 波纹扁球壳的大挠度方程. 钱伟长教授九十华诞祝寿文集. 上海：上海大学出版社，2003：15-22.

[76] 王璠，刘人怀. 波纹扁球壳的各向异性参数. 暨南大学学报（自然科学版），2003，(3)：10-15.

[77] Liu R H，Wang F. Nonlinear stability of corrugated shallow spherical shell. Int. J. of Applied Mechanics and Engineering，2005，10 (2)：295-309.

[78] 王璠，刘人怀. 波纹扁球壳的非线性动态屈曲. 振动工程学报，2005，(4)：426-432.

[79] 王璠，刘人怀，潘燕环. 正弦波纹扁球壳的非线性强迫振动. 力学季刊，2006，(2)：175-183.

[80] Панов Д Ю. О больщих пртибах круглых мембран со слабым гофром. Прикладная Математика и Механика，1941，5 (2)：303.

[81] Феодосьев В И. О вольших прогибах и устойчивости круглой мембраны с мелкой гофрировкой. Приклалная Математика и Механика，1945，9 (5)：389.

[82] Феодосьев В И. 精密仪器弹性软件的理论和计算. 卢文达，等译. 北京：科学出版社，1965.

[83] Grover H J，Bell J C. Some evaluations of stresses in aneroid capsules. Proceedings of the Society of Experimental Stress Analysis，1948：125-131.

[84] Way S. Bending of circular plates with large deflection. Trans. of the ASME，1934，56：627.

[85] Stange K. Der spannungszusang einer kreisringschall. Ingenieur-Archiv，1931，2：47-91.

[86] Timoshenko S. Theory of Plates and Shells. New York：Mcgraw-Hill，1940.

[87] Wildhack W A，Dressler R F，Lloyd E C. Investigations of the properties of corrugated diaphragm. Tran. of the ASME，1957，(1)：65-82.

[88] Dressler R F. Bending and stretching of corrugated diaphragms. J. of Basic Engineering，1959，81：651-659.

[89] Аксельрад Э Д. Уравнеия деформации оболочек вращения иизгиба тонкостенных стержней при больших упругих перемещениях. Механика и Машиностроение，Иэвестия АН СССР，1960，(4)：84-92.

[90] Аксельад Э Л. Нэгиб и потеря устойчвостн тонкостенных труб при лруб п гидростатическом давлении. Механика и Машиностроение，Иэвестия АН СССР，1962，(1)：98-114.

[91] Аксельрад Э Л. Расчет гофрированной мембраны как непологой оболчки. Механика и Машиностроение，Нэвестия АН СССР，1963，(5)：67-76.

[92] Аксельрад Э Л. Вольшие прогибы гофрированной мембраны как непологой оболочки. Механика и Машиностроение，Нэвестия АН СССР，1964，(1)：46-53.

[93] Аксельрад Э Л. периодпческие рещения осесимметричой задачи теории оболочек. Механика Твердого Тела，1966，(2)：77-83.

[94] Hamada M，Seguchi Y. On the accuracy of the v. Kármán equations for axisymmetric non-linear bending for circular plates. Trans. Japan Soc. Aero. Space Sci.，1965，8 (12)：6-14.

[95] Hamada M，Seguchi Y. On the strength of a disk spring. Bulletin of JSME，1966，9 (35)：492-502.

[96] Hamada M. Numerical method for nonlinear axisymmetric bending of arbitrary shells of revolution and large deflection analyses of corrugated diaphragm and bellows. Bulletin of JSME，1968，11 (43)：24-33.

[97] Bihari I，Elbert A. Deformation of circular corrugated plates and shells. Periodica Polytechnica Mechanical Engineering，1978，22 (2)：123-143.

[98] 陈山林. 浅正弦波纹圆板在均布载荷下的大挠度弹性特征. 应用数学和力学，1980，(2)：261-272.

[99] Andryewa L E. Elastic elements of instruments. Moscow：Masgiz，1981.

[100] 周哲伟，王蜀. 圆弧形波纹膜片的矩阵联乘法. 应用数学和力学，1985，(6)：551-566.

[101] 钱伟长，郑思梁. 轴对称圆环壳的一般解. 应用数学和力学，1980，(3)：287-300.

[102] 宋卫平，叶开沅. 中心集中载荷作用下波纹圆板的变形应力和稳定性研究. 中国科学，A辑，1989，(1)：40-47.

[103] Liu R H，Yuan H. Nonlinear bending of corrugated annular plate with large boundary corrugation. Applied Mechanics and Engineering，1997，2 (3) 353-367.

[104] 袁鸿，刘人怀. 均布载荷作用下带边缘大波纹膜片的非线性弯曲. 力学学报，2003，(1)：14-20.

[105] 袁鸿，刘人怀. 复合载荷作用下带边缘大波纹膜片的非线性弯曲. 应用数学和力学，2003，(4)：367-372.

[106] 袁鸿，张湘伟. 集中载荷作用下具有光滑中心波纹膜片的非线性分析. 力学季刊，2003，(1)：124-128.

[107] 袁鸿，刘人怀. 复合载荷作用下具有光滑中心波纹膜片的非线性分析. 应用力学学报，2003，

(1)：27-30.

[108] 袁鸿，刘人怀. 均布载荷作用下具有光滑中心波纹膜片的非线性分析. 应用力学学报，2004，(1)：117-120.

[109] 袁鸿，张湘伟. 波纹壳的格林函数方法. 应用数学和力学，2005，(7)：763-769.

[110] 袁鸿，张湘伟，刘人怀. 波纹膜片的非线性稳定. 工程力学，2005，(6)：202-206.

[111] 袁鸿，刘人怀. 均布载荷作用下波纹扁壳的非线性振动. 应用数学和力学，2007，(5)：514-520.

[112] 袁鸿. 复合载荷作用下波纹扁壳的非线性振动. 振动与冲击，2007，(12)：28-31.

[113] 袁鸿. 波纹扁壳非线性振动的格林函数方法. 中国科学 G 辑，2008，(5)：592-599.

[114] 袁鸿. 带刚性硬中心波纹扁壳的非线性振动. 暨南大学学报（自然科学版），2008，(3)：262-267.

[115] 袁鸿. 波纹壳的摄动解法. 应用力学学报，1999，(1)：144-148.

[116] 袁鸿，龚胜海，吴立彬. 浅正弦波纹圆板大挠度问题的摄动解法. 中北大学学报（自然科学版），2010，(4)：331-337.

[117] Wang Y G，Gao D，Wang X Z. On the nonlinear vibration of heated corrugated circular plates with shallow sinusoidal corrugations. International Journal of Mechanical Sciences，2008，50 (6)：1082-1089.

[118] Chernopiskii D I. Stressed state of thick-walled corrugated spherical shells. Translated from Prikladanaya Mekhanika，1979，15 (10)：128-133.

[119] Гузъ А Н，НЕМИЩ Ю Н. Методы возмущениий в пространстренных задачах теории упрутости. киев，1982.

[120] Liew K M，Peng L X，Kitipornchai S. Nonlinear analysis of corrugated plates using a FSDT and a meshfree method. Computer Methods in Applied Mechanics and Engineering，2007，196 (21-24)：2358-2376.

[121] Gulgazaryan G R，Gulgazaryan L G. Vibrations of a corrugated orthotropic cylindrical shell with free edges. International Applied Mechanics，2006，42 (12)：1398-1413.

[122] Вольмир А С. 柔韧板与柔韧壳. 卢文达，等译. 北京：科学出版社，1963.

[123] Пономарев С Л，Андреева Л Е. Расчет упругих злементов машин и приборов. Москва：Машиностроение，1980.

[124] 刘广玉，庄肇康. 仪表弹性元件. 北京：国防工业出版社，1981.

[125] 翁善臣，林友德，徐振远，等. 仪表弹性元件设计基础. 北京：机械工业出版社，1982.

[126] Liu R H. Study on Nonlinear Mechanics of Plates and Shells. New York，Beijing，Guangzhou：Science Press and Jinan University，1998.

[127] 刘人怀. 精密仪器仪表弹性元件设计的力学原理. 广州：暨南大学出版社，2006.

[128] 钱伟长，樊大钧，黄黔. 环壳理论与直交异性板理论在计算三圆弧波纹膜片上的比较. 应用数学和力学，1984，(1)：41-48.

[129] 胡寿镛. 波纹管几何参数的选定. 仪器制造，1983，(6)：16-21.

[130] 刘岩，段玫，张道伟. 波纹管应力分析研究进展. 管道技术与设备，2006，(4)：31-33，36.

[131] Feely F J，Goryl W M. Stress studies on piping expansion bellows. Journal of Applied Mechanics，1950，17 (2)：136-141.

[132] Murphy G. Analysis of stresses and displacements in heat exchanger expansion joints. Transactions of ASME，1952，74 (3)：397-402.

[133] Turner C E. Stress and deflection studies of flat plate and toroidal expansion bellows，subjected to axil，eccentric or internal pressure loading. J. Mech. Eng. Sci.，1959：113-143.

[134] Hamada M，Takezono S. Strength of U-shaped bellows，Bull. JSME，1965：525-531；1966：513-523.

[135] Anderson W F. Atomics International Division of North American Aviation. NAA-SR-4527，1964；1965.

[136] Clark R A. An expansion bellows problem. Trans. ASME，Ser. E，J. Appl. Mech，1970，37（1）：61-69.

[137] 钱伟长，郑思梁. 轴对称圆环壳的复变量方程和轴对称细环壳的一般解. 清华大学学报（自然科学版），1979，（1）：27-47.

[138] 钱伟长，郑思梁. 轴对称圆环壳的一般解. 应用数学和力学，1980，（3）：287-300.

[139] 钱伟长，吴明德. U型波纹管的非线性特性摄动法计算. 应用数学和力学，1983，（5）：595-608.

[140] 徐志翘，刘燕，杨嘉实，等. 变厚度U型波纹管大挠度问题的摄动解. 清华大学学报（自然科学版），1985，（1）：39-51.

[141] 刘人怀. U型波纹管非线性变形的刚度和应力分析理论. 第九届全国弹性元件学术会议，广州，1992.

[142] 刘人怀，王志伟. 变厚度U型波纹管非线性变形分析. 管道技术与设备，1996，（1）：1-4.

[143] Liu R H，Wang Z W. Nonlinear deformation analysis of a U-shaped bellows with varying thickness. Archive of Applied Mechanics，2000，70：366-376.

[144] 朱卫平. U型波纹管及相关结构环向屈曲的有限元分析——基本方程及环板的屈曲. 应用力学学报，2002（4）：19-25，158-159.

[145] 于长波，王建军，李楚林，等. 多层U型波纹管的疲劳寿命有限元分析. 压力容器，2008，（2）：23-27.

[146] 李建国. Ω形膨胀节. 化工设备与管道，1998，（5）：13-17.

[147] 黎廷新，李添祥. Ω形膨胀节圆环的椭圆度对应力的影响. 压力容器，1996，（6）：19-22，3.

[148] Clark R A. On theory of thin elastic toroidal shells. Journal of Mathematics Physicals，1950，29（1/4）：146-177.

[149] 吴培媛，黎廷新. Ω形波纹管圆环开口量对应力和刚度的影响. 石油化工设备，1997，（1）：11-14.

[150] 易南伟，黎廷新，岑汉钊，等. 椭圆Ω截面形膨胀节的应力分析. 石油化工设备，1998，（3）：4-7.

[151] 苏旭. Ω形波纹管刚度研究. 广东造船，2004，（4）：36-38.

[152] 陈山林，黄骏. 圆环壳初参数积分方程的应用——S型波纹管问题. 重庆建筑工程院学报，1989，（2）：44-52.

[153] 陈山林. 轴对称圆环壳的初参数积分方程及其应用. 力学学报，1989，（6）：714-721.

[154] 陈山林. S型波纹管的轴对称应力和位移. 成都科技大学学报，1982，（3）：49-58.

[155] 陈民. S型波纹管纯弯曲应力和位移的研究. 成都科技大学学报，1989，（5）：127-132.

[156] 陈民. C型波纹管纯弯曲刚度设计公式. 成都科技大学学报，1990，（5）：63-66，74.

[157] 朱卫平. 用初参数法解C型波纹管在子午面内整体弯曲. 力学季刊，2000，（3）：311-315.

[158] 朱卫平，黄黔，郭平. 柔性圆环壳在子午面内整体弯曲的复变量方程及细环壳的一般解. 应

用数学和力学，1999，(9)：889-895.

[159] 王永岗，戴诗亮，吕英民. C型波纹管的非轴对称自振频率. 清华大学学报（自然科学版），2002，(2)：216-219.

[160] 朱卫平，黄黔. 中细柔性圆环壳整体弯曲的一般解及在波纹管计算中的应用（Ⅲ）——C型波纹管的计算. 应用数学和力学，2002，(10)：1025-1034.

[161] 曹红芍，杨方，李德雨，等. 矩形波纹管应力分析. 压力容器，2009，(6)：24-27.

第二章　单层板壳非线性力学

在内边缘均布力矩作用下中心开孔圆底扁球壳的非线性稳定问题①

本文在前两个问题[1,2]的基础上对在内边缘均布力矩作用下中心开孔圆底扁球壳的非线性稳定问题进行了研究。由于此问题在中心有奇异性，求解时不很容易，我们用修正迭代法容易地克服了这个困难。据作者所知，尚未有人研究过此问题。

今设有一厚度为 h，中曲面半径为 R，内、外缘半径分别为 b，a 的中心开孔圆底扁球壳，在内边缘上有均布力矩（$-M$）作用，如图 1 所示。

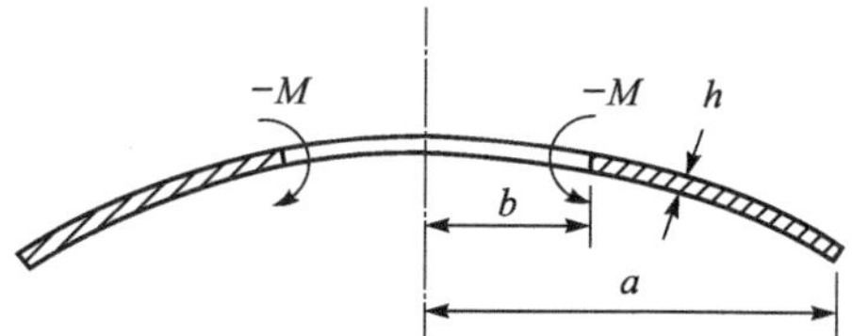

图 1　中心开孔圆底扁球壳

为了求解简单起见，引入下列无量纲量

$$\left.\begin{aligned}&\rho=\frac{r}{a},\alpha=\frac{b}{a},y=\sqrt{12(1-\nu^2)}\,\frac{W}{h},\\&\theta=-\frac{\mathrm{d}y}{\mathrm{d}\rho},\quad s_r=\frac{12(1-\nu^2)a^2}{Eh^3}N_r,\\&s=\rho s_r,\quad k=\sqrt{12(1-\nu^2)}\,\frac{a^2}{Rh}\\&m=\frac{12(1-\nu^2)\ \sqrt{12(1-\nu^2)}a^2}{Eh^4}M\end{aligned}\right\}\tag{1}$$

其中，ν 是泊松比，E 是弹性模量。

于是，挠度 W 和径向薄膜力 N_r 所满足的本问题的基本方程和边界条件[3]的无量纲形式为

$$\left.\begin{aligned}L(\rho\theta)&=s\left(-k+\frac{\theta}{\rho}\right)\\L(\rho s)&=\theta\left(k-\frac{\theta}{2\rho}\right)\end{aligned}\right\}\tag{2a,b}$$

---

① 本文原载《科学通报》，1965，(3)：253-255.（本文未阐明的符号与文献 [1] 相同）.

$$
\left.\begin{aligned}
&\text{内边缘悬空} \\
&\text{当}\ \rho=\alpha\ \text{时，}\quad \frac{\mathrm{d}\theta}{\mathrm{d}\rho}+\nu\frac{\theta}{\rho}=-m, \\
&\qquad s=0 \\
&\text{外边缘简单支承} \\
&\text{当}\ \rho=1\ \text{时，}\quad \frac{\mathrm{d}\theta}{\mathrm{d}\rho}+\nu\frac{\theta}{\rho}=0, \\
&\qquad s=0
\end{aligned}\right\}\tag{3}
$$

其中
$$
L(\cdots)=\frac{\mathrm{d}}{\mathrm{d}\rho}\frac{1}{\rho}\frac{\mathrm{d}}{\mathrm{d}\rho}(\cdots)
$$

我们用修正迭代法来求解非线性边值问题（2）、（3）。我们选取壳外缘转角 β 作为迭代参数。

在一次近似中，略去方程（2a）右端中含 s 的项，有下列边值问题

$$
\left.\begin{aligned}
&L(\rho\theta_1)=0 \\
&L(\rho s_1)=\theta_1\left(k-\frac{\theta_1}{2\rho}\right)
\end{aligned}\right\}\tag{4}
$$

$$
\left.\begin{aligned}
&\text{当}\ \rho=\alpha\ \text{时，}\quad \frac{\mathrm{d}\theta_1}{\mathrm{d}\rho}+\nu\frac{\theta_1}{\rho}=-m, \\
&\qquad s_1=0 \\
&\text{当}\ \rho=1\ \text{时，}\quad \frac{\mathrm{d}\theta_1}{\mathrm{d}\rho}+\nu\frac{\theta_1}{\rho}=0, \\
&\qquad s_1=0
\end{aligned}\right\}\tag{5}
$$

解线性边值问题（4）、（5），并利用条件

$$
\text{当}\ \rho=1\ \text{时},\theta_1=\beta \tag{6}
$$

我们可得到一次近似解

$$
\left.\begin{aligned}
&\theta_1=f_1(\rho,\beta) \\
&s_1=g_1(\rho,k,\beta)
\end{aligned}\right\}\tag{7}
$$

在二次近似中，我们有下列边值问题

$$
L(\rho\theta_2)=s_1\left(-k+\frac{\theta_1}{\rho}\right)\tag{8}
$$

$$
\left.\begin{aligned}
&\text{当}\ \rho=\alpha\ \text{时},\frac{\mathrm{d}\theta_2}{\mathrm{d}\rho}+\nu\frac{\theta_2}{\rho}=-m \\
&\text{当}\ \rho=1\ \text{时},\frac{\mathrm{d}\theta_2}{\mathrm{d}\rho}+\nu\frac{\theta_2}{\rho}=0
\end{aligned}\right\}\tag{9}
$$

解线性边值问题（8）、（9），并利用条件

$$
\text{当}\ \rho=1\ \text{时},\theta_2=\beta \tag{10}
$$

我们得到内边缘力矩 m 的二次近似解析解

$$
m=(\alpha_1+\alpha_2k^2)\beta+\alpha_3k\beta^2+\alpha_4\beta^3 \tag{11}
$$

其中

$$\left.\begin{aligned}
\alpha_1 &= \frac{(1-\alpha^2)\lambda_1\lambda_2}{2\alpha^2} \\
\alpha_2 &= \frac{1}{2(1-\alpha^2)\alpha^2}\Big(\frac{2+\nu}{96}\lambda_1\alpha^8 - \frac{1}{16}\lambda_1\lambda_2\alpha^6\ln\alpha + \frac{1+4\nu+\nu^2}{24}\alpha^6 - \frac{1}{4}\lambda_2^2\alpha^4\ln^2\alpha \\
&\quad - \frac{2+5\nu+\nu^2}{16}\alpha^4 + \frac{1}{16}\lambda_1\lambda_2\alpha^2\ln\alpha + \frac{1+4\nu+\nu^2}{24}\alpha^2 + \frac{2+\nu}{96}\lambda_1\Big) \\
\alpha_3 &= -\frac{1}{2(1-\alpha^2)\alpha^2}\Big(\frac{2+\nu}{128}\lambda_1^2\alpha^8 - \frac{1-5\nu}{64}\lambda_1\lambda_2\alpha^6\ln\alpha + \frac{7+24\nu+\nu^2}{256}\lambda_1\alpha^6 \\
&\quad - \frac{5-11\nu}{32}\lambda_2^2\alpha^4\ln^2\alpha - \frac{4+3\nu+5\nu^2}{32}\lambda_2\alpha^4\ln\alpha - \frac{15+36\nu+3\nu^2}{256}\lambda_1\alpha^4 \\
&\quad - \frac{1-7\nu}{32}\lambda_2^2\alpha^2\ln^2\alpha + \frac{5+7\nu^2}{32}\lambda_2\alpha^2\ln\alpha - \frac{7-\nu}{256}\lambda_1^2\alpha^2 - \frac{1}{64}\lambda_1\lambda_2^2\ln\alpha + \frac{11+6\nu-\nu^2}{256}\lambda_1\Big) \\
\alpha_4 &= \frac{1}{2(1-\alpha^2)\alpha^2}\Big(\frac{2+\nu}{768}\lambda_1^3\alpha^8 + \frac{1}{64}\nu\lambda_1^2\lambda_2\alpha^6\ln\alpha + \frac{7+16\nu+\nu^2}{768}\lambda_1^2\alpha^6 \\
&\quad - \frac{1-7\nu}{64}\lambda_1\lambda_2^2\alpha^4\ln^2\alpha - \frac{7+8\nu+9\nu^2}{128}\lambda_1\lambda_2\alpha^4\ln\alpha - \frac{1-8\nu-22\nu^2-3\nu^4}{256}\alpha^4 \\
&\quad - \frac{1-2\nu+5\nu^2}{32}\lambda_2^2\alpha^2\ln^2\alpha + \frac{3+\nu+4\nu^2}{64}\lambda_1\lambda_2\alpha^2\ln\alpha \\
&\quad - \frac{23+34\nu+72\nu^2+46\nu^3+17\nu^4}{768}\alpha^2 \\
&\quad + \frac{1}{64}\lambda_1\lambda_2^3\ln^2\alpha + \frac{1+3\nu}{128}\lambda_1\lambda_2^2\ln\alpha + \frac{17+13\nu+27\nu^2+31\nu^3+8\nu^4}{768}\Big) \\
\lambda_1 &= 1-\nu \\
\lambda_2 &= 1+\nu
\end{aligned}\right\} \tag{12}$$

由极值条件$\frac{\mathrm{d}m}{\mathrm{d}\beta}=0$解得：$\beta^* = \frac{-\alpha_3 k \pm \sqrt{(\alpha_3^2 - 3\alpha_2\alpha_4)\ k^2 - 3\alpha_1\alpha_4}}{3\alpha_4}$ (13)

将β^*代入式（11）中，得临界力矩公式

$$m^* = (\alpha_1 + \alpha_2 k^2)\beta^* + \alpha_3 k\beta^{*2} + \alpha_4\beta^{*3} \tag{14}$$

其中对应于式（13）负号的m^*为上临界力矩，正号的m^*为下临界力矩。

确定临界点的公式为

$$k_0 = \sqrt{\frac{3\alpha_1\alpha_4}{\alpha_3^2 - 3\alpha_2\alpha_4}} \tag{15}$$

图 2 绘出了各种α值下的稳定性曲线。通过计算得出两点结论：

（1）中心开孔太大或太小，壳体的稳定性能越好；

（2）$\alpha=0.40$的中心开孔圆底扁球壳是对失稳反应最灵敏的壳体，此时，壳体的几何参数k达到最小值

$$(k_0)_{\min} = 12.65$$

我们使式（11）中的$k=0$，还附带得到了外缘简支、内缘承受均布力矩的环形薄板大挠度问题的刚度设计公式。

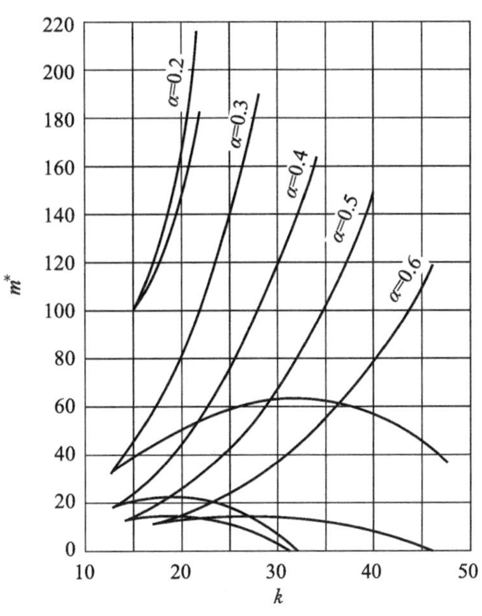

图 2 各种 α 值下的稳定曲线（ν=0.3）

本文所用的方法不难推广到中心开孔圆底扁球壳在其他边缘条件和其他载荷下的非线性稳定问题中。

参 考 文 献

［1］ 叶开沅，刘人怀，平庆元，等. 在对称线布载荷下的圆底扁薄球壳的非线性稳定问题. 科学通报，1965 (2)：142.

［2］ 叶开沅，刘人怀，张传智，等. 圆底扁薄球壳在边缘力矩作用下的非线性稳定问题. 科学通报，1965 (2)：145.

［3］ Chien W. Z.，The intrinsic theory of thin shells and plates. Quart Appl. Math.，1944，1 (4)：297-327.

在边缘载荷作用下中心开孔圆底扁薄球壳的轴对称稳定性①

一、前言

近代工程技术的发展，要求解决薄壳理论中核算壳体稳定性的问题。这对于选择材料及结构尺寸有意义。本文讨论的问题对于精密仪器中的弹性元件及建筑工程中的壳体屋顶等的设计具有实际意义。

文献［1］首先指出了壳体屈曲的非线性特点，但由于非线性在数学上的困难，以后的进展缓慢。由于工程技术的迫切需要，近年来国内外对此问题的研究兴趣一直未减少。目前大家多从事最简单的壳体稳定性问题的研究，特别是扁球壳的对称屈曲问题。作者[2]曾用修正迭代法克服了非线性数学和扁球壳中心开孔所带来的困难。方法的特点是使用逐次逼近法的程序和摄动法的参数（挠度或转角）。在取相同参数的情况下，修正迭代法和摄动法所得的结果相同，但前者程序比较简单，能获得解析公式，结果精确度较高。本文继续用修正迭代法得到了在边缘载荷作用下中心开孔圆底扁薄球壳临界载荷的二次近似解析公式（此公式仅适合于轴对称失稳情形），并将计算结果绘成了图表。

二、基本方程

在均布载荷 q 的作用下，圆底扁薄球壳的轴对称大挠度弯曲方程为[3]

$$-\frac{D}{r}\frac{\mathrm{d}}{\mathrm{d}r}r\frac{\mathrm{d}}{\mathrm{d}r}\frac{1}{r}\frac{\mathrm{d}}{\mathrm{d}r}r(r\vartheta)+\frac{1}{r}\frac{\mathrm{d}}{\mathrm{d}r}[rN_r(\theta+\vartheta)]=-q \tag{1a}$$

$$\frac{r}{Eh}\frac{\mathrm{d}}{\mathrm{d}r}\frac{1}{r}\frac{\mathrm{d}}{\mathrm{d}r}(r^2N_r)+\vartheta(\theta+\frac{1}{2}\vartheta)=0 \tag{1b}$$

其中

$$\vartheta=\frac{\mathrm{d}w}{\mathrm{d}r},\ \theta=\frac{r}{R},\ D=\frac{Eh^3}{12(1-\nu^2)} \tag{2}$$

将方程（1a）两端乘以 $r\mathrm{d}r$，积分一次后，得

$$-Dr\frac{\mathrm{d}}{\mathrm{d}r}\frac{1}{r}\frac{\mathrm{d}}{\mathrm{d}r}(r\vartheta)+rN_r(\theta+\vartheta)=F(r) \tag{3}$$

其中

$$F(r)=C-\int qr\,\mathrm{d}r \tag{4}$$

另外，关于 u、M_r、Q，还有公式：

$$u=\frac{r}{Eh}\left[(1-\nu)N_r+r\frac{\mathrm{d}N_r}{\mathrm{d}r}\right],$$

① 本文原载《力学学报》，1977，(3)：206-212.

$$M_r = -D\left(\frac{d\vartheta}{dr} + \frac{\nu}{r}\vartheta\right), \tag{5a ～ c}$$

$$Q = -D\frac{d}{dr}\frac{1}{r}\frac{d}{dr}(r\vartheta)$$

应用公式（5c），可把方程（3）改写为

$$r[Q + N_r(\theta + \vartheta)] = F(r) \tag{6}$$

由于我们要研究如图 1 所示的中心开孔圆底扁薄球壳的非线性稳定问题，故使用基本方程（3）和（1b）时，需要确定载荷函数 F（r）。此时，内边缘承受了线布载荷 p，壳体表面无均布载荷。F（r）可以由壳体部分的平衡条件来确定。这个壳体部分是用圆锥曲面沿半径为 r 的圆周切割出来的（图 2）。将作用在壳体部分的全部力投影到对称轴上，便有

$$2\pi r[Q\cos(\theta + \vartheta) + N_r\sin(\theta + \vartheta)] + 2\pi bp = 0 \tag{7}$$

根据角（$\theta+\vartheta$）微小的条件[3]，式（7）可简化为

$$r[Q + N_r(\theta + \vartheta)] = -bp \tag{8}$$

将式（8）同式（6）比较，就得

$$F(r) = -bp \tag{9}$$

应用结果（9），方程（3）成为

$$D\frac{d}{dr}\frac{1}{r}\frac{d}{dr}(r\vartheta) - N_r(\theta + \vartheta) = \frac{b}{r}p \tag{10}$$

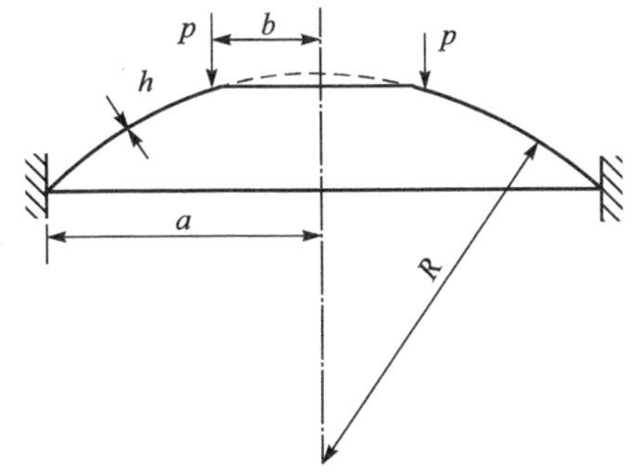

图 1　中心开孔圆底扁薄球壳

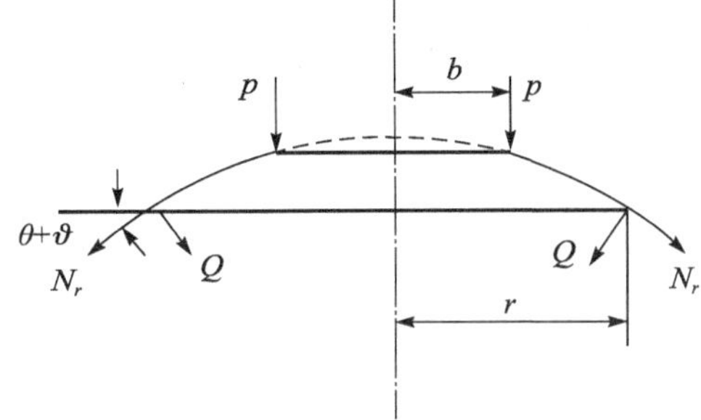

图 2　载荷函数 F（r）的确定

于是，方程（10）和（1b）就组成内边缘承受线布载荷的中心开孔圆底扁薄球壳的轴对称大挠度弯曲方程。与此方程有关的边界条件，我们仅讨论下面一种情况。

当 $r=a$ 时，边缘夹紧：

$$w = 0, \quad \vartheta = 0, \quad u = 0 \tag{11}$$

当 $r=b$ 时，边缘悬空：

$$M_r = 0, \quad N_r = 0 \tag{12}$$

为了使求解简单，我们所给的条件（12）中的第二式是一个近似的条件，球壳中心孔越小，此条件的精确度越好。

为了简化计算，我们引进下列无量纲量：

$$\rho=\frac{r}{b},\quad \alpha=\frac{a}{b},\quad y=\sqrt{12\lambda_1\lambda_2}\,\frac{w}{h}+\frac{1}{2}k\rho^2,$$

$$\varphi=-\frac{\mathrm{d}y}{\mathrm{d}\rho},\quad S=\frac{12\lambda_1\lambda_2 b^2}{Eh^3}\rho N_r,\quad k=\sqrt{12\lambda_1\lambda_2}\,\frac{b^2}{Rh},\tag{13}$$

$$P=\frac{\sqrt{12\lambda_1\lambda_2}\,b^3}{Dh}p$$

其中，$\lambda_1=1-\nu$，$\lambda_2=1+\nu$。

于是，由基本方程（10）、（1b）和边界条件（11）、（12）得到了无量纲边值问题

$$L(\rho\varphi)=S\varphi-P,$$

$$L(\rho S)=\frac{1}{2}(k^2\rho^2-\varphi^2)\tag{14a,b}$$

$$\text{当}\ \rho=\alpha\ \text{时，}\ y=\frac{1}{2}k\rho^2,\ \varphi=-k\rho,\ \rho\frac{\mathrm{d}S}{\mathrm{d}\rho}-\nu S=0\tag{15}$$

$$\text{当}\ \rho=1\ \text{时，}\ \frac{\mathrm{d}\varphi}{\mathrm{d}\rho}+\nu\frac{\varphi}{\rho}=-\lambda_2 k,\ S=0\tag{16}$$

其中

$$L(\cdots)=\rho\frac{\mathrm{d}}{\mathrm{d}\rho}\frac{1}{\rho}\frac{\mathrm{d}}{\mathrm{d}\rho}(\cdots)$$

这样，我们的问题便化为在边界条件（15）和（16）下求解非线性微分方程组（14）。

三、边值问题的求解

本文与前文[2]的不同之处是取内缘挠度作为迭代参数。这时非线性边值问题转化为一系列线性边值问题。

在一次近似中，略去方程（14a）中的非线性项 $S\varphi$，得到线性微分方程组

$$L(\rho\varphi_1)=-P,$$

$$L(\rho S_1)=\frac{1}{2}(k^2\rho^2-\varphi_1^2)\tag{17a,b}$$

相应的边界条件为

$$\text{当}\ \rho=\alpha\ \text{时，}\quad y_1=\frac{1}{2}k\rho^2,\quad \varphi_1=-k\rho,\quad \rho\frac{\mathrm{d}S}{\mathrm{d}\rho}-\nu S_1=0\tag{18}$$

$$\text{当}\ \rho=1\ \text{时，}\quad \frac{\mathrm{d}\varphi_1}{\mathrm{d}\rho}+\nu\frac{\varphi_1}{\rho}=-\lambda_2 k,\ S_1=0\tag{19}$$

将方程（17a）积分，利用边界条件（18）和（19），得环形薄板小挠度理论的解：

$$\varphi_1=-\frac{1}{2}P\rho\ln\rho+\left[\frac{1}{2}(\lambda_1\alpha^2\ln\alpha+1)\beta_1 P-k\right]\rho$$

$$+\frac{1}{2}(\lambda_2\ln\alpha-1)\alpha^2\beta_1 P\rho^{-1}\tag{20}$$

此处 $\beta_1=(\lambda_1\alpha^2+\lambda_2)^{-1}$。

我们以中心开孔圆底扁薄球壳的无量纲内边缘挠度 Y_m 为迭代参数：

$$Y_m = \sqrt{12\lambda_1\lambda_2}\,\frac{w}{h}\bigg|_{r=b} \tag{21}$$

再应用式（13）和（18），可得

$$Y_m = \int_1^{\alpha}(\varphi + k\rho)\mathrm{d}\rho \tag{22}$$

在一次近似中有

$$\begin{aligned} Y_m &= \int_1^{\alpha}(\varphi_1 + k\rho)\mathrm{d}\rho \\ &= \left(\frac{1}{8}\lambda_1\alpha^4 + \frac{1}{2}\lambda_2\alpha^2\ln^2\alpha - \alpha^2\ln\alpha + \frac{1}{4}\lambda_2\alpha^2 - \frac{3+\nu}{8}\right)\beta_1 P \end{aligned} \tag{23}$$

$$P = 2\beta_1^{-1}\beta_2 Y_m \tag{24}$$

其中

$$\beta_2 = \left(\frac{1}{4}\lambda_1\alpha^4 + \lambda_2\alpha^2\ln^2\alpha - 2\alpha^2\ln\alpha + \frac{1}{2}\lambda_2\alpha^2 - \frac{3+\nu}{4}\right)^{-1}$$

将式（24）代入式（20），有

$$\begin{aligned} \varphi_1 =& -\beta_1^{-1}\beta_2 Y_m\rho\ln\rho + [(\lambda_1\alpha^2\ln\alpha + 1)\beta_2 Y_m - k]\rho \\ & + (\lambda_2\ln\alpha - 1)\alpha^2\beta_2 Y_m\rho^{-1} \end{aligned} \tag{25}$$

利用式（25）再将方程（17b）积分，并利用边界条件（18）和（19），得到 S 的一次近似解为

$$S_1 = f(\rho, Y_m) \tag{26}$$

在二次近似中，将一次近似的解 φ_1、S_1 形式地代入方程（14a），便得关于 φ_2 的线性微分方程

$$L(\rho\varphi_2) = S_1\varphi_1 - P \tag{27}$$

相应的边界条件为

$$\text{当 } \rho=\alpha \text{ 时，} y_2 = \frac{1}{2}k\rho^2,\ \varphi_2 = -k\rho \tag{28}$$

$$\text{当 } \rho=1 \text{ 时，} \frac{\mathrm{d}\varphi_2}{\mathrm{d}\rho} + \nu\frac{\varphi_2}{\rho} = -\lambda_2 k \tag{29}$$

将方程（27）积分，并利用边界条件（28）和（29），得 φ 的二次近似解为

$$\varphi_2 = g(\rho, P, Y_m) \tag{30}$$

利用式（22），在二次近似中有

$$Y_m = \int_1^{\alpha}(\varphi_2 + k\rho)\mathrm{d}\rho \tag{31}$$

将上式整理后，便得边缘载荷的二次近似解析解

$$P = (\alpha_1 + \alpha_2 k^2)Y_m + \alpha_3 k Y_m^2 + \alpha_4 Y_m^3 \tag{32}$$

其中

$$\begin{aligned} \alpha_1 =& 2\beta_1^{-1}\beta_2, \\ \alpha_2 =& \beta_1\beta_2^2\left[\frac{83-29\nu}{13824}\lambda_1^2\alpha^{12} + \frac{31-13\nu}{576}\lambda_1\lambda_2\alpha^{10}\ln\alpha\right. \\ & - \frac{195-214\nu+55\nu^2}{2304}\lambda_1\alpha^{10} - \frac{17-37\nu-29\nu^2+25\nu^3}{96}\alpha^8\ln^2\alpha \end{aligned}$$

$$
\begin{aligned}
&+\frac{43-221\nu+155\nu^2-13\nu^3}{288}\alpha^6\ln\alpha+\frac{391-955\nu+989\nu^2-353\nu^3}{4608}\alpha^8\\
&-\frac{1}{2}\lambda_1\lambda_2^2\alpha^6\ln^4\alpha+(1-2\nu)\lambda_2\alpha^6\ln^3\alpha\\
&-\frac{21-87\nu-21\nu^2-25\nu^3}{48}\alpha^6\ln^2\alpha-\frac{5+12\nu+75\nu^2}{72}\alpha^6\ln\alpha\\
&+\frac{417+45\nu-105\nu^2+379\nu^3}{3456}\alpha^6+\frac{23-46\nu-25\nu^2}{96}\lambda_2\alpha^4\ln^2\alpha\\
&-\frac{83-341\nu-197\nu^2-13\nu^3}{288}\alpha^4\ln\alpha-\frac{321+359\nu+839\nu^2+353\nu^3}{4608}\alpha^4\\
&-\frac{71+60\nu+13\nu^2}{576}\lambda_2\alpha^2\ln\alpha-\frac{83-329\nu-275\nu^2-55\nu^3}{2304}\alpha^2\\
&-\frac{293+178\nu+29\nu^2}{13824}\lambda_2\Big],
\end{aligned}
$$

$$
\begin{aligned}
\alpha_3=&-\beta_1\beta_2^3\Big(\frac{95-41\nu}{18432}\lambda_1^3\alpha^{14}+\frac{73-37\nu}{768}\lambda_1^2\lambda_2\alpha^{12}\ln\alpha\\
&-\frac{2587-2778\nu+731\nu^2}{18432}\lambda_1^2\alpha^{12}\\
&+\frac{15-9\nu}{32}\nu\lambda_1\lambda_2\alpha^{10}\ln^3\alpha-\frac{71+41\nu-125\nu^2+49\nu^3}{128}\lambda_1\alpha^{10}\ln^2\alpha\\
&+\frac{213-275\nu+195\nu^2-37\nu^3}{256}\lambda_1\alpha^{10}\ln\alpha\\
&-\frac{947+1577\nu-2759\nu^2+1027\nu^3}{6144}\lambda_1\alpha^{10}\\
&-\frac{3-9\nu}{8}\lambda_1\lambda_2^2\alpha^8\ln^5\alpha-\frac{3+27\nu-45\nu^2+3\nu^3}{8}\lambda_2\alpha^8\ln^4\alpha\\
&+\frac{62-53\nu-285\nu^2+45\nu^3-9\nu^4}{32}\alpha^8\ln^3\alpha\\
&-\frac{89-1064\nu-96\nu^2-12\nu^3+147\nu^4}{128}\alpha^8\ln^2\alpha\\
&-\frac{469+1280\nu+138\nu^2-844\nu^3+37\nu^4}{384}\alpha^8\ln\alpha\\
&+\frac{20229-17364\nu+2454\nu^2+7036\nu^3-5875\nu^4}{18432}\alpha^8\\
&+\frac{3-9\nu}{8}\lambda_2^3\alpha^6\ln^5\alpha-\frac{9-48\nu-3\nu^2}{8}\lambda_2^2\alpha^6\ln^4\alpha\\
&-\frac{10+441\nu+60\nu^2+9\nu^3}{32}\lambda_2\alpha^6\ln^3\alpha\\
&+\frac{419+1598\nu+474\nu^2+498\nu^3+147\nu^4}{128}\alpha^6\ln^2\alpha\\
&-\frac{991-206\nu+2112\nu^2+994\nu^3+37\nu^4}{384}\alpha^6\ln\alpha
\end{aligned}
$$

$$+\frac{2901+15990\nu-1608\nu^2+8266\nu^3+5875\nu^4}{18432}\alpha^6$$

$$+\frac{21+9\nu}{32}\nu\lambda_2^2\alpha^4\ln^3\alpha-\frac{43+449\nu+287\nu^2+49\nu^3}{128}\lambda_2\alpha^4\ln^2\alpha$$

$$+\frac{107+1514\nu+1272\nu^2+382\nu^3+37\nu^4}{256}\alpha^4\ln\alpha$$

$$-\frac{5587+3960\nu+5146\nu^2+4152\nu^3+1027\nu^4}{6144}\alpha^4$$

$$-\frac{353+467\nu+223\nu^2+37\nu^3}{768}\lambda_2\alpha^2\ln\alpha$$

$$-\frac{155-9266\nu-11232\nu^2-4846\nu^3-731\nu^4}{18432}\alpha^2$$

$$-\frac{881+867\nu+315\nu^2+41\nu^3}{18432}\lambda_2\Big),$$

$$\alpha_4=-\beta_1\beta_2^4\Big(\frac{353-191\nu}{331776}\lambda_1^4\alpha^{16}+\frac{445+162\nu-283\nu^2}{13824}\lambda_1^3\alpha^{14}\ln\alpha$$

$$-\frac{3869-4452\nu+1231\nu^2}{82944}\lambda_1^3\alpha^{14}+\frac{9+2\nu-7\nu^2}{32}\nu\lambda_1^2\alpha^{12}\ln^3\alpha$$

$$-\frac{371+425\nu-1235\nu^2+439\nu^3}{2304}\lambda_1^2\alpha^{12}\ln^2\alpha$$

$$+\frac{829-1580\nu+1196\nu^2-283\nu^3}{3456}\lambda_1^2\alpha^{12}\ln\alpha$$

$$-\frac{2377+9535\nu-16075\nu^2+6431\nu^3}{82944}\lambda_1^2\alpha^{12}$$

$$-\frac{1-5\nu}{12}\lambda_1^2\lambda_2^2\alpha^{10}\ln^6\alpha-\frac{3+10\nu-5\nu^2}{8}\lambda_1^2\lambda_2\alpha^{10}\ln^5\alpha$$

$$+\frac{19-10\nu+86\nu^3+\nu^4}{24}\lambda_1\alpha^{10}\ln^4\alpha$$

$$-\frac{85+257\nu+930\nu^2+69\nu^3+63\nu^4}{144}\lambda_1\alpha^{10}\ln^3\alpha$$

$$+\frac{869+3844\nu+1086\nu^2+40\nu^3-439\nu^4}{576}\lambda_1\alpha^{10}\ln^2\alpha$$

$$-\frac{30167+45508\nu-5766\nu^2-19052\nu^3+1415\nu^4}{13824}\lambda_1\alpha^{10}\ln\alpha$$

$$+\frac{90801-43782\nu-3936\nu^2+16390\nu^3-15409\nu^4}{82944}\lambda_1\alpha^{10}$$

$$+\frac{1-6\nu+17\nu^2}{12}\lambda_2^3\alpha^8\ln^6\alpha-\frac{1+17\nu^2}{2}\lambda_2^2\alpha^8\ln^5\alpha$$

$$+\frac{18+117\nu+288\nu^2+11\nu^3+\nu^4}{12}\lambda_2\alpha^8\ln^4\alpha$$

$$-\frac{439+2328\nu+2717\nu^2+180\nu^3+24\nu^4}{72}\alpha^8\ln^3\alpha$$

$$
\begin{aligned}
&+\frac{4357+12113\nu+3006\nu^2+550\nu^3-1155\nu^4-439\nu^5}{384}\alpha^8\ln^2\alpha\\
&-\frac{14438+13611\nu+9631\nu^2-7701\nu^3-3951\nu^4}{1728}\alpha^8\ln\alpha\\
&+\frac{303111+268779\nu-203526\nu^2+3682\nu^3-19065\nu^4-40645\nu^5}{165888}\alpha^8\\
&-\frac{1-5\nu}{12}\lambda_2^4\alpha^6\ln^6\alpha-\frac{1+30\nu+5\nu^2}{8}\lambda_2^3\alpha^6\ln^5\alpha\\
&+\frac{53+297\nu+63\nu^2-\nu^3}{24}\lambda_2^2\alpha^6\ln^4\alpha\\
&-\frac{937+2929\nu+732\nu^2+147\nu^3+63\nu^4}{144}\lambda_2\alpha^6\ln^3\alpha\\
&+\frac{3613+7811\nu+3502\nu^2+4822\nu^3+2789\nu^4+439\nu^5}{576}\alpha^6\ln^2\alpha\\
&-\frac{14321-44561\nu+100070\nu^2+107690\nu^3+27337\nu^4+1415\nu^5}{13824}\alpha^6\ln\alpha\\
&-\frac{131499-99453\nu-120714\nu^2-55378\nu^3-46817\nu^4-15409\nu^5}{82944}\alpha^6\\
&+\frac{27+26\nu+7\nu^2}{32}\nu\lambda_2^2\alpha^4\ln^3\alpha\\
&-\frac{2659+11640\nu+11086\nu^2+3984\nu^3+439\nu^4}{2304}\lambda_2\alpha^4\ln^2\alpha\\
&+\frac{5503+26716\nu+29773\nu^2+14293\nu^3+3136\nu^4+283\nu^5}{3456}\alpha^4\ln\alpha\\
&-\frac{105773+138209\nu+111644\nu^2+80072\nu^3+36095\nu^4+6431\nu^5}{82944}\alpha^4\\
&-\frac{4661+8410\nu+6144\nu^2+2102\nu^3+283\nu^4}{13824}\lambda_2\alpha^2\ln\alpha\\
&+\frac{3095+32895\nu+49954\nu^2+32402\nu^3+10023\nu^4+1231\nu^5}{82944}\alpha^2\\
&\left.-\frac{8087+11204\nu+6402\nu^2+1748\nu^3+191\nu^4}{331776}\lambda_2\right)
\end{aligned}
$$

下面以 $\alpha=3$，$\nu=0.3$ 的情况为例，按照公式（32），在图 3 上绘出了不同 k 值下的特征曲线。由图看出，当 k 很小时，P-Y_m 曲线单调上升，说明壳体具有平板性质，无跳跃现象产生。当 k 较大时，P-Y_m 曲线出现了迴形线状态，这时壳体便产生了稳定问题，即跳跃现象。精密仪器中正是利用这种突然的跳跃来获得控制的信号，而建筑中的屋盖则应避免这种破坏现象发生。对应于 P-Y_m 曲线极大值一点的载荷称作上临界载荷，对应于曲线极小值一点

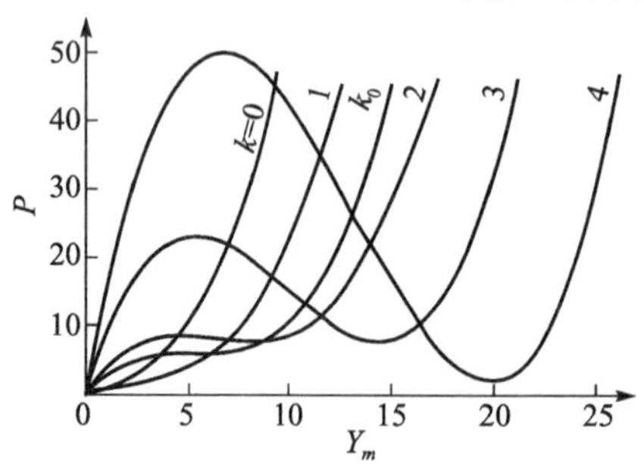

图 3　各种 k 值下的特征曲线（$\alpha=3$，$\nu=0.3$）

的载荷称作下临界载荷。对公式（32）应用极值条件

$$\frac{\mathrm{d}P}{\mathrm{d}Y_m}=0 \tag{33}$$

我们得到产生跳跃现象时的无量纲内边缘临界挠度的公式

$$Y_m^*=\frac{-\alpha_3 k\pm\sqrt{(\alpha_3^2-3\alpha_2\alpha_4)k^2-3\alpha_1\alpha_4}}{3\alpha_4} \tag{34}$$

将 Y_m^* 代入式（32），便得二次近似的临界载荷公式

$$P^*=(\alpha_1+\alpha_2 k^2)Y_m^*+\alpha_3 k Y_m^{*2}+\alpha_4 Y_m^{*3} \tag{35}$$

其中对应于式（34）负号的 P^* 是上临界载荷，正号的 P^* 是下临界载荷。

按照公式（34）和（35），我们得到了不同 α 值的上、下临界载荷数据，图 4 绘出了 $\nu=0.3$ 时的稳定曲线。由图看出，随着 α 值的增大，即圆底扁球壳中心孔减小，稳定曲线越来越低，表明壳体在几何参数 k 和载荷 P 都小时就可以失稳。

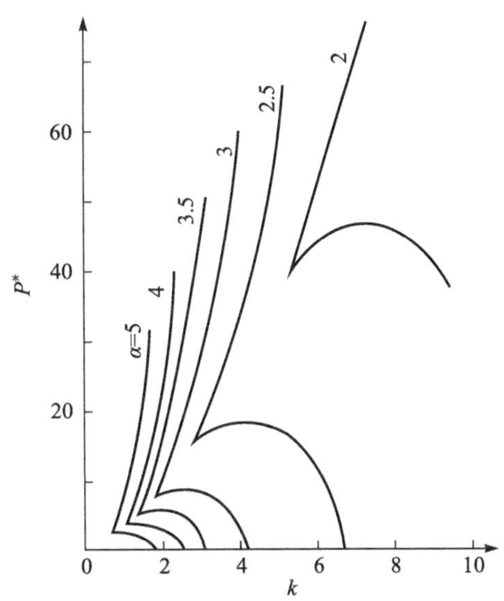

图 4　各种 α 值下的稳定曲线（$\nu=0.3$）

下面求临界点即上、下临界载荷相重点的公式。记临界点的壳体几何参数为 k_0，临界载荷为 P_0^*。二次方程（33）有重根的条件为方程的判别式等于零：

$$(\alpha_3^2-3\alpha_2\alpha_4)k^2-3\alpha_1\alpha_4=0 \tag{36}$$

由此解得

$$k_0=\sqrt{\frac{3\alpha_1\alpha_4}{\alpha_3^2-3\alpha_2\alpha_4}} \tag{37}$$

可见，当 $k<k_0$ 时，壳体不会发生失稳现象。当 $k\geqslant k_0$ 时，壳体就会出现失稳现象。因此，几何参数 k_0 是区分壳体失稳与否的分界点。图 5 与图 6 分别给出了 k_0-α 和 P_0^*-α 的关系曲线。由图看出，随着 α 的增加，即圆底扁球壳中心孔减小，k_0 和 P_0^* 开始减小很快，后来下降缓慢，逐渐趋近于横轴。这表明壳体对失稳的反应随 α 的增加而变得

灵敏起来，在 α 较大时，只要很小的 k_0 和 P_0^*，壳体就会出现失稳现象。

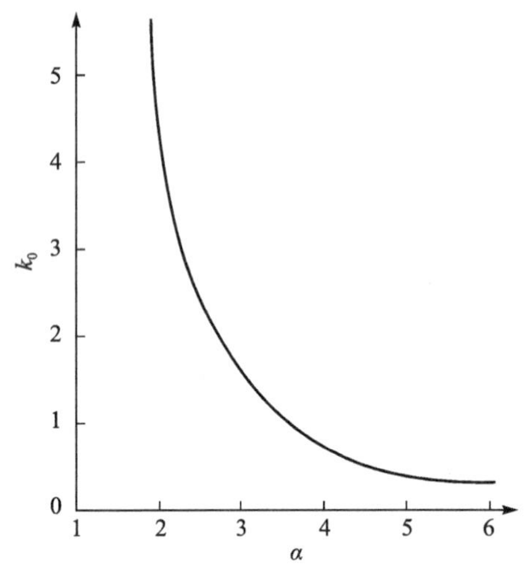

图 5　临界几何参数 k_0 与 α 的关系曲线（ν=0.3）

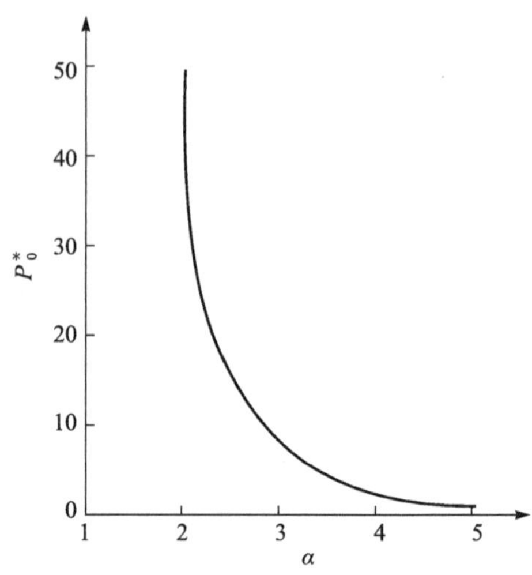

图 6　临界点载荷 P_0^* 与 α 的关系曲线（ν=0.3）

本文的数值计算工作由原中国科学院兰州地质研究所计算室刘玉茹同志协助进行。

符　号

r	扁球壳的中曲面点至对称中心轴的距离
R	扁球壳的中曲面半径
h	扁球壳的厚度
a,b	扁球壳的外、内边缘半径
w	横向挠度
u	径向薄膜位移
θ	扁球壳经线方向弧的倾斜角
ϑ	扁球壳经线方向弧的旋转角
N_r	径向薄膜内力
M_r	径向弯矩
Q	径向剪力
q	均布载荷
p	作用在扁球壳内边缘的线布载荷
$F(r)$	载荷函数
E	弹性模量
ν	泊松比
D	抗弯刚度
C	积分常数
ρ	无量纲径向坐标

α	扁球壳的无量纲外边缘半径
k	扁球壳的无量纲几何参数
k_0	扁球壳的临界点的无量纲几何参数
$\varphi, \varphi_1, \varphi_2$	经线方向弧的新无量纲旋转角及其一、二次近似值
S, S_1	无量纲径向薄膜内力及其一次近似值
y, y_1, y_2	无量纲挠度及其一、二次近似值
Y_m	无量纲内边缘挠度
Y_m^*	无量纲内边缘的临界挠度
P	无量纲内边缘线布载荷
P^*	无量纲临界载荷
P_1^*, P_2^*	无量纲上、下临界载荷
P_0^*	临界点的无量纲临界载荷
$f(\rho, Y_m), g(\rho, P, Y_m)$	S_1 与 φ_2 表达式的缩写
$\lambda_1, \lambda_2, \beta_1, \beta_2, \alpha_1, \alpha_2, \alpha_3, \alpha_4$	辅助量
L	微分算子

参考文献

[1] von Kármán, T. and Tsien, H. S.(钱学森). The buckling of spherical shells by external pressure. J. Aeronaut. Sci., 1939, 7 (2): 43.

[2] 刘人怀. 在内边缘均布力矩作用下中心开孔圆底扁球壳的非线性稳定问题. 科学通报, 1965 (3): 253.

[3] B. И. 费奥多谢夫. 精密仪器弹性元件的理论与计算. 北京: 科学出版社, 1963: 186.

集中载荷作用下开顶扁球壳的非线性稳定问题①

一、引言

在建筑和精密仪器的弹性元件等工程中，常使用具有硬中心的边缘固定的开顶扁球壳。这种壳体在中心集中载荷作用时，在一定条件下会丧失稳定性。对于建筑工程，需要防止这种现象发生；对于精密仪器，则应利用失稳所产生的跳跃来作为自动控制的信号。由于本问题涉及了非线性的数学问题，结构亦较复杂，所以给研究带来了巨大的困难，因而至今尚无人讨论过。

关于开顶扁球壳的非线性稳定问题，以往仅有刘人怀[1-4]、Tillman[5]研究过，他们涉及的是无硬中心的开顶扁球壳。

本文使用修正迭代法来求解本问题的非线性微分方程的边值问题，获得了较精确的解析解。这一方法是叶开沅和刘人怀于 1965 年在研究扁球壳的非线性稳定问题中提出的[1,6,7]，随后应用于板壳的一系列非线性问题，均获得十分满意的结果。

本文所获得的结果对于工程设计来说有现实意义。

二、基本方程

考虑一个在中心集中载荷 P 作用下，厚度为 h，中曲面半径为 R，内、外缘半径分别为 b、a 的具有硬中心的开顶扁球壳（图 1）。

按照在均布载荷 q 作用下的扁球壳的非线性弯曲理论[1,2]，我们有下列关于挠度 w 和径向薄膜内力 N_r 所满足的基本方程：

$$\frac{D}{r}\frac{\mathrm{d}}{\mathrm{d}r}r\frac{\mathrm{d}}{\mathrm{d}r}\frac{1}{r}\frac{\mathrm{d}}{\mathrm{d}r}(r\vartheta)-\frac{1}{r}\frac{\mathrm{d}}{\mathrm{d}r}[rN_r(\theta+\vartheta)]=q,$$

$$\frac{r}{Eh}\frac{\mathrm{d}}{\mathrm{d}r}\frac{1}{r}\frac{\mathrm{d}}{\mathrm{d}r}(r^2N_r)+\vartheta\left(\theta+\frac{1}{2}\vartheta\right)=0 \qquad (2.1\text{a,b})$$

其中，r 是扁球壳中曲面点至对称中心轴的距离，ϑ 和 θ 分别是壳体经线方向弧的旋转角和倾斜角，E 是弹性模量，ν 是泊松比，D 是抗弯刚度。

$$\vartheta=\frac{\mathrm{d}w}{\mathrm{d}r},\theta=\frac{r}{R},D=\frac{Eh^3}{12(1-\nu^2)} \qquad (2.2)$$

求得 w 和 N_r 后，便可按下列公式计算径向薄膜位移 u，径向弯矩 M_r 和径向剪力 Q_r

$$u=\frac{r}{Eh}\left[(1-\nu)N_r+r\frac{\mathrm{d}N_r}{\mathrm{d}r}\right],M_r=-D\left(\frac{\mathrm{d}\vartheta}{\mathrm{d}r}+\frac{\nu}{r}\vartheta\right),$$

$$Q_r=-D\frac{\mathrm{d}}{\mathrm{d}r}\frac{1}{r}\frac{\mathrm{d}}{\mathrm{d}r}(r\vartheta) \qquad (2.3\text{a}\sim\text{c})$$

先将方程（2.1 a）两端乘以 $r\mathrm{d}r$，然后积分一次，得

---

① 本文原载《应用数学和力学》，1988，9（2）：95-106. 作者：刘人怀，成振强.

$$Dr\frac{\mathrm{d}}{\mathrm{d}r}\frac{1}{r}\frac{\mathrm{d}}{\mathrm{d}r}(r\vartheta)-rN_r(\theta+\vartheta)=F(r) \tag{2.4}$$

其中

$$F(r)=\int qr\,\mathrm{d}r+C \tag{2.5}$$

应用式（2.3c），方程（2.4）成为

$$r[Q_r+N_r(\theta+\vartheta)]=-F(r) \tag{2.6}$$

现在，我们需要确定载荷函数 $F(r)$。为此，用圆锥曲面沿半径为 r 的圆周从壳体中切割出如图 2 所示的部分壳体，将作用在这部分壳体上的全部力投影到对称轴上，便有

$$2\pi r[Q_r\cos(\theta+\vartheta)+N_r\sin(\theta+\vartheta)]+P=0 \tag{2.7}$$

根据角（$\theta+\vartheta$）微小的条件，式（2.7）可简化为

$$r[Q_r+N_r(\theta+\vartheta)]=-\frac{P}{2\pi} \tag{2.8}$$

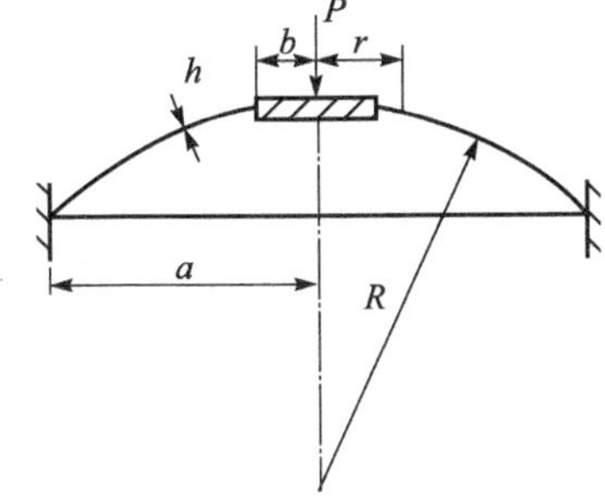

图 1　具有硬中心的开顶扁球壳

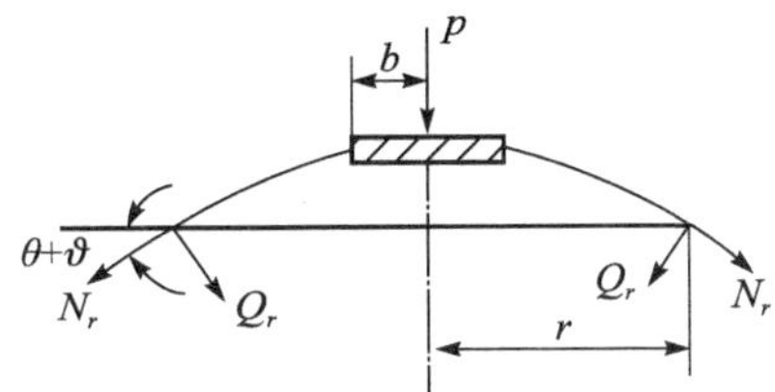

图 2　载荷函数 $F(r)$ 的确定

将式（2.8）同式（2.6）比较，就得

$$F(r)=\frac{P}{2\pi} \tag{2.9}$$

应用这一结果，方程（2.4）成为

$$Dr\frac{\mathrm{d}}{\mathrm{d}r}\frac{1}{r}\frac{\mathrm{d}}{\mathrm{d}r}(r\vartheta)-rN_r(\theta+\vartheta)=\frac{P}{2\pi} \tag{2.10}$$

于是，方程（2.10）和（2.1b）就组成具有硬中心的开顶扁球壳在中心集中载荷作用下的轴对称大挠度弯曲方程。关于相应的边界条件如下。

当 $r=a$ 时，外边缘夹紧固定：

$$w=0,\quad \vartheta=0,\quad u=0 \tag{2.11}$$

当 $r=b$ 时，内边缘被固定在可上下移动的不变形的硬中心上：

$$\vartheta=0,\quad u=0 \tag{2.12}$$

为了简化计算，我们引入下列无量纲量：

$$\rho=\frac{r}{a},\alpha=\frac{b}{a},y=\sqrt{12\lambda_1\lambda_2}\,\frac{w}{h},\varphi=-\frac{\mathrm{d}y}{\mathrm{d}\rho},$$

$$S=\frac{a^2}{D}\rho N_r,k=\sqrt{12\lambda_1\lambda_2}\,\frac{a^2}{Rh},Q=\sqrt{12\lambda_1\lambda_2}\,\frac{a^2P}{2\pi Dh} \tag{2.13}$$

其中

$$\lambda_1 = 1-\nu,\quad \lambda_2 = 1+\nu \tag{2.14}$$

利用这些无量纲量，方程（2.10）、（2.1b）和边界条件（2.11）、（2.12）简化为

$$L(\rho\varphi) = S(\varphi - k\rho) - Q,$$
$$L(\rho S) = \varphi\left(k\rho - \frac{1}{2}\varphi\right) \tag{2.15a,b}$$

$$当\ \rho=1\ 时，\quad y=0,\quad \varphi=0,\quad \rho\frac{\mathrm{d}S}{\mathrm{d}\rho}-\nu S=0 \tag{2.16}$$

$$当\ \rho=\alpha\ 时，\quad \varphi=0,\quad \rho\frac{\mathrm{d}S}{\mathrm{d}\rho}-\nu S=0 \tag{2.17}$$

其中

$$L(\cdots) = \rho\frac{\mathrm{d}}{\mathrm{d}\rho}\frac{1}{\rho}\frac{\mathrm{d}}{\mathrm{d}\rho}(\cdots) \tag{2.18}$$

这样，我们的问题便化为在边界条件（2.16）和（2.17）下求解非线性微分方程组（2.15）。

三、边值问题的求解

我们用修正迭代法解边值问题（2.15）～（2.17）。在一次近似中，我们略去方程（2.15a）中含 S 的项，便得到如下的线性边值问题：

$$L(\rho\varphi_1) = -Q,$$
$$L(\rho S_1) = \varphi_1\left(k\rho - \frac{1}{2}\varphi_1\right) \tag{3.1a,b}$$

$$当\ \rho=1\ 时，y_1=0,\ \varphi_1=0,\ \rho\frac{\mathrm{d}S_1}{\mathrm{d}\rho}-\nu S_1=0 \tag{3.2a～c}$$

$$当\ \rho=\alpha\ 时，\varphi_1=0,\ \rho\frac{\mathrm{d}S_1}{\mathrm{d}\rho}-\nu S_1=0 \tag{3.3a, b}$$

将方程（3.1a）积分，利用边界条件（3.2b）和（3.3a），得小挠度理论的解

$$\varphi_1 = \frac{Q}{2(1-\alpha^2)}\left[(\alpha^2-1)\rho\ln\rho - \alpha^2\ln\alpha\left(\rho-\frac{1}{\rho}\right)\right] \tag{3.4}$$

我们以扁球壳的无量纲内边缘挠度 Y_m 为迭代参数：

$$Y_m = \sqrt{12\lambda_1\lambda_2}\,\frac{w}{h}\bigg|_{r=b} \tag{3.5}$$

再应用式（2.13）和（3.2a），可得

$$Y_m = \int_\alpha^1 \varphi\,\mathrm{d}\rho \tag{3.6}$$

将解（3.4）代入上式，得

$$Q = 2(1-\alpha^2)\beta Y_m \tag{3.7}$$

其中

$$\beta = \left(\frac{1}{4}\alpha^4 - \alpha^2\ln^2\alpha - \frac{1}{2}\alpha^2 + \frac{1}{4}\right)^{-1} \tag{3.8}$$

将式（3.7）代入式（3.4），有

$$\varphi_1 = \beta Y_m\left[(\alpha^2-1)\rho\ln\rho - \alpha^2\ln\alpha\left(\rho-\frac{1}{\rho}\right)\right] \tag{3.9}$$

利用式（3.9）以及边界条件（3.2c）和（3.3b），我们积分方程（3.1b），得 S 的一次近似解为

$$\begin{aligned} S_1 = {} & k\beta Y_m\Big[\frac{1}{8}(\alpha^2-1)\left(\rho^3\ln\rho-\frac{3}{4}\rho^3\right) \\ & -\frac{1}{2}\alpha^2\ln\alpha\left(\frac{1}{4}\rho^3-\rho\ln\rho\right)+A_1\rho+B_1\frac{1}{\rho}\Big] \\ & -\beta^2Y_m^2\Big[\frac{1}{16}(\alpha^2-1)^2\left(\rho^3\ln^2\rho-\frac{3}{2}\rho^3\ln\rho+\frac{7}{8}\rho^3\right) \\ & -\frac{1}{4}(\alpha^2-1)\alpha^2\ln\alpha\left(\frac{1}{2}\rho^3\ln\rho-\frac{3}{8}\rho^3-\rho\ln^2\rho+\rho\ln\rho\right) \\ & +\frac{1}{2}\alpha^4\ln^2\alpha\left(\frac{1}{8}\rho^3-\rho\ln\rho-\frac{1}{2}\frac{1}{\rho}\ln\rho\right)-A_2\rho-B_2\frac{1}{\rho}\Big] \end{aligned} \tag{3.10}$$

其中

$$\begin{aligned} A_1 = {} & \frac{1}{2\lambda_2(\alpha^2-1)}\Big(\frac{5-3\nu}{16}\alpha^6-\lambda_1\alpha^4\ln^2\alpha-\frac{1}{4}\lambda_2\alpha^4\ln\alpha-\frac{5-3\nu}{16}\alpha^4 \\ & +\frac{1}{4}\lambda_2\alpha^2\ln\alpha-\frac{5-3\nu}{16}\alpha^2+\frac{5-3\nu}{16}\Big), \\ B_1 = {} & \frac{1}{2\lambda_2(\alpha^2-1)}\Big(\frac{5-3\nu}{16}\alpha^6-\lambda_1\alpha^4\ln^2\alpha-\frac{5-3\nu}{8}\alpha^4+\frac{5-3\nu}{16}\alpha^2\Big), \\ A_2 = {} & \frac{1}{2\lambda_1(\alpha^2-1)}\Big(\frac{9-7\nu}{64}\alpha^8-\frac{1}{2}\lambda_1\alpha^6\ln^3\alpha-\frac{1}{2}\lambda_1\alpha^6\ln^2\alpha \\ & -\frac{3}{16}\lambda_2\alpha^6\ln\alpha-\frac{9-7\nu}{32}\alpha^6+\nu\alpha^4\ln^3\alpha+\frac{1}{2}\lambda_1\alpha^4\ln^2\alpha+\frac{3}{8}\lambda_2\alpha^4\ln\alpha \\ & -\frac{3}{16}\lambda_2\alpha^2\ln\alpha+\frac{9-7\nu}{32}\alpha^2-\frac{9-7\nu}{64}\Big), \\ B_2 = {} & \frac{1}{2\lambda_2(\alpha^2-1)}\Big(\frac{9-7\nu}{64}\alpha^8-\frac{1}{2}\lambda_1\alpha^6\ln^3\alpha+\frac{5}{8}\lambda_2\alpha^6\ln^2\alpha-\frac{27-21\nu}{64}\alpha^6 \\ & +\nu\alpha^4\ln^3\alpha-\frac{5}{8}\lambda_2\alpha^4\ln^2\alpha+\frac{27-21\nu}{64}\alpha^4-\frac{9-7\nu}{64}\alpha^2\Big) \end{aligned} \tag{3.11a ~ d}$$

在二次近似中，我们有下述线性边值问题：

$$L(\rho\varphi_2) = S_1(\varphi_1-k\rho)-Q \tag{3.12}$$

$$当\ \rho=1\ 时, y_2=0, \varphi_2=0 \tag{3.13a,b}$$

$$当\ \rho=\alpha\ 时, \varphi_2=0 \tag{3.14}$$

将方程（3.12）积分，并利用边界条件（3.13b）和（3.14），得 φ 的二次近似解为

$$\begin{aligned} \varphi_2 = {} & \frac{Q}{2(1-\alpha^2)}\left[(\alpha^2-1)\rho\ln\rho-\alpha^2\ln\alpha\left(\rho-\frac{1}{\rho}\right)\right] \\ & -k^2\beta Y_m\Big[\frac{1}{192}(\alpha^2-1)\left(\rho^5\ln\rho-\frac{1}{6}\rho^5\right)-\frac{1}{16}\alpha^2\ln\alpha\left(\frac{1}{12}\rho^5-\rho^3\ln\rho+\frac{3}{4}\rho^3\right) \end{aligned}$$

$$
\begin{aligned}
&+\frac{1}{8}A_1\rho^3+\frac{1}{2}B_1\rho\ln\rho-C_2\rho-D_2\frac{1}{\rho}\Big]\\
&+k\beta^2Y_m^2\Big\{\frac{1}{128}(\alpha^2-1)^2\Big(\rho^5\ln^2\rho-\frac{11}{6}\rho^5\ln\rho+\frac{35}{36}\rho^5\Big)\\
&-\frac{1}{32}(\alpha^2-1)\alpha^2\ln\alpha\Big(\frac{1}{2}\rho^5\ln\rho-\frac{11}{24}\rho^5-3\rho^3\ln^2\rho+5\rho^3\ln\rho-\frac{21}{8}\rho^3\Big)\\
&+\frac{1}{8}\alpha^4\ln^2\alpha\Big(\frac{1}{16}\rho^5-\rho^3\ln\rho+\frac{5}{8}\rho^3+\frac{1}{2}\rho\ln^2\rho-\frac{1}{2}\rho\ln\rho\Big)\\
&+\frac{1}{2}A_1\Big[\frac{1}{4}(\alpha^2-1)\Big(\rho^3\ln\rho-\frac{3}{4}\rho^3\Big)-\alpha^2\ln\alpha\Big(\frac{1}{4}\rho^3-\rho\ln\rho\Big)\Big]\\
&+\frac{1}{2}B_1\Big[\frac{1}{2}(\alpha^2-1)(\rho\ln^2\rho-\rho\ln\rho)-\alpha^2\ln\alpha\Big(\rho\ln\rho+\frac{1}{\rho}\ln\rho\Big)\Big]\\
&-\frac{1}{8}A_2\rho^3-\frac{1}{2}B_2\rho\ln\rho+C_3\rho+D_3\frac{1}{\rho}\Big\}-\beta^3Y_m^3\Big\{\frac{1}{384}(\alpha^2\\
&-1)^3\Big(\rho^5\ln^3\rho-\frac{11}{4}\rho^5\ln^2\rho+\frac{35}{12}\rho^5\ln\rho-\frac{71}{72}\rho^5\Big)\\
&-\frac{1}{32}(\alpha^2-1)^2\alpha^2\ln\alpha\Big(\frac{1}{4}\rho^5\ln^2\rho-\frac{11}{24}\rho^5\ln\rho+\frac{35}{144}\rho^5-\rho^3\ln^3\rho\\
&+3\rho^3\ln^2\rho-\frac{27}{8}\rho^3\ln\rho+\frac{25}{16}\rho^3\Big)+\frac{1}{16}(\alpha^2-1)\alpha^4\ln^2\alpha\Big(\frac{1}{8}\rho^5\ln\rho\\
&-\frac{11}{96}\rho^5-\frac{3}{2}\rho^3\ln^2\rho+\frac{5}{2}\rho^3\ln\rho-\frac{21}{16}\rho^3-\rho\ln^2\rho+\rho\ln\rho\Big)\\
&-\frac{1}{16}\alpha^6\ln^3\alpha\Big(\frac{1}{24}\rho^5-\rho^3\ln\rho+\frac{5}{8}\rho^3+\rho\ln^2\rho-\rho\ln\rho-\frac{1}{\rho}\ln^2\rho-\frac{1}{\rho}\ln\rho\Big)\\
&-\frac{1}{2}A_2\Big[\frac{1}{4}(\alpha^2-1)\Big(\rho^3\ln\rho-\frac{3}{4}\rho^3\Big)-\alpha^2\ln\alpha\Big(\frac{1}{4}\rho^3-\rho\ln\rho\Big)\Big]\\
&-\frac{1}{2}B_2\Big[\frac{1}{2}(\alpha^2-1)(\rho\ln^2\rho-\rho\ln\rho)-\alpha^2\ln\alpha\Big(\rho\ln\rho+\frac{1}{\rho}\ln\rho\Big)\Big]\\
&-C_4\rho-D_4\frac{1}{\rho}\Big\}
\end{aligned}
\tag{3.15}
$$

其中

$$
\begin{aligned}
C_2=&\frac{1}{4\lambda_1\lambda_2(\alpha^2-1)^2}\Big(\frac{31-13\nu}{576}\lambda_2\alpha^{10}+\frac{1-15\nu+8\nu^2}{24}\alpha^8\ln\alpha-\frac{17+\nu}{576}\lambda_2\alpha^8\\
&-\lambda_1^2\alpha^6\ln^3\alpha-\frac{1}{4}\lambda_1\lambda_2\alpha^6\ln^2\alpha-\frac{17-54\nu+25\nu^2}{48}\alpha^6\ln\alpha-\frac{26-17\nu}{144}\lambda_2\alpha^6\\
&+\frac{1}{4}\lambda_1\lambda_2\alpha^4\ln^2\alpha+\frac{14-9\nu+\nu^2}{24}\alpha^4\ln\alpha+\frac{26-17\nu}{144}\lambda_2\alpha^4\\
&-\frac{13-7\nu}{48}\lambda_2\alpha^2\ln\alpha+\frac{17+\nu}{576}\lambda_2\alpha^2-\frac{31-13\nu}{576}\lambda_2\Big),
\end{aligned}
$$

$$
D_2=-\frac{1}{4\lambda_1\lambda_2(\alpha^2-1)^2}\Big(\frac{31-13\nu}{576}\lambda_2\alpha^{10}+\frac{1-15\nu+8\nu^2}{24}\alpha^8\ln\alpha
$$

$$
-\frac{31-13\nu}{288}\lambda_2\alpha^8-\lambda_1^2\alpha^6\ln^3\alpha-\frac{1-15\nu+8\nu^2}{12}\alpha^6\ln\alpha
$$

$$
+\frac{1-15\nu+8\nu^2}{24}\alpha^4\ln\alpha+\frac{31-13\nu}{288}\lambda_2\alpha^4-\frac{31-13\nu}{576}\lambda_2\alpha^2\Big),
$$

$$
C_3=\frac{1}{16\lambda_1\lambda_2(\alpha^2-1)^2}\Big[\frac{73-37\nu}{288}\lambda_2\alpha^{12}-\frac{5+12\nu-9\nu^2}{8}\alpha^{10}\ln^2\alpha
$$

$$
-\frac{10+63\nu-43\nu^2}{24}\alpha^{10}\ln\alpha-\frac{37-13\nu}{96}\lambda_2\alpha^{10}-(1-7\nu)\lambda_1\alpha^8\ln^4\alpha
$$

$$
+\frac{5+14\nu-7\nu^2}{2}\alpha^8\ln^3\alpha-(1+2\nu)\lambda_1\alpha^8\ln^2\alpha-\frac{17-348\nu+211\nu^2}{48}\alpha^8\ln\alpha
$$

$$
-\frac{71-59\nu}{96}\lambda_2\alpha^8-(5-11\nu)\lambda_1\alpha^6\ln^4\alpha-\frac{5+12\nu-9\nu^2}{2}\alpha^6\ln^3\alpha
$$

$$
+\frac{31+52\nu-59\nu^2}{8}\alpha^6\ln^2\alpha+\frac{57-39\nu}{16}\lambda_1\alpha^6\ln\alpha+\frac{251-179\nu}{144}\lambda_2\alpha^6
$$

$$
-\nu\lambda_2\alpha^4\ln^3\alpha-\frac{9+16\nu-17\nu^2}{4}\alpha^4\ln^2\alpha-\frac{211-36\nu-55\nu^2}{48}\alpha^4\ln\alpha
$$

$$
-\frac{71-59\nu}{96}\lambda_2\alpha^4+\frac{77-47\nu}{48}\lambda_2\alpha^2\ln\alpha-\frac{37-13\nu}{96}\lambda_2\alpha^2+\frac{73-37\nu}{288}\lambda_2\Big],
$$

$$
D_3=-\frac{1}{16\lambda_1\lambda_2(\alpha^2-1)}\Big[\frac{73-37\nu}{288}\lambda_2\alpha^{12}-\frac{5+12\nu-9\nu^2}{8}\alpha^{10}\ln^2\alpha
$$

$$
-\frac{35+132\nu-95\nu^2}{48}\alpha^{10}\ln\alpha-\frac{73-37\nu}{96}\lambda_2\alpha^{10}-(1-7\nu)\lambda_1\alpha^8\ln^4\alpha
$$

$$
+(4+7-5\nu^2)\alpha^8\ln^3\alpha+\frac{15-4\nu-3\nu^2}{8}\alpha^8\ln^2\alpha
$$

$$
+\frac{35+132\nu-95\nu^2}{16}\alpha^8\ln\alpha+\frac{73-37\nu}{144}\lambda_2\alpha^8
$$

$$
-(5-11\nu)\lambda_1\alpha^6\ln^4\alpha-(4+7\nu-5\nu^2)\alpha^6\ln^3\alpha
$$

$$
-\frac{15-44\nu+21\nu^2}{8}\alpha^6\ln^2\alpha-\frac{35+132\nu-95\nu^2}{16}\alpha^6\ln\alpha
$$

$$
+\frac{73-37\nu}{144}\lambda_2\alpha^6+\frac{5-28\nu+15\nu^2}{8}\alpha^4\ln^2\alpha+\frac{35+132\nu-95\nu^2}{48}\alpha^4\ln\alpha-\frac{73-37\nu}{96}\lambda_2\alpha^4
$$

$$
+\frac{73-37\nu}{288}\lambda_2\alpha^2\Big],
$$

$$
C_4=\frac{1}{8\lambda_1\lambda_2(\alpha^2-1)^2}\Big(\frac{445-283\nu}{13824}\lambda_2\alpha^{14}-\frac{9+20\nu-21\nu^2}{64}\alpha^{12}\ln^2\alpha
$$

$$
-\frac{179+216\nu-251\nu^2}{576}\alpha^{12}\ln\alpha-\frac{1051-565\nu}{13824}\lambda_2\alpha^{12}+\nu\lambda_1\alpha^{10}\ln^5\alpha
$$

$$
+\frac{9+17\nu}{8}\lambda_1\alpha^{10}\ln^4\alpha+\frac{12-3\nu}{8}\lambda_2\alpha^{10}\ln^3\alpha-\frac{83-71\nu}{192}\lambda_2\alpha^{10}\ln^2\alpha
$$

$$
+\frac{19+33-34\nu^2}{24}\alpha^{10}\ln\alpha-\frac{325-379\nu}{4608}\lambda_2\alpha^{10}-\frac{1-4\nu+11\nu^2}{2}\alpha^8\ln^5\alpha
$$

$$+\frac{4-3\nu}{2}\lambda_2\alpha^8\ln^4\alpha-\frac{24-9\nu}{8}\lambda_2\alpha^8\ln^3\alpha+\frac{137+132\nu-197\nu^2}{64}\alpha^8\ln^2\alpha$$
$$-\frac{17+504\nu-377\nu^2}{288}\alpha^8\ln\alpha+\frac{5065-4255\nu}{13824}\lambda_2\alpha^8-\frac{1-5\nu}{2}\lambda_1\alpha^6\ln^5\alpha$$
$$-\frac{25+12\nu-29\nu^2}{8}\alpha^6\ln^4\alpha+\frac{12-9\nu}{8}\lambda_2\alpha^6\ln^3\alpha$$
$$-\frac{155+172\nu-239\nu^2}{64}\alpha^6\ln^2\alpha-\frac{211-108\nu-31\nu^2}{144}\alpha^6\ln\alpha$$
$$-\frac{5065-4255\nu}{13824}\lambda_2\alpha^6+\frac{3}{8}\nu\lambda_2\alpha^4\ln^3\alpha+\frac{41+48\nu-65\nu^2}{48}\alpha^4\ln^2\alpha$$
$$+\frac{287+24\nu-167\nu^2}{192}\alpha^4\ln\alpha+\frac{325-379\nu}{4608}\lambda_2\alpha^4-\frac{65-47\nu}{144}\lambda_2\alpha^2\ln\alpha$$
$$+\frac{1051-565\nu}{13824}\lambda_2\alpha^2-\frac{445-283\nu}{13824}\lambda_2\Big),$$

$$D_4=-\frac{1}{8\lambda_1\lambda_2(\alpha^2-1)^2}\Big(\frac{445-283\nu}{13824}\lambda_2\alpha^{14}-\frac{9+20\nu-21\nu^2}{64}\alpha^{12}\ln^2\alpha$$
$$-\frac{439+450\nu-565\nu^2}{1152}\alpha^{12}\ln\alpha-\frac{445-283\nu}{3456}\lambda_2\alpha^{12}+\nu\lambda_1\alpha^{10}\ln^5\alpha$$
$$+\frac{11+19\nu}{8}\lambda_1\alpha^{10}\ln^4\alpha+\frac{109-55\nu}{48}\lambda_2\alpha^{10}\ln^3\alpha+\frac{9-7\nu}{16}\lambda_2\alpha^{10}\ln^2\alpha$$
$$+\frac{439+450\nu-565\nu^2}{288}\alpha^{10}\ln\alpha+\frac{2225-1415\nu}{13824}\lambda_2\alpha^{10}$$
$$-\frac{1-4\nu+11\nu^2}{2}\alpha^8\ln^5\alpha+\frac{7}{4}\lambda_1\lambda_2\alpha^8\ln^4\alpha-\frac{109-55\nu}{24}\lambda_2\alpha^8\ln^3\alpha$$
$$-\frac{27-21\nu}{32}\lambda_1\alpha^8\ln^2\alpha-\frac{439+450\nu-565\nu^2}{192}\alpha^8\ln\alpha-\frac{1-5\nu}{2}\lambda_1\alpha^6\ln^5\alpha$$
$$-\frac{25+33\nu}{8}\lambda_1\alpha^6\ln^4\alpha+\frac{109-55\nu}{48}\lambda_2\alpha^6\ln^3\alpha+\frac{9-34\nu+21\nu^2}{16}\alpha^6\ln^2\alpha$$
$$+\frac{439+450\nu-565\nu^2}{288}\alpha^6\ln\alpha-\frac{2225-1415\nu}{13824}\lambda_2\alpha^6$$
$$-\frac{9-52\nu+35\nu^2}{64}\alpha^4\ln^2\alpha-\frac{439+450\nu-565\nu^2}{1152}\alpha^4\ln\alpha$$
$$+\frac{445-283\nu}{3456}\lambda_2\alpha^4-\frac{445-283\nu}{13824}\lambda_2\alpha^2\Big) \qquad (3.16a\sim f)$$

将式（3.15）代入式（3.6），得此壳体的二次似特征关系式

$$Q=(\alpha_1+\alpha_2k^2)Y_m+\alpha_3kY_m^2+\alpha_4Y_m^3 \qquad (3.17)$$

其中

$$\alpha_1=2(1-\alpha^2)\beta,$$
$$\alpha_2=\frac{\beta^2}{2\lambda_1\lambda_2(1-\alpha^2)}\Big(\frac{83-29\nu}{6912}\lambda_2\alpha^{12}-\frac{31-13\nu}{288}\lambda_2\alpha^{10}\ln\alpha$$
$$+\frac{71-162\nu+55\nu^2}{1152}\alpha^{10}-\frac{17-54\nu+25\nu^2}{48}\alpha^8\ln^2\alpha$$

$$+\frac{31-13\nu}{144}\lambda_2\alpha^8\ln\alpha-\frac{817-1134\nu+353\nu^2}{2304}\alpha^8$$

$$+\lambda_1^2\alpha^6\ln^4\alpha+\frac{17-54\nu+25\nu^2}{24}\alpha^6\ln^2\alpha+\frac{971-1242\nu+379\nu^2}{1728}\alpha^6$$

$$-\frac{17-54\nu+25\nu^2}{48}\alpha^4\ln^2\alpha-\frac{31-13\nu}{144}\lambda_2\alpha^4\ln\alpha$$

$$-\frac{817-1134\nu+353\nu^2}{2304}\alpha^4+\frac{31-13\nu}{288}\lambda_2\alpha^2\ln\alpha$$

$$+\frac{71-162\nu+55\nu^2}{1152}\alpha^2+\frac{83-29\nu}{6912}\lambda_2\Big),$$

$$\alpha_3=\frac{\beta^3}{8\lambda_1\lambda_2(1-\alpha^2)}\Big[\frac{95-41\nu}{2304}\lambda_2\alpha^{14}-\frac{73-37\nu}{96}\lambda_2\alpha^{12}\ln\alpha$$

$$+\frac{835-1890\nu+731\nu^2}{2304}\alpha^{12}+\frac{15-9\nu}{4}\nu\alpha^{10}\ln^3\alpha$$

$$+\frac{19+66\nu-49\nu^2}{16}\alpha^{10}\ln^2\alpha+\frac{73-37\nu}{32}\lambda_2\alpha^{10}\ln\alpha$$

$$-\frac{1835-2898\nu+1027\nu^2}{768}\alpha^{10}+(3-9\nu)\lambda_1\alpha^8\ln^5\alpha$$

$$-(6-3\nu)\lambda_2\alpha^8\ln^4\alpha-\frac{15-9\nu}{4}\nu\alpha^8\ln^3\alpha-\frac{57+198\nu-147\nu^2}{16}\alpha^8\ln^2\alpha$$

$$-\frac{73-37\nu}{48}\lambda_2\alpha^8\ln\alpha+\frac{11675-17010\nu+5875\nu^2}{2304}\alpha^8$$

$$+(3-9\nu)\lambda_1\alpha^6\ln^5\alpha+(6-3\nu)\lambda_2\alpha^6\ln^4\alpha-\frac{15-9\nu}{4}\nu\alpha^6\ln^3\alpha$$

$$+\frac{57+198\nu-147\nu^2}{16}\alpha^6\ln^2\alpha-\frac{73-37\nu}{48}\lambda_2\alpha^6\ln\alpha$$

$$-\frac{11675-17010\nu+5875\nu^2}{2304}\alpha^6+\frac{15-9\nu}{4}\nu\alpha^4\ln^3\alpha$$

$$-\frac{19+66\nu-49\nu^2}{16}\alpha^4\ln^2\alpha+\frac{73-37\nu}{32}\lambda_2\alpha^4\ln\alpha$$

$$+\frac{1835-2898\nu+1027\nu^2}{768}\alpha^4-\frac{73-37\nu}{96}\lambda_2\alpha^2\ln\alpha$$

$$-\frac{835-1890\nu+731\nu^2}{2304}\alpha^2-\frac{95-41\nu}{2304}\lambda_2\Big],$$

$$\alpha_4=\frac{\beta^4}{4\lambda_1\lambda_2(1-\alpha^2)}\Big(\frac{353-191\nu}{82944}\lambda_2\alpha^{16}-\frac{445-283\nu}{3456}\lambda_2\alpha^{14}\ln\alpha$$

$$+\frac{1199-2754\nu+1231\nu^2}{20736}\alpha^{14}+\frac{9-7\nu}{8}\nu\alpha^{12}\ln^3\alpha$$

$$+\frac{601-439\nu}{576}\lambda_2\alpha^{12}\ln^2\alpha+\frac{445-283\nu}{864}\lambda_2\alpha^{12}\ln\alpha$$

$$-\frac{8959-15714\nu+6431\nu^2}{20736}\alpha^{12}+\frac{1-5\nu}{3}\lambda_1\alpha^{10}\ln^6\alpha-\frac{5}{2}\lambda_1\lambda_2\alpha^{10}\ln^5\alpha$$

$$
\begin{aligned}
&-\frac{8+\nu}{6}\lambda_2\alpha^{10}\ln^4\alpha-\frac{9-7\nu}{4}\nu\alpha^{10}\ln^3\alpha-\frac{601-439\nu}{144}\lambda_2\alpha^{10}\ln^2\alpha\\
&-\frac{2225-1415\nu}{3456}\lambda_2\alpha^{10}\ln\alpha+\frac{23633-38718\nu+15409\nu^2}{20736}\alpha^{10}\\
&+\frac{1-6\nu+17\nu^2}{3}\alpha^8\ln^6\alpha+\frac{8+\nu}{3}\lambda_2\alpha^8\ln^4\alpha+\frac{601-439\nu}{96}\lambda_2\alpha^8\ln^2\alpha\\
&-\frac{63845-102870\nu+40645\nu^2}{41472}\alpha^8+\frac{1-5\nu}{3}\lambda_1\alpha^6\ln^6\alpha+\frac{5}{2}\lambda_1\lambda_2\alpha^6\ln^5\alpha\\
&-\frac{8+\nu}{6}\lambda_2\alpha^6\ln^4\alpha+\frac{9-7\nu}{4}\nu\alpha^6\ln^3\alpha-\frac{601-439\nu}{144}\lambda_2\alpha^6\ln^2\alpha\\
&+\frac{2225-1415\nu}{3456}\lambda_2\alpha^6\ln\alpha+\frac{23633-38718\nu+15409\nu^2}{20736}\alpha^6\\
&-\frac{9-7\nu}{8}\nu\alpha^4\ln^3\alpha+\frac{601-439\nu}{576}\lambda_2\alpha^4\ln^2\alpha-\frac{445-283\nu}{864}\lambda_2\alpha^4\ln\alpha\\
&-\frac{8959-15714\nu+6431\nu^2}{20736}\alpha^4+\frac{445-283\nu}{3456}\lambda_2\alpha^2\ln\alpha\\
&+\frac{1199-2754\nu+1231\nu^2}{20736}\alpha^2+\frac{353-191\nu}{82944}\lambda_2\Big)
\end{aligned}
\tag{3.18a ~ d}
$$

我们以 $\alpha=0.3$，$\nu=0.3$ 的情况为例，按照公式（3.17）在图 3 上绘出了不同几何参数 k 值下的特征曲线。由图看出，当 k 很小时，$Q—Y_m$ 曲线单调上升，说明壳体具有平板性质，无跳跃现象产生。当 k 较大时，$Q—Y_m$ 曲线出现了迴形线状态，这时壳体便产生了跳跃现象。

对式（3.17）应用极值条件

$$\frac{\mathrm{d}Q}{\mathrm{d}Y_m}=0 \tag{3.19}$$

我们得到产生跳跃现象时的无量纲内边缘临界挠度的公式：

$$Y_m^*=\frac{-\alpha_3 k\pm\sqrt{(\alpha_3^2-3\alpha_2\alpha_4)k^2-3\alpha_1\alpha_4}}{3\alpha_4} \tag{3.20}$$

将此 Y_m^* 值代入式（3.17），得二次近似的临界载荷公式

$$Q^*=(\alpha_1+\alpha_2 k^2)Y_m^*+\alpha_3 kY_m^{*2}+\alpha_4 Y_m^{*3} \tag{3.21}$$

其中对应于式（3.20）负号的 Q^* 是上临界载荷，正号的 Q^* 是下临界载荷。它们分别是 $Q—Y_m$ 曲线上的极大值点和极小值点。

按照公式（3.21）进行数值计算，我们得到了不同 α 值的上、下临界载荷数据，结果给在图 4 中。由图看到，上临界载荷随 k 的增加而增加；下临界载荷在很小的范围内随 k 的增加而增加，之后随 k 的增加而减少。当 k 值相当大时，下临界载荷为负值，这表明此时若无反向载荷作用，扁壳本身无能力恢复至原来形状。

下面，我们来讨论上、下临界载荷曲线的相重点，亦即壳体屈曲临界点。记临界点的壳体几何参数为 k_0，临界载荷为 Q_0^*。众所周知，二次方程有重根的条件应为方程的判别式等于零，于是由式（3.19）得

$$(\alpha_3^2-3\alpha_2\alpha_4)k^2-3\alpha_1\alpha_4=0 \tag{3.22}$$

解此方程，即得

$$k_0 = \sqrt{\frac{3\alpha_1\alpha_4}{\alpha_3^2 - 3\alpha_2\alpha_4}} \tag{3.23}$$

显然，当 $k<k_0$ 时，壳体不会发生失稳现象。当 $k\geqslant k_0$ 时，壳体就会出现失稳现象。因此，几何参数 k_0 是区分壳体失稳与否的分界点。

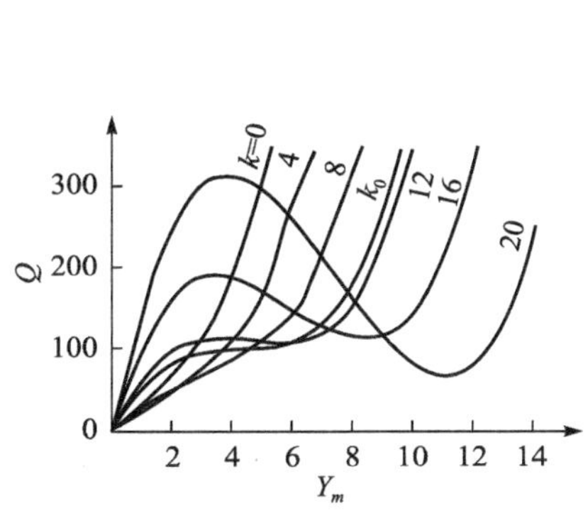

图 3　各种 k 值下的特征曲线

（α=0.3，ν=0.3）

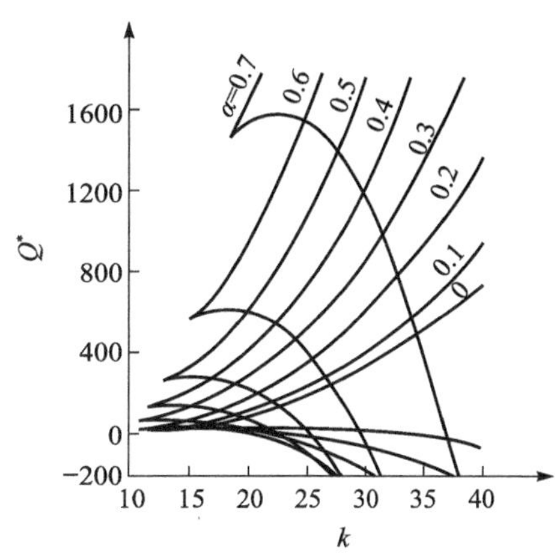

图 4　各种 α 值下的稳定曲线

（ν=0.3）

图 5 与图 6 分别给出了 k_0—α 和 Q_0^*—α 的关系曲线。由图 5 可知，此曲线存在 k_0 的一个极小值。相应于这一极小值的壳体是对失稳反应最灵敏的壳体。为了得到此值，我们令式（3.23）的关于 α 的导数为零

$$\frac{\mathrm{d}k_0}{\mathrm{d}\alpha} = 0 \tag{3.24}$$

于是得到

$$\alpha = 0.2436 \tag{3.25}$$

将此值代入式（3.23），便得 k_0 的极小值

$$(k_0)_{\min} = 11.05 \tag{3.26}$$

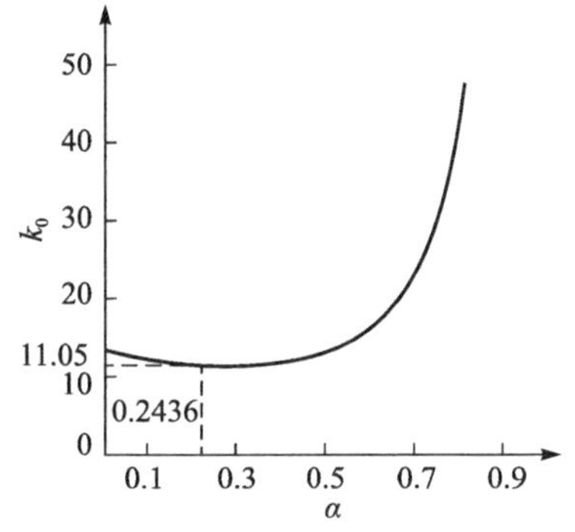

图 5　临界几何参数 k_0 与 α 的关系曲线（ν=0.3）

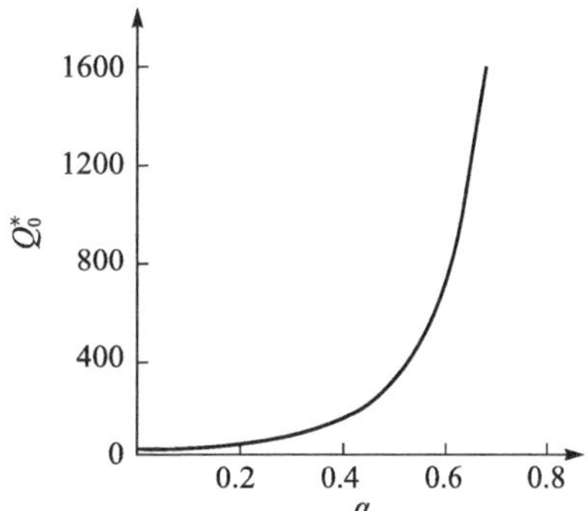

图 6　临界点载荷 Q_0^* 与 α 的关系曲线（ν=0.3）

参考文献

[1] 刘人怀．在内边缘均布力矩作用下中心开孔圆底扁球壳的非线性稳定问题．科学通报，1965，(3)：253.

[2] 刘人怀．在边缘载荷作用下中心开孔圆底扁薄球壳的轴对称稳定性．力学学报，1977，(3)：206.

[3] 刘人怀．双层金属中心开孔扁球壳的非线性热稳定问题．中国科学技术大学学报，1981，11(1)：84.

[4] Liu Ren-huai. Nonlinear thermal stability of bimetallic shallow shells of revolution. International Journal of Non-Linear Mechanics，1983，18 (5)：409.

[5] Tillman. S. C. On the buckling behaviour of shallow spherical caps under a uniform pressure load. International Journal of Solids and Structures，1970，6 (1)：37.

[6] 叶开沅，刘人怀，平庆元，等．在对称线布载荷作用下的圆底扁薄球壳的非线性稳定问题．科学通报，1965，(2)：142；兰州大学学报，1965，(2)：10.

[7] 叶开沅，刘人怀，张传智，等．圆底扁薄球壳在边缘力矩作用下的非线性稳定问题．科学通报，1965，(2)：145.

均布载荷作用下开顶扁球壳的非线性稳定问题①

一、引言

在工程结构中，大量使用薄壁壳体。按照设计要求，需要核算它的稳定性。这对于选择材料以及结构的尺寸有着重要的意义。

近几十年来，薄壳非线性稳定理论的研究一直是引人注意的课题。但是，这项课题的研究不仅在理论上，而且在实验上都遇到了巨大的困难，所以解决的范围较窄、问题较少，大都只讨论较简单的扁球壳、圆柱壳和扁锥壳，而对于扁球壳开顶等较复杂的情况，则研究甚少。刘人怀[1-4]、罗祖道等[5]、Tillman[6]等先后对开顶扁球壳的非线性稳定问题作了研究，获得了一些有益的结果。然而，在这一领域内，迄今还有一些亟待解决的问题尚未被讨论过，本问题即属此范围。

我们考虑一个具有刚性中心的开顶扁球壳，外边缘完全固定，承受均布载荷作用。我们采用叶开沅和刘人怀于 1965 年所提出的修正迭代法[1,7,8]，克服了非线性数学和壳体开顶的困难，获得了临界载荷的解析公式。这一方法由于吸收了钱伟长[9]的以中心挠度做摄动参数的摄动法以及逐次逼近法的优点，因而程序简单明确，计算量少，而且结果精度较高。本文所获得的结果可供工程设计部门参考使用。

二、边值问题的建立

考虑一个均布载荷作用下具有硬中心的开顶扁球壳，如图 1 所示。壳体厚度为 h，中曲面半径为 R，内、外边缘半径分别为 b、a，r 是扁球壳中曲面点至对称中心轴的距离。壳体边缘是夹紧的。

根据均布载荷 q 作用下的扁球壳非线性弯曲理论[1,2,10]，本文所涉及的壳体的挠度 w 和径向薄膜力 N_r 所满足的基本方程是

$$\frac{D}{r}\frac{\mathrm{d}}{\mathrm{d}r}r\frac{\mathrm{d}}{\mathrm{d}r}\frac{1}{r}\frac{\mathrm{d}}{\mathrm{d}r}(r\vartheta)-\frac{1}{r}\frac{\mathrm{d}}{\mathrm{d}r}[rN_r(\theta+\vartheta)]=q,$$

$$\frac{r}{Eh}\frac{\mathrm{d}}{\mathrm{d}r}\frac{1}{r}\frac{\mathrm{d}}{\mathrm{d}r}(r^2N_r)+\vartheta\left(\theta+\frac{1}{2}\vartheta\right)=0 \tag{2.1a,b}$$

其中，ϑ 和 θ 分别是壳体经线方向弧的旋转角和倾斜角，E 为弹性模量，ν 为泊松比，D 是抗弯刚度。

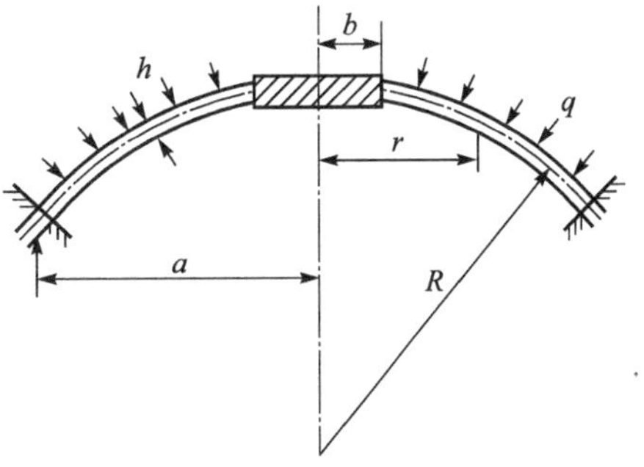

图 1　均布载荷作用下的开顶扁球壳

① 本文原载《应用数学和力学》，1988，9 (3)：205-217. 作者：刘人怀，李东.

$$\vartheta = \frac{\mathrm{d}w}{\mathrm{d}r}, \theta = \frac{r}{R}, D = \frac{Eh^3}{12(1-\nu^2)} \tag{2.2}$$

求出 w 和 N_r 后，我们可由下列公式计算径向薄膜位移 u，径向剪力 Q_r 以及径向弯矩 M_r

$$\begin{aligned} u &= \frac{r}{Eh}\left[r\frac{\mathrm{d}N_r}{\mathrm{d}r} + (1-\nu)N_r\right], \\ Q_r &= -D\frac{\mathrm{d}}{\mathrm{d}r}\frac{1}{r}\frac{\mathrm{d}}{\mathrm{d}r}(r\vartheta), \\ M_r &= -D\left(\frac{\mathrm{d}\vartheta}{\mathrm{d}r} + \frac{\nu}{r}\vartheta\right) \end{aligned} \tag{2.3a ~ c}$$

为简化计算，先将方程（2.1a）积分得

$$Dr\frac{\mathrm{d}}{\mathrm{d}r}\frac{1}{r}\frac{\mathrm{d}}{\mathrm{d}r}(r\vartheta) - rN_r(\theta+\vartheta) = F(r) \tag{2.4}$$

其中 $F(r)$为载荷函数，C 为积分常数，

$$F(r) = \int qr\,\mathrm{d}r + C \tag{2.5}$$

然后，应用式（2.3b），方程（2.4）化为

$$r[Q_r + N_r(\theta+\vartheta)] = -F(r) \tag{2.6}$$

现在，我们来确定载荷函数 $F(r)$。为此，用圆锥曲面从半径为 r 处切割壳体，如图 2 所示。将作用在这部分壳体上的所有力向对称轴投影，并注意到 $\theta+\vartheta$ 的微小条件，这样就得到

$$2\pi r[Q_r + N_r(\theta+\vartheta)] + \pi q(r^2 - b^2) = 0 \tag{2.7}$$

将式（2.7）与式（2.6）比较，就得到

$$F(r) = \frac{1}{2}q(r^2 - b^2) \tag{2.8}$$

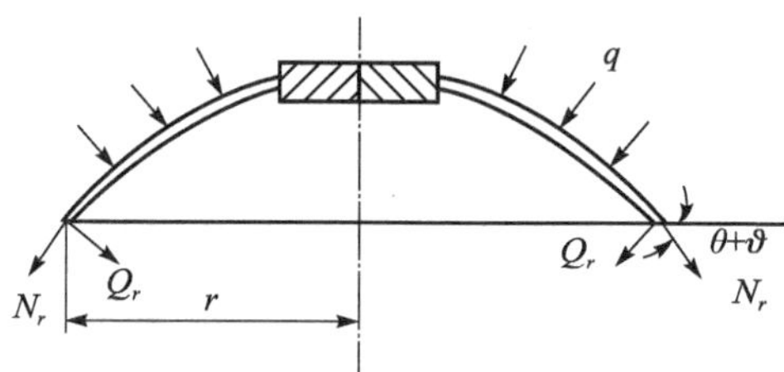

图 2 载荷函数 $F(r)$ 的确定

这样，方程（2.4）最后化为

$$Dr\frac{\mathrm{d}}{\mathrm{d}r}\frac{1}{r}\frac{\mathrm{d}}{\mathrm{d}r}(r\vartheta) - rN_r(\theta+\vartheta) = \frac{1}{2}q(r^2 - b^2) \tag{2.9}$$

因此，方程（2.9）和（2.1b）就组成具有硬中心的开顶扁球壳在均布载荷作用下的轴对称大挠度方程组。

我们在此仅讨论边缘夹紧情况。

当 $r=a$ 时，外边缘夹紧：

$$w=0,\ \vartheta=0,\ u=0 \tag{2.10}$$

当 $r=b$ 时，内边缘被固定在可上、下移动的无变形的一块硬中心上：

$$\vartheta=0, u=0 \tag{2.11}$$

为了计算方便，引入下列无量纲量

$$\begin{aligned}&\rho=\frac{r}{a},\ \alpha=\frac{b}{a},\ y=\sqrt{12\lambda_1\lambda_2}\,\frac{w}{h}+\frac{1}{2}k\rho^2,\\&\varphi=-\frac{\mathrm{d}y}{\mathrm{d}\rho},\ S=\frac{a^2}{D}\rho N_r,\ k=\sqrt{12\lambda_1\lambda_2}\,\frac{a^2}{Rh},\\&Q=\sqrt{3\lambda_1\lambda_2}\,\frac{a^4q}{Dh}\end{aligned} \tag{2.12}$$

其中

$$\lambda_1=1-\nu,\ \lambda_2=1+\nu \tag{2.13}$$

利用这些无量纲量，方程（2.9）、（2.1b）及边界条件（2.10）、（2.11）简化为

$$\begin{aligned}&L(\rho\varphi)=S\varphi+Q(\alpha^2-\rho^2),\\&L(\rho S)=\frac{1}{2}(k^2\rho^2-\varphi^2)\end{aligned} \tag{2.14a,b}$$

当 $\rho=1$ 时，$y=\frac{1}{2}k$，$\varphi=-k$，$\frac{\mathrm{d}S}{\mathrm{d}\rho}-\nu S=0$　　(2.15)

当 $\rho=\alpha$ 时，$\varphi=-k\alpha$，$\rho\frac{\mathrm{d}S}{\mathrm{d}\rho}-\nu S=0$　　(2.16)

其中 L 是微分算子，

$$L(\cdots)=\rho\frac{\mathrm{d}}{\mathrm{d}\rho}\frac{1}{\rho}\frac{\mathrm{d}}{\mathrm{d}\rho}(\cdots) \tag{2.17}$$

这样一来，问题就化为在边界条件（2.15）、（2.16）下求解非线性微分方程组（2.14）～（2.16）。

三、边值问题的求解

我们使用修正迭代法解此问题。在第一次近似中，我们略去方程（2.14a）中的非线性项 $S\varphi$，便得到如下的线性边值问题

$$\begin{aligned}&L(\rho\varphi_1)=Q(\alpha^2-\rho^2),\\&L(\rho S_1)=\frac{1}{2}(k^2\rho^2-\varphi_1^2)\end{aligned} \tag{3.1a,b}$$

当 $\rho=1$ 时，$y_1=\frac{1}{2}k$，$\varphi_1=-k$，$\frac{\mathrm{d}S_1}{\mathrm{d}\rho}-\nu S_1=0$　　(3.2a～c)

当 $\rho=\alpha$ 时，$\varphi_1=-k\alpha$，$\rho\frac{\mathrm{d}S_1}{\mathrm{d}\rho}-\nu S_1=0$　　(3.3a,b)

应用式（3.2b）和（3.3a），积分方程（3.1a），就得到小挠度理论的解

$$\begin{aligned}\varphi_1=&-\frac{1}{8(1-\alpha^2)}\{\alpha^2(1-\alpha^2+4\alpha^2\ln\alpha)Q\rho^{-1}-[(Q-8k)+8k\alpha^2\\&-Q\alpha^4+4Q\alpha^4\ln\alpha]\rho-4\alpha^2(1-\alpha^2)Q\rho\ln\rho+(1-\alpha^2)Q\rho^3\}\end{aligned} \tag{3.4}$$

以扁球壳的无量纲内边缘挠度 W_m 作迭代参数

$$W_m = \sqrt{12(1-\nu^2)}\,\frac{w}{h}\bigg|_{r=b} \tag{3.5}$$

利用式（2.12）及（3.2a），可得到

$$W_m = \int_\alpha^1 (\varphi + k\rho)\mathrm{d}\rho \tag{3.6}$$

再将式（3.4）代入上式，即得到一次近似下壳体的特征关系

$$Q = 8\beta_1^{-1}\beta_2 W_m \tag{3.7}$$

其中

$$\begin{aligned}\beta_1 &= (1-\alpha^2)^{-1} \\ \beta_2 &= 4(1-5\alpha^2+4\alpha^2\ln\alpha+7\alpha^4-4\alpha^4\ln\alpha+16\alpha^4\ln^2\alpha-4\alpha^6)^{-1}\end{aligned} \tag{3.8}$$

将式（3.7）代入式（3.4）中，得到

$$\begin{aligned}\varphi_1 = -k\rho-\beta_2 W_m[&(\alpha^2-\alpha^4+4\alpha^4\ln\alpha)\rho^{-1}-(1-\alpha^4 \\ &+4\alpha^4\ln\alpha)\rho-4\alpha^2(1-\alpha^2)\rho\ln\rho+(1-\alpha^2)\rho^3]\end{aligned} \tag{3.9}$$

应用式（3.9）及边界条件（3.2c）、（3.3b），积分方程（3.1b），便得到无量纲薄膜内力的近似解为

$$\begin{aligned}S_1 = k\beta_1\beta_2 W_m\Big[&B\rho^{-1}+A\rho-\frac{1}{2}\alpha^2(1-2\alpha^2+4\alpha^2\ln\alpha+\alpha^4 \\ &-4\alpha^4\ln\alpha)\rho\ln\rho+\frac{1}{8}(1-4\alpha^2+5\alpha^4+4\alpha^4\ln\alpha-2\alpha^6-4\alpha^6\ln\alpha)\rho^3 \\ &+\frac{1}{2}\alpha^2(1-\alpha^2)^2\rho^3\ln\rho-\frac{1}{24}(1-\alpha^2)^2\rho^5\Big] \\ &+\beta_1\beta_2^2 W_m^2\Big[D\rho^{-1}+\frac{1}{4}\alpha^4(1-3\alpha^2+8\alpha^2\ln\alpha+3\alpha^4-16\alpha^4\ln\alpha \\ &+16\alpha^4\ln^2\alpha-\alpha^6+8\alpha^6\ln\alpha-16\alpha^6\ln^2\alpha)\rho^{-1}\ln\rho+C\rho \\ &+\frac{1}{2}\alpha^2(1-4\alpha^2+4\alpha^2\ln\alpha+6\alpha^4-8\alpha^4\ln\alpha-4\alpha^6 \\ &+4\alpha^6\ln\alpha+16\alpha^6\ln^2\alpha+\alpha^8-16\alpha^8\ln^2\alpha)\rho\ln\rho \\ &+\alpha^4(1-3\alpha^2+4\alpha^2\ln\alpha+3\alpha^4-8\alpha^4\ln\alpha \\ &-\alpha^6+4\alpha^6\ln\alpha)\rho\ln^2\rho-\frac{1}{16}(1-5\alpha^2+18\alpha^4+16\alpha^4\ln\alpha \\ &-34\alpha^6-48\alpha^6\ln\alpha+29\alpha^8+48\alpha^8\ln\alpha+16\alpha^8\ln^2\alpha \\ &-9\alpha^{10}-16\alpha^{10}\ln\alpha-16\alpha^{10}\ln^2\alpha)\rho^3-\frac{1}{2}\alpha^2(1-5\alpha^2+9\alpha^4+4\alpha^4\ln\alpha-7\alpha^6 \\ &-8\alpha^6\ln\alpha+2\alpha^8+4\alpha^8\ln\alpha)\rho^3\ln\rho-\alpha^4(1-\alpha^2)^3\rho^3\ln^2\rho \\ &+\frac{1}{72}(3-11\alpha^2+15\alpha^4+12\alpha^4\ln\alpha-9\alpha^6-24\alpha^6\ln\alpha \\ &+2\alpha^8+12\alpha^8\ln\alpha)\rho^5+\frac{1}{6}\alpha^2(1-\alpha^2)^3\rho^5\ln\rho-\frac{1}{96}(1-\alpha^2)^3\rho^7\Big]\end{aligned} \tag{3.10}$$

其中

$$A=-\frac{1}{\lambda_1}\Big(\frac{2-\nu}{12}-\frac{11-4\nu}{12}\alpha^2+\frac{7-\nu}{8}\alpha^4-\nu\alpha^4\ln\alpha$$
$$+\frac{4-5\nu}{12}\alpha^6+\nu\alpha^6\ln\alpha+2\lambda_1\alpha^6\ln^2\alpha-\frac{11-7\nu}{24}\alpha^8\Big),$$

$$B=-\frac{1}{\lambda_2}\Big(\frac{2-\nu}{12}\alpha^2-\frac{19-11\nu}{24}\alpha^4+\frac{\lambda_1}{2}\alpha^4\ln\alpha+\frac{13-8\nu}{12}\alpha^6$$
$$-\frac{\lambda_1}{2}\alpha^6\ln\alpha+2\lambda_1\alpha^6\ln^2\alpha-\frac{11-7\nu}{24}\alpha^8\Big),$$

$$C=\frac{1}{\lambda_1}\Big[\frac{5-3\nu}{96}-\frac{73-23\nu}{144}\alpha^2+\frac{679-209\nu}{288}\alpha^4+\frac{5-19\nu}{12}\alpha^4\ln\alpha-\frac{275-67\nu}{72}\alpha^6$$
$$+\frac{1+13\nu}{3}\alpha^6\ln\alpha+(1-5\nu)\alpha^6\ln^2\alpha+\frac{461+125\nu}{288}\alpha^8-\frac{23+47\nu}{12}\alpha^8\ln\alpha$$
$$-(2-6\nu)\alpha^8\ln^2\alpha-8\nu\alpha^8\ln^3\alpha+\frac{151-197\nu}{144}\alpha^{10}+\frac{7}{6}\lambda_1\alpha^{10}\ln\alpha$$
$$+\lambda_1\alpha^{10}\ln^2\alpha+4\lambda_1\alpha^{10}\ln^3\alpha-\frac{211-173\nu}{288}\alpha^{12}\Big],$$

$$D=\frac{1}{\lambda_2}\Big[\frac{5-3\nu}{96}\alpha^2+\frac{1+4\nu}{18}\alpha^4+\frac{1-3\nu}{4}\alpha^4\ln\alpha+\frac{73-311\nu}{288}\alpha^6$$
$$+\frac{11+5\nu}{3}\alpha^6\ln\alpha+(1-5\nu)\alpha^6\ln^2\alpha-\frac{155-213\nu}{96}\alpha^8$$
$$-\frac{97+13\nu}{12}\alpha^8\ln\alpha+7\lambda_2\alpha^8\ln^2\alpha-8\nu\alpha^8\ln^3\alpha$$
$$+\frac{143-139\nu}{72}\alpha^{10}+\frac{25+\nu}{6}\alpha^{10}\ln\alpha-(8+2\nu)\alpha^{10}\ln^2\alpha$$
$$+4\lambda_1\alpha^{10}\ln^3\alpha-\frac{211-173\nu}{288}\alpha^{12}\Big]$$

在二次近似中，我们有下述边值问题

$$L(\rho\varphi_2)=S_1\varphi_1+Q(\alpha^2-\rho^2) \tag{3.11}$$

当 $\rho=1$ 时，$y_2=\frac{1}{2}k$，$\varphi_2=-k$　　(3.12a，b)

当 $\rho=\alpha$ 时，$\varphi_2=-k\alpha$　　(3.13)

将方程（3.11）积分，并利用边界条件（3.12b）及（3.13），就得到 φ 的二次近似解

$$\begin{aligned}\varphi_2=\varphi_1&+k^2\beta_1\beta_2W_m[\phi_1(\rho)+A_1\rho+B_1\rho^{-1}]+k\beta_1\beta_2^2W_m^2[\phi_2(\rho)\\&+A_2\rho+B_2\rho^{-1}]+\beta_1\beta_2^3W_m^3[\phi_3(\rho)+A_3\rho+B_3\rho^{-1}]\end{aligned} \tag{3.14}$$

其中

$$A_n=\frac{\alpha\,\phi_n(\alpha)-\phi_n(1)}{1-\alpha^2},$$

$$B_n = \frac{\alpha^2\phi_n(1) - \alpha\phi_n(\alpha)}{1-\alpha^2} \quad (n=1,2,3)$$

$$\begin{aligned}
\phi_1(\rho) = & -\frac{B}{2}\rho\ln\rho - \frac{A}{8}\rho^3 \\
& + \frac{1}{64}\alpha^2(1-2\alpha^2+4\alpha^2\ln\alpha+\alpha^4-4\alpha^4\ln\alpha)(4\rho^3\ln\rho-3\rho^3) \\
& - \frac{1}{192}(1-4\alpha^2+5\alpha^4+4\alpha^4\ln\alpha-2\alpha^6-4\alpha^6\ln\alpha)\rho^5 \\
& - \frac{1}{576}\alpha^2(1-\alpha^2)^2(12\rho^5\ln\rho-5\rho^5) + \frac{1}{1152}(1-\alpha^2)^2\rho^7,
\end{aligned}$$

$$\begin{aligned}
\phi_2(\rho) = & -A\Big[\frac{1}{2}\alpha^2(1-\alpha^2+4\alpha^2\ln\alpha)\rho\ln\rho - \frac{1}{8}(1-3\alpha^2+2\alpha^4+4\alpha^4\ln\alpha)\rho^3 \\
& - \frac{1}{2}\alpha^2(1-\alpha^2)\rho^3\ln\rho + \frac{1}{24}(1-\alpha^2)\rho^5\Big] + B\Big[\frac{1}{2}\alpha^2(1-\alpha^2+4\alpha^2\ln\alpha)\rho^{-1}\ln\rho \\
& + \frac{1}{2}(1-2\alpha^2+\alpha^4+4\alpha^4\ln\alpha)\rho\ln\rho + \alpha^2(1-\alpha^2)\rho\ln^2\rho - \frac{1}{8}(1-\alpha^2)\rho^3\Big] \\
& - \frac{1}{2}D\rho\ln\rho - \frac{1}{8}C\rho^3 - \frac{1}{64}\alpha^2(1-5\alpha^2+4\alpha^2\ln\alpha+9\alpha^4-12\alpha^4\ln\alpha-7\alpha^6+12\alpha^6\ln\alpha \\
& + 16\alpha^6\ln^2\alpha + 2\alpha^8 - 4\alpha^8\ln\alpha - 16\alpha^8\ln^2\alpha)\rho^3 + \frac{1}{1152}(9-37\alpha^2+72\alpha^4+104\alpha^4\ln\alpha \\
& - 84\alpha^6 - 280\alpha^6\ln\alpha + 55\alpha^8 + 248\alpha^8\ln\alpha + 144\alpha^8\ln^2\alpha - 15\alpha^{10} - 72\alpha^{10}\ln\alpha \\
& - 144\alpha^{10}\ln^2\alpha)\rho^5 - \frac{1}{3456}(15-62\alpha^2+96\alpha^4+60\alpha^4\ln\alpha-66\alpha^6-120\alpha^6\ln\alpha+17\alpha^8 \\
& + 60\alpha^8\ln\alpha)\rho^7 + \frac{1}{1536}(1-\alpha^2)^3\rho^9 + \frac{1}{16}\alpha^4(1-3\alpha^2+8\alpha^2\ln\alpha+3\alpha^4-16\alpha^4\ln\alpha \\
& + 16\alpha^4\ln^2\alpha - \alpha^6 + 8\alpha^6\ln\alpha - 16\alpha^6\ln^2\alpha)(\rho\ln^2\rho - \rho\ln\rho) - \frac{1}{64}\alpha^2(2-5\alpha^2+8\alpha^2\ln\alpha \\
& + 3\alpha^4 - 4\alpha^4\ln\alpha + \alpha^6 - 16\alpha^6\ln\alpha + 32\alpha^6\ln^2\alpha - \alpha^8 + 12\alpha^8\ln\alpha \\
& - 32\alpha^8\ln^2\alpha)(4\rho^3\ln\rho - 3\rho^3) + \frac{1}{576}\alpha^2(4-15\alpha^2+4\alpha^2\ln\alpha+21\alpha^4 \\
& + 4\alpha^4\ln\alpha - 13\alpha^6 - 20\alpha^6\ln\alpha + 3\alpha^8 + 12\alpha^8\ln\alpha)(12\rho^5\ln\rho - 5\rho^5) \\
& - \frac{5}{6912}\alpha^2(1-\alpha^2)^3(24\rho^7\ln\rho - 7\rho^7) - \frac{3}{64}\alpha^4(1-3\alpha^2 \\
& + 4\alpha^2\ln\alpha + 3\alpha^4 - 8\alpha^4\ln\alpha - \alpha^6 + 4\alpha^6\ln\alpha)(8\rho^3\ln^2\rho - 12\rho^3\ln\rho + 7\rho^3) \\
& + \frac{1}{576}\alpha^4(1-\alpha^2)^3(72\rho^5\ln^2\rho - 60\rho^5\ln\rho + 19\rho^5),
\end{aligned}$$

$$\begin{aligned}
\phi_3(\rho) = & -C\Big[\frac{1}{2}\alpha^2(1-\alpha^2+4\alpha^2\ln\alpha)\rho\ln\rho - \frac{1}{8}(1-3\alpha^2+2\alpha^4+4\alpha^4\ln\alpha)\rho^3 \\
& - \frac{1}{2}\alpha^2(1-\alpha^2)\rho^3\ln\rho + \frac{1}{24}(1-\alpha^2)\rho^5\Big] \\
& + D\Big[\frac{1}{2}\alpha^2(1-\alpha^2+4\alpha^2\ln\alpha)\rho^{-1}\ln\rho + \frac{1}{2}(1-2\alpha^2+\alpha^4+4\alpha^4\ln\alpha)\rho\ln\rho
\end{aligned}$$

$$
\begin{aligned}
&+\alpha^2(1-\alpha^2)\rho\ln^2\rho-\frac{1}{8}(1-\alpha^2)\rho^3\Big]+\frac{1}{128}\alpha^2(1-6\alpha^2+4\alpha^2\ln\alpha+23\alpha^4\\
&-4\alpha^4\ln\alpha-52\alpha^6+8\alpha^6\ln\alpha+64\alpha^6\ln^2\alpha+63\alpha^8-40\alpha^8\ln\alpha-176\alpha^8\ln^2\alpha-38\alpha^{10}\\
&+52\alpha^{10}\ln\alpha+160\alpha^{10}\ln^2\alpha+9\alpha^{12}-20\alpha^{12}\ln\alpha+64\alpha^{10}\ln^3\alpha-48\alpha^{12}\ln^2\alpha-64\alpha^{12}\ln^3\alpha)\rho^3\\
&-\frac{1}{3456}(9-39\alpha^2+125\alpha^4+204\alpha^4\ln\alpha-209\alpha^6-676\alpha^6\ln\alpha+51\alpha^8+984\alpha^8\ln\alpha\\
&+816\alpha^8\ln^2\alpha+247\alpha^{10}-936\alpha^{10}\ln\alpha\\
&-2064\alpha^{10}\ln^2\alpha-265\alpha^{12}+604\alpha^{12}\ln\alpha\\
&+1680\alpha^{12}\ln^2\alpha+81\alpha^{14}-180\alpha^{14}\ln\alpha\\
&+576\alpha^{12}\ln^3\alpha-432\alpha^{14}\ln^2\alpha-576\alpha^{14}\ln^3\alpha)\rho^5\\
&+\frac{1}{13824}(30-149\alpha^2+450\alpha^4\\
&+396\alpha^4\ln\alpha-910\alpha^6-1460\alpha^6\ln\alpha\\
&+1070\alpha^8+2004\alpha^8\ln\alpha+480\alpha^8\ln^2a-645\alpha^{10}\\
&-1212\alpha^{10}\ln\alpha-960\alpha^{10}\ln^2\alpha+154\alpha^{12}\\
&+272\alpha^{12}\ln\alpha+480\alpha^{12}\ln^2\alpha)\rho^7-\frac{1}{4608}(3-13\alpha^2\\
&+22\alpha^4+12\alpha^4\ln\alpha-18\alpha^6-36\alpha^6\ln\alpha\\
&+7\alpha^8+36\alpha^8\ln\alpha-\alpha^{10}-12\alpha^{10}\ln\alpha)\rho^9\\
&+\frac{1}{11520}(1-\alpha^2)^4\rho^{11}+\frac{1}{15360}\alpha^2(1-\alpha^2)^4(9\rho^9-40\rho^9\ln\rho)\\
&+\frac{1}{16}\alpha^6(1-4\alpha^2+12\alpha^2\ln\alpha+6\alpha^4-36\alpha^4\ln\alpha+48\alpha^4\ln^2\alpha\\
&-4\alpha^6+36\alpha^6\ln\alpha-96\alpha^6\ln^2\alpha+64\alpha^6\ln^3\alpha+\alpha^8\\
&-12\alpha^8\ln\alpha+48\alpha^8\ln^2\alpha-64\alpha^8\ln^3\alpha)(\rho^{-1}\ln\rho+\rho^{-1}\ln^2\rho)\\
&-\frac{1}{16}\alpha^4(1-7\alpha^2+8\alpha^2\ln\alpha+18\alpha^4-44\alpha^4\ln\alpha\\
&+16\alpha^4\ln^2\alpha-22\alpha^6+84\alpha^6\ln\alpha-48\alpha^6\ln^2\alpha\\
&+13\alpha^8-68\alpha^8\ln\alpha+48\alpha^8\ln^2\alpha-3\alpha^{10}\\
&+20\alpha^{10}\ln\alpha+64\alpha^8\ln^3\alpha-16\alpha^{10}\ln^2\alpha\\
&-64\alpha^{10}\ln^3\alpha)(\rho\ln^2\rho-\rho\ln\rho)-\frac{1}{3456}\alpha^2(15-76\alpha^2+24\alpha^2\ln\alpha+169\alpha^4\\
&+32\alpha^4\ln\alpha-216\alpha^6-312\alpha^6\ln\alpha\\
&+96\alpha^6\ln^2\alpha+169\alpha^8+504\alpha^8\ln\alpha-48\alpha^8\ln^2\alpha\\
&-76\alpha^{10}-320\alpha^{10}\ln\alpha-192\alpha^{10}\ln^2\alpha\\
&+15\alpha^{12}+72\alpha^{12}\ln\alpha+144\alpha^{12}\ln^2\alpha)(12\rho^5\ln\rho\\
&-5\rho^5)+\frac{1}{20736}\alpha^2(15-77\alpha^2+158\alpha^4\\
&+60\alpha^4\ln\alpha-162\alpha^6-180\alpha^6\ln\alpha+83\alpha^8\\
&+180\alpha^8\ln\alpha-17\alpha^{10}-60\alpha^{10}\ln\alpha)(24\rho^7\ln\rho-7\rho^7)
\end{aligned}
$$

$$+\frac{3}{64}\alpha^4(1-4\alpha^2+4\alpha^2\ln\alpha+6\alpha^4-8\alpha^4\ln\alpha-4\alpha^6+16\alpha^6\ln^2 a+\alpha^8$$
$$+8\alpha^8\ln\alpha-32\alpha^8\ln^2\alpha-4\alpha^{10}\ln\alpha+16\alpha^{10}\ln^2\alpha)(8\rho^3\ln^2\rho-12\rho^3\ln\rho+7\rho^3)$$
$$-\frac{1}{1728}\alpha^4(4-19\alpha^2+4\alpha^2\ln\alpha+36\alpha^4-34\alpha^6-24\alpha^6\ln\alpha+16\alpha^8$$
$$+32\alpha^8\ln\alpha-3\alpha^{10}-12\alpha^{10}\ln\alpha)(72\rho^5\ln^2\rho-60\rho^5\ln\rho+19\rho^5)$$
$$+\frac{5}{41472}\alpha^4(1-\alpha^2)^4(288\rho^7\ln^2\rho-168\rho^7\ln\rho+37\rho^7)$$
$$+\frac{1}{64}\alpha^6(1-4\alpha^2+4\alpha^2\ln\alpha$$
$$+6\alpha^4-12\alpha^4\ln\alpha-4\alpha^6+12\alpha^6\ln\alpha+\alpha^8$$
$$-4\alpha^8\ln\alpha)(32\rho^3\ln^3\rho-72\rho^3\ln^2\rho+84\rho^3\ln\rho-45\rho^3)$$
$$-\frac{1}{1728}\alpha^6(1-\alpha^2)^4(288\rho^5\ln^3\rho-360\rho^5\ln^2\rho+288\rho^5\ln\rho-65\rho^5)$$
$$+\frac{1}{128}\alpha^2(2-7\alpha^2+8\alpha^2\ln\alpha+2\alpha^4-8\alpha^4\ln\alpha$$
$$+22\alpha^6-40\alpha^6\ln\alpha+48\alpha^6\ln^2\alpha-38\alpha^8$$
$$+88\alpha^8\ln\alpha-32\alpha^8\ln^2\alpha+25\alpha^{10}-64\alpha^{10}\ln\alpha$$
$$-80\alpha^{10}\ln^2\alpha-6\alpha^{12}+16\alpha^{12}\ln\alpha$$
$$+128\alpha^{10}\ln^3\alpha+64\alpha^{12}\ln^2\alpha-128\alpha^{12}\ln^3\alpha)(4\rho^3\ln\rho-3\rho^3)$$

将式（3.14）代入式（3.5），便得到壳体的二次近似特征关系式

$$Q=(a_1+a_2k^2)W_m+a_3kW_m^2+a_4W_m^3 \tag{3.15}$$

其中

$$a_1=8\beta_1^{-1}\beta_2,$$
$$a_2=-\lambda_1^{-1}\lambda_2^{-1}\beta_1\beta_2^2\Big[\frac{227+144\nu-83\nu^2}{3456}\alpha^{14}-\frac{35+20\nu-15\nu^2}{48}\alpha^{12}\ln\alpha$$
$$+\frac{557-888\nu+283\nu^2}{1152}\alpha^{12}-\frac{13-48\nu+23\nu^2}{6}\alpha^{10}\ln^2\alpha$$
$$+\frac{37-10\nu+\nu^2}{24}\alpha^{10}\ln\alpha-\frac{3025-3456\nu+1007\nu^2}{1152}\alpha^{10}+8\lambda_1^2\alpha^8\ln^4\alpha$$
$$-2\lambda_1^2\alpha^8\ln^3\alpha+(5-17\nu+8\nu^2)\alpha^8\ln^2\alpha$$
$$-\frac{17-552\nu+295\nu^2}{144}\alpha^8\ln\alpha+\frac{14945-17784\nu+5287\nu^2}{3456}\alpha^8$$
$$+2\lambda_1^2\alpha^6\ln^3\alpha-\frac{7-20\nu+9\nu^2}{2}\alpha^6\ln^2\alpha-\frac{77+232\nu-133\nu^2}{48}\alpha^6\ln\alpha$$
$$-\frac{1155-1744\nu+557\nu^2}{384}\alpha^6+\frac{2-3\nu+\nu^2}{3}\alpha^4\ln^2\alpha$$
$$+\frac{25+46\nu-27\nu^2}{24}\alpha^4\ln\alpha+\frac{811-2376\nu+845\nu^2}{1152}\alpha^4$$
$$-\frac{19+12\nu-7\nu^2}{144}\alpha^2\ln\alpha+\frac{299+1440\nu-587\nu^2}{3456}\alpha^2$$

$$-\frac{35+24\nu-11\nu^2}{1152}\Big],$$

$$a_3=-\lambda_1^{-1}\lambda_2^{-1}\beta_1\beta_2^3\Big[\frac{2989+1610\nu-1379\nu^2}{17280}\alpha^{18}-\frac{2407+1132\nu-1275\nu^2}{576}\alpha^{16}\ln\alpha$$

$$+\frac{7599-10625\nu+3496\nu^2}{2880}\alpha^{16}+\frac{74\nu-46\nu^2}{3}\alpha^{14}\ln^3\alpha$$

$$+\frac{289+332\nu-333\nu^2}{24}\alpha^{14}\ln^2\alpha+\frac{2637+3820\nu-2273\nu^2}{576}\alpha^{14}\ln\alpha$$

$$-\frac{44671-54310\nu+17059\nu^2}{2880}\alpha^{14}+(24-96\nu$$

$$+72\nu^2)\alpha^{12}\ln^5\alpha-(68-40\nu+20\nu^2)\alpha^{12}\ln^4\alpha$$

$$+\frac{237-304\nu+155\nu^2}{6}\alpha^{12}\ln^3\alpha-\frac{479+656\nu-595\nu^2}{12}\alpha^{12}\ln^2\alpha$$

$$+\frac{4945-852\nu-1573\nu^2}{192}\alpha^{12}\ln\alpha+\frac{45682-61485\nu+20813\nu^2}{1440}\alpha^{12}$$

$$+(24-96\nu+72\nu^2)\alpha^{10}\ln^5\alpha+(48+24\nu-24\nu^2)\alpha^{10}\ln^4\alpha$$

$$-\frac{151-24\nu-7\nu^2}{2}\alpha^{10}\ln^3\alpha+\frac{1127+1996\nu-1595\nu^2}{24}\alpha^{10}\ln^2\alpha$$

$$-\frac{37654+5380\nu-16137\nu^2}{576}\alpha^{10}\ln\alpha$$

$$-\frac{21879-38335\nu+14636\nu^2}{720}\alpha^{10}+(20-64\nu$$

$$+44\nu^2)\alpha^8\ln^4\alpha+\frac{195+176\nu-139\nu^2}{6}\alpha^8\ln^3\alpha$$

$$-\frac{523+1436\nu-975\nu^2}{24}\alpha^8\ln^2\alpha$$

$$+\frac{33735+10300\nu-16523\nu^2}{576}\alpha^8\ln\alpha+\frac{17953-56530\nu+24817\nu^2}{1440}\alpha^8$$

$$+\frac{21-92\nu+55\nu^2}{6}\alpha^6\ln^3\alpha+\frac{13+118\nu-65\nu^2}{6}\alpha^6\ln^2\alpha$$

$$-\frac{12693+6588\nu-7257\nu^2}{576}\alpha^6\ln\alpha$$

$$+\frac{351+49850\nu-25021\nu^2}{2880}\alpha^6+\frac{13-52\nu+23\nu^2}{24}\alpha^4\ln^2\alpha$$

$$+\frac{1613+1580\nu-1185\nu^2}{576}\alpha^4\ln\alpha-\frac{2202+6355\nu-3467\nu^2}{1440}\alpha^4$$

$$-\frac{75+44\nu-31\nu^2}{576}\alpha^2\ln\alpha$$

$$+\frac{1897+3290\nu-1727\nu^2}{5760}\alpha^2-\frac{227+145\nu-82\nu^2}{8640}\Big],$$

$$a_4=-\lambda_1^{-1}\lambda_2^{-1}\beta_1\beta_2^4\Big[\frac{238954+98800\nu-140154\nu^2}{2073600}\alpha^{22}-\frac{81726+26720\nu-55006\nu^2}{17280}\alpha^{20}\ln\alpha$$

$$
\begin{aligned}
&+\frac{249384-349552\nu+119912\nu^2}{82944}\alpha^{20}+\frac{157\nu-127\nu^2}{3}\alpha^{18}\ln^3\alpha\\
&+\frac{5949-1839\nu-1848\nu^2}{108}\alpha^{18}\ln^2\alpha-\frac{62768-162720\nu+68272\nu^2}{8640}\alpha^{18}\ln\alpha\\
&-\frac{3879361-5326680\nu+1879079\nu^2}{207360}\alpha^{18}+\frac{64-384\nu+320\nu^2}{3}\alpha^{16}\ln^6\alpha\\
&-(184-144\nu-40\nu^2)\alpha^{16}\ln^5\alpha+\frac{224-612\nu-92\nu^2}{3}\alpha^{16}\ln^4\alpha\\
&-\frac{187+192\nu-1027\nu^2}{9}\alpha^{16}\ln^3\alpha-\frac{32069-2524\nu-13137\nu^2}{144}\alpha^{16}\ln^2\alpha\\
&+\frac{1000932-577640\nu-118412\nu^2}{8640}\alpha^{16}\ln\alpha\\
&+\frac{8666966-14289720\nu+5900914\nu^2}{207360}\alpha^{16}\\
&+\frac{64-384\nu+1088\nu^2}{3}\alpha^{14}\ln^6\alpha-(24-48\nu+216\nu^2)\alpha^{14}\ln^5\alpha\\
&+\frac{74+816\nu+310\nu^2}{3}\alpha^{14}\ln^4\alpha\\
&+\frac{61-1310\nu-563\nu^2}{6}\alpha^{14}\ln^3\alpha+\frac{37290+11139\nu-21291\nu^2}{108}\alpha^{14}\ln^2\alpha\\
&-\frac{5279806-1813360\nu-1321646\nu^2}{17280}\alpha^{14}\ln\alpha\\
&-\frac{4221089-10798200\nu+5492071\nu^2}{103680}\alpha^{14}\\
&+\frac{64-384\nu+320\nu^2}{3}\alpha^{12}\ln^6\alpha+(184-48\nu+56\nu^2)\alpha^{12}\ln^5\alpha\\
&-\frac{790-48\nu+218\nu^2}{3}\alpha^{12}\ln^4\alpha+\frac{1721+4812\nu+211\nu^2}{18}\alpha^{12}\ln^3\alpha\\
&-\frac{110445+107076\nu-94521\nu^2}{432}\alpha^{12}\ln^2\alpha\\
&+\frac{6486112-1257360\nu-2127472\nu^2}{17280}\alpha^{12}\ln\alpha\\
&+\frac{3280898-50424600\nu+32440552\nu^2}{518400}\alpha^{12}\\
&+(24-144\nu+120\nu^2)\alpha^{10}\ln^5\alpha+(154-32\nu-42\nu^2)\alpha^{10}\ln^4\alpha\\
&-\frac{817+382\nu-107\nu^2}{6}\alpha^{10}\ln^3\alpha+\frac{3455+7397\nu-4662\nu^2}{36}\alpha^{10}\ln^2\alpha\\
&-\frac{2106857-53560\nu-864417\nu^2}{8640}\alpha^{10}\ln\alpha\\
&+\frac{4709378+12210480\nu-9934418\nu^2}{207360}\alpha^{10}
\end{aligned}
$$

$$
\begin{aligned}
&+(10-52\nu+42\nu^2)\alpha^8\ln^4\alpha+\frac{299-52\nu-79\nu^2}{6}\alpha^8\ln^3\alpha \\
&-\frac{3307+9500\nu-5471\nu^2}{144}\alpha^8\ln^2\alpha+\frac{88726+20785\nu-47421\nu^2}{1080}\alpha^8\ln\alpha \\
&-\frac{4456264+5005680\nu-4876504\nu^2}{207360}\alpha^8 \\
&+\frac{4-23\nu+17\nu^2}{3}\alpha^6\ln^3\alpha+\frac{528+573\nu-495\nu^2}{108}\alpha^6\ln^2\alpha \\
&-\frac{37099+26120\nu-28259\nu^2}{2880}\alpha^6\ln\alpha \\
&+\frac{3437218+2918640\nu-2937778\nu^2}{414720}\alpha^6 \\
&+\frac{5-28\nu+15\nu^2}{48}\alpha^4\ln^2\alpha+\frac{8123+8920\nu-7843\nu^2}{8640}\alpha^4\ln\alpha \\
&-\frac{599344+585200\nu-497984\nu^2}{414720}\alpha^4-\frac{77+40\nu-37\nu^2}{2880}\alpha^2\ln\alpha \\
&+\frac{119227+147800\nu-101027\nu^2}{1036800}\alpha^2-\frac{173+100\nu-73\nu^2}{34560}\Bigg]
\end{aligned}
$$

我们从 $\alpha=0.3$，$\nu=0.3$ 的情况为例，按公式（3.15）在图 3 上绘出不同几何参数 k 值下的特征曲线。由图 3 看出，只有当 k 大于某一值 k_0 后，曲线才呈现出迴形线形态，这时壳体出现了失稳现象。

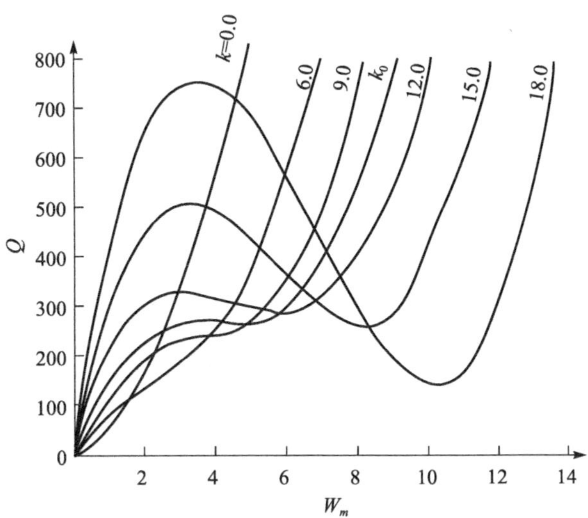

图 3　各种 k 值下的特征曲线（$\nu=0.3$，$\alpha=0.3$）

对式（3.15）应用极值条件

$$\frac{\mathrm{d}Q}{\mathrm{d}W_m}=0 \tag{3.16}$$

就得到产生失稳时的临界无量纲内缘挠度公式

$$W_m^* = \frac{-a_3k \pm \sqrt{(a_2^3 - 3a_2a_4)k^2 - 3a_1a_4}}{3a_4} \tag{3.17}$$

将此 W_m^* 代入式（3.15），就得到二次近似的临界载荷公式

$$Q^* = (a_1 + a_2k^2)W_m^* + a_3kW_m^{*2} + a_4W_m^{*3} \tag{3.18}$$

这里，对应于式（3.17）中的正、负号，所得到的 Q^* 分别是下、上临界载荷，它们分别对应于特征曲线上的极小值和极大值点。

按照式（3.18）进行数值计算，我们得到了几个不同 α 值的上、下临界载荷数据，并将结果绘在图 4 中。从图上看，上临界载荷是几何参数 k 的单调上升函数，而下临界载荷只在一段范围内是 k 的增值函数，随后随着 k 的增大而迅速下降。当 k 值较大时，下临界载荷变为负值，表明若无反向作用，则壳体不能恢复原状。

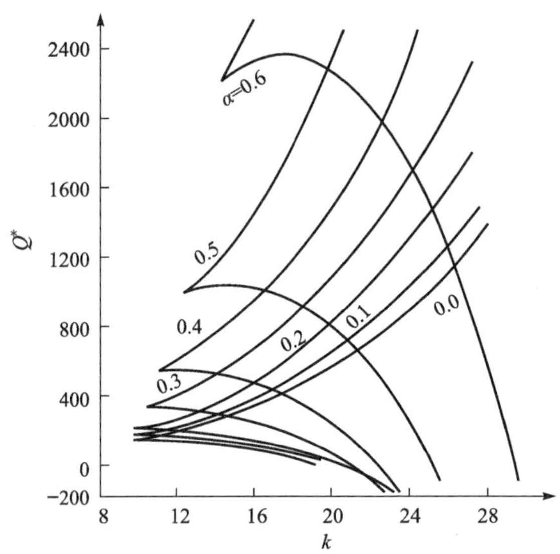

图 4　各种 α 值下的稳定曲线（$\nu=0.3$）

我们现在研究壳体的屈曲临界点，也即上、下临界载荷相重合的点。记此时壳体的几何参数为 k_0，相应的临界载荷为 Q_0^*。由根的判别式为零，得到决定临界几何参数 k_0 的方程

$$(a_3^2 - 3a_2a_4)k_0^2 - 3a_1a_4 = 0 \tag{3.19}$$

由此求出

$$k_0 = \sqrt{\frac{3a_1a_4}{a_3^2 - 3a_2a_4}} \tag{3.20}$$

显而易见，当 $k<k_0$ 时，壳体不会产生失稳。当 $k\geqslant k_0$ 时，壳体就出现失稳。所以，k_0 是确定壳体失稳与否的分界点。图 5 及图 6 给出了 $k_0-\alpha$ 以及 $Q_0^*-\alpha$ 的关系曲线。

最后，在式（3.18）中令 $\alpha\to 0$，便得到封顶扁球壳在均布载荷作用下的临界载荷计算公式。在 $\nu=0.3$ 情况下，其计算结果绘在图 7 中，并与实验结果及以前的理论分析结果作一比较。从图上看，我们的理论结果与实验结果吻合较好。

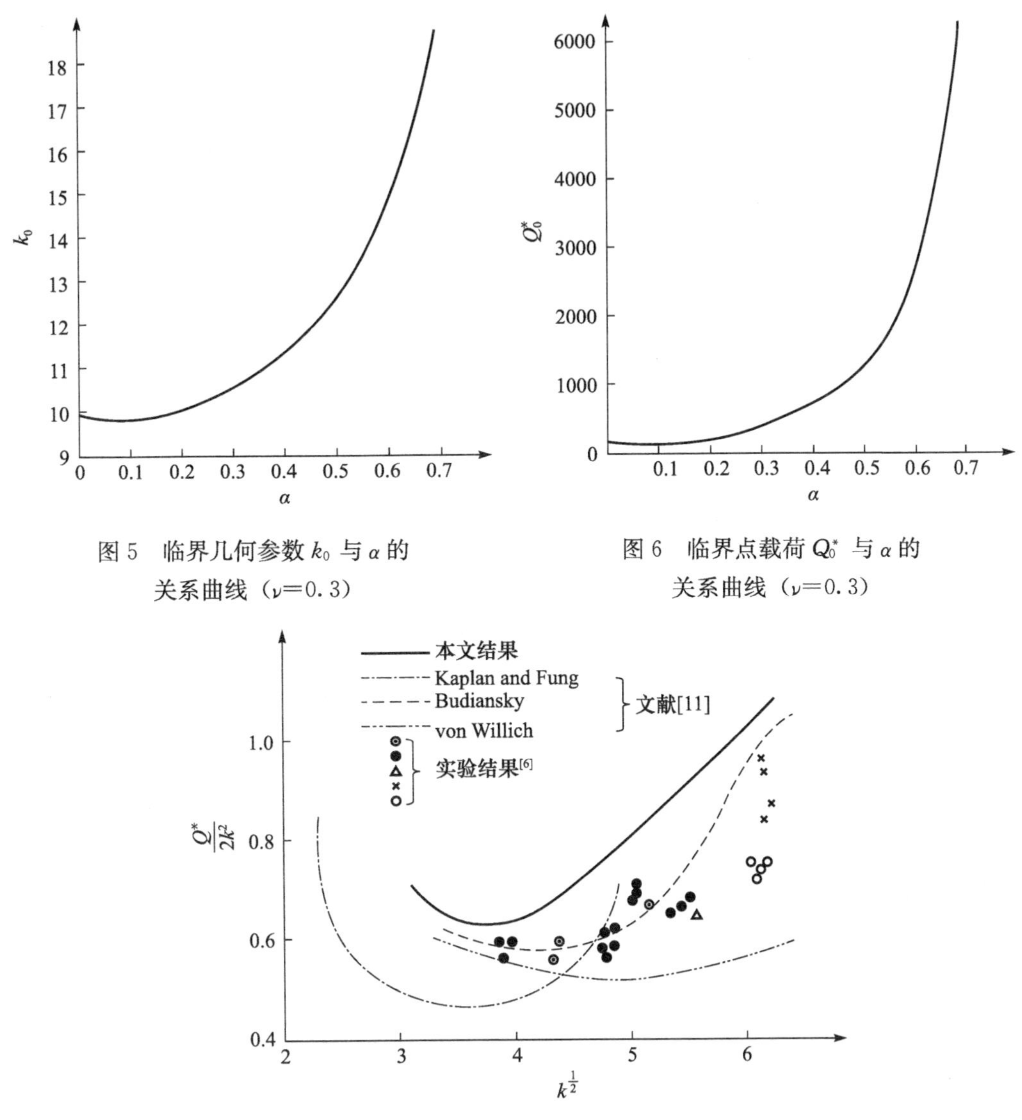

图 5　临界几何参数 k_0 与 α 的关系曲线（$\nu=0.3$）

图 6　临界点载荷 Q_0^* 与 α 的关系曲线（$\nu=0.3$）

图 7　理论结果与实验结果的比较（$\nu=0.3$）

参 考 文 献

[1]　刘人怀. 在内边缘均布力矩作用下中心开孔圆底扁球壳的非线性稳定问题. 科学通报，1965，(3)：253.

[2]　刘人怀. 在边缘载荷作用下中心开孔圆底扁薄球壳的轴对称稳定性. 力学学报，1977，(3)：206.

[3]　刘人怀. 双层金属中心开孔扁球壳的非线性热稳定问题. 中国科学技术大学学报，1981，11 (1)：84.

[4]　Liu Ren-huai. Nonlinear thermal stability of bimetallic shallow shells of revolution. International Journal of Non-Linear Mechanics，1983，18 (5)：409.

[5] 罗祖道，聂德耀，刘汉东，等. 双层金属球面扁壳的热稳定问题. 力学学报，1966，9 (1)：1.

[6] Tillman，S. C. On the buckling behaviour of shallow spherical caps under a uniform pressure load. International Journal of Solids and Structures，1970，6 (1)：37.

[7] 叶开沅，刘人怀，平庆元，等. 在轴对称线布载荷作用下圆底扁球壳的非线性稳定问题. 兰州大学学报，1965，(2)：10；科学通报，1965，(2)：142.

[8] 叶开沅，刘人怀，张传智，等. 圆底扁薄球壳在边缘力矩作用下的非线性稳定问题. 科学通报，1965，(2)：145.

[9] Chien Wei-zang. Large deflection of a circular clamped plate under uniform pressure. Chinese Journal of Physics，1947，7 (2)：102.

[10] Феодосьев，В. И. *Упругие Элеменмы Точного Приборосмроеhия*. Москва：Оборонгиз，1979.

[11] Fung，Y. C. and Sechler，E. E. Thin-Shell Structures：Theory，Experiment，and Design. New Jersey：Prentice-Hall，Inc. Engle-wood Cliffs，1974.

On the nonlinear stability of a truncated shallow spherical shell under axisymmetrically distributed load①

中文摘要

本文研究了轴对称分布载荷作用下具有硬中心的开顶扁球壳的非线性稳定问题，此问题尚无人研究过。应用修正迭代法，克服了非线性数学和壳体开顶的困难，获得了临界载荷的解析公式，可供工程设计部门参考使用。

1. Introduction

The investigation of the nonlinear stability of thin shell is of great importance to aeronautical, astronautical, naval, civil, automatic control, and precision instrument manufacturing engineering, etc. It has, therefore, received a great deal of attention. But the investigation of these problems is rather difficult both theoretically and experimentally because the features of such stability problems are nonlinear. For this reason, the study in recent years has mainly focussed on some simpler cases, such as the problems of shallow spherical shells, cylindrical shells and shallow conical shells. There are only a few works about more complicated problems on the nonlinear stability of truncated shallow spherical shells. Liu Renhuai[1-6], Lo Zhudao[7] and Tillman[8] have discussed these problems and obtained some optimistic results. But many problems of this kind have not yet been studied, such as the problem presented here.

In this paper, we consider a truncated shallow spherical shell with a nondeformable rigid body at the center under an axisymmetrically distributed load for the cases when the outer edge of the shell is loosely clamped or rigidly clamped. We have applied Yeh Kai-yuan and Liu Renhuai's modified iteration method[1,9,10] to overcome difficulties due to the nonlinear differential equations and the truncation of the shell, and obtained an analytic solution for the critical load. As a special case, we discuss in detail the case when the load varies according to the linear law. The present results can be applied directly to engineering design.

2. Nonlinear Boundary Value Problems

We consider a truncated shallow spherical shell with a nondeformable rigid body at the center subjected to the action of an axisymmetrically distributed load $q(r)$ as shown in Fig.1. Here h is the shell thickness, R is the radius of curvature of the middle surface, b and a are the upper and lower radius respectively, and r is the distance from a

① Reprinted from *Solid Mechanics Archives*, 1989, 14 (2): 81-102. Authors: Lin Renhuai and He Linghui.

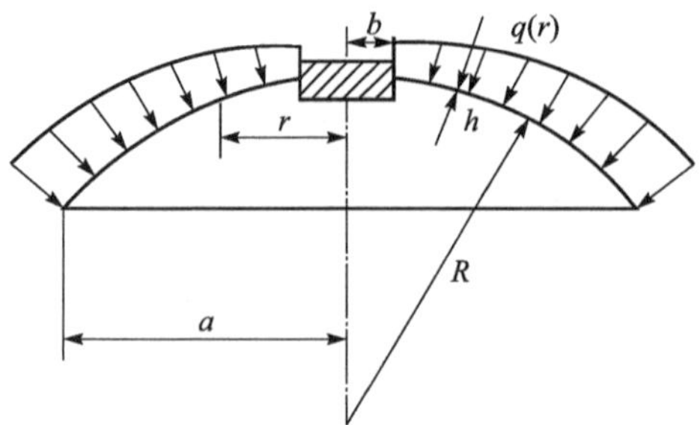

Fig. 1　An axisymmetric distributed loaded truncated shallow spherical shell

point on the middle surface of the shell to the axis of the shell. The inner edge of the shell is rigidly clamped to the nondeformable rigid body which can move, without rotation, along the axis of the shell, and the outer edge is loosely clamped or rigidly clamped.

We can easily obtain the basic equations satisfied by the deflection w and the radial membrane stress N_r as follows[1-6,11]

$$\begin{aligned} &\frac{D}{r}\frac{\mathrm{d}}{\mathrm{d}r}r\frac{\mathrm{d}}{\mathrm{d}r}\frac{1}{r}\frac{\mathrm{d}}{\mathrm{d}r}(r\Theta)-\frac{1}{r}\frac{\mathrm{d}}{\mathrm{d}r}[rN_r(\theta+\Theta)]=q(r)\\ &\frac{r}{Eh}\frac{\mathrm{d}}{\mathrm{d}r}\frac{1}{r}\frac{\mathrm{d}}{\mathrm{d}r}(r^2N_r)+\Theta\left(\theta+\frac{1}{2}\Theta\right)=0 \end{aligned} \tag{2.1a, b}$$

where Θ and θ are the angle of rotation and the slope angle of the tangent to a meridian respectively, E is the elastic modulus. ν is Poisson's ratio and D is the flexural rigidity

$$\Theta=\frac{\mathrm{d}w}{\mathrm{d}r},\ \theta=\frac{r}{R},\ D=\frac{Eh^3}{12(1-\nu^2)} \tag{2.2}$$

Having w and N_r, the radial membrane displacement u, the radial transverse force Q_r and the radial moment M_r can be derived from the following expressions

$$\begin{aligned} &u=\frac{r}{Eh}\left[r\frac{\mathrm{d}N_r}{\mathrm{d}r}+(1-\nu)N_r\right]\\ &Q_r=-D\frac{\mathrm{d}}{\mathrm{d}r}\frac{1}{r}\frac{\mathrm{d}}{\mathrm{d}r}(r\Theta)\\ &M_r=-D\left(\frac{\mathrm{d}\Theta}{\mathrm{d}r}+\frac{\nu}{r}\Theta\right) \end{aligned} \tag{2.3a,b,c}$$

In order to simplify the following calculation, integrating (2.1a), we have

$$Dr\frac{\mathrm{d}}{\mathrm{d}r}\frac{1}{r}\frac{\mathrm{d}}{\mathrm{d}r}(r\Theta)-rN_r(\theta+\Theta)=F(r) \tag{2.4}$$

where $F(r)$ is the load function given by

$$F(r)=\int q(r)\mathrm{d}r+c \tag{2.5}$$

where c is the integral constant.

Using (2.3b), Eq. (2.4) is reduced to

$$r[Q_r+N_r(\theta+\Theta)]=-F(r) \tag{2.6}$$

Considering the condition of vertical equilibrium of a part shell as shown in Fig. 2, all forces satisfy

$$r[Q_r+N_r(\theta+\Theta)]=-\int_b^r q(\xi)\xi\mathrm{d}\xi \tag{2.7}$$

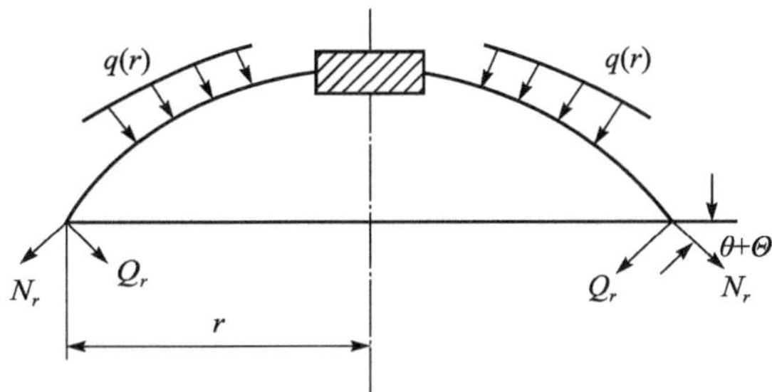

Fig. 2　Determination of the load function $F(r)$

Comparing (2.7) and (2.6), we obtain the following expression

$$F(r)=\int_b^r q(\xi)\xi \mathrm{d}\xi \tag{2.8}$$

Thus, the reduction of Eq. (2.4) is completed as

$$Dr\frac{\mathrm{d}}{\mathrm{d}r}\frac{1}{r}\frac{\mathrm{d}}{\mathrm{d}r}(r\Theta)-rN_r(\theta+\Theta)=\int_b^r q(\xi)\xi \mathrm{d}\xi \tag{2.9}$$

Then Eqs. (2.9) and (2.1b) are the basic equations for the large deflection of the truncated shallow spherical shell with a nondeformable rigid body at the center. Corresponding to Eqs. (2.9) and (2.1b), the boundary conditions are given as follows

(1) Loosely clamped along the outer edge

$$w=0,\ \Theta=0,\ N_r=0 \quad \text{at } r=a;$$
$$\Theta=0,\ u=0 \quad \text{at } r=b \tag{2.10}$$

(2) Rigidly clamped along the outer edge

$$w=0,\ \Theta=0,\ u=0 \quad \text{at } r=a;$$
$$\Theta=0,\ u=0 \quad \text{at } r=b \tag{2.11}$$

Making the calculation simpler, nondimensional parameters are introduced as follows

$$\rho=\frac{r}{a},\ \alpha=\frac{b}{a},\ y=\sqrt{12\lambda_1\lambda_2}\frac{w}{h}+\frac{1}{2}\rho^2$$
$$\phi=-\frac{\mathrm{d}y}{\mathrm{d}p},\ S=\frac{a^2}{D}\rho N_r,\ k=\sqrt{12\lambda_1\lambda_2}\frac{a^2}{Rh} \tag{2.12}$$
$$Q=\sqrt{3\lambda_1\lambda_2}\frac{a^4q(r)}{Dh}$$

where

$$\lambda_1=1-\nu,\ \lambda_2=1+\nu \tag{2.13}$$

Therefore, Eqs. (2.9) and (2.1b) become

$$L(\rho\phi)=S\phi-2\int_\alpha^\rho Q(\eta)\eta \mathrm{d}\eta$$
$$L(\rho S)=\frac{1}{2}(k^2\rho^2-\phi^2) \tag{2.14a,b}$$

where

$$L(\,) = \rho \frac{\mathrm{d}}{\mathrm{d}\rho} \frac{1}{\rho} \frac{\mathrm{d}}{\mathrm{d}\rho}(\,) \tag{2.15}$$

and the boundary conditions (2. 10) and (2. 11) become

(1) Loosely clamped along the outer edge

$$\begin{aligned} & y = \frac{k}{2},\ \phi = -k,\ S = 0 \qquad \text{at } \rho = 1 \\ & \phi = -k,\ \rho \frac{\mathrm{d}S}{\mathrm{d}\rho} - \nu S = 0 \qquad \text{at } \rho = \alpha \end{aligned} \tag{2.16}$$

(2) Rigidly clamped along the outer edge

$$\begin{aligned} & y = \frac{k}{2},\ \phi = -k,\ \frac{\mathrm{d}S}{\mathrm{d}\rho} - \nu S = 0 \qquad \text{at } \rho = 1 \\ & \phi = -k\alpha,\ \rho \frac{\mathrm{d}S}{\mathrm{d}\rho} - \nu S = 0 \qquad \text{at } \rho = \alpha \end{aligned} \tag{2.17}$$

Thus the problem is reduced to the calculation of the nonlinear differential Eqs. (2. 14) with boundary conditions (2. 16) or (2. 17) .

3. Solution of Nonlinear Boundary Value Problems

Let

$$Q(\rho) = Q_0 G(\rho) \tag{3.1}$$

in which Q_0 is the load parameter, $G(\rho)$ is a consecutive function defined in the closed interval $[\alpha,\ 1]$, which can reflect distributed characteristics of the load.

Substituting (3. 1) into (2. 14a), we have

$$L(\rho\phi) = S\phi - 2Q_0 \int_\alpha^\rho G(\eta)\eta \mathrm{d}\eta \tag{3.2}$$

According to the Weierstrass' theorem, when n is sufficiently large, the integration $\int_\alpha^\rho G(\eta)\eta \mathrm{d}\eta$ in (3. 2) can be uniformly approximated with arbitrary small error by a Bernstein's multinomial

$$B_n(\rho) = \sum_{i=1}^{n} \left[\frac{C_n^i}{(1-\alpha)^n} \int_\alpha^{i/n} G(\eta)\eta \mathrm{d}\eta \right] (\rho - \alpha)^i (1-\rho)^{n-i}$$

In order to simplify the calculation, we arrange $B_n(\rho)$ in form as $\sum_{i=0}^{\infty} \eta \overline{C}_i \rho^i$. We can maintain

$$\int_\alpha^\rho G(\eta)\eta \mathrm{d}\eta = \sum_{n=0}^{\infty} \overline{C}_n \rho^n \tag{3.3}$$

where $\overline{C}_n$ is a coefficient of ρ^n in $B_i(\rho)$ when $i \to \infty$.

Thus, the Eq. (2. 14a) becomes

$$L(\rho\phi) = S\phi - 2Q_0 \sum_{n=0}^{\infty} \overline{C}_n \rho^n \tag{3.4}$$

Now we solve two nonlinear boundary value problems (3. 4), (2. 14b), (2. 16) and (3.4), (2.14b), (2.17) by the modified iteration method. In the first approximation, omitting $S\phi$ in Eq. (3.4) leads to the linear boundary value problems as follows

$$L(\rho\phi_1) = -2Q_0 \sum_{n=0}^{\infty} \overline{C}_n \rho^n$$

$$L(\rho S_1) = \frac{1}{2}(k^2\rho^2 - \phi_1^2) \tag{3.5a,b}$$

(1) Loosely clamped along the outer edge

$$y_1 = \frac{k}{2},\ \phi_1 = -k,\ S_1 = 0 \quad \text{at } \rho = 1; \tag{3.6a,b,c}$$

$$\phi_1 = -k\alpha,\ \rho\frac{dS}{d\rho} - \nu S_1 = 0 \quad \text{at } \rho = \alpha \tag{3.7a,b}$$

(2) Rigidly clamped along the outer edge

$$y_1 = \frac{k}{2},\ \phi_1 = -k,\ \frac{dS_1}{d\rho} - \nu S_1 = 0 \quad \text{at } \rho = 1; \tag{3.8a,b,c}$$

$$\phi_1 = -k\alpha,\ \rho\frac{dS_1}{d\rho} - \nu S_1 = 0 \quad \text{at } \rho = \alpha \tag{3.9a,b}$$

For the first problem (3.5) — (3.7), using the boundary conditions (3.6b) and (3.7a) and integrating equation (3.8a), we obtain

$$\phi_1 = -k\rho + Q_0\left[a_1\rho^{-1} + a_2\rho + a_3\rho\ln\rho - 2\sum_{n=1}^{\infty}\frac{\overline{C}_n}{n(n+2)}\rho^{n+1}\right] \tag{3.10}$$

where

$$a_1 = \frac{1}{1-\alpha^2}\left[\overline{C}_0\alpha^2\ln\alpha - 2\sum_{n=1}^{\infty}\frac{\overline{C}_n\alpha^2(1-\alpha^n)}{n(n+2)}\right]$$

$$a_2 = \frac{1}{1-\alpha^2}\left[-\overline{C}_0\alpha^2\ln\alpha + 2\sum_{n=1}^{\infty}\frac{\overline{C}_n(1-\alpha^{n+2})}{n(n+2)}\right] \tag{3.11}$$

$$a_3 = -\overline{C}_0$$

Let the nondimensional inner edge deflection W_m of the shell be an iteration parameter, which is given by

$$W_m = \sqrt{12(1-\nu^2)}\left.\frac{w}{h}\right|_{\rho=\alpha} \tag{3.12}$$

From (2.12) and (3.6a), we have

$$W_m = \int_\alpha^1 (\phi + k\rho)d\rho \tag{3.13}$$

Substituting (3.10) into (3.13), the characteristic relation of first approximation is obtained as

$$Q_0 = \beta W_m \tag{3.14}$$

in which

$$\beta = \left[\frac{2a_2 - a_3}{4} - a_1\ln\alpha - \frac{2a_2 - a_3}{4}\alpha^2 - \frac{a_3}{2}\alpha^2\ln\alpha - 2\sum_{n=1}^{\infty}\frac{C_n(1-\alpha^{n+2})}{n^2(n+2)^2}\right]^{-1} \tag{3.15}$$

Substituting (3.15) into (3.10), we obtain

$$\phi_1 = -k\rho + \beta W_m\left[a_1\rho^{-1} + a_2\rho + a_3\rho\ln\rho - 2\sum_{n=1}^{\infty}\frac{\overline{C}_n}{n(n+2)}\rho^{n+1}\right] \tag{3.16}$$

Obviously, for the second problem (3.5), (3.8) and (3.9), the first approximational solution ϕ_1 is just the same as (3.16).

Substituting the solution (3.16) into Eq. (3.5b) and integrating, we obtain the first approximational solution for the nondimensional radial membrane stress

$$\begin{aligned} S_1 = {}& k\beta W_m[A\rho^{-1} + B\rho + b_1\rho\ln\rho + b_2\rho^3 + b_3\rho^3\ln\rho \\ & + \sum_{n=1}^{\infty}d_n\rho^{n+3}] + \beta^2W_m^2[C\rho^{-1} + D\rho + b_4\rho^{-1}\ln\rho + b_5\rho\ln\rho \\ & + b_6\rho\ln^2\rho + b_7\rho^3 + b_8\rho^3\ln^3\rho + b_9\rho^3\ln^2\rho + \sum_{n=1}^{\infty}e_n\rho^{n+1} \\ & + \sum_{n=1}^{\infty}f_n\rho^{n+3} + \sum_{n=1}^{\infty}g_n\rho^{n+3}\ln\rho + \sum_{n=1}^{\infty}h_n\rho^{n+4}] \end{aligned} \tag{3.17}$$

where

$$\begin{aligned} & b_1 = \frac{a_1}{2},\ b_2 = \frac{4a_2 - 3a_3}{32},\ b_3 = \frac{a_3}{8},\ b_4 = \frac{a_1^2}{4},\ b_5 = -\frac{2a_1a_2 - a_1a_3}{4}, \\ & b_6 = -\frac{a_1a_3}{4},\ b_7 = -\frac{8a_2^2 - 12a_2a_3 + 7a_3^2}{128},\ b_8 = -\frac{4a_2a_3 - 3a_3^2}{32},\ b_9 = -\frac{a_3^2}{16}, \\ & d_n = -\frac{2\overline{C}_n}{n(n+2)^2(n+4)},\ e_n = \frac{2a_1\overline{C}_n}{n^2(n+2)^2}, \\ & f_n = \frac{2(n+2)(n+4)a_2 - 4(n+3)a_3}{n(n+2)^3(n+4)^2}\overline{C}_n,\ g_n = \frac{2a_3\overline{C}_n}{n(n+2)^2(n+4)}, \\ & h_n = -\frac{2}{(n+3)(n+5)}\sum_{i=1}^{n}\frac{\overline{C}_i\overline{C}_{n-i+1}}{i(i+2)(n-i+1)(n-i+3)} \end{aligned} \tag{3.18}$$

For the first problem

$$\begin{aligned} A = {}& \frac{1}{\lambda_2 + \lambda_1\alpha^2}\{[b_1 - (1-\nu)b_2]\alpha^2 + (1-\nu)b_1\alpha^2\ln\alpha \\ & + [(3-\nu)b_2 + b_3]\alpha^4 + (3-\nu)b_3\alpha^4\ln\alpha \\ & - \sum_{n=1}^{\infty}\alpha^2[(1-\nu) - (n+3-\nu)\alpha^{n+2}]d_n\} \\ B = {}& \frac{1}{\lambda_2 + \lambda_1\alpha^2}\{-(1+\nu)b_2 - b_1\alpha^2 - (1-\nu)b_1\alpha^2\ln\alpha \\ & - [(3-\nu)b_2 + b_3]\alpha^4 + (3-\nu)b_3\alpha^4\ln\alpha \\ & - \sum_{n=1}^{\infty}[(1-\nu) + (n+3-\nu)\alpha^{n+4}]d_n\} \\ C = {}& \frac{1}{\lambda_2 + \lambda_1\alpha^2}\{b_4 - (1+\nu)b_4\ln\alpha \end{aligned}$$

$$+[b_5-(1-\nu)b_7]\alpha^2+[(1-\nu)b_5+2b_6]\alpha^2\ln\alpha$$
$$+(1-\nu)b_6\alpha^2\ln^2\alpha+[(3-\nu)b_7+b_8]\alpha^4$$
$$+[(3-\nu)b_8+2b_9]\alpha^4\ln\alpha+(3-\nu)b_9\alpha^4\ln^2\alpha$$
$$-\sum_{n=1}^{\infty}(1-\nu)(e_n+f_n+g_n)\alpha^2+\sum_{n=1}^{\infty}(n+1-\nu)e_n\alpha^{n+2}$$
$$-\sum_{n=1}^{\infty}[(n+3-\nu)f_n+g_n]\alpha^{n+4}+\sum_{n=1}^{\infty}(n+3-\nu)g_n\alpha^{n+4}\ln\alpha$$
$$+\sum_{n=1}^{\infty}(n+4-\nu)h_n\alpha^{n+5}\}$$

$$D=\frac{1}{\lambda_2+\lambda_1\alpha^2}\{-b_4-(1+\nu)b_7+(1+\nu)b_4\ln\alpha$$
$$-b_5\alpha^2-[(1-\nu)b_5+2b_6]\alpha^2\ln\alpha$$
$$-(1-\nu)b_6\alpha^2\ln^2\alpha-[(3-\nu)b_7+b_8]\alpha^4$$
$$-[(3-\nu)b_8+2b_9]\alpha^4\ln\alpha-(3-\nu)b_9\alpha^4\ln^2\alpha$$
$$-\sum_{n=1}^{\infty}(1+\nu)(e_n+f_n+g_n)-\sum_{n=1}^{\infty}(n+1-\nu)e_n\alpha^{n+2}$$
$$-\sum_{n=1}^{\infty}[(n+3-\nu)f_n+g_n]\alpha^{n+4}-\sum_{n=1}^{\infty}(n+3$$
$$-\nu)g_n\alpha^{n+4}\ln\alpha-\sum_{n=1}^{\infty}(n+4-\nu)h_n\alpha^{n+5}\}\tag{3.19}$$

For the second problem

$$A=\frac{1}{\lambda_2(1-\alpha^2)}\{-[(3-\nu)b_2+b_3]\alpha^2+(1-\nu)b_1\alpha^2\ln\alpha$$
$$+[(3-\nu)b_2+b_3]\alpha^4+(3-\nu)b_3\alpha^4\ln\alpha$$
$$-\sum_{n=1}^{\infty}(n+3-\nu)d_n\alpha^2(1-\alpha^{n+2})\}$$

$$B=\frac{1}{\lambda_1(1-\alpha^2)}\{-b_1-(3-\nu)b_2-b_3+b_1\alpha^2$$
$$+(1-\nu)b_1\alpha^2\ln\alpha+[(3-\nu)b_2+b_3]\alpha^4$$
$$+(3-\nu)b_3\alpha^4\ln\alpha-\sum_{n=1}^{\infty}(n+3-\nu)d_n(1-\alpha^{n+4})\}$$

$$C=\frac{1}{\lambda_2(1-\alpha^2)}\{b_4-(1+\nu)b_4\ln\alpha-[b_4+(3-\nu)b_7$$
$$+b_8]\alpha^2+[(1-\nu)b_5+2b_6]\alpha^2\ln\alpha+(1-\nu)b_6\alpha^2\ln^2\alpha$$
$$+[(3-\nu)b_7+b_8]\alpha^4+[(3-\nu)b_8+2b_9]\alpha^4\ln\alpha$$
$$+(3-\nu)b_9\alpha^4\ln^2\alpha-\sum_{n=1}^{\infty}(n+1-\nu)e_n\alpha^2(1-\alpha^2)$$

$$
-\sum_{n=1}^{\infty}[(n+3-\nu)f_n+g_n]\alpha^2(1-\alpha^{n+2})+\sum_{n=1}^{\infty}(n+3
$$

$$
-\nu)g_n\alpha^{n+4}\ln\alpha-\sum_{n=1}^{\infty}(n+4-\nu)h_n\alpha^2(1-\alpha^{n+3})\}
$$

$$
D=\frac{1}{\lambda_1(1-\alpha^2)}\{-b_5-(3-\nu)b_7-b_8-(1+\nu)b_4\ln\alpha
$$

$$
+b_5\alpha^2+[(1-\nu)b_5+2b_6]\alpha^2\ln\alpha+(1-\nu)b_6\alpha^2\ln^2\alpha
$$

$$
+[(3-\nu)b_7+b_8]\alpha^4+[(3-\nu)b_8+2b_9]\alpha^4\ln\alpha
$$

$$
+(3-\nu)b_9\alpha^4\ln^2\alpha-\sum_{n=1}^{\infty}(n+1-\nu)e_n(1-\alpha^{n+2})
$$

$$
-\sum_{n=1}^{\infty}[(n+3-\nu)f_n+g_n](1-\alpha^{n+4})+\sum_{n=1}^{\infty}(n+3
$$

$$
-\nu)g_n\alpha^{n+4}\ln\alpha-\sum_{n=1}^{\infty}(n+4-\nu)h_n(1-\alpha^{n+5})\} \tag{3.20}
$$

In the second approximation, for the above two problems, we have the linear boundary value problem as follows

$$
L(\rho\phi_2)=S_1\phi_1-2Q_0\sum_{n=1}^{\infty}C_n\rho^n \tag{3.21}
$$

$$
y_2=\frac{k}{2},\ \phi_2=-k \quad \text{at} \quad \rho=1 \tag{3.22a,b}
$$

$$
\phi_2=-k\alpha \quad \text{at} \quad \rho=\alpha \tag{3.23}
$$

Substituting (3.10) and (3.17) into (3.21) and integrating it under boundary conditions (3.22b) and (3.23), the second approximational solution of ϕ is obtained

$$
\begin{aligned}
\phi_2=&\phi_1+k^2\beta W_m[\Psi_1(\rho)+A_1\rho+B_1\rho^{-1}]\\
&+k\beta^2W_m^2[\Psi_2(\rho)+A_2\rho+B_2\rho^{-1}]\\
&+\beta^3W_m^3[\Psi_3(\rho)+A_3\rho+B_3\rho^{-1}]
\end{aligned} \tag{3.24}
$$

in which

$$
A_n=\frac{\alpha\Psi_n(\alpha)-\Psi_n(1)}{1-\alpha^2}
$$

$$
B_n=\frac{\alpha^2\Psi_n(1)-\alpha\Psi_n(\alpha)}{1-\alpha^2} \qquad (n=1,2,3)
$$

$$
\Psi_1(\rho)=-\frac{A}{2}\rho\ln\rho+c_1\rho^3+c_2\rho^3\ln\rho+c_3\rho^5+c_4\rho^5\ln\rho+\sum_{n=1}^{\infty}i_n\ln\rho^{n+5}
$$

$$
\begin{aligned}
\Psi_2(\rho)=&c_5\rho^{-1}\ln\rho+c_6\rho\ln\rho+c_7\rho\ln^2\rho+c_8\rho^3+c_9\rho^3\ln\rho\\
&+c_{10}\rho^3\ln^2\rho+c_{11}\rho^5+c_{12}\rho^5\ln\rho+c_{13}\rho^5\ln^2\rho\\
&-\sum_{n=1}^{\infty}\frac{A}{a_1}e_n\rho^{n+1}+\sum_{n=1}^{\infty}j_n\rho^{n+3}+\sum_{n=1}^{\infty}b_1d_n\rho^{n+3}\ln\rho
\end{aligned}
$$

$$+\sum_{n=1}^{\infty}k_n\rho^{n+5}+\sum_{n=1}^{\infty}l_n\rho^{n+5}\ln\rho+\sum_{n=1}^{\infty}m_n\rho^{n+6}$$

$$\begin{aligned}\Psi_3(\rho)=&c_{14}\rho^{-1}\ln\rho+c_{15}\rho^{-1}\ln^2\rho+c_{16}\rho\ln\rho+c_{17}\rho\ln^2\rho+c_{18}\rho\ln^3\rho\\&+c_{19}\rho^3+c_{20}\rho^3\ln\rho+c_{21}\rho^3\ln^2\rho+c_{22}\rho^3\ln^3\rho+c_{23}\rho^5\\&+c_{24}\rho^5\ln\rho+c_{25}\rho^5\ln^2\rho+c_{26}\rho^5\ln^3\rho+\sum_{n=1}^{\infty}o_n\rho^{n+1}\\&+\sum_{n=1}^{\infty}p_n\rho^{n+1}\ln\rho+\sum_{n=1}^{\infty}q_n\rho^{n+4}+\sum_{n=1}^{\infty}r_n\rho^{n+3}\ln\rho+\sum_{n=1}^{\infty}s_n\rho^{n+3}\ln^2\rho\\&+\sum_{n=1}^{\infty}t_n\rho^{n+4}+\sum_{n=1}^{\infty}u_n\rho^{n+5}+\sum_{n=1}^{\infty}v_n\rho^{n+5}\ln\rho+\sum_{n=1}^{\infty}w_n\rho^{n+5}\ln^2\rho\\&+\sum_{n=1}^{\infty}x_n\rho^{n+6}+\sum_{n=1}^{\infty}y_n\rho^{n+6}\ln\rho+\sum_{n=1}^{\infty}z_n\rho^{n+7}\end{aligned}$$

$$c_1=-\frac{4B-3b_1}{32}\qquad c_2=-\frac{b_1}{8}\qquad c_3=-\frac{12b_2-5b_3}{288}$$

$$c_4=-\frac{b_3}{24}\qquad c_5=-\frac{a_1A}{2}$$

$$c_6=\frac{(2a_2-a_3)A+2a_1B-2C-a_1b_1+b_4}{4}$$

$$c_7=\frac{a_1b_1+a_3A-b_4}{4}$$

$$c_8=\frac{8a_1b_2-6a_1b_3-6a_2b_1+7a_3b_1+6b_5-7b_6+(8a_2-6a_3)B-8D}{64}$$

$$c_9=\frac{2a_1b_3+2a_2b_1-3a_3b_1-2b_5+3b_6+2a_3B}{16}\qquad c_{10}=\frac{a_3b_1-b_6}{8}$$

$$c_{11}=\frac{72a_2b_2-30a_2b_3-30a_3b_2+19a_3b_3-72b_7+30b_8-19b_9}{1728}$$

$$c_{12}=\frac{6a_2b_3+6a_3b_2-5a_3b_3-6b_8+5b_9}{144}$$

$$c_{13}=\frac{a_3b_3-b_9}{24}\qquad c_{14}=-\frac{2a_1C+a_1b_4}{4}\qquad c_{15}=-\frac{a_1b_4}{4}$$

$$c_{16}=\frac{(2a_2-a_3)C+2a_1D-a_1b_5+a_1b_6-a_2b_4+a_3b_4}{4}$$

$$c_{17}=\frac{a_1b_5-a_1b_6+a_2b_4-a_3b_4+a_3C}{4}\qquad c_{18}=\frac{a_1b_6+a_3b_4}{6}$$

$$\begin{aligned}c_{19}=&\frac{1}{256}[32a_1b_7-24a_1b_8+28a_1b_9-24a_2b_5+28a_2b_6\\&+28a_3b_5-45a_3b_6+(32a_2-24a_3)D]\end{aligned}$$

$$c_{20}=\frac{8a_1b_8-12a_1b_9+8a_2b_5-12a_2b_6-12a_3b_5+21a_3b_6+8a_3D}{64}$$

$$c_{21}=\frac{4a_1b_9+4a_2b_6+4a_3b_5-9a_3b_6}{32}\quad c_{22}=\frac{a_3b_6}{8}$$

$$c_{23}=\frac{288a_2b_7-120a_2b_8+76a_2b_9-120a_3b_7+76a_3b_8-65a_3b_9}{6912}$$

$$c_{24}=\frac{24a_2b_8-20a_2b_9+24a_3b_7-20a_3b_8+19a_3b_9}{576}$$

$$c_{25}=\frac{4a_2b_9+4a_3b_8-5a_3b_9}{96}\quad c_{26}=\frac{a_3b_9}{24}$$

$$i_n=-\frac{d_n}{(n+4)(n+6)}$$

$$j_n=\frac{a_1d_n-e_n}{(n+2)(n+4)}-\frac{2(n+2)(n+4)B-4(n+3)b_1}{n(n+2)^3(n+4)^2}\overline{C}_n$$

$$\begin{aligned}k_n=&\frac{[(n+4)(n+6)a_2-2(n+5)a_3]d_n}{(n+4)^2(n+6)^2}\\&-\frac{(n+4)(n+6)f_n-2(n+5)g_n}{(n+4)^2(n+6)^2}\\&-\frac{2(n+4)(n+6)b_2-4(n+5)b_3}{n(n+2)(n+4)^2(n+6)^2}\overline{C}_n\end{aligned}$$

$$l_n=\frac{a_3d_n-g_n}{(n+4)(n+6)}-\frac{2b_3\overline{C}_n}{n(n+2)(n+4)(n+6)}$$

$$m_n=-\frac{2}{(n+5)(n+7)}\sum_{i=1}^{n}\frac{\overline{C}_id_{n-i+1}}{i(i+2)}-\frac{h_n}{(n+5)(n+7)}$$

$$o_n=\frac{a_1e_n}{n(n+2)}-\frac{2n(n+2)\overline{C}-4(n+1)b_4}{n^3(n+2)^3}\overline{C}_n$$

$$p_n=-\frac{2b_4\overline{C}_n}{n^2(n+2)^2}$$

$$\begin{aligned}q_n=&\frac{a_1f_n}{n(n+2)}-\frac{2(n+3)a_1g_n}{(n+2)^2(n+4)^2}+\frac{(n+2)(n+4)a_2-2(n+3)a_3}{(n+2)^2(n+4)^2}e_n\\&-\frac{2}{n(n+2)^4(n+4)^3}\{(n+2)^2(n+4)^2D\\&-2(n+2)(n+3)(n+4)b_5+2[(n+2)^2\\&+(n+2)(n+4)+(n+4)^2]b_6\}\overline{C}_n\end{aligned}$$

$$r_n=\frac{a_1g_n+a_3e_n}{(n+2)(n+4)}-\frac{2(n+2)(n+4)b_5-8(n+3)b_6}{n(n+2)^3(n+4)^2}\overline{C}_n$$

$$s_n=-\frac{2b_6\overline{C}_n}{n(n+2)^2(n+4)}$$

$$t_n=\frac{a_1h_n}{(n+3)(n+5)}-\frac{2}{(n+3)(n+5)}\sum_{n=1}^{\infty}\frac{\overline{C}_ie_{n-i+1}}{i(i+2)}$$

$$u_n=\frac{(n+4)(n+6)a_2-2(n+5)a_3}{(n+4)^2(n+6)^2}f_n$$

$$
\begin{aligned}
&-\frac{1}{(n+4)^3(n+6)^3}\{2(n+4)(n+5)(n+6)a_2\\
&-2[(n+4)^2+(n+4)(n+6)+(n+6)^2]a_3\}g_n\\
&-\frac{2}{n(n+2)(n+4)^3(n+6)^3}\{(n+4)^2(n+6)^2b_7\\
&-2(n+4)(n+5)(n+6)b_8+2[(n+4)^2\\
&+(n+4)(n+6)+(n+6)^2]b_9\}\overline{C}_n\\
v_n=&\frac{a_3f_n}{(n+4)(n+6)}+\frac{(n+4)(n+6)a_2-4(n+5)a_3}{(n+4)^2(n+6)^2}g_n\\
&-\frac{2(n+4)(n+6)b_8-8(n+5)b_9}{n(n+2)(n+4)^2(n+6)^2}\overline{C}_n\\
w_n=&\frac{a_3g_n}{(n+4)(n+6)}-\frac{2b_9\overline{C}_n}{n(n+2)(n+4)(n+6)}\\
x_n=&\frac{(n+5)(n+7)a_2-2(n+6)a_3}{(n+5)^2(n+7)^2}h_n\\
&-\frac{2}{(n+5)(n+7)}\sum_{i=1}^{n}\frac{\overline{C}_if_{n-i+1}}{i(i+2)}\\
&+\frac{4(n+6)}{(n+5)^2(n+7)^2}\sum_{i=1}^{n}\frac{\overline{C}_ig_{n-i+1}}{i(i+2)}\\
y_n=&\frac{a_3h_n}{(n+5)(n+7)}-\frac{2}{(n+5)(n+7)}\sum_{i=1}^{n}\frac{\overline{C}_ig_{n-i+1}}{i(i+2)}\\
z_n=&-\frac{2}{(n+6)(n+8)}\sum_{i=1}^{n}\frac{\overline{C}_ih_{n-i+1}}{i(i+2)}
\end{aligned}
\tag{3.25}
$$

Substituting (3.24) into (3.13), the characteristic relation for second approximation is obtained as

$$
Q_0=(\alpha_1+\alpha_2k^2)W_m+\alpha_3kW_m^2+\alpha_4W_m^3 \tag{3.26}
$$

in which

$$
\begin{aligned}
\alpha_1=&\beta\\
\alpha_2=&\beta^2\Big[-\frac{A+4A_1}{8}(1-\alpha^2)+B_1\ln\alpha-\frac{A}{4}\alpha^2\ln\alpha\\
&-\frac{4c_1-c_2}{16}(1-\alpha^4)+\frac{c_2}{4}\alpha^4\ln\alpha-\frac{6c_3-c_4}{36}(1-\alpha^6)\\
&+\frac{c_4}{6}\alpha^6\ln\alpha-\sum_{n=1}^{\infty}\frac{i_n}{n+6}(1-\alpha^{n+6})\Big]\\
\alpha_3=&\beta^3\Big[B_2\ln\alpha+\frac{c_5}{2}\ln^2\alpha-\frac{2A_2-c_6+c_7}{4}(1-\alpha^2)+\frac{c_6-c_7}{2}\alpha^2\ln\alpha\\
&+\frac{c_7}{2}\alpha^2\ln^2\alpha-\frac{8c_8-2c_9+c_{10}}{32}(1-\alpha^4)+\frac{2c_9-c_{10}}{8}\alpha^4\ln\alpha\\
&+\frac{c_{10}}{4}\alpha^4\ln^2\alpha-\frac{18c_{11}-3c_{12}+c_{13}}{108}(1-\alpha^6)+\frac{2c_{12}-c_{13}}{18}\alpha^6\ln\alpha
\end{aligned}
$$

$$+\frac{c_{13}}{6}\alpha^6\ln^2\alpha+\sum_{n=1}^{\infty}\frac{Ae_n}{a_1(n+2)}(1-\alpha^{n+2})$$

$$-\sum_{n=1}^{\infty}\frac{(n+4)j_n-b_1d_n}{(n+4)^2}(1-\alpha^{n+4})$$

$$+\sum_{n=1}^{\infty}\frac{b_1d_n}{n+4}\alpha^{n+4}\ln\alpha-\sum_{n=1}^{\infty}\frac{(n+6)k_n-l_n}{(n+6)^2}(1-\alpha^{n+6})$$

$$+\sum_{n=1}^{\infty}\frac{l_n}{n+6}\alpha^{n+6}\ln\alpha-\sum_{n=1}^{\infty}\frac{m_n}{n+7}(1-\alpha^{n+7})\Big]$$

$$\alpha_4=\beta^4\Big[B_3\ln\alpha+\frac{c_{14}}{2}\ln^2\alpha+\frac{c_{15}}{3}\ln^3\alpha-\frac{4A_3-2c_{16}+2c_{17}-3c_{18}}{8}(1-\alpha^2)$$

$$+\frac{2c_{16}-2c_{17}+3c_{18}}{4}\alpha^2\ln\alpha+\frac{2c_{17}-3c_{18}}{4}\alpha^2\ln^2\alpha+\frac{c_{18}}{2}\alpha^2\ln^3\alpha$$

$$-\frac{32c_{19}-8c_{20}+4c_{21}-3c_{22}}{128}(1-\alpha^4)+\frac{8c_{20}-4c_{21}+3c_{22}}{32}\alpha^4\ln\alpha$$

$$+\frac{4c_{21}-3c_{22}}{16}\alpha^4\ln^2\alpha+\frac{c_{22}}{4}\alpha^4\ln^3\alpha-\frac{36c_{23}-6c_{24}+2c_{25}-c_{26}}{216}(1-\alpha^6)$$

$$+\frac{6c_{24}-2c_{25}+c_{26}}{36}\alpha^6\ln\alpha+\frac{2c_{25}-c_{26}}{12}\alpha^6\ln^2\alpha$$

$$+\frac{c_{26}}{6}\alpha^6\ln^3\alpha-\sum_{n=1}^{\infty}\frac{(n+2)o_n-p_n}{(n+2)^2}(1-\alpha^{n+2})+\sum_{n=1}^{\infty}\frac{p_n}{n+2}\alpha^{n+2}\ln\alpha$$

$$-\sum_{n=1}^{\infty}\frac{(n+4)^2q_n-(n+4)r_n+2s_n}{(n+4)^3}(1-\alpha^{n+4})$$

$$+\sum_{n=1}^{\infty}\frac{(n+4)r_n-2s_n}{(n+4)^2}\alpha^{n+4}\ln\alpha+\sum_{n=1}^{\infty}\frac{s_n}{n+4}\alpha^{n+4}\ln^2\alpha$$

$$-\sum_{n=1}^{\infty}\frac{t_n}{n+5}(1-\alpha^{n+5})-\sum_{n=1}^{\infty}\frac{(n+6)^2u_n-(n+6)v_n+2w_n}{(n+6)^3}(1-\alpha^{n+6})$$

$$+\sum_{n=1}^{\infty}\frac{(n+6)v_n-2w_n}{(n+6)^2}\alpha^{n+6}\ln\alpha+\sum_{n=1}^{\infty}\frac{w_n}{n+6}\alpha^{n+6}\ln^2\alpha$$

$$-\sum_{n=1}^{\infty}\frac{(n+7)x_n-y_n}{(n+7)^2}(1-\alpha^{n+7})$$

$$+\sum_{n=1}^{\infty}\frac{y_n}{n+7}\alpha^{n+7}\ln\alpha-\sum_{n=1}^{\infty}\frac{z_n}{n+8}(1-\alpha^{n+8})\Big] \quad (3.27)$$

When the geometrical parameter k is large according to the nonlinear characteristic relation (3.26) the characteristic curves between Q_0 and W_m will become gyratory, and the shell may become stable. Applying the following extremal condition

$$\frac{\mathrm{d}Q_0}{\mathrm{d}W_m}=0 \quad (3.28)$$

we obtain the nondimensional critical deflection W_m^* of the shell when snap-through buckling of the shell occurs

$$W_m^* = \frac{-\alpha_3 k \pm \sqrt{(\alpha_3^2 - 3\alpha_2\alpha_4)k^2 - 3\alpha_1\alpha_4}}{3\alpha_4} \tag{3.29}$$

Substituting W_m^* into (3.26), we obtain the following formula of the nondimensional critical buckling load of second approximation

$$Q_0^* = (\alpha_1 + \alpha_2 k^2)W_m^* + \alpha_3 k W_m^{*2} + \alpha_4 W_m^{*3} \tag{3.30}$$

where Q_0^* with respect to the positive and negative signs of (3.29) are the nondimensional lower and upper critical buckling loads, respectively.

The point at which the upper and lower critical buckling loads coincide gives the critical geometry of the shell. From the condition of the multiple root of the Eq. (3.28), we obtain the formula of the geometrical parameter k_0 for the critical point

$$k_0 = \sqrt{3\alpha_1\alpha_4/\alpha_3^2 - 3\alpha_2\alpha_4} \tag{3.31}$$

Obviously, when $k \geqslant k_0$ the shell can fail by buckling, and when $k < k_0$ no buckling occurs. Thus, k_0 is used to distinguish between buckling and no buckling for the shell.

4. Numerical Example

As an application of the results of this paper, we consider a clamped truncated shallow spherical shell under an axisymmetrical linear distributed load as follows

$$q = q_0\left(1 - \frac{r}{a}\right) \tag{4.1}$$

From the expressions (2.12), (3.1) and (3.4), we have

$$F(\rho) = 1 - \rho \tag{4.2}$$

$$\bar{c}_0 = -\frac{\alpha^2}{2} + \frac{\alpha^3}{3}$$

$$\bar{c}_1 = 0 \qquad \bar{c}_2 = \frac{1}{2}$$

$$\bar{c}_3 = -\frac{1}{3} \qquad \bar{c}_n = 0 \quad (n = 4, 5, 6, \cdots) \tag{4.3}$$

According to the expression (3.26), in one special case for $\alpha = 0.3$, $\nu = 0.3$, the characteristic curves for several values of the geometrical parameter k are drawn in Fig. 3. It is seen that only when $k \geqslant k_0$, the curves become gyratory, implying that the shell now becomes unstable.

Through numerical computation from (3.30), values of the critical buckling load corresponding to various values of k and α are given in Fig.4. From this figure, it can be seen that curves of the upper critical buckling load increase monotonically with the increase in the values of k. Meanwhile curves of the lower critical buckling load increase with the increase of k in a small range, but subsequently fall monotonically, and it is also observed that when the values of k are too large, values of the lower critical buckling load will become negative, implying that the shell now cannot restore its initial form unless an opposite load is applied.

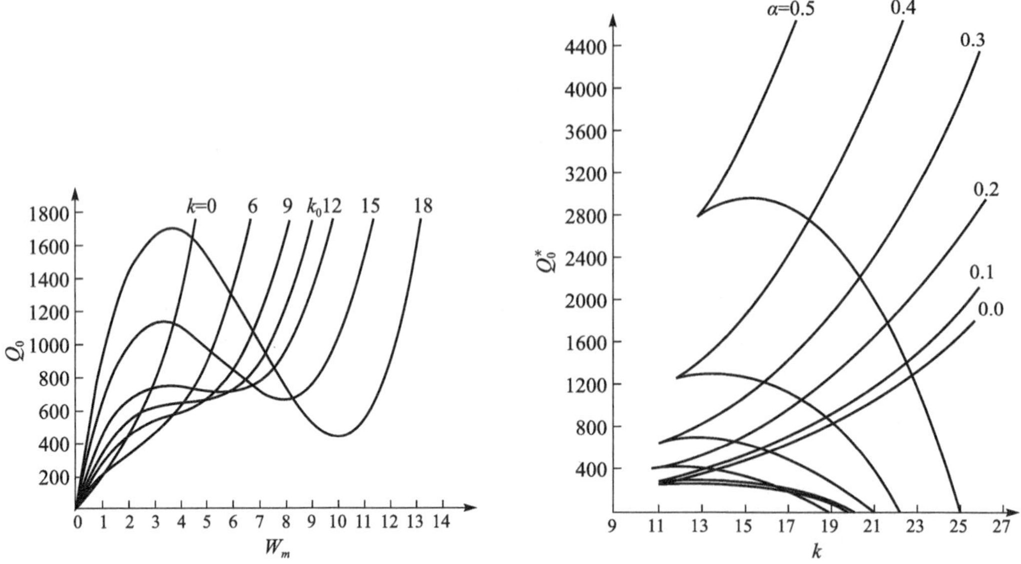

Fig. 3 Characteristic curves for various values of k ($\nu=0.3$, $\alpha=0.3$)

Fig. 4 Curves of the nondimensional critical buckling loads ($\nu=0.3$)

At the critical point of the shell, applying the expressions (3.30) and (3.31), we can obtain the relationships between k_0 and α, and Q_0^* and α, as shown in Fig.5 and Fig.6, respectively. When α is determined, k_0 is also determined. When $k<k_0$ no buckling of the shell occurs, and when $k\geqslant k_0$ the shell may become unstable.

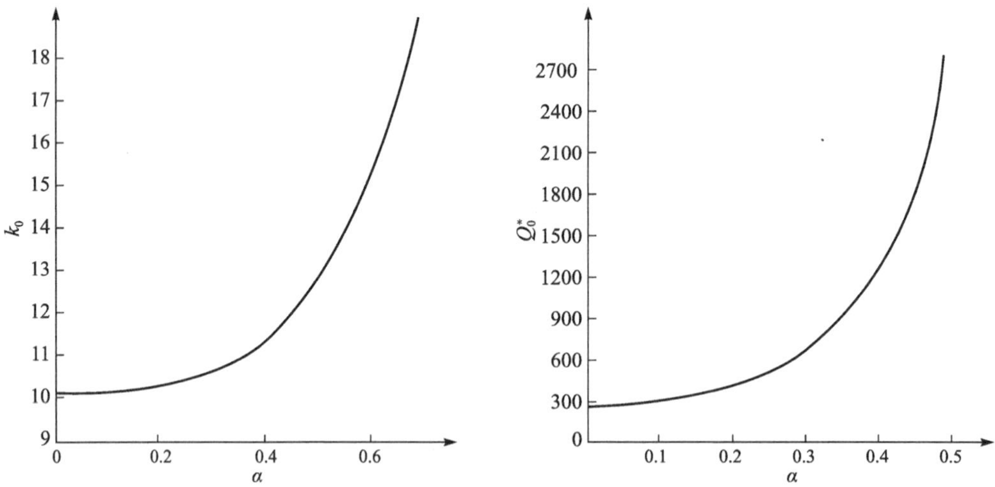

Fig. 5 Relationship between k_0 and α ($\nu=0.3$)

Fig. 6 Relationship between Q_0^* and α ($\nu=0.3$)

References

[1] Liu Renhuai, Nonlinear stability of a circular shallow spherical shell with a circular hole in the

center under the action of uniform moment at the inner edge, *Bulletin of the Sciences*, 3 (1965), 253 (in Chinese).

[2] Liu Renhuai, Axisymmetrical buckling of thin circular shallow spherical shell with a circular hole at the center by shearing forces uniformly distributed along the inner edge, *Mechanics*, 3 (1977), 206 (in Chinese).

[3] Liu Renhuai, Nonlinear thermal stability of a bimetallic shallow spherical shell with a circular hole at the center, *Journal of the China University of Science and Technology*, 11, 1 (1981), 84 (in Chinese).

[4] Liu Renhuai, Nonlinear thermal stability of bimetallic shallow shells of revolution, *International Journal of Nonlinear Mechanics*, 18, 5 (1983), 409.

[5] Liu Renhuai, Cheng Zhen-qiang, On the nonlinear stability of a truncated shallow spherical shell under a concentrated load, *Applied Mathematics and Mechanics*, 9, 2 (1988), 101.

[6] Liu Renhuai, Li Dong, On the nonlinear stability of a truncated shallow spherical shell under a uniformly distributed load, *Applied Mathematics and Mechanics*, 9, 3 (1988), 227.

[7] Lo Zhu-dau. Thermal stability of a bimetallic shallow shell, *Acta Mechanics Sinica*, 9, 1 (1966), 1 (in Chinese).

[8] Tillman, S. C., On the buckling behaviour of shallow spherical caps under a uniform pressure load, *International Journal of Solids and Structures*, 6, 1 (1970), 37.

[9] Yeh Kai-yuan, Liu Renhuai, Ping Qing-yuan, et al. Nonlinear stability of a thin circular shallow spherical shell under actions of axisymmetrical uniformly distributed line loads, *Journal of Lanzhou University*, 2 (1965), 10; *Bulletin of Sciences*, 2 (1965), 142 (in Chinese).

[10] Yeh Kai-yuan, Liu Renhuai, Zhang Chuan-zhe et al. Nonlinear stability of a thin circular shallow spherical shell under the action of uniform edge moment, *Bulletin of Sciences*, 2 (1965), 145 (in Chinese).

[11] Feodosev, V. I., Elastic Elements of Precision-Instruments Manufacture, Oborongiz, Moscow, 1949 (in Russian).

均布载荷作用下具有硬中心的开顶扁球壳的非线性屈曲①

一、引言

薄壳稳定性的研究对航天、航空、航海、精密仪器仪表、建筑等工程有着重大的实际意义，因此，薄壳稳定性理论的研究一直是国际前沿研究课题。特别是对于扁球壳的非线性屈曲问题，研究者更多。然而，对于扁球壳开顶较复杂情况，则研究较少。刘人怀等[1−9]、罗祖道等[10]和 Tillman[11]等先后对开顶扁球壳的非线性屈曲问题进行了研究，获得了一系列有益的结果。

本文是作者以往工作的继续，研究了均布载荷作用下，具有硬中心，外缘夹紧固定的开顶扁球壳的非线性屈曲问题。据我们所知，此问题尚未有人讨论。应用作者和叶开沅等[1,12,13]于1965年创立的修正迭代法，使非线性数学和壳体开顶的困难得以克服，获得了临界载荷解析公式。本文所得的结果可供工程设计部门直接使用。

二、非线性边值问题的建立与求解

考虑如图1所示的均布载荷 q 作用在壳体和硬中心上的开顶扁球壳，其厚度为 h，中曲面半径为 R，内外半径分别为 b 和 a，壳体中曲面上的点到对称中心轴的距离为 r。壳体的内边缘被固定在可上、下移动的不变形的硬中心上，而外边缘则是夹紧固定的。

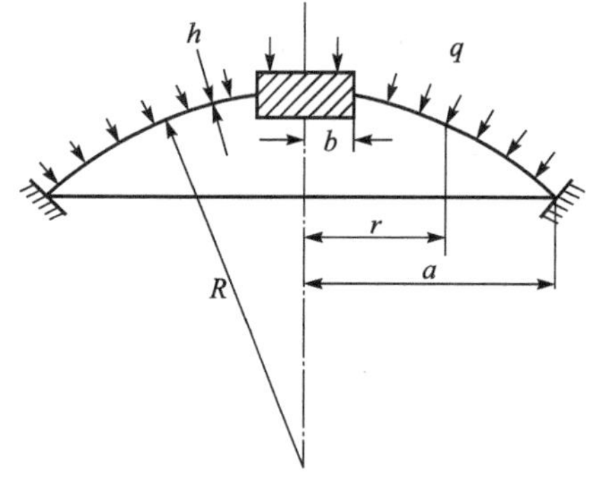

图1　开顶扁球壳

在此情况下，壳体的挠度 w 和径向薄膜内力 N_r 所满足的大挠度弯曲方程为[1,2]

$$Dr\frac{\mathrm{d}}{\mathrm{d}r}\frac{1}{r}\frac{\mathrm{d}}{\mathrm{d}r}(r\vartheta)-rN_r\left(\vartheta+\frac{r}{R}\right)=\frac{1}{2}qr^2,$$

$$\frac{r}{Eh}\frac{\mathrm{d}}{\mathrm{d}r}\frac{1}{r}\frac{\mathrm{d}}{\mathrm{d}r}(r^2N_r)+\vartheta\left(\frac{1}{2}\vartheta+\frac{r}{R}\right)=0 \tag{1a,b}$$

其中 ϑ 是壳体经线方向弧的旋转角，E 是弹性模量，ν 是泊松比，D 是抗弯刚度，

$$\vartheta=\frac{\mathrm{d}w}{\mathrm{d}r},D=\frac{Eh^3}{12(1-\nu^2)} \tag{2}$$

相应的边界条件为

当 $r=a$ 时，$w=0$，$\vartheta=0$，$u=0$

① 本文原载《工程力学》，1995，3（增刊）：1839-1844. 作者：刘人怀，张小果.

当 $r=b$ 时，$\vartheta=0$，$u=0$ (3)

其中 u 是径向位移，

$$u=\frac{r}{Eh}\left[r\frac{\mathrm{d}N_r}{\mathrm{d}r}+(1-\nu)N_r\right] \tag{4}$$

为求解简单，引入下列无量纲量：

$$\begin{aligned}&\rho=\frac{r}{a},\ \alpha=\frac{b}{a},\ y=\sqrt{12\lambda_1\lambda_2}\,\frac{w}{h}+\frac{1}{2}k\rho^2,\ \varphi=-\frac{\mathrm{d}y}{\mathrm{d}\rho},\ S=\frac{a^2}{D}\rho N_r,\\&k=\sqrt{12\lambda_1\lambda_2}\,\frac{a^2}{Rh},\ Q=\sqrt{3\lambda_1\lambda_2}\,\frac{a^4}{Dh}q,\ \lambda_1=1-\nu,\lambda_2=1+\nu\end{aligned} \tag{5}$$

于是，非线性边值问题（1）和（3）转化为无量纲形式：

$$\begin{aligned}L(\rho\varphi)&=S\varphi-Q\rho^2,\\L(\rho S)&=\frac{1}{2}(k^2\rho^2-\varphi^2)\end{aligned} \tag{6a,b}$$

当 $\rho=1$ 时，$y=\frac{1}{2}k\rho^2$，$\varphi=-k\rho$，$\rho\frac{\mathrm{d}S}{\mathrm{d}\rho}-\nu S=0$ (7a～c)

当 $\rho=\alpha$ 时，$\varphi=-k\rho,\rho=\frac{\mathrm{d}S}{\mathrm{d}\rho}-\nu S=0$ (8a，b)

其中 L 是微分算子，

$$L(\cdots)=\rho\frac{\mathrm{d}}{\mathrm{d}\rho}\frac{1}{\rho}\frac{\mathrm{d}}{\mathrm{d}\rho}(\cdots) \tag{9}$$

现在，我们应用修正迭代法求解无量纲非线性边值问题（6)～(8)。以扁球壳的无量纲内边缘挠度 Y_m 作为迭代参数，应用式（7a），便有

$$Y_m=\sqrt{12\lambda_1\lambda_2}\,\frac{w}{h}\bigg|_{r=b}=\int_\alpha^1(\varphi+k\rho)\mathrm{d}\rho \tag{10}$$

对于一次近似，由式（6)～(8）可得下列线性边值问题

$$\begin{aligned}L(\rho\varphi_1)&=-Q\rho^2,\\L(\rho S_1)&=\frac{1}{2}(k^2\rho^2-\varphi_1^2)\end{aligned} \tag{11}$$

$$\begin{aligned}&\text{当 }\rho=1\text{ 时，}y_1=\frac{1}{2}k\rho^2,\varphi_1=-k\rho,\ \rho=\frac{\mathrm{d}S_1}{\mathrm{d}\rho}-\nu S_1=0\\&\text{当 }\rho=\alpha\text{ 时，}\varphi_1=-k\rho,\ \rho\frac{\mathrm{d}S_1}{\mathrm{d}\rho}-\nu S_1=0\end{aligned} \tag{12}$$

应用式（10)，这一边值问题的解为

$$\varphi_1=f_1(\rho,Y_m),S_1=f_2(\rho,Y_m) \tag{13}$$

对于二次近似，由式（6)～(8）可得关于 φ_2 的线性边值问题

$$L(\rho\varphi_2)=S_1\varphi_1-Q\rho^2 \tag{14}$$

当 $\rho=1$ 时，$y_2=\frac{1}{2}k\rho^2$，$\varphi_2=-k\rho$

当 $\rho=\alpha$ 时，$\varphi_2=-k\rho$ (15)

此边值问题的解为

$$\varphi_2=f_3(\rho,Q,Y_m) \tag{16}$$

将此解代入式（10），便得开顶扁球壳的非线性特征关系式

$$Q=(\alpha_0+k^2\alpha_1)Y_m-k\alpha_2Y_m^2+\alpha_3Y_m^3 \tag{17}$$

其中

$$\alpha_0=32\beta_1,$$

$$\begin{aligned}\alpha_1=&\frac{32\beta_1^2\beta_2^2}{\lambda_1\lambda_2}\Big(\frac{3-4\nu}{64}\lambda_2+\frac{45-288\nu+83\nu^2}{768}\alpha^2+\frac{4-\nu}{24}\lambda_2\alpha^2\ln\alpha\\&-\frac{153-130\nu+43\nu^2}{384}\alpha^4-\frac{173+64\nu-77\nu^2}{288}\alpha^4\ln\alpha-\frac{3-\nu}{4}\lambda_1\alpha^4\ln^2\alpha\\&+\frac{97+1200\nu-49\nu^2}{2304}\alpha^6+\frac{77+312\nu-197\nu^2}{288}\alpha^6\ln\alpha\\&-\frac{1}{12}\lambda_1\lambda_2\alpha^6\ln^2\alpha-\lambda_1^2\alpha^6\ln^3\alpha+\frac{175-234\nu+23\nu^2}{576}\alpha^8\\&-\frac{115+1980\nu-47\nu^2}{288}\alpha^8\ln\alpha+\frac{1}{2}\lambda_1^2\alpha^8\ln^2\alpha\\&-\frac{299-672\nu+39\nu^2}{2304}\alpha^{10}+\frac{67+132\nu-97\nu^2}{288}\alpha^{10}\ln\alpha-\frac{5-\nu}{24}\lambda_1\alpha^{10}\ln^2\alpha\\&-\frac{31-11\nu}{384}\lambda_2\alpha^{12}-\frac{5-\nu}{48}\lambda_2\alpha^{12}\ln\alpha+\frac{53-5\nu}{192}\lambda_2\alpha^{14}\Big),\end{aligned}$$

$$\begin{aligned}\alpha_2=&\frac{32\beta_1^3\beta_2^2}{\lambda_1\lambda_2}\Big[\frac{89+162\nu+24\nu^2}{576}-\frac{4387+303\nu+3639\nu_2}{3456}\alpha^2\\&+\frac{23+18\nu-9\nu^2}{32}\alpha^2\ln\alpha+\frac{36556-22375\nu-78906\nu^2}{86400}\alpha^4\\&-\frac{296-369\nu-95\nu^2}{144}\alpha^4\ln\alpha-\frac{15-26\nu+7\nu^2}{4}\alpha^4\ln^2\alpha\\&-\frac{5851+52400\nu+70849\nu^2}{86400}\alpha^6+\frac{1573-10690\nu+3897\nu^2}{2880}\alpha^6\ln\alpha\\&-\frac{2713-611\nu+162\nu^2}{24}\alpha^6\ln^2\alpha-2\lambda_1^2\alpha^6\ln^3\alpha-\frac{31873-400\nu-55853\nu^2}{28800}\alpha^8\\&+\frac{2713+90\nu-1683\nu^2}{720}\alpha^8\ln\alpha+\frac{24-22\nu+\nu^2}{3}\alpha^8\ln^2\alpha-(5-7\nu)\lambda_1\alpha^8\ln^3\alpha\\&-\frac{2747+286\nu-1981\nu^2}{3456}\alpha^{10}-\frac{209-334\nu+297\nu^2}{288}\alpha^{10}\ln\alpha\\&+\frac{995-1080\nu-59\nu^2}{144}\alpha^{10}\ln^2\alpha+\frac{1581-118\nu+1881\nu^2}{1728}\alpha^{12}\\&+\frac{323+67\nu-86\nu^2}{144}\alpha^{12}\ln\alpha-\frac{43-56\nu+5\nu^2}{12}\alpha^{12}\ln^2\alpha\\&-\frac{15162-28195\nu+18438\nu^2}{172800}\alpha^{14}-\frac{3505+2190\nu-343\nu^2}{2800}\alpha^{14}\ln\alpha\\&-\frac{13-\nu}{12}\lambda_1\alpha^{14}\ln^2\alpha+\frac{30999+37500\nu-9299\nu^2}{86400}\alpha^{16}-\frac{16-5\nu}{180}\lambda_2\alpha^{16}\ln\alpha\\&+\frac{2111+1600\nu-49\nu^2}{90225}\alpha^{18}\Big],\end{aligned}$$

$$
\begin{aligned}
\alpha_3 =128\beta_1^4\beta_2^2\Big[& \frac{26+18\nu-5\nu^2}{288}-\frac{703-608\nu+225\nu^2}{384}\alpha^2+\frac{4-\nu}{12}\lambda_2\alpha^2\ln\alpha \\
& +\frac{4717-8145\nu+3338\nu^2}{4320}\alpha^4-\frac{75-112\nu-23\nu^2}{96}\alpha^4\ln\alpha-\frac{3-\nu}{2}\lambda_1\alpha^4\ln^2\alpha \\
& -\frac{31075-3708\nu-36183\nu^2}{17280}\alpha^6-\frac{460-364\nu+78\nu^2}{96}\alpha^6\ln\alpha \\
& -\frac{3-96\nu+44\nu^2}{24}\alpha^6\ln^2\alpha+\frac{1-18\nu+11\nu^2}{3}\alpha^6\ln^3\alpha \\
& +\frac{131081-196512\nu+36511\nu^2}{34560}\alpha^8+\frac{1471+2000\nu-2221\nu^2}{720}\alpha^8\ln\alpha \\
& -\frac{13-64\nu-81\nu^2}{12}\alpha^8\ln^2\alpha+\frac{1+25\nu^2}{2}\alpha^8\ln^3\alpha \\
& -\frac{3767+3984\nu-22751\nu^2}{6912}\alpha^{10}-\frac{2845-1950\nu+3194\nu^2}{720}\alpha^{10}\ln\alpha \\
& +\frac{23+4\nu+47\nu^2}{6}\alpha^{10}\ln^2\alpha+\frac{13-6\nu+19\nu^2}{3}\alpha^{10}\ln^3\alpha-\frac{678+335\nu+572\nu^2}{720}\alpha^{12} \\
& -\frac{5121-8100\nu-1781\nu^2}{1440}\alpha^{12}\ln\alpha+\frac{11-73\nu+18\nu^2}{12}\alpha^{12}\ln^2\alpha \\
& -\frac{11557-32964\nu+13619\nu^2}{17280}\alpha^{14}+\frac{776+568\nu+235\nu^2}{1440}\alpha^{14}\ln\alpha \\
& -(13-10\nu+5\nu^2)\alpha^{14}\ln^2\alpha-\frac{749-2904\nu-1101\nu^2}{2304}\alpha^{16} \\
& -\frac{165+32\nu-73\nu^2}{288}\alpha^{16}\ln\alpha+\frac{9\nu}{4}\alpha^{16}\ln^2\alpha+\frac{3437-11430\nu-167\nu^2}{17280}\alpha^{18} \\
& -\frac{57+460\nu+23\nu^2}{1440}\alpha^{18}\ln\alpha-\frac{241-2461\nu-93\nu^2}{8640}\alpha^{20}+\frac{33+2\nu}{720}\lambda_2\alpha^{20}\ln\alpha \\
& -\frac{383+420\nu-35\nu^2}{8640}\alpha^{22}\Big]
\end{aligned}
\tag{18}
$$

对式（17）应用极值条件：

$$\frac{\mathrm{d}Q}{\mathrm{d}Y_m}=0 \tag{19}$$

便得壳体发生屈曲时的无量纲内边缘临界挠度：

$$Y_m^*=\frac{k\alpha_2\pm\sqrt{(\alpha_2^2-3\alpha_1\alpha_3)k^2-3\alpha_0\alpha_3}}{3\alpha_3} \tag{20}$$

将此式代入式（17），我们得到临界载荷公式

$$Q^*=(\alpha_0+k^2\alpha_1)Y_m^*-k\alpha_2Y_m^{*2}+\alpha_3Y_m^{*3} \tag{21}$$

其中对应于式（20）负号的 Q^* 是上临界载荷，正号的 Q^* 是下临界载荷。

现在，我们来讨论上、下临界载荷的相重点，亦即区分壳体失稳与否的分界点，记此点的壳体几何参数为 k_0。显而易见，我们可以从二次方程（19）的判别式为零的条件求得 k_0 值：

$$k_0=\sqrt{\frac{3\alpha_0\alpha_3}{\alpha_2^2-3\alpha_1\alpha_3}} \tag{22}$$

三、结果与讨论

以泊松比 $\nu=0.3$ 为例，按照公式（21），进行了数值计算。我们将所得结果给在图 2 中。由图看出，上临界载荷随着壳体几何参数 k 的增加而增加；下临界载荷在很小的范围内随 k 的增加而增加，之后就随 k 值的增加而减少。而且，临界载荷随着壳体开孔的增大而增大。换句话说，硬中心半径越小，扁球壳对屈曲的反应越灵敏。

图 3 给出了 k_0-α 的关系曲线。由图看出，随着硬中心半径的增大，临界几何参数 k_0 值在增加，也就是说，壳体的稳定性能变得更好。

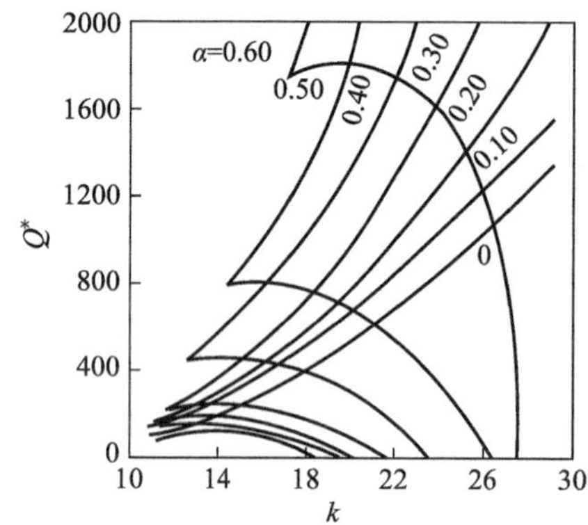

图 2　各种 α 值下的稳定曲线（$\nu=0.3$）

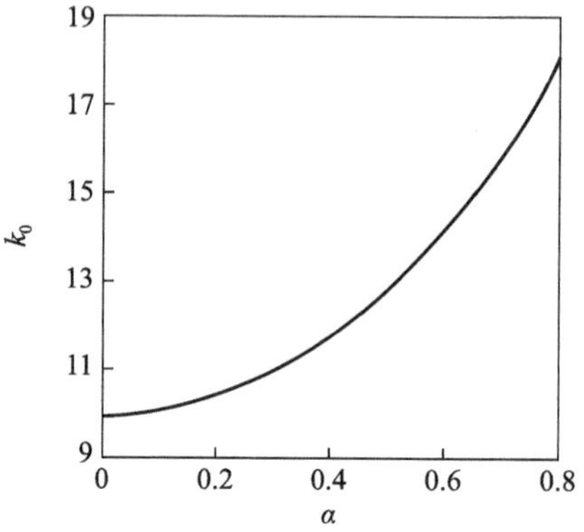

图 3　临界几何参数 k_0 与 α 的关系曲线（$\nu=0.3$）

参考文献

[1]　刘人怀. 在内边缘均布力矩作用下中心开孔圆底扁球壳的非线性稳定问题. 科学通报，1965，(3)：253.

[2]　刘人怀. 在边缘载荷作用下中心开孔圆底扁薄球壳的轴对称稳定性. 力学学报，1977，(3)：206.

[3]　刘人怀. 双层金属中心开孔扁球壳的非线性热稳定问题. 中国科学技术大学学报，1981，11 (1)：84.

[4]　Liu Ren-huai. Non-linear thermal stability of bimetallic shallow shells of revolution. International Journal of Non-Linear Mechanics，1983，18 (5)：409.

[5]　刘人怀，成振强. 集中载荷作用下开顶扁球壳的非线性稳定问题. 应用数学和力学，1988，9 (2)：95.

[6]　刘人怀，李东. 均布载荷作用下开顶扁球壳的非线性稳定问题. 应用数学和力学，1988，9 (3)：205.

[7]　Liu Ren-huai and He Ling-hui. On the nonlinear stability of a truncated shallow spherical shell under axisymmetrically distributed load. Solid Mechanics Archives，1989，14 (2)：81.

[8]　刘人怀. 新型跳跃薄膜的研究. 仪表技术与传感器. 1991，(3)：10.

[9]　刘人怀，钟诚. 考虑横向剪切的对称层合圆柱正交异性中心开孔扁球壳的非线性屈曲. 暨南大学学报（自然科学与医学版），1994，15 (1)：1.

[10] 罗祖道，聂德耀，刘汉东，等．双层金属球面扁壳的热稳定性．力学学报，1966，9 (1)：2.

[11] Tillman，S. C. On the buckling behaviour of shallow spherical caps under a uniform pressure load. International Journal of Solids and Structures，1970，6 (1)：37.

[12] 叶开沅，刘人怀，平庆元，等. 在对称线布载荷作用下的圆底扁薄球壳的非线性稳定问题. 科学通报，1965，(2)：142.

[13] 叶开沅，刘人怀，张传智，等. 圆底扁薄球壳在边缘力矩作用下的非线性稳定问题. 科学通报，1965，(1)：145.

中心开孔扁球壳在均布载荷作用下的非线性屈曲①

一、引言

在现代工程中，薄壳结构得到了广泛的应用。因此，研究薄壳的稳定性具有十分重要的意义。在开顶扁球壳方面，已有刘人怀等[1-10]、罗祖道等[11]、Tillman[12]等进行了研究。然而，尚有一些问题未被讨论过，本问题即属此范围。

二、边值问题的求解

考虑如图1所示的在均布载荷 q 作用下的中心开孔圆底扁薄球壳，内边缘悬空，外边缘夹紧固定。这里，R 为中曲面半径，r 为中曲面点至对称中心轴的距离，b、a 分别为内、外边缘半径，h 为厚度。

按照扁球壳非线性弯曲理论[1,2]，本问题所涉及壳体的非线性边值问题为

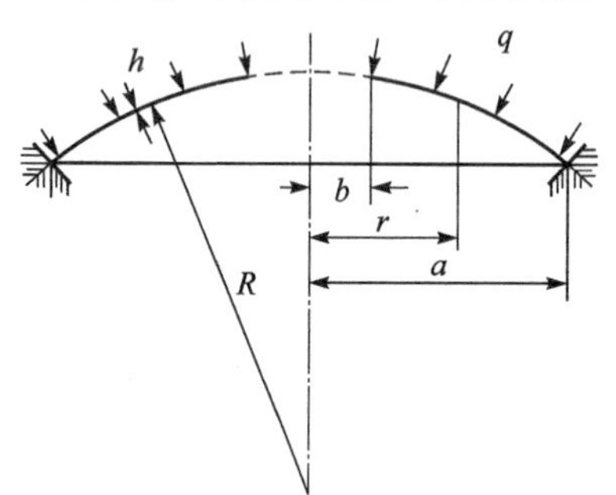

图1　中心开孔圆底扁薄球壳

$$
\begin{aligned}
&Dr\frac{\mathrm{d}}{\mathrm{d}r}\frac{1}{r}\frac{\mathrm{d}}{\mathrm{d}r}(r\vartheta)-rN_r(\theta+\vartheta)=\frac{1}{2}q(r^2-b^2),\\
&\frac{r}{Eh}\frac{\mathrm{d}}{\mathrm{d}r}\frac{1}{r}\frac{\mathrm{d}}{\mathrm{d}r}(r^2N_r)+\vartheta\left(\frac{1}{2}\vartheta+\theta\right)=0
\end{aligned}
\tag{1}
$$

$$
\begin{aligned}
&\text{当 } r=a \text{ 时，} w=0,\ \vartheta=0,\ u=0\\
&\text{当 } r=b \text{ 时，} N_r=0,\ M_r=0
\end{aligned}
\tag{2}
$$

其中 u 是径向薄膜位移，ϑ 和 θ 分别是经线方向弧的旋转角和倾斜角，w 是横向挠度，N_r 是径向薄膜内力，M_r 是径向弯矩，E 是弹性模量，ν 是泊松比，D 是抗弯刚度，

$$
\begin{aligned}
&u=\frac{r}{Eh}\left[r\frac{\mathrm{d}N_r}{\mathrm{d}r}+(1-\nu)N_r\right],M_r=-D\left(\frac{\mathrm{d}\vartheta}{\mathrm{d}r}+\frac{\nu}{r}\vartheta\right),\\
&\vartheta=\frac{\mathrm{d}w}{\mathrm{d}r},\theta=\frac{r}{R},D=\frac{Eh^3}{12(1-\nu^2)}
\end{aligned}
\tag{3}
$$

为简化求解，引入以下无量纲量

$$
\begin{aligned}
&\rho=\frac{r}{a},\alpha=\frac{b}{a},y=\sqrt{12\lambda_1\lambda_2}\,\frac{w}{h}+\frac{1}{2}k\rho^2,\varphi=-\frac{\mathrm{d}y}{\mathrm{d}\rho},\\
&S=\frac{a^2}{D}\rho N_r,k=\sqrt{12\lambda_1\lambda_2}\,\frac{a^2}{Rh},Q=\sqrt{3\lambda_1\lambda_2}\,\frac{a^4q}{Dh},\\
&\lambda_1=1-\nu,\lambda_2=1+\nu
\end{aligned}
\tag{4}
$$

则由问题（1）和（2）得到无量纲非线性边值问题

① 本文原载《暨南大学学报》，1996，17（5）：1-7. 作者：刘人怀，梅魁银．

$$L(\rho\varphi)=S\varphi+Q(\alpha^2-\rho^2),\quad L(\rho S)=\frac{1}{2}(k^2\rho^2-\varphi^2) \tag{5a, b}$$

当 $\rho=1$ 时，$y=\frac{1}{2}k$，$\varphi=-k$，$\frac{\mathrm{d}S}{\mathrm{d}\rho}-\nu S=0$ (6a～c)

当 $\rho=\alpha$ 时，$\frac{\mathrm{d}\varphi}{\mathrm{d}\rho}+\frac{\nu}{\rho}\varphi=-\lambda_2 k$，$S=0$ (7a, b)

其中 L 是微分算子，

$$L(\cdots)=\rho\frac{\mathrm{d}}{\mathrm{d}\rho}\frac{1}{\rho}\frac{\mathrm{d}}{\mathrm{d}\rho}(\cdots) \tag{8}$$

应用叶开沅和刘人怀的修正迭代法[1]求解问题（5）～（7）。此时，以扁球壳的无量纲内边缘挠度 W_m 作为迭代参数。利用式（4）和式（6a），便有

$$W_m=\sqrt{12\lambda_1\lambda_2}\frac{w}{h}\bigg|_{r=b}=\int_\alpha^1(\varphi+k\rho)\mathrm{d}\rho \tag{9}$$

对于一次近似，有如下线性边值问题

$$L(\rho\varphi_1)=Q(\alpha^2-\rho^2),\quad L(\rho S_1)=\frac{1}{2}(k^2\rho^2-\varphi_1^2) \tag{10}$$

当 $\rho=1$ 时，$y_1=\frac{1}{2}k$，$\varphi_1=-k$，$\frac{\mathrm{d}S_1}{\mathrm{d}\rho}-\nu S_1=0$ (11)

当 $\rho=\alpha$ 时，$\frac{\mathrm{d}\varphi_1}{\mathrm{d}\rho}+\frac{\nu}{\rho}\varphi_1=-\lambda_2 k$，$S_1=0$ (12)

利用式（9），问题（10）～（12）的解为

$$\varphi_1=W_m(a_1\rho^3+a_2\rho\ln\rho+a_3\rho+a_4\rho^{-1})-k\rho \tag{13a}$$

$$\begin{aligned}S_1=&\frac{1}{48}kW_m\Big[2a_1\rho^5+6a_2\rho^3\ln\rho+3\Big(2a_3-\frac{3}{2}a_2\Big)\rho^3\\&+24a_4\rho\ln\rho+kA_1\rho+kA_2\rho^{-1}\Big]\\&-\frac{1}{48}W_m^2\Big[\frac{1}{2}a_1^2\rho^7+2a_1a_2\rho^5\ln\rho+a_1\Big(2a_3-\frac{5}{6}a_2\Big)\rho^5+3a_2^2\rho^3\ln^2\rho\\&+3a_2\Big(2a_3-\frac{3}{2}a_2\Big)\rho^3\ln\rho+3\Big(2a_1a_4+a_3^2-\frac{3}{2}a_2a_3+\frac{7}{8}a_2^2\Big)\rho^3\\&+12a_2a_4\rho\ln^2\rho+12a_4(2a_3-a_2)\rho\ln\rho\\&+A_3\rho+12a_4^2\rho^{-1}\ln\rho+A_4\rho^{-1}\Big]\end{aligned} \tag{13b}$$

其中

$$\begin{aligned}a_1&=-4\beta_1\beta_2,\\a_2&=16\beta_1\beta_2\alpha^2,\\a_3&=-4\beta_2(4\lambda_2\alpha^4\ln\alpha+\lambda_1\alpha^4-\lambda_1),\\a_4&=4\beta_2(4\lambda_2\alpha^4\ln\alpha+\lambda_1\alpha^4+\lambda_2\alpha^2),\end{aligned}$$

$$
\begin{aligned}
\beta_1 &= \lambda_2\alpha^2 + \lambda_1, \\
\beta_2 &= [(7+3\nu)\alpha^6 - 16\lambda_2\alpha^4\ln^2\alpha - 4(5-\nu)\alpha^4\ln\alpha \\
&\quad - (1+7\nu)\alpha^4 - 4(1+\nu)\alpha^2\ln\alpha - (7-5\nu)\alpha^2 + \lambda_1]^{-1}, \\
A_1 &= -\beta_1^{-1}\Big[2(\lambda_2\alpha^6 + 5 - \nu)a_1 + \frac{3}{2}(4\lambda_2\alpha^4\ln\alpha \\
&\quad - 3\lambda_2\alpha^4 - 5 + 3\nu)a_2 + 6(\lambda_1\alpha^2 + 3 - \nu)a_3 \\
&\quad + 12(2\lambda_2\alpha^2\ln\alpha - \lambda_2\alpha^2 + \lambda_2 + \beta_1)a_4\Big], \\
A_2 &= -2\beta_1^{-1}\alpha^2\Big[(\lambda_1\alpha^4 - 5 + \nu)a_1 + \frac{3}{4}(4\lambda_1\alpha^2\ln\alpha \\
&\quad - 3\lambda_1\alpha^2 + 5 - 3\nu)a_2 + 3(\lambda_1\alpha^2 - 3 + \nu)a_3 \\
&\quad + 12(\lambda_1\ln\alpha - 1)a_4\Big], \\
A_3 &= \beta_1^{-1}\Big[\frac{1}{2}(\lambda_2\alpha^8 + 7 - \nu)a_1^2 + \frac{3}{8}(8\lambda_2\alpha^4\ln^2\alpha \\
&\quad - 12\lambda_2\alpha^4\ln\alpha + 7\lambda_2\alpha^4 + 9 - 7\nu)a_2^2 \\
&\quad + 3(\lambda_2\alpha^4 + 3 - \nu)a_3^2 - 12(\lambda_2\ln\alpha + 1)a_4^2 \\
&\quad + \frac{1}{6}(12\lambda_2\alpha^6\ln\alpha - 5\lambda_2\alpha^6 - 13 + 5\nu)a_1a_2 \\
&\quad + 2(\lambda_2\alpha^6 + 5 - \nu)a_1a_3 + 6(\lambda_2\alpha^4 + 3 - \nu)a_1a_4 \\
&\quad + \frac{3}{2}(4\lambda_2\alpha^4\ln\alpha - 3\lambda_2\alpha^4 - 5 + 3\nu)a_2a_3 \\
&\quad + 6(2\lambda_2\alpha^2\ln^2\alpha - 2\lambda_2\alpha^2\ln\alpha + \lambda_2\alpha^2 - \beta_1 - \lambda_2)a_2a_4 \\
&\quad + 12(2\lambda_2\alpha^2\ln\alpha - \lambda_2\alpha^2 + \beta_1 + \lambda_2)a_3a_4)\Big], \\
A_4 &= \beta_1^{-1}\Big[\frac{1}{2}(\lambda_1\alpha^6 - 7 + \nu)\alpha^2a_1^2 + \frac{3}{8}(8\lambda_1\alpha^2\ln^2\alpha \\
&\quad - 12\lambda_1\alpha^2\ln\alpha + 7\lambda_1\alpha^2 - 9 + 7\nu)\alpha^2a_2^2 \\
&\quad + 3(\lambda_1\alpha^2 - 3 + \nu)\alpha^2a_3^2 + 12(\alpha^2 - \lambda_1\ln\alpha)a_4^2 \\
&\quad + \frac{1}{6}(12\lambda_1\alpha^4\ln\alpha - 5\lambda_1\alpha^4 + 13 - 5\nu)\alpha^2a_1a_2 \\
&\quad + 2(\lambda_1\alpha^4 - 5 + \nu)\alpha^2a_1a_3 + 6(\lambda_1\alpha^2 - 3 + \nu)\alpha^2a_1a_4 \\
&\quad + \frac{3}{2}(4\lambda_1\alpha^2\ln\alpha - 3\lambda_1\alpha^2 + 5 - 3\nu)\alpha^2a_2a_3 \\
&\quad + 12(\lambda_1\ln^2\alpha - \lambda_1\ln\alpha + 1)\alpha^2a_2a_4 + 24(\lambda_1\ln\alpha - 1)\alpha^2a_3a_4\Big]
\end{aligned} \tag{14}
$$

对于二次近似，可得关于 φ_2 的如下线性边值问题

$$L(\rho\varphi_2) = S_1\varphi_1 + Q(\alpha^2 - \rho^2) \tag{15}$$

当 $\rho=1$ 时，$y_2=\frac{1}{2}k$，$\varphi_2=-k$ (16)

当 $\rho=\alpha$ 时，$\dfrac{d\varphi_2}{d\rho}+\dfrac{\nu}{\rho}\varphi_2=-\lambda_2 k$ (17)

应用解 (13)，解上述问题 (15)～(17)，得到解 φ_2。然后再应用式 (9)，便得壳体的二次近似特征关系式

$$Q=(\alpha_1+\alpha_2 k^2)W_m+\alpha_3 k W_m^2+\alpha_4 W_m^3 \tag{18}$$

其中

$$\alpha_1=32\beta_1\beta_2,$$

$$\begin{aligned}\alpha_2=&-\frac{\beta_2}{3}\Big\{\frac{a_1}{12}\Big[\Big(\frac{\beta_1}{8}+(7+\nu)\beta_3\Big)\alpha^8-\frac{\beta_1}{8}-\beta_4\Big]+\frac{a_2}{2}\Big[\Big(\frac{\beta_1}{6}+(5+\nu)\beta_3\Big)\alpha^6\ln\alpha\\&-\frac{1}{3}\Big(\frac{2}{3}\beta_1+\frac{29+7\nu}{2}\beta_3\Big)\alpha^6+\frac{2}{9}\beta_1+\frac{7}{6}\beta_4\Big]\\&+\frac{a_3}{2}\Big[\Big(\frac{\beta_1}{6}+(5+\nu)\beta_3\Big)\alpha^6-\frac{\beta_1}{6}-\beta_4\Big]\\&+6a_4\Big[\Big(\frac{\beta_1}{4}+(3+\nu)\beta_3\Big)\alpha^4\ln\alpha-\frac{1}{4}(\beta_1+(5+3\nu)\beta_3)\alpha^4+\frac{1}{4}(\beta_1+3\beta_4)\Big]\\&+\frac{A_1}{4}\Big[\Big(\frac{\beta_1}{4}+(3+\nu)\beta_3\Big)\alpha^4-\frac{\beta_1}{4}-\beta_4\Big]\\&+A_2\Big[\Big(\frac{\beta_1}{2}+\lambda_2\beta_3\Big)\alpha^2\ln\alpha-\frac{1}{2}(\beta_1-\lambda_1\beta_3)\alpha^2+\frac{1}{2}(\beta_1+\beta_4)\Big]\Big\},\end{aligned}$$

$$\begin{aligned}\alpha_3=&\frac{\beta_2}{3}\Big\{\frac{a_1^2}{16}\Big[\Big(\frac{\beta_1}{10}+(9+\nu)\beta_3\Big)\alpha^{10}-\frac{\beta_1}{10}-\beta_4\Big]+\frac{3}{4}a_2^2\Big[\Big(\frac{\beta_1}{6}+(5+\nu)\beta_3\Big)\alpha^6\ln^2\alpha\\&-\frac{1}{6}\Big(\frac{13}{6}\beta_1+(43+11\nu)\beta_3\Big)\alpha^6\ln\alpha+\frac{2}{9}\Big(\beta_1+\frac{109+35\nu}{8}\beta_3\Big)\alpha^6-\frac{2}{9}\Big(\beta_1+\frac{35}{8}\beta_4\Big)\Big]\\&+\frac{1}{12}(9a_3^2+a_1A_1)\Big[\Big(\frac{\beta_1}{6}+(5+\nu)\beta_3\Big)\alpha^6-\frac{\beta_1}{6}-\beta_4\Big]\\&+\Big(6a_4^2+\frac{1}{2}a_2A_2\Big)\Big[\Big(\frac{\beta_1}{2}+\lambda_2\beta_3\Big)\alpha^2\ln^2\alpha\\&-(\beta_1-\lambda_1\beta_3)\alpha^2\ln\alpha+\frac{1}{2}\Big(\frac{3}{2}\beta_1-\lambda_1\beta_3\Big)\alpha^2-\frac{1}{2}\Big(\frac{3}{2}\beta_1+\beta_4\Big)\Big]\\&+\frac{a_1a_2}{12}\Big[5\Big(\frac{\beta_1}{8}+(7+\nu)\beta_3\Big)\alpha^8\ln\alpha\\&-\frac{1}{8}\Big(\frac{19}{4}\beta_1+(191+33\nu)\beta_3\Big)\alpha^8+\frac{1}{8}\Big(\frac{19}{4}\beta_1+33\beta_4\Big)\Big]+\frac{5}{12}a_1a_3\Big[\Big(\frac{\beta_1}{8}+(7+\nu)\beta_3\Big)\alpha^8\\&-\frac{\beta_1}{8}-\beta_4\Big]+a_1a_4\Big[2\Big(\frac{\beta_1}{6}+(5+\nu)\beta_3\Big)\alpha^6\ln\alpha\\&-\frac{1}{6}\Big(\frac{\beta_1}{2}-(7-\nu)\beta_3\Big)\alpha^6+\frac{1}{6}\Big(\frac{\beta_1}{2}+\beta_4\Big)\Big]\\&+\frac{3}{2}a_2a_3\Big[\Big(\frac{\beta_1}{6}+(5+\nu)\beta_3\Big)\alpha^6\ln\alpha\\&-\frac{1}{12}\Big(\frac{13}{6}\beta_1+(43+11\nu)\beta_3\Big)\alpha^6+\frac{1}{12}\Big(\frac{13}{6}\beta_1+11\beta_4\Big)\Big]\end{aligned}$$

$$+3a_2a_4\Big[3\Big(\frac{\beta_1}{4}+(3+\nu)\beta_3\Big)\alpha^4\ln^2\alpha-\frac{1}{2}\Big(\frac{13}{4}\beta_1+2(9+5\nu)\beta_3\Big)\alpha^4\ln\alpha$$

$$+\frac{1}{8}\Big(\frac{17}{2}\beta_1+(23+21\nu)\beta_3\Big)\alpha_4-\frac{1}{8}\Big(\frac{17}{2}\beta_1+21\beta_4\Big)\Big]$$

$$+a_3a_4\Big[6\Big(\frac{\beta_1}{2}+2(3+\nu)\beta_3\Big)\alpha^4\ln\alpha-\frac{3}{2}\Big(\frac{7}{4}\beta_1$$

$$+(7+5\nu)\beta_3\Big)\alpha^4+\frac{3}{2}\Big(\frac{7}{2}\beta_1+5\beta_4\Big)\Big]$$

$$+\frac{1}{4}(a_1A_2+a_3A_1-A_3)\Big[\Big(\frac{\beta_1}{4}+(3+\nu)\beta_3\Big)\alpha^4-\frac{\beta_1}{4}-\beta_4\Big]$$

$$+\frac{1}{4}a_2A_1\Big[\Big(\frac{\beta_1}{4}+(3+\nu)\beta_3\Big)\alpha^4\ln\alpha$$

$$-\frac{1}{4}(\beta_1+(5+3\nu)\beta_3)\alpha^4+\frac{1}{4}(\beta_1+3\beta_4)\Big]$$

$$+(a_3A_2+a_4A_1-A_4)\Big[\Big(\frac{\beta_1}{2}+\lambda_2\beta_3\Big)\alpha^2\ln\alpha$$

$$-\frac{1}{2}(\beta_1-\lambda_1\beta_3)\alpha^2+\frac{1}{2}(\beta_1+\beta_4)\Big]-a_4A_2\Big(\frac{\beta_1}{2}\ln^2\alpha-\lambda_1\beta_3\ln\alpha+\beta_3\Big)\Big\},$$

$$\alpha_4=-\frac{2}{3}\beta_2\Big\{\frac{a_1^3}{240}\Big[\Big(\frac{\beta_1}{12}+(11+\nu)\beta_3\Big)\alpha^{12}-\frac{\beta_1}{12}-\beta_4\Big]$$

$$+\frac{a_2^3}{8}\Big[\Big(\frac{\beta_1}{6}+(5+\nu)\beta_3\Big)\alpha^6\ln^3\alpha-\frac{1}{4}\Big(\frac{13}{6}\beta_1+(43+11\nu)\beta_3\Big)\alpha^6\ln^2\alpha$$

$$+\frac{1}{3}\Big(2\beta_1+\frac{109+35\nu}{4}\beta_3\Big)\alpha^6\ln\alpha-\frac{1}{72}\Big(\frac{119}{6}\beta_1$$

$$+(145+71\nu)\beta_3\Big)\alpha^6+\frac{1}{72}\Big(\frac{119}{6}\beta_1+71\beta_4\Big)\Big]$$

$$+\frac{a_3^3}{8}\Big[\Big(\frac{\beta_1}{6}+(5+\nu)\beta_3\Big)\alpha^6-\frac{\beta_1}{6}-\beta_4\Big]$$

$$+3a_4^3\Big[\frac{\beta_1}{3}\ln^3\alpha+\Big(\frac{\beta_1}{2}-\lambda_1\beta_3\Big)\ln^2\alpha+\lambda_2\beta_3\ln\alpha+\beta_3\Big]$$

$$+\frac{a_1^2a_2}{32}\Big[\Big(\frac{\beta_1}{10}+(9+\nu)\beta_3\Big)\alpha^{10}\ln\alpha-\frac{1}{120}\Big(\frac{79}{10}\beta_1+(483+67\nu)\beta_3\Big)\alpha^{10}$$

$$+\frac{1}{120}\Big(\frac{79}{10}\beta_1+67\beta_4\Big)\Big]+\frac{a_1^2a_3}{32}\Big[\frac{1}{10}(\beta_1+(9+\nu)\beta_3)\alpha^{10}-\frac{\beta_1}{10}-\beta_4\Big]$$

$$+\frac{1}{48}\Big(\frac{13}{2}a_1^2a_4+5a_1a_3^2\Big)\Big[\Big(\frac{\beta_1}{8}+(7+\nu)\beta_3\Big)\alpha^8-\frac{\beta_1}{8}-\beta_4\Big]$$

$$+\frac{a_1a_2^2}{48}\Big[5\Big(\frac{\beta_1}{8}+(7+\nu)\beta_3\Big)\alpha^8\ln^2\alpha$$

$$-\frac{1}{4}\Big(\frac{19}{4}\beta_1+(191+33\nu)\beta_3\Big)\alpha^8\ln\alpha+\frac{1}{96}\Big(\frac{577}{8}\beta_1$$

$$+(2449+463\nu)\beta_3\Big)\alpha^8-\frac{1}{96}\Big(\frac{577}{8}\beta_1+463\beta_4\Big)\Big]$$

$$
\begin{aligned}
&+\frac{3}{8}a_2^2a_3\left[\left(\frac{\beta_1}{6}+(5+\nu)\beta_3\right)\alpha^6\ln^2\alpha\right.\\
&-\frac{1}{6}\left(\frac{13}{6}\beta_1+(43+11\nu)\beta_3\right)\alpha^6\ln\alpha\\
&\left.+\frac{2}{9}\left(\beta_1+\frac{109+35\nu}{8}\beta_3\right)\alpha^6-\frac{2}{9}\left(\beta_1+\frac{35}{8}\beta_4\right)\right]\\
&+\frac{3}{2}a_2^2a_4\left[\left(\frac{\beta_1}{4}+(3+\nu)\beta_3\right)\alpha^4\ln^3\alpha\right.\\
&-3\left(\frac{5}{16}\beta_1+(2+\nu)\beta_3\right)\alpha^4\ln^2\alpha+\frac{3}{8}\left(\frac{7}{2}\beta_1\right.\\
&\left.+(11-5\nu)\beta_3\right)\alpha^4\ln\alpha-\frac{1}{16}\left(\frac{23}{2}\beta_1+(21+25\nu)\beta_3\right)\alpha^4\\
&\left.+\frac{1}{16}\left(\frac{23}{2}\beta_1+25\beta_4\right)\right]+\frac{3}{8}a_2a_3^2\left[\left(\frac{\beta_1}{6}\right.\right.\\
&\left.+(5+\nu)\beta_3\right)\alpha^6\ln\alpha-\frac{1}{12}\left(\frac{13}{6}\beta_1+(43+11\nu)\beta_3\right)\alpha^6\\
&\left.+\frac{1}{12}\left(\frac{13}{6}\beta_1+11\beta_4\right)\right]+3a_3^2a_4\left[\left(\frac{\beta_1}{4}\right.\right.\\
&\left.+(3+\nu)\beta_3\right)\alpha^4\ln\alpha-\frac{1}{8}\left(\frac{7}{4}\beta_1+(7+5\nu)\beta_3\right)\alpha^4\\
&\left.+\frac{1}{8}\left(\frac{7}{4}\beta_1+5\beta_4\right)\right]-\frac{3}{2}a_1a_4^2\left[\left(\frac{\beta_1}{4}+(3+\nu)\beta_3\right)\alpha^4\ln\alpha\right.\\
&-\frac{1}{4}\left(\frac{3}{2}\beta_1+(11+5\nu)\beta_3\right)\alpha^4\\
&\left.+\frac{1}{4}\left(\frac{3}{2}\beta_1+5\beta_4\right)\right]-3a_4^2(a_2-a_3)\left[\left(\frac{\beta_1}{2}\right.\right.\\
&\left.+\lambda_2\beta_3\right)\alpha^2\ln^2\alpha-(\beta_1-\lambda_1\beta_3)\alpha^2\ln\alpha+\frac{1}{2}\left(\frac{3}{2}\beta_1-\lambda_1\beta_3\right)\alpha^2\\
&\left.-\frac{1}{2}\left(\frac{3}{2}\beta_1+\beta_4\right)\right]+\frac{1}{24}a_1a_2a_3\left[\frac{5}{8}\beta_1\right.\\
&+5(7+\nu)\alpha^8\ln\alpha-\frac{1}{8}\left(\frac{19}{4}\beta_1+(191+33\nu)\beta_3\right)\alpha^8\\
&\left.+\frac{1}{8}\left(\frac{19}{4}\beta_1+33\beta_4\right)\right]+\frac{1}{2}a_1a_2a_4\left[\frac{\beta_1}{6}\right.\\
&+(5+\nu)\beta_3\alpha^6\ln^2\alpha-\frac{1}{2}\left(\frac{\beta_1}{2}+\frac{23+7\nu}{3}\beta_3\right)\alpha^6\ln\alpha\\
&+\frac{1}{6}\left(\frac{7}{12}\beta_1+(3+2\nu)\beta_3\right)\alpha^6-\frac{1}{3}\left(\frac{27}{4}\beta_1\right.\\
&\left.\left.+\beta_4\right)\right]+a_1a_3a_4\left[\left(\frac{\beta_1}{6}+(5+\nu)\beta_3\right)\alpha^6\ln\alpha\right.\\
&\left.-\frac{1}{12}\left(\frac{\beta_1}{2}-(7-\nu)\beta_3\right)\alpha^6+\frac{1}{12}\left(\frac{\beta_1}{2}+\beta_4\right)\right]
\end{aligned}
$$

$$
\begin{aligned}
&+\frac{3}{2}a_2a_3a_4\Big[3\Big(\frac{\beta_1}{4}+(3+\nu)\beta_3\Big)\alpha^4\ln^2\alpha\\
&-\Big(\frac{13}{8}\beta_1+(9+5\nu)\beta_3\Big)\alpha^4\ln\alpha+\frac{1}{8}\Big(\frac{17}{2}\beta_1+(23\\
&+21\nu)\beta_3)\alpha^4-\frac{1}{8}\Big(\frac{17}{2}\beta_1+21\beta_4\Big)\Big]\\
&-\frac{a_1A_3}{24}\Big[\Big(\frac{\beta_1}{6}+(5+\nu)\beta_3\Big)\alpha^6-\frac{\beta_1}{6}-\beta_4\Big]-\frac{1}{8}(a_1A_4\\
&+a_3A_3)\Big[\Big(\frac{\beta_1}{4}+(3+\nu)\beta_3\Big)\alpha^4-\frac{\beta_1}{4}-\beta_4\Big]\\
&-\frac{a_2A_3}{8}\Big[\Big(\frac{\beta_1}{4}+(3+\nu)\beta_3\Big)\alpha^4\ln\alpha-\frac{1}{4}(\beta_1\\
&+(5+3\nu)\beta_3)\alpha^4+\frac{1}{4}(\beta_1+3\beta_4)\Big]\\
&-\frac{a_2A_4}{2}\Big[\Big(\frac{\beta_1}{2}+\lambda_2\beta_3\Big)\alpha^2\ln^2\alpha-(\beta_1-\lambda_1\beta_3)\alpha^2\ln\alpha+\frac{1}{2}\Big(\frac{3}{2}\beta_1\\
&-\lambda_1\beta_3\Big)\alpha^2-\frac{1}{2}\Big(\frac{3}{2}\beta_1+\beta_4\Big)\Big]\\
&-\frac{1}{2}(a_3A_4+a_4A_3)\Big[\Big(\frac{\beta_1}{2}+\lambda_2\beta_3\Big)\alpha^2\ln\alpha-\frac{1}{2}(\beta_1-\lambda_1\beta_3)\alpha^2\\
&+\frac{1}{2}(\beta_1+\beta_4)\Big]+\frac{a_4A_4}{4}\Big(\frac{\beta_1}{2}\ln^2\alpha-\lambda_1\beta_3\ln\alpha+\beta_3\Big)\Big\},
\end{aligned}
$$

$$
\beta_3=-\frac{1}{2}\alpha^2+\ln\alpha+\frac{1}{2},
$$

$$
\beta_4=\lambda_2\alpha^2\ln\alpha+\frac{\lambda_1}{2}(\alpha^2-1) \tag{19}
$$

对式（18）使用极值条件

$$
\frac{\mathrm{d}Q}{\mathrm{d}W_m}=0 \tag{20}
$$

便得壳体失稳时的无量纲内边缘挠度

$$
W_m^*=\frac{1}{3\alpha_4}\left[-\alpha_3k\pm\sqrt{(\alpha_3^2-3\alpha_2\alpha_4)k^2-3\alpha_1\alpha_4}\right] \tag{21}
$$

将式（21）代入式（18），便得壳体的临界载荷公式

$$
Q^*=(\alpha_1+\alpha_2k^2)W_m^*+\alpha_3kW_m^{*2}+\alpha_4W_m^{*3} \tag{22}
$$

其中对应于式（21）中的正、负号所得到的 Q^* 分别是下、上临界载荷。

按照式（22）进行了数值计算，其结果给在图 2 中。由图看出，上临界载荷是几何参数 k 的单调上升函数，而下临界载荷只在一定范围内是 k 的增值函数，随后将随 k 的增加而迅速减小。并且，对于同一壳体而言，临界载荷随着开孔的增大而增加。

最后，由式（20）的判别式为零，可得壳体的屈曲临界点的几何参数 k_0。它是区分壳体失稳与否的几何参数值，其公式为

$$k_0 = \sqrt{\frac{3\alpha_1\alpha_4}{\alpha_3^2 - 3\alpha_2\alpha_4}} \tag{23}$$

我们在图 3 中给出了 k_0-α 曲线。由图看出，k_0 随着壳体开孔的增大而增加。

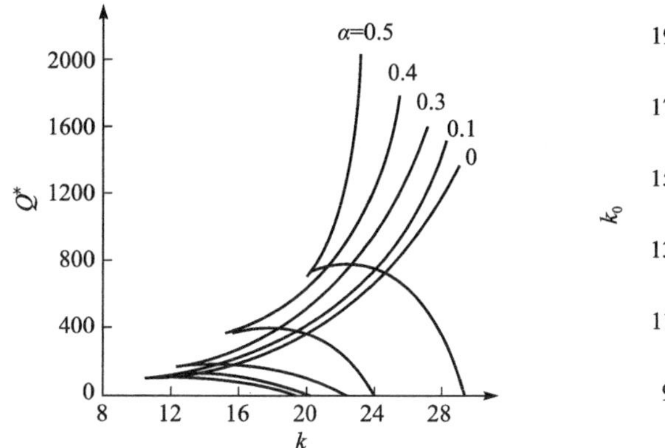

图 2　各种 α 值下的稳定曲线（ν=0.3）

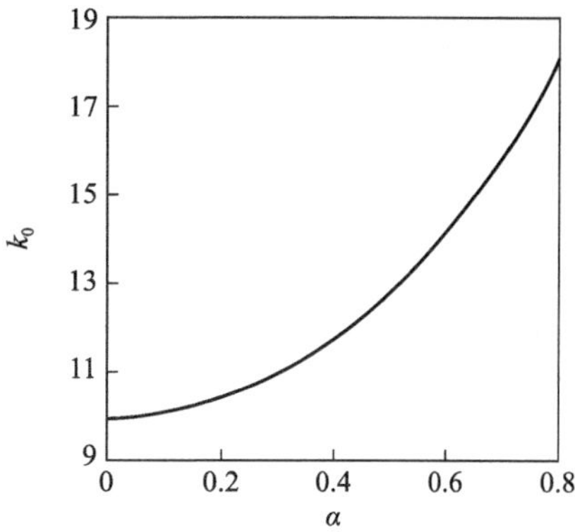

图 3　临界几何参数 k_0 与 α 的关系曲线（ν=0.3）

参 考 文 献

[1]　刘人怀. 在内边缘均布力矩作用下中心开孔圆底扁球壳的非线性稳定问题. 科学通报，1965，(3)：253.

[2]　刘人怀. 在边缘载荷作用下中心开孔圆底扁薄球壳的轴对称稳定性. 力学，1977，(3)：206.

[3]　刘人怀. 双层金属中心开孔扁球壳的非线性热稳定问题. 中国科学技术大学学报，1981，11 (1)：84.

[4]　Liu Renhuai. Non-linear thermal stability of bimetallic shallow shells of revolution. International Journal of Non-Linear Mechanics，1983，18 (5)：409.

[5]　刘人怀，成振强. 集中载荷作用下开顶扁球壳的非线性稳定问题. 应用数学和力学，1988，9 (2)：95.

[6]　刘人怀，李东. 均布载荷作用下开顶扁球壳的非线性稳定问题. 应用数学和力学，1988，9 (3)：205.

[7]　Liu Renhuai，He Linghui. On the nonlinear stability of a truncated shallow spherical shell under axisymmetrically distrinbuted load. Solid Mechanics Archives，1989，14 (2)：81.

[8]　刘人怀，钟诚. 考虑横向剪切的对称层合圆柱正交异性中心开孔扁球壳的非线性屈曲. 暨南大学学报（自然科学版），1994，15 (1)：1.

[9]　刘人怀，张小果. 均布载荷作用下具有硬中心的开顶扁球壳的非线性屈曲. 工程力学，1995，增刊：1839.

[10]　Liu Renhuai. On non-linear buckling of symmetrically laminated，cylindrically orthotropic，truncated，shallow，spherical shells under uniform pressure including shear effects. International Journal of Non-Linear Mechanics，1996，31 (1)：101.

[11]　罗祖道，聂德耀，刘汉东，等. 双层金属球面扁壳的热稳定性. 力学学报，1966，9 (1)：1.

[12]　Tillman S C. On the buckling behaviour of shallow spherical caps under a uniform pressure load. International Journal of Solids and Structures，1970，6 (1)：37.

A new approach to nonlinear vibration of orthotropic thin circular plates①

中文摘要

本文为板的非线性振动提供了一种新的解析法。应用哈密尔顿原理，导出了决定振动基频的非线性特征值方程。使用修正迭代法，求得了方程的解析解。对于纤维铺设成的圆柱正交各向异性薄圆板的非线性自由振动情况，用现有新方法来处理，最终可导出板的振幅-频率之间的解析关系，方法既简单，又有很好的精确度。

1. Introduction

The increasing application of plates, especially laminated composite plates in engineering has led to extensive studies of the mechanical behavior of plates. Much attention has been received from investigating nonlinear flexural vibrations of plates[1,2] in recent years. The mathematical equations of the problem involve coupled nonlinearities arising from large deformation of plates, which result in the difficulty of the problem that many investigators have been endeavoring to overcome. Remarkable advances have been made in studies of the problem with the advent of computers and FEM, as well as other numerical techniques to tackle nonlinear differential equations.

Despite the increasing favor towards using numerical methods to solve vibration problems of plates, many efforts have also been made to push the development of analytic studies of the problems[3]. However, aside from the traditional Ritz', Galerkin's, and perturbation methods, other new analytic methods are rarely developed. The proposal of new methods serves to facilitate the development of analytic investigation of the problems, which now lags behind numerical studies of the same problems.

The present paper presents a new analytic approach to the nonlinear flexural vibration of plates. Hamilton's principle is employed to derive the nonlinear eigenvalue equations determining the fundamental frequencies of vibrations, a modified iterative method is applied to solve the analytic solution of the equations. This iterative method absorbs the advantage of rapid convergence of the modified iteration method used to analyze the nonlinear bending and stability of plates and shallow shells[4-6]. The case of the nonlinear flexural free vibration of fiber-reinforced cylindrically orthotropic thin circular plates is reconsidered with this new method as an illustration. An analytic relation for the amplitude-frequency response of the plates is finally derived, and the numerical

① Reprinted from Proceedings of the International Conference on Vibration Engineering, Edited by Wen, B. C., Northeastern University Press, Shengyang, 1998, 248-252. Authors: Liu Renhuai and Li Dong.

results are obtained by the point collocation method. The present solution method possesses high accuracy. It is expected that the present method, when incorporated with numerical techniques, can be extended to studies of the nonlinear free or forced vibration of various plates, such as laminated composite plates with irregular plane forms.

2. Basic Equations

An orthotropic thin circular plate with uniform thickness h and radius a is considered here. w is the deflection of the plate, u is the radial displacement, r is the radial coordinate, t is the time variable, N_r and N_θ are the radial and circumferential membrane forces, m is the mass per unit area of the plate, E_r and E_θ are the radial and circumferential Young's moduli, V_r and V_θ are the radial and circumferential Poisson's ratios.

Neglecting the in-plane inertia and considering the case of axisymmetric free vibration, we have the in-plane equilibrium equation as follows

$$N_\theta - (rN_r)_{,r} = 0 \tag{1}$$

where $()_{,r}$ is the partial derivative with respect to r.

The boundary conditions for immovably or movably clamped outer edge of the plate are

$$w \text{ is finite},\ w_{,r} = 0,\ \lim_{r\to 0}(rw_{,rrr} + w_{,rr} - \beta^2 r^{-1} w_{,r}) = 0,$$

$$rN_r = 0 \qquad \text{at} \quad r = 0 \tag{2}$$

$$w = 0, \quad w_{,r} = 0,$$

$$(rN_r)_{,r} - \lambda N_r = 0 \qquad \text{at} \quad r = a \tag{3}$$

where $\beta^2 = E_\theta/E_r = \nu_\theta/\nu_r$ is the orthotropic parameter, $\lambda = \nu_\theta$ or $\lambda \to \infty$ indicates the immovably or movably clamped edge conditions respectively.

Hamilton's principle demands that

$$\delta \int_{t_1}^{t_2} (T - V)\mathrm{d}t = 0 \tag{4}$$

in which δ indicates variation, T and V are the total kinetic and potential energies of the plate, t_1 and t_2 are two arbitrary different times of the motion. Generally, an exact solution of Eq. (4) is unavailable, a loose expression of (4) as

$$\int_0^{2\pi/\omega^*} \delta(T - V)\mathrm{d}t = 0 \tag{5}$$

will yield a good approximate solution, in which ω^* is the circular frequency of vibrations.

Besides, we have the compatible equation as follows[7]

$$r^2(rN_r)_{,rr} + r(rN_r)_{,r} - \beta^2 rN_r + \frac{1}{2} rhE_\theta w_{,r}^2 = 0 \tag{6}$$

The following dimensionless quantities are introduced

$$x = \frac{r}{a},\ \overline{N} = \frac{a}{D} rN_r,\ \overline{W} = \frac{w}{h},$$

$$\tau = t\left(\frac{D}{ma^4}\right)^{\frac{1}{2}}, \ \omega = \omega^* \left(\frac{ma^4}{D}\right)^{\frac{1}{2}} \tag{7}$$

where $D = E_\theta h^3/12\ (\beta^2 - \nu_\theta^2)$ is the flexural rigidity.

Using these dimensionless quantities in Eqs. (5), (6) and conditions (2), (3) leads to the following dimensionless governing equations and boundary conditions

$$\int_0^{2\pi/\omega} \delta(T - V)\mathrm{d}\tau = 0 \tag{8}$$

$$x^2 \overline{N}'' + x\overline{N}' - \beta^2 \overline{N} + 6(\beta^2 - \nu_\theta^2) x \overline{W}'^2 = 0$$

$$\overline{W} \text{ is finite}, \ \overline{W}' = 0, \tag{9}$$

$$\lim_{x \to 0^+} (x\overline{W}''' + \overline{W}'' - \beta^2 x^{-1} \overline{W}' = 0, \ \overline{N} = 0 \qquad \text{at } x = 0 \tag{10}$$

$$\overline{W} = 0, \ \overline{W}' = 0, \ \overline{N}' - \lambda \overline{N} = 0 \qquad \text{at } x = 1 \tag{11}$$

where $()'$ indicates the derivative with respect to x, ω is the dimensionless frequency as defined in (7).

We assume that the solutions of Eqs. (8), (9) have the following time-mode forms

$$\overline{W} = W(x)\cos\omega\tau, \ \overline{N} = N(x)\cos^2\omega\tau \tag{12}$$

where $\overline{W}\ (x)$ and $N\ (x)$ are unspecified functions.

Substituting (12) into Eq. (8) and using (1), (10) and (11), after integrating the equation by part over interval $[0, 2\pi/\omega]$, we finally arrive at

$$\int_0^1 [W^{(4)} + 2x^{-1}W''' - \beta^2 x^{-2} W'' + \beta^2 x^{-3} W'$$

$$- \omega^2 W - \frac{3}{4} x^{-1} (NW')'] \delta W x \mathrm{d}x = 0$$

Since δW is an arbitrary independent function, the above equation reduces to

$$W^{(4)} + 2x^{-1}W''' - \beta^2 x^{-2} W'' + \beta^2 x^{-3} W' - \omega^2 W - \frac{3}{4} x^{-1} (NW')' = 0 \tag{13}$$

Substitution of (12) into (9)-(11) yields

$$x^2 N'' + xN' - \beta^2 N + 6(\beta^2 - \upsilon_\theta^2) x W'^2 = 0 \tag{14}$$

$$W = W_m, \ W' = 0$$

$$\lim_{x \to 0^+} (xW''' + W'' - \beta^2 \chi^{-1} W') = 0, \ N = 0 \qquad \text{at } x = 0 \tag{15}$$

$$W = 0, \ W' = 0, \ N' - \lambda N = 0 \qquad \text{at } x = 1 \tag{16}$$

where $W_m = W\ (x)\ |_{x=0}$ is the dimensionless vibration amplitude of the center of the plate.

3. Analytic Solutions

We begin with the following linearized eigenvalue equations

$$W_1^{(4)} + 2x^{-1} W_1''' - \beta^2 x^{-2} W_1'' + \beta^2 x^{-3} W_1' - \omega_L^2 W_1 = 0 \tag{17}$$

$$x^2 N_1'' + xN_1' - \beta^2 N_1 + 6(\beta^2 - \nu_\theta^2) x W_1'^2 = 0 \tag{18}$$

$$W_1 = W_m,\ W_1' = 0$$

$$\lim_{x\to 0^+}(xW_1''' + W_1'' - \beta^2 x^{-1} W_1') = 0,\ N_1 = 0 \qquad \text{at } x = 0 \tag{19}$$

$$W_1 = 0,\ W_1' = 0,\ N_1' - \lambda N_1 = 0 \qquad \text{at } x = 1 \tag{20}$$

where the subscript indicates the order of iteration.

The solution of Eq. (17) is obtained as

$$W_1 = W_m\left(\sum_{j=0}^{\infty} A_j^{(1)} x^{4j} + \varepsilon \sum_{j=0}^{\infty} B_j^{(1)} x^{4j+1+\beta}\right) \tag{21}$$

in which ε is an unspecified coefficient, and

$$A_0^{(1)} = B_0^{(1)} = 1$$

$$A_{j+1}^{(1)} = \frac{\omega_L^2}{8(2j+1)(j+1)(4j+3+\beta)(4j+3-\beta)} A_j^{(1)}$$

$$B_{j+1}^{(1)} = \frac{\omega_L^2}{8(j+1)(2j+2+\beta)(4j+3+\beta)(4j+5+\beta)} B_j^{(1)}$$

$$(j = 0, 1, \cdots)$$

Solution (21) obviously satisfies the center condition in (19). Substituting (21) into condition (20) yields

$$\sum_{j=0}^{\infty} A_j^{(1)} + \varepsilon \sum_{j=0}^{\infty} B_j^{(1)} = 0$$

$$\sum_{j=0}^{\infty} 4jA_j^{(1)} + \varepsilon \sum_{j=0}^{\infty} (4j+1+\beta) B_j^{(1)} = 0 \tag{22}$$

Elimination ε from Eq. (22), we obtain the following algebraic equation

$$\sum_{j=0}^{\infty} A_j^{(1)} \left[\sum_{j=0}^{\infty} (4j+1+\beta) B_j^{(1)}\right] - \sum_{j=0}^{\infty} B_j^{(1)} \left[\sum_{j=0}^{\infty} 4jA_j^{(1)}\right] = 0 \tag{23}$$

The value of ω_L is computed from Eq. (23), then values of $A_j^{(1)}$ and $B_j^{(1)}$ are specified via the recursion formulae listed above, finally, ε is obtained from Eq. (22) as

$$\varepsilon = -\frac{\sum_{j=0}^{\infty} A_j^{(1)}}{\sum_{j=0}^{\infty} B_j^{(1)}}$$

Now putting solution (21) into Eq. (18), using integration and considering the boundary conditions (19), (20), we have the solution of Eq. (18)

$$N_1 = W_m^2\left(\sum_{j=0}^{\infty} C_j^{(1)} x^{4j-1} + \sum_{j=0}^{\infty} D_j^{(1)} x^{4j+\beta} + \sum_{j=0}^{\infty} E_j^{(1)} x^{4j+2\beta+1}\right) \tag{24}$$

in which

$$C_0^{(1)} = C_1^{(1)} = 0$$

$$D_0^{(1)} = (\lambda-\beta)^{-1}\Big[\sum_{j=0}^{\infty} (4j-1-\lambda) C_j^{(1)}$$

$$+ \sum_{j=0}^{\infty} (4j+\beta-\lambda) D_j^{(1)} + \sum_{j=0}^{\infty} (4j+2\beta+1-\lambda) E_j^{(1)}\Big]$$

$$C_j^{(1)}=-\frac{96(\beta^2-\nu_\theta^2)\sum_{j=0}^{j}i(j-i)A_i^{(1)}A_{j-i}^{(1)}}{(4j-i)^2-\beta^2}\qquad(j\geqslant 2)$$

$$D_j^{(1)}=-\frac{48(\beta^2-\nu_\theta^2)\varepsilon\sum_{i=0}^{j}i(4j-4i+1+\beta)A_i^{(1)}B_{j-i}^{(1)}}{(4j+\beta)^2-\beta^2}\qquad(j\geqslant 1)$$

$$E_j^{(1)}=-\left[6(\beta^2-\nu_\theta^2)\varepsilon^2\sum_{i=0}^{j}(4j+1+\beta)(4j-4i+1+\beta)B_i^{(1)}B_{j-i}^{(1)}\right]\cdot[(4j+2\beta+1)^2-\beta^2]^{-1}\qquad(j=1,2,\cdots)$$

In the second step of iteration, we have the following modified linear eigenvalue equation

$$W_2^{(4)}+2x^{-1}W_2'''-\beta^2x^{-2}W_2''+\beta^2x^{-3}W_2' -\omega_{NL}^2W_2-\frac{3}{4}x^{-1}(N_1W_1')'=0 \tag{25}$$

$$W_2=W_m,\ W_2'=0,\quad \lim_{x\to 0^+}(xW_2'''+W_2''-\beta^2x^{-1}W_2')=0,\qquad \text{at } x=0 \tag{26}$$

$$W_2=0,\ W_2'=0,\qquad \text{at } x=1 \tag{27}$$

where ω_{NL} is nonlinear frequency of vibrations obtained via second iteration.

Considering condition (26) and setting solutions (21), (24) in (25), we can obtain the solution of Eq. (25) as

$$W_2=W_m\Big(\sum_{j=0}^{\infty}A_j^{(2)}x^{4j}+\eta\sum_{j=0}^{\infty}B_j^{(2)}x^{4j+1+\beta}\Big) +W_m^3\Big(\sum_{j=0}^{\infty}C_j^{(2)}x^{4j-4}+\sum_{j=0}^{\infty}D_j^{(2)}x^{4j+\beta-3} +\sum_{j=0}^{\infty}E_j^{(2)}x^{4j+2\beta-2}+\sum_{j=0}^{\infty}F_j^{(2)}x^{4j+3\beta-1}\Big) \tag{28}$$

in which

$$A_0^{(2)}=B_0^{(2)}=C_0^{(2)}=D_0^{(2)}=E_0^{(2)}=F_0^{(2)}=0$$

$$C_1^{(2)}=D_1^{(2)}=0$$

$$A_{j+1}^{(2)}=\frac{\omega_{NL}^2}{8(2j+1)(j+1)(4j+3+\beta)(4j+3-\beta)}A_j^{(2)}$$

$$B_{j+1}^{(2)}=\frac{\omega_{NL}^2}{8(j+1)(2j+2+\beta)(4j+3+\beta)(4j+5+\beta)}B_j^{(2)}$$

$$C_{j+1}^{(2)}=\frac{\omega_{NL}^2C_j^{(2)}+6(2j-1)\sum_{i=0}^{j}iA_i^{(1)}C_{j-i}^{(1)}}{8j(2j-1)(4j-1-\beta)(4j-1+\beta)}\qquad(j\geqslant 1)$$

$$D_{j+1}^{(2)}=\{4\omega_{NL}^2D_j^{(2)}+3(4j+\beta-1)$$

$$
\begin{aligned}
&\cdot \sum_{i=0}^{j}\left[(4i+1+\beta)\varepsilon B_i^{(1)}D_{j-i}^{(1)}+4iA_i^{(1)}D_{j-i}^{(1)}\right]\Big\}\\
&\cdot \left[32j(2i+\beta)(4j+1+\beta)(4j-1+\beta)\right]^{-1}\quad (j\geqslant 1)\\
E_{j+1}^{(2)} &= \{2\omega_{NL}^2E_j^{(2)}+3(2j+\beta)\\
&\cdot \sum_{i=0}^{j}\left[(4i+1+\beta)\varepsilon B_i^{(1)}D_{j-i}^{(1)}+4iA_i^{(1)}E_{j-i}^{(1)}\right]\Big\}\\
&\cdot \left[8(2j+\beta)(2j+1+\beta)(4j+1+\beta)(4j+1+3\beta)\right]^{-1}\\
F_{j+1}^{(2)} &= \{4\omega_{NL}^2F_j^{(2)}+3(4j+3\beta+1)\\
&\cdot \sum_{i=0}^{j}(4i+1+\beta)\varepsilon B_i^{(1)}E_{j-i}^{(1)}\Big\}\left[16(2j+\beta+1)\right.\\
&\cdot \left.(2j+2\beta+1)(4j+3\beta+1)(4j+3\beta+3)\right]^{-1}\quad (j=0,1,\cdots)
\end{aligned}
\tag{29}
$$

The unspecified coefficient η will be determined by boundary conditions (27). Placing solution (28) into conditions (27), we have

$$
\begin{aligned}
&W_m[f_1(\omega_{NL})+\eta f_3(\omega_{NL})]+W_m^3f_6(\omega_{NL})=0\\
&W_m[f_4(\omega_{NL})+\eta f_2(\omega_{NL})]+W_m^3f_5(\omega_{NL})=0
\end{aligned}
\tag{30}
$$

where

$$
\begin{aligned}
f_1(\omega_{NL}) &= \sum_{j=0}^{\infty}A_j^{(2)},\\
f_2(\omega_{NL}) &= \sum_{j=0}^{\infty}(4j+1+\beta)B_j^{(2)},\\
f_3(\omega_{NL}) &= \sum_{j=0}^{\infty}B_j^{(2)},\\
f_4(\omega_{NL}) &= \sum_{j=0}^{\infty}4jA_j^{(2)},\\
f_5(\omega_{NL}) &= \sum_{j=0}^{\infty}[4(j-1)C_j^{(2)}+(4j+\beta-3)D_j^{(2)}\\
&\quad +(4j+2\beta-2)E_j^{(2)}+(4j+3\beta-1)F_j^{(2)}],\\
f_6(\omega_{NL}) &= \sum_{j=0}^{\infty}[C_j^{(2)}+D_j^{(2)}+E_j^{(2)}+F_j^{(2)}]
\end{aligned}
$$

Obviously, the above $f_1(\omega_{NL})\sim f_6(\omega_{NL})$ are fast-converged infinite power series for ω_{NL}. Again eliminating η from Eq. (30) leads to

$$
W_m^2=\frac{f_1(\omega_{NL})f_2(\omega_{NL})-f_3(\omega_{NL})f_4(\omega_{NL})}{f_3(\omega_{NL})f_5(\omega_{NL})-f_2(\omega_{NL})f_6(\omega_{NL})}\tag{31}
$$

The values of nonlinear frequencies ω_{NL} are calculated from (31) vs. certain values of the amplitude W_m, then coefficients $A_j^{(2)}$, $B_j^{(2)}$, …, $F_j^{(2)}$ are specified via recursion formulae (29), finally, η is determined from Eq. (30)

$$\eta = -\frac{\sum_{j=0}^{\infty} A_j^{(2)} + w_m^2 \sum_{j=0}^{\infty} [C_j^{(2)} + D_j^{(2)} + E_j^{(2)} + F_j^{(2)}]}{\sum_{j=0}^{\infty} B_j^{(2)}}$$

Thus the solution of Eqs. (13), (14) up to second order iteration is completed.

In actual computations, we truncate the infinite power series in (21), (24) and (28) by taking $j=0$ up to k, the value or vibration frequencies obtained from (31) is $w_{NL}^{(k)}$. Increasing the value of k until the relative error $|w_{NL}^{(k+1)} - w_{NL}^{(k)}| / w_{NL}^{(k+1)}$ is less than 10^{-8}. The computation can also be performed with a small calculator.

From relation (31), values for vibration frequencies of immovably or movably clamped orthotropic thin circular plates are given in Figs. 1 and 2. Our analytic solutions agree well with those obtained by the orthogonal point collocation method[7]. Besides, from these figures, we can conclude that the in-plane restraint of outer edges of clamped plates tends to strengthen the nonlinearity of vibrations.

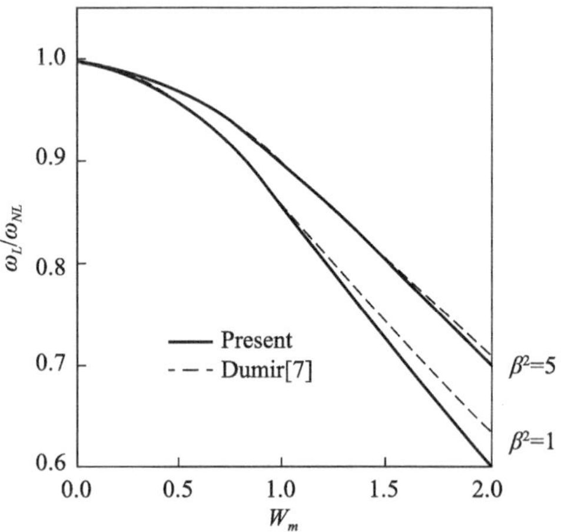

Fig. 1 Amplitude-frequency response for immovably clamped orthotropic thin circular plates ($\lambda=\nu_\theta=0.3$)

4. Remarks

The present iterative method can be extended to studies of nonlinear vibrations of laminated composite plates with irregular planforms and boundary conditions. Harmonic time-mode solutions are substituted into Eq. (5), nonlinear eigenvalue equations thus derived are linearized as described in this paper, in each step of iteration, the linearized equations are solved numerically if analytic solutions are unavailable. Boundary conditions are used to generate amplitude-frequency relations.

Further use of the iterative method in studies of the nonlinear vibration of annular plates and shallow shells is in progress.

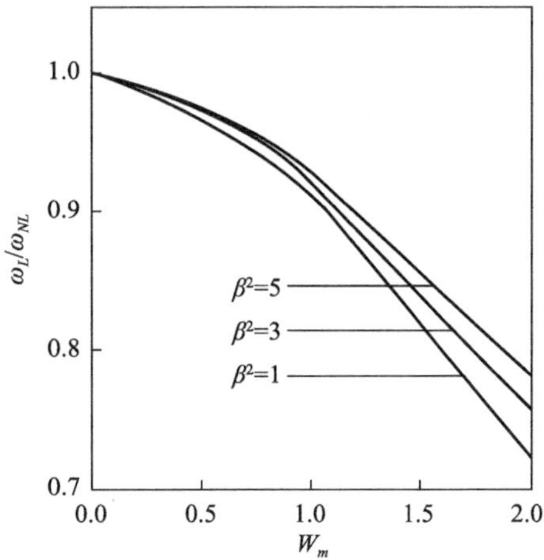

Fig. 2 Amplitude-frequency response for movably clamped orthotropic thin circular plates ($\lambda=\infty$, $\nu_\theta=0.3$)

References

[1] Leissa, A. W., Nonlinear analysis of plate and shell vibrations, Pros. Int. Conf. Recent Adv. Struct. Dynamics 1, 241-260, 1984, University of Southampton, UK.

[2] Reddy, J. N., A review of the literature on finite-element modeling of laminated composite plates, Shock Vib. Digest, 17 (4), 3-8, 1985.

[3] Sathyamoorthy, M., Nonlinear vibration analysis of plates: A review and survey of current developments, Appl. Mech. Rev., 40 (11), 1553-1561, 1987.

[4] Liu Renhuai, Nonlinear thermal stability of bimetallic shallow shells of revolution, Int. J. Nonlinear Mech., 18 (5), 409-429, 1983.

[5] Liu Renhuai, Large deflection of corrugated circular plate with a plane central region under the action of concentrated loads at the center, Int. J. Nonlinear Mech., 19 (5), 409-419, 1984.

[6] Liu Renhuni, and Li, D., On the nonlinear stability of truncated shallow spherical shell under a uniformly distributed load, Appl. Math. Mech., 9 (3), 227-240, 1988.

[7] Dumir, P. C., Nonlinear axisymmetric vibration of orthotropic thin circular plates with elastically restrained edges, Comput. & Struct., 22 (4), 677-686, 1986.

正交异性扁薄球壳的非线性轴对称振动①

一、引言

由于工程技术和科学技术的迅速发展，扁球壳的非线性振动问题一直受到研究者的关注。Singh 等[1]、Ramachandran[2,3] 和 Yasuda 等[4] 已经分别研究过各向同性扁球壳的非线性轴对称自由振动和强迫振动。Dumir[5] 首先研究了正交异性开顶扁薄球壳的非线性轴对称自由振动，得到了一些有意义的结果。然而，此问题的研究遇到了求解一对非线性数学方程的困难。Sinharay 等[6] 提出了研究此问题的一种修正方法，从而简化了对这一问题的分析。但是，Sinharay 等的方法是基于一些来自经验的假设，而且对具有拱高相对厚度和大振幅有较大比值的壳体来说，其方程未含有振动的耦合特征关系。接着，刘人怀和王璠等[7,8] 研究了层合复合材料封闭与开顶扁球壳的非线性强迫振动问题，获得了解析解。

本文是以往工作的继续，其目的是提出一种新的研究扁薄球壳非线性振动的方法。在假设壳体的挠度和应力函数为与时间分离形式的基础上，为更精确地描述壳体的振动特性，我们引入一个小量，以便模仿壳体的非对称弹性特征关系，同时导出一组非线性耦合的代数方程和微分方程。在作者[9-13] 应用修正迭代法[14-16] 处理圆板和波纹圆板非线性振动问题的基础上，进一步发展了这种方法。最后，得到了本问题的解析解。

二、非线性本征值问题

我们研究一个如图 1 所示的圆柱正交异性扁薄球壳。这里，h 是厚度，a 是半径，H^* 是拱高。同时，以壳体的中曲面为坐标曲面，r 是径向坐标，θ 是环向坐标。

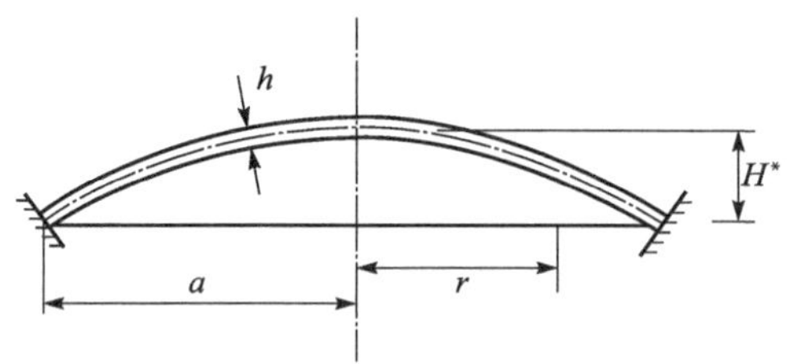

图 1　圆柱正交异性扁薄球壳

这个扁薄球壳的非线性轴对称自由振动问题的 von Kármán-Donnel 型的控制方程组已经在文献［5］等中给出。为了便于随后的分析，我们宁愿采用 Hamilton 原理推出的变分形式的平衡方程

① 本文原载《振动工程学报》，2005，18（4）：395-405. 作者：刘人怀，李东，袁鸿.

$$\int_{t_1}^{t_2}\int_0^a\left\{DL_1(w^*)-\frac{1}{r}\frac{\partial}{\partial r}\left[\psi^*\left(\frac{\partial w^*}{\partial r}+\frac{2H^*}{a^2}r\right)\right]+\gamma h\frac{\partial^2 w^*}{\partial t^2}\right\}\delta w^* r\mathrm{d}r\mathrm{d}t=0 \tag{1}$$

和相容方程

$$L_2(\psi^*)+\frac{E_\theta h}{2}\frac{\partial w^*}{\partial r}\left(\frac{\partial w^*}{\partial r}+\frac{4H^*}{a^2}r\right)=0 \tag{2}$$

其中 t 是时间变量，γ 是质量密度，w^* 是挠度，ψ^* 是应力函数，β 是正交异性参数，D 是抗弯刚度，E_r 和 E_θ 分别是径向和环向的弹性模量，ν_r 和 ν_θ 分别是径向和环向的泊松比，N_r 和 N_θ 分别是径向和环向的薄膜内力，L_1 和 L_2 是偏微分算子，

$$N_r=\frac{\psi^*}{r},N_\theta=\frac{\partial\psi^*}{\partial r},\beta=\frac{E_\theta}{E_r}=\frac{\nu_\theta}{\nu_r},D=\frac{E_\theta h^3}{12(\beta-\nu_\theta^2)},$$

$$L_1(\cdots)=\frac{\partial^4(\cdots)}{\partial r^4}+\frac{2}{r}\frac{\partial^3(\cdots)}{\partial r^3}-\frac{\beta}{r^2}\frac{\partial^2(\cdots)}{\partial r^2}+\frac{\beta}{r^3}\frac{\partial(\cdots)}{\partial r},$$

$$L_2(\cdots)=r\frac{\partial^2(\cdots)}{\partial r^2}+\frac{\partial(\cdots)}{\partial r}-\frac{\beta}{r}(\cdots) \tag{3}$$

相应的边界条件为

在 $r=0$ 时，w^* 有限，$\dfrac{\partial w^*}{\partial r}=0$，$\psi^*=0$，

$$\lim_{r\to0}r\left(\frac{\partial^3 w^*}{\partial r^3}+\frac{1}{r}\frac{\partial^2 w^*}{\partial r^2}-\frac{\beta}{r^2}\frac{\partial w^*}{\partial r}\right)=0; \tag{4}$$

在 $r=a$ 时，$w^*=0$，$\dfrac{\partial^2 w^*}{\partial r^2}+\dfrac{\lambda}{r}\dfrac{\partial w^*}{\partial r}=0$，$\dfrac{\partial\psi^*}{\partial r}-\nu\dfrac{\psi^*}{r}=0$

这里，取 λ 和 ν 的不同数值，便有如下四种常用的边界条件：

(1) 铰支，$\lambda=\nu=\nu_\theta$；

(2) 简支，$\lambda=\nu_\theta$，$\nu=\infty$；　(5a～d)

(3) 滑动固定，$\lambda=\nu=\infty$；

(4) 夹紧固定，$\lambda=\infty$，$\nu=\nu_\theta$。

引入以下的无量纲量

$$w=\frac{w^*}{h},\psi=\frac{a}{D}\psi^*,x=\frac{r}{a},$$

$$\tau=t\sqrt{\frac{D}{\gamma ha^4}},H=\frac{H^*}{h},m=\sqrt{\beta} \tag{6}$$

于是，方程 (1)、(2) 和边界条件 (4) 转化为无量纲形式

$$\int_{\tau_1}^{\tau_2}\int_0^1\left\{\frac{\partial^2 w}{\partial\tau^2}+\overline{L}_1(w)-\frac{1}{x}\frac{\partial}{\partial x}\left[\psi\left(\frac{\partial w}{\partial x}+2xH\right)\right]\right\}\delta wx\,\mathrm{d}x\mathrm{d}\tau=0,$$

$$\overline{L}_2(x^m\psi)+6(\beta-\nu_\theta^2)\frac{\partial w}{\partial x}\left(\frac{\partial w}{\partial x}+4xH\right)=0 \tag{7a,b}$$

在 $x=0$ 时，w 有限，$\dfrac{\partial w}{\partial x}=0$，

$$\lim_{x\to0}\left(x\frac{\partial^3 w}{\partial x^3}+\frac{\partial^2 w}{\partial x^2}-\frac{\beta}{x}\frac{\partial w}{\partial x}\right)=0,\psi=0; \tag{8}$$

在 $x=1$ 时，$w=0$，$\dfrac{\partial^2 w}{\partial x^2}+\lambda\dfrac{\partial w}{\partial x}=0$，$\dfrac{\partial\psi}{\partial x}-\nu\psi=0$

其中

$$\overline{L}_1(\cdots)=\frac{1}{x}\frac{\partial}{\partial x}x^m\frac{\partial}{\partial x}x^{-(2m-1)}\frac{\partial}{\partial x}x^m\frac{\partial}{\partial x}(\cdots),$$

$$\overline{L}_2(\cdots)=x^m\frac{\partial}{\partial x}x^{-(2m-1)}\frac{\partial}{\partial x}(\cdots) \tag{9}$$

设方程（7）的解为

$$w(x,\tau)=W(x)(\xi+\cos\omega\tau),$$

$$\psi(x,\tau)=N(x)(\xi+\cos\omega\tau)+T(x)(\xi+\cos\omega\tau)^2 \tag{10a,b}$$

其中 ξ 是一个反映壳体弹性特征关系的非对称性的参数，从物理上讲，可解释为描述非线性振荡器[17]漂移大小的一个内禀量。ω 是无量纲振动的固有频率。$W(x)$、$N(x)$ 和 $T(x)$ 是待定函数。

将式（10）代入方程（7a），应用边界条件（8），在一个振动周期$\left[0,\dfrac{2\pi}{\omega}\right]$内对方程积分后，便得

$$\int_0^1\Big\{L_3(W)-\frac{\omega^2}{2\xi^2+1}W-\frac{1}{x}\frac{\mathrm{d}}{\mathrm{d}x}\Big[2xHN+\frac{\xi(2\xi^2+3)}{2\xi^2+1}\Big(2xHT+N\frac{\mathrm{d}W}{\mathrm{d}x}\Big)$$

$$+\frac{8\xi^4+24\xi^2+3}{4(2\xi^2+1)}T\frac{\mathrm{d}W}{\mathrm{d}x}\Big]\Big\}\delta Wx\,\mathrm{d}x+\Big\{\int_0^1\Big(N\frac{\mathrm{d}W}{\mathrm{d}x}+2xHT\Big)\frac{\mathrm{d}W}{\mathrm{d}x}\mathrm{d}x$$

$$+\xi\int_0^1\Big[2xL_3(W)W+4xHN\frac{\mathrm{d}W}{\mathrm{d}x}+3T\Big(\frac{\mathrm{d}W}{\mathrm{d}x}\Big)^2\Big]\mathrm{d}x+2\xi^2\int_0^1\Big(N\frac{\mathrm{d}W}{\mathrm{d}x}$$

$$+2xHT\Big)\frac{\mathrm{d}W}{\mathrm{d}x}\mathrm{d}x+2\xi^3\int_0^1 T\Big(\frac{\mathrm{d}W}{\mathrm{d}x}\Big)^2\mathrm{d}x\Big\}\frac{\delta\xi}{(2\xi^2+1)}=0 \tag{11}$$

其中

$$L_3(\cdots)=\frac{1}{x}\frac{\mathrm{d}}{\mathrm{d}x}x^m\frac{\mathrm{d}}{\mathrm{d}x}x^{-(2m-1)}\frac{\mathrm{d}}{\mathrm{d}x}x^m\frac{\mathrm{d}}{\mathrm{d}x}(\cdots) \tag{12}$$

由于 δW 和 $\delta\xi$ 的任意性，于是有如下方程成立

$$L_3(W)-\frac{\omega^2}{2\xi^2+1}W=\frac{1}{x}\frac{\mathrm{d}}{\mathrm{d}x}\Big[2xHN+f(\xi)\Big(2xHT+N\frac{\mathrm{d}W}{\mathrm{d}x}\Big)$$

$$+g(\xi)T\frac{\mathrm{d}W}{\mathrm{d}x}\Big], \tag{13a,b}$$

$$\alpha_0+\alpha_1\xi+\alpha_2\xi^2+\alpha_3\xi^3=0$$

其中 $f(\xi)$ 和 $g(\xi)$ 是两个函数，

$$f(\xi)=\frac{\xi(2\xi^2+3)}{2\xi^2+1},g(\xi)=\frac{8\xi^4+24\xi^2+3}{4(2\xi^2+1)},$$

$$\alpha_0=\int_0^1\Big(N\frac{\mathrm{d}W}{\mathrm{d}x}+2xHT\Big)\frac{\mathrm{d}W}{\mathrm{d}x}\mathrm{d}x,$$

$$\alpha_1 = \frac{3}{2}\alpha_3 + 2\int_0^1 \left[xL_3(W)W + 2xHT\frac{\mathrm{d}W}{\mathrm{d}x} \right]\mathrm{d}x, \tag{14}$$
$$\alpha_2 = 2\alpha_0,$$
$$\alpha_3 = 2\int_0^1 T\left(\frac{\mathrm{d}W}{\mathrm{d}x}\right)^2 \mathrm{d}x$$

再将式（10）代入方程（7b）和边界条件（8），便有

$$L_4(x^m N) = 24Hx(\nu_\theta^2 - \beta)\frac{\mathrm{d}W}{\mathrm{d}x},$$
$$L_4(x^m T) = 6(\nu_\theta^2 - \beta)\left(\frac{\mathrm{d}W}{\mathrm{d}x}\right)^2 \tag{15a,b}$$

在 $x=0$ 时，$W=W_0$，$\frac{\mathrm{d}W}{\mathrm{d}x}=0$，

$$\lim_{x\to 0}\left(x\frac{\mathrm{d}^3W}{\mathrm{d}x^3} + \frac{\mathrm{d}^2W}{\mathrm{d}x^2} - \frac{\beta}{x}\frac{\mathrm{d}W}{\mathrm{d}x}\right) = 0,$$
$$N = 0, T = 0 \tag{16}$$

在 $x=1$ 时，$W=0$，$\frac{\mathrm{d}^2W}{\mathrm{d}x^2}+\lambda\frac{\mathrm{d}W}{\mathrm{d}x}=0$，

$$\frac{\mathrm{d}N}{\mathrm{d}x} - \nu N = 0, \frac{\mathrm{d}T}{\mathrm{d}x} - \nu T = 0 \tag{17}$$

其中

$$L_4(\cdots) = x^m \frac{\mathrm{d}}{\mathrm{d}x}x^{-(2m-1)}\frac{\mathrm{d}}{\mathrm{d}x}(\cdots) \tag{18}$$
$$W_0 = W(x)\big|_{x=0} \tag{19}$$

值得指出，W_0 的关系式对壳体振动时中心朝内的振幅 W_m 将取正值，而对中心朝外的振幅$\overline{W}_m$ 则取负值。由式（10a），可得

$$W_m = W_0(\xi+1), \overline{W}_m = W_0(\xi-1) \tag{20a,b}$$

显而易见，参数 ξ 是指壳体振幅的非对称的程度。

这样一来，正交异性扁薄球壳的非线性轴对称自由振动问题便归结为求解非线性本征值方程（13）、（15）和边界条件（16）、（17）。

三、渐进解

与文献［9］～［13］一样，我们使用改进的修正近代法。在一次近似中，我们由方程（13a）、（15）和边界条件（16）、（17）得到如下线性的边值问题

$$L_3(W_1) - \omega_L^2 W_1 = \frac{2H}{x} + \frac{\mathrm{d}}{\mathrm{d}x}(xN_1),$$
$$L_4(x^m N_1) = 24Hx(\nu_\theta^2 - \beta)\frac{\mathrm{d}W_1}{\mathrm{d}x},$$
$$L_4(x^m T_1) = 6(\nu_\theta^2 - \beta)\left(\frac{\mathrm{d}W_1}{\mathrm{d}x}\right)^2 \tag{21a$\sim$c}$$

在 $x=0$ 时，$W_1=W_0$，$\frac{\mathrm{d}W_1}{\mathrm{d}x}=0$，

$$\lim_{x\to 0}\left(x\frac{\mathrm{d}^3W_1}{\mathrm{d}x^3}+\frac{\mathrm{d}^2W_1}{\mathrm{d}x^2}-\frac{\beta}{x}\frac{\mathrm{d}W_1}{\mathrm{d}x}\right)=0,$$

$$N_1=0,T_1=0 \qquad (22\mathrm{a}\sim\mathrm{e})$$

在 $x=1$ 时，$W_1=0$，$\dfrac{\mathrm{d}^2W_1}{\mathrm{d}x^2}+\lambda\dfrac{\mathrm{d}W_1}{\mathrm{d}x}=0$，

$$\frac{\mathrm{d}N_1}{\mathrm{d}x}-\nu N_1=0,\frac{\mathrm{d}T_1}{\mathrm{d}x}-\nu T_1=0 \qquad (23\mathrm{a}\sim\mathrm{d})$$

其中，ω_L 是壳体振动的线性固有频率。

用无穷幂级数的形式，可求得方程（21a，b）的解。注意式（22a～d），便得

$$N_1=W_0\left(\zeta\sum_{j=0}^{\infty}A_jx^{4j+m}+\eta\sum_{j=0}^{\infty}B_jx^{4j+m+2}+\sum_{j=0}^{\infty}C_jx^{4j+5}\right),$$

$$W_1=W_0\left(\zeta\sum_{j=0}^{\infty}D_jx^{4j+m+3}+\eta\sum_{j=0}^{\infty}E_jx^{4j+m+1}+\sum_{j=0}^{\infty}F_jx^{4j}\right) \qquad (24\mathrm{a,b})$$

其中 ζ 和 η 是待定系数，A_j、B_j、C_j、D_j、E_j、F_j 由以下循环公式确定

$$A_0=B_0=F_0=1,C_0=\frac{12(\nu_\theta^2-\beta)H}{(9-m^2)(25-m^2)}\omega_L^2,$$

$$A_{j+1}=\frac{6(\nu_\theta^2-\beta)H^2(4j+m+1)(4j+m-1)+j(2j+m)\omega_L^2}{4(2j+m+2)(2j+m+1)(4j+m+1)(4j+m-1)(2j+1)(j+1)}A_j,$$

$$B_{j+1}=\frac{12(\nu_\theta^2-\beta)H^2(4j+m+3)(4j+m+1)+(2j+1)(2j+m+1)\omega_L^2}{8(2j+m+3)(2j+m+2)(4j+m+3)(4j+m+1)(2j+3)(j+1)}B_j,$$

$$C_{j+1}=\frac{384(\nu_\theta^2-\beta)H^2(2j+3)(j+1)+(4j+m+5)(4j-m+5)\omega_L^2}{8(4j+m+9)(4j+m+7)(4j-m+9)(4j-m+7)(2j+3)(j+1)}C_j,$$

$$D_j=\frac{(j+1)(2j+m+2)}{3(\nu_\theta^2-\beta)H(4j+m+3)}A_{j+1},$$

$$E_j=\frac{(2j+1)(2j+m+1)}{6(\nu_\theta^2-\beta)H(4j+m+1)}B_j,$$

$$F_{j+1}=\frac{(4j+m+5)(4j-m+5)}{96(\nu_\theta^2-\beta)H(j+1)}C_j \qquad (25)$$

将解（24）代入相应的边界条件（23a～c），便得到一组代数方程。线性频率 ω_L 和系数 ζ、η 的值可通过求解这些代数方程而得到。

令

$$A_j^{(1)}=\zeta A_j,B_j^{(1)}=\eta B_j,C_j^{(1)}=C_j,D_j^{(1)}=\zeta D_j,E_j^{(1)}=\eta E_j,F_j^{(1)}=F_j,$$

$$G_j^{(1)}=(4j+m+3)D_j^{(1)},H_j^{(1)}=(4j+m+1)E_j^{(1)},I_j^{(1)}=4(j+1)F_{j+1}^{(1)},$$

$$K_j^{(1)}=\sum_{i=0}^{j}G_i^{(1)}G_{j-i}^{(1)},L_j^{(1)}=\sum_{i=0}^{j}H_i^{(1)}H_{j-i}^{(1)},M_j^{(1)}=\sum_{i=0}^{j}I_i^{(1)}I_{j-i}^{(1)},$$

$$N_j^{(1)}=2\sum_{i=0}^{j}G_i^{(1)}I_{j-i}^{(1)},O_j^{(1)}=2\sum_{i=0}^{j}H_i^{(1)}I_{j-i}^{(1)},P_j^{(1)}=2\sum_{i=0}^{j}G_i^{(1)}H_{j-i}^{(1)},$$

$$Q_j^{(1)}=\sum_{i=0}^{j}D_i^{(1)}D_{j-i}^{(1)},R_j^{(1)}=\sum_{i=0}^{j}E_i^{(1)}E_{j-i}^{(1)},S_j^{(1)}=\sum_{i=0}^{j}F_i^{(1)}F_{j-i}^{(1)},$$

$$T_j^{(1)}=2\sum_{i=0}^{j}D_i^{(1)}E_{j-i}^{(1)},U_j^{(1)}=2\sum_{i=0}^{j}D_i^{(1)}F_{j-i}^{(1)},V_j^{(1)}=2\sum_{i=0}^{j}E_i^{(1)}E_{j-i}^{(1)} \tag{26}$$

这些公式在随后的分析中将是有用的。

将解（24b）代入方程（21c），然后进行积分，同时考虑条件（22e）和（23d），便得

$$\begin{aligned}T_1=&W_0^2\Big(\overline{A}x^m+\sum_{j=0}^{\infty}A_j^{(2)}x^{4j+2m+5}+\sum_{j=0}^{\infty}B_j^{(2)}x^{4j+2m+1}+\sum_{j=0}^{\infty}C_j^{(2)}x^{4j+7}\\&+\sum_{j=0}^{\infty}D_j^{(2)}x^{4j+m+6}+\sum_{j=0}^{\infty}E_j^{(2)}x^{4j+m+4}+\sum_{j=0}^{\infty}F_j^{(2)}x^{4j+2m+3}\Big)\end{aligned} \tag{27}$$

其中

$$\begin{aligned}&A_j^{(2)}=\frac{6(\nu_\theta^2-\beta)K_j^{(1)}}{(4j+2m+5)^2-\beta},\quad B_j^{(2)}=\frac{6(\nu_\theta^2-\beta)L_j^{(1)}}{(4j+2m+1)^2-\beta},\\&C_j^{(2)}=\frac{6(\nu_\theta^2-\beta)M_j^{(1)}}{(4j+7)^2-\beta},\quad D_j^{(2)}=\frac{6(\nu_\theta^2-\beta)N_j^{(1)}}{(4j+m+6)^2-\beta},\\&E_j^{(2)}=\frac{6(\nu_\theta^2-\beta)O_j^{(1)}}{(4j+m+4)^2-\beta},\quad F_j^{(2)}=\frac{6(\nu_\theta^2-\beta)P_j^{(1)}}{(4j+2m+3)^2-\beta},\\&\overline{A}=(\nu-m)^{-1}\Big[\sum_{j=0}^{\infty}(4j+2m+5-\nu)A_j^{(2)}+\sum_{j=0}^{\infty}(4j+2m+1-\nu)B_j^{(2)}\\&\quad+\sum_{j=0}^{\infty}(4j+7-\nu)C_j^{(2)}+\sum_{j=0}^{\infty}(4j+2m+6-\nu)D_j^{(2)}\\&\quad+\sum_{j=0}^{\infty}(4j+m+4-\nu)E_j^{(2)}+\sum_{j=0}^{\infty}(4j+2m+3-\nu)F_j^{(2)}\Big]\end{aligned} \tag{28}$$

将解（24）和式（27）代入式（14），可得 α_0、α_1、α_2 和 α_3 的近似公式如下

$$\begin{aligned}&\alpha_0=W_0^3\bar{\alpha}_0,\alpha_1=3W_0^4\bar{\alpha}_3+2\omega_L^2W_0^2\bar{\alpha}_1,\\&\alpha_2=2W_0^3\bar{\alpha}_0,\alpha_3=2W_0^4\bar{\alpha}_3\end{aligned} \tag{29}$$

其中

$$\begin{aligned}\bar{\alpha}_0=&\sum_{j=0}^{\infty}\Bigg\{\frac{2H\sum\limits_{i=0}^{j}G_i^{(1)}A_{j-i}^{(2)}}{4j+3m+9}+\frac{\sum\limits_{i=0}^{j}[B_i^{(1)}K_{j-i}^{(1)}+2H(G_i^{(1)}F_{j-i}^{(2)}+H_i^{(1)}A_{j-i}^{(2)})]}{4j+3m+7}\\&+\frac{\sum\limits_{i=0}^{j}[B_i^{(1)}P_{j-i}^{(1)}+A_i^{(1)}K_{j-i}^{(1)}+2H(G_i^{(1)}B_{j-i}^{(2)}+H_i^{(1)}F_{j-i}^{(2)})]}{4j+3m+5}\\&+\frac{\sum\limits_{i=0}^{j}(B_i^{(1)}L_{j-i}^{(1)}+A_i^{(1)}P_{j-i}^{(1)}+2HH_i^{(1)}B_{j-i}^{(2)})}{4j+3m+3}\\&+\frac{\sum\limits_{i=0}^{j}A_i^{(1)}L_{j-i}^{(1)}}{4j+3m+1}+\frac{\sum\limits_{i=0}^{j}[C_i^{(1)}K_{j-i}^{(1)}+2H(G_i^{(1)}D_{j-i}^{(2)}+I_i^{(1)}A_{j-i}^{(2)})]}{2(2j+m+5)}\\&+\frac{\sum\limits_{i=0}^{j}[B_i^{(1)}N_{j-i}^{(1)}+C_i^{(1)}P_{j-i}^{(1)}+2H(G_i^{(1)}E_{j-i}^{(2)}+H_i^{(1)}D_{j-i}^{(2)}+I_i^{(1)}F_{j-i}^{(2)})]}{2(2j+m+4)}\end{aligned}$$

$$
+\frac{\sum_{i=0}^{j}[A_i^{(1)}N_{j-i}^{(1)}+B_i^{(1)}O_{j-i}^{(1)}+C_i^{(1)}L_{j-i}^{(1)}+2H(H_i^{(1)}E_{j-i}^{(2)}+I_i^{(1)}B_{j-i}^{(2)})]}{2(2j+m+3)}
$$

$$
+\frac{\sum_{i=0}^{j}A_i^{(1)}O_{j-i}^{(1)}+2H\overline{A}G_j^{(1)}}{2(2j+m+2)}+\frac{H\overline{A}H_j^{(1)}}{2j+m+1}
$$

$$
+\frac{\sum_{i=0}^{j}[C_i^{(1)}N_{j-i}^{(1)}+2H(G_i^{(1)}C_{j-i}^{(2)}+I_i^{(1)}D_{j-i}^{(2)})]}{4j+m+11}
$$

$$
+\frac{\sum_{i=0}^{j}[B_i^{(1)}M_{j-i}^{(1)}+C_i^{(1)}O_{j-i}^{(1)}+2H(H_i^{(1)}C_{j-i}^{(2)}+I_i^{(1)}E_{j-i}^{(2)})]}{4j+m+9}
$$

$$
\left.+\frac{\sum_{i=0}^{j}A_i^{(1)}M_{j-i}^{(1)}}{4j+m+7}+\frac{2H\overline{A}I_j^{(1)}}{4j+m+5}+\frac{\sum_{i=0}^{j}(C_i^{(1)}M_{j-i}^{(1)}+2HI_i^{(1)}C_{j-i}^{(2)})}{4(j+3)}\right\},
$$

$$
\overline{\alpha}_1=\sum_{j=0}^{\infty}\left\{\frac{Q_j^{(1)}}{2(2j+m+4)}+\frac{T_j^{(1)}}{2(2j+m+3)}+\frac{R_j^{(1)}}{2(2j+m+2)}\right.
$$

$$
\left.+\frac{U_j^{(1)}}{4j+m+5}+\frac{V_j^{(1)}}{4j+m+3}+\frac{S_j^{(1)}}{2(2j+1)}\right\},
$$

$$
\overline{\alpha}_3=\sum_{j=0}^{\infty}\left\{\frac{\sum_{i=0}^{j}(K_i^{(1)}A_{j-i}^{(2)}}{2(2j+2m+5)}+\frac{\sum_{i=0}^{j}(K_i^{(1)}F_{j-i}^{(1)}+P_i^{(1)}A_{j-i}^{(2)})}{4(j+m+2)}\right.
$$

$$
+\frac{\sum_{i=0}^{j}(K_i^{(1)}B_{j-i}^{(2)}+L_i^{(1)}A_{j-i}^{(2)}+P_i^{(1)}F_{j-i}^{(2)})}{2(2j+2m+3)}+\frac{\sum_{i=0}^{j}(L_i^{(1)}F_{j-i}^{(2)}+P_i^{(1)}B_{j-i}^{(2)})}{4(j+m+1)}
$$

$$
+\frac{\sum_{i=0}^{j}L_i^{(1)}B_{j-i}^{(2)}}{2(2j+2m+1)}+\frac{\sum_{i=0}^{j}(K_i^{(1)}D_{j-i}^{(2)}+N_i^{(1)}A_{j-i}^{(2)})}{4j+3m+11}
$$

$$
+\frac{\sum_{i=0}^{j}(K_i^{(1)}E_{j-i}^{(2)}+N_i^{(1)}F_{j-i}^{(2)}+O_i^{(1)}A_{j-i}^{(2)}+P_i^{(1)}D_{j-i}^{(2)})}{4j+3m+9}
$$

$$
+\frac{\sum_{i=0}^{j}(L_i^{(1)}D_{j-i}^{(2)}+N_i^{(1)}B_{j-i}^{(2)}+O_i^{(1)}F_{j-i}^{(2)}+P_i^{(1)}E_{j-i}^{(2)})}{4j+3m+7}
$$

$$
+\frac{\sum_{i=0}^{j}(L_i^{(1)}E_{j-i}^{(2)}+O_i^{(1)}B_{j-i}^{(2)})+\overline{A}K_i^{(1)}}{4j+3m+5}+\frac{\overline{A}P_j^{(1)}}{4j+3m+3}+\frac{\overline{A}L_j^{(1)}}{4j+3m+1}
$$

$$
+\frac{\sum_{i=0}^{j}(K_i^{(1)}C_{j-i}^{(2)}+M_i^{(1)}A_{j-i}^{(2)}+N_i^{(1)}D_{j-i}^{(2)})}{2(2j+m+6)}
$$

$$+\frac{\sum_{i=0}^{j}(M_i^{(1)}F_{j-i}^{(2)}+N_i^{(1)}E_{j-i}^{(2)}+O_i^{(1)}D_{j-i}^{(2)}+P_i^{(1)}C_{j-i}^{(2)})}{2(2j+m+5)}$$

$$+\frac{\sum_{i=0}^{j}(L_i^{(1)}C_{j-i}^{(2)}+M_i^{(1)}B_{j-i}^{(2)}+O_i^{(1)}E_{j-i}^{(2)})}{2(2j+m+4)}$$

$$+\frac{\overline{A}N_j^{(1)}}{2(2j+m+3)}+\frac{\overline{A}O_j^{(1)}}{2(2j+m+2)}$$

$$+\frac{\sum_{i=0}^{j}(M_i^{(1)}D_{j-i}^{(2)}+N_i^{(1)}C_{j-i}^{(2)})}{4j+m+13}+\frac{\sum_{i=0}^{j}(M_i^{(1)}E_{j-i}^{(2)}+O_i^{(1)}C_{j-i}^{(2)})}{4j+m+11}$$

$$\left.+\frac{\overline{A}M_j^{(1)}}{4j+m+7}+\frac{\sum_{i=0}^{j}M_i^{(1)}C_{j-i}^{(2)}}{2(2j+7)}\right\} \tag{30}$$

于是，通过代数方程（13b），ξ 的渐进值被指定为相对于所给定值 W_0 的 ξ_1，并且由式（20a）得到

$$W_m=W_0(\xi_1+1) \tag{31}$$

为了找到非线性振动的固有频率 ω_{NL}，我们进一步求解以下改进的迭代方程和相应的边界条件

$$L_3(W_2)-\frac{\omega_{NL}^2}{(2\xi_1^2+1)}W_2=\frac{1}{x}\frac{\mathrm{d}}{\mathrm{d}x}\left[2xHN_2+f(\xi_1)\left(2xHT_1+N_1\frac{\mathrm{d}W_1}{\mathrm{d}x}\right)+g(\xi_1)T_1\frac{\mathrm{d}W_1}{\mathrm{d}x}\right],$$

$$L_4(x^mN_2)=24H(\nu_\theta^2-\beta)x\frac{\mathrm{d}W_2}{\mathrm{d}x} \tag{32a,b}$$

在 $x=0$ 时，$W_2=W_0$，$\frac{\mathrm{d}W_2}{\mathrm{d}x}=0$，

$$\lim_{x\to 0}\left(x\frac{\mathrm{d}^3W_2}{\mathrm{d}x^3}+\frac{\mathrm{d}^2W_2}{\mathrm{d}x^2}-\frac{\beta}{x}\frac{\mathrm{d}W_2}{\mathrm{d}x}\right)=0,N_2=0 \tag{33a ~ d}$$

在 $x=1$ 时，$W_2=0$，$\frac{\mathrm{d}^2W_2}{\mathrm{d}x^2}+\lambda\frac{\mathrm{d}W_2}{\mathrm{d}x}=0$，$\frac{\mathrm{d}N_2}{\mathrm{d}x}-\nu N_2=0$　(34a~c)

将解（24）、（27）和 ξ_1 的值代入方程（32a），应用条件（33）后，便得到方程（32）的精确解如下

$$N_2=W_0\left(\overline{\zeta}\sum_{j=0}^{\infty}A_j^{(3)}x^{4j+m}+\overline{\eta}\sum_{j=0}^{\infty}B_j^{(3)}x^{4j+m-2}+\sum_{j=0}^{\infty}C_j^{(3)}x^{4j+5}\right)$$

$$+W_0^2\left(\sum_{j=0}^{\infty}D_j^{(3)}x^{4j+m+4}+\sum_{j=0}^{\infty}E_j^{(3)}x^{4j+m+6}+\sum_{j=0}^{\infty}F_j^{(3)}x^{4j+2m+3}\right.$$

$$\left.+\sum_{j=0}^{\infty}G_j^{(3)}x^{4j+2m+5}+\sum_{j=0}^{\infty}H_j^{(3)}x^{4j+11}\right)+W_0^3\left(\sum_{j=0}^{\infty}I_j^{(3)}x^{4j+m+6}\right.$$

$$+\sum_{j=0}^{\infty}J_j^{(3)}x^{4j+m+12}+\sum_{j=0}^{\infty}K_j^{(3)}x^{4j+2m+3}+\sum_{j=0}^{\infty}L_j^{(3)}x^{4j+2m+5}$$

$$+\sum_{j=0}^{\infty}M_j^{(3)}x^{4j+3m+4}+\sum_{j=0}^{\infty}N_j^{(3)}x^{4j+3m+6}+\sum_{j=0}^{\infty}P_j^{(3)}x^{4j+13}\Bigg), \tag{35a}$$

$$W_2=W_0\left(\bar{\zeta}\sum_{j=0}^{\infty}\overline{A}_j^{(3)}x^{4j+m+3}+\bar{\eta}\sum_{j=0}^{\infty}\overline{B}_j^{(3)}x^{4j+m+1}+\sum_{j=0}^{\infty}\overline{C}_j^{(3)}x^{4j}\right)$$

$$+W_0^2\Bigg(\sum_{j=0}^{\infty}\overline{D}_j^{(3)}x^{4j+m+3}+\sum_{j=0}^{\infty}\overline{E}_j^{(3)}x^{4j+m+5}+\sum_{j=0}^{\infty}\overline{F}_j^{(3)}x^{4j+2m+2}$$

$$+\sum_{j=0}^{\infty}\overline{G}_j^{(3)}x^{4j+2m+4}+\sum_{j=0}^{\infty}\overline{H}_j^{(3)}x^{4j+10}\Bigg)+W_0^3\Bigg(\sum_{j=0}^{\infty}\overline{I}_j^{(3)}x^{4j+m+5}$$

$$+\sum_{j=0}^{\infty}\overline{J}_j^{(3)}x^{4j+m+11}+\sum_{j=0}^{\infty}\overline{K}_j^{(3)}x^{4j+2m+2}+\sum_{j=0}^{\infty}\overline{L}_j^{(3)}x^{4j+2m+4}$$

$$+\sum_{j=0}^{\infty}\overline{M}_j^{(3)}x^{4j+3m+3}+\sum_{j=0}^{\infty}\overline{N}_j^{(3)}x^{4j+3m+5}+\sum_{j=0}^{\infty}\overline{P}_j^{(3)}x^{4j+12}\Bigg) \tag{35b}$$

其中$\bar{\zeta}$和$\bar{\eta}$是待定系数，$A_j^{(3)}$，$\overline{A}_j^{(3)}$，…，$P_j^{(3)}$，$\overline{P}_j^{(3)}$ 类似于式（25），表达在 ω_{NL} 的有限幂级数中，为简单起见，这些系数列在附录中。

由式（35）的系数递推公式可知，仅 $\overline{A}_0^{(3)}$、$\overline{B}_0^{(3)}$ 和 $\overline{C}_0^{(3)}$ 是未知的。将式（35）代入边界条件（34a～c），便得一组代数方程

$$\begin{bmatrix}\alpha_{11} & \alpha_{12} & (\alpha_{13}+W_0\alpha_{14}+W_0^2\alpha_{15})\\ \alpha_{21} & \alpha_{22} & (\alpha_{23}+W_0\alpha_{24}+W_0^2\alpha_{25})\\ \alpha_{31} & \alpha_{32} & (\alpha_{33}+W_0\alpha_{34}+W_0^2\alpha_{35})\end{bmatrix}\begin{bmatrix}\bar{\xi}\\ \bar{\eta}\\ 1\end{bmatrix}=0 \tag{36}$$

其中矩阵元素 $\alpha_{ij}(i=1,2,3;j=1,2,3,4,5)$表达在 ω_{NL} 的幂级数中，为简洁起见，我们将它们给在附录中。

方程（36）中的向量 $[\bar{\xi},\ \bar{\eta},\ 1]^{\mathrm{T}}$ 的非零需要由方程（36）中的矩阵的行列式为零来确定，由此有

$$\sum_{j=0}^{\infty}a_j\omega_{NL}^{2j}+W_0\sum_{j=0}^{\infty}b_j\omega_{NL}^{2j}+W_0^2\sum_{j=0}^{\infty}c_j\omega_{NL}^{2j}=0 \tag{37}$$

这里的系数 a_j、b_j 和 c_j 容易由计算机确定。实际计算中，可取 $\overline{A}_0^{(3)}$、$\overline{B}_0^{(3)}$ 和 $\overline{C}_0^{(3)}$ 为 1。将式（31）代入方程（37），便可导出壳体振动的振幅—频率响应关系式（即 W_m-ω_{NL}/ω_L 关系式）。将式（29）代入方程（13b），便可导出壳体振动的 W_0 和非对称的内禀量 ξ 的关系式。最后，根据式（31）得到壳体振动的振幅和非对称的内禀量 ξ 的关系。

四、结果和讨论

我们首先研究 Sinharay 等[6]所处理的一种情况，即具有滑动固定和夹紧固定外边界的各向同性扁薄球壳的振动问题。数值结果表明，本文结果与 Sinharay 等结果大部分是一致的（图 2），两个结果的差异随着具有较大几何参数值（$H=1$）的壳体的振幅 W_m 的增加而增加。如前面引言中所述，Sinharay 等的非耦合方程存在缺陷，故不能期

望更满意的结果。

对于固定和铰支两种边界条件下壳体振动的数值结果已给在图 3～图 6 中。显而易见，几何参数 H 对振动响应行为有很大影响，它可从变硬型到变软型来改变非线性。随着振幅 W_m 的增加，变软型的非线性将返回到变硬型，并且初始变软的程度先随着几何参数的增加而增加，然后随着几何参数的增加而减小（图 5）。正交异性参数 β 对振动的影响与几何参数和振幅是相应的，对于具有较小几何参数和较小振幅的壳体来说，影响是轻微的，而对于具有较大几何参数和振幅的壳体而言，影响则很大。在影响较大时，壳体振动的非线性和初始变软则随着参数 β 的增加而减小。还需要指出，描述振幅非对称的内禀量 ξ 先随着几何参数的增加而增加，最终随着几何参数的增加而减小（图 6），而且也随着各向异性参数的增加而减小。ξ 值先随着振幅的增加而增加，但是随后在达到最大值以后就衰微了。

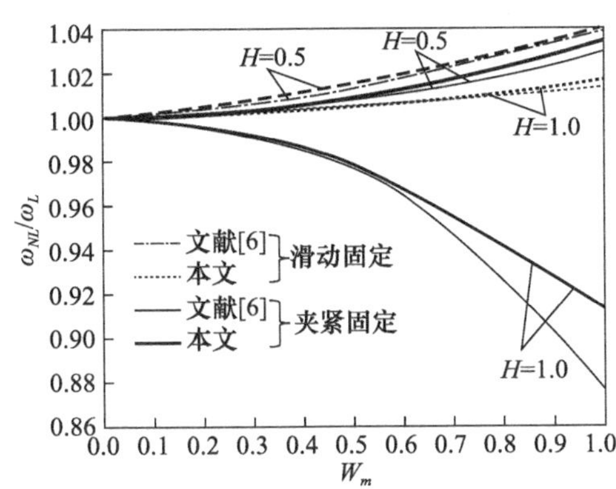

图 2　固定边各向同性扁薄球壳理论结果间的比较（$\beta=1$，$\nu_\theta=0.3$）

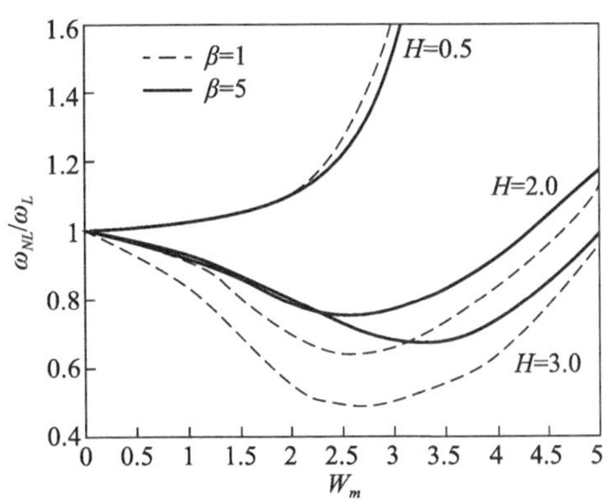

图 3　几何参数和正交异性参数对固定边球壳振频的影响（$\nu_\theta=0.3$）

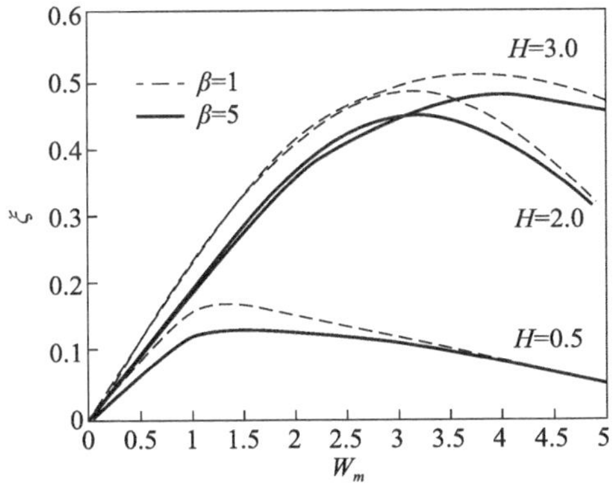

图 4　几何参数和正交异性参数对固定边球壳振幅的非对称性影响（$\nu_\theta=0.3$）

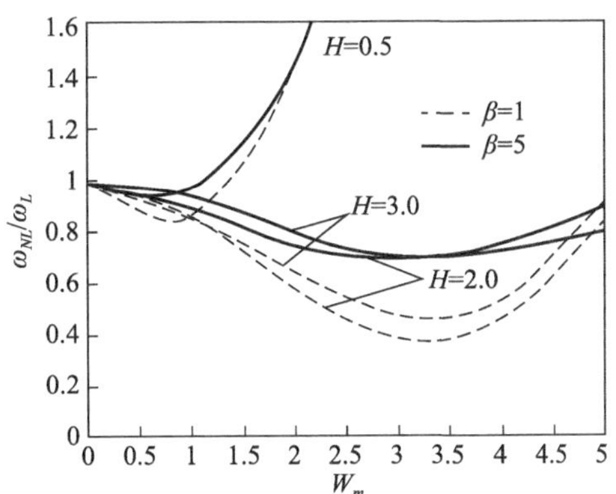

图 5　几何参数和正交异性参数对铰支边球壳振动频率的影响（$\nu_\theta=0.3$）

相应地可以断定，具有较小几何参数 H 值的壳体的非线性振动将类似于圆板，它们的振动频率将随着振幅的增加而增加，而且振幅的非对称程度很轻微。对于具有几何参数较大值的壳体，它们的振动频率先随着振幅的增加而减小，而且其振幅的初始非对称程度随着振幅的增加而变得更大。随后，频率与振幅回到不变的状态，而且振幅的非对称性衰退，再次显出平板振动的行为。还要指出一种有趣的现象，即振幅的非对称性将随着非线性的最初变软的程度来增加和减小。

最后，我们讨论四种常用的边界条件对扁球壳的影响，结果给在图 7 中。可以指出，铰支边壳体的振动具有最强的非线性和相应的初始变软，滑动固定边壳体的振动的非线性和相应的初始变软则是最弱的。至于具有最小几何参数值的壳体，固定支承下的振动较简支情况下显示了更强的非线性。另外，一般来说，固定支承下壳体的振动较简支情况下显示了更强的非线性和初始的变软。

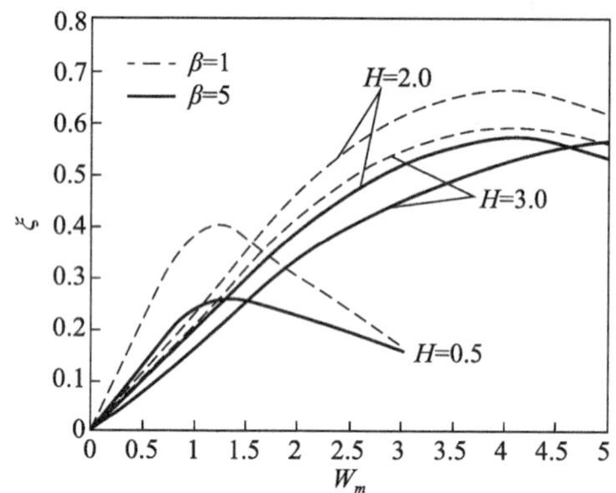

图 6　几何参数和正交异性参数对铰支边球壳振幅的非对称性影响（ν_θ=0.3）

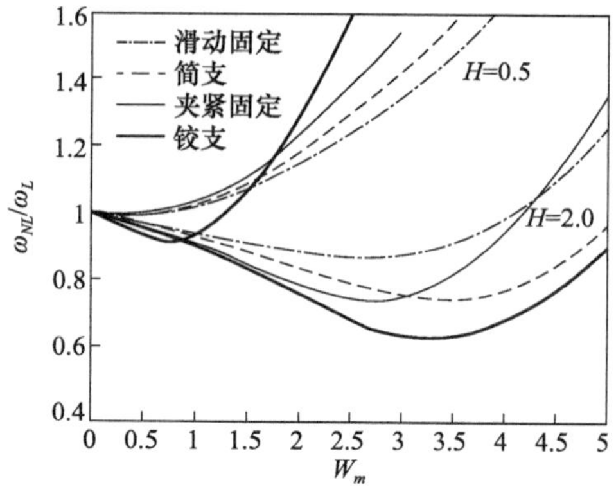

图 7　不同边界条件对扁球壳振动频率的影响（ν_θ=0.3，β=3）

参考文献

[1] Singh P N, Sundararajan V, Das Y C. Large amplitude axisymmetric vibrations of moderately thick spherical caps. J. Sound Vib., 1972, 20 (3): 269-276.

[2] Ramachandran J. Vibration of shallow spherical shells at large amplitudes. J. Appl. Mech., 1974, 41 (3): 811-812.

[3] Ramachandran J. Large amplitude vibrations of shallow spherical shell with concentrated mass. J. Appl. Mech., 1976, 43: 363-365.

[4] Yasuda K, Kushida G. Nonlinear forced oscillations of a shallow spherical shell. Bull. JSME, 1984, 27 (232): 2233-2240.

[5] Dumir P C. Nonlinear axisymmetric response of orthotropic thin truncated conical and spherical caps. Acta. Mech., 1986, 60: 121-132.

[6] Sinharay G C., Banerjee B. Large amplitude free vibration of shallow spherical and cylindrical shell—A new approach. Inter. J. Non-Linear Mech., 1985, 20: 69-78.

[7] 刘人怀，王璠. 复合材料层合扁球壳的非线性强迫振动. 力学学报，1997，29 (2): 236-241.

[8] 刘人怀，王璠，肖潭. 阻尼对层合复合材料中心开孔扁球壳非线性振动的影响. 应用力学研究与实践，刘人怀主编，广州：暨南大学出版社，2000，35-43.

[9] Liu Renhuai. Study on Nonlinear Mechanics of Plates and Shells. New York, Beijing and Guan-

gzhou: Science Press and Jinan University Press, 1998.

[10] Li Dong and Liu Renhuai. Application of the modified iteration method to nonlinear vibration of corrugated circular plates. Appl. Math. Mech., 1990, 11 (1): 13-22.

[11] Li Dong and Liu Renhuai. Nonlinear vibration of thin circular plates. Advances in Applied Mathematics and Mechanics in China. Vol. 3 edited by Chien Wei-zang, Oxford: International Academic Publishers and Pergamon Press, 1991, 15-23.

[12] Li Dong. Large amplitude vibration of thin annular plates. Appl. Math. Mech., 1991, 12 (6): 583-593.

[13] Liu Renhuai and Li dong. A new approach to nonlinear vibration of orthotropic thin circular plates. Proceeding of the International Conference on Vibration Engineering, Shenyang: Northeastern University Press, 1998: 248-252.

[14] 叶开沅，刘人怀，平庆元，等. 在对称线布荷载作用下的圆底扁薄球壳的非线性稳定性. 科学通报，1965，(2)：142-145.

[15] 叶开沅，刘人怀，张传智，等. 圆底薄球壳在边缘力矩作用下的非线性稳定问题. 科学通报，1965，(2)：145-147.

[16] 刘人怀，在内边缘荷载作用下中心开孔圆底扁球壳的非线性稳定问题. 科学通报，1965，(3)：253-255.

[17] Nayfeh A H, Mook D T. Nonlinear Oscillations. New York: John Wiley & Sons Inc., 1979.

附　录

为使用本文公式方便起见，我们将式（35）的系数公式给在下面：

$$A_{j+1}^{(3)}=\frac{3H(\nu_0^2-\beta)(4j+m+3)}{(2j+m+2)(j+1)}\overline{A}_j^{(3)},$$

$$B_j^{(3)}=\frac{6H(\nu_0^2-\beta)(4j+m+1)}{(2j+m+1)(2j+1)}\overline{B}_j^{(3)},$$

$$C_j^{(3)}=\frac{96H(\nu_0^2-\beta)(j+1)}{(4j+5)^2-m^2}\overline{C}_{j+1}^{(3)},$$

$$D_j^{(3)}=\frac{3H(\nu_0^2-\beta)(4j+m+3)}{(2j+m+2)(j+1)}\overline{D}_j^{(3)},$$

$$E_j^{(3)}=\frac{6H(\nu_0^2-\beta)(4j+m+5)}{(2j+m+3)(2j+3)}\overline{E}_j^{(3)},$$

$$F_j^{(3)}=\frac{48H(\nu_0^2-\beta)(2j+m+1)}{(4j+3m+3)(4j+m+3)}\overline{F}_j^{(3)},$$

$$G_j^{(3)}=\frac{48H(\nu_0^2-\beta)(2j+m+2)}{(4j+3m+5)(4j+m+5)}\overline{G}_j^{(3)},$$

$$H_j^{(3)}=\frac{48H(\nu_0^2-\beta)(2j+5)}{(4j+11)^2-m^2}\overline{H}_j^{(3)},$$

$$I_j^{(3)}=\frac{6H(\nu_0^2-\beta)(4j+m+5)}{(2j+m+3)(2j+3)}\overline{I}_j^{(3)},$$

$$J_j^{(3)}=\frac{3H(\nu_0^2-\beta)(4j+m+11)}{(2j+m+6)(j+3)}\overline{J}_j^{(3)},$$

$$K_j^{(3)}=\frac{48H(\nu_0^2-\beta)(2j+m+1)}{(4j+3m+3)(4j+m+3)}\overline{K}_j^{(3)},$$

$$L_j^{(3)}=\frac{48H(\nu_0^2-\beta)(2j+m+2)}{(4j+3m+5)(4j+m+5)}\overline{L}_j^{(3)},$$

$$M_j^{(3)} = \frac{3H(\nu_\theta^2 - \beta)(4j+3m+3)}{(2j+m+2)(j+m+1)}\overline{M}_j^{(3)},$$

$$N_j^{(3)} = \frac{6H(\nu_\theta^2 - \beta)(4j+3m+5)}{(2j+2m+3)(2j+m+3)}\overline{N}_j^{(3)},$$

$$P_j^{(3)} = \frac{96H(\nu_\theta^2 - \beta)(j+3)}{(4j+13)^2 - m^2}\overline{P}_j^{(3)},$$

$$\overline{A}_{j+1}^{(3)} = \frac{\mu\omega_{NL}^2 \overline{A}_j^{(3)} + 2H(4j+m+5)A_{j+1}^{(3)}}{4(2j+m+3)(4j+m+7)(4j+m+5)(2j+3)},$$

$$\overline{A}_0^{(3)} = \frac{H}{2(m+3)(m+1)}A_0^{(3)},$$

$$\overline{B}_{j+1}^{(3)} = \frac{\mu\omega_{NL}^2 \overline{B}_j^{(3)} + 2H(4j+m+3)B_j^{(3)}}{8(2j+m+2)(4j+m+5)(4j+m+3)(j+1)},$$

$$\overline{C}_{j+2}^{(3)} = \frac{\mu\omega_{NL}^2 \overline{C}_{j+1}^{(3)} + 4H(2j+3)C_j^{(3)}}{8[(4j+7)^2 - m^2](2j+3)(j+2)},$$

$$\overline{C}_1^{(3)} = -\frac{\mu\omega_{NL}^2}{8(m^2-9)}\overline{C}_0^{(3)},$$

$$\overline{D}_{j+1}^{(3)} = \frac{1}{4(2j+m+3)(4j+m+7)(4j+m+5)(2j+3)}\Big\{\mu\omega_{NL}^2 \overline{D}_j^{(3)} + 2H(4j+m+5)[D_j^3 + f(\xi_1)E_j^{(2)}] + \eta f(\xi_1)(4j+m+5) \cdot \Big[\sum_{i+n=j+1} 4nB_iF_n + \sum_{i+n=j}(4n+m+1)C_iE_n\Big]\Big\},$$

$$\overline{D}_0^{(3)} = \frac{Hf(\xi_1)}{2(m+3)(m+1)}\overline{A},$$

$$\overline{E}_{j+1}^{(3)} = \frac{1}{8(2j+m+4)(4j+m+9)(4j+m+7)(j+2)}\Big\{\mu\omega_{NL}^2 \overline{E}_j^{(3)} + 2H(4j+m+7)E_j^{(3)} + f(\xi_1)(4j+m+7)\Big[2HD_j^{(2)} + \zeta\sum_{i+n=j}(4n+m+3)C_iD_n + \zeta\sum_{i+n=j+2}4nA_iF_n\Big]\Big\},$$

$$\overline{E}_0^{(3)} = \frac{\zeta f(\xi_1)}{2(m+5)(m+2)}A_0F_1,$$

$$\overline{F}_{j+1}^{(3)} = \frac{1}{4(2j+m+3)(2j+m+2)(4j+3m+5)(4j+m+5)}\Big\{\mu\omega_{NL}^2 \overline{F}_j^{(3)} + 4H(2j+m+2)[F_j^{(3)} + f(\zeta_1)F_j^{(2)}] + 2\eta\zeta f(\xi_1)(2j+m+2)\Big[\sum_{i+n=j+1}(4n+m+1)A_iE_n + \sum_{i+n=j}(4n+m+3)B_iD_n\Big]\Big\},$$

$$\overline{F}_0^{(3)} = \frac{\eta\zeta f(\xi_1)}{2(3m+1)(m+1)}A_0E_0,$$

$$\overline{G}_{j+1}^{(3)} = \frac{1}{4(2j+m+4)(2j+m+3)(4j+3m+7)(4j+m+7)}\Big\{\mu\omega_{NL}^2 \overline{G}_j^{(3)} + 4H(2j+m+3)G_j^{(3)} + 4Hf(\xi_1)(2j+m+3) \cdot (A_j^{(2)} + B_{j+1}^{(2)}) + 2f(\xi_1)(2j+m+3)\sum_{i+n=j+1}\Big[\zeta^2(4n+m+3)A_iD_n + \eta^2(4n+m+1)B_iE_n\Big]\Big\},$$

$$\overline{G}_0^{(3)}=\frac{1}{6(m+3)(m+2)(m+1)}\Big\{2Hf(\xi_1)B_0^{(2)}$$

$$+\zeta^2 f(\xi_1)(m+3)A_0D_0+\eta^2 f(\xi_1)(m+1)B_0E_0\Big\},$$

$$\overline{H}_{j+1}^{(3)}=\frac{1}{8[(4j+13)^2-m^2](2j+7)(j+3)}\Big\{\mu\omega_{\mathrm{NL}}^2\overline{H}_j^{(3)}+8H(j+3)H_j^{(3)}$$

$$+8Hf(\xi_1)(j+3)C_{j+1}^{(2)}+16f(\xi_1)\sum_{i+n=j+2}n(j+3)C_iF_n\Big\},$$

$$\overline{H}_0^{(3)}=-\frac{1}{5(m^2-81)}[Hf(\xi_1)C_0^{(2)}+2C_0F_1],$$

$$\overline{I}_{j+1}^{(3)}=\frac{1}{8(2j+m+4)(4j+m+9)(4j+m+7)(j+2)}\Big\{\mu\omega_{\mathrm{NL}}^2\overline{I}_j^{(3)}$$

$$+2H(4j+m+7)I_j^{(3)}+g(\xi_1)(4j+m+7)\Big[4\overline{A}(j+2)F_{j+2}$$

$$+\eta\sum_{i+n=j}(4n+m+1)C_i^{(2)}E_n+\sum_{i+n=j+1}4nE_i^{(2)}F_n\Big]\Big\},$$

$$\overline{I}_0^{(3)}=\frac{\overline{A}g(\xi_1)}{2(m+5)(m+2)}F_1,$$

$$\overline{J}_{j+1}^{(3)}=\frac{1}{4(2j+m+7)(4j+m+15)(4j+m+13)(2j+7)}\Big\{\mu\omega_{\mathrm{NL}}^2\overline{J}_j^{(3)}$$

$$+(4j+m+13)\Big[2HJ_j^{(3)}+\zeta g(\xi_1)\sum_{i+n=j+1}(4n$$

$$+m+3)C_i^{(2)}D_n+4g(\xi_1)\sum_{i+n=j+2}nD_i^{(2)}F_n\Big]\Big\},$$

$$\overline{J}_0^{(3)}=\frac{g(\xi_1)}{20(m+11)(m+5)}[\zeta(m+3)C_0^{(2)}D_0+4D_0^{(2)}F_1],$$

$$\overline{K}_{j+2}^{(3)}=\frac{1}{4(2j+m+5)(2j+m+4)(4j+3m+9)(4j+m+9)}\Big\{\mu\omega_{\mathrm{NL}}^2\overline{K}_{j+1}^{(3)}$$

$$+4H(2j+m+4)K_{j+1}^{(3)}+g(\xi_1)(2j+m+4)\Big[2\eta\overline{A}(4j+m+9)E_{j+2}$$

$$+2\zeta\sum_{i+n=j}(4n+m+3)D_i^{(2)}D_n+2\sum_{i+n=j+1}\Big((4n+m+1)\eta E_i^{(2)}E_n$$

$$+4nA_i^{(2)}F_n\Big)+8\sum_{i+n=j+2}nB_i^{(2)}F_n\Big]\Big\},$$

$$\overline{K}_0^{(3)}=\frac{\eta\overline{A}g(\xi_1)}{2(3m+1)(m+1)}E_0,$$

$$\overline{K}_1^{(3)}=\frac{1}{4(3m+5)(m+5)(m+3)(m+2)}\Big\{\mu\omega_{\mathrm{NL}}^2\overline{K}_0^{(3)}$$

$$+4H(m+2)K_0^{(3)}+2g(\xi_1)(m+2)\Big[\eta\overline{A}(m+5)E_1$$

$$+4B_0^{(2)}F_1+\eta(m+1)E_0^{(2)}E_0\Big]\Big\},$$

$$\overline{L}_{j+1}^{(3)}=\frac{1}{4(2j+m+4)(2j+m+3)(4j+3m+7)(4j+m+7)}\Big\{\mu\omega_{\mathrm{NL}}^2\overline{L}_j^{(3)}$$

$$+4H(2j+m+3)L_j^{(3)}+2g(\xi_1)(2j+m+3)\Big[\zeta\overline{A}(4j+m+7)D_{j+1}$$

$$+\sum_{i+n=j}\Big(\zeta(4n+m+3)E_i^{(2)}D_n+\eta(4n+m+1)D_i^{(2)}E_n\Big)+\sum_{i+n=j+1}4nF_i^{(2)}F_n\Big]\Big\},$$

$$\overline{L}_0^{(3)}=\frac{\zeta\overline{A}g(\xi_1)}{6(m+2)(m+1)}D_0,$$

$$\overline{M}_{j+1}^{(3)}=\frac{1}{4(2j+2m+3)(2j+m+3)(4j+3m+7)(4j+3m+5)}\Big\{\mu\omega_{NL}^2\overline{M}_j^{(3)}+2H(4j+3m+5)M_j^{(3)}+g(\xi_1)(4j+3m+5)\Big[\sum_{i+n=j}\Big(\zeta(4n+m+3)F_i^{(2)}D_n+\eta(4n+m+1)A_i^{(2)}E_n\Big)+\eta\sum_{i+n=j+1}(4n+m+1)B_i^{(2)}E_n\Big]\Big\},$$

$$\overline{M}_0^{(3)}=\frac{\eta g(\xi_1)}{12(2m+1)(m+1)}B_0^{(2)}E_0,$$

$$\overline{N}_{j+1}^{(3)}=\frac{1}{8(j+m+2)(2j+m+4)(4j+3m+9)(4j+3m+7)}\Big\{\mu\omega_{NL}^2\overline{N}_j^{(3)}+2H(4j+3m+7)N_j^{(3)}+g(\xi_1)(4j+3m+7)\Big[\zeta\sum_{i+n=j}(4n+m+3)A_i^{(2)}D_n+\sum_{i+n=j+1}\Big(\zeta(4n+m+3)B_i^{(2)}D_n+\eta(4n+m+1)F_i^{(2)}E_n\Big)\Big]\Big\},$$

$$\overline{N}_0^{(3)}=\frac{g(\xi_1)}{8(3m+5)(m+2)(m+1)}\big[\zeta(m+3)B_0^{(2)}D_0+\eta(m+1)F_0^{(2)}E_0\big],$$

$$\overline{P}_{j+1}^{(3)}=\frac{1}{8[(4j+15)^2-m^2](2j+7)(j+4)}\Big\{\mu\omega_{NL}^2\overline{P}_j^{(3)}+4H(2j+7)P_j^{(3)}+8g(\xi_1)\sum_{i+n=j+2}n(2j+7)C_i^{(2)}F_n\Big\},$$

$$\overline{P}_0^{(3)}=-\frac{g(\xi_1)}{3(m^2-121)}C_0^{(2)}F_1,$$

$$\mu=\frac{1}{2\xi_1^2+1}\tag{附 1}$$

最后，再将式（36）的矩阵元素的公式给在下面：

$$\alpha_{11}=\sum_{j=0}^{\infty}\overline{A}_j^{(3)},$$

$$\alpha_{12}=\sum_{j=0}^{\infty}\overline{B}_j^{(3)},$$

$$\alpha_{13}=\sum_{j=0}^{\infty}\overline{C}_j^{(3)},$$

$$\alpha_{14}=\sum_{j=0}^{\infty}\Big(\overline{D}_j^{(3)}+\overline{E}_j^{(3)}+\overline{F}_j^{(3)}+\overline{G}_j^{(3)}+\overline{H}_j^{(3)}\Big),$$

$$\alpha_{15}=\sum_{j=0}^{\infty}(\overline{I}_j^{(3)}+\overline{J}_j^{(3)}+\overline{K}_j^{(3)}+\overline{L}_j^{(3)}+\overline{M}_j^{(3)}+\overline{N}_j^{(3)}+\overline{P}_j^{(3)}),$$

$$\alpha_{21}=\sum_{j=0}^{\infty}\overline{A}_j^{(3)}(4j+m+3)(4j+m+2+\lambda),$$

$$\alpha_{22}=\sum_{j=0}^{\infty}\overline{B}_j^{(3)}(4j+m+1)(4j+m+\lambda),$$

$$\alpha_{23}=\sum_{j=0}^{\infty}\overline{C}_j^{(3)}4j(4j-1+\lambda),$$

$$\alpha_{24}=\sum_{j=0}^{\infty}\overline{D}_j^{(3)}(4j+m+3)(4j+m+2+\lambda)$$

$$
\begin{aligned}
&+\overline{E}_j^{(3)}(4j+m+5)(4j+m+4+\lambda)\\
&+2\overline{F}_j^{(3)}(2j+m+1)(4j+2m+1+\lambda)\\
&+2\overline{G}_j^{(3)}(2j+m+1)(4j+2m+3+\lambda)\\
&+2\overline{H}_j^{(3)}(2j+5)(4j+9+\lambda),
\end{aligned}
$$

$$
\begin{aligned}
\alpha_{25}=\sum_{j=0}^{\infty}\Big[&\overline{I}_j^{(3)}(4j+m+5)(4j+m+4+\lambda)\\
&+\overline{J}_j^{(3)}(4j+m+1)(4j+m+\lambda)\\
&+2\overline{K}_j^{(3)}(2j+m+1)(4j+2m+1+\lambda)\\
&+2\overline{L}_j^{(3)}(2j+m+2)(4j+2m+3+\lambda)\\
&+\overline{M}_j^{(3)}(4j+3m+3)(4j+3m+2+\lambda)\\
&+\overline{N}_j^{(3)}(4j+3m+5)(4j+3m+4+\lambda)\\
&+4\overline{P}_j^{(3)}(j+3)(4j+11+\lambda)\Big],
\end{aligned}
$$

$$\alpha_{31}=\sum_{j=0}^{\infty}A_j^{(3)}(4j+m-\nu),$$

$$\alpha_{32}=\sum_{j=0}^{\infty}B_j^{(3)}(4j+m-2-\nu),$$

$$\alpha_{33}=\sum_{j=0}^{\infty}C_j^{(3)}(4j+5-\nu),$$

$$
\begin{aligned}
\alpha_{34}=\sum_{j=0}^{\infty}\Big[&D_j^{(3)}(4j+m+4-\nu)+E_j^{(3)}(4j+m+6-\nu)\\
&+F_j^{(3)}(4j+2m+3-\nu)+G_j^{(3)}(4j+m+5-\nu)\\
&+H_j^{(3)}(4j+11-\nu)\Big],
\end{aligned}
$$

$$
\begin{aligned}
\alpha_{35}=\sum_{j=0}^{\infty}\Big[&H_j^{(3)}(4j+m+6-\nu)+J_j^{(3)}(4j+m+12-\nu)\\
&+K_j^{(3)}(4j+2m+3-\nu)+L_j^{(3)}(4j+2m+5-\nu)\\
&+M_j^{(3)}(4j+3m+4-\nu)+N_j^{(3)}(4j+3m+6-\nu)\\
&+P_j^{(3)}(4j+13-\nu)\Big]
\end{aligned}
\qquad(附 2)
$$

第三章　波纹板壳非线性力学

波纹圆板的特征关系式[①]

一、前言

波纹圆板在精密仪器的灵敏弹性元件中起着很重要的作用。然而，这种弹性元件的理论研究至今还很不充分。造成这种状况的主要原因是波纹圆板本身形状复杂，参数很多，特别是大挠度非线性微分方程组在数学上求解极困难。以往的理论计算大致可分为两类。第一类是采用壳体的大挠度方程来求解，如文献［1］用 Галёркин 方法求得了浅正弦截面波纹圆板的特征关系式。其优点是能探讨波纹圆板的跳跃现象。缺点是仅讨论了浅正弦波纹一种形状，而且所涉及的挠度范围很小（厚度的 1～5 倍范围内）。此后，文献［2］用同样方法处理了较深波纹圆板的大挠度问题。第二类是将波纹圆板看成是结构上的各向异性圆平板，采用各向异性圆板大挠度方程来进行研究，如文献［3］用 Галёркин 方法获得了适应各种波纹圆板的特征关系式。其优点是能同时讨论各种形状的波纹圆板，涉及的挠度范围较大（厚度的 10 倍左右）。缺点是解法本身限制了结果的精确度和适应范围。文献［4］也用能量法处理了正弦截面波纹圆板，获得了类似的结果。

本文注意到第二类方法的优点和不足之处，期望从非线性微分方程的求解上来获得更好的结果，以适应工程设计的需要。我们使用各向异性圆板的大挠度方程，并采用修正迭代法[5]来求解。需要指出，这一解法只是改动了摄动法的程序，其结果与摄动法是相同的。本文所得到的波纹圆板的特征关系式简便适用，提高了精确度和适用范围，可供精密仪器弹性元件设计时参考。

二、边值问题的建立与求解

波纹圆板在径向和周向有很不相同的刚度，这从结构上讲，类似于各向异性圆平板。所以我们以一各向异性圆板来代替波纹圆板，这板在径向和周向的刚度对应于波纹圆板的刚度，而厚度、半径及材料是相同的。在此替换下，半径为 R，厚度为 h，承受均布载荷 q 的各向异性圆板（即波纹圆板）的中面挠度 w 和径向薄膜力 T_1 所满足的大挠度方程为[3]

$$r\frac{\mathrm{d}^3w}{\mathrm{d}r^3}+\frac{\mathrm{d}^2w}{\mathrm{d}r^2}-\frac{k_2k_2'}{r}\frac{\mathrm{d}w}{\mathrm{d}r}=\frac{k_2}{D}\left(T_1r\frac{\mathrm{d}w}{\mathrm{d}r}+\frac{1}{2}qr^2\right),$$

---

① 本文原载《力学学报》，1978，(1)：47-52.

$$r\frac{\mathrm{d}^2(T_1 r)}{\mathrm{d}r^2}+\frac{\mathrm{d}(T_1 r)}{\mathrm{d}r}-k_1k_2T_1=-\frac{1}{2}Ehk_2\left(\frac{\mathrm{d}w}{\mathrm{d}r}\right)^2 \tag{1}$$

我们讨论两种常用的边界条件：

1）夹紧固定情况

当 $r=0$ 时，$\frac{\mathrm{d}w}{\mathrm{d}r}=0$，$T_1$ 有限

当 $r=R$ 时，$w=0$，$\frac{\mathrm{d}w}{\mathrm{d}r}=0$，$u=0$ (2)

2）滑动固定情况

当 $r=0$ 时，$\frac{\mathrm{d}w}{\mathrm{d}r}=0$，$T_1$ 有限

当 $r=R$ 时，$w=0$，$\frac{\mathrm{d}w}{\mathrm{d}r}=0$，$T_1=0$ (3)

这里 r 是板的径向坐标，E 是弹性模量，ν 是泊松比，k_1,k_2,k_2'是与径向和周向刚度有关的参数[3]，抗弯刚度 $D=\frac{Eh^3}{12\ (1-\nu^2/k_2k_2')}$，径向位移 $u=\frac{r}{k_2Eh}\left[\frac{\mathrm{d}(T_1r)}{\mathrm{d}r}-\nu T_1\right]$。

为简化以后的计算，我们引入下列无量纲量：

$$\rho=\frac{r}{R},y=\frac{w}{h},\ \varphi=\frac{\mathrm{d}y}{\mathrm{d}\rho},S=-\frac{k_2Rr}{D}T_1,$$

$$P=\frac{k_2R^4}{2Dh}q,\beta_1^2=k_2k_2',\beta_2^2=k_1k_2,\beta_3=\frac{Eh^3k_2^2}{2D} \tag{4}$$

将式 (4) 代入式 (1)～(3)，便得波纹圆板大挠度的无量纲非线性的基本方程和边界条件：

$$\frac{\mathrm{d}^2\varphi}{\mathrm{d}\rho^2}+\frac{1}{\rho}\frac{\mathrm{d}\varphi}{\mathrm{d}\rho}-\frac{\beta_1^2}{\rho^2}\varphi=-\frac{1}{\rho}S\varphi+P\rho,$$

$$\frac{\mathrm{d}^2S}{\mathrm{d}\rho^2}+\frac{1}{\rho}\frac{\mathrm{d}S}{\mathrm{d}\rho}-\frac{\beta_2^2}{\rho^2}S=\frac{\beta_3}{\rho}\varphi^2 \tag{5a,b}$$

1）夹紧固定情况

当 $\rho=0$ 时，$\varphi=0$，$S=0$

当 $\rho=1$ 时，$y=0$，$\varphi=0$，$\frac{\mathrm{d}S}{\mathrm{d}\rho}-\frac{\nu}{\rho}S=0$ (6)

2）滑动固定情况

当 $\rho=0$ 时，$\varphi=0$，$S=0$

当 $\rho=1$ 时，$y=0$，$\varphi=0$，$S=0$ (7)

首先，我们求解夹紧固定下的边值问题 (5) 和 (6)。在一次近似中，略去方程 (5a) 右端含 S 的非线性项，并以一次近似解 φ_1 代替式 (5b) 右端的 φ，再应用边界条件 (6)，便得下列线性边值问题

$$\frac{\mathrm{d}^2\varphi_1}{\mathrm{d}\rho^2}+\frac{1}{\rho}\frac{\mathrm{d}\varphi_1}{\mathrm{d}\rho}-\frac{\beta_1^2}{\rho^2}\varphi_1=P\rho,$$

$$\frac{\mathrm{d}^2S_1}{\mathrm{d}\rho^2}+\frac{1}{\rho}\frac{\mathrm{d}S_1}{\mathrm{d}\rho}-\frac{\beta_2^2}{\rho^2}S_1=\frac{\beta_3}{\rho}\varphi_1^2 \tag{8a,b}$$

当 $\rho=0$ 时，$\varphi_1=0$，$S_1=0$ (9a，b)

当 $\rho=1$ 时，$y_1=0$，$\varphi_1=0$，$\frac{\mathrm{d}s_1}{\mathrm{d}\rho}-\frac{\nu}{\rho}S_1=0$ (10a～c)

在应用式（9a)，(10b）后，方程（8a）的通解为

$$\varphi_1=\frac{P}{\beta_1^2-9}(\rho^{\beta_1}-\rho^3) \tag{11}$$

由式（11)，(10a)，还可得无量纲挠度

$$y_1=\frac{P}{\beta_1+3}\left[\frac{1}{\beta_1-3}\left(\frac{1}{\beta_1+1}\rho^{\beta_1+1}-\frac{1}{4}\rho^4\right)+\frac{1}{4(\beta_1+1)}\right] \tag{12}$$

令 y_0 为板中心的无量纲挠度，即

当 $\rho=0$ 时，$y=y_0$ (13)

应用此关系，在式（12）中令 $\rho=0$，便得

$$P=4(\beta_1+1)(\beta_1+3)y_0 \tag{14}$$

此即波纹圆板的小挠度特征关系式。将此式代入式（11)，就有

$$\varphi_1=m_1(\rho^{\beta_1}-\rho^3)y_0 \tag{15}$$

其中

$$m_1=4(\beta_1+1)/(\beta_1-3)$$

我们再求解方程（8b)。将式（15）代入方程（8b）的右端，并应用式（9b)，(10c)，得到方程（8b）的解为

$$S_1=\left\{m_2\rho^{\beta_2}+\beta_3m_1^2\left[\frac{1}{(2\beta_1+1)^2-\beta_2^2}\rho^{2\beta_1+1}\right.\right.$$
$$\left.\left.-\frac{2}{(\beta_1+4)^2-\beta_2^2}\rho^{\beta_1+4}-\frac{1}{\beta_2^2-49}\rho^7\right]\right\}y_0^2 \tag{16}$$

其中

$$m_2=\frac{\beta_3m_1^2}{\beta_2-\nu}\left[\frac{7-\nu}{\beta_2^2-49}+\frac{2(\beta_1+4-\nu)}{(\beta_1+4)^2-\beta_2^2}-\frac{2\beta_1+1-\nu}{(2\beta_1+1)^2-\beta_2^2}\right]$$

现在求二次近似解。由于我们只进行到第二次近似，故只求解关于 φ_2 的边值问题就可以了。将已得的式（15)、(16）形式地代入方程（5a）的右端，并应用式（6)，便得关于 φ_2 的线性边值问题：

$$\frac{\mathrm{d}^2\varphi_2}{\mathrm{d}\rho^2}+\frac{1}{\rho}\frac{\mathrm{d}\varphi_2}{\mathrm{d}\rho}-\frac{\beta_1^2}{\rho_2}\varphi_2=-\frac{1}{\rho}S_1\varphi_1+P\rho \tag{17}$$

当 $\rho=0$ 时，$\varphi_2=0$ (18)

当 $\rho=1$ 时，$y_2=0$，$\varphi_2=0$ (19a，b)

应用式（18)，(19b)，方程（17）的解为

$$\varphi_2=\frac{P}{\beta_1^2-9}(\rho^{\beta_1}-\rho^3)+\left\{(\beta_1+1)(m_3+m_4)\rho^{\beta_1}\right.$$
$$-m_1m_2\left[\frac{1}{(\beta_2+1)(2\beta_1+\beta_2+1)}\rho^{\beta_1+\beta_2+1}+\frac{1}{\beta_1^2-(\beta_2+4)^2}\rho^{\beta_2+4}\right]$$
$$+\frac{\beta_3m_1^3}{2}\left[\frac{1}{(2\beta_1+1)^2-\beta_2^2}\left(\frac{2}{(\beta_1+5)(3\beta_1+5)}\rho^{2\beta_1+5}\right.\right.$$

$$
\begin{aligned}
&-\frac{1}{2(\beta_1+1)(2\beta_1+1)}\rho^{3\beta_1+2}\Big) \\
&+\frac{2}{(\beta_1+4)^2-\beta_2^2}\left(\frac{2}{(\beta_1+5)(3\beta_1+5)}\rho^{2\beta_1+5}-\frac{1}{8(\beta_1+4)}\rho^{\beta_1+8}\right) \\
&+\frac{1}{\beta_1^2-49}\left(\frac{1}{8(\beta_1+4)}\rho^{\beta_1+8}+\frac{2}{\beta_1^2-121}\rho^{11}\right)\Big]\Big\}y_0^3 \qquad (20)
\end{aligned}
$$

其中

$$
\begin{aligned}
m_3&=\frac{(\beta_1-3)(\beta_1+2\beta_2+5)m_1m_2}{(\beta_1+1)(\beta_2+1)(2\beta_1+\beta_2+1)[\beta_1^2-(\beta_2+4)^2]}, \\
m_4&=\frac{\beta_3m_1^3}{4(\beta_1+1)}\left\{\frac{3\beta_1^2+4\beta_1-39}{2(\beta_1+4)(\beta_1+5)(3\beta_1+5)[(\beta_1+4)^2-\beta_2^2]}\right. \\
&\quad-\frac{5\beta_1^2-8\beta_1-21}{(\beta_1+1)(\beta_1+5)(2\beta_1+1)(3\beta_1+5)[(2\beta_1+1)^2-\beta_2^2]} \\
&\quad\left.-\frac{\beta_1^2+16\beta_1-57}{4(\beta_1+4)(\beta_1^2-121)(\beta_2^2-49)}\right\}
\end{aligned}
$$

由式 (20)，(19a)，可得无量纲挠度

$$
\begin{aligned}
y_2=&\frac{P}{4(\beta_1+1)(\beta_1^2-9)}[\beta_1-3+4\rho^{\beta_1+1})-(\beta_1+1)\rho^4] \\
&+\Big\{m_5+m_6+(m_3+m_4)(\rho^{\beta_1+1}-1) \\
&-m_1m_2\Big[\frac{1}{(\beta_2+1)(\beta_1+\beta_2+2)(2\beta_1+\beta_2+1)}\rho^{\beta_1+\beta_2+2} \\
&+\frac{1}{(\beta_2+5)[\beta_1^2-(\beta_2+4)^2]}\rho^{\beta_2+5}\Big]+\frac{\beta_3m_1^3}{2}\Big[\frac{1}{(2\beta_1+1)^2+\beta_2^2} \\
&\times\Big(\frac{1}{(\beta_1+3)(\beta_1+5)(3\beta_1+5)}\rho^{2(\beta_1+3)} \\
&-\frac{1}{6(\beta_1+1)^2(2\beta_1+1)}\rho^{3(\beta_1+1)}\Big) \\
&+\frac{2}{(\beta_1+4)^2-\beta_2^2}\Big(\frac{1}{(\beta_1+3)(\beta_1+5)(3\beta_1+5)}\rho^{2(\beta_1+3)} \\
&-\frac{1}{8(\beta_1+4)(\beta_1+9)}\rho^{\beta_1+9}\Big) \\
&+\frac{1}{\beta_2^2-49}\Big(\frac{1}{8(\beta_1+4)(\beta_1+9)}\rho^{\beta_1+9}+\frac{1}{6(\beta_1^2-121)}\rho^{12}\Big)\Big]\Big\}y_0^3 \qquad (21)
\end{aligned}
$$

其中

$$
\begin{aligned}
m_5&=m_1m_2\left\{\frac{1}{(\beta_2+1)(\beta_1+\beta_2+2)(2\beta_1+\beta_2+1)}\right. \\
&\quad\left.+\frac{1}{(\beta_2+5)[(\beta_1^2-(\beta_2+4)^2]}\right\}, \\
m_6&=\frac{\beta_3m_1^3}{2}\Big\{\frac{1}{(2\beta_1+1)^2-\beta_2^2}\Big[\frac{1}{6(\beta_1+1)^2(2\beta_1+1)}
\end{aligned}
$$

$$-\frac{1}{(\beta_1+3)(\beta_1+5)(3\beta_1+5)}\Big]+\frac{2}{(\beta_1+4)^2-\beta_2^2}$$

$$\times\Big[\frac{1}{8(\beta_1+4)(\beta_1+9)}-\frac{1}{(\beta_1+3)(\beta_1+5)(3\beta_1+5)}\Big]$$

$$-\frac{1}{\beta_2^2-49}\Big[\frac{1}{6(\beta_1^2-121)}+\frac{1}{8(\beta_1+4)(\beta_1+9)}\Big]\Big\}$$

将式（21）应用于定义（13），便得到波纹圆板在夹紧固定边界条件下的二次近似特征关系式

$$P=4(\beta_1+1)(\beta_1+3)[y_0+(m_3+m_4-m_5-m_6)y_0^3] \tag{22}$$

为便于实际应用，将式（22）转化为有量纲形式

$$q=m_7[h^2w_0+(m_3+m_4-m_5-m_6)w_0^3] \tag{23}$$

其中，w_0 为板中心挠度值，

$$m_7=4Ehk_2(\beta_1+1)(\beta_1+3)/R^4\beta_3$$

对于滑动固定边界条件情况，我们容易从上述解推得。比较两种边界条件式（6）和（7），仅有最后一个条件不同。若将式（6）的最后一个条件改写为

$$\text{当}\ \rho=1\ \text{时},\frac{1}{\nu}\frac{\mathrm{d}S}{\mathrm{d}\rho}-\frac{1}{\rho}S=0 \tag{24}$$

并令 $\nu\to\infty$，那么就得

$$S=0 \tag{25}$$

此即式（7）的最后一个条件。而且，在边值问题（5），（6）中，仅有式（24）显含 ν，故我们可在边值问题的最后结果式（23）中令 $\nu\to\infty$，就可得到波纹圆板在滑动固定边界条件下的二次近似特征关系式。实际计算表明，参数 m_1，m_3，m_4，m_5，m_6，m_7 的表达式和特征关系式（23）与夹紧固定边界条件情况相似，仅 m_2 改为

$$m_2=\beta_3m_1^2\Big[\frac{2}{(\beta_1+4)^2-\beta_2^2}+\frac{1}{\beta_2^2-49}-\frac{1}{(2\beta_1+1)^2-\beta_2^2}\Big] \tag{26}$$

三、实例计算

［例 1］　绘制无光滑中心的正弦截面波纹圆板（图 1）的特征曲线。

此波纹圆板的周边为滑动固定边界条件，其他的数据为 $E=1\times10^4\,\mathrm{kg/mm^2}$，$\nu=0.3$，$R=76\mathrm{mm}$，$H=1.20\mathrm{mm}$，$h=0.33\mathrm{mm}$，$l=25\frac{1}{3}\mathrm{mm}$。将这些值代入式（23），便得此正弦截面波纹圆板的特征关系式

$$q=0.0304w_0+0.00235w_0^3 \tag{27}$$

在此式中，w_0 的单位是 mm，q 的单位是 $\mathrm{kg/cm^2}$。

由公式（27）绘制的特征曲线见图 2。由图看出，本文计算曲线与实验值在已有的 $w_0<9h$ 的实验值范围内非常一致。而文献［1］的计算曲线在 $w_0>5h$ 的范围却与实验值偏差很大。以 $w_0=8.79h$ 的情况为例，本文结果与实验值误差为 0.90%，而文献［1］的结果与实验值误差竟达 39.0%。

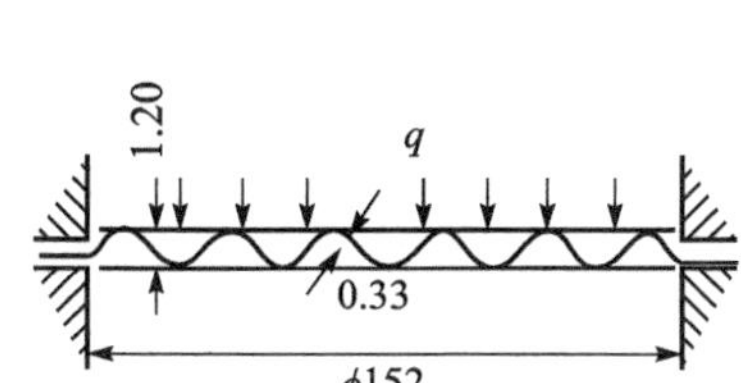

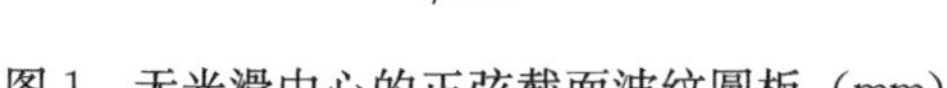
图 1　无光滑中心的正弦截面波纹圆板（mm）

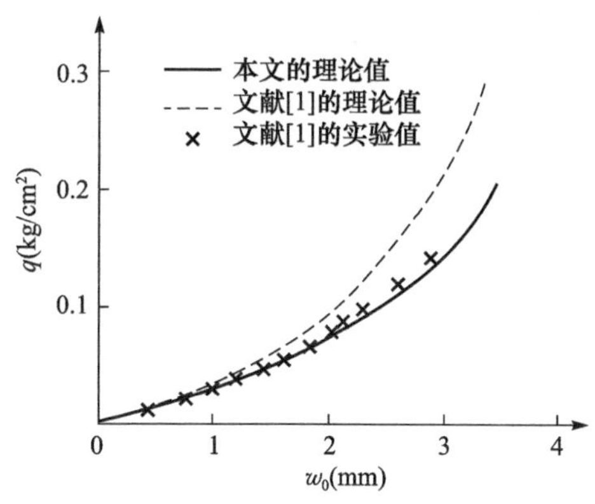

图 2　计算值和实验值的比较

[例 2]　绘制具有光滑中心的锯齿形截面波纹圆板（图 3）的特征曲线。

此波纹圆板的周边为夹紧固定边界条件，其他的数据为 $E=1.35\times10^4\text{kg/mm}^2$，$\nu=0.3$，$R=24\text{mm}$，$H=0.414\text{mm}$，$h=0.101\text{mm}$，$\theta_0=8°45'$，$l=5.4\text{mm}$。将这些值代入式（23），便得此锯齿形截面波纹圆板的特征关系式

$$q=0.106w_0+0.147w_0^3 \tag{28}$$

在此式中，w_0 的单位是 mm，q 的单位是 kg/cm^2。

由公式（28）绘制的特征曲线见图 4。由图看出，在已有的 $w_0<18h$ 的实验值范围内，本文计算曲线与实验值非常一致，而文献［3］的计算曲线在 $w_0>10h$ 的范围与实验值误差较大。以 $w_0=17.8h$ 的情况为例，本文结果与实验值误差为 4.6%，而文献［3］的结果与实验值的误差却达 17.2%。

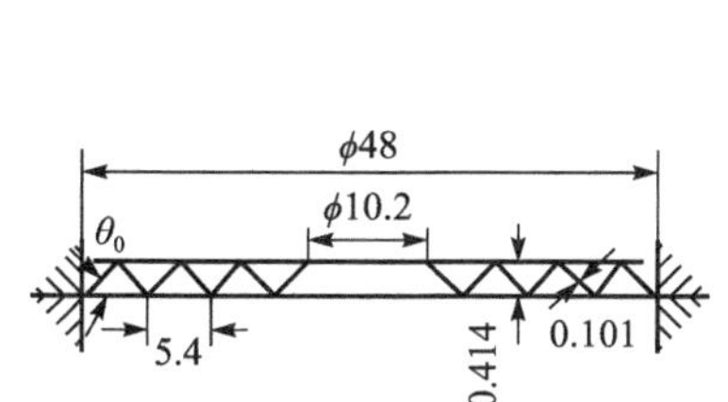

图 3　有光滑中心的锯齿形截面波纹圆板（mm）

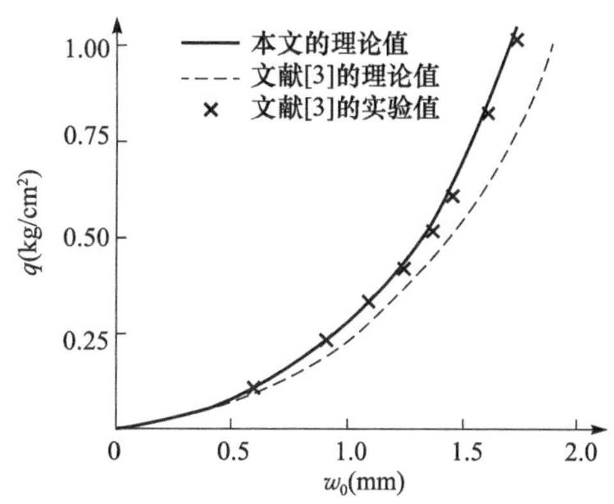

图 4　计算值和实验值的比较

四、结语

由上面两种不同波纹、不同边界条件的实例计算可以看到，在 $w_0<18h$ 的范围内，本文获得的特征关系式与实验值很好符合。有理由相信，本文公式适用范围还可适当扩大。由于许多精密仪器中的波纹圆板是在 $w_0>10h$ 情况下工作，故本文公式（23）

可供工程设计部门参考使用。

特征关系式（23）对于深、浅的不同形状（如正弦、锯齿形、梯形等）波纹圆板均能适用。自然，波纹越密集，即 l/R 越小时，波纹圆板就越接近于各向异性圆平板情况，本文公式结果的精确度就越高。实例 1 属于波纹不密集（$l/R=1/3$）的情况，亦即不大接近于各向异性圆板状态，但本文结果的精确度仍很高。由此说明，对于波纹较少情况，本文公式仍能满足工程设计的要求。

显而易见，对于无光滑中心情况，本文公式是较精确的。但在精密仪器中，为了便于和适当的仪器零件焊接在一起，波纹圆板中心需留下一光滑部分，这就给本文公式的精确度带来了影响。可是，一般说来，这光滑中心是较小的，因此带来的误差也是小的。

参考文献

[1] 费奥多谢夫，В. И. 精密仪器弹性元件的理论与计算. 卢文达，熊大達，译. 北京：科学出版社，1963.

[2] Аксельрад，Э. Л. Вольшие прогибы гофрированной мембраны как неполoгой оболочки. Изв. *АН СССР*，*Мех.*－*Маш.*，1964，(1)：46-53.

[3] Анлреева，Л. Е. Расчёт гофрированных мембран，как анизотропных пластинок. *Инженерный Сборник*，1955，21：138-141.

[4] 赤坂隆. Corrngated diaphragm の弹性特性について. 日本航空学会誌，1995，3（22～23）：279-288.

[5] 刘人怀. 在内边缘均布力矩作用下中心开孔圆底扁球壳的非线性稳定问题. 科学通报，1965，(3)：253-255.

具有光滑中心的波纹圆板的特征关系式①

一、前言

灵敏弹性元件在精密仪器中起着很重要的作用，而波纹圆板在弹性元件中，又占有十分突出的地位。文献［1］研究了无光滑中心的波纹圆板，所得的特征关系式对于具有光滑中心的波纹圆板来说，仅能近似应用。然而，在精密仪器中，为了便于和适当的仪器零件焊接在一起，波纹圆板中心需留下一光滑部分。而波纹圆板有了光滑中心后，就大大增加了求解的难度。因此，一些文献[1-7]只研究无光滑中心的简单情况。据作者所知，对于具有光滑中心情况，目前仅有正弦波纹圆板解[8]，从文献［1］看到，其公式精确度是比较差的。鉴于此，求解具有光滑中心的含各种波纹形状的波纹圆板的大挠度问题，无论从理论和工程上讲，都是十分必要的。本文用修正迭代法进行研究，使这一问题得到了解决，获得了具有夹紧固定和滑动固定边界条件下的特征关系式。最后，通过实例计算，比较了前文和本文公式的数值误差。计算表明，在一般情况下，使用前文的简捷公式进行设计计算是在允许误差范围以内。

二、边值问题

具有光滑中心的波纹圆板如图 1 所示，其光滑中心和波纹部分的中面挠度 w 和径向薄膜力 T_1 在各向同性和各向异性圆板的大挠度理论中分别满足的方程为

当 $0\leqslant r\leqslant R_1$ 时，

$$\begin{aligned} & r\frac{\mathrm{d}^3 w_a}{\mathrm{d}r^3}+\frac{\mathrm{d}^2 w_a}{\mathrm{d}r^2}-\frac{1}{r}\frac{\mathrm{d}w_a}{\mathrm{d}r}=\frac{1}{D_a}\left(T_{1,a}r\frac{\mathrm{d}w_a}{\mathrm{d}r}+\frac{1}{2}qr^2\right), \\ & r\frac{\mathrm{d}^2(T_{1,a}r)}{\mathrm{d}r^2}+\frac{\mathrm{d}(T_{1,a}r)}{\mathrm{d}r}-T_{1,a}=-\frac{1}{2}Eh\left(\frac{\mathrm{d}w_a}{\mathrm{d}r}\right)^2 \end{aligned} \tag{1}$$

当 $R_1<r\leqslant R$ 时，

$$\begin{aligned} & r\frac{\mathrm{d}^3 w_b}{\mathrm{d}r^3}+\frac{\mathrm{d}^2 w_b}{\mathrm{d}r^2}-\frac{k_2k_2'}{r}\frac{\mathrm{d}w_b}{\mathrm{d}r}=\frac{k_2}{D_b}\left(T_{1,b}r\frac{\mathrm{d}w_b}{\mathrm{d}r}+\frac{1}{2}qr^2\right), \\ & r\frac{\mathrm{d}^2(T_{1,b}r)}{\mathrm{d}r^2}+\frac{\mathrm{d}(T_{1,b}r)}{\mathrm{d}r}-k_1k_2T_{1,b}=-\frac{1}{2}Ehk_2\left(\frac{\mathrm{d}w_b}{\mathrm{d}r}\right)^2 \end{aligned} \tag{2}$$

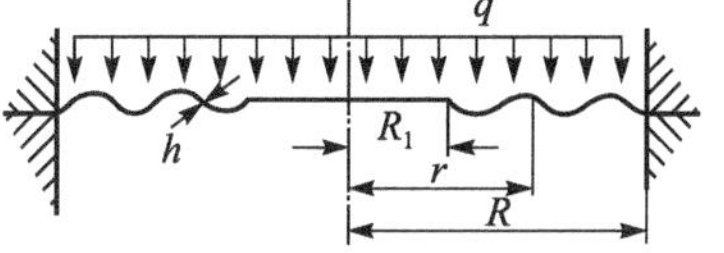

图 1　具有光滑中心的波纹圆板

其中，r 是板的径向坐标，R 是波纹圆板半径，R_1 是光滑中心部分的半径，h 是厚度，E 是弹性模量，ν 是泊松比，q 是均布载荷，k_1，k_2，k_2'是与径向和周向刚度有关的参数（其公式见附录），下角标 a，b 分别表示光滑中心和波纹部分的量。抗弯刚度 D、径向

① 本文原载《中国科学技术大学学报》，1979，9（2）：75-86.

位移 u 和径向弯矩 M_1 的公式为

$$D_a=\frac{Eh^3}{12(1-\nu^2)},\qquad D_b=\frac{Eh^3}{12\left(1-\frac{\nu^2}{k_2k_2'}\right)},$$

$$u_a=\frac{r}{Eh}\left[\frac{\mathrm{d}(T_{1,a}r)}{\mathrm{d}r}-\nu T_{1,a}\right],\qquad u_b=\frac{r}{k_2Eh}\left[\frac{\mathrm{d}(T_{1,b}r)}{\mathrm{d}r}-\nu T_{1,b}\right],$$

$$M_{1,a}=-D_a\left(\frac{\mathrm{d}^2w_a}{\mathrm{d}r^2}+\frac{\nu}{r}\frac{\mathrm{d}w_a}{\mathrm{d}r}\right),\qquad M_{1,b}=-\frac{D_b}{k_2}\left(\frac{\mathrm{d}^2w_b}{\mathrm{d}r^2}+\frac{\nu}{r}\frac{\mathrm{d}w_b}{\mathrm{d}r}\right)\tag{3}$$

我们讨论两种常用的边界条件：

1）夹紧固定情况

当 $r=0$ 时，$\frac{\mathrm{d}w_a}{\mathrm{d}r}=0$，$T_{1,a}$有限

$$\text{当 } r=R \text{ 时，} w_b=0,\ \frac{\mathrm{d}w_b}{\mathrm{d}r}=0,\ u_b=0\tag{4}$$

2）滑动固定情况

当 $r=0$ 时，$\frac{\mathrm{d}w_a}{\mathrm{d}r}=0$，$T_{1,a}$有限

$$\text{当 } r=R \text{ 时，} w_b=0,\ \frac{\mathrm{d}w_b}{\mathrm{d}r}=0,\ T_{1,b}=0\tag{5}$$

在光滑中心和波纹部分的连接处，尚需满足下面的连续条件：

$$\text{当 } r=R_1 \text{ 时，} w_a=w_b,\ \frac{\mathrm{d}w_a}{\mathrm{d}r}=\frac{\mathrm{d}w_b}{\mathrm{d}r},\ M_{1,a}=M_{1,b},$$

$$T_{1,a}=T_{1,b},\ u_a=u_b\tag{6}$$

引用下列无量纲量：

$$\rho=\frac{r}{R},\ y_a=\frac{w_a}{h},\ y_b=\frac{w_b}{h},\varphi_a=\frac{\mathrm{d}y_a}{\mathrm{d}\rho},$$

$$\varphi_b=\frac{\mathrm{d}y_b}{\mathrm{d}\rho},\ S_a=-\frac{Rr}{D_a}T_{1,a},\ S_b=-\frac{k_2Rr}{D_b}T_{1,b},\ P=\frac{R^4}{2D_ah}q,$$

$$\beta_0=6(1-\nu^2),\ \beta_1^2=k_2k_2',\ \beta_2^2=k_1k_2,\ \beta_3=\frac{Eh^3k_2^2}{2D_b},$$

$$\beta_4=\frac{D_b}{D_ak_2},\ \beta_5=\nu(1-\beta_4),\ \beta_6=\nu\left(1-\frac{1}{k_2}\right),\ \beta_7=\frac{\beta_4}{k_2}\tag{7}$$

利用这些无量纲量，方程（1）、（2），边界条件（4）、（5）和连续条件（6）成为

当 $0\leqslant\rho\leqslant\rho_1$ 时，

$$L(\rho\varphi_a)=-\frac{1}{\rho}S_a\varphi_a+P\rho,$$

$$L(\rho S_a)=\frac{\beta_0}{\rho}\varphi_a^2\tag{8a,b}$$

当 $\rho_1<\rho\leqslant1$ 时，

$$\frac{\mathrm{d}^2\varphi_b}{\mathrm{d}\rho^2}+\frac{1}{\rho}\frac{\mathrm{d}\varphi_b}{\mathrm{d}\rho}-\frac{\beta_1^2}{\rho^2}\varphi_b=-\frac{1}{\rho}S_b\varphi_b+\frac{P}{\beta_4}\rho,$$

$$\frac{d^2 S_b}{d\rho^2}+\frac{1}{\rho}\frac{dS_b}{d\rho}-\frac{\beta_2^2}{\rho^2}S_b=-\frac{\beta_3}{\rho}\varphi_b^2 \tag{9a,b}$$

1）夹紧固定情况

当 $\rho=0$ 时，$\varphi_a=0$，$S_a=0$

当 $\rho=1$ 时，$y_b=0$，$\varphi_b=0$，$\frac{dS_b}{d\rho}-\frac{\nu}{\rho}S_b=0$　(10a ～e)

2）滑动固定情况

当 $\rho=0$ 时，$\varphi_a=0$，$S_a=0$

当 $\rho=1$ 时，$y_b=0$，$\varphi_b=0$，$S_b=0$　(11)

连续条件：

当 $\rho=\rho_1$ 时，$y_a=y_b$，$\varphi_a=\varphi_b$，$\frac{d\varphi_a}{d\rho}+\frac{\beta_5}{\rho}\varphi_a=\beta_4\frac{d\varphi_b}{d\rho}$，

$$S_a=\beta_4 S_b,\frac{dS_a}{d\rho}-\frac{\beta_6}{\rho}S_a=\beta_7\frac{dS_b}{d\rho} \tag{12}$$

其中

$$\rho_1=\frac{R_1}{R},$$
$$L(\cdots)=\frac{d}{d\rho}\frac{1}{\rho}\frac{d}{d\rho}(\cdots) \tag{13}$$

三、边值问题的求解

我们首先求解夹紧固定边界条件下的非线性边值问题（8）、（9）、（10）、（12）。与前文[1]一样，我们使用修正迭代法求解。

在一次近似中，由非线性边值问题（8）、（9）、（10）、（12）得下列线性边值问题：

当 $0\leqslant\rho\leqslant\rho_1$ 时，

$$L(\rho\varphi_{a,1})=P\rho,$$
$$L(\rho S_{a,1})=\frac{\beta_0}{\rho}\varphi_{a,1}^2 \tag{14a,b}$$

当 $\rho_1<\rho\leqslant 1$ 时，

$$\frac{d^2\varphi_{b,1}}{d\rho^2}+\frac{1}{\rho}\frac{d\varphi_{b,1}}{d\rho}-\frac{\beta_1^2}{\rho^2}\varphi_{b,1}=\frac{P}{\beta_2}\rho,$$
$$\frac{d^2 S_{b,1}}{d\rho^2}+\frac{1}{\rho}\frac{dS_{b,1}}{d\rho}-\frac{\beta_2^2}{\rho^2}S_{b,1}=\frac{\beta_3}{\rho}\varphi_{b,1}^2 \tag{15a,b}$$

当 $\rho=0$ 时，$\varphi_{a,1}=0$，$S_{a,1}=1$　(16a，b)

当 $\rho=1$ 时，$y_{b,1}=0$，$\varphi_{b,1}=0$，

$$\frac{dS_{b,1}}{d\rho}-\frac{\nu}{\rho}S_{b,1}=0 \tag{17a ～ c}$$

当 $\rho=\rho_1$ 时，$y_{a,1}=y_{b,1}$，$\varphi_{a,1}=\varphi_{b,1}$，

$$\frac{d\varphi_{a,1}}{d\rho}+\frac{\beta_5}{\rho}\varphi_{a,1}=\beta_4\frac{d\varphi_{b,1}}{d\rho},S_{a,1}=\beta_4 S_{b,1},$$

$$\frac{\mathrm{d}S_{a,1}}{\mathrm{d}\rho}-\frac{\beta_6}{\rho}S_{a,1}=\beta_7\frac{\mathrm{d}S_{b,1}}{\mathrm{d}\rho} \tag{18a ～ e}$$

应用式（16a)、(17b)、(18b，c)、方程（14a）和（15a）的解为

$$\varphi_{a,1}=(a_1\rho+a_2\rho^3)P,$$
$$\varphi_{b,1}=(b_1\rho^{-\beta_1}+b_2\rho^{\beta_1}+b_3\rho^3)P \tag{19}$$

其中

$$a_1=b_3\left\{\lambda_1\left[\frac{\beta_4}{4}(\beta_1^2+3)-\beta_5-1\right](\rho_1^{\beta_1+2}-\rho^{-(\beta_1-2)})\right.$$
$$\left.-\lambda_1(\beta_1\beta_4-\beta_5-1)(\rho_1^{2\beta-1}-\rho_1^{-1})+\left[\frac{\beta_4}{8}(\beta_1^2-9)+1\right]\rho_1^2\right\},a_2=\frac{1}{8},$$
$$b_1=\lambda_1b_3\left\{(\beta_1\beta_4-\beta_5-1)\rho_1^{\beta_1}-\left[\frac{\beta_4}{4}(\beta_1^2+3)-\beta_5-1\right]\rho_1^3\right\},$$
$$b_2=-(b_1+b_3),$$
$$b_3=-\frac{1}{\beta_4(\beta_1^2-9)},$$
$$\lambda_1=-\frac{1}{(\beta_1\beta_4-\beta_5-1)\rho_1^{\beta_1}+(\beta_1\beta_4+\beta_5+1)\rho_1^{-\beta_1}}$$

由式（19)、(17a)、(18a)，还可得无量纲挠度

$$y_{a,1}=\left(\frac{1}{2}a_1\rho^2+\frac{1}{4}a_2\rho^4+\frac{1}{\alpha_1}\right)P,$$
$$y_{b,1}=-\left(\frac{b_1}{\beta_1-1}\rho^{-(\beta_1-1)}-\frac{b_2}{\beta_1+1}\rho^{\beta_1+1}-\frac{b_3}{4}\rho^4+\lambda_2\right)P \tag{20a,b}$$

其中

$$\alpha_1=\frac{1}{b_3}\left\{\frac{\lambda_1}{2(\beta_1^2-1)}\left[\left(\frac{\beta_4}{4}(\beta_1^2+3)-\beta_5-1\right)\left((\beta_1+1)^2\rho_1^{-(\beta_1-4)}\right.\right.\right.$$
$$\left.-(\beta_1-1)^2\rho_1^{\beta_1+4}-4\beta_1\rho_1^3\right)+(\beta_1\beta_4-\beta_5-1)$$
$$\left.\times\left((\beta_1-1)^2\rho_1^{2\beta_1+1}+4\beta_1\rho_1^{\beta_1}\right)-(\beta_1+1)^2(\beta_1\beta_4-\beta_5-1)\rho_1\right]$$
$$\left.+\frac{\beta_1-1}{2(\beta_1+1)}\rho_1^{\beta_1+1}-\frac{1}{4}\left(1-\frac{1}{8b_3}\right)\rho_1^4-\frac{\beta_1-3}{4(\beta_1+1)}\right\}^{-1},$$
$$\lambda_2=\frac{1}{\beta_1+1}\left[\frac{b_\beta}{4}(\beta_1-3)-\frac{2\beta_1b_1}{\beta_1-1}\right] \tag{21}$$

令 y_0 为波纹圆板中心的无量纲挠度，即

当 $\rho=0$ 时，$y=y_0$　(22)

应用此关系式，在式（20a）中令 $\rho=0$，便得

$$P=\alpha_1y_0 \tag{23}$$

此即波纹圆板的小挠度特征关系式。将式（23）代入式（19），就有

$$\varphi_{a,1}=\alpha_1y_0(a_1\rho+a_2\rho^3),$$
$$\varphi_{b,1}=\alpha_1y_0(b_1\rho^{-\beta_1}+b_2\rho^{\beta_1}+b_3\rho^3) \tag{24}$$

我们再求解方程（14b）和（15b）。将式（24）代入它们的右端，并应用式（16b）、（17c）、（18d，e），便得方程（14b）和（15b）的解

$$
\begin{aligned}
S_{a,1} &= \alpha_1^2 y_0^2 \sum_{i=0}^{3} c_{i+1}\rho^{2i+1}, \\
S_{b,1} &= \alpha_1^2 y_0^2 (d_1\rho^{-\beta_2} + d_2\rho^{-(2\beta_1-1)} + d_3\rho + d_4\rho^{2\beta_1+1} \\
&\quad + d_5\rho^{-(\beta_1-4)} + d_6\rho^{\beta_1+4} + d_7\rho^7 + d_8\rho^{\beta_2})
\end{aligned} \tag{25}
$$

其中

$$
\begin{aligned}
c_1 &= \beta_4(d_1\rho_1^{-(\beta_2+1)} + d_2\rho_1^{-2\beta_1} + d_3 + d_4\rho_1^{2\beta_1} \\
&\quad + d_5\rho_1^{-(\beta_1-3)} + d_6\rho_1^{\beta_1+3} + d_7\rho_1^6 + d_8\rho_1^{\beta_2-1}), \\
c_2 &= \frac{1}{8}\beta_0 a_1^2, \\
c_3 &= \frac{1}{12}\beta_0 a_1 a_2, \\
c_4 &= \frac{1}{48}\beta_0 a_2^2, \\
d_1 &= \frac{\beta_2-\nu}{(\beta_2+\nu)(\beta_2\beta_7+\beta_4\beta_6-\beta_4)\rho_1^{\beta_2}-(\beta_2-\nu)(\beta_2\beta_7-\beta_4\beta_6+\beta_4)\rho_1^{-\beta_2}} \\
&\quad \times\{2c_2\rho_1^3 + 4c_3\rho_1^5 + [6c_4 - d_7(\beta_4\beta_6-\beta_4+7\beta_7)]\rho_1^7 \\
&\quad + d_2(2\beta_1\beta_7-\beta_4\beta_6+\beta_4-\beta_7)\rho_1^{-(2\beta_1-1)} - d_3(\beta_4\beta_6-\beta_4+\beta_7)\rho_1 \\
&\quad - d_4(2\beta_1\beta_7+\beta_4\beta_6-\beta_4+\beta_7)\rho_1^{2\beta_1+1} \\
&\quad + d_5(\beta_1\beta_7-\beta_4\beta_6+\beta_4-4\beta_7)\rho_1^{-(\beta_1-4)} \\
&\quad - d_6(\beta_1\beta_7+\beta_4\beta_6-\beta_4+4\beta_7)\rho_1^{\beta_1+4} \\
&\quad - \lambda_3(\beta_2\beta_7+\beta_4\beta_6-\beta_4)\rho_1^{\beta_2}\}, \\
d_2 &= \frac{\beta_3 b_1^2}{(2\beta_1-1)^2-\beta_2^2}, \\
d_3 &= -\frac{2\beta_3 b_1 b_2}{\beta_2^2-1}, \\
d_4 &= \frac{\beta_3 b_2^2}{(2\beta_1+1)^2-\beta_2^2}, \\
d_5 &= \frac{2\beta_3 b_1 b_3}{(\beta_1-4)^2-\beta_2^2}, \\
d_6 &= \frac{2\beta_3 b_2 b_3}{(\beta_1+4)^2-\beta_2^2}, \\
d_7 &= -\frac{\beta_3 b_3^2}{\beta_2^2-49}, \\
d_8 &= \frac{\beta_2+\nu}{\beta_2-\nu}d_1 + \lambda_3, \\
\lambda_3 &= \frac{1}{\beta_2-\nu}[(2\beta_1-1+\nu)d_2 - (1-\nu)d_3 - (2\beta_1+1-\nu)d_4
\end{aligned}
$$

$$+(\beta_1-4+\nu)d_5-(\beta_1+4-\nu)d_6-(7-\nu)d_7]$$

现在求二次近似解。由于我们只进行到二次近似，故只需求解关于 φ_2 的边值问题。应用方程（8a）、（9a）、边界条件（10a，c，d）以及式（24）、（25），便得关于 φ_2 的线性边值问题：

当 $0\leqslant\rho\leqslant\rho_1$ 时，

$$L(\rho\varphi_{a,2})=-\frac{1}{\rho}S_{a,1}\varphi_{a,1}+P\rho \tag{26}$$

当 $\rho_1<\rho\leqslant1$ 时，

$$\frac{\mathrm{d}^2\varphi_{b,2}}{\mathrm{d}\rho^2}+\frac{1}{\rho}\frac{\mathrm{d}\varphi_{b,2}}{\mathrm{d}\rho}-\frac{\beta_1^2}{\rho^2}\varphi_{b,2}=-\frac{1}{\rho}S_{b,1}\varphi_{b,1}+\frac{P}{\beta_4}\rho \tag{27}$$

当 $\rho=0$ 时，$\varphi_{a,2}=0$　　(28)

当 $\rho=1$ 时，$y_{b,2}=0$，$\varphi_{b,2}=0$　　(29a，b)

当 $\rho=\rho_1$ 时，$y_{a,2}=y_{b,2}$，$\varphi_{a,2}=\varphi_{b,2}$，

$$\frac{\mathrm{d}\varphi_{a,2}}{\mathrm{d}\rho}+\frac{\beta_5}{\rho}\varphi_{a,2}=\beta_4\frac{\mathrm{d}\varphi_{b,2}}{\mathrm{d}\varphi} \tag{30a～c}$$

应用式（28）、（29b）、（30b，c），方程（26）和（27）的解为

$$\begin{aligned}\varphi_{a,2}&=g_1\rho+\frac{1}{8}P\rho^3-\alpha_1^3y_0^3\sum_{i=1}^{5}e_i\rho^{2i+1},\\ \varphi_{b,2}&=g_2\rho^{-\beta_1}+g_3\rho^{\beta_1}+b_3P\rho^3+\alpha_1^3y_0^3(f_1\rho^{-(\beta_1+\beta_2-1)}+f_2\rho^{\beta_1+\beta_2+1}\\ &\quad+f_3\rho^{-(\beta_1-\beta_2-1)}+f_4\rho^{\beta_1-\beta_2+1}+f_5\rho^{-(3\beta_1-2)}+f_6\rho^{3\beta_1+2}\\ &\quad+f_7\rho^{-(2\beta_1-5)}+f_8\rho^{2\beta_1+5}+f_9\rho^{-(\beta_1-8)}+f_{10}\rho^{\beta_1+8}+f_{11}\rho^{-(\beta_1-2)}\\ &\quad+f_{12}\rho^{\beta_1+2}+f_{13}\rho^{-(\beta_2-4)}+f_{14}\rho^{\beta_2+4}+f_{15}\rho^5+f_{16}\rho^{11})\end{aligned} \tag{31}$$

其中

$$e_1=\frac{1}{8}a_1c_1,\qquad e_2=\frac{1}{24}(a_1c_2+a_2c_1),$$

$$e_3=\frac{1}{48}(a_1c_3+a_2c_2),\qquad e_4=\frac{1}{80}(a_1c_4+a_2c_3),$$

$$e_5=\frac{1}{120}a_2c_4,\qquad f_1=-\frac{b_1d_1}{(\beta_2-1)(2\beta_1+\beta_2-1)},$$

$$f_2=-\frac{b_2d_8}{(\beta_2+1)(2\beta_1+\beta_2+1)},\quad f_3=\frac{b_1d_8}{(\beta_2+1)(2\beta_1-\beta_2-1)},$$

$$f_4=\frac{b_2d_1}{(\beta_2-1)(2\beta_1-\beta_2+1)},\quad f_5=-\frac{b_1d_2}{4(\beta_1-1)(2\beta_1-1)},$$

$$f_6=-\frac{b_2d_4}{4(\beta_1+1)(2\beta_1+1)},\quad f_7=-\frac{b_1d_5+b_3d_2}{(\beta_1-5)(3\beta_1-5)},$$

$$f_8=-\frac{b_2d_6+b_3d_4}{(\beta_1+5)(3\beta_1+5)},\quad f_9=\frac{b_1d_7+b_3d_5}{16(\beta_1-4)},$$

$$f_{10}=-\frac{b_2d_7+b_3d_6}{16(\beta_1+4)},\quad f_{11}=\frac{b_1d_3+b_2d_2}{4(\beta_1-1)},$$

$$f_{12}=-\frac{b_1d_4+b_2d_3}{4(\beta_1+1)},\quad f_{13}=\frac{b_3d_1}{\beta_1^2-(\beta_2-4)^2},$$

$$f_{14}=\frac{b_3d_8}{\beta_1^2-(\beta_2+4)^2},\quad f_{15}=\frac{b_1d_6+b_2d_5+b_3d_3}{\beta_1^2-25},$$

$$f_{16}=\frac{b_3d_7}{\beta_1^2-121},\quad g_1=\lambda_4P+\lambda_5\alpha_1^3y_0^3,$$

$$g_2=b_1P-\lambda_1\lambda_6\alpha_1^3y_0^3,\quad g_3=b_2P+\lambda_7\alpha_1^3y_0^3,$$

$$\lambda_4=b_1\rho_1^{-(\beta_1+1)}+b_2\rho_1^{\beta_1-1}+\left(b_3-\frac{1}{8}\right)\rho_1^2,$$

$$\begin{aligned}\lambda_5=&f_1\rho^{-(\beta_1+\beta_2)}+f_2\rho_1^{\beta_1+\beta_2}+f_3\rho_1^{-(\beta_1-\beta_2)}+f_4\rho_1^{\beta_1-\beta_2}\\&+f_5\rho_1^{-(3\beta_1-1)}+f_6\rho_1^{3\beta_1+1}+f_7\rho_1^{-2(\beta_1-2)}+f_8\rho_1^{2(\beta_1+2)}\\&+f_9\rho_1^{-(\beta_1-7)}+f_{10}\rho_1^{\beta_1+7}+f_{11}\rho_1^{-(\beta_1-1)}+f_{12}\rho_1^{\beta_1+1}\\&+\lambda_7\rho^{\beta_1-1}-\lambda_1\lambda_6\rho_1^{-(\beta_1+1)}+f_{13}\rho_1^{-(\beta_2-3)}+f_{14}\rho_1^{\beta_2+3}+e_1\rho_1^2\\&+(e_2+f_{15})\rho_1^4+e_3\rho_1^6+e_4\rho_1^8+(e_5+f_{16})\rho_1^{10},\end{aligned}$$

$$\begin{aligned}\lambda_6=&-f_1[\beta_4(\beta_1+\beta_2-1)+\beta_5+1]\rho_1^{-(\beta_1+\beta_2-1)}\\&+f_2[\beta_4(\beta_1+\beta_2+1)-\beta_5-1]\rho_1^{\beta_1+\beta_2+1}\\&-f_3[\beta_4(\beta_1-\beta_2-1)+\beta_5+1]\rho_1^{-(\beta_1-\beta_2-1)}\\&+f_4[\beta_4(\beta_1-\beta_2+1)-\beta_5-1]\rho_1^{\beta_1-\beta_2+1}\\&-f_5[\beta_4(3\beta_1-2)+\beta_5+1]\rho_1^{-(3\beta_1-2)}\\&+f_6[\beta_4(3\beta_1+2)-\beta_5-1]\rho_1^{3\beta_1+2}\\&-f_7[\beta_4(2\beta_1-5)+\beta_5+1]\rho_1^{-(2\beta_1-5)}\\&+f_8[\beta_4(2\beta_1+5)-\beta_5-1]\rho_1^{2\beta_1+5}\\&-f_9[\beta_4(\beta_1-8)+\beta_5+1]\rho_1^{-(\beta_1-8)}\\&+f_{10}[\beta_4(\beta_1+8)-\beta_5-1]\rho_1^{\beta_1+8}\\&-f_{11}[\beta_4(\beta_1-2)+\beta_5+1]\rho_1^{-(\beta_1-2)}\\&+f_{12}[\beta_4(\beta_1+2)-\beta_5-1]\rho_1^{\beta_1+2}\\&-\lambda_8(\beta_1\beta_4-\beta_5-1)\rho_1^{\beta_1}\\&-f_{13}[\beta_4(\beta_2-4)+\beta_5+1]\rho_1^{-(\beta_2-4)}\\&+f_{14}[\beta_4(\beta_2+4)-\beta_5-1]\rho_1^{\beta_2+4}\\&+2e_1\rho_1^3+[4e_2+f_{15}(5\beta_4-\beta_5-1)]\rho_1^5+6e_3\rho_1^7\\&+8e_4\rho_1^9+[10e_5+f_{16}(11\beta_4-\beta_5-1)]\rho_1^{11},\end{aligned}$$

$$\lambda_7=\lambda_1\lambda_6-\lambda_8,\qquad \lambda_8=\sum_{i=1}^{16}f_i$$

由式 (31)、(29a)、(30a)，还可得无量纲挠度的二次近似公式

$$y_{a,2}=\left(\frac{1}{\alpha_1}+\frac{1}{32}\rho^4\right)P+\frac{1}{2}g_1\rho^2-\alpha_1^3y_0^3\left[\lambda_9+\sum_{i=1}^{5}\frac{e_i}{2(i+1)}\rho^{2(i+1)}\right],$$

$$
\begin{aligned}
y_{b,2} =& \left[\frac{1}{4\beta_4(\beta_1+1)(\beta_1+3)}+\frac{2\beta_1 b_1}{\beta_1^2-1}+\frac{1}{4}b_3\rho^4\right]P-\frac{g_2}{\beta_1-1}\rho^{-(\beta_1-1)} \\
&+\frac{g_3}{\beta_1+1}\rho^{\beta_1+1}-a_1^3 y_0^3\left[\lambda_{10}+\frac{f_1}{\beta_1+\beta_2-2}\rho^{-(\beta_1+\beta_2-2)}\right. \\
&-\frac{f_2}{\beta_1+\beta_2+2}\rho^{\beta_1+\beta_2+2}+\frac{f_3}{\beta_1-\beta_2-2}\rho^{-(\beta_1-\beta_2-2)} \\
&-\frac{f_4}{\beta_1-\beta_2+2}\rho^{\beta_1-\beta_2+2}+\frac{f_5}{3(\beta_1-1)}\rho^{-3(\beta_1-1)} \\
&-\frac{f_6}{3(\beta_1+1)}\rho^{3(\beta_1+1)}+\frac{f_7}{2(\beta_1-3)}\rho^{-2(\beta_1-3)} \\
&-\frac{f_8}{2(\beta_1+3)}\rho^{2(\beta_1+3)}+\frac{f_9}{\beta_1-9}\rho^{-(\beta_1-9)}-\frac{f_{10}}{\beta_1+9}\rho^{\beta_1+9} \\
&+\frac{f_{11}}{\beta_1-3}\rho^{-(\beta_1-3)}-\frac{f_{12}}{\beta_1+3}\rho^{\beta_1+3}+\frac{f_{13}}{\beta_2-5}\rho^{-(\beta_2-5)} \\
&\left.-\frac{f_{14}}{\beta_2+5}\rho^{\beta_2+5}-\frac{1}{6}f_{15}\rho^6-\frac{1}{12}f_{16}\rho^{12}\right] \qquad \text{(32a,b)}
\end{aligned}
$$

其中

$$
\begin{aligned}
\lambda_9 =& \frac{f_1}{\beta_1+\beta_2-2}\rho_1^{-(\beta_1+\beta_2-2)}-\frac{f_2}{\beta_1+\beta_2+2}\rho_1^{\beta_1+\beta_2+2} \\
&+\frac{f_3}{\beta_1-\beta_2-2}\rho_1^{-(\beta_1-\beta_2-2)}-\frac{f_4}{\beta_1-\beta_2+2}\rho_1^{\beta_1-\beta_2+2} \\
&+\frac{f_5}{3(\beta_1-1)}\rho_1^{-3(\beta_1-1)}-\frac{f_6}{3(\beta_1+1)}\rho_1^{3(\beta_1+1)} \\
&+\frac{f_7}{2(\beta_1-3)}\rho_1^{-2(\beta_1-3)}-\frac{f_8}{2(\beta_1+3)}\rho_1^{2(\beta_1+3)} \\
&+\frac{f_9}{\beta_1-9}\rho_1^{-(\beta_1-9)}-\frac{f_{10}}{\beta_1+9}\rho_1^{\beta_1+9}+\frac{f_{11}}{\beta_1-3}\rho_1^{-(\beta_1-3)} \\
&-\frac{f_{12}}{\beta_1+3}\rho_1^{\beta_1+3}-\frac{\lambda_1\lambda_6}{\beta_1-1}\rho_1^{-(\beta_1-1)} \\
&-\frac{\lambda_7}{\beta_1+1}\rho_1^{\beta_1+1}+\frac{f_{13}}{\beta_2-5}\rho_1^{-(\beta_2-5)}-\frac{f_{14}}{\beta_2+5}\rho_1^{\beta_2+5} \\
&+\lambda_{10}+\frac{1}{2}\lambda_5\rho_1^2-\frac{1}{4}e_1\rho_1^4-\frac{1}{6}(e_2+f_{15})\rho_1^6 \\
&-\frac{1}{8}e_3\rho_1^8-\frac{1}{10}e_4\rho_1^{10}-\frac{1}{12}(e_5+f_{16})\rho_1^{12},
\end{aligned}
$$

$$
\begin{aligned}
\lambda_{10} =& \frac{\lambda_1\lambda_6}{\beta_1-1}+\frac{\lambda_7}{\beta_1+1}-\frac{f_1}{\beta_1+\beta_2-2}+\frac{f_2}{\beta_1+\beta_2+2}-\frac{f_3}{\beta_1-\beta_2-2} \\
&+\frac{f_4}{\beta_1-\beta_2+2}-\frac{f_5}{3(\beta_1-1)}+\frac{f_6}{3(\beta_1+1)}-\frac{f_7}{2(\beta_1-3)} \\
&+\frac{f_8}{2(\beta_1+3)}-\frac{f_9}{\beta_1-9}+\frac{f_{10}}{\beta_1+9}-\frac{f_{11}}{\beta_1-3}+\frac{f_{12}}{\beta_1+3}
\end{aligned}
$$

$$-\frac{f_{13}}{\beta_2-5}+\frac{f_{14}}{\beta_2+5}+\frac{1}{6}f_{15}+\frac{1}{12}f_{16}$$

将式（32a）应用于定义（22），便得具有光滑中心的波纹圆板在夹紧固定边界条件下的二次近似特征关系式

$$P=\alpha_1 y_0+\alpha_3 y_0^3 \tag{33}$$

其中

$$\alpha_3=\lambda_9\alpha_1^4 \tag{34}$$

为便于应用，我们将式（33）转化为有量纲形式

$$q=\frac{2D_a\alpha_1}{R^4}\left(w_0+\frac{\lambda_9\alpha_1^3}{h^2}w_0^3\right) \tag{35}$$

其中 w_0 为波纹圆板中心的挠度值。

下面给出滑动固定边界条件下的二次近似特征关系式。将夹紧固定和滑动固定的边值问题进行比较，我们看到，仅有式（10）和（11）中的最后一个边界条件不同。若将式（10）的最后一个条件改写为

$$\text{当}\ \rho=1\ \text{时，}\frac{1}{\nu}\frac{\mathrm{d}S_b}{\mathrm{d}\rho}-\frac{1}{\rho}S_b=0 \tag{36}$$

并令 $\nu\to\infty$，就得式（11）的最后一个条件

$$S_b=0 \tag{37}$$

此外，我们还看到，在边界夹紧固定下的边值问题（8）、（9）、（10）、（12）中，仅有式（10）的最后一个条件显含 ν。所以，只要在式（35）中令 $\nu\to\infty$，便可得波纹圆板在滑动固定边界条件下的二次近似特征关系式。经过实际计算，除符号 d_1，d_8 和 λ_3 外，式（35）和其他符号与前形式相同，仅 d_1，d_8 和 λ_3 改为

$$\begin{aligned}
d_1=&-\frac{1}{(\beta_2\beta_7+\beta_4\beta_6-\beta_4)\rho_1^{\beta_2}+(\beta_2\beta_7-\beta_4\beta_6+\beta_4)\rho_1^{-\beta_2}}\{2c_2\rho_1^3+4c_3\rho_1^5\\
&+[6c_4-d_7(\beta_4\beta_6-\beta_4+7\beta_7)]\rho_1^7+d_2(2\beta_1\beta_7-\beta_4\beta_6+\beta_4\\
&-\beta_7)\rho_1^{-(2\beta_1-1)}-d_3(\beta_4\beta_6-\beta_4+\beta_7)\rho_1-d_4(2\beta_1\beta_7+\beta_4\beta_6-\beta_4\\
&+\beta_7)\rho_1^{2\beta_1+1}+d_5(\beta_1\beta_7-\beta_4\beta_6+\beta_4-4\beta_7)\rho_1^{-(\beta_1-4)}-d_6(\beta_1\beta_7\\
&+\beta_4\beta_6-\beta_4+4\beta_7)\rho_1^{\beta_1+4}-\lambda_3(\beta_2\beta_7+\beta_4\beta_6-\beta_4)\rho_1^{\beta_2}\},
\end{aligned} \tag{38}$$

$$d_8=-d_1+\lambda_3,$$

$$\lambda_3=-\sum_{i=2}^{7}d_i$$

四、实例计算

例：绘制具有光滑中心的锯齿形截面波纹圆板（图 2）的特征曲线。

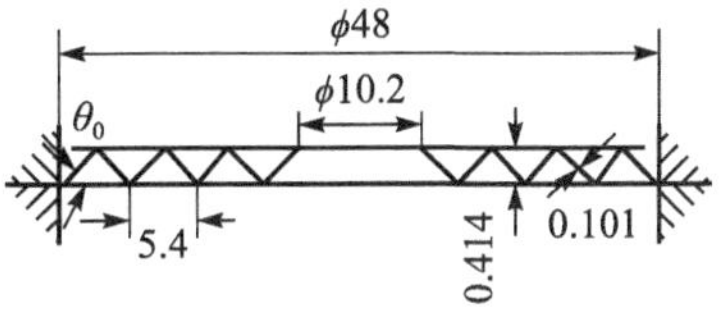

图 2　锯齿形截面波纹圆板（mm）

此波纹圆板的周边为夹紧固定边界条件。其余数据为

$E=1.35\times10^4\,\mathrm{kg/mm^2}$，　　$\nu=0.3$，　　$H=0.414\mathrm{mm}$，

$h=0.101\text{mm}$,　　　　　$R=24\text{mm}$,　　$R_1=5.1\text{mm}$,

$\theta_0=8°45'$,　　　　　　$l=5.4\text{mm}$

将这些数值代入式（35），便得此波纹圆板的特征关系式

$$q=0.103w_0+0.129w_0^3 \tag{39}$$

这里，w_0 的单位是 mm，q 的单位是 kg/cm^2。

由式（39）绘制的特征曲线见图 3。显而易见，本文与前文[1]的特征曲线很接近，它们之间的误差很小。这一事实说明，小的光滑中心$\left(\frac{R_1}{R}=0.21\right)$对特征曲线影响较小。与文献［7］相比，它们最接近实验值。以 $w_0=17.8h$ 为例，本文结果与实验值的相对误差为 6.0%，前文[1]为−4.6%，文献［7］为 17.2%。本文和前文[1]的理论值从上下两面逼近实验值的现象不难由波纹圆板的力学性质解释，全波纹板与具有光滑中心的波纹板相较，刚度要高些。

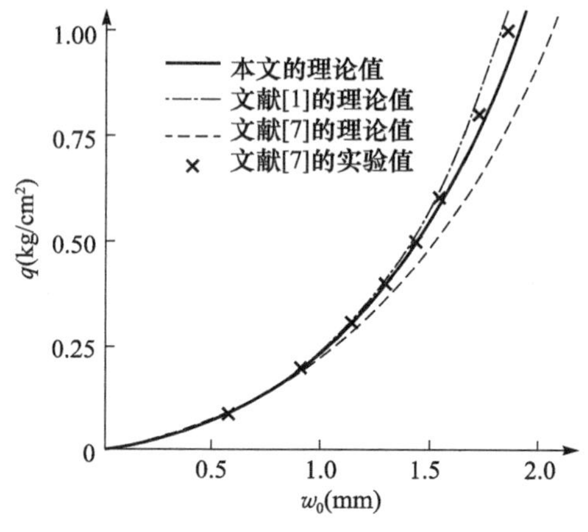

图 3　理论值和实验值的比较

五、结语

本文的特征关系式（35）适用于具有光滑中心的含不同深、浅形状（如正弦、锯齿形、梯形等）的波纹圆板。由实例计算看到，公式的精确度较高，适于工程应用。随着波纹数的增加，波纹圆板越接近各向异性圆板情况，本文公式的精确度就越高。显而易见，前文[1]的公式是本文公式在光滑中心半径 R_1 为零时的特殊情况。由于本文公式（35）较前文[1]公式复杂，加之波纹圆板光滑中心的尺寸在仪表中较小，其半径一般小于板半径的 30%，对特征曲线影响较小，故在设计时，使用前文[1]的简捷公式进行计算一般也是可以的。

参考文献

［1］　刘人怀. 波纹圆板的特征关系式. 力学学报，1978，(1)：47.

［2］　刘人怀. 在边缘力矩作用下夹层圆板的非线性轴对称弯曲问题. 中国科学技术大学学报，1980，10 (2)：56.

[3]　刘人怀. 在边缘载荷作用下中心开孔圆底扁薄球壳的轴对称稳定性. 力学，1977，(3)：206.

[4]　叶开沅，刘人怀，平庆元，等. 在对称线布载荷作用下的圆底扁薄球壳的非线性稳定问题. 兰州大学学报（自然科学），1965，(2)：10；科学通报，1965，(2)：142.

[5]　叶开沅，刘人怀，张传智，等. 圆底扁薄球壳在边缘力矩作用下的非线性稳定问题. 科学通报，1965，(2)：145.

[6]　刘人怀. 在内边缘均布力矩作用下中心开孔圆底扁球壳的非线性稳定问题. 科学通报，1965 (3)：253.

[7]　Anapeeва，Л. E. Расчёт Гофрированных мембран，kak анизотропных пластинок. *Инженерный Сборник*，1955，21：128.

[8]　费奥多谢夫，B. H. 精密仪器弹性元件的理论与计算. 卢文达，熊大逵，译. 北京：科学出版社，1963：288.

附　录

为使用本文公式方便起见，我们把与径向、周向刚度有关的参数 k_1，k_2 和 k_2' 的一般公式列出[7]：

$$k_1 = \frac{12}{h^2 l}\int_0^l \frac{z^2}{\cos\theta}\mathrm{d}x + \frac{1}{l}\int_0^l \cos\theta\mathrm{d}x,$$

$$k_2 = \frac{1}{l}\int_0^l \frac{1}{\cos\theta}\mathrm{d}x,$$

$$k_2' = \frac{12}{h^2 l}\int_0^l \frac{z^2}{\cos\theta}\mathrm{d}x + \frac{1}{l}\int_0^l \frac{1}{\cos^3\theta}\mathrm{d}x \tag{附 1}$$

其中，l 是波长，θ 是任意点的切线倾斜角，x 和 z 是坐标轴，它们的意义示于附图 1 中。

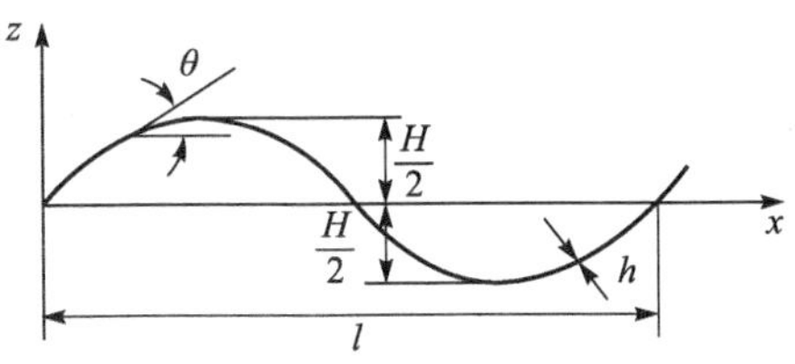

附图 1　一个波纹截面

由式 (1)，我们给出精密仪器设计中最通常遇到的波纹截面的 k_1，k_2 和 k_2' 的公式。

1) 正弦截面（图 1）

文献 [7] 仅给出了浅波纹截面（$\theta \leqslant 15°$）情况的公式。我们为使应用范围扩大，对 $\cos\theta$，$\frac{1}{\cos\theta}$ 和 $\frac{1}{\cos^3\theta}$ 的级数展开式选取前两项计算。对展开式而言，在 $\theta \leqslant 30°$ 范围内，与精确值的误差不超过 4%。这时的 k_1，k_2 和 k_2' 的公式为

$$k_1 = 1 + \frac{3H^2}{2h^2} + \frac{\pi^2 H^2}{4l^2}\left(\frac{3H^2}{4h^2} - 1\right),$$

$$k_2 = 1 + \frac{\pi^2 H^2}{4l^2},$$

$$k_2' = 1 + \frac{3H^2}{2h^2} + \frac{3\pi^2 H^2}{4l^2}\left(1 + \frac{H^2}{4h^2}\right) \tag{附 2}$$

其中，H 是波幅。

2) 锯齿形截面（正文中的图 2）。

$$k_1 = \frac{H^2}{h^2 \cos\theta_0} + \cos\theta_0,$$

$$k_2 = \frac{1}{\cos\theta_0},$$

$$k_2' = \frac{H^2}{h^2 \cos\theta_0} + \frac{1}{\cos^3\theta_0} \tag{附 3}$$

其中，θ_0 是截线与 x 轴的夹角。

3）梯形截面（附图 2）

$$k_1 = \frac{H^2}{lh^2}\left(\frac{l-2a}{\cos\theta_0} + 6a\right) + \left(1 - \frac{2a}{l}\right)\cos\theta_0 + \frac{2a}{l},$$

$$k_2 = \frac{1}{l}\left(\frac{l-2a}{\cos\theta_0} + 2a\right),$$

$$k_2' = \frac{H^2}{lh^2}\left(\frac{l-2a}{\cos\theta_0} + 6a\right) + \frac{1}{l}\left(\frac{l-2a}{\cos^3\theta_0} + 2a\right) \tag{附 4}$$

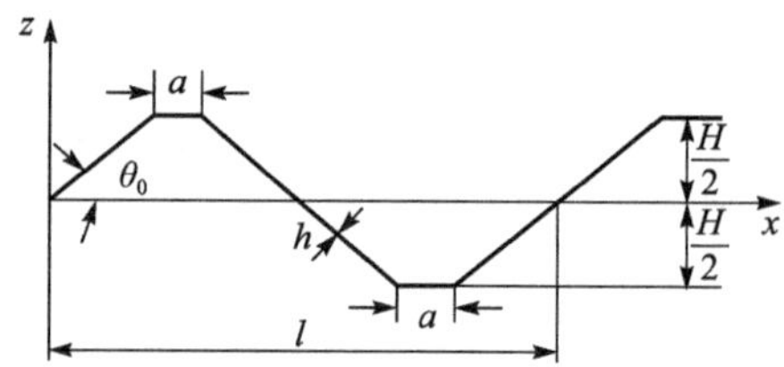

附图 2　一个梯形波纹截面

波纹环形板的非线性弯曲①

一、引言

本文所讨论的波纹圆板，在技术上应用范围很广，尤其是精密仪器的弹性元件方面的应用。因此，这种板的非线性弯曲理论在近年来得到了迅速发展，出现了一些新的论文[1-4]。一般说来，这些理论分析可分为两类：一类是使用扁薄壳的非线性弯曲理论进行分析。Панов[5]，Феодосьев[6,7]，Аксельрад[8]和陈山林[9]等进行了这方面的工作。但其解的分析较复杂，因而迄今仅仅讨论了一种特殊情况，即具有正弦波纹的波纹圆板。而且在挠度较大时，理论结果与实验值之间还存在较大差别。

在实际情况下，因为波纹圆板的波纹分布得均匀致密，所以可将它视为各向异性圆板。于是，我们应用各向异性圆板的非线性弯曲理论来进行分析。在文献［1］～［4］中，Андреева[10,11]，赤坂隆[12]，Haringx[13,14]，Бурмистров[15]和张其浩[16]等也进行了这方面的探讨。应用这种分析，所得的解适用于具有不同形状波纹的波纹圆板。但是只有少数是关于波纹环形板的工作。Андреева[11]和赤坂隆[12]分别用迦辽金法和能量法讨论过具有硬中心的波纹环形板。然而，所得的结果尚不能令人满意。因此，进一步讨论这种波纹环形板的工作是很有意义的。

为了得到更令人满意的解，在本文中我们采用修正迭代法来研究波纹环形板。这一方法早在 1965 年由文献［17］～［19］提出，后来在一系列板壳的非线性弯曲和稳定问题中得到了应用[1-4,20-27]。此法吸收了文献［28］的摄动法和常用的逐次逼近法的优点，迭代程序明确、简单、精确度高，具有较广泛的适用范围。

二、基本方程和边界条件

图 1 所示为一个在集中载荷 p 作用下的具有硬中心的波纹环形板。外边界夹紧固定，内边界固定在一个硬中心上，且可以上下移动。令 h 是厚度，a 是外半径，b 是内半径，r 是径向坐标，L 是波长，H 是波幅。

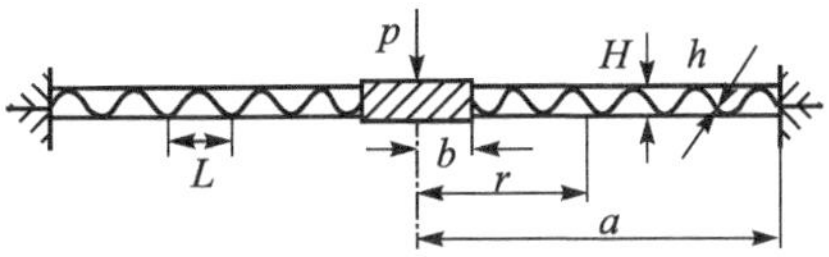

图 1　波纹环形板

从承受均匀载荷的各向异性圆板的非线性弯曲的一般理论[1-3,10,11]出发，我们易于得到以下在集中载荷 p 作用下具有硬中心的波纹环形板的由挠度 w 和径向薄膜内力 N_r 所满足的基本方程和边界条件

$$r\frac{\mathrm{d}^3w}{\mathrm{d}r^3}+\frac{\mathrm{d}^2w}{\mathrm{d}r^2}-k_t\frac{\mathrm{d}w}{\mathrm{d}r}\left(\frac{1}{D}rN_r+\frac{k_t'}{r}\right)=\frac{k_tp}{2\pi D},$$

① 本文原载《中国科学》，A辑，1984，(3)：247-253.

$$r\frac{\mathrm{d}^2w}{\mathrm{d}r^2}(rN_r)+\frac{\mathrm{d}}{\mathrm{d}r}(rN_r)-k_rk_tN_r+\frac{Ehk_t}{2}\left(\frac{\mathrm{d}w}{\mathrm{d}r}\right)^2=0 \tag{1}$$

当 $r=a$ 时，$w=0$，$\frac{\mathrm{d}w}{\mathrm{d}r}=0$，$u=0$

当 $r=b$ 时，$\frac{\mathrm{d}w}{\mathrm{d}r}=0$，$u=0$　　(2)

式中 E 是弹性模量，ν 是泊松比，D 是抗弯刚度，

$$D=\frac{Eh^3}{12\left(1-\frac{\nu^2}{k_tk_t'}\right)} \tag{3}$$

u 是径向位移，

$$u=\frac{r}{k_tEh}\left[\frac{\mathrm{d}}{\mathrm{d}r}(rN_r)-\nu N_r\right] \tag{4}$$

k_r，k_t 和 k_t' 分别是与径向和周向刚度有关的参数。

（1）对于正弦波纹（图 2(a)）

$$k_r=1+\frac{3H^2}{2h^2}+\frac{\pi^2H^2}{4L^2}\left(\frac{3H^2}{4h^2}-1\right),$$
$$k_t=1+\frac{\pi^2H^2}{4L^2},$$
$$k_t'=1+\frac{3H^2}{2h^2}+\frac{3\pi^2H^2}{4L^2}\left(\frac{H^2}{4h^2}+1\right) \tag{5}$$

（2）对于锯齿形波纹（图 2(b)）

$$k_r=\frac{H^2}{h^2\cos\theta_0}+\cos\theta_0,$$
$$k_t=\frac{1}{\cos\theta_0},$$
$$k_t'=\frac{H^2}{h^2\cos\theta_0}+\frac{1}{\cos^3\theta_0} \tag{6}$$

（3）对于梯形波纹（图 2(c)）

$$k_r=\frac{H^2}{Lh^2}\left(\frac{L-2L_0}{\cos\theta_0}+6L_0\right)+\left(1-\frac{2L_0}{L}\right)\cos\theta_0+\frac{2L_0}{L},$$
$$k_t=\frac{1}{L}\left(\frac{L-2L_0}{\cos\theta_0}+2L_0\right),$$
$$k_t'=\frac{H^2}{Lh^2}\left(\frac{L-2L_0}{\cos\theta_0}+6L_0\right)+\frac{1}{L}\left(\frac{L-2L_0}{\cos^3\theta_0}+2L_0\right) \tag{7}$$

我们首先引入下列符号

$$y=\frac{r}{a},\alpha=\frac{b}{a},W=\frac{w}{h},\varphi=\frac{\mathrm{d}W}{\mathrm{d}y},$$
$$S=-\frac{ak_tr}{D}N_r,P=\frac{a^2k_t}{2\pi hD}p,\beta_1^2=k_tk_t',$$
$$\beta_2^2=k_rk_t,\beta_3=\frac{Eh^3k_t^2}{2D} \tag{8}$$

以便将非线性边值问题（1）和（2）变换成无量纲的形式。用了这些符号后，问题（1）和（2）分别化为

$$y\frac{d^2\varphi}{dy^2}+\frac{d\varphi}{dy}-\frac{\beta_1^2}{y}\varphi+S\varphi=P,$$

$$y\frac{d^2S}{dy^2}+\frac{dS}{dy}-\frac{\beta_2^2}{y}S-\beta_3\varphi^2=0 \tag{9}$$

当 $y=1$ 时，$W=0$，$\varphi=0$，$\frac{dS}{dy}-\nu\frac{S}{y}=0$

当 $y=\alpha$ 时，$\varphi=0$，$\frac{dS}{dy}-\nu\frac{S}{y}=0$　(10)

再引入两个微分算子

$$L_1(\cdots)=y^{\beta_1}\frac{d}{dy}y^{1-2\beta_1}\frac{d}{dy}(\cdots),$$

$$L_2(\cdots)=y^{\beta_2}\frac{d}{dy}y^{1-2\beta_2}\frac{d}{dy}(\cdots) \tag{11}$$

方程（9）可表示成下列简化形式

$$L_1(y^{\beta_1}\varphi)=-S\varphi+P,$$

$$L_2(y^{\beta_2}S)=\beta_3\varphi^2 \tag{12}$$

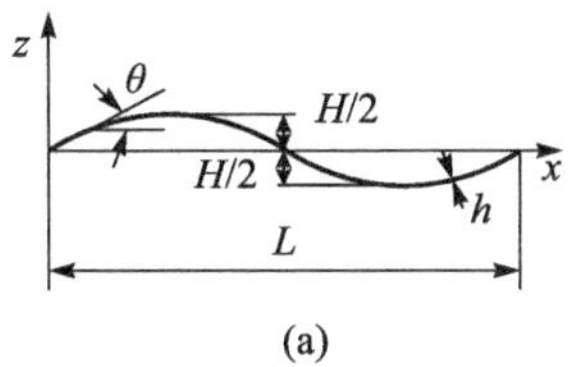

(a)

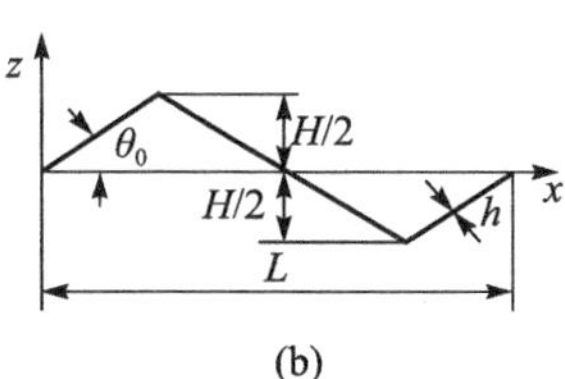

(b)

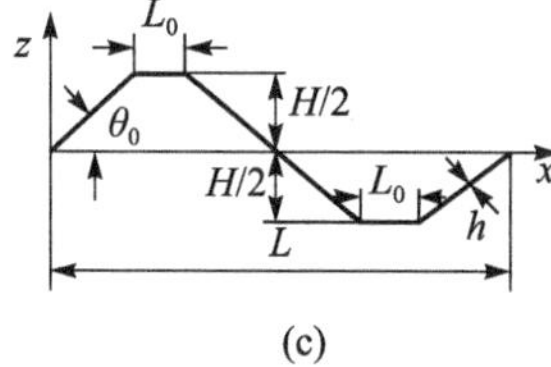

(c)

图 2　波纹环形板的径向截面上的一个波纹

(a) 正弦波纹；(b) 锯齿形波纹；(c) 梯形波纹

这样，在集中载荷 p 作用下具有硬中心的波纹环形板的非线性弯曲问题，便归结为计算无量纲非线性边值问题（12）和（10）。因为这个问题包含了两个非线性项 S_φ 和 φ^2，所以求解它是很困难的。

三、求解无量纲非线性边值问题

为了得到无量纲非线性边值问题（12）和（10）的更令人满意的解，我们使用了修正迭代法。

在方程（12）中，取非线性项 $S\varphi=0$ 和 $\varphi^2=\varphi_1^2$，得到的第一次近似的边值问题如下

$$L_1(y^{\beta_1}\varphi_1)=P,$$

$$L_2(y^{\beta_2}S_1)=\beta_3\varphi_1^2 \tag{13a,b}$$

当 $y=1$ 时，$W_1=0$，$\varphi_1=0$，$\frac{dS_1}{dy}-\nu\frac{S_1}{y}=0$　(14a～c)

当 $y=\alpha$ 时，$\varphi_1=0$，$\frac{dS_1}{dy}-\nu\frac{S_1}{y}=0$　(15a，b)

求解问题（13a）、（14b）和（15a），得到

$$\varphi_1 = \frac{P}{\mu}(a_1 y^{\beta_1} + a_2 y^{-\beta_1} - y) \tag{16}$$

式中

$$a_1 = \frac{\alpha - \alpha^{-\beta_1}}{\alpha^{\beta_1} - \alpha^{-\beta_1}}, a_2 = \frac{\alpha^{\beta_1} - \alpha}{\alpha^{\beta_1} - \alpha^{-\beta_1}},$$

$$\mu = \beta_1^2 - 1 \tag{17}$$

假定迭代参数是波纹环形板在 $y=\alpha$ 处的无量纲挠度 W_m。在应用式（8）的第三个和边界条件（10）的第一个以后，则可将 W_m 表示成如下形式

$$W_m = -\int_\alpha^1 \varphi \mathrm{d}y \tag{18}$$

将式（16）代入式（18）中，进行积分后，便得在集中载荷 p 作用下具有硬中心的波纹环形板的小挠度特征关系式

$$P = \lambda_1 \mu W_m \tag{19}$$

式中

$$\lambda_1 = -\left[\frac{a_1(1-\alpha^{1+\beta_1})}{1+\beta_1} + \frac{a_2(1-\alpha^{1-\beta_1})}{1-\beta_1} + \frac{1}{2}(\alpha^2 - 1)\right]^{-1} \tag{20}$$

现在利用式（19），便可将式（16）的斜度 φ_1 表为下面形式

$$\varphi_1 = \lambda_1 W_m (a_1 y^{\beta_1} + a_2 y^{-\beta_1} - y) \tag{21}$$

应用已得的解（21），关于无量纲径向薄膜力的问题（13b），（14c）和（15b）的解是

$$\begin{aligned} S_1 = {} & \beta_3 \lambda_1^2 W_m^2 (b_1 y^{1+2\beta_1} + b_2 y^{1-2\beta_1} + b_3 y^{2+\beta_1} + b_4 y^{2-\beta_1} \\ & + b_5 y^{\beta_2} + b_6 y^{-\beta_2} + b_7 y^3 + b_8 y) \end{aligned} \tag{22}$$

式中

$$\begin{aligned} & b_1 = \frac{a_1^2}{(1+2\beta_1)^2 - \beta_2^2}, \qquad b_2 = \frac{a_2^2}{(1-2\beta_1)^2 - \beta_2^2}, \\ & b_3 = -\frac{2a_1}{(2+\beta_1)^2 - \beta_2^2}, \qquad b_4 = -\frac{2a_2}{(2-\beta_1)^2 - \beta_2^2}, \\ & b_5 = \frac{\omega_2 \alpha - \omega_1 \alpha^{-\beta_2}}{(\beta_2 - \nu)(\alpha^{\beta_2} - \alpha^{-\beta_2})}, \qquad b_6 = \frac{\omega_2 \alpha - \omega_1 \alpha^{\beta_2}}{(\beta_2 + \nu)(\alpha^{\beta_2} - \alpha^{-\beta_2})}, \\ & b_7 = \frac{1}{9 - \beta_2^2}, \qquad b_8 = \frac{2a_1 a_2}{1 - \beta_2^2}, \\ & \omega_1 = -[b_1(1+2\beta_1-\nu) + b_2(1-2\beta_1-\nu) + b_3(2+\beta_1-\nu) \\ & \qquad + b_4(2-\beta_1-\nu) + b_7(3-\nu) + b_8(1-\nu)], \\ & \omega_2 = -[b_1(1+2\beta_1-\nu)\alpha^{2\beta_1} + b_2(1-2\beta_1-\nu)\alpha^{-2\beta_1} \\ & \qquad + b_3(2+\beta_1-\nu)\alpha^{1+\beta_1} + b_4(2-\beta_1-\nu)\alpha^{1-\beta_1} \\ & \qquad + b_7(3-\nu)\alpha^2 + b_8(1-\nu)] \end{aligned} \tag{23}$$

对于二次近似，我们限于研究斜度 φ，于是有如下的线性边值问题

$$L_1(y^{\beta_1}\varphi_2) = -S_1\varphi_1 + P \tag{24}$$

当 $y=1$ 时，$W_2=0$，$\varphi_2=0$ (25a，b)

当 $y=\alpha$ 时，$\varphi_2=0$ (26)

应用解（21）和（22），得到问题（24）、(25b）和（26）的解

$$\varphi_2=\frac{P}{\mu}(a_1y^{\beta_1}+a_2y^{-\beta_1}-y)-\beta_3\lambda_1^3W_m^3[F(y)+c_{17}y^{\beta_1}+c_{18}y^{-\beta_1}], \tag{27}$$

式中

$$\begin{aligned}F(y)=&c_1y^{2+3\beta_1}+c_2y^{2-3\beta_1}+c_3y^{3+2\beta_1}+c_4y^{3-2\beta_1}\\&+c_5y^{4+\beta_1}+c_6y^{4-\beta_1}+c_7y^{2+\beta_1}+c_8y^{2-\beta_1}+c_9y^{1+\beta_1+\beta_2}\\&+c_{10}y^{1+\beta_1-\beta_2}+c_{11}y^{1-\beta_1+\beta_2}+c_{12}y^{1-\beta_1-\beta_2}+c_{13}y^{2+\beta_2}\\&+c_{14}y^{2-\beta_2}+c_{15}y^5+c_{16}y^3,\end{aligned}$$

$$c_1=\frac{a_1b_1}{4(1+\beta_1)(1+2\beta_1)},\quad c_2=\frac{a_2b_2}{4(1-\beta_1)(1-2\beta_1)},$$

$$c_3=\frac{a_1b_3-b_1}{3(1+\beta_1)(3+\beta_1)},\quad c_4=\frac{a_2b_4-b_2}{3(1-\beta_1)(3-\beta_1)},$$

$$c_5=\frac{a_1b_7-b_3}{8(2+\beta_1)},\quad c_6=\frac{a_2b_7-b_4}{8(2-\beta_1)},$$

$$c_7=\frac{a_1b_8+a_2b_1}{4(1+\beta_1)},\quad c_8=\frac{a_1b_2+a_2b_8}{4(1-\beta_1)},$$

$$c_9=\frac{a_1b_5}{(1+\beta_2)(1+2\beta_1+\beta_2)},\quad c_{10}=\frac{a_1b_6}{(1-\beta_2)(1+2\beta_1-\beta_2)},$$

$$c_{11}=\frac{a_2b_5}{(1+\beta_2)(1-2\beta_1+\beta_2)},\quad c_{12}=\frac{a_2b_6}{(1-\beta_2)(1-2\beta_1-\beta_2)},$$

$$c_{13}=\frac{b_5}{\beta_1^2-(2+\beta_2)^2},\quad c_{14}=\frac{b_6}{\beta_1^2-(2-\beta_2)^2},$$

$$c_{15}=-\frac{b_7}{25-\beta_1^2},\quad c_{16}=\frac{a_1b_4+a_2b_3-b_8}{9-\beta_1^2},$$

$$c_{17}=\frac{F(1)\alpha^{-\beta_1}-F(\alpha)}{\alpha^{\beta_1}-\alpha^{-\beta_1}},\quad c_{18}=-\frac{F(1)\alpha^{\beta_1}-F(\alpha)}{\alpha^{\beta}-\alpha^{-\beta_1}} \tag{28}$$

将解（27）代入式（18）中，进行积分后，得到在集中载荷 p 作用下具有硬中心的波纹环形板的特征关系式

$$P=\mu(\lambda_1W_m+\lambda_3W_m^3) \tag{29}$$

式中

$$\begin{aligned}\lambda_3=\beta_3\lambda_1^4\Big[&\frac{c_1}{3(1+\beta_1)}(\alpha^{3(1+\beta_1)}-1)+\frac{c_2}{3(1-\beta_1)}(\alpha^{3(1-\beta_1)}-1)\\&+\frac{c_3}{2(2+\beta_1)}(\alpha^{2(2+\beta_1)}-1)+\frac{c_4}{2(2-\beta_1)}(\alpha^{2(2-\beta_1)}-1)\\&+\frac{c_5}{5+\beta_1}(\alpha^{5+\beta_1}-1)+\frac{c_6}{5-\beta_1}(\alpha^{5-\beta_1}-1)\end{aligned}$$

$$
+\frac{c_7}{3+\beta_1}(\alpha^{3+\beta_1}-1)+\frac{c_8}{3-\beta_1}(\alpha^{3-\beta_1}-1)
$$

$$
+\frac{c_9}{2+\beta_1+\beta_2}(\alpha^{2+\beta_1+\beta_2}-1)+\frac{c_{10}}{2+\beta_1-\beta_2}(\alpha^{2+\beta_1-\beta_2}-1)
$$

$$
+\frac{c_{11}}{2-\beta_1+\beta_2}(\alpha^{2-\beta_1+\beta_2}-1)+\frac{c_{12}}{2-\beta_1-\beta_2}(\alpha^{2-\beta_1-\beta_2}-1)
$$

$$
+\frac{c_{13}}{3+\beta_2}(\alpha^{3+\beta_2}-1)+\frac{c_{14}}{3-\beta_2}(\alpha^{3-\beta_2}-1)+\frac{c_{15}}{6}(\alpha^6-1)
$$

$$
+\frac{c_{16}}{4}(\alpha^4-1)+\frac{c_{17}}{1+\beta_1}(\alpha^{1+\beta_1}-1)+\frac{c_{18}}{1-\beta_1}(\alpha^{1-\beta_1}-1)\Big] \tag{30}
$$

为便于实际应用，我们将此特征关系式转回到有量纲的形式

$$
p=\lambda_0\mu(h^2\lambda_1 w_m+\lambda^3 w_m^3) \tag{31}
$$

式中，w_m 是波纹环形板在 $r=b$ 处的挠度，

$$
\lambda_0=\frac{\pi Ehk_t'}{6a^2(\beta_1^2-\nu^2)} \tag{32}
$$

四、数例

我们以具有正弦波纹的波纹环形板为例（图 1）。这一波纹环形板的尺寸是：外半径 $a=78$mm，内半径 $b=18$mm，厚度 $h=0.4$mm，波长 $L=12$mm，波幅 $H=1.340$mm，弹性模量 $E=1.345\times10^4$kg/mm^2，泊松比 $\nu=0.3$。

对于这样一种情况，我们由特征关系式（31）得到

$$
p=4.744w_m+0.3168w_m^3 \tag{33}
$$

式中 w_m 的单位是 mm，p 的单位是 kg。

按照式（33），图 3 给出了特征曲线。为了便于比较，我们把用 Андреева[11] 和赤坂隆[12] 的公式进行计算所得的结果以及赤坂隆的实验值都一并给在图 3 上。由此图看到，本文结果与实验值完全一致，较 Андреева 和赤坂隆的公式有更高的精确度。

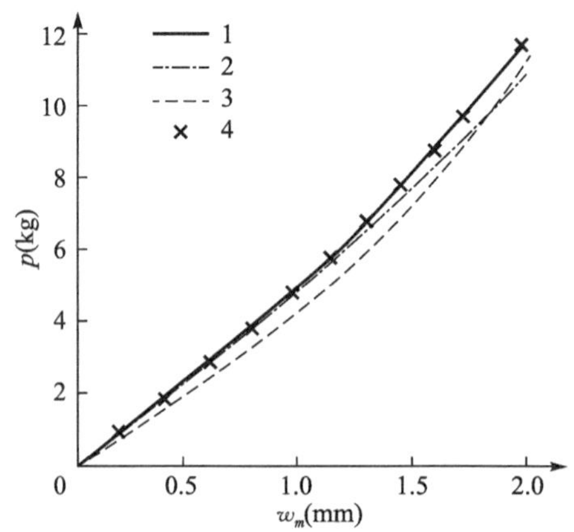

图 3　理论与实验结果之间的比较

1. 本文，2. 赤坂隆[12]，3. Андреева[11]，4. 实验值[12]

五、结语

本文是文献［1］～［4］工作的继续。使用修正迭代法，得到了在集中载荷作用下具有硬中心的波纹环形板的非线性弯曲问题的解析解。这解有较高的精确度，并且对于深、浅不同形状波纹的波纹环形板均能适用。

作者对于西德鲁尔大学的 Zerna 教授的热情接待表示衷心的感谢。

参 考 文 献

［1］ 刘人怀. 波纹圆板的特征关系式. 力学学报，1978，(1)：47.

［2］ 刘人怀. 具有光滑中心的波纹圆板的特征关系式. 中国科学技术大学学报，1979，9 (2)：75.

［3］ Liu Ren-huai. Large deflection of corrugated circular plate with plane boundary region. Solid Mechanics Archives，1984，9 (4)：383.

［4］ Liu Ren-huai. Large deflection of corrugated circular plate with a plane central region under the action of concentrated loads at the center. International Journal of Non-Linear Mechanics，1984，19 (5)：409.

［5］ Панов，Д. Ю. О больших прогибах круглых мембран со слабым гофром. ПММ.，1941，5 (2)：303.

［6］ Феодосьев，В. И. О больших прогибах и устойчивости круглой мембраны с Мелкой гофрировкой. ПММ.，1945，9 (2)：389.

［7］ Феодосьев，В. И. Упругие Элементы Точного Приборостроения（中译本：精密仪器弹性元件的理论与计算，卢文达和熊大達译. 北京：科学出版社，1963）. Москва：Оборонгиз，1949.

［8］ Аксельрад，Э. Л. Ъольшие прогибы гофрированной мембраны как непологой оболочки. Изв. АН СССР Мех. Маw.，1964，(1)：46.

［9］ 陈山林. 浅正弦波纹圆板在均布载荷下的大挠度弹性特征. 应用数学和力学，1980，1 (2)：261.

［10］ Андреева，Л. Е. расгет гофрированных мембран，как анизотриных лдастпнок. Иижененерный Сборник，1955，21：128.

［11］ Андреева，Л. Е. Упрусие Элемнты Приборов. Москва：Машгиз，1962.

［12］ 赤坂隆. Corrngated diaphragm の弾性特性について. 日本航空学会誌，1955，3 (22～23)：279.

［13］ Haringx，J. A. Nonlinearity of corrugated diaphragms. Appl. Sci. Res.，Ser. A，1956，16 (1)：55.

［14］ Haringx，J. A. Design of corrugated diaphragms. Trans. ASME，Ser. A，1957，79 (1)：55.

［15］ Бурмистров，Е. Ф. СиммЕтричный изгпб неоднородных и однородных ортотропных оболочек вращения с учегом ьольших прогибов и неравномерного температурного поля. Инженерный Сборник. 1960，27：185.

［16］ 张其浩. 波纹圆板的特征关系的研讨. 力学与实践，1980，2 (3)：64.

［17］ 叶开沅，刘人怀，张传智，等. 圆底扁薄球壳在边缘力矩作用下的非线性稳定问题. 科学通报，1965，(2)：145.

［18］ 叶开沅，刘人怀，平庆元，等. 在对称线布载荷作用下的圆底扁薄球壳的非线性稳定问题.

兰州大学学报，1965，(2)：10.
[19] 刘人怀. 在内边缘均布力矩作用下中心开孔圆底扁薄球壳的非线性稳定问题. 科学通报，1965，(3)：253.
[20] 刘人怀. 在边缘载荷作用下中心开孔圆底扁薄球壳的轴对称稳定性. 力学学报，1977，(3)：206.
[21] 叶开沅，刘人怀，张传智，等. 弹性圆底扁球壳在边缘均布力矩作用下的非线性稳定问题. 应用数学与力学，1980，1 (1)：71.
[22] 刘人怀. 在边缘力矩作用下夹层圆板的非线性轴对称弯曲问题. 中国科学技术大学学报，1980，10 (2)：56.
[23] 叶开沅，顾淑贤. 均布载荷作用下圆底扁薄球壳的非线性稳定性. 1980 年全国计算力学会议文集，北京：北京大学出版社，1980：280.
[24] 刘人怀. 双层金属中心开孔扁球壳的非线性热稳定问题. 中国科学技术大学学报，1981，11 (1)：84.
[25] 刘人怀. 双层金属截头偏锥壳的热稳定性. 力学学报，特刊，1981，172.
[26] 刘人怀. 夹层圆板的非线性弯曲. 应用数学与力学，1981，2 (2)：173.
[27] Liu Ren-huai. Non-linear thermal stability of bimetallic shallow shells of revolution. International Journal of Non-Linear Mechanics，1983，18 (5)：409.
[28] Chien Wei-zang. Large deflection of a circular clamped plate under uniform pressure. Chinese Journal of Physics，1947，7 (2)：102.

Large deflection of corrugated circular plate with a plane central region under the action of concentrated loads at the center①

中文摘要

本文研究了中心载荷作用下具有光滑中心的波纹圆板的大挠度问题。由于理论分析相当困难，至今还没有人研究过这种波纹圆板。我们再次应用修正迭代法，并借助各向异性圆板的非线性弯曲理论，研究了这种波纹圆板，给出了一个相当精确的解，可直接用在测量仪器弹性元件设计里。

1. Introduction

Corrugated circular plates are often adopted as elastic elements in measuring instruments which are applied widely in engineering. The practical importance of the plates has led to a number of special investigations. So far as we know, in most cases, investigators[1-14] discussed only the problem of large deflections of a uniformly loaded corrugated circular plate. It is unfortunate that there are only a few works about the problem of large deflections of a corrugated circular plate under a concentrated load. Akasaka[5] has investigated a simpler problem, the nonlinear bending of a corrugated circular plate with a nondeformable rigid body at the center. However, in practical instruments, a corrugated circular plate usually does not have a nondeformable rigid body at the center but a plane region. Obviously, the theoretical analysis of the plate is very difficult. The difficulties in the investigation are due to more nonlinear differential equations and a more complicated shape of the plate. Therefore up to now the plate has not yet been studied. Proceeding as in the author's papers[11-13] we can once more apply the modified iteration method[15-17] to study the plate by means of the nonlinear bending theory for anisotropic circular plates. This method was suggested by Yeh and the author in 1965 and then applied successfully in a system of problems of the nonlinear theory of shells and plates[18-22]. This paper gives a quite accurate solution which may be applied directly to the design of elastic elements in measuring instruments.

2. Basic Equations

Consider a corrugated circular plate with a plane central region under the action of a concentrated load P at the center as shown in Fig. 1. Here r is the radial coordinate, R_1 and R are radii of the plane central region and the boundary respectively, h is the thickness, H is the amplitude, L is the wavelength. θ is the slope angle of the tangent to

---

① Reprinted from *International Journal of Non-Linear Mechanics*, 1984, 19 (5): 409-419.

one diametral section.

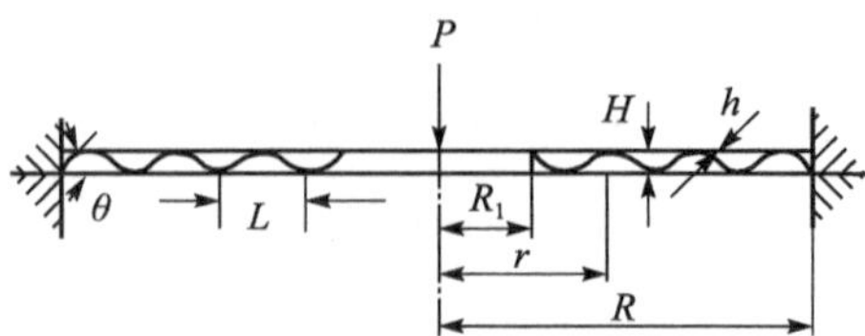

Fig. 1　A corrugated circular plate with a plane central region of concentrated loads at the center

In order to keep our formulas simpler, we now shall distinguish between some magnitudes of two regions of the plate. For this purpose, we indicate these magnitudes of the plane central region and the corrugated region by the subscripts a and b respectively.

According to large-deflection theories of isotropic and anisotropic circular plates[14,11-13,23-25], we obtain the following basic equations of the plate.

for $0\leqslant r\leqslant R_1$,

$$
\begin{aligned}
& r\frac{\mathrm{d}^3 w_a}{\mathrm{d}r^3}+\frac{\mathrm{d}^2 w_a}{\mathrm{d}r^2}-\frac{\mathrm{d}w_a}{\mathrm{d}r}\left(\frac{1}{D_a}rN_{r,a}+\frac{1}{r}\right)=\frac{P}{2\pi D_a},\\
& r\frac{\mathrm{d}^2}{\mathrm{d}r^2}(rN_{r,a})+\frac{\mathrm{d}}{\mathrm{d}r}(rN_{r,a})-N_{r,a}+\frac{Eh}{2}\left(\frac{\mathrm{d}w_a}{\mathrm{d}r}\right)^2=0;
\end{aligned}
\tag{1}
$$

for $R_1\leqslant r\leqslant R$,

$$
\begin{aligned}
& r\frac{\mathrm{d}^3 w_b}{\mathrm{d}r^3}+\frac{\mathrm{d}^2 w_b}{\mathrm{d}r^2}-k_t\frac{\mathrm{d}w_b}{\mathrm{d}r}\left(\frac{1}{D_b}rN_{r,b}+\frac{k_t'}{r}\right)=\frac{k_t}{2\pi D_b}P,\\
& r\frac{\mathrm{d}^2}{\mathrm{d}r^2}(rN_{r,b})+\frac{\mathrm{d}}{\mathrm{d}r}(rN_{r,b})-k_rk_tN_{r,b}+\frac{1}{2}Ehk_t\left(\frac{\mathrm{d}w_b}{\mathrm{d}r}\right)^2=0,
\end{aligned}
\tag{2}
$$

where w is the deflection, N_r is the radial membrane stress, E is Young's modulus, ν is Poisson's ratio, D is the flexural rigidity,

$$
D_a=\frac{Eh^3}{12(1-\nu^2)},\ D_b=\frac{Eh^3}{12\left(1-\dfrac{\nu^2}{k_tk_t'}\right)},
\tag{3}
$$

and k_r, k_t and k_t' are parameters for the radial and circumferential rigidity respectively,

$$
\begin{aligned}
k_r &=\frac{12}{h^2L}\int_0^L\frac{z^2}{\cos\theta}\mathrm{d}x+\frac{1}{L}\int_0^L\cos\theta\mathrm{d}x,\\
k_t &=\frac{1}{L}\int_0^L\frac{1}{\cos\theta}\mathrm{d}x,\\
k_t' &=\frac{12}{h^2L}\int_0^L\frac{z^2}{\cos\theta}\mathrm{d}x+\frac{1}{L}\int_0^L\frac{1}{\cos^3\theta}\mathrm{d}x.
\end{aligned}
\tag{4}
$$

For the sinusoidal corrugation (Fig. 2a),

$$
\begin{aligned}
k_r &\approx 1+\frac{3H^2}{2h^2}+\frac{\pi^2H^2}{4L^2}\left(\frac{3H^2}{4h^2}-1\right),\\
k_t &\approx 1+\frac{\pi^2H^2}{4L^2},
\end{aligned}
$$

$$k_t' \approx 1+\frac{3H^2}{2h^2}+\frac{3\pi^2 H^2}{4L^2}\left(\frac{H^2}{4h^2}+1\right); \tag{5}$$

For the toothed corrugation (Fig. 2b),

$$k_r=\frac{H^2}{h^2\cos\theta_0}+\cos\theta_0,$$
$$k_t=\frac{1}{\cos\theta_0},$$
$$k_t'=\frac{H^2}{h^2\cos\theta_0}+\frac{1}{\cos^3\theta_0}; \tag{6}$$

and for the trapezoidal corrugation (Fig. 2c),

$$k_r=\frac{H^2}{Lh^2}\left(\frac{L-2L_0}{\cos\theta_0}+6L_0\right)+\left(1-\frac{2L_0}{L}\right)\cos\theta_0+\frac{2L_0}{L},$$
$$k_t=\frac{1}{L}\left(\frac{L-2L_0}{\cos\theta_0}+2L_0\right),$$
$$k_t'=\frac{H^2}{Lh^2}\left(\frac{L-2L_0}{\cos\theta_0}+6L_0\right)+\frac{1}{L}\left(\frac{L-2L_0}{\cos^3\theta_0}+2L_0\right). \tag{7}$$

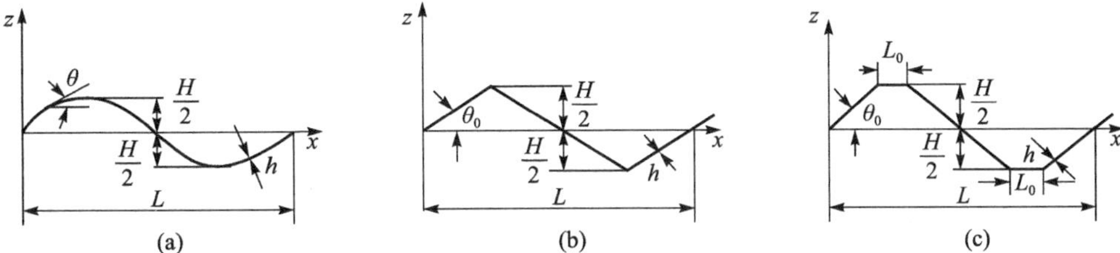

Fig. 2 A corrugation in one diametral section of the corrugated circular plate
(a) The sinusoidal corrugation; (b) the toothed corrugation;
(c) the trapezoidal corrugation

The nonlinear equations (1) and (2) will be solved for the following customary two types of boundary conditions:

1) Edge rigidly clamped

$$\frac{dw_a}{dr}=0,\ N_{r,a}\ \text{finite} \qquad \text{at } r=0,$$
$$w_b=0,\ \frac{dw_b}{dr}=0,\quad u_b=0 \qquad \text{at } r=R; \tag{8}$$

2) Edge clamped but free to slip

$$\frac{dw_a}{dr}=0\quad N_{r,a}\ \text{finite} \qquad \text{at } r=0,$$
$$w_b=0,\ \frac{dw_b}{dr}=0,\quad N_{r,b}=0 \qquad \text{at } r=R, \tag{9}$$

where u is the radial displacement,

$$u_a=\frac{r}{Eh}\left[\frac{d}{dr}(rN_{r,a})-\nu N_{r,a}\right],$$

$$u_b = \frac{r}{k_t E h}\left[\frac{\mathrm{d}}{\mathrm{d}r}(rN_{r,b}) - \nu N_{r,b}\right]. \tag{10}$$

Lastly, we must also consider the following conditions of continuity between the two regions

$$w_a = w_b,\ \frac{\mathrm{d}w_a}{\mathrm{d}r} = \frac{\mathrm{d}w_b}{\mathrm{d}r},\ M_{r,a} = M_{r,b},$$

$$N_{r,a} = N_{r,b},\ u_a = u_b \quad \text{at } r = R_1, \tag{11}$$

where M_r is the radial moment,

$$M_{r,a} = -D_a\left(\frac{\mathrm{d}^2 w_a}{\mathrm{d}r^2} + \frac{\nu}{r}\frac{\mathrm{d}w_a}{\mathrm{d}r}\right),\ M_{r,b} = -\frac{D_b}{k_t}\left(\frac{\mathrm{d}^2 w_b}{\mathrm{d}r^2} + \frac{\nu}{r}\frac{\mathrm{d}w_b}{\mathrm{d}r}\right). \tag{12}$$

Defining the nondimensional parameters as

$$y = \frac{r}{R},\ y_1 = \frac{R_1}{R},\ W = \frac{w}{h},\ \phi = \frac{\mathrm{d}W}{\mathrm{d}y},$$

$$S_a = \frac{Rr}{D_a}N_{r,a},\ S_b = \frac{k_t Rr}{D_b}N_{r,b},\ p = \frac{R^2}{2\pi h D_a}P,$$

$$\beta_0 = 6(1-\nu^2),\ \beta_1^2 = k_t k_t',\ \beta_2^2 = k_r k_t,\ \beta_3 = \frac{D_b}{D_a k_t},$$

$$\beta_4 = \frac{\beta_3}{k_t},\ \beta_5 = \nu(1-\beta_3),\ \beta_6 = \nu\left(1-\frac{1}{k_t}\right). \tag{13}$$

Then equations (1) and (2) transform to

for $0 \leqslant y \leqslant y_1$

$$L_0(y\phi_a) = S_a\phi_a + p,$$

$$L_0(yS_a) = -\beta_0\phi_a^2; \tag{14a,b}$$

for $y_1 \leqslant y \leqslant 1$

$$y_1(h^{\beta_1}\phi_b) = S_b\phi_b + \frac{p}{\beta_3},$$

$$L_2(y^{\beta_2}S_b) = -\frac{\beta_0}{\beta_4}\phi_b^2, \tag{15a,b}$$

in which L_i are the differential operators defined by

$$L_0() = y\frac{\mathrm{d}}{\mathrm{d}y}\frac{1}{y}\frac{\mathrm{d}}{\mathrm{d}y}(),$$

$$L_1() = y^{\beta_1}\frac{\mathrm{d}}{\mathrm{d}y}y^{-(2\beta_1-1)}\frac{\mathrm{d}}{\mathrm{d}y}(),$$

$$L_2() = y^{\beta_2}\frac{\mathrm{d}}{\mathrm{d}y}y^{-(2\beta_2-1)}\frac{\mathrm{d}}{\mathrm{d}y}(). \tag{16}$$

Similarly boundary conditions (8) and (9) can be written as:

1) Edge rigidly clamped

$$\phi_a = 0,\quad S_a = 0 \qquad \text{at } y = 0,$$

$$W_b = 0,\quad \phi_b = 0,\quad \frac{\mathrm{d}S_b}{\mathrm{d}y} - \nu\frac{S_b}{y} = 0 \qquad \text{at } y = 1; \tag{17}$$

2) Edge clamped but free to slip

$$\phi_a = 0, \quad S_a = 0 \qquad \text{at } y = 0,$$
$$W_b = 0, \quad \phi_b = 0, \quad S_b = 0 \qquad \text{at } y = 1, \tag{18}$$

and the continuity conditions as

$$W_a = W_b, \quad \phi_a = \phi_b, \quad \frac{d\phi_a}{dy} + \beta_5 \frac{\phi_a}{y} = \beta_3 \frac{d\phi_b}{dy},$$
$$S_a = \beta_3 S_b, \quad \frac{dS_a}{dy} - \beta_6 \frac{S_a}{y} = \beta_4 \frac{dS_b}{dy} \quad \text{at } y = y_1. \tag{19}$$

The four nonlinear differential equations (14) and (15), the five boundary conditions (17) or (18), and five continuity conditions (19) constitute a nondimensional nonlinear boundary value problem governing the large deflection of a corrugated circular plate with a plane central region under the action of concentrated loads at the center.

3. Analytical Solution

We shall begin with the nondimensional nonlinear boundary value problem (14), (15), (17), and (19) of a corrugated circular plate with a rigidly clamped edge. This problem will be solved by the modified iteration method.

For the first approximation, by neglecting nonlinear terms in Eqs. (14a) and (15a), and again introducing $\phi_{a,1}$ and $\phi_{b,1}$ into the right of Eqs. (14b) and (15b) respectively, we find the following linear boundary value problem

for $0 \leqslant y \leqslant y_1$,

$$L_0(y\phi_{a,1}) = p,$$
$$L_0(yS_{a,1}) = -\beta_0 \phi_{a,1}^2; \tag{20a,b}$$

for $y_1 \leqslant y \leqslant 1$,

$$L_1(y^{\beta_1}\phi_{b,1}) = \frac{p}{\beta_3},$$
$$L_2(y^{\beta_2}S_{b,1}) = -\frac{\beta_0}{\beta_4}\phi_{b,1}^2; \tag{21a,b}$$

$$\phi_{a,1} = 0, \ S_{a,1} = 0 \qquad \text{at } y = 0; \tag{22a,b}$$

$$W_{b,1} = 0, \ \phi_{b,1} = 0, \ \frac{dS_{b,1}}{dy} - \nu \frac{S_{b,1}}{y} = 0 \qquad \text{at } y = 1; \tag{23a-c}$$

$$W_{a,1} = W_{b,1}, \ \phi_{a,1} = \phi_{b,1}, \ \frac{d\phi_{a,1}}{dy} + \frac{\beta_5}{y}\phi_{a,1} = \beta_3 \frac{d\phi_{b,1}}{dy},$$
$$S_{a,1} = \beta_3 S_{b,1}, \ \frac{dS_{a,1}}{dy} - \frac{\beta_6}{y}S_{a,1} = \beta_4 \frac{dS_{b,1}}{dy} \qquad \text{at } y = y_1. \tag{24a-e}$$

Solutions of the problem (20a), (21a), (22a), (23a, b) and (24a-c) are

$$\phi_{a,1} = p(a_1 y\ln y + a_2 y),$$
$$\phi_{b,1} = p(b_1 y^{\beta_1} + b_2 y^{-\beta_1} + b_3 y); \tag{25}$$

and

$$W_{a,1} = p\left[\frac{a_1}{2}y^2\ln y + \frac{1}{2}\left(a_2 - \frac{a_1}{2}\right)y^2 + \frac{1}{\alpha_1}\right],$$

$$W_{b,1}=p\Big[\frac{b_1}{\beta_1+1}(y^{\beta_1+1}-1)-\frac{b_2}{\beta_1-1}(y^{-(\beta_1-1)}-1)+\frac{b_3}{2}(y^2-1)\Big]. \quad (26a,b)$$

where

$$a_1=\frac{1}{2},$$

$$a_2=\frac{1}{\mu_1}\Big\{\frac{1}{2}(\beta_1\beta_3-\beta_5-1)y_1^{\beta_1}\ln y_1+\Big[\frac{1}{\beta_1+1}-\frac{1}{4}(\beta_1\beta_3-\beta_5+1)\Big]y_1^{\beta_1}+\frac{1}{2}(\beta_1\beta_3+\beta_5+1)y_1^{-\beta_1}\ln y_1+\Big[\frac{1}{\beta_1-1}-\frac{1}{4}(\beta_1\beta_3+\beta_5-1)\Big]y_1^{-\beta_1}+2\beta_1\beta_3 b_3 y_1^{-1}\Big\}-\frac{1}{4},$$

$$b_1=\frac{1}{\mu_1}\Big\{b_3(\beta_1\beta_3+\beta_5+1)y_1^{-\beta_1}+\Big[b_3(\beta_3-\beta_5-1)-\frac{1}{2}\Big]y_1\Big\},$$

$$b_2=\frac{1}{\mu_1}\Big\{b_3(\beta_1\beta_3-\beta_5-1)y_1^{-\beta_1}-\Big[b_3(\beta_3-\beta_5-1)-\frac{1}{2}\Big]y_1\Big\},$$

$$b_3=-\frac{1}{\beta_3(\beta_1^2-1)},$$

$$\mu_1=-\big[(\beta_1\beta_3-\beta_5-1)y_1^{\beta_1}+(\beta_1\beta_3+\beta_5+1)y_1^{-\beta_1}\big], \quad (27)$$

$$\alpha_1=\Big[\frac{b_1}{\beta_1+1}(y_1^{\beta_1+1}-1)-\frac{b_2}{\beta_1-1}(y_1^{-(\beta_1-1)}-1)-\frac{1}{4}y_1^2\ln y_1+\frac{1}{2}\Big(b_3-a_2+\frac{1}{4}\Big)y_1^2-\frac{b_3}{2}\Big]^{-1}. \quad (28)$$

We introduce the notation W_0 for the nondimensional center deflection as an iteration parameter

$$W_0=W\,|_{y=0}. \quad (29)$$

Substituting the solution (26a) into this expression, we find

$$p=\alpha_1 W_0. \quad (30)$$

This is the characteristic relation of small deflection of a corrugated circular plate with a plane central region. Under such a case, we rewrite the above solution (25) for the slope ϕ in these forms

$$\phi_{a,1}=\alpha_1 W_0(a_1 y\ln y+a_2 y),$$

$$\phi_{b,1}=\alpha_1 W_0(b_1 y^{\beta_1}+b_2 y^{-\beta_1}+b_3 y). \quad (31)$$

Using obtained solution (31), solutions of the problem (20b), (21b), (22b), (23c) and (24 d, e) for the nondimensional radial membrane stress S are

$$S_{a,1}=\alpha_1^2\beta_0 W_0^2(c_1 y^3\ln^2 y+c_2 y^3\ln y+c_3 y^3+c_4 y),$$

$$S_{b,1}=\alpha_1^2\beta_0 W_0^2\Big[\frac{1}{\beta_4}(d_1 y^{2\beta_1+1}+d_2 y^{-(2\beta_1-1)}+d_3 y^{\beta_1+2}+d_4 y^{-(\beta_1-2)}+d_5 y^3+d_6 y)+d_7 y^{\beta_2}+d_8 y^{-\beta_2}\Big], \quad (32)$$

where

$$c_1=-\frac{1}{32},\ c_2=-\frac{1}{8}\left(a_2-\frac{3}{8}\right),$$

$$c_3=-\frac{1}{8}\left(a_2^2-\frac{3}{4}a_2+\frac{7}{32}\right),$$

$$c_4=-\frac{1}{\mu_2}\{\beta_2\beta_4\lambda_2[(\beta_2+\nu)y_1^{\beta_2-1}-(\beta_2-\nu)y^{-(\beta_2+1)}]$$
$$+\beta_3\lambda_3[(\beta_2+\nu)y_1^{\beta_2}+(\beta_2-\nu)y_1^{-\beta_2}]-2\beta_2\beta_3\beta_4\lambda_1 y_1^{-1}\},$$

$$d_1=-\frac{b_1^2}{(2\beta_1+1)^2-\beta_2^2},\ d_2=-\frac{b_2^2}{(2\beta_1-1)^2-\beta_2^2},$$

$$d_3=-\frac{2b_1b_3}{(\beta_1+2)^2-\beta_2^2},\ d_4=-\frac{2b_2b_3}{(\beta_1-2)^2-\beta_2^2},$$

$$d_5=\frac{b_3^2}{\beta_2^2-9},\ d_6=\frac{2b_1b_2}{\beta_2^2-1},$$

$$d_7=\frac{1}{\mu_2}\{\lambda_1[\beta_2\beta_4-\beta_3(\beta_6-1)]y_1^{-\beta_2}$$
$$-(\beta_2+\nu)[\lambda_3y_1-\lambda_2(\beta_6-1)]\},$$

$$d_8=\frac{1}{\mu_2}\{\lambda_1[\beta_2\beta_4+\beta_3(\beta_6-1)]y_1^{\beta_2}$$
$$-(\beta_2-\nu)[\lambda_3y_1-\lambda_2(\beta_6-1)]\},$$

$$\lambda_1=-\frac{1}{\beta_4}[d_1(2\beta_1+1-\nu)-d_2(2\beta_1-1+\nu)+d_3(\beta_1+2-\nu)$$
$$-d_4(\beta_1-2+\nu)+d_5(3-\nu)+d_6(1-\nu)],$$

$$\lambda_2=-(c_1\ln^2y_1+c_2\ln y_1+c_3)y_1^3+\frac{\beta_3}{\beta_4}(d_1y_1^{2\beta_1+1}$$
$$+d_2y_1^{-(2\beta_1-1)}+d_3y_1^{\beta_1+2}+d_4y_1^{-(\beta_1-2)}+d_5y_1^3+d_6y_1),$$

$$\lambda_3=\frac{\beta_6-3}{32}y_1^2\ln^2y_1-\left[c_2(\beta_6-3)+\frac{1}{16}\right]y_1^2\ln y_1$$
$$+[c_2-c_3(\beta_6-3)]y_1^2-d_1(2\beta_1+1)y_1^{2\beta_1}$$
$$+d_2(2\beta_1-1)y_1^{-2\beta_1}-d_3(\beta_1+2)y_1^{\beta_1+1}$$
$$+d_4(\beta_1-2)y_1^{-(\beta_1-1)}-3d_5y_1^2-d_6,$$

$$\mu_2=-(\beta_2+\nu)[\beta_2\beta_4+\beta_3(\beta_6-1)]y_1^{\beta_2}$$
$$+(\beta_2-\nu)[\beta_2\beta_4-\beta_3(\beta_6-1)]y_1^{-\beta_2}. \tag{33}$$

For simplicity, the second approximation is restricted to the determination of the slope ϕ only. We rewrite the nonlinear term $S\phi$ of Eqs. (14a) and (15a) in $S_1\phi_1$. Then we obtain the linear boundary value problem for ϕ as follows

for $0\leqslant y\leqslant y_1$,

$$L_0(y\phi_{a,2})=S_{a,1}\phi_{a,1}+p; \tag{34}$$

for $y_1\leqslant y\leqslant 1$,

$$L_1(y^{\beta_1}\phi_{b,2}) = S_{b,1}\phi_{b,1} + \frac{p}{\beta_3}; \tag{35}$$

$$\phi_{a,2} = 0 \qquad \text{at } y = 0; \tag{36}$$

$$W_{b,2} = 0, \quad \phi_{b,2} = 0, \qquad \text{at } y = 1; \tag{37}$$

$$W_{a,2} = W_{b,2}, \quad \phi_{a,2} = \phi_{b,2},$$

$$\frac{\mathrm{d}\phi_{a,2}}{\mathrm{d}y} + \beta_5\frac{\phi_{a,2}}{y} + \beta_3\frac{\mathrm{d}\phi_{b,2}}{y} \qquad \text{at } y = y_1. \tag{38}$$

Using solutions (31) and (32), the solutions to the problem are

$$\phi_{a,2} = p(a_1 y\ln y + a_2 y) + \alpha_1^3\beta_0 W_0^3[F_1(y) + e_7 y],$$

$$\phi_{b,2} = p(b_1 y^{\beta_1} + b_2 y^{-\beta_1} + b_3 y) + \alpha_1^3\beta_0 W_0^3[F_2(y) + f_{17}y^{\beta_1} + f_{18}y^{-\beta_1}]; \tag{39}$$

and

$$W_{a,2} = p\left[\frac{1}{4}y^2\ln y + \frac{1}{2}\left(a_2 - \frac{1}{4}\right)y^2 + \frac{1}{\alpha_1}\right] + \alpha_1^3\beta_0 W_0^3[G_1(y) - G_1(y_1) + G_2(y_1) - G_2(1)],$$

$$W_{b,2} = p\left[\frac{b_1}{\beta_1+1}(y^{\beta_1+1} - 1) - \frac{b_2}{\beta_1-1}(y^{-(\beta_1-1)} - 1) + \frac{b_3}{2}(y^2 - 1)\right] + \alpha_1^3\beta_0 W_0^3[G_2(y) - G_2(1)], \tag{40a,b}$$

where

$$F_1(y) = \frac{1}{8}(e_1 y^5\ln^3 y + e_2 y^5\ln^2 y + e_3 y^5\ln y + e_4 y^5 + e_5 y^3\ln y + e_6 y^3),$$

$$F_2(y) = f_1 y^{3\beta_1+2} + f_2 y^{-(3\beta_1-2)} + f_3 y^{2\beta_1+3} + f_4 y^{-(2\beta_1-3)} + f_5 y^{\beta_1+4} \\ f_6 y^{-(\beta_1-4)} + f_7 y^{\beta_1+2} + f_8 y^{-(\beta_1-2)} + f_9 y^{\beta_1+\beta_2+1} \\ + f_{10} y^{\beta_1-\beta_2+1} + f_{11} y^{-(\beta_1-\beta_2-1)} + f_{12} y^{-(\beta_1+\beta_2-1)} + f_{13} y^{\beta_2+2} \\ + f_{14} y^{-(\beta_2-2)} + f_{15} y^5 + f_{16} y^3,$$

$$G_1(y) = \frac{1}{48}\left[e_1\ln^3 y - \left(\frac{e_1}{2} - e_2\right)\ln^2 y + \left(\frac{e_1}{6} - \frac{e_2}{3} + e_3\right)\ln y - \left(\frac{e_1}{36} - \frac{e_2}{18} + \frac{e_3}{6} - e_4\right)\right]y^6 + \frac{1}{32}\left[e_5\ln y - \left(\frac{e_5}{4} - e_6\right)\right]y^4 + \frac{e_7}{2}y^2,$$

$$G_2(y) = \frac{f_1}{3(\beta_1+1)}y^{3(\beta_1+1)} - \frac{f_2}{3(\beta_1-1)}y^{-3(\beta_1-1)} + \frac{f_3}{2(\beta_1+2)}y^{2(\beta_1+2)} \\ - \frac{f_4}{2(\beta_1-2)}y^{-2(\beta_1-2)} + \frac{f_5}{\beta_1+5}y^{(\beta_1+5)} - \frac{f_6}{\beta_1-5}y^{-(\beta_1-5)} \\ + \frac{f_7}{\beta_1+3}y^{\beta_1+3} - \frac{f_8}{\beta_1-3}y^{-(\beta_1-3)} + \frac{f_9}{\beta_1+\beta_2+2}y^{\beta_1+\beta_2+2} \\ + \frac{f_{10}}{\beta_1-\beta_2+2}y^{\beta_1-\beta_2+2} - \frac{f_{11}}{\beta_1-\beta_2-2}y^{-(\beta_1-\beta_2-2)}$$

$$-\frac{f_{12}}{\beta_1+\beta_2-2}y^{-(\beta_1+\beta_2-2)}+\frac{f_{13}}{\beta_2+3}y^{\beta_2+3}-\frac{f_{14}}{\beta_2-3}y^{-(\beta_2-3)}$$

$$+\frac{f_{15}}{6}y^6+\frac{f_{16}}{4}y^4+\frac{f_{17}}{\beta_1+1}y^{\beta_1+1}-\frac{f_{18}}{\beta_1-1}y^{-(\beta_1-1)},$$

$$e_1=-\frac{1}{192},\ e_2=\frac{1}{6}\left(c_2-\frac{a_2}{16}+\frac{5}{128}\right),$$

$$e_3=\frac{1}{3}\left[a_2\left(c_2+\frac{5}{192}\right)-\frac{5}{12}c_2+\frac{c_3}{2}-\frac{19}{1536}\right],$$

$$e_4=-\frac{1}{3}\left[a_2\left(\frac{5}{12}c_2-c_3+\frac{19}{2304}\right)-\frac{1}{24}\left(\frac{19}{6}c_2-5c_3+\frac{65}{768}\right)\right],$$

$$e_5=\frac{1}{2}c_4,\ e_6=c_4\left(a_2-\frac{3}{8}\right),$$

$$e_7=-\frac{1}{\mu_1}\Big[\beta_1\beta_3\omega_2(y_1^{\beta_1-1}+y_1^{-(\beta_1+1)})$$

$$-\omega_3(y_1^{\beta_1}-y_1^{-\beta_1})+2\beta_1\beta_3\omega_1y_1^{-1}\Big],$$

$$f_1=\frac{b_1d_1}{4\beta_4(\beta_1+1)(2\beta_1+1)},\ f_2=\frac{b_2d_2}{4\beta_4(\beta_1-1)(2\beta_1-1)},$$

$$f_3=\frac{b_1d_3+b_3d_1}{3\beta_4(\beta_1+1)(\beta_1+3)},\ f_4=\frac{b_2d_4+b_3d_2}{3\beta_4(\beta_1-1)(\beta_1-3)},$$

$$f_5=\frac{b_1d_5+b_3d_3}{8\beta_4(\beta_1+2)},\ f_6=-\frac{b_2d_5+b_3d_4}{8\beta_4(\beta_1-2)},$$

$$f_7=\frac{b_1d_6+b_2d_1}{4\beta_4(\beta_1+1)},\ f_8=-\frac{b_1d_2+b_2d_6}{4\beta_4(\beta_1-1)},$$

$$f_9=\frac{b_1d_7}{(\beta_2+1)(2\beta_1+\beta_2+1)},\ f_{10}=-\frac{b_1d_8}{(\beta_2-1)(2\beta_1-\beta_2+1)},$$

$$f_{11}=-\frac{b_2d_7}{(\beta_2+1)(2\beta_1-\beta_2-1)},\ f_{12}=\frac{b_2d_8}{(\beta_2-1)(2\beta_1+\beta_2-1)},$$

$$f_{13}=-\frac{b_3d_7}{\beta_1^2-(\beta_2+2)^2},\ f_{14}=-\frac{b_3d_8}{\beta_1^2-(\beta_2-2)^2},$$

$$f_{15}=-\frac{b_3d_5}{\beta_4(\beta_1^2-25)},\ f_{16}=-\frac{b_1d_4+b_2d_3+b_3d_6}{\beta_4(\beta_1^2-9)},$$

$$f_{17}=-\frac{1}{\mu_1}[\omega_1(\beta_1\beta_3+\beta_5+1)y_1^{-\beta_1}-\omega_3y_1+\omega_2(\beta_5+1)],$$

$$f_{18}=-\frac{1}{\mu_1}[\omega_1(\beta_1\beta_3-\beta_5-1)y_1^{\beta_1}+\omega_3y_1-\omega_2(\beta_5+1)],$$

$$\omega_1=-F_2(1),\omega_2=F_2(y_1)-F_1(y_1),$$

$$\omega_3=\beta_3[f_1(3\beta_1+2)y_1^{3\beta_1+1}-f_2(3\beta_1-2)y_1^{-(3\beta_1-1)}$$

$$+f_3(2\beta_1+3)y_1^{2(\beta_1+1)}-f_4(2\beta_1-3)y_1^{-2(\beta_1-1)}$$

$$+f_5(\beta_1+4)y_1^{\beta_1+3}-f_6(\beta_1-4)y_1^{-(\beta_1-3)}$$

$$+f_7(\beta_1+2)y_1^{\beta_1+1}-f_8(\beta_1-2)y_1^{-(\beta_1-1)}$$

$$+f_9(\beta_1+\beta_2+1)y_1^{\beta_1+\beta_2}+f_{10}(\beta_1-\beta_2+1)y_1^{\beta_1-\beta_2}$$
$$-f_{11}(\beta_1-\beta_2-1)y_1^{(-\beta_1-\beta_2)}-f_{12}(\beta_1+\beta_2-1)y_1^{-(\beta_1+\beta_2)}$$
$$+f_{13}(\beta_2+2)y_1^{\beta_2+1}-f_{14}(\beta_2-2)y_1^{-(\beta_2-1)}$$
$$+5f_{15}y_1^4+3f_{16}y_1^2]-\frac{1}{8}\{e_1(\beta_5+5)y_1^4\ln^3y_1$$
$$+[3e_1+e_2(\beta_5+5)]y_1^4\ln^2y_1+[2e_2+e_3(\beta_5+5)]y_1^4\ln y_1$$
$$+[e_3+e_4(\beta_5+5)]y_1^4+e_5(\beta_5+3)y_1^2\ln y_1$$
$$+[e_5+e_6(\beta_5+3)]y_1^2\}. \tag{41}$$

Substituting the solution (40a) into the above expression (29) for the nondimensional center deflection, we obtain the characteristic relation of the corrugated circular plate with a plane central region

$$p=\alpha_1W_0+\alpha_3W_0^3, \tag{42}$$

where

$$\alpha_3=\alpha_1^4\beta_0[G_1(y_1)-G_2(y_1)+G_2(1)]. \tag{43}$$

In order to facilitate application, we may rewrite the relation (42) in the dimensional form

$$P=\alpha_0(\alpha_1h^2w_0+\alpha_3w_0^3), \tag{44}$$

where w_0 is the center deflection of the plate,

$$\alpha_0=\frac{\pi Eh}{\beta_0R^2}. \tag{45}$$

For an edge clamped but free to slip, the solution of the nondimensional nonlinear boundary value problem (14), (15), (18) and (19) can readily be found from the relation (44). For this purpose, the last of the boundary conditions (17) is used and expressed as

$$\frac{1}{\nu}\frac{dS_b}{dy}-\frac{S_b}{y}=0 \qquad \text{at } y=1, \tag{46}$$

and let $\nu\to\infty$, the condition (46) becomes

$$S_b=0 \qquad \text{at } y=1. \tag{47}$$

This is the last condition of the boundary condition (18). It is seen that in (17) and (18) only the last condition is different. Hence, let $\nu\to\infty$ in the relation (44), and we can at once obtain the characteristic relation of a corrugated circular plate with a plane central region for an edge clamped but free to slip. The present relation and most parameters coincide with the above results in appearance, except that c_4, d_7, d_8, λ_1 and μ_2 become

$$c_4=-\frac{1}{\mu_2}[\beta_2\beta_4\lambda_2(y_1^{\beta_2-1}+y_1^{-(\beta_2+1)})+\beta_3\lambda_3(y_1^{\beta_2}-y_1^{-\beta_2})-2\beta_2\beta_3\beta_4\lambda_1y_1^{-1}],$$

$$d_7=\frac{1}{\mu_2}\{\lambda_1[\beta_2\beta_4-\beta_3(\beta_6-1)]y_1^{-\beta_2}-\lambda_3y_1+\lambda_2(\beta_6-1)\},$$

$$d_8=\frac{1}{\mu_2}\{\lambda_1[\beta_2\beta_4+\beta_3(\beta_6-1)]y_1^{\beta_2}+\lambda_3 y_1-\lambda_2(\beta_6-1)\},$$

$$\lambda_1=\frac{1}{\beta_4}\sum_{i=1}^{6}d_i,$$

$$\mu_2=-\{[\beta_2\beta_4+\beta_3(\beta_6-1)]y_1^{\beta_2}+[\beta_2\beta_4-\beta_3(\beta_6-1)]y_1^{-\beta_2}\}. \tag{48}$$

Especially in the limiting case where R_1 is infinitely small, we can at once obtain the characteristic relation of the large deflection of a corrugated circular plate with full corrugations under the action of concentrated loads at the center. The relation is still given by a formula of the type (44). But it is remarkably simple. Here for an edge rigidly clamped, the coefficients α_1 and α_3 are

$$\alpha_1=2\beta_3(\beta_1+1)^2,$$

$$\alpha_3=\alpha_1^4\beta_0\beta_4^{-1}b_3^3\left[\frac{m_1}{3(\beta_1+1)}+\frac{m_2}{2(\beta_1+2)}+\frac{m_3}{\beta_1+5}+\frac{m_4}{\beta_1+\beta_2+2}+\frac{m_5}{\beta_2+3}+\frac{1}{6}m_6+\frac{m_7}{\beta_1+1}\right], \tag{49}$$

where

$$m_0=\frac{1}{\beta_2-\nu}\left[\frac{2(\beta_1+2-\nu)}{(\beta_1+2)^2-\beta_2^2}-\frac{2\beta_1+1-\nu}{(2\beta_1+1)^2-\beta_2^2}+\frac{3-\nu}{\beta_2^2-9}\right],$$

$$m_1=\frac{1}{4(\beta_1+1)(2\beta_1+1)[(2\beta_1+1)^2-\beta_2^2]},$$

$$m_2=-\frac{1}{3(\beta_1+1)(\beta_1+3)}\left[\frac{1}{(2\beta_1+1)^2-\beta_2^2}+\frac{2}{(\beta_1+2)^2-\beta_2^2}\right],$$

$$m_3=-\frac{1}{8(\beta_1+2)}\left[\frac{1}{\beta_2^2-9}-\frac{2}{(\beta_1+2)^2-\beta_2^2}\right],$$

$$m_4=\frac{m_0}{(\beta_2+1)(2\beta_1+\beta_2+1)},$$

$$m_5=\frac{m_0}{\beta_1^2-(\beta_2+2)^2},$$

$$m_6=-\frac{1}{(\beta_1^2-25)(\beta_2^2-9)},$$

$$m_7=-\sum_{i=1}^{6}m_i, \tag{50}$$

and for an edge clamped but free to slip, coefficients α_1 and α_3 as well as expressions for the parameters m_i ($i=1$, 2, …, 7) again coincide with expressions (49) and (50) in appearance, except that m_0 becomes

$$m_0=\frac{2}{(\beta_1+2)^2-\beta_2^2}-\frac{1}{(2\beta_1+1)^2-\beta_2^2}+\frac{1}{\beta_2^2-9}. \tag{51}$$

4. Numerical Example

Now let us consider a corrugated circular plate with toothed corrugations for the

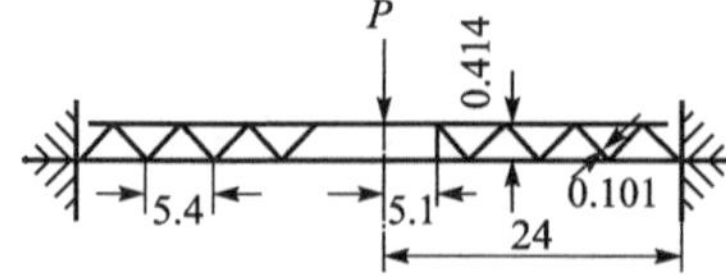

Fig. 3　A corrugated circular plate with toothed corrugations

case of a rigidly clamped edge as shown in Fig. 3. Assume $E = 1.35 \times 10^4 \mathrm{kg\ mm}^{-2}$, $\nu = 0.3$, $R = 24\mathrm{mm}$, $R_1 = 5.1\mathrm{mm}$, $H = 0.414\mathrm{mm}$, $h = 0.101\mathrm{mm}$, $\theta_0 = 8°45'$ and $L = 5.4\mathrm{mm}$. Then, from the characteristic relation (44), we obtain a numerical formula for the load-deflection relation

$$P = 0.4761 w_0 + 0.5958 w_0^3, \tag{52}$$

where w_0 is in mm, P is in kg.

This result is shown in Fig. 4. It is seen from this figure that the characteristic curve rises monotonically.

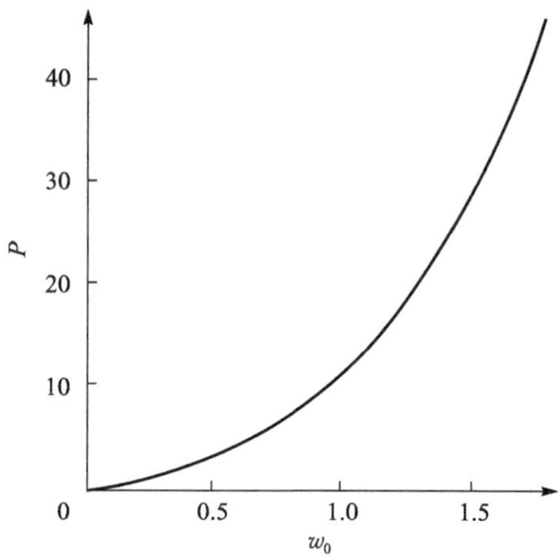

Fig. 4　The characteristic curve

5. Conclusions

We summarize the results obtained in the following:

(1) Analytical solutions of large deflections of a corrugated circular plate with a plane central region under the action of concentrated loads at the center are obtained for a rigidly clamped edge and an edge clamped but free to slip.

(2) Analytical solutions of large deflections of a corrugated circular plate with full corrugations under the action of concentrated loads at the center are obtained for a rigidly clamped edge and a clamped edge but free to slip.

(3) The present solutions can be applied to manifold cases for corrugated circular plates with deep or shallow corrugations of various types.

Acknowledgements

The work was completed during the time the author was an Alexander von Humboldt-Foundation fellow in Ruhr-University Bochum, West Germany. The author gratefully acknowledges Professor W. Zerna for his support and hearty hospitality.

References

[1] D. Y. Panov. On large deflection of circular membranes with weak corrugation, *Prikl. Mat. Mekh.* 5, 308 (1941) (in Russian).

[2] V. I. Feodosev. On large deflection and stability of a circular membrane with fine corrugation, *Prikl, Mat. Mekh.* 9, 389 (1945) (in Russian).

[3] V. I. Feodosev. *Elastic Elements of Precision-Instruments Manufacture*, Oborongiz, Moscow (1949) (in Russian).

[4] L. E. Andryewa. The calculation of a corrugated membrane as an anisotropic plate, *Engineer's Collection* 21, 128 (1955) (in Russian).

[5] T. Akasaka. On the elastic properties of the corrugated diaphragm, *J. Jap. Soc. Aeronaut. Eng.*, 3, 279 (1955) (in Japanese).

[6] J. A. Haringx. Nonlinearity of corrugated diaphragms, *Appl. Scient. Res.* Ser. A, 16, (1956).

[7] J. A. Haringx. Design of corrugated diaphragms, *Trans. Am. Soc. Mech. Engrs.* Ser. A, 79. (1) (1957).

[8] E. V. Burmistrov. Axisymmetrical bending of isotropic and anisotropic shells of revolution under large deflection and non-uniform temperature field, *Engineer's Collection* 27, 185 (1960) (in Russian).

[9] L. E. Andryewa. *Elastic Elements of Instruments*. Masgiz, Moscow (1962) (in Russian).

[10] E. L. Axelrad. Large deflection of a corrugated membrane as a non-shallow shell, *Izv. Akad. Nouk.* SSSR, 1, 46 (1964) (in Russian).

[11] Liu Renhuai. The characteristic relations of corrugated circular plates, *Acta Mech. Sin.* (1), 47 (1978) (in Chinese).

[12] Liu Renhuai. The characteristic relations of corrugated circular plates with plane central region, *J. Chin. Univ. Sci. Technol.* 9, 75 (1979) (in Chinese).

[13] Liu Renhuai. Large deflection of corrugated circular plate with plane boundary region, *Solid Mechanics Archives* 9, 383 (1984).

[14] S. L. Chen. Elastic behavior of uniformly loaded circular corrugated plate with sine-shaped shallow waves in large deflection, *Appl. Math. Mech.* 1, 261 (1980).

[15] K. Y. Yeh, Liu Renuai, Q. Y. Ping and S. L. Li. Nonlinear stability of thin circular shallow spherical shell under actions of axisymmetric uniform distributed line loads, *J. Lanzhou Univ.* (2), 10 (1965); *Bull. Sci.* (2), 142 (1965) (in Chinese).

[16] K. Y. Yeh, Liu Renhuai, C. Z. Zhang and Y. F. Xu. Nonlinear stability of thin circular shallow spherical shell under the action of uniform edge moment, *Bull. Sci.* (2), 145 (1965) (in Chinese).

[17] Liu Renhuai. Nonlinear stability of circular shallow spherical shell with a hole in the center under the action of uniform moment at the inner edge, *Bull. Sci.* (3), 253 (1965) (in Chinese).

[18] Liu Renhuai. Axisymmetrical buckling of thin circular shallow spherical shell with a circular hole at the center by shearing forces uniformly distributed along the inner edge, *Mechanics* (3), 206 (1977) (in Chinese).

[19] K. Y. Yeh, Liu Renhuai. C. Z. Zhang and Y. F. Xu. Nonlinear stability of thin elastic circular

shallow spherical shell under the action of uniform edge moment, *Appl. Math. Mech.* 1, 71 (1980).

[20] Liu Renhuai. Nonlinear axisymmetrical bending of circular sandwich plates under the action of uniform edge moment, *J. Chin. Univ. Sci. Technol.* 10, 56 (1980) (in Chinese).

[21] Liu Renhuai. Nonlinear bending of circular sandwich plates, *Appl. Math. Mech.* 2, 189 (1981).

[22] Liu Renhuni. Nonlinear thermal stability of bimetallic shallow shells of revolution, *Int. J. Nonlinear Mechanics* 18, 409-429 (1983).

[23] T. von Kármán. Festigkeitsprobleme im Mashinenbau, *Encykl. Math. Wiss.* 4, 348 (1910).

[24] W. C. Chien. Large deflection of a circular clamped plate under uniform pressure, *Chin. J. Plys.* 7, 102 (1947).

[25] S. Timoshenko and S. Woinowsky-Krieger. *Theory of Plates and Shells*, 2nd ed. McGraw-Hill, NewYork (1959).

Large deflection of corrugated circular plate with plane boundary region①

中文摘要

本文研究了带有光滑中心和平面边缘区域的波纹圆板的大挠度问题，由于困难太大，此问题一直无人研究。我们使用修正迭代法，获得了相当精确的解析解，可直接应用于精密仪器弹性元件的设计。

1. Introduction

Corrugated circular plates are an important type of elastic elements in precision instruments. Therefore, the problem of large deflection of the plates has been attracting the attention of numerous investigators. In general, we may distinguish between two kinds of previous theoretical analyses. In the case of the first kind, corrugated circular plates are regarded as cylindrically anisotropic circular plates. The theory for large deflection of cylindrically anisotropic circular plates was applied by Liu[1,2], Andryewa[3,4], Akasaka[5], Haringx[6,7], Burmistrov[8], etc. This procedure is characterized by the fact that solutions can be applied to corrugated circular plates with corrugations of various types. But the theoretical analysis of the plates is rather difficult and it is avoided frequently by considering the simplest case, a corrugated circular plate with full corrugations. Liu[2] discussed a general case of practical importance, i. e., a corrugated circular plate with a plane central region, and obtained a quite accurate solution of this problem by means of the modified iteration method. The method was suggested by Yeh and Liu[9-11] in 1965, and applied in a system problems of nonlinear theory of shells and plates[12-15].

In the case of the second kind, the theory for large deflection of thin shallow shells was used by Panov[16], Feodosev[17,18], Axelrad[19] and Chen[20], etc. Unfortunately, so far only a corrugated circular plate with sinusoidal corrugation has been considered, and there are still serious discrepancies between the theoretical and experimental results when the deflection is large.

This paper is a continuation of the previous papers[1,2]. The type of corrugated circular plate which we discuss here is the most important one: the plate with a plane boundary region and a plane central region. As we know, the problem for large deflection of the plate has not yet been studied. The difficulties in the investigation are due to a number of nonlinear differential equations and more complicated shape of the plate.

---

① Reprinted from *Solid Mechanics Archives*, 1984, 9 (4): 383-406.

To overcome these difficulties, we still use the modified iteration method to solve this problem. A quite accurate solution is obtained, and it can be applied directly to design of elastic elements of precision instruments.

2. Fundamental Equations

Let us consider a corrugated circular plate with a plane boundary region and a plane central region under uniform pressure q (Fig. 1). Here h is the thickness, R_1, R_2 and R are radiuses of the plane central region, the corrugated region and the boundary respectively, r is the radial coordinate, L is the wavelength, H is the amplitude.

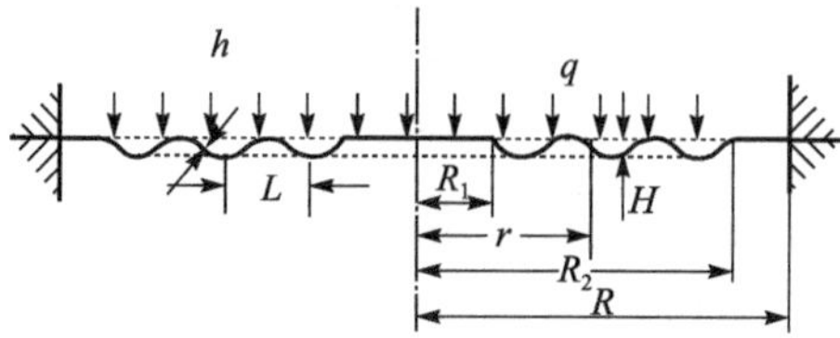

Fig. 1 A uniformly loaded corrugated circular plate with a plane boundary region

In order to simplify the following presentation, we introduce a simple notation for some magnitudes of three regions of the plate. We shall indicate these magnitudes of the plane central region, the corrugated region and the plane boundary region by subscripts a, b and c respectively.

To analyse parameters for rigidities, we consider a corrugation in one diametral section of the plate (Fig. 2a). Here x and z are coordinates, θ is the slope angle of the tangent to one diametral section. Using the theory of cylindrically anisotropic circular plate[1-4], we have the definition of the following parameters for rigidities:

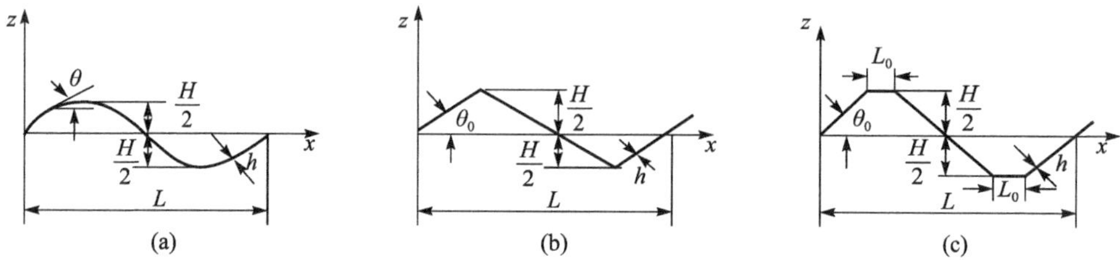

Fig. 2 A corrugation in one diamecral section of the corrugated circular plate:
(a) The sinusoidal corrugation; (b) The toothed corrugation; (c) The trapezoidal corrugation

parameter for the radial rigidity:

$$k_r = \frac{12}{h^2 L}\int_0^L \frac{z^2}{\cos\theta}dx + \frac{1}{L}\int_0^L \cos\theta dx, \tag{1a}$$

parameters for the circumferential rigidity:

$$k_t = \frac{1}{L}\int_0^L \frac{1}{\cos\theta}dx, \tag{1b}$$

$$k_t' \frac{12}{h^2 L}\int_0^L \frac{z^2}{\cos\theta}\mathrm{d}x + \frac{1}{L}\int_0^L \frac{1}{\cos^3\theta}\mathrm{d}x. \tag{1c}$$

From (1), several cases of practical importance are derived and are given by the following formulas:

(1) Sinusoidal corrugation (Fig. 2a)

$$k_r \doteq 1 + \frac{3H^2}{2h^2} + \frac{\pi^2 H^2}{4L^2}\left(\frac{3H^2}{4h^2} - 1\right),$$

$$k_t \doteq 1 + \frac{\pi^2 H^2}{4L^2},$$

$$k_t' \doteq 1 + \frac{3H^2}{2h^2} + \frac{3\pi^2 H^2}{4L^2}\left(\frac{H^2}{4h^2} + 1\right); \tag{2}$$

(2) Toothed corrugation (Fig. 2b)

$$k_r = \frac{H^2}{h^2\cos\theta_0} + \cos\theta_0,$$

$$k_t = \frac{1}{\cos\theta_0},$$

$$k_t' = \frac{H^2}{h^2\cos\theta_0} + \frac{1}{\cos^3\theta_0}; \tag{3}$$

(3) Trapezoidal corrugation (Fig. 2c)

$$k_r = \frac{H^2}{Lh^2}\left(\frac{L-2L_0}{\cos\theta_0} + 6L_0\right) + \left(1 - \frac{2L_0}{L}\right)\cos\theta_0 + \frac{2L_0}{L},$$

$$k_t = \frac{1}{L}\left(\frac{L-2L_0}{\cos\theta_0} + 2L_0\right),$$

$$k_t' = \frac{H^2}{Lh^2}\left(\frac{L-2L_0}{\cos\theta_0} + 6L_0\right) + \frac{1}{L}\left(\frac{L-2L_0}{\cos\theta_0} + 2L_0\right). \tag{4}$$

Having the above formulas for k_r, k_t and k_t', according to large-deflection theories of isotropic and anisotropic circular plates[1-4,21-23], we now obtain fundamental equations satisfied by the deflection w and the radial membrane stress N_r of the corrugated circular plate with a plane boundary region and a plane central region as follows:

for $0 \leqslant r \leqslant R_1$,

$$\frac{\mathrm{d}}{\mathrm{d}r}\frac{1}{r}\frac{\mathrm{d}}{\mathrm{d}r}r\frac{\mathrm{d}w_a}{\mathrm{d}r} - \frac{1}{D_a}N_{r,a}\frac{\mathrm{d}w_a}{\mathrm{d}r} = \frac{1}{2D_a}qr,$$

$$r\frac{\mathrm{d}}{\mathrm{d}r}\frac{1}{r}\frac{\mathrm{d}}{\mathrm{d}r}(r^2 N_{r,a}) + \frac{Eh}{2}\left(\frac{\mathrm{d}w_a}{\mathrm{d}r}\right)^2 = 0; \tag{5}$$

for $R_1 \leqslant r \leqslant R_2$,

$$r\frac{\mathrm{d}^3 w_b}{\mathrm{d}r^3} + \frac{\mathrm{d}^2 w_b}{\mathrm{d}r^2} - k_t\frac{\mathrm{d}w_b}{\mathrm{d}r}\left(\frac{1}{D_b}rN_{r,b} + \frac{k_t'}{r}\right) = \frac{k_t}{2D_b}qr^2,$$

$$r\frac{\mathrm{d}^2}{\mathrm{d}r^2}(rN_{r,b}) + \frac{\mathrm{d}}{\mathrm{d}r}(rN_{r,b}) - k_r k_t N_{r,b} + \frac{1}{2}Ehk_t\left(\frac{\mathrm{d}w_b}{\mathrm{d}r}\right)^2 = 0; \tag{6}$$

for $R_2 \leqslant r \leqslant R$,

$$\frac{\mathrm{d}}{\mathrm{d}r}\frac{1}{r}\frac{\mathrm{d}}{\mathrm{d}r}r\frac{\mathrm{d}w_c}{\mathrm{d}r}-\frac{1}{D_c}N_{r,c}\frac{\mathrm{d}w_c}{\mathrm{d}r}=\frac{1}{2D_c}qr,$$

$$r\frac{\mathrm{d}}{\mathrm{d}r}\frac{1}{r}\frac{\mathrm{d}}{\mathrm{d}r}(r^2N_{r,c})+\frac{Eh}{2}\left(\frac{\mathrm{d}w_c}{\mathrm{d}r}\right)^2=0, \tag{7}$$

where E is Young's modulus, ν is Poisson's ratio, D is the flexural rigidity,

$$D_a=D_c=\frac{Eh^3}{12(1-\nu^2)},D_b=\frac{Eh^3}{12\left(1-\frac{\nu^2}{k_tk_t'}\right)} \tag{8}$$

These nonlinear equations will be solved under the following customary edge conditions for rigidly clamped edge:

$$w_c=0,\quad \frac{\mathrm{d}w_c}{\mathrm{d}r}=0,\ u_c=0 \qquad \text{at } r=R;$$

$$\frac{\mathrm{d}w_a}{\mathrm{d}r}=0,\ N_{r,a}\ \text{finite} \qquad \text{at } r=0 \tag{9}$$

where u is the radial displacement,

$$u_a=\frac{r}{Eh}\left[\frac{\mathrm{d}}{\mathrm{d}r}(rN_{r,a})-\nu N_{r,a}\right],$$

$$u_b=\frac{r}{k_tEh}\left[\frac{\mathrm{d}}{\mathrm{d}r}(rN_{r,b})-\nu N_{r,b}\right],$$

$$u_c=\frac{r}{Eh}\left[\frac{\mathrm{d}}{\mathrm{d}r}(rN_{r,c})-\nu N_{r,c}\right] \tag{10}$$

We must also consider the following conditions of continuity between regions:

$$w_a=w_b,\ \frac{\mathrm{d}w_a}{\mathrm{d}r}=\frac{\mathrm{d}w_b}{\mathrm{d}r},\ M_{r,a}=M_{r,b},\ N_{r,a}=N_{r,b},$$

$$u_a=u_b \quad \text{at } r=R_1; \tag{11}$$

$$w_b=w_c,\ \frac{\mathrm{d}w_b}{\mathrm{d}r}=\frac{\mathrm{d}w_c}{\mathrm{d}r},\ M_{r,b}=M_{r,c},\ N_{r,b}=N_{r,c},$$

$$u_b=u_c \quad \text{at } r=R_2, \tag{12}$$

where M_r is the radial moment;

$$M_{r,a}=-D_a\left(\frac{\mathrm{d}^2w_a}{\mathrm{d}r^2}+\frac{\nu}{r}\frac{\mathrm{d}w_a}{\mathrm{d}r}\right),$$

$$M_{r,b}=-\frac{D_b}{k_t}\left(\frac{\mathrm{d}^2w_b}{\mathrm{d}r^2}+\frac{\nu}{r}\frac{\mathrm{d}w_b}{\mathrm{d}r}\right),$$

$$M_{r,c}=-D_c\left(\frac{\mathrm{d}^2w_c}{\mathrm{d}r^2}+\frac{\nu}{r}\frac{\mathrm{d}w_c}{\mathrm{d}r}\right). \tag{13}$$

In order to simplify the following calculations, let us introduce the following dimensionless notations:

$$y=\frac{r}{R},\ y_1=\frac{R_1}{R},\ y_2=\frac{R_2}{R},\ W=\frac{w}{h},\ \phi=\frac{\mathrm{d}W}{\mathrm{d}y},$$

$$S_a=-\frac{Rr}{D_a}N_{r,a},\ S_b=-\frac{k_tRr}{D_b}N_{r,b},\ S_c=-\frac{Rr}{D_c}N_{r,c},$$

$$Q=\frac{R^4}{2D_ah}q, \beta_0=6\lambda_1\lambda_2, \beta_1^2=k_tk_t', \beta_2^2=k_rk_t,$$

$$\beta_3=\frac{D_b}{D_ak_t}, \beta_4=\frac{\beta_0}{\beta_5}, \beta_5=\frac{\beta_3}{k_t}, \lambda_1=1-\nu, \lambda_2=1+\nu. \tag{14}$$

With the help of these notations, the nonlinear boundary value problem (5-7), (9), (11), and (12) may be rewritten in the following, more convenient dimensionless form:

for $0\leqslant y\leqslant y_1$,

$$L_0(y\phi_a)=-S_a\phi_a+Qy^2,$$
$$L_0(yS_a)=\beta_0\phi_a^2; \tag{15a,b}$$

for $y_1\leqslant y\leqslant y_2$,

$$L_1(y^{\beta_1}\phi_b)=-S_b\phi_b+\frac{Q}{\beta_3}y^2,$$
$$L_2(y^{\beta_2}S_b)=\beta_4\phi_b^2; \tag{16a,b}$$

for $y_2\leqslant y\leqslant 1$,

$$L_0(y\phi_c)=-S_c\phi_c+Qy^2,$$
$$L_0(yS_c)=\beta_0\phi_c^2; \tag{17a,b}$$

$$W_c=0, \quad \phi_c=0, \quad \frac{dS_c}{dy}-\nu\frac{S_c}{y}=0 \quad \text{at } y=1; \tag{18}$$

$$\phi_a=0, \quad S_a=0 \quad \text{at } y=0; \tag{19}$$

$$W_a=W_b, \quad \phi_a=\phi_b, \quad \frac{d\phi_a}{dy}+\nu\frac{\phi_a}{y}=\beta_3\left(\frac{d\phi_b}{dy}+\nu\frac{\phi_b}{y}\right),$$

$$S_a=\beta_3S_b, \quad \frac{dS_a}{dy}-\nu\frac{S_a}{y}=\beta_5\left(\frac{dS_b}{dy}-\nu\frac{S_b}{y}\right) \quad \text{at } y=y_1; \tag{20}$$

$$W_b=W_c, \quad \phi_b=\phi_c, \quad \beta_3\left(\frac{d\phi_b}{dy}+\nu\frac{\phi_b}{y}\right)=\frac{d\phi_c}{dy}+\nu\frac{\phi_c}{y},$$

$$\beta_3S_b=S_c, \quad \beta_5\left(\frac{dS_b}{dy}-\nu\frac{S_b}{y}\right)=\frac{dS_c}{dy}-\nu\frac{S_c}{y} \quad \text{at } y=y_2, \tag{21}$$

where

$$L_0(\cdots)=y\frac{d}{dr}\frac{1}{y}\frac{d}{dy}(\cdots),$$

$$L_1(\cdots)=y^{\beta_1}\frac{d}{dy}y^{-(2\beta_1-1)}\frac{d}{dy}(\cdots),$$

$$L_2(\cdots)=y^{\beta_2}\frac{d}{dy}y^{-(2\beta_2-1)}\frac{d}{dy}(\cdots). \tag{22}$$

3. Analytical Solution

Since the dimensionless nonlinear boundary value problem (15)-(21) contains more equations and conditions, it is quite difficult to solve them. We can overcome this difficulty by the modified iteration method.

For first approximation, by neglecting nonlinear terms in Eqs. (15a), (16a) and

(17a), and again introducing $\phi_{a,1}$, $\phi_{b,1}$ and $\phi_{c,1}$ into the right of Eqs. (15b), (16b) and (17b) respectively, we have the linear boundary value problem as follows:

for $0\leqslant y\leqslant y_1$,

$$L_0(y\phi_{a,1}) = Qy^2,$$
$$L_0(yS_{a,1}) = \beta_0\phi_{a,1}^2; \qquad (23a,b)$$

for $y_1\leqslant y\leqslant y_2$,

$$L_1(y^{\beta_1}\phi_{b,1}) = \frac{Q}{\beta_3}y^2,$$
$$L_2(y^{\beta_2}S_{b,1}) = \beta_4\phi_{b,1}^2; \qquad (24a,b)$$

for $y_2\leqslant y\leqslant 1$,

$$L_0(y\phi_{c,1}) = Qy^2,$$
$$L_0(yS_{c,1}) = \beta_0\phi_{c,1}^2; \qquad (25a,b)$$

$$W_{c,1} = 0,\quad \phi_{c,1} = 0,\quad \frac{dS_{c,1}}{dy} - \nu\frac{S_{c,1}}{y} = 0 \qquad \text{at } y = 1; \qquad (26a,b,c)$$

$$\phi_{a,1} = 0,\qquad S_{a,1} = 0 \qquad \text{at } y = 0; \qquad (27a,b)$$

$$W_{a,1} = W_{b,1},\quad \phi_{a,1} = \phi_{b,1}, \qquad (28a,b)$$

$$\frac{d\phi_{a,1}}{dy} + \nu\frac{\phi_{a,1}}{dy} = \beta_3\left(\frac{d\phi_{b,1}}{dy} + \nu\frac{\phi_{b,1}}{y}\right), \qquad (28c)$$

$$S_{a,1} = \beta_3 S_{b,1}, \qquad (28d)$$

$$\frac{dS_{a,1}}{dy} - \nu\frac{S_{a,1}}{dy} = \beta_5\left(\frac{dS_{b,1}}{dy} - \nu\frac{S_{b,1}}{y}\right) \quad \text{at } y = y; \qquad (28e)$$

$$W_{b,1} = W_{c,1},\quad \phi_{b,1} = \phi_{c,1}, \qquad (29a,b)$$

$$\beta_3\left(\frac{d\phi_{b,1}}{dy} + \nu\frac{\phi_{b,1}}{y}\right) = \frac{d\phi_{c,1}}{dy} + \nu\frac{\phi_{c,1}}{y}, \qquad (29c)$$

$$\beta_3 S_{b,1} = S_{c,1}, \qquad (29d)$$

$$\beta_5\left(\frac{dS_{b,1}}{dy} - \nu\frac{S_{b,1}}{y}\right) = \frac{dS_{c,1}}{dy} - \nu\frac{S_{c,1}}{y} \qquad \text{at } y = y_2. \qquad (29e)$$

Under conditions (26a, b), (27a), (28a, b, c) and (29a, b, c), solutions of Eqs. (23a), (24a) and (25a) are

$$\phi_{a,1} = Q(A_{a,3}y^3 + A_{a,1}y),$$
$$\phi_{b,1} = Q(A_{b,3}y^{\beta_1} + A_{b,2}y^{-\beta_1} + A_{b,1}y^3),$$
$$\phi_{c,1} = Q(A_{c,3}y^3 + A_{c,1}y + A_{c,-1}y^{-1}), \qquad (30)$$

where

$$A_{a,1} = A_{b,3}y_1^{\beta_1-1} + A_{b,2}y_1^{-(\beta_1+1)} + \left(A_{b,1} - \frac{1}{8}\right)y_1^2,$$

$$A_{a,3} = \frac{1}{8},$$

$$A_{b,1} = -\frac{1}{\beta_3(\beta_1^2 - 9)},$$

$$A_{b,2}=a_0\{a_1[(a_3-\lambda_2)y_2-(a_3+\lambda_1)y_2^{-1}]y_1^2y_2^{\beta_1}-a_2(a_3-\lambda_2)y_1^{\beta_1-1}\}y_2^{-1},$$

$$A_{b,3}=a_0\{a_4[(a_4+\lambda_2)y_2-(a_4-\lambda_1)y_2^{-1}]y_1^2y_2^{-\beta_1}-a_2(a_4+\lambda_2)y^{-(\beta_1+1)}\}y_2^{-1},$$

$$A_{c,-1}=-\left(A_{c,1}+\frac{1}{8}\right),$$

$$A_{c,1}=a_0\left\{\left[(3+\nu)\left(\beta_3A_{b,1}-\frac{1}{8}\right)y_2^2-\frac{\lambda_1}{8}y_2^{-2}\right][(a_3-\lambda_2)y_1^{\beta_1}y_2^{-\beta_1}+(a_4+\lambda_2)y_1^{-\beta_1}y_2^{\beta_1}]-\left[\left(A_{b,1}-\frac{1}{8}\right)y_2^2+\frac{1}{8}y_2^{-2}\right][a_3(a_4+\lambda_2)y_1^{-\beta_1}y_2^{\beta_1}-a_4(a_3-\lambda_2)y_1^{\beta_1}y_2^{-\beta_1}]+2\beta_1\beta_3a_1y_1^3y_2^{-1}\right\}y_1^{-1},$$

$$A_{c,3}=\frac{1}{8},$$

$$a_0=\{(a_3-\lambda_2)[(a_4+\lambda_2)y_2-(a_4-\lambda_1)y_2^{-1}]y_1^{\beta_1}y_2^{-\beta_1}-(a_4+\lambda_2)[(a_3-\lambda_2)y_2-(a_3+\lambda_1)y_2^{-1}]y_1^{-\beta_1}y_2^{\beta_1}\}^{-1}y_1y_2,$$

$$a_1=-A_{b,1}[\beta_3(3+\nu)-\lambda_2]+\frac{1}{4},$$

$$a_2=a_1y_2^4+\left\{A_{b,1}[\beta_3(3+\nu)+\lambda_1]-\frac{1}{2}\right\}y_2^2+\frac{1}{4},$$

$$a_3=\beta_3(\beta_1+\nu),$$

$$a_4=\beta_3(\beta_1-\nu).\tag{31}$$

Now multiplying (30) by dy and integrating it, and using (26a), (28a) and (29a), we obtain the following formulas for the deflection to first approximation:

$$W_{a,1}=Q\left(\frac{A_{a,3}}{4}y^4+\frac{A_{a,1}}{2}y^2+\frac{1}{\alpha_1}\right),\tag{32a}$$

$$W_{b,1}=Q\left[\frac{A_{b,3}}{\beta_1+1}(y^{\beta_1+1}-y_2^{\beta_1+1})-\frac{A_{b,2}}{\beta_1-1}(y^{-(\beta_1-1)}-y_2^{-(\beta_1-1)})+\frac{A_{b,1}}{4}(y^4-y_2^4)+\frac{A_{c,3}}{4}(y_2^4-1)+\frac{A_{c,1}}{2}(y_2^2-1)+A_{c,-1}\ln y_2\right],\tag{32b}$$

$$W_{c,1}=Q\left[\frac{A_{c,3}}{4}(y^4-1)+\frac{A_{c,1}}{2}(y^2-1)+A_{c,-1}\ln y\right],\tag{32c}$$

where

$$\alpha_1=\left[\frac{A_{b,3}}{\beta_1+1}(y_1^{\beta_1+1}-y_2^{\beta_1+1})-\frac{A_{b,2}}{\beta_1-1}(y_1^{-(\beta_1-1)}-y_2^{-(\beta_1-1)})+\frac{1}{4}\left(A_{b,1}-\frac{1}{8}\right)(y_1^4-y_2^4)-\frac{1}{2}(A_{a,1}y_1^2-A_{c,1}y_2^2)+A_{c,-1}\ln y_2-\frac{1}{2}A_{c,1}-\frac{1}{32}\right]^{-1}.\tag{33}$$

When we introduce, as an iteration parameter, a notation of the dimensionless center deflection

$$W_0 = W_a \mid_{y=0}, \tag{34}$$

the expression (32a) becomes

$$Q = \alpha_1 W_0. \tag{35}$$

This is the characteristic relation of small deflection of the corrugated circular plate with a plane boundary region and a plane central region. Substituting this relation into (30), we obtain

$$\begin{aligned}
\phi_{a,1} &= \alpha_1 (A_{a,3} y^3 + A_{a,1} y) W_0, \\
\phi_{b,1} &= \alpha_1 (A_{b,3} y^{\beta_1} + A_{b,2} y^{-\beta_1} + A_{b,1} y^3) W_0, \\
\phi_{c,1} &= \alpha_1 (A_{c,3} y^3 + A_{c,1} y + A_{c,-1} y^{-1}) W_0.
\end{aligned} \tag{36}$$

Using the solution (36), under conditions (26c), (27b), (28d, e) and (29d, e), solutions of Eqs. (23b), (24b) and (25b) are

$$\begin{aligned}
S_{a,1} &= \alpha_1^2 \beta_4 W_0^2 \sum_{i=0}^{3} B_{a,2i+1} y^{2i+1}, \\
S_{b,1} &= \alpha_1^2 \beta_4 (B_{b,8} y^{2\beta_1+1} + B_{b,7} y^{-(2\beta_1-1)} + B_{b,6} y^{\beta_1+4} \\
&\quad + B_{b,5} y^{-(\beta_1-4)} + B_{b,4} y^{\beta_2} + B_{b,3} y^{-\beta_2} + B_{b,2} y^7 + B_{b,1} y) W_0^2, \\
S_{c,1} &= \alpha_1^2 \beta_4 \Big[\sum_{i=0}^{4} B_{c,2i-1} y^{2i-1} + (B_{c,2} y + B_{c,0} y^{-1}) \ln y \Big] W_0^2,
\end{aligned} \tag{37}$$

where

$$\begin{aligned}
B_{a,1} &= \beta_3 (B_{b,8} y_1^{2\beta_1} + B_{b,7} y_1^{-2\beta_1} + B_{b,6} y_1^{\beta_1+3} + B_{b,5} y_1^{-(\beta_1-3)} \\
&\quad + B_{b,4} y_1^{\beta_2-1} + B_{b,3} y_1^{-(\beta_2+1)} + B_{b,2} y_1^6 + B_{b,1}) - \sum_{i=1}^{3} B_{a,2i+1} y_1^{2i},
\end{aligned}$$

$$B_{a,3} = \frac{\beta_5}{8} A_{a,1}^2, \qquad B_{a,5} = \frac{\beta_5}{96} A_{a,1},$$

$$B_{a,7} = \frac{\beta_5}{3072}, \qquad B_{b,1} = -\frac{2A_{b,2} A_{b,3}}{\beta_2^2 - 1},$$

$$B_{b,2} = -\frac{A_{b,1}^2}{\beta_2^2 - 49},$$

$$\begin{aligned}
B_{b,3} &= b_0 \Big\{ b_3 \Big[\lambda_3 \beta_3 (y_2 - y_2^{-1}) - b_5 \Big(y_2 + \frac{\lambda_1}{\lambda_2} y_2^{-1} \Big) \Big] y_2^{\beta_1} \\
&\quad - (\lambda_1 \beta_3 - b_5) \Big[\lambda_1 b_1 (y_2 - y_2^{-1}) - b_2 \Big(y_2 + \frac{\lambda_1}{\lambda_2} y_2^{-1} \Big) \Big] y_1^{\beta_2} \Big\},
\end{aligned}$$

$$\begin{aligned}
B_{b,4} &= b_0 \Big\{ \lambda_1 \beta_3 + b_4) \Big[\lambda_1 b_1 (y_2 - y_2^{-1}) - b_2 \Big(y_2 + \frac{\lambda_1}{\lambda_2} y_2^{-1} \Big) \Big] y_1^{-\beta_2} \\
&\quad - b_3 \Big[\lambda_1 \beta_3 (y_2 - y_2^{-1}) + b_4 \Big(y_2 + \frac{\lambda_1}{\lambda_2} y_2^{-1} \Big) \Big] y_2^{-\beta_2} \Big\},
\end{aligned}$$

$$B_{b,5} = \frac{2A_{b,1} A_{b,2}}{(\beta_1 - 4)^2 - \beta_2^2}, \qquad B_{b,6} = \frac{2A_{b,1} A_{b,3}}{(\beta_1 + 4)^2 - \beta_2^2},$$

$$B_{b,7}=\frac{A_{b,2}^2}{(2\beta_1-1)^2-\beta_2^2},\qquad B_{b,8}=\frac{A_{b,3}^2}{(2\beta_1+1)^2-\beta_2^2},$$

$$B_{c,-1}=\frac{1}{\lambda_2}\Big[\sum_{i=0}^{3}B_{c,2i+1}(2i+\lambda_1)+B_{c,2}+B_{c,0}\Big],$$

$$B_{c,0}=-\frac{\beta_5}{2}A_{c,-1}^2,$$

$$B_{c,1}=b_0[(\lambda_1\beta_3-b_5)(b_1b_4-\beta_3b_2)y_1^{\beta_2}y_2^{-\beta_2}$$

$$+(\lambda_1\beta_3+b_4)(b_1b_5-\beta_3b_2)y_1^{-\beta_2}y_2^{\beta_2}-2\beta_2\beta_3\beta_5b_3]-\frac{1}{2}B_{c,2},$$

$$B_{c,2}=\beta_5A_{c,1}A_{c,-1},\qquad B_{c,3}=\frac{\beta_5}{8}\left(A_{c,1}^2+\frac{1}{4}A_{c,-1}\right),$$

$$B_{c,5}=\frac{\beta_5}{96}A_{c,1},\qquad B_{c,7}=B_{a,7},$$

$$b_0=\left\{(\lambda_1\beta_3+b_4)\left[\lambda_1\beta_3(y_2-y_2^{-1})-b_5\left(y_2+\frac{\lambda_1}{\lambda_2}y_2^{-1}\right)\right]y_1^{-\beta_2}y_2^{\beta_2}\right.$$

$$\left.-(\lambda_1\beta_3-b_5)\left[\lambda_1\beta_3(y_2-y_2^{-1})+b_4\left(y_2+\frac{\lambda_1}{\lambda_2}y_2^{-1}\right)\right]y^{\beta_2}y^{-\beta_2}\right\}^{-1},$$

$$b_1=\sum_{i=1}^{3}B_{c,2i+1}\left[y_2^{2i+1}+\frac{1}{\lambda_2}(2i+\lambda_1)y_2^{-1}\right]$$

$$+B_{c,2}\left[\left(\ln y_2-\frac{1}{2}\right)y_2+\frac{1}{2}y_2^{-1}\right]+B_{c,0}\left(\ln y_2+\frac{1}{\lambda_2}\right)y_2^{-1}$$

$$-\beta_3(B_{b,8}y_2^{2\beta_1+1}+B_{b,7}y^{-(2\beta_1-1)}+B_{b,6}y_2^{\beta_1+4}+B_{b,5}y_2^{-(\beta_1-4)}$$

$$+B_{b,2}y_2^7+B_{b,1}y_2),$$

$$b_2=\sum_{i=1}^{3}B_{c,2i+1}(2i+\lambda_1)(y_2^{2i+1}-y_2^{-1})$$

$$+B_{c,2}\left(\lambda_1\ln y_2+\frac{1}{2}\lambda_2\right)y_2-\lambda_2\left(B_{c,0}\ln y_2+\frac{1}{2}B_{c,2}\right)y_2^{-1}$$

$$-\beta_5[B_{b,8}(2\beta_1+\lambda_1)y_2^{2\beta_1+1}-B_{b,7}(2\beta_1-\lambda_1)y^{-(2\beta_1-1)}$$

$$+B_{b,6}(\beta_1+4-\nu)y_2^{\beta_1+4}-B_{b,5}(\beta_1-4+\nu)y_2^{-(\beta_1-4)}$$

$$+B_{b,2}(7-\nu)y_2^7+\lambda_1B_{b,1}y_2],$$

$$b_3=B_{b,8}[\beta_5(2\beta_1+\lambda_1)-\lambda_1\beta_3]y_1^{2\beta_1+1}$$

$$-B_{b,7}[\beta_5(2\beta_1-\lambda_1)+\lambda_1\beta_3]y_1^{-(2\beta_1-1)}$$

$$+B_{b,6}[\beta_5(\beta_1+4-\nu)-\lambda_1\beta_3]y_1^{\beta_1+4}$$

$$-B_{b,5}[\beta_5(\beta_1-4+\nu)+\lambda_1\beta_3]y_1^{-(\beta_1-4)}$$

$$+B_{b,2}[\beta_5(7-\nu)-\lambda_1\beta_3]y_1^7+\lambda_1B_{b,1}(\beta_5-\beta_3)y_1$$

$$-\sum_{i=1}^{3}2iB_{a,2i+1}y_1^{2i+1},$$

$$b_4=\beta_5(\beta_2+\nu),\qquad b_5=\beta_5(\beta_2-\nu).\tag{38}$$

For second approximation, we have the linear boundary value problem for ϕ as follows:

for $0\leqslant y\leqslant y_1$,

$$L_0(y\phi_{a,2})=-S_{a,1}\phi_{a,1}+Qy^2;\tag{39a}$$

for $y_1\leqslant y\leqslant y_2$,

$$L_1(y^{\beta_1}\phi_{b,2})=-S_{b,1}\phi_{b,1}+\frac{Q}{\beta_3}y^2;\tag{39b}$$

for $y_2\leqslant y\leqslant 1$,

$$L_0(y\phi_{c,2})=-S_{c,1}\phi_{c,1}+Qy^2;\tag{39c}$$

$$W_{c,2}=0,\quad \phi_{c,2}=0\qquad \text{at } y=1;\tag{40a,b}$$

$$\phi_{a,2}=0,\qquad \text{at } y=0;\tag{41}$$

$$W_{a,2}=W_{b,2},\quad \phi_{a,2}=\phi_{b,2},$$
$$\frac{d\phi_{a,2}}{dy}+\nu\frac{\phi_{a,2}}{y}=\beta_3\left(\frac{d\phi_{b,2}}{dy}+\nu\frac{\phi_{b,2}}{y}\right)\qquad \text{at } y=y_1;\tag{42a,b,c}$$

$$W_{b,2}=W_{c,2},\quad \phi_{b,2}=\phi_{c,2},$$
$$\beta_3\left(\frac{d\phi_{b,2}}{dy}+\nu\frac{\phi_{b,2}}{y}\right)=\frac{d\phi_{c,2}}{dy}+\nu\frac{\phi_{c,2}}{y}\qquad \text{at } y=y_2.\tag{43a,b,c}$$

Using solutions (36) and (37), under conditions (40b), (41), (42b, c) and (43b, c), solutions of Eqs. (39) are

$$\begin{aligned}
\phi_{a,2}=&\frac{1}{8}Qy^3+(A_{a,1}Q+\alpha_1^3\beta_4c_1W_0^3)y-\frac{1}{8}\alpha_1^3\beta_4F_1(y)W_0^3,\\
\phi_{b,2}=&A_{b,1}Qy^3+(A_{b,3}Q+\alpha_1^3\beta_4c_3W_0^3)y^{\beta_1}+(A_{b,2}Q\\
&+\alpha_1^3\beta_4c_2W_0^3)y^{-\beta_1}-\alpha_1^3\beta_4F_2(y)W_0^3,\\
\phi_{c,2}=&\frac{1}{8}Qy^3+(A_{c,1}Q+\alpha_1^3\beta_4c_5W_0^3)y+(A_{c,-1}Q+\alpha_1^3\beta_4c_4W_0^3)y^{-1}\\
&-\frac{1}{2}\alpha_1^3\beta_4F_3(y)W_0^3,
\end{aligned}\tag{44}$$

where

$$\begin{aligned}
F_1(y)=&\sum_{i=1}^{5}C_{a,2i+1}y^{2i+1},\\
F_2(y)=&C_{b,16}y^{3\beta_1+2}+C_{b,15}y^{-(3\beta_1-2)}+C_{b,14}y^{2\beta_1+5}\\
&+C_{b,13}y^{-(2\beta_1-5)}+C_{b,12}y^{\beta_1+8}+C_{b,11}y^{-(\beta_1-8)}\\
&+C_{b,10}y^{\beta_1+2}+C_{b,9}y^{-(\beta_1-2)}+C_{b,8}y^{\beta_1+\beta_2+1}+C_{b,7}y^{-(\beta_1+\beta_2-1)}\\
&+C_{b,6}y^{\beta_1-\beta_2+1}+C_{b,5}y^{-(\beta_1-\beta_2-1)}+C_{b,4}y^{\beta_2+4}\\
&+C_{b,3}y^{-(\beta_2-4)}+C_{b,2}y^{11}+C_{b,1}y^5,\\
F_3(y)=&\sum_{i=1}^{5}C_{c,2i+1}y^{2i+1}+\sum_{i=0}^{3}C_{c,2i}y^{2i-1}\ln y+(C_{c,10}y+C_{c,8}y^{-1})\ln^2 y,
\end{aligned}$$

$$C_{a,3}=A_{a,1}B_{a,1},\qquad C_{a,5}=\frac{1}{3}\left(\frac{1}{8}B_{a,1}+A_{a,1}B_{a,3}\right),$$

$$C_{a,7}=\frac{1}{6}\left(\frac{1}{8}B_{a,3}+A_{a,1}B_{a,5}\right),$$

$$C_{a,9}=\frac{1}{10}\left(\frac{1}{8}B_{a,5}+A_{a,1}B_{a,7}\right),\quad C_{a,11}=\frac{B_{a,7}}{120},$$

$$C_{b,1}=-\frac{A_{b,3}B_{b,5}+A_{b,2}B_{b,6}+A_{b,1}B_{b,1}}{\beta_1^2-25},$$

$$C_{b,2}=-\frac{A_{b,1}B_{b,2}}{\beta_1^2-121},\quad C_{b,3}=-\frac{A_{b,1}B_{b,3}}{\beta_1^2-(\beta_2-4)^2},$$

$$C_{b,4}=-\frac{A_{b,1}B_{b,4}}{\beta_1^2-(\beta_2+4)^2},\quad C_{b,5}=-\frac{A_{b,2}B_{b,4}}{(\beta_2+1)(2\beta_1-\beta_2-1)},$$

$$C_{b,6}=-\frac{A_{b,3}B_{b,3}}{(\beta_2-1)(2\beta_1-\beta_2+1)},$$

$$C_{b,7}=\frac{A_{b,2}B_{b,3}}{(\beta_2-1)(2\beta_1+\beta_2-1)},$$

$$C_{b,8}=\frac{A_{b,3}B_{b,4}}{(\beta_2+1)(2\beta_1+\beta_2+1)},\quad C_{b,9}=-\frac{A_{b,3}B_{b,7}+A_{b,2}B_{b,1}}{4(\beta_1-1)},$$

$$C_{b,10}=\frac{A_{b,3}B_{b,1}+A_{b,2}B_{b,8}}{4(\beta_1+1)},\quad C_{b,11}=-\frac{A_{b,2}B_{b,2}+A_{b,1}B_{b,5}}{16(\beta_1-4)},$$

$$C_{b,12}=\frac{A_{b,3}B_{b,2}+A_{b,1}B_{b,6}}{16(\beta_1+4)},\quad C_{b,13}=\frac{A_{b,2}B_{b,5}+A_{b,1}B_{b,7}}{(\beta_1-5)(3\beta_1-5)},$$

$$C_{b,14}=\frac{A_{b,3}B_{b,6}+A_{b,1}B_{b,8}}{(\beta_1+5)(3\beta_1+5)},\quad C_{b,15}=\frac{A_{b,2}B_{b,7}}{4(\beta_1-1)(2\beta_1-1)},$$

$$C_{b,16}=\frac{A_{b,3}B_{b,8}}{4(\beta_1+1)(2\beta_1+1)},\quad C_{c,0}=C_{c,8}-A_{c,-1}B_{c,-1},$$

$$C_{c,1}=-\frac{1}{2}C_{c,2},\quad C_{c,2}=A_{c,1}B_{c,-1}+A_{c,-1}B_{c,1}-C_{c,10},$$

$$C_{c,3}=-\frac{1}{4}\left(3C_{c,4}-\frac{1}{8}B_{c,-1}-A_{c,1}B_{c,1}-A_{c,-1}B_{c,3}\right),$$

$$C_{c,4}=\frac{1}{4}\left(\frac{1}{8}B_{c,0}+A_{c,1}B_{c,2}\right),$$

$$C_{c,5}=-\frac{1}{12}\left(5C_{c,6}-\frac{1}{8}B_{c,1}-A_{c,1}B_{c,3}-A_{c,-1}B_{c,5}\right),$$

$$C_{c,6}=\frac{1}{96}B_{c,2},$$

$$C_{c,7}=\frac{1}{24}\left(\frac{1}{8}B_{c,3}+A_{c,1}B_{c,5}+A_{c,-1}B_{c,7}\right),$$

$$C_{c,8}=-\frac{1}{2}A_{c,-1}B_{c,0},\quad C_{c,9}=\frac{1}{40}\left(\frac{1}{8}B_{c,5}+A_{c,1}B_{c,7}\right),$$

$$C_{c,10}=\frac{1}{2}(A_{c,1}B_{c,0}+A_{c,-1}B_{c,2}),$$

$$C_{c,11}=\frac{1}{480}B_{c,7},$$

$$c_1=c_3y_1^{\beta_1-1}+c_2y_1^{-(\beta_1+1)}+\frac{1}{y_1}\left[\frac{1}{8}F_1(y_1)-F_2(y_1)\right],$$

$$
\begin{aligned}
c_2 =& -a_0\{(a_3-\lambda_2)[d_1(y_2-y_2^{-1})-d_2(\lambda_2 y_2+\lambda_1 y_2^{-1})]y_1^{\beta_1} \\
& +d_3[(a_3-\lambda_2)y_2-(a_3+\lambda_1)y_2^{-1}]y_2^{\beta_1}\}y_1^{-1}y_2^{-1}, \\
c_3 =& -a_0\{(a_4+\lambda_2)[d_1(y_2-y_2^{-1})-d_2(\lambda_2 y_2+\lambda_1 y_2^{-1})]y_1^{-\beta_1} \\
& +d_3[(a_4+\lambda_2)y_2-(a_4-\lambda_1)y_2^{-1}]y_2^{-\beta_1}\}y_1^{-1}y_2^{-1}, \\
c_4 =& \frac{1}{2}\sum_{i=0}^{5}C_{c,2i+1}-c_5, \\
c_5 =& -a_0\{(a_3-\lambda_2)(d_1+a_4 d_2)y_1^{\beta_1}y_2^{-\beta_1} \\
& +(a_4+\lambda_2)(d_1-a_3 d_2)y_1^{-\beta_1}y_2^{\beta_1}+2\beta_1\beta_3 d_3\}y_1^{-1}y_2^{-1}, \\
d_1 =& \beta_3[C_{b,16}(3\beta_1+2+\nu)y_2^{3\beta_1+2}-C_{b,15}(3\beta_1-2-\nu)y_2^{-(3\beta_1-2)} \\
& +C_{b,14}(2\beta_1+5+\nu)y_2^{2\beta_1+5}-C_{b,13}(2\beta_1-5-\nu)y_2^{-(2\beta_1-5)} \\
& +C_{b,12}(\beta_1+8+\nu)y_2^{\beta_1+8}-C_{b,11}(\beta_1-8-\nu)y_2^{-(\beta_1-8)} \\
& +C_{b,10}(\beta_1+2+\nu)y_2^{\beta_1+2}-C_{b,9}(\beta_1-2-\nu)y_2^{-(\beta_1-2)} \\
& +C_{b,8}(\beta_1+\beta_2+\lambda_2)y_2^{\beta_1+\beta_2+1}-C_{b,7}(\beta_1+\beta_2-\lambda_2)y_2^{-(\beta_1+\beta_2-1)} \\
& +C_{b,6}(\beta_1-\beta_2+\lambda_2)y_2^{\beta_1-\beta_2+1}-C_{b,5}(\beta_1-\beta_2-\lambda_2)y_2^{-(\beta_1-\beta_2-1)} \\
& +C_{b,4}(\beta_2+4+\nu)y_2^{\beta_2+4}-C_{b,3}(\beta_2-4-\nu)y_2^{-(\beta_2-4)} \\
& +C_{b,2}(11+\nu)y_2^{11}+C_{b,1}(5+\nu)y_2^{5}] \\
& -\frac{1}{2}\left\{\sum_{i=0}^{5}C_{c,2i+1}[(2i+\lambda_2)y_2^{2i+1}+\lambda_1 y_2^{-1}]\right. \\
& +\sum_{i=0}^{3}C_{c,2i}[(2i-\lambda_1)\ln y_2+1]y_2^{2i-1} \\
& \left.+C_{c,10}(\lambda_2\ln y_2+2)y_2\ln y_2-C_{c,8}(\lambda_1\ln y_2-2)y_2^{-1}\ln y_2\right\}, \\
d_2 =& F_2(y_2)-\frac{1}{2}F_3(y_2)+\frac{1}{2}\sum_{i=0}^{5}C_{c,2i+1}y_2^{-1}, \\
d_3 =& \frac{1}{8}\sum_{i=1}^{5}2iC_{a,2i+1}y_1^{2i+1}-C_{b,16}[\beta_3(3\beta_1+2+\nu)-\lambda_2]y_1^{3\beta_1+2} \\
& +C_{b,15}[\beta_3(3\beta_1-2-\nu)+\lambda_2]y_1^{-(3\beta_1-2)} \\
& -C_{b,14}[\beta_3(2\beta_1+5+\nu)-\lambda_2]y_1^{2\beta_1+5} \\
& +C_{b,13}[\beta_3(2\beta_1-5-\nu)+\lambda_2]y_1^{-(2\beta_1-5)} \\
& -C_{b,12}[\beta_3(\beta_1+8+\nu)-\lambda_2]y_1^{\beta_1+8} \\
& +C_{b,11}[\beta_3(\beta_1-8-\nu)+\lambda_2]y_1^{-(\beta_1-8)} \\
& -C_{b,10}[\beta_3(\beta_1+2+\nu)-\lambda_2]y_1^{\beta_1+2} \\
& +C_{b,9}[\beta_3(\beta_1-2-\nu)+\lambda_2]y_1^{-(\beta_1-2)} \\
& -C_{b,8}[\beta_3(\beta_1+\beta_2+\lambda_2)-\lambda_2]y_1^{\beta_1+\beta_2+1} \\
& +C_{b,7}[\beta_3(\beta_1+\beta_2-\lambda_2)+\lambda_2]y_1^{-(\beta_1+\beta_2-1)} \\
& -C_{b,6}[\beta_3(\beta_1-\beta_2+\lambda_2)-\lambda_2]y_1^{\beta_1-\beta_2+1} \\
& +C_{b,5}[\beta_3(\beta_1-\beta_2-\lambda_2)+\lambda_2]y_1^{-(\beta_1-\beta_2-1)} \\
& -C_{b,4}[\beta_3(\beta_2+4+\nu)-\lambda_2]y_1^{\beta_2+4}
\end{aligned}
$$

$$+C_{b,3}[\beta_3(\beta_2-4-\nu)+\lambda_2]y_1^{-(\beta_2-4)}$$
$$-C_{b,2}[\beta_3(11+\nu)-\lambda_2]y_1^{11}-C_{b,1}[\beta_3(5+\nu)-\lambda_2]y_1^5. \tag{45}$$

Multiplying (44) by dy and integrating it, and using conditions (40a), (42a) and (43a), we obtain the following solutions for the deflection to the second approximation

$$W_{a,2}=QG_1(y)+\alpha_1^3\beta_4G_2(y)W_0^3+e_1, \tag{46a}$$
$$W_{b,2}=QG_3(y)+\alpha_1^3\beta_4G_4(y)W_0^3+e_2, \tag{46b}$$
$$W_{c,2}=QG_5(y)+\alpha_1^3\beta_4G_6(y)W_0^3+e_3, \tag{46c}$$

where

$$G_1(y)=\frac{1}{2}\left(\frac{1}{2}A_{a,3}y^4+A_{a,1}y^2\right),$$

$$G_2(y)=\frac{1}{2}c_1y^2-\sum_{i=2}^{6}\frac{C_{a,2i-1}}{16i}y^{2i},$$

$$G_3(y)=\frac{A_{b,3}}{\beta_1+1}y^{\beta_1+1}-\frac{A_{b,2}}{\beta_1-1}y^{-(\beta_1-1)}+\frac{1}{4}A_{b,1}y^4,$$

$$\begin{aligned}G_4(y)=&\frac{c_3}{\beta_1+1}y^{\beta_1+1}-\frac{c_2}{\beta_1-1}y^{-(\beta_1-1)}-\frac{C_{b,16}}{3(\beta_1+1)}y^{3(\beta_1+1)}\\&+\frac{C_{b,15}}{3(\beta_1-1)}y^{-3(\beta_1-1)}-\frac{C_{b,14}}{2(\beta_1+3)}y^{2(\beta_1+3)}\\&+\frac{C_{b,13}}{2(\beta_1-3)}y^{-2(\beta_1-3)}-\frac{C_{b,12}}{\beta_1+9}y^{\beta_1+9}\\&+\frac{C_{b,11}}{\beta_1-9}y^{-(\beta_1-9)}-\frac{C_{b,10}}{\beta_1+3}y^{\beta_1+3}\\&+\frac{C_{b,9}}{\beta_1-3}y^{-(\beta_1-3)}-\frac{C_{b,8}}{\beta_1+\beta_2+2}y^{\beta_1+\beta_2+2}\\&+\frac{C_{b,7}}{\beta_1+\beta_2-2}y^{-(\beta_1+\beta_2-2)}-\frac{C_{b,6}}{\beta_1-\beta_2+2}y^{\beta_1-\beta_2+2}\\&+\frac{C_{b,5}}{\beta_1-\beta_2-2}y^{-(\beta_1-\beta_2-2)}-\frac{C_{b,4}}{\beta_2+5}y^{\beta_2+5}\\&+\frac{C_{b,3}}{\beta_2-5}y^{-(\beta_2-5)}-\frac{1}{12}C_{b,2}y^{12}-\frac{1}{6}C_{b,1}y^6,\end{aligned}$$

$$C_5(y)=\frac{1}{4}A_{c,3}y^4+\frac{1}{2}A_{c,1}y^2+A_{c,-1}\ln y,$$

$$\begin{aligned}C_6(y)=&\frac{1}{2}c_5y^2+c_4\ln y-\frac{1}{4}\left[\sum_{i=1}^{6}\frac{C_{c,2i-1}}{i}y^{2i}+\sum_{i=1}^{3}\frac{C_{c,2i}}{i}\left(\ln y-\frac{1}{2i}\right)y^{2i}\right.\\&\left.+C_{c,10}\left(\ln^2y-\ln y+\frac{1}{2}\right)y^2+\frac{2}{3}C_{c,8}\ln^3y+C_{c,0}\ln^2y\right],\end{aligned}$$

$$\begin{aligned}e_1&=Q[G_3(y_1)-G_1(y_1)]+\alpha_1^3\beta_4[G_4(y_1)-G_2(y_1)]W_0^3+e_2,\\e_2&=Q[G_5(y_2)-G_3(y_2)]+\alpha_1^3\beta_4[G_6(y_2)-G_4(y_2)]W_0^3+e_3,\\e_3&=-QG_5(1)-\alpha_1^3\beta_4G_6(1)W_0^3.\end{aligned} \tag{47}$$

Substituting the solution (46a) into the expression (34) for the dimensionless center deflection, we have

$$Q = \alpha_1 W_0 + \alpha_3 W_0^3, \tag{48}$$

where

$$\alpha_3 = \alpha_1^4 \beta_4 [G_6(1) - G_6(y_2) + G_4(y_2) - G_4(y_1) + G_2(y_1)]. \tag{49}$$

The expression (48) is the characteristic relation of corrugated circular plate with a plane boundary region and a plane central region for rigidly clamped edge. In order to facilitate application, the relation (48) can be written in dimensional form

$$q = \alpha_0 (\alpha_1 h^2 w_0 + \alpha_3 w_0^3), \tag{50}$$

where w_0 is the center deflection,

$$\alpha_0 = \frac{Eh}{\beta_0 R^4}. \tag{51}$$

Especially, if we put $R_2 = R$ in (50), we can obtain the characteristic relation of large deflection of corrugated circular plate only with a plane central region under uniform pressure which has been studied by the author[2] in 1979.

If we put $R_2 = R$ and $R_1 = 0$ again in (50), we can yet obtain the characteristic relation of large deflection of corrugated circular plate with full corrugations under uniform pressure which had been studied by the author[1] in 1978. Here the relation is remarkably simple.

4. Numerical Examples

Now let us consider the following two numerical examples:

Example 1. A corrugated circular plate with sinusoidal corrugations (Fig. 1).

Let us apply the above theory to a numerical example. Assume $E = 10^4 \text{kg/mm}^2$, $\nu = 0.3$, $R = 40\text{mm}$, $R_1 = 3.8\text{mm}$, $R_2 = 30.2\text{mm}$, $H = 0.75\text{mm}$, $h = 0.22\text{mm}$, $L = 6.6\text{mm}$. Substituting these values into the relation (50), we find

$$q = 0.09415 w_0 + 0.06962 w_0^3, \tag{52}$$

where w_0 is in (mm), q is in (kg/cm^2).

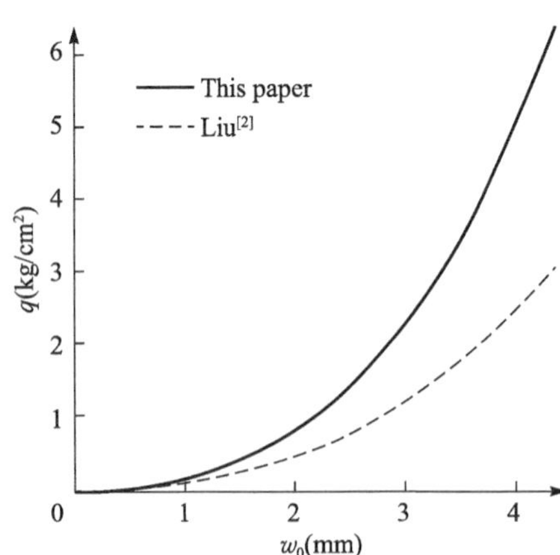

Fig. 3 Comparison between the approximate solution and the present solution

Using (52), we obtain useful results. Fig. 3 indicates the characteristic curve. This is shown as the full line. If we neglect the effect for the plane boundary region of the plate, i. e., we use the formula of (2), we can at once obtain an approximate solution. These numerical results are shown as the dashed line in Fig. 3. From this figure it is seen that, the approximate solution is far away from the present solution in a range of $w_0 > 4h$. When $w_0 = 20h$, the relative error between the approximate solution and the present solution is about 52.4%.

Example 2. A corrugated circular plate with toothed corrugations (Fig. 4).

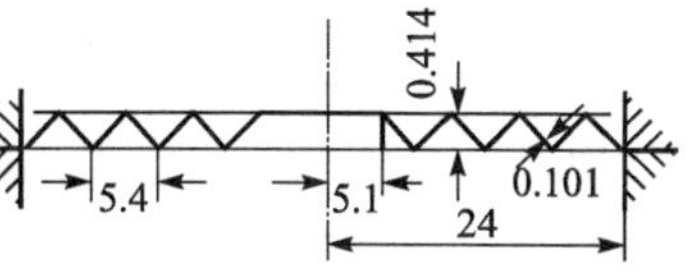

Fig. 4

Let us take a second example, a particular case in which $R_2=R$, and $E=1.35\times 10^4$ kg/mm^2, $\nu=0.3$, $R=24$mm, $R_1=5.1$mm, $H=0.414$mm, $h=0.101$mm, $\theta_0=8°45'$ and $L=5.4$mm. Then, from the characteristic relation (50), we find

$$q=0.1029w_0+0.1295w_0^3, \tag{53}$$

where w_0 is in (mm), q is in (kg/cm^2).

According to the relation (53), results are calculated and shown in Fig. 5, which is the characteristic curve. For the purpose of comparison, Fig. 5 also includes author's[1] and Andryewa's[3] results showing the results obtained from the approximate theory in which the effect for the plane central region of the plate is neglected. From this figure it may be seen that, reasonable agreement between present results and experimental results[3] in the wide range of $w_0\leqslant 18h$. But Andryewa's theoretical results are far away from experimental results in the range of $w_0>10h$. When $w_0=17.8h$, the relative error between the present and experimental results is 6.0%; for the results of [1] and [3], the errors are −4.6% and 17.2% respectively.

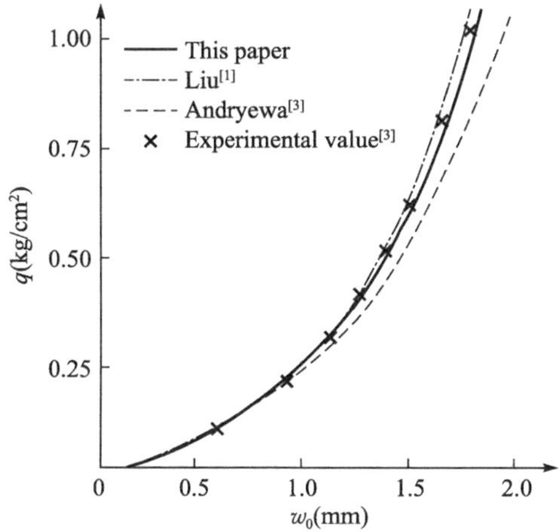

Fig. 5 Comparison between theoretical and experimental results

5. Discussion and Conclusion

In this paper, the analytical solution of large deflection of corrugated circular plate with a plane boundary region and a plane central region has been obtained by the modified iteration method. The solution can be applied to many cases for the plate with deep or shallow corrugations of various types. From the second example in the previous section, it is seen that even for $w_0=17.8h$, which corresponds to very flexible plate, the accuracy of the characteristic relation (50) is still very high.

Thus it can be expected that this paper will be satisfactory in practical cases. In general, this accuracy depends on the magnitude of $L/(R_2-R_1)$. The error increases with the increase of $L/(R_2-R_1)$. Namely, if corrugations of the plate are very close, the value of $L/(R_2-R_1)$ becomes small, and the corrugated region of the plate approaches the case of cylindrically anisotropic annular plate, then the accuracy of the relation (50) is higher.

Our calculations show yet that for a corrugated circular plate, in which R_1 and $(R-R_2)$ are low numbers, sometimes their effects on bending are negligible.

Acknowledgements

The work of this paper was completed when the author was an Alexander von Humboldt-Foundation fellow in West Germany. The author wishes to thank Professor W. Zerna, Ruhr University, for his help and warm hospitality.

References

[1] Liu Renhuai, The characteristic relations of corrugated circular plates, *Acta Mechanics Sinica*, No. 1, 47, 1978 (in Chinese).

[2] Liu Renhuai, The characteristic relations of corrugated circular plates with plane central region, *Journal of China University of Science and Technology*, 9, No. 2, 75, 1979 (in Chinese).

[3] Andryewa, L. E., The calculation of a corrugated membrane as an anisotropic plate, *The Engineer's Collection*. 21, 128, 1955 (in Russian).

[4] Andryewa, L. E., *Elastic Elements of Instruments*, Moscow: Masgiz, 1962 (in Russian).

[5] Akasaka. T., On the elastic properties of the corrugated diaphragm, *Journal of the Japan Society of Aeronautical Engineering*, 3, No. 22-23. 279, 1955 (in Japanese).

[6] Haringx, J. A., Nonlinearity of corrugated diaphragms, *Appl. Sci. Res.*, Ser. A. 16, 1956.

[7] Haringx. J. A., Design of corrugated diaphragms, *Trans. ASME*, Ser. A, 79, No. 1, 1957.

[8] Burmistrov, E. V., Axisymmetrical bending of isotropic and anisotropic shells of revolution under large deflection and non-uniform temperature field, *The Engineer's Collection*, 27, 185, 1960 (in Russian).

[9] Yeh, K. Y., Liu Renhuai, Ping, Q. Y. and Li, S. L., Nonlinear stability of thin circular shallow spherical shell under actions of axisymmetric uniform distributed line loads, *The Journal of Lanzhou University*, No. 2, 10, 1965; *Bulletin of Sciences*, No. 2, 142, 1965 (in Chinese).

[10] Yeh. K. Y., Liu Renhuai, Zhang, C, Z. and Xu, Y. F., Nonlinear stability of thin circular shallow spherical shell under the action of uniform edge moment, *Bulletin of Sciences*, No. 2, 145, 1965 (in Chinese).

[11] Liu Renhuai, Nonlinear stability of circular shallow spherical shell with a hole in the center under the action of uniform moment at the inner edge, *Bulletin of Sciences*, No. 3, 253, 1965 (in

Chinese).

[12] Liu Renhuai, Axisymmetrical buckling of thin circular shallow spherical shell with a circular hole at the center by shearing forces uniformly distributed along the inner edge, *Mechanics*, No. 3, 206, 1977 (in Chinese).

[13] Yeh, K. Y., Liu Renhuai, Zhang. C. Z. and Xu, Y. F., Nonlinear stability of thin elastic circular shallow spherical shell under the action of uniform edge moment, *Applied Mathematics and Mechanics*, 1, No. 1, 71, 1980.

[14] Liu Renhuai. Nonlinear axisymmetrical bending of circular sandwich plates under the action of uniform edge moment, *Journal of China University of Science and Technology*, 10, No. 2, 56, 1980 (in Chinese).

[15] Liu Renhuai, Nonlinear bending of circular sandwich plates, *Applied Mathematics and Mechanics*, 2, No. 2, 189, 1981.

[16] Panov, D. Y., On large deflection of circular membranes with weak corrugation. *Prikl. Mat. Mekh.*, 5, No. 2, 308, 1941 (in Russian).

[17] Feodosev. V. I., On large deflections and stability of a circular membrane with fine corrugation, *Prikl. Mat. Mekh.*, 9, No. 5, 389, 1945 (in Russian).

[18] Feodosev, V. I., *Elastic Elements of Precision-Instruments Manufacture*, Moscow: Oborongiz, 1949 (in Russian).

[19] Axelrad, E. L., Large deflection of a corrugated membrane as a non-shallow shell, *Izv. AN SSSR*, *Mec. Mas.*, 1, 46, 1964 (in Russian).

[20] Chen, S. L., Elastic behavior of uniformly loaded circular corrugated plate with sine-shaped shallow waves in large deflection, *Applied Mathematics and Mechanics*, 1, No. 2, 261, 1980.

[21] von Kármán, T. Festigkeitsprobleme im mashinenbau, *Encykl. Math. Wiss.*, 4, No. 4, 348, 1910.

[22] Chien, W. C., Large deflection of a circular clamped plate under uniform pressure, *Chinese Journal of Physics*, 7, No. 2, 102, 1947.

[23] Timoshenko, S. and Woinowsky-Krieger. S., *Theory of Plates and Shells*, 2nd edition, McGraw-Hill Book Co., New York, 1959.

在复合载荷作用下波纹环形板的非线性弯曲①

一、引言

波纹圆板是一类具有非常复杂形状的壳体。这样的板被用作弹性元件，广泛地应用在测量仪表里。因此，对于工程说来，研究这种板的非线性弯曲问题十分重要[1-3]。到现在，已经有许多人研究了这一问题，例如 Панов[4]，Феодосьев[5,6]，Андреева[3,7]，Haringx[8,9]，Бурмистров[10]，Аксельрад[11]，刘人怀[12−15]，陈山林[16]和张其浩[17]等。但是，仅有少数人进行了波纹环形板的非线性弯曲问题的研究工作，例如 Андреева[3]，赤坂隆[18]和刘人怀[1,2]。然而，由于非线性数学和复合载荷的困难，一个有实际意义的而且更加困难的问题，亦即在均布压力和中心集中载荷共同作用下，具有硬中心的波纹环形板的非线性弯曲，却始终还未被人研究过。按照正交各向异性圆板的非线性弯曲理论，我们使用修正迭代法[19−21]克服了上述困难，求解了这个问题。

二、基本方程

考虑一个在均布压力 q 和中心集中载荷 p 共同作用下，具有硬中心的波纹环形板（图 1），令 R 是外半径，R_1 是内半径，h 是厚度，L 是波纹，H 是波幅，r 是径向坐标。

按照正交各向异性圆板理论[7,13]，对波纹环形板的刚度参数定义如下。

对于径向刚度参数：

$$k_r=\frac{12}{h^2L}\int_0^L\frac{z^2}{\cos\theta}\mathrm{d}x+\frac{1}{L}\int_0^L\cos\theta\mathrm{d}x, \tag{2.1a}$$

对于周向刚度参数：

$$k_t=\frac{1}{L}\int_0^L\frac{1}{\cos\theta}\mathrm{d}x,$$

$$k_t'=\frac{12}{h^2L}\int_0^L\frac{z^2}{\cos\theta}\mathrm{d}x+\frac{1}{L}\int_0^L\frac{1}{\cos^3\theta}\mathrm{d}x \tag{2.1b,c}$$

其中，x 和 z 是坐标，θ 是径向截面的切线倾斜角（图 2）。

特别地，我们给出以下有实际重要性的三种情况。

（1）正弦波纹（图 2）

$$k_r\doteq 1+\frac{3H^2}{2h^2}+\frac{\pi^2H^2}{4L^2}\left(\frac{3H^2}{4h^2}-1\right),$$

$$k_t\doteq 1+\frac{\pi^2H^2}{4L^2},$$

① 本文原载《中国科学》，A 辑，1985，(6)：537-545.

$$k_t' \doteq 1 + \frac{3H^2}{2h^2} + \frac{3\pi^2 H^2}{4L^2}\left(\frac{H^2}{4h^2} + 1\right) \tag{2.2}$$

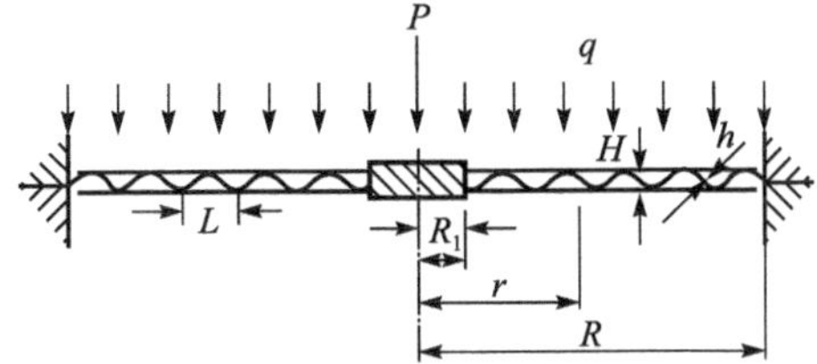

图 1　具有硬中心的波纹环形板

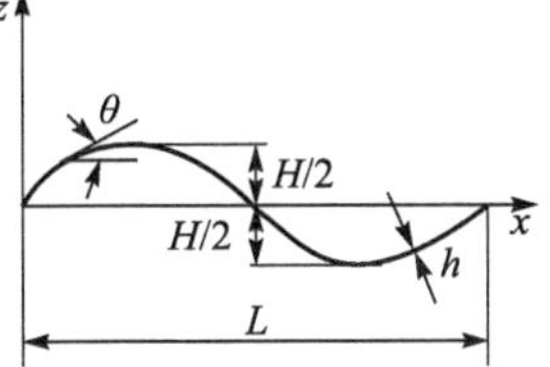

图 2　在波纹环形板径向截面上的一个波纹

(2) 锯齿形波纹（图 3(a)）

$$k_r = \frac{H^2}{h^2\cos\theta_0} + \cos\theta_0,$$

$$k_t = \frac{1}{\cos\theta_0},$$

$$k_t' = \frac{H^2}{h^2\cos\theta_0} + \frac{1}{\cos^3\theta_0} \tag{2.3}$$

(3) 梯形波纹（图 3(b)）

$$k_r = \frac{H^2}{Lh^2}\left(\frac{L-2L_0}{\cos\theta_0} + 6L_0\right) + \left(1 - \frac{2L_0}{L}\right)\cos\theta_0 + \frac{2L_0}{L},$$

$$k_t = \frac{1}{L}\left(\frac{L-2L_0}{\cos\theta_0} + 2L_0\right),$$

$$k_t' = \frac{H^2}{Lh^2}\left(\frac{L-2L_0}{\cos\theta_0} + 6L_0\right) + \frac{1}{L}\left(\frac{L-2L_0}{\cos^3\theta_0} - 2L_0\right) \tag{2.4}$$

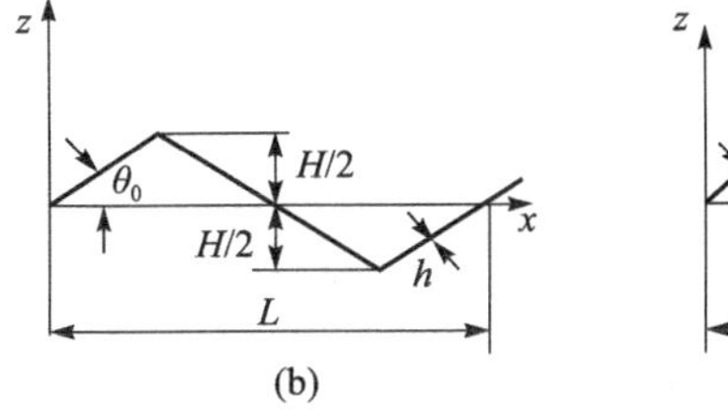

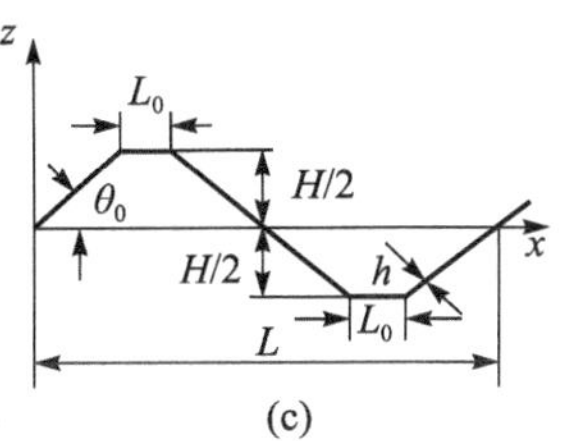

图 3　锯齿形波纹和梯形波纹

有了上述刚度参数公式，应用正交各向异性圆板的非线性弯曲理论[1-3]，就能得到这种波纹环形板的基本方程。由此，可得

$$r\frac{d^3w}{dr^3} + \frac{d^2w}{dr^2} - k_t\frac{dw}{dr}\left(\frac{1}{D}rN_r + \frac{k_t'}{r}\right) = \frac{k_t}{2D}\left(qr^2 + \frac{p}{\pi}\right),$$

$$r\frac{d^2}{dr^2}(rN_r) + \frac{d}{dr}(rN_r) - k_r k_t N_r + \frac{Ehk_t}{2}\left(\frac{dw}{dr}\right)^2 = 0 \tag{2.5}$$

其中，w 是挠度，N_r 是径向薄膜内力，E 是弹性模量，ν 是泊松比，D 是抗弯刚度，且有

$$D = \frac{Eh^3}{12\left(1 - \dfrac{\nu^2}{k_t k_t'}\right)} \tag{2.6}$$

这些非线性微分方程将在相应的边界条件下进行求解。在一般情况下，这种波纹环形板的外边界是夹紧固定或滑动固定，而内边界被固定在可上下移动的不变形的硬中心上。因此，边界条件如下。

(1) 沿外边界夹紧固定

当 $r=R_1$ 时，$\frac{\mathrm{d}w}{\mathrm{d}r}=0$，$u=0$。

当 $r=R$ 时，$w=0$，$\frac{\mathrm{d}w}{\mathrm{d}r}=0$，$u=0$。　　(2.7)

(2) 沿外边界滑动固定

当 $r=R_1$ 时，$\frac{\mathrm{d}w}{\mathrm{d}r}=0$，$u=0$。

当 $r=R$ 时，$w=0$，$\frac{\mathrm{d}w}{\mathrm{d}r}=0$，$N_r=0$。　　(2.8)

式中，u 是径向位移，且有

$$u=\frac{r}{k_tEh}\left[\frac{\mathrm{d}}{\mathrm{d}r}(rN_r)-\nu N_r\right] \tag{2.9}$$

为使最终的解更一般化，我们引入下列无量纲量：

$$\begin{aligned}
&y=\frac{r}{R},\quad y_1=\frac{R_1}{R},\quad W=\frac{w}{h},\\
&\phi=\frac{\mathrm{d}W}{\mathrm{d}y},\quad S=-\frac{Rk_tr}{D}N_r\quad U=\frac{2\beta_3R}{h^2}u,\\
&\beta_1^2=k_tk_t',\quad \beta_2^2=k_rk_t,\quad \beta_3=\frac{Eh^3k_t^2}{2D}\\
&Q=\frac{R^4k_t}{2Dh}q,\quad P=\frac{R^2k_t}{2\pi hD}p
\end{aligned} \tag{2.10}$$

进行代换，基本微分方程 (2.5) 化成无量纲形式

$$\begin{aligned}
&L_1(y^{\beta_1}\phi)=-S\phi+Qy^2+P,\\
&L_2(y^{\beta_2}S)=\beta_3\phi^2
\end{aligned} \tag{2.11a,b}$$

其中

$$\begin{aligned}
&L_1(\cdots)=y^{\beta_1}\frac{\mathrm{d}}{\mathrm{d}y}y^{1-2\beta_1}\frac{\mathrm{d}}{\mathrm{d}y}(\cdots),\\
&L_2(\cdots)=y^{\beta_2}\frac{\mathrm{d}}{\mathrm{d}y}y^{1-2\beta_2}\frac{\mathrm{d}}{\mathrm{d}y}(\cdots)
\end{aligned} \tag{2.12}$$

同样地，可将方程 (2.9) 表示为

$$U=-\left(y\frac{\mathrm{d}S}{\mathrm{d}y}-\nu S\right) \tag{2.13}$$

并且，边界条件 (2.7) 和 (2.8) 化为

(1) 沿外边界夹紧固定

当 $y=y_1$ 时，$\phi=0$，$y\frac{\mathrm{d}S}{\mathrm{d}y}-\nu S=0$。　　(2.14)

当 $y=1$ 时，$w=0$，$\phi=0$，$y\dfrac{\mathrm{d}S}{\mathrm{d}y}-\nu S=0$ (2.15)

(2) 沿外边界滑动固定

当 $y=y_1$ 时，$\phi=0$，$y\dfrac{\mathrm{d}S}{\mathrm{d}y}-\nu S=0$ (2.16)

当 $y=1$ 时，$W=0$，$\phi=0$，$S=0$ (2.17)

显然，把方程（2.11a）中的无量纲均布压力 Q 和无量纲集中载荷 P 表示为单一载荷参数 G 的量将是方便的。于是，令

$$Q=\mu_1 G,\quad P=\mu_2 G, \tag{2.18}$$

式中，μ_1 和 μ_2 是常数。在这种情况下，方程（2.11a）化为

$$L_1=(y^{\beta_1}\phi)=-S\phi+(\mu_1 y^2+\mu_2)G \tag{2.19}$$

方程（2.19）、（2.11b）、（2.14）和（2.15）或方程（2.19），（2.11b）、（2.16）和（2.17）便组成了在均布压力和中心集中载荷共同作用下，具有硬中心的波纹环形板的非线性弯曲的无量纲非线性边值问题。

三、解析解

我们首先讨论具有夹紧固定外边界的波纹环形板的无量纲非线性边值问题（2.19）、（2.11b）、（2.14）和（2.15）。为了简化具有滑动固定外边界的波纹环形板的第二个无量纲非线性边值问题（2.19）、（2.11b）、（2.16）和（2.17）的计算，我们暂时将边界条件式（2.15）的最后一个条件中的 ν 记为 ν'。

现在，应用修正迭代法求解第一个问题。

对于第一次近似，我们有线性边值问题如下：

$$L_1(y^{\beta_1}\phi_1)=(\mu_1 y^2+\mu_2)G,$$
$$L_2(y^{\beta_2}S_1)=\beta_3\phi_1^2 \tag{3.1a,b}$$

当 $y=y_1$ 时，$\phi_1=0$，$y\dfrac{\mathrm{d}S_1}{\mathrm{d}y}-\nu S_1=0$ (3.2a,b)

当 $y=1$ 时，$W_1=0$，$\phi_1=0$，$y\dfrac{\mathrm{d}S_1}{\mathrm{d}y}-\nu' S_1=0$ (3.3a～c)

满足边界条件（3.2a）和（3.3b）的方程（3.1a）的解是

$$\phi_1=G(a_1 y^{\beta_1}+a_2 y^{-\beta_1}+a_3 y^3+a_4 y) \tag{3.4}$$

其中

$$\begin{aligned}
a_1&=\frac{1}{y_1^{\beta_1}-y_1^{-\beta_1}}[y_1^{-\beta_1}(a_3+a_4)-y_1(a_3 y_1^2+a_4)],\\
a_2&=-\frac{1}{y_1^{\beta_1}-y_1^{-\beta_1}}[y_1^{\beta_1}(a_3+a_4)-y_1(a_3 y_1^2+a_4)],\\
a_3&=\frac{\mu_1}{9-\beta_1^2},\quad a_4=\frac{\mu_2}{1-\beta_1^2}
\end{aligned} \tag{3.5}$$

以在波纹环形板 $y=y_1$ 处的无量纲挠度 W（y_1）作为迭代参数，并记为 W_m。应用式（2.10)的第四式和边界条件（2.15）的第一式，便有

$$W_m = -\int_{y_1}^{1} \phi \, \mathrm{d}y \tag{3.6}$$

将式（3.4）代入式（3.6），即得波纹环形板的小挠度特征关系式如下

$$G = \lambda_1 W_m \tag{3.7}$$

其中

$$\lambda_1 = \left[\frac{a_1}{1+\beta_1}(y_1^{1+\beta_1} - 1) + \frac{a_2}{1-\beta_1}(y_1^{1-\beta_1} - 1) + \frac{a_3}{4}(y_1^4 - 1) + \frac{a_4}{2}(y_1^2 - 1) \right]^{-1} \tag{3.8}$$

于是，式（3.4）化为

$$\phi_1 = \lambda_1 W_m (a_1 y^{\beta_1} + a_2 y^{-\beta_1} + a_3 y^3 + a_4 y) \tag{3.9}$$

应用此解，我们求得问题（3.1b），（3.2b）和（3.3c）的解

$$\begin{aligned} S_1 = \beta_3 \lambda_1^2 W_m^2 (& b_1 y^{1+2\beta_1} + b_2 y^{1-2\beta_1} + b_3 y^{4+\beta_1} \\ & + b_4 y^{4-\beta_1} + b_5 y^{2+\beta_1} + b_6 y^{2-\beta_1} + b_7 y^{\beta_2} \\ & + b_8 y^{-\beta_2} + b_9 y^7 + b_{10} y^5 + b_{11} y^3 + b_{12} y) \end{aligned} \tag{3.10}$$

其中

$$\begin{aligned}
& b_1 = \frac{a_1^2}{(1+2\beta_1)^2 - \beta_2^2}, \qquad b_2 = \frac{a_2^2}{(1-2\beta_1)^2 - \beta_2^2}, \\
& b_3 = \frac{2a_1 a_3}{(4+\beta_1)^2 - \beta_2^2}, \qquad b_4 = \frac{2a_2 a_3}{(4-\beta_1)^2 - \beta_2^2}, \\
& b_5 = \frac{2a_1 a_4}{(2+\beta_1)^2 - \beta_2^2}, \qquad b_6 = \frac{2a_2 a_4}{(2-\beta_1)^2 - \beta_2^2}, \\
& b_7 = \frac{\omega_2(\beta_2 + \nu') - \omega_1(\beta_2 + \nu) y_1^{-\beta_2}}{(\beta_2 + \nu')(\beta_2 - \nu) y_1^{\beta_2} - (\beta_2 + \nu)(\beta_2 - \nu') y_1^{-\beta_2}}, \\
& b_8 = \frac{\omega_2(\beta_2 - \nu') - \omega_1(\beta_2 - \nu) y_1^{\beta_2}}{(\beta_2 + \nu')(\beta_2 - \nu) y_1^{\beta_2} - (\beta_2 + \nu)(\beta_2 - \nu') y_1^{-\beta_2}}, \\
& b_9 = \frac{a_3^2}{49 - \beta_2^2}, \qquad b_{10} = \frac{2a_3 a_4}{25 - \beta_2^2}, \\
& b_{11} = \frac{a_4^2}{9 - \beta_2^2}, \qquad b_{12} = \frac{2a_1 a_2}{1 - \beta_2^2}, \\
& \omega_1 = -[b_1(1+2\beta_1 - \nu') + b_2(1-2\beta_1 - \nu') + b_3(4+\beta_1 - \nu') \\
& \qquad + b_4(4-\beta_1 - \nu') + b_5(2+\beta_1 - \nu') + b_6(2-\beta_1 - \nu') \\
& \qquad + b_9(7-\nu') + b_{10}(5-\nu') + b_{11}(3-\nu') + b_{12}(1-\nu')], \\
& \omega_2 = -[b_1(1+2\beta_1 - \nu) y_1^{1+2\beta_1} + b_2(1-2\beta_1 - \nu) y_1^{1-2\beta_1} \\
& \qquad + b_3(4+\beta_1 - \nu) y_1^{4+\beta_1} + b_4(4-\beta_1 - \nu) y_1^{4-\beta_1} \\
& \qquad + b_5(2+\beta_1 - \nu) y_1^{2+\beta_1} + b_6(2-\beta_1 - \nu) y_1^{2-\beta_1} \\
& \qquad + b_9(7-\nu) y_1^7 + b_{10}(5-\nu) y_1^5 \\
& \qquad + b_{11}(3-\nu) y_1^3 + b_{12}(1-\nu) y_1]
\end{aligned} \tag{3.11}$$

对于第二次近似，关于 ϕ 的线性边值问题如下

$$L_1(y^{\beta_1}\phi_2) = -S_1\phi_1 + (\mu_1 y^2 + \mu_2)G \tag{3.12}$$

$$当\ y = y_1\ 时，\phi_2 = 0 \tag{3.13}$$

$$当\ y = 1\ 时，W_2 = 0, \phi_2 = 0 \tag{3.14}$$

应用已得的解（3.9）和（3.10），这一问题的解是

$$\begin{aligned}\phi_2 =& G(a_1 y^{\beta_1} + a_2 y^{-\beta_1} + a_3 y^3 + a_4 y) \\ & - \beta_3\lambda_1^3 W_m^3[F(y) + c_{28} y^{\beta_1} + c_{29} y^{-\beta_1}],\end{aligned} \tag{3.15}$$

其中

$$c_1 = \frac{a_1 b_1}{4(1+\beta_1)(1+2\beta_1)}, \qquad c_2 = \frac{a_2 b_2}{4(1-\beta_1)(1-2\beta_1)},$$

$$c_3 = \frac{a_1 b_3 + a_3 b_1}{(5+\beta_1)(5+3\beta_1)}, \qquad c_4 = \frac{a_2 b_4 + a_3 b_2}{(5-\beta_1)(5-3\beta_1)},$$

$$c_5 = \frac{a_1 b_5 + a_4 b_1}{3(1+\beta_1)(3+\beta_1)}, \qquad c_6 = \frac{a_2 b_6 + a_4 b_2}{3(1-\beta_1)(3-\beta_1)},$$

$$c_7 = \frac{a_1 b_9 + a_3 b_3}{16(4+\beta_1)}, \qquad c_8 = \frac{a_2 b_9 + a_3 b_4}{16(4-\beta_1)},$$

$$c_9 = \frac{a_1 b_{10} + a_3 b_5 + a_4 b_3}{12(3+\beta_1)}, \qquad c_{10} = \frac{a_2 b_{10} + a_3 b_6 + a_4 b_4}{12(3-\beta_1)},$$

$$c_{11} = \frac{a_1 b_{11} + a_4 b_5}{8(2+\beta_1)}, \qquad c_{12} = \frac{a_2 b_{11} + a_4 b_6}{8(2-\beta_1)},$$

$$c_{13} = \frac{a_1 b_{12} + a_2 b_1}{4(1+\beta_1)}, \qquad c_{14} = \frac{a_1 b_2 + a_2 b_{12}}{4(1-\beta_1)},$$

$$c_{15} = \frac{a_1 b_7}{(1+\beta_2)(1+2\beta_1+\beta_2)}, \qquad c_{16} = \frac{a_1 b_8}{(1-\beta_2)(1+2\beta_1-\beta_2)},$$

$$c_{17} = \frac{a_2 b_7}{(1+\beta_2)(1-2\beta_1+\beta_2)}, \qquad c_{18} = \frac{a_2 b_8}{(1-\beta_2)(1-2\beta_1-\beta_2)},$$

$$c_{19} = \frac{a_3 b_7}{(4+\beta_2)^2-\beta_1^2}, \qquad c_{20} = \frac{a_3 b_8}{(4-\beta_2)^2-\beta_1^2},$$

$$c_{21} = \frac{a_4 b_7}{(2+\beta_2)^2-\beta_1^2}, \qquad c_{22} = \frac{a_4 b_8}{(2-\beta_2)^2-\beta_1^2},$$

$$c_{23} = \frac{a_3 b_9}{121-\beta_1^2}, \qquad c_{24} = \frac{a_3 b_{10} + a_4 b_9}{81-\beta_1^2},$$

$$c_{25} = \frac{a_3 b_{11} + a_4 b_{10}}{49-\beta_1^2}, \qquad c_{26} = \frac{a_1 b_4 + a_2 b_3 + a_3 b_{12} + a_4 b_{11}}{25-\beta_1^2},$$

$$c_{27} = \frac{a_1 b_6 + a_2 b_5 + a_4 b_{12}}{9-\beta_1^2}, \qquad c_{28} = \frac{F(1) y_1^{-\beta_1} - F(y_1)}{y_1^{\beta_1} - y_1^{-\beta_1}},$$

$$c_{29} = \frac{F(y_1) - y_1^{\beta_1} F(1)}{y_1^{\beta_1} - y_1^{-\beta_1}},$$

$$\begin{aligned}F(y) =& c_1 y^{2+3\beta_1} + c_2 y^{2-3\beta_1} + c_3 y^{5+2\beta_1} + c_4 y^{5-2\beta_1} + c_5 y^{3+2\beta_1} \\ & + c_6 y^{3-2\beta_1} + c_7 y^{8+\beta_1} + c_8 y^{8-\beta_1} + c_9 y^{6+\beta_1} + c_{10} y^{6-\beta_1} \\ & + c_{11} y^{4+\beta_1} + c_{12} y^{4-\beta_1} + c_{13} y^{2+\beta_1} + c_{14} y^{2-\beta_1} + c_{15} y^{1+\beta_1+\beta_2}\end{aligned}$$

$$+c_{16}y^{1+\beta_1-\beta_2}+c_{17}y^{1-\beta_1+\beta_2}+c_{18}y^{1-\beta_1-\beta_2}+c_{19}y^{4+\beta_2}$$
$$+c_{20}y^{4-\beta_2}+c_{21}y^{2+\beta_2}+c_{22}y^{2-\beta_2}+c_{23}y^{11}$$
$$+c_{24}y^{9}+c_{25}y^{7}+c_{26}y^{5}+c_{27}y^{3} \tag{3.16}$$

将解（3.15）代入式（3.6），得波纹环形板大挠度的特征关系式如下

$$G=\lambda_1 W_m+\lambda_3 W_m^3 \tag{3.17}$$

其中

$$\begin{aligned}\lambda_3=\beta_3\lambda_1^4\Big[&\frac{c_1}{3(1+\beta_1)}(y_1^{3(1+\beta_1)}-1)+\frac{c_2}{3(1-\beta_1)}(y_1^{3(1-\beta_1)}-1)\\
&+\frac{c_3}{2(3+\beta_1)}(y_1^{2(3+\beta_1)}-1)+\frac{c_4}{2(3-\beta_1)}(y_1^{2(3-\beta_1)}-1)\\
&+\frac{c_5}{2(2+\beta_1)}(y_1^{2(2+\beta_1)}-1)+\frac{c_6}{2(2-\beta_1)}(y_1^{2(2-\beta_1)}-1)\\
&+\frac{c_7}{9+\beta_1}(y_1^{9+\beta_1}-1)+\frac{c_8}{9-\beta_1}(y_1^{9-\beta_1}-1)\\
&+\frac{c_9}{7+\beta_1}(y_1^{7+\beta_1}-1)+\frac{c_{10}}{7-\beta_1}(y_1^{7-\beta_1}-1)\\
&+\frac{c_{11}}{5+\beta_1}(y_1^{5+\beta_1}-1)+\frac{c_{12}}{5-\beta_1}(y_1^{5-\beta_1}-1)\\
&+\frac{c_{13}}{3+\beta_1}(y_1^{3+\beta_1}-1)+\frac{c_{14}}{3-\beta_1}(y_1^{3-\beta_1}-1)\\
&+\frac{c_{15}}{2+\beta_1+\beta_2}(y_1^{2+\beta_1+\beta_2}-1)+\frac{c_{16}}{2+\beta_1-\beta_2}(y_1^{2+\beta_1-\beta_2}-1)\\
&+\frac{c_{17}}{2-\beta_1+\beta_2}(y_1^{2-\beta_1+\beta_2}-1)+\frac{c_{18}}{2-\beta_1-\beta_2}(y_1^{2-\beta_1-\beta_2}-1)\\
&+\frac{c_{19}}{5+\beta_2}(y_1^{5+\beta_2}-1)+\frac{c_{20}}{5-\beta_2}(y_1^{5-\beta_2}-1)\\
&+\frac{c_{21}}{3+\beta_2}(y_1^{3+\beta_2}-1)+\frac{c_{22}}{3-\beta_2}(y_1^{3-\beta_2}-1)+\frac{c_{23}}{12}(y_1^{12}-1)\\
&+\frac{c_{24}}{10}(y_1^{10}-1)+\frac{c_{25}}{8}(y_1^{8}-1)+\frac{c_{26}}{6}(y_1^{6}-1)+\frac{c_{27}}{4}(y_1^{4}-1)\\
&+\frac{c_{28}}{1+\beta_1}(y_1^{1+\beta_1}-1)+\frac{c_{29}}{1-\beta_1}(y_1^{1-\beta_1}-1)\Big]\end{aligned} \tag{3.18}$$

为了实际需要，现将此式化为有量纲形式

$$g=\alpha(h^2\lambda_1 w_m+\lambda_3 w_m^3) \tag{3.19}$$

其中，w_m 是波纹环形板 $r=R_1$ 处的挠度，而 g，α，μ_1 和 μ_2 有下列意义。

（1）取均布压力 q 作为主要载荷

$$g=q,\quad \alpha=\frac{2D}{R^4h^2k_t},\ \mu_1=1,\ \mu_2=\frac{p}{\pi R^2 q} \tag{3.20}$$

（2）取集中载荷 p 作为主要载荷

$$g=p,\quad \alpha=\frac{2\pi D}{R^2h^2k_t},\ \mu_1=\frac{\pi R^2 q}{p},\ \mu_2=1 \tag{3.21}$$

下面，考虑在均布压力和中心集中载荷共同作用下，具有硬中心和滑动固定外边界的波纹环形板的无量纲非线性边值问题（2.19）、（2.11b）、（2.16）和（2.17）。实际上，这一波纹环形板大挠度的特征关系式，能够容易地直接从前面关于具有夹紧固定外边界的波纹环形板的特征关系式（3.19）得到。比较上述两个无量纲非线性边值问题，显而易见，仅仅边界条件（2.15）和（2.17）的最后一个条件不相同。如果我们将边界条件（2.15）的最后一式表述为下式

$$\text{当 } y=1 \text{ 时，} \frac{y}{\nu'}\frac{\mathrm{d}S}{\mathrm{d}y}-S=0 \tag{3.22}$$

然后令 $\nu'\to\infty$，则此式化为

$$\text{当 } y=1 \text{ 时，} S=0 \tag{3.23}$$

此即条件（2.17）的最后一式。这一事实简化了第二个问题的分析。在此情况下，令特征关系式（3.19）的 $\nu'\to\infty$，则容易得到具有硬中心和滑动固定外边界的波纹环形板的大挠度特征关系式。计算表明，除 b_7、b_8 和 ω_1 改为

$$b_7=\frac{\omega_2-\omega_1(\beta_2+\nu)y_1^{-\beta_2}}{(\beta_2-\nu)y_1^{\beta_2}+(\beta_2+\nu)y_1^{-\beta_2}},$$

$$b_8=-\frac{\omega_1(\beta_2-\nu)y_1^{\beta_2}+\omega_2}{(\beta_2-\nu)y_1^{\beta_2}+(\beta_2+\nu)y_1^{-\beta_2}},$$

$$\omega_1=\sum_{i=1}^{6}b_i+\sum_{i=9}^{12}b_j \tag{3.24}$$

以外，特征关系式和其他参数都与第一个问题中的结果相同。

特别地，在 $R_1=0$ 的极限情况下，具有硬中心的波纹环形板将化为波纹圆板。于是，从上述公式就给出在均布压力和中心集中载荷共同作用下，具有夹紧固定或滑动固定边界的波纹圆板的大挠度特征关系式。据我们所知，因为非线性数学和复合载荷的困难，这样的波纹圆板过去尚未被研究过。

四、算例

作为一个特例，我们考虑一个在硬中心和板的表面上同时有均布压力作用下，具有硬中心和夹紧固定外边界的锯齿形波纹环形板（图 4）的非线性弯曲问题。这一波纹环形板的尺寸是：$R=24.75\text{mm}$，$R_1=4.75\text{mm}$，$H=0.4\text{mm}$，$h=0.135\text{mm}$，$\theta_0=23°45'$，并且，$E=13500\text{kg/mm}^2$，$\nu=0.3$。在这种情况下，

$$p=0 \tag{4.1}$$

于是，由式（3.20）知，

$$\mu_2=0 \tag{4.2}$$

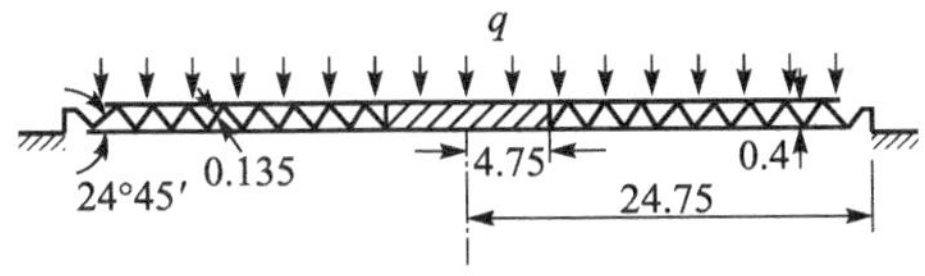

图 4　具有硬中心和夹紧固定外边界的锯齿形波纹环形板（mm）

现在，应用式（3.19）和（3.20）计算载荷—挠度关系，即得

$$q = 0.1453 w_m + 0.2445 w_m^3 \tag{4.3}$$

式中 w_m 的单位是 mm，q 的单位是 kg/cm^2。

这些结果见图 5。为了便于比较，图 5 也给出了 Андреева[3] 的实验值以及用 Галеркин 方法所得的理论结果。由此图可知，本文结果与实验值十分符合，较 Андреева 的理论结果有较高的精确度。

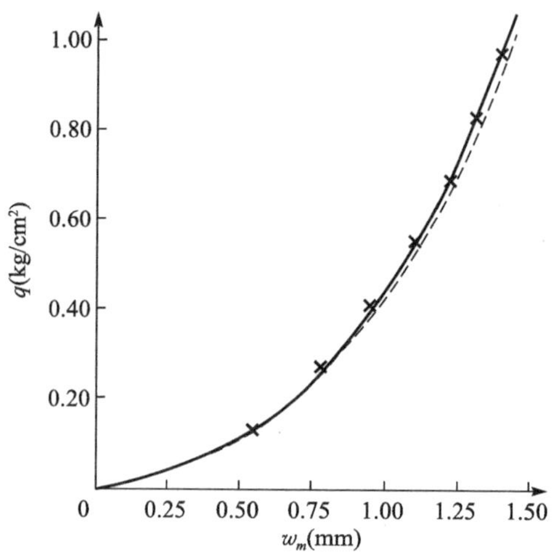

图 5　理论和实验结果之间的比较

(—— 本文，---- Андреева，× 实验值[3])

参考文献

[1]　刘人怀. 波纹环形板的非线性弯曲. 中国科学，A 辑，1984，(3)：247.

[2]　Liu Ren-huai（刘人怀）. On large deflection of corrugated annular plates under uniform pressure. The Advances of Applied Mathematics and Mechanics in China. vol. 1，Beijing：China Academic Publishers，1987，138.

[3]　Андреева，Л. Е. упpyzue Элеменmы Приборов. Москва：Машгиз，1962.

[4]　Панов，Д. Ю. О больших прогибах круглых мембран со слабым гофром，1941，5（2）：303.

[5]　Феодосьев，В. И. О больших прогибах и устойчивости круглой мембраны с мелкой гофрировкой. ПММ，1945，9（5）：389.

[6]　Феодосьев，В. И. Упpyrue Элеmehmы Точного Приборосrроения（中译本：精密仪器弹性元件的理论与计算. 卢文达，熊大遒译. 北京：科学出版社，1963）. Москва：Оборонгиз，1949.

[7]　Андреева，Л. Е. расёт гофрированных мембран，как анизотропных пластинок. Инженерный Сборнuк，1955，21：128.

[8]　Haringx，J. A. Nonlinearity of corrugated diaphragms. Appl. Scient. Res.，Ser. A，1956，16：45.

[9]　Haringx，J. A. Design of corrugated diaphragms. Trans. ASME，Ser A，1957，79（1）：55.

[10]　Бурмистров，Е. Ф. Симметричный изгиб неоднородных и однородных ортотропных оболочек

вращения с учегом ъольших прогибов и неравномерного температурного поля. Инженерный Сборник，1960，27：185.

［11］ Аксельрад，Э. Л. Ъольшие прогибы гофрированной мембраны как непологой оболочки. Изв. АН СССР，Mex. —Maw.，1964，(1)：46.

［12］ 刘人怀. 波纹圆板的特征关系式. 力学学报，1978，(1)：47.

［13］ 刘人怀. 具有光滑中心的波纹圆板的特征关系式. 中国科学技术大学学报，1979，(2)：75.

［14］ Liu Ren-huai（刘人怀）. Large deflection of corrugated circular plate with plane boundary region. Solid Mechanics Archives，1984，9 (4)：383.

［15］ Liu Ren-huai（刘人怀）. Large deflection of corrugated circular plate with a plane central region under the action of concentrated loads at the center. Int. J. Non-Linear Mechanics，1984，19 (5)：409.

［16］ 陈山林. 浅正弦波纹圆板在均布载荷下的大挠度弹性特征. 应用数学和力学，1980，1 (2)：261.

［17］ 张其浩. 波纹圆板的特征关系的研讨. 力学与实践，1980，2 (3)：64.

［18］ 赤坂隆. Corrugated diaphragmの弹性特性について. 日本航空学会誌，1955，3 (22～23)：279.

［19］ 叶开沅，刘人怀，平庆元，等. 在对称线布载荷作用下的圆底扁薄球壳的非线性稳定问题. 兰州大学学报，1965，(2)：10；科学通报，1965，(2)：142.

［20］ 叶开沅，刘人怀，张传智，等. 圆底扁薄球壳在边缘力矩作用下的非线性稳定问题. 科学通报，1965，(2)：145.

［21］ 刘人怀. 在内边缘均布力矩作用下中心开孔圆底扁球壳的非线性稳定问题. 科学通报，1965，(3)：253.

On large deflection of corrugated annular plates under uniform pressure①

中 文 摘 要

本文研究了具有硬中心的波纹环形板承受均布载荷的大挠度问题。由于研究的困难，此问题尚无人研究过。我们仍使用修正迭代法处理这一问题，获得了相当精确的解析解，可直接应用到弹性元件的设计中。

1. Introduction

The nonlinear bending of the corrugated annular plate is one of the important problems in the design of elastic elements of precision instruments which are applied widely in various engineering fields. However, this problem was only studied by a very few because of the difficulties of nonlinear mathematical problem involved and the more complicated shape of the plate. Akasaka[1] applied the energy method to study a corrugated annular plate with a nondeformable rigid body at the center by means of the nonlinear bending theory for anisotropic circular plates. He calculated two cases of concentrated load and uniform pressure, respectively. Afterwards, Andryewa[2] discussed the same problem by applying Galerkin's method. But their results are not satisfactory yet.

Recently, Liu Renhuai[3] obtained a quite accurate analytic solution of the abovementioned plate under a concentrated load by means of the modified iteration method. This method was suggested by Yeh Kai-yuan and Liu Renhuai in 1965[4-6], then applied successfully to solve a system of problems of nonlinear bending and stability of plates and shells[7-19]. This method incorporates the advantages of Chien Wei-zang's perturbation method[20] and usual successive approximations, has a broad validity scope, involves certain iteration process, and is simple and highly accurate.

This paper is a further work of the previous paper[3]. We have studied a nonlinear bending problem of practical interest for a corrugated annular plate with a nondeformable rigid body at the center under uniform pressure. We have used the modified iteration method to solve the problem. Analytic solutions to the problem obtained may be applied directly to the design of elastic elements.

2. The Nonlinear Boundary Value Problem

Let us consider the case of a corrugated annular plate with a nondeformable rigid

① Reprinted from *The Advances of Applied Mathematics and Mechanics in China*, Edited by Chien, W. Z. and Fu, Z. Z., Vol. 1, China Academic Publishers, Beijing, 1987, 138-152.

body at the center under uniform pressure q as shown in Fig. 1. The outer edge is rigidly clamped or movably clamped, and the inner edge is clamped in the nondeformable rigid body which can be moved up and down. Here h is the thickness, a and b are the outer and inner radii respectively, r is the radial coordinate, L is the wavelength, and H is the amplitude.

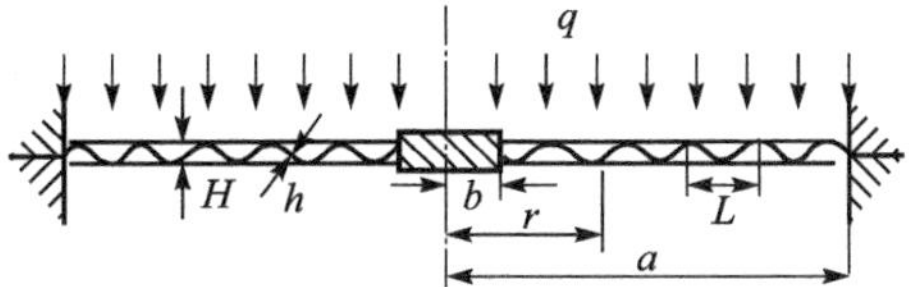

Fig. 1 A corrugated annular plate with a nondeformable rigid body at the center

According to the theory of the anisotropic circular plate[2,8], the definition of parameters for rigidities of a corrugated annular plate can be represented by the following formulas:

parameter for the radial rigidity:

$$k_r = \frac{12}{h^2 L}\int_0^L \frac{z^2}{\cos\theta}\mathrm{d}x + \frac{1}{L}\int_0^L \cos\theta \mathrm{d}x, \tag{1a}$$

parameters for the circumferential rigidity:

$$k_t = \frac{1}{L}\int_0^L \frac{1}{\cos\theta}\mathrm{d}x,$$

$$k_t' = \frac{12}{h^2 L}\int_0^L \frac{z^2}{\cos\theta}\mathrm{d}x + \frac{1}{L}\int_0^L \frac{1}{\cos^3\theta}\mathrm{d}x, \tag{1b,c}$$

where x and z are coordinates, and θ is the slope angle of the tangent to one diametral section (Fig. 2).

Usual values of the above parameters in some cases of practical interest are given below.

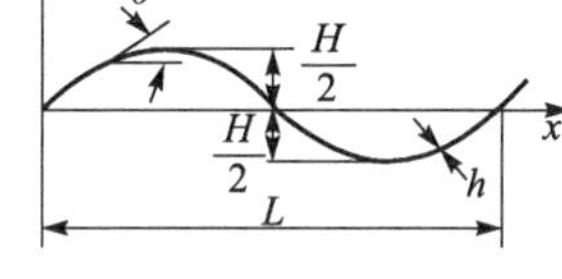

Fig. 2a A corrugation in one diametral section of the corrugated annular plate

1) Sinusoidal corrugation (Fig. 2a)

$$k_r \approx 1 + \frac{3H^2}{2h^2} + \frac{\pi^2 H^2}{4L^2}\left(\frac{3H^2}{4h^2} - 1\right),$$

$$k_t \approx 1 + \frac{\pi^2 H^2}{4L^2},$$

$$k_t' \approx 1 + \frac{3H^2}{2h^2} + \frac{3\pi^2 H^2}{4L^2}\left(\frac{H^2}{4h^2} + 1\right). \tag{2}$$

2) Toothed corrugation (Fig. 2b)

$$k_r = \frac{H^2}{h^2\cos\theta_0} + \cos\theta_0,$$

$$k_t = \frac{1}{\cos\theta_0},$$

$$k_t' = \frac{H^2}{h^2\cos\theta_0} + \frac{1}{\cos^3\theta_0}. \tag{3}$$

3) Trapezoidal corrugation (Fig. 2c)

$$k_r = \frac{H^2}{Lh^2}\left(\frac{L-2L_0}{\cos\theta_0}+6L_0\right)+\left(1-\frac{2L_0}{L}\right)\cos\theta_0+\frac{2L_0}{L},$$

$$k_t = \frac{1}{L}\left(\frac{L-2L_0}{\cos\theta_0}+2L_0\right),$$

$$k_t' = \frac{H^2}{Lh^2}\left(\frac{L-2L_0}{\cos\theta_0}+6L_0\right)+\frac{1}{L}\left(\frac{L-2L_0}{\cos^3\theta_0}+2L_0\right). \tag{4}$$

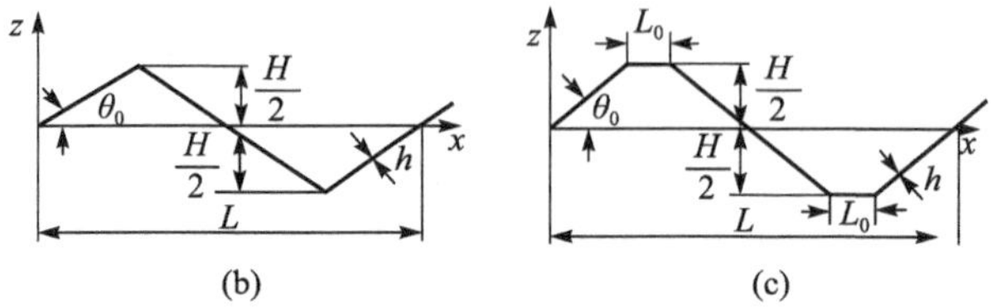

(b) (c)

Fig. 2b, c A corrugation in one diametral section of the corrugated annular plate

Having the formula (1), and proceeding from the nonlinear bending theory for anisotropic circular plates, we obtain fundamental equations for large deflection satisfied by the deflection w and the radial membrane stress N_r of the corrugated annular plate with a nondeformable rigid body at the center as follows

$$r\frac{d^3w}{dr^3}+\frac{d^2w}{dr^2}-k_t\frac{dw}{dr}\left(\frac{1}{D}rN_r+\frac{k_t'}{r}\right)=\frac{k_tq}{2D}r^2,$$

$$r\frac{d^2}{dr^2}(rN_r)+\frac{d}{dr}(rN_r)-k_rk_tN_r+\frac{Ehk_t}{2}\left(\frac{dw}{dr}\right)^2=0. \tag{5}$$

where E is the Young's modulus, D is the flexural rigidity,

$$D=\frac{Eh^3}{12\left(1-\frac{\nu^2}{k_tk_t'}\right)}, \tag{6}$$

and ν is the Poisson's ratio, respectively.

These nonlinear equations will be solved for the following two customary types of edge conditions:

1) Rigidly clamped along the outer edge

$$\frac{dw}{dr}=0,\quad u=0,\qquad \text{at } r=b;$$

$$w=0,\quad \frac{dw}{dr}=0,\quad u=0,\qquad \text{at } r=a, \tag{7}$$

2) Movably clamped along the outer edge

$$\frac{dw}{dr}=0,\quad u=0,\qquad \text{at } r=b;$$

$$w=0,\quad \frac{dw}{dr}=0,\quad N_r=0,\qquad \text{at } r=a, \tag{8}$$

where u is the radial displacement,

$$u=\frac{r}{k_tEh}\left[\frac{\mathrm{d}}{\mathrm{d}r}(rN_r)-\nu N_r\right]. \tag{9}$$

To express Eqs. (5) and the edge conditions (7)-(8) in a simpler dimensionless form let us introduce the following notations:

$$y=\frac{r}{a},\quad \alpha=\frac{b}{a},\quad W=\frac{w}{h},\quad \phi=\frac{\mathrm{d}W}{\mathrm{d}y},\quad S=-\frac{ak_tr}{D}N_r,$$
$$P=\frac{a^4k_t}{2Dh}q,\ \beta_1^2=k_tk_t',\ \beta_2^2=k_rk_t,\ \beta_3=\frac{Eh^3k_t^2}{2D}. \tag{10}$$

With these notations, Eqs. (5) become

$$L_1(y^{\beta_1}\phi)=-S\phi+Py^2,$$
$$L_2(y^{\beta_2}S)=\beta_3\phi^2, \tag{11}$$

and the conditions (7) - (8) become:

1) Rigidly clamped along the outer edge

$$\phi=0,\ y\frac{\mathrm{d}S}{\mathrm{d}y}-\nu S=0\qquad \text{at } y=\alpha; \tag{12a,b}$$

$$W=0,\ \phi=0,\ y\frac{\mathrm{d}S}{\mathrm{d}y}-\nu S=0\qquad \text{at } y=1. \tag{13a-c}$$

2) Movably clamped along the outer edge

$$\phi=0,\ y\frac{\mathrm{d}S}{\mathrm{d}y}-\nu S=0\qquad \text{at } y=\alpha; \tag{14a,b}$$

$$W=0,\ \phi=0,\ S=0\qquad \text{at } y=1, \tag{15a-c}$$

where

$$L_1(\cdots)=y^{\beta_1}\frac{\mathrm{d}}{\mathrm{d}y}y^{1-2\beta_1}\frac{\mathrm{d}}{\mathrm{d}y}(\cdots),$$
$$L_2(\cdots)=y^{\beta_2}\frac{\mathrm{d}}{\mathrm{d}y}y^{1-2\beta_2}\frac{\mathrm{d}}{\mathrm{d}y}(\cdots). \tag{16}$$

Eqs. (11)-(13) or Eqs. (11), (14) and (15) are the dimensionless nonlinear boundary value problem for discussing a corrugated annular plate with a nondeformable rigid body at the center under uniform pressure.

It is seen that in problems (11)-(13) for rigidly clamped edge and problems (11), (14) and (15) for movably clamped edge only the last one of the edge conditions is different. By using a simple method, i. e., let $\nu\to\infty$ in the condition (13c), we get the problems (11), (14), and (15) from the problems (11)-(13). In order to distinguish between the Poisson's ratios in two conditions (12b) and (13c), we indicate the ratio ν in the condition (13c) by ν'. Therefore, the solution for the problems (11), (14) and (15) can readily be obtained by setting $\nu'\to\infty$ from the solution for the problems (11)-(13).

3. Solution of the Problem

We shall begin with the dimensionless nonlinear boundary value problem (11)-(13) for the plate with a rigidly clamped outer edge. This problem will be solved by the modified iteration method.

For the first approximation, by taking $S\phi=0$ and $\phi^2=\phi_1^2$ for nonlinear terms in Eqs. (11), we have the following linear boundary value problem

$$L_1(y^{\beta_1}\phi_1)=Py^2,$$
$$L_2(y^{\beta_2}S_1)=\beta_3\phi_1^2, \tag{17a,b}$$

$$\phi_1=0,\ y\frac{dS_1}{dy}-\nu S_1=0 \qquad \text{at } y=\alpha, \tag{18a,b}$$

$$W_1=0,\ \phi_1=0,\ y\frac{dS_1}{dy}-\nu' S_1=0 \qquad \text{at } y=1. \tag{19a-c}$$

The solution of the problem (17a), (18a) and (19b) is

$$\phi_1=\frac{P}{\mu}(a_1y^{\beta_1}+a_2y^{-\beta_1}-y^3), \tag{20}$$

where

$$a_1=\frac{\alpha^3-\alpha^{-\beta_1}}{\alpha^{\beta_1}-\alpha^{-\beta_1}},\ a_2=\frac{\alpha^{-\beta_1}-\alpha^3}{\alpha^{\beta_1}-\alpha^{-\beta_1}},$$
$$\mu=\beta_1^2-9. \tag{21}$$

We introduce a notation W_m of the dimensionless deflection at $y=\alpha$ for the plate as an iteration parameter

$$W_m=W\mid_{y=\alpha}. \tag{22}$$

Then using the fourth in expressions (10) and the edge condition (13a), we obtain the relation between W_m and ϕ in the followimg

$$W_m=-\int_\alpha^1\phi dy. \tag{23}$$

Substituting the solution (20) into expression (23), after integrating we obtain the characteristic relation for the small deflection of the corrugated annular plate with a nondeformable rigid body at the center in the following form:

$$P=\lambda_1\mu W_m, \tag{24}$$

where

$$\lambda_1=-\left[\frac{a_1}{1+\beta_1}(1-\alpha^{1+\beta_1})+\frac{a_2}{1-\beta_1}(1-\alpha^{1-\beta_1})-\frac{1}{4}(1-\alpha^4)\right]^{-1}. \tag{25}$$

Making use of the relation (24), we can represent the solution (20) in the form

$$\phi_1=\lambda_1W_m(a_1y^{\beta_1}+a_2y^{-\beta_1}-y^3). \tag{26}$$

Substituting this solution into Eq. (17b), we find the following solution of the problem (17b), (18b) and (19c)

$$S_1=\beta_1\lambda_1^2W_m^2(b_1y^{1+2\beta_1}+b_2y^{1-2\beta_1}+b_3y^{4+\beta_1}+b_4y^{4-\beta_1}+b_5y^{\beta_2}+b_6y^{-\beta_2}+b_7y^7+b_8y), \tag{27}$$

where

$$b_1 = \frac{a_1^2}{(1+2\beta_1)^2 - \beta_2^2}, \qquad b_2 = \frac{a_2^2}{(1-2\beta_1)^2 - \beta_2^2},$$

$$b_3 = -\frac{2a_1}{(4+\beta_1)^2 - \beta_2^2}, \qquad b_4 = -\frac{2a_2}{(4-\beta_1)^2 - \beta_2^2},$$

$$b_5 = \frac{\omega_2(\beta_2 + \nu') - \omega_1(\beta_2 + \nu)\alpha^{-\beta_2}}{(\beta_2 + \nu')(\beta_2 - \nu)\alpha^{\beta_2} - (\beta_2 + \nu)(\beta_2 - \nu')\alpha^{-\beta_2}},$$

$$b_6 = \frac{\omega_2(\beta_2 - \nu') - \omega_1(\beta_2 - \nu)\alpha^{\beta_2}}{(\beta_2 + \nu')(\beta_2 - \nu)\alpha^{\beta_2} - (\beta_2 + \nu)(\beta_2 - \nu')\alpha^{-\beta_2}},$$

$$b_7 = \frac{1}{49 - \beta_2^2}, \qquad b_8 = \frac{2a_1 a_2}{1 - \beta_2^2},$$

$$\begin{aligned} \omega_1 = &-[b_1(1+2\beta_1 - \nu') + b_2(1-2\beta_1 - \nu') + b_3(4+\beta_1 - \nu') \\ &+ b_4(4-\beta_1 - \nu') + b_7(7-\nu') + b_8(1-\nu')], \\ \omega_2 = &-[b_1(1+2\beta_1 - \nu)\alpha^{1+2\beta_1} + b_2(1-2\beta_1 - \nu)\alpha^{1-2\beta_1} \\ &+ b_3(4+\beta_1 - \nu)\alpha^{4+\beta_1} + b_4(4-\beta_1 - \nu)\alpha^{4-\beta_1} \\ &+ b_7(7-\nu)\alpha^7 + b_8(1-\nu)\alpha], \end{aligned} \tag{28}$$

or, with $\nu' = \nu$,

$$b_5 = \frac{\omega_2 - \omega_1\alpha^{-\beta_2}}{(\beta_2 - \nu)(\alpha^{\beta_2} - \alpha^{-\beta_2})}, \qquad b_6 = \frac{\omega_2 - \omega_1\alpha^{\beta_2}}{(\beta_2 + \nu)(\alpha^{\beta_2} - \alpha^{-\beta_2})},$$

$$\begin{aligned} \omega_1 = &-[b_1(1+2\beta_1 - \nu) + b_2(1-2\beta_1 - \nu) + b_3(4+\beta_1 - \nu) \\ &+ b_4(4-\beta_1 - \nu) + b_7(7-\nu) + b_8(1-\nu)]. \end{aligned} \tag{29}$$

For the second approximation, we have the linear boundary value problem for ϕ as follows

$$L_1(y^{\beta_1}\phi_2) = -S_1\phi_1 + Py^2, \tag{30}$$

$$\phi_2 = 0 \qquad \text{at } y = \alpha, \tag{31}$$

$$W_2 = 0, \quad \phi_2 = 0 \qquad \text{at } y = 1. \tag{32a,b}$$

Substituting solutions (26) and (27) into Eq. (30). We find the following solution of the problem (30), (31) and (32b)

$$\begin{aligned} \phi_2 = &\frac{P}{\mu}(a_1 y^{\beta_1} + a_2 y^{-\beta_1} - y^3) \\ &- \beta_3\lambda_1^3 W_m^3 [F(y) + c_{17} y^{\beta_1} + c_{18} y^{-\beta_1}], \end{aligned} \tag{33}$$

where

$$\begin{aligned} F(y) = &c_1 y^{2+3\beta_1} + c_2 y^{2-3\beta_1} + c_3 y^{5+2\beta_1} + c_4 y^{5-2\beta_1} + c_5 y^{8+\beta_1} \\ &+ c_6 y^{8-\beta_1} + c_7 y^{2+\beta_1} + c_8 y^{2-\beta_1} + c_9 y^{1+\beta_1+\beta_2} + c_{10} y^{1+\beta_1-\beta_2} \\ &+ c_{11} y^{1-\beta_1+\beta_2} + c_{12} y^{1-\beta_1-\beta_2} + c_{13} y^{4+\beta_2} + c_{14} y^{4-\beta_2} + c_{15} y^{11} + c_{16} y^5, \end{aligned}$$

$$c_1 = \frac{a_1 b_1}{4(1+\beta_1)(1+2\beta_1)}, \qquad c_2 = \frac{a_2 b_2}{4(1-\beta_1)(1-2\beta_1)},$$

$$c_3 = \frac{a_1 b_3 - b_1}{(5+\beta_1)(5+3\beta_1)}, \qquad c_4 = \frac{a_2 b_4 - b_2}{(5-\beta_1)(5-3\beta_1)},$$

$$c_5 = \frac{a_1 b_7 - b_3}{16(4+\beta_1)}, \qquad c_6 = \frac{a_2 b_7 - b_4}{16(4-\beta_1)},$$

$$c_7 = \frac{a_1 b_8 + a_2 b_1}{4(1+\beta_1)}, \qquad c_8 = \frac{a_1 b_2 + a_2 b_8}{4(4-\beta_1)},$$

$$c_9 = \frac{a_1 b_5}{(1+\beta_2)(1+2\beta_1+\beta_2)},$$

$$c_{10} = \frac{a_1 b_6}{(1-\beta_2)(1+2\beta_1-\beta_2)},$$

$$c_{11} = \frac{a_2 b_5}{(1+\beta_2)(1-2\beta_1+\beta_2)},$$

$$c_{12} = \frac{a_2 b_6}{(1-\beta_2)(1-2\beta_1-\beta_2)},$$

$$c_{13} = \frac{b_5}{\beta_1^2 - (4+\beta_2)^2}, \qquad c_{14} = \frac{b_6}{\beta_1^2 - (4-\beta_2)^2},$$

$$c_{15} = -\frac{b_7}{121-\beta_1^2}, \qquad c_{16} = \frac{a_1 b_4 + a_2 b_3 - b_8}{25-\beta_1^2},$$

$$c_{17} = -\frac{F(\alpha) - F(1)\alpha^{-\beta_1}}{\alpha^{\beta_1} - \alpha^{-\beta_1}},$$

$$c_{18} = \frac{F(\alpha) - F(1)\alpha^{\beta_1}}{\alpha^{\beta_1} - \alpha^{-\beta_1}}. \tag{34}$$

Substituting this solution into expression (23), after integrating we obtain

$$P = \mu(\lambda_1 W_m + \lambda_3 W_m^3), \tag{35}$$

where

$$\begin{aligned}
\lambda_3 = \beta_3 \lambda_1^4 \Big\{ & \frac{c_1}{3(1+\beta_1)}(\alpha^{3(1+\beta_1)} - 1) + \frac{c_2}{3(1-\beta_1)}(\alpha^{3(1-\beta_1)} - 1) \\
& + \frac{c_3}{2(3+\beta_1)}(\alpha^{2(3+\beta_1)} - 1) + \frac{c_4}{2(3-\beta_1)}(\alpha^{2(3-\beta_1)} - 1) \\
& + \frac{c_5}{9+\beta_1}(\alpha^{9+\beta_1} - 1) + \frac{c_6}{9-\beta_1}(\alpha^{9-\beta_1} - 1) + \frac{c_7}{3+\beta_1}(\alpha^{3+\beta_1} - 1) \\
& + \frac{c_8}{3-\beta_1}(\alpha^{3-\beta_1} - 1) + \frac{c_9}{2+\beta_1+\beta_2}(\alpha^{2+\beta_1+\beta_2} - 1) \\
& + \frac{c_{10}}{2+\beta_1-\beta_2}(\alpha^{2+\beta_1-\beta_2} - 1) + \frac{c_{11}}{2-\beta_1+\beta_2}(\alpha^{2-\beta_1+\beta_2} - 1) \\
& + \frac{c_{12}}{2-\beta_1-\beta_2}(\alpha^{2-\beta_1-\beta_2} - 1) + \frac{c_{13}}{5+\beta_2}(\alpha^{5+\beta_2} - 1) \\
& + \frac{c_{14}}{5-\beta_2}(\alpha^{5-\beta_2} - 1) + \frac{c_{15}}{12}(\alpha^{12} - 1) + \frac{c_{16}}{6}(\alpha_6 - 1) \\
& + \frac{c_{17}}{1+\beta_1}(\alpha^{1+\beta_1} - 1) + \frac{c_{18}}{1-\beta_1}(\alpha^{1-\beta_1} - 1) \Big\}.
\end{aligned} \tag{36}$$

This is the nonlinear characteristic relation of the corrugated annular plate with a nondeformable rigid body at the center and a rigidly clamped outer edge under uniform pressure.

In order to facilitate the application, we may rewrite the relation (35) into dimensional form

$$q=\lambda_0\mu(h^2\lambda_1 w_m+\lambda_3 w_m^3),\tag{37}$$

where w_m is the deflection at $r=b$ for the plate,

$$\lambda_0=\frac{Ehk_t'}{6a^4(\beta_1^2-\nu^2)}.\tag{38}$$

Finally, let $\nu'\to\infty$ in the relation (37), we can at once obtain the nonlinear characteristic relation of the corrugated annular plate with a nondeformable rigid body at the center and a movably clamped outer edge under uniform pressure. At this moment the present relation and most parameters coincide with the above results in appearance, only b_5, b_6 and ω_1 become

$$b_5=\frac{\omega_2-\omega_1(\beta_2+\nu)\alpha^{-\beta_2}}{(\beta_2-\nu)\alpha^{\beta_2}+(\beta_2+\nu)\alpha^{-\beta_2}},\ b_6=-\frac{\omega_2+\omega_1(\beta_2-\nu)\alpha^{\beta_2}}{(\beta_2-\nu)\alpha^{\beta_2}+(\beta_2+\nu)\alpha^{-\beta_2}},$$
$$\omega_1=b_1+b_2+b_3+b_4+b_7+b_8.\tag{39}$$

Specifically, in the limiting case where the inner radius b is infinitely small, from the above relations we can obtain the nonlinear characteristic relation of a corrugated circular plate with a rigidly clamped edge or a movably clamped edge under uniform pressure which had been studied by the author[8] in 1978.

4. Applied Examples

Example 1. We consider a corrugated annular plate with sinusoidal corrugations in the case of a rigidly clamped outer edge (Fig. 1). The known values are given: $a=78$mm, $b=18$mm, $h=0.2$mm, $L=12$mm, $H=1.354$mm, $E=1.345\times10^4$kg/mm^2, $\nu=0.3$.

Substituting these data into the relation (37), we find

$$q=0.02092w_m+0.001005w_m^3,\tag{40}$$

where w_m is in (mm), q is in (kg/cm^2).

According to expression (40), the characteristic curve is shown in Fig. 3. It is seen that the curve rises monotonically.

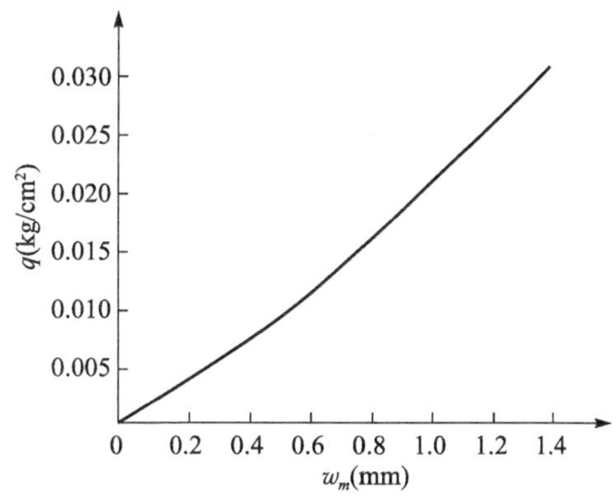

Fig. 3　The characteristic curve

Example 2. We consider now a particular example of a corrugated circular plate with sinusoidal corrugations in the case of a movably clamped edge (Fig. 4).

Assume $a=76$mm, $b=0$, $h=0.33$mm, $L=25\frac{1}{3}$mm, $H=1.20$mm, $E=1\times10^4$kg/mm^2, $\nu=0.3$.

Using the above relation in which $b=0$, we find

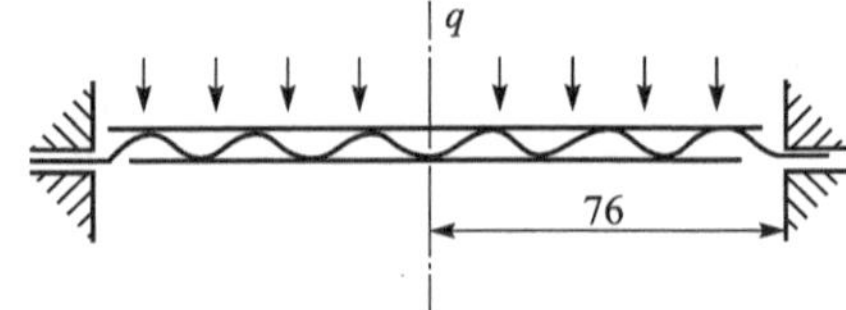

Fig. 4 A corrugated circular plate

$$q = 0.03036 w_m + 0.002347 w_m^3, \qquad (41)$$

where the center deflection w_m is in (mm), q is in (kg/cm^2).

This result is shown in Fig. 5. For the purpose of comparison, Fig. 5 also includes Feodosev's[21] experimental value and theoretical results obtained by means of Galerkin's method. From this figure it may be seen that, reasonable agreement between present results and experimental value in the wide range of $w_m < 9h$. But, Feodosev's theoretical results are far away from experimental value in the range of $w_m > 5h$. When $w_m = 8.79h$, the relative error between the present results and experimental value is only 0.90%, but for Feodosev's results, the error is about 39.0%.

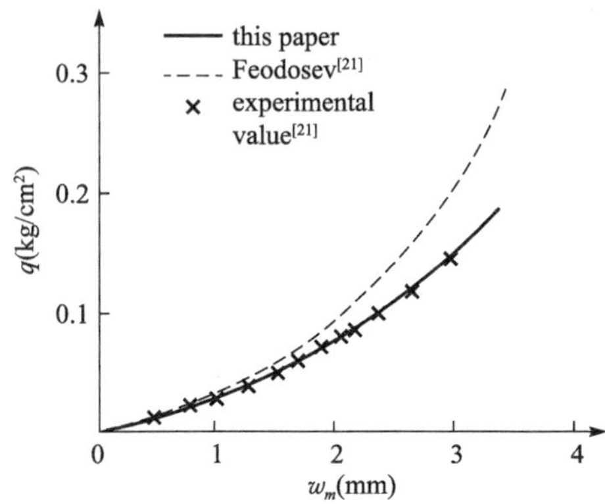

Fig. 5 Comparison between theoretical and experimental results

5. Conclusion

By means of the nonlinear bending theory for anisotropic circular plates, this paper gives analytic solutions for large deflection of corrugated annular plates with a nondeformable rigid body at the center under uniform pressure by the modified iteration method. Two marked advantages of solving the problem with our method are that:

1) Solutions have a simple form and very satisfactory accuracy.

2) Solutions are suitable for manifold cases of corrugated annular plates with deep or shallow corrugations of various types.

References

[1] Akasaka T., On the elastic properties of the corrugated diaphragm, *Journal of the Japan Society of Aeronautical Engineering*, Vol. 3, No. 22-23, 279 (1995) (in Japanese).

[2] Andryewa, L. E., *Elastic Elements of Instruments*. Moscow: Masgiz (1962) (in Russian).

[3] Liu Renhuai, Nonlinear bending of corrugated annular plates, Science in China, Ser. A, Vol.

27, No. 6, 640 (1984).

[4] Yeh Kaiyuan, Liu Renhuai, Ping Qingyuan and Li Silai, Nonlinear stability of thin circular shallow spherical shell under actions of axisymmetric uniform distributed line loads, *Journal of Lanzhou University*. No. 2, 10 (1965); *Bulletin of Sciences*; No. 2, 142 (1965) (in Chinese).

[5] Yeh Kaiyuan, Liu Renhuai, Zhang Chuanzhi and Xu Yifan, Nonlinear stability of thin circular shallow spherical shell under the action of uniform edge moment, *Bulletin of Sciences*, No. 2, 145 (1965) (in Chinese).

[6] Liu Renhuai, Nonlinear stability of circular shallow spherical shell with a hole in the center under the action of uniform moment at the inner edge, *Bulletin of Sciences*, No. 3, 253 (1965) (in Chinese).

[7] Liu Renhuni, Axisymmetrical buckling of thin circular shallow spherical shell with a circular hole at the center by shearing forces uniformly distributed along the inner edge, *Mechanics*, No. 3, 206 (1977) (in Chinese).

[8] Liu Renhuai, The characteristic relations of corrugated circular plates, *Acta Mechanics Sinica*, No. 1, 47 (1978) (in Chinese).

[9] Liu Renhuai. The characteristic relations of corrugated circular plates with plane center region, *Journal of China University of Science and Technology*, Vol. 9, No. 2, 75 (1979) (in Chinese).

[10] Yeh Kaiyuan, Liu Renhuai, Zhang Chuanzhi and Xu Yifan, Nonlinear stability of thin elastic circular shallow spherical shell under the action of uniform edge moment, *Applied Mathematics and Mechanics*, Vol. 1, No. 1, 71 (1980).

[11] Liu Renhuai, Nonlinear axisymmetrical bending of circular sandwich plates under the action of uniform edge moment. *Journal of China University of Science and Technology*. Vol. 10, No. 2, 56 (1980) (in Chinese).

[12] Yeh Kaiyuan and Gu Shuxian, Nonlinear stability of a spherical shallow shell under uniformly distributed pressure. *Symposium of the First Chinese National Congress on Computational Structural Mechanics*, 280 (1980) (in Chinese).

[13] Liu, Renhuai, The thermal stability of bimetallic truncated shallow conical shells, *Acta Mechanica Sinica*, *Special Issue*, 172 (1981) (in Chinese).

[14] Liu Renhuai, Nonlinear bending of circular sandwich plates, *Applied Mathematics and Mechanics*, Vol. 2, No. 2, 189 (1981).

[15] Liu Renhuai, Nonlinear thermal stability of bimetallic shallow spherical shells with a circular hole at the center, *Journal of China University of Science and Technology*, Vol. 11, No. 1, 84 (1981) (in Chinese).

[16] Liu Renhaui, Nonlinear thermal stability of bimetallic shallow shells of revolution, *International Journal of Nonlinear Mechanics*, *Vol*. 18, No. 5, 409 (1983).

[17] Liu Renhusi, Large deflection of corrugated circular plate with plane boundary region, *Solid Mechanics Archives*, *Vol*. 9, No. 4, 383 (1984).

[18] Liu Renhuai, Nonlinear bending of circular sandwich plates under the action of axisymmetric uniform distributed line loads, Progress in Applied Mechanics, Edited by Yeh, K. Y., Martinus Nijhoff Publishers, Dordrecht, 293 (1987).

[19] Liu Renhuai, Large deflection of corrugated circular plate under the action of concentrated loads at the center, International Journal of Nonlinear Mechanics, Vol. 19, No. 5, 409 (1984).

[20] Chien Weizang, Large deflection of a circular clamped plate under uniform pressure, *Chinese Journal of Physics*, Vol. 7, No, 2, 102 (1947).

[21] Feodosev, V. I., *Elastic Elements of Precision-Instruments Manufacture*, Oborongiz (1949) (in Russian).

复合载荷下波纹圆板的非线性分析①

一、引言

众所周知，在精密仪器里广泛使用着一种名为波纹圆板的弹性元件。因此，研究波纹圆板的大挠度问题在理论和实践上都有着极其重要的意义。因为这一问题牵涉到此种板的复杂形状和非线性微分方程组，所以问题十分复杂。于是，大家通常研究一些比较简单的问题。

最初，Панов[1]，Феодосьев[2,3]以及后来的 Аксельрад[4]和陈山林[5]等都从壳体的大挠度理论出发，讨论了具有光滑中心的波纹圆板。然而这板的理论分析是那样困难，以至于他们仅讨论了正弦波纹和均布载荷一种情形，而且结果也不精确。因此，研究另一种更有效的方法是十分必要的。

为此，Андреева[6,8]，赤坂隆[7]，Haringx[9,10]和 Бурмистров[11]等进行了新的处理，将波纹圆板视为各向异性圆板。这样，便对具有各种形状的深、浅波纹的波纹圆板都能适用。在这种情况下，他们使用各向异性圆板的大挠度理论讨论了两个比较简单的问题，即承受均布载荷的具有全波纹的波纹圆板以及承受均布载荷或中心集中载荷的具有刚性中心的波纹环形板。然而，他们的结果仍然不能令人满意。最近，刘人怀[12,13]应用修正迭代法成功地重新考虑了这些板，得到了十分精确的解析解。这一方法是在 1965 年研究圆底扁薄球壳的非线性稳定问题中，首先由叶开沅和刘人怀[14-16]提出来的。它结合了钱伟长教授的摄动法[17]和通常的逐次逼近法的优点，对求解板壳非线性问题十分有效。

此外，刘人怀还讨论了几个更困难的情况：均布载荷下具有光滑中心的波纹圆板[18]，中心集中载荷下具有光滑中心的波纹圆板[19]，均布载荷下具有光滑中心和平面边缘区域的波纹圆板[20]，均布载荷下具有刚性中心的波纹环形板[21]以及均布载荷和中心集中载荷共同作用下的具有刚性中心的波纹环形板[22]。

本文是作者上述工作的继续。我们研究了一个有实际意义的问题，即具有光滑中心的波纹圆板在均布和中心集中载荷联合作用下的非线性弯曲。显然，此板的大挠度理论分析将比我们以往所处理过的问题都更加困难。产生分析上的困难的原因是此板带有一个光滑中心并承受了复合载荷。如同以前一样，我们使用修正迭代法克服了这些困难，获得了可供精密仪器弹性元件设计时所需要的公式。

二、方程

现在考虑图 1 所示的具有光滑中心的波纹圆板。这里，R 和 R_1 分别是边缘和光滑中心的半径，h 是厚度，L 是波长，H 是波幅，r 是径向坐标，q 是均布载荷，p 是集

---

① 本文原载《应用数学和力学》，1988，9 (8)：661-674.

中载荷。

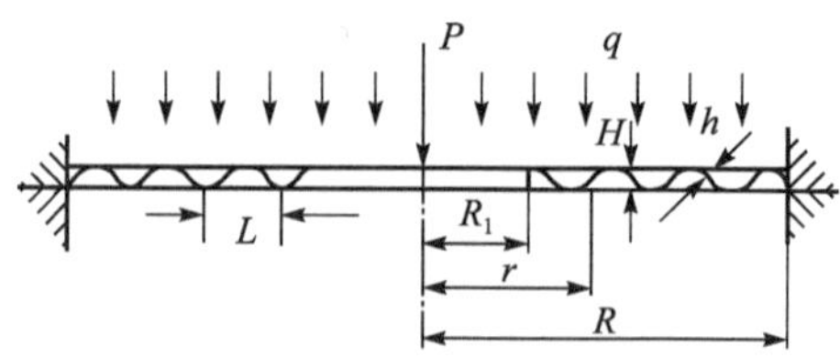

图 1　具有光滑中心的波纹圆板

这一波纹圆板由光滑中心和环形波纹区域所组成。我们将用角标 a 和 b 分别表示光滑中心和环形波纹区域的量。

在研究时，我们假定波纹圆板的波纹分布是均匀而致密的。于是，我们可将此板的波纹环形区域视为一个各向异性的环形板。应用这种分析方法，按照各向同性和各向异性圆板的大挠度理论[6,12,13,23]，我们得到这个问题的基本方程如下。

当 $0\leqslant r\leqslant R_1$ 时，

$$r\frac{\mathrm{d}^3w_a}{\mathrm{d}r^3}+\frac{\mathrm{d}^2w_a}{\mathrm{d}r_2}-\frac{\mathrm{d}w_a}{\mathrm{d}r}\left(\frac{1}{D_a}rN_{r,a}+\frac{1}{r}\right)=\frac{1}{2D_a}\left(qr^2+\frac{p}{\pi}\right),$$

$$r\frac{\mathrm{d}^2}{\mathrm{d}r^2}(rN_{r,a})+\frac{\mathrm{d}}{\mathrm{d}r}(rN_{r,a})-N_{r,a}+\frac{Eh}{2}\left(\frac{\mathrm{d}w_a}{\mathrm{d}r}\right)^2=0 \tag{2.1}$$

当 $R_1\leqslant r\leqslant R$ 时，

$$r\frac{\mathrm{d}^3w_b}{\mathrm{d}r^3}+\frac{\mathrm{d}^2w_b}{\mathrm{d}r_2}-k_t\frac{\mathrm{d}w_b}{\mathrm{d}r}\left(\frac{1}{D_b}rN_{r,b}+\frac{k_t'}{r}\right)=\frac{k_t}{2D_b}\left(qr^2+\frac{p}{\pi}\right),$$

$$r\frac{\mathrm{d}^2}{\mathrm{d}r^2}(rN_{r,b})+\frac{\mathrm{d}}{\mathrm{d}r}(rN_{r,b})-k_rk_tN_{r,b}+\frac{1}{2}Ehk_t\left(\frac{\mathrm{d}w_b}{\mathrm{d}r}\right)^2=0 \tag{2.2}$$

其中，w 是挠度，N_r 是径向薄膜内力，θ 是径向截面的切线倾斜角，E 是弹性模量，ν 是泊松比，D 是抗弯刚度，

$$D_a=\frac{Eh^3}{12(1-\nu^2)},\quad D_b=\frac{Eh^3}{12(1-\nu^2/k_tk_t')}$$

k_r，k_t 和 k_t' 分别是径向和周向刚度参数，

$$k_r=\frac{12}{h^2L}\int_0^L\frac{z^2}{\cos\theta}\mathrm{d}x+\frac{1}{L}\int_0^L\cos\theta\mathrm{d}x,$$

$$k_t=\frac{1}{L}\int_0^L\frac{1}{\cos\theta}\mathrm{d}x,$$

$$k_t'=\frac{12}{h^2L}\int_0^L\frac{z^2}{\cos\theta}\mathrm{d}x+\frac{1}{L}\int_0^L\frac{1}{\cos^3\theta}\mathrm{d}x \tag{2.3}$$

特别，对于正弦波纹（图 2（a））

$$k_r\doteq1+\frac{3H^2}{2h^2}+\frac{\pi^2H^2}{4L^2}\left(\frac{3H^2}{4h^2}-1\right),$$

$$k_t\doteq1+\frac{\pi^2H^2}{4L^2},$$

$$k_t' \doteq 1+\frac{3H^2}{2h^2}+\frac{3\pi^2 H^2}{4L^2}\left(\frac{H^2}{4h^2}+1\right) \tag{2.4}$$

对于锯齿形波纹（图 2（b）），

$$\begin{aligned} k_r &= \frac{H^2}{h^2\cos\theta_0}+\cos\theta_0, \\ k_t &= \frac{1}{\cos\theta_0}, \\ k_t' &= \frac{H^2}{h^2\cos\theta_0}+\frac{1}{\cos^3\theta_0} \end{aligned} \tag{2.5}$$

对于梯形波纹（图 2（c）），

$$\begin{aligned} k_r &= \frac{H^2}{Lh^2}\left(\frac{L-2L_0}{\cos\theta_0}+6L_0\right)+\left(1-\frac{2L_0}{L}\right)\cos\theta_0+\frac{2L_0}{L}, \\ k_t &= \frac{1}{L}\left(\frac{L-2L_0}{\cos\theta_0}+2L_0\right), \\ k_t' &= \frac{H^2}{Lh^2}\left(\frac{L-2L_0}{\cos\theta_0}+6L_0\right)+\frac{1}{L}\left(\frac{L-2L_0}{\cos^3\theta_0}+2L_0\right) \end{aligned} \tag{2.6}$$

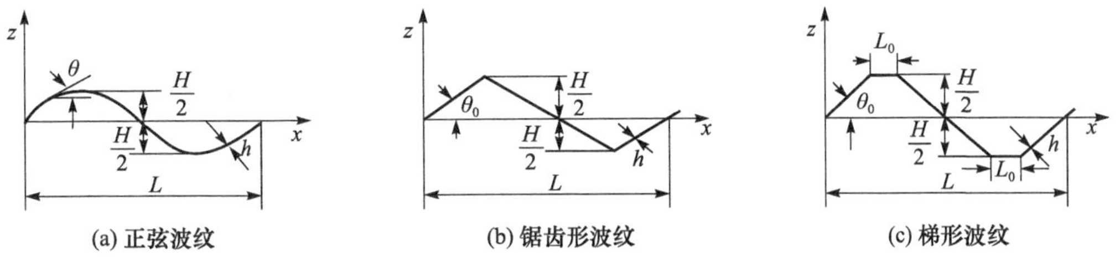

图 2　波纹圆板的径向截面上的一个波纹

我们讨论下面两种常用的边界条件。

（1）夹紧固定边界

$$\begin{aligned} &\text{当 } r=0 \text{ 时，} \frac{\mathrm{d}w_a}{\mathrm{d}r}=0，N_{r,a}\text{有限} \\ &\text{当 } r=R \text{ 时，} w_b=0，\frac{\mathrm{d}w_b}{\mathrm{d}r}=0，u_b=0 \end{aligned} \tag{2.7}$$

（2）滑动固定边界

$$\begin{aligned} &\text{当 } r=0 \text{ 时，} \frac{\mathrm{d}w_a}{\mathrm{d}r}=0，N_{r,a}\text{有限} \\ &\text{当 } r=R \text{ 时，} w_b=0，\frac{\mathrm{d}w_b}{\mathrm{d}r}=0，N_{r,b}=0 \end{aligned} \tag{2.8}$$

其中 u 是径向位移，

$$\begin{aligned} u_a &= \frac{r}{Eh}\left[\frac{\mathrm{d}}{\mathrm{d}r}(rN_{r,a})-\nu N_{r,a}\right], \\ u_b &= \frac{r}{k_t Eh}\left[\frac{\mathrm{d}}{\mathrm{d}r}(rN_{r,b})-\nu N_{r,b}\right] \end{aligned} \tag{2.9}$$

此外，在两个区域的连接处，尚需有下面的连续条件：

当 $r=R_1$ 时，$w_a=w_b$，$\frac{\mathrm{d}w_a}{\mathrm{d}r}=\frac{\mathrm{d}w_b}{\mathrm{d}r}$，

$$M_{r,a}=M_{r,b}，N_{r,a}=N_{r,b}，u_a=u_b \tag{2.10}$$

其中 M_r 是径向弯矩，

$$M_{r,a}=-D_a\left(\frac{\mathrm{d}^2w_a}{\mathrm{d}r^2}+\frac{\nu}{r}\frac{\mathrm{d}w_a}{\mathrm{d}r}\right),$$

$$M_{r,b}=-\frac{D_b}{k_t}\left(\frac{\mathrm{d}^2w_b}{\mathrm{d}r^2}+\frac{\nu}{r}\frac{\mathrm{d}w_b}{\mathrm{d}r}\right) \tag{2.11}$$

为使解更便于应用，我们引进下列无量纲量：

$$\begin{aligned}
&y=\frac{r}{R},\quad y_1=\frac{R_1}{R},\quad W=\frac{w}{h},\\
&\phi=\frac{\mathrm{d}W}{\mathrm{d}y},\quad S_a=\frac{Rr}{D_a}N_{r,a},\quad S_b=\frac{k_tRr}{D_b}N_{r,b},\\
&Q=\frac{R^4}{2D_ah}q,\quad P=\frac{R^2}{2\pi hD_a}p,\\
&\beta_0=6(1-\nu^2),\quad \beta_1^2=k_tk_t',\quad \beta_2^2=k_rk_t,\quad \beta_3=\frac{D_b}{D_ak_t},\\
&\beta_4=\frac{\beta_4}{k_t},\quad \beta_5=\nu(1-\beta_3),\quad \beta_6=\nu\left(1-\frac{1}{k_t}\right)
\end{aligned} \tag{2.12}$$

应用这些量，可把基本方程（2.1）和（2.2）化为无量纲形式：

当 $0\leqslant y\leqslant y_1$ 时，

$$L_0(y\phi_a)=S_a\phi_a+y^2Q+P,$$

$$L_0(yS_a)=-\beta_0\phi_a^2 \tag{2.13a,b}$$

当 $y_1<y\leqslant 1$ 时，

$$L_1(y^{\beta_1}\phi_b)=S_b\phi_b+\frac{1}{\beta_3}(y^2Q+P),$$

$$L_2(y^{\beta_2}S_b)=-\frac{\beta_0}{\beta_4}\phi_b^2 \tag{2.14a,b}$$

类似地，可把边界条件（2.7）和（2.8）化为

（1）夹紧固定边界

当 $y=0$ 时，$\phi_a=0$，$S_a=0$

当 $y=1$ 时，$W_b=0$，$\phi_b=0$，$\frac{\mathrm{d}S_b}{\mathrm{d}y}-\nu\frac{S_b}{y}=0$　　(2.15)

（2）滑动固定边界

当 $y=0$ 时，$\phi_a=0$，$S_a=0$

当 $y=1$ 时，$W_b=0$，$\phi_b=0$，$S_b=0$　　(2.16)

而连续条件（2.10）成为

当 $y=y_1$ 时，

$$W_a=W_b,\phi_a=\phi_b,\frac{\mathrm{d}\phi_a}{\mathrm{d}y}+\beta_5\frac{\phi_a}{y}=\beta_3\frac{\mathrm{d}\phi_b}{\mathrm{d}y},$$

$$S_a = \beta_3 S_b, \frac{\mathrm{d}S_a}{\mathrm{d}y} - \beta_6 \frac{S_a}{y} = \beta_4 \frac{\mathrm{d}S_b}{\mathrm{d}y} \tag{2.17}$$

其中

$$L_0(\cdots) = y\frac{\mathrm{d}}{\mathrm{d}y}\frac{1}{y}\frac{\mathrm{d}}{\mathrm{d}y}(\cdots),$$

$$L_1(\cdots) = y^{\beta_1}\frac{\mathrm{d}}{\mathrm{d}y}y^{-(2\beta_1-1)}\frac{\mathrm{d}}{\mathrm{d}y}(\cdots),$$

$$L_2(\cdots) = y^{\beta_2}\frac{\mathrm{d}}{\mathrm{d}y}y^{-(2\beta_2-1)}\frac{\mathrm{d}}{\mathrm{d}y}(\cdots) \tag{2.18}$$

为了便于研究，我们引入单载荷参数 G，于是便有

$$Q = \mu_1 G,$$

$$P = \mu_2 G \tag{2.19}$$

其中 μ_1 和 μ_2 是常数。在这种情况下，方程（2.13a）和（2.14a）化为

$$L_0(y\phi_a) = S_a\phi_a + G(\mu_1 y^2 + \mu_2) \tag{2.20}$$

$$L_1(y^{\beta_1}\phi_b) = S_b\phi_b + \frac{G}{\beta_3}(\mu_1 y^2 + \mu_2) \tag{2.21}$$

现在，我们的问题便在于定出满足控制方程（2.20）、（2.13b）、（2.21）和（2.14b），边界条件（2.15）或（2.16）以及连续条件（2.17）的 S 和 ϕ。

三、解答

首先，我们讨论关于在均布及中心集中载荷联合作用下，具有光滑中心和夹紧固定边界的波纹圆板的非线性弯曲的无量纲非线性边值问题（2.20）、(2.13b)、(2.21)、(2.14b)、(2.15）和（2.17）。使用修正迭代法，便能获得此问题的解析解。为此，我们取无量纲中心挠度 W_0 作为迭代参数

$$W_0 = W_a\big|_{y=0} \tag{3.1}$$

对第一次近似，我们忽略方程（2.20）和（2.21）中的非线性项，便得到下面的线性边值问题：

当 $0 \leqslant y \leqslant y_1$ 时，

$$L_0(y\phi_{a,1}) = G(\mu_1 y^2 + \mu_2),$$

$$L_0(yS_{a,1}) = -\beta_0\phi_{a,1}^2 \tag{3.2a,b}$$

当 $y_1 < y \leqslant 1$ 时，

$$L_1(y^{\beta_1}\phi_{b,1}) = \frac{G}{\beta_3}(\mu_1 y^2 + \mu_2),$$

$$L_2(y^{\beta_2}S_{b,1}) = -\frac{\beta_0}{\beta_4}\phi_{b,1}^2 \tag{3.3a,b}$$

当 $y=0$ 时，$\phi_{a,1}=0$，$S_{a,1}=0$　　(3.4a,b)

当 $y=1$ 时，$W_{b,1}=0$，$\phi_{b,1}=0$，$\dfrac{\mathrm{d}S_{b,1}}{\mathrm{d}y} - \nu\dfrac{S_{b,1}}{y} = 0$　　(3.5a～c)

当 $y=y_1$ 时，$W_{a,1}=W_{b,1}$，$\phi_{a,1}=\phi_{b,1}$，

$$\frac{\mathrm{d}\phi_{a,1}}{\mathrm{d}y} + \beta_5\frac{\phi_{a,1}}{y} = \beta_3\frac{\mathrm{d}\phi_{b,1}}{\mathrm{d}y},$$

$$S_{a,1}=\beta_3 S_{b,1}, \frac{\mathrm{d}S_{a,1}}{\mathrm{d}y}-\beta_6\frac{S_{a,1}}{y}=\beta_4\frac{\mathrm{d}S_{b,1}}{\mathrm{d}y} \tag{3.6a ~ e}$$

边值问题（3.2a）、（3.3a）、（3.4a）、（3.5a，b）、（3.6a~c）的解是

$$\begin{aligned}\phi_{a,1}&=G(a_1y^3+a_2y\ln y+a_3y),\\ \phi_{b,1}&=G(b_1y^{\beta_1}+b_2y^{-\beta_1}+b_3y^3+b_4y)\end{aligned} \tag{3.7}$$

和

$$\begin{aligned}W_{a,1}&=G\left[\frac{a_1}{4}y^4+\frac{a_2}{2}y^2\ln y+\frac{1}{2}\left(a_3-\frac{a_2}{2}\right)y^2+\frac{1}{\alpha_1}\right],\\ W_{b,1}&=G\left[\frac{b_1}{\beta_1+1}(y^{\beta_1+1}-1)-\frac{b_2}{\beta_1-1}(y^{-(\beta_1-1)}-1)\right.\\ &\left.\quad+\frac{b_3}{4}(y^4-1)+\frac{b_4}{2}(y^2-1)\right]\end{aligned} \tag{3.8a,b}$$

其中

$$\begin{aligned}a_1&=\frac{\mu_1}{8}, a_2=\frac{\mu_2}{2},\\ a_3&=\frac{1}{\lambda_1}\left\{\mu_1\left[\frac{1}{\beta_1+3}+\frac{1}{8}(\beta_1\beta_3-\beta_5-3)\right]y_1^{\beta_1+2}\right.\\ &\quad+\mu_1\left[\frac{1}{\beta_1-3}+\frac{1}{8}(\beta_1\beta_3+\beta_5+3)\right]y_1^{-(\beta_1-2)}\\ &\quad+\frac{\mu_2}{2}(\beta_1\beta_3-\beta_5-1)y_1^{\beta_1}\ln y_1\\ &\quad+\frac{\mu_2}{2}(\beta_1\beta_3+\beta_5+1)y_1^{-\beta_1}\ln y_1\\ &\quad+\mu_2\left[\frac{1}{\beta_1+1}-\frac{1}{4}(\beta_1\beta_3-\beta_5+\beta_1)\right]y_1^{\beta_1}\\ &\quad+\mu_2\left[\frac{1}{\beta_1-1}-\frac{1}{4}(\beta_1\beta_3+\beta_5-1)\right]y_1^{-\beta_1}\\ &\quad\left.-2\beta_1\left(\frac{\mu_1}{\beta_1^2-9}+\frac{\mu_2}{\beta_1^2-1}\right)y_1^{-1}\right\}-\frac{\mu_2}{4},\\ b_1&=-\frac{1}{\lambda_1}\left\{\left(\frac{\beta_5+1}{\beta_3}+\beta_1\right)\left(\frac{\mu_1}{\beta_1^2-9}+\frac{\mu_2}{\beta_1^2-1}\right)y_1^{-\beta_1}\right.\\ &\quad-\mu_1\left[\frac{1}{\beta_1^2-9}\left(\frac{\beta_5+1}{\beta_3}-3\right)-\frac{1}{4}\right]y_1^3\\ &\quad\left.+\mu_2\left[\frac{1}{\beta_1^2-1}\left(1-\frac{\beta_5+1}{\beta_3}\right)+\frac{1}{2}\right]y_1\right\},\\ b_2&=\frac{1}{\lambda_1}\left\{\left(\frac{\beta_5+1}{\beta_3}-\beta_1\right)\left(\frac{\mu_1}{\beta_1^2-9}+\frac{\mu_2}{\beta_1^2-1}\right)y_1^{\beta_1}\right.\\ &\quad-\mu_1\left[\frac{1}{\beta_1^2-9}\left(\frac{\beta_5+1}{\beta_3}-3\right)-\frac{1}{4}\right]y_1^3\\ &\quad\left.-\mu_2\left[\frac{1}{\beta_1^2-1}\left(\frac{\beta_5+1}{\beta_3}-1\right)-\frac{1}{2}\right]y_1\right\},\end{aligned}$$

$$b_3 = -\frac{\mu_1}{\beta_3(\beta_1^2 - 9)}, \qquad b_4 = -\frac{\mu_2}{\beta_3(\beta_1^2 - 1)},$$

$$\lambda_1 = -[(\beta_1\beta_3 - \beta_5 - 1)y_1^{\beta_1} + (\beta_1\beta_3 + \beta_5 + 1)y_1^{-\beta_1}], \tag{3.9}$$

$$\alpha_1 = \left[\frac{b_1}{\beta_1+1}(y_1^{\beta_1+1} - 1) - \frac{b_2}{\beta_1 - 1}(y_1^{-(\beta_1-1)} - 1) + \frac{b_3}{4}(y_1^4 - 1)\right.$$
$$\left. + \frac{b_4}{2}(y_1^2 - 1) - \frac{a_1}{4}y_1^4 - \frac{a_2}{2}y_1^2\ln y_1 - \frac{1}{2}\left(a_3 - \frac{a_2}{2}\right)y_1^2\right]^{-1} \tag{3.10}$$

将式（3.8a）代入式（3.1），便得

$$G = \alpha_1 W_0 \tag{3.11}$$

此即小挠度的特征关系式。将此式代入式（3.7）得

$$\phi_{a,1} = \alpha_1 W_0(a_1 y^3 + a_2 y\ln y + a_3 y),$$

$$\phi_{b,1} = \alpha_1 W_0(b_1 y^{\beta_1} + b_2 y^{-\beta_1} + b_3 y^3 + b_4 y) \tag{3.12}$$

应用解（3.12），可得问题（3.2b）、（3.3b）、（3.4b）、（3.5c）和（3.6d，e）的解

$$S_{a,1} = \alpha_1^2\beta_0 W_0^2(c_1 y^7 + c_2 y^5\ln y + c_3 y^5 + c_4 y^3\ln^2 y$$
$$+ c_5 y^3\ln y + c_6 y^3 + c_7 y),$$

$$S_{b,1} = \alpha_1^2\beta_0 W_0^2\left[\frac{1}{\beta_4}(d_1 y^{2\beta_1+1} + d_2 y^{-(2\beta_1-1)} + d_3 y^{\beta_1+4}\right.$$
$$+ d_4 y^{-(\beta_1-4)} + d_5 y^{\beta_1+2} + d_6 y^{-(\beta_1-2)} + d_7 y^7$$
$$\left. + d_8 y^5 + d_9 y^3 + d_{10} y) + d_{11} y^{\beta_2} + d_{12} y^{-\beta_2}\right] \tag{3.13}$$

其中

$$c_1 = -\frac{a_1^2}{48}, c_2 = -\frac{a_1 a_2}{12}, c_3 = \frac{a_1}{12}\left(\frac{5}{12}a_2 - a_3\right), c_4 = -\frac{a_2^2}{8}$$

$$c_5 = \frac{a_2}{4}\left(\frac{3}{4}a_2 - a_3\right), c_6 = \frac{1}{8}\left(\frac{3}{2}a_2 a_3 - \frac{7}{8}a_2^2 - a_3^2\right),$$

$$c_7 = \frac{1}{\lambda_2}\{\beta_2\beta_4\psi_2[(\beta_2 + \nu)y_1^{\beta_2 - 1} - (\beta_2 - \nu)y_1^{-(\beta_2+1)}]$$
$$- \beta_3\psi_3[(\beta_2 + \nu)y_1^{\beta_2} + (\beta_2 - \nu)y_1^{-\beta_2}] - 2\beta_2\beta_3\beta_4\psi_1 y_1^{-1}\},$$

$$d_1 = -\frac{b_1^2}{(2\beta_1 + 1)^2 - \beta_2^2}, \qquad d_2 = -\frac{b_2^2}{(2\beta_1 - 1)^2 - \beta_2^2},$$

$$d_3 = -\frac{2b_1 b_3}{(\beta_1 + 4)^2 - \beta_2^2}, \qquad d_4 = -\frac{2b_2 b_3}{(\beta_1 - 4)^2 - \beta_2^2},$$

$$d_5 = -\frac{2b_1 b_4}{(\beta_1 + 2)^2 - \beta_2^2}, \qquad d_6 = -\frac{2b_2 b_4}{(\beta_1 - 2)^2 - \beta_2^2},$$

$$d_7 = \frac{b_3^2}{\beta_2^2 - 49}, \qquad d_8 = \frac{2b_3 b_4}{\beta_2^2 - 25},$$

$$d_9 = \frac{b_4^2}{\beta_2^2 - 9}, \qquad d_{10} = \frac{2b_1 b_2}{\beta_2^2 - 1},$$

$$d_{11}=\frac{1}{\lambda_2}\{\psi_1[\beta_3(\beta_6-1)-\beta_2\beta_4)]y_1^{-\beta_2}$$
$$-(\beta_2+\nu)[\psi_3y_1+\psi_2(\beta_6-1)]\},$$
$$d_{12}=-\frac{1}{\lambda_2}\{\psi_1[\beta_3(\beta_6-1)+\beta_2\beta_4)]y_1^{\beta_2}$$
$$+(\beta_2-\nu)[\psi_3y_1+\psi_2(\beta_6-1)]\},$$
$$\lambda_2=(\beta_2+\nu)[\beta_3(\beta_6-1)+\beta_2\beta_4]y_1^{\beta_2}$$
$$+(\beta_2-\nu)[\beta_3(\beta_6-1)-\beta_2\beta_4]y_1^{-\beta_2},$$
$$\psi_1=-\frac{1}{\beta_4}[d_1(2\beta_1+1-\nu)-d_2(2\beta_1-1+\nu)+d_3(\beta_1+4-\nu)$$
$$-d_4(\beta_1-4+\nu)+d_5(\beta_1+2-\nu)-d_6(\beta_1-2+\nu)$$
$$+d_7(7-\nu)+d_8(5-\nu)+d_9(3-\nu)+d_{10}(1-\nu)],$$
$$\psi_2=-(c_1y_1^7+c_2y_1^5\ln y+c_3y_1^5+c_4y_1^3\ln^2y_1+c_5y_1^3\ln y_1+c_6y_1^3)$$
$$+\frac{\beta_3}{\beta_4}(d_1y_1^{2\beta_1+1}+d_2y_1^{-(2\beta_1-1)}+d_3y_1^{\beta_1+4}+d_4y_1^{-(\beta_1-4)}$$
$$+d_5y_1^{\beta_1+2}+d_6y_1^{-(\beta_1-2)}+d_7y_1^7+d_8y_1^5+d_9y_1^3+d_{10}y_1),$$
$$\psi_3=c_1(\beta_6-7)y_1^6+c_2(\beta_6-5)y_1^4\ln y_1+[c_3(\beta_6-5)-c_2]y_1^4$$
$$+c_4(\beta_6-3)y_1^2\ln^2y_1+[c_5(\beta_6-3)-2c_4]y_1^2\ln y_1$$
$$+[c_6(\beta_6-3)-c_5]y_1^2+d_1(2\beta_1+1)y_1^{2\beta_1}-d_2(2\beta_1-1)y_1^{-2\beta_1}$$
$$+d_3(\beta_1+4)y_1^{\beta_1+3}-d_4(\beta_1-4)y_1^{-(\beta_1-3)}+d_5(\beta_1+2)y_1^{\beta_1+1}$$
$$-d_6(\beta_1-2)y_1^{-(\beta_1-1)}+7d_7y_1^6+5d_8y_1^4+3d_9y_1^2+d_{10}\tag{3.14}$$

对第二次近似，我们有如下的关于 ϕ 的线性边值问题：

当 $0\leqslant y\leqslant y_1$ 时，

$$L_0(y\phi_{a,2})=S_{a,1}\phi_{a,1}+G(\mu_1y^2+\mu_2)\tag{3.15}$$

当 $y_1<y\leqslant 1$ 时，

$$L_1(y^{\beta_1}\phi_{b,2})=S_{b,1}\phi_{b,1}+\frac{G}{\beta_3}(\mu_1y^2+\mu_2)\tag{3.16}$$

当 $y=0$ 时，$\phi_{a,2}=0$ (3.17)

当 $y=1$ 时，$W_{b,2}=0$，$\phi_{b,2}=0$ (3.18)

当 $y=y_1$ 时，$W_{a,2}=W_{b,2}$，$\phi_{a,2}=\phi_{b,2}$，

$$\frac{d\phi_{a,2}}{dy}+\beta_5\frac{\phi_{a,2}}{y}=\beta_3\frac{d\phi_{b,2}}{dy}\tag{3.19}$$

解此问题，便得

$$\phi_{a,2}=G(a_1y^3+a_2y\ln y+a_3y)+\alpha_1^3\beta_0W_0^3[F_1(y)+e_{13}y],$$
$$\phi_{b,2}=G(b_1y^{\beta_1}+b_2y^{-\beta_1}+b_3y^3+b_4y)$$
$$+\alpha_1^3\beta_0W_0^3[F_2(y)+f_{28}y^{\beta_1}+f_{29}y^{-\beta_1}]\tag{3.20}$$

和

$$W_{a,2}=G\Big[\frac{a_1}{4}y^4+\frac{a_2}{2}y^2\ln y+\frac{1}{2}\Big(a_3-\frac{a^2}{2}\Big)y^2+\alpha_1^{-1}\Big]$$

$$
\begin{aligned}
&+\alpha_1^3\beta_0 W_0^3[H_1(y)-H_1(y_1)+H_2(y_1)-H_2(1)],\\
W_{b,2}=&G\Big[\frac{b_1}{\beta_1+1}(y^{\beta_1+1}-1)-\frac{b_2}{\beta_1-1}(y^{-(\beta_1-1)}-1)\\
&+\frac{b_3}{4}(y^4-1)+\frac{b_4}{2}(y^2-1)\Big]\\
&+\alpha_1^3\beta_0 W_0^3[H_2(y)-H_2(1)]
\end{aligned}
\tag{3.21a,b}
$$

其中

$$
\begin{aligned}
F_1(y)=&e_1y^{11}+e_2y^9\ln y+e_3y^9+e_4y^7\ln^2 y+e_5y^7\ln y+e_6y^7\\
&+e_7y^5\ln^3 y+e_8y^5\ln^2 y+e_9y^5\ln y+e_{10}y^5+e_{11}y^3\ln y+e_{13}y^3,\\
F_2(y)=&f_1y^{3\beta_1+2}+f_2y^{-(3\beta_1-2)}+f_3y^{2\beta_1+5}+f_4y^{-(2\beta_1-5)}+f_5y^{2\beta_1+3}\\
&+f_6y^{-(2\beta_1-3)}+f_7y^{\beta_1+8}+f_8y^{-(\beta_1-8)}+f_9y^{\beta_1+6}+f_{10}y^{-(\beta_1-6)}\\
&+f_{11}y^{\beta_1+4}+f_{12}y^{-(\beta_1-4)}+f_{13}y^{\beta_1+2}+f_{14}y^{-(\beta_1-2)}+f_{15}y^{\beta_1+\beta_2+1}\\
&+f_{16}y^{\beta_1-\beta_2+1}+f_{17}y^{-(\beta_1-\beta_2-1)}+f_{18}y^{-(\beta_1+\beta_2-1)}+f_{19}y^{\beta_2+4}\\
&+f_{20}y^{-(\beta_2-4)}+f_{21}y^{\beta_2+2}+f_{22}y^{-(\beta_2-2)}+f_{23}y^{11}\\
&+f_{24}y^9+f_{25}y^7+f_{26}y^5+f_{27}y^3,\\
H_1(y)=&\frac{e_1}{12}y^{12}+\frac{1}{10}\left(e_2\ln y-\frac{e_2}{10}+e_3\right)y^{10}\\
&+\frac{1}{8}\left[e_4\ln^2 y-\left(\frac{e_4}{4}-e_5\right)\ln y+\frac{e_4}{32}-\frac{e_5}{8}+e_6\right]y^8\\
&+\frac{1}{6}\left[e_7\ln^3 y-\left(\frac{e_7}{2}-e_8\right)\ln^2 y+\left(\frac{e_7}{6}-\frac{e_8}{3}+e_9\right)\ln y\right.\\
&\left.-\frac{e_7}{36}+\frac{e_8}{18}-\frac{e_9}{6}+e_{10}\right]y^6+\frac{1}{4}\left(e_{11}\ln y-\frac{e_{11}}{4}+e_{12}\right)y^4+\frac{e_{13}}{2}y^2,\\
H_2(y)=&\frac{f_1}{3(\beta_1+1)}y^{3(\beta_1+1)}-\frac{f_2}{3(\beta_1-1)}y^{-3(\beta_1-1)}\\
&+\frac{f_3}{2(\beta_1+3)}y^{2(\beta_1+3)}-\frac{f_4}{2(\beta_1-3)}y^{-2(\beta_1-3)}\\
&+\frac{f_5}{2(\beta_1+2)}y^{2(\beta_1+2)}-\frac{f_6}{2(\beta_1-2)}y^{-2(\beta_1-2)}\\
&+\frac{f_7}{\beta_1+9}y^{\beta_1+9}-\frac{f_8}{\beta_1-9}y^{-(\beta_1-9)}+\frac{f_9}{\beta_1+7}y^{\beta_1+7}\\
&-\frac{f_{10}}{\beta_1-7}y^{-(\beta_1-7)}+\frac{f_{11}}{\beta_1+5}y^{\beta_1+5}-\frac{f_{12}}{\beta_1-5}y^{-(\beta_1-5)}\\
&+\frac{f_{13}}{\beta_1+3}y^{\beta_1+3}-\frac{f_{14}}{\beta_1-3}y^{-(\beta_1-3)}+\frac{f_{15}}{\beta_1+\beta_2+2}y^{\beta_1+\beta_2+2}\\
&+\frac{f_{16}}{\beta_1-\beta_2+2}y^{\beta_1-\beta_2+2}-\frac{f_{17}}{\beta_1-\beta_2-2}y^{-(\beta_1-\beta_2-2)}\\
&-\frac{f_{18}}{\beta_1+\beta_2-2}y^{-(\beta_1+\beta_2-2)}+\frac{f_{19}}{\beta_2+5}y^{\beta_2+5}
\end{aligned}
$$

$$-\frac{f_{20}}{\beta_2-5}y^{-(\beta_2-5)}+\frac{f_{21}}{\beta_2+3}y^{\beta_2+3}-\frac{f_{22}}{\beta_2-3}y^{-(\beta_2-3)}$$
$$+\frac{f_{23}}{12}y^{12}+\frac{f_{24}}{10}y^{10}+\frac{f_{25}}{8}y^{8}+\frac{f_{26}}{6}y^{6}+\frac{f_{27}}{4}y^{4}$$
$$+\frac{f_{28}}{\beta_1+1}y^{\beta_1+1}-\frac{f_{29}}{\beta_1-1}y^{-(\beta_1-1)},$$

$$e_1=\frac{a_1c_1}{120},e_2=\frac{1}{80}(a_1c_2+a_2c_1),$$

$$e_3=\frac{1}{80}\left[a_1c_3+a_3c_1-\frac{9}{40}(a_1c_2+a_2c_1)\right],e_4=\frac{1}{48}(a_1c_4+a_2c_2),$$

$$e_5=\frac{1}{48}\left[a_1c_5+a_2c_3+a_3c_2-\frac{7}{12}(a_1c_4+a_2c_2)\right],$$

$$e_6=\frac{1}{48}\left[a_1c_6+a_3c_3+\frac{37}{288}(a_1c_4+a_2c_2)-\frac{7}{24}(a_1c_5+a_2c_3+a_3c_2)\right],$$

$$e_7=\frac{a_2c_4}{24},\qquad e_8=\frac{1}{24}\left[a_2c_5+c_4\left(a_3-\frac{5}{4}a_2\right)\right],$$

$$e_9=\frac{1}{24}\left[a_2\left(c_6+\frac{19}{24}c_4\right)+a_3c_5-\frac{5}{6}(a_2c_5+a_3c_4)\right],$$

$$e_{10}=\frac{1}{24}\left[a_1c_7+a_3c_6-\frac{65}{288}a_2c_4+\frac{19}{72}(a_2c_5+a_3c_4)-\frac{5}{12}(a_2c_6+a_3c_5)\right],$$

$$e_{11}=\frac{a_2c_7}{8},\qquad e_{12}=\frac{c_7}{8}\left(a_3-\frac{3}{4}a_2\right),$$

$$e_{13}=-\frac{1}{\lambda_1}[\beta_1\beta_3\omega_2(y_1^{\beta_1-1}+y_1^{-(\beta_1+1)})-\omega_3(y_1^{\beta_1}-y_1^{-\beta_1})+2\beta_1\beta_3\omega_1y_1^{-1}],$$

$$f_1=\frac{b_1d_1}{4\beta_4(\beta_1+1)(2\beta_1+1)},\qquad f_2=\frac{b_2d_2}{4\beta_4(\beta_1-1)(2\beta_1-1)},$$

$$f_3=\frac{b_1d_3+b_3d_1}{\beta_4(\beta_1+5)(3\beta_1+5)},\qquad f_4=\frac{b_2d_4+b_3d_2}{\beta_4(\beta_1-5)(3\beta_1-5)},$$

$$f_5=\frac{b_1d_5+b_4d_1}{3\beta_4(\beta_1+1)(\beta_1+3)},\qquad f_6=\frac{b_2d_6+b_4d_2}{3\beta_4(\beta_1-1)(\beta_1-3)},$$

$$f_7=\frac{b_1d_7+b_3d_3}{16\beta_4(\beta_1+4)},\qquad f_8=-\frac{b_2d_7+b_3d_4}{16\beta_4(\beta_1-4)},$$

$$f_9=\frac{b_1d_8+b_3d_5+b_4d_3}{12\beta_4(\beta_1+3)},\qquad f_{10}=-\frac{b_2d_8+b_3d_6+b_4d_4}{12\beta_4(\beta_1-3)},$$

$$f_{11}=\frac{b_1d_9+b_4d_5}{8\beta_4(\beta_1+2)},\qquad f_{12}=-\frac{b_2d_9+b_4d_6}{8\beta_4(\beta_1-2)},$$

$$f_{13}=\frac{b_1d_{10}+b_2d_1}{4\beta_4(\beta_1+1)},\qquad f_{14}=-\frac{b_1d_2+b_2d_{10}}{4\beta_4(\beta_1-1)},$$

$$f_{15}=\frac{b_1d_{11}}{(\beta_2+1)(2\beta_1+\beta_2+1)},\qquad f_{16}=-\frac{b_1d_{12}}{(\beta_2-1)(2\beta_1-\beta_2+1)},$$

$$f_{17}=-\frac{b_2d_{11}}{(\beta_2+1)(2\beta_1-\beta_2+1)},\qquad f_{18}=\frac{b_2d_{12}}{(\beta_2-1)(2\beta_1+\beta_2-1)},$$

$$f_{19}=\frac{b_3d_{11}}{(\beta_2+4)^2-\beta_1^2},\qquad f_{20}=\frac{b_3d_{12}}{(\beta_2-4)^2-\beta_1^2},$$

$$f_{21}=\frac{b_4d_{11}}{(\beta_2+2)^2-\beta_1^2},\qquad f_{22}=\frac{b_4d_{12}}{(\beta_2-2)^2-\beta_1^2},$$

$$f_{23}=-\frac{b_3d_7}{\beta_4(\beta_1^2-121)},\qquad f_{24}=-\frac{b_3d_8+b_4d_7}{\beta_4(\beta_1^2-81)},$$

$$f_{25}=-\frac{b_3d_9+b_4d_8}{\beta_4(\beta_1^2-49)},\qquad f_{26}=-\frac{b_1d_4+b_2d_3+b_3d_{10}+b_4d_9}{\beta_4(\beta_1^2-25)},$$

$$f_{27}=-\frac{b_1d_6+b_2d_5+b_4d_{10}}{\beta_4(\beta_1^2-9)},$$

$$f_{28}=-\frac{1}{\lambda_1}[\omega_1(\beta_1\beta_3+\beta_5+1)y_1^{-\beta_1}+\omega_2(\beta_5+1)-\omega_3y_1],$$

$$f_{29}=-\frac{1}{\lambda_1}[\omega_1(\beta_1\beta_3-\beta_5-1)y_1^{\beta_1}-\omega_2(\beta_5+1)+\omega_3y_1],$$

$$\omega_1=-F_2(1),\qquad \omega_2=F_2(y_1)-F_1(y_1),$$

$$\begin{aligned}\omega_3=&\beta_3[f_1(3\beta_1+2)y_1^{3\beta_1+1}-f_2(3\beta_1-2)y_1^{-(3\beta_1-1)}+f_3(2\beta_1+5)y_1^{2(\beta_1+2)}\\&-f_4(2\beta_1-5)y_1^{-2(\beta_1-2)}+f_5(2\beta_1+3)y_1^{2(\beta_1+1)}\\&-f_6(2\beta_1-3)y_1^{-(\beta_1-1)}+f_7(\beta_1+8)y_1^{\beta_1+7}-f_8(\beta_1-8)y_1^{-(\beta_1-7)}\\&+f_9(\beta_1+6)y_1^{\beta_1+5}-f_{10}(\beta_1-6)y_1^{-(\beta_1-5)}+f_{11}(\beta_1+4)y_1^{\beta_1+3}\\&-f_{12}(\beta_1-4)y_1^{-(\beta_1-3)}+f_{13}(\beta_1+2)y_1^{\beta_1+1}-f_{14}(\beta_1-2)y_1^{-(\beta_1-1)}\\&+f_{15}(\beta_1+\beta_2+1)y_1^{\beta_1+\beta_2}+f_{16}(\beta_1-\beta_2+1)y_1^{\beta_1-\beta_2}\\&-f_{17}(\beta_1-\beta_2-1)y_1^{-(\beta_1-\beta_2)}-f_{18}(\beta_1+\beta_2-1)y_1^{-(\beta_1+\beta_2)}\\&+f_{19}(\beta_2+4)y_1^{\beta_2+3}-f_{20}(\beta_2-4)y_1^{-(\beta_2-3)}\\&+f_{21}(\beta_2+2)y_1^{\beta_2+1}-f_{22}(\beta_2-2)y_1^{-(\beta_2-1)}]\\&+[11\beta_3f_{23}-e_1(\beta_5+11)]y_1^{10}-e_2(\beta_5+9)y_1^8\ln y_1\\&+[9\beta_3f_{24}-e_2-e_3(\beta_5+9)]y_1^8-e_4(\beta_5+7)y_1^6\ln^2y_1\\&-[e_5(\beta_5+7)+2e_4]y_1^6\ln y_1+[7\beta_3f_{25}-e_5-e_6(\beta_5+7)]y_1^6\\&-e_7(\beta_5+5)y_1^4\ln^3y_1-[e_8(\beta_5+5)+3e_7]y_1^4\ln^2y_1\\&-[e_9(\beta_5+5)]+2e_8]y_1^4\ln y_1+[5\beta_3f_{26}-e_9-e_{10}(\beta_5+5)]y_1^4\\&-e_{11}(\beta_5+3)y_1^2\ln y_1+[3\beta_3f_{27}-e_{11}-e_{12}(\beta_5+3)]y_1^2\end{aligned}\tag{3.22}$$

将解（3.21a）代入式（3.1），便得波纹圆板的非线性特征关系式

$$G=\alpha_1W_0+\alpha_3W_0^3\tag{3.23}$$

其中

$$\alpha_3=\alpha_1^4\beta_0[H_1(y_1)-H_2(y_1)+H_2(1)]\tag{3.24}$$

为便于应用，我们将式（3.23）改写为有量纲形式

$$g=\alpha_1^*w_0+\alpha_3^*w_0^3\tag{3.25}$$

其中，w_0 为波纹圆板的中心挠度，而 g，α_1^*，α_3^*，μ_1 和 μ_2 有如下意义。

（1）以均布载荷 q 为主要载荷

$$g = q, \alpha_1^* = \frac{2D_a}{R^4}\alpha_1, \alpha_3^* = \frac{2D_a}{R^4h^2}\alpha_3, \mu_1 = 1, \mu_2 = \frac{p}{\pi R^2 q} \tag{3.26}$$

（2）以中心集中载荷 p 为主要载荷

$$g = p, \alpha_1^* = \frac{2\pi D_a}{R^2}\alpha_1, \alpha_3^* = \frac{2\pi D_a}{R^2h^2}\alpha_3, \mu_1 = \frac{\pi R^2 q}{p}, \mu_2 = 1 \tag{3.27}$$

其次，我们讨论具有光滑中心和滑动固定边界的波纹圆板在均布和中心集中载荷联合作用下的无量纲非线性边值问题（2.20）、（2.13b）、（2.21）、（2.14b）、（2.16）和（2.17）。因为上述两个无量纲非线性边值问题十分相似，所以我们易于直接从式（3.25）得到现在问题的非线性特征关系式。将它们进行比较后可知，仅仅只有边界条件（2.15）和（2.16）的最末一式子是不相同的。值得注意，若将式（2.15）的最末一个条件改写为

$$当\ y=1\ 时，\frac{1}{\nu}\frac{\mathrm{d}S_b}{\mathrm{d}y} - \frac{S_b}{y} = 0 \tag{3.28}$$

并在式（3.28）中令 $\nu \to \infty$，便得式（2.16）的最末一个条件

$$当\ y=1\ 时，S_b = 0 \tag{3.29}$$

因此，在式（3.25）中令 $\nu \to \infty$，我们立即得到第二个问题的非线性特征关系式，可以看出，除 c_7，d_{11}，d_{12}，λ_2 和 ψ_1 改变为

$$\begin{aligned}
c_7 &= \frac{1}{\lambda_2}[\beta_2\beta_4\psi_2(y_1^{\beta_2-1} + y_1^{-(\beta_2+1)}) - \beta_3\psi_3(y_1^{\beta_2} - y_1^{-\beta_2}) \\
&\quad - 2\beta_2\beta_3\beta_4\psi_1 y_1^{-1}], \\
d_{11} &= \frac{1}{\lambda_2}\{\psi_1[\beta_3(\beta_6 - 1) - \beta_2\beta_4]y_1^{-\beta_2} - \psi_3 y_1 - \psi_2(\beta_6 - 1)\}, \\
d_{12} &= -\frac{1}{\lambda_2}\{\psi_1[\beta_3(\beta_6 - 1) + \beta_2\beta_4]y_1^{\beta_2} - \psi_3 y_1 - \psi_2(\beta_6 - 1)\}, \\
\lambda_2 &= [\beta_3(\beta_6 - 1) + \beta_2\beta_4]y_1^{\beta_2} - \beta_3(\beta_6 - 1) - \beta_2\beta_4]y_1^{-\beta_2}, \\
\psi_1 &= \frac{1}{\beta_4}\sum_{i=1}^{10} d_i
\end{aligned} \tag{3.30}$$

以外，特征关系式和其余参数均与第一个问题结果相同。

在 $R_1 \to 0$ 的特殊情况下，我们还能得到没有光滑中心的波纹圆板在均布和中心集中载荷联合作用下的非线性特征关系式。计算表明，特征关系式仍是式（3.25）的形式，对于夹紧固定边界情况，仅需将系数 α_1，α_3，b_1，d_{11} 和 f_{26} 改成为

$$\begin{aligned}
\alpha_1 &= -\left(\frac{b_1}{\beta_1 + 1} + \frac{b_3}{4} + \frac{b_4}{2}\right)^{-1}, \\
\alpha_3 &= -\frac{a_1^4\beta_0}{\beta_1 + 1}\Big[\frac{2}{3}f_1 + \frac{\beta_1 + 5}{2(\beta_1 + 3)}f_3 + \frac{\beta_1 + 3}{2(\beta_1 + 2)}f_5 \\
&\quad + \frac{8}{\beta_1 + 9}f_7 + \frac{6}{\beta_1 + 7}f_9 + \frac{4}{\beta_1 + 5}f_{11} + \frac{\beta_2 + 1}{\beta_1 + \beta_2 + 2}f_{15} \\
&\quad - \frac{\beta_1 - \beta_2 - 4}{\beta_2 + 5}f_{19} - \frac{\beta_1 - \beta_2 - 2}{\beta_2 + 3}f_{21} - \frac{\beta_1 - 11}{12}f_{23}
\end{aligned}$$

$$-\frac{\beta_1-9}{10}f_{24}-\frac{\beta_1-7}{8}f_{25}-\frac{\beta_1-5}{6}f_{26}\Big],$$

$$b_1=-(b_3+b_4),$$

$$d_{11}=-\frac{1}{\beta_4(\beta_2-\nu)}[d_1(2\beta_1+1-\nu)+d_3(\beta_1+4-\nu)$$
$$+d_5(\beta_1+2-\nu)+d_7(7-\nu)+d_8(5-\nu)+d_9(3-\nu)],$$

$$f_{26}=-\frac{b_4d_9}{\beta_4(\beta_1^2-25)} \tag{3.31}$$

而对于滑动固定边界情况，除 d_{11} 改为

$$d_{11}=-\frac{1}{\beta_4}(d_1+d_3+d_5+d_7+d_8+d_9) \tag{3.32}$$

以外，其余公式与式（3.31）相同。

最后，如果在式（3.25）中再令 $R_1=R$，我们还能得到在均布和中心集中载荷联合作用下各向同性圆板的非线性特征关系式。

四、算例

为了说明上面特征关系式的计算过程，我们考虑如图 3 所示的一个非常简单的例子，即在均布和中心集中载荷联合作用下，具有滑动固定边界的无光滑中心的正弦形波纹圆板。已知：$R=76\text{mm}$，$R_1=0$，$h=0.33\text{mm}$，$H=1.20\text{mm}$，$L=25\frac{1}{3}\text{mm}$，$E=1\times10^4\text{kg/mm}^2$ 和 $\nu=0.3$。我们取均布载荷 q 作为主要载荷，于是由式（3.26）有

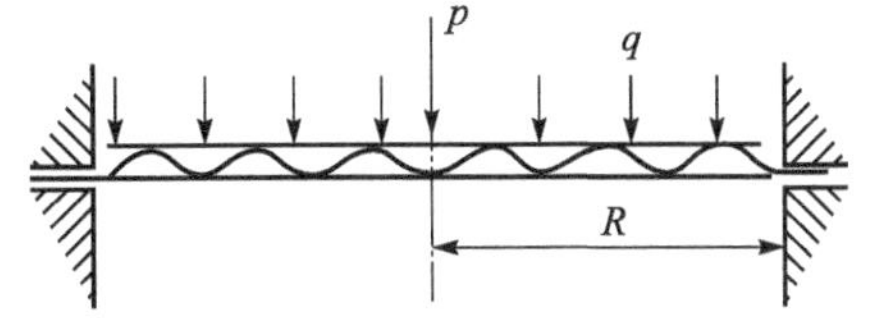

图 3　无光滑中心的正弦截面波纹圆板

$$\mu_1=1 \tag{4.1}$$

将上述值代入式（3.25），便得

$$q=\alpha_1^*w_0+\alpha_3^*w_0^3 \tag{4.2}$$

其中，q 的单位是 kg/cm^2，w_0 的单位是 mm，而对几个 μ_2 值计算所得的系数 α_1^* 和 α_3^* 的数值给在表 1 中。

表 1　系数 α_1^* 和 α_3^* 的值

μ_2	0	0.25	0.50	0.75	1.00
α_1^*	0.03038	0.01810	0.01288	0.01000	0.008176
α_3^*	0.002350	0.001097	0.0007099	0.0005246	0.0004161

图 4 示出了对于几个 μ_2 从值的特征曲线。由图看到，随着 μ_2 的增大，特征曲线降低。显然，这是符合物理意义的。特别当 $\mu_2=0$ 时，我们便得到仅承受均布载荷 q 的波纹圆板的特征曲线。同时，我们还把 Феодосьев[3] 的理论和实验结果也给在此图中。显而易见，本文结果与实验值相当吻合，而 Феодосьев 的理论结果却在 $w_0>4.5h$ 的范围内与实验值相差很大。以 $w_0=8.79h$ 情况为例，本文结果与实验值的相对误差仅为

0.97％，而 Феодосьев 的结果的相对误差竟高达 39.0％。

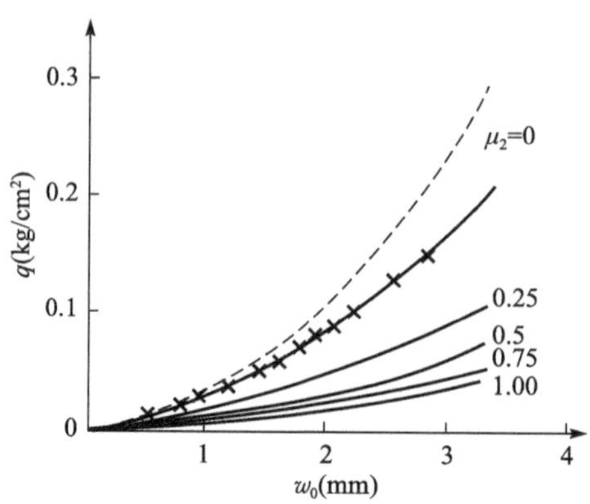

图 4　特征曲线

(—— 本文结果，- - - - $\mu_2=0$ 时的 Феодосьев[3] 的结果，× $\mu_2=0$ 时的实验值[3])

五、结语

本文对承受均布和中心集中载荷联合作用的波纹圆板的非线性弯曲问题进行了研究。应用各向同性和各向异性的圆板大挠度理论，我们讨论了两种板，即具有光滑中心的波纹圆板和无光滑中心的波纹圆板。关于板的边界条件，我们也考虑了两种情况，即夹紧固定边界和滑动固定边界。使用修正迭代法，求得了上述问题的解析解。这些解对于具有深或浅的不同形状的波纹圆板均能适用。

由上节例子可见，甚至当 $\omega_0=8.79h$，亦即板产生很大弯曲情况时，特征关系式 (3.25) 的精确度仍然是非常高的。因此，我们可以预料，本文对实验情况将是适用的。一般说来，这一精确度依赖于 $L/(R-R_1)$ 的大小，其误差随着 $L/(R-R_1)$ 的增加而增加。即是说，如果板的波纹非常密，$L/(R-R_1)$ 的值变得很小，因而波纹区域接近正交各向异性环形板情况，则解的精确度较高。

本文结果对于精密仪器弹性元件的设计将是有益的。

参 考 文 献

[1] Панов, Д. Ю. О больших прогибах круглых мембран со слабым гофром. *ПММ*, 1941, 5 (2): 303-318.

[2] Феодосьев, В. И. О больших прогибах и устойчивости круглой мембраны с мелкой гофрировкой. *ПММ*, 1945, 9 (5): 389-412.

[3] Феодосьев, В. И. *Упругие Элементы Точного Приборостроения*. Москва: Оборонгиз, 1949.

[4] Аксельрад, Э. Л. Ъольшие прогибы гофрированной мембраны как непологой оболочки. *Изв. АН СССР, Мех. —Маш.*, 1964, (1): 46-53.

[5] 陈山林. 浅正弦波纹圆板在均布载荷下的大挠度弹性特征. 应用数学和力学, 1980, 1 (2): 261-272.

[6] Андреева, Л. Е. расёт гофрированных мембран, как анизотропных пластинок. Инженерный Сборник, 1955, 21: 128-141.

[7] 赤坂隆. Corrugated diaphragmの弾性特性について, 日本航空学会誌, 1955, 3 (22-23): 279-288.

[8] Андреева, Л. Е. *Уиругие Элементы Приборов*. Москва: Машгиз, 1962.

[9] Haringx, J. A. Nonlinearity of corrugated diaphragms. Appl. Scient. Res., Ser. A, 1956, 16: 45.

[10] Haringx, J. A. Design of corrugated diaphragms. Trans. ASME, Ser. A, 1957, 79 (1): 55.

[11] Бурмистров, Е. Ф. Снмметричный изгиб неоднородных и однородных ортотропных оболочек вращения с учетом больших прогибов и неравномерного температурного поля. Инженерный Сборник, 1960, 27: 185-199.

[12] 刘人怀. 波纹圆板的特征关系式. 力学学报, 1978, (1): 47-52.

[13] 刘人怀. 波纹环形板的非线性弯曲. 中国科学, A辑, 1984, (3): 247-253.

[14] 叶开沅, 刘人怀, 平庆元, 等. 在对称线布载荷作用下的圆底扁薄球壳的非线性稳定问题. 兰州大学学报, 1965, (2): 10～33; 科学通报, 1965, (2): 142-145.

[15] 叶开沅, 刘人怀, 张传智, 等. 圆底扁薄球壳在边缘力矩作用下的非线性稳定问题. 科学通报, 1965, (2): 145-147.

[16] 刘人怀. 在内边缘均布力矩作用下中心开孔圆底扁球壳的非线性稳定问题. 科学通报, 1965, (3): 253-255.

[17] Chien Wei-zang. Large deflection of a circular clamped plate under uniform pressure. Chinese Journal of Physics, 1947, 7 (2): 102-113.

[18] 刘人怀. 具有光滑中心的波纹圆板的特征关系式. 中国科学技术大学学报, 1979, 9 (2); 75-86.

[19] Liu Ren-huai. Large deflection of corrugated circular plate with a plane central region under the action of concentrated loads at the center. International Journal of Non-Linear Mechanics, 1984, 19 (5): 409-419.

[20] Liu Ren-huai. Large deflection of corrugated circular plate with plane boundary region. Solid Mechanics Archives, 1984, 9 (4): 383-406.

[21] Liu Ren-huai. On large deflection of corrugated annular plates under uniform pressure. The Advances of Applied Mathematics and Mechanics in China, 1987, 1: 138-152.

[22] 刘人怀. 在复合载荷作用下波纹环形板的非线性弯曲. 中国科学, A 辑, 1985, (6); 537-545.

[23] von Kármán T. Festigkeitsprobleme im Mashinenbau. Encykl. Math. Wiss., 1910, 4 (4): 348-351.

On the non-linear bending and vibration of corrugated circular plates①

中文摘要

本文将波纹圆板等效为正交各向异性圆板，研究波纹圆板的非线性弯曲和振动。首先，利用哈密尔顿原理导出问题的基本方程、边界条件和连续条件，然后使用迦辽金方法获得了波纹圆板的特征关系式以及波纹圆板在自由振动下的非线性固有周期和振幅的解析关系式。所获结果不仅再次证实了我们用修正迭代法所获结果的可靠性，而且第一次获得了波纹圆板非线性振动的公式和数据，对精密仪器弹性元件的设计有指导性意义。

1. Introduction

Corrugated circular plates play an important role in the elastic elements of precision instruments, but problems involving them are quite difficult because of the complicated shape as well as the techniques to handle nonlinear simultaneous differential equations with variable coefficients. Previous studies were mainly limited to the static case of large deflections of corrugated circular plates, and were carried on in two ways. In the first way, the nonlinear bending theory of thin shallow shells had been used by Panov[1], Feodosev[2,3], Axelrad[4], Chen[5], et al., to treat the problem, and the analysis of the stiffness, the strength and the stability of the plates can be achieved simultaneously. But difficulties due to the nonlinear mathematical analysis are rather great, so far only the case for sinusoidal corrugated circular plates under actions of uniform pressure has been considered, and the results are not yet satisfactory compared with experimental results. In the second approach another treatment was suggested by assuming a corrugated circular plate as an anisotropic circular plate and large deflection theory of thin anisotropic circular plates had been adopted to treat the same problem by Andryewa[6,7], Akasaka[8], Haringx[9,10], Burmistrov[11], et al. The latter analysis technique is applicable to corrugated circular plates with deep or shallow corrugations of various types. However, results are still not good owing to the lack of improvement in skills to solve nonlinear differential equations. For this purpose, Liu Renhuai[12-19] applied the modified iteration method[20-22] and effectively solved a succession of more difficult problems of large deflection of corrugated circular and annular plates under actions of various loads. The modified iteration method was first forwarded by Yeh Kai-yuan and Liu Renhuai in 1965, then applied successfully to the solution of a series of prob-

---

① Reprinted from *International Journal of Non-Linear Mechanics*, 1989, 24 (3): 165-176. Authors: Liu Renhuai and Li dong.

lems of nonlinear bending and stability of thin plates and shells.

Problems of the nonlinear vibration of corrugated circular plates are much more difficult when the time variable is involved; there is only one existing study[23] in which an attempt was made to deal with the problem of nonlinear free vibrations of corrugated circular plates with full corrugations, but regretfully, no numerical results and discussions are given. As we know, corrugated circular plates with central plane regions are more practically and widely applied in precision instruments than those with full corrugations, hence investigations on the former case are important both theoretically and practically.

In this paper, the authors follow the second approach in studying the non-linear bending and vibration of corrugated circular plates with plane central regions. The governing equations and corresponding boundary continuity conditions are derived by using Hamilton's principle, then, Galerkin's method is applied to find the elastic characteristic relation for uniformly loaded corrugated circular plates and the nonlinear period-amplitude relation for free vibrational corrugated circular plates. Extensive numerical results are given in the figures and effects of various geometrical parameters of the plates on the bending and vibration behaviours of the plates are discussed. From these results it can be concluded that the present characteristic curves are in good agreement with Andryewa's experimental values and Liu Renhuai's theoretical results obtained by means of the modified iteration method. Results presented here are instructively significant in the design of elastic elements of precision instruments.

2. Governing Equations

Let us now consider a corrugated circular plate with a plane central region under uniform pressure q as shown in Fig. 1. Here a and b are the radii of the boundary and the plane central region respectively, r is the radial coordinate, h is the thickness, E is Young's modulus, ν is Poisson's ratio, m is mass density per unit area of the plate, w and u are the deflection and radial displacement respectively, N_r and N_θ are the radial and circumferential membrane stress respectively, ε_r and ε_θ are the radial and circumferential membrane strain respectively, M_r and M_θ are the radial and circumferential moments, and K_r and K_θ are the variation of radial and circumferential curvature respectively. It is noted that the deformation of the plate is axisymmetrical, hence displacement components and stresses of the plate are functions of the variables r and t, where t indicates time. Assuming the corrugated circular plate as a plate composed of a circular plate and an anisotropic annular plate, their physical magnitudes are indicated by the subscripts 1 and 2 respectively. The total potential energy of the plate is expressed as

$$V = V_1 + V_2 + V_3 \tag{2.1}$$

in which V_1, V_2 and V_3 are the bending, stretching and load potential energy respectively,

$$
\begin{aligned}
V_1 &= \int_0^b \frac{1}{2}(M_{r,1}K_{r,1} + M_{\theta,1}K_{\theta,1})2\pi r \mathrm{d}r \\
&\quad + \int_b^a \frac{1}{2}(M_{r,2}K_{r,2} + M_{\theta,2}K_{\theta,2})2\pi r \mathrm{d}r \\
V_2 &= \int_0^b \frac{1}{2}(N_{r,1}\varepsilon_{r,1} + N_{\theta,1}\varepsilon_{\theta,1})2\pi r \mathrm{d}r \\
&\quad + \int_b^a \frac{1}{2}(N_{r,2}\varepsilon_{r,2} + N_{\theta,2}\varepsilon_{\theta,2})2\pi r \mathrm{d}r \\
V_3 &= -\int_0^b q w_1 2\pi r \mathrm{d}r - \int_b^a q w_2 2\pi r \mathrm{d}r
\end{aligned} \tag{2.2a-c}
$$

and the kinetic energy of the plate is

$$
T = \int_0^b \frac{1}{2} m_1 \left(\frac{\partial w_1}{\partial t}\right)^2 2\pi r \mathrm{d}r + \int_b^a \frac{1}{2} m_2 \left(\frac{\partial w_2}{\partial t}\right)^2 2\pi r \mathrm{d}r \tag{2.3}
$$

From [3, 7], we have

$$
\begin{aligned}
M_{r,1} &= -D_1\left(\frac{\partial^2 w_1}{\partial r^2} + \frac{\nu}{r}\frac{\partial w_1}{\partial r}\right), \\
M_{\theta,1} &= -D_1\left(\frac{1}{r}\frac{\partial w_1}{\partial r} + \nu\frac{\partial w_1}{\partial r^2}\right),
\end{aligned} \tag{2.4a,b}
$$

$$
\begin{aligned}
M_{r,2} &= -D_2 k_t^{-1}\left(\frac{\partial^2 w_2}{\partial r^2} + \frac{\nu}{r}\frac{\partial w_2}{\partial r}\right), \\
M_{\theta,2} &= -D_2 k_t^{-1}\left(\frac{k_t k_t'}{r}\frac{\partial w_2}{\partial r} + \nu\frac{\partial^2 w_2}{\partial r^2}\right),
\end{aligned} \tag{2.5a,b}
$$

$$
\begin{aligned}
\varepsilon_{r,1} &= -\frac{1}{Eh}(N_{r,1} - \nu N_{\theta,1}) = \frac{\partial u_1}{\partial r} + \frac{1}{2}\left(\frac{\partial w_1}{\partial r}\right)^2, \\
\varepsilon_{\theta,1} &= \frac{1}{Eh}(N_{\theta,1} - \nu N_{r,1}) = \frac{u_1}{r}
\end{aligned} \tag{2.6a,b}
$$

$$
\begin{aligned}
\varepsilon_{r,2} &= \frac{1}{Ehk_t}(k_r k_t N_{r,2} - \nu N_{\theta,2}) = \frac{\partial u_2}{\partial r} + \frac{1}{2}\left(\frac{\partial w_2}{\partial r}\right)^2, \\
\varepsilon_{\theta,2} &= \frac{1}{Ehk_t}(N_{\theta,2} - \nu N_{r,2}) = \frac{u_2}{r}
\end{aligned} \tag{2.7a,b}
$$

with

$$
\begin{aligned}
K_{r,1} &= -\frac{\partial^2 w_1}{\partial r^2},\ K_{r,2} = -\frac{\partial^2 w_2}{\partial r^2}, \\
K_{\theta,1} &= -\frac{1}{r}\frac{\partial w_1}{\partial r},\ K_{\theta,2} = -\frac{1}{r}\frac{\partial w_2}{\partial r}
\end{aligned} \tag{2.8a-d}
$$

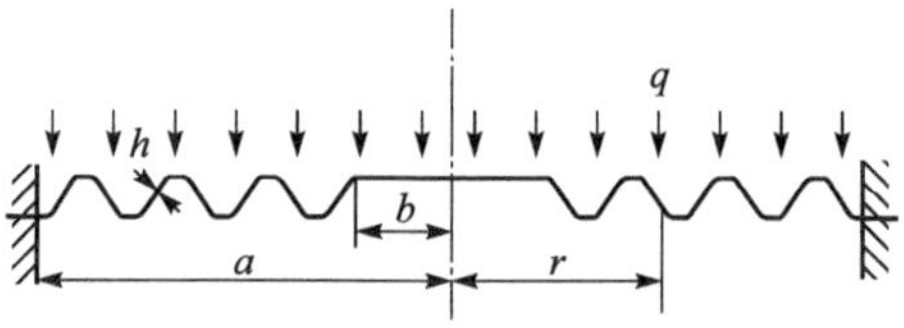

Fig. 1　A uniformly loaded corrugated circular plate with a plane central region

where k_r, k_t and k_t' are parameters[14,15] for the radial and circumferential rigidity respectively, and D is the flexural rigidity

$$D_1=\frac{Eh^3}{12(1-\nu^2)},\quad D_2=\frac{Eh^3}{12\left(1-\dfrac{\nu^2}{k_tk_t'}\right)}$$

According to Hamilton's principle[24], the solutions of the kinetic problem satisfy the following functional stationary condition

$$\delta\Pi=0 \tag{2.9}$$

in which t_1 and t_2 indicate different times of the motion, and

$$\Pi=\int_{t_1}^{t_2}(T-V)\mathrm{d}t \tag{2.10}$$

Substituting expressions (2.1)-(2.5) and (2.8) into (2.10), we have

$$\begin{aligned}\Pi=&\int_{t_1}^{t_2}\int_0^b\left\{\frac{1}{2}m_1\left(\frac{\partial w_1}{\partial t}\right)^2+qw_1-\frac{1}{2}D_1\left[\left(\frac{\partial^2 w_1}{\partial r^2}\right)^2\right.\right.\\&\left.+\frac{2\nu}{r}\frac{\partial w_1}{\partial r}\frac{\partial^2 w_1}{\partial r^2}+\frac{1}{r^2}\left(\frac{\partial w_1}{\partial r}\right)^2\right]-\frac{1}{2}(N_{r,1}\varepsilon_{r,1}\\&\left.+N_{\theta,1}\varepsilon_{\theta,1})\right\}2\pi r\mathrm{d}r\mathrm{d}t+\int_{t_1}^{t_2}\int_b^a\left\{\frac{1}{2}m_2\left(\frac{\partial w_2}{\partial t}\right)^2+qw_2\right.\\&-\frac{1}{2}D_2k_t^{-1}\left[\left(\frac{\partial^2 w_2}{\partial r^2}\right)^2+\frac{2\nu}{r}\frac{\partial w_2}{\partial r}\frac{\partial^2 w_2}{\partial r^2}+\frac{k_tk_t'}{r^2}\left(\frac{\partial w_2}{\partial r}\right)^2\right]\\&\left.-\frac{1}{2}(N_{r,2}\varepsilon_{r,2}+N_{\theta,2}\varepsilon_{\theta,2})\right\}2\pi r\mathrm{d}r\mathrm{d}t\end{aligned}$$

Substituting this expression into condition (2.9), and integrating by parts, we can obtain

$$\begin{aligned}&\int_{t_1}^{t_2}\int_0^b\left[q-m_1\frac{\partial^2 w_1}{\partial t^2}-D_1\left(\frac{\partial^4 w_1}{\partial r^4}+\frac{2}{r}\frac{\partial^3 w_1}{\partial r^3}-\frac{1}{r^2}\frac{\partial^2 w_1}{\partial r^2}\right.\right.\\&\left.\left.+\frac{1}{r^3}\frac{\partial w_1}{\partial r}\right)+\frac{1}{r}\frac{\partial}{\partial r}\left(rN_{r,1}\frac{\partial w_1}{\partial r}\right)\right]\delta w_1 r\mathrm{d}r\mathrm{d}t\\&+\int_{t_1}^{t_2}\int_0^b\left[\frac{\partial}{\partial r}(rN_{r,1})-N_{\theta,1}\right]\delta u_1\mathrm{d}r\mathrm{d}t+\int_{t_1}^{t_2}\int_b^a\left[q-m_2\frac{\partial^2 w_2}{\partial r^2}\right.\\&-D_2k_t^{-1}\left(\frac{\partial^4 w_2}{\partial r^4}+\frac{2}{r}\frac{\partial^2 w_2}{\partial r^3}-\frac{k_tk_t'}{r^2}\frac{\partial^2 w_2}{\partial r^2}+\frac{k_tk_t'}{r^3}\frac{\partial w_2}{\partial r}\right)\\&\left.+\frac{1}{r}\frac{\partial}{\partial r}\left(rN_{r,2}\frac{\partial w_2}{\partial r}\right)\right]\delta w_2 r\mathrm{d}r\mathrm{d}t+\int_{t_1}^{t_2}\int_b^a\left[\frac{\partial}{\partial r}(rN_{r,2})\right.\\&\left.-N_{\theta,2}\right]\delta u_2\mathrm{d}r\mathrm{d}t+\int_{t_1}^{t_2}\left\{\left[D_1\left(\frac{\partial^3 w_1}{\partial r^3}+\frac{1}{r}\frac{\partial^2 w_1}{\partial r^2}-\frac{1}{r^2}\frac{\partial w_1}{\partial r}\right)\right.\right.\\&\left.-N_{r,1}\frac{\partial w_1}{\partial r}\right]r\delta w_1\Big|_{r=0}^{r=b}-(rN_{r,1})\delta u_1\Big|_{r=0}^{r=b}+\left[D_2k_t^{-1}\left(\frac{\partial^3 w_2}{\partial r^3}\right.\right.\\&\left.\left.+\frac{1}{r}\frac{\partial^2 w_2}{\partial r^2}-\frac{k_tk_t'}{r^2}\frac{\partial w_2}{\partial r}\right)-N_{r,2}\frac{\partial w_2}{\partial r}\right]r\delta w_2\Big|_{r=b}^{r=a}\end{aligned}$$

$$- (rN_{r,2})\delta u_2 \Big|_{r=b}^{r=a} - D_1\left(\frac{\partial^2 w_1}{\partial r^2} + \frac{v}{r}\frac{\partial w_1}{\partial r}\right) r\delta\left(\frac{\partial w_1}{\partial r}\right)\Big|_{r=0}^{r=b}$$

$$- D_2 k_t^{-1}\left(\frac{\partial^2 w_2}{\partial r^2} + \frac{\nu}{r}\frac{\partial w_2}{\partial r}\right) r\delta\left(\frac{\partial w_2}{\partial r}\right)\Big|_{r=b}^{r=a}\Bigg\} dt$$

$$+ \left[\int_0^b m_1 \frac{\partial w_1}{\partial t}\delta w_1 r dr + \int_b^a m_2 \frac{\partial w_2}{\partial t}\delta w_2 r dr\right]\Big|_{t=t_1}^{t=t_2} = 0 \qquad (2.11)$$

Hamilton's principle requires that

$$\delta w_1 \Big|_{t=t_1} = \delta w_1 \Big|_{t=t_2} = \delta w_2 \Big|_{t=t_1} = \delta w_2 \Big|_{t=t_2} = 0$$

so the last two integrals in the above equation vanish, and since δw_1, δw_2, δu_1, δu_2 are arbitrary independent functions, it is easy to get the basic equations of the problem from (2.11).

For $0 \leqslant r \leqslant b$

$$m_1 \frac{\partial^2 w_1}{\partial t^2} + D_1\left(\frac{\partial^4 w_1}{\partial r^4} + \frac{2}{r}\frac{\partial^3 w_1}{\partial r^3}\right.$$

$$\left. - \frac{1}{r^2}\frac{\partial^2 w_1}{\partial r^2} + \frac{1}{r^3}\frac{\partial w_1}{\partial r}\right) - \frac{1}{r}\frac{\partial}{\partial r}\left(rN_{r,1}\frac{\partial w_1}{\partial r}\right) = q,$$

$$\frac{\partial}{\partial r}(rN_{r,1}) - N_{\theta,1} = 0 \qquad (2.12a,b)$$

For $b < r \leqslant a$

$$m_2 \frac{\partial^2 w_2}{\partial t^2} + D_2 k_t^{-1}\left(\frac{\partial^4 w_2}{\partial r^4} + \frac{2}{r}\frac{\partial^3 w_2}{\partial r^3} - \frac{k_t k_t'}{r^2}\frac{\partial^2 w_2}{\partial t^2}\right.$$

$$\left. + \frac{k_t k_t'}{r^3}\frac{\partial w_2}{\partial r}\right) - \frac{1}{r}\frac{\partial}{\partial r}\left(rN_{r,2}\frac{\partial w_2}{\partial r}\right) = q,$$

$$\frac{\partial}{\partial r}(rN_{r,2}) - N_{\theta,2} = 0 \qquad (2.13a,b)$$

From equation (2.11), various boundary conditions of the plates can be obtained, however, we are here only interested in one usual type of boundary condition, i.e., the plate is rigidly clamped along the outer edge

$$\frac{\partial w_1}{\partial r} = 0,\ N_{r,1} \text{ is finite} \qquad \text{at } r = 0 \qquad (2.14a,b)$$

$$\frac{\partial w_2}{\partial r} = 0,\ w_2 = 0,\ u_2 = 0 \qquad \text{at } r = a \qquad (2.15a\text{-}c)$$

Besides, Eq. (2.11) yields also the continuity conditions as follows

$$u_1 = u_2,\quad N_{r,1} = N_{r,2},\quad w_1 = w_2,\quad \frac{\partial w_1}{\partial r} = \frac{\partial w_2}{\partial r},$$

$$D_1\left(\frac{\partial^2 w_1}{\partial r^2} + \frac{\nu}{r}\frac{\partial w_1}{\partial r}\right) = D_2 k_t^{-1}\left(\frac{\partial^2 w_2}{\partial r^2} + \frac{\nu}{r}\frac{\partial w_2}{\partial r}\right),$$

$$D_1\left(\frac{\partial^3 w_1}{\partial r^3} + \frac{1}{r}\frac{\partial^2 w_1}{\partial r^2} - \frac{1}{r^2}\frac{\partial w_1}{\partial r}\right)$$

$$= D_2 k_1^{-1}\left(\frac{\partial^3 w_2}{\partial r^3} + \frac{1}{r}\frac{\partial^2 w_2}{\partial r^2} - \frac{k_t k_t'}{r^2}\frac{\partial w_2}{\partial r}\right) \qquad \text{at } r = b \qquad (2.16)$$

Applying expressions (2.6), (2.7) and Eqs. (2.12b), (2.13b), we have the following compatibility equations

for $0 \leqslant r \leqslant b$

$$r\frac{\partial^2}{\partial r^2}(rN_{r,1})+\frac{\partial}{\partial r}(rN_{r,1})-N_{r,1}=-\frac{1}{2}Eh\left(\frac{\partial w_1}{\partial r}\right)^2 \tag{2.17a}$$

for $b \leqslant r \leqslant a$

$$r\frac{\partial^2}{\partial r^2}(rN_{r,2})+\frac{\partial}{\partial r}(rN_{r,2})-k_rk_tN_{r,2}=-\frac{1}{2}k_tEh\left(\frac{\partial w_2}{\partial r}\right)^2 \tag{2.17b}$$

Using (2.6b), (2.7b) and (2.12b), (2.13b) we still have

$$u_1=\frac{r}{Eh}\left[\frac{\partial}{\partial r}(rN_{r,1})-\nu N_{r,1}\right],$$

$$u_2=\frac{r}{Ehk_t}\left[\frac{\partial}{\partial r}(rN_{r,2})-\nu N_{r,2}\right] \tag{2.18a,b}$$

To make the resulting solution simpler, the following non-dimensional parameters are introduced

$$x=\frac{r}{a},\quad \alpha=\frac{b}{a},\quad \tau=t\left(\frac{D_1}{m_1a^4}\right)^{1/2},\quad W_1=\frac{w_1}{h},$$

$$W_2=\frac{w_2}{h},\quad N_1=-\frac{a}{D_1}rN_{r,1},\quad N_2=-\frac{a}{D_2}k_trN_{r,2},$$

$$Q=\frac{a^4}{D_1h}q,\quad m^2=k_tk_t',\quad n^2=k_rk_t,\quad \lambda_1=6(1-\nu^2),$$

$$\lambda_2=6k_t^2\left(1-\frac{\nu^2}{k_tk_t'}\right),\quad \lambda_3=\frac{D_2}{D_1k_t},\quad \lambda_4=\nu(1-\lambda_3),$$

$$\lambda_5=\nu\left(1-\frac{1}{k_t}\right),\quad \lambda_6=\frac{\lambda_1}{\lambda_2}$$

Upon applying these parameters and expressions (2.18a, b), the governing equations (2.12a), (2.13a) and (2.17a, b), the boundary conditions (2.14), (2.15) as well as the continuity conditions (2.16) for a motional corrugated circular plate with a plane central region under uniform pressure are transformed into the following non-dimensional forms.

For $0 \leqslant x \leqslant \alpha$

$$\frac{\partial^2W_1}{\partial\tau^2}+\frac{\partial^4W_1}{\partial x^4}+\frac{2}{x}\frac{\partial^3W_1}{\partial x^3}-\frac{1}{x^2}\frac{\partial^2W_1}{\partial x^2}+\frac{1}{x^3}\frac{\partial W_1}{\partial x}$$

$$+\frac{1}{x}\frac{\partial}{\partial x}\left(N_1\frac{\partial W_1}{\partial x}\right)=Q,$$

$$x^2\frac{\partial^2N_1}{\partial x^2}+x\frac{\partial N_1}{\partial x}-N_1=\lambda_1x\left(\frac{\partial W_1}{\partial x}\right)^2 \tag{2.19a,b}$$

For $\alpha < x \leqslant 1$

$$k_t\frac{\partial^2W_2}{\partial\tau^2}+\lambda_3\left[\frac{\partial^4W_2}{\partial x^4}+\frac{2}{x}\frac{\partial^3W_2}{\partial x^3}-\frac{m^2}{x^2}\frac{\partial^2W_2}{\partial x^2}\right.$$

$$+\frac{m^2}{x^3}\frac{\partial W_2}{\partial x}+\frac{1}{x}\frac{\partial}{\partial x}\left(N_2\frac{\partial W_2}{\partial x}\right)\Big]=Q,$$

$$x^2\frac{\partial^2 N_2}{\partial x^2}+x\frac{\partial N_2}{\partial x}-n^2 N_2=\lambda_2 x\left(\frac{\partial W_2}{\partial x}\right)^2 \tag{2.20a,b}$$

with

$$\frac{\partial W_1}{\partial x}=0,\qquad N_1=0\qquad \text{at } x=0 \tag{2.21a,b}$$

$$W_2=0,\quad \frac{\partial W_2}{\partial x}=0,\quad \frac{\partial N_2}{\partial x}-\nu N_2=0\quad \text{at } x=1 \tag{2.22a-c}$$

$$W_1=W_2,\ \frac{\partial W_1}{\partial x}=\frac{\partial W_2}{\partial x},\ \alpha\frac{\partial^2 W_1}{\partial x^2}+\lambda_4\frac{\partial W_1}{\partial x}=\alpha\lambda_3\frac{\partial^2 W_2}{\partial x^2},$$

$$\alpha^2\frac{\partial^3 W_1}{\partial x^3}+\alpha\frac{\partial^2 W_1}{\partial x^2}-\frac{\partial W_1}{\partial x}=\lambda_3\left(\alpha^2\frac{\partial^3 W_2}{\partial x^3}+\alpha\frac{\partial^2 W_2}{\partial x^2}-m^2\frac{\partial W_2}{\partial x}\right),$$

$$N_1=\lambda_3 N_2,\ \alpha\frac{\partial N_1}{\partial x}-\lambda_5 N_1=\alpha\lambda_6\frac{\partial N_2}{\partial x}\quad \text{at } x=\alpha \tag{2.23a-f}$$

where the relation $m_2=k_t m_1$ has been used.

Finally, the problem of nonlinear bending and vibration of corrugated circular plates with plane central regions is reduced to solving the non-dimensional nonlinear equations (2.19) and (2.20) under boundary conditions (2.21), (2.22) and continuity condition (2.23).

3. Analytical Solutions

Galerkin's method is applied to solve the above problem (2.19)-(2.23). The solutions of the small deflection theory for uniformly loaded circular plates and anisotropic annular plates[25] are chosen to model nonlinear vibration of corrugated circular plates with plane central regions by assuming

for $0\leqslant x\leqslant\alpha$, $W_1(x,\tau)=f(\tau)(A_1x^4+A_2x^2+A_3)$,

for $\alpha<x\leqslant 1$, $W_2(x,\tau)=f(\tau)(B_1x^{m+1}+B_2x^{-m+1}+B_3x^4+B_4x^2+B_5)$ (3.1a, b)

where f is an unknown function of τ

$$A_1=\frac{1}{32},\qquad B_3=\frac{1}{4\lambda_3(9-m^2)}$$

Since W_1 and W_2 are chosen to satisfy boundary conditions (2.21a), (2.22a, b) and continuity conditions (2.23a-d), we can easily determine that

$$A_2=\frac{B_2}{2}(1-m)(\alpha^{-m-1}-\alpha^{m-1})+2B_3(\alpha^2-\alpha^{m-1})-2A_1\alpha^2,$$

$$A_3=B_5-A_2\alpha^2+(B_3-A_1)\alpha^4+B_2\alpha^{-m+1}+B_1\alpha^{m+1},$$

$$B_1=(m+1)^{-1}[(m-1)B_2-4B_3],$$

$$B_2=\frac{4\alpha^m(m\lambda_3-\lambda_4-1)+[4(1+\lambda_4)-\lambda_3(3+m^2)]\alpha^3}{4\lambda_3(1-m)(9-m^2)[(1+\lambda_4-m\lambda_3)\alpha^m-(1+\lambda_4+m\lambda_3)\alpha^{-m}]},$$

$$B_4=0,\quad B_5=(m+1)^{-1}[(3-m)B_3-2mB_2]$$

Substituting expressions (3.1a, b) into Eqs. (2.19b) and (2.20b), using boundary conditions (2.21b), (2.22c) and continuity conditions (2.23e, f), the solutions for membrane stresses can be obtained as

$$
\begin{aligned}
N_1 &= f^2(\tau)(C_1x^7 + C_2x^5 + C_3x^3 + C_4x),\\
N_2 &= f^2(\tau)(E_1x^{2m+1} + E_2x^{-2m+1} + E_3x^{m+4} + E_4x^{-m+4}\\
&\quad + E_5x^7 + E_6x + E_7x^n + E_8x^{-n})
\end{aligned}
\tag{3.2a,b}
$$

where

$$
\begin{aligned}
C_1 &= \frac{1}{3}\lambda_1A_1^2, \quad C_2 = \frac{2}{3}\lambda_1A_1A_2, \quad C_3 = \frac{1}{2}\lambda_1A_2^2,\\
C_4 &= \lambda_3(E_1\alpha^{2m} + E_2\alpha^{-2m} + E_3\alpha^{m+3} + E_4\alpha^{-m+3} + E_5\alpha^6 + E_6\\
&\quad + E_7\alpha^{n-1} + E_8\alpha^{-n-1}) - C_1\alpha^6 - C_2\alpha^4 - C_3\alpha^2,\\
E_1 &= \frac{\lambda_2(m+1)^2}{(2m+1)^2 - n^2}B_1^2, \quad E_2 = \frac{\lambda_2(m-1)^2}{(2m-1)^2 - n^2}B_2^2,\\
E_3 &= \frac{8\lambda_2(m+1)}{(m+4)^2 - n^2}B_1B_3, \quad E_4 = -\frac{8\lambda_2(m-1)}{(m-4)^2 - n^2}B_2B_3,\\
E_5 &= \frac{16\lambda_2}{49 - n^2}B_3^2, \quad E_6 = \frac{2(m^2-1)}{n^2-1}\lambda_2B_1B_2,\\
E_7 &= (n-\nu)^{-1}[(\nu-2m-1)E_1 + (\nu+2m-1)E_2 + (\nu-m-4)E_3\\
&\quad + (\nu+m-4)E_4 + (\nu-7)E_5 + (\nu-1)E_6 + (n+\nu)E_8],\\
E_8 &= [(n+\nu)(n\lambda_6 + \lambda_3\lambda_5 - \lambda_3)\alpha^n - (n-\nu)(n\lambda_6 - \lambda_3\lambda_5\\
&\quad + \lambda_3)\alpha^{-n}]^{-1}(n-\nu)\{E_1(\lambda_3 - \lambda_3\lambda_5 - \lambda_6 - 2m\lambda_6)\alpha^{2m+1}\\
&\quad + E_2(\lambda_3 - \lambda_3\lambda_5 - \lambda_6 + 2m\lambda_6)\alpha^{-2m+1} + E_3(\lambda_3 - m\lambda_6 - \lambda_3\lambda_5\\
&\quad - 4\lambda_6)\alpha^{m+4} + E_4(\lambda_3 + m\lambda_6 - \lambda_3\lambda_5 - 4\lambda_6)\alpha^{-m+4}\\
&\quad + (n-\nu)^{-1}(\lambda_3 - n\lambda_6 - \lambda_3\lambda_5)[(\nu-2m-1)E_1\\
&\quad + (\nu+2m-1)E_2 + (\nu-m-4)E_3 + (\nu+m-4)E_4\\
&\quad + (\nu-7)E_5 + (\nu-1)E_6]\alpha^n + [6C_1 + E_5(\lambda_3 - \lambda_3\lambda_5\\
&\quad - 7\lambda_6)]\alpha^7 + 4C_2\alpha^5 + 2C_3\alpha^3 + E_6(\lambda_3 - \lambda_3\lambda_5 - \lambda_6)\alpha\}
\end{aligned}
$$

Now, putting (3.1), (3.2) into the following Galerkin equation

$$
\begin{aligned}
&\int_0^\alpha\Big[\frac{\partial^2W_1}{\partial\tau^2} + \frac{\partial^4W_1}{\partial x^4} + \frac{2}{x}\frac{\partial^3W_1}{\partial x^3} - \frac{1}{x^2}\frac{\partial^2W_1}{\partial x^2} + \frac{1}{x^3}\frac{\partial W_1}{\partial x}\\
&\quad + \frac{1}{x}\frac{\partial}{\partial x}\Big(N_1\frac{\partial W_1}{\partial x}\Big) - Q\Big]xW_1\,\mathrm{d}x + \int_\alpha^1\Big\{k_t\frac{\partial^2W_2}{\partial\tau^2}\\
&\quad + \lambda_3\Big[\frac{\partial^4W_2}{\partial x^4} + \frac{2}{x}\frac{\partial^3W_2}{\partial x^3} - \frac{m^2}{x^2}\frac{\partial^2W_2}{\partial x^2} + \frac{m^2}{x^3}\frac{\partial W_2}{\partial x}\\
&\quad + \frac{1}{x}\frac{\partial}{\partial x}\Big(N_2\frac{\partial W_2}{\partial x}\Big)\Big] - Q\Big\}xW_2\,\mathrm{d}x = 0
\end{aligned}
$$

and integrating by parts, we have

$$
\mu_1\frac{\mathrm{d}^2f}{\mathrm{d}\tau^2} + \mu_2f + \mu_3f^3 = \mu_4Q \tag{3.3}
$$

If we define

$$F_{(j)} = \frac{4A_1}{j+4}\alpha^{j+4} + \frac{2A_2}{j+2}\alpha^{j+2}$$

$$G_{(j)} = \frac{m+1}{m+1+j}B_1(1-\alpha^{j+1+m}) + \frac{m-1}{m-1-j}B_2(1 - \alpha^{j+1-m}) + \frac{4B_3}{j+4}(1-\alpha^{j+4})$$

then the coefficients μ_1, μ_2, μ_3 and μ_4 can be expressed as follows

$$\mu_1 = \frac{A_1^2}{10}\alpha^{10} + \frac{A_1A_2}{4}\alpha^8 + \left(\frac{A_1A_3}{3} + \frac{A_2^2}{6}\right)\alpha^6 + \frac{A_2A_3}{3}\alpha^4 + \frac{A_3^2}{2}\alpha^2 + k_t\left[\frac{B_3^2}{10}(1-\alpha^{10}) + \frac{B_3B_5}{3}(1-\alpha^6) + \frac{B_1B_2}{2}(1-\alpha^4) + \frac{B_5^2}{2}(1-\alpha^2) + \frac{B_1^2}{4+2m}(1-\alpha^{4+2m}) + \frac{2B_1B_3}{7+m}(1-\alpha^{7+m}) + \frac{2B_1B_5}{3+m}(1-\alpha^{3+m}) + \frac{2B_2B_3}{7-m}(1-\alpha^{7-m}) + \frac{2B_2B_5}{3-m}(1-\alpha^{3-m}) + \frac{B_2^2}{4-2m}(1-\alpha^{4-2m})\right],$$

$$\mu_2 = 4\lambda_3B_3(m^2-9)\left[\frac{B_1(1+m)(1-\alpha^{3+m})}{3+m} + \frac{B_2(1-m)(1-\alpha^{3-m})}{3-m} + \frac{2}{3}B_3(1-\alpha^6)\right] - 32A_1F_{(2)},$$

$$\mu_3 = -\lambda_3\{[(1+m)B_1E_3 + 4B_3E_1]G_{(4+2m)} + (1+m)B_1E_1G_{(1+3m)} + [(1-m)B_2E_4 + 4B_3E_2]G_{(4-2m)} + (1-m)B_2E_2G_{(1-3m)} + (1+m)B_1E_7G_{(m+n)} + (1+m)B_1E_8G_{(m-n)} + (1-m)B_2E_7G_{(n-m)} + (1-m)B_2E_8G_{(-m-n)} + [(1+m)B_1E_5 + 4B_3E_3]G_{(7+m)} + [(1-m)B_2E_5 + 4B_3E_4]G_{(7-m)} + [(1+m)B_1E_6 + (1-m)B_2E_1]G_{(1+m)} + [(1+m)B_1E_2 + (1-m)B_2E_6]G_{(1-m)} + 4B_3E_5G_{(10)} + [(1+m)B_1E_4 + (1-m)B_2E_3 + 4B_3E_6]G_{(4)} + 4B_3E_7G_{(3+n)} + 4B_3E_8G_{(3-n)}\} - 2A_2C_4F_{(2)} - 2(A_2C_3 + 2A_1C_4)F_{(4)} - 2(A_2C_2 + 2A_1C_3)F_{(6)} - 2(A_2C_1 + 2A_1C_2)F_{(8)} - 4A_1C_1F_{(10)},$$

$$\mu_4 = \frac{A_1}{6}\alpha^6 + \frac{A_2}{4}\alpha^4 + \frac{A_3}{2}\alpha^2 + \frac{B_3}{6}(1-\alpha^6) + \frac{B_5}{2}(1-\alpha^2) + \frac{B_1}{3+m}(1-\alpha^{3+m}) + \frac{B_2}{3-m}(1-\alpha^{3-m})$$

It is observed that the non-dimensional center deflection of the corrugated circular plate is

$$W_0(\tau) = W_1(x,\tau)\mid_{x=0} = A_3 f(\tau)$$

then equation (3.3) can be written in the form

$$\frac{d_2 W_0}{d\tau^2} + \left(\frac{\mu_2}{\mu_1}\right)W_0 + \left(\frac{\mu_3}{\mu_1 A_3^2}\right)W_0^3 = \left(\frac{\mu_4 A_3}{\mu_1}\right)Q \tag{3.4}$$

If we set

$$\frac{d^2 W_0}{d\tau^2} = 0$$

in equation (3.4), then the characteristic relation for a uniformly loaded corrugated circular plate with a plane central region can be obtained. We may retrace this relation in the following dimensional form to facilitate application

$$q = \left(\frac{\mu_2 D_1}{\mu_4 a^4 A_3}\right)w_0 + \left(\frac{\mu_3 D_1}{\mu_4 a^4 A_3^3 h^2}\right)w_0^3 \tag{3.5}$$

in which w_0 is the dimensional center deflection of the plate.

If we put $Q=0$ in Eq. (3.4), the equation for nonlinear free vibration of the corrugated circular plate with a plane central region can be written as

$$\frac{d^2 W_0}{d\tau^2} + \omega_L^2 (W_0 + \varepsilon W_0^3) = 0 \tag{3.6}$$

in which ω_L is the natural circular frequency for linear free vibration of the plate, and ε is the nonlinearity parameter:

$$\omega_L^2 = \frac{\mu_2}{\mu_1}, \quad \varepsilon = \frac{\mu_3}{\mu_2 A_3^2} \tag{3.7a,b}$$

Equation (3.6) is Duffin's equation, hence the ratio of nonlinear period T_{NL} to linear period T_L is given by [26]

$$\frac{T_{NL}}{T_L} = \frac{2K(S)}{\pi\sqrt{1+\varepsilon W_m^2}} \tag{3.8}$$

where W_m is the non-dimensional center amplitude, and K is the complete elliptic integral of the first kind:

$$T_L = \frac{2\pi}{\omega_L}, \quad S = \left[\frac{\varepsilon W_m^2}{2(1+\varepsilon W_m^2)}\right]^{1/2}$$

4. Results and Discussion

For convenience to check on the accuracy of the theoretical results obtained above, first, we consider the nonlinear bending problem of corrugated circular plates. An example studied in [6] is reconsidered here, in which the plate has toothed corrugations, as shown in Fig. 2. The dimensions of the plate are

$H=0.414$mm, $h=0.101$mm, $a=24.00$mm,

$b=5.10$mm, $\theta_0=8°45'$, $l=5.40$mm,

$E=1.35\times10^4$kg/mm^2, $\nu=0.3$

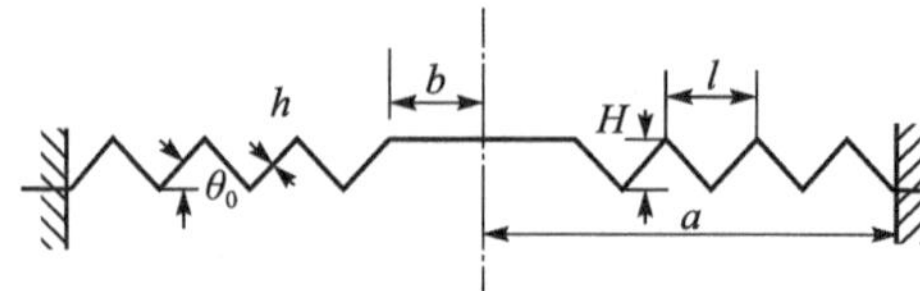

Fig. 2 Corrugated circular plate with toothed corrugations

Substituting these values into relation (3.5), we have

$$q = 0.1029 w_0 + 0.1343 w_0^3 \tag{4.1}$$

where w_0 is in (mm), q is in (kg/cm^2).

From formula (4.1), results are presented in Fig. 3, which demonstrates very good agreement with previous experimental results[6] and Liu Renhuai's theoretical results[13] obtained by means of the modified iteration method. It can be observed that Andryewa's theoretical results[6] by applying Galerkin's method are far from the experimental values when the deflection becomes larger, owing to roughness in the choice of deflection function.

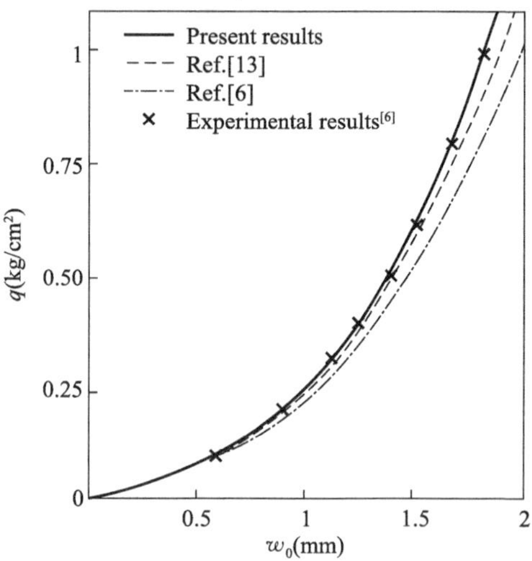

Fig. 3 Comparison between theoretical and experimental results

Next, we shall further discuss the large deflection and large amplitude free vibration of corrugated circular plates with toothed corrugations as shown in Fig. 2. The effect of the size of the plane central region on mechanical behaviour of the plate is first investigated. We have studied the case

$H=0.3$mm, $h=0.1$mm, $a=25.0$mm, $\theta_0=6°51'$,

$l=5.0$mm, $E=1.2\times10^4$kg/mm^2, $\nu=0.3$

and $b=0.0$, 2.5, 7.5, 12.5, 25.0 mm respectively.

Numerical results computed from formulae (3.5) and (3.8) are depicted in Figs. 4 and 5. From these results, it can be concluded that when the size of the plane central region is small ($\alpha<0.5$), the characteristic curves and the curves of the vibration period for various values of α are slightly different. This is of practical interest, for in applications the size of the plane central regions of corrugated circular plates is usually small ($\alpha<0.3$), hence the characteristic relation and the formula of the vibration period of corrugated circular plates with full corrugations can be used directly on the design of corrugated circular plates with small plane central regions. Also, only small errors will be made. Noticeably there is a sensitive value of α^*, when $\alpha=\alpha^*$, the corresponding characteristic curve is the flattest one, indicating that under the same load q, the corresponding plate has the largest deflection, or, this type of corrugated circular plate may be applied as the most sensitive elastic element. From Fig. 5, it may be seen that when $\alpha<0.5$, and the central amplitude reaches three times the thickness of the plates, the decrease of nonlinear period is smaller than 15%, therefore it is more convenient to apply linear theory to analyse the nonlinear vibrations of corrugated circular plates with small plane central regions in practical design. Meanwhile, we see that for curves of vibration period there is also a sensitive value of α^*, when $\alpha=\alpha^*$, the nonlinear effect of vibration is the weakest. This value of α^* is determined by using the extremum condition

$$\left.\frac{\mathrm{d}\varepsilon}{\mathrm{d}\alpha}\right|_{\alpha=\alpha^*}=0 \tag{4.2}$$

eventually, for the plates discussed above, we have $\alpha^*=0.3322$.

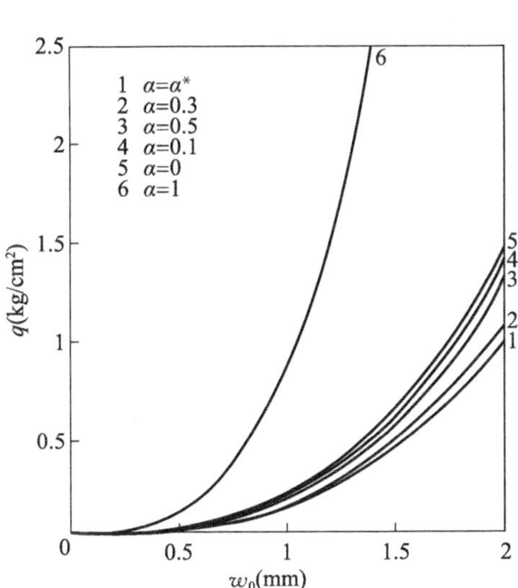

Fig. 4　The effect of the size of plane central regions on the rigidity of corrugated circular plates

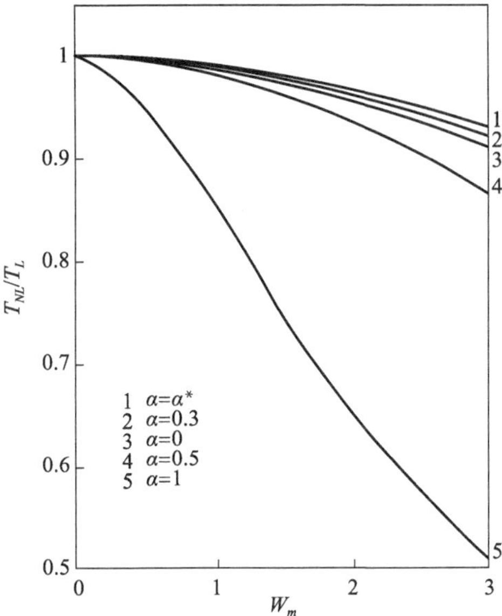

Fig. 5　The effect of the size of plane central regions on the vibration period of corrugated circular plates

Finally, let us discuss the effect of the depth of corrugations on mechanical behaviour of corrugated circular plates. We still study the corrugated plate with toothed corrugations as shown in Fig. 2, as an example, assuming $E=1.2\times10^4\,\text{kg/mm}^2$, $\nu=0.3$, $h=0.1\text{mm}$, $a=25\text{mm}$, $b=7\text{mm}$, $l=4\text{mm}$, and

(1) $H=0.15\text{mm}$, $\theta_0=4°17'$; (2) $H=0.25\text{mm}$, $\theta_0=7°8'$;

(3) $H=0.35\text{mm}$, $\theta_0=9°56'$; (4) $H=0.45\text{mm}$, $\theta_0=12°41'$;

(5) $H=h=0.1\text{mm}$, $\theta_0=0°$.

According to formulae (3.5) and (3.8), through numerical computation, results are presented in Figs. 6 and 7. It is noted that for shallower corrugations the nonlinear effect is stronger; when corrugations become deeper, the nonlinear effect however becomes weaker, especially in the case of vibration of corrugated circular plates.

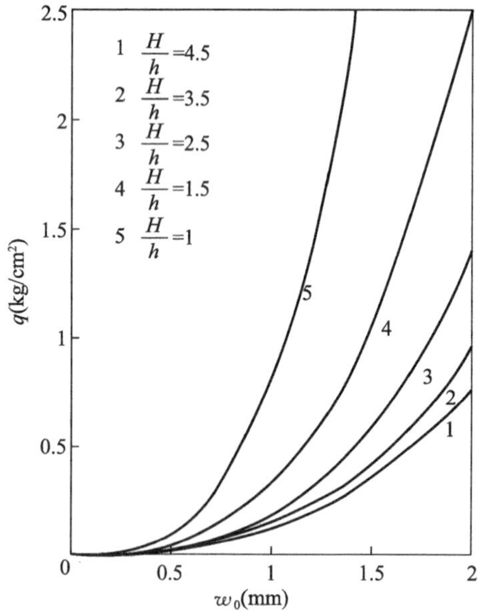

Fig. 6 The effect of the depth of corrugations on the rigidity of corrugated circular plates

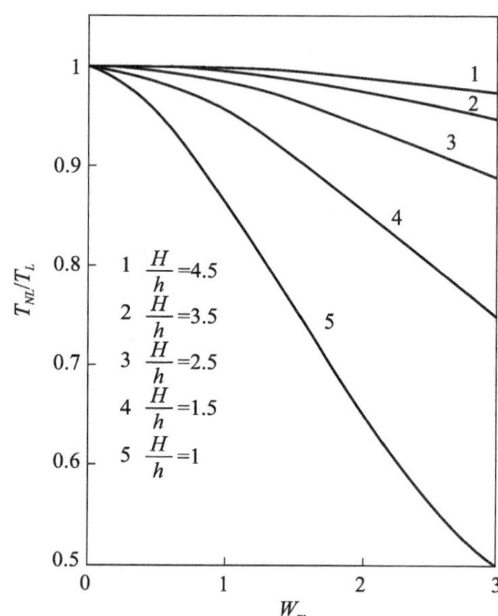

Fig. 7 The effect of the depth of corrugations on the period of vibration of corrugated circular plates

References

[1] D. Y. Panov, On large deflection of circular membranes with weak corrugations. *Prikl. Mat. Mekh.* 5, 308 (1941) (in Russian).

[2] V. I. Feodosev, On large deflection and stability of a circular membrane with fine corrugations. *Prikl. Mat. Mekh.* 9, 389 (1945) (in Russian).

[3] V. I. Feodosev, *Elastic Elements of Precision-Instruments Manufacture*. Oborongiz, Moscow (1949) (in Russian).

[4] E. L. Axelrad, Large deflection of a corrugated membrane as a non-shallow shell. *Izv. AN SSSR, Mekh. Mas.* 1, 46 (1964) (in Russian).

[5] S. L. Chen, Elastic behavior of uniformly loaded circular corrugated plate with sine-shaped shallow waves in large deflection. *Appl. Math. Mech.* 1, 261 (1980).

[6] L. E. Andryewa, The calculation of a corrugated membrane as an anisotropic plate. *Engineer's Collection* 21, 128 (1955) (in Russian).

[7] L. E. Andryewa, *Elastic Elements of Instruments*. Masgiz, Moscow (1962) (in Russian).

[8] T. Akasaka, On the elastic properties of the corrugated diaphragm. *J. Jap. Soc. Aeronaut. Eng.* 3, 279 (1955) (in Japanese).

[9] J. A. Haringx, Nonlinearity of corrugated diaphragms. *Appl. Sei. Res.*, *Ser.* A, 16, 45 (1956).

[10] J. A. Haringx, Design of corrugated diaphragms. *Trans. ASME*, *Ser.* A, 79, 55 (1957).

[11] E. V. Burmistrov. Axismmetrical bending of isotropic and anisotropic shells of revolution under large deflection and non-uniform temperature field. *Engineer's Collection* 27, 185 (1960) (in Russian).

[12] Liu Renhuai, The characteristic relations of corrugated circular plates. *Acta Mech. Sin.* 1, 47 (1978) (in Chinese).

[13] Liu Renhuai, The characteristic relations of corrugated circular plates with plane central regions. *J. China Univ. Sci. Technol.* 9, 75 (1979) (in Chinese).

[14] Liu Renhuai, Large deflection of corrugated circular plate with a plane central region under the action of concentrated loads at the center. *Int. J. Nonlinear Mech.* 19, 409 (1984).

[15] Lin Renhuai, Large deflection of corrugated circular plate with plane boundary region. *Solid Mech. Archs.* 9, 383 (1984).

[16] Liu Renhuai, Nonlinear bending of corrugated annular plates. *Sci. Sin. Ser.* A, 27, 640 (1984).

[17] Liu Renhuai, Nonlinear analysis of corrugated annular plates under compound load. *Sci. Sin. Ser.* A, 28, 959 (1985).

[18] Liu Renhuai, On large deflection of corrugated annular plates under uniform pressure. *Adv. Appl. Math. Mech. China* 1, 138 (1987).

[19] Liu Renhuai, Nonlinear analysis of a corrugated circular plate under combined lateral loading. *Appl. Math. Mech.* 9, 8 (1988).

[20] K. Y. Yeh, Liu Renhuai, SL Li, et al., Nonlinear stability of thin circular shallow spherical shell under actions of axisymmetric uniformly distributed line loads. *J. Lanzhou Univ.* 2, 10; *Bull. Sci.* 2, 142 (1965) (in Chinese).

[21] K. Y. Yeh, Liu Renhuai, C Z Zhang, et al., Nonlinear stability of thin circular shallow spherical shell under the action of uniform edge moment. *Bull. Sci.* 2, 145 (1965) (in Chinese).

[22] Liu Renhuai, Nonlinear stability of circular shallow spherical shell with a hole in the center under the action of uniform moment at the inner edge. *Bull. Sci.* 3, 253 (1965) (in Chinese).

[23] X. Z. Wang, L. X. Wang, X. F. Hu, Nonlinear vibration of circular corrugated plates, *Appl. Math. Mech.* 8, 3 (1987).

[24] R. Courant and D. Hilbert, *Methods of Mathematical Physics*, Vol. 1. New York (1953).

[25] S. G. Lekhnitzky, *Anisotropic Plates*. Gostekhizdat, Moscow (1957) (in Russian).

[26] C. Y. Chia, *Nonlinear Analysis of Plates*, McGraw-Hilll, New York (1980).

U 型波纹管非线性变形的刚度和应力分析理论①

U 型波纹管是一种典型的弹性元件，在仪表工业中有着广泛的应用，其力学性能的研究已有四十多年的历史。三十年前钱伟长等[1,2]给出的轴对称圆环壳的一般解为用解析方法精细地分析圆环壳进行了开创性的工作并奠定了基础。此后，许多中国学者应用这一结果对 U 型波纹管进行了一系列研究，形成了一条具有我国特色的波纹管分析、计算的路子。钱伟长等[3]考虑了连接内、外圆环壳的环板大挠度。徐志翘等[4]分析了变壁厚对刚度的影响。本节将沿着这条思路再作进一步的发展。在以往的工作中，只考虑了环板的非线性变形，忽略了圆环壳的非线性变形。显然是不够完善的。本节不仅考虑了圆环壳的非线性变形，而且计及了压缩角的影响。将原视为环板的部分改作为变厚度截头扁锥壳，从而使计算模型更接近实际的 U 型波纹管。我们从 Reissner[5]的轴对称壳体大挠度问题方程组出发，分别导出圆环壳、变厚度扁锥壳大挠度问题的非线性方程组。选取无量纲半波轴向相对位移为摄动参数，导出各级线性的摄动方程，求出各级摄动方程的通解，其中圆环壳的各级方程可借用钱伟长的轴对称圆环壳一般解。最后，使用连接条件和摄动条件确定通解中的待定系数。所获的解用实验结果进行了验证，令人满意。我们的计算结果可直接供设计部门参考使用。

一、变厚度截头扁锥壳的大挠度解

图 1 是典型的 U 型波纹管的子午线。由于周期性和对称性，我们只需考虑半个波 $ABCD$，其中 AB 和 CD 段分别代表等厚度外、内圆环壳，BC 段代表变厚度截头扁锥壳。φ_B 是压缩角，一般小于 5°，它是联接内、外圆环壳的截头扁锥壳的斜角。我们采用圆柱坐标系（$r\theta\xi$），ξ 轴置于波纹管轴线上。

按照旋转壳大挠度问题的 Reissner 方程[5]，我们可导得变厚度截头扁锥壳大挠度问题的控制方程如下

$$
\begin{aligned}
&\frac{\mathrm{d}}{\mathrm{d}r}\frac{1}{r}\frac{\mathrm{d}}{\mathrm{d}r}(r\beta)+\frac{\mathrm{d}(\ln D)}{\mathrm{d}r}\left(\frac{\mathrm{d}\beta}{\mathrm{d}r}+\nu\frac{\beta}{r}\right)=\frac{1}{Dr}[(\varphi_{\mathrm{B}}+\beta)\psi_{\mathrm{H}}-\psi_{\mathrm{V}}],\\
&\frac{\mathrm{d}}{\mathrm{d}r}\frac{1}{r}\frac{\mathrm{d}}{\mathrm{d}r}(r\psi_{\mathrm{H}})-\frac{\mathrm{d}(\ln C)}{\mathrm{d}r}\left(\frac{\mathrm{d}\psi_{\mathrm{H}}}{\mathrm{d}r}-\nu\frac{\psi_{\mathrm{H}}}{r}\right)=-\frac{C\beta}{r}\left(\varphi_{\mathrm{B}}+\frac{\beta}{2}\right),\quad(r_{\mathrm{C}}\leqslant r\leqslant r_{\mathrm{B}})\qquad(1)
\end{aligned}
$$

其中，β 是壳体中曲面法线转角，ψ_{H} 和 ψ_{V} 分别是水平向和竖直向内力，D 是抗弯刚度，C 是抗拉刚度，h 是壳体厚度，E 是弹性模量，ν 是泊松比，Q 是波纹管均布内压力，N_{ξ} 是径向内力，

$$
\psi_{\mathrm{H}}=rN_{\xi},
$$

$$
\psi_{\mathrm{V}}=\psi_{\mathrm{V}}(r_{\mathrm{C}})-\frac{r^2-r_{\mathrm{C}}^2}{2}Q,
$$

① 本文是全国第九届弹性元件学术会议论文，广州，1992 年 3 月 20 日；压力容器和压力管道的分析与计算，北京：科学出版社，2014，272—287. 作者：刘人怀，胡俍.

$$D=\frac{Eh^3}{12(1-\nu^2)},$$

$$C=Eh. \tag{2}$$

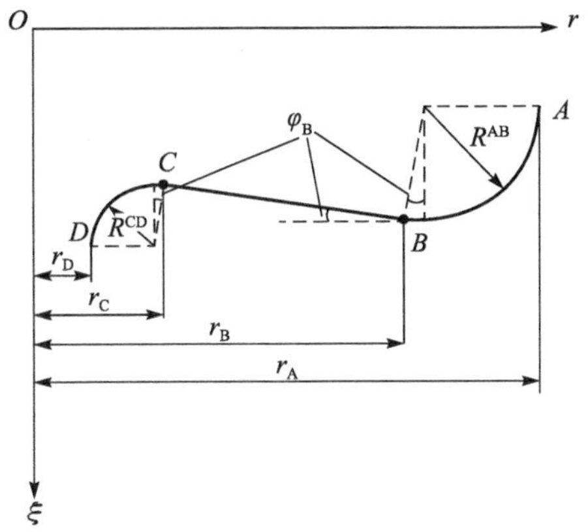

图 1　U 型波纹管的半个波段

相应的径向位移 u、竖向的挠度 w、径向内力 N_ξ、环向内力 N_θ、径向弯矩 M_ξ 和环向弯矩 M_θ 的公式（图 2）为

$$u=\frac{1}{C}\left(r\frac{\mathrm{d}\psi_{\mathrm{H}}}{\mathrm{d}r}-\nu\psi_{\mathrm{H}}\right),$$

$$\frac{\mathrm{d}w}{\mathrm{d}r}=\beta-\frac{\nu}{C}(\varphi_{\mathrm{B}}+\beta)\frac{\mathrm{d}\psi_{\mathrm{H}}}{\mathrm{d}r}+\frac{\varphi_{\mathrm{B}}+\beta}{Cr}\psi_{\mathrm{H}},$$

$$N_\xi=\frac{\psi_{\mathrm{H}}}{r},$$

$$N_\theta=\frac{\mathrm{d}\psi_{\mathrm{H}}}{\mathrm{d}r},$$

$$M_\xi=-D\left(\frac{\mathrm{d}\beta}{\mathrm{d}r}+\nu\frac{\beta}{r}\right),$$

$$M_\theta=-D\left(\nu\frac{\mathrm{d}\beta}{\mathrm{d}r}+\frac{\beta}{r}\right). \tag{3}$$

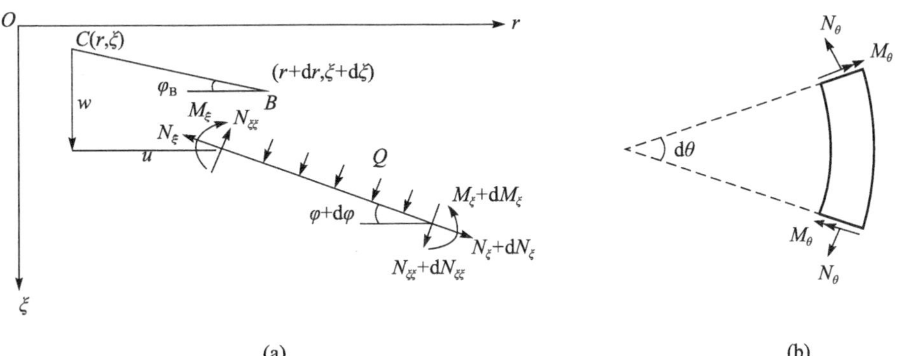

图 2　截头扁锥壳

根据 Андреева[6] 的研究，对于波纹不太深的液压成形波纹管，可设其厚度变化为

$$h=\frac{h_c r_c}{r} \tag{4}$$

徐志翘等[4] 的研究表明，上述壁厚变化与实测结果一致。我们也将实测结果与此公式比较，结果也基本一致。

引入下列无量纲量：

$$\begin{aligned}
&\rho=\frac{r}{a}-\rho_1,\alpha=\frac{a\sqrt{12(1-\nu^2)}}{h_c(\rho_1-1)}\varphi_B,\bar{\beta}=\frac{a\sqrt{12(1-\nu^2)}}{h_c(\rho_1-1)}\beta,\bar{\psi}_H=\frac{a}{D_c(\rho_1-1)^3}\psi_H,\\
&\overline{P}=-\frac{a^2\sqrt{12-(1-\nu^2)}}{D_c h_c(\rho_1-1)^4[\omega_P^{BC}+\omega_Q^{BC}(\rho^2+2\rho_1\rho+2\rho_1-1)]}\psi_V,a=\frac{1}{2}(r_B-r_C),\\
&b=\frac{1}{2}(r_B+r_C),\rho_1=\frac{b}{a},D_c=\frac{Eh_c^3}{12(1-\nu^2)},
\end{aligned} \tag{5}$$

其中，$\overline{P}$ 为无量纲载荷，由无量纲参数 ω_P^{BC} 和 ω_Q^{BC} 控制。

（1）仅有轴向集中力 P 作用，波纹管处于拉伸变形状态：

$$\omega_P^{BC}=1,\omega_Q^{BC}=0,P=\frac{2\pi D_c h_c(\rho_1-1)^4}{a^2\sqrt{12(1-\nu^2)}}\overline{P} \tag{6}$$

（2）仅有均布内压力 Q 作用，波纹管处于膨胀状态：

$$\omega_P^{BC}=\left(\frac{2r_c}{r_B-r_C}\right)^2,\omega_Q^{BC}=1,Q=\frac{2D_c h_c(\rho_1-1)^4}{a^4\sqrt{12(1-\nu^2)}}\overline{P} \tag{7}$$

应用这些无量纲量，方程组（1）转化为下面无量纲形式：

$$\begin{aligned}
&L_1(\bar{\beta})-\alpha(\rho+\rho_1)^2\bar{\psi}_H=(\rho+\rho_1)^2\bar{\beta}\bar{\psi}_H+(\rho+\rho_1)^2[\omega_P^{BC}\\
&\qquad+(\rho^2+2\rho_1\rho+2\rho_1-1)\omega_Q^{BC}]\overline{P},\\
&L_2(\bar{\psi}_H)+\frac{\alpha}{(\rho+\rho_1)^2}\bar{\beta}=-\frac{1}{2(\rho+\rho_1)^2}\bar{\beta}^2,\quad(-1\leqslant\rho\leqslant 1)
\end{aligned} \tag{8}$$

其中

$$\begin{aligned}
&L_1(\cdots)=\frac{d^2}{d\rho^2}(\cdots)-\frac{2}{\rho+\rho_1}\frac{d}{d\rho}(\cdots)-\frac{1+\nu}{(\rho+\rho_1)^2}(\cdots),\\
&L_2(\cdots)=\frac{d^2}{d\rho^2}(\cdots)+\frac{2}{\rho+\rho_1}\frac{d}{d\rho}(\cdots)-\frac{1+\nu}{(\rho+\rho_1)^2}(\cdots)
\end{aligned} \tag{9}$$

公式（3）成为下面形式：

$$\begin{aligned}
&u=\frac{h_c^2(\rho_1-1)^2}{12a(1-\nu^2)}\left[(\rho+\rho_1)^2\frac{d\bar{\psi}_H}{d\rho}-\nu(\rho+\rho_1)\bar{\psi}_H\right],\\
&\omega=\int_{-1}^{\rho}\left\{\frac{h_c(\rho_1-1)}{\sqrt{12(1-\nu^2)}}\bar{\beta}-\nu a\left[\frac{h_c(\rho_1-1)}{a\sqrt{12(1-\nu^2)}}\right]^3(\rho+\rho_1)(\alpha+\bar{\beta})\right.\\
&\qquad\left.\times\frac{d\bar{\psi}_H}{d\rho}+a\left[\frac{h_c(\rho_1-1)}{a\sqrt{12(1-v^2)}}\right]^3(\alpha+\bar{\beta})\bar{\psi}_H\right\}d\rho,\\
&N_\xi=\frac{D_c(\rho_1-1)^3}{a^2(\rho+\rho_1)}\bar{\psi}_H,
\end{aligned}$$

$$N_\theta = \frac{D_c(\rho_1-1)^2}{a^2}\frac{d\overline{\psi}_H}{d\rho},$$

$$M_\xi = -\frac{D_c h_c(\rho_1-1)^4}{a^2\sqrt{12(1-\nu^2)}(\rho+\rho_1)^3}\left(\frac{d\overline{\beta}}{d\rho}+\frac{\nu}{\rho+\rho_1}\overline{\beta}\right),$$

$$M_\theta = -\frac{D_c h_c(\rho_1-1)^4}{a^2\sqrt{12(1-\nu^2)}(\rho+\rho_1)^3}\left(\nu\frac{d\overline{\beta}}{d\rho}+\frac{1}{\rho+\rho_1}\overline{\beta}\right) \tag{10}$$

方程组（8）中出现了非线性项 $\overline{\beta}$、$\overline{\psi}_H$ 和 $\overline{\beta}^2$，这给求解带来了困难，我们使用摄动法克服了求解困难。现在取无量纲小参数 ε 作为摄动参数，并将下面几个物理量表示成渐近展开式：

$$\overline{\beta}(\rho,\varepsilon) = \sum_{i=1}^{\infty}\overline{\beta}_i^{BC}(\rho)\varepsilon^i,$$

$$\overline{\psi}_H(\rho,\varepsilon) = \sum_{i=1}^{\infty}\psi_i^{BC}(\rho)\varepsilon^i,$$

$$\overline{P}(\varepsilon) = \sum_{i=1}^{\infty}p_i\varepsilon^i \tag{11}$$

其中，p_i 是待定常数。

将式（11）代入方程组（8），比较 ε 的同次幂项，便得各级摄动方程：

$$L_1(\beta_i^{BC}) - \alpha(\rho+\rho_1)^2\psi_i^{BC} = (\rho+\rho_1)^2\sum_{k=1}^{i-1}\beta_k^{BC}\psi_{i-k}^{BC} + (\rho+\rho_1)^2 \times[\omega_P^{BC} + (\rho^2+2\rho_1\rho+2\rho_1-1)\omega_Q^{BC}]p_i,$$

$$L_2(\psi_i^{BC}) + \frac{\alpha}{(\rho+\rho_1)^2}\beta_i^{BC} = -\frac{1}{2(\rho+\rho_1)^2}\sum_{k=1}^{i-1}\beta_k^{BC}\beta_{i-k}^{BC}.$$

$$(i=1,2,3,\cdots,\quad -1\leqslant\rho\leqslant 1) \tag{12}$$

其中，当求和号 $\sum$ 的终值小于初值时，就不再求和。

显而易见，问题便归结为求解如下形式的变系数二阶常微分方程组的通解：

$$L_1(f) - \alpha(\rho+\rho_1)^2 g = F(\rho),$$

$$L_2(g) + \frac{\alpha}{(\rho+\rho_1)^2}f = G(\rho)$$

$$(-1\leqslant p<1,\quad F(\rho),G(\rho)\in C^1[-1,1]) \tag{13}$$

上面方程组的系数和右端项均可展开成收敛半径为 ρ_1 的幂级数，可记为

$$F(\rho) = \rho_*\sum_{i=0}^{\infty}F_i\rho^i,$$

$$G(\rho) = \rho_*\sum_{i=0}^{\infty}G_i\rho^i \tag{14}$$

按照常微分方程理论，方程组（13）的解也可表示成收敛半径为 ρ_1 的如下形式的幂级数：

$$f(\rho) = \sum_{m=0}^{\infty}[\overline{A}a_m(1,0,0,0,0,0) + \overline{B}a_m(0,1,0,0,0,0)$$

$$
\begin{aligned}
&+\overline{C}a_m(0,0,1,0,0,0)+\overline{D}a_m(0,0,0,1,0,0)\\
&+p_*a_m(0,0,0,0,F_i,G_i)]\rho^m,
\end{aligned}
$$

$$
\begin{aligned}
g(\rho)=\sum_{m=0}^{\infty}&[\overline{A}c_m(1,0,0,0,0,0)+\overline{B}c_m(0,1,0,0,0,0)\\
&+\overline{C}c_m(0,0,1,0,0,0)+\overline{D}c_m(0,0,0,1,0,0)\\
&+p_*c_m(0,0,0,0,F_i,G_i)]\rho^m \qquad (15)
\end{aligned}
$$

这两个式子中的前四项是齐次通解，第五项是非齐次特解，$\overline{A}$、$\overline{B}$、$\overline{C}$、$\overline{D}$ 是待定常数，而系数 a_m 和 c_m 则可由下列递推公式确定：

$$
\begin{aligned}
&a_0(a_0,a_1,c_0,c_1,F_i,G_i)=a_0,\\
&a_1(a_0,a_1,c_0,c_1,F_i,G_i)=a_1,\\
&a_2(a_0,a_1,c_0,c_1,F_i,G_i)=\frac{1+3\nu}{2}\rho_1^{-2}a_0+\rho_1^{-1}a_1+\frac{\alpha}{2}\rho_1^2c_0+\frac{1}{2}F_0,\\
&a_3(a_0,a_1,c_0,c_1,F_i,G_i)=-\frac{1+3\nu}{3}\rho_1^{-3}a_0-\frac{1-3\nu}{6}\rho_1^{-2}a_1+\frac{2}{3}\rho_1^{-1}a_2(a_0,a_1,c_0,c_1,F_i,G_i)\\
&\qquad+\frac{\alpha}{3}\rho_1c_0+\frac{\alpha}{6}\rho_1^2c_1+\frac{1}{6}F_1,
\end{aligned}
$$

$$
\begin{aligned}
a_{m+2}(a_0,a_1,c_0,c_1,F_i,G_i)=&\frac{1}{(m+1)(m+2)}\sum_{n=0}^{m}(-1)^{m-n}\rho_1^{-(m-n+2)}[m-3n+1\\
&+3(m-n+1)\nu]a_n(a_0,a_1,c_0,c_1,F_i,G_i)\\
&+\frac{2}{m+2}\rho_1^{-1}a_{m+1}(a_0,a_1,c_0,c_1,F_i,G_i)\\
&+\frac{\alpha}{(m+1)(m+2)}[c_{m-2}(a_0,a_1,c_0,c_1,F_i,G_i)\\
&+2\rho_1c_{m-1}(a_0,a_1,c_0,c_1,F_i,G_i)\\
&+\rho_1^2C_m(a_0,a_1,c_0,c_1,F_i,G_i)]\\
&+\frac{1}{(m+1)(m+2)}F_m,\quad (m\geqslant 2),
\end{aligned}
$$

$$
\begin{aligned}
&c_0(a_0,a_1,c_0,c_1,F_i,G_i)=c_0,\\
&c_1(a_0,a_1,c_0,c_1,F_i,G_i)=c_1,
\end{aligned}
$$

$$
\begin{aligned}
c_{m+2}(a_0,a_1,c_0,c_1,F_i,G_i)=&\frac{1}{(m+1)(m+2)}\sum_{n=0}^{m}(-1)^{m-n}\rho_1^{-(m-n+2)}[m+n+1\\
&+(m-n+1)\nu]c_n(a_0,a_1,c_0,c_1,F_i,G_i)\\
&-\frac{2}{m+2}\rho_1^{-1}c_{m+1}(a_0,a_1,c_0,c_1,F_i,G_i)\\
&-\frac{\alpha}{(m+1)(m+2)}\sum_{n=0}^{m}(-1)^{m-n}\rho_1^{-(m-n+2)}(m-n+1)\\
&\times a_n(a_0,a_1,c_0,c_1,F_i,G_i)+\frac{1}{(m+1)(m+2)}G_m,\quad (m\geqslant 0)
\end{aligned}
\qquad (16)
$$

在实际计算时，对式（15）总是截取前 M 项，即 $0\leqslant m\leqslant M$。由于 $\rho_1>1$，所以上

述解在 $|\rho|\leqslant 1$ 范围内是有效的。

将各级摄动方程组的右端项展开成 ρ 的幂级数，利用上述结果易得各级摄动方程组的通解 $\beta_i^{BC}=(\rho,\ \overline{A}_i,\ \overline{B}_i,\ \overline{C}_i,\ \overline{D}_i,\ p_i)$，$\psi_i^{BC}(\rho,\ \overline{A}_i,\ \overline{B}_i,\ \overline{C}_i,\ \overline{D}_i,\ p_i)$ $(i=1,\ 2,\ 3,\ \cdots)$。然后由式（10）得

$$
\begin{aligned}
u(\rho) &= \sum_{i=1}^{\infty} u_i^{BC}(\rho,\overline{A}_i,\overline{B}_i,\overline{C}_i,\overline{D}_i,p_i)\varepsilon^i,\\
w(\rho) &= \sum_{i=1}^{\infty} w_i^{BC}(\rho,\overline{A}_i,\overline{B}_i,\overline{C}_i,\overline{D}_i,p_i)\varepsilon^i,\\
M_\xi(\rho) &= \sum_{i=1}^{\infty} M_{\xi i}^{BC}(\rho,\overline{A}_i,\overline{B}_i,\overline{C}_i,\overline{D}_i,p_i)\varepsilon^i
\end{aligned}
\tag{17}
$$

其中，$\overline{A}_i$，$\overline{B}_i$，$\overline{C}_i$，$\overline{D}_i$，p_i 由连接条件和摄动条件确定。

二、等厚度圆环壳的大挠度解

与前面一样，再按照旋转壳大挠度问题的 Reissner 方程[5]，我们又可得到如下的等厚度圆环壳大挠度问题的控制方程

$$
\begin{aligned}
&\frac{1+\lambda\sin\varphi}{\sin\varphi}\frac{d^2\beta}{d\varphi^2}+\lambda\cot\varphi\frac{d\beta}{d\varphi}-\frac{\lambda^2\cos^2\varphi}{(1+\lambda\sin\varphi)\sin\varphi}\beta-\nu\lambda\beta-\frac{R\lambda}{D}\psi_H\\
&=-\left[\frac{3\lambda^2\cos\varphi}{2(1+\lambda\sin\varphi)}-\frac{\nu\lambda}{2}\cot\varphi\right]\beta^2+\frac{R\lambda}{D}\cot\varphi\beta\psi_H-\frac{R\lambda}{D}\cot\varphi\psi_V,\\
&\frac{1+\lambda\sin\varphi}{\sin\varphi}\frac{d^2\psi_H}{d\varphi^2}+\lambda\cot\varphi\frac{d\psi_H}{d\varphi}-\frac{\lambda^2\cos^2\varphi}{(1+\lambda\sin\varphi)\sin\varphi}\psi_H+\nu\lambda\psi_H+CR\lambda\beta\\
&=-\left(\frac{2\lambda^2\cos\varphi}{1+\lambda\sin\varphi}+\nu\lambda\cot\varphi\right)\beta\psi_H-\nu\lambda\frac{dB}{dr}\psi_H-\frac{CR\lambda}{2}\cot\varphi\beta^2\\
&\quad+\left(\frac{\lambda^2\cos\varphi}{1+\lambda\sin\varphi}+\nu\lambda\cot\varphi\right)\psi_V-[2R^2\cos\varphi(1+\lambda\sin\varphi)+r_0R(1+\lambda\sin\varphi)^2\cot\varphi]Q,\\
&\qquad\qquad(\varphi_0\leqslant\varphi\leqslant\varphi_1)
\end{aligned}
\tag{18}
$$

其中，内力与位移的定义和正值方向见图 3，

$$
\begin{aligned}
\psi_H &= r_0(1+\lambda\sin\varphi)(\cos\varphi N_\xi-\sin\varphi N_{\xi\xi}-\sin\varphi\beta N_\xi-\cos\varphi\beta N_{\xi\xi}),\\
\psi_V &= r_0(1+\lambda\sin\varphi)(\sin\varphi N_\xi+\cos\varphi N_{\xi\xi}+\cos\varphi\beta N_\xi-\sin\varphi\beta N_{\xi\xi})\\
&= \psi_V(\varphi_0)+r_0R\left[\sin\varphi\left(1+\frac{\lambda}{2}\sin\varphi\right)-\sin\varphi_0\left(1+\frac{\lambda}{2}\sin\varphi_0\right)\right]Q,\\
\lambda &= \frac{R}{r_0}
\end{aligned}
\tag{19}
$$

位移和内力的公式为

$$
\begin{aligned}
u &= \frac{1}{C}\left[\frac{1+\lambda\sin\varphi}{\lambda}\frac{d\psi_H}{d\varphi}-\nu(\cos\varphi-\sin\varphi\beta)\psi_H-\nu\sin\varphi\psi_V+r_0^2(1+\lambda\sin\varphi)^2\sin\varphi Q\right],\\
\frac{dw}{d\varphi} &= \frac{1}{C}\left\{-\nu(\sin\varphi+\cos\varphi\beta)\frac{d\psi_H}{d\varphi}+\frac{\lambda}{1+\lambda\sin\varphi}[\sin\varphi\cos\varphi+(\cos^2\varphi-\sin^2\varphi)\beta]\psi_H\right.\\
&\quad\left.+\frac{\lambda\sin^2\varphi}{1+\lambda\sin\varphi}\psi_V-\nu r_0R(1+\lambda\sin\varphi)\sin^2\varphi Q\right\}-\frac{1}{2}R\sin\varphi\beta^2+R\cos\varphi\beta,
\end{aligned}
$$

$$
\begin{aligned}
N_\xi &= \frac{1}{r_0(1+\lambda\sin\varphi)}[(\cos\varphi-\sin\varphi\beta)\psi_H+\sin\varphi\psi_V],\\
N_\theta &= \frac{1}{R}\frac{d\psi_H}{d\varphi}+r_0\sin\varphi(1+\lambda\sin\varphi)Q,\\
M_\xi &= -\frac{D}{R}\left[\frac{d\beta}{d\varphi}-\frac{\nu\lambda\sin\varphi}{2(1+\lambda\sin\varphi)}\beta^2+\frac{\nu\lambda\cos\varphi}{1+\lambda\sin\varphi}\beta\right],\\
M_\theta &= -\frac{D}{R}\left[\nu\frac{d\beta}{d\varphi}-\frac{\lambda\sin\varphi}{2(1+\lambda\sin\varphi)}\beta^2+\frac{\lambda\cos\varphi}{1+\lambda\sin\varphi}\beta\right].
\end{aligned}
\tag{20}
$$

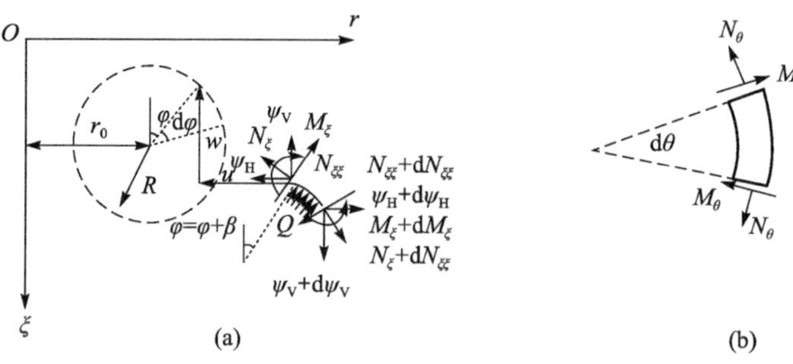

图 3　圆环壳

引入下列无量纲量：

$$
2\mu=\frac{R\lambda}{r}\sqrt{12(1-\nu^2)},\tilde{\beta}=\beta,\tilde{\psi}_H=\frac{\sqrt{12(1-\nu^2)}}{Eh^2}\psi_H,
$$

$$
\overline{P}=\frac{\sqrt{12(1-\nu^2)}}{Eh^2\left\{\omega_P+\omega_Q\left[\sin\varphi\left(1+\frac{1}{2}\lambda\sin\varphi\right)-\sin\varphi_0\left(1+\frac{1}{2}\lambda\sin\varphi_0\right)\right]\right\}}\psi_V \tag{21}
$$

其中，$\overline{P}$ 即是前面的无量纲载荷，由无量纲参数 ω_P 和 ω_Q 控制。

（1）仅有轴向集中为力 P 作用，波纹管处于拉伸变形状态.

$$
\omega_P^{AB}=\frac{[h_c(\rho_1-1)]^4}{12h_A^2a^2(1-\nu^2)},\omega_Q^{AB}=0; \tag{22}
$$

$$
\omega_P^{CD}=-\frac{h_c^2(\rho_1-1)^4}{12a^2(1-\nu^2)},\omega_Q^{CD}=0. \tag{23}
$$

（2）仅有均匀内压力 Q 作用，波纹管处于膨胀状态：

$$
\omega_P^{AB}=\frac{r_A^2h_c^4(\rho_1-1)^4}{12h_A^2a^4(1-\nu^2)},\omega_Q^{AB}=\frac{2R_{AB}h_c^4(r_A-R_{AB})(\rho_1-1)^4}{12h_A^2a^4(1-\nu^2)}; \tag{24}
$$

$$
\omega_P^{CD}=-\frac{r_D^2h_c^2(\rho_1-1)^4}{12a^4(1-\nu^2)},\omega_Q^{CD}=-\frac{2R_{CD}h_c^2(r_D-R_{CD})(\rho_1-1)^4}{12a^4(1-\nu^2)}. \tag{25}
$$

于是，方程（18）化为如下无量纲形式

$$
L_3(\tilde{\beta})-\lambda\nu\tilde{\beta}-2\mu\tilde{\psi}_H=-\frac{\lambda}{2}\left(\frac{3\lambda\cos\varphi}{1+\lambda\sin\varphi}-\nu\cot\varphi\right)\tilde{\beta}^2+2\mu\cot\varphi\tilde{\beta}\tilde{\psi}_H
$$

$$
-2\mu\cot\varphi\left\{\omega_P-\left[\sin\varphi_0\left(1+\frac{1}{2}\lambda\sin\varphi_0\right)\right.\right.
$$

$$- \sin\varphi\left(1+\frac{1}{2}\lambda\sin\varphi\right)\Big]\omega_{Q}\Big\}\overline{P},$$

$$\begin{aligned} L_3(\tilde{\psi}_H)+\lambda\nu\tilde{\psi}_H+2\mu\tilde{\beta} = & -\lambda\left(\frac{2\lambda\cos\varphi}{1+\lambda\sin\varphi}+\nu\cot\varphi\right)\tilde{\beta}\tilde{\psi}_H-\nu\lambda\frac{d\tilde{\beta}}{d\varphi}\tilde{\psi}_H \\ & -\mu\cot\varphi\tilde{\beta}^2+\left\{\lambda\left(\frac{\lambda\cos\varphi}{1+\lambda\sin\varphi}+\nu\cot\varphi\right)\omega_P\right. \\ & +\left\{\lambda\left(\frac{\lambda\cot\varphi}{1+\lambda\sin\varphi}+\nu\cot\varphi\right)\left[\sin\varphi\left(1+\frac{1}{2}\lambda\sin\varphi\right)\right.\right. \\ & \left.-\sin\varphi_0\left(1+\frac{1}{2}\lambda\sin\varphi_0\right)\right]-2\lambda\cos\varphi(1+\lambda\sin\varphi) \\ & \left.\left.-\cot\varphi(1+\lambda\sin\varphi)^2\right\}\omega_Q\right\}\overline{P} \end{aligned} \tag{26}$$

其中

$$L_3(\cdots)=\frac{1+\lambda\sin\varphi}{\sin\varphi}\frac{d^2}{d\varphi^2}(\cdots)+\lambda\cot\varphi\frac{d}{d\varphi}(\cdots)-\frac{\lambda^2\cos^2\varphi}{\sin\varphi(1+\lambda\sin\varphi)}(\cdots) \tag{27}$$

位移和内力成为

$$\begin{aligned} u = & \frac{h}{\sqrt{12(1-\nu^2)}}\left\{\nu\sin\varphi\tilde{\beta}\tilde{\psi}_H+\frac{1+\lambda\sin\varphi}{\lambda}\frac{d\tilde{\psi}_H}{d\varphi}-\nu\cos\varphi\tilde{\psi}_H\right. \\ & -\left\{\nu\omega_P\sin\varphi-\left[\nu\sin\varphi\left(\sin\varphi_0\left(1+\frac{1}{2}\lambda\sin\varphi_0\right)\right.\right.\right. \\ & \left.\left.\left.\left.-\sin\varphi\left(1+\frac{1}{2}\lambda\sin\varphi\right)\right)+\frac{\sin\varphi}{\lambda}(1+\lambda\sin\varphi)^2\right]\omega_Q\right\}\overline{P}\right\}, \end{aligned}$$

$$\begin{aligned} w = & \frac{h}{\sqrt{12(1-\nu^2)}}\int_{\varphi_0}^{\varphi}\left\{-\nu\cos\varphi\tilde{\beta}\frac{d\tilde{\psi}_H}{d\varphi}+\frac{\lambda(\cos^2\varphi-\sin^2\varphi)}{1+\lambda\sin\varphi}\tilde{\beta}\tilde{\psi}_H\right. \\ & -\frac{R\sqrt{12(1-\nu^2)}}{2h}\sin\varphi\tilde{\beta}^2+\frac{R\cos\varphi\sqrt{12(1-\nu^2)}}{h}\tilde{\beta} \\ & -\nu\sin\varphi\frac{d\tilde{\psi}_H}{d\varphi}+\frac{\lambda\sin\varphi\cos\varphi}{1+\lambda\sin\varphi}\tilde{\psi}_H \\ & +\left\{\frac{\lambda\sin^2\varphi}{1+\lambda\sin\varphi}\left[\omega_P+\left(\sin\varphi\left(1+\frac{1}{2}\lambda\sin\varphi\right)\right.\right.\right. \\ & \left.\left.-\sin\varphi_0\left(1+\frac{1}{2}\lambda\sin\varphi_0\right)\right)\omega_Q\right] \\ & \left.\left.-\nu\sin^2\varphi(1+\lambda\sin\varphi)\omega_Q\right\}\overline{P}\right\}d\varphi, \end{aligned}$$

$$\begin{aligned} N_\xi = & \frac{Eh^2}{r_0(1+\lambda\sin\varphi)\sqrt{12(1-\nu^2)}}\left\{(\cos\varphi-\sin\varphi\tilde{\beta})\tilde{\psi}_H+\sin\varphi\left\{\omega_P\right.\right. \\ & \left.\left.+\left[\sin\varphi\left(1+\frac{1}{2}\lambda\sin\varphi\right)-\sin\varphi_0\left(1+\frac{1}{2}\lambda\sin\varphi_0\right)\right]\omega_Q\right\}\overline{P}\right\}, \end{aligned}$$

$$N_\theta=\frac{Eh^2}{R\sqrt{12(1-\nu^2)}}\left[\frac{d\tilde{\psi}_H}{d\varphi}+\sin\varphi(1+\lambda\sin\varphi)\omega_Q\overline{P}\right],$$

$$M_\xi = \frac{D}{R}\left[\frac{\nu\lambda\sin\varphi}{2(1+\lambda\sin\varphi)}\tilde{\beta}^2 - \frac{\mathrm{d}\tilde{\beta}}{\mathrm{d}\varphi} - \frac{\nu\lambda\cos\varphi}{1+\lambda\sin\varphi}\tilde{\beta}\right],$$

$$M_\theta = \frac{D}{R}\left[\frac{\lambda\sin\varphi}{2(1+\lambda\sin\varphi)}\tilde{\beta}^2 - \nu\frac{\mathrm{d}\tilde{\beta}}{\mathrm{d}\varphi} - \frac{\lambda\cos\varphi}{1+\lambda\sin\varphi}\tilde{\beta}\right] \tag{28}$$

现在，仍用摄动法求解非线性微分方程组（26）。设

$$\bar{\beta}(\varphi,\varepsilon) = \sum_{i=1}^{\infty}\beta_i(\varphi)\varepsilon^i,$$

$$\bar{\psi}_{\mathrm{H}}(\varphi,\varepsilon) = \sum_{i=1}^{\infty}\psi_i(\varphi)\varepsilon^i \tag{29}$$

将式（29）代入方程组（26），比较 ε 的同次幂项，便得各级摄动方程

$$\begin{aligned}
&L_3(\beta_i) - \nu\lambda\beta_i - 2\mu\psi_i \\
&= -\frac{\lambda}{2}\left(\frac{3\lambda\cos\varphi}{1+\lambda\sin\varphi} - \nu\cot\varphi\right)\sum_{k=1}^{i-1}\beta_k\beta_{i-k} + 2\mu\cot\varphi\sum_{k=1}^{i-1}\beta_k\beta_{i-k} - 2\mu\cot\varphi\Big\{\omega_{\mathrm{P}} \\
&\quad - \left[\sin\varphi_0\left(1+\frac{1}{2}\lambda\sin\varphi_0\right) - \sin\varphi\left(1+\frac{1}{2}\lambda\sin\varphi\right)\right]\omega_{\mathrm{Q}}\Big\}p_i, \\
&L_3(\psi_i) + \nu\lambda\psi_i + 2\mu\beta_i \\
&= -\lambda\left(\frac{2\lambda\cos\varphi}{1+\lambda\sin\varphi} + \nu\cot\varphi\right)\sum_{k=1}^{i-1}\beta_k\psi_{i-k} - \nu\lambda\sum_{k=1}^{i-1}\frac{\mathrm{d}\beta_k}{\mathrm{d}\varphi}\psi_{i-k} - \mu\cot\varphi\sum_{k=1}^{i-1}\beta_k\beta_{i-k} \\
&\quad + \Big\{\lambda\left(\frac{\lambda\cos\varphi}{1+\lambda\sin\varphi} + \nu\cot\varphi\right)\omega_{\mathrm{P}} + \Big\{\lambda\left(\frac{\lambda\cos\varphi}{1+\lambda\sin\varphi} + \nu\cot\varphi\right) \\
&\quad \times\Big[\sin\varphi\left(1+\frac{1}{2}\lambda\sin\varphi\right) \\
&\quad - \sin\varphi_0\left(1+\frac{1}{2}\lambda\sin\varphi_0\right)\Big] - 2\lambda\cos\varphi(1+\lambda\sin\varphi) - \cot\varphi(1+\lambda\sin\varphi)^2\Big\}\omega_{\mathrm{Q}}\Big\}p_i, \\
&\qquad (i = 1,2,\cdots,\quad \varphi_0 \leqslant \varphi \leqslant \varphi_1)
\end{aligned} \tag{30}$$

在 Love-Kirchhoff 薄壳假设的精度范围内，与 1 相比忽略 $O(\lambda/2\mu)$[1]，从而记

$$\begin{aligned}
x_i(\varphi) &= (1+\lambda\sin\varphi)\left\{\beta_i + \left[i\sqrt{1+\left(\frac{\nu\lambda}{2\mu}^2\right)} + \frac{\nu\lambda}{2\mu}\right]\psi_i\right\} \\
&\approx (1+\lambda\sin\varphi)(\beta_i + i\psi_i) \qquad (i = 1,2,\cdots)
\end{aligned} \tag{31}$$

再将方程（30）化成关于复变量 $x_i(\varphi)$ 的方程

$$\begin{aligned}
L_4(x_i) = &(1+\lambda\sin\varphi)\Big\{2\mu\cos\varphi\sum_{k=1}^{i-1}\beta_k\psi_{i-k} - 2\mu\cos\varphi\Big\{\omega_{\mathrm{P}} - \Big[\sin\varphi_0\left(1+\frac{1}{2}\lambda\sin\varphi_0\right) \\
&- \sin\varphi\left(1+\frac{1}{2}\lambda\sin\varphi\right)\Big]\omega_{\mathrm{Q}}\Big\}p_i - i\mu\cos\varphi\sum_{k=1}^{i-1}\beta_k\beta_{i-k} - i\nu\lambda\sin\varphi\sum_{k=1}^{i-1}\frac{\mathrm{d}\beta_k}{\mathrm{d}\varphi}\psi_{i-k} \\
&- i\cos\varphi\omega_{\mathrm{Q}}p_i\Big\}, \qquad (i = 1,2\cdots,\varphi_0 \leqslant \varphi \leqslant \varphi_1)
\end{aligned} \tag{32}$$

这里，当求和号 $\sum$ 的终值小于初值时，就不再求和，

$$L_4(\cdots) = (1+\lambda\sin\varphi)\frac{\mathrm{d}^2}{\mathrm{d}\varphi^2}(\cdots) - \lambda\cos\varphi\frac{\mathrm{d}}{\mathrm{d}\varphi}(\cdots) + i2\mu\sin\varphi(\cdots) \tag{33}$$

至此，问题归结为求下面变系数二阶复常微分方程的通解

$$L_4[h(\varphi)] = H(\varphi), \tag{34}$$

其中，$H(\varphi)$ 是解析函数。

我们注意到，尽管引入的复变量不同，但上述方程与 Новожинов[7] 的圆环壳方程相比，只是右端项不同。由 $H(\varphi)$ 的连续性，我们可在区间 $[-\pi, \pi]$ 上将它展开成 Fourier 级数，这种级数表达式在区间 $[\varphi_0, \varphi_1]$ 上显然收敛到 $H(\varphi)$。同时，若圆环壳的几何形状和外载荷关于 $\varphi=\frac{\pi}{2}$ 平面是上下对称的，则 $H(\varphi)$ 的 Fourier 级数展开式为

$$H(\varphi) = q_* \sum_{n=1}^{\infty} H_n \sin\left[n\left(\frac{\pi}{2}-\varphi\right)\right] \tag{35}$$

钱伟长等[2] 曾给出了圆环壳 Новожинов 方程的一般解，陈山林[8] 又将其推广到方程右端项是 Fourier 级数的情形，借用此解，方程 (34) 的通解为

$$\begin{aligned} h(\varphi) =& \widetilde{B}_1 e^{(\gamma+i\delta)\varphi}[f_a(\varphi)+if_b(\varphi)] + \widetilde{B}_2 e^{-(\gamma+i\delta)\varphi}[f_a(\pi-\varphi)+if_b(\pi-\varphi)] \\ &+ q_* \sum_{n=1}^{\infty} d_n \sin\left[n\left(\frac{\pi}{2}-\varphi\right)\right], \end{aligned} \tag{36}$$

其中，$\widetilde{B}_1$、$\widetilde{B}_2$ 是复待定常数，前面两项是齐次方程的通解，第三项是非齐次方程的特解，而 γ、δ、b_n、b_{-n} 和 d_n 由下面递推关系式 (38) 确定：

$$\begin{aligned} f_a(\varphi) &= 1 - \sum_{n=1}^{\infty}[\mathrm{Re}(b_n+b_{-n})\cos(n\varphi) - \mathrm{Im}(b_n-b_{-n})\sin(n\varphi)], \\ f_b(\varphi) &= -\sum_{n=1}^{\infty}[\mathrm{Im}(b_n+b_{-n})\cos(n\varphi) + \mathrm{Re}(b_n-b_{-n})\sin(n\varphi)]; \end{aligned} \tag{37}$$

$$\begin{aligned} & b_0 = 1, \\ & \frac{b_n}{b_{n-1}} = \frac{\mu - \frac{i\lambda}{2}[\alpha+i(\delta+n-1)][\alpha+i(\delta+n-2)]}{[\alpha+i(\delta+n)]^2 - \left\{\mu - \frac{i\lambda}{2}[\alpha+i(\delta+n+1)][\alpha+i(\delta+n+2)]\right\}\frac{b_{n+1}}{b_n}}, \\ & \frac{b_{-n}}{b_{-n+1}} = \frac{\mu - \frac{i\lambda}{2}[\alpha+i(\delta-n+1)][\alpha+i(\delta-n+2)]}{[\alpha+i(\delta-n)]^2 - \left\{\mu - \frac{i\lambda}{2}[\alpha+i(\delta-n-1)][\alpha+i(\delta-n-2)]\right\}\frac{b_{-n-1}}{b_{-n}}}, \\ & \left\{\mu - \frac{i\lambda}{2}[\alpha+i(\delta-1)][\alpha+i(\delta-2)]\right\}\frac{b_{-1}}{b_0} + (\alpha+i\delta)^2 - \left\{\mu - \frac{i\lambda}{2}[\alpha+i(\delta+1)]\right. \\ & \left.\times[\alpha+i(\delta+2)]\right\}\frac{b_1}{b_0} = 0, \\ & d_0 = 0, \\ & \left[i\mu - \frac{\lambda}{2}(n-1)(n-2)\right]d_{n-1}^2 + n^2 d_n + \left[i\mu - \frac{\lambda}{2}(n+1)(n+2)\right]d_{n+1} = -H_n, \\ & \qquad (n>0) \end{aligned} \tag{38}$$

实际计算时，式 (38) 总是截取 N 项，即 $0<n<N$。

对于 $\overset{\frown}{AB}$ 圆环壳，有

$$\varphi_0 = \frac{\pi}{2}, \qquad \varphi_1 = \pi + \varphi_B \tag{39}$$

再展开各级复变量方程的右端项，利用上述结果，并取

$$\begin{aligned} \theta &= \varphi - \frac{\pi}{2}, \\ \widetilde{B}_1 &= -(I_i + \mathrm{i}J_i)\mathrm{e}^{-\gamma(\pi+\varphi_B)-\mathrm{i}\frac{\pi\delta}{2}}, \\ \widetilde{B}_2 &= (I_i + \mathrm{i}J_i)\mathrm{e}^{-\gamma\varphi_B+\mathrm{i}\frac{\pi\delta}{2}}, \end{aligned} \tag{40}$$

便得到各级摄动方程的通解 $\beta_i^{AB}(\theta, I_i, J_i, p_i)$ 和 $\psi_i^{AB}(\theta, I_i, J_i, p_i)$ $(i=1,2,\cdots)$.

对于$\overset{\frown}{CD}$圆环壳，有

$$\varphi_0 = -\frac{\pi}{2}, \qquad \varphi_1 = \varphi_B. \tag{41}$$

并取

$$\begin{aligned} \theta &= \varphi + \frac{\pi}{2}, \\ \widetilde{B}_1 &= (K_i + \mathrm{i}L_i)\mathrm{e}^{-\gamma\varphi_B+\mathrm{i}\frac{\pi\delta}{2}}, \\ \widetilde{B}_2 &= -(K_i + \mathrm{i}L_i)\mathrm{e}^{-\gamma(\pi+\varphi_B)-\mathrm{i}\frac{\pi\delta}{2}}. \end{aligned} \tag{42}$$

又得到各级摄动方程的通解 $\beta_i^{CD}(\theta, K_i, L_i, p_i)$ 和 $\psi_i^{CD}(\theta, K_i, L_i, p_i)$ $(i=1, 2, \cdots)$.

最后，由式（29）和（28）得

$$\begin{aligned} u^{AB}(\theta) &= \sum_{i=1}^{\infty} u_i^{AB}(\theta, I_i, J_i, p_i), \\ w^{AB}(\theta) &= \sum_{i=1}^{\infty} w_i^{AB}(\theta, I_i, J_i, p_i), \\ M_\xi^{AB}(\theta) &= \sum_{i=1}^{\infty} M_{\xi i}^{AB}(\theta, I_i, J_i, p_i); \end{aligned} \tag{43}$$

$$\begin{aligned} u^{CD}(\theta) &= \sum_{i=1}^{\infty} u_i^{CD}(\theta, K_i, L_i, p_i), \\ w^{CD}(\theta) &= \sum_{i=1}^{\infty} w_i^{CD}(\theta, K_i, L_i, p_i), \\ M_\xi^{CD}(\theta) &= \sum_{i=1}^{\infty} M_{\xi i}^{CD}(\theta, K_i, L_i, p_i). \end{aligned} \tag{44}$$

其中，I_i，J_i，K_i，L_i 和 p_i 由连接条件和摄动条件确定。

三、连接条件

在$\overset{\frown}{AB}$圆环壳和$\overline{BC}$锥壳以及$\overline{BC}$锥壳和$\overset{\frown}{CD}$圆环壳交接的地方，是自动相连的，因此，位移和内力应该满足以下的连续条件：

在 B 点处，

$$\begin{aligned} M_\xi^{AB}(\theta_B) &= -M_\xi^{BC}(\rho_B), \\ \psi_H^{AB}(\theta_B) &= -\psi_H^{BC}(\rho_B), \\ u^{AB}(\theta_B) &= u^{BC}(\rho_B), \\ \beta^{AB}(\theta_B) &= \beta^{BC}(\rho_B) \end{aligned} \tag{45}$$

在 C 点处，

$$
\begin{aligned}
M_\xi^{CD}(\theta_C) &= M_\xi^{BC}(\rho_C),\\
\psi_H^{CD}(\theta_C) &= \psi_H^{BC}(\rho_C),\\
u^{CD}(\theta_C) &= u^{BC}(\rho_C),\\
\beta^{CD}(\theta_C) &= \beta^{BC}(\rho_C)
\end{aligned} \tag{46}
$$

其中，$\theta_B=\frac{\pi}{2}+\varphi_B$，$\theta_C=\frac{\pi}{2}+\varphi_B$，$\rho_B=1$，$\rho_C=-1$。

另外，我们取 A、D 两点的无量纲轴向相对位移作为摄动参数：

$$
\varepsilon=\frac{\Delta w}{r_D}=\frac{w^{CD}(\theta_D)-w^{CD}(\theta_C)+w^{BC}(\rho_C)-w^{BC}(\rho_B)+w^{AB}(\theta_B)-w^{AB}(\theta_A)}{r_D}, \tag{47}
$$

其中，$\theta_A=\theta_D=0$。

由式（47），便有以下摄动条件

$$
w^{CD}(\theta_D)-w^{CD}(\theta_C)+w^{BC}(\rho_C)-w^{BC}(\rho_B)+w^{AB}(\theta_B)-w^{AB}(\theta_A)=r_D\varepsilon \tag{48}
$$

我们看到，2 小节和 3 小节的摄动解中，总共有 9 个待定常数 A、B、C、D、I、J、K、L 和 p，而现在总共有 9 个条件，即式（45）、（46）和（48），故是可解的。也就是说，对于每级摄动解，均要求解一个九阶线性代数方程组。一般说来，摄动法中取 $i=3$，其数值结果已较令人满意。

四、算例

为了检验理论，下面用两个算例进行比较。

例 1　一支铍青铜的波纹管，承受均匀内压力，其有关数据如下：

$r_D=12.75\text{mm}$，　$r_A=19\text{mm}$，　$R^{AB}=R^{CD}=0.75\text{mm}$，　波数 $n=8$，

$h=0.12\text{mm}$，　$\varphi_B=0°$，　$E=1.35\times10^6\text{kg/cm}^2$，　$\nu=0.33$

使用上述数据，经前面的公式计算后，将数值结果给在图 4 中。同时也给出了 Андреева[6] 的实验值和用数值方法所得的理论值。由图看到，我们的计算结果与实验值完全吻合。

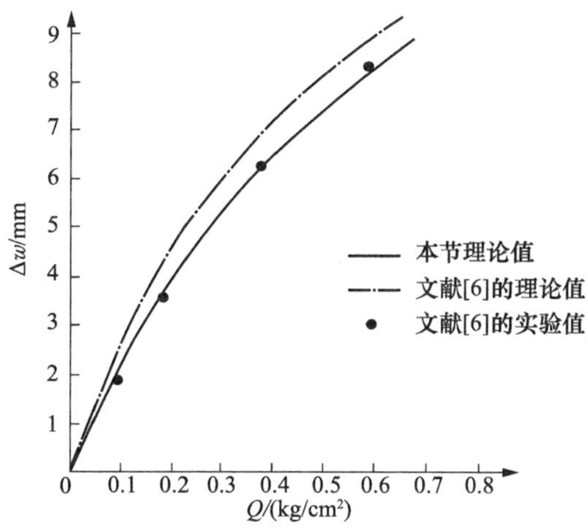

图 4　波纹管特性的理论值与实验值比较

例 2　一支承受轴向集中力压缩的波纹管，其有关数据如下：

$r_A=28\text{cm}$，　$r_D=18\text{cm}$，　$R^{AB}=R^{CD}=2.5\text{cm}$，　$\varphi_B=0°$，　$h=0.15\text{cm}$，$E=2.1\times 10^6\text{kg/cm}^2$，　$\nu=0.3$，$P=100\text{kg}$

戴福隆[9]测量了一支波纹管内，外表面的径向和环向应力，实验做得十分精细，其结果与我们的计算结果一并给在图 5 中。显然，计算结果与实验值吻合。

通过以上两个算例可以看出，由本节的理论公式所给的理论值与实验值符合，故可供工程设计部门使用。

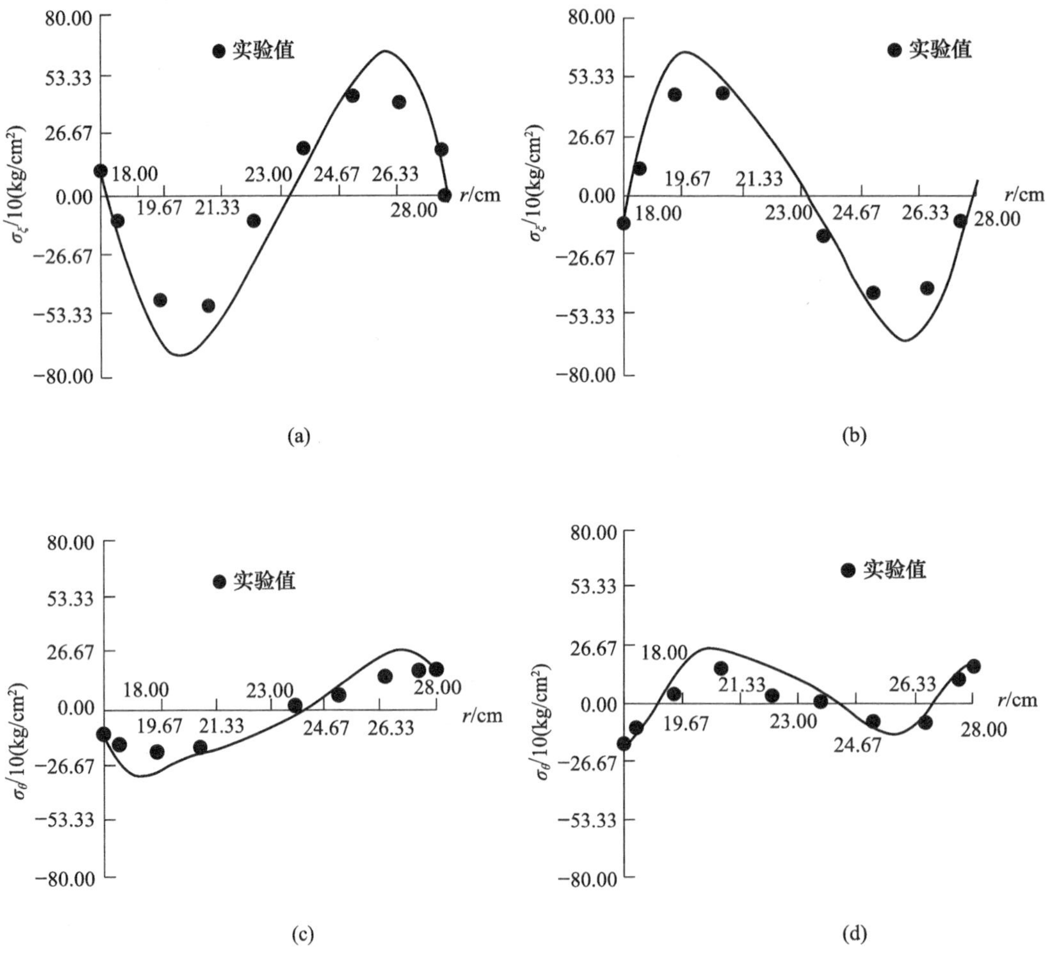

图 5　波纹管应力的理论值与实验值比较

（a）外表面径向应力分布；（b）内表面径向应力分布；（c）外表面环向应力分布；（d）内表面环向应力分布

参 考 文 献

[1]　钱伟长，郑思梁. 轴对称圆环壳的复变量方程和轴对称细环壳的一般解. 清华大学学报（自然科学版），1979，19（1）：27-47.

[2]　钱伟长，郑思梁. 轴对称圆环壳的一般解. 应用数学和力学，1980，1（3）：287-300.

[3] 钱伟长，吴明德．U型波纹管的非线性特性摄动法计算．应用数学和力学，1983，4 (5)：595-608.

[4] 徐志翘，刘燕，杨嘉实，等．变厚度U型波纹壳大挠度问题的摄动解．清华大学学报（自然科学版），1985，25 (1)：39-51.

[5] Reissner E. On axisymmetrical deformations of thin shells of revolution. Proc. Symp. Appl. Math. (Elasticity)，1950，3：27-52.

[6] Андреева. 波纹管的计算与设计．翁善臣，等译．北京：国防工业出版社，1982.

[7] Новожинов. 薄壳理论．白鹏飞，译．北京：科学出版社，1959，287-300.

[8] 陈山林．圆环壳在一般荷载下的轴对称问题．应用数学和力学，1986，7 (5)：425-434.

[9] 戴福隆．波纹壳的光弹性贴片法应力测定．固体力学学报，1984，5 (2)：224-230.

Non-linear bending of a corrugated annular plate with a plane boundary region and a non-deformable rigid body at the center under compound load①

中 文 摘 要

由于非线性数学和复合载荷的困难，一个有实际意义的而且更加困难的问题，亦即复合载荷作用下具有刚性中心和光滑边缘的波纹环形板的非线性弯曲问题，却始终还未被人研究过。通过修正迭代法的应用，并基于各向异性圆板非线性弯曲理论，本文成功地求解了这个问题。本文结果可直接地用于精密仪器弹性元件的设计。

1. Introduction

Corrugated circular plates are a kind of shell with a complex shape. They are an important type of elastic elements, and are applied widely in modern engineering, especially in measuring instruments. Therefore, the nonlinear problem of these plates has been attracting increasingly the attention of numerous investigators in recent years. Liu Renhuai[1-9], Panov[10], Feodosev[11,12], Andryewa[13,14], Akasaka[15], Haringx[16,17], Burmistrov[18], Chen[19], et al., have already made theoretical analysis on the problem for circular plates with corrugations of various types. Nevertheless, because of the difficulty in nonlinear mathematics and compound loads, a problem with more practical significance and high level of difficulty has never been attempted, i. e. the nonlinear bending problem of a corrugated annular plate with a plane boundary region and a non-deformable rigid body at the center under compound load. Through application of the modified iteration method, on the basis of the nonlinear bending theory for anisotropic circular plates, this paper has successfully solved the problem. The method was originally suggested by Yeh Kaiyuan and Liu Renhuai in 1965[20-22], and successfully applied in a system problem of nonlinear theory of plates and shells. The present results can be applied directly to the design of elastic elements of precision instruments.

2. Fundamental Equations

Let us consider a corrugated annular plate with a plane boundary region and a non-deformable rigid body at the center under the combined action of a uniform pressure q and a concentrated load p at the center, as shown in Fig. 1. Here R_1, R_2 and R are radii of the non-deformable rigid body, the corrugated region and the plane boundary region, respectively, r is the radial coordinate, L is the wavelength, H is the amplitude and h is

① Reprinted from *International Journal of Non-Linear Mechanics*, 1993, 28 (3): 353-364. Authors: Liu Renhuai and Zhou Renpo.

the thickness.

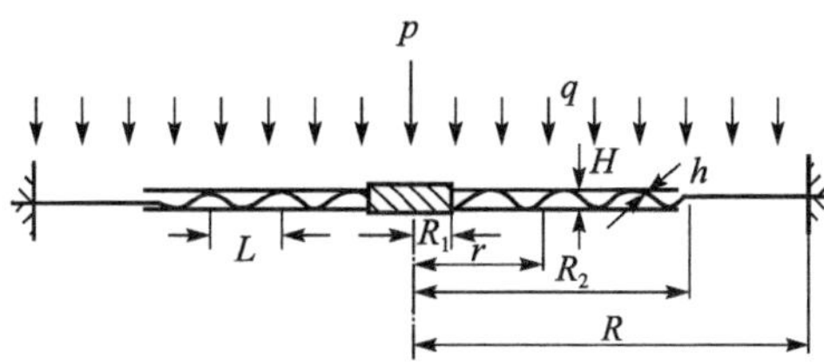

Fig. 1　A corrugated annular plate with a plane boundary region and a non-deformable rigid body at the center under compound load

In order to simplify the following presentation, we introduce a simple notation for some magnitudes of different regions of the plate. We shall indicate these magnitudes of the corrugated and the plane boundary regions by subscripts a and b, respectively. In this paper, the corrugations of the annular plate are assumed to be closely and evenly spaced. Then we may consider the corrugated region of the plate as an anisotropic annular plate. Using this analysis, according to nonlinear bending theories of isotropic and anisotropic circular plates, we obtain the fundamental equations of the plate as follows:

for $R_1 \leqslant r \leqslant R_2$,

$$r\frac{\mathrm{d}^3 w_a}{\mathrm{d}r^3}+\frac{\mathrm{d}^2 w_a}{\mathrm{d}r^2}-k_t\frac{\mathrm{d}w_a}{\mathrm{d}r}\left(\frac{1}{D_a}rN_{r,a}+\frac{k_t'}{r}\right)=\frac{k_t}{2D_a}\left(qr^2+\frac{p}{\pi}\right),$$

$$r\frac{\mathrm{d}^2}{\mathrm{d}r^2}(rN_{r,a})+\frac{\mathrm{d}}{\mathrm{d}r}(rN_{r,a})-k_rk_tN_{r,a}+\frac{Ehk_t}{2}\left(\frac{\mathrm{d}w_a}{\mathrm{d}r}\right)^2=0; \tag{2.1}$$

for $R_2 \leqslant r \leqslant R$,

$$r\frac{\mathrm{d}^3 w_b}{\mathrm{d}r^3}+\frac{\mathrm{d}^2 w_b}{\mathrm{d}r^2}-\frac{\mathrm{d}w_b}{\mathrm{d}r}\left(\frac{1}{D_b}rN_{r,b}+\frac{1}{r}\right)=\frac{1}{2D_b}\left(qr^2+\frac{p}{\pi}\right),$$

$$r\frac{\mathrm{d}^2}{\mathrm{d}r^2}(rN_{r,b})+\frac{\mathrm{d}}{\mathrm{d}r}(rN_{r,b})-N_{r,b}+\frac{Eh}{2}\left(\frac{\mathrm{d}w_b}{\mathrm{d}r}\right)^2=0 \tag{2.2}$$

where w is the deflection, N_r is the radial membrane stress, k_r, k_t and k_t' are parameters[1-4] for the radial and circumferential rigidity, respectively, E is Young's modulus, ν is Poisson's ratio, and D is the flexural rigidity:

$$D_a=\frac{Eh^3}{12\left(1-\frac{\nu^2}{k_tk_t'}\right)},\qquad D_b=\frac{Eh^3}{12(1-\nu^2)}. \tag{2.3}$$

These nonlinear differential equations will be solved in conjunction with appropriate boundary conditions. In general, the outer edge of the plate is either rigidly clamped or loosely clamped, and the inner edge of the plate is clamped in the non-deformable rigid body which can be moved up and down. Hence, the boundary conditions are given as follows:

(1) Rigidly clamped along the outer edge:

$$\frac{dw_a}{dr}=0,\quad u_a=0\qquad \text{at } r=R_1,$$

$$w_b=0,\quad \frac{dw_b}{dr}=0,\quad u_b=0\qquad \text{at } r=R. \tag{2.4}$$

(2) Loosely clamped along the outer edge:

$$\frac{dw_a}{dr}=0,\quad u_a=0\qquad \text{at } r=R_1,$$

$$w_b=0,\ \frac{dw_b}{dr}=0,\quad N_{r,b}=0\qquad \text{at } r=R \tag{2.5}$$

where u is the radial displacement given by

$$u_a=\frac{r}{k_tEh}\left[\frac{d}{dr}(rN_{r,a})-\nu N_{r,a}\right],$$

$$u_b=\frac{r}{Eh}\left[\frac{d}{dr}(rN_{r,b})-\nu N_{r,b}\right]. \tag{2.6}$$

Again, the continuous conditions between two regions are given below:

$$w_a=w_b,\quad \frac{dw_a}{dr}=\frac{dw_b}{dr},\quad M_{r,a}=M_{r,b},$$

$$N_{r,a}=N_{r,b}\quad u_a=u_b\quad \text{at } r=R_2 \tag{2.7}$$

where M is the radial moment given by

$$M_{r,a}=-\frac{D_a}{k_t}\left(\frac{d^2w_a}{dr^2}+\frac{\nu}{r}\frac{dw_a}{dr}\right),$$

$$M_{r,b}=-D_b\left(\frac{d^2w_b}{dr^2}+\frac{\nu}{r}\frac{dw_b}{dr}\right). \tag{2.8}$$

In order to simplify the following calculations, we introduce the following nondimensional parameters:

$$y=\frac{r}{R},\ y_1=\frac{R_1}{R},\ y_2=\frac{R_2}{R},$$

$$W_a=\frac{w_a}{h},\ W_b=\frac{w_b}{h},$$

$$\phi_a=\frac{dw_a}{dy},\ \phi_b=\frac{dw_b}{dy},$$

$$S_a=-\frac{k_tRr}{D_a}N_{r,a},\ S_b=-\frac{Rr}{D_b}N_{r,b},$$

$$Q=\frac{R^4k_t}{2D_ah}q,\ P=\frac{R^2k_t}{2\pi D_ah}p,$$

$$\beta_0=6(1-\nu)(1+\nu),\ \beta_1^2=k_tk_t',\ \beta_2^2=k_rk_t,$$

$$\beta_3=\frac{Eh^3k_t^2}{2D_a},\ \beta_4=\frac{D_a}{k_tD_b},\ \beta_5=\frac{\beta_4}{k_t}. \tag{2.9}$$

Then equations (2.1) and (2.2), boundary conditions (2.4) and (2.5), and continuous conditions (2.7) are transformed to the non-dimensional form

for $y_1\leqslant y\leqslant y_2$,

$$L_1(y^{\beta_1}\phi_a) = -S_a\phi_a + Qy^2 + P, \tag{2.10a}$$

$$L_2(y^{\beta_2}S_a) = \beta_3\phi_a^2; \tag{2.10b}$$

for $y_2 \leqslant y \leqslant 1$,

$$L_0(y\phi_b) = -S_b\phi_b + \beta_4(Qy^2 + P), \tag{2.11a}$$

$$L_0(yS_b) = \beta_0\phi_b^2. \tag{2.11b}$$

(1) Rigidly clamped along the outer edge:

$$\phi_a = 0, \quad y\frac{\mathrm{d}S_a}{\mathrm{d}y} - \nu S_a = 0 \qquad \text{at } y = y_1, \tag{2.12a,b}$$

$$W_b = 0, \quad \phi_b = 0, \quad y\frac{\mathrm{d}S_b}{\mathrm{d}y} - \nu S_b = 0 \qquad \text{at } y = 1. \tag{2.13a-c}$$

(2) Loosely clamped along the outer edge:

$$\phi_a = 0, \quad y\frac{\mathrm{d}S_a}{\mathrm{d}y} - \nu S = 0 \qquad \text{at } y = y_1, \tag{2.14a,b}$$

$$W_b = 0, \quad \phi_b = 0, \quad S_b = 0 \qquad \text{at } y = 1, \tag{2.15a-c}$$

and the continuous conditions as

$$W_a = W_b, \quad \phi_a = \phi_b, \quad \beta_4 S_a = S_b,$$

$$\beta_4\left(\frac{\mathrm{d}\phi_a}{\mathrm{d}y} + \frac{\nu}{y}\phi_a\right) = \frac{\mathrm{d}\phi_a}{\mathrm{d}y} + \frac{\nu}{y}\phi_b,$$

$$\beta_5\left(\frac{\mathrm{d}S_a}{\mathrm{d}y} - \frac{\nu}{y}S_a\right) = \frac{\mathrm{d}S_b}{\mathrm{d}y} - \frac{\nu}{y}S_b \qquad \text{at } y = y_2 \tag{2.16}$$

where

$$L_0(\ ) = y\frac{\mathrm{d}}{\mathrm{d}r}\frac{1}{y}\frac{\mathrm{d}}{\mathrm{d}r}(\ ),$$

$$L_1(\ ) = y^{\beta_1}\frac{\mathrm{d}}{\mathrm{d}y}y^{1-2\beta_1}\frac{\mathrm{d}}{\mathrm{d}y}(\ ),$$

$$L_2(\ ) = y^{\beta_2}\frac{\mathrm{d}}{\mathrm{d}y}y^{1-2\beta_2}\frac{\mathrm{d}}{\mathrm{d}y}(\ ). \tag{2.17}$$

It will be less difficult if we use the single load parameter G to replace the non-dimensional loads Q and P in equations (2.10a) and (2.11a), which are

$$Q = \mu_1 G, \quad P = \mu_2 G \tag{2.18}$$

where μ_1 and μ_2 are constants.

Then equations (2.10a) and (2.11a) can be written as

$$L_1(y^{\beta_1}\phi_a) = -S_a\phi_a + (\mu_1 y^2 + \mu_2)G \tag{2.19}$$

$$L_0(y\phi_b) = -S_b\phi_b + \beta_4(\mu_1 y^2 + \mu_2)G. \tag{2.20}$$

3. Analytical Solution

First of all, we shall deal with the non-dimensional nonlinear boundary value problem (2.19), (2.10b), (2.20), (2.11b), (2.12), (2.13) and (2.17) of a corrugated annular plate with a plane boundary region, a non-deformable rigid body at the center and a rigidly clamped outer edge under the combined action of uniform pressure and a single concentrated load at the center. To make it easier to solve the second non-dimensional nonlinear boundary value problem (2.19), (2.10b), (2.20) (2.11b), (2.14) (2.15) and (2.17) for the plate with a loosely clamped outer edge, we temporarily write Poisson's ν in (2.13c) by ν'. The first problem will be solved by the modified iteration method.

For the first approximation, by neglecting nonlinear terms in (2.19) and (2.20), and again introducing ϕ_{a1} and ϕ_{b1} into the right-hand side of equations (2.10b) and (2.11b) respectively, we find the linear boundary value problem as follows:

for $y_1 \leqslant y \leqslant y_2$,

$$L_1(y^{\beta_1}\phi_{a1}) = (\mu_1 y^2 + \mu_2)G, \tag{3.1a}$$

$$L_1(y^{\beta_2}S_{a1}) = \beta_3\phi_{a1}^2; \tag{3.1b}$$

for $y_2 \leqslant y \leqslant 1$,

$$L_0(y\phi_{b1}) = \beta_4(\mu_1 y^2 + \mu_2)G, \tag{3.2a}$$

$$L_0(yS_{b1}) = \beta_0\phi_{b1}^2, \tag{3.2b}$$

$$\phi_{a1} = 0,\ y = \frac{dS_{a1}}{dy} - \nu S_{a1} = 0 \quad \text{at } y = y_1, \tag{3.3a,b}$$

$$W_{b1} = 0,\ \phi_{b1} = 0,\ y = \frac{dS_{b1}}{dy} - \nu' S_{b1} = 0 \quad \text{at } y = 1, \tag{3.4a-c}$$

$$W_{a1} = W_{b1},\ \phi_{a1} = \phi_{b1},\ \beta_4 S_{a1} = S_{b1},$$

$$\beta_4\left(\frac{d\phi_{a1}}{dy} + \frac{\nu}{y}\phi_{a1}\right) = \frac{d\phi_{b1}}{dy} + \frac{\nu}{y}\phi_{b1},$$

$$\beta_5\left(\frac{dS_{a1}}{dy} - \frac{\nu}{y}S_{a1}\right) = \frac{dS_{b1}}{dy} - \frac{\nu}{y}S_{b1} \quad \text{at } y = y_2. \tag{3.5a-e}$$

Solving the problems (3.1a), (3.2a), (3.3a), (3.4b), (3.5b) and (3.5d), we obtain

$$\phi_{a1} = (a_1 y^{\beta_1} + a_2 y^{-\beta_1} + a_3 y^3 + a_4 y)G, \tag{3.6a}$$

$$\phi_{b1} = (b_1 y^3 + b_2 y\ln y + b_3 y + b_4 y^{-1})G \tag{3.6b}$$

where

$$a_1 = \frac{I_1 t_1 - I_2 t_2}{t_1 t_3 - \beta_4 t_2 t_4},\ a_2 = -a_1 y_1^{2\beta_1} - a_3 y_1^{3+\beta_1} - a_4 y_1^{1+\beta_1},$$

$$a_3 = \frac{\mu_1}{9-\beta_1^2},\ a_4 = \frac{\mu_2}{1-\beta_1^2},$$

$$b_1 = \frac{\beta_4\mu_1}{8},\ b_2 = \frac{\beta_4\mu_2}{2},$$

$$b_3 = \frac{\beta_4 I_1 t_4 - I_2 t_3}{t_1 t_3 - \beta_4 t_2 t_4},\ b_4 = -b_1 - b_3,$$

$$I_1 = a_3 y_1^{3+\beta_1} y_2^{-\beta_1} + a_4 y_1^{1+\beta_1} y_2^{-\beta_1} - a_3 y_2^3 - a_4 y_2 + b_1 y_2^3 + b_2 y_2 \ln y_2 - b_1 y_2^{-1},$$

$$I_2 = -\beta_4[(\beta_1-\nu)(a_3 y_1^{3+\beta_1} y_2^{-\beta_1} + a_4 y_1^{1+\beta_1} y_2^{-\beta_1}) + (3+\nu)a_3 y_2^3 + (1+\nu)a_4 y_2] + (3+\nu)b_1 y_2^3 + (1+\nu)b_2 y_2 \ln y_2 + b_2 y_2 + (1-\nu)b_1 y_2,$$

$$t_1 = (1+\nu)y_2 + (1-\nu)y_2^{-1},\quad t_2 = y_2 - y_2^{-1},$$

$$t_3 = y_2^{\beta_1} - y_1^{2\beta_1} y_2^{-\beta_1},$$

$$t_4 = (\beta_1+\nu)y_2^{\beta_1} + (\beta_1-\nu)y_1^{2\beta_1} y_2^{-\beta_1}. \tag{3.7}$$

Now multiplying (3.6a) and (3.6b) by dy and integrating them, respectively, and using (3.4a) and (3.5a), we obtain the following formulas for the deflection to the first approximation:

$$W_{a1} = [F_1(y) - F_1(y_2) + F_2(y_2) - F_2(1)]G, \tag{3.8a}$$

$$W_{b1} = [F_2(y) - F_2(1)]G \tag{3.8b}$$

where

$$F_1(y) = \frac{a_1}{1+\beta_1} y^{1+\beta_1} + \frac{a_2}{1-\beta_1} y^{1-\beta_1} + \frac{a_3}{4} y^4 + \frac{a_4}{2} y^2,$$

$$F_2(y) = \frac{b_1}{4} y^4 + b_2\left(\frac{1}{2}y^2 \ln y - \frac{1}{4}y^2\right)\frac{b_3}{2}y^2 + b_4 \ln y. \tag{3.9}$$

When we introduce, as an iteration parameter, a notation of the non-dimensional deflection at $y=y_1$ for the plate

$$W_m = W_a|_{y=y_1} \tag{3.10}$$

the expression (3.8a) becomes

$$G = \lambda_1 W_m \tag{3.11}$$

where

$$\lambda_1 = [F_1(y_1) - F_1(y_2) + F_2(y_2) - F_2(1)]^{-1}. \tag{3.12}$$

This is the characteristic relation of small deflection of the corrugated annular plate. In such a case, we rewrite the above solution (3.6) for the slope ϕ in these forms

$$\phi_{a1} = \lambda_1 W_m (a_1 y^{\beta_1} + a_2 y^{-\beta_1} + a_3 y^3 + a_4 y), \tag{3.13a}$$

$$\phi_{b1} = \lambda_1 W_m (b_1 y^3 + b_2 y \ln y + b_3 y + b_4 y^{-1}). \tag{3.13b}$$

Using the obtained solution (3.13), solutions of the problem (3.1b), (3.2b), (3.3b), (3.4c), (3.5c) and (3.5e) for the non-dimensional radial membrane strese S are

$$S_{a1} = \beta_3 \lambda_1^2 W_m^2 (c_1 y^{1+2\beta_1} + c_2 y^{1-2\beta_1} + c_3 y^{4+\beta_1} + c_4 y^{4-\beta_1} + c_5 y^{2+\beta_1} + c_6 y^{2-\beta_1} + c_7 y^{\beta_2} + c_8 y^{-\beta_2} + c_9 y^7 + c_{10} y^5 + c_{11} y^3 + c_{12} y),$$

$$S_{b1} = \beta_0 \lambda_1^2 W_m^2 (d_1 y^3 \ln^2 y + d_2 y \ln^2 y + d_3 y^5 \ln y + d_4 y^3 \ln y + d_5 y \ln y + d_6 y^{-1} \ln y + d_7 y^7 + d_8 y^5 + d_9 y^3 + d_{10} y + d_{11} y^{-1}) \tag{3.14}$$

where

$$c_1=\frac{a_1^2}{(1+2\beta_1)^2-\beta_2^2},\quad c_2=\frac{a_2^2}{(1-2\beta_1)^2-\beta_2^2},\quad c_3=\frac{2a_1a_3}{(4+\beta_1)^2-\beta_2^2},$$

$$c_4=\frac{2a_2a_3}{(4-\beta_1)^2-\beta_2^2},\quad c_5=\frac{2a_1a_4}{(2+\beta_1)^2-\beta_2^2},\quad c_6=\frac{2a_2a_4}{(2-\beta_1)^2-\beta_2^2},$$

$$c_7=\frac{I_3t_5-I_4t_6}{\beta_3(\beta_4t_5t_7-\beta_5t_6t_8)},\quad c_8=\frac{[J_1+(\beta_2-\nu)c_7y_1^{\beta_2}]y_1^{\beta_2}}{\beta_2+\nu},\quad c_9=\frac{a_3^2}{49-\beta_2^2},$$

$$c_{10}=\frac{2a_3a_4}{25-\beta_2^2},\quad c_{11}=\frac{a_4^2}{9-\beta_2^2},\quad c_{12}=\frac{2a_1a_2}{1-\beta_2^2},$$

$$d_1=\frac{b_2^2}{8},\quad d_2=\frac{b_2b_4}{2},\quad d_3=\frac{b_1b_2}{2},\quad d_4=\frac{4b_2b_3-3b_2^2}{16},$$

$$d_5=\frac{2b_3b_4-b_2b_4}{2},\quad d_6=-\frac{b_4^2}{2},\quad d_7=\frac{b_1^2}{48},$$

$$d_8=\frac{12b_1b_3-5b_1b_2}{144},\quad d_9=\frac{16b_1b_4-12b_2b_3+7b_2^2+8b_3^2}{64},$$

$$d_{10}=\frac{(1+\nu')(\beta_5I_3t_8-\beta_4I_4t_7)}{\beta_0(\beta_4t_5t_7-\beta_5t_6t_8)},\quad d_{11}=\frac{J_2+(1-\nu')d_{10}}{1+\nu'},$$

$$\begin{aligned}J_1=&(1+2\beta_1-\nu)c_1y_1^{1+2\beta_1}+(1-2\beta_1-\nu)c_2y_1^{1-2\beta_1}\\&+(4+\beta_1-\nu)c_3y_1^{4+\beta_1}+(4-\beta_1-\nu)c_4y_1^{4-\beta_1}\\&+(2+\beta_1-\nu)c_5y_1^{2+\beta_1}+(2-\beta_1-\nu)c_6y_1^{2-\beta_1}\\&+(7-\nu)c_9y_1^7+(5-\nu)c_{10}y_1^5\\&+(3-\nu)c_{11}y_1^3+(1-\nu)c_{12}y_1,\end{aligned}$$

$$J_2=d_3+d_4+d_5+d_6+(7-\nu')d_7+(5-\nu')d_8+(3-\nu')d_9,$$

$$\begin{aligned}J_3=&-\beta_3\beta_4(c_1y_2^{1+2\beta_1}+c_2y_2^{1-2\beta_1}+c_3y_2^{4+\beta_1}+c_4y_2^{4-\beta_1}+c_5y_2^{2+\beta_1}\\&+c_6y_2^{2-\beta_1}+c_9y_2^7+c_{10}y_2^5+c_{11}y_2^3+c_{12}y_2)\\&+\beta_0(d_1y_2^3\ln^2y_2+d_2y_2\ln^2y_2+d_3y_2^5\ln y_2+d_4y_2^3\ln y_2\\&+d_5y_2\ln y_2+d_6y_2^{-1}\ln y_2+d_7y_2^7+d_8y_2^5+d_9y_2^3),\end{aligned}$$

$$\begin{aligned}J_4=&-\beta_3\beta_5[(1+2\beta_1-\nu)c_1y_2^{1+2\beta_1}+(1-2\beta_1-\nu)c_1y_2^{1-2\beta_1}\\&+(4+\beta_1-\nu)c_3y_2^{4+\beta_1}+(4-\beta_1-\nu)c_4y_2^{4-\beta_1}\\&+(2+\beta_1-\nu)c_5y_2^{2+\beta_1}+(2-\beta_1-\nu)c_6y_2^{2-\beta_1}\\&+(7-\nu)c_9y_2^7+(5-\nu)c_{10}y_2^5+(3-\nu)c_{11}y_2^3+(1-\nu)c_{12}y_2)]\\&+\beta_0\{[(3-\nu)y_2^3\ln^2y_2+2y_2^3\ln y_2]d_1\\&+[(1-\nu)y_2\ln^2y_2+2y_2\ln y_2]d_2+[(5-\nu)y_2^5\ln y_2+y_2^5]d_3\\&+[(3-\nu)y_2^3\ln y_2+y_2^3]d_4+[(1-\nu)y_2\ln y_2+y_2]d_5\\&+[-(1+\nu)y_2^{-1}\ln y_2+y_2^{-1}]d_6+(7-\nu)d_7y_2^7+(5-\nu)d_8y_2^5\\&+(3-\nu)d_9y_2^3],\end{aligned}$$

$$I_3=J_3-\frac{\beta_3\beta_4J_1y_1^{\beta_1}y_2^{-\beta_2}}{\beta_2+\nu}+\frac{\beta_0J_2y_2^{-1}}{1+\nu'},$$

$$I_4=J_4+\beta_3\beta_5J_1y_1^{\beta_2}y_2^{-\beta_2}-\frac{\beta_0(1+\nu)J_2y_2^{-1}}{1+\nu'},$$

$$t_5 = (1-\nu)(1+\nu')y_2 - (1+\nu)(1-\nu')y_2^{-1},$$
$$t_6 = (1+\nu')y_2 + (1-\nu')y_2^{-1},$$
$$t_7 = y_2^{\beta_2} + \frac{\beta_2 - \nu}{\beta_2 + \nu} y_1^{2\beta_2} y_1^{-\beta_2},$$
$$t_8 = (\beta_2 - \nu)(y_2^{\beta_2} - y_1^{2\beta_2} y_2^{-\beta_2}). \tag{3.15}$$

For the second approximation, by confining ourselves to the slope ϕ, we have the following linear boundary value problem:

for $y_1 \leqslant y \leqslant y_2$,

$$L_1(y^{\beta_1}\phi_{a2}) = -S_{a1}\phi_{a1} + (\mu_1 y^2 + \mu_2)G, \tag{3.16}$$

for $y_2 \leqslant y \leqslant 1$,

$$L_0(y\phi_{b2}) = -S_{b1}\phi_{b1} + \beta_4(\mu_1 y^2 + \mu_2)G, \tag{3.17}$$
$$\phi_{a2} = 0 \quad \text{at } y = y_1, \tag{3.18}$$
$$W_{b2} = 0, \quad \phi_{b2} = 0 \quad \text{at } y = 1, \tag{3.19a,b}$$
$$W_{a2} = W_{b2}, \quad \phi_{a2} = \phi_{b2},$$
$$\beta_4\left(\frac{d\phi_{a2}}{dy} + \frac{\nu}{y}\phi_{a2}\right) = \frac{d\phi_{b2}}{dy} + \frac{\nu}{y}\phi_{b2} \quad \text{at } y = y_2. \tag{3.20a-c}$$

Using solutions (3.13) and (3.14), the solution of this problem is

$$\begin{aligned}\phi_{a2} = &\lambda_1 W_m(a_1 y^{\beta_1} + a_2 y^{-\beta_1} + a_3 y^3 + a_4 y)\\ &-\beta_3\lambda_1^3 W_m^3[F_3(y) + e_{28} y^{\beta_1} + e_{29} y^{-\beta_1}],\\ \phi_{b2} = &\lambda_1 W_m(b_1 y^3 + b_2 y\ln y + b_3 y + b_4 y^{-1})\\ &-\beta_0\lambda_1^3 W_m^3[F_4(y) + f_{21} y + f_{22} y^{-1}]\end{aligned} \tag{3.21}$$

where

$$\begin{aligned}F_3(y) = &e_1 y^{2+3\beta_1} + e_2 y^{2-3\beta_1} + e_3 y^{5+2\beta_1} + e_4 y^{5-2\beta_1}\\ &+ e_5 y^{3+2\beta_1} + e_6 y^{3-2\beta_1} + e_7 y^{8+\beta_1} + e_8 y^{8-\beta_1}\\ &+ e_9 y^{6+\beta_1} + e_{10} y^{6-\beta_1} + e_{11} y^{4+\beta_1} + e_{12} y^{4-\beta_1}\\ &+ e_{13} y^{2+\beta_1} + e_{14} y^{2-\beta_1} + e_{15} y^{1+\beta_1+\beta_2} + e_{16} y^{1+\beta_1-\beta_2}\\ &+ e_{17} y^{1-\beta_1+\beta_2} + e_{18} y^{1-\beta_1-\beta_2} + e_{19} y^{4+\beta_2} + e_{20} y^{4-\beta_2}\\ &+ e_{21} y^{2+\beta_2} + e_{22} y^{2-\beta_2} + e_{23} y^{11} + e_{24} y^9\\ &+ e_{25} y^7 + e_{26} y^5 + e_{27} y^3,\\ F_4(y) = &f_1 y^5 \ln^3 y + f_2 y^3 \ln^3 y + f_3 y \ln^3 y + f_4 y^7 \ln^2 y\\ &+ f_5 y^5 \ln^2 y + f_6 y^3 \ln^2 y + f_7 y \ln^2 y + f_8 y^{-1} \ln^2 y\\ &+ f_9 y^9 \ln y + f_{10} y^7 \ln y + f_{11} y^5 \ln y + f_{12} y^3 \ln y\\ &+ f_{13} y \ln y + f_{14} y^{-1} \ln y + f_{15} y^{11} + f_{16} y^9\\ &+ f_{17} y^7 + f_{18} y^5 + f_{19} y^3 + f_{20} y,\\ J_5 = &-\beta_3 F_3(y_2) + \beta_0 F_4(y_2),\\ J_6 = &\beta_0\{(5+\nu) f_1 y_2^5 \ln^3 y_2 + (3+\nu) f_2 y_2^3 \ln^3 y_2\\ &+ (1+\nu) f_3 y_2 \ln^3 y_2 + (7+\nu) f_4 y_2^7 \ln^2 y_2\end{aligned}$$

$$
\begin{aligned}
&+[(5+\nu)f_5+3f_1]y_2^5\ln^2 y_2+[(3+\nu)f_6+3f_2]y_2^3\ln^2 y_2\\
&+[(1+\nu)f_7+3f_3]y_2\ln^2 y_2-(1-\nu)f_8 y_2^{-1}\ln^2 y_2\\
&+(9+\nu)f_9 y_2^9\ln y_2+[(7+\nu)f_{10}+2f_4]y_2^7\ln y_2\\
&+[(5+\nu)f_{11}+2f_5]y_2^5\ln y_2+[(3+\nu)f_{12}+2f_6]y_2^3\ln y_2\\
&+[(1+\nu)f_{13}+2f_7]y_2\ln y_2+[(-1+\nu)f_{14}+2f_8]y_2^{-1}\ln y_2\\
&+(11+\nu)f_{15}y_2^{11}+[(9+\nu)f_{16}+f_9]y_2^9\\
&+[(7+\nu)f_{17}+f_{10}]y_2^7+[(5+\nu)f_{18}+f_{11}]y_2^5\\
&+[(3+\nu)f_{19}+f_{12}]y_2^3+[(1+\nu)f_{20}+f_{13}]y_2+f_{14}y_2^{-1}\}\\
&-\beta_3\beta_4[(2+3\beta_1+\nu)e_1 y_2^{2+3\beta_1}+(2-3\beta_1+\nu)e_2 y_2^{2-3\beta_1}\\
&+(5+2\beta_1+\nu)e_3 y_2^{5+2\beta_1}+(5-2\beta_1+\nu)e_4 y_2^{5-2\beta_1}+(3+2\beta_1\\
&+\nu)e_5 y_2^{3+2\beta_1}+(3-2\beta_1+\nu)e_6 y_2^{3-2\beta_1}+(8+\beta_1+\nu)e_7 y_2^{8+\beta_1}\\
&+(8-\beta_1+\nu)e_8 y_2^{8-\beta_1}+(6+\beta_1+\nu)e_9 y_2^{6+\beta_1}\\
&+(6-\beta_1+\nu)e_{10}y_2^{6-\beta_1}+(4+\beta_1+\nu)e_{11}y_2^{4+\beta_1}\\
&+(4-\beta_1+\nu)e_{12}y_2^{4-\beta_1}+(2+\beta_1+\nu)e_{13}y_2^{2+\beta_1}+(2-\beta_1\\
&+\nu)e_{14}y_2^{2-\beta_1}+(1+\beta_1+\beta_2+\nu)e_{15}y_2^{1+\beta_1+\beta_2}\\
&+(1+\beta_1-\beta_2+\nu)e_{16}y_2^{1+\beta_1-\beta_2}+(1-\beta_1+\beta_2+\nu)e_{17}y_2^{1-\beta_1+\beta_2}\\
&+(1-\beta_1-\beta_2+\nu)e_{18}y_2^{1-\beta_1-\beta_2}+(4+\beta_2+\nu)e_{19}y_2^{4+\beta_2}\\
&+(4-\beta_2+\nu)e_{20}y_2^{4-\beta_2}+(2+\beta_2+\nu)e_{21}y_2^{2+\beta_2}\\
&+(2-\beta_2+\nu)e_{22}y_2^{2-\beta_2}+(11+\nu)e_{23}y_2^{11}+(9+\nu)e_{24}y_2^9\\
&+(7+\nu)e_{25}y_2^7+(5+\nu)e_{26}y_2^5+(3+\nu)e_{27}y_2^3],
\end{aligned}
$$

$$I_5=J_5+\beta_3F_3(y_1)y_1^{\beta_1}y_2^{-\beta_1}-\beta_0F_4(1)y_2^{-1},$$

$$I_6=J_6-\beta_3\beta_4(\beta_1-\nu)F_3(y_1)y_1^{\beta_1}y_2^{-\beta_1}+\beta_0(1-\nu)F_4(1)y_2^{-1},$$

$$e_1=\frac{a_1c_1}{4(1+\beta_1)(1+2\beta_1)},\quad e_2=\frac{a_2c_2}{4(1-\beta_1)(1-2\beta_1)},$$

$$e_3=\frac{a_1c_3+a_3c_1}{(5+\beta_1)(5+3\beta_1)},\quad e_4=\frac{a_2c_4+a_3c_2}{(5-\beta_1)(5-3\beta_1)},$$

$$e_5=\frac{a_1c_5+a_4c_1}{3(1+\beta_1)(3+\beta_1)},\quad e_6=\frac{a_2c_6+a_4c_2}{3(1-\beta_1)(3-\beta_1)},$$

$$e_7=\frac{a_1c_9+a_3c_3}{16(4+\beta_1)},\qquad e_8=\frac{a_2c_9+a_3c_4}{16(4-\beta_1)},$$

$$e_9=\frac{a_1c_{10}+a_3c_5+a_4c_3}{12(3+\beta_1)},\quad e_{10}=\frac{a_2c_{10}+a_3c_6+a_4c_4}{12(3-\beta_1)},\quad e_{11}=\frac{a_1c_{11}+a_4c_5}{8(2+\beta_1)},$$

$$e_{12}=\frac{a_2c_{11}+a_4c_6}{8(2-\beta_1)},\quad e_{13}=\frac{a_1c_{12}+a_2c_1}{4(1+\beta_1)},\quad e_{14}=\frac{a_1c_2+a_2c_{12}}{4(1-\beta_1)},$$

$$e_{15}=\frac{a_1c_7}{(1+\beta_2)(1+2\beta_1+\beta_2)}\qquad e_{16}=\frac{a_1c_8}{(1-\beta_2)(1+2\beta_1-\beta_2)},$$

$$e_{17}=\frac{a_2c_7}{(1+\beta_2)(1-2\beta_1+\beta_2)},\quad e_{18}=\frac{a_2c_8}{(1-\beta_2)(1-2\beta_1-\beta_2)},$$

$$e_{19}=\frac{a_3c_7}{(4+\beta_2)^2-\beta_1^2},\quad e_{20}=\frac{a_3c_8}{(4-\beta_2)^2-\beta_1^2},$$

$$e_{21}=\frac{a_4c_7}{(2+\beta_2)^2-\beta_1^2},\quad e_{22}=\frac{a_4c_8}{(2-\beta_2)^2-\beta_1^2},$$

$$e_{23}=\frac{a_3c_9}{121-\beta_1^2},\quad e_{24}=\frac{a_3c_{10}+a_4c_9}{81-\beta_1^2},\quad e_{25}=\frac{a_3c_{11}+a_4c_{10}}{49-\beta_1^2},$$

$$e_{26}=\frac{a_1c_4+a_2c_3+a_3c_{12}+a_4c_{11}}{25-\beta_1^2},\quad e_{27}=\frac{a_1c_6+a_2c_5+a_4c_{12}}{9-\beta_1^2},$$

$$e_{28}=\frac{I_5t_1-I_6t_2}{\beta_3(t_1t_3-\beta_4t_2t_4)},\quad e_{29}=-y_1^{\beta_1}F_3(y_1)-e_{28}y_1^{2\beta_1},$$

$$f_1=\frac{b_2d_1}{24},\quad f_2=\frac{b_2d_2}{8},\quad f_3=\frac{b_2d_6+b_4d_2}{6},\quad f_4=\frac{b_1d_1+b_2d_3}{48},$$

$$f_5=\frac{4(b_1d_2+b_2d_4+b_3d_1)-5b_2d_1}{96},\quad f_6=\frac{4(b_2d_5+b_3d_2+b_4d_1)-9b_2d_2}{32},$$

$$f_7=\frac{(b_2d_{11}+b_3d_6+b_4d_5)-(b_2d_6+b_4d_2)}{4},\quad f_8=-\frac{b_4d_6}{4},$$

$$f_9=\frac{b_1d_3+b_2d_7}{80},\quad f_{10}=\frac{12(b_1d_4+b_2d_8+b_3d_3)-7(b_1d_1+b_2d_3)}{576},$$

$$f_{11}=\frac{1}{1728}[72(b_1d_5+b_2d_9+b_3d_4+b_4d_3)-60(b_1d_2+b_2d_4+b_3d_1)+57b_2d_1],$$

$$f_{12}=\frac{1}{64}[8(b_1d_6+b_2d_{10}+b_3d_5+b_4d_4)-12(b_2d_5+b_3d_2+b_4d_1)+21b_2d_2],$$

$$f_{13}=\frac{1}{4}[2(b_3d_{11}+b_4d_{10})-(b_2d_{11}+b_3d_6+b_4d_5)+(b_2d_6+b_4d_2],$$

$$f_{14}=-\frac{2b_4d_{11}+b_4d_6}{4},\quad f_{15}=\frac{b_1d_7}{120},$$

$$f_{16}=\frac{40(b_1d_8+b_3d_7)-9(b_1d_3+b_2d_7)}{3200},$$

$$f_{17}=\frac{1}{13824}[288(b_1d_9+b_3d_8+b_4d_7)-84(b_1d_4+b_2d_8+b_3d_3)+37(b_1d_1+b_2d_3],$$

$$f_{18}=\frac{1}{6912}[288(b_1d_{10}+b_3d_9+b_4d_8)-120(b_1d_5+b_2d_9+b_3d_4+b_4d_3)+76(b_1d_2+b_2d_4+b_3d_1)-65b_2d_1],$$

$$f_{19}=\frac{1}{256}[32(b_1d_{11}+b_3d_{10}+b_4d_9)-24(b_1d_6+b_2d_{10}+b_3d_5+b_4d_4)+28(b_2d_5+b_3d_2+b_4d_1)-45b_2d_2],$$

$$f_{20}=\frac{1}{8}[(b_2d_{11}+b_3d_6+b_4d_5)-2(b_3d_{11}+b_4d_{10})-(b_2d_6+b_4d_2)],$$

$$f_{21}=\frac{\beta_4I_5t_4-I_6t_3}{\beta_0(t_1t_3-\beta_4t_2t_4)},\quad f_{22}=-F_4(1)-f_{21}. \tag{3.22}$$

Multiplying (3.21) by dy and integrating them, and using conditions (3.19a) and

(3. 20a), we obtain the following solutions for the deflection to the second approximation:

$$W_{a2} = F_1(y)G + \beta_3\lambda_1^3 W_m^3\{[F_5(y) - F_2(y_2)] + \frac{\beta_0}{\beta_3}[F_6(y_2) - F_6(1)]\}, \tag{3.23a}$$

$$W_{b2} = F_2(y)G + \beta_0\lambda_1^3 W_m^3 [F_6(y) - F_6(1)] \tag{3.23b}$$

where

$$\begin{aligned} F_5(y) = &\frac{e_1}{3(1+\beta_1)}y^{3(1+\beta_1)} + \frac{e_2}{3(1-\beta_1)}y^{3(1-\beta_1)} \\ &+ \frac{e_3}{2(3+\beta_1)}y^{2(3+\beta_1)} + \frac{e_4}{2(3-\beta_1)}y^{2(3-\beta_1)} \\ &+ \frac{e_5}{2(2+\beta_1)}y^{2(2+\beta_1)} + \frac{e_6}{2(2-\beta_1)}y^{2(2-\beta_1)} \\ &+ \frac{e_7}{9+\beta_1}y^{9+\beta_1} + \frac{e_8}{9-\beta_1}y^{9-\beta_1} + \frac{e_9}{7+\beta_1}y^{7+\beta_1} \\ &+ \frac{e_{10}}{7-\beta_1}y^{7-\beta_1} + \frac{e_{11}}{5+\beta_1}y^{5+\beta_1} + \frac{e_{12}}{5-\beta_1}y^{5-\beta_1} \\ &+ \frac{e_{13}}{3+\beta_1}y^{3+\beta_1} + \frac{e_{14}}{3-\beta_1}y^{3-\beta_1} + \frac{e_{15}}{2+\beta_1+\beta_2}y^{2+\beta_1+\beta_2} \\ &+ \frac{e_{16}}{2+\beta_1-\beta_2}y^{2+\beta_1-\beta_2} + \frac{e_{17}}{2-\beta_1+\beta_2}y^{2-\beta_1+\beta_2} \\ &+ \frac{e_{18}}{2-\beta_1-\beta_2}y^{2-\beta_1-\beta_2} + \frac{e_{19}}{5+\beta_2}y^{5+\beta_2} + \frac{e_{20}}{5-\beta_2}y^{5-\beta_2} \\ &+ \frac{e_{21}}{3+\beta_2}y^{3+\beta_2} + \frac{e_{22}}{3-\beta_2}y^{3-\beta_2} + \frac{e_{23}}{12}y^{12} + \frac{e_{24}}{10}y^{10} + \frac{e_{25}}{8}y^8 \\ &+ \frac{e_{26}}{6}y^6 + \frac{e_{27}}{4}y^4 + \frac{e_{28}}{1+\beta_1}y^{1+\beta_1} + \frac{e_{29}}{1-\beta_1}y^{1-\beta_1}, \end{aligned}$$

$$\begin{aligned} F_6(y) = &\frac{f_1}{6}y^6\ln^3 y + \frac{f_2}{4}y^4\ln^3 y + \frac{f_3}{2}y^2\ln^3 y + \frac{f_8}{3}\ln^3 y \\ &+ \frac{f_4}{8}y^8\ln^2 y - \frac{f_1-2f_5}{12}y^6\ln^2 y - \frac{3f_2-4f_6}{16}y^4\ln^2 y \\ &- \frac{3f_3-2f_7}{4}y^2\ln^2 y + \frac{f_{14}}{2}\ln^2 y + \frac{f_9}{10}y^{10}\ln y \\ &- \frac{f_4-4f_{10}}{32}y^8\ln y + \frac{f_1-2f_5+6f_{11}}{36}y^6\ln y \\ &+ \frac{3f_2-4f_6+8f_{12}}{32}y^4\ln y + \frac{3f_3-2f_7+2f_{13}}{4}y^2\ln y + f_{22}\ln y \\ &+ \frac{f_{15}}{12}y^{12} - \frac{f_9-10f_{16}}{100}y^{10} + \frac{f_4-4f_{10}+32f_{17}}{256}y^8 \\ &- \frac{f_1-2f_5+6f_{11}-36f_{13}}{216}y^6 - \frac{3f_2-4f_6+8f_{12}-32f_{19}}{128}y^4 \end{aligned}$$

$$-\frac{3f_3-2f_7+2f_{13}-4f_{20}-4f_{21}}{8}y^2. \tag{3.24}$$

Substituting solution (3.23a) into the expression (3.10), we have

$$G=\lambda_1 W_m+\lambda_3 W_m^3 \tag{3.25}$$

where

$$\lambda_3=\lambda_1^4\{\beta_3[F_5(y_1)-F_5(y_2)]+\beta_0[F_6(y_2)-F_6(1)]\}. \tag{3.26}$$

Expression (3.25) is the characteristic relation of a corrugated annular plate with a plane boundary region, a non-deformable rigid body at the center and a rigidly clamped outer edge. In order to facilitate application, relation (3.25) can be written in dimensional form as

$$g=\alpha(h^2\lambda_1 w_m+\lambda_3 w_m^3) \tag{3.27}$$

where the w_m is the deflection at $r=R_1$ for the plate and g, α, μ_1 and μ_2 have the following meaning:

(1) The main load is a uniform pressure q:

$$g=q,\ \alpha=\frac{2D_a}{R^4h^2k_t},\ \mu_1=1,\ \mu_2=\frac{p}{\pi R^2 q}. \tag{3.28}$$

(2) The main load is a concentrated load p:

$$g=p,\ \alpha=\frac{2\pi D_a}{R^2h^2k_t},\ \mu_1=\frac{\pi R^2 q}{p},\ \mu_2=1. \tag{3.29}$$

Now we shall study another non-dimensional nonlinear boundary value problem (2.19), (2.10b), (2.20), (2.11b), (2.14), (2.15) and (2.17) of the corrugated annular plate with a loosely clamped outer edge. Actually, the characteristic relation of large deflection of this corrugated annular plate is easy to obtain from the above results (3.27) obtained for the rigidly clamped outer edge. Comparing the difference between the two kinds of boundary conditions, only (2.13c) differs from (2.15c). We can rewrite (2.13c) as

$$\frac{y}{\nu'}\frac{\mathrm{d}S}{\mathrm{d}y}-S=0 \qquad \text{at } y=1 \tag{3.30}$$

and let $\nu'\to\infty$, we obtain

$$S=0 \qquad \text{at } y=1. \tag{3.31}$$

There, (3.31) is just the same as (2.15c). For this reason, the analysis of the second problem can be simplified. Let $\nu'\to\infty$ and, from the characteristic relation (3.27), we can directly obtain the characteristic relation of the corrugated annular plate with a plane boundary region, a non-deformable rigid body at the center and a loosely clamped outer edge. After solving the problem, we know that only c_7, d_{10}, d_{11}, I_3, I_4, J_2 and t_5 need to be reformed and the other parameters remain the same as the results of the first problem. At this time, we have

$$c_7=\frac{I_3t_5-I_4t_2}{\beta_3(\beta_4t_5t_7-\beta_5t_2t_8)},\quad d_{10}=\frac{\beta_4I_3t_7-\beta_5I_4t_8}{\beta_0(\beta_4t_5t_7-\beta_5t_2t_8)},$$

$$d_{11}=J_2-d_{10},\quad I_3=J_3-\frac{\beta_3\beta_4J_1}{\beta_2+\nu}y_1^{\beta_2}y_2^{-\beta_2}+\beta_0y_2^{-1}J_2,$$
$$I_4=J_4+\beta_3\beta_5y_1^{\beta_2}y_2^{-\beta_2}J_1-\beta_0(1+\nu)y_2^{-1}J_2,$$
$$J_2=d_7+d_8+d_9,\quad t_5=(1-\nu)y_2+(1+\nu)y_2^{-1}. \tag{3.32}$$

Obviously, the above formulas are suitable for manifold cases of corrugated annular plates with deep or shallow corrugations of various types.

4. Numerical Examples

As a particular example, we consider a corrugated annular plate with toothed corrugations for the case of a rigidly clamped outer edge under the pressure uniformly distributed over surfaces of the plate and the body as shown in Fig. 2. Assume $E=13\,500\text{kg/mm}^2$, $\nu=0.3$, $R=30.75\text{mm}$, $R_1=4.75\text{mm}$, $R_2=24.75\text{mm}$, $H=0.4$ mm, $\theta=23°45'$.

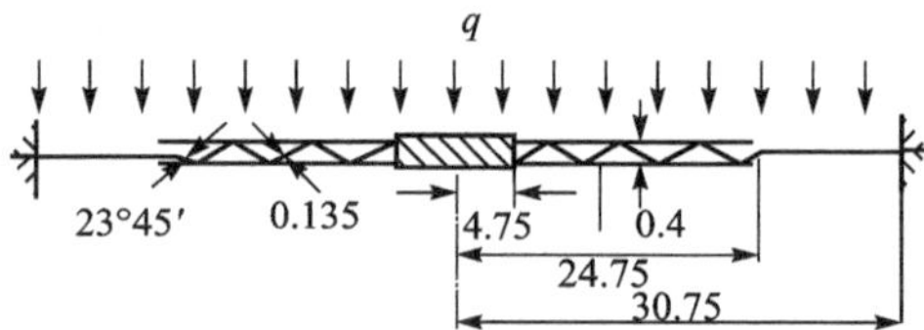

Fig. 2　A corrugated annular plate with toothed corrugations for the case of a rigidly clamped outer edge under uniform pressure

Then we have

$$p=0 \tag{4.1}$$

and, from (3.28), we obtain

$$\mu_2=0. \tag{4.2}$$

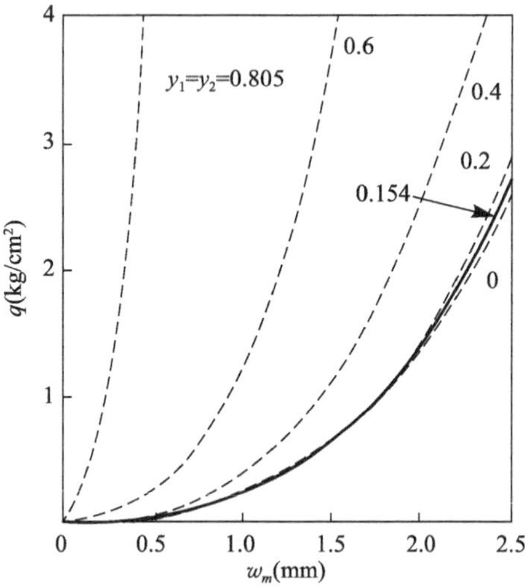

Fig. 3　The effect of the size of the non-dimensional inner radius on the rigidity of corrugated annular plates

Substituting these values into formulas (3.27) and (3.28), we obtain

$$q=0.06570w_m+0.1666w_m^3 \tag{4.3}$$

where w_m is in mm and q is in kg/cm^2.

Fig. 3 indicates the different characteristic curves of the plate when the non-dimensional inner radius y_1 is changed from 0 to $y_1=y_2$, with the other parameters unchanged. From this figure, it may be seen that the rigidity of the plate increases with an increase of y_1.

Fig. 4 indicates the different characteristic curves of this plate when the radius y_2 of the corrugated region is changed from 0.2 to $y_2=1$ with the other parameters unchanged. It is seen that the rigidity of the plate decreases with the increase of y_2.

Fig. 5 indicates the characteristic curve of this plate in the particular case of $R=R_2=24.75\text{mm}$. To make it easier to compare with other results, we give Andryevw's experiment results and theoretical results[14] from Galerkin's method. From this figure, it can be concluded that the present results are in good agreement with the experimental values.

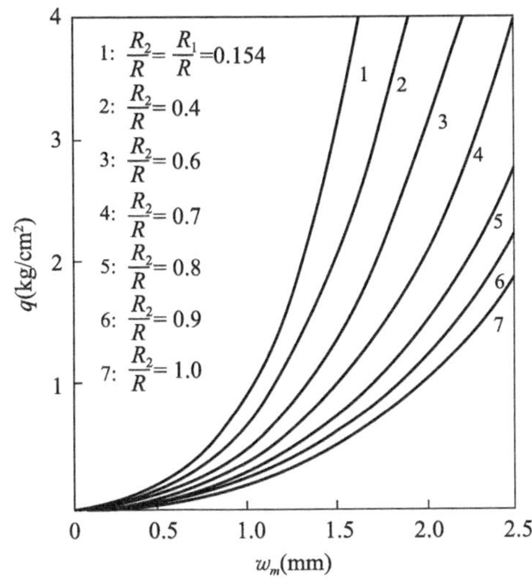

Fig. 4 The effect of the size of the radius of the corrugated region on the rigidity of corrugated annular plates

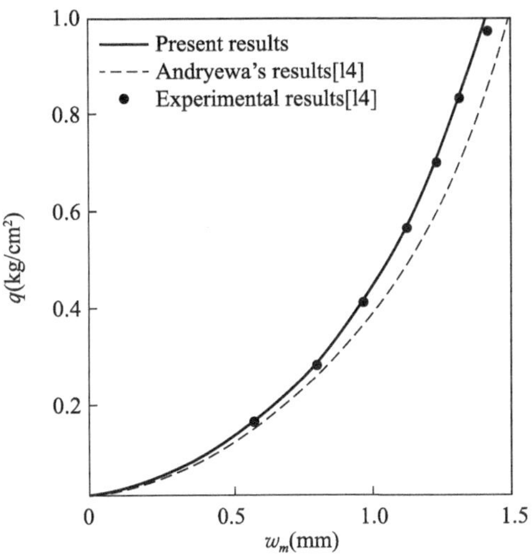

Fig. 5 Comparison between the oretical and experimental results

References

[1] Liu Renhuai, The characteristic relations of corrugated circular plates. *Acta Mech.* sin. 1, 47 (1978) (in Chinese).

[2] Liu Renhuai, The characteristic relations of corrugated circular plates with plane central regions. *J. China Univ. Sci. Technol.* 9, 75 (1979) (in Chinese).

[3] Liu Renhuai, Large deflection of corrugated circular plate with a plane central region under the action of concentrated loads at the center. *Int. J. Nonlinear Mech.* 19, 409 (1984).

[4] Liu Renhuai, Large deflection of corrugated circular plate with plane boundary region. *Solid Mech. Archs.* 9, 383 (1984).

[5] Liu Renhuai, Nonlinear bending of corrugated annular plates. *Sci. Sin. Ser.* A, 27, 640 (1984).

[6] Liu Renhuai, Nonlinear analysis of corrugated annular plates under compound load. *Sci. Sin. Ser.* A, 28, 959 (1985).

[7] Liu Renhuai, On large deflection of corrugated annular plates under uniform pressure, *Adv, Appl. Math. Mech. China*, 1. 138 (1987).

[8] Liu Renhuai, Nonlinear analysis of a corrugated circular plate under combined lateral loading. *Appl. Math. Mech.* 9, 711 (1988).

[9] Liu Renhuai and D. Li, On the nonlinear bending and vibration of corrugated circular plates. *Int. J. Nonlinear Mech.* 24, 165 (1989).

[10] D. Y. Panov, On large deflection of circular membranes with weak corrugations. *Prikl. Math. Mekh.* 5, 308 (1941) (in Russian).

[11] V. I. Feodosev, On large deflection and stability of a circular membrane with fine corrugations. *Prikl. Math. Mekh.* 9, 389 (1945) (in Russian).

[12] V. I. Feodosev, *Elastic Elements of Precision-Instruments Manufacture*. Oborongiz, Moscow (1949) (in Russian).

[13] L. E. Andryeva, The calculation of a corrugated membrane as an anisotropic plate. *Engrs Collection*, 21, 128 (1955) (in Russian).

[14] L. E. Andryeva, *Elastic Elements of Instruments*. Masgiz, Moscow (1962) (in Russian).

[15] T. Akasaka, On the elastic properties of the corrugated diaphragm. *J. Jap. Soc. Aeronaut, Eng.* 3, 279 (1955) (in Japanese).

[16] J. A. Haringx, Nonlinearity of corrugated diaphragms. *Appl. Sci. Res. Ser.* A, 16, 45 (1956).

[17] J. A. Haringx, Design of corrugated diaphragms. *Trans. ASME*, Ser. A, 79, 55 (1957).

[18] E. V. Burmistrov, Axisymmetrical bending of isotropic and anisotropic shells of revolution under large deflection and non-uniform temperature field. *Engrs Collection*, 27, 185 (1960) (in Russian).

[19] S. L. Chen, Elastic behavior of uniformly loaded circular corrugated plate with sine-shaped shallow waves in large deflection. *Appl. Math. Mech.* 1, 26, (1980).

[20] K. Y. Yeh, Liu Renhuai, Q. Y. Ping, et al. Nonlinear stability of thin circular shallow spherical shell under actions of axisymmetric uniformly distributed line loads. *J. Lanzhou Univ.* 2, 10; *Bull. Sci.* 2, 142 (1965) (in Chinese).

[21] K. Y. Yeh, Liu Renhuai, C. Z. Zhang, et al., Nonlinear stability of thin circular shallow spherical shell under action of uniform edge moment. *Bull. Sci.* 2, 145 (1965) (in Chinese).

[22] Liu Renhuai, Nonlinear stabilitity of circular shallow spherical shell with hole in the center under the action of uniform moment at the inner edge. *Bull. Sci.* 3, 253 (1965) (in Chinese).

均布载荷作用下波纹环形板的非线性弯曲①

在精密仪器弹性元件中，波纹圆板占有十分重要的地位。因此，研究波纹圆板的非线性弯曲问题就成了精密仪器弹性元件开发研究工作的重要组成部分。这类问题的研究是从最简单的全波纹圆板开始的。但在实际上，为了便于和适当的零件焊接在一起，需要在板的中心留下一块刚性部分，于是便开始了波纹环形板的研究工作，如Андреева[1]，赤坂隆[2]和刘人怀[3-5]等。显然，有了刚性中心后，求解问题的难度大大增加了。本文在以前工作[3-5]的基础上，考虑了一种形状更加复杂的波纹环形板，即不但有一块刚性中心，而且有平面边缘。据我们所知，这种板在理论上尚无人研究过。应用正交各向异性和各向同性圆板的非线性弯曲理论及修正迭代法，我们得到了均布载荷作用下具有刚性中心和平面边缘的波纹环形板的非线性特征关系式。所得结果可供精密仪器弹性元件设计时应用。

一、边值问题的建立

考虑承受均布压力 q 的具有刚性中心和平面边缘的波纹环形板（如图 1 所示）。它的挠度 w 和径向薄膜力内 N_r 所满足的正交各向异性和各向同性圆板的大挠度方程为[3-7]

当 $R_1 < r \leqslant R_2$ 时，

$$
\begin{aligned}
& r\frac{\mathrm{d}^3 w_a}{\mathrm{d}r^3}+\frac{\mathrm{d}^2 w_a}{\mathrm{d}r^2}-k_t\frac{\mathrm{d}w_a}{\mathrm{d}r}\left(\frac{1}{D_a}rN_{r,a}+\frac{k_t'}{r}\right)=\frac{k_t}{2D_a}qr^2,\\
& r\frac{\mathrm{d}^2}{\mathrm{d}r^2}(rN_{r,a})+\frac{\mathrm{d}}{\mathrm{d}r}(rN_{r,a})-k_rk_tN_{r,a}+\frac{Ehk_t}{2}\left(\frac{\mathrm{d}w_a}{\mathrm{d}r}\right)^2=0
\end{aligned}
\tag{1}
$$

当 $R_2 \leqslant r \leqslant R$ 时，

$$
\begin{aligned}
& r\frac{\mathrm{d}^3 w_b}{\mathrm{d}r^3}+\frac{\mathrm{d}^2 w_b}{\mathrm{d}r^2}-\frac{\mathrm{d}w_b}{\mathrm{d}r}\left(\frac{1}{D_b}rN_{r,b}+\frac{1}{r}\right)=\frac{1}{2D_b}qr^2,\\
& r\frac{\mathrm{d}^2}{\mathrm{d}r^2}(rN_{r,b})+\frac{\mathrm{d}}{\mathrm{d}r}(rN_{r,b})-N_{r,b}+\frac{Eh}{2}\left(\frac{\mathrm{d}w_b}{\mathrm{d}r}\right)^2=0
\end{aligned}
\tag{2}
$$

其中，r 是径向坐标，R 是外半径，R_1 是内半径，R_2 是波纹区域的半径，h 是厚度，L 是波长，H 是波幅，k_r、k_t 和 k_t' 分别是波纹环形板的径向和周向刚度参数，E 是弹性模量，ν 是泊松比，D 是抗弯刚度，

$$
D_a=\frac{Eh^3}{12\left(1-\dfrac{\nu^2}{k_tk_t'}\right)},\quad D_b=\frac{Eh^3}{12(1-\nu^2)}
\tag{3}
$$

① 本文原载《华南理工大学学报》，1994，22（增刊）：1-10. 作者：刘人怀，翟赏中.

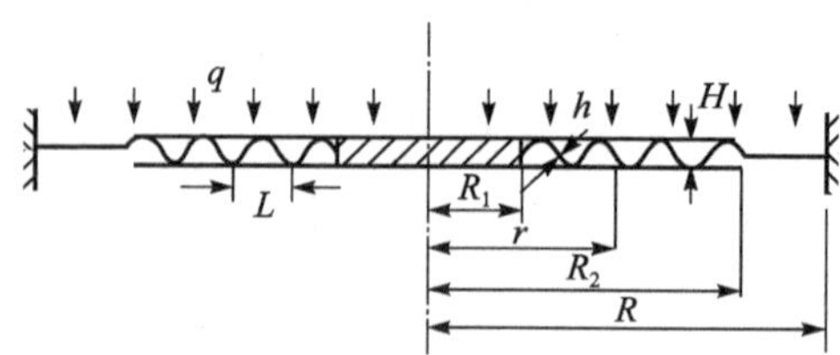

图 1　具有刚性中心和平面边缘的波纹环形板

我们讨论下面两种常用的边界条件：

(1) 沿外边界夹紧固定。

当 $r=R_1$ 时，$\dfrac{\mathrm{d}w_a}{\mathrm{d}r}=0$，$u_a=0$

$$\text{当 } r=R \text{ 时，} w_b=0,\ \frac{\mathrm{d}w_b}{\mathrm{d}r}=0,\ u_b=0 \tag{4}$$

(2) 沿外边界滑动固定。

当 $r=R_1$ 时，$\dfrac{\mathrm{d}w_a}{\mathrm{d}r}=0$，$u_a=0$

$$\text{当 } r=R \text{ 时，} w_b=0,\ \frac{\mathrm{d}w_b}{\mathrm{d}r}=0,\ N_{r,b}=0 \tag{5}$$

其中，u 是径向位移，

$$u_a=\frac{r}{k_tEh}\left[\frac{\mathrm{d}}{\mathrm{d}r}(rN_{r,a}-\nu N_{r,a})\right],$$

$$u_b=\frac{\nu}{Eh}\left[\frac{\mathrm{d}}{\mathrm{d}r}(rN_{r,b}-\nu N_{r,b})\right] \tag{6}$$

在波纹区域和平面边缘的连接处，还需要满足以下连续条件：

当 $r=R_2$ 时，$w_a=w_b$，$\dfrac{\mathrm{d}w_a}{\mathrm{d}r}=\dfrac{\mathrm{d}w_b}{\mathrm{d}r}$，

$$M_{r,a}=M_{r,b},\ N_{r,a}=N_{r,b},\ u_a=u_b \tag{7}$$

其中，M_r 是径向弯矩，

$$M_{r,a}=-\frac{D_a}{k_t}\left(\frac{\mathrm{d}^2w_a}{\mathrm{d}r^2}+\frac{\nu}{r}\frac{\mathrm{d}w_a}{\mathrm{d}r}\right),$$

$$M_{r,b}=-D_b\left(\frac{\mathrm{d}^2w_a}{\mathrm{d}r^2}+\frac{\nu}{r}\frac{\mathrm{d}w_a}{\mathrm{d}r}\right) \tag{8}$$

于是，我们的问题便归结为求解非线性边值问题（1）、（2）、（4）、（7）或（1）、（2）、（5），（7）。为求解简单，我们引入下列无量纲量：

$$\rho=\frac{r}{R},\quad y_a=\frac{w_a}{h},\quad y_b=\frac{w_b}{h},\quad \varphi_a=\frac{\mathrm{d}y_a}{\mathrm{d}\rho},$$

$$\varphi_b=\frac{\mathrm{d}y_b}{\mathrm{d}\rho},\quad \rho_1=\frac{R_1}{R},\quad \rho_2=\frac{R_2}{R},\quad S_a=-\frac{Rk_tr}{D_a}N_{r,a},$$

$$S_b=-\frac{Rr}{D_b}N_{r,b},\quad P=\frac{R^4}{2D_bh}q,\quad \beta_0=b(1-\nu^2),\quad \beta_1^2=k_tk_t',$$

$$\beta_2^2=k_r k_t,\quad \beta_3=\frac{Eh^3k_t^2}{2D_a},\quad \beta_4=\frac{D_a}{k_tD_b},\quad \beta_5=\nu(1-\beta_4),$$

$$\beta_6=\nu\left(1-\frac{1}{k_t}\right),\quad \beta_7=\frac{\beta_4}{k_t}\tag{9}$$

进行代换后，方程（1）、(2)、(4)、(5) 和 (7) 成为下述无量纲形式：

当 $\rho_1\leqslant\rho\leqslant\rho_2$ 时，$L_1\ (\rho^{\beta_1}\varphi_a)\ =-S_a\varphi_a+\dfrac{P}{\beta_4}\rho^2$，

$$L_2\ (\rho^{\beta_2}S_a)\ =\beta_3\varphi_a^2\tag{10}$$

当 $\rho_2\leqslant\rho\leqslant 1$ 时，$L_3\ (\rho\varphi_b)\ =-S_b\varphi_b+P\rho^2$，

$$L_3\ (\rho S_b)\ =\beta_0\varphi_b^2\tag{11}$$

（1）沿外边界夹紧固定。

当 $\rho=\rho_1$ 时，$\varphi_a=0$，$\dfrac{\mathrm{d}S_a}{\mathrm{d}\rho}-\dfrac{\nu}{\rho}S_a=0$；

当 $\rho=1$ 时，$y_b=0$，$\varphi_b=0$，$\dfrac{\mathrm{d}S_b}{\mathrm{d}\rho}-\dfrac{\nu'}{\rho}S_b=0$　　(12a～e)

（2）沿外边界滑动固定。

当 $\rho=\rho_1$ 时，$\varphi_a=0$，$\dfrac{\mathrm{d}S_a}{\mathrm{d}\rho}-\dfrac{\nu}{\rho}S_a=0$

当 $\rho=1$ 时，$y_b=0$，$\varphi_b=0$，$S_b=0$　　(13a～e)

连续条件：

当 $\rho=\rho_2$ 时，$y_a=y_b$，$\varphi_a=\varphi_b$，$\beta_4\dfrac{\mathrm{d}\varphi_a}{\mathrm{d}\rho}=\dfrac{\mathrm{d}\varphi_b}{\mathrm{d}\rho}+\dfrac{\beta_5}{\rho}\varphi_b$，

$$\beta_4S_a=S_b,\beta_7\frac{\mathrm{d}S_a}{\mathrm{d}\rho}=\frac{\mathrm{d}S_b}{\mathrm{d}\rho}-\frac{\beta_6}{\rho}S_b\tag{14}$$

这里，为了以后求解方便，我们将边界条件（12e）中的泊松比 ν 暂记为 ν'。另外

$$L_1(\cdots)=\rho^{\beta_1}\frac{\mathrm{d}}{\mathrm{d}\rho}\rho^{1-2\beta_1}\frac{\mathrm{d}}{\mathrm{d}\rho}(\cdots)$$

$$L_2(\cdots)=\rho^{\beta_2}\frac{\mathrm{d}}{\mathrm{d}\rho}\rho^{1-2\beta_2}\frac{\mathrm{d}}{\mathrm{d}\rho}(\cdots)$$

$$L_3(\cdots)=\rho\frac{\mathrm{d}}{\mathrm{d}\rho}\frac{1}{\rho}\frac{\mathrm{d}}{\mathrm{d}\rho}(\cdots)\tag{15}$$

二、边值问题的求解

我们用修正迭代法先求解沿外边界夹紧固定的波纹环形板的无量纲非线性边值问题(10)、(11)、(12) 和 (14)。在第一次近似中，我们有线性边值问题如下：

当 $\rho_1\leqslant\rho\leqslant\rho_2$ 时，　$L_1\ (\rho^{\beta_1}\varphi_{a1})\ =\dfrac{P}{\beta_4}\rho^2$，

$$L_2\ (\rho^{\beta_2}S_{a1})\ =\beta_3\varphi_{a1}^2\tag{16a,b}$$

当 $\rho_2<\rho\leqslant 1$ 时，　$L_3\ (\rho\varphi_{b1})\ =P\rho^2$，

$$L_3\ (\rho S_{b1})\ =\beta_0\varphi_{b1}^2 \tag{17a,b}$$

当 $\rho=\rho_1$ 时，$\varphi_{a,1}=0$，$\dfrac{\mathrm{d}S_{a1}}{\mathrm{d}\rho}-\dfrac{\nu}{\rho}S_{a1}=0$ (18a～c)

当 $\rho=1$ 时，$y_{b1}=0$，$\varphi_{b1}=0$，$\dfrac{\mathrm{d}S_{b1}}{\mathrm{d}\rho}-\dfrac{\nu'}{\rho}S_{b1}=0$ (19a～c)

当 $\rho=\rho_2$ 时，$y_{a1}=y_{b1}$，$\varphi_{a1}=\varphi_{b1}$，$\beta_4\dfrac{\mathrm{d}\varphi_{a1}}{\mathrm{d}\rho}=\dfrac{\mathrm{d}\varphi_{b1}}{\mathrm{d}\rho}+\dfrac{\beta_5}{\rho}\varphi_{b1}$，

$$\beta_4 S_{a1}=S_{b1},\ \beta_7\frac{\mathrm{d}S_{a1}}{\mathrm{d}\rho}=\frac{\mathrm{d}S_{b1}}{\mathrm{d}\rho}-\frac{\beta_6}{\rho}S_{b1} \tag{20a～e}$$

应用条件（18a)、(19a，b）和（20a，b，c)，方程（16a）和（17a）的解为

$$\begin{aligned}
\varphi_{a1}&=(a_1\rho^{-\beta_1}+a_2\rho^{\beta_1}+a_3\rho^3)P,\\
\varphi_{b1}&=(b_1\rho^{-1}+b_2\rho+b_3\rho^3)P,
\end{aligned} \tag{21a,b}$$

$$\begin{aligned}
y_{a1}&=-\Big[\frac{1}{2}\Big(b_2+\frac{b_3}{2}\Big)-b_1\ln\rho_2+\frac{a_1}{1-\beta_1}\rho_2^{1-\beta_1}+\frac{a_2}{1+\beta_1}\rho_2^{1+\beta_1}-\frac{1}{2}b_2\rho_2^2\\
&\quad-\frac{b_3-a_3}{4}\rho_2^4-\frac{a_1}{1-\beta_1}\rho_1^{1-\beta_1}-\frac{a_2}{1+\beta_1}\rho_1^{1+\beta_1}-\frac{a_3}{4}\rho^4\Big]P,\\
y_{b1}&=-\Big[\frac{1}{2}\Big(b_2+\frac{b_3}{2}\Big)-b_1\ln\rho_2-\frac{1}{2}b_2\ln^2\rho-\frac{1}{4}b_3\rho^4\Big]P
\end{aligned} \tag{22a,b}$$

其中

$$\begin{aligned}
a_1&=\frac{\lambda_3\lambda_5-\lambda_2\lambda_6}{\lambda_1\lambda_5-\lambda_2\lambda_4},\qquad a_2=-a_3\rho_1^{3-\beta_1}-a_1\rho_1^{-2\beta_1},\\
a_3&=\frac{1}{\beta_4(9-\beta_1^2)},\qquad b_1=\frac{\lambda_1\lambda_6-\lambda_3\lambda_4}{\lambda_1\lambda_5-\lambda_2\lambda_4},\\
b_2&=-\frac{1}{8}-b_1,\qquad b_3=\frac{1}{8},\\
\lambda_1&=\rho_2^{-\beta_1}-\rho_1^{-2\beta_1}\rho_1^{\beta_1},\qquad \lambda_2=-\rho_2^{-1}+\rho_2,\\
\lambda_3&=a_3(\rho_1^{3-\beta_1}-1)\rho_2^{\beta_1},\qquad \lambda_4=-\beta_1\beta_4(\rho_2^{-\beta_1}+\rho_2^{-2\beta_1}\rho_2^{\beta_1})\rho_2^{-1},\\
\lambda_5&=(1-\beta_5)\rho_2^{-2}+1+\beta_5,\\
\lambda_6&=\frac{1}{9-\beta_1^2}(\beta_1\rho_1^{3-\beta_1}\rho_2^{-1+\beta_1}-3\rho_2^2)+\frac{1}{8}[(3+\beta_5)\rho_2^2-1-\beta_5]
\end{aligned} \tag{23}$$

令 y_m 为波纹环形板内边缘的无量纲挠度：

当 $\rho=\rho_1$ 时，　$y_a=y_m$ (24)

将式（22a）代入上式，便得波纹环形板的小挠度特征关系式

$$P=\alpha_1 y_m \tag{25}$$

其中

$$\begin{aligned}
\alpha_1&=\Big[b_1\ln\rho_2+\frac{a_1}{1-\beta_1}(\rho_1^{1-\beta_1}-\rho_2^{1-\beta_1})+\frac{a_2}{1+\beta_1}(\rho_1^{1+\beta_1}-\rho_2^{1+\beta_1})\\
&\quad+\frac{b_2}{2}(\rho_2^2-1)+\frac{b_3}{4}(\rho_2^4-1)+\frac{a_3}{4}(\rho_1^4-\rho_2^4)\Big]^{-1}
\end{aligned} \tag{26}$$

再将式（25）代入式（21)，又得

$$\varphi_{a1}=\alpha_1 y_m(a_1\rho^{-\beta_1}+a_2\rho^{\beta_1}+a_3\rho^3),$$
$$\varphi_{b1}=\alpha_1 y_m(b_1\rho^{-1}+b_2\rho+b_3\rho^3) \tag{27}$$

将式 (27) 代入方程 (16b) 和 (17b) 的右端，应用条件 (18b)，(19c) 和 (20d,e)，便得到方程 (16b) 和 (17b) 的解为

$$S_{a1}=\alpha_1^2 y_m^2(c_1\rho^{-\beta_2}+c_2\rho^{\beta_2}+c_3\rho^{1-2\beta_1}+c_4\rho^{1+2\beta_1}+c_5\rho^{4-\beta_1}+c_6\rho^{4+\beta_1}+c_7\rho+c_8\rho^7),$$
$$S_{b1}=\alpha_1^2 y_m^2(d_1\rho^{-1}+d_2\rho^{-1}\ln\rho+d_3\rho+d_4\rho\ln\rho+d_5\rho^3+d_6\rho^5+d_7\rho^7) \tag{28a,b}$$

其中

$$c_1=\frac{\lambda_9\lambda_{11}-\lambda_8\lambda_{12}}{\lambda_7\lambda_{11}-\lambda_8\lambda_{10}},$$

$$\begin{aligned}c_2=\frac{\rho_1^{1-\beta_2}}{\beta_2-\nu}\big[&(\beta_2-\nu)c_1\rho_1^{-1-\beta_2}-(1-2\beta_1-\nu)c_3\rho_1^{-2\beta_1}\\&-(1+2\beta_1-\nu)c_4\rho_1^{2\beta_1}-(4-\beta_1-\nu)c_5\rho_1^{3-\beta_1}\\&-(4+\beta_1-\nu)c_6\rho_1^{3+\beta_1}-(1-\nu)c_7-(7-\nu)c_8\rho_1^6\big],\end{aligned}$$

$$c_3=\frac{\beta_3 a_1^2}{(1-2\beta_1)^2-\beta_2^2},\qquad c_4=\frac{\beta_3 a_2^2}{(1+2\beta_1)^2-\beta_2^2},$$

$$c_5=\frac{2\beta_3 a_2 a_3}{(4-\beta_1)^2-\beta_2^2},\qquad c_6=\frac{2\beta_3 a_2 a_3}{(4+\beta_1)^2-\beta_2^2},$$

$$c_7=\frac{2\beta_3 a_1 a_2}{1-\beta_2^2},\qquad c_8=\frac{\beta_3 a_3^2}{49-\beta_2^2},$$

$$d_1=\frac{\lambda_7\lambda_{12}-\lambda_9\lambda_{10}}{\lambda_7\lambda_{11}-\lambda_8\lambda_{10}},\quad d_2=-\frac{\beta_0}{2}b_1^2,$$

$$\begin{aligned}d_3=\frac{1}{1-\nu'}\big[&(1+\nu')d_1-d_2-d_4-(3-\nu')d_5\\&-(5-\nu')d_6-(7-\nu')d_7\big],\end{aligned}$$

$$d_4=\beta_0 b_1 b_2,\qquad d_5=\frac{\beta_0}{8}(b_2^2+2b_1b_3),$$

$$d_6=\frac{\beta_0}{12}b_2b_3,\qquad d_7=\frac{\beta_0}{48}b_3^2,$$

$$\lambda_7=\beta_4\left(\rho_2^{-\beta_2}+\frac{\beta_2+\nu}{\beta_2-\nu}\rho_2^{-2\beta_2}\rho_2^{\beta_2}\right),\lambda_8=-\left(\rho_2^{-1}+\frac{1+\nu'}{1-\nu'}\rho_2\right),$$

$$\begin{aligned}\lambda_9=\frac{\beta_4\rho_2^{\beta_2}}{\beta_2-\nu}\big[&(1-2\beta_1-\nu)c_3\rho_1^{1-2\beta_1-\beta_2}+(1+2\beta_1-\nu)c_4\rho_1^{1+2\beta_1-\beta_2}\\&+(4-\beta_1-\nu)c_5\rho_1^{4-\beta_1-\beta_2}+(4+\beta_1-\nu)c_6\rho_1^{4+\beta_1-\beta_2}\\&+(1-\nu)c_7\rho_1^{1-\beta_2}+(7-\nu)c_8\rho_1^{7-\beta_2}\big]\\&-\beta_4(c_3\rho_2^{1-2\beta_1}+c_4\rho_2^{1+2\beta_1}+c_5\rho_2^{4-\beta_1}+c_6\rho_2^{4+\beta_2}+c_7\rho_2+c_8\rho_2^7)\\&-\frac{\rho_2}{1-\nu'}\big[d_2+d_4+(3-\nu')d_5+(5-\nu')d_6+(7-\nu')d_7\big]\\&+d_2\rho_2^{-1}\ln\rho_2+d_4\rho_2\ln\rho_2+d_5\rho_2^3+d_6\rho_2^5+d_7\rho_2^7,\end{aligned}$$

$$\lambda_{10}=\beta_2\beta_7\rho_2^{-1}\left(\frac{\beta_2+\nu}{\beta_2-\nu}\rho_1^{-2\beta_2}\rho_2^{\beta_2}-\rho_2^{-2\beta_2}\right),$$

$$\lambda_{11}=\frac{1+\beta_6}{\rho_2^2}-\frac{1+\nu'}{1-\nu'}(1-\beta_6),$$

$$\begin{aligned}\lambda_{12}=&\frac{\beta_2\beta_7\rho_2^{-1+\beta_2}}{\beta_2-\nu}[(1-2\beta_1-\nu)c_3\rho_1^{1-2\beta_1-\beta_2}+(1+2\beta_1-\nu)c_4\rho_1^{1+2\beta_1-\beta_2}\\&+(4-\beta_1-\nu)c_5\rho_1^{4-\beta_1-\beta_2}+(4+\beta_1-\nu)c_6\rho_1^{4+\beta_1-\beta_2}\\&+(1-\nu)c_7\rho_1^{1-\beta_2}+(7-\nu)c_8\rho^{7-\beta_2}]\\&-\beta_7[(1-2\beta_1)c_3\rho_2^{-2\beta_1}+(1+2\beta_1)c_4\rho_2^{2\beta_1}+(4-\beta_1)c_5\rho_2^{3-\beta_1}\\&+(4+\beta_1)c_6\rho_2^{3+\beta_1}+c_7+7c_8\rho_2^6]+d_2\rho_2^{-2}-(1+\beta_6)d_2\rho_2^{-2}\ln\rho_2\\&-\frac{1-\beta_6}{1-\nu'}[d_2+d_4+(3-\nu')d_5+(5-\nu')d_6+(7-\nu')d_7]\\&+d_4-(1-\beta_6)d_4\ln\rho_2+(3-\beta_6)d_5\rho_2^2\\&+(5-\beta_6)d_6\rho_2^4+(7-\beta_6)d_7\rho_2^7\end{aligned}\tag{29}$$

在第二次近似中，我们有如下的关于 φ_2 的线性边值问题：

当 $\rho_1\leqslant\rho\leqslant\rho_2$ 时， $L_1\ (\rho^{\beta_1}\varphi_{a2})\ =-S_{a1}\varphi_{a1}+\dfrac{P}{\beta_4}\rho^2$ (30)

当 $\rho_2\leqslant\rho\leqslant1$ 时， $L_3\ (\rho\varphi_{b2})\ =-S_{b1}\varphi_{b1}+P\rho^2$ (31)

当 $\rho=\rho_1$ 时，$\varphi_{a2}=0$ (32)

当 $\rho=1$ 时，$y_{b2}=0$，$\varphi_{b2}=0$ (33)

当 $\rho=\rho_2$ 时，$y_{a2}=y_{b2}$，$\varphi_{a2}=\varphi_{b2}$；$\beta_4\dfrac{d\varphi_{a2}}{d\rho}=\dfrac{d\varphi_{b2}}{d\rho}+\dfrac{\beta_5}{\rho}\varphi_{b2}$ (34)

解此边值问题，得解为

$$\begin{aligned}\varphi_{a2}=&(a_1\rho^{-\beta_1}+a_2\rho^{\beta_1}+a_3\rho^3)P\\&-\alpha_1^3y_m^3[f_1\rho^{-\beta_1}+f_2\rho^{\beta_1}+F_1(\rho)],\\\varphi_{b2}=&(b_1\rho^{-1}+b_2\rho+b_3\rho^3)P-\alpha_1^3y_m^3[g_1\rho^{-1}+g_2\rho+G_1(\rho)];\end{aligned}\tag{35}$$

$$\begin{aligned}y_{a2}=&y_{a1}-\alpha_1^3y_m^3\Big[g_1\ln\rho_2+\frac{1}{2}g_2(\rho_2^2-1)+\frac{f_1}{1-\beta_1}(\rho^{1-\beta_1}-\rho_2^{1-\beta_1})\\&+\frac{f_2}{1+\beta_1}(\rho^{1+\beta_1}-\rho_2^{1+\beta_1})+F_2(\rho)-F_2(\rho_2)\\&+G_2(\rho_2)-G_2(1)\Big],\\y_{b2}=&y_{b1}-\alpha_1^3y_m^3\Big[g_1\ln\rho+\frac{1}{2}g_2(\rho^2-1)+G_2(\rho)-G_2(1)\Big]\end{aligned}\tag{36a,b}$$

其中

$$\begin{aligned}F_1(\rho)=&f_3\rho^{1-\beta_1-\beta_2}+f_4\rho^{1-\beta_1+\beta_2}+f_5\rho^{1+\beta_1-\beta_2}\\&+f_6\rho^{1+\beta_1+\beta_2}+f_7\rho^{2-\beta_1}+f_8\rho^{2+\beta_1}+f_9\rho^{2-3\beta_1}\\&+f_{10}\rho^{2+3\beta_1}+f_{11}\rho^{4-\beta_2}+f_{12}\rho^{4+\beta_2}+f_{13}\rho^{5-2\beta_1}\\&+f_{14}\rho^{5+2\beta_1}+f_{15}\rho^{8-\beta_1}+f_{16}\rho^{8+\beta_1}+f_{17}\rho^5+f_{18}\rho^{11},\end{aligned}$$

$$F_2(\rho)=\frac{f_3}{2-\beta_1-\beta_2}\rho^{2-\beta_1-\beta_2}+\frac{f_4}{2-\beta_1+\beta_2}\rho^{2-\beta_1+\beta_2}$$
$$+\frac{f_5}{2+\beta_1-\beta_2}\rho^{2+\beta_1-\beta_2}+\frac{f_6}{2+\beta_1+\beta_2}\rho^{2+\beta_1+\beta_2}$$
$$+\frac{f_7}{3-\beta_1}\rho^{3-\beta_1}+\frac{f_8}{3+\beta_1}\rho^{3+\beta_1}+\frac{f_9}{3(1-\beta_1)}\rho^{3-3\beta_1}$$
$$+\frac{f_{10}}{3(1+\beta_1)}\rho^{3+3\beta_1}+\frac{f_{11}}{5-\beta_2}\rho^{5-\beta_2}+\frac{f_{12}}{5+\beta_2}\rho^{5+\beta_2}$$
$$+\frac{f_{13}}{2(3-\beta_1)}\rho^{6-2\beta_1}+\frac{f_{14}}{2(3+\beta_1)}\rho^{6+2\beta_1}+\frac{f_{15}}{9-\beta_1}\rho^{9-\beta_1}$$
$$+\frac{f_{16}}{9+\beta_1}\rho^{9+\beta_1}+\frac{f_{17}}{6}\rho^6,+\frac{f_{18}}{12}\rho^{12},$$

$$G_1(\rho)=g_3\rho^{-1}\ln\rho+g_4\rho^{-1}\ln^2\rho+g_5\rho\ln\rho+g_6\rho\ln^2\rho+g_7\rho^3$$
$$+g_8\rho^3\ln\rho+g_9\rho^5+g_{10}\rho^5\ln\rho+g_{11}\rho^7+g_{12}\rho^9+g_{13}\rho^{11},$$

$$G_2(\rho)=\frac{1}{2}g_3\ln^2\rho+\frac{1}{3}g_4\ln^3\rho+\frac{1}{4}(g_6-g_5)\rho^2$$
$$+\frac{1}{2}(g_5-g_6)\rho^2\ln\rho+\frac{1}{2}g_6\rho^2\ln^2\rho+\frac{1}{16}(4g_7-g_8)\rho^4$$
$$+\frac{1}{4}g_8\rho^4\ln\rho+\frac{1}{36}(6g_9-g_{10})\rho^6+\frac{1}{6}g_{10}\rho^6\ln\rho$$
$$+\frac{1}{8}g_{11}\rho^8+\frac{1}{10}g_{12}\rho^{10}+\frac{1}{12}g_{13}\rho^{12},$$

$$f_1=\frac{\lambda_5\lambda_{13}-\lambda_2\lambda_{14}}{\lambda_1\lambda_5-\lambda_2\lambda_4},\qquad f_2=-\rho_1^{-\beta_1}[F_1(\rho_1)+f_1\rho_1^{-\beta_1}],$$

$$f_3=\frac{a_1c_1}{(1-\beta_1-\beta_2)^2-\beta_1^2},\qquad f_4=\frac{a_1c_2}{(1-\beta_1+\beta_2)^2-\beta_1^2},$$

$$f_5=\frac{a_2c_1}{(1+\beta_1-\beta_2)^2-\beta_1^2},\qquad f_6=\frac{a_2c_2}{(1+\beta_1+\beta_2)^2-\beta_1^2},$$

$$f_7=\frac{a_1c_7+a_2c_3}{(2-\beta_1)^2-\beta_1^2},\qquad f_8=\frac{a_1c_4+a_2c_7}{(2+\beta_1)^2-\beta_1^2},$$

$$f_9=\frac{a_1c_3}{(2-3\beta_1)^2-\beta_1^2},\qquad f_{10}=\frac{a_2c_4}{(2+3\beta_1)^2-\beta_1^2},$$

$$f_{11}=\frac{a_3c_1}{(4-\beta_2)^2-\beta_1^2},\qquad f_{12}=\frac{a_3c_2}{(4+\beta_2)^2-\beta_1^2},$$

$$f_{13}=\frac{a_1c_5+a_3c_3}{(5-2\beta_1)^2-\beta_1^2},\qquad f_{14}=\frac{a_2c_6+a_3c_4}{(5+2\beta_1)^2-\beta_1^2},$$

$$f_{15}=\frac{a_1c_8+a_3c_5}{(8-\beta_1)^2-\beta_1^2},\qquad f_{16}=\frac{a_2c_8+a_3c_6}{(8+\beta_1)^2-\beta_1^2},$$

$$f_{17}=\frac{a_1c_6+a_2c_5+a_3c_7}{25-\beta_1^2},\qquad f_{18}=\frac{a_3c_8}{121-\beta_1^2},$$

$$g_1=\frac{\lambda_1\lambda_{14}-\lambda_4\lambda_{13}}{\lambda_1\lambda_5-\lambda_2\lambda_4},\quad g_2=-[G_1(1)+g_1],$$

$$g_3=-\frac{b_1}{4}(2d_1+d_2),\quad g_4=-\frac{1}{4}b_1d_2,$$

$$g_5=\frac{1}{4}(2b_1d_3+2b_2d_1-b_1d_4-b_2d_2),\quad g_6=\frac{1}{4}(b_1d_4+b_2d_2),$$

$$g_7=\frac{1}{32}[4b_1d_5+b_2(4d_3-3d_4)+b_3(4d_1-3d_2)],$$

$$g_8=\frac{1}{8}(b_2d_4+b_3d_2),$$

$$g_9=\frac{1}{288}[12b_1d_6+12b_2d_5+b_3(12d_3-5d_4)],$$

$$g_{10}=\frac{1}{24}b_3d_4,$$

$$g_{11}=\frac{1}{48}(b_1d_7+b_2d_6+b_3d_5),\qquad g_{12}=\frac{1}{80}(b_2d_7+b_3d_6),$$

$$g_{13}=\frac{1}{120}b_3d_7,$$

$$\lambda_{13}=\rho_1^{-\beta_1}\rho_2^{\beta_1}F_1(\rho_1)-F_1(\rho_2)+G_1(\rho_2)-\rho_2G_1(1),$$

$$\begin{aligned}\lambda_{14}=&\beta_1\beta_4\rho_1^{-\beta_1}\rho_2^{-1+\beta_1}F_1(\rho_1)-\beta_4\frac{\mathrm{d}F_1(\rho_2)}{\mathrm{d}\rho}+\frac{\mathrm{d}G_1(\rho_2)}{\mathrm{d}\rho}\\&+\beta_5\rho_2^{-1}G_1(\rho_2)-(1+\beta_5)G_1(1)\end{aligned}\tag{37}$$

将解（36a）代入式（24），便得波纹环形板的非线性特征关系式

$$P=\alpha_1y_m+\alpha_3y_m^3\tag{38}$$

其中

$$\begin{aligned}\alpha_3=\Big[&\frac{f_1}{1-\beta_1}(\rho_1^{1-\beta_1}-\rho_2^{1-\beta_1})+\frac{f_2}{1+\beta_1}(\rho_1^{1+\beta_1}-\rho_2^{1+\beta_1})+g_1\ln\rho_2\\&-\frac{1}{2}(g_2-g_3\rho_2^2)+F_2(\rho_1)-F_2(\rho_2)+G_2(\rho_2)-G_2(1)\Big]\alpha_1^4\end{aligned}\tag{39}$$

为使用方便，我们将式（38）转换成有量纲形式：

$$q=\frac{2D_b}{R_4}\left(\alpha_1w_m+\frac{\alpha_3}{h^2}w_m^3\right)\tag{40}$$

其中，w_m 为波纹环形板内边缘的挠度。

下面，我们求解波纹环形板在滑动固定外边界条件下的无量纲非线性边值问题（10）、（11）、（13）和（14）。将此问题与前一问题相比较，我们发现仅有边界条件（12e）和（13e）不相同。若将边界条件（12e）改写为

$$\text{当}\ \rho=1\ \text{时，}\ \frac{1}{\nu'}\frac{\mathrm{d}S_b}{\mathrm{d}\rho}-\frac{S_b}{\rho}=0\tag{41}$$

并令 $\nu'\to\infty$，即得与式（13e）相同的形式。鉴于此，我们可直接地利用式（40），只要令$\nu'\to\infty$，就可得到现有问题的解。经过计算，在保留原有公式的形式下，仅改变下面的公式即可：

$$c_1 = \frac{\lambda_{15}\lambda_{16} - \lambda_2\lambda_{17}}{\lambda_7\lambda_{16} - \lambda_2\lambda_{10}}, \quad d_1 = \frac{\lambda_7\lambda_{17} - \lambda_{10}\lambda_{15}}{\lambda_7\lambda_{16} - \lambda_2\lambda_{10}},$$

$$d_3 = -d_1 - d_5 - d_6 - d_7,$$

$$\begin{aligned}
\lambda_{15} = & \frac{\beta_4\rho_2^{\beta_2}}{\beta_2 - \nu}[(1-2\beta_1-\nu)c_3\rho_1^{1-2\beta_1-\beta_2} + (1+2\beta_1-\nu)c_4\rho_1^{1+2\beta_1-\beta_2} \\
& + (4-\beta_1-\nu)c_5\rho_1^{4-\beta_1-\beta_2} + (4+\beta_1-\nu)]c_6\rho_1^{4+\beta_1-\beta_2} \\
& + (1-\nu)c_7\rho_1^{1-\beta_2} + (7-\nu)c_8\rho_1^{7-\beta_2}] \\
& - \beta_4(c_3\rho_1^{1-2\beta_1} + c_4\rho_2^{1+2\beta_1} + c_5\rho_2^{4-\beta_1} \\
& + c_6\rho_2^{4+\beta_1} + c_7\rho_2 + c_8\rho_2^7) + d_2\rho_2^{-1}\ln\rho_2 \\
& - \rho_2(d_5 + d_6 + d_7) + d_4\rho_2\ln\rho_2 + d_5\rho_2^3 + d_6\rho_2^5 + d_7\rho_2^7,
\end{aligned}$$

$$\lambda_{16} = \rho_2^{-2}(1+\beta_6) + 1 - \beta_6,$$

$$\begin{aligned}
\lambda_{17} = & \frac{\beta_2\beta_7\rho_2^{-1+\beta_2}}{\beta_2 - \nu}[(1-2\beta_1-\nu)c_3\rho_1^{1-2\beta_1-\beta_2} \\
& + (1+2\beta_1-\nu)c_4\rho_1^{1+2\beta_1-\beta_2} + (4-\beta_1-\nu)c_5\rho_1^{4-\beta_1-\beta_2} \\
& + (4+\beta_1-\nu)c_6\rho_1^{4+\beta_1-\beta_2} + (1-\nu)c_7\rho_1^{1-\beta_2} + (7-\nu)c_8\rho_1^{7-\beta_2})] \\
& - \beta_7[(1-2\beta_1)c_3\rho_2^{1-2\beta_1} + (1+2\beta_1)c_4\rho_2^{2\beta_1} + (4-\beta_1)c_5\rho_2^{3-\beta_1} \\
& + (4+\beta_1)c_6\rho_2^{3+\beta_1} + c_7 + 7c_8\rho_2^6] + d_2\rho^{-2} - (1+\beta_6)d_2\rho_2^{-2}\ln\rho_2 \\
& - (1-\beta_6)(d_5 + d_6 + d_7) + d_4 + (1-\beta_6)d_4\ln\rho_2 \\
& + (3-\beta_6)d_5\rho_2^2 + (5-\beta_6)d_6\rho_2^4 + (7-\beta_7)d_7\rho_2^6
\end{aligned}$$

三、算例和讨论

现在，我们来绘制承受均布压力 q、沿外边缘夹紧固定的具有刚性中心和平面边缘的正弦波纹环形板的特征曲线。已知数据为

$$h=0.1\text{mm},\ H=0.4\text{mm},\ R=24\text{mm},\ L=5.4\text{mm},$$
$$E=1.35\times10^6\text{kg/cm}^2,\ \nu=0.3$$

我们讨论两种情况。对于第一种情况，

设

$$R_2 = 19.2\text{mm},\ \rho_1 = 0.1, 0.2, 0.3$$

将上述数据代入特征关系式（40），经过计算，所得结果给在图 2 中。由图看出，随着刚性中心半径的增大，波纹环形板的非线性效应逐渐减弱。

对于第二种情况，

设

$$R_1 = 4.8\text{mm}, \quad \rho_2 = 0.5, 0.6, 0.7, 0.8, 0.9, 1.0$$

将这些数据代入式（40），所得计算结果给在图 3 中。由图看出，随着波纹区域半径的增大，亦即平面边缘区域的缩小，波纹环形板的非线性效应逐渐增强。

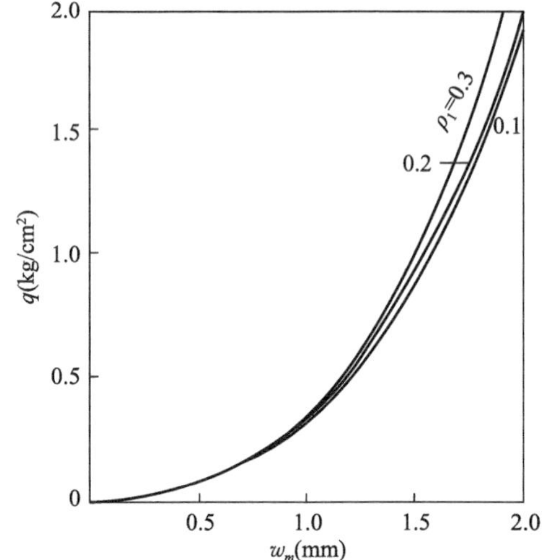

图 2　刚性中心半径的大小对特征曲线的影响

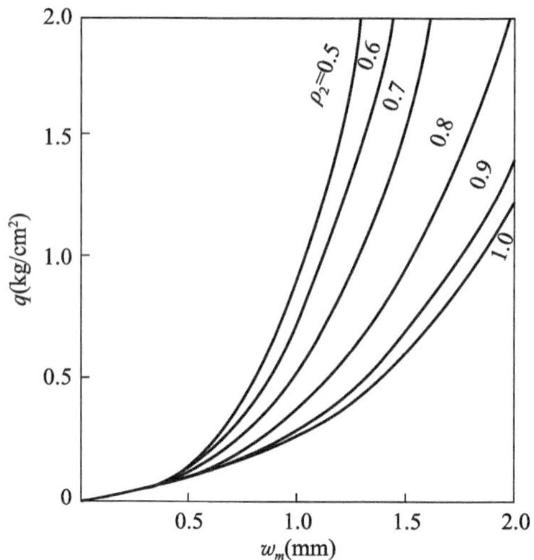

图 3　波纹区域半径的大小对特征曲线的影响

参考文献

[1]　Андреева，Л. Е. Ynpyzue Элементы Приборов. Москва：Машгиз，1992.

[2]　赤坂隆. Corrugated diaphragmの弹性特性について. 日本航空学会誌，1955，3 (22-23)：279.

[3]　刘人怀. 波纹环形板的非线性弯曲. 中国科学，A 辑，1984 (3)：247.

[4]　刘人怀. 在复合载荷作用下波纹环形板的非线性弯曲. 中国科学，A 辑，1985 (6)：537.

[5]　Liu Renhuai. On large deflection of corrugated annular plates under uniform pressure. The Advances of Applied Mathematics and Mechanics In China，vol. 1，Beijing：China Academic Publishers，1987，138.

[6]　Liu Renhuai. Large deflection of corrugated circular plate with plane boundary region. Solid Mechanics Archives，1984，9 (4)：383.

[7]　Liu Renhuai. Large deflection of corrugated circular plate with a plane central region under the action of concentrated loads at the center. International Journal of Non-Linear Mechanics，1984，19 (5)：409.

波纹管的制造与理论研究概况①

波纹管是一种圆柱型的薄壁金属或非金属壳体，沿着圆周有波纹的折皱（图 1）。

波纹管的尺寸在非常宽阔的范围内变化，其直径小到 2mm，大到 10m 左右。

波纹管的功能：在轴向力、内压力、外压力和弯矩的单独或联合作用下，能产生相应的位移。这种功能使波纹管在各种技术领域里有广泛的应用，成为液压气压敏感元件、密封元件、热膨胀接头元件和柔性元件的主要形式，在仪器仪表、自动化系统、国防军工、石油化工、核电站和航天航空等各种工业部门中得到了重要的应用[1-8]。

一、制造历史

最早是在 1844 年，在欧洲，人们提出了波纹管的设想，即一种理论上的构造形状。当时，欧洲需要为火车锅炉的蒸汽设计一种测压计，共提出了三种测压元件：波纹管，波纹板和弹簧管。后两种设计成功，而波纹管则由于加工制造的困难没有设计成功，半途而废。

直到 1881 年，普鲁士的一家工厂提出了三块波纹板组成的厘式测压计，把膜盒测压计提高到一个新水平，这是世界上第一支波纹管，不过，这仅是波纹管的一种雏型（图 2）。

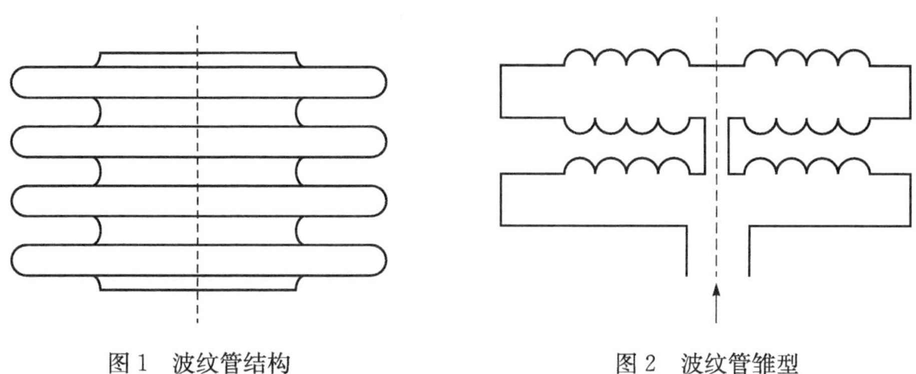

图 1　波纹管结构　　　　图 2　波纹管雏型

1903 年，英国由于气象工作的需要，终于正式生产了波纹管，有好几种型号，其材料主要是铜和铜合金。

直到第二次世界大战初期，即 1940 年左右，由于飞机潜艇和气象的要求，波纹管的生产规模才有了较大发展。英国有 5 个专业厂，美国有 7 个工厂，德国也有专业厂。

从 1960 年开始，随着石油化学、航天航空和原子能等工业的迅猛发展，波纹管的生产规模迅速扩张。在美国、英国、德国和日本，增设了许多制造厂。伴随工业和科

① 本文是全国首届管道技术与设备学术会议的大会特邀报告，南京，1994 年 10 月 14 日；压力容器和压力管道的分析与计算，北京：科学出版社，2014，267-272，299-301.

技的发展，波纹管的制造技术也渐渐趋于成熟。

我国在中华人民共和国成立前，已在上海新新仪表厂和新成仪表厂试制过波纹管。中华人民共和国成立后，这两个厂合并成为上海仪表厂。现在的上海弹性元件厂就是从上海仪表厂分出来的，这样，我国才有了自己的制造波纹管的工厂。到 20 世纪 90 年代，全国大型专业生产波纹管的工厂有 19 个。据不完全统计，全国已有 500 余个制造厂。

二、波纹管的应用

按照功能划分，可将波纹管的应用分为以下五个方面。

(1) 作为仪表敏感元件。

在民用方面，可用于各种控制仪表（调温等）和测量仪表（测压、测温、测压差、测温差）以及制冷机、制氦机等控调机构。

在军用方面，可用作鱼雷、炸弹等的触发元件，降落伞的自动开伞装置，发动机增压器、指示器等。

(2) 在密封系统中作密封元件。

(3) 用于轴位移和角位移的弹性传递。

(4) 作介质隔离器。

(5) 作管道热膨胀的补偿器和远距离液压传递元件。

重点介绍一下第五个方面，现代工程的一个重要特点是管道化程度越来越高。管道作为水、石油、蒸汽、煤气等的输送通路，几乎遍布各种工程、遍布所有城市以至农村。

对管道的热胀冷缩和防震，需要进行补偿，这是管道工程中的重要技术之一。

传统的补偿技术是采用方形补偿器，我们称为第一代补偿器，即是在直线管道中增加“Ⅱ”形管道，形成一定的弹性，以解决补偿问题，它主要有以下三个缺点：一是补偿能力差，因为管道壁厚较大，故不可能具备很大的补偿量；只好增加数量，那么就会大大增加工程造价。若在地下还需增加观察井，造价会更大；二是占地面积大，一般占地面积要3～4m^2，在城市中很难实施；三是应力大，容易损坏管道。

为了解决上述问题，出现了第二代补偿器，即套筒式补偿器，或称滑动套筒装置，如图 3 所示。这种补偿器采用填料密封，让两端带有动密封的一段套筒在管道中滑动，以实现补偿。这一代补偿器的优点是，补偿量可达到 300～400mm，而且占地面积也不大。但是它有两个致命弱点：一是密封不可靠，易泄漏；二是安装很困难，稍微有点倾斜，就很容易卡住，实现不了补偿。

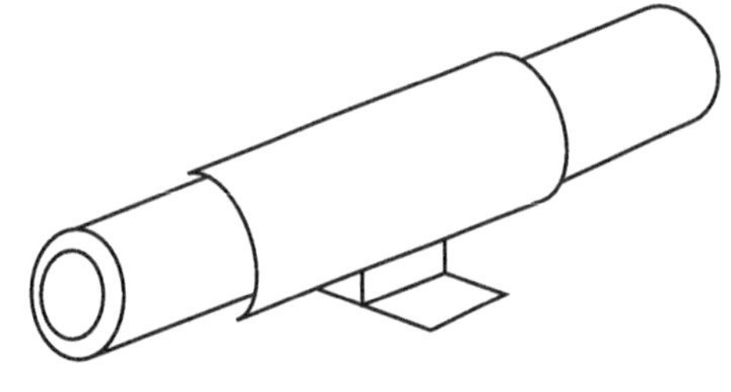

图 3　套筒式补偿器

由于这些缺点，这一代产品在国外未得到广泛应用，在国内采用的也不多。

鉴于此，便出现了第三代补偿器，即波纹膨胀节，这种补偿器弥补了上述产品的不足，占地面积小，便于安装，密封性能好，工作状况可靠。特别应该指出，前两代补偿器只能吸收与管道轴线方向一致的运动，故使用效果受到限制，而波纹膨胀节不同，不需保养就能吸收较小空间的多种运动。由于它具有许多突出的优点，因此十余年来，在国际上被迅速推广到几乎所有新建工程中，形成了管道补偿中的一个热点，而在我国，从 20 世纪 80 年代末起，短短几年就已有 200 余个工厂生产这种产品。

但是，波纹膨胀节也存在两个缺点：①由于许多企业多采用单层波纹管作为补偿器件，因此可靠性差，这便增加了工程的造价；②出于耐腐蚀的考虑，一般采用价格昂贵的不锈钢波纹管，因此波纹膨胀节的价格明显高于套筒式。

上述缺点造成工程造价的提高，对于国情而言，不容轻视。因此，改进膨胀节，势在必行。现已有多层式波纹膨胀节，3～12 层，内外层采用耐腐蚀的不锈钢，夹层用普通碳素钢，降低了成本，提高了可靠性。此外，还出现了复合材料波纹膨胀节，单层式或多层式，也改善了性能。

波纹膨胀节的波型常见有以下四种。

（1）U 型（图 4（a））。

特点是伸缩性好，耐压能力强，制造容易，补偿能力较强。

（2）Ω 型（图 4（b））。

特点是耐压能力最大，伸缩性小，制造较复杂，不易产生应力集中，补偿能力差。

（3）V 型（图 4（c））。

特点是耐压能力最小，伸缩性好，制造复杂，补偿能力强。

（4）S 型（图 4（d））。

特点是耐压能力强，制造复杂，补偿能力较差。

因此，工程上最常应用 U 型膨胀节。

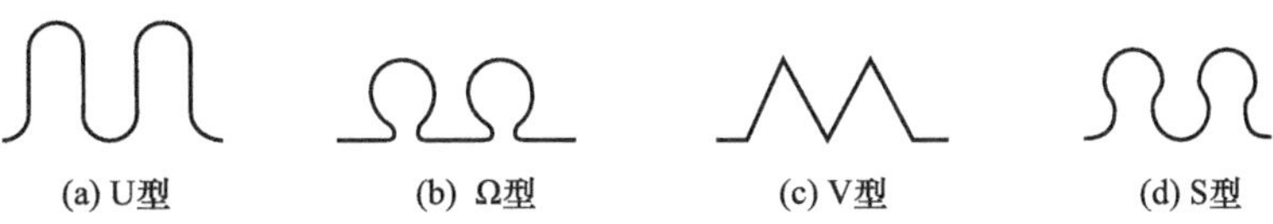

图 4　波纹膨胀节的波型

三、我国波纹管产业状况和问题

长期以来，我国波纹管产业只是依赖于其他产业的一个行业。20 世纪 80 年代以前，国内制造波纹管的工厂除了有限的几个厂外，大多是仪表厂的一个车间。80 年代以后，这些车间升格成为分厂，基本上不再是仪表厂的附属单位。同时，一大批专业化波纹管厂相继诞生，如电冰箱温控器波纹管生产厂、波纹膨胀节生产厂等。

20 世纪 80 年代以前，全国波纹管一年生产量在 50 万只左右，产值 1000 多万元。1988 年，总产量超过 1000 万只，产值 1.8 亿元。按国家统计，1980 年仪表工业 14 类产品中，有 10 类的年产值都低于 1.8 亿元，即低于波纹管一个单独产品 1988 年的产值。目前，波纹管的年产值几乎每年都以 30％的高速度在增长。1993 年，年产值已达 5 亿元以上。

由于波纹管最先应用在仪表上，因此，一般将它划归仪表元件。这种划分，在1985年以前没有问题。1985年以后，波纹管的应用领域以惊人的速度急剧扩大，石油、化工、轻工、电子、供热、电力、造船、航天、航空、民用等工业对它们需要量急剧增长，大大超过了仪表工业自身的需要。1985年以前，波纹管90%都用于仪表工业。到1988年，这一比例降到30%，即70%的波纹管都用在非仪表工业中，在这些领域，波纹管已经不是仪表元件，而是管件、结构件，成为一种横跨多种工业的独立性产业产品。

但是，我国在这一领域中仍存在以下几个问题：

（1）在国际上的现代化国家，波纹管已经趋于普及。但在我们国家，仍然在初期，与发达国家约有20年差距。

（2）我国的波纹管工艺技术陈旧，分散生产，技术力量弱，产品质量不高。

（3）理论研究与实际工程结合得不好。

显然，上述问题都应尽快解决。

四、波纹管的理论研究情况

研究波纹管理论的学者很多。

在早期，Feely 等[9]和 Turner[10]等借助一维的梁理论，将波纹管视为梁来研究。这种方法的特点是，忽略波纹管的周向作用，将其子午线近似地看作为一根曲梁，这种方法只能给出很粗糙的结果。

此外，一些学者则应用能量方法，即采用最小势能原理来研究波纹管。这一方法的特点是，首先将波纹管的总势能用其子午线的变形转角来表示，再将此角近似地用含有待定常数的公式来表达，最后，使总势能最小，从而确定待定常数。Феодосвев[2]给出了波纹管的刚度公式，Dahl[11]研究了 Ω 型波纹管，Turner 等[12]研究了 S 型等波纹管。这种方法所给出的结果误差很大。

以前，人们认为波纹管由环形板和圆环壳组成，研究的最大难点是寻求圆环壳的解。于是，学者采用薄壳线性理论来研究圆环壳。

首先 Reissner[13]和 Meissner[14]建立了圆环壳一般轴对称问题的基本微分方程组，包含两个方程和两个变量（子午线转角和剪力），为波纹管的研究打下理论基础，然后 Törker[15]，Clark[16]和诺沃日洛夫[17]以前述理论为基础，进行了复变量化的简化工作，推导出圆环壳的一个复变量的二阶变系数非齐次常微分工程，他们所得的方程形式各不相同。1979年，钱伟长等[18]用统一的复变量化过程，分别导出了前述学者的复变量方程，并证明了它们的差异都在 Love-Kirchhoff 薄壳假定的容许范围以内。

下面将介绍圆环壳的线弹性方程的求解情况。

（1）解析方法

Wissler[19]首先用幂级数的方法求解了轴对称圆环壳方程的齐次式。Tao[20]对幂级数解进行了收敛证明，但这证明有错误。直到1979年，钱伟长[21]才进行了正确的分析，纠正了 Tao 的错误，澄清了长期以来关于收敛区域的争议。Chang[22]、Ota 等[23]和 Hamada 等[24]也用幂级数方法研究了细环壳问题。

在使用渐近积分方法方面，Clark[16]给出了圆环壳方程的渐近解，并与 Wissler 的幂级数解进行了比较，接着，他又继续给出了轴向力作用下的渐近解[25]。Jenssen[26]给出了一个与齐次解精度相当的特解。Anderson[27]进一步改进了 Clark 的工作。

在使用三角级数方面，诺沃日洛夫[17]首先给出了用三角级数表示的圆环壳非齐次微分方程的特解。接着，钱伟长等[18,28]用这方法成功地获得了圆环壳的一般解，并证明了解的收敛性，同时，还继续进行了许多有意义的研究[29-32]。徐志翘等[33]、陈山林等[34,35]和 Turner[10]等也曾使用这一方法求解。

此外，董明德[36]使用积分变换法求解圆环壳的基本方程，获得了一个新的级数解，但未进行数值计算。

（2）数值方法

谢志成等[37]使用有限元法给出了圆环壳的数值解，并与钱伟长等[31]的理论解和 Turner 等[12]的实验结果进行了比较，结果一致。

Булгаков[38]、Лаупа 等[39]、Миткевич 等[40]、Сухарев[41]、Grubitzsch[42]和 Hampl[43]等采用有限差分法对波纹管进行了数值计算，也给出了许多有用的数值结果。

几十年来，学者采用多种途径，多种方法研究波纹管，由于限于上述线性理论研究，理论结果始终与实验结果差距甚大，根本无法用于工程设计。鉴于此，学者便开始用板壳非线性理论来研究波纹管，但是，由于困难太大，进展十分缓慢。

安德列娃[3]使用差分法得到了波纹管的非线性微分方程的相当复杂的数值解。钱伟长等[32]在解析方法上迈出了成功的一步，采用非线性分析理论，并使用摄动法处理了波纹管中的环形板，而对圆环壳仍采用原来所得的线性理论解。接着，徐志翘等[33]采用钱伟长方法，对变厚度的 U 型波纹管进行了处理。虽然上述理论研究有了进步，但仍与实验结果存在严重的差异。最近，Axelrad[44]使用三角级数方法和摄动法研究了波纹管的非线性理论，但计算较繁。

1986 年，在钱伟长先生的关怀下，刘人怀等[45]承接了国家“七五”科技攻关波纹管和膜片设计项目，并于 1990 年完成，该年 8 月 17 日获国家科技成果鉴定。研究时，我们注意到前人研究的不足之处：只考虑了环形板的非线性变形，忽略了重要的圆环壳的非线性变形。于是，我们计及了压缩角的影响，将原视作环形板的部分改为变厚度截头扁锥壳，对圆环壳和变厚度截头扁锥壳皆使用非线性理论进行分析，应用摄动法求解，获得了波纹管整体非线性理论的解析解，与已有的实验值十分符合。最后，采用 Fortran-77 计算机语言编制了计算程序，绘制了图表，其公式和程序已供有关工程设计部门使用。

参考文献

[1] 钱伟长. 波纹管的制造、设计、实验和理论. 第五届全国弹性元件学术会议论文，上海，1978 年 12 月.

[2] 费奥多谢夫 B И. 精密仪器弹性元件的理论与计算. 卢文达，熊大逵，译. 北京：科学出版社，1963：135-185.

[3] 安德列娃 П E. 波纹管的计算与设计. 翁善臣，等译. 北京：国防工业出版社，1982.

[4] 刘广玉，庄肇康. 仪表弹性元件. 北京：国防工业出版社，1981：168-198.

[5] 中国石油化工总公司，中华人民共和国化学工业部，中华人民共和国机械工业部. 钢制石油化工压力容器设计规定. 北京：全国压力容器标准化技术委员会，1985：183-191.

[6] 中国石油化工总公司，中华人民共和国化学工业部，中华人民共和国机械工业部. 钢制石油化工压力容器设计规定，一九八五年编制说明. 全国压力容器标准化技术委员会，1985：241-262.

[7] 钟培滋. 波形膨胀节国内外新发展综述. 第九届全国弹性文件学术会议论文集，上海，1991，178-186.

[8] 樊大钧. 波纹管设计学. 北京：北京理工大学出版社，1988.

[9] Feely F J，Goryl W M. Stress studies on piping expansion bellows. J of Applied Mechanics，1950，17（2）：135-141.

[10] Turner C E. Stress and deflection studies of flat-plate and toroidal expansion bellows，subjected to axial，eccentric or internal pressure loading. J Mech Eng Sci，1959，1（2）：113-129；130.

[11] Dahl N C. Toroidal-Shell expansion joints. J of Applied Mechanics，ASME，1953，20（4）：497-503.

[12] Turner C E，Ford H. Stress and deflection studies of pipeline expansion bellows. Proc Inst Mech Engrs，1957，（15）：526-550.

[13] Reissner H. Spannungen in Kugelschalen（Kuppeln）. Müller-Breslau-Festschrift，Leipzig，1912，1：181-193.

[14] Meissner E. Das Elastizitätsproblem für dünne Schalen von Ringflächen-Kugel-und Kegel-form. Phys Z，1913，14：343-349.

[15] Törker F. Zur Integration der Differentialgleichungen der drehsymmetrisch belasteten Rotationsschate bei beliebiger Wandstärke. Ing-Arch. 1938，9：282-288.

[16] Clark R A. On the theory of thin elastic toroidal shells. J Math Phys，1950，29：146-178.

[17] 诺沃日洛夫 B B. 薄壳理论. 北京石油学院材料力学教研组，译. 北京：科学出版社，1959：287-300.

[18] 钱伟长，郑思梁. 轴对称圆环壳的复变量方程和轴对称细环壳的一般解. 清华大学学报（自然科学版），1979，19（1）：27-47.

[19] Wissler H. Festigkeitsberechnung von Ringflachenschalen. Dissertation，Technisch Hochschule in Zurich，1916.

[20] Tao L N. On toroidal shells. J Math Phys，1959，38（2）：130-134.

[21] 钱伟长. 环壳方程级数解的收敛性问题及其有关收敛定理的研究. 兰州大学学报，力学专号，1979：1-38.

[22] Chang W. The state of stress in toroidal and similar shells with alimental rings under torsionally symmetrically stress. 清华大学工程学报（A），1949，5：289-349.

[23] Ota T，Hamada M. On the strength of toroidal shells：lst report，a proposition on the solution. Bulletin of JSME，1963，6（24）：638-654.

[24] Hamada M，Takezono S. Strength of U-shaped bellows（1st report and second report，Case of axial loading）. Bulletin of JSME，1965，8（32）：525-531；1966，9（35）：502-513.

[25] Clark R A. An expansion bellows problem. Trans JSME，Ser E，J Appl Mech，1970，37（1）：61-69.

[26] Jenssen O. Asymptotic integration of the differential equation for a special case of symmetrically

loaded toroidal shells. J Math Phys，1960，39（1）：1-17.

[27] Anderson W F. Atomics International Division of North American Aviation. NAA-SR-4527 Pt. I，1964；Pt. II，1965.

[28] 钱伟长，郑思梁. 轴对称圆环壳的一般解. 应用数学和力学，1980，1（3）：287-300.

[29] 钱伟长. 细环壳极限方程的非齐次解及其在仪器仪表上的应用. 仪器仪表学报，1980，1（1）：92-115.

[30] 钱伟长. 半圆弧波纹管的计算：细环壳理论的应用. 清华大学学报（自然科学版），1979，19（1）：84-99.

[31] 钱伟长，郑思梁. 半圆弧波纹管的计算：环壳一般解的应用. 应用数学和力学，1981，2（1）：97-111.

[32] 钱伟长，吴明德. U型波纹管的非线性特性摄动法计算. 应用数学和力学，1983，4（5）：595-608.

[33] 徐志翘，刘燕，杨嘉实，等. 变厚度U型波纹壳大挠度问题的摄动解. 清华大学学报（自然科学版），1985，25（1）：39-51.

[34] 陈山林. 圆环壳在一般荷载下的轴对称问题. 应用数学和力学，1986，7（5）：425-434.

[35] 陈山林，王玳瑜，邹定祺. 圆环壳在离心力作用下的应力和位移. 应用数学和力学，1986，7（6）：545-552.

[36] 董明德. Novozhilov环壳方程的新解. 应用数学和力学，1985，6（5）：401-413.

[37] 谢志成，付承诵，郑思梁. 有曲率突变的轴对称壳（波纹壳）的有限元解. 应用数学和力学，1981，2（1）：113-130.

[38] Булгаков В Н. Напряжения и перемещения сильфонов. Численные методыв прнкладной теории упругости. Кнев，Наукова Думка，1968，211-248.

[39] Лаупа А，ВейлН А. Расчет компенсаторов с U-оьразными гафрами. Приклал ная Механика，1962，29（1）：130-139.

[40] Миткевич В М，шулика А К. К численному решению краевых задач статики осесимметричных оболочек методом сведения к задачам коши. Прикладная Механика，1972，（5）：34-40.

[41] Сухарев В А. Расчет сильфонов. Численные методы в прикладиой теории упругости. Киев. Наукова Думка，1968，176-210.

[42] Grubitzsch W. Festigkeitsberechnung von kompensatoren. Maschinenbautechnik，1962，11（12）：663-667.

[43] Hampl M. Anuloidove skorepiny a vlnove kompensatory pro potrubi. Nakladatelstvi ceskoslovenske Akademie VED，Praha，1958，98.

[44] Axelrad E L. Theory of Flexible Shells. Amsterdam：Elsevier Science，1987.

[45] 刘人怀，胡偎. U型波纹管非线性变形的刚度和应力分析理论. 全国第九届弹性元件学术会议论文. 广州，1992.

中心受载下具有平面边缘区域的固支波纹环形板的非线性分析①

波纹圆板以其独特的性能在各种技术领域里被广泛使用，尤其是作为敏感元件在精密仪器仪表和传感器中更占突出地位。因此，对它的非线性理论与实验的研究备受各国科学家的重视。1955 年，Андреева[1] 和赤坂隆[2] 开创性地采用圆柱正交各向异性圆板大挠度理论进行了研究。接着，Haringx[3,4] 和 Бурмистров[5] 也进行了研究工作。但是，由于波纹圆板本身形状复杂以及问题的非线性性质，理论的研究进展一直很缓慢，始终仅停留在研究最简单的问题上，且所获结果的精度也不高，远不能满足工程实际迅速发展的需要。1978 年，刘人怀[6] 使用叶开沅和刘人怀等[7-9] 早年共同提出的修正迭代法研究了这一问题，获得了既精确又简单的解析解。此文在 20 世纪 60 年代中期完成，令人遗憾的是整整推迟了十余年才得以公诸于世。此后，刘人怀又获得了一系列结果[10-21]。本文是前面工作的继续。我们对中心集中载荷作用下具有平面边缘区域和刚性中心的波纹环形板进行了非线性分析，仍使用修正迭代法进行求解，获得了较精确的解析解。据我们所知，目前尚无人研究过这一问题。本文所获得公式可供工程设计直接应用。

一、基本方程

考虑一个在集中载荷 p 作用下，具有刚性中心和平面边缘区域的波纹环形板（图 1），外半径为 R，中心部分的半径为 R_1，波纹部分的半径为 R_2，厚度为 h，波长为 L，波幅为 H，选取 r 为径向坐标。

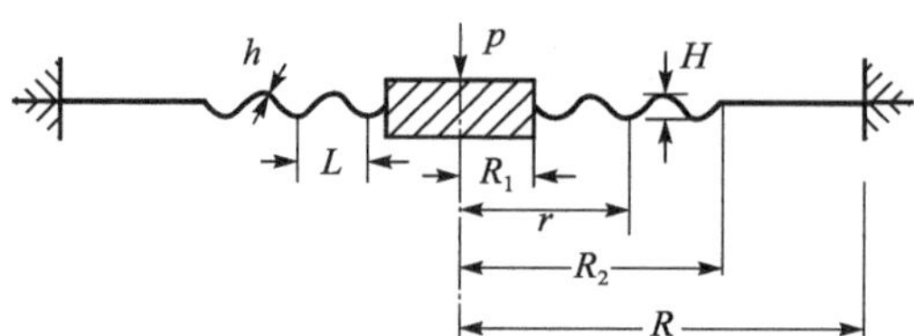

图 1　具有刚性中心和平面边缘区域的波纹环形板

应用正交各向异性圆板和各向同性圆板的大挠度弯曲理论，我们易于得到在集中载荷作用下具有刚性中心和平面边缘区域的波纹环形板的基本方程如下[10-21]。

对于 $R_1 \leqslant r \leqslant R_2$，

$$r\frac{\mathrm{d}^3 w_a}{\mathrm{d}r^3}+\frac{\mathrm{d}^2 w_a}{\mathrm{d}r^2}-\frac{k_t k_t'}{r}\frac{\mathrm{d}w_a}{\mathrm{d}r}=\frac{k_t}{D_a}\left(rN_{r,a}\frac{\mathrm{d}w_a}{\mathrm{d}r}+\frac{p}{2\pi}\right),$$

① 本文原载《暨南大学学报》，1995，16 (1)：1-13. 作者：刘人怀，徐加初.

$$r\frac{\mathrm{d}^2(rN_{r,a})}{\mathrm{d}r^2}+\frac{\mathrm{d}(rN_{r,a})}{\mathrm{d}r}-k_rk_tN_{r,a}=-\frac{1}{2}Ehk_t\left(\frac{\mathrm{d}w_a}{\mathrm{d}r}\right)^2 \tag{1}$$

对于 $R_2<r\leqslant R$，

$$r\frac{\mathrm{d}^3w_b}{\mathrm{d}r^3}+\frac{\mathrm{d}^2w_b}{\mathrm{d}r^2}-\frac{1}{r}\frac{\mathrm{d}w_b}{\mathrm{d}r}=\frac{1}{D_b}\left(rN_{r,b}\frac{\mathrm{d}w_b}{\mathrm{d}r}+\frac{p}{2\pi}\right),$$

$$r\frac{\mathrm{d}^2(rN_{r,b})}{\mathrm{d}r^2}+\frac{\mathrm{d}(rN_{r,b})}{\mathrm{d}r}-N_{r,b}=-\frac{1}{2}Eh\left(\frac{\mathrm{d}w_b}{\mathrm{d}r}\right)^2 \tag{2}$$

其中，w 是挠度，N_r 是径向薄膜内力，E 是弹性模量，ν 是泊松比，下标“a”表示板的波纹部分，下标“b”表示板的平面边缘区域部分，D 是抗弯刚度，

$$D_a=\frac{Eh^3}{12\left(1-\dfrac{\nu^2}{k_tk_t'}\right)},\quad D_b=\frac{Eh^3}{12(1-\nu^2)} \tag{3}$$

k_r、k_t 和 k_t' 分别是波纹部分的径向和周向等效刚度有关的参数，其定义可参看文献[10]。

非线性微分方程组（1）和（2）将在相应的边界条件下求解。在一般情况下，这种波纹环形板的外边界具有夹紧固定和滑动固定，而内边界则固定在可以上下移动的刚性中心上。因此，波纹环形板的通常边界条件如下。

（1）沿外边界夹紧固定：

当 $r=R_1$ 时，$\dfrac{\mathrm{d}w_a}{\mathrm{d}r}=0$，$u_a=0$

当 $r=R$ 时，$w_b=0$，$\dfrac{\mathrm{d}w_b}{\mathrm{d}r}=0$，$u_b=0$ (4)

（2）沿外边界滑动固定：

当 $r=R_1$ 时，$\dfrac{\mathrm{d}w_a}{\mathrm{d}r}=0$，$u_a=0$

当 $r=R$ 时，$w_b=0$，$\dfrac{\mathrm{d}w_b}{\mathrm{d}r}=0$，$N_{r,b}=0$ (5)

此外，在波纹部分与平面边缘区域部分的连接处，还满足如下连续条件：

当 $r=R_2$ 时，$w_a=w_b$，$\dfrac{\mathrm{d}w_a}{\mathrm{d}r}=\dfrac{\mathrm{d}w_b}{\mathrm{d}r}$，$M_{r,a}=M_{r,b}$，

$$N_{r,a}=N_{r,b},u_a=u_b \tag{6}$$

其中 M_r 为径向弯矩，u 为径向位移，

$$M_{r,a}=-\frac{D_a}{k_t}\left(\frac{\mathrm{d}^2w_a}{\mathrm{d}r^2}+\frac{\nu}{r}\frac{\mathrm{d}w_a}{\mathrm{d}r}\right),$$

$$M_{r,b}=-D_b\left(\frac{\mathrm{d}^2w_b}{\mathrm{d}r^2}+\frac{\nu}{r}\frac{\mathrm{d}w_b}{\mathrm{d}r}\right),$$

$$u_a=\frac{r}{k_tEh}\left[\frac{\mathrm{d}(rN_{r,a})}{\mathrm{d}r}-\nu N_{r,a}\right],$$

$$u_b=\frac{r}{Eh}\left[\frac{\mathrm{d}(rN_{r,b})}{\mathrm{d}r}-\nu N_{r,b}\right] \tag{7}$$

为了简化运算，且使最后的解更一般化，引入下列无量纲参量：

$$
\begin{aligned}
&y=\frac{r}{R},\quad y_1=\frac{R_1}{R},\quad y_2=\frac{R_2}{R},\quad W_a=\frac{w_a}{h},\\
&W_b=\frac{w_b}{h},\quad \varphi_a=\frac{\mathrm{d}w_a}{\mathrm{d}y},\quad \varphi_b=\frac{\mathrm{d}w_b}{\mathrm{d}y},\quad S_a=-\frac{k_tRr}{D_a}N_{r,a},\\
&S_b=-\frac{Rr}{D_b}N_{r,b},\quad P=\frac{R^2k_t}{2\pi hD_a}p,\quad \beta_0=6(1-\nu^2),\quad \beta_1^2=k_tk_t',\\
&\beta_2^2=k_rk_t,\quad \beta_3=\frac{Eh^3k_t^2}{2D_a},\quad \beta_4=\frac{D_a}{D_bk_t},\quad \beta_5=\nu(1-\beta_4),\\
&\beta_6=\nu\left(1-\frac{1}{k_t}\right),\quad \beta_7=\frac{\beta_4}{k_t}
\end{aligned}
\tag{8}
$$

引入上述无量纲参量后，非线性边值问题（1）～(6）变换成无量纲形式如下：

对于 $y_1\leqslant y\leqslant y_2$，

$$
\begin{aligned}
&L_1(y^{\beta_1}\varphi_a)=-S_a\varphi_a+P,\\
&L_2(y^{\beta_2}S_a)=\beta_3\varphi_a^2;
\end{aligned}
\tag{9a,b}
$$

对于 $y_2\leqslant y\leqslant 1$，

$$
\begin{aligned}
&L_0(y\varphi_b)=-S_b\varphi_b+\beta_4P,\\
&L_0(yS_b)=\beta_0\varphi_b^2
\end{aligned}
\tag{10a,b}
$$

其中

$$
\begin{aligned}
&L_0(\cdots)=y\frac{\mathrm{d}}{\mathrm{d}y}\frac{1}{y}\frac{\mathrm{d}}{\mathrm{d}y}(\cdots),\\
&L_1(\cdots)=y^{\beta_1}\frac{\mathrm{d}}{\mathrm{d}y}y^{1-2\beta_1}\frac{\mathrm{d}}{\mathrm{d}y}(\cdots),\\
&L_2(\cdots)=y^{\beta_2}\frac{\mathrm{d}}{\mathrm{d}y}y^{1-2\beta_2}\frac{\mathrm{d}}{\mathrm{d}y}(\cdots)
\end{aligned}
\tag{11}
$$

(1）沿外边界夹紧固定：

当 $y=y_1$ 时，　$\varphi_a=0$，　$\dfrac{\mathrm{d}S_a}{\mathrm{d}y}-\dfrac{\nu}{y}S_a=0$　　(12a,b)

当 $y=1$ 时，　$W_b=0$，　$\varphi_b=0$，　$\dfrac{\mathrm{d}S_b}{\mathrm{d}y}-\dfrac{\nu}{y}S_b=0$　　(13a～c)

(2）沿外边界滑动固定：

当 $y=y_1$ 时，　$\varphi_a=0$，　$\dfrac{\mathrm{d}S_a}{\mathrm{d}y}-\dfrac{\nu}{y}S_a=0$，　　(14a,b)

当 $y=1$ 时，　$W_b=0$，　$\varphi_b=0$，　$S_b=0$　　(15a～c)

连续条件：

当 $y=y_2$ 时，　$W_a=W_b$，　$\varphi_a=\varphi_b$，　$\dfrac{\mathrm{d}\varphi_b}{\mathrm{d}y}=\dfrac{\beta_5}{y}\varphi_b=\beta_4\dfrac{\mathrm{d}\varphi_a}{\mathrm{d}y}$

$$\frac{\mathrm{d}S_b}{\mathrm{d}y}-\frac{\beta_6}{y}S_b=\beta_7\frac{\mathrm{d}S_a}{\mathrm{d}y},\ S_b=\beta_4S_a \tag{16a～e}$$

方程（9)～(16）构成本文所要求解的两类边界条件下的无量纲非线性边值问题。

二、边值问题的求解

由于无量纲非线性边值问题式（9）～式（16）包含多个方程、边界条件及连续条件，因此求解它们非常困难。我们采用修正迭代法来克服这一困难。

我们首先求解具有夹紧固定外边界的波纹环形板的非线性边值问题（9）、（10）、（12）、（13）和（16）。

在一次近似中，略去方程（9a）和（10a）中的非线性项，我们有线性边值问题如下。

对于 $y_1 \leqslant y \leqslant y_2$，

$$L_1(y^{\beta_1}\varphi_{a,1}) = P,$$
$$L_2(y^{\beta_2}S_{a,1}) = \beta_3\varphi_{a,1}^2 \tag{17a,b}$$

对于 $y_2 \leqslant y \leqslant 1$，

$$L_0(y\varphi_{b,1}) = \beta_4 P,$$
$$L_0(yS_{b,1}) = \beta_0\varphi_{b,1}^2 \tag{18a,b}$$

当 $y = y_1$ 时，$\quad \varphi_{a,1} = 0, \quad \dfrac{\mathrm{d}S_{a,1}}{\mathrm{d}y} - \dfrac{\nu}{y}S_{a,1} = 0 \qquad (19\mathrm{a,b})$

当 $y = 1$ 时，$\quad W_{b,1} = 1, \quad \varphi_{b,1} = 0, \quad \dfrac{\mathrm{d}S_{b,1}}{\mathrm{d}y} - \dfrac{\nu}{y}S_{b,1} = 0 \qquad (20\mathrm{a}\sim\mathrm{c})$

当 $y = y_2$ 时，$\quad W_{a,1} = W_{b,1}, \quad \varphi_{a,1} = \varphi_{b,1}, \quad \dfrac{\mathrm{d}\varphi_{b,1}}{\mathrm{d}y} + \dfrac{\beta_5}{y}\varphi_{b,1} = \beta_4\dfrac{\mathrm{d}\varphi_{a,1}}{\mathrm{d}y},$

$$\frac{\mathrm{d}S_{b,1}}{\mathrm{d}y} - \frac{\beta_6}{y}S_{b,1} = \beta_7\frac{\mathrm{d}S_{a,1}}{\mathrm{d}y}, \ S_{b,1} = \beta_4 S_{a,1} \tag{21a～e}$$

求解边值问题（17a)、(18a)、(19a)、(20b）和（21b，c)，我们得到

$$\varphi_{a,1} = P(a_1 y^{\beta_1} + a_2 y^{-\beta_1} + a_3 y),$$
$$\varphi_{b,1} = P(b_1 y + b_2 y^{-1} + b_3 y\ln y) \tag{22}$$

其中

$$a_1 = \frac{t_3 t_5 - t_2 t_6}{t_1 t_5 - t_2 t_4}, \qquad b_1 = \frac{t_1 t_6 - t_3 t_4}{t_1 t_5 - t_2 t_4},$$
$$a_2 = -a_3 y_1^{1+\beta_1} - a_1 y_1^{2\beta_1}, \qquad b_2 = -b_1,$$
$$a_3 = \frac{1}{1-\beta_1^2}, \qquad b_3 = \frac{1}{2}\beta_4,$$
$$t_1 = y_2^{\beta_1} - y_1^{2\beta_1} y_2^{-\beta_1}, \qquad t_2 = y_2^{-1} - y_2,$$
$$t_3 = a_3(y_1^{1+\beta_1} y_2^{-\beta_1} - y_2) + b_3 y_2 \ln y_2,$$
$$t_4 = \beta_1\beta_4(y_2^{\beta_1 - 1} + y_1^{2\beta_1} y_2^{-\beta_1 - 1}),$$
$$t_5 = \beta_5(y_2^{-2} - 1) - y_2^{-2} - 1,$$
$$t_6 = b_3(\ln y_2 + \beta_5 \ln y_2 + 1) - a_3\beta_4(\beta_1 y_1^{1+\beta_1} y_2^{-\beta_1 - 1} + 1) \tag{23}$$

将式（22）再积分一次，并利用条件（20a）和（21a)，就得到一次近似的无量纲挠度如下

$$W_{a,1}=P\left(\frac{a_1}{1+\beta_1}y^{1+\beta_1}+\frac{a_2}{1-\beta_1}y^{1-\beta_1}+\frac{1}{2}a_3y^2+a_4\right),$$
$$W_{b,1}=P\left[\frac{1}{2}b_1y^2+b_2\ln y+\frac{1}{2}b_3y^2\left(\ln y-\frac{1}{2}\right)+b_4\right] \tag{24a,b}$$

其中

$$a_4=\frac{1}{2}b_1y_2^2+b_2\ln y_2+\frac{1}{2}b_3y_2^2\left(\ln y_2-\frac{1}{2}\right)$$
$$+b_4-\frac{a_1}{1+\beta_1}y_2^{1+\beta_1}-\frac{a_2}{1-\beta_1}y_2^{1-\beta_1}-\frac{1}{2}a_3y_2^2,$$
$$b_4=\frac{1}{4}b_3-\frac{1}{2}b_1 \tag{25}$$

以在波纹环形板 $y=y_1$ 处的无量纲挠度 $W_a(y_1)$ 作为迭代参数，并记为 W_m，则有

$$W_m=W_a\big|_{y=y_1} \tag{26}$$

将式（24a）代入式（26），得到波纹环形板的小挠度特征关系式

$$P=\lambda_1W_m \tag{27}$$

其中

$$\lambda_1=\left(\frac{a_1}{1+\beta_1}y_1^{1+\beta_1}+\frac{a_2}{1-\beta_1}y_1^{1-\beta_1}+\frac{1}{2}a_3y_1^2+a_4\right)^{-1} \tag{28}$$

将式（27）代入式（24），有

$$\varphi_{a,1}=\lambda_1W_m(a_1y^{\beta_1}+a_2y^{-\beta_1}+a_3y),$$
$$\varphi_{b,1}=\lambda_1W_m(b_1y+b_2y^{-1}+b_3y\ln y) \tag{29a,b}$$

应用已得的解（29），并利用条件（19b）、（20c）和（21d，e），我们得到方程（17b）和（18b）的解

$$S_{a,1}=\beta_3\lambda_1^2W_m^2[F_1(y)+c_7y^{\beta_2}+c_8y^{-\beta_2}],$$
$$S_{b,1}=\beta_0\lambda_1^2W_m^2[F_2(y)+f_7y+f_8y^{-1}], \tag{30a,b}$$

其中

$$F_1(y)=c_1y^{1+2\beta_1}+c_2y^{1-2\beta_1}+c_3y^{2+\beta_1}+c_4y^{2-\beta_1}+c_5y^3+c_6y,$$
$$F_2(y)=f_1y^3\ln^2y+f_2y^3\ln y+f_3y^3+f_4\ln^2y+f_5y\ln y+f_6y^{-1}\ln y,$$
$$F_1'(y)=\frac{\mathrm{d}F_1(y)}{\mathrm{d}y},\qquad F_2'(y)=\frac{\mathrm{d}F_2(y)}{\mathrm{d}y},$$
$$c_1=\frac{a_1^2}{(1+2\beta_1-\beta_2)(1+2\beta_1+\beta_2)},$$
$$c_2=\frac{a_2^2}{(1-2\beta_1-\beta_2)(1-2\beta_1+\beta_2)},$$
$$c_3=\frac{2a_1a_3}{(2+2\beta_1-\beta_2)(2+\beta_1+\beta_2)},\quad c_4=\frac{2a_2a_3}{(2-\beta_1-\beta_2)(2-\beta_1+\beta_2)},$$
$$c_5=\frac{a_3^2}{9-\beta_2^2},\qquad c_6=\frac{2a_1a_2}{1-\beta_2^2},$$
$$c_7=\frac{t_7t_{12}-t_9t_{10}}{t_7t_{11}-t_8t_{10}},$$

$$c_8=\frac{\beta_2-\nu}{\beta_2+\nu}y^{2\beta_2}c_7-\frac{y_1^{1+\beta_2}}{\beta_2+\nu}\left[\frac{\nu}{y_1}F_1(y_1)+F_1'(y_1)\right],$$

$$f_1=\frac{1}{8}b_3^2,\qquad f_2=-\frac{3}{16}b_3^2+\frac{1}{4}b_1b_3,$$

$$f_3=\frac{1}{8}b_1^2+\frac{7}{64}b_3^2-\frac{3}{16}b_1b_3,\qquad f_4=\frac{1}{2}b_2b_3,$$

$$f_5=b_1b_2-\frac{1}{2}b_2b_3,\qquad f_6=-\frac{1}{2}b_2^2,$$

$$f_7=\frac{t_9t_{11}-t_8t_{12}}{t_7t_{11}-t_8t_{10}},$$

$$f_8=\frac{1-\nu}{1+\nu}f_7-\frac{1}{1+\nu}[\nu F_2(1)-F_2'(1)],$$

$$t_7=y_2+\frac{1-\nu}{1+\nu}y_2^{-1},\qquad t_8=-\frac{\beta_3\beta_4}{\beta_0}\left(y_2^{\beta_2}+\frac{\beta_2-\nu}{\beta_2+\nu}y_1^{2\beta_2}y_2^{-\beta_2}\right),$$

$$t_9=\frac{\beta_3\beta_4}{\beta_0}\left\{F_1(y_2)-\frac{1}{\beta_2+\nu}y_1^{1+\beta_2}y_2^{-\beta_2}\left[\frac{\nu}{y_1}F_1(y_1)-F_1'(y_1)\right]\right\}$$
$$-F_2(y_2)+\frac{1}{(1+\nu)y_2}[\nu F_2(1)-F_2'(1)],$$

$$t_{10}=1-\beta_6-\frac{1-\nu}{1+\nu}(1+\beta_6)y_2^{-2},$$

$$t_{11}=-\frac{\beta_2\beta_3\beta_7}{\beta_0}\left(y_2^{-1+\beta_2}-\frac{\beta_2-\nu}{\beta_2+\nu}y_1^{2\beta_2}y_2^{-1-\beta_2}\right),$$

$$t_{12}=\frac{\beta_3\beta_7}{\beta_0}\left\{F_1'(y_2)+\frac{\beta_2}{\beta_2+\nu}y_1^{1+\beta_2}y_2^{-1-\beta_2}\left[\frac{\nu}{y_1}F_1(y_1)-F_1'(y_1)\right]\right\}$$
$$-\frac{1+\beta_6}{1+\nu}y_2^{-2}[\nu F_2(1)-F_2'(1)]-F_2'(y_2)+\frac{\beta_6}{y_2}F_2(y_2)\tag{31}$$

对于二次近似，我们只考虑关于 φ 的边值问题。由方程（9）、（10）和边界条件（12）、（13）及（16），便有线性边值问题如下

对于 $y_1\leqslant y\leqslant y_2$，

$$L_1(y^{\beta_1}\varphi_{a,2})=-S_{a,1}\varphi_{a,1}+P\tag{32}$$

对于 $y_2\leqslant y\leqslant 1$，

$$L_0(y\varphi_{b,2})=-S_{b,1}\varphi_{b,1}+\beta_4P\tag{33}$$

当 $y=y_1$ 时，　$\varphi_{a,2}=0$　(34)

当 $y=1$ 时，　$W_{b,2}=0$，　$\varphi_{b,2}=0$　(35a,b)

当 $y=y_2$ 时，　$W_{a,2}=W_{b,2}$，　$\varphi_{a,2}=\varphi_{b,2}$，

$$\frac{\mathrm{d}\varphi_{b,2}}{\mathrm{d}y}+\frac{\beta_5}{y}\varphi_{b,2}=\beta_4\frac{\mathrm{d}\varphi_{a,2}}{\mathrm{d}y}\tag{36a～c}$$

利用解（29）和（30），得到边值问题（32）、（33）、（34）、（35b）和（36b，c）的解为

$$\varphi_{a,2}=P(a_1y^{\beta_1}+a_2y^{-\beta_1}+a_3y)-\beta_3\lambda_1^3W_m^3[F_3(y)+d_{17}y^{\beta_1}+d_{18}y^{-\beta_1}],$$
$$\varphi_{b,2}=P(b_1y+b_2y^{-1}+b_3y\ln y)-\beta_0\lambda_1^3W_m^3[F_4(y)+h_{14}y+h_{15}y^{-1}]\tag{37}$$

其中

$$F_3(y)=d_1y^{2+3\beta_1}+d_2y^{2-3\beta_1}+d_3y^{3+2\beta_1}+d_4y^{3-2\beta_1}+d_5y^{4+\beta_1}+d_6y^{4-\beta_1}+d_7y^{2+\beta_1}+d_8y^{2-\beta_1}+d_9y^{1+\beta_1+\beta_2}+d_{10}y^{1+\beta_1-\beta_2}+d_{11}y^{1-\beta_1+\beta_2}+d_{12}y^{1-\beta_1-\beta_2}+d_{13}y^{2+\beta_2}+d_{14}y^{2-\beta_2}+d_{15}y^5+d_{16}y^3,$$

$$F_4(y)=h_1y^5\ln^3y+h_2y^5\ln^2y+h_3y^5\ln y+h_4y^5+h_5y^3\ln^3y+h_6y^3\ln^2y+h_7y^3\ln y+h_8y^3+h_9\ln^3y+h_{10}y\ln^2y+h_{11}y\ln y+h_{12}y^{-1}\ln^2y+h_{13}y^{-1}\ln y,$$

$$F_3'(y)=\frac{dF_3(y)}{dy},\quad F_4'(y)=\frac{dF_4(y)}{dy},$$

$$d_1=\frac{a_1c_1}{4(1+\beta_1)(1+2\beta_1)},\quad d_2=\frac{a_2c_2}{4(1-\beta_1)(1-2\beta_1)},$$

$$d_3=\frac{a_1c_3+a_3c_1}{3(3+\beta_1)(1+\beta_1)},\quad d_4=\frac{a_2c_4+a_3c_2}{3(3-\beta_1)(1-\beta_1)},$$

$$d_5=\frac{a_1c_5+a_3c_3}{8(2+\beta_1)},\quad d_6=\frac{a_2c_5+a_3c_4}{8(2-\beta_1)},$$

$$d_7=\frac{a_1c_6+a_2c_1}{4(1+\beta_1)},\quad d_8=\frac{a_1c_2+a_2c_6}{4(1-\beta_1)},$$

$$d_9=\frac{a_1c_7}{(1+\beta_2)(1+2\beta_1+\beta_2)},\quad d_{10}=\frac{a_1c_8}{(1-\beta_2)(1+2\beta_1-\beta_2)},$$

$$d_{11}=\frac{a_2c_7}{(1+\beta_2)(1-2\beta_1+\beta_2)},\quad d_{12}=\frac{a_2c_8}{(1-\beta_2)(1-2\beta_1-\beta_2)},$$

$$d_{13}=\frac{a_3c_7}{(2+\beta_1+\beta_2)(2-\beta_1+\beta_2)},\quad d_{14}=\frac{a_3c_8}{(2+\beta_1-\beta_2)(2-\beta_1-\beta_2)},$$

$$d_{15}=\frac{a_3c_5}{25-\beta_1^2},\quad d_{16}=\frac{a_1c_4+a_2c_3+a_3c_6}{9-\beta_1^2}$$

$$h_1=\frac{1}{24}f_1b_3,\quad h_2=-\frac{5}{95}f_1b_3+\frac{1}{24}(f_1b_1+f_2b_3),$$

$$h_3=\frac{19}{576}f_1b_3-\frac{5}{144}(f_1b_1+f_2b_3)+\frac{1}{24}(f_2b_1+f_3b_3),$$

$$h_4=-\frac{65}{6912}f_1b_3+\frac{19}{1728}(f_1b_1+f_2b_3)-\frac{5}{288}(f_2b_1+f_3b_3)+\frac{1}{24}f_3b_1,$$

$$h_5=\frac{1}{8}f_4b_3,$$

$$h_6=-\frac{9}{32}f_4b_3+\frac{1}{8}(f_4b_1+f_1b_2+f_5b_3),$$

$$h_7=\frac{21}{64}f_4b_3-\frac{3}{16}(f_4b_1+f_1b_2+f_5b_3)+\frac{1}{8}(f_2b_2+f_7b_3+f_5b_1),$$

$$h_8=-\frac{45}{256}f_4b_3+\frac{7}{64}(f_4b_1+f_1b_2+f_5b_3)$$

$$-\frac{3}{32}(f_5b_1+f_2b_2+f_7b_3)+\frac{1}{8}(f_7b_1+f_3b_2),$$

$$h_9=\frac{1}{6}(f_4b_2+f_6b_3),$$

$$h_{10}=-\frac{1}{4}(f_4b_2+f_6b_3)+\frac{1}{4}(f_6b_1+f_8b_3+f_5b_2),$$

$$h_{11}=\frac{1}{4}(f_4b_2+f_6f_3)-\frac{1}{4}(f_6b_1+f_8b_3+f_5b_2)+\frac{1}{2}(f_7b_2+f_8b_1)$$

$$h_{12}=-\frac{1}{4}f_6b_2,\qquad h_{13}=-\frac{1}{2}f_8b_2-\frac{1}{4}f_6b_2,$$

$$d_{17}=\frac{t_{13}t_{18}-t_{15}t_{16}}{t_{13}t_{17}-t_{14}t_{16}},\qquad h_{14}=\frac{t_{15}t_{17}-t_{14}t_{18}}{t_{13}t_{17}-t_{14}t_{16}},$$

$$d_{18}=-F_3(y_1)y_1^{\beta_1}-d_{17}y_1^{2\beta_1},\qquad h_{15}=-F_4(1)-h_{14},$$

$$t_{13}=\beta_0(y_2-y_2^{-1}),\qquad t_{14}=\beta_3(y_1^{2\beta_1}y_2^{-\beta_1}-y_2^{\beta_1}),$$

$$t_{15}=\beta_3[F_3(y_2)-F_3(y_1)y_1^{\beta_1}y_2^{-\beta_1}]+\beta_0[F_4(1)y_2^{-1}-F_4(y_2)],$$

$$t_{16}=\beta_0[\beta_5(y_2^{-2}-1)-y_2^{-2}-1],$$

$$t_{17}=\beta_1\beta_3\beta_4(y_2^{\beta_1-1}+y_1^{2\beta_1}y_2^{-1-\beta_1}),$$

$$t_{18}=\beta_0[F_4'(y_2)+F_4(1)y_2^{-2})]+\beta_0\beta_5[F_4(y_2)y_2^{-1}-F_4(1)y_2^{-2}]$$
$$-\beta_3\beta_4[F_3'(y_2)+\beta_1y_1^{\beta_1}y_2^{-1-\beta_1}F_3(y_1)]\tag{38}$$

将式（37）对 y 积分一次，并利用条件（35a）和（36a），我们得到二次近似的挠度公式

$$W_{a,2}=P\left(\frac{a_1}{1+\beta_1}y^{1+\beta_1}+\frac{a_2}{1-\beta_1}y^{1-\beta_1}+\frac{1}{2}a_3y^2+a_4\right)$$
$$-\beta_3\lambda_1^3W_m^3[F_5(y)+d_{19}],$$
$$W_{b,2}=P\left[\frac{1}{2}b_1y^2+b_2\ln y+\frac{1}{2}b_3y^2\left(\ln y-\frac{1}{2}\right)+64\right]$$
$$-\beta_0\lambda_1^3W_m^3[F_6(y)+h_{16}]\tag{39a,b}$$

其中

$$F_5(y)=\frac{d_1}{3(1+\beta_1)}y^{3(1+\beta_1)}+\frac{d_2}{3(1-\beta_1)}y^{3(1-\beta_1)}$$
$$+\frac{d_3}{2(2+\beta_1)}y^{2(2+\beta_1)}+\frac{d_4}{2(2-\beta_1)}y^{2(2-\beta_1)}$$
$$+\frac{d_5}{5+\beta_1}y^{5+\beta_1}+\frac{d_6}{5-\beta_1}y^{5-\beta_1}$$
$$+\frac{d_7}{3+\beta_1}y^{3+\beta_1}+\frac{d_8}{3-\beta_1}y^{3-\beta_1}$$
$$+\frac{d_9}{2+\beta_1+\beta_2}y^{2+\beta_1+\beta_2}+\frac{d_{10}}{2+\beta_1-\beta_2}y^{2+\beta_1-\beta_2}$$
$$+\frac{d_{11}}{2-\beta_1+\beta_2}y^{2-\beta_1+\beta_2}+\frac{d_{12}}{2-\beta_1-\beta_2}y^{2-\beta_1-\beta_2}$$

$$+\frac{d_{13}}{3+\beta_2}y^{3+\beta_2}+\frac{d_{14}}{3-\beta_2}y^{3-\beta_2}+\frac{1}{6}d_{15}y^6$$

$$+\frac{1}{4}d_{16}y^4+\frac{d_{17}}{1+\beta_1}y^{1+\beta_1}+\frac{d_{18}}{1-\beta_1}y^{1-\beta_1},$$

$$F_6(y)=\frac{1}{6}h_1y^6\ln^3y+\frac{1}{12}(-h_1+2h_2)y^6\ln^2y$$

$$+\frac{1}{36}(h_1-2h_2+6h_3)y^6\ln y$$

$$-\frac{1}{216}(h_1-2h_2+6h_3-36h_4)y^6+\frac{1}{4}h_5y^4\ln^3y$$

$$-\frac{1}{16}(3h_5-4h_6)y^4\ln^2y+\frac{1}{32}(3h_5-4h_6+8h_7)y^4\ln y$$

$$-\frac{1}{128}(3h_5-4h_6+8h_7-32h_8)y^4+\frac{1}{2}h_9y^2\ln^3y$$

$$-\frac{1}{4}(3h_9-2h_{10})y^2\ln^2y+\frac{1}{4}(3h_9-2h_{10}+2h_{11})y^2\ln y$$

$$-\frac{1}{8}(3h_9-2h_{10}+2h_{11}-4h_{14})y^2+\frac{1}{3}h_{12}\ln^3y$$

$$+\frac{1}{2}h_{13}\ln^2y+h_{15}\ln y$$

$$h_{16}=-F_6(1),$$

$$h_{19}=\frac{1}{\beta_3}\{\beta_0[F_6(y_2)+h_{16}]-\beta_3F_5(y_2)\} \tag{40}$$

将解（39）代入表达式（26），我们得到波纹环形板的大挠度特征关系式

$$P=\lambda_1W_m+\lambda_3W_m^3 \tag{41}$$

其中

$$\lambda_3=\beta_3\lambda_1^4[F_5(y_1)+d_{19}] \tag{42}$$

表达式（41）就是夹紧固定外边界下具有刚性中心和平面边缘区域的波纹环形板在集中载荷作用下的大挠度特征关系式。为了实际工程的需要，将表达式（41）写成有量纲的形式

$$p=\lambda_0(\lambda_1h^2w_m+\lambda_3w_m^3) \tag{43}$$

其中，w_m 是波纹环形板内边缘的挠度值，

$$\lambda_0=\frac{2\pi D_a}{R^2k_th^2} \tag{44}$$

下面考虑滑动固定外边界情况下的具有刚性中心和平面边缘区域的波纹环形板在集中载荷作用下的无量纲非线性边值问题（9）、（10）、（14）、（15）和（16）的求解。实际上，这一波纹环形板的特征关系式可以很容易地从前一个问题的特征关系式（43）而得到。比较这两个问题的边界条件，我们发现只有边界条件（15c）不同于（13c）。因此，与文献［11］方法相似，只要在式（43）中令 $\nu\to\infty$，即将第一个问题中式（31）的下列几个参数改为

$$f_8=-f_7-F_2(1),$$

$$
\begin{aligned}
t_7 &= y_2 - y_2^{-1} \\
t_9 &= \frac{\beta_3\beta_4}{\beta_0}\left\{F_1(y_2) - \frac{1}{\beta_2+\nu}y_2^{-\beta_2}y_1^{1+\beta_2}\left[\frac{\nu}{y_1}F_1(y_1) - F_1'(y_1)\right]\right\} \\
&\quad - F_2(y_2) + \frac{1}{y_2}F_2(1), \\
t_{10} &= 1 - \beta_6 + (1+\beta_6)y_2^{-2}, \\
t_{12} &= \frac{\beta_3\beta_7}{\beta_0}\left\{F_1'(y_2) + \frac{\beta_2}{\beta_2+\nu}y_1^{1+\beta_2}y_2^{-1-\beta_2}\left[\frac{\nu}{y_1}F_1(y_1) - F_1'(y_1)\right]\right\} \\
&\quad - F_2'(y_2) - (1+\beta_6)y_2^{-2}F_2(y_1) + \beta_6 y_2^{-1}F_2(y_2)
\end{aligned}
\tag{45}
$$

那么，特征关系式和其他参数都与夹紧固定外边界情况的结果相同。

三、数值结果与讨论

我们以具有正弦波形的波纹环形板在外边界夹紧固定的情况为例，讨论这一波纹环形板的非线性弯曲问题。这一波纹环形板的有关参数如下：

外半径 R=78mm，厚度 h=0.4mm，波长 L=12mm，波幅 H=1.34mm，弹性模量$E=1.345\times10^4\,\text{kg/mm}^2$，泊松比 ν=0.3。

我们分以下两种情况进行讨论。

第一种情况：设波纹环形板的波纹部分半径 R_2=72mm，刚性中心的无量纲半径 y_1 分别取值为 0.1，0.2 和 0.3。将上述数据代入有关公式，经过计算，将得到的结果绘于图 2 中。

从图上可以看出，随着刚性中心区域的增大，波纹环形板的非线性特征逐渐减弱。

第二种情况：设刚性中心的半径 R_1=18mm，波纹区域的无量纲半径 y_2 分别取值 0.6，0.7，0.8 和 1。图 3 给出上述数据的数值结果。从图 3 可以看出，当挠度较大时，

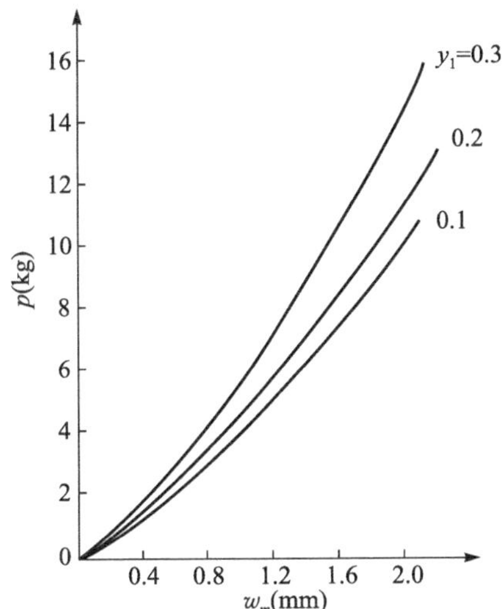

图 2　刚性中心的大小对波纹环形板特征的影响

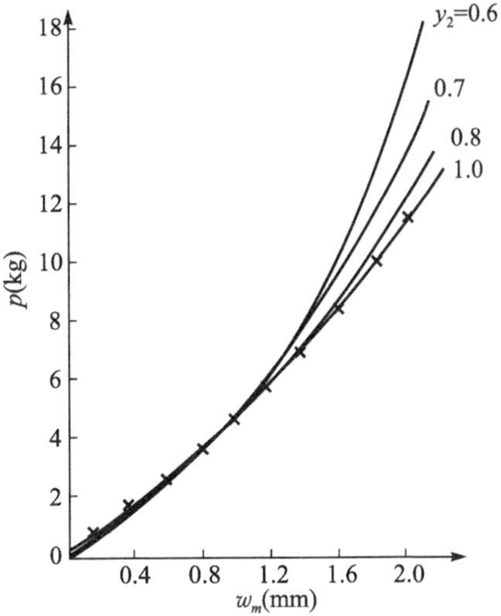

图 3　波纹区域的大小对波纹环形板特征的影响

随着波纹区域的增大，波纹环形板的非线性特征逐渐增强：当挠度较小时，随波纹区域的增大，波纹环形板的非线性特征反而略为减弱。

特别地，当 $y_2=1$ 时，平面边缘区域部分完全退化掉，即得到只有刚性中心的波纹环形板。本文在这一特例情况下的结果与文献[13]的结果完全相同，与文献[2]的实验值也符合得非常好。

参考文献

[1] Андреева. Л. Е Расчёт гофрированных мембран，как анизотропных пластинок，Инженерный Сборник，1955，21：128.

[2] 赤坂隆. Corrugated diaphragmの弾性特性について，日本航空学会誌，1955，3（22-23）：279.

[3] Haringx，J. A. Nonlinearity of corrugated diaphragms. Appl. Sci. Res. Ser. A，1956，16：45.

[4] Haringx，J. A. Design of corrugated diaphragms. Trans. ASME，Ser. A，1957，79（1）；55.

[5] Бурмистров，Е. Ф. Симметричный изгиб неоднородных и однородных ортотропных оболочек вращения с учетом больших прогибов и неравномерного температурного поля，*Инженерный Сборник*，1960，27：185.

[6] 刘人怀. 波纹圆板的特征关系式. 力学学报，1978（1）：47.

[7] 叶开沅，刘人怀，平庆元，等. 在对称线布载荷下的圆底扁薄球壳的非线性稳定问题. 兰州大学学报（自然科学版），1965（2）：10；科学通报，1965（2）：142.

[8] 叶开沅，刘人怀，张传智，等. 圆底扁薄球壳在边缘力矩作用下的非线性稳定问题. 科学通报，1965（2）：145.

[9] 刘人怀. 在内边缘均布力矩作用下中心开孔圆底扁球壳的非线性稳定问题. 科学通报，1965（3）：253.

[10] 刘人怀. 具有光滑中心的波纹圆板的特征关系式. 中国科学技术大学学报，1979，9（2）：75.

[11] Liu Ren-huai. Large deflection of corrugated circular plate under the action of concentrated loads at the center. International Journal of Non-Linear Mechanics，1984，19（5）：409.

[12] Liu Ren-huai. Large deflection of corrugated circular plate with plane boundary region. Solid Mechanics Archives，1984，9（4）：383.

[13] 刘人怀. 波纹环形板的非线性弯曲. 中国科学，A辑，1984（3）：247.

[14] 刘人怀. 在复合载荷作用下波纹环形板的非线性弯曲. 中国科学，A辑，1985（6）：537.

[15] Liu Ren-huai. On large deflection of corrugated annular plates under uniform pressure. The Advances of Applied Mathematics and Mechanics in China，Beijing：China Academic Publishers，1978，138.

[16] Liu Ren-huai. Nonlinear analysis of a corrugated circular plate under combined lateral loading. Applied Mathematics and Mechanics，1988，9（8）：711.

[17] Liu Ren-huai，Li Dong. On the non-linear bending and vibration of corrugated circular plates. International Journal of Non-Linear Mechanics，1989，24（3）：165.

[18] Liu Ren-huai，Li Dong. Application of the modified iteration method to nonlinear vibration of corrugated circular plates. Applied Mathematics and Mechanics，1990，11（1）：13.

[19] 刘人怀，翟赏中．均布载荷作用下波纹环形板的非线性弯曲．华南理工大学学报（自然科学版），1994，12：1.

[20] 刘人怀．波纹膜片理论的研究．仪表技术与传感器，1991，(5)：9；(6)：9.

[21] Liu Ren-huai，Zou Renpo. Non-linear bending of a corrugated annular plate with a plane boundary region and a non-deformable rigid body at the center under compound load. International Journal of Non-Linear Mechanics，1993，28 (3)：353.

Nonlinear bending of corrugated annular plate with large boundary corrugation[①]

中 文 摘 要

本文研究了具有刚性中心和边缘大波纹的波纹环形板的非线性弯曲问题，使用格林函数方法把简化的 Reissner 方程化为积分方程，成功地进行了研究，得到了这种板的应力和位移，其结果对测量仪表弹性元件的设计有益。

1. Introduction

Corrugated circular plates are widely applied in engineering, especially as elastic elements in measuring instruments. Thus, the investigation of nonlinear bending of the plates is very important. The nonlinear bending theory of thin shallow shells has been used by Panov (1941), Feodosev (1949), Chen (1980), Song and Yeh (1989), et al., to treat the problem only for a particular case of corrugated circular plates with sinusoidal corrugations. For the plate with uniformly distributed and dense corrugations, by assuming a corrugated circular plate as an orthotropic circular plate, a new approach of theoretical analysis for the plates was suggested by Haringx (1950), Andryewa (1955, 1981), Akasaka (1955), et al., to treat the same problem. Based on the theory of orthotropic circular plates, numerous results in succession have been obtained by Liu (1978, 1979, 1984a, 1984b, 1984c, 1985, 1987, 1988, 1991) and Liu and Zou (1993). Simultaneously, the nonlinear vibration of the plates is also investigated by Liu and Li (1989). In engineering practice, a corrugated circular plate with a large boundary corrugation is often encountered. It is unfortunate that there are few items in the literature about the plate because of difficulties with a more complicated shape and nonlinear mathematics. Hamada et al. (1968) solved this problem of a plate with a flat center by utilizing the finite difference method. Bihari and Elbert (1978) solved the multipoint problem of a plate and obtained characteristic curves of the plate by directly solving 6 coupled differential equations of the first order. But their results are not yet satisfactory and the distribution of stress in the annular plate has not yet obtained. In this paper, we apply the method of Green's function and reduce the simplified Reissner's equations to the integral equations to carry out the investigation successfully. The stresses and displacements in the plate are obtained. Results presented here are available for the design of elastic elements of measuring instruments.

---

① Reprinted from *Applied Mechanics and Engineering*, 1997, 2 (3): 353-367. Authors: Liu Renhuai and Yuan Hong.

2. Fundamental Equations and Boundary Conditions

Let (r,θ,z) denote a set of circular cylindrical coordinates in a fixed, righthanded Cartesian frame $Oxyz$. A shell of revolution is generated by rotating a plane curve about a fixed z-axis. See Fig. 1.

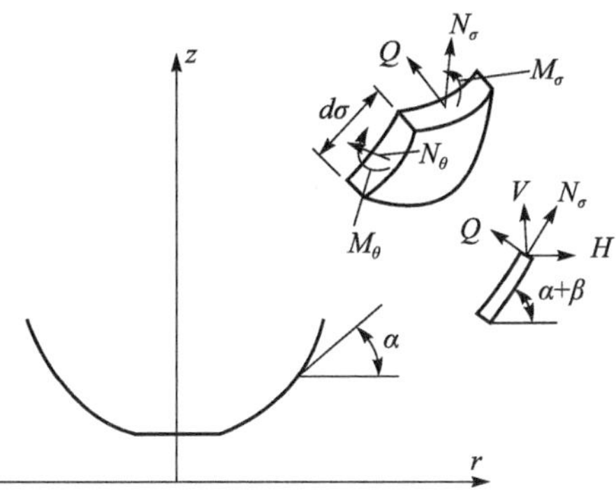

Fig. 1　Geometry of a shell of revolution

Moderate rotation theory of simplified Reissner's equations of axisymmetric shells of revolution under the action of a concentrated load at the center can be written as follows (Libai and Simmonds, 1988)

$$D[(r\beta')' - r^{-1}\beta] - F(\sin\alpha + \beta\cos\alpha) + \frac{Q_0}{2\pi}(\cos\alpha - \beta\sin\alpha) = 0, \tag{2.1}$$

$$A[(rF')' - r^{-1}F] + \beta\sin\alpha + \frac{1}{2}\beta^2\cos\alpha = 0 \tag{2.2}$$

where Q_0 is the central concentrated load. σ is the arc length along meridional direction, α is the slope angle of the tangent to one diametral section,

$$D = \frac{Eh^3}{12(1-\nu^2)}, \quad A = \frac{1}{Eh}, \quad (\ )' = \frac{d}{d\sigma}(\ ).$$

The relations of inner forces and displacements and the stress function F and the rotating angle β are

$$H = r^{-1}F, \tag{2.3}$$

$$N_\theta = F', \tag{2.4}$$

$$N_\sigma = r^{-1}F(\cos\alpha - \beta\sin\alpha) + \frac{Q_0}{2p}(\sin\alpha + \beta\cos\alpha), \tag{2.5}$$

$$M_\sigma = D\left[\beta' + \nu r^{-1}\left(\beta\cos\alpha - \frac{1}{2}\beta^2\sin\alpha\right)\right], \tag{2.6}$$

$$M_\theta = D\left[r^{-1}\left(\beta\cos\alpha - \frac{1}{2}\beta^2\sin\alpha\right) + \nu\beta'\right], \tag{2.7}$$

$$u = Ar\left\{F' - \nu\left[r^{-1}F(\cos\alpha - \beta\sin\alpha) + \frac{Q_0}{2\pi r}(\sin\alpha + \beta\cos\alpha)\right]\right\}, \tag{2.8}$$

$$w = w(0) + \int_0^\sigma \left(\beta\cos\alpha + e_\sigma\sin\alpha - \frac{1}{2}\beta^2\sin\alpha\right)d\sigma. \tag{2.9}$$

e_σ is the membrane strain along meridional direction in expression (2.9)

$$e_\sigma = A[N_\sigma - \nu N_\theta]. \tag{2.10}$$

Consider a corrugated circular annular plate with a large boundary corrugation and a nondeformable rigid body at the center as shown in Fig. 2. Here a is the outer radius, b is the inner radius. It is clear that we have:

$$\beta = 0, \qquad u = 0 \qquad \text{at } r = b. \tag{2.11}$$

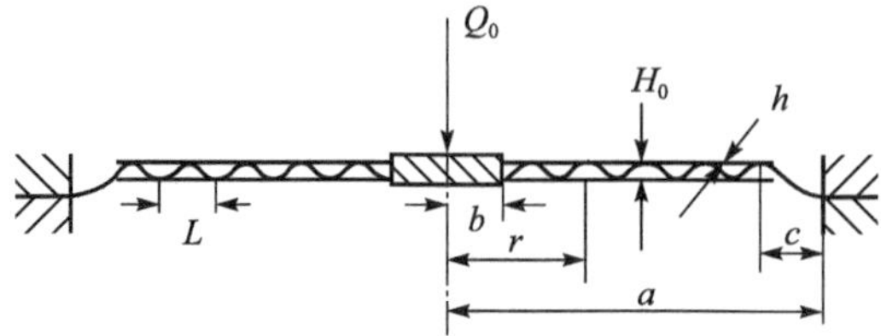

Fig. 2 A corrugated circular annular plate with a large boundary corrugation

Discussing the following customary two types of outer boundary conditions:

1) Edge rigidly clamped

$$w = 0, \quad \beta = 0, \quad u = 0 \qquad \text{at } r = a. \tag{2.12}$$

2) Edge loosely clamped

$$w = 0, \quad \beta = 0, \quad H = 0 \qquad \text{at } r = a. \tag{2.13}$$

Making the resulting solution simple, we introduce nondimensional parameters as follows

$$x = \frac{\sigma}{a}, \quad R = \frac{r}{a}, \quad g = \frac{aF}{D}, \quad W = \frac{w}{h},$$

$$\lambda^2 = \frac{AD}{a^2}, \quad Q = \frac{Q_0 a}{2\pi D}, \quad \rho = \frac{b}{a} \tag{2.14}$$

and introducing the change of variable

$$\xi = \int_0^x R^{-1} \mathrm{d}x \tag{2.15}$$

then Eqs. (2.1) (2.3) and boundary conditions (2.1)-(2.13) transform to:

$$\frac{\mathrm{d}^2\beta}{\mathrm{d}\xi^2} - \beta = f_\beta, \tag{2.16}$$

$$\frac{\mathrm{d}^2 g}{\mathrm{d}\xi^2} - g = f_g \tag{2.17}$$

where

$$f_\beta = gR(\sin\alpha + \beta\cos\alpha) - QR(\cos\alpha - \beta\sin\alpha), \tag{2.18}$$

$$f_g = -\frac{1}{\lambda^2}\left(\beta\sin\alpha + \frac{1}{2}\beta^2\cos\alpha)\right)R \tag{2.19}$$

and

$$\beta = 0, \quad \frac{\mathrm{d}g}{\mathrm{d}\xi} - \nu g\cos\alpha = \nu Q\sin\alpha \qquad \text{at } R = \rho, \tag{2.20}$$

for a rigidly clamped edge

$$w=0,\quad \beta=0,\quad \frac{\mathrm{d}g}{\mathrm{d}\xi}-\nu g\cos\alpha=\nu Q\sin\alpha \qquad \text{at } R=1, \tag{2.21}$$

for a loosely clamped edge

$$w=0,\quad \beta=0,\quad g=0 \qquad \text{at } R=1. \tag{2.22}$$

3. Integral Equation and Its Solution

By applying the method of Green's function, the nondimensional nonlinear boundary value problem (2.16), (2.17), (2.20) and (2.21) of a corrugated circular plate with a rigidly clamped edge reduces to the following integral equations:

$$\beta=\int_0^{\xi_N}G(\xi,\eta)f_\beta\mathrm{d}\eta, \tag{3.1}$$

$$g=\int_0^{\xi_N}G_2(\xi,\eta)f_g\mathrm{d}\eta+(C_1e^{\xi}+C_2e^{-\xi})Q \tag{3.2}$$

where C_1 and C_2 are known constants

$$C_1=\frac{\nu\sin\alpha_N(1+\nu\cos\alpha_0)-\nu\sin\alpha_0(1+\nu\cos\alpha_N)e^{-\xi_N}}{(1+\nu\cos\alpha_0)(1-\nu\cos\alpha_N)e^{-\xi_N}-(1-\nu\cos\alpha_0)(1+\nu\cos\alpha_N)e^{-\xi_N}}, \tag{3.3}$$

$$C_2=\frac{\nu\sin\alpha_0(1+\nu\cos\alpha_N)e^{-\xi_N}-\nu\sin\alpha_N(1-\nu\cos\alpha_0)}{(1+\nu\cos\alpha_0)(1-\nu\cos\alpha_N)e^{-\xi_N}-(1-\nu\cos\alpha_0)(1+\nu\cos\alpha_N)e^{-\xi_N}} \tag{3.4}$$

where ξ_N is the value of ξ at $R=1$, α_0 and α_N are values of α at $R=\rho$ and $R=1$ respectively, $G(\xi,\eta)$ and $G_2(\xi,\eta)$ are Green's functions

$$\begin{aligned}G_2(\xi,\eta)&=\frac{1}{4}(e^{|\xi-\eta|}-e^{-|\eta-\xi|})\\&\quad+\frac{1}{4}\frac{(e^{\xi-\xi_N}-e^{\xi_N-\xi})(e^{-\eta}-e^{\eta})+(e^{\xi_N-\eta}-e^{\eta-\xi_N})(e^{\xi}-e^{-\xi})}{e^{-\xi_N}-e^{\xi_N}}\end{aligned} \tag{3.5}$$

$$\begin{aligned}\Delta\cdot G_2(\xi,\eta)&=\Delta\cdot\frac{1}{4}(e^{|\xi-\eta|}-e^{-|\eta-\xi|})\\&\quad-\frac{1}{2}(1+\nu\cos\alpha_0)(1+\nu\cos\alpha_N)e^{-\xi_N+\eta+\xi}\\&\quad-\frac{1}{2}(1-\nu\cos\alpha_0)(1-\nu\cos\alpha_N)e^{\xi_N-\eta-\xi}\\&\quad-\frac{1}{4}[(1-\nu\cos\alpha_0)(1+\nu\cos\alpha_N)e^{-\xi_N}\\&\quad+(1+\nu\cos\alpha_0)(1-\nu\cos\alpha_N)e^{\xi_N}](e^{\eta-\xi}+e^{\xi-\eta})\end{aligned} \tag{3.6}$$

where

$$\Delta=(1+\nu\cos\alpha_0)(1-\nu\cos\alpha_N)e^{\xi_N}-(1+\nu\cos\alpha_N)(1-\nu\cos\alpha)e^{-\xi_N}. \tag{3.7}$$

Similarly, we can obtain the integral equations of the nondimensional nonlinear boundary value problem (2.16), (2.17), (2.20) and (2.22) of a corrugated circular plate with a loosely clamped edge. In this case (3.1) and (3.5) are still valid, but (3.2)-(3.4), (3.6) and (3.7) are substituted by (3.8)-(3.12) respectively:

$$g=\int_0^{\xi_N}\overline{G}_2(\xi,\eta)f_g\mathrm{d}\eta+(\overline{C}_1e^{\xi}+\overline{C}_2e^{-\xi})Q, \tag{3.8}$$

$$\overline{C}_1 = \frac{\nu \sin\alpha_0 e^{-\xi_N}}{(1-\nu\cos\alpha_0)e^{\xi_N} + (1+\nu\cos\alpha_0)e^{\xi_N}}, \tag{3.9}$$

$$\overline{C}_2 = -\frac{\nu \sin\alpha_0 e^{\xi_N}}{(1-\nu\cos\alpha_0)e^{-\xi_N} + (1+\nu\cos\alpha_0)e^{\xi_N}}, \tag{3.10}$$

$$\begin{aligned}\overline{\Delta}\cdot\overline{G}_2(\xi,\eta) =& \overline{\Delta}\cdot\frac{1}{4}(e^{|\xi-\eta|} - e^{-|\xi-\eta|}) \\ &+ \frac{1}{2}(1+\nu\cos\alpha_0)e^{-\xi_N+\eta+\xi} - \frac{1}{2}(1-\nu\cos\alpha_0)e^{\xi_N-\eta-\xi} \\ &+ \frac{1}{4}[(1-\nu\cos\alpha_0)e^{-\xi_N} - (1+\nu\cos\alpha_0)e^{\xi_N}](e^{\eta-\xi}+e^{\xi-\eta})\end{aligned} \tag{3.11}$$

$$\overline{\Delta} = (1-\nu\cos\alpha_0)e^{-\xi_N} + (1+\nu\cos\alpha_0)e^{\xi_N}. \tag{3.12}$$

Eqs. (3.1) and (3.2) can reduce to a nonlinear integral equation with only one unknown function

$$\begin{aligned}\beta_\tau =& -\frac{1}{\lambda^2}\int_0^{\xi_N}\int_0^{\xi_N} G(\tau,\xi)G_2(\xi,\eta)R_\xi(\sin\alpha_\xi + \beta_\xi\cos\alpha_\xi)\beta_\eta(\sin\alpha_\eta \\ &+ \frac{1}{2}\beta_\eta\cos\alpha_\eta)R_\eta \mathrm{d}\xi\mathrm{d}\eta + Q\int_G^{\xi_N} G(\tau,\xi)[C_1e^{\xi} + C_2e^{-\xi}) \\ &\cdot R_\xi(\sin\alpha_\xi + \beta_\xi\cos\alpha_\xi) - R_\xi(\cos\alpha_\xi - \beta_\eta\sin\alpha_\xi)]\mathrm{d}\xi\end{aligned} \tag{3.13}$$

where subscripts ξ, η, τ denote the functions of ξ, η, τ respectively. $\beta_\tau = \beta(\tau)$, $\alpha_\xi = \alpha(\xi)$, etc.

We can solve the integral Eq. (3.13) by the following iterative format

$$\begin{aligned}\beta_\tau^{*(m)} =& -\frac{1}{\lambda^2}\int_0^{\xi_N}\int_0^{\xi_N} G(\tau,\xi)G_2(\xi,\eta)R_\xi R_\eta\beta_\tau^{*(m)} \\ &\cdot(\sin\alpha_\xi + \beta_\xi^{(m-1)}\cos\alpha_\xi)\left(\sin\alpha_\eta + \frac{1}{2}\beta_\eta^{(m-1)}\cos\alpha_\eta\right)\mathrm{d}\xi\mathrm{d}\eta \\ &+ Q\int_0^{\xi_N} G(\tau,\xi)[C_1e^{\xi} + C_2e^{-\xi}) \\ &\cdot R_\xi(\sin\alpha_\xi + \beta_\xi^{(m-1)}\cos\alpha_\xi) - R_\xi(\cos\alpha_\xi - \beta_\xi^{(m-1)}\sin\alpha_\xi)]\mathrm{d}\xi.\end{aligned} \tag{3.14}$$

In the above expression we take $m=1$ and assume $\beta^{(0)}=0$, then $\beta^*(1)$ obtained from the first iterative will be the linear solution. For the purpose of convergence of the computed results, we take

$$\beta^{(m)} = \overline{\overline{\lambda}}\beta^{*(m)} + (1+\overline{\overline{\lambda}})\beta^{(m-1)} \qquad 0 < \overline{\overline{\lambda}} < 1 \tag{3.15}$$

where $\overline{\overline{\lambda}}$ is a so-called interpolated parameter important to prevent divergence or to obtain a better convergence rate. Computation shows that when load Q is small, any value of $\overline{\overline{\lambda}}$ can assure the convergence of iteration. $\overline{\overline{\lambda}}$ equal or almost equal to 1 yields a faster convergence rate; when load Q is large, $\overline{\overline{\lambda}}$ cannot be taken too large in order to assure convergence.

Take $\overline{\overline{\lambda}} = 0.8$ in practical computation discretize (3.14) by subdividing the interval of arc length into 50 parts. Give Q a small enough value and iterate according to (3.14) and (3.15) until all points reach.

$$\frac{|\beta^{*(m)} - \beta^{(m-1)}|}{1+|\beta^{*(m)}|} \leqslant 0.0001 \tag{3.16}$$

Thus, the nonlinear solution associated with this Q is obtained. Then, give load Q an increment and take the convergence solution associated with the old load as the initial value of iterative process associated with the new load. In this way, we obtain solutions associated with any loads.

Substituting $G_2(\xi,\eta)$ with $\overline{G}_2(\xi,\eta)$, C_1 and C_2 with $\overline{C}_1$ and $\overline{C}_2$ in Eq. (3.14), we can obtain an iterative format of the integral equation associated with the case of loosely champed edge. Then, we acquire g from Eq. (3.2). Thus, we can obtain inner forces and displacements from Eqs. (2.3)-(2.9). According to the theory of thin shell, we calculate the total stress components of the upper and lower surfaces which are important in engineering.

$$\sigma_\sigma\left(\pm\frac{h}{2}\right)=\frac{N_\sigma}{h}\mp\frac{6M_\sigma}{h^2} \tag{3.17}$$

$$\sigma_\theta\left(\pm\frac{h}{2}\right)=\frac{N_\theta}{h}\mp\frac{6M_\theta}{h^2} \tag{3.18}$$

The total stress value at any internal point between the upper and lower surfaces can be obtained by linear interpolation of the total stress of the upper and lower surfaces which are given by Eqs. (3.17) and (3.18).

4. Calculation Examples and Discussions

Example 1

For convenience, to check on the accuracy of the computation results obtained above, we consider the nonlinear bending problem of a corrugated annular plate as shown in Fig. 3. The dimensions of the plate are

$E=1.345\times10^5\,\text{N/mm}^2$, $\nu=0.3$, $h=0.4\text{mm}$,

$a=78\text{mm}$, $b=18\text{mm}$, $L=12\text{mm}$, $H_0=1.340\text{mm}$.

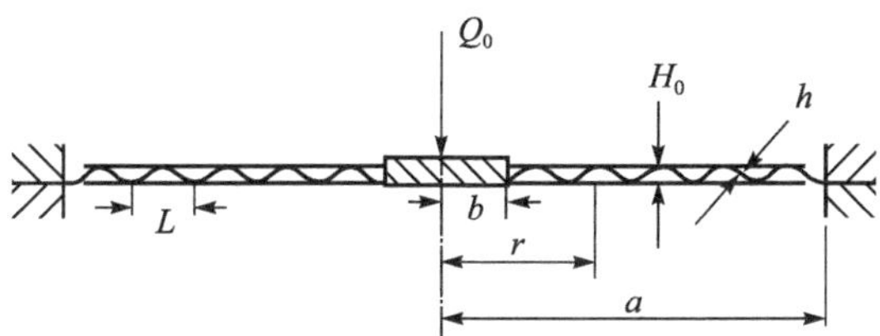

Fig. 3　A corrugated circular annular plate

Numerical results from formulas (3.14) and (2.9), are presented in Fig. 4. It can be observed that results in this paper (curve 1) show a very good agreement with Akasaka's experimental results (Akasaka, 1955) and Liu RenHuai's theoretical results (Liu, 1978) (curve 2).

Example 2

Let us take as the second example the corrugated annular plate as shown in Fig. 2. The function $\alpha(\sigma)$ is continuous since the meridian is built up of a sinusoidal curve and a large boundary circular corrugation with the curvature radius r_0. We study the case:

$E=1.345\times10^5\,\text{N/mm}^2$,　$\nu=0.3$,　$h=0.4\text{mm}$,　$a=78\text{mm}$,

$b=18\text{mm}$,　$L=12\text{mm}$,　$H_0=1.340\text{mm}$,　$c=15\text{mm}$,　$r_0=17.32\text{mm}$

and the edge may be rigidly clamped or loosely clamped.

Fig. 5 indicates the characteristic curves. Curve 1 shows the case of a rigidly clamped edge. Curve 2 shows the case of a loosely clamped edge. It can be concluded that when the load is small, the curves 1 and 2 are slightly different. The deflection associated with a loosely clamped edge is larger with the increment of load.

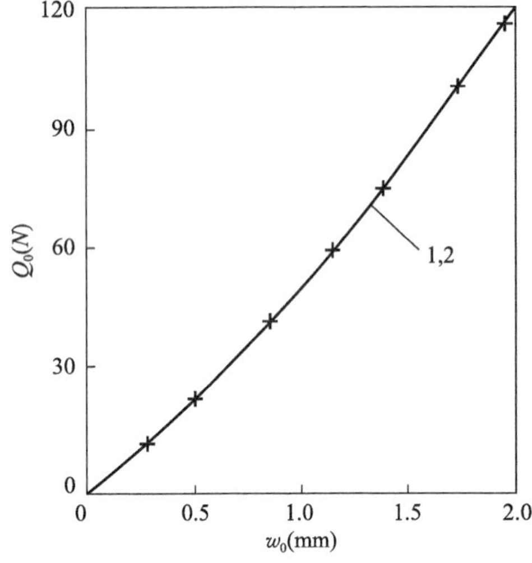

Fig. 4　Comparision between theoretical and experimental results

Fig. 5　The characteristic curves

Figs. 6-9 show circumferential and meridional total stress distribution of the upper and lower surfaces associated with a rigidly clamped edge when nondimensional load $Q=2$. Nondimensional total stresses σ_θ^L, σ_θ^u, σ_σ^L, σ_σ^u are defined as:

$$\sigma_\theta^L=\sigma_\theta\left(-\frac{h}{2}\right)\frac{12(1-\nu^2)a^2}{Eh^2},\qquad \sigma_\theta^u=\sigma_\theta\left(\frac{h}{2}\right)\frac{12(1-\nu^2)a^2}{Eh^2}.$$

$$\sigma_\sigma^L=\sigma_\sigma\left(-\frac{h}{2}\right)\frac{12(1-\nu^2)a^2}{Eh^2},\qquad \sigma_\sigma^u=\sigma_\sigma\left(\frac{h}{2}\right)\frac{12(1-\nu^2)a^2}{Eh^2},$$

The meridional total stress distributions of the upper and lower surfaces σ_σ^u and σ_σ^L are approximately asymmetric about the middle surface. The reason is that the meridional membrane stress in a corrugated shell is much smaller than the meridional bending stress. Circumferential membrane and bending stress are the same order, so circumferential total stress distributions of the upper and lower surfaces σ_θ^u and σ_θ^L are not approximately symmetric. For the corrugated annular plate as shown in Fig. 4, the maximum stress occurs in the neighborhood of the inner edge.

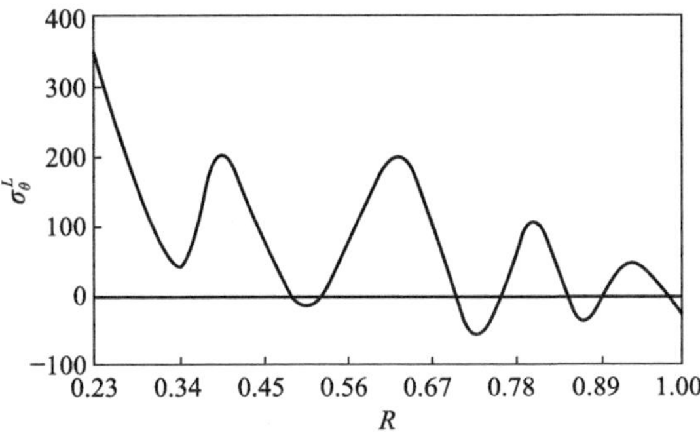

Fig. 6 Circumferential total stress of the lower surface

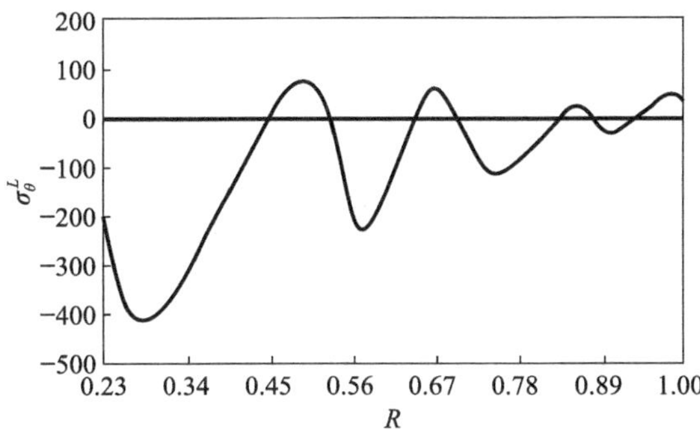

Fig. 7 Circumferential total stress of the upper surface

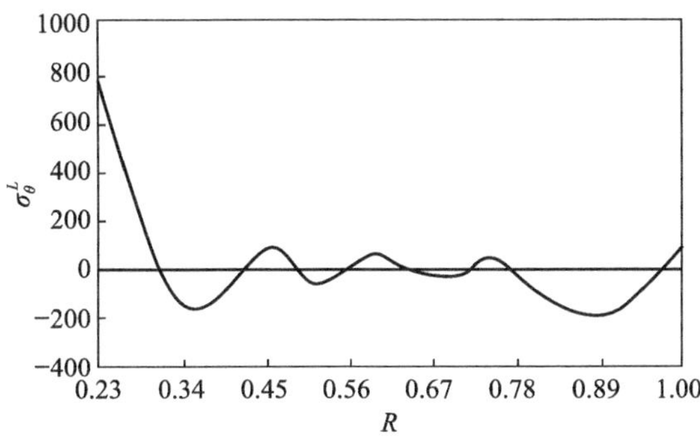

Fig. 8 Meridional total stress of the lower surface

5. Conclusion

Based on the simplified Reissner's equations, the numerical solutions of the nonlinear bending problem of a corrugated annular plate with a nondeformable body at the center and a large boundary corrugation under the action of concentrated load are obtained. Both rigidly and loosely clamped edges are discussed. The obtained characteristic curves and stress distributions are available for reference to design. The solution method in this paper can be applied to corrugated shells of arbitrary diametral sections.

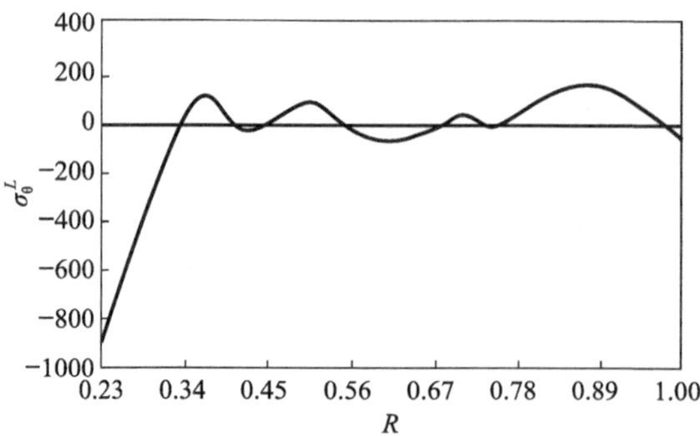

Fig. 9　Meridional total strees of the upper surface

Acknowledgements

This work was supported by Guangdong Natural Science Foundation.

Nomenclature

a, b —outer and inner radii, respectively

A —stretching compliance

c —geometrical size of large boundary corrugation

$C_1, C_2, \overline{C}_1, \overline{C}_2$ —constants

D —flexural rigidity

e_σ —membrane strain along meridional direction

E —Young's modulus

F —stress function

g —nondimensional stress function

$G(\xi,\eta), G_2(\xi,\eta)$ —Green's function

$\overline{G}(\xi,\eta), \overline{G_2}(\xi,\eta)$ —Green's function

h —thickness

H, V —horizontal and vertical stress resultants

H_0 —amplitude

L —wavelength

m —iterative number

M_σ, M_θ —meridional and circumferential moments, respectively

N_σ, N_θ —meridional and circumferential stress resultants, respectively.

Q —nondimensional central concentrated load

Q_0 —central concentrated load

r, θ, z —circular cylindrical coordinates

r_0 —radius of large boundary corrugation

u —radial displacement

w —deflection

W —nondimensional deflection

x,ρ,R —nondimensional parameter of σ,b and r, respectively

α —the slope angle of the tangent to one diametrial section

α_0,α_N —values of α at $r=a$ and $r=b$, respectively

β —rotating angle

λ^2 —nondimensional parameter

$\bar{\bar{\lambda}}$ —interpolated parameter

ν —Poisson's ratio

ξ —new variable

ξ_N —value of ξ at $R=1$

σ —arc length along meridional direction

$\sigma_\theta^L,\sigma_\theta^u$ —circumferential total stresses of lower and upper surfaces, respectively

$\sigma_\sigma^L,\sigma_\sigma^u$ —meridional total stresses of lower and upper surfaces, respectively

References

Akasaka T. (1955): On the elastic properties of the corrugated diaphragm. -J. Jap. Soc. Aeronaut. Engng., v. 3, p. 279, (in Japanese).

Andryewa L. E. (1955); The calculation of a corrugated membrane as an anisotropic plate. -Engr's Collection, v. 21, p. 128, (in Russian).

Andryewa L. E. (1981): Elastic Elements of Instruments. -Masgiz, Moscow (in Russian).

Bihari I. and Elbert A. (1978): Deformation of circular corrugated plates and shells. Periodica Polytech. -Mech. Eng. Mas., v. 22, p. 123.

Chen S. L. (1980): Elastic behavior of uniformly loaded circular corrugated plate with sine-shaped shallow waves in large deflection. -Appl. Math. Mech. v. 1, p. 261.

Feodosev V. I. (1949): Elastic Elements of Precision-Instruments Manufacture. -Moscow: Oborongiz, (in Russian).

Haringx J. A. (1950): The rigidity of corrugated diaphragms. -Appl. Sei. Res. Ser. A2, p. 299.

Hamada M., Seguchi Y, Ito S., et al. (1968): Numerical method for nonlinear axisymmetric bending of arbitrary shells of revolution and large deflection analyses of corrugated diaphragm and bellows. Bull. JSME. v. 11, p. 24.

Li D. and Liu Renhuai. (1990): Application of the modified iteration method to nonlinear vibration of corrugated circular plates. -Appl. Math. Mech. v. 11, p. 13.

Libai A. and Simmonds J. G. (1988): The Nonlinear Theory of Elastic Shells of One Spatial Dimension. -Academic Press, Baston.

Liu Renhuai. (1978): The characteristic relations of corrugated circular plates. -Acta Mech. Sin. v. 1, p. 47, (in Chinese).

Liu Renhuai. (1979): The characteristic relations of corrugated circular plates with plane central region. -J. Chin. Univ. Sci. Technol., v. 9, p. 75, (in Chinese).

Liu Renhuai. (1984a): Nonlinear bending of corrugated annular plates. -Sci. Sin. Ser. A. v, 27,

p. 640.

Liu Renhuai (1984b): Large deflection of corrugated circular plate with a plane central region under the action of concentrated loads at the center. -Int. J. Nonlinear Mech., v. 19, p. 409.

Liu Renhuai. (1984c): Large deflection of corrugated circular plate with plane boundary region Solid Mech. Archs. v. 9, p. 383.

Liu Renhuai. (1985): Nonlinear analysis of corrugated annular plates under compound load. -Sci. Sin. Ser. A, v. 28, p. 959.

Liu Renhuai. (1987): On large deflection of corrugaled annular plates under uniform pressure. -Adv. Appl. Math. Mech. China, v. 1, p. 138.

Liu Renhuai. (1988): Nonlinear analysis of a corrugated circular plate under combined lateral loading. -Appl. Math. Mech., v. 9, p. 711.

Liu Renhuai. (1991): The study on corrugated circular plates theory. -Instrument Technique and Sensor, v. 5, p. 9; (in Chinese).

Liu Renhuai and Zou Renpo. (1993): Non-linear bending of a corrugated annular plate with a plane boundary region and a non-deformable rigid body at the center under compound load. *-Int. J. Nonlinear Mech*, v. 28, p. 353.

Liu Renhuai. and Li D. (1989): On the nonlinear bending and vibration of corrugated circular plates-Int. J. Nonlinear Mech., v. 24, p. 165.

Panov D. Y. (1941): On large deflection of circular membranes with weak corrugations. *-Prik. Mat. Mekh.*, v. 5, p. 308, (in Russian).

Song W. P. and Yeh K. Y. (1989): Study of deformation stress and stability of corrugated circular plate under the action of concentrated loads at the center. *-Sci. Sin. Ser.* A, v. 32, p. 40.

Nonlinear deformation analysis of a U-shaped bellows with varying thickness①

中文摘要

本文将U型波纹管视为一个组合元件，即由圆环壳和带压缩角的截头扁锥壳所组成，且截头扁锥壳的壁厚沿径向坐标而改变。对这一变厚度U型波纹管使用积分方程方法，应用积分方程迭代法和梯度法，给出了相当精确的数值解，可以直接应用于波纹管的设计。

1. Introduction

U-shaped bellows are often adopted as an important pressure-measuring element in precision instruments and an expansion joint in pipeline engineering. With sufficient corrugations, a U-shaped bellow allows practically unlimited elastic displacements and rotations. It can sustain large external and internal pressure. In previous analyses, U-shaped bellows are generally simplified as composite members, which consist of circular ring shells and annular plates. The linear deformation of bellows was investigated in many papers [1-4]. In recent years, a few works have been published on the problem of nonlinear deformation of bellows, also based on the composite model of circular ring shells and annular plates.

A considerably more complicated numerical solution of the nonlinear differential equations for bellows was obtained using the finite difference method and Newton method [5]. References [6] and [7] respectively, applied the perturbation method to study the nonlinear problem of bellows with uniform wall thickness and varying wall thickness. Still, the nonlinear analyses were limited to the annular plates. That is to say, the nonlinear solution of bellows was obtained by the nonlinear solution of annular plates in combination with the linear solution of circular ring shells. The bellows' nonlinear deformation problem was also investigated by means of the trigonometric-series method and perturbation method [8]. There are serious discrepancies between the above theoretical results and experimental results. Thus, it might be profitable to discuss the bellows further.

In a practical case, the profile of U-shaped bellows is more complicated, and the wall thickness varies along its profile. Obviously, a more accurate analysis depends largely on the proper description of the profile and the wall thickness of a real bellows,

① Reprinted from *Archive of Applied Mechanics*, 2000, 70: 366-376. Authors: Liu Renhuai and Wang Zhiwei.

and is rather difficult. In order to make the model close to real bellows, a U-shaped bellow is treated as a composite member consisting of circular ring shells and truncated shallow conical shells with compressed angle θ_0; the wall thickness of the truncated shallow conical shells is considered to vary with the radial coordinate in this paper. Then, the integral-equation method for U-shaped bellows with compressed angle and varying wall thickness is developed. The nonlinear integral equations of circular ring shells and truncated shallow conical shells of a U-shaped bellows are respectively derived, and the iteration procedure for nonlinear analysis is suggested by means of the integral equation iteration in conjunction with the gradient method. Numerical results are obtained and compared with previous theories and experiments. This paper gives a more accurate numerical solution, which may be applied directly to the design of bellows.

2. Fundamental Equations of U-shaped Bellows

Let us consider the geometrically nonlinear axisymmetric deformation for U-shaped bellows under axial compression force P_0 and internal pressure Q_0. The geometry of a U-shaped bellows is shown in Fig. 1. Because of the axisymmetric and periodical deformation, we only need to take a typical part $ABCD$ of a bellow into consideration in Fig. 2. Here, the part AB is the outside circular ring shell of the radius R_{AB} with uniform wall thickness; the part BC is the truncated shallow conical shell with varying wall thickness; the part CD is the inside circular ring shell of the radius R_{CD} with uniform wall thickness. Let r and z be the radial coordinate, and the axial coordinate respectively, and θ_0 be called the compressed angle; in general, $|\theta_0|<5°$.

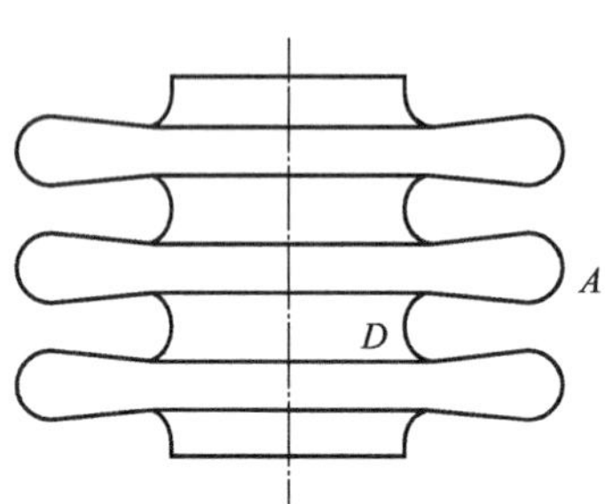

Fig. 1 Geometry of a U-shaped bellow

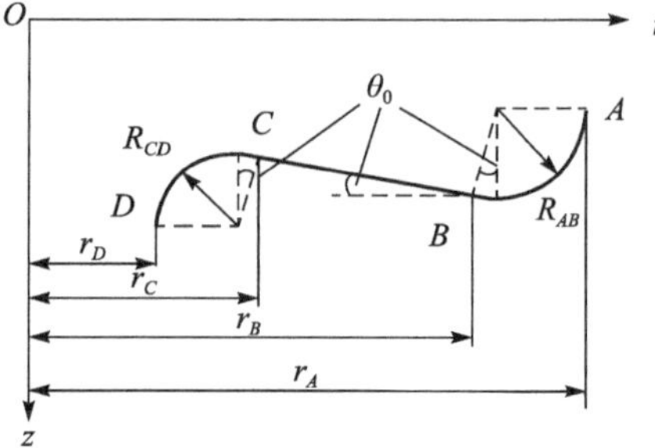

Fig. 2 Typical part of a bellow

According to the simplified Reissner equations of the large deflection theory for revolution shell [9-11], we obtain the governing equations of circular ring shells AB and CD as follows:

$$D\left[\frac{\mathrm{d}}{\mathrm{d}\sigma}\left(r\frac{\mathrm{d}\beta}{\mathrm{d}\sigma}\right)-\frac{\sin\beta}{r}\right]-\psi_H\sin(\alpha+\beta)+\psi_V\cos(\alpha+\beta)=0,$$

$$A\left[\frac{\mathrm{d}}{\mathrm{d}\sigma}\left(r\frac{\mathrm{d}\psi_H}{\mathrm{d}\sigma}\right)-\frac{\psi_H}{r}\right]+\cos\alpha-\cos(\alpha+\beta)+A\left[\frac{\mathrm{d}}{\mathrm{d}\sigma}(r^2P_H)+\nu rP_T\right]=0, \tag{1}$$

where ψ_H is the force function, β is the angle of rotation of a tangent to a meridian, α is the angle between r and the tangent to a meridian, h is the thickness of the shell, $d\sigma$ is the length of an infinitesimal element of a meridian, E is Young's modulus, ν is Poisson's ratio, D is the flexural rigidity, H and ψ_V/r are, respectively, the radial and axial components of forces in the shells, and P_H and P_T are, respectively, the radial and tangent loads

$$\psi_H = rH,\quad \psi_V = \frac{P_0}{2\pi} - \frac{1}{2}Q_0 r^2,\quad P_H = -Q_0\sin(\alpha+\beta),$$

$$P_T = 0\quad D = \frac{Eh^3}{12(1-\nu^2)},\quad A = \frac{1}{Eh}. \tag{2}$$

Having β and ψ_H, the meridional force N_σ, the hoop force N_θ, the meridional transverse force Q_σ, the meridional moment M_σ, the hoop moment M_θ, the radial displacement u and the axial displacement w can be derived as

$$N_\sigma = r^{-1}\psi_H\cos(\alpha+\beta) + r^{-1}\psi_V\sin(\alpha+\beta),$$

$$N_\theta = \frac{d\psi_H}{d\sigma} + rP_H,$$

$$Q_\sigma = -r^{-1}\psi_H\sin(\alpha+\beta) + r^{-1}\psi_V\cos(\alpha+\beta),$$

$$M_\sigma = -D\left\{\frac{d\beta}{d\sigma} + \nu r^{-1}[\sin(\alpha+\beta) - \sin\alpha]\right\},$$

$$M_\theta = -D\left\{r^{-1}[\sin(\alpha+\beta) - \sin\alpha] + \nu\frac{d\beta}{d\sigma}\right\},$$

$$u = \frac{r}{Eh}(N_\theta - \nu N_\sigma),$$

$$\frac{dw}{d\sigma} = \sin(\alpha+\beta) - \sin\alpha + A(N_\sigma - \nu N_\theta)\sin(\alpha+\beta). \tag{3a-g}$$

The nonlinear equation (1) will be solved for the following boundary conditions:

1) for the outside circular ring shell

$$\beta = 0,\quad \psi_H = 0\quad \text{at } r = r_A, \tag{4}$$

$$\beta = \beta^B,\quad \psi_H = \psi_H^B\quad \text{at } r = r_B, \tag{5}$$

2) for the inside circular ring shell

$$\beta = \beta^C,\quad \psi_H = \psi_H^C\quad \text{at } r = r_C, \tag{6}$$

$$\beta = 0,\quad \psi_H = 0\quad \text{at } r = r_D. \tag{7}$$

The fundamental equations of large deflections for the truncated shallow conical shell BC with varying thickness are

$$\frac{d}{dr}\left[\frac{1}{r}\frac{d}{dr}(r\beta)\right] + \frac{d\ln D}{dr}\left(\frac{d\beta}{dr} + \frac{\nu}{r}\beta\right) = \frac{\theta_0+\beta}{Dr}\psi_H - \frac{1}{Dr}\psi_V,$$

$$\frac{d}{dr}\left[\frac{1}{r}\frac{d}{dr}(r\psi_H)\right] + \frac{d\ln A}{dr}\left(\frac{d\psi_H}{dr} - \frac{\nu}{r}\beta\right) = -\frac{\beta}{Ar}\left(\theta_0 + \frac{\beta}{2}\right). \tag{8}$$

The corresponding boundary conditions can be written as

$$\beta = \beta^B,\quad \psi_H = \psi_H^B\quad \text{at } r = r_B, \tag{9}$$

$$\beta = \beta^C,\quad \psi_H = \psi_H^C\quad \text{at } r = r_C. \tag{10}$$

Expressions for the internal forces and displacements of the shell BC may be written as

$$\begin{aligned}
N_\sigma &= r^{-1}\psi_H = H,\\
N_\theta &= \frac{\mathrm{d}\psi_H}{\mathrm{d}r},\\
Q_\sigma &= r^{-1}\psi_V,\\
M_\sigma &= -D\left(\frac{\mathrm{d}\beta}{\mathrm{d}r}+\frac{\nu}{r}\beta\right),\\
M_\theta &= -D\left(\nu\frac{\mathrm{d}\beta}{\mathrm{d}r}+\frac{\beta}{r}\right),\\
u &= A\left(r\frac{\mathrm{d}\psi_H}{\mathrm{d}r}-\nu\psi_H\right),\\
\frac{\mathrm{d}w}{\mathrm{d}r} &= \beta - A\nu(\theta_0+\beta)\frac{\mathrm{d}\psi_H}{\mathrm{d}r}+A\frac{\theta_0+\beta}{r}\psi_H.
\end{aligned} \tag{11a-g}$$

The thickness of the truncated shallow conical shell BC in the case of a seamless bellows varies along its profile. This variation is determined by manufacturing technology, and is more complex. According to the investigation [5], it may be approximately described by

$$h = \frac{h_c r_c}{r}, \tag{12}$$

where h_c is the thickness of the shell at the point C. Equation (12) is also in good agreement with the author's experimental measurements.

3. Integral Equation Forms of Fundamental Equations

For the inside circular ring shell CD, by introducing the following dimensionless variables:

$$\begin{aligned}
&x = \frac{\sigma}{r_D},\quad R = \frac{r}{r_D},\quad g = \frac{r_D\psi_H}{D},\\
&\lambda^2 = \frac{AD}{r_D^2},\quad Q = \frac{Q_0 r_D^3}{D},\quad P = \frac{P_0 r_D}{2\pi D},
\end{aligned} \tag{13}$$

and the relationship

$$\xi = \int_0^x R^{-1}\mathrm{d}x, \tag{14}$$

Eqs. (1) and conditions (6)-(7) become

$$\begin{aligned}
&\frac{\mathrm{d}^2\beta}{\mathrm{d}\xi^2}-\beta = f_\beta,\\
&\frac{\mathrm{d}^2 g}{\mathrm{d}\xi^2}-g = f_g,
\end{aligned} \tag{15}$$

$$\beta = \beta^c,\quad g = g^c = \frac{r_D\psi_H^C}{D}\qquad \text{at } \xi = \xi_C, \tag{16}$$

$$\beta = 0,\quad g = 0\qquad \text{at } \xi = \xi_D, \tag{17}$$

where

$$f_\beta = gR\sin(\alpha+\beta) - (PR - 0.5QR^3)\cos(\alpha+\beta) + \sin\beta - \beta,$$
$$f_g = -\frac{R}{\lambda^2}[\cos\alpha - \cos(\alpha+\beta)] + Q\frac{\mathrm{d}}{\mathrm{d}\xi}[R^2\sin(\alpha+\beta)]. \tag{18}$$

Taking f_β and f_g as known functions, the nonlinear boundary value problem (15)-(17) is of the Sturm-Liouville type [12]. Following Ref. [12], the associated Green's function of the nonlinear boundary value problem (15)-(17) can be derived as

$$G(\xi,\eta) = \begin{cases} \dfrac{[e^{\xi-\xi_D} - e^{-(\xi-\xi_D)}][e^{\eta-\xi_C} - e^{-(\eta-\xi_C)}]}{2[e^{\xi_C-\xi_D} - e^{-(\xi_C-\xi_D)}]}, & \xi_D \leqslant \zeta \leqslant \eta, \\ \dfrac{[e^{\xi-\xi_C} - e^{-(\xi-\xi_C)}][e^{\eta-\xi_D} - e^{-(\eta-\xi_D)}]}{2[e^{\xi_C-\xi_D} - e^{-(\xi_C-\xi_D)}]}, & \eta \leqslant \zeta \leqslant \xi_C. \end{cases} \tag{19}$$

Then, the problem (15)-(17) can be transformed into the following set of nonlinear integral equations[12]:

$$\beta(\xi) = \phi_\beta(\xi) + \int_{\xi_D}^{\xi_C} G(\xi,\eta)[f_\beta(\eta) + \phi_\beta(\eta)]\mathrm{d}\eta,$$
$$g(\xi) = \phi_g(\xi) + \int_{\xi_D}^{\xi_C} G(\xi,\eta)[f_g(\eta) + \phi_g(\eta)]\mathrm{d}\eta, \tag{20a-b}$$

where

$$\phi_\beta(\xi) = \frac{\beta^C}{\xi_C - \xi_D}(\xi - \xi_D), \quad \phi_g(\xi) = \frac{g^C}{\xi_C - \xi_D}(\xi - \xi_D), \tag{21}$$

Substituting expression (20b) into (20a), we obtain the integral equation containing only β

$$\begin{aligned} \beta(\xi) =& \phi_\beta(\xi) + \int_{\xi_D}^{\xi_C} G(\xi,\eta)\{\phi_\beta(\eta) + \phi_g(\eta)R(\eta)\sin[\alpha(\eta)+\beta(\eta)] - [PR(\eta) \\ &- 0.5QR^3(\eta)] \times \cos[\alpha(\eta)+\beta(\eta)] + \sin[\beta(\eta)] - \beta(\eta)\}\mathrm{d}\eta \\ &+ \int_{\xi_D}^{\xi_C}\int_{\xi_D}^{\xi_C} G(\xi,\eta)G(\eta,\gamma)\{\phi_g(\gamma) - \frac{R(\gamma)}{\lambda^2}\{\cos[\alpha(\gamma)] \\ &- \cos[\alpha(\gamma)+\beta(\gamma)]\}\}R(\eta)\sin[\alpha(\eta)+\beta(\eta)]\mathrm{d}\eta\mathrm{d}\gamma \\ &- \int_{\xi_D}^{\xi_C}\int_{\xi_D}^{\xi_C} G(\xi,\eta)G_{,\gamma}(\eta,\gamma)QR(\eta)\sin[\alpha(\eta)+\beta(\eta)]R^2(\gamma)\sin[\alpha(\gamma) \\ &+ \beta(\gamma)]\mathrm{d}\eta\mathrm{d}\gamma. \end{aligned} \tag{22}$$

In the case of a moderate rotation, $\beta^2 \ll 1$, we can simplify Eq. (22) as

$$\begin{aligned} \beta(\xi) =& \phi_\beta(\xi) + \int_{\xi_D}^{\xi_C} G(\xi,\eta)\{\phi_\beta(\eta) + \phi_g(\eta)R(\eta)[\sin(\alpha(\eta)) + \beta(\eta)\cos(\alpha(\eta))] \\ &- [PR(\eta) - 0.5QR^3(\eta)][\cos(\alpha(\eta)) - \beta(\eta)\sin(\alpha(\eta))]\}\mathrm{d}\eta \\ &+ \int_{\xi_D}^{\xi_C}\int_{\xi_D}^{\xi_C} G(\xi,\eta)G,(\eta,\gamma)\{\phi_g(\gamma) - \frac{R(\gamma)}{\lambda^2}\beta(\gamma)[\sin(\alpha(\gamma)) + \frac{1}{2}\beta(\gamma)\cos(\alpha(\gamma))]\} \\ &\times R(\eta)[\sin(\alpha(\eta)) + \beta(\eta)\cos(\alpha(\eta))]\mathrm{d}\eta\mathrm{d}\gamma \\ &- \int_{\xi_D}^{\xi_C}\int_{\xi_D}^{\xi_C} G,(\xi,\eta)G_{,\gamma}(\eta,\gamma)QR(\eta)[\sin(\alpha(\eta)) + \beta(\eta)\cos(\alpha(\eta))]R^2(\gamma) \\ &\times [\sin(\alpha(\gamma)) + \beta(\gamma)\cos(\alpha(\gamma))]\mathrm{d}\eta\mathrm{d}\gamma. \end{aligned} \tag{23}$$

In the same manner, we can also derive the integral equations for the outside circular ring shell AB. They are analogous to Eqs. (20) and (22).

Let us now discuss the truncated shallow conical shell BC with varying wall thickness. By defining the dimensionless variables as

$$\rho=\frac{r}{a}-\rho_1,\quad \rho_1=\frac{b}{a},\quad \bar{\theta}_0=\frac{a\sqrt{12(1-\nu^2)}}{h_C(\rho_0-1)}\theta_0,$$

$$\bar{\beta}=\frac{a\sqrt{12(1-\nu^2)}}{h_C(\rho_1-1)}\beta,\quad \bar{\psi}_H=\frac{a}{D_C(\rho_0-1)}\psi_H,$$

$$\bar{P}=\frac{a^2\sqrt{12(1-\nu^2)}}{2\pi D_C h_C(\rho_1-1)^4}P_0,\quad \bar{Q}=-\frac{a^4\sqrt{12(1-\nu^2)}}{2D_C h_C(\rho_1-1)^4}Q_0,\tag{24}$$

where

$$a=\frac{r_B-r_C}{2},\quad b=\frac{r_B+r_C}{2},\quad D_C=\frac{Eh_C^3}{12(1-\nu^2)},\tag{25}$$

Eqs. (8) and conditions (9)-(10) can be transformed into

$$\frac{\mathrm{d}^2\bar{\beta}}{\mathrm{d}\rho^2}-\frac{2}{\rho+\rho_1}\frac{\mathrm{d}\bar{\beta}}{\mathrm{d}\rho}-\frac{1+3\nu}{(\rho+\rho_1)^2}\bar{\beta}=(\rho+\rho_1)^2\{(\bar{\theta}_0+\bar{\beta})\bar{\psi}_H+[\bar{P}-\bar{Q}(\rho+\rho_1)^2]\},$$

$$\frac{\mathrm{d}^2\bar{\psi}_H}{\mathrm{d}\rho^2}+\frac{2}{\rho+\rho_1}\frac{\mathrm{d}\bar{\psi}_H}{\mathrm{d}\rho}-\frac{1+\nu}{(\rho+\rho_1)^2}\bar{\psi}_H=-\frac{1}{(\rho+\rho_1)^2}\left(\bar{\theta}_0+\frac{\bar{\beta}}{2}\right)\bar{\beta},\tag{26}$$

and

$$\bar{\beta}=\bar{\beta}^B=\frac{a\sqrt{12(1-\nu^2)}}{h_C(\rho_1-1)}\beta^B,\quad \bar{\psi}_H=\bar{\psi}_H^B=\frac{a}{D_C(\rho_1-1)^3}\psi_H^B\qquad \text{at } \rho=1,\tag{27}$$

$$\bar{\beta}=\bar{\beta}^C=\frac{a\sqrt{12(1-\nu^2)}}{h_C(\rho_1-1)}\beta^C,\quad \bar{\psi}_H=\bar{\psi}_H^C=\frac{a}{D_C(\rho_1-1)^3}\psi_H^C\qquad \text{at } \rho=-1.\tag{28}$$

Taking the right sides of Eq. (26) as known functions, the above boundary value problem is also of the Sturm-Liouville type [12]. As in Ref. [12], the associated Green's functions of the nonlinear boundary value problem (26)-(28) can also be derived as

$$G_1(\xi,\eta)=\begin{cases}[(\xi+\rho_1)^{r_1}(\rho_1-1)^{r_2}-(\xi+\rho_1)^{r_2}(\rho_1-1)^{r_1}]\\ \quad\times[(\eta+\rho_1)^{r_1}(\rho_1+1)^{r_2}-(\eta+\rho_1)^{r_2}(\rho_1+1)^{r_1}]/K_I,\ -1\leqslant\zeta\leqslant\eta,\\ [(\xi+\rho_1)^{r_1}(\rho_1+1)^{r_2}-(\xi+\rho_1)^{r_2}(\rho_1+1)^{r_1}]\\ \quad\times[(\eta+\rho_1)^{r_1}(\rho_1-1)^{r_2}-(\eta+\rho_1)^{r_2}(\rho_1-1)^{r_1}]/K_I,\ \eta\leqslant\zeta\leqslant1,\end{cases}$$

$$G_2(\xi,\eta)=\begin{cases}[(\xi+\rho_1)^{r_3}(\rho_1-1)^{r_4}-(\xi+\rho_1)^{r_4}(\rho_1-1)^{r_3}]\\ \quad\times[(\eta+\rho_1)^{r_3}(\rho_1+1)^{r_4}-(\eta+\rho_1)^{r_4}(\rho_1+1)^{r_3}]/K_{II},\ -1\leqslant\zeta\leqslant\eta,\\ [(\xi+\rho_1)^{r_3}(\rho_1+1)^{r_4}-(\xi+\rho_1)^{r_4}(\rho_1+1)^{r_3}]\\ \quad\times[(\eta+\rho_1)^{r_3}(\rho_1-1)^{r_4}-(\eta+\rho_1)^{r_4}(\rho_1-1)^{r_3}]/K_{II},\ \eta\leqslant\zeta\leqslant1,\end{cases}\tag{29}$$

where

$$K_I = (r_1 - r_2)(\rho_1 - 1)^2[(\rho_1 - 1)^{r_2}(\rho_1 + 1)^{r_1} - (\rho_1 - 1)^{r_1}(\rho_1 + 1)^{r_2}],$$

$$K_{II} = \frac{(r_3 - r_4)}{(\rho_1 - 1)^2}[(\rho_1 - 1)^{r_4}(\rho_1 + 1)^{r_3} - (\rho_1 - 1)^{r_3}(\rho_1 + 1)^{r_4}],$$

$$r_{1,2} = \frac{3 \pm \sqrt{13 + 12\nu}}{2}, \qquad r_{3,4} = \frac{-1 \pm \sqrt{5 + 4\nu}}{2}. \tag{30}$$

Then the problem (26)-(28) can be transformed into the following nonlinear integral equations

$$\bar{\beta}(\rho) = \phi_1(\rho) + \int_{-1}^{1} G_1(\rho,\xi) h_1(\xi)\mathrm{d}\xi,$$

$$\bar{\psi}_H(\rho) = \phi_2(\rho) + \int_{-1}^{1} G_2(\rho,\xi) h_2(\xi)\mathrm{d}\xi, \tag{31a-b}$$

where

$$\phi_1(\rho) = \frac{1}{2}[\bar{\beta}^B + \bar{\beta}^C + \rho(\bar{\beta}^B - \bar{\beta}^C)],$$

$$\phi_2(\rho) = \frac{1}{2}[\bar{\psi}_H^B + \bar{\psi}_H^C + \rho(\bar{\psi}_H^B - \bar{\psi}_H^C)],$$

$$h_1(\xi) = (\rho_1 - 1)^2\Big[(\bar{\theta}_0 + \bar{\beta}(\xi))\bar{\psi}_H(\xi) + \bar{P} - \bar{Q}(\xi + \rho_1)^2 + \frac{\bar{\beta}^B - \bar{\beta}^C}{(\xi + \rho_1)^3} + \frac{1 + 3\nu}{(\xi + \rho_1)^4}\phi_1(\xi)\Big],$$

$$h_2(\xi) = -\frac{1}{(\rho_1 - 1)^2}\Big[\Big(\bar{\beta}(\xi)(\bar{\theta}_0 + \frac{1}{2}\bar{\beta}(\xi)\Big) + (\xi + \rho_1)(\bar{\psi}_H^B - \bar{\psi}_H^C) - (1 + \nu)\phi_2(\xi)\Big]. \tag{32}$$

Substituting expression (31b) into (31a), the integral equation containing only $\bar{\beta}$ is obtained as

$$\begin{aligned}\bar{\beta}(\rho) = {} & \phi_1(\rho) + \int_{-1}^{1} G_1(\rho,\xi)(\rho_1 - 1)^2\Big\{\bar{P} - \bar{Q}(\xi + \rho_1)^2 \\ & + \frac{\bar{\beta}^B - \bar{\beta}^C}{(\xi + \rho_1)^3} + \frac{1 + 3\nu}{(\xi + \rho_1)^4}\phi_1(\xi) + \phi_2(\xi)[\bar{\theta}_0 + \bar{\beta}(\xi)]\Big\}\mathrm{d}\xi \\ & - \int_{-1}^{1}\int_{-1}^{1} G_1(\rho,\xi)G_2(\xi,\eta)[\bar{\theta}_0 + \bar{\beta}(\xi)]\Big\{\bar{\beta}(\eta)\Big[\bar{\theta}_0 + \frac{1}{2}\bar{\beta}(\eta)\Big] \\ & + (\bar{\psi}_H^B - \bar{\psi}_H^C)(\eta + \rho_1) - (1 + \nu)\phi_2(\eta)\Big\}\mathrm{d}\xi\mathrm{d}\eta. \end{aligned} \tag{33}$$

4. Junction Conditions and Displacement of U-shaped Bellows

In practice, the following junction conditions at B and C must be satisfied in order to solve above integral equations:

$$\psi_H^{AB}(\xi_B) = \psi_H^{BC}(1),\ \psi_V^{AB}(\xi_B) = \psi_V^{BC}(1),\ \beta^{AB}(\xi_B) = \beta^{BC}(1),\ M_\sigma^{AB}(\xi_B) = M_\sigma^{BC}(1),$$
$$u^{AB}(\xi_B) = u^{BC}(1),\ w^{AB}(\xi_B) = w^{BC}(1), \tag{34a-f}$$

$$\psi_H^{BC}(-1) = \psi_H^{CD}(\xi_C),\ \psi_V^{BC}(-1) = \psi_V^{CD}(\xi_C),\ \beta^{BC}(-1) = \beta^{CD}(\xi_C),\ M_\sigma^{BC}(-1) = M_\sigma^{CD}(\xi_C),$$
$$u^{BC}(-1) = u^{CD}(\xi_C),\ w^{BC}(-1) = w^{CD}(\xi_C). \tag{35a-f}$$

In the above junction conditions, conditions (34a-c) and (35a-c) are automatically satisfied, and conditions (34f) and (35f) will be applied to determine the value of the axial displacement of a bellow. Then the remaining conditions at the points B and C which must be satisfied are

$$M_{\sigma}^{AB}(\xi_B) = M_{\sigma}^{BC}(1), \; u^{AB}(\xi_B) = u^{BC}(1), \tag{36}$$

$$M_{\sigma}^{BC}(-1) = M_{\sigma}^{CD}(\xi_C), \; u^{BC}(-1) = u^{CD}(\xi_C). \tag{37}$$

By integrating expressions (3g) and (11g), and considering conditions (34f) and (35f), we yield the axial relative displacement of point A and D of the typical part AB-CD of the bellow

$$\begin{aligned}\Delta w = w_A - w_D = & \int_{\xi_D}^{\xi_C} r[\sin(\alpha+\beta) - \sin\alpha + e_{\sigma}\sin(\alpha+\beta)]\mathrm{d}\xi \\ & + \int_{\xi_B}^{\xi_A} r[\sin(\alpha+\beta) - \sin\alpha + e_{\sigma}\sin(\alpha+\beta)]\mathrm{d}\xi \\ & + \int_{-1}^{1} a\left[\beta - A\nu(\theta_0+\beta)\frac{\mathrm{d}\psi_H}{\mathrm{d}r} + A\frac{\theta_0+\beta}{r}\psi_H\right]\mathrm{d}\rho,\end{aligned} \tag{38}$$

where

$$e_{\sigma} = A(N_{\sigma} - \nu N_{\theta}). \tag{39}$$

In the case of a moderate rotation, the above expression may be simplifed as

$$\begin{aligned}\Delta w = & \int_{\xi_D}^{\xi_C} r[\beta\cos\alpha + e_{\sigma}(\sin\alpha + \beta\cos\alpha)]\mathrm{d}\xi \\ & + \int_{\xi_B}^{\xi_A} r[\beta\cos\alpha + e_{\sigma}(\sin\alpha + \beta\cos\alpha)]\mathrm{d}\xi \\ & + \int_{-1}^{1} a\left[\beta - A\nu(\theta_0+\beta)\frac{\mathrm{d}\psi_H}{\mathrm{d}r} + A\frac{\theta_0+\beta}{r}\psi_H\right]\mathrm{d}\rho.\end{aligned} \tag{40}$$

5. Procedure for Numerical Calculations

The integral equations (23), (20b), (33) and (31b) for β, g, $\bar{\beta}$ and $\bar{\psi}_H$ include four unknown parameters β^B, β^C, ψ_H^B, and ψ_H^C. The junction conditions (36) and (37) will determine them.

The steps for numerical calculations are as follows:

(1) Assume some initial approximate values β_i^B, β_i^C, ψ_{Hi}^B and ψ_{Hi}^C for β^B, β^C, ψ_H^B, and ψ_H^C, respectively.

(2) Obtain the approximate values β_i and ψ_{Hi} for β and ψ_H in a bellow, by using iterations in the integral Eqs. (23), (20b), (33) and (31b).

(3) Solve the nonlinear Eqs. (36) and (37) to obtain the next approximate values β_{i+1}^B, β_{i+1}^C, ψ_{Hi+1}^B and ψ_{Hi+1}^C for β^B, β^C, ψ_H^B and ψ_H^C by using the gradient method.

(4) Take the values β_{i+1}^B, β_{i+1}^C, ψ_{Hi+1}^B and ψ_{Hi+1}^C as the initial approximate values β_i^B, β_i^C, ψ_{Hi}^B and ψ_{Hi}^C, respectively, and repeat steps (2)-(4).

6. Examples and Discussion

According to the above iterative procedure, we obtain the numerical results for four typical examples.

Example 1: Model in Ref. [8]

The parameters of the bellow are r_D=6.65mm, r_A=11.72mm, R_{CD}=0.15mm, R_{AB}=0.15mm, h_C=0.10mm, θ_0=−1.20°, E=130kN/mm^2, ν=0.30.

Example 2: Model in Ref. [5]

The parameters of the bellow are r_D=12.75mm, r_A=19.00mm, R_{CD}=0.75mm, R_{AB}=0.75mm, h=0.12mm, θ_0=0°, E=13 500kg/mm^2, ν=0.33.

Example 3: Model in Ref. [13]

The parameters of the bellow are r_D=18.0cm, r_A=28.0cm, R_{CD}=2.5cm, R_{AB}=2.5cm, h=0.15cm, θ_0=0°, E=2.1×10^{6}kg/cm^2, ν=0.30.

Example 4: Actual bellows *H*100×75×0.2×9.5

The parameters of the bellows are r_D=37.50mm, r_A=50.00mm, R_{CD}=1.79mm, R_{AB}=2.15mm, h_C=0.20mm, θ_0=4.28°, E=13 500kg/mm^2, ν=0.30.

In Example 1, the value of compressed angle is calculated by the given parameters of the bellow [8]. The nonlinear relation between axial force and extension of the bellows is obtained and given in Fig. 3. The present results are in good agreement with the theory and experiments by using the trigonometric-series method and perturbation method [8].

In Example 2 or Example 3, the wall thickness of a bellow is uniform, but the present theory assumes a bellow with varying wall thickness. We take respectively h_C=0.141mm, 0.168mm in the calculation so that the average wall thickness of the truncated shallow conical shell *BC* equals the uniform thickness *h*.

The relation between internal pressure and displacement in Example 2 is given in Fig. 4. It is compared with the experiment and theory [5]. The present results are in good agreement with the experiments and better than the theoretical results [5].

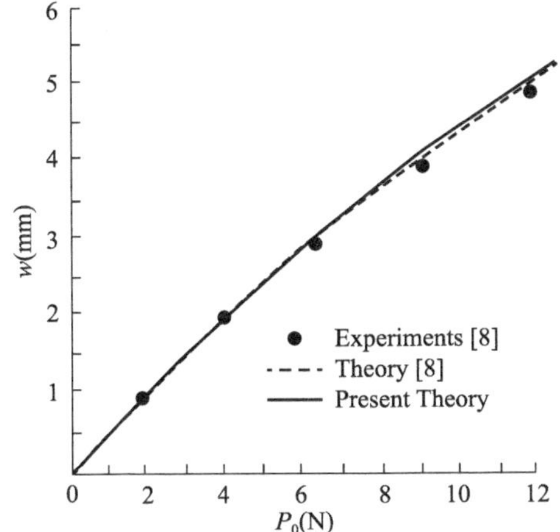

Fig. 3　The nonlinear relation between axial force and extension in Example 1 (9 waves)

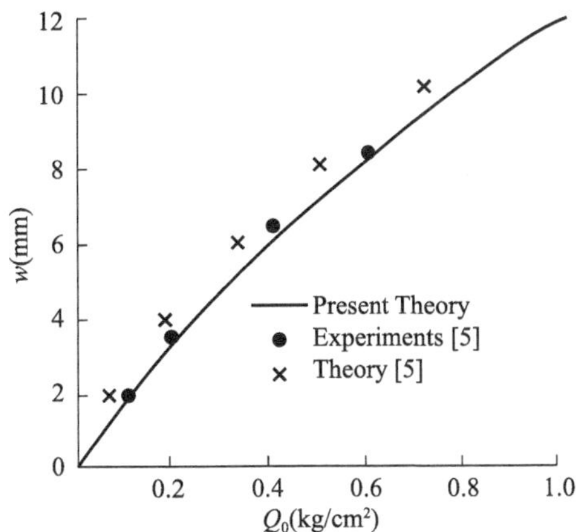

Fig. 4.　The nonlinear relation between internal pressure and displacement in Example 2 (8 waves)

The stress distributions in Example 3 are given in Fig. 5 when $P_0=100\text{kg}$, which checks well with the experiments [13].

Fig. 6 makes a comparison between numerical stresses and experimental stresses for the bellows in Example 4 when $Q_0=0.001\text{kg/mm}^2$. Both results coincide well.

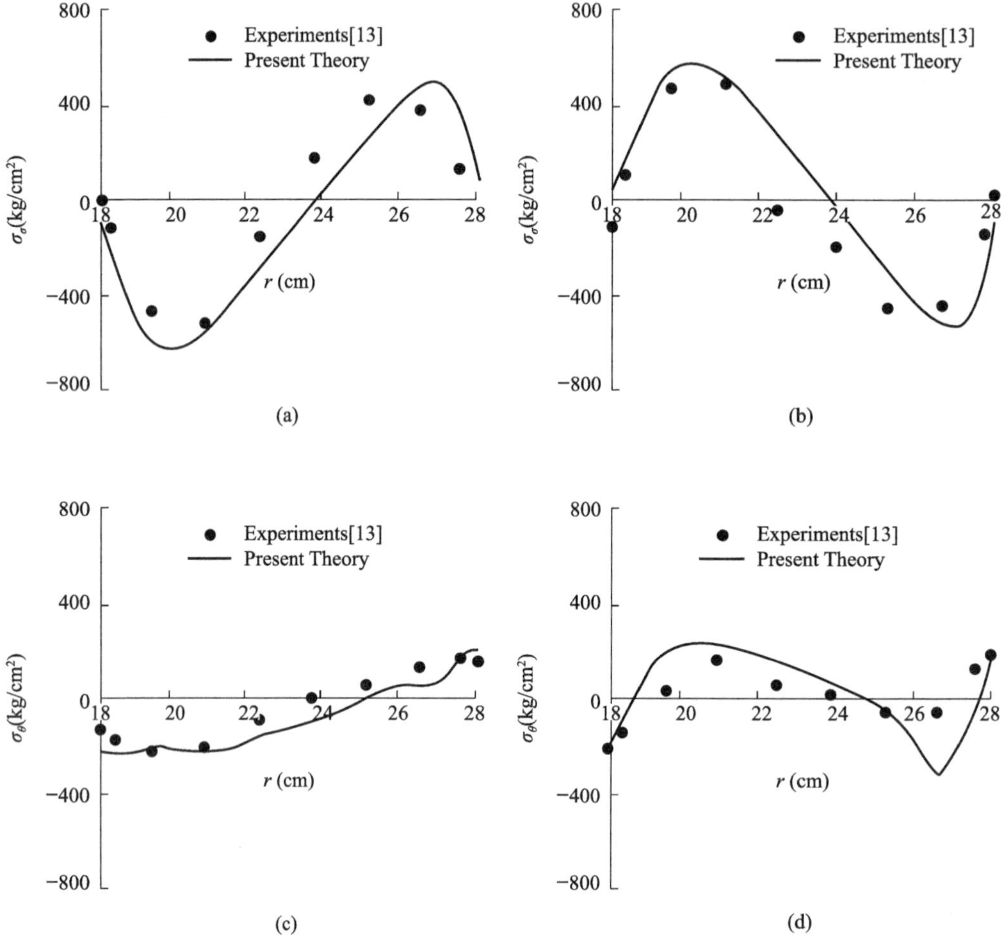

Fig. 5a-d. The stress distributions in Example 3. (a) σ_σ along the outer surface of the bellows; (b) σ_σ along the inner surface of the bellows; (c) σ_θ along the outer surface of the bellows; (d) σ_θ along the inner surface of the bellows

The iteration number and time for convergence are obviously relevant to the number of iteration points taken in the calculation. For the above examples, if we respectively divide the parts AB, BC and CD of bellows into two segments and take three Gauss integral points as the iteration points in every segment, the iteration procedure converges fast. Table 1 gives an insight into the iteration procedure, The results are obtained by a Pentium-S type PC. If we divide the parts AB, BC, and CD into more segments, the iteration number and time for convergence increase largely but the iteration results change little.

Our numerical results indicate that the present theory is more appropriate for a real U-shaped bellow, and that the suggested iteration procedure converges fast and is of high precision.

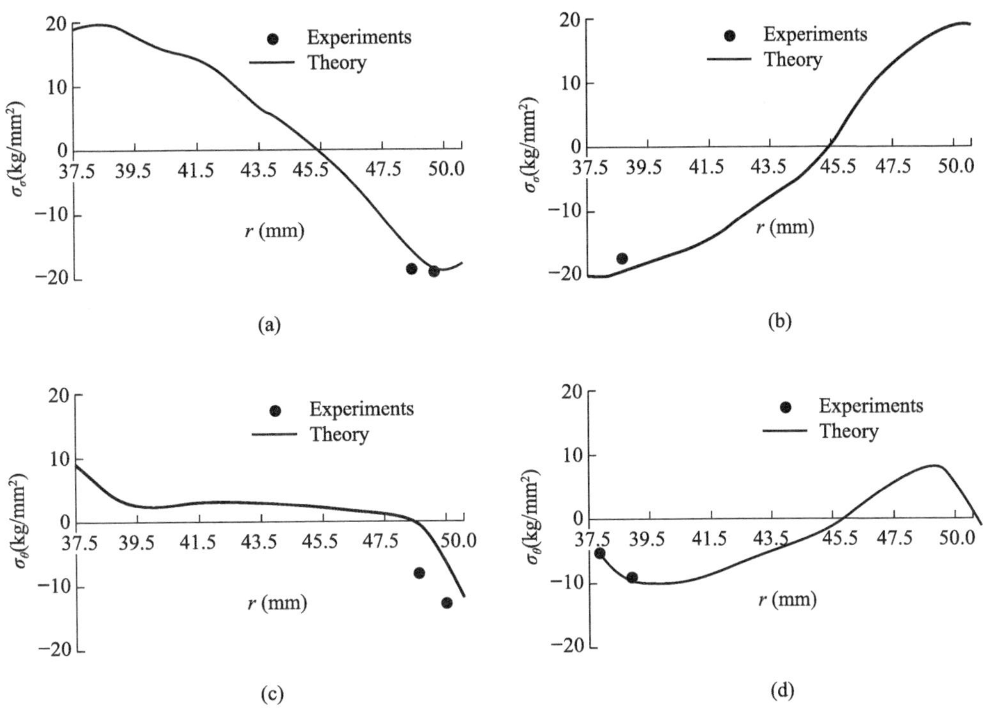

Fig. 6a-d. The stress distributions in Example 4. (a) σ_σ along the outer surface of the bellows; (b) σ_σ along the inner surface of the bellows; (c) σ_θ along the outer surface of the bellows; (d) σ_θ along the inner surface of the bellows

Table 1. Iteration number and time for convergence

Examples	Iteration number	Iteration time (s)
Example 1	21	31
Example 2	46	49
Example 3	19	29
Example 4	20	30

References

[1] Turner, C. E.: Stress and deflection studies of flat-plate and toroidal expansion bellows, subjected to axial, eccentric or internal pressure loading. J. Mech. Eng. Sci. I (1959) 130-143; II (1959) 113-129.

[2] Hamada, M.; Takezono, S.: Strength of U-shaped bellows. Bull. JSME 8 (1965) 525-531;

9 (1966) 513-523.

[3] Anderson, W. F.: Atomics International Division of North American Aviation. NAA-SR-4527 Pt. I (1964); Pt. II (1965).

[4] Clark, R. A.: An expansion bellows problem. Trans ASME, Ser. E, J. Appl. Mech. 37 (1970) 61-69.

[5] Andryewa, L. E.: The Computation and Design of Bellows. Moscow, Mashgiz 1975 (in Russian).

[6] Chien, W. Z.; Wu, M. D.: The nonlinear characteristics of U-shaped bellows: calculations by the method of perturbation. Appl. Math. Mech. 4 (1983) 649-665.

[7] Xu, Z. Q.; Liu, Y.; Yong, J. S.; et al.: Large deflection of a U-shaped bellows: with varying thickness. Journal of Qinghua University. 25 (1985) 39-51 (in Chinese).

[8] Axelrad, E. L.: Theory of Flexible Shells. Amsterdam, Elsevier Science. 1987.

[9] Reissner, E.: On axisymmetrical deformations of thin shells of revolution. Proc. Symp. Appl. Math. (Elasticity) (1950) 27-52.

[10] Simmonds, J. G.; Libai, A.: A simplified version of Reissner's non-linear equations for a first-approximation theory of shells of revolution. Comput. Mech. 2 (1987) 1-5.

[11] Koiter, W. T.: The intrinsic equation of shell theory with some applications. Mech. Today 5 (1980) 139-154.

[12] Birkhoff, G.; Rota, G. C.: Ordinary Differential Equations, 277-314. New York, John Wiley & Sons. 1969 (Second Edition).

[13] Dai, F. L.: Measurement of stress distribution in corrugated tube by photoelastic coating method. Acta Mech Solida Sinica. 2 (1984) 224-230 (in Chinese).

波纹扁球壳的大挠度方程①

一、引言

波纹壳是精密仪器仪表弹性元件中的一类重要形式。由于形状复杂和参数众多，故对其进行非线性弯曲分析将是十分困难的。这类壳体包括三种形式：第一种形式是波纹圆板，其非线性问题已被 Panov[1]、Feodosef[2]、Andryewa[3,4]、Akasaka[5]、Haringx[6]、刘人怀[7-21]等研究过；第二种形式是波纹管，其非线性问题已被钱伟长等[22]、徐志翘等[23]、Axelrad[24]和刘人怀等[25]研究过；第三种形式是有别于前两种的其他形式的波纹壳，非线性问题更加复杂，研究极少，Salashiling[26]曾应用 Andryewa[27]提出的波纹圆板等效正交异性弹性常数讨论了圆柱波纹块壳的非线性弯曲问题。本文中，我们考虑一种有重要应用价值的壳体，即承受均布压力的波纹扁球壳，将这种壳体视为结构上的圆柱正交异性扁球壳，并采用 Salashiling[26]方法，本文建立了波纹扁球壳的大挠度方程。

二、基本方程

考虑如图 1（a）所示的均布载荷 q 作用下的波纹扁球壳，其拱高为 f，厚度为 h，外边缘半径为 a，曲率半径为 R，波幅为 H，波长为 l。

假定波纹扁球壳的波纹呈均匀和致密分布，则可将此波纹扁球壳用一个等效的圆柱正交异性扁球壳来代替（图 1（b））。我们在等效壳中面上设置圆柱坐标系，r 是径向坐标，θ 是环向坐标。

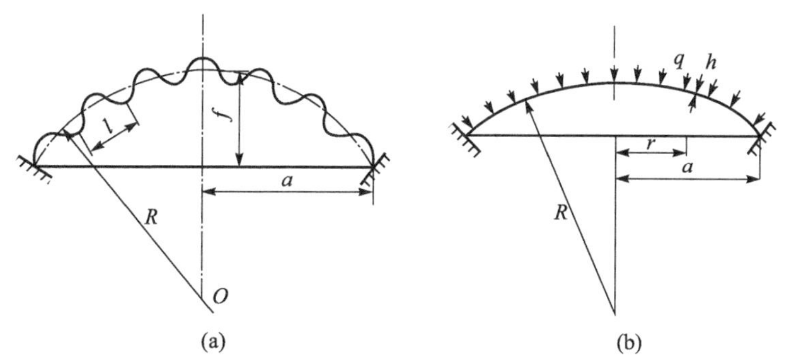

图 1　波纹扁球壳

在轴对称条件下，圆柱正交异性扁球壳与壳中面相距为 z 的任一点的径向、环向和轴向位移 u、v 和 w 可表达为如下形式：

① 本文原载《钱伟长教授九十华诞祝寿文集》，上海：上海大学出版社，2003，15-22. 作者：刘人怀，王璠.

$$u = u_0 - z\frac{\mathrm{d}w_0}{\mathrm{d}r} + \frac{h}{\pi}\gamma_{rz}^0 \sin\left(\frac{\pi z}{h}\right), v = 0, w = w_0 \tag{1}$$

其中，u_0 和 w_0 分别是壳体中曲面上点的径向位移和挠度，γ_{rz}^0 是壳体中曲面上点的横向剪切应变。

壳体的几何方程为

$$\varepsilon_r = \frac{\partial u}{\partial r} + \frac{r}{R}\frac{\partial w}{\partial r} + \frac{1}{2}\left(\frac{\partial w}{\partial r}\right)^2, \varepsilon_\theta = \frac{u}{r}$$

$$\gamma_{rz} = \frac{\partial u}{\partial z} + \frac{\partial w}{\partial r}, \gamma_{r\theta} = \gamma_{\theta z} = \varepsilon_z = 0 \tag{2}$$

其中，ε_r、ε_θ、ε_z、$\gamma_{r\theta}$、$\gamma_{\theta z}$ 和 γ_{rz} 是壳体中一点的应变分量。

将式（1）代入式（2），可得

$$\varepsilon_r = \varepsilon_r^0 + z\kappa_r, \varepsilon_\theta = \varepsilon_\theta^0 + z\kappa_\theta, \gamma_{rz} = \gamma_{rz}^0 \cos\left(\frac{\pi}{h}z\right) \tag{3a-c}$$

其中，ε_r^0 和 ε_θ^0 分别是壳体中曲面上点的径向和环向应变，κ_r 和 κ_θ 分别是壳体中曲面上点的径向和环向曲率：

$$\varepsilon_r^0 = \frac{\mathrm{d}u_0}{\mathrm{d}r} + \frac{r}{R}\frac{\mathrm{d}w}{\mathrm{d}r} + \frac{1}{2}\left(\frac{\mathrm{d}w}{\mathrm{d}r}\right)^2, \varepsilon_\theta^0 = \frac{1}{r}u_0$$

$$\kappa_r = -\frac{\mathrm{d}^2 w}{\mathrm{d}r^2} + \frac{h}{\pi z}\frac{\mathrm{d}\gamma_{rz}^0}{\mathrm{d}r}\sin\left(\frac{\pi z}{h}\right), \kappa_\theta = -\frac{1}{r}\left[\frac{\mathrm{d}w}{\mathrm{d}r} - \frac{h}{z\pi}\gamma_{rz}^0 \sin\left(\frac{\pi z}{h}\right)\right] \tag{4a-d}$$

按照 Andryewa[27]方法，我们可将正应变 ε_r 和 ε_θ 写为如下形式

$$\varepsilon_r = \varepsilon_{r1} + \varepsilon_{r2}, \varepsilon_\theta = \varepsilon_{\theta 1} + \varepsilon_{\theta 2} \tag{5}$$

其中，ε_{r1} 和 $\varepsilon_{\theta 1}$ 分别是由拉伸引起的壳体的径向和环向应变，ε_{r2} 和 $\varepsilon_{\theta 2}$ 分别是由弯曲引起的径向和环向应变：

$$\varepsilon_{r1} = \varepsilon_r^0, \varepsilon_{r2} = z\kappa_r, \varepsilon_{\theta 1} = \varepsilon_\theta^0, \varepsilon_{\theta 2} = z\kappa_\theta \tag{6}$$

由于等效扁球壳具有双重正交异性性质，则可将应力与应变关系分别表示为

对于拉伸有

$$\varepsilon_{r1} = \frac{\sigma_{r1}}{E_{r1}} - \nu_{\theta r1}\frac{\sigma_{\theta 1}}{E_{\theta 1}}, \varepsilon_{\theta 1} = \frac{\sigma_{\theta 1}}{E_{\theta 1}} - \nu_{r\theta 1}\frac{\sigma_{r1}}{E_{r1}} \tag{7}$$

对于弯曲有

$$\varepsilon_{r2} = \frac{\sigma_{r2}}{E_{r2}} - \nu_{\theta r2}\frac{\sigma_{\theta 2}}{E_{\theta 2}}, \varepsilon_{\theta 2} = \frac{\sigma_{\theta 2}}{E_{\theta 2}} - \nu_{r\theta 2}\frac{\sigma_{r2}}{E_{r2}} \tag{8}$$

其中，σ_{r1} 和 $\sigma_{\theta 1}$ 是由拉伸引起的壳体的径向和环向应力，σ_{r2} 和 $\sigma_{\theta 2}$ 是由弯曲引起的壳体的径向和环向应力，E_{r1} 和 $E_{\theta 1}$ 分别是拉伸时壳体材料的弹性模量，E_{r2} 和 $E_{\theta 2}$ 分别是弯曲时壳体材料的弹性模量，$\nu_{\theta r1}$ 和 $\nu_{\theta r2}$ 分别是拉伸和弯曲时 θ 方向伸缩时决定 r 方向缩伸时的泊松比，$\nu_{r\theta 1}$ 和 $\nu_{r\theta 2}$ 分别是拉伸和弯曲时 r 方向伸缩时决定 θ 方向缩伸时的泊松比：

$$\nu_{\theta r1}E_{r1} = \nu_{r\theta 1}E_{\theta 1}, \nu_{\theta r2}E_{r2} = \nu_{r\theta 2}E_{\theta 2} \tag{9}$$

显然，在目前情况下，应有

$$\nu_{\theta r1} = \nu_{\theta r2} = \nu \tag{10}$$

其中，ν 是波纹壳材料的泊松比。

等效圆柱正交异性扁球壳的弹性模量表示为如下形式：

$$E_{r1}=\frac{E}{\kappa_{r1}},E_{\theta1}=\kappa_{\theta1}E,E_{r2}=\frac{E}{\kappa_{r2}},E_{\theta2}=\kappa_{\theta2}E \tag{11}$$

其中，E 是波纹壳材料的弹性模量，κ_{r1}、$\kappa_{\theta1}$、κ_{r2} 和 $\kappa_{\theta2}$ 是与径向和环向刚度有关的参数。通过对波纹扁球壳和等效圆柱正交异性扁球壳的刚度比较、计算，我们有

$$\kappa_{r1}=\frac{\int_0^s \frac{y^2}{J_r^0}\mathrm{d}s+\int_0^s \frac{\cos^2\varphi_1}{bh}\mathrm{d}s}{\int_0^l \frac{y^2}{J_r^0}\mathrm{d}l+\int_0^l \frac{\cos^2\varphi_2}{bh}\mathrm{d}l},\kappa_{\theta1}=\frac{\int_0^b \frac{y^2}{J_r''}\mathrm{d}l+\int_0^b \frac{\cos^2\varphi_3}{lh}\mathrm{d}l}{\int_0^b \frac{y^2}{J_r'}\mathrm{d}l+\int_0^b \frac{\cos^2\varphi_3}{sh}\mathrm{d}l}$$

$$\kappa\kappa_{r2}=\frac{s}{l},\kappa_{\theta2}=\frac{h\int_0^s y^2\,\mathrm{d}s+\frac{h^3}{12}\int_0^s\cos^2\varphi_1\,\mathrm{d}s}{h\int_0^l y^2\,\mathrm{d}l+\frac{h^3}{12}\int_0^l\cos^2\varphi_2\,\mathrm{d}l} \tag{12}$$

其中，s 是波纹弧长，b 是壳元环向宽度，φ_1 和 φ_2 分别是波纹壳和等效壳的径向切线，φ_3 是壳体环向切线斜角，y 是纵向坐标，J_r^0、J_r'、J_r''为惯性力矩。

$$J_r^0=h\int_0^b y^2\,\mathrm{d}l+\frac{h^3}{12}\int_0^b\cos^2\varphi_3\,\mathrm{d}l,J_r'=h\int_0^s y^2\,\mathrm{d}s+\frac{h^3}{12}\int_0^s\cos^2\varphi_2\,\mathrm{d}s$$

$$J_r''=h\int_0^l y^2\,\mathrm{d}l+\frac{h^3}{12}\int_0^l\cos^2\varphi_2\,\mathrm{d}l \tag{13}$$

应用式（6）、（9）和（11），由式（7）和（8）可得

$$\begin{aligned}\sigma_{r1}&=Q'_{11}\varepsilon_r^0+Q'_{12}\varepsilon_\theta^0, &\sigma_{\theta1}&=Q'_{12}\varepsilon_r^0+Q'_{22}\varepsilon_\theta^0\\ \sigma_{r2}&=z(Q'_{11}\kappa_r+Q''_{12}\kappa_\theta), &\sigma_{\theta2}&=z(Q''_{12}\kappa_r+Q''_{22}\kappa_\theta)\end{aligned} \tag{14}$$

其中

$$Q'_{11}=\frac{E}{\kappa_{r1}\left(1-\frac{\nu^2}{\kappa_{r1}\kappa_{\theta1}}\right)},Q'_{12}=\frac{E\nu}{\kappa_{r1}\left(1-\frac{\nu^2}{\kappa_{r1}\kappa_{\theta1}}\right)},Q'_{22}=\frac{E\kappa_{\theta1}}{\left(1-\frac{\nu^2}{\kappa_{r1}\kappa_{\theta1}}\right)}$$

$$Q''_{11}=\frac{E}{\kappa_{r2}\left(1-\frac{\nu^2}{\kappa_{r2}\kappa_{\theta2}}\right)},Q''_{12}=\frac{E\nu}{\kappa_{r2}\left(1-\frac{\nu^2}{\kappa_{r2}\kappa_{\theta2}}\right)},Q''_{22}=\frac{E\kappa_{\theta2}}{\left(1-\frac{\nu^2}{\kappa_{r2}\kappa_{\theta2}}\right)} \tag{15}$$

等效壳的横向剪力由下式决定：

$$\tau_{rz}=G_{rz}\gamma_{rz} \tag{16}$$

其中，G_{rz}是决定 r 和 z 方向之间夹角变化的剪切模量。

将式（3c）代入式（16），便得

$$\tau_{rz}=G_{rz}\gamma_{rz}^0\cos\left(\frac{\pi}{h}z\right) \tag{17}$$

由此可见，横向剪应力在壳体中面上为 $G_{rz}\gamma_{rz}^0$，在壳体上下表面上为零。于是，我们可按 Salashiling[26] 的方法，令

$$G_{rz}=\frac{G}{\kappa_{\theta1}} \tag{18}$$

其中，G 是材料的剪切模量。

现在，用相邻经线和两平行圆从等效壳中截出一个微元（图 2）。这里，N_r 和 N_θ 分别是径向和环向薄膜力，M_r 和 M_θ 分别是径向和环向弯矩，Q_r 是径向剪力，其表达式为

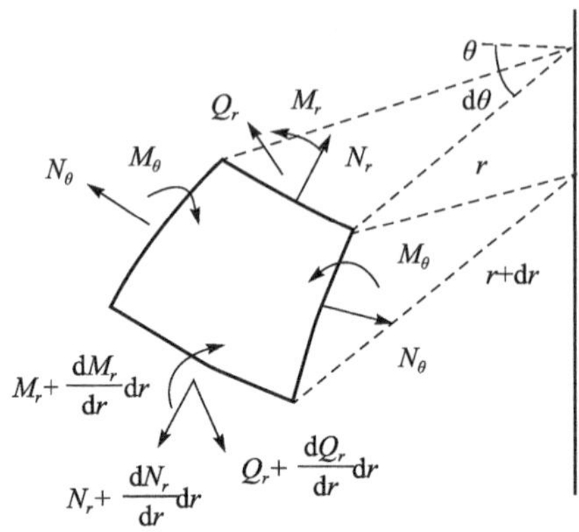

图 2　内力

$$N_r=\int_{-h/2}^{h/2}(\sigma_{r1}+\sigma_{r2})\mathrm{d}z,N_\theta=\int_{-h/2}^{h/2}(\sigma_{\theta1}+\sigma_{\theta2})\mathrm{d}z,M_r=\int_{-h/2}^{h/2}(\sigma_{r1}+\sigma_{r2})z\mathrm{d}z$$

$$M_\theta=\int_{-h/2}^{h/2}(\sigma_{\theta1}+\sigma_{\theta2})z\mathrm{d}z,Q_r=\int_{-h/2}^{h/2}\tau_{r\theta}\mathrm{d}z \tag{19}$$

将式（14）和（17）代入式（19），便得

$$N_r=A_{11}\varepsilon_r^0+A_{12}\varepsilon_\theta^0,N_\theta=A_{12}\varepsilon_r^0+A_{22}\varepsilon_\theta^0 \tag{20}$$

$$M_r=-\left(D_{11}\frac{\mathrm{d}^2w}{\mathrm{d}r^2}+D_{12}\frac{1}{r}\frac{\mathrm{d}w}{\mathrm{d}r}+D_{11}'\frac{\mathrm{d}\gamma_{rz}^0}{\mathrm{d}r}+D_{12}'\frac{\gamma_{rz}^0}{r}\right)$$

$$M_\theta=-\left(D_{12}\frac{\mathrm{d}^2w}{\mathrm{d}r^2}+D_{22}\frac{1}{r}\frac{\mathrm{d}w}{\mathrm{d}r}+D_{12}'\frac{\mathrm{d}\gamma_{rz}^0}{\mathrm{d}r}+D_{22}'\frac{\gamma_{rz}^0}{r}\right) \tag{21}$$

$$Q_r=G_0\gamma_{rz}^0 \tag{22}$$

其中，A_{ij}、D_{ij} 和 D_{ij}' 分别是拉伸和弯曲刚度，G_0 是剪切刚度：

$$A_{ij}=hQ_{ij}',D_{ij}=\frac{h^3}{12}Q_{ij}'',D_{ij}'=-\frac{2h^3}{\pi^3}Q_{ij}'',(i,j=1,2) \tag{23}$$

$$G_0=\frac{2Gh}{\pi\kappa_{\theta1}} \tag{24}$$

由壳体的微元平衡，便得大挠度下这一等效壳体的平衡方程如下：

$$\frac{\mathrm{d}(rN_r)}{\mathrm{d}r}-N_\theta=0,\frac{\mathrm{d}(rM_r)}{\mathrm{d}r}-M_\theta-rQ_r=0$$

$$\frac{\mathrm{d}}{\mathrm{d}r}\left[rN_r\left(\frac{r}{R}+\frac{\mathrm{d}w}{\mathrm{d}r}\right)+rQ_r\right]+qr=0 \tag{25a-c}$$

由式（4a,b）两个方程消去 u_0 便得

$$\varepsilon_r^0=\frac{\mathrm{d}}{\mathrm{d}r}(r\varepsilon_\theta^0)+\frac{r}{R}\frac{\mathrm{d}w}{\mathrm{d}r}+\frac{1}{2}\left(\frac{\mathrm{d}w}{\mathrm{d}r}\right)^2 \tag{26}$$

按式（20），可得

$$\varepsilon_r^0 = A_1 N_r - A_2 N_\theta, \varepsilon_\theta^0 = -A_2 N_r + A_3 N_\theta \tag{27}$$

其中

$$A_0 = \frac{1}{A_{11}A_{22} - A_{12}^2}, A_1 = A_0 A_{22}, A_2 = A_0 A_{12}, A_3 = A_0 A_{11} \tag{28}$$

将式（27）代入式（26），便得到协调方程

$$\frac{\mathrm{d}}{\mathrm{d}r}[r(A_3 N_\theta - A_2 N_r)] - A_1 N_r + A_2 N_\theta + \frac{1}{2}\left(\frac{\mathrm{d}w}{\mathrm{d}r}\right)^2 + \frac{r}{R}\frac{\mathrm{d}w}{\mathrm{d}r} = 0 \tag{29}$$

由方程（25a），可得

$$N_\theta = \frac{\mathrm{d}(rN_r)}{\mathrm{d}r} \tag{30}$$

将此式代入方程（29），便有

$$A_3 \frac{\mathrm{d}}{\mathrm{d}r} r \frac{\mathrm{d}(rN_r)}{\mathrm{d}r} - A_1 N_r + \frac{1}{2}\left(\frac{\mathrm{d}w}{\mathrm{d}r}\right)^2 + \frac{r}{R}\frac{\mathrm{d}w}{\mathrm{d}r} = 0 \tag{31}$$

应用式（22），方程（25c）成为

$$\frac{\mathrm{d}}{\mathrm{d}r}\left[rN_r\left(\frac{r}{R} + \frac{\mathrm{d}w}{\mathrm{d}r}\right) + G_0 r\gamma_{rz}^0\right] + qr = 0 \tag{32}$$

将式（21）和（22）代入方程（25b），便得

$$D_{11}\frac{\mathrm{d}}{\mathrm{d}r} r \frac{\mathrm{d}^2 w}{\mathrm{d}r^2} - D_{22}\frac{1}{r}\frac{\mathrm{d}w}{\mathrm{d}r} + D_{11}'\frac{\mathrm{d}}{\mathrm{d}r} r \frac{\mathrm{d}\gamma_{rz}^0}{\mathrm{d}r} - D_{22}'\frac{\gamma_{rz}^0}{r} + G_0 r\gamma_{rz}^0 = 0 \tag{33}$$

于是，方程（32）、（33）和（31）就组成波纹扁球壳在均布压力作用下的轴对称大挠度方程组。

参考文献

[1] Panov D. Y., On large deflection of circular membranes with weak corrugation, Prikl. Mat. Mekh. 1941, (5): 308 (in Russian)

[2] Feodosef V. I., *Elastic Elements of Precision-Instruments Manufacture*, Oborongiz, Moscow, 1949 (in Russian)

[3] Andryewa. L. E., The calculation of a corrugated membrane as an anisotropic plate, Engineer's Collection, 21, 1955, 128-141 (in Russian)

[4] Andryewa, L. E., *Elastic Elements of Instruments*. Moscow, Masgiz, 1962 (in Russian)

[5] Akasaka T., On the elastic properties of the corrugated diaphragm, *J. Jap. Soc. Aeronant. Eng.*, 3, 1955, 279 (in Japanese)

[6] Haringx J. A., Nonlinearity of corrugated diaphragms, *Appl. Sci. Res. Ser.* A, 1956, 16.

[7] 刘人怀. 波纹圆板的特征关系式. 力学学报，1978 (1): 47-52.

[8] 刘人怀. 具有光滑中心的波纹圆板的特征关系式. 中国科学技术大学学报，1979，9 (2): 75-86.

[9] 刘人怀. 波纹环形板的非线性弯曲. 中国科学（A 辑），1984 (3): 247-253.

[10] Liu Ren-huai, Large deflection of corrugated circular plate under the action of concentrated loads at the center, *International Journal of Nor-linear Mechanics*, 1984, 19 (5): 409-419.

[11] Liu Ren-huai, Large deflection of corrugated circular plate with plane boundary region, *Solid Mechanics Archives*, 1984, 9 (4): 383-406.

[12] 刘人怀. 在复合载荷作用下波纹环形板的非线性弯曲. 中国科学，1985，6：537-545.

[13] Liu Ren-huai，On large deflection of corrugated annular plates under uniform pressure pressure，*The Advances of Applied Mathematical and Mechanics in China*，China Academic Publishers，1987，1：138-152.

[14] 刘人怀. 复合载荷下波纹圆板的非线性分析. 应用数学和力学，1988，9（8）：661-674.

[15] Liu Ren-huai and Li Dong，On the non-linear bending and vibration of corrugated circular plate，*International Journal of Non-Linear Mechanics*，1989，24（3）：165-176.

[16] 李东，刘人怀，修正迭代法在波纹板圆板非线性振动问题中的应用. 应用数学和力学，1990，11（1）：13-21.

[17] Liu Ren-Huai，Zou Ren-Po，Non-linear bending of a corrugated annular plate with a plane boundary region and a non-deformable rigid body at the center under compound load，*International Journal of Non-linear Mechanics*，1993，28（3），353-364.

[18] 刘人怀，翟赏中. 均布载荷作用下波纹环形板的非线性弯曲. 华南理工大学学报，1994. 22（增刊）：1-10.

[19] 刘人怀，徐加初. 中心受载下具有平面边缘区域的固支波纹环形板的非线性分析. 暨南大学学报（自然科学版），1995，16（1）：1-13.

[20] Liu Ren-huai，Yuan Hong，Nonlinear bending of corrugated annular plate with large boundary corrugation，*Journal of Applied Mechanics and Engineering*，1997，2（3）：353-367.

[21] Liu Ren-huai，*Study on Nonlinear Mechanics of Plates and Shells*，New York，Beijing and Guangzhou：Science Press and Jinan University Press，1998.

[22] 钱伟长，吴明德，U 型波纹管的非线性特性摄动法计算. 应用数学和力学，1983，4（5）：595-608.

[23] 徐志翘，刘燕，杨嘉实，等，变厚度 U 型波纹壳大挠度非线性问题的摄动法. 清华大学学报，1985，25（1）：39-51.

[24] Axelrad E. L.，*Theory of Flexible Shells*，Amsterdam：Elsevier Science Publishers，1987.

[25] Liu Ren-huai，Wang Zhi-wei，Nonlinear deformation analysis of a U-shaped bellows with varying thickness. *Archive Applied Mechanics*，2000，70：366-376.

[26] Salashiling V. I.，The calculation of a cylindrical corrugated shell，*The Collection of Construction Frame*，Moscow：Steroigiz，1965，339-345（in Russian）

[27] Andryewa L. E.，*Elastic Elements of Instruments*，Moscow：Masgiz，1962（in Russian）

Nonlinear stability of corrugated shallow spherical shell①

1. Introduction

Corrugated shells are important elastic elements, which can be widely applied in precise instruments and other engineering. It is difficult to analyse nonlinear problems of corrugated shells because of their complicated shape and too many parameters. There are three kinds of corrugated shells. The first kind is the corrugated circular plate which has been studied by Panov (1941), Feodosev (1949), Andryewa (1955; 1962), Akasaka (1955), Haringx (1956), Liu (1978; 1984a; 1984b), Liu and Li (1989), Liu and Zou (1993), Liu and Yuan (1997), Liu (1998). The second kind is the corrugated cylindrical shells, especially bellows. Its nonlinear problem was studied by Andryewa (1975), Chien and Wu (1983), Xu et al. (1985), Axelrad (1987), Galishin (1999; 2003), Liu and Wang (2000), and Semenyuk (2002). The third one is the corrugated shallow shell. However, this shell was only studied by a very few due to its great difficulty. Salashiling (1965) studied a nonlinear problem of a cylindrical corrugated panel using Andryewa's (1962) equivalent orthotropic elastic constants of the corrugated circular plate. In this paper, a corrugated shallow spherical shell was discussed. As we know, the nonlinear stability problem of the shell has not yet been studied. By assuming a corrugated shallow spherical shell as a cylindrically orthotropic shallow spherical shell, the nonlinear bending theory of the shell is established. Using this theory, nonlinear stability problems of a corrugated shallow spherical shell with a rigidly clamped edge under uniform pressure are investigated, and critical buckling pressure is given.

2. Fundamental Equations

Now consider a corrugated circular shallow spherical shell of thickness h, radius a and radius of curvature R under the action of uniform pressure q as shown in Fig. la. Here l is the wavelength, H is the amplitude, f is the center height.

In this study, the corrugations of the corrugated circular shallow spherical shell are assumed to be closely and evenly spaced. Then, we may consider the shell as a cylindrically orthotropic shallow spherical shell (Fig. 1b). We choose the middle surface of the cylindrically orthotropic shallow spherical shell as the coordinate surface. Here r

① Reprinted from *International Journal of Applied Mechanics and Engineering*, 2005, 10 (2): 295-309. Authors: Liu Renhuai and Wang Fan.

is the radial coordinate, θ is the circumferential coordinate. In the case of a cylindrically orthotropic shallow spherical shell, we must consider the effect of transverse shear deformation. It can be concluded from the axisymmetric condition that the radial, circumferential and axial displacements u, v, w at an arbitrary point of the shell at a distance z from the middle surface of the shell may be written as follows

$$u = u_0 - z\frac{\mathrm{d}w_0}{\mathrm{d}r} + \frac{h}{\pi}\sin\left(\frac{\pi z}{h}\right)\gamma_{rz}^0, \tag{2.1a}$$

$$v = 0, \tag{2.1b}$$

$$w = w_0 \tag{2.1c}$$

where u_0 and w_0 are the radial displacement and the deflection of a point on the middle surface, respectively. γ_{rz}^0 is the transverse shear strain of the middle surface.

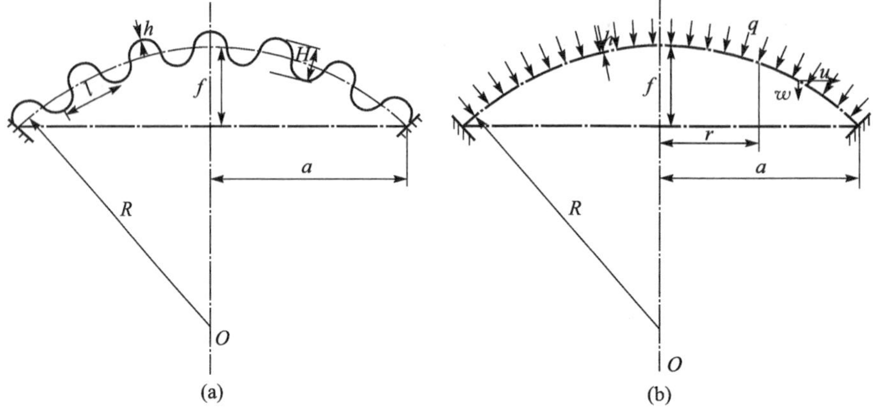

Fig. 1 A corrugated circular shallow spherical shell.

The geometrical equations of the shell are

$$\varepsilon_r = \frac{\partial u}{\partial r} + \frac{r}{R}\frac{\mathrm{d}w}{\mathrm{d}r} + \frac{1}{2}\left(\frac{\mathrm{d}w}{\mathrm{d}r}\right)^2, \ \varepsilon_\theta = \frac{u}{r},$$

$$\gamma_{rz} = \frac{\partial u}{\partial z} + \frac{\mathrm{d}w}{\mathrm{d}r}, \ \gamma_{r\theta} = \gamma_{\theta z} = \varepsilon_z = 0 \tag{2.2}$$

where ε_r, ε_θ, ε_z, $\gamma_{r\theta}$, $\gamma_{\theta z}$ and γ_{rz} are strain components at a point in the shell.

Substituting Eqs. (2.1a, b, c) into (2.2), we obtain

$$\varepsilon_r = \varepsilon_r^0 + z\kappa_r, \tag{2.3a}$$

$$\varepsilon_\theta = \varepsilon_\theta^0 + z\kappa_\theta, \tag{2.3b}$$

$$\gamma_{rz} = \gamma_{rz}^0\cos\left(\frac{\pi}{h}z\right) \tag{2.3c}$$

where ε_r^0 and ε_θ^0 are the radial and circumferential strain of the middle surface respectively, κ_r, and κ_θ are the radial and circumferential curvatures of the middle surface respectively

$$\varepsilon_r^0 = \frac{\mathrm{d}u_0}{\mathrm{d}r} + \frac{r}{R}\frac{\mathrm{d}w}{\mathrm{d}r} + \frac{1}{2}\left(\frac{\mathrm{d}w}{\mathrm{d}r}\right)^2, \tag{2.4a}$$

$$\varepsilon_\theta^0 = \frac{u_0}{r}, \tag{2.4b}$$

$$\kappa_r = -\frac{d^2 w}{dr^2} + \frac{h}{\pi z}\sin\left(\frac{\pi z}{h}\right)\frac{d\gamma_{rz}^0}{dr}, \tag{2.4c}$$

$$\kappa_\theta = -\frac{1}{r}\left[\frac{dw}{dr} - \frac{h}{\pi z}\sin\left(\frac{\pi}{h}z\right)\gamma_{rz}^0\right]. \tag{2.4d}$$

According to Andryewa's method (1962), now we have

$$\varepsilon_r = \varepsilon_{r1} + \varepsilon_{r2}, \quad \varepsilon_\theta = \varepsilon_{\theta 1} + \varepsilon_{\theta 2} \tag{2.5}$$

where ε_{r1} and $\varepsilon_{\theta 1}$ are the radial and circumferential strains of the shell due to the tension respectively. ε_{r2} and $\varepsilon_{\theta 2}$ are the radial and circumferential strains of the shell due to the flexure respectively

$$\varepsilon_{r1} = \varepsilon_r^0, \qquad \varepsilon_{r2} = z\kappa_r,$$

$$\varepsilon_{\theta 1} = \varepsilon_\theta^0, \qquad \varepsilon_{\theta 2} = z\kappa_\theta. \tag{2.6}$$

In such a case the stress-strain relations of the shell may be separated as follows

$$\varepsilon_{r1} = \frac{\sigma_{r1}}{E_{r1}} - \nu_{\theta r1}\frac{\sigma_{\theta 1}}{E_{\theta 1}}, \qquad \varepsilon_{\theta 1} = \frac{\sigma_{\theta 1}}{E_{\theta 1}} - \nu_{r\theta 1}\frac{\sigma_{r1}}{E_{r1}},$$

$$\varepsilon_{r2} = \frac{\sigma_{r2}}{E_{r2}} - \nu_{\theta r2}\frac{\sigma_{\theta 2}}{E_{\theta 2}}, \qquad \varepsilon_{\theta 2} = \frac{\sigma_{\theta 2}}{E_{\theta 2}} - \nu_{r\theta 2}\frac{\sigma_{r2}}{E_{r2}} \tag{2.7}$$

where σ_{r1} and $\sigma_{\theta 1}$ are the radial and circumferential stress of the shell due to the tension respectively, σ_{r2} and $\sigma_{\theta 2}$ are the radial and circumferential stress of the shell due to the flexure respectively, E_{r1} and $E_{\theta 1}$ are Young's moduli for the tension in the r and θ direction respectively, E_{r2} and $E_{\theta 2}$ are Young's moduli for the flexure in the r and θ direction respectively, $\nu_{\theta r1}$ and $\nu_{\theta r2}$ are Poisson's ratios for the tension and flexure characterizing contraction in the r direction during tension applied in the θ direction respectively, $\nu_{r\theta 1}$ and $\nu_{r\theta 2}$ are Poisson's ratios for the tension and flexure characterizing contraction in the θ direction during tension applied in the r direction respectively.

$$\nu_{\theta r1}E_{r1} = \nu_{r\theta 1}E_{\theta 1}, \qquad \nu_{\theta r2}E_{r2} = \nu_{r\theta 2}E_{\theta 2}. \tag{2.8}$$

Obviously, in the present case we have (Andryewa, 1962)

$$\nu_{\theta r1} = \nu_{\theta r2} = \nu \tag{2.9}$$

where ν is Poisson's ratio of the material.

At the same time, the relations between Young's moduli of the corrugated circular shallow spherical shell and the equivalent cylindrically orthotropic shallow spherical shell can be written as

$$E_{r1} = \frac{E}{k_{r1}}, \qquad E_{\theta 1} = k_{\theta 1}E,$$

$$E_{r2} = \frac{E}{k_{r2}}, \qquad E_{\theta 2} = k_{\theta 2}E, \tag{2.10}$$

in which E is Young's modulus of the material, k_{r1}, $k_{\theta 1}$, k_{r2} and $k_{\theta 2}$ are parameters of the radial and circumferential rigidities respectively.

The parameters k_{r1}, $k_{\theta 1}$, k_{r2} and $k_{\theta 2}$ can be obtained by comparing rigidities of a corrugated shallow spherical shell and a equivalent cylindrically orthotripic shallow spherical shell. The result is

$$k_{r1}=\frac{\int_0^s \frac{y^2}{J_r^0}\mathrm{d}s+\int_0^s \frac{\cos^2\varphi_1}{bh}\mathrm{d}s}{\int_0^l \frac{y^2}{J_r^0}\mathrm{d}l+\int_0^l \frac{\cos^2\varphi_2}{bh}\mathrm{d}l},\quad k_{\theta 1}=\frac{\int_0^b \frac{y^2}{J_r''}\mathrm{d}l+\int_0^b \frac{\cos^2\varphi_3}{lh}\mathrm{d}l}{\int_0^b \frac{y^2}{J_r'}\mathrm{d}l+\int_0^s \frac{\cos^2\varphi_3}{sh}\mathrm{d}l},$$

$$k_{r2}=\frac{s}{l},\qquad k_{\theta 2}=\frac{h\int_0^s y^2\,\mathrm{d}s+\frac{h^3}{12}\int_0^s \cos^2\varphi_1\,\mathrm{d}s}{h\int_0^l y^2\,\mathrm{d}l+\frac{h^3}{12}\int_0^l \cos^2\varphi_2\,\mathrm{d}l}, \tag{2.11}$$

$$J_r^0=h\int_0^b y^2\,\mathrm{d}l+\frac{h^3}{12}\int_0^b \cos^2\varphi_3\,\mathrm{d}l,$$

$$J_r'=h\int_0^s y^2\,\mathrm{d}s+\frac{h^3}{12}\int_0^s \cos^2\varphi_2\,\mathrm{d}s$$

$$J_r''=h\int_0^l y^2\,\mathrm{d}l+\frac{h^3}{12}\int_0^l \cos^2\varphi_2\,\mathrm{d}l \tag{2.12}$$

where s is the length of a wave, b is the width of a shell element, φ_1 and φ_2 are slope angles of the tangent to a diametral section, φ_3 is the slope angle of the tangent to a circumferential section.

From relation (2.7), and using expressions (2.6), (2.8), (2.9) and (2.10), we obtain

$$\begin{aligned}\sigma_{r1}&=Q_{11}'\varepsilon_r^0+Q_{12}'\varepsilon_\theta^0, &\sigma_{\theta 1}&=Q_{12}'\varepsilon_r^0+Q_{22}'\varepsilon_\theta^0\\ \sigma_{r2}&=z(Q_{11}''\kappa_r+Q_{12}''\kappa_\theta), &\sigma_{\theta 2}&=z(Q_{12}''\kappa_r+Q_{22}''\kappa_\theta)\end{aligned} \tag{2.13}$$

where

$$\begin{aligned}Q_{11}'&=\frac{E}{k_{r1}\left(1-\frac{\nu^2}{k_{r1}k_{\theta 1}}\right)}, &Q_{12}'&=\frac{E\nu}{k_{r1}\left(1-\frac{\nu^2}{k_{r1}k_{\theta 1}}\right)},\\ Q_{22}'&=\frac{Ek_{\theta 1}}{1-\frac{\nu^2}{k_{r1}k_{\theta 1}}}, &Q_{11}''&=\frac{E}{k_{r2}\left(1-\frac{\nu^2}{k_{r2}k_{\theta 2}}\right)},\\ Q_{12}''&=\frac{E\nu}{k_{r2}\left(1-\frac{\nu^2}{k_{r2}k_{\theta 2}}\right)}, &Q_{22}''&=\frac{Ek_{\theta 2}}{1-\frac{\nu^2}{k_{r2}k_{\theta 2}}}.\end{aligned} \tag{2.14}$$

The transverse shear stress τ_{rz} of the shell is defined by

$$\tau_{rz}=G_{rz}\gamma_{rz} \tag{2.15}$$

where G_{rz} is the shear modulus characterizing change of the angle in the rz plane.

Substituting Eq. (2.3c) into (2.15), we find

$$\tau_{rz} = G_{rz}\gamma_{rz}^0 \cos\left(\frac{\pi}{h}z\right). \tag{2.16}$$

This stress at the middle surface of the shell is $G_{rz}\gamma_{rz}^0$ and at the upper and lower faces it is zero. According to Salashiling's suggestion (1965), we take

$$G_{rz} = \frac{G}{k_{\theta 1}} \tag{2.17}$$

where G is the shear modulus of the material.

Let us consider an element cut out of the shell by two meridians and two parallel circles as shown in Fig. 2. Here N_r and N_θ are the radial and circumferential membrane forces respectively, M_r and M_θ are the radial and circumferential moments respectively, Q_r is the radial shear force. They are given by the expressions

$$N_r = \int_{-h/2}^{h/2} (\sigma_{r1} + \sigma_{r2})\mathrm{d}z, \qquad N_\theta = \int_{-h/2}^{h/2} (\sigma_{\theta 1} + \sigma_{\theta 2})\mathrm{d}z,$$

$$M_r = \int_{-h/2}^{h/2} (\sigma_{r1} + \sigma_{r2})z\mathrm{d}z, \qquad M_\theta = \int_{-h/2}^{h/2} (\sigma_{\theta 1} + \sigma_{\theta 2})z\mathrm{d}z,$$

$$Q_r = \int_{-h/2}^{h/2} \tau_{rz}\,\mathrm{d}z. \tag{2.18}$$

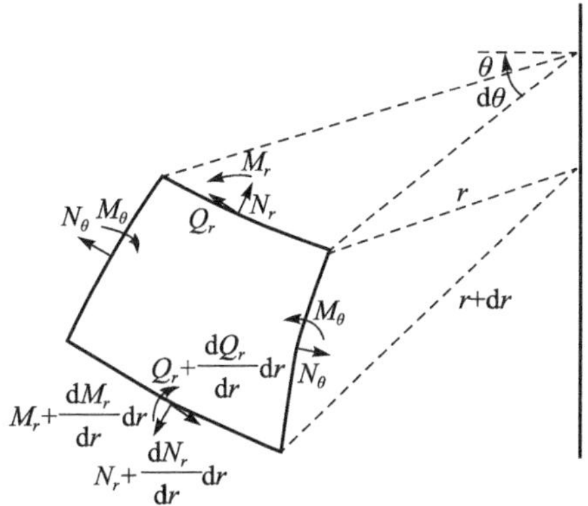

Fig. 2 Internal forces acting on a shell element.

Substituting expressions (2.13) and (2.16) into (2.18), we obtain

$$N_r = A_{11}\varepsilon_r^0 + A_{12}\varepsilon_\theta^0, \quad N_\theta = A_{12}\varepsilon_r^0 + A_{22}\varepsilon_\theta^0, \tag{2.19}$$

$$M_r = -\left(D_{11}\frac{\mathrm{d}^2 w}{\mathrm{d}r^2} + D_{12}\frac{1}{r}\frac{\mathrm{d}w}{\mathrm{d}r} + D_{11}'\frac{\mathrm{d}\gamma_{rz}^0}{\mathrm{d}r} + D_{12}'\frac{\gamma_{rz}^0}{r}\right),$$

$$M_\theta = -\left(D_{12}\frac{\mathrm{d}^2 w}{\mathrm{d}r^2} + D_{22}\frac{1}{r}\frac{\mathrm{d}w}{\mathrm{d}r} + D_{12}'\frac{\mathrm{d}\gamma_{rz}^0}{\mathrm{d}r} + D_{22}'\frac{\gamma_{rz}^0}{r}\right), \tag{2.20}$$

$$Q_r = G_0\gamma_{rz}^0 \tag{2.21}$$

where A_{ij}, D_{ij} and D_{ij}' are the extensional and flexural rigidities respectively, G_0 is the shear rigidity

$$A_{ij} = hQ_{ij}', \qquad D_{ij} = \frac{h^3}{12}Q_{ij}'',$$

$$D_{ij}'' = -\frac{2h^3}{\pi^3}Q_{ij}'', \qquad (i,j=1,2), \tag{2.22}$$

$$G_0 = \frac{2Gh}{\pi\kappa_{\theta 1}}. \tag{2.23}$$

From the equilibrium of differential element of the shell, we obtain the equilibrium equations of the shell with large deflection

$$\frac{\mathrm{d}(rN_r)}{\mathrm{d}r} - N_\theta = 0, \tag{2.24a}$$

$$\frac{\mathrm{d}(rM_r)}{\mathrm{d}r} - M_\theta - rQ_r = 0, \tag{2.24b}$$

$$\frac{\mathrm{d}}{\mathrm{d}r}\left[rN_r\left(\frac{r}{R} + \frac{\mathrm{d}w}{\mathrm{d}r}\right) + rQ_r\right] + qr = 0. \tag{2.24c}$$

Eliminating u_0 from expressions (2.4a, b), we have

$$\varepsilon_r^0 = \frac{\mathrm{d}}{\mathrm{d}r}(r\varepsilon_\theta^0) + \frac{r}{R}\frac{\mathrm{d}w}{\mathrm{d}r} + \frac{1}{2}\left(\frac{\mathrm{d}w}{\mathrm{d}r}\right)^2. \tag{2.25}$$

The strains ε_r^0 and ε_θ^0 can be expressed in terms of N_r and N_θ by appling expressions (2.19). This gives

$$\varepsilon_r^0 = A_1 N_r - A_2 N_\theta, \qquad \varepsilon_\theta^0 = -A_2 N_r + A_3 N_\theta \tag{2.26}$$

where

$$A_0 = \frac{1}{A_{11}A_{22} - A_{12}^2}, \quad A_1 = A_0 A_{22}, \quad A_2 = A_0 A_{12}, \quad A_3 = A_0 A_{11}. \tag{2.27}$$

Substituting expressions (2.26) into (2.25), we obtain a compatibility equation

$$\frac{\mathrm{d}}{\mathrm{d}r}[r(A_3 N_\theta - A_2 N_r)] - A_1 N_r + A_2 N_\theta + \frac{1}{2}\left(\frac{\mathrm{d}w}{\mathrm{d}r}\right)^2 + \frac{r}{R}\frac{\mathrm{d}w}{\mathrm{d}r} = 0. \tag{2.28}$$

Equations (2.24) and (2.28) are the fundamental equations for the large deflection of the corrugated circular shallow spherical shell under the action of uniform pressure.

From Eq. (2.24a), we have

$$N_\theta = \frac{\mathrm{d}(rN_r)}{\mathrm{d}r}. \tag{2.29}$$

Sustituting this equation into (2.28), we find

$$A_3 \frac{\mathrm{d}}{\mathrm{d}r} r \frac{\mathrm{d}(rN_r)}{\mathrm{d}r} - A_1 N_r + \frac{1}{2}\left(\frac{\mathrm{d}w}{\mathrm{d}r}\right)^2 + \frac{r}{R}\frac{\mathrm{d}w}{\mathrm{d}r} = 0. \tag{2.30}$$

Using Eq. (2.21), Eq. (2.24c) becomes

$$\frac{\mathrm{d}}{\mathrm{d}r}\left[rN_r\left(\frac{r}{R} + \frac{\mathrm{d}w}{\mathrm{d}r}\right) + G_0 r\gamma_{rz}^0\right] + qr = 0. \tag{2.31}$$

Substituting Eqs. (2.20) and (2.21) into (2.24b), we have

$$D_{11}\frac{\mathrm{d}}{\mathrm{d}r} r \frac{\mathrm{d}^2 w}{\mathrm{d}r^2} - D_{22}\frac{1}{r}\frac{\mathrm{d}w}{\mathrm{d}r} + D_{11}'\frac{\mathrm{d}}{\mathrm{d}r} r \frac{\mathrm{d}\gamma_{rz}^0}{\mathrm{d}r} - D_{22}'\frac{\gamma_{rz}^0}{r} + G_0 r\gamma_{rz}^0 = 0. \tag{2.32}$$

Thus the problem of nonlinear stability of the corrugated circular shallow spherical shell under the action of uniform pressure is reduced to the integration of the three Eqs. (2.31), (2.32) and (2.30).

Eqs. (2.31), (2.32) and (2.30) will be solved under the following customary boundary conditions for rigidly clamped edge

$$w=0,\quad \frac{\mathrm{d}w}{\mathrm{d}r}=0,\quad u_0=0 \qquad \text{at } r=a,$$
$$\gamma_{rz}^0=0,\quad N_r \text{ finite} \qquad \text{at } r=0 \tag{2.33}$$

where

$$u_0=r\left[A_3\frac{\mathrm{d}(rN_r)}{\mathrm{d}r}-A_2N_r\right]. \tag{2.34}$$

In order to simplify the following calculations, let us introduce the following dimensionless variables

$$\rho=\frac{r}{a},\quad W=\frac{w}{h},\quad \Phi=k\rho+\frac{\mathrm{d}w}{\mathrm{d}\rho},\quad S=\frac{a}{D_{11}}rN_r,$$
$$\psi=\frac{a}{h}\gamma_{rz}^0,\quad P=\frac{a^4}{D_{11}h}q,\quad k=\frac{a^2}{Rh},\quad m=\frac{a^2}{D_{11}}G_0,$$
$$\beta_1^2=\frac{D_{22}}{D_{11}},\quad \beta_2^2=\frac{A_1}{A_3},\quad \beta_3=\frac{D_{11}'}{D_{11}},$$
$$\beta_4=\frac{D_{22}'}{D_{11}},\quad \beta_5=\frac{h^2}{2A_3D_{11}},\quad \beta_6=\frac{A_2}{A_3}. \tag{2.35}$$

With the help of these variables, the nonlinear boundary value problem (2.31), (2.32), (2.30) and (2.33) may be rewritten in a more convenient dimensionless form

$$m\frac{\mathrm{d}(\rho\psi)}{\mathrm{d}\rho}=-\frac{\mathrm{d}(S\Phi)}{\mathrm{d}\rho}-P\rho, \tag{2.36a}$$

$$\rho^{\beta_1}\frac{\mathrm{d}}{\mathrm{d}\rho}\rho^{1-2\beta_1}\frac{\mathrm{d}}{\mathrm{d}\rho}(\rho^{\beta_1}\Phi)=-\beta_3\frac{\mathrm{d}}{\mathrm{d}\rho}\rho\frac{\mathrm{d}\psi}{\mathrm{d}\rho}-m\rho\psi+\beta_4\frac{\psi}{\rho}+k(1-\beta_1^2), \tag{2.36b}$$

$$\rho^{\beta_2}\frac{\mathrm{d}}{\mathrm{d}\rho}\rho^{1-2\beta_2}\frac{\mathrm{d}}{\mathrm{d}\rho}(\rho^{\beta_2}S)=-\beta_5(\Phi^2-k^2\rho^2), \tag{2.36c}$$

$$W=0,\quad \Phi=k\rho,\quad \frac{\mathrm{d}S}{\mathrm{d}\rho}-\beta_6\frac{S}{\rho}=0 \quad \text{at} \quad \rho=1,$$
$$\psi=0,\quad S=0 \qquad \text{at} \qquad \rho=0. \tag{2.37}$$

3. Analytical Solution

We apply modified iteration method (Yeh and Liu) to solve the dimensionless nonlinear boundary value problem (2.36) and (2.37). Let the dimensionless center deflection as an iteration parameter, be denoted by W_m

$$W_m=-\int_0^1(\Phi-k\rho)\mathrm{d}\rho. \tag{3.1}$$

In the first approximation, by neglecting the nonlinear term $S\Phi$ in Eq. (2.36a), we obtain the following linear boundary value problem

$$m\frac{\mathrm{d}(\rho\psi_1)}{\mathrm{d}\rho}=-P\rho,$$

$$\rho^{\beta_1}\frac{\mathrm{d}}{\mathrm{d}\rho}\rho^{1-2\beta_1}\frac{\mathrm{d}}{\mathrm{d}\rho}(\rho^{\beta_1}\Phi_1)=-\beta_3\frac{\mathrm{d}}{\mathrm{d}\rho}\rho\frac{\mathrm{d}\psi_1}{\mathrm{d}\rho}-m\rho\psi_1+\beta_4\frac{\psi_1}{\rho}+k(1-\beta_1^2), \tag{3.2}$$

$$\rho^{\beta_2}\frac{\mathrm{d}}{\mathrm{d}\rho}\rho^{1-2\beta_2}\frac{\mathrm{d}}{\mathrm{d}\rho}(\rho^{\beta_2}S_1)=-\beta_5(\Phi_1^2-k^2\rho^2),$$

$$W_1=0,\quad \Phi_1=k\rho,\quad \frac{\mathrm{d}S_1}{\mathrm{d}\rho}-\beta_6\frac{S_1}{\rho}=0\quad \text{at}\quad \rho=1, \tag{3.3}$$

$$\psi_1=0,\quad S_1=0\quad \text{at}\quad \rho=0. \tag{3.4}$$

Solving the boundary value problem (3.2), (3.3), (3.4) and using expression (3.1), we obtain the following solutions for the first approximation

$$\psi_1=-\frac{W_m}{2m}\rho,$$

$$\Phi_1=c_0W_m\left\{-\frac{\beta_2+\beta_3}{m(1-\beta_1^2)}\rho+\frac{1}{3^2-\beta_1^2}\rho^3+\left[\frac{\beta_2+\beta_3}{m(1-\beta_1^2)}-\frac{1}{3^2-\beta_1^2}\right]\rho^{\beta_1}\right\}+k\rho$$

$$\begin{aligned}S_1=&-\frac{1}{2}\beta_5W_m^2\left[\frac{c_1^2}{3^2-\beta_4^2}\rho^3+\frac{2c_1c_2}{5^2-\beta_4^2}\rho^5+\frac{c_2^2}{7^2-\beta_4^2}\rho^7\right.\\&\left.+\frac{2c_1c_3}{(2+\beta_1)^2-\beta_4^2}\rho^{2+\beta_1}+\frac{2c_1c_3}{(4+\beta_1)^2-\beta_4^2}\rho^{4+\beta_1}+\frac{c_3^2}{(1+2\beta_1)^2-\beta_4^2}\rho^{1+2\beta_1}+c_4\rho^{\beta_4}\right]\\&-\beta_5kW_m\left[\frac{c_1}{3^2-\beta_4^2}\rho^3+\frac{c_2}{5^2-\beta_4^2}\rho^5+\frac{c_3}{(2+\beta_1)^2-\beta_4^2}\rho^{2+\beta_1}+c_5\rho^{\beta_4}\right]\end{aligned} \tag{3.5}$$

where

$$\begin{aligned}c_0=&\left\{\frac{\beta_2+\beta_3}{2m(1-\beta_1^2)}-\frac{1}{4(3^2-\beta_1^2)}+\frac{1}{1+\beta_1}\left[\frac{1}{3^2-\beta_1^2}-\frac{\beta_2+\beta_3}{m(1-\beta_1^2)}\right]\right\}^{-1},\\c_1=&-\frac{c_0(\beta_2+\beta_3)}{m(1-\beta_1^2)},c_2=\frac{c_0}{3^2-\beta_1^2},c_3=c_0\left[\frac{\beta_2+\beta_3}{m(1-\beta_1^2)}-\frac{1}{3^2-\beta_1^2}\right],\\c_4=&\frac{1}{\beta_6-\beta_4}\left[\frac{c_1^2}{3^2-\beta_4^2}\left(\frac{1}{3}-\beta_6\right)+\frac{2c_1c_2}{5^2-\beta_4^2}\left(\frac{1}{5}-\beta_6\right)\right.\\&+\frac{c_2^2}{7^2-\beta_4^2}\left(\frac{1}{7}-\beta_6\right)+\frac{c_3^2}{(1+2\beta_1)^2-\beta_4^4}\left(\frac{1}{1+2\beta_1}-\beta_6\right)\\&\left.+\frac{2c_1c_3}{(2+\beta_1)^2-\beta_4^4}\left(\frac{1}{2+\beta_1}-\beta_6\right)+\frac{2c_2c_3}{(4+\beta_1)^2-\beta_4^4}\left(\frac{1}{4+\beta_1}-\beta_6\right)\right],\\c_5=&\frac{1}{\beta_6-\beta_4}\left[\frac{c_1}{3^2-\beta_4^2}\left(\frac{1}{3}-\beta_6\right)+\frac{c_2}{5^2-\beta_4^2}\left(\frac{1}{5}-\beta_6\right)\right.\\&\left.+\frac{c_3}{(2+\beta_1)^2-\beta_4^2}\left(\frac{1}{2+\beta_1}-\beta_6\right)\right].\end{aligned} \tag{3.6}$$

For the second approximation, we have the following linear boundary value problem

$$m\frac{\mathrm{d}(\rho\psi_2)}{\mathrm{d}\rho}=-\frac{\mathrm{d}(S_1\psi_1)}{\mathrm{d}\rho}-P\rho,$$

$$\rho^{\beta_1}\frac{\mathrm{d}}{\mathrm{d}\rho}\rho^{1-2\beta_1}\frac{\mathrm{d}}{\mathrm{d}\rho}(\rho^{\beta_1}\Phi_2)=-\beta_3\frac{\mathrm{d}}{\mathrm{d}\rho}\rho\frac{\mathrm{d}\psi_2}{\mathrm{d}\rho}-m\rho\psi_2+\beta_4\frac{\psi_2}{\rho}+k(1-\beta_1^2), \tag{3.7}$$

$$\rho^{\beta_2}\frac{\mathrm{d}}{\mathrm{d}\rho}\rho^{1-2\beta_2}\frac{\mathrm{d}}{\mathrm{d}\rho}(\rho^{\beta_2}S_2)=-\beta_5(\Phi_2^2-k^2\rho^2),$$

$$W_2=0,\quad \Phi_2=k\rho,\quad \frac{\mathrm{d}S_2}{\mathrm{d}\rho}-\beta_6\frac{S_2}{\rho}=0\quad \text{at}\quad \rho=1,$$

$$\psi_2=0,\quad S_2=0\quad \text{at}\quad \rho=0. \tag{3.8}$$

Solving this problem, we obtain the solution for the second approximation. Using expression (3.1), we have the nonlinear characteristic relation of the shell

$$P=(k^2\alpha_0+\alpha_1)W_m+k\alpha_2W_m^2+\alpha_3W_m^2 \tag{3.9}$$

where

$$\alpha_0=-\frac{1}{A'},\qquad \alpha_1=-\frac{1}{A'}\left[\int_0^1 f_1(\rho)\mathrm{d}y-\frac{f_1(1)}{1+\beta_1}\right],$$

$$\alpha_2=-\frac{1}{A'}\left[\int_0^1 f_2(\rho)\mathrm{d}y-\frac{f_2(1)}{1+\beta_1}\right],\qquad \alpha_3=-\frac{1}{A'}\left[\int_0^1 f_3(\rho)\mathrm{d}y-\frac{f_3(1)}{1+\beta_1}\right],$$

$$A'=\frac{\beta_1-3}{4(9-\beta_1^2)(1+\beta_1)}-\frac{\beta_2+\beta_3}{\mu(1-\beta_1^2)}\left(\frac{1}{2}-\frac{1}{1+\beta_1}\right),$$

$$\begin{aligned}f_1(\rho)=&-\frac{1}{m}\left\{d_8\rho^3\frac{3\beta_2+\beta_3}{3^2-\beta_1^2}+d_9\rho^5\frac{5\beta_2+\beta_3}{5^2-\beta_1^2}+d_{10}\rho^{2+\beta_1}\frac{\beta_2(2+\beta_1)+\beta_3}{(2+\beta_1)^2-\beta_1^2}\right.\\&\left.+d_{11}\rho^{\beta_4}\frac{\beta_4\beta_2+\beta_3}{\beta_4^2-\beta_1^2}\right\}+d_8\frac{\rho^5}{5^2-\beta_1^2}+d_9\frac{\rho^7}{7^2-\beta_1^2}\\&+d_{10}\frac{\rho^{4+\beta_1}}{(4+\beta_1)^2-\beta_1^2}+d_{11}\frac{y^{4+\beta_4}}{(2+\beta_4)^2-\beta_1^2},\end{aligned}$$

$$\begin{aligned}f_2(\rho)=&-\frac{1}{m}\left\{(c_1d_8+d_1)\frac{(3\beta_2+\beta_3)}{3^2-\beta_1^2}\rho^3+(c_1d_9+c_2d_8+d_5)\frac{5\beta_2+\beta_3}{5^2-\beta_1^2}\rho^5\right.\\&+(c_1d_{10}+c_3c_8+d_4)\frac{\beta_2(2+\beta_1)+\beta_3}{(2+\beta_1)^2-\beta_1^2}\rho^{2+\beta_1}+(c_1d_{11}+d_7)\frac{\beta_4\beta_2+\beta_3}{\beta_4^2-\beta_1^2}\rho^{\beta_4}\\&+(c_2d_9+d_2)\frac{7\beta_2+\beta_3}{7^2-\beta_1^2}\rho^7+(c_2d_{10}+d_9+d_6)\frac{\beta_2(4+\beta_1)+\beta_3}{(4+\beta_1)^2-\beta_1^2}\rho^{4+\beta_1}\\&+c_2d_{11}\frac{\beta_2(2+\beta_4)+\beta_3}{(2+\beta_4)^2-\beta_1^2}\rho^{\beta_4+2}+(c_3d_8+d_4)\frac{\beta_2(2+\beta_1)+\beta_3}{(2+\beta_1)^2-\beta_1^2}\rho^{2+\beta_1}\\&\left.+(c_3d_{10}+d_3)\frac{\beta_2(1+2\beta_1)+\beta_3}{(2\beta_1+1)^2-\beta_1^2}\rho^{2\beta_1+1}+c_3d_{11}\frac{\beta_2(\beta_4+\beta_1-1)+\beta_3}{(\beta_4+\beta_1-1)^2-\beta_1^2}\rho^{\beta_4+\beta_1-1}\right\}\\&+(c_1d_8+d_1)\frac{\rho^5}{5^2-\beta_1^2}+(c_1d_9+d_8+d_5)\frac{\rho^7}{7^2-\beta_1^2}+(c_1d_{10}+d_4)\frac{\rho^{4+\beta_1}}{(4+\beta_1)^2-\beta_1^2}\\&+(c_1d_{11}+d_7)\frac{\rho^{2+\beta_4}}{(2+\beta_4)^2-\beta_1^2}+c_2d_9\frac{\rho^8}{8^2-\beta_1^2}+(c_2d_{10}+d_9)\frac{\rho^{6+\beta_1}}{(6+\beta_1)^2-\beta_1^2}\\&+c_2d_{11}\frac{\rho^{4+\beta_4}}{(4+\beta_4)^2-\beta_1^2}+(c_3d_{10}+d_3)\frac{\rho^{3+2\beta_1}}{(3+2\beta_1)^2-\beta_1^2}\\&+d_2\frac{\rho^9}{9^2+\beta_1^2}+c_3d_{11}\frac{\rho^{\beta_4+\beta_1+1}}{(\beta_4+\beta_1+1)^2-\beta_1^2},\end{aligned}$$

$$
\begin{aligned}
f_3(\rho) = & -\frac{1}{m}\left\{c_1 d_1 \rho^3 \frac{3\beta_2+\beta_3}{3^2-\beta_1^2} + (c_1 d_2 + c_2 d_5)\rho^7 \frac{7\beta_2+\beta_3}{7^2-\beta_1^2}\right. \\
& + (c_1 d_3 + c_3 d_4)\rho^{2\beta_1+1} \frac{\beta_2(1+2\beta_1)+\beta_3}{(1+2\beta_1)^2-\beta_1^2} + (c_1 d_4 + c_3 d_1)\rho^{2+\beta_1} \frac{\beta_2(2+\beta_1)+\beta_3}{(2+\beta_1)^2-\beta_1^2} \\
& + (c_1 d_5 + c_2 d_1)\rho^5 \frac{5\beta_2+\beta_3}{5^2-\beta_1^2} + (c_1 d_6 + c_2 d_4 + c_3 d_5)\rho^{4+\beta_1} \frac{\beta_2(4+\beta_1)+\beta_3}{(4+\beta_1)^2-\beta_1^2} \\
& + c_1 d_7 \rho^{\beta_4} \frac{\beta_4\beta_2+\beta_3}{\beta_4^2-\beta_1^2} + c_2 d_2 \rho^9 \frac{9\beta_2+\beta_3}{9^2-\beta_1^2} + (c_2 d_3 + c_3 d_6)\rho^{2\beta_1+3} \frac{\beta_2(3+2\beta_1)+\beta_3}{(3+2\beta_1)^2-\beta_1^2} \\
& + (c_2 d_6 + c_3 d_2)\rho^{6+\beta_1} \frac{\beta_2(6+\beta_1)+\beta_3}{(6+\beta_1)^2-\beta_1^2} + c_2 d_7 \rho^{2+\beta_4} \frac{\beta_2(2+\beta_4)+\beta_3}{(2+\beta_4)^2-\beta_1^2} \\
& \left. + c_3 d_1 \rho^{2+\beta_1} \frac{\beta_2(2+\beta_1)+\beta_3}{(2+\beta_1)^2-\beta_1^2} + c_3 d_7 \rho^{\beta_1+\beta_4-1} \frac{(\beta_1+\beta_4-1)+\beta_3}{(\beta_1+\beta_4-1)^2-\beta_1^2}\right\} \\
& + c_1 d_1 \frac{\rho^5}{5^2-\beta_1^2} + (c_1 d_2 + c_2 d_5)\frac{\rho^9}{9^2-\beta_1^2} + (c_1 d_3 + c_3 d_4)\frac{\rho^{2\beta_1+3}}{(2\beta_1+3)^2-\beta_1^2} \\
& + (c_1 d_4 + c_3 d_1)\frac{\rho^{\beta_1+4}}{(\beta_1+4)^2-\beta_1^2} + (c_1 d_5 + c_2 d_1)\frac{\rho^7}{7^2-\beta_1^2} \\
& + (c_1 d_6 + c_2 d_4)\frac{\rho^{\beta_1+6}}{(\beta_1+6)^2-\beta_1^2} + c_1 d_7 \frac{\rho^{\beta_4+2}}{(\beta_4+2)^2-\beta_1^2} + c_2 d_2 \frac{\rho^{11}}{11^2-\beta_1^2} \\
& + (c_2 d_3 + c_3 d_6)\frac{\rho^{5+2\beta_1}}{(5+2\beta_1)^2-\beta_1^2} + (c_2 d_6 + c_3 d_2)\frac{\rho^{8+\beta_1}}{(8+\beta_1)^2-\beta_1^2} \\
& + c_2 d_7 \frac{\rho^{4+\beta_4}}{(4+\beta_4)^2-\beta_1^2} + c_3 d_3 \frac{\rho^{2+3\beta_1}}{(2+3\beta_1)^2-\beta_1^2} \\
& + c_3 d_4 \frac{\rho^{3+2\beta_1}}{(3+2\beta_1)^2-\beta_1^2} + c_3 d_7 \frac{\rho^{\beta_1+\beta_4+1}}{(\beta_1+\beta_4+1)^2-\beta_1^2},
\end{aligned}
$$

$$
d_6 = -\beta_5 \frac{c_2 c_3}{2[(4+\beta_1)^2-\beta_4^2]},\quad d_7 = -\beta_5 \frac{c_4}{2},\quad d_8 = -\beta_5 \frac{k c_1}{3^2-\beta_4^2},
$$

$$
d_9 = -\beta_5 k \frac{c_2}{5^2-\beta_4^2},\quad d_{10} = -\beta_5 k \frac{c_3}{(2+\beta_1)^2-\beta_4^2},\quad d_{11} = -\beta_5 k c_5. \tag{3.10}
$$

The critical dimensionless load P^* of the shell can be solved by the extreme condition

$$
\frac{\mathrm{d}P}{\mathrm{d}W_m} = 0. \tag{3.11}
$$

From this condition, when buckling appears, the dimensionless center deflection can be written in the form

$$
W_m^* = \frac{1}{3\alpha_3}\left[-k\alpha_2 \pm \sqrt{k^2(\alpha_2^2 - 3\alpha_0\alpha_3) - 3\alpha_1\alpha_3}\right]. \tag{3.12}
$$

Substituting W_m^* into Eq. (3.9), the dimensionless critical load can be obtained

$$
P^* = (k^2\alpha_0 + \alpha_1)W_m^* + k\alpha_2 W_m^{*2} + \alpha_3 W_m^{*3}. \tag{3.13}
$$

4. Numerical Examples

Now let us consider an example of a sinusoidal corrugated shallow spherical shell. Assume: a=80mm, h=0.4mm, l=12mm, H=1.35mm, E=1.345×10^{4}kg/mm^2,

$\nu=0.3$，$R=130$mm.

Equivalent elastic coefficients of sinusoidal corrugated shallow spherical shells under different curvature radius R are shown in Table 1.

Table 1　Equivalent elastic coefficients of sinusoidal corrugated shallow spherical shells

R	k_{r1}	$k_{\theta 1}$	k_{r2}	$k_{\theta 2}$
80	6.717	1.0354	1.0353	11.665
100	8.513	1.0343	1.0343	13.254
120	10.067	1.0336	1.0336	14.385
130	11.362	1.0331	1.0331	15.194
160	12.422	1.0327	1.0327	15.783
180	13.284	1.0324	1.0324	16.221
200	13.986	1.0322	1.0322	16.552
220	14.559	1.0320	1.0320	16.808
240	15.030	1.0318	1.0318	17.010
260	15.421	1.0317	1.0317	17.171
280	15.741	1.0316	1.0316	17.302
300	16.020	1.0315	1.0315	17.408
340	16.449	1.0313	1.0313	17.571
380	16.765	1.0312	1.0312	17.688
400	16.892	1.0311	1.0311	17.733
450	17.145	1.0310	1.0310	17.823
500	17.330	1.0309	1.0309	17.889
550	17.471	1.0309	1.0309	17.938
600	17.579	1.0308	1.0308	17.975
650	17.680	1.0307	1.0307	18.009
700	17.733	1.0307	1.0307	18.027
750	17.788	1.0306	1.0306	18.046
800	17.834	1.0306	1.0306	18.061
900	17.904	1.0306	1.0306	18.085
1000	17.955	1.0305	1.0305	18.102

Table 1 shows that $k_{\theta 1}=k_{r2}\approx 1$ when curvature radii R increase the value of k_{r1} and $k_{\theta 2}$ gradually approximates. When $R\rightarrow\infty$, corrugated shell becomes corrugated plate, $k_{r1}=k_{\theta 2}$, $k_{\theta 1}=k_{r2}$. The results are coincident with Andryewa's (1962).

Substituting these values into formula (3.13), the dimensionless critical load can be obtained

$$P^* = 1.305W_m^{*3} - 0.7912W_m^{*2} + 8.635W_m^*. \tag{4.1}$$

According to formula (4.1), values of the upper and lower critical buckling loads for several values of shear rigidity m are given in Fig. 3. It can be seen that the upper critical buckling load increases with the increases of k or m, and the lower critical buckling load increase only with the increase of a small range of k, then it decreases with the increase of k.

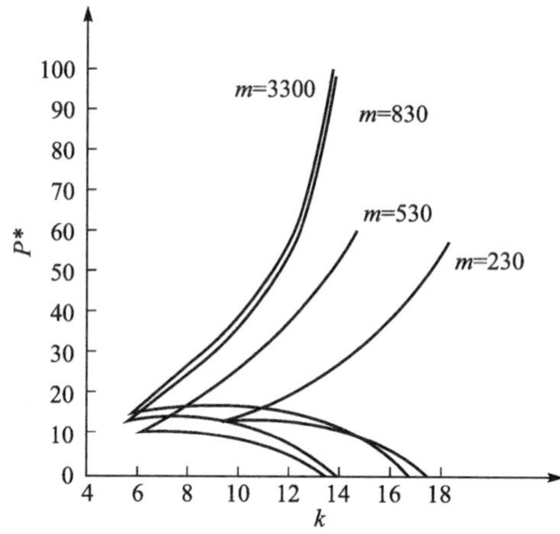

Fig. 3　The characteristic curves.

Nomenclature

f —center height of the corrugated circular shallow spherical shell

h —thickness of the corrugated circular shallow spherical shell

H —the amplitude of corrugate

k_{r1} —parameters of the radial rigidities due to the tension

$k_{\theta 1}$ —parameters of the circumferential rigidities due to the tension.

k_{r2} —parameters of the radial rigidities due to the flexure.

$k_{\theta 2}$ —parameters of the circumferential rigidities due to the flexure

l —wavelength of corrugate

R —radius of curvature

q —uniform pressure

ε_{r1} —radial strains of the shell due to the tension

$\varepsilon_{\theta 1}$ —circumferential strains of the shell due to the tension

ε_{r2} —radial strains of the shell due to the flexure

$\varepsilon_{\theta 2}$ —circumferential strains of the shell due to the flexure

References

Akasaka T. (1955): *On the elastic properties of the corrugated diaphragm*. -J. Jap. Soc. Aeronaut. Eng., vol. 3, pp. 22-23 (in Japanese).

Andryewa L. E. (1955): *The calculation of a corrugated membrane as an anisotropic plate*. -Engineer's Collection, vol. 21, pp. 128-141 (in Russian).

Andryewa L. E. (1962): *Elastic Elements of Instruments*. -Moscow: Masgiz (in Russian).

Andryewa L. E. (1975): *The Computation and Design of Bellow*. -Moscow: Masgiz (in Russian).

Axelrad E. L (1987): *Theory of Flexible Shells*. -Amsterdam: Elsevier Science.

Chien W. Z. and Wu M. D. (1983): *The nonlinear characteristics of U-shaped bellows-Calculations by the method of perturbation*. -Appl. Math. Mech., vol. 4, No. 5, pp. 649-665.

Feodosev V. I. (1949): *Elastic Elements of Precision-Instruments Manufacture*. -Moscow: Oborongiz, (in Russian).

Galishin A. Z. (1999): *Determination of the axisymmetric geometrically nonlinear thermoviscoelastoplastic stress-strain state of shell of revolution*. -International Applied Mechanics, vol. 35, No. 12, pp. 1229-1237.

Galishin A. Z. (2003): *Determination of the axisymmetric geometrically nonlinear thermoviscoelastoplastic state of laminated orthotropic shells*. -International Applied Mechanics, vol. 39, No. 1, pp. 56-63.

Haringx J. A. (1956): *Nonlinearity of corrugated diaphragms*. -Appl. Sci. Res., Ser. A, vol 16, pp. 45-47.

Liu R. H. (1965): *Nonlinear stability of circular shallow spherical shell with a hole in the center under the action of uniform moment at the inner edge*. -Bull. Sci., No. 3, pp. 253-255 (in Chinese).

Liu R. H. (1978): *The characteristic relations of corrugated circular plates*. -Acta Mechanica Sinica., No. 1, pp. 47-52 (in Chinese).

Liu R. H. (1984a): *Large deflection of corrugated circular plate with a plane central region under the action of concentrated loads at the center*. -Int. J. Non-linear Mech., vol. 19, No. 5, pp. 409-419.

Liu R. H. (1984b): *Large deflection of corrugated circular plate boundary region*. -Solid Mech. Archs., vol. 10, No. 9, pp. 383-406.

Liu R. H. (1998): *Study on Nonlinear Mechanics of Plates and Shells*. -New York: Science Press and Jinan University Press.

Liu R. H. and Li D. (1989): *On non-linear bending and vibration of corrugated circular plate*. -Int. J. Non-Linear Mech., vol. 24. No. 3, pp. 165-176.

Liu R. H. and Wang Z. W. (2000): *Nonlinear deformation analysis of a U-shaped bellows with varying thickness*. -Arch. Appl. Mech., vol. 70, No. 4, pp. 366-376.

Liu R. H. and Yuan H. (1997): *Non-linear bending of corrugated annular plate with large boundary corrugation*. -Appl. Mech. Eng. vol. 2, No. 2, pp. 353-367.

Liu R. H. and Zou R. P. (1993): *Nonlinear bending of a corrugated annular plate with a plane boundary region and a non-deformable rigid body at the center under compound load*. -Int. J. Non-linear Mech., vol. 28, No. 4, pp. 353-364.

Panov D. Y. (1941): *On large deflection of circular membranes with weak corrugation*. -Prikl. Mat.

Mekh. , vol. 5, pp. 308-318 (in Russian).

Salashiling V. I. (1965): *The Calculation of a Cylindrical Corrugated Shell*. -The Collection of Construction Frame, pp. 339-345 (in Russian).

Semenyuk N. P. (2002): *On design models in stability problems for corrugated cylindrical shells*. -International Applied Mechanics, vol. 38, No. 10, pp. 1245-1252.

Semenyuk N. P. (2002): *Stability of orthotropic cylindrical shells under axial compression*. -Mechanics of Composite Materials, vol. 38, No. 3, pp. 243-252.

Xu Z. Q. , Liu Y. , Yong J. S. , et al. (1985): *Large deflection of a U-shaped bellows with varying thickness*. -Journal of Qinghua University, vol. 15, No. 1, pp. 39-51 (in Chinese).

Yeh K. Y. , Liu R. H. , Pin Q. Y. , et al. (1965a): *Nonlinear stability of thin circular shallow spherical shell under actions of axisymmetric uniform distribution line loads*. -Bull. Sci, No. 2, pp. 142-145 (in Chinese).

Yeh K. Y. , Liu R. H. , Zhang C. Z. , et al. (1965b): *Nonlinear stability of thin circular shallow spherical shell under the actions of uniform moment*. -Bull. Sci. , No. 2, pp. 145-147 (in Chinese).

第四章　双层扁壳非线性力学

双层金属截头扁锥壳的热稳定性①

符　号

r　壳体坐标参考面的点至对称轴的距离

φ　壳子午线方向弧的新无量纲旋转角

h_0　壳坐标参考面至内表面的距离

a，b　截头扁锥壳大、小端的半径

y_0　截头扁锥壳小端的无量纲挠度

f　扁锥壳拱高

E_1，E_2　壳内、外层弹性模量，在一般温度变化范围内，视为常量

ν_1，ν_2　壳内、外层泊松比，在一般温度变化范围内，视为常量

α_1，α_2　壳内、外层的热膨胀系数

α，μ　相当一次膨胀系数、二次膨胀系数

C，D　双层壳体的抗拉刚度、抗弯刚度

T_1^*，T_2^*　无量纲上、下临界温度

t^*　临界温度

ϑ　壳子午线方向弧的旋转角

$\overset{*}{t^c}$　平均温度

L　无量纲跳跃幅度

l　跳跃幅度

$\left.\begin{matrix}\beta_0,\ \beta_1\\ \beta_2,\ \beta_3\\ \lambda_1,\ \lambda_2\end{matrix}\right\}$　辅助量

ρ_1　截头扁锥壳小端的无量纲半径

h_1，h_2　壳内、外层厚度

h　壳总厚度

S　无量纲径向薄膜力

y　无量纲挠度

---

① 本文原载《力学学报》，特刊，1981，(2)：172-180.

θ 扁锥壳斜角

y_0^* 截头扁锥壳小端的无量纲临界挠度

$y_{0,1}^*$，$y_{0,2}^*$ 截头扁锥壳小端的无量纲上、下临界挠度

k 无量纲几何参数

k_0 无量纲临界几何参数

f_0 临界拱高

T 无量纲温度改变量

T^* 无量纲临界温度

u，w 壳坐标参考面点的径向位移、横向挠度

T_c^* 无量纲平均温度

N_r 径向薄膜力

M_r 径向弯矩

t 温度改变量

ρ 无量纲径向坐标

H 微分算子

一、引言

在精密仪器的构造里，弹性元件起着十分重要的作用。在这些弹性元件中，有一类采用双层金属扁壳做成的热敏弹性元件，被用于自动控制仪表。这类元件具有这样一种特性，在均匀加热的情况下，由于两层金属的热膨胀系数不同，将产生弯曲，以致发生跳跃现象。自动控制仪表便应用这跳跃现象作为控制信号。

以前，已有 Панов[9]、Wittrick[10]、Григолюк[11]、罗祖道[12]等进行过双层金属扁球壳方面的理论研究。然而，在双层金属扁锥壳方面，据作者所知，尚无人研究。

本文应用 Korolev[13] 和 Radkowski[14] 所建议的选择多层壳体坐标参考面的方法，将双层扁锥壳的基本方程表示成单层壳体理论的形式，从而建立了在均匀温度作用下双层金属扁锥壳的大挠度弯曲理论。由于问题的非线性性质，因而在数学上存在巨大的困难。本文应用我们以往所提出的修正迭代法[1-8]，克服了非线性的数学困难，获得了双层金属截头扁锥壳的热稳定问题的解析解。这一方法的特点是使用迭代法的程序和摄动法的参数，程序比较简单，结果精确度较高，在取相同参数的情况下，所得结果与摄动法结果相同。最后，我们还将计算结果绘制成曲线，以供精密仪器热敏弹性元件设计时应用。

二、双层金属扁锥壳的大挠度弯曲方程

按照通常的方法，设双层旋转扁壳的各层泊松比相等：

$$\nu_1=\nu_2=\nu \tag{1}$$

于是，双层旋转扁壳在均匀温度作用下的轴对称大挠度弯曲方程为[12-14]

$$
\left.\begin{aligned}
&Dr - \frac{\mathrm{d}}{\mathrm{d}r}\frac{1}{r}\frac{\mathrm{d}}{\mathrm{d}r}(r\vartheta) - rN_r(\theta+\vartheta) = 0\\
&\frac{1}{C(1-\nu^2)}r\frac{\mathrm{d}}{\mathrm{d}r}\frac{1}{r}\frac{\mathrm{d}}{\mathrm{d}r}(r^2N_r) + \vartheta\left(\theta+\frac{\vartheta}{2}\right) = 0
\end{aligned}\right\}\tag{2}
$$

其中

$$
\left.\begin{aligned}
&D=\frac{E_1[h_0^3-(h_0-h_1)^3]+E_2[(h-h_0)^3+(h_0-h_1)^3]}{3(1-\nu^2)}\\
&C=\frac{E_1h_1+E_2h_2}{1-\nu^2}\\
&h_0=\frac{E_1h_1^2+2E_2h_1h_2+E_2h_2^2}{2(E_1h_1+E_2h_2)},\quad \vartheta=-\frac{\mathrm{d}w}{\mathrm{d}r}
\end{aligned}\right\}\tag{3}
$$

另外，关于 N_r 和 M_r，还有公式

$$
\left.\begin{aligned}
&N_r = C\left[\frac{\mathrm{d}u}{\mathrm{d}r}+\nu\frac{u}{r}+\vartheta\left(\theta+\frac{\vartheta}{2}\right)\right]-\alpha t\\
&M_r = D\left(\frac{\mathrm{d}\vartheta}{\mathrm{d}r}+\nu\frac{\vartheta}{r}\right)+\mu t
\end{aligned}\right\}\tag{4a,b}
$$

这里，t 是壳体相对于没有热应力时的初始温度的温度改变量，

$$
\alpha=\frac{\alpha_1E_1h_1+\alpha_2E_2h_2}{1-\nu},\quad \mu=\frac{E_1h_1(2h_0-h_1)}{2(1-\nu)}(\alpha_1-\alpha_2)\tag{5}
$$

本文研究如图 1 所示的双层金属截头扁锥壳的热稳定性问题。为了将上述双层旋转扁壳的方程转化为双层扁锥壳的方程，只需将确定壳体形状的子午线方向弧的倾斜角 θ 用扁锥壳的相应角代替。在扁锥壳中，θ 是一常数，为

$$
\theta=\tan\theta=\frac{f}{a}\tag{6}
$$

将式 (6) 代入方程 (2)，便得到双层金属扁锥壳的大挠度弯曲方程。

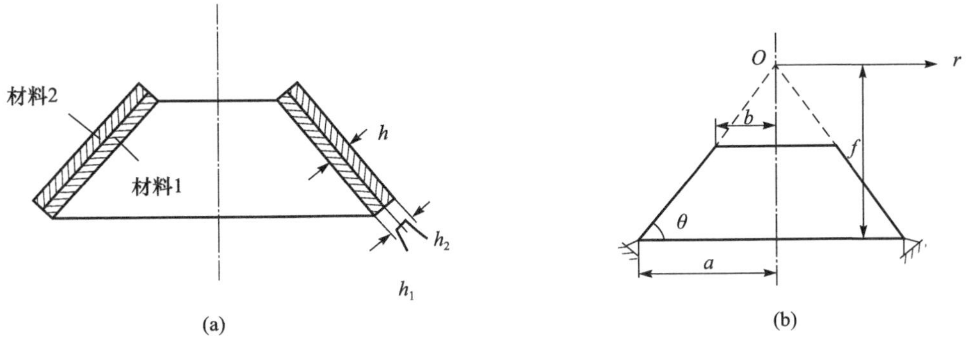

图 1　双层截头扁锥壳

在热敏元件的实际情况下，截头扁锥壳的大、小端边界条件是简单自由支承，即

$$
\left.\begin{aligned}
&\text{当}\quad r=a\text{ 时},\quad w=0,\quad M_r=0,\quad N_r=0\\
&\text{当}\quad r=b\text{ 时},\quad M_r=0,\quad N_r=0
\end{aligned}\right\}\tag{7}
$$

于是，基本方程 (2) 和边界条件 (7) 就组成在均匀温度作用下的双层截头扁锥壳的大挠度弯曲问题的非线性边值问题。

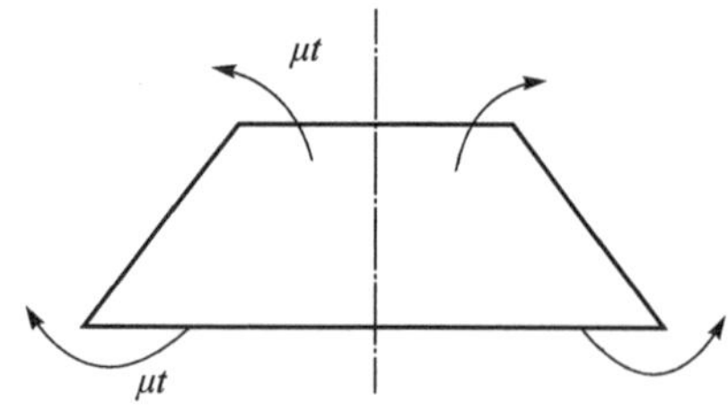

图 2　在均布弯矩作用下的截头扁锥壳

按照式（4b），边界条件 $M_r=0$，可以写为

$$D\left(\frac{\mathrm{d}\vartheta}{\mathrm{d}r}+\nu\frac{\vartheta}{r}\right)=-\mu t \tag{8}$$

由此可见，双层截头扁锥壳的温度改变所造成的效果与截头扁锥壳大小端承受相当于（$-\mu t$）的均布弯矩作用的效果是完全相同的（图 2）。

为了简化计算，我们引进下列无量纲量：

$$\left.\begin{aligned}&\rho=\frac{r}{a},\quad \rho_1=\frac{b}{a},\quad y=-\sqrt{\frac{C(1-\nu^2)}{2D}}w,\quad \varphi=-\left(\frac{\mathrm{d}y}{\mathrm{d}\rho}+k\right)\\&S=\frac{a^2\rho}{D}N_r,\quad k=\sqrt{\frac{C(1-\nu^2)}{2D}}f,\quad T=\frac{\mu a^2}{D}\sqrt{\frac{C(1-\nu^2)}{2D}}t\end{aligned}\right\} \tag{9}$$

将式（9）代入式（2）和（7），并应用式（6）和（8），便得双层金属截头扁锥壳的无量纲非线性边值问题：

$$\begin{aligned}&H(\rho\varphi)=S\varphi+\frac{k}{\rho}\\&H(\rho S)=k^2-\varphi^2\end{aligned} \tag{10a,b}$$

当 $\rho=1$ 时，
$$y=0,\quad \frac{\mathrm{d}\varphi}{\mathrm{d}\rho}+\frac{\nu}{\rho}(\varphi+k)=T,\quad S=0 \tag{11a,b,c}$$

当 $\rho=\rho_1$ 时，
$$\frac{\mathrm{d}\varphi}{\mathrm{d}\rho}+\frac{\nu}{\rho}(\varphi+k)=T,\quad S=0 \tag{12a,b}$$

其中

$$H(\cdots)=\rho\frac{\mathrm{d}}{\mathrm{d}\rho}\frac{1}{\rho}\frac{\mathrm{d}}{\mathrm{d}\rho}(\cdots) \tag{13}$$

三、非线性边值问题的解

采用修正迭代法，并取截头扁锥壳小端挠度作为迭代参数，我们便将非线性边值问题（10）—（12）化为一系列易于求解的线性边值问题。在一次近似中，略去方程（10a）右端的非线性项 $S\varphi$，并以一次近似解 φ_1 代替方程（10b）右端的 φ，再应用边界条件（11）和（12），便得下列线性边值问题：

$$H(\rho\varphi_1)=\frac{k}{\rho},\quad H(\rho S_1)=k^2-\varphi_1^2 \tag{14a,b}$$

当 $\rho=1$ 时，
$$y_1=0,\quad \frac{\mathrm{d}\varphi_1}{\mathrm{d}\rho}+\frac{\nu}{\rho}(\varphi_1+k)=T,\quad S_1=0 \tag{15a,b,c}$$

当 $\rho=\rho_1$ 时，
$$\frac{\mathrm{d}\varphi_1}{\mathrm{d}\rho}+\frac{\nu}{\rho}(\varphi_1+k)=T,\quad S_1=0 \tag{16a,b}$$

首先，将方程（14a）积分，并应用边界条件（15b）和（16a），便得双层环形圆板的小挠度理论解：

$$\varphi_1=\frac{T}{1+\nu}\rho-k \tag{17}$$

我们以双层截头扁锥壳小端的无量纲挠度 y_0 为迭代参数

$$y_0=-\sqrt{\frac{C(1-\nu^2)}{2D}}\omega\Big|_{r=b} \tag{18}$$

则应用式（9）和边界条件（11a），有

$$y_0=\int_{\rho_1}^{1}(\varphi+k)\mathrm{d}\rho \tag{19}$$

在一次近似中，将式（17）代入上式，得

$$T=\beta_0 y_0 \tag{20}$$

其中

$$\beta_0=\frac{2\lambda_1}{1-\rho_1^2},\quad \lambda_1=1+\nu \tag{21}$$

将式（20）代入式（17），得

$$\varphi_1=\frac{1}{\lambda_1}\beta_0 y_0\rho-k \tag{22}$$

利用式（22）再将方程（14b）积分，并应用边界条件（15c）和（16b），得 S 的一次近似解：

$$\begin{aligned}S_1=&\frac{\rho_1^2}{\lambda_1}\left[\frac{2k}{3(1+\rho_1)}\beta_0 y_0-\frac{1}{8\lambda_1}\beta_0^2 y_0^2\right]\frac{1}{\rho}-\frac{1}{\lambda_1}\left[\frac{2k(1+\rho_1+\rho_1^2)}{3(1+\rho_1)}\beta_0 y_0\right.\\&\left.-\frac{1+\rho_1^2}{8\lambda_1}\beta_0^2 y_0^2\right]\rho+\frac{2k}{3\lambda_1}\beta_0 y_0\rho^2-\frac{1}{8\lambda_1^2}\beta_0^2 y_0^2\rho^3\end{aligned} \tag{23}$$

在二次近似中，将一次近似解（22）和（23）代入方程（10a）的右端，并应用相应的边界条件（11a,b）和（12a），便得关于 φ_2 的线性边值问题：

$$H(\rho\varphi_2)=S_1\varphi_1+\frac{k}{\rho} \tag{24}$$

当 $\rho=1$ 时，　$y_2=0$，　$\dfrac{\mathrm{d}\varphi_2}{\mathrm{d}\rho}+\dfrac{\nu}{\rho}(\varphi_2+k)=T$　(25a,b)

当 $\rho=\rho_1$ 时，　$\dfrac{\mathrm{d}\varphi_2}{\mathrm{d}\rho}+\dfrac{\nu}{\rho}(\varphi_2+k)=T$　(26)

应用边界条件（25b）和（26），得方程（24）的解：

$$\begin{aligned}\varphi_2=&\frac{k\rho_1^2}{\lambda_1}\left[\frac{2k}{3(1+\rho_1)}\beta_0 y_0-\frac{1}{8\lambda_1}\beta_0^2 y_0^2\right]-k-\frac{\rho_1^2}{\lambda_1\lambda_2(1-\rho_1^2)}\left\{\frac{k^2}{3}\left[\frac{7+5\nu}{12}\right.\right.\\&\left.-\frac{4+2\nu}{3}\rho_1+\frac{3+\nu}{4}\rho_1^2-\frac{2\nu}{1+\rho_1}\rho_1+\frac{4+8\nu}{3(1+\rho_1)}\rho_1^2-\frac{4+2\nu}{3(1+\rho_1)}\rho_1^3\right]\beta_0 y_0\\&-\frac{k}{3\lambda_1}\left(\frac{22+13\nu}{60}-\frac{1+2\nu}{4}\rho_1+\frac{1+2\nu}{4}\rho_1^2-\frac{22+13\nu}{60}\rho_1^3\right.\\&\left.\left.+\frac{1}{1+\rho_1}\lambda_1\rho_1^2\ln\rho_1\right)\beta_0^2 y_0^2+\frac{1}{16\lambda_1^2}\left(\frac{2+\nu}{6}+\lambda_1\rho_1^2\ln\rho_1-\frac{2+\nu}{6}\rho_1^4\right)\beta_0^3 y_0^3\right\}\frac{1}{\rho}\\&+\frac{1}{\lambda_1}\left\{T-\frac{1}{\lambda_1(1-\rho_1^2)}\left[\frac{k^2}{3}\left(\frac{7+5\nu}{12}+\frac{4+8\nu}{3}\rho_1^2-\frac{8+10\nu}{3}\rho_1^3+\frac{3+\nu}{4}\rho_1^4\right.\right.\right.\end{aligned}$$

$$- \frac{2\nu}{1+\rho_1}\rho_1^3 + \frac{4+8\nu}{3(1+\rho_1)}\rho_1^4 - \frac{4+2\nu}{3(1+\rho_1)}\rho_1^5\Big)\beta_0 y_0 - \frac{k}{3\lambda_1}\Big(\frac{22+13\nu}{60}$$

$$+ \frac{3\nu}{4}\rho_1^2 - \frac{3\nu}{4}\rho_1^3 - \frac{22+13\nu}{60}\rho_1^5 + \frac{1}{1+\rho_1}\lambda_1\rho_1^4\ln\rho_1\Big)\beta_0^2 y_0^2 + \frac{1}{16\lambda_1^2}\Big(\frac{2+\nu}{6}$$

$$- \frac{1}{4}\lambda_2\rho_1^2 + \frac{1}{4}\lambda_2\rho_1^4 + \lambda_1\rho_1^4\ln p_1 - \frac{2+\nu}{6}\rho_1^6\Big)\beta_0^3 y_0^3\Big]\Big\}\rho + \frac{\rho_1^2}{\lambda_1^2}\Big[\frac{k}{3(1+\rho_1)}\beta_0^2 y_0^2$$

$$- \frac{1}{16\lambda_1}\beta_0^3 y_0^3\Big]\rho\ln\rho + \frac{k}{3\lambda_1}\Big[\frac{2k(1+\rho_1+\rho_1^2)}{3(1+\rho_1)}\beta_0 y_0 - \frac{1+\rho_1^2}{8\lambda_1}\beta_0^2 y_0^2\Big]\rho^2$$

$$- \frac{1}{8\lambda_1}\Big[\frac{2k^2}{3}\beta_0 y_0 + \frac{2k(1+\rho_1+\rho_1^2)}{3\lambda_1(1+\rho_1)}\beta_0^2 y_0^2 - \frac{1+\rho_1^2}{8\lambda_1^2}\beta_0^3 y_0^3\Big]\rho^3$$

$$+ \frac{19k}{360\lambda_1^2}\beta_0^2 y_0^2 \rho^4 - \frac{1}{192\lambda_1^3}\beta_0^3 y_0^3 \rho^5 \tag{27}$$

其中

$$\lambda_2 = 1-\nu \tag{28}$$

将式（27）代入式（19），得双层金属截头扁锥壳的非线性特征关系式：

$$T = (\beta_0 + k^2\beta_1)y_0 - k\beta_2 y_0^2 + \beta_3 y_0^3 \tag{29}$$

其中

$$\left.\begin{aligned}
\beta_1 = {} & \frac{\beta_0^3}{8\lambda_1^3\lambda_2(1+\rho_1)}\Big[\frac{19+7\nu}{108}\lambda_2 + \frac{19+7\nu}{108}\lambda_2\rho_1 - \frac{9+5\nu}{4}\lambda_2\rho_1^2 \\
& - \frac{7+5\nu}{9}\lambda_1\rho_1^2\ln\rho_1 + \frac{205+121\nu}{108}\lambda_2\rho_1^3 + (1+3\nu)\lambda_1\rho_1^3\ln\rho_1 \\
& + \frac{205+121\nu}{108}\lambda_2\rho_1^4 - (1+3\nu)\lambda_1\rho_1^4\ln\rho_1 - \frac{9+5\nu}{4}\lambda_2\rho_1^5 \\
& + \frac{7+5\nu}{9}\lambda_1\rho_1^5\ln\rho_1 + \frac{19+7\nu}{108}\lambda_2\rho_1^6 + \frac{19+7\nu}{108}\lambda_2\rho_1^7\Big] \\
\beta_2 = {} & \frac{\beta_0^4}{4\lambda_1^4\lambda_2(1+\rho_1)}\Big(\frac{133+43\nu}{1800}\lambda_2 + \frac{133+43\nu}{1800}\lambda_2\rho_1 \\
& - \frac{28+13\nu}{50}\lambda_2\rho_1^2 - \frac{22+13\nu}{90}\lambda_1\rho_1^2\ln\rho_1 - \frac{133+43\nu}{1800}\lambda_2\rho_1^3 \\
& - \frac{7-17\nu}{90}\lambda_1\rho_1^3\ln\rho_1 + \frac{35+17\nu}{36}\lambda_2\rho_1^4 - \frac{2}{3}\lambda_1^2\rho_1^4\ln^2\rho_1 \\
& - \frac{133+43\nu}{1800}\lambda_2\rho_1^5 + \frac{7-17\nu}{90}\lambda_1\rho_1^5\ln\rho_1 - \frac{28+13\nu}{50}\lambda_2\rho_1^6 \\
& + \frac{22+13\nu}{90}\lambda_1\rho_1^6\ln\rho_1 + \frac{133+43\nu}{1800}\lambda_2\rho_1^7 + \frac{133+43\nu}{1800}\lambda_2\rho_1^8\Big) \\
\beta_3 = {} & \frac{\beta_0^5}{32\lambda_1^5\lambda_2}\Big(\frac{17+5\nu}{144}\lambda_2 - \frac{5+2\nu}{9}\lambda_2\rho_1^2 - \frac{2+\nu}{6}\lambda_1\rho_1^2\ln\rho_1 \\
& + \frac{7+3\nu}{8}\lambda_2\rho_1^4 - \lambda_1^2\rho_1^4\ln^2\rho_1 - \frac{5+2\nu}{9}\lambda_2\rho_1^6 \\
& + \frac{2+\nu}{6}\lambda_1\rho_1^6\ln\rho_1 + \frac{17+5\nu}{144}\lambda_2\rho_1^8\Big)
\end{aligned}\right\} \tag{30}$$

四、主要结果

以 $\rho_1=0.4$，$\nu=0.3$ 的情况为例，按式（29），在图 3 上绘出了不同 k 值下的 T-y_0 曲线。工程上感兴趣的是 T-y_0 特征曲线出现极值情况，此时壳体产生了丧失稳定问题，出现了跳跃现象。精密仪器正是利用这种跳跃现象来作为控制信号。为确定 T-y_0 曲线的极值点，对式（29）应用极值条件

$$\frac{\mathrm{d}T}{\mathrm{d}y_0}=3\beta_3y_0^2-2k\beta_2y_0+(\beta_0+k^2\beta_1)=0 \tag{31}$$

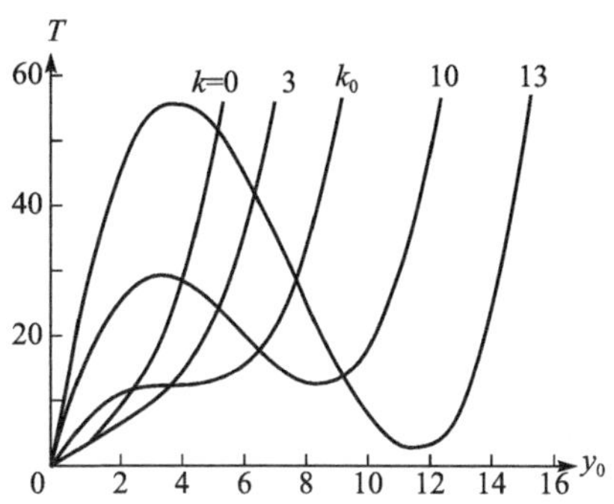

图 3　不同 k 值下的特征曲线

（$\rho_1=0.4$，$\nu=0.3$）

解此二次代数方程，得产生跳跃现象时的截头扁锥壳小端的无量纲临界挠度

$$y_{s0}^*=\frac{k\beta_2\pm\sqrt{k^2(\beta_2^2-3\beta_1\beta_3)-3\beta_0\beta_3}}{3\beta_3} \tag{32}$$

这里，对应于负号的 y_0^* 是上临界挠度，对应于正号的 y_0^* 是下临界挠度，分别记为 $y_{0,1}^*$，$y_{0,2}^*$。

将式（32）代入式（29），得双层金属截头扁锥壳的无量纲临界温度

$$T^*=(\beta_0+k^2\beta_1)y_0^*-k\beta_2y_0^{*2}+\beta_3y_0^{*3} \tag{33}$$

这里，对应于 $y_{0,1}^*$的 T^* 是上临界温度，对应于 $y_{0,2}^*$的 T^* 是下临界温度，分别记为 T_1^*，T_2^*。

求得 T^* 后，可按下式计算壳体的临界温度

$$t^*=\frac{D}{\mu a^2}\sqrt{\frac{2D}{C(1-\nu^2)}}T^* \tag{34}$$

按式（33），在图 4 上绘出了不同 ρ_1 值下的无量纲临界温度曲线。每一条曲线的上、下支分别相应于上、下临界温度。我们称上、下临界温度的相重点为临界点，并记此点的 k 值为 k_0。由图看出，当 $k<k_0$ 时，壳体未具有跳跃的本能，而当 $k\geqslant k_0$ 时，壳体便具有跳跃的本能。因此，k_0 值是判别壳体能否出现跳跃现象的分界点。令二次方程（31）的判别式等于零：

$$4k^2\beta_2^2-12\beta_3(\beta_0+k^2\beta_1)=0 \tag{35}$$

则得计算无量纲临界几何参数 k_0 的公式

$$k_0=\sqrt{\frac{3\beta_0\beta_3}{\beta_2^2-3\beta_1\beta_3}} \tag{36}$$

求得 k_0 后，可按下式计算壳体的临界拱高

$$f_0=\sqrt{\frac{2D}{C(1-\nu^2)}}k_0 \tag{37}$$

按式（36），在图 5 上绘出了壳体的无量纲临界几何参数 k_0 与双层截头扁锥壳小端的无量纲半径 ρ_1 的关系曲线。由图看出，曲线呈凹形。当 ρ_1 较小时，k_0 随着 ρ_1 的增

大而减小；而当 ρ_1 较大时，这一关系恰恰相反，k_0 随着 ρ_1 的增大而增加。亦即，k_0—ρ_1 曲线有一极小值存在，相应于这一极小值的壳体其拱高为最小，它是对失稳反应最灵敏的壳体。对式（36）应用极值条件

$$\frac{\mathrm{d}k_0}{\mathrm{d}\rho_1}=0 \tag{38}$$

并用迭代法解此方程，得 k_0 取最小值的壳体为

$$\rho_1=0.18794$$

再由式（36）和（33），得相应的 $(k_0)_{\min}$和 T^* 值：

$$(k_0)_{\min}=5.4718,\quad T^*=12.807$$

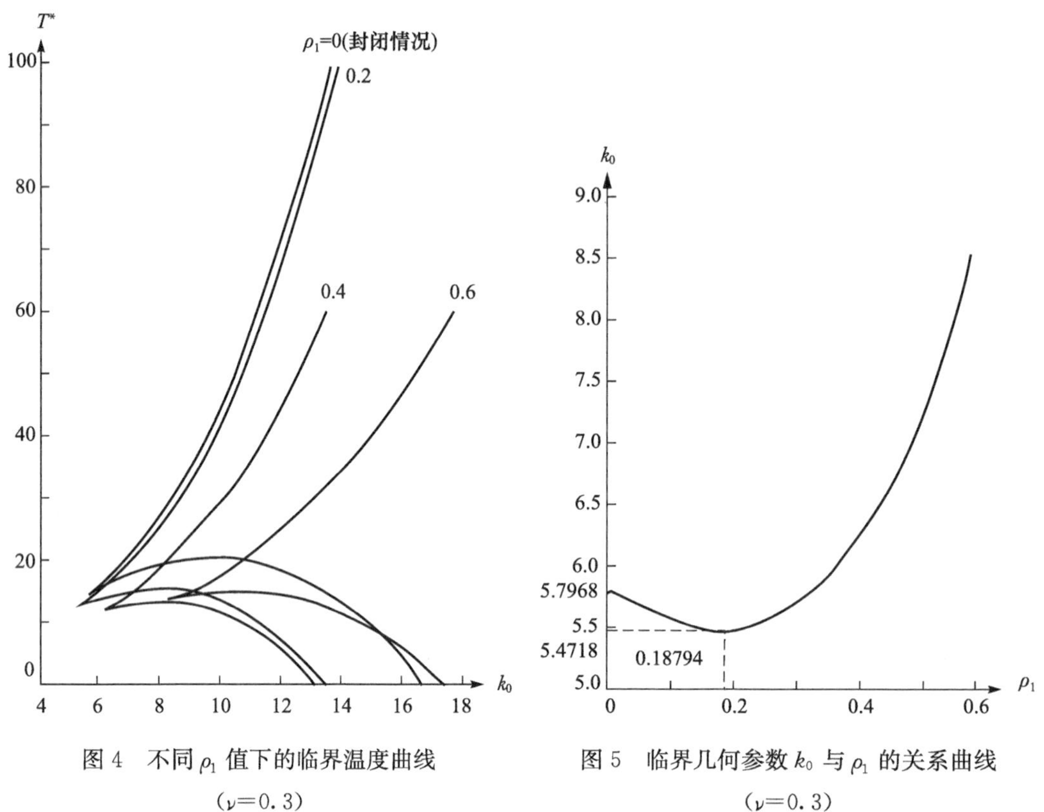

图 4　不同 ρ_1 值下的临界温度曲线（ν=0.3）

图 5　临界几何参数 k_0 与 ρ_1 的关系曲线（ν=0.3）

此外，在热敏元件设计时，还需要应用平均温度这个参数。平均温度就是上、下临界温度的平均值。在无量纲情况下，其公式为

$$T_c^*=\frac{T_1^*+T_2^*}{2}=\frac{k\beta_2}{3\beta_3}\left[\beta_0+k^2\left(\beta_1-\frac{2\beta_2^2}{9\beta_3}\right)\right] \tag{39}$$

求得 T_c^* 后，可按下式计算壳体的平均温度

$$t_c^*=\frac{D}{\mu a^2}\sqrt{\frac{2D}{C(1-\nu^2)}}T_c^* \tag{40}$$

按式（39），在图 6 上绘出了不同 ρ_1 值下的无量纲平均温度 T_c^* 的曲线。显然，T_c^*-k 曲线单调上升，而且，对于具有相同 k 值的壳体，平均温度随着截头扁锥壳小端

的无量纲半径 ρ_1 的增大而减小。

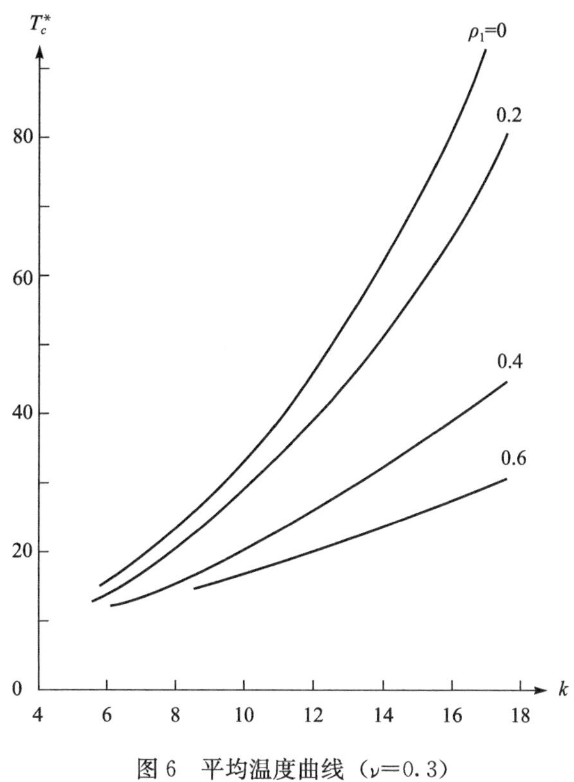

图 6　平均温度曲线（$\nu=0.3$）

最后，我们给出热敏元件设计时所需要的跳跃幅度的公式。跳跃幅度就是壳体丧失稳定时跳跃的距离，其意义可参见图 7。由图 7 看到，无量纲跳跃幅度 L 的公式为

$$L = y_0' - y_{0,1}^* \tag{41}$$

显然，此式中的 y_0' 应是下面三次代数方程

$$\beta_3 y_0^3 - k\beta_2 y_0^2 + (\beta_0 + k^2\beta_1) y_0 - T_1^* = 0 \tag{42}$$

的一个根。解此方程，得此根为

$$y_0' = \frac{k\beta_2}{3\beta_3} + 2\sqrt[3]{\frac{T_1^* - T_2^*}{4\beta_3}} \tag{43}$$

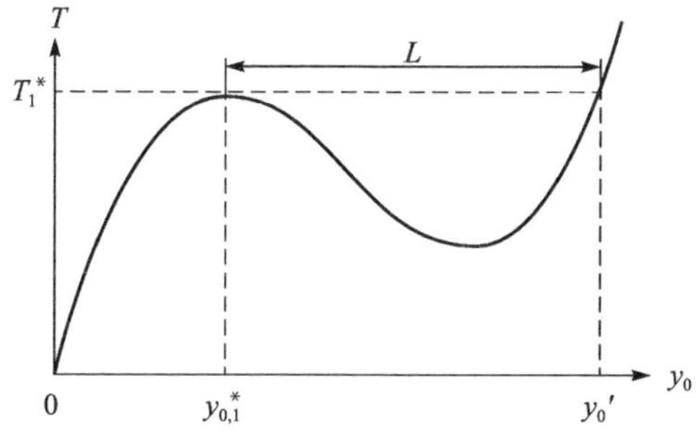

图 7　扁锥壳的跳跃幅度

求得 L 后，可按下式计算壳体的跳跃幅度

$$l=\sqrt{\frac{2D}{C(1-\nu^2)}}L \tag{44}$$

按式（41），在图 8 上绘出了不同 ρ_1 值下的无量纲跳跃幅度 L 的曲线。由图看出，L-k 曲线也是单调上升的，而且，在较大的范围内，对于具有相同 k 值的壳体，跳跃幅度随着截头扁锥壳小端的无量纲半径 ρ_1 的增大而减小。

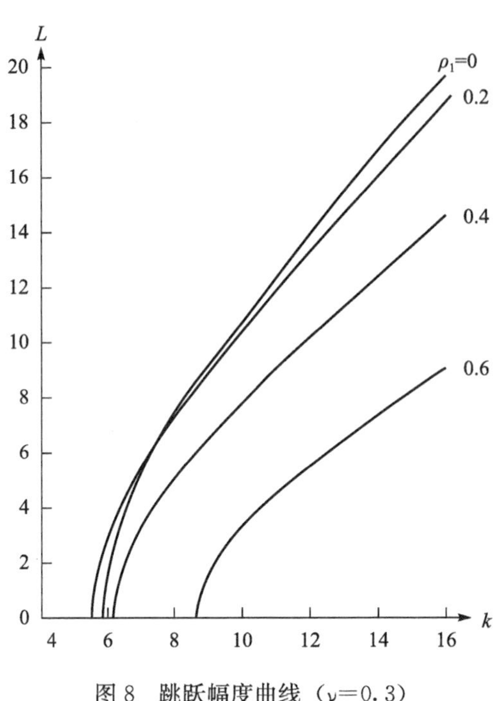

图 8　跳跃幅度曲线（ν=0.3）

参 考 文 献

［1］　叶开沅，刘人怀，平庆元，等．在对称线布载荷作用下的圆底扁薄球壳的非线性稳定问题．科学通报，1965，（2）：142-145.

［2］　叶开沅，刘人怀，张传智，等．圆底扁薄球壳在边缘力矩作用下的非线性稳定问题．科学通报，1965，（2）：145-147.

［3］　刘人怀．在内边缘均布力矩作用下中心开孔圆底扁球壳的非线性稳定问题．科学通报，1965，（3）：253-255.

［4］　刘人怀．在边缘载荷作用下中心开孔圆底扁薄球壳的轴对称稳定性．力学学报，1977，（3）：206-212.

［5］　刘人怀．波纹圆板的特征关系式．力学学报，1978，（1）：47-52.

［6］　刘人怀．具有光滑中心的波纹圆板的特征关系式．中国科学技术大学学报（纪念校庆 20 周年文集），1979，9（2）：75-86.

［7］　叶开沅，刘人怀，张传智，等．弹性圆底扁球壳在边缘均布力矩作用下的非线性稳定问题．应用数学和力学，1980，1（1）：71-87.

[8] 刘人怀. 在边缘力矩作用下夹层圆板的非线性轴对称弯曲问题. 中国科学技术大学学报，1980，10 (2)：56-67.

[9] Панов，Д Ю. Об устойчивости биметаллической оболочки при нагреве. ПММ，1947，11 (6)：603.

[10] Wittrick，W. H.，Myers D. M.，Blunden W. R. Stability of a bimetallic disk. Quarterly Journal of Mech. and Appl. Math.，1953，6 (1)：15-31.

[11] Григолюк，Э. И. Тонкие ъимталлические оболочки и пластииы. Инж. сборим，1953：18.

[12] 罗祖道，聂德耀，刘汉东，等. 双层金属球面扁壳的热稳定性. 力学学报，1966，9 (1)：1-13.

[13] KopoleB，B. И. Thin two-layer plates and shells. Engrs'Colln，1955，22：98 (in Russian).

[14] Radkowsky，P. P. Buckling of thin single-and multi-layer conical and cylindrical shells with rotationally symmetric stresses. Proc，3rd U. S. natn. Congr. appl. Mech.，1958，443.

双层金属中心开孔扁球壳的非线性热稳定问题①

一、前言

双层金属旋转扁壳的非线性热稳定理论是近年来精密仪器热敏弹性元件设计和壳体稳定理论中受到关注的问题之一。在这一问题中，有两种壳体，即双层金属扁球壳和双层金属扁锥壳受到重视。关于双层金属截头扁锥壳的情况，本文作者[1]首先进行了理论探讨，得到了令人满意的结果。至于较简单的双层金属扁球壳情况，已有多人作了理论工作，如Панов[2]，Wittrick[3]，Григолюк[4]，罗祖道[5]等．其中文献［2］—［4］仅对最简单的双层金属封顶扁球壳作了计算，而文献［5］对具有一般意义的双层金属中心开孔扁球壳进行了分析研究。但是，此问题的解决还不能完全令人满意。本文试图就此作出努力，进行探讨。最后，给出了双层金属中心开孔扁球壳的非线性热稳定问题的相当精确的解析解，可供精密仪器设计部门设计时使用。

二、双层金属扁球壳的大挠度方程和边界条件

考虑如图1所示的内、外层厚度分别为 h_1 和 h_2 的双层金属中心开孔扁球壳在均匀温度作用下的非线性热稳定问题。

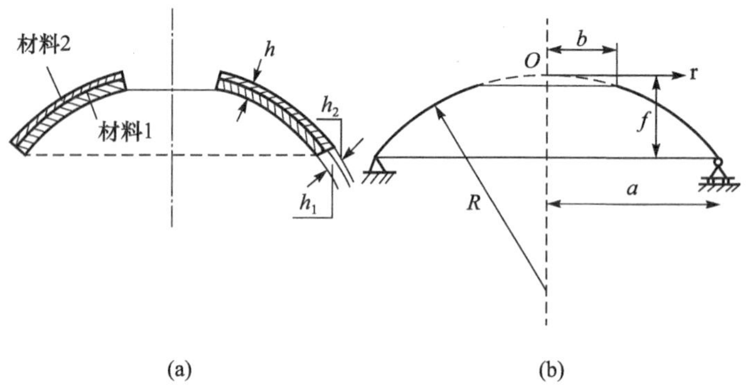

图1　双层金属中心开孔扁球壳

若选取离壳体内表面距离为

$$h_0 = \frac{E_1 h_1^2 + 2E_2 h_1 h_2 + E_2 h_2^2}{2(E_1 h_1 + E_2 h_2)} \tag{1}$$

的平行面作为坐标参考面，并设壳内外层的泊松比相等，即

$$\nu_1 = \nu_2 = \nu, \tag{2}$$

则双属金属扁球壳在均匀温度作用下的轴对称大挠度方程为[1,5]

① 本文原载《中国科学技术大学学报》，1981，11（1）：84-99.

$$
\left.\begin{aligned}
&Dr\frac{\mathrm{d}}{\mathrm{d}r}\frac{1}{r}\frac{\mathrm{d}}{\mathrm{d}r}r\frac{\mathrm{d}w}{\mathrm{d}r}-rN_r\left(\frac{\mathrm{d}w}{\mathrm{d}r}-\frac{2f}{a^2}r\right)=0,\\
&\frac{1}{C(1-\nu^2)}r\frac{\mathrm{d}}{\mathrm{d}r}\frac{1}{r}\frac{\mathrm{d}}{\mathrm{d}r}(r^2N_r)+\frac{\mathrm{d}w}{\mathrm{d}r}\left(\frac{1}{2}\frac{\mathrm{d}w}{\mathrm{d}r}-\frac{2f}{a^2}r\right)=0,
\end{aligned}\right\}\tag{3}
$$

其中 E_1 和 E_2 分别为壳内、外层材料的弹性模量。在一般温度变化范围内，可将 E_1、E_2 和 ν 视为常数，R 为壳半径，f 为拱高，h 为壳总厚度，a、b 分别为壳外、内缘半径，r 为壳坐标参考面上的点至对称轴的距离，w 为挠度，N_r 为径向薄膜内力，C 为抗拉刚度，D 为抗弯刚度，

$$
\left.\begin{aligned}
&C=\frac{E_1h_1+E_2h_2}{1-\nu^2},\\
&D=\frac{E_1[h_0^3-(h_0-h_1)^3]+E_2[(h-h_0)^3+(h_0-h_1)^3]}{3(1-\nu^2)}
\end{aligned}\right\}\tag{4}
$$

在热敏弹性元件的实际情况下，双层金属中心开孔扁球壳的内、外缘是简单自由支承，其边界条件为

当 $r=a$ 时，　　$w=0,\quad M_r=0,\quad N_r=0.$　　(5)

当 $r=b$ 时，　　$M_r=0,\quad N_r=0.$　　(6)

这里，径向弯矩 M_r 与挠度 w 以及壳体的相对于没有热应力时的初始温度的温度改变量 t 存在下面关系

$$
M_r=-D\left(\frac{\mathrm{d}^2w}{\mathrm{d}r^2}+\frac{\nu}{r}\frac{\mathrm{d}w}{\mathrm{d}r}\right)+\mu t\tag{7}
$$

而相当二次膨胀系数 μ 与壳体内、外层的热膨胀系数 α_1、α_2 的关系为

$$
\mu=\frac{E_1h_1(2h_0-h_1)}{2(1-\nu)}-(\alpha_1-\alpha_2)\tag{8}
$$

于是，基本方程（3）和边界条件（5）、（6）就组成在均匀温度作用下双层金属中心开孔扁球壳的热稳定问题的非线性边值问题。

显然，边界条件 $M_r=0$ 可以由式（7）改写为下面形式

$$
-D\left(\frac{\mathrm{d}^2w}{\mathrm{d}r^2}+\frac{\nu}{r}\frac{\mathrm{d}w}{\mathrm{d}r}\right)=-\mu t\tag{9}
$$

这说明，由于壳体两层金属的热膨胀系数不同，亦即 μ 不等于零，双层金属中心开孔扁球壳的温度改变所造成的效果相当于壳内外边界承受均布弯矩作用的效果（图 2）。

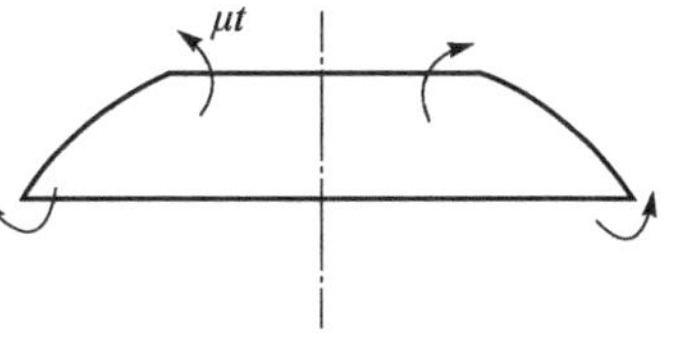

图 2　承受均布弯矩作用的双层金属中心开孔扁球壳

引入下列符号：

$$
\left.\begin{aligned}
&\rho=\frac{r}{a},\quad \rho_1=\frac{b}{a},\quad y=-\sqrt{\frac{C(1-\nu^2)}{2D}}w,\\
&\varphi=-\left(\frac{\mathrm{d}y}{\mathrm{d}\rho}+k\rho\right),\quad S=\frac{a^2\rho}{D}N_r,\quad k=\sqrt{\frac{2C(1-\nu^2)}{D}}f,\\
&T=\frac{\mu a^2}{\lambda_1D}\sqrt{\frac{C(1-\nu^2)}{2D}}t,\quad \lambda_1=1+\nu,
\end{aligned}\right\}\tag{10}
$$

则非线性边值问题（3）、（5）和、（6）简化成如下的无量纲形式

$$\left.\begin{aligned}H(\rho\varphi)&=S\varphi,\\H(\rho S)&=-(\varphi^2-k^2\rho^2),\end{aligned}\right\}\tag{11a,b}$$

当 $\rho=1$ 时，　$y=0$，　$\dfrac{\mathrm{d}\varphi}{\mathrm{d}\rho}+\dfrac{\nu}{\rho}\varphi=\lambda_1(T-k)$，　$S=0$　　(12a,b,c)

当 $\rho=\rho_1$ 时，　$\dfrac{\mathrm{d}\varphi}{\mathrm{d}\rho}+\dfrac{\nu}{\rho}\varphi=\lambda_1(T-k)$，　$S=0$　　(13a,b)

其中

$$H(\cdots)=\rho\frac{\mathrm{d}}{\mathrm{d}\rho}\frac{1}{\rho}\frac{\mathrm{d}}{\mathrm{d}\rho}(\cdots)\tag{14}$$

在壳体中心孔的半径 b 趋于零的极限情况下，上述边值问题将化为较简单的在均匀温度作用下双层金属封顶扁球壳的热稳定问题的非线性边值问题。此时，仅边界条件（6）化为如下的中心条件：

当 $r=0$ 时，　$\dfrac{\mathrm{d}w}{\mathrm{d}r}=0$，　N_r 有限　　(15)

相应的无量纲形式为

当 $\rho=0$ 时，　$\varphi=0$，　$S=0$　　(16)

三、双层金属中心开孔扁球壳的热稳定问题的解

因为边值问题式（11）～式（13）含有一些非线性项，所以极难求解。与前文［1］一样，我们使用修正迭代法克服了求解的困难。

在一次近似中，忽略方程（11a）右端的非线性项 $S\varphi$，则有如下的线性边值问题：

$$\left.\begin{aligned}H(\rho\varphi_1)&=0,\\H(\rho S_1)&=-(\varphi_1^2-k^2\rho^2),\end{aligned}\right\}\tag{17a,b}$$

当 $\rho=1$ 时，　$y_1=0$，　$\dfrac{\mathrm{d}\varphi_1}{\mathrm{d}\rho}+\dfrac{\nu}{\rho}\varphi_1=\lambda_1(T-k)$，　$S_1=0$　　(18a,b,c)

当 $\rho=\rho_1$ 时，　$\dfrac{\mathrm{d}\varphi_1}{\mathrm{d}\rho}+\dfrac{\nu}{\rho}\varphi_1=\lambda_1(T-k)$，　$S_1=0$　　(19a,b)

首先，应用边界条件（18b）和（19a），积分微分方程（17a），得

$$\varphi_1=(T-k)\rho\tag{20}$$

我们采用壳内边缘的无量纲挠度 y_0 作为迭代参数，即

$$y_0=-\sqrt{\frac{C(1-\nu^2)}{2D}}w\bigg|_{r=b},\tag{21}$$

那么，在应用式（10）和边界条件（12a）后，有

$$y_0=\int_{\rho_1}^{1}(\varphi+k\rho)\mathrm{d}\rho\tag{22}$$

将式（20）代入此式，经过整理，得双层金属中心开孔扁球壳的小挠度理论的特征关系式

$$T = \beta_0 y_0, \tag{23}$$

其中

$$\beta_0 = \frac{2}{1-\rho_1^2}, \tag{24}$$

再将式（23）代入式（20），有

$$\varphi_1 = (\beta_0 y_0 - k)\rho, \tag{25}$$

然后，应用式（25）和边界条件（18c）、（19b），积分微分方程（17b），得

$$S_1 = \frac{1}{8}(2k\beta_0 y_0 - \beta_0^2 y_0^2)[\rho_1^2\rho^{-1} - (1+\rho_1^2)\rho + \rho^3] \tag{26}$$

在二次近似中，有下面关于 φ 的线性边值问题：

$$H(\rho\varphi_2) = S_1\varphi_1, \tag{27}$$

当 $\rho=1$ 时，　$y_2=0$，　$\frac{d\varphi_2}{d\rho}+\frac{\nu}{\rho}\varphi_2=\lambda_1(T-k)$　(28a,b)

当 $\rho=\rho_1$ 时，　$\frac{d\varphi_2}{d\rho}+\frac{\nu}{\rho}\varphi_2=\lambda_1(T-k)$　(29)

应用式（25）、（26）和边界条件（28b）、（29），积分微分方程（27），得

$$\begin{aligned}\varphi_2 =& (T-k)\rho - \frac{1}{16}(2k^2\beta_0 y_0 - 3k\beta_0^2 y_0^2 + \beta_0^3 y_0^3)\left\{\frac{\beta_0\rho_1^2}{2\lambda_2}\left(\frac{2+\nu}{6} + \lambda_1\rho_1^2\ln\rho_1\right.\right. \\ &\left. - \frac{2+\nu}{6}\rho_1^4\right)\rho^{-1} + \frac{1}{2}\left[\frac{\beta_0}{\lambda_1}\left(\frac{2+\nu}{6} + \frac{1+3\nu}{4}\rho_1^2 - \frac{1+3\nu}{4}\rho_1^4 + \lambda_1\rho_1^4\ln\rho_1\right.\right. \\ &\left.\left.\left. - \frac{2+\nu}{6}\rho_1^6\right) - \rho_1^2\right]\rho + \rho_1^2\rho\ln\rho - \frac{1}{4}(1+\rho_1^2)\rho^3 + \frac{1}{12}\rho^5\right\},\end{aligned} \tag{30}$$

其中

$$\lambda_2 = 1-\nu \tag{31}$$

最后，将式（30）代入式（22），便得双层金属中心开孔扁球壳的大挠度理论的特征关系式

$$T = \beta_1[(\beta_1^{-1} + 2k^2)\beta_0 y_0 - 3k\beta_0^2 y_0^2 + \beta_0^3 y_0^3], \tag{32}$$

其中

$$\begin{aligned}\beta_1 =& \frac{\beta_0^2}{32\lambda_1\lambda_2}\left(\frac{17+5\nu}{144}\lambda_2 - \frac{5+2\nu}{9}\lambda_2\rho_1^2 - \frac{2+\nu}{6}\lambda_1\rho_1^2\ln\rho_1 + \frac{7+3\nu}{8}\lambda_2\rho_1^4 - \lambda_1^2\rho_1^4\ln^2\rho_1\right. \\ &\left. - \frac{5+2\nu}{9}\lambda_2\rho_1^6 + \frac{2+\nu}{6}\lambda_1\rho_1^6\ln\rho_1 + \frac{17+5\nu}{144}\lambda_2\rho_1^8\right),\end{aligned} \tag{33}$$

当 $\nu=0.3$ 时，式（33）为

$$\begin{aligned}\beta_1 =& \frac{\beta_0^2}{29.12}(0.089\,930\,6 - 0.435\,556\rho_1^2 - 0.498\,333\rho_1^2\ln\rho_1 + 0.691\,25\rho_1^4 - 1.69\rho_1^4\ln^2\rho_1 \\ & - 0.435\,556\rho_1^6 + 0.498\,333\rho_1^6\ln\rho_1 + 0.089\,930\,6\rho_1^8)\end{aligned} \tag{34}$$

在壳体的无量纲内边缘半径 ρ_1 取不同值的情况下，β_0 和 β_1 的数值如表 1 所示。

表 1 $\nu=0.3$ 时的 β_0 和 β_1 的值

ρ_1	0	0.05	0.10	0.15	0.20
β_0	2	2.005 01	2.020 20	2.046 04	2.083 33
β_1	0.0123 531	0.012 767 5	0.013 485 5	0.014 182 9	0.014 698 0
ρ_1	0.25	0.30	0.40	0.50	0.60
β_0	2.133 33	2.197 80	2.380 95	2.666 67	3.125 00
β_1	0.014 945 0	0.014 885 1	0.013 836 9	0.011 721 6	0.008 893 13

四、双层金属封顶扁球壳的热稳定问题的解

对于上面的双层金属中心开孔扁球壳的情况，我们仅得到二次近似解，为了说明这一解的精确度，我们以 $\rho_1=0$，即壳体封顶的特殊情况为例，进行了更高次近似的计算。

在一次近似中，与前面类似，我们有如下的线性边值问题：

$$\left.\begin{aligned} &H(\rho\rho_1)=0,\\ &H(\rho S_1)=-(\varphi_1^2-k^2\rho^2),\end{aligned}\right\} \tag{35}$$

当 $\rho=1$ 时，$$y_1=0,\ \frac{d\varphi_1}{d\rho}+\frac{v}{\rho}\varphi_1=\lambda_1(T-k),\quad S_1=0 \tag{36}$$

当 $\rho=0$ 时，$$\varphi_1=0,\quad S_1=0 \tag{37}$$

现在，我们采用壳中心的无量纲挠度 y_0 作为迭代参数，则应用式（10）和边界条件（12a）后，有下面公式成立：

$$y_0=\int_0^1(\varphi+k\rho)d\rho \tag{38}$$

与前面求解相似，在应用式（38）后，可得边值问题（35）～(37）的解如下

$$\left.\begin{aligned} &\varphi_1=(2y_0-k)\rho,\\ &S_1=-\frac{1}{2}(ky_0-y_0^2)(\rho-\rho^3).\end{aligned}\right\} \tag{39}$$

在二次近似中，有下面的线性边值问题：

$$\left.\begin{aligned} &H(\rho\varphi_2)=S_1\varphi_1,\\ &H(\rho S_2)=-(\varphi_2^2-k^2\rho^2),\end{aligned}\right\} \tag{40a,b}$$

当 $\rho=1$ 时，$$y_2=0,\ \frac{d\varphi_2}{d\rho}+\frac{\nu}{\rho}\varphi_2=\lambda_1(T-k),\quad S_2=0 \tag{41a,b,c}$$

当 $\rho=0$ 时，$$\varphi_2=0,\quad S_2=0 \tag{42a,b}$$

应用式（39）以及边界条件（41b）、(42a)，积分微分方程（40a)，得

$$\varphi_2=(T-k)\rho-\frac{1}{8}(k^2y_0-3ky_0^2+2y_0^3)\left(\frac{2+\nu}{3\lambda_1}\rho-\frac{1}{2}\rho^3+\frac{1}{6}\rho^5\right) \tag{43}$$

再将此式代入式（38)，则得双层金属封顶扁球壳的二次近似特征关系式

$$T=\frac{\gamma_1}{\lambda_1}[(2\lambda_1\gamma_1^{-1}+k^2)y_0-3ky_0^2+2y_0^3], \tag{44}$$

其中

$$\gamma_1 = \frac{17+5\nu}{288} \tag{45}$$

值得指出，式（44）与（32）在壳内边缘半径 ρ_1 趋于零的极限情况下的形式是一致的。

将式（44）代入式（43），有

$$\varphi_2 = (2y_0 - k)\rho - \frac{1}{16}(k^2 y_0 - 3ky_0^2 + 2y_0^3)\left(\frac{7}{18}\rho - \rho^3 + \frac{1}{3}\rho^5\right) \tag{46}$$

应用式（46）和边界条件（41c）、(42b)，积分微分方程式（40b)，我们便得到无量纲径向薄膜内力 S 的二次近似解

$$\begin{aligned} S_2 = -\frac{1}{2}\Big[&(ky_0 - y_0^2)(\rho - \rho^3) + \frac{1}{288}(2y_0 - k)(k^2 y_0 - 3ky_0^2 + 2y_0^3)\left(\rho - \frac{7}{2}\rho^3 + 3\rho^5 - \frac{1}{2}\rho^7\right) \\ &- \frac{1}{27\,648}(k^2 y_0 - 3ky_0^2 + 2y_0^3)^2\left(\frac{23}{20}\rho - \frac{49}{12}\rho^7 + 7\rho^5 - \frac{17}{3}\rho^7 + \frac{9}{5}\rho^9 - \frac{1}{5}\rho^{11}\right)\Big] \end{aligned} \tag{47}$$

在三次近似中，有如下关于 φ_3 的线性边值问题

$$H(\rho\varphi_3) = S_2\varphi_2 \tag{48}$$

当 $\rho=1$ 时， $y_3=0$， $\dfrac{d\varphi_3}{d\rho} + \dfrac{v}{\rho}\varphi_3 = \lambda_1(T-k)$ (49a,b)

当 $\rho=0$ 时， $\varphi_3=0$ (50)

应用式（46)、(47）和边界条件（49b)、(50)，积分微分方程（48)，得

$$\begin{aligned} \varphi_a = &(T-k)\rho + \frac{1}{16}(2y_0 - k)(ky_0 - y_0^2)\left(\frac{4+2\nu}{3\lambda_1}\rho - \rho^3 + \frac{1}{3}\rho^5\right) \\ &+ \frac{1}{4608}(k^2 y_0 - 3ky_0^2 + 2y_0^3)\Big[(2y_0 - k)^2\left(\frac{13+17\nu}{60\lambda_1}\rho - \rho^3 + \frac{7}{6}\rho^5 - \frac{1}{2}\rho^7 + \frac{1}{20}\rho^9\right) \\ &- (ky_0 - y_0^2)\left(\frac{29+31\nu}{15\lambda_1}\rho - 7\rho^3 + \frac{25}{3}\rho^5 - 4\rho^7 + \frac{3}{5}\rho^9\right)\Big] \\ &- \frac{1}{1\,105\,920}(2y_0 - k)(k^2 y_0 - 3ky_0^2 + 2y_0^3)^2\left(\frac{3627+2197\nu}{840\lambda_1}\rho - \frac{209}{24}\rho^3\right. \\ &\left.+ \frac{365}{24}\rho^5 - \frac{185}{12}\rho^7 + \frac{191}{24}\rho^9 - \frac{9}{5}\rho^{11} + \frac{1}{7}\rho^{13}\right) + \frac{1}{637\,009\,920}(k^2 y_0 - 3ky_0^2 \\ &+ 2y_0^3)^3\left(\frac{11\,617+11\,747\nu}{1260\lambda_1}\rho - \frac{161}{4}\rho^3 + \frac{2957}{36}\rho^5 - \frac{647}{6}\rho^7 + \frac{1141}{12}\rho^9 - \frac{261}{5}\rho^{11}\right. \\ &\left.+ \frac{113}{7}\rho^{13} - \frac{18}{7}\rho^{15} + \frac{1}{6}\rho^{17}\right) \end{aligned} \tag{51}$$

最后，将此式代入式（38)，经过计算和整理，我们便得到双层金属封顶扁球壳的三次近似特征关系式

$$T = \frac{1}{\lambda_1}\sum_{i=1}^{9} m_i y_0^i, \tag{52}$$

其中

$$
\left.\begin{aligned}
m_1 &= 2\lambda_1 + \gamma_1 k^2 + \gamma_2 k^4,\\
m_2 &= -3\gamma_1 k - (7\gamma_2 + \gamma_3)k^3 + \gamma_4 k^5,\\
m_3 &= 2\gamma_1 + 2(9\gamma_2 + 2\gamma_3)k^2 - 8\gamma_4 k^4 + \gamma_5 k^6,\\
m_4 &= -5(4\gamma_2 + \gamma_3)k + 25\gamma_4 k^3 - 9\gamma_5 k^5,\\
m_5 &= 2(4\gamma_2 + \gamma_3) - 38\gamma_4 k^2 + 33\gamma_5 k^4,\\
m_6 &= 28\gamma_4 k - 63\gamma_5 k^3,\\
m_7 &= -8\gamma_4 + 66\gamma_5 k^2,\\
m_8 &= -36\gamma_5 k,\\
m_9 &= 8\gamma_5,
\end{aligned}\right\} \tag{53}
$$

$$
\left.\begin{aligned}
\gamma_1 &= \frac{17 + 5\nu}{288},\\
\gamma_2 &= \frac{17 - 103\nu}{8\,294\,400},\\
\gamma_3 &= -\frac{149 + 209\nu}{2\,073\,600},\\
\gamma_4 &= -\frac{21\,971 + 6956\nu}{9\,754\,214\,400},\\
\gamma_5 &= -\frac{48\,581 + 51\,311\nu}{16\,855\,282\,483\,200},
\end{aligned}\right\} \tag{54}
$$

当 $\nu=0.3$ 时，式（53）为

$$
\left.\begin{aligned}
m_1 &= 2.6 + 0.064\,236\,1k^2 - 1.675\,83\times10^{-6}k^4,\\
m_2 &= -0.192\,708k + 1.138\,24\times10^{-4}k^3 - 2.466\,40\times10^{-6}k^5,\\
m_3 &= 0.128\,472 - 4.385\,37\times10^{-4}k^2 + 1.973\,12\times10^{-5}k^4 - 3.795\,50\times10^{-9}k^6,\\
m_4 &= 5.439\,82\times10^{-4}k - 6.166\,00\times10^{-5}k^3 + 3.415\,95\times10^{8}k^5,\\
m_5 &= -2.175\,93\times10^{-4} + 9.372\,32\times10^{-5}k^2 - 1.252\,52\times10^{-7}k^4,\\
m_6 &= -6.905\,92 - 10^{-5}k + 2.391\,17\times10^{-7}k^3,\\
m_7 &= 1.973\,12\times10^{-5} - 2.505\,03\times10^{-7}k^2,\\
m_8 &= 1.366\,38\times10^{-7}k,\\
m_9 &= -3.036\,40\times10^{-3}
\end{aligned}\right\} \tag{55}
$$

五、主要结果

1. 临界温度

以 $\rho_1=0$，$\nu=0.3$ 的情况为例，按照三次近似特征关系式（52），我们在图 3 上绘制了 $T-y_0$ 特征曲线。与前文［1］一样，我们关心的是这些特征曲线有极值点的情形。相应于这些极值点，壳体将丧失稳定，出现跳跃现象，从而给予仪表以控制信号。

对特征关系式应用极值条件

$$\frac{\mathrm{d}T}{\mathrm{d}y_0}=0, \tag{56}$$

我们即可确定壳体出现跳跃现象时的临界挠度值。

在壳体中心开孔的情况下，应将式（32）代入式（56）。于是，得关于壳内边缘挠度 y_0 的二次代数方程：

$$3\beta_0^2 y_0^2-6k\beta_0 y_0+(\beta_1^{-1}+2k^3)=0, \tag{57}$$

解此方程，得临界挠度

$$y_0^*=\frac{1}{\beta_0}\left[k\pm\sqrt{\frac{1}{3}(k^2-\beta_1^{-1})}\right] \tag{58}$$

此式中的负号相应于上临界挠度，而正号则相应于下临界挠度。我们分别将它们记为 $y_{0,1}^*$ 和 $y_{0,2}^*$。

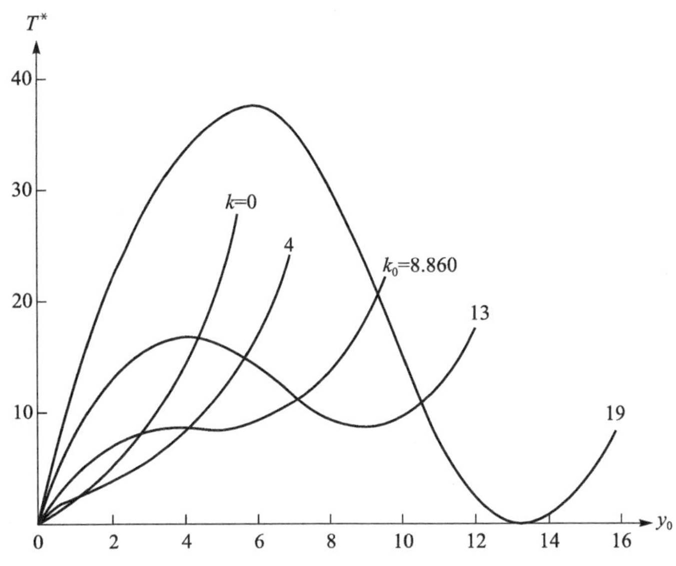

图 3　不同 k 值下的特征曲线（$\rho_1=0$，$\nu=0.3$）

将式（58）代入式（32），便得双层金属中心开孔扁球壳的无量纲临界温度

$$T^*=\beta_1[(\beta_1^{-1}+2k^2)\beta_0 y_0^*-3k\beta_0^2 y_0^{*2}+\beta_0^3 y_0^{*3}], \tag{59}$$

这里，相应于 $y_{0,1}^*$ 的 T^* 是上临界温度，而相应于 $y_{0,2}^*$ 的 T^* 则是下临界温度，我们分别将它们记为 T_1^* 和 T_2^*。

根据公式（59），我们很容易算出壳体临界温度的数值。为了方便和明显，我们把求得的结果表示在图 4 上。每一条曲线由两条分支曲线组成，上面一支相应于上临界温度，下面一支相应于下临界温度。

在壳体封顶的情况下，对于二次近似的临界温度，可以直接使用公式（59）进行计算。但为了读者使用方便，我们给出直接计算它的相当简单的公式

$$T^*=\frac{\gamma_1}{\lambda_1}[(2\lambda_1\gamma_1^{-1}+k^2)y_0^*-3ky_0^{*2}+2y_0^{*3}], \tag{60}$$

其中

$$y_0^* = \frac{1}{2}\left[k \pm \sqrt{\frac{1}{3}(k^2 - 4\lambda_1\gamma_1^{-1})}\right], \tag{61}$$

这里的符号规则与前面相同。

对于三次近似情况，则应将式（52）代入极值条件（56），便得关于壳中心挠度 y_0 的八次代数方程

$$\sum_{i=1}^{9} i m_i y_0^{i-1} = 0, \tag{62}$$

使用简化牛顿法，并选取由式（61）计算所得的临界挠度值为初值，我们便不难由方程（62）求得三次近似中的临界挠度 y_0^* 值。将所得结果代入式（52），便得双层金属封顶扁球壳的无量纲临界温度

$$T^* = \frac{1}{\lambda_1}\sum_{i=1}^{9} m_i y_0^{*\,i} \tag{63}$$

按照式（60）和（61）所得的计算结果都给在图 5 中，相当精确的三次近似结果用实线表示，二次近似结果用虚线表示。显然，两条曲线相当吻合。

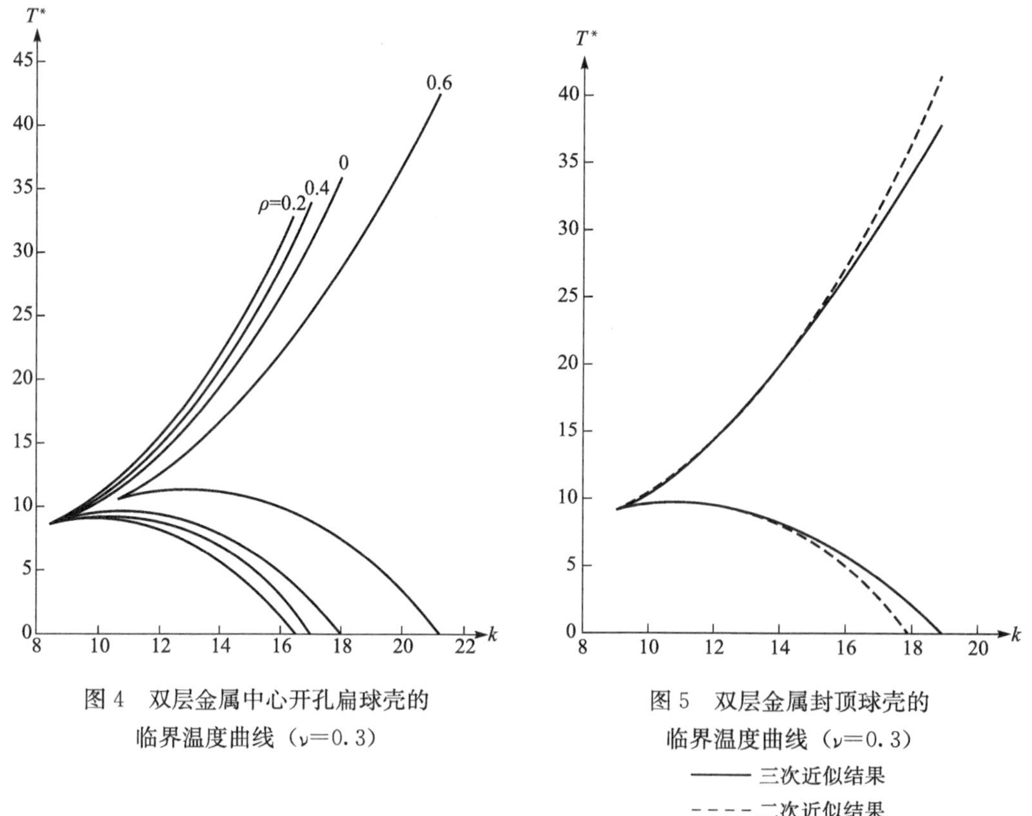

图 4　双层金属中心开孔扁球壳的临界温度曲线（$\nu=0.3$）

图 5　双层金属封顶球壳的临界温度曲线（$\nu=0.3$）
—— 三次近似结果
- - - - 二次近似结果

得到壳体的无量纲临界的温度 T^* 后，可由式（10）得到双层金属扁球壳的临界温度

$$t^* = \frac{\lambda_1 D}{\mu a^2}\sqrt{\frac{2D}{C(1-\nu^2)}}\,T^* \tag{64}$$

2. 平均温度

现在，我们引进热敏弹性元件设计中一个重要参数，即平均温度 t_C^* 的概念，其定义为

$$t_C^* = \frac{\lambda_1 D}{\mu a^2}\sqrt{\frac{2D}{C(1-\nu^2)}}T_C^*, \tag{65}$$

其中 T_C^* 是壳体的无量纲平均温度，即

$$T_C^* = \frac{1}{2}(T_1^* + T_2^*). \tag{66}$$

在壳体中心开孔的情况下，将式（59）代入式（66），得

$$T_C^* = k. \tag{67}$$

在壳体封顶的情况下，计算表明，二次和三次近似解也都满足公式（67）。

这就是说，一个双层金属扁球壳的无量纲平均温度 T_C^* 的值，恰恰就等于它自身的无量纲几何参数 k 的值。

将式（67）代入式（65），得平均温度

$$t_C^* = \eta\frac{f}{a^2} \tag{68}$$

其中

$$\eta = \frac{4\{E_1[h_0^3-(h_0-h_1)^3]+E_2[(h-h_0)^3+(h_0-h_1)^3]\}}{3E_1h_1(2h_0-h_1)(\alpha_1-\alpha_2)} \tag{69}$$

显而易见，双层金属中心开孔扁球壳的平均温度 t_C^* 既与开孔无关，又与壳体材料的泊松比无关。

3. 临界拱高

对于元件而言，确定图 4—图 5 中的上、下临界温度曲线的交点，即临界点是十分重要的，我们把临界点的壳体无量纲几何参数 k 的值称为临界几何参数 k_0，则 k_0 值就是区分壳体能否具有跳跃本能的分界点。

在壳体中心开孔的情况下，为使二次代数方程（57）具有重根，应令其判别式为零，即

$$36k^2\beta_0^4 - 12\beta_0^4(\beta_1^{-1}+2k^2) = 0, \tag{70}$$

由此方程解得

$$k_0 = \sqrt{\beta_1^{-1}} \tag{71}$$

在壳体封顶的情况下，对于二次近似，可直接由上式导出计算公式

$$k_0 = 2\sqrt{\lambda_1\gamma_1^{-1}} \tag{72}$$

对于三次近似，则需使用式（52），求解下面非线性方程组：

$$\left.\begin{aligned}\frac{\mathrm{d}T}{\mathrm{d}y_0} &= 0,\\ \frac{\mathrm{d}^2T}{\mathrm{d}y_0^2} &= 0\end{aligned}\right\} \tag{73}$$

亦即

$$\left.\begin{aligned}&\sum_{i=1}^{9} i m_i y_0^{i-1}=0,\\&\sum_{i=2}^{9} i(i-1) m_i y_0^{i-2}=0\end{aligned}\right\} \tag{74}$$

使用牛顿法，选取二次近似中的 k_0 值和相应的 y_0^* 值为初值，我们就不难求出方程组（74）的解。

得到临界几何参数 k_0 后，便可按式（10）计算双层金属球壳的临界拱高：

$$f_0=\sqrt{\frac{D}{2C(1-\nu^2)}}k_0 \tag{75}$$

值得注意的是，根据式（67），我们立即可知：若一个壳体的临界几何参数已知，则其临界点的临界温度 T_0^* 值就等于该值，亦即我们有下面公式成立：

$$T_0^*=k_0 \tag{76}$$

下面，我们给出在 $\nu=0.3$ 的情况下双层金属封顶扁球壳的临界几何参数 k_0 以及相应的临界挠度、临界温度值：

对于二次近似：　$k_0=8.997$，　$y_0^*=4.499$，　$T_0^*=8.997$.

对于三次近似：　$k_0=8.860$，　$y_0^*=4.430$，　$T_0^*=8.860$.

关于双层金属中心开孔扁球壳的临界几何参数 k_0 的值，则用图 6 表示，由图看出，曲线呈凹形，这说明 k_0 有一个极小值 $(k_0)_{\min}$ 存在，相应于此值的壳体的拱高为最小，对失稳的反应最灵敏。为了求得 $(k_0)_{\min}$ 值，我们对式（71）使用极值条件

$$\frac{\mathrm{d}k_0}{\mathrm{d}\rho_1}=0, \tag{77}$$

应用迭代法解此方程，可得具有 $(k_0)_{\min}$ 值的壳体为

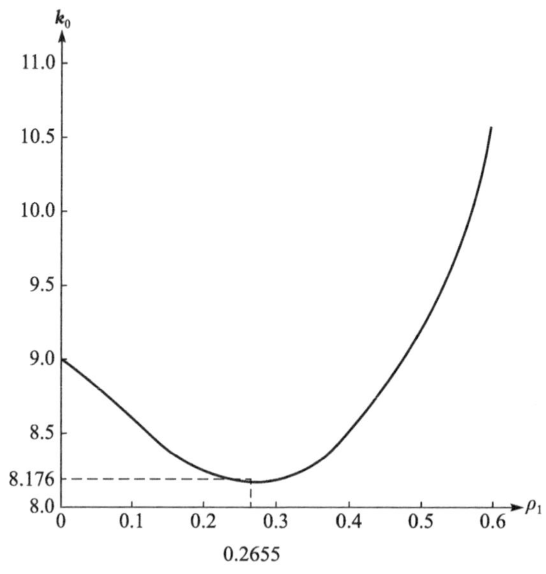

图 6　临界几何参数 k_0 与 ρ_1 的关系曲线（$\nu=0.3$）

$$\rho_2 = 0.2655$$

将此值代入式（71）和（76），即得 $(k_0)_{\min}$ 和相应的 T_0^* 值

$$(k_0)_{\min} = 8.176, \qquad T_0^* = 8.176.$$

4. 跳跃幅度

最后，我们再引进热敏弹性元件设计中另一个重要参数，即跳跃幅度的概念。它的定义为

$$l = \sqrt{\frac{2D}{C(1-\nu^2)}} L, \tag{78}$$

其中 L 是壳体的无量纲跳跃幅度（图 7），即

$$L = y_0' - y_{0,1}^* \tag{79}$$

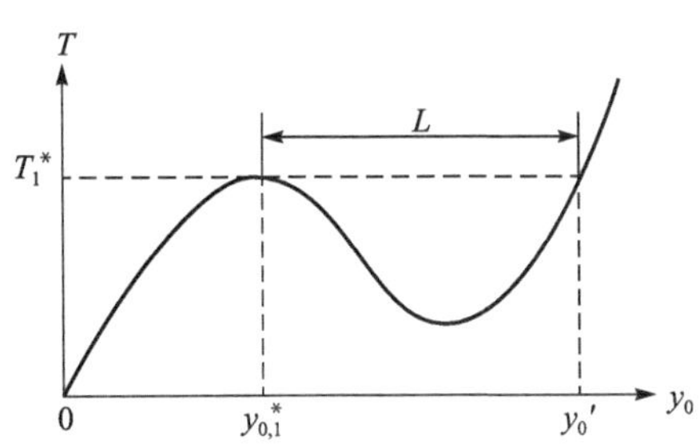

图 7　双层金属扁球壳的跳跃幅度

在壳体中心开孔的情况下，式（79）中的 y_0' 是下面三次代数方程

$$\beta_0^3 y_0^3 - 3k\beta_0^2 y_0^2 + (\beta_1^{-1} + 2k^2)\beta_0 y_0 - \beta_1^{-1} T_1^* = 0 \tag{80}$$

的一个根，为

$$y_0' = \beta_0^{-1}\left[k + \sqrt[3]{2\beta_1^{-1}(T_1^* - T_2^*)}\right] \tag{81}$$

在壳体封顶的情况下，对于二次近似，可直接由上式导出计算公式

$$y_0' = \frac{k}{2} + \sqrt[3]{\lambda_1 \gamma_1^{-1}(T_1^* - T_2^*)} \tag{82}$$

对于三次近似，则式（79）中的 y_0' 应是下面九次代数方程

$$\frac{1}{\lambda_1}\sum_{i=1}^{9} m_i y_0^i - T_1^* = 0 \tag{83}$$

的一个根，使用简化牛顿法，选取二次近似中的 y_0' 值为初值，我们不难求得方程（83）的根。

下面，我们将所得的数值结果绘制在图 8 和图 9 上。由图 9 看到，跳跃幅度的二次近似曲线与相当精确的三次近似曲线是相当一致的。

六、讨论

应用修正迭代法，本文获得了双层金属中心开孔扁球壳的非线性热稳定问题的解析解。在壳体封顶的特殊情况下，还获得了相当精确的三次近似解。在图 10 和图 11 上，我们分别给出了双层金属封顶扁球壳的临界温度和跳跃幅度的二次和三次近似结果的相对误差。由图看出，在很大的区域内，两次近似结果的相对误差不超过±10%。至于双层金属封顶扁球壳的临界几何参数 k_0 值，其二次和三次近似结果的相对误差也仅为−1.5%。通过上述比较，说明本文公式具有较高的精确度，在一般情况下，因为二次近似公式相当简单，所以使用它既方便，且又能满足精确度的要求。

将本文结果与文献［5］的结果进行比较，可知平均温度 t_C^* 的公式是相同的。在双层金属封顶扁球壳的特殊情况下，本文结果与文献［3］，［5］的结果也是趋于一致的（参见表 2）。

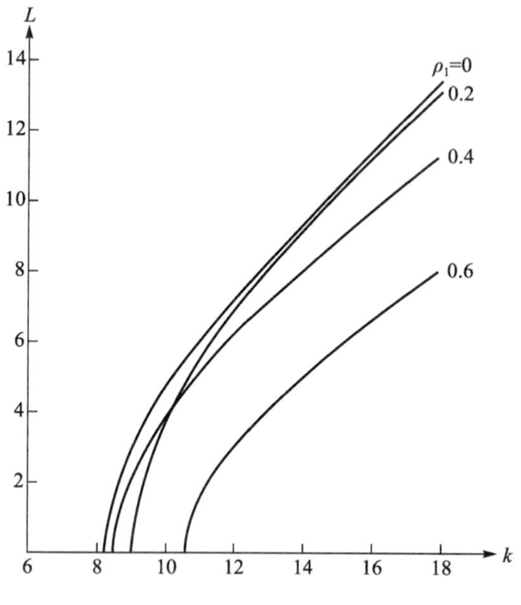

图 8　双层金属中心开孔扁球壳的跳跃幅度曲线（ν=0.3）

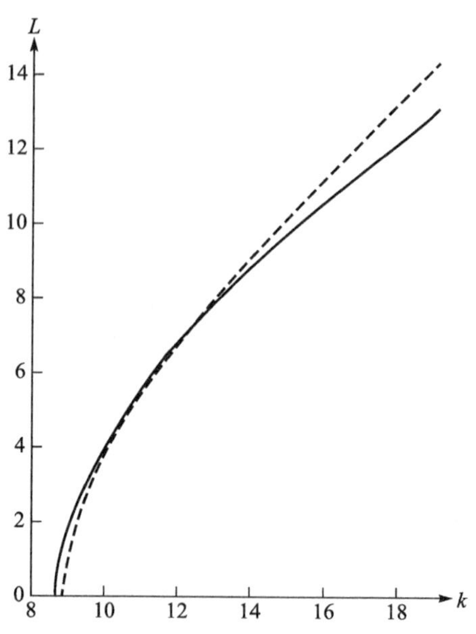

图 9　双层金属封顶扁球壳的跳跃幅度曲线（ν=0.3）

——三次近似结果

- - - - 二次近似结果

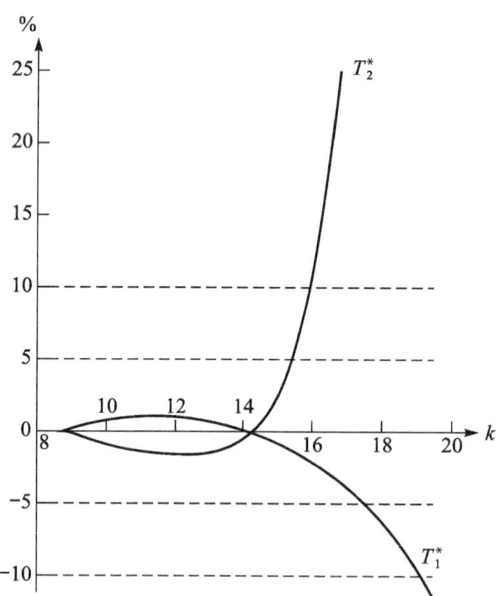

图 10　双层金属封顶扁球壳的临界温度的二次和三次近似结果的相对误差（ν=0.3）

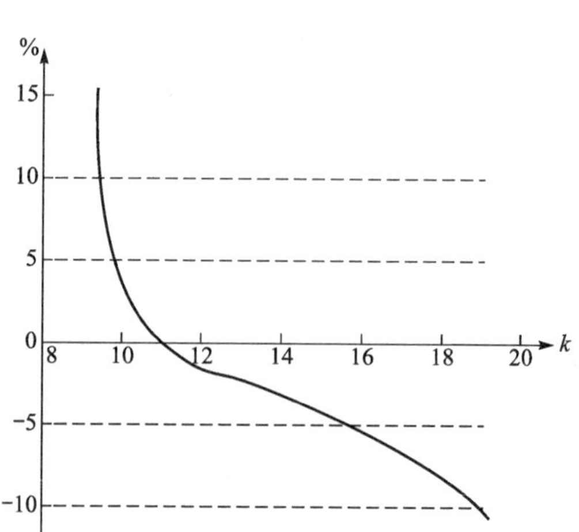

图 11　双层金属封顶扁球壳的跳跃幅度的二次和三次近似结果的相对误差（ν=0.3）

表 2　双层金属封顶扁球壳的上临界温度 T_1^* 的比较 ($\nu=1/3$)

k	9.164	10.21	11.31	12.47	14.32	15.62
文献 [3]	9.297	10.50	13.13	15.72	20.46	24.21
文献 [5]	9.175	10.98	12.83	15.38	20.52	24.85
本文二次近似结果	9.174	10.69	12.76	15.41	20.68	25.24
本文三次近似结果	9.207	10.79	12.91	15.58	20.72	24.93

参考文献

[1] 刘人怀. 双层金属截头扁锥壳的热稳定性. 1980 年中国力学学会弹塑性力学学术交流会论文集，重庆 (1980)；力学学报，特刊，1981，(2)：172.

[2] Панов Д Ю. Об устойчивости биметаллической оболочки при нагреве. ПММ，1947，11 (6)：603.

[3] Wittrick W H，Myers D M，Blunden W R. Stability of a bimetallic disk. Quarterly Journal of Mech. and Appl. Math.，1953，6 (1)：15.

[4] Григолюк Э N. Тонкие ъиметаллические оболочки и пластииы. Nнж. сборник，1953，18.

[5] 罗祖道，聂德耀，刘汉东，等. 双层金属球面扁壳的热稳定性. 力学学报，1966，9 (1)：1.

Non-linear thermal stability of bimetallic shallow shells of revolution①

中文摘要

本文建立了双层金属旋转扁壳在均匀温度作用下的非线性热稳定性理论。应用修正迭代法，对于双层金属扁球壳和双层金属扁锥壳两种特殊情况进行了求解，得到了很好的结果。特别是对于双层金属扁锥壳情况，以前尚无人研究过。本文对工程中仪表的弹性元件设计有益。

1. Introduction

Bimetallic shallow shells of revolution are important elastic elements for thermal sensitivity in precision instruments. So far as we know, there are two kinds of bimetallic shallow shells of revolution which are very attractive for design purposes. The first kind of shell is the bimetallic shallow spherical shell which has been studied by Panov[1], Wittrick[2], Grigolyuk[3], Lo[4], etc. But their results are not yet satisfactory. The second one is the bimetallic shallow conical shell. Unfortunately, the shell has not yet been studied because of the difficulties of nonlinear mathematics.

In this paper, general theories for thermal stability of the bimetallic shallow spherical shell with a circular hole at the center and the bimetallic truncated shallow conical shell under a uniform temperature field are presented. According to the selection of reference surface of coordinates as suggested by Korolev[5] and Radkovsky[6], basic equations for a double-layered shell are simplified into a form similar to that of classical shell theory. By analogy, the thermal effect is replaced by an equivalent edge moment loaded uniformly along both boundaries of shell. The problems are then solved according to the modified iteration method as suggested by Yeh and the author[7-9] in 1965. This method incorporates advantages of Chien's perturbation method[10] and usual successive approximations. It is an effective, simple, accurate method of solving nonlinear differential equations. We had applied the method to solve several practical problems with excellent results[11-16]. The paper gives some rather interesting results, which may be valuable in designing elastic elements of various instruments in engineering.

2. Basic Equations of Nonlinear Bending Theory for Bimetallic Shallow Shells of Revolution

Now consider an open bimetallic shallow shell of revolution of upper radius b, lower radius a, and height f as show in Fig. 1. In the shell we take a surface parallel to the

① Reprinted from *International Journal of Non-Linear Mechanics*, 1983, 18 (5): 409-429.

internal surface and at a distance h_0 from the internal surface as the reference suface of coordinates.

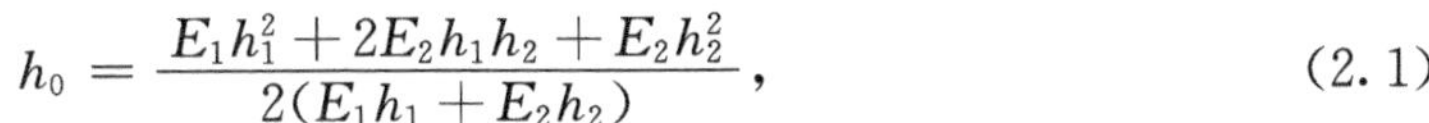

$$h_0 = \frac{E_1 h_1^2 + 2E_2 h_1 h_2 + E_2 h_2^2}{2(E_1 h_1 + E_2 h_2)}, \tag{2.1}$$

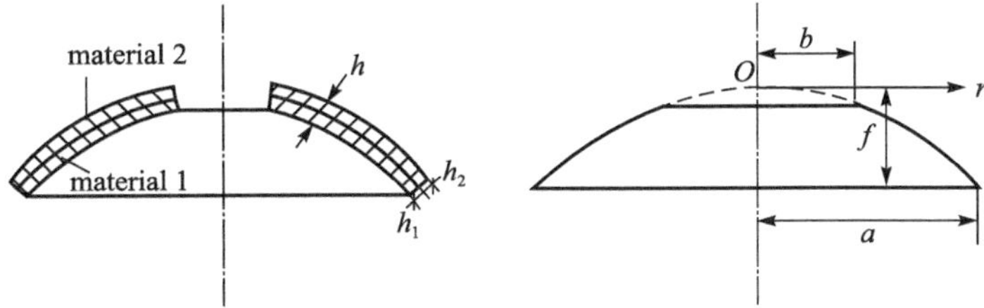

Fig. 1　An open himetallic shallow shell of revolution

Here h_1 and h_2 are thicknesses of the inner and outer layer respectively, E_1 and E_2 are Young's moduli of the inner and outer layer respectively. As usual, we assume that Young's modulus E_1, and Poisson's ratio ν_1 of the inner layer and Young's modulus E_2, and Poisson's ratio ν_2 of the outer layer do not vary in the region of the general temperature, and ν_1 is equal to ν_2

$$\nu_1 = \nu_2 = \nu. \tag{2.2}$$

Then, basic equations of large deflections for the bimetallic shallow shell of revolution under a uniform temperature field are [4-6]

$$\begin{gathered} Dr \frac{\mathrm{d}}{\mathrm{d}r} \frac{1}{r} \frac{\mathrm{d}}{\mathrm{d}r} r \frac{\mathrm{d}w}{\mathrm{d}r} - rN_r\left(\frac{\mathrm{d}w}{\mathrm{d}r} - \theta\right) = 0, \\ \frac{1}{C(1-\nu^2)} r \frac{\mathrm{d}}{\mathrm{d}r} \frac{1}{r} \frac{\mathrm{d}}{\mathrm{d}r}(r^2 N_r) + \frac{\mathrm{d}w}{\mathrm{d}r}\left(\frac{1}{2}\frac{\mathrm{d}w}{\mathrm{d}r} - \theta\right) = 0, \end{gathered} \tag{2.3}$$

where r is the distance from a point on the reference surface of coordinates to the axis of the shell, h is the total thickness of the shell, θ is the angle between a normal to the shell and its axis of revolution, w is the deflection of a point on the reference surface of coordinates, N_r is the radial membrane stress, C is the extensional rigidity and D is the flexural rigidity,

$$\begin{aligned} C &= \frac{E_1 h_1 + E_2 h_2}{1-\nu^2}, \\ D &= \frac{E_1[h_0^3 - (h_0 - h_1)^3] + E_2[(h-h_0)^3 + (h_0 - h_1)^3]}{3(1-\nu^2)}. \end{aligned} \tag{2.4}$$

The radial membrane stress N_r and the radial moment M_r may be calculated as follows

$$\begin{aligned} N_r &= C\left[\frac{\mathrm{d}u}{\mathrm{d}r} + \nu\frac{u}{r} + \frac{\mathrm{d}w}{\mathrm{d}r}\left(\frac{1}{2}\frac{\mathrm{d}w}{\mathrm{d}r} - \theta\right)\right] - \alpha t, \\ M_r &= -D\left(\frac{\mathrm{d}^2 w}{\mathrm{d}r^2} + \frac{\nu}{r}\frac{\mathrm{d}w}{\mathrm{d}r}\right) + \mu t, \end{aligned} \tag{2.5a,b}$$

where u is the radial displacement of a point on the reference surface of coordinates, α_1

and α_2 are coefficients of linear expansion of the inner and outer layer respectively, t is the variation of the temperature with respect to the initial temperature without thermal stress, α and μ are appropriate coefficients of linear expansion,

$$\alpha = \frac{\alpha_1 E_1 h_1 + \alpha_2 E_2 h_2}{1-\nu}, \quad \mu = \frac{E_1 h_1 (2h_0 - h_1)}{2(1-\nu)}(\alpha_1 - \alpha_2). \tag{2.6}$$

Equations (2.3) will be solved under the following customary boundary conditions for simply supported edges

$$\begin{aligned} w=0,\ M_r = 0,\ N_r = 0 \quad & \text{at } r = a; \\ M_r = 0,\ N_r = 0 \quad & \text{at } r = b. \end{aligned} \tag{2.7}$$

According to (2.5b), the boundary condition $M_r=0$ may be written

$$-D\left(\frac{\mathrm{d}^2 w}{\mathrm{d}r^2} + \frac{\nu}{r}\frac{\mathrm{d}w}{\mathrm{d}r}\right) = -\mu t. \tag{2.8}$$

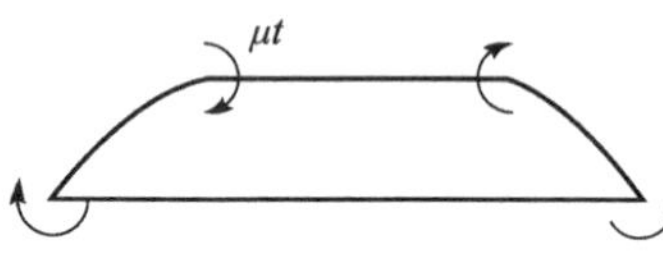

Fig. 2

From this it may be seen that the thermal effect is replaced by an equivalent edge moment ($-\mu t$) acting uniformly along both boundaries of an open bimetallic shallow shell of revolution as shown in Fig. 2.

In order to simplify the following calculations, let us introduce the following dimensionless variables

$$\begin{aligned} &\rho = \frac{r}{a},\ \rho_b = \frac{b}{a},\ W = -\sqrt{\frac{C(1-\nu^2)}{2D}}w,\ S = \frac{a^2 \rho}{D}N_r, \\ &K = \frac{a}{f}k,\ k = \sqrt{\frac{C(1-\nu^2)}{2D}}f,\ T = \frac{\mu a^2}{D}\sqrt{\frac{C(1-\nu^2)}{2D}}t. \end{aligned} \tag{2.9}$$

Using these notations, the nonlinear boundary value problem (2.3) and (2.7) becomes

$$\begin{aligned} &H\left(\rho\frac{\mathrm{d}W}{\mathrm{d}\rho}\right) = S\left(\frac{\mathrm{d}W}{\mathrm{d}\rho} + K\theta\right), \\ &H(\rho S) = -\frac{\mathrm{d}W}{\mathrm{d}\rho}\left(\frac{\mathrm{d}W}{\mathrm{d}\rho} + 2K\theta\right); \end{aligned} \tag{2.10}$$

$$\begin{aligned} W = 0,\ \frac{\mathrm{d}^2 W}{\mathrm{d}\rho^2} + \frac{\nu}{\rho}\frac{\mathrm{d}W}{\mathrm{d}\rho} = -T,\ S = 0 \quad & \text{at } \rho = 1; \\ \frac{\mathrm{d}^2 W}{\mathrm{d}\rho^2} + \frac{\nu}{\rho}\frac{\mathrm{d}W}{\mathrm{d}\rho} = -T,\ S = 0 \quad & \text{at } \rho = \rho_b, \end{aligned} \tag{2.11}$$

where

$$H(\cdots) = \rho\frac{\mathrm{d}}{\mathrm{d}\rho}\frac{1}{\rho}\frac{\mathrm{d}}{\mathrm{d}\rho}(\cdots). \tag{2.12}$$

We shall solve this nonlinear boundary value problem by the modified iteration method for two particular cases, the bimetallic shallow spherical shell and the bimetallic shallow conical shell.

3. Solution For Bimetallic Closed Shallow Spherical Shell

We shall begin the theory of nonlinear thermal stability of bimetallic shallow shells of revolution with the simple problem of the stability of a bimetallic closed shallow spherical shell. We can therefore transform the problem (2.10) and (2.11) into the nonlinear boundary value problem for the bimetallic closed shallow spherical shell by introducing, instead of θ, the parameter of the shallow spherical shell.

$$\theta \approx \frac{2f}{a^2} r. \tag{3.1}$$

Then we obtain

$$\begin{aligned} &H(\rho\phi) = S\phi, \\ &H(\rho S) = -(\phi^2 - 4k^2\rho^2); \end{aligned} \tag{3.2a,b}$$

$$W = 0,\ \frac{\mathrm{d}\phi}{\mathrm{d}\rho} + \nu\frac{\phi}{\rho} = T - 2\lambda_1 k,\ S = 0 \quad \text{at } \rho = 1; \tag{3.3a,b,c}$$

$$\phi = 0,\ S = 0 \quad \text{at } \rho = 0, \tag{3.4a,b}$$

where we employ the notations

$$\begin{aligned} &\phi = -\left(\frac{\mathrm{d}W}{\mathrm{d}\rho} + 2k\rho\right), \\ &\lambda_1 = 1 + \nu. \end{aligned} \tag{3.5a,b}$$

To obtain a satisfactory solution, the modified iteration method can be used.

For the first approximation, by neglecting the nonlinear term $S\phi$ in Eq. (3.2a), and again introducing ϕ_1 into the right side of Eq. (3.2b), we find the following linear boundary value problem

$$\begin{aligned} &H(\rho\phi_1) = 0, \\ &H(\rho S_1) = -(\phi_1^2 - 4k^2\rho^2); \end{aligned} \tag{3.6a,b}$$

$$W_1 = 0,\ \frac{\mathrm{d}\phi_1}{\mathrm{d}\rho} + \nu\frac{\phi_1}{\rho} = T - 2\lambda_1 k,\ S_1 = 0 \quad \text{at } \rho = 1; \tag{3.7a,b,c}$$

$$\phi_1 = 0,\ S_1 = 0 \qquad \text{at } \rho = 0. \tag{3.8a,b}$$

The solution of the problem (3.6a), (3.7b) and (3.8a) is

$$\phi_1 = \frac{1}{\lambda_1}(T - 2\lambda_1 k)\rho. \tag{3.9}$$

When we introduce a notation of the dimensionless center deflection as an iteration parameter

$$W_0 = -\sqrt{\frac{C(1-\nu^2)}{2D}}\, W\big|_{r=0}. \tag{3.10}$$

Then using (3.3a) and (3.5a), we have

$$W_0 = \int_0^1 (\phi + 2k\rho)\,\mathrm{d}\rho. \tag{3.11}$$

Substituting (3.9) into (3.11), we obtain

$$T = 2\lambda_1 W_0. \tag{3.12}$$

This is the solution for small deflections. Substituting this solution into (3.9), we find

$$\phi_1 = 2(W_0 - k)\rho. \tag{3.13}$$

Substituting this value of ϕ_1 into Eq. (3.6b), under the boundary conditions (3.7c) and (3.8b), the solution of Eq. (3.6b) is

$$S_1 = -\left(kW_0 - \frac{1}{2}W_0^2\right)(\rho - \rho^3). \tag{3.14}$$

For the second approximation, we have the following linear boundary value problem

$$\begin{aligned} H(\rho\phi_2) &= S_1\phi_1, \\ H(\rho S_2) &= -(\phi_2^2 - 4k^2\rho^2); \end{aligned} \tag{3.15a,b}$$

$$W_2 = 0,\ \frac{\mathrm{d}\phi_2}{\mathrm{d}\rho} + \nu\frac{\phi_2}{\rho} = T - 2\lambda_1 k,\ S_2 = 0 \quad \text{at } \rho = 1; \tag{3.16a,b,c}$$

$$\phi_2 = 0,\ S_2 = 0 \quad \text{at } \rho = 0. \tag{3.17a,b}$$

Using (3.13) and (3.14), the solution of the problem (3.15a), (3.16b) and (3.17a) gives

$$\phi_2 = \frac{1}{\lambda_1}(T - 2\lambda_1 k)\rho - \frac{1}{4}(2k^2W_0 - 3kW_0^2 + W_0^3)\left(\frac{2+\nu}{3\lambda_1}\rho - \frac{1}{2}\rho^3 + \frac{1}{6}\rho^5\right). \tag{3.18}$$

Substituting (3.18) into (3.11), we obtain a second approximation characteristic relation of the bimetallic closed shallow spherical shell

$$T = \gamma_1\left[\left(\frac{\lambda_1}{\gamma_1} + 2k^2\right)W_0 - 3kW_0^2 + W_0^3\right], \tag{3.19}$$

where

$$\gamma_1 = \frac{17 + 5\nu}{144}. \tag{3.20}$$

Using (3.19), the solution (3.18) becomes

$$\phi_2 = 2(W_0 - k)\rho - \frac{1}{8}(2k^2W_0 - 3kW_0^2 + W_0^3)\left(\frac{7}{18}\rho - \rho^3 + \frac{1}{3}\rho^5\right). \tag{3.21}$$

Using (3.21), the solution of the problem (3.15b), (3.16c) and (3.17b) gives

$$\begin{aligned} S_2 = -\frac{1}{2}\Big[&(2kW_0 - W_0^2)(\rho - \rho^3) + \frac{1}{72}(W_0 - k)(2k^2W_0 - 3kW_0^3)\Big(\rho \\ &- \frac{7}{2}\rho^3 + 3\rho^5 - \frac{1}{2}\rho^7\Big) - \frac{1}{13824}(2k^2W_0 - 3kW_0^2 + W_0^3)^2\Big(\frac{23}{20}\rho \\ &- \frac{49}{12}\rho^3 + 7\rho^5 - \frac{17}{3}\rho^7 + \frac{9}{5}\rho^9 - \frac{1}{5}\rho^{11}\Big)\Big]. \end{aligned} \tag{3.22}$$

For the third approximation, we have the following linear boundary value problem for ϕ

$$H(\rho_3) = S_2\phi_2; \tag{3.23}$$

$$W_3 = 0,\ \frac{\mathrm{d}\phi_3}{\mathrm{d}\rho} + \nu\frac{\phi_3}{\rho} = T - 2\lambda_1 k \quad \text{at } \rho = 1; \tag{3.24a,b}$$

$$\phi_3 = 0 \quad \text{at } \rho = 0. \tag{3.25}$$

Using (3.21) and (3.22), the solution of the problem (3.23), (3.24b) and (3.25) gives

$$
\begin{aligned}
\phi_3 =& \frac{1}{\lambda_1}(T-2\lambda_1 k)\rho+\frac{1}{8}(W_0-k)(2kW_0-W_0^2)\left(\frac{4+2\nu}{3\lambda_1}\rho-\rho^3+\frac{1}{3}\rho^5\right) \\
&+\frac{1}{2304}(2k^2W_0-3kW_0^2+W_0^3)\Big[(W_0-k)^2\Big(\frac{13+17\nu}{15\lambda_1}\rho \\
&-4\rho^3+\frac{14}{3}\rho^5-2\rho^7+\frac{1}{5}\rho^9\Big) \\
&-(2kW_0-W_0^2)\left(\frac{29+31\nu}{15\lambda_1}\rho-7\rho^3+\frac{25}{3}\rho^5-4\rho^7+\frac{3}{5}\rho^9\right)\Big] \\
&-\frac{1}{138240}(W_0-k)(2k^2W_0-3kW_0^2+W_0^3)^2\Big(\frac{3627+2197\nu}{840\lambda_1}\rho \\
&-\frac{209}{24}\rho^3+\frac{365}{24}\rho^5-\frac{185}{12}\rho^7+\frac{191}{24}\rho^9-\frac{9}{5}\rho^{11}+\frac{1}{7}\rho^{13}\Big) \\
&+\frac{1}{79626240}(2k^2W_0-3kW_0^2+W_0^3)^3\Big(\frac{11167+11747\nu}{1260\lambda_1}\rho \\
&-\frac{161}{4}\rho^3+\frac{2957}{36}\rho^5-\frac{647}{6}\rho^7+\frac{1141}{12}\rho^9 \\
&-\frac{261}{5}\rho^{11}+\frac{113}{7}\rho^{13}-\frac{18}{7}\rho^{15}+\frac{1}{6}\rho^{17}\Big).
\end{aligned}
\tag{3.26}
$$

Substituting (3.26) into (3.11), we obtain the third approximation of the characteristic relation of the bimetallic closed shallow spherical shell

$$
T=\sum_{i=1}^{9} m_i W_0^i, \tag{3.27}
$$

where

$$
\begin{aligned}
m_1 &= 2(\lambda_1+\gamma_1 k^2+\gamma_2 k^4), \\
m_2 &= -3\gamma_1 k-(7\gamma_2+\gamma_3)k^3+4\gamma_4 k^5, \\
m_3 &= \gamma_1+(9\gamma_2+2\gamma_3)k^2-16\gamma_4 k^4+8\gamma_5 k^6, \\
m_4 &= -5\left(\gamma_2+\frac{1}{4}\gamma_3\right)k+25\gamma_4 k^3-36\gamma_5 k^5, \\
m_5 &= \left(\gamma_2+\frac{1}{4}\gamma_3\right)-19\gamma_4 k^2+66\gamma_5 k^4, \\
m_6 &= 7\gamma_4 k-63\gamma_5 k^3, \\
m_7 &= -\gamma_4+33\gamma_5 k^2, \\
m_8 &= -9\gamma_5 k, \\
m_9 &= \gamma_5, \\
\gamma_2 &= \frac{17-103\nu}{1036800}, \\
\gamma_3 &= -\frac{149+209\nu}{259200},
\end{aligned}
\tag{3.28}
$$

$$\gamma_4 = -\frac{21971 + 6956\nu}{1219276800},$$

$$\gamma_5 = -\frac{48581 + 51311\nu}{2106910310400}. \tag{3.29}$$

The calculations have been carried out and we have seen that snap-through buckling of the shell occurs under an elevated temperature. In order to find the critical buckling temperature of the shell, we differentiate expressions (3.19) and (3.27) with respect to W_0 and set the derivative equal to zero

$$\frac{dT}{dW_0} = 0, \tag{3.30}$$

which gives:
for the second approximation,

$$W_0^2 - 2kW_0 + \frac{1}{3}\left(\frac{\lambda_1}{\gamma_1} + 2k^2\right) = 0; \tag{3.31a}$$

for the third approximation,

$$\sum_{i=1}^{9} im_i W_0^{i-1} = 0. \tag{3.31b}$$

Solving the quadratic Eq. (3.31a), the critical center deflection W_0 for the second approximation gives

$$W_0^* = k \pm \sqrt{\frac{1}{3}\left(k^2 - \frac{\lambda_1}{\gamma_1}\right)}, \tag{3.32}$$

where W_0^* with respect to the negative sign is the upper critical center deflection $W_{0,1}^*$ and W_0^* with respect to the positive sign is the lower critical center deflection $W_{0,2}^*$. Solving Eq. (3.31b), the critical center deflection W_0^* for the third approximation can be obtained by Newton's method. Here for given values of Poisson's ratio ν and the geometrical parameter k, the initial value of W_0^* may be taken on the basis of the expression (3.32).

Substituting W_0^* into (3.19) and (3.27) respectively, we obtain the following formulas of the dimensionless critical buckling temperature:
for the second approximation,

$$T^* = \gamma_1\left[\left(\frac{\lambda_1}{\gamma_1} + 2k^2\right)W_0^* - 3kW_0^{*2} + W_0^{*3}\right], \tag{3.33a}$$

for the third approximation,

$$T^* = \sum_{i=1}^{9} m_i W_0^{*i}, \tag{3.33b}$$

where T^* with respect to $W_{0,1}^*$ is the dimensionless upper critical buckling temperature T_1^*, and T^* with respect to $W_{0,2}^*$ is the dimensionless lower critical buckling temperature T_2^*.

Calculated results of the dimensionless critical buckling temperature for $\nu=0.3$ are shown in Fig. 3. The more exact results for the third approximation are shown as the full line. The results for the second approximation are shown as the dashed line. Each curve contains two branch curves. The upper and lower branch curves are called the curves for the upper and lower critical bucking temperature respectively. From this figure it may be seen that reasonable agreement between the dashed line and the full line exists. For convenience the relative error is shown in Fig. 4. From these two figures it can be concluded that the second approximation can be applied with sufficient accuracy in a large range of the geometrical parameter k.

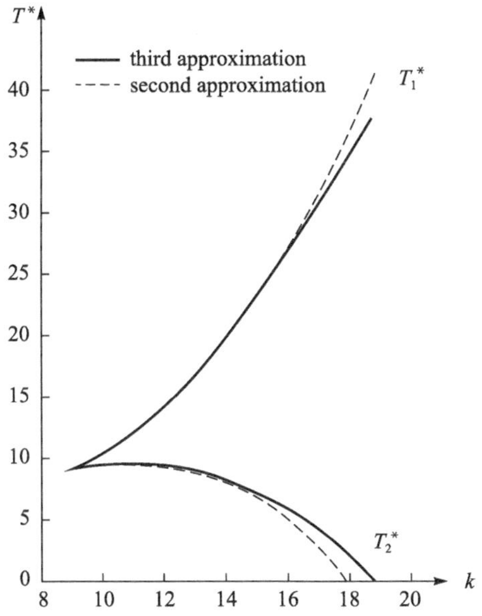

Fig. 3 Curves of the dimensionless critical buckling temperature (ν=0. 3)

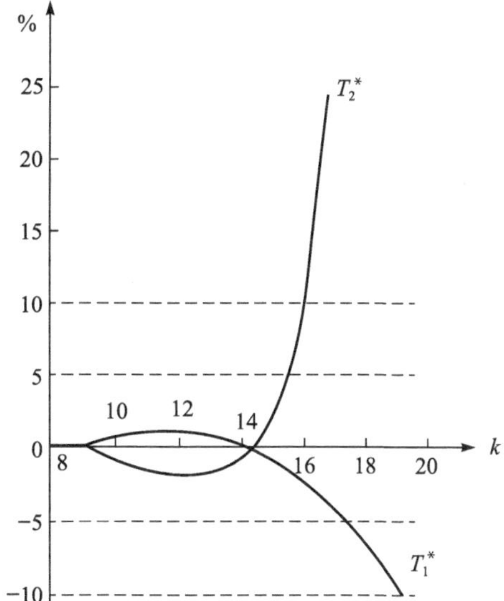

Fig. 4 Relative error between results of the dimensionless critical buckling temperature for second approximation and third approximation (ν=0. 3)

Having T^*, we now obtain the critical buckling temperature t^* by using (2. 9) from which we find

$$t^* = \frac{D}{\mu a^2}\sqrt{\frac{2D}{C(1-\nu^2)}}T^*. \tag{3. 34}$$

We shall now derive the equation for the critical point, i. e. , the point at which the upper and lower critical buckling temperatures are coincided. Then, if the value of k at the point is denoted by k_0, it can be seen that when the geometrical parameter $k < k_0$ no buckling takes place for the shell and when $k \geqslant k_0$ buckling occurs. Thus the critical geometrical parameter k_0 is used to distinguish between buckling and nobuckling for the shell.

For the second approximation, let the discriminant of the quadratic equation (3. 31a) be zero, then we obtain the formula for the critical point

$$k_0 = \sqrt{\frac{2\lambda_1}{\gamma_1}}. \tag{3.35}$$

From the formula (3. 35), it is seen that the dimensionless critical geometrical parameter k_0 depends only on the magnitude of Poisson's ratio ν. For $\nu=0.3$, we obtain

$$k_0 = 4.499.$$

For the third approximation, we have following nonlinear equations for k_0 and W_0

$$\begin{aligned} \frac{\mathrm{d}T}{\mathrm{d}W_0} &= \sum_{i=1}^{9} im_i W_0^{i-1} = 0, \\ \frac{\mathrm{d}^2 T}{\mathrm{d}W_0^2} &= \sum_{i=2}^{9} i(i-1)m_i W_0^{i-2} = 0. \end{aligned} \tag{3.36}$$

Numnerical solution of this system of equations can be obtained by Newton's method. Here for $\nu=0.3$ initial values of k_0 and W_0 may be taken on the basis of results for the second appoximation. Then, we find

$$k_0 = 4.430.$$

We observe that the relative error between these two k_0 is only -1.6%.

Having k_0, we can readily calculate the critical arch height f_0 of the shell by the expression

$$f_0 = \sqrt{\frac{2D}{C(1-\nu^2)}}k_0. \tag{3.37}$$

Next, we introduce the notation for the dimensionless average temperature

$$T_c^* = \frac{1}{2}(T_1^* + T_2^*). \tag{3.38}$$

This is an important parameter in the design. Using the solution (3. 33), we find

$$T_c^* = 2\lambda_1 k. \tag{3.39}$$

Having T_c^*, we can calculate the average temperature

$$t_c^* = \frac{D}{\mu a^2}\sqrt{\frac{2D}{C(1-\nu^2)}}T_c^* = \eta\frac{f}{a^2}, \tag{3.40}$$

where

$$\eta = \frac{4\{E_1[h_0^3 - (h_0 - h_1)^3] + E_2[(h-h_0)^3 + (h_0 - h_1)^3]\}}{3E_1 h_1(2h_0 - h_1)(\alpha_1 - \alpha_2)}. \tag{3.41}$$

It is seen from the formula (3. 40) the average temperature t_c^* of a bimetallic closed shallow spherical shell is independent of Poisson's ν.

Finally, we introduce the notation for the dimensionless snapping deflection of instability (see Fig. 5), i. e., we put

$$L = W_0' - W_{0,1}^*, \tag{3.42}$$

where W_0' is a root of the following equations:

for the second approximation,

$$W_0^3 - 3kW_0^2 + \left(\frac{\lambda_1}{\gamma_1} + 2k^2\right)W_0 - \frac{T_1^*}{\gamma_1} = 0; \tag{3.43a}$$

for the third approximation,

$$\sum_{i=1}^{9} m_i W_0^i - T_1^* = 0. \qquad (3.43b)$$

Solving the cubic equation (3.43a), we have

$$W_0' = k + \sqrt[3]{\frac{2(T_1^* - T_2^*)}{\gamma_1}}. \qquad (3.44)$$

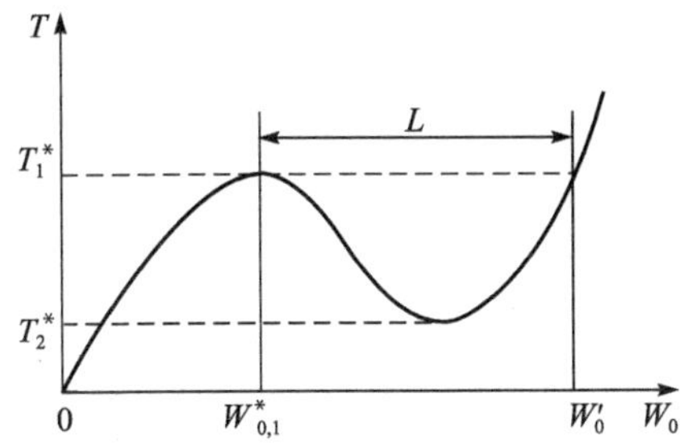

Fig. 5　The dimensionless snapping deflection L

Solving equation (3.43b) by Newton's method, we can easily obtain a more exact solution for W_0'. Then using (3.42), we have numerical results of these two solutions for $\nu=0.3$ (Fig. 6). The relative error is shown in Fig. 7. From these two figures, it can be concluded once again that the second approximation has a very satisfactory accuracy.

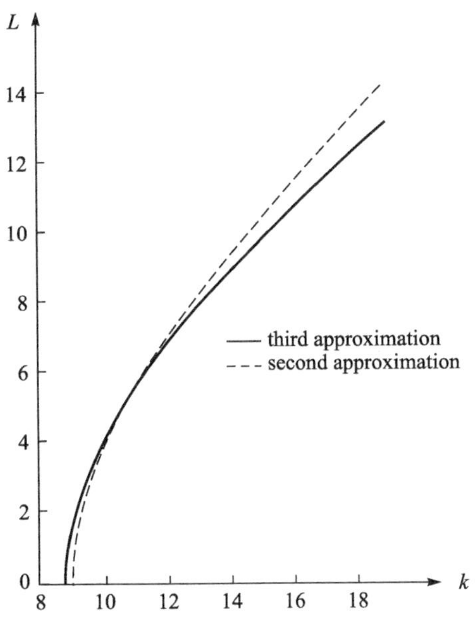

Fig. 6　Curves of the dimensionless snapping distance L ($\nu=0.3$).

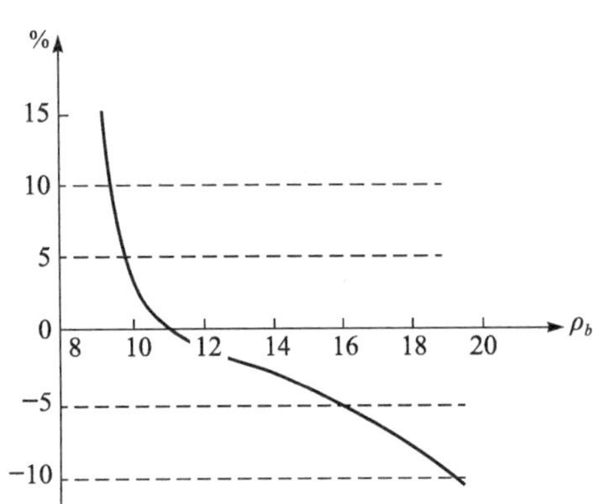

Fig. 7　Relative error between results of the dimensionless snapping distance for second approximation and third approximation ($\nu=0.3$)

Having the expression (3.42) for the dimensionless snapping deflection L, we can obtain the snapping deflection

$$l = \sqrt{\frac{2D}{C(1-\nu^2)}} L. \qquad (3.45)$$

To compare the present results with Wittrick[2] and Lo et al.[4], we now consider the upper critical buckling temperature T_1^*. From Table 1 it may be seen that the present results appear to be in good agreement with their results. However, the present formulas are much simpler than the formulae in [2] and [4].

Table 1 Comparison between the upper critical buckling temperature T_1^* in this paper and references [2] and [4]

$$\nu=\frac{1}{3}$$

	k	4.582	5.105	5.657	6.237	7.159	7.810
T_1^*	[2]	12.09	13.65	17.06	20.43	26.60	31.47
	[4]	11.93	14.27	16.68	19.99	26.68	32.30
	second approximation	11.93	13.90	16.59	20.03	26.88	32.81
	third approximation	11.97	14.03	16.79	20.25	26.93	32.41

From the discussion of this section, it follows that in the investigation of nonlinear thermal stability of a bimetallic shallow shell although the accuracy of the solution for the third approximation is very high, in even the simplest case it is also more complicated and the solution for the second approximation is not only more simple but also more accurate. Hence, for practical purposes we shall find the formulas of the following shells by limiting the solution to the second approximation.

4. Solution for Bimetallic Shallow Spherical Shell With a Circular Hole at the Center

Now consider a bimetallic shallow spherical shell with a circular hole at the center. Using the preceding nonlinear boundary value problems (2.10-2.11) and (3.2-3.4), we easily obtain the following nonlinear boundary value problem for the shell

$$\begin{aligned} &H(\rho\phi)=S\phi, \\ &H(\rho S)=-(\phi^2-4k^2\rho^2); \end{aligned} \tag{4.1}$$

$$W=0,\ \frac{d\phi}{d\rho}+\nu\frac{\phi}{\rho}=T-2\lambda_1 k,\ S=0 \quad \text{at } \rho=1, \tag{4.2a,b,c}$$

$$\frac{d\phi}{d\rho}+\nu\frac{\phi}{\rho}=T-2\lambda_1 k,\ S=0 \quad \text{at } \rho=\rho_b. \tag{4.3a,b}$$

As before, by using the modified iteration method we can also obtain the solution of the problem.

Now, let us introduce, as an iteration parameter, a notation of the dimensioness inner edge deflection

$$W_b=-\sqrt{\frac{C(1-\nu^2)}{2D}}w\,|_{r=b}. \tag{4.4}$$

Then using (3.5a) and (4.2a), we find

$$W_b=\int_{\rho_b}^{1}(\phi+2k\rho)\,d\rho. \tag{4.5}$$

For the first approximation, we have the linear boundary value problem as follows

$$\begin{aligned} &H(\rho\phi_1)=0, \\ &H(\rho S_1)=-(\phi_1^2-4k^2\rho^2); \end{aligned} \tag{4.6}$$

$$W_1=0,\ \frac{\mathrm{d}\phi_1}{\mathrm{d}\rho}+\nu\frac{\phi_1}{\rho}=T-2\lambda_1 k,\ S_1=0 \quad \text{at } \rho=1;$$

$$\frac{\mathrm{d}\phi}{\mathrm{d}\rho}+\nu\frac{\phi_1}{\rho}=T-2\lambda_1 k,\ S_1=0 \quad \text{at } \rho=\rho_b. \tag{4.7}$$

Using (4.5), the solutions of the problem are

$$\phi_1=(\beta_0 W_b-2k)\rho,$$

$$S_1=\frac{1}{8}(4k\beta_0 W_b-\beta_0^2 W_b^2)[\rho_b^2\rho^{-1}-(1+\rho_b^2)\rho+\rho^3], \tag{4.8}$$

where

$$\beta_0=\frac{2}{1-\rho_b^2}. \tag{4.9}$$

For the second approximation, we have the following linear boundary value problem for ϕ

$$H(\rho\phi_2)=S_1\phi_1; \tag{4.10}$$

$$W_2=0,\ \frac{\mathrm{d}\phi_2}{\mathrm{d}\rho}+\nu\frac{\phi_2}{\rho}=T-2\lambda_1 k \quad \text{at } \rho=1; \tag{4.11a,b}$$

$$\frac{\mathrm{d}\phi_2}{\mathrm{d}\rho}+\nu\frac{\phi_2}{\rho}=T-2\lambda_1 k \quad \text{at } \rho=\rho_b. \tag{4.12}$$

Using (4.8), under the boundary conditions (4.11b) and (4.12), the solution of Eq. (4.10) is

$$\begin{aligned}\phi_2=&\frac{1}{\lambda_1}(T-2\lambda_1 k)\rho-\frac{1}{16}(8k^2\beta_0 W_b-6k\beta_0^2 W_b^2\\&+\beta_0^3 W_b^3)\left\{\frac{\beta_0\rho_b^2}{2\lambda_2}\left(\frac{2+\nu}{6}+\lambda_1\rho_b^2\ln\rho_b-\frac{2+\nu}{6}\rho_b^4\right)\rho^{-1}\right.\\&+\frac{1}{2}\left[\frac{\beta_0}{\lambda_1}\left(\frac{2+\nu}{6}+\frac{1+3\nu}{4}\rho_b^2-\frac{1+3\nu}{4}\rho_b^4+\lambda_1\rho_b^4\ln\rho_b-\frac{2+\nu}{6}\rho_b^6\right)-\rho_b^2\right]\rho\\&\left.+\rho_b^2\rho\ln\rho-\frac{1}{4}(1+\rho_b^2)\rho^3+\frac{1}{12}\rho^5\right\},\end{aligned} \tag{4.13}$$

where

$$\lambda_2=1-\nu. \tag{4.14}$$

Substituting (4.13) into (4.5), we obtain the nonlinear characteristic relation of the bimetallic shallow spherical shell with a circular hole at the center

$$T=\beta_1\left[\left(\frac{\lambda_1}{\beta_1}+8k^2\right)\beta_0 W_b-6k\beta_0^2 W_b^2+\beta_0^3 W_b^3\right], \tag{4.15}$$

where

$$\begin{aligned}\beta_1=&\frac{\beta_0^2}{32\lambda_2}\left(\frac{17+5\nu}{144}\lambda_2-\frac{5+2\nu}{9}\lambda_1\rho_b^2-\frac{2+\nu}{6}\lambda_1\rho_b^2\ln\rho_b+\frac{7+3\nu}{8}\lambda_2\rho_b^4\right.\\&\left.-\lambda_1^2\rho_b^4\ln^2\rho_b-\frac{5+2\nu}{9}\lambda_2\rho_b^6+\frac{2+\nu}{6}\lambda_1\rho_b^6\ln\rho_b+\frac{17+5\nu}{144}\lambda_2\rho_b^8\right).\end{aligned} \tag{4.16}$$

As ρ_b becomes infinitely small, the relation (4.15) coincides with the relation (3.19) of the bimetallic closed shallow spherical shell.

We obtain the necessary equation for determining the critical inner edge deflection W_b by equating the derivative of expression (4.15) with respect to W_b to zero, which gives

$$3\beta_0^2 W_b^2 - 12k\beta_0 W_b + \left(\frac{\lambda_1}{\beta_1} + 8k^2\right) = 0. \tag{4.17}$$

Solving this quadratic equation, we obtain

$$W_b^* = \frac{1}{\beta_0}\left[2k \pm \sqrt{\frac{1}{3}\left(4k^2 - \frac{\lambda_1}{\beta_1}\right)}\right], \tag{4.18}$$

where the meaning of the signs is the same as those in the formula (3.33).

Substituting (4.18) into (4.15), we have the following formula for the dimensionless critical buckling temperature

$$T^* = \beta_1\left[\left(\frac{\lambda_1}{\beta_1} + 8k^2\right)\beta_0 W_b^* - 6k\beta_0^2 W_b^{*2} + \beta_0^3 W_b^{*3}\right]. \tag{4.19}$$

Calculated results of the dimensionless critical buckling temperature for $\nu=0.3$ and several values of k are shown in Fig. 8.

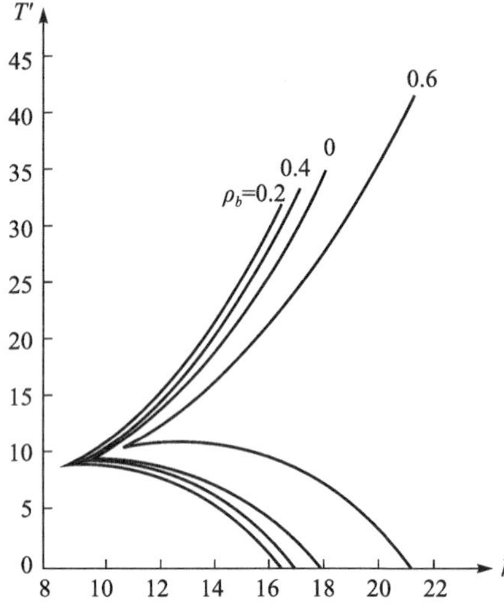

Fig. 8 Curves of the dimensionless critical buckling temperature ($\nu=0.3$)

Now, let us derive the equation for the critical point. Let the discriminant of the quadratic equation (4.17) be zero as before. We obtain the formula for the critical point

$$k_0 = \frac{1}{2}\sqrt{\frac{\lambda_1}{\beta_1}}. \tag{4.20}$$

It can be seen from the formula (4.20) that if we know the dimensionless upper radius ρ_b and Poisson's ratio ν of a bimetallic shallow spherical shell with a circular hole at the center, we can at once obtain the critical geometrical parameter k_0 of the shell. These values of k_0 for $\nu=0.3$ are graphically represented in Fig. 9. It is seen that the curve is bent convex downward. Then this curve has a minimum value of k_0. A corresponding shell for this value is most sensitive to instability. To find the minimum numerical value of k_0, we equate to zero the derivative with respect to ρ_b of the expression (4.20)

$$\frac{\mathrm{d}k_0}{\mathrm{d}\rho_b} = 0. \tag{4.21}$$

By the application of the iteration method we find the numerical solution of Eq. (4.21) for $\nu=0.3$

$$\rho_b = 0.2655.$$

Substituting this value into (4.20), we obtain a minimum value of k_0

$$(k_0)_{\min} = 4.088.$$

A corresponding critical buckling temperature is

$$T_0^* = 10.63.$$

To find the average temperature of the shell, we proceed as was explained in the preceding section. The results are exactly the same with formulae (3.39) and (3.40). It may be seen that the average temperature t_c^* of a shell is independent of Poisson's ratio ν and the extent of the central opening of the shell.

By using the same method we can obtain also the formula for the dimensionless snapping deflection L of the shell.

$$L = W_b{}' - W_{b,1}^*, \tag{4.22}$$

where

$$W_b{}' = \frac{1}{\beta_0}\left[2k + \sqrt[3]{\frac{2}{\beta_1}(T_1^* - T_2^*)}\right]. \tag{4.23}$$

This means that we can plot a set of curves giving L in terms of k_0 for $\nu=0.3$, each curve in the set corresponding to a particular value of ρ_b. Such curves are given in Fig. 10. It is seen that these curves rise monotonically.

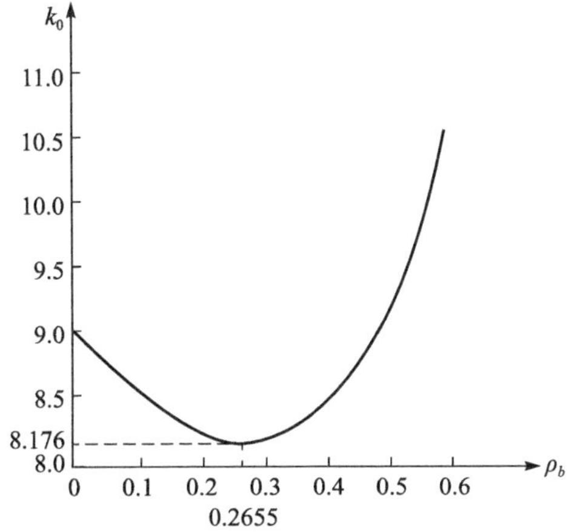

Fig. 9 Curve of the critical geometrical parameter k_0 ($\nu=0.3$)

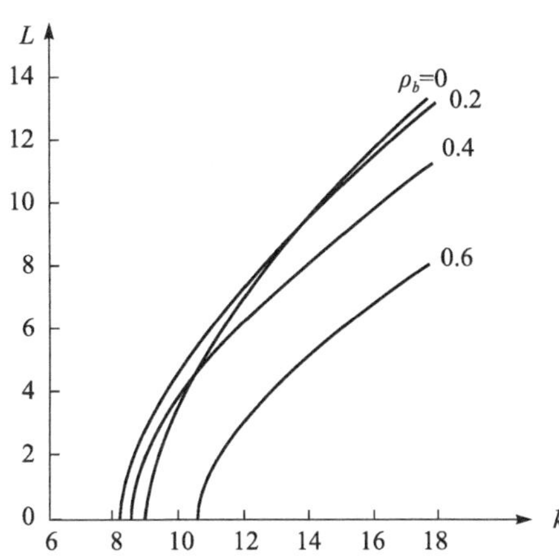

Fig. 10 Curves of the dimensionless snapping distance L ($\nu=0.3$)

5. Solution for Bimetallic Truncated Shallow Conical Shell

Now consider a bimetallic truncated shallow conical shell. We may transform the problem (2.10) and (2.11) into the dimensionless nonlinear boundary value problem for the bimetallic truncated shallow conical shell by introducing, instead of θ, the slope angle of the shallow conical shell

$$\theta \approx \frac{f}{a}. \tag{5.1}$$

Then we obtain

$$H(\rho\psi)=S\psi+\frac{k}{\rho},\quad H(\rho S)=k^2-\psi^2;\tag{5.2a,b}$$

$$W=0,\ \frac{\mathrm{d}\psi}{\mathrm{d}\rho}+\frac{\nu}{\rho}(\psi+k)=T,\ S=0\quad \text{at } \rho=1;\tag{5.3a,b,c}$$

$$\frac{\mathrm{d}\psi}{\mathrm{d}\rho}+\frac{\nu}{\rho}(\psi+k)=T,\ S=0\quad \text{at } \rho=\rho_b,\tag{5.4a,b}$$

where we employ the notation

$$\psi=-\left(\frac{\mathrm{d}W}{\mathrm{d}\rho}+k\right).\tag{5.5}$$

A more accurate solution of this problem is given by means of the modified iteration method. As an iteration parameter, introducing the notation for the dimensionless top edge deflection of the shell

$$W_b=-\sqrt{\frac{C(1-\nu^2)}{2D}}w\,|_{r=b},\tag{5.6}$$

then using (5.3a) and (5.5), we represent the formula for W_b in the following form:

$$W_b=\int_{\rho_b}^{1}(\psi+k)\mathrm{d}\rho.\tag{5.7}$$

For the first approximation, we have the following linear boundary value problem

$$H(\rho\psi_1)=\frac{k}{\rho},\quad H(\rho S_1)=k^2-\psi_1^2;\tag{5.8}$$

$$W_1=0,\ \frac{\mathrm{d}\psi_1}{\mathrm{d}\rho}+\frac{\nu}{\rho}(\psi_1+k)=T,\ S_1=0\quad \text{at } \rho=1;$$

$$\frac{\mathrm{d}\psi_1}{\mathrm{d}\rho}+\frac{\nu}{\rho}(\psi_1+k)=T,\ S_1=0\quad \text{at } \rho=\rho_b.\tag{5.9}$$

Introducing the notation

$$\delta_0=\frac{2}{1-\rho_b^2}\tag{5.10}$$

and using (5.7), the solution of the problem (5.8) and (5.9) can be written in the following form

$$\begin{aligned}\psi_1&=\frac{1}{\lambda_1}\delta_0W_b\rho-k,\\ S_1&=\frac{\rho_b^2}{\lambda_1}\left[\frac{2k}{3(1+\rho_b)}\delta_0W_b-\frac{1}{8\lambda_1}\delta_0^2W_b^2\right]\frac{1}{\rho}\\ &\quad-\frac{1}{\lambda_1}\left[\frac{2k(1+\rho_b+\rho_b^2)}{3(1+\rho_b)}\delta_0W_b-\frac{1+\rho_b^2}{8\lambda_1}\delta_0^2W_b^2\right]\rho\\ &\quad+\frac{2k}{3\lambda_1}\delta_0W_b\rho^2-\frac{1}{8\lambda_1^2}\delta_0^2W_b^2\rho^3.\end{aligned}\tag{5.11}$$

For the second approximation, we have the following linear boundary value problem for ψ

$$H(\rho\psi_2)=S_1\psi_1+\frac{k}{\rho};\tag{5.12}$$

$$W_2=0,\ \frac{\mathrm{d}\psi_2}{\mathrm{d}\rho}+\frac{\nu}{\rho}(\psi_2+k)=T\quad \text{at } \rho=1;\tag{5.13a,b}$$

$$\frac{\mathrm{d}\psi_2}{\mathrm{d}\rho}+\frac{\nu}{\rho}(\psi_2+k)=T\qquad \text{at } \rho=\rho_b.\tag{5.14}$$

Using (5.11), the solution of the problem (5.12), (5.13b) and (5.14) gives

$$\begin{aligned}\psi_2=&\frac{k\rho_b^2}{\lambda_1}\left[\frac{2k}{3(1+\rho_b)}\delta_0W_b-\frac{1}{8\lambda_1}\delta_0^2W_b^2\right]-k-\frac{\rho_b^2}{\lambda_1\lambda_2(1-\rho_b^2)}\left\{\frac{k^2}{3}\left[\frac{7+5\nu}{12}-\frac{4+2\nu}{3}\rho_b\right.\right.\\&\left.+\frac{3+\nu}{4}\rho_b^2-\frac{2\nu}{1+\rho_b}\rho_b+\frac{4+8\nu}{3(1+\rho_b)}\rho_b^2-\frac{4+2\nu}{3(1+\rho_b)}\rho_b^3\right]\delta_0W_b\\&-\frac{k}{3\lambda_1}\left(\frac{22+13\nu}{60}-\frac{1+2\nu}{4}\rho_b+\frac{1+2\nu}{4}\rho_b^2-\frac{22+13\nu}{60}\rho_b^3+\frac{1}{1+\rho_b}\lambda_1\rho_b^2\ln\rho_b\right)\delta_0^2W_b^2\\&\left.+\frac{1}{16\lambda_1^2}\left(\frac{2+\nu}{6}+\lambda_1\rho_b^2\ln\rho_b-\frac{2+\nu}{6}\rho^4\right)\delta_0^3W_b^3\right\}\frac{1}{\rho}\\&+\frac{1}{\lambda_1}\left\{T-\frac{1}{\lambda_1+\rho_b^2}\left[\frac{k^2}{3}\left(\frac{7+5\nu}{12}+(2+4\nu)\rho_b^2-\frac{8+10\nu}{3}\rho_b^3+\frac{3+\nu}{4}\rho_b^4-\frac{2\nu}{1+\rho_b}\rho_b^3\right.\right.\right.\\&\left.+\frac{4+8\nu}{3(1+\rho_b)}\rho_b^4-\frac{4+2\nu}{3(1+\rho_b)}\rho_b^5\right)\delta_0W_b\\&-\frac{k}{3\lambda_1}\left(\frac{22+13\nu}{60}+\frac{3\nu}{4}\rho_b^3-\frac{3\nu}{4}\rho_b^3-\frac{22+13\nu}{60}\rho_b^5+\frac{1}{1+\rho_b}\lambda_1\rho_b^4\ln\rho_b\right)\delta_0^2W_b^2\\&\left.\left.+\frac{1}{16\lambda_1^2}\left(\frac{2+\nu}{6}-\frac{1}{4}\lambda_2\rho_b^2+\frac{1}{4}\lambda_2\rho_b^4+\lambda_1\rho_b^4\ln\rho_b-\frac{2+\nu}{6}\rho_b^6\right)\delta_0^3W_b^3\right]\right\}\rho\\&+\frac{\rho_b^2}{\lambda_1^2}\left[\frac{k}{3(1+\rho_b)}\delta_0^2W_b^2-\frac{1}{16\lambda_1}\delta_0^3W_b^3\right]\rho\ln\rho\\&+\frac{k}{3\lambda_1}\left[\frac{2k(1+\rho_b+\rho_b^2)}{3(1+\rho_b)}\delta_0W_b-\frac{1+\rho_b^2}{8\lambda_1}\delta_0^2W_b^2\right]\rho^2\\&-\frac{1}{8\lambda_1}\left[\frac{2k^2}{3}\delta_0W_b+\frac{2k(1+\rho_b+\rho_b^2)}{3\lambda_1(1+\rho_b)}\delta_0^2W_b^2-\frac{1+\rho_b^2}{8\lambda_1^2}\delta_0^3W_b^3\right]\rho^3\\&+\frac{19k}{360\lambda_1^2}\delta_0^2W_b^2\rho^4-\frac{1}{192\lambda_1^3}\delta_0^3W_b^3\rho^5.\end{aligned}\tag{5.15}$$

Substituting (5.15) into (5.7), we obtain the nonlinear characteristic relation of the bimetallic truncated shallow conical shell

$$T=(\delta_0+k^2\delta_1)W_b-k\delta_2W_b^2+\delta_3W_b^3,\tag{5.16}$$

where

$$\delta_1=\frac{\delta_0^3}{8\lambda_1^3\lambda_2(1+\rho_b)}\left[\frac{19+7\nu}{108}\lambda_2+\frac{19+7\nu}{108}\lambda_2\rho_b-\frac{9+5\nu}{4}\lambda_2\rho_b^2\right.$$

$$-\frac{7+5\nu}{9}\lambda_1\rho_b^2\ln\rho_b+\frac{205+121\nu}{108}\lambda_2\rho_b^3+(1+3\nu)\lambda_1\rho_b^3\ln\rho_b+\frac{205+121\nu}{108}\lambda_2\rho_b^4$$
$$-(1+3\nu)\lambda_1\rho_b^4\ln\rho_b-\frac{9+5\nu}{4}\lambda_2\rho_b^5$$
$$+\frac{7+5\nu}{9}\lambda_1\rho_b^5\ln\rho_b+\frac{19+7\nu}{108}\lambda_2\rho_b^6+\frac{19+7\nu}{108}\lambda_2\rho_b^7,$$

$$\delta_2=\frac{\delta_0^4}{4\lambda_1^4\lambda_2(1+\rho_b)}\Big(\frac{133+43\nu}{1800}\lambda_2+\frac{133+43\nu}{1800}\lambda_2\rho_b-\frac{28+13\nu}{50}\lambda_2\rho_b^2$$
$$-\frac{22+13\nu}{90}\lambda_1\rho_b^2\ln\rho_b-\frac{133+43\nu}{1800}\lambda_2\rho_b^3$$
$$-\frac{7-17\nu}{90}\lambda_1\rho_b^3\ln\rho_b+\frac{35+17\nu}{36}\lambda_2\rho_b^4$$
$$-\frac{2}{3}\lambda_1^2\rho_b^4\ln^2\rho_b-\frac{133+43\nu}{1800}\lambda_2\rho_b^5$$
$$+\frac{7-17\nu}{90}\lambda_1\rho_b^5\ln\rho_b-\frac{28+13\nu}{50}\lambda_2\rho_b^6+\frac{22+13\nu}{90}\lambda_1\rho_b^6\ln\rho_b$$
$$+\frac{133+43\nu}{1800}\lambda_2\rho_b^7+\frac{133+43\nu}{1800}\lambda_2\rho_b^8\Big),$$

$$\delta_3=\frac{\delta_0^5}{32\lambda_1^5\lambda_2}\Big(\frac{17+5\nu}{144}\lambda_2-\frac{5+2\nu}{9}\lambda_2\rho_b^2-\frac{2+\nu}{6}\lambda_1\rho_b^2\ln\rho_b+\frac{7+3\nu}{8}\lambda_2\rho_b^4$$
$$-\lambda_1^2\rho_b^4\ln^2\rho_b-\frac{5+2\nu}{9}\lambda_2\rho_b^6+\frac{2+\nu}{6}\lambda_1\rho_b^6\ln\rho_b+\frac{17+5\nu}{144}\lambda_2\rho_b^8\Big). \tag{5.17}$$

In the limiting case where ρ_b is infinitely small, formulae (5.10) and (5.17) simplify

$$\delta_0=2\lambda_1,\ \delta_1=\frac{19+7\nu}{108},\ \delta_2=\frac{133+43\nu}{450},\ \delta_3=\frac{17+5\nu}{144}. \tag{5.18}$$

Substituting (5.18) into (5.16), we obtain the nonlinear characteristic relation of the bimetallic closed shallow conical shell.

As before, in order to find the critical buckling temperature of the shell, we differentiate expression (5.16) with respect to W_b and set the derivative equal to zero, i.e.

$$3\delta_3W_b^2-2k\delta_2W_b+(\delta_0+k^2\delta_1)=0. \tag{5.19}$$

Solving this quadratic equation, the dimensionless critical top edge deflection W_b gives

$$W_b^*=\frac{1}{3\delta_3}\left[k\delta_2\pm\sqrt{k_2(\delta_2^2-3\delta_1\delta_3)-3\delta_0\delta_3}\right], \tag{5.20}$$

where the meaning of the signs is also the same as in the formula (3.32).

Substituting (5.20) into (5.16), we have the following formula for the dimensionless critical buckling temperature

$$T^*=(\delta_0+k^2\delta_1)W_b^*-k\delta_2W_b^{*2}+\delta_3W_b^{*3}. \tag{5.21}$$

Calculated results of the dimensional critical buckling temperature for $\nu=0.3$ and several values of k are shown in Fig. 11.

Now, let us derive the equation for the critical point. Let the discriminant of the quadratic equation (5.19) be zero, we obtain the formula for the critical point

$$k_0 = \sqrt{\frac{3\delta_0\delta_3}{\delta_2^2 - 3\delta_1\delta_3}}. \tag{5.22}$$

The values of k_0 are given by curve in Fig. 12. It is seen that the curve has a minimun value of k_0. To find the minimum numerical value of k_0, as before, we equal to zero the derivative with respect to ρ_b of the expression (5.22)

$$\frac{\mathrm{d}k_0}{\mathrm{d}\rho_b} = 0. \tag{5.23}$$

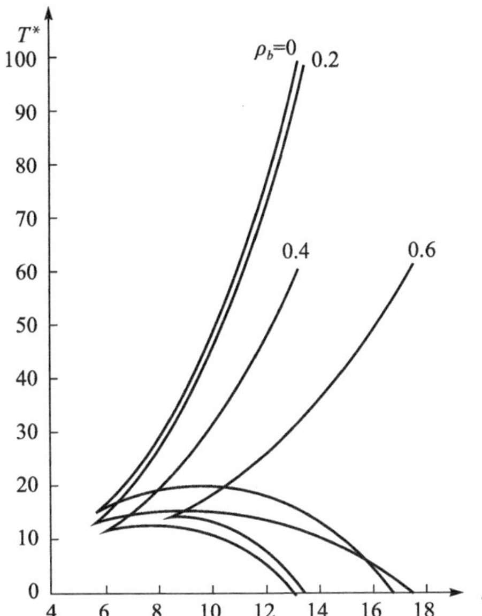

Fig. 11　Curves of the dimensionless critical buckling temperature (ν=0.3)

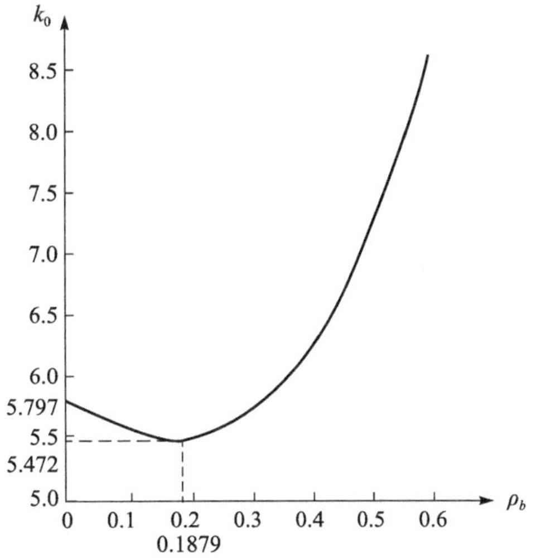

Fig. 12　Curves of the critical geometrical parameter k_0 (ν=0.3)

By the application of the iteration method we find the numerical solution of equation (5.23)

$$\rho_b = 0.1879.$$

Substituting this value into (5.22), we obtain a minimum value of k_0

$$(k_0)_{\min} = 5.472.$$

A corresponding critical buckling temperature is

$$T_0^* = 12.81.$$

Next, as before, we find the dimensionless average temperature T_c^* of the shell

$$T_c^* = \frac{k\delta_2}{3\delta_3}\left[\delta_0 + k^2\left(\delta_1 - \frac{2\delta_2^2}{9\delta_3}\right)\right]. \tag{5.24}$$

Now we can plot a set of curves giving T_c^* in terms of k for ν=0.3, each curve in the set corresponding to a particular value of ρ_b. Such curves are given in Fig. 13. It is seen from this figure that these curves rise monotonically, and for the same value of k the dimensionless average temperature T_c^* of a shell with large ρ_b is low. Finally, we obtain the dimensionless snapping deflection L of the shell

$$L = W_b' - W_{b,1}^*, \tag{5.25}$$

where

$$W_b' = \frac{k\delta_2}{3\delta_3} + 2\sqrt[3]{\frac{T_1^* - T_2^*}{4\delta_3}}. \tag{5.26}$$

The dimensionless snapping deflection L obtained for $\nu=0.3$ and several values of ρ_b are shown in Fig. 14. It is seen that these curves rise monotonically, and for the same value of k, the dimensionless snapping deflection induced in a shell diminishes mostly with the increase of ρ_b.

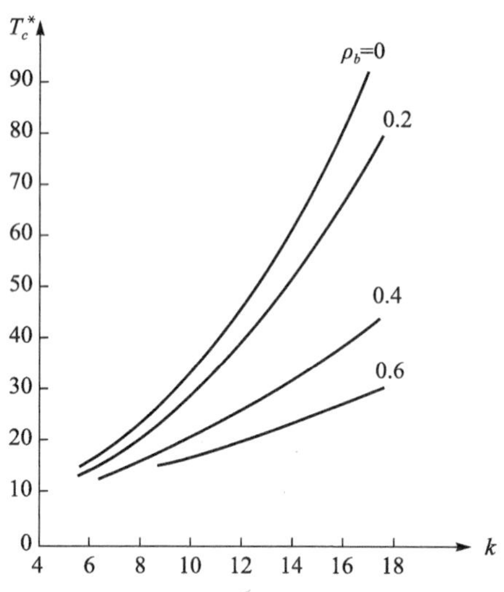

Fig. 13　Curves of the dimensionless average temperature ($\nu=0.3$)

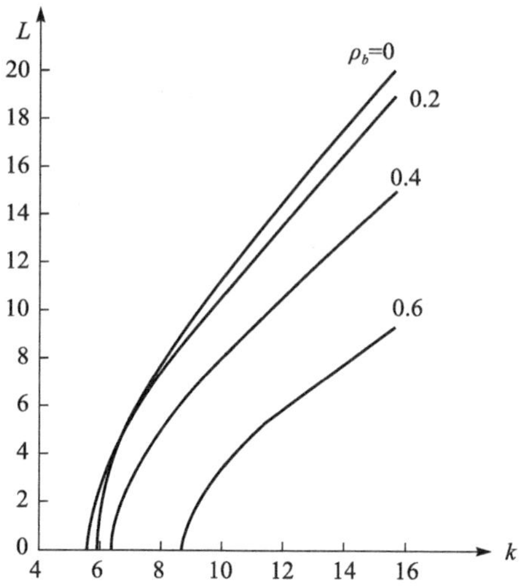

Fig. 14　Curves of the dimensionless snapping distance L ($\nu=0.3$)

Acknowledgements

This paper was written in Ruhr-University Bochum, Federal Republic of Germany. The author is grateful to the Alexander Von Humboldt Foundation, Ruhr-University Bochum and Professor W. Zerna for his help and hearty hospitality.

References

[1] D. Y. Panov, On the stability of a bimetallic membrane on heating, *Prikl. Mat. Mekh.* 11, 603 (in Russian) (1947).

[2] W. H. Wittrick, D. M. Myers, W. R. Blunden, Stability of a bimetallic disk, *Q. J. Mech. Appl. Math.*, 6, 15 (1953).

[3] E. I. Grigolyuk, Thin bimetallic shells and plates, *Engineer's Collection* 18, (in Russian) (1953).

[4] Z. D. Lo, D. Y. Nie, H. D. Liu, et al., Thermal stability of bimetallic shallow spherical shell, *Acta Mech. Sin.*, 9, 1 (in Chinese) (1966).

[5] V. I. Korolev, Thin two-layer plates and shells, *Engineer's Collection*, 22, 98 (in Russian)

(1955).

[6] P. P. Radkovsky, Buckling of thin single-and multi-layer conical and cylindrical shells with rotationally symmetric stresses, *Proc. 3rd U. S. Nati. Congr. Appl. Mech.*, p. 443 (1958).

[7] K. Y. Yeh, Liu Renhuai, Q. Y. Ping, et al., Nonlinear stability of thin circular shallow spherical shell under actions of axisymmetric uniform distributed line loads, *Bull. Sci.*, 2, 142 (in Chinese) (1965).

[8] K. Y. Yeh, Liu Renhuai, C. Z. Zhang, et al., Nonlinear stability of thin circular shallow spherical shell under the action of uniform edge moment, *Bull. Sci.*, 2, 145 (in Chinese) (1965).

[9] Liu Renhuai, Nonlinear stability of circular shallow spherical shell with a hole in the center under the action of uniform moment at the inner edge, *Bull. Sci.*, 3, 253 (in Chinese) (1965).

[10] W. C. Chien, Large deflection of a circular clamped plate under uniform pressure, *Chin. J. Phys.*, 7, 102 (1947).

[11] Liu Renhuai, Axisymmetrical buckling of thin circular shallow spherical shell with a circular hole at the center by shearing forces uniformly distributed along the inner edge, *Mechanics*, 3, 206 (in Chinese) (1977).

[12] Liu Renhuai, The characteristic relations of corrugated circular plates, *Acta Mech. Sin.*, 1, 47 (in Chinese) (1978).

[13] Liu Renhuai, The characteristic relations of corrugated circular plates with plane center region, *J. China Univ. Sci. Technol.*, 9, 75 (in Chinese) (1979).

[14] K. Y. Yeh, Liu Renhuai, C. Z. Zhang, et al., Nonlinear stability of thin elastic circular shallow spherical shell under the action of uniform edge moment, *Appl. Math. Mech.*, 1, 71 (1980).

[15] Liu Renhuai, Nonlinear axisymmetrical bending of circular plates under the action of uniform edge moment, *J. China Univ. Sci. Technol.*, 10, 56 (in Chinese) (1980).

[16] Liu Renhuai, Nonlinear bending of circular sandwich plates, *Appl. Math. Mech.*, 2 (2), 189 (1981).

刘人怀院士文集

刘人怀◎著

第二卷

科学出版社

北京

内 容 简 介

本文集由著者在力学和管理科学领域60多年中所发表的文章汇编而成，分为上、下篇。上篇是力学部分，结合工程需要，对单层板壳、波纹板壳、双层板壳、夹层板壳、复合材料层合板壳和网格扁壳等六类板壳的非线性弯曲、稳定和振动问题，以及厚、薄板壳弯曲问题进行理论探索。下篇是管理科学部分，联合实际，对工程管理、公共管理、工商管理、科技管理和教育管理的问题进行了研究。

本文集可作为高校力学和管理科学专业，以及相关专业老师、研究生、本科生的参考书，适合政府、高等学校、科研院所、科技社团、企业等领域的领导、管理人员和科研人员参考阅读，又可作为航天、航空、航海、机械、建筑和交通等工程设计师与工程师的设计制造指导书。

GS京（2025）0637号

图书在版编目（CIP）数据

刘人怀院士文集 / 刘人怀著. -- 北京 : 科学出版社，2025. 4. -- ISBN 978-7-03-080897-4

Ⅰ. O3-53；C93-53

中国国家版本馆CIP数据核字第2024JH1723号

责任编辑：陈会迎/ 责任校对：姜丽策

责任印制：张 伟/ 封面设计：有道设计

科 学 出 版 社出版

北京东黄城根北街16号

邮政编码：100717

http://www.sciencep.com

北京中科印刷有限公司印刷

科学出版社发行　　各地新华书店经销

*

2025年4月第 一 版　开本：787×1092 1/16

2025年4月第一次印刷　印张：153 3/4

字数：3 646 000

定价：698.00元（全五卷）

（如有印装质量问题，我社负责调换）

目　录

第五章　夹层板壳非线性力学 ······ 399

在边缘力矩作用下夹层圆板的非线性轴对称弯曲问题 ······ 399

夹层圆板的非线性弯曲 ······ 410

夹层圆板大挠度问题的精确解 ······ 428

Nonlinear bending of circular sandwich plates under the action of axisymmetric uniformly distributed line loads ······ 441

夹层圆板大挠度问题的进一步研究 ······ 464

夹层矩形板的非线性振动 ······ 472

简支夹层矩形板的非线性弯曲 ······ 485

On the non-linear buckling of circular shallow spherical sandwich shells under the action of uniform edge moments ······ 500

Nonlinear vibration of shallow conical sandwich shells ······ 512

复合材料面层夹层板中转动一致有效理论 ······ 525

夹层环形板的非线性弯曲 ······ 531

Nonlinear theory of sandwich shells part Ⅰ: Exact kinematics of moderately thick shells ······ 541

Nonlinear theory of sandwich shells part Ⅱ: Approximate theories ······ 558

On nonlinear stability of shallow conical sandwich shells ······ 575

考虑横向剪应力连续的复合材料面层夹层壳非线性一致有效理论 ······ 589

Large deflection of annular sandwich plates ······ 599

第六章　复合材料层合板壳非线性力学 ······ 611

四边简支对称正交层合矩形板的非线性弯曲问题 ······ 611

对称圆柱正交异性层合扁球壳的非线性稳定问题 ······ 618

考虑横向剪切的对称层合圆柱正交异性扁球壳的非线性稳定问题 ······ 626

A simple theory for non-linear bending of laminated composite rectangular plates including higher-order effects ······ 636

考虑横向剪切的对称层合圆柱正交异性中心开孔扁球壳的非线性屈曲 ······ 647

Non-linear buckling of symmetrically laminated, cylindrically orthotropic, shallow, conical shells considering shear ······ 660

On non-linear buckling of symmetrically laminated, cylindrically orthotropic, truncated, shallow, spherical shells under uniform pressure including shear effects ······ 672

复合材料层合扁锥壳的非线性稳定问题 ······ 687

复合材料层合扁球壳的非线性强迫振动 ······ 693

Large deflection bending of symmetrically laminated rectilinearly orthotropic elliptical plates including transverse shear …… 699
Nonlinear dynamic buckling of symmetrically laminated cylindrically orthotropic shallow spherical shells …… 715
Nonlinear stability of symmetrically laminated cylindrically orthotropic truncated shallow conical shells including transverse shear …… 725
复合材料层合板壳非线性力学的研究进展 …… 741
第七章　网格扁壳非线性力学 …… 769
网格扁壳的非线性弯曲理论 …… 769
网格扁壳的非线性自由振动分析 …… 776
Non-linear buckling of squarely-latticed shallow spherical shells …… 781
双层正交正放网格扁壳结构的非线性弯曲理论 …… 802
矩形底双层网格扁壳的非线性弯曲 …… 807
矩形底双层网格扁壳的非线性屈曲 …… 812
第八章　厚、薄板壳弯曲分析 …… 818
高压聚乙烯反应器厚壁筒体径向开孔的应力计算 …… 818
尿素合成塔底部球形封头开孔的应力计算 …… 831
铂重整装置反应器椭球封头中心开孔接管的强度问题 …… 842
500 万吨/年常减压装置减压塔下端部分壳体的应力分析 …… 854
在轴向压力与均匀外压力共同作用下薄壁截头圆锥形壳的稳定性 …… 872
加氢反应器顶部厚壁壳体的应力分析 …… 879
双锥密封中的内力分析 …… 897
厚壁圆柱壳轴对称变形近似理论的应力公式 …… 900
双层套箍式厚壁压力容器环沟部位的应力状态 …… 902
厚壁球壳的弯曲理论及其在高压容器上的应用 …… 916
厚管板的设计 …… 927
固定式厚管板的弯曲问题 …… 945
层合圆薄板的轴对称弯曲问题 …… 958
焦炭塔鼓胀与开裂变形机理及疲劳断裂寿命预测的研究进展 …… 964
膜盒基体的理论与设计 …… 974

第五章　夹层板壳非线性力学

在边缘力矩作用下夹层圆板的非线性轴对称弯曲问题[①]

符　号

r，θ，z	圆柱坐标系
a	夹层圆板的半径
t	表层厚度
h_0	夹层圆板的总厚度
E，ν	表层材料的弹性模量和泊松比
G_2	夹心剪切模量
D	夹层圆板的抗弯刚度
M	边缘均布力矩
u_i，ν_i，$w_{i(i=1,2,3)}$	上表层、夹心和下表层的径向、环向和法向位移
u，w	夹层圆板中面上点的径向位移和挠度
ψ	夹心中面法线在 rz 平面内的转角
ε_{ri}，$\varepsilon_{\theta i}$，ε_{zi}，$\gamma_{r\theta i}$，$\gamma_{\theta zi}$，$\gamma_{rzi(i=1,2,3)}$	上表层、夹心和下表层点的伸长和剪切应变分量
σ_{ri}，$\sigma_{\theta i}$，σ_{zi}，$\tau_{r\theta i}$，$\tau_{\theta zi}$，$\tau_{rzi(i=1,2,3)}$	上表层、夹心和下表层点的正应力和剪应力分量
σ_{r0}，$\sigma_{\theta 0}$	夹层圆板中面内的径向和环向应力
$U_{i(i=1,2,3)}$	上表层、夹心和下表层的应变能
V	外力功
U	夹层圆板的总势能
M_r，Q_r	夹层圆板的径向弯矩和横向力
φ	应力函数
ρ	无量纲径向坐标
k	夹层圆板的无量纲特征参数
W，W_0	夹层圆板中面上点的无量纲挠度及其中心值
S_r，S_r（0）	夹层圆板中面内的无量纲径向应力及其中心值
S_θ，S_θ（0），S_θ（1）	夹层圆板中面内的无量纲环向应力及其中心和边缘值
m	无量纲边缘均布力矩

① 本文原载《中国科学技术大学学报》，1980，10（2）：56-67.

$\alpha_1, \cdots, \alpha_9, \lambda$　辅助量

L　微分算子

一、引言

在航空工业中，夹层板是使用广泛且很重要的结构元件。在夹层圆板方面，已有一些人研究[1-4]。但关于其非线性弯曲方面的问题，却很少有人探讨。这主要是因为非线性数学问题的求解存在较大困难。

本文考虑由两层平面刚度较大、厚度较小的各向同性表层和一层材料较软、厚度较大的各向同性夹心所组成的夹层圆形薄板。应用 Reissner 的夹层板理论[5,6]中的假定，我们导出了这种夹层圆板在边缘力矩作用下的非线性轴对称弯曲理论的平衡方程和边界条件，并使用修正迭代法[7-11]进行了求解，获得了较精确的解析公式。所得结果可作为工程设计的有益参考。

二、基本方程

今考虑一总厚度为 h_0（由于表层很薄，故 h_0 以两表层中面间的距离来量度）、半径为 a 的夹层面板，在边缘上有均布力矩 M 作用，如图 1 所示。

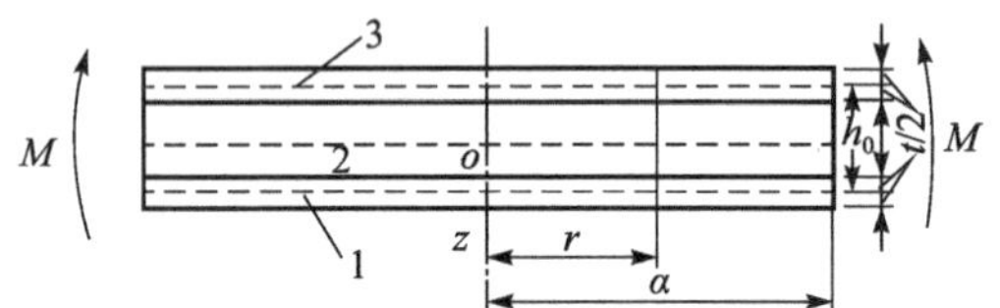

图 1　在边缘力矩作用下的夹层圆板

采用 Reissner[5,6]的假定：

（1）材料服从于胡克定律。

（2）夹心横向不可压缩。

（3）夹心沿板面方向不能承受载荷。

（4）表层处于薄膜应力状态。

（5）夹心中面法线在变形后保持直线。

在轴对称受载和上述假定下，夹层圆板中任一点的位移为

上表层

$$u_1 = u + \frac{1}{2}h_0\psi, v_1 = 0, w_1 = w \tag{1}$$

下表层

$$u_3 = u - \frac{1}{2}h_0\psi, v_3 = 0, w_3 = w \tag{2}$$

夹心

$$u_2 = u + z\psi, v_2 = 0, w_2 = w \tag{3}$$

将式（1）～式（3）分别代入下述夹层圆板的几何方程：

$$\begin{aligned} &\varepsilon_{ri} = \frac{\mathrm{d}u_i}{\mathrm{d}r} + \frac{1}{2}\left(\frac{\mathrm{d}w_i}{\mathrm{d}r}\right)^2, \\ &\varepsilon_{\theta i} = \frac{u_i}{r}, \\ &\varepsilon_{zi} = \gamma_{r\theta i} = \gamma_{\theta zi} = \gamma_{rzi} = 0; i = 1,3 \end{aligned} \tag{4}$$

$$\gamma_{rz2}=\frac{\partial u_2}{\partial z}+\frac{\mathrm{d}w_2}{\mathrm{d}r},$$

$$\varepsilon_{r2}=\varepsilon_{\theta 2}=\varepsilon_{z2}=\gamma_{r\theta 2}=\gamma_{\theta z2}=0 \tag{5}$$

便得

上表层

$$\varepsilon_{r1}=\frac{\mathrm{d}u}{\mathrm{d}r}+\frac{h_0}{2}\frac{\mathrm{d}\psi}{\mathrm{d}r}+\frac{1}{2}\left(\frac{\mathrm{d}w_i}{\mathrm{d}r}\right)^2,$$

$$\varepsilon_{\theta 1}=\frac{u}{r}+\frac{h_0}{2r}\psi \tag{6}$$

下表层

$$\varepsilon_{r3}=\frac{\mathrm{d}u}{\mathrm{d}r}-\frac{h_0}{2}\frac{\mathrm{d}\psi}{\mathrm{d}r}+\frac{1}{2}\left(\frac{\mathrm{d}w_i}{\mathrm{d}r}\right)^2,$$

$$\varepsilon_{\theta 3}=\frac{u}{r}-\frac{h_0}{2r}\psi \tag{7}$$

夹心

$$\gamma_{rz2}=\psi+\frac{\mathrm{d}w}{\mathrm{d}r} \tag{8}$$

将式（6）~式（8）分别代入下述胡克定律：

$$\sigma_{ri}=\frac{E}{1-\nu^2}(\varepsilon_{ri}+\nu\varepsilon_{\theta i}),$$

$$\sigma_{\theta i}=\frac{E}{1-\nu^2}(\varepsilon_{\theta i}+\nu\varepsilon_{ri}), \tag{9}$$

$$\sigma_{zi}=\tau_{r\theta i}=\tau_{\theta zi}=\tau_{rzi}=0;i=1,3$$

$$\tau_{rz2}=G_2\gamma_{rz2},$$

$$\sigma_{r2}=\sigma_{\theta 2}=\sigma_{z2}=\tau_{r\theta 2}=\tau_{\theta z2}=0 \tag{10}$$

便得

上表层

$$\sigma_{r1}=\sigma_{r0}+\frac{Eh_0}{2(1-\nu^2)}\left(\frac{\mathrm{d}\psi}{\mathrm{d}r}+\frac{\nu}{r}\psi\right),$$

$$\sigma_{\theta 1}=\sigma_{\theta 0}+\frac{Eh_0}{2(1-\nu^2)}\left(\frac{\psi}{r}+\nu\frac{\mathrm{d}\psi}{\mathrm{d}r}\right) \tag{11}$$

下表层

$$\sigma_{r3}=\sigma_{r0}-\frac{Eh_0}{2(1-\nu^2)}\left(\frac{\mathrm{d}\psi}{\mathrm{d}r}+\frac{\nu}{r}\psi\right),$$

$$\sigma_{\theta 3}=\sigma_{\theta 0}-\frac{Eh_0}{2(1-\nu^2)}\left(\frac{\psi}{r}+\nu\frac{\mathrm{d}\psi}{\mathrm{d}r}\right) \tag{12}$$

夹心

$$\tau_{rz2}=G_2\left(\psi+\frac{\mathrm{d}w}{\mathrm{d}r}\right) \tag{13}$$

其中

$$\sigma_{r0}=\frac{E}{1-\nu^2}\left[\frac{\mathrm{d}u}{\mathrm{d}r}+\frac{\nu}{r}u+\frac{1}{2}\left(\frac{\mathrm{d}w}{\mathrm{d}r}\right)^2\right],$$

$$\sigma_{\theta 0}=\frac{E}{1-\nu^2}\left[\frac{u}{r}+\nu\frac{\mathrm{d}u}{\mathrm{d}r}+\frac{\nu}{2}\left(\frac{\mathrm{d}w}{\mathrm{d}r}\right)^2\right] \tag{14}$$

由弹性体应变能的公式：

$$U_i=\frac{1}{2}\iiint_{V_i}(\sigma_r\varepsilon_r+\sigma_\theta\varepsilon_\theta+\sigma_z\varepsilon_z+\tau_{r\theta}\gamma_{r\theta}+\tau_{\theta z}\gamma_{\theta z}+\tau_{rz}\gamma_{rz})r\mathrm{d}r\mathrm{d}\theta\mathrm{d}z,i=1,2,3 \tag{15}$$

得到表层和夹心的应变能公式

$$U_i=\frac{1}{2E}\iiint_{V_i}[(\sigma_{ri}+\sigma_{\theta i})^2-2(1+\nu)\sigma_{ri}\sigma_{\theta i}]r\mathrm{d}r\mathrm{d}\theta\mathrm{d}z,i=1,2$$

$$U_i=\frac{1}{2G_2}\iiint_{V_2}\tau_{rz2}^2 r\mathrm{d}r\mathrm{d}\theta\mathrm{d}z \tag{16}$$

将式（11）～式（13）代入式（16），并对 z 进行积分，便得

$$U_1=\frac{t}{2E}\iint_{s_1}[(\sigma_{r0}+\sigma_{\theta 0})^2-2(1+\nu)\sigma_{r0}\sigma_{\theta 0}]r\mathrm{d}r\mathrm{d}\theta$$

$$+\frac{th_0}{2}\iint_{s_1}\left(\sigma_{r0}\frac{\mathrm{d}\psi}{\mathrm{d}r}+\frac{1}{r}\sigma_{\theta 0}\psi\right)r\mathrm{d}r\mathrm{d}\theta+\frac{D}{4}\iint_{s_1}\left[\left(\frac{\mathrm{d}\psi}{\mathrm{d}r}+\frac{\psi}{r}\right)^2-2(1-\nu)\frac{\psi}{r}\frac{\mathrm{d}\psi}{\mathrm{d}r}\right]r\mathrm{d}r\mathrm{d}\theta,$$

$$U_3=\frac{t}{2E}\iint_{s_3}[(\sigma_{r0}+\sigma_{\theta 0})^2-2(1+\nu)\sigma_{r0}\sigma_{\theta 0}]r\mathrm{d}r\mathrm{d}\theta$$

$$-\frac{th_0}{2}\iint_{s_3}\left(\sigma_{r0}\frac{\mathrm{d}\psi}{\mathrm{d}r}+\frac{1}{r}\sigma_{\theta 0}\psi\right)r\mathrm{d}r\mathrm{d}\theta+\frac{D}{4}\iint_{s_3}\left[\left(\frac{\mathrm{d}\psi}{\mathrm{d}r}+\frac{\psi}{r}\right)^2-2(1-\nu)\frac{\psi}{r}\frac{\mathrm{d}\psi}{\mathrm{d}r}\right]r\mathrm{d}r\mathrm{d}\theta,$$

$$U_2=\frac{G_2h_0}{2}\iint_{s_2}\left(\psi+\frac{\mathrm{d}w}{\mathrm{d}r}\right)^2r\mathrm{d}r\mathrm{d}\theta \tag{17}$$

其中

$$D=\frac{Eth_0^2}{2(1-\nu^2)} \tag{18}$$

边缘均布力矩 M 的外力功为

$$V=\int_0^{2\pi}M\psi a\,\mathrm{d}\theta=2\pi aM\psi \tag{19}$$

这样，夹层圆板的总势能为

$$U=U_1+U_2+U_3-V=\frac{t}{E}\iint_{s}[(\sigma_{r0}+\sigma_{\theta 0})^2-2(1+\nu)\sigma_{r0}\sigma_{\theta 0}]r\mathrm{d}r\mathrm{d}\theta$$

$$+\frac{D}{2}\iint_{s}\left[\left(\frac{\mathrm{d}\psi}{\mathrm{d}r}+\frac{\psi}{r}\right)^2-2(1-\nu)\frac{\psi}{r}\frac{\mathrm{d}\psi}{\mathrm{d}r}\right]r\mathrm{d}r\mathrm{d}\theta$$

$$+\frac{G_2h_0}{2}\iint_{s}\left(\psi+\frac{\mathrm{d}w}{\mathrm{d}r}\right)^2r\mathrm{d}r\mathrm{d}\theta-2\pi aM\psi \tag{20}$$

根据势能原理，以 u、w、ψ 作自变量，对总势能变分为零，有

$$\delta U = 0 \tag{21}$$

将式（20）代入，经部分积分后，可得在边缘均布力矩 M 作用下夹层圆板的大挠度理论的平衡方程和边界条件，这里仅列出简支边界条件：

$$\sigma_{\theta 0} - \frac{\mathrm{d}}{\mathrm{d}r}(r\sigma_{r0}) = 0,$$

$$2t\frac{\mathrm{d}}{\mathrm{d}r}\left(r\sigma_{r0}\frac{\mathrm{d}w}{\mathrm{d}r}\right) + G_2h_0\frac{\mathrm{d}}{\mathrm{d}r}\left[r\left(\psi + \frac{\mathrm{d}w}{\mathrm{d}r}\right)\right] = 0,$$

$$D\frac{\mathrm{d}}{\mathrm{d}r}\frac{1}{r}\frac{\mathrm{d}}{\mathrm{d}r}(r\psi) - G_2h_0\left(\psi + \frac{\mathrm{d}w}{\mathrm{d}r}\right) = 0 \tag{22a,b,c}$$

在 $r=a$ 时，

$$w = 0, \sigma_{r0} = 0, M_r = D\left(\frac{\mathrm{d}\psi}{\mathrm{d}r} + \nu\frac{\psi}{r}\right) = M \tag{23}$$

在 $r=0$ 时，

$$\psi = 0, Q_r + 2t\sigma_{r0}\frac{\mathrm{d}w}{\mathrm{d}r} = G_2h_0\left(\psi + \frac{\mathrm{d}w}{\mathrm{d}r}\right) + 2t\sigma_{r0}\frac{\mathrm{d}w}{\mathrm{d}r}\text{有限}, \sigma_{r0}\text{有限} \tag{24a,b,c}$$

关于应变协调方程，可由方程（14）消去位移 u 求得

$$\sigma_{r0} - \frac{\mathrm{d}}{\mathrm{d}r}(r\sigma_{\theta 0}) - \frac{E}{2}\left(\frac{\mathrm{d}w}{\mathrm{d}r}\right)^2 = 0 \tag{25}$$

就这样，我们得到了一组对于函数 σ_{r0}、$\sigma_{\theta 0}$、w 和 ψ 的四个微分方程和相应的边界条件。下面，我们化简这组非线性边值问题。

引入应力函数 φ：

$$\sigma_{r0} = \frac{1}{r}\frac{\mathrm{d}\varphi}{\mathrm{d}r}, \quad \sigma_{\theta 0} = \frac{\mathrm{d}^2\varphi}{\mathrm{d}r^2} \tag{26}$$

则方程（22a）自动满足，而方程（22b）和（25）成为

$$2t\frac{\mathrm{d}}{\mathrm{d}r}\left(\frac{\mathrm{d}\varphi}{\mathrm{d}r}\frac{\mathrm{d}w}{\mathrm{d}r}\right) + G_2h_0\frac{\mathrm{d}}{\mathrm{d}r}\left[r\left(\psi + \frac{\mathrm{d}w}{\mathrm{d}r}\right)\right] = 0,$$

$$\frac{\mathrm{d}}{\mathrm{d}r}\frac{1}{r}\frac{\mathrm{d}}{\mathrm{d}r}r\frac{\mathrm{d}\varphi}{\mathrm{d}r} + \frac{E}{2r}\left(\frac{\mathrm{d}w}{\mathrm{d}r}\right)^2 = 0 \tag{27a,b}$$

将方程（27a）乘以 $\mathrm{d}r$，积分一次，并应用式（24b），得

$$\psi = -\left(1 + \frac{2t}{G_2h_0r}\frac{\mathrm{d}\varphi}{\mathrm{d}r}\right)\frac{\mathrm{d}w}{\mathrm{d}r} \tag{28}$$

将此式代入方程（22c），得

$$Dr\frac{\mathrm{d}}{\mathrm{d}r}\frac{1}{r}\frac{\mathrm{d}}{\mathrm{d}r}r\frac{\mathrm{d}w}{\mathrm{d}r} - 2t\left(1 - \frac{D}{G_2h_0}r\frac{\mathrm{d}}{\mathrm{d}r}\frac{1}{r}\frac{\mathrm{d}}{\mathrm{d}r}\right)\left(\frac{\mathrm{d}\varphi}{\mathrm{d}r}\frac{\mathrm{d}w}{\mathrm{d}r}\right) = 0 \tag{29}$$

则四个方程简化为仅含 w 和 φ 的两个基本方程（27b）和（29），相应的边界条件为式（23）和（24a,c）。为了计算方便，我们以 w 和 σ_{r0} 为未知函数，并引进下列无量纲量：

$$\rho = \frac{r}{a}, W = \sqrt{2(1-\nu^2)}\frac{w}{h_0}, S_r = \frac{2ta^2}{D}\sigma_{r0},$$

$$k = \frac{D}{G_2h_0a^2}, m = \frac{a^2\sqrt{2(1-\nu^2)}}{Dh_0}M \tag{30}$$

则边值问题（29）、(27b)、(23) 和（24a，c）成为

$$L\left(\rho\frac{\mathrm{d}W}{\mathrm{d}\rho}\right)=\left(\frac{1}{\rho}-kL\right)\left(\rho S_r\frac{\mathrm{d}W}{\mathrm{d}\rho}\right),$$

$$L(\rho^2S_r)=-\frac{1}{\rho}\left(\rho\frac{\mathrm{d}W}{\mathrm{d}\rho}\right)^2 \tag{31a,b}$$

在 $\rho=1$ 时，

$$W=0, S_r=0, \frac{\mathrm{d}^2W}{\mathrm{d}\rho^2}+\frac{\nu}{\rho}\frac{\mathrm{d}W}{\mathrm{d}\rho}=-\left[k\frac{W}{d\rho}\left(S_r\frac{\mathrm{d}W}{\mathrm{d}\rho}\right)+m\right] \tag{32a,b,c}$$

在 $\rho=0$ 时，

$$\frac{\mathrm{d}W}{\mathrm{d}\rho}=-kS_r\frac{\mathrm{d}W}{\mathrm{d}\rho},\ S_r\ \text{有限} \tag{33}$$

其中

$$L(\cdots)=\frac{\mathrm{d}}{\mathrm{d}\rho}\frac{1}{\rho}\frac{\mathrm{d}}{\mathrm{d}\rho}(\cdots) \tag{34}$$

显而易见，方程（31）和边界条件（32c）、(33a) 都是非线性形式。我们应用修正迭代法，克服了求解这组非线性边值问题的困难。

三、边值问题的求解

在一次近似中，略去方程（31a）和边界条件（32c）、(33a) 右端的非线性项，并将由方程（31a）所得的解 W_1 代入方程（31b），便得下列线性边值问题：

$$L\left(\rho\frac{\mathrm{d}W_1}{\mathrm{d}\rho}\right)=0,$$

$$L(\rho^2S_{r1})=-\frac{1}{\rho}\left(\rho\frac{\mathrm{d}W_1}{\mathrm{d}\rho}\right)^2 \tag{35a,b}$$

在 $\rho=1$ 时，

$$W_1=0, S_{r1}=0, \frac{\mathrm{d}^2W}{\mathrm{d}\rho^2}+\frac{\nu}{\rho}\frac{\mathrm{d}W_1}{\mathrm{d}\rho}=-m \tag{36a,b,c}$$

在 $\rho=0$ 时，

$$\frac{\mathrm{d}W_1}{\mathrm{d}\rho}=0, S_r\ \text{有限} \tag{37a,b}$$

应用边界条件（36a，c）和（37a），方程（35a）的解是

$$W_1=\frac{1}{2\lambda}(1-\rho^2)m \tag{38}$$

其中

$$\lambda=1+\nu$$

我们用无量纲中心挠度 W_0 作为迭代参数：

$$W|_{\rho=0}=W_0 \tag{39}$$

再应用式（38），便得一次近似中的边缘力矩和中心挠度的关系式

$$m=\alpha_1W_0 \tag{40}$$

其中

$$\alpha_1 = 2\lambda$$

将式（40）代入式（38），得一次近似挠度公式

$$W_1 = (1-\rho^2)W_0 \tag{41}$$

将式（41）代入方程（35b）的左端，应用边界条件（36b）和（37b），方程（35b）的解是

$$S_{r1} = \frac{1}{2}(1-\rho^2)W_0^2 \tag{42}$$

此即一次近似中的径向应力公式。

在二次近似中，由非线性边值问题（31）～(33）得下列线性边值问题：

$$L\left(\rho\frac{\mathrm{d}W_2}{\mathrm{d}\rho}\right)=\left(\frac{1}{\rho}-kL\right)\left(\rho S_{r1}\frac{\mathrm{d}W_1}{\mathrm{d}\rho}\right),$$

$$L(\rho^2 S_{r2}) =-\frac{1}{\rho}\left(\rho\frac{\mathrm{d}W_2}{\mathrm{d}\rho}\right)^2 \tag{43a,b}$$

在 $\rho=1$ 时，

$$W_2 = 0, S_{r2} = 0, \frac{\mathrm{d}^2W_2}{\mathrm{d}\rho^2}+\frac{\nu}{\rho}\frac{\mathrm{d}W_2}{\mathrm{d}\rho} =-\left[k\frac{\mathrm{d}}{\mathrm{d}\rho}\left(S_{r1}\frac{\mathrm{d}W_1}{\mathrm{d}\rho}\right)+m\right] \tag{44a,b,c}$$

在 $\rho=0$ 时，

$$\frac{\mathrm{d}W_2}{\mathrm{d}\rho} =-kS_{r1}\frac{\mathrm{d}W_1}{\mathrm{d}\rho},\ S_{r2}\text{ 有限} \tag{45a,b}$$

应用式（41）和（42），在边界条件式（44a，c）和式（45a）情况下，方程（43a）的解是

$$\begin{aligned}W_2 =&\frac{1}{2\lambda}\left[m-\frac{1}{2}\left(\lambda k+\frac{17+5\nu}{72}\right)W_0^3\right]\\&-\frac{1}{2\lambda}\left[m-\left(\lambda k+\frac{2+\nu}{12}\right)W_0^3\right]\rho^2\\&-\frac{1}{4}\left(k+\frac{1}{8}\right)W_0^3\rho^4+\frac{1}{144}W_0^3\rho^6\end{aligned} \tag{46}$$

应用定义式（39），由式（46）得二次近似中的边缘力矩和中心挠度的关系式

$$m = \alpha_1 W_0 + \alpha_3 W_0^3 \tag{47}$$

其中

$$\alpha_3 = \frac{1}{2}\left(\lambda k+\frac{17+5\nu}{72}\right)$$

将式（47）代入式（46），便得二次近似的挠度公式

$$W_2 = W_0-\left[W_0-\frac{1}{4}\left(k+\frac{7}{72}\right)W_0^3\right]\rho^2-\frac{1}{4}\left(k+\frac{1}{8}\right)W_0^3\rho^4+\frac{1}{144}W_0^3\rho^6 \tag{48}$$

将式（48）代入方程（43b）的右端，应用边界条件（44b）和（45b），方程（43b）的解是

$$\begin{aligned}S_{r2} =&\frac{1}{2}\left[W_0^2-\frac{1}{6}\left(k+\frac{1}{12}\right)W_0^4+\frac{1}{12}\left(\frac{1}{4}k^2+\frac{2}{45}k+\frac{23}{11520}\right)W_0^6\right]\\&-\frac{1}{2}\left[W_0^2-\frac{1}{2}\left(k+\frac{7}{72}\right)W_0^4+\frac{1}{16}\left(k^2+\frac{7}{36}k+\frac{49}{5184}\right)W_0^6\right]\rho^2\end{aligned}$$

$$-\frac{1}{6}\Big[\Big(k+\frac{1}{8}\Big)W_0^4-\frac{1}{4}\Big(k^2+\frac{2}{9}k+\frac{7}{576}\Big)W_0^6\Big]\rho^4$$

$$+\frac{1}{48}\Big[\frac{1}{6}W_0^4-\Big(k^2+\frac{7}{24}k+\frac{17}{864}\Big)W_0^6\Big]\rho^6$$

$$+\frac{1}{960}\Big(k+\frac{1}{8}\Big)W_0^6\rho^8-\frac{1}{69120}W_0^6\rho^{10} \tag{49}$$

此即二次近似的径向应力公式。

在三次近似中，关于挠度 W 有下列线性边值问题：

$$L\Big(\rho\frac{\mathrm{d}W_3}{\mathrm{d}\rho}\Big)=\Big(\frac{1}{\rho}-kL\Big)\Big(\rho S_{r2}\frac{\mathrm{d}W_2}{\mathrm{d}\rho}\Big) \tag{50}$$

在 $\rho=1$ 时，

$$W_3=0,\frac{\mathrm{d}^2W_3}{\mathrm{d}\rho^2}+\frac{\nu}{\rho}\frac{\mathrm{d}W_3}{\mathrm{d}\rho}=-\Big[k\frac{\mathrm{d}}{\mathrm{d}\rho}\Big(S_{r2}\frac{\mathrm{d}W_2}{\mathrm{d}\rho}\Big)+m\Big] \tag{51}$$

在 $\rho=0$ 时，

$$\frac{\mathrm{d}W_3}{\mathrm{d}\rho}=-kS_{r2}\frac{\mathrm{d}W_2}{\mathrm{d}\rho} \tag{52}$$

应用式（48）和（49），在边界条件（51）和（52）情况下，方程（50）的解是

$$W_3=\frac{m}{2\lambda}-\frac{1}{4\lambda}\Big[\Big(\lambda k+\frac{17+5\nu}{72}\Big)W_0^3-\frac{1}{36}\Big(5\lambda k^2+\frac{11}{16}\lambda k+\frac{11+26\nu}{1200}\Big)W_0^5$$

$$+\frac{1}{36}\Big(\lambda k^3+\frac{634+409\nu}{1800}k^2+\frac{29+13\nu}{768}k+\frac{21971+6956\nu}{16934400}\Big)W_0^7$$

$$-\frac{1}{720}\Big(\lambda k^4+\frac{917}{2880}\lambda k^3+\frac{94273+94903\nu}{2540160}k^2+\frac{10085+10309\nu}{5419008}k$$

$$+\frac{48581+51311\nu}{1463132160}\Big)W_0^9\Big]-\frac{1}{2\lambda}\Big[m-\Big(\lambda k+\frac{2+\nu}{12}\Big)W_0^3$$

$$+\frac{1}{6}\Big(\frac{5}{2}\lambda k^2+\frac{1}{3}\lambda k-\frac{7+8\nu}{960}\Big)W_0^5-\frac{1}{16}\Big(\lambda k^3+\frac{269+239\nu}{1080}k^2$$

$$+\frac{2561+1961\nu}{129600}k+\frac{3627+2197\nu}{7257600}\Big)W_0^7+\frac{1}{920}\Big(\lambda k^4+\frac{37}{120}\lambda k^3$$

$$+\frac{15403+15418\nu}{453600}k^2+\frac{33437+33557\nu}{21772800}k+\frac{11617+11747\nu}{522547200}\Big)W_0^9\Big]\rho^2$$

$$-\frac{1}{4}\Big[\Big(k+\frac{1}{8}\Big)W_0^3-\frac{1}{4}\Big(5k^2+\frac{3}{4}k+\frac{11}{576}\Big)W_0^5$$

$$+\frac{1}{48}\Big(13k^3+\frac{71}{24}k^2+\frac{559}{2280}k+\frac{209}{69120}\Big)W_0^7$$

$$-\frac{1}{192}\Big(5k^4+\frac{289}{180}k^3+\frac{259}{1440}k^2+\frac{2461}{311040}k+\frac{161}{1658880}\Big)W_0^9\Big]\rho^4$$

$$+\frac{1}{36}\Big[\frac{1}{4}W_0^3-\Big(5k^2+\frac{17}{16}k+\frac{13}{384}\Big)W_0^5$$

$$+\frac{1}{2}\Big(5k^3+\frac{371}{288}k^2+\frac{35}{384}k+\frac{73}{55296}\Big)W_0^7$$

$$-\frac{1}{16}\left(5k^4+\frac{251}{144}k^3+\frac{10733}{51840}k^2+\frac{769}{82944}k+\frac{2957}{29859840}\right)W_0^9\Big]\rho^6$$

$$+\frac{1}{192}\left[\frac{1}{4}\left(\frac{13}{3}k+\frac{1}{4}\right)W_0^5-\left(5k^3+\frac{43}{24}k^2+\frac{137}{864}k+\frac{37}{13824}\right)W_0^7\right.$$

$$\left.+\frac{1}{4}\left(5k^4+\frac{289}{144}k^3+\frac{115}{432}k^2+\frac{15977}{1244160}k+\frac{647}{4976640}\right)W_0^9\right]\rho^8$$

$$-\frac{1}{480}\left[\frac{1}{60}W_0^5-\frac{1}{8}\left(\frac{29}{5}k^2+\frac{11}{12}k+\frac{191}{8640}\right)W_0^7\right.$$

$$\left.+\left(k^4+\frac{89}{160}k^3+\frac{1061}{11520}k^2+\frac{289}{55296}k+\frac{1141}{19906560}\right)W_0^9\right]\rho^{10}$$

$$-\frac{1}{2880}\left[\frac{1}{5}\left(\frac{1}{3}k+\frac{1}{64}\right)W_0^7-\frac{1}{2}\left(k^3+\frac{47}{160}k^2+\frac{199}{8640}k+\frac{29}{92160}\right)W_0^9\right]\rho^{12}$$

$$+\frac{1}{1354752}\left[\frac{1}{10}W_0^7-\left(\frac{67}{10}k^2+k+\frac{113}{5760}\right)W_0^9\right]\rho^{14}$$

$$+\frac{1}{6193152}\left(\frac{1}{3}k+\frac{1}{80}\right)W_0^9\rho^{16}-\frac{1}{8599633920}W_0^9\rho^{18} \tag{53}$$

应用定义（39），由式（53）得三次近似中的边缘力矩和中心挠度的关系式

$$m=\sum_{n=0}^{4}\alpha_{2n+1}W_0^{2n+1} \tag{54}$$

其中

$$\alpha_5=-\frac{1}{72}\left(5\lambda k^2+\frac{11}{16}\lambda k+\frac{11+26\nu}{1200}\right),$$

$$\alpha_7=\frac{1}{72}\left(\lambda k^3+\frac{634+409\nu}{1800}k^2+\frac{29+13\nu}{768}k+\frac{21971+6956\nu}{16934400}\right),$$

$$\alpha_9=-\frac{1}{1440}\left(\lambda k^4+\frac{917}{2880}\lambda k^3+\frac{94273+94903\nu}{2540160}k^2\right.$$

$$\left.+\frac{10085+10309\nu}{5419008}k+\frac{48581+51311\nu}{1463132160}\right)$$

最后，引入环向应力 $\sigma_{\theta b}$ 的无量纲量：

$$S_\theta=\frac{2ta^2}{D}\sigma_{\theta b} \tag{55}$$

再应用式（22a）和（49），便得二次近似的环向应力公式

$$S_{\theta 2}=\frac{1}{2}\left[W_0^2-\frac{1}{6}\left(k+\frac{1}{12}\right)W_0^4+\frac{1}{12}\left(\frac{1}{4}k^2+\frac{2}{45}k+\frac{23}{11520}\right)W_0^6\right]$$

$$-3\left[\frac{1}{2}W_0^2-\frac{1}{4}\left(k+\frac{7}{72}\right)W_0^4+\frac{1}{32}\left(k^2+\frac{7}{36}k+\frac{49}{5184}\right)W_0^6\right]\rho^2$$

$$-\frac{5}{24}\left[4\left(k+\frac{1}{8}\right)W_0^4-\left(k^2+\frac{2}{9}k+\frac{7}{576}\right)W_0^6\right]\rho^4$$

$$+\frac{7}{48}\left[\frac{1}{6}W_0^4-\left(k^2+\frac{7}{24}k+\frac{17}{864}\right)W_0^6\right]\rho^6$$

$$+\frac{3}{320}\left(k+\frac{1}{8}\right)W_0^6\rho^8-\frac{11}{69120}W_0^6\rho^{10} \tag{56}$$

由式（49）和（56），我们可得到夹层圆板中面内的径向应力和环向应力在中心和边缘的公式

$$S_r(0)=S_\theta(0)=\frac{1}{2}\left[W_0^2-\frac{1}{6}\left(k+\frac{1}{12}\right)W_0^4+\frac{1}{12}\left(\frac{1}{4}k^2+\frac{2}{45}k+\frac{23}{11520}\right)W_0^6\right],$$

$$S_\theta(1)=-\left[W_0^2+\frac{1}{6}\left(k+\frac{1}{12}\right)W_0^4+\frac{1}{48}\left(k^2+\frac{7}{45}k+\frac{53}{8640}\right)W_0^6\right] \tag{57}$$

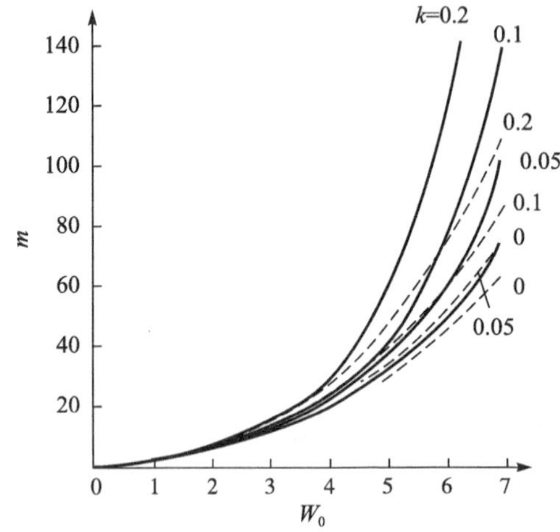

图 2　各种 k 值下的边缘力矩与中心挠度关系曲线（$\nu=0.3$）
—— 三次近似解
---- 二次近似解

由上述公式，我们得到了下面一些有用的结果。图 2 示出了几个不同的 k 值下的边缘力矩 m 与中心挠度 W_0 的关系曲线。较精确的三次近似结果用实线表示，二次近似结果用虚线表示。显而易见，随着中心挠度的增大，板的刚度也增大，并且对于相同的 m 值，具有较小 k 值的板将产生较大的挠度。由图看出，在中心挠度不太大的范围内，实线和虚线异常接近，而在中心挠度比较大时，实线和虚线相差就比较大。对于工程设计来说，只要中心挠度不是特别大，使用简单的二次近似解公式进行计算，将是十分恰当的。但是，随着中心挠度的增大，二次近似解将出现越来越大的误差，当误差超过工程允许范围时，我们就必须使用精确的三次近似解。

在图 3、图 4 内，我们还给出了各种 k 值下夹层圆板中心和边缘的径向与环向应力曲线。由图看出，中心的径向和环向应力值较小，而且都是拉应力；边缘的环向应力值较大，始终是压应力。

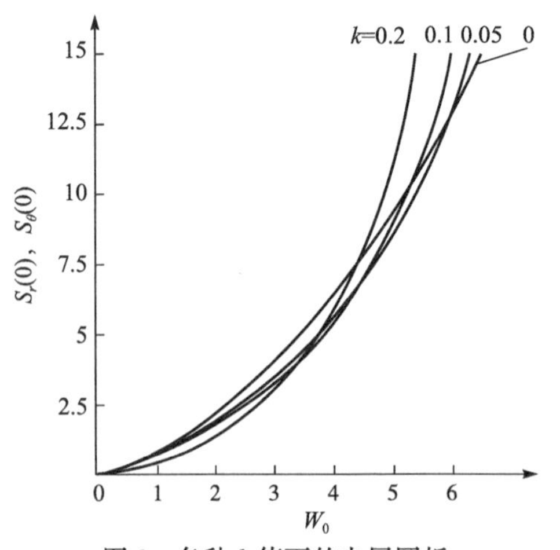

图 3　各种 k 值下的夹层圆板中心的径向和环向应力

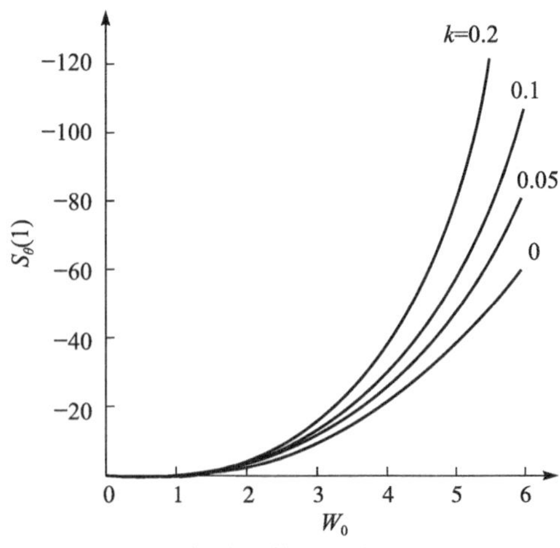

图 4　各种 k 值下的夹层圆板边缘的环向应力

参考文献

[1] Zaid M. Symmetrical bending of circular sandwich plates. *Proceedings of the Second U. S. National Congress of Applied Mechanics*, ASME, New York, 1955: 413.

[2] Bruun E R. Thermal deflection of a circular sandwich plate. *AIAA J.*, 1963, 1 (5): 1213.

[3] Huang J C, Ebcioglu I K. Circular sandwich plate under radial compression and thermal gradient. *AIAA J.*, 1965, 3 (6): 1146.

[4] Amato A J, Ebcioglu I K. Axisymmetric buckling of annular sandwich panels. *AIAA J.*, 1972, 10 (10): 1351.

[5] Reissner E. Finite deflections of sandwich plates. *J. Aeron. Sci.*, 1948, 15 (7): 435; 1950, 17 (2): 125.

[6] 沃耳密尔 A C. 柔韧板与柔韧壳. 卢文达，黄择言，卢鼎霍，译. 北京：科学出版社，1959.

[7] 叶开沅，刘人怀，平庆元，等. 在对称线布载荷作用下的圆底扁薄球壳的非线性稳定问题. 科学通报，1965 (2): 142.

[8] 叶开沅，刘人怀，张传智，等. 圆底扁薄球壳在边缘力矩作用下的非线性稳定问题. 科学通报，1965 (2): 145.

[9] 刘人怀. 在内边缘均布力矩作用下中心开孔圆底扁球壳的非线性稳定问题. 科学通报，1965 (3): 253.

[10] 刘人怀. 在边缘载荷作用下中心开孔圆底扁薄球壳的轴对称稳定性. 力学学报，1977 (3): 206.

[11] 刘人怀. 波纹圆板的特征关系式. 力学学报，1978 (1): 47.

夹层圆板的非线性弯曲[①]

符　号

$r,\ \theta,\ z$　圆柱坐标系

a　夹层圆板的半径

$t,\ h$　表板和夹心厚度

h_0　上、下表板中面间的距离

$E,\ \nu$　表层材料的弹性模量和泊松比

G_2　夹心剪切模量

D_f　表板的抗弯刚度

$D,\ C$　夹层圆板的抗弯和抗剪刚度

q　均布横向载荷

$u_i,\ v_i,\ w_{i(i=1,2,3)}$　上表板、夹心和下表板的径向、环向和法向位移

$u,\ w$　夹层圆板中面上点的径向位移和挠度

ψ　上、下表板中面上任意点在变形后的连线与夹层圆板变形前法线的夹角

$\varepsilon_{ri},\ \varepsilon_{\theta i},\ \varepsilon_{zi},\ \gamma_{r\theta i},\ \gamma_{\theta zi},\ \gamma_{rzi(i=1,2,3)}$　上表板、夹心和下表板点的伸长和剪切应变分量

$\sigma_{ri},\ \sigma_{\theta i},\ \sigma_{zi},\ \tau_{r\theta i},\ \tau_{\theta zi},\ \tau_{rzi(i=1,2,3)}$　上表板、夹心和下表板点的正应力和剪应力分量

$\sigma_{r0},\ \sigma_{\theta 0}$　夹层圆板中面内的径向和环向应力

$U_{i(i=1,2,3)}$　上表板、夹心和下表板的应变能

V　外力功

U　夹层圆板的总势能

$M_r,\ Q_r$　夹层圆板的径向弯矩和横向力

m_r　表板的径向弯径

φ　应力函数

ρ　无量纲径向坐标

k　夹层圆板的无量纲特征参数

$W,\ W_0$　夹层圆板中面上点的无量纲挠度及其中心值

$S_r,\ S_r(0)$　夹层圆板中面内的无量纲径向应力及其中心值

$S_\theta,\ S_\theta(0),\ S_\theta(1)$　夹层圆板中面内的无量纲环向应力及其中心和边缘值

---

① 本文原载《应用数学和力学》，1981，2（2）：173-190.

P　无量纲横向载荷

A_2，A_3，B_2，B_3，α_1，…，α_9，λ_1，λ_2，$l_{1,1}$，…，$l_{11,3}$，m_1，…，m_{33}，$n_{0,2}$，…，$n_{22,6}$，R_1，…，E_{33}　辅助量

L　微分算子

一、前言

夹层板是航空、宇航和船舶制造等工业部门中的重要结构元件，它由三层，即一块厚夹心和两块薄板所构成，因为这种板具有高的刚度和轻的重量的特性，所以在工业中应用十分广泛。因此，夹层圆板已经成为一些研究者的对象[1-4]。但是对于其非线性弯曲方面的问题，则由于非线性数学的困难而仅有少数人探讨过，作者[5]使用修正迭代法[6-11]曾对边缘力矩作用下的夹层圆板的非线性弯曲进行过研究，获得了相当精确的三次近似解析解。中国科学院北京力学研究所固体力学研究室板壳组[12]使用摄动法对均布横向载荷作用下的夹层圆板的非线性弯曲求解，得到了简支和固定边界条件下的二次近似解。

本文所研究的夹层板，是上、下表板厚度相等、材料各向同性，夹心材料横观各向同性的薄圆板。我们使用变分法导出了这种夹层圆板的非线性轴对称弯曲理论的平衡方程和边界条件。当表板很薄时，还导出了简化方程。最后，使用修正迭代法对承受均布横向载荷作用的具有滑动固定边界条件的夹层圆板进行了研究，获得了相当精确的三次近似解析解。所得结果可供工程设计时参考应用。

二、基本方程

在推导基本方程和边界条件时，引进下列假设：①材料服从于胡克定律。②夹心横向不可压缩。③夹心沿板面方向不能承受载荷。④表板为直法线假设，夹心中面法线在变形后保持直线。

现在，考虑在均布横向载荷作用下半径为 a 的夹层圆板，其坐标、尺寸如图 1 所示。在轴对称和上述假设下，夹层圆板中任意一点的位移为

上表板$\left[\dfrac{h}{2}\leqslant z\leqslant\dfrac{h}{2}+t\right]$

$$u_1=u+\frac{h_0}{2}\psi-\left(z-\frac{h_0}{2}\right)\frac{\mathrm{d}w_1}{\mathrm{d}r},v_1=0,w_1=w \tag{2.1}$$

下表板$\left[-\left(\dfrac{h}{2}+t\right)\leqslant z\leqslant-\dfrac{h}{2}\right]$

$$u_3=u-\frac{h_0}{2}\psi-\left(z+\frac{h_0}{2}\right)\frac{\mathrm{d}w_3}{\mathrm{d}r},v_3=0,w_3=w \tag{2.2}$$

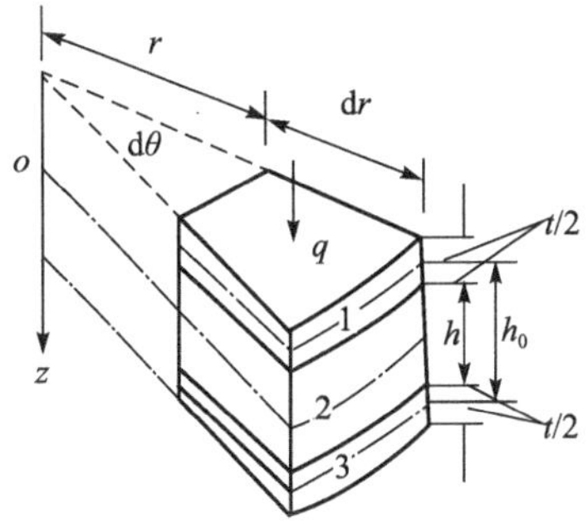

图 1　夹层圆板的坐标和尺寸

夹心 $\left[-\frac{h}{2}\leqslant z\leqslant\frac{h}{2}\right]$

$$u_2=u+\frac{z}{h}\left(h_0\psi+t\frac{\mathrm{d}w_2}{\mathrm{d}r}\right),v_2=0,w_2=w \tag{2.3}$$

将式（2.1）、(2.2）和（2.3）分别代入下述夹层圆板的几何方程：

$$\varepsilon_{r1}=\frac{\partial u_1}{\partial r}+\frac{1}{2}\left(\frac{\mathrm{d}w_1}{\mathrm{d}r}\right)^2,\varepsilon_{\theta i}=\frac{u_i}{r},\varepsilon_{zi}=\gamma_{r\theta i}=\gamma_{\theta zi}=\gamma_{rzi}=0\quad i=1,3 \tag{2.4}$$

$$\gamma_{rz2}=\frac{\partial u_2}{\partial z}+\frac{\mathrm{d}w_2}{\mathrm{d}r},\varepsilon_{r2}=\varepsilon_{\theta 2}=\varepsilon_{z2}=\gamma_{r\theta 2}=\gamma_{\theta z2}=0 \tag{2.5}$$

便得：

上表板

$$\begin{aligned}\varepsilon_{r1}&=\frac{\mathrm{d}u}{\mathrm{d}r}+\frac{h_0}{2}\frac{\mathrm{d}\psi}{\mathrm{d}r}-\left(z-\frac{h_0}{2}\right)\frac{\mathrm{d}^2w}{\mathrm{d}r^2}+\frac{1}{2}\left(\frac{\mathrm{d}w}{\mathrm{d}r}\right)^2,\\ \varepsilon_{\theta 1}&=\frac{u}{r}+\frac{h_0}{2r}\psi-\left(z-\frac{h_0}{2}\right)\frac{1}{r}\frac{\mathrm{d}w}{\mathrm{d}r}\end{aligned} \tag{2.6}$$

下表板

$$\begin{aligned}\varepsilon_{r3}&=\frac{\mathrm{d}u}{\mathrm{d}r}-\frac{h_0}{2}\frac{\mathrm{d}\psi}{\mathrm{d}r}-\left(z+\frac{h_0}{2}\right)\frac{\mathrm{d}^2w}{\mathrm{d}r^2}+\frac{1}{2}\left(\frac{\mathrm{d}w}{\mathrm{d}r}\right)^2,\\ \varepsilon_{\theta 3}&=\frac{u}{r}-\frac{h_0}{2r}\psi-\left(z+\frac{h_0}{2}\right)\frac{1}{r}\frac{\mathrm{d}w}{\mathrm{d}r}\end{aligned} \tag{2.7}$$

夹心

$$\gamma_{rz2}=\frac{h_0}{h}\left(\psi+\frac{\mathrm{d}w}{\mathrm{d}r}\right) \tag{2.8}$$

将式（2.6）、(2.7）和（2.8）分别代入下述胡克定律：

$$\begin{aligned}\sigma_{ri}&=\frac{E}{1-\nu^2}(\varepsilon_{ri}+\nu\varepsilon_{\theta i}),\sigma_{\theta i}=\frac{E}{1-\nu^2}(\varepsilon_{\theta i}+\nu\varepsilon_{ri}),\\ \sigma_{zi}&=\tau_{r\theta i}=\tau_{\theta zi}=\tau_{rzi}=0,i=1,3\end{aligned} \tag{2.9}$$

$$\tau_{rz2}=G_2\gamma_{rz2},\sigma_{r2}=\sigma_{\theta 2}=\sigma_{z2}=\tau_{r\theta 2}=\tau_{\theta z2}=0 \tag{2.10}$$

便得：

上表板

$$\begin{aligned}\sigma_{r1}&=\sigma_{r0}+\frac{Eh_0}{2(1-\nu^2)}\left(\frac{\mathrm{d}\psi}{\mathrm{d}r}+\frac{\nu}{r}\psi\right)-\frac{E}{1-\nu^2}\left(z-\frac{h_0}{2}\right)\left(\frac{\mathrm{d}^2w}{\mathrm{d}r^2}+\frac{\nu}{r}\frac{\mathrm{d}w}{\mathrm{d}r}\right),\\ \sigma_{\theta 1}&=\sigma_{\theta 0}+\frac{Eh_0}{2(1-\nu^2)}\left(\frac{\psi}{r}+\nu\frac{\mathrm{d}\psi}{\mathrm{d}r}\right)-\frac{E}{1-\nu^2}\left(z-\frac{h_0}{2}\right)\left(\frac{1}{r}\frac{\mathrm{d}w}{\mathrm{d}r}+\nu\frac{\mathrm{d}^2w}{\mathrm{d}r^2}\right)\end{aligned} \tag{2.11}$$

下表板

$$\begin{aligned}\sigma_{r3}&=\sigma_{r0}-\frac{Eh_0}{2(1-\nu^2)}\left(\frac{\mathrm{d}\psi}{\mathrm{d}r}+\frac{\nu}{r}\psi\right)-\frac{E}{1-\nu^2}\left(z+\frac{h_0}{2}\right)\left(\frac{\mathrm{d}^2w}{\mathrm{d}r^2}+\frac{\nu}{r}\frac{\mathrm{d}w}{\mathrm{d}r}\right),\\ \sigma_{\theta 3}&=\sigma_{\theta 0}-\frac{Eh_0}{2(1-\nu^2)}\left(\frac{\psi}{r}+\nu\frac{\mathrm{d}\psi}{\mathrm{d}r}\right)-\frac{E}{1-\nu^2}\left(z+\frac{h_0}{2}\right)\left(\frac{1}{r}\frac{\mathrm{d}w}{\mathrm{d}r}+\nu\frac{\mathrm{d}^2w}{\mathrm{d}r^2}\right)\end{aligned} \tag{2.12}$$

夹心

$$\tau_{rz2}=\frac{G_2h_0}{h}\left(\psi+\frac{\mathrm{d}w}{\mathrm{d}r}\right) \tag{2.13}$$

其中

$$\sigma_{r0}=\frac{E}{1-\nu^2}\left[\frac{\mathrm{d}u}{\mathrm{d}r}+\frac{\nu}{r}u+\frac{1}{2}\left(\frac{\mathrm{d}w}{\mathrm{d}r}\right)^2\right],$$

$$\sigma_{\theta0}=\frac{E}{1-\nu^2}\left[\frac{u}{r}+\nu\frac{\mathrm{d}u}{\mathrm{d}r}+\frac{\nu}{2}\left(\frac{\mathrm{d}w}{\mathrm{d}r}\right)^2\right] \tag{2.14}$$

由弹性体应变能的公式：

$$U_i=\frac{1}{2}\iiint\limits_V(\sigma_r\varepsilon_r+\sigma_\theta\varepsilon_\theta+\sigma_z\varepsilon_z+\tau_{r\theta}\gamma_{r\theta}+\tau_{\theta z}\gamma_{\theta z}+\tau_{rz}\gamma_{rz})r\mathrm{d}r\mathrm{d}\theta\mathrm{d}z, i=1,2,3 \tag{2.15}$$

得到表板和夹心的应变能公式

$$U_i=\frac{1}{2E}\iiint\limits_{V_i}[(\sigma_{ri}+\sigma_{\theta i})^2-(1-\nu)\sigma_{ri}\sigma_{\theta i}]r\mathrm{d}r\mathrm{d}\theta\mathrm{d}z,\quad i=1,3$$

$$U_2=\frac{1}{2G_2}\iiint\limits_{V_2}\tau_{rz2}^2r\mathrm{d}r\mathrm{d}\theta\mathrm{d}z \tag{2.16}$$

将式（2.11）、（2.12）和（2.13）代入式（2.16），并对 z 进行积分，便得

$$\begin{aligned}U_1=&\frac{t}{2E}\iint\limits_{s_1}[(\sigma_{r0}+\sigma_{\theta0})^2-2(1+\nu)\sigma_{r0}\sigma_{\theta0}]r\mathrm{d}r\mathrm{d}\theta\\&+\frac{th_0}{2}\iint\limits_{s_1}\left(\sigma_{r0}\frac{\mathrm{d}\psi}{\mathrm{d}r}+\frac{1}{r}\sigma_{\theta0}\psi\right)r\mathrm{d}r\mathrm{d}\theta\\&+\frac{D_f}{2}\iint\limits_{s_1}\left[\left(\frac{\mathrm{d}^2w}{\mathrm{d}r^2}+\frac{1}{r}\frac{\mathrm{d}w}{\mathrm{d}r}\right)^2-2(1-\nu)\frac{1}{r}\frac{\mathrm{d}w}{\mathrm{d}r}\frac{\mathrm{d}^2w}{\mathrm{d}r^2}\right]r\mathrm{d}r\mathrm{d}\theta\\&+\frac{D}{4}\iint\limits_{s_1}\left[\left(\frac{\mathrm{d}\psi}{\mathrm{d}r}+\frac{\psi}{r}\right)^2-2(1-\nu)\frac{\psi}{r}\frac{\mathrm{d}\psi}{\mathrm{d}r}\right]r\mathrm{d}r\mathrm{d}\theta,\end{aligned}$$

$$\begin{aligned}U_3=&\frac{t}{2E}\iint\limits_{s_3}[(\sigma_{r0}+\sigma_{\theta0})^2-2(1+\nu)\sigma_{r0}\sigma_{\theta0}]r\mathrm{d}r\mathrm{d}\theta\\&+\frac{th_0}{2}\iint\limits_{s_3}\left(\sigma_{r0}\frac{\mathrm{d}\psi}{\mathrm{d}r}+\frac{1}{r}\sigma_{\theta0}\psi\right)r\mathrm{d}r\mathrm{d}\theta\\&+\frac{D_f}{2}\iint\limits_{s_3}\left[\left(\frac{\mathrm{d}^2w}{\mathrm{d}r^2}+\frac{1}{r}\frac{\mathrm{d}w}{\mathrm{d}r}\right)^2-2(1-\nu)\frac{1}{r}\frac{\mathrm{d}w}{\mathrm{d}r}\frac{\mathrm{d}^2w}{\mathrm{d}r^2}\right]r\mathrm{d}r\mathrm{d}\theta\\&+\frac{D}{4}\iint\limits_{s_3}\left[\left(\frac{\mathrm{d}\psi}{\mathrm{d}r}+\frac{\psi}{r}\right)^2-2(1-\nu)\frac{\psi}{r}\frac{\mathrm{d}\psi}{\mathrm{d}r}\right]r\mathrm{d}r\mathrm{d}\theta,\end{aligned}$$

$$U_2=\frac{C}{2}\iint\limits_{s_2}\left(\psi+\frac{\mathrm{d}w}{\mathrm{d}r}\right)^2r\mathrm{d}r\mathrm{d}\theta \tag{2.17}$$

其中

$$D_f=\frac{Et^3}{12(1-\nu^2)},D=\frac{Eth_0^2}{2(1-\nu^2)},C=\frac{G_2h_0^2}{h} \tag{2.18}$$

均布横向载荷 q 的外力功为

$$V=\iint_s qwr\,\mathrm{d}r\mathrm{d}\theta \tag{2.19}$$

这样，夹层圆板的总势能为

$$\begin{aligned}U&=U_1+U_2+U_3-V\\&=\frac{t}{E}\iint_s\left[(\sigma_{r0}+\sigma_{\theta 0})^2-2(1+\nu)\sigma_{r0}\sigma_{\theta 0}\right]r\mathrm{d}r\mathrm{d}\theta\\&\quad+D_f\iint_s\left[\left(\frac{\mathrm{d}^2w}{\mathrm{d}r^2}+\frac{1}{r}\frac{\mathrm{d}w}{\mathrm{d}r}\right)^2-2(1-\nu)\frac{1}{r}\frac{\mathrm{d}w}{\mathrm{d}r}\frac{\mathrm{d}^2w}{\mathrm{d}r^2}\right]r\mathrm{d}r\mathrm{d}\theta\\&\quad+\frac{D}{2}\iint_s\left[\left(\frac{\mathrm{d}\psi}{\mathrm{d}r}+\frac{\psi}{r}\right)^2-2(1-\nu)\frac{\psi}{r}\frac{\mathrm{d}\psi}{\mathrm{d}r}\right]r\mathrm{d}r\mathrm{d}\theta\\&\quad+\frac{C}{2}\iint_s\left(\psi+\frac{\mathrm{d}w}{\mathrm{d}r}\right)^2r\mathrm{d}r\mathrm{d}\theta-\iint_s qwr\,\mathrm{d}r\mathrm{d}\theta\end{aligned}\tag{2.20}$$

根据势能原理，以 u、w、ψ 作自变量，对总势能变分为零，有

$$\delta U=0 \tag{2.21}$$

将式（2.20）代入，经部分积分后，可得夹层圆板在均布横向载荷 q 作用下的大挠度理论的平衡方程和边界条件：

$$\begin{aligned}&\sigma_{\theta 0}-\frac{\mathrm{d}}{\mathrm{d}r}(r\sigma_{r0})=0,\\&2D_f\frac{\mathrm{d}}{\mathrm{d}r}r\frac{\mathrm{d}}{\mathrm{d}r}\frac{1}{r}\frac{\mathrm{d}}{\mathrm{d}r}r\frac{\mathrm{d}w}{\mathrm{d}r}-2t\frac{\mathrm{d}}{\mathrm{d}r}\left(r\sigma_{r0}\frac{\mathrm{d}w}{\mathrm{d}r}\right)\\&-C\frac{\mathrm{d}}{\mathrm{d}r}\left[r\left(\psi+\frac{\mathrm{d}w}{\mathrm{d}r}\right)\right]-qr=0,\\&D\frac{\mathrm{d}}{\mathrm{d}r}\frac{1}{r}\frac{\mathrm{d}}{\mathrm{d}r}(r\psi)-C\left(\psi+\frac{\mathrm{d}w}{\mathrm{d}r}\right)=0\end{aligned}\tag{2.22a-c}$$

当 $r=0$ 时或当 $r=a$ 时，

$$\begin{aligned}&r\sigma_{r0}=0\text{ 或 }\delta u=0,\\&2rD_f\frac{\mathrm{d}}{\mathrm{d}r}\frac{1}{r}\frac{\mathrm{d}}{\mathrm{d}r}r\frac{\mathrm{d}w}{\mathrm{d}r}-2rt\sigma_{r0}\frac{\mathrm{d}w}{\mathrm{d}r}-rC\left(\psi+\frac{\mathrm{d}w}{\mathrm{d}r}\right)=0\text{ 或 }\delta w=0,\\&rM_r=Dr\left(\frac{\mathrm{d}\psi}{\mathrm{d}r}+\frac{\nu}{r}\psi\right)=0\text{ 或 }\delta\psi=0,\\&rm_r=-D_fr\left(\frac{\mathrm{d}^2w}{\mathrm{d}r^2}+\frac{\nu}{r}\frac{\mathrm{d}w}{\mathrm{d}r}\right)=0\text{ 或 }\frac{\mathrm{d}}{\mathrm{d}r}(\delta w)=0\end{aligned}\tag{2.23}$$

关于应变协调方程，可由方程（2.14）消去位移 u 求得

$$\sigma_{r0}-\frac{\mathrm{d}}{\mathrm{d}r}(r\sigma_{\theta 0})-\frac{E}{2}\left(\frac{\mathrm{d}w}{\mathrm{d}r}\right)^2=0 \tag{2.24}$$

方程（2.22）和（2.24）就是夹层圆板大挠度理论的基本方程，而边界条件（如简支、夹紧固定、滑动固定和悬空等）由式（2.23）决定。

引入应力函数 φ：

$$\sigma_{r0}=\frac{1}{r}\frac{\mathrm{d}\varphi}{\mathrm{d}r},\sigma_{\theta 0}=\frac{\mathrm{d}^2\varphi}{\mathrm{d}r^2} \tag{2.25}$$

则方程（2.22a）自动满足，而方程（2.22b，c）和（2.24）成为

$$\begin{aligned}&2D_f\frac{\mathrm{d}}{\mathrm{d}r}r\frac{\mathrm{d}}{\mathrm{d}r}\frac{1}{r}\frac{\mathrm{d}}{\mathrm{d}r}r\frac{\mathrm{d}w}{\mathrm{d}r}-2t\frac{\mathrm{d}}{\mathrm{d}r}\left(\frac{\mathrm{d}\varphi}{\mathrm{d}r}\frac{\mathrm{d}w}{\mathrm{d}r}\right)\\&-C\frac{\mathrm{d}}{\mathrm{d}r}\left[r\left(\psi+\frac{\mathrm{d}w}{\mathrm{d}r}\right)\right]-qr=0,\\&D\frac{\mathrm{d}}{\mathrm{d}r}\frac{1}{r}\frac{\mathrm{d}}{\mathrm{d}r}(r\psi)-C\left(\psi+\frac{\mathrm{d}w}{\mathrm{d}r}\right)=0,\\&\frac{\mathrm{d}}{\mathrm{d}r}\frac{1}{r}\frac{\mathrm{d}}{\mathrm{d}r}r\frac{\mathrm{d}\varphi}{\mathrm{d}r}+\frac{E}{2r}\left(\frac{\mathrm{d}w}{\mathrm{d}r}\right)^2=0\end{aligned} \tag{2.26a-c}$$

将方程（2.26a）乘以 $\mathrm{d}r$，积分一次，可解得

$$\psi=\frac{2D_f}{C}\frac{\mathrm{d}}{\mathrm{d}r}\frac{1}{r}\frac{\mathrm{d}}{\mathrm{d}r}r\frac{\mathrm{d}w}{\mathrm{d}r}-\left(\frac{2t}{Cr}\frac{\mathrm{d}\varphi}{\mathrm{d}r}+1\right)\frac{\mathrm{d}w}{\mathrm{d}r}-\frac{1}{2C}qr \tag{2.27}$$

将此式代入方程（2.26b），得

$$\begin{aligned}&\frac{2DD_f}{C}\frac{\mathrm{d}}{\mathrm{d}r}\frac{1}{r}\frac{\mathrm{d}}{\mathrm{d}r}r\frac{\mathrm{d}}{\mathrm{d}r}\frac{1}{r}\frac{\mathrm{d}}{\mathrm{d}r}r\frac{\mathrm{d}w}{\mathrm{d}r}-(D+2D_f)\frac{\mathrm{d}}{\mathrm{d}r}\frac{1}{r}\frac{\mathrm{d}}{\mathrm{d}r}r\frac{\mathrm{d}w}{\mathrm{d}r}\\&-\frac{2tD}{C}\frac{\mathrm{d}}{\mathrm{d}r}\frac{1}{r}\frac{\mathrm{d}}{\mathrm{d}r}\left(\frac{\mathrm{d}\varphi}{\mathrm{d}r}\frac{\mathrm{d}w}{\mathrm{d}r}\right)+\frac{2t}{r}\frac{\mathrm{d}\varphi}{\mathrm{d}r}\frac{\mathrm{d}w}{\mathrm{d}r}+\frac{1}{2}qr=0\end{aligned} \tag{2.28}$$

于是，基本方程组（2.22）和（2.24）被简化为（2.28）和（2.26c）两个方程。

由方程（2.28）、（2.26c）和边界条件（2.23）不难看出，当夹心剪切模量 G_2 无限增强（即 $C\to\infty$）时，它们将转化为单层圆板（厚度为 $h+2t$）的基本方程和边界条件。

三、方程的简化

若表板很薄，即 $t\ll h$，那么上述方程可被简化。此时式（2.1）～（2.3）成为

上表板

$$u_1=u+\frac{1}{2}h_0\psi,v_1=0,w_1=w \tag{3.1}$$

下表板

$$u_3=u-\frac{1}{2}h_0\psi,v_3=0,w_3=w \tag{3.2}$$

夹心

$$u_2=u+z\psi,v_2=0,w_2=w \tag{3.3}$$

这时，板总厚度可用两表板中面间的距离 h_0 来量度。

几何方程和胡克定律仍为式（2.4）、（2.5）和（2.9）、（2.10）。经过类似的运算，得表板和夹心的应变和应力公式

$$\varepsilon_{r1}=\frac{\mathrm{d}u}{\mathrm{d}r}+\frac{h_0}{2}\frac{\mathrm{d}\psi}{\mathrm{d}r}+\frac{1}{2}\left(\frac{\mathrm{d}w}{\mathrm{d}r}\right)^2,\varepsilon_{\theta 1}=\frac{u}{r}+\frac{h_0}{2r}\psi, \tag{3.4}$$

$$\varepsilon_{r3}=\frac{\mathrm{d}u}{\mathrm{d}r}-\frac{h_0}{2}\frac{\mathrm{d}\psi}{\mathrm{d}r}+\frac{1}{2}\left(\frac{\mathrm{d}w}{\mathrm{d}r}\right)^2,\varepsilon_{\theta 3}=\frac{u}{r}-\frac{h_0}{2r}\psi, \tag{3.5}$$

$$\gamma_{rz2}=\psi+\frac{\mathrm{d}w}{\mathrm{d}r}, \tag{3.6}$$

$$\sigma_{r1}=\sigma_{r0}+\frac{Eh_0}{2(1-\nu^2)}\left(\frac{\mathrm{d}\psi}{\mathrm{d}r}+\frac{\nu}{r}\psi\right),$$

$$\sigma_{\theta 1}=\sigma_{\theta 0}+\frac{Eh_0}{2(1-\nu^2)}\left(\frac{\psi}{r}+\nu\frac{\mathrm{d}\psi}{\mathrm{d}r}\right), \tag{3.7}$$

$$\sigma_{r3}=\sigma_{r0}-\frac{Eh_0}{2(1-\nu^2)}\left(\frac{\mathrm{d}\psi}{\mathrm{d}r}+\frac{\nu}{r}\psi\right),$$

$$\sigma_{\theta 3}=\sigma_{\theta 0}-\frac{Eh_0}{2(1-\nu^2)}\left(\frac{\psi}{r}+\nu\frac{\mathrm{d}\psi}{\mathrm{d}r}\right), \tag{3.8}$$

$$\tau_{rz2}=G_2\left(\psi+\frac{\mathrm{d}w}{\mathrm{d}r}\right), \tag{3.9}$$

这里，σ_{r0}和$\sigma_{\theta 0}$仍与式（2.14）相同。

将式（3.7）～（3.9）代入应变能公式（2.16），并应用式（2.19），得总势能

$$\begin{aligned}U=&\frac{t}{E}\iint_s[(\sigma_{r0}+\sigma_{\theta 0})^2-2(1+\nu)\sigma_{r0}\sigma_{\theta 0}]r\mathrm{d}r\mathrm{d}\theta\\&+\frac{D}{2}\iint_s\left[\left(\frac{\mathrm{d}\psi}{\mathrm{d}r}+\frac{\psi}{r}\right)^2-2(1-\nu)\frac{\psi}{r}\frac{\mathrm{d}\psi}{\mathrm{d}r}\right]r\mathrm{d}r\mathrm{d}\theta\\&+\frac{G_2h_0}{2}\iint_s\left(\psi+\frac{\mathrm{d}w}{\mathrm{d}r}\right)^2r\mathrm{d}r\mathrm{d}\theta-\iint_s qwr\mathrm{d}r\mathrm{d}\theta\end{aligned} \tag{3.10}$$

其中D仍由式（2.18）的第二式表示。

对总势能变分为零，即

$$\delta U=0 \tag{3.11}$$

推得简化情况下的平衡方程和边界条件：

$$\begin{gathered}\sigma_{\theta 0}-\frac{\mathrm{d}}{\mathrm{d}r}(r\sigma_{r0})=0,\\2t\frac{\mathrm{d}}{\mathrm{d}r}\left(r\sigma_{r0}\frac{\mathrm{d}w}{\mathrm{d}r}\right)+G_2h_0\frac{\mathrm{d}}{\mathrm{d}r}\left[r\left(\psi+\frac{\mathrm{d}w}{\mathrm{d}r}\right)\right]+qr=0,\\Dr\frac{\mathrm{d}}{\mathrm{d}r}\frac{1}{r}\frac{\mathrm{d}}{\mathrm{d}r}(r\psi)-G_2h_0r\left(\psi+\frac{\mathrm{d}w}{\mathrm{d}r}\right)=0\end{gathered} \tag{3.12a-c}$$

当$r=0$时或当$r=a$时，

$$\begin{aligned}r\sigma_{r0}&=0\text{ 或 }\delta u=0,\\rQ_r&=G_2h_0r\left(\psi+\frac{\mathrm{d}w}{\mathrm{d}r}\right)-2tr\sigma_{r0}\frac{\mathrm{d}w}{\mathrm{d}r}\text{ 或 }\delta w=0,\\rM_r&=Dr\left(\frac{\mathrm{d}\psi}{\mathrm{d}r}+\frac{\nu}{r}\psi\right)=0\text{ 或 }\delta\psi=0\end{aligned} \tag{3.13}$$

而应变协调方程仍为方程（2.24）。

引入如式（2.25）的应力函数φ，则方程（3.12a）自动满足。

将方程（3.12b）乘以 $\mathrm{d}r$，积分一次，可解得

$$\psi=-\frac{1}{G_2h_0}\left(\frac{2t}{r}\frac{\mathrm{d}\varphi}{\mathrm{d}r}\frac{\mathrm{d}w}{\mathrm{d}r}+\frac{1}{2}qr\right)-\frac{\mathrm{d}w}{\mathrm{d}r} \tag{3.14}$$

将式（3.14）代入方程（3.12c），得

$$Dr\frac{\mathrm{d}}{\mathrm{d}r}\frac{1}{r}\frac{\mathrm{d}}{\mathrm{d}r}r\frac{\mathrm{d}w}{\mathrm{d}r}-2t\left(1-\frac{D}{G_2h_0}r\frac{\mathrm{d}}{\mathrm{d}r}\frac{1}{r}\frac{\mathrm{d}}{\mathrm{d}r}\right)\left(\frac{\mathrm{d}\varphi}{\mathrm{d}r}\frac{\mathrm{d}w}{\mathrm{d}r}\right)-\frac{1}{2}qr^2=0 \tag{3.15}$$

则方程（3.15）和（2.26c）就是表板很薄情况下的基本方程组。显而易见，此简化方程较前大为简单。由于实际情况中夹层板大多属于这种情况，所以这组方程非常实用。若将 Reissner[13] 所得的夹层矩形板的大挠度方程进行极坐标变换，所得方程与此简化方程完全相同。

四、在滑动固定边界条件下夹层圆板的非线性轴对称弯曲问题的求解

现在，应用简化方程，研究如图 2 所示的夹层圆板的非线性轴对称弯曲问题。若以 σ_{r0} 和 w 为未知数，由方程（3.15）、（2.26c）和（3.13）得边值问题：

$$D\frac{\mathrm{d}}{\mathrm{d}r}\frac{1}{r}\frac{\mathrm{d}}{\mathrm{d}r}r\frac{\mathrm{d}w}{\mathrm{d}r}=2t\left(\frac{1}{r}-\frac{D}{G_2h_0}\frac{\mathrm{d}}{\mathrm{d}r}\frac{1}{r}\frac{\mathrm{d}}{\mathrm{d}r}\right)\left(r\sigma_{r0}\frac{\mathrm{d}w}{\mathrm{d}r}\right)+\frac{1}{2}qr,$$

$$\frac{\mathrm{d}}{\mathrm{d}r}\frac{1}{r}\frac{\mathrm{d}}{\mathrm{d}r}(r^2\sigma_{r0})=-\frac{E}{2r}\left(\frac{\mathrm{d}w}{\mathrm{d}r}\right)^2 \tag{4.1}$$

在 $r=a$ 时，

$$w=0,\psi=-\frac{1}{G_2h_0}\left(2t\sigma_{r0}\frac{\mathrm{d}w}{\mathrm{d}r}+\frac{1}{2}qr\right)-\frac{\mathrm{d}w}{\mathrm{d}r}=0,\sigma_{r0}=0 \tag{4.2}$$

在 $r=0$ 时，

$$\psi=-\frac{1}{G_2h_0}\left(2t\sigma_{r0}\frac{\mathrm{d}w}{\mathrm{d}r}+\frac{1}{2}qr\right)-\frac{\mathrm{d}w}{\mathrm{d}r}=0,\sigma_{r0}\text{ 有限} \tag{4.3}$$

为计算简单起见，引入无量纲量

$$\rho=\frac{r}{a},W=\sqrt{2(1-\nu^2)}\frac{w}{h_0},S_r=\frac{2ta^2}{D}\sigma_{r0},$$

$$k=\frac{D}{G_2h_2a^2},P=\frac{\sqrt{2(1-\nu^2)}a^4}{2h_0D}q \tag{4.4}$$

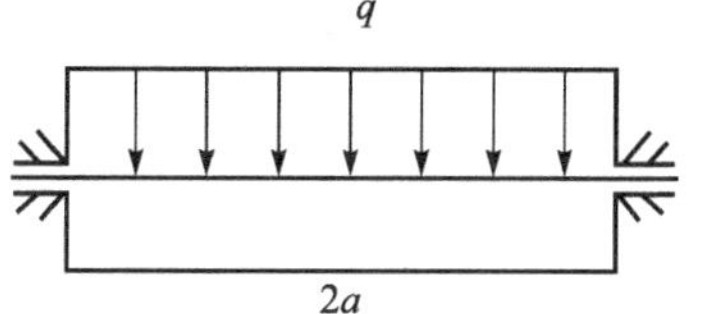

图 2　具有滑动固定边界条件的夹层圆板

则边值问题（4.1）～（4.3）成为

$$L\left(\rho\frac{\mathrm{d}W}{\mathrm{d}\rho}\right)=\left(\frac{1}{\rho}-kL\right)\left(\rho S_r\frac{\mathrm{d}W}{\mathrm{d}\rho}\right)+P\rho,$$

$$L(\rho^2S_r)=-\frac{1}{\rho}\left(\frac{\mathrm{d}W}{\mathrm{d}\rho}\right)^2, \tag{4.5a,b}$$

在 $\rho=1$ 时，

$$W=0,\frac{\mathrm{d}W}{\mathrm{d}\rho}=-k\left(S_r\frac{\mathrm{d}W}{\mathrm{d}\rho}+P\rho\right),S_r=0 \tag{4.6a-c}$$

在 $\rho=0$ 时，

$$\frac{\mathrm{d}W}{\mathrm{d}\rho}=-kS_r\frac{\mathrm{d}W}{\mathrm{d}\rho},S_r\text{ 有限} \tag{4.7a-c}$$

其中

$$L(\cdots)=\frac{\mathrm{d}}{\mathrm{d}\rho}\frac{1}{\rho}\frac{\mathrm{d}}{\mathrm{d}\rho}(\cdots) \tag{4.8}$$

我们使用修正迭代法求解此组非线性边值问题。在一次近似中，先略去方程（4.5a）和边界条件（4.6b）、（4.7a）中的非线性项，然后将由方程（4.5a）所得的解 W_1 代入方程（4.5b）的右端，便得下列线性边值问题

$$L\left(\rho\frac{\mathrm{d}W_1}{\mathrm{d}\rho}\right)=P\rho,$$

$$L(\rho^2 S_{r1})=-\frac{1}{\rho}\left(\frac{\mathrm{d}W_1}{\mathrm{d}\rho}\right) \tag{4.9a,b}$$

在 $\rho=1$ 时，

$$W_1=0,\frac{\mathrm{d}W_1}{\mathrm{d}\rho}=-kP\rho,S_{r1}=0 \tag{4.10a-c}$$

在 $\rho=0$ 时，

$$\frac{\mathrm{d}W_1}{\mathrm{d}\rho}=0,S_{r1}\text{ 有限} \tag{4.11a,b}$$

应用边界条件（4.10a，b）和（4.11a），方程（4.9a）的解是

$$W_1=\frac{1}{8}\left(\frac{1}{\lambda_1}-\frac{1}{2}\lambda_2\rho^2+\frac{1}{4}\rho^4\right)P \tag{4.12}$$

其中

$$\lambda_1=\frac{4}{1+16k},\ \lambda_2=1+8k \tag{4.13}$$

我们以无量纲中心挠度作为迭代参数：

$$W_0=W|_{\rho=0} \tag{4.14}$$

应用式（4.12），得一次近似中的载荷和中心挠度的关系式

$$P=\alpha_1 W_0 \tag{4.15}$$

其中

$$\alpha_1=8\lambda_1 \tag{4.16}$$

显而易见，式（4.15）就是夹层圆板小挠度理论的解。

将式（4.15）代入式（4.12），有

$$W_1=\left(1-\frac{1}{2}\lambda_1\lambda_2\rho^2+\frac{1}{4}\lambda_1\rho^4\right)W_0 \tag{4.17}$$

再将这个 W_1 值代入方程（4.9b），应用边界条件（4.10c）和（4.11b），方程（4.9b）的解是

$$S_{r1}=\frac{\lambda_1^2}{4}\left[\left(\frac{1}{2}\lambda_2^2-\frac{1}{3}\lambda_2+\frac{1}{12}\right)-\frac{1}{2}\lambda_2^2\rho^2+\frac{1}{3}\lambda_2\rho^4-\frac{1}{12}\rho^6\right]W_0^2 \tag{4.18}$$

此即一次近似中的径向应力公式。

在二次近似中，有下面的边值问题

$$L\left(\rho\frac{\mathrm{d}W_2}{\mathrm{d}\rho}\right)=\left(\frac{1}{\rho}-kL\right)\left(\rho S_{r1}\frac{\mathrm{d}W_1}{\mathrm{d}\rho}\right)+P\rho,$$

$$L(\rho S_{r2}) = -\frac{1}{\rho}\left(\frac{\mathrm{d}W_2}{\mathrm{d}\rho}\right)^2 \tag{4.19a,b}$$

在 $\rho=1$ 时，

$$W_2 = 0, \frac{\mathrm{d}W_2}{\mathrm{d}\rho} = -k\left(S_{r1}\frac{\mathrm{d}W_1}{\mathrm{d}\rho} + P\rho\right), S_{r2} = 0 \tag{4.20a-c}$$

在 $\rho=0$ 时，

$$\frac{\mathrm{d}W_2}{\mathrm{d}\rho} = -kS_{r1}\frac{\mathrm{d}W_1}{\mathrm{d}\rho}, S_{r2}\text{ 有限} \tag{4.21a,b}$$

应用式（4.17）和（4.18），在边界条件（4.20a，b）和（4.21 a）情况下，方程（4.19a）的解是

$$\begin{aligned} W_2 = & B_2 + \frac{1}{2}A_2\rho^2 + \frac{1}{32}P\rho^4 - \frac{\lambda_1^3}{64}\left[\frac{1}{4}\left(\lambda_2^4 + \lambda_2^3 - \frac{7}{3}\lambda_2^2 + \lambda_2 - \frac{1}{6}\right)\rho^4\right. \\ & - \frac{1}{9}\left(3\lambda_2^3 - 2\lambda_2^2 - \frac{1}{3}\lambda_2 + \frac{1}{12}\right)\rho^6 + \frac{5\lambda_2}{48}\left(\frac{4}{3}\lambda_2 - 1\right)\rho^8 \\ & \left. - \frac{1}{60}\left(\frac{3}{2}\lambda_2 - 1\right)\rho^{10} + \frac{1}{1080}\rho^{12}\right]W_0^3 \end{aligned} \tag{4.22a}$$

其中

$$\begin{aligned} A_2 &= -\frac{\lambda_2}{8}P + \frac{\lambda_1^3}{64}\left(\lambda_2^4 - \lambda_2^3 + \frac{1}{9}\lambda_2^2 + \frac{5}{36}\lambda_2 - \frac{2}{45}\right)W_0^3, \\ B_2 &= \frac{1}{8\lambda_1}P - \frac{\lambda_1^3}{128}\left(\frac{1}{2}\lambda_2^4 - \frac{5}{6}\lambda_2^3 + \frac{5}{9}\lambda_2^2 - \frac{191}{1080}\lambda_2 + \frac{1}{45}\right)W_0^3 \end{aligned} \tag{4.22b}$$

利用定义（4.14），由式（4.22a，b）得二次近似中的载荷和中心挠度的关系式

$$P = \alpha_1 W_0 + \alpha_3 W_0^3 \tag{4.23}$$

其中

$$\alpha_3 = \frac{\lambda_1^4}{16}\left(\frac{1}{2}\lambda_2^4 - \frac{5}{6}\lambda_2^3 + \frac{5}{9}\lambda_2^2 - \frac{191}{1080}\lambda_2 + \frac{1}{45}\right) \tag{4.24}$$

将式（4.23）代入式（4.22a，b），得二次近似挠度公式

$$\begin{aligned} W = & W_0 - \frac{\lambda_1}{2}\left[\lambda_2 W_0 - \frac{\lambda_1^3}{128}\left(\frac{1}{2}\lambda_2^5 - \frac{2}{3}\lambda_2^4 + \frac{1}{18}\lambda_2^3 + \frac{281}{1080}\lambda_2^2 - \frac{49}{360}\lambda_2 + \frac{1}{45}\right)W_0^3\right]\rho^2 \\ & + \frac{\lambda_1}{4}\left[W_0 - \frac{\lambda_1^3}{128}\left(\lambda_2^5 - 2\lambda_2^3 + \frac{29}{18}\lambda_2^2 - \frac{529}{1080}\lambda_2 + \frac{11}{180}\right)W_0^3\right]\rho^4 \\ & + \frac{\lambda_1^3}{192}\left[\left(\lambda_2^3 - \frac{2}{3}\lambda_2^2 - \frac{1}{9}\lambda_2 + \frac{1}{36}\right)\rho^6 - \frac{5\lambda_2}{16}\left(\frac{4}{3}\lambda_2 - 1\right)\rho^8\right. \\ & \left. + \frac{1}{20}\left(\frac{3}{2}\lambda_2 - 1\right)\rho^{10} - \frac{1}{360}\rho^{12}\right]W_0^3 \end{aligned} \tag{4.25}$$

将此式代入方程（4.19b），应用边界条件（4.20c）和（4.21b），得方程（4.19b）的解，即二次近似径向应力公式

$$\begin{aligned} S_{r2} = & (n_{0,2}W_0^2 + n_{0,4}W_0^4 + n_{0,6}W_0^6) - (n_{2,2}W_0^2 + n_{2,4}W_0^4 + n_{2,6}W_0^6)\rho^2 \\ & - (n_{4,2}W_0^2 + n_{4,4}W_0^4 + n_{4,6}W_0^6)\rho^4 - (n_{6,2}W_0^2 + n_{6,4}W_0^4 + n_{6,6}W_0^6)\rho^6 \\ & - (n_{8,4}W_0^4 + n_{8,6}W_0^6)\rho^8 - (n_{10,4}W_0^4 + n_{10,6}W_0^6)\rho^{10} \end{aligned}$$

$$
\begin{aligned}
&-(n_{12,4}W_0^4+n_{12,6}W_0^6)\rho^{12}-(n_{14,4}W_0^4+n_{14,6}W_0^6)\rho^{14}\\
&-(n_{16,6}\rho^{16}+n_{18,6}\rho^{18}+n_{20,6}\rho^{20}+n_{22,6}\rho^{22})W_0^6
\end{aligned}
\tag{4.26}
$$

其中

$$
\begin{aligned}
n_{0,2}&=-\frac{\lambda_1^2}{4}\left(\frac{1}{2}\lambda_2^2-\frac{1}{3}\lambda_2+\frac{1}{12}\right),\\
n_{0,4}&=\frac{\lambda_1^5}{1536}\left(\frac{1}{2}\lambda_2^6-\lambda_2^5+\frac{11}{15}\lambda_2^4-\frac{11}{40}\lambda_2^3+\frac{593}{7560}\lambda_2^2-\frac{71}{3024}\lambda_2+\frac{1}{240}\right),\\
n_{0,6}&=-\frac{\lambda_1^8}{786432}\left(\frac{1}{2}\lambda_2^{10}-\frac{26}{15}\lambda_2^9+\frac{12}{5}\lambda_2^8-\frac{6229}{3780}\lambda_2^7+\frac{2153}{3780}\lambda_2^6-\frac{2164}{14175}\lambda_2^5\right.\\
&\qquad\left.+\frac{64741}{453600}\lambda_2^4-\frac{186223}{1603800}\lambda_2^3+\frac{4317389}{89812800}\lambda_2^2-\frac{5363}{534600}\lambda_2+\frac{6431}{7484400}\right),\\
n_{2,2}&=\frac{1}{8}\lambda_1^2\lambda_2^2,\\
n_{2,4}&=-\frac{\lambda_1^5\lambda_2}{512}\left(\frac{1}{2}\lambda_2^5-\frac{2}{3}\lambda_2^4+\frac{1}{18}\lambda_2^3+\frac{281}{1080}\lambda_2^2-\frac{49}{360}\lambda_2+\frac{1}{45}\right),\\
n_{2,6}&=\frac{\lambda_1^8}{131072}\left(\frac{1}{4}\lambda_2^{10}-\frac{2}{3}\lambda_2^9+\frac{1}{2}\lambda_2^8+\frac{67}{360}\lambda_2^7-\frac{311}{648}\lambda_2^6+\frac{2261}{9720}\lambda_2^5\right.\\
&\qquad\left.+\frac{26761}{1166400}\lambda_2^4-\frac{13289}{194400}\lambda_2^3+\frac{11699}{388800}\lambda_2^2-\frac{49}{8100}\lambda_2+\frac{1}{2025}\right),\\
n_{4,2}&=-\frac{1}{12}\lambda_1^2\lambda_2,\\
n_{4,4}&=\frac{\lambda_1^5}{1536}\left(\lambda_2^6+\frac{1}{2}\lambda_2^5-\frac{8}{3}\lambda_2^4+\frac{5}{3}\lambda_2^3-\frac{31}{135}\lambda_2^2-\frac{3}{40}\lambda_2+\frac{1}{45}\right),\\
n_{4,6}&=-\frac{\lambda_1^8}{393216}\left(\lambda_2^{10}-\frac{4}{3}\lambda_2^9-\frac{17}{9}\lambda_2^8+\frac{2591}{540}\lambda_2^7-\frac{3383}{1080}\lambda_2^6-\frac{167}{1620}\lambda_2^5\right.\\
&\qquad\left.+\frac{101}{81}\lambda_2^4-\frac{452309}{583200}\lambda_2^3+\frac{15341}{64800}\lambda_2^2-\frac{3733}{97200}\lambda_2+\frac{11}{4050}\right),\\
n_{6,2}&=\frac{1}{48}\lambda_1^2,\\
n_{6,4}&=-\frac{\lambda_1^5}{9216}\left(9\lambda_2^5-7\lambda_2^4-\frac{14}{3}\lambda_2^3+\frac{16}{3}\lambda_2^2-\frac{559}{360}\lambda_2+\frac{11}{60}\right),\\
n_{6,6}&=\frac{\lambda_1^8}{2359296}\left(3\lambda_2^{10}+6\lambda_2^9-27\lambda_2^8+21\lambda_2^7+\frac{1823}{180}\lambda_2^6-\frac{1118}{45}\lambda_2^5\right.\\
&\qquad\left.+\frac{2705}{162}\lambda_2^4-\frac{2386}{405}\lambda_2^3+\frac{462781}{388800}\lambda_2^2-\frac{1091}{8100}\lambda_2+\frac{3}{400}\right),\\
n_{8,4}&=\frac{\lambda_1^4}{3840}\left(\frac{14}{3}\lambda_2^3-\frac{13}{4}\lambda_2^2-\frac{1}{3}\lambda_2+\frac{1}{12}\right),\\
n_{8,6}&=-\frac{\lambda_1^7}{491520}\left(3\lambda_2^8-\frac{7}{6}\lambda_2^7-\frac{581}{72}\lambda_2^6+\frac{1063}{108}\lambda_2^5-\frac{11861}{3240}\lambda_2^4-\frac{401}{4320}\lambda_2^3\right.\\
&\qquad\left.+\frac{4957}{12960}\lambda_2^2-\frac{1153}{12960}\lambda_2+\frac{11}{2160}\right),
\end{aligned}
$$

$$n_{10,4}=-\frac{\lambda_1^4\lambda_2}{23040}\left(\frac{49}{6}\lambda_2-6\right),$$

$$n_{10,6}=\frac{\lambda_1^7}{368640}\left(\frac{7}{3}\lambda_2^7-\frac{105}{32}\lambda_2^6-\frac{25}{48}\lambda_2^5+\frac{2441}{864}\lambda_2^4-\frac{81031}{51840}\lambda_2^3+\frac{5327}{17280}\lambda_2^2-\frac{97}{8640}\lambda_2-\frac{29}{8640}\right),$$

$$n_{12,4}=\frac{\lambda_1^4}{32256}\left(\frac{47}{30}\lambda_2-1\right),$$

$$n_{12,6}=-\frac{\lambda_1^7}{24772608}\left(89\lambda_2^6-\frac{2387}{15}\lambda_2^5+\frac{3128}{45}\lambda_2^4+\frac{1787}{90}\lambda_2^3-\frac{108443}{5400}\lambda_2^2+\frac{3841}{900}\lambda_2-\frac{161}{450}\right),$$

$$n_{14,4}=-\frac{1}{645120}\lambda_1^4,$$

$$n_{14,6}=\frac{\lambda_1^7}{49545216}\left(\frac{914}{15}\lambda_2^5-\frac{697}{6}\lambda_2^4+\frac{1391}{20}\lambda_2^3-\frac{1219}{120}\lambda_2^2-\frac{488}{225}\lambda_2+\frac{43}{150}\right),$$

$$n_{16,6}=-\frac{\lambda_1^6}{53084160}\left(27\lambda_2^3-\frac{147}{4}\lambda_2^2+\frac{221}{18}\lambda_2+\frac{1}{18}\right),$$

$$n_{18,6}=\frac{\lambda_1^6}{53084160}\left(\frac{113}{36}\lambda_2^2-\frac{11}{3}\lambda_2+1\right),$$

$$n_{20,6}=-\frac{\lambda_1^6}{486604800}\left(\frac{3}{2}\lambda_2-1\right),$$

$$n_{22,6}=\frac{1}{17517772800}\lambda_1^6$$

在三次近似中，关于挠度 W 的边值问题为

$$L\left(\rho\frac{\mathrm{d}W_3}{\mathrm{d}\rho}\right)=\left(\frac{1}{\rho}-kL\right)\left(\rho S_{r2}\frac{\mathrm{d}W_2}{\mathrm{d}\rho}\right)+P\rho \tag{4.27}$$

在 $\rho=1$ 时，

$$W_3=0,\frac{\mathrm{d}W_3}{\mathrm{d}\rho}=-k\left(S_{r2}\frac{\mathrm{d}W_2}{\mathrm{d}\rho}+P\rho\right) \tag{4.28}$$

在 $\rho=0$ 时，

$$\frac{\mathrm{d}W_3}{\mathrm{d}\rho}=-kS_{r2}\frac{\mathrm{d}W_2}{\mathrm{d}\rho} \tag{4.29}$$

应用式（4.25）和（4.26），在边界条件（4.28）和（4.29）情况下，方程（4.27）的解是

$$\begin{aligned}W_3=&B_3+\frac{1}{2}A_3\rho^2+\frac{1}{4}\left(\frac{1}{8}P+\frac{m_1}{8}-km_3\right)\rho^4+\frac{1}{6}\left(\frac{m_3}{24}-km_5\right)\rho^6\\&+\frac{1}{8}\left(\frac{m_5}{48}-km_7\right)\rho^8+\frac{1}{10}\left(\frac{m_7}{80}-km_9\right)\rho^{10}+\frac{1}{12}\left(\frac{m_9}{120}-km_{11}\right)\rho^{12}\\&+\frac{1}{14}\left(\frac{m_{11}}{168}-km_{13}\right)\rho^{14}+\frac{1}{16}\left(\frac{m_{13}}{224}-km_{15}\right)\rho^{16}+\frac{1}{18}\left(\frac{m_{15}}{288}-km_{17}\right)\rho^{18}\\&+\frac{1}{20}\left(\frac{m_{17}}{360}-km_{19}\right)\rho^{20}+\frac{1}{22}\left(\frac{m_{19}}{440}-km_{21}\right)\rho^{22}+\frac{1}{24}\left(\frac{m_{21}}{528}-km_{23}\right)\rho^{24}\\&+\frac{1}{26}\left(\frac{m_{23}}{624}-km_{25}\right)\rho^{26}+\frac{1}{28}\left(\frac{m_{25}}{728}-km_{27}\right)\rho^{28}+\frac{1}{30}\left(\frac{m_{27}}{840}-km_{29}\right)\rho^{30}\end{aligned}$$

$$+\frac{1}{32}\left(\frac{m_{29}}{960}-km_{31}\right)\rho^{32}+\frac{1}{34}\left(\frac{m_{31}}{1088}-km_{33}\right)\rho^{34}+\frac{m_{33}}{44064}\rho^{36} \qquad (4.30)$$

其中

$$A_3=-\left(\frac{\lambda_2}{8}P+\frac{\lambda_2 m_1}{8}+\frac{m_3}{24}+\frac{m_5}{48}+\frac{m_7}{80}+\frac{m_9}{120}+\frac{m_{11}}{168}+\frac{m_{13}}{224}+\frac{m_{15}}{288}\right.$$
$$\left.+\frac{m_{17}}{360}+\frac{m_{19}}{440}+\frac{m_{21}}{528}+\frac{m_{23}}{624}+\frac{m_{25}}{728}+\frac{m_{27}}{840}+\frac{m_{29}}{960}+\frac{m_{31}}{1088}+\frac{m_{33}}{1224}\right),$$

$$B_3=-\left[\frac{1}{2}A_3+\frac{1}{32}P+\frac{m_1}{32}+\frac{m_3}{4}\left(\frac{1}{36}-k\right)\right.$$
$$+\frac{m_5}{6}\left(\frac{1}{64}-k\right)+\frac{m_7}{8}\left(\frac{1}{100}-k\right)+\frac{m_9}{10}\left(\frac{1}{144}-k\right)+\frac{m_{11}}{12}\left(\frac{1}{196}-k\right)$$
$$+\frac{m_{13}}{14}\left(\frac{1}{256}-k\right)+\frac{m_{15}}{16}\left(\frac{1}{324}-k\right)+\frac{m_{17}}{18}\left(\frac{1}{400}-k\right)+\frac{m_{19}}{20}\left(\frac{1}{484}-k\right)$$
$$+\frac{m_{21}}{22}\left(\frac{1}{576}-k\right)+\frac{m_{23}}{24}\left(\frac{1}{676}-k\right)+\frac{m_{25}}{26}\left(\frac{1}{784}-k\right)$$
$$+\frac{m_{27}}{28}\left(\frac{1}{900}-k\right)+\frac{m_{29}}{30}\left(\frac{1}{1024}-k\right)+\frac{m_{31}}{32}\left(\frac{1}{1156}-k\right)$$
$$\left.+\frac{m_{33}}{34}\left(\frac{1}{1296}-k\right)\right],$$

$$m_1=l_{1,1}n_{0,2}W_0^3+(l_{1,1}n_{0,4}+l_{1,3}n_{0,2})W_0^5+(l_{1,1}n_{0,6}+l_{1,3}n_{0,4})W_0^7$$
$$+l_{1,3}n_{0,6}W_0^9,$$

$$m_3=(l_{3,1}n_{0,2}-l_{1,1}n_{2,2})W_0^3+(l_{3,1}n_{0,4}+l_{3,3}n_{0,2}-l_{1,1}n_{2,4}-l_{1,3}n_{2,2})W_0^5$$
$$+(l_{3,1}n_{0,6}+l_{3,3}n_{0,4}-l_{1,1}n_{2,6}-l_{1,3}n_{2,4})W_0^7+(l_{3,3}n_{0,6}-l_{1,3}n_{2,6})W_0^9,$$

$$m_5=-(l_{3,1}n_{2,2}+l_{1,1}n_{4,2})W_0^3+(l_{5,3}n_{0,2}-l_{3,1}n_{2,4}-l_{3,3}n_{2,2}-l_{1,1}n_{4,4}-l_{1,3}n_{4,2})W_0^5$$
$$+(l_{5,3}n_{0,4}-l_{3,1}n_{2,6}-l_{3,3}n_{2,4}-l_{1,1}n_{4,6}-l_{1,3}n_{4,4})W_0^7$$
$$+(l_{5,3}n_{0,6}-l_{3,3}n_{2,6}-l_{1,3}n_{4,6})W_0^9,$$

$$m_7=-(l_{3,1}n_{4,2}+l_{1,1}n_{6,2})W_0^3+(l_{7,3}n_{0,2}-l_{5,3}n_{2,2}-l_{3,1}n_{4,4}$$
$$-l_{3,3}n_{4,2}-l_{1,1}n_{6,4}-l_{1,3}n_{6,2})W_0^5$$
$$+(l_{7,3}n_{0,4}-l_{5,3}n_{2,4}-l_{3,1}n_{4,6}-l_{3,3}n_{4,4}-l_{1,1}n_{6,6}-l_{1,3}n_{6,4})W_0^7$$
$$+(l_{7,3}n_{0,8}-l_{5,3}n_{2,6}-l_{3,3}n_{4,6}-l_{1,3}n_{6,6})W_0^9,$$

$$m_9=-l_{3,1}n_{6,2}W_0^3+(l_{9,3}n_{0,2}-l_{7,3}n_{2,2}-l_{5,3}n_{4,2}-l_{3,1}n_{6,4}-l_{3,3}n_{6,2}$$
$$-l_{1,1}n_{8,4})W_0^5+(l_{9,3}n_{0,4}-l_{7,3}n_{2,4}-l_{5,3}n_{4,4}-l_{3,1}n_{6,6}$$
$$-l_{3,3}n_{6,4}-l_{1,1}n_{8,6}-l_{1,3}n_{8,4})W_0^7+(l_{9,3}n_{0,6}-l_{7,3}n_{2,6}$$
$$-l_{5,3}n_{4,6}-l_{3,3}n_{6,6}-l_{1,3}n_{8,6})W_0^9,$$

$$m_{11}=(l_{11,3}n_{0,2}-l_{9,3}n_{2,2}-l_{7,3}n_{4,2}-l_{5,3}n_{6,2}-l_{3,1}n_{8,4}-l_{1,1}n_{10,4})W_0^5$$
$$+(l_{11,3}n_{0,4}-l_{9,3}n_{2,4}-l_{7,3}n_{4,4}-l_{5,3}n_{6,4}-l_{3,1}n_{8,6}-l_{3,3}n_{8,4}$$
$$-l_{1,1}n_{10,6}-l_{1,3}n_{10,4})W_0^7+(l_{11,3}n_{0,6}-l_{9,3}n_{2,6}-l_{7,3}n_{4,6}$$
$$-l_{5,3}n_{6,6}-l_{3,3}n_{8,6}-l_{1,3}n_{10,6})W_0^9,$$

$$
\begin{aligned}
m_{13} =& -[(l_{11,3}n_{2,2}+l_{9,3}n_{4,2}+l_{7,3}n_{6,3}+l_{3,1}n_{10,4}+l_{1,1}n_{12,4})W_0^5 \\
&+(l_{11,3}n_{2,4}+l_{9,3}n_{4,4}+l_{7,3}n_{6,4}+l_{5,3}n_{8,4}+l_{3,1}n_{10,6}+l_{3,3}n_{10,4} \\
&+l_{1,1}n_{12,6}+l_{1,3}n_{12,4})W_0^7+(l_{11,3}n_{2,6}+l_{9,3}n_{4,6}+l_{7,3}n_{6,6} \\
&+l_{5,3}n_{8,6}+l_{3,3}n_{10,6}+l_{1,3}n_{12,6})W_0^9],
\end{aligned}
$$

$$
\begin{aligned}
m_{15} =& -[(l_{11,3}n_{4,2}+l_{9,3}n_{6,2}+l_{3,1}n_{12,4}+l_{1,1}n_{14,4})W_0^5 \\
&+(l_{11,3}n_{4,4}+l_{9,3}n_{6,4}+l_{7,3}n_{8,4}+l_{5,3}n_{10,4}+l_{3,1}n_{12,6}+l_{3,3}n_{12,4}+l_{1,1}n_{14,6} \\
&+l_{1,3}n_{14,4})W_0^7+(l_{11,3}n_{4,6}+l_{9,3}n_{6,6}+l_{7,3}n_{8,6}+l_{5,3}n_{10,6} \\
&+l_{3,3}n_{12,6}+l_{1,3}n_{14,6})W_0^9],
\end{aligned}
$$

$$
\begin{aligned}
m_{17} =& -[(l_{11,3}n_{6,2}+l_{3,1}n_{14,4})W_0^5+(l_{11,3}n_{6,4}+l_{9,3}n_{8,4}+l_{7,3}n_{10,4} \\
&+l_{5,3}n_{12,4}+l_{3,1}n_{14,6}+l_{3,3}n_{14,4}+l_{1,1}n_{16,6})W_0^7+(l_{11,3}n_{6,6} \\
&+l_{9,3}n_{8,6}+l_{7,3}n_{10,6}+l_{5,3}n_{12,6}+l_{3,3}n_{14,6}+l_{1,3}n_{16,6})W_0^9],
\end{aligned}
$$

$$
\begin{aligned}
m_{19} =& -[(l_{11,3}n_{8,4}+l_{9,3}n_{10,4}+l_{7,3}n_{12,4}+l_{5,3}n_{14,4}+l_{3,1}n_{16,6}+l_{1,1}n_{18,6})W_0^7 \\
&+(l_{11,3}n_{8,6}+l_{9,3}n_{10,6}+l_{7,3}n_{12,6}+l_{5,3}n_{14,6}+l_{3,3}n_{16,6}+l_{1,3}n_{18,6})W_0^9],
\end{aligned}
$$

$$
\begin{aligned}
m_{21} =& -[(l_{11,3}n_{10,4}+l_{9,3}n_{12,4}+l_{7,3}n_{14,4}+l_{3,1}n_{18,6}+l_{1,1}n_{20,6})W_0^7 \\
&+(l_{11,3}n_{10,6}+l_{9,3}n_{12,6}+l_{7,3}n_{14,6}+l_{5,3}n_{16,6}+l_{3,3}n_{18,6}+l_{1,3}n_{20,6})W_0^9],
\end{aligned}
$$

$$
\begin{aligned}
m_{23} =& -[(l_{11,3}n_{12,4}+l_{9,3}n_{14,4}+l_{3,1}n_{20,6}+l_{1,1}n_{22,6})W_0^7 \\
&+(l_{11,3}n_{12,6}+l_{9,3}n_{14,6}+l_{7,3}n_{16,6}+l_{5,3}n_{18,6}+l_{3,3}n_{20,6}+l_{1,3}n_{22,6})W_0^9],
\end{aligned}
$$

$$
\begin{aligned}
m_{25} =& -[(l_{11,3}n_{14,4}+l_{3,1}n_{22,6})W_0^7 \\
&+(l_{11,3}n_{14,6}+l_{9,3}n_{16,6}+l_{7,3}n_{18,6}+l_{5,3}n_{20,6}+l_{3,3}n_{22,6})W_0^9],
\end{aligned}
$$

$$m_{27} = -(l_{11,3}n_{16,6}+l_{9,3}n_{18,6}+l_{7,3}n_{20,6}+l_{5,3}n_{22,6})W_0^9,$$

$$m_{29} = -(l_{11,3}n_{18,6}+l_{9,3}n_{20,6}+l_{7,3}n_{22,6})W_0^9,$$

$$m_{31} = -(l_{11,3}n_{20,6}+l_{9,3}n_{22,6})W_0^9,$$

$$m_{33} = -l_{11,3}n_{22,6}W_0^9,$$

$$l_{1,1} = -\lambda_1\lambda_2,$$

$$l_{1,3} = \frac{\lambda_1^4}{128}\left(\frac{1}{2}\lambda_2^5-\frac{2}{3}\lambda_2^4+\frac{1}{18}\lambda_2^3+\frac{281}{1080}\lambda_2^2-\frac{49}{360}\lambda_2+\frac{1}{45}\right),$$

$$l_{3,1} = \lambda_1,$$

$$l_{3,3} = -\frac{\lambda_1^4}{128}\left(\lambda_2^5-2\lambda_2^3+\frac{29}{18}\lambda_2^2-\frac{529}{1080}\lambda^2+\frac{11}{180}\right),$$

$$l_{5,3} = \frac{\lambda_1^3}{96}\left(3\lambda_2^3-2\lambda_2^2-\frac{1}{3}\lambda_2+\frac{1}{12}\right),$$

$$l_{7,3} = -\frac{5\lambda_1^3\lambda_2}{384}\left(\frac{4}{3}\lambda_2-1\right),$$

$$l_{9,3} = \frac{\lambda_1^3}{384}\left(\frac{3}{2}\lambda_2-1\right),$$

$$l_{11,3} = -\frac{\lambda_1^3}{5760}$$

利用定义（4.14），由式（4.30）得三次近似中的载荷和中心挠度的九次方关系式

$$P=\sum_{i=0}^{4}\alpha_{2i+1}W_0^{2i+1} \tag{4.31}$$

其中

$$\begin{aligned}\alpha_5=&-8\lambda_1[R_1(l_{1,1}n_{0,4}+l_{1,3}n_{0,2})\\&+R_3(l_{3,1}n_{0,4}+l_{3,3}n_{0,2}-l_{1,1}n_{2,4}-l_{1,3}n_{2,2})\\&+R_5(l_{5,3}n_{0,2}-l_{3,1}n_{2,4}-l_{3,3}n_{2,2}-l_{1,1}n_{4,4}-l_{1,3}n_{4,2})\\&+R_7(l_{7,3}n_{0,2}-l_{5,3}n_{2,2}-l_{3,1}n_{4,4}-l_{3,3}n_{4,2}-l_{1,1}n_{6,4}-l_{1,3}n_{6,2})\\&+R_9(l_{9,3}n_{0,2}-l_{7,3}n_{2,2}-l_{5,3}n_{4,2}-l_{3,1}n_{6,4}-l_{3,3}n_{6,2}-l_{1,1}n_{8,4})\\&+R_{11}(l_{11,3}n_{0,2}-l_{9,3}n_{2,2}-l_{7,3}n_{4,2}-l_{5,3}n_{6,2}-l_{3,1}n_{8,4}-l_{1,1}n_{10,4})\\&-R_{13}(l_{11,3}n_{2,2}+l_{9,3}n_{4,2}+l_{7,3}n_{6,2}+l_{3,1}n_{10,4}+l_{1,1}n_{12,4})\\&-R_{15}(l_{11,3}n_{4,2}+l_{9,3}n_{6,2}+l_{3,1}n_{12,4}\\&+l_{1,1}n_{14,4})-R_{17}(l_{11,3}n_{6,2}+l_{3,1}n_{14,4})],\end{aligned} \tag{4.32a}$$

$$\begin{aligned}\alpha_7=&-8\lambda_1[R_1(l_{1,1}n_{0,6}+l_{1,3}n_{0,4})\\&+R_3(l_{3,1}n_{0,6}+l_{3,3}n_{0,4}-l_{1,1}n_{2,6}-l_{1,3}n_{2,4})\\&+R_5(l_{5,3}n_{0,4}-l_{3,1}n_{2,6}-l_{3,3}n_{2,4}-l_{1,1}n_{4,6}-l_{1,3}n_{4,4})\\&+R_7(l_{7,3}n_{0,4}-l_{5,3}n_{2,4}-l_{3,1}n_{4,6}-l_{3,3}n_{4,4}-l_{1,1}n_{6,6}-l_{1,3}n_{6,4})\\&+R_9(l_{9,3}n_{0,4}-l_{7,3}n_{2,4}-l_{5,3}n_{4,4}-l_{3,1}n_{6,6}-l_{3,3}n_{6,4}-l_{1,1}n_{8,6}-l_{1,3}n_{8,4})\\&+R_{11}(l_{11,3}n_{0,4}-l_{9,3}n_{2,4}-l_{7,3}n_{4,4}-l_{5,3}n_{6,4}-l_{3,1}n_{8,6}\\&-l_{3,3}n_{8,4}-l_{1,1}n_{10,6}-l_{1,3}n_{10,4})\\&-R_{13}(l_{11,3}n_{2,4}+l_{9,3}n_{4,4}+l_{7,3}n_{6,4}+l_{5,3}n_{8,4}+l_{3,1}n_{10,6}\\&+l_{3,3}n_{10,4}+l_{1,1}n_{12,6}+l_{1,3}n_{12,4})\\&-R_{15}(l_{11,3}n_{4,4}+l_{9,3}n_{6,4}+l_{7,3}n_{8,4}+l_{5,3}n_{10,4}\\&+l_{3,1}n_{12,6}+l_{3,3}n_{12,4}+l_{1,1}n_{14,6}+l_{1,3}n_{14,4})\\&-R_{17}(l_{11,3}n_{6,4}+l_{9,3}n_{8,4}+l_{7,3}n_{10,4}+l_{5,3}n_{12,4}\\&+l_{3,1}n_{14,6}+l_{3,3}n_{14,4}+l_{1,1}n_{16,6})\\&-R_{19}(l_{11,3}n_{8,4}+l_{9,3}n_{10,4}+l_{7,3}n_{12,4}+l_{5,3}n_{14,4}+l_{3,1}n_{16,6}+l_{1,1}n_{18,6})\\&-R_{21}(l_{11,3}n_{10,4}+l_{9,3}n_{12,4}+l_{7,3}n_{14,4}+l_{3,1}n_{18,6}+l_{1,1}n_{20,6})\\&-R_{23}(l_{11,3}n_{12,4}+l_{9,3}n_{14,4}+l_{3,1}n_{20,6}+l_{1,1}n_{22,6})\\&-R_{25}(l_{11,3}n_{14,4}+l_{3,1}n_{22,6})],\end{aligned} \tag{4.32b}$$

$$\begin{aligned}\alpha_9=&-8\lambda_1[R_1l_{1,3}n_{0,6}+R_3(l_{3,3}n_{0,6}-l_{1,3}n_{2,6})\\&+R_5(l_{5,3}n_{0,6}-l_{3,3}n_{2,6}-l_{1,3}n_{4,6})\\&+R_7(l_{7,3}n_{0,6}-l_{5,3}n_{2,6}-l_{3,3}n_{4,6}-l_{1,3}n_{6,6})\\&+R_9(l_{9,3}n_{0,6}-l_{7,3}n_{2,6}-l_{5,3}n_{4,6}-l_{3,3}n_{6,6}-l_{1,3}n_{8,6})\\&+R_{11}(l_{11,3}n_{0,6}-l_{9,3}n_{2,6}-l_{7,3}n_{4,6}-l_{5,3}n_{6,6}-l_{3,3}n_{8,6}-l_{1,3}n_{10,6})\\&-R_{13}(l_{11,3}n_{2,6}+l_{9,3}n_{4,6}+l_{7,3}n_{6,6}+l_{5,3}n_{8,6}+l_{3,3}n_{10,6}+l_{1,3}n_{12,6})\end{aligned}$$

$$-R_{15}(l_{11,3}n_{4,6}+l_{9,3}n_{6,6}+l_{7,3}n_{8,6}+l_{5,3}n_{10,6}+l_{3,3}n_{12,6}+l_{1,3}n_{14,6})$$
$$-R_{17}(l_{11,3}n_{6,6}+l_{9,3}n_{8,6}+l_{7,3}n_{10,6}+l_{5,3}n_{12,6}+l_{3,3}n_{14,6}+l_{1,3}n_{16,6})$$
$$-R_{19}(l_{11,3}n_{8,6}+l_{9,3}n_{10,6}+l_{7,3}n_{12,6}+l_{5,3}n_{14,6}+l_{3,3}n_{16,6}+l_{1,3}n_{18,6})$$
$$-R_{21}(l_{11,3}n_{10,6}+l_{9,3}n_{12,6}+l_{7,3}n_{14,6}+l_{5,3}n_{16,6}+l_{3,3}n_{18,6}+l_{1,3}n_{20,6})$$
$$-R_{23}(l_{11,3}n_{12,6}+l_{9,3}n_{14,6}+l_{7,3}n_{16,6}+l_{5,3}n_{18,6}+l_{3,3}n_{20,6}+l_{1,3}n_{22,6})$$
$$-R_{25}(l_{11,3}n_{14,6}+l_{9,3}n_{16,6}+l_{7,3}n_{18,6}+l_{5,3}n_{20,6}+l_{3,3}n_{22,6})$$
$$-R_{27}(l_{11,3}n_{16,6}+l_{9,3}n_{18,6}+l_{7,3}n_{20,6}+l_{5,3}n_{22,6})$$
$$-R_{29}(l_{11,3}n_{18,6}+l_{9,3}n_{20,6}+l_{7,3}n_{22,6})$$
$$-R_{31}(l_{11,3}n_{20,6}+l_{9,3}n_{22,6})-R_{33}l_{11,3}n_{22,6}\Big], \tag{4.32c}$$

$$R_1=\frac{1}{8\lambda_1},\quad R_3=\frac{1}{4}\left(\frac{1}{18}+k\right),\quad R_5=\frac{1}{6}\left(\frac{3}{64}+k\right),$$
$$R_7=\frac{1}{8}\left(\frac{1}{25}+k\right),\quad R_9=\frac{1}{10}\left(\frac{5}{144}+k\right),\quad R_{11}=\frac{1}{12}\left(\frac{3}{98}+k\right),$$
$$R_{13}=\frac{1}{14}\left(\frac{7}{256}+k\right),\quad R_{15}=\frac{1}{16}\left(\frac{2}{81}+k\right),\quad R_{17}=\frac{1}{18}\left(\frac{9}{400}+k\right),$$
$$R_{19}=\frac{1}{20}\left(\frac{5}{242}+k\right),\quad R_{21}=\frac{1}{22}\left(\frac{11}{576}+k\right),\quad R_{23}=\frac{1}{24}\left(\frac{3}{169}+k\right),$$
$$R_{25}=\frac{1}{26}\left(\frac{13}{784}+k\right),\quad R_{27}=\frac{1}{28}\left(\frac{7}{450}+k\right),\quad R_{29}=\frac{1}{30}\left(\frac{15}{1024}+k\right),$$
$$R_{31}=\frac{1}{32}\left(\frac{4}{289}+k\right),\quad R_{33}=\frac{1}{34}\left(\frac{17}{1296}+k\right)$$

最后，引入环向应力 $\sigma_{\theta b}$ 的无量纲量：

$$S_\theta=\frac{2ta^2}{D}\sigma_{\theta b} \tag{4.33}$$

再应用式（4.26），可由方程（2.12）得二次近似无量纲环向应力公式

$$\begin{aligned}S_{\theta 2}=&(n_{0,2}W_0^2+n_{0,4}W_0^4+n_{0,6}W_0^6)-3(n_{2,2}W_0^2+n_{2,4}W_0^4+n_{2,6}W_0^6)\rho^2\\&-5(n_{4,2}W_0^2+n_{4,4}W_0^4+n_{4,6}W_0^6)\rho^4\\&-7(n_{6,2}W_0^2+n_{6,4}W_0^4+n_{6,6}W_0^6)\rho^6\\&-9(n_{8,4}W_0^4+n_{8,6}W_0^6)\rho^8\\&-11(n_{10,4}W_0^4+n_{10,6}W_0^6)\rho^{10}\\&-13(n_{12,4}W_0^4+n_{12,6}W_0^6)\rho^{12}\\&-15(n_{14,4}W_0^4+n_{14,6}W_0^6)\rho^{14}\\&-(17n_{16,6}\rho^{16}+19n_{18,6}\rho^{18}+21n_{20,6}\rho^{20}+23n_{22,6}\rho^{22})W_0^6\end{aligned} \tag{4.34}$$

由式（4.26）和（4.34），我们还可写出夹层圆板中心和边缘的应力公式

$$S_r(0)=S_\theta(0)=n_{0,2}W_0^2+n_{0,4}W_0^4+n_{0,6}W_0^6,$$

$$S_\theta(1)=-\frac{\lambda_1^2}{2}\Big[\left(\frac{1}{2}\lambda_2^2-\frac{2}{3}\lambda_2+\frac{1}{4}\right)W_0^2+\frac{\lambda_1^3}{192}\left(\frac{1}{4}\lambda_2^6-\frac{3}{4}\lambda_2^5+\frac{13}{15}\lambda_2^4\right.$$
$$\left.-\frac{113}{240}\lambda_2^3+\frac{104}{945}\lambda_2^2-\frac{17}{10080}\lambda_2-\frac{5}{2016}\right)W_0^4+\frac{\lambda_1^6}{49152}\left(\frac{1}{8}\lambda_2^{10}\right.$$

$$-\frac{17}{30}\lambda_2^9+\frac{17}{15}\lambda_2^8-\frac{4021}{3024}\lambda_2^7+\frac{577}{560}\lambda_2^6-\frac{32051}{56700}\lambda_2^5+\frac{1252103}{5443200}\lambda_2^4-\frac{3178631}{44906400}\lambda_2^3+\frac{382043}{23950080}\lambda_2^2-\frac{59}{25200}\lambda_2+\frac{353}{2138400}\Big)W_0^6\Big]$$

应用公式（4.23）、（4.31）和（4.35），我们得到了下面一些有用的结果。图 3 示出了几个不同 k 值下的载荷 P 与中心挠度 W_0 的关系曲线，较精确的三次近似结果用实线表示，二次近似结果用虚线表示。显而易见，随着中心挠度的增大，板的刚度也增大。并且，对于相同的 P 值，具有较大 k 值的板将产生较大的中心挠度。由图看出，二次和三次近似曲线异常接近，这说明解的精确度很高。为了便于应用，我们用图 4 表示本文二次与三次近似解的相对误差。由此图可知，随着特征参数 k 的增大，它们的相对误差在增加。一般来说，由于二次近似解非常简单，对于工程设计来说，采用它是比较方便的。

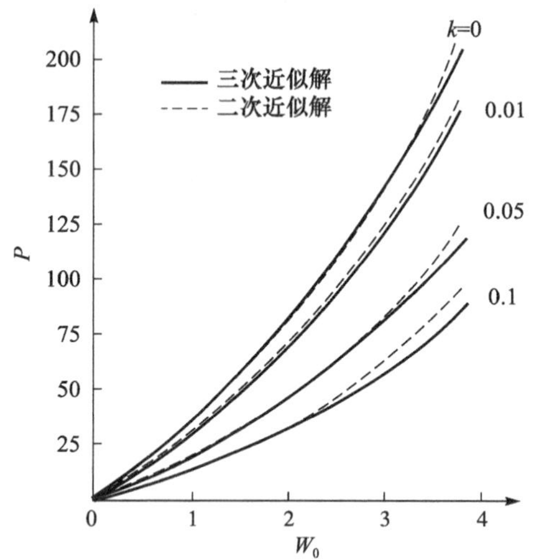

图 3　各种 k 值下的载荷与中心挠度关系曲线

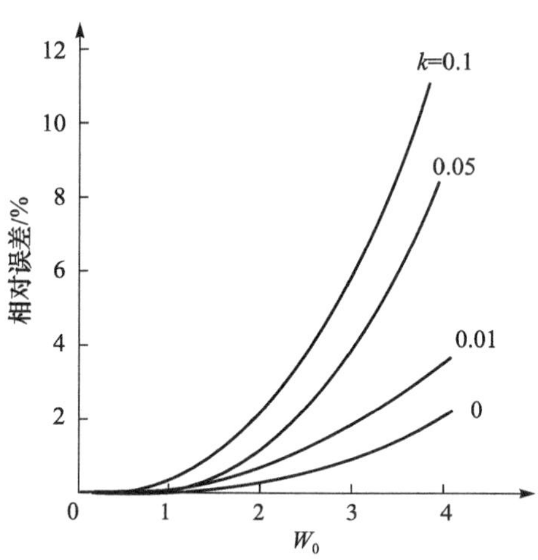

图 4　二次与三次近似解的相对误差

在图 5、图 6 内，给出了各种 k 值下的夹层圆板中心和边缘的径向与环向应力曲线。由图看出，中心的径向和环向应力始终是正值，边缘的环向应力始终是负值，亦即夹层圆板的边缘区域是承受压应力的。而且，对于相同的中心挠度值，具有较大 k 值的夹层圆板将有较大的边缘环向应力以及较小的中心径向和环向应力。

五、结论

（1）本文给出了具有软夹心的夹层圆板在均布横向载荷作用下的非线性轴对称弯曲理论的基本方程和边界条件。

（2）本文给出了这种夹层圆板在表板很薄时的上述方程和边界条件的简化形式。

（3）应用简化方程，使用修正迭代法求解了滑动固定边界条件下的夹层圆板，获得了挠度的二次和三次近似解和应力的二次近似解。由二次和三次近似数值结果非常一致的事实，说明本文所得到的解的精确度是相当高的。对于工程设计说来，一般使用二次近似解就能满足精确度要求。

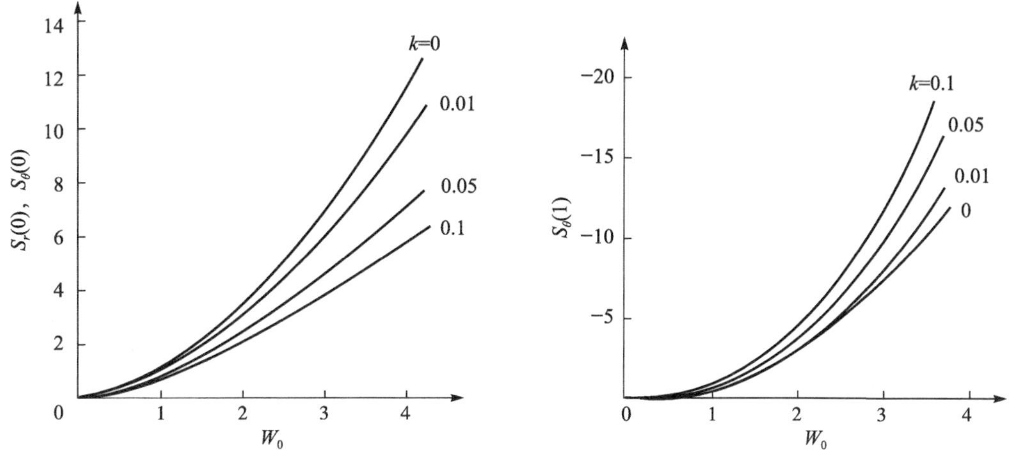

图 5 各种 k 值下的夹层圆板中心的径向和环向应力　　图 6 各种 k 值下的夹层圆板边缘的环向应力

参 考 文 献

[1] Zaid M. Symmetrical bending of circular sandwich plates. *Proceedings of the Second United States National Congress of Applied Mechanics*, *ASME*, New York, 1955: 413.

[2] Bruun E R. Thermal deflection of a circular sandwich plate. *AIAA J.*, 1963, 1 (5): 1213.

[3] Huang J C, Ebcioglu I K. Circular sandwich plate under radial compression and thermal gradient. *AIAA J.*, 1965, 3 (6): 1146.

[4] Amato A J, Ebcioglu I K. Axisymmetric buckling of annular sandwich panels. *AIAA J.*, 1972, 10 (10): 1351.

[5] 刘人怀. 在边缘力矩作用下夹层圆板的非线性轴对称弯曲问题. 中国科学技术大学学报, 1980, 10 (2): 56.

[6] 叶开沅, 刘人怀, 平庆元, 等. 在对称线布载荷作用下的圆底扁薄球壳的非线性稳定问题. 兰州大学学报, 1965, (2): 10; 科学通报, 1965, (2): 142.

[7] 叶开沅, 刘人怀, 张传智, 等. 圆底扁薄球壳在边缘力矩作用下的非线性稳定问题. 科学通报, 1965, (2): 145.

[8] 刘人怀. 在内边缘均布力矩作用下中心开孔圆底扁球壳的非线性稳定问题. 科学通报, 1965, (3): 253.

[9] 刘人怀. 在边缘载荷作用下中心开孔圆底扁薄球壳的轴对称稳定性. 力学学报, 1977, (3): 206.

[10] 刘人怀. 波纹圆板的特征关系式. 力学学报, 1978, (1): 47.

[11] 刘人怀. 具有光滑中心的波纹圆板的特征关系式. 中国科学技术大学学报, 1979, 9 (2): 75.

[12] 中国科学院北京力学研究所固体力学研究室板壳组. 夹层板壳的弯曲、稳定和振动. 北京: 科学出版社, 1977: 42.

[13] Reissner E. Finite deflections of sandwich plates. *J. Aeron. Sci.*, 1948, 15 (7): 435; 1950, 17 (2): 125.

夹层圆板大挠度问题的精确解①

一、引言

夹层板的大挠度问题是一个很有实际意义的问题，已经引起许多人专门研究。Reissner[1]首先建立了具有软夹心和极薄表板的夹层矩形板的大挠度理论。但是，对于夹层圆板的大挠度问题说来，却只有极少数人研究过。近来，中国科学院力学研究所板壳组[2]研究了这方面的一个问题。接着，刘人怀[3,4]进一步做了一些工作。本文是前文[4]工作的继续。应用幂级数方法，我们求解了在均布载荷作用下的具有不同边界条件的夹层圆板的大挠度问题，获得了精确解。我们所处理的四种边界条件是：①固定；②滑动固定；③简支；④铰支。最后，我们还将此精确解与刘人怀[4]用修正迭代法所得的解析解进行了比较。由比较可知：以前的解析解的精确性是十分令人满意的。

二、基本方程和边界条件

今考虑一半径为 a，承受均布载荷 q 的夹层圆板。为了简便起见，本文使用的主要符号与文献［4］所用的相同。应用文献［4］所给的简化方程，即得由夹层圆板中面上的挠度 w 和径向应力 σ_{r0} 所满足的大挠度方程组：

$$D\frac{\mathrm{d}}{\mathrm{d}r}\frac{1}{r}\frac{\mathrm{d}}{\mathrm{d}r}r\frac{\mathrm{d}w}{\mathrm{d}r}+\frac{2tD}{G_2h_0}\frac{\mathrm{d}}{\mathrm{d}r}\frac{1}{r}\frac{\mathrm{d}}{\mathrm{d}r}\left(r\sigma_{r0}\frac{\mathrm{d}w}{\mathrm{d}r}\right)-2t\sigma_{r0}\frac{\mathrm{d}w}{\mathrm{d}r}-\frac{1}{2}qr=0,$$

$$\frac{\mathrm{d}}{\mathrm{d}r}\frac{1}{r}\frac{\mathrm{d}}{\mathrm{d}r}(r^2\sigma_{r0})+\frac{E}{2r}\left(\frac{\mathrm{d}w}{\mathrm{d}r}\right)^2=0 \tag{2.1}$$

其中，D 是夹层圆板的抗弯刚度，为

$$D=\frac{Eth_0^2}{2(1-\nu^2)} \tag{2.2}$$

非线性方程组（2.1）将分别在下列四种常用的边界条件下求解。

（1）固定：

$$\begin{aligned}&\text{当 } r=a \text{ 时},w=0,\psi=0,u=0;\\&\text{当 } r=0 \text{ 时},\psi=0,\sigma_{r0}\text{ 有限}\end{aligned} \tag{2.3}$$

（2）滑动固定：

$$\begin{aligned}&\text{当 } r=a \text{ 时},w=0,\psi=0,\sigma_{r0}=0;\\&\text{当 } r=0 \text{ 时},\psi=0,\sigma_{r0}\text{ 有限}\end{aligned} \tag{2.4}$$

（3）简支：

$$\text{当 } r=a \text{ 时},w=0,M_r=0,\sigma_{r0}=0;$$

① 本文原载《应用数学和力学》，1982，3（1）：11-23. 作者：刘人怀，施云方.

$$当\ r=0\ 时, \psi=0, \sigma_{r0}\ 有限 \tag{2.5}$$

(4) 铰支：

$$当\ r=a\ 时, w=0, M_r=0, u=0;$$
$$当\ r=0\ 时, \psi=0, \sigma_{r0}\ 有限 \tag{2.6}$$

这里，u 是夹层圆板中面上点的径向位移，ψ 是夹层圆板中面法线在径向平面内的转角，M_r 是夹层圆板的径向弯矩，$\sigma_{\theta 0}$是夹层圆板中面内的环向应力，

$$u=\frac{r}{E}(\sigma_{\theta 0}-\nu\sigma_{r0})$$

$$\psi=-\frac{2t}{G_2h_0}\sigma_{r0}\frac{\mathrm{d}w}{\mathrm{d}r}-\frac{\mathrm{d}w}{\mathrm{d}r}-\frac{qr}{2G_2h_0},$$

$$M_r=D\left(\frac{\mathrm{d}\psi}{\mathrm{d}r}+\nu\frac{\psi}{r}\right)$$

$$\sigma_{\theta 0}=\frac{\mathrm{d}}{\mathrm{d}r}(r\sigma_{r0}) \tag{2.7a-d}$$

首先，将非线性方程组（2.1）和边界条件（2.3)～(2.6）转化为无量纲形式。为此，我们引入下列符号：

$$\rho=\frac{r}{a},\quad W=\sqrt{2(1-\nu^2)}\frac{w}{h_0},\quad \phi=\frac{\mathrm{d}W}{\mathrm{d}\rho},$$

$$s_r=\frac{2ta^2}{D}\sigma_{r0},\quad S_\theta=\frac{2ta^2}{D}\sigma_{\theta 0},\quad k=\frac{D}{G_2h_0a^2},$$

$$P=\frac{\sqrt{2(1-\nu^2)}a^4}{2h_0D}q \tag{2.8}$$

应用这些符号和式（2.7）以后，方程组（2.1）化为

$$L[\rho(kS_r+1)\phi]-S_r\phi-P\rho=0,$$

$$L(\rho^2S_r)+\frac{\phi^2}{\rho}=0 \tag{2.9}$$

边界条件（2.3)～(2.6）化为

(1) 固定：

$$当\ \rho=1\ 时, W=0, (kS_r+1)\phi+kP=0, \frac{\mathrm{d}}{\mathrm{d}\rho}(\rho S_r)-\nu S_r=0;$$

$$当\ \rho=0\ 时, (kS_r+1)\phi=0, S_r\ 有限 \tag{2.10}$$

(2) 滑动固定：

$$当\ \rho=1\ 时, W=0, (kS_r+1)\phi+kP=0, S_r=0;$$
$$当\ \rho=0\ 时, (kS_r+1)\phi=0, S_r\ 有限 \tag{2.11}$$

(3) 简支：

$$当\ \rho=1\ 时, W=0,$$

$$\frac{\mathrm{d}}{\mathrm{d}\rho}[(kS_r+1)\phi]+\nu(kS_r+1)\phi+k(1+\nu)P=0, S_r=0;$$

$$当\ \rho=0\ 时, (kS_r+1)\phi=0, S_r\ 有限 \tag{2.12}$$

(4) 铰支：

当 $\rho=1$ 时, $W=0$,

$$\frac{\mathrm{d}}{\mathrm{d}\rho}[(kS_r+1)\phi]+\nu(kS_r+1)\phi+k(1+\nu)P=0,$$

$$\frac{\mathrm{d}}{\mathrm{d}\rho}(\rho S_r)-\nu S_r=0; \tag{2.13}$$

当 $\rho=0$ 时, $(kS_r+1)\phi=0$, S_r 有限

其中

$$L(\cdots)=\frac{\mathrm{d}}{\mathrm{d}\rho}\frac{1}{\rho}\frac{\mathrm{d}}{\mathrm{d}\rho}(\cdots) \tag{2.14}$$

有了 S_r 和 ϕ，就能得到所有的应力和位移计算公式。无量纲环向应力 S_θ 由式（2.7d）求得。积分 ϕ，并注意到 $\rho=1$ 时的边界条件 $W=0$，则可得到无量挠度 W。于是，有

$$S_\theta=\frac{\mathrm{d}}{\mathrm{d}\rho}(\rho S_r),$$

$$W=\int_1^\rho \phi\,\mathrm{d}\rho \tag{2.15}$$

三、用幂级数方法求解

应用幂级数方法，我们可求得上述非线性方程组和边界条件的精确解。假定 S_r 和 ϕ 分别是 ρ 的对称函数和反对称函数，在用幂级数表示后，为

$$S_r=\sum_{i=0}^{\infty}a_{2i}\rho^{2i}$$

$$\phi=\sum_{i=0}^{\infty}b_{2i+1}\rho^{2i+1} \tag{3.1a,b}$$

其中 a_{2i}，b_{2i+1}（$i=0$，1，2，…）是待定的常数。显然，这些数已满足夹层圆板的两个中心条件。

将式（3.1）代入式（2.15），有

$$S_\theta=\sum_{i=0}^{\infty}(2i+1)a_{2i}\rho^{2i}$$

$$W=\sum_{i=0}^{\infty}\frac{b_{2i+1}}{2(i+1)}[\rho^{2(i+1)}-1] \tag{3.2a,b}$$

这说明，如果知道了常数 a_{2i} 和 b_{2i+1}（$i=0,1,2,\cdots$），则夹层圆板的所有应力和位移都能确定。

将级数（3.1）代入方程（2.9），并注意到方程（2.9）必须对任何 ρ 值都成立，于是我们得到常数 a_{2i} 和 b_{2i+1} 之间的关系式

$$a_{2i}=-\frac{1}{4i(i+1)}\sum_{m=0}^{i-1}b_{2m+1}b_{2i-2m-1},\ i=1,2,3,\cdots$$

$$b_{2i+1}=\frac{1}{ka_0+1}\left[\frac{1}{4i(i+1)}\sum_{m=0}^{i-1}a_{2m}b_{2i-2m-1}-k\sum_{m=0}^{i}a_{2m}b_{2i-2m+1}\right],\ i=2,3,4,\cdots$$

$$b_3 = \frac{1}{8(ka_0+1)}[(a_0 - 8ka_2)b_1 + P] \tag{3.3}$$

由此看到，只要知道了 a_0 和 b_1 的值，其他常数便可以依次确定。这样一来，我们便可用级数（3.1a）和（3.2）确定板中所有点的 S_r，S_θ 和 W 的值。

为了确定常数 a_0 和 b_1，我们需要应用前面每种边界条件中尚未用过的两个条件。将式（3.1）代入这两个边界条件，便得关于待定常数 a_0 和 b_1 的非线性方程组：

（1）对于固定

$$\sum_{i=0}^{\infty}(2i+1-\nu)a_{2i} = 0,$$

$$\sum_{i=0}^{\infty}(k\sum_{m=0}^{i}a_{2m}b_{2i-2m+1} + b_{2i+1}) + kP = 0 \tag{3.4}$$

（2）对于滑动固定

$$\sum_{i=0}^{\infty}a_{2i} = 0,$$

$$\sum_{i=0}^{\infty}(k\sum_{m=0}^{i}a_{2m}b_{2i-2m+1} + b_{2i+1}) + kP = 0 \tag{3.5}$$

（3）对于简支

$$\sum_{i=0}^{\infty}a_{2i} = 0,$$

$$\sum_{i=0}^{\infty}(2i+1+\nu)(k\sum_{m=0}^{i}a_{2m}b_{2i-2m+1} + b_{2i+1}) + k(1+\nu)P = 0 \tag{3.6}$$

（4）对于铰支

$$\sum_{i=0}^{\infty}(2i+1-\nu)a_{2i} = 0,$$

$$\sum_{i=0}^{\infty}(2i+1+\nu)(k\sum_{m=0}^{i}a_{2m}b_{2i-2m+1} + b_{2i+1}) + k(1+\nu)P = 0 \tag{3.7}$$

使用牛顿法，可求得这些方程组关于 a_0 和 b_1 的数值解。对于给定的泊松比 ν、无量纲特征参数 k 和无量纲横向载荷 P 的值，本文采用一种简单的方法来选取 a_0 和 b_1 的初始值。我们从很小的载荷，如 $P=P_0$ 开始，这时可令 a_0 和 b_1 的初始值为零。求解上面的方程组，便得在 $P=P_0$ 情况下 a_0 和 b_1 的数值解。然后，对于十分接近 P_0 值的另一载荷值情况，我们便可将刚才所得的解作为此时的初始值。于是依次类推，按照载荷 P 的由小到大的顺序，便能毫无困难地依次给出 a_0 和 b_1 的其他初始值。为了简化以后的计算，在确定 a_0 和 b_1 的初始值时，使用插值法是有益的。这样，上述方程组的牛顿迭代程序能够迅速收敛。

在 DJS-8 机上完成了繁冗的数值计算，结果给在表 1 中。

计算表明，级数（3.1）收敛很慢。当 $k\leqslant 0.01$ 时，我们必须取级数前面 25 项才能得到足够精确的结果。随着 k 值的增大，级数收敛变得更慢，因此级数中必须计算的项数越来越多。当 $k=0.10$ 时，这些级数至少要取 80 项。

应用式（3.2b），可得下列关于无量纲中心挠度 W_0 的公式

$$W_0 = W|_{\rho=0} = -\sum_{i=0}^{\infty} \frac{b_{2i+1}}{2(i+1)} \tag{3.8}$$

表 1(a)　固定夹层圆板的 a_0 和 b_1 的值（$\nu=0.3$）

$k=$	0		0.01		0.05		0.10	
P	a_0	b_1	a_0	b_1	a_0	b_1	a_0	b_1
5	0.047429	−0.62302	0.061064	−0.67170	0.13751	−0.85953	0.27434	−1.0695
10	0.18696	−1.2345	0.23905	−1.3245	0.51388	−1.6389	0.93262	−1.9088
15	0.41099	−1.8245	0.52006	−1.9431	1.0526	−2.3021	1.7441	−2.5211
20	0.70863	−2.3856	0.88555	−2.5181	1.6844	−2.8526	2.5953	−2.9802
25	1.0676	−2.9133	1.3164	−3.0456	2.3631	−3.3095	3.4439	−3.3412
30	1.4759	−3.4058	1.7955	−3.5259	3.0621	−3.6930	4.2754	−3.6373
35	1.9226	−3.8631	2.3088	−3.9616	3.7665	−4.0198	5.0853	−3.8883
40	2.3984	−4.2865	2.8456	−4.3567	4.4682	−4.3025	5.8730	−4.1066
45	2.8958	−4.6781	3.3976	−4.7155	5.1629	−4.5506	6.6393	−4.3004
50	3.4085	−5.0405	3.9590	−5.0421	5.8484	−4.7711	7.3856	−4.4751
55	3.9320	−5.3760	4.5254	−5.3405	6.5236	−4.9694	8.1133	−4.6345
60	4.4624	−5.6872	5.0937	−5.6140	7.1881	−5.1494	8.8238	−4.7816
65	4.9970	−5.9764	5.6615	−5.8658	7.8420	−5.3142	9.5184	−4.9183
70	5.5335	−6.2457	6.2274	−6.0983	8.4854	−5.4663	10.198	−5.0463
75	6.0704	−6.4971	6.7902	−6.3138	9.1187	−5.6075	10.864	−5.1668
80	6.6062	−6.7322	7.3490	−6.5142	9.7423	−5.7395	11.518	−5.2809
85	7.1401	−6.9527	7.9035	−6.7013	10.356	−5.8633	12.160	−5.3893
90	7.6714	−7.1598	8.4532	−6.8765	10.962	−5.9801	12.790	−5.4927
95	8.1995	−7.3549	8.9978	−7.0410	11.559	−6.0907	13.410	−5.5916
100	8.7240	−7.5390	9.5374	−7.1959	12.147	−6.1958	14.021	−5.6865

表 1(b)　滑动固定夹层圆板的 a_0 和 b_1 的值

$k=$	0		0.01		0.05		0.10	
P	a_0	b_1	a_0	b_1	a_0	b_1	a_0	b_1
5	0.024367	−0.62422	0.029851	−0.67372	0.057756	−0.86949	0.10526	−1.1058
10	0.096910	−1.2438	0.11841	−1.3399	0.22521	−1.7076	0.39513	−2.1112
15	0.21601	−1.8543	0.26282	−1.9916	0.48712	−2.4904	0.81323	−2.9688
20	0.37911	−2.4519	0.45873	−2.6230	0.82373	−3.2047	1.3062	−3.6813
25	0.58296	−3.0332	0.70077	−3.2295	1.2155	−3.8472	1.8372	−4.2724
30	0.82382	−3.5957	0.98307	−3.8081	1.6457	−4.4209	2.3842	−4.7684
35	1.0977	−4.1373	1.2997	−4.3569	2.1013	−4.9323	2.9349	−5.1909
40	1.4006	−4.6569	1.6449	−4.8753	2.5726	−5.3888	3.4826	−5.5568

续表

$k=$	0		0.01		0.05		0.10	
P	a_0	b_1	a_0	b_1	a_0	b_1	a_0	b_1
45	1.7286	−5.1537	2.0136	−5.3635	3.0528	−5.7980	4.0236	−5.8784
50	2.0780	−5.6278	2.4012	−5.8224	3.5370	−6.1666	4.5562	−6.1647
55	2.4454	−6.0792	2.8036	−6.2533	4.0219	−6.5004	5.0796	−6.4228
60	2.8279	−6.5086	3.2175	−6.6577	4.5052	−6.8045	5.5936	−6.6576
65	3.2226	−6.9166	3.6401	−7.0372	4.9853	−7.0829	6.0983	−6.8732
70	3.6274	−7.3042	4.0690	−7.3937	5.4613	−7.3393	6.5939	−7.0726
75	4.0400	−7.6723	4.5022	−7.7288	5.9325	−7.5765	7.0808	−7.2583
80	4.4588	−8.0219	4.9381	−8.0442	6.3985	−7.7969	7.5594	−7.4322
85	4.8823	−8.3541	5.3756	−8.3412	6.8590	−8.0028	8.0300	−7.5959
90	5.3091	−8.6699	5.8134	−8.6215	7.3141	−8.1957	8.4932	−7.7506
95	5.7382	−8.9702	6.2509	−8.8863	7.7636	−8.3773	8.9491	−7.8974
100	6.1686	−9.2560	6.6872	−9.1368	8.2076	−8.5486	9.3983	−8.0373

表 1(c)　简支夹层圆板的 a_0 和 b_1 的值（$\nu=0.3$）

k	0		0.01		0.05		0.10	
P	a_0	b_1	a_0	b_1	a_0	b_1	a_0	b_1
5	0.22980	−1.5457	0.24501	−1.5874	0.30821	−1.7446	0.39023	−1.9171
10	0.82268	−2.8940	0.86536	−2.9444	1.0281	−3.1094	1.2096	−3.2477
15	1.6072	−3.9899	1.6682	−4.0208	1.8854	−4.0991	2.1045	−4.1316
20	2.4650	−4.8684	2.5310	−4.8674	2.7564	−4.8337	2.9726	−4.7642
25	3.3365	−5.5807	3.3974	−5.5448	3.6015	−5.4034	3.7950	−5.2502
30	4.1956	−6.1686	4.2446	−6.0994	4.4106	−5.8631	4.5727	−5.6440
35	5.0314	−6.6628	5.0645	−6.5633	5.1835	−6.2463	5.3106	−5.9755
40	5.8403	−7.0850	5.8549	−6.9588	5.9227	−6.5742	6.0139	−6.2627
45	6.6217	−7.4510	6.6167	−7.3015	6.6314	−6.8606	6.6871	−6.5167
50	7.3766	−7.7721	7.3512	−7.6026	7.3129	−7.1149	7.3341	−6.7452
55	8.1064	−8.0568	8.0607	−7.8703	7.9700	−7.3439	7.9580	−6.9535
60	8.8129	−8.3117	8.7469	−8.1108	8.6052	−7.5524	8.5614	−7.1453
65	9.4977	−8.5417	9.4119	−8.3287	9.2206	−7.7440	9.1465	−7.3234
70	10.163	−8.7508	10.057	−8.5278	9.8182	−7.9215	9.7152	−7.4900
75	10.809	−8.9420	10.685	−8.7109	10.400	−8.0870	10.269	−7.6468

续表

k	0		0.01		0.05		0.10	
P	a_0	b_1	a_0	b_1	a_0	b_1	a_0	b_1
80	11.438	−9.1179	11.296	−8.8803	10.966	−8.2424	10.809	−7.7950
85	12.051	−9.2806	11.891	−9.0378	11.519	−8.3888	11.337	−7.9358
90	12.649	−9.4318	12.472	−9.1851	12.059	−8.5276	11.853	−8.0701
95	13.234	−9.5729	13.041	−9.3234	12.588	−8.6594	12.359	−8.1985
100	13.806	−9.7050	13.596	−9.4538	13.106	−8.7852	12.855	−8.3217

表 1(d) 铰支夹层圆板的 a_0 和 b_1 的值（$\nu=0.3$）

$k=$	0		0.01		0.05		0.10	
P	a_0	b_1	a_0	b_1	a_0	b_1	a_0	b_1
5	0.59429	−1.4118	0.62887	−1.4369	0.76263	−1.5222	0.91577	−1.6006
10	1.6835	−2.3378	1.7457	−2.3479	1.9683	−2.3723	2.1964	−2.3828
15	2.7969	−2.9666	2.8693	−2.9565	3.1223	−2.9152	3.3739	−2.8712
20	3.8579	−3.4356	3.9329	−3.4072	4.1941	−3.3125	4.4533	−3.2303
25	4.8611	−3.8084	4.9352	−3.7646	5.1956	−3.6276	5.4564	−3.5178
30	5.8128	−4.1178	5.8844	−4.0610	6.1398	−3.8901	6.3996	−3.7600
35	6.7199	−4.3824	6.7882	−4.3145	7.0367	−4.1162	7.2945	−3.9707
40	7.5887	−4.6137	7.6532	−4.5365	7.8940	−4.3158	8.1492	−4.1584
45	8.4242	−4.8194	8.4847	−4.7343	8.7176	−4.4951	8.9701	−4.3283
50	9.2305	−5.0048	9.2870	−4.9129	9.5120	−4.6583	9.7618	−4.4841
55	10.011	−5.1738	10.063	−5.0761	10.281	−4.8086	10.528	−4.6284
60	10.768	−5.3291	10.817	−5.2264	11.027	−4.9481	11.271	−4.7630
65	11.505	−5.4730	11.550	−5.3660	11.753	−5.0785	11.995	−4.8896
70	12.223	−5.6071	12.264	−5.4965	12.460	−5.2012	12.700	−5.0091
75	12.924	−5.7329	12.962	−5.6191	13.151	−5.3173	13.389	−5.1225
80	13.609	−5.8513	13.644	−5.7349	13.827	−5.4274	14.063	−5.2305
85	14.280	−5.9633	14.312	−5.8446	14.489	−5.5324	14.723	−5.3337
90	14.938	−6.0696	14.966	−5.9490	15.138	−5.6328	15.370	−5.4327
95	15.583	−6.1709	15.609	−6.0486	15.775	−5.7291	16.005	−5.5278
100	16.217	−6.2676	16.240	−6.1440	16.401	−5.8216	16.629	−5.6195

如欲求夹层圆板的中心和边缘应力，我们就必须用级数（3.1 a）和（3.2a），其结果为

$$S_r(0) = S_r|_{\rho=0} = a_0,$$

$$S_\theta(0) = S_\theta|_{\rho=0} = a_0,$$

$$S_r(1) = S_r|_{\rho=1} = \sum_{i=0}^{\infty} a_{2i},$$

$$S_\theta(1) = S_\theta|_{p=0} = \sum_{i=0}^{\infty} (2i+1) a_{2i} \tag{3.9}$$

由边界条件（2.11）和（2.12）可知，在滑动固定和简支两种边界情况下，夹层圆板的无量纲边缘径向应力 S_r（1）恒等于零。

最后，我们将夹层圆板的挠度和应力的计算结果给在图 1～图 4 中。

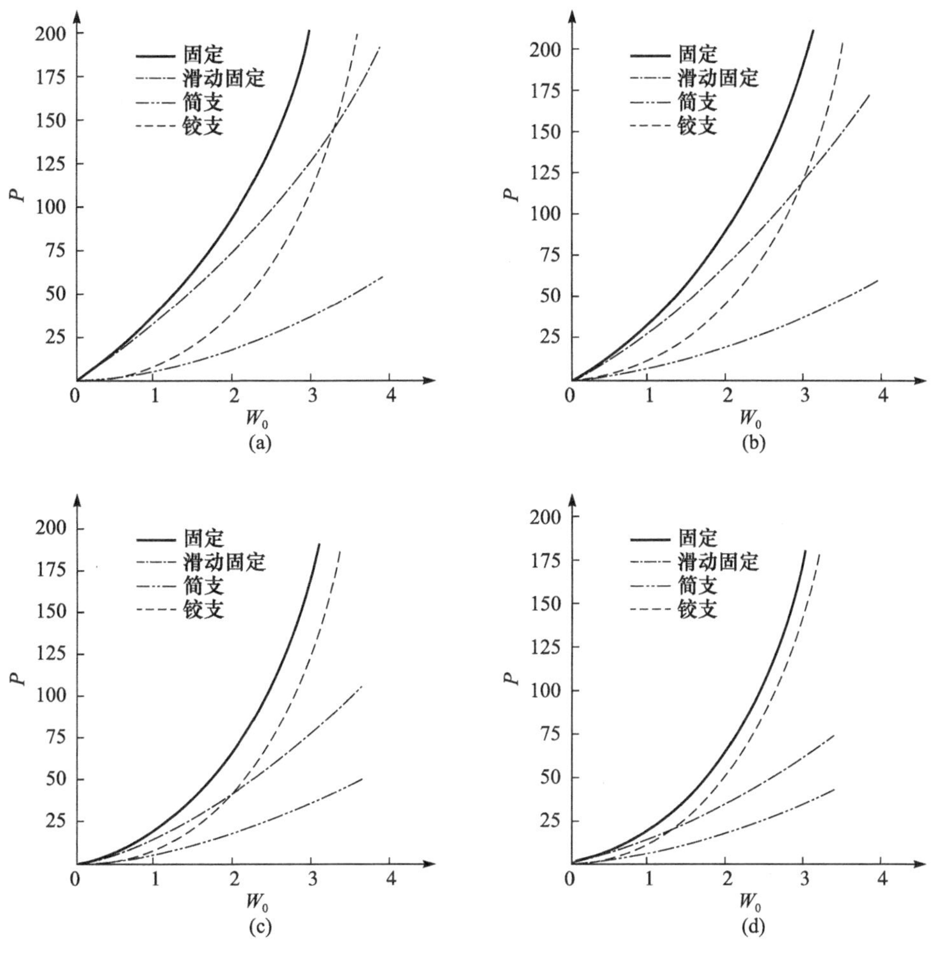

图 1　不同边界条件下的载荷与中心挠度的关系曲线

(a) $\nu=0.3$，$k=0$；(b) $\nu=0.3$，$k=0.01$；

(c) $\nu=0.3$，$k=0.05$；(d) $\nu=0.3$，$k=0.10$.

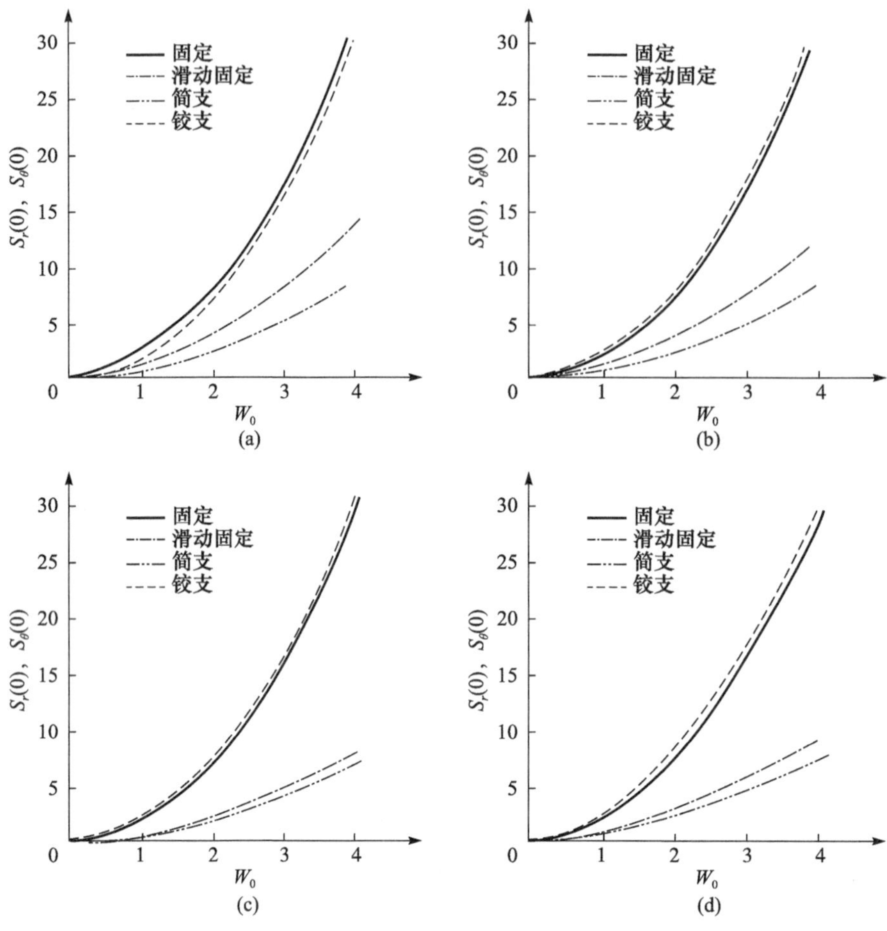

图 2　不同边界条件下的夹层圆板中心的径向应力和环向应力

(a) ν=0.3，k=0；(b) ν=0.3，k=0.01；(c) ν=0.3，k=0.05；(d) ν=0.3，k=0.10.

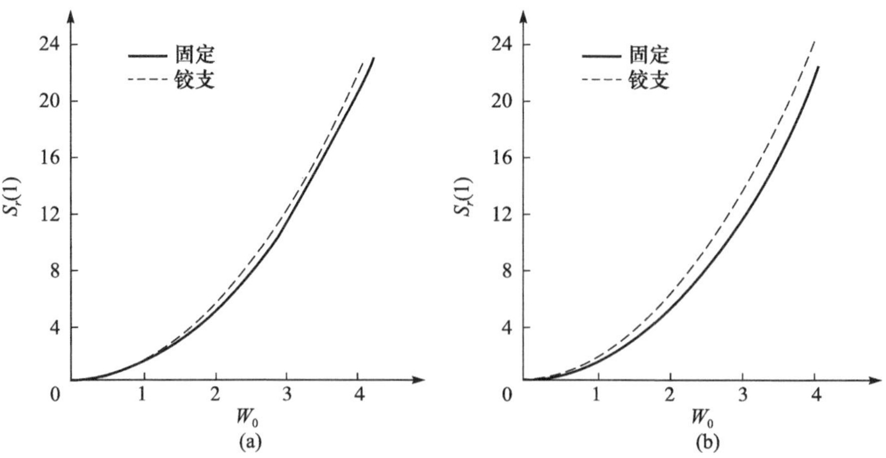

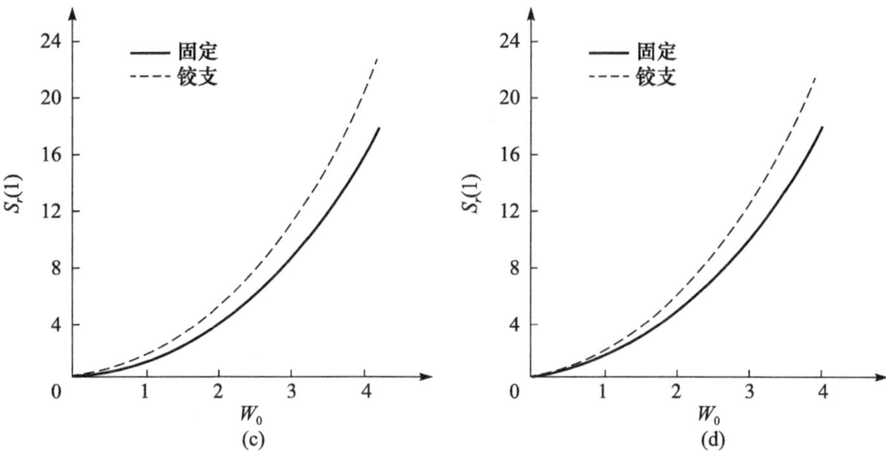

图 3　不同边界条件下的夹层圆板边缘的径向应力

(a) ν=0.3，k=0；(b) ν=0.3，k=0.01；(c) ν=0.3，k=0.05；(d) ν=0.3，k=0.10

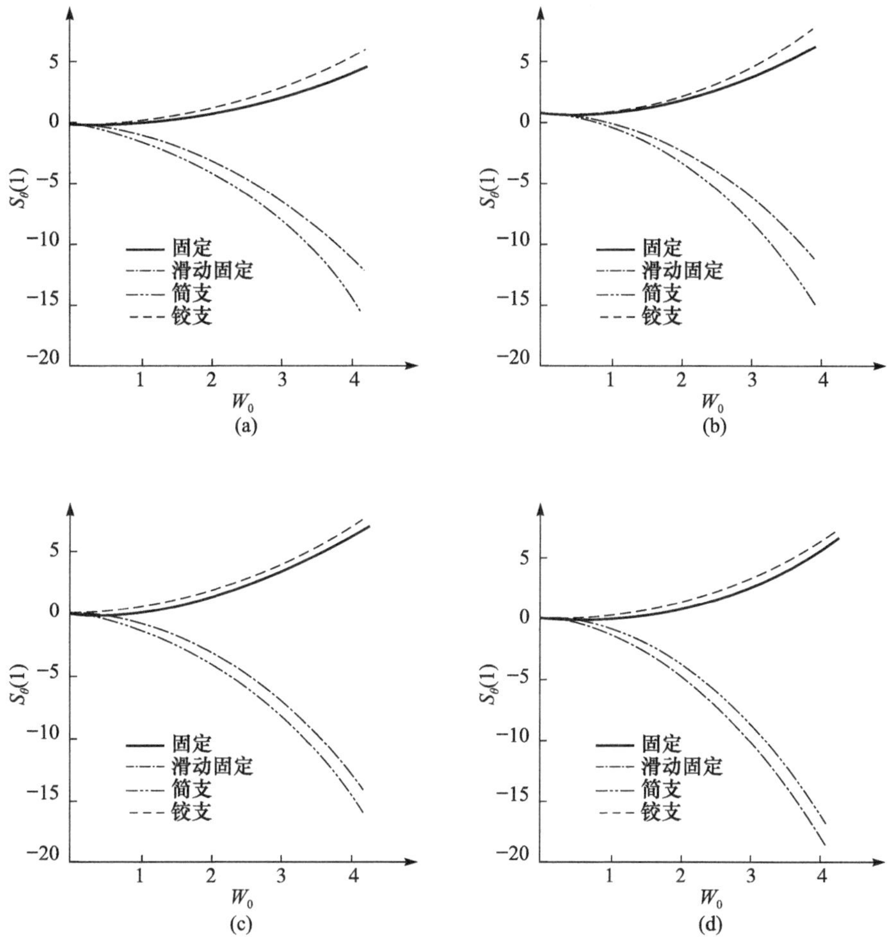

图 4　不同边界条件下的夹层圆板边缘的环向应力

(a) ν=0.3，k=0；(b) ν=0.3，k=0.01；(c) ν=0.3，k=0.05；(d) ν=0.3，k=0.10

四、讨论

本文使用幂级数方法求得了夹层圆板大挠度问题的精确解。虽然此法的数值计算量相当大，但我们使用电子计算机容易地克服了这一困难。应用本文的精确解，我们能验证文献［4］中所给的解析解的精确度。以滑动固定夹层圆板为例，我们将这两个解的数值结果及其相对误差给在表 2 中。为简单起见，今将文献［4］中的不同近似的解析解加以区分。为此，我们分别用下标 2 和 3 表示二次近似解和三次近似解。由表可知：

（1）三次近似解与精确解几乎重合，它们的相对误差小于 0.5%；

（2）二次近似解便有令人满意的精确性。一般来说，它与精确解之相对误差小于 10%。这说明，对夹层圆板的大挠度问题而言，修正迭代法[4]能给出十分精确的解，即使只做到二次近似，就能得到满意的近似解。

表 2(a)　文献［4］的解析解与本文精确解的比较

$k=0$

W_0			0.46535	1.0544	1.4587	1.9438	2.4723	3.0157	3.4847	3.9625
P	本文		15	35	50	70	95	125	155	190
	文献［4］	P_2	15.000	35.007	50.036	70.144	95.455	126.16	157.27	194.08
		相对误差（%）	0	−0.02	−0.07	−0.2	−0.5	−0.9	−1.5	−2.1
	文献［4］	P_3	15.00	35.00	50.000	69.999	94.997	125.00	155.05	190.22
		相对误差（%）	0	0	0	0.001	0.003	0	−0.03	−0.1
$S_r(0)$	本文		0.21601	1.0977	2.0780	3.6274	5.7382	8.3184	10.842	13.667
	文献［4］	S_{r2}（0）	0.21600	1.0974	2.0761	3.6175	5.7003	8.2074	10.607	13.221
		相对误差（%）	0.005	0.03	0.09	0.3	0.7	1.3	2.2	3.3
$S_\theta(1)$	本文		−0.14458	−0.74671	−1.4388	−2.5816	−4.2350	−6.4074	−8.6920	−11.433
	文献［4］	$S_{\theta 2}$（1）	−0.14458	−0.74686	−1.4398	−2.5871	−4.2579	−6.4810	−8.8638	−11.797
		相对误差（%）	0	0.02	0.07	0.2	0.5	1.1	2.0	3.2

表 2(b)　文献［4］的解析解与本文精确解的比较

$k=0.01$

W_0			0.53828	1.0475	1.5122	2.0600	2.5369	2.9575	3.5085	3.9882
P	本文		15	30	45	65	85	105	135	165
	文献［4］	P_2	15.000	30.011	45.068	65.301	85.812	106.67	138.66	171.52
		相对误差（%）	0	−0.04	−0.2	−0.5	−1.0	−1.6	−2.7	−4.0

续表

W_0			0.53828	1.0475	1.5122	2.0600	2.5369	2.9575	3.5085	3.9882
P	文献[4]	P_3	15.000	30.000	44.999	64.993	84.978	104.96	134.96	165.08
		相对误差(%)	0	0	0.002	0.01	0.03	0.04	0.03	−0.05
$S_r(0)$	本文		0.26282	0.98307	2.0136	3.6401	5.3756	7.1220	9.6776	12.121
	文献[4]	S_{r2} (0)	0.26281	0.98261	2.0097	3.6175	5.3054	6.9652	9.3100	11.452
		相对误差(%)	0.004	0.05	0.2	0.6	1.3	2.2	3.8	5.5
$S_\theta(1)$	本文		−0.19551	−0.74562	−1.5693	−2.9573	−4.5570	−6.2900	−9.0467	−11.924
	文献[4]	$S_{\theta 2}$ (1)	−0.19551	−0.74588	−1.5717	−2.9719	−4.6068	−6.4120	−9.3754	−12.612
		相对误差(%)	0	0.03	0.2	0.5	1.1	1.9	3.6	5.8

表 2(c)　文献 [4] 的解析解与本文精确解的比较

k=0.05

W_0			0.55524	1.0734	1.5387	1.9528	2.4950	2.9646	3.5094	3.9866
P	本文		10	20	30	40	55	70	90	110
	文献[4]	P_2	10.001	20.021	30.118	40.365	56.135	72.468	95.174	118.93
		相对误差(%)	−0.01	−0.1	−0.4	−0.9	−2.1	−3.5	−5.7	−8.1
	文献[4]	P_3	10.000	19.999	29.995	39.976	54.907	69.792	89.663	109.79
		相对误差(%)	0	0.005	0.02	0.06	0.2	0.3	0.4	0.2
$S_r(0)$	本文		0.22521	0.82373	1.6457	2.5726	4.0219	5.4613	7.3141	9.0797
	文献[4]	S_{r2} (0)	0.22518	0.82238	1.6353	2.5345	3.8874	5.1541	6.6730	8.0374
		相对误差(%)	0.01	0.2	0.6	1.5	3.3	5.6	8.8	11.5
$S_\theta(1)$	本文		−0.22712	−0.86149	−1.8057	−2.9717	−5.0088	−7.2875	−10.588	−14.110
	文献[4]	$S_{\theta 2}$ (1)	−0.22714	−0.86263	−1.8152	−3.0099	−5.1661	−7.7093	−11.687	−16.364
		相对误差(%)	0.009	0.1	0.5	1.3	3.1	5.8	10.4	16.0

表 2(d)　文献 [4] 的解析解与本文精确解的比较

k=0.10

W_0			0.40275	1.1415	1.4649	2.0293	2.5084	2.9258	3.4693	3.9416
P	本文		5	15	20	30	40	50	65	80
	文献[4]	P_2	5.0002	15.034	20.113	30.515	41.332	52.607	70.370	89.093
		相对误差(%)	−0.004	−0.2	−0.6	−1.7	−3.3	−5.2	−8.3	−11.4

续表

W_0			0.40275	1.1415	1.4649	2.0293	2.5084	2.9258	3.4693	3.9416
P	文献[4]	P_3	5.0000	14.998	19.993	29.949	39.853	49.738	64.732	80.311
		相对误差(%)	0	0.01	0.04	0.2	0.4	0.5	0.4	−0.4
$S_r(0)$	本文		0.10526	0.81323	1.3062	2.3843	3.4826	4.5562	6.0983	7.5594
	文献[4]	S_{r2} (0)	0.10525	0.80991	1.2930	2.3118	3.2823	4.1650	5.3596	6.5113
		相对误差(%)	0.01	0.4	1.0	3.0	5.8	8.6	12.1	13.9
$S_\theta(1)$	本文		−0.12922	−1.0702	−1.7990	−3.5998	−5.7209	−8.0654	−11.885	−15.974
	文献[4]	$S_{\theta 2}$ (1)	−0.12923	−1.0740	−1.8152	−3.7065	−6.0773	−8.9146	−14.090	−20.469
		相对误差(%)	0.008	0.4	0.9	3.0	6.2	1.5	18.6	28.1

致谢：刘怀玉同志参加了本文一部分数值计算工作，特此致谢。

参考文献

[1] Reissner E. Finite deflections of sandwich plates. J. Aeron. Sci.，1948，15（7）：435；1950，17（2）：125.

[2] 中国科学院北京力学研究所固体力学研究室板壳组. 夹层板壳的弯曲、稳定和振动. 北京：科学出版社，1977：42.

[3] 刘人怀. 在边缘力矩作用下夹层圆板的非线性轴对称弯曲问题. 中国科学技术大学学报，1980，10（2）：56.

[4] 刘人怀. 夹层圆板的非线性弯曲. 应用数学和力学，1981，2（2）：173.

Nonlinear bending of circular sandwich plates under the action of axisymmetric uniformly distributed line loads①②

中文摘要

本文研究了在对称线布载荷作用下具有夹紧固定边界的夹层圆板的非线性弯曲问题。因为问题相当困难故一直未有人研究。困难来源于非线性微分方程和载荷的不连续分布。引入 Heaviside 函数，并使用钱伟长先生提出的摄动法，获得了令人满意的解。

1. Introduction

In recent years, many investigators have paid great attention to sandwich plates because of the importance of such plates in engineering. But few studies have considered the nonlinear bending problems of circular sandwich plates. This paper is an extension of the author's previous papers[1-3]. Here, we will consider a more difficult problem, i. e. , the nonlinear bending problems of the plates with rigidly clamped edges under the action of axisymmetric uniformly distributed line loads. As we know, the problem has not yet been studied because of serious difficulties. It is apparent that the difficulties are due to the nonlinear differential equations and the discontinuity of the loads. To conquer these difficulties the Heaviside function will be introduced, and the perturbation method proposed by Chien[4] will be used. The obtained solution of this problem may be applied directly to engineering design.

2. Fundamental Equations

Let us consider a rigidly clamped circular sandwich plate in which the load p is uniformly distributed along a circle of radius b as shown in Fig. 1. In order to simplify the following relation, the system of notation for the displacements and the stresses in this paper is the same as that used in the discussion of [2]. Proceeding by the simplified equations in [2], we can easily obtain the fundamental equations of nonlinear bending for the plate. The equations satisfied by the deflection w and the radial stress σ_{r0} of the middle plane of the plate are

at $0\leqslant r<b$,

---

① Reprinted from *Progress in Applied Mechanics*, Edited by Yeh, K. Y. , Martinus Nijhoff Publishers, Dordrecht, 1987, 293-321.

② This paper was completed at Ruhr-University, Bochum, Federal Republic of Germany. The author wishes to express his thanks to the Alexander von Humboldt Foundation, Ruhr-University, Bochum, and especially to W. Zerna for his warm hospitality.

$$Dr\frac{\mathrm{d}}{\mathrm{d}r}\frac{1}{r}\frac{\mathrm{d}}{\mathrm{d}r}r\frac{\mathrm{d}w}{\mathrm{d}r}+2t\left(\frac{D}{G_2h_0}r\frac{\mathrm{d}}{\mathrm{d}r}\frac{1}{r}\frac{\mathrm{d}}{\mathrm{d}r}-1\right)\left(r\sigma_{r0}\frac{\mathrm{d}w}{\mathrm{d}r}\right)=0, \tag{2.1a}$$

at $b<r\leqslant a$,

$$Dr\frac{\mathrm{d}}{\mathrm{d}r}\frac{1}{r}\frac{\mathrm{d}}{\mathrm{d}r}r\frac{\mathrm{d}w}{\mathrm{d}r}+2t\left(\frac{D}{G_2h_0}r\frac{\mathrm{d}}{\mathrm{d}r}\frac{1}{r}\frac{\mathrm{d}}{\mathrm{d}r}-1\right)\left(r\sigma_{r0}\frac{\mathrm{d}w}{\mathrm{d}r}\right)-pb=0, \tag{2.1b}$$

and

$$r\frac{\mathrm{d}}{\mathrm{d}r}\frac{1}{r}\frac{\mathrm{d}}{\mathrm{d}r}(r^2\sigma_{r0})=-\frac{E}{2}\left(\frac{\mathrm{d}w}{\mathrm{d}r}\right)^2. \tag{2.2}$$

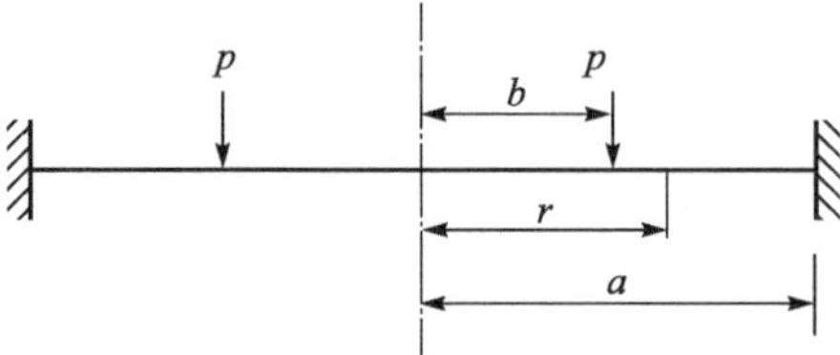

Fig. 1　A circular sandwich plate under the action of axisymmetric uniformly distributed line loads

To simplify the calculations, we now make use of the Heaviside function

$$H(r-b)=\begin{cases}1 & \text{at } r\geqslant b,\\ 0 & \text{at } r<b,\end{cases} \tag{2.3}$$

thus (2.1a) and (2.1b) can be reduced to one:

$$Dr\frac{\mathrm{d}}{\mathrm{d}r}\frac{1}{r}\frac{\mathrm{d}}{\mathrm{d}r}r\frac{\mathrm{d}w}{\mathrm{d}r}+2t\left(\frac{D}{G_2h_0}\frac{\mathrm{d}}{\mathrm{d}r}\frac{1}{r}\frac{\mathrm{d}}{\mathrm{d}r}-1\right)\left(r\sigma_{r0}\frac{\mathrm{d}w}{\mathrm{d}r}\right)-H(r-b)pb=0. \tag{2.4}$$

In this way the solution of (2.2) and (2.4) can automatically satisfy the jump condition for the shearing force and continuity conditions at the place of the action of the load.

The boundary conditions for a rigidly clamped edge are

$$w=0,\ \psi=0,\ u=0\quad \text{for } r=a;$$
$$\psi=0,\ \sigma_{r0}\ \text{finite}\qquad \text{for } r=0. \tag{2.5}$$

Using expressions for the rotation ψ of a normal to the middle plane of the plate in the diametral plane and the radial displacement u of the middle plane of the plate, we can represent the boundary conditions in the following form:

$$w=0,\ \left(\frac{2t}{G_2h_0}\sigma_{r0}+1\right)\frac{\mathrm{d}w}{\mathrm{d}r}+\frac{pb}{G_2h_0a}=0,$$
$$\frac{\mathrm{d}}{\mathrm{d}r}(r\sigma_{r0})-\nu\sigma_{r0}=0\quad \text{for } r=a;$$
$$\left(\frac{2t}{G_2h_0}\sigma_{r0}+1\right)\frac{\mathrm{d}w}{\mathrm{d}r}=0,\ \sigma_{r0}\quad \text{finite for } r=0. \tag{2.6}$$

Equations (2.4), (2.2), and the boundary conditions (2.6) containing the two unknown functions w and σ_{r0} will be used in solving the problem. Having the solution for the nonlinear boundary value problem, we can find the tangential stress σ_{r0} of the middle plane of the plate:

$$\sigma_{\theta 0} = \frac{\mathrm{d}(r\sigma_{r0})}{\mathrm{d}r}. \tag{2.7}$$

To represent these equations (2.4), (2.2) and the conditions (2.6) in a simpler dimensionless form, let us introduce the following notations:

$$y = \frac{r}{a},\ \beta = \frac{b}{a},\ W = \frac{\sqrt{2(1-\nu^2)}}{h_0}w,\ S_r = \frac{2ta^2}{D}\sigma_{r0},$$

$$S_\theta = \frac{2ta^2}{D}\sigma_{\theta 0},\ k = \frac{D}{G_2 h_0 a^2},\ P = \frac{a^3\sqrt{2(1-\nu^2)}}{h_0 D}p. \tag{2.8}$$

With these notations, (2.4), (2.2) and the conditions (2.6) become, respectively,

$$L\left(y\frac{\mathrm{d}W}{\mathrm{d}y}\right) = -(kL-1)\left(yS_r\frac{\mathrm{d}W}{\mathrm{d}y}\right) + H(y-\beta)\beta P,$$

$$L(y^2 S_r) = -\left(\frac{\mathrm{d}W}{\mathrm{d}y}\right)^2, \tag{2.9}$$

$$W = 0,\ (kS_r+1)\frac{\mathrm{d}W}{\mathrm{d}y} + k\beta P = 0,\ \frac{\mathrm{d}S_r}{\mathrm{d}y} + \lambda_1 S_r = 0 \quad \text{for } y = 1;$$

$$(kS_r+1)\frac{\mathrm{d}W}{\mathrm{d}y} = 0,\ S_r \text{ finite} \quad \text{for } y = 0, \tag{2.10}$$

where

$$L(\cdots) = y\frac{\mathrm{d}}{\mathrm{d}y}\frac{1}{y}\frac{\mathrm{d}}{\mathrm{d}y}(\cdots), \tag{2.11}$$

$$\lambda_1 = 1-\nu, \tag{2.12}$$

and (2.7) becomes

$$S_\theta = \frac{\mathrm{d}(yS_r)}{\mathrm{d}y}. \tag{2.13}$$

Equations (2.9) and (2.10) are the dimensionless nonlinear boundary value problem for discussing a rigidly clamped circular sandwich plate under the action of axisymmetric uniformly distributed line load. It should be noted that the difficulty in solving this problem is great. It is only just possible to find the solution by the Heaviside function. However, even in this case, the results are still very complicated.

3. Solving the Nonlinear Boundary Value Problem by the Perturbation Method

To obtain a satisfactory solution of the above nonlinear boundary value problem, it is necessary to use the perturbation method. Assuming that the perturbation parameter is the dimensionless center deflection W_0 for the circular sandwich plate, we represent dimensionless deflection W, dimensionless radial stress S_r, and dimensionless axisymmetric uniformly distributed line load P by the following power series for W_0:

$$W = w_1(y)W_0 + w_3(y)W_0^3 + \cdots,$$

$$S_r = S_2(y)W_0^2 + S_4(y)W_0^4 + \cdots,$$

$$P = p_1 W_0 + p_3 W_0^3 + \cdots, \tag{3.1}$$

in which w_1, w_3, $\cdots S_2$, S_4, $\cdots$and p_1, p_3, $\cdots$are functions and constants to be determined later, respectively. Substituting these series in the problems (2.9) and (2.10), and observing that these equations must be satisfied for any power of W_0, we find a system of linear boundary value problems for w_1, S_2, p_1, w_3, S_4, p_3, $\cdots$. For w_1 we have the linear boundary value problem as follows

$$L\left(y\frac{\mathrm{d}w_1}{\mathrm{d}y}\right)=H(y-\beta)\beta p_1, \tag{3.2}$$

$$w_1=0,\ \frac{\mathrm{d}w_1}{\mathrm{d}y}=-k\beta p_1\quad \text{for } y=1;$$

$$w_1=1,\ \frac{\mathrm{d}w_1}{\mathrm{d}y}=0\quad \text{for } y=0. \tag{3.3}$$

The solution of this problem is

$$\begin{aligned} w_1=&-\frac{\beta p_1}{4}\{\alpha y^2-\left(\beta^2\ln\beta-\frac{1}{2}\beta^2+2k+\frac{1}{2}\right)\\ &-H(y-\beta)[y^2\ln y-(\ln\beta+1)y^2+\beta^2\ln y-\beta^2(\ln\beta-1)]\}, \end{aligned} \tag{3.4}$$

where

$$p_1=\frac{8}{\beta(2\beta^2\ln\beta-\beta^2+4k+1)}, \tag{3.5}$$

$$\alpha=\frac{1}{2}\beta^2-\ln\beta+2k-\frac{1}{2}. \tag{3.6}$$

For S_2 we have the linear boundary value problem as follows

$$L(y^2S_2)=-\left(\frac{\mathrm{d}w_1}{\mathrm{d}y}\right)^2, \tag{3.7}$$

$$\frac{\mathrm{d}S_2}{\mathrm{d}y}+\lambda_1S_2=0\quad \text{for } y=1;$$

$$S_2 \text{ finite}\quad \text{for } y=0. \tag{3.8}$$

The solution of this problem is

$$\begin{aligned} S_2=&-\frac{\beta^2p_1^2}{32}\left\{\alpha^2y^2-\frac{1}{\lambda_1}\left[\frac{1}{4}\lambda_2\beta^6-\frac{1-7\nu}{2}\beta^4\ln\beta\right.\right.\\ &+\left(\lambda_2k-\frac{13-19\nu}{8}\right)\beta^4-2\lambda_1\beta^2\ln^2\beta\\ &+4\lambda_1(2k+1)\beta^2\ln\beta-\left(2(3-\nu)k-\frac{1-7\nu}{4}\right)\beta^2\\ &\left.+4(3-\nu)k^2+(5-3\nu)k+\frac{9-7\nu}{8}\right]\\ &+H(y-\beta)\left[y^2\ln^2y-\left(\beta^2+4k+\frac{3}{2}\right)y^2\ln y\right.\\ &+\left(\beta^2\ln\beta+\frac{5}{4}\beta^2-\ln^2\beta+(4k-1)\ln\beta+5k+\frac{5}{8}\right)y^2\\ &+2\beta^2\ln^2y-2\beta^2(\beta^2+4k+1)\ln y+\beta^2(2\beta^2\ln\beta-\beta^2-2\ln^2\beta\\ &+4(2k+1)\ln\beta-4k+1)-\beta^4y^{-2}\ln y\\ &\left.\left.-\beta^4\left(\frac{1}{4}\beta^2-\frac{3}{2}\ln\beta+k+\frac{3}{8}\right)y^{-2}\right]\right\}, \end{aligned} \tag{3.9}$$

where

$$\lambda_2 = 1 + \nu. \tag{3.10}$$

For w_3, we have the linear boundary value problem as follows

$$L\left(y\frac{\mathrm{d}w_3}{\mathrm{d}y}\right) = -(kL-1)\left(yS_2\frac{\mathrm{d}w_1}{\mathrm{d}y}\right) + H(y-\beta)\beta p_3, \tag{3.11}$$

$$w_3 = 0,\ \frac{\mathrm{d}w_3}{\mathrm{d}y} = -k\left(S_2\frac{\mathrm{d}w_1}{\mathrm{d}y} + \beta p_3\right) \quad \text{for } y = 1;$$

$$w_3 = 0,\ \frac{\mathrm{d}w_3}{\mathrm{d}y} = -kS_2\frac{\mathrm{d}w_1}{\mathrm{d}y} \quad \text{for } y = 0. \tag{3.12}$$

The solution of this problem is

$$\begin{aligned} w_3 = {} & \frac{\beta^3 p_1^3}{512}\left[\sum_{i=1}^{3} f_{2i}y^{2i} + H(y-\beta)\sum_{i=-1}^{3}\sum_{j=0}^{3} g_{2i,j}y^{2i}\ln^j y\right] \\ & - \frac{\beta p_3}{4}\{\alpha y^2 - H(y-\beta)[y^2\ln y - (\ln\beta + 1)y^2 \\ & + \beta^2\ln y - \beta^2(\ln\beta - 1)]\}, \end{aligned} \tag{3.13}$$

where

$$\begin{aligned} p_3 = {} & -\frac{\beta^3 p_1^4}{1024\lambda_1}\left\{\frac{1+5\nu}{24}\beta^{10}\ln\beta - \frac{5+31\nu}{144}\beta^{10} + \frac{13-31\nu}{12}\beta^8\ln^2\beta\right. \\ & - \left(\frac{1+17\nu}{6}k + \frac{139-229\nu}{48}\right)\beta^8\ln\beta - \left(\frac{91-127\nu}{36}k - \frac{79}{36}\lambda_1\right)\beta^8 \\ & - 3\lambda_1\beta^6\ln^3\beta + \left(\frac{4+68\nu}{3}k + \frac{86}{9}\lambda_1\right)\beta^6\ln^2\beta \\ & + \left(\frac{20-68\nu}{3}k^2 + \frac{92-152\nu}{3}k - \frac{601-493\nu}{54}\right)\beta^6\ln\beta \\ & - \left(\frac{221}{9}\lambda_1 k^2 + \frac{391-463\nu}{18}k - \frac{5213-2135\nu}{2592}\right)\beta^6 + 12\lambda_1 k\beta^4\ln^3\beta \\ & - 24\lambda_1 k(3k+2)\beta^4\ln^2\beta + \left[32\lambda_1 k^3 + (206-122\nu)k^2 + (79-51\nu)k + \frac{63-49\nu}{16}\right]\beta^4\ln\beta \\ & - \left[(60-28\nu)k^3 + (65-23\nu)k^2 + \frac{73+269\nu}{24}k + \frac{1607-1193\nu}{288}\right]\beta^4 \\ & - k[(96-32\nu)k^2 + (40-24\nu)k + 9 - 7\nu]\beta^2\ln\beta \\ & + \left[(152-56\nu)k^3 + \frac{352-178\nu}{3}k^2 + \frac{280-181\nu}{9}k + \frac{445-283\nu}{288}\right]\beta^2 \\ & - 32(5-\nu)k^4 - \frac{920-272\nu}{9}k^3 \\ & \left. - \frac{250-106\nu}{9}k^2 - \frac{275-149\nu}{72}k - \frac{353-191\nu}{2592}\right\}, \end{aligned} \tag{3.14}$$

$$\begin{aligned} f_2 = {} & -\frac{1}{\lambda_1}\left\{\frac{1+5\nu}{96}\beta^{10} + \frac{19-61\nu}{48}\beta^8\ln\beta - \left(\frac{13+5\nu}{24}k + \frac{491-617\nu}{576}\right)\beta^8\right. \\ & - \frac{25-61\nu}{12}\beta^6\ln^2\beta + \left(\frac{10+8\nu}{3}k + \frac{146-173\nu}{18}\right)\beta^6\ln\beta \end{aligned}$$

$$
\begin{aligned}
& -\left(\frac{13+11\nu}{3}k^2+\frac{13+5\nu}{18}k+\frac{3463-3301\nu}{864}\right)\beta^6 \\
& +\frac{3}{2}\lambda_1\beta^4\ln^3\beta-[(3+9\nu)k+6\lambda_1]\beta^4\ln^2\beta \\
& -\left[(12-44\nu)k^2-(7-3\nu)k-\frac{123-81\nu}{16}\right]\beta^4\ln\beta \\
& -\left[8\lambda_2k^3-(24-19\nu)k^2+\frac{49-67\nu}{8}k-\frac{263-323\nu}{96}\right]\beta^4 \\
& -8\lambda_1k\beta^2\ln^3\beta+12\lambda_1k(4k+1)\beta^2\ln^2\beta \\
& -\left[64\lambda_1k^3+(52-28\nu)k^2-(4-12\nu)k+\frac{9-7\nu}{8}\right]\beta^2\ln\beta \\
& +\left[(24-8\nu)k^3-(13-19\nu)k^2+\frac{43}{6}\lambda_1k+\frac{341-251\nu}{144}\right]\beta^2 \\
& +k\left[(48-16\nu)k^2+(20-12\nu)k+\frac{9-7\nu}{2}\right]\ln\beta \\
& -32(3-\nu)k^4-\frac{80-56\nu}{3}k^3-\frac{20-11\nu}{3}k^2 \\
& \left.+\frac{4-13\nu}{18}k-\frac{445-283\nu}{1728}\right\},
\end{aligned}
$$

$$
\begin{aligned}
f_4= & -\frac{1}{\lambda_1}\left\{\frac{1}{32}\lambda_2\beta^8-\frac{\nu}{2}\beta^6\ln\beta+\left(\frac{1}{2}k-\frac{15-17\nu}{64}\right)\beta^6\right. \\
& -\frac{3-9\nu}{8}\beta^4\ln^2\beta-\left(\frac{1+3\nu}{2}k-\frac{27-21\nu}{32}\right)\beta^4\ln\beta \\
& +\left(\frac{7-5\nu}{2}k^2-\frac{39-33\nu}{16}k+\frac{15-33\nu}{64}\right)\beta^4+\frac{1}{2}\lambda_1\beta^2\ln^3\beta \\
& -\frac{3}{4}\lambda_1\beta^2\ln^2\beta-\left(8\lambda_1k^2-\frac{11-9\nu}{2}k+\frac{9-15\nu}{16}\right)\beta^2\ln\beta \\
& +\left(12\lambda_1k^3-\frac{15-13\nu}{2}k^2+\frac{9}{4}\lambda_1k+\frac{7}{64}\lambda_2\right)\beta^2 \\
& -2\lambda_1k\ln^3\beta+3\lambda_1k(4k-1)\ln^2\beta \\
& -\left[24\lambda_1k^3-(9-11\nu)k^2+\frac{11-9\nu}{4}k+\frac{9-7\nu}{32}\right]\ln\beta \\
& \left.+16\lambda_1k^4-2(3-5\nu)k^3+4\lambda_1k^2-\frac{5-3\nu}{16}k-\frac{9-7\nu}{64}\right\},
\end{aligned}
$$

$$f_6=\frac{1}{18}\alpha^3,$$

$$g_{-2,0}=k\beta^6\left(\frac{1}{2}\beta^2-3\ln\beta+2k+\frac{17}{4}\right),$$

$$g_{-2,1}=2k\beta^6,$$

$$g_{-2,2}=g_{-2,3}=0,$$

$$g_{0,0}=-\frac{\beta^2}{\lambda_1}\left\{\frac{1+5\nu}{48}\beta^8\ln\beta-\frac{4+23\nu}{144}\beta^8+\frac{13-31\nu}{24}\beta^6\ln^2\beta-\left(\frac{1+17\nu}{12}k\right.\right.$$

$$+\frac{59-101\nu}{32}\Big)\beta^6\ln\beta-\Big(\frac{8-53\nu}{36}k-\frac{1051-1213\nu}{576}\Big)\beta^6$$

$$-\frac{3}{2}\lambda_1\beta^4\ln^3\beta+\Big(\frac{2+34\nu}{3}k+\frac{55}{9}\lambda_1\Big)\beta^4\ln^2\beta$$

$$+\Big[\frac{10-34\nu}{3}k^2+(7-23\nu)k-\frac{4045-3883\nu}{432}\Big]\beta^4\ln\beta$$

$$-\Big(\frac{107-251\nu}{18}k^2+\frac{203-365\nu}{36}k-\frac{14107-12649\nu}{2592}\Big)\beta^4$$

$$+6\lambda_1 k\beta^2\ln\beta-12\lambda_1 k(3k+2)\beta^2\ln^2\beta$$

$$+\Big[16\lambda_1 k^3+(91-73\nu)k^2+\frac{131-105\nu}{4}k+\frac{9-7\nu}{32}\Big]\beta^2\ln\beta$$

$$-\Big(22\lambda_1 k^3+\frac{107-69\nu}{2}k^2+\frac{45+9\nu}{8}k+\frac{27-21\nu}{64}\Big)\beta^2$$

$$-k\Big[(48-16\nu)k^2+(20-12\nu)k+\frac{9-7\nu}{2}\Big]\ln\beta$$

$$+k\Big[(48-16\nu)k^2+(20-12\nu)k+\frac{9-7\nu}{2}\Big]\Big\},$$

$$g_{0,1}=\frac{\beta^2}{\lambda_1}\Big\{\frac{1+5\nu}{48}\beta^8+\frac{19-37\nu}{24}\beta^6\ln\beta-\Big(\frac{1+17\nu}{12}k+\frac{401-455\nu}{288}\Big)\beta^6$$

$$-\frac{8}{3}\lambda_1\beta^4\ln^2\beta+\Big(\frac{14+22\nu}{3}k+\frac{65}{9}\lambda_1\Big)\beta^4\ln\beta$$

$$+\Big(\frac{10-34\nu}{3}k^2+\frac{32-68\nu}{9}k-\frac{2437-2185\nu}{432}\Big)\beta^4$$

$$-24\lambda_1 k^2\beta^2\ln\beta+\Big[16\lambda_1 k^3+(37-19\nu)k^2+\frac{17+9\nu}{4}k+\frac{9-7\nu}{32}\Big]\beta^2$$

$$-k\Big[(48-16\nu)k^2+(20-12\nu)k+\frac{9-7\nu}{2}\Big]\Big\},$$

$$g_{0,2}=-\beta^4\Big[\frac{1}{4}\beta^4-\frac{3}{2}\beta^2\ln\beta+\Big(4k+\frac{17}{8}\Big)\beta^2-6k\ln\beta+k\Big(12k+\frac{21}{2}\Big)\Big],$$

$$g_{0,3}=-\frac{1}{3}\beta^6,$$

$$g_{2,0}=-\frac{1}{\lambda_1}\Big\{\frac{\nu}{2}\beta^8\ln\beta+\frac{3}{32}\lambda_1\beta^8+\frac{3-15\nu}{4}\beta^6\ln^2\beta+\Big(2\lambda_1 k-\frac{33-39\nu}{8}\Big)\beta^6\ln\beta$$

$$-\Big(\frac{3+5\nu}{4}k+\frac{3}{16}\lambda_1\Big)\beta^6-\frac{3}{2}\lambda_1\beta^4\ln^3\beta+[(3+9\nu)k+6\lambda_1]\beta^4\ln^2\beta$$

$$+\Big[(24-32\nu)k^2+\frac{1+9\nu}{2}k-\frac{45-33\nu}{8}\Big]\beta^4\ln\beta$$

$$-\Big(\frac{13+3\nu}{2}k^2-\frac{45-57\nu}{4}k+\frac{15}{4}\lambda_1\Big)\beta^4+8\lambda_1 k\beta^2\ln^3\beta-12\lambda_1 k(4k+1)\beta^2\ln^2\beta$$

$$+\Big[64\lambda_1 k^3+(52-28\nu)k^2-(4-12\nu)k+\frac{9-7\nu}{8}\Big]\beta^2\ln\beta-k\Big[8\lambda_1 k^2$$

$$-(46-30\nu)k-\frac{27-3\nu}{4}\Big]\beta^2-k\Big[(48-16\nu)k^2+(20-12\nu)k+\frac{9-7\nu}{2}\Big]\ln\beta$$

$$-k\Big[(48-16\nu)k^2+(20-12\nu)k+\frac{9-7\nu}{2}\Big]\Big\},$$

$$g_{2,1}=\frac{1}{\lambda_1}\Big\{\frac{\nu}{2}\beta^8-3\nu\beta^6\ln\beta+\Big(2\lambda_1 k-\frac{15-18\nu}{4}\Big)\beta^6+(12k+3\lambda_1)\beta^4\ln\beta$$

$$+\Big[(24-32\nu)k^2+(2-9\nu)k-\frac{12-9\nu}{2}\Big]\beta^4$$

$$+\Big[64\lambda_1 k^3+(76-52\nu)k^2+(23-15\nu)k+\frac{9-7\nu}{8}\Big]\beta^2$$

$$-k\Big[(48-16\nu)k^2+(20-12\nu)k+\frac{9-7\nu}{2}\Big]\Big\},$$

$$g_{2,2}=3\beta^2\Big[\frac{1}{4}\beta^4-\frac{1}{2}\beta^2\ln\beta-\Big(3k-\frac{7}{8}\Big)\beta^2-2k(8k+3)\Big],$$

$$g_{2,3}=8k\beta^2,$$

$$g_{4,0}=\frac{1}{\lambda_1}\Big\{\frac{3-5\nu}{16}\beta^6\ln\beta-\frac{3\nu}{16}\beta^6-\frac{3-9\nu}{8}\beta^4\ln^2\beta+\Big(\frac{1-3\nu}{4}k+\frac{39-9\nu}{32}\Big)\beta^4\ln\beta$$

$$-\Big(\frac{6-3\nu}{4}k-\frac{99-117\nu}{64}\Big)\beta^4+\frac{1}{2}\lambda_1\beta^2\ln^3\beta-\frac{3}{4}\lambda_1\beta^2\ln^2\beta$$

$$-\Big(8\lambda_1 k^2-\frac{11-9\nu}{2}k+\frac{9-15\nu}{16}\Big)\beta^2\ln\beta-\Big(\frac{27}{2}\lambda_1 k^2-\frac{19-13\nu}{4}k$$

$$-\frac{51-33\nu}{32}\Big)\beta^2-2\lambda_1 k\ln^3\beta+3\lambda_1 k(4k-1)\ln^2\beta$$

$$-\Big[24\lambda_1 k^3-(9-11\nu)k^2+\frac{11-9\nu}{4}k+\frac{9-7\nu}{32}\Big]\ln\beta$$

$$-30\lambda_1 k^3-\frac{19-13\nu}{2}k^2-\frac{25-19\nu}{8}k-\frac{27-21\nu}{64}\Big\},$$

$$g_{4,1}=-\frac{1}{\lambda_1}\Big[\frac{3-5\nu}{16}\beta^6+\frac{3}{8}\lambda_2\beta^4\ln\beta+\Big(\frac{1-3\nu}{4}k+\frac{57-63\nu}{32}\Big)\beta^4$$

$$-\Big(8\lambda_1 k^2-\frac{5-3\nu}{2}k-\frac{45-39\nu}{16}\Big)\beta^2-24\lambda_1 k^3$$

$$-(21-19\nu)k^2-\frac{21-19\nu}{4}k-\frac{9-7\nu}{32}\Big],$$

$$g_{4,2}=\frac{3}{4}\beta^4+\frac{15}{4}\beta^2-12k^2-\frac{9}{2}k,$$

$$g_{4,3}=-\frac{1}{2}\beta^2+2k,$$

$$g_{6,0}=\frac{1}{6}\Big[\frac{1}{4}\beta^4\ln\beta+\frac{19}{48}\beta^4-\frac{1}{2}\beta^2\ln^2\beta+\Big(2k-\frac{1}{2}\Big)\beta^2\ln\beta+\Big(\frac{19}{6}k+\frac{13}{24}\Big)\beta^2$$

$$+\frac{1}{3}\ln^3\beta-\Big(2k-\frac{1}{2}\Big)\ln^2\beta+\Big(4k^2-2k+\frac{1}{4}\Big)\ln\beta+\frac{19}{3}k^2+\frac{13}{6}k+\frac{16}{27}\Big],$$

$$g_{6,1}=-\frac{1}{3}\left[\frac{1}{8}\beta^4+\left(k+\frac{13}{24}\right)\beta^2+2k^2+\frac{13}{6}k+\frac{2}{3}\right],$$

$$g_{6,2}=\frac{1}{3}\left(\frac{1}{4}\beta^2+k+\frac{13}{24}\right),$$

$$g_{6,3}=-\frac{1}{18}. \tag{3.15}$$

For S_4 we have the linear boundary value problem as follows

$$L(y^2S_4)=-2\frac{\mathrm{d}w_1}{\mathrm{d}y}\frac{\mathrm{d}w_3}{\mathrm{d}y}, \tag{3.16}$$

$$\frac{\mathrm{d}S_4}{\mathrm{d}y}+\lambda_1 S_4=0 \quad \text{for } y=1;$$

$$S_4 \quad \text{finite} \quad \text{for } y=0. \tag{3.17}$$

The solution of this problem is

$$S_4=\frac{\beta^4 p_1^4}{512}\left[\sum_{i=0}^{3}m_{2i}y^{2i}-H(y-\beta)\sum_{i=-2}^{3}\sum_{j=0}^{4}n_{2i,j}y^{2i}\ln^j y\right], \tag{3.18}$$

where

$$m_0=\frac{1}{\lambda_1}\left\{\sum_{i=-2}^{3}[n_{2i,1}+(2i+1-\nu)n_{2i,0}]-\sum_{i=1}^{3}(2i+1-\nu)m_{2i}\right\},$$

$$m_2=\frac{\alpha}{4}\left(f_2-\frac{128}{\beta^2 p_1^3}p_3\alpha\right),$$

$$m_4=\frac{1}{6}\alpha f_4,$$

$$m_6=\frac{1}{144}\alpha^4,$$

$$n_{-4,0}=-\frac{k\beta^8}{8}\left(\frac{1}{2}\beta^2-3\ln\beta+2k+\frac{19}{4}\right),$$

$$n_{-4,1}=-\frac{1}{1}k\beta^8,$$

$$n_{-4,2}=n_{-4,3}=n_{-4,4}=0,$$

$$\begin{aligned}n_{-2,0}=&-\frac{\beta^5 p_1}{16\lambda_1}\left\{\frac{1-2\nu}{576}\beta^{12}\ln\beta+\frac{1+18\nu}{1152}\beta^{12}-\frac{3+20\nu}{288}\beta^{10}\ln^2\beta\right.\\&-\left(\frac{18-7\nu}{144}k-\frac{633-1469\nu}{13824}\right)\beta^{10}\ln\beta+\left(\frac{59-27\nu}{144}k-\frac{3813-5177\nu}{27648}\right)\beta^{10}\\&+\frac{1+51\nu}{96}\beta^8\ln^3\beta-\left(\frac{89-93\nu}{72}k-\frac{925-1365\nu}{1152}\right)\beta^8\ln^2\beta-\left(\frac{109-89\nu}{72}k^2\right.\\&\left.+\frac{2437+231\nu}{1728}k-\frac{7098-7225\nu}{1152}\right)\beta^8\ln\beta+\left(\frac{643-393\nu}{144}k^2+\frac{13943-8747\nu}{6912}k\right.\\&\left.-\frac{299981-289049\nu}{82994}\right)\beta^8-\frac{35}{144}\lambda_1\beta^6\ln^4\beta+\lambda_1\left(\frac{53}{8}k-\frac{683}{1152}\right)\beta^6\ln^3\beta\\&-\left(\frac{331}{36}\lambda_1 k^2+\frac{1141-2525\nu}{288}k-\frac{49555-43363\nu}{13824}\right)\beta^6\ln^2\beta-\left(\frac{37}{6}\lambda_1 k^3\right.\end{aligned}$$

$$+\frac{16673-14201\nu}{864}k^2+\frac{47553-13361\nu}{6912}k+\frac{1450375-1393351\nu}{331776}\Big)\beta^6\ln\beta$$

$$+\Big(\frac{159-155\nu}{9}k^3+\frac{44355-41963\nu}{1728}k^2+\frac{606815-604223\nu}{41472}k$$

$$+\frac{3284315-3394619\nu}{663552}\Big)\beta^6-\frac{47}{9}\lambda_1 k\beta^4\ln^4\beta+\lambda_1\Big(\frac{76}{3}k^2+\frac{803}{72}k+\frac{289}{288}\Big)\beta^4\ln^3\beta$$

$$-\Big(\frac{44}{3}\lambda_1 k^3-\frac{1229-1389\nu}{36}k^2-\frac{18287-15023\nu}{1728}k+\frac{5227-6419\nu}{2304}\Big)\beta^4\ln^2\beta$$

$$-\Big(\frac{80}{9}\lambda_1 k^4+\frac{1679-1659\nu}{18}\lambda_1 k^3+\frac{36151-38343\nu}{432}k^2+\frac{263473-234337\nu}{20736}k$$

$$-\frac{53989-78189\nu}{27648}\Big)\beta^4\ln\beta+\Big(\frac{259}{9}\lambda_1 k^4+\frac{41011-40315\nu}{432}k^3+\frac{120725-129269\nu}{3456}k^2$$

$$-\frac{640589-654893\nu}{55296}k-\frac{436891-760123\nu}{663552}\Big)\beta^4-\frac{64}{9}\lambda_1 k(4k+1)\beta^2\ln^3\beta$$

$$+\frac{2}{9}\lambda_1 k(768k^2+604k+103)\beta^2\ln^2\beta-\Big(\frac{448}{3}\lambda_1 k^4+\frac{3290-3192\nu}{9}k^3$$

$$+\frac{13847-14042\nu}{108}k^2+\frac{2765-4593\nu}{432}k-\frac{5405-3131\nu}{6912}\Big)\beta^2\ln\beta$$

$$+\Big(\frac{128}{9}\lambda_1 k^5+\frac{1120-1104\nu}{9}k^4+\frac{4321-4071\nu}{27}k^3-\frac{5539-4147\nu}{648}k^2$$

$$-\frac{18797-16511\nu}{1296}k-\frac{3193-1669\nu}{5184}\Big)\beta^2+\Big(24\lambda_1 k^4+\frac{58-76\nu}{3}k^3+\frac{23-32\nu}{3}k^2$$

$$+\frac{49-103\nu}{96}k-\frac{353-191\nu}{3456}\Big)\ln\beta-16\lambda_1 k^5-\frac{398}{9}\lambda_1 k^4-\frac{281-326\nu}{9}k^3$$

$$-\frac{1529-1889\nu}{144}k^2-\frac{4495-6925\nu}{5184}k+\frac{1627-817\nu}{20736}\Big\},$$

$$n_{-2,1}=-\frac{\beta^5 p_1}{8\lambda_1}\Big\{\frac{1+8\nu}{288}\beta^{10}+\frac{1}{8}\lambda_1\beta^8\ln^2\beta+\Big(\frac{1}{2}\lambda_1 k-\frac{85-31\nu}{288}\Big)\beta^8\ln\beta+\Big(\frac{61-25\nu}{144}k$$

$$-\frac{17-15\nu}{128}\Big)\beta^8-\frac{7}{12}\lambda_1\beta^6\ln^3\beta-\lambda_1\Big(k-\frac{95}{36}\Big)\beta^6\ln^2\beta+\Big(4\lambda_1 k^2-\frac{217-163\nu}{72}k$$

$$-\frac{155-161\nu}{48}\Big)\beta^6\ln\beta+\Big(\frac{76-67\nu}{18}k^2+\frac{649-703\nu}{288}k+\frac{2854-3259\nu}{2592}\Big)\beta^6$$

$$-3\lambda_1 k\beta^4\ln^3\beta-\lambda_1\Big(6k^2-\frac{46}{3}k+\frac{2}{3}\Big)\beta^4\ln^2\beta+\Big(8\lambda_1 k^3-\frac{40}{3}\lambda_1 k^2-\frac{1637-1583\nu}{72}k$$

$$+\frac{385-439\nu}{288}\Big)\beta^4\ln\beta+\Big(\frac{55}{3}\lambda_1 k^3+\frac{223}{18}\lambda_1 k^2-\frac{1163-1649\nu}{864}k-\frac{149-203\nu}{216}\Big)\beta^4$$

$$-3\lambda_1 k(16k^2-1)\beta^2\ln\beta+\Big[32\lambda_1 k^4+(37-35\nu)k^3-\frac{164-155\nu}{6}k^2-\frac{3041-2807\nu}{288}k$$

$$-\frac{91-55\nu}{288}\Big]\beta^2-8\lambda_1 k^4-\frac{58-76\nu}{9}k^3-\frac{23-32\nu}{9}k^2-\frac{49-103\nu}{288}k+\frac{353-191\nu}{10368}\Big\},$$

$$n_{-2,2}=\frac{\beta^6}{4}\Big[\frac{1}{4}\beta^4-\frac{3}{2}\beta^2\ln\beta+\Big(3k+\frac{21}{8}\Big)\beta^2-12k\ln\beta+8k^2+21k\Big],$$

$$n_{-2,3}=\frac{\beta^6}{3}\left(\frac{1}{4}\beta^2+2k\right),$$

$$n_{-2,4}=0,$$

$$\begin{aligned}n_{0,0}=\frac{\beta^3 p_1}{16\lambda_1}\Big\{&\frac{37+125\nu}{1728}\beta^{12}\ln\beta-\frac{37+125\nu}{3456}\beta^{12}+\frac{13-31\nu}{48}\beta^{10}\ln^3\beta-\Big(\frac{13-7\nu}{12}k\\
&+\frac{183-201\nu}{288}\Big)\beta^{10}\ln^2\beta+\Big(\frac{269-143\nu}{72}k-\frac{35-71\nu}{576}\Big)\beta^{10}\ln\beta-\Big(\frac{1367-665\nu}{864}k\\
&-\frac{305-62\nu}{1728}\Big)\beta^{10}-\frac{25-79\nu}{24}\beta^8\ln^4\beta+\Big(\frac{23-17\nu}{4}k+\frac{2051-3293\nu}{288}\Big)\beta^8\ln^3\beta\\
&-\Big(\frac{35-29\nu}{3}k^2+\frac{1735-1105\nu}{72}k+\frac{6451-7477\nu}{576}\Big)\beta^8\ln^2\beta+\Big(\frac{1405-1171\nu}{36}k^2\\
&+\frac{1789-1357\nu}{72}k+\frac{113497-140551\nu}{20736}\Big)\beta^8\ln\beta-\Big(\frac{403-373\nu}{24}k^2+\frac{406-469\nu}{144}k\\
&+\frac{73105-78775\nu}{41472}\Big)\beta^8-\frac{5}{6}\lambda_1\beta^6\ln^5\beta-\lambda_1\Big(4k-\frac{223}{108}\Big)\beta^6\ln^4\beta+\Big(\frac{110}{3}\lambda_1k^2+\frac{1109-515\nu}{36}k\\
&+\frac{11+25\nu}{9}\Big)\beta^6\ln^3\beta-\Big(\frac{136}{3}\lambda_1k^3+\frac{3367-3169\nu}{18}k^2+\frac{23693-12083\nu}{432}k+\frac{2615+8023\nu}{1728}\Big)\\
&\beta^6\ln^2\beta+\Big(\frac{4114-4006\nu}{27}k^3+\frac{7837-7621\nu}{36}k^2-\frac{9641-16283\nu}{1296}k-\frac{77779-124921\nu}{15552}\Big)\\
&\beta^6\ln\beta-\Big(\frac{3733-3571\nu}{54}k^3+\frac{380-425\nu}{12}k^2-\frac{190597-194971\nu}{10368}k-\frac{397807-519793\nu}{124416}\Big)\beta^6\\
&-\lambda_1\Big(24k^2+\frac{5}{3}k-\frac{13}{12}\Big)\beta^4\ln^4\beta+\Big(96\lambda_1k^3+\frac{445-463\nu}{3}k^2+\frac{2297-2135\nu}{108}k\\
&-\frac{3173-3659\nu}{864}\Big)\beta^4\ln^3\beta-\Big(64\lambda_1k^4+\frac{1684-1600\nu}{3}k^3+\frac{2977-3193\nu}{9}k^2\\
&+\frac{1541-3431\nu}{216}k-\frac{6025-5431\nu}{864}\Big)\beta^4\ln^2\beta+\Big(\frac{752}{3}\lambda_1k^4+\frac{1939-1897\nu}{3}k^3\\
&+\frac{17807-22397\nu}{216}\Big)k^2-\frac{62341-53755\nu}{648}k-\frac{98303-59585\nu}{20736}\Big)\beta^4\ln\beta\\
&-\Big(\frac{3520}{27}\lambda_1k^4+\frac{5063-4847\nu}{54}k^3-\frac{89131-83461\nu}{1296}k^2-\frac{42377-9815\nu}{7776}k\\
&-\frac{20255+141583\nu}{124416}\Big)\beta^4-6\lambda_1k(4k+1)\beta^2\ln^4\beta+6\lambda_1k(32k^2+28k+5)\beta^2\ln^3\beta\\
&-\Big[448\lambda_1k^4+\frac{2242-2134\nu}{3}k^3+(278-268\nu)k^2+\frac{3655-3673\nu}{144}k\\
&-\frac{565-187\nu}{1728}\Big]\beta^2\ln^2\beta+\Big[128\lambda_1k^5+(568-552\nu)k^4+\frac{2675-2297\nu}{9}k^3-\frac{451-601\nu}{12}k^2\\
&-\frac{5503-5809\nu}{288}k+\frac{8741-9227\nu}{10368}\Big]\beta^2\ln\beta-\Big[96\lambda_1k^5+(58-46\nu)k^4\\
&-\frac{4337-4355\nu}{36}k^3+\frac{561-403\nu}{16}k^2+\frac{10463-8801\nu}{384}k+\frac{81289-55207\nu}{41472}\Big]\beta^2\end{aligned}$$

$$+\left(32\lambda_1 k^4+\frac{232-304\nu}{9}k^3+\frac{92-128\nu}{9}k^2+\frac{49-103\nu}{72}k-\frac{353-191\nu}{2592}\right)\ln^2\beta$$
$$-\left(128\lambda_1 k^5+\frac{1888}{9}\lambda_1 k^4+\frac{1204-1240\nu}{9}k^3+\frac{701-737\nu}{18}k^2+\frac{5021-5507\nu}{1296}k\right.$$
$$\left.-\frac{77+85\nu}{5184}\right)\ln\beta+64\lambda_1 k^5+\frac{512}{9}\lambda_1 k^4+\frac{254-164\nu}{9}k^3+\frac{149+31\nu}{36}k^2$$
$$\left.+\frac{2375+55\nu}{2592}k+\frac{2041-1231\nu}{10368}\right\},$$

$$n_{0,1}=-\frac{\beta^3 p_1}{8\lambda_1}\left\{\frac{1+8\nu}{144}\beta^{12}+\frac{1}{4}\lambda_1\beta^{10}\ln^2\beta-\left(\frac{13-7\nu}{24}k+\frac{83+7\nu}{288}\right)\beta^{10}\ln\beta+\left(\frac{79-37\nu}{48}k\right.\right.$$
$$\left.-\frac{59}{144}\lambda_1\right)\beta^{10}-\frac{7}{6}\lambda_1\beta^8\ln^3\beta+\left(\frac{77-59\nu}{12}k+\frac{469-523\nu}{144}\right)\beta^8\ln^2\beta-\left(\frac{35-29\nu}{6}k^2\right.$$
$$\left.+\frac{601-391\nu}{48}k+\frac{1589-1571\nu}{576}\right)\beta^8\ln\beta+\left(\frac{53-44\nu}{3}k^2+\frac{1393-1303\nu}{288}k\right.$$
$$\left.+\frac{5291-8045\nu}{10368}\right)\beta^8-\frac{\lambda_1}{6}(32k-1)\beta^6\ln^3\beta+\left(\frac{128}{3}\lambda_1 k^2+\frac{119-101\nu}{6}k-\frac{61-115\nu}{48}\right)$$
$$\beta^6\ln^2\beta-\left(\frac{68}{3}\lambda_1 k^3+\frac{1711-1657\nu}{18}k^2+\frac{2759-1553\nu}{216}k-\frac{5045-7367\nu}{1728}\right)\beta^6\ln\beta$$
$$+\left(\frac{631-613\nu}{9}k^3+\frac{459-447\nu}{8}k^2-\frac{13055-12569\nu}{1296}k-\frac{1985-3929\nu}{5184}\right)\beta^6$$
$$-6\lambda_1 k(4k+1)\beta^4\ln^3\beta+\lambda_1\left(96k^3+\frac{292}{3}k^2+\frac{38}{3}k-\frac{2}{3}\right)\beta^4\ln^2\beta$$
$$-\left(32\lambda_1 k^4+\frac{718-706\nu}{3}k^3+\frac{943-979\nu}{9}k^2-\frac{344+203\nu}{144}k\right.$$
$$\left.-\frac{1513-1063\nu}{576}\right)\beta^4\ln\beta+\left(\frac{352}{3}\lambda_1 k^4+\frac{1403-1349\nu}{9}k^3-\frac{587-425\nu}{54}k^2\right.$$
$$\left.-\frac{3109-2758\nu}{216}k-\frac{296-53\nu}{864}\right)\beta^4-\left(96\lambda_1 k^4+\frac{248-224\nu}{3}k^3\right.$$
$$\left.+\frac{95-80\nu}{3}k^2+\frac{181-145\nu}{36}k+\frac{445-283\nu}{1728}\right)\beta^2\ln\beta$$
$$+\left[64\lambda_1 k^5+(84-76\nu)k^4+\frac{223-169\nu}{9}k^3+\frac{169-91\nu}{24}k^2\right.$$
$$\left.+\frac{143-107\nu}{36}k+\frac{1759-1435\nu}{2592}\right]\beta^2-64\lambda_1 k^5-\frac{800}{9}\lambda_1 k^4-2(27-26\nu)k^3$$
$$\left.-\frac{517-481\nu}{36}k^2-\frac{4139-3653\nu}{2592}k-\frac{629-467\nu}{10368}\right\},$$

$$n_{0,2}=\frac{\beta^3 p_1}{16\lambda_1}\left\{\frac{11+7\nu}{48}\beta^{10}\ln\beta-\frac{29-11\nu}{288}\beta^{10}-\frac{43+11\nu}{24}\beta^8\ln^2\beta+\left(\frac{85-67\nu}{12}k\right.\right.$$
$$\left.+\frac{67+203\nu}{288}\right)\beta^8\ln\beta-\left(\frac{7-25\nu}{18}k+\frac{317-227\nu}{576}\right)\beta^8+\frac{1}{3}\lambda_1\beta^6\ln^3\beta+\lambda_1\left(\frac{10}{3}k-1\right)\beta^6\ln^2\beta$$
$$+\left(\frac{146}{3}\lambda_1 k^2-\frac{139+239\nu}{36}k-\frac{164+25\nu}{108}\right)\beta^6\ln\beta+\left(\frac{193-103\nu}{18}k^2+\frac{557-179\nu}{144}k\right.$$

$$+\frac{4259-3125\nu}{5184}\Big)\beta^6-12\lambda_1 k\beta^4\ln^3\beta+4\lambda_1\Big(6k^2+\frac{23}{3}k-\frac{1}{3}\Big)\beta^4\ln^2\beta$$

$$+\Big(96\lambda_1 k^3+\frac{19-\nu}{3}k^2-\frac{287-233\nu}{36}k+\frac{311-473\nu}{288}\Big)\beta^4\ln\beta$$

$$+\Big(\frac{320}{3}\lambda_1 k^3+\frac{1157-1319\nu}{18}k^2-\frac{5005-4519\nu}{216}k+\frac{925-331\nu}{1728}\Big)\beta^4$$

$$-6\lambda_1 k(16k^2+8k+1)\beta^2\ln\beta+\Big[256\lambda_1 k^4+(318-298\nu)k^3+\frac{266}{3}\lambda_1 k^2$$

$$+\frac{47-353\nu}{144}k-\frac{485-251\nu}{576}\Big]\beta^2-32\lambda_1 k^4-\frac{232-304\nu}{9}k^3-\frac{92-128\nu}{9}k^2$$

$$-\frac{49-103\nu}{72}k+\frac{353-191\nu}{2592}\Big\},$$

$$n_{0,3}=\beta^4\Big[\frac{1}{8}\beta^4+\frac{1}{4}\beta^2\ln\beta-\Big(\frac{5}{2}k+\frac{1}{48}\Big)\beta^2+2k\ln\beta-12k^2-\frac{13}{2}k\Big],$$

$$n_{0,4}=-\beta^4\Big(\frac{1}{8}\beta^2-k\Big),$$

$$n_{2,0}=-\frac{\beta p_1}{32\lambda_1}\Big\{\frac{1-19\nu}{48}\beta^{12}\ln^2\beta-\frac{17-143\nu}{576}\beta^{12}\ln\beta-\frac{17+253\nu}{1152}\beta^{12}-\frac{1-7\nu}{2}\beta^{10}\ln^3\beta$$

$$-\Big(\frac{11+19\nu}{12}k-\frac{59-93\nu}{16}\Big)\beta^{10}\ln^2\beta-\Big(\frac{1-43\nu}{48}k+\frac{135-345\nu}{128}\Big)\beta^{10}\ln\beta$$

$$-\Big(\frac{73-33\nu}{16}k-\frac{3263-3029\nu}{2304}\Big)\beta^{10}+\frac{19-73\nu}{12}\beta^8\ln^4\beta-\Big(\frac{41-95\nu}{6}k$$

$$+\frac{641-767\nu}{48}\Big)\beta^8\ln^3\beta-\Big(\frac{52-40\nu}{3}k^2-\frac{607-544\nu}{18}k-\frac{6895-4465\nu}{864}\Big)\beta^8\ln^2\beta$$

$$+\Big[(2-6\nu)k^2+\frac{425-119\nu}{36}k+\frac{29585-54047\nu}{10368}\Big]\beta^8\ln\beta$$

$$-\Big(\frac{1165-1063\nu}{24}k^2+\frac{1333-1543\nu}{192}k+\frac{51283-72181\nu}{20736}\Big)\beta^8$$

$$+\lambda_1\Big(\frac{8}{3}k+\frac{49}{9}\Big)\beta^6\ln^4\beta-\Big(\frac{40}{3}\lambda_1 k^2+\frac{499-283\nu}{9}k+\frac{1306-631\nu}{108}\Big)\beta^6\ln^3\beta$$

$$-\Big(\frac{256}{3}\lambda_1 k^3-\frac{791-449\nu}{9}k^2+\frac{1319-2075\nu}{108}k+\frac{19007-21599\nu}{2592}\Big)\beta^6\ln^2\beta$$

$$-\Big(\frac{194-50\nu}{9}k^3-\frac{5849-5759\nu}{36}k^2-\frac{88891-74797\nu}{1296}k+\frac{10511-46151\nu}{10368}\Big)\beta^6\ln\beta$$

$$-\Big(\frac{1603}{9}\lambda_1 k^3+\frac{4520-4799\nu}{36}k^2-\frac{33083-34541\nu}{1296}k-\frac{25697-45218\nu}{10368}\Big)\beta^6$$

$$+4\lambda_1 k\beta^4\ln^5\beta-\lambda_1\Big(8k^2-6k+\frac{3}{2}\Big)\beta^4\ln^4\beta-\Big[(90-102\nu)k^2+\frac{25}{2}\lambda_1 k$$

$$-\frac{57-63\nu}{16}\Big]\beta^4\ln^3\beta-\Big[128\lambda_1 k^4-(212-228\nu)k^3+(104-74\nu)k^2+\frac{3307-2809\nu}{24}k$$

$$+\frac{249-143\nu}{32}\Big]\beta^4\ln^2\beta-\Big[240\lambda_1 k^4-(378-346\nu)k^3-\frac{4751-4685\nu}{12}k^2$$

$$
\begin{aligned}
&+\frac{365-301\nu}{16}k+\frac{2227-2137\nu}{384}\Big]\beta^4\ln\beta-\Big[224\lambda_1k^4+(375-350\nu)k^3-\frac{541-475\nu}{12}k^2\\
&-\frac{1438-811\nu}{48}k-\frac{1133+694\nu}{1152}\Big]\beta^4+8\lambda_1k(4k+1)\beta^2\ln^4\beta-8\lambda_1k(32k^2+12k+1)\beta^2\ln^3\beta\\
&+\Big[640\lambda_1k^4+\frac{856-712\nu}{3}k^3-(70-86\nu)k^2-\frac{505-523\nu}{18}k+\frac{41-95\nu}{432}\Big]\beta^2\ln^2\beta\\
&-\Big[512\lambda_1k^5+(160-96\nu)k^4-\frac{3196-3160\nu}{9}k^3-\frac{79-130\nu}{6}k^2\\
&+\frac{2707-2275\nu}{72}k+\frac{32147-21941\nu}{10368}\Big]\beta^2\ln\beta-\Big(208\lambda_1k^4+\frac{22+14\nu}{9}k^3\\
&+\frac{163-115\nu}{4}k^2+\frac{203-165\nu}{8}k+\frac{13169-10217\nu}{5184}\Big)\beta^2\\
&-\Big(32\lambda_1k^4+\frac{232-304\nu}{9}k^3+\frac{91-128\nu}{9}k^2+\frac{49-103\nu}{72}k-\frac{353-191\nu}{2592}\Big)\ln^2\beta\\
&+\Big(128\lambda_1k^5+\frac{1024}{9}\lambda_1k^4+\frac{508-328\nu}{9}k^3+\frac{149+31\nu}{18}k^2+\frac{2375+55\nu}{1296}k\\
&+\frac{2041-1231\nu}{5184}\Big)\ln\beta+160\lambda_1k^5+\frac{1820}{9}\lambda_1k^4+\frac{1070-980\nu}{9}k^3\\
&+\frac{2125-1765\nu}{72}k^2+\frac{9245-6815\nu}{2592}k+\frac{2455-1645\nu}{10368}\Big\},
\end{aligned}
$$

$$
\begin{aligned}
n_{2,1}=&\frac{\beta p_1}{32\lambda_1}\Big\{\frac{1-19\nu}{48}\beta^{12}\ln\beta-\frac{7-25\nu}{288}\beta^{12}+\frac{7+47\nu}{24}\beta^{10}\ln^2\beta-\Big(\frac{11+19\nu}{12}k-\frac{815-1265\nu}{288}\Big)\beta^{10}\ln\beta\\
&-\Big(\frac{81-45\nu}{36}k+\frac{289-451\nu}{576}\Big)\beta^{10}-\frac{1}{3}\lambda_1\beta^8\ln^3\beta-\Big[(17-23\nu)k+\frac{21-51\nu}{8}\Big]\beta^8\ln^2\beta\\
&-\Big(\frac{52-40\nu}{3}k^2-\frac{1333-1063\nu}{36}k-\frac{8809-8431\nu}{864}\Big)\beta^8\ln\beta-\Big(\frac{149-95\nu}{6}k^2+\frac{1567-973\nu}{72}k\\
&+\frac{10253-9443\nu}{5184}\Big)\beta^8-\frac{9}{2}\lambda_1\beta^6\ln^3\beta-\Big(\frac{416}{3}\lambda_1k^2+\frac{508-454\nu}{9}k-\frac{26-53\nu}{12}\Big)\beta^6\ln^2\beta\\
&-\Big(\frac{256}{3}\lambda_1k^3-\frac{409-319\nu}{3}k^2-\frac{2803+275\nu}{108}k+\frac{1243-2755\nu}{288}\Big)\beta^6\ln\beta\\
&-\Big[\frac{1196-1052\nu}{9}k^3+(87-82\nu)k^2-\frac{25223-26681\nu}{648}k-\frac{2927-3710\nu}{864}\Big]\beta^6\\
&+6\lambda_1k(8k+3)\beta^4\ln^3\beta-12\lambda_1k(24k^2+25k+6)\beta^4\ln^2\beta\\
&-\Big[128\lambda_1k^4-(168-184\nu)k^3-(61-83\nu)k^2+\frac{535-427\nu}{12}k\\
&+\frac{2093-1679\nu}{288}\Big]\beta^4\ln\beta-\Big[400\lambda_1k^4+(424-436\nu)k^3-\frac{32+58\nu}{3}k^2\\
&-\frac{443-366\nu}{8}k-\frac{101+142\nu}{288}\Big]\beta^4+\Big(\frac{64-16\nu}{3}k^3+\frac{46-16\nu}{3}k^2\\
&+\frac{73-37\nu}{18}k+\frac{445-283\nu}{864}\Big)\beta^2\ln\beta-\Big[512\lambda_1k^5+(800-736\nu)k^4+\frac{5036-4532\nu}{9}k^3\\
&+\frac{614-566\nu}{3}k^2+\frac{2197+2035\nu}{72}k+\frac{1669-1507\nu}{129}6\Big]\beta^2
\end{aligned}
$$

$$+128\lambda_1 k^5+\frac{1744}{9}\lambda_1 k^4+\frac{1088}{9}\lambda_1 k^3+\frac{203}{6}\lambda_1 k^2+\frac{1145}{324}\lambda_1 k+\frac{23}{432}\lambda_1\Big\},$$

$$n_{2,2}=-\frac{\beta p_1}{32\lambda_1}\Big\{\frac{19-37\nu}{24}\beta^{10}\ln\beta-\frac{59-95\nu}{144}\beta^{10}-\frac{5-59\nu}{12}\beta^8\ln^2\beta$$
$$-\Big(\frac{61-43\nu}{6}k-\frac{401-527\nu}{48}\Big)\beta^8\ln\beta+\Big(\frac{143-179\nu}{36}k-\frac{193-211\nu}{72}\Big)\beta^8-3\lambda_1\beta^6\ln^3\beta$$
$$-\Big(\frac{86}{3}\lambda_1 k+\frac{7}{36}\lambda_1\Big)\beta^6\ln^2\beta-\Big(\frac{376}{3}\lambda_1 k^2-\frac{8+46\nu}{3}k-\frac{2725-997\nu}{432}\Big)\beta^6\ln\beta$$
$$+\Big(\frac{193-265\nu}{9}k^2+\frac{257-365\nu}{18}k+\frac{784-2161\nu}{1296}\Big)\beta^6+12\lambda_1 k\beta^4\ln^3\beta-24\lambda_1 k(3k+2)\beta^4\ln^2\beta$$
$$-\Big[288\lambda_1 k^3+(198-186\nu)k^2+\frac{123}{2}\lambda_1 k+\frac{51-57\nu}{16}\Big]\beta^4\ln\beta$$
$$-\Big[164\lambda_1 k^3+(39-57\nu)k^2-\frac{679-598\nu}{12}k-\frac{317-272\nu}{144}\Big]\beta^4$$
$$-\Big[640\lambda_1 k^4+(744-712\nu)k^3+\frac{860-842\nu}{3}k^2+\frac{278-287\nu}{9}k-\frac{121-31\nu}{288}\Big]\beta^2$$
$$+32\lambda_1 k^4+\frac{232-304\nu}{9}k^3+\frac{92-128\nu}{9}k^2+\frac{49-103\nu}{72}k-\frac{353-191\nu}{2592}\Big\},$$

$$n_{2,3}=\beta^2\Big[\frac{3}{16}\beta^4-\frac{3}{8}\beta^2\ln\beta-\Big(\frac{13}{4}k-\frac{17}{32}\Big)\beta^2-16k^2-7k\Big],$$

$$n_{2,4}=2k\beta^2,$$

$$n_{4,0}=-\frac{1}{24\lambda_1}\Big\{\frac{1-3\nu}{4}\beta^8\ln\beta-\frac{7+55\nu}{96}\beta^8-\frac{3-17\nu}{4}\beta^6\ln^2\beta-\Big(4\nu k-\frac{61-3\nu}{16}\Big)\beta^6\ln\beta$$
$$-\Big(\frac{57+5\nu}{12}k-\frac{660-823\nu}{144}\Big)\beta^6+\frac{5-11\nu}{2}\beta^4\ln^3\beta-\Big[(1-15\nu)k$$
$$+\frac{33-15\nu}{8}\Big]\beta^4\ln^2\beta-\Big[(28-20\nu)k^2-\frac{133-59\nu}{4}k+\frac{277-587\nu}{96}\Big]\beta^4\ln\beta$$
$$-\Big(\frac{331-269\nu}{6}k^2-\frac{2209-2163\nu}{72}k-\frac{8063-6113\nu}{1152}\Big)\beta^4$$
$$-2\lambda_1\beta^2\ln^4\beta+2\lambda_1\beta^2\ln^3\beta+\Big[56\lambda_1 k^2-(34-30\nu)k+\frac{15-21\nu}{4}\Big]\beta^2\ln^2\beta$$
$$-\Big[160\lambda_1 k^3-(108-100\nu)k^2+30\lambda_1 k-\frac{1-15\nu}{8}\Big]\beta^2\ln\beta$$
$$-\Big(232\lambda_1 k^3-\frac{101-8\nu}{9}k^2-\frac{583-117\nu}{72}k-\frac{1487-1415\nu}{432}\Big)\beta^2$$
$$+8\lambda_1 k\ln^4\beta-16\lambda_1 k(4k-1)\ln^3\beta+\Big[192\lambda_1 k^3-(84-92\nu)k^2+(17-15\nu)k$$
$$+\frac{9-7\nu}{8}\Big]\ln^2\beta-\Big[256\lambda_1 k^4-(144-176\nu)k^3+56\lambda_1 k^2-\frac{9-7\nu}{2}k-\frac{9-7\nu}{8}\Big]\ln\beta$$
$$-\frac{992}{3}\lambda_1 k^4-\frac{458-334\nu}{3}k^3-\frac{2983-2517\nu}{36}k^2-\frac{1453-1243\nu}{108}k-\frac{423-329\nu}{1152}\Big\},$$

$$n_{4,1}=\frac{1}{24\lambda_1}\left\{\frac{1-3\nu}{4}\beta^8+\frac{3}{2}\lambda_2\beta^6\ln\beta-(4\nu k-\frac{255-329\nu}{48})\beta^6+\lambda_2\left(6k\right.\right.$$

$$\left.+\frac{19}{8}\right)\beta^4\ln\beta-\left[(28-20\nu)k^2-\frac{295-321\nu}{12}k-\frac{541-531\nu}{32}\right]\beta^4$$

$$-\left[160\lambda_1k^3+\frac{224-248\nu}{3}k^2-\frac{163-101\nu}{6}k-\frac{2069-1763\nu}{144}\right]\beta^2$$

$$\left.-256\lambda_1k^4-32(11-10\nu)k^3-\frac{533-471\nu}{3}k^2-\frac{1259-1109\nu}{36}k-\frac{171-133\nu}{96}\right\},$$

$$n_{4,2}=-\frac{1}{24\lambda_1}\left\{\frac{9-11\nu}{4}\beta^6+\frac{3}{2}\lambda_2\beta^4\ln\beta+\left[(7-9\nu)k+\frac{107-113\nu}{8}\right]\beta^4-\left[56\lambda_1k^2-(15-11\nu)k\right.\right.$$

$$\left.\left.-\frac{35-32\nu}{2}\right]\beta^2-192\lambda_1k^3-4(41-39\nu)k^2-\frac{119-113\nu}{3}k-\frac{9-7\nu}{8}\right\},$$

$$n_{4,3}=\frac{1}{6}\left(\beta^4+\frac{55}{24}\beta^2-16k^2-\frac{19}{3}k\right),$$

$$n_{4,4}=-\frac{1}{3}\left(\frac{1}{4}\beta^2-k\right),$$

$$n_{6,0}=-\frac{1}{48}\left[\frac{1}{6}\beta^6\ln\beta+\frac{71}{288}\beta^6-\frac{1}{2}\beta^4\ln^2\beta+\left(2k-\frac{1}{2}\right)\beta^4\ln\beta+\left(\frac{71}{24}k+\frac{49}{128}\right)\beta^4\right.$$

$$+\frac{2}{3}\beta^2\ln^3\beta-(4k-1)\beta^2\ln^2\beta+\left(8k^2-4k+\frac{1}{2}\right)\beta^2\ln\beta+\left(\frac{71}{6}k^2+\frac{49}{16}k+\frac{5243}{6912}\right)\beta^2$$

$$-\frac{1}{3}\ln^4\beta+\left(\frac{8}{3}k-\frac{2}{3}\right)\ln^3\beta-\left(8k^2-4k+\frac{1}{2}\right)\ln^2\beta+\left(\frac{32}{3}k^3-8k^2+2k-\frac{1}{6}\right)\ln\beta$$

$$\left.+\frac{142}{9}k^3+\frac{49}{8}k^2+\frac{5243}{1728}k+\frac{23921}{82944}\right],$$

$$n_{6,1}=\frac{1}{48}\left[\frac{1}{6}\beta^6+(2k+\frac{47}{48})\beta^4+\left(8k^2+\frac{47}{6}k+\frac{65}{32}\right)\beta^2+\frac{32}{3}k^3+\frac{47}{3}k^2+\frac{65}{8}k+\frac{4667}{3456}\right],$$

$$n_{6,2}=-\frac{1}{48}\left[\frac{1}{2}\beta^4+\left(4k+\frac{47}{24}\right)\beta^2+8k^2+\frac{47}{6}k+\frac{65}{32}\right],$$

$$n_{6,3}=\frac{1}{144}\left(2\beta^2+8k+\frac{47}{12}\right),$$

$$n_{6,4}=-\frac{1}{144}. \tag{3.19}$$

4. Major Results

From the solution in the previous section, we obtain the relation between the load and the center deflection as

$$P=p_1W_0+p_3W_0^3, \tag{4.1}$$

and the formula for the radial stress

$$S_r=S_2W_0^2+S_4W_0^4. \tag{4.2}$$

Substituting (4.2) into (2.13), we obtain the formula for the tangential stress

$$S_\theta=S_2'W_0^2+S_4'W_0^4, \tag{4.3}$$

where

$$S_2' = -\frac{\beta^2 p_1^2}{32}\Big\{3\alpha^2 y^2 - \frac{1}{\lambda_1}\Big[\frac{1}{4}\lambda_1\beta^6 + \frac{1-7\nu}{2}\beta^4\ln\beta + \Big(\lambda_2 k - \frac{13-19\nu}{8}\Big)\beta^4 - 2\lambda_1\beta^2\ln^2\beta$$

$$+4\lambda_1(2k+1)\beta^2\ln\beta - \Big(2(3-\nu)k - \frac{1-7\nu}{4}\Big)\beta^2 + 4(3-\nu)k^2 + (5-3\nu)k$$

$$+\frac{9-7\nu}{8}\Big] + H(y-\beta)\Big[3y^2\ln^2 y - \Big(3\beta^2 + 12k + \frac{5}{2}\Big)y^2\ln y$$

$$+\Big(3\beta^2\ln\beta + \frac{11}{4}\beta^2 - 3\ln^2\beta + 3(4k-1)\ln\beta + 11k + \frac{3}{8}\Big)y^2 + 2\beta^2\ln^2 y$$

$$-2\beta^2(\beta^2 + 4k - 1)\ln y + 2\beta^4\ln\beta - 3\beta^4 - 2\beta^2\ln^2\beta + 4(2k+1)\beta^2\ln\beta$$

$$-(12k+1)\beta^2 + \beta^4 y^{-2}\ln y + \beta^4\Big(\frac{1}{4}\beta^2 - \frac{3}{2}\ln\beta + k + \frac{5}{8}\Big)y^{-2}\Big]\Big\}, \tag{4.4a}$$

$$S_4' = \frac{\beta^4 p_1^4}{512}\Big\{\sum_{i=0}^{3} m_{2i}(2i+1)y^{2i} - H(y-\beta)\sum_{i=-2}^{3}\sum_{j=0}^{4} n_{2i,j}[(2i+1)\ln^j y + j\ln^{j-1} y]y^{2i}\Big\}, \tag{4.4b}$$

For a particular position at the center of the plate, substituting $y=0$ in expressions (4.2) and (4.3), we obtain the stresses at the center of the plate

$$\begin{aligned} S_r(0) &= S_\theta(0) \\ &= S_2(0)W_0^2 + S_4(0)W_0^4, \end{aligned} \tag{4.5}$$

where

$$S_2(0) = \frac{\beta^2 p_1^2}{32\lambda_1}\Big\{\frac{1}{4}\lambda_2\beta^6 + \frac{1-7\nu}{2}\beta^4\ln\beta + \Big(\lambda_2 k - \frac{13-19\nu}{8}\Big)\beta^4$$

$$-2\lambda_1\beta^2\ln^2\beta + 4\lambda_1(2k+1)\beta^2\ln\beta - \Big[2(3-\nu)k - \frac{1-7\nu}{4}\Big]\beta^2$$

$$+4(3-\nu)k^2 + (5-3\nu)k + \frac{9-7\nu}{8}\Big\}, \tag{4.6a}$$

$$S_4(0) = \frac{\beta^4 p_1^4}{512} m_0. \tag{4.6b}$$

For a particular position at the edge of the plate, substituting $y=1$ in expressions (4.2) and (4.3), we obtain the stresses at the edge of the plate

$$\begin{aligned} S_r(1) &= S_2(1)W_0^2 + S_4(1)W_0^4, \\ S_\theta(1) &= \nu S_r(1), \end{aligned} \tag{4.7}$$

where

$$S_2(1) = \frac{\beta^2 p_1^2}{32\lambda_1}\Big[\frac{1}{2}\beta^6 - 3\beta^4\ln\beta + \Big(2k + \frac{3}{4}\Big)\beta^4 - \Big(4k + \frac{3}{2}\Big)\beta^2 + 8k^2 + 2k + \frac{1}{4}\Big],$$

$$S_4(1) = \frac{\beta^4 p_1^4}{512}\Big(\sum_{i=0}^{3} m_{2i} - \sum_{i=-2}^{3} n_{2i,0}\Big). \tag{4.8}$$

Numerical values of the coefficients p_1, p_3, $S_2(0)$, $S_4(0)$, $S_2(1)$, and $S_4(1)$ have been computed for various dimensionless characteristic parameters k and dimensionless radius β by assuming Poisson's ratio $\nu=0.3$. The calculations have been completed by PRIME computer and are given in Tables 1-6.

Table 1　Numerical values of p_1. ν=0.3

k \ β	0.05	0.20	0.35	0.50	0.65	0.80
0	162.84	48.121	36.849	39.660	57.660	134.45
0.01	156.47	45.911	34.617	36.083	48.553	87.431
0.02	150.59	43.896	32.639	33.097	41.936	64.777
0.03	145.12	42.050	30.876	30.568	36.906	51.447
0.04	140.04	40.353	29.293	28.398	32.953	42.666
0.05	135.30	38.788	27.865	26.515	29.765	36.446
0.06	130.88	37.340	26.569	24.867	27.140	31.809
0.07	126.73	35.996	25.389	23.411	24.940	28.219
0.08	122.84	34.745	24.309	22.117	23.070	25.356
0.09	119.18	33.578	23.317	20.958	21.461	23.0222
0.10	115.73	32.487	22.403	19.915	20.062	21.080

Table 2　Numerical values of p_3. ν=0.3

k \ β	0.05	0.20	0.35	0.50	0.65	0.80
0	13.617	4.5554	3.8972	5.7758	47.776	4111.5
0.01	15.096	4.9248	4.0735	5.3604	26.778	743.25
0.02	16.416	5.2543	4.2349	5.1222	17.287	229.64
0.03	17.611	5.5528	4.3848	4.9890	12.487	95.721
0.04	18.705	5.8266	4.5254	4.9198	9.8449	48.792
0.05	19.716	6.0803	4.6582	4.8908	8.2918	28.913
0.06	20.657	6.3174	4.7843	4.8872	7.3292	19.290
0.07	21.541	6.5405	4.9042	4.9000	6.7061	14.146
0.08	22.375	6.7516	5.0187	4.9233	6.2880	11.174
0.09	23.167	6.9522	5.1282	4.9533	5.9988	9.3454
0.10	23.922	7.1436	5.2329	4.9872	5.7934	8.1610

Table 3　Numerical values of $S_2(0)$. ν=0.3

k \ β	0.05	0.20	0.35	0.50	0.65	0.80
0	2.3957	2.1757	2.0441	1.9708	1.9338	1.9208
0.01	2.3254	2.1173	1.9905	1.9173	1.8782	1.8709
0.02	2.2642	2.0679	1.9480	1.8803	1.8507	1.8671
0.03	2.2106	2.0260	1.9140	1.8543	1.8370	1.8709
0.04	2.1636	1.9903	1.8868	1.8358	1.8307	1.8760
0.05	2.1222	1.9597	1.8648	1.8228	1.8284	1.8808
0.06	2.0856	1.9335	1.8469	1.8136	1.8284	1.8849
0.07	2.0532	1.9108	1.8325	1.8073	1.8297	1.8886
0.08	2.0245	1.8913	1.8207	1.8030	1.8318	1.8917
0.09	1.9988	1.8744	1.8111	1.8002	1.8343	1.8944
0.10	1.9760	1.8597	1.8034	1.7986	1.8370	1.8968

Table 4　Numerical values of $S_4(0)$. ν=0.3

k \ β	0.05	0.20	0.35	0.50	0.65	0.80
0	0.010777	−0.0053845	0.0036663	1.0433	162.01	86327
0.01	−0.0025789	−0.013954	−0.030390	−0.10583	42.329	10552
0.02	−0.013588	−0.020560	−0.048084	−0.54886	7.9979	2477.8
0.03	−0.022645	−0.025619	−0.056489	−0.68391	−2.1399	823.80
0.04	−0.030075	−0.029448	−0.059398	−0.68571	−4.7358	340.03
0.05	−0.036142	−0.032285	−0.059090	−0.63313	−4.9110	162.65
0.06	−0.041060	−0.034320	−0.056928	−0.56209	−4.3606	86.577
0.07	−0.045009	−0.035694	−0.053718	−0.48881	−3.6630	49.972
0.08	−0.048133	−0.036527	−0.049956	−0.42023	−3.0095	30.733
0.09	−0.050559	−0.036908	−0.045922	−0.35882	−2.4522	19.890
0.10	−0.052386	−0.036914	−0.041807	−0.30506	−1.9944	13.423

Table 5　Numerical values of $S_2(1)$. ν=0.3

k \ β	0.05	0.20	0.35	0.50	0.65	0.80
0	0.72901	0.82269	0.93330	1.0472	1.1617	1.2761
0.01	0.72965	0.82127	0.92981	1.0419	1.1561	1.2804
0.02	0.73219	0.82243	0.93033	1.0435	1.1633	1.3013
0.03	0.73620	0.82555	0.93365	1.0491	1.1751	1.3194
0.04	0.74137	0.83014	0.93893	1.0571	1.1882	1.3335
0.05	0.74743	0.83586	0.94555	1.0664	1.2012	1.3447
0.06	0.75419	0.84240	0.95309	1.0763	1.2136	1.3536
0.07	0.76147	0.84956	0.96123	1.0864	1.2251	1.3609
0.08	0.76914	0.85715	0.96973	1.0965	1.2357	1.3669
0.09	0.77709	0.86505	0.97843	1.1065	1.2454	1.3719
0.10	0.78525	0.87314	0.98720	1.1161	1.2543	1.3762

Table 6　Numerical values of $S_4(1)$. ν=0.3

k \ β	0.05	0.20	0.35	0.50	0.65	0.80
0	−0.0045683	0.00089932	0.019810	1.4897	246.67	132708
0.01	−0.024509	0.0019225	−0.0010191	0.41535	94.585	20043
0.02	−0.00017936	0.0035071	−0.010424	−0.054294	40.872	5508.7
0.03	0.0022370	0.0055180	−0.013229	−0.24659	19.248	2068.8
0.04	0.0047843	0.0078532	−0.012211	−0.30950	9.6591	940.78
0.05	0.0074469	0.010440	−0.0089951	−0.31241	5.0818	486.76
0.06	0.010212	0.013220	−0.0045519	−0.28864	2.7690	276.37
0.07	0.013063	0.016153	0.00054737	−0.25446	1.5481	168.29
0.08	0.015991	0.019202	0.0059514	−0.21774	0.88165	108.24
0.09	0.018980	0.022344	0.011466	−0.18211	0.50906	72.737
0.10	0.022024	0.025555	0.016958	−0.14915	0.29779	50.670

Knowing the coefficients, we can easily calculate the deflection and the stresses of the circular sandwich plate from formulae (4.1), (4.5), and (4.7). The results of these calculations are represented graphically in Figs. 2-4.

Fig. 2 indicates the relation between the dimensionless load P and the dimensionless center deflection W_0 for several values of k and β. It is obvious that the curves rise monotonically. When β is smaller, for the same value of P, the center deflection induced in a plate with large β is large. But when β is larger, the result turns out to be contrary to the above conclusions.

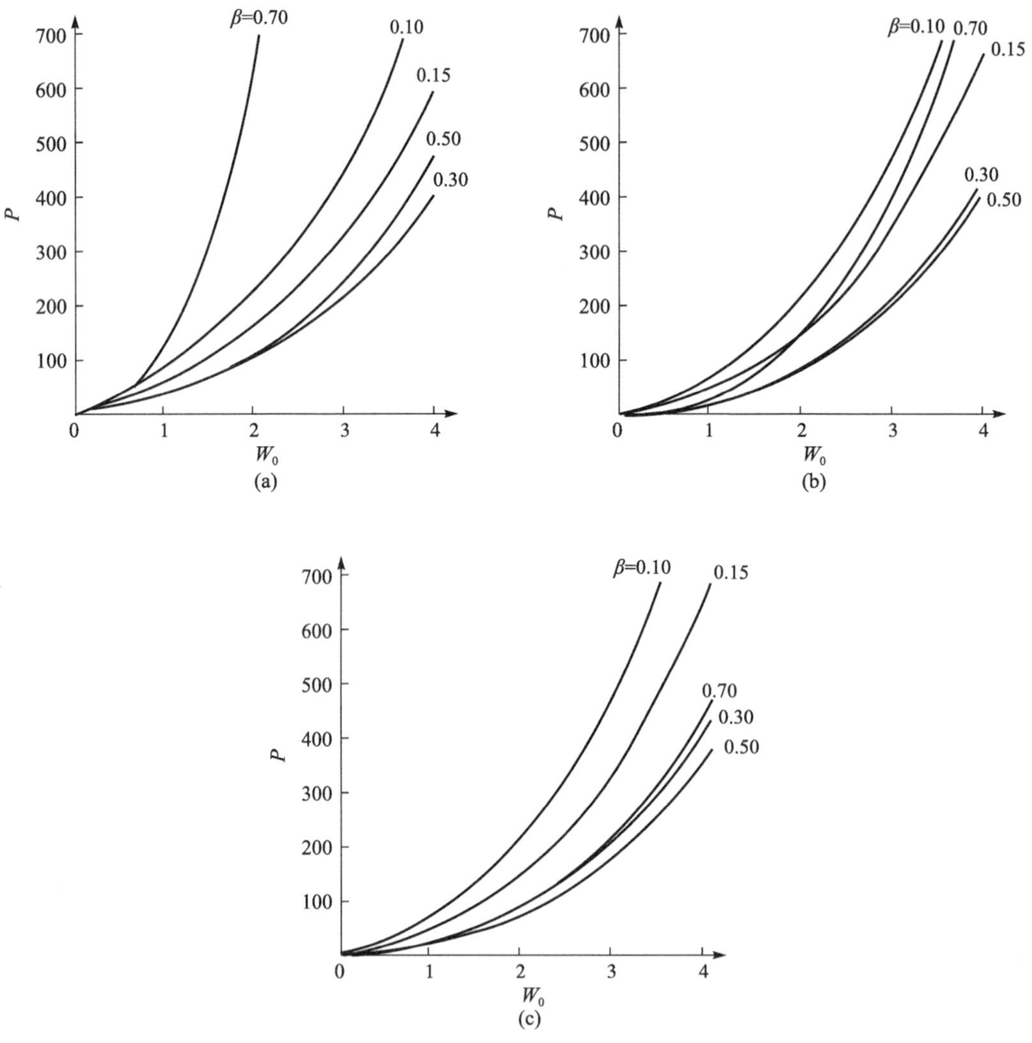

Fig. 2 Variation of the load with center deflection (ν=0.3). (a) k=0.01. (b) k=0.05. (c) k=0.10

In Fig. 3, curves are given for the dimensionless center radial stress $S_r(0)$ and the dimensionless center tangential stress $S_\theta(0)$. It can be seen from the figures that, for small k, the stresses are positive. However, for larger k, when the center deflection is larger, the stresses induced in a plate with large β are negative.

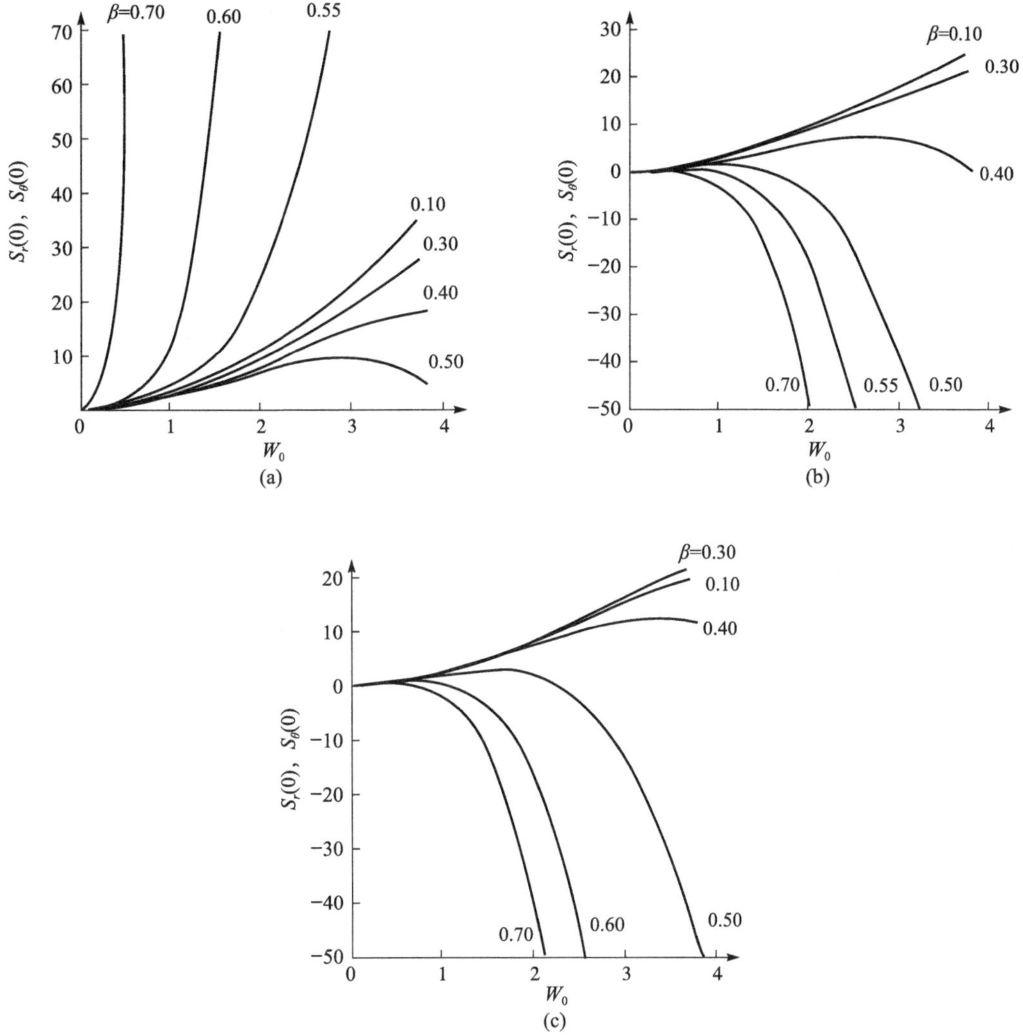

Fig. 3　The radial stress and the tangential stress at the center of the circular sandwich plate (ν=0.3). (a) k=0.01. (b) k=0.05. (c) k=0.10

Fig. 4 shows the curves for the dimensionless edge radial stress $S_r(1)$. From the figures, we see that, for small k the curves rise monotonically, and the stresses are positive. However for large k, when the center deflection is larger, the stresses in a plate with middle β are negative.

In the limiting case where β is infinitely small, from the above solution, we can obtain the solution for a concentrated load acting at the center of the circular sandwich plate with a rigidly clamped edge. But the solution does not hold near the point of application of a concentrated load, since the deflection and the stresses approach infinity *as r* approaches zero. The accurate theory in [2] must be considered if we want the deflection and the stresses near the point of application of the load.

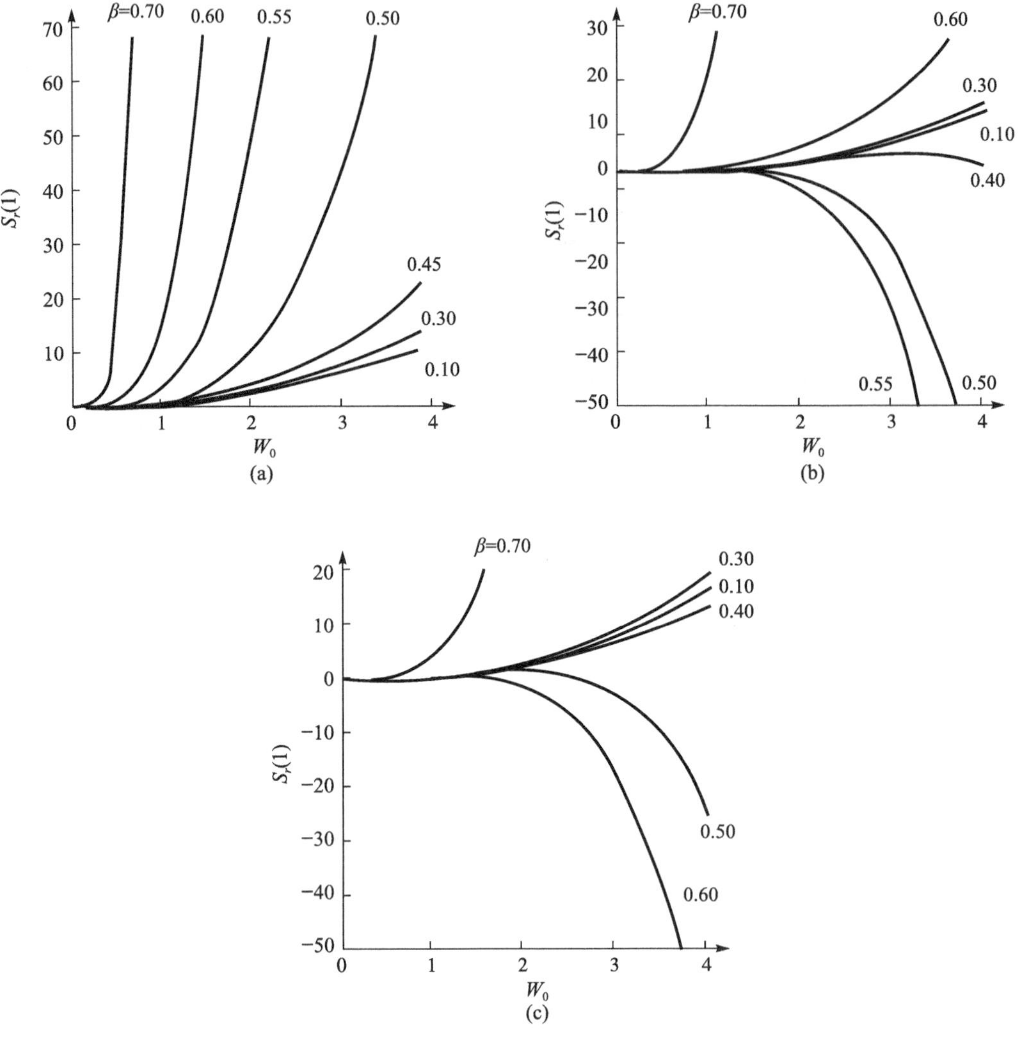

Fig. 4 The radial stress at the edge of the circular sandwich plate (ν=0.3) (a) k=0.01. (b) k=0.05. (c) k=0.10

Acknowledgements

The author would like to thank Mr. Hu Si-yi for his help in the numerical calculations.

References

[1] Liu Renhuai, Nonlinear axisymmetrical bending of circular sandwich plates under the action of uniform edge moment (in Chinese), *Journal of the China University of Science and Technology* 10, 2 (1980), 56.

[2] Liu Renhuai, Nonlinear bending of circular sandwich plates, *Applied Mathematics and Mechan-*

ics 2，2（1981），189.

[3] Liu Renhuai and Shi Yun-fang，Exact solution for circular sandwich plate with large deflection，*Applied Mathematics and Mechanics* 3，1（1982），11.

[4] Chien Wei-zang，Large deflection of a circular clamped plate under uniform pressure，*Chinese Journal of physics*7，2（1947），102.

夹层圆板大挠度问题的进一步研究①

符　号

a　夹层圆板的半径

h　夹心厚度

t　表层厚度

h_0　厚度 $h_0 = h + t$

r　径向坐标

E_f　表板的弹性模量

ν_f　表板的泊松比

G^c　夹心的剪切模量

C　夹心的抗弯刚度 $C = \dfrac{G^c h_0^2}{h}$

D　夹层圆板的抗弯刚度 $D = \dfrac{E_f t h_0^2}{2(1-\nu_f^2)}$

D_f　表板的抗弯刚度 $D_f = \dfrac{E_f t^3}{12(1-\nu_f^2)}$

k　剪切参数 $k = \dfrac{D}{G^c h_0 a^2}$

q　横向均布载荷

w　夹层板的 z 方向位移

σ_r^0　表板的 r 方向应力

σ_θ^0　表板的 θ 方向应力

一、引言

自 20 世纪 40 年代以来，已发表了一些关于夹层矩形板的论文，其中包括 Reissner[1]、Hoff[2]、Libore 和 Batdorf[3]提出的著名理论以及 Kan 和 Huang[4]、Bhimaraddi 和 Chandrashekhara[5]等所获得的解。同时，夹层圆板的研究也引起一些人的兴趣[6-9]，但大多数均局限于线性分析，仅有很少几篇论文涉及夹层圆板的大挠度问题。在 1980 年和 1981 年，刘人怀[10,11]首次建立了在均布载荷和边缘力矩作用下夹层圆板非线性弯曲的基本方程，并用叶开沅和刘人怀等[12-14]1965 年提出的修正迭代法进行了求解。接着，刘人怀和施云方[15]应用幂级数方法得到了承受均布载荷的夹层圆板的精确解。刘人怀[16]还用摄动法讨论了更困难的承受同心圆载荷的夹层圆板。然而，上述工作均将

① 本文原载《应用数学和力学》，1989，10 (12)：1041-1047. 作者：刘人怀，朱高秋.

夹层板的表板视为薄膜，即忽略了表板的抗弯刚度。对于不同类型的大挠度的夹层板而言，这一近似的适用性便成为一个疑问。为此，刘人怀[11]推导了计及表板抗弯刚度的夹层圆板非线性弯曲的一般方程。由于数学上的复杂性，尚未得到这些方程的解。本文试图寻求计及表板抗弯刚度的夹层板的解。为此，我们提出了修正幂级数方法，得到了在均布载荷作用下，具有滑动固定边界条件的夹层圆板的解，并与刘人怀和施云方[15]的结果进行了比较。本文结果可作为工程应用中一个精确的解。

二、修正幂级数方法和数值结果

考虑一承受均布载荷的等厚夹层圆板。在计及表板抗弯刚度的情况下，刘人怀[11]建立了此板的非线性弯曲方程。为了将这些方程无量纲化，我们引入：

$$P=\frac{\sqrt{2(1-\nu_f^2)}a^4}{2h_0D}q, W=\sqrt{2(1-\nu_f^2)}\frac{w}{h_0}, S_r=\frac{2ta^2}{D}\sigma_r^0,$$

$$S_\theta=\frac{2ta^2}{D}\sigma_\theta^0, \rho=\frac{r}{a}, \phi=\frac{\mathrm{d}W}{\mathrm{d}\rho}, \lambda=\frac{D}{Ca^2},$$

$$\varepsilon=\frac{2D_f}{Ca^2}, L=\frac{\mathrm{d}}{\mathrm{d}\rho}\frac{1}{\rho}\frac{\mathrm{d}}{\mathrm{d}\rho}\rho \tag{2.1}$$

于是，基本方程化为[11]

$$\varepsilon L^2\phi-\left(1+\frac{\varepsilon}{\lambda}\right)L\phi-\lambda L(S_r\phi)+S_r\phi+P\rho=0,$$

$$L(\rho S_r)+\frac{1}{\rho}\phi^2=0 \tag{2.2}$$

而滑动固定边界条件成为

当 $\rho=1$ 时，

$$W=0, \varepsilon L\phi-(\lambda S_r+1)\phi-\lambda P=0, S_r=0, \phi=0;$$

当 $\rho=0$ 时，

$$\varepsilon L\phi-(\lambda S_r+1)\phi=0, S_r<\infty \tag{2.3}$$

这里，S_θ 由下式决定：

$$S_\theta=\frac{\mathrm{d}}{\mathrm{d}\rho}(\rho S_r) \tag{2.4}$$

我们注意到

$$\frac{\varepsilon}{\lambda}=\frac{2D_f}{D}=\frac{1}{3}\left(\frac{t}{h_0}\right)^2 \tag{2.5}$$

对于工程中常用的夹层板，我们知道

$$\lambda<1, t/h_0\ll 1$$

所以

$$\varepsilon\ll 1$$

显然，因为最高阶导数前有一个小参数，故这是一个边界层型问题。众所周知，因为这些方程阶数高且为非线性，故用奇异摄动理论[17]将是十分困难的。此外，若对这类方程采用传统的幂级数方法[15]，则因基本方程最高阶导数系数为一小参数而无法求解。

为此，我们提出一种修正幂级数方法。假设：

$$\phi = \sum_{i=0}^{\infty} b_{2i+1}\rho^{2i+1}, S_r = \sum_{i=0}^{\infty} a_{2i}\rho^{2i} \tag{2.6}$$

于是，由式（2.1）和（2.4），可得下列表达式

$$W = \sum_{i=0}^{\infty} \frac{b_{2i+2}}{2(i+1)}(\rho^{2i+2}-1), S_\theta = \sum_{i=0}^{\infty}(2i+1)a_{2i}\rho^{2i} \tag{2.7}$$

显然，式（2.6）和（2.7）已满足以下边界条件

当 $\rho=1$ 时，

$$\varepsilon L\phi-(\lambda S_r+1)\phi=0，S_r<\infty$$

当 $\rho=1$ 时，

$$W=0$$

将式（2.6）代入基本方程（2.2），在展开并归并 ρ 的同次幂系数后，可得

$$192\varepsilon b_5-8\left(1+\frac{\varepsilon}{\lambda}\right)b_3-8\lambda\alpha_3+\alpha_1+P=0,$$

$$16\varepsilon i^2(i^2-1)b_{2i+1}-4\left(1+\frac{\varepsilon}{\lambda}\right)i(i-1)b_{2i-1}-4\lambda i(i-1)\alpha_{2i-1}$$
$$+\alpha_{2i-3}=0，i=3,4,5,\cdots \tag{2.8}$$

再将式（2.6）代入边界条件（2.3），又得

$$\sum_{i=0}^{\infty}4\varepsilon i(i+1)b_{2i+1}-\sum_{i=0}^{\infty}b_{2i+1}-\lambda P=0,$$
$$\sum_{i=0}^{\infty}a_{2i}=0,$$
$$\sum_{i=0}^{\infty}b_{2i+1}=0 \tag{2.9a-c}$$

其中

$$\alpha_{2i+1}=\sum_{k=0}^{i}a_{2(i-k)}b_{2k+1}, \beta_{2i+1}=\sum_{k=0}^{i}b_{2(i-k)+1}b_{2k+1},$$
$$a_{2i}=-\frac{\beta_{2i-1}}{4i(i+1)}，i=1,2,3,\cdots \tag{2.10}$$

然而，代表夹层板表板的边界条件，即

$$当\ \rho=1\ 时, \phi=0$$

的方程（2.9c）在以前的近似理论[15]中并未出现过，故可以予以放弃。但剩余的方程和边界条件仍比文献［15］的理论更为精确。

这样一来，按照方程（2.8）、（2.9）和（2.10），并吸收奇异摄动法[17]的思想，我们提出如下的修正迭代公式

$$a_0^{(n+1)}=-\sum_{i=1}^{\infty}a_{2i}^{(n)},$$

$$b_1^{(n+1)}=\sum_{i=1}^{\infty}[4\varepsilon i(i+1)-1]b_{2i+1}^{(n)}-\lambda P,$$

$$b_3^{(n+1)} = \frac{1}{8\left(1+\frac{\varepsilon}{\lambda}+a_0\lambda\right)}\{192\varepsilon b_5^{(n)}+\lambda[b^{(n+1)}]^3+a_0^{(n+1)}b_1^{(n+1)}+P\},$$

$$\alpha_{2i+1}^{(n+1)} = \sum_{k=0}^{i} a_{2(i-k)}^{(n+1)} b_{2k+1}^{(n+1)},$$

$$\beta_{2i+1}^{(n+1)} = \sum_{k=0}^{i} b_{2(i-k)}^{(n+1)} b_{2k+1}^{(n+1)},$$

$$a_{2i}^{(n+1)} = -\frac{\beta_{2i-1}^{(n+1)}}{4i(i+1)},$$

$$b_{2i-1}^{(n+1)} = \frac{1}{4i(i-1)\left(1+\frac{\varepsilon}{\lambda}+a_0\lambda\right)}\left[16\varepsilon i^2(i^2-1)b_{2i+1}^{(n)} + \alpha_{2i-3}^{n+1} - 4\lambda i(i-1)\sum_{k=0}^{i-2} a_{2(i-k-1)}^{(n+1)} b_{2k+1}^{(n+1)}\right],$$

$$i = 3,4,5,\cdots; n = 0,1,2,\cdots \tag{2.11}$$

由式（2.6）和（2.7），我们得到夹层圆板的中心挠度、中心应力和边缘应力的计算公式

$$W(0) = -\sum_{i=0}^{\infty}\frac{b_{2i+1}}{2(i+1)}, S_r(0) = S_r|_{\rho=0} = a_0,$$

$$S_\theta(0) = S_\theta|_{\rho=0} = a_0, S_\theta(1) = S_\theta|_{\rho=0} = \sum_{i=0}^{\infty}(2i+1)a_{2i} \tag{2.12}$$

然后，我们以刘人怀等[15]的结果作为迭代初值，按上面的迭代公式进行求解，其结果给在表 1～表 8 中。同时，还与文献［15］的结果进行了比较，其中 $k=D/(G^c h_0 a^2)$ 是文献［15］中的剪切参数。

表 1　$k=0.01$，$t/h=0.05$

P		5.0	10.0	15.0	20.0	25.0	30.0	35.0
$W(0)$	本文	0.17972	0.35824	0.53443	0.70735	0.87621	1.0404	1.1997
	文献[15]	0.18104	0.36085	0.53828	0.71235	0.88228	1.0475	1.2077
	相对误差	0.74%	0.73%	0.72%	0.71%	0.69%	0.68%	0.66%
$S_r(0)$	本文	0.029531	0.11716	0.26011	0.45416	0.69405	0.97405	1.2883
	文献[15]	0.029850	0.11841	0.26282	0.45873	0.70077	0.98306	1.2997
	相对误差	1.08%	1.07%	1.04%	1.01%	0.97%	0.93%	0.88%
$S_\theta(0)$	本文	−0.021727	−0.086401	−0.19256	−0.33795	−0.51977	−0.73846	−0.97999
	文献[15]	−0.022065	−0.087736	−0.19551	−0.34306	−0.52750	−0.74561	−0.99410
	相对误差	1.55%	1.54%	1.53%	1.51%	1.49%	1.46%	1.44%

表 2　$k=0.01$，$t/h=0.10$

P		5.0	10.0	15.0	20.0	25.0	30.0	35.0
$W(0)$	本文	0.17829	0.35540	0.53026	0.70192	0.86963	1.0328	1.1911
	文献[15]	0.18104	0.36085	0.53828	0.71235	0.88228	1.0475	1.2077
	相对误差	1.54%	1.53%	1.51%	1.49%	1.46%	1.42%	1.39%

续表

P		5.0	10.0	15.0	20.0	25.0	30.0	35.0
$S_r(0)$	本文	0.029167	0.11573	0.25702	0.44892	0.68634	0.96367	1.2752
	文献[15]	0.029850	0.11841	0.26282	0.45873	0.70077	0.98306	1.2997
	相对误差	2.34%	2.31%	2.26%	2.19%	2.10%	2.01%	1.92%
$S_\theta(0)$	本文	−0.021368	−0.084980	−0.18943	−0.33253	−0.51155	−0.72344	−0.96505
	文献[15]	−0.022065	−0.087736	−0.19551	−0.34306	−0.52750	−0.74561	−0.99410
	相对误差	3.26%	3.24%	3.21%	3.17%	3.12%	3.06%	3.01%

表 3　$k=0.01$，$t/h=0.15$

P		5.0	10.0	15.0	20.0	25.0	30.0	35.0
$W(0)$	本文	0.17679	0.35245	0.52592	0.69627	0.86277	1.0249	1.1822
	文献[15]	0.18104	0.36085	0.53828	0.71235	0.88228	1.0475	1.2077
	相对误差	2.40%	2.38%	2.35%	2.31%	2.26%	2.21%	2.15%
$S_r(0)$	本文	0.028775	0.11420	0.25369	0.44326	0.67800	0.95244	1.2610
	文献[15]	0.029850	0.11841	0.26282	0.45873	0.70077	0.98306	1.2997
	相对误差	3.74%	3.68%	3.60%	3.49%	3.36%	3.22%	3.07%
$S_\theta(0)$	本文	−0.020998	−0.083521	−0.18621	−0.32695	−0.50311	−0.71173	−0.94973
	文献[15]	−0.02265	−0.087736	−0.19551	−0.34306	−0.52750	−0.74561	−0.99410
	相对误差	5.08%	5.05%	4.99%	4.93%	4.85%	4.76%	4.67%

表 4　$k=0.01$，$t/h=0.20$

P		5.0	10.0	15.0	20.0	25.0	30.0	35.0
$W(0)$	本文	0.17527	0.34944	0.52149	0.69051	0.85578	1.0168	1.1731
	文献[15]	0.18104	0.36085	0.53828	0.71235	0.88228	1.0475	1.2077
	相对误差	3.29%	3.26%	3.22%	3.16%	3.10%	3.02%	2.95%
$S_r(0)$	本文	0.028369	0.11261	0.25023	0.43739	0.66932	0.94072	1.2462
	文献[15]	0.029850	0.11841	0.26282	0.45873	0.70077	0.98306	1.2997
	相对误差	5.22%	5.15%	5.03%	4.88%	4.70%	4.50%	4.29%
$S_\theta(0)$	本文	−0.020627	−0.082055	−0.18297	−0.32135	−0.49464	−0.69997	−0.93435
	文献[15]	−0.022065	−0.087736	−0.19551	−0.34306	−0.52750	−0.74561	−0.99410
	相对误差	6.97%	6.92%	6.85%	6.75%	6.64%	6.52%	6.39%

表 5　$k=0.05$，$t/h=0.05$

P		5.0	10.0	15.0	20.0	25.0	30.0	35.0
$W(0)$	本文	0.27422	0.54346	0.80368	1.0523	1.2881	1.5111	1.7217
	文献[15]	0.28031	0.55524	0.82049	1.0734	1.3128	1.5387	1.7517
	相对误差	2.22%	2.17%	2.09%	2.00%	1.91%	1.83%	1.74%
$S_r(0)$	本文	0.055751	0.21770	0.47182	0.79975	1.1830	1.6053	2.0542
	文献[15]	0.057756	0.22521	0.48711	0.82373	1.2155	1.6457	2.1013
	相对误差	3.60%	3.45%	3.24%	3.00%	2.75%	2.51%	2.29%

续表

P		5.0	10.0	15.0	20.0	25.0	30.0	35.0
$S_\theta(0)$	本文	−0.054894	−0.21645	−0.47625	−0.82288	−1.2442	−1.7289	−2.2673
	文献[15]	−0.057641	−0.22711	−0.49920	−0.86148	−1.3010	−1.8057	−2.3654
	相对误差	5.00%	4.93%	4.82%	4.69%	4.56%	4.44%	4.33%

表 6　k=0.05，t/h=0.10

P		5.0	10.0	15.0	20.0	25.0	30.0	35.0
$W(0)$	本文	0.26834	0.53207	0.78739	1.0318	1.2642	1.4843	1.6924
	文献[15]	0.28031	0.55524	0.82049	1.0734	1.3128	1.5387	1.7517
	相对误差	4.46%	4.36%	4.20%	4.03%	3.84%	3.67%	3.50%
$S_r(0)$	本文	0.053822	0.21045	0.45700	0.77639	1.1511	1.5657	2.0077
	文献[15]	0.057756	0.22521	0.48711	0.82373	1.2155	1.6457	2.1013
	相对误差	7.31%	7.02%	6.59%	6.10%	5.60%	5.11%	4.66%
$S_\theta(0)$	本文	−0.052327	−0.20647	−0.45476	−0.78668	−1.1909	−1.6568	−2.1750
	文献[15]	−0.57641	−0.22711	−0.49920	−0.86148	−1.3010	−1.8057	−2.3654
	相对误差	10.16%	10.00%	9.77%	9.51%	9.24%	8.99%	8.75%

表 7　k=0.05，t/h=0.15

P		5.0	10.0	15.0	20.0	25.0	30.0	35.0
$W(0)$	本文	0.26271	0.52114	0.77174	1.0121	1.2411	1.4583	1.6641
	文献[15]	0.28031	0.55524	0.82049	1.0734	1.3128	1.5387	1.7517
	相对误差	6.70%	6.54%	6.31%	6.05%	5.78%	5.51%	5.27%
$S_r(0)$	本文	0.051985	0.20352	0.44278	0.75387	1.1203	1.5271	1.9623
	文献[15]	0.57756	0.22521	0.48711	0.82373	1.2155	1.6457	2.1013
	相对误差	11.10%	10.66%	10.01%	9.27%	8.50%	7.76%	7.08%
$S_\theta(0)$	本文	−0.049942	−0.19719	−0.43475	−0.75292	−1.1412	−1.5894	−2.0887
	文献[15]	−0.057641	−0.22711	−0.49920	−0.86148	−1.3010	−1.8057	−2.3654
	相对误差	15.42%	15.18%	14.82%	14.42%	14.01%	13.61%	13.25%

表 8　k=0.05，t/h=0.20

P		5.0	10.0	15.0	20.0	25.0	30.0	35.0
$W(0)$	本文	0.25734	0.51070	0.75678	0.99326	1.2189	1.4334	1.6368
	文献[15]	0.28031	0.55524	0.82049	1.0734	1.3128	1.5387	1.7517
	相对误差	8.93%	8.72%	8.42%	8.07%	7.70%	7.35%	7.02%
$S_r(0)$	本文	0.050247	0.19695	0.42923	0.73230	1.0906	1.4898	1.9182
	文献[15]	0.057756	0.22521	0.48711	0.82373	1.2155	1.6457	2.1013
	相对误差	14.94%	14.35%	13.49%	12.49%	11.45%	10.46%	9.55%
$S_\theta(0)$	本文	−0.047733	−0.18859	−0.41617	−0.72153	−1.0948	−1.5265	−2.0080
	文献[15]	−0.057641	−0.22711	−0.49920	−0.86148	−1.3010	−1.8057	−2.3654
	相对误差	20.76%	20.42%	19.95%	19.40%	18.83%	18.29%	17.80%

三、讨论和结论

在上一节，应用我们的修正幂级数方法，首次解决了计及表板抗弯刚度的夹层圆板的非线性弯曲问题。由前述分析和结果，我们可知：

（1）在本文的分析中，计及了表板的抗弯刚度。我们所作的唯一近似是忽略了表板的一个边界条件，而在前述理论［15］中它也是未考虑的。故其他方程和边界条件以及所得到的结果仍然比文献［15］精确。我们的工作可用于各种工程问题。

（2）在综合了传统的幂级数方法和奇异摄动法的优点后，我们提出了修正幂级数方法。对于求解一类最高阶导数项含小参数的问题，此法具有一定的有效性。

（3）当 t/h 较小时，文献［15］的理论结果与本文解接近。随着 t/h 的增加，由忽略表板抗弯刚度所带来的相对误差逐渐增大。当 $t/h<0.20$ 和 $t/h<0.10$ 时，夹层板的挠度和应力的误差分别不超过 10%。因此，在 t/h 较小的情况下，对于一般的工程问题而言，可应用文献［15］的近似理论。

（4）边缘应力的相对误差大于中心的应力和挠度的相对误差。这是因为在我们的理论中计及了表板的抗弯刚度，出现了所谓的应力边界层的结果。

（5）当 k 较小时，文献［15］的近似理论解与我们的结果接近。这表明表板抗弯刚度的影响，不仅随 t/h 变化，而且也与夹层板的剪切参数有关。

致谢：感谢薛大为教授和王秀喜副教授的宝贵意见。

参考文献

［1］ Reissner E. Finite deflections of sandwich plates. J. Aero. Sci.，1948，15（7）：435.

［2］ Hoff N J. Bending and buckling of sandwich plates. NACA TN，1950：2225.

［3］ Libove C，Batdorf S B. A general small-deflection theory for flat sandwich plates. NACA TN，1948：1526.

［4］ Kan H P，Huang J C. Large deflection of rectangular sandwich plates. AIAA J.，1967，5（9）：1706.

［5］ Bhimaraddi A，Chandrashekhara K. Comparison of elasticity and sandwich theories for a rectangular sandwich plates. Aero. J.，1984，88（876）：229.

［6］ Zaid M. Symmetrical bending of circular sandwich plates. Proceedings of the Second United States National Congress of Applied. Mechanics，ASME，New York，1955：413.

［7］ Bruun E R. Thermal deflection of a circular sandwich plate. AIAA J.，1963，1（5）：1213.

［8］ Huang J C，Ebcioglu I K. Circular sandwich plate under radial compression and thermal gradient. AIAA J.，1965，3（6）：1146.

［9］ Amato A J，Ebcioglu I K. Axisymmetric buckling of annular sandwich panels. AIAA J.，1972，10（10）：1351.

［10］ 刘人怀. 在边缘力矩作用下夹层圆板的非线性轴对称弯曲问题. 中国科学技术大学学报，1980，10（2）：56.

［11］ 刘人怀. 夹层圆板的非线性弯曲. 应用数学和力学，1980，2（2）：173.

［12］ 叶开沅，刘人怀，平庆元，等. 在对称线布载荷作用下的圆底扁薄球壳的非线性稳定问题.

科学通报，1965，(2)：142；兰州大学学报，1965，(2)：10.

[13] 叶开沅，刘人怀，张传智，等. 圆底扁薄球壳在边缘力矩作用下的非线性稳定问题. 科学通报，1965，(2)：145.

[14] 刘人怀. 在内边缘均布力矩作用下中心开孔圆底扁球壳的非线性稳定问题. 科学通报，1965，(3)：253.

[15] 刘人怀，施云方. 夹层圆板大挠度问题的精确解. 应用数学和力学，1982，3 (1)：11.

[16] Liu Ren-huai. Nonlinear bending of circular sandwich plates under the action of axisymmetric uniformly distributed line loads. Progress in Applied Mechanics，(ed. Yeh Kai-yuan)，Martinus Nijhoff Publishers，Dordrecht，1987：293.

[17] Nayfeh A H. Perturbation Methods. New York：John Wiley & Sons，1973.

夹层矩形板的非线性振动①

符　号

x,y,z	直角坐标
t	时间
a,b	夹层矩形板边长的一半
h_1	每块表层的厚度
h_0	上、下表层中面间的距离
E,ν	表层材料的弹性模量和泊松比
G_2	夹心的剪切模量
D	夹层矩形板的抗弯刚度
q	横向载荷
$u_i,v_i,w_i(i=1,2,3)$	上表层、夹心和下表层的点沿 x,y 和 z 方向的位移
u,v,w	夹层矩形板中面上点沿 x,y 和 z 方向位移
ψ_x,ψ_y	夹心中面法线在 xz 和 yz 平面内的转角
$\varepsilon_{xi},\varepsilon_{yi},\varepsilon_{zi},\gamma_{xyi},\gamma_{yzi},\gamma_{zxi}(i=1,2,3)$	上表层、夹心和下表层的应变分量
$\sigma_{xi},\sigma_{yi},\sigma_{zi},\tau_{xyi},\tau_{yzi},\tau_{zxi}(i=1,2,3)$	上表层、夹心和下表层的应力分量
$\sigma_{x0},\sigma_{y0},\tau_{xy0}$	夹层矩形板中面内的应力分量
$U_i(i=1,2,3)$	上表层、夹心和下表层的应变能
V	外力功
U	总势能
$\overline{T}$	动能
ρ	夹层矩形板的面密度
ρ_1,ρ_2	表层和夹心的体积密度
M_x,M_y,M_{xy}	夹层矩形板的弯矩和扭矩
Q_x,Q_y	夹层矩形板的横向力
φ	应力函数
λ	夹层矩形板边长比
ξ,η	无量纲直角坐标
k	夹层矩形板的无量纲特征参数

① 本文原载《中国科学》，A辑，1991，(10)：1075-1086. 作者：刘人怀，吴建成.

τ 无量纲时间
W_m 无量纲最大振幅
ω_L 线性振动的固有频率
T_L, T_{NL} 线性和非线性周期

一、引言

近年来，随着航天、航空与航海等工业部门的飞跃发展，具有刚度高、重量轻等优良特性的夹层板得到了更广泛的应用和更大的关注。但由于结构复杂和数学分析的困难，以往人们大多研究较简单的夹层板的线性力学问题。显然，这种研究状况远不能满足工程实际的需要。早在 40 年前，Reissner[1] 首先建立了具有软夹心和极薄表层的夹层矩形板的大挠度理论。然而由于非线性数学的困难，直到 1967 年，才有 Kan 和 Huang[2] 用摄动法讨论了在均布载荷作用下的矩形板大挠度问题。对于具有软夹心和极薄表板的夹层圆板的大挠度问题，文献［3］～［6］先后用修正迭代法、摄动法和幂级数方法进行了研究，文献［7］也用摄动法讨论过这个问题。对于计及表层抗弯刚度的夹层圆板的更加复杂的问题，文献［3］采用变分原理建立了非线性理论，并用修正幂级数方法[8]进行了具体的求解，获得了满意的结果。但是，就我们所知，对于更加困难的有很大理论和实际意义的夹层矩形板的非线性振动问题，至今尚无人研究过。

本文所考虑的夹层板，是上、下表层很薄且厚度相等、材料各向同性，夹心较厚、较软、材料横观各向同性的薄矩形板。我们先使用 Hamilton 原理导出了这种夹层矩形板的非线性振动问题的运动方程和边界条件，借助于文献［9］的方法，对基本方程和边界条件进行了简化。最后，我们使用 Галёркин 方法求解了简支和铰支边界夹层矩形板的非线性自由振动问题，获得了非线性周期与振幅的关系式。

二、基本方程

考虑如图 1 所示的承受横向载荷 $q(x,y,t)$ 作用的夹层矩形板，并采用下列假设：

(1) 材料服从于胡克定律；
(2) 夹心横向不可压缩；
(3) 夹心沿板面方向不能承受载荷；
(4) 表层处于薄膜应力状态；
(5) 夹心中面法线在变形后仍为直线。

基于上述假设，夹层矩形板中任一点的位移为

上表层

$$u_1 = u + \frac{1}{2}h_0\psi_x, \quad v_1 = v + \frac{1}{2}h_0\psi_y, \quad w_1 = w \tag{1}$$

下表层

$$u_3 = u - \frac{1}{2}h_0\psi_x, \quad v_3 = v - \frac{1}{2}h_0\psi_y, \quad w_3 = w \tag{2}$$

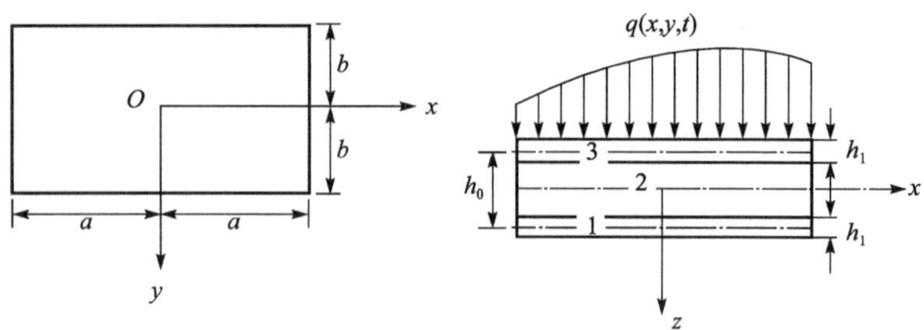

图 1　夹层矩形板的坐标和几何尺寸

夹心

$$u_2 = u + z\psi_x, \quad v_2 = v + z\psi_y, \quad w_2 = w \tag{3}$$

将式 (1)～式 (3) 代入下述几何方程

$$\varepsilon_{xi} = \frac{\partial u_i}{\partial x} + \frac{1}{2}\left(\frac{\partial w_i}{\partial x}\right)^2, \quad \varepsilon_{yi} = \frac{\partial v_i}{\partial y} + \frac{1}{2}\left(\frac{\partial w_i}{\partial y}\right)^2,$$

$$\gamma_{xyi} = \frac{\partial v_i}{\partial x} + \frac{\partial u_i}{\partial y} + \frac{\partial w_i}{\partial x}\frac{\partial w_i}{\partial y}, \quad \varepsilon_{zi} = \gamma_{yzi} = \gamma_{zxi}, \quad i = 1,3 \tag{4}$$

$$\gamma_{zx2} = \frac{\partial u_2}{\partial z} + \frac{\partial w_2}{\partial x}, \quad \gamma_{yz2} = \frac{\partial v_2}{\partial z} + \frac{\partial w_2}{\partial y}, \quad \varepsilon_{x2} = \varepsilon_{y2} = \varepsilon_{z2} = \gamma_{xy2} = 0 \tag{5}$$

便得

上表层

$$\varepsilon_{x1} = \frac{\partial u}{\partial x} + \frac{h_0}{2}\frac{\partial \psi_x}{\partial x} + \frac{1}{2}\left(\frac{\partial w}{\partial x}\right)^2, \quad \varepsilon_{y1} = \frac{\partial v}{\partial y} + \frac{h_0}{2}\frac{\partial \psi_y}{\partial y} + \frac{1}{2}\left(\frac{\partial w}{\partial y}\right)^2,$$

$$\gamma_{xy1} = \frac{\partial v}{\partial x} + \frac{\partial u}{\partial y} + \frac{h_0}{2}\left(\frac{\partial \psi_y}{\partial x} + \frac{\partial \psi_x}{\partial y}\right) + \frac{\partial w}{\partial x}\frac{\partial w}{\partial y} \tag{6}$$

下表层

$$\varepsilon_{x3} = \frac{\partial u}{\partial x} - \frac{h_0}{2}\frac{\partial \psi_x}{\partial x} + \frac{1}{2}\left(\frac{\partial w}{\partial x}\right)^2, \quad \varepsilon_{y3} = \frac{\partial v}{\partial y} - \frac{h_0}{2}\frac{\partial \psi_y}{\partial y} + \frac{1}{2}\left(\frac{\partial w}{\partial y}\right)^2,$$

$$\gamma_{xy3} = \frac{\partial v}{\partial x} + \frac{\partial u}{\partial y} - \frac{h_0}{2}\left(\frac{\partial \psi_y}{\partial x} + \frac{\partial \psi_x}{\partial y}\right) + \frac{\partial w}{\partial x}\frac{\partial w}{\partial y} \tag{7}$$

夹心

$$\gamma_{zx2} = \psi_x + \frac{\partial w}{\partial x}, \quad \gamma_{yz2} = \psi_y + \frac{\partial w}{\partial y} \tag{8}$$

将式 (6)～式 (8) 代入下述胡克定律

$$\sigma_{xi} = \frac{E}{1-\nu^2}(\varepsilon_{xi} + \nu\varepsilon_{yi}), \quad \sigma_{yi} = \frac{E}{1-\nu^2}(\varepsilon_{yi} + \nu\varepsilon_{xi}),$$

$$\tau_{xyi} = \frac{E}{2(1+\nu)}\gamma_{xyi}, \quad \sigma_{zi} = \tau_{yzi} = \tau_{zxi} = 0, \quad i = 1,3 \tag{9}$$

$$\tau_{zx2} = G_2\gamma_{zx2}, \quad \tau_{yz2} = G_2\gamma_{yz2},$$

$$\sigma_{x2} = \sigma_{y2} = \sigma_{z2} = \tau_{xy2} = 0 \tag{10}$$

便得

上表层

$$\sigma_{x1}=\sigma_{x0}+\frac{Eh_0}{2(1-\nu^2)}\left(\frac{\partial\psi_x}{\partial x}+\nu\frac{\partial\psi_y}{\partial y}\right),$$

$$\sigma_{y1}=\sigma_{y0}+\frac{Eh_0}{2(1-\nu^2)}\left(\frac{\partial\psi_y}{\partial y}+\nu\frac{\partial\psi_x}{\partial x}\right),$$

$$\tau_{xy1}=\tau_{xy0}+\frac{Eh_0}{4(1+\nu)}\left(\frac{\partial\psi_y}{\partial x}+\frac{\partial\psi_x}{\partial y}\right) \tag{11}$$

下表层

$$\sigma_{x3}=\sigma_{x0}-\frac{Eh_0}{2(1-\nu^2)}\left(\frac{\partial\psi_x}{\partial x}+\nu\frac{\partial\psi_y}{\partial y}\right),$$

$$\sigma_{y3}=\sigma_{y0}-\frac{Eh_0}{2(1-\nu^2)}\left(\frac{\partial\psi_y}{\partial y}+\nu\frac{\partial\psi_x}{\partial x}\right),$$

$$\tau_{xy3}=\tau_{xy0}-\frac{Eh_0}{4(1+\nu)}\left(\frac{\partial\psi_y}{\partial x}+\frac{\partial\psi_x}{\partial y}\right) \tag{12}$$

夹心

$$\tau_{zx2}=G_2\left(\psi_x+\frac{\partial w}{\partial x}\right),\quad \tau_{yz2}=G_2\left(\psi_y+\frac{\partial w}{\partial y}\right) \tag{13}$$

其中

$$\sigma_{x0}=\frac{E}{1-\nu^2}\left[\frac{\partial u}{\partial x}+\nu\frac{\partial v}{\partial y}+\frac{1}{2}\left(\frac{\partial w}{\partial x}\right)^2+\frac{\nu}{2}\left(\frac{\partial w}{\partial y}\right)^2\right],$$

$$\sigma_{y0}=\frac{E}{1-\nu^2}\left[\frac{\partial v}{\partial y}+\nu\frac{\partial u}{\partial x}+\frac{1}{2}\left(\frac{\partial w}{\partial y}\right)^2+\frac{\nu}{2}\left(\frac{\partial w}{\partial x}\right)^2\right],$$

$$\tau_{xy0}=\frac{E}{2(1+\nu)}\left(\frac{\partial v}{\partial x}+\frac{\partial u}{\partial y}+\frac{\partial w}{\partial x}\frac{\partial w}{\partial y}\right) \tag{14}$$

由弹性应变能公式

$$U_i=\iiint_{V_i}\frac{1}{2E}(\sigma_x\varepsilon_x+\sigma_y\varepsilon_y+\sigma_z\varepsilon_z+\tau_{xy}\gamma_{xy}+\tau_{yz}\gamma_{yz}+\tau_{zx}\gamma_{zx})\mathrm{d}x\mathrm{d}y\mathrm{d}z \tag{15}$$

可得表层和夹心的应变能公式

$$U_i=\iiint_{V_i}\frac{1}{2E}[(\sigma_{xi}+\sigma_{yi})^2+2(1+\nu)(\tau_{xyi}^2-\sigma_{xi}\sigma_{yi})]\mathrm{d}x\mathrm{d}y\mathrm{d}z,\quad i=1,3$$

$$U_2=\iiint_{V_2}\frac{1}{2G_2}(\tau_{zx2}^2+\tau_{yz2}^2)\mathrm{d}x\mathrm{d}y\mathrm{d}z \tag{16}$$

将式 (11)～式 (13) 代入上式，并对 z 进行积分，便得

$$\begin{aligned}U_1=&\frac{h_1}{2E}\iint_{S_1}[(\sigma_{x0}+\sigma_{y0})^2+2(1+\nu)(\tau_{xy0}^2-\sigma_{x0}\sigma_{y0})]\mathrm{d}x\mathrm{d}y\\&+\frac{h_0h_1}{2}\iint_{S_1}\left[\sigma_{x0}\frac{\partial\psi_x}{\partial x}+\sigma_{y0}\frac{\partial\psi_y}{\partial y}+\tau_{xy0}\left(\frac{\partial\psi_y}{\partial x}+\frac{\partial\psi_x}{\partial y}\right)\right]\mathrm{d}x\mathrm{d}y\\&+\frac{D}{4}\iint_{S_1}\left[\left(\frac{\partial\psi_x}{\partial x}\right)^2+\left(\frac{\partial\psi_y}{\partial y}\right)^2+2\nu\frac{\partial\psi_x}{\partial x}\frac{\partial\psi_y}{\partial y}+\frac{1-\nu}{2}\left(\frac{\partial\psi_y}{\partial x}+\frac{\partial\psi_x}{\partial y}\right)^2\right]\mathrm{d}x\mathrm{d}y,\end{aligned}$$

$$U_3 = \frac{h_1}{2E}\iint\limits_{S_1}[(\sigma_{x0}+\sigma_{y0})+2(1+\nu)(\tau_{xy0}^2-\sigma_{x0}\sigma_{y0})]\mathrm{d}x\mathrm{d}y$$
$$-\frac{h_0h_1}{2}\iint\limits_{S_1}\left[\sigma_{x0}\frac{\partial\psi_x}{\partial x}+\sigma_{y0}\frac{\partial\psi_y}{\partial y}+\tau_{xy0}\left(\frac{\partial\psi_y}{\partial x}+\frac{\partial\psi_x}{\partial y}\right)\right]\mathrm{d}x\mathrm{d}y$$
$$+\frac{D}{4}\iint\limits_{S_1}\left[\left(\frac{\partial\psi_x}{\partial x}\right)^2+\left(\frac{\partial\psi_y}{\partial y}\right)^2+2\nu\frac{\partial\psi_x}{\partial x}\frac{\partial\psi_y}{\partial y}+\frac{1-\nu}{2}\left(\frac{\partial\psi_y}{\partial x}+\frac{\partial\psi_x}{\partial y}\right)^2\right]\mathrm{d}x\mathrm{d}y,$$
$$U_2 = \frac{G_2h_0}{2}\iint\limits_{S_2}\left[\left(\psi_x+\frac{\partial w}{\partial x}\right)^2+\left(\psi_y+\frac{\partial w}{\partial y}\right)^2\right]\mathrm{d}x\mathrm{d}y \tag{17}$$

其中

$$D=\frac{Eh_0^2h_1}{2(1-\nu^2)} \tag{18}$$

横向载荷的外力功为

$$V=\iint\limits_S qw\,\mathrm{d}x\mathrm{d}y \tag{19}$$

因此，夹层矩形板的总势能为

$$U = U_1+U_2+U_3-V$$
$$=\frac{h_1}{E}\iint\limits_S[(\sigma_{x0}+\sigma_{y0})^2+2(1+\nu)(\tau_{xy0}^2-\sigma_{x0}\sigma_{y0})]\mathrm{d}x\mathrm{d}y$$
$$+\frac{D}{2}\iint\limits_S\left[\left(\frac{\partial\psi_x}{\partial x}\right)^2+\left(\frac{\partial\psi_y}{\partial y}\right)^2+2\nu\frac{\partial\psi_x}{\partial x}\frac{\partial\psi_y}{\partial y}+\frac{1-\nu}{2}\left(\frac{\partial\psi_y}{\partial x}+\frac{\partial\psi_x}{\partial y}\right)^2\right]\mathrm{d}x\mathrm{d}y$$
$$+\frac{G_2h_0}{2}\iint\limits_S\left[\left(\psi_x+\frac{\partial w}{\partial x}\right)^2+\left(\psi_y+\frac{\partial w}{\partial y}\right)^2\right]\mathrm{d}x\mathrm{d}y-\iint\limits_S qw\,\mathrm{d}x\mathrm{d}y \tag{20}$$

至于夹层板的动能，我们有如下表达式

$$\overline{T}=\iint\limits_S\frac{1}{2}\rho\left(\frac{\partial w}{\partial t}\right)^2\mathrm{d}x\mathrm{d}y$$
$$=\frac{1}{2}\iint\limits_S[\rho_2(h_0-h_1)+2h_1\rho_1]\left(\frac{\partial w}{\partial t}\right)^2\mathrm{d}x\mathrm{d}y \tag{21}$$

按照 Hamilton 原理，动力学问题的真实解应使泛函 Π 取驻值，即

$$\delta\Pi=0 \tag{22}$$

而 Π 应为

$$\Pi=\int_{t_1}^{t_2}(\overline{T}-U)\mathrm{d}t \tag{23}$$

其中 t_1 和 t_2 表示任意的时刻。

将式（20）和（21）代入式（22），经分部积分后，可得夹层矩形板非线性振动的运动方程和边界条件

$$\frac{\partial\sigma_{x0}}{\partial x}+\frac{\partial\tau_{xy0}}{\partial y}=0,$$

$$\frac{\partial\tau_{xy0}}{\partial x}+\frac{\partial\sigma_{x0}}{\partial y}=0,$$

$$G_2h_0\left(\frac{\partial\psi_x}{\partial x}+\frac{\partial\psi_y}{\partial y}+\nabla^2 w\right)=\rho\frac{\partial^2 w}{\partial t^2}-q-2h_1\left(\sigma_{x0}\frac{\partial^2 w}{\partial x^2}+\sigma_{y0}\frac{\partial^2 w}{\partial y^2}+2\tau_{xy0}\frac{\partial^2 w}{\partial x\partial y}\right),$$

$$\frac{D}{G_2h_0}\left(\frac{\partial^2\psi_x}{\partial x^2}+\frac{1-\nu}{2}\frac{\partial^2\psi_x}{\partial y^2}+\frac{1+\nu}{2}\frac{\partial^2\psi_y}{\partial x\partial y}\right)-\left(\psi_x+\frac{\partial w}{\partial x}\right)=0,$$

$$\frac{D}{G_2h_0}\left(\frac{1+\nu}{2}\frac{\partial^2\psi_x}{\partial x\partial y}+\frac{1-\nu}{2}\frac{\partial^2\psi_y}{\partial x^2}+\frac{\partial^2\psi_y}{\partial y^2}\right)-\left(\psi_y+\frac{\partial w}{\partial y}\right)=0 \tag{24a-e}$$

当 $x=\pm a$ 时，

$$\sigma_{x0}=0 \quad 或 \quad \delta u=0,$$

$$\tau_{xy0}=0 \quad 或 \quad \delta v=0,$$

$$Q_x=G_2h_0\left(\psi_x+\frac{\partial w}{\partial x}\right)=-2h_1\left(\sigma_{x0}\frac{\partial w}{\partial x}+\tau_{xy0}\frac{\partial w}{\partial y}\right) \quad 或 \quad \delta w=0,$$

$$M_x=D\left(\frac{\partial\psi_x}{\partial x}+\nu\frac{\partial\psi_y}{\partial y}\right)=0 \quad 或 \quad \delta\psi_x=0,$$

$$M_{xy}=\frac{D}{2}(1-\nu)\left(\frac{\partial\psi_y}{\partial x}+\frac{\partial\psi_x}{\partial y}\right)=0 \quad 或 \quad \delta\psi_y=0 \tag{25}$$

当 $y=\pm b$ 时，

$$\tau_{xy0}=0 \quad 或 \quad \delta u=0,$$

$$\sigma_{y0}=0 \quad 或 \quad \delta v=0,$$

$$Q_y=G_2h_0\left(\psi_y+\frac{\partial w}{\partial y}\right)=-2h_1\left(\tau_{xy0}\frac{\partial w}{\partial x}+\sigma_{y0}\frac{\partial w}{\partial y}\right) \quad 或 \quad \delta w=0,$$

$$M_{xy}=\frac{D}{2}(1-\nu)\left(\frac{\partial\psi_y}{\partial x}+\frac{\partial\psi_x}{\partial y}\right)=0 \quad 或 \quad \delta\psi_x=0,$$

$$M_y=D\left(\frac{\partial\psi_y}{\partial y}+\nu\frac{\partial\psi_x}{\partial x}\right)=0 \quad 或 \quad \delta\psi_y=0 \tag{26}$$

对于应变协调方程，可由方程（14）消去位移 u 和 v 求得

$$\frac{\partial^2\sigma_{x0}}{\partial y^2}+\frac{\partial^2\sigma_{y0}}{\partial x^2}-\nu\left(\frac{\partial^2\sigma_{x0}}{\partial x^2}+\frac{\partial^2\sigma_{y0}}{\partial y^2}\right)-2(1+\nu)\frac{\partial^2\tau_{xy0}}{\partial x\partial y}=E\left[\left(\frac{\partial^2 w}{\partial x\partial y}\right)^2-\frac{\partial^2 w}{\partial x^2}\frac{\partial^2 w}{\partial y^2}\right] \tag{27}$$

方程组（24）和（27）就是夹层矩形板非线性振动理论的基本方程，而边界条件由式（25）和（26）决定。显然，这组方程相当复杂。为此，我们应用文献［9］的方法简化。

设 ψ_x，ψ_y 可由另外两个函数 γ，f 表示

$$\psi_x=\frac{\partial\gamma}{\partial x}+\frac{\partial f}{\partial y},\quad \psi_y=\frac{\partial\gamma}{\partial y}-\frac{\partial f}{\partial x} \tag{28}$$

将式（28）代入方程（24d，e），可得

$$\frac{\partial}{\partial x}\left(\frac{D}{G_2h_0}\nabla^2\gamma-\gamma-w\right)+\frac{\partial}{\partial y}\left(\frac{1-\nu}{2}\frac{D}{G_2h_0}\nabla^2 f-f\right)=0,$$

$$\frac{\partial}{\partial y}\left(\frac{D}{G_2h_0}\nabla^2\gamma-\gamma-w\right)-\frac{\partial}{\partial x}\left(\frac{1-\nu}{2}\frac{D}{G_2h_0}\nabla^2 f-f\right)=0 \tag{29}$$

显然，此方程为 Cauchy-Riemann 方程，它的解可以用一个复变函数 g（$x+iy$）表示

如下

$$\frac{1-\nu}{2}\frac{D}{G_2h_0}\nabla^2 f-f+i\left(\frac{D}{G_2h_0}\nabla^2\gamma-\gamma-w\right)=g(x+iy) \tag{30}$$

非齐次微分方程（30）的解可表示为任一特解与相应的齐次方程的解之和。其特解 f_1，γ_1 和 w_1 可取为

$$f_1+i\gamma_1=-g_1(x+iy),w_1=0 \tag{31}$$

而相应的齐次方程为

$$\frac{1-\nu}{2}\frac{D}{G_2h_0}\nabla^2 f-f=0,$$

$$\frac{D}{G_2h_0}\nabla^2\gamma-\gamma-w=0 \tag{32a,b}$$

由于特解（31）代表刚性运动，不产生挠度，也不影响 ψ_x，ψ_y 的值，因此这组特解可以略去。于是，只要 f，γ 和 w 满足齐次方程（32）就可以了。由方程（32b）可得

$$w=\frac{D}{G_2h_0}\nabla^2\gamma-\gamma \tag{33}$$

将式（28）和（33）代入方程（24c），便得

$$D\nabla^4\gamma=\left[\rho\frac{\partial^2}{\partial t^2}-2h_1\left(\sigma_{x0}\frac{\partial^2}{\partial x^2}+\sigma_{y0}\frac{\partial^2}{\partial y^2}+2\tau_{xy0}\frac{\partial^2}{\partial x\partial y}\right)\right]\left(\frac{D}{G_2h_0}\nabla^2\gamma-\gamma\right)-q \tag{34}$$

引入下列应力函数 φ（x，y）

$$\sigma_{x0}=\frac{\partial^2\varphi}{\partial y^2},\sigma_{y0}=\frac{\partial^2\varphi}{\partial x^2},\tau_{xy0}=-\frac{\partial^2\varphi}{\partial x\partial y} \tag{35}$$

这样，方程（24a，b）自然满足，而方程（34）和（27）成为

$$D\nabla^4\gamma=\rho\frac{\partial^2 w}{\partial t^2}-2h_1L(w,\varphi)-q,$$

$$\frac{1}{E}\nabla^4\varphi=-\frac{1}{2}L(w,w) \tag{36}$$

其中 L 为微分算子，

$$L(w,\varphi)=\frac{\partial^2 w}{\partial x^2}\frac{\partial^2\varphi}{\partial y^2}+\frac{\partial^2 w}{\partial y^2}\frac{\partial^2\varphi}{\partial x^2}-2\frac{\partial^2 w}{\partial x\partial y}\frac{\partial^2\varphi}{\partial x\partial y} \tag{37}$$

于是，夹层矩形板非线性振动问题的基本方程（24）和（27）被简化成方程组（36）和（32）。

三、解析解

现在，我们讨论简支和铰支夹层矩形板的非线性自由振动问题。此时，边界条件为

（1）简支边界

当 $x=\pm a$ 时，

$$\gamma=0,\quad \frac{\partial^2\gamma}{\partial x^2}=0,\quad \frac{\partial^2\varphi}{\partial y^2}=0,\quad \frac{\partial^2\varphi}{\partial x\partial y}=0,\quad f=0;$$

当 $x=\pm b$ 时，

$$\gamma=0,\quad \frac{\partial^2\gamma}{\partial y^2}=0,\quad \frac{\partial^2\varphi}{\partial x^2}=0,\quad \frac{\partial^2\varphi}{\partial x\partial y}=0,\quad f=0 \tag{38}$$

(2) 铰支边界

当 $x=\pm a$ 时，

$$\gamma=0,\quad \frac{\partial^2\gamma}{\partial x^2}=0,\quad \int_0^x\left[\frac{1}{E}\left(\frac{\partial^2\varphi}{\partial y^2}-\nu\frac{\partial^2\varphi}{\partial x^2}\right)-\frac{1}{2}\left(\frac{\partial w}{\partial x}\right)^2\right]\mathrm{d}x=0,\quad \frac{\partial^2\varphi}{\partial x\partial y}=0,\quad f=0;$$

当 $x=\pm b$ 时，

$$\gamma=0,\quad \frac{\partial^2\gamma}{\partial y^2}=0,\quad \int_0^y\left[\frac{1}{E}\left(\frac{\partial^2\varphi}{\partial x^2}-\nu\frac{\partial^2\varphi}{\partial y^2}\right)-\frac{1}{2}\left(\frac{\partial w}{\partial y}\right)^2\right]\mathrm{d}y=0,\quad \frac{\partial^2\varphi}{\partial x\partial y}=0,\quad f=0 \tag{39}$$

由于 f 在边界上为零，且满足方程 (32a)，所以在简支和铰支边界条件下，它在整个区域内都等于零。

为简化上述问题的求解，我们引入下列无量纲量

$$\lambda=\frac{a}{b},\quad \xi=\frac{x}{a},\quad \eta=\frac{y}{b},\quad \Gamma=\frac{2\sqrt{1-\nu^2}}{h_0}\gamma,\quad \phi=\frac{4(1-\nu^2)}{Eh_0}\varphi,$$

$$W=\frac{2\sqrt{1-\nu^2}}{h_0}w,\quad k=\frac{D}{G_2h_0a^2},\quad \tau=\sqrt{\frac{D}{a^4\rho}}t \tag{40}$$

利用这些无量纲量，方程组 (33)、(36) 和边界条件 (38)、(39) 成为

$$L_1^2\Gamma=\frac{\partial^2 W}{\partial\tau^2}-\lambda^2L_2(\phi,W),$$

$$L_1^2\phi=-\frac{1}{2}\lambda^2L_2(W,W) \tag{41a,b}$$

(1) 简支边界

当 $\xi=\pm1$ 时，

$$\Gamma=0,\quad \frac{\partial^2\Gamma}{\partial\xi^2}=0,\quad \frac{\partial^2\varphi}{\partial\eta^2}=0,\quad \frac{\partial^2\phi}{\partial\xi\partial\eta}=0 \tag{42a-d}$$

当 $\eta=\pm1$ 时，

$$\Gamma=0,\quad \frac{\partial^2\Gamma}{\partial\eta^2}=0,\quad \frac{\partial^2\varphi}{\partial\xi^2}=0,\quad \frac{\partial^2\phi}{\partial\xi\partial\eta}=0 \tag{43a-d}$$

(2) 铰支边界

当 $\xi=\pm1$ 时，

$$\Gamma=0,\quad \frac{\partial^2\Gamma}{\partial\xi^2}=0,\quad \int_0^\xi\left[\lambda^2\frac{\partial^2\varphi}{\partial\eta^2}-\nu\frac{\partial^2\varphi}{\partial\xi^2}-\frac{1}{2}\left(\frac{\partial W}{\partial\xi}\right)^2\right]\mathrm{d}\xi=0,\quad \frac{\partial^2\varphi}{\partial\xi\partial\eta}=0 \tag{44a-d}$$

当 $\eta=\pm1$ 时，

$$\Gamma=0,\quad \frac{\partial^2\Gamma}{\partial\eta^2}=0,\quad \int_0^\eta\left[\frac{\partial^2\varphi}{\partial\xi^2}-\nu\lambda^2\frac{\partial^2\varphi}{\partial\eta^2}-\frac{1}{2}\left(\frac{\partial W}{\partial\eta}\right)^2\right]\mathrm{d}\eta=0,\quad \frac{\partial^2\varphi}{\partial\xi\partial\eta}=0 \tag{45a-d}$$

其中

$$W = kL_1\Gamma - \Gamma,$$

$$L_1\phi = \frac{\partial^2\varphi}{\partial\xi^2} + \lambda^2\frac{\partial^2\varphi}{\partial\eta^2},$$

$$L_2(W,\phi) = \frac{\partial^2 W}{\partial\xi^2}\frac{\partial^2\varphi}{\partial\eta^2} + \frac{\partial^2 W}{\partial\eta^2}\frac{\partial^2\phi}{\partial\xi^2} - 2\frac{\partial^2 W}{\partial\xi\partial\eta}\frac{\partial^2\varphi}{\partial\xi\partial\eta} \tag{46a-c}$$

下面，应用 Галёркин 方法首先求解简支夹层矩形板的无量纲非线性振动问题 (41)～(43)。为此设解为

$$\Gamma = F(\tau)\cos\frac{\pi}{2}\xi\cos\frac{\pi}{2}\eta \tag{47}$$

其中 $F(\tau)$ 是 τ 的一个待定函数。

显然，解（47）已满足边界条件（42a，b）和（43a，b）。将解（47）代入式（46a)，便得

$$W = e\Gamma \tag{48}$$

其中

$$e = -\left[\frac{1}{4}k\pi^2(1+\lambda^2)+1\right] \tag{49}$$

应用式（48)，方程（41b）的解为

$$\phi = \sum_{m=0}^{\infty}\sum_{n=0}^{\infty} d_{mn}\cos m\pi\xi\cos n\pi\eta + \frac{1}{2}(c_1\xi^2 + c_2\eta^2) \tag{50}$$

其中

$$d_{mn} = b_{mn} + c_{mn},$$

$$b_{10} = -\frac{\lambda^2}{32}e^2F^2,\quad b_{01} = -\frac{1}{32\lambda^2}e^2F^2,\quad \text{其余 } b_{mn} = 0,$$

$$c_{mn} = \frac{4\lambda}{\pi(m^2+\lambda^2n^2)^2}\left[\frac{m(-1)^n\varepsilon_n\mathrm{sh}^2(m\pi/\lambda)}{m\pi/\lambda + \mathrm{sh}(m\pi/\lambda)\mathrm{ch}(m\pi/\lambda)}A_m + \frac{n(-1)^m\varepsilon_m\mathrm{sh}^2(n\pi\lambda)}{n\pi\lambda + \mathrm{sh}(n\pi\lambda)\mathrm{ch}(n\pi\lambda)}B_n\right],$$

$$\varepsilon_0 = \frac{1}{2},\quad \varepsilon_n = 1\quad (m,n = 0,1,2,\cdots) \tag{51}$$

应用边界条件式（42c，d）和式（43c，d)，得到

$$c_1 = c_2 = 0,$$

$$B_n + \sum_{p=1}^{\infty}(-1)^pA_pH\left(\frac{p}{\lambda},n\right) = -n^2\lambda^2\sum_{q=0}^{\infty}(-1)^qb_{qn},$$

$$A_n + \sum_{p=1}^{\infty}(-1)^pB_pH(p\lambda,n) = -n^2\sum_{q=0}^{\infty}(-1)^qb_{qn} \tag{52}$$

其中

$$H(\zeta,j) = \frac{4(-1)^j\zeta j^2}{\pi(\zeta^2+j^2)^2}\frac{\mathrm{sh}^2\zeta\pi}{\zeta\pi + \mathrm{sh}\zeta\pi\mathrm{ch}\zeta\pi} \tag{53}$$

将式（47)，(48）和（50）代入方程（41a)，并用 Галёркин 方程，便得

$$\int_0^1\int_0^1\left(\frac{\partial^2 W}{\partial\tau^2}-\lambda^2 L_2(\phi,W)-L_1^2\Gamma\right)\Gamma\mathrm{d}\xi\mathrm{d}\eta=0 \tag{54}$$

经过推算，上式成为

$$\frac{\mathrm{d}^2F}{\mathrm{d}\tau^2}+\omega_L^2F+\alpha^2F^3=0 \tag{55}$$

其中 ω_L 是夹层矩形板的线性振动的固有频率，

$$\omega_L^2=\frac{\pi^4(1+\lambda^2)^2}{4[\pi^2k(1+\lambda^2)+4]},\quad \alpha^2=\frac{\pi^4\lambda^2\theta}{16}\left[\frac{1}{4}\pi^2k(1+\lambda^2)+1\right]^2,$$

$$\theta=\frac{2(d_{10}+d_{01})}{e^2F^2} \tag{56}$$

方程（55）是著名的 Duffin 方程。若初始条件为

$$F(0)=A,\frac{\mathrm{d}F(0)}{\mathrm{d}\tau}=0 \tag{57}$$

并注意到板的无量纲最大振幅发生在板中心处，且为

$$W_m=eF \tag{58}$$

则得方程（55）的解为

$$F=A\ \mathrm{cn}(\mu\tau,\beta) \tag{59}$$

其中 cn 为椭圆余弦，

$$\mu=\sqrt{\omega_L^2+\left(\frac{\alpha}{e}\right)^2W_m^2},\quad \beta=\sqrt{\frac{\alpha^2W_m^2}{2(e^2\omega_L^2+\alpha^2W_m^2)}} \tag{60}$$

于是，我们得到非线性周期与线性周期的比值

$$\frac{T_{NL}}{T_L}=\frac{2\omega_LK(\beta)}{\pi\sqrt{\omega_L^2+(\alpha/e)^2W_m^2}} \tag{61}$$

其中 K 是第一类完全的椭圆积分。

最后，我们求解铰支夹层矩形板的无量纲非线性振动问题（41），（44）和（45），其步骤与前面一样。除下述公式

$$c_1=\frac{\pi^2(\nu+\lambda^2)}{32(1-\nu^2)}e^2F^2,\quad c_2=\frac{\pi^2(1+\nu\lambda^2)}{32\lambda^2(1-\nu^2)}e^2F^2,$$

$$A_m=B_n=0,\qquad c_{mn}=0,\quad (m,n=0,1,2,\cdots)$$

$$\omega_L^2=\frac{\pi^4(1+\lambda^2)^2}{4[\pi^2k(1+\lambda^2)+4]},$$

$$\alpha^2=\frac{\pi^4[(3-\nu^2)(1+\lambda^4)+4\nu\lambda^2]}{256[1-\nu^2]}e^2 \tag{62}$$

以外，其余公式形式与前面相同。

四、结果与讨论

首先，我们讨论简支夹层矩形板的非线性振动周期与振幅的关系。在式（52）中，我们分别取 $n=1$，2，3，…，12，可求得 A_n 和 B_n 的值。按照我们的计算，n 取到 8，精确度已满足要求。在表 1 中列出了 A_n 和 B_n 的若干值。再按照式（56），可求得 θ 值，其结果给在表 2 中。

最后，按照式（61），我们在图 2～图 5 中给出了各种 k 值和 λ 值下的夹层矩形板的非线性振动周期与振幅的关系曲线。由这些图可知：

（1）随着夹层矩形板无量纲特征参数 k 的增大，非线性的影响增大。

（2）随着夹层矩形板边长比 λ 的增大，非线性的影响减小。当板狭长时，非线性的影响很小。

表 1　式（52）中 A_n 和 B_n 的值

λ	1	2	3	4	6	8	10
A_1/e^2F^2	0.024237	0.12609	0.28221	0.50054	1.1252	2.0001	3.1250
A_2/e^2F^2	0.0039377	−0.0021686	−0.0024141	−0.0016074	−0.00065218	−0.00030179	−0.00015991
A_3/e^2F^2	−0.0018592	0.0025135	0.0031098	0.0024204	0.0011792	0.00059446	0.00032935
A_4/e^2F^2	0.00094895	−0.0025178	−0.0032217	−0.0027910	−0.0016079	−0.00089111	−0.00052103
A_5/e^2F^2	−0.00052595	0.0023816	0.0030873	0.0028574	0.0018823	0.0011435	0.00070824
A_6/e^2F^2	0.00031153	−0.0021942	−0.0028727	−0.0027702	−0.0020179	−0.0013305	−0.00087235
A_7/e^2F^2	−0.00019448	0.0019974	0.0026422	0.0026188	0.0020538	0.0014518	0.0010039
A_8/e^2F^2	0.00012663	−0.0018095	−0.0024201	−0.0024471	−0.0020270	−0.0015173	−0.0011009
B_1/e^2F^2	0.024237	−0.0074830	−0.022038	−0.027743	−0.031120	−0.031558	−0.031428
B_2/e^2F^2	0.0039377	0.014852	0.017796	0.018152	0.017350	0.016583	0.016133
B_3/e^2F^2	−0.0018592	−0.0078481	−0.0090379	−0.0088325	−0.0080056	−0.0074864	−0.0072217
B_4/e^2F^2	0.00094895	0.0049242	0.0054851	0.0052034	0.0045778	0.0042375	0.0040733
B_5/e^2F^2	−0.00052595	−0.0033935	−0.0036741	−0.0034168	−0.0029545	−0.0027204	−0.0026103
B_6/e^2F^2	0.00031153	0.0024806	0.0026257	0.0024096	0.0020616	0.0018924	0.0018140
B_7/e^2F^2	−0.00019448	−0.0018900	−0.0019660	−0.0017878	−0.0015192	−0.0013918	−0.0013333
B_8/e^2F^2	0.00012663	0.0014857	0.0015250	0.0013779	0.0011654	0.0010663	0.0010211

表 2　式（56）中 θ 的值

λ	1	2	3	4	6	8	10
θ	0.06492	0.03531	0.01818	0.01087	0.005106	0.002944	0.001908

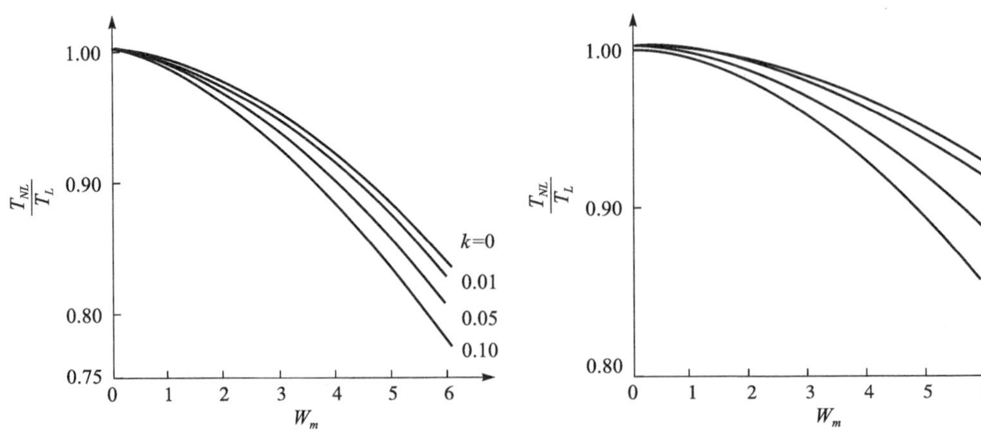

图 2　振动周期和振幅的关系（$\lambda=1$）

图 3　振动周期和振幅的关系（$\lambda=2$）

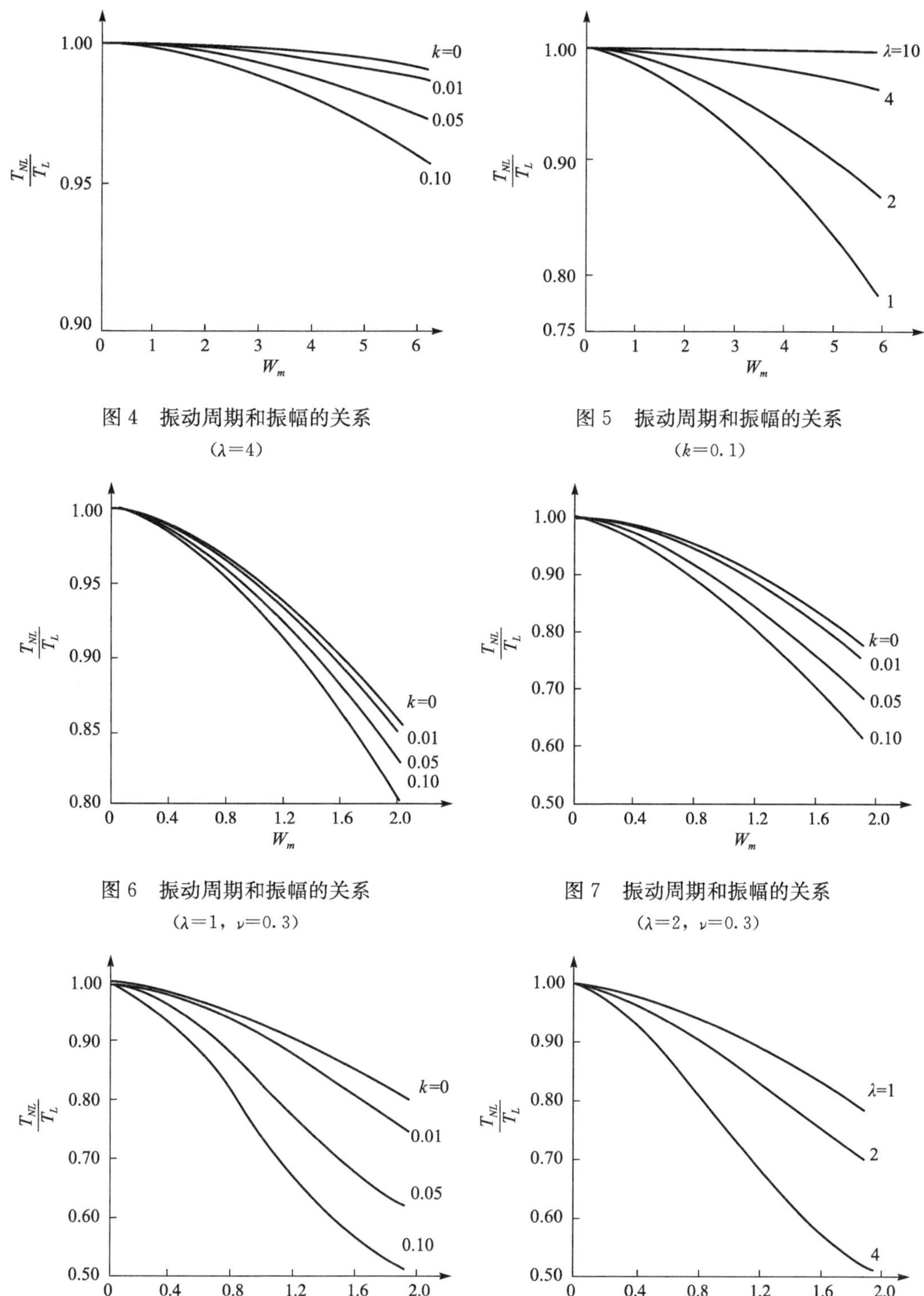

图 4　振动周期和振幅的关系

(λ=4)

图 5　振动周期和振幅的关系

(k=0.1)

图 6　振动周期和振幅的关系

(λ=1，ν=0.3)

图 7　振动周期和振幅的关系

(λ=2，ν=0.3)

图 8　振动周期和振幅的关系

(λ=4，ν=0.3)

图 9　振动周期和振幅的关系

(k=0.1，ν=0.3)

下面，我们再讨论铰支夹层矩形板的振动周期与振幅的关系，其结果给在图 6～图 9 中。我们由这些图得出以下结论：

（1）随着夹层矩形板无量纲特征参数 k 的增大，非线性的影响增大。

（2）随着夹层矩形板边长比 λ 的增大，非线性的影响也增大。

最后，我们将上述两种不同边界条件下的结果进行比较。由图 10 显然可见，铰支的振动周期受非线性影响大，而简支的振动周期受影响较小。

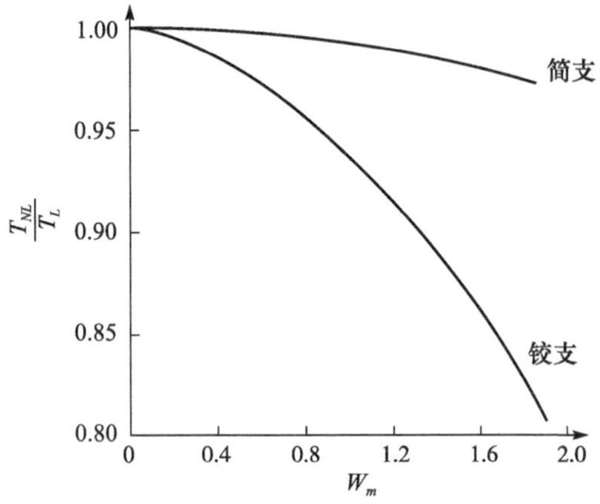

图 10　两种不同边界振动周期与振幅的关系
（λ=1，k=0.1，ν=0.3）

参考文献

[1] Reissner E. Finite deflections of sandwich plates. J. Aeron. Sci., 1948, 15 (7): 435; 1950, 17 (2): 125.

[2] Kan H P, Huang J C. Large deflection of rectangular sandwich plates. AIAA J., 1967, 5 (9): 1706.

[3] 刘人怀. 在边缘力矩作用下夹层圆板的非线性轴对称弯曲问题. 中国科学技术大学学报, 1980, 10 (2): 56.

[4] 刘人怀. 夹层圆板的非线性弯曲. 应用数学和力学, 1981, 2 (2): 173.

[5] 刘人怀, 施云方. 夹层圆板大挠度问题的精确解. 应用数学和力学, 1982, 3 (1): 11.

[6] Liu Ren-huai. Progress in Appied Mechanics (Ed. Yeh, K. Y.). Martinus Nijhoff Publishers, Dordrecht, 1987: 291.

[7] 中国科学院北京力学研究所固体力学研究室板壳组. 夹层板壳的弯曲、稳定和振动. 北京: 科学出版社, 1977.

[8] 刘人怀, 朱高秋. 夹层圆板大挠度问题的进一步研究. 应用数学和力学, 1989, 10 (12): 1041.

[9] 胡海昌. 各向同性夹层板反对称小挠度的若干问题. 力学学报, 1963, 6 (1): 53.

简支夹层矩形板的非线性弯曲①

一、引言

作为结构元件的夹层板在航空、宇航和航海工程中得到了广泛的应用。因此近年来，许多研究者对这种板进行了研究。但是，因为面临非线性微分方程和夹层结构复杂的巨大困难，仅有少数人研究了夹层板的非线性问题。首先，Reissner[1]建立了具有软夹心的夹层矩形板的非线性弯曲理论。此时，视表层如薄膜一样，忽略了表层的抗弯刚度。然后，刘人怀[2]进一步建立了计及表层抗弯刚度的具有软夹心的夹层圆板的更为精确的非线性弯曲理论，并且给出了忽略表层抗弯刚度的简化理论。刘人怀等[2-7]、Kan 和 Huang[8]、Alwan[9]和 Kamiya[10]等先后讨论了夹层圆板和夹层矩形板的非线性弯曲和振动问题。然而，所得的简支夹层矩形板的非线性弯曲问题的结果尚不能令人满意。所以，进一步研究这一问题是很有意义的。

值得指出，应用文献［11］的方程来求解夹层矩形板的实际问题是不大方便的。为此，须推导这种板的新方程。在忽略表层抗弯刚度的情况下，我们使用变分法导出了均布横向载荷作用下具有软夹心的夹层矩形板的非线性弯曲的基本方程和边界条件。然后，对上述方程和边界条件进行了简化。最后，使用摄动法研究了均布横向载荷作用下的简支夹层矩形板，获得了相当精确的解析解。本文所得结果可供工程设计时参考应用。

二、基本方程

考虑在任意横向载荷 $q(x,y)$ 作用下的夹层矩形薄板，如图 1 所示。这里 x，y 和 z 为直角坐标，$2a$ 和 $2b$ 为边长，t 为表层厚度，h_0 为上下表层中面间的距离，1，2 和 3 分别表示上表层、夹心和下表层。

在推导基本方程和边界条件时，采用 Reissner[1]的假定：

（1）材料服从于胡克定律。

（2）夹心横向不可压缩。

（3）夹心沿板面方向不能承受载荷。

（4）表层处于薄膜应力状态。

（5）夹心中面法线在变形后保持直线。

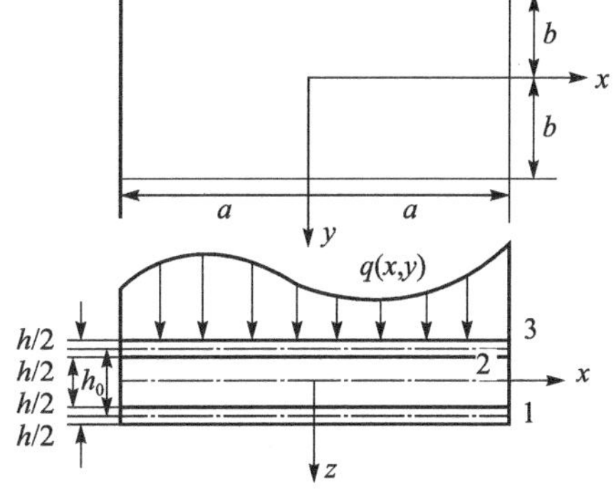

图 1　夹层矩形板的几何形状

在这些假定下，夹层矩形板中任意一点的位移为

① 本文原载《应用数学和力学》，1993，14（3）：203-218. 作者：刘人怀，成振强.

上表层$\left[\frac{1}{2}(h_0-t)\leqslant z\leqslant\frac{1}{2}(h_0+t)\right]$

$$u_1=u+\frac{1}{2}h_0\psi_x,\quad v_1=v+\frac{1}{2}h_0\psi_y,\quad w_1=w \tag{2.1}$$

下表层$\left[-\frac{1}{2}(h_0+t)\leqslant z\leqslant-\frac{1}{2}(h_0-t)\right]$

$$u_3=u-\frac{1}{2}h_0\psi_x,\quad v_3=v-\frac{1}{2}h_0\psi_y,\quad w_3=w \tag{2.2}$$

夹心$\left[-\frac{1}{2}(h_0-t)\leqslant z\leqslant\frac{1}{2}(h_0-t)\right]$

$$u_2=u+z\psi_x,\quad v_2=v+z\psi_y,\quad w_2=w \tag{2.3}$$

其中u_i，v_i和w_i（i=1，2，3）分别为上表层、夹心和下表层在x，y和z轴方向的位移，u，v和w分别为板中面在x，y和z方向的位移，ψ_x和ψ_y分别为夹心中面法线在xz和yz平面内的转角。

设ε_{xi}，ε_{yi}，ε_{zi}，γ_{xyi}，γ_{yzi}和γ_{xzi}（i=1，2，3）分别为上表层、夹心和下表层的应变分量，则将式（2.1）～（2.3）分别代入下述夹层矩形板的几何方程

$$\begin{aligned}
\varepsilon_{xi}&=\frac{\partial u_i}{\partial x}+\frac{1}{2}\left(\frac{\partial w_i}{\partial x}\right)^2,\\
\varepsilon_{yi}&=\frac{\partial v_i}{\partial y}+\frac{1}{2}\left(\frac{\partial w_i}{\partial y}\right)^2,\\
\gamma_{xyi}&=\frac{\partial v_i}{\partial x}+\frac{\partial u_i}{\partial y}+\frac{\partial w_i}{\partial x}\frac{\partial w_i}{\partial y},\\
\varepsilon_{zi}&=\gamma_{xzi}=\gamma_{yzi}=0,\quad (i=1,3)
\end{aligned} \tag{2.4}$$

$$\begin{aligned}
\gamma_{xz2}&=\frac{\partial u_2}{\partial z}+\frac{\partial w_2}{\partial x},\\
\gamma_{yz2}&=\frac{\partial v_2}{\partial z}+\frac{\partial w_2}{\partial y},\\
\varepsilon_{x2}&=\varepsilon_{y2}=\varepsilon_{z2}=\gamma_{xy2}=0
\end{aligned} \tag{2.5}$$

便得：

上表层

$$\begin{aligned}
\varepsilon_{x1}&=\frac{\partial u}{\partial x}+\frac{h_0}{2}\frac{\partial\psi_x}{\partial x}+\frac{1}{2}\left(\frac{\partial w}{\partial x}\right)^2,\\
\varepsilon_{y1}&=\frac{\partial v}{\partial y}+\frac{h_0}{2}\frac{\partial\psi_y}{\partial y}+\frac{1}{2}\left(\frac{\partial w}{\partial y}\right)^2,\\
\gamma_{xy1}&=\frac{\partial v}{\partial x}+\frac{\partial u}{\partial y}+\frac{h_0}{2}\left(\frac{\partial\psi_y}{\partial x}+\frac{\partial\psi_x}{\partial y}\right)+\frac{\partial w}{\partial x}\frac{\partial w}{\partial y}
\end{aligned} \tag{2.6}$$

下表层

$$\begin{aligned}
\varepsilon_{x3}&=\frac{\partial u_i}{\partial x}-\frac{h_0}{2}\frac{\partial\psi_x}{\partial x}+\frac{1}{2}\left(\frac{\partial w}{\partial x}\right)^2,\\
\varepsilon_{y3}&=\frac{\partial v}{\partial y}-\frac{h_0}{2}\frac{\partial\psi_y}{\partial y}+\frac{1}{2}\left(\frac{\partial w}{\partial y}\right)^2,
\end{aligned}$$

$$\gamma_{xy3}=\frac{\partial v}{\partial x}+\frac{\partial u}{\partial y}-\frac{h_0}{2}\left(\frac{\partial\psi_y}{\partial x}+\frac{\partial\psi_x}{\partial y}\right)+\frac{\partial w}{\partial x}\frac{\partial w}{\partial y} \tag{2.7}$$

夹心

$$\gamma_{xz2}=\psi_x+\frac{\partial w}{\partial x},$$
$$\gamma_{yz2}=\psi_y+\frac{\partial w}{\partial y} \tag{2.8}$$

设 σ_{xi}，σ_{yi}，σ_{zi}，τ_{xyi}，τ_{yzi} 和 τ_{xzi}（$i=1$，2，3）分别为上表层、夹心和下表层点的应力分量，E，ν 和 G 分别为表层材料的弹性模量、泊松比和剪切模量，G_2 为夹心的剪切模量，将式（2.6）～(2.8) 分别代入下述胡克定律：

$$\begin{aligned}
\sigma_{xi}&=\frac{E}{1-\nu^2}(\varepsilon_{xi}+\nu\varepsilon_{yi}),\\
\sigma_{yi}&=\frac{E}{1-\nu^2}(\varepsilon_{yi}+\nu\varepsilon_{xi}),\\
\tau_{xyi}&=G\gamma_{xyi},\\
\sigma_{zi}&=\tau_{xzi}=\tau_{yzi}=0,\quad i=1,3
\end{aligned} \tag{2.9}$$

$$\begin{aligned}
\tau_{xz2}&=G_2\gamma_{xz2},\\
\tau_{yz2}&=G_2\gamma_{yz2},\\
\sigma_{x2}&=\sigma_{y2}=\sigma_{z2}=\tau_{xy2}=0
\end{aligned} \tag{2.10}$$

便得：

上表层

$$\begin{aligned}
\sigma_{x1}&=\sigma_{x0}+\frac{Eh_0}{2(1-\nu^2)}\left(\frac{\partial\psi_x}{\partial x}+\nu\frac{\partial\psi_y}{\partial y}\right),\\
\sigma_{y1}&=\sigma_{y0}+\frac{Eh_0}{2(1-\nu^2)}\left(\frac{\partial\psi_y}{\partial y}+\nu\frac{\partial\psi_x}{\partial x}\right),\\
\tau_{xy1}&=\tau_{xy0}+\frac{Gh_0}{2}\left(\frac{\partial\psi_y}{\partial x}+\frac{\partial\psi_x}{\partial y}\right),
\end{aligned} \tag{2.11}$$

下表层

$$\begin{aligned}
\sigma_{x3}&=\sigma_{x0}-\frac{Eh_0}{2(1-\nu^2)}\left(\frac{\partial\psi_x}{\partial x}+\nu\frac{\partial\psi_y}{\partial y}\right),\\
\sigma_{y3}&=\sigma_{y0}-\frac{Eh_0}{2(1-\nu^2)}\left(\frac{\partial\psi_y}{\partial y}+\nu\frac{\partial\psi_x}{\partial x}\right),\\
\tau_{xy3}&=\tau_{xy0}-\frac{Gh_0}{2}\left(\frac{\partial\psi_y}{\partial x}+\frac{\partial\psi_x}{\partial y}\right),
\end{aligned} \tag{2.12}$$

夹心

$$\begin{aligned}
\tau_{xz2}&=G_2\left(\psi_x+\frac{\partial w}{\partial x}\right),\\
\tau_{yz2}&=G_2\left(\psi_y+\frac{\partial w}{\partial y}\right)
\end{aligned} \tag{2.13}$$

其中，σ_{x0}，σ_{y0} 和 τ_{xy0} 分别为夹层矩形板中面内的应力，

$$\sigma_{x0}=\frac{E}{1-\nu^2}\left[\frac{\partial u}{\partial x}+\nu\frac{\partial v}{\partial y}+\frac{1}{2}\left(\frac{\partial w}{\partial x}\right)^2+\frac{\nu}{2}\left(\frac{\partial w}{\partial y}\right)^2\right],$$

$$\sigma_{y0}=\frac{E}{1-\nu^2}\left[\frac{\partial v}{\partial y}+\nu\frac{\partial u}{\partial x}+\frac{1}{2}\left(\frac{\partial w}{\partial y}\right)^2+\frac{\nu}{2}\left(\frac{\partial w}{\partial y}\right)^2\right],$$

$$\tau_{xy0}=G\left(\frac{\partial v}{\partial x}+\frac{\partial u}{\partial y}+\frac{\partial w}{\partial x}\frac{\partial w}{\partial y}\right) \tag{2.14}$$

根据弹性体应变能公式：

$$U_i=\frac{1}{2}\iiint\limits_{V_i}(\sigma_x\varepsilon_x+\sigma_y\varepsilon_y+\sigma_z\varepsilon_z+\tau_{xy}\gamma_{xy}+\tau_{yz}\gamma_{yz}+\tau_{xz}\gamma_{xz})\mathrm{d}x\mathrm{d}y\mathrm{d}z,\quad i=1,2,3 \tag{2.15}$$

我们得到表层和夹心的应变能公式：

$$U_i=\frac{1}{2E}\iiint\limits_{V_i}[(\sigma_{xi}+\sigma_{yi})^2+2(1+\nu)(\tau_{xyi}^2-\sigma_{xi}\sigma_{yi})]\mathrm{d}x\mathrm{d}y\mathrm{d}z,\quad i=1,3$$

$$U_2=\frac{1}{2G_2}\iiint\limits_{V_2}(\tau_{xz2}^2+\tau_{yz2}^2)\mathrm{d}x\mathrm{d}y\mathrm{d}z \tag{2.16}$$

将式（2.11）～（2.13）代入式（2.16），并对 z 进行积分，便得

$$\begin{aligned}U_1=&\frac{t}{2E}\iint\limits_{S_1}[(\sigma_{x0}+\sigma_{y0})^2+2(1+\nu)(\tau_{xy0}^2-\sigma_{x0}\sigma_{y0})]\mathrm{d}x\mathrm{d}y\\&+\frac{th_0}{2}\iint\limits_{S_1}\left[\sigma_{x0}\frac{\partial\psi_x}{\partial x}+\sigma_{y0}\frac{\partial\psi_y}{\partial y}+\tau_{xy0}\left(\frac{\partial\psi_y}{\partial x}+\frac{\partial\psi_x}{\partial y}\right)\right]\mathrm{d}x\mathrm{d}y\\&+\frac{D}{4}\iint\limits_{S_1}\left[\left(\frac{\partial\psi_x}{\partial x}\right)^2+\left(\frac{\partial\psi_y}{\partial y}\right)^2+2\nu\frac{\partial\psi_x}{\partial x}\frac{\partial\psi_y}{\partial y}+\frac{1-\nu}{2}\left(\frac{\partial\psi_y}{\partial x}+\frac{\partial\psi_x}{\partial y}\right)^2\right]\mathrm{d}x\mathrm{d}y,\end{aligned}$$

$$\begin{aligned}U_3=&\frac{t}{2E}\iint\limits_{S_3}[(\sigma_{x0}+\sigma_{y0})^2+2(1+\nu)(\tau_{xy0}^2-\sigma_{x0}\sigma_{y0})]\mathrm{d}x\mathrm{d}y\\&-\frac{th_0}{2}\iint\limits_{S_3}\left[\sigma_{x0}\frac{\partial\psi_x}{\partial x}+\sigma_{y0}\frac{\partial\psi_y}{\partial y}+\tau_{xy0}\left(\frac{\partial\psi_y}{\partial x}+\frac{\partial\psi_x}{\partial y}\right)\right]\mathrm{d}x\mathrm{d}y\\&+\frac{D}{4}\iint\limits_{S_3}\left[\left(\frac{\partial\psi_x}{\partial x}\right)^2+\left(\frac{\partial\psi_y}{\partial y}\right)^2+2\nu\frac{\partial\psi_x}{\partial x}\frac{\partial\psi_y}{\partial y}+\frac{1-\nu}{2}\left(\frac{\partial\psi_y}{\partial x}+\frac{\partial\psi_x}{\partial y}\right)^2\right]\mathrm{d}x\mathrm{d}y,\end{aligned}$$

$$U_2=\frac{G_2h_0}{2}\iint\limits_{S_2}\left[\left(\psi_x+\frac{\partial w}{\partial x}\right)^2+\left(\psi_y+\frac{\partial w}{\partial y}\right)^2\right]\mathrm{d}x\mathrm{d}y \tag{2.17}$$

其中，D 为夹层矩形板的抗弯刚度

$$D=\frac{Eth_0^2}{2(1-\nu^2)} \tag{2.18}$$

任意横向载荷 $q(x,y)$ 的外力功为

$$V=\iint\limits_{S}qw\mathrm{d}x\mathrm{d}y \tag{2.19}$$

这样，夹层矩形板的总势能为

$$
\begin{aligned}
U &= U_1 + U_2 + U_3 - V \\
&= \frac{t}{E}\iint_S [(\sigma_{x0}+\sigma_{y0})^2 + 2(1+\nu)(\tau_{xy0}^2 - \sigma_{x0}\sigma_{y0})]\mathrm{d}x\mathrm{d}y \\
&\quad + \frac{D}{2}\iint_S \left[\left(\frac{\partial\psi_x}{\partial x}\right)^2 + \left(\frac{\partial\psi_y}{\partial y}\right)^2 + 2\nu\frac{\partial\psi_x}{\partial x}\frac{\partial\psi_y}{\partial y} + \frac{1-\nu}{2}\left(\frac{\partial\psi_y}{\partial x}+\frac{\partial\psi_x}{\partial y}\right)^2\right]\mathrm{d}x\mathrm{d}y \\
&\quad + \frac{G_2h_0}{2}\iint_{S_2}\left[\left(\psi_x+\frac{\partial w}{\partial x}\right)^2 + \left(\psi_y+\frac{\partial w}{\partial y}\right)^2\right]\mathrm{d}x\mathrm{d}y - \iint_S qw\mathrm{d}x\mathrm{d}y
\end{aligned} \tag{2.20}
$$

按照势能原理，并以 u，v，w，ψ_x 和 ψ_y 作为自变量，便有

$$\delta U = 0 \tag{2.21}$$

将式（2.20）代入此式，经部分积分后，可得在任意横向载荷 $q(x,y)$ 作用下夹层矩形板大挠度理论的平衡方程和边界条件：

$$
\begin{aligned}
&\frac{\partial\sigma_{x0}}{\partial_x} + \frac{\partial\tau_{xy0}}{\partial_y} = 0, \\
&\frac{\partial\tau_{xy0}}{\partial_x} + \frac{\partial\sigma_{y0}}{\partial_y} = 0, \\
&G_2h_0\left(\frac{\partial\psi_x}{\partial x}+\frac{\partial\psi_y}{\partial y}+\nabla^2 w\right) = -q - 2t\left(\sigma_{x0}\frac{\partial^2 w}{\partial x^2} + \sigma_{y0}\frac{\partial^2 w}{\partial y^2} + 2\tau_{xy0}\frac{\partial^2 w}{\partial x\partial y}\right), \\
&\frac{D}{G_2h_0}\left(\frac{\partial^2\psi_x}{\partial x^2} + \frac{1-\nu}{2}\frac{\partial^2\psi_x}{\partial y^2} + \frac{1+\nu}{2}\frac{\partial^2\psi_y}{\partial x\partial y}\right) - \left(\psi_x+\frac{\partial w}{\partial x}\right) = 0, \\
&\frac{D}{G_2h_0}\left(\frac{1+\nu}{2}\frac{\partial^2\psi_x}{\partial x\partial y} + \frac{1-\nu}{2}\frac{\partial^2\psi_y}{\partial x^2} + \frac{\partial^2\psi_y}{\partial y^2}\right) - \left(\psi_y+\frac{\partial w}{\partial y}\right) = 0
\end{aligned} \tag{2.22a-e}
$$

当 $x=\pm a$ 时，

$$
\begin{aligned}
&\sigma_{x0} = 0, \text{或 } \delta u = 0; \\
&\tau_{xy0} = 0, \text{或 } \delta v = 0; \\
&G_2h_0\left(\psi_x+\frac{\partial w}{\partial x}\right) + 2t\left(\sigma_{x0}\frac{\partial w}{\partial x} + \tau_{xy0}\frac{\partial w}{\partial y}\right) = 0, \quad \text{或 } \delta w = 0; \\
&D\left(\frac{\partial\psi_x}{\partial x} + \nu\frac{\partial\psi_y}{\partial y}\right) = 0, \quad \text{或 } \delta\psi_x = 0; \\
&\frac{D(1-\nu)}{2}\left(\frac{\partial\psi_y}{\partial x}+\frac{\partial\psi_x}{\partial y}\right) = 0, \quad \text{或 } \delta\psi_y = 0
\end{aligned} \tag{2.23}
$$

当 $y=\pm b$ 时，

$$
\begin{aligned}
&\tau_{xy0} = 0, \text{或 } \delta u = 0; \\
&\sigma_{y0} = 0, \text{或 } \delta v = 0; \\
&G_2h_0\left(\psi_y+\frac{\partial w}{\partial y}\right) + 2t\left(\frac{\partial w}{\partial x} + \sigma_{y0}\frac{\partial w}{\partial y}\right) = 0, \text{或 } \delta w = 0; \\
&D\left(\frac{\partial\psi_y}{\partial y} + \nu\frac{\partial\psi_x}{\partial x}\right) = 0, \text{或 } \psi_y = 0; \\
&\frac{D(1-\nu)}{2}\left(\frac{\partial\psi_y}{\partial x}+\frac{\partial\psi_x}{\partial y}\right) = 0, \text{或 } \psi_x = 0
\end{aligned} \tag{2.24}
$$

其中∇^2是二阶拉普拉斯算子，

$$\nabla^2(\cdots) = \left(\frac{\partial^2}{\partial x^2} + \frac{\partial^2}{\partial y^2}\right)(\cdots) \tag{2.25}$$

至于弯矩 M_x 和 M_y，扭矩 M_{xy}，横向剪力 Q_x 和 Q_y 则有公式：

$$\begin{aligned}
M_x &= D\left(\frac{\partial \psi_x}{\partial x} + \nu \frac{\partial \psi_y}{\partial y}\right), \\
M_y &= D\left(\frac{\partial \psi_y}{\partial y} + \nu \frac{\partial \psi_x}{\partial x}\right), \\
M_{xy} &= \frac{D(1-\nu)}{2}\left(\frac{\partial \psi_y}{\partial x} + \frac{\partial \psi_x}{\partial y}\right), \\
Q_x &= G_2 h_0\left(\psi_x + \frac{\partial w}{\partial x}\right), \\
Q_y &= G_2 h_0\left(\psi_y + \frac{\partial w}{\partial y}\right)
\end{aligned} \tag{2.26a-e}$$

由方程（2.14）消去位移 u 和 v' 便得应变协调方程：

$$\frac{\partial^2 \sigma_{x0}}{\partial y^2} + \frac{\partial^2 \sigma_{y0}}{\partial x^2} - \nu\left(\frac{\partial^2 \sigma_{x0}}{\partial x^2} + \frac{\partial^2 \sigma_{y0}}{\partial y^2}\right) - 2(1+\nu)\frac{\partial^2 \tau_{xy0}}{\partial x \partial y} = E\left[\left(\frac{\partial^2 w}{\partial x \partial y}\right)^2 - \frac{\partial^2 w}{\partial x^2}\frac{\partial^2 w}{\partial y^2}\right] \tag{2.27}$$

于是，我们得到了夹层矩形板大挠度理论的基本方程和边界条件（2.22），（2.27），（2.23）和（2.24）。显然此问题相当繁杂，因而进一步简化是十分必要的。

应用文献［12］的处理夹层板小挠度方程的方法，我们可将 ψ_x，ψ_y 用两个新函数 ω，f 来表示：

$$\psi_x = \frac{\partial \omega}{\partial x} + \frac{\partial f}{\partial y}, \psi_y = \frac{\partial \omega}{\partial y} - \frac{\partial f}{\partial x} \tag{2.28}$$

将式（2.28）代入方程（2.22d，e）便得

$$\begin{aligned}
&\frac{\partial}{\partial x}\left(\frac{D}{G_2 h_0}\nabla^2 \omega - \omega - w\right) + \frac{\partial}{\partial y}\left[\frac{D(1-\nu)}{2G_2 h_0}\nabla^2 f - f\right] = 0, \\
&\frac{\partial}{\partial y}\left(\frac{D}{G_2 h_0}\nabla^2 \omega - \omega - w\right) - \frac{\partial}{\partial x}\left[\frac{D(1-\nu)}{2G_2 h_0}\nabla^2 f - f\right] = 0
\end{aligned} \tag{2.29}$$

显而易见，若把方程（2.29）中括号里的量看作两个独立的函数，则此方程就是 Cauchy-Riemann 方程，因而它的解可表示为

$$\frac{D(1-\nu)}{2G_2 h_0}\nabla^2 f - f + i\left(\frac{D}{G_2 h_0}\nabla^2 \omega - \omega - w\right) = F(x+iy) \tag{2.30}$$

这是一个关于 f，ω 和 w 的非齐次偏微分方程，其通解为方程（2.30）的任一特解与相应的齐次方程的通解之和。由于 $F(x+iy)$ 的实部和虚部都是调和函数，所以方程（2.30）的特解 f_1，ω_1，w_1 可取为

$$f_1 + i\omega_1 = -F(x+iy), \quad w_1 = 0 \tag{2.31}$$

相应的齐次方程为

$$\begin{aligned}
&\frac{D(1-\nu)}{2G_2 h_0}\nabla^2 f - f = 0, \\
&\frac{D}{G_2 h_0}\nabla^2 \omega - \omega - w = 0
\end{aligned} \tag{2.32a,b}$$

特解（2.31）既不产生挠度，也不影响 ψ_x 和 ψ_y 的值，因此这组特解可以略去。这样，f，ω 和 w 只要理解为满足齐次方程（2.32）的函数便可以了。由方程（2.32b）得到

$$w = \frac{D}{G_2 h_0} \nabla^2 \omega - \omega \tag{2.33}$$

将式（2.28）和（2.33）代入方程（2.22c），得到 ω 需满足的方程如下：

$$D\nabla^4 \omega = -q - 2t\left(\sigma_{x0}\frac{\partial^2}{\partial x^2} + \sigma_{y0}\frac{\partial^2}{\partial y^2} + 2\tau_{xy0}\frac{\partial^2}{\partial x \partial y}\right)\left(\frac{D}{G_2 h_0}\nabla^2\omega - \omega\right) \tag{2.34}$$

这样一来，夹层矩形板的挠度 w 和转角 ψ_x，ψ_y 便通过式（2.28）和（2.33）用两个函数 ω，f 来表示，而 ω，f 则应分别满足方程（2.34）和（2.32a）。

下面，我们引入如下的应力函数 φ（x，y）：

$$\sigma_{x0} = \frac{\partial^2 \varphi}{\partial y^2},\quad \sigma_{y0} = \frac{\partial^2 \varphi}{\partial x^2},\quad \tau_{xy0} = -\frac{\partial^2 \varphi}{\partial x \partial y} \tag{2.35}$$

这样，方程（2.22a，b）已被满足，而方程（2.34）和（2.27）成为

$$D\nabla^4\omega = -q - 2tL(w,\varphi),$$
$$\frac{1}{E}\nabla^4\omega = -\frac{1}{2}(w,w) \tag{2.36a,b}$$

其中 w 的表达式如式（2.33）所示，

$$L(w,\varphi) = \frac{\partial^2 w}{\partial x^2}\frac{\partial^2 \varphi}{\partial y^2} + \frac{\partial^2 w}{\partial y^2}\frac{\partial^2 \varphi}{\partial x^2} - 2\frac{\partial^2 w}{\partial x \partial y}\frac{\partial^2 \varphi}{\partial x \partial y} \tag{2.37}$$

由于引进了函数 ω，f 以及应力函数 φ，基本方程组大大简化。但是必须指出，对于大多数具体问题，在边界条件的表达式中，ω 和 f 是耦合的，因此还必须联立求解。在个别问题的边界条件中，ω 与 f 不同时出现，这时问题就更为简化，ω 与 f 便可以分别独立地求解了。

今以四边简支夹层矩形板作为一个例子。由式（2.23）和（2.24）知，在这种情况下，它的边界条件为

当 $x=\pm a$ 时，$\sigma_{x0}=0$，$\tau_{xy0}=0$，$w=0$，$\psi_y=0$，$M_x=0$　(2.38)

当 $y=\pm b$ 时，$\sigma_{y0}=0$，$\tau_{xy0}=0$，$w=0$，$\psi_x=0$，$M_y=0$　(2.39)

现在，我们将这些边界条件表示为 ω 和 φ 的显式。为此，考虑一块简支多边形板。于是在任一边界 l 上有

$$\sigma_{n0}=0,\quad \tau_{nl0}=0,\quad w=0,\quad \psi_l=0,\quad M_n=0 \tag{2.40a-e}$$

这里 n 为边界法向，

$$M_n = D\left(\frac{\partial \psi_n}{\partial n} + \nu\frac{\partial \psi_l}{\partial l}\right) \tag{2.41}$$

显然，式（2.40e）可简化为如下形式：

$$\frac{\partial \psi_n}{\partial n} = 0 \tag{2.42}$$

由式（2.28），有

$$\psi_n = \frac{\partial w}{\partial n} + \frac{\partial f}{\partial l},\quad \psi_l = \frac{\partial w}{\partial l} - \frac{\partial f}{\partial n} \tag{2.43}$$

应用式（2.33），（2.42）和（2.43），边界条件（2.40c～e）化为

$$\frac{D}{G_2h_0}\nabla^2\omega-\omega=0,\quad \frac{\partial w}{\partial l}-\frac{\partial f}{\partial n}=0,\quad \frac{\partial^2\omega}{\partial n^2}+\frac{\partial^2 f}{\partial n\partial l}=0 \tag{2.44}$$

这些边界条件可进一步简化，最后化成如下形式：

$$\omega=0,\quad \frac{\partial f}{\partial n}=0,\quad \frac{\partial^2\omega}{\partial n^2}=0 \tag{2.45a-c}$$

这里，函数 f 还满足方程（2.32a）。

将矢量场的 Gauss 定理应用于平面情形，有

$$\iint_S \nabla \boldsymbol{v}\,\mathrm{d}S=\oint_l \boldsymbol{v}\cdot\boldsymbol{n}\,\mathrm{d}l \tag{2.46}$$

其中 $\boldsymbol{v}$ 是平面矢量，S 为多边形板面区域，

$$\nabla(\cdots)=\left(i\frac{\partial}{\partial x}+j\frac{\partial}{\partial x}\right)(\cdots) \tag{2.47}$$

取 $\boldsymbol{v}=\nabla f$，则式（2.46）成为

$$\iint_S \nabla^2 f\,\mathrm{d}S=\oint_l \nabla f\cdot\boldsymbol{n}\,\mathrm{d}l=\oint_l\frac{\partial f}{\partial n}\mathrm{d}l \tag{2.48}$$

应用式（2.32a）和（2.45b），我们由上式得到

$$\iint_S f\,\mathrm{d}S=0 \tag{2.49}$$

由 S 的任意性可知

$$f=0 \tag{2.50}$$

因此，四边简支夹层矩形板的边界条件（2.38）和（2.39）化为

$$\text{当 } x=\pm a \text{ 时},\quad \omega=0,\quad \frac{\partial^2\omega}{\partial x^2}=0,\quad \frac{\partial^2\varphi}{\partial y^2}=0,\quad \frac{\partial^2\varphi}{\partial x\partial y}=0 \tag{2.51}$$

$$\text{当 } y=\pm b \text{ 时},\quad \omega=0,\quad \frac{\partial^2\omega}{\partial y^2}=0,\quad \frac{\partial^2\varphi}{\partial x^2}=0,\quad \frac{\partial^2\varphi}{\partial x\partial y}=0 \tag{2.52}$$

为简单起见，本文仅讨论在均布载荷 q_0 作用下简夹层矩形板的大挠度问题。于是，问题归结为求解非线性边值问题（2.36）、（2.51）和（2.52）。

为了计算方便，引入下列无量纲量：

$$\lambda=\frac{a}{b},\quad \xi=\frac{x}{a},\quad \eta=\frac{y}{b},\quad \Omega=\frac{2\sqrt{1-\nu^2}}{h_0}\omega,\quad \Phi=\frac{4(1-\nu^2)}{Eh_0}\varphi,$$

$$W=\frac{2\sqrt{1-\nu^2}}{h_0}w,\quad k=\frac{D}{G_2h_0a^2},\quad Q=\frac{2a^4\sqrt{1-\nu^2}}{Dh_0}q_0 \tag{2.53}$$

利用这些无量纲量，非线性边值问题（2.36），（2.51）和（2.52）化为下面的无量纲形式：

$$L_1^2\Omega=-Q-\lambda^2L_2(W,\Phi),$$

$$L_1^2\Phi=-\frac{\lambda^2}{2}L_2(W,W) \tag{2.54a,b}$$

$$\text{当 } \xi=\pm 1 \text{ 时},\quad \Omega=0,\quad \frac{\partial^2\Omega}{\partial\xi^2}=0,\quad \frac{\partial^2\Phi}{\partial\eta^2}=0,\quad \frac{\partial^2\Phi}{\partial\xi\partial\eta}=0 \tag{2.55}$$

$$\text{当}\ \eta = \pm 1\ \text{时}, \quad \Omega = 0, \quad \frac{\partial^2 \Omega}{\partial \eta^2} = 0, \quad \frac{\partial^2 \Phi}{\partial \xi^2} = 0, \quad \frac{\partial^2 \Phi}{\partial \xi \partial \eta} = 0 \tag{2.56}$$

其中

$$W = kL_1\Omega - \Omega \tag{2.57}$$

$$L_1(\cdots) = \left(\frac{\partial^2}{\partial \xi^2} + \lambda^2 \frac{\partial^2}{\partial \eta^2}\right)(\cdots),$$

$$L_2(W,\Phi) = \frac{\partial^2 W}{\partial \xi^2}\frac{\partial^2 \Phi}{\partial \eta^2} + \frac{\partial^2 W}{\partial \eta^2}\frac{\partial^2 \Phi}{\partial \xi^2} - 2\frac{\partial^2 W}{\partial \xi \partial \eta}\frac{\partial^2 \Phi}{\partial \xi \partial \eta} \tag{2.58}$$

三、非线性边值问题的解

我们用摄动法求解无量纲非线性边值问题（2.54）～(2.56)。设夹层矩形板的无量纲中心挠度 W（0，0）记为 W_0，则可将 Q，Ω，Φ 和 W 展成如下形式的升幂摄动级数：

$$\begin{aligned} Q &= \alpha_1 W_0 + \alpha_3 W_0^2 + \cdots, \\ \Omega &= \Omega_1(\xi,\eta) W_0 + \Omega_3(\xi,\eta) W_0^3 + \cdots, \\ \Phi &= \Phi_2(\xi,\eta) W_0^2 + \Phi_4(\xi,\eta) W_0^4 + \cdots, \\ W &= W_1(\xi,\eta) W_0 + W_3(\xi,\eta) W_0^3 + \cdots \end{aligned} \tag{3.1}$$

依照定义给出

$$W_1(0,0) = 1, \quad W_{2i+1}(0,0) = 0, \quad i = 1,2,3,\cdots \tag{3.2}$$

将式（3.1）代入边值问题(2.54)～(2.56)，使 W_0 的同次幂相等，便得到以下一系列线性边值问题：

对于一次近似，

$$L_1^2 \Omega_1 = -\alpha_1 \tag{3.3}$$

$$\xi = \pm 1\ \text{时}, \quad \Omega_1 = 0, \quad \frac{\partial^2 \Omega_1}{\partial \xi^2} = 0;$$

$$\eta = \pm 1\ \text{时}, \quad \Omega_1 = 0, \quad \frac{\partial^2 \Omega_1}{\partial \eta^2} = 0 \tag{3.4}$$

对于二次近似，

$$L_1^2 \Phi_2 = -\frac{\lambda^2}{2} L_2(W_1, W_1) \tag{3.5}$$

$$\xi = \pm 1\ \text{时}, \quad \frac{\partial^2 \Phi_2}{\partial \eta^2} = 0, \quad \frac{\partial^2 \Phi_2}{\partial \xi \partial \eta} = 0;$$

$$\eta = \pm 1\ \text{时}, \quad \frac{\partial^2 \Phi_2}{\partial \xi^2} = 0, \quad \frac{\partial^2 \Phi_2}{\partial \xi \partial \eta} = 0 \tag{3.6}$$

对于三次近似，

$$L_1^2 \Omega_3 = -\alpha_3 - \lambda^2 L^2(W_1, \Phi_2) \tag{3.7}$$

$$\xi = \pm 1\ \text{时}, \quad \Omega_3 = 0, \quad \frac{\partial^2 \Omega_3}{\partial \xi^2} = 0;$$

$$\eta = \pm 1\ \text{时}, \quad \Omega_3 = 0, \quad \frac{\partial^2 \Omega_3}{\partial \eta^2} = 0 \tag{3.8}$$

类似地，我们还能得到更高阶近似的线性边值问题。

由式（2.57），我们还有

$$W_i = kL_1\Omega_i - \Omega_i, \quad i=1,3,\cdots \tag{3.9}$$

为了得到一阶近似的解，我们取满足边界条件（3.4）的双重 Fourier 级数：

$$\Omega_1 = \sum_{i=1}^{\infty}\sum_{j=1}^{\infty}\Omega_{ij}^{(1)}\cos\left(i-\frac{1}{2}\right)\pi\xi\cos\left(j-\frac{1}{2}\right)\pi\eta \tag{3.10}$$

将此式代入方程（3.3），并在方程两端同乘以 $\cos(i-1/2)\pi\xi\cos(j-1/2)\pi\eta$，且对 ξ 和 η 在各自区间内积分，得

$$\Omega_{ij}^{(1)} = \frac{(-1)^{i+j+1}4\alpha_1}{\pi^6\left(i-\frac{1}{2}\right)\left(j-\frac{1}{2}\right)\left[\left(i-\frac{1}{2}\right)^2+\lambda^2\left(j-\frac{1}{2}\right)^2\right]^2} \tag{3.11}$$

将式（3.10）代入式（3.9），可得

$$W_1 = \alpha_1\sum_{i=1}^{\infty}\sum_{j=1}^{\infty}W_{ij}^{(1)}\cos\left(i-\frac{1}{2}\right)\pi\xi\cos\left(j-\frac{1}{2}\right)\pi\eta \tag{3.12}$$

其中

$$W_{ij}^{(1)} = (-1)^{i+j}\frac{4\pi^2\left[\left(i-\frac{1}{2}\right)^2+\lambda^2\left(j-\frac{1}{2}\right)^2\right]k+4}{\pi^6\left(i-\frac{1}{2}\right)\left(j-\frac{1}{2}\right)\left[\left(i-\frac{1}{2}\right)^2+\lambda^2\left(j-\frac{1}{2}\right)^2\right]^2} \tag{3.13}$$

应用式（3.12）和式（3.2），便有

$$\alpha_1 = \beta^{-1} \tag{3.14}$$

其中

$$\beta = \sum_{i=1}^{\infty}\sum_{j=1}^{\infty}W_{ij}^{(1)} \tag{3.15}$$

于是，式（3.12）可写为如下形式：

$$W_1 = \beta^{-1}\sum_{i=1}^{\infty}\sum_{j=1}^{\infty}W_{ij}^{(1)}\cos\left(i-\frac{1}{2}\right)\pi\xi\cos\left(j-\frac{1}{2}\right)\pi\eta \tag{3.16}$$

对于二次阶近似，假定方程（3.5）的解为

$$\Phi_2 = \sum_{m=1}^{\infty}\sum_{n=1}^{\infty}\Phi_{mn}^{(2)}X_m(\xi)Y_n(\eta) \tag{3.17}$$

它满足边界条件（3.6），而且 $X_m(\xi)$ 和 $Y_n(\eta)$ 是由下式给定的梁本征函数：

$$X_m(\xi) = \frac{\mathrm{ch}\lambda_m\xi}{\mathrm{ch}\lambda_m} - \frac{\cos\lambda_m\xi}{\cos\lambda_m},$$

$$Y_n(\eta) = \frac{\mathrm{ch}\lambda_n\eta}{\mathrm{ch}\lambda_n} - \frac{\cos\lambda_n\eta}{\cos\lambda_n}, \quad m,n=1,2,\cdots \tag{3.18}$$

其中 λ_m 为下述超越方程的根：

$$\mathrm{th}\lambda_m + \mathrm{tg}\lambda_m = 0, \quad m=1,2,\cdots \tag{3.19}$$

且给在表 1 中。

表 1　λ_m 的值

m	1	2	3	4	$m>4$
λ_m	2.3650	5.4978	8.6394	11.7810	π（$m-0.25$）

函数 $X_m(\xi)$ 和 $Y_n(\eta)$ 满足下列正交关系

$$\int_{-1}^{1} X_m X_n \mathrm{d}\xi = \begin{cases} 0, & m \neq n \\ 2, & m = n \end{cases}$$

$$\int_{-1}^{1} Y_m Y_n \mathrm{d}\eta = \begin{cases} 0, & m \neq n \\ 2, & m = n \end{cases} \tag{3.20}$$

将式 (3.16) 和 (3.17) 代入方程 (3.5)，得

$$\begin{aligned} &\sum_{m=1}^{\infty}\sum_{n=1}^{\infty}\left[(\lambda_m^4 + \lambda^4\lambda_n^4)X_m Y_n + 2\lambda^2 \frac{\mathrm{d}^2 X_m}{\mathrm{d}\xi^2}\frac{\mathrm{d}^2 Y_n}{\mathrm{d}\eta^2}\right]\Phi_{mn}^{(1)} \\ &= \pi^4\beta^{-2}\lambda^2 \sum_{i=1}^{\infty}\sum_{j=1}^{\infty}\sum_{m=1}^{\infty}\sum_{n=1}^{\infty}\left[\left(i-\frac{1}{2}\right)\left(j-\frac{1}{2}\right)\left(m-\frac{1}{2}\right)\left(n-\frac{1}{2}\right)\right. \\ &\quad \cdot \sin\left(i-\frac{1}{2}\right)\pi\xi\sin\left(m-\frac{1}{2}\right)\pi\xi\sin\left(j-\frac{1}{2}\right)\pi\eta\sin\left(n-\frac{1}{2}\right)\pi\eta \\ &\quad - \left(i-\frac{1}{2}\right)^2\left(n-\frac{1}{2}\right)^2\cos\left(i-\frac{1}{2}\right)\pi\xi\cos\left(m-\frac{1}{2}\right)\pi\xi \\ &\quad \left.\cdot\cos\left(j-\frac{1}{2}\right)\pi\eta\cos\left(n-\frac{1}{2}\right)\pi\eta\right]W_{ij}^{(1)}W_{mn}^{(1)} \end{aligned} \tag{3.21}$$

将此方程两端乘以 $X_r(\xi)Y_s(\eta)$，对 ξ 和 η 在各自区间内积分，并且利用式 (3.20)，我们便导出下面一组线性代数方程：

$$\begin{aligned} &(\lambda_r^4 + \lambda^4\lambda_s^4)\Phi_{rs}^{(2)} + 2\lambda^2\sum_{m=1}^{\infty}\sum_{n=1}^{\infty}\lambda_m^2\lambda_n^2 K^{rm}H^{sn}\Phi_{mn}^{(2)} \\ &= \frac{\pi^4\lambda^2}{4\beta^2}\sum_{i=1}^{\infty}\sum_{j=1}^{\infty}\sum_{m=1}^{\infty}\sum_{n=1}^{\infty}\left[\left(i-\frac{1}{2}\right)\left(j-\frac{1}{2}\right)\left(m-\frac{1}{2}\right)\left(n-\frac{1}{2}\right)M_4^{rim}N_4^{sjn}\right. \\ &\quad \left.-\left(i-\frac{1}{2}\right)^2\left(n-\frac{1}{2}\right)M_1^{rim}N_1^{sjn}\right]W_{ij}^{(1)}W_{mn}^{(1)}, \quad r,s = 1,2,3,\cdots \end{aligned} \tag{3.22}$$

这里

$$K^{rm} = \frac{1}{2\lambda_m^2}\int_{-1}^{1} X_r \frac{\mathrm{d}^2 X_m}{\mathrm{d}\xi^2}\mathrm{d}\xi = \begin{cases} -\dfrac{4\lambda_r}{\lambda_r^4 - \lambda_4^4}(\lambda_r \mathrm{tg}\lambda_r - \lambda_m \mathrm{tg}\lambda_m), & r \neq m \\ -\mathrm{tg}^2\lambda_r - \dfrac{\mathrm{tg}\lambda_r}{\lambda_r}, & r = m \end{cases}$$

$$H^{sn} = \frac{1}{2\lambda_n^2}\int_{-1}^{1} Y_s \frac{\mathrm{d}^2 Y_n}{\mathrm{d}\eta^2}\mathrm{d}\eta = \begin{cases} -\dfrac{4\lambda_s^2}{\lambda_s^4 - \lambda_4^4}(\lambda_s \mathrm{tg}\lambda_s - \lambda_n \mathrm{tg}\lambda_n), & s \neq n \\ -\mathrm{tg}^2\lambda_s - \dfrac{\mathrm{tg}\lambda_s}{\lambda_s}, & s = n \end{cases}$$

$$M^{mri} = \int_{-1}^{1} X_m \cos\left(r-\frac{1}{2}\right)\pi\xi\cos\left(i-\frac{1}{2}\right)\pi\xi\mathrm{d}\xi = -T_{m,r+i-1} - T_{m,r-i},$$

$$N^{nsj} = \int_{-1}^{1} Y_n \cos\left(s-\frac{1}{2}\right)\pi\eta\cos\left(j-\frac{1}{2}\right)\pi\eta\mathrm{d}\eta = -T_{n,s+j-1} - T_{n,s-j},$$

$$M^{rim} = \int_{-1}^{1} X_r \sin\left(i-\frac{1}{2}\right)\pi\xi\sin\left(m-\frac{1}{2}\right)\pi\xi\mathrm{d}\xi = T_{r,i+m-1} - T_{r,i-m},$$

$$N^{sjn} = \int_{-1}^{1} Y_s \sin\left(j-\frac{1}{2}\right)\pi\eta\sin\left(n-\frac{1}{2}\right)\pi\eta\mathrm{d}\eta = -T_{s,j+n-1} - T_{s,j-n},$$

$$T_{m,n}=(-1)^n\frac{2\lambda_m^3\operatorname{tg}\lambda_m}{\lambda_m^4-n^4\pi^4}\tag{3.23}$$

联立求解这些代数方程，得到 $\Phi_{mn}^{(2)}$，由此 Φ_2 得以确定。

对于三阶近似，我们取双重 Fourier 级数：

$$\Omega_3=\sum_{m=1}^{\infty}\sum_{n=1}^{\infty}\Omega_{mn}^{(3)}\cos\left(m-\frac{1}{2}\right)\pi\xi\cos\left(n-\frac{1}{2}\right)\pi\eta\tag{3.24}$$

它满足边界条件（3.8）。现在，我们将此式连同式（3.16）和式（3.17）一起代入方程（3.7），便得

$$\begin{aligned}&\pi^4\sum_{m=1}^{\infty}\sum_{n=1}^{\infty}\left[\left(m-\frac{1}{2}\right)^2+\lambda^2\left(n-\frac{1}{2}\right)^2\right]^2\Omega_{mn}^{(3)}\cos\left(m-\frac{1}{2}\right)\pi\xi\cos\left(n-\frac{1}{2}\right)\pi\eta\\&=-\alpha_3+\frac{\pi^2\lambda^2}{\beta}\sum_{i=1}^{\infty}\sum_{j=1}^{\infty}\sum_{m=1}^{\infty}\sum_{n=1}^{\infty}\left[\left(i-\frac{1}{2}\right)^2X_m\frac{\mathrm{d}^2Y_n}{\mathrm{d}\eta^2}\cos\left(i-\frac{1}{2}\right)\pi\xi\right.\\&\quad\cdot\cos\left(j-\frac{1}{2}\right)\pi\eta+\left(j-\frac{1}{2}\right)^2Y_n\frac{\mathrm{d}^2X_m}{\mathrm{d}\xi^2}\cos\left(i-\frac{1}{2}\right)\pi\xi\\&\quad\cdot\cos\left(j-\frac{1}{2}\right)\pi\eta+2\left(i-\frac{1}{2}\right)\left(j-\frac{1}{2}\right)\frac{\mathrm{d}X_m}{\mathrm{d}\xi}\frac{\mathrm{d}Y_n}{\mathrm{d}\eta}\sin\left(i-\frac{1}{2}\right)\pi\xi\\&\quad\left.\cdot\sin\left(j-\frac{1}{2}\right)\pi\eta\right]W_{ij}^{(1)}\Phi_{mn}^{(2)}\end{aligned}\tag{3.25}$$

将此方程两端同乘以 $\cos(r-1/2)\pi\xi\cos(s-1/2)\pi\eta$，并对 ξ 和 η 在各自区间内积分，即得

$$\begin{aligned}\Omega_{rs}^{(3)}&=(-1)^{r+s+1}\frac{4\alpha_3}{\pi^6\left(r-\frac{1}{2}\right)\left(s-\frac{1}{2}\right)\left[\left(r-\frac{1}{2}\right)^2+\lambda^2\left(s-\frac{1}{2}\right)^2\right]^2}\\&\quad+\frac{\lambda^2}{\pi^2\left[\left(r-\frac{1}{2}\right)^2+\lambda^2\left(s-\frac{1}{2}\right)^2\right]^2\beta}\\&\quad\cdot\sum_{i=1}^{\infty}\sum_{j=1}^{\infty}\sum_{m=1}^{\infty}\sum_{n=1}^{\infty}\left[\left(i-\frac{1}{2}\right)^2M_1^{mri}N_2^{nsj}+\left(j-\frac{1}{2}\right)^2M_2^{mri}N_1^{nsj}\right.\\&\quad\left.+2\left(i-\frac{1}{2}\right)\left(j-\frac{1}{2}\right)M_3^{mir}N_3^{njs}\right]W_{ij}^{(1)}\Phi_{mn}^{(2)}\end{aligned}\tag{3.26}$$

其中

$$\begin{aligned}M_2^{mri}&=\int_{-1}^{1}\frac{\mathrm{d}^2X_m}{\mathrm{d}\xi^2}\cos\left(r-\frac{1}{2}\right)\pi\xi\cos\left(i-\frac{1}{2}\right)\pi\xi\mathrm{d}\xi\\&=\pi^2[(r+i-1)^2T_{m,r+i-1}+(r-i)^2T_{m,r-i}],\\N_2^{nsj}&=\int_{-1}^{1}\frac{\mathrm{d}^2Y_n}{\mathrm{d}\eta^2}\cos\left(s-\frac{1}{2}\right)\pi\eta\cos\left(j-\frac{1}{2}\right)\pi\eta\mathrm{d}\eta\\&=\pi^2[(s+j-1)^2T_{n,s+j-1}+(s-j)^2T_{n,s-j}],\\M_3^{mir}&=\int_{-1}^{1}\frac{\mathrm{d}X_m}{\mathrm{d}\xi}\sin\left(i-\frac{1}{2}\right)\pi\xi\cos\left(r-\frac{1}{2}\right)\pi\xi\mathrm{d}\xi\\&=\pi[(r+i-1)T_{m,r+i-1}-(r-i)T_{m,r-i}],\end{aligned}$$

$$N_3^{njs} = \int_{-1}^{1} \frac{dY_n}{d\eta} \sin\left(j - \frac{1}{2}\right)\pi\eta\cos\left(s - \frac{1}{2}\right)\pi\eta d\eta$$
$$= \pi[(s+j-1)T_{n,s+j-1} - (s-j)T_{n,s-j}], \tag{3.27}$$

将式（3.24）代入式（3.9），便得

$$W_3 = \sum_{r=1}^{\infty}\sum_{s=1}^{\infty}\left\{\alpha_3 W_{rs}^{(1)} - \frac{\pi^4\lambda^2}{4\beta}\sum_{i=1}^{\infty}\sum_{j=1}^{\infty}\sum_{m=1}^{\infty}\sum_{n=1}^{\infty}(-1)^{r+s}\left(r-\frac{1}{2}\right)\left(j-\frac{1}{2}\right)\right.$$
$$\cdot\left[\left(i-\frac{1}{2}\right)^2 M_1^{mri}N_2^{nsj} + \left(j-\frac{1}{2}\right)^2 M_2^{mri}N_1^{nsj} + 2\left(i-\frac{1}{2}\right)\left(j-\frac{1}{2}\right)\right.$$
$$\left.\left.\cdot M_3^{mir}N_3^{njs}\right]W_{rs}^{(1)}W_{ij}^{(1)}\Phi_{mn}^{(2)}\right\}\cos\left(r-\frac{1}{2}\right)\pi\xi\cos\left(s-\frac{1}{2}\right)\pi\eta \tag{3.28}$$

应用式（3.2），由上式可得

$$\alpha_3 = \frac{\pi^4\lambda^2}{4\beta^2}\sum_{r=1}^{\infty}\sum_{s=1}^{\infty}\sum_{i=1}^{\infty}\sum_{j=1}^{\infty}\sum_{m=1}^{\infty}\sum_{n=1}^{\infty}(-1)^{r+s}$$
$$\cdot\left(r-\frac{1}{2}\right)\left(s-\frac{1}{2}\right)\left[\left(i-\frac{1}{2}\right)^2 M_1^{mri}N_2^{nsj} + \left(j-\frac{1}{2}\right)^2 M_2^{mri}N_1^{nsj}\right.$$
$$\left. + 2\left(i-\frac{1}{2}\right)\left(j-\frac{1}{2}\right)M_3^{mir}N_3^{njs}\right]W_{rs}^{(1)}W_{ij}^{(1)}\Phi_{mn}^{(1)} \tag{3.29}$$

四、结果与讨论

由上一节的解（3.1），我们得到夹层矩形板的载荷与中心挠度的特征关系式为

$$Q = \alpha_1 W_0 + \alpha_3 W_0^3 \tag{4.1}$$

应用这一公式，我们得到了下面一些有用的数值结果。对边长比 λ 和特征参数 k 的不同值，均布载荷 Q 对中心挠度 W_0 的曲线给在图 2(a)-(d) 中。由图看到，对于相同的 Q 值，具有较大 k 值或较小 λ 值的夹层矩形板将产生较大的中心挠度。

特别地，在 $k=0$，$t=\dfrac{h}{2}$ 和 $h_0=\dfrac{h}{\sqrt{3}}$ 的情况下，上述夹层矩形板的边值问题就转化为厚度为 h 的单层矩形板的边值问题。因此，前面的公式也给出了均布载荷作用下简支

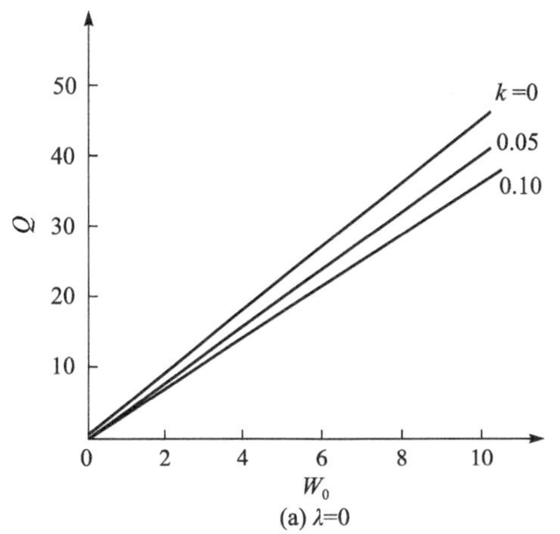

(a) $\lambda=0$

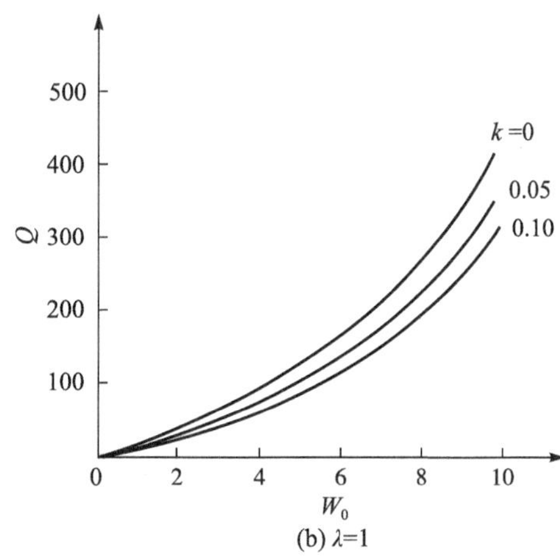

(b) $\lambda=1$

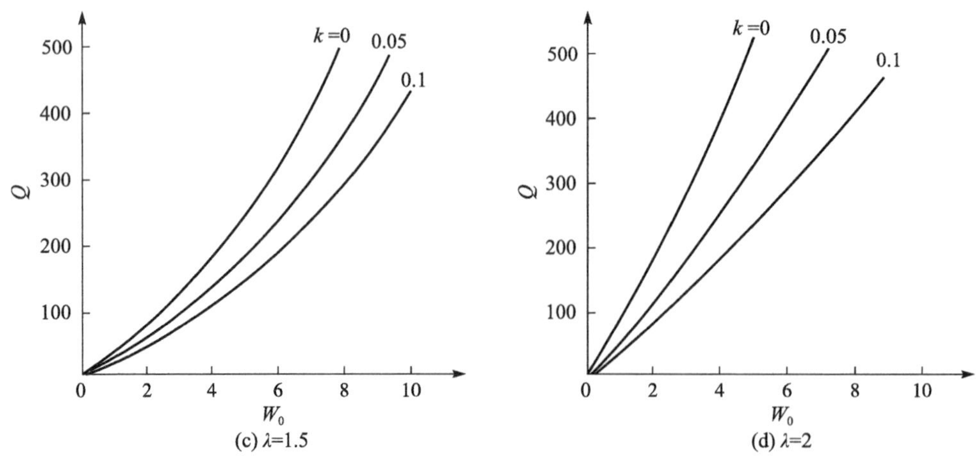

图 2　夹层矩形板各种 k 和 λ 值下的载荷—挠度曲线

矩形板的大挠度问题的解。为了说明本文解的正确性，我们在图 3 中给出了简支正方形板（$\lambda=1$）情况下的本文结果以及文献［13］和［14］的理论值与实验值。由图看到，本文结果与实验值十分吻合。由此可得出结论：本文公式的精确性是十分令人满意的。

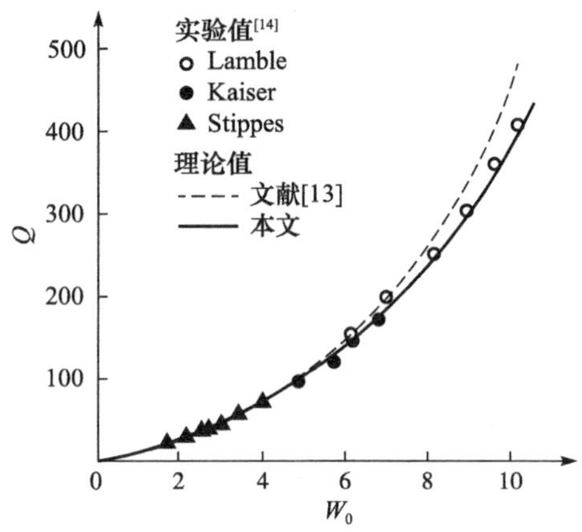

图 3　简支正方形板的理论和实验结果之间的比较

参 考 文 献

［1］　Reissner E. Finite deflections of sandwich plates. J. Aeron. Sci.，1948，15（7）：435；1950，17（2）：125.

［2］　刘人怀．夹层圆板的非线性弯曲．应用数学和力学，1981，2（2）：173.

［3］　刘人怀．在边缘力矩作用下夹层圆板的非线性轴对称弯曲问题．中国科学技术大学学报，1980，10（2）：56.

［4］　刘人怀，施云方．夹层圆板大挠度问题的精确解．应用数学和力学，1982，3（1）：11.

［5］　Liu Ren-huai. Nonlinear bending of circular sandwich plates under the action of axisymmetric uniformly distributed line loads. Progress in Applied Mechanics (edited by Yeh Kai-yuan), Martinus Nijhoff Publishers, Dordrecht, 1987: 293.

[6] 刘人怀，朱高秋. 夹层圆板大挠度问题的进一步研究 . 应用数学和力学，1989，10 (12): 1041.

[7] 刘人怀，吴建成. 夹层矩形板的非线性振动. 中国科学，A 辑，1991，(10): 1075.

[8] Kan H P, Huang J C. Large deflection of rectangular sandwich plates. AIAA Journal, 1967, 5 (9): 1706.

[9] Alwan A M. Bending of sandwich plates with large deflections. Journal of the Engineering Mechanics Division, Proceedings of the American Society of Civil Engineers, 1967, 93 (3): 83.

[10] Kamiya N. Governing equations for large deflections of sandwich plates. AIAA Journal, 1967, 5 (9): 1706.

[11] 中国科学院北京力学研究所固体力学研究室板壳组. 夹层板壳的弯曲、稳定和振动. 北京：科学出版社，1977.

[12] 胡海昌. 各向同性夹层板反对称小挠度的若干问题. 力学学报，1963，6 (1): 53.

[13] Sundara Raja Iyengar K T, Naqvi M M. Large deflections of rectangular plates. Int. J. Non-Linear Mech., 1966, 1 (2): 109.

[14] Brown J C, Harvey J M. Large deflections of rectangular plates subjected to uniform lateral pressure and compressive edge loading. J. Mech. Eng. Sci., 1969, 11 (3): 305.

On the non-linear buckling of circular shallow spherical sandwich shells under the action of uniform edge moments①

中文摘要

本文所研究的夹层扁球壳是由两层薄的各向同性表层和一层厚的各向同性软夹心所组成。使用变分法导出这种夹层扁球壳在边缘均布力矩作用下的非线性轴对称弯曲理论的基本方程和边界条件。使用修正迭代法求解这一边值问题，获得了相当精确的临界屈曲载荷解析解。本问题尚无人研究过。

1. Introduction

In recent years, with the essential advantages of light weight and high rigidity, sandwich plates and shells have been used as an important pattern of structural elements in aeronautical, astronautical, and naval engineering. However, nonlinear problems for sandwich plates and shells are only investigated by a few because of the difficulties of nonlinear mathematical problems. Reissner[1], Yu[2], Alwan[3], Kan and Huang[4], Kamiya[5], Liu Renhuai[6-13], Rao and Valsarajan[14,15], Ng and Das[16] and others have made some investigations in this field.

This paper is concerned with a circular, planform, shallow, spherical, sandwich shell consisting of two thin isotropic faces and a thicker, soft, isotropic core. Fundamental equations and boundary conditions of nonlinear axisymmetric bending theory for the shell under the action of uniform edge moments have been first derived by means of calculus of variations. We have solved the nonlinear boundary value problem for the shell by using Yeh Kay-yuan and Liu Renhuai's modified iteration method[17-19] and obtained a more accurate analytic solution. To the author's knowledge, there is no analysis of the study of this problem so far. Results obtained in this paper will be valuable in engineering design.

2. Fundamental Equations and Boundary Conditions

Consider a circular shallow spherical sandwich shell under the action of uniform edge moment M, as shown in Fig. 1. Here r is the radial coordinate, θ is the circumferential coordinate, z is the distance from any point in the shell to the middle surface of the shell, t is the thickness of the face, h_0 is the distance between middle surfaces of the upper and lower faces, R is the radius of curvature, a is the radius and subscripts

① Reprinted from *International Journal of Non-Linear Mechanics*, 1995, 30 (1): 33-43. Authors: Liu Renhuai and Cheng Zhenqiang.

1, 2 and 3 refer to the upper face, core, and lower face respectively.

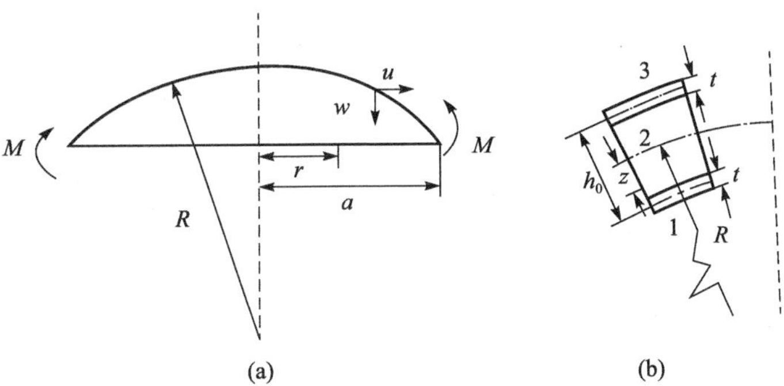

Fig. 1 Geometry and the coordinate of a circular shallow, spherical, sandwich shell

We derive the fundamental equations and boundary conditions of the shallow, spherical, sandwich shell on the basis of the following assumptions which are similar to Reissner's assumptions[1] for the large deflection theory of a sandwich plate:

(a) The material is elastic and follows Hooke's law.

(b) The core in the lateral direction is incompressible.

(c) The core is not subjected to loads in the middle surface direction.

(d) The faces are treated as membranes.

(e) The line normal to the middle surface of the core remains straight during bending.

Assuming that u_i, v_i and w_i $(i=1,2,3)$ are the radial, circumferential and axial displacements of the upper face, core and lower face, respectively, and u and w are the radial displacement and the deflection of the middle surface of the shell, respectively, on the above assumptions it can be concluded from the axisymmetrical condition that displacements at a point of the shallow, spherical, sandwich shell may be written as follows:

the upper face:

$$u_1 = u + \frac{1}{2}h_0\psi, \quad v_1 = 0, \quad w_1 = w; \tag{2.1}$$

the lower face:

$$u_3 = u - \frac{1}{2}h_0\psi, \quad v_3 = 0, \quad w_3 = w; \tag{2.2}$$

the core:

$$u_2 = u + z\psi, \quad v_2 = 0, \quad w_2 = w; \tag{2.3}$$

where ψ represents rotation of the line normal to the undeformed middle surface.

Assuming that ε_{ri}, $\varepsilon_{\theta i}$, ε_{zi}, $\gamma_{r\theta i}$, $\gamma_{\theta zi}$ and γ_{rzi} $(i=1,2,3)$ are strains at a point in the upper face, core, and lower face, respectively, and substituting formulas (2.1)-(2.3) into the following geometrical equations:

$$
\begin{aligned}
\varepsilon_{ri} &= \frac{\mathrm{d}u_i}{\mathrm{d}r} + \frac{\mathrm{d}w_i}{\mathrm{d}r}\left(\frac{r}{R} + \frac{1}{2}\frac{\mathrm{d}w_i}{\mathrm{d}r}\right) \\
\varepsilon_{\theta i} &= \frac{u_i}{r} \\
\varepsilon_{zi} &= \gamma_{r\theta i} = \gamma_{\theta z i} = \gamma_{rzi} = 0, \quad i = 1,3
\end{aligned} \tag{2.4}
$$

$$
\begin{aligned}
\gamma_{rz2} &= \frac{\partial u_2}{\partial z} + \frac{\mathrm{d}w_2}{\mathrm{d}r} \\
\varepsilon_{r2} &= \varepsilon_{\theta 2} = \varepsilon_{z2} = \gamma_{r\theta 2} = \gamma_{\theta z2} = 0,
\end{aligned} \tag{2.5}
$$

we obtain for:

the upper face:

$$
\begin{aligned}
\varepsilon_{r1} &= \frac{\mathrm{d}u}{\mathrm{d}r} + \frac{h_0}{2}\frac{\mathrm{d}\psi}{\mathrm{d}r} + \frac{\mathrm{d}w}{\mathrm{d}r}\left(\frac{r}{R} + \frac{1}{2}\frac{\mathrm{d}w}{\mathrm{d}r}\right), \\
\varepsilon_{\theta 1} &= \frac{u}{r} + \frac{h_0}{2r}\psi;
\end{aligned} \tag{2.6}
$$

the lower face:

$$
\begin{aligned}
\varepsilon_{r3} &= \frac{\mathrm{d}u}{\mathrm{d}r} - \frac{h_0}{2}\frac{\mathrm{d}\psi}{\mathrm{d}r} + \frac{\mathrm{d}w}{\mathrm{d}r}\left(\frac{r}{R} + \frac{1}{2}\frac{\mathrm{d}w}{\mathrm{d}r}\right), \\
\varepsilon_{\theta 3} &= \frac{u}{r} - \frac{h_0}{2r}\psi;
\end{aligned} \tag{2.7}
$$

the core:

$$
\gamma_{rz2} = \psi + \frac{\mathrm{d}w}{\mathrm{d}r}. \tag{2.8}
$$

Assuming that $\sigma_{ri}, \sigma_{\theta i}, \sigma_{zi}, \tau_{r\theta i}, \tau_{\theta zi}$ and τ_{rzi} ($i=1,2,3$) are stresses at a point in the upper face, core and lower face, respectively, E and ν are Young's modulus and Poisson's ratio of the face, respectively, G_2 is the shear modulus of the core, and substituting formulas (2.6)-(2.8) into the following Hooke's law:

$$
\begin{aligned}
\sigma_{ri} &= \frac{E}{1-\nu^2}(\varepsilon_{ri} + \nu\varepsilon_{\theta i}) \\
\sigma_{\theta i} &= \frac{E}{1-\nu^2}(\varepsilon_{\theta i} + \nu\varepsilon_{ri}) \\
\sigma_{zi} &= \tau_{r\theta i} = \tau_{\theta zi} = \tau_{rzi} = 0, \quad i = 1,3
\end{aligned} \tag{2.9}
$$

$$
\begin{aligned}
\tau_{rz2} &= G_2\gamma_{rz2} \\
\sigma_{r2} &= \sigma_{\theta 2} = \sigma_{z2} = \tau_{r\theta 2} = \tau_{\theta z2} = 0,
\end{aligned} \tag{2.10}
$$

we obtain for:

the upper face:

$$
\begin{aligned}
\sigma_{r1} &= \sigma_{r0} + \frac{Eh_0}{2(1-\nu^2)}\left(\frac{\mathrm{d}\psi}{\mathrm{d}r} + \frac{\nu}{r}\psi\right) \\
\sigma_{\theta 1} &= \sigma_{\theta 0} + \frac{Eh_0}{2(1-\nu^2)}\left(\frac{\psi}{r} + \nu\frac{\mathrm{d}\psi}{\mathrm{d}r}\right);
\end{aligned} \tag{2.11}
$$

the lower face:

$$\sigma_{r3}=\sigma_{r0}-\frac{Eh_0}{2(1-\nu^2)}\left(\frac{\mathrm{d}\psi}{\mathrm{d}r}+\frac{\nu}{r}\psi\right)$$

$$\sigma_{\theta 3}=\sigma_{\theta 0}-\frac{Eh_0}{2(1-\nu^2)}\left(\frac{\psi}{r}+\nu\frac{\mathrm{d}\psi}{\mathrm{d}r}\right);\tag{2.12}$$

the core:

$$\tau_{rz2}=G_2\left(\psi+\frac{\mathrm{d}w}{\mathrm{d}r}\right)\tag{2.13}$$

where σ_{r0} and $\sigma_{\theta 0}$ are the radial and circumferential stresses of the middle surface of the shallow, spherical, sandwich shell, respectively,

$$\sigma_{r0}=\frac{E}{1-\nu^2}\left[\frac{\mathrm{d}u}{\mathrm{d}r}+\frac{\nu}{r}u+\frac{\mathrm{d}w}{\mathrm{d}r}\left(\frac{r}{R}+\frac{1}{2}\frac{\mathrm{d}w}{\mathrm{d}r}\right)\right]$$

$$\sigma_{\theta 0}=\frac{E}{1-\nu^2}\left[\frac{u}{r}+\nu\frac{\mathrm{d}u}{\mathrm{d}r}+\nu\frac{\mathrm{d}w}{\mathrm{d}r}\left(\frac{r}{R}+\frac{1}{2}\frac{\mathrm{d}w}{\mathrm{d}r}\right)\right].\tag{2.14}$$

The strain energy of an elastic body, denoted by U, is

$$U=\frac{1}{2}\iiint_V(\sigma_r\varepsilon_r+\sigma_\theta\varepsilon_\theta+\sigma_z\varepsilon_z+\tau_{r\theta}\gamma_{r\theta}+\tau_{\theta z}\gamma_{\theta z}+\tau_{rz}\gamma_{rz})r\mathrm{d}r\mathrm{d}\theta\mathrm{d}z.\tag{2.15}$$

Thus from this expression the strain energies of the faces and the core can be written as

$$U_i=\frac{1}{2E}\iiint_{V_i}[(\sigma_{ri}+\sigma_{\theta i})^2-2(1+\nu)\sigma_{ri}\sigma_{\theta i}]r\mathrm{d}r\mathrm{d}\theta\mathrm{d}z,i=1,3$$

$$U_2=\frac{1}{2G_2}\iiint_{V_2}\tau_{rz2}^2r\mathrm{d}r\mathrm{d}\theta\mathrm{d}z.\tag{2.16}$$

Substituting formulas (2.11)-(2.13) into these expressions and integrating them with respect to z, we find

$$\begin{aligned}U_1=&\frac{t}{2E}\iint_{s_1}[(\sigma_{r0}+\sigma_{\theta 0})^2-2(1+\nu)\sigma_{r0}\sigma_{\theta 0}]r\mathrm{d}r\mathrm{d}\theta\\&+\frac{th_0}{2}\iint_{s_1}\left(\sigma_{r0}\frac{\mathrm{d}\psi}{\mathrm{d}r}+\frac{1}{r}\sigma_{\theta 0}\psi\right)r\mathrm{d}r\mathrm{d}\theta\\&+\frac{D}{4}\iint_{s_1}\left[\left(\frac{\mathrm{d}\psi}{\mathrm{d}r}+\frac{\psi}{r}\right)^2-2(1-\nu)\frac{\psi}{r}\frac{\mathrm{d}\psi}{\mathrm{d}r}\right]r\mathrm{d}r\mathrm{d}\theta\\U_3=&\frac{t}{2E}\iint_{s_3}[(\sigma_{r0}+\sigma_{\theta 0})^2-2(1+\nu)\sigma_{r0}\sigma_{\theta 0}]r\mathrm{d}r\mathrm{d}\theta\\&-\frac{th_0}{2}\iint_{s_3}\left(\sigma_{r0}\frac{\mathrm{d}\psi}{\mathrm{d}r}+\frac{1}{r}\sigma_{\theta 0}\psi\right)r\mathrm{d}r\mathrm{d}\theta\\&+\frac{D}{4}\iint_{s_3}\left[\left(\frac{\mathrm{d}\psi}{\mathrm{d}r}+\frac{\psi}{r}\right)^2-2(1-\nu)\frac{\psi}{r}\frac{\mathrm{d}\psi}{\mathrm{d}r}\right]r\mathrm{d}r\mathrm{d}\theta\\U_2=&\frac{G_2h_0}{2}\iint_{s_2}\left(\psi+\frac{\mathrm{d}w}{\mathrm{d}r}\right)^2r\mathrm{d}r\mathrm{d}\theta,\end{aligned}\tag{2.17}$$

where D represents the flexural rigidity of the shallow, spherical, sandwich shell,

$$D=\frac{Eth_0^2}{2(1-\nu^2)}.\tag{2.18}$$

The work of the uniform edge moment M per unit length, due to the rotation ψ, is

$$V=\int_0^{2\pi} M\psi a\,\mathrm{d}\theta = 2\pi a M\psi. \tag{2.19}$$

Then the total potential energy of the shallow, spherical, sandwich shell is

$$U=U_1+U_2+U_3-V=\frac{t}{E}\iint_s\left[(\sigma_{r0}+\sigma_{\theta 0})^2-2(1+\nu)\sigma_{r0}\sigma_{\theta 0}\right]r\mathrm{d}r\mathrm{d}\theta$$

$$+\frac{D}{2}\iint_s\left[\left(\frac{\mathrm{d}\psi}{\mathrm{d}r}+\frac{\psi}{r}\right)^2-2(1-\nu)\frac{\psi}{r}\frac{\mathrm{d}\psi}{\mathrm{d}r}\right]r\mathrm{d}r\mathrm{d}\theta$$

$$+\frac{G_2h_0}{2}\iint_s\left(\psi+\frac{\mathrm{d}w}{\mathrm{d}r}\right)^2 r\mathrm{d}r\mathrm{d}\theta-2\pi a M\psi. \tag{2.20}$$

Applying the principle of potential energy, we have

$$\delta U=0. \tag{2.21}$$

Substituting expression (2.20) into (2.21) and integrating by parts, we obtain the equilibrium equations and boudary conditions for the large deflection theory of the shallow, spherical, sandwich shell under the action of the edge moment. In the case of a simply-supported edge, these equilibrium equations and boundary conditions can be written as

$$\sigma_{\theta 0}-\frac{\mathrm{d}}{\mathrm{d}r}(r\sigma_{r0})=0 \tag{2.22a}$$

$$2t\frac{\mathrm{d}}{\mathrm{d}r}\left[r\sigma_{r0}\left(\frac{\mathrm{d}w}{\mathrm{d}r}+\frac{r}{R}\right)\right]+G_2h_0\frac{\mathrm{d}}{\mathrm{d}r}\left[r\left(\frac{\mathrm{d}w}{\mathrm{d}r}+\psi\right)\right]=0 \tag{2.22b}$$

$$Dr\frac{\mathrm{d}}{\mathrm{d}r}\frac{1}{r}\frac{\mathrm{d}}{\mathrm{d}r}(r\psi)-G_2h_0r\left(\frac{\mathrm{d}w}{\mathrm{d}r}+\psi\right)=0 \tag{2.22c}$$

$$w=0,\quad \sigma_{r0}=0,\quad M_r=M\quad \text{at } r=a;$$

$$\psi=0,\quad \sigma_{r0}\ \text{finite}\quad \text{at } r=0 \tag{2.23}$$

in which M_r and Q_r are the radial moment and the radial shearing force of the shallow, spherical, sandwich shell respectively,

$$M_r=D\left(\frac{\mathrm{d}\psi}{\mathrm{d}r}+\nu\frac{\psi}{r}\right)$$

$$Q_r=G_2h_0\left(\frac{\mathrm{d}w}{\mathrm{d}r}+\psi\right). \tag{2.24}$$

Eliminating the displacement u from the two equations (2.14), the compatibility equation is given by

$$\sigma_{r0}-\frac{\mathrm{d}}{\mathrm{d}r}(r\sigma_{\theta 0})-E\frac{\mathrm{d}w}{\mathrm{d}r}\left(\frac{1}{2}\frac{\mathrm{d}w}{\mathrm{d}r}+\frac{r}{R}\right)=0. \tag{2.25}$$

Thus the problem of the large deflection of a shallow, spherical, sandwich shell having a simply-supported edge and loaded by the moment uniformly distributed along the edge is reduced to the integration of equations (2.22) and (2.25) in conjunction with boundary conditions (2.23). Now we begin the simplification of the nonlinear boundary value problems (2.22), (2.25) and (2.23). As usual, a stress function ϕ is defined by

$$\sigma_{r0}=\frac{1}{r}\frac{\mathrm{d}\phi}{\mathrm{d}r},\quad \sigma_{\theta0}=\frac{\mathrm{d}^2\phi}{\mathrm{d}r^2}. \tag{2.26}$$

Thus the stress function satisfies equation (2.22a) exactly and (2.22b) and (2.25) become

$$2t\frac{\mathrm{d}}{\mathrm{d}r}\left[\frac{\mathrm{d}\varphi}{\mathrm{d}r}\left(\frac{\mathrm{d}w}{\mathrm{d}r}+\frac{r}{R}\right)\right]+G_2h_0\frac{\mathrm{d}}{\mathrm{d}r}\left[r\left(\frac{\mathrm{d}w}{\mathrm{d}r}+\psi\right)\right]=0 \tag{2.27a}$$

$$r\frac{\mathrm{d}}{\mathrm{d}r}\frac{1}{r}\frac{\mathrm{d}}{\mathrm{d}r}\left(r\frac{\mathrm{d}\varphi}{\mathrm{d}r}\right)+E\frac{\mathrm{d}w}{\mathrm{d}r}\left(\frac{1}{2}\frac{\mathrm{d}w}{\mathrm{d}r}+\frac{r}{R}\right)=0. \tag{2.27b}$$

By integrating (2.27a) we obtain

$$\psi=-\left[\frac{\mathrm{d}w}{\mathrm{d}r}+\frac{2t}{G_2h_0r}\frac{\mathrm{d}\varphi}{\mathrm{d}r}\left(\frac{\mathrm{d}w}{\mathrm{d}r}+\frac{r}{R}\right)\right]. \tag{2.28}$$

Substituting this expression into (2.22c), we find

$$Dr\frac{\mathrm{d}}{\mathrm{d}r}\frac{1}{r}\frac{\mathrm{d}}{\mathrm{d}r}r\frac{\mathrm{d}w}{\mathrm{d}r}-2t\left(1-\frac{D}{G_2h_0}r\frac{\mathrm{d}}{\mathrm{d}r}\frac{1}{r}\frac{\mathrm{d}}{\mathrm{d}r}\right)\left[\frac{\mathrm{d}\varphi}{\mathrm{d}r}\left(\frac{\mathrm{d}w}{\mathrm{d}r}+\frac{r}{R}\right)\right]=0. \tag{2.29}$$

In this way the nonlinear boundary value problem of (2.22),(2.25) and (2.23) can be replaced by the simplified problem of (2.27b),(2.29) and (2.23) .

Let us introduce the non-dimensional parameters

$$\rho=\frac{r}{a},\quad W=2\sqrt{1-\nu^2}\frac{w}{h_0},\quad \phi=\frac{\mathrm{d}w}{\mathrm{d}\rho}+k_1\rho,\quad S=\frac{2ta}{D}r\sigma_{r0},$$

$$m=\frac{2a^2\sqrt{1-\nu^2}}{Dh_0}M,\quad k_1=2\sqrt{1-\nu^2}\frac{a^2}{Rh_0},\quad k_2=\frac{D}{G_2h_0a^2}. \tag{2.30}$$

Upon substitution, the nonlinear boudary value problem of (2.29), (2.27b) and (2.23) transform to

$$L(\rho\phi)=-(k_2L-1)(S\phi)$$
$$L(\rho S)=-\frac{1}{2}(\phi^2-k_1^2\rho^2) \tag{2.31}$$

$$W=0,\quad \frac{\mathrm{d}\phi}{\mathrm{d}\rho}+\nu\frac{\phi}{\rho}=-k_2\frac{\mathrm{d}}{\mathrm{d}\rho}\left(\frac{S\phi}{\rho}\right)-m+k_1(1+\nu),$$
$$S=0\quad \text{at } \rho=1, \tag{2.32a-c}$$
$$\phi=0,\quad S=0\quad \text{at } \rho=0, \tag{2.33a,b}$$

where L is a differential operator defined by

$$L(\cdots)=\rho\frac{\mathrm{d}}{\mathrm{d}\rho}\frac{1}{\rho}\frac{\mathrm{d}}{\mathrm{d}\rho}(\cdots). \tag{2.34}$$

3. Solution of Non-linear Boundary Value Problem

An approximate solution of the non-dimensional nonlinear boundary value problem of (2.31)-(2.33) is formulated by the modified iteration method. Thus, in the first approximation, the linear boundary value problem is given by

$$L(\rho\phi_1)=0$$

$$L(\rho S_1)=-\frac{1}{2}(\phi_1^2-k_1^2\rho^2) \tag{3.1a,b}$$

$$W_1=0,\quad \frac{d\phi_1}{d\rho}+\nu\frac{\phi_1}{\rho}=-m+k_1(1+\nu),$$

$$S_1=0\quad \text{at } \rho=1, \tag{3.2a-c}$$

$$\phi_1=0,\quad S_1=0\quad \text{at } \rho=0. \tag{3.3a,b}$$

Integrating (3.1a) and using boundary conditions (3.2a,b) and (3.3a), the solution of (3.1a) is obtained:

$$\phi_1=\left(k_1-\frac{m}{\lambda}\right)\rho \tag{3.4}$$

where

$$\lambda=1+\nu. \tag{3.5}$$

We introduce the non-dimensional center deflection W_0 as an iteration parameter

$$W_0=W|_{\rho=0} \tag{3.6}$$

Then, using (2.30) and (2.32a), we have

$$W_0=-\int_0^1(\phi-k_1\rho)\,d\rho. \tag{3.7}$$

Substituing the solution (3.4) into (3.7), we obtain

$$m=2\lambda W_0. \tag{3.8}$$

Substituting this relation into (3.4), we obtain the solution of the first approximation

$$\phi_1=(k_1-2W_0)\rho. \tag{3.9}$$

Using this solution, a solution of (3.1b) satisfying boundary conditions (3.2a) and (3.3b) can be written as

$$S_1=\frac{1}{4}W_0(k_1-W_0)(\rho^3-\rho). \tag{3.10}$$

For the second approximation, we have following linear boundary value problem:

$$L(\rho\phi_2)=-(k_2L-1)(S_1\phi_1)$$

$$L(\rho S_2)=-\frac{1}{2}(\phi_1^2-k_1^2\rho^2) \tag{3.11a,b}$$

$$W_2=0,\quad \frac{d\phi_2}{d\rho}+\nu\frac{\phi_2}{\rho}=-k_2\frac{d}{d\rho}\left(\frac{S_1\phi_1}{\rho}\right)-m+k_1(1+\nu),$$

$$S_2=0\quad \text{at } \rho=1, \tag{3.12a-c}$$

$$\phi_2=0,\quad S_2=0\quad \text{at } \rho=0. \tag{3.13a,b}$$

Applying solutions (3.9), (3.10) and boundary conditions (3.12a, b) and (3.13a), the solution of (3.11a) is given by

$$\phi_2=\frac{1}{8}W_0(k_1-W_0)(k_1-2W_0)\left[\frac{1}{12}\rho^5-2\left(k_2+\frac{1}{8}\right)\rho^3+2\left(k_2+\frac{2+\nu}{12\lambda}\right)\rho\right]$$
$$+\left(k_1-\frac{m}{\lambda}\right)\rho. \tag{3.14}$$

Introducing this solution in (3.7), we obtain the characteristic relation of the second approximation as

$$m = \alpha_1 W_0 + \alpha_2 W_0^2 + \alpha_3 W_0^3 \tag{3.15}$$

where

$$\alpha_1 = 2\lambda_1 + \beta_1 k_1^2, \quad \alpha_2 = -3\beta_1 k_1, \quad \alpha_3 = 2\beta_1, \quad \beta_1 = \frac{1}{8}\left(\lambda k_2 + \frac{17+5\nu}{72}\right). \tag{3.16}$$

Substituting (3.15) into (3.14), we obtain

$$\phi_2 = \frac{W_0}{16}(k_1 - W_0)(k_1 - 2W_0)\left[\frac{1}{6}\rho^5 - 4\left(k_2 + \frac{1}{8}\right)\rho^3 + 2\left(k_2 + \frac{7}{72}\right)\rho\right] + (k_1 - 2W_0)\rho. \tag{3.17}$$

By means of solution (3.17) and solving (3.11b), we obtain the following solution satisfying boundary conditions (3.12c) and (3.13b)

$$\begin{aligned} S_2 = &-\frac{1}{256}W_0^2(k_1 - W_0)^2(k_1 - 2W_0)^2\left[\frac{1}{8640}\rho^{11} - \frac{1}{120}\left(k_2 + \frac{1}{8}\right)\rho^9\right. \\ &+ \frac{1}{6}\left(k_2^2 + \frac{7}{24}k_2 + \frac{17}{864}\right)\rho^7 - \frac{1}{3}\left(k_2^3 + \frac{2}{9}k_2 + \frac{7}{576}\right)\rho^5 + \frac{1}{4}\left(k^2 + \frac{7}{72}\right)^2\rho^3 \\ &\left. - \frac{1}{3}\left(\frac{1}{4}k_2^2 + \frac{2}{45}k_2 + \frac{23}{11520}\right)\rho\right] - \frac{1}{32}W_0(k_1 - W_0)(k_1 - 2W_0)^2\left[\frac{1}{144}\rho^7\right. \\ &\left. - \frac{1}{3}\left(k_2 + \frac{1}{8}\right)\rho^5 + \frac{1}{2}\left(k^2 + \frac{7}{72}\right)\rho^3 - \frac{1}{6}\left(k_2 + \frac{1}{12}\right)\rho\right] \\ &+ \frac{1}{4}W_0(k_1 - W_0)(\rho^3 - \rho). \end{aligned} \tag{3.18}$$

For the third approximation, we have the following linear boundary value problem of the non-dimensional deflection W

$$L(\rho\phi_3) = -(k_2 L - 1)(S_2\phi_2) \tag{3.19}$$

$$W_3 = 0, \quad \frac{\mathrm{d}\phi_3}{\mathrm{d}\rho} + \nu\frac{\phi_3}{\rho} = -k_2\frac{\mathrm{d}}{\mathrm{d}\rho}\left(\frac{S_2\phi_2}{\rho}\right) - m + k_1(1+\nu) \quad \text{at } \rho = 1 \tag{3.20a,b}$$

$$\phi_3 = 0 \quad \text{at } \rho = 0. \tag{3.21}$$

By virtue of solutions (3.17) and (3.18), the solution of (3.19) satisfying boundary conditions (3.20b) and (3.21) is

$$\begin{aligned} \phi_3 = &-\frac{1}{6144}W_0^3(k_1 - W_0)^3(k_1 - 2W_0)^3\left[\frac{1}{9953280}\rho^{17} - \frac{1}{8064}\left(\frac{1}{3}k_2 + \frac{1}{80}\right)\rho^{15}\right. \\ &+ \frac{1}{2016}\left(\frac{67}{10}k_2^2 + k_2 + \frac{113}{5760}\right)\rho^{13} - \frac{1}{10}\left(k_2^3 + \frac{47}{160}k_2^2 + \frac{199}{8640}k_2 + \frac{29}{92160}\right)\rho^{11} \\ &+ \left(k_2^4 + \frac{89}{160}k_2^3 + \frac{1061}{11520}k_2^2 + \frac{289}{55296}k_2 + \frac{1141}{19906560}\right)\rho^9 \\ &- \frac{1}{2}\left(5k_2^4 + \frac{289}{144}k_2^3 + \frac{115}{432}k_2^2 + \frac{15977}{1244160}k_2 + \frac{647}{4976640}\right)\rho^7 \\ &+ \frac{1}{2}\left(5k_2^4 + \frac{251}{144}k_2^3 + \frac{10733}{51840}k_2^2 + \frac{769}{82944}k_2 + \frac{2957}{29859840}\right)\rho^5 \end{aligned}$$

$$-\frac{1}{4}\left(5k_2^4+\frac{289}{180}k_2^3+\frac{259}{1440}k_2^2+\frac{2461}{311040}k_2+\frac{161}{1658880}\right)\rho^3$$

$$+\frac{1}{4}\left(k_2^4+\frac{37}{120}k_2^3+\frac{15403+15418\nu}{453600\lambda}k_2^2+\frac{33437+33557\nu}{21772800\lambda}k_2+\frac{11617+11747\nu}{522547200\lambda}\right)\rho\Big]$$

$$-\frac{1}{256}W_0^2(k_1-W_0)^2(k_1-2W_0)^3\Big[\frac{1}{241920}\rho^{13}-\frac{1}{300}\left(\frac{1}{3}k_2+\frac{1}{64}\right)\rho^{11}$$

$$+\frac{1}{96}\left(\frac{29}{5}k_2^2+\frac{11}{12}k_2+\frac{191}{8640}\right)\rho^9-\frac{1}{6}\left(5k_2^3+\frac{43}{24}k_2^2+\frac{137}{864}k_2+\frac{37}{13824}\right)\rho^7$$

$$+\frac{1}{3}\left(5k_2^3+\frac{371}{288}k_2^2+\frac{35}{384}k_2+\frac{73}{55296}\right)\rho^5-\frac{1}{4}\left(\frac{13}{3}k_2^3+\frac{71}{72}k_2^2+\frac{559}{8640}k_2+\frac{209}{207360}\right)\rho^3$$

$$+\frac{1}{4}\left(k_2^3+\frac{269+239\nu}{1080\lambda}k_2^2+\frac{2561+1961\nu}{129600\lambda}k_2+\frac{3627+2197\nu}{7257600\lambda}\right)\rho\Big]$$

$$-\frac{1}{32}W_0(k_1-W_0)(k_1-2W_0)^3\Big[\frac{1}{11520}\rho^9-\frac{1}{72}\left(k_2+\frac{1}{16}\right)\rho^7$$

$$+\left(\frac{1}{3}k_2^2+\frac{1}{16}k_2+\frac{7}{3456}\right)\rho^5-\frac{1}{2}\left(k_2^2+\frac{5}{36}k_2+\frac{1}{288}\right)\rho^3+\frac{1}{6}\left(k_2^2+\frac{1}{8}k_2+\frac{13+17\nu}{5760\lambda}\right)\rho\Big]$$

$$+\frac{1}{16}W_0^2(k_1-W_0)^2(k_1-2W_0)\Big[\frac{1}{1920}\rho^9-\frac{1}{16}\left(k_2+\frac{1}{18}\right)\rho^7+\left(k_2^2+\frac{11}{48}k_2+\frac{25}{3456}\right)\rho^5$$

$$-\frac{1}{2}\left(3k_2^2+\frac{17}{36}k_2+\frac{7}{576}\right)\rho^3+\frac{1}{2}\left(k_2^2+\frac{5}{36}k_2+\frac{29+31\nu}{8640\lambda}\right)\rho\Big]$$

$$+\frac{1}{4}W_0(k_1-W_0)(k_1-2W_0)\Big[\frac{1}{24}\rho^5-\left(k_2+\frac{1}{8}\right)\rho^3$$

$$+\left(k_2+\frac{2+\nu}{12\lambda}\right)\rho\Big]+\left(k_1-\frac{m}{\lambda}\right)\rho. \tag{3.22}$$

Substituting solution (3.22) into (3.7), we obtain the characteristic relation of the third approximation

$$m=\sum_{n=1}^{9}\omega_n W_0^n \tag{3.23}$$

where

$$\omega_1=2\lambda_1-\beta_3k_1^4+\beta_1k_1^2$$

$$\omega_2=-\beta_4k_1^5+(\beta_2+7\beta_3)k_1^3-3\beta_1k_1$$

$$\omega_3=-\beta_5k_1^6+8\beta_4k_1^4-2(2\beta_2+9\beta_3)k_1^2+2\beta_1$$

$$\omega_4=9\beta_5k_1^5-25\beta_4k_1^3+5(\beta_2+4\beta_3)k_1$$

$$\omega_5=-33\beta_5k_1^4+38\beta_4k_1^2-2(\beta_2+4\beta_3)$$

$$\omega_6=63\beta_5k_1^3-28\beta_4k_1$$

$$\omega_7=-66\beta_5k_1^2+8\beta_4$$

$$\omega_8=36\beta_5k_1$$

$$\omega_9=-8\beta_5$$

$$\beta_2=\frac{1}{192}\left(\lambda k_2^2+\frac{7}{48}\lambda k_2+\frac{149+209\nu}{43200}\right)$$

$$\beta_3=\frac{1}{1152}\left(\lambda k_2^2+\frac{1}{8}\lambda k_2+\frac{17-103\nu}{28800}\right)$$

$$\beta_4=\frac{1}{4608}\left(\lambda k_2^3+\frac{634+409\nu}{1800}k_2^2+\frac{29+13\nu}{768}k_2+\frac{21971+6956\nu}{16934400}\right)$$

$$\beta_5=\frac{1}{184320}\left(\lambda k_2^4+\frac{917}{2800}\lambda k_2^3+\frac{94273+94903\nu}{2540160}k_2^2\right.$$
$$\left.+\frac{10085+10309\nu}{5419008}k_2+\frac{48581+51311\nu}{1463132160}\right). \tag{3.24}$$

If necessary, a higher-order approximation can be obtained in a similar manner.

The use of the relation (3.23) will now be illustrated by a numerical example. A shallow spherical sandwich shell has the parameters $k_2=0.05$ and $\nu=0.3$. In such a case, results are calculated and shown in Fig. 2, which is a set of characteristic curves for several values of the geometrical parameter k_1.

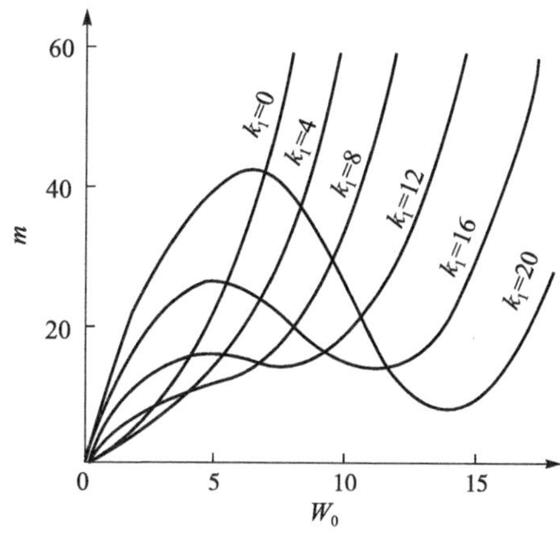

Fig. 2　Characteristic curves for several values of k_1 ($k_2=0.05$, $\nu=0.3$)

It is observed from Fig. 2 that when k_1 is smaller the curves increase monotonically, this means that no buckling takes place for the shell. When k_1 is larger the curves will become gyratory, and the shell may undergo buckling. In a similar manner, from the relation (3.15) of the second approximation, we can also obtain a similar conclusion to the above. Therefore, in order to find the critical buckling moment of the shell, we apply the following extremal condition:

$$\frac{\mathrm{d}m}{\mathrm{d}W_0}=0. \tag{3.25}$$

For the second approximation, (3.25) can be written as

$$3\alpha_3W_0^2+2\alpha_2W_0+\alpha_1=0. \tag{3.26}$$

Solving this quadratic equation, the non-dimensional critical center deflection W_0^* is given by

$$W_0^*=\frac{-\alpha_2\pm\sqrt{\alpha_2^2-3\alpha_1\alpha_3}}{3\alpha_3}. \tag{3.27}$$

Substituting expression (3.27) into the relation (3.15), we obtain the following formula of the non-dimensional critical buckling moment m^* as follows:

$$m^*=\alpha_1W_0^*+\alpha_2W_0^{*2}+\alpha_3W_0^{*3} \tag{3.28}$$

in which m^* with respect to the positive and negative signs of expression (3.27) are the non-dimensional lower and upper critical buckling moments, respectively.

For the third approximation, Eq. (3.25) can be written as

$$\sum_{n=1}^{9} n\omega_n W_0^{n-1} = 0; \tag{3.29}$$

this is an eighth-degree equation for W_0. Solving this equation, we can obtain the non-dimensional critical center deflection W_0^* using Newton's method. Here, for a given Poisson's ratio ν, a given geometrical parameter k_1 and a given physical parameter k_2, the initial value of W_0^* may be taken on the basis of expression (3.27). Substituting this value of W_0^* into the relation (3.23), we obtain the non-dimensional critical buckling moment m^* for the third approximation:

$$m^* = \sum_{n=1}^{9} \omega_n W_0^{*n} \tag{3.30}$$

Numerical results are obtained by taking relations (3.28) and (3.30) for $\nu=0.3$ and shown in Fig. 3. Each curve corresponds to a particular value of k_2, and it contains two branch curves. The upper and lower branch curves are called the curves for the upper and lower critical buckling moment, respectively. In practical engineering, it is sufficient only to consider the upper critical buckling moment. The more exact results for the third approximation are shown as the solid line. The results for the second approximation are shown as the dashed line. It is seen that a good agreement is found between the full line and the dashed line in a wide range of values of k_1. Thus, in general, the results of the second approximation can be applied with sufficient accuracy in most engineering designs.

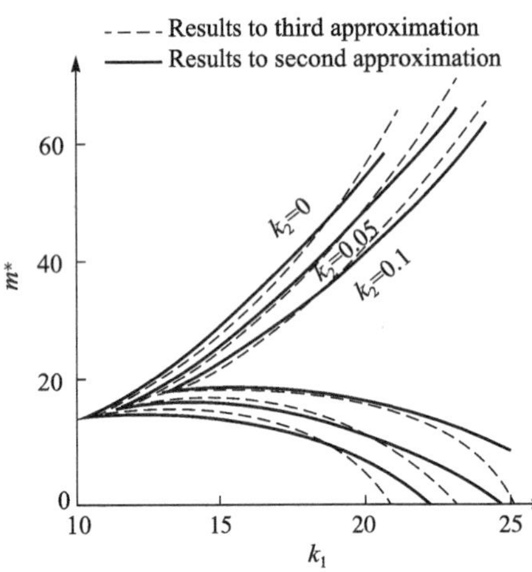

Fig. 3　Curves of the non-dimensional critical buckling moment ($\nu=0.3$)

References

[1] E. Reissner, Finite deflections of sandwich plates. *J. Aeronaut. Sci.* 15, 435 (1948); 17, 125 (1950).

[2] Y. Y. Yu, Nonlinear flexural vibrations of sandwich plates. *J. Acoust, Soc. Amer.* 34, 1176 (1962).

[3] A. M. Alwan, Bending of sandwich plates with large deflections. *Proc. ASCE J. Engng Mech. Div.* 93, 83 (1967).

[4] H. P. Kan and J. C. Huang, Large deflection of rectangular sandwich plates. *AIAA J.* 5, 1706 (1967).

[5] N. Kamiya, Governing equations for large deflections of sandwich plates. *AIAA J.* 14, 250 (1976).

[6] Liu Renhuai, Nonlinear axisymmetrical bending of circular sandwich plates under the action of uniform edge moment. *J. Chin. Univ. Sci. Technol.* 10, 56 (1980) (in Chinese).

[7] Liu Renhuai, Nonlinear bending of circular sandwich plates. *Appl. Math. Mech.* 2, 189 (1981).

[8] Liu Renhuai and Y. F. Shi, Exact solution for circular sandwich plate with large deflection. *Appl. Math. Mech.* 3, 11 (1982).

[9] Liu Renhuai, Nonlinear bending of circular sandwich plates under the action of axisymmetric uniformly distributed line loads. In *Progress in Applied Mechanics* (Edited by K. Y. Yeh), 293, Martinus Nijhoff, Dordrecht (1987).

[10] Liu Renhuai and G. Q. Zhu, Further study on large deflection of circular sandwich plates. *Appl. Math. Mech.* 10, 1099 (1989).

[11] Liu Renhuai and J. C. Wu, Nonlinear vibration of rectangular sandwich plates. *Sci. China A*, 35, 472 (1992).

[12] Liu Renhuai and J. F. Zhu, *Nonlinear Theory of Sandwich Shells*. Press of Mechanical Industry, Beijing (1993) (in Chinese).

[13] Liu Renhuai and Z. Q. Cheng, Nonlinear bending of simply supported rectangular sandwich plates. *Appl. Math. Mech.* 14, 217 (1993).

[14] N. R. Rao and K. V. Valsarajan, Large deflection analysis of clamped skew sandwich plates by parametric differentiation. *Comput. Struct.* 17, 599 (1983).

[15] N. R. Rao and K. V. Valsarajan, An integral equation solution for the finite deflection of clamped skew sandwich plates. *Comput. Struct.* 22, 665 (1986).

[16] S. F. Ng and B. Das, Finite deflection of skew sandwich plates on elastic foundations by the Galerkin method. *J. Struct. Mech.* 14, 355 (1986).

[17] K. Y. Yeh, Liu Renhuai, Q. Y. Ping and S. L. Li, Nonlinear stability of thin circular shallow spherical shell under actions of axisymmetric uniform distribution line loads. *Bull. Sci.* 2, 142 (1965) (in Chinese).

[18] K. Y. Yeh, Liu Renhuai, C. Z. Zhang and U. F. Xu, Nonlinear stability of thin circular shallow spherical shell under the action of uniform edge moment. *Bull. Sci.* 2, 145 (1965) (in Chinese).

[19] Liu Renhuai, Nonlinear stability of circular shallow spherical shell with a hole in the center under the action of uniform moment at the inner edge. *Bull. Sci.* 3, 253 (1965) (in Chinese).

Nonlinear vibration of shallow conical sandwich shells①

中文摘要

本文建立了夹层扁锥壳的非线性振动理论，使用迦辽金方法求解这种壳体的非线性自由振动问题，所获结果能直接用在工程设计中。此问题尚无人研究过。

1. Introduction

In recent years, the use of sandwich plates and shells has increased considerably because of a number of beneficial properties. They are found in various industries such as aircraft, missile, storage tank, and packing structure manufacture. Therefore, a considerable amount of research has been completed. In most cases, however, as the theoretical analysis is very difficult, investigators discussed only the problem of the linear theory for the sandwich plates and shells. Obviously, this cannot satisfy the needs of the rapid development of modern technology. So far as we know, there are only a few works about the nonlinear problem of these sandwich plates and shells. Reissner[1], Kan and Huang[2], Liu Renhuai[3-9] and Struk[10], et al., discussed nonlinear bending and buckling for sandwich plates and shells, respectively. Recently, Liu Renhuai and Wu Jian-cheng[11] studied nonlinear vibration for rectangular sandwich plates; however, nonlinear vibration for shallow conical sandwich shells has not yet been investigated. In this paper, a theory of nonlinear vibration of shallow conical sandwich shells is developed and the problem of nonlinear free vibration for the shells with two types of boundary conditions is solved by Galerkin's method. All results are expressed in curves which may be applied directly to engineering design.

2. Basic Equations

Let us consider a shallow conical sandwich shell of radius a, thickness h_0, face thickness t, core thickness h, slope angle α, radial coordinate r, circumferential coordinate θ and with the z-axis in the thickness direction (Fig. 2. 1).

In discussing deformation of the shallow conical sandwich shell, the following assumptions are adopted:

(1) The material follows Hooke's law.

(2) The core is transversely incompressible.

(3) The core in the direction of the surface of the shell is not subjected to loads.

① Reprinted from *International Journal of Non-Linear Mechanics*, 1995, 30 (2): 97-109. Authors: Liu Renhuai and Li Jun.

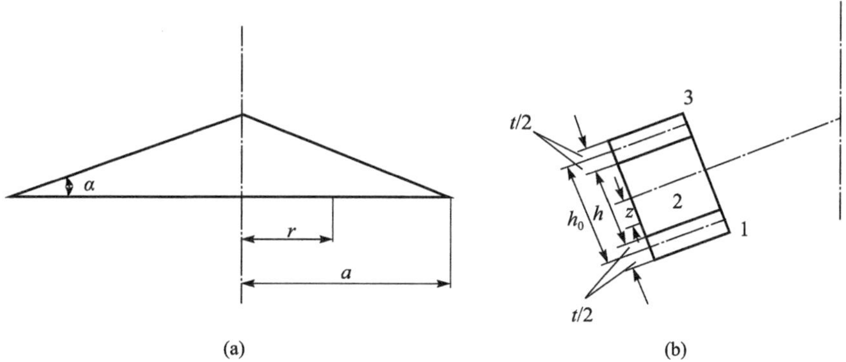

Fig. 2. 1 Geometry of shallow conical sandwich shell

(4) The face is in the membrance stress state.

(5) The normal line to the middle surface of the core remains straight during bending.

In the case when the shell is deformed axisymmetrically with respect to the axis of the shell, using the above assumptions, displacements at a point of the shell can be expressed in the form:

the upper face$\left[\dfrac{h}{2}\leqslant z\leqslant\dfrac{h}{2}+t\right]$:

$$u_1=u+\frac{h_0}{2}\psi,\quad v_1=0,\quad w_1=w \tag{2.1}$$

the core$\left[-\dfrac{h}{2}\leqslant z\leqslant\dfrac{h}{2}\right]$:

$$u_2=u+z\psi,\quad v_2=0,\quad w_2=w \tag{2.2}$$

the lower face$\left[-\left(\dfrac{h}{2}+t\right)\leqslant z\leqslant-\dfrac{h}{2}\right]$:

$$u_3=u-\frac{h_0}{2}\psi,\quad v_3=0,\quad w_3=w \tag{2.3}$$

in which u_i, v_i, w_i ($i=1, 2, 3$) are the radial, circumferential and vertical displacements of the upper face, the core and the lower face of the shell, respectively; u, w are the radial displacement and the deflection of the middle surface of the shell, respectively, and ψ is the rotation of the normal to the middle surface.

The strain-displacement relations of the shell are defined by

$$\begin{aligned}
\varepsilon_{ri}&=\frac{\partial u_i}{\partial r}+\frac{\partial w_i}{\partial r}\left(\frac{1}{2}\frac{\partial w_i}{\partial r}+\alpha\right),\\
\varepsilon_{\theta i}&=\frac{u_i}{r},\\
\varepsilon_{zi}&=\gamma_{r\theta i}=\gamma_{\theta zi}=\gamma_{rzi}=0\quad(i=1,3)
\end{aligned} \tag{2.4}$$

$$\begin{aligned}
\gamma_{rz2}&=\frac{\partial u_2}{\partial z}+\frac{\partial w_2}{\partial r},\\
\varepsilon_{r2}&=\varepsilon_{\theta 2}=\varepsilon_{z2}=\gamma_{r\theta 2}=\gamma_{\theta z2}=0
\end{aligned} \tag{2.5}$$

in which $\varepsilon_{ri}, \varepsilon_{\theta i}, \varepsilon_{zi}, \gamma_{r\theta i}, \gamma_{\theta zi}, \gamma_{rzi}$ ($i=1, 2, 3$) are strains of the upper face, the core and the lower face, respectively.

Introduction of Eqs. (2.1)-(2.3) into (2.4) and (2.5) yields:

the upper face,

$$\varepsilon_{r1} = \frac{\partial u}{\partial r} + \frac{h_0}{2}\frac{\partial \psi}{\partial r} + \frac{\partial w}{\partial r}\left(\frac{1}{2}\frac{\partial w}{\partial r} + \alpha\right),$$

$$\varepsilon_{\theta 1} = \frac{1}{r}\left(u + \frac{h_0}{2}\psi\right), \tag{2.6}$$

the core,

$$\gamma_{rz2} = \psi + \frac{\partial w}{\partial r}, \tag{2.7}$$

the lower face,

$$\varepsilon_{r3} = \frac{\partial u}{\partial r} - \frac{h_0}{2}\frac{\partial \psi}{\partial r} + \frac{\partial w}{\partial r}\left(\frac{1}{2}\frac{\partial w}{\partial r} + \alpha\right),$$

$$\varepsilon_{\theta 3} = \frac{1}{r}\left(u - \frac{h_0}{2}\psi\right). \tag{2.8}$$

Hooke's law for the shell can be written in the form:

$$\sigma_{ri} = \frac{E}{1-\nu^2}(\varepsilon_{ri} + \nu\varepsilon_{\theta i}),$$

$$\sigma_{\theta i} = \frac{E}{1-\nu^2}(\varepsilon_{\theta i} + \nu\varepsilon_{ri}), \tag{2.9}$$

$$\sigma_{zi} = \tau_{r\theta i} = \tau_{\theta zi} = \tau_{rzi} = 0 \quad i = 1,3$$

$$\tau_{rz2} = G_2\gamma_{rz2},$$

$$\sigma_{r2} = \sigma_{\theta 2} = \sigma_{z2} = \tau_{r\theta 2} = \tau_{\theta z2} = 0 \tag{2.10}$$

in which $\sigma_{ri}, \sigma_{\theta i}, \sigma_{zi}, \tau_{r\theta i}, \tau_{\theta zi}$ and τ_{rzi} ($i=1, 2, 3$) are stresses of the upper face, the core and the lower face, respectively, E is Young's moudlus of the face, ν is Poisson's ratio of the face and G_2 is the shear modulus of the core.

Substitution of Eqs. (2.6)-(2.8) into (2.9) and (2.10) leads to:

for the upper face,

$$\sigma_{r1} = \sigma_{r0} + \frac{Eh_0}{2(1-\nu^2)}\left(\frac{\partial \psi}{\partial r} + \nu\frac{\psi}{r}\right),$$

$$\sigma_{\theta 1} = \sigma_{\theta 0} + \frac{Eh_0}{2(1-\nu^2)}\left(\frac{\psi}{r} + \nu\frac{\partial \psi}{\partial r}\right) \tag{2.11}$$

for the core,

$$\tau_{rz2} = G_2\left(\psi + \frac{\partial w}{\partial r}\right) \tag{2.12}$$

for the lower face,

$$\sigma_{r3} = \sigma_{r0} - \frac{Eh_0}{2(1-\nu^2)}\left(\frac{\partial \psi}{\partial r} + \nu\frac{\psi}{r}\right),$$

$$\sigma_{\theta 3}=\sigma_{\theta 0}-\frac{Eh_0}{2(1-\nu^2)}\left(\frac{\psi}{r}+\nu\frac{\partial\psi}{\partial r}\right) \tag{2.13}$$

where σ_{r0}, $\sigma_{\theta 0}$ are the radial and circumferential stresses of the middle surface, and:

$$\sigma_{r0}=\frac{E}{1-\nu^2}\left[\frac{\partial u}{\partial r}+\nu\frac{u}{r}+\frac{\partial w}{\partial r}\left(\frac{1}{2}\frac{\partial w}{\partial r}+\alpha\right)\right]$$

$$\sigma_{\theta 0}=\frac{E}{1-\nu^2}\left[\frac{u}{r}+\nu\frac{\partial u}{\partial r}+\nu\frac{\partial w}{\partial r}\left(\frac{1}{2}\frac{\partial w}{\partial r}+\alpha\right)\right]. \tag{2.14}$$

The strain energy of the elastic body, denoted by U, is

$$U=\frac{1}{2}\iiint_V(\sigma_r\varepsilon_r+\sigma_\theta\varepsilon_\theta+\sigma_z\varepsilon_z+\tau_{r\theta}\gamma_{r\theta}+\tau_{\theta z}\gamma_{\theta z}+\tau_{rz}\gamma_{rz})r\mathrm{d}r\mathrm{d}\theta\mathrm{d}z. \tag{2.15}$$

By virtue of this expression, the strain energies of the face and the core are:

$$U_i=\frac{1}{2E}\iiint_{V_i}[(\sigma_{ri}+\sigma_{\theta i})^2-2(1+\nu)\sigma_{ri}\sigma_{\theta i}]r\mathrm{d}r\mathrm{d}\theta\mathrm{d}z,\quad i=1,3$$

$$U_2=\frac{1}{2G_2}\iiint_{V_2}\tau_{rz2}^2r\mathrm{d}r\mathrm{d}\theta\mathrm{d}z. \tag{2.16}$$

Substituting Eqs. (2.11)-(2.13) into (2.16) and integrating once with respect to z, we obtain:

$$\begin{aligned}U_1=&\frac{t}{2E}\iint_{S_1}[(\sigma_{r0}+\sigma_{\theta 0})^2-2(1+\nu)\sigma_{r0}\sigma_{\theta 0}]r\mathrm{d}r\mathrm{d}\theta\\&+\frac{h_0t}{2}\iint_{S_1}\left(\sigma_{r0}\frac{\partial\psi}{\partial r}+\frac{1}{r}\sigma_{\theta 0}\psi\right)r\mathrm{d}r\mathrm{d}\theta\\&+\frac{D}{4}\iint_{S_1}\left[\left(\frac{\partial\psi}{\partial r}+\frac{\psi}{r}\right)^2-2(1-\nu)\frac{\psi}{r}\frac{\partial\psi}{\partial r}\right]r\mathrm{d}r\mathrm{d}\theta\\U_2=&\frac{G_2h_0}{2}\iint_{S_2}\left(\psi+\frac{\partial w}{\partial r}\right)^2r\mathrm{d}r\mathrm{d}\theta\\U_3=&\frac{t}{2E}\iint_{S_3}[(\sigma_{r0}+\sigma_{\theta 0})^2-2(1+\nu)\sigma_{r0}\sigma_{\theta 0}]r\mathrm{d}r\mathrm{d}\theta\\&-\frac{h_0t}{2}\iint_{S_3}\left(\sigma_{r0}\frac{\partial\psi}{\partial r}+\frac{1}{r}\sigma_{\theta 0}\psi\right)r\mathrm{d}r\mathrm{d}\theta\\&+\frac{D}{4}\iint_{S_3}\left[\left(\frac{\partial\psi}{\partial r}+\frac{\psi}{r}\right)^2-2(1-\nu)\frac{\psi}{r}\frac{\partial\psi}{\partial r}\right]r\mathrm{d}r\mathrm{d}\theta\end{aligned} \tag{2.17}$$

where D is the flexural rigidity of the shell:

$$D=\frac{Eh_0^2t}{2(1-\nu^2)}. \tag{2.18}$$

Then the total strain energy of the shallow conical sandwich shell is

$$\begin{aligned}U=U_1+U_2+U_3=&\frac{t}{E}\iint_S[(\sigma_{r0}+\sigma_{\theta 0})^2-2(1+\nu)\sigma_{r0}\sigma_{\theta 0}]r\mathrm{d}r\mathrm{d}\theta\\&+\frac{D}{2}\iint_S\left[\left(\frac{\partial\psi}{\partial r}+\frac{\psi}{r}\right)^2-2(1-\nu)\frac{\psi}{r}\frac{\partial\psi}{\partial r}\right]r\mathrm{d}r\mathrm{d}\theta\\&+\frac{G_2h_0}{2}\iint_S\left(\psi+\frac{\partial w}{\partial r}\right)^2r\mathrm{d}r\mathrm{d}\theta\end{aligned} \tag{2.19}$$

The kinetic energy of the shallow conical sandwich shell is

$$T=\frac{1}{2}\iint_{S} m\left(\frac{\partial w}{\partial t}\right)^{2} r\mathrm{d}r\mathrm{d}\theta \tag{2.20}$$

in which m, m_1, m_2 are the mass densities per unit area of the shell, face and core, respectively, where:

$$m=2m_1+m_2 \tag{2.21}$$

Using Hamilton's principle, we have

$$\delta\Pi=0 \tag{2.22}$$

where Π is the functional, t is the time,

$$\Pi=\int_{t_1}^{t_2}(T-U)\mathrm{d}t. \tag{2.23}$$

Taking variations of u, w and ψ independently and simultaneously, and integrating them by parts, we obtain the following equations of flexural motion and the associated boundary conditions for the nonlinear free vibration of the shallow conical sandwich shell:

$$\begin{aligned}
&\sigma_{\theta 0}-\frac{\partial}{\partial r}(r\sigma_{r0})=0\\
&G_2h_0\frac{\partial}{\partial r}\left[r\left(\psi+\frac{\partial w}{\partial r}\right)\right]+2t\frac{\partial}{\partial r}\left[r\sigma_{r0}\left(\frac{\partial w}{\partial r}+\alpha\right)\right]-mr\frac{\partial^2 w}{\partial t^2}=0\\
&Dr\frac{\partial}{\partial r}\frac{1}{r}\frac{\partial}{\partial r}(r\psi)-G_2h_0r\left(\psi+\frac{\partial w}{\partial r}\right)=0
\end{aligned} \tag{2.24a-c}$$

$$\begin{aligned}
&\text{At } r=0 \quad \text{or} \quad r=a:\\
&r\sigma_{r0}=0 \quad \text{or} \quad \delta u=0\\
&rQr+2tr\sigma_{r0}\left(\frac{\partial w}{\partial r}+\alpha\right)=0 \quad \text{or} \quad \delta w=0\\
&rMr=0 \quad \text{or} \quad \delta\psi=0,
\end{aligned} \tag{2.25}$$

in which Q_r is the shearing force and M_r is the radial moment:

$$\begin{aligned}
Q_r&=G_2h_0\left(\psi+\frac{\partial w}{\partial r}\right)\\
M_r&=D\left(\frac{\partial\psi}{\partial r}+\nu\frac{\psi}{r}\right)
\end{aligned} \tag{2.26}$$

At the same time eliminating u from Eq. (2.14), we find a compatibility equation:

$$\sigma_{r0}-\frac{\partial}{\partial r}(r\sigma_{\theta 0})-E\frac{\partial w}{\partial r}\left(\frac{1}{2}\frac{\partial w}{\partial r}+\alpha\right)=0 \tag{2.27}$$

Now it is convenient to introduce a stress function as one of the dependent variables for the governing Eqs. (2.24) and (2.27). This stress function ϕ is defined by:

$$\sigma_{r0}=\frac{1}{r}\frac{\partial\varphi}{\partial r},\quad \sigma_{\theta 0}=\frac{\partial^2\phi}{\partial r^2}. \tag{2.28}$$

Obviously, the stress function satisfies Eq. (2.24a) exactly, and Eqs. (2.24b) and (2.27) are simplified to yield:

$$G_2 h_0 \frac{\partial}{\partial r}\left[r\left(\psi+\frac{\partial w}{\partial r}\right)\right]+2t\frac{\partial}{\partial r}\left[\frac{\partial\varphi}{\partial r}\left(\frac{\partial w}{\partial r}+\alpha\right)\right]-mr\frac{\partial^2 w}{\partial t^2}=0$$

$$r\frac{\partial}{\partial r}\frac{1}{r}\frac{\partial}{\partial r}\left(r\frac{\partial\varphi}{\partial r}\right)+E\frac{\partial w}{\partial r}\left(\frac{1}{2}\frac{\partial w}{\partial r}+\alpha\right)=0 \tag{2.29}$$

Equations (2.29) and (2.24c) are now the fundamental equations for the nonlinear free vibration of the shallow conical sandwich shell. Using boundary conditions (2.25), these equations will be solved for the following two types of boundary conditions usually encountered in engineering:

(1) Rigidly clamped edge

$$w=0,\quad \psi=0, u=0 \quad \text{at } r=a$$
$$\psi=0,\quad \sigma_{r0} \text{ finite} \quad \text{at } r=0. \tag{2.30}$$

(2) Loosely clamped edge

$$w=0,\quad \psi=0,\quad \sigma_{r0}=0 \quad \text{at } r=a$$
$$\psi=0,\quad \sigma_{r0} \text{ finite} \quad \text{at } r=0. \tag{2.31}$$

Using Eqs. (2.14) and (2.24a), the radial displacement u can be expressed as

$$u=\frac{r}{E}\left[\frac{\partial}{\partial r}(r\sigma_{r0}-\nu\sigma_{r0})\right] \tag{2.32}$$

Let us introduce the following nondimensional parameters

$$y=\frac{r}{a},\quad W=\frac{2\sqrt{1-\nu^2}}{h_0}w,\ \Psi=\frac{2\sqrt{1-\nu^2}}{h_0}a\psi,\quad S=\frac{2ta}{P}\frac{\partial\phi}{\partial r},$$

$$\tau=\sqrt{\frac{D}{ma^4}}t,\quad k_1=\frac{2\sqrt{1-\nu^2}}{h_0}a\alpha,\quad k_2=\frac{D}{G_2 h_0 a^2}. \tag{2.33}$$

With these quantities, Eqs. (2.29), (2.24c) and boundary conditions (2.30), (2.31) become:

$$L(W,\Psi,S)=\frac{1}{k_2}\frac{\partial}{\partial y}\left[y\left(\Psi+\frac{\partial W}{\partial y}\right)\right]$$
$$+\frac{\partial}{\partial y}\left[S\left(\frac{\partial W}{\partial y}+k_1\right)\right]-y\frac{\partial^2 W}{\partial\tau^2}=0 \tag{2.34a}$$

$$\frac{\partial}{\partial y}\frac{1}{y}\frac{\partial}{\partial y}(y\Psi)-\frac{1}{k_2}\Psi=\frac{1}{k_2}\frac{\partial W}{\partial y} \tag{2.34b}$$

$$y\frac{\partial}{\partial y}\frac{1}{y}\frac{\partial}{\partial y}(yS)=-\frac{\partial W}{\partial y}\left(\frac{1}{2}\frac{\partial W}{\partial y}+k_1\right) \tag{2.34c}$$

$$W=0,\quad \Psi=0,\quad \frac{\partial S}{\partial y}-\lambda\frac{S}{y}=0 \quad \text{at } y=1 \tag{2.35a-c}$$

$$\Psi=0,\quad S=0 \quad \text{at } y=0 \tag{2.36a,b}$$

in which the numerical values of the factor are,

for a rigidly clamped edge,

$$\lambda=\nu \tag{2.37}$$

and for a loosely clamped edge,

$$\lambda=\infty. \tag{2.38}$$

3. Analytical Solution

We shall begin to solve the nondimensional nonlinear boundary value problem (2. 34)-(2. 36) for the shallow conical sandwich shell with a rigidly clamped outer edge or a loosely clamped outer edge by Galenkin's method.

According to the solution obtained from the small deflection theory for the shallow conical sandwich shell, boundary condition (2. 35a) is satisfied by assuming that the nondimensional deflection W is in the separable form

$$W=\frac{1}{4}f(\tau)(y^4-2\beta_1 y^2+2\beta_1-1) \tag{3. 1}$$

in which $f(\tau)$ is an unknown function of τ

$$\beta_1=8k_2+1 \tag{3. 2}$$

Substituting (3. 1) into the compatibility equation (2. 33c) yields:

$$y\frac{\partial}{\partial y}\frac{1}{y}\frac{\partial}{\partial y}(yS)=-f(\tau)(y^3-\beta_1 y)\left[\frac{1}{2}f(\tau)(y^3-\beta_1 y)+k_1\right]. \tag{3. 3}$$

Integrating Eq. (3. 3) twice and using conditions (2. 35c) and (2. 36b), whose solution is:

$$S=-\frac{1}{8}f^2(\tau)\left(\frac{1}{12}y^7-\frac{1}{3}\beta_1 y^5+\frac{1}{2}\beta_1^2 y^3+\beta_3 y\right)-\frac{k_1}{3}f(\tau)\left(\frac{1}{5}y^4-\beta_1 y^2+\beta_2 y\right) \tag{3. 4}$$

where

$$\beta_2=\left[\frac{1}{1-\lambda}(2-\lambda)\beta_1-\frac{1}{5}(4-\lambda)\right],$$

$$\beta_3=-\frac{1}{1-\lambda}\left[\frac{1}{2}(3-\lambda)\beta_1^2-\frac{1}{3}(5-\lambda)\beta_1+\frac{1}{12}(7-\lambda)\right]. \tag{3. 5}$$

Inserting expression (3. 1) into Eq. (2. 34b) leads to

$$\frac{\partial}{\partial y}\frac{1}{y}\frac{\partial}{\partial y}(y\Psi)-\frac{1}{k_2}\Psi=\frac{1}{k_2}f(\tau)(y^3-\beta_1 y). \tag{3. 6}$$

The rotation is assumed to be of the separable form

$$\Psi=f(\tau)Y(y). \tag{3. 7}$$

Substituting this expression into (3. 6), we find that the function $Y(y)$ satisfies the following ordinary differential equation:

$$\frac{\mathrm{d}^2Y}{\mathrm{d}y^2}+\frac{1}{y}\frac{\mathrm{d}Y}{\mathrm{d}y}-\left(\frac{1}{y^2}+\frac{1}{k_2}\right)Y=\frac{1}{k_2}(y^3-\beta_1 y). \tag{3. 8}$$

We take the general soultion of this equation in the form:

$$Y=Y^{(0)}+Y^{(*)} \tag{3. 9}$$

in which $Y^{(0)}$ is the general solution of the homogeneous equation:

$$\frac{\mathrm{d}^2Y}{\mathrm{d}y^2}+\frac{1}{y}\frac{\mathrm{d}Y}{\mathrm{d}y}-\left(\frac{1}{y^2}+\frac{1}{k_2}\right)Y=0 \tag{3. 10}$$

and $Y^{(*)}$ is a particular solution of Eq. (3. 8) .

A solution of Eq. (3. 10) can be taken in the form of a power series

$$Y=\sum_{i=0}^{\infty}a_i y^i \tag{3.11}$$

where a_i are constants to be determined later.

Substituting this series into Eq. (3.10) and equating the coefficients for each power of y to zero, we obtain the following relation between the coefficients:

$$a_i=\frac{1}{(i^2-1)k_2}a_{i-2} \quad i=3,5,7,\cdots.$$

$$a_i=0 \quad i=0,2,4,\cdots. \tag{3.12}$$

These relations show that when the constant a_1 is assigned, all the other constants can be determined successively. Then, these coefficients may be represented by the expression:

$$a_{2n+1}=\frac{1}{2^{2n}n!(n+1)!k_2^n}a_1 \quad n=1,2,3\cdots. \tag{3.13}$$

With these relations, series (3.11) becomes

$$Y_1=a_1 y\left[1+\sum_{n=1}^{\infty}\frac{1}{n!(n+1)!k_2^n}\left(\frac{y}{2}\right)^{2n}\right]. \tag{3.14}$$

It can be shown, by Dalembert discriminate, that the series is convergent for any value of y. Thus, it can be used to represent one of the basic solutions of Eq. (3.10). Here the solution contains one arbitrary constant a_1.

Following the general procedure described in the theory of the differential equation, and using the first basic solution (3.14), the second basic solution can be written in the form:

$$Y_2=Y_1\int\frac{e^{-\int P(y)dy}}{Y_1^2}dy \tag{3.15}$$

in which $P(y)$ is the coefficient of the first derivative of the function Y in Eq. (3.10),

$$P(y)=\frac{1}{y}. \tag{3.16}$$

Having solutions (3.14) and (3.15), we conclude that the general solution of equation (3.10) is:

$$Y^{(0)}=c_{10}Y_1+c_{20}Y_1\int\frac{e^{-\int\frac{1}{y}dy}}{Y_1^2}dy=c_{10}Y_1+c_{20}Y_1\int\frac{1}{yY_1^2}dy \tag{3.17}$$

where c_{10} and c_{20} are undetermined constants.

From consideration of the right-hand side of Eq. (3.8), we conclude that a particular solution of the equation can be represented by the following multinomial:

$$Y^{(*)}=b_3y^3+b_2y^2+b_1y+b_0 \tag{3.18}$$

in which b_i $(i=0,1,2,3)$ are undetermined coefficients.

Substituting this multinomial into (3.8) and using (3.2), we find

$$b_3=-1,\quad b_2=b_0=0,\quad b_1=1. \tag{3.19}$$

Then, the particular solution (3.18) becomes

$$Y^{(*)}=-y^3+y. \tag{3.20}$$

Using expression (3.17) and (3.20), the general solution of Eq. (3.8) is

$$Y = c_{10}Y_1 + c_{20}Y_1\int \frac{1}{yY_1^2}\mathrm{d}y - y^3 + y. \tag{3.21}$$

Having solution (3.21), and using solution (3.13), the solution (3.7) of Eq. (3.6) can be written in the form:

$$\begin{aligned}\Psi = f(\tau)\Bigg\{ & y\left[1+\sum_{n=1}^{\infty}\frac{a_n}{k_2^n}\left(\frac{y}{2}\right)^{2n}\right]\Bigg[c_1 \\ & + c_2\int \frac{1}{y^3\left[1+\sum_{n=1}^{\infty}\frac{a_n}{k_2^n}\left(\frac{y}{2}\right)^{2n}\right]^2}\mathrm{d}y\Bigg] - y^3 + y\Bigg\}\end{aligned} \tag{3.22}$$

in which c_1 and c_2 are undetermined constants

$$a_n = \frac{1}{n!(n+1)!}. \tag{3.23}$$

Applying conditions (2.34b) and (2.35a), we obtain

$$c_1 = c_2 = 0. \tag{3.24}$$

Then solution (3.22) becomes

$$\Psi = -f(\tau)(y^3 - y). \tag{3.25}$$

Instead of satisfying Eq. (2.33a), the Galerkin's procedure is applied to this equation. This leads to the following condition on W, Ψ and S:

$$\int_0^1 L(W,\Psi,S)W\mathrm{d}y = 0. \tag{3.26}$$

Substituting solutions (3.1), (3.4) and (3.25) into this condition and integrating it, we obtain a nonlinear ordinary differential equation for the time function $f(\tau)$:

$$\frac{\mathrm{d}^2 f}{\mathrm{d}\tau^2} + \omega_0^2(f + \alpha_2 f^2 + \alpha_3 f^3) = 0 \tag{3.27}$$

or

$$\frac{\mathrm{d}^2 f}{\mathrm{d}\tau^2} = -\omega_0^2(f + \alpha_2 f^2 + \alpha_3 f^3) \tag{3.28}$$

where ω_0 is the linear natural circular frequency,

$$\omega_0 = \sqrt{\frac{u_1}{\mu_0}}$$

$$\begin{aligned}\alpha_2 = & -\frac{k_1}{\mu_1}\Big[\frac{1}{20}\Big(\frac{19}{3}\beta_1^3 - \frac{67}{7}\beta_1^2 + \frac{83}{18}\beta_1\Big) - \frac{1}{3}\beta_2\Big(\beta_1^2 - \frac{4}{3}\beta_1 + \frac{1}{2}\Big) \\ & + \frac{1}{2}\beta_3\Big(\frac{1}{3}\beta_1 - \frac{1}{5}\Big) - \frac{37}{1320}\Big]\end{aligned}$$

$$\alpha_3 = \frac{1}{8\mu_1}\Big[\frac{1}{3}\beta_1^4 - \frac{2}{3}\beta_1^3 + \frac{1}{2}\beta_1^2 - \frac{1}{6}\beta_1 + \beta_3\Big(\beta_1^2 - \frac{4}{3}\beta_1 + \frac{1}{2}\Big) + \frac{1}{42}\Big]$$

$$u_0 = -\frac{1}{6}\Big(\beta_1^2 - \frac{5}{4}\beta_1 + \frac{2}{5}\Big)$$

$$\mu_1 = \frac{k_1^2}{3}\left[\beta_1^2 - \frac{4}{5}\beta_1 - 2\beta_2\left(\frac{1}{3}\beta_1 - \frac{2}{5}\right) + \frac{1}{10}\right] - \frac{1}{k_2}(\beta_1 - 1)\left(\beta_1 - \frac{2}{3}\right). \tag{3.29}$$

The corresponding initial conditions are:

$$f = A, \quad \frac{\mathrm{d}f}{\mathrm{d}\tau} = 0 \quad \text{at } \tau = 0 \tag{3.30}$$

where A is the non-dimensional inward amplitude.

Multiplying both sides of Eq. (3.28) by $\mathrm{d}f$, this equation can be written in the following form:

$$\frac{\mathrm{d}f}{\mathrm{d}\tau}\mathrm{d}\left(\frac{\mathrm{d}f}{\mathrm{d}\tau}\right) = -\omega_0^2(f + \alpha_2 f^2 + \alpha_3 f^3)\mathrm{d}f. \tag{3.31}$$

Integrating once and using condition (3.30) leads to:

$$\left(\frac{\mathrm{d}f}{\mathrm{d}\tau}\right)^2 = 2\omega_0^2(A - f)\left[\frac{1}{2}(A + f) + \frac{\alpha_2}{3}(A^2 + f^2 + Af) + \frac{\alpha_3}{4}(A + f)(A^2 + f^2)\right]. \tag{3.32}$$

In the case of nonlinear free vibration, the nonlinear period T of the shallow conical sandwich shell is:

$$T = 2\int_{\tau_1}^{\tau_2}\mathrm{d}\tau = 2\int_{A^*}^{A}\frac{\mathrm{d}f}{\mathrm{d}f/\mathrm{d}\tau} \tag{3.33}$$

in which τ_1, τ_2 are two moments when the shell is located at the nondimensional inward and outward amplitude A, A^*, respectively, and A^* is the negative root of the right-hand side of Eq. (3.32). Then we have

$$\frac{1}{2}(A + A^*) + \frac{\alpha_2}{3}(A^2 + f^2 + Af) + \frac{\alpha_3}{4}(A + f)(A^2 + f^2) = 0$$

or

$$\frac{\alpha_3}{4}A^{*3} + \left(\frac{\alpha_3}{4}A + \frac{\alpha_2}{3}\right)A^{*2} + \left(\frac{\alpha_3}{4}A^2 + \frac{\alpha_2}{3}A + \frac{1}{2}\right)A^* + \frac{\alpha_3}{4}A^3 + \frac{\alpha_2}{3}A^2 + \frac{1}{2}A = 0 \tag{3.34}$$

which is a cubic equation for A^*. The root is easy to find using the known analytical formula or other numerical methods.

The corresponding linear period T_0 is

$$T_0 = \frac{2\pi}{\omega_0}. \tag{3.35}$$

Hence using Eqs. (3.32), (3.33) and (3.35), the ratio of nonlinear period T to linear period T_0 is given by

$$\frac{T}{T_0} = \frac{1}{\sqrt{2}\pi}\int_{A^*}^{A}\left\{(A - f)\left[\frac{1}{2}(A + f) + \frac{\alpha_2}{3}(A^2 + f^2 + Af) + \frac{\alpha_3}{4}(A + f)(A^2 + f^2)\right]\right\}^{-1/2}\mathrm{d}f. \tag{3.36}$$

Numerical computations can be performed for this integral.

4. Results and Discussion

In Table 1, according to formula (2.30), we first give the numerical values of the coefficients α_2, α_3 and the linear natural circular frequency ω_0 for $k_2=0.05$, $\nu=0.3$ and several values of k_1. It is observed that the coefficients α_2 and α_3 are small, and the values of ω_0 and $|\alpha_2/\alpha_3|$ increase with increasing k_1, but the value of α_3 decreases with increasing k_1.

Table 1 Values of α_2, α_3, and ω_0 ($k_2=0.05$, $\nu=0.3$)

k_1	Rigidly clamped				Loosely clamped			
	ω_0	$100\alpha_2$	$100\alpha_3$	$\|\alpha_2/\alpha_3\|$	ω_0	$100\alpha_2$	$100\alpha_3$	$\|\alpha_2/\alpha_3\|$
0	7.596	−0.000	1.354	0.000	7.596	0.000	0.282	0.000
1	7.926	−6.737	1.243	5.418	7.669	−1.525	0.277	5.513
2	8.840	−10.830	0.999	10.837	7.884	−2.886	0.262	11.027
3	10.184	−12.242	0.753	16.255	8.230	−3.973	0.240	16.540
4	11.811	−12.136	0.560	21.673	8.691	−4.750	0.215	22.053
5	13.620	−11.407	0.421	27.091	9.251	−5.242	0.190	27.567
6	15.547	−10.505	0.323	32.510	9.891	−5.502	0.166	33.080
7	17.555	−9.613	0.253	37.928	10.598	−5.591	0.145	38.593
8	19.618	−8.797	0.203	43.346	11.360	−5.561	0.126	44.106
9	21.720	−8.074	0.166	48.764	12.166	−5.456	0.110	49.620

In order to have some knowledge of the outward vibration amplitude A^* of the shallow conical sandwich shells, we obtain the values of A^* for $k_1=2$, $k_2=0.05$ and $\nu=0.3$ by means of Newton's method. The values are given in Table 2. It is seen from this table that the absolute value of A^* is less throughout than the inward vibration amplitude A. Only when $k_1=0$, namely, when the shallow conical sandwich shells are transformed into the circular sandwich plates, is the absolute value of A^* equal A.

Table 2 Values of A^* ($k_1=2$, $k_2=0.05$, $\nu=0.3$)

A	A^*	
	Rigidly clamped	Loosely clamped
1.5	−1.3547	−1.4581
3.0	−2.4822	−2.8392
4.5	−3.4640	−4.1549
6.0	−4.3656	−5.4178
7.5	−5.2394	−6.6404
9.0	−6.1255	−7.8343

Then, from expression (3.36), the ratio of nonlinear period T to linear period T_0 is plotted against the nondimensional amplitude A in Fig. 4.1 for rigidly clamped shallow conical sandwich shells, in Fig. 4.2 for loosely clamped shallow conical sandwich shells, and in Fig. 4. 3 for shallow concial sandwich shells with various boundary conditions. From these figures, it is seen that:

(1) for rigidly clamped shells, the nonlinear effect increases with increasing geometrical parameter k_1;

(2) for loosely clamped shells, the nonlinear effect diminishes with increasing geometrical parameter k_1;

(3) the nonlinear effect increases with the increase of physical parameter k_2;

(4) for rigidly clamped shells the nonlinear effect on the period of vibration is stronger, but is weaker for loosely clamped shells.

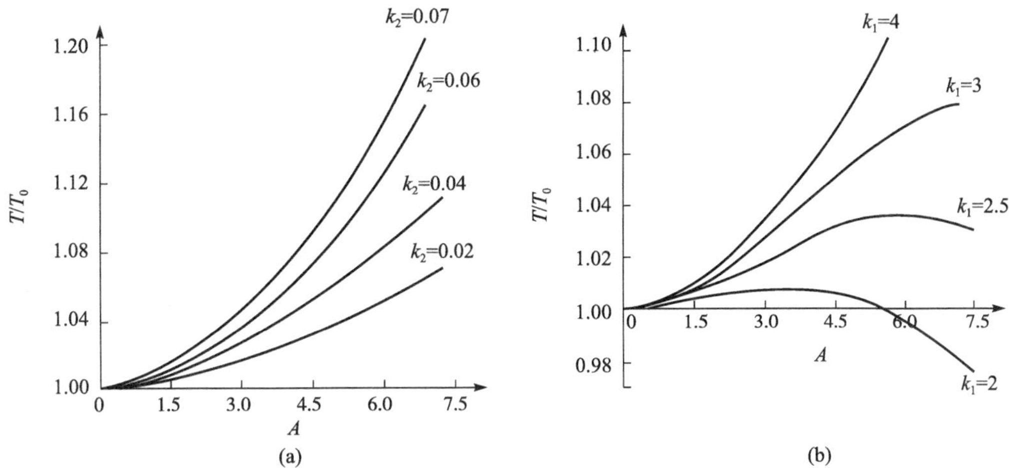

Fig. 4. 1　Period of vibration of rigidly clamped shallow conical sandwich shells

(a) $k_1=7$, $\nu=0.3$, (b) $k_2=0.05$, $\nu=0.3$

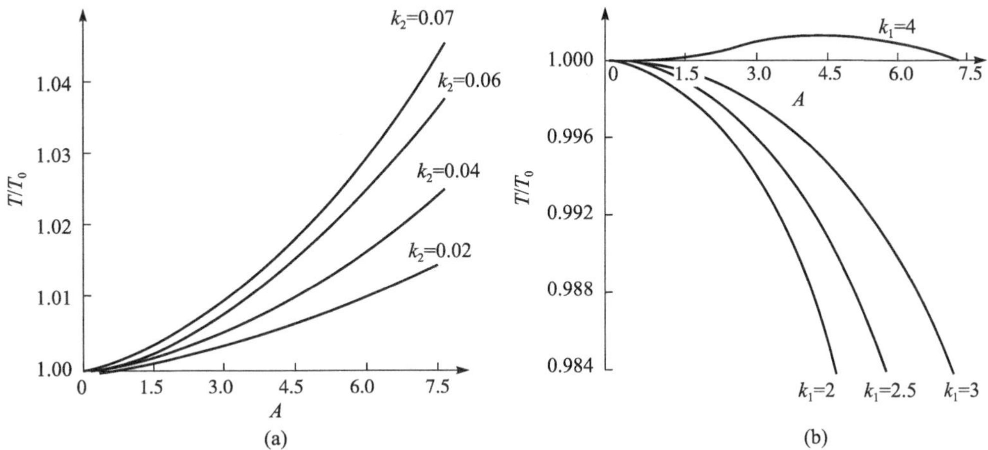

Fig. 4. 2　Period of vibration of loosely clamped shallow conical sandwich shells

(a) $k_1=7$, $\nu=0.3$, (b) $k_2=0.05$, $\nu=0.3$

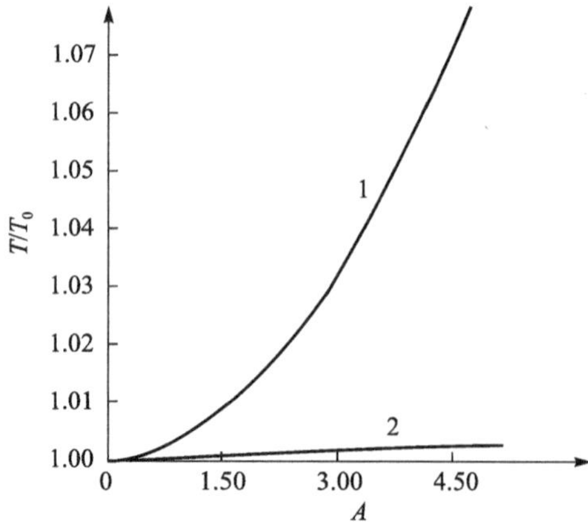

Fig. 4. 3 Period of vibration of shallow conical sandwich shells for various boundary conditions

1. rigidly clamped edge; 2. loosely clamped edge

($k_1=4$, $k_2=0.05$, $\nu=0.3$)

References

[1] E. Reissner, Finite deflection of sandwich plates. *J. Aeronaut. Sci.* 15 (7), 435-440 (1948); 17 (2), 125 (1950).

[2] H. P. Kan and J. C. Huang, Large deflection of rectangular sandwich plates. *J. AIAA* 5 (9), 1706-1708 (1967).

[3] Liu Renhuai, Nonlinear bending of circular sandwich plates. *Appl. Math. Mech.* 2 (2), 189-207 (1981).

[4] Liu Renhuai, Nonlinear axisymmetrical bending of circular sandwich plates under the action of uniform edge moment. *J. China University of Science and Technology* 10 (2), 56-67 (1981) (in Chinese).

[5] Liu Renhuai and Y. F. Shi, Exact solution for circular sandwich plate with large deflection. *Appl. Math. Mech.* 3 (1), 11-24 (1982).

[6] Liu Renhuai, Nonlinear bending of circular sandwich plates under the action of axisymmetric uniformly distributed line loads. In *Progress in Applied Mechanics*, pp. 293-321. Martinus Nighoff, Dordrecht (1987).

[7] Liu Renhuai and G. Q. Zhu, Further study on large deflection of circular sandwich plates. *Appl. Math. Mech.* 10 (12), 1099-1106 (1989).

[8] Liu Renhuai and Z. Q. Cheng, Nonlinear bending of simply supported rectangular sandwich plates. *Appl. Math. Mech.* 14 (3), 217-234 (1993).

[9] Liu Renhuai and J. F. Zhu, *Nonlinear Theory of Sandwich Shells. Press of Mechanical Industry*, Beijing (1993) (in Chinese).

[10] R. Struk, Nonlinear stability problem of an open conical sandwich shell under extreme pressure and compression. *Int. J. Nonlinear Mech.* 14 (3), 217-233 (1984).

[11] Liu Renhuai and J. C Wu, Nonlinear vibration of rectangular sandwich plates. *Science in China* (*Series A*) 35 (4), 472-486 (1992).

复合材料面层夹层板中转动一致有效理论①

一、引言

复合材料在近代工业部门的广泛应用，导致了复合材料板壳理论的产生和发展。复合材料面层夹层板是指面层为复合材料的夹层板，其面层一般由复合材料单向层板层合而成，芯层一般为各向异性材料。这类夹层板构造上的复杂性，形成了其宏观上各向同性夹层板无法企及的许多优越性能，如比强度和比刚度高、可设计性好、成型工艺简单、抗振性能好等，但也造成了理论分析上的巨大困难。

早在 20 世纪 40 年代末，Reissner[1] 考虑了薄面层和软芯的夹层板大挠度问题。Librescu[2] 在研究夹层板线性和非线性行为时，考虑了两面层对称布置的各向异性夹层板。Ebcioǧlu[3] 讨论了很一般的夹层板非线性问题：夹层板面层厚度任意、各向异性，芯层为各向异性硬核。Ebcioǧlu 同时考虑了面层和芯层的横向剪切变形以及芯层的横向伸展效应，所得控制方程含有 10 个未知位移变量。刘人怀等[4-10] 则研究了夹层圆板和夹层矩形板的非线性弯曲和非线性振动问题。

复合材料面层夹层板的研究甚少，且基本限于线性屈曲问题[11-13]，理论模型也是沿用了各向同性夹层板模型。

在上述文献中，一般认为夹层板面层变形符合薄膜假设或符合克希霍夫-拉甫假设。这对各向同性夹层板的变形描述是合适的，但若将其用于复合材料面层夹层板，则将导致两个方面的理论缺陷。第一，无法描述面层的横向剪切变形。我们知道，对于复合材料面层夹层板。其面层和芯层的横向剪切模量都是很低的，需要同时考虑面层及芯层的横向剪切变形。第二，无法比较精确地得到工程上颇为重要的层间应力。

为消除上述理论缺陷，精确地反映复合材料面层夹层板的位移和应力分布状况，本文假设夹层板中任一点面内位移分别在面层和芯层中为横向坐标的逐段三次函数。引入夹层板层间（面层与芯层粘结处，下同）和表面应力协调条件，导出了位移场的修正形式。根据能量误差一致原则，通过严格的量阶分析并弃去不完全项，得到了应变场的表达式。然后应用最小位能原理建立了复合材料面层夹层板的控制方程和边界条件。

二、位移场

设复合材料面层夹层板的面层和芯层厚度均匀，总厚度为 h，上、下面层厚度分别为 h'' 和 h'，芯层厚度为 h_c，$h=h'+h''+h_c$（如图 1）。记夹层板上、下表面分别为 $\underset{(3)}{\omega}$ 和 $\underset{(0)}{\omega}$，层间依次为 $\underset{(1)}{\omega}$ 和 $\underset{(2)}{\omega}$，取坐标面 x_1ox_2 与 $\underset{(0)}{\omega}$ 重合，x_3 坐标线与 $\underset{(0)}{\omega}$ 垂直，则 $\underset{(m)}{\omega}$（$m=$

---

① 本文原载《上海力学》，1996，17（3）：222-228. 作者：刘人怀，王志伟.

0，1，2，3）可由 $x_3 = \underset{(m)}{x_3}$ 表示。本文采用哑指标求和约定，希腊指标取值 1 和 2，拉丁指标取值 1，2 和 3。设夹层板各层材料具有平行于坐标面 x_1ox_2 的弹性对称面，应力张量 S^{ij} 和应变张量 e_{ij} 服从本构关系

$$S^{\alpha\beta} = Q^{\alpha\beta\rho\sigma} e_{\rho\sigma}, S^{\alpha 3} = Q^{\alpha 3\rho 3} e_{\rho 3} \tag{1}$$

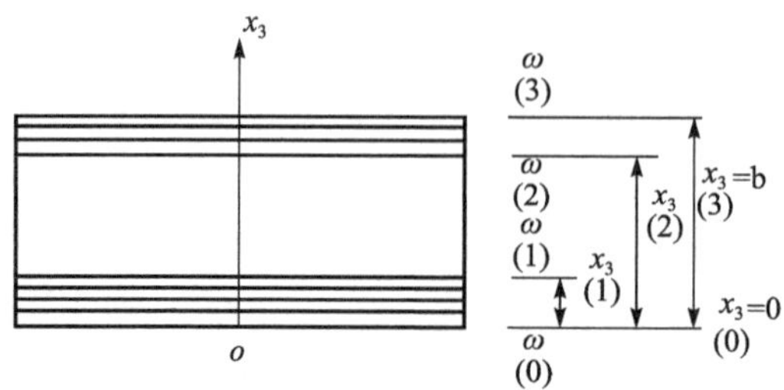

图 1　复合材料面层夹层板

其中，Q^{ijkl} 为弹性常数。这里，略去了夹层板横向法应力的效应。

应变张量 e_{ij} 与位移 v_i 的关系为

$$e_{ij} = \frac{1}{2}(v_i|_j + v_j|_i + v_r|_i v_r|_j) \tag{2}$$

其中“|”表示协变导数。

令

$$\eta_{ij} = \frac{1}{2}(v_i|_j + v_j|_i),$$

$$\omega_{ij} = \frac{1}{2}(v_i|_j - v_j|_i) \tag{3}$$

则式（2）可写成

$$e_{ij} = \eta_{ij} + \frac{1}{2}\omega_{ki}\omega_{kj} + \frac{1}{2}(\eta_{kj}\omega_{ki} + \eta_{ki}\omega_{kj}) + \frac{1}{2}\eta_{kj}\eta_{kj} \tag{4}$$

考虑小应变中转动变形，有

$$e_{ij} = o(\varepsilon^2),\quad \eta_{ij} = o(\varepsilon^2),\quad \omega_{3\alpha} = o(\varepsilon),\quad \omega_{\alpha\beta} = o(\varepsilon^2) \tag{5}$$

其中 ε 为一小量。

考虑式（5）中的量阶关系，由式（4）得

$$e_{\alpha\beta} = \eta_{\alpha\beta} + \frac{1}{2}\omega_{3\alpha}\omega_{3\beta} + \frac{1}{2}(\eta_{3\alpha}\omega_{3\beta} + \eta_{3\beta}\omega_{3\alpha}) + o(\varepsilon^4),$$

$$e_{\alpha 3} = \eta_{\alpha 3} + o(\varepsilon^3) \tag{6a,b}$$

夹层板应变能密度为

$$W = \frac{1}{2}Q^{\alpha\beta\rho\sigma}e_{\alpha\beta}e_{\rho\sigma} + Q^{\alpha 3\rho 3}e_{\alpha 3}e_{\rho 3} = o[Q(e_{\alpha\beta})^2 + G(e_{\alpha 3})^2] \tag{7}$$

其中 $Q = o(Q^{\alpha\beta\rho\sigma})$，$G = o(Q^{\alpha 3\rho 3})$。

本文取 W 的误差界为 $o(Q\varepsilon^6)$，则 $e_{\alpha\beta}$ 只需取至量阶为 ε^3 的项。对复合材料面层夹层板，由于面层和芯层的横向剪切模量较低，即 G/Q 很小，$e_{\alpha 3}$ 只需取至量阶为 ε^2 的项。所以，在式（6a,b）中，可分别略去 $o(\varepsilon^4)$ 项和 $o(\varepsilon^3)$ 项。

设复合材料面层夹层板位移场为

$$v_\alpha = u_\alpha + u_\alpha^{(1)} x_3 + u_\alpha^{(2)} (x_3)^2 + u_\alpha^{(3)} (x_3)^3 + \sum_{m=1}^{2} \underset{(m)}{u_\alpha} (x_3 - \underset{(m)}{x_3}) H(x_3 - \underset{(m)}{x_3}), \tag{8}$$

$$v_3 = u_3$$

式中，v_α 分别在面层和芯层中为 x_3 的逐段三次光滑曲线，且在层间连续，

$$H(x_3 - \underset{(m)}{x_3}) = \begin{cases} 1, & x_3 \geqslant \underset{(m)}{x_3} \\ 0, & x_3 < \underset{(m)}{x_3} \end{cases} \tag{9}$$

由式（6b）得夹层板横向剪应变为

$$e_{\alpha 3} = \frac{1}{2}(u_{3,\alpha} + u_\alpha^{(1)}) + u_\alpha^{(2)} x_3 + \frac{3}{2} u_\alpha^{(3)} (x_3)^2 + \frac{1}{2} \sum_{m=1}^{2} \underset{(m)}{u_\alpha} H(x_3 - \underset{(m)}{x_3}) \tag{10}$$

限定夹层板上下表面无切向力作用，则夹层板表面剪应力协调条件为

$$S^{\alpha 3}(x_\alpha, 0) = 0, \quad S^{\alpha 3}(x_\alpha, h) = 0 \tag{11}$$

夹层板层间横向剪应力连续条件为

$$S^{\alpha 3}(x_\alpha, \underset{(1)}{x_3} - 0) = S^{\alpha 3}(x_\alpha, \underset{(1)}{x_3} + 0),$$

$$S^{\alpha 3}(x_\alpha, \underset{(2)}{x_3} - 0) = S^{\alpha 3}(x_\alpha, \underset{(2)}{x_3} + 0) \tag{12}$$

将式（1）和式（10）代入式（11）和式（12），便得

$$u_{3,\alpha} + u_\alpha^{(1)} = 0,$$

$$2h u_\alpha^{(2)} + h^2 u_\alpha^{(3)} + \sum_{m=1}^{2} \underset{(m)}{u_\alpha} = 0,$$

$$({}_{f'}Q^{\alpha 3\rho 3} - {}_cQ^{\alpha 3\rho 3})[2 \underset{(1)}{x_3} u_\rho^{(2)} + 3(\underset{(1)}{x_3})^2 u_\rho^{(3)}] = {}_cQ^{\alpha 3\rho 3} \underset{(1)}{u_\rho},$$

$$({}_cQ^{\alpha 3\rho 3} - {}_{f''}Q^{\alpha 3\rho 3})[2 \underset{(2)}{x_3} u_\rho^{(2)} + 3(\underset{(2)}{x_3})^2 u_\rho^{(3)} + \underset{(1)}{u_\rho}] = {}_{f''}Q^{\alpha 3\rho 3} \underset{(2)}{u_\rho} \tag{13}$$

式中${}_{f'}Q^{\alpha 3\rho 3}$，${}_{f''}Q^{\alpha 3\rho 3}$分别为下表层中和上表层中与芯层粘结的铺层的弹性常数，${}_cQ^{\alpha 3\rho 3}$为芯层的弹性常数。

记 $u_\alpha^{(2)} = \varphi_\alpha$，由式（13）可解得

$$u_\alpha^{(1)} = - u_{3,\alpha},$$

$$\underset{(m)}{u_\alpha} = \underset{(m)}{\alpha}{}_\alpha^\lambda \varphi_\lambda,$$

$$u_\alpha^{(3)} = c_\alpha^\lambda \varphi_\lambda \tag{14}$$

从而得修正位移场为

$$v_\alpha = u_\alpha - u_{3,\alpha} x_3 + \varphi_\alpha (x_3)^2 + c_\alpha^\lambda \varphi_\lambda (x_3)^3 + \sum_{m=1}^{2} \underset{(m)}{\alpha}{}_\alpha^\lambda \varphi_\lambda (x_3 - \underset{(m)}{x_3}) H(x_3 - \underset{(m)}{x_3})$$

$$v_3 = u_3 \tag{15}$$

三、应变场

将位移表达式（15）代入式（6），并弃去不完全项后得

$$e_{\alpha\beta}=\sum_{n=0}^{3}\overset{(n)}{\underset{(o)}{e}}_{\alpha\beta}(x_3)^n+\sum_{m=1}^{2}\overset{(1)}{\underset{(m)}{e}}_{\alpha\beta}(x_3-\underset{(m)}{x_3})H(x_3-\underset{(m)}{x_3}),$$

$$e_{\alpha 3}=\sum_{n=1}^{2}\overset{(n)}{\underset{(0)}{e}}_{\alpha 3}(x_3)^n+\sum_{m=1}^{2}\overset{(0)}{\underset{(m)}{e}}_{\alpha 3}H(x_3-\underset{(m)}{x_3}) \tag{16}$$

其中

$$\overset{(0)}{\underset{(0)}{e}}_{\alpha\beta}=\overset{(0)}{\underset{(0)}{\eta}}_{\alpha\beta}+\frac{1}{2}\overset{(0)}{\underset{(0)}{\omega}}_{3\alpha}\overset{(0)}{\underset{(0)}{\omega}}_{3\beta},$$

$$\overset{(1)}{\underset{(0)}{e}}_{\alpha\beta}=\overset{(1)}{\underset{(0)}{\eta}}_{\alpha\beta}+\frac{1}{2}(\overset{(0)}{\underset{(0)}{\omega}}_{3\alpha}\overset{(1)}{\underset{(0)}{\omega}}_{3\beta}+\overset{(1)}{\underset{(0)}{\omega}}_{3\alpha}+\overset{(0)}{\underset{(0)}{\omega}}_{3\beta})+\frac{1}{2}(\overset{(1)}{\underset{(0)}{\eta}}_{\alpha 3}\overset{(0)}{\underset{(0)}{\omega}}_{3\beta}+\overset{(1)}{\underset{(0)}{\eta}}_{\beta 3}+\overset{(0)}{\underset{(0)}{\omega}}_{3\alpha}),$$

$$\begin{aligned}\overset{(2)}{\underset{(0)}{e}}_{\alpha\beta}=&\overset{(1)}{\underset{(0)}{\eta}}_{\alpha\beta}+\frac{1}{2}(\overset{(0)}{\underset{(0)}{\omega}}_{3\alpha}\overset{(2)}{\underset{(0)}{\omega}}_{3\beta}+\overset{(1)}{\underset{(0)}{\omega}}_{3\alpha}\overset{(1)}{\underset{(0)}{\omega}}_{3\beta}+\overset{(2)}{\underset{(0)}{\omega}}_{3\alpha}\overset{(0)}{\underset{(0)}{\omega}}_{3\beta})\\&+\frac{1}{2}(\overset{(2)}{\underset{(0)}{\eta}}_{\alpha 3}\overset{(0)}{\underset{(0)}{\omega}}_{3\beta}+\overset{(2)}{\underset{(0)}{\eta}}_{\beta 3}\overset{(0)}{\underset{(0)}{\omega}}_{3\alpha}+\overset{(1)}{\underset{(0)}{\eta}}_{\alpha 3}\overset{(1)}{\underset{(0)}{\omega}}_{3\beta}+\overset{(1)}{\underset{(0)}{\eta}}_{\beta 3}\overset{(1)}{\underset{(0)}{\omega}}_{3\alpha}),\end{aligned} \tag{17}$$

$$\overset{(3)}{\underset{(0)}{e}}_{\alpha\beta}=\overset{(3)}{\underset{(0)}{\eta}}_{\alpha\beta},\overset{(1)}{\underset{(m)}{e}}_{\alpha\beta}=\overset{(1)}{\underset{(m)}{\eta}}_{\alpha\beta},\overset{(1)}{\underset{(0)}{e}}_{\alpha 3}=\overset{(1)}{\underset{(0)}{\eta}}_{\alpha 3},\overset{(2)}{\underset{(0)}{e}}_{\alpha 3}=\overset{(2)}{\underset{(0)}{\eta}}_{\alpha 3},$$

$$\overset{(0)}{\underset{(m)}{e}}_{\alpha 3}=\overset{(0)}{\underset{(m)}{\eta}}_{\alpha 3};\overset{(0)}{\underset{(0)}{\eta}}_{\alpha\beta}=\frac{1}{2}(u_\alpha|_\beta+u_\beta|_\alpha),$$

$$\overset{(1)}{\underset{(0)}{\eta}}_{\alpha\beta}=-u_{3,\alpha\beta},\overset{(2)}{\underset{(0)}{\eta}}_{\alpha\beta}=\frac{1}{2}(\varphi_\alpha|_\beta+\varphi_\beta|_\alpha),$$

$$\overset{(3)}{\underset{(0)}{\eta}}_{\alpha\beta}=\frac{1}{2}[(c_\alpha^\lambda\varphi_\lambda)|_\beta+(c_\beta^\lambda\varphi_\lambda)|_\alpha],$$

$$\overset{(1)}{\underset{(m)}{\eta}}_{\alpha\beta}=\frac{1}{2}[(\underset{(m)}{a_\alpha^\lambda}\varphi_\lambda)|_\beta+(\underset{(m)}{a_\beta^\lambda}\varphi_\lambda)|_\alpha],$$

$$\overset{(1)}{\underset{(0)}{\eta}}_{\alpha 3}=\varphi_\alpha,\overset{(2)}{\underset{(0)}{\eta}}_{\alpha 3}=\frac{3}{2}c_\alpha^\lambda\varphi_\lambda,\overset{(0)}{\underset{(m)}{\eta}}_{\alpha 3}=\frac{1}{2}\underset{(m)}{a_\alpha^\lambda}\varphi_\lambda,\overset{(0)}{\underset{(0)}{\omega}}_{3\alpha}=u_{3,\alpha},$$

$$\overset{(1)}{\underset{(0)}{\omega}}_{3\alpha}=-\varphi_\alpha,\overset{(2)}{\underset{(0)}{\omega}}_{3\alpha}=-\frac{3}{2}c_\alpha^\lambda\varphi_\lambda,\overset{(0)}{\underset{(m)}{\omega}}_{3\alpha}=-\frac{1}{2}\underset{(m)}{a_\alpha^\lambda}\varphi_\lambda \tag{18}$$

若有 $h\varphi_\alpha=o(\varepsilon^2)$，则将式（17）和式（18）代入式（16），并在 $e_{\alpha\beta}$ 表达式中略去量阶为 $o(\varepsilon^4)$ 的项后得应变位移关系

$$\begin{aligned}e_{\alpha\beta}=&\frac{1}{2}(u_\alpha|_\beta+u_\beta|_\alpha+u_{3,\alpha}u_{3,\beta})-u_{3,\alpha\beta}x_3+\frac{1}{2}(\varphi_\alpha|_\beta+\varphi_\beta|_\alpha)(x_3)^2\\&+\frac{1}{2}[(c_\alpha^\lambda\varphi_\lambda)|_\beta+(c_\beta^\lambda\varphi_\lambda)|_\alpha](x_3)^3+\frac{1}{2}\sum_{m=1}^{2}[(\underset{(m)}{\alpha_\alpha^\lambda}\varphi_\lambda)|_\beta+(\underset{(m)}{\alpha_\beta^\lambda}\varphi_\lambda)|_\alpha]H(x_3-\underset{(m)}{x_3}),\end{aligned}$$

$$e_{\alpha 3}=\varphi_\alpha x_3+\frac{3}{2}c_\alpha^\lambda\varphi_\lambda(x_3)^2+\frac{1}{2}\sum_{m=1}^{2}\underset{(m)}{\alpha_\alpha^\lambda}\varphi_\lambda H(x_3-\underset{(m)}{x_3}) \tag{19}$$

四、平衡方程和边界条件

夹层板总应变能为

$$U_1=\iint_{\underset{(0)}{\omega}}\int_0^h\left(\frac{1}{2}Q^{\alpha\beta\rho\omega}e_{\alpha\beta}e_{\rho\omega}+Q^{\alpha 3\rho 3}e_{\alpha 3}e_{\rho 3}\right)\mathrm{d}\sigma\mathrm{d}x_3, \tag{20}$$

载荷位能为

$$U_2 = -\int_{\underset{(0)}{\omega}} \overset{*}{p}{}^3 v_3 \mathrm{d}\sigma - \int_{s\Omega} n_\alpha \overset{*}{s}{}^{\alpha i} v_i \mathrm{d}\Omega$$

$$= -\int_{\underset{(0)}{\omega}} \overset{*}{p}{}^3 v_3 \mathrm{d}\sigma - \int_{\Gamma_s} \left[\int_0^h (\overset{*}{S}{}^{\alpha\beta} v_\beta + \overset{*}{s}{}^{\alpha 3} v_3) \mathrm{d}x_3 \right] n_\alpha \mathrm{d}s \tag{21}$$

式中 $\overset{*}{s}{}^{\alpha i}$ 为夹层板边界侧面 Ω 上给定的外加应力，$\overset{*}{p}{}^3$ 为外加横向载荷，Γ_s 为应力边界侧面 $s\Omega$ 与 $\underset{(0)}{\omega}$ 面的交线，n_α 为 Ω 的单位外法线矢量分量，$\mathrm{d}s$ 为 Γ_s 上的线元。

根据最小位能原理

$$\delta(U_1 + U_2) = 0 \tag{22}$$

将式（15）和（19）代入式（22），完成变分操作，并经分部积分和整理后得

$$\begin{aligned}
&-\int_{\underset{(0)}{\omega}} \Big\{ \underset{(0)}{\overset{(0)}{L}}{}^{\alpha\beta}|_\beta \delta u_\alpha + \big[(u_{3,\alpha} \underset{(0)}{\overset{(0)}{L}}{}^{\alpha\beta} + \underset{(0)}{\overset{(1)}{L}}{}^{\beta\alpha}|_\alpha)|_\beta + \overset{*}{p}{}^3 \big] \delta u_3 \\
&+ \Big[\underset{(0)}{\overset{(2)}{L}}{}^{\alpha\beta}|_\beta + c_\lambda^\alpha \underset{(0)}{\overset{(3)}{L}}{}^{\lambda\beta}|_\beta + \sum_{m=1}^{2} \underset{(m)}{a}{}_\lambda^\alpha \underset{(m)}{\overset{(1)}{L}}{}^{\lambda\beta}|_\beta - 2\underset{(0)}{\overset{(1)}{N}}{}^{\alpha 3} \\
&- 3c_\lambda^\alpha \underset{(0)}{\overset{(2)}{N}}{}^{\lambda 3} - \sum_{m=1}^{2} \underset{(m)}{a}{}_\lambda^\alpha \underset{(m)}{\overset{(0)}{N}}{}^{\lambda 3} \Big] \delta\varphi_\alpha \Big\} \mathrm{d}\sigma \\
&+ \int_{\Gamma_s} \Big[(\underset{(0)}{\overset{(0)}{L}}{}^{\alpha\beta} - \underset{(0)}{\overset{*(0)}{L}}{}^{\alpha\beta}) \delta u_\alpha + (u_{3,\alpha} \underset{(0)}{\overset{(0)}{L}}{}^{\alpha\beta} + \underset{(0)}{\overset{(1)}{L}}{}^{\alpha\beta}|_\alpha - \underset{(0)}{\overset{*(0)}{N}}{}^{\beta 3}) \delta u_3 \\
&- (\underset{(0)}{\overset{(1)}{L}}{}^{\alpha\beta} - \underset{(0)}{\overset{*(1)}{L}}{}^{\alpha\beta}) \delta u_{3,\alpha} + (\underset{(0)}{\overset{(2)}{L}}{}^{\alpha\beta} + c_\lambda^\alpha \underset{(0)}{\overset{(3)}{L}}{}^{\lambda\beta} + \sum_{m=1}^{2} \underset{(m)}{a}{}_\lambda^\alpha \underset{(m)}{\overset{(1)}{L}}{}^{\lambda\beta} - \underset{(0)}{\overset{*(2)}{L}}{}^{\alpha\beta} \\
&- c_\lambda^\alpha \underset{(0)}{\overset{*(3)}{L}}{}^{\lambda\beta} - \sum_{m=1}^{2} \underset{(m)}{a}{}_\lambda^\alpha \underset{(m)}{\overset{*(1)}{L}}{}^{\lambda\beta}) \delta\varphi_\alpha \Big] n_\beta \mathrm{d}s = 0
\end{aligned} \tag{23}$$

这里，$\underset{(m)}{\overset{*(n)}{L}}{}^{\alpha\beta}$，$\underset{(m)}{\overset{*(n)}{N}}{}^{\alpha 3}$的表达式同式（24），只需将 $s^{\alpha\beta}$ 和 $s^{\alpha 3}$ 分别换成 $\overset{*}{s}{}^{\alpha\beta}$ 和 $\overset{*}{s}{}^{\alpha 3}$ 即可。此时，我们定义

$$\begin{aligned}
\underset{(m)}{\overset{(n)}{L}}{}^{\alpha\beta} &= \int_0^h s^{\alpha\beta} (x_3 - \underset{(m)}{x_3})^n H(x_3 - \underset{(m)}{x_3}) \mathrm{d}x_3, \\
\underset{(m)}{\overset{(n)}{N}}{}^{\alpha 3} &= \int_0^h s^{\alpha 3} (x_3 - \underset{(m)}{x_3})^n H(x_3 - \underset{(m)}{x_3}) \mathrm{d}x_3
\end{aligned} \tag{24}$$

这样，我们便得到复合材料面层夹层板的平衡方程和边界条件如下

平衡方程

$$\begin{gathered}
\underset{(0)}{\overset{(0)}{L}}{}^{\alpha\beta}|_\beta = 0, \\
(u_{3,\alpha} \underset{(0)}{\overset{(0)}{L}}{}^{\alpha\beta} + \underset{(0)}{\overset{(1)}{L}}{}^{\beta\alpha}|_\alpha)_\beta + \overset{*}{p}{}^3 = 0, \\
\underset{(0)}{\overset{(2)}{L}}{}^{\alpha\beta}|_\beta + c_\lambda^\alpha \underset{(0)}{\overset{(3)}{L}}{}^{\lambda\beta}|_\beta + \sum_{m=1}^{2} \underset{(m)}{a}{}_\lambda^\alpha \underset{(m)}{\overset{(1)}{L}}{}^{\lambda\beta}|_\beta - 2\underset{(0)}{\overset{(1)}{N}}{}^{\alpha 3} - 3c_\lambda^\alpha \underset{(0)}{\overset{(2)}{N}}{}^{\lambda 3} - \sum_{m=1}^{2} \underset{(m)}{a}{}_\lambda^\alpha \underset{(m)}{\overset{(0)}{N}}{}^{\lambda 3} = 0
\end{gathered} \tag{25}$$

力边界条件

$$
\begin{aligned}
&n_\beta \overset{(0)}{\underset{(0)}{L}}{}^{\alpha\beta} = n_\beta \overset{*(0)}{\underset{(0)}{L}}{}^{\alpha\beta},\\
&n_\beta \overset{(1)}{\underset{(0)}{L}}{}^{\alpha\beta} = n_\beta \overset{*(1)}{\underset{(0)}{L}}{}^{\alpha\beta},\\
&n_\beta \left(u_{3,\alpha} \overset{(0)}{\underset{(0)}{L}}{}^{\alpha\beta} + \overset{(1)}{\underset{(0)}{L}}{}^{\alpha\beta}\big|_\alpha\right) = n_\beta \overset{*(0)}{\underset{(0)}{N}}{}^{\beta 3},\\
&n_\beta \left(\overset{(2)}{\underset{(0)}{L}}{}^{\alpha\beta} + c_\lambda^\alpha \overset{(3)}{\underset{(0)}{L}}{}^{\lambda\beta} + \sum_{m=1}^{2} \underset{(m)}{a_\lambda^\alpha} \overset{(1)}{\underset{(m)}{L}}{}^{\lambda\beta}\right)\\
&= n_\beta \left(\overset{*(2)}{\underset{(0)}{L}}{}^{\alpha\beta} + c_\lambda^\alpha \overset{*(3)}{\underset{(0)}{L}}{}^{\lambda\beta} + \sum_{m=1}^{2} \underset{(m)}{a_\lambda^\alpha} \overset{*(1)}{\underset{(m)}{L}}{}^{\lambda\beta}\right)
\end{aligned}
\tag{26}
$$

几何边界条件

$$
\begin{aligned}
&u_\alpha = \overset{*}{u}_\alpha, u_3 = \overset{*}{u}_3,\\
&u_{3,\alpha} = \overset{*}{u}_{3,\alpha}, \varphi_\alpha = \overset{*}{\varphi}_\alpha
\end{aligned}
\tag{27}
$$

参 考 文 献

[1] Reissner E. Finite deflections of sandwich plates. J. Aeron Sci., 1948, 15 (7): 435-440; 1950, 17 (2): 125.

[2] Librescu L. Elastostatics and Kinetics of Anisotropic and Heterogeneous Shell-Type Structures. Noordhoff Int. Pub., Leyden, 1973.

[3] Ebcioğlu I K. A general nonlinear theory of sandwich panels. Int. J. Engng. Sci., 1989, 27 (8): 865.

[4] 刘人怀. 在边缘力矩作用下夹层圆板的非线性轴对称弯曲问题. 中国科学技术大学学报, 1980, 10 (2): 56-67.

[5] 刘人怀. 夹层圆板的非线性弯曲. 应用数学和力学, 1981, 2 (2): 173-190.

[6] 刘人怀, 施云方. 夹层圆板大挠度问题的精确解. 应用数学和力学, 1982, 3 (1): 11-23.

[7] Liu Ren-huai. Nonlinear bending of circular sandwich plates under the action of axisymmetric uniformly distributed line loads. Progress in Applied Mechanics (edited by Yeh, K. Y.), Martinus Nijhoff Publishers, Dordrecht, 1987: 293-321.

[8] 刘人怀, 朱高秋. 夹层圆板大挠度问题的进一步研究. 应用数学和力学, 1989, 10 (12): 1041-1047.

[9] 刘人怀, 吴建成. 夹层矩形板的非线性振动. 中国科学 (A辑), 1991, (10): 1075-1086.

[10] 刘人怀, 成振强. 简支夹层矩形板的非线性弯曲. 应用数学和力学, 1993, 14 (3): 203-218.

[11] Pearce T R A, Webber J P H. Buckling of sandwich panels with laminated face plates. Aeronautical Quarterly, 1972, 23: 148-160.

[12] Koganti M R. Buckling analysis of FRP faced anisotropic sandwich plates. J. AIAA, 1985, (23): 1247.

[13] Koganti M R, Meyer-Piening H R. Analysis of sandwich plates using a hybrid-stress finite element. J. AIAA, 1991, 29 (9): 1498.

夹层环形板的非线性弯曲①

一、引言

近几十年来，由于夹层板壳具有重量轻、强度高和刚度大等优良特性，所以在航天、航空和船舶制造工程等领域得到了越来越广泛的应用。但是，因为结构复杂的关系，大多研究夹层板壳的线性问题，远不能满足工程结构设计的要求，作者[1-10]在以往一系列夹层板壳非线性问题研究的基础上，进一步讨论了在内边缘均布剪力作用下夹层环形板的非线性弯曲问题。就我们所知，这一问题尚无人研究过。本文应用摄动法，获得了此问题的解析解。最后，将所得结果绘制成曲线，便于工程设计时应用。

二、边值问题的建立与求解

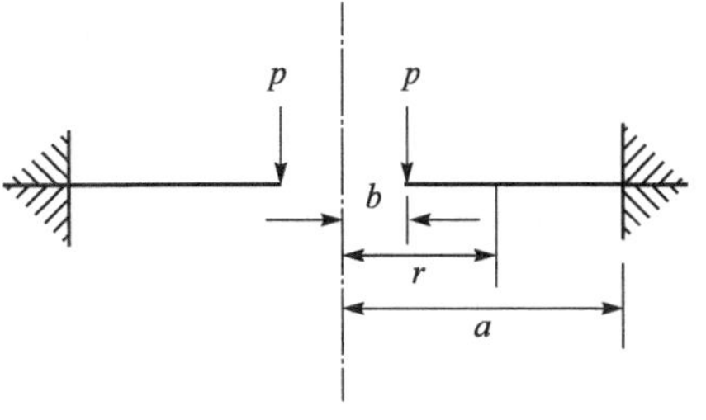

图 1　具有固定边界条件的夹层环形板

考虑如图 1 所示的在内边缘均布剪力 p 作用下具有薄表层和软夹心的夹层环形板。这里，a、b 分别为外、内缘半径。为了简单起见，本文所使用的主要符号与文献［9］相同。应用文献［9］的简化方程，我们得到此夹层环形板中面上的挠度 w 和径向应力 σ_{r0} 所满足的大挠度方程组：

$$Dr\frac{\mathrm{d}}{\mathrm{d}r}\frac{1}{r}\frac{\mathrm{d}}{\mathrm{d}r}r\frac{\mathrm{d}w}{\mathrm{d}r}-2t\Big(1-\frac{D}{G_2h_0}r\frac{\mathrm{d}}{\mathrm{d}r}\frac{1}{r}\frac{\mathrm{d}}{\mathrm{d}r}\Big)\Big(r\sigma_{r0}\frac{\mathrm{d}w}{\mathrm{d}r}\Big)-pb=0,$$

$$r\frac{\mathrm{d}}{\mathrm{d}r}\frac{1}{r}\frac{\mathrm{d}}{\mathrm{d}r}(r^2\sigma_{r0})+\frac{E}{2}\Big(\frac{\mathrm{d}w}{\mathrm{d}r}\Big)^2=0 \tag{2.1}$$

其中 D 为抗弯刚度，

$$D=\frac{Eth_0^2}{2(1-\nu^2)} \tag{2.2}$$

方程组（2.1）将在下述边界条件下求解。

当 $r=a$ 时，边缘夹紧固定：

$$w=0,\quad \psi=-\Big(\frac{2t}{G_2h_0}\sigma_{r0}+1\Big)\frac{\mathrm{d}w}{\mathrm{d}r}-\frac{pb}{G_2h_0r}=0,$$

$$u=\frac{r}{E}\Big[\frac{\mathrm{d}}{\mathrm{d}r}(r\sigma_{r0})-\nu\sigma_{r0}\Big]=0 \tag{2.3}$$

当 $r=b$ 时，边缘悬空：

$$M_r=D\Big(\frac{\mathrm{d}\psi}{\mathrm{d}r}+\nu\frac{\psi}{r}\Big)=0,\quad \sigma_{r0}=0 \tag{2.4}$$

---

① 本文原载《暨南大学学报》，1997，18（1）：1-11. 作者：刘人怀，朱金福，张小果.

为使求解简单，引入下列无量纲量：

$$x=\frac{r}{a},\quad \alpha=\frac{b}{a},W=\frac{\sqrt{2(1-\nu^2)}}{h_0}w,\quad S_r=\frac{2ta^2}{D}\sigma_{r0},$$

$$k=\frac{D}{G_2h_0a^2},\quad P=a^3\frac{\sqrt{2(1-\nu^2)}}{Dh_0}p,\quad \lambda_1=1-\nu \tag{2.5}$$

于是，非线性边值问题（2.1）、（2.3）和（2.4）转化为下述无量纲形式

$$L\left(x\frac{\mathrm{d}W}{\mathrm{d}x}\right)=-(kL-1)\left(xS_r\frac{\mathrm{d}W}{\mathrm{d}x}\right)+\alpha P,$$

$$L(x^2S_r)=-\left(\frac{\mathrm{d}W}{\mathrm{d}x}\right)^2 \tag{2.6}$$

当 $x=1$ 时，

$$W=0,\quad \frac{\mathrm{d}S_r}{\mathrm{d}x}+\lambda_1S_r=0,\quad (kS_r+1)\frac{\mathrm{d}W}{\mathrm{d}x}+k\alpha P=0 \tag{2.7}$$

当 $x=\alpha$ 时，

$$S_r=0,\frac{\mathrm{d}^2W}{\mathrm{d}x^2}+\left(k\frac{\mathrm{d}S_r}{\mathrm{d}x}+\frac{1-\lambda_1}{\alpha}\right)\frac{\mathrm{d}W}{\mathrm{d}x}-\frac{k\lambda_1}{\alpha}P=0 \tag{2.8}$$

其中，L 是微分算子，

$$L(\cdots)=x\frac{\mathrm{d}}{\mathrm{d}x}\frac{1}{x}\frac{\mathrm{d}}{\mathrm{d}x}(\cdots) \tag{2.9}$$

我们使用摄动法求解上述非线性边值问题（2.6）～（2.8）。假定夹层环形板的无量纲内边缘挠度为 W_m，将夹层环形板的 W、S_r 和 P 展开为 W_m 的幂级数

$$W=W_1(x)W_m+W_3(x)W_m^3+\cdots,$$

$$S_r=S_{r2}(x)W_m^2+S_{r4}(x)W_m^4+\cdots,$$

$$P=P_1W_m+P_3W_m^3+\cdots \tag{2.10}$$

将展开式（2.10）代入问题（2.6）～（2.8）中，使 W_m 的同此幂相等，我们就得到一系列线性边值问题。

对于 W_m，我们有关于 W_1 的边值问题

$$L\left(x\frac{\mathrm{d}W_1}{\mathrm{d}x}\right)=\alpha P_1 \tag{2.11}$$

当 $x=1$ 时，

$$W_1=0,\quad \frac{\mathrm{d}W_1}{\mathrm{d}x}+k\alpha P_1=0 \tag{2.12}$$

当 $x=\alpha$ 时，

$$W_1=1,\quad \frac{\mathrm{d}^2W_1}{\mathrm{d}x^2}+\frac{1-\lambda_1}{\alpha}\frac{\mathrm{d}W_1}{\mathrm{d}x}-\frac{k\lambda_1}{\alpha}P_1=0 \tag{2.13}$$

此问题的解为

$$W_1=\alpha P_1\left[\frac{1}{4}x^2\ln x+\left(a_1-\frac{1}{8}\right)x^2+a_2\ln x+a_3\right] \tag{2.14}$$

其中

$$a_1=\frac{(\lambda_1-2)\alpha^2\ln\alpha-\alpha^2}{4[(2-\lambda_1)\alpha^2+\lambda_1]},$$

$$a_2=\frac{(2-\lambda_1)\alpha^2\ln\alpha+[2k(\lambda_1-2)+1]\alpha^2-2k\lambda_1}{8[(2-\lambda_1)\alpha^2+\lambda_1]},\tag{2.15}$$

$$a_3=\frac{2(2-\lambda_1)\alpha^2\ln\alpha+(4-\lambda_1)\alpha^2+\lambda_1}{8[(2-\lambda_1)\alpha^2+\lambda_1]},$$

$$P_1=\frac{1}{\alpha\left[\frac{1}{4}\alpha^2\ln\alpha+a_2\ln\alpha+(1-\alpha^2)a_3\right]}\tag{2.16}$$

对于W_m^2，我们有关于S_{r2}的边值问题

$$L(x^2S_{r2})=-\left(\frac{\mathrm{d}W_1}{\mathrm{d}x}\right)^2\tag{2.17}$$

$$当\ x=1\ 时,\quad \frac{\mathrm{d}S_{r2}}{\mathrm{d}x}+\lambda_1S_{r2}=0\tag{2.18}$$

$$当\ x=\alpha\ 时,\quad S_{r2}=0\tag{2.19}$$

利用解（2.14），我们求解上面问题（2.17）～(2.19)，得

$$S_{r2}=-\alpha^2P_1^2(b_1x^2\ln^2x+b_2x^2\ln x+b_3x^2+b_4\ln^2x+b_5\ln x+b_6x^{-2}\ln x+b_7x^{-2}+b_8)\tag{2.20}$$

其中

$$b_1=\frac{1}{32},\quad b_2=\frac{1}{4}\left(a_1-\frac{3}{16}\right),\quad b_3=\frac{1}{2}\left(a_1^2-\frac{3}{8}a_1+\frac{7}{128}\right),$$

$$b_4=\frac{1}{4}a_2,\quad b_5=\left(2a_1-\frac{1}{4}\right)a_2,\quad b_6=-\frac{1}{2}a_2^2,$$

$$b_7=(\lambda_1+\lambda_2\alpha^2)^{-1}[(b_2+(2+\lambda_1)b_3+b_5+b_6)\alpha^2-\lambda_1(b_1\alpha^4\ln^2\alpha$$
$$+b_2\alpha^4\ln\alpha+b_3\alpha^4+b_4\alpha^2\ln^2\alpha+b_5\alpha^2\ln\alpha+b_6\ln\alpha)],$$

$$b_8=-(\lambda_1+\lambda_2\alpha^2)^{-1}[b_2+(2+\lambda_1)b_3+b_5+b_6+\lambda_2(b^1\alpha^4\ln^2\alpha$$
$$+b_2\alpha^4\ln\alpha+b_3\alpha^4+b_4\alpha^2\ln^2\alpha+b_5\alpha^2\ln\alpha+b_6\ln\alpha)],$$

$$\lambda_2=1+\nu\tag{2.21}$$

对于W_m^3，我们有关于W_3的边值问题

$$L\left(x\frac{\mathrm{d}W_3}{\mathrm{d}x}\right)=-(kL-1)\left(xS_{r2}\frac{\mathrm{d}W_1}{\mathrm{d}x}\right)\alpha P_3\tag{2.22}$$

当$x=1$时，

$$W_3=0,\quad \frac{\mathrm{d}W_3}{\mathrm{d}x}+kS_{r2}\frac{\mathrm{d}W_1}{\mathrm{d}x}+k\alpha P_3=0\tag{2.23}$$

当$x=\alpha$时，

$$W_3=0,\quad \frac{\mathrm{d}^2W_3}{\mathrm{d}x^2}+\frac{1-\lambda_1}{\alpha}\frac{\mathrm{d}W_3}{\mathrm{d}x}+k\frac{\mathrm{d}S_{r2}}{\mathrm{d}x}\frac{\mathrm{d}W_1}{\mathrm{d}x}-\frac{k\lambda_1}{\alpha}P_3=0\tag{2.24}$$

利用解（2.14）和（2.20），边值问题（2.21）～(2.23）的解为

$$W_3=\alpha P_3\left(\frac{1}{4}x^2\ln x-e_1x^2+e_2\ln x+e_1\right)$$

$$+\alpha^3 P_1^3(c_1x^6\ln^3x+c_2x^6\ln^2x+c_3x^6\ln x+c_4x^6+c_5x^4\ln^3x+c_6x^4\ln^2x$$
$$+c_7x^4\ln x+c_8x^4+c_9x^2\ln^3x+c_{10}x^2\ln^2x+c_{11}x^2\ln x$$
$$+c_{12}x^2+c_{13}\ln^3x+c_{14}\ln^2x+c_{15}\ln x+c_{16}x^{-2}\ln x+c_{17}x^{-2}+c_{18}) \tag{2.25}$$

其中

$$P_3=-\alpha^3P_1^4(c_1\alpha^6\ln^3\alpha+c_2\alpha^6\ln^2\alpha+c_3\alpha^6\ln\alpha+c_4\alpha^6+c_5\alpha^4\ln^3\alpha+c_6\alpha^4\ln^2\alpha$$
$$+c_7\alpha^4\ln\alpha+c_8\alpha^4+c_9\alpha^2\ln^3\alpha+c_{10}\alpha^2\ln^2\alpha+c_{11}\alpha^2\ln\alpha$$
$$+c_{12}\alpha^2+c_{13}\ln^3\alpha+c_{14}\ln^2\alpha+c_{16}\alpha^{-2}\ln\alpha+c_{17}\alpha^{-2}+c_{18}) \tag{2.26}$$

$$c_1=\frac{d_1}{6},$$
$$c_2=-\frac{1}{6}\left(\frac{d_1}{2}-d\right),$$
$$c_3=\frac{1}{6}\left(\frac{d_1}{6}-\frac{d_2}{3}+d_3\right),$$
$$c_4=-\frac{1}{6}\left(\frac{d_1}{36}-\frac{d_2}{18}+\frac{d_3}{6}-d_4\right),$$
$$c_5=\frac{d_5}{4},$$
$$c_6=-\frac{1}{4}\left(\frac{3}{4}d_5-d_6\right),$$
$$c_7=\frac{1}{4}\left(\frac{3}{8}d_5-\frac{d_6}{2}+d_8\right),$$
$$c_8=-\frac{1}{4}\left(\frac{3}{32}d_5-\frac{d_6}{8}+\frac{d_8}{4}-d_{10}\right),$$
$$c_9=\frac{d_7}{2},$$
$$c_{10}=-\frac{1}{2}\left(\frac{3}{2}d_7-d_9\right),$$
$$c_{11}=\frac{1}{2}\left(\frac{3}{2}d_7-d_9+d_{11}\right),$$
$$c_{12}=-\frac{1}{2}(c_{11}-d_{19}),$$
$$c_{13}=\frac{d_{12}}{3},$$
$$c_{14}=\frac{d_{13}}{2},$$
$$c_{15}=k(2a_1b_7+a_2b_8),$$
$$c_{16}=-\frac{d_{14}}{2},$$
$$c_{17}=-\frac{1}{2}\left(\frac{d_{14}}{2}+d_{15}\right),$$
$$c_{18}=-(c_4+c_8+c_{12}+c_{17}),$$

$$d_1 = -\frac{b_1}{48},$$

$$d_2 = -\frac{1}{192}(16a_1b_1 - 5b_1 + 4b_2),$$

$$d_3 = \frac{1}{1152}(80a_1b_1 - 96a_1b_2 - 19b_1 + 20b_2 - 24b_3),$$

$$d_4 = -\frac{1}{13824}(340a_1b_1 - 480a_1b_2 + 1152a_1b_3 - 65b_1 + 76b_2 - 120b_3),$$

$$d_5 = \frac{1}{2}\left(kb_1 - \frac{b_4}{8}\right),$$

$$d_6 = k\left(2a_1b_1 + \frac{b_2}{2}\right) - \frac{1}{64}(16a_1b_4 + 8a_2b_1 - 9b_4 + 4b_5),$$

$$d_7 = \frac{1}{2}\left[kb_4 - \frac{1}{3}\left(a_2b_4 + \frac{b_6}{2}\right)\right],$$

$$d_8 = k\left(2a_1b_2 + \frac{b_3}{2}\right) + \frac{1}{640}(384a_1b_4 - 160a_1b_5 + 192a_2b_1 - 80a_2b_2 - 105b_4 + 96b_5 - 40b_8),$$

$$d_9 = k\left(2a_1b_4 + a_2b_1 + \frac{b_5}{2}\right) - \frac{1}{4}\left(2a_1b_6 - a_2b_4 + a_2b_5 - \frac{b_6}{2} + \frac{b_7}{2}\right),$$

$$d_{10} = 2ka_1b_3 - \frac{1}{152}(112a_1b_4 - 96a_1b_5 + 128a_1b_8 + 56a_2b_1 - 48a_2b_2 + 64a_2b_3 - 45b_4 + 28b_5 - 24b_8),$$

$$d_{11} = k\left(2a_1b_5 + a_2b_2 + \frac{b_8}{2}\right) + \frac{1}{4}\left(2a_1b_6 - 4a_1b_7 - a_2b_4 + a_2b_5 - 2a_2b_8 - \frac{b_6}{2} + \frac{b_7}{2}\right),$$

$$d_{12} = k\left(a_2b_4 + \frac{b_6}{2}\right) + \frac{1}{4}a_2b_6,$$

$$d_{13} = k\left(2a_1b_6 + a_2b_5 + \frac{b_7}{2}\right) + \frac{a_2}{2}\left(\frac{b_6}{2} + b_7\right),$$

$$d_{14} = ka_2b_6,$$

$$d_{15} = ka_2b_7,$$

$$d_{16} = k(2a_1b_8 + a_2b_3) - \frac{1}{24}\left(6a_1b_6 - 12a_1b_7 - 3a_2b_4 + 3a_2b_5 - 6a_2b_8 - \frac{3}{2}b_6 + \frac{3}{2}b_7\right),$$

$$e_1 = \frac{f_1(f_2 + \lambda_1 f_3\alpha^{-1}) - (f_4 + \lambda_1 f_5\alpha^{-1})}{2f_1(\lambda_2\alpha + \lambda_1\alpha^{-1})},$$

$$e_2 = \frac{f_1(f_2 - \lambda_2 f_3\alpha) - (f_4 - \lambda_2 f_5\alpha)}{f_1(\lambda_2\alpha + \lambda_1\alpha^{-1})},$$

$$f_1 = f_6[2(\lambda_2\alpha + \lambda_1\alpha^{-1})(c_1\alpha^6\ln^3\alpha + c_2\alpha^6\ln^2\alpha + c_3\alpha^6\ln\alpha + c_4\alpha^6 + c_5\alpha^4\ln^3\alpha + c_6\alpha^4\ln^2\alpha + c_7\alpha^4\ln\alpha + c_8\alpha^4 + c_9\alpha^2\ln^3\alpha + c_{10}\alpha^2\ln^2\alpha + c_{11}\alpha^2\ln\alpha + c_{12}\alpha^2 + c_{13}\ln^3\alpha$$

$$+c_{14}\ln^2\alpha+c_{15}\ln\alpha+c_{16}\alpha^{-2}\ln\alpha+c_{17}\alpha^{-2}+c_{18})+f_7],$$

$$f_2=\frac{\alpha}{2}\left[\lambda_2\ln\alpha+\frac{1}{2}(4-\lambda_1)\right]-k\lambda_1\alpha^{-1},$$

$$f_3=k+\frac{1}{4},$$

$$f_4=(1-\lambda_1)f_8+f_9\alpha-kf_{10},$$

$$f_5=k^2(b_3+b_7+b_8)+f_{11},$$

$$f_6=\left[\lambda_2\alpha\ln^2\alpha+2(1-k\lambda_2)\alpha\ln\alpha-2k\lambda_1\alpha^{-1}\ln\alpha+\frac{\lambda_1}{4}\alpha^{-1}\right]^{-1},$$

$$f_7=-f_4(\alpha^2-2\ln\alpha-1)-f_5(2\lambda_2\alpha\ln\alpha+\lambda_1\alpha-\lambda_1\alpha^{-1}),$$

$$f_8=d_1\alpha^5\ln^3\alpha+d_2\alpha^5\ln^2\alpha+d_3\alpha^5\ln\alpha+d_4\alpha^5+d_5\alpha^3\ln^3\alpha+d_6\alpha^3\ln^2\alpha+d_8\alpha^3\ln\alpha$$
$$+d_{10}\alpha^3+d_7\alpha\ln^3\alpha+d_9\alpha\ln^2\alpha+d_{11}\alpha\ln\alpha+d_{16}\alpha+d_{12}\alpha^{-1}\ln^2\alpha+d_{13}\alpha^{-1}\ln\alpha$$
$$+d_{15}\alpha^{-1}+d_{14}\alpha^{-3}\ln\alpha+d_{15}\alpha^{-3},$$

$$f_9=5d_1\alpha^4\ln^3\alpha+(3d_1+5d_2)\alpha^4\ln^2\alpha+(2d_2+5d_3)\alpha^4\ln\alpha+(d_3+5d_4)\alpha^4$$
$$+3d_5\alpha^2\ln^3\alpha+3(d_5+d_6)\alpha^2\ln^2\alpha+(2d_6+3d_8)\alpha^2\ln\alpha+(d_8+3d_{10})\alpha^2$$
$$+d_7\ln^3\alpha+(3d_7+d_9)\ln^2\alpha+(2d_9+d_{11})\ln\alpha$$
$$-2d_{12}\alpha^{-2}\ln^2\alpha+(2d_{12}-d_{13})\alpha^{-2}\ln\alpha-(c_{15}-d_{13})\alpha^{-2}-3d_{14}\alpha^{-4}\ln\alpha$$
$$+(d_{14}-3d_{15})\alpha^{-4}+d_{11}+d_{16},$$

$$f_{10}=\left(\frac{1}{2}\alpha^2\ln\alpha+2a_1\alpha^2+a_2\right)[2b_1\alpha\ln^2\alpha+2(b_1+b_2)\alpha\ln\alpha$$
$$+(b_2+2b_3)\alpha+2b_4\alpha^{-1}\ln\alpha+b_5\alpha^{-1}-2b_6\alpha^{-3}\ln\alpha+(b_6-2b_7)\alpha^{-3}],$$

$$f_{11}=c_3+6c_4+c_7+4c_8+c_{11}+2c_{12}+c_{15}+c_{16}-2c_{17} \tag{2.27}$$

对于 W_m^4，我们有关 S_{r4} 的边值问题

$$L(x^2S_{r4})=-2\left(\frac{\mathrm{d}W_1}{\mathrm{d}x}\right)\left(\frac{\mathrm{d}W_3}{\mathrm{d}x}\right) \tag{2.28}$$

$$\text{当 } x=1 \text{ 时，}\quad \frac{\mathrm{d}S_{r4}}{\mathrm{d}x}+\lambda_1S_{r4}=0 \tag{2.29}$$

$$\text{当 } x=\alpha \text{ 时，}\quad S_{r4}=0 \tag{2.30}$$

利用解（2.14）和（2.25），求解问题（2.28）～（2.30），得

$$S_{r4}=-\alpha^4P_1^4(g_1x^6\ln^4x+g_2x^6\ln^3x+g_3x^6\ln^2x+g_4x^6\ln x+g_5x^6$$
$$+g_6x^4\ln^4x+g_7x^4\ln^3x+g_8x^4\ln^2x+g_9x^4\ln x+g_{10}x^4+g_{11}x^2\ln^4x+g_{12}x^2\ln x$$
$$+g_{13}x^2\ln^2x+g_{14}x^2\ln x+g_{15}x^2+g_{16}\ln^4x+g_{17}\ln^3x+g_{18}\ln^2x+g_{19}\ln x$$
$$+g_{20}x^{-2}\ln^3x+g_{21}x^{-2}\ln^2x+g_{22}x^{-2}\ln x$$
$$+g_{23}x^{-2}+g_{24}x^{-4}\ln x+g_{25}x^{-4}+g_{26}) \tag{2.31}$$

其中

$$g_1=\frac{h_1}{48},$$

$$g_2=\frac{1}{48}\left(h_2-\frac{7}{6}h_1\right),$$

$$g_3=\frac{1}{48}\left(h_3-\frac{7}{8}h_2+\frac{37}{48}h_1\right),$$

$$g_4=\frac{1}{48}\left(h_4-\frac{7}{12}h_3+\frac{37}{96}h_2-\frac{175}{576}h_1\right),$$

$$g_5=\frac{1}{48}\left(h_5-\frac{7}{24}h_4+\frac{37}{288}h_3-\frac{175}{2304}h_2+\frac{781}{13824}h_1\right),$$

$$g_6=\frac{h_6}{24},$$

$$g_7=\frac{1}{24}\left(h_7-\frac{5}{3}h_6\right),$$

$$g_8=\frac{1}{24}\left(h_8-\frac{5}{4}h_7+\frac{19}{12}h_6\right),$$

$$g_9=\frac{1}{24}\left(h_9-\frac{5}{6}h_8+\frac{19}{24}h_7-\frac{65}{72}h_6\right),$$

$$g_{10}=\frac{1}{24}\left(h_{10}-\frac{5}{12}h_9+\frac{19}{72}h_8-\frac{65}{288}h_7+\frac{211}{864}h_6\right),$$

$$g_{11}=\frac{h_{11}}{8},$$

$$g_{12}=\frac{1}{8}(h_{12}-3h_{11}),$$

$$g_{13}=\frac{1}{8}\left(h_{13}-\frac{9}{4}h_{12}+\frac{21}{4}h_{11}\right),$$

$$g_{14}=\frac{1}{8}\left(h_{14}-\frac{3}{2}h_{13}+\frac{21}{8}h_{12}-\frac{45}{8}h_{11}\right),$$

$$g_{15}=\frac{1}{8}\left(h_{15}-\frac{3}{4}h_{14}+\frac{7}{8}h_{13}-\frac{45}{32}h_{12}+\frac{93}{32}h_{11}\right),$$

$$g_{16}=\frac{h_{16}}{8},$$

$$g_{17}=\frac{1}{2}\left(\frac{h_{17}}{3}-\frac{h_{16}}{2}\right),$$

$$g_{18}=\frac{1}{4}\left(h_{18}-h_{17}+\frac{3}{2}h_{16}\right),$$

$$g_{19}=\frac{1}{2}\left(h_{19}-\frac{h_{18}}{2}+\frac{h_{17}}{2}-\frac{3}{4}h_{16}\right),$$

$$g_{20}=-\frac{h_{20}}{4},$$

$$g_{21}=-\frac{1}{4}\left(h_{21}+\frac{3}{2}h_{20}\right),$$

$$g_{22}=-\frac{1}{2}\left(h_{22}+\frac{h_{21}}{2}+\frac{3}{4}h_{20}\right),$$

$$g_{23}=\frac{m_2-\lambda_1 m_1}{\lambda_2+\lambda_1\alpha^{-2}},$$

$$g_{24}=\frac{h_{23}}{8},$$

$$g_{25}=\frac{1}{8}\left(h_{24}+\frac{3}{4}h_{23}\right),$$

$$g_{26}=-\frac{\lambda_2 m_1+m_2\alpha^{-2}}{\lambda_2+\lambda_1\alpha^{-2}},$$

$$h_1=d_1,$$

$$h_2=4a_1d_1+d_2,$$

$$h_3=4a_1d_2+d_3,$$

$$h_4=4a_1d_3+d_4,$$

$$h_5=4a_1d_4,$$

$$h_6=d_5,$$

$$h_7=4a_1d_5+2a_2d_1+d_6,$$

$$h_8=4a_1d_6+2a_2d_2+d_8,$$

$$h_9=4a_1d_8+2a_2d_3+d_{10},$$

$$h_{10}=2(2a_1d_{10}+a_2d_4),$$

$$h_{11}=d_7,$$

$$h_{12}=4a_1d_7+2a_2d_5+d_9,$$

$$h_{13}=4a_1d_9+2a_2d_5+d_{11}-\frac{f_1}{2},$$

$$h_{14}=2a_1(2d_{11}-f_1)+2a_2d_8+d_{16}-f_1m_3,$$

$$h_{15}=2[2a_1(d_{16}-f_1m_3)+a_2d_{10}],$$

$$h_{16}=2a_2d_7+d_{12},$$

$$h_{17}=4a_1d_{12}+2a_2d_9+d_{13},$$

$$h_{18}=4a_1d_{13}+a_2(2d_{11}-f_1)-t_2f_1+m_4,$$

$$h_{19}=-2[2a_1(e_2f_1-m_4)-a_2(d_{16}-f_1m_3)],$$

$$h_{20}=2a_2d_{12}+d_{13},$$

$$h_{21}=4a_1d_{14}+2a_2d_{13}+d_{15},$$

$$h_{22}=2[2a_1d_{15}-a_2(e_2f_1-m_4)],$$

$$h_{23}=2a_2d_{14},$$

$$h_{24}=2a_2d_{15},$$

$$\begin{aligned}m_1=&g_1\alpha^6\ln^4\alpha+g_2\alpha^6\ln^3\alpha+g_3\alpha^6\ln^2\alpha+g_4\alpha^6\ln\alpha+g_5\alpha^6+g_6\alpha^4\ln^4\alpha\\&+g_7\alpha^4\ln^3\alpha+g_8\alpha^4\ln^2\alpha+g_9\alpha^4\ln\alpha+g_{10}\alpha^4+g_{11}\alpha^2\ln^4\alpha+g_{12}\alpha^2\ln^3\alpha\\&+g_{13}\alpha^2\ln^2\alpha+g_{14}\alpha^2\ln\alpha+g_{15}\alpha^2+g_{16}\ln^4\alpha+g_{17}\ln^3\alpha+g_{18}\ln^2\alpha+g_{19}\ln\alpha\\&+g_{20}\alpha^{-2}\ln^3\alpha+g_{21}\alpha^{-2}\ln^2\alpha+g_{22}\alpha^{-2}\ln\alpha+g_{24}\alpha^{-4}\ln\alpha+g_{25}\alpha^{-4},\end{aligned}$$

$$\begin{aligned}m_2=&g_4+6g_5+g_9+4g_{10}+g_{14}+2g_{15}+g_{19}+g_{22}\\&+g_{24}-4g_{25}+\lambda_1(g_5+g_{10}+g_{15}+g_{25}),\end{aligned}$$

$$m_3 = -\left(2e_1 - \frac{1}{4}\right),$$

$$m_4 = 2a_1 b_7 + a_2 b_8 \tag{2.32}$$

如果继续往前求解，我们还可得到高阶近似解。

三、主要结果和结论

根据上述结果，我们有

$$P = P_1 W_m + P_3 W_m^3, \tag{3.1}$$

$$S_r = S_{r2} W_m^2 + S_{r4} W_m^4 \tag{3.2}$$

对于夹层环形板中面上的环向应力 $\sigma_{\theta 0}$，可按下面公式计算

$$\sigma_{\theta 0} = \frac{\mathrm{d}(r\sigma_{\theta 0})}{\mathrm{d}r} \tag{3.3}$$

设无量纲环向应力 S_θ 为

$$S_\theta = \frac{2ta^2}{D}\sigma_{\theta 0} \tag{3.4}$$

则有

$$S_\theta = \frac{\mathrm{d}(xS_r)}{\mathrm{d}x} \tag{3.5}$$

将式（3.2）代入上式，得

$$S_\theta = \frac{\mathrm{d}}{\mathrm{d}x}(xS_{r2})W_m^2 + \frac{\mathrm{d}}{\mathrm{d}x}(xS_{r4})W_m^4 \tag{3.6}$$

由式（3.2）和（3.6），可得到夹层环形板外边缘的无量纲径向应力 $S_r(1)$ 和内边缘的无量纲环向应力 $S_\theta(\alpha)$ 的计算公式

$$S_r(1) = S_{r2}(1)W_m^2 + S_{r4}(1)W_m^4, \tag{3.7}$$

$$S_\theta(\alpha) = \left[S_{r2}(\alpha) + \alpha \frac{\mathrm{d}S_{r2}}{\mathrm{d}x}\Big|_{x=\alpha}\right]W_m^2 + \left[S_{r4}(\alpha) + \alpha \frac{\mathrm{d}S_{r4}}{\mathrm{d}x}\Big|_{x=\alpha}\right]W_m^4 \tag{3.8}$$

其中

$$S_{r2}(1) = -\alpha^2 P_1^2(b_3 + b_7 + b_8),$$

$$S_{r4}(1) = -\alpha^4 P_1^4(g_5 + g_{10} + g_{15} + g_{23} + g_{25} + g_{26}) \tag{3.9}$$

按照式（3.1）、（3.7）和（3.8），对于 $\nu=0.3$ 和 $k=0.05$ 情况，我们进行了数值计算，所得的数值结果分别给在图 2～图 4 中。

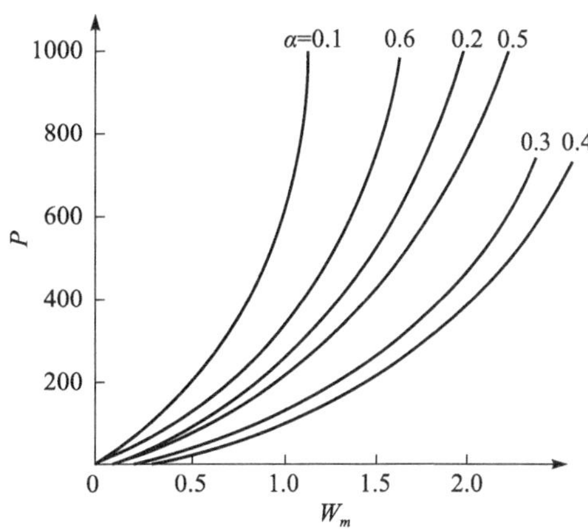

图 2　夹层环形板的特征曲线
（$\nu=0.3$，$k=0.05$）

由图 2 看到，夹层环形板的特征曲线呈单调上升，且在中心孔较小时，在同一载荷下，挠度 W_m 随孔半径 α 的增大而增加，而当中心孔较大时，结果则相反。

由图 3 可知，夹层环形板外边缘的径

向应力是负值，其绝对值随着 W_m 的增大而增加。在中心孔较小时，对于同一 W_m 值，其径向应力绝对值随着孔半径 α 的增大而减少，然而在中心孔较大时，情况则相反。

最后，从图 4 看出，夹层环形板内边缘的环向应力在中心孔较小或较大时，随挠度 W_m 的增大而增加，而在中心孔中等大小时，其绝对值随着 W_m 的增大而增加。

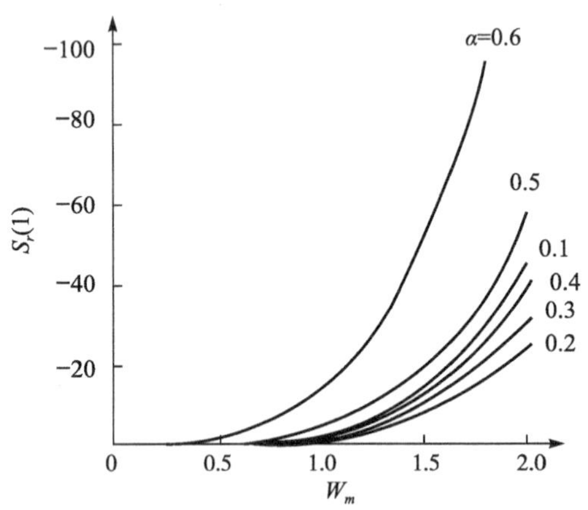

图 3　夹层环形板外边缘的径向应力

(ν=0.3，k=0.05)

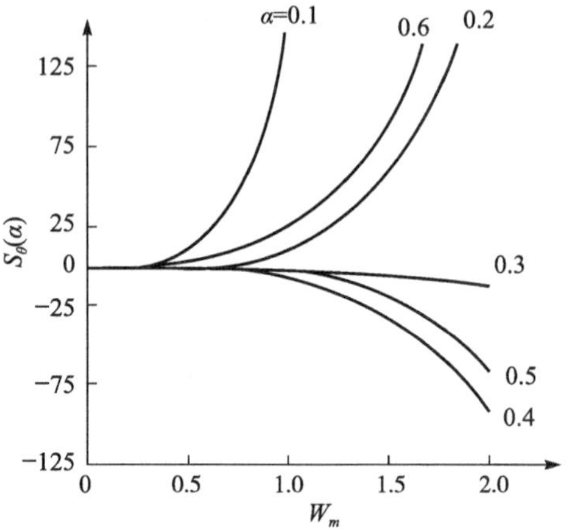

图 4　夹层环形板内边缘的环向应力

(ν=0.3，k=0.05)

参 考 文 献

[1] Liu Ren-huai，Cheng Zhenqiang. On the non-linear buckling of circular shallow spherical sandwich shells under the action of uniform edge moments. International Journal of Non-Linear Mechanics，1995，30 (1)：33.

[2] Liu Ren-huai，Li Jun. Non-linear vibration of shallow conical sandwich shells. International Journal of Non-Linear Mechanics，1995，30 (2)：97.

[3] 刘人怀，朱金福. 夹层壳非线性理论. 北京：机械工业出版社，1993.

[4] 刘人怀，成振强. 简支夹层矩形板的非线性弯曲. 应用数学和力学，1993，14 (3)：203.

[5] 刘人怀，吴建成. 夹层矩形板的非线性振动. 中国科学，A 辑，1991，(10)：1075.

[6] 刘人怀，朱高秋. 夹层圆板大挠度问题的进一步研究. 应用数学和力学，1989，10 (12)：1041.

[7] Liu Ren-huai. Nonlinear bending of circular sandwich plates under the action of axisymmetric uniformly distributed line loads. Progress in Applied Mechanics，Martinus Nijhoff Publishers，Dordrecht，1987：293.

[8] 刘人怀，施云方. 夹层圆板大挠度问题的精确解. 应用数学和力学，1982，3 (1)：11.

[9] 刘人怀. 夹层圆板的非线性弯曲. 应用数学和力学，1981，2 (2)：173.

[10] 刘人怀. 在边缘力矩作用下夹层圆板的非线性轴对称弯曲问题. 中国科学技术大学学报，1980，10 (2)：56.

Nonlinear theory of sandwich shells
part Ⅰ: Exact kinematics of moderately thick shells①

中文摘要

夹层壳由两个薄的表层和其中间的一个较厚的夹心所组成。由于具有强度与重量和刚度与重量的高比值，夹层壳越来越广泛地用在航空、航天、航海、土木和汽车制造等工程中。因此，夹层板壳的研究越来越受到关注。本文考虑了夹心的横向剪切刚度对夹层壳力学特性影响，将夹心考虑为一个中厚壳，为此建立中厚壳的非线性几何理论。

1. Introduction

Sandwich shells consist of two layers of thin faces and one layer of a relatively thick core that is between the faces. Because of their high rates of strength to weight and stiffness to weight, sandwich shells are used more and more widely in aeronautics and astronautics, naval offshore and civil engineering and automobile industry, etc. Thus the literature on the analysis and computation of sandwich plates and shells is growing constantly (Adi Murthy and Alwar, 1976; Group of Plates and Shells, 1977; Habip, 1965; Wu, 1991; Liu et al., 1980, 1981, 1982, 1987, 1989, 1991, 1993, 1995; Reissner, 1949; Schmidt, 1964; Zhu, 1989).

The instability of sandwich shells is an essential problem in most cases, and the stability analysis must use the nonlinear theory. Although there have been some large deflection equations (Liu and Cheng, 1995; Roman and Kao, 1975; Wampner, 1967; Wang, 1950) of sandwich shells, a systematic study on the nonlinear theory of them is still lacking.

We know that the effect of the transverse shear stiffness of its core on the mechanical characteristics of the sandwich shell should be taken into account. Therefore the core is considered as a moderately thick shell and the nonlinear geometric theory of moderately thick shells is established in this paper. A novel analysis is used to derive the concise expressions of the important geometric quantities of deformed shells. The exact compatibility equations that are the developments of Koiter's equations of thin shells are obtained.

In the follow-up paper (Liu and Zhu, 1997), the authors will simplify the equations obtained in this paper under the condition of small strain associated with moderate

① Reprinted from *Applied Mechanics and Engineering*. 1997, 2 (2): 213-240. Authors: Liu Renhuai and Zhu Jinfu.

rotation. Then the associated physical equations, including the constitutive equations, the strain energy expression, the equilibrium equations and the boundary conditions of the sandwich shells, will be derived.

2. The Geometry of Moderately Thick Shells

Now consider a moderately thick shell $\mathfrak{H}$ (Fig. 1) . The initial state of the shell is taken as the reference configuration $\mathfrak{I}$, its middle surface in the reference configuration $\mathfrak{I}$ is the reference surface $\mathfrak{R}$. On the reference surface $\mathfrak{R}$, a local surface coordinate (θ^α, z) is set up. Here, the coordinate axis z is normal to the reference surface $\mathfrak{R}$.

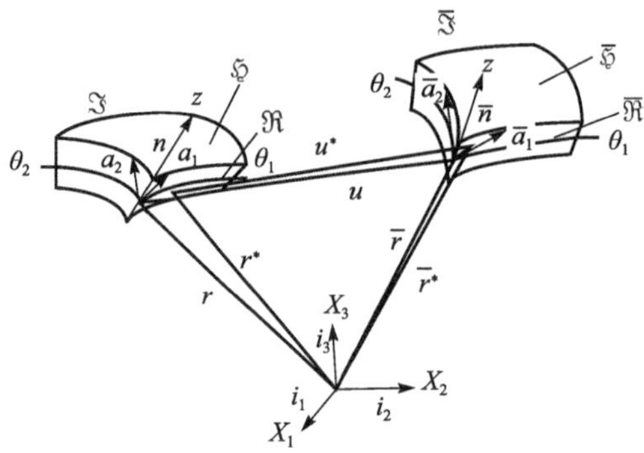

Fig. 1 Motion of a moderately thick shell

At some instant t, the shell moves to a new place $\bar{\mathfrak{H}}$ in the current configuration $\bar{\mathfrak{I}}$, the original middle surface $\mathfrak{R}$ occupies place $\bar{\mathfrak{R}}$. $\bar{\mathfrak{R}}$ need not be the middle surface of $\bar{\mathfrak{H}}$. But in order to discuss conveniently, we still call $\bar{\mathfrak{R}}$ the present middle surface. According to Chien (1944), we take (θ^α, z) as the comoving coordinate, the coordinate line z in the present configuration $\bar{\mathfrak{I}}$ need not be a straight line and not be normal to the present middle surface $\bar{\mathfrak{R}}$.

The position vectors of a typical Particle X on $\mathfrak{R}$ and $\bar{\mathfrak{R}}$ are assumed to be r and $\bar{r}$, respectively. Thus, the base vectors of $\mathfrak{R}$ and $\bar{\mathfrak{R}}$ are:

$$a_\alpha = r_{,\alpha}, \tag{2. 1a}$$

$$\bar{a}_\alpha = \bar{r}_{,\alpha} \tag{2. 1b}$$

and the outer unit normals to $\mathfrak{R}$ and $\bar{\mathfrak{R}}$ are:

$$n = \frac{1}{2} \in^{\alpha\beta} a_\alpha \times a\beta, \tag{2. 2a}$$

$$\bar{n} = \frac{1}{2} \bar{\in}^{\alpha\beta} \bar{a}_\alpha \times \bar{a}\beta \tag{2. 2b}$$

where $\in^{\alpha\beta}$and $\bar{\in}^{\alpha\beta}$ are Eddington tensors of 2-space on $\mathfrak{R}$ amd $\bar{\mathfrak{R}}$, respectively.

Ler r^* and $\bar{r}^*$ be the position vectors of a typical particle X of $\mathfrak{H}$ in $\mathfrak{I}$ and $\bar{\mathfrak{I}}$, respectively, then the base vectors at X in $\mathfrak{I}$ and $\bar{\mathfrak{I}}$ are:

$$g_\alpha = r^*_{,\alpha}, \quad \bar{g}_\alpha = \bar{r}_{,\alpha}, \tag{2.3a,b}$$

$$g_3 = r^*_{,3}, \quad \bar{g}_3 = \bar{r}^*_{,3}. \tag{2.3c,d}$$

Here and in what follows:

$$(\ \ )_{,\alpha} = \frac{\partial}{\partial \theta^\alpha}, (\ \ )_{,3} = \frac{\partial}{\partial z}.$$

Obviously, we have:

$$a_\alpha = g_\alpha |_z = 0, \quad \bar{a}_\alpha = \bar{g}_\alpha |_z = 0,$$

$$a_3 = g_3 = n, \quad \bar{a}_3 = \bar{g}_3 |_z = 0$$

where $(\ \ )|_{z=0}$ means the value of the quantity in the brackets when $z=0$.

We recall that the first fundamental forms of $\mathfrak{R}$ and $\bar{\mathfrak{R}}$ are:

$$a_{\alpha\beta} = a_\alpha \cdot a_\beta, \tag{2.4a}$$

$$\bar{a}_{\alpha\beta} = \bar{a}_\alpha \cdot a_\beta. \tag{2.4b}$$

Furthermore:

$$a_{\alpha 3} = 0, \qquad a_{33} = 1, \tag{2.5a,b}$$

$$\bar{a}_{\alpha 3} = \bar{a}_\alpha \cdot \bar{a}_3, \quad \bar{a}_{33} = \bar{a}_3 \cdot \bar{a}_3. \tag{2.5c,d}$$

The second fundamental forms of $\mathfrak{R}$ and $\bar{\mathfrak{R}}$ are:

$$b_{\alpha\beta} = a_{\alpha,\beta} \cdot n = - n_{,\alpha} \cdot a_\beta, \tag{2.6a}$$

$$\bar{b}_{\alpha\beta} = \bar{a}_{\alpha,\beta} \cdot \bar{n} = - \bar{n}_{,\alpha} \cdot \bar{a}_\beta. \tag{2.6b}$$

It is easy to prove that $a_{\alpha\beta}$, $\bar{a}_{\alpha\beta}$, $b_{\alpha\beta}$ and $\bar{b}_{\alpha\beta}$ are the symmetric tensors (Koiter, 1966).

In $\mathfrak{R}$, the outer unit normal n coincides with the coordinate line z, hence there exists the relation (Naghdi, 1972):

$$r^* = r + zn.$$

According to Eqs. (2.1a) and (2.6a), we have:

$$g_\alpha = r^*_{,\alpha} = a_\alpha - z b^\beta_\alpha a_\beta = \mu^\beta_\alpha a_\beta \tag{2.7}$$

$$\mu^\beta_\alpha = \delta^\beta_\alpha - z b^\beta_\alpha \tag{2.8}$$

where δ^β_α is the Kronecker symbol.

The space metric tensor $g_{\alpha\beta}$ can be given by:

$$g_{\alpha\beta} = g_\alpha \cdot g_\beta = \mu^\gamma_\alpha \mu^\pi_\beta a_{\gamma\pi} = a_{\alpha\beta} - 2z b_{\alpha\beta} + z^2 c_{\alpha\beta}, \tag{2.9a}$$

$$g_{\alpha 3} = 0, \tag{2.9b}$$

$$g_{33} = 1 \tag{2.9c}$$

in which $c_{\alpha\beta}$ is the third fundamental form of $\mathfrak{R}$:

$$c_{\alpha\beta} = b^\lambda_\alpha b_{\lambda\beta}. \tag{2.10}$$

Let a and g be the determiants of the surface metric tensor $a_{\alpha\beta}$ and the space metric tensor $g_{\alpha\beta}$, b the determinant of $b_{\alpha\beta}$, then from (2.9a), we find:

$$g = \mu^2 a \tag{2.11}$$

where

$$\mu = 1 - 2Hz + Kz^2, \tag{2.12a}$$

$$H = b^\alpha_\alpha / 2, \tag{2.12b}$$

$$K = \frac{1}{2} \in^{\alpha\lambda} \in \beta\mu b^\beta_\alpha b^\mu_\lambda. \tag{2.12c}$$

Hence

$$(g/a)^{1/2} = \mu = 1 - 2Hz + Kz^2. \tag{2.13}$$

The reciprocal base vectors of a_α and g_α are:

$$a^\alpha = a^{\alpha\beta} a_\beta, \tag{2.14a}$$

$$g^\alpha = g^{\alpha\beta} g_\beta \tag{2.14b}$$

and we have:

$$a^{\alpha\beta} = \in^{\alpha\lambda} \in^{\beta\mu} a_{\lambda\mu}, \tag{2.15a}$$

$$a^{\alpha\lambda} a_{\lambda\beta} = \delta^\alpha_\beta, \tag{2.15b}$$

$$g^{\alpha\lambda} g_{\lambda\beta} = \delta^\alpha_\beta. \tag{2.15c}$$

Similary to Eq. (2.7), we can write:

$$g^\alpha = \overset{-1}{\mu^\alpha_\beta} a^\beta \tag{2.16}$$

where $\overset{-1}{\mu^\alpha_\beta}$ is just the inverse of μ^α_β:

$$\overset{-1}{\mu^\alpha_\beta} = (\mu^\alpha_\beta)^{-1} = (\delta^\alpha_\beta - zb^\alpha_\beta)^{-1} = \delta^\alpha_\beta + zb^\alpha_\beta + z^2 b^\alpha_\gamma b^\gamma_\beta + \cdots. \tag{2.17}$$

In fact, we know that:

$$\delta^\alpha_\beta = g^\alpha \cdot g_\beta = \overset{-1}{\mu^\alpha} \gamma a^\gamma \cdot \mu^\pi_\beta a_\pi = \overset{-1}{\mu^\alpha_\gamma} \mu^\gamma_\beta.$$

Hence

$$g^{\alpha\beta} = g^\alpha \cdot g^\beta = \overset{-1}{\mu^\alpha} \gamma \overset{-1}{\mu^\beta_\pi} a^{\gamma\pi} = a^{\alpha\beta} + 2zb^{\alpha\beta} + 3z^2 b^\alpha_\gamma b^{\gamma\beta} + \cdots. \tag{2.18}$$

The reciprocal base vector of $(\bar{a}_\alpha, \bar{n})$ is:

$$\bar{a}^\alpha = \bar{a}^{\alpha\beta} \bar{a}_\beta \tag{2.19}$$

where

$$\bar{a}^{\alpha\beta} = \overline{\in}^{\alpha\lambda} \overline{\in}^{\beta\mu} \bar{a}^{\lambda\mu}. \tag{2.20}$$

This gives:

$$\frac{1}{2} \overline{\in}^{\alpha\lambda} \overline{\in}^{\beta\mu} \bar{a}_{\lambda\mu} \bar{a}_{\alpha\beta} = 1. \tag{2.21}$$

For late convenience, we define another set of base vectors $(\bar{a}_\alpha, \bar{a}_3)$ on $\mathfrak{R}$. Their reciprocal base vectors are $(\bar{a}^{*\alpha}, \bar{a}^{*3})$. At present, because $\bar{a}_3$ is not the normal to $\mathfrak{R}$, $(\bar{a}_\alpha, \bar{a}_3)$ is not a 2-D base vector, but is a 3-D base vector defined on $\mathfrak{R}$. In addition, $\bar{a}^{*3}$ is orthogonal to $\bar{a}_\alpha$, thus we can write:

$$\bar{a}^{*3} = A\bar{n}. \tag{2.22}$$

Furthermore:

$$\bar{a}^{*i} \cdot \bar{a}_j = \delta^i_j, \tag{2.23a}$$

$$\bar{a}^{\alpha}\cdot\bar{a}_{\beta}=\delta^{\alpha}_{\beta} \tag{2.23b}$$

but

$$\bar{a}^{\alpha}\cdot\bar{a}_{3}\neq 0. \tag{2.24}$$

Meanwhile, using Eqs. (2.20) and (2.21), $\bar{a}^{*ij}$ can be expressed as:

$$\begin{aligned}\bar{a}^{*\alpha\beta}&=\frac{1}{2}\bar{\in}^{*\alpha lm}\,\bar{\in}^{*\beta kn}\bar{a}_{lk}\bar{a}_{mn}\\&=\frac{\bar{a}}{\bar{a}^{*}}\bar{\in}^{\alpha\lambda}\,\bar{\in}^{\beta\mu}\bar{a}_{\lambda\mu}\bar{a}_{33}-\frac{\bar{a}}{\bar{a}^{*}}\bar{\in}^{\alpha\lambda}\,\bar{\in}^{\beta\mu}\bar{a}_{3\lambda}\bar{a}_{3\mu}\\&=\frac{\bar{a}}{\bar{a}^{*}}\bar{\in}^{\alpha\beta}\bar{a}_{33}-\frac{\bar{a}}{\bar{a}^{*}}\bar{\in}^{\alpha\lambda}\,\bar{\in}^{\beta\mu}\bar{a}_{3\lambda}\bar{a}_{3\mu},\end{aligned} \tag{2.25}$$

$$\begin{aligned}\bar{a}^{*\alpha 3}&=\bar{a}^{*3\alpha}=-\frac{1}{2}\bar{\in}^{*\alpha lm}\,\bar{\in}^{*3\beta\mu}\bar{a}_{l\beta}\bar{a}_{m\mu}\\&=\frac{\bar{a}}{\bar{a}^{*}}\bar{\in}^{\alpha\gamma}\,\bar{\in}^{\beta\mu}\bar{a}_{\gamma\beta}\bar{a}_{3\mu}=-\frac{\bar{a}}{\bar{a}^{*}}\bar{a}^{\alpha\mu}\bar{a}_{3\mu},\end{aligned} \tag{2.26}$$

$$\begin{aligned}\bar{a}^{*33}&=\frac{1}{2}\bar{\in}^{*3\alpha\gamma}\,\bar{\in}^{*3\beta\mu}\bar{a}_{\alpha\beta}\bar{a}_{\gamma\mu}\\&=\frac{\bar{a}}{2\bar{a}^{*}}\bar{\in}^{\alpha\gamma}\,\bar{\in}^{\beta\mu}\bar{a}_{\alpha\beta}\bar{a}_{\gamma\mu}=\frac{\bar{a}}{\bar{a}^{*}}\end{aligned} \tag{2.27}$$

where $\bar{\in}^{*ijk}$ is the 3-D Eddington tensor with base vector $(\bar{a}_{\alpha},\bar{a}_{3})$ and

$$\bar{a}=\det(\bar{a}_{\alpha\beta}),\quad \bar{a}^{*}=\det(\bar{a}_{ij}).$$

From Eq. (2.22), we obtain:

$$\bar{a}^{*33}=\bar{a}^{*3}\cdot\bar{a}^{*3}=A^{2}. \tag{2.28}$$

Comparing this formula with (2.27), we have:

$$A=(\bar{a}/\bar{a}^{*})^{1/2}. \tag{2.29}$$

Moreover, we recall that:

$$\bar{a}^{*3}=\bar{a}^{*3\alpha}\bar{a}_{\alpha}+\bar{a}^{*33}\bar{a}_{3}$$

and dot multiplying both sides of it by $\bar{n}$, we obtain:

$$A=1/(\bar{a}_{3}\cdot\bar{n}), \tag{2.30a}$$

$$\bar{a}_{3}\cdot\bar{n}=(\bar{a}^{*}/\bar{a})^{1/2}, \tag{2.30b}$$

$$\bar{a}^{*3}=\bar{n}/(\bar{a}_{3}\cdot\bar{n}), \tag{2.30c}$$

$$\bar{a}^{*3}=(\bar{a}^{*}/\bar{a})^{1/2}\bar{n}. \tag{2.30d}$$

Obviously, $\bar{a}^{*3}$ expressed by Eqs. (2.30c,d) satisfies:

$$\begin{aligned}&\bar{a}^{*3}\cdot\bar{a}_{3}=1,\\&\bar{a}^{*3}\cdot\bar{a}_{\alpha}=0.\end{aligned} \tag{2.31}$$

We will discuss the geometric quantities $\bar{a}^{*3}$, $\bar{a}^{*3\alpha}$ and A, etc. further at the end of the next section.

3. Displacements and Strains of Moderately Thick Shells

In this section, the displacements and strains of the moderately thick shell are discussed. Let u be the displacement vector of a typical particle X on $\mathfrak{R}$ at time t (Fig. 1), the position vector of the particle may be written:

$$\bar{r}=r+u. \tag{3.1}$$

The position vector of a typical particle X^* of the shell in the present configuration $\bar{\mathfrak{I}}$ should be:

$$\bar{r}^*=r^*+u^*=r+zn+u^* \tag{3.2a}$$

where u^* is the displacement vector of it.

On the other hand, we can write:

$$\bar{r}^*=\bar{r}+z\bar{a}_3+z^2\overset{2}{u}+\sum_m \overset{m}{u}z^m. \tag{3.2b}$$

Comparing (3.2a) and (3.2b), using (3.1), we have:

$$u^*=u+z(\bar{a}_3-n)+z^2\overset{2}{u}+\sum_m \overset{m}{u}z^m, \tag{3.3}$$

which is the general expression of the displacement of a typical particle X^* in $\bar{\mathfrak{I}}$. In practice, only the first few terms of (3.3) are taken into account. The more terms are taken, the higher the accuracy of the theory, but the more tedious it is. For the classic theory of shells (either Love's or Reissner/Mindlin's theory), only the first two terms are taken.

In order to observe the effect of each term of Eq. (3.3) on the strains of the shells, wc take the first three terms from (3.3) temporarily, that is, we assume:

$$u^*=u+z\beta+z^2\overset{2}{u} \tag{3.4}$$

where β is the total rotation of the transverse cross-section of the shell:

$$\beta=\bar{a}_3-n.$$

The roration β can be decomposed into two parts: the local rotation vector ω and the transverse shear angle γ, i.e.:

$$\beta=\omega+\gamma, \tag{3.5}$$

from which

$$\bar{a}_3=n+\beta=n+\omega+\gamma=\bar{n}+\gamma, \tag{3.6a}$$

$$\bar{n}=n+\omega. \tag{3.6b}$$

With the use of (2.1b), (2.2b), (3.1) and (3.4), we obtain:

$$\bar{a}_\alpha=\bar{r}_{,\alpha}+u_{,\alpha}=a_\alpha+u_{,\alpha}, \tag{3.7a}$$

$$\bar{g}_\alpha=\bar{a}_\alpha+z\bar{a}_{3,\alpha}+z^2\overset{2}{u}_{,\alpha} \tag{3.7b}$$

Let:

$$\bar{a}_{3,\alpha}=-(B_{\alpha\beta}\bar{a}^{*3}+B_3\bar{a}^{*3})=-(B_\alpha^\beta\bar{a}_\beta^*+B_\alpha^3\bar{a}_3^*), \tag{3.8a}$$

$$\overset{2}{u}_{,\alpha}=D_{\alpha\beta}\bar{a}^{*\beta}+D_{3\alpha}\bar{a}^{*3}=D_\alpha^\beta\bar{a}_\beta+D_\alpha^3\bar{a}_3. \tag{3.8b}$$

Then we find:

$$B_{\alpha\beta}=-\bar{a}_{3,\alpha}\cdot\bar{a}_\beta,$$

$$B_\alpha^\beta=-\bar{a}_{3,\alpha}\cdot\bar{a}^{*\beta},$$

$$B_{3\alpha}=-\bar{a}_{3,\alpha}\cdot\bar{a}_3,$$
$$B_\alpha^3=-\bar{a}_{3,\alpha}\cdot\bar{a}^{*3} \tag{3.9}$$

and

$$D_{\alpha\beta}=\overset{2}{u}_{3,\alpha}\cdot\bar{a}_\beta,$$
$$D_\alpha^\beta=\overset{2}{u}_{3,\alpha}\cdot\bar{a}^{*\beta},$$
$$D_{3\alpha}=\overset{2}{u}_{3,\alpha}\cdot\bar{a}_3,$$
$$D_\alpha^3=\overset{2}{u}_{3,\alpha}\cdot\bar{a}^{*3}, \tag{3.10}$$

Substituting (3.8) in (3.7b), we have:

$$\bar{g}_\alpha=\bar{\mu}_\alpha^\beta\bar{a}_\beta+z^2D_\alpha^\beta\bar{a}_\beta+(D_\alpha^3z^2-B_\alpha^3z)\bar{a}_3=\bar{\mu}_\alpha^{*\beta}\bar{a}_\beta+\mu_\alpha^{*3}\bar{a}_3 \tag{3.11}$$

where

$$\bar{\mu}_\alpha^\beta=\delta_\alpha^\beta-B_\alpha^\beta z, \tag{3.12a}$$
$$\bar{\mu}_\alpha^{*\beta}=\bar{\mu}_\alpha^\beta-D_\alpha^\beta z^2, \tag{3.12b}$$
$$\bar{\mu}_\alpha^{*3}=-B_\alpha^3z+D_\alpha^3z^2. \tag{3.12c}$$

In addition, we have:

$$\bar{g}_3=\bar{r}_{,\alpha}^*=\bar{a}_\alpha+2\overset{2}{u}z. \tag{3.13}$$

Using the Lagrangian description and letting

$$u=u^\alpha a_\alpha+u_3n,$$
$$\beta=\beta^\alpha a_\alpha+\beta_3n, \tag{3.14a,b}$$
$$\omega=\omega^\alpha a_\alpha+\omega_3n,$$
$$\gamma=\gamma^\alpha a_\alpha+\gamma_3n, \tag{3.14c,d}$$
$$\overset{2}{u}=\Omega^\alpha a_\alpha+\Omega_3n, \tag{3.14e}$$

we obtain

$$u_{,\alpha}=u^\gamma\|_\alpha a_\gamma+\phi_\alpha n, \tag{3.15a}$$
$$\begin{aligned}\bar{a}_{3,\alpha}=&[(\omega^\gamma+\gamma^\gamma)|_\alpha-b_\alpha^\gamma(1+\omega_3+\gamma_3)]a_\gamma\\&+[(\omega_3+\gamma_3)_{,\alpha}+b_\alpha^\gamma(\omega_\gamma+\gamma_\gamma)]n,\end{aligned} \tag{3.15b}$$
$$\overset{2}{u}_{,\alpha}=(\Omega^\gamma|_\alpha-b_\alpha^\gamma\Omega_3)a_\gamma+(\Omega_{3,\alpha}-b_\alpha^\gamma\Omega_\gamma)n \tag{3.15c}$$

where

$$u^\gamma\|_\alpha=u^\gamma|_\alpha-b_\alpha^\gamma u_3, \tag{3.16a}$$
$$\phi_\alpha=u_{3,\alpha}+b_\alpha^\gamma u_\gamma. \tag{3.16b}$$

From (3.7a), we find:

$$\bar{a}_\alpha=l_\alpha^\gamma a_\gamma+\phi_\alpha n \tag{3.17}$$

where

$$l_\alpha^\gamma=\delta_\alpha^\gamma+u^\gamma\|_\alpha. \tag{3.18}$$

Substituting (3.17) in (2.4b) and (2.5c,d), we obtain

$$\bar{a}_{\alpha\beta}=l_\alpha^\gamma l_\beta^\mu a_{\gamma\mu}+\phi_\alpha\phi_\beta=a_{\alpha\beta}+2\theta_{\alpha\beta}+u^\gamma\|_\alpha u_\gamma\|_\alpha+\phi_\alpha\phi_\beta, \tag{3.19a}$$
$$\bar{a}_{\alpha3}=\phi_\alpha+\beta_\alpha+\beta^\gamma u_\gamma\|_\alpha+\beta_3\phi_\alpha, \tag{3.19b}$$

$$\bar{a}_{33} = 1 + 2\beta_3 + \beta^\alpha\beta_\alpha + \beta_3\beta_3 \tag{3.19c}$$

where $\theta_{\alpha\beta}$ is the symmetric part of $u_\alpha \|_\beta$.

Let $\psi_{\alpha\beta}$ be its antisymmetric part, then $\psi_{\alpha\beta}$ is equivalent to a scalar ϕ, i. e. :

$$\theta_{\alpha\beta} = \frac{1}{2}(u_\alpha \|_\beta + u_\beta \|_\alpha) = \frac{1}{2}(u_\alpha |_\beta + u_\beta |_\alpha) - b_{\alpha\beta}u_3, \tag{3.20a}$$

$$\psi_{\alpha\beta} = \frac{1}{2}(u_\alpha \|_\beta - u_\beta \|_\alpha) = \frac{1}{2}(u_\alpha |_\beta - u_\beta |_\alpha), \tag{3.20b}$$

$$\phi = \frac{1}{2} \in^{\alpha\beta}\psi_{\alpha\beta} = \frac{1}{2} \in^{\alpha\beta} u_\alpha |_\beta. \tag{3.20c}$$

By the use of (3.9), (3.10) together with (3.15b,c) and (3.17), we can deduce that:

$$\begin{aligned} B_{\alpha\beta} &= b_{\alpha\beta} - (\omega_\beta + \gamma_\beta)|_\alpha - (\omega^\gamma + \gamma^\gamma)|_\alpha u_\gamma \|_\beta + b_{\alpha\beta}(\omega_3 + \gamma_3) \\ &\quad + b_\alpha^\gamma(1 + \omega_3 + \gamma_3)u_\gamma \|_\beta - [(\omega_3 + \gamma_3)_{,\alpha} + b_\alpha^\gamma(\omega_\gamma + \gamma_\gamma)]\phi_\beta \\ &= b_{\alpha\beta} + K_{\alpha\beta}, \end{aligned} \tag{3.21a}$$

$$B_{3\alpha} = -\bar{a}_{3,\alpha} \cdot \bar{a}_3 = -(\bar{a}_{33})_{,\alpha}/2 = -e_{33,\alpha}, (\text{see Eq. (3.29c)}) \tag{3.21b}$$

$$D_{\alpha\beta} = (\Omega_\gamma |_\alpha - b_{\gamma\alpha}\Omega_3)l_\beta^\gamma + (\Omega_3 |_\alpha - b_\alpha^\gamma\Omega_\gamma)\phi_\beta, \tag{3.21c}$$

$$D_{3\alpha} = (\Omega_\gamma |_\alpha - b_\alpha^\gamma\Omega_3)(\omega^\gamma + \gamma^\gamma) + (\Omega_3 |_\alpha - b_\alpha^\gamma\Omega_\gamma)(1 + \omega_3 + \gamma_3) \tag{3.21d}$$

where

$$\begin{aligned} K_{\alpha\beta} = B_{\alpha\beta} - b_{\alpha\beta} &= -(\omega_\beta + \gamma_\beta)|_\alpha - (\omega^\gamma + \gamma^\gamma)|_\alpha u_\gamma \|_\beta \\ &\quad + b_{\alpha\beta}(\omega_3 + \gamma_3) + b_\alpha^\gamma(1 + \omega_3 + \gamma_3)u_\gamma \|_\beta \\ &\quad - [(\omega_3 + \gamma_3)_{,\alpha} + b_\alpha^\gamma(\omega_\gamma + \gamma_\gamma)]\phi_\beta. \end{aligned} \tag{3.22}$$

From Eq. (2.6b), we have

$$\bar{b}_{\alpha\beta} = b_{\alpha\beta} + \chi_{\alpha\beta} \tag{3.23}$$

where

$$\begin{aligned} \chi_{\alpha\beta} &= \bar{b}_{\alpha\beta} - b_{\alpha\beta} \\ &= -\omega_\beta |_\alpha - \omega^\gamma |_\beta u_\gamma \|_\beta + b_{\alpha\beta}\omega_3 \\ &\quad + b_\alpha^\gamma(1 + \omega_3)u_\gamma \|_\beta - (\omega_{3,\alpha} + b_\alpha^\gamma\omega_\gamma)\phi_\beta. \end{aligned} \tag{3.24}$$

Introducting the notation:

$$\lambda_{\alpha\beta} = -\gamma_\beta |_\alpha - \gamma_\alpha^\gamma u_\gamma \|_\beta + b_{\alpha\beta}\gamma_3 + b_\alpha^\gamma\gamma_3 u_\gamma \|_\beta - (\gamma_{3,\alpha} + b_\alpha^\gamma\gamma_\gamma)\phi_\beta, \tag{3.25}$$

we can rewrite Eq. (3.22) as:

$$K_{\alpha\beta} = \chi_{\alpha\beta} + \lambda_{\alpha\beta} \tag{3.26}$$

In general, $\chi_{\alpha\beta}$ is the symmetric tensor, but $\lambda_{\alpha\beta}$ is not, and neither is $K_{\alpha\beta}$.

Using (3.11) and (3.13), we can obtain the space metric tensor of the shell in $\bar{\mathfrak{I}}$:

$$\begin{aligned} \bar{g}_{\alpha\beta} = \bar{g}_\alpha \cdot \bar{g}_\beta &= \bar{a}_{\alpha\beta} + z(\bar{a}_{3,\alpha} \cdot \bar{a}_{3,\beta} + \bar{a}_{3,\beta} \cdot \bar{a}_\alpha) \\ &\quad + z^2(\bar{a}_{3,\alpha} \cdot \bar{a}_{3,\beta} + \bar{a}_\beta \cdot \overset{2}{u}_{,\alpha} + \bar{a}_\alpha + \bar{a}_\alpha \overset{2}{u}_{,\beta}) \\ &\quad + z^3(\bar{a}_{3,\beta} \cdot \overset{2}{u}_{,\alpha} + \bar{a}_{3,\alpha}\overset{2}{u}_{,\beta}) + z^4 \overset{2}{u}_{,\alpha} \cdot \overset{2}{u}_{,\beta}, \end{aligned} \tag{3.27a}$$

$$\bar{g}_{\alpha 3} = \bar{g}_{3\alpha} = \bar{g}_{\alpha} \cdot \bar{g}_{3} = \bar{a}_{\alpha 3} + z(\bar{a}_{3,\alpha} \cdot \bar{a}_{3} + 2\overset{2}{u} \cdot \bar{a}_{\alpha})$$
$$+ z^{2}(\bar{a}_{3} \cdot \overset{2}{u}_{,\alpha} + 2\bar{a}_{3,\alpha} \cdot \overset{2}{u}) + 2z^{3}\overset{2}{u}_{,\alpha} \cdot \overset{2}{u}, \tag{3.27b}$$

$$\bar{g}_{33} = \bar{g}_{3} \cdot \bar{g}_{3} = \bar{a}_{33} + 4z\bar{a}_{3} \cdot \overset{2}{u} + 4z^{2}\overset{2}{u} \cdot \overset{2}{u}. \tag{3.27c}$$

The Green strain tensor is defined by:

$$\gamma_{ij} = \frac{1}{2}(\bar{g}_{ij} - g_{ij}).$$

Substituting (3.27) and (2.9) into the above formula and using (3.16)-(3.26), we have:

$$\gamma_{\alpha\beta} = e_{\alpha\beta} - zK_{(\alpha\beta)} + z^{2}U_{\alpha\beta} - \frac{1}{2}(B_{\alpha}^{\gamma}D_{\gamma\alpha} + B_{\beta}^{\gamma}D_{\gamma\alpha} + D_{\alpha}^{3}B_{3\beta} + D_{\beta}^{3}B_{3\alpha})z^{3}$$
$$+ \frac{1}{2}(D_{\beta}^{\gamma}D_{\gamma\alpha} + D_{\alpha}^{3}D_{3\beta})z^{4}, \tag{3.28a}$$

$$\gamma_{3\alpha} = e_{3\alpha} + z\overset{1}{e}3\alpha + z^{2}\overset{2}{e}_{3\alpha} + z^{3}\{D_{\alpha}^{\beta}(\Omega_{\gamma}l_{\beta}^{\gamma} + \Omega_{3}\phi_{\alpha})$$
$$+ D_{\alpha}^{3}[\Omega^{\gamma}(\omega_{\gamma} + \gamma_{\gamma}) + \Omega_{3}(1 + \omega_{3} + \gamma_{3})]\}, \tag{3.28b}$$

$$\gamma_{33} = e_{33} + z\overset{1}{e}_{33} + z^{2}\overset{2}{e}_{33} \tag{3.28c}$$

where $e_{\alpha\beta}$ is the membrane strain tensor, $e_{3\alpha}$ and e_{33} are the transverse shear strain and normal strain, respectively, $K_{(\alpha\beta)}$ is the symmetric part of $K_{\alpha\beta}$, Eq. (3.22):

$$e_{\alpha\beta} = \frac{1}{2}(\bar{a}_{\alpha\beta} - a_{\alpha\beta}) = \theta_{\alpha\beta} + \frac{1}{2}(u^{\gamma}\|_{\alpha}u_{\gamma}\|_{\beta} + \phi_{\alpha}\phi_{\beta}), \tag{3.29a}$$

$$e_{\alpha 3} = \frac{1}{2}(\bar{a}_{\alpha 3} - a_{\alpha 3}) = \frac{1}{2}(\phi_{\alpha} + \beta_{\alpha} + \beta^{\gamma}u_{\gamma}\|_{\alpha} + \beta_{3}\phi_{\alpha}), \tag{3.29b}$$

$$e_{33} = \frac{1}{2}(\bar{a}_{33} - a_{33}) = \beta_{3} + \frac{1}{2}(\beta^{\alpha}\beta_{\alpha} + \beta^{3}\beta_{3}), \tag{3.29c}$$

$$U_{\alpha\beta} = \frac{1}{2}(b_{\alpha}^{\gamma}K_{\gamma\beta} + b_{\beta}^{\gamma}K_{\gamma\alpha} + K_{\alpha}^{\gamma}K_{\gamma\beta} + D_{\alpha\beta} + D_{\beta\alpha} + \bar{a}_{33}B_{\alpha}^{3}B_{\beta}^{3}). \tag{3.29d}$$

and

$$\overset{1}{e}_{3\alpha} = \Omega_{\gamma}l_{\alpha}^{\gamma} + \Omega_{3}\phi_{\alpha} + \frac{1}{2}e_{33,\alpha}, \tag{3.30a}$$

$$\overset{2}{e}_{3\alpha} = \frac{1}{2}D_{3\alpha} - B_{\alpha}^{\beta}(\Omega_{\gamma}l_{\beta}^{\gamma} + \Omega_{3}\phi_{\beta}) - B_{\alpha}^{3}[\Omega^{\gamma}(\omega_{\gamma} + \gamma_{\gamma}) + \Omega_{3}(1 + \omega_{3} + \gamma_{3})], \tag{3.30b}$$

$$\overset{1}{e}_{33} = 2[\Omega^{\gamma}(\omega_{\gamma} + \gamma_{\gamma}) + \Omega_{3}(1 + \omega_{3} + \gamma_{3})], \tag{3.30c}$$

$$\overset{2}{e}_{33} = z(\Omega^{\alpha}\Omega_{\alpha} + \Omega_{3}\Omega_{3}). \tag{3.30d}$$

It can be known from (2.2b) and (3.6b) that ω is not an independent kinematic quantity and can be expressed by the derivative of the displacement u. There are two kinds of methods to find ω. One is exact, it can be shown as follows.

With the use of (2.2b), the unit normal $\bar{n}$ to $\bar{\mathfrak{R}}$ can be expressed as:

$$\bar{n} = \frac{1}{2}\bar{\in}^{\alpha\beta}\bar{a}_{\alpha} \times \bar{a}_{\beta} = n_{\alpha}a^{\alpha} + n_{3}n$$

where

$$n_\alpha = -j^{-1}\phi_\lambda[(1+\theta_\sigma^\sigma)\delta_\alpha^\lambda + u^\lambda\|_\alpha], \tag{3.31a}$$

$$n_3 = j^{-1}(1+\theta_\sigma^\sigma+\bar{\theta}/\alpha+\phi^2), \tag{3.31b}$$

$$j = \sqrt{\bar{a}/a}, \tag{3.31c}$$

$$\bar{\theta} = \det(\theta_{\alpha\beta}). \tag{3.31d}$$

From (3.6b), we have

$$\omega_\alpha = n_\alpha, \tag{3.32a}$$

$$\omega_3 = n_3 - 1. \tag{3.32b}$$

Another method is approximate. Using the conditions:

$$\bar{n}\cdot\bar{a}_\alpha = 0, \quad \bar{n}\cdot\bar{n} = 1$$

we obtain:

$$\phi_\alpha + \omega_\alpha + \omega^\gamma u_\gamma\|_\alpha + \phi_\alpha\omega_3 = 0,$$

$$(\omega_3)^2 + 2\omega_3 + \omega^\alpha\omega_\alpha = 0.$$

Solving these equations approximately, we obtain:

$$\omega_\alpha = -\phi_\alpha + \phi u_\gamma\|_\alpha + \cdots, \tag{3.33a}$$

$$\omega_3 = -\phi^\alpha\phi_\alpha. \tag{3.33b}$$

We can see obviously that Eqs. (3.33) are really the approximate expressions of Eqs. (3.32) .

By the use of (3.32), Eqs. (3.29b,c) can be rewritten as:

$$e_{\alpha 3} = \frac{1}{2}(\gamma_\alpha + \gamma^\gamma u_\gamma\|_\alpha + \gamma_3\phi_\alpha), \tag{3.34a}$$

$$e_{33} = \gamma_3 + \gamma_3\omega_3 + \gamma^\alpha\omega_\alpha + \frac{1}{2}(\gamma^\alpha\gamma_\alpha + \gamma^3\gamma_3). \tag{3.34b}$$

In effect, Eqs. (3.34) can be obtained simply by the last equation of (3.6a):

$$e_{\alpha 3} = \frac{1}{2}\bar{a}_{\alpha 3} = \frac{1}{2}\bar{n}_\alpha\cdot(\bar{n}+\gamma) = \frac{1}{2}\bar{a}_\alpha\cdot\gamma = \frac{1}{2}(\gamma_\alpha + \gamma^\gamma u_\gamma\|_\alpha + \gamma_3\phi_\alpha),$$

$$e_{33} = \frac{1}{2}(\bar{a}_{33} - a_{33}) = \frac{1}{2}[(\bar{n}+\gamma)\cdot(\bar{n}+\gamma) - 1] = \gamma\cdot\bar{n} + \frac{1}{2}\gamma\cdot\gamma$$

$$= \gamma_3 + \gamma_3\omega_3 + \gamma_\alpha\omega_\alpha + \frac{1}{2}(\gamma^\alpha\gamma_\alpha + \gamma^3\gamma_3).$$

Now we have obtained all the expressions of strains of the shell and other relevant kinematic quantities. If the expression for the displacement u^* includes infinite terms, expressions of strains would also have infinite terms.

It can be seen from (3.28) that the fourth power, the cube and the square of z appear in the expressions of $\gamma_{\alpha\beta}$, $\gamma_{3\alpha}$ and γ_{33}, respectively when the displacement u^* still contains the square term of z. But one can see that the higher order terms of them are incomplete if these expressions are analysed. Because the square term of the expression for $\gamma_{\alpha\beta}$ and the first power term of the expressions for $\gamma_{\alpha 3}$, and γ_{33} contain already the square term $\overset{2}{u}$ of the displacement u^*. We can infer that, in general, the mth power term ($m \geqslant 2$) of the displacement will appear in the mth power term of $\gamma_{\alpha\beta}$ and in the ($m-1$)th power term of $\gamma_{\alpha 3}$ and $\gamma_{\alpha 33}$, therefore, only up to the mth power term for $\gamma_{\alpha\beta}$, the ($m-1$)th power terms for $\gamma_{\alpha 3}$ and γ_{33} are required if up to the mth power term is included in the displacement. Consequently, Eqs. (3.28) should be simplified as:

$$\gamma_{\alpha\beta} = e_{\alpha\beta} - zK_{(\alpha\beta)} + z^2 U_{\alpha\beta}, \tag{3.35a}$$

$$\gamma_{3\alpha} = e_{3\alpha} + z\overset{1}{e}_{3\alpha}, \tag{3.35b}$$

$$\gamma_{33} = e_{33} + z\overset{1}{e}_{33}. \tag{3.35c}$$

In the following, we reconsider Eqs. (2.25)-(2.30) in the preceding section. We try to simplify the expressions of $\bar{a}^{*\alpha 3}$, $\bar{a}^{*33}$ and A with the strains given above. We recall that:

$$\bar{a}^{*3} = \bar{a}^{*3\alpha}\bar{a}_{\alpha} + \bar{a}^{*33}\bar{a}_3 = \left(\bar{a}^{*3\alpha} + \frac{\bar{a}}{\bar{a}^*}\bar{\gamma}^{\alpha}\right)\bar{a}_{\alpha} + \frac{\bar{a}}{\bar{a}^*}(1+\bar{\gamma}_n)\bar{n}, \tag{3.36}$$

in which

$$\bar{\gamma}_n = \gamma \cdot \bar{n} = \gamma_3 + \gamma_3\omega_3 + \gamma^{\alpha}\omega_{\alpha}, \tag{3.37a}$$

$$\bar{\gamma}^{\alpha} = \gamma \cdot \bar{a}^{\alpha} = \bar{a}^{\alpha\beta}\gamma \cdot \bar{a}_{\beta} = \bar{a}^{\alpha\beta}(\gamma_{\beta} + \gamma^{\gamma}u_{\gamma}\|_{\beta} + \gamma_3\phi_{\beta}). \tag{3.37b}$$

We know that $\bar{a}^{*3}$ is parallel to $\bar{n}$, from (3.36) we have

$$\bar{a}^{*3\alpha} = -\frac{\bar{a}}{\bar{a}^*}\bar{\gamma}^{\alpha} = -A^2\gamma^{\alpha}, \tag{3.38}$$

which is the concise form of $\bar{a}^{*3\alpha}$ by γ^{α} and A.

Once again, from (3.36) we obtain:

$$\bar{a}^{*3} = \frac{\bar{a}}{\bar{a}^*}(1+\bar{\gamma}_n)\bar{n}. \tag{3.39}$$

Hence:

$$\frac{\bar{a}}{\bar{a}^*}(1+\bar{\gamma}_n) = A$$

from which

$$A^{-1} = 1 + \bar{\gamma}_n = 1 + \gamma_3 + \gamma^{\alpha}\omega_{\alpha} + \gamma_3\omega_3. \tag{3.40}$$

Because from (2.30a) we have:

$$A^{-1} = \bar{a}_3 \cdot \bar{n} = (\bar{n} + \gamma) \cdot \bar{n} = 1 + \bar{\gamma}_n,$$

which is the same as (3.40), then:

$$(\bar{a}^*/\bar{a})^{1/2} = A^{-1} = 1 + \gamma_3 + \gamma^{\alpha}\omega_{\alpha} + \gamma_3\omega_3 \approx 1 + e_{33}, \tag{3.41a}$$

$$\bar{a}^{*33} = \bar{a}/\bar{a}^* = A^2 = 1 - 2\gamma_3 - 2(\gamma^{\alpha}\omega_{\alpha} + \gamma_3\omega_3) + 3\gamma_3\gamma_3 + \cdots \approx 1 - 2e_{33}. \tag{3.41b}$$

From (2.25) together with (3.29b), we have:

$$\bar{a}^{*\alpha\beta} = \bar{a}^{*33}a_{33}\bar{a}^{\alpha\beta} - 4\bar{a}^{*33}\,\bar{\epsilon}^{\alpha\gamma}\,\bar{\epsilon}^{\beta\mu}e_{3\gamma}e_{3\mu}. \tag{3.42}$$

If the strain is small, the second term on the right-hand side of (3.42) is a small quantity of 2nd order higher than the first term, so that:

$$\bar{a}^{*\alpha\beta} = \bar{a}^{*33}\bar{a}_{33}\bar{a}^{\alpha\beta} \approx (1-e_{33})(1+2e_{33})\bar{a}^{\alpha\beta}$$

$$=(1-4e_{33}e_{33})\bar{a}^{\alpha\beta}\approx\bar{a}^{\alpha\beta} \tag{3.43}$$

which means that $a^{*\alpha\beta}$ is different from $\bar{a}^{\alpha\beta}$ only by a smal quantity of 2nd order. In addition, from (3.38) and (3.41b) we have:

$$\bar{a}^{*3\alpha}=-(2-2e_{33})\bar{\gamma}^{\alpha}, \tag{3.44a}$$

$$\bar{a}^{*3}=(1-e_{33})\bar{n}, \tag{3.44b}$$

which shows that $\bar{a}^{*3\alpha}$ is different from $\bar{\gamma}^{\alpha}$ and $\bar{a}^{*3}$ from $\bar{n}$ only by a small quantity of higher order.

4. Christoffel Symbols and Compatibility Equations

For the covariant derivative of 2-space, e.g. $u_{\alpha}|_{\beta}$ etc., the Chrictoffel symbol is defined as:

$$\Gamma_{\alpha\beta\gamma}=a_{\beta,\gamma}\cdot a_{\alpha}=\frac{1}{2}(a_{\alpha\gamma,\beta}+a_{\alpha\beta,\gamma}-a_{\beta\gamma,\alpha}) \tag{4.1}$$

The first subscript α in (4.1) can be raised or lowered by the metric tensor $a_{\alpha\beta}$ or $a^{\alpha\beta}$:

$$\Gamma^{\alpha}_{\beta\gamma}=a^{\alpha\pi}\Gamma_{\pi\beta\gamma}=a_{\beta,\gamma}\cdot a^{\alpha} \tag{4.2}$$

The last two subscripts β and γ of the Christoffel symbol are symmetric.

The compatibility equations of the reference surface $\mathfrak{R}$ are the well-known Gauss-Codazzi conditions:

$$R^{\pi}_{\gamma\alpha\beta}+b_{\gamma\alpha}b^{\pi}_{\beta}-b_{\gamma\beta}b^{\pi}_{\alpha}=0, \tag{4.3a}$$

$$b_{\gamma\beta}|_{\alpha}-b_{\gamma\alpha}|_{\beta}=0 \tag{4.3b}$$

where $R^{\pi}_{\gamma\alpha\beta}$ is the Riemann-Christoffel tensor of curvature which is expressed as:

$$R^{\pi}_{\gamma\alpha\beta}=\Gamma^{\pi}_{\gamma\beta,\alpha}-\Gamma^{\pi}_{\gamma\alpha,\beta}+\Gamma^{\sigma}_{\gamma\beta}\Gamma^{\pi}_{\sigma\alpha}-\Gamma^{\sigma}_{\gamma\alpha}\Gamma^{\pi}_{\sigma\beta}.$$

Thus the Gauss-Codazzi conditions can be rewritten as:

$$\in^{\alpha\beta}\in^{\gamma\mu}(\Gamma_{\alpha\beta\mu,\lambda}+\Gamma^{\pi}_{\alpha\mu}\Gamma_{\pi\beta\lambda}+b_{\alpha\mu}b_{\beta\lambda})=0, \tag{4.4a}$$

$$\in^{\alpha\beta}\in^{\lambda\mu}b_{\beta\mu}|_{\lambda}=0. \tag{4.4b}$$

The key problem is to obtain the compatibility equations and Christoffel symbols in the present configuration $\bar{\mathfrak{J}}$ for our purpose. To this end, we define a vector function on $\bar{\mathfrak{R}}$:

$$V=V^{\gamma}\bar{a}_{\gamma}+V^{3}\bar{a}_{3}=V_{\gamma}\bar{a}^{*\gamma}+V_{3}\bar{a}^{*3}. \tag{4.5}$$

We suppose that the vector can be differentiated more than twice, then the compatibility equations can be determined by the following formula:

$$V_{,\alpha\beta}-V_{,\beta\alpha}=0. \tag{4.6}$$

On the other hand, we observe that:

$$V_{,\alpha}=V_{\gamma;\alpha}\bar{a}^{*\gamma}+V_{3;\alpha}\bar{a}^{*3} \tag{4.7}$$

where $V_{\gamma,\alpha}$ and $V_{3,\alpha}$ are 3-D covariant derivatives of V_{γ} and V_{3} based on $(\bar{a}_{\alpha},\bar{a}_{3})$ respectively:

$$V_{\gamma;\alpha}=V_{\gamma:\alpha}-V_{3}\bar{\Gamma}^{*3}_{\gamma\alpha}, \tag{4.8a}$$

$$V_{3;\alpha}=V_{3:\alpha}-V_{\gamma}\bar{\Gamma}^{*\gamma}_{3\alpha}, \tag{4.8b}$$

$$V_{\gamma:\alpha} = V_{\gamma,\alpha} - V_\pi \overline{\Gamma}^{*\pi}_{\gamma\alpha}, \tag{4.8c}$$

$$V_{3:\alpha} = V_{3,\alpha} - V_3 \overline{\Gamma}^{*3}_{3\alpha}. \tag{4.8d}$$

Making the second derivation on (4.7), we obtain:

$$V_{,\alpha\beta} = V_{\gamma;\alpha\beta}\bar{a}^{*\gamma} + V_{3;\alpha\beta}\bar{a}^{*3} \tag{4.9}$$

where

$$\begin{aligned} V_{\gamma;\alpha\beta} = & V_{\gamma:\alpha\beta} - V_{3,\alpha}\overline{\Gamma}^{*3}_{\gamma\beta} - V_{3,\beta}\overline{\Gamma}^{*3}_{\gamma\alpha} + V_\pi \overline{\Gamma}^{*\pi}_{3\alpha}\Gamma^{*3}_{\gamma\beta} \\ & - V_3(\overline{\Gamma}^{*3}_{\gamma\alpha,\beta} - \overline{\Gamma}^{*\pi}_{\gamma\beta}\overline{\Gamma}^{*3}_{\pi\alpha} - \overline{\Gamma}^{*3}_{\gamma\beta}\overline{\Gamma}^{*3}_{3\alpha}), \end{aligned} \tag{4.10a}$$

$$\begin{aligned} V_{3;\alpha\beta} = & V_{3:\alpha\beta} - V_{\gamma,\alpha}\overline{\Gamma}^{*\gamma}_{3\beta} - V_{\gamma,\beta}\overline{\Gamma}^{*\gamma}_{3\alpha} + V_3 \overline{\Gamma}^{*3}_{\gamma\alpha}\overline{\Gamma}^{*\gamma}_{3\beta} \\ & - V_\pi(\overline{\Gamma}^{*\pi}_{3\alpha,\beta} - \overline{\Gamma}^{*\pi}_{3\alpha}\overline{\Gamma}^{*3}_{3\beta} - \overline{\Gamma}^{*\pi}_{\gamma\alpha}\overline{\Gamma}^{*\gamma}_{3\beta}), \end{aligned} \tag{4.10b}$$

$$V_{\gamma:\alpha\beta} = V_{\gamma,\alpha\beta} - V_{\pi,\beta}\overline{\Gamma}^{*\pi}_{\gamma\alpha} - V_{\pi,\alpha}\overline{\Gamma}^{*\pi}_{\gamma\beta} - V_\pi(\overline{\Gamma}^{*\pi}_{\gamma\alpha,\beta} - \overline{\Gamma}^{*\xi}_{\gamma\beta}\overline{\Gamma}^{*\pi}_{\xi\alpha}), \tag{4.10c}$$

$$V_{3:\alpha\beta} = V_{3,\alpha\beta} - V_{3,\beta}\overline{\Gamma}^{*3}_{3\alpha} - V_{3,\alpha}\overline{\Gamma}^{*3}_{3\beta} - V_3(\overline{\Gamma}^{*3}_{3\alpha,\beta} - \overline{\Gamma}^{*3}_{3\beta}\overline{\Gamma}^{*3}_{3\alpha}). \tag{4.10d}$$

In Eq. (4.10), $\overline{\Gamma}^{*i}_{jk}$ is the Christoffel symbol based on $(\bar{a}^{*\alpha}, \bar{a}^{*3})$, their expressions will be given later in this section.

Exchanging the subscripts α and β of Eqs. (4.9) and (4.10), we can obtain the corresponding expressions for $V_{,\beta\alpha}$. Then substituting them in Eq. (4.6) and eliminating the symmetric terms related to the subscripts α and β, we have finally:

$$V_{,\alpha\beta} - V_{,\beta\alpha} = \overline{R}^{*i}_{j\alpha\beta} V_i \bar{a}^{*j} = \overline{R}^{*}_{ij\alpha\beta} V^i \bar{a}^{*j} = 0,$$

which is equivalent to:

$$\overline{R}^{*i}_{j\alpha\beta} = 0 \tag{4.11}$$

or

$$\overline{R}^{*}_{ij\alpha\beta} = 0 \tag{4.12}$$

where $\overline{R}^{*i}_{j\alpha\beta}$ and $\overline{R}^{*}_{ij\alpha\beta}$ are the Riemann-Christoffel tensor in $\mathfrak{I}$:

$$\overline{R}^{*i}_{j\alpha\beta} = \overline{\Gamma}^{*i}_{j\beta,\alpha} - \overline{\Gamma}^{*i}_{j\alpha,\beta} + \overline{\Gamma}^{*i}_{\alpha k}\overline{\Gamma}^{*k}_{j\beta} - \overline{\Gamma}^{*i}_{\beta k}\overline{\Gamma}^{*k}_{j\alpha}, \tag{4.13a}$$

$$\overline{R}^{*}_{ij\alpha\beta} = \bar{a}_{il}\overline{R}^{*l}_{i\alpha\beta} = \overline{\Gamma}^{*}_{ij\beta,\alpha} - \overline{\Gamma}^{*}_{ij\alpha,\beta} + \overline{\Gamma}^{*}_{kj\alpha}\overline{\Gamma}^{*k}_{i\beta} - \overline{\Gamma}^{*}_{kj\beta}\overline{\Gamma}^{*k}_{i\alpha}. \tag{4.13b}$$

Formulae (4.11) and (4.12) are the compatibility equations of the shell deformation. The formulae are simple but not convenient for application. To use them conveniently, with (4.13a,b), we first alter them into the following forms:

$$\in^{\alpha\beta}\in^{\lambda\mu}(\overline{\Gamma}^{*}_{\alpha\beta\mu,\lambda} + \overline{\Gamma}^{*\pi}_{\alpha\mu}\overline{\Gamma}^{*}_{\pi\beta\lambda} + \overline{\Gamma}^{*3}_{\alpha\mu}\overline{\Gamma}^{*}_{3\beta\lambda}) = 0, \tag{4.14a}$$

$$\in^{\alpha\beta}\in^{\lambda\mu}(\overline{\Gamma}^{*3}_{\beta\mu,\lambda} + \overline{\Gamma}^{*\xi}_{\lambda\beta}\overline{\Gamma}^{*3}_{\alpha\xi} + \overline{\Gamma}^{*3}_{3\alpha}\overline{\Gamma}^{*3}_{\beta\lambda}) = 0, \tag{4.14b}$$

$$\in^{\alpha\beta}\in^{\lambda\mu}(\overline{\Gamma}^{*}_{\lambda3\beta,\alpha} + \overline{\Gamma}^{*\xi}_{\lambda\beta}\overline{\Gamma}^{*}_{\xi3\alpha} + \overline{\Gamma}^{*}_{33\alpha}\overline{\Gamma}^{*3}_{\beta\lambda}) = 0, \tag{4.14c}$$

$$\in^{\alpha\beta}\in^{\lambda\mu}(\overline{\Gamma}^{*}_{33\beta,\alpha} + \overline{\Gamma}^{*\xi}_{3\beta}\overline{\Gamma}^{*}_{\xi3\alpha}) = 0, \tag{4.14d}$$

which is not applicable directly yet. We represent $\overline{\Gamma}^{*}_{\alpha\beta\mu}$ with $\Gamma_{\alpha\beta\mu}$ and strains etc. as follows:

$$\overline{\Gamma}^{*}_{ijk} = \bar{a}_{j,k} \cdot \bar{a}_i = \frac{1}{2}(\bar{a}_{ij,k} + \bar{a}_{ik,j} - \bar{a}_{jk,i}) \tag{4.15a}$$

$$\overline{\Gamma}^{*i}_{jk} = \bar{a}^{*il}\overline{\Gamma}^{*}_{ljk} = \bar{a}_{j,k} \cdot \bar{a}^{*i} \tag{4.15b}$$

and define:

$$\bar{\Gamma}^{\alpha}_{\beta\gamma} = \bar{a}_{\beta,\gamma} \cdot \bar{a}_{\alpha}, \tag{4.16}$$

which is the Christoffel symbol of 2-space based on $(\bar{a}_{\alpha}, \bar{n})$. Introducing of:

$$\bar{a}_{ij} = a_{ij} + 2e_{ij}$$

into Eq. (4.15a), we have:

$$\bar{\Gamma}^{*}_{\alpha\beta\gamma} = \bar{a}_{\alpha\sigma}\Gamma^{\sigma}_{\beta\gamma} + e_{\alpha\beta\gamma}, \tag{4.17a}$$

$$\bar{\Gamma}^{*}_{3\beta\gamma} = \bar{a}_{3\sigma}\Gamma^{\sigma}_{\beta\gamma} + e_{3\beta\gamma} + b_{\beta\gamma}, \tag{4.17b}$$

where

$$e_{i\beta\gamma} = e_{i\beta}|_{\gamma} + e_{i\gamma}|_{\beta} - e_{\beta\gamma}|_{i}, \tag{4.18a}$$

$$e_{\beta\gamma}|_{3} = \gamma_{\beta\gamma}|_{3}|_{z=0}. \tag{4.18b}$$

Obviously, $e_{i\beta\gamma}$ is symmetric in relation to subscripts β and γ.

Eq. (4.17) can be combined to the form:

$$\bar{\Gamma}^{*}_{i\beta\gamma} = \bar{a}_{i\sigma}\Gamma^{\sigma}_{\beta\gamma} + e_{i\beta\gamma} + \bar{a}^{*3}_{i} b_{\beta\gamma}. \tag{4.19}$$

Using (4.15b) together with (4.19), we can deduce that:

$$\begin{aligned}
\bar{\Gamma}^{*\alpha}_{\beta\gamma} &= \bar{a}^{*\alpha i}\bar{\Gamma}^{*}_{i\beta\gamma} \\
&= \Gamma^{\alpha}_{\beta\gamma} + \bar{a}^{*3\alpha} b_{\beta\gamma} + \bar{a}^{*\alpha i} e_{i\beta\gamma} \\
&= \Gamma^{\alpha}_{\beta\gamma} + \bar{a}^{*3\alpha}(b_{\beta\gamma} + e_{3\beta\gamma}) + \bar{a}^{*\alpha\xi} e_{\xi\beta\gamma}.
\end{aligned} \tag{4.20a}$$

$$\begin{aligned}
\bar{\Gamma}^{*3}_{\beta\gamma} &= \bar{a}^{*3i}\bar{\Gamma}^{*}_{i\beta\gamma} \\
&= A^{2}(b_{\beta\gamma} + e_{3\beta\gamma})\bar{a}^{*3\xi} e_{\xi\beta\gamma}.
\end{aligned} \tag{4.20b}$$

We will find other expressions of $\bar{\Gamma}^{*}_{3\beta\mu}$, $\bar{\Gamma}^{*3}_{\beta\mu}$ and $\bar{\Gamma}^{\alpha *}_{\beta\mu}$ in another way and obtain two interesting and useful results. We first recall that:

$$\begin{aligned}
\bar{\Gamma}^{*}_{3\beta\gamma} &= \bar{a}_{\beta,\gamma} \cdot \bar{a}_{3} = (\bar{\Gamma}^{\alpha}_{\beta\gamma}\bar{a}_{\alpha} + b_{\beta\gamma}\bar{n}) \cdot \bar{a}_{3} \\
&= \bar{a}_{3\mu}\Gamma^{\mu}_{\beta\gamma} + \bar{a}^{*\mu\sigma}\bar{a}_{3\sigma}e_{\mu\beta\gamma} + \bar{b}_{\beta\gamma}/A,
\end{aligned} \tag{4.21a}$$

$$\bar{\Gamma}^{*3}_{\beta\gamma} = \bar{a}_{\beta,\gamma} \cdot \bar{a}^{*3} = A\bar{a}_{\beta,\gamma} \cdot \bar{n} = A\bar{b}_{\beta\gamma}. \tag{4.21b}$$

Then comparing (4.21a) with (4.17b), we obtain:

$$\bar{b}_{\beta\gamma} = A(b_{\beta\gamma} + e_{3\beta\gamma} - \bar{a}^{\mu\sigma}\bar{a}_{3\sigma}e_{\mu\beta\gamma}). \tag{4.22}$$

And comparing (4.21b) with (4.20b), we find:

$$\bar{b}_{\beta\gamma} = A(b_{\beta\gamma} + e_{3\beta\gamma}) + A^{-1}\bar{a}^{*3\xi}e_{\xi\beta\gamma}. \tag{4.23}$$

Furthermore, comparing (4.22) with (4.23), we obtain:

$$\bar{a}^{*3\xi} = -A^{2}\bar{a}^{\xi\sigma}\bar{a}_{\sigma 3},$$

which is just Eq. (2.26) . This has proved the correctness of Eqs. (4.22) and (4.23). Eq. (4.22) and (4.23) are the development of Koiter's tendors of the curvature changes (Koiter, 1966).

Substituting (4.22) into (4.20a), we obtain another expression of $\bar{\Gamma}^{*\alpha}_{\beta\mu}$:

$$\bar{\Gamma}^{*\alpha}_{\beta\mu} = \Gamma^{\alpha}_{\beta\mu} + A^{-1}\bar{a}^{*3\alpha}\bar{b}_{\beta\mu} + \bar{a}^{*\alpha\sigma}(\delta^{\pi}_{\sigma} + \xi\bar{a}^{*3\pi}\bar{a}_{3\sigma})e_{\pi\beta\mu} \tag{4.24}$$

in which:

$$\bar{a}^{\alpha\sigma} = \xi \bar{\alpha}^{*\alpha\sigma}. \tag{4.25}$$

From (3.43), we know that:

$$\xi = 1 + O(e_{33}^2). \tag{4.26}$$

In addition, we have:

$$\bar{\Gamma}^*_{\alpha 3\beta} = \bar{a}_{3,\beta} \cdot \bar{a}^*_\alpha = -B_{\alpha\beta}, \tag{4.27a}$$

$$\bar{\Gamma}^{*\lambda}_{3\alpha} = \bar{a}_{3,\alpha} \cdot \bar{a}^{*\lambda} = -B^\lambda_\alpha, \tag{4.27b}$$

$$\bar{\Gamma}^{*3}_{3\alpha} = \bar{a}_{3,\alpha} \cdot \bar{a}^{*3} = -B^3_\alpha, \tag{4.27c}$$

$$\bar{\Gamma}^*_{33\beta} = \bar{a}_{3,\beta} \cdot \bar{a}_3 = -B_{3\beta}. \tag{4.27d}$$

Substituting these formulae into (4.14) and going through a tedious algebraic operations, we obtain:

$$Ke^\alpha_\alpha + \in^{\alpha\beta} \in^{\lambda\mu} \Big\{ e_{\alpha\mu} |_{\beta\lambda} b_{\alpha\mu} \chi_{\beta\lambda} + \frac{1}{2} [\chi_{\beta\lambda} \chi_{\alpha\mu} + \bar{a}^{*\pi\sigma} (\delta^\xi_\sigma + \xi \bar{a}^{*3\xi} a_{3\sigma}) e_{\xi\alpha\mu} e_{\pi\beta\lambda}] \Big\} = 0, \tag{4.28a}$$

$$\in^{\alpha\beta} \in^{\lambda\mu} \{ \chi_{\lambda\beta} |_\alpha + a^{*\pi\sigma} (\delta^\xi_\sigma + \xi \bar{a}^{*3\xi} a_{3\sigma}) b_{\xi\alpha} e_{\pi\beta\lambda} + b_{\beta\lambda} [a^{*3\sigma} (A^{-1} b_{\sigma\alpha} - B_{\sigma\alpha}) + A_{,\alpha}/A - A^2 B_{3\alpha}] \} = 0, \tag{4.28b}$$

$$\in^{\alpha\beta} \in^{\lambda\mu} \{ \lambda_{\lambda\beta} |_\alpha + a^{*\pi\sigma} (\delta^\xi_\sigma + \xi a^{*3\xi} a_{3\sigma}) \lambda_{\xi\alpha} e_{\pi\beta\lambda} + b_{\beta\lambda} [A^{-1} a^{*3\xi} \lambda_{\xi\alpha} + A(1+A) B_{3\alpha} - A_{,\alpha}/A] \} = 0, \tag{4.28c}$$

$$\in^{\alpha\beta} \in^{\lambda\mu} (B_{3\beta} |_\alpha - a^{*3\xi} B_{\xi\alpha} B_{3\beta}) = 0 \tag{4.28d}$$

where K is defined by (2.12c).

Equations (4.28a-d) are the intrinsic forms (Chien, 1944) of the compatibility conditions of the moderately thick shells. If Kirchhoff-Love's assumption is used. Eqs. (4.28a-d) can be simplified as:

$$Ke^\alpha_\alpha + \in^{\alpha\beta} \in^{\lambda\mu} \left[e_{\alpha\mu} |_{\beta\lambda} + b_{\alpha\mu} \chi_{\beta\lambda} + \frac{1}{2} (\chi_{\beta\lambda} \chi_{\alpha\mu} + \bar{a}^{\pi\sigma} e_{\sigma\alpha\mu} e_{\pi\beta\lambda}) \right] = 0, \tag{4.29a}$$

$$\in^{\alpha\beta} \in^{\lambda\mu} (\chi_{\lambda\beta} |_\alpha + \bar{a}^{\pi\sigma} \bar{b}_{\sigma\alpha} e_{\pi\beta\lambda}) = 0, \tag{4.29b}$$

which are the same as Koiter's equations (Koiter, 1966) except that we use different notations from Koiter's. Thus Eqs. (4.28) are the developments of Koiter's equations.

Since Eqs. (4.28) are very complicated, we will simplify them in the next paper (Liu and Zhu, 1996).

Nomenclature

$a_\alpha, \bar{a}_\alpha$	—	basic vectors of the middle surface in reference and current configuration, respectively
$a_{\alpha\beta}, \bar{a}_{\alpha\beta}$	—	metric tensors of the middle surface in reference and current configuration, respectively
$b_{\alpha\beta}, \bar{b}_{\alpha\beta}$	—	curvature tensors in reference and current configuration, respectively
$e_{i\beta\gamma}$	—	$e_{i\beta} \|_\gamma + e_{i\gamma} \|_\gamma - e_{\beta\gamma} \|_\gamma$

$e_{\alpha\beta}$, $\chi_{\alpha\beta}$	—	strain and curvature changing tensors
g_α, $\bar{g}_\alpha$	—	basic vectors in reference and current configuration, respectively
$g_{\alpha\beta}$, $\bar{g}_{\alpha\beta}$	—	metric tensors in reference and current configuration, respectively
$R_{\alpha\beta\gamma\mu}$, $\bar{R}_{\alpha\beta\gamma\mu}$	—	Riemman-Christoffel curvature tensors in reference and current configuration, respectively
$\theta_{\alpha\beta}$	—	the linear part of $e_{\alpha\beta}$
$\lambda_{\alpha\beta}$	—	the contribution to curvature changing due to transverse shear
Γ_{ijk}, $\bar{\Gamma}_{ijk}$	—	Christoffel symbols in reference and current configuration, respectively
$\in_{\alpha\beta}$	—	the alternating tensor
ϕ_α, ϕ	—	out-plane and in-plane rotations at the middle surface
γ_α	—	the transvers shear deformation
ω_α	—	the total out-plane rotation at the middle surface

References

Adi Murthy N. K. and Alwar R. S. (1976): *Nonlinear dynamic buckling of sandwich panels*. -J. Appl. Mech., v. 43, pp. 459-463.

Akkas N. (1972): *On the buckling and initial postbuckling behavior of shallow spherical and conical sandwich shells*. -J. Appl. Mech., v. 39, pp. 163-171.

Bauld N. R, Jr. (1974): *Imperfecton sensitivity of axially compressed stringer reinforced cylindrical sandwich panels*. -Int. J. Solids Structures, v. 10, pp. 883-902.

Chien W. Z. (1944): *The intrinsic theory of thin shells and plates, part I. General theory*. -Quart. Appl. Math., v. 1, pp. 297-327.

Group of Plates and Shells (1977): *Bending, Stability and Vibration of Sandwich Plates and Shells*. -Beijing: China Academic Publishers (in Chinese).

Habip L. M. (1965): *A survey of modern developments in the analysis of sandwich structures*. -AMR, v. 18, pp. 93-101.

Kan H-P and Huang J-C. (1967): *Large deflection of rectangular sandwich plates*. -J AIAA, v. 5, pp. 1706-1707.

Koiter W. T. (1966): *On the nonlinear theory of thin elastic shells* (Ⅰ, Ⅱ, Ⅲ). -Proc. Kon. Ned. Ak. Wet., v. B69, pp. 1-60.

Liu Renhuai. (1987): *Nonlinear bending of circular sandwich plates under the action of axisymmetric uniformly distributed line loads*, In: Progress in Applied Mechanics (K. Y. Yeh Ed.), pp. 293-321. Dordrecht: Martinus Nijhoff Publishers.

Liu Renhuai. (1981): *Nonlinear bending of circular sandwich plates*. -Appl. Math. Mech., v. 2, pp. 189-208.

Liu Renhuai. (1980): *Nonlinear axisymmetrical bending of circular sandwich plates under the action of uniform edge moments*. -J. Chin. Univ. Sci. Technol., v. 10, pp. 56-67 (in Chinese).

Liu Renhuai and Cheng Z. Q. (1995): *On the nonlinear buckling of circular shallow spherical sandwich shells under the action of uniform edge moments*. -Int. J. Nonlinear Mechanics, v. 30, pp. 33-43.

Liu Renhuai. and Cheng Z. Q. (1993): *Nonlinear bending of simply supported rectangular sandwich plates*. -Appl. Math. Mech., v. 14, pp. 217-234.

Liu Renhuai. and Li J. (1995): *Nonlinear vibration of shallow conical sandwich shells*. -Int. J. Nonlinear Mechanics, v. 30, pp. 97-109.

Liu Renhuai. and Shi Y. F. (1982): *Exact solution for circular sandwich plates with large deflection*. -Appl. Math. Mech., v. 3, pp. 11-24.

Liu Renhuai. and Wu J. C. (1991): *Nonlinear vibration of rectangular sandwich plates*. -Science in China (Series A), v. 35, pp. 472-486.

Liu Renhuai. and Zhu G. Q. (1989): *Further study on large deflection of circular sandwich plates*. -Appl. Math. Mech., v. 10, pp. 1099-1106.

Liu Renhuai. and Zhu J. F. (1997): *Nonlinear theory of sandwich shells, part* Ⅱ*-Approximate theories*. -Applied Mech. Engng., v. 2, pp. 241-269.

Naghdi P. M. (1972): *The theory of shells and plates*, In: Mechanics of Solids Ⅱ (C. Truesdell ed.), Berlin: Springer-Verlag.

Rajagopal S. V., Singh G. and Sandsiva Rao Y. U. K (1987): *Nonlinear analysis of sandwich plates.*, -Int. J. Nonlinear Mech., v. 22, pp. 161-174.

Reese C. D. and Bert C. W. (1974): *Buckling of orthotropic sandwich cylinders under axial compression and bending*. -J. Aircraft, v. 11, pp. 207-212.

Reissner E. (1949): *Small bending and stretching of sandwich-type shells*. -NACA TN1832.

Roman J. G. and Kao J. S. (1975): *Nonlinear equations for shallow sandwich shells with orthotropic cores*. -J. AIAA, v. 13, pp. 961-968.

Schmidt R. (1964): *Sandwich shells of arbitrary shape*. -J. Appl. Mech., v. 31, pp. 239-245.

Wampner G. A. (1967): *Theory for moderately large deflections of sandwich shells with dissimilar facing*. -Int. J. Solids Struct., v. 3, pp. 367-375.

Wang C-T (1950): *Principle and application of complementary energy method for thin homogeneous and sandwich plates and shells with finite deflections*. -NACA TN2620.

Wang Z. M. and Dai F. L. (1983): *Bending, stability and vibration of orthotropic multilayer, sandwich and stiffened shallow spherical shells*. -Acta Mechanica, No. 5, pp. 480-492.

Wand Z. M., Dai F. L. and Lu M. S. (1984): *Stability and vibration of multilayer, sandwich and sti ffened composite circular cylindrical panels*. -Acta Mech. Solids, No. 4, pp. 517-530.

Zhu J. F. (1989): *Large Deflection Theories of Sandwich Shells and Their Applications*. -Shanghai: Ph. D. Dissertation (in Chinese), Shanghai University of Technology.

Nonlinear theory of sandwich shells part Ⅱ: Approximate theories①

中文摘要

本文按照“能量误差一致”原则，使用量阶分析对已求得的中厚壳的非线性几何理论的方程进行化简，获得了便于应用的夹层壳非线性理论的平衡方程和协调条件。

1. Introduction

The authors have established the exact kinematics of the moderately thick shells in Liu and Zhu (1997). These equations are rather complicated. The complexity is caused by some small quantities in most cases. If these small quantities are identified and neglected, the equations will be reduced to simpler ones. Simplification of the equations is so important that we can obtain simple but almost accurate equations and save a lot of unnecessary labor, meanwhile the error estimates and the application range of the equations can be pointed out.

The order analysis of magnitude (simply OAM) (Budiansky, 1968; Danielson, 1970; John, 1965; Koiter, 1966, 1973, 1980) is the main tool for filtering small quantities in the equations. In accordance with the principle of energy error of consistency (Koiter, 1980; Koiter and Simmonds, 1973), we can decide which small quantities may be neglected. Pietraszkiewicz (1979, 1983) and Schmidt (1985) et al. have classified shell theories systematically. It makes it more clear to understand large deflection deformations of shells.

It is worthy to mention that almost fifty years ago Chien (1944) set up the intrinsic theory of thin shells and simplified it systematically. Although the concepts of large deflection motion of shells were not so clear at that time as they are now, his immortal work has profound effect the development of shell theories.

According to the principle of energy error consistency, the exact equations obtained by Liu and Zhu (1997) will be simplified with the use of OAM. The notations used in this paper are the same as those of Liu and Zhu (1997). Therefore no explanation will be given unless new symbols appear.

2. Method and Steps of Simplifying The Theory of Shells

We assume that the sandwich shells are thin and the strains in the shells are small even though the deflection is large. Here the large deflection is mainly caused by the

① Reprinted from *Applied Mechanics and Engineering*, 1997, 2 (2): 241-269. Authors: Liu Renhuai and Zhu Jinfu.

relative large local rotation. Then the theory of shells can be simplified in accordance with the following methods and steps.

According to Pietraszkiewicz (1979, 1983), we define a small parameter θ:

$$\theta = \max(h/L, h/L^*, h/d, \sqrt{h/R}, \sqrt{n}) \tag{2.1}$$

where:

h—the thickness of the shells,

L—the characteristic length of the deformation,

L^*—the characteristic length of the curvature model,

d—the distance from a typical particle of the shells to the edge of the shells,

R—the minimum curvature radius of the reference surface.

The small parameter θ plays a key role in OAM.

Next we decide the kind of shell motion. Pietraszkiewicz (1979, 1983) has shown that the local rotation vector has the same order of magnitude as the vector ($\in^{\alpha\beta}\phi_\beta a_\alpha + \phi^n$). In terms of the magnitude order of ϕ_α and ϕ relative to θ, the small strain motion of the shells can be classified into the following several kinds:

1. $\phi_\alpha = O(\theta^2)$, $\phi = O(\theta^2)$: small/small rotation,
2. $\phi_\alpha = O(\theta)$, $\phi = O(\theta^2)$: moderate/small rotation,
3. $\phi_\alpha = O(\theta)$, $\phi = O(\theta)$: moderate/moderate rotation,
4. $\phi_\alpha = O(\sqrt{\theta})$, $\phi = O(\theta^2)$: large/small rotation,
5. $\phi_\alpha = O(\sqrt{\theta})$, $\phi = O(\theta)$: large/moderate rotation,
6. $\phi_\alpha = O(\sqrt{\theta})$, $\phi = O(\sqrt{\theta})$: large/large rotation,
7. $\phi_\alpha \geqslant O(1)$: finite rotation,

in which the small/small rotation theory is the well-known small deflection theory, i.e., the linear theory. Because the stiffness of the out-plane is smaller than that of the in-plane, in general there exists the relation:

$$\phi \leqslant O(\phi_\alpha).$$

The large and finite rotation theories of shells are very complicated. The deformation of shells in most cases belongs to the range of moderate rotation except for high elastic structures. Therefore, the moderate rotation is considered in this paper.

Finally, to obtain the simplified equations, we carry on OAM for every term of the equations and neglect the small quantities of higher order in accordance with the principle of energy error of consistency.

In order to analyze the magnitude order of the derivative of any quantity, it is necessary to define a length parameter (Koiter and Simmonds, 1973; Koiter, 1980):

$$\lambda = \min(L, L^*, h/\theta). \tag{2.2}$$

For any quantity T associated with the theory of thin shells, we have:

$$T|_\alpha = O(T\lambda^{-1}) \tag{2.3}$$

Let E be the elastic parameter on the reference surface $\mathfrak{R}$, then the elastic parameter E^* of the shell space $\mathfrak{H}$ can be expressed as

$$E^* = O(E + \theta^2 E + \theta^4 E + \cdots).$$

In accordance with the analysis results obtained by Naghdi et al. (1982, 1983) for the moderate rotation deformation accompanied by small strain, there is the following relation

$$\theta_{\alpha\beta} = O(\theta^2) = O(\eta).$$

Hence

$$e^{\alpha\beta} = O(\eta + \theta\eta + \theta^2 \eta)$$

If

$$hK_{\alpha\beta} = O(e_{\alpha\beta})$$

then the strain energy density is

$$\sum = O(E\eta^2 + 2\theta E\eta^2 + 4\theta^2 E\eta^2 + 4\theta^3 E\eta^2 + 5\theta^4 E\eta^2 + \cdots).$$

For the first order theory, the relative error bound of the strain energy is $O(\theta)$, so that we can take

$$E^* = O(E) \quad \text{and} \quad e_{\alpha\beta} = O(\eta) = O(\theta^2).$$

For the second order theory, the relative error bound of the strain energy is $Q(\theta^3)$, so that we can take

$$E^* = O(E + \theta^2 E) \quad \text{and} \quad e_{\alpha\beta} = O(\eta) = O(\eta + \theta\eta + \theta^2 \eta).$$

Stresses in the shells have the following magnitude order relation (John, 1965)

$$\sigma^{3\alpha} = O(\theta \sigma^{\alpha\beta}),$$
$$\sigma^{33} = O(\theta^2 \sigma^{\alpha\beta}).$$

Even for the second order theory, σ^{33} may be often neglected.

The Kirchhoff-Love kinematic assumption (simply the K-L assumption) for the theory of thin shells is equivalent to supposing that the transverse shear moduli $G_{3\alpha} = \infty$, so that the transverse shear strains are zero. In effect for the thin shells with isotropic material, there is $G_{3\alpha} = O(E)$, hence, $e_{3\alpha} = O(\theta\eta)$, neglecting $e_{3\alpha}$ is allowed for the first order theory. But for the shells which are not isotropic, the K-L assumption may not be proper. e. g. for sandwich shells which have relatively soft cores: $G_c/G_f = O(q)$, then $e_{3\alpha}$ will be the same order as in-plane strains $e_{\alpha\beta}$. Neglecting it will bring a larger error of deflection, so that the influence of the transverse shear stiffnesses of cores on deflections of sandwich shells must be included. In terms of the structural features of sandwich shells, we define two additional parameters

$$\theta_1 = G_c/G_f \tag{2.4a}$$

$$\theta_2 = t/h_c \tag{2.4b}$$

where G_c is the greater transverse shear modulus of the core, G_f is the in-plane shear modulus of the faces, t and h_c are the thickness of the faces and the core, respectively.

We have

$$h = h_c + 2t.$$

The small parameters θ_1 and θ_2 are important for sandwich shells. It will be seen that the essential mechanical performances of the sandwich shells depend upon the magnitude order of θ_1 and θ_2.

3. Second Order Moderate/Moderate Rotation Theory

At first, for the deformation and the structure of the sandwich shells, we assume that:

1) the strain is small, i. e. $\eta \ll O(1)$;

2) the core is relatively thick, and the normal to the reference surface $\mathfrak{R}$ is not the normal to the instance middle surface $\bar{\mathfrak{R}}$, but is almost a straight line after deformation, that is to assume $\overset{2}{u} \leqslant O(\beta/R)$ (Liu and Zhu, 1997), thus $\overset{2}{u}$ may be neglected;

3) the thicknesses of the faces are the same, and the faces are thinner than the core, i. e. $\theta_2 < 1$;

4) the sandwich shell is thin, i. e. $h/R \ll O(1)$;

5) the materials of the faces are the same and the core is softer than the faces, i. e. $\theta_1 < 1$;

Based on the above assumptions, the geometric relations are first simplified as follows.

3. 1. Displacements of the Sandwich Shells

With the assumptions (1-4) and the condition of displacement continuity between the layers, the displacements of the sandwich shells can be expressed as

—in the upper face, $\left(\frac{h_c}{2} \leqslant z \leqslant \frac{h_c}{2} + t\right)$:

$$_{f_+}u_\alpha^* = u_\alpha + \frac{1}{2}(h_c + t)\gamma_\alpha + \omega_\alpha{}^z - \frac{1}{2}(h_c + t)\left(z - \frac{h_c}{2}\right) b_\alpha^\lambda(\omega_\lambda + \gamma_\lambda), \tag{3. 1a}$$

$$_{f_+}u_3 = w, \tag{3. 1b}$$

—in the core, $\left(-\frac{h_c}{2} \leqslant z \leqslant \frac{h_c}{2}\right)$:

$$_c u_\alpha^* = u_\alpha + (\omega_\alpha + \gamma_\alpha) z, \tag{3. 2a}$$

$$_c u_3 = w, \tag{3. 2b}$$

—in the lower face, $\left(-\frac{h_c}{2} - t \leqslant z \leqslant -\frac{h_c}{2}\right)$:

$$_{f_-}u_\alpha^* = u_\alpha - \frac{1}{2}(h_c + t)\gamma_\alpha + \omega_\alpha{}^z + \frac{1}{2}(h_c + t)\left(z - \frac{h_c}{2}\right) b_\alpha^\lambda(\omega_\lambda + \gamma_\lambda), \tag{3. 3a}$$

$$_{f_-}u_3 = w, \tag{3. 3b}$$

in which the left subscripts f_+, f_- and c mean the upper and lower faces and the core, respectively.

The displacements given by (3. 1)-(3. 3) are consistent and continuous on the inter-face of the faces and the core.

3.2. Strain Relation of the Sandwich Shells

In accordance with the principle of energy error of consistency, formula (3.29) in Liu and Zhu (1997) has to remain unchanged in the case of moderate/moderate rotation, that is:

$$e_{\alpha\beta} = \theta_{\alpha\beta} + \frac{1}{2}(u^{\gamma}\|_{\alpha}u_{\gamma}\|_{\beta} + \phi_{\alpha}\phi_{\beta}). \tag{3.4}$$

Even if for the second order theory, it is unnecessary to hold the nonlinear terms in the expressions of the curvature change tensor and the transverse shear strain tensor, etc. Otherwise, the nonlinear terms of the moments and the shear resultants will be led into the equilibrium equations, which is valueless from the engineering view. Therefore, we have

$$\chi_{\alpha\beta} = \frac{1}{2}(\phi_{\alpha}|_{\beta} + \phi_{\beta}|_{\alpha} + b_{\alpha}^{\gamma}u_{\gamma}\|_{\beta} + b_{\beta}^{\gamma}u_{\gamma}\|_{\alpha}), \tag{3.5}$$

$$\lambda_{\alpha\beta} = -\gamma_{\beta}|_{\alpha} \tag{3.6}$$

$$U_{\alpha\beta} = \frac{1}{2}(b_{\alpha}^{\gamma}\chi_{\gamma\beta} + b_{\beta}^{\gamma}\chi_{\gamma\alpha}), \tag{3.7}$$

$$e_{3\alpha} = \gamma_{\alpha}/2, \tag{3.8}$$

$$\omega_{\alpha} = -\phi_{\alpha}, \tag{3.9}$$

$$K_{\alpha\beta} = \chi_{\alpha\beta} + \lambda_{\alpha\beta}, \tag{3.10}$$

$$\overset{1}{e}_{3\alpha} \approx 0 \quad \text{and} \quad \overset{1}{e}_{33} \approx 0. \tag{3.11}$$

The strain in the upper face is

$$_{f_+}\gamma_{\alpha\beta} = e_{\alpha\beta} - \frac{1}{2}(h_c + t)\lambda_{\alpha\beta} - z\chi_{\alpha\beta} + \frac{1}{2}z^2(b_{\alpha}^{\gamma}\chi_{\gamma\beta} + b_{\beta}^{\gamma}\chi_{\gamma\alpha}). \tag{3.12}$$

The strain in the lower face is:

$$_{f_-}\gamma_{\alpha\beta} = e_{\alpha\beta} + \frac{1}{2}(h_c + t)\lambda_{\alpha\beta} - z\chi_{\alpha\beta} + \frac{1}{2}z^2(b_{\alpha}^{\gamma}\chi_{\gamma\beta} + b_{\beta}^{\gamma}\chi_{\gamma\alpha}). \tag{3.13}$$

The strains in the core are

$$_{c}\gamma_{\alpha\beta} = e_{\alpha\beta} - zK_{\alpha\beta} + \frac{1}{2}z^2(b_{\alpha}^{\gamma}\chi_{\gamma\beta} + b_{\beta}^{\gamma}\chi_{\gamma\alpha}), \tag{3.14}$$

$$\gamma_{3\beta} = e_{3\alpha} = \gamma_{\alpha}/2, \tag{3.15}$$

$$K_{\alpha\beta} = \chi_{\alpha\beta} + \lambda_{\alpha\beta}. \tag{3.16}$$

In Eqs. (3.12)-(3.16), $e_{\alpha\beta}$, $\chi_{\alpha\beta}$ and $\lambda_{\alpha\beta}$ are given by (3.4)-(3.6).

We have obtained the simplified geometric relations of the second order moderate/moderate rotation theory of the sandwich shells. In what follows, we try to build up the physical equations.

3.3. Constitutive Equations

In terms of the assumption (5) given previously, the second Piola-Kirchhoff stress tensor of the sandwich shells can be expressed as

$$\sigma_{ij} = L^{ijkl}\gamma_{kl} \tag{3.17}$$

where L^{ijkl} is the elastic tensor of the material of the faces or the core. According to John (1965), even for the second order theory the hypothesis $\sigma^{33}=0$ is still suitable to the sandwich shells, thus:

$$\gamma^{33} = -\frac{L^{33\alpha\beta}}{L^{3333}}\gamma_{\alpha\beta}$$

and

$$\sigma^{\alpha\beta} = H^{\alpha\beta\gamma\mu}\gamma_{\gamma\mu}, \tag{3.18a}$$

$$\sigma^{3\alpha} = 2L^{3\alpha3\beta}\gamma_{3\beta} \tag{3.18b}$$

where

$$H^{\alpha\beta\gamma\mu} = L^{\alpha\beta\gamma\mu} - \frac{L^{33\gamma\mu}}{L^{3333}}L^{\alpha\beta33}. \tag{3.19}$$

For homogeneous materials, $H^{\alpha\beta\gamma\mu}$ and $L^{3\alpha3\beta}$ can be expressed as the series of z

$$H^{\alpha\beta\gamma\mu} = H_0^{\alpha\beta\gamma\mu} + zH_1^{\alpha\beta\gamma\mu} + O(h^2R^{-2}H_0^{\alpha\beta\gamma\mu}), \tag{3.20a}$$

$$L^{3\alpha3\beta} = L_0^{3\alpha3\beta} + zL_1^{3\alpha3\beta} + O(h^2R^{-2}L_0^{3\alpha3\beta}), \tag{3.20b}$$

and there are the following magnitude order relations:

$$H_1^{\alpha\beta\gamma\mu} = O(R^{-1}H_0^{\alpha\beta\gamma\mu}), \tag{3.21a}$$

$$L^{3\alpha3\beta} = O(R^{-1}L_0^{3\alpha3\beta}). \tag{3.21b}$$

We define the nominal resultants of stresses and the nominal resultants of stress moments as follows

$$N^{\alpha\beta} = \int_{-h/2}^{h/2}\sigma^{\alpha\beta}\mu\,\mathrm{d}z \tag{3.22a}$$

$$Q^{\alpha} = \int_{-h/2}^{h/2}\sigma^{\alpha3}\mu\,\mathrm{d}z = \int_{-h/2}^{h/2}c\sigma^{\alpha3}\mu\,\mathrm{d}z \tag{3.22b}$$

$$M^{\alpha\beta} = \int_{-h/2}^{h/2}\sigma^{\alpha\beta}\mu z\,\mathrm{d}z. \tag{3.22c}$$

Obviously $N^{\alpha\beta}$ and $M^{\alpha\beta}$ are symmetrical tensors. In accordance with the principle of energy error consistency for the second order theory, we should take (see Eq. (2.12) in Liu and Zhu (1997))

$$\mu = 1 - 2Hz. \tag{3.23}$$

Substituting (3.18), (3.20) and (3.23) into (3.22) and through tedious algebraic operations, we find:

$$\begin{aligned} N^{\alpha\beta} &= E^{\alpha\beta\gamma\mu}e_{\gamma\mu} + 2HD_1^{\alpha\beta\gamma\mu}\chi_{\gamma\mu}, \\ M^{\alpha\beta} &= -D^{\alpha\beta\gamma\mu}(\chi_{\gamma\mu} + \lambda_{\gamma\mu}) - D_f^{\alpha\beta\gamma\mu}\chi_{\gamma\mu} - 2HD_2^{\alpha\beta\gamma\mu}e_{\gamma\mu}, \\ Q^{\alpha} &= h_c L_0^{3\alpha3\beta}\gamma_\beta \end{aligned} \tag{3.24}$$

where

$$E^{\alpha\beta\gamma\mu} = 2t_f H_0^{\alpha\beta\gamma\mu} + h_c H_0^{\alpha\beta\gamma\mu}, \tag{3.25a}$$

$$2HD_1^{\alpha\beta\gamma\mu} = 2H_fD_1^{\alpha\beta\gamma\mu} + 2H_cD_1^{\alpha\beta\gamma\mu} + 2H_sD_1^{\alpha\beta\gamma\mu}, \tag{3.25b}$$

$$2H_fD_1^{\alpha\beta\gamma\mu} = \frac{t^3}{6}[{}_fH_0^{\alpha\beta\gamma\mu}(2H\delta_\sigma^\gamma + b_\sigma^\gamma) - {}_fH_1^{\alpha\beta\gamma\mu}], \tag{3.26a}$$

$$2H_cD_1^{\alpha\beta\gamma\mu} = \frac{h_c^3}{12}[{}_cH_0^{\alpha\beta\gamma\mu}(2H\delta_\sigma^\gamma + b_\sigma^\gamma) - {}_cH_1^{\alpha\beta\gamma\mu}], \tag{3.26b}$$

$$2H_sD_1^{\alpha\beta\gamma\mu} = 2tz_1^2[{}_fH_0^{\alpha\beta\gamma\mu}(2H\delta_\sigma^\gamma + b_\sigma^\gamma) - {}_fH_1^{\alpha\beta\gamma\mu}], \tag{3.26c}$$

$$D^{\alpha\beta\gamma\mu} = 2tz_1^2\, {}_fH_0^{\alpha\beta\gamma\mu} + \frac{h_c^3}{12}\, {}_cH_0^{\alpha\beta\gamma w}, \tag{3.27a}$$

$$D_f^{\alpha\beta\gamma\mu} = \frac{t^3}{6}\, {}_fH_0^{\alpha\beta\gamma m}, \tag{3.27b}$$

$$2HD_2^{\alpha\beta\gamma\mu} = 2H_fD_2^{\alpha\beta\gamma\mu} + 2H_cD_2^{\alpha\beta\gamma\mu} + 2H_sD_2^{\alpha\beta\gamma\mu}, \tag{3.27c}$$

$$2H_fD_2^{\alpha\beta\gamma\mu} = \frac{t^3}{6}(2H_fD_0^{\alpha\beta\gamma\mu} - {}_fH_1^{\alpha\beta\gamma\mu}), \tag{3.28a}$$

$$2H_cD_2^{\alpha\beta\gamma\mu} = \frac{h_c^3}{12}(2H_cH_0^{\alpha\beta\gamma\mu} - {}_cH_1^{\alpha\beta\gamma\mu}), \tag{3.28b}$$

$$2H_sD_2^{\alpha\beta\gamma\mu} = 2tz_1^2(2H_fH_0^{\alpha\beta\gamma\mu} - {}_fH_1^{\alpha\beta\gamma\mu}), \tag{3.28c}$$

and we have

$$z_1^2 = \frac{1}{2}(h_c + t).$$

In Eqs. (3.25)-(3.28), the left subscripts f and c denote the faces and the core, respectively.

It should be noted from (3.24) that for the second order theory, the couple terms between $N^{\alpha\beta}$ and $M^{\alpha\beta}$ exist, which makes the constitutive Eq. (3.24) more complicated.

If the materials of the sandwich shells are isotropic, we recall that

$$H^{\alpha\beta\gamma\mu} = \frac{E}{2(1+\nu)}\left(g^{\alpha\gamma}g^{\beta\mu} + g^{\alpha\mu}g^{\beta\gamma} + \frac{2\nu}{1-\nu}g^{\alpha\beta}g^{\gamma\mu}\right), \tag{3.29a}$$

$$L^{3\alpha3\beta} = \frac{E}{2(1+\nu)}g^{\alpha\beta} \tag{3.29b}$$

where E is Young's modulus and ν is Poisson's ratio.

In order to analyse the contributions of every layer of the sandwich shells to the stiffnesses, we introduce (3.29) in (3.25) and (3.27), and write the results in the following form:

$$D^{\alpha\beta\gamma\mu} = 2tz_1^2\left(1 + \frac{1}{6}\theta_1\theta_2^{-1}\right){}_fH_0^{\alpha\beta\gamma\mu}, \tag{3.30a}$$

$$E^{\alpha\beta\gamma\mu} = 2t\left(1 + \frac{1}{2}\theta_1\theta_2^{-1}\right){}_fH_0^{\alpha\beta\gamma\mu}, \tag{3.30b}$$

$$2HD_1^{\alpha\beta\gamma\mu} = 2tz_1^2\left[1 + \frac{1}{6}(\theta_1\theta_2^{-1} + 2\theta_2^2)\right][{}_fH_0^{\alpha\beta\gamma\mu}(2H\delta_\sigma^\gamma + b_\sigma^\gamma) - {}_fH_1^{\alpha\beta\gamma\mu}], \tag{3.31a}$$

$$2HD_2^{\alpha\beta\gamma\mu} = 2tz_1^2\left[1 + \frac{1}{6}(\theta_1\theta_2^{-1} + 2\theta_2^2)\right][{}_fH_0^{\alpha\beta\gamma\mu} - {}_fH_1^{\alpha\beta\gamma\mu}]. \tag{3.31b}$$

It is shown from (3.30) and (3.31) that only when $\theta_1\theta_2^{-1} < O(\theta^2)$, can the effect of the core on the tension stiffness and the bending stiffness of the sandwich shells be neglected, and only when $\theta_2 < O(\theta)$, can the contribution of the bending stiffness of the faces to the bending stiffness of the sandwich shells be ignored.

In accordance with Chien (1944), we can get the expressions of real resultants of the stresses and real resultants of stress moments as follows

$$N_t^{\alpha\beta} = \int_{-h/2}^{h/2} \sigma^{\beta\gamma} \mu_{\gamma}^{\alpha} \mu \, dz = N^{\alpha\beta} - b_{\gamma}^{\alpha} M^{\beta\gamma}, \tag{3.32a}$$

$$M_t^{\alpha\beta} = \int_{-h/2}^{h/2} \sigma^{\beta\gamma} \mu_{\gamma}^{\alpha} \mu z \, dz = M^{\alpha\beta} - b_{\gamma}^{\alpha} K_t^{\beta\gamma} \tag{3.32b}$$

where

$$K_t^{\beta\gamma} = \int_{-h/2}^{h/2} \sigma^{\beta\gamma} \mu z^2 \, dz = D^{\beta\gamma\sigma\lambda} e_{\sigma\lambda} + O(h^3 t R^{-1} {}_f E h \chi). \tag{3.33}$$

It can be seen from (3.32) that $N_t^{\alpha\beta}$ and $M_t^{\alpha\beta}$ are asymmetric. For the second order theory, $N_t^{\alpha\beta}$ and $M_t^{\alpha\beta}$ cannot be equivalent to $N^{\alpha\beta}$ and $M^{\alpha\beta}$.

3.4. Strain Energy

At first, we introduce an improved resultant of the stress moment:

$$\overline{M}^{\alpha\beta} = M^{\alpha\beta} - \frac{1}{2}(b_{\sigma}^{\alpha} K_t^{\sigma\beta} + b_{\sigma}^{\alpha} K_t^{\sigma\alpha}) = \frac{1}{2}(M_t^{\alpha\beta} + M_t^{\beta\alpha}), \tag{3.34}$$

which can be expressed as:

$$\overline{M}^{\alpha\beta} = - D^{\alpha\beta\gamma\mu}(\chi_{\gamma\mu} + \lambda_{(\gamma\mu)}) - D_f^{\alpha\beta\gamma\mu} \chi_{\gamma\mu} - 2H\overline{D}_2^{\alpha\beta\gamma\mu} e_{\gamma\mu} \tag{3.35}$$

where

$$2H\overline{D}_2^{\alpha\beta\gamma\mu} = 2H_f \overline{D}_2^{\alpha\beta\gamma\mu} + 2H_c \overline{D}_2^{\alpha\beta\gamma\mu} + 2H_s \overline{D}_2^{\alpha\beta\gamma\mu}, \tag{3.36}$$

$$2H_f \overline{D}_2^{\alpha\beta\gamma\mu} = 2H_f D_2^{\alpha\beta\gamma\mu} + \frac{1}{2}(b_{\sigma}^{\alpha} D_f^{\sigma\beta\gamma\mu} + b_{\sigma}^{\beta} D_f^{\sigma\alpha\gamma\mu}), \tag{3.37a}$$

$$2H_c \overline{D}_2^{\alpha\beta\gamma\mu} + 2H_s \overline{D}_2^{\alpha\beta\gamma\mu} = 2H_c D_2^{\alpha\beta\gamma\mu} + 2H_s D_2^{\alpha\beta\gamma\mu} + \frac{1}{2}(b_{\sigma}^{\alpha} D^{\sigma\beta\gamma\mu} + b_{\sigma}^{\beta} D^{\sigma\alpha\gamma\mu}). \tag{3.37b}$$

In addition we have:

$$N_t^{\alpha\beta} = N^{\alpha\beta} - b_{\gamma}^{\alpha} \overline{M}^{\gamma\beta}, \tag{3.38a}$$

$$M_t^{\alpha\beta} = \overline{M}^{\alpha\beta} - \frac{1}{2}(b_{\gamma}^{\alpha} K_t^{\gamma\beta} - b_{\gamma}^{\beta} K_t^{\gamma\alpha}) \tag{3.38b}$$

We will take $\overline{M}^{\alpha\beta}$ instead of $M^{\alpha\beta}$ in the following analysis.

The strain energy of the sandwich shells is

$$U_1 = \int_{\Omega_0} \widetilde{\sum} \, d\Omega_0 \tag{3.39}$$

where Ω_0 is the volume of the sandwich shells in the reference configuration $\Im$, and

$$\widetilde{\sum} = \frac{1}{2} H^{\alpha\beta\gamma\mu} \gamma_{\alpha\beta} \gamma_{\gamma\mu} + 2L^{3\alpha3\beta} \gamma_{3\alpha} \gamma_{3\beta}. \tag{3.40}$$

Because

$$d\Omega_0 = \mu dz dA_0,$$

here A_0 is the area of the reference surface $\mathfrak{R}$, thus:

$$U_1 = \int_{A_0} \left(\int_h \widetilde{\sum} \mu dz\right) dA_0 = \int_{A_0} \sum dA_0 \tag{3.41}$$

where

$$\sum = \int_h \widetilde{\sum} \mu dz, \tag{3.42}$$

or

$$\sum = \frac{1}{2} E^{\alpha\beta\gamma\mu} e_{\alpha\beta} e_{\gamma\mu} + \frac{1}{2} D_f^{\alpha\beta\gamma\mu} \chi_{\alpha\beta} \chi_{\gamma\mu} + \frac{1}{2} D^{\alpha\beta\gamma\mu} (\chi_{\alpha\beta} + \lambda_{\alpha\beta})(\chi_{\gamma\mu} + \lambda_{\gamma\mu})$$

$$\overbrace{+ 2HD_1^{\alpha\beta\gamma\mu} e_{\alpha\beta} \chi_{\gamma\mu} + 2H\overline{D}_2^{\alpha\beta\gamma\mu} \chi_{\alpha\beta} e_{\gamma\mu}}^{\alpha\beta \text{ cannot be exchanged with } \gamma\mu} + \frac{h_2}{2} c L_0^{3\alpha 3\beta} \gamma_\alpha \gamma_\beta. \tag{3.43}$$

It is obvious from (3.43), (3.24) and (3.34) that the following conjugate relations exist

$$N^{\alpha\beta} = -\frac{\partial \sum}{\partial e_{\alpha\beta}},$$

$$\overline{M}^{\alpha\beta} = \frac{\partial \sum}{\partial \chi_{\alpha\beta}},$$

$$Q^\alpha = \frac{\partial \sum}{\partial \gamma_\alpha}. \tag{3.44}$$

3.5. Equilibrium Equations and Boundary Conditions

In order to discuss the issue clearly and conveniently, we introduce:

$$m^{\alpha\beta} = -\frac{\partial \sum}{\partial \lambda_{(\alpha\beta)}} = -D^{\alpha\beta\gamma\mu}(\chi_{\gamma\mu} + \lambda_{(\gamma\mu)}) \tag{3.45}$$

and

$$t^{\alpha\beta} = \overline{M}^{\alpha\beta} - m^{\alpha\beta} = -D_f^{\alpha\beta\gamma\mu} \chi_{\gamma\mu} - 2H\overline{D}_2^{\alpha\beta\gamma\mu} e_{\gamma\mu} \tag{3.46}$$

The latter is the resultant of stress moment in the faces related to their own middle surfaces and the former is the resultant of stress moment of the sandwich shells when the latter can be considered as zero.

If the sandwich shell-load system is conservative, its potential energy function is

$$U = \int_{A_0} \left(\sum - \bar{p} \cdot u\right) dA_0 - \int_{s_0} (q \cdot u + T \cdot \beta + t \cdot \omega) ds_0 \tag{3.47}$$

where we suppose that there is no distributed moment action on the reference surface, $\bar{p}$ is the distributed load vector on the reference surface, q is the distributed load vector on the boundary, T and t are the distributed moments on the boundary corresponding to $m^{\alpha\beta}$ and $t^{\alpha\beta}$, respectively, s_0 is the arc of the boundary in the reference surface.

In terms of the minimum potential principle, i. e. $\delta U=0$, from (3.47) we obtain the equilibrium equations and the boundary conditions as follows

equilibrium equations:

$$[l_\alpha^\gamma N^{\alpha\beta}-b_\alpha^\gamma\overline{M}^{\alpha\beta}]|_\beta-b_\alpha^\gamma(\phi_\beta N^{\alpha\beta}+\overline{M}^{\alpha\beta}|_\beta)+\bar{p}^\gamma=0, \tag{3.48a}$$

$$[\phi_\beta N^{\alpha\beta}-\overline{M}_\beta^{\alpha\beta}|_\beta]|_\alpha+b_\beta^\gamma(l_\alpha^\gamma N^{\alpha\beta}-b_\alpha^\gamma\overline{M}^{\alpha\beta})+\bar{p}^3=0, \tag{3.48b}$$

$$m^{\alpha\beta}|_\beta-Q^\alpha=0, \tag{3.48c}$$

boundary conditions:

$$[l_\alpha^\gamma N^{\alpha\beta}-b_\alpha^\gamma\overline{M}^{\alpha\beta}]\overset{0}{v}_\beta=q^\gamma \quad \text{or} \quad u_\alpha=u_\alpha^0, \tag{3.49a}$$

$$[\phi_\alpha N^{\alpha\beta}+\overline{M}^{\alpha\beta}|_\alpha]\overset{0}{v}_\beta=q^3 \quad \text{or} \quad w=w^0, \tag{3.49b}$$

$$m^{\alpha\beta}\overset{0}{v}_\beta=T^\alpha \quad \text{or} \quad \beta_\alpha=\beta_\alpha^0, \tag{3.49c}$$

$$t^{\alpha\beta}\overset{0}{v}_\beta=t^\alpha \quad \text{or} \quad \omega_\alpha=\omega_\alpha^0. \tag{3.49d}$$

For l_α^γ and ϕ_α see Liu and Zhu (1997), v_β is the outer unit normal to the edge in the reference surface, $u_\alpha^0, w^0, \beta_\alpha^0$ and ϕ_α^0 are the specified values at the boundary corresponding to $u_\alpha, w, \beta_\alpha$ and ϕ_α, respectively.

We have developed the second order theory of sandwich shells. It is rather complicated and not convenient for application. But in the case where the first order theory is not accurate enough, it is useful.

4. First Order Moderate/Moderate Rotation Theory

In terms of the principle of energy error consistency stated in Section 2, the small quantities of order higher than $O(\theta\eta)$ in the strains can be ignored for the first order theory. Therefore, for the purpose of this section, what we need to do is to neglect the small quantities of order higher than $O(\theta\eta)$ in the formulae of the second order theory developed in last section.

When the second order theory is used to solve the problem in practice, the displacement method is usually used so that the compatibility conditions may not be required. But for the first order theory, the mixed method may be applied, hence, the compatibility conditions will be discussed in this section.

4.1. Strain and Compatibility Conditions

In terms of the formulae obtained in the preceding section and neglecting the small quantities of higher order, we have

$$e_{\alpha\beta}=\theta_{\alpha\beta}+\frac{1}{2}(\alpha_{\alpha\beta}\phi^2+\phi_\alpha\phi_\beta), \tag{4.1}$$

$$\chi_{\alpha\beta}=\frac{1}{2}(\phi_\alpha|_\beta+\phi_\beta|_\alpha)-\frac{1}{2}(\in_{\gamma\beta}b_\alpha^\gamma+\in_{\gamma\alpha}b_\beta^\gamma)\phi, \tag{4.2}$$

$$\lambda_{\alpha\beta}=-\gamma_\beta|_\alpha, \tag{4.3}$$

$$U_{\alpha\beta}=O(\theta\lambda^{-1}R^{-1}), \tag{4.4}$$

$$e_{3\alpha}=\gamma_\alpha/2, \tag{4.5}$$

$$\phi_{\alpha} = u_{3,\alpha} + b_{\alpha}^{\gamma} u_{\gamma} = w_{,\alpha} + b_{\alpha}^{\gamma} u_{\gamma}, \tag{4.6}$$

$$\phi = \frac{1}{2} \in^{\alpha\beta} \psi_{\alpha\beta} = \frac{1}{2} \in^{\alpha\beta} u_{\beta} |_{\alpha}. \tag{4.7}$$

The strain of the sandwich shells is

$$_{f_+}\gamma_{\alpha\beta} = e_{\alpha\beta} - \frac{1}{2}(h_c + t)\lambda_{\alpha\beta} - z\chi_{\alpha\beta}, \tag{4.8}$$

$$_{f_-}\gamma_{\alpha\beta} = e_{\alpha\beta} + \frac{1}{2}(h_c + t)\lambda_{\alpha\beta} - z\chi_{\alpha\beta}, \tag{4.9}$$

$$_{c}\gamma_{\alpha\beta} = e_{\alpha\beta} - z\chi_{\alpha\beta} + \lambda_{(\alpha\beta)}, \tag{4.10a}$$

$$_{c}\gamma_{3\beta} = e_{3\alpha} = \gamma_{\alpha}/2. \tag{4.10b}$$

In Eqs. (4.8)-(4.10), $e_{\alpha\beta}$, $\chi_{\alpha\beta}$ and $\lambda_{\alpha\beta}$ are given by (4.1)-(4.3) .

The exact compatibility equations for the moderately thick shells were given in Liu and Zhu (1997). In order to simplify these equations under the condition of moderate rotation, it should be noted first that the conjugate relation based on the precision demands between the equilibrium equations and the compatibility equations really exists although there is not an exact static-geometric analogy in our nonlinear theory of sandwich shells. Equation (4.28d) in Liu and Zhu (1997) is the compatibility condition on the transverse normal strain e_{33} which is not an independent quantity if Koiter's plane stress assumption (i. e. $\sigma^{33}=0$) is accepted, so that e_{33} is not contained in our nonlinear theory and Eq. (4.28d) in Liu and Zhu (1997) is considered to be satisfied naturally. On account of the conjugate relation of precision, the equilibrium equation of moment normal to the reference surface will not be introduced neither and can be satisfied naturally.

By OAM and the principle of energy error consistency and using the strain relation given previously, we have

$$\in^{\alpha\beta} \in^{\lambda\mu} \left(e_{\alpha\mu} |_{\beta\lambda} + b_{\alpha\mu}\chi_{\beta\lambda} + \frac{1}{2}\chi_{\alpha\mu}\chi_{\beta\lambda} \right) = 0, \tag{4.11a}$$

$$\in^{\alpha\beta} \in^{\lambda\mu} [\chi_{\alpha\mu} |_{\beta\lambda} + a^{\xi\sigma}(b_{\alpha\xi} + \chi_{\alpha\xi}) e_{\sigma\beta\lambda}] = 0 \tag{4.11b}$$

$$\in^{\alpha\beta} \in^{\lambda\mu} [\lambda_{[\alpha\mu]} |_{\beta\lambda} + a^{\xi\sigma}(b_{\alpha\xi} + \chi_{\alpha\xi}) e_{\sigma\beta\lambda}] = 0 \tag{4.11c}$$

where

$$\lambda_{[\alpha\mu]} = \frac{1}{2}(\lambda_{\alpha\mu} - \lambda_{\mu\alpha}).$$

In most cases (except the membrane state), the second terms of Eqs. (4.11b, c) can be neglected, hence

$$\in^{\alpha\beta} \in^{\lambda\mu} \chi_{\beta\lambda} |_{\alpha} = 0, \tag{4.11d}$$

$$\in^{\alpha\beta} \in^{\lambda\mu} \lambda_{[\beta\lambda]} |_{\alpha} = 0, \tag{4.11e}$$

If we can consider $\lambda_{\alpha\beta}$ as symmetric approximately, Eq. (4.11e) is satisfied, then the compatibility equations are reduced to (4.11a) and (4.11d).

4.2. Constitutive Equations

It is noted that for the first order theory we can take

$$H^{\alpha\beta\gamma\mu}=H_0^{\alpha\beta\gamma\mu}, \tag{4.12a}$$

$$L^{3\alpha3\beta}=L_0^{3\alpha3\beta}, \tag{4.12b}$$

$$\mu=1. \tag{4.12c}$$

Using the same method as that in the previous section, we obtain:

$$N^{\alpha\beta}=E^{\alpha\beta\gamma\mu}e_{\gamma\mu}, \tag{4.13a}$$

$$M^{\alpha\beta}=-D^{\alpha\beta\gamma\mu}(\chi_{\gamma\mu}+\lambda_{(\gamma\mu)})-D_f^{\alpha\beta\gamma\mu}\chi_{\gamma\mu}, \tag{4.13b}$$

$$Q^{\alpha}=h_c L_0^{3\alpha3\beta}\gamma_\beta \tag{4.13c}$$

and

$$M_t^{\alpha\beta}=\overline{M}^{\alpha\beta}=M^{\alpha\beta}, \tag{4.14a}$$

$$N_t^{\alpha\beta}=N^{\alpha\beta}, \tag{4.14b}$$

i.e. for the first order theory, the nominal resultants of stress and stress moment can be considered as the real ones.

For isotropic sandwich shells, Eqs. (4.13) can be rewritten as a simple form

$$N^{\alpha\beta}=\widetilde{E}[(1-\nu_f)e^{\alpha\beta}+\nu_f a^{\alpha\beta}e_\gamma^\gamma], \tag{4.15a}$$

$$M^{\alpha\beta}=-\widetilde{D}[(1-\nu_f)(\chi^{\alpha\beta}+\lambda^{(\alpha\beta)})+\nu_f a^{\alpha\beta}(\chi_\gamma^\gamma+\lambda_\gamma^\gamma)]$$
$$-D_f[(1-\nu_f)\chi^{\alpha\beta}+\nu_f a^{\alpha\beta}\chi_\gamma^\gamma], \tag{4.15b}$$

$$Q^{\alpha}=h_c G_c\gamma^\alpha \tag{4.15c}$$

where G_c is the transverse shear modulus of the core, ν_f is Poisson's ratio of the faces and E_f is Young's modulus of the faces, and

$$\widetilde{E}=\hat{E}\left(1+\frac{1}{2}\theta_1\theta_2^{-1}\right), \tag{4.16a}$$

$$\widetilde{D}=D\left(1+\frac{1}{6}\theta_1\theta_2^{-1}\right), \tag{4.16b}$$

$$D_f=\frac{E_f t^3}{6(1-\nu_f^2)}, \tag{4.16c}$$

$$\hat{E}=\frac{2tE_f}{1-\nu_f^2}, \tag{4.17a}$$

$$D=\frac{t(h_c+t)^2E_f}{2(1-\nu_f^2)}. \tag{4.17b}$$

It can be seen from (4.15) that for the first order theory, only when $\theta_1\theta_2^{-1}<O(1)$, can the tension stiffness and bending stiffness of the core be ignored. In this case, $\widetilde{E}=\hat{E}$ and $\widetilde{D}=D$. It is also true that only when $\theta_2<0(\sqrt{\theta})$, can the bending stiffness of the faces be neglected, that is $D_f=0$.

4. 3. Strain Energy

Neglecting the higher order terms in (3. 43), we have for the first order theory

$$\sum = \frac{1}{2}E^{\alpha\beta\gamma\mu}e_{\alpha\beta}e_{\gamma\mu} + \frac{1}{2}D^{\alpha\beta\gamma\mu}(\chi_{\alpha\beta} + \lambda_{\alpha\beta})(\chi_{\gamma\mu} + \lambda_{\gamma\mu})$$

$$+ \frac{h_c}{2}\,{}_cL_0^{3\alpha3\beta}\gamma_\alpha\gamma_\beta + \frac{1}{2}D_f^{\alpha\beta\gamma\mu}\chi_{\alpha\beta}\chi_{\gamma\mu}. \tag{4. 18}$$

Obviously, we have:

$$N^{\alpha\beta} = \frac{\partial\sum}{\partial e_{\alpha\beta}}, \tag{4. 19a}$$

$$\overline{M}^{\alpha\beta} = -\frac{\partial\sum}{\partial\chi_{\alpha\beta}}, \tag{4. 19b}$$

$$Q^\alpha = \frac{\partial\sum}{\partial\gamma_\alpha}, \tag{4. 19c}$$

$$m^{\alpha\beta} = -\frac{\partial\sum}{\partial\lambda_{\alpha\beta}} = -D^{\alpha\beta\gamma\mu}(\chi_{\gamma\mu} + \lambda_{(\gamma\mu)}), \tag{4. 19d}$$

$$t^{\alpha\beta} = M^{\alpha\beta} - m^{\alpha\beta} = -D_f^{\alpha\beta\gamma\mu}\chi_{\gamma\mu}. \tag{4. 19e}$$

4. 4. Equilibrium Equations and Boundary Conditions

Similar to (3. 47), we have

$$U = \int_{A_0}(\sum - p\cdot u)\mathrm{d}A_0 - \int_{S_0}(q\cdot u + T\cdot\beta + t\cdot\omega)\mathrm{d}s_0.$$

Using the minimum potential principle $\delta U=0$, we obtain the equilibrium equations:

$$\left[N^{\alpha\beta} + \frac{1}{2}\in^{\beta\alpha}\phi N_\gamma^\gamma - \frac{1}{2}(b_\pi^\alpha M^{\pi\beta} - b_\pi^\beta M^{\pi\alpha})\right]\Big|_\beta$$

$$- b_\beta^\alpha(\phi_\gamma N^{\beta\gamma} + M^{\beta\gamma}|_\gamma) + p^\alpha = 0, \tag{4. 20a}$$

$$[\phi_\beta N^{\alpha\beta} + M^{\alpha\beta}|_\beta]|_\alpha + b_{\alpha\beta}N^{\alpha\beta} + p^3 = 0, \tag{4. 20b}$$

$$m^{\alpha\beta}|_\beta - Q^\alpha = 0 \tag{4. 20c}$$

and the boundary conditions

$$\left[N^{\alpha\beta} + \frac{1}{2}\in^{\beta\alpha}\phi N_\gamma^\gamma - \frac{1}{2}(b_\pi^\alpha M^{\pi\beta} - b_\pi^\beta M^{\pi\alpha})\right]\overset{0}{\nu}_\beta = q^\alpha \quad \text{or} \quad u_\alpha = u_\alpha^0 \tag{4. 21a}$$

$$[\phi_\alpha N^{\alpha\beta} + M^{\alpha\beta}|_\alpha]\overset{0}{\nu}_\beta = q^3 \quad \text{or} \quad w = w^0 \tag{4. 21b}$$

$$m^{\alpha\beta}\overset{0}{\nu}_\beta = T^\alpha \quad \text{or} \quad \beta_\alpha = \beta_\alpha^0 \tag{4. 21c}$$

$$t^{\alpha\beta}\overset{0}{\nu}_\beta = t^\alpha \quad \text{or} \quad \phi_\alpha = \phi_\alpha^0 \tag{4. 21d}$$

This system of equations is the development of the famous Sander-Koiter equations of thin elastic shells (Koiter, 1966; Yamaki, 1984) .

If the bending stiffness of the faces related to their own middle surface is neglected, i. e. $D_f^{\alpha\beta\gamma\mu}=0$, then $m^{\alpha\beta}=M^{\alpha\beta}$ and the boundary condition (4. 21d) should be abandoned.

5. First Order Moderate/Small Rotation Theory

Because the moderate/small rotation motion of the shells satisfies the condition

$$\phi_\alpha = O(\theta), \phi = O(\theta^2),$$

for the first order theory, ϕ is negligible. Therefore, what we need to do is to neglect the terms containing ϕ in the equations of Section 4 in this paper.

5.1. Strain Relations and Compatibility Conditions

The strains are

$$e_{\alpha\beta} = \theta_{\alpha\beta} + \frac{1}{2}\phi_\alpha\phi_\beta, \tag{5.1}$$

$$\chi_{\alpha\beta} = \frac{1}{2}(\phi_\alpha |_\beta + \phi_\beta |_\alpha), \tag{5.2}$$

$$\lambda_{\alpha\beta} = \gamma_\beta |_\alpha. \tag{5.3}$$

The compatibiliy equations are

$$\in^{\alpha\beta} \in^{\lambda\mu} \left(e_{\alpha\mu} |_{\beta\lambda} + b_{\alpha\mu}\chi_{\beta\lambda} + \frac{1}{2}\chi_{\alpha\mu}\chi_{\beta\lambda} \right) = 0 \tag{5.4a}$$

$$\in^{\alpha\beta} \in^{\lambda\mu}\chi_{\lambda\beta} |_\alpha = 0, \tag{5.4b}$$

$$\in^{\alpha\beta} \in^{\lambda\mu}\lambda_{[\lambda\beta]} |_\alpha = 0. \tag{5.4c}$$

5.2. Constitutive Equations

$$N^{\alpha\beta} = E^{\alpha\beta\gamma\mu} e_{\gamma\mu}, \tag{5.5a}$$

$$M^{\alpha\beta} = - D^{\alpha\beta\gamma\mu}(\chi_{\gamma\mu} + \lambda_{(\gamma\mu)}) - D_f^{\alpha\beta\gamma\mu}\chi_{\gamma\mu}, \tag{5.5b}$$

$$Q^\alpha = h_\alpha L^{3\alpha3\beta}\gamma_\beta \tag{5.5c}$$

for isotropic sandwich shells

$$N^{\alpha\beta} = \widetilde{E}[(1-\nu_f)e^{\alpha\beta} + \nu_f a^{\alpha\beta} e_\gamma^\gamma], \tag{5.6a}$$

$$\begin{aligned} M^{\alpha\beta} = &- \widetilde{D}[(1-\nu_f)(\chi^{\alpha\beta} + \lambda^{(\alpha\beta)}) + \nu_f a^{\alpha\beta}(\chi_\gamma^\gamma + \lambda_\gamma^\gamma)] \\ &- D_f[(1-\nu_f)\chi^{\alpha\beta} + \nu_f a^{\alpha\beta}\chi_\gamma^\gamma], \end{aligned} \tag{5.6b}$$

$$Q^\alpha = h_c G_c \gamma^\alpha, \tag{5.6c}$$

Eqs. (5.5) and (5.6) are the same as (4.13) and (4.15), but the expressions of strains are different. Here strain formulae (5.1) and (5.2) should be used.

5.3. Equilibrium Equations and Boundary Conditions

The equilibrium equations are

$$N^{\alpha\beta} |_\beta - b_\beta^\alpha(\phi_\gamma N^{\beta\gamma} + M^{\beta\gamma} |_\gamma) + p^\alpha = 0, \tag{5.7a}$$

$$[\phi_\beta N^{\alpha\beta} + M^{\alpha\beta} |_\beta] |_\alpha + b_{\alpha\beta}N^{\alpha\beta} + p^3 = 0 \tag{5.7b}$$

$$m^{\alpha\beta} |_\beta - Q^\alpha = 0 \tag{5.7c}$$

and the boundary conditions are

$$N^{\alpha\beta}\overset{0}{\nu}_\beta = q^\alpha \quad \text{or} \quad u_\alpha = \overset{0}{u\alpha} \tag{5.8a}$$

$$[\phi_\alpha N^{\alpha\beta} + M^{\alpha\beta} |_\alpha]\overset{0}{\nu}_\beta = q^3 \quad \text{or} \quad w = w^0, \tag{5.8b}$$

$$m^{\alpha\beta}\overset{0}{\nu}_\beta = T^\alpha \quad \text{or} \quad \beta_\alpha = \beta_\alpha^0, \tag{5.8c}$$

$$t^{\alpha\beta}\overset{0}{\nu}_{\beta} = t^{\alpha} \qquad \text{or} \quad \phi_{\alpha} = \phi_{\alpha}^{0}, \tag{5.8d}$$

Eqs. (5.7) are consistent, but they still contain some megligible terms in most cases, e.g., $b_{\beta}^{\alpha}(\phi_{\gamma}N^{\beta\gamma}+M^{\beta\gamma}|_{\gamma})$ in Eq. (5.7a). Hence (5.7) can be reduced to

$$N^{\alpha\beta}|_{\beta}+p^{\alpha}=0, \tag{5.9a}$$

$$m^{\alpha\beta}|_{\beta\alpha}+t^{\alpha\beta}|_{\beta\alpha}+(b_{\alpha\beta}+\chi_{\alpha\beta})N^{\alpha\beta}-\phi_{\alpha}p^{\alpha}+p^{3}=0, \tag{5.9b}$$

$$m^{\alpha\beta}|_{\beta}-Q^{\alpha}=0. \tag{5.9c}$$

Meanwhile, we take

$$\phi_{\alpha}=w_{,\alpha}=w|_{\alpha}, \tag{5.10a}$$

$$\chi_{\alpha\beta}=w|_{\alpha\beta}. \tag{5.10b}$$

On the other hand, for the shallow shells, whose deformations fall in the range of small strain accompanied by moderate/small rotation, we can take

$$\gamma_{\alpha}=w_{s,\alpha} \tag{5.11}$$

where w_s is the additional deflection due to the transverse shear of the core.

Thus

$$\gamma_{\alpha\beta}=-w_{s}|_{\alpha\beta}, \tag{5.12}$$

which is symmetric.

Formulae (5.10) and (5.12) satisfy the compatibility conditions (5.4b,c). Only compatibility condition (5.4a) remains to be discussed.

5.4. Further Simplification of the Equations

Substituting (5.10), (5.11) and (5.12) into (5.5b,c), we obtain:

$$M^{\alpha\beta}=-D^{\alpha\beta\gamma\mu}w_{b}|_{\gamma\mu}-D_{f}^{\alpha\beta\gamma\mu}w|_{\gamma\mu}=m^{\alpha\beta}+t^{\alpha\beta}, \tag{5.13a}$$

$$Q^{\alpha}=h_{\alpha}L_{0}^{\alpha3\beta3}w_{s}|_{\beta} \tag{5.13b}$$

where

$$w_{b}=w-w_{s} \tag{5.14}$$

and we further have

$$\sum=\frac{1}{2}E^{\alpha\beta\gamma\mu}e_{\alpha\beta}e_{\gamma\mu}+\frac{1}{2}D^{\alpha\beta\gamma\mu}w_{b}|_{\alpha\beta}w_{b}|_{\alpha\beta}$$
$$+\frac{h_{c}}{2}cL_{0}^{3\alpha3\beta}w_{s}|_{\alpha}w_{s}|_{\beta}+\frac{1}{2}D_{f}^{\alpha\beta\gamma\mu}w|_{\alpha\beta}w|_{\gamma\mu}. \tag{5.15}$$

Suppose p^{α} has potential, then

$$p^{\alpha}=-\psi|^{\alpha}. \tag{5.16}$$

Let us introduce the resultant function F which satisfies:

$$N^{\alpha\beta}=\in^{\alpha\gamma}\in^{\beta\mu}F|_{\gamma\mu}+a^{\alpha\beta}\psi, \tag{5.17}$$

then Eq. (5.9a) is satisfied.

By the use of (5.13), (5.16) and (5.17), Eqs. (5.4a) and (5.9b,c) can be reduced into

$$D^{\alpha\beta\gamma\mu}w_{b}|_{\alpha\beta\gamma\mu}+D_{f}^{\alpha\beta\gamma\mu}w|_{\alpha\beta\gamma\mu}-\in^{\alpha\gamma}\in^{\beta\mu}(b_{\alpha\beta}+w|_{\alpha\beta})F|_{\gamma\mu}$$
$$-2(2H+w|_{\alpha}^{\alpha})\psi-w|_{\alpha}\psi|^{\alpha}-p^{3}=0, \tag{5.18a}$$

$$\in^{\alpha\beta} \in^{\lambda\mu}\left(\in^{\gamma\sigma} \in^{\xi\pi} C_{\alpha\mu\gamma\xi} F \mid_{\sigma\pi\beta\gamma} + b_{\alpha\mu} w \mid_{\beta\lambda} + \frac{1}{2} w \mid_{\alpha\mu} w \mid_{\beta\lambda} + C_{\alpha\mu\gamma\xi} a^{\gamma\xi} \psi \mid_{\beta\lambda}\right) = 0, \tag{5.18b}$$

$$D^{\alpha\beta\gamma\mu} w_b \mid_{\gamma\mu} + h_c L_0^{\alpha 3\beta 3} (w - w_b) = f^{\alpha\beta} \tag{5.18c}$$

where $C_{\alpha\beta\gamma\mu}$ is the reciprocal tensor of $E^{\alpha\beta\gamma\mu}$, $f^{\alpha\beta}$ is a constant tensor.

Equations (5.18) contain three unknown functions F, w, w_b. Together with proper boundary conditions, they are a definite problem.

For isotropic sandwich shells, we have the constitutive equations

$$e^{\alpha\beta} = \frac{1}{E(1-\nu_f^2)}[(1+\nu_f) \in^{\alpha\gamma} \in^{\beta\mu} F \mid_{\gamma\mu} + a^{\alpha\beta}\psi - \nu_f a^{\alpha\beta}(\nabla^2 F + 2\psi)] \tag{5.19a}$$

$$m^{\alpha\beta} = -\widetilde{D}[(1-\nu_f) w_b \mid^{\alpha\beta} + \nu_f a^{\alpha\beta} \nabla^2 w_b], \tag{5.19b}$$

$$t^{\alpha\beta} = -D_f[(1-\nu_f) w \mid^{\alpha\beta} + \nu_f a^{\alpha\beta} \nabla^2 w], \tag{5.19c}$$

$$Q^{\alpha} = h_c G_c w_s \mid^{\alpha}. \tag{5.19d}$$

the equilibrium equations and compatibility condition

$$\widetilde{D}\nabla^4 w_b + D_f \nabla^4 w - \in^{\alpha\gamma} \in^{\beta\mu} (b_{\alpha\beta} + w \mid_{\alpha\beta}) F \mid_{\gamma\mu} - (2H + \nabla^2 w)\psi - w \mid_{\alpha} \psi^{\alpha} - p^3 = 0, \tag{5.20a}$$

$$\nabla^4 F + \widetilde{E}(1-\nu_f^2) \in^{\alpha\beta} \in^{\gamma\mu}\left(b_{\beta\mu} w \mid_{\alpha\gamma} + \frac{1}{2} w \mid_{\beta\mu} w \mid_{\alpha\gamma}\right) + (1-\nu_f) \nabla^2 \psi = 0, \tag{5.20b}$$

$$w = \left(1 - \frac{\widetilde{D}}{h_c G_c} \nabla^2\right) w_b + \text{const} \tag{5.20c}$$

where

$$\nabla^2 = (\ ) \mid_{\alpha}^{\alpha},$$

$$\nabla^4 = \nabla^2 \nabla^2 = (\ ) \mid_{\alpha\beta}^{\alpha\beta}.$$

In Eq. (5.20c), the constant may be zero. In addition, in Eqs. (5.19) and (5.20a), D_f (hence $t^{\alpha\beta}$) can be ignored in most cases.

Acknowledgments

The authors wish to express their great gratitude to Professor W. Z. Chien for his encouragement and beneficial discussions.

Nomenclature

$D_f^{\alpha\beta\gamma\mu}, D^{\alpha\beta\gamma\mu}, E^{\alpha\beta\gamma\mu}$	—	the bending stiffness of face and plate, and stretching stiffness tensor, respectively
$e_{\alpha\beta}, \chi_{\alpha\beta}$	—	strain and curvature changing tensors
h_c, t, h	—	the core, face, and total thickness, respectively
$N^{\alpha\beta}, M^{\alpha\beta}, Q^{\alpha}$	—	the stress resultant tensor, stress moment tensor and transverse shear forces
$t^{\alpha\beta}, m^{\alpha\beta}$	—	the stress moment tensor of face and core, respectively
u_{α}, w	—	in-plane displacements and deflection at the middle surface

$\theta_{\alpha\beta}$ — the linear part of $e_{\alpha\beta}$

ϕ_α, ϕ — the out-plane and in-plane rotation at the middle surface

γ_α — the transverse shear deformation

ω_α — the total out-plane rotation at the middle surface

$\lambda_{\alpha\beta}$ — the contribution to curvatmre at the middle sunface

σ^{ij} — the stress tensor

$\sum$ — the strain enerav

References

Budiansky B. (1968): *Notes on nonlinear shell theory*. J. Appl. Mech. v. 35, pp. 393-401.

Chien W. Z. (1944a): *The intrinsic theory of thin shells and plates*, Part Ⅰ, General theory. Quart. Appl. Math., No. 1, pp. 297-327.

Chien W. Z. (1944b): *The intrinsic theory of thin shells and plates*, Part Ⅱ, Application to the plates. Quart. Appl. Math., No. 2, pp. 43-57.

Chien W. Z. (1944c): *The intrinsic theory of thin shells and plates*, Part Ⅲ, *Application to thin shells*. Quart Appl. Math., No. 2, pp. 120-135.

Danielson D. A. (1970): *Simplified intrinsic equations for arbitrary elastic shells*. Int. J. Engng. Sci., v. 8, pp. 251-259.

John F. (1965): *Estimates for the derivatives of the stresses in a thin shell and interior shell equations*. Commun. Pure and App. Math., v. 18, pp. 235-267.

Koiter W. T. (1966): *On the nonlinear theory of thin elastic shells* (Ⅰ, Ⅱ, Ⅲ). Proc. Kon. Ned. AK. Wet., v. B69, pp. 1-60.

Koiter W. T. (1980): *The intrinsic equations of shell theory with some applications*, In: Mechanics Today: 139-154.

Koiter W. T. and Simmonds J. G. (1973): *Foundation of shell theory*, In: Proc. 13th Int. Congr. Theor. Appl. Mech. (E. Becker and G. K. Mikhaiv Eds.). Berlin: Springer-Verkag.

Liu Renhuai. and Zhu J. F. (1997): *Nonlinear theory of sandwich shells, part I-Exact kinematics of moderately thick shells*. Applied Mech. and Engi., v. 2, No. 2, pp. 213-240.

Naghdi P. M. and Vongsarnpigoon L. (1982): *Small strain accompanied by moderate rotation*. Arch. Rational Mech. Anal., v. 80. pp. 263-294.

Naghdi P. M. and Vongsarnpigoon L. (1983): *A theory of shells with small strain accompanied by moderate rotation*. Arch. Rational Mech. Anal., v. 83, pp. 245-283.

Pietraszkiewicz W. (1979): *Finite Rotations and Lagrangean Description in the Nonlinear Theory of Shells*. Institute of Fluid-Flow Machinery, Polish Academy of Science.

Pietraszkiewicz W. and Badur J. (1983): *Fintite rotation in the description of continuum deformation*. Int. J. Engng. Sci., v. 21, pp. 1097-1110.

Schmidt R. (1985): *A current trend in shell theory: constrained geometrically nonlinear Kirchhoff Love type theories based on polar decomposition of strains and rotations*. Comp. Struct., v. 20, pp. 265-280.

Yamaki Y. (1984): *Elastic Stability of Circular Cylindrical Shells*. Amsterdam: North-Holland.

On nonlinear stability of shallow conical sandwich shells①

中文摘要

在本文中，使用变分法推导出了在轴对称分布横向载荷作用下夹层扁锥壳非线性弯曲理论的基本方程和边界条件。使用修正迭代法，对于四种边界条件，获得了壳体在均匀外压力作用下的非线性稳定的解析解。

1. Introduction

In recent years, sandwich structures have become more widely utilized in aeronautical and mechanical engineering. Many papers have been published internationally. Most of them are confined to linearity. Reissner (1948) first presented the large deflection theory of a rectangular sandwich plate with two thin faces and a soft core. On account of the nonlinear analysis, until 1967, this problem had just been discussed by Kan and Huang (1967). Liu (1981) established the nonlinear bending theory of circular sandwich plates that takes into account faces with resistance to bending and solved several large deflection problems (Liu, 1980, 1981, 1987; Liu and Shi, 1982; Liu and Zhu, 1989; Liu et al., 1997) of circular sandwich plates using the modified iteration method as suggested by Yeh and Liu. (1965a, 1965b), perturbation method and power series method respectively. Recently, Liu and Wu (1992), Liu and Zhu (1993), Liu and Cheng (1993) also studied nonlinear bending and vibration of rectangular sandwich plates. Liu and Wang (1996), Wang and Liu (1997, 1998a, 1998b) developed the theory of consistent moderate rotation of sandwich plates and shells faced with laminated composites and investigated a few interesting problems. Comparatively, the investigation of nonlinear problems for conical sandwich shells received less attention. Struk (1984) discussed the nonlinear stability of conical sandwich shells under the combined action of external pressure and axial compression by Ritz's procedure. Afterward, Liu and Zhu (1993, 1997a, 1997b) established the general nonlinear bending theory of sandwich shells and studied nonlinear stability and vibration for sandwich shells.

In this paper, similar to Reissner's assumptions for the rectangular sandwich plate theory, we derive fundamental equations and boundary conditions of the nonlinear bending theory for shallow conical sandwich shells under axisymmetrical distributed lateral

---

① Reprinted from *Applied Mechanics and Engineering*, 2000, 5 (2): 367-387. Authors: Liu Renhuai and Li Jun.

load by the method of calculus of variations. Concerning four types of boundary conditions, the analytic solutions of nonlinear stability for the shells under uniform pressure have been obtained by the modified iteration method. As we know, these solutions have not been obtained before.

2. Fundamental Equations

Consider a shallow conical sandwich shell under the axisymmetrical distributed lateral load $q(r)$ as shown in Fig 1. Here h_0 is the thickness, t is the thickness of the face, h is the thickness of the core, a is the radius, r is the radial coordinate, θ is the circumferential coordinate, α is the slope angle, z is the coordinate in the thickness direction.

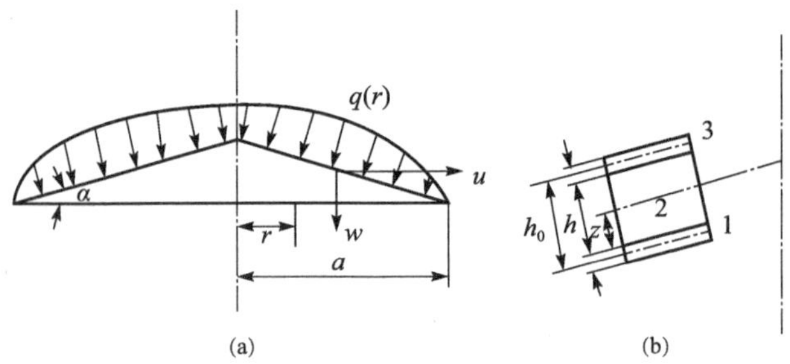

Fig. 1　Geometry of shallow conical sandwich shell.

To derive the fundamental equations and corresponding boundary conditions, we introduce the following assumptions:

(1) Material is submitted to Hooke's law,

(2) The core is transversely uncompressible,

(3) The core in the direction of the surface of the shell is not subjected to loads,

(4) The face is in the membrane stress state,

(5) The normal line to the middle surface of the core remains straight during bending.

Let u_i, v_i, w_i $(i=1,2,3)$ be the radial, circumferential and vertical displacements of the upper face, core, and lower face respectively, u, w the radial displacement and deflection of the middle surface of the shell respectively, and ψ the rotation of the normal to the middle surface of the shell. Using the above assumptions and the condition of symmetry, displacements at a point of the shell can be written as:

The upper face $\left[\frac{h}{2}\leqslant z\leqslant\frac{h}{2}+t\right]$:

$$u_1 = u+\frac{h_0}{2}\psi,\quad v_1 = 0,\quad w_1 = w \tag{2.1}$$

the core $\left[-\frac{h}{2}\leqslant z\leqslant\frac{h}{2}\right]$:

$$u_2 = u+z\psi,\quad v_2 = 0,\quad w_2 = w \tag{2.2}$$

the lower face $\left[-\left(\frac{h}{2}+t\right)\leqslant z\leqslant-\frac{h}{2}\right]$:

$$u_3 = u - \frac{h_0}{2}\psi, \quad v_3 = 0, \quad w_3 = w \tag{2.3}$$

Now let $\varepsilon_{ri}, \varepsilon_{\theta i}, \varepsilon_{zi}, \gamma_{r\theta i}, \gamma_{\theta z i}, \gamma_{rzi}$ $(i=1,2,3)$ be the strain components of the upper face, core, and lower face respectively, then the strain-displacement relations can be expressed in the form:

$$\begin{aligned} &\varepsilon_{ri} = \frac{\mathrm{d}u_i}{\mathrm{d}r} + \frac{\mathrm{d}w_i}{\mathrm{d}r}\left(\frac{1}{2}\frac{\mathrm{d}w_i}{\mathrm{d}r} + \alpha\right), \quad \varepsilon_{\theta i} = \frac{u}{r} \\ &\varepsilon_{zi} = \gamma_{r\theta i} = \gamma_{\theta z i} = \gamma_{rzi} = 0 \quad i = 1,3, \end{aligned} \tag{2.4}$$

$$\begin{aligned} &\gamma_{rz2} = \frac{\partial u_2}{\partial z} + \frac{\mathrm{d}w_2}{\mathrm{d}r}, \\ &\varepsilon_{r2} = \varepsilon_{\theta 2} = \varepsilon_{z2} = \gamma_{r\theta 2} = \gamma_{\theta z 2}. \end{aligned} \tag{2.5}$$

Introduction of Eqs. (2.1)-(2.3) into (2.4) and (2.5) yields the upper face:

$$\begin{aligned} &\varepsilon_{r1} = \frac{\mathrm{d}u}{\mathrm{d}r} + \frac{h_0}{2}\frac{\mathrm{d}\psi}{\mathrm{d}r} + \frac{\mathrm{d}w}{\mathrm{d}r}\left(\frac{1}{2}\frac{\mathrm{d}w}{\mathrm{d}r} + \alpha\right), \\ &\varepsilon_{\theta 1} = \frac{1}{r}\left(u + \frac{h_0}{2}\psi\right) \end{aligned} \tag{2.6}$$

the core:

$$\gamma_{rz2} = \psi + \frac{\mathrm{d}w}{\mathrm{d}r} \tag{2.7}$$

the lower face:

$$\begin{aligned} &\varepsilon_{r3} = \frac{\mathrm{d}w}{\mathrm{d}r} - \frac{h_0}{2}\frac{\mathrm{d}\psi}{\mathrm{d}r} + \frac{\mathrm{d}w}{\mathrm{d}r}\left(\frac{1}{2}\frac{\mathrm{d}w}{\mathrm{d}r} + \alpha\right), \\ &\varepsilon_{\theta 3} = \frac{1}{r}\left(u - \frac{h_0}{2}\psi\right). \end{aligned} \tag{2.8}$$

Let $\sigma_{ri}, \sigma_{\theta i}, \sigma_{zi}, \tau_{r\theta i}, \tau_{\theta z i}, \tau_{rzi}$ $(i=1,2,3)$ be the stress components of the upper face, core, lower face respectively, E Yong's modulus, ν Poisson's ratio of the face and G_2 the shear modulus of the core. Then Hook's law is given by:

$$\begin{aligned} &\sigma_{ri} = \frac{E}{1-\nu^2}(\varepsilon_{ri} + \nu\varepsilon_{\theta i}), \quad \sigma_{\theta i} = \frac{E}{1-\nu^2}(\varepsilon_{\theta i} + \nu\varepsilon_{ri}), \\ &\sigma_{zi} = \tau_{r\theta i} = \tau_{\theta z i} = \tau_{rzi} = 0 \quad i = 1,3, \end{aligned} \tag{2.9}$$

$$\begin{aligned} &\tau_{rz2} = G_2\gamma_{rz2}, \\ &\sigma_{r2} = \sigma_{\theta 2} = \sigma_{z2} = \tau_{r\theta 2} = \tau_{\theta z 2} = 0. \end{aligned} \tag{2.10}$$

Inserting Eqs. (2.6)-(2.8) into (2.9) and (2.10), we obtain the upper face:

$$\begin{aligned} &\sigma_{r1} = \sigma_{r0} + \frac{Eh_0}{2(1-\nu^2)}\left(\frac{\mathrm{d}\psi}{\mathrm{d}r} + \nu\frac{\psi}{r}\right), \\ &\sigma_{\theta 1} = \sigma_{\theta 0} + \frac{Eh_0}{2(1-\nu^2)}\left(\frac{\psi}{r} + \nu\frac{\mathrm{d}w}{\mathrm{d}r}\right), \end{aligned} \tag{2.11}$$

the core:

$$\tau_{rz2} = G_2\left(\psi + \frac{\mathrm{d}w}{\mathrm{d}r}\right), \tag{2.12}$$

the lower face:

$$\sigma_{r3} = \sigma_{r0} - \frac{Eh_0}{2(1-\nu^2)}\left(\frac{d\psi}{dr} + \nu\frac{\psi}{r}\right),$$

$$\sigma_{\theta 3} = \sigma_{\theta 0} - \frac{Eh_0}{2(1-\nu^2)}\left(\frac{\psi}{r} + \nu\frac{d\psi}{dr}\right) \tag{2.13}$$

where σ_{r0}, $\sigma_{\theta 0}$ are the radial and circumferential stresses of the middle surface of the shell, respecticely,

$$\sigma_{r0} = \frac{E}{1-\nu^2}\left[\frac{du}{dr} + \nu\frac{u}{r} + \frac{dw}{dr}\left(\frac{1}{2}\frac{dw}{dr} + \alpha\right)\right],$$

$$\sigma_{\theta 0} = \frac{E}{1-\nu^2}\left[\frac{u}{r} + \nu\frac{du}{dr} + \nu\frac{dw}{dr}\left(\frac{1}{2}\frac{dw}{dr} + \alpha\right)\right]. \tag{2.14}$$

The strain energy of the elastic body, denoted by U, is:

$$U = \frac{1}{2}\iiint_V (\sigma_r\varepsilon_r + \sigma_\theta\varepsilon_\theta + \sigma_z\varepsilon_z + \tau_{r\theta}\gamma_{r\theta} + \tau_{\theta z}\gamma_{\theta z} + \tau_{rz}\gamma_{rz})r\mathrm{d}r\mathrm{d}\theta\mathrm{d}z. \tag{2.15}$$

According to this formula, the strain energies of the face and core are:

$$U_i = \frac{1}{2E}\iiint_{V_i}[(\sigma_{ri} + \sigma_{\theta i})^2 - 2(1+\nu)\sigma_{ri}\sigma_{\theta i}]r\mathrm{d}r\mathrm{d}\theta\mathrm{d}z, \quad i = 1,3,$$

$$U_2 = \frac{1}{2G_2}\iiint_{V_2}\tau_{rz2}^2 r\mathrm{d}r\mathrm{d}\theta\mathrm{d}z. \tag{2.16}$$

Substituting Eqs. (2.11)-(2.13) into (2.16) and integrating them with respect to z, we obtain:

$$\begin{aligned}
U_1 =& \frac{t}{2E}\iint_S[(\sigma_{r0} + \sigma_{\theta 0})^2 - 2(1+\nu)\sigma_{r0}\sigma_{\theta 0}]r\mathrm{d}r\mathrm{d}\theta \\
&+ \frac{h_0 t}{2}\iint_{S_1}\left(\sigma_{r0}\frac{d\psi}{dr} + \frac{1}{r}\sigma_{\theta 0}\psi\right)r\mathrm{d}r\mathrm{d}\theta \\
&+ \frac{D}{4}\iint_{S_1}\left[\left(\frac{d\psi}{dr} + \frac{\psi}{r}\right)^2 - 2(1-\nu)\frac{\psi}{r}\frac{d\psi}{dr}\right]r\mathrm{d}r\mathrm{d}\theta, \\
U_2 =& \frac{G_2 h_0}{2}\iint_{S_2}\left(\psi + \frac{dw}{dr}\right)^2 r\mathrm{d}r\mathrm{d}\theta, \\
U_3 =& \frac{t}{2E}\iint_{S_3}[(\sigma_{r0} + \sigma_{\theta 0})^2 - 2(1+\nu)\sigma_{r0}\sigma_{\theta 0}]r\mathrm{d}r\mathrm{d}\theta \\
&- \frac{h_0 t}{2}\iint_{S_3}\left(\sigma_{r0}\frac{d\psi}{dr} + \frac{1}{r}\sigma_{\theta 0}\psi\right)r\mathrm{d}r\mathrm{d}\theta \\
&+ \frac{D}{4}\iint_{S_3}\left[\left(\frac{d\psi}{dr} + \frac{\psi}{r}\right)^2 - 2(1-\nu)\frac{\psi}{r}\frac{d\psi}{dr}\right]r\mathrm{d}r\mathrm{d}\theta,
\end{aligned} \tag{2.17}$$

in which D is the flexural rigidity of the shell:

$$D = \frac{Eh_0^2 t}{2(1-\nu^2)}. \tag{2.18}$$

The potential energy of the external load $q(r)$ is:

$$V = \iint_S qwr\,\mathrm{d}r\mathrm{d}\theta. \tag{2.19}$$

Thus the total potential energy of the shallow conical sandwich shell is obtained

$$U = U_1 + U_2 + U_3 - V = \frac{t}{E}\iint_S [(\sigma_{r0} + \sigma_{\theta0})^2 - 2(1+v)\sigma_{r0}\sigma_{\theta0}] r\mathrm{d}r\mathrm{d}\theta$$

$$+ \frac{D}{2}\iint_S \left[\left(\frac{\mathrm{d}\psi}{\mathrm{d}r} + \frac{\psi}{r}\right)^2 - 2(1-\nu)\frac{\psi}{r}\frac{\mathrm{d}\psi}{\mathrm{d}r}\right] r\mathrm{d}r\mathrm{d}\theta + \frac{G_2 h_0}{2}\iint_S \left(\psi + \frac{\mathrm{d}w}{\mathrm{d}r}\right)^2 r\mathrm{d}r\mathrm{d}\theta - \iint_S qwr\,\mathrm{d}r\mathrm{d}\theta. \tag{2.20}$$

Applying the principle of the potential energy to expression (2.20), the following equations of equlilbrium and boundary conditions of the nonlinear bending theory for the shallow conical sandwich shell are obtained:

$$\begin{aligned} &\sigma_{\theta0} - \frac{\mathrm{d}}{\mathrm{d}r}(r\sigma_{r0}) = 0, \\ &2t\frac{\mathrm{d}}{\mathrm{d}r}\left[r\sigma_{r0}\left(\frac{\mathrm{d}w}{\mathrm{d}r} + \alpha\right)\right] + G_2 h_0 \frac{\mathrm{d}}{\mathrm{d}r}[r(\psi + \frac{\mathrm{d}w}{\mathrm{d}r})] + qr = 0. \\ &Dr\frac{\mathrm{d}}{\mathrm{d}r}\frac{1}{r}\frac{\mathrm{d}}{\mathrm{d}r}(r\psi) - G_2 h_0 r\left(\psi + \frac{\mathrm{d}w}{\mathrm{d}r}\right) = 0. \end{aligned} \tag{2.21a,b,c}$$

$$\begin{aligned} &\text{at } r = 0 \quad \text{or} \quad \text{at } t = a, \\ &r\sigma_{r0} = 0, \quad \text{or} \quad \delta u = 0, \\ &rQ_r + 2tr\sigma_{r0}\left(\frac{\mathrm{d}w}{\mathrm{d}r} + \alpha\right) = 0, \quad \text{or} \quad \delta w = 0, \\ &rM_r = 0 \quad \text{or} \quad \delta\psi = 0 \end{aligned} \tag{2.22}$$

where M_r is the radial moment, Q_r is the shearing force:

$$\begin{aligned} M_r &= D\left(\frac{\mathrm{d}\psi}{\mathrm{d}r} + \nu\frac{\psi}{r}\right), \\ Q_r &= G_2 h_0\left(\psi + \frac{\mathrm{d}w}{\mathrm{d}r}\right). \end{aligned} \tag{2.23a,b}$$

Eliminating the displacement u from Eq. (2.14), we obtain the following compatibility equation:

$$\sigma_{r0} - \frac{\mathrm{d}}{\mathrm{d}r}(r\sigma_{\theta0}) - E\frac{\mathrm{d}w}{\mathrm{d}r}(\frac{1}{2}\frac{\mathrm{d}w}{\mathrm{d}r} + \alpha) = 0. \tag{2.24}$$

The system of Eqs. (2.21) and (2.24) governs nonlinear bending of the shallow conical sandwich shell. The fundamental Eqs. (2.21) and (2.24) can be simplified. As usual, a stress function φ is defined by:

$$\sigma_{r0} = \frac{1}{r}\frac{\mathrm{d}\varphi}{\mathrm{d}r}, \quad \sigma_{\theta0} = \frac{\mathrm{d}^2\varphi}{\mathrm{d}r^2}. \tag{2.25}$$

The stress function thus satisfies Eq. (2.21a) exactly, and Eqs. (2.21b) and (2.24) become:

$$2t\frac{d}{dr}\left[\frac{d\varphi}{dr}\left(\frac{dw}{dr}+\alpha\right)\right]+G_2h_0\frac{d}{dr}\left[r\left(\psi+\frac{dw}{dr}\right)\right]+qr=0,$$

$$r\frac{d}{dr}\frac{1}{r}\frac{d}{dr}\left(r\frac{d\varphi}{dr}\right)+E\frac{dw}{dr}\left(\frac{1}{2}\frac{dw}{dr}+\alpha\right)=0. \qquad (2.26a,b)$$

Multiplying Eq. (2.2a) by dr and integrating once, we find:

$$\psi=-\frac{1}{G_2h_0}\left[\frac{2t}{r}\frac{d\varphi}{dr}(\frac{dw}{dr}+\alpha)+\frac{1}{r}\int_0^r qr\,dr\right]-\frac{dw}{dr}. \qquad (2.27)$$

Substituting expression (2.27) into (2.21c) leads to:

$$Dr\frac{d}{dr}\frac{1}{r}\frac{d}{dr}(r\frac{dw}{dr})+2t(\frac{D}{G_2h_0}r\frac{d}{dr}\frac{1}{r}\frac{d}{dr}-1)[\frac{d\varphi}{dr}(\frac{dw}{dr}+\alpha)]$$

$$+\frac{D}{G_2h_0}r\frac{dq}{dr}-\int_0^r qr\,dr=0. \qquad (2.28)$$

Then Eqs. (2.28) and (2.26b) constitute a system of equations of nonlinear bending for the shallow conical sandwich shell.

3. Solution of Nonlinear Boundary Value Problem

In the case of uniform pressure q_0, the governing Eqs. (2.28) and (2.26b) satisfied by w and σ_{r0} are simplified to yield:

$$Dr\frac{d}{dr}\frac{1}{r}\frac{d}{dr}r\frac{dw}{dr}=-2t\left(\frac{D}{G_2h_0}r\frac{d}{dr}\frac{1}{r}\frac{d}{dr}-1\right)\left[r\sigma_{r0}\left(\frac{dw}{dr}+\alpha\right)\right]+\frac{1}{2}q_0r^2,$$

$$r\frac{d}{dr}\frac{1}{r}\frac{d}{dr}(r^2\sigma_{r0})=-E\frac{dw}{dr}\left(\frac{1}{2}\frac{dw}{dr}+\alpha\right). \qquad (3.1)$$

By virtue of boundary condition (2.22), four types of boundary conditions usually encountered in engineering are presented in the following:

1) Rigidly clamped edge

$$w=0,\quad \psi=0,\quad u=0\quad \text{at } r=a,$$

$$\psi=0,\quad \sigma_{r0}\text{ finite}\quad \text{at } r=0. \qquad (3.2)$$

2) Loosely clamped edge

$$w=0,\quad \psi=0,\quad \sigma_{r0}=0\quad \text{at } r=a,$$

$$\psi=0,\quad \sigma_{r0}\text{ finite}\quad \text{at } r=0. \qquad (3.2)$$

3) Simply supported edge

$$w=0,\quad M_r=0,\quad \sigma_{r0}=0\quad \text{at } r=a,$$

$$\psi=0,\quad \sigma_{r0}\text{ finite}\quad \text{at } r=0. \qquad (3.4)$$

4) Hinged edge

$$w=0,\quad M_r=0,\quad u=0\quad \text{at } r=a,$$

$$\psi=0,\quad \sigma_{r0}\text{ finite}\quad \text{at } r=0, \qquad (3.5)$$

in which using expressions (2.14), (2.23a) and (2.27), u,ψ and M_r can be written as:

$$u=\frac{r}{E}\left[\frac{d}{dr}[r\sigma_{r0}]-\nu\sigma_{r0}\right],$$

$$\psi = -\frac{1}{G_2 h_0}\left[2t\sigma_{r0}\left(\frac{\mathrm{d}w}{\mathrm{d}r}+\alpha\right)+\frac{1}{2}q_0 r\right]-\frac{\mathrm{d}w}{\mathrm{d}r}, \tag{3.6}$$

$$M_r = -D\left(\frac{\mathrm{d}^2 w}{\mathrm{d}r^2}+\frac{\nu}{r}\frac{\mathrm{d}w}{\mathrm{d}r}\right)-\frac{D}{G_2 h_0}\left\{2t\left(\frac{\mathrm{d}}{\mathrm{d}r}+\frac{\nu}{r}\right)\left[\sigma_{r0}\left(\frac{\mathrm{d}w}{\mathrm{d}r}+\alpha\right)\right]+\frac{1+\nu}{2}q_0\right\}.$$

To make the resulting solution more general, we introduce the following new variables:

$$y = \frac{r}{a},\quad W = 2\sqrt{1-\nu^2}\frac{w}{h_0},\quad \Phi = \frac{\mathrm{d}W}{\mathrm{d}y}+k_1,$$

$$S = \frac{2ta}{D}r\sigma_{r0},\quad Q = \frac{\sqrt{1-\nu^2}a^4}{Dh_0}q_0. \tag{3.7}$$

Upon substituting, the governing Eqs. (3.1) are transformed to the nondimensional form:

$$L(y\Phi) = -(k_2 L-1)(S\Phi)+Qy^2-\frac{k_1}{y},$$

$$L(yS) = -\frac{1}{2}(\Phi^2-k_1^2) \tag{3.8a,b}$$

where

$$L(\cdots) = y\frac{\mathrm{d}}{\mathrm{d}y}\frac{1}{y}\frac{\mathrm{d}}{\mathrm{d}y}(\cdots),$$

$$k_1 = \frac{2\sqrt{1-\nu^2}}{h_0}a\alpha,\quad k_2 = \frac{D}{G_2 h_0 a^2}. \tag{3.9}$$

For convenience, four types of boundary conditions (3.2)-(3.5) are combined in a nondimensional form:

$$W = 0,\quad \left(\frac{\mathrm{d}}{\mathrm{d}y}+\frac{\lambda_1}{y}\right)\left(k_2\frac{S\Phi}{y}+\Phi\right)+k_2(1+\lambda_1)Q-\frac{\lambda_1 k_1}{y} = 0,$$

$$\frac{\mathrm{d}S}{\mathrm{d}y}-\lambda_2\frac{S}{y} = 0\quad \text{at } y = 1, \tag{3.10a,b,c}$$

$$k_2\frac{S\Phi}{y}+\Phi-k_1 = 0,\quad S = 0\quad \text{at } y = 0 \tag{3.11a,b}$$

in which λ_1 and λ_2 are parameters to be determined by the prescribed boundary conditions:

1) Rigidly clamped edge

$$\lambda_1 = \infty,\quad \lambda_2 = \nu. \tag{3.12}$$

2) Loosely clamped edge

$$\lambda_1 = \infty,\quad \lambda_2 = \infty. \tag{3.13}$$

3) Simply supported edge

$$\lambda_1 = \nu,\quad \lambda_2 = \infty \tag{3.14}$$

4) Hinged edge

$$\lambda_1 = \nu,\quad \lambda_2 = \nu. \tag{3.15}$$

The nondimensional nonlinear boundary value problem Eqs. (3.8), (3.10) and

(3.11) will be solved by the modified iteration method. We introduce the notation W_0 for the nondimensional central deflection of the shallow conical sanwich shell as iteration parameter:

$$w_0 = 2\sqrt{1-\nu^2}\left.\frac{w}{h_0}\right|_{r=0} \tag{3.16}$$

Using the third equation of expression (3.7) and the boundary condition Eq. (3.10a), we have:

$$W_0 = -\int_0^1 (\Phi - k_1)\mathrm{d}y. \tag{3.17}$$

For the first approximation, by neglecting the nonlinear terms $S\Phi$ in Eq. (3.8a) and boundary conditions (3.10b), (3.11a), and again introducing Φ_1 into the right of Eq. (3.8b), we find the following linear boundary value problem:

$$L(y\Phi_1) = Qy^2 - \frac{k_1}{y},$$

$$L(yS_1) = -\frac{1}{2}(\Phi_1^2 - k_1^2), \tag{3.18a,b}$$

$$W_1 = 0,\quad \frac{\mathrm{d}\Phi_1}{\mathrm{d}y} + \lambda_1\frac{\Phi_1}{y} = -k_2(1+\lambda_1)Q + \frac{\lambda_1 k_1}{y},$$

$$\frac{\mathrm{d}S_1}{\mathrm{d}y} - \lambda_2\frac{S_1}{y} = 0 \quad \text{at } y = 1, \tag{3.19a,b,c}$$

$$\Phi_1 = k_1,\quad S_1 = 0 \quad \text{at } y = 0. \tag{3.20a,b}$$

The solution to the problem given by Eqs. (3.18a), (3.19a, b) and (3.20a) is:

$$\Phi_1 = \frac{Q}{8}(y^3 - \beta_1 y) + k_1 \tag{3.21}$$

where

$$\beta_1 = \frac{3+\lambda_1}{1+\lambda_1} + 8k_2. \tag{3.22}$$

Substituting this solution into expression (3.17), we obtain the characteristic relation of small deflection for the shell as follows:

$$Q = 8\beta_2 W_0 \tag{3.23}$$

where

$$\beta_2 = \frac{4}{2\beta_1 - 1} \tag{3.24}$$

With this relation the solution (3.21) becomes:

$$\Phi_1 = \beta_2 W_0(y^3 - \beta_1 y) + k_1. \tag{3.25}$$

By using this solution, the solution to the problem given by Eqs. (3.18b), (3.19c) and (3.20b) is:

$$S_1 = -\frac{1}{8}\beta_2^2 W_0^2\left(\frac{1}{12}y^7 - \frac{1}{3}\beta_1 y^5 + \frac{1}{2}\beta_1^2 y^3 + \beta_4 y\right)$$

$$-\frac{1}{3}k_1\beta_2 W_0\left(\frac{1}{5}y^4 - \beta_1 y^2 + \beta_3 y\right) \tag{3.26}$$

where

$$\beta_3 = \frac{2-\lambda_2}{1-\lambda_2}\beta_1 - \frac{4-\lambda_2}{5(1-\lambda_2)},$$

$$\beta_4 = -\frac{3-\lambda_2}{2(1-\lambda_2)}\beta_1^2 + \frac{5-\lambda_2}{3(1-\lambda_2)}\beta_1 - \frac{7-\lambda_2}{12(1-\lambda_2)}. \tag{3.27}$$

For simplicity the second approximation is restricted to the determination of Φ only. We rewrite the nonlinear term $S\Phi$ of the nonlinear boundary value problem Eqs. (3.8a), (3.10a,b) and (3.11a) in $S_1\Phi_1$. Then we have the following linear boundary value problem for Φ:

$$L(y\Phi_2) = -(k_2L-1)(S_1\Phi_1) + Qy^2 - \frac{k_1}{y}, \tag{3.28}$$

$$W_2 = 0,\quad \frac{d\Phi_2}{dy} + \lambda_1\frac{\Phi_2}{y} = -k_2\left(\frac{d}{dy}+\frac{\lambda_1}{y}\right)\left(\frac{S_1\Phi_1}{y}\right)$$

$$-k_2(1+\lambda_1)Q + \frac{\lambda_1 k_1}{y}\quad \text{at } y=1, \tag{3.29a,b}$$

$$\Phi_2 = -k_2\frac{S_1\Phi_1}{y} + k_1 \quad \text{at } y=0. \tag{3.30}$$

Utilizing solutions of Eqs. (3.25) and (3.26), the solution to the above problem is:

$$\begin{aligned}
\Phi_2 = &-\frac{1}{8}\beta_2^3W_0^3\left\{\frac{1}{1440}y^{11} - \frac{1}{12}\left(\frac{1}{16}\beta_1 + k_2\right)y^9\right.\\
&+\frac{5}{12}\beta_1\left(\frac{1}{24}\beta_1 + k_2\right)y^7 - \frac{1}{6}\left(\frac{1}{8}\beta_1^3 + 5\beta_1^2k_2 - \frac{1}{4}\beta_4\right)y^5\\
&\left.+\left[k_2\left(\frac{1}{2}\beta_1^3 - \beta_4\right) - \frac{1}{8}\beta_1\beta_4\right]y^3 + \beta_7 y\right\}\\
&-\frac{1}{8}k_1\beta_2^2W_0^2\left[\frac{37}{3780}y^8 - \frac{1}{35}\left(\frac{53}{15}\beta_1 + \frac{259}{12}k_2\right)y^6 + \frac{1}{9}\beta_3y^5\right.\\
&+\frac{1}{15}\beta_1\left(\frac{19}{6}\beta_1 + 53k_2\right)y^4 - \frac{1}{3}\beta_3(\beta_1+8\beta_2)y^3\\
&\left.-\frac{1}{3}\left(\frac{19}{2}k_2\beta_1^2 - \beta_4\right)y^2 + 8\beta_6 y - k_2\beta_4\right]\\
&-\frac{1}{3}k_1^2\beta_2W_0\left[\frac{1}{120}y^5 - \frac{1}{8}\left(\beta_1 + \frac{8}{5}k_2\right)y^3\right.\\
&\left.+\frac{1}{3}\beta_3y^2 + \beta_5 y - k_2\beta_3\right] + \frac{Q}{8}(y^3 - \beta_1 y) + k_1
\end{aligned} \tag{3.31}$$

where

$$\beta_5 = \frac{1}{8}\beta_1^2 - \frac{2-\lambda_1}{3(1+\lambda_1)}\beta_3 - \frac{5+\lambda_1}{120(1+\lambda_1)},$$

$$\begin{aligned}
\beta_6 = &\frac{1}{8}\left\{\frac{1}{3} - \left[\beta_3 - \frac{19(4+\lambda_1)}{30(1+\lambda_1)}\right]\beta_1^2 + \frac{53(6+\lambda_1)}{525(1+\lambda_1)}\beta_1\right.\\
&\left.-\frac{5+\lambda_1}{9(1+\lambda_1)}\beta_3 - \frac{2+\lambda_1}{3(1+\lambda_1)}\beta_4 - \frac{37(8+\lambda_1)}{3780(1+\lambda_1)}\right\},
\end{aligned} \tag{3.32}$$

$$\beta_7 = \frac{1}{8}\left\{\frac{5+\lambda_1}{6(1+\lambda_1)}\beta_1^3 + \left[\beta_4 - \frac{5(7+\lambda_1)}{36(1+\lambda_1)}\right]\beta_1^2 + \frac{9+\lambda_1}{24(1+\lambda_1)}\beta_1 - \frac{5+\lambda_1}{3(1+\lambda_1)}\beta_4 - \frac{11+\lambda_1}{180(1+\lambda_1)}\right\}.$$

Introduction solution Eq. (3.31) into expression (3.17), we obtain the following characteristic relation of nonlinear bending for the shell:

$$Q = (\alpha_0 + k_1^2\alpha_1) + k_1\alpha_2 W_0^2 + \alpha_3 W_0^3 \tag{3.33}$$

where

$$\begin{aligned}
\alpha_0 &= 8\beta_2,\\
\alpha_1 &= \frac{8}{3}\beta_2^2\left[\frac{1}{32}\beta_1 + \left(k_2 - \frac{1}{9}\right)\beta_3 - \frac{1}{2}\beta_5 + \frac{1}{20}k_2 - \frac{1}{720}\right],\\
\alpha_2 &= \beta_2^3\left[\frac{19}{18}\left(k_2 - \frac{1}{25}\right)\beta_1^2 + \frac{1}{3}\left(\frac{1}{4}\beta_3 - \frac{53}{25}k_2 + \frac{53}{1225}\right)\beta_1\right.\\
&\quad \left. + \frac{1}{3}\left(2k_2 - \frac{1}{18}\right)\beta_3 + \left(k_2 - \frac{1}{9}\right)\beta_4 - 4\beta_6 + \frac{37}{420}k_2 - \frac{37}{34020}\right],\\
\alpha_3 &= -\frac{1}{2}\beta_2^4\left[\frac{1}{4}\left(k_2 - \frac{1}{36}\right)\beta_1^3 - \frac{5}{18}\left(k_2 - \frac{1}{64}\right)\beta_1^2 - \frac{1}{16}\left(\beta_4 - \frac{5}{3}k_2 + \frac{1}{60}\right)\beta_1\right.\\
&\quad \left. - \frac{1}{2}\left(k_2 - \frac{1}{36}\right)\beta_4 + \beta_7 - \frac{1}{60}k_2 + \frac{1}{8640}\right].
\end{aligned} \tag{3.34}$$

Especially if we put $k_1 = 0$ in relation (3.33), we can obtain the characteristic relation of large deflection for the circular sandwich plate under uniform pressure, which has been studied by Liu (1981).

To illustrate the foregoing results, let us consider a shallow conical sandwich shell with a rigidly clamped edge for $k_2 = 0.05$ and $\nu = 0.3$. The relation between load Q and central deflection W_0 of the shell is shown in Fig. 2 for various values of the geometric parameters k_1. Evidently, when k_1 is larger, curves become gyratory. At this time, the shell occurs snap-through. In order to get the critical load, let us use the following extremal condition

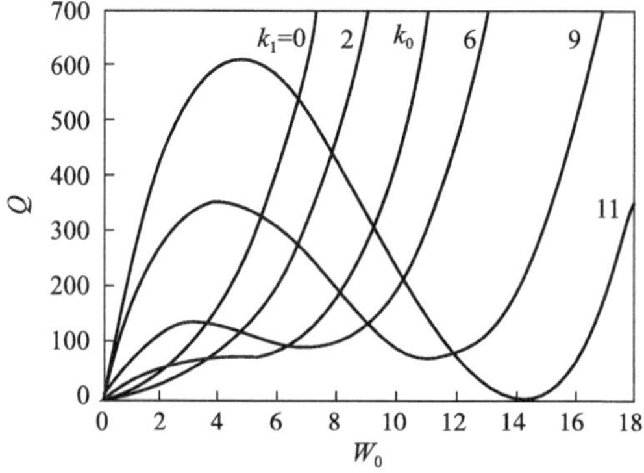

Fig. 2 The characteristic curves ($k_2 = 0.05$, $\nu = 0.3$).

$$\frac{\mathrm{d}Q}{\mathrm{d}w_0} = 0. \tag{3.35}$$

Then we have

$$3\alpha_3 W_0^2 + 2k_1\alpha_2 W_0 + \alpha_0 + k_1^2\alpha_1 = 0. \tag{3.36}$$

Solving this quadratic equation, we obtain

$$W_0^* = \frac{1}{3\alpha_3}\left[-\alpha_2 k_1 \pm \sqrt{(\alpha_2^2 - 3\alpha_1\alpha_3)k_1^2 - 3\alpha_0\alpha_3}\right] \tag{3.37}$$

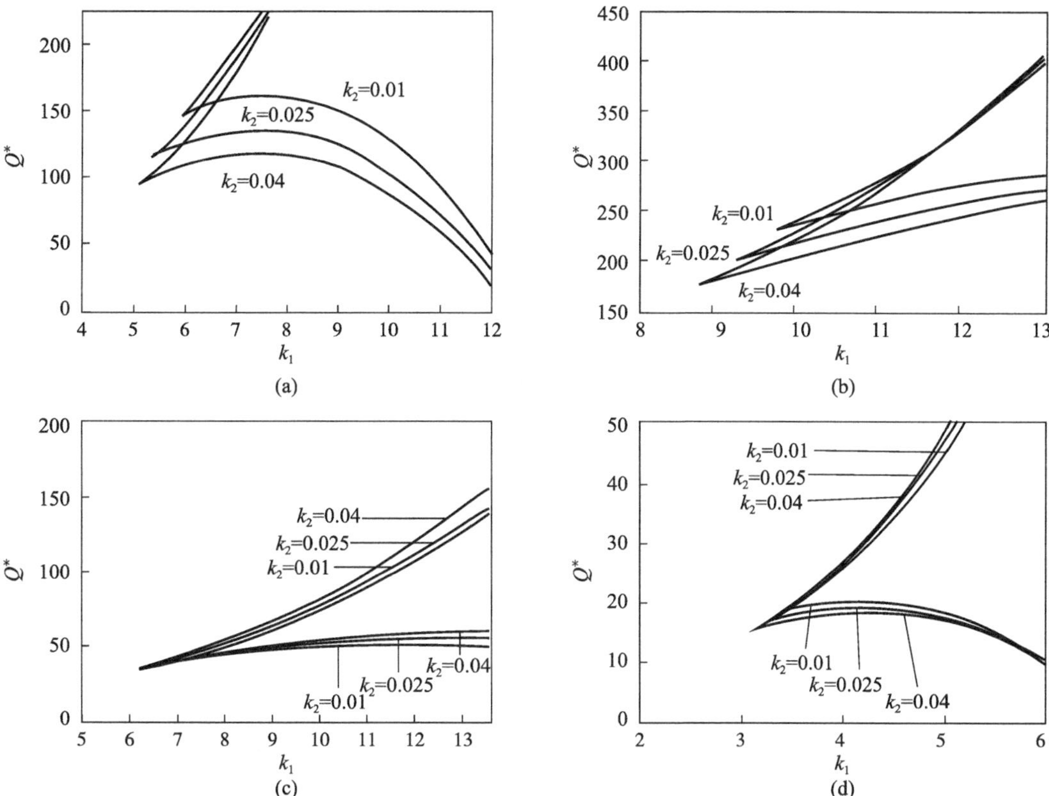

Fig. 3　Curves of nondimensional critical loads for various boundary conditions ($\nu=0.3$): (a) rigidly clamped edge, (b) loosely clamped edge, (c) simply supported edge, (d) hinged edge.

Substituting this expression into relation (3.33), we obtain the following formula for the nondimensional critical load

$$Q^* = (\alpha_0 + k_1^2\alpha_1)W_0^* + k_1\alpha_2 W_0^{*2} + \alpha_3 W_0^{*3}, \tag{3.38}$$

in which Q^* with respect to W_0^* with the negative and positive signs in expression (3.37) are the upper and lower critical loads, respectively.

The numerical results obtained from formula (3.38) for various boundary conditions and different values of the physical parameter k_2 by taking $\nu=0.3$ are shown in Fig. 3. These figures indicate that the upper critical loads increase with increasing k_1, and in large range of k_1, the lower critical loads diminish as k_1 increases, In general, we discuss only the upper critical loads because of the engineering practice.

Now, we shall derive the equation for the critical point at which the upper and lower critical loads coincide. Let the discriminant of the quadratic Eq. (3. 36) be zero, we then obtain the formula for the critical point

$$k_0 = \sqrt{\frac{3\alpha_0\alpha_3}{\alpha_2^2 - 3\alpha_1\alpha_3}}. \tag{3. 39}$$

Obviously, when $k_1 < k_0$, no snap-through occurs for the shell, and when $k_1 \geqslant k_0$, snap-through can occur. Calculated results of formula (3. 39) for various boundary conditions and $\nu=0.3$ are shown in Fig. 4. It may be seen that the critical geometrical parameter k_0 diminishes as k_2 increases. Having k_0 and using formula (3. 38), we can at once calculate the critical load Q_0^* of the critical point for various boundary conditions and $\nu=0.3$. These results are given in Fig. 5. It will be noted that all curves drop monotonically.

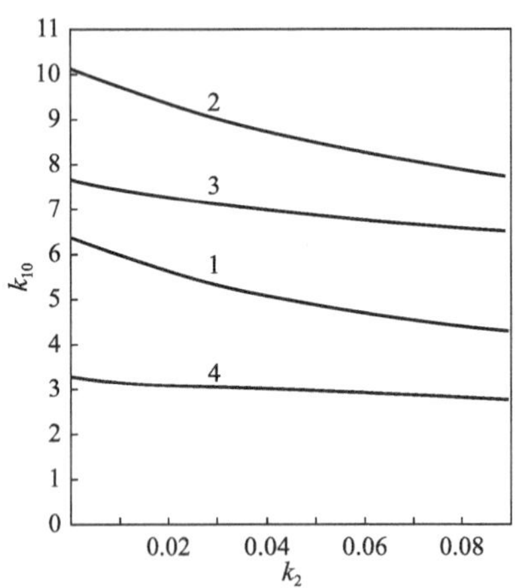

Fig. 4 Curves of the critical geometrical parameter k_0 for various boundary condition ($\nu=0.3$):
1-rigidly clamped edge; 2-loosely clamped edge; 3-simply supported edge; 4-hinged edge.

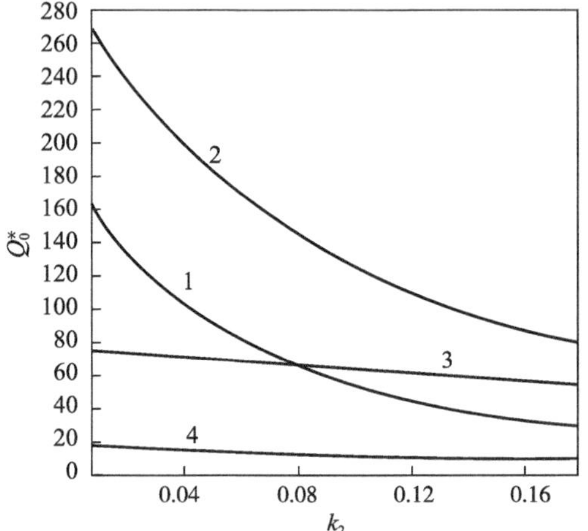

Fig. 5 Curves of nondimensional critical loads at the critical point for various boundary condition ($\nu=0.3$):
1-rigidly clamped edge; 2-loosely clamped edge; 3-simply supported edge; 4-hinged edge.

References

Kan H. P. and Huang J. C. (1967): *Large deflections of rectangular sandwich plates*. -AIAA Journal, v. 5, pp. 1706-1708.

Liu R. H. (1965): *Nonlinear stability of circular shallow spherical shell with a hole in the center under the action of uniform moment at the inner edge*. -Bulletin of Sciences, No. 3, pp. 253-255 (in Chinese).

Liu R. H. (1980): *Nonlinear axisymmetrical bending of circular sandwich plates under the action of uniform edge moment*. -J. of China University of Science and Technology, v. 10, pp. 56-67 (in Chinese).

Liu R. H. (1981): *Nonlinear bending of circular sandwich plates*. -Applied Mathematics and Mechanics, v. 2, pp. 189-206.

Liu R. H. (1987): *Nonlinear bending of circular sandwich plate under the action of axisymmetric uniformly distributed line loads*. -Progress in Applied Mechanics, ed. by K. Y. Yeh, Martinus Nijhoff Publishers, Dordrecht, pp. 293-321.

Liu R. H. and Shi Y. F. (1982): *Exact solution for circular sandwich plate with large deflection*. -Applied Mathematics and Mechanics, v. 3, pp. 11-24.

Liu R. H. and Zhu G. Q. (1989): *Further study on large deflection of circular sandwich plates*. -Applied Mathematics and Mechanics, v. 10, pp. 1099-1106.

Liu R. H. and Wu J. C. (1992): *Nonlinear vibration of rectangular sandwich plates*. -Science in China, Series A, v. 35. pp. 472-486.

Liu R. H. and Cheng Z. Q. (1993): *Nonlinear bending of simply supported rectangular sandwich plates*. -Applied Mathematics and Mechanics, v. 14, pp. 217-234.

Liu R. H. and Cheng Z. Q. (1995): *On the non-linear buckling of circular shallow spherical sandwich shells under the action of uniform edge moments*. -Int. J. of Non-Linear Mechanics, v. 30, pp. 33-43.

Liu R. H. and Li J. (1995): *Non-linear vibration of shallow conical sandwich shells*. -Int. J. of Non-Linear Mechanics, v. 30, pp. 97-109.

Liu R. H. and Wang Z. W. (1996): *The theory of consistent moderate rotation of sandwich plates faced with laminated composites*. -Shanghai J. of Mechanics, v. 17, pp. 222-228 (in Chinese).

Liu R. H., Zhu J. F. and Zhang X. G. (1997): *Nonlinear bending of annular sandwich plates*. -J. of Jinan University, v. 18, pp. 1-11 (in Chinese).

Liu R. H. and Zhu J. F. (1993): *Nonlinear Theory of Sandwich Shells*. -Press of Mechanical Industry, Beijing (in Chinese).

Liu R. H. and Zhu J. F. (1997a): *Nonlinear theory of sandwich shells, part I-Exact kinematics of moderately thick shells*. -Applied Mech. and Engi., v. 2, pp. 213-240.

Liu R. H. and Zhu J. F. (1997b): *Nonlinear theory of sandwich shells, part II-Approximate theories*. -Applied Mech. and Engi., v. 2, pp. 241-269.

Reissner E. (1948): *Finite deflection of sandwich plates*. -J. Aeron. Sci., v. 15, pp. 435-440; v. 17, pp. 125.

Struk R. (1984): *Nonlinear stability problem of an open conical sandwich shell under external pressure and compression*. -Int. J. of Non-Linear Mechanics, v. 14, pp. 217-233.

Wang Z. W. and Liu R. H. (1997): *Nonlinear analysis of sandwich plates considering continuity of transverse shear stresses between layers*. -Modern Mechanics and Development of Science and Technology, ed. by F. G. Zhuang, Qinghua University Press, Beijing, pp. 1469-1472 (in Chinese).

Wang Z. W. and Liu R. H. (1998a): *A refined nonlinear theory for sandwich plates faced with orthotropic laminated composites and its application*. -Chinese J. of Applied Mechanics, v. 15, pp. 18-24 (in Chinese).

Wang Z. W. and Liu R. H. (1998b): *Nonlinear buckling for sandwich shallow shells accounting for continuity conditions of stresses at interfaces*. -Proceedings of the 3rd International Conference on Nonlinear Mechanics, Shanghai University Press, Shanghai, pp. 381-386.

Yeh K. Y.，Liu R. H.，Ping Q.，et al. (1965a)：*Nonlinear stability of thin circular shallow spherical shell under actions of axisymmetric uniform distributed line loads*. -Bulletin of Sciences，No. 2，pp. 142-145 (in Chinese).

Yeh K. Y.，Liu，R. H.，Zhang C. Z.，et al. (1965b)：*Nonlinear stability of thin circular shallow spherical shell under the action of uniform edge moment*. -Bulletin of Sciences，No. 2，pp. 145-147 (in Chinese).

考虑横向剪应力连续的复合材料面层夹层壳非线性一致有效理论①

一、引言

夹层壳在许多工程领域已得到了广泛应用，其理论研究始于 20 世纪 40 年代末，Reissner[1]在面层为薄膜和夹芯为软芯假设的基础上，建立了完整的小挠度理论，Wang[2]采用 Reissner 方法，将 Reissner 小挠度理论推广到了夹层壳大挠度问题，Reissner-Wang 理论仅适用于面层很薄、夹芯很软的各向同性夹层壳。为此，Schmidt[3]修正 Reissner 面层薄膜假设，视面层为薄壳，克希霍夫-拉甫假设适用，建立了考虑面层自身抗弯刚度的任意形状夹层壳体的小挠度 Schmidt 理论。

较为复杂的夹层壳已有所研究，Wempner[4]建立了面层不等厚也不同质的夹层壳中等大挠度理论，Reese 和 Bert[5]考虑了面层和芯层皆为正交各向异性的夹层壳，Ronan 和 Kao[6]则推导了芯层是正交各向异性的夹层壳非线性方程。刘人怀和朱金福[7-9]在面层为薄壳、芯层为中厚壳、各层材料为正交各向异性的假设的基础上，根据“能量误差一致”原则系统地建立了夹层壳的小应变中转动一致有效理论，包括：小应变中转动一阶近似理论，小应变中转动二阶近似理论，并运用所建立的中转动理论研究了夹层圆柱壳的非线性稳定问题，分析了初始几何缺陷对夹层壳非线性稳定的影响，刘人怀等[10-13]还研究了夹层扁壳的非线性屈曲和振动问题。

复合材料面层夹层壳的研究还甚少[14-16]，王震鸣等[17]建立了计及面层自身抗弯刚度和芯层面内刚度的复合材料面层夹层扁壳的有限挠度方程。但对这类夹层壳的研究文献，大多还限于特定壳体的线性屈曲问题，夹层壳模型仍沿用经典的 Reissner 模型。

夹层壳面层变形用薄膜假设或克希霍夫-拉甫假设，这对较薄的各向同性材料面层变形描述是合适的，但若将其用于复合材料层合面层，则将导致两方面的理论缺陷：第一，无法描述面层的横向剪切变形。我们知道，对于复合材料面层夹层壳体，其芯层和面层的横向剪切模量都是很低的，需要同时考虑芯层和面层的横向剪切变形。第二，无法比较精确地得到工程上颇为重要的层间应力（本文“层间”特指面层与芯层的粘接处）。

本文首先假设复合材料面层夹层壳中任一点面内位移分别在面层和芯层中为横向坐标的逐段三次函数，通过考虑夹层壳层间和表面横向剪应力协调条件，导出了位移场的修正表达式。根据“能量误差一致”原则，通过严格的量阶分析并弃去不完全项，得到了小应变中转动理论的应变位移关系及本构方程，然后应用最小位能原理建立了复合材料面层夹层壳中转动理论的控制方程和边界条件。

① 本文原载《应用力学研究与实践》，广州：暨南大学出版社，2000，1-12. 作者：刘人怀，王志伟.

二、未变形夹层壳几何及基本量阶关系

设复合材料面层夹层壳的面层和芯层厚度均匀，总厚度为 h（图 1）。记夹层壳上、下表面分别为 $\underset{(3)}{\omega}$ 和 $\underset{(0)}{\omega}$，层间依次为 $\underset{(1)}{\omega}$ 和 $\underset{(2)}{\omega}$，取 $\underset{(0)}{\omega}$ 为参考曲面，在 $\underset{(0)}{\omega}$ 上取一曲面坐标系 $\{\theta^\alpha\}(\alpha=1,2)$，并取 θ^3 坐标线与 $\underset{(0)}{\omega}$ 垂直。建立坐标系 $\{\theta^i\}(i=1,2,3)$ 后，曲面 $\underset{(m)}{\omega}$ $(m=0,1,2,3)$ 可由 $\theta^3=\underset{(m)}{\theta^3}$ 描述。

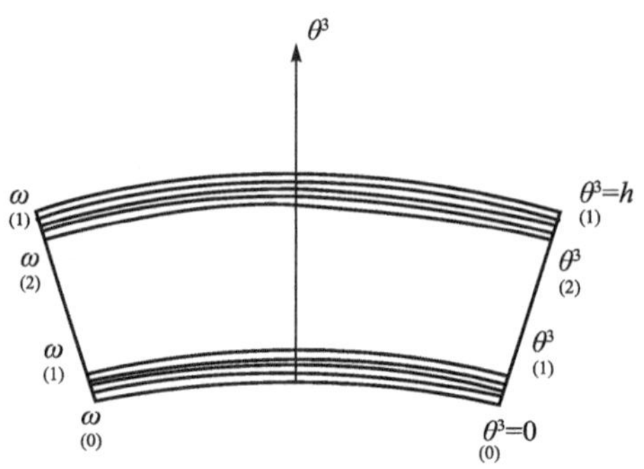

图 1　复合材料面层夹层壳体

$\underset{(0)}{\omega}$ 的度量张量是

$$a_{\alpha\beta}=\boldsymbol{a}_\alpha\cdot\boldsymbol{a}_\beta,\quad a_{\alpha3}=\boldsymbol{a}_\alpha\cdot\boldsymbol{a}_3=0,\quad a_{33}=\boldsymbol{a}_3\cdot\boldsymbol{a}_3=1,$$
$$a^{\alpha\beta}=\boldsymbol{a}^\alpha\cdot\boldsymbol{a}^\beta,\quad a^{\alpha3}=\boldsymbol{a}^\alpha\cdot\boldsymbol{a}^3=0,\quad a^{33}=\boldsymbol{a}^3\cdot\boldsymbol{a}^3=1 \tag{1}$$

这里 $\boldsymbol{a}_i$ 和 $\boldsymbol{a}^i$ 分别表示 $\underset{(0)}{\omega}$ 上的协变和逆变基矢，且 $\boldsymbol{a}_3=\boldsymbol{a}^3=\boldsymbol{n}$。

夹层壳的空间度量张量为

$$g_{\alpha\beta}=\boldsymbol{g}_\alpha\cdot\boldsymbol{g}_\beta=\mu_\alpha^\delta\mu_\beta^\lambda a_{\delta\lambda},\quad g_{\alpha3}=\boldsymbol{g}_\alpha\cdot\boldsymbol{g}_3=0,\quad g_{33}=\boldsymbol{g}_3\cdot\boldsymbol{g}_3=1 \tag{2}$$

其中，δ_α^β为 Kronecker 符号，b_α^β 为参考曲面 $\underset{(0)}{\omega}$ 的混合曲率张量，$\boldsymbol{g}_i$ 为空间协变基矢，

$$\mu_\alpha^\beta=\delta_\alpha^\beta-\theta^3 b_\alpha^\beta \tag{3}$$

μ_α^β 的逆记为 $\overset{-1}{\mu}{}_\alpha^\beta$，满足

$$\overset{-1}{\mu}{}_\lambda^\alpha\mu_\beta^\lambda=\mu_\beta^\lambda\overset{-1}{\mu}{}_\lambda^\alpha=\delta_\beta^\alpha \tag{4}$$

μ_α^β 的行列式记为 μ，即

$$\mu=\det(\mu_\alpha^\beta)=1-2H\theta^3+K(\theta^3)^2 \tag{5}$$

其中

$$H=\frac{1}{2}b_\alpha^\alpha,\quad K=\det(b_\beta^\alpha)$$

另外，用“$\|$”表示张量分量关于 g_{ij} 的协变导数，“$|$”表示关于 $a_{\alpha\beta}$ 的协变导数，$\overline{T}^i$ 表示关于 $\boldsymbol{a}_i$ 的矢量分量，而 T^i 表示关于 $\boldsymbol{g}_i$ 的矢量分量。于是，我们可得到以下有用公式

$$T_\alpha\|_\beta=\mu_\alpha^\lambda(\overline{T}_\lambda|_\beta-b_{\lambda\beta}\overline{T}_3),\quad T_\alpha\|_3=\mu_\alpha^\lambda\overline{T}_{\lambda,3},$$

$$T_3\parallel_\alpha=\overline{T}_{3,\alpha}+b_\alpha^\lambda\overline{T}_\lambda,\quad T_3\parallel_3=\overline{T}_{3,3} \tag{6}$$

本文采用哑标求和约定，希腊指标取值 1 和 2，拉丁指标取值 1，2 和 3。

设夹层壳各层材料具有平行于 $\theta^3=0$ 的弹性对称面，第二类 Piola-Kirchhoff 应力张量 S^{ij} 和 Green 应变张量 e_{ij} 服从本构关系

$$S^{\alpha\beta}=Q^{\alpha\beta\omega\rho}e_{\omega\rho},\quad S^{\alpha3}=Q^{\alpha3\omega3}e_{\omega3} \tag{7}$$

其中 Q^{ijkl} 为空间弹性常数。这里略去了夹层壳横向法应力的效应。

Green 应变张量 e_{ij} 与位移 V_i 的关系为

$$e_{ij}=\frac{1}{2}(V_i\parallel_j+V_j\parallel_i+V^r\parallel_iV_r\parallel_j) \tag{8}$$

令

$$\eta_{ij}=\frac{1}{2}(V_i\parallel_j+V_j\parallel_i),\quad \omega_{ij}=\frac{1}{2}(V_i\parallel_j-V_j\parallel_i) \tag{9}$$

则式（8）可以写成

$$e_{ij}=\eta_{ij}+\frac{1}{2}\omega_{ki}\omega_j^k+\frac{1}{2}(\eta_{kj}\omega_i^k+\eta_{ki}\omega_j^k)+\frac{1}{2}\eta_{ki}\eta_j^k \tag{10}$$

考虑小应变中转动变形，有

$$\begin{aligned}&e_{ij}=o(\varepsilon^2),\quad \eta_{ij}=o(\varepsilon^2),\\&\omega_{3\alpha}=o(\varepsilon),\quad \omega_{\alpha\beta}=o(\varepsilon^2)\end{aligned} \tag{11}$$

式中 ε 为一小量。

考虑式（11）中的量阶关系，由式（10）可得

$$\begin{aligned}&e_{\alpha\beta}=\eta_{\alpha\beta}+\frac{1}{2}\omega_{3\alpha}\omega_{3\beta}+\frac{1}{2}(\eta_{3\alpha}\omega_{3\beta}+\eta_{3\beta}\omega_{3\alpha})+o(\varepsilon^4)\\&e_{\alpha3}=\eta_{\alpha3}+o(\varepsilon^3)\end{aligned} \tag{12a,b}$$

记

$$\begin{aligned}&Q=o(Q^{\alpha\beta\gamma\nu}),\quad G=o(Q^{\alpha3\gamma3}),\\&\overline{Q}=o(\overline{Q}^{\alpha\beta\gamma\nu}),\quad \overline{G}=o(\overline{Q}^{\alpha3\gamma3})\end{aligned} \tag{13}$$

式中 $\overline{Q}^{ijkl}$ 为曲面弹性常数。

空间弹性常数与曲面弹性常数有下列量阶关系

$$\begin{aligned}&Q=o[(1+\varepsilon^2+\varepsilon^4+\cdots)\overline{Q}],\\&G=o[(1+\varepsilon^2+\varepsilon^4+\cdots)\overline{G}]\end{aligned} \tag{14}$$

夹层壳体应变能密度为

$$\begin{aligned}W&=\frac{1}{2}Q^{\alpha\beta\omega\rho}e_{\alpha\beta}e_{\omega\rho}+2Q^{\alpha3\omega3}e_{\alpha3}e_{\omega3}\\&=o[Q(e_{\alpha\beta})^2+G(e_{\alpha3}{}^2)]\end{aligned} \tag{15}$$

取 W 的误差界为 $o(\overline{Q}\varepsilon^6)$，则 $e_{\alpha\beta}$ 只需取至量阶为 ε^3 的项，Q 只需取 $o(\overline{Q})$。对复合材料面层夹层壳，由于面层和芯层的横向剪切模量均较低，即 G/Q 很小，故 $e_{\alpha3}$ 只需取至量阶为 ε^3 的项，G 只需取 $o(\overline{G})$。所以，在式（12a，b）中可分别略去 $o(\varepsilon^4)$ 项和 $o(\varepsilon^3)$ 项。

三、位移场

夹层壳变形后，假设位移矢 V 分别在面层和芯层中为 θ^3 的逐段光滑曲线。于是，可取满足这一逐段光滑位移假设近似位移场为

$$\begin{aligned} &V=\overline{V}_\alpha \boldsymbol{a}^\alpha+\overline{V}_3 \boldsymbol{a}^3, \\ &\overline{V}_\alpha=u_\alpha+u_\alpha^{(1)}\theta^3+u_\alpha^{(2)}(\theta^3)^2+u_\alpha^{(3)}(\theta^3)^3+\sum_{m=1}^{2}\underset{(m)}{u}_\alpha(\theta^3-\underset{(m)}{\theta}{}^3)H(\theta^3-\underset{(m)}{\theta}{}^3), \\ &\overline{V}_3=u_3 \end{aligned} \tag{16}$$

式中

$$H(\theta^3-\underset{(m)}{\theta}{}^3)=\begin{cases}1, & \theta^3\geqslant \underset{(m)}{\theta}{}^3\\ 0, & \theta^3<\underset{(m)}{\theta}{}^3\end{cases} \tag{17}$$

由式（12b）得夹层壳横向剪应变为

$$\begin{aligned} e_{\alpha3}=&\frac{1}{2}(u_\alpha^{(1)}+u_{3,\alpha}+b_\alpha^\lambda u_\lambda)+u_\alpha^{(2)}\theta^3+\frac{1}{2}(3u_\alpha^{(3)}-b_\alpha^\lambda u_\lambda^{(2)})(\theta^3)^2\\ &-b_\alpha^\lambda u_\lambda^{(3)}(\theta^3)^3+\frac{1}{2}\sum_{m=1}^{2}(\underset{(m)}{u}_\alpha-\underset{(m)}{\theta}{}^3 b_\alpha^\lambda \underset{(m)}{u}_\lambda)H(\theta^3-\underset{(m)}{\theta}{}^3) \end{aligned} \tag{18}$$

由于式（16）为位移场的近似表达式，故式（18）中含有不完全项。事实上，在式（18）中，θ^3 的二次方项系数已出现 $u_\alpha^{(3)}$，那么 θ^3 的三次方项系数中还应含有 $u_\alpha^{(4)}$，因此式（18）中 θ^3 的三次方项为不完全项，可弃去。从而可得

$$\begin{aligned} e_{\alpha3}=&\frac{1}{2}(u_\alpha^{(1)}+u_{3,\alpha}+b_\alpha^\lambda u_\lambda)+u_\alpha^{(2)}\theta^3+\frac{1}{2}(3u_\alpha^{(3)}-b_\alpha^\lambda u_\lambda^{(2)})(\theta^3)^2\\ &+\frac{1}{2}\sum_{m=1}^{2}(\underset{(m)}{u}_\alpha-\underset{(m)}{\theta}{}^3 b_\alpha^\lambda \underset{(m)}{u}_\lambda)H(\theta^3-\underset{(m)}{\theta}{}^3) \end{aligned} \tag{19}$$

限定夹层壳上下面无切向力作用，则夹层壳表面剪应力协调条件为

$$S^{\alpha3}(\theta^\alpha,0)=0,\quad S^{\alpha3}(\theta^\alpha,h)=0 \tag{20}$$

夹层壳层间横向剪应力连续条件为

$$\begin{aligned} &S^{\alpha3}(\theta^\alpha,\underset{(1)}{\theta}{}^3-0)=S^{\alpha3}(\theta^\alpha,\underset{(1)}{\theta}{}^3+0),\\ &S^{\alpha3}(\theta^\alpha,\underset{(2)}{\theta}{}^3-0)=S^{\alpha3}(\theta^\alpha,\underset{(2)}{\theta}{}^3+0) \end{aligned} \tag{21}$$

将式（7）和（18）代入式（20）和（21），便得

$$\begin{aligned} &u_\alpha^{(1)}+u_{3,\alpha}+b_\alpha^\lambda u_\lambda=0,\\ &2hu_\alpha^{(2)}+3h^2u_\alpha^{(3)}-h^2b_\alpha^\lambda u_\lambda^{(2)}+\sum_{m=1}^{2}(\underset{(m)}{u}_\alpha-\underset{(m)}{\theta}{}^3 b_\alpha^\lambda \underset{(m)}{u}_\lambda)=0,\\ &({}_{f'}Q^{\alpha3\rho3}-{}_{c}Q^{\alpha3\rho3})[2\underset{(1)}{\theta}{}^3u_\rho^{(2)}-u_\lambda^{(2)}+3(\underset{(1)}{\theta}{}^3)^2u_\rho^{(2)}]={}_{c}Q^{\alpha3\rho3}[\underset{(1)}{u}_\rho-\underset{(1)}{\theta}{}^3 b_\rho^\lambda \underset{(1)}{u}_\lambda],\\ &({}_{c}Q^{\alpha3\rho3}-{}_{f''}Q^{\alpha3\rho3})[2\underset{(2)}{\theta}{}^3u_\rho^{(2)}-(\underset{(2)}{\theta}{}^3)^2b_\rho^\lambda u_\lambda^{(2)}+3(\underset{(2)}{\theta}{}^3)^2u_\rho^{(3)}\\ &+\underset{(1)}{u}_\rho-\underset{(1)}{\theta}{}^3 b_\rho^\lambda \underset{(1)}{u}_\lambda]={}_{f''}Q^{\alpha3\beta3}[\underset{(2)}{u}_\rho-\underset{(2)}{\theta}{}^3 b_\rho^\lambda \underset{(2)}{u}_\lambda] \end{aligned} \tag{22}$$

式中${}_{f'}Q^{\alpha3\beta3}$，${}_{f''}Q^{\alpha3\beta3}$分别为下表层中和上表层中与芯层粘结的铺层的弹性常数，$Q^{\alpha3\rho3}$为芯层的弹性常数。

记$u_\alpha^{(2)}=\varphi_\alpha$，由式（22）可解得

$$
\begin{aligned}
u_\alpha^{(1)} &= -u_{3,\alpha}-b_\alpha^\lambda u_\lambda,\\
\underset{(m)}{u}_\alpha &= \underset{(m)}{a}{}_\alpha^\lambda\varphi_\lambda,\\
u_\alpha^{(3)} &= c_\alpha^\lambda\varphi_\lambda
\end{aligned}
\tag{23}
$$

从而得到修正的位移场为

$$
\begin{aligned}
\overline{V}_\alpha &= u_\alpha-(u_{3,\alpha}+b_\alpha^\lambda u_\lambda)\theta^3+\varphi_\alpha(\theta^3)^2+c_\alpha^\lambda\varphi_\lambda(\theta^3)^3\\
&\quad+\sum_{m=1}^{2}\underset{(m)}{a}{}_\alpha^\lambda\varphi_\lambda(\theta^3-\underset{(m)}{\theta}{}^3)H(\theta^3-\underset{(m)}{\theta}{}^3),\\
\overline{V} &= u_3
\end{aligned}
\tag{24}
$$

四、应变场

令

$$
\begin{aligned}
&\phi_\omega^3=u_{3,\omega}+b_\omega^\lambda u\lambda,\quad \phi_{\alpha\beta}=u_\alpha\,|_\beta-b_{\alpha\beta}u_3,\\
&\underset{(0)(1)}{\phi_\omega^3}=-b_\omega^\lambda(u_{3,\lambda}+b_\lambda^\delta u_\delta),\quad \underset{(0)(1)}{\phi_{\alpha\beta}}=-\phi_\alpha^3\,|_\beta,\\
&\underset{(0)(2)}{\phi_\omega^3}=b_\omega^\lambda\varphi_\lambda,\quad \underset{(0)(2)}{\phi_{\alpha\beta}}=\phi_\alpha\,|_\beta,\\
&\underset{(0)(3)}{\phi_\omega^3}=b_\omega^\lambda c_\lambda^\delta\varphi_\delta,\quad \underset{(0)(3)}{\phi_{\alpha\beta}}=(c_\alpha^\lambda\varphi_\lambda)\,|_\beta,\\
&\underset{(m)(1)}{\phi_\omega^3}=b_\omega^\lambda\underset{(m)}{a}{}_\lambda^\delta\varphi_\delta,\quad \underset{(m)(1)}{\phi_{\alpha\beta}}=(\underset{(m)}{a}{}_\alpha^\lambda\varphi_\lambda)\,|_\beta
\end{aligned}
\tag{25}
$$

将位移表达式（24）代入式（12），并弃去不完全项后，可得

$$
\begin{aligned}
e_{\alpha\beta} &= \sum_{n=0}^{3}\overset{(n)}{\underset{(0)}{e_{\alpha\beta}}}(\theta^3)^n+\sum_{m=1}^{2}\overset{(1)}{\underset{(m)}{e_{\alpha\beta}}}(\theta^3-\underset{(m)}{\theta}{}^3)H(\theta^3-\underset{(m)}{\theta}{}^3),\\
e_{\alpha3} &= \sum_{n=1}^{2}\overset{(n)}{\underset{(0)}{e_{\alpha3}}}(\theta^3)^n+\sum_{m=1}^{2}\overset{(0)}{\underset{(m)}{e_{\alpha3}}}H(\theta^3-\underset{(m)}{\theta}{}^3)
\end{aligned}
\tag{26}
$$

其中

$$
\begin{aligned}
&\overset{(0)}{\underset{(0)}{e_{\alpha\beta}}}=\overset{(0)}{\underset{(0)}{\eta_{\alpha\beta}}}+\frac{1}{2}\overset{(0)}{\underset{(0)}{\omega_{3\alpha}}}\overset{(0)}{\underset{(0)}{\omega_{3\beta}}},\\
&\overset{(1)}{\underset{(0)}{e_{\alpha\beta}}}=\overset{(1)}{\underset{(0)}{\eta_{\alpha\beta}}}+\frac{1}{2}(\overset{(0)}{\underset{(0)}{\omega_{3\alpha}}}\overset{(1)}{\underset{(0)}{\omega_{3\beta}}}+\overset{(1)}{\underset{(0)}{\omega_{3\alpha}}}+\overset{(0)}{\underset{(0)}{\omega_{3\beta}}})+\frac{1}{2}(\overset{(1)}{\underset{(0)}{\eta_{\alpha3}}}\overset{(0)}{\underset{(0)}{\omega_{3\beta}}}+\overset{(1)}{\underset{(0)}{\eta_{\beta3}}}\overset{(0)}{\underset{(0)}{\omega_{3\alpha}}}),\\
&\overset{(2)}{\underset{(0)}{e_{\alpha\beta}}}=\overset{(2)}{\underset{(0)}{\eta_{\alpha\beta}}}+\frac{1}{2}(\overset{(0)}{\underset{(0)}{\omega_{3\alpha}}}\overset{(2)}{\underset{(0)}{\omega_{3\beta}}}+\overset{(1)}{\underset{(0)}{\omega_{3\alpha}}}\overset{(0)}{\underset{(0)}{\omega_{3\beta}}}+\overset{(2)}{\underset{(0)}{\omega_{3\alpha}}}\overset{(0)}{\underset{(0)}{\omega_{3\beta}}})\\
&\quad+\frac{1}{2}(\overset{(2)}{\underset{(0)}{\eta_{\alpha3}}}\overset{(0)}{\underset{(0)}{\omega_{3\beta}}}+\overset{(2)}{\underset{(0)}{\eta_{\beta3}}}\overset{(0)}{\underset{(0)}{\omega_{3\alpha}}}+\overset{(1)}{\underset{(0)}{\eta_{\alpha3}}}\overset{(1)}{\underset{(0)}{\omega_{3\beta}}}+\overset{(1)}{\underset{(0)}{\eta_{\beta3}}}\overset{(1)}{\underset{(0)}{\omega_{3\alpha}}},\\
&\overset{(3)}{\underset{(0)}{e_{\alpha\beta}}}=\overset{(3)}{\underset{(0)}{\eta_{\alpha\beta}}},\\
&\overset{(1)}{\underset{(m)}{e_{\alpha\beta}}}=\overset{(1)}{\underset{(m)}{\eta_{\alpha\beta}}},
\end{aligned}
\tag{27}
$$

$$
\begin{aligned}
&\underset{(0)}{\overset{(1)}{e}}_{\alpha 3} = \underset{(0)}{\overset{(1)}{\eta}}_{\alpha 3}, \\
&\underset{(0)}{\overset{(2)}{e}}_{\alpha 3} = \underset{(0)}{\overset{(2)}{\eta}}_{\alpha 3}, \\
&\underset{(m)}{\overset{(0)}{e}}_{\alpha 3} = \underset{(m)}{\overset{(0)}{\eta}}_{\alpha 3}, \\
&\underset{(0)}{\overset{(0)}{\eta}}_{\alpha\beta} = \frac{1}{2}(\phi_{\alpha\beta} + \phi_{\beta\alpha}), \\
&\underset{(0)}{\overset{(1)}{\eta}}_{\alpha\beta} = \frac{1}{2}(\underset{(0)(1)}{\phi_{\alpha\beta}} + \underset{(0)(1)}{\phi_{\beta\alpha}} - b_\alpha^\lambda \phi_{\lambda\beta} - b_\beta^\lambda \phi_{\lambda\alpha}), \\
&\underset{(0)}{\overset{(2)}{\eta}}_{\alpha\beta} = \frac{1}{2}(\underset{(0)(2)}{\phi_{\alpha\beta}}\ \underset{(0)(2)}{\phi_{\beta\alpha}} - b_\alpha^\lambda \underset{(0)(1)}{\phi_{\lambda\beta}} - b_\beta^\lambda \underset{(0)(1)}{\phi_{\lambda\alpha}}), \\
&\underset{(0)}{\overset{(3)}{\eta}}_{\alpha\beta} = \frac{1}{2}(\underset{(0)(3)}{\phi_{\alpha\beta}} + \underset{(0)(3)}{\phi_{\beta\alpha}} - b_\beta^\lambda \underset{(0)(2)}{\phi_{\lambda\beta}} - \underset{(0)(2)}{\phi_{\lambda\alpha}}), \\
&\underset{(m)}{\overset{(1)}{\eta}}_{\alpha\beta} = \frac{1}{2}[\underset{(m)(1)}{\phi_{\alpha\beta}} + \underset{(m)(1)}{\phi_{\beta\alpha}} - \underset{(m)}{\theta^3}(b_\alpha^\lambda \underset{(m)(1)}{\phi_{\lambda\beta}} + b_\beta^\lambda \underset{(m)(1)}{\phi_{\lambda\alpha}})], \\
&\underset{(0)}{\overset{(1)}{\eta}}_{\alpha 3} = \varphi_\alpha, \\
&\underset{(0)}{\overset{(2)}{\eta}}_{\alpha 3} = \frac{1}{2}(3c_\alpha^\lambda - b_\alpha^\lambda)\varphi_\lambda, \\
&\underset{(m)}{\overset{(0)}{\eta}}_{\alpha 3} = \frac{1}{2}\underset{(m)}{\alpha_\nu^\lambda}(\delta_\alpha^\nu - \underset{(m)}{\theta^3} b_\alpha^\nu)\varphi_\lambda, \\
&\underset{(0)}{\overset{(0)}{\omega}}_{\alpha 3} = \phi_\alpha^3, \\
&\underset{(0)}{\overset{(1)}{\omega}}_{3\alpha} = -\varphi_\alpha - b_\alpha^\lambda \phi_\lambda^3, \\
&\underset{(0)}{\overset{(2)}{\omega}}_{3\alpha} = \frac{3}{2}(b_\alpha^\lambda - c_\alpha^\lambda)\varphi_\lambda, \\
&\underset{(m)}{\overset{(0)}{\omega}}_{3\alpha} = -\frac{1}{2}\underset{(m)}{\alpha_\nu^\lambda}(\delta_\alpha^\nu - \underset{(m)}{\theta^3} b_\alpha^\nu)\varphi_\lambda
\end{aligned}
\tag{28}
$$

若有 $h\varphi_\alpha = o\ (\varepsilon^2)$，则可将式（27）和（28）代入式（26），并进行量阶分析略去小量后，便得应变位移关系

$$
\begin{aligned}
e_{\alpha\beta} =& \frac{1}{2}(\phi_{\alpha\beta} + \phi_{\beta\alpha} + \phi_\alpha^3 \phi_\beta^3) + \frac{1}{2}(\underset{(0)(1)}{\phi_{\alpha\beta}} + \underset{(0)(1)}{\phi_{\beta\alpha}})\theta^3 \\
& + \frac{1}{2}(\underset{(0)(2)}{\phi_{\alpha\beta}} + \underset{(0)(2)}{\phi_{\beta\alpha}})(\theta^3)^2 + \frac{1}{2}(\underset{(0)(3)}{\phi_{\alpha\beta}} + \underset{(0)(3)}{\phi_{\alpha\beta}})(\theta^3)^3 \\
& \frac{1}{2}\sum_{m=1}^{2}(\underset{(m)(1)}{\phi_{\alpha\beta}} + \underset{(m)(1)}{\phi_{\beta\alpha}})(\theta^3 - \underset{(m)}{\theta^3})H(\theta^3 - \underset{(m)}{\theta^3}),
\end{aligned}
\tag{29}
$$

$$
e_{\alpha 3} = \varphi_\alpha \theta^3 + \frac{3}{2}c_\alpha^\lambda \varphi_\lambda (\theta^3)^2 + \frac{1}{2}\sum_{m=1}^{2}\underset{(m)}{\alpha_\alpha^\lambda}\varphi_\lambda H(\theta^3 - \underset{(m)}{\theta^3})
$$

五、本构方程

按照本文假设，我们只需取

$$Q=\bar{Q},\quad \mu=1,\quad \mu_\beta^\alpha=\delta_\beta^\alpha,\quad \overset{-1}{\mu}{}_\beta^\alpha=\delta_\beta^\alpha \tag{30}$$

故可定义

$$\begin{aligned}
\underset{(m)}{\overset{(n)}{L}}{}^{\rho\alpha}&=\int_0^h \mu\mu_\omega^\rho S^{\alpha\omega}(\theta^3-\underset{(m)}{\theta}{}^3)^n H(\theta^3-\underset{(m)}{\theta}{}^3)\mathrm{d}\theta^3\\
&=\int_0^h S^{\alpha\rho}(\theta^3-\underset{(m)}{\theta}{}^3)^n H(\theta^3-\underset{(m)}{\theta}{}^3)\mathrm{d}\theta^3,\\
\underset{(m)}{\overset{(n)}{N}}{}^{\alpha3}&=\int_0^h \mu S^{\alpha3}(\theta^3-\underset{(m)}{\theta}{}^3)^n H(\theta^3-\underset{(m)}{\theta}{}^3)\mathrm{d}\theta^3\\
&=\int_0^h S^{\alpha3}(\theta^3-\underset{(m)}{\theta}{}^3)^n H(\theta^3-\underset{(m)}{\theta}{}^3)\mathrm{d}\theta^3
\end{aligned} \tag{31}$$

将应变位移关系式（29）代入上式，得本构方程

$$\begin{aligned}
\underset{(m)}{\overset{(n)}{L}}{}^{\rho\alpha}&=\sum_{q=0}^{3}\underset{(m)}{\overset{(n)}{}}D_{(0)}^{(q)\,\rho\alpha\gamma\nu}\underset{(0)}{\overset{(q)}{e}}{}_{\gamma\nu}+\sum_{p=1}^{2}\underset{(m)}{\overset{(n)}{}}D_{(p)}^{(1)\,\rho\alpha\gamma\nu}\underset{(p)}{\overset{(1)}{e}}{}_{\gamma\nu},\\
\underset{(m)}{\overset{(n)}{N}}{}^{\alpha3}&=2\sum_{q=1}^{2}\underset{(m)}{\overset{(n)}{}}D_{(0)}^{(q)\,\alpha3\gamma3}\underset{(0)}{\overset{(q)}{e}}{}_{\gamma3}+2\sum_{p=1}^{2}\underset{(m)}{\overset{(n)}{}}D_{(p)}^{(0)\,\alpha3\gamma3}\underset{(p)}{\overset{(0)}{e}}{}_{\gamma3},
\end{aligned} \tag{32}$$

其中

$$\begin{aligned}
\underset{(0)}{\overset{(0)}{e}}{}_{\alpha\beta}&=\frac{1}{2}(\phi_{\alpha\beta}+\phi_{\beta\alpha}+\phi_\alpha^3\phi_\beta^3),\\
\underset{(0)}{\overset{(1)}{e}}{}_{\alpha\beta}&=\frac{1}{2}(\underset{(0)(1)}{\phi_{\alpha\beta}}+\underset{(0)(1)}{\phi_{\beta\alpha}}),\\
\underset{(0)}{\overset{(2)}{e}}{}_{\alpha\beta}&=\frac{1}{2}(\underset{(0)(2)}{\phi_{\alpha\beta}}+\underset{(0)(2)}{\phi_{\beta\alpha}}),\\
\underset{(0)}{\overset{(3)}{e}}{}_{\alpha\beta}&=\frac{1}{2}(\underset{(0)(3)}{\phi_{\alpha\beta}}+\underset{(0)(3)}{\phi_{\beta\alpha}}),\\
\underset{(m)}{\overset{(1)}{e}}{}_{\alpha\beta}&=\frac{1}{2}(\underset{(m)(1)}{\phi_{\alpha\beta}}+\underset{(m)(1)}{\phi_{\beta\alpha}}),\\
\underset{(0)}{\overset{(1)}{e}}{}_{\alpha3}&=\varphi_\alpha,\\
\underset{(0)}{\overset{(2)}{e}}{}_{\alpha3}&=\frac{3}{2}c_\alpha^\lambda\varphi_\lambda,\\
\underset{(m)}{\overset{(0)}{e}}{}_{\alpha3}&=\frac{1}{2}\underset{(m)}{\alpha_\alpha^\lambda}\phi_\lambda,
\end{aligned} \tag{33}$$

$$\underset{(m)}{\overset{(n)}{}}D_{(p)}^{(q)\,ijkl}=\int_0^h \bar{Q}^{ijkl}(\theta^3-\underset{(m)}{\theta}{}^3)^n(\theta^3-\underset{(p)}{\theta}{}^3)^q H(\theta^3-\underset{(m)}{\theta}{}^3)H(\theta^3-\underset{(p)}{\theta}{}^3)\mathrm{d}\theta^3 \tag{34}$$

六、平衡方程和边界条件

夹层壳总应变能为

$$U_1=\iint\limits_{\underset{(0)}{\omega}}\int_0^h W\mu\,\mathrm{d}\sigma\mathrm{d}\theta^3 \tag{35}$$

记

$$\overline{W}=\int_0^h W\mu\,\mathrm{d}\theta^3 \tag{36}$$

按本文近似理论，有

$$\overline{W}=\frac{1}{2}\int_0^h Q^{\alpha\beta\omega\lambda}e_{\alpha\beta}e_{\omega\lambda}\,\mathrm{d}\theta^3+2\int_0^h Q^{\alpha3\omega3}e_{\alpha3}e_{\omega3}\,\mathrm{d}\theta^3 \tag{37}$$

将应变位移关系式（29）代入上式，可发现下列“匹配”关系

$$\begin{aligned}\overset{(n)}{\underset{(m)}{L}}{}^{\alpha\beta}&=\frac{\partial\overline{W}}{\partial\overset{(n)}{\underset{(m)}{e}}{}_{\alpha\beta}}\\ \overset{(n)}{\underset{(m)}{N}}{}^{\alpha3}&=\frac{1}{2}\frac{\partial\overline{W}}{\partial\overset{(n)}{\underset{(m)}{e}}{}_{\alpha3}}\end{aligned} \tag{38}$$

载荷位能为

$$\begin{aligned}U_2&=-\int\limits_{\underset{(0)}{\omega}}\overset{*}{P}{}^3V_3\,\mathrm{d}\sigma-\int\limits_{{}_s\Omega}\nu_\alpha\overset{*}{S}{}^{\alpha i}V_i\,\mathrm{d}\Omega\\ &=-\int\limits_{\underset{(0)}{\omega}}\overset{*}{P}{}^3u_3\,\mathrm{d}\sigma-\int_{\Gamma_s}\mathrm{d}S_0\nu_\alpha\int_0^h(\overset{*}{S}{}^{\alpha\beta}\overline{V}_\beta+\overset{*}{S}{}^{\alpha3}\overline{V}_3)\,\mathrm{d}\theta^3\end{aligned} \tag{39}$$

式中 $\overset{*}{S}{}^{\alpha i}$ 为夹层壳边界侧面Ω 上给定的外加应力，$\overset{*}{P}{}^3$ 为外加横向载荷，Γ_s 为应力边界侧面${}_s\Omega$ 与 $\underset{(0)}{\omega}$ 面的交线，ν_α 为Ω 的单位外法线矢量分量，${}_0\nu_\alpha$ 为 ν_α 在$\theta^3=0$ 时的对应值。

将位移表达式（24）代入上式，得

$$\begin{aligned}U_2=&-\int\limits_{\underset{(0)}{\omega}}\overset{*}{P}{}^3u_3\,\mathrm{d}\sigma-\int_{\Gamma_s}\Big[(\overset{\overset{*}{(0)}}{\underset{(0)}{L}}{}^{\alpha\beta}-b^\beta_\lambda\overset{\overset{*}{(1)}}{\underset{(0)}{L}}{}^{\alpha\lambda})u_\beta-\overset{\overset{*}{(1)}}{\underset{(0)}{L}}{}^{\alpha\beta}u_{3,\beta}\\ &+(\overset{\overset{*}{(2)}}{\underset{(0)}{L}}{}^{\alpha\beta}+C^\beta_\lambda\overset{\overset{*}{(3)}}{\underset{(0)}{L}}{}^{\alpha\lambda}+\sum_{m=1}^{2}\underset{(m)}{a^\beta_\lambda}\overset{\overset{*}{(1)}}{\underset{(m)}{L}}{}^{\alpha\lambda})\varphi_\beta+\overset{\overset{*}{(0)}}{\underset{(0)}{N}}{}^{\alpha3}u_3\Big]{}_0\nu_\alpha\,\mathrm{d}S\end{aligned} \tag{40}$$

根据最小位能原理

$$\delta U=\delta(U_1+U_2)=0 \tag{41}$$

有

$$\begin{aligned}&\iint\limits_{\underset{(0)}{\omega}}\Big[\sum_{m,n}\frac{\partial\overline{W}}{\partial\overset{(n)}{\underset{(m)}{e}}{}_{\alpha\beta}}\delta\overset{(n)}{\underset{(m)}{e}}{}_{\alpha\beta}+\sum_{m,n}\frac{\partial\overline{W}}{\partial\overset{(n)}{\underset{(m)}{e}}{}_{\alpha3}}\delta\overset{(n)}{\underset{(m)}{e}}{}_{\alpha3}-\overset{*}{P}{}^3\delta u_3\Big]\mathrm{d}\sigma\\ &-\int_{\Gamma_s}\Big[(\overset{\overset{*}{(0)}}{\underset{(0)}{L}}{}^{\alpha\beta}-b^\beta_\lambda\overset{\overset{*}{(1)}}{\underset{(0)}{L}}{}^{\alpha\lambda})\delta u_\beta-\overset{\overset{*}{(0)}}{\underset{(0)}{L}}{}^{\alpha\beta}\delta u_{3,\beta}\\ &+(\overset{\overset{*}{(2)}}{\underset{(0)}{L}}{}^{\alpha\beta}+c^\beta_\lambda\overset{\overset{*}{(3)}}{\underset{(0)}{L}}{}^{\alpha\lambda}+\sum_{m=1}^{2}\underset{(m)}{a^\beta_\lambda}\overset{\overset{*}{(1)}}{\underset{(m)}{L}}{}^{\alpha\lambda})\delta\varphi_\beta+\overset{\overset{*}{(0)}}{\underset{(0)}{N}}{}^{\alpha3}\delta u_3\Big]{}_0\nu_\alpha\,\mathrm{d}S=0\end{aligned} \tag{42}$$

将式（33）代入上式，注意到式（38）及（25），并经过分部积分和整理后可得

$$
\begin{aligned}
& -\int_{\underset{(0)}{\omega}} \{[\overset{(0)}{\underset{(0)}{\hat{L}}}{}^{\alpha\beta}|_\beta - b^\alpha_\beta(\phi^3_\lambda \overset{(0)}{\underset{(0)}{\hat{L}}}{}^{\lambda\beta} + \overset{(1)}{\underset{(0)}{\hat{L}}}{}^{\beta\lambda}|_\lambda)]\delta u_\alpha + [(\phi^3_\alpha \overset{(0)}{\underset{(0)}{\hat{L}}}{}^{\alpha\beta} + \overset{(0)}{\underset{(0)}{\hat{L}}}{}^{\beta\alpha})|_\beta + b_{\alpha\beta}\overset{(0)}{\underset{(0)}{\hat{L}}}{}^{\alpha\beta} \\
& + \overset{*}{p}{}^3]\delta u_3 + [\overset{(2)}{\underset{(0)}{\hat{L}}}{}^{\alpha\beta}|_\beta + c^\alpha_\lambda \overset{(3)}{\underset{(0)}{\hat{L}}}{}^{\lambda\beta}|_\beta + \sum_{m=1}^{2} \underset{(m)}{a^\alpha_\lambda} \overset{(1)}{\underset{(m)}{\hat{L}}}{}^{\lambda\beta}|_\beta - 2\overset{(1)}{\underset{(0)}{N}}{}^{\alpha 3} - 3C^\alpha_\lambda \overset{(2)}{\underset{(0)}{N}}{}^{\lambda 3} \\
& - \sum_{m=1}^{2} \underset{(m)}{a^\alpha_\lambda} \overset{(0)}{\underset{(m)}{N}}{}^{\lambda 3}]\delta\varphi_\alpha\}\mathrm{d}\sigma + \int_{\Gamma_3}\{[(\overset{(0)}{\underset{(0)}{\hat{L}}}{}^{\alpha\beta} - b^\alpha_\lambda \overset{(1)}{\underset{(0)}{\hat{L}}}{}^{\lambda\beta}) - (\overset{*}{\overset{(0)}{\underset{(0)}{\hat{L}}}}{}^{\alpha\beta} - b^\alpha_\lambda \overset{(1)}{\underset{(0)}{\hat{L}}}{}^{\lambda\beta})]\delta u_\alpha \\
& + [\phi^3_\alpha \overset{(0)}{\underset{(0)}{\hat{L}}}{}^{\alpha\beta} + \overset{(1)}{\underset{(0)}{\hat{L}}}{}^{\beta\alpha}|_\alpha - \overset{*}{\overset{(0)}{\underset{(0)}{N}}}{}^{\beta 3}]\delta u_3 - [\overset{(0)}{\underset{(0)}{\hat{L}}}{}^{\alpha\beta} - \overset{(1)}{\underset{(0)}{\hat{L}}}{}^{\alpha\beta}]\delta u_{3,\alpha} + [\overset{(2)}{\underset{(0)}{\hat{L}}}{}^{\alpha\beta} + c^\alpha_\lambda \overset{(3)}{\underset{(0)}{\hat{L}}}{}^{\lambda\beta} \\
& + \sum_{m=1}^{2} \underset{(m)}{a^\alpha_\lambda} \overset{(1)}{\underset{(m)}{\hat{L}}}{}^{\lambda\beta} - (\overset{*}{\overset{(2)}{\underset{(0)}{\hat{L}}}}{}^{\alpha\beta} + c^\alpha_\lambda \overset{*}{\overset{(3)}{\underset{(0)}{\hat{L}}}}{}^{\lambda\beta} + \sum_{m=1}^{2} \underset{(m)}{a^\alpha_\lambda} \overset{*}{\overset{(1)}{\underset{(m)}{\hat{L}}}}{}^{\lambda\beta})]\delta\varphi_\alpha\}{}_0\nu_\alpha \mathrm{d}S = 0
\end{aligned} \tag{43}
$$

从而得平衡方程和边界条件如下。

平衡方程

$$
\begin{aligned}
& \overset{(0)}{\underset{(0)}{\hat{L}}}{}^{\alpha\beta}|_\beta - b^\alpha_\beta(\phi^3_\lambda \overset{(0)}{\underset{(0)}{\hat{L}}}{}^{\beta\lambda} + \overset{*}{\overset{(0)}{\underset{(0)}{\hat{L}}}}{}^{\beta\lambda}|_\lambda) = 0, \\
& (\phi^3_\alpha \overset{(0)}{\underset{(0)}{\hat{L}}}{}^{\alpha\beta} + \overset{(1)}{\underset{(0)}{\hat{L}}}{}^{\beta\lambda}|_\alpha)|_\beta + b_{\alpha\beta}\overset{(0)}{\underset{(0)}{\hat{L}}}{}^{\alpha\beta} + \overset{*}{p}{}^3 = 0, \\
& \overset{(2)}{\underset{(0)}{\hat{L}}}{}^{\alpha\beta}|_\beta + c^\alpha_\lambda \overset{(3)}{\underset{(0)}{L}}{}^{\lambda\beta}|_\beta + \sum_{m=1}^{2} \underset{(m)}{a^\alpha_\lambda} \overset{(1)}{\underset{(0)}{\hat{L}}}{}^{\lambda\beta}|_\beta - 2\overset{(1)}{\underset{(0)}{N}}{}^{\lambda 3} - 3C^\alpha_\lambda \overset{(2)}{\underset{(0)}{N}}{}^{\lambda 3} - \sum_{m=1}^{2} \underset{(m)}{a^\alpha_\lambda} \overset{(0)}{\underset{(m)}{N}}{}^{\lambda 3} = 0
\end{aligned} \tag{44}
$$

静力边界条件

$$
\begin{aligned}
& {}_0\nu_\beta(\overset{(0)}{\underset{(0)}{\hat{L}}}{}^{\alpha\beta} - b^\alpha_\beta \overset{(1)}{\underset{(0)}{\hat{L}}}{}^{\lambda\beta}) = {}_0\nu_\beta(\overset{*}{\overset{(0)}{\underset{(0)}{\hat{L}}}}{}^{\alpha\beta} - b^\alpha_\lambda \overset{*}{\overset{(1)}{\underset{(0)}{\hat{L}}}}{}^{\lambda\beta}), \\
& {}_0\nu_\beta \overset{(0)}{\underset{(0)}{\hat{L}}}{}^{\alpha\beta} = {}_0\nu_\beta \overset{*}{\overset{(1)}{\underset{(0)}{\hat{L}}}}{}^{\alpha\beta}, \\
& {}_0\nu_\beta(\phi^3_\alpha \overset{(0)}{\underset{(0)}{\hat{L}}}{}^{\alpha\beta} + \overset{(1)}{\underset{(0)}{\hat{L}}}{}^{\beta\alpha}|_\alpha) = {}_0\nu_\beta \overset{*}{\overset{(0)}{\underset{(0)}{N}}}{}^{\beta 3} \\
& {}_0\nu_\beta(\overset{(2)}{\underset{(0)}{\hat{L}}}{}^{\alpha\beta} + c^\alpha_\lambda \overset{(3)}{\underset{(0)}{\hat{L}}}{}^{\lambda\beta} + \sum_{m=1}^{2} \underset{(m)}{a^\alpha_\lambda} \overset{(1)}{\underset{(m)}{\hat{L}}}{}^{\lambda\beta}) = {}_0V_\beta(\overset{*}{\overset{(2)}{\underset{(0)}{L}}}{}^{\alpha\beta} + c^\alpha_\lambda \overset{*}{\overset{(3)}{\underset{(0)}{\hat{L}}}}{}^{\lambda\beta} + \sum_{m=1}^{2} \underset{(m)}{a^\alpha_\lambda} \overset{*}{\overset{(1)}{\underset{(m)}{\hat{L}}}}{}^{\lambda\beta})
\end{aligned} \tag{45}
$$

几何边界条件

$$
\begin{aligned}
& u_\alpha = \overset{*}{u}_\alpha, \quad u_{3,\alpha} = \overset{*}{u}_{3,\alpha}, \\
& u_3 = \overset{*}{u}_3, \quad \varphi_\alpha = \overset{*}{\varphi}_\alpha
\end{aligned}
$$

参 考 文 献

[1] Reissner E. Small bending and stretching of sandwich-type shells. NACA，1950：975.

[2] Wang C T. Principle and application of complementary energy method for thin homogeneous and sandwich plates and shells with finite deflections. NACA TN，1952：2620.

[3] Schmidt R. Sandwich shells of arbitrary shape. J. Appl. Mech.，1964，31 (2)：239.

[4] Wempner G A. Theory for moderately large deflections of sandwich shells with dissimilar facings. Int. J. Solids Struct.，1967，3 (3)：367.

[5] Reese C D, Bert C W. Buckling of orthotropic sandwich cylinders under axial compression and bending. J. Aircraft, 1974, 11: 207.

[6] Ronan J G, Kao J S. Nonlinear equations for shallow sandwich shells with orthotropic cores. J. AIAA, 1975, 13 (7): 961.

[7] 刘人怀，朱金福. 夹层壳非线性理论. 北京：机械工业出版社，1993.

[8] Liu Ren-huai and Zhu, J. F. Nonlinear theory of sandwich shells, Part Ⅰ—Exact kinematics of moderately thick shells. Appl. Mech. Eng., 1997, 2 (2): 213.

[9] Liu Ren-huai and Zhu, J. F. Nonlinear theory of sandwich shells, Part Ⅱ—Approximate theories. Appl. Meeh Eng. 1997, 2 (2): 241.

[10] Liu Ren-huai and Cheng, Z. Q. On the non-linear buckling of circular shallow spherical sandwich shells under the action of uniform edge moments. Int. J. Non-Linear Mech. 1995, 30 (1): 33.

[11] Liu Ren-huai and Li, J. Non-linear vibration of shallow conical sandwich shells. Int. J. Non-Linear Mech., 1995, 30 (2): 97.

[12] Wang, Z. W. and Liu Ren-huai. Nonlinear buckling for sandwich shallow shells accounting for continuity conditions of stresses at interfaces. Proc. 3rdInt. Conf. Nonlinear Mech., Shanghai University Press, Shanghai, 1998: 381.

[13] Liu Ren-huai. Study on Nonlinear Mechanics of Plates and Shells. New York, Beijing, Guang zhou: Science Press and Jinan University Press, 1998.

[14] Schmit, L. A. Jr. and Monforton, G. R. Finite deflection discrete element analysis of sandwich plates and cylindrical shells with laminated faces. J. AIAA, 1970, 8 (8): 1454.

[15] Rao, K. M. Buckling analysis of FRP-faced anisotropic cylindrical sandwich panel. J. Eng. Mech., 1985, 111: 529.

[16] Koganti, M. R. Critical shear loading of curved sandwich panels faced with fiberreinforced plastic. J. AIAA, 1986, 24: 1531.

[17] 王震鸣，刘国玺，吕明身. 各向异性多层扁壳的大挠度方程. 应用数学和力学，1982，(1)：49.

Large deflection of annular sandwich plates①

中文摘要

本文应用修正迭代法，获得了在均布压力作用下具有刚性中心的夹层环形板大挠度问题的解析解，可直接应用于工程设计。

1. Introduction

It is common knowledge that the sandwich plate is applied widely in aeronautical and astronautical engineering because of high rigidity and lightweight features. Therefore, it is of great importance both theoretically and practically to study nonlinear bending problems for the plate. At first, Reissner[1] established a nonlinear bending theory of a rectangular sandwich plate with a soft core and two very thin faces. Scholars[2] of the Chinese Research Institute of Mechanics solved a large deflection problem for the circular sandwich plate with a soft core and two very thin faces under the action of uniform lateral load by the perturbation method. The author[3] solved a nonlinear bending problem of the plate under uniform edge moment, and a more accurate third approximation solution was obtained using the modified iteration method. This method was suggested by Yeh Kai-yuan and Liu Ren-huai[4-6] in 1965. The method incorporates the advantages of Chien Wei-zang's perturbation method[7] and usual successive approximations, and it is an effective, simple, accurate method for solving nonlinear differential equations.

After that, the author[8] presented a more accurate nonlinear bending theory of a circular sandwich plate with a soft core, taking into account the bending rigidity of the faces, and also gave a simplified theory of the plate in the case of neglecting the bending rigidity of the faces. In the case of including the bending rigidity of the faces, the authors[9] first discussed the nonlinear bending problem for the plate using the modified power series method. Unfortunately, such studies are few yet because of quite complication of the problem.

So far as we know, nonlinear problems of bending and vibration for circular and annular sandwich plates with very thin faces behaving as membranes were studied by Liu Ren-huai, Du Guojun, Yang Jingning, Ho Chaosheng, Zhang Xiuli and Kirichok, et al. [8,10-26]

This paper is a further work of the authors' previous papers[12,13]. A large deflection

① Reprinted from *Journal of Mechanics and MEMS*, 2009, 1 (2): 145-156.

problem of an annular sandwich plate with a nondeformable rigid body at the center under uniform pressure is studied. We still use the modified iteration method to solve this problem. Analysic solutions presented here may be applied directly to the engineering design.

2. Fundamental Equations

Now consider an annular sandwich plate with a nondeformable rigid body at the center under the action of uniform pressure q as shown in Fig. 1. The outer edge of the plate is rigidly clamped and the inner edge is fixed on the nondeformable rigid body which can be moved up down. Here a is the outer radius, b is the inner radius, r is the radial coordinate.

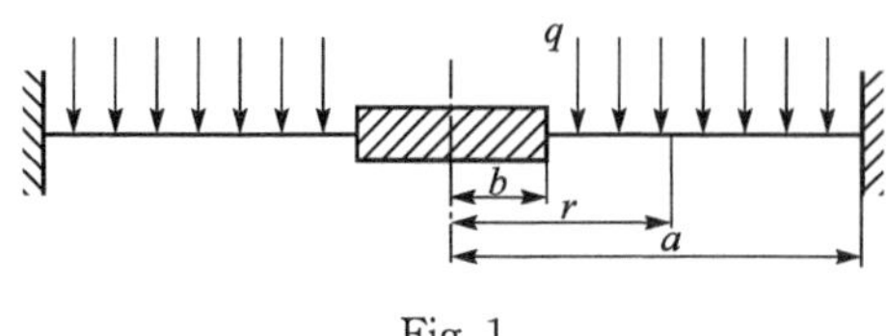

Fig. 1

Using the same system of notation and the simplified equations of Ref. [8], we can easily obtain the fundamental equations of large deflection of the plate as follows.

$$Dr\frac{\mathrm{d}}{\mathrm{d}r}\frac{1}{r}\frac{\mathrm{d}}{\mathrm{d}r}r\frac{\mathrm{d}w}{\mathrm{d}r}-2t\left(1-\frac{D}{G_2h_0}r\frac{\mathrm{d}}{\mathrm{d}r}\frac{1}{r}\frac{\mathrm{d}}{\mathrm{d}r}\right)\left(r\sigma_{r0}\frac{\mathrm{d}w}{\mathrm{d}r}\right)-\frac{1}{2}q(r^2-b^2)=0,$$

$$r\frac{\mathrm{d}}{\mathrm{d}r}\frac{1}{r}\frac{\mathrm{d}}{\mathrm{d}r}(r^2\sigma_{r0})+\frac{E}{2}\left(\frac{\mathrm{d}w}{\mathrm{d}r}\right)^2=0 \tag{1}$$

where w is the deflection of the middle plane of the plate, σ_{r0} is the radial stress of the middle plane of the plate, D is the flexural rigidity of the plate, E is Young's modulus of the face, ν is Poisson's ratio of the face, G_2 is the shear modulus of the core, t is the thickness of the face, h_0 is the distance from middle of thickness of the lower face to middle of thickness of the upper face,

$$D=\frac{Eth_0^2}{2(1-\nu^2)} \tag{2}$$

Equations (1) will be solved under the following boundary conditions:

$$w=0,\ \psi=0,\ u=0 \text{ at } r=a$$
$$\psi=0,\ u=0 \text{ at } r=b \tag{3}$$

where ψ is the rotation of a normal to the middle plane of the plate in the diametral plane, u is the radial displacement of the middle plane of the plate,

$$\psi=-\frac{1}{G_2h_0r}\left[2tr\sigma_{r0}\frac{\mathrm{d}w}{\mathrm{d}r}+\frac{q}{2}(r^2-b^2)\right]-\frac{\mathrm{d}w}{\mathrm{d}r},$$

$$u=\frac{r}{E}\left[\frac{\mathrm{d}}{\mathrm{d}r}(r\sigma_{r0})-\nu\sigma_{r0}\right] \tag{4}$$

In order to simplify the calculations, let us introduce the following nondimensional variables

$$\rho=\frac{r}{a},\ \alpha=\frac{b}{a},\ W=\sqrt{2(1-\nu^2)}\frac{w}{h_0},\ \phi=\frac{\mathrm{d}w}{\mathrm{d}\rho},\ S_r=\frac{2ta^2}{D}\sigma_{r0}$$

$$P=\frac{\sqrt{2(1-\nu^2)}}{2h_0D}a^4q,\ k=\frac{D}{G_2h_0a^2} \tag{5}$$

Using these nondimensional variables, the fundamental equations (1) and boundary conditions (3) become

$$L(\rho\phi) = (1-KL)(\rho S_r\phi) + P(\rho^2-\alpha^2),$$
$$L(\rho^2 S_r) = -\phi^2 \tag{6a,b}$$

$$W=0,\ \phi = -k[S_r\phi - P(\alpha^2-1)],\ \frac{d}{d\rho}(\rho S_r) - \nu S_r = 0 \quad \text{at } \rho = 1; \tag{7a,b,c}$$

$$\phi = -kS_r\phi,\ \frac{d}{d\rho}(\rho S_r) - \nu S_r = 0 \quad \text{at } \rho = \alpha \tag{8a,b}$$

where L is a differential operator

$$L(\cdots) = \rho\frac{d}{d\rho}\frac{1}{\rho}\frac{d}{d\rho}(\cdots) \tag{9}$$

Thus Equations (6) and boundary conditions (7) and (8) constitute a nondimensional nonlinear boundary value problem for ϕ and S_r of the annular sandwich plate with a nondeformable rigid body at the center under uniform pressure.

3. Analytical Solution

The nondimensional nonlinear boundary value problem (6), (7) and (8) will be solved by the modified iteration method. At first, We introduce a notation W_m of the nondimensional inner edge deflection as an iteration parameter

$$W_m = W\,|_{\rho=\alpha} \tag{10}$$

Using the fourth equation of expressions (5) and the boundary condition (7a), We obtain

$$W_m = -\int_\alpha^1 \phi d\rho \tag{11}$$

For the first approximation, neglecting the nonlinear term $S_r\phi$ in Eq. (6a) and conditions (7b) and (8a) leads to the linear boundary value problem as follows

$$L(\rho\phi_1) = P(\rho^2-\alpha^2),$$
$$L(\rho^2 S_{r1}) = -\phi_1^2 \tag{12a,b}$$

$$W_1 = 0,\ \phi_1 = kP(\alpha^2-1),\ \frac{d}{d\rho}(\rho S_{r1}) - \nu S_{r1} = 0 \quad \text{at } \rho = \alpha \tag{13a,b,c}$$

$$\phi_1 = 0,\ \frac{d}{d\rho}(\rho S_{r1}) - \nu S_{r1} = 0 \quad \text{at } \rho = \alpha \tag{14a,b}$$

Eq. (12a) can be solved easily by direct integration in conjunction with the corresponding boundary conditions (13b) and (14a). Then we obtain

$$\phi_1 = P(a_1\rho^3 + a_2\rho\ln\rho + a_3\rho + a_4\rho^{-1}) \tag{15}$$

where

$$\begin{aligned}
a_1 &= \frac{1}{8},\\
a_2 &= -\frac{1}{2}\alpha^2,\\
a_3 &= \frac{1}{8(\alpha^2-1)}[4\alpha^4\ln\alpha - \alpha^4 + 1 - 8k(\alpha^2-1)],\\
a_4 &= \frac{\alpha^2}{8(\alpha^2-1)}[4\alpha^2\ln\alpha - \alpha^2 + 1 - 8k(\alpha^2-1)]
\end{aligned} \tag{16}$$

Substituting the solution (15) into (11), the linear characteristic relation is obtained as

$$P = \alpha_1 W_m \tag{17}$$

where

$$\alpha_1 = 4[a_1(\alpha^4 - 1) + a_2(2\alpha^1 \ln\alpha - \alpha^2 + 1) + 2\alpha(\alpha^2 - 1) + 4\alpha\ln\alpha]^{-1} \tag{18}$$

Using relation (17), the solution (15) may be written as

$$\phi_1 = \alpha_1 W_m (a_1\rho^3 + a_2\rho\ln\rho + a_3\rho + a_4\rho^{-1}) \tag{19}$$

Using the solution (19) and integrating Eq. (12b) twice under conditions (13c) and (14b), the solution of Eq. (12b) is

$$\begin{aligned} S_{r1} =& -\alpha_1^2 W_m^2 (b_1\rho^6 + b_2\rho^4\ln\rho + b_3\rho^4 + b_4\rho^2\ln^2\rho + b_5\rho^2\ln\rho \\ & + b_6\rho^2 + b_7\ln^2\rho + b_8\ln\rho + b_9 + b_{10}\rho^{-2}\ln\rho + b_{11}\rho^{-2}) \end{aligned} \tag{20}$$

where

$$\begin{aligned} b_1 &= \frac{1}{48}a_1^2, \\ b_2 &= \frac{1}{12}a_1 a_2, \\ b_3 &= \frac{a_1}{144}(12a_3 - 5a_2), \\ b_4 &= \frac{1}{8}a_2^2, \\ b_5 &= \frac{a_2}{16}(4a_3 - 3a_2), \\ b_6 &= \frac{1}{64}(7a_2^2 + 8a_3^2 + 16a_1a_4 - 12a_2a_3), \\ b_7 &= \frac{1}{2}a_2a_4, \\ b_8 &= \frac{a_4}{2}(2a_3 - a_2), \\ b_9 &= -\frac{1}{\lambda_1(\alpha^2 - 1)}\{b_1(7-\nu)(\alpha^8 - 1) + b_2[(5-\nu)\alpha^6\ln\alpha + \alpha^6 - 1] + b_3(5-\nu)(\alpha^6 - 1) \\ & \quad + b_4\alpha^4\ln\alpha[(3-\nu)\ln\alpha + 2] + b_5[(3-\nu)\alpha^4\ln\alpha + \alpha^4 - 1] \\ & \quad + b_6(3-\nu)(\alpha^4 - 1) + b_7\alpha^2\ln\alpha(\lambda_1\ln\alpha + 2) + b_8(\lambda_1\alpha^2\ln\alpha + \alpha^2 - 1) + \lambda_2\lambda_{10}\ln\alpha\}, \\ b_{10} &= -\frac{1}{2}a_4^2, \\ b_{11} &= -\frac{1}{\lambda_2(\alpha^2 - 1)}\{b_1\alpha^2(7-\nu)(\alpha^6 - 1) + b_2\alpha^2[(5-\nu)\alpha^4\ln\alpha + \alpha^4 - 1] \\ & \quad + b_3\alpha^2(5-\nu)(\alpha^4 - 1) + b_4\alpha^4\ln\alpha[(3-\nu)\ln\alpha + 2] + b_5\alpha^2[(3-\nu)\alpha^2\ln\alpha + \alpha^2 - 1] \\ & \quad + b_6\alpha^2(3-\nu)(\alpha^2 - 1) + b_7\alpha^2\ln\alpha(\lambda_1\ln\alpha + 2) + \lambda_1 b_8\alpha^2\ln\alpha - b_{10}(\alpha^2 + \lambda_2\ln\alpha - 1)\}, \\ \lambda_1 &= 1 - \nu, \\ \lambda_2 &= 1 + \nu \end{aligned} \tag{21}$$

For the second approximation, from the problem (6)-(8) the following linear boundary value problem for ϕ is obtained

$$L(\rho\phi_2) = (1-kL)(\rho S_{r1}\phi_1) + P(\rho^2-\alpha^2) \tag{22}$$

$$W_2 = 0,\ \phi_2 = -k[S_{r1}\phi_1 - P(\alpha^2-1)] \quad \text{at } \rho = 1; \tag{23a,b}$$

$$\phi_2 = -kS_{r1}\phi_1 \quad \text{at } \rho = \alpha \tag{24}$$

Using solutions (19) and (20), the solution of this problem is

$$\begin{aligned}\phi_2 = &P(a_1\rho^3 + a_2\rho\ln\rho + a_3\rho + a_4\rho^{-1}) + \alpha_1^3 W_m^3(c_1\rho^{11} + c_2\rho^9\ln\rho + c_3\rho^9 + c_4\rho^7\ln^2\rho \\ &+ c_5\rho^7\ln\rho + c_6\rho^7 + c_7\rho^5\ln^3\rho + c_8\rho^5\ln^2\rho + c_9\rho^5\ln\rho + c_{10}\rho^5 + c_{11}\rho^3\ln^3\rho \\ &+ c_{12}\rho^3\ln^2\rho + c_{13}\rho^3\ln\rho + c_{14}\rho^3 + c_{15}\rho\ln^2\rho + c_{16}\rho\ln^2\rho + c_{17}\rho\ln\rho + c_{18}\rho \\ &+ c_{19}\rho^{-1}\ln^2\rho + c_{20}\rho^{-1}\ln\rho + c_{21}\rho^{-1} + c_{22}\rho^{-1} + c_{22}\rho^{-3}\ln\rho + c_{23}\rho^{-3})\end{aligned} \tag{25}$$

where

$$c_1 = -\frac{1}{120}a_1b_1,$$

$$c_2 = -\frac{1}{80}(a_1b_2 + a_2b_1),$$

$$c_3 = \frac{9}{3200}(a_1b_2 + a_2b_1) - \frac{1}{80}(a_1b_3 + a_3b_1) + ka_1b_1,$$

$$c_4 = -\frac{1}{48}(a_1b_4 + a_2b_2),$$

$$c_5 = \frac{7}{576}(a_1b_4 + a_2b_2) - \frac{1}{48}(a_1b_5 + a_2b_3 + a_3b_2) + k(a_1b_2 + a_2b_1),$$

$$\begin{aligned}c_6 = &-\frac{37}{13824}(a_1b_4 + a_2b_2) + \frac{7}{1152}(a_1b_5 + a_2b_3 + a_3b_2) \\ &- \frac{1}{48}(a_1b_6 + a_3b_3 + a_4b_1) + k(a_1b_3 + a_3b_1),\end{aligned}$$

$$c_7 = -\frac{1}{24}a_2b_4,$$

$$c_8 = \frac{5}{96}a_2b_4 - \frac{1}{24}(a_1b_7 + a_2b_5 + a_3b_4) + k(a_1b_4 + a_2b_2),$$

$$\begin{aligned}c_9 = &-\frac{19}{576}a_2b_4 + \frac{5}{144}(a_1b_7 + a_2b_5 + a_3b_4) - \frac{1}{24}(a_1b_8 \\ &+ a_2b_6 + a_3b_5 + a_4b_2) + k(a_1b_5 + a_2b_3 + a_3b_2),\end{aligned}$$

$$\begin{aligned}c_{10} = &\frac{65}{6912}a_2b_4 - \frac{19}{1728}(a_1b_7 + a_2b_5 + a_3b_4) + \frac{5}{288}(a_1b_8 + a_2b_6 + a_3b_5 + a_4b_2) \\ &- \frac{1}{24}(a_1b_9 + a_3b_6 + a_4b_3) + k(a_1b_6 + a_3b_3 + a_4b_1),\end{aligned}$$

$$c_{11} = -\frac{1}{8}a_2b_7 + ka_2b_4,$$

$$c_{12} = \frac{9}{32}a_2b_7 - \frac{1}{8}(a_2b_8 + a_3b_7 + a_4b_4) + k(a_1b_7 + a_2b_5 + a_3b_4),$$

$$c_{13} = -\frac{21}{64}a_2b_7 + \frac{3}{16}(a_2b_8 + a_3b_7 + a_4b_4) - \frac{1}{8}(a_1b_{10} + a_2b_9 + a_3b_8 + a_4b_5)$$

$$
\begin{aligned}
&+k(a_1b_8+a_2b_6+a_3b_5+a_4b_2),\\
c_{14}&=\frac{45}{256}a_2b_7-\frac{7}{64}(a_2b_8+a_3b_7+a_4b_4)+\frac{3}{32}(a_1b_{10}+a_2b_9+a_3b_8+a_4b_5)\\
&\quad-\frac{1}{8}(a_1b_{11}+a_3b_9+a_4b_6)+k(a_1b_9+a_3b_6+a_4b_3),\\
c_{15}&=-\frac{1}{6}(a_2b_{10}+a_4b_7)+ka_2b_7,\\
c_{16}&=\frac{1}{4}(a_2b_{10}-a_2b_{11}-a_3b_{10}+a_4b_7-a_4b_8)+k(a_2b_8+a_3b_7+a_4b_4),\\
c_{17}&=-\frac{1}{4}(a_2b_{10}-a_2b_{11}-a_3b_{10}+2a_3b_{11}+a_4b_7-a_4b_8+2a_4b_9)\\
&\quad+k(a_1b_{10}+a_2b_9+a_3b_8+a_4b_5),\\
c_{18}&=d_1-\frac{d_3-\alpha d_4}{\alpha^2-1},\\
c_{19}&=\frac{1}{4}a_4b_{10}+k(a_2b_{10}+a_4b_7),\\
c_{20}&=\frac{1}{4}a_4(b_{10}+2b_{11})+k(a_2b_{11}+a_3b_{10}+a_4b_8),\\
c_{21}&=d_2+\frac{\alpha(\alpha d_3-d_4)}{\alpha^2-1},\\
c_{22}&=ka_4b_{10},\\
c_{23}&=ka_4b_{11},\\
d_1&=\frac{1}{8}(a_2b_{10}-a_2b_{11}-a_3b_{10}+2a_3b_{11}+a_4b_7-a_4b_8+2a_4b_9)+k(a_1b_{11}+a_3b_9+a_4b_6),\\
d_2&=k(a_3b_{11}+a_4b_9),\\
d_3&=-(c_1+c_3+c_6+c_{10}+c_{14}+c_{23}+d_1+d_2)\\
&\quad+k(a_1+a_3+a_4)(b_1+b_3+b_6+b_9+b_{11}),\\
d_4&=-c_1\alpha^{11}-c_2\alpha^9\ln\alpha+(ka_1b_1-c_3)\alpha^9-c_4\alpha^7\ln^2\alpha-[c_5-k(a_1b_2+a_2b_1)]\alpha^7\ln\alpha\\
&\quad-[c_6-k(a_1b_3+a_3b_1)]\alpha^7-c_7\alpha^5\ln^3\alpha-[c_8-k(a_1b_4+a_2b_2)]\alpha^5\ln^2\alpha\\
&\quad-[c_9-k(a_1b_5+a_2b_3+a_3b_2)]\alpha^5\ln\alpha-[c_{10}-k(a_1b_6+a_3b_3+a_4b_1)]\alpha^5\\
&\quad-(c_{11}-ka_2b_4)\alpha^3\ln^3\alpha-[c_{12}-k(a_1b_7+a_2b_5+a_3b_4)]\alpha^3\ln^2\alpha\\
&\quad-[c_{13}-k(a_1b_8+a_2b_6+a_3b_5+a_4b_2)]\alpha^3\ln\alpha-[c_{14}-k(a_1b_9+a_3b_6+a_4b_3)]\alpha^3\\
&\quad-(c_{15}-ka_2b_7)\alpha\ln^3\alpha-[c_{16}-k(a_2b_8+a_3b_7+a_4b_4)]\alpha\ln^2\alpha\\
&\quad-[c_{17}-k(a_1b_{10}+a_2b_9+a_3b_8+a_4b_5)]\alpha\ln\alpha-[d_1-k(a_1b_{11}+a_3b_9+a_4b_6)]\alpha\\
&\quad-\frac{1}{4}a_4b_{10}\alpha^{-1}\ln^2\alpha-\frac{1}{4}a_4(b_{10}+2b_{11})\alpha^{-1}\ln\alpha
\end{aligned}
\tag{26}
$$

Substituting the solution (25) into (11), we obtain the nonlinear characteristic relation of the annular sandwich plate

$$
P=\alpha_1W_m+\alpha_3W_m^3 \tag{27}
$$

where

$$\alpha_3 = -\alpha_1^4\Big[\frac{c_1}{12}(\alpha^{12}-1)+\frac{c_2}{100}(10\alpha^{10}\ln\alpha-\alpha^{10}+1)+\frac{c_3}{10}(\alpha^{10}-1)$$
$$+\frac{c_4}{256}(32\alpha^8\ln^2\alpha-8\alpha^8\ln\alpha+\alpha^8-1)+\frac{c_5}{64}(8\alpha^8\ln\alpha-\alpha^8+1)$$
$$+\frac{c_6}{8}(\alpha^8-1)+\frac{c_7}{216}(36\alpha^6\ln^3\alpha-18\alpha^6\ln^2\alpha+6\alpha^6\ln\alpha-\alpha^6+1)$$
$$+\frac{c_8}{108}(18\alpha^6\ln^2\alpha-6\alpha^6\ln\alpha+\alpha^6-1)+\frac{c_9}{36}(6\alpha^6\ln\alpha-\alpha^6+1)+\frac{c_{10}}{6}(\alpha^6-1)$$
$$+\frac{c_{11}}{128}(32\alpha^4\ln^3\alpha-24\alpha^4\ln^2\alpha+12\alpha^4\ln\alpha-3\alpha^4+3)+\frac{c_{12}}{32}(8\alpha^4\ln^2\alpha-4\alpha^4\ln\alpha+\alpha^4-1)$$
$$+\frac{c_{13}}{16}(4\alpha^4\ln\alpha-\alpha^4+1)+\frac{c_{14}}{4}(\alpha^4-1)+\frac{c_{15}}{8}(4\alpha^2\ln^3\alpha-6\alpha^2\ln^2\alpha+6\alpha^2\ln\alpha-3\alpha^2+3)$$
$$+\frac{c_{16}}{4}(2\alpha^2\ln^2\alpha-2\alpha^2\ln\alpha+\alpha^2-1)+\frac{c_{17}}{2}(2\alpha^2\ln\alpha-\alpha^2+1)+\frac{c_{18}}{2}(\alpha^2-1)+\frac{c_{19}}{3}\ln^3\alpha$$
$$+\frac{c_{20}}{2}\ln^2\alpha+c_{21}\ln\alpha-\frac{c_{22}}{4}(2\alpha^{-2}\ln\alpha+\alpha^{-2}-1)-\frac{c_{23}}{2}(\alpha^{-2}-1)\Big] \tag{28}$$

4. Results and Discussion

Now let us introduce the nondimensional variable S_θ for the tangential stress $\sigma_{\theta 0}$ of the middle plane of the annular sandwich plate

$$S_\theta = \frac{2ta^2}{D}\sigma_{\theta 0} \tag{29}$$

where

$$\sigma_{\theta 0} = \frac{\mathrm{d}}{\mathrm{d}r}(r\sigma_{r0}) \tag{30}$$

Using Eq. (30) and (5), expression (29) becomes

$$S_\theta = \frac{\mathrm{d}}{\mathrm{d}\rho}(\rho S_r) \tag{31}$$

Substituting solution (20) into Eq. (31), we obtain the following formula of the nondimensional tangential stress for the first approximation

$$S_\theta = -\alpha_1^2 W_m^2[7b_1\rho^6+b_2\rho^4(5\ln\rho+1)+5b_3\rho^4+b_4\rho^2\ln\rho(3\ln\rho+2)$$
$$+b_5\rho^2(3\ln\rho+1)+3b_6\rho^2+b_7\ln\rho(\ln\rho+2)+b_8(\ln\rho+1)+b_9$$
$$-b_{10}\rho^{-2}(\ln\rho-1)-b_{11}\rho^{-2}] \tag{32}$$

Finally, from formulas (20) and (32) we obtain the stresses at the inner and outer edges of the annular sandwich plate

$$S_r(1) = -\alpha_1^2 W_m^2(b_1+b_3+b_6+b_9+b_{11}),$$
$$S_\theta(1) = -\alpha_1^2 W_m^2(7b_1+b_2+5b_3+b_5+3b_6+b_8+b_9+b_{10}-b_{11}),$$
$$S_r(\alpha) = -\alpha_1^2 W_m^2(b_1\alpha^6+b_2\alpha^4\ln\alpha+b_3\alpha^4+b_4\alpha^2\ln^2\alpha+b_5\alpha^2\ln\alpha+b_6\alpha^2+b_7\ln^2\alpha$$
$$+b_8\ln\alpha+b_9+b_{10}\alpha^{-2}\ln\alpha+b_{11}\alpha^{-2}),$$
$$S_\theta(\alpha) = -\alpha_1^2 W_m^2[7b_1\alpha^6+b_2\alpha^4(5\ln\alpha+1)+5b_3\alpha^4+b_4\alpha^2\ln\alpha(3\ln\alpha+2)+b_5\alpha^2(3\ln\alpha+1)$$
$$+3b_6\alpha^2+b_7\ln\alpha(\ln\alpha+2)+b_8(\ln\alpha+1)+b_9-b_{10}\alpha^{-2}(\ln\alpha-1)-b_{11}\alpha^{-2}] \tag{33}$$

According to the above formulae (27), (20), (32), and (33), the numerical results of nonlinear bending of the annular sandwich plate for nondimensional characteristic parameter k and nondimensional inner radius α by assuming Poisson's ratio $\nu=0.3$ are represented graphically in Figs. 2-8.

Figs. 2-3 indicate relations between the nondimensional uniform pressure P and the nondimensional inner edge deflection W_m for several values of k and α, respectively. It is obvious that these curves rise monotonically. For the same value of P, the nondimensional inner edge deflection W_m of an annular sandwich plate with small α is larger, and W_m of the plate with small k is low.

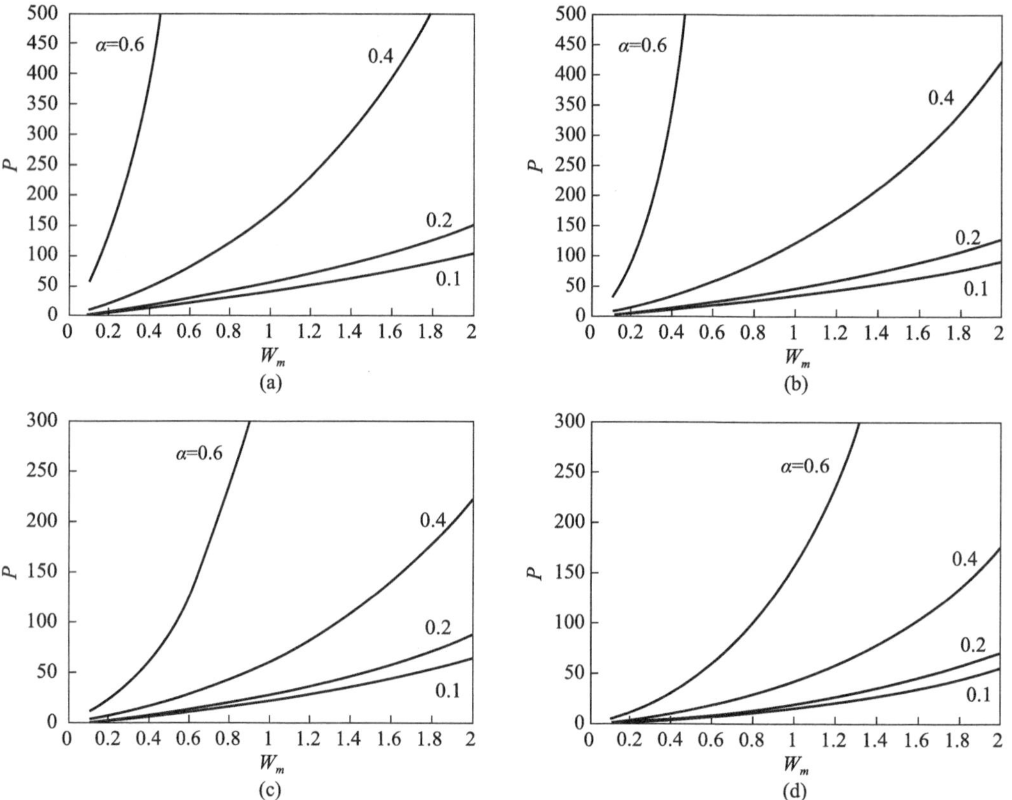

Fig. 2　Variation of the pressure P with the inner edge deflection W_m for several values of α ($\nu=0.3$).

(a) $k=0$, (b) $k=0.01$, (c) $k=0.05$, (d) $k=0.1$

The distributions of the nondimensional radial and tangential stresses S_r and S_θ along the nondimensional radius ρ in the case of $\alpha=0.2$ and $k=0.05$ are shown in Fig. 4a, b respectively. Obviously, the maximum stress of the annular sandwich plate is the radial stress and is located at the inner edge of the plate.

Fig. 5 shows the curves for the nondimensional inner edge radial stress $S_r(\alpha)$. It can be seen that the curves rise monotonically, and for the same value of the inner edge deflection, the radial stress $S_r(\alpha)$ at the inner edge induced in a plate with larger α is high.

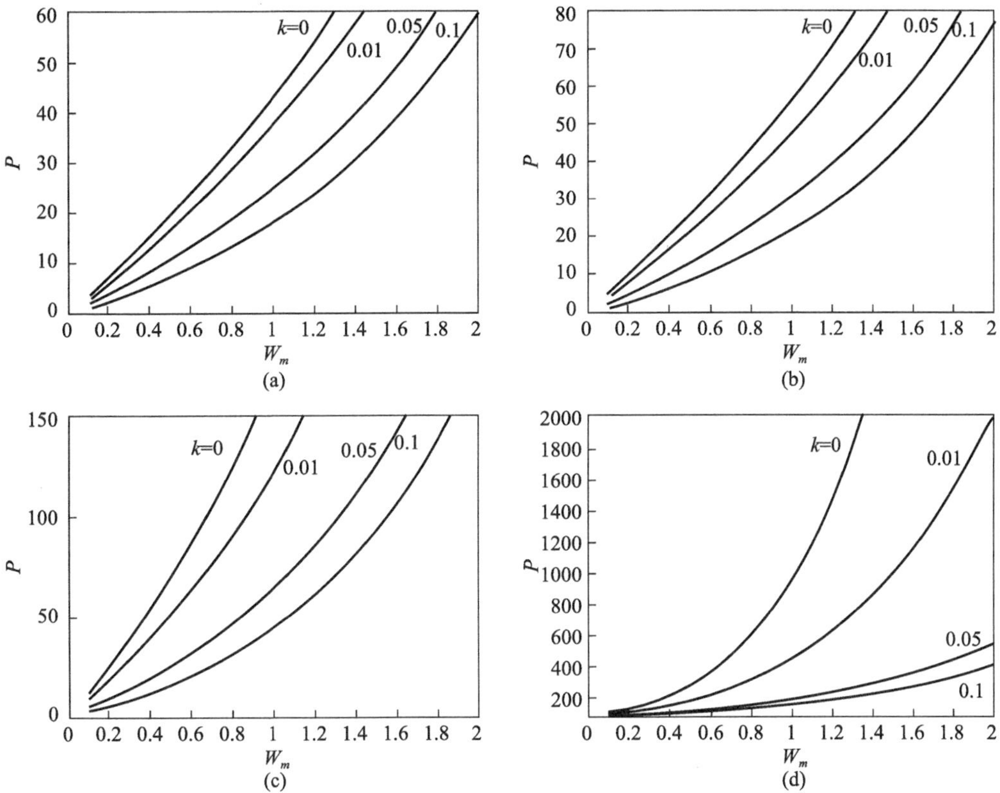

Fig. 3　Variation of the pressure P with the inner edge deflection W_m for several values of k ($\nu=0.3$).

(a) $\alpha=0.1$, (b) $\alpha=0.2$, (c) $\alpha=0.3$, (d) $\alpha=0.6$

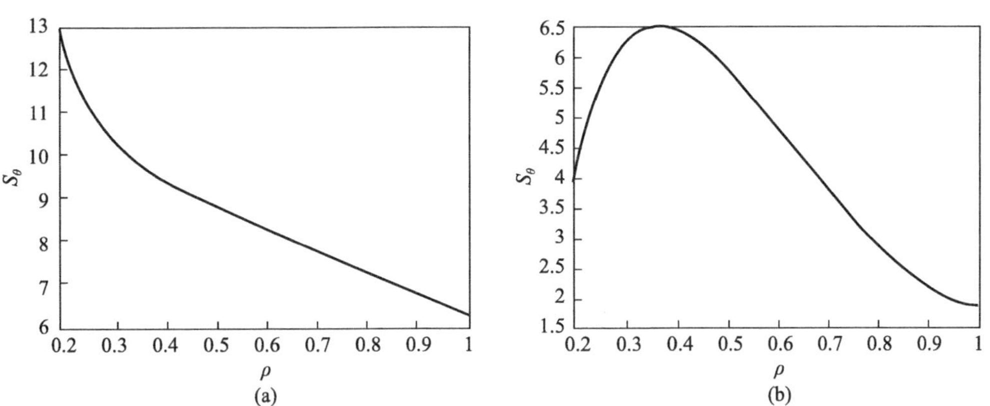

Fig. 4　The variation of the radial and tangential stress S_r, S_θ alone the radius ρ ($\nu=0.3$, $\alpha=0.2$, $k=0.05$)

(a) S_r, (b) S_θ

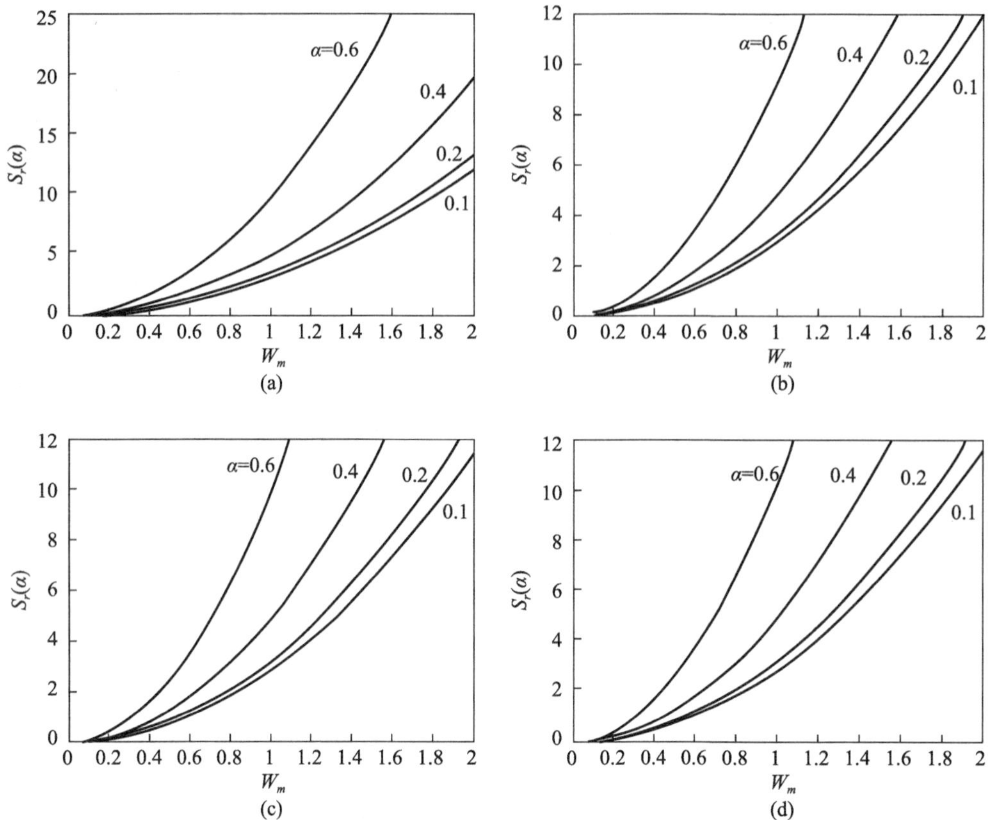

Fig. 5 The radial stress $S_r(\alpha)$ at the inner edge of the annular sandwich plate for several values of α (ν=0.3)
(a) k=0, (b) k=0.01, (c) k=0.05, (d) k=0.1

Finally, the results of numerical calculation for the stresses $S_r(1)$, $S_\theta(\alpha)$ and $S_\theta(1)$ of the annular sandwich plate for the several values of α are given in Figs. 6-8 respectively.

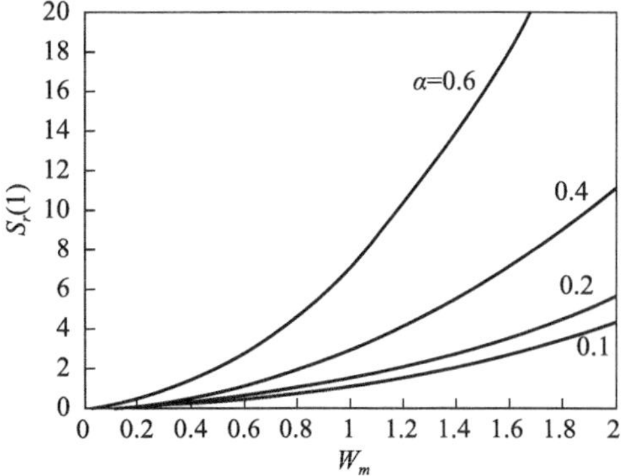

Fig. 6 The radial stress $S_r(1)$ at the outer edge of the annular sandwich plate for several values of α (ν=0.3, k=0.05)

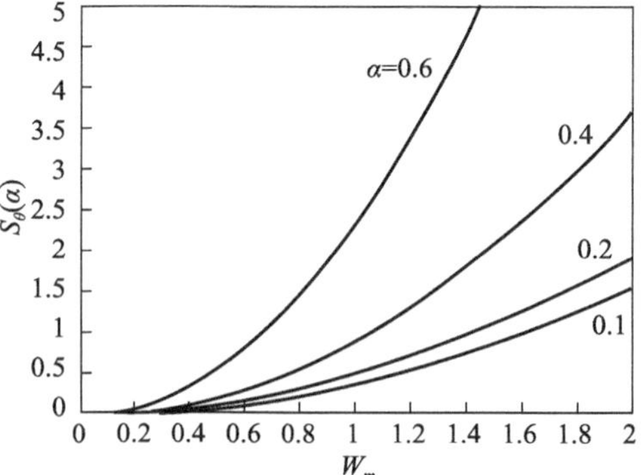

Fig. 7 The tangential stress $S_\theta(\alpha)$ at the inner edge of the annular sandwich plate for several values of α (ν=0.3, k=0.05)

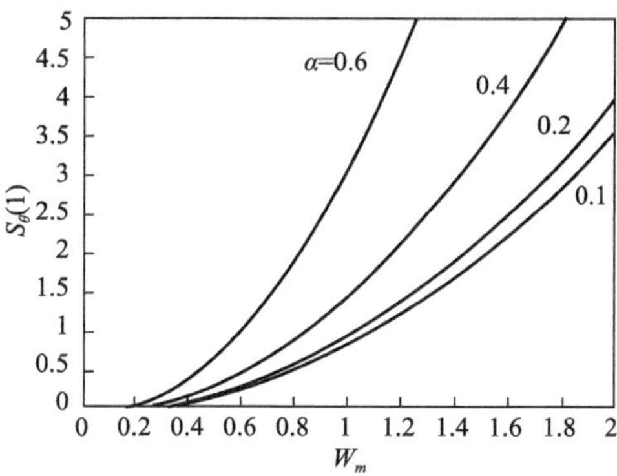

Fig. 8　The tangential stress $S_\theta(1)$ at the outer edge of the annular sandwich plate for several values of α ($\nu=0.3$, $k=0.05$)

Acknowledgements

The author would like to thank Mr. Wang Bo for his help in the numerical calculations.

References

[1] Reissner, E. Finite deflections of sandwich plates, *Journal of the Aeronautical Sciences*, 1948, 15 (7): 435-440; 1950, 17 (2): 125-130.

[2] Chinese Research Institute of Mechanics. Bending, Stability and Vibration of Sandwich Plates and Shells, Science Press, Beijing, 1972, 42-47. (in Chinese).

[3] Liu Ren-huai. Nonlinear axisymmetrical bending of circular sandwich plates under the action of uniform edge moment, *Journal of the China University of Science and Technology*, 1980, 10 (2): 56-67. (in Chinese).

[4] Yeh Kai-yuan, Liu Ren-huai, Ping Qing-yuan and Li Si-lai. Nonlinear stability of thin circular shallow spherical shell under actions of axisymmetric uniform distributed line loads, *Bulletin of Science*, 1965, 2: 142-145. (in Chinese).

[5] Yeh Kai-yuan, Liu Ren-huai, Zhang Chuan-zhe and Xu Yi-fan. Nonlinear stability of thin circular shallow spherical shell under the action of uniform edge moment, *Bulletin of Science*, 1965, 2: 145-147. (in Chinese).

[6] Liu Ren-huai. Nonlinear stability of circular shallow spherical shell with a hole in the center under the action of uniform moment at the inner edge, *Bulletin of Science*, 1965, 3: 253-255. (in Chinese).

[7] Chien Wei-zang. Large deflection of a circular clamped plate under uniform pressure, *Chinese Journal of Physics*, 1947, 7 (2): 102-107.

[8] Liu Ren-huai. Nonlinear bending of circular sandwich plates, *Applied Mathematics and Mechanics*, 1981, 2 (2): 189-208.

[9] Liu Ren-huai and Zhu Gao-qiu. Further study on large deflection of circular sandwich plates, *Applied Mathematics and Mechanics*, 1989, 10 (12): 1099-1106.

[10] Liu Ren-huai and Shi Yun-fan. Exact solution for circular sandwich plate with large deflection,

Applied Mathematics and Mechanics, 1982, 3 (1): 11-24.

[11] Liu Ren-huai. Nonlinear bending of circular sandwich plates under the action of axisymmetric uniformly distributed line loads, *Progress in Applied Mechanics (Edited by Yeh Kai-yuan)*, Martinus Nijhoff Publishers, Dordrecht, 1987, 293-321.

[12] Liu Ren-huai, Zhu Jin-fu and Zhang Xiao-guo. Nonlinear bending of annular sandwich plates, *Journal of Jinan University*, 1997, 18 (1): 1-11. (in Chinese).

[13] Xu Jia-chu, Wang Cheng and Liu Ren-huai. Nonlinear bending of annular sandwich plates with variable thickness, *Engineering Mechanics*, 2001, 18 (4): 28-37. (in Chinese).

[14] Liu Ren-huai. Study on Nonlinear Mechanics of Plates and Shells, Science *Press and Jinan University Press*, New York, Beijing and Guangzhou, 1998.

[15] Liu Ren-huai. Nonlinear Theory and Analysis of Sandwich Plates and Shells, *Jinan University Press*, Guangzhou, 2007. (in Chinese).

[16] Du Guojun and Li Huijian. Nonlinear vibration of circular sandwich plate under the uniformed load, *Applied Mathematics and Mechanics*, 2000, 21 (2): 217-226.

[17] Du Guojun and Ma Jianqing. Nonlinear vibration of circular sandwich plates under circumjacent load, *Applied Mathematics and Mechanics*, 2006, 27 (10): 1417-1424.

[18] Du Guojun and Ma Jianqing. Nonlinear vibration and buckling of circular sandwich plates under complex load, *Applied Mathematics and Mechanics*, 2007, 28 (8): 1081-1091.

[19] Du Guojun, Zhang Xiuli and Hu Yuda. Nonlinear vibration and solution stability, *Journal of Vibration and Shock*, 2007, 26 (11): 156-159. (in Chinese).

[20] Du Guojun, Hu Yuda and Zhang Xiuli. Time domain characteristic analysis of large amplitude vibration on circular sandwich plate under static load, *Engineering Mechanics*, 2008, 25 (4): 39-44. (in Chinese).

[21] Yang Jingning, Zhao Yonggang and Ma Liansheng. Numerical solution for the shooting method of the large deflection of sandwich plates, *Journal of Lanzhou Railway University*, 2001, 20 (6): 111-114. (in Chinese).

[22] Yang Jingning, Zhao Yonggang, Qin Ping and Liu Caixue. Geometrically nonlinear deformation of circular sandwich plates under transversely non-uniform temperature rise, *Key Engineering Materials*, 2007, 353-358 (2): 1161-1164.

[23] Hou Chaosheng, Zhang Shoukai and Lin Feng. Cublic spline solutions of axisymmetrical nonlinear bending and buckling of circular sandwich plates, *Applied Mathematics and Mechanics*, 2005, 26 (1): 131-138.

[24] Hou Chaosheng, Zhou Chengxiang and Yang Lijun. Axisymmetrical nonlinear bending and buckling of annular sandwich plates, *Journal of Tianjin University (Science and Technology)*, 2006, 39 (11): 1293-1298. (in Chinese).

[25] Zhang Xiuli. The effect of initial deflection in nonlinear vibration property of circular sandwich plate, *Science and Technology Information*, 2007, 29 (64): 83-84. (in Chinese).

[26] Kirichok, I. F., Karnaukhov, M. V. Single-frequency vibrations and vibrational heating of a piezoelectric circular sandwich plate under monoharmonic electromechanical loading, *International Applied Mechanics*, 2008, 44 (1): 65-72.

第六章　复合材料层合板壳非线性力学

四边简支对称正交层合矩形板的非线性弯曲问题①

一、引言

复合材料层合板壳结构元件，具有轻质高强、可设计性好等许多其他结构元件无法企及的优异性能，已经在近代工业的许多重要部门得到了广泛应用。因此，对复合材料层合板壳的力学分析便显得十分必要。

关于层合板线性问题的早期研究结果，在 Лехницкий1947 年的著作[1]中就有过介绍，但基于 von Kármán 型板理论基础上的非线性问题的较普遍的理论，则是到 20 世纪 60 年代中后期才由 Stavsky[2]，Whitney 和 Leissa[3]等建立起来。此后，一些研究者对工程上最常用的复合材料层合矩形板的非线性弯曲问题进行了不少有益的研究。Turvey 和 Wittrick[4]采用动力松弛法研究了简支斜交层合板的大挠度和后屈曲问题。Chia 和 Prabhakara[5,6]采用摄动法求解了对称正交或斜交层合板在四边夹紧固定和滑动固定边界条件下，承受均布载荷时的非线性弯曲问题。Zaghloul 和 Kennedy[7]对对称正交或斜交层合板在四边夹紧固定或简支两种边界条件下的大挠度问题进行了实验研究，并给出了相应的有限差分解。Prabhakara 和 Chia[8-10]采用了广义多重 Fourier 级数方法，研究了横向和面内载荷联合作用下的四边简支或四边滑动固定非对称各向异性层合板的非线性弯曲问题，以及简支非对称正交层合板在横向载荷或均匀边缘弯矩作用下的非线性弯曲问题。Chandra[11]使用迦辽金法分析了一个四边固定反对称层合板的非线性弯曲问题。周次青[12-14]采用奇异摄动法讨论了多种边界条件下各向异性层合板的非线性弯曲问题，导出了挠度和力函数的 N 阶一致有效渐近展开式。近年，Chia[15,16]又采用了广义多重 Fourier 级数方法研究了非对称斜交层合板在对转动具有弹性约束的边界条件下，或在固支和简支混合边界条件下的非线性弯曲问题。

本文研究了工程中常用的四边简支对称正交层合矩形板在任意横向载荷和面内载荷联合作用下的非线性弯曲问题。我们采用双重 Fourier 级数方法，将用挠度和力函数表示的 Kármán 型非线性微分方程组转化为一组非线性代数方程组，这就从理论上给出了一种获得该问题的非常精确的解析解的途径。最后，针对板承受正弦分布载荷和均布载荷的情况进行了具体的分析计算，导出了载荷与板中心挠度关系的近似公式。

---

① 本文原载《应用数学和力学》，1990，11 (9)：753-759. 作者：刘人怀，何陵辉.

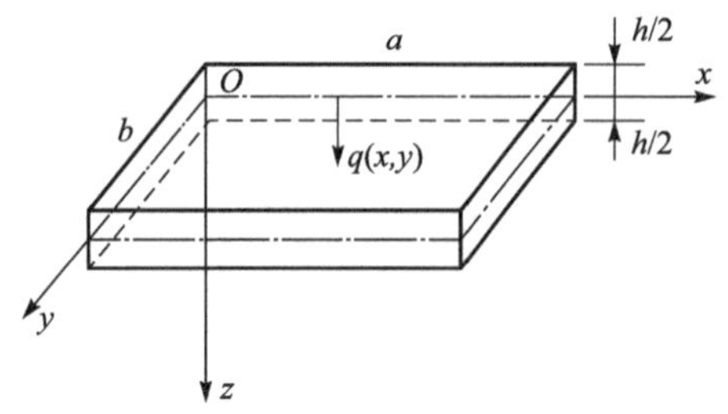

图 1 横向载荷作用下的层合矩形板

二、边值问题的建立

在笛卡儿坐标系 $O\text{-}xyz$ 中，考虑一块在 x 轴和 y 轴方向长度分别为 a 和 b，z 轴方向厚度为 h 的对称正交复合材料层合薄板（图 1）。此时板的中面与 xOy 坐标面重合。

为了简单起见，本文采用文献［17］的公式和符号。于是，此板的第 k 层材料的应力应变关系为

$$\begin{Bmatrix} \sigma_x^{(k)} \\ \sigma_y^{(k)} \\ \sigma_{xy}^{(k)} \end{Bmatrix} = \begin{bmatrix} C_{11}^{(k)} & C_{12}^{(k)} & 0 \\ C_{12}^{(k)} & C_{22}^{(k)} & 0 \\ 0 & 0 & C_{66}^{(k)} \end{bmatrix} \begin{Bmatrix} \varepsilon_x \\ \varepsilon_y \\ \varepsilon_{xy} \end{Bmatrix} \tag{2.1}$$

其中，$C_{ij}^{(k)}$ 为第 k 层材料的弹性常数。

此板的薄膜力、横向力、弯矩和扭矩分别为

$$[N_x, N_y, N_{xy}] = \int_{-h/2}^{h/2} [\sigma_x^{(k)}, \sigma_y^{(k)}, \sigma_{xy}^{(k)}] \mathrm{d}z, \tag{2.2}$$

$$[Q_s, Q_y] = \int_{-h/2}^{h/2} [\sigma_{xy}^{(k)}, \sigma_{yz}^{(k)}] \mathrm{d}z, \tag{2.3}$$

$$[M_x, M_y, M_{zy}] = \int_{-h/2}^{h/2} [\sigma_x^{(k)}, \sigma_y^{(k)}, \sigma_{xy}^{(k)}] z \mathrm{d}z \tag{2.4}$$

我们引入力函数 ψ，使得

$$N_x = \psi_{,yy}, N_y = \psi_{,xx}, N_{xy} = -\psi_{,xy} \tag{2.5a-c}$$

则此板在横向载荷 $q(x,y)$ 作用下的非线性弯曲问题的基本方程为

$$\begin{aligned} & D_{11} w_{,xxxx} + 2(D_{12} + 2D_{66}) w_{,xxyy} + D_{22} w_{,yyyy} \\ & \quad = q + w_{,xx}\psi_{,yy} + w_{,yy}\psi_{,xx} - 2w_{,xy}\psi_{,xy}, \\ & A_{22}^* \psi_{,xxxx} + (2A_{12}^* + A_{66}^*)\psi_{,xxyy} + A_{11}^*\psi_{,yyyy} = w_{,xy}^2 - w_{,xx} w_{,yy} \end{aligned} \tag{2.6a,b}$$

其中，w 为挠度，A_{ij} 和 D_{ij} 分别为板的拉压刚度矩阵和弯曲刚度矩阵的元素，A_{ij}^* 为拉压刚度矩阵的逆矩阵的元素，且有

$$A_{ij} = \int_{-h/2}^{h/2} C_{ij}^{(k)} \mathrm{d}z, \quad D_{ij} = \int_{-h/2}^{h/2} C_{ij}^{(k)} z^2 \mathrm{d}z \tag{2.7}$$

我们考虑航空工程中常用的具有刚性边框的简支板，板的边缘在变形后仍为直线，因此，其边界条件可写成

$$\begin{aligned} & w(0,y) = w(a,y) = M_x(0,y) = M_x(a,y) = 0, \\ & w(x,0) = w(x,b) = M_y(x,0) = M_y(x,b) = 0, \\ & \frac{1}{b}\int_0^b N_x \mathrm{d}y = \overline{N}_x, \quad \frac{1}{a}\int_0^a N_y \mathrm{d}x = \overline{N}_y, \\ & u(a,y) - u(0,y) = \delta_x, \quad v(x,b) - v(x,0) = \delta_y \end{aligned} \tag{2.8a-d}$$

根据式（2.4）和（2.5），这些条件可由 w 和 ψ 表示成

$$w(0,y) = w(a,y) = w_{,xx}(0,y) = w_{,xx}(a,y) = 0,$$

$$
\begin{aligned}
& w(x,0)=w(x,b)=x_{,yy}(x,0)=w_{,yy}(x,b)=0,\\
& \frac{1}{b}\int_0^b \psi_{,yy}\mathrm{d}y=\overline{N}_x,\quad \frac{1}{a}\int_0^a \psi_{,xx}\mathrm{d}x=\overline{N}_y,\\
& \int_0^a \left(A_{11}^{*}\psi_{,yy}+A_{12}^{*}\psi_{,xx}-\frac{1}{2}w_{,x}^{2}\right)\mathrm{d}x=\delta_x,\\
& \int_0^b \left(A_{12}^{*}\psi_{,yy}+A_{22}^{*}\psi_{,xx}-\frac{1}{2}w_{,y}^{2}\right)\mathrm{d}y=\delta_y
\end{aligned}
\tag{2.9a-e}
$$

其中 $\overline{N}_x$ 和 $\overline{N}_y$ 为单位长度板边受到的 x 和 y 轴方向的平均法向合力，δ_x 和 δ_y 为板在 x 轴和 y 轴方向的伸长。

于是，本文要讨论的问题就被归结为一组边值问题（2.6）和（2.9），求解这组边值问题，得出 w 和 ψ，便可由式（2.5a-c）算得 N_x，N_y 和 N_{xy}，进而由以下关系求出 M_x，M_y，M_{xy}，ε_x，ε_y 及 ε_{xy}：

$$
\begin{aligned}
& M_x=-D_{11}w_{,xx}-D_{12}w_{,yy},\\
& M_y=-D_{12}w_{,xx}-D_{22}w_{,yy},\\
& M_{xy}=-2D_{66}w_{,xy}
\end{aligned}
\tag{2.10a-c}
$$

$$
\begin{aligned}
& \varepsilon_x=A_{11}^{*}N_x+A_{12}^{*}N_y-zw_{,xx},\\
& \varepsilon_y=A_{12}^{*}N_x+A_{22}^{*}N_y-zw_{,yy},\\
& \varepsilon_{xy}=A_{66}^{*}N_{xy}-2zw_{,xy}
\end{aligned}
\tag{2.11a-c}
$$

还可由式（2.11）求出各点应力。

三、边值问题的求解

为求解边值问题（2.6）和（2.9），将 w，ψ 和 q 都展开成为双重 Fourier 级数形式

$$
\begin{aligned}
& w=\sum_{m=1}^{\infty}\sum_{n=1}^{\infty}w_{mn}\sin\frac{m\pi x}{a}\sin\frac{n\pi y}{b},\\
& \psi=\frac{\overline{N}_x}{2}y^2+\frac{\overline{N}_y}{2}x^2+\sum_{m=0}^{\infty}\sum_{n=0}^{\infty}\psi_{mn}\cos\frac{m\pi x}{a}\cos\frac{n\pi y}{b},\\
& q=\sum_{m=1}^{\infty}\sum_{n=1}^{\infty}q_{mn}\sin\frac{m\pi x}{a}\sin\frac{n\pi y}{b}
\end{aligned}
\tag{3.1a-c}
$$

其中

$$
q_{mn}=\frac{4}{ab}\int_0^b\int_0^a q(x,y)\sin\frac{m\pi x}{a}\sin\frac{n\pi y}{b}\mathrm{d}x\mathrm{d}y \tag{3.2}
$$

显然，式（3.1a，b）已满足边界条件（2.9a-c）。将其代入方程（2.6b），同时方程两边同乘以 $\cos\dfrac{i\pi x}{a}\cos\dfrac{j\pi y}{b}$ 并对 x 从 0 到 a，y 从 0 到 b 积分，得到

$$
\psi_{ij}=a_{ij}\sum_{k=1}^{\infty}\sum_{l=1}^{\infty}\sum_{r=1}^{\infty}\sum_{s=1}^{\infty}e_{klrs}^{ij}w_{kl}w_{rs} \tag{3.3}
$$

其中 i，j 不同时为零，且

$$a_{ij} = \frac{1}{4a^2b^2}\left[i^4\frac{A_{22}^*}{a^4} + i^2j^2\frac{2A_{12}^* + A_{66}^*}{a^2b^2} + j^4\frac{A_{11}^*}{b^4}\right]^{-1},$$

$$\begin{aligned} e_{klrs}^{ij} = & klrs[\delta_{i,(k+r)} + \delta_{i,|k-r|}][\delta_{j,(l+s)} + \delta_{j,|l-s|}] \\ & -\frac{1}{2}(k^2s^2 + l^2r^2)[\delta_{i,(k+r)} - \delta_{i,|k-r|}][\delta_{j,(l+s)} - \delta_{j,|l-s|}] \end{aligned} \tag{3.4}$$

$\delta_{i,j}$ 为 Kronecker 记号

$$\delta_{i,j} = \begin{cases} 1, & \text{当 } i = j \text{ 时} \\ 0, & \text{当 } i \neq j \text{ 时} \end{cases} \tag{3.5}$$

再将式（3.1a，b）代入式（2.9d，e），利用式（3.3）和（3.4）可得

$$\begin{aligned} & aA_{11}^*\overline{N}_x + aA_{12}^*\overline{N}_y - \frac{\pi^2}{8a}\sum_{m=1}^{\infty}\sum_{n=1}^{\infty} m^2 w_{mn}^2 = \delta_x, \\ & bA_{12}^*\overline{N}_x + bA_{22}^*\overline{N}_y - \frac{\pi^2}{8b}\sum_{m=1}^{\infty}\sum_{n=1}^{\infty} n^2 w_{mn}^2 = \delta_y \end{aligned} \tag{3.6}$$

或

$$\begin{aligned} \overline{N}_x &= \frac{1}{A_{12}^{*2} - A_{11}^*A_{22}^*}\left[A_{12}^*\frac{\delta_y}{b} - A_{22}^*\frac{\delta_x}{a} - \frac{\pi^2}{8}\sum_{m=1}^{\infty}\sum_{n=1}^{\infty}\left(A_{22}^*\frac{m^2}{a^2} - A_{12}^*\frac{n^2}{b^2}\right)w_{mn}^2\right], \\ \overline{N}_y &= \frac{1}{A_{12}^{*2} - A_{11}^*A_{22}^*}\left[A_{12}^*\frac{\delta_x}{a} - A_{11}^*\frac{\delta_y}{b} + \frac{\pi^2}{8}\sum_{m=1}^{\infty}\sum_{n=1}^{\infty}\left(A_{12}^*\frac{m^2}{a^2} - A_{11}^*\frac{n^2}{b^2}\right)w_{mn}^2\right] \end{aligned} \tag{3.7}$$

最后，将式（3.1a），（3.1b）和（3.1c）代入方程（2.6a），同时两边同乘以 $\sin\frac{i\pi x}{a}\cdot\sin\frac{j\pi y}{b}$，并对 x、y 分别从 0 到 a 和从 0 到 b 积分，可得

$$b_{ij}w_{ij} + \pi^2\left(i^2\frac{\overline{N}_x}{a^2} + j^2\frac{\overline{N}_y}{b^2}\right)w_{ij} + \frac{\pi^4}{2a^2b^2}\sum_{k=1}^{\infty}\sum_{l=1}^{\infty}\sum_{r=0}^{\infty}\sum_{s=0}^{\infty} f_{klrs}^{ij}w_{kl}\psi_{rs} = q_{ij} \tag{3.8}$$

其中

$$\begin{aligned} b_{ij} = & \pi^4\left[i^4\frac{D_{11}}{a^4} + 2i^2j^2\frac{D_{12} + 2D_{66}}{a^2b^2} + j^4\frac{D_{22}}{b^4}\right], \\ f_{klrs}^{ij} = & klrs[\delta_{i,(k+r)} + \delta_{i,(r-k)} - \delta_{i,(k-r)}][\delta_{j,(l+s)} + \delta_{j,(s-l)} - \delta_{j,(l-s)}] \\ & -\frac{1}{2}(k^2s^2 + l^2r^2)[\delta_{j,(k+r)} + \delta_{i,(k-r)} - \delta_{i,(r-k)}][\delta_{j,(l+s)} + \delta_{j,(l-s)} - \delta_{j,(s-l)}] \end{aligned} \tag{3.9}$$

至此，我们已将一组边值问题式（2.6）、式（2.8）及式（2.9）转化为一组非线性代数方程（3.3）、（3.7）和（3.8）。对于给定的 $q(x, y)$，截取式（3.1a，b）中级数的前有限项，并将与之相应的式（3.3）和（3.7）代入方程（3.8），便得到关于 w_{ij} 的三次方程组，方程的个数与未知数 w_{ij} 的个数相等。求得这些方程的数值解，即能由式（3.3）算出系数 ψ_{ij}，进而得到所有与板的应力和应变有关的数据。所得解的精确性，可通过引入计算的系数 w_{ij} 数目逐步增加时所得数值结果的变化情况来判断。

四、算例

例 1　正弦分布载荷作用下边缘载荷为零的正交层合矩形板的大挠度特征。

设

$$q = q_0 \sin\frac{\pi x}{a}\sin\frac{\pi y}{b} \tag{4.1}$$

取式（3.1a）级数的首项，则由式（3.2）、（3.3）和（3.8），可得 q_0 与板中心挠度 w_m 近似关系为

$$q_0 = \frac{D_{11}b^4 + 2(D_{12}+2D_{66})a^2b^2 + D_{22}a^4}{a^4b^4}\pi^4 w_m + \frac{a^4A_{11}^* + b^4A_{22}^*}{16a^4b^4A_{11}^*A_{22}^*}\pi^4 w_m^3 \tag{4.2}$$

以玻璃纤维/环氧复合材料层合正方形板为例，设其特征常数为

$a = b = 15.24\text{cm}$，　$h = 0.2438\text{cm}$，

$A_{11} = A_{22} = 3.321\times10^4\text{kgf/cm}$　$(32.55\times10^6\text{N/m})$，

$A_{12} = 0.8036\times10^4\text{kgf/cm}$　$(7.875\times10^6\text{N/m})$，

$A_{66} = 0.5357\times10^4\text{kgf/cm}$　$(5.25\times10^6\text{N/m})$，

$D_{11} = D_{22} = 1.630\times10^2\text{kgf}\cdot\text{cm}$　$(15.98\text{N}\cdot\text{m})$，

$D_{12} = 0.3908\times10^2\text{kgf}\cdot\text{cm}$　$(3.830\text{N}\cdot\text{m})$，

$D_{66} = 0.2594\times10^2\text{kgf}\cdot\text{cm}$　$(2.542\text{N}\cdot\text{m})$

则由式（4.2）可以给出板的无量纲载荷-中心挠度关系曲线，如图 2 所示。

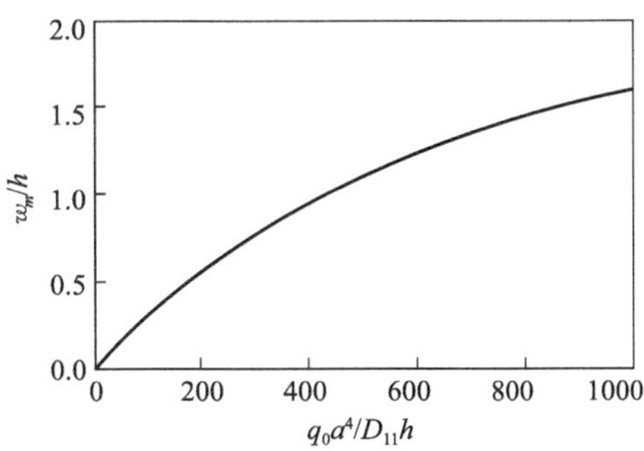

图 2　正弦载荷作用下简支层合正方形板的载荷与中心挠度关系曲线

例 2　均布载荷作用下，相对位移 δ_x 和 δ_y 为零的层合矩形板的大挠度特征。

仍取级数（3.1a）的首项，由式（3.3）、（3.7）和（3.8），即得板的载荷—中心挠度关系为

$$q = \frac{D_{11}b^4 + 2(D_{12}+2D_{66})a^2b^2 + D_{12}a^4}{16a^4b^4}\pi^6 w_m + \frac{4a^2b^2A_{11}^*A_{12}^*A_{22}^* + (a^4A_{11}^* + b^4A_{22}^*)(A_{12}^{*2} - 3A_{11}^*A_{22}^*)}{256a^4b^4A_{11}^*A_{22}^*(A_{12}^{*2} - A_{11}^*A_{22}^*)}\pi^6 w_m^3 \tag{4.3}$$

我们考虑碳纤维/环氧复合材料层合板，其弹性常数为

$E_1 = 14.08\times10^5\text{kgf/cm}^2(138.0\times10^9\text{N/m}^2)$，

$E_2 = E_3 = 1.479\times10^5\text{kgf/cm}^2(14.49\times10^9\text{N/m}^2)$，

$G_{12} = G_{23} = G_{31} = 0.5985\times10^5\text{kgf/cm}^2(5.865\times10^9\text{N/m}^2)$，

$$\nu_{12}=\nu_{23}=\nu_{31}=0.21$$

对照式（2.1）确定$C_{ij}^{(k)}$后，通过数值计算，我们得到单层和三层［0°/90°/0°］层合板的无量纲载荷—中心挠度曲线（图3）。虽然只取了w展开式的首项，但本文结果几乎与文献［12］完全一样，而且本文公式更为简单。

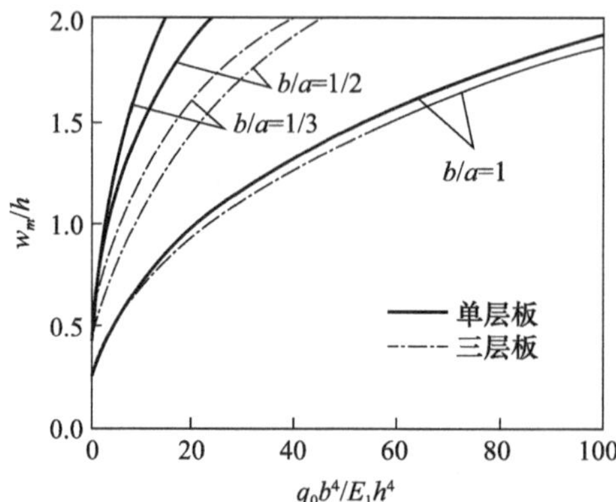

图3　均布载荷作用下简支层合矩形板的载荷与中心挠度关系曲线

参考文献

［1］ Лехницкий С Г. *Анизотропные Пластинки*. Гостехиздат，Москва（1947）. 中译本：列赫尼茨基С Г. 各向异性板. 北京：科学出版社，1963.

［2］ Stavsky Y. On the general theory of heterogeneous aeolotropic plates. Aeronaut，Q.，1964，15：29.

［3］ Whitney J M，Leissa A W. Analysis of heterogeneous anisotropic plates，J. Appl. Mech. 1969，36：261.

［4］ Turvey G J，Wittrick W H. The large deflection and postbuckling behaviour of some laminated plates. Aeronaut. Q.，1973，24：77.

［5］ Chia C Y. Large deflections of heterogeneous anisotropic rectangular plates. Int. J. Solids Struct.，1974，10：965.

［6］ Chia C Y，Prabhakara M K. Large deflection of unsymmetric cross-ply and angle-ply plates. J. Mech. Eng. Sci.，1976，18：179.

［7］ Zaghloul S A，Kennedy J B. Nonlinear behaviour of symmetrically laminated plates. J. Appl. Mech.，1975，42：234.

［8］ Prabhakara M K，Chia C Y. Finite deflections of unsymmetrically layered anisotropic rectangular plates subjected to the combined action of transverse and inplane loads. J. Appl. Mech.，1975，42：517.

［9］ Prabhakara M K，Chia C Y. Nonlinear analysis of laminated cross-ply plates. J. Eng. Mech. Div.，Proc. 1977，103（4）：749.

［10］ Prabhakara M K. Finite deflections of unsymmetric angle-ply anisotropic rectangular plates under edge moments，J. Appl. Mech.，1977，44：171.

［11］ Chandra R. Nonlinear bending of antisymmetric angle-ply laminated plates. Fib. Sci. and

Tech.，1977，10：123.

[12]　周次青. 对称正交铺设矩形叠层板的非线性弯曲. 应用数学和力学，1985，6 (9)：819.

[13]　周次青. 不对称的各向异性叠层矩形板的非线性弯曲. 应用数学和力学，1986，7 (11)：1003.

[14]　周次青. 对称铺设各向异性叠层矩形板的非线性弯曲. 应用数学和力学，1988，9 (3)：267.

[15]　Chia C Y. Nonlinear bending of unsymmetric angle-ply plates with edges elastically restrained against rotation. Acta Mech.，1984，53：201.

[16]　Chia C Y. Large deflection of unsymmetric laminates with mixed boundary conditions. Int. J. Non-Linear Mech.，1985，20：273.

[17]　Chia C Y. Nonlinear Analysis of Plates. New York：McGraw-Hill，1980.

对称圆柱正交异性层合扁球壳的非线性稳定问题①

一、引言

近年来，使用正交异性层合复合材料作为结构元件与日俱增。因此，许多研究者对这样的元件给予了极大的重视。因为它们的结构不仅十分复杂，而且其数学问题又常常是非线性的，所以这些元件的研究在理论和实验上都存在极大的困难。鉴于此，以往对薄层合壳的非线性稳定问题的研究主要集中在一些较简单的情况，例如层合圆柱壳的问题，十分遗憾，对于更加困难的薄层合复合材料扁球壳的非线性稳定问题却没有什么研究。本文将研究均布压力作用下具有夹紧固定边界的对称圆柱正交异性层合扁球壳的非线性稳定问题。我们应用叶开沅和刘人怀的修正迭代法[1-3]求解了这一问题，获得了临界载荷的解析解。这一方法综合了钱伟长的摄动法[4]和常用的逐次逼近法的优点。对于求解非线性微分方程来说，它是一种有效的、简单的、很精确的方法。本文结果可直接用于工程设计。

二、非线性边值问题

现考虑如图1所示的均布压力 q 作用下薄层合复合材料扁球壳。我们取壳体的中曲面作为坐标曲面。这里，r 是径向坐标，h 是壳体厚度，R 是中曲面的曲率半径，a 是半径，H 是拱高。每层与中曲面相距情况如图2所示。

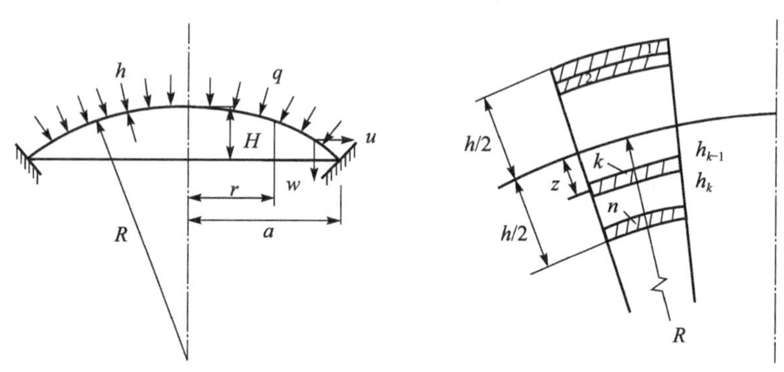

图1　扁球壳　　　　图2　与中曲面相距情况

对于大挠度分析中与壳体变形相应的基本假定为：

（1）壳体的材料是线弹性的，壳体的应力和应变服从胡克定律；

（2）挠度是与壳体厚度同阶大小的量，但与壳体其他尺寸比仍然是小量；

（3）每层的厚度和弹性常数是相同的，但各层可能有不同的厚度和弹性常数；

① 本文原载《应用数学和力学》，1991，12（3）：251-258.

(4) 壳体是对称圆柱正交异性铺设层合壳体；

(5) 壳体在变形前的中曲面法线在变形后仍是一样的；

(6) 壳体在横向是不可压缩的；

(7) 壳体在横向中的法向应力可以忽略。

应用上述假定以及轴对称条件，与壳体中曲面距离为 z 的任一点的径向、环向和轴向位移 u，v，w 为

$$u = u_0 - z\frac{\mathrm{d}w}{\mathrm{d}r}, v = 0, w = w_0 \tag{2.1}$$

其中 u_0 和 w_0 分别是中曲面上点的径向位移和挠度。

壳体的几何方程为

$$\varepsilon_r = \frac{\partial u}{\partial r} + \frac{r}{R}\frac{\mathrm{d}w}{\mathrm{d}r} + \frac{1}{2}\left(\frac{\mathrm{d}w}{\mathrm{d}r}\right)^2, \quad \varepsilon_\theta = \frac{u}{r}, \quad \gamma_{r\theta} = 0 \tag{2.2}$$

其中 ε_r，ε_θ 和 $\gamma_{r\theta}$ 是壳体中点的应变分量。

壳体 k 层的胡克定律为

$$\sigma_r^{(k)} = Q_{11}^{(k)}\varepsilon_r + Q_{12}^{(k)}\varepsilon_\theta, \quad \sigma_\theta^{(k)} = Q_{12}^{(k)}\varepsilon_r + Q_{22}^{(k)}\varepsilon_\theta, \quad \tau_{r\theta}^{(k)} = 0 \tag{2.3}$$

其中 $\sigma_r^{(k)}$，$\sigma_\theta^{(k)}$ 和 $\tau_{r\theta}^{(k)}$ 是壳体 k 层中点的应力分量，$E_r^{(k)}$ 和 $E_\theta^{(k)}$ 分别是 k 层 r 和 θ 方向的弹性模量，$\nu_{r\theta}^{(k)}$ 是 k 层 r 方向伸长时决定 θ 方向收缩的泊松比，$\nu_{\theta r}^{(k)}$ 是 k 层 θ 方向伸长时决定 r 方向收缩的泊松比，

$$\begin{aligned} &Q_{11}^{(k)} = \frac{E_r^{(k)}}{1-\nu_{r\theta}^{(k)}\nu_{\theta r}^{(k)}}, \quad Q_{12}^{(k)} = \frac{\nu_r^{(k)}E_\theta^{(k)}}{1-\nu_{r\theta}^{(k)}\nu_{\theta r}^{(k)}}, \\ &Q_{22}^{(k)} = \frac{E_\theta^{(k)}}{1-\nu_{r\theta}^{(k)}\nu_{\theta r}^{(k)}} \end{aligned} \tag{2.4}$$

而且弹性常数间满足下列关系

$$\nu_{r\theta}^{(k)}E_\theta^{(k)} = \nu_{\theta r}^{(k)}E_r^{(k)} \tag{2.5}$$

将式 (2.1) 代入式 (2.2)，得

$$\varepsilon_r = \varepsilon_r^0 + z\kappa_r, \quad \varepsilon_\theta = \varepsilon_\theta^0 + z\kappa_\theta \tag{2.6}$$

式中 ε_r^0 和 ε_θ^0 是由下式决定的中曲面的径向应变：

$$\varepsilon_r^0 = \frac{\mathrm{d}u_0}{\mathrm{d}r} + \frac{r}{R}\frac{\mathrm{d}w}{\mathrm{d}r} + \frac{1}{2}\left(\frac{\mathrm{d}w}{\mathrm{d}r}\right)^2, \quad \varepsilon_\theta^0 = \frac{u_0}{r} \tag{2.7}$$

而 κ_r 的 κ_θ 是由下式决定的中曲面的径向和环向曲率：

$$\kappa_r = -\frac{\mathrm{d}^2 w}{\mathrm{d}r^2}, \quad \kappa_\theta = -\frac{1}{r}\frac{\mathrm{d}w}{\mathrm{d}r} \tag{2.8}$$

因为在经典壳体理论中，内力为

$$\begin{aligned} &N_r = \sum_{k=1}^{n}\int_{h_{k-1}}^{h_k}\sigma_r^{(k)}\,\mathrm{d}z, \quad N_\theta = \sum_{k=1}^{n}\int_{h_{k-1}}^{h_k}\sigma_\theta^{(k)}\,\mathrm{d}z, \\ &M_r = \sum_{k=1}^{n}\int_{h_{k-1}}^{h_k}\sigma_r^{(k)}z\,\mathrm{d}z, \quad M_\theta = \sum_{k=1}^{n}\int_{h_{k-1}}^{h_k}\sigma_\theta^{(k)}z\,\mathrm{d}z, \\ &Q_r = \sum_{k=1}^{n}\int_{h_{k-1}}^{h_k}\tau_{rz}^{(k)}\,\mathrm{d}z \end{aligned} \tag{2.9}$$

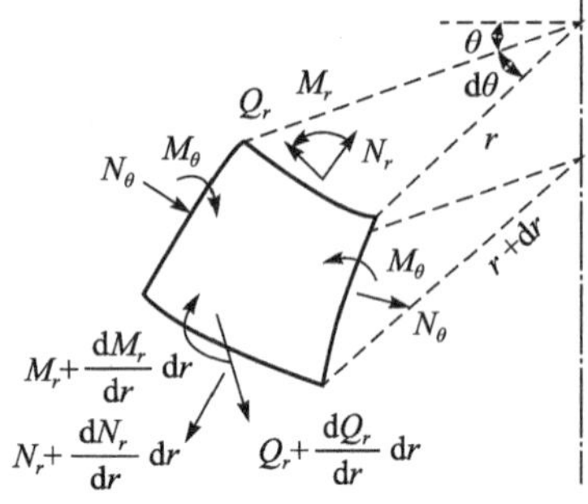

图 3　作用在壳体微元上的内力

其中 n 是层数，N_r 和 N_θ 是薄膜力，Q_r 是径向剪力，M_r 和 M_θ 是弯矩。这些内力示于图 3 中。

将式（2.3）代入式（2.9），得

$$N_r = A_{11}\varepsilon_r^0 + A_{12}\varepsilon_\theta^0,$$
$$N_\theta = A_{12}\varepsilon_r^0 + A_{22}\varepsilon_\theta^0 \tag{2.10}$$
$$M_r = D_{11}\kappa_r + D_{12}\kappa_\theta,$$
$$M_\theta = D_{12}\kappa_r + D_{22}\kappa_\theta \tag{2.11}$$

其中 A_{11}，A_{12} 和 A_{22} 是抗拉刚度，D_{11}，D_{12} 和 D_{22} 是弯曲刚度，

$$A_{11} = \sum_{k=1}^{n} Q_{11}^{(k)}(h_k - h_{k-1}),\quad A_{12} = \sum_{k=1}^{n} Q_{12}^{(k)}(h_k - h_{k-1}),$$
$$A_{22} = \sum_{k=1}^{n} Q_{22}^{(k)}(h_k - h_{k-1}) \tag{2.12}$$
$$D_{11} = \frac{1}{3}\sum_{k=1}^{n} Q_{11}^{(k)}(h_k^3 - h_{k-1}^3),\quad D_{12} = \frac{1}{3}\sum_{k=1}^{n} Q_{12}^{(k)}(h_k^3 - h_{k-1}^3),$$
$$D_{22} = \frac{1}{3}\sum_{k=1}^{n} Q_{22}^{(k)}(h_k^3 - h_{k-1}^3) \tag{2.13}$$

现在让我们讨论壳体微元的平衡（图 3）。这壳体的平衡方程组为

$$\frac{d(rN_r)}{dr} - N_\theta = 0,$$
$$\frac{d(rM_r)}{dr} - M_\theta - rQ_r = 0, \tag{2.14a-c}$$
$$\frac{d}{dr}\left[rN_r\left(\frac{r}{R} + \frac{dw}{dr}\right) + rQ_r\right] + rq = 0$$

借助式（2.10），可将应变 ε_r^0 和 ε_θ^0 写为

$$\varepsilon_r^0 = A_1 N_r - A_2 N_\theta,\quad \varepsilon_\theta^0 = A_3 N_\theta - A_2 N_r \tag{2.15}$$

其中

$$A_0 = \frac{1}{A_{11}A_{22} - A_{12}^2},$$
$$A_1 = A_0 A_{22},\quad A_2 = A_0 A_{12},\quad A_3 = A_0 A_{11} \tag{2.16}$$

由式（2.7）的两个方程消去 u_0，便得

$$\varepsilon_r^0 = \frac{d}{dr}(r\varepsilon_\theta^0) + \frac{r}{R}\frac{dw}{dr} + \frac{1}{2}\left(\frac{dw}{dr}\right)^2 \tag{2.17}$$

将式（2.15）代入式（2.17），并应用方程（2.14a），便得协调方程

$$A_3\frac{d}{dr}r\frac{d(rN_r)}{dr} - A_1 N_r + \frac{1}{2}\left(\frac{dw}{dr}\right)^2 + \frac{r}{R}\frac{dw}{dr} = 0 \tag{2.18}$$

由方程（2.14b，c）消去 Q_r，并应用式（2.8）和（2.11），得

$$\frac{d}{dr}r^{\beta_1}\frac{d}{dr}r^{1-2\beta_1}\frac{d}{dr}\left(r^{\beta_1}\frac{dw}{dr}\right) - \frac{1}{D_{11}}\frac{d}{dr}\left[rN_r\left(\frac{r}{R} + \frac{dw}{dr}\right)\right] - \frac{1}{D_{11}}rq = 0 \tag{2.19}$$

其中

$$\beta_1^2 = \frac{D_{22}}{D_{11}} \tag{2.20}$$

对 r 积分，方程（2.19）成为

$$r^{\beta_1}\frac{\mathrm{d}}{\mathrm{d}r}r^{1-2\beta_1}\frac{\mathrm{d}}{\mathrm{d}r}\left(r^{\beta_1}\frac{\mathrm{d}w}{\mathrm{d}r}\right)-\frac{1}{D_{11}}rN_r\left(\frac{r}{R}+\frac{\mathrm{d}w}{\mathrm{d}r}\right)-\frac{1}{2D_{11}}r^2q=0 \tag{2.21}$$

方程（2.18）和（2.21）便组成了均布压力 q 作用下对称圆柱正交异性层合扁球壳大挠度理论的方程组。

如果壳体沿边界夹紧固定，则相应的边界条件为

$$\begin{aligned}&\text{当 } r=a \text{ 时，} w=0，\frac{\mathrm{d}w}{\mathrm{d}r}=0，u_0=0\\&\text{当 } r=0 \text{ 时，} \frac{\mathrm{d}w}{\mathrm{d}r}=0，N_r \text{ 有限}\end{aligned} \tag{2.22}$$

其中

$$u_0 = r\left[A_3\frac{\mathrm{d}(rN_r)}{\mathrm{d}r}-A_2N_r\right] \tag{2.23}$$

控制方程（2.18）和（2.21）便在边界条件（2.22）下被求解。

为使最终的解更一般化，我们引入以下无量纲量：

$$\begin{gathered}y=\frac{r}{a},W=\frac{w}{h},\Phi=-\left(\frac{\mathrm{d}W}{\mathrm{d}y}+ky\right),S=\frac{a}{D_{11}}rN_r\\k=\frac{a^2}{Rh},P=\frac{a^4}{2D_{11}h}q,\beta_2^2=\frac{A_1}{A_3},\beta_3=\frac{h^2}{A_3D_{11}},\beta_4=\frac{A_2}{A_3}\end{gathered} \tag{2.24}$$

代入后，控制方程（2.18）、（2.21）和边界条件（2.22）转化为下述无量纲形式

$$\begin{aligned}&L_1(y^{\beta_1}\Phi)=S\Phi-Py^2+k(\beta_1^2-1)\\&L_2(y^{\beta_2}S)=-\beta_3(\Phi+ky)\left[\frac{1}{2}(\Phi+ky)-ky\right]\end{aligned} \tag{2.25a,b}$$

$$\text{当 } y=1 \text{ 时}, W=0,\Phi=-k,\frac{\mathrm{d}S}{\mathrm{d}y}-\beta_4\frac{S}{y}=0 \tag{2.26}$$

$$\text{当 } y=0 \text{ 时}, \Phi=0,S=0 \tag{2.27}$$

其中 L_1 和 L_2 是线性微分算子，

$$\begin{aligned}&L_1(\cdots)=y^{\beta_1}\frac{\mathrm{d}}{\mathrm{d}y}y^{1-2\beta_1}\frac{\mathrm{d}}{\mathrm{d}y}(\cdots),\\&L_2(\cdots)=y^{\beta_2}\frac{\mathrm{d}}{\mathrm{d}y}y^{1-2\beta_2}\frac{\mathrm{d}}{\mathrm{d}y}(\cdots)\end{aligned} \tag{2.28}$$

三、非线性边值问题的解

我们现在用修正迭代法来求解无量纲非线性边值问题（2.25）～（2.27），在第一次近似中，忽略方程（2.25a）的 $S\Phi$，便得如下线性边值问题：

$$L_1(y^{\beta_1}\Phi_1)=-Py^2+k(\beta_1^2-1),$$

$$L_2(y^{\beta_2}S_1)=-\beta_3(\Phi_1+ky)\left[\frac{1}{2}(\Phi_1+ky)-ky\right] \tag{3.1a,b}$$

当 $y=1$ 时，$W_1=0$，$\Phi_1=-k$，$\dfrac{\mathrm{d}S_1}{\mathrm{d}y}-\beta_4\dfrac{S_1}{y}=0$ (3.2a-c)

当 $y=0$ 时，$\Phi_0=0$，$S_1=0$ (3.3a，b)

应用边界条件（3.2b）和（3.3a），并积分方程（3.1a），我们得到

$$\Phi_1=-[\lambda_0P(y^{\beta_1}-y^3)+ky] \tag{3.4}$$

其中

$$\lambda_0=\frac{1}{\beta_1^2-9} \tag{3.5}$$

我们取壳体的无量纲中心挠度 W_m 作为迭代参数：

$$W_m=W|_{y=0} \tag{3.6}$$

由式（2.24）和边界条件（3.2a），无量纲中心挠度 W_m 为

$$W_m=\int_0^1(\Phi+ky)\mathrm{d}y \tag{3.7}$$

将解（3.4）代入式（3.7），便得线性特征关系式如下

$$P=\alpha_1W_m \tag{3.8}$$

其中

$$\alpha_1=4(\beta_1+1)(\beta_1+3) \tag{3.9}$$

考虑到式（3.8），便可将解（3.4）改写为

$$\Phi_1=-[\lambda_0\alpha_1W_m(y^{\beta_1}-y^3)+ky] \tag{3.10}$$

应用解（3.10），并在边界条件（3.2c）和（3.3b）下积分方程（3.1b），便得方程（3.1b）的解为

$$\begin{aligned}S_1=\beta_3\Big[&\frac{1}{2}\alpha_1^2W_m^2(b_1y^{2\beta_1+1}+b_2y^{\beta_1+4}+b_3y^7+b_4y^{\beta_2})\\&+k\alpha_1W_m(b_5y^{\beta_1+2}+b_6y^5+b_7y^{\beta_2})\Big]\end{aligned} \tag{3.11}$$

其中

$$\begin{aligned}&b_1=-\frac{\lambda_0^2}{(2\beta_1+1)^2-\beta_2^2},\quad b_2=\frac{2\lambda_0^2}{(\beta_1+4)^2-\beta_2^2},\quad b_3=\frac{\lambda_0^2}{\beta_2^2-49},\\&b_4=-\frac{1}{\beta_2-\beta_4}[b_1(2\beta_1-\beta_4+1)+b_2(\beta_1-\beta_4+4)-b_3(\beta_4-7)],\\&b_5=-\frac{\lambda_0}{(\beta_1+2)^2-\beta_2^2},\quad b_6=-\frac{\lambda_0}{\beta_2^2-25},\\&b_7=-\frac{1}{\beta_2-\beta_4}[b_5(\beta_1-\beta_4+2)-b_6(\beta_4-5)]\end{aligned} \tag{3.12}$$

在第二次近似中，由问题（2.25）～（2.27）得到如下关于 Φ 的线性边值问题

$$L_1(y^{\beta_1}\Phi_2)=S_1\Phi_1-Py^2+k(\beta_1^2-1) \tag{3.13}$$

当 $y=1$ 时，$W_2=0$，$\Phi_2=-k$ (3.14a，b)

当 $y=0$ 时，$\Phi_2=0$ (3.15)

应用解（3.10）和式（3.11），问题（3.13）、（3.14b）和（3.15）的解是

$$\begin{aligned}\Phi_2 =& -[\lambda_0 P(y^{\beta_1} - y^3) + ky] + \beta_3\Big[\frac{1}{2}\alpha_1^3 W_m^3 (c_1 y^{3\beta_1+2} + c_2 y^{2\beta_1+5} \\ &+ c_3 y^{\beta_1+8} + c_4 y^{\beta_1+\beta_2+1} + c_5 y^{\beta_2+4} + c_6 y^{11} + c_7 y^{\beta_1}) \\ &+ k\alpha_1^2 W_m^2 (c_8 y^{2\beta_1+3} + c_9 y^{\beta_1+6} + c_{10} y^{\beta_1+\beta_2+1} + c_{11} y^{\beta_2+4} \\ &+ c_{12} y^{\beta_2+2} + c_{13} y^9 + c_{14} y^{\beta_1}) + k^2 \alpha_1 W_m (c_{15} y^{\beta_1+4} \\ &+ c_{16} y^{\beta_2+2} + c_{17} y^7 + c_{18} y^{\beta_1})\Big]\end{aligned} \tag{3.16}$$

其中

$$c_1 = -\frac{\lambda_0 b_1}{4(2\beta_1+1)(\beta_1+1)}, \qquad c_2 = \frac{\lambda_0(b_1 - b_2)}{(3\beta_1+5)(\beta_1+5)},$$

$$c_3 = \frac{\lambda_0(b_2 - b_3)}{16(\beta_1+4)}, \qquad c_4 = -\frac{\lambda_0 b_4}{(2\beta_1+\beta_2+1)(\beta_2+1)},$$

$$c_5 = \frac{\lambda_0 b_4}{(\beta_2+4)^2 - \beta_1^2}, \qquad c_6 = -\frac{\lambda_0 b_3}{\beta_1^2 - 121},$$

$$c_7 = -\sum_{i=1}^{6} c_i, \qquad c_8 = -\frac{2\lambda_0 b_4 + b_1}{6(\beta_1+3)(\beta_1+1)},$$

$$c_9 = -\frac{2\lambda_0(b_6 - b_5) + b_2}{24(\beta_1+3)}, \qquad c_{10} = -\frac{\lambda_0 b_7}{(2\beta_1+\beta_2+1)(\beta_2+1)},$$

$$c_{11} = \frac{\lambda_0 b_7}{(\beta_2+4)^2 - \beta_1^2}, \qquad c_{12} = \frac{b_4}{2[\beta_1^2 - (\beta_2+2)^2]},$$

$$c_{13} = -\frac{2\lambda_0 b_6 - b_3}{2(\beta_1^2 - 81)}, \qquad c_{14} = -\sum_{i=8}^{13} c_i,$$

$$c_{15} = -\frac{b_5}{8(\beta_1+2)}, \qquad c_{16} = \frac{b_7}{\beta_1^2 - (\beta_2+2)^2},$$

$$c_{17} = \frac{b_6}{\beta_1^2 - 49}, \qquad c_{18} = -(c_{15} + c_{16} + c_{17}) \tag{3.17}$$

将解（3.16）代入式（3.7），我们得到对称圆柱正交异性层合扁球壳的非线性特征关系式

$$P = (\alpha_1 + k^2\alpha_2) W_m + k\alpha_3 W_m^2 + \alpha_4 W_m^3 \tag{3.18}$$

其中

$$\begin{aligned}\alpha_2 &= -\alpha_1^2\beta_3\left(\frac{c_{15}}{\beta_1+5} + \frac{c_{16}}{\beta_2+3} + \frac{c_{17}}{8} + \frac{c_{18}}{\beta_1+1}\right), \\ \alpha_3 &= -\alpha_1^3\beta_3\Big[\frac{c_8}{2(\beta_1+2)} + \frac{c_9}{\beta_1+7} + \frac{c_{10}}{\beta_1+\beta_2+2} + \frac{c_{11}}{\beta_2+5} \\ &\quad + \frac{c_{12}}{\beta_2+3} + \frac{c_{13}}{10} + \frac{c_{14}}{\beta_1+1}\Big], \\ \alpha_4 &= -\frac{1}{2}\alpha_1^4\beta_3\Big[\frac{c_1}{3(\beta_1+1)} + \frac{c_2}{2(\beta_1+3)} + \frac{c_3}{\beta_1+9} + \frac{c_4}{\beta_1+\beta_2+2} \\ &\quad + \frac{c_5}{\beta_2+5} + \frac{c_6}{12} + \frac{c_7}{\beta_1+1}\Big]\end{aligned} \tag{3.19}$$

为得到临界载荷，我们使用极值条件

$$\frac{\mathrm{d}P}{\mathrm{d}W_m}=0 \tag{3.20}$$

由此得

$$3\alpha_4 W_m^2+2k\alpha_3 W_m+\alpha_1+k^2\alpha_2=0 \tag{3.21}$$

解此代数方程，便得壳体失稳时的无量纲临界挠度 W_m^*

$$W_m^*=-\frac{\alpha_3 k+\sqrt{(\alpha_3^2-3\alpha_2\alpha_4)k^2-3\alpha_1\alpha_4}}{3\alpha_4} \tag{3.22}$$

将此式代入式（3.18），得如下形式的无量纲临界载荷公式：

$$P^*=(\alpha_1+k^2\alpha_2)W_m^*+k\alpha_3 W_m^{*2}+\alpha_4 W_m^{*3} \tag{3.23}$$

现在回到方程（3.21），令其判别式为零。于是，得到壳体临界几何参数 k_0

$$k_0=\sqrt{\frac{3\alpha_1\alpha_4}{\alpha_3^2-3\alpha_2\alpha_4}} \tag{3.24}$$

显然，k_0 被用来区分壳体屈曲与否。当 $k\geqslant k_0$ 时，壳体将会出现屈曲；当 $k<k_0$ 时，壳体不会发生屈曲。

四、算例

考虑一个对称圆柱正交异性层合扁球壳。为了简化计算，作为一个例子，我们考虑一特殊情况，壳体的各层具有同样的厚度和弹性常数。于是，由式（2.20）和式（2.24）可知

$$\beta_1^2=\beta_2^2=\frac{E_\theta}{E_r},\quad \beta_3=12(\beta_1^2-\beta_4^2),\quad \beta_4=\upsilon_{r\theta}\beta_1^2 \tag{4.1}$$

由式（3.23）显而易见，无量纲临界载荷 P^* 仅依赖于壳体的几何参数 k，泊松比 $\nu_{r\theta}$ 以及 θ 和 r 方向的弹性模量之比值 E_θ/E_r。这说明可作出一簇用 k 表示临界载荷 P^* 的曲线。对于 $\nu_{r\theta}$ 或 E_θ/E_r 的一个给定值，每一曲线对应着一个 E_θ/E_r 或 $\nu_{r\theta}$ 值。假定 $\nu_{r\theta}=0.25$ 和 $E_\theta/E_r=1.5$，这样的曲线分别给在图 4 和图 5 中。显然，临界载荷 P^* 的值随着 k 和 E_θ/E_r 的增加而增加，同时又随着 $\nu_{r\theta}$ 的增加而减小。

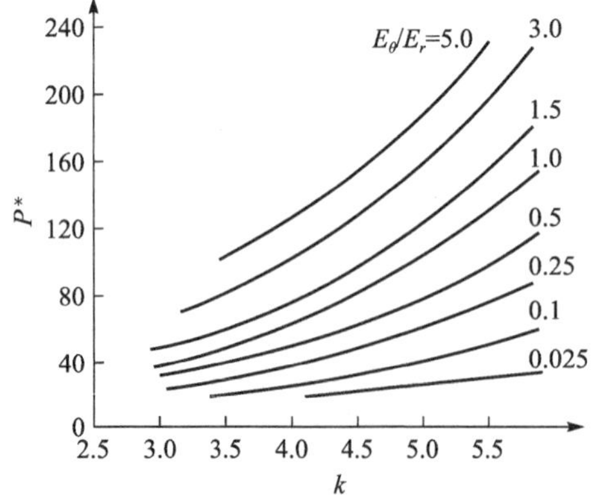

图 4　各种 E_θ/E_r 值下的稳定曲线（$\nu_{r\theta}=0.25$）

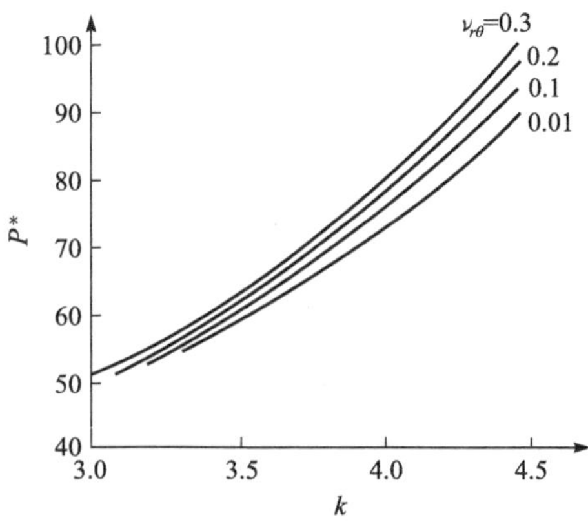

图 5　各种 $\nu_{r\theta}$ 值下的稳定曲线（$E_\theta/E_r=1.5$）

最后，我们将壳体的临界几何参数 k_0 的数值计算结果给在图 6 和图 7 中。由图 6 可知，当 $\nu_{r\theta}=0.25$ 时，在 E_θ/E_r 很小的情况下，k_0 值随着 E_θ/E_r 的增加而迅速减小，然后又随着 E_θ/E_r 的增加而增加。此外，由图 7 可看出，对于 $E_\theta/E_r=1.5$，k_0 值随着 $\nu_{r\theta}$ 的增加而减小。

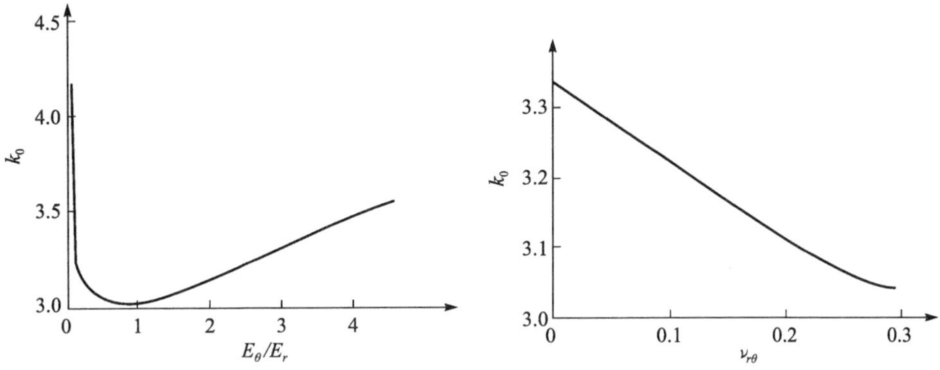

图 6　k_0 与 E_θ/E_r 的关系曲线（$\nu_{r\theta}=0.25$）　　图 7　k_0 与 $\nu_{r\theta}$ 的关系曲线（$E_\theta/E_r=1.5$）

参 考 文 献

[1]　叶开沅，刘人怀，平庆元，等. 在轴对称线布载荷作用下圆底扁球壳的非线性稳定问题. 兰州大学学报，1965 (2)：10；科学通报，1965 (2)：142.

[2]　叶开沅，刘人怀，张传智，等. 圆底扁薄球壳在边缘力矩作用下的非线性稳定问题. 科学通报，1965 (2)：145.

[3]　刘人怀. 在内边缘均布力矩作用下中心开孔圆底扁球壳的非线性稳定问题. 科学通报，1965 (3)：253.

[4]　Chien Wei-zang. Large deflection of a circular clamped plate under uniform pressure. Chinese Journal of Physics，1947，7 (2)：102.

考虑横向剪切的对称层合圆柱正交异性扁球壳的非线性稳定问题①

一、引言

众所周知，复合材料薄层合板壳在工程中应用十分广泛。因此，研究这种板壳的非线性问题有着重大的理论和实际意义，有关这种课题和文献的情况，可参阅文献[1]～[7]。但是，由于非线性数学的困难，对于相当重要的复合材料薄层合扁球壳的非线性问题，还未见研究过。

本文建立了横向剪切变形影响的对称层合圆柱正交异性扁球壳的大挠度理论，得到了这一壳体理论的基本方程。依靠文献［8］～［10］的修正迭代法，我们求解了均布压力作用下对称层合圆柱正交异性固定边扁球壳的非线性稳定问题。修正迭代法吸收文献［11］的摄动法和通常的逐次逼近法的优点，是求解非线性微分方程的一种有效、简单且较精确的方法。它已成功地在一系列板壳非线性理论的问题中得到应用，例如，薄扁球壳的非线性稳定分析[12-18]。最后，通过数值例子，讨论了横向剪切变形对临界荷载的影响。所得的解既精确又可直接用于工程设计。

二、基本方程

在讨论复合材料的薄层合扁球壳的变形和应力时，为了避免繁冗，我们采用如下的假设。

（1）材料是弹性的且服从胡克定律。壳体每层的厚度和弹性常数是相同的，但各层可以有不同的厚度和弹性常数；（2）壳体横向不可压缩；（3）壳体是对称层合圆柱正交铺层；（4）壳体的中曲面的法线在弯曲中保持为直线；（5）壳体的横向正应力忽略不计。

我们考虑如图 1 所示的在均布压力 q 作用下，厚度为 h，半径为 a，曲率半径为 R 和拱高为 H 的层合扁球壳。我们以壳体的中曲面作为坐标曲面。这里，r 是径向坐标，θ 是环向坐标。每层与中曲面的距离如图 2 所示。

应用上述假设，在轴对称条件下，与壳体中曲面相距为 z 的任一点的径向、环向和轴向位移可表达为如下形式

$$u = u_0 + z\psi, \quad v = 0, \quad w = w_0 \tag{1}$$

其中 u_0 和 w_0 分别是壳体中曲面上点的径向位移和挠度，ψ 为壳体变形前的中曲面法线的转角。

将这些表达式代入以下的壳体的几何方程：

① 本文原载《中国科学》，A 辑，1991，(7)：742-751.

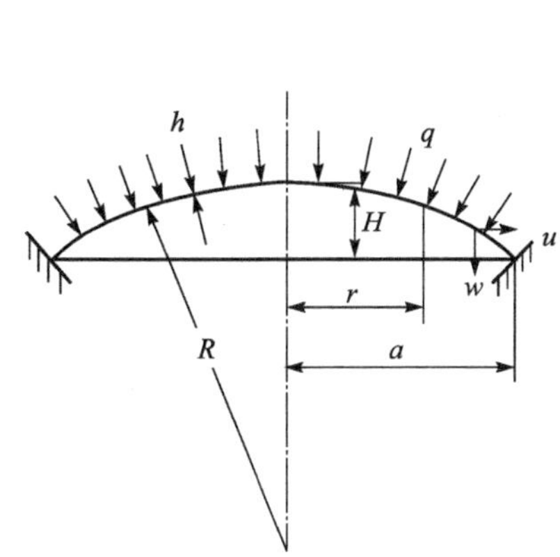

图 1　在均布压力 q 作用下的层合扁球壳

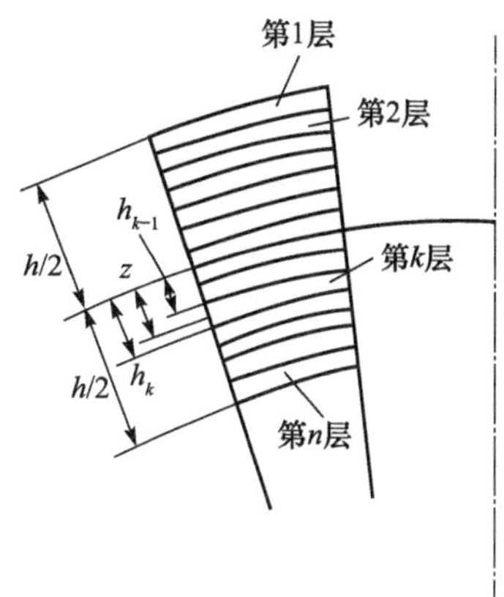

图 2　每层与中曲面的距离

$$
\begin{aligned}
&\varepsilon_r=\frac{\partial u}{\partial r}+\frac{r}{R}\frac{\mathrm{d}w}{\mathrm{d}r}+\frac{1}{2}\left(\frac{\mathrm{d}w}{\mathrm{d}r}\right)^2,\quad \varepsilon_\theta=\frac{u}{r},\\
&\gamma_{rz}=\frac{\partial u}{\partial z}+\frac{\mathrm{d}w}{\mathrm{d}r},\gamma_{r\theta}=\gamma_{\theta z}=\varepsilon_z=0
\end{aligned}
\tag{2}
$$

便得

$$
\varepsilon_r=\varepsilon_r^0+z\kappa_r,\quad \varepsilon_\theta=\varepsilon_\theta^0+z\kappa_\theta,\quad \gamma_{rz}=\psi+\frac{\mathrm{d}w}{\mathrm{d}r}
\tag{3a-c}
$$

其中 ε_r^0 和 ε_θ^0 分别是中曲面上的径向和环向应变，κ_r 和 κ_θ 分别是中曲面上的径向和环向曲率，

$$
\varepsilon_r^0=\frac{\mathrm{d}u_0}{\mathrm{d}r}+\frac{r}{R}\frac{\mathrm{d}w}{\mathrm{d}r}+\frac{1}{2}\left(\frac{\mathrm{d}w}{\mathrm{d}r}\right)^2,\quad \varepsilon_\theta^0=\frac{u_0}{r},
\tag{4}
$$

$$
\kappa_r=\frac{\mathrm{d}\psi}{\mathrm{d}r},\quad \kappa_\theta=\frac{\psi}{r}
\tag{5}
$$

在圆柱正交异性扁球壳的情况下，壳体第 k 层的胡克定律为

$$
\begin{aligned}
\sigma_r^{(k)}&=Q_{11}^{(k)}\varepsilon_r+Q_{12}^{(k)}\varepsilon_\theta,\\
\sigma_\theta^{(k)}&=Q_{12}^{(k)}\varepsilon_r+Q_{22}^{(k)}\varepsilon_\theta,\\
\tau_{rz}^{(k)}&=G_{rz}^{(k)}\gamma_{rz},\quad \tau_{r\theta}^{(k)}=\tau_{\theta z}^{(k)}=\sigma_z^{(k)}=0
\end{aligned}
\tag{6a-d}
$$

其中 $E_r^{(k)}$ 和 $E_\theta^{(k)}$ 分别是 k 层 r 和 θ 方向的弹性模量，$\nu_{r\theta}^{(k)}$ 是 k 层 r 方向伸缩时决定 θ 方向缩伸时的泊松比，$\nu_{\theta r}^{(k)}$ 是 k 层 θ 方向伸缩时决定 r 方向缩伸时的泊松比，$G_{rz}^{(k)}$ 是 k 层的决定 r 和 z 方向之间夹角变化的剪切模量，

$$
Q_{11}^{(k)}=\frac{E_r^{(k)}}{1-\nu_{r\theta}^{(k)}\nu_{\theta r}^{(k)}},\quad Q_{12}^{(k)}=\frac{\nu_{r\theta}^{(k)}E_\theta^{(k)}}{1-\nu_{r\theta}^{(k)}\nu_{\theta r}^{(k)}},\quad Q_{22}^{(k)}=\frac{E_\theta^{(k)}}{1-\nu_{r\theta}^{(k)}\nu_{\theta r}^{(k)}}
\tag{7}
$$

并且在弹性常数之间有下式成立

$$
\nu_{r\theta}^{(k)}E_\theta^{(k)}=\nu_{\theta r}^{(k)}E_r^{(k)}
\tag{8}
$$

如图 3 所示，用二相邻经线及二平行圆截取壳的一个单元体。在图中，我们给出了作用在单元体上的内力。这里，N_r 和 N_θ 分别是径向和环向薄膜力，M_r 和 M_θ 分别是径向和环向弯矩，Q_r 是径向剪力。它们由下列表达式给出：

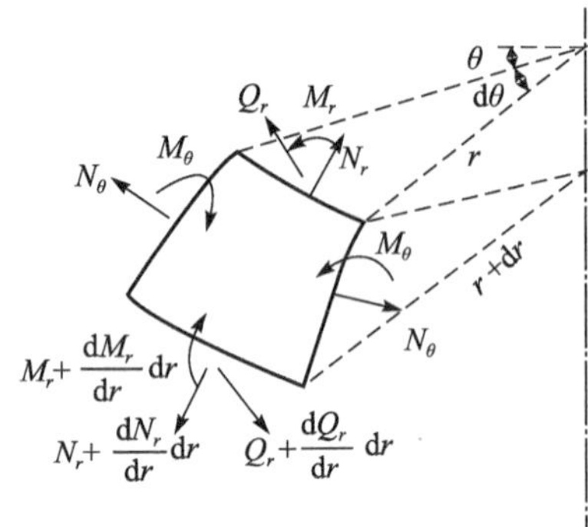

图 3 用二相邻经线及二平行圆截取壳的一个单元体

$$N_r = \sum_{k=1}^{n}\int_{h_{k-1}}^{h_k}\sigma_r^{(k)}\,\mathrm{d}z,\quad N_\theta = \sum_{k=1}^{n}\int_{h_{k-1}}^{h_k}\sigma_\theta^{(k)}\,\mathrm{d}z \tag{9}$$

$$M_r = \sum_{k=1}^{n}\int_{h_{k-1}}^{h_k}\sigma_r^{(k)}z\,\mathrm{d}z,\quad M_\theta = \sum_{k=1}^{n}\int_{h_{k-1}}^{h_k}\sigma_\theta^{(k)}z\,\mathrm{d}z \tag{10}$$

$$Q_r = \sum_{k=1}^{n}\int_{h_{k-1}}^{h_k}\tau_{rz}^{(k)}\,\mathrm{d}z \tag{11}$$

其中 n 是层数。

将式（6）代入式（9）和（10），便得

$$N_r = A_{11}\varepsilon_r^0 + A_{12}\varepsilon_\theta^0,\quad N_\theta = A_{12}\varepsilon_r^0 + A_{22}\varepsilon_\theta^0 \tag{12}$$

$$M_r = D_{11}\kappa_r + D_{12}\kappa_\theta,\quad M_\theta = D_{12}\kappa_r + D_{22}\kappa_\theta \tag{13}$$

其中 A_{ij} 和 D_{ij} 分别是抗拉刚度和弯曲刚度，

$$A_{ij} = \sum_{k=1}^{n}Q_{ij}^{(k)}(h_k - h_{k-1}),\quad D_{ij} = \frac{1}{3}\sum_{k=1}^{n}Q_{ij}^{(k)}(h_k^3 - h_{k-1}^3),\quad i,j = 1,2 \tag{14}$$

应用式（3c），式（6c）变为

$$\tau_{rz}^{(k)} = G_{rz}^{(k)}\left(\psi + \frac{\mathrm{d}w}{\mathrm{d}r}\right) \tag{15}$$

将式（15）代入式（11），可得

$$Q_r = G\left(\psi + \frac{\mathrm{d}w}{\mathrm{d}r}\right) \tag{16}$$

其中 G 是剪切刚度，

$$G = \sum_{k=1}^{n}G_{rz}^{(k)}(h_k - h_{k-1}) \tag{17}$$

由式（15）显而易见，剪应力在层间不连续，在壳体表面上不等于零，而且在每一层中，沿壳厚也不按抛物线规律分布。所以，这是不符合实际情况的。为此，我们进行一些修正，采用

$$\tau_{rz} = \frac{3Q_r}{2h}\left[1 - 4\left(\frac{z}{h}\right)^2\right] \tag{18}$$

然后，分别按式（16）和（18）计算余能。令它们相等后，便得到代替式（17）的一个更精确的关于剪切刚度 G 的公式

$$G = \frac{4h^2}{9\sum_{k=1}^{n}\frac{1}{G_{rz}^{(k)}}\left[h_k - h_{k-1} - \frac{8}{3h^2}(h_k^3 - h_{k-1}^3) + \frac{16}{5h^4}(h_k^5 - h_{k-1}^5)\right]} \tag{19}$$

由这个壳体的微元体的平衡，便得有限挠度下壳体的平衡方程为

$$\frac{\mathrm{d}(rN_r)}{\mathrm{d}r}-N_\theta=0,$$
$$\frac{\mathrm{d}(rM_r)}{\mathrm{d}r}-M_\theta-rQ_r=0, \tag{20a-c}$$
$$\frac{\mathrm{d}}{\mathrm{d}r}\left[rN_r\left(\frac{r}{R}+\frac{\mathrm{d}w}{\mathrm{d}r}\right)+rQ_r\right]+rq=0$$

由式（4）的两个方程消去 u_0，便得

$$\varepsilon_r^0=\frac{\mathrm{d}}{\mathrm{d}r}(r\,\varepsilon_\theta^0)+\frac{r}{R}\frac{\mathrm{d}w}{\mathrm{d}r}+\frac{1}{2}\left(\frac{\mathrm{d}w}{\mathrm{d}r}\right)^2 \tag{21}$$

应用式（12），将应变 ε_r^0 和 ε_θ^0 用 N_r 和 N_θ 表示，便有

$$\varepsilon_r^0=A_1N_r-A_2N_\theta,\quad \varepsilon_\theta^0=A_3N_\theta-A_2N_r \tag{22}$$

其中

$$A_0=\frac{1}{A_{11}A_{22}-A_{12}^2},\quad A_1=A_0A_{22},$$
$$A_2=A_0A_{12},\qquad A_3=A_0A_{11} \tag{23}$$

将式（22）代入（21），就得到协调方程

$$\frac{\mathrm{d}}{\mathrm{d}r}[r(A_3N_\theta-A_2N_r)]-A_1N_r+A_2N_\theta+\frac{1}{2}\left(\frac{\mathrm{d}w}{\mathrm{d}r}\right)^2+\frac{r}{R}\frac{\mathrm{d}w}{\mathrm{d}r}=0 \tag{24}$$

应用式（5）、（13）和（16），方程（20b）成为

$$D_{11}\frac{\mathrm{d}}{\mathrm{d}r}r\frac{\mathrm{d}\psi}{\mathrm{d}r}-\left(Gr+\frac{D_{22}}{r}\right)\psi-Gr\frac{\mathrm{d}w}{\mathrm{d}r}=0 \tag{25}$$

将式（16）代入方程（20c），然后用 dr 乘这个方程并积分，就得到

$$Gr\frac{\mathrm{d}w}{\mathrm{d}r}+rN_r\left(\frac{r}{R}+\frac{\mathrm{d}w}{\mathrm{d}r}\right)+Gr\psi+\frac{1}{2}qr^2=0 \tag{26}$$

由方程（20a）得

$$N_\theta=\frac{\mathrm{d}(rN_r)}{\mathrm{d}r} \tag{27}$$

将此式代入方程（24），便得

$$A_3\frac{\mathrm{d}}{\mathrm{d}r}r\frac{\mathrm{d}(rN_r)}{\mathrm{d}r}-A_1N_r+\frac{1}{2}\left(\frac{\mathrm{d}w}{\mathrm{d}r}\right)^2+\frac{r}{R}\frac{\mathrm{d}w}{\mathrm{d}r}=0 \tag{28}$$

于是，在均布压力 q 的作用下，考虑横向剪切的对称层合圆柱正交异性扁球壳的非线性稳定问题便归结为求解三个方程（25）、（26）和（28）。方程组（25）、（26）和（28）将在以下通常的夹紧固定边界条件下求解：

当 $r=a$ 时，　$w=0$，　$\psi=0$，　$u_0=0$；

当 $r=0$ 时，　$\psi=0$，　N_r 有限　　(29)

其中

$$u_0=r\left[A_3\frac{\mathrm{d}(rN_r)}{\mathrm{d}r}-A_2N_r\right] \tag{30}$$

为了简化以后的计算，我们引入下列无量纲量：

$$y=\frac{r}{a},\qquad W=\frac{w}{h},\qquad \Phi=ky+\frac{\mathrm{d}w}{\mathrm{d}y},\qquad \Psi=\frac{a}{h}\psi,$$

$$S=\frac{a}{D_{11}}rN_r,\quad P=\frac{a^4}{2D_{11}h}q,\quad k=\frac{a^2}{Rh},\quad m=\frac{a^2}{D_{11}}G,\tag{31}$$

$$\beta_1^2=\frac{D_{22}}{D_{11}},\quad \beta_2^2=\frac{A_1}{A_3},\quad \beta_3=\frac{h^2}{A_2D_{11}},\quad \beta_4=\frac{A_2}{A_3}$$

借助这些量，非线性边值问题（25）、（26）、（28）和（29）化简为如下的形式：

$$\begin{aligned}&my\Phi=-[S\Phi+my\Phi+(P-km)y^2],\\&L_1(y^{\beta_1}\Psi)=my(\Psi+\Phi-ky),\\&L_2(y^{\beta_2}S)=-\beta_3(\Phi-ky)\left[\frac{1}{2}(\Phi-ky)+ky\right]\end{aligned}\tag{32a-c}$$

$$\begin{aligned}&\text{当 } y=1\text{ 时},\quad W=0,\quad \Psi=0,\quad \frac{\mathrm{d}S}{\mathrm{d}y}-\beta_4\frac{S}{y}=0;\\&\text{当 } y=0\text{ 时},\quad \Psi=0,\quad S=0\end{aligned}\tag{33}$$

其中

$$L_1(\cdots)=y^{\beta_1}\frac{\mathrm{d}}{\mathrm{d}y}y^{1-2\beta_1}\frac{\mathrm{d}}{\mathrm{d}y}(\cdots),\quad L_2(\cdots)=y^{\beta_2}\frac{\mathrm{d}}{\mathrm{d}y}y^{1-2\beta_2}\frac{\mathrm{d}}{\mathrm{d}y}(\cdots)\tag{34}$$

三、解析解

我们使用修正迭代法求解无量纲非线性边值问题（32）和（33）。此时，选取无量纲中心挠度

$$W_m=-\int_0^1(\Phi-ky)\mathrm{d}y\tag{35}$$

作为迭代参数。

对于第一次近似，忽略方程（32a）的非线性项 $S\Phi$，便得下述的线性边值问题：

$$\begin{aligned}&my\Phi_1=-[my\Psi_1+(P-km)y^2],\\&L_1(y^{\beta_1}\Psi_1)=my(\Psi_1+\Phi_1-ky),\\&L_2(y^{\beta_2}S_1)=-\beta_3(\Phi_1-ky)\left[\frac{1}{2}(\Phi_1-ky)+ky\right]\end{aligned}\tag{36}$$

$$\begin{aligned}&\text{当 } y=1\text{ 时},\quad W_1=0,\quad \Psi_1=0,\quad \frac{\mathrm{d}S_1}{\mathrm{d}y}-\beta_4\frac{S_1}{y}=0;\\&\text{当 } y=0\text{ 时},\quad \Psi_1=0,\quad S_1=0\end{aligned}\tag{37}$$

求解问题（36）和（37），并应用式（35），便得一次近似解

$$\Phi_1=f_1(y,k,W_m),\quad \Psi_1=f_2(y,W_m),\quad S_1=f_3(y,k,W_m)\tag{38}$$

对于第二次近似，有下述线性边值问题：

$$\begin{aligned}&my\Phi_2=-[S_1\Phi_1+my\Psi_2+(P-km)y^2],\\&L_1(y^{\beta_1}\Psi_2)=my(\Psi_2+\Phi_2-ky)\end{aligned}\tag{39}$$

$$\begin{aligned}&\text{当 } y=1\text{ 时},\quad W_2=0,\quad \Psi_2=0;\\&\text{当 } y=0\text{ 时},\quad \Psi_2=0\end{aligned}\tag{40}$$

求解问题（39）和（40），可得二次近似解 Φ_2 和 Ψ_2，应用式（35），得到对称层合圆柱正交异性扁球壳的非线性特征关系式

$$P=(\alpha_1+k^2\alpha_2)W_m+k\alpha_3W_m^2+\alpha_4W_m^3 \tag{41}$$

其中

$$\alpha_1=\frac{4m(\beta_1+3)(\beta_1+1)}{2(\beta_1+3)(\beta_1+1)+m},$$

$$\alpha_2=-\alpha_1^2\beta_3\Big[\frac{c_{25}}{\beta_1+5}+\frac{c_{26}}{\beta_2+3}+\frac{c_{27}}{8}+\frac{c_{28}}{6}+\frac{c_{29}}{\beta_1+1}+\frac{1}{m}\Big(\frac{b_8}{\beta_1+3}+\frac{b_9}{6}+\frac{b_{10}}{4}+\frac{b_{11}}{\beta_2+1}\Big)\Big],$$

$$\begin{aligned}\alpha_3=&-\alpha_1^3\beta_3\Big\{\frac{c_{15}}{2(\beta_1+2)}+\frac{c_{16}}{\beta_1+7}+\frac{c_{17}}{\beta_1+5}+\frac{c_{18}}{\beta_1+\beta_2+2}+\frac{c_{19}}{\beta_2+5}\\&+\frac{c_{20}}{\beta_2+3}+\frac{c_{21}}{10}+\frac{c_{22}}{8}+\frac{c_{23}}{6}+\frac{c_{24}}{\beta_1+1}+\frac{1}{m}\Big[\frac{1}{2(\beta_1+1)}\Big(\lambda_0b_8+\frac{b_1}{2}\Big)\\&+\frac{1}{\beta_1+5}\Big(\lambda_0b_9-\lambda_0b_8+\frac{b_2}{2}\Big)+\frac{1}{\beta_1+3}\Big(\lambda_0b_{10}-\frac{b_8}{m}+\frac{b_3}{2}\Big)\\&+\frac{\lambda_0b_{11}}{\beta_1+\beta_2}-\frac{\lambda_0b_{11}}{\beta_2+3}-\frac{1}{\beta_2+1}\Big(\frac{b_{11}}{m}-\frac{b_7}{2}\Big)-\frac{1}{8}\Big(\lambda_0b_9-\frac{b_4}{2}\Big)\\&-\frac{1}{6}\Big(\lambda_0b_{10}+\frac{b_9}{m}-\frac{b_5}{2}\Big)-\frac{1}{4}\Big(\frac{b_{10}}{m}-\frac{b_6}{2}\Big)\Big]\Big\},\end{aligned}$$

$$\begin{aligned}\alpha_4=&-\frac{1}{2}\alpha_1^4\beta_3\Big\{\frac{c_1}{3(\beta_1+1)}+\frac{c_2}{2(\beta_1+3)}+\frac{c_3}{2(\beta_1+2)}+\frac{c_4}{\beta_1+9}\\&+\frac{c_5}{\beta_1+7}+\frac{c_6}{\beta_1+5}+\frac{c_7}{\beta_1+\beta_2+2}+\frac{c_8}{\beta_2+5}+\frac{c_9}{\beta_2+3}+\frac{c_{10}}{2}\\&+\frac{c_{11}}{10}+\frac{c_{12}}{8}+\frac{c_{13}}{6}+\frac{c_{14}}{\beta_1+1}+\frac{1}{m}\Big[\frac{\lambda_0b_1}{3\beta_1+1}+\frac{\lambda_0(b_2-b_1)}{2(\beta_1+2)}\\&+\frac{1}{2(\beta_1+1)}\Big(\lambda_0b_3-\frac{b_1}{m}\Big)+\frac{\lambda_0(b_4-b_2)}{\beta_1+7}+\frac{1}{\beta_1+5}\Big(\lambda_0b_5-\lambda_0b_3-\frac{b_2}{m}\Big)\\&+\frac{1}{\beta_1+3}\Big(\lambda_0b_6-\frac{b_3}{m}\Big)+\lambda_0b_7\Big(\frac{1}{\beta_1+\beta_2}-\frac{1}{\beta_2+3}\Big)-\frac{b_7}{m(\beta_2+1)}\\&-\frac{1}{10}\lambda_0b_4-\frac{1}{8}\Big(\lambda_0b_5+\frac{b_4}{m}\Big)-\frac{1}{6}\Big(\lambda_0b_6+\frac{b_5}{m}\Big)-\frac{b_6}{4m}\Big]\Big\};\end{aligned}$$

$$b_1=-\frac{\lambda_0^2}{(2\beta_1+1)^2-\beta_2^2},\qquad b_2=\frac{2\lambda_0^2}{(\beta_1+4)^2-\beta_2^2},$$

$$b_3=-\frac{2\lambda_0}{m[(\beta_1+2)^2-\beta_2^2]},\qquad b_4=\frac{\lambda_0^2}{\beta_2^2-49},$$

$$b_5=\frac{2\lambda_0}{m(\beta_2^2-25)},\qquad b_6=\frac{1}{m^2(\beta_2^2-9)},$$

$$\begin{aligned}b_7=&-\frac{1}{\beta_2-\beta_4}[b_1(2\beta_1-\beta_4+1)+b_2(\beta_1-\beta_4+4)+b_3(\beta_1-\beta_4+2)\\&-b_4(\beta_4-7)-b_5(\beta_4-5)-b_6(\beta_4-3)],\end{aligned}$$

$$b_8=-\frac{\lambda_0}{(\beta_1+2)^2-\beta_2^2},\quad b_9=-\frac{\lambda_0}{\beta_2^2-25},\quad b_{10}=-\frac{1}{m(\beta_2^2-9)},$$

$$b_{11}=-\frac{1}{\beta_2-\beta_4}[b_8(\beta_1-\beta_4+2)-b_9(\beta_4-5)-b_{10}(\beta_4-3)];$$

$$c_1=-\frac{\lambda_0 b_1}{4(2\beta_1+1)(\beta_1+1)},\qquad c_2=-\frac{\lambda_0(b_2-b_1)}{(3\beta_1+5)(\beta_1+5)},$$

$$c_3=-\frac{\lambda_0 b_3 m-b_1}{3m(\beta_1+3)(\beta_1+1)},\qquad c_4=-\frac{\lambda_0(b_4-b_2)}{16(\beta_1+4)},$$

$$c_5=-\frac{\lambda_0 m(b_5-b_3)-b_2}{12m(\beta_1+3)},\qquad c_6=-\frac{\lambda_0 b_6 m-b_3}{8m(\beta_1+2)},$$

$$c_7=-\frac{\lambda_0 b_7}{(2\beta_1+\beta_2+1)(\beta_2+1)},\qquad c_8=-\frac{\lambda_0 b_7}{\beta_1^2-(\beta_2+4)^2},$$

$$c_9=-\frac{b_7}{m[\beta_1^2-(\beta_2+2)^2]},\qquad c_{10}=-\frac{\lambda_0 b_4}{\beta_1^2-121},$$

$$c_{11}=-\frac{\lambda_0 b_5 m+b_4}{m(\beta_1^2-81)},\qquad c_{12}=-\frac{\lambda_0 b_6 m+b_5}{m(\beta_1^2-49)},$$

$$c_{13}=-\frac{b_6}{m(\beta_1^2-25)},\qquad c_{14}=-\sum_{i=1}^{13}c_i,$$

$$c_{15}=-\frac{2\lambda_0 b_8+b_1}{6(\beta_1+3)(\beta_1+1)},\qquad c_{16}=-\frac{2\lambda_0(b_9-b_8)+b_2}{24(\beta_1+3)},$$

$$c_{17}=-\frac{2m\lambda_0 b_{10}-2b_8+b_3 m}{16m(\beta_1+2)},\qquad c_{18}=-\frac{\lambda_0 b_{11}}{(2\beta_1+\beta_2+1)(\beta_2+1)},$$

$$c_{19}=-\frac{\lambda_0 b_{11}}{\beta_1^2-(\beta_2+4)^2},\qquad c_{20}=-\frac{2b_{11}-b_7 m}{2m[\beta_1^2-(\beta_2+2)^2]},$$

$$c_{21}=-\frac{2\lambda_0 b_9-b_4}{2(\beta_1^2-81)},\qquad c_{22}=-\frac{2\lambda_0 b_{10}m+2b_9-b_5 m}{2m(\beta_1^2-49)},$$

$$c_{23}=-\frac{2b_{10}-b_6 m}{2m(\beta_1^2-25)},\qquad c_{24}=-\sum_{i=15}^{23}c_i,$$

$$c_{25}=-\frac{b_8}{8(\beta_1+2)},\qquad c_{26}=-\frac{b_{11}}{\beta_1^2-(\beta_2+2)^2},$$

$$c_{27}=\frac{b_9}{\beta_1^2-49},\qquad c_{28}=\frac{b_{10}}{\beta_1^2-25},$$

$$c_{29}=-\sum_{i=25}^{28}c_i,\qquad \lambda_0=\frac{1}{\beta_1^2-9} \tag{42}$$

为了得到壳体的临界荷载 P^*，我们应用下面的极值条件：

$$\frac{\mathrm{d}P}{\mathrm{d}W_m}=0 \tag{43}$$

于是，得到屈曲发生时的壳体无量纲临界中心挠度

$$W_m^*=-\frac{\alpha_3 k+\sqrt{(\alpha_3^2-3\alpha_2\alpha_4)k^2-3\alpha_1\alpha_4}}{3\alpha_4} \tag{44}$$

将 W_m^* 代入式（41），得无量纲临界荷载

$$P^* = (\alpha_1 + k^2\alpha_2)W_m^* + k\alpha_3 W_m^{*2} + \alpha_4 W_m^{*3} \tag{45}$$

令二次方程（43）的判别式为零，可得用来区分壳体屈曲与否的临界几何参数 k_0 的公式

$$k_0 = \sqrt{\frac{3\alpha_1\alpha_4}{\alpha_3^2 - 3\alpha_2\alpha_4}} \tag{46}$$

特别地，在式（45）中的剪切刚度 G 趋近于无限大，即 $m\to\infty$ 的情况下，还可得到对称层合圆柱正交异性扁球壳的无量纲临界荷载的公式。此时，横向剪切对弯曲的影响已完全被忽略。

四、数值例子

今考虑下面两个例子。

例 1　对称层合圆柱正交异性扁球壳。

为使计算简单，假设壳体各层具有相同的厚度和弹性常数，且有

$$\frac{E_\theta}{E_r} = 1.5, \quad \nu_{r\theta} = 0.2$$

在这样的情况下，可得

$$\beta_1^2 = \beta_2^2 = \frac{E_\theta}{E_r} = 1.5,$$
$$\beta_3 = 12(\beta_1^2 - \beta_4^2) = 16.92,$$
$$\beta_4 = \frac{\nu_{r\theta}E_\theta}{E_r} = 0.3$$

按照式（45），我们计算了临界荷载，其数值结果给在图 4 中。由图可见，稳定曲线单调上升，壳体的临界荷载随着无量纲剪切刚度 m 的增加而增加，且在 m 趋近于无限大时，临界荷载将取最大值。

应用式（45）和（46），还可得到如图 5 和图 6 所示的 k_0 和 m 以及 P_0^* 和 k_0 的关系曲线。显而易见，临界几何参数 k_0 随着无量纲剪切刚度 m 的增加而增加，且当 m 为无限大时，相应的 k_0 为极大值：

$$k_0 = 3.122$$

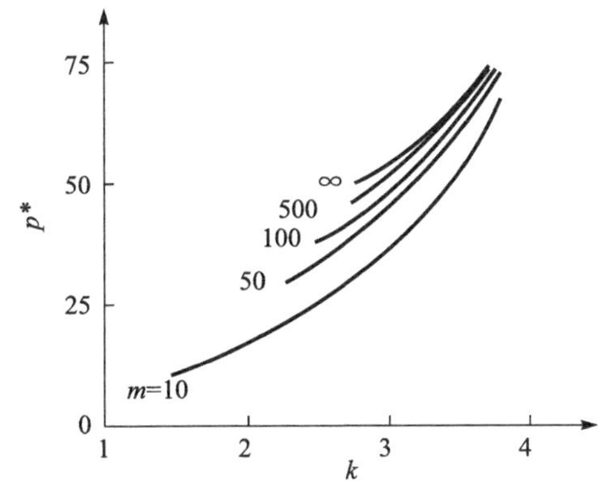

图 4　剪切刚度对临界荷载的影响

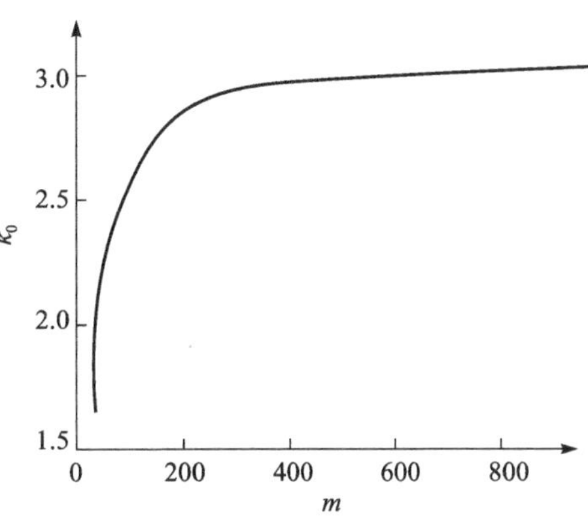

图 5　k_0-m 曲线

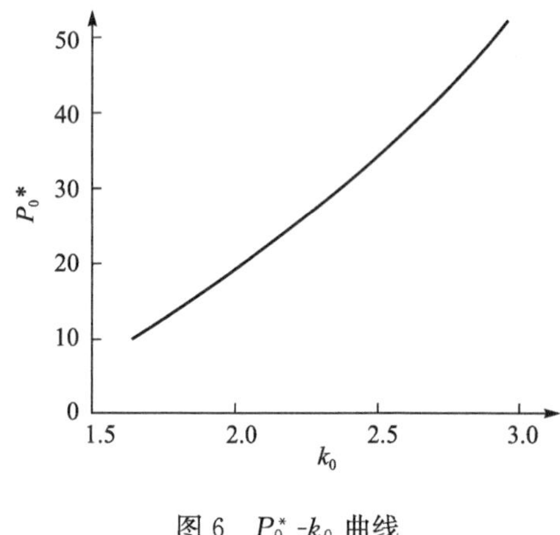

图 6 P_0^*-k_0 曲线

例 2 圆柱正交异性扁球壳。

为了便于比较，我们考虑一种特殊情况，即不计及横向剪切的单层圆柱正交异性扁球壳。假设 $n=1$ 和 $\nu_{\theta r}=\frac{1}{3}$。然后应用式（45），我们得到有用的结果。图 7 给出了各向同性（$E_\theta/E_r=1$）和圆柱正交异性扁球壳的稳定曲线，用实线表示。为了比较，也在图 7 中给出了 Varadan[19] 和 Hyman[20] 的结果，由图看出，本文结果与他们的结果是一致的。此外，在图 8 中，还给出了本文的单层各向同性扁球壳（$\nu_{\theta r}=0.3$）的数值结果以及文献［21］，［22］的实验和理论值。由图上看到，我们的理论结果与实验值吻合。根据我们的分析，由于实验时边界条件的影响，实验值偏低了一些。

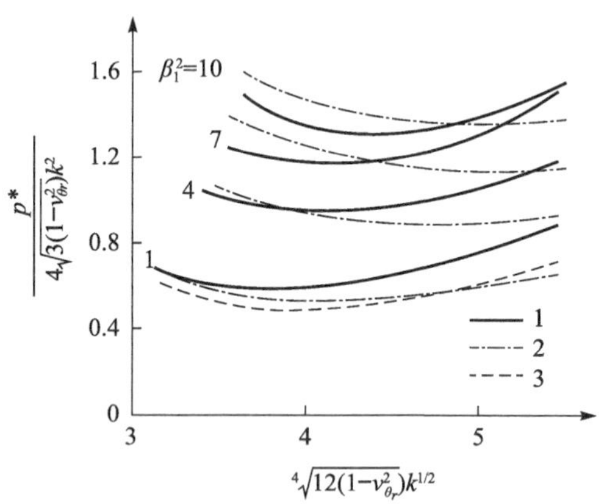

图 7 理论结果之间的比较$\left(\nu_{\theta r}=\frac{1}{3}\right)$

（1—本文结果，2—Varadan[19]，3—Hyman[20]）

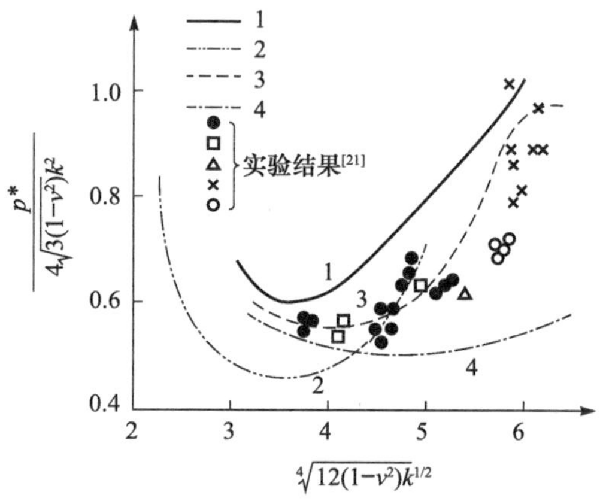

图 8 理论和实验结果之间的比较

（1—本文结果，2—Kaplan 和 Fung[22]，3—Budiansky 和 Huang[22]，4—von Willich[22]. $\nu=\nu_{\theta r}=0.3$；$E=E_\theta=E_r$）

本文主要部分是在加拿大卡尔加里大学土木工程系访问期间完成的。作者对卡尔加里大学贾春元教授的热情接待和对本文的有益讨论；上海市应用数学和力学研究所吴建成博士研究生在数值计算方面给予的协助一并表示谢意。

参考文献

［1］ Chia, C. Y. Non-Linear Analysis of Plates. McGraw-Hill, New York (1980).

［2］ Chia, C. Y. Nonlinear vibration and postbuckling of unsymmetrically laminated imperfect shallow cylindrical panels with maxed boundary conditions resting on elastic foundation. Int. J. Eng. Sci., 25, 427 (1987).

[3] Chia, C. Y. Nonlinear free vibration and post-buckling of symmetrically laminated orthotropic imperfect shallow cylindrical panels with two adjacent edges simply supported and the orther edges clamped. Int. J. Solids Struct, 23, 1123 (1987).

[4] Chia, C. Y. Nonlinear analysis of doubly curved symmetrically laminated shallow shells with rectangular planform. Ingenieur-Archiv, 58, 252 (1988).

[5] Lu, V. P. & Chia, C. Y. Nonlinear vibration and postbuckling of unsymmetric cross-ply circular cylindrical shells. International Journal of Solids and Structures, 24, 195 (1988).

[6] Saigal, S. et al. Geometrically nonlinear finite element analysis of imperfect laminated shells. J. Composite Materials, 20, 197 (1986).

[7] Zhang, Y. & Mattheus, F. L. Postbuckling beharioure of curred panels of generally layered composite materials. J. Composite Structures, 1, 115 (1983).

[8] 叶开沅，刘人怀，平庆元，等. 在对称线布载荷作用下的圆底扁薄球壳的非线性稳定性. 兰州大学学报，(2)，10 (1965)；科学通报，(2)，142 (1965).

[9] 叶开沅，刘人怀，张传智，等. 圆底扁薄球壳在边缘力矩作用下的非线性稳定问题. 科学通报，(2)，145 (1965).

[10] 刘人怀. 在内边缘均布力矩作用下中心开孔圆底扁薄球壳的非线性稳定问题. 科学通报，(3)，253 (1965).

[11] Chien Weizang, Large deflection of a circular clamped plate under uniform pressure. Chinese Journal of Physics, 7 (2), 102 (1947).

[12] 刘人怀. 在边缘载荷作用下中心开孔圆底扁薄球壳的轴对称稳定性. 力学学报，(3)，206 (1977).

[13] Yeh Kaiyuan & Liu Ren-huai et al. Nonlinear stability of thin elastic circular shallow spherical shell under the action of uniform edge moment. Applied Mathematics and Mechanics, 1 (1), 71 (1980).

[14] 刘人怀. 双层金属中心开孔扁球壳的非线性热稳定问题. 中国科学技术大学学报，11 (1)，84 (1981).

[15] Liu Renhuai. Non-linear thermal stability of bimetallic shallow shells of revolution. International Journal of Non-Linear Mechanics, 18 (5), 409 (1983).

[16] 刘人怀，成振强. 集中载荷作用下开顶扁球壳的非线性稳定问题. 应用数学和力学，9 (2)，95 (1988).

[17] 刘人怀，李东. 均布载荷作用下开顶扁球壳的非线性稳定问题. 应用数学和力学，9 (3)，205 (1988).

[18] Liu Renhuai & He Linghui, On the nonlinear stability of a truncated shallow spherical shell under axisymmetrically distributed load. Solid Mechanics Archives, 14 (2), 81 (1989).

[19] Varadan T. K. Snap-buckling of orthatropic shallow spherical shells. J. Appl. Mech., 45, 445 (1978).

[20] Hyman B. I. Snap-through of shallow clamped spherical caps under uniform pressure. Int. J. Non-Linear Mechanics, (6): 55 (1971).

[21] Tillman S. C. On the buckling behaviour of shallow spherical caps under a uniform pressure load. International Journal of Solids and Structures, 6 (1), 37 (1970).

[22] Kaplan A. Buckling of spherical shells. Thin-Shell Structures, Eds. Fung, Y. C. & Sechler, E. E., Prentice-Hall, New Jersey, 247 (1974).

A simple theory for non-linear bending of laminated composite rectangular plates including higher-order effects①

中文摘要

本文应用虚位移原理，系统地提出了一个计及高阶影响的复合材料层合板的非线性弯曲理论的简化理论。同时，作为一个特殊情况，本文研究了在均匀横向载荷作用下对称正交层合板的非线性弯曲问题。本文所得到的结果与由三维理论所得的结果相当一致。

1. Introduction

In recent years, geometric nonlinear problems of composite laminated structures have attracted increasing attention. Under Kirchhoff's hypothesis, some researchers have studied the nonlinear mechanical behaviour of anisotropic laminated plates, and obtained many optimistic results, including analyses for some typical problems[1-7]. But these results are only applicable to very thin plates, because Kirchhoff's hypothesis neglects transverse shear deformation effects.

As is well known, for fiber-reinforced composite laminated plates, the ratios of in-plane elastic modulus to transverse shear modulus are relatively large. Thus the transverse shear deformation effects are considerable. In view of this fact, there is the need to establish a theory which can account for transverse deformations. In 1966, Habip[8] formulated a general theory of heterogeneous anisotropic plates based on the three-dimensional elasticity theory. The theory includes transverse shear and normal strain, acceleration, and a temperature field. Later, Wu and Vinson[9,10] studied nonlinear oscillations of laminated anisotropic plates under the Berger approximation. The limitation of the Berger approximation is that it could be used only for plates with immovable in-plane boundary conditions. Using the finite element method, Reddy[11-17], Hinrichsen[18], et al. have studied some nonlinear problems of laminated plates such as bending, free vibration, and transient response. In addition, Sivakumaran and Chia[19-21] have developed a refined theory on the basis of the Reissner-Mindlin-type theory, and investigated nonlinear oscillations of laminated rectangular plates using the Galerkin procedure and principle of harmonic balance. However, the Reissner-Mindlin-type plate theories do not satisfy the condition of zero transverse shear stress on the top and

① Reprinted from *International Journal of Non-Linear Mechanics*, 1991, 26 (5): 537-545. Authors: Liu Renhuai and He Linghui.

bottom surfaces of the plate, and the three-dimensional theories are intractable to analyse concrete problems. Thus, it is necessary to develop a simple theory which is not only more accurate but also simpler.

In this study, a simple theory for nonlinear bending problems of laminated composite plates including higher-order effects is formulated by using the principle of virtual displacements. Through introducing a displacement field obtained by modifying Reddy's theory[22] for linear problems, the present theory accounts for a parabolic variation of the transverse shear strains through thickness, so that it does not require using shear correction factors in computing the shear stress. Moreover, because the total deflection of a plate is decomposed into a deflection due to bending and a deflection due to shear, solution of the governing equations for present theory becomes simpler. In addition, nonlinear bending of a symmetrically laminated cross-ply plate under a uniform transverse load is studied as a special case. The results given in this paper are in good agreement with those obtained from three-dimensional theory[17].

2. Governing Equations and Boundary Conditions

Consider a laminated rectangular plate of constant thickness h in the z direction, length a in the x direction and width b in the y direction. The reference plane $z=0$ is located in the undeformed mid-plane and two adjacent edges of the plate are assumed to coincide with x-and y-axes respectively. The plate is constructed of finite homogeneous orthotropic layers perfectly bonded together. Each layer is of arbitrary thickness, elastic properties and orientation of orthotropic axes with respect to the plate axes. Under the action of a transverse load $q(x,y)$, the displacement field of the plate is assumed similar to [22] as follows

$$
\begin{aligned}
u(x,y,z) &= u_0(x,y) - zw^b_{,x}(x,y) + z^2\xi_x(x,y) + z^3\xi_x(x,y),\\
v(x,y,z) &= v_0(x,y) - zw^b_{,y}(x,y) + z^2\xi_y(x,y) + z^3\xi_y(x,y),\\
w(x,y) &= w^b(x,y) + w^s(x,y)
\end{aligned}
\tag{2.1a-c}
$$

where $u(x,y,z)$, $v(x,y,z)$ and $w(x,y)$ denote, respectively, displacement components of a point (x,y,z) in directions x, y and z, $u_0(x,y)$ and $v_0(x,y)$ are values of $u(x,y,z)$ and $v(x,y,z)$ on the mid-plane, w^b and w^s represent deflections of the plate due to bending and shear respectively, and unknown functions $\xi_x(x,y)$, $\xi_y(x,y)$, $\zeta_x(x,y)$ and $\zeta_y(x,y)$ will be determined using the condition that the transverse shear stress σ_{zx} and σ_{zy} vanish on the plate top and bottom surface

$$
\sigma_{zx}(x,y,\pm h/2) = 0,\quad \sigma_{zy}(x,y,\pm h/2) = 0 \tag{2.2a,b}
$$

For orthotropic plate or plates laminated of orthotropic layers, the conditions (2.2a,b) are equivalent to indicating that the corresponding strains $\varepsilon_{zx}(x,y,z)$ and $\varepsilon_{zy}(x,y,z)$ be zero on these surfaces

$$
\varepsilon_{zx}(x,y,\pm h/2) = 0,\quad \varepsilon_{zy}(x,y,\pm h/2) = 0 \tag{2.3a,b}
$$

Due to

$$\varepsilon_{zx} = w^s_{,x} + 2z\xi_x + 3z^2\zeta_x,$$
$$\varepsilon_{zy} = w^s_{,y} + 2z\xi_y + 3z^2\zeta_y \qquad (2.4\text{a,b})$$

from (2.3a,b) we obtain

$$\xi_x = \xi_y = 0, \quad \zeta_x = -\frac{4}{3h^2}w^s_{,x} \quad \zeta_y = -\frac{4}{3h^2}w^s_{,y} \qquad (2.5\text{a-d})$$

Thus, the strain components in a von Kármán sense corresponding to (2.1a-c) can be written as:

$$\varepsilon_x = \varepsilon^0_x - zw^b_{,xx} - \frac{4z^3}{3h^2}w^s_{,xx},$$
$$\varepsilon_y = \varepsilon^0_y - zw^b_{,yy} - \frac{4z^3}{3h^2}w^s_{,yy},$$
$$\varepsilon_z = 0,$$
$$\varepsilon_{xy} = \varepsilon^0_{xy} - 2zw^b_{,xy} - \frac{8z^3}{3h^2}w^s_{,xy}, \qquad (2.6\text{a-f})$$
$$\varepsilon_{zx} = \left(1 - \frac{4z^2}{h^2}\right)w^s_{,x},$$
$$\varepsilon_{zy} = \left(1 - \frac{4z^2}{h^2}\right)w^s_{,y}$$

in which ε^0_x, ε^0_y and ε^0_{xy} are strains on the mid-plane

$$\varepsilon^0_x = u^0_{,x} + \frac{1}{2}w^2_{,x},$$
$$\varepsilon^0_y = v^0_{,y} + \frac{1}{2}w^2_{,y},$$
$$\varepsilon^0_{xy} = u^0_{,y} + y^0_{,x} + w_{,x}w_{,y} \qquad (2.7\text{a-c})$$

and satisfy the following relation

$$\varepsilon^0_{x,yy} + \varepsilon^0_{y,xx} - \varepsilon^0_{xy,xy} = w^2_{,xy} - w_{,xx} + w_{,yy} \qquad (2.8)$$

Now, we assume that the constitutive equations for the kth layer are expressed as

$$\begin{Bmatrix} \sigma^{(k)}_x \\ \sigma^{(k)}_y \\ \sigma^{(k)}_{zy} \\ \sigma^{(k)}_{zx} \\ \sigma^{(k)}_{xy} \end{Bmatrix} = \begin{bmatrix} C^{(k)}_{11} & C^{(k)}_{12} & 0 & 0 & C^{(k)}_{16} \\ C^{(k)}_{12} & C^{(k)}_{22} & 0 & 0 & C^{(k)}_{26} \\ 0 & 0 & C^{(k)}_{44} & C^{(k)}_{45} & 0 \\ 0 & 0 & C^{(k)}_{45} & C^{(k)}_{55} & 0 \\ C^{(k)}_{16} & C^{(k)}_{26} & 0 & 0 & C^{(k)}_{66} \end{bmatrix} \begin{Bmatrix} \varepsilon_x \\ \varepsilon_y \\ \varepsilon_{zy} \\ \varepsilon_{zx} \\ \varepsilon_{xy} \end{Bmatrix} \qquad (2.9)$$

where $C^{(k)}_{ij}$ are the material stiffness coefficients of the layer which can be reduced from the engineering constants of the layer (see [23]).

Defining stress resultants and stress couples by

$$(N_x, N_y, N_{xy}) = \int_{-h/2}^{h/2} (\sigma^{(k)}_x, \sigma^{(k)}_y, \sigma^{(k)}_{xy})\,\mathrm{d}z,$$
$$(M_x, M_y, M_{xy}) = \int_{-h/2}^{h/2} (\sigma^{(k)}_x, \sigma^{(k)}_y, \sigma^{(k)}_{xy})\,z\,\mathrm{d}z,$$

$$(P_x, P_y, P_{xy}) = \int_{-h/2}^{h/2} (\sigma_x^{(k)}, \sigma_y^{(k)}, \sigma_{xy}^{(k)}) z^3 \mathrm{d}z,$$

$$(Q_x, Q_y) = \int_{-h/2}^{h/2} (\sigma_{zx}^{(k)}, \sigma_{zy}^{(k)}) \mathrm{d}z,$$

$$(R_x, R_y) = \int_{-h/2}^{h/2} (\sigma_{zx}^{(k)}, \sigma_{zy}^{(k)}) z^2 \mathrm{d}z, \tag{2.10a-e}$$

the following relations are obtained by substituting (2.9) into (2.10a-e)

$$\begin{Bmatrix} \{\varepsilon^0\} \\ \{M\} \\ \{P\} \end{Bmatrix} = \begin{bmatrix} [A^*] & [B^*] & [E^*] \\ [B^*]^{\mathrm{T}} & [D^*] & [F^*] \\ [E^*]^{\mathrm{T}} & [F^*]^{\mathrm{T}} & [H^*] \end{bmatrix} \begin{Bmatrix} \{N\} \\ \{K^b\} \\ \{K^s\} \end{Bmatrix},$$

$$\begin{Bmatrix} Q_y \\ Q_x \end{Bmatrix} = \begin{bmatrix} A_{44} - \frac{4}{h^2} D_{44} & A_{45} - \frac{4}{h^2} D_{45} \\ A_{45} - \frac{4}{h^2} D_{55} & A_{55} - \frac{4}{h^2} D_{55} \end{bmatrix} \begin{Bmatrix} w^s_{,y} \\ w^s_{,x} \end{Bmatrix},$$

$$\begin{Bmatrix} R_y \\ R_x \end{Bmatrix} = \begin{bmatrix} D_{44} - \frac{4}{h^2} F_{44} & D_{45} - \frac{4}{h^2} F_{45} \\ D_{45} - \frac{4}{h^2} F_{45} & D_{55} - \frac{4}{h^2} F_{55} \end{bmatrix} \begin{Bmatrix} w^s_{,y} \\ w^s_{,x} \end{Bmatrix} \tag{2.11a-c}$$

in which

$$\{\varepsilon^0\} = \begin{Bmatrix} \varepsilon_x^0 \\ \varepsilon_y^0 \\ \varepsilon_{xy}^0 \end{Bmatrix}, \quad \{M\} = \begin{Bmatrix} M_x \\ M_y \\ M_{xy} \end{Bmatrix}, \quad \{P\} = \begin{Bmatrix} P_x \\ P_y \\ P_{xy} \end{Bmatrix},$$

$$\{N\} = \begin{Bmatrix} N_x \\ N_y \\ N_{xy} \end{Bmatrix}, \quad \{K^b\} = \begin{Bmatrix} w^b_{,xx} \\ w^b_{,yy} \\ 2w^b_{,xy} \end{Bmatrix}, \quad \{K^s\} = \frac{4}{3h^2} \begin{Bmatrix} w^s_{,xx} \\ w^s_{,yy} \\ 2w^s_{,xy} \end{Bmatrix},$$

$$[A^*] = \begin{bmatrix} A_{11}^* & A_{12}^* & A_{16}^* \\ A_{21}^* & A_{22}^* & A_{26}^* \\ A_{61}^* & A_{62}^* & A_{66}^* \end{bmatrix}, \quad [B^*] = \begin{bmatrix} B_{11}^* & B_{12}^* & B_{16}^* \\ B_{21}^* & B_{22}^* & B_{26}^* \\ B_{61}^* & B_{62}^* & B_{66}^* \end{bmatrix},$$

$$[D^*] = \begin{bmatrix} D_{11}^* & D_{12}^* & D_{16}^* \\ D_{21}^* & D_{22}^* & D_{26}^* \\ D_{61}^* & D_{62}^* & D_{66}^* \end{bmatrix}, \quad [E^*] = \begin{bmatrix} E_{11}^* & E_{12}^* & E_{16}^* \\ E_{21}^* & E_{22}^* & E_{26}^* \\ E_{61}^* & E_{62}^* & E_{66}^* \end{bmatrix},$$

$$[F^*] = \begin{bmatrix} F_{11}^* & F_{12}^* & F_{16}^* \\ F_{21}^* & F_{22}^* & F_{26}^* \\ F_{61}^* & F_{62}^* & F_{66}^* \end{bmatrix}, \quad [H^*] = \begin{bmatrix} H_{11}^* & H_{12}^* & H_{16}^* \\ H_{21}^* & H_{22}^* & H_{26}^* \\ H_{61}^* & H_{62}^* & H_{66}^* \end{bmatrix},$$

$$[A] = \begin{bmatrix} A_{11} & A_{12} & A_{16} \\ A_{12} & A_{22} & A_{26} \\ A_{16} & A_{26} & A_{66} \end{bmatrix}, \quad [B] = \begin{bmatrix} B_{11} & B_{12} & B_{16} \\ B_{12} & B_{22} & B_{26} \\ B_{16} & B_{26} & B_{66} \end{bmatrix},$$

$$[D]=\begin{bmatrix}D_{11} & D_{12} & D_{16}\\ D_{12} & D_{22} & D_{26}\\ D_{16} & D_{26} & D_{66}\end{bmatrix},\quad [E]=\begin{bmatrix}E_{11} & E_{12} & E_{16}\\ E_{12} & E_{22} & E_{26}\\ E_{16} & E_{26} & E_{66}\end{bmatrix},$$

$$[F]=\begin{bmatrix}F_{11} & F_{12} & F_{16}\\ F_{12} & F_{22} & F_{26}\\ F_{16} & F_{26} & F_{66}\end{bmatrix},\quad [H]=\begin{bmatrix}H_{11} & H_{12} & H_{16}\\ H_{12} & H_{22} & H_{26}\\ H_{16} & H_{26} & H_{66}\end{bmatrix},$$

$$[A^*]=[A]^{-1},[B^*]=[A]^{-1}[B],$$

$$[D^*]=[B][A]^{-1}[B]-[D],$$

$$[E^*]=[A]^{-1}[E],\ [F^*]=[B][A]^{-1}[E]-[F],$$

$$[H^*]=[E][A]^{-1}[E]-[H],$$

$$(A_{ij},B_{ij},D_{ij},E_{ij},F_{ij},H_{ij})=\int_{-h/2}^{h/2}C_{ij}^{(k)}(1,z,z^2,z^3,z^4,z^6)\mathrm{d}z \tag{2.12}$$

Using Eqs. (2.1a-c), (2.6a-f) and (2.10a-e) in the principle of virtual displacements, and taking variations with respect to u_0, v_0 and w, the equilibrium equations of the plate are obtained

$$N_{x,x}+N_{xy,y}=0,$$

$$N_{xy,x}+N_{y,y}=0,$$

$$M_{x,xx}+2M_{xy,xy}+M_{y,yy}+N_xw_{,xx}+2N_{xy}w_{,xy}+N_yw_{,yy}+q=0,$$

$$\frac{4}{3h}(P_{x,xx}+2P_{xy,xy}+P_{y,yy})+(N_xw_{,xx}+2N_{xy}w_{,xy}+N_yw_{,yy})$$

$$+(Q_{x,x}+Q_{y,y})-\frac{4}{h^2}(R_{x,x}+R_{y,y})+q=0 \tag{2.13a-d}$$

and the boundary conditions are of the form: specify

$$\left.\begin{array}{lll}N_x & \text{or} & u_0\\ N_{xy} & \text{or} & v_0\\ Q_{1x} & \text{or} & w^b\\ Q_{2x} & \text{or} & w^s\\ M_x & \text{or} & w^b_{,x}\\ P_x & \text{or} & w^s_{,x}\\ M_{xy} & \text{or} & w^b_{,y}\\ P_{xy} & \text{or} & w^s_{,y}\end{array}\right\}\quad \text{along } x=0 \text{ or } a, \tag{2.14a}$$

$$\left.\begin{array}{lll}N_y & \text{or} & v_0\\ N_{xy} & \text{or} & u_0\\ Q_{1y} & \text{or} & w^b\\ Q_{2y} & \text{or} & w^s\\ M_y & \text{or} & w^b_{,y}\\ P_y & \text{or} & w^s_{,y}\\ M_{xy} & \text{or} & w^b_{,x}\\ P_{xy} & \text{or} & w^s_{,x}\end{array}\right\}\quad \text{along } y=0 \text{ or } b, \tag{2.14b}$$

where

$$Q_{1x} = N_x \frac{\partial w}{\partial x} + N_{xy} \frac{\partial w}{\partial y} + \frac{\partial M_x}{\partial x} + \frac{\partial M_{xy}}{\partial y},$$

$$Q_{1y} = N_y \frac{\partial w}{\partial x} + N_{xy} \frac{\partial w}{\partial y} + \frac{\partial M_y}{\partial y} + \frac{\partial M_{xy}}{\partial y},$$

$$Q_{2x} = N_x \frac{\partial w}{\partial x} + N_{xy} \frac{\partial w}{\partial y} + \frac{4}{3h^2} \frac{\partial P_x}{\partial x} + \frac{4}{3h^2} \frac{\partial P_{xy}}{\partial y} + Q_x - \frac{4}{h^2} R_x,$$

$$Q_{2y} = N_y \frac{\partial w}{\partial y} + N_{xy} \frac{\partial w}{\partial x} + \frac{4}{3h^2} \frac{\partial P_y}{\partial y} + \frac{4}{3h^2} \frac{\partial P_{xy}}{\partial x} + Q_y - \frac{4}{h^2} R_y \tag{2.15a-d}$$

Introducing a force function $\varphi(x, y)$ such that

$$N_x = \varphi_{,yy} \quad N_y = \varphi_{,xx} \quad N_{xy} = -\varphi_{,xy} \tag{2.16a-c}$$

Obviously, Eqs. (2.13a, b) hold voluntarily in this case. Substituting (2.11a-c) and (2.16a-c) into (2.8) and (2.13c, d), the governing equations for the problem discussed in this paper can be written in terms of w^b, w^s and φ as follows

$$L_{11}(w^b) + L_{12}(w^s) + L_{13}(\varphi) = \frac{1}{2} L(w, w),$$

$$L_{21}(w^b) + L_{22}(w^s) + L_{23}(\varphi) = L(\varphi, w) - q,$$

$$L_{31}(w^b) + L_{32}(w^s) + L_{33}(\varphi) = L(\varphi, w) - q \tag{2.17a-c}$$

where

$$L_{11}() = B_{21}^*()_{,xxxx} + (2B_{26}^* - B_{61}^*)()_{,xxxy} + (B_{11}^* + B_{22}^* - 2B_{66}^*)()_{,xxyy} + (2B_{16}^* - B_{62}^*)()_{,xyyy} + B_{12}^*()_{,yyyy},$$

$$L_{12}() = \frac{4}{3h^2}[E_{21}^*()_{,xxxx} + (2E_{26}^* - E_{61}^*)()_{,xxxy} + (E_{11}^* + E_{22}^* - 2E_{66}^*)()_{,xxyy} + (2E_{16}^* - E_{62}^*)()_{,xyyy} + E_{12}^*()_{,yyyy}],$$

$$L_{13}() = A_{22}^*()_{,xxxx} - (A_{26}^* + A_{62}^*)()_{,xxxy} + (A_{12}^* + A_{21}^* + A_{66}^*)()_{,xxyy} - (A_{16}^* + A_{61}^*)()_{,xyyy} + A_{11}^*()_{,yyyy},$$

$$L_{21}() = D_{11}^*()_{,xxxx} + 2(D_{16}^* + D_{61}^*)()_{,xxxy} + (D_{12}^* + D_{21}^* + 4D_{66}^*)()_{,xxyy} + 2(D_{26}^* + D_{62}^*)()_{,xyyy} + D_{22}^*()_{,yyyy},$$

$$L_{22}() = \frac{4}{3h^2}[F_{11}^*()_{,xxxx} + (2F_{16}^* + F_{61}^*)()_{,xxxy} + (F_{12}^* + F_{21}^* + 4F_{66}^*)()_{,xxyy} + 2(F_{26}^* + F_{62}^*)()_{,xyyy} + F_{22}^*()_{,yyyy}],$$

$$L_{23}() = L_{11}(),$$

$$L_{31}() = L_{22}(),$$

$$L_{32}() = \frac{16}{9h^4}[H_{11}^*()_{,xxxx} + 2(H_{16}^* + H_{61}^*)()_{,xxxy} + (H_{12}^* + H_{21}^* + 4H_{66}^*)()_{,xxyy} + 2(H_{26}^* + H_{62}^*)()_{xyyy} + H_{22}^*()_{,yyyy}] + (A_{55} - \frac{8}{h^2}D_{55} + \frac{16}{h^4} + \frac{16}{h^4}F_{55})()_{,xx}$$

$$+(2A_{45}-\frac{16}{h^2}D_{45}+\frac{32}{h^4}F_{45})()_{,xy}+(A_{44}-\frac{8}{h^2}D_{44}+\frac{16}{h^4}F_{44})()_{,yy},$$

$$L_{33}()=L_{12}(),$$

$$L(\varphi,w)=2\varphi_{,xy}w_{,xy}-\varphi_{,xx}w_{,yy}-\varphi_{,yy}w_{,xx} \tag{2.18}$$

In this paper, we consider the following simply supported boundary conditions

$$w^b=w^s=w^b_{,xx}=w^s_{,xx}=w^b_{,y}=w^s_{,y}=v_0=\varphi_{,yy}=0$$

$$\text{at } x=0,a; \tag{2.19a}$$

$$w^b=w^s=w^b_{,yy}=w^s_{,yy}=w^b_{,x}=w^s_{,x}=u_0=\varphi_{,xx}=0$$

$$\text{at } y=0,b \tag{2.19b}$$

3. Solution for a Symmetrically Laminated Rectangular Plate

As an application of the theory presented in this paper, a symmetrically laminated cross-ply rectangular plate under simply supported boundary conditions (2.19a) and (2.19b) is considered. In this case, some of the material stiffness coefficients of the plate vanish:

$$C_{16}^{(k)}=C_{26}^{(k)}=C_{45}^{(k)}=0 \tag{3.1}$$

and further, from (2.12) we have

$$[B]=[E]=0,\ [D^*]=-[D],\ [F^*]=-[F],\ [H^*]=-[H] \tag{3.2}$$

For simplification, the following non-dimensional parameters are introduced

$$\xi=\frac{x}{a},\quad \eta=\frac{y}{b},\quad W^b=\frac{w^b}{h},\quad W^s=\frac{w^s}{h},\quad \Phi=\frac{\varphi}{A_{22}h^2}$$

$$Q=\frac{qa^4}{A_{22}h^3},\quad \lambda=\frac{a}{b},\quad a_1=\frac{2A_{12}^*+A_{66}^*}{A_{22}^*},$$

$$a_2=\frac{A_{11}^*}{A_{22}^*},\quad a_3=\frac{1}{2A_{22}^*A_{22}},$$

$$b_1=\frac{2(D_{12}+2D_{66})}{D_{11}},\quad b_2=\frac{D_{22}}{D_{11}},$$

$$b_3=\frac{4F_{11}}{3D_{11}h^2},\quad b_4=\frac{8(F_{12}+2F_{66})}{3D_{11}h^2},$$

$$b_5=\frac{4F_{22}}{3D_{11}h^2},\quad b_6=\frac{A_{22}h^2}{D_{11}},\quad c_1=\frac{2(F_{12}+2F_{66})}{F_{11}},$$

$$c_2=\frac{F_{22}}{F_{11}},\quad c_3=\frac{4H_{11}}{3F_{11}h^2},$$

$$c_4=\frac{8(H_{12}+2H_{66})}{3F_{11}h^2},\quad c_5=\frac{4H_{22}}{3F_{11}h^2},$$

$$c_6=\frac{3a^2}{4F_{11}}(A_{55}h^2-8D_{55}+\frac{16}{h^2}F_{55}),$$

$$c_7=\frac{3a^2}{4F_{11}}(A_{44}h^2-8D_{44}+\frac{16}{h^2}F_{44}),\quad c_8=\frac{3A_{22}h^4}{4F_{11}},$$

$$d_1=A_{11}^*A_{22},\quad d_2=A_{22}^*A_{22},\quad d_3=A_{12}^*A_{22} \tag{3.3}$$

Thus, the governing equations (2.17a-c) are simplified as follows

$$\mathscr{L}_{11}(\Phi)=a_3\lambda^2\mathscr{L}(W,W),$$
$$\mathscr{L}_{21}(W^b)+\mathscr{L}_{22}(W^s)=-b_6\lambda^2\mathscr{L}(\Phi,W)+b_6Q,$$
$$\mathscr{L}_{31}(W^b)+\mathscr{L}_{32}(W^s)=-c_8\lambda^2\mathscr{L}(\Phi,W)+c_8Q \tag{3.4a-c}$$

where

$$\mathscr{L}_{11}()=()_{,\xi\xi\xi\xi}+a_1\lambda^2()_{,\xi\xi\eta\eta}+a_2\lambda^4()_{,\eta\eta\eta\eta},$$
$$\mathscr{L}_{21}()=()_{,\xi\xi\xi\xi}+b_1\lambda^2()_{,\xi\xi\eta\eta}+b_2\lambda^4()_{,\eta\eta\eta\eta},$$
$$\mathscr{L}_{22}()=b_3()_{,\xi\xi\xi\xi}+b_4\lambda^2()_{,\xi\xi\eta\eta}+b_5\lambda^4()_{,\eta\eta\eta\eta},$$
$$\mathscr{L}_{31}()=()_{,\xi\xi\xi\xi}+c_1\lambda^2()_{,\xi\xi\eta\eta}+c_2\lambda^4()_{,\eta\eta\eta\eta},$$
$$\mathscr{L}_{32}()=c_3()_{,\xi\xi\xi\xi}+c_4\lambda^2()_{,\xi\xi\eta\eta}+c_5\lambda^4()_{,\eta\eta\eta\eta}$$
$$-c_6()_{,\xi\xi}-c_7\lambda^2()_{,\eta\eta}$$
$$\mathscr{L}(\Phi,W)=2\Phi_{,\xi\eta}W_{,\xi\eta}-\Phi_{,\xi\xi}W_{,\eta\eta}-\Phi_{,\eta\eta}W_{,\xi\xi} \tag{3.5a-f}$$

and the boundary conditions (2.17) and (2.21) become

$$W^b=W^s=W^b_{,\xi\xi}=W^s_{,\xi\xi}=W^b_{,\eta}=W^s_{,\eta}=\Phi_{,\eta\eta}$$
$$=\int_0^1\left(d_3\lambda^2\Phi_{,\eta\eta}+d_2\Phi_{,\eta\eta}-\frac{1}{2}\lambda^2W^2_{,\eta}\right)\mathrm{d}\eta=0 \qquad \text{at } \xi=0,1 \tag{3.6a-f}$$
$$W^b=W^s=W^b_{,\eta\eta}=W^s_{,\eta\eta}=W^b_{,\xi}=W^s_{,\xi}=\Phi_{,\xi\xi}$$
$$=\int_0^1\left(d_1\lambda^2\Phi_{,\eta\eta}+d_3\Phi_{,\xi\xi}-\frac{1}{2}W^2_{,\xi}\right)\mathrm{d}\xi=0 \qquad \text{at } \eta=0,1 \tag{3.7a-f}$$

In order to solve the above boundary value problems (3.4)-(3.7), we express w^b, w^s and Φ in the following double Fourier series forms

$$W^b=\sum_{m=1}^{\infty}\sum_{n=1}^{\infty}W^b_{mn}\sin m\pi\xi\sin n\pi\eta,$$
$$W^s=\sum_{m=1}^{\infty}\sum_{n=1}^{\infty}W^s_{mn}\sin m\pi\xi\sin n\pi\eta,$$
$$\Phi=\sum_{m=1}^{\infty}\sum_{n=1}^{\infty}\Phi_{mn}\sin m\pi\xi\sin n\pi\eta,$$
$$Q=\sum_{m=1}^{\infty}\sum_{n=1}^{\infty}Q_{mn}\sin m\pi\xi\sin n\pi\eta \tag{3.8a-d}$$

in which

$$Q_{mn}=4\int_0^1\int_0^1 Q\sin m\pi\xi\sin n\pi\eta\,\mathrm{d}\xi\mathrm{d}\eta. \tag{3.9}$$

Clearly, expressions (3.8a-d) satisfy all of the conditions given in (3.6)-(3.7). Substituting them into Eqs. (3.4), multiplying both sides of the three equations by sin $p(\pi\xi)$ sin $q(\pi\eta)$ and integrating over their respective intervals provided that term-by-term integrations are permissible, the following nonlinear algebraic equations are obtained

$$\Phi_{pq}=\frac{2a_3\lambda^2}{p^4+a_1\lambda^2p^2q^2+a_2\lambda^4q^4}\sum_{k=1}^{\infty}\sum_{l=1}^{\infty}\sum_{r=1}^{\infty}\sum_{s=1}^{\infty}e^{pq}_{klrs}W_{kl}W_{rs},$$

$$(p^4+b_1\lambda^2p^2q^2+b_2\lambda^4q^4)W^b_{pq}+(b_3p^4+b_4\lambda^2p^2q^2+b_5\lambda^4q^4)W^s_{pq}$$

$$=-2b_6\lambda^2\sum_{k=1}^{\infty}\sum_{l=1}^{\infty}\sum_{r=1}^{\infty}\sum_{s=1}^{\infty}e^{pq}_{klrs}\Phi_{kl}W_{rs}+\frac{b_6}{\pi^4}Q_{pq},$$

$$(p^4+c_1\lambda^2p^2q^2+c_2\lambda^4q^4)W^b_{pq}+(c_3p^4+c_4\lambda^2p^2q^2+c_5\lambda^4q^4+\frac{c_6p^2}{\pi^2}+\frac{\lambda^2c_7q^2}{\pi^2})W^s_{pq}$$

$$=-2c_8\lambda^2\sum_{k=1}^{\infty}\sum_{l=1}^{\infty}\sum_{r=1}^{\infty}\sum_{s=1}^{\infty}e^{pq}_{klrs}\Phi_{kl}W_{rs}+\frac{c_8}{\pi^4}Q_{pq}\tag{3.10a-c}$$

where

$$\begin{aligned}e^{pq}_{klrs}=&klrs\int_a^1[\cos(k+r)\pi\xi+\cos(k-r)\pi\xi]\sin p\pi\xi d\xi\\&\cdot\int_0^1[\cos(1+s)\pi\eta+\cos(1-s)\pi\eta]\sin q\pi\eta d\eta\\&-\frac{k^2s^2+l^2r^2}{2}\int_0^1[\cos(k+r)\pi\xi-\cos(k-r)\pi\xi]\sin p\pi\xi d\xi\\&\cdot\int_0^1[\cos(1+s)\pi\eta-\cos(l-s)\pi\eta]\sin q\pi\eta d\eta\end{aligned}\tag{3.11}$$

Combining Eqs. (3.10b) and (3.10c), we obtain

$$W^s_{pq}=\Delta(p,q;\lambda)W^b_{pq}\tag{3.12}$$

where

$$\begin{aligned}\Delta(p,q;\lambda)=&[c_8(p^4+b_1\lambda^2p^2q^2+b_2\lambda^4q^4)-b_6(p^4+c_1\lambda^2p^2q^2\\&+c_2\lambda^4q^4][b_6(c_3p^4+c_4\lambda^2p^2q^2+c_5\lambda^4q^4\\&+\frac{c_6p^2+\lambda^2c_7q^2}{\pi})-c_8(b_3p^4+b_4\lambda^2p^2+b_5\lambda^4q^4)]^{-1}\end{aligned}\tag{3.13a}$$

and further, from (2.1c) we have

$$W^b_{pq}=\frac{1}{1+\Delta(p,q;\lambda)}W_{pq}\tag{3.13b}$$

By substituing (3.12), (3.13a, b) and (3.10a) into (3.10b), the relations between Q and W_{pq} are obtained

$$\begin{aligned}&\frac{(p^4+b_1\lambda^2p^2q^2+b_2\lambda^4q^4)+\Delta(p,q;\lambda)(b_3p^4+b_4\lambda^2p^2q^2+b_5\lambda^4q^4)}{1+\Delta(p,q;\lambda)}\\&+4a_3b_6\lambda^4\sum_{i=1}^{\infty}\sum_{j=1}^{\infty}\sum_{k=1}^{\infty}\sum_{l=1}^{\infty}\sum_{m=1}^{\infty}\sum_{n=1}^{\infty}\sum_{r=1}^{\infty}\sum_{s=1}^{\infty}\frac{e^{pq}_{ijkl}e^{ij}_{mnrs}}{i^4+a_1\lambda^2i^2j^4+a_2\lambda^4j^4}W_{kl}W_{mn}W_{rs}\\&=\frac{b_6Q_{pq}}{\pi^4}\end{aligned}\tag{3.14}$$

For prescribed Q, the coefficients Q_{pq} are also determined. Truncating the first finite terms in series (3.8a-d), and substituting them in (3.4), a set of cubic equations for W_{pq}, in which number of equations is equal to number of unknowns, are obtained. As soon as the unknowns W_{pq} are determined, the values of F_{pq} can be calculated from (3.10a), and then the stress resultants and stress couples are also derived from (2.16a-c) and (2.11a-c).

4. Numerical Results

Here we present numerical results for cross-ply (0°/90°/0°, equal thickness layers) square plates under uniformly distributed load q_0. The material properties of the plates are given in terms of the engineering constants as follows

$$E_1 = 1.724 \times 10^8 \mathrm{kN/m}(25 \times 10^6 \mathrm{psi})$$
$$E_2 = E_3 = 6.89 \times 10^6 \mathrm{kN/m}(10^6 \mathrm{psi})$$
$$G_{12} = G_{13} = 3.45 \times 10^6 \mathrm{kN/m}(0.5 \times 10^6 \mathrm{psi})$$
$$G_{23} = 13.78 \times 10^6 \mathrm{kN/m}(0.2 \times 10^6 \mathrm{psi})$$
$$\nu_{12} = \nu_{13} = \nu_{32} = 0.25 \tag{4.1}$$

Taking only the first term ($p=q=1$) in series (3.8a,b) and the first three terms ($p,q=1,3$) in series (3.8c), through numerical computations, the relations between nondimensional load $Q_0\left(=\dfrac{q_0a^2}{E_2h^4}\right)$ and central deflection $W_m(=W_{11})$ of the plates are obtained from equation (3.14)

$$Q_0 = 103.0W_m + 17.75W_m^3 \quad \text{for} \quad a/h = 10 \tag{4.2a}$$

$$Q_0 = 25.76W_m + 17.75W_m^3 \quad \text{for} \quad a/h = 4 \tag{4.2b}$$

The load-deflection curves corresponding to (4.2a) and (4.2b) are illustrated and compared with those obtained from the shear deformable plate theory (SDPT) and the three-dimensional elasticity theory (3-DET) [17] in Fig. 1. It is clear to see that the present results are in good agreement with those of the 3-DET, and are more accurate than those of the SDPT. Moreover, from equation (3.13) we know the error of deflection due to neglect transverse shear deformation effects will be 36% for $a/h=10$ and 89% for $a/h=4$.

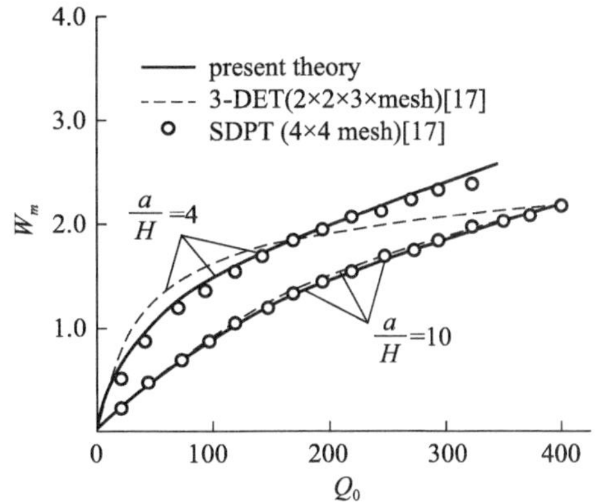

Fig. 1 Load-deflection curves for cross-ply (0°/90°/0°) square plates under uniformly distributed load.

References

[1] C. Y. Chia, *Nonlinear Analysis of Plates*. McGraw-Hill, New York (1980).

[2] C. Y. Chia, Nonlinear bending of unsymmetric angle-ply plates with edges elastically restrained against rotation. *Acta Mech*. 53, 201-212 (1984).

[3] C. Y. Chia, Large deflection of unsymmetric laminates with mixed boundary conditions. *Int. J. Nonlinear Mech*. 20, 273-282 (1985).

[4] C. Y. Chia, Postbuckling of composite plates under inplane compressive and shear loading having edges elastically restrained against rotation, in *Proc. 3rd Int. Conf. Compos. Struct.*, pp.

278-289. Elsevier Applied Science, London (1985).

[5] D. Hui, Imperfection sensitivity of axially compressed laminated flat plates due to bending-stretching coupling. *Int. J. Solids Struct.* 22, 13-22 (1986).

[6] T. R. Tauchert and N. N. Huang, Thermal buckling and postbuckling behavior of antisymmetric angle-ply laminates, in *Proc. Int. Symp. Compos. Mat. Struct.*, pp, 357-363, Beijing, Technomic. Lancaster, PA (1986).

[7] Liu Renhuai and L. H. He, Nonlinear bending of simply supported symmetric laminated cross-ply rectangular plates. *Appl. Math. Mech.* 11, 801-807 (1990).

[8] L. M. Habip, Theory of elastic plates in the reference state. *Int. J. Solids Struct.* 2, 157-166 (1966).

[9] C. I. Wu and J. R. Vinson, On the nonlinear oscillations of plates composed of composite materials, J. *Compos. Mater.* 3, 548-561 (1969).

[10] C. I. Wu and J. R. Vinson, Nonlinear oscillations of laminated specially orthotropic plates with clamped and simply supported edges. *J. Acoust. Soc.* Am. 49, 1561-1567 (1971).

[11] J. N. Reddy, Free vibration of antisymmetric, angle-ply laminated plates including transverse shear deformation by the finite element method. *J. Sound Vib.* 66, 565-576 (1979).

[12] J. N. Reddy and C. Chao, Nonlinear bending of thick rectangular, laminated composite plates. *Int. J. Nonlinear Mech.* 16, 291-301 (1981).

[13] J. N. Reddy, Nonlinear oscillations of laminated, anisotropic rectangular plates. *ASME J. Appl. Mech.* 49, 396-402 (1982).

[14] J. N. Reddy, Large amplitude flexural vibration of layered composite plates with cutouts. *J. Sound Vib.* 83, 1-10 (1983).

[15] J. N. Reddy, Geometrically nonlinear transient analysis of laminated composite plates. *AIAA J.* 21, 621-629 (1983).

[16] N. S, Putcha and J. N. Reddy, A refined mixed shear flexible finite element for the nonlinear analysis of laminated plates. *Comput. Struct.* 22, 529-538 (1986).

[17] T. Kuppusamy and J. N. Reddy, A three-dimensional nonlinear analysis of cross-ply rectangular composite plates. *Comput. Struct.* 18, 263-272 (1984).

[18] R. L. Hinrichsen and A. N. Palazotto, Nonlinear finite element analysis of thick composite plates using cubic spline functions. *AIAA J.* 24, 1836-1842 (1986).

[19] K. S. Sivakumaran, A refined theory of generally laminated plates. This is submitted to the Faculty of Graduate Studies, The requirements for the degree of Doctor of Philosophy (January 1983).

[20] K. S. Sivakumaran and C. Y. Chia, Nonlinear vibration of generally laminated anisotropic thick plates. *Ing. Arch.* 54, 220-231 (1984).

[21] K. S. Sivakumaran and C. Y. Chia, Large-amplitude oscillations of unsymmetrically laminated anisotropic rectangular plates including shear, rotatory inertia, and transverse normal stress. *ASME J. Appl. Mech.* 52, 536-542 (1985).

[22] J. N. Reddy, A simple higher-order theory for laminated composite plates. *ASME J, Appl. Mech.* 51, 745-752 (1984).

[23] R. M. Jones, *Mechanics of Composite Materials*. McGraw-Hill, New York (1975).

考虑横向剪切的对称层合圆柱正交异性中心开孔扁球壳的非线性屈曲①

一、引言

随着科学技术的迅猛发展，复合材料日益广泛应用于航天、航空、航海等工程结构中，因此，研究复合材料板壳的非线性力学行为不仅具有非常重要的理论意义，还具有非常重要的工程实际意义。众所周知，由于非线性数学问题和结构复杂的困难，对复合材料层合壳体非线性问题的研究至今仍限制在少量的壳型范围内[1-3]，而对相当重要的层合扁球壳，仅有作者最近才研究过[4,5]。

本文在前文[5]的理论基础上，建立考虑横向剪切的对称层合圆柱正交异性中心开孔扁球壳的非线性屈曲理论。最后，应用我们于 1965 年提出的修正迭代法[6-8]，研究了在均布压力作用下考虑横向剪切的对称层合圆柱正交异性中心开孔扁球壳的非线性轴对称屈曲问题。

二、基本方程

考虑如图 1 所示的在均布压力 q 作用下，厚度为 h，内、外边缘半径分别为 b 和 a，径向坐标为 r，曲率半径为 R 和拱高为 H 的对称层合圆柱正交异性中心开孔扁球壳。

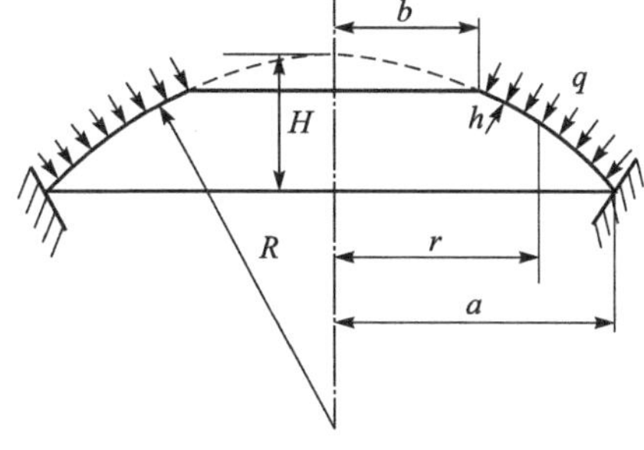

图 1　在均布压力 q 作用下的中心开孔扁球壳

按照前文[5]，并使用相同的符号，在均布压力 q 作用下，考虑横向剪切的对称层合圆柱正交异性封顶扁球壳的由挠度 w、壳体变形前的中曲面法线转角 ψ 和径向薄膜力 N_r 所满足的轴对称大挠度弯曲方程为

$$\frac{\mathrm{d}}{\mathrm{d}r}\left[rN_r\left(\frac{r}{R}+\frac{\mathrm{d}w}{\mathrm{d}r}\right)+Gr\left(\psi+\frac{\mathrm{d}w}{\mathrm{d}r}\right)\right]+rq=0$$

$$D_{11}\frac{\mathrm{d}}{\mathrm{d}r}r\frac{\mathrm{d}\psi}{\mathrm{d}r}-\left(Gr+\frac{D_{22}}{r}\right)\psi-Gr\frac{\mathrm{d}w}{\mathrm{d}r}=0 \qquad \text{(1a-c)}$$

$$A_3\frac{\mathrm{d}}{\mathrm{d}r}r\frac{\mathrm{d}}{\mathrm{d}r}(rN_r)-A_1N_r+\frac{1}{2}\left(\frac{\mathrm{d}w}{\mathrm{d}r}\right)^2+\frac{r}{R}\frac{\mathrm{d}w}{\mathrm{d}r}=0$$

用 $\mathrm{d}r$ 乘方程（1a），从 b 到 r 积分一次后，得

$$Gr\frac{\mathrm{d}w}{\mathrm{d}r}+rN_r\left(\frac{r}{R}+\frac{\mathrm{d}w}{\mathrm{d}r}\right)+Gr\psi+\frac{1}{2}q(r^2-b^2)=0 \qquad (2)$$

于是，方程（2）和（1b，c）就组成在均布压力作用下考虑横向剪切的对称层合

① 本文原载《暨南大学学报》，1994，15（1），1-12. 作者：刘人怀，钟诚.

圆柱正交异性中心开孔扁球壳轴对称大挠度弯曲方程，至于相关的边界条件，我们讨论下面较常见的一种情况。

当 $r=a$ 时，边缘夹紧固定：

$$w=0,\psi=0,u=0 \tag{3}$$

当 $r=b$ 时，边缘悬空：

$$M_r=0,N_r=0 \tag{4}$$

其中 u 为径向位移，M_r 为径向弯矩，

$$u=r\left[A_3\frac{\mathrm{d}}{\mathrm{d}r}(rN_r)-A_2N_r\right],$$

$$M_r=D_{11}\frac{\mathrm{d}\psi}{\mathrm{d}r}+\frac{D_{12}}{r}\psi \tag{5}$$

为了计算简单起见，引入以下无量纲量：

$$\begin{aligned}
&y=\frac{r}{a}, && \alpha=\frac{b}{a}, && W=\frac{w}{h},\\
&\Phi=ky+\frac{\mathrm{d}w}{\mathrm{d}y}, && \Psi=\frac{a}{h}\psi, && S=\frac{a}{D_{11}}rN_r,\\
&P=\frac{a^4}{2D_{11}h}q, && k=\frac{a^2}{Rh}, && m=\frac{a^2}{D_{11}}G,\\
&\beta_1^2=\frac{D_{22}}{D_{11}}, && \beta_2^2=\frac{A_1}{A_3}, && \beta_3=\frac{h^2}{A_3D_{11}},\\
&\beta_4=\frac{A_2}{A_3}, && \beta_5=\frac{D_{12}}{D_{11}}
\end{aligned} \tag{6}$$

再应用方程（2）、（1b，c）、（3）和（4），我们得到如下的无量纲非线性边值问题：

$$\begin{aligned}
&my\Phi=-\left[S\Phi+my\Psi-kmy^2+P(y^2-\alpha^2)\right],\\
&L_1(y^{\beta_1}\Psi)=my(\Psi+\Phi-ky),\\
&L_2(y^{\beta_2}S)=-\beta_3(\Phi-ky)\left[\frac{1}{2}(\Phi-ky)+ky\right]
\end{aligned} \tag{7a-c}$$

$$\text{当 } y=1 \text{ 时，}\quad W=0,\quad \Psi=0,\quad \frac{\mathrm{d}S}{\mathrm{d}y}-\beta_4\frac{S}{y}=0 \tag{8a-c}$$

$$\text{当 } y=\alpha \text{ 时，}\quad S=0,\quad \frac{\mathrm{d}\Psi}{\mathrm{d}y}+\beta_5\frac{\Psi}{y}=0 \tag{9}$$

其中

$$\begin{aligned}
L_1&=y^{\beta_1}\frac{\mathrm{d}}{\mathrm{d}y}y^{1-2\beta_1}\frac{\mathrm{d}}{\mathrm{d}y}(\cdots),\\
L_2&=y^{\beta_2}\frac{\mathrm{d}}{\mathrm{d}y}y^{1-2\beta_2}\frac{\mathrm{d}}{\mathrm{d}y}(\cdots)
\end{aligned} \tag{10}$$

三、边值问题的求解

现在，应用修正迭代法求解上述无量纲非线性边值问题（7）、（8）和（9）。在第一次近似中，首先略去方程（7a）中的非线性项 $S\Phi$，便得如下的线性边值问题：

$$my\Phi_1 = -[my\Psi_1 - kmy^2 + P(y^2 - \alpha^2)],$$
$$L_1(y^{\beta_1}\Psi_1) = my(\Psi_1 + \Phi_1 - ky), \tag{11a-c}$$
$$L_2(y^{\beta_2}S_1) = -\beta_3(\Phi_1 - ky)\left[\frac{1}{2}(\Phi_1 - ky) + ky\right]$$

$$当\ y = 1\ 时，\quad W_1 = 0，\quad \Psi_1 = 0，\quad \frac{\mathrm{d}S_1}{\mathrm{d}y} - \beta_4\frac{S_1}{y} = 0 \tag{12a-c}$$

$$当\ y = \alpha\ 时，\quad \frac{\mathrm{d}\Psi_1}{\mathrm{d}y} + \beta_5\frac{\Psi_1}{y} = 0，\quad S_1 = 0 \tag{13a,b}$$

将方程（11a）代入（11b），再利用边界条件（12b）和（13a），方程（11b）的解是

$$\Psi_1 = -(a_1y^3 + a_2y + a_3y^{\beta_1} + a_4y^{-\beta_1})P \tag{14}$$

其中

$$a_1 = \frac{1}{9-\beta_1^2},\quad a_2 = \frac{\alpha^2}{1-\beta_1^2},$$
$$a_3 = \delta_1[F_0(1)(\beta_5 - \beta_1)\alpha^{-1-\beta_1} - F_0'(\alpha) - \beta_5\overline{F}_0(\alpha)],$$
$$a_4 = -F_0(1) + a_3,$$
$$\delta_1 = [(\beta_5 - \beta_1)\alpha^{-1-\beta_1} - (\beta_5 + \beta_1)\alpha^{-1+\beta_1}]^{-1}, \tag{15}$$
$$F_0(y) = -(a_1y^3 + a_2y),$$
$$F_0'(y) = \frac{\mathrm{d}F_0(y)}{\mathrm{d}y},$$
$$\overline{F}_0(y) = \frac{F_0(y)}{y}$$

将解（14）代入方程（11a），可得

$$\Phi_1 = \left(a_1y^3 + a_5y + a_3y^{\beta_1} + a_6y^{-\beta_1} + \frac{\alpha^2}{my}\right)P + ky \tag{16}$$

其中

$$a_5 = a_2 - \frac{1}{m},\quad a_6 = \frac{\alpha^2}{m} \tag{17}$$

应用边界条件（8a），我们以层合扁球壳的无量纲内边缘挠度

$$W_m = -\int_\alpha^1(\Phi - ky)\mathrm{d}y \tag{18}$$

作为迭代参数。于是，将解（16）代入此式，便得第一次近似下的特征关系：

$$P = \alpha_1W_m \tag{19}$$

式中

$$\alpha_1 = -\left[\frac{a_1}{4}(1-\alpha^4) + \frac{1}{2m}(a_2m - 1)(1-\alpha^2) + \frac{a_3}{1+\beta_1}(1-\alpha^{1+\beta_1}) - \frac{1}{m}\alpha^2\ln\alpha + \frac{a_4}{1-\beta_1}(1-\alpha^{1-\beta_1})\right]^{-1} \tag{20}$$

应用式（19），解（16）成为

$$\Phi_1 = \alpha_1(a_1y^3 + a_5y + a_3y^{\beta_1} + a_6y^{-\beta_1})W_m + ky \tag{21}$$

将解（21）代入方程（11c），再利用边界条件（12c）、（13b），可求得

$$S_1 = [F_1(y) + \alpha_2 y^{\beta_2} + \alpha_4 y^{-\beta_1}]W_m^2 + k[F_2(y) + \alpha_3 y^{\beta_2} + \alpha_5 y^{-\beta_2}]W_m \tag{22}$$

其中

$$F_1(y) = \alpha_1^2(b_1 y^7 + b_2 y^5 + b_3 y^3 + b_4 y + b_5 y^{-1} + b_6 y^{1+2\beta_1} + b_7 y^{1-2\beta_1} + b_8 y^{4+\beta_1} + b_9 y^{4-\beta_1} + b_{10} y^{2+\beta_1} + b_{11} y^{2-\beta_1} + b_{12} y^{\beta_1}\ln y + b_{13} y^{\beta_2} + b_{14} y^{\beta_2}\ln y),$$

$$F_2(y) = \alpha_1(b_{15} y^5 + b_{16} y^3 + b_{17} y + b_{18} y^{2+\beta_1} + b_{19} y^{2-\beta_1}),$$

$$\alpha_i = -\delta_2[F'_{i-1}(1) - \beta_4 \overline{F}_{i-1}(1) + F_{i-1}(\alpha)(\beta_2 + \beta_4)\alpha^{\beta_2}], \quad i = 2,3$$

$$\alpha_i = \delta_2\{[F'_{i-3}(1) - \beta_4 \overline{F}_{i-3}(1)]\alpha^{\beta_2} + F_{i-3}(\alpha)(\beta_2 + \beta_4)\}, \quad i = 4,5$$

$$\overline{F}_{i-1}(y) = \frac{F_{i-1}(y)}{y}, \quad F'_{i-1}(y) = \frac{\mathrm{d}F_{i-1}(y)}{\mathrm{d}y},$$

$$\delta_2 = [\beta_2 - \beta_4 + (\beta_2 + \beta_4)\alpha^{2\beta_2}]^{-1},$$

$$b_1 = -\frac{a_1^2\beta_3}{2(49-\beta_2^2)}, \quad b_2 = \frac{a_1 a_5\beta_3}{25-\beta_2^2},$$

$$b_3 = -\frac{(a_5^2 + 2a_1 a_6)\beta_3}{2(9-\beta_2^2)}, \quad b_4 = \frac{(a_5 a_6 - a_3 a_4)\beta_3}{1-\beta_2^2},$$

$$b_5 = \frac{a_6^2\beta_3}{2(1-\beta_2^2)}, \quad b_6 = -\frac{a_3^2\beta_3}{2[(1+2\beta_1)^2-\beta_2^2]},$$

$$b_7 = -\frac{a_4^2\beta_3}{2[(1-2\beta_1)^2-\beta_2^2]}, \quad b_8 = -\frac{a_1 a_3\beta_3}{(4+\beta_1)^2-\beta_2^2},$$

$$b_9 = -\frac{a_1 a_4\beta_3}{(4-\beta_1)^2-\beta_2^2}, \quad b_{10} = \frac{a_3 a_5\beta_3}{(2+\beta_1)^2-\beta_2^2},$$

$$b_{11} = \frac{a_4 a_5\beta_3}{(2-\beta_1)^2-\beta_2^2}, \quad b_{12} = -\frac{a_3 a_6\beta_3}{2\beta_2},$$

$$b_{13} = \frac{a_3 a_6\beta_3}{4\beta_2^2}, \quad b_{14} = \frac{a_4 a_6\beta_3}{2\beta_2},$$

$$b_{15} = -\frac{a_1\beta_3}{25-\beta_2^2}, \quad b_{16} = \frac{a_5\beta_2}{9-\beta_2^2},$$

$$b_{17} = -\frac{a_6\beta_2}{1-\beta_2^2}, \quad b_{18} = -\frac{a_3\beta_3}{(2+\beta_1)^2-\beta_2^2},$$

$$b_{19} = -\frac{a_4\beta_3}{(2-\beta_1)^2+\beta_2^2} \tag{23}$$

在第二次近似中，我们有如下线性边值问题：

$$my\Phi_2 = -[S_1\Phi_1 + my\Psi_2 - kmy^2 + P(y^2 - \alpha^2)],$$

$$L_1(y^{\beta_1}\Psi_2) = my(\Psi_2 + \Phi_2 - ky) \tag{24a,b}$$

$$当\ y = 1\ 时，\quad W_2 = 0，\quad \Psi_2 = 0 \tag{25a,b}$$

$$当\ y = \alpha\ 时，\frac{\mathrm{d}\psi_2}{\mathrm{d}y} + \beta_5\frac{\Psi_2}{y} = 0 \tag{26}$$

将方程（24a）代入方程（24b），再利用边界条件（25b）、（26），方程（24b）的解是

$$\Psi_2 = [-\alpha_1^3 F_3(y) + \alpha_6 y^{\beta_1} + \alpha_{10} y^{-\beta_1}]W_m^3 + k[-\alpha_1^2 F_4(y) + \alpha_7 y^{\beta_1} + \alpha_{11} y^{-\beta_1}]W_m^2$$
$$+ k^2[-\alpha_1 F_5(y) + \alpha_8 y^{\beta_1} + a_{12} y^{-\beta_1}]W_m + [-F_6(y) + \alpha_9 y^{\beta_1} + \alpha_{13} y^{-\beta_1}]P \quad (27)$$

其中

$$F_3(y) = c_1 y^{11} + c_2 y^9 + c_3 y^7 + c_4 y^5 + c_5 y^3 + c_6 y + c_7 y^{-1} + c_8 y^{8+\beta_1}$$
$$+ c_9 y^{8-\beta_1} + c_{10} y^{6+\beta_1} + c_{11} y^{6-\beta_1} + c_{12} y^{4+\beta_1} + c_{13} y^{4-\beta_1}$$
$$+ c_{14} y^{2+\beta_1} + c_{15} y^{2-\beta_1} + c_{16} y^{\beta_1} + c_{17} y^{5+2\beta_1} + c_{18} y^{5-2\beta_1}$$
$$+ c_{19} y^{3+2\beta_1} + c_{20} y^{3-2\beta_1} + c_{21} y^{1+2\beta_1} + c_{22} y^{1-2\beta_1} + c_{23} y^{2+3\beta_1}$$
$$+ c_{24} y^{2-3\beta_1} + c_{25} y^{4+\beta_1} \ln y + c_{26} y^{4-\beta_1} \ln y + c_{27} y^{2+\beta_1} \ln y$$
$$+ c_{28} y^{2-\beta_1} \ln y + c_{29} y^{\beta_1} \ln y + c_{30} y^{-\beta_1} \ln y + c_{31} y^{1+2\beta_1} \ln y$$
$$+ c_{32} y^{1-2\beta_1} \ln y + c_{33} y \ln y + c_{34} y^{\beta_1} \ln^2 y + c_{35} y^{-\beta_1} \ln^2 y,$$

$$F_4(y) = c_{36} y^9 + c_{37} y^7 + c_{38} y^5 + c_{39} y^3 + c_{40} y + c_{41} y^{6+\beta_1} + c_{42} y^{6-\beta_1}$$
$$+ c_{43} y^{4+\beta_1} + c_{44} y^{4-\beta_1} + c_{45} y^{2+\beta_1} + c_{46} y^{2-\beta_1} + c_{47} y^{3+2\beta_1}$$
$$+ c_{48} y^{3-2\beta_1} + c_{49} y^{1+2\beta_1} + c_{50} y^{1-2\beta_1} + c_{51} y^{\beta_1} + c_{52} y^{2+\beta_1} \ln y$$
$$+ c_{53} y^{2-\beta_1} \ln y + c_{54} y^{\beta_1} \ln y + c_{55} y^{-\beta_1} \ln y,$$

$$F_5(y) = c_{56} y^7 + c_{57} y^5 + c_{58} y^3 + c_{59} y^{4+\beta_1} + c_{60} y^{4-\beta_1} + c_{61} y^{2+\beta_1} + c_{62} y^{2-\beta_1},$$

$$F_6(y) = c_{63} y^3 + c_{64} y,$$

$$\alpha_i = \delta_1\{[F_{i-3}(1)(\beta_5 - \beta_1)\alpha^{-1-\beta_1} - F'_{i-3}(\alpha) - \beta_5 \overline{F}_{i-3}(\alpha)]\alpha_1^{9-i}\}, \quad i = 6 \sim 9$$

$$\alpha_i = F_{i-7}(1)\alpha_1^{13-i} - \alpha_{i-4}, \quad i = 10 \sim 13$$

$$c_1 = \frac{a_1 b_1}{121 - \beta_1^2}, \qquad c_2 = \frac{a_1 b_2 - a_5 b_1}{81 - \beta_1^2},$$

$$c_3 = \frac{a_1 b_3 - a_5 b_2 + a_6 b_1}{49 - \beta_1^2},$$

$$c_4 = \frac{a_1 b_4 - a_5 b_3 + a_6 b_2 + a_3 b_9 + a_4 b_8}{25 - \beta_1^2},$$

$$c_5 = \frac{a_1 b_5 - a_5 b_4 + a_6 b_3 + a_3 b_{11} + a_4 b_{10}}{9 - \beta_1^2},$$

$$c_6 = -\frac{a_5 b_5 - a_6 b_4 - a_4 b_{13} - a_3 \alpha_4 - a_4 \alpha_2}{1 - \beta_1^2} - \frac{2(a_3 b_{14} + a_4 b_{12})}{(1 - \beta_1^2)^2},$$

$$c_7 = \frac{a_6 b_5}{1 - \beta_1^2}, \qquad c_8 = \frac{a_1 b_8 + a_3 b_1}{(8 + \beta_1)^2 - \beta_1^2},$$

$$c_9 = \frac{a_1 b_9 + a_4 b_1}{(8 - \beta_1)^2 - \beta_1^2}, \qquad c_{10} = \frac{a_1 b_{10} - a_5 b_8 + a_3 b_2}{(6 + \beta_1)^2 - \beta_1^2},$$

$$c_{11} = \frac{a_1 b_{11} - a_5 b_9 + a_4 b_2}{(6 - \beta_1)^2 - \beta_1^2},$$

$$c_{12} = \frac{a_1 b_{13} - a_5 b_{10} + a_6 b_8 + a_3 b_{13} + a_1 \alpha_2}{(4 + \beta_1)^2 - \beta_1^2} - \frac{2 a_1 b_{12}(4 + \beta_1)}{[(4 + \beta_1)^2 - \beta_1^2]^2},$$

$$c_{13}=-\frac{a_5b_{11}-a_6b_9-a_4b_3-a_1\alpha_4}{(4-\beta_1)^2-\beta_1^2}-\frac{2a_1b_{14}(4-\beta_1)}{[(4-\beta_1)^2-\beta_1^2]^2},$$

$$c_{14}=\frac{a_6b_{10}-a_5b_{13}+a_6b_4+a_4b_6-a_5\alpha_2}{(2+\beta_1)^2-\beta_1^2}-\frac{2a_5b_{12}(2+\beta_1)}{[(2+\beta_1)^2-\beta_1^2]^2},$$

$$c_{15}=\frac{a_6b_{11}+a_3b_7+a_4b_4-a_5\alpha_4}{(2-\beta_1)^2-\beta_1^2}-\frac{2a_5b_{14}(2-\beta_1)}{[(2-\beta_1)^2-\beta_1^2]^2},$$

$$c_{16}=-\frac{a_6b_{13}+a_3b_5+a_6\alpha_2}{4\beta_1^2}-\frac{a_5b_{12}-a_6b_{12}}{8\beta_1^2},$$

$$c_{17}=\frac{a_1b_6+a_3b_8}{(5+2\beta_1)^2-\beta_1^2},\qquad c_{18}=\frac{a_1b_7+a_4b_9}{(5-2\beta_1)^2-\beta_1^2},$$

$$c_{19}=-\frac{a_5b_6-a_3b_{10}}{(3+2\beta_1)^2-\beta_1^2},\qquad c_{20}=-\frac{a_5b_7-a_4b_{11}}{(3-2\beta_1)^2-\beta_1^2},$$

$$c_{21}=\frac{a_6b_6+a_3b_{13}+a_3\alpha_2}{(1+2\beta_1)^2-\beta_1^2}-\frac{2a_3b_{12}(1+2\beta_1)}{[(1+2\beta_1)^2-\beta_1^2]^2},$$

$$c_{22}=\frac{a_6b_7+a_4\alpha_4}{(1-2\beta_1)^2-\beta_1^2}-\frac{2a_4b_{14}(1-2\beta_1)}{[(1-2\beta_1)^2-\beta_1^2]^2},$$

$$c_{23}=\frac{a_3b_6}{(2+3\beta_1)^2-\beta_1^2},\qquad c_{24}=\frac{a_4b_7}{(2-3\beta_1)^2-\beta_1^2},$$

$$c_{25}=\frac{a_1b_{12}}{(4+\beta_1)^2-\beta_1^2},\qquad c_{26}=\frac{a_1b_{14}}{(4-\beta_1)^2-\beta_1^2},$$

$$c_{27}=-\frac{a_5b_{12}}{(2+\beta_1)^2-\beta_1^2},\qquad c_{28}=-\frac{a_5b_{14}}{(2-\beta_1)^2-\beta_1^2},$$

$$c_{29}=\frac{a_6b_{13}+a_3b_5+a_6\alpha_2}{2\beta_1}-\frac{a_6b_{12}}{4\beta_1^2},$$

$$c_{30}=-\frac{a_6b_5-a_6\alpha_4}{2\beta_1}-\frac{a_6b_{14}}{4\beta_1^2},$$

$$c_{31}=\frac{a_3b_{12}}{(1+2\beta_1)^2-\beta_1^2},\qquad c_{32}=\frac{a_4b_{14}}{(1-2\beta_1)^2-\beta_1^2},$$

$$c_{33}=\frac{a_3b_{14}+a_4b_{12}}{1-\beta_1^2},\qquad c_{34}=\frac{a_6b_{12}}{4\beta_1},$$

$$c_{35}=-\frac{a_6b_{14}}{4\beta_1},\qquad c_{36}=\frac{b_1+a_1b_{15}}{81-\beta_1^2},$$

$$c_{37}=\frac{b_2+a_1b_{16}-a_5b_{15}}{49-\beta_1^2},\qquad c_{38}=\frac{b_3+a_1b_{17}-a_5b_{16}+a_6b_{15}}{25-\beta_1^2},$$

$$c_{39}=\frac{b_4-a_5b_{17}+a_6b_{16}+a_3b_{19}+a_4b_{18}}{9-\beta_1^2},$$

$$c_{40}=\frac{b_5+a_6b_{17}+a_3\alpha_5+a_4\alpha_3}{1-\beta_1^2},\qquad c_{41}=\frac{b_8+a_1b_{18}+a_3b_{15}}{(6+\beta_1)^2-\beta_1^2},$$

$$c_{42}=\frac{b_9+a_1b_{19}+a_4b_{15}}{(6-\beta_1)^2-\beta_1^2},\qquad c_{43}=\frac{b_{10}-a_5b_{18}+a_3b_{16}+a_1\alpha_3}{(4+\beta_1)^2-\beta_1^2},$$

$$c_{44}=\frac{b_{11}-a_5b_{19}+a_4b_{16}+a_1\alpha_5}{(4-\beta_1)^2-\beta_1^2},$$

$$c_{45}=\frac{b_{13}+a_6b_{18}+a_3b_{17}+\alpha_2-a_5\alpha_3}{(2+\beta_1)^2-\beta_1^2}-\frac{2b_{12}(2+\beta_1)}{[(2+\beta_1)^2-\beta_1^2]^2},$$

$$c_{46}=\frac{a_6b_{19}+a_4b_{17}+\alpha_4-a_5\alpha_5}{(2-\beta_1)^2-\beta_1^2}-\frac{2b_{14}(2-\beta_1)}{[(2-\beta_1)^2-\beta_1^2]^2},$$

$$c_{47}=\frac{b_6+a_3b_{18}}{(3+2\beta_1)^2-\beta_1^2},\qquad c_{48}=\frac{b_7+a_4b_{19}}{(3-2\beta_1)^2-\beta_1^2},$$

$$c_{49}=\frac{a_3\alpha_3}{(1+2\beta_1)^2-\beta_1^2},\qquad c_{50}=\frac{a_4\alpha_5}{(1-2\beta_1)^2-\beta_1^2},$$

$$c_{51}=-\frac{a_6\alpha_3}{4\beta_1^2},\qquad c_{52}=\frac{b_{12}}{(2+\beta_1)^2-\beta_1^2},$$

$$c_{53}=\frac{b_{14}}{(2-\beta_1)^2-\beta_1^2},\qquad c_{54}=\frac{a_6\alpha_3}{2\beta_1},$$

$$c_{55}=-\frac{a_6\alpha_5}{2\beta_1},\qquad c_{56}=\frac{b_{15}}{49-\beta_1^2},$$

$$c_{57}=\frac{b_{16}}{25-\beta_1^2},\qquad c_{58}=\frac{b_{17}}{9-\beta_1^2},$$

$$c_{59}=\frac{b_{18}}{(4+\beta_1)^2-\beta_1^2},\qquad c_{60}=\frac{b_{19}}{(4-\beta_1)^2-\beta_1^2},$$

$$c_{61}=\frac{\alpha_3}{(2+\beta_1)^2-\beta_1^2},\qquad c_{62}=\frac{\alpha_5}{(2-\beta_1)^2-\beta_1^2},$$

$$c_{63}=\frac{1}{9-\beta_1^2},\qquad c_{64}=-\frac{\alpha^2}{1-\beta_1^2}\tag{28}$$

将解（27）代入式（24a），并应用式（18），便得这一层合扁球壳的非线性特征关系式：

$$P=\lambda_0W_m^2+k\lambda_1W_m^2+(k^2\lambda_2+\lambda_3)W_m\tag{29}$$

其中

$$\lambda_i=-\frac{F_{i+7}(1)-F_{i+7}(\alpha)}{F_{10}(1)-F_{10}(\alpha)},\quad i=0\sim2$$

$$\lambda_3=\frac{1}{F_{(10)}(1)-F_{10}(\alpha)},$$

$$\begin{aligned}F_7(y)=\alpha_1^3(&d_1y^{12}+d_2y^{10}+d_3y^8+d_4y^6+d_5y^4\\&+d_6y^2+d_7y^{-2}+d_8y^{9+\beta_1}+d_9y^{9-\beta_1}+d_{10}y^{7+\beta_1}\\&+d_{11}y^{7-\beta_1}+d_{12}y^{5+\beta_1}+d_{13}y^{5-\beta_1}+d_{14}y^{3+\beta_1}+d_{15}y^{3-\beta_1}\\&+d_{16}y^{1+\beta_1}+d_{17}y^{1-\beta_1}+d_{18}y^{-1+\beta_1}+d_{19}y^{-1-\beta_1}\\&+d_{20}y^{6+2\beta_1}+d_{21}y^{6-2\beta_1}+d_{22}y^{4+2\beta_1}+d_{23}y^{4-2\beta_1}+d_{24}y^{2+2\beta_1}\\&+d_{25}y^{2-2\beta_1}+d_{26}y^{2\beta_1}+d_{27}y^{-2\beta_1}+d_{28}y^{3+3\beta_1}+d_{29}y^{3-3\beta_1}\\&+d_{30}y^{1+3\beta_1}+d_{31}y^{1-3\beta_1}+d_{32}y^{5+\beta_1}\ln y+d_{33}y^{5-\beta_1}\ln y\\&+d_{34}y^{3+\beta_1}\ln y+d_{35}y^{3-\beta_1}\ln y+d_{36}y^{1+\beta_1}\ln y+d_{37}y^{1-\beta_1}\ln y\end{aligned}$$

$$+d_{38}y^{-1+\beta_1}\ln y+d_{39}y^{-1-\beta_1}\ln y+d_{40}y^{2+2\beta_1}\ln y+d_{41}y^{2-2\beta_1}\ln y$$
$$+d_{42}y^{2\beta_1}\ln y+d_{43}y^{-2\beta_1}\ln y+d_{44}y^2\ln y+d_{45}y^{1+\beta_1}\ln^2 y$$
$$+d_{46}y^{1-\beta_1}\ln^2 y+d_{47}\ln^2 y+d_{48}\ln y)+d_{49}y^{1+\beta_1}+d_{50}y^{1-\beta_1},$$

$$F_8(y)=\alpha_1^2(d_{51}y^{10}+d_{52}y^8+d_{53}y^6+d_{54}y^4$$
$$+d_{55}y^2+d_{56}y^{7+\beta_1}+d_{57}y^{7-\beta_1}+d_{58}y^{5+\beta_1}+d_{59}y^{5-\beta_1}$$
$$+d_{60}y^{3+\beta_1}+d_{61}y^{3-\beta_1}+d_{62}y^{1+\beta_1}+d_{63}y^{1-\beta_1}+d_{64}y^{-1+\beta_1}$$
$$+d_{65}y^{-1-\beta_1}+d_{66}y^{4+2\beta_1}+d_{67}y^{4-2\beta_1}+d_{68}y^{2+2\beta_1}+d_{69}y^{2-2\beta_1}$$
$$+d_{70}y^{2\beta_1}+d_{71}y^{-2\beta_1}+d_{72}y^{3+\beta_1}\ln y+d_{73}y^{3-\beta_1}\ln y+d_{74}y^{1+\beta_1}\ln y$$
$$+d_{75}y^{1-\beta_1}\ln y+d_{76}\ln y)+d_{77}y^{1+\beta_1}+d_{78}y^{1-\beta_1},$$

$$F_9(y)=\alpha_1(d_{79}y^8+d_{80}y^6+d_{81}y^4+d_{82}y^2+d_{83}y^{5+\beta_1}$$
$$+d_{84}y^{5-\beta_1}+d_{85}y^{3+\beta_1}+d_{86}y^{3-\beta_1}+d_{87}y^{1+\beta_1}$$
$$+d_{88}y^{1-\beta_1})+d_{89}y^{1+\beta_1}+d_{90}y^{1-\beta_1},$$

$$F_{10}(y)=d_{91}y^4+d_{92}y^2+d_{93}y^{1+\beta_1}+d_{94}y^{1-\beta_1}+d_{95}\ln y,$$

$$d_1=-\frac{c_1}{12},\qquad d_2=\frac{a_1b_1-mc_2}{10m},$$

$$d_3=\frac{a_1b_2-a_5b_1-mc_3}{8m},\qquad d_4=\frac{a_1b_3-a_5b_2+a_6b_1-mc_4}{6m},$$

$$d_5=\frac{a_1b_4-a_5b_3+a_6b_2+a_3b_9+a_4b_8-mc_5}{4m},$$

$$d_6=\frac{a_1b_5-a_5b_4+a_6b_3+a_3b_{11}+a_4b_{10}-mc_6}{2m}-\frac{c_{31}}{4},$$

$$d_7=-\frac{a_6b_5}{2m},\qquad d_8=-\frac{c_8}{9+\beta_1},$$

$$d_9=-\frac{c_9}{9-\beta_1},\qquad d_{10}=\frac{a_1b_8+a_3b_1-mc_{10}}{(7+\beta_1)m},$$

$$d_{11}=\frac{a_1b_9+a_4b_1-mc_{11}}{(7-\beta_1)m},$$

$$d_{12}=\frac{a_1b_{10}-a_5b_8+a_3b_2-mc_{12}}{(5+\beta_1)m}+\frac{c_{25}}{(5+\beta_1)^2},$$

$$d_{13}=\frac{a_1b_{11}-a_5b_9+a_4b_2-mc_{13}}{(5-\beta_1)m}+\frac{c_{26}}{(5-\beta_1)^2},$$

$$d_{14}=\frac{a_1b_{13}-a_5b_{10}+a_6b_8+a_3b_3+a_1\alpha_2-mc_{14}}{(3+\beta_1)m}-\frac{a_1b_{12}-mc_{27}}{(3+\beta_1)^2m},$$

$$d_{15}=-\frac{a_5b_{11}-a_6b_9-a_4b_3-a_1\alpha_4+mc_{15}}{(3-\beta_1)m}-\frac{a_1b_{14}-mc_{28}}{(3-\beta_1)^2m},$$

$$d_{16}=\frac{a_6b_{10}-a_5b_{13}+a_3b_4+a_4\alpha_6-a_5\alpha_2-mc_{16}}{(1+\beta_1)m}-\frac{a_5b_{12}-mc_{29}}{(1+\beta_1)^2m}-\frac{2c_{34}}{(1+\beta_1)^3},$$

$$d_{17}=\frac{a_6b_{11}+a_3b_7+a_4b_4-a_5\alpha_4}{(1-\beta_1)m}-\frac{a_5b_{14}+mc_{30}}{(1-\beta_1)^2m}-\frac{2c_{35}}{(1-\beta_1)^3},$$

$$d_{18}=-\frac{a_6b_{13}+a_3b_5+a_6\alpha_2}{(1-\beta_1)m}+\frac{a_6b_{12}}{(1-\beta_1)^2},$$

$$d_{19}=-\frac{a_4b_5+a_6\alpha_4}{(1+\beta_1)m}+\frac{a_6b_{14}}{(1+\beta_1)^2},\quad d_{20}=-\frac{c_{17}}{2(3+\beta_1)},$$

$$d_{21}=-\frac{c_{18}}{2(3-\beta_1)},\quad d_{22}=\frac{a_1b_6+a_3b_8-mc_{19}}{2(2+\beta_1)m},$$

$$d_{23}=\frac{a_1b_7+a_4b_9-mc_{20}}{2(2-\beta_1)m},$$

$$d_{24}=-\frac{a_5b_6-a_3b_{10}+mc_{21}}{2(1+\beta_1)m}+\frac{c_{31}}{4(1+\beta_1)^2},$$

$$d_{25}=-\frac{a_5b_7-a_4b_{11}+mc_{22}}{2(1-\beta_1)m}+\frac{c_{32}}{4(1-\beta_1)^2},$$

$$d_{26}=\frac{a_6b_6+a_3b_{13}+a_3\alpha_2}{2\beta_1m}-\frac{a_3b_{12}}{4\beta_1^2m},\quad d_{27}=\frac{a_6b_7+a_4\alpha_4}{2\beta_1m}-\frac{a_4b_{14}}{4\beta_1^2m},$$

$$d_{28}=-\frac{c_{23}}{3(1+\beta_1)},\quad d_{29}=-\frac{c_{24}}{3(1-\beta_1)},$$

$$d_{30}=\frac{a_3b_6}{(1+3\beta_1)m},\quad d_{31}=\frac{a_4b_7}{(1-3\beta_1)m},$$

$$d_{32}=-\frac{c_{25}}{5+\beta_1},\quad d_{33}=-\frac{c_{26}}{5-\beta_1},$$

$$d_{34}=\frac{a_1b_{12}-mc_{27}}{(3+\beta_1)m},\quad d_{35}=\frac{a_1b_{14}-mc_{28}}{(3-\beta_1)m},$$

$$d_{36}=-\frac{a_5b_{12}+mc_{29}}{(1+\beta_1)m}+\frac{2c_{34}}{(1+\beta_1)^2},$$

$$d_{37}=-\frac{a_5b_{14}+mc_{30}}{(1-\beta_1)m}+\frac{2c_{35}}{(1-\beta_1)^2},$$

$$d_{38}=-\frac{a_6b_{12}}{(1-\beta_1)m},\quad d_{39}=-\frac{a_6b_{14}}{(1+\beta_1)m},$$

$$d_{40}=-\frac{c_{31}}{2(1+\beta_1)},\quad d_{41}=-\frac{c_{32}}{2(1-\beta_1)},$$

$$d_{42}=\frac{a_3b_{12}}{2\beta_1m},\quad d_{43}=-\frac{a_4b_{14}}{2\beta_1m},$$

$$d_{44}=-\frac{c_{33}}{2},\quad d_{45}=-\frac{c_{34}}{1+\beta_1},$$

$$d_{46}=-\frac{c_{35}}{1-\beta_1},\quad d_{47}=\frac{a_3b_{14}+a_4b_{12}}{2m},$$

$$d_{48}=-\frac{a_5b_5-a_6b_4-a_4b_{13}-a_3\alpha_4-a_4\alpha_2+mc_7}{m},$$

$$d_{49}=\frac{\alpha_6}{1+\beta_1},\quad d_{50}=\frac{\alpha_{10}}{1-\beta_1},$$

$$d_{51}=-\frac{c_{36}}{10},\quad d_{52}=\frac{b_1+a_1b_{15}-mc_{37}}{8m},$$

$$d_{53}=\frac{b_2+a_1b_{16}-a_5b_{15}-mc_{38}}{6m},$$

$$d_{54}=\frac{b_3+a_1b_{17}-a_5b_{16}+a_6b_{15}-mc_{39}}{4m},$$

$$d_{55}=\frac{b_4-a_5b_{17}+a_6b_{16}+a_3b_{19}+a_4b_{18}-mc_{40}}{2m},$$

$$d_{56}=-\frac{c_{41}}{7+\beta_1},\qquad d_{57}=-\frac{c_{42}}{7-\beta_1},$$

$$d_{58}=\frac{b_8+a_1b_{18}+a_3b_{15}-mc_{43}}{(5+\beta_1)m},$$

$$d_{59}=\frac{b_9+a_1b_{19}+a_4b_{15}-mc_{44}}{(5-\beta_1)m},$$

$$d_{60}=\frac{b_{10}-a_5b_{18}+a_3b_{16}+a_1\alpha_3-mc_{45}}{(3+\beta_1)m}+\frac{c_{52}}{(3+\beta_1)^2},$$

$$d_{61}=\frac{b_{11}-a_5b_{19}+a_4b_{16}+a_1\alpha_5-mc_{46}}{(3-\beta_1)m}+\frac{c_{53}}{(3-\beta_1)^2},$$

$$d_{62}=\frac{b_{13}+a_6b_{18}+a_3b_{17}+\alpha_2-a_5\alpha_3-mc_{51}}{(1+\beta_1)m}-\frac{b_{12}-mc_{54}}{(1+\beta_1)^2m},$$

$$d_{63}=\frac{a_6b_{19}+a_4b_{17}+\alpha_4-a_5\alpha_5}{(1-\beta_1)m}+\frac{b_{14}+mc_{55}}{(1-\beta_1)^2m},$$

$$d_{64}=-\frac{a_6\alpha_3}{(1-\beta_1)m},\qquad d_{65}=-\frac{a_6\alpha_5}{(1+\beta_1)m},$$

$$d_{66}=-\frac{c_{47}}{2(2+\beta_1)m},\qquad d_{67}=-\frac{c_{48}}{2(2-\beta_1)},$$

$$d_{68}=\frac{b_6+a_3b_{18}-mc_{49}}{2(1+\beta_1)m},\qquad d_{69}=\frac{b_7+a_4b_{19}-mc_{50}}{2(1-\beta_1)m},$$

$$d_{70}=\frac{a_3\alpha_3}{2\beta_1m},\qquad d_{71}=\frac{a_4\alpha_5}{2\beta_1m},$$

$$d_{72}=-\frac{c_{52}}{3+\beta_1},\qquad d_{73}=-\frac{c_{53}}{3-\beta_1},$$

$$d_{74}=\frac{b_{12}-mc_{54}}{(1+\beta_1)m},\qquad d_{75}=-\frac{b_{14}+mc_{55}}{(1-\beta_1)m},$$

$$d_{76}=\frac{b_5+a_6b_{17}+a_3\alpha_5+a_4\alpha_3}{m},\qquad d_{77}=\frac{\alpha_7}{1+\beta_1},$$

$$d_{78}=\frac{\alpha_{11}}{1-\beta_1},\qquad d_{79}=-\frac{c_{56}}{8},$$

$$d_{80}=\frac{b_{15}-mc_{57}}{6m},\qquad d_{81}=\frac{b_{16}-mc_{58}}{4m},$$

$$d_{82}=\frac{b_{17}}{2m},\qquad d_{83}=-\frac{c_{59}}{5+\beta_1},$$

$$d_{84}=-\frac{c_{60}}{5-\beta_1},\qquad d_{85}=\frac{b_{18}-mc_{61}}{(3+\beta_1)m},$$

$$d_{86}=-\frac{b_{19}-mc_{62}}{(3-\beta_1)m},\qquad d_{87}=\frac{\alpha_3}{(1+\beta_1)m},$$

$$d_{88}=\frac{\alpha_5}{(1-\beta_1)m},\qquad d_{89}=\frac{\alpha_8}{1+\beta_1},$$
$$d_{90}=\frac{\alpha_{12}}{1-\beta_1},\qquad d_{91}=-\frac{c_{63}}{4},$$
$$d_{92}=\frac{1-mc_{64}}{2m},\qquad d_{93}=\frac{\alpha_9}{1+\beta_1},$$
$$d_{94}=\frac{\alpha_{13}}{1-\beta_1},\qquad d_{95}=-\frac{\alpha^2}{m} \tag{30}$$

对式（29）求导，并令其为零，则有

$$3\lambda_0 W_m^2+2k\lambda_1 W_m+(k^2\lambda_2+\lambda_3)=0 \tag{31}$$

解此方程，便得壳体屈曲时的无量纲临界内边缘挠度

$$W_m^*=-\frac{k\lambda_1+\sqrt{k^2(\lambda_1^2-3\lambda_0\lambda_2)-3\lambda_0\lambda_3}}{3\lambda_0} \tag{32}$$

应用此式，则由式（29）得到无量纲临界荷载

$$P^*=\lambda_0 W_m^{*3}+k\lambda_1 W_m^{*2}+(k^2\lambda_2+\lambda_3)W_m^* \tag{33}$$

令方程（31）的判别式为零，可得如下临界几何参数 k_0 的计算公式

$$k_0=\sqrt{\frac{3\lambda_0\lambda_3}{\lambda_1^2-3\lambda_0\lambda_2}} \tag{34}$$

显然，当 $k\geqslant k_0$ 时，壳体将发生屈曲；当 $k<k_0$ 时，壳体不会屈曲。

四、算例分析

现在，我们应用前面所获得的对称层合圆柱正交异性中心开孔扁球壳的非线性临界荷载的计算公式进行实例计算。为计算简单，且不失一般性，我们假设壳体各层厚度相等、弹性常数相同，且有

$$E_\theta/E_r=1.5,\quad \nu_{\theta r}=0.3$$

于是，便有

$$\beta_1^2=\beta_2^2=1.5,\quad \beta_3=16.92,$$
$$\beta_4=\beta_5=0.3$$

我们将不同的无量纲中心孔半径 α、无量纲剪切刚度 m 以及几何参数 k 值下的临界荷载计算结果绘成曲线，如图 2～图 7 所示，而无量纲中心孔半径 α 下的临界几何参数 k_0 的计算结果则如图 8、图 9 所示。值得指出，在 α 趋于零，即壳体封顶情况下，本文结果与前文[5]的结果一致。

从图 2～图 7 中，我们可以看出：

（1）临界荷载随壳体无量纲几何参数 k 的增大而增大。

（2）在同一无量纲中心孔半径的情况下，临界荷载在 k 较小时，随 k 的增大而增大，而当 k 较大时，则随 k 的增大而减小。

（3）临界荷载随无量纲中心孔半径 α 的增大而减小。

由图 8、图 9 看出，临界几何参数 k_0 随着壳体中心孔半径的增大而减小，而且，一般说来，它又随着无量纲剪切刚度 m 的增加而增加。

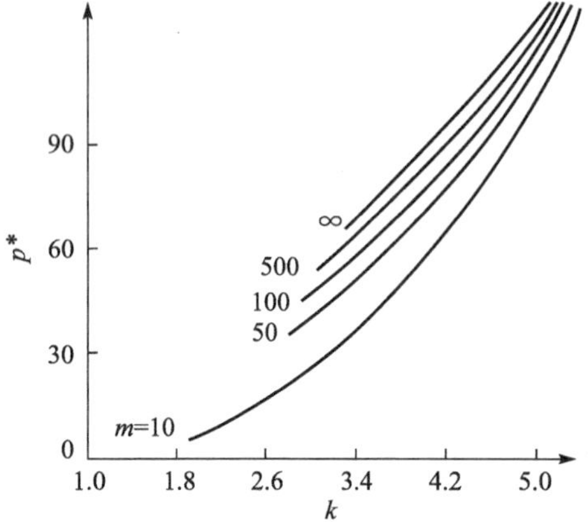

图 2　剪切刚度对临界荷载的影响
（$\alpha=0$）

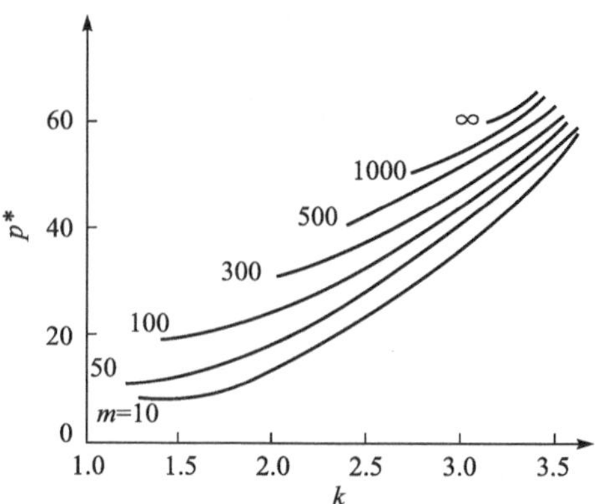

图 3　剪切刚度对临界荷载的影响
（$\alpha=0.1$）

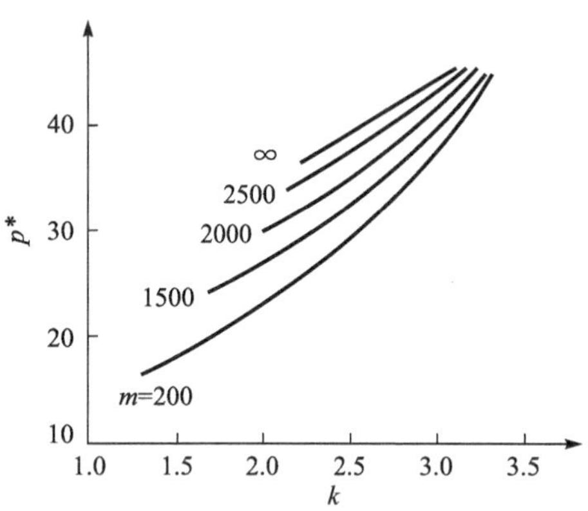

图 4　剪切刚度对临界荷载的影响
（$\alpha=0.2$）

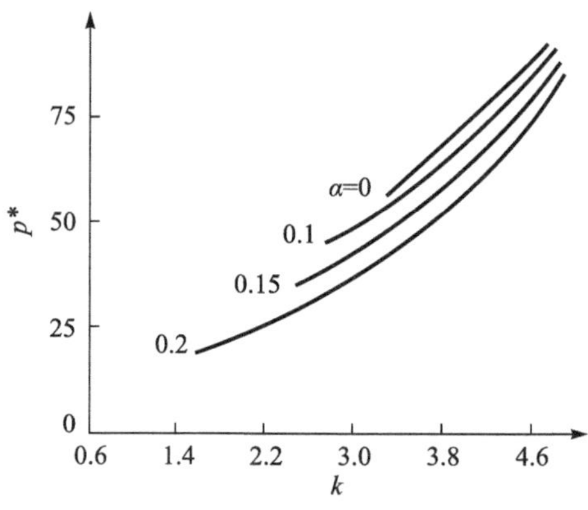

图 5　开孔半径对临界荷载的影响
（$m=10^3$）

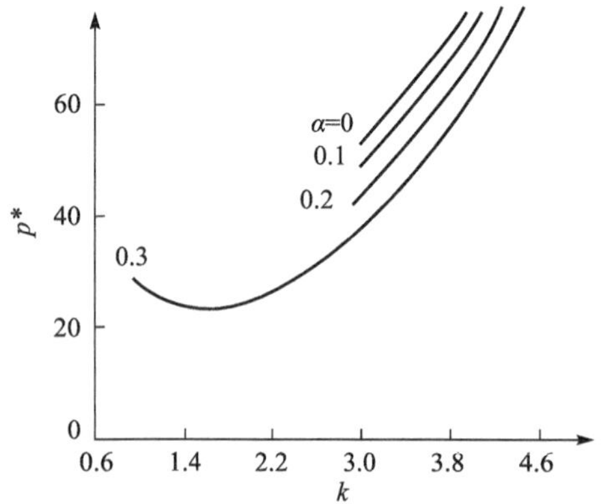

图 6　开孔半径对临界荷载的影响
（$m=10^4$）

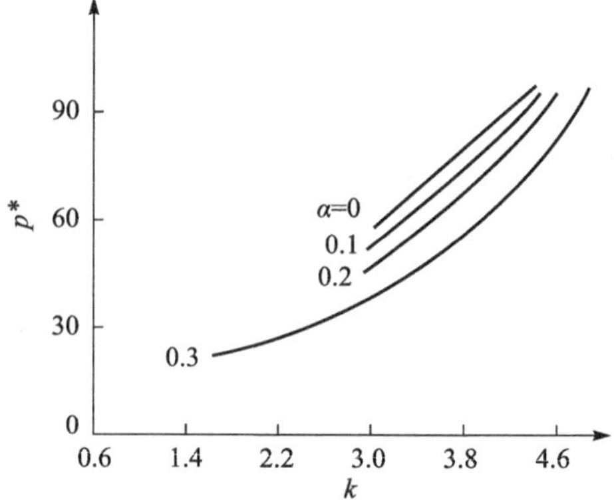

图 7　开孔半径对临界荷载的影响
（$m=1.5\times10^4$）

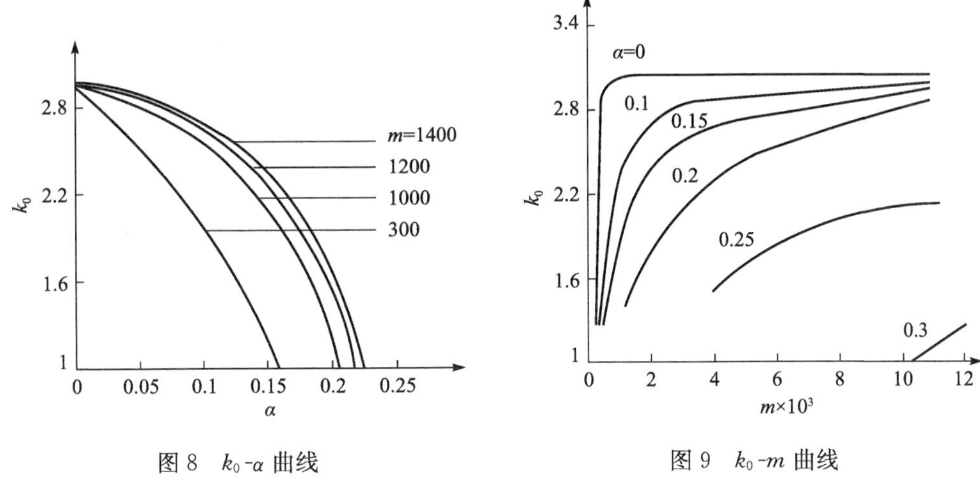

图 8　k_0-α 曲线　　　　图 9　k_0-m 曲线

参考文献

[1] Alfutov N A，Popov B G. Nonlinear analysis of composite structures// Herakkovich C T，et al. Handbook of Composites，Vol. 2-Structure and Design Amsterdam：Elsevier Sci，Pub，1989：115.

[2] Iu V P，Chia C Y. Non-linear vibration and postbuckling of unsymmetric cross-ply circular cylindrical shells. International Journal of Solids and Structures，1988，24 (2)：195.

[3] 王震鸣，游绍建，杨明. 复合材料圆柱曲板在轴压下的非线性弹性稳定问题. 复合材料学报，1988，5 (2)：37.

[4] 刘人怀. 对称圆柱正交异性层合扁球壳的非线性稳定性问题. 应用数学和力学，1991，12 (3)：251.

[5] 刘人怀. 考虑横向剪切对称层合圆柱正交异性扁球壳的非线性稳定问题. 中国科学，A 辑，1991，(7)：742.

[6] 叶开沅，刘人怀，平庆元，等. 在轴对称线布载荷作用下圆底扁球壳的非线性稳定性问题. 科学通报，1965，(2)：142.

[7] 叶开沅，刘人怀，张传智，等. 圆底扁球薄球壳在边缘力矩作用下的非线性稳定性问题. 科学通报，1965，(2)：145.

[8] 刘人怀. 在内边缘均布力矩作用下中心开孔圆底扁球壳的非线性稳定问题. 科学通报，1965，(3)：253.

Non-linear buckling of symmetrically laminated, cylindrically orthotropic, shallow, conical shells considering shear[①]

中文摘要

本文建立了在轴对称分布载荷作用下考虑横向剪切影响的对称层合圆柱正交异性扁锥壳的非线性弯曲理论。使用修正迭代法，求解了在均布压力作用下具有夹紧固定边界的这一壳体的非线性屈曲问题，得到了这一壳体临界压力的相当精确的解，可直接应用到结构元件的设计里，本问题尚无人研究过。

1. Introduction

In recent years laminated composite plates and shells have become important structural members in various industries such as aeronautical, astronautical, naval, civil and vessel engineering. Thus, a great number of investigators studied bending, buckling, and vibration of these plates and shells[1-11]. However, to date a more important problem, namely, nonlinear buckling of symmetrically laminated, cylindrically orthotropic, shallow conical shells that take into account the effect of the transverse shear, has not been discussed because of difficulties due to nonlinear mathematics and complicated structures. For this reason, a theory for nonlinear bending of the symmetrically laminated, cylindrically orthotropic, shallow, conical shells subjected to an axisymmetrically distributed load including transverse shear effects is established in this paper. Basic equations of the theory are derived and solved for the nonlinear buckling problem of the shells with a rigid clamped edge under uniform lateral pressure using the modified iteration method as suggested by Yeh and Liu[12-14]. A quite accurate analytical solution of the critical buckling pressure for the shells is obtained, and can be applied directly to the design of structural members.

2. Basic Equations

The present investigation is concerned with nonlinear buckling of a symmetrically laminated, cylindrically orthotropic, shallow, conical shell subjected to an axisymmetrically distributed load $q(r)$. The geometry and coordinate system of the shell is shown in Fig. 1. Here, h is the total thickness of the shell, n is the number of layers, a is the radius, α is the slope angle, r is the radial coordinate, θ is the circumferential coordinate and z is the coordinate in the thickness direction.

① Reprinted from *International Journal of Non-Linear Mechanics*, 1996, 31 (1): 89-99.

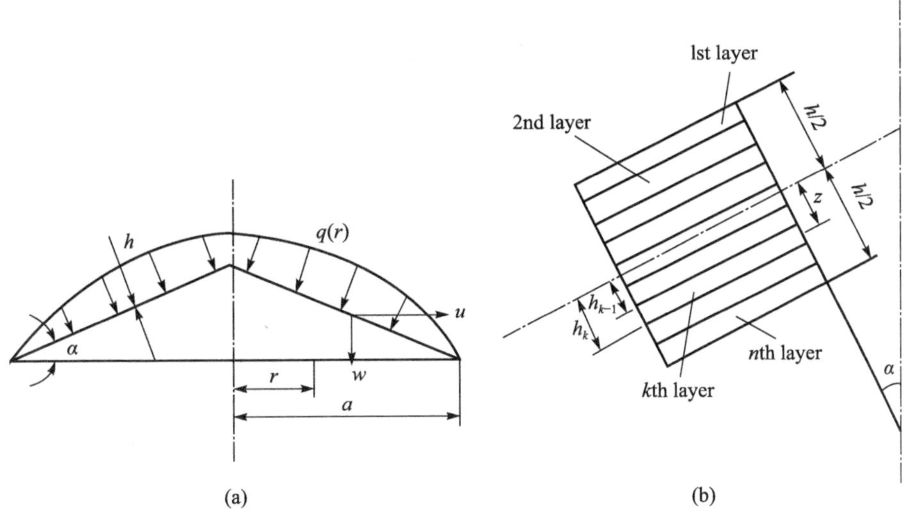

Fig. 1 (a) and (b). Geometry of symmetrically laminated, cylindrically orthotropic, shallow, conical shell

In the following discussion on the stresses and deformations in the shell, let us introduce the following assumptions:

(1) the material is linearly elastic;

(2) any layer of the shell has uniform thickness, and each layer may have different thickness and elastic constants, but must be symmetrical with respect to the middle surface of the shell;

(3) the shell is transversely incompressible;

(4) the normal line to the middle surface of the shell remains straight during bending;

(5) the transverse normal stress of the shell can be neglected.

Under these assumptions and the condition of symmetry, the radial, circumferential and vertical displacements at an arbitrary point of the shell may be approximated by

$$u = u_0 - z\psi, \quad v = 0, \quad w = w_0 \tag{2.1}$$

where u_0 and w_0 are the values of u and w at the middle surface, and ψ is the rotation of the normal to the middle surface of the shell in the r direction.

For the case of large deflection, the strain-displacement relations at an arbitrary point of the shell can be written in the form

$$\varepsilon_r = \frac{\partial u}{\partial r} + \frac{\mathrm{d}w}{\mathrm{d}r}\left(\alpha + \frac{1}{2}\frac{\mathrm{d}w}{\mathrm{d}r}\right), \quad \varepsilon_\theta = \frac{u}{r},$$

$$r_{rz} = \frac{\partial u}{\partial z} + \frac{\mathrm{d}w}{\mathrm{d}r}, \quad \varepsilon_z = \gamma_{r\theta} = \gamma_{\theta z} = 0. \tag{2.2}$$

Substituting Eqs. (2.1) into (2.2), we obtain

$$\varepsilon_r = \varepsilon_r^0 + z\kappa_r, \quad \varepsilon_\theta = \varepsilon_\theta^0 + z\kappa_\theta, \quad \gamma_{rz} = \frac{\mathrm{d}w}{\mathrm{d}r} - \psi \tag{2.3}$$

where ε_r^0 and ε_θ^0 are the strains of the middle surface of the shell, and κ_r and κ_θ are the curvatures:

$$\varepsilon_r^0 = \frac{du_0}{dr} + \frac{dw}{dr}\left(\alpha + \frac{1}{2}\frac{dw}{dr}\right), \quad \varepsilon_\theta^0 = \frac{u_0}{r},$$

$$\kappa_r = -\frac{d\psi}{dr}, \quad \kappa_\theta = -\frac{\psi}{r}. \tag{2.4a-d}$$

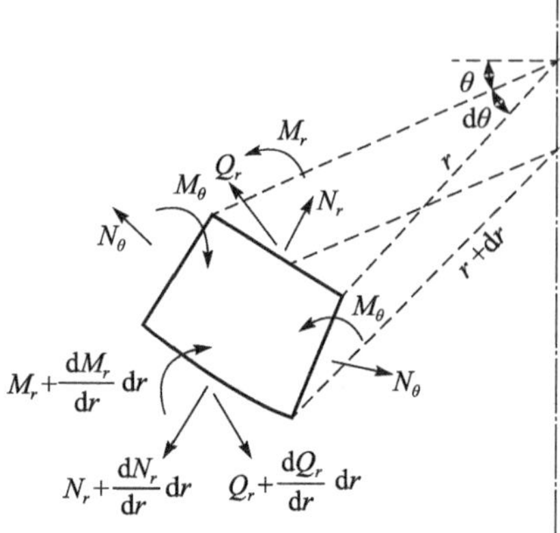

Fig. 2 An element of a symmetrically laminated, cylindrically orthotropic, shallow, conical shell

Hooke's laws for the kth layer of the shell are given by

$$\sigma_r^{(k)} = Q_{11}^{(k)}\varepsilon_r + Q_{12}^{(k)}\varepsilon_\theta, \quad \sigma_\theta^{(k)} = Q_{12}^{(k)}\varepsilon_r + Q_{22}^{(k)}\varepsilon_\theta,$$

$$\tau_{rz}^{(k)} = G_{rz}^{(k)}\gamma_{rz}, \quad \sigma_z^{(k)} = \tau_{r\theta}^{(k)} = \tau_{\theta z}^{(k)} = 0 \tag{2.5}$$

in which $E_r^{(k)}$ and $E_\theta^{(k)}$ are Young's moduli of the kth layer along the r and θ directions, respectively, $\nu_{r\theta}^{(k)}$ and $\nu_{\theta r}^{(k)}$ are Poisson's ratios of the kth layer, with the first subscript indicating the direction of the tensile force and the second indicating the direction of contraction, $G_{rz}^{(k)}$ is the transverse shear modulus of the kth layer, characterizing the change of the angle in the rz plane, and $Q_{ij}^{(k)}$ are the reduced stiffnesses of the kth layer,

$$Q_{11}^{(k)} = \frac{E_r^{(k)}}{1-\nu_{r\theta}^{(k)}\nu_{\theta r}^{(k)}}, \quad Q_{12}^{(k)} = \frac{\nu_{\theta r}^{(k)}E_r^{(k)}}{1-\nu_{r\theta}^{(k)}\nu_{\theta r}^{(k)}}, \quad Q_{22}^{(k)} = \frac{E_\theta^{(k)}}{1-\nu_{r\theta}^{(k)}\nu_{\theta r}^{(k)}}, \tag{2.6}$$

and these elastic constants satisfy the condition of elastic symmetry

$$\nu_{r\theta}^{(k)}E_\theta^{(k)} = \nu_{\theta r}^{(k)}E_r^{(k)}. \tag{2.7}$$

Now let us consider an element cut from the shell by two adjacent generatrix planes and two sections perpendicular to the generatrices (Fig. 2) . Forces and moments per unit length acting on each side of the element are the membrane forces N_r, N_θ, the bending moments M_r, M_θ, and the transverse shear force Q_r, which are defined by

$$N_r = \sum_{k=1}^{n}\int_{h_{k-1}}^{h_k}\sigma_\theta^{(k)}dz, \quad N_\theta = \sum_{k=1}^{n}\int_{h_{k-1}}^{h_k}\sigma_\theta^{(k)}dz,$$

$$M_r = \sum_{k=1}^{n}\int_{h_{k-1}}^{h_k}\sigma_\theta^{(k)}z\,dz, \quad M_\theta = \sum_{k=1}^{n}\int_{h_{k-1}}^{h_k}\sigma_\theta^{(k)}z\,dz,$$

$$Q_r = \sum_{k=1}^{n}\int_{h_{k-1}}^{h_k}\tau_{rz}^{(k)}dz. \tag{2.8a-e}$$

Substituting expressions (2.5) into (2.8), we have

$$N_r = A_{11}\varepsilon_r^0 + A_{12}\varepsilon_\theta^0, \quad N_\theta = A_{12}\varepsilon_r^0 + A_{22}\varepsilon_\theta^0, \tag{2.9}$$

$$M_r = D_{11}\kappa_r + D_{12}\kappa_\theta, \quad M_\theta = D_{12}\kappa_r + D_{22}\kappa_\theta, \tag{2.10}$$

$$Q_r = G\left(\frac{dw}{dr} - \psi\right) \tag{2.11}$$

in which A_{ij}, D_{ij} and G are the extensional, bending and shear rigidities, respectively,

$$A_{ij}=\sum_{k=1}^{n}Q_{ij}^{(k)}(h_k-h_{k-1}),\ D_{ij}=\frac{1}{3}\sum_{k=1}^{n}Q_{ij}^{(k)}(h_k^3-h_{k-1}^3)\quad i,j=1,2 \tag{2.12}$$

$$G=\sum_{k=1}^{n}Q_{rz}^{(k)}(h_k-h_{k-1}). \tag{2.13}$$

From expressoins (2.5) and (2.3), it is seen that the transverse shear stress varies across the thickness of the shell in a different way from the actual conditions. Therefore, we apply a necessary correction to the transverse shear stress. To obtain a simple satisfactory result, we take the transverse shear stress of the shell in the form

$$\tau_{rz}=\frac{3Q_r}{2h}\left[1-\left(\frac{2z}{h}\right)^2\right]. \tag{2.14}$$

Then by equating the complementary energy obtained by the transverse shear stress (2.14) to the complementary energy obtained by the transverse shear force (2.11), a more exact expression, instead of expression (2.13), is given by

$$G=\frac{4h^2}{9}\left\{\sum_{k=1}^{n}\frac{1}{G_{rz}^{(k)}}\left[h_k-h_{k-1}-\frac{8}{3h^2}(h_k^3-h_{k-1}^3)+\frac{16}{5h^4}(h_k^5-h_{k-1}^5)\right]\right\}^{-1} \tag{2.15}$$

Equations of equilibrium of the shell can be derived by considering the equilibrium of an element of the shell such as that shown in Fig. 2. Hence, we find

$$\frac{\mathrm{d}(rN_r)}{\mathrm{d}r}-N_\theta=0,$$

$$\frac{\mathrm{d}(rM_r)}{\mathrm{d}r}-M_\theta-rQ_r=0,$$

$$\frac{\mathrm{d}}{\mathrm{d}r}\left[rN_r\left(\alpha+\frac{\mathrm{d}w}{\mathrm{d}r}\right)+rQ_r\right]+rq=0. \tag{2.16a-c}$$

To derive the compatibility equation of the shell, Eqs. (2.9) may be rewritten in the following form:

$$\varepsilon_r^0=A_1N_r-A_2N_\theta,\quad \varepsilon_\theta^0=A_3N_\theta-A_2N_r \tag{2.17}$$

where

$$A_0=\frac{1}{A_{11}A_{22}-A_{12}^2},\ A_1=A_0A_{22},\ A_2=A_0A_{12},\ A_3=A_0A_{11}. \tag{2.18}$$

Eliminating u_0 in Eqs. (2.4a, b), and using equations (2.17), we obtain the following compatibility equation:

$$\frac{\mathrm{d}}{\mathrm{d}r}[r(A_3N_\theta-A_2N_r)]-A_1N_r+A_2N_\theta+\frac{\mathrm{d}w}{\mathrm{d}r}\left(\alpha+\frac{1}{2}\frac{\mathrm{d}w}{\mathrm{d}r}\right)=0. \tag{2.19}$$

From Eq. (2.16a), we have

$$N_\theta=\frac{\mathrm{d}(rN_r)}{\mathrm{d}r}. \tag{2.20}$$

By virtue of this equation, the compatibility equation becomes

$$A_3\frac{\mathrm{d}}{\mathrm{d}r}r\frac{\mathrm{d}(rN_r)}{\mathrm{d}r}-A_1N_r+\frac{1}{2}\left(\frac{\mathrm{d}w}{\mathrm{d}r}\right)^2+\alpha\frac{\mathrm{d}w}{\mathrm{d}r}=0. \tag{2.21}$$

Substituting Eqs. (2.10) and (2.11) into (2.16b), and using Eqs. (2.4c, d), we obtain

$$D_{11}\frac{\mathrm{d}}{\mathrm{d}r}r\frac{\mathrm{d}\psi}{\mathrm{d}r}-\left(Gr+\frac{D_{22}}{r}\right)\psi+Gr\frac{\mathrm{d}w}{\mathrm{d}r}=0. \tag{2.22}$$

Introduction of Eq. (2.11) in (2.16c) yields

$$\frac{\mathrm{d}}{\mathrm{d}r}\left[rN_r\left(\alpha+\frac{\mathrm{d}w}{\mathrm{d}r}\right)+Gr\left(\frac{\mathrm{d}w}{\mathrm{d}r}-\psi\right)\right]+rq=0. \tag{2.23}$$

Thus the investigation of nonlinear buckling of the symmetrically laminated, cylindrically orthotropic, shallow, conical shell subjected to axisymmetrically distributed load reduces to the solution of the three nonlinear differential equations (2.21), (2.22) and (2.23).

In the case of uniform lateral pressure q_0, multiplying Eq. (2.23) by dr and integrating, we have

$$rN_r\left(\alpha+\frac{\mathrm{d}w}{\mathrm{d}r}\right)+Gr\left(\frac{\mathrm{d}w}{\mathrm{d}r}-\psi\right)+\frac{1}{2}q_0r^2=0. \tag{2.24}$$

Now the basic equations (2.21), (2.22) and (2.24) will be solved under the following customary boundary conditions for a rigidly clamped edge:

$$\begin{aligned} &w=0,\quad \psi=0,\quad u_0=0 \quad \text{at } r=a\\ &\psi=0,\quad N_r \text{ finite} \quad \text{at } r=0, \end{aligned} \tag{2.25}$$

where

$$u_0=r\left[A_3\frac{\mathrm{d}(rN_r)}{\mathrm{d}r}-A_2N_r\right]. \tag{2.26}$$

Making the resulting solution more general, the following non-dimensional parameters are introduced:

$$\begin{aligned} &y=\frac{r}{a},\ W=\frac{w}{h},\ \varphi=\frac{\mathrm{d}W}{\mathrm{d}y}+k,\ \Psi=\frac{a}{h}\psi,\\ &S=\frac{a}{D_{11}}rN_r,\ P=\frac{a^4}{2D_{11}h}q_0,\ k=\frac{a}{h}\alpha,\ m=\frac{a^2}{D_{11}}G,\\ &\beta_1^2=\frac{D_{22}}{D_{11}},\ \beta_2^2=\frac{A_1}{A_3},\ \beta_3=\frac{h^2}{A_3D_{11}},\ \beta_4=\frac{A_2}{A_3}. \end{aligned} \tag{2.27}$$

Upon substitution, the nonlinear boundary-value problem (2.21), (2.22), (2.24) and (2.25) is transformed to the non-dimensional form

$$\begin{aligned} &my(\varphi-k)=-S\varphi+my\Psi-Py^2,\\ &L_1(y^{\beta_1}\Psi)=my(\Psi-\varphi+k),\\ &L_2(y^{\beta_2}S)=-\frac{\beta_3}{2}(\varphi^2-k^2). \end{aligned} \tag{2.28a-c}$$

$$W=0,\ \Psi=0,\ \frac{\mathrm{d}S}{\mathrm{d}y}-\beta_4\frac{S}{y}=0 \quad \text{at } y=1 \tag{2.29a-c}$$

$$\Psi=0,\ S=0 \quad \text{at } y=0 \tag{2.30a,b}$$

where L_1 and L_2 are the differential operators defined by

$$L_1(\cdots) = y^{\beta_1}\frac{d}{dy}y^{1-2\beta_1}\frac{d}{dy}(\cdots)$$

$$L_2(\cdots) = y^{\beta_2}\frac{d}{dy}y^{1-2\beta_2}\frac{d}{dy}(\cdots). \tag{2.31}$$

3. Formula of Critical Buckling Pressure

Now the non-dimensional, nonlinear boundary-value problem as stated above is solved by the modified iteration method. At first, we take the non-dimensional center deflection W_m as an iteration parameter. Then using the third expression of (2.27) and condition (2.29a), we have

$$W_m = -\int_0^1 (\varphi - k)\,dy. \tag{3.1}$$

For first approximation, by neglecting a nonlinear term $S\varphi$ in Eq. (2.28a), from problem (2.28)-(2.30) we obtain the non-dimensional linear boundary-value problem as follows:

$$my(\varphi_1 - k) = my\Psi_1 - Py^2,$$

$$L_1(y^{\beta_1}\Psi_1) = my(\Psi_1 - \varphi_1 + k),$$

$$L_2(y^{\beta_2}S_1) = -\frac{\beta_3}{2}(\varphi_1^2 - k^2), \tag{3.2a-c}$$

$$W_1 = 0,\quad \Psi_1 = 0,\quad \frac{dS_1}{dy} - \beta_4\frac{S_1}{y} = 0 \quad \text{at } y = 1 \tag{3.3a-c}$$

$$\Psi_1 = 0,\quad S_1 = 0, \quad \text{at } y = 0 \tag{3.4a,b}$$

From Eq. (3.2a), we find

$$\varphi_1 = \Psi_1 - \frac{P}{m}y + k. \tag{3.5}$$

Substituting this expression into the right-hand side of Eq. (3.2b) yields

$$L_1(y^{\beta_1}\Psi_1) = Py^2. \tag{3.6}$$

Using conditions (3.3b) and (3.4a), the solution of Eq. (3.6) is

$$\Psi_1 = \lambda P(y^{\beta_1} - y^3) \tag{3.7}$$

where

$$\lambda = -\frac{1}{9 - \beta_1^2}. \tag{3.8}$$

Inserting solution (3.7) into (3.5) leads to

$$\varphi_1 = P\left(\lambda y^{\beta_1} - \lambda y^3 - \frac{1}{m}y\right) + k. \tag{3.9}$$

By substitution and integration, expression (3.1) yields

$$P = \alpha_1 W_m \tag{3.10}$$

where

$$\alpha_1 = \frac{4m(1+\beta_1)(3+\beta_1)}{2(1+\beta_1)(3+\beta_1)+m}. \tag{3.11}$$

This is the characteristic relation of small deflection of a symmetrically laminated, cylindrically orthotropic, shallow, conical shell with a rigidly clamped edge.

Using relation (3.10), solutions (3.7) and (3.9) are rewritten in the form

$$\Psi_1 = \lambda\alpha_1 W_m(y^{\beta_1} - y^3)$$

$$\varphi_1 = \alpha_1 W_m\left(\lambda y^{\beta_1} - \lambda y^3 - \frac{1}{m}y\right) + k. \qquad (3.12\text{a,b})$$

Under conditions (3.3b) and (3.4b), and using solution (3.12b), the solution of Eq. (3.2c) is

$$S_1 = -\frac{\beta_3}{2}[\alpha_1^2 W_m^2(b_1 y^{1+2\beta_1} + b_2 y^{4+\beta_1} + b_3 y^{2+\beta_1} + b_4 y^{\beta_2} + b_5 y^7 + b_6 y^5 + b_7 y^3) + 2k\alpha_1 W_m(b_8 y^{1+\beta_1} + b_9 y^{\beta_2} + b_{10} y^4 + b_{11} y^2)] \qquad (3.13)$$

where

$$b_1 = \frac{\lambda^2}{(1+2\beta_1)^2 - \beta_2^2}, \qquad b_2 = -\frac{2\lambda^2}{(4+\beta_1)^2 - \beta_2^2},$$

$$b_3 = -\frac{2\lambda}{m[(2+\beta_1)^2 - \beta_2^2]},$$

$$b_4 = \frac{1}{\beta_4 - \beta_2}[b_1(1+2\beta_1 - \beta_4) + b_2(4+\beta_1 - \beta_4) + b_3(2+\beta_1 - \beta_4) + b_5(7-\beta_4) + b_6(5-\beta_4) + b_7(3-\beta_4)],$$

$$b_5 = \frac{\lambda^2}{49 - \beta_2^2}, \qquad b_6 = \frac{2\lambda}{m(25 - \beta_2^2)},$$

$$b_7 = \frac{1}{m^2(9 - \beta_2^2)}, \qquad b_8 = \frac{\lambda}{(1+\beta_1^2) - \beta_2^2},$$

$$b_9 = \frac{1}{\beta_4 - \beta_2}[b_8(1+\beta_1 - \beta_4) + b_{10}(4-\beta_4) + b_{11}(2-\beta_4)],$$

$$b_{10} = -\frac{\lambda}{16 - \beta_2^2}, \qquad b_{11} = -\frac{1}{m(4 - \beta_2^2)}. \qquad (3.14)$$

For the second approximation, according to problem (2.28)-(2.30), we have the non-dimensional linear boundary-value problem for φ_2 and Ψ_2 as follows:

$$my(\varphi_2 - k) = -S_1\varphi_1 + my\Psi_2 - Py^2,$$

$$L_1(y^{\beta_1}\Psi_2) = my(\Psi_2 - \varphi_2 + k), \qquad (3.15\text{a,b})$$

$$W_2 = 0, \quad \Psi_2 = 0 \qquad \text{at } y = 1 \qquad (3.16\text{a,b})$$

$$\Psi_2 = 0 \qquad \text{at } y = 0. \qquad (3.17)$$

From Eq. (3.15a), we obtain

$$\varphi_2 = -\frac{1}{my}S_1\varphi_1 + \Psi_2 - \frac{P}{m}y + k. \qquad (3.18)$$

Introducing this expression into the right-hand side of Eq. (3.15b), we find

$$L_1(y^{\beta_1}\Psi_2) = S_1\varphi_1 + py^2. \qquad (3.19)$$

Using the solutions (3.12b), (3.13) obtained, and the boundary conditions (3.16b), (3.17), the solution of Eq. (3.15b) can be expressed as

$$\Psi_2 = \lambda P(y^{\beta_1} - y^3) - \frac{\beta_3}{2}[\alpha_1^3 W_m^3(c_1 y^{2+3\beta_1} + c_2 y^{5+2\beta_1}$$

$$+c_3y^{3+2\beta_1}+c_4y^{8+\beta_1}+c_5y^{6+\beta_1}+c_6y^{4+\beta_1}+c_7y^{1+\beta_1+\beta_2}+c_8y^{4+\beta_2}+c_9y^{2+\beta_2}+c_{10}y^{11}$$
$$+c_{11}y^9+c_{12}y^7+c_{13}y^5+c_{14}y^{\beta_1})+k\alpha_1^2W_m^2(c_{15}y^{2+2\beta_1}+c_{16}y^{5+\beta_1}+c_{17}y^{3+\beta_1}$$
$$+c_{18}y^{1+\beta_1+\beta_2}+c_{19}y^{4+\beta_2}+c_{20}y^{2+\beta_2}+c_{21}y^{1+\beta_2}+c_{22}y^8+c_{23}y^6+c_{24}y^4+c_{25}y^{\beta_1})$$
$$+2k^2\alpha_1W_m(c_{26}y^{2+\beta_1}+c_{27}y^{1+\beta_2}+c_{28}y^5+c_{29}y^3+c_{30}y^{\beta_1})] \tag{3.20}$$

where

$$c_1=\frac{\lambda b_1}{4(1+\beta_1)(1+2\beta_1)},\qquad c_2=\frac{\lambda(b_2-b_1)}{(5+\beta_1)(5+3\beta_1)},$$
$$c_3=\frac{\lambda b_3-b_1m^{-1}}{3(1+\beta_1)(3+\beta_1)},\qquad c_4=\frac{\lambda(b_5-b_2)}{16(4+\beta_1)},$$
$$c_5=\frac{\lambda b_6-\lambda b_3-b_2m^{-1}}{12(3+\beta_1)},\qquad c_6=\frac{\lambda b_7-b_3m^{-1}}{8(2+\beta_1)},$$
$$c_7=\frac{\lambda b_4}{(1+\beta_2)(1+2\beta_1+\beta_2)},\qquad c_8=\frac{\lambda b_4}{\beta_1^2-(4+\beta_2)^2},$$
$$c_9=\frac{b_4}{m[\beta_1^2-(2+\beta_2)^2]},\qquad c_{10}=\frac{\lambda b_5}{\beta_1^2-121},$$
$$c_{11}=\frac{\lambda b_6+b_5m^{-1}}{\beta_1^2-81},\qquad c_{12}=\frac{\lambda b_7+b_6m^{-1}}{\beta_1^2-49},$$
$$c_{13}=\frac{b_7}{m(\beta_1^2-25)},\qquad c_{14}=-\sum_{i=1}^{13}c_i,$$
$$c_{15}=\frac{b_1+2\lambda b_8}{(2+\beta_1)(2+3\beta_1)},\qquad c_{16}=\frac{b_2+2\lambda b_{10}-2\lambda b_8}{5(5+2\beta_1)},$$
$$c_{17}=\frac{b_3+2\lambda b_{11}-2b_8m^{-1}}{3(3+2\beta_1)},\qquad c_{18}=\frac{2\lambda b_9}{(1+\beta_2)(1+2\beta_1+\beta_2)},$$
$$c_{19}=\frac{2\lambda b_9}{\beta_1^2-(4+\beta_2)^2},\qquad c_{20}=\frac{2b_9}{m[\beta_1^2-(2+\beta_2)^2]},$$
$$c_{21}=-\frac{b_4}{\beta_1^2-(1+\beta_2)^2},\qquad c_{22}=-\frac{b_5-2\lambda b_{10}}{\beta_1^2-64},$$
$$c_{23}=-\frac{b_6-2b_{10}m^{-1}-2\lambda b_{11}}{\beta_1^2-36},\qquad c_{24}=-\frac{b_7-2b_{11}m^{-1}}{\beta_1^2-16},$$
$$c_{25}=-\sum_{i=15}^{24}c_i,\qquad c_{26}=\frac{b_8}{4(1+\beta_1)},$$
$$c_{27}=-\frac{b_9}{\beta_1^2-(1+\beta_2)^2},\qquad c_{28}=-\frac{b_{10}}{\beta_1^2-25},$$
$$c_{29}=-\frac{b_{11}}{\beta_1^2-9},\qquad c_{30}=-\sum_{i=26}^{29}c_i. \tag{3.21}$$

Substituting expression (3.18) into (3.1), using solutions (3.12b), (3.13) and (3.20), and performing the integration, we finally obtain the following nonlinear characteristic relation of the shell:

$$P=(\alpha_1+k^2\alpha_0)W_m+k\alpha_2W_m^2+\alpha_3W_m^3 \tag{3.22}$$

where

$$\alpha_0=\alpha_1^2\beta_3\left[\frac{1}{m}\left(\frac{b_8}{1+\beta_1}+\frac{b_9}{\beta_2}+\frac{b_{10}}{4}+\frac{b_{11}}{2}\right)-\frac{c_{26}}{3+\beta_1}-\frac{c_{27}}{2+\beta_2}-\frac{c_{28}}{6}-\frac{c_{29}}{4}-\frac{c_{30}}{1+\beta_1}\right],$$

$$\alpha_2=\frac{1}{2}\alpha_1^3\beta_3\left\{\frac{1}{m}\left[\frac{b_1+2\lambda b_8}{1+2\beta_1}+\frac{b_2+2\lambda b_{10}-2\lambda b_8}{4+\beta_1}+\frac{b_3+2\lambda b_{11}-2b_8m^{-1}}{2+\beta_1}+\frac{2\lambda b_9}{\beta_1+\beta_2}\right.\right.$$
$$-\frac{2\lambda b_9}{3+\beta_2}-\frac{2b_9}{m\ (1+\beta_2)}+\frac{b_4}{\beta_2}+\frac{1}{7}\ (b_5-2\lambda b_{10})\ +\frac{1}{5}\left(b_6-\frac{2b_{10}}{m}-2\lambda b_{11}\right)$$
$$\left.+\frac{1}{3}\left(b_7-\frac{2b_{11}}{m}\right)\right]-\frac{c_{15}}{3+2\beta_1}-\frac{c_{16}}{6+\beta_1}-\frac{c_{17}}{4+\beta_1}-\frac{c_{18}}{2+\beta_1+\beta_2}-\frac{c_{19}}{5+\beta_2}$$
$$\left.-\frac{c_{20}}{3+\beta_2}-\frac{c_{21}}{2+\beta_2}-\frac{c_{22}}{9}-\frac{c_{23}}{7}-\frac{c_{24}}{5}-\frac{c_{25}}{1+\beta_1}\right\},$$

$$\alpha_3=\frac{1}{2}\alpha_1^4\beta_3\left\{\frac{1}{m}\left[\frac{\lambda b_1}{1+3\beta_1}+\frac{\lambda\ (b_2-b_1)}{2\ (2+\beta_1)}+\frac{\lambda b_3-b_1m^{-1}}{2\ (1+\beta_1)}+\frac{\lambda\ (b_5-b_2)}{7+\beta_1}\right.\right.$$
$$+\frac{\lambda b_6-\lambda b_3-b_2m^{-1}}{5+\beta_1}+\frac{\lambda b_7-b_3m^{-1}}{3+\beta_1}+\frac{\lambda b_4}{\beta_1+\beta_2}-\frac{\lambda b_4}{3+\beta_2}-\frac{b_4}{m\ (1+\beta_2)}$$
$$\left.-\frac{1}{10}\lambda b_5-\frac{1}{8}\left(\lambda b_6+\frac{b_5}{m}\right)-\frac{1}{6}\left(\lambda b_7+\frac{b_6}{m}\right)-\frac{b_7}{4m}\right]-\frac{c_1}{3\ (1+\beta_1)}$$
$$-\frac{c_2}{2\ (3+\beta_1)}-\frac{c_3}{2\ (2+\beta_1)}-\frac{c_4}{9+\beta_1}-\frac{c_5}{7+\beta_1}-\frac{c_6}{5+\beta_1}-\frac{c_7}{2+\beta_1+\beta_2}$$
$$\left.-\frac{c_8}{5+\beta_2}-\frac{c_9}{3+\beta_2}-\frac{c_{10}}{12}-\frac{c_{11}}{10}-\frac{c_{12}}{8}-\frac{c_{13}}{6}-\frac{c_{14}}{1+\beta_1}\right\}. \tag{3.23}$$

To obtain the critical buckling pressure of the shell we use the extremal condition

$$\frac{\mathrm{d}P}{\mathrm{d}W_m}=0, \tag{3.24}$$

from which

$$3\alpha_3W_m^2+2k\alpha_2W_m+\alpha_1+k^2\alpha_0=0. \tag{3.25}$$

Solving this quadratic equation, we find

$$W_m^*=\frac{1}{3\alpha_3}\left[-k\alpha_2\pm\sqrt{k^2(\alpha_2^2-3\alpha_0\alpha_3)-3\alpha_1\alpha_3}\right]. \tag{3.26}$$

Upon substitution, relation (3.22) becomes

$$P^*=(\alpha_1+k^2\alpha_0)W_m^*+k\alpha_2W_m^{*2}+\alpha_3W_m^{*3}. \tag{3.27}$$

This is the formula of the non-dimensional, critical buckling pressure for the symmetrically laminated, cylindrically orthotropic, shallow, conical shell with a rigidly clamped edge under uniform lateral pressure including transverse shear effects. The non-dimensional critical buckling pressure P^* with respect to the non-dimensional critical center deflection W_m^* with the negative and positive signs are called the upper and lower critical buckling pressure, respectively. Obviously, in practical applications, only the upper critical buckling pressure arouses the interest of all. Thus, in the following discussion, we consider only the upper critical buckling pressure.

For the critical point at which the upper and lower critical buckling pressure coincide, it is used to distinguish between buckling and no buckling for the shell. k_0 corresponding to such a point is called the critical geometrical parameter. Obviously, let the discriminant of the quadratic equation (3.25) be zero; the formula for the critical point is given by

$$k_0=\sqrt{\frac{3\alpha_1\alpha_3}{\alpha_2^2-3\alpha_0\alpha_3}} \tag{3.28}$$

4. Numerical Examples

In order to simplify the following calculations and to make the final results general, as a particular example, a symmetrically laminated, cylindrically orthotropic, shallow, conical shell under consideration consists of n layers of thin, orthotropic sheets all of the same thickness and elastic properties. The elastic constants of the composite material are

$$\frac{E_\theta}{E_r}=1.5,\quad \nu_{\theta r}=0.3.$$

Hence from expressions (2.27), several parameters can be calculated as follows:

$$\beta_1^2=\beta_2^2=1.5,\quad \beta_3=16.92,\quad \beta_4=0.3.$$

Substituting these values into relation (3.27), we obtain interesting numerical results. Fig. 3 indicates the calculated results of the non-dimensional critical buckling pressure for several values of the non-dimensional shear rigidity m. From this figure, it may be seen that curves of the critical buckling pressure for the shell rise monotonically, and, for the same value of the geometrical parameter k, the critical buckling pressure of the shell with larger shear rigidity m is high. Especially in the limiting case $m=\infty$, implying that the effect of the transverse shear deformation is neglected, the corresponding critical buckling pressure is the maximum value. Obviously, the effect of the transverse shear deformation on buckling of the shell is very important.

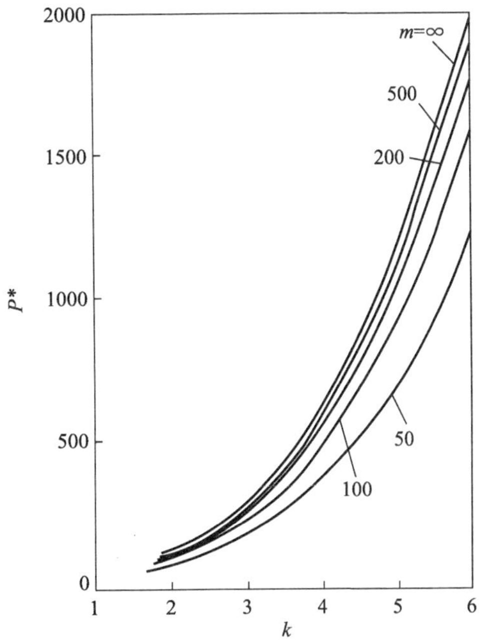

Fig. 3　Curves of the non-dimensional critical buckling pressure for several values of the non-dimensional shear rigidity m

According to formula (3.28), results are calculated and shown in Fig. 4, which is the curve of the critical geometrical parameter k_0. It is seen from this figure that k_0 increases with the non-dimensional shear rigidity m, and when $m>100$, k_0 increases very slowly. In other words, a shell with small transverse shear modulus is more sensitive

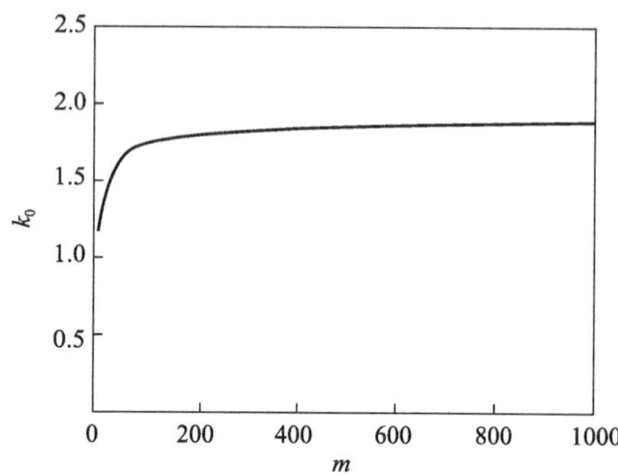

Fig. 4　Curve of the critical geometrical parameter k_0

to instability.

Acknowledgements

This paper was written at the Institute of Statics and Dynamics, Ruhr-University Bochum, Germany. The author is grateful to the Alexander von Humboldt Foundation, Ruhr-University Bochum, and Professor Wilfried B. Krätzig for his support and hearty hospitality. At the same time, this work was supported by the National Natural Science Foundation of China.

References

[1] C, Y. Chia, *Nonlinear Analysis of Plates*. McGraw-Hill, New York (1980).

[2] Liu Renhuai, Nonlinear stability of symmetrically laminated, cylindrically orthotropic shallow spherical shells including transverse shear. *Science in China (Series A)* 35, 734-746 (1992).

[3] Liu Renhuai and L. H. He, A simple theory for nonlinear bending of laminated composite rectangular plates including higher-order effects, *Int. J. Nonlinear Mech.* 26, 537-545 (1991).

[4] Liu Renhuai, On the nonlinear stability of symmetrically laminated cylindrically orthotropic shallow spherical shells. *Appl, Math. Mech.* 12, 271 -279 (1991).

[5] Liu Renhuai, Large deflection equations of symmetrically laminated composite, cylindrically orthotropic, shallow, spherical shells. In *Applied Mathematics and Mechanics*, pp. 279-284. Science Press, Beijing (1993).

[6] Liu Renhuai and C. Zhong, Nonlinear buckling of symmetrically laminated, cylindrically orthotropic, shallow, spherical shells with a circular hole at the center including transverse shear. *J. Jinan Univ. (Natural Science)* 15, 1-12 (1994) (in Chinese).

[7] Liu Renhuai and L. H. He, Nonlinear bending of simply supported symmetric laminated cross-ply rectangular plates. *Appl. Math. Mech.* 11, 801-807 (1990).

[8] Liu Renhuai and L. H. He, Axisymmetrical bending of laminated circular plates. *J. Jiangxi Polytech. Univ.* 13, 199-204 (1991) (in Chinese).

[9] F. Gu and Z. D. Lo, Large deflection bending of composite circular plates exhibiting rectilinear orthotropic. *Acta Materiae Composite Sinica* 5 (2), 11-16 (1988) (in Chinese).

[10] Y. Basar, Y. Ding and R. Schultz, Refined shear-deformation models for composite laminates with finite rotations. *Int. J. Solids Structures* 30, 2611-2638 (1993).

[11] J. N. Reddy, A refined nonlinear theory of plates with transverse shear deformation. *Int. J. Solids Structures* 20, 881-896 (1984).

[12] K. Y. Yeh, Liu Renhuai, Q. Y. Ping and S. L. Li, Nonlinear stability of thin circular shallow spherical shell under actions of axisymmetric uniform distributed line loads. *Bull. Sci.* 2, 142-145 (1965); *J. Lanzhou Univ.* 2, 10-33 (1965) (in Chinese).

[13] K. Y. Yeh, Liu Renhuai, Zhang C. Z. and Xu, Y. F. Nonlinear stability of thin circular shallow spherical shell under the action of uniform edge moment. *Bull. Sci.* 2, 145-147 (1965) (in Chinese).

[14] Liu Renhuai, Nonlinear stability of circular shallow spherical shell with a hole in the center under the action of uniform moment at the inner edge. *Bull. Sci* 3, 253-255 (1965) (in Chinese).

On non-linear buckling of symmetrically laminated, cylindrically orthotropic, truncated, shallow, spherical shells under uniform pressure including shear effects①

中文摘要

本文研究了考虑横向剪切影响的并具有夹紧固定外边界和刚性中心的对称层合圆柱正交异性开顶扁球壳在均布压力作用下的非线性屈曲问题。使用修正迭代法，获得了临界压力的解析解，可直接用于工程设计。这个问题尚未有人研究过。

1. Introduction

It is common knowledge that the use laminated anisotropic composites as structural members, particularly thin plates and shells, has increased considerably. This arises from the fact that, by taking advantage of its anisotropic material properties and light weight with high strength, the materials can be used efficiently. Therefore, there is already a considerable amount of research[1-12] on nonlinear problems of plates and shells. However, the nonlinear buckling problems for laminated, anisotropic, shallow, spherical shells have been investigated by relatively few researchers because of the difficulties of nonlinear mathematical problems and more complicated structures. The present author[2] first derived basic equations of nonlinear bending theory for symmetrically laminated, cylindrically orthotropic, shallow, spherical shells and solved several practical buckling problems[2-5]. This paper is an extension of the previous papers. The type of shells which we discuss here is the more important one: the symmetrically laminated, cylindrically orthotropic, truncated, shallow, spherical shell with a rigidly clamped outer edge and a non-deformable rigid body at the center under uniform pressure including the effect of the transverse shear. As we know, this problem has not yet been studied. We use the modified iteration method which was suggested by the author and Yeh Kaiyuan[13-15] in 1965 to successfully solve this problem. The analytical solution for the critical buckling pressure is obtained, and can be applied directly to engineering design.

2. Basic Equations

Fig. 1 shows a symmetrically laminated, cylindrically orthotropic, truncated, shallow, spherical shell with a non-deformable rigid body at the center under the action of uniform pressure q. The outer edge is rigidly clamped and the inner edge is fixed on the non-deformable rigid body which can be moved up and down. Here h is the thick-

① Reprinted from *International Journal of Non-Linear Mechanics*, 1996, 31 (1): 101-115.

ness, a is the outer radius, b is the inner radius, r is the radial coordinate, R is the radius of curvature and n is the total number of layers.

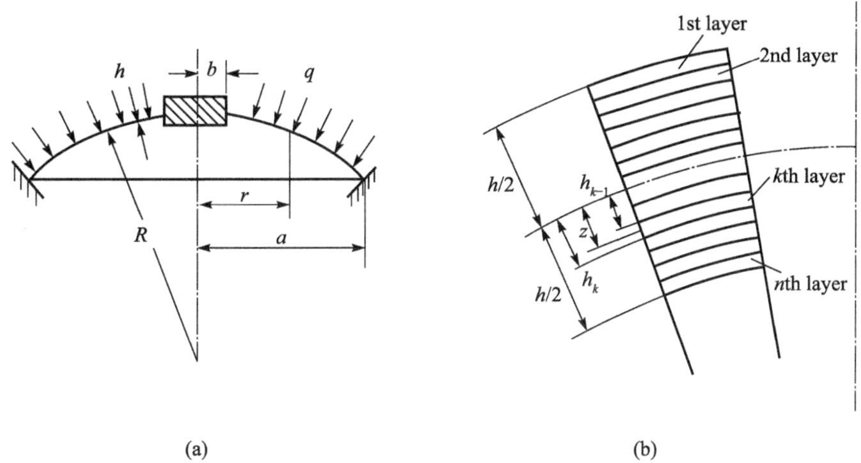

Fig. 1a, b　Geometry of symmetrically laminated, cylindrically orthotropic, truncated, shallow, spherical shell

In the case of rotationally axisymmetric deformations, starting with the general theory for nonlinear bending of symmetrically laminated, cylindrically orthotropic, shallow, spherical shells[2], we can easily obtain basic equations and boundary conditions of the nonlinear bending theory satisfied by the deflection w, the rotation ψ of the normal to the middle surface of the shell and the radial membrane force N_r of the symmetrically laminated, cylindrically orthotropic, truncated, shallow, spherical shell with a non-deformable rigid body at the center under uniform pressure q as follows:

$$D_{11}\frac{\mathrm{d}}{\mathrm{d}r}r\frac{\mathrm{d}\psi}{\mathrm{d}r}-\left(Gr+\frac{D_{22}}{r}\right)\psi-Gr\frac{\mathrm{d}w}{\mathrm{d}r}=0,$$

$$rN_r\left(\frac{r}{R}+\frac{\mathrm{d}w}{\mathrm{d}r}\right)+Gr\left(\frac{\mathrm{d}w}{\mathrm{d}r}+\psi\right)=-\frac{1}{2}q\ (r^2-b^2),$$

$$A_3\frac{\mathrm{d}}{\mathrm{d}r}r\frac{\mathrm{d}}{\mathrm{d}r}(rN_r)-A_1N_r+\frac{1}{2}\left(\frac{\mathrm{d}w}{\mathrm{d}r}\right)^2+\frac{r}{R}\frac{\mathrm{d}w}{\mathrm{d}r}=0. \tag{2.1}$$

$$w=0,\qquad \psi=0,\qquad u=0\qquad \text{at } r=a \tag{2.2}$$

$$\psi=0,\qquad u=0\qquad \text{at } r=b \tag{2.3}$$

in which u is the radial displacement, $E_r^{(k)}$ and $E_\theta^{(k)}$ are moduli of elasticity of the kth layer along the principal material axes r and θ, respectively, $\nu_{r\theta}^{(k)}$ and $\nu_{\theta r}^{(k)}$ are Poisson's ratios of the kth layer with the first subscript indicating the direction of the tensile force and the second indicating the direction of contraction, $G_{rz}^{(k)}$ is the transverse shear modulus of the kth layer characterizing the change of the angle in the rz plane, $Q_{ij}^{(k)}$ are the reduced stiffnesses of the kth layer, A_{ij} are the extensional rigidities, D_{ij} are the bending rigidities, and G is the shear rigidity,

$$u=r\left[A_3\frac{\mathrm{d}\ (rN_r)}{\mathrm{d}r}-A_2N_r\right].\tag{2.4}$$

$$Q_{11}^{(k)}=\frac{E_r^{(k)}}{1-\nu_{r\theta}^{(k)}\nu_{\theta r}^{(k)}},\qquad Q_{12}^{(k)}=\frac{\nu_{r\theta}^{(k)}E_\theta^{(k)}}{1-\nu_{r\theta}^{(k)}\nu_{\theta r}^{(k)}},\qquad Q_{22}^{(k)}=\frac{E_\theta^{(k)}}{1-\nu_{r\theta}^{(k)}\nu_{\theta r}^{(k)}}.\tag{2.5}$$

$$A_{ij}=\sum_{k=1}^{n}Q_{ij}^{(k)}(h_k-h_{k-1}),$$

$$D_{ij}=\frac{1}{3}\sum_{k=1}^{n}Q_{ij}^{(k)}(h_k^3-h_{k-1}^3)\quad i,j=1,2\tag{2.6}$$

$$G=\frac{4h^2}{9}\left\{\sum_{k=1}^{n}\frac{1}{G_{rz}^{(k)}}\left[h_k-h_{k-1}-\frac{8}{3h^2}(h_k^3-h_{k-1}^3)+\frac{16}{5h^4}(h_k^5-h_{k-1}^5)\right]\right\}^{-1}\tag{2.7}$$

$$A_1=A_0A_{22},\qquad A_2=A_0A_{12},\qquad A_3=A_0A_{11},$$

$$A_0=\frac{1}{A_{11}A_{22}-A_{12}^2}.\tag{2.8}$$

In order to simplify the following calculations, let us introduce the following non-dimensional notations:

$$y=\frac{r}{a},\quad \alpha=\frac{b}{a},\quad W=\frac{w}{h},\quad \varphi=\frac{\mathrm{d}W}{\mathrm{d}y}+ky,\quad \Psi=\frac{a}{h}\psi,$$

$$S=\frac{a}{D_{11}}rN_r,\quad P=\frac{a^4}{2D_{11}h}q,\quad k=\frac{a^2}{Rh},\quad m=\frac{a^2}{D_{11}}G,$$

$$\beta_1^2=\frac{D_{22}}{D_{11}},\quad \beta_2^2=\frac{A_1}{A_3},\quad \beta_3=\frac{h^2}{A_3D_{11}},\quad \beta_4=\frac{A_2}{A_3}\tag{2.9}$$

With the help of these notations, and using expression (2.4), the nonlinear boundary value problem (2.1)-(2.3) becomes

$$my(\varphi-ky)=-[S\varphi+my\Psi+P(y^2-\alpha^2)],$$

$$L_1(y^{\beta_1}\Psi)=my(\Psi+\varphi-ky),$$

$$L_2(y^{\beta_2}S)=-\frac{\beta_3}{2}(\varphi^2-k^2y^2)\tag{2.10a-c}$$

$$W=0,\quad \Psi=0,\quad y\frac{\mathrm{d}S}{\mathrm{d}y}-\beta_4S=0\quad \text{at } y=1\tag{2.11a-c}$$

$$\Psi=0,\ y\frac{\mathrm{d}S}{\mathrm{d}y}-\beta_4S=0\qquad \text{at } y=\alpha\tag{2.12a,b}$$

where L_1 and L_2 are the differential operators defined by

$$L_1(\cdots)=y^{\beta_1}\frac{\mathrm{d}}{\mathrm{d}y}y^{1-2\beta_1}\frac{\mathrm{d}}{\mathrm{d}y}(\cdots),$$

$$L_2(\cdots)=y^{\beta_2}\frac{\mathrm{d}}{\mathrm{d}y}y^{1-2\beta_2}\frac{\mathrm{d}}{\mathrm{d}y}(\cdots),\tag{2.13}$$

Equations (2.10) and boundary conditions (2.11) and (2.12) constitute a non-dimensional nonlinear boundary-value problem for φ, Ψ and S of the symmetrically laminated, cylindrically orthotropic, truncated, shallow, spherical shell with a non-deformable rigid body at the center subjected to uniform pressure.

3. Solution of the Problem

Now we use the modified iteration method to solve the preceding non-dimensional, nonlinear boundary-value problem (2.10)-(2.12). First, let us introduce as an iteration parameter a notation of the non-dimensional inner edge deflection

$$W_m = \frac{w}{h}\bigg|_{r=b} \tag{3.1}$$

Using the boundary condition (2.11a) and the fourth equation of expressions (2.9), we obtain

$$W_m = -\int_a^1 (\varphi - ky)\,\mathrm{d}y. \tag{3.2}$$

For the first approximation, neglecting the nonlinear term S_φ in Eq. (2.10a), and introducing φ_1 into the right-hand sides of Eqs. (2.10b, c), we obtain the following linear boundary-value problem:

$$\begin{aligned} &my(\varphi_1 - ky) = -[my\Psi_1 + P(y^2 - \alpha^2)],\\ &L_1(y^{\beta_1}\Psi_1) = my(\Psi_1 + \varphi_1 - ky),\\ &L_2(y^{\beta_2}S_1) = -\frac{\beta_3}{2}(\varphi_1^2 - k^2y^2) \end{aligned} \tag{3.3a-c}$$

$$W_1 = 0,\quad \Psi_1 = 0,\quad y\frac{\mathrm{d}S_1}{\mathrm{d}y} - \beta_4 S_1 = 0 \quad \text{at } y = 1 \tag{3.4a-c}$$

$$\Psi_1 = 0,\quad y\frac{\mathrm{d}S_1}{\mathrm{d}y} - \beta_4 S_1 = 0 \quad \text{at } y = \alpha. \tag{3.5a,b}$$

Substituting Eq. (3.3a) into (3.3b), we find

$$L_1(y^{\beta_1}\Psi_1) = -P(y^2 - \alpha^2) \tag{3.6}$$

This equation can be solved easily by direct integration in conjunction with the corresponding boundary conditions (3.4b) and (3.5a). We then obtain

$$\Psi_1 = P(a_1 y^{\beta_1} + a_2 y^{-\beta_1} + a_3 y^3 + a_4 y) \tag{3.7}$$

where

$$\begin{aligned} &a_1 = a_0[(a_3 + a_4)\alpha^{-\beta_1} + 8a_3a_4\alpha],\\ &a_2 = -a_0[(a_3 + a_4)\alpha^{\beta_1} + 8a_3a_4\alpha],\\ &a_3 = -\frac{1}{9 - \beta_1^2},\\ &a_4 = \frac{\alpha^2}{1 - \beta_1^2},\\ &a_0 = \frac{1}{\alpha^{\beta_1} - \alpha^{-\beta_1}}. \end{aligned} \tag{3.8}$$

Inserting solution (3.7) into Eq. (3.3a) and using expression (3.2) leads to

$$P = \alpha_1 W_m. \tag{3.9}$$

This is the characteristic relation for small deflection. Here

$$\alpha_1=\left\{\frac{a_1}{1+\beta_1}(1-\alpha^{1+\beta_1})+\frac{a_2}{1-\beta_1}(1-\alpha^{1-\beta_1})+\frac{a_3}{4}(1-\alpha^4)\right.$$
$$\left.+\frac{a_4}{2}(1-\alpha^2)+\frac{1}{m}\left[\frac{1}{2}(1-\alpha^2)+\alpha^2\ln\alpha\right]\right\}^{-1}. \tag{3.10}$$

Introducing relation (3.9) into solution (3.7) yields

$$\Psi_1=\alpha_1 W_m(a_1 y^{\beta_1}+a_2 y^{-\beta_1}+a_3 y^3+a_4 y). \tag{3.11}$$

Using solution (3.11) and boundary conditions (3.4c) and (3.5b), the solution of Eq. (3.3b) is

$$S_1=-\beta_3\left[\frac{1}{2}\alpha_1^2 W_m^2(b_1 y^{1+2\beta_1}+b_2 y^{1-2\beta_1}+b_3 y^{4+\beta_1}+b_4 y^{4-\beta_1}+b_5 y^{2+\beta_1}+b_6 y^{2-\beta_1}\right.$$
$$+b_7 y^{\beta_1}+b_8 y^{-\beta_1}+b_9 y^{\beta_2}+b_{10} y^{-\beta_2}+b_{11} y^7+b_{12} y^5+b_{13} y^3+b_{14} y+b_{15} y^{-1})$$
$$\left.-k\alpha_1 W_m(b_{16} y^{2+\beta_1}+b_{17} y^{2-\beta_1}+b_{18} y^{\beta_2}+b_{19} y^{-\beta_2}+b_{20} y^5+b_{21} y^3+b_{22} y)\right] \tag{3.12}$$

where

$$b_1=\frac{a_1^2}{(1+2\beta_1)^2-\beta_2^2},\qquad b_2=\frac{a_2^2}{(1-2\beta_1)^2-\beta_2^2},$$
$$b_3=\frac{2a_1 a_3}{(4+\beta_1)^2-\beta_2^2},\qquad b_4=\frac{2a_2 a_3}{(4-\beta_1)^2-\beta_2^2},$$
$$b_5=\frac{2a_1(a_4+m^{-1})}{(2+\beta_1)^2-\beta_2^2},\qquad b_6=\frac{2a_2(a_4+m^{-1})}{(2-\beta_1)^2-\beta_2^2},$$
$$b_7=\frac{2a_1\alpha^2}{m(\beta_2^2-\beta_1^2)},\qquad b_8=\frac{2a_2\alpha^2}{m(\beta_2^2-\beta_1^2)},$$
$$b_9=\frac{\lambda_1\alpha^{-\beta_2}-\lambda_3}{(\beta_2-\beta_4)(\alpha^{\beta_2}-\alpha^{-\beta_2})},\qquad b_{10}=\frac{\lambda_1\alpha^{\beta_2}-\lambda_3}{(\beta_2+\beta_4)(\alpha^{\beta_2}-\alpha^{-\beta_2})},$$
$$b_{11}=\frac{a_3^2}{49-\beta_2^2},\qquad b_{12}=\frac{2a_3(a_4+m^{-1})}{25-\beta_2^2},$$
$$b_{13}=\frac{(a_4+m^{-1})^2-2a_3\alpha^2 m^{-1}}{9-\beta_2^2},\qquad b_{14}=\frac{2[a_1 a_2-\alpha^2 m^{-1}(a_4+m^{-1})]}{1-\beta_2^2},$$
$$b_{15}=\frac{4}{m^2(1-\beta_2^2)},\qquad b_{16}=\frac{a_1}{(2+\beta_1)^2-\beta_2^2},$$
$$b_{17}=\frac{a_2}{(2-\beta_1)^2-\beta_2^2},\qquad b_{18}=\frac{\lambda_2\alpha^{-\beta_2}-\lambda_4}{(\beta_2-\beta_4)(\alpha^{\beta_2}-\alpha^{-\beta_2})},$$
$$b_{19}=\frac{\lambda_2\alpha^{\beta_2}-\lambda_4}{(\beta_2+\beta_4)(\alpha^{\beta_2}-\alpha^{-\beta_2})},\qquad b_{20}=\frac{a_3}{25-\beta_2^2},$$
$$b_{21}=\frac{a_4+m^{-1}}{9-\beta_2^2},\qquad b_{22}=-\frac{\alpha^2}{m(1-\beta_2^2)},$$
$$\lambda_1=b_1(1+2\beta_1-\beta_4)+b_2(1-2\beta_1-\beta_4)+b_3(4+\beta_1-\beta_4)+b_4(4-\beta_1-\beta_4)$$
$$+b_5(2+\beta_1-\beta_4)+b_6(2-\beta_1-\beta_4)+b_7(\beta_1-\beta_4)-b_8(\beta_1+\beta_4)+b_{11}(7-\beta_4)$$
$$+b_{12}(5-\beta_4)+b_{13}(3-\beta_4)+b_{14}(1-\beta_4)-b_{15}(1+\beta_4),$$
$$\lambda_2=b_{16}(2+\beta_1-\beta_4)+b_{17}(2-\beta_1-\beta_4)+b_{20}(5-\beta_4)+b_{21}(3-\beta_4)+b_{22}(1-\beta_4),$$
$$\lambda_3=b_1(1+2\beta_1-\beta_4)\alpha^{1+2\beta_1}+b_2(1-2\beta_1-\beta_4)\alpha^{1-2\beta_1}+b_3(4+\beta_1-\beta_4)\alpha^{4+\beta_1}$$

$$+b_4(4-\beta_1-\beta_4)\alpha^{4-\beta_1}+b_5(2+\beta_1-\beta_4)\alpha^{2+\beta_1}+b_6(2-\beta_1-\beta_4)\alpha^{2-\beta_1}$$
$$+b_7(\beta_1-\beta_4)\alpha^{\beta_1}-b_8(\beta_1+\beta_4)\alpha^{-\beta_1}+b_{11}(7-\beta_4)\alpha^7+b_{12}(5-\beta_4)\alpha^5$$
$$+b_{13}(3-\beta_4)\alpha^3+b_{14}(1-\beta_4)\alpha-b_{15}(1+\beta_4)\alpha^{-1},$$
$$\lambda_4=b_{16}(2+\beta_1-\beta_4)\alpha^{2+\beta_1}+b_{17}(2-\beta_1-\beta_4)\alpha^{2-\beta_1}+b_{20}(5-\beta_4)\alpha^5$$
$$+b_{21}(3-\beta_4)\alpha^3+b_{22}(1-\beta_4)\alpha. \tag{3.13}$$

For the second approximation, it is restricted to the determination of φ and Ψ only. We rewrite the nonlinear term $S\varphi$ of Eq. (2.10a) in $S_1\varphi_1$. We then have the following linear boundary-value problem:

$$my(\varphi_2-ky)=-[S_1\varphi_1+my\Psi_2+P(y^2-\alpha^2)],$$
$$L_1(y^{\beta_1}\Psi_2)=my(\Psi_2+\varphi_2-ky) \tag{3.14a,b}$$
$$W_2=0,\qquad \Psi_2=0\qquad \text{at } y=1 \tag{3.15a,b}$$
$$\Psi_2=0,\qquad \text{at } y=\alpha. \tag{3.16}$$

Introducing Eq. (3.14a) into (3.14b), we find

$$L_1(y^{\beta_1}\Psi_2)=-[S_1\varphi_1+P(y^2-\alpha^2)]. \tag{3.17}$$

Using solutions (3.11) and (3.12), the solution of Eq. (3.17) satisfying boundary conditions (3.15b) and (3.16) is given by

$$\Psi_2=P(a_1y^{\beta_1}+a_2y^{-\beta_1}+a_3y^3+a_4y)-\beta_3\Big\{\frac{1}{2}\alpha_1^3W_m^3[F_3(y)+e_3y^{\beta_1}+f_3y^{-\beta_1}]$$
$$-k\alpha_1^2W_m^2[F_2(y)+e_2y^{\beta_1}+f_2y^{-\beta_1}]+k^2\alpha_1W_m[F_1(y)+e_1y^{\beta_1}+f_1y^{-\beta_1}]\Big\} \tag{3.18}$$

where

$$F_3(y)=c_1y^{2+3\beta_1}+c_2y^{2-3\beta_1}+c_3y^{5+2\beta_1}+c_4y^{5-2\beta_1}+c_5y^{3+2\beta_1}+c_6y^{3-2\beta_1}+c_7y^{1+2\beta_1}$$
$$+c_8y^{1-2\beta_1}+c_9y^{1+\beta_1+\beta_2}+c_{10}y^{1+\beta_1-\beta_2}+c_{11}y^{1-\beta_1+\beta_2}+c_{12}y^{1-\beta_1-\beta_2}+c_{13}y^{8+\beta_1}$$
$$+c_{14}y^{8-\beta_1}+c_{15}y^{6+\beta_1}+c_{16}y^{6-\beta_1}+c_{17}y^{4+\beta_1}+c_{18}y^{4-\beta_1}+c_{19}y^{2+\beta_1}+c_{20}y^{2-\beta_1}$$
$$+c_{21}y^{\beta_1}\ln y+c_{22}y^{-\beta_1}\ln y+c_{23}y^{\beta_1}+c_{24}y^{4+\beta_2}+c_{25}y^{4-\beta_2}+c_{26}y^{2+\beta_2}+c_{27}y^{2-\beta_2}$$
$$+c_{28}y^{\beta_2}+c_{29}y^{-\beta_2}+c_{30}y^{11}+c_{31}y^9+c_{32}y^7+c_{33}y^5+c_{34}y^3+c_{35}y+c_{36}y^{-1},$$
$$F_2(y)=c_{37}y^{3+2\beta_1}+c_{38}y^{3-2\beta_1}+c_{39}y^{1+\beta_1+\beta_2}+c_{40}y^{1+\beta_1-\beta_2}+c_{41}y^{1-\beta_1+\beta_2}+c_{42}y^{1-\beta_1-\beta_2}$$
$$+c_{43}y^{6+\beta_1}+c_{44}y^{6-\beta_1}+c_{45}y^{4+\beta_1}+c_{46}y^{4-\beta_1}+c_{47}y^{2+\beta_1}+c_{48}y^{2-\beta_1}$$
$$+c_{49}y^{4+\beta_2}+c_{50}y^{4-\beta_2}+c_{51}y^{2+\beta_2}+c_{52}y^{2-\beta_2}+c_{53}y^{\beta_2}+c_{54}y^{-\beta_2}$$
$$+c_{55}y^9+c_{56}y^7+c_{57}y^5+c_{58}y^3+c_{59}y,$$
$$F_1(y)=c_{60}y^{4+\beta_1}+c_{61}y^{4-\beta_1}+c_{62}y^{2+\beta_2}+c_{63}y^{2-\beta_2}+c_{64}y^7+c_{65}y^5+c_{66}y^3,$$
$$e_i=a_0[F_i(1)\alpha^{-\beta_1}-F_i(\alpha)],\qquad f_i=-a_0[F_i(1)\alpha^{\beta_1}-F_i(\alpha)],\ i=1,2,3$$
$$c_1=\frac{d_1}{4(1+\beta_1)(1+2\beta_1)},\qquad c_2=\frac{d_2}{4(1-\beta_1)(1-2\beta_1)},$$
$$c_3=\frac{d_3}{(5+\beta_1)(5+3\beta_1)},\qquad c_4=\frac{d_4}{(5-\beta_1)(5-3\beta_1)},$$
$$c_5=\frac{d_5}{3(1+\beta_1)(3+\beta_1)},\qquad c_6=\frac{d_6}{3(1-\beta_1)(3-\beta_1)},$$

$$c_7=\frac{d_7}{(1+\beta_1)(1+3\beta_1)},\qquad c_8=\frac{d_8}{(1-\beta_1)(1-3\beta_1)},$$

$$c_9=\frac{d_9}{(1+\beta_2)(1+2\beta_1+\beta_2)},\qquad c_{10}=\frac{d_{10}}{(1-\beta_2)(1+2\beta_1-\beta_2)},$$

$$c_{11}=\frac{d_{11}}{(1+\beta_2)(1-2\beta_1+\beta_2)},\qquad c_{12}=\frac{d_{12}}{(1-\beta_2)(1-2\beta_1-\beta_2)},$$

$$c_{13}=\frac{d_{13}}{16(4+\beta_1)},\qquad c_{14}=\frac{d_{14}}{16(4-\beta_1)},$$

$$c_{15}=\frac{d_{15}}{12(3+\beta_1)},\qquad c_{16}=\frac{d_{16}}{12(3-\beta_1)},$$

$$c_{17}=\frac{d_{17}}{8(2+\beta_1)},\qquad c_{18}=\frac{d_{18}}{8(2-\beta_1)},$$

$$c_{19}=\frac{d_{19}}{4(1+\beta_1)},\qquad c_{20}=\frac{d_{20}}{4(1-\beta_1)},$$

$$c_{21}=\frac{d_{21}}{2\beta_1},\qquad c_{22}=-\frac{d_{22}}{2\beta_1},$$

$$c_{23}=-\frac{d_{21}}{4\beta_1^2},\qquad c_{24}=\frac{d_{23}}{(4+\beta_2)^2-\beta_1^2},$$

$$c_{25}=\frac{d_{24}}{(4-\beta_2)^2-\beta_1^2},\qquad c_{26}=\frac{d_{25}}{(2+\beta_2)^2-\beta_1^2},$$

$$c_{27}=\frac{d_{26}}{(2-\beta_2)^2-\beta_1^2},\qquad c_{28}=\frac{d_{27}}{\beta_2^2-\beta_1^2},$$

$$c_{29}=\frac{d_{28}}{\beta_2^2-\beta_1^2},\qquad c_{30}=\frac{d_{29}}{121-\beta_1^2},$$

$$c_{31}=\frac{d_{30}}{81-\beta_1^2},\qquad c_{32}=\frac{d_{31}}{49-\beta_1^2},$$

$$c_{33}=\frac{d_{32}}{25-\beta_1^2},\qquad c_{34}=\frac{d_{33}}{9-\beta_1^2},$$

$$c_{35}=\frac{d_{34}}{1-\beta_1^2},\qquad c_{36}=\frac{d_{35}}{1-\beta_1^2},$$

$$c_{37}=\frac{d_{36}}{3(1+\beta_1)(3+\beta_1)},\qquad c_{38}=\frac{d_{37}}{3(1-\beta_1)(3-\beta_1)},$$

$$c_{39}=\frac{d_{38}}{(1+\beta_2)(1+2\beta_1+\beta_2)},\qquad c_{40}=\frac{d_{39}}{(1-\beta_2)(1+2\beta_1-\beta_2)},$$

$$c_{41}=\frac{d_{40}}{(1+\beta_2)(1-2\beta_1+\beta_2)},\qquad c_{42}=\frac{d_{41}}{(1-\beta_2)(1-2\beta_1-\beta_2)},$$

$$c_{43}=\frac{d_{42}}{12(3+\beta_1)},\qquad c_{44}=\frac{d_{43}}{12(3-\beta_1)},$$

$$c_{45}=\frac{d_{44}}{8(2+\beta_1)},\qquad c_{46}=\frac{d_{45}}{8(2-\beta_1)},$$

$$c_{47}=\frac{d_{46}}{4(1+\beta_1)},\qquad c_{48}=\frac{d_{47}}{4(1-\beta_1)},$$

$$c_{49}=\frac{d_{48}}{(4+\beta_2)^2-\beta_1^2},\qquad c_{50}=\frac{d_{49}}{(4-\beta_2)^2-\beta_1^2},$$

$$c_{51}=\frac{d_{50}}{(2+\beta_2)^2-\beta_1^2},\qquad c_{52}=\frac{d_{51}}{(2-\beta_2)^2-\beta_1^2},$$

$$c_{53}=\frac{d_{52}}{\beta_2^2-\beta_1^2},\qquad c_{54}=\frac{d_{53}}{\beta_2^2-\beta_1^2},$$

$$c_{55}=\frac{d_{54}}{81-\beta_1^2},\qquad c_{56}=\frac{d_{55}}{49-\beta_1^2},$$

$$c_{57}=\frac{d_{56}}{25-\beta_1^2},\qquad c_{58}=\frac{d_{57}}{9-\beta_1^2},$$

$$c_{59}=\frac{d_{58}}{1-\beta_1^2},\qquad c_{60}=\frac{b_{16}}{8(2+\beta_1)},$$

$$c_{61}=\frac{b_{17}}{8(2-\beta_1)},\qquad c_{62}=\frac{b_{18}}{(2+\beta_2)^2-\beta_1^2},$$

$$c_{63}=\frac{b_{19}}{(2-\beta_2)^2-\beta_1^2},\qquad c_{64}=\frac{b_{20}}{49-\beta_1^2},$$

$$c_{65}=\frac{b_{21}}{25-\beta_1^2},\qquad c_{66}=\frac{b_{22}}{9-\beta_1^2},$$

$$d_1=a_1b_1,\quad d_2=a_2b_2,\quad d_3=a_1b_3+a_3b_1,$$

$$d_4=a_2b_4+a_3b_2,\quad d_5=a_1b_5+b_1\left(a_4+\frac{1}{m}\right),$$

$$d_6=a_2b_6+b_2\left(a_4+\frac{1}{m}\right),$$

$$d_7=a_1b_7-\frac{b_1}{m}\alpha^2,\quad d_8=a_2b_8-\frac{b_2}{m}\alpha^2,$$

$$d_9=a_1b_9,\quad d_{10}=a_1b_{10},\quad d_{11}=a_2b_9,\quad d_{12}=a_2b_{10},$$

$$d_{13}=a_1b_{11}+a_3b_3,\quad d_{14}=a_2b_{11}+a_3b_4,$$

$$d_{15}=a_1b_{12}+a_3b_5+b_3\left(a_4+\frac{1}{m}\right),$$

$$d_{16}=a_2b_{12}+a_3b_6+b_4\left(a_4+\frac{1}{m}\right),$$

$$d_{17}=a_1b_{13}+a_3b_7+a_4b_5+\frac{1}{m}(b_5-b_3\alpha^2),$$

$$d_{18}=a_2b_{13}+a_3b_8+a_4b_6+\frac{1}{m}(b_6-b_4\alpha^2),$$

$$d_{19}=a_1b_{14}+a_2b_1+a_4b_7+\frac{1}{m}(b_7-b_5\alpha^2),$$

$$d_{20}=a_1b_2+a_2b_{14}+a_4b_8+\frac{1}{m}(b_8-b_6\alpha^2),$$

$$d_{21}=a_1b_{15}-\frac{b_7}{m}\alpha^2,\qquad d_{22}=a_2b_{15}-\frac{b_8}{m}\alpha^2,$$

$$d_{23}=a_3b_9,\quad d_{24}=a_3b_{10},\quad d_{25}=b_9\left(a_4+\frac{1}{m}\right),$$

$$d_{26}=b_{10}\left(a_4+\frac{1}{m}\right),\qquad d_{27}=-\frac{b_9}{m}\alpha^2,$$

$$d_{28}=-\frac{b_{10}}{m}\alpha^2, \qquad d_{29}=a_3b_{11},$$

$$d_{30}=a_3b_{12}+b_{11}\left(a_4+\frac{1}{m}\right),$$

$$d_{31}=a_3b_{13}+a_4b_{12}+\frac{1}{m}(b_{12}-b_{11}\alpha^2),$$

$$d_{32}=a_1b_4+a_2b_3+a_3b_{14}+a_4b_{13}+\frac{1}{m}(b_{13}-b_{12}\alpha^2),$$

$$d_{33}=a_1b_6+a_2b_5+a_3b_{15}+a_4b_{14}+\frac{1}{m}(b_{14}-b_{13}\alpha^2),$$

$$d_{34}=a_1b_8+a_2b_7+a_4b_{15}+\frac{1}{m}(b_{15}-b_{14}\alpha^2),$$

$$d_{35}=-\frac{b_{15}}{m}\alpha^2, \qquad d_{36}=a_1b_{16}+\frac{b_1}{2},$$

$$d_{37}=a_2b_{17}+\frac{b_2}{2}, \qquad d_{38}=a_1b_{18},$$

$$d_{39}=a_1b_{19}, \quad d_{40}=a_2b_{18}, \quad d_{41}=a_2b_{19},$$

$$d_{42}=a_2b_{20}+a_3b_{16}+\frac{b_3}{2},$$

$$d_{43}=a_2b_{20}+a_3b_{17}+\frac{b_4}{2},$$

$$d_{44}=a_1b_{21}+b_{16}\left(a_4+\frac{1}{m}\right)+\frac{b_5}{2},$$

$$d_{45}=a_2b_{21}+b_{17}\left(a_4+\frac{1}{m}\right)+\frac{b_6}{2},$$

$$d_{46}=a_1b_{22}-\frac{b_{16}}{m}\alpha^2+\frac{b_7}{2},$$

$$d_{47}=a_2b_{22}-\frac{b_{17}}{m}\alpha^2+\frac{b_8}{2}, \quad d_{48}=a_3b_{18}, \quad d_{49}=a_3b_{19},$$

$$d_{50}=b_{18}\left(a_4+\frac{1}{m}\right)+\frac{b_9}{2}, \qquad d_{51}=b_{19}\left(a_4+\frac{1}{m}\right)+\frac{b_{10}}{2},$$

$$d_{52}=-\frac{b_{18}}{m}\alpha^2, \quad d_{53}=-\frac{b_{19}}{m}\alpha^2, \quad d_{54}=a_3b_{20}+\frac{b_{11}}{2},$$

$$d_{55}=a_3b_{21}+b_{20}\left(a_4+\frac{1}{m}\right)+\frac{b_{12}}{2},$$

$$d_{56}=a_3b_{22}+a_4b_{21}+\frac{1}{m}(b_{21}-b_{20}\alpha^2)+\frac{b_{13}}{2},$$

$$d_{57}=a_1b_{17}+a_2b_{16}+a_4b_{22}+\frac{1}{m}(b_{22}-b_{21}\alpha^2)+\frac{b_{14}}{2},$$

$$d_{58}=\frac{b_{15}}{2}-\frac{b_{22}}{m}\alpha^2. \tag{3.19}$$

Substituting solution (3.18) into (3.14a), and using expression (3.2), we obtain the nonlinear characteristic relation

$$P=(k^2\alpha_0+\alpha_1)W_m-k\alpha_2W_m^2+\alpha_3W_m^3 \tag{3.20}$$

where

$$\alpha_0=\alpha_1^2\beta_3[G_1(1)-G_1(\alpha)-G_4(1)+G_4(\alpha)],$$
$$\alpha_2=\alpha_1^3\beta_3[G_2(1)-G_2(\alpha)-G_5(1)+G_5(\alpha)],$$
$$\alpha_3=\frac{1}{2}\alpha_1^4\beta_3[G_3(1)-G_3(\alpha)-G_6(1)+G_6(\alpha)],$$

$$G_1(y)=\frac{e_1}{1+\beta_1}y^{1+\beta_1}+\frac{f_1}{1-\beta_1}y^{1-\beta_1}+\frac{c_{60}}{5+\beta_1}y^{5+\beta_1}+\frac{c_{61}}{5-\beta_1}y^{5-\beta_1}+\frac{c_{62}}{3+\beta_2}y^{3+\beta_2}+\frac{c_{63}}{3-\beta_2}y^{3-\beta_2}+\frac{c_{64}}{8}y^8+\frac{c_{65}}{6}y^6+\frac{c_{66}}{4}y^4,$$

$$\begin{aligned}G_2(y)=&\frac{e_2}{1+\beta_1}y^{1+\beta_1}+\frac{f_2}{1-\beta_1}y^{1-\beta_1}+\frac{c_{37}}{2(2+\beta_1)}y^{4+2\beta_1}+\frac{c_{38}}{2(2-\beta_1)}y^{4-2\beta_1}\\&+\frac{c_{39}}{2+\beta_1+\beta_2}y^{2+\beta_1+\beta_2}+\frac{c_{40}}{2+\beta_1-\beta_2}y^{2+\beta_1-\beta_2}+\frac{c_{41}}{2-\beta_1+\beta_2}y^{2-\beta_1+\beta_2}\\&+\frac{c_{42}}{2-\beta_1-\beta_2}y^{2-\beta_1-\beta_2}+\frac{c_{43}}{7+\beta_1}y^{7+\beta_1}+\frac{c_{44}}{7-\beta_1}y^{7-\beta_1}+\frac{c_{45}}{5+\beta_1}y^{5+\beta_1}\\&+\frac{c_{46}}{5-\beta_1}y^{5-\beta_1}+\frac{c_{47}}{3+\beta_1}y^{3+\beta_1}+\frac{c_{48}}{3-\beta_1}y^{3-\beta_1}+\frac{c_{49}}{5+\beta_2}y^{5+\beta_2}+\frac{c_{50}}{5-\beta_2}y^{5-\beta_2}\\&+\frac{c_{51}}{3+\beta_2}y^{3+\beta_2}+\frac{c_{52}}{3-\beta_2}y^{3-\beta_2}+\frac{c_{53}}{1+\beta_2}y^{1+\beta_2}+\frac{c_{54}}{1-\beta_2}y^{1-\beta_2}+\frac{c_{55}}{10}y^{10}\\&+\frac{c_{56}}{8}y^8+\frac{c_{57}}{6}y^6+\frac{c_{58}}{4}y^4+\frac{c_{59}}{2}y^2,\end{aligned}$$

$$\begin{aligned}G_3(y)=&\frac{e_3}{1+\beta_1}y^{1+\beta_1}+\frac{f_3}{1-\beta_1}y^{1-\beta_1}+\frac{c_1}{3(1+\beta_1)}y^{3+3\beta_1}+\frac{c_2}{3(1-\beta_1)}y^{3-3\beta_1}\\&+\frac{c_3}{2(3+\beta_1)}y^{6+2\beta_1}+\frac{c_4}{2(3-\beta_1)}y^{6-2\beta_1}+\frac{c_5}{2(2+\beta_1)}y^{4+2\beta_1}+\frac{c_6}{2(2-\beta_1)}y^{4-2\beta_1}\\&+\frac{c_7}{2(1+\beta_1)}y^{2+2\beta_1}+\frac{c_8}{2(1-\beta_1)}y^{2-2\beta_1}+\frac{c_9}{2+\beta_1+\beta_2}y^{2+\beta_1+\beta_2}\\&+\frac{c_{10}}{2+\beta_1-\beta_2}y^{2+\beta_1-\beta_2}+\frac{c_{11}}{2-\beta_1+\beta_2}y^{2-\beta_1+\beta_2}+\frac{c_{12}}{2-\beta_1-\beta_2}y^{2-\beta_1-\beta_2}\\&+\frac{c_{13}}{9+\beta_1}y^{9+\beta_1}+\frac{c_{14}}{9-\beta_1}y^{9-\beta_1}+\frac{c_{15}}{7+\beta_1}y^{7+\beta_1}+\frac{c_{16}}{7-\beta_1}y^{7-\beta_1}+\frac{c_{17}}{5+\beta_1}y^{5+\beta_1}\\&+\frac{c_{18}}{5-\beta_1}y^{5-\beta_1}+\frac{c_{19}}{3+\beta_1}y^{3+\beta_1}+\frac{c_{20}}{3-\beta_1}y^{3-\beta_1}+\frac{c_{21}}{1+\beta_1}y^{1+\beta_1}\left(\ln y-\frac{1}{1+\beta_1}\right)\\&+\frac{c_{22}}{1-\beta_1}y^{1-\beta_1}\left(\ln y-\frac{1}{1-\beta_1}\right)+\frac{c_{23}}{1+\beta_1}y^{1+\beta_1}+\frac{c_{24}}{5+\beta_2}y^{5+\beta_2}+\frac{c_{25}}{5-\beta_2}y^{5-\beta_2}\\&+\frac{c_{26}}{3+\beta_2}y^{3+\beta_2}+\frac{c_{27}}{3-\beta_2}y^{3-\beta_2}+\frac{c_{28}}{1+\beta_2}y^{1+\beta_2}+\frac{c_{29}}{1-\beta_2}y^{1-\beta_2}+\frac{c_{30}}{12}y^{12}+\frac{c_{31}}{10}y^{10}\\&+\frac{c_{32}}{8}y^8+\frac{c_{33}}{6}y^6+\frac{c_{34}}{4}y^4+\frac{c_{35}}{2}y^2+c_{36}\ln y,\end{aligned}$$

$$G_4(y)=\frac{1}{m}\Big(\frac{b_{16}}{3+\beta_1}y^{3+\beta_1}+\frac{b_{17}}{3-\beta_1}y^{3-\beta_1}+\frac{b_{18}}{1+\beta_2}y^{1+\beta_2}+\frac{b_{19}}{1-\beta_2}y^{1-\beta_2}$$
$$+\frac{b_{20}}{6}y^6+\frac{b_{21}}{4}y^4+\frac{b_{22}}{2}y^2\Big),$$

$$G_5(y)=\frac{1}{m}\Big(\frac{d_{36}}{2(1+\beta_1)}y^{2+2\beta_1}+\frac{d_{37}}{2(1-\beta_1)}y^{2-2\beta_1}+\frac{d_{38}}{\beta_1+\beta_2}y^{\beta_1+\beta_2}+\frac{d_{39}}{\beta_1-\beta_2}y^{\beta_1-\beta_2}$$
$$-\frac{d_{40}}{\beta_1-\beta_2}y^{-\beta_1+\beta_2}-\frac{d_{41}}{\beta_1+\beta_2}y^{-\beta_1-\beta_2}+\frac{d_{42}}{5+\beta_1}y^{5+\beta_1}+\frac{d_{43}}{5-\beta_1}y^{5-\beta_1}+\frac{d_{44}}{3+\beta_1}y^{3+\beta_1}$$
$$+\frac{d_{45}}{3-\beta_1}y^{3-\beta_1}+\frac{d_{46}}{1+\beta_1}y^{1+\beta_1}+\frac{d_{47}}{1-\beta_1}y^{1-\beta_1}+\frac{d_{48}}{3+\beta_2}y^{3+\beta_2}+\frac{d_{49}}{3-\beta_2}y^{3-\beta_2}$$
$$+\frac{d_{50}}{1+\beta_2}y^{1+\beta_2}+\frac{d_{51}}{1-\beta_2}y^{1-\beta_2}-\frac{d_{52}}{1-\beta_2}y^{-1+\beta_2}-\frac{d_{53}}{1+\beta_2}y^{-1-\beta_2}+\frac{d_{54}}{8}y^8$$
$$+\frac{d_{55}}{6}y^6+\frac{d_{56}}{4}y^4+\frac{d_{57}}{2}y^2+d_{58}\ln y\Big),$$

$$G_6(y)=\frac{1}{m}\Big(\frac{d_1}{1+3\beta_1}y^{1+3\beta_1}+\frac{d_2}{1-3\beta_1}y^{1-3\beta_1}+\frac{d_3}{2(2+\beta_1)}y^{4+2\beta_1}+\frac{d_4}{2(2-\beta_1)}y^{4-2\beta_1}$$
$$+\frac{d_5}{2(1+\beta_1)}y^{2+2\beta_1}+\frac{d_6}{2(1-\beta_1)}y^{2-2\beta_1}+\frac{d_7}{2\beta_1}y^{2\beta_1}-\frac{d_8}{2\beta_1}y^{-2\beta_1}+\frac{d_9}{\beta_1+\beta_2}y^{\beta_1+\beta_2}$$
$$+\frac{d_{10}}{\beta_1-\beta_2}y^{\beta_1-\beta_2}-\frac{d_{11}}{\beta_1-\beta_2}y^{-\beta_1+\beta_2}-\frac{d_{12}}{\beta_1+\beta_2}y^{-\beta_1-\beta_2}+\frac{d_{13}}{7+\beta_1}y^{7+\beta_1}$$
$$+\frac{d_{14}}{7-\beta_1}y^{7-\beta_1}+\frac{d_{15}}{5+\beta_1}y^{5+\beta_1}+\frac{d_{16}}{5-\beta_1}y^{5-\beta_1}+\frac{d_{17}}{3+\beta_1}y^{3+\beta_1}+\frac{d_{18}}{3-\beta_1}y^{3-\beta_1}$$
$$+\frac{d_{19}}{1+\beta_1}y^{1+\beta_1}+\frac{d_{20}}{1-\beta_1}y^{1-\beta_1}-\frac{d_{21}}{1-\beta_1}y^{-1+\beta_1}-\frac{d_{22}}{1+\beta_1}y^{-1-\beta_1}+\frac{d_{23}}{3+\beta_2}y^{3+\beta_2}$$
$$+\frac{d_{24}}{3-\beta_2}y^{3-\beta_2}+\frac{d_{25}}{1+\beta_2}y^{1+\beta_2}+\frac{d_{26}}{1-\beta_2}y^{1-\beta_2}-\frac{d_{27}}{1-\beta_2}y^{-1+\beta_2}-\frac{d_{28}}{1+\beta_2}y^{-1-\beta_2}$$
$$+\frac{d_{29}}{10}y^{10}+\frac{d_{30}}{8}y^8+\frac{d_{31}}{6}y^6+\frac{d_{32}}{4}y^4+\frac{d_{33}}{2}y^2+d_{34}\ln y-\frac{d_{35}}{2}y^{-2}\Big). \quad (3.21)$$

According to relation (3.20), a simple calculation yields the conclusion that buckling of the shell occurs for large values of pressure q. In order to find the critical buckling pressure of the shell, we apply the following extremal condition:

$$\frac{\mathrm{d}P}{\mathrm{d}W_m}=0 \quad (3.22)$$

from which

$$3\alpha_3 W_m^2-2k\alpha_2 W_m+k^2\alpha_0+\alpha_1=0. \quad (3.23)$$

Solutions of this quadratic equation are

$$W_m^*=\frac{1}{3\alpha_3}\left[k\alpha_2\pm\sqrt{k^2(\alpha_2^2-3\alpha_0\alpha_3)-3\alpha_1\alpha_3}\right]. \quad (3.24)$$

Substituting W_m^* into relation (3.20), we obtain the following formula of the nondimensional critical buckling pressure:

$$P^*=(k_1^2\alpha_0+\alpha_1)W_m^*-k\alpha_2 W_m^{*2}+\alpha_3 W_m^{*3} \quad (3.25)$$

in which for W_m^* with the negative and positive signs in expression (3.24), P^* are the upper and lower critical buckling pressures, respectively. It is pointed out that we are interested only in the upper critical buckling pressure when considering engineering practice.

Let the discriminant of the quadratic equation (3.23) be zero; the formula for the critical point, which is used to distinguish between buckling and no buckling for the shell, is given by

$$k_0 = \sqrt{\frac{3\alpha_1\alpha_3}{\alpha_2^2 - 3\alpha_0\alpha_3}}. \tag{3.26}$$

Especially in the limiting case $b=0$, the symmetrically laminated, cylindrically orthotropic, truncated, shallow, spherical shell with a non-deformable rigid body at the center becomes a symmetrically laminated, cylindrically orthotropic, shallow, spherical shell. Thus, the preceding formulae give the critical buckling pressure of the symmetrically laminated, cylindrically orthotropic, truncated, shallow, spherical shell with a rigidly clamped edge under uniform pressure which has been studied by the author[2]。

4. Numerical Examples

For the purpose of simple calculation and without affecting the generality of results, let us take as a particular example a symmetrically laminated, cylindrically orthotropic, truncated, shallow, spherical shell. Each layer of the shell has the same thickness and elastic properties. Assume

$$\frac{E_\theta}{E_r} = 30, \quad \nu_{\theta r} = 0.25.$$

Then from expressions (2.9), we first obtain

$$\beta_1^2 = \beta_2^2 = 30, \quad \beta_3 = 359.25, \quad \beta_4 = 0.25.$$

Using these values and formula (3.25), we obtain useful results, Curves of the non-dimensional critical buckling pressure P^* for $m=100$ and $\alpha=0.1$ are shown in Figs. 2 and 3, respectively. It can be seen that these curves rise monotonically and for the same value of k the non-dimensional critical buckling pressure P^* of a shell with small α and m is low.

Finally, from formula (3.26), results are calculated and shown in Fig. 4. It is seen that the critical geometrical parameter k_0 increases with the increase of the non-dimensional shear rigidity m, and k_0 diminishes with the increase of the non-dimensional inner radius α. It indicates that a shell with small m and larger α is more sensitive to instability.

For the purpose of comparison, we consider a second example, the special case of one-layer, cylindrically orthotropic, shallow, spherical shells without a non-deformable rigid body at the center, and not including the effect of the transverse shear. Assume $n=1$ and $\nu_{\theta r}=1/3$. Then, according to formula (3.25), stability curves of isotropic ($E_\theta/E_r=1$) and cylindrically orthotropic, shallow, spherical shells are indicated by solid lines in Fig. 5.

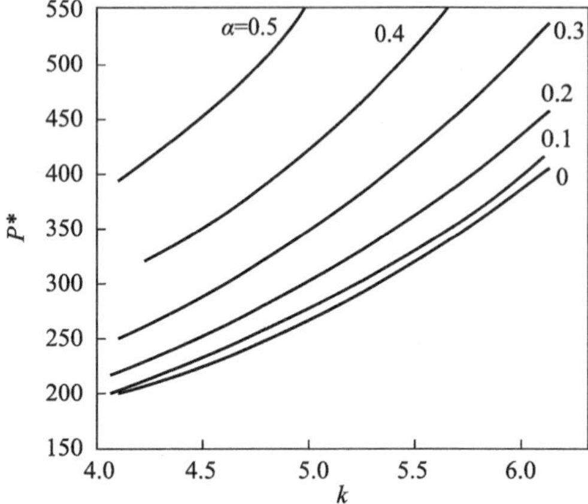

Fig. 2　Curves of the non-dimensional critical buckling pressure for several values of α ($m=100$)

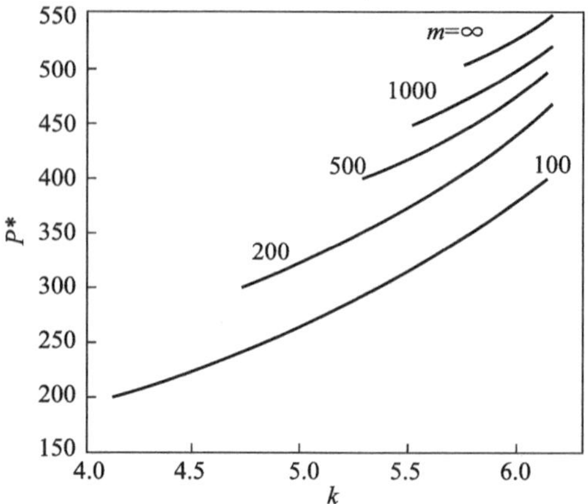

Fig. 3　Curves of the non-dimensional critical buckling pressure for several values of m ($\alpha=0.1$)

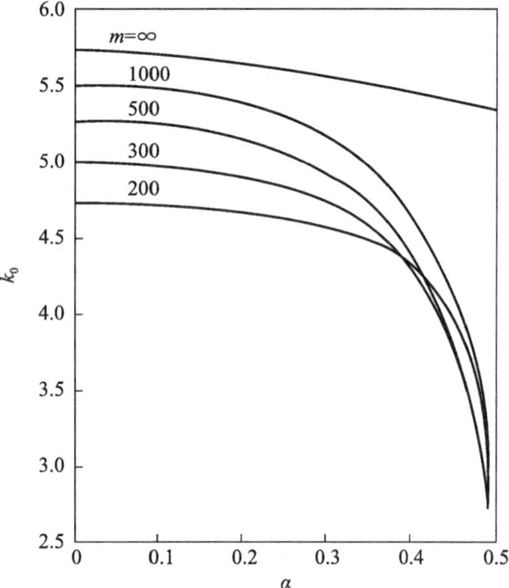

Fig. 4　Curves of the critical geomnetrical parameter k_0

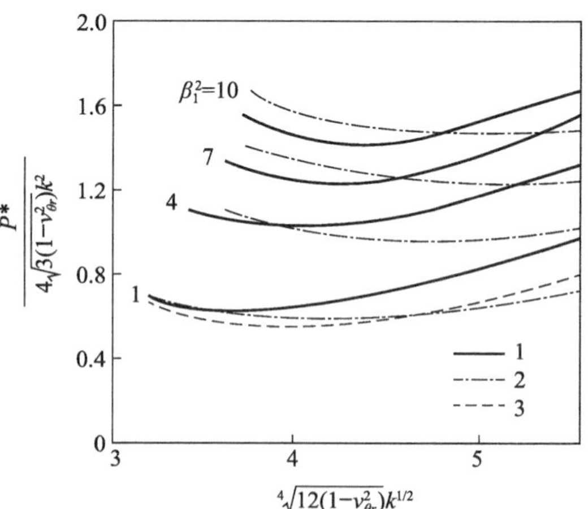

Fig. 5　Comparison between theoretical results ($\nu_{\theta r}=1/3$):

1. present results; 2. Varadan[16]; 3. Hyman[17]

The results of Varadan[16] and Hyman[17] are also included in the figure. They show that our results are in good agreement with other authors' results. Moreover, Fig. 6 shows the present numerical results for a one-layer, isotropic, shallow, spherical shell as well as the experimental and theoretical values[18,19] obtained by other authors. The figure illustrates that our results are in better agreement with the experimental ones. In accordance with our analysis, the experimental values are on the low side because of the effect of the boundary condition in the experiment.

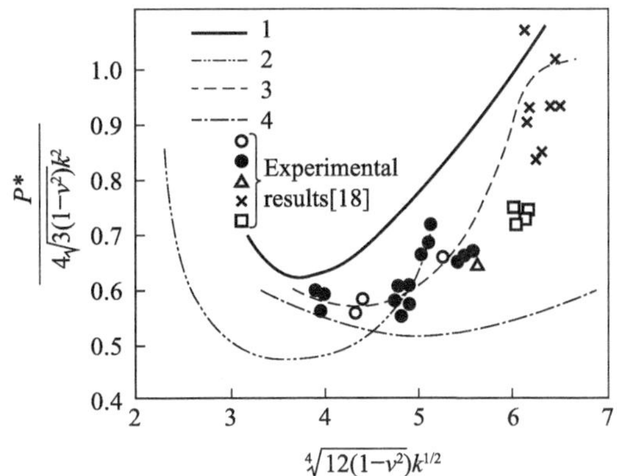

Fig. 6 Comparison between theoretical and experimental results

($\nu=\nu_{\theta r}=0.3$, $E=E_\theta=E_r$)

1. present results; 2. Kaplan and Fung; 3. Budiansky and Huang; 4. von Willich

Acknowledgements

This paper was written at the Institute of Statics and Dynamics, Ruhr-University Bochum, Germany. The author is grateful to the Alexander von Humboldt Foundation, Ruhr-University Bochum, and Professor Wilfried B. Krätzig for his support and hearty hospitality. At the same time, this work was supported by the National Natural Science Foundation of China.

References

[1] C. Y. Chia, *Nonlinear Analysis of plates*. McGraw-Hill, New York (1980).

[2] Liu Renhuai, Nonlinear stability of symmetrically laminated cylindrically orthotropic shallow spherical shells including transverse shear. *Science in China* (*Series A*) 35, 734-746 (1992).

[3] Liu Renhuai, On the nonlinear stability of symmetrically laminated cylindrically orthotropic shallow spherical shells. *Appl. Math. Mech.* 12, 271-279 (1991).

[4] Liu Renhuai, Large deflection equations of symmetrically laminated composite cylindrically orthotropic shallow spherical shells. In *Applied Mathematics and Mechanics*, pp. 279-284. Science Press, Beijing (1993).

[5] Liu Renhuai and C. Zhong, Nonlinear buckling of symmetrically laminated, cylindrically orthotropic shallow spherical shells with a circular hole at the center including transverse shear. *J. Jinan Univ.* (*Natural Science*) 15, 1-12 (1994) (in Chinese).

[6] Liu Renhuai and L. H. He, A simple theory for nonlinear bending of laminated composite rectangular plates including higher-order effects. *Int. J. Nonlinear Mechanics* 26, 537-545 (1991).

[7] Liu Renhuai and L. H. He, Axisymmetrical bending of laminated circular plates. *J. Jiangxi Polytech. Univ.* 13, 199-204 (1991) (in Chinese).

[8] Liu Renhuai and L. H. He, Nonlinear bending of simply supported symmetric laminated cross-ply rectangular plates. *Appl. Math. Mech.* 11, 801-807 (1990).

[9] Y. Başar, Y. Ding and R. Schultz, Refined shear-deformation models for composite laminates with finite rotations, *Int. J. Solids Structures*. 30, 2611-2638 (1993).

[10] C. Y. Chia, Effect of transverse shear on nonlinear vibration of anti-symmetric cross-ply shallow spherical shell with rectangular planform. In *Proceedings of the Second International Symposi-*

um on Composite Materials and Structures, pp. 555-560. Peking University Press, Beijing (1992).

[11] P. C. Dumir and A. Bhaskar, Nonlinear forced vibration of orthotropic thin rectangular plates. *Int. J. Mech. Sci.* 30, 371-380 (1988).

[12] J. N, Reddy, A refined nonlinear theory of plates with transverse shear deformation. *Int. J. Solids Structures* 20, 881-896 (1984).

[13] K. Y. Yeh, Liu Renhuai, Q. Y. Ping and S. L. Li, Nonlinear stability of thin circular shallow spherical shell under actions of axisymmetric uniform distributed line loads. *Bull. Sci.* 2, 142-145 (1965); *J. Lanzhou Univ.* 2, 10-33 (1965) (in Chinese).

[14] K. Y. Yeh, Liu Renhuai, C. Z. Zhang and Y. F. Xu, Nonlinear stability of thin circular shallow spherical shell under the action of uniform edge moment. *Bull. Sci.* 2, 145-147 (1965) (in Chinese).

[15] Liu Renhuai, Nonlinear stability of circular shallow spherical shell with a hole in the center under the action of uniform moment at the inner edge. *Bull. Sci.* 3, 253-255 (1965) (in Chinese).

[16] T. K. Varadan, Snap-buckling of orthotropic shallow spherical shells. *J. Appl. Mech.* 45, 445-447 (1978).

[17] B. I. Hyman, Snap-through of shallow clamped spherical caps under uniform pressure. *Int. J. Nonlinear Mechanics* 6, 55-67 (1971).

[18] S. C. Tillman, On the buckling behavior of shallow caps under a uniform pressure load. *Int. J. Solids Structures* 6, 37-52 (1970).

[19] A. Kaplan, Buckling of spherical shells. In *Thin-shell Structures* (edited by Y. C. Fung and E. E. Sechler), pp. 247-288. Prentice-Hall, Englewood Cliffs, NJ (1974).

复合材料层合扁锥壳的非线性稳定问题①

一、引言

近年来，复合材料层合扁壳在航空、航天、航海、土木和压力容器等工程中得到了广泛应用。因此，研究这类板壳的弯曲、稳定和振动的学者日益增加。作者[1-10]在以往研究的基础上，进一步研究了复合材料层合扁锥壳的非线性稳定问题。由于结构的复杂和非线性数学的困难，此问题尚未被讨论过。本文建立的理论和所获公式可供工程设计时参考应用。

二、基本方程

考虑如图1所示的均布压力 q 作用下的层合扁锥壳，并以壳体的中面作为坐标曲面。这里，r、θ 分别是径向和环向坐标，h 为厚度，a 为半径，α 为斜角，n 为层数。

采用如下假设：①材料是线弹性的；②每层的厚度和弹性常数相同，但各层可以有不同的厚度和弹性常数且与中曲面对称，呈圆柱正交异性；③壳体横向不可压缩；④直法线假定；⑤壳体的横向正应力忽略不计。

于是，壳体内与中曲面相距为 z 的任一点的径向、环向和轴向位移为

$$u = u_0 - z\frac{\mathrm{d}w}{\mathrm{d}r},\quad v = 0,\quad w = w_0 \tag{2.1}$$

其中，u_0 和 w_0 分别是中曲面上点的径向位移和挠度。

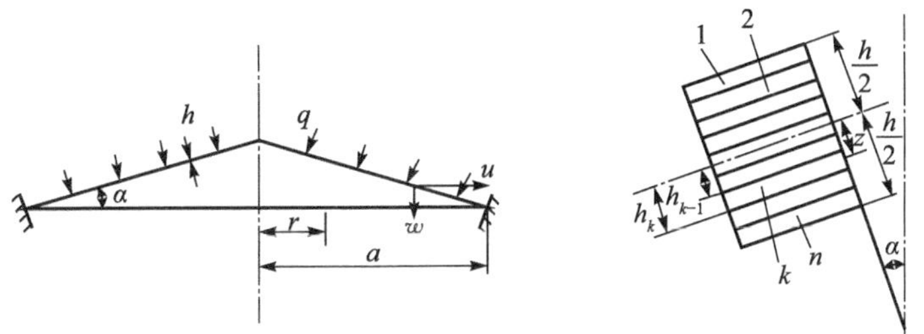

图1　对称层合圆柱正交异性扁锥壳

壳体内任一点的几何方程为

$$\varepsilon_r = \frac{\partial u}{\partial r} + \frac{\mathrm{d}w}{\mathrm{d}r}\left(\alpha + \frac{1}{2}\frac{\mathrm{d}w}{\mathrm{d}r}\right),\quad \varepsilon_\theta = \frac{u}{r},\quad \gamma_{r\theta} = 0 \tag{2.2}$$

将式（1）代入上式，有

$$\varepsilon_r = \varepsilon_r^0 + z\kappa_r,\quad \varepsilon_\theta = \varepsilon_\theta^0 + z\kappa_\theta \tag{2.3}$$

① 本文原载《第九届全国复合材料学术会议论文集》，北京：世界图书出版公司，1996，249-255.

其中 ε_r^0 和 ε_θ^0 是壳体中曲面上的应变，κ_r 和 κ_θ 是中曲面上的曲率，

$$\varepsilon_r^0 = \frac{\mathrm{d}u_0}{\mathrm{d}r} + \frac{\mathrm{d}w}{\mathrm{d}r}\left(\alpha + \frac{1}{2}\frac{\mathrm{d}w}{\mathrm{d}r}\right), \qquad \varepsilon_\theta^0 = \frac{u_0}{r}, \tag{2.4}$$

$$\kappa_r = -\frac{\mathrm{d}^2 w}{\mathrm{d}r^2}, \qquad \kappa_\theta = -\frac{1}{r}\frac{\mathrm{d}w}{\mathrm{d}r} \tag{2.5}$$

壳体第 k 层的胡克定律为

$$\sigma_r^{(k)} = Q_{11}^{(k)}\varepsilon_r + Q_{12}^{(k)}\varepsilon_\theta, \quad \sigma_\theta^{(k)} = Q_{12}^{(k)}\varepsilon_r + Q_{22}^{(k)}\varepsilon_\theta, \quad \tau_{r\theta}^{(k)} = 0 \tag{2.6}$$

其中 $E_r^{(k)}$ 和 $E_\theta^{(k)}$ 是弹性模量，$\nu_{r\theta}^{(k)}$ 和 $\nu_{\theta r}^{(k)}$ 是泊松比，$Q_{ij}^{(k)}$ 是 k 层的刚度系数，

$$Q_{11}^{(k)} = \frac{E_r^{(k)}}{1-\nu_{r\theta}^{(k)}\nu_{\theta r}^{(k)}}, \quad Q_{12}^{(k)} = \frac{\nu_{\theta r}^{(k)}E_r^{(k)}}{1-\nu_{r\theta}^{(k)}\nu_{\theta r}^{(k)}},$$

$$Q_{22}^{(k)} = \frac{E_\theta^{(k)}}{1-\nu_{r\theta}^{(k)}\nu_{\theta r}^{(k)}} \tag{2.7}$$

而且

$$\nu_{r\theta}^{(k)}E_\theta^{(k)} = \nu_{\theta r}^{(k)}E_r^{(k)} \tag{2.8}$$

设径向和环向薄膜力为 N_r 和 N_θ，径向和环向弯矩为 M_r 和 M_θ，径向剪力为 Q_r，其定义为

$$N_r = \sum_{k=1}^{n}\int_{h_{k-1}}^{h_k}\sigma_r^{(k)}\mathrm{d}z, \quad N_\theta = \sum_{k=1}^{n}\int_{h_{k-1}}^{h_k}\sigma_\theta^{(k)}\mathrm{d}z, \quad M_r = \sum_{k=1}^{n}\int_{h_{k-1}}^{h_k}\sigma_r^{(k)}z\,\mathrm{d}z,$$

$$M_\theta = \sum_{k=1}^{n}\int_{h_{k-1}}^{h_k}\sigma_\theta^{(k)}z\,\mathrm{d}z, \quad Q_r = \sum_{k=1}^{n}\int_{h_{k-1}}^{h_k}\tau_{rz}^{(k)}\mathrm{d}z \tag{2.9}$$

将式（2.6）代入上式，可得

$$N_r = A_{11}\varepsilon_r^0 + A_{12}\varepsilon_\theta^0, \quad N_\theta = A_{12}\varepsilon_r^0 + A_{22}\varepsilon_\theta^0$$

$$M_r = D_{11}\kappa_r + D_{12}\kappa_\theta, \quad M_\theta = D_{12}\kappa_r + D_{22}\kappa_\theta \tag{2.10a-d}$$

其中 A_{ij} 是抗拉刚度，D_{ij} 是弯曲刚度，

$$A_{ij} = \sum_{k=1}^{n}Q_{ij}^{(k)}(h_k - h_{k-1}), \quad D_{ij} = \frac{1}{3}\sum_{k=1}^{n}Q_{ij}^{(k)}(h_k^3 - h_{k-1}^3) \tag{2.11}$$

由壳体微元体的平衡，可得平衡方程组：

$$\frac{\mathrm{d}(rN_r)}{\mathrm{d}r} - N_\theta = 0,$$

$$\frac{\mathrm{d}(rM_r)}{\mathrm{d}r} - M_\theta - rQ_r = 0, \tag{2.12a-c}$$

$$\frac{\mathrm{d}}{\mathrm{d}r}\left[rN_r\left(\alpha + \frac{\mathrm{d}w}{\mathrm{d}r}\right) + rQ_r\right] + rq = 0$$

由式（2.4）的两个方程消去 u_0，并应用式（2.10a,b）和方程（2.12a），便得协调方程：

$$A_3\frac{\mathrm{d}}{\mathrm{d}r}r\frac{\mathrm{d}(rN_r)}{\mathrm{d}r} - A_1N_r + \frac{1}{2}\left(\frac{\mathrm{d}w}{\mathrm{d}r}\right)^2 + \alpha\frac{\mathrm{d}w}{\mathrm{d}r} = 0 \tag{2.13}$$

其中

$$A_1 = A_0A_{22}, \quad A_3 = A_0A_{11}, A_0 = (A_{11}A_{22} - A_{12}^2)^{-1} \tag{2.14}$$

由方程（2.12b,c）消去 Q_r，并应用式（2.10c,d）和（2.5），对 r 积分一次后，

可得

$$r^{\beta_1}\frac{\mathrm{d}}{\mathrm{d}r}r^{1-2\beta_1}\frac{\mathrm{d}}{\mathrm{d}r}\left(r^{\beta_1}\frac{\mathrm{d}w}{\mathrm{d}r}\right)-\frac{rN_r}{D_{11}}\left(\alpha+\frac{\mathrm{d}w}{\mathrm{d}r}\right)-\frac{r^2q}{2D_{11}}=0 \tag{2.15}$$

其中

$$\beta_1^2=\frac{D_{22}}{D_{11}} \tag{2.16}$$

于是，方程（2.15）和（2.13）便组成了均布压力 q 作用下对称层合圆柱正交异性扁锥壳大挠度理论的控制方程组。

我们讨论常用的夹紧固定边界条件：

当 $r=a$ 时，$w=0$，$\frac{\mathrm{d}w}{\mathrm{d}r}=0$，$u_0=0$

当 $r=0$ 时，$\frac{\mathrm{d}w}{\mathrm{d}r}=0$，$N_r$ 有限　　(2.17)

其中

$$u_0=r\left[A_3\frac{\mathrm{d}(rN_r)}{\mathrm{d}r}-A_2N_r\right],\quad A_2=A_0A_{12} \tag{2.18}$$

三、解析解

为了便于求解上述非线性边值问题（2.15）、（2.13）和（2.17），我们首先引入如下的无量纲量：

$$\begin{aligned}&y=\frac{r}{a},\quad W=\frac{w}{h},\quad \Phi=\frac{\mathrm{d}W}{\mathrm{d}y}+k,\quad S=\frac{a}{D_{11}}rN_r,\quad k=\frac{a}{h}\alpha,\\&P=\frac{a^4}{2D_{11}h}q,\quad \beta_2^2=\frac{A_1}{A_3},\quad \beta_3=\frac{h^2}{A_3D_{11}},\quad \beta_4=\frac{A_2}{A_3}\end{aligned} \tag{3.1}$$

应用这些量，非线性边值问题（2.15）、（2.16）和（2.17）转化为下面的无量纲形式

$$\begin{aligned}&L_1(y^{\beta_1}\Phi)=S\Phi+Py^2-k\beta_1^2y^{-1},\\&L_2(y^{\beta_2}S)=-\frac{\beta_3}{2}(\Phi^2-k^2)\end{aligned} \tag{3.2}$$

$$\text{当 } y=1 \text{ 时},W=0,\Phi=k,\frac{\mathrm{d}S}{\mathrm{d}y}-\beta_4\frac{S}{y}=0 \tag{3.3}$$

$$\text{当 } y=0 \text{ 时},\Phi=k,S=0 \tag{3.4}$$

其中，L_1 和 L_2 是线性微分算子，

$$\begin{aligned}&L_1(\cdots)=y^{\beta_1}\frac{\mathrm{d}}{\mathrm{d}y}y^{1-2\beta_1}\frac{\mathrm{d}}{\mathrm{d}y}(\cdots),\\&L_2(\cdots)=y^{\beta_2}\frac{\mathrm{d}}{\mathrm{d}y}y^{1-2\beta_2}\frac{\mathrm{d}}{\mathrm{d}y}(\cdots)\end{aligned} \tag{3.5}$$

应用叶开沅和刘人怀提出的修正迭代法[8-10]，易于求解上述无量纲非线性边值问题。此时，我们取壳体的无量纲中心挠度 W_m 作为迭代参数：

$$W_m=W\mid_{y=0} \tag{3.6}$$

由式（3.1）和（3.3a），无量纲中心挠度 W_m 为

$$W_m = -\int_0^1 (\Phi - k)\mathrm{d}y \tag{3.7}$$

对于一次近似，有线性边值问题如下

$$\begin{aligned} &L_1(y^{\beta_1}\Phi_1) = Py^2 - k\beta_1^2 y^{-1}, \\ &L_2(y^{\beta_2}S_1) = -\frac{\beta_3}{2}(\Phi^2 - k^2) \end{aligned} \tag{3.8}$$

$$\begin{aligned} &\text{当 } y = 1 \text{ 时}, W_1 = 0, \Phi_1 = k, \frac{\mathrm{d}S_1}{\mathrm{d}y} - \beta_4\frac{S_1}{y} = 0 \\ &\text{当 } y = 0 \text{ 时}, \Phi_1 = k, S_1 = 0 \end{aligned} \tag{3.9}$$

应用式（3.7），这一边值问题的解为

$$\begin{aligned} &\Phi_1 = \lambda\alpha_1 W_m(y^{\beta_1} - y^3) + k, \\ &S_1 = -\beta_3\Big[\frac{1}{2}\alpha_1^2 W_m^2(b_1 y^{1+2\beta_1} + b_2 y^{4+\beta_1} + b_3 y^{\beta_2} + b_4 y^7) \\ &\qquad + k\alpha_1 W_m(b_5 y^{1+\beta_1} + b_6 y^{\beta_2} + b_7 y^4)\Big] \end{aligned} \tag{3.10}$$

其中

$$\begin{aligned} &\alpha_1 = 4(1+\beta_1)(3+\beta_1), \quad \lambda = -\frac{1}{9-\beta_1^2}, \\ &b_1 = \frac{\lambda^2}{(1+2\beta_1)^2 - \beta_2^2}, \quad b_2 = \frac{2\lambda^2}{(4+\beta_1)^2 - \beta_2^2}, \\ &b_3 = \frac{b_1(1+2\beta_1-\beta_4) + b_2(4+\beta_1-\beta_4) + b_4(7-\beta_4)}{\beta_4 - \beta_2}, \\ &b_4 = \frac{\lambda^2}{49-\beta_2^2}, \quad b_5 = \frac{\lambda}{(1+\beta_1)^2 - \beta_2^2}, \\ &b_6 = \frac{b_5(1+\beta_1-\beta_4) + b_7(4-\beta_4)}{\beta_4 - \beta_2}, \quad b_7 = -\frac{\lambda}{16-\beta_2^2} \end{aligned} \tag{3.11}$$

对于二次近似，有关于 Φ 的线性边值问题

$$L_1(y^{\beta_1}\Phi_2) = S_1\Phi_1 + Py^2 - k\beta_1^2 y^{-1} \tag{3.12}$$

$$\text{当 } y = 1 \text{ 时}, W_2 = 0, \Phi_2 = k \tag{3.13}$$

$$\text{当 } y = 0 \text{ 时}, \Phi_2 = k \tag{3.14}$$

应用解（3.10），上述边值问题的解为

$$\begin{aligned} \Phi_2 = {} & \lambda P(y^{\beta_1} - y^3) + k - \frac{\beta_3}{2}[\alpha_1^3 W_m^3(c_1 y^{2+3\beta_1} + c_2 y^{5+2\beta_1} + c_3 y^{8+\beta_1} \\ & + c_4 y^{1+\beta_1+\beta_2} + c_5 y^{4+\beta_2} + c_6 y^{11} + c_7 y^{\beta_1}) + k\alpha_1^2 W_m^2(c_8 y^{2+2\beta_1} \\ & + c_9 y^{5+\beta_1} + c_{10} y^{1+\beta_1+\beta_2} + c_{11} y^{4+\beta_2} + c_{12} y^{1+\beta_2} + c_{13} y^8 \\ & + c_{14} y^{\beta_1}) + 2k^2\alpha_1 W_m(c_{15} y^{2+\beta_1} + c_{16} y^{1+\beta_2} + c_{17} y^5 + c_{18} y^{\beta_1})] \end{aligned} \tag{3.15}$$

其中

$$c_1 = \frac{\lambda b_1}{4(1+\beta_1)(1+2\beta_1)}, \qquad c_2 = \frac{\lambda(b_2 - b_1)}{(5+\beta_1)(5+3\beta_1)},$$

$$
\begin{aligned}
&c_3 = \frac{\lambda(b_4 - b_2)}{16(4+\beta_1)}, && c_4 = \frac{\lambda b_3}{(1+\beta_2)(1+2\beta_1+\beta_2)}, \\
&c_5 = -\frac{\lambda b_3}{(4+\beta_2)^2 - \beta_1^2}, && c_6 = -\frac{\lambda b_4}{121 - \beta_1^2}, \\
&c_7 = -\sum_{i=1}^{6} c_i, && c_8 = \frac{b_1 + 2\lambda b_5}{(2+\beta_1)(2+3\beta_1)}, \\
&c_9 = \frac{b_2 + 2\lambda b_7 - 2\lambda b_5}{5(5+2\beta_1)}, && c_{10} = \frac{2\lambda b_6}{(1+\beta_2)(1+2\beta_1+\beta_2)}, \\
&c_{11} = -\frac{2\lambda b_6}{(4+\beta_2)^2 - \beta_1^2}, && c_{12} = \frac{b_3}{(1+\beta_2)^2 - \beta_1^2}, \\
&c_{13} = \frac{b_4 - 2\lambda b_7}{b_4 - \beta_1^2}, && c_{14} = -\sum_{i=8}^{13} c_i, \\
&c_{15} = \frac{b_5}{4(1+\beta_1)}, && c_{16} = \frac{b_6}{(1+\beta_2)^2 - \beta_1^2}, \\
&c_{17} = \frac{b_7}{25 - \beta_1^2}, && c_{18} = -\sum_{i=15}^{17} c_i
\end{aligned}
\tag{3.16}
$$

将解（3.15）代入式（3.7），我们得到对称层合圆柱正交异性扁锥壳的非线性特征关系式

$$
P = (\alpha_1 + k^2\alpha_0)W_m + k\alpha_2 W_m^2 + \alpha_3 W_m^3 \tag{3.17}
$$

其中

$$
\begin{aligned}
\alpha_0 &= -\alpha_1^2\beta_3\left(\frac{c_{15}}{3+\beta_1} + \frac{c_{16}}{2+\beta_2} + \frac{c_{17}}{6} + \frac{c_{18}}{1+\beta_1}\right), \\
\alpha_2 &= -\frac{1}{2}\alpha_1^3\beta_3\left(\frac{c_8}{3+2\beta_1} + \frac{c_9}{6+\beta_1} + \frac{c_{10}}{2+\beta_1+\beta_2} + \frac{c_{11}}{5+\beta_2}\right. \\
&\quad \left. + \frac{c_{12}}{2+\beta_1} + \frac{c_{13}}{9} + \frac{c_{14}}{1+\beta_1}\right), \\
\alpha_3 &= -\frac{1}{2}\alpha_1^4\beta_3\left[\frac{c_1}{3(1+\beta_1)} + \frac{c_2}{2(3+\beta_1)} + \frac{c_3}{9+\beta_1} + \frac{c_4}{2+\beta_1+\beta_2}\right. \\
&\quad \left. + \frac{c_5}{5+\beta_2} + \frac{c_6}{12} + \frac{c_7}{1+\beta_1}\right]
\end{aligned}
\tag{3.18a-c}
$$

应用极值条件：

$$
\frac{\mathrm{d}P}{\mathrm{d}W_m} = 0 \tag{3.19}
$$

得到壳体失稳时的无量纲临界挠度 W_m^*

$$
W_m^* = \frac{1}{3\alpha_3}\left[-k\alpha_2 \pm \sqrt{k^2(\alpha_2^2 - 3\alpha_0\alpha_3) - 3\alpha_1\alpha_3}\right] \tag{3.20}
$$

将此式代入式（3.17），便得无量纲临界载荷

$$
P^* = (\alpha_1 + k^2\alpha_0)W_m^* + k\alpha_2 W_m^{*2} + \alpha_3 W_m^{*3} \tag{3.21}
$$

令方程（3.19）判别式为零，便得壳体的临界几何参数

$$
k_0 = \sqrt{\frac{3\alpha_1\alpha_3}{\alpha_2^2 - 3\alpha_0\alpha_3}} \tag{3.22}
$$

k_0 被用来区分壳体失稳与否。当 $k \geqslant k_0$ 时，壳体出现失稳；当 $k < k_0$ 时，壳体没有失稳。

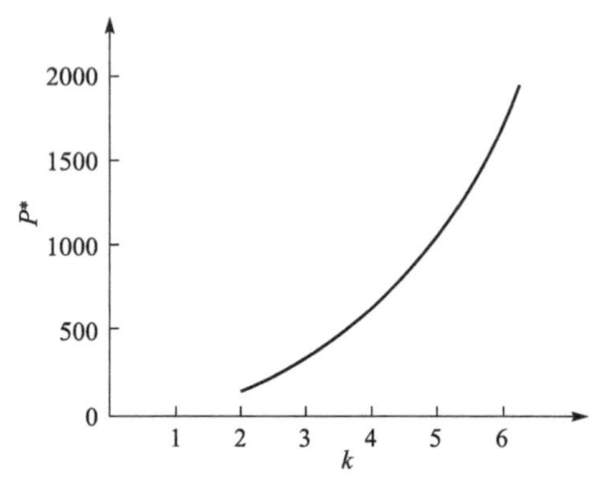

图 2　无量纲临界载荷 P^* 曲线

四、数例

为使计算简单且不失一般性，我们考虑一个特例。假设壳体各层具有相同的厚度和相同的弹性常数，且有

$$\frac{E_\theta}{E_r} = 1.5, \quad \nu_{\theta r} = 0.3$$

按照式（3.21），我们计算了临界载荷。由于实际工程中仅有上临界载荷有意义，故只应用式（3.20）中的负号所对应的 W_m^* 值。所得的上临界载荷值绘在图 2 中。由图可见，临界载荷曲线单调上升。

参考文献

[1] 刘人怀. 考虑横向剪切的对称层合圆柱正交异性扁球壳的非线性稳定问题. 中国科学（A 辑），1991（7）：742-751.

[2] Liu Ren-huai and He Ling-hui. A simple theory for non-linear bending of laminated composite rectangular plates including higher-order effects. Int. J. Non-Linear Mech.，1991，26（5）：537-545.

[3] 刘人怀. 对称圆柱正交异性层合扁球壳的非线性稳定问题. 应用数学和力学，1991，12（3）：251-258.

[4] 刘人怀，何陵辉. 层合圆板的轴对称弯曲问题. 江西工业大学学报，1991，13（2）：199-205.

[5] 刘人怀，何陵辉. 四边简支对称正交层合矩形板的非线性弯曲问题. 应用数学和力学，1990，11（9）：753-759.

[6] 刘人怀. 对称层合圆柱正交异性复合材料扁球壳的大挠度方程. 应用数学和力学：钱伟长八十诞辰祝寿文集. 北京：科学出版社和重庆出版社，1993：279-284.

[7] 刘人怀，钟诚. 考虑横向剪切的对称层合圆柱正交异性中心开孔扁球壳的非线性屈曲. 暨南大学学报（自然科学版），1994，15（1）：1-12.

[8] 叶开沅，刘人怀，平庆元，等. 在对称线布截荷作用下的圆底扁球壳的非线性稳定问题. 科学通报，1965，(2)：142-145.

[9] 叶开沅，刘人怀，张传智，等. 圆底扁球壳在边缘力矩作用下的非线性稳定问题. 科学通报，1965，(2)：145-147.

[10] 刘人怀. 在内边缘均布力矩作用下的中心开孔圆底扁球壳的非线性稳定问题. 科学通报，1965，(3)：253-255.

复合材料层合扁球壳的非线性强迫振动①

一、引言

复合材料优越的特性，使得复合材料层合板壳在航空航天等各种工程结构中得到广泛应用。因此，研究复合材料层合板壳的非线性特性、大挠度变形、非线性屈曲和非线性振动有着十分重要的意义。在 20 世纪 80 年代，Chia[1]就对复合材料层合板的后屈曲特性和非线性弯曲振动进行研究。以后，刘人怀[2-9]建立了考虑横向剪切的对称层合圆柱正交异性扁球壳和扁锥壳的大挠度理论以及考虑层间位移和横向剪应力连续条件的层合板理论，并对这些板壳的非线性弯曲和稳定问题进行了求解。Wang[10]对复合材料锥壳的有限挠度问题进行了研究。Alwar 等[11]对复合材料层合环形球壳进行了轴对称的非线性分析。Muc[12]对复合材料层合扁球壳在外压力作用下的屈曲和过屈曲行为进行了分析。Xu[13]对在弹性基础上的对称层合复合材料扁球壳的动态非线性问题进行了研究，得到非线性自由振动的 Fourier-Bessel 级数解。Qatu[14]对复合材料层合板壳的非线性剪切变形理论的有效范围进行了讨论。

本文在过去工作的基础上，进一步研究了复合材料层合扁球壳的非线性强迫振动问题。

二、非线性强迫振动微分方程

考虑对称层合圆柱正交异性扁球壳，受均布强迫荷载 q（r，t）的作用，厚度为 h，半径为 a，曲率半径为 R，拱高为 H。

采用文献［2］提出的位移场和修正的剪切刚度 G 的表达式，引入动荷载及 D'Alembert 惯性力，进行无量纲化，得到如下的复合材料层合扁球壳的无量纲非线性强迫振动微分方程

$$
\begin{aligned}
&\frac{1}{y}\frac{\partial}{\partial y}[S\Phi+my\Psi+my\Phi-kmy^2]+P-\bar{\rho}\frac{\partial^2 W}{\partial t^2}=0,\\
&y^{\beta_1}\frac{\partial}{\partial y}y^{1-2\beta_1}\frac{\partial}{\partial y}(y^{\beta_1}\Psi)=my(\Psi+\Phi-ky),\\
&y^{\beta_2}\frac{\partial}{\partial y}y^{1-2\beta_2}\frac{\partial}{\partial y}(y^{\beta_2}S)=-\beta_3(\Phi-ky)\left[\frac{1}{2}(\Phi-ky)+ky\right]
\end{aligned}
\tag{1}
$$

和夹紧固定的边界条件

$$
\begin{aligned}
&\text{当 } y=1 \text{ 时}, W=0, \Psi=0, \frac{\mathrm{d}S}{\mathrm{d}y}-\beta_4\frac{S}{y}=0\\
&\text{当 } y=0 \text{ 时}, \Psi=0, S=0
\end{aligned}
\tag{2}
$$

其中 ψ 为壳体变形前的中曲面法线的转角，N_r 为径向薄膜力，w 为挠度，ρ 为密度，

① 本文原载《力学学报》，1997，29（2）：236-241. 作者：刘人怀，王璠.

A_i，D_{ij}（i，j=1，2）与文献［2］相同，

$$
\begin{aligned}
&y=\frac{r}{a},W=\frac{w}{h},\Phi=ky+\frac{\partial w}{\partial y},\Psi=\frac{a}{h}\eta\psi,S=\frac{a}{D_{11}}rN_r,\\
&P=\frac{a^4}{D_{11}h}q(r,t),k=\frac{a^2}{Rh},m=\frac{a^2}{D_{11}}G,\beta_1^2=\frac{D_{22}}{D_{11}},\\
&\beta_2^2=\frac{A_1}{A_3},\beta_3=\frac{h^2}{A_3D_{11}},\beta_4=\frac{A_2}{A_3},\bar{\rho}=\frac{ha^4}{D_{11}}\rho
\end{aligned}
\tag{3}
$$

设 $W(y, t)$ 是时间和空间的可分离函数

$$W(y,t)=f(t)W_1(y) \tag{4}$$

其中

$$W_1(y)=\frac{(y^4-1)}{4(y-\beta_1^2)}-\frac{(y^{\beta_1+1}-1)}{(9-\beta_1^2)(1+\beta_1)}-\frac{1}{2m}(y^2-1) \tag{5}$$

将式（4）代入方程（1），并考虑虚功方程和边界条件（2），我们得到如下方程

$$
\begin{aligned}
&\int_0^1\Big\{f^3(t)H_1(y)H_{14}(y)+f^2(t)[H_2(y)H_{14}(y)+kyH_1(y)]\\
&+kyH_2(y)f(t)-PH_9(y)+\frac{\mathrm{d}^2f(t)}{\mathrm{d}t^2}H_{10}(y)-f^3(t)H_{11}(y)\\
&-f^2(t)H_{12}(y)-f(t)H_{13}(y)+myH_{14}(y)f(t)+P\int_0^y y\mathrm{d}y\\
&-\int_0^y\bar{\rho}\frac{\partial^2W}{\partial t^2}y\mathrm{d}y\Big\}W_1(y)y\mathrm{d}y=0
\end{aligned}
\tag{6}
$$

其中

$$
\begin{aligned}
&H_1(y)=F_1(y)+a_{10}y^{\beta_2},\qquad H_2(y)=F_2(y)+a_{11}y^{\beta_2},\\
&H_3(y)=(F_1+a_{10}y^{\beta_2})\left[a_0(y^3-y^{\beta_1})-\frac{y}{m}\right],\\
&H_4(y)=(F_2+a_{11}y^{\beta_2})\left[a_0(y^3-y^{\beta_1})-\frac{y}{m}\right]+H_1(y)ky,\\
&H_5(y)=ky(F_2+a_{11}y^{\beta_2}),\qquad H_6(y)=y^{\beta_1}\frac{\mathrm{d}}{\mathrm{d}y}y^{1-2\beta_1}\frac{\mathrm{d}}{\mathrm{d}y}[y^{\beta_1}H_3(y)],\\
&H_7(y)=y^{\beta_1}\frac{\mathrm{d}}{\mathrm{d}y}y^{1-2\beta_1}\frac{\mathrm{d}}{\mathrm{d}y}[y^{\beta_1}H_4(y)],\\
&H_8(y)=y^{\beta_1}\frac{\mathrm{d}}{\mathrm{d}y}y^{1-2\beta_1}\frac{\mathrm{d}}{\mathrm{d}y}[y^{\beta_1}H_5(y)],\quad H_9(y)=\frac{my(y^3-1)}{2(9-\beta_1^2)},\\
&H_{10}(y)=my\bar{\rho}[b_1(y^2-1)+b_2(y^3-1)+b_3(y^{4+\beta_1}-1)+b_4(y^5-1)],\\
&H_{11}(y)=my[H_6(y)-H_6(1)],\quad H_{12}(y)=my[H_7(y)-H_7(1)],\\
&H_{13}(y)=my[H_8(y)-H_8(1)],\quad H_{14}(y)=a_0(y^3-y^{\beta_1})-\frac{y}{m},\\
&F_1=a_1y^7+a_2y^{4+\beta_1}+a_3y^{1+2\beta_1}+a_4y^5+a_5y^{2+\beta_1}+a_6y^3,\\
&F_2=a_7y^5+a_8y^{2+\beta_1}+a_9y^3,\\
&F_i'(y)=\frac{\mathrm{d}F_i}{\mathrm{d}y},\quad \overline{F}_i(y)=\frac{F_i(y)}{y},\quad i=1,2
\end{aligned}
$$

$$a_0=\frac{1}{9-\beta_1^2},\quad a_1=-\frac{\beta_3 a_0^2}{2(49-\beta_2^2)},\quad a_2=\frac{\beta_3 a_0^2}{(4+\beta_1)^2-\beta_2^2},$$

$$a_3=-\frac{\beta_3 a_0^2}{2[(1+2\beta_1)^2-\beta_2^2]},\quad a_4=\frac{\beta_3 a_0}{m(25-\beta_2^2)},$$

$$a_5=-\frac{\beta_3 a_0}{m[(2+\beta_1)^2-\beta_2^2]},\quad a_6=-\frac{\beta_3}{2m^2(9-\beta_2^2)},\quad a_7=\frac{\beta_3 k a_0}{(25-\beta_2^2)},$$

$$a_8=-\frac{\beta_3 k a_0}{(2+\beta_1)^2-\beta_2},\quad a_9=-\frac{\beta_3 k}{m(9-\beta_2^2)},$$

$$a_{10}=\frac{[F_1'(1)-\beta_4\overline{F}_1(1)]}{\beta_4-\beta_2},\quad a_{11}=\frac{[F_2'(1)-\beta_4\overline{F}_2(1)]}{\beta_4-\beta_2},$$

$$b_1=\frac{a_0}{24(49-\beta_1^2)},$$

$$b_2=-\frac{a_0}{8(9-\beta_1^2)}+\frac{a_0}{2(1+\beta_1)(9-\beta_1^2)}+\frac{1}{4m(9-\beta_1^2)},$$

$$b_3=-\frac{a_0}{(1+\beta_1)(3+\beta_1)[(4+\beta_1)^2-\beta_3^2]},\quad b_4=-\frac{1}{4m(25-\beta_1^2)}\tag{7}$$

由方程（6）求得壳体关于 f（t）的具有二次和三次非线性的强迫振动微分方程

$$\frac{\mathrm{d}^2 f(t)}{\mathrm{d}t^2}+\omega_0^2 f(t)+\varepsilon[\lambda f^2(t)+\lambda f^3(t)]+b\cos\Omega t=0\tag{8}$$

其中

$$\omega_0^2=\frac{c_4}{c_1},\quad \varepsilon=\frac{c_2}{c_1},\quad \lambda=\frac{c_3}{c_2},\quad P=q_0\cos\Omega t,\quad b=\frac{c_5}{c_1}q_0,$$

$$c_1=\int_0^1 W_1(y)y\left[-\int_0^y\bar{\rho}yW_1(y)\mathrm{d}y+H_{10}(y)\right]\mathrm{d}y,$$

$$c_2=\int_0^1 W_1(y)y[H_1(y)H_{14}(y)-H_{11}(y)]\mathrm{d}y,$$

$$c_3=\int_0^1 W_1(y)y[H_2(y)H_{14}(y)+kyH_1(y)-H_{12}(y)]\mathrm{d}y,\tag{9}$$

$$c_4=\int_0^1 W_1(y)y[kyH_2(y)-H_{13}(y)+myH_{14}(y)]\mathrm{d}y,$$

$$c_5=\int_0^1 W_1(y)y\left[-H_9(y)+\int_0^y y\mathrm{d}y\right]\mathrm{d}y$$

三、非线性强迫振动的周期解

对方程（8）所示的非自治系统，我们采用小参数法进行求解。要求$|c_2|<|c_1|$[15]。当 ω_0 与 Ω 既不接近，又不能公约时，我们得到方程（8）的非共振周期解如下

$$f(t)=\frac{b}{\omega_0^2-\Omega^2}\cos\Omega t+\varepsilon\left[-\frac{\lambda}{2\omega_0^2}\left(\frac{b}{\omega_0^2-\Omega^2}\right)^2-\frac{3b^3}{4(\omega_0^2-\Omega^2)^4}\cos\Omega t\right.$$
$$\left.-\frac{\lambda}{2(\omega_0^2-4\Omega^2)}\left(\frac{b}{\omega_0^2-\Omega^2}\right)^2\cos2\Omega t-\frac{1}{4(\omega_0^2-9\Omega^2)}\left(\frac{b}{\omega_0^2-\Omega^2}\right)^3\cos3\Omega t\right]$$
$$+\varepsilon^2\left\{\frac{1}{\omega_0^2}\left[\frac{3\lambda}{2\omega_0^2}\left(\frac{b}{\omega_0^2-\Omega^2}\right)^4+\frac{3\lambda}{8(\omega_0^2-4\Omega^2)}\left(\frac{b}{\omega_0^2-\Omega^2}\right)^4\right]\right.$$

$$+\frac{1}{\omega_0^2-\Omega^2}\left[\frac{\lambda^2}{\omega_0^2}\left(\frac{b}{\omega_0^2-\Omega^2}\right)^3+\frac{\lambda^2}{2(\omega_0^2-4\Omega^2)}\left(\frac{b}{\omega_0^2-\Omega^2}\right)^3\right.$$

$$\left.+\frac{15}{8(\omega_0^2-9\Omega^2)}\left(\frac{b}{\omega_0^2-\Omega^2}\right)^5\right]\cos\Omega t+\frac{1}{(\omega_0^2-4\Omega^2)}\left[\frac{3\lambda}{2\omega_0^2}\left(\frac{b}{\omega_0^2-\Omega^2}\right)^4\right.$$

$$\left.+\frac{\lambda}{\omega_0^2-4\Omega^2}\left(\frac{b}{\omega_0^2-\Omega^2}\right)^4+\frac{3}{8(\omega_0^2-9\Omega^2)}\left(\frac{b}{\omega_0^2-\Omega^2}\right)^5\right]\cos 2\Omega t$$

$$+\frac{1}{\omega_0^2-9\Omega^2}\left[\frac{\lambda^2}{2(\omega_0^2-4\Omega^2)}\left(\frac{b}{\omega_0^2-\Omega^2}\right)^3+\frac{9}{16(\omega_0^2-\Omega^2)}\left(\frac{b}{\omega_0^2-\Omega^2}\right)^5\right]\cos 3\Omega t$$

$$+\frac{1}{\omega_0^2-16\Omega^2}\left[\frac{\lambda^2}{4(\omega_0^2-9\Omega^2)}\left(\frac{b}{\omega_0^2-\Omega^2}\right)^4+\frac{3\lambda}{8(\omega_0^2-4\Omega^2)}\left(\frac{b}{\omega_0^2-\Omega^2}\right)^4\right]\cos 4\Omega t$$

$$\left.+\frac{1}{16(\omega_0^2-9\Omega^2)(\omega_0^2-25\Omega^2)}\left(\frac{b}{\omega_0^2-\Omega^2}\right)^5\cos 5\Omega t\right\}+O(\varepsilon^3) \tag{10}$$

当

$$\omega_0^2=\Omega^2+\varepsilon\sigma \tag{11}$$

$$b=\varepsilon M \tag{12}$$

时，我们得到如下共振周期解

$$f(t)=A_0\cos\Omega t+\varepsilon\left(A_1\cos\Omega t+\frac{\lambda A_0^2}{6\Omega^2}\cos 2\Omega t+\frac{A_0^3}{32}\cos 3\Omega t-\frac{\lambda A_0^2}{2\Omega^2}\right)$$

$$+\varepsilon^2\left[\left(\frac{\lambda\sigma}{2\Omega^4}A_0^2+2\lambda A_0A_1+\frac{15\lambda A_0^4}{12\Omega^4}\right)+A_2\cos\Omega t\right.$$

$$+\frac{1}{3\Omega^2}\left(\frac{\lambda\sigma A_0^2}{6\Omega^2}+2\lambda A_0A_1+\frac{15\lambda A_0^4}{32\Omega^2}\right)\cos 2\Omega t$$

$$+\frac{1}{8\Omega^2}\left(\frac{\sigma A_0^3}{32\Omega^2}+\frac{3}{8}A_0^2A_1+\frac{\lambda^2A_0^3}{6\Omega^2}+\frac{3A_0^5}{64\Omega^2}\right)\cos 3\Omega t$$

$$\left.+\frac{1}{15\Omega^2}\left(\frac{5A_0^4}{32\Omega^2}\right)\cos 4\Omega t+\frac{1}{24\Omega^2}\left(\frac{3A_0^5}{128\Omega^2}\right)\cos 5\Omega t\right]+O(\varepsilon^3) \tag{13}$$

这里，A_0 和 A_1 分别由下列方程确定

$$\sigma A_0+\frac{3}{4}A_0^3-M=0 \tag{14}$$

$$A_1=\left(\frac{5}{6}\lambda^2A_0^3-\frac{3}{4}\,\frac{A_0^3}{32\Omega^2}\right)\frac{1}{\sigma+\frac{9}{8}A_0^2} \tag{15}$$

四、幅频特性曲线

对于该非自治系统，在非线性主共振时的特性，由幅频特性曲线来体现。

将式（11）代入方程（14），得到下列共振曲线方程

$$(\omega_0^2-\Omega^2)A_0+\frac{3}{4}\varepsilon A_0^3-b=0 \tag{16}$$

现在进行实例计算，为计算简单，且不失一般性，我们假设壳体各层厚度相等，弹性常数相同，且有 $E_\theta/E_r=1.5$，$\nu=0.2$。于是，由方程（16），我们得到如图 1 和图 2 所示的幅频特性曲线。

图 3 和图 4 是在其他参数不变，改变剪切刚度时所得到的幅频特性曲线。由图看到，当改变剪切刚度时，幅频特性曲线各点切线的斜率发生变化，随着剪切刚度 m 的增加，曲线上各点切线的斜率变大。

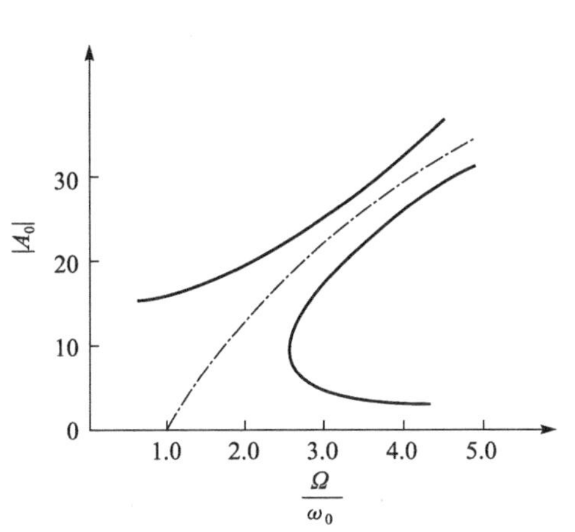

图 1　$P=15$，$k=1.4$，$m=80$，$\bar{\rho}=1.2$

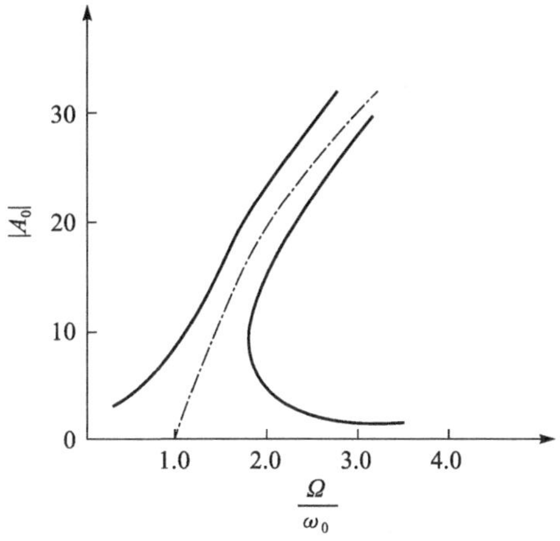

图 2　$P=5$，$k=1.4$，$m=80$，$\bar{\rho}=1.2$

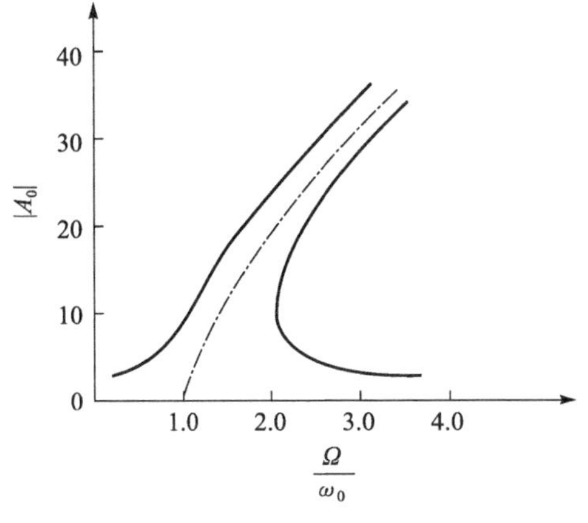

图 3　$m=800$，$P=10$

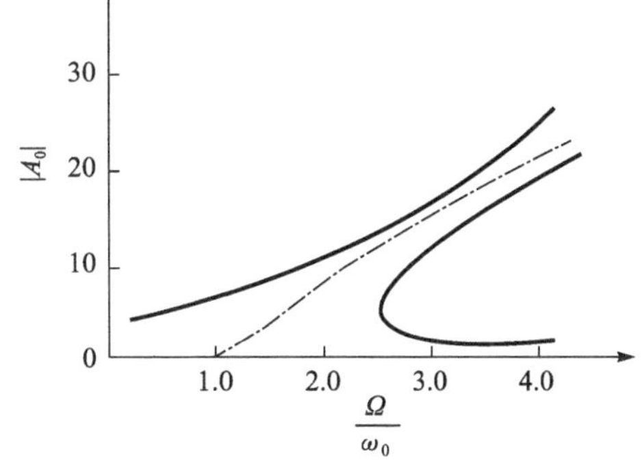

图 4　$m=8$，$P=10$

参 考 文 献

[1]　Chia C Y. Nonlinear Analysis of Plates. New York：McGraw-Hill，1980.

[2]　刘人怀. 考虑横向剪切的对称层合圆柱正交异性扁球壳的非线性稳定问题. 中国科学（A 辑），1991（7）：742-751.

[3]　Liu Ren-huai. Non-linear buckling of symmetrically laminated，cylindrically orthotropic，shallow，conical shells considering shear. Int. J. Non-Linear Mechanics，1996，31（1）：89-99.

[4]　Liu Ren-huai. On non-linear buckling of symmetrically laminated，cylindrically orthotropic，truncated，shallow，spherical shells under uniform pressure including shear effects. Int. J. Non-Linear Mechanics，1996，31（1）：101-115.

[5]　刘人怀，钟诚. 考虑横向剪切的对称层合圆柱正交异性中心开孔扁球壳的非线性屈曲. 暨南大

学学报（自然科学版），1994，15（1）：1-12.

[6] Liu Ren-huai，He Ling-hui. A simple theory for non-linear bending of laminated composite rectangular plates including higher-order effects. Int. J. Non-Linear Mechanics，1991，26（5）：537-545.

[7] 刘人怀，何陵辉. 四边简支对称正交层合矩形板的非线性弯曲问题. 应用数学和力学，1990，11（9）：753-759.

[8] 刘人怀，何陵辉. 层合圆薄板的轴对称弯曲问题. 江西工业大学学报，1991，13（2）：199-205.

[9] 何陵辉，刘人怀. 一种考虑层间位移和横向剪应力连续条件的层合板理论. 固体力学学报，1994，15（4）：319-326.

[10] Wang H，Wang Tsun-kuei. Donnell type theory of finite deflection of stiffened thin conical shells composed of composite materials. Appl. Math. Mech.，1990，11（9）：857-868.

[11] Alwar R S，Narasimhan M C. Axisymmetric non-linear analysis of laminated orthotropic annular spherical shells. Int. J. Non-Linear Mechanics，1992，27（4）：611-622.

[12] Muc A. Buckling and post-buckling behaviour of laminated shallow spherical shells subjected to external pressure. Int. J. Non-Linear Mechanics，1992，27（3）：465-476.

[13] Xu C S. Multi-mode nonlinear vibration and large deflection of symmetrically laminated imperfect spherical caps on elastic foundations. Int. J. Mech. Sci.，1992，34（6）：459-474.

[14] Qatu M S. On the validity of nonlinear shear deformation theories for laminated plates and shells. Composite Structure，1994，27（4）：395-401.

[15] 陈予恕，唐云，等. 非线性动力学中的现代分析方法. 北京：科学出版社，1992.

Large deflection bending of symmetrically laminated rectilinearly orthotropic elliptical plates including transverse shear①

中 文 摘 要

本文考虑了横向剪切的影响，导出了对称层合直线型正交异性椭圆板的非线性弯曲的基本方程组。应用迦辽金方法获得了承受均布压力的夹紧固定的这种板的解析解。作为层合椭圆板的特殊情况，对层合圆板也进行了研究。

1. Introduction

The research on plates and shells composed of laminated composite material has been developing rapidly in recent years, due to their unique advantages of high strength and light weight, and the fact that they can be tailored for specific applications. Thus, they have been widely applied in aeronautical, astronautical, naval, and civil engineering.

As for the nonlinear bending problem of laminated composite elliptical plates, as well as its particular case of laminated composite circular plates, there are only a few works available, such as references[1-5]. Unfortunately, the above papers disregard the effect of transverse shear deformation on large-deflection bending of the plates. In such a case, however, the effect of shear becomes very important, and obviously it is no longer negligible. This paper continues the authors' previous research[6-13]. We derive basic equations of nonlinear bending for symmetrically laminated rectilinearly orthotropic elliptical plates by including transverse shear and using Galerkin's method. We obtain the analytical solution of the plates with a rigidly clamped edge under uniform lateral pressure. As a particular case of the laminated elliptical plate, the laminated circular plate is also studied. The results obtained may be valuable in designing structural elements in engineering.

2. Fundamental Equations

Consider a symmetrically laminated elliptical plate of rectilinear orthotropy with its filaments parallel to the major axis as shown in Fig. 1. The axes of the plate are $2a$ in the x direction and $2b$ in the y direction. The thickness is h, the total numbers of layers is n. The origin of the coordinate system is chosen to coincide with the center of the midplane of the undeformed plate. The plate is assumed to be subjected to uniform lateral pressure q.

---

① Reprinted from *Archive of Applied Mechanics*. 1997, 67: 507-520. Authors: Liu Renhuai, Xu Jiachu and Zhai Shangzhong.

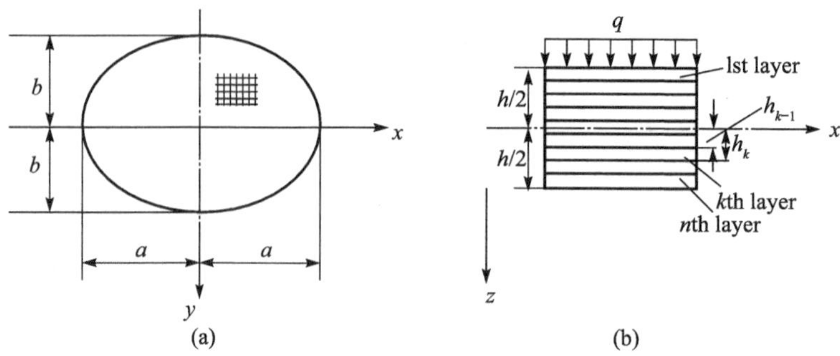

Fig. 1a，b　Geometry and elastic direction of symmetrically laminated rectilinearly orthotropic elliptical plate

In the following discussion of the deformation and stresses of the plate, to avoid excessive complexity, we adopt the following assumptions:

(1) The material of the plate is linearly elastic. Any layer of the plate has a uniform thickness, and each layer may have different thickness and elastic constants, but must be symmetrical with respect to the midplane of the plate.

(2) The plate is transversely incompressible.

(3) The normal line to the midplane of the plate remains straight during bending.

(4) The effect of the transverse normal stress in the plate on the deformation can be neglected.

Under the above assumptions, the displacements u, v, w at an arbitrary point of the plate in the x, y, z directions may be written in form

$$u = u_0(x,y) + z\psi_x(x,y),$$
$$v = v_0(x,y) + z\psi_y(x,y),$$
$$w = w_0(x,y), \tag{1}$$

in which u_0, v_0, w_0 are the displacements at a corresponding point of the midplane of the plate in the x, y, z directions, ψ_x and ψ_y are the rotations of the normal to the midplane of the plate in the x and y directions, respectively.

For the case of large deflection, the strain-displacement relations of the plate are

$$\varepsilon_x = \varepsilon_x^0 + z\kappa_x, \qquad \gamma_{xy} = \gamma_{xy}^0 + z\kappa_{xy},$$
$$\varepsilon_y = \varepsilon_y^0 + z\kappa_y, \qquad \gamma_{yz} = \frac{\partial w}{\partial y} + \psi_y,$$
$$\varepsilon_z = 0, \qquad \gamma_{zx} = \frac{\partial w}{\partial x} + \psi_x, \tag{2}$$

where $\varepsilon_x^0, \varepsilon_y^0, \gamma_{xy}^0$ are the strains of the midplane of the plate, κ_x, κ_y and κ_{xy} are the curvatures and twisting curvatures, respectively,

$$\varepsilon_x^0 = \frac{\partial u_0}{\partial x} + \frac{1}{2}\left(\frac{\partial w}{\partial x}\right)^2, \qquad \kappa_x = \frac{\partial \psi_x}{\partial x},$$

$$\varepsilon_y^0 = \frac{\partial v_0}{\partial y} + \frac{1}{2}\left(\frac{\partial w}{\partial y}\right)^2, \qquad \kappa_y = \frac{\partial \psi_y}{\partial y},$$

$$\gamma_{xy}^0 = \frac{\partial u_0}{\partial y} + \frac{\partial v_0}{\partial y} + \frac{\partial w}{\partial x}\frac{\partial w}{\partial y}, \quad \kappa_{xy} = \frac{\partial \psi_x}{\partial y} + \frac{\partial \psi_y}{\partial x}. \tag{3}$$

The Hooke's law for the kth layer of the plate can be written in the form

$$\begin{aligned} \sigma_x^{(k)} &= Q_{11}^{(k)}\varepsilon_x + Q_{12}^{(k)}\varepsilon_y, & \tau_{xy}^{(k)} &= Q_{66}^{(k)}\gamma_{xy}, \\ \sigma_y^{(k)} &= Q_{12}^{(k)}\varepsilon_x + Q_{22}^{(k)}\varepsilon_y, & \tau_{yz}^{(k)} &= Q_{44}^{(k)}\gamma_{yz}, \\ \sigma_z^{(k)} &= 0, & \tau_{zx}^{(k)} &= Q_{55}^{(k)}\gamma_{zx}, \end{aligned} \tag{4}$$

in which $Q_{ij}^{(k)}$ are the reduced stiffnesses of the kth layer, $E_x^{(k)}$ and $E_y^{(k)}$ are Young's moduli along x and y direction, $\nu_{xy}^{(k)}$ and $\nu_{yx}^{(k)}$ are Poisson's ratios characterizing contraction in y or x direction during tension applied in x or y direction, $G^{(k)}$ is the in-plane shear modulus, $G_{yz}^{(k)}$ and $G_{zx}^{(k)}$ are the transverse shear moduli characterizing changes of angles in the yz and zx planes, respectively,

$$\begin{aligned} Q_{11}^{(k)} &= \frac{E_x^{(k)}}{1-\nu_{xy}^{(k)}\nu_{yx}^{(k)}}, & Q_{44}^{(k)} &= G_{yz}^{(k)}, \\ Q_{12}^{(k)} &= \frac{\nu_{xy}^{(k)}E_y^{(k)}}{1-\nu_{xy}^{(k)}\nu_{yx}^{(k)}}, & Q_{55}^{(k)} &= G_{zx}^{(k)}, \\ Q_{22}^{(k)} &= \frac{E_y^{(k)}}{1-\nu_{xy}^{(k)}\nu_{yx}^{(k)}}, & Q_{66}^{(k)} &= G^{(k)}, \end{aligned} \tag{5}$$

and the elastic constants satisfy the condition of elastic symmetry

$$\nu_{xy}^{(k)}E_y^{(k)} = \nu_{yx}^{(k)}E_x^k. \tag{6}$$

Let us consider an element cut out of the plate by two pairs of planes parallel to the xz and yz planes, as shown in Fig. 2. According to the classical plate theory, the membrane forces N_x, N_y, N_{xy}, the bending moments M_x, M_y, the twisting moment M_{xy}, and the transverse shear forces Q_x, Q_y are defined by

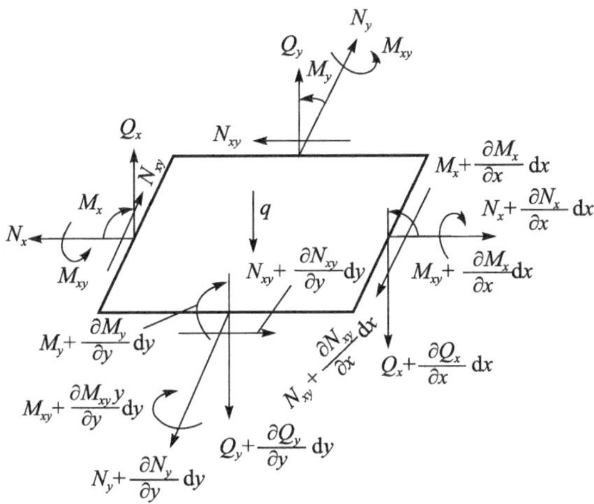

Fig. 2　An element of symmetrically laminated rectilinearly orthotropic elliptical plate

$$[N_x, N_y, N_{xy}] = \int_{-\frac{h}{2}}^{\frac{h}{2}} [\sigma_x^{(k)}, \sigma_y^{(k)}, \tau_{xy}^{(k)}] \mathrm{d}z, \tag{7}$$

$$[M_x, M_y, M_{xy}] = \int_{-\frac{h}{2}}^{\frac{h}{2}} [\sigma_x^{(k)}, \sigma_y^{(k)}, \tau_{xy}^{(k)}] z \mathrm{d}z, \tag{8}$$

$$[Q_x, Q_y] = \int_{-\frac{h}{2}}^{\frac{h}{2}} [\tau_{zx}^{(k)}, \tau_{yz}^{(k)}] \mathrm{d}z. \tag{9}$$

Substituting expressions (4) into (7) and (8), and using expressions (2), we have

$$\begin{aligned} N_x &= A_{11}\left[\frac{\partial u_0}{\partial x} + \frac{1}{2}\left(\frac{\partial w}{\partial x}\right)^2\right] + A_{12}\left[\frac{\partial v_0}{\partial y} + \frac{1}{2}\left(\frac{\partial w}{\partial y}\right)^2\right], \\ N_y &= A_{12}\left[\frac{\partial u_0}{\partial x} + \frac{1}{2}\left(\frac{\partial w}{\partial x}\right)^2\right] + A_{22}\left[\frac{\partial v_0}{\partial y} + \frac{1}{2}\left(\frac{\partial w}{\partial y}\right)^2\right], \\ N_{xy} &= A_{66}\left(\frac{\partial u_0}{\partial y} + \frac{\partial v_0}{\partial x} + \frac{\partial w}{\partial x}\frac{\partial w}{\partial y}\right), \end{aligned} \tag{10}$$

$$\begin{aligned} M_x &= D_{11}\frac{\partial \psi_x}{\partial x} + D_{12}\frac{\partial \psi_y}{\partial x}, \\ M_y &= D_{12}\frac{\partial \psi_x}{\partial x} + D_{22}\frac{\partial \psi_y}{\partial x}, \\ M_{xy} &= D_{66}\left(\frac{\partial \psi_x}{\partial y} + \frac{\partial \psi_y}{\partial x}\right), \end{aligned} \tag{11}$$

where A_{ij} and D_{ij} are the extensional and bending rigidities, respectively,

$$A_{ij} = \sum_{k=1}^{n} Q_{ij}^{(k)}(h_k - h_{k-1}), \quad D_{ij} = \frac{1}{3}\sum_{k=1}^{n} Q_{ij}^{(k)}(h_k^3 - h_{k-1}^3). \tag{12}$$

Introduction of expressions (4) into (9) yields

$$Q_x = C_{55}\left(\frac{\partial w}{\partial x} + \psi_x\right), \quad Q_y = C_{44}\left(\frac{\partial w}{\partial y} + \psi_y\right), \tag{13}$$

where C_{44} and C_{55} are the shear rigidities,

$$C_{44} = \sum_{k=1}^{n} Q_{44}^{(k)}(h_k - h_{k-1}), \quad C_{55} = \sum_{k=1}^{n} Q_{55}^{(k)}(h_k - h_{k-1}). \tag{14}$$

From expressions (2) and (4), it is seen that the transverse shear stresses are discontinuous between layers, and they are not zero at the upper and lower surfaces of the plate, while their distribution in every layer through the thickness of the plate does not follow the parabolic law. Obviously, there is no agreement between the present stress distribution and the actual one. In such a case, some corrections need to be introduced. In accordance with Reissner's theory, we assume a parabolic law for the distribution of the transverse shear stresses τ_{zx} and τ_{yz} through the thickness of the plate. In this manner, we have

$$\tau_{zx} = \frac{3Q_x}{2h}\left[1 - \left(\frac{2z}{h}\right)^2\right], \quad \tau_{yz} = \frac{3Q_y}{2h}\left[1 - \left(\frac{2z}{h}\right)^2\right], \tag{15}$$

Then, calculating complementary energies separately from expressions (15) and (13) and equating the two energies, we find, instead of C_{44} and C_{55} in expressions (14), the following more exact expressions:

$$C_{44}=\frac{4h^2}{9\sum_{k=1}^{n}\frac{1}{Q_{44}^{(k)}}\left[h_k-h_{k-1}-\frac{8}{3h^2}(h_k^3-h_{k-1}^3)+\frac{16}{5h^4}(h_k^5-h_{k-1}^5)\right]},$$

$$C_{55}=\frac{4h^2}{9\sum_{k=1}^{n}\frac{1}{Q_{55}^{(k)}}\left[h_k-h_{k-1}-\frac{8}{3h^2}(h_k^3-h_{k-1}^3)+\frac{16}{5h^4}(h_k^5-h_{k-1}^5)\right]}. \tag{16}$$

Upon discussing the conditions of equilibrium of the element of the plate (Fig. 2), equations of equilibrium for large deflections of the plate can be written in the form

$$\frac{\partial N_x}{\partial x}+\frac{\partial N_{xy}}{\partial y}=0,\quad \frac{\partial N_{xy}}{\partial x}+\frac{\partial N_y}{\partial y}=0,$$

$$\frac{\partial^2 M_x}{\partial x^2}+2\frac{\partial^2 M_{xy}}{\partial x\partial y}+\frac{\partial^2 M_y}{\partial y^2}+N_x\frac{\partial^2 w}{\partial x^2}+2N_{xy}\frac{\partial^2 w}{\partial x\partial y}+N_y\frac{\partial^2 w}{\partial y^2}+q=0,$$

$$\frac{\partial M_x}{\partial x}+\frac{\partial M_{xy}}{\partial y}-Q_x=0,\quad \frac{\partial M_{xy}}{\partial x}+\frac{\partial M_y}{\partial y}-Q_y=0. \tag{17}$$

Inserting expressions (10),(11) and (13) into the above equations leads to the following equations

$$A_{11}\frac{\partial^2 u_0}{\partial x^2}+A_{66}\frac{\partial^2 u_0}{\partial y^2}+(A_{12}+A_{66})\frac{\partial^2 v_0}{\partial x\partial y}+\left(A_{11}\frac{\partial^2 w}{\partial x^2}+A_{66}\frac{\partial^2 w}{\partial y^2}\right)\frac{\partial w}{\partial x}$$
$$+(A_{12}+A_{66})\frac{\partial^2 w}{\partial x\partial y}\frac{\partial w}{\partial y}=0,$$

$$(A_{12}+A_{66})\frac{\partial^2 u_0}{\partial x\partial y}+A_{66}\frac{\partial^2 v_0}{\partial x^2}+A_{22}\frac{\partial^2 v_0}{\partial y^2}+\left(A_{66}\frac{\partial^2 w}{\partial x^2}+A_{22}\frac{\partial^2 w}{\partial y^2}\right)\frac{\partial w}{\partial y}$$
$$+(A_{12}+A_{66})\frac{\partial^2 w}{\partial x\partial y}\frac{\partial w}{\partial x}=0,$$

$$D_{11}\frac{\partial^3\psi_x}{\partial x^3}+(D_{12}+2D_{66})\left(\frac{\partial^3\psi_x}{\partial x\partial y^2}+\frac{\partial^3\psi_y}{\partial x^2\partial y}\right)+D_{22}\frac{\partial^3\psi_y}{\partial y^3}$$
$$+\left[\frac{\partial u_0}{\partial x}+\frac{1}{2}\left(\frac{\partial w}{\partial x}\right)^2\right]\left(A_{11}\frac{\partial^2 w}{\partial x^2}+A_{12}\frac{\partial^2 w}{\partial y^2}\right)$$
$$+\left[\frac{\partial v_0}{\partial y}+\frac{1}{2}\left(\frac{\partial w}{\partial y}\right)^2\right]\left(A_{12}\frac{\partial^2 w}{\partial x^2}+A_{22}\frac{\partial^2 w}{\partial y^2}\right)$$
$$+2A_{66}\left(\frac{\partial u_0}{\partial y}+\frac{\partial v_0}{\partial x}+\frac{\partial w}{\partial x}\frac{\partial w}{\partial y}\right)\frac{\partial^2 w}{\partial x\partial y}+q=0,$$

$$D_{11}\frac{\partial^2\psi_x}{\partial x^2}+D_{66}\frac{\partial^2\psi_x}{\partial y^2}+(D_{12}+D_{66})\frac{\partial^2\psi_y}{\partial x\partial y}-C_{55}\left(\frac{\partial w}{\partial x}+\psi_x\right)=0,$$

$$(D_{12}+D_{66})\frac{\partial^2\psi_x}{\partial x\partial y}+D_{66}\frac{\partial^2\psi_y}{\partial x^2}+D_{22}\frac{\partial^2\psi_y}{\partial y^2}-C_{44}\left(\frac{\partial w}{\partial y}+\psi_y\right)=0. \tag{18}$$

Thus, the investigation of large deflections of symmetrically laminated rectilinearly orthotropic elliptical plates under uniform lateral pressure reduces to the solution of the five nonlinear differential Eqs. (18).

Now, these nonlinear equations will be solved together with the following boundary conditions for a rigidly clamped edge

$$u_0 = v_0 = w = \psi_x = \psi_y = 0 \quad \text{at the boundary } \frac{x^2}{a^2} + \frac{y^2}{b^2} - 1 = 0. \tag{19}$$

Defining the nondimensional parameters as

$$\begin{aligned}&\xi = \frac{x}{a},\ \eta = \frac{y}{b},\ \lambda = \frac{a}{b},\ U = \frac{a}{h^2}u_0,\ V = \frac{a}{h^2}v_0,\ W = \frac{w}{h},\ \Psi_\xi = \frac{a}{h}\psi_x,\\ &\Psi_\eta = \frac{a}{h}\psi_y,\ Q = \frac{a^4}{D_{11}h}q,\ a_{ij} = \frac{A_{ij}}{A_{11}},\ d_{ij} = \frac{D_{ij}}{D_{11}},\ \beta_1 = \frac{A_{11}h^2}{D_{11}},\ \beta_2 = \frac{A_{12}h^2}{D_{11}},\\ &\beta_3 = \frac{A_{22}h^2}{D_{11}},\ \beta_4 = \frac{2A_{66}h^2}{D_{11}},\ \beta_5 = \frac{C_{55}a^2}{D_{11}},\ \beta_6 = \frac{C_{44}a^2}{D_{11}}.\end{aligned} \tag{20}$$

Eqs. (18) and boundary conditions (19) transform to

$$\begin{aligned}L_1(U,V,W,\Psi_\xi,\Psi_\eta) =& \frac{\partial^2 U}{\partial \xi^2} + \lambda^2 a_{66}\frac{\partial^2 U}{\partial \eta^2} + \lambda(a_{12}+a_{66})\frac{\partial^2 V}{\partial\xi\partial\eta} + \left(\frac{\partial^2 W}{\partial\xi^2}\right.\\ &\left.+\lambda^2 a_{66}\frac{\partial^2 W}{\partial\eta^2}\right)\frac{\partial W}{\partial\xi} + \lambda^2(a_{12}+a_{66})\frac{\partial W}{\partial\eta}\frac{\partial^2 W}{\partial\xi\partial\eta} = 0,\\ L_2(U,V,W,\Psi_\xi,\Psi_\eta) =& \lambda(a_{12}+a_{66})\frac{\partial^2 U}{\partial\xi\partial\eta} + a_{66}\frac{\partial^2 V}{\partial\xi^2} + \lambda^2 a_{22}\frac{\partial^2 V}{\partial\eta^2} + \lambda\left(a_{66}\frac{\partial^2 W}{\partial\xi^2}\right.\\ &\left.+\lambda^2 a_{22}\frac{\partial^2 W}{\partial\eta^2}\right)\frac{\partial W}{\partial\eta} + \lambda(a_{12}+a_{66})\frac{\partial W}{\partial\xi}\frac{\partial^2 W}{\partial\xi\partial\eta} = 0\\ L_3(U,V,W,\Psi_\xi,\Psi_\eta) =& \frac{\partial^3\Psi_\xi}{\partial\xi^3} + (d_{12}+2d_{66})\left(\lambda^2\frac{\partial^3\Psi_\xi}{\partial\xi\partial\eta^2} + \lambda\frac{\partial^3\Psi_\eta}{\partial\xi^2\partial\eta}\right) + \lambda^3 d_{22}\frac{\partial^3\Psi_\eta}{\partial\eta^3}\\ &+\left[\frac{\partial U}{\partial\xi} + \frac{1}{2}\left(\frac{\partial W}{\partial\xi}\right)^2\right]\left(\beta_1\frac{\partial^2 W}{\partial\xi^2} + \lambda^2\beta_2\frac{\partial^2 W}{\partial\eta^2}\right) + \left[\lambda\frac{\partial V}{\partial\eta} + \frac{1}{2}\lambda^2\left(\frac{\partial W}{\partial\eta}\right)^2\right]\\ &\left(\beta_2\frac{\partial^2 W}{\partial\xi^2} + \lambda^2\beta_3\frac{\partial^2 W}{\partial\eta^2}\right) + \lambda\beta_4\left(\lambda\frac{\partial U}{\partial\eta} + \frac{\partial V}{\partial\xi} + \lambda\frac{\partial W}{\partial\xi}\frac{\partial W}{\partial\eta}\right)\frac{\partial^2 W}{\partial\xi\partial\eta} + Q = 0,\\ L_4(U,V,W,\Psi_\xi,\Psi_\eta) =& \frac{\partial^2\Psi_\xi}{\partial\xi^2} + \lambda^2 d_{66}\frac{\partial^2\Psi_\xi}{\partial\eta^2} + \lambda(d_{12}+d_{66})\frac{\partial^2\Psi_\eta}{\partial\xi\partial\eta} - \beta_5\left(\frac{\partial W}{\partial\xi} + \Psi_\xi\right) = 0,\\ L_5(U,V,W,\Psi_\xi,\Psi_\eta) =& \lambda(d_{12}+d_{66})\frac{\partial^2\Psi_\xi}{\partial\xi\partial\eta} + d_{66}\frac{\partial^2\Psi_\eta}{\partial\xi^2} + \lambda^2 d_{22}\frac{\partial^2\Psi_\eta}{\partial\eta^2} - \beta_6\left(\lambda\frac{\partial W}{\partial\eta} + \Psi_\eta\right) = 0,\end{aligned} \tag{21}$$

and

$$U = V = W = \Psi_\xi = \Psi_\eta = 0 \quad \text{along} \quad \xi^2 + \eta^2 - 1 = 0. \tag{22}$$

3. Solution of the Nonlinear Boundary Value Problem

Obviously, it is very difficult to solve the above nondimensional nonlinear boundary value problem (21) and (22) because the problem contains some nonlinear terms. To conquer this difficulty, we use Galerkin's method. The boundary conditions (22) and symmetry conditions are satisfied by taking

$$U = \xi(1-\xi^2-\eta^2)\sum_{m=0}^{\infty}\sum_{n=0}^{\infty} b_{mn}\xi^{2m}\eta^{2n},$$

$$V=\eta(1-\xi^2-\eta^2)\sum_{m=0}^{\infty}\sum_{n=0}^{\infty}c_{mn}\xi^{2m}\eta^{2n},$$

$$W=(1-\xi^2-\eta^2)\sum_{m=0}^{\infty}\sum_{n=0}^{\infty}e_{mn}\xi^{2m}\eta^{2n},$$

$$\Psi_\xi=\xi(1-\xi^2-\eta^2)\sum_{m=0}^{\infty}\sum_{n=0}^{\infty}f_{mn}\xi^{2m}\eta^{2n},$$

$$\Psi_\eta=\eta(1-\xi^2-\eta^2)\sum_{m=0}^{\infty}\sum_{n=0}^{\infty}g_{mn}\xi^{2m}\eta^{2n}. \tag{23}$$

In general, it is impossible to determine all the coefficients b_{mn}, C_{mn}, e_{mn}, f_{mn} and g_{mn} in the series (23) by Galerkin's method. Thus, we take only a term as an approximate solution

$$\begin{aligned}U&=b_{00}\xi(1-\xi^2-\eta^2),\\V&=c_{00}\eta(1-\xi^2-\eta^2),\\W&=W_0(1-\xi^2-\eta^2),\\\Psi_\xi&=f_{00}\xi(1-\xi^2-\eta^2),\\\Psi_\eta&=g_{00}\eta(1-\xi^2-\eta^2),\end{aligned} \tag{24}$$

in which W_0 is the nondimensional central deflection of the plate.

Applying the Galerkin procedure to Eq. (21), the following conditions are obtained:

$$\int_0^1\int_0^{\sqrt{1-\eta^2}}L_1(U,V,W,\Psi_\xi,\Psi_\eta)\xi(1-\xi^2-\eta^2)\mathrm{d}\xi\mathrm{d}\eta=0,$$

$$\int_0^1\int_0^{\sqrt{1-\eta^2}}L_2(U,V,W,\Psi_\xi,\Psi_\eta)\eta(1-\xi^2-\eta^2)\mathrm{d}\xi\mathrm{d}\eta=0,$$

$$\int_0^1\int_0^{\sqrt{1-\eta^2}}L_3(U,V,W,\Psi_\xi,\Psi_\eta)(1-\xi^2-\eta^2)\mathrm{d}\xi\mathrm{d}\eta=0,$$

$$\int_0^1\int_0^{\sqrt{1-\eta^2}}L_4(U,V,W,\Psi_\xi,\Psi_\eta)\xi(1-\xi^2-\eta^2)\mathrm{d}\xi\mathrm{d}\eta=0,$$

$$\int_0^1\int_0^{\sqrt{1-\eta^2}}L_5(U,V,W,\Psi_\xi,\Psi_\eta)\eta(1-\xi^2-\eta^2)\mathrm{d}\xi\mathrm{d}\eta=0. \tag{25}$$

Inserting expressions (24) into (25) and performing the integration, we obtain the nonlinear characteristic relation of the symmetrically laminated rectilinearly orthotropic elliptical plate with a rigidly clamped edge under uniform lateral pressure

$$Q=\alpha_1W_0+\alpha_3W_0^3, \tag{26}$$

where

$$\alpha_1=2\omega_3[3+\lambda^2(d_{12}+2d_{66})]+2\lambda\omega_1(d_{12}+2d_{66}+3\lambda^2d_{22}),$$

$$\alpha_3=\frac{2}{3}[(1+\omega_1)(\beta_1+\lambda^2\beta_2)]+\lambda(1+\omega_2)(\beta_2+\lambda^2\beta_3),$$

$$\omega_1=2\,\frac{(1+\lambda a_{66})(a_{66}+3\lambda^2a_{22})-\lambda(a_{66}+\lambda^2a_{22})(a_{12}+a_{66})}{(3+\lambda^2a_{66})(a_{66}+3\lambda^2a_{22})-\lambda^2(a_{12}+a_{66})^2},$$

$$\omega_2 = 2\frac{(3+\lambda a_{22})(a_{66}+\lambda^2 a_{22})-\lambda(a_{12}+a_{22})(a_{12}+\lambda a_{66})}{(3+\lambda^2 a_{66})(a_{66}+3\lambda^2 a_{22})-\lambda^2(a_{12}+a_{66})^2},$$

$$\omega_3 = \frac{\beta_5(3\lambda^2 d_{22}+d_{66}+\frac{13}{64}\beta_5)-\lambda\beta_5(d_{12}+d_{66})}{(3+\lambda^2 d_{66}+\frac{13}{64}\beta_5)(3\lambda^2 d_{22}+d_{66}+\frac{13}{64}\beta_5)-\lambda^2(d_{12}+d_{66})^2},$$

$$\omega_4 = \frac{\beta_6(3+3\lambda^2 d_{66}+\frac{13}{64}\beta_5)-\lambda\beta_5(d_{12}+d_{66})}{(3+\lambda^2 d_{66}+\frac{13}{64}\beta_5)(3\lambda^2 d_{22}+d_{66}+\frac{13}{64}\beta_6)-\lambda^2(d_{12}+d_{66})^2}. \tag{27}$$

At the same time, we have

$$b_{00}=\omega_1 W_0^2,\ c_{00}=\omega_2 W_0^2,\ f_{00}=\omega_3 W_0,\ g_{00}=\omega_4 W_0, \tag{28}$$

Thus, solution (24) becomes

$$\begin{aligned}
U &= \omega_1 W_0^2 \xi(1-\xi^2-\eta^2),\\
V &= \omega_2 W_0^2 \eta(1-\xi^2-\eta^2),\\
W &= W_0(1-\xi^2-\eta^2),\\
\Psi_\xi &= \omega_3 W_0 \xi(1-\xi^2-\eta^2),\\
\Psi_\eta &= \omega_4 W_0 \eta(1-\xi^2-\eta^2).
\end{aligned} \tag{29}$$

The stress at a general point within the plate can be found from the above formulas (10) and (11) for the forces and moments. Of most interest are the membrane stresses σ_x^m, σ_y^m and τ_{xy}^m in the middle surface, the maximum transverse shear stresses τ_{zx}^0 and τ_{yz}^0 at $z=0$, and the extreme-fibre bending stresses σ_x^b, σ_y^b and τ_{xy}^b. Let us introduce the following nondimensional stresses

$$\begin{aligned}
(\sigma_\xi^m, \sigma_\eta^m, \tau_{\xi\eta}^m) &= \frac{a^2}{A_{11}h}(\sigma_x^m, \sigma_y^m, \tau_{xy}^m),\\
(\sigma_\xi^b, \sigma_\eta^b, \tau_{\xi\eta}^b) &= \frac{a^2}{A_{11}h}(\sigma_x^b, \sigma_y^b, \tau_{xy}^b),\\
(\tau_{\zeta\xi}^0, \tau_{\eta\zeta}^0) &= \frac{a^2}{A_{11}h}(\tau_{zx}^0, \tau_{yz}^0).
\end{aligned} \tag{30}$$

Using expressions (10), (11) and (13), these nondimensional stresses can be expressed as follows:

$$\begin{aligned}
\sigma_\xi^m &= \frac{\partial U}{\partial \xi}+\frac{1}{2}\left(\frac{\partial W}{\partial \xi}\right)^2+\lambda a_{12}\left[\frac{\partial V}{\partial \eta}+\frac{1}{2}\lambda\left(\frac{\partial W}{\partial \eta}\right)^2\right],\\
\sigma_\eta^m &= a_{12}\left[\frac{\partial U}{\partial \xi}+\frac{1}{2}\left(\frac{\partial W}{\partial \xi}\right)^2\right]+\lambda a_{22}\left[\frac{\partial V}{\partial \eta}+\frac{1}{2}\lambda\left(\frac{\partial W}{\partial \eta}\right)^2\right],\\
\tau_{\xi\eta}^m &= a_{66}\left(\lambda\frac{\partial U}{\partial \eta}+\frac{\partial V}{\partial \xi}+\lambda\frac{\partial W}{\partial \xi}\frac{\partial W}{\partial \eta}\right),\\
\sigma_\xi^b &= \frac{6}{\beta_1}\left(\frac{\partial \Psi_\xi}{\partial \xi}+\lambda d_{12}\frac{\partial \Psi_\eta}{\partial \eta}\right),\\
\sigma_\eta^b &= \frac{6}{\beta_1}\left(d_{12}\frac{\partial \Psi_\xi}{\partial \xi}+\lambda d_{22}\frac{\partial \Psi_\eta}{\partial \eta}\right),
\end{aligned}$$

$$\tau_{\xi\eta}^{b}=\frac{6}{\beta_1}d_{66}\left(\lambda\frac{\partial\Psi_\xi}{\partial\eta}+\frac{\partial\Psi_\eta}{\partial\xi}\right),$$

$$\tau_{\zeta\xi}^{0}=\beta_7\left(\frac{\partial W}{\partial\xi}+\Psi_\xi\right),$$

$$\tau_{\eta\zeta}^{0}=\beta_8\left(\lambda\frac{\partial W}{\partial\eta}+\Psi_\eta\right),\tag{31}$$

where

$$\beta_7=\frac{3C_{55}a}{2A_{11}h},\quad \beta_8=\frac{3C_{44}a}{2A_{11}h}.\tag{32}$$

Substituting solutions (29) into (31), we have

$$\sigma_\xi^m=W_0^2\{\omega_1(1-3\xi^2-\eta^2)+2\xi^2+\lambda a_{12}[\omega_2(1-\xi^2-3\eta^2)+2\eta^2]\},$$

$$\sigma_\eta^m=W_0^2\{a_{12}[\omega_1(1-3\xi^2-\eta^2)+2\xi^2]+\lambda a_{22}[\omega_2(1-\xi^2-3\eta^2)+2\eta^2]\},$$

$$\tau_{\xi\eta}^m=-2a_{66}W_0^2(\lambda\omega_1+\omega_2-2)\xi\eta,$$

$$\sigma_\xi^b=\frac{6}{\beta_1}W_0[\omega_3(1-3\xi^2-\eta^2)+\lambda d_{12}\omega_4(1-\xi^2-3\eta^2)],$$

$$\sigma_\eta^b=\frac{6}{\beta_1}W_0[d_{12}\omega_3(1-3\xi^2-\eta^2)+\lambda d_{22}\omega_4(1-\xi^2-3\eta^2)],$$

$$\tau_{\xi\eta}^b=-\frac{12}{\beta_1}d_{66}W_0(\lambda\omega_3+\omega_4)\xi\eta,$$

$$\tau_{\zeta\xi}^0=\beta_7[\omega_3(1-\xi^2-\eta^2)-2]\xi,$$

$$\tau_{\eta\zeta}^0=\beta_8[\omega_4(1-\xi^2-\eta^2)-2]\eta.\tag{33}$$

Especially if we put $a=b$ into (26) and (33), we can obtain the characteristic relation and nondimensional stresses of the symmetrically laminated rectilinearly orthotropic circular plate. Here, the relation and stress formulas are remarkably simple.

4. Numerical Examples

In order to simplify the following numerical calculation, let us apply the above theory to two more particular examples. For such two cases, the layers of symmetrically laminated rectilinearly orthotropic elliptical and circular plates are considered to be of the same thickness and elastic constants, and the elastic constants of the boron-epoxy, graphite-epoxy, and glass-epoxy composite materials are given in Table 1, Here,

$$G_{yz}=G_{zx}=G_s.$$

Table 1　Numerical values of elastic constants[14]

Material	E_y (MPa)	E_x/E_y	G/E_y	ν_{xy}
boron-epoxy	20.7×10^5	10	0.33	0.22
graphite-epoxy	5.175×10^6	40	0.6	0.25
glass-epoxy	17.94×10^4	3	0.6	0.25

Example 1: A symmetrically laminated rectilinearly orthotropic elliptical plate. The dimensions of the elliptical plate are: $a=10$cm, $b=5$cm, $h=2$cm.

From the expression (26), calculated results of the load-deflection relation for different composite materials and several values of G_s, n are obtained and shown in Figs. 3-7. Figs. 3-5 indicate that the rigidity of the plate increases with the transverse shear modulus G_s. In the limiting case where G_s approaches infinity, which corresponds to the disregard of the transverse shear deformation, the rigidity of the plate is maximum. Obviously, in the present case, the effect of the transverse shear deformation becomes very important and must be taken into consideration in analysing laminated composite plates. For the purpose of comparison, Fig. 5 also shows the theoretical results[1] obtained by using Kirchhoff's hypothesis in a particular case of an elliptical orthotropic plate. From this figure, it is seen that the present results are in good agreement with the results[1].

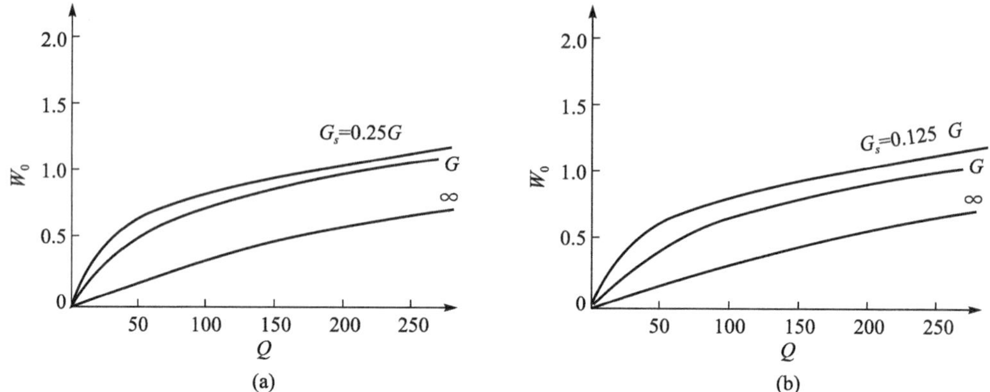

Fig. 3a, b　The characteristic curves for different composite materials and several values of G_s ($n=9$)

(a) graphite-epoxy; (b) boron-epoxy

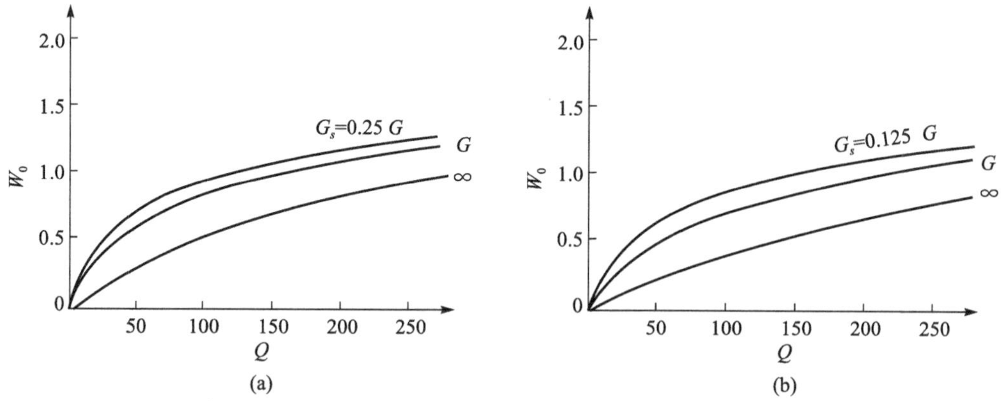

Fig. 4a, b　The characteristic curves for different composite materials and several values of G_s ($n=5$)

(a) graphite-epoxy; (b) boron-epoxy

From Figs. 6 and 7, it may be seen that the rigidity of the plate increases with the increase in the number of layers, but for $n \geqslant 9$ the rigidity increases quite slowly.

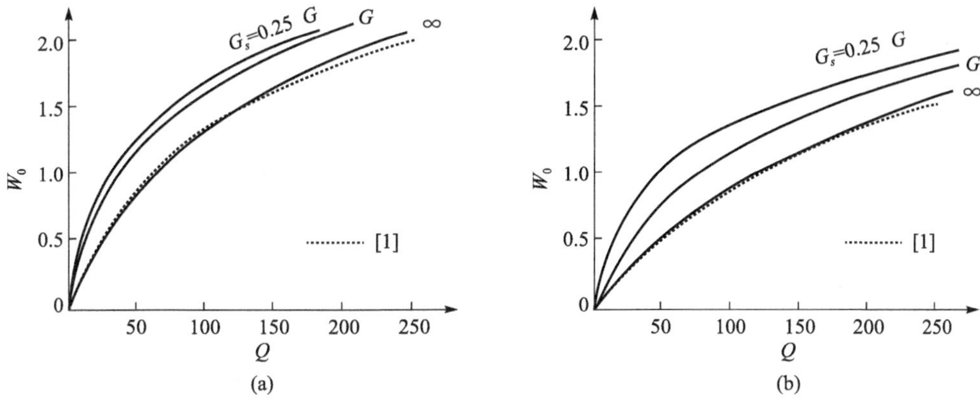

Fig. 5a, b The characteristic curves for different composite materials and several values of G_s ($n=1$)

(a) graphite-epoxy; (b) boron-epoxy

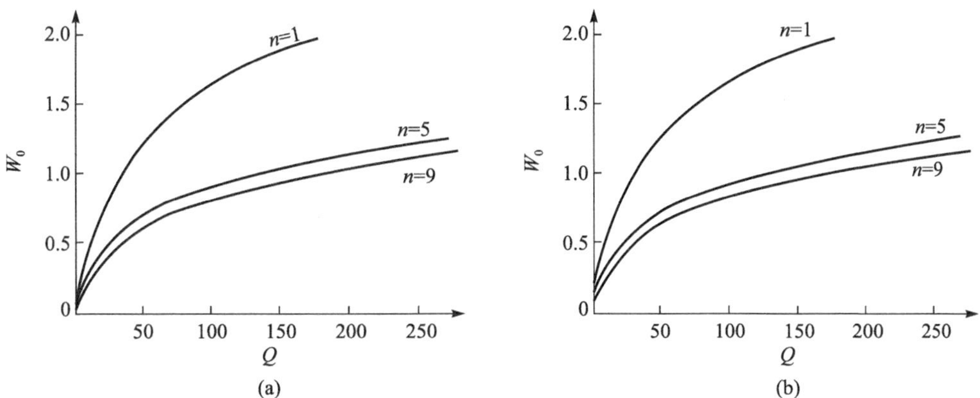

Fig. 6a, b Effect of number of layers on rigidity for synmetrically laminated rectilinearly orthotropic graphite-epoxy elliptical plate

(a) $G_s=0.25G$; (b) $G_s=G$

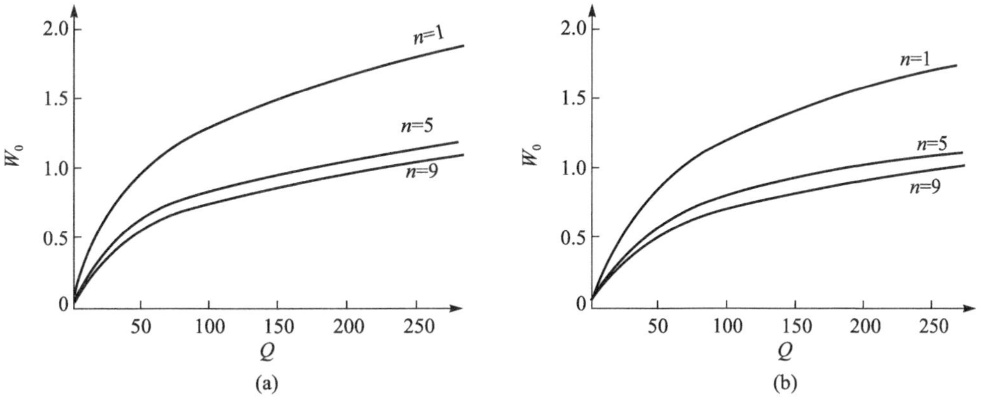

Fig. 7a, b Effect of number of layers on rigidity for symmetrically laminated rectilinearly orthotropic graphite-epoxy elliptical plate

(a) $G_s=0.125G$; (b) $G_s=0.5G$

According to formulas (33), we obtain some useful results for the nondimensional bending and membrane stresses at the center and the ends of the major and minor axes of the boron-epoxy plate having nine layers. Such results are given by curves in Figs. 8-10. It may be observed that at the ends of the major axis of the plate, the membrane stresses and extreme-fibre bending stresses in the direction of the major axis increase rapidly with the central deflection, and these stresses in the direction of the minor axis are slightly affected. Similarly, at the ends of the minor axis, the stresses in the direction of the minor axis increase rapidly with the central deflection, whereas these stresses in the direction of the major axis change only slightly. Fig. 10 indicates the value of the central stresses in the direction of the major axis is much higher than that in the direction of the minor axis.

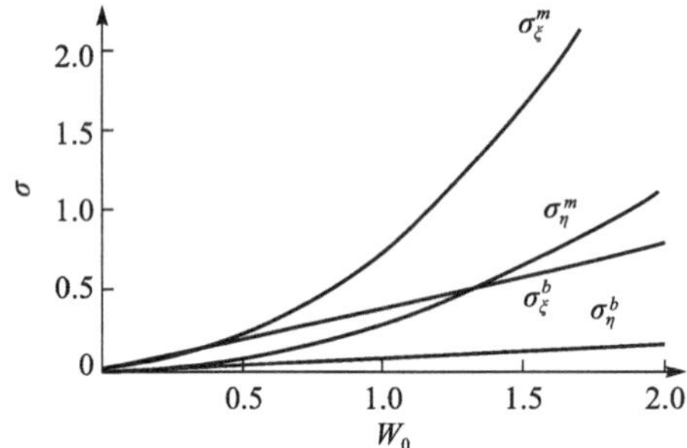

Fig. 8　Membrane and extreme-fibre bending stresses at the center of a symmetrically laminated rectilinearly orthotropic boron-epoxy elliptical plate ($n=9$)

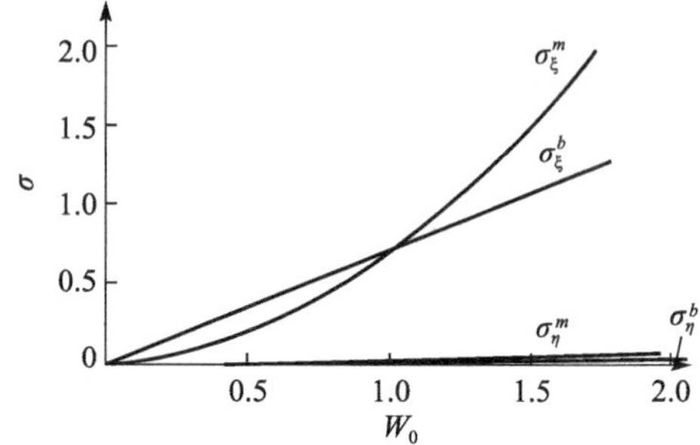

Fig. 9　Membrane and extreme-fibre bending stresses at the ends of the major axis of a symmetrically laminated rectilinearly orthotropic boron-epoxy elliptical plate ($n=9$)

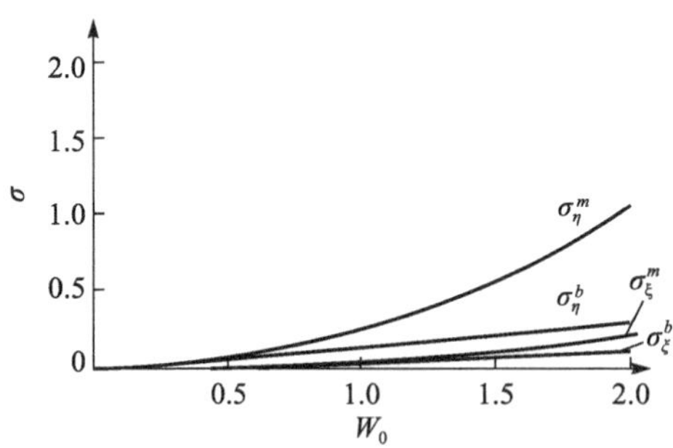

Fig. 10　Membrane and extreme-fibre bending stresses at the ends of the minor axis of a symmetrically laminated rectilinearly orthotropic baron-epoxy elliptical plate ($n=9$)

Example 2: A symmetrically laminated rectilinearly orthotropic circular plate.

For comparison, we consider a special case of a symmetrically laminated rectilinearly orthotropic circular plate in which $a=b$. In such a case, we have $\lambda=1$.

Using (26) and (33), we obtain useful numerical results. The characteristic curves for graphite-epoxy circular plates having different number of layers are shown in Fig. 11. It is seen from Fig. 11 that the effect of the number of layers on bending of a laminated circular plate is similar to that of a laminated elliptical plate.

Fig. 12a, b, c show the nondimensional stress curves for a boron-epoxy circular plate at the center and the ends of x, y axes, respectively. It may be observed that the membrane and extreme-fibre bending central stresses in the direction of the x axis increase rapidly with the central deflection, but these kinds of stresses in the direction of the y axis are only slightly affected. This is to be expected in view of the fact the reinforced fibres are parallel to the x axis. Again, at the end of the x axis, the membrane and extreme-fibre bending stresses in the direction of the x axis increase rapidly with the increase of the central deflection, but the stresses in that direction of the y axis change slightly only. Similarly, at the end of the y axis, the stresses in the direction of the y axis increase rapidly with the central deflection, but the stresses in the direction of the axis increase very slowly.

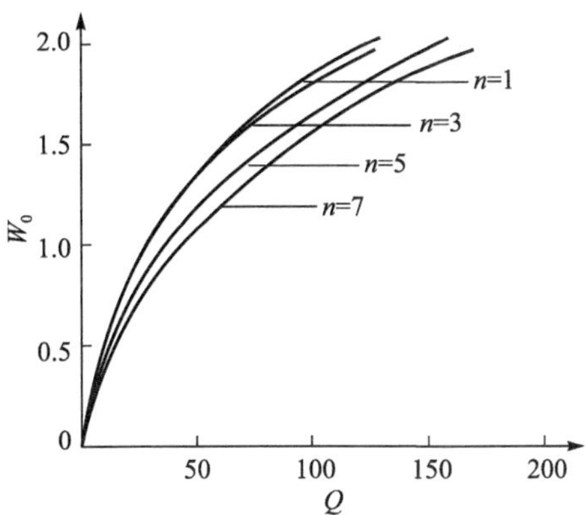

Fig. 11 Effect of the number of layers on bending of a graphite-epoxy laminated circular plate ($G_s=G$, $a/h=10$)

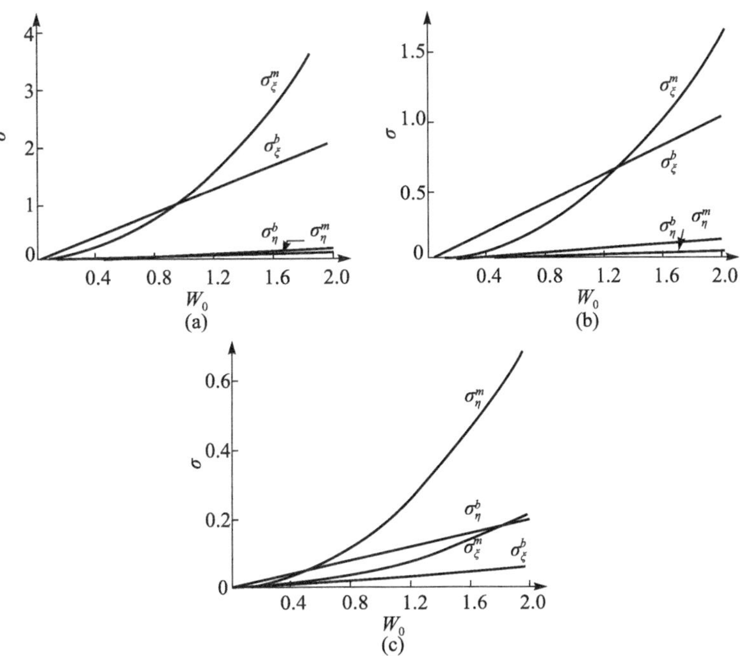

Fig. 12a-c The membrane and extreme-fibre bending stresses of a boron-epoxy circular plate ($G_s=G$, $a/h=10$)

(a) at the center; (b) at the end of the x axis; (c) at the end of the y axis

The distribution of the membrane and extreme-fibre bending stresses along x and y axes for a boron-epoxy plate in the case of $W_0=2$ are shown in Figs. 13a, b, respectively. It is seen from the figures that along the x axis the maximum stress is the membrane stress in the direction of the x axis, and is located at the end of the x axis. On the other hand, along the y axis, the maximum stress is still the membrane bending stress in the direction of the x axis, but is located at the center of the plate.

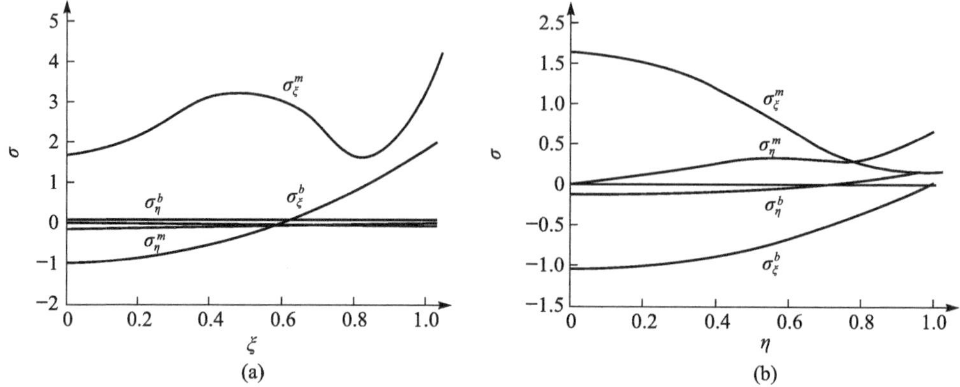

Fig. 13a, b　The variation of the membrane and extreme-fibre bending stresses along coordinate axes of a boron-epoxy circular plate ($G_s=G$, $a/h=10$, $W_0=2$)
(a) along the x axis, (b) along the y axis

Fig. 14 indicates characteristic curves for a graphite-epoxy circular plate having different transverse shear moduli. Here $E_x/E_y=20$, $G/E_y=0.5$, $\nu_{xy}=0.3$, $a/h=10$, $G_{yz}=G_{zx}=G_s$. Obviously, the effect of the transverse shear deformation on bending of the plate is very important, and the rigidity of the plate increases with the increasing transverse shear modulus. The dashed curve corresponds to the disregarded transverse shear deformation.

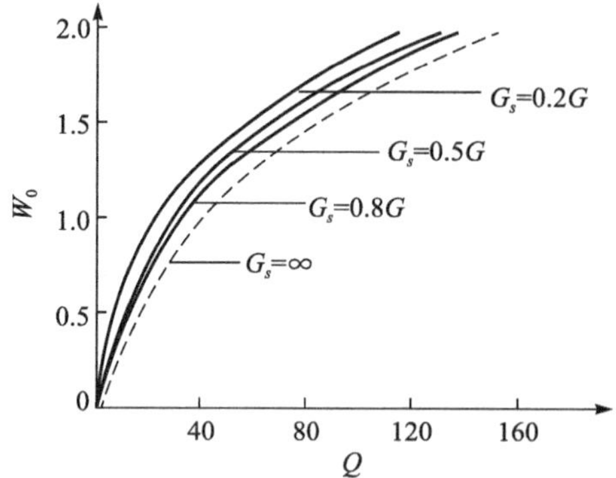

Fig. 14　Effect of the transverse shear modulus on bending of a graphite-epoxy laminated circular plate ($a/h=10$)

Finally, for comparison, we consider a limiting case where the transverse shear moduli G_{yz}, G_{zx} approach infinity corresponds to the disregarding of the transverse shear deformation. In the case of $G_s\to\infty$, and the number of layers $n=1$, substituting those values in Table 1 for a glass-epoxy material and a graphite-epoxy material, respectively, into the relation (26), gives useful results as presented in Table 2. For comparison, Table 2 also gives the theoretical results[4]. It shows good agreement between the present results and the earlier results.

Table 2 Comparison between two theoretical results

material / $\frac{qa^4}{E_y h^4}$ / W_0	glass-epoxy		graphite-epoxy	
	present-results	Wang[4]	present-results	Wang[4]
0.2	2.058	2.059	17.06	17.13
0.5	5.600	5.629	46.54	46.91
1.0	14.46	14.69	120.8	122.6
1.5	29.82	30.54	250.6	254.6
2.0	54.96	56.61	463.6	468.5
3.0	147.6	151.3	1250	1248

References

[1] Prabhakara, M.K., Chia, C.Y., Bending of elliptical orthotropic plates with large deflection, *Acta Mech.*, 21, (1975), 29-40.

[2] Ou, G.X., Liang, J.G., Zhou, L., Nonlinear bending of laminated elliptical plates, *The Second National Conference on Composite Material*, China 1982 (in Chinese).

[3] Gu, F., Lo, Z.D., Large deflection bending of composite circular plates exhibiting rectilinear orthotropy, *Acta Mat. Compositae Sinica*, 5, (1998), 11-16 (in Chinese).

[4] Wang, R.J. Large deflection of symmetrically laminated orthotropic composite circular plates, *Acta Mat. Compositae Sinica*, 4, (1988), 84-89 (in Chinese).

[5] Sathyamoorthy, M., Nonlinear vibration of orthotopic circular plates-A comparison, *Fibre Science and Technology*, 16, (1982), 111-117 (in Chinese).

[6] Liu Renhuai, He, L.H., A simple theory for nonlinear bending of laminated composite rectangular plates including higher-order effects, *Int. J. Non-Linear Mech.*, 26, (1991), 537-545.

[7] Liu Renhuai, He, L.H., Axisymmetrical bending of laminated circular plates, *Journal of Jiangxi Polytechnic University*, 2, (1991), 199-205 (in Chinese).

[8] Liu Renhuai, He, L.H., Nonlinear bending of simply supported symmetric laminated crossply rectangular plates, *Appl. Math. Mech.*, 11, (1990), 801-807.

[9] Liu Renhuai, On the nonlinear stability of symmetrically laminated cylindrically orthotropic shallow spherical shells, *Appl. Math. Mech.*, 12, (1991), 271-279.

[10] Liu Renhuai, Nonlinear stability of symmetrically laminated cylindrically orthotropic shallow spherical shells including transverse shear, *Science in China* (Series A), 35, (1992), 734-746.

[11] Liu Renhuai, Zhong, C., Nonlinear buckling of symmetrically laminated cylindrically orthotropic shallow spherical shells with a circular hole at the center including transverse shear, *Journal*

of Jinan University，15，(1994)，1-12 (in Chinese).

[12] Liu Renhuai，Nonlinear buckling of symmetrically laminated，cylindrically orthotropic，shallow，conical shells considering shear，*Int. J. Non-Linear Mech.*，31，(1991)，89-99.

[13] Liu Renhuai，On nonlinear buckling of symmetrically laminated cylindrically orthotropic，truncated，shallow，spherical shells under uniform pressure including shear effects，*Int. J. Nonlinear Mech.*，31，(1996)，101-105.

[14] Jones，R. M.，*Mechanics of Composite Materials*，New York，McGraw-Hill，1975.

Nonlinear dynamic buckling of symmetrically laminated cylindrically orthotropic shallow spherical shells①

中文摘要

本文应用突变理论提出了承受轴对称载荷的对称层合圆柱正交异性扁球壳的非线性动态屈曲的尖点突变模型。我们计及了横向剪切影响，讨论了材料的剪切模量、几何尺寸以及有关参数对于非线性动态屈曲的影响。

1. Introduction

Composite materials have been used widely for their superior properties. Laminated composite plates and shells have become important structural members in various industries such as aeronautical, astronautical, naval, civil, and vessel engineering. The study of nonlinear problems of laminated composite plates and shells has important theoretical and practical significance. The nonlinear behaviour of laminated plates in [1, 2] was already investigated in the 1980s. Geometrically nonlinear laminated shells were discussed in [3, 4], while nonlinear vibration and postbuckling of laminated plates were studied in [5]. Later, in [6], the author discussed the nonlinear bending of laminated cross-ply rectangular plates, and in [7-9], they investigated the nonlinear buckling of laminated composite shallow spherical shells and presented a simple theory for nonlinear bending of laminated composite rectangular plates including high-order effects. Soon after, an axisymmetric nonlinear analysis of laminated orthotropic annular spherical shells in [10] was carried out. In 1994, a study was made in [11] on the problem of unsymmetrically laminated moderately thick shallow spherical shells. Recently, nonlinear buckling of symmetrically laminated cylindrically orthotropic shallow conical shells considering shear was investigated in [12]. The nonlinear buckling of laminated composite truncated in [13] shallow spherical shells was considered under uniform pressure including shear effects. The nonlinear vibration of laminated truncated conical shells was discussed in [14].

The catastrophe theory was formulated more than thirty years ago and used in many fields. It is natural to study buckling problems by means of the catastrophe theory because the instability of an elastic structure is due to the vanishing of a minimal extreme value of its potential energy. This paper presents a model of a cusped catastrophe

① Reprinted from *Archive of Applied Mechanics*, 1998, 68: 375-384. Authors: Liu Renhuai and Wang Fan.

at nonlinear dynamic buckling of symmetrically laminated cylindrically orthotropic shallow spherical shells subjected to an axisymmetrical load by using the catastrophe theory. Effects of transverse shear are taken into account. Effects of the shear modulus, geometry, and material parameters on the nonlinear dynamic buckling are discussed.

2. Fundamental Equations

The present investigation will concern with a symmetrically laminated cylindrically orthotropic shallow spherical shell. The geometry and coordinate system of the shell is shown in Fig. 1, in which H is the shell rise, a is the base radius, h is the shell thickness, R is the radius of curvature, $q(r,t)$ is load, r is the radial coordinate, θ is the circumferential coordinate and z is the coordinate in the thickness direction.

In the case of axisymmetric deformation of the shallow spherical shell, the displacement field and the modified shear rigidity G may be written in the form [13]

$$u = u_0 + z\psi, \quad v = 0, \quad w = w_0, \tag{1}$$

$$G = \frac{4h^2}{9\sum_{k=1}^{n}\frac{1}{G_{rz}^{(k)}}\left[h_k - h_{k-1} - \frac{8}{3h_2}(h_k^3 - h_{k-1}^3) + \frac{16}{5h^4}(h_k^5 - h_{k-1}^5)\right]}. \tag{2}$$

in which $u_0(r,t), w_0(r,t)$, are the radial and transverse displacements of the middle surface, respectively. $\psi(r,t)$ is the rotation of a normal to the underformed middle surface. u, v, w are the radial, circumferential and transverse displacements at an arbitrary point of the layer, respectively. The thickness and the transverse shear modules of the kth layer are denoted by h_k and $G_{rz}^{(k)}$, respectively.

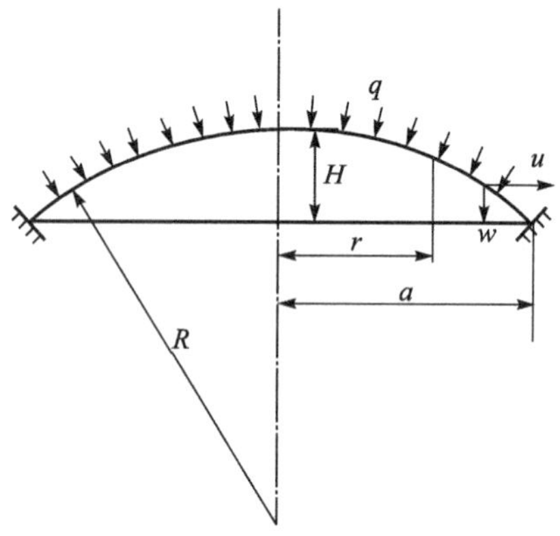

Fig. 1　Geometry of laminated shallow spherical shell

Considering the nonlinear term of the deformation, the compatibility equation of strains must be included in the differential equations of the nonlinear vibration of the shell. We shall first expand the definition of w, ψ, N_r in differential equations [8] to the time domain, then include the forced load and D'Alembert inertial force. The differential equations of the transverse motion of the shell can be given by

$$\frac{1}{r}\frac{\partial}{\partial r}\left\{rN_r(r,t)\left[\frac{r}{R} + \frac{\partial w(r,t)}{\partial r}\right] + rG\left[\psi(r,t) + \frac{\partial w(r,t)}{\partial r}\right]\right\} + q(r,t) - \rho h\frac{\partial^2 w(r,t)}{\partial t^2} = 0,$$

$$D_{11}\frac{\partial}{\partial r}r\frac{\partial\psi(r,t)}{\partial r} - \left(Gr + \frac{D_{22}}{r}\right)\psi(r,t) - Gr\frac{\partial w(r,t)}{\partial r} = 0,$$

$$A_3\frac{\partial}{\partial r}r\frac{\partial[rN_r(r,t)]}{\partial r} - A_1N_r(r,t) + \frac{1}{2}\left[\frac{\partial w(r,t)}{\partial r}\right]^2 + \frac{r}{R}\frac{\partial w(r,t)}{\partial r} = 0. \tag{3}$$

The basic equations (3) will be solved under the following boundary conditions for a rigidly clamped edge:

$$w=0,\quad \psi=0,\quad u=0 \qquad \text{at } r=a,$$
$$\psi=0,\quad N_r \ \text{ finite} \qquad \text{at } r=0. \tag{4}$$

where N_r is the radial membrane force, $E_r^{(k)}$ and $E_\theta^{(k)}$ are moduli of elasticity of the kth layer along the principal material axes r and θ respectively, $\nu_{r\theta}^{(k)}$, $\nu_{\theta r}^{(k)}$ ate Poisson's ratios of the kth layer with the first subscript indicating the direction of the tensile force and the second one indicating the direction of contraction, $Q_{ij}^{(k)}$ is the reduced stiffness of the kth layer, A_{ij} is the extensional rigidity, D_{ij} is the bending rigidity, while

$$u=r\left[A_3\frac{\partial(rN_r)}{\partial r}-A_2N_r\right],$$
$$Q_{11}^{(k)}=\frac{E_r^{(k)}}{1-\nu_{r\theta}^{(k)}\nu_{\theta r}^{(k)}},\ Q_{12}^{(k)}=\frac{\nu_{\theta r}^{(k)}E_\theta^{(k)}}{1-\nu_{r\theta}^{(k)}\nu_{\theta r}^{(k)}},\ Q_{12}^{(k)}=\frac{E_\theta^{(k)}}{1-\nu_{r\theta}^{(k)}\nu_{\theta r}^{(k)}},$$
$$A_{ij}=\sum_{k=1}^{n}Q_{ij}^{(k)}(h_k-h_{k-1}),\ D_{ij}=\frac{1}{3}\sum_{k=1}^{n}Q_{ij}^{(k)}(h_k^3-h_{k-1}^3)\quad i,j=1,2,$$
$$A_1=A_0A_{22},\ A_2=A_0A_{12},\ A_3=A_0A_{11},\ A_0=\frac{1}{A_{11}A_{22}-A_{12}^2}. \tag{5}$$

In order to simplify the following calculations, the dimensionless parameters will be defined as:

$$y=\frac{r}{a},\ W=\frac{w}{h},\ \Psi=\frac{a}{h}\psi,\ S=\frac{a}{D_{11}}rN_r,\ P=\frac{a^4}{D_{11}h}q(r,t),\ k=\frac{a^2}{Rh},$$
$$m=\frac{a^2}{D_{11}}G,\ \beta_1^2=\frac{D_{22}}{D_{11}},\ \beta_2^2=\frac{A_1}{A_3},\ \beta_3=\frac{h^2}{A_3D_{11}},\ \beta_4=\frac{A_2}{A_3},\ \bar{\rho}=\frac{\rho}{D_{11}}ha^4. \tag{6}$$

Upon substitution, Eqs. (3) and conditions (4) will be transformed into the dimensionless form

$$\frac{1}{y}\frac{\partial}{\partial y}\left[S\left(ky+\frac{\partial W}{\partial y}\right)+my\left(\Psi+\frac{\partial W}{\partial y}\right)\right]+p-\bar{\rho}\frac{\partial^2 W}{\partial t^2}=0, \tag{7}$$

$$y^{\beta_1}\frac{\partial}{\partial y}y^{1-2\beta_1}\frac{\partial}{\partial y}(y^{\beta_1}\Psi)=my\left(\Psi+\frac{\partial W}{\partial y}\right), \tag{8}$$

$$y^{\beta_2}\frac{\partial}{\partial y}y^{1-2\beta_2}\frac{\partial}{\partial y}(y^{\beta_2}S)=-\beta_3\frac{\partial W}{\partial y}\left(\frac{1}{2}\frac{\partial W}{\partial y}+ky\right). \tag{9}$$

$$W=0,\ \Psi=0,\ \frac{\partial S}{\partial y}-\beta_4\frac{S}{y}=0\quad \text{at } y=1, \tag{10a-c}$$

$$\Psi=0,\ S=0 \qquad \text{at } y=0. \tag{11a,b}$$

We shall choose the deflection function W in the particular form which satisfies the boundary condition (10a),

$$W(y,t)=f(t)W_1(y), \tag{12}$$

where $W_1(y)$ is the solution for small deflection of the shell

$$W_1(y)=\frac{y^4-1}{4(9-\beta_1^2)}-\frac{y^{\beta_1+1}-1}{(9-\beta_1^2)(1+\beta_1)}-\frac{1}{2m}(y^2-1). \tag{13}$$

Substituting expression (12) into (9), (10c) and (11b), we obtain

$$S=-\frac{1}{2}\beta_3 f^2(t)a_0^2\left[\frac{y^7}{(49-\beta_1^2)}-\frac{2y^{4+\beta_1}}{(4+\beta_1)^2-\beta_2^2}+\frac{y^{1+2\beta_1}}{(1+2\beta_1)^2-\beta_2^2}\right]$$
$$+\frac{1}{2}\beta_3 f^2(t)\left\{\frac{2a_0}{m}\left[\frac{y^5}{(25-\beta_2^2)}-\frac{y^{2+\beta_1}}{(2+\beta_1)^2-\beta_2^2}\right]+\frac{y^3}{m^2(9-\beta_2^2)}\right\}$$
$$-\beta_3 f(t)k\left\{a_0^2\left[\frac{y^5}{(25-\beta_2^2)}-\frac{y^{2+\beta_1}}{(2+\beta_1)^2-\beta_1^2}\right]-\frac{y^3}{m(9-\beta_1^2)}\right\}+Cy^{\beta_2}. \tag{14}$$

where

$$C=f^2(t)\frac{F_1'(1)-\beta_4\overline{F}_1(1)}{\beta_4-\beta_2}+f(t)\frac{F_2'(1)-\beta_4\overline{F}_2(1)}{\beta_4-\beta_2},$$
$$F_1(y)=a_1y^7+a_2y^{4+\beta_1}+a_3y^{1+2\beta_1}+a_4y^5+a_5y^{2+\beta_1}+a_6y^3,$$
$$F_2(y)=a_7y^5+a_8y^{2+\beta_1}+a_9y^3,$$
$$F_i'(y)=\frac{\mathrm{d}F_i}{\mathrm{d}y},\overline{F}_i(y)=\frac{F_i(y)}{y}\quad(i=1,2),$$
$$a_0=\frac{1}{(49-\beta_1^2)},\ a_1=-\frac{\beta_3a_0^2}{2(49-\beta_2^2)},\ a_2=\frac{\beta_3a_0^2}{(4+\beta_1)^2-\beta_2^2},$$
$$a_3=\frac{\beta_3a_0^2}{2[(1+2\beta_1)^2-\beta_2^2]},\ a_4=\frac{\beta_3a_0}{m(25-\beta_2^2)},$$
$$a_5=-\frac{\beta_3a_0}{m[(2+\beta_1)^2-\beta_2^2]},\ a_6=-\frac{\beta_3}{2m^2(9-\beta_2^2)},$$
$$a_7=\frac{\beta_3ka_0}{(25-\beta_2^2)},\ a_8=-\frac{\beta_3ka_0}{(2+\beta_1)^2-\beta_2^2},\ a_9=-\frac{\beta_3k}{m(9-\beta_2^2)} \tag{15}$$

Substituting solutions (12) and (14) into Eq. (7), the expression for $(\Psi+\partial W/\partial y)$ can be given.

Then, using Eq. (8) and conditions (10b), (11a) we can obtain

$$\Psi=\bar{\rho}f''(t)[b_1(y^7-1)+b_2(y^3-1)+b_3(y^{4+\beta_1}-1)+b_4(y^5-1)]$$
$$-f^3(t)[H_6(y)-H_6(1)]-f^2(t)[H_7(y)-H_7(1)]$$
$$-f(t)[H_8(y)-H_8(1)]-\frac{P}{2(9-\beta_1^2)}(y^3-1). \tag{16}$$

Here

$$b_1=\frac{a_0}{24(49-\beta_1^2)},\ b_2=-\frac{a_0}{8(9-\beta_1^2)}+\frac{a_0}{2(1+\beta_1)(9-\beta_1^2)}+\frac{1}{4m(9-\beta_1^2)},$$
$$b_3=-\frac{a_0}{(1+\beta_1)(3+\beta_1)[(4+\beta_1)^2-\beta_2^2]},\ b_4=-\frac{1}{4m(25-\beta_1^2)},$$
$$a_{10}=\frac{F_1'(1)-\beta_4\overline{F}_1(1)}{\beta_4-\beta_2},\ a_{11}=\frac{F_2'(1)-\beta_3\overline{F}_2(1)}{\beta_4-\beta_2},$$
$$H_1(y)=F_1(y)+a_{10}y^{\beta_2},\ H_2(y)=F_2(y)+a_{11}y^{\beta_2},$$
$$H_3(y)=H_1(y)\left[a_0(y^3-y^{\beta_1})-\frac{y}{m}\right],$$

$$H_4(y)=H_2(y)\left[a_0(y^3-y^{\beta_1})-\frac{y}{m}\right]+kyH_1(y),$$

$$H_5(y)=ky[F_2(y)+a_{11}y^{\beta_2}],\ H_6(y)=y^{\beta_1}\frac{\mathrm{d}}{\mathrm{d}y}y^{1-2\beta_1}\frac{\mathrm{d}}{\mathrm{d}y}[y^{\beta_1}H_3(y)],$$

$$H_7(y)=y^{\beta_1}\frac{\mathrm{d}}{\mathrm{d}y}y^{1-2\beta_1}\frac{\mathrm{d}}{\mathrm{d}y}[y^{\beta_1}H_4(y)],\ H_8(y)=y^{\beta_1}\frac{\mathrm{d}}{\mathrm{d}y}y^{1-2\beta_1}\frac{\mathrm{d}}{\mathrm{d}y}[y^{\beta_1}H_5(y)].\quad(17)$$

Thus, the expressions of $W(y,t)$, $\Psi(y,t)$ and $S(y,t)$ with respect to the unknown $f(t)$ and the variable y are all given.

3. Catastrophe Model of Dynamic Buckling

In terms of the catastrophe theory[15,16], the manifold consisting of all critical points of the potential energy of the system can be determined by variation of the potential energy with regard to the state variable.

Using the above method, the following equation is obtained:

$$\int_0^1\Big\{f^3(t)H_1(y)H_{14}(y)+f^2(t)[H_2(y)H_{14}(y)+kyH_1(y)]+kyH_2(y)f(t)$$
$$-PH_9(y)+f''(t)H_{10}(y)-f^3(t)H_{11}(y)-f^2(t)H_{12}(y)-f(t)H_{13}(y)$$
$$+myH_{14}(y)f(t)+\int_0^y Py\mathrm{d}y-\int_0^y\bar{\rho}\frac{\partial^2W}{\partial t^2}y\mathrm{d}y\Big\}W_1(y)y\mathrm{d}y=0,\quad(18)$$

where

$$H_9(y)=\frac{my(y^3-1)}{2(9-\beta_1^2)},$$
$$H_{10}(y)=my\bar{\rho}[b_1(y^2-1)+b_2(y^3-1)+b_3(y^{4+\beta_1}-1)+b_4(y^5-1)]$$
$$H_{11}(y)=my[H_6(y)-H_6(1)],\ H_{12}(y)=my[H_7(y)-H_7(1)],$$
$$H_{13}(y)=my[H_8(y)-H_8(1)],\ H_{14}(y)=a_0(y-y^{\beta_1})-\frac{y}{m}.\quad(19)$$

Let

$$P=P_0\sin\theta t,\ f(t)=f_0\sin\theta t.\quad(20)$$

Substituting these expressions into (18), we can obtain the manifold of the cusped catastrophe from the following bifurcation equation:

$$f_{01}^3-x_1f_{01}-x_2=0.\quad(21)$$

Here f_{01} is the state variable, x_1 and x_2 are the control variables. They depend on quantities c_i ($i=1,2,\cdots5$) which are functions of the dimensionless shear rigidity m, geometrical parameter k, parameters of material $\beta_1,\beta_2,\beta_3,\beta_4$ and dimensionless density $\bar{\rho}$, defined as follows

$$f_{01}=f_0+\frac{C_3}{3C_2},$$

$$x_1=-\frac{4}{3}\left[\frac{3C_2C_4-C_3^2}{3C_2^2}-\frac{C_1}{C_2}\theta^2\right],x_2=-\frac{4}{3}\left[\frac{2C_3^3-9C_2C_3C_4}{27C_2^3}+P_0\frac{C_5}{C_2}\right],$$

where

$$
\begin{aligned}
C_1 &= \int_0^1 yW_1(y)\left[H_{10}(y) - \int_0^y \bar{\rho} yW_1(y)\mathrm{d}y\right]\mathrm{d}y,\\
C_2 &= \int_0^1 yW_1(y)[H_1(y)H_{14}(y) - H_{11}(y)]\mathrm{d}y,\\
C_3 &= \int_0^1 yW_1(y)[H_2(y)H_{14}(y) + kyH_1(y) - H_{12}(y)]\mathrm{d}y,\\
C_4 &= \int_0^1 yW_1(y)[kyH_2(y) - H_{13}(y) + myH_{14}(y)]\mathrm{d}y,\\
C_5 &= \int_0^1 yW_1(y)\left[\int_0^y y\mathrm{d}y - H_9(y)\right]\mathrm{d}y.
\end{aligned} \tag{22}
$$

The equation satisfying the bifurcation set of the catastrophe manifold (21) can be represented as

$$
48\left[\frac{2C_2^3 - 9C_2C_3C_4}{27C_2^3} + P_0\frac{C_5}{C_2}\right]^2 - \frac{256}{27}\left[\frac{3C_2C_4 - C_3^2}{3C_2^2} - \frac{C_1}{C_2}\theta^2\right]^3 = 0. \tag{23}
$$

Thus, Eq. (23) is the critical equation of the system. When all parameters satisfy Eq. (23), the cusped catastrophe will occur.

4. Results and Discussion

Considering Eqs. (21) and (23), we shall analyze the manifold of the cusped catastrophe with control variables P_0, θ, and the state variable f_{01}, shown in Fig. 2.

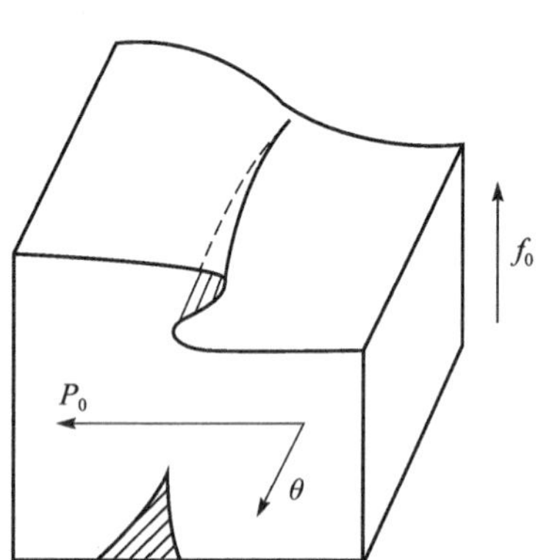

Fig. 2 Manifold of cusped catastrophe
(m=80, k=1.6, $\bar{\rho}$=1.2, θ=4.2)

Obviously, when

$$\theta^2 < \frac{3C_2C_4 - C_3^2}{3C_1C_2},$$

Eq. (21) will have only one real root. As P_0 changes, the amplitude of the vibration of the shell changes smoothly.

When

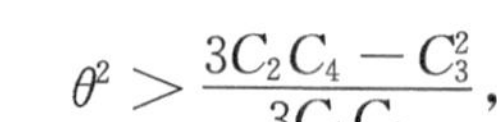

$$\theta^2 > \frac{3C_2C_4 - C_3^2}{3C_1C_2},$$

the number of real roots of Eq. (21) is different. The amplitude of vibration of the shell may suddenly change. Then, there will be the following three cases:

(1) When θ and P_0 are outside of the diamond domain enclosed by the bifurcation set, Eq. (21) will only have one root. It will correspond to the state of stable equilibrium and f_{01} will be on the upper and the lower sheet of the catastrophe manifold.

(2) When θ and P_0 are within the diamond domain, Eq. (21) will have three real roots, two of them corresponding to the state of stable equilibrium and the solution placed on the upper and lower sheet of the catastrophe manifold, respectively. Another real root will correspond to the state of the unstable equilibrium and is located in the middle sheet of the catastrophe manifold. However, this case is not possible in practice.

(3) When θ and P_0 are on the boundary line of the diamond domain, Eq. (21) will have two real roots. Both of them will be corresponding to the state of stable equilibrium. However, a slight change of θ and P_0 can cause the catastrophe of the amplitude f_{01}. The positions and properties of the catastrophe will be related to the route in which θ and P_0 change. If P_0 increases gradually, the sudden jump will occur at point P_1 and cause the sudden rise of f_{01}, shown in Fig. 3. If P_0 decreases gradually, the sudden jump will occur at point P_2 and cause the sudden drop of f_{01}.

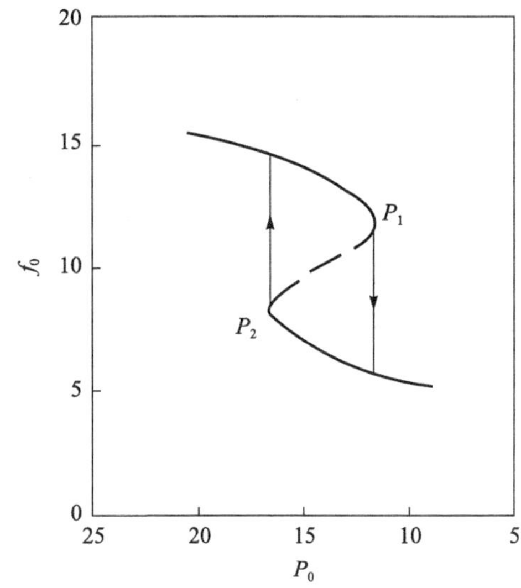

Fig. 3　Position of the critical points (m=80, k=1.6, $\bar{\rho}$=1.2, θ=4.2)

At the same time, we may discuss the effect of several parameters on the dynamic buckling. As an example, without affecting the generality of results we shall consider a symmetrically laminated cylindrically orthotropic shallow spherical shell, which consists of n layers of thin, orthotropic sheets of the same thickness and elastic properties. Assume

$$\frac{E_\theta}{E_r} = 1.5,\ \nu_{r\theta} = 0.3,$$

Then, from Eq. (6), several parameters can be calculated as

$$\beta_1^2 = 1.5,\ \beta_2^2 = 1.5,\ \beta_3 = 16.92,\ \beta_4 = 0.3.$$

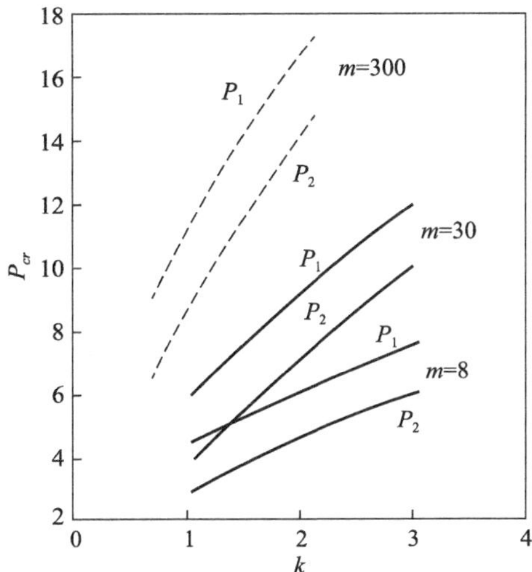

Fig. 4　Effect of dimensionless shear rigidity on the critical load ($\bar{\rho}$=1.2, θ=4.2)

(1) The effect of the dimensionless shear rigidity on the dynamic buckling

a. *The effect of the dimensionless shear rigidity on the critical load*

The amplitude P_0 of the load causing the sudden jump is called the critical load. P_1 and P_2 shown in Fig. 3 are critical loads. From Fig. 4, it may be seen that the critical loads P_1 and P_2 increase with the dimensionless shear rigidity m.

b. *The effect of the dimensionless shear rigidity on the diamond domain enclosed by the bifurcation set*

Fig. 5 clearly shows that the field angle of the diamond domain decreases with the in-

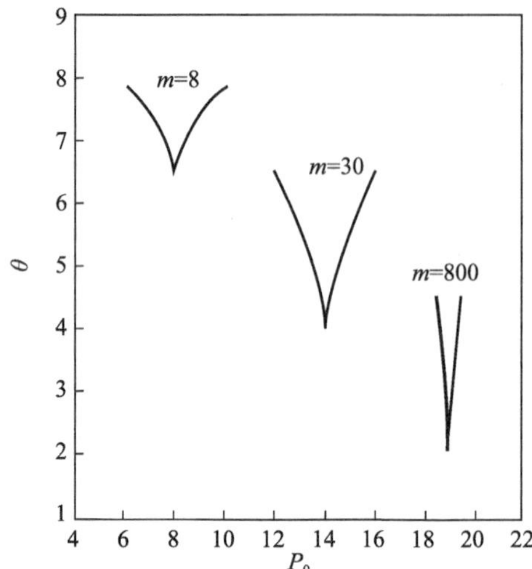

Fig. 5 Effect of dimensionless shear rigidity on the bifurcation set ($k=1.6$, $\bar{\rho}=1.2$)

crease of the dimensionless shear rigidity m. Especially in the limiting case $m=\infty$, the diamond domain will become a straight line.

c. *The effect of the dimensionless shear rigidity on the curve of the amplitude of the load to the amplitude of vibration*

Fig. 6 gives several curves of the amplitude of the load to the amplitude of vibration when dimensionless shear rigidity m is changed from 8 to 800. From Fig. 6, it can be seen that $P_1=P_2$ when $m\to\infty$.

(2) The effect of the geometrical parameter on the dynamic buckling

Fig. 7 shows several curves of θ to P_0 for different values of the geometrical parameter k. It is seen from Fig. 7 that the vertex of the diamond domain will vary with the geometrical parameter k. When the geometrical parameter k increases, the vertex value for θ of the diamond domain drops.

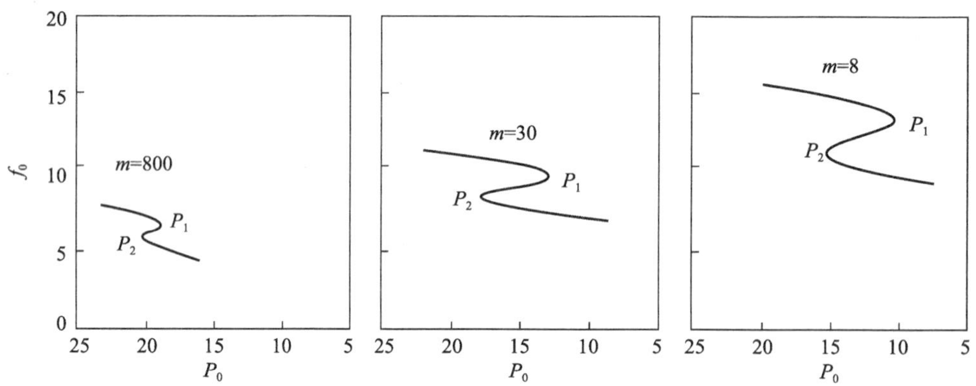

Fig. 6 Effect of dimensionless shear rigidity on the curve of load amplitude ($k=1.6$, $\bar{\rho}=1.2$, $\theta=4.2$)

(3) The effect of the orthotropic parameter β_1 of the material on the dynamic buckling

Fig. 8 shows several curves of θ to P_0 for various values of the orthotropic parameter β_1 of the material. It is seen that values of θ and P_0 of the vertex of the bifurcation set increase with the decrease of β_1 and the diamond domain becomes wider.

(4) The effect of the dimensionless density of the material on the dynamic bucking

Fig. 9 gives several curves of θ to P_0 for various values of dimensionless density $\bar{\rho}$ of the material. It may be seen that the value for θ of the vertex of the bifurcation set increases as the dimensionless density $\bar{\rho}$ of the material decreases, but the value for P_0 does not change substantially.

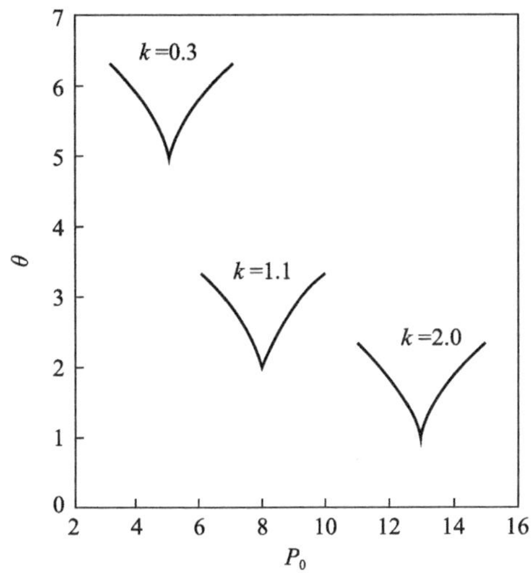

Fig. 7　Effect of the geometrical parameter on the bifurcation set ($\bar{\rho}=1.2$, $m=8$)

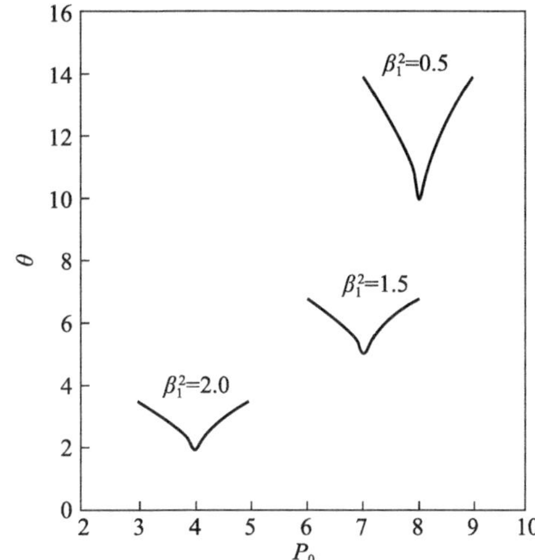

Fig. 8　Effect of the orthotropic parameter of the material on the bifurcation set ($k=1.4$, $\bar{\rho}=1.8$, $m=8$)

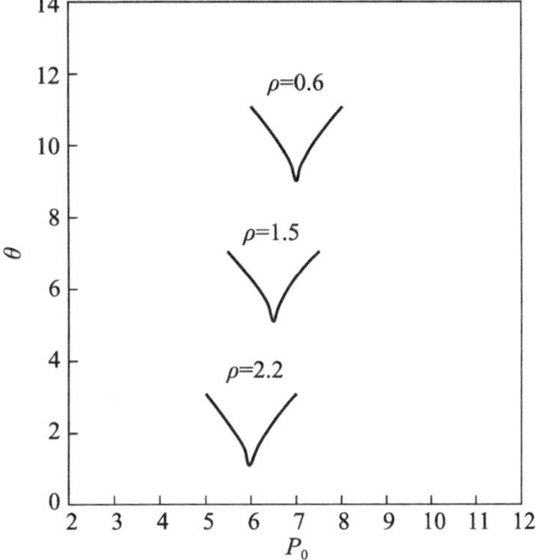

Fig. 9　Effect of dimensionless density of the material on the bifurcation set ($k=1.3$, $m=10$)

References

[1]　Chia, C. Y.: Nonlinear Analysis of Plates. New York: McGraw-Hill, 1980.

[2]　Chia, C. Y.: Geometrically nonlinear behaviour of composite plate: a review. Appl. Mech. Rev. 41 (1988) 439-451.

[3] Reddy, J. N.; Chandrashekhara, K.: Geometrically nonlinear analysis of laminated shells including transverse shear strains. AIAA J. 23 (1985) 440-441.

[4] Reddy, J. N.; Liu, C. F.: A higher-order shear deformation theory of laminated elastic shells. Int. J. Eng. Sci. 23 (1985) 319-330.

[5] Fu, Y. M.; Chia, C. Y.: Multi-mode non-linear vibration and postbuckling of anti-symmetric imperfect angle-ply cylindrical thick panels. Int. J. Non-Linear Mech. 24 (1989) 365-381.

[6] Liu Renhuai; He, L. H.: Nonlinear bending of simply supported symmetric laminated cross-ply rectangular plates. App. Math. Mech. 11 (1990) 801-807.

[7] Liu Renhuai: On the nonlinear stability of symmetrically laminated cylindrically orthotropic shallow spherical shells. Appl. Math. Mech. 12 (1991) 271-279.

[8] Liu Renhuai: Nonlinear stability of symmetrically laminated cylindrically orthotropic shallow spherical shells including transverse shear. Science in China (Series A) 35 (1992) 734-746.

[9] Liu Renhuai; He, L. H.: A simple theory for nonlinear bending of laminated composite rectangular plates including higher-order effects. Int. J. Non-Linear Mech. 26 (1991) 537-545.

[10] Alwar, R. S.; Narasimhan, M. C.: Axisymmetric nonlinear analysis of laminated orthotropic annular spherical shells. Int. J. Non-Linear Mech. 27 (1992) 611-622.

[11] Xu, C. S.; Chia, C. Y.: Non-linear analysis of unsymmetrically laminated moderately thick shallow spherical shells. Int. J. Non-Linear Mech. 29 (1994) 247-260.

[12] Liu Renhuai: Non-linear buckling of symmetrically laminated, cylindrically orthotropic, shallow, conical shells considering shear. Int. Non-Linear Mech. 31 (1996) 89-99.

[13] Liu Renhuai: On nonlinear buckling of symmetrically laminated, cylindrically orthotropic, truncated, shallow, spherical shells under uniform pressure including shear effects. Int. J. Non-Linear Mech. 31 (1996) 101-115.

[14] Xu, C. S.; Xia, Z; Chia, C. Y.: Nonlinear theory and vibration analysis of laminated truncated, thick, conical shells. Int. J. Non-Linear Mech. 31 (1996) 139-154.

[15] Poston, T.; Stewart, I. N.: Catastrophe Theory and Catastrophes. London: Pitman 1978.

[16] Zeeman, E. C: Catastrophe Theory and Application, New York: Halstol Press 1978.

Nonlinear stability of symmetrically laminated cylindrically orthotropic truncated shallow conical shells including transverse shear[①]

中文摘要

本文使用修正迭代法研究了考虑横向剪切的对称层合圆柱正交异性具有硬中心的截头扁锥壳在均布压力作用下的非线性稳定问题，获得了临界载荷的解析公式。本文结果可直接应用于工程设计。

1. Introduction

Recently, composite materials have been applied extensively for their essential feature of light weight and high strength. Therefore, they have been studied by many investigators. However, the nonlinear problems for laminated composite shallow shells of revolution are only investigated by few because of the difficulties of nonlinear mathematical problems and complicated structures. The author (1991, 1992) first studied a nonlinear stability problem for a symmetrically laminated cylindrically orthotropic shallow spherical shell and has succeeded in solving a succession of nonlinear problems of the shell (Liu, 1993, 1996; Liu and Zhong, 1994; Liu and Wang, 1997; Wang et al., 2000; Wang and Liu, 2001; Zhu et al., 2008). After that, we further studied the nonlinear stability problem for a symmetrically laminated cylindrically orthotropic shallow conical shell using the modified iteration method (Liu, 1996). The method was originally suggested by Yeh Kai-Yuan and Liu Ren-Huai in 1965, and successfully applied in a system problems of nonlinear problem of plates and shells.

Nevertheless, because of the difficulty in nonlinear mathematics and truncated shells, a problem with more practical significance and high level of difficulty has never been attempted, i.e., the nonlinear stability problem of a symmetrically laminated cylindrically orthotropic truncated shallow conical shell with a non-deformable rigid body at the center under uniform pressure including transverse shear. As before, we have used the modified iteration method to overcome these difficulties and obtained an analytical solution of the critical buckling applicable to engineering design.

2. Nonlinear Boundary Value Problem Formulation

Let us consider a symmetrically laminated cylindrically orthotropic truncated shallow conical shell with a non-deformable rigid body at the center subjected to the action

① Reprinted from *International Journal of Applied Mechanics and Engineering*, 2009, 14 (3): 769-790. Authors: Liu Renhuai and Su Wei.

of a uniformly distributed load q_0 as shown in Fig. 1. The outer edge is rigidly clamped and the inner edge is fixed on the non-deformable rigid body which can be moved up and down. Here h is the total thickness of the shell, n is the number of layers, α is the slope angle, b and a are the upper and lower radii, respectively, r is the radical coordinate, θ is the circumferential coordinate and z is the coordinate in the thickness direction.

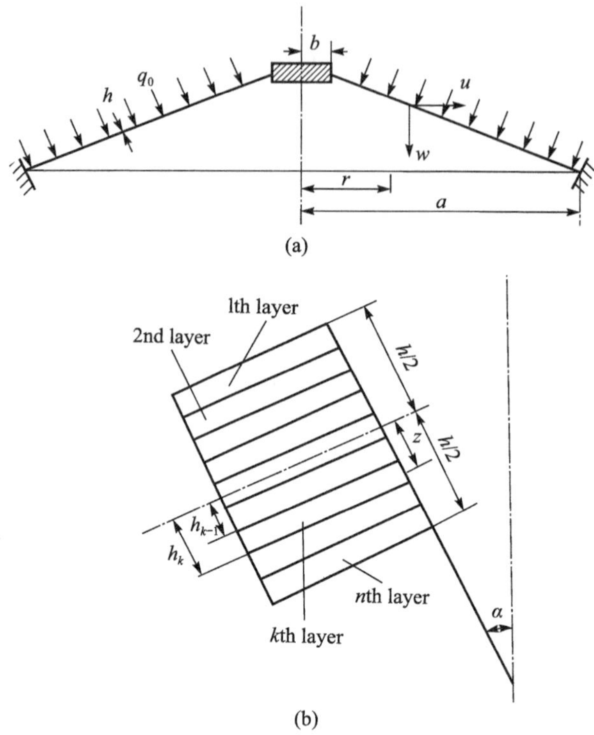

Fig. 1　A symmetrically laminated cylindrically orthotropic truncated shallow conical shell

According to the nonlinear bending theory (Liu, 1996b) for symmetrically laminated cylindrically orthotropic shallow conical shells, we can easily obtain the basic equations satisfied by the deflection w, the rotation ψ of the normal to the middle surface of the shell and the radical membrane force N_r of the symmetrically laminated cylindrically orthotropic truncated shallow conical shell with a non-deformable rigid body at the center under the action of a uniformly distributed load q_0 as follows

$$\begin{aligned}
&\frac{\mathrm{d}}{\mathrm{d}r}\left[rN_r\left(\alpha+\frac{\mathrm{d}w}{\mathrm{d}r}\right)+\mathrm{Gr}\left(\frac{\mathrm{d}w}{\mathrm{d}r}-\psi\right)+q_0 r\right]=0,\\
&A_3\frac{\mathrm{d}}{\mathrm{d}r}r\frac{\mathrm{d}(rN_r)}{\mathrm{d}r}-A_1N_r+\frac{1}{2}\left(\frac{\mathrm{d}w}{\mathrm{d}r}\right)^2+\alpha\frac{\mathrm{d}w}{\mathrm{d}r}=0,\\
&D_{11}\frac{\mathrm{d}}{\mathrm{d}r}r\frac{\mathrm{d}\psi}{\mathrm{d}r}-\left(\mathrm{Gr}+\frac{D_{22}}{r}\right)\psi+\mathrm{Gr}\frac{\mathrm{d}w}{\mathrm{d}r}=0
\end{aligned}\qquad(2.1\text{a-c})$$

where A_{ij}, D_{ij} and G are the extensional, flexural and shear rigidities, respectively,

$Q_{ij}^{(k)}$ are the reduced stiffnesses of the kth layer, $E_r^{(k)}$ and $E_\theta^{(k)}$ are moduli of elasticity of the kth layer along the principal material axes r and θ, respectively, $\nu_{r\theta}^{(k)}$ and $\nu_{\theta r}^{(k)}$ are Poisson's ratios of the kth layer with the first subscript indicating the direction of the tensile force and the other indicating the direction of contradiction, $G_{rz}^{(k)}$ is the transverse shear modulus of the kth layer characterizing the change of the angle in the rz plane

$$\begin{aligned}
&A_{ij}=\sum_{k=1}^{n}Q_{ij}^{(k)}(h_k-h_{k-1}),\quad D_{ij}=\frac{1}{3}\sum_{k=1}^{n}Q_{ij}^{(k)}(h_k^3-h_{k-1}^3),(i,j=1,2),\\
&G=\frac{4h^2}{9}\left\{\sum_{k=1}^{n}\frac{1}{G_{rz}^{(k)}}\left[h_k-h_{k-1}-\frac{8}{3h^2}(h_k^3-h_{k-1}^3)+\frac{16}{5h^4}(h_k^5-h_{k-1}^5)\right]\right\}^{-1},\\
&Q_{11}^{(k)}=\frac{E_r^{(k)}}{1-\nu_{r\theta}^{(k)}\nu_{\theta r}^{(k)}},\quad Q_{12}^{(k)}=\frac{\nu_{\theta r}^{(k)}E_r^{(k)}}{1-\nu_{r\theta}^{(k)}\nu_{\theta r}^{(k)}},\quad Q_{22}^{(k)}=\frac{E_\theta^{(k)}}{1-\nu_{r\theta}^{(k)}\nu_{\theta r}^{(k)}},\\
&A_0=\frac{1}{A_{11}A_{22}-A_{12}^2},\quad A_1=A_0A_{22},\quad A_2=A_0A_{12},\quad A_3=A_0A_{11}.
\end{aligned}\tag{2.2}$$

Having w, ψ and N_r, the radical displacement u_0, the transverse shear force Q_r and the radial moment M_r can be derived from the following expressions

$$\begin{aligned}
&u_0=r\left[A_3\frac{\mathrm{d}(rN_r)}{\mathrm{d}r}-A_2N_r\right],\\
&Q_r=G\left(\frac{\mathrm{d}w}{\mathrm{d}r}-\psi\right),\\
&M_r=-\left(D_{11}\frac{\mathrm{d}\psi}{\mathrm{d}r}+D_{12}\frac{\psi}{r}\right).
\end{aligned}\tag{2.3a-c}$$

Making the following calculation simpler, integrating Eq. (2.1a), we have

$$rN_r\left(\alpha+\frac{\mathrm{d}w}{\mathrm{d}r}\right)+\mathrm{Gr}\left(\frac{\mathrm{d}w}{\mathrm{d}r}-\psi\right)=F(r)\tag{2.4}$$

where F (r) is the load function, and c is the integral constant

$$F(r)=-\frac{1}{2}q_0r^2+c.\tag{2.5}$$

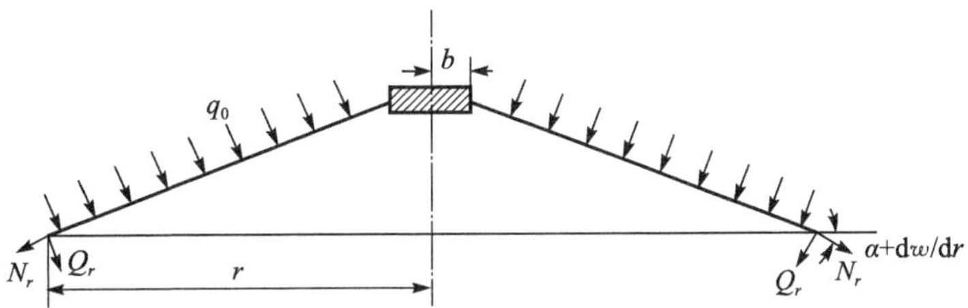

Fig. 2 Determination of the load function F (r)

Considering the equilibrium condition of the section of the shell as shown in Fig. 2, and observing that for the shell the angle $\alpha+\frac{\mathrm{d}w}{\mathrm{d}r}$ is small, along the central vertical axis, all forces satisfy the following equation

$$2\pi r\left[Q_r + N_r\left(\alpha + \frac{dw}{dr}\right) + \pi q_0(r^2 - b^2)\right] = 0. \tag{2.6}$$

Using expression (2. 3b), Eq. (2. 4) becomes

$$r\left[Q_r + N_r\left(\alpha + \frac{dw}{dr}\right)\right] = F(r). \tag{2.7}$$

A comparison of Eq. (2. 6) and Eq. (2. 7) leads to the following expression

$$F(r) = -\frac{1}{2}q_0(r^2 - b^2). \tag{2.8}$$

Using the above expression, Eq. (2. 4) becomes

$$rN_r\left(\alpha + \frac{dw}{dr}\right) + Gr\left(\frac{dw}{dr} - \psi\right) = -\frac{1}{2}q_0(r^2 - b^2). \tag{2.9}$$

Then, Eq. (2. 9) and (2. 1b,c) constitute the basic equations for the large deflection of the symmetrically laminated cylindrically orthotropic truncated shallow conical shell with a non-deformable rigid body at the center under a uniformly distributed load q_0.

If the shell is rigidly clamped along its lower edge, the appropriate boundary conditions can be expressed as

$$w = 0,\quad \psi = 0,\quad u_0 = 0 \quad \text{at} \quad r = a, \tag{2.10}$$

$$\psi = 0,\quad u_0 = 0 \quad \text{at} \quad r = b. \tag{2.11}$$

In order to simplify the following calculations, let us introduce the following non-dimensional parameters

$$y = \frac{r}{a},\quad \eta = \frac{b}{a},\quad W = \frac{w}{h},\quad \varphi = \frac{dW}{dy} + k,\quad \Psi = \frac{a}{h}\psi,\quad S = \frac{a}{D_{11}}rN_r,\quad P = \frac{a^4 q_0}{2D_{11}h}$$

$$k = \frac{a}{h}\alpha,\quad m = \frac{a^2}{D_{11}}G,\quad \beta_1^2 = \frac{D_{22}}{D_{11}},\quad \beta_2^2 = \frac{A_1}{A_3},\quad \beta_3 = \frac{h^2}{A_3 D_{11}},\quad \beta_4 = \frac{A_2}{A_3}. \tag{2.12}$$

Using these non-dimensional parameters, nonlinear boundary value problem (2. 9), (2. 1b,c) becomes

$$\begin{aligned} &my(\varphi - k) = -S\varphi + my\Psi + P(\eta^2 - y^2),\\ &L_1(y^{\beta_1}\Psi) = my(\Psi - \varphi + k),\\ &L_2(y^{\beta_2}S) = -\frac{1}{2}\beta_3(\varphi^2 - k^2), \end{aligned} \tag{2.13a-c}$$

$$W = 0,\quad \Psi = 0,\quad y\frac{dS}{dy} - \beta_4 S = 0 \quad \text{at} \quad y = 1, \tag{2.14a-c}$$

$$\Psi = 0,\quad y\frac{dS}{dy} - \beta_4 S = 0 \quad \text{at} \quad y = \eta \tag{2.15a,b}$$

where L is a differential operator

$$\begin{aligned} L_1(\cdots) &= y^{\beta_1}\frac{d}{dy}y^{1-2\beta_1}\frac{d}{dy}(\cdots),\\ L_2(\cdots) &= y^{\beta_2}\frac{d}{dy}y^{1-2\beta_2}\frac{d}{dy}(\cdots). \end{aligned} \tag{2.16}$$

3. Solution of the Nonlinear Boundary Value Problem

Now we solve the above non-dimensional nonlinear boundary value problem

(2.13)-(2.15) resorting to the modified iteration method. Let the non-dimensional upper edge deflection W_m of the shell be an iteration parameter, which is given by

$$W_m = \frac{w}{h}\bigg|_{r=b}. \tag{3.1}$$

From Eqs (2.12) and (2.14a) we have

$$W_m = -\int_{\eta}^{1}(\varphi - k)\mathrm{d}y. \tag{3.2}$$

For the first approximation, by neglecting the nonlinear term $S\varphi$ in Eq. (2.13a) and again introducing φ_1 into right sides of Eqs (3.13b,c), we find the following non-dimensional linear boundary value problem

$$\begin{aligned}
&my(\varphi_1 - k) = my\Psi_1 + P(\eta^2 - y^2),\\
&L_1(y^{\beta_1}\Psi_1) = my(\Psi_1 - \varphi_1 + k),\\
&L_2(y^{\beta_2}S_1) = -\frac{1}{2}\beta_3(\varphi_1^2 - k^2),
\end{aligned} \tag{3.3a-c}$$

$$W_1 = 0,\quad \Psi_1 = 0,\quad y\frac{\mathrm{d}S_1}{\mathrm{d}y} - \beta_4 S_1 = 0,\quad \text{at}\quad y = 1, \tag{3.4a-c}$$

$$\Psi_1 = 0,\quad y\frac{\mathrm{d}S_1}{\mathrm{d}y} - \beta_4 S_1 = 0\quad \text{at}\quad y = \eta. \tag{3.5a,b}$$

Substituting Eq. (3.3a) into the right side of Eq. (3.3b), we have

$$L_1(y^{\beta_1}\Psi_1) = -P(\eta^2 - y^2). \tag{3.6}$$

Under the boundary conditions (3.4b) and (3.5a), the solution of Eq. (3.6) is

$$\Psi_1 = P(a_1 y^{\beta_1} + a_2 y^{-\beta_1} + a_3 y^3 + a_4 y) \tag{3.7}$$

where

$$\begin{aligned}
a_1 &= a_0[(a_3 + a_4)\eta^{-\beta_1} - 8a_3a_4\eta],\\
a_2 &= -a_0[(a_3 + a_4)\eta^{\beta_1} - 8a_3a_4\eta],\\
a_3 &= \frac{1}{9 - \beta_1^2},\\
a_4 &= \frac{\eta^2}{1 - \beta_1^2},\\
a_0 &= -\frac{1}{\eta^{-\beta_1} - \eta^{\beta_1}}.
\end{aligned} \tag{3.8}$$

Substituting expression (3.7) into Eq. (3.3a), and using expression (3.2), we obtain

$$P = \eta_1 W_m \tag{3.9}$$

where

$$\eta_1 = \left\{\frac{a_1}{1+\beta_1}(\eta^{1+\beta_1} - 1) + \frac{a_2}{1-\beta_1}(\eta^{1-\beta_1} - 1) + \frac{a_3}{4}(\eta^4 - 1) + \frac{a_4}{2}(\eta^2 - 1) + \frac{1}{m}\left[\frac{1}{2}(1 - \eta^2) + \eta^2\ln\eta\right]\right\}^{-1}. \tag{3.10}$$

Substituting this relation into expression (3.7), we find

$$\Psi_1 = \eta_1 W_m (a_1 y^{\beta_1} + a_2 y^{-\beta_1} + a_3 y^3 + a_4 y). \tag{3.11}$$

By integrating Eq. (3.3c) under boundary conditions (3.4c) and (3.5b), the solution of S for the first approximation can be written as

$$S_1 = -\frac{\beta_3}{2}\Big[\eta_1^2 W_m^2 (b_1 y^{(1+2\beta_1)} + b_2 y^{(1-2\beta_1)} + b_3 y^{(4+\beta_1)} + b_4 y^{(4-\beta_1)} + b_5 y^{(2+\beta_1)} + b_6 y^{(2-\beta_1)} + b_7 y^7 + b_8 y^5 + b_9 y^3 + b_{10} y + b_{11} y^{-1} + b_{12} y^{\beta_2} \ln y + b_{13} y^{-\beta_2} \ln y + b_{14} y^{\beta_2} + b_{15} y^{-\beta_2}) + 2k\eta_1 W_m (b_{16} y^{(1+\beta_1)} + b_{17} y^{(1-\beta_1)} + b_{18} y^4 + b_{19} y^2 + b_{20} + b_{21} y^{\beta_2} + b_{22} y^{-\beta_2})\Big] \tag{3.12}$$

For the second approximation, we have the non-dimensional linear boundary value problem for φ and ψ as follows

$$my(\varphi_2 - k) = -S_1\varphi_1 + my\Psi_2 + P(\eta^2 - y^2),$$
$$L_1(y^{\beta_1}\Psi_2) = my(\Psi_2 - \varphi_2 + k), \tag{3.13a,b}$$
$$W_2 = 0, \quad \Psi_2 = 0 \quad \text{at} \quad y = 1, \tag{3.14a,b}$$
$$\Psi_2 = 0 \quad \text{at} \quad y = \eta. \tag{3.15}$$

Substituting Eq. (3.13a) into Eq. (3.13b), we have

$$L_1(y^{\beta_1}\Psi_2) = S_1\varphi_1 + P(y^2 - \eta^2). \tag{3.16}$$

Integrating Eq. (3.16), and using boundary conditions (3.14b) and (3.15), we find the solution of the second approximation for Ψ_2

$$\Psi_2 = P(a_1 y^{\beta_1} + a_2 y^{-\beta_1} + a_3 y^3 + a_4 y) - \beta_3\Big\{\frac{1}{2}\eta_1^3 W_m^3 [F_3(y) + e_3 y^{\beta_1} + f_3 y^{-\beta_1}] + k\eta_1^2 W_m^2 [F_2(y) + e_2 y^{\beta_1} + f_2 y^{-\beta_1}] + k^2 \eta_1 W_m [F_1(y) + e_1 y^{\beta_1} + f_1 y^{-\beta_1}]\Big\} \tag{3.17}$$

Substituting the solution (3.17) into Eq. (3.13a), and using expression (3.2), the characteristic relation for the second approximation is obtained as

$$P = (k^2\eta_0 + \eta_1)W_m + k\eta_2 W_m^2 + \eta_3 W_m^3 \tag{3.18}$$

Appling the following extremal condition

$$\frac{\mathrm{d}P}{\mathrm{d}W_m} = 0, \tag{3.19}$$

we obtain

$$3\eta_3 W_m^2 + 2k\eta_2 W_m + k^2\eta_0 + \eta_1 = 0. \tag{3.20}$$

Solving this equation, the non-dimensional critical upper edge deflection W_m^* of the shell for the second approximation gives

$$W_m^* = \frac{1}{3\eta_3}\Big[-k\eta_2 \pm \sqrt{k^2(\eta_2^2 - 3\eta_0\eta_3) - 3\eta_1\eta_3}\Big]. \tag{3.21}$$

Substituting W_m^* into expression (3.18), the following non-dimensional critical buckling load is obtained

$$P^* = (k^2\eta_0 + \eta_1)W_m^* + k\eta_2 W_m^{*2} + \eta_3 W_m^{*3} \tag{3.22}$$

where P^* with respect to the positive and negative of expression (3.21) are the non-dimensional lower and upper critical buckling loads, respectively. In practice, we are only interested in the non-dimensional upper critical bucking load.

Let the discriminant of the quadratic Eq. (3.20) be zero, the formula for the critical point is given by

$$k_0 = \sqrt{\frac{3\eta_1\eta_3}{\eta_2^2 - 3\eta_0\eta_3}}. \tag{3.23}$$

The critical geometrical parameter k_0 is used to distinguish buckling and no buckling for the shell.

4. Numerical Examples

To simplify calculations and without affecting the generality of results, as a particular example, each layer of the symmetrically laminated cylindrically orthotropic truncated shallow conical shell has the same thickness and elastic properties, and

$$\frac{E_\theta}{E_r} = 1.5, \quad \nu_{\theta r} = 0.3.$$

According to expression (2.12), we obtain

$$\beta_1^2 = \beta_2^2 = 1.5, \quad \beta_3 = 16.92, \quad \beta_4 = 0.3.$$

Let us now apply relation (3.18) to a numerical example. Assume $m=100$ and $\eta=0.1$. Characteristic curves for several values of the geometrical parameter k are shown in Fig. 3. It is seen from Fig. 3 that when k is small, the curves of $P \sim W_m$ increase monotonically. This means that no snapping takes place for the shell. When k is larger, the curves of $P \sim W_m$ are gyratory, and snapping can occur.

From Eq. (3.22), the non-dimensional critical buckling loads have been calculated, and the numerical results for $m=100$ and $\eta=0.1$ are shown in Figs. 4 and 5, respectively. It is seen that these curves rise monotonically, and for the same value of the

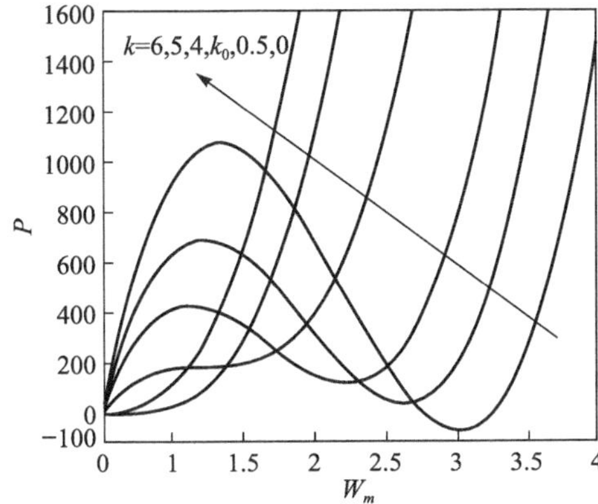

Fig. 3　Curves for several values of k ($m=100$, $\eta=0.1$)

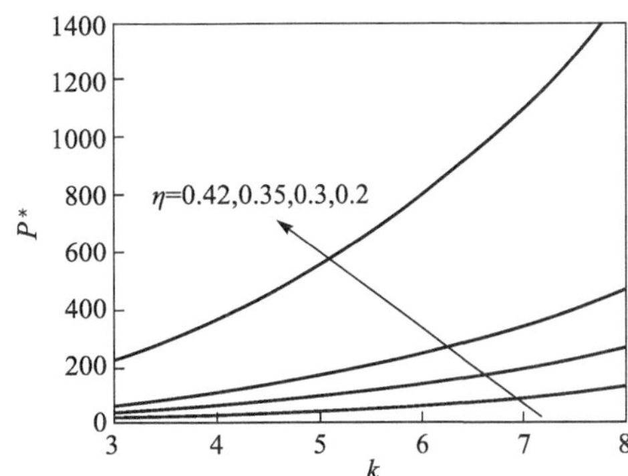

Fig. 4　Curves of the non-dimensional critical load for several values of η ($m=100$)

geometrical parameter k, the non-dimensional critical buckling load P^* of a shell with larger non-dimensional upper radius η and non-dimensional shear rigidity m is low.

Through numerical computation from expression (3.23), values of the critical geometrical parameter k_0 corresponding to various values of the non-dimensional shear rigidity m and the non-dimensional upper radius η are given in Fig. 6. From this figure it can be seen that k_0 increases with m and η, and when $m>100$, k_0 increases very slowly. Obviously, a shell with small transverse shear modulus and upper radius is more sensitive to instability.

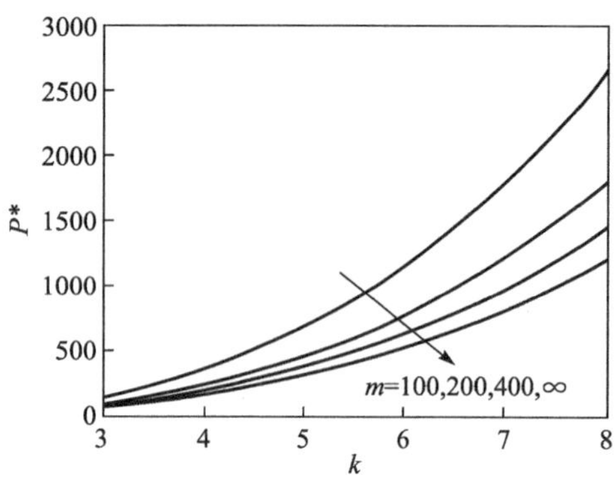

Fig. 5 Curves of the non-dimensional critical load for several values of m (η=0.1)

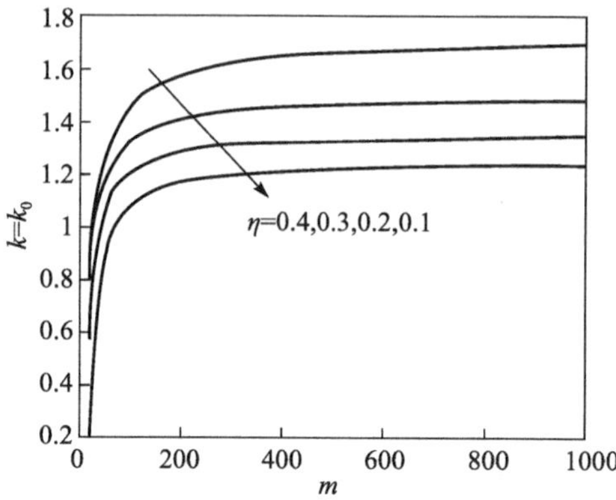

Fig. 6 Curves of the geometrical parameter k_0

Acknowledgements

The authors are grateful to the Key Laboratory of Disaster Forecast and Control in Engineering, Ministry of Education of P. R. China and the Key Laboratory of Diagnosis of Fault in Engineering Structures of Guangdong Province of P. R. China for the support.

Nomenclature

A_{ij}, D_{ij}, G—extensional, flexural and shear rigidities

a, b—upper and lower radii of conical shell

$E_r^{(k)}$, $E_\theta^{(k)}$—moduli of elasticity of the kth layer along the principal material axes r and θ

$G_{rz}^{(k)}$—transverse shear modulus of the kth layer characterizing the change of the angle in the rz plane

h—thickness of the shallow conical shell

N_r—radical membrane force

n—number of layers

$Q_r^{(k)}$—reduced stiffnesses of the kth layer

q_0—uniformly distributed load

α—slope angle

Appendix

Coefficients b_i ($i=1,2,\cdots,22$) are as follows

$$b_1=\frac{a_1^2}{(1+2\beta_1)^2-\beta_2^2},\quad b_2=\frac{a_2^2}{(1-2\beta_1)^2-\beta_2^2},\quad b_3=\frac{2a_1a_3}{(4+\beta_1)^2-\beta_2^2},$$

$$b_4=\frac{2a_2a_3}{(4-\beta_1)^2-\beta_2^2},\quad b_5=\frac{2a_1(a_4-m^{-1})}{(2+\beta_1)^2-\beta_2^2},\quad b_6=\frac{2a_2(a_4-m^{-1})}{(2-\beta_1)^2-\beta_2^2},$$

$$b_7=\frac{a_3^2}{49-\beta_2^2},\quad b_8=\frac{2a_3(a_4-m^{-1})}{25-\beta_2^2},\quad b_9=\frac{(a_4-m^{-1})^2+2a_3\eta^2m^{-1}}{9-\beta_2^2},$$

$$b_{10}=\frac{2[a_1a_2+\eta^2m^{-1}(a_4-m^{-1})]}{1-\beta_2^2},\quad b_{11}=\frac{\eta^4}{m^2(1-\beta_2^2)},\quad b_{12}=\frac{\eta^2a_1}{m\beta_2},$$

$$b_{13}=-\frac{\eta^2a_2}{m\beta_2},\quad b_{14}=\frac{(\lambda_1+\lambda_7)\eta^{-\beta_2}+\lambda_8\eta^{\beta_2}-\lambda_3}{(\beta_2-\beta_4)(\eta^{\beta_2}-\eta^{-\beta_2})},\quad b_{15}=\frac{(\lambda_1+\lambda_6)\eta^{\beta_2}+\lambda_5\eta^{-\beta_2}-\lambda_3}{(\beta_2+\beta_4)(\eta^{\beta_2}-\eta^{-\beta_2})},$$

$$b_{16}=\frac{a_1}{(1+\beta_1)^2-\beta_2^2},\quad b_{17}=\frac{a_2}{(1-\beta_1)^2-\beta_2^2},\quad b_{18}=\frac{a_3}{16-\beta_2^2},\quad b_{19}=\frac{a_4-m^{-1}}{4-\beta_2^2},$$

$$b_{20}=-\frac{\eta^2}{m\beta_2^2},\quad b_{21}=\frac{\lambda_2\eta^{-\beta_2}-\lambda_4}{(\beta_2-\beta_4)(\eta^{\beta_2}-\eta^{-\beta_2})},\quad b_{22}=\frac{\lambda_2\eta^{\beta_2}-\lambda_4}{(\beta_2+\beta_4)(\eta^{\beta_2}-\eta^{-\beta_2})},$$

$$\begin{aligned}
\lambda_1&=b_1(1+2\beta_1-\beta_4)+b_2(1-2\beta_1-\beta_4)+b_3(4+\beta_1-\beta_4)+b_4(4-\beta_1-\beta_4)\\
&\quad+b_5(2+\beta_1-\beta_4)+b_6(2-\beta_1-\beta_4)+b_7(7-\beta_4)+b_8(5-\beta_4)+b_9(3-\beta_4)\\
&\quad+b_{10}+(1-\beta_4)-b_{11}(1+\beta_4),\\
\lambda_2&=b_{16}(1+\beta_1-\beta_4)+b_{17}(1-\beta_1-\beta_4)+b_{18}(4-\beta_4)+b_{19}(2-\beta_4)-b_{20}\beta_4,\\
\lambda_3&=b_1(1+2\beta_1-\beta_4)\eta^{1+2\beta_1}+b_2(1-2\beta_1-\beta_4)\eta^{1-2\beta_1}+b_3(4+\beta_1-\beta_4)\eta^{4+\beta_1}+b_4\\
&\quad+(4-\beta_1-\beta_4)\eta^{4-\beta_1}+b_5(2+\beta_1-\beta_4)\eta^{2+\beta_1}+b_6(2-\beta_1-\beta_4)\eta^{2-\beta_1}\\
&\quad+b_7(7-\beta_4)\eta^7+b_8(5-\beta_4)\eta^5+b_9(3-\beta_4)\eta^3+b_{10}(1-\beta_4)\eta-b_{11}(1+\beta_4)\eta^{-1},\\
\lambda_4&=b_{16}(1+\beta_1-\beta_4)\eta^{1+\beta_1}+b_{17}(1-\beta_1-\beta_4)\eta^{1-\beta_1}\\
&\quad+b_{18}(4-\beta_4)\eta^4+b_{19}(2-\beta_4)\eta^2-b_{20}\beta_4,\\
\lambda_5&=[(\beta_2+\beta_4)\ln\eta-1]b_{13},\\
\lambda_6&=b_{13}-b_{12}(\beta_2-\beta_4)\ln\eta,\\
\lambda_7&=b_{12}+b_{13}(\beta_2+\beta_4)\ln\eta,\\
\lambda_8&=-[(\beta_2-\beta_4)\ln\eta+1]b_{12}.
\end{aligned}\tag{A.1}$$

Auxiliary coefficients for Eq. (3.17)

$$F_1(y)=c_{73}y^{2+\beta_1}+c_{74}y^{2-\beta_1}+c_{75}y^{1+\beta_2}+c_{76}y^{1-\beta_2}+c_{77}y^5+c_{78}y^3+c_{79}y,$$

$$\begin{aligned}
F_2(y)=&c_{45}y^{2+2\beta_1}+c_{46}y^{2-2\beta_1}+c_{47}y^{1+\beta_1+\beta_2}+c_{48}y^{1+\beta_1-\beta_2}+c_{49}y^{1-\beta_1+\beta_2}\\
&+c_{50}y^{1-\beta_1-\beta_2}+c_{51}y^{5+\beta_1}+c_{52}y^{5-\beta_1}+c_{53}y^{3+\beta_1}+c_{54}y^{3-\beta_1}+c_{55}y^{1+\beta_1}+c_{56}y^{1-\beta_1}\\
&+c_{57}y^{4+\beta_2}+c_{58}y^{4-\beta_2}+c_{59}y^{2+\beta_2}+c_{60}y^{2-\beta_2}+c_{61}y^8+c_{62}y^6+c_{63}y^4\\
&+c_{64}y^2+c_{65}+c_{66}y^{\beta_1}\ln y+c_{67}y^{-\beta_1}\ln y+c_{68}y^{\beta_1}+c_{69}y^{1+\beta_2}\ln y\\
&+c_{70}y^{1-\beta_2}\ln y+c_{71}y^{1+\beta_2}+c_{72}y^{1-\beta_2},
\end{aligned}$$

$$
\begin{aligned}
F_3(y) = & c_1 y^{2+3\beta_1} + c_2 y^{2-3\beta_1} + c_3 y^{5+2\beta_1} + c_4 y^{5-2\beta_1} + c_5 y^{3+2\beta_1} + c_6 y^{3-2\beta_1} \\
& + c_7 y^{1+2\beta_1} + c_8 y^{1-2\beta_1} + c_9 y^{1+\beta_1+\beta_2} + c_{10} y^{1+\beta_1-\beta_2} + c_{11} y^{1-\beta_1+\beta_2} + c_{12} y^{1-\beta_1-\beta_2} \\
& + c_{13} y^{8+\beta_1} + c_{14} y^{8-\beta_1} + c_{15} y^{6+\beta_1} + c_{16} y^{6-\beta_1} + c_{17} y^{4+\beta_1} + c_{18} y^{4-\beta_1} + c_{19} y^{2+\beta_1} \\
& + c_{20} y^{2-\beta_1} + c_{21} y^{\beta_1} \ln y + c_{22} y^{-\beta_1} \ln y + c_{23} y^{\beta_1} + c_{24} y^{4+\beta_2} + c_{25} y^{4-\beta_2} \\
& + c_{26} y^{2+\beta_2} + c_{27} y^{2-\beta_2} + c_{28} y^{11} + c_{29} y^9 + c_{30} y^7 + c_{31} y^5 + c_{32} y^3 + c_{33} y \\
& + c_{34} y^{-1} + c_{35} y^{1+\beta_1+\beta_2} \ln y + c_{36} y^{1+\beta_1-\beta_2} \ln y + c_{37} y^{1-\beta_1+\beta_2} \ln y + c_{38} y^{1-\beta_1-\beta_2} \ln y \\
& + c_{39} y^{4+\beta_2} \ln y + c_{40} y^{4-\beta_2} \ln y + c_{41} y^{2+\beta_2} \ln y + c_{42} y^{2-\beta_2} \ln y + c_{43} y^{\beta_1} \ln^2 y \\
& + c_{44} y^{-\beta_1} \ln^2 y,
\end{aligned}
$$

$$
e_i = a_0[F_i(1)\eta^{-\beta_1} - F_i(\eta)],
$$

$$
f_i = -a_0[F_i(1)\eta^{\beta_1} - F_i(\eta)], \quad i = 1,2,3,
$$

$$
c_1 = \frac{d_1}{4(1+\beta_1)(1+2\beta_1)}, \quad c_2 = \frac{d_2}{4(1-\beta_1)(1-2\beta_1)}, \quad c_3 = \frac{d_3}{(5+\beta_1)(5+3\beta_1)},
$$

$$
c_4 = \frac{d_4}{(5-\beta_1)(5-3\beta_1)}, \quad c_5 = \frac{d_5}{3(1+\beta_1)(3+\beta_1)}, \quad c_6 = \frac{d_6}{3(1-\beta_1)(3-\beta_1)},
$$

$$
c_7 = \frac{d_7}{(1+\beta_1)(1+3\beta_1)}, \quad c_8 = \frac{d_8}{(1-\beta_1)(1-3\beta_1)},
$$

$$
c_9 = \frac{d_9}{(1+\beta_2)(1+2\beta_1+\beta_2)} - \frac{2(1+\beta_1+\beta_2)d_{10}}{(1+\beta_2)^2(1+2\beta_1+\beta_2)^2},
$$

$$
c_{10} = \frac{d_{11}}{(1-\beta_2)(1+2\beta_1-\beta_2)} - \frac{2(1+\beta_1-\beta_2)d_{12}}{(1-\beta_2)^2(1+2\beta_1-\beta_2)^2},
$$

$$
c_{11} = \frac{d_{13}}{(1+\beta_2)(1-2\beta_1+\beta_2)} - \frac{2(1-\beta_1+\beta_2)d_{14}}{(1+\beta_2)^2(1-2\beta_1+\beta_2)^2},
$$

$$
c_{12} = \frac{d_{15}}{(1-\beta_2)(1-2\beta_1-\beta_2)} - \frac{2(1-\beta_1-\beta_2)d_{16}}{(1-\beta_2)^2(1-2\beta_1-\beta_2)^2}
$$

$$
c_{13} = \frac{d_{17}}{16(4+\beta_1)}, \quad c_{14} = \frac{d_{18}}{16(4-\beta_1)}, \quad c_{15} = \frac{d_{19}}{12(3+\beta_1)}, \quad c_{16} = \frac{d_{20}}{12(3-\beta_1)},
$$

$$
c_{17} = \frac{d_{21}}{8(2+\beta_1)}, \quad c_{18} = \frac{d_{22}}{8(2-\beta_1)}, \quad c_{19} = \frac{d_{23}}{4(1+\beta_1)}, \quad c_{20} = \frac{d_{24}}{4(1-\beta_1)},
$$

$$
c_{21} = \frac{1}{2\beta_1}\left(d_{25} - \frac{d_{26}}{2\beta_1}\right), \quad c_{22} = -\frac{1}{2\beta_1}\left(d_{27} + \frac{d_{28}}{2\beta_1}\right), \quad c_{23} = \frac{1}{4\beta_1^2}\left(\frac{d_{26}}{2\beta_1} - d_{25}\right),
$$

$$
c_{24} = \frac{d_{29}}{(4+\beta_2)^2 - \beta_1^2} - \frac{2(4+\beta_2)d_{30}}{[(4+\beta_2)^2 - \beta_1^2]^2}, \quad c_{25} = \frac{d_{31}}{(4-\beta_2)^2 - \beta_1^2} - \frac{2(4-\beta_2)d_{32}}{[(4-\beta_2)^2 - \beta_1^2]^2},
$$

$$
c_{26} = \frac{d_{33}}{(2+\beta_2)^2 - \beta_1^2} - \frac{2(2+\beta_2)d_{34}}{[(2+\beta_2)^2 - \beta_1^2]^2}, \quad c_{27} = \frac{d_{35}}{(2-\beta_2)^2 - \beta_1^2} - \frac{2(2-\beta_2)d_{36}}{[(2-\beta_2)^2 - \beta_1^2]^2},
$$

$$
c_{28} = \frac{d_{37}}{121 - \beta_1^2}, \quad c_{29} = \frac{d_{38}}{81 - \beta_1^2}, \quad c_{30} = \frac{d_{39}}{49 - \beta_1^2}, \quad c_{31} = \frac{d_{40}}{25 - \beta_1^2}, \quad c_{32} = \frac{d_{41}}{9 - \beta_1^2},
$$

$$
c_{33} = \frac{d_{42}}{1 - \beta_1^2}, \quad c_{34} = \frac{d_{43}}{1 - \beta_1^2}, \quad c_{35} = \frac{d_{10}}{(1+\beta_2)(1+\beta_2+2\beta_1)},
$$

$$c_{36}=\frac{d_{12}}{(1-\beta_2)(1-\beta_2+2\beta_1)},\quad c_{37}=\frac{d_{14}}{(1+\beta_2)(1+\beta_2-2\beta_1)},$$

$$c_{38}=\frac{d_{16}}{(1-\beta_2)(1-\beta_2-2\beta_1)},\quad c_{39}=\frac{d_{30}}{(4+\beta_2)^2-\beta_1^2},\quad c_{40}=\frac{d_{32}}{(4-\beta_2)^2-\beta_1^2},$$

$$c_{41}=\frac{d_{34}}{(2+\beta_2)^2-\beta_1^2},\quad c_{42}=\frac{d_{36}}{(2-\beta_2)^2-\beta_1^2},\quad c_{43}=\frac{d_{26}}{4\beta_1},\quad c_{44}=-\frac{d_{28}}{4\beta_1},$$

$$c_{45}=\frac{d_{44}}{(2+\beta_1)(3\beta_1+2)},\quad c_{46}=\frac{d_{45}}{(2-3\beta_1)(2-\beta_1)},\quad c_{47}=\frac{d_{46}}{(1+\beta_2)(1+\beta_2+2\beta_1)},$$

$$c_{48}=\frac{d_{47}}{(1-\beta_2)(1-\beta_2+2\beta_1)},\quad c_{49}=\frac{d_{48}}{(1+\beta_2)(1+\beta_2-2\beta_1)},$$

$$c_{50}=\frac{d_{49}}{(1-\beta_2)(1-\beta_2-2\beta_1)},\quad c_{51}=\frac{d_{50}}{5(5+2\beta_1)},\quad c_{52}=\frac{d_{51}}{5(5-2\beta_1)},$$

$$c_{53}=\frac{d_{52}}{3(3+2\beta_1)},\quad c_{54}=\frac{d_{53}}{3(3-2\beta_1)},\quad c_{55}=\frac{d_{54}}{1+2\beta_1},\quad c_{56}=\frac{d_{55}}{1-2\beta_1},$$

$$c_{57}=\frac{d_{56}}{(4+\beta_2)^2-\beta_1^2},\quad c_{58}=\frac{d_{57}}{(4-\beta_2)^2-\beta_1^2},\quad c_{59}=\frac{d_{58}}{(2+\beta_2)^2-\beta_1^2},$$

$$c_{60}=\frac{d_{59}}{(2-\beta_2)^2-\beta_1^2},\quad c_{61}=\frac{d_{60}}{64-\beta_1^2},\quad c_{62}=\frac{d_{61}}{36-\beta_1^2},\quad c_{63}=\frac{d_{62}}{16-\beta_1^2},\quad c_{64}=\frac{d_{63}}{4-\beta_1^2},$$

$$c_{65}=-\frac{d_{64}}{\beta_1^2},\quad c_{66}=\frac{d_{65}}{2\beta_1},\quad c_{67}=-\frac{d_{66}}{2\beta_1},\quad c_{68}=-\frac{d_{65}}{4\beta_1^2},\quad c_{69}=\frac{b_{12}}{(1+\beta_2)^2-\beta_1^2},$$

$$c_{70}=\frac{b_{13}}{(1-\beta_2)^2-\beta_1^2},\quad c_{71}=\frac{b_{14}}{(1+\beta_2)^2-\beta_1^2}-\frac{2(1+\beta_2)b_{12}}{[(1+\beta_2)^2-\beta_1^2]^2},$$

$$c_{72}=\frac{b_{15}}{(1-\beta_2)^2-\beta_1^2}-\frac{2(1-\beta_2)b_{13}}{[(1-\beta_2)^2-\beta_1^2]^2},\quad c_{73}=\frac{b_{16}}{4(1+\beta_1)},\quad c_{74}=\frac{b_{17}}{4(1-\beta_1)},$$

$$c_{75}=\frac{b_{21}}{(1+\beta_2)^2-\beta_1^2},\quad c_{76}=\frac{b_{22}}{(1-\beta_2)^2-\beta_1^2},\quad c_{77}=\frac{b_{18}}{25-\beta_1^2},\quad c_{78}=\frac{b_{19}}{9-\beta_1^2},$$

$$c_{79}=\frac{b_{20}}{1-\beta_1^2},$$

$$d_1=a_1b_1,\quad d_2=a_2b_2,\quad d_3=a_1b_3+a_3b_1,\quad d_4=a_2b_4+a_3b_2,$$

$$d_5=a_1b_5+b_1\left(a_4-\frac{1}{m}\right),\quad d_6=a_2b_6+b_2\left(a_4-\frac{1}{m}\right),$$

$$d_7=\frac{b_1}{m}\eta^2,\quad d_8=\frac{b_2}{m}\eta^2,\quad d_9=a_1b_{14},\quad d_{10}=a_1b_{12},$$

$$d_{11}=a_1b_{15},\quad d_{12}=a_1b_{13},\quad d_{13}=a_2b_{14},\quad d_{14}=a_2b_{12},$$

$$d_{15}=a_2b_{15},\quad d_{16}=a_2b_{13},\quad d_{17}=a_1b_7+a_3b_3,\quad d_{18}=a_3b_4+a_2b_7,$$

$$d_{19}=a_1b_8+a_3b_5+b_3\left(a_4-\frac{1}{m}\right),\quad d_{20}=a_2b_8+a_3b_6+b_4\left(a_4-\frac{1}{m}\right),$$

$$d_{21}=a_1b_9+a_4b_5+\frac{1}{m}(\eta^2b_3-b_5),\quad d_{22}=a_2b_9+a_4b_6+\frac{1}{m}(\eta^2b_4-b_6),$$

$$d_{23}=a_1b_{10}+a_2b_1+\frac{\eta^2}{m}b_5,\quad d_{24}=a_1b_2+a_2b_{10}+\frac{\eta^2}{m}b_6,$$

$$d_{25}=a_1b_{11}+\frac{\eta^2}{m}b_{14},\quad d_{26}=\frac{\eta^2}{m}b_{12},\quad d_{27}=a_2b_{11}+\frac{\eta^2}{m}b_{15},\quad d_{28}=\frac{\eta^2}{m}b_{13},$$

$$d_{29}=a_3b_{14},\quad d_{30}=a_3b_{12},\quad d_{31}=a_3b_{15},\quad d_{32}=a_3b_{13},$$

$$d_{33}=b_{14}\left(a_4-\frac{1}{m}\right),\quad d_{34}=b_{12}\left(a_4-\frac{1}{m}\right),\quad d_{35}=b_{15}\left(a_4-\frac{1}{m}\right),$$

$$d_{36}=b_{13}\left(a_4-\frac{1}{m}\right),\quad d_{37}=a_3b_7,\quad d_{38}=a_3b_8+b_7\left(a_4-\frac{1}{m}\right),$$

$$d_{39}=a_3b_9+a_4b_8+\frac{1}{m}(\eta^2b_7-b_8),\quad d_{40}=a_1b_4+a_2b_3+a_3b_{10}+a_4b_9+\frac{1}{m}(\eta^2b_8-b_9),$$

$$d_{41}=a_1b_6+a_2b_5+a_3b_{11}+a_4b_{10}+\frac{1}{m}(\eta^2b_9-b_{10}),\quad d_{42}=a_4b_{11}+\frac{1}{m}(\eta^2b_{10}-b_{11}),$$

$$d_{43}=\frac{\eta^2}{m}b_{11},\quad d_{44}=a_1b_{16}+\frac{b_1}{2},\quad d_{45}=a_2b_{17}+\frac{b_2}{2},\quad d_{46}=a_1b_{21},\quad d_{47}=a_1b_{22},$$

$$d_{48}=a_2b_{21},\quad d_{49}=a_2b_{22},\quad d_{50}=a_1b_{18}+a_3b_{16}+\frac{b_3}{2},\quad d_{51}=a_2b_{18}+a_3b_{17}+\frac{b_4}{2},$$

$$d_{52}=a_1b_{19}+b_{16}\left(a_4-\frac{1}{m}\right)+\frac{b_5}{2},\quad d_{53}=a_2b_{19}+b_{17}\left(a_4-\frac{1}{m}\right)+\frac{b_6}{2},$$

$$d_{54}=a_1b_{20}+\eta^2\frac{b_{16}}{m},\quad d_{55}=a_2b_{20}+\eta^2\frac{b_{17}}{m},\quad d_{56}=a_3b_{21},\quad d_{57}=a_3b_{22},$$

$$d_{58}=b_{21}\left(a_4-\frac{1}{m}\right),\quad d_{59}=b_{22}\left(a_4-\frac{1}{m}\right),\quad d_{60}=a_3b_{18}+\frac{b_7}{2},$$

$$d_{61}=a_3b_{19}+b_{18}\left(a_4-\frac{1}{m}\right)+\frac{b_8}{2},\quad d_{62}=a_3b_{20}+a_4b_{19}+\frac{1}{m}(\eta^2b_{18}-b_{19})+\frac{b_9}{2},$$

$$d_{63}=a_1b_{17}+a_2b_{16}+a_4b_{20}+\frac{1}{m}(\eta^2b_{19}-b_{20})+\frac{b_{10}}{2},\quad d_{64}=\frac{\eta^2}{m}b_{20}+\frac{b_{11}}{2},$$

$$d_{65}=\frac{\eta^2}{m}b_{21},\quad d_{66}=\frac{\eta^2}{m}b_{22}. \tag{A. 2}$$

Coefficients η_i are as follows

$$\eta_0=\eta_1^2\beta_3[G_1(1)-G_1(\eta)-G_4(1)+G_4(\eta)],$$

$$\eta_2=\eta_1^3\beta_3[G_2(1)-G_2(\eta)-G_5(1)+G_5(\eta)],$$

$$\eta_3=\frac{1}{2}\eta_1^4\beta_3[G_3(1)-G_3(\eta)-G_6(1)+G_6(\eta)],$$

$$G_1(y)=\frac{e_1y^{1+\beta_1}}{1+\beta_1}+\frac{f_1y^{1-\beta_1}}{1-\beta_1}+\frac{c_{73}y^{3+\beta_1}}{3+\beta_1}+\frac{c_{74}y^{3-\beta_1}}{3-\beta_1}+\frac{c_{75}y^{2+\beta_2}}{2+\beta_2}+\frac{c_{76}y^{2-\beta_2}}{2-\beta_2}+c_{77}\frac{y^6}{6}$$
$$+c_{78}\frac{y^4}{4}+c_{79}\frac{y^2}{2},$$

$$G_2(y)=\frac{e_2y^{1+\beta_1}}{1+\beta_1}+\frac{f_2y^{1-\beta_1}}{1-\beta_1}+c_{45}\frac{y^{3+2\beta_1}}{3+2\beta_1}+c_{46}\frac{y^{3-2\beta_1}}{3-2\beta_1}+c_{47}\frac{y^{2+\beta_2+\beta_1}}{2+\beta_1+\beta_2}$$

$$+c_{48}\frac{y^{2-\beta_2+\beta_1}}{2+\beta_1-\beta_2}+c_{49}\frac{y^{2+\beta_2-\beta_1}}{2-\beta_1+\beta_2}+c_{50}\frac{y^{2-\beta_2-\beta_1}}{2-\beta_1-\beta_2}+c_{51}\frac{y^{6+\beta_1}}{6+\beta_1}+c_{52}\frac{y^{6-\beta_1}}{6-\beta_1}$$

$$+c_{53}\frac{y^{4+\beta_1}}{4+\beta_1}+c_{54}\frac{y^{4-\beta_1}}{4-\beta_1}+c_{55}\frac{y^{2+\beta_1}}{2+\beta_1}+c_{56}\frac{y^{2-\beta_1}}{2-\beta_1}+c_{57}\frac{y^{5+\beta_2}}{5+\beta_2}+c_{58}\frac{y^{5-\beta_2}}{5-\beta_2}$$

$$+c_{59}\frac{y^{3+\beta_2}}{3+\beta_2}+c_{60}\frac{y^{3-\beta_2}}{3-\beta_2}+c_{61}\frac{y^9}{9}+c_{62}\frac{y^7}{7}+c_{63}\frac{y^5}{5}+c_{64}\frac{y^3}{3}+c_{65}y$$

$$+c_{66}\left(\frac{y^{1+\beta_1}\ln y}{1+\beta_1}-\frac{y^{1+\beta_1}}{(1+\beta_1)^2}\right)+c_{67}\left(\frac{y^{1-\beta_1}\ln y}{1-\beta_1}-\frac{y^{1-\beta_1}}{(1-\beta_1)^2}\right)$$

$$+c_{68}\frac{y^{1+\beta_1}}{1+\beta_1}+c_{69}\left(\frac{y^{2+\beta_2}\ln y}{2+\beta_2}-\frac{y^{2+\beta_2}}{(2+\beta_2)^2}\right)+c_{70}\left(\frac{y^{2-\beta_2}\ln y}{2-\beta_2}-\frac{y^{2-\beta_2}}{(2-\beta_2)^2}\right)$$

$$+c_{71}\frac{y^{2+\beta_2}}{2+\beta_2}+c_{72}\frac{y^{2-\beta_2}}{2-\beta_2},$$

$$G_3(y)=\frac{e_3y^{1+\beta_1}}{1+\beta_1}+\frac{f_3y^{1-\beta_1}}{1-\beta_1}+c_1\frac{y^{3+3\beta_1}}{3+3\beta_1}+c_2\frac{y^{3-3\beta_1}}{3-3\beta_1}+c_3\frac{y^{6+2\beta_1}}{6+2\beta_1}+c_4\frac{y^{6-2\beta_1}}{6-2\beta_1}$$

$$+c_5\frac{y^{4+2\beta_1}}{4+2\beta_1}+c_6\frac{y^{4-2\beta_1}}{4-2\beta_1}+c_7\frac{y^{2+2\beta_1}}{2+2\beta_1}+c_8\frac{y^{2-2\beta_1}}{2-2\beta_1}+c_9\frac{y^{2+\beta_1+\beta_2}}{2+\beta_1+\beta_2}$$

$$+c_{10}\frac{y^{2+\beta_1-\beta_2}}{2+\beta_1-\beta_2}+c_{11}\frac{y^{2-\beta_1+\beta_2}}{2-\beta_1+\beta_2}+c_{12}\frac{y^{2-\beta_1-\beta_2}}{2-\beta_1-\beta_2}+c_{13}\frac{y^{9+\beta_1}}{9+\beta_1}+c_{14}\frac{y^{9-\beta_1}}{9-\beta_1}$$

$$+c_{15}\frac{y^{7+\beta_1}}{7+\beta_1}+c_{16}\frac{y^{7-\beta_1}}{7-\beta_1}+c_{17}\frac{y^{5+\beta_1}}{5+\beta_1}+c_{18}\frac{y^{5-\beta_1}}{5-\beta_1}+c_{19}\frac{y^{3+\beta_1}}{3+\beta_1}+c_{20}\frac{y^{3-\beta_1}}{3-\beta_1}$$

$$+c_{21}\left(\frac{y^{1+\beta_1}\ln y}{1+\beta_1}-\frac{y^{1+\beta_1}}{(1+\beta_1)^2}\right)+c_{22}\left(\frac{y^{1-\beta_1}\ln y}{1-\beta_1}-\frac{y^{1-\beta_1}}{(1-\beta_1)^2}\right)+c_{23}\frac{y^{1+\beta_1}}{1+\beta_1}$$

$$+c_{24}\frac{y^{5+\beta_2}}{5+\beta_2}+c_{25}\frac{y^{5-\beta_2}}{5-\beta_2}+c_{26}\frac{y^{3+\beta_2}}{3+\beta_2}+c_{27}\frac{y^{3-\beta_2}}{3-\beta_2}+c_{28}\frac{y^{12}}{12}+c_{29}\frac{y^{10}}{10}+c_{30}\frac{y^8}{8}$$

$$+c_{31}\frac{y^6}{6}+c_{32}\frac{y^4}{4}+c_{33}\frac{y^2}{2}+c_{34}\ln y+c_{35}\left(\frac{y^{2+\beta_1+\beta_2}\ln y}{2+\beta_1+\beta_2}-\frac{y^{2+\beta_1+\beta_2}}{(2+\beta_1+\beta_2)^2}\right)$$

$$+c_{36}\left(\frac{y^{2+\beta_1-\beta_2}\ln y}{2+\beta_1-\beta_2}-\frac{y^{2+\beta_1-\beta_2}}{(2+\beta_1-\beta_2)^2}\right)+c_{37}\left(\frac{y^{2-\beta_1+\beta_2}\ln y}{2-\beta_1+\beta_2}-\frac{y^{2-\beta_1+\beta_2}}{(2-\beta_1+\beta_2)^2}\right)$$

$$+c_{38}\left(\frac{y^{2-\beta_1-\beta_2}\ln y}{2-\beta_1-\beta_2}-\frac{y^{2-\beta_1-\beta_2}}{(2-\beta_1-\beta_2)^2}\right)+c_{39}\left(\frac{y^{5+\beta_2}\ln y}{5+\beta_2}-\frac{y^{5+\beta_2}}{(5+\beta_2)^2}\right)$$

$$+c_{40}\left(\frac{y^{5-\beta_2}\ln y}{5-\beta_2}-\frac{y^{5-\beta_2}}{(5-\beta_2)^2}\right)+c_{41}\left(\frac{y^{3+\beta_2}\ln y}{3+\beta_2}-\frac{y^{3+\beta_2}}{(3+\beta_2)^2}\right)$$

$$+c_{42}\left(\frac{y^{3-\beta_2}\ln y}{3-\beta_2}-\frac{y^{3-\beta_2}}{(3-\beta_2)^2}\right)+c_{43}\left(\frac{y^{1+\beta_1}\ln^2 y}{1+\beta_1}-2\frac{y^{1+\beta_1}\ln y}{(1+\beta_1)^2}+2\frac{y^{1+\beta_1}}{(1+\beta_1)^3}\right)$$

$$+c_{44}\left(\frac{y^{1-\beta_1}\ln^2 y}{1-\beta_1}-2\frac{y^{1-\beta_1}\ln y}{(1-\beta_1)^2}+2\frac{y^{1-\beta_1}}{(1-\beta_1)^3}\right),$$

$$G_4(y)=\frac{1}{m}\left(\frac{b_{16}y^{1+\beta_1}}{1+\beta_1}+\frac{b_{17}y^{1-\beta_1}}{1-\beta_1}+\frac{b_{18}y^4}{4}+\frac{b_{19}y^2}{2}+b_{20}\ln y+b_{21}\frac{y^{\beta_2}}{\beta_2}-b_{22}\frac{y^{-\beta_2}}{\beta_2}\right),$$

$$\begin{aligned}G_5(y)=\frac{1}{m}\Big\{&\frac{d_{44}y^{1+2\beta_1}}{1+2\beta_1}+\frac{d_{45}y^{1-2\beta_1}}{1-2\beta_1}+\frac{d_{46}y^{\beta_1+\beta_2}}{\beta_1+\beta_2}-\frac{d_{49}y^{-\beta_1-\beta_2}}{\beta_1+\beta_2}+(d_{47}+d_{48})\ln y\\&+\frac{d_{50}y^{4+\beta_1}}{4+\beta_1}+\frac{d_{51}y^{4-\beta_1}}{4-\beta_1}+\frac{d_{52}y^{2+\beta_1}}{2+\beta_1}+\frac{d_{53}y^{2-\beta_1}}{2-\beta_1}+\frac{d_{54}y^{\beta_1}}{\beta_1}-\frac{d_{55}y^{-\beta_1}}{\beta_1}+\frac{d_{56}y^{3+\beta_2}}{3+\beta_2}\\&+\frac{d_{57}y^{3-\beta_2}}{3-\beta_2}+\frac{d_{58}y^{1+\beta_2}}{1+\beta_2}+\frac{d_{59}y^{1-\beta_2}}{1-\beta_2}+\frac{d_{60}y^7}{7}+\frac{d_{61}y^5}{5}+\frac{d_{62}y^3}{3}\\&+d_{63}y-d_{64}y^{-1}-\frac{d_{65}y^{-1+\beta_2}}{1-\beta_2}-\frac{d_{66}y^{-1-\beta_2}}{1+\beta_2}+\frac{b_{14}\beta_2-b_{12}}{\beta_2^2}\\&+y^{\beta_2}-\frac{b_{15}\beta_2+b_{13}}{\beta_2^2}y^{-\beta_2}+\frac{b_{12}\ln y}{\beta_2}y^{\beta_2}-\frac{b_{13}\ln y}{\beta_2}y^{-\beta_2}\Big\},\end{aligned}$$

$$\begin{aligned}G_6(y)=\frac{1}{m}\Big\{&\frac{d_1y^{1+3\beta_1}}{1+3\beta_1}+\frac{d_2y^{1-3\beta_1}}{1-3\beta_1}+\frac{d_3y^{4+2\beta_1}}{2(2+\beta_1)}+\frac{d_4y^{4-2\beta_1}}{2(2-\beta_1)}+\frac{d_5y^{2+2\beta_1}}{2(1+\beta_1)}+\frac{d_6y^{2-2\beta_1}}{2(1-\beta_1)}\\&+\frac{d_7y^{2\beta_1}}{2\beta_1}-\frac{d_8y^{-2\beta_1}}{2\beta_1}+\frac{d_9(\beta_1+\beta_2)-d_{10}}{(\beta_1+\beta_2)^2}y^{\beta_1+\beta_2}-\frac{d_{15}(\beta_1+\beta_2)+d_{16}}{(\beta_1+\beta_2)^2}y^{-\beta_1-\beta_2}\\&+\frac{d_{17}y^{7+\beta_1}}{7+\beta_1}+\frac{d_{18}y^{7-\beta_1}}{7-\beta_1}+\frac{d_{19}y^{5+\beta_1}}{5+\beta_1}+\frac{d_{20}y^{5-\beta_1}}{5-\beta_1}+\frac{d_{21}y^{3+\beta_1}}{3+\beta_1}+\frac{d_{22}y^{3-\beta_1}}{3-\beta_1}+\frac{d_{23}y^{1+\beta_1}}{1+\beta_1}\\&+\frac{d_{24}y^{1-\beta_1}}{1-\beta_1}+\frac{a_1b_{11}}{-1+\beta_1}y^{-1+\beta_1}-\frac{a_2b_{11}}{1+\beta_1}y^{-1-\beta_1}+\frac{d_{29}(3+\beta_2)-d_{30}}{(3+\beta_2)^2}y^{3+\beta_2}\\&+\frac{d_{31}(3-\beta_2)-d_{32}}{(3-\beta_2)^2}y^{3-\beta_2}+\frac{d_{33}(1+\beta_2)-d_{34}}{(1+\beta_2)^2}y^{1+\beta_2}+\frac{d_{35}(1-\beta_2)-d_{36}}{(1-\beta_2)^2}y^{1-\beta_2}\\&+\frac{\eta^2m^{-1}b_{14}(\beta_2-1)-d_{26}}{(\beta_2-1)^2}y^{-1+\beta_2}-\frac{\eta^2m^{-1}b_{15}(\beta_2+1)+d_{28}}{(\beta_2+1)^2}y^{-1-\beta_2}+\frac{d_{37}}{10}y^{10}\\&+\frac{d_{38}}{8}y^8+\frac{d_{39}}{6}y^6+\frac{d_{40}}{4}y^4+\frac{d_{41}}{2}y^2-\frac{d_{43}}{2}y^{-2}+(d_{11}+d_{13}+d_{42})\ln y+\frac{d_{34}y^{1+\beta_2}\ln y}{1+\beta_2}\\&-\frac{d_{28}y^{-1-\beta_2}\ln y}{1+\beta_2}+\frac{d_{36}y^{1-\beta_2}\ln y}{1-\beta_2}-\frac{d_{26}y^{\beta_2-1}\ln y}{1-\beta_2}+\frac{d_{10}y^{\beta_1+\beta_2}\ln y}{\beta_1+\beta_2}-\frac{d_{16}y^{-\beta_1-\beta_2}\ln y}{\beta_1+\beta_2}\\&+(d_{12}+d_{14})\frac{\ln^2 y}{2}+\frac{d_{30}}{3+\beta_2}y^{3+\beta_2}\ln y+\frac{d_{32}}{3-\beta_2}y^{3-\beta_2}\ln y\Big\}.\end{aligned}\qquad (A.3)$$

References

An N., Wang F. and Liu R. H. (2001): Nonlinear bending of symmetrically laminated cylindrically orthotropic shallow spherical shells under linear load. —Journal of Jinan University, vol. 27, No. 3, pp. 398-405 (in Chinese).

Liu R. H. (1965): Nonlinear stability of circular shallow spherical shell with a hole in the center under the action of uniform moment at the inner edge. —Bulletin of Sciences, No. 3, pp. 253-255 (in Chinese).

Liu R. H. (1991): On the nonlinear stability of symmetrically laminated cylindrically orthotropic shallow spherical shells. —Applied Mathematics and Mechanics, vol. 12, No. 3, pp. 271-279.

Liu R. H. (1992): Nonlinear stability of symmetrically laminated cylindrically orthotropic shallow spherical shells including transverse shear. —Science in China, Series A, vol. 35, No. 6, pp. 734-746.

Liu R. H. (1993): Large deflection equations of symmetrically laminated composite cylindrically orthotropic shallow spherical shells. — Applied Mathematics and Mechanics, Wei-Zang Chien Eightieth Anniversary Volume (Yeh K. Y. , Ed.), Beijing and Chongqing: Science Press and Chongqing Publishing House, pp. 355-361.

Liu R. H. (1996): On non-linear buckling of symmetrically laminated, cylindrically orthotropic, truncated shallow, spherical shells under uniform pressure including shear effects. — International Journal of Non-Linear Mechanics. , vol. 31, No. 1, pp. 101-115.

Liu R. H. (1996): Non-linear buckling of symmetrically laminated, cylindrically orthotropic, shallow, conical considering shear. — International Journal of Non-Linear Mechanics. , vol. 31, No. 1, pp. 89-99.

Liu R. H. (1996): Nonlinear stability of laminated composite shallow conical shells. —Proceedings of the Ninth National Conference of Composite Materials (Zhang Z. M. , Ed.), Beijing: Publishing Company of World Books, pp. 249-255 (in Chinese).

Liu R. H. (1998): Study on NonLinear Mechanics of Plates and Shells. —New York, Beijing Guangzhou: Science Press and Jinan University Press.

Liu R. H. and Wang F. (1997): Nonlinear forced vibration of laminated composite shallow spherical shells. —Acta Mechanica Sinica, vol. 29, No. 2, pp. 236-241 (in Chinese).

Liu R. H. and Wang F. (1998): Nonlinear dynamic buckling of symmetrically cylindrically orthotropic shallow spherical shells. —Archive of Applied Mechanics, No. 68, pp. 375-384.

Liu R. H. and Zhong C. (1994): Nonlinear buckling of symmetrically laminated cylindrically orthotropic shallow spherical shells with a circular hole at the center including transverse shear. —Journal of Jinan University, vol. 15, No. 1, pp. 1-12 (in Chinese).

Wang F. and Liu R. H. (2001): Nonlinear dynamic buckling of symmetrically laminated truncated shallow spherical shell. —Acta Mechanica Sinica, vol. 22, No. 3, pp. 309-314 (in Chinese).

Wang F. , Xiao T. and Liu R. H. (2000): The effect of damping on the nonlinear vibration of a laminated composite truncated shallow spherical shell. — Research and Practice in Applied Mechanics (Liu R. H. Ed.), Guangzhou: Jinan University Press, pp. 35-43 (in Chinese).

Yeh K. Y. , Liu R. H. , Pin Q. Y. and Li S. L. (1965a): Nonlinear stability of thin circular shallow spherical shell under actions of axisymmetric uniform distribution line loads. — Bulletin of Sciences, No. 2, pp. 142-145 (in Chinese).

Yeh K. Y. , Liu R. H. , Zhang C. Z. and Xu U. F. (1965): Nonlinear stability of thin circular shallow spherical shell under the actions of uniform moment. —Bulletin of Sciences, No. 2, pp. 145-147 (in Chinese).

Zhu Y. A., Wang F. and Liu R. H. (2008): Thermal buckling of axisymmetrically laminated cylindrically orthotropic shallow spherical shells including transverse shear. —Applied Mathematics and Mechanics, vol. 29, No. 3, pp. 291-300.

复合材料层合板壳非线性力学的研究进展①

一、引言

随着世界各国在航天、航空、石油化工、机械工程、海洋工程、交通运输、能源领域的不断发展，传统的单一材料已不能满足日益增加的对材料高强度、高刚度、高断裂韧性和低比重等综合优良性能的要求，寻求先进复合材料已成为材料技术发展过程中的关键。我国政府对先进材料的研究和开发大力支持，在“863 计划”与“973 计划”中，先进材料技术分别为其中的 7 个与 6 个重点领域之一。近期实施的重大基础研究攀登计划中，30 个重大课题有 7 个与材料有直接关系。国家自然科学基金委资助的研究课题，与材料有关的约占 1/4。综上所述，先进材料技术已成为我国当前经济建设中的一项关键技术，因此针对先进复合材料广泛开展的研究工作具有重要的意义。

关于复合材料，广义上来讲，它是由两种或者两种以上具有不同性质的材料，经过物理或化学的方法复合而形成的一种具有新性能的多相固体材料。层合板壳结构则是将多层复合材料的单层板壳粘合在一起组成的结构体系。由于复合材料层合板壳是由多种组分材料组合而成的，与单一材料的板壳结构相比，无明确的材料主方向，各层间材料间断和不连续，具有明显的几何非线性和材料非线性等新的特点。其失效也具有如基体开裂、脱胶、分层、分层裂纹偏转、多分层以及分层传播等多种模式，因此需要建立与层合板壳相适应的几何、本构模型来描述层合板壳的结构整体性能和力学性能。本文将从结构非线性力学的角度对复合材料层合板壳的研究情况进行回顾和综述。

二、复合材料层合板壳非线性理论

在板壳弯曲理论中[1,2]，当层合板受到一般载荷作用时，总可以把每一个载荷分解为两个分载荷：一个是作用在结构中面之内的纵向载荷，另一个是垂直于中面的横向载荷。纵向载荷和横向载荷所引起的应力、应变和位移分别按平面应力问题和板壳弯曲问题进行计算。层合板的平衡微分方程为

$$\left.\begin{aligned}&\frac{\partial N_x}{\partial x}+\frac{\partial N_{xy}}{\partial y}=0\\&\frac{\partial N_{xy}}{\partial x}+\frac{\partial N_y}{\partial y}=0\\&\frac{\partial^2 M_x}{\partial x^2}+2\frac{\partial^2 M_{xy}}{\partial x\partial y}+\frac{\partial^2 M_y}{\partial y^2}+N_x\frac{\partial^2 w}{\partial x^2}+2N_{xy}\frac{\partial^2 w}{\partial x\partial y}+N_y\frac{\partial^2 w}{\partial y^2}+q=0\end{aligned}\right\}\tag{1}$$

其中，M_x，M_y，M_{xy} 为层合板横向的弯曲内力，N_x，N_y，N_{xy} 为作用于中面内的薄膜内力，q 为横向载荷。

① 本文原载《力学学报》，2017，49 (3)：487-506. 作者：刘人怀，薛江红.

根据层合板的应力应变关系和有关假设[3]，层合板的本构方程为

$$\left.\begin{aligned}\begin{bmatrix}N_x\\N_y\\N_{xy}\end{bmatrix}&=\begin{bmatrix}A_{11}&A_{12}&A_{16}\\A_{12}&A_{22}&A_{26}\\A_{16}&A_{26}&A_{66}\end{bmatrix}\begin{bmatrix}\varepsilon_x\\\varepsilon_y\\\varepsilon_{xy}\end{bmatrix}+\begin{bmatrix}B_{11}&B_{12}&B_{16}\\B_{12}&B_{22}&B_{26}\\B_{16}&B_{26}&B_{66}\end{bmatrix}\begin{bmatrix}\chi_x\\\chi_y\\\chi_{xy}\end{bmatrix}\\\begin{bmatrix}M_x\\M_y\\M_{xy}\end{bmatrix}&=\begin{bmatrix}B_{11}&B_{12}&B_{16}\\B_{12}&B_{22}&B_{26}\\B_{16}&B_{26}&B_{66}\end{bmatrix}\begin{bmatrix}\varepsilon_x\\\varepsilon_y\\\varepsilon_{xy}\end{bmatrix}+\begin{bmatrix}D_{11}&D_{12}&D_{16}\\D_{12}&D_{22}&D_{26}\\D_{16}&D_{26}&D_{66}\end{bmatrix}\begin{bmatrix}\chi_x\\\chi_y\\\chi_{xy}\end{bmatrix}\end{aligned}\right\}\tag{2}$$

其中，ε_x，ε_y，ε_{xy}是层合板的中面应变，χ_x，χ_y，χ_{xy}是其中面曲率、扭曲率，A_{ij}为层合板的抗拉刚度，表示的是薄膜内力与中面应变的刚度关系，D_{ij}为弯曲刚度，表示的是弯曲内力矩与曲率及扭曲率有关的刚度系数；B_{ij}为耦合刚度，表示弯曲、拉压之间的耦合关系。若令E_1，E_2，μ_{12}，μ_{21}为正交各向异性单层板主方向的材料常数，$Q_{ij}^{(k)}$和θ分别为第k层的弹性刚度和该层坐标系与材料主方向的夹角，则有

$$\left.\begin{aligned}A_{ij}&=\sum_{k=1}^{n}Q_{ij}^{(k)}(z_k-z_{k-1})\\B_{ij}&=\frac{1}{2}\sum_{k=1}^{n}Q_{ij}^{(k)}(z_k^2-z_{k-1}^2)\\D_{ij}&=\frac{1}{3}\sum_{k=1}^{n}Q_{ij}^{(k)}(z_k^3-z_{k-1}^3)\end{aligned}\right\}\tag{3}$$

$$\begin{bmatrix}Q_{11}^{(k)}\\Q_{12}^{(k)}\\Q_{22}^{(k)}\\Q_{16}^{(k)}\\Q_{26}^{(k)}\\Q_{66}^{(k)}\end{bmatrix}=\begin{bmatrix}c^4&2c^2s^2&s^4&4c^2s^2\\c^2s^2&c^4+s^4&c^2s^2&-4c^2s^2\\s^4&2c^2s^2&c^4&4c^2s^2\\c^3s&cs^3-c^3s&-cs^3&-2cs(c^2-s^2)\\cs^3&c^3s-cs^3&-c^3s&2cs(c^2-s^2)\\c^2s^2&-2c^2s^2&c^2s^2&(c^2-s^2)^2\end{bmatrix}\begin{bmatrix}Q_{11}\\Q_{12}\\Q_{22}\\Q_{66}\end{bmatrix}\tag{4}$$

$$\left.\begin{aligned}&Q_{11}=\frac{E_1}{1-\mu_{12}\mu_{21}},Q_{22}=\frac{E_2}{1-\mu_{12}\mu_{21}}\\&Q_{12}=\mu_{12}Q_{22}=\mu_{21}Q_{11},Q_{66}=G_{12}\\&E_2=\mu_{21}E_1,c=\cos\theta,s=\sin\theta\end{aligned}\right\}\tag{5}$$

在纵向载荷和横向载荷的作用下，若板内产生的应变量和转动分量都远远小于1，且应变量与转动分量为等量级的微小量时，由线性理论知

$$\left.\begin{aligned}&\varepsilon_x=\frac{\partial u}{\partial x},\varepsilon_y=\frac{\partial v}{\partial x},\varepsilon_{xy}=\frac{1}{2}\left(\frac{\partial u}{\partial y}+\frac{\partial v}{\partial x}\right)\\&\chi_x=-\frac{\partial^2 w}{\partial x^2},\chi_y=-\frac{\partial^2 w}{\partial y^2},\chi_{xy}=-\frac{\partial^2 w}{\partial x\partial y}\end{aligned}\right\}\tag{6}$$

当作用在复合材料层合板上的外载荷较大，所引起的面外变形与厚度之比不再是小量时，面内应变就必须考虑，此时薄膜力在弯曲平衡时起了很大的作用，形成几何非线性问题。1910年，von Kármán[4]首先给出了平板大挠度非线性方程，此方程中考虑了应变分量的二阶小项及薄膜应力在变形后位形上对平衡方程的影响，从此奠定了

板壳几何非线性分析的基础[5,6]。现有的复合材料层合板壳的非线性理论是从上述理论发展起来的，可分为以下 5 种：经典大挠度弯曲理论，一阶剪切变形理论，高阶剪切变形理论，锯齿理论和广义分层理论。

1. 经典大挠度弯曲理论

经典大挠度弯曲理论是依据基尔霍夫（Kirchhoff）直法线假设，忽略了横向剪切变形而建立的。根据 Kirchhoff 假定，变形前垂直于中面的直线在变形后仍然垂直于中面，在层合板内距中面为 z 的点，其位移沿厚度做线性变化，即

$$\left.\begin{aligned} u^{(z)} &= u(x,y) - z\frac{\partial w}{\partial x} \\ v^{(z)} &= v(x,y) - z\frac{\partial w}{\partial y} \\ w^{(z)} &= w(x,y) \end{aligned}\right\} \tag{7}$$

根据基尔霍夫的应变假设在板内任一点处有 $\varepsilon_{zx}=\varepsilon_{zy}=\varepsilon_z=0$。由上式得结构内任意点的应变表达式

$$\left.\begin{aligned} \varepsilon_x^{(z)} &= \varepsilon_x - z\frac{\partial^2 w}{\partial x^2} \\ \varepsilon_y^{(z)} &= \varepsilon_y - z\frac{\partial^2 w}{\partial y^2} \\ \gamma_{xy}^{(z)} &= \gamma_{xy} - z\frac{\partial^2 w}{\partial x\partial y} \end{aligned}\right\} \tag{8}$$

其中，ε_{ij} 为板中面的应变。

在小应变（ε_x，ε_y，$\varepsilon_{xy}\ll 1$）、大转动的非线性问题中，由于板的挠度 $w\gg u$，v，根据中等大挠度的 von Kármán 非线性几何公式，层合板的中面应变、曲率改变量的表达式为

$$\left.\begin{aligned} \varepsilon_x &= \frac{\partial u}{\partial x} + \frac{1}{2}\left(\frac{\partial w}{\partial x}\right)^2, & \chi_x &= -\frac{\partial^2 w}{\partial x^2} \\ \varepsilon_y &= \frac{\partial v}{\partial x} + \frac{1}{2}\left(\frac{\partial w}{\partial y}\right)^2, & \chi_y &= -\frac{\partial^2 w}{\partial y^2} \\ \varepsilon_{xy} &= \frac{\partial u}{\partial y} + \frac{\partial v}{\partial x} + \frac{\partial w}{\partial x}\frac{\partial w}{\partial y}, & \chi_{xy} &= -\frac{\partial^2 w}{\partial x\partial y} \end{aligned}\right\} \tag{9}$$

在 20 世纪 60 年代中后期 Stavsky[7]，Whitney 和 Leissa[8] 最先建立了基于 von Kármán 非线性几何公式的适用于各向异性材料的非线性理论。Chia 等[9-17] 采用摄动法和广义多重 Fourier 级数方法，对斜交层合板、非对称各向异性层合板在不同边界条件下的非线性弯曲问题进行了研究。然而使用幂级数解法和其他方法去处理层合壳体非线性问题，异常困难，解的精确度常常不能满足要求。为此，20 世纪 90 年代初，叶开沅和刘人怀等在其前期板壳非线性研究成果基础上[18-39]，刘人怀又创造性地将修正迭代法引入到层合结构的非线性问题分析中，率先研究了层合矩形板的非线性弯曲[40]，以及极难求解的层合扁球壳和层合扁锥壳的非线性稳定[41,42] 等问题，解决了复合材料层合板壳领域的一大难题。他们不仅求解了非线性控制微分方程，获得了临界载荷的解析解，而且他们的解法结合了摄动法和逐次逼近法的优点，不仅程序简单、计算量

小，而且收敛快、所获解析解的精度高。

此外，Turvey 和 Wittrick[43]采用动力松弛法研究了简支斜交层合板的大挠度和后屈曲问题。Zaghloul 和 Kennedy[44]对对称正交或斜交层合板在四边夹紧固定或简支两种边界条件下的大挠度问题进行了实验研究，并给出了相应的有限差分解。Chandra[45]使用伽辽金法分析了一个四边固定反对称层合板的非线性弯曲问题。周次青[46-48]采用奇异摄动法讨论了多种边界条件下各向异性层合板的非线性弯曲问题，导出了挠度和应力函数的 N 阶一致有效渐近展开式。

2. 一阶剪切变形理论

层合板/壳的经典大挠度弯曲理论是在经典层合板理论基础上，考虑了层合板中面的薄膜变形以及面外的横向大挠度对薄膜变形的影响。但经典层合板壳理论却忽略了横向剪切变形，只考虑了横向剪力的作用。由于层合板/壳是由很多各向异性的叠片按一定规律堆砌而成的，其等效横向剪切模量远小于面内纤维方向的等效弹性模量，因此层合板对横向剪切变形很敏感。基于此考虑产生了一阶剪切变形理论。在一阶变形理论中，位移场取为

$$\left.\begin{aligned} u^{(z)} &= u_0(x,y) - z\psi_x \\ v^{(z)} &= v_0(x,y) - z\psi_y \\ w^{(z)} &= w_0(x,y) \end{aligned}\right\} \tag{10}$$

其中，u_0，v_0 和 w_0 为板中面的变形，ψ_x 和 ψ_y 为板内任一点处法线对板中面的转角。

一阶剪切变形理论基于 Reissner-Mindlin 中厚板理论[49,50]，考虑了横向剪切变形的影响，最早由 Stavsky[51]针对各向同性层合板提出，之后 Yang 等[52]将该理论推广到各向异性板，在 20 世纪七八十年代被 Ambartsumyan[53]，Bert[54]和 Whitney[55]引入到各向异性板和层合板中。Dong 和 Tso[56]建立了层合圆柱壳的一阶剪切变形理论。之后，Reddy 和他的合作者[57-59]将 von Kármán 非线性几何方程引入到 Reissner-Mindlin 剪切变形理论中，考虑了小应变、大转动的几何非线性，建立了各向异性板考虑中面薄膜影响的非线性一阶剪切变形理论，并基于该理论应用有限元算法求解问题[60,61]。由于模型简单，变量少，理论值较符合实验值，一阶剪切变形理论得到了广泛的应用。Iu 等[62-66]研究了层合圆柱壳、圆锥壳的后屈曲和非线性振动，Fares 等[67-70]运用 Reissner 的混合变分原理，提出了一种考虑热效应的精细一阶剪切变形理论，使得应力在上下边界处得以连续。Whitney 和 Pagano[71]分析了对称和反对称铺设的正交及斜交矩形板受正弦分布载荷的弯曲与自由振动。Fortier 和 Rossettos[72]以及 Sinha 和 Rath[73]分别对非对称斜交铺设的矩形厚板的自由振动和屈曲进行了研究。Bert 和 Chen[74]给出了反对称斜交铺设的简支矩形板自由振动的近似解。

在层合板壳计及一阶剪切变形非线性理论领域中，刘人怀等创造性地将一阶剪切变形理论引入到各种复合材料层合薄壁板壳结构中，结合修正迭代法，率先涉及复杂的复合材料层合旋转壳体领域，建立了计及横向剪切影响的对称层合扁球壳和扁锥壳的非线性稳定和振动理论，系统地分析了对称层合圆柱正交异性扁球壳的非线性稳定问题[75]、非线性强迫振动[76]与动态屈曲[77]、层合圆板的大挠度弯曲[78]与受迫振动[79]、

椭圆板的大挠度弯曲[80]、开顶扁球壳的非线性稳定性[81-82]、非线性强迫振动[83]和动态屈曲[84]、扁锥壳的非线性屈曲[85]。这些问题的研究结果可直接用于工程设计，刘人怀等得出了重要的结论：若使用不计及横向剪切影响的经典理论进行工程设计，那将是保守的。同时刘人怀和王志伟还率先研究了考虑面层和芯层的横向剪切变形的复合材料面层夹层板壳的非线性一致有效理论[86-87]，给出了复合材料面层夹层矩形板的非线性弯曲[88]和夹层扁壳的非线性弯曲与屈曲的解答[88-90]。接着，徐加初等[91]研究了具有夹层的正交异性复合材料圆锥壳体的非线性轴对称屈曲问题。

由于一阶剪切变形理论假设恒定的横向剪切应力，不满足在自由边界无面力的边界条件。为了克服这个问题，提出了各种解决的办法。一种办法是引入一个剪切校正因子，文献［92-97］给出了几种情况下的剪切校正因子；另一种办法是对一阶剪切变形理论进行细化。

3. 高阶剪切变形理论

一阶剪切变形理论在预测层合结构的整体响应，即挠度、固有频率和屈曲载荷方面提供了良好的结果，但在研究沿厚度方向的应力响应方面，一阶剪切变形理论给出的结果都远远偏离了结构的实际情况。为了克服一阶剪切形变理论的局限，出现了各种高阶剪切变形理论。在二阶和三阶变形理论中，位移分量被展开成层合板厚度的高阶多项式（二次和三次），其中，以 Reddy[98,99]在 1984 年提出的三阶剪切理论影响最为广泛。在 Reddy 的三阶剪切变形理论中包含了更多的位移函数，其位移表达式具有如下的形式

$$\left.\begin{aligned} u(x,y,z) &= u_0(x,y) + z\psi_x(x,y) + z^2\phi_x(x,y) + z^3\theta_x(x,y) \\ v(x,y,z) &= v_0(x,y) + z\psi_y(x,y) + z^2\phi_y(x,y) + z^3\theta_y(x,y) \\ w(x,y) &= w_0(x,y) \end{aligned}\right\} \tag{11}$$

其中，u_0，v_0，w_0，ψ_x，ψ_y，ϕ_x，ϕ_y，θ_x，θ_y 是坐标（x，y）的未知函数，且 u_0，v_0，w_0，ψ_x，ψ_y 的含义与一阶变形理论一致。在此基础上，Reddy[100]进一步引入 von Kármán 非线性几何方程，建立了非线性的高阶剪切形变理论，Reddy 和 Liu[101]建立了层合壳的高级剪切变形理论。根据 Reddy 理论得到的横向剪切应变沿厚度方向呈抛物线分布，因此该理论不需要剪切修正，被大量用来计算层合板壳的挠度、自然频率和屈曲载荷[102-109]。除了 Reddy 的理论，Kant 和 Swaminathan[110-112]建立了考虑横向应变的三阶理论，其中面内位移和横向位移都为横向坐标的三次多项式，并分析了复合材料层合板/夹层板自由振动与弯曲问题。Barut 等[113]发展了三阶理论分析多层夹层板弯曲问题，其中面内位移为三次多项式而横向位移为二次多项式。

与国际研究同步，国内也开展对高阶剪切变形理论的研究。Reissner-Mindlin 型板理论不能满足在板的上、下表面横向剪切应力为零的条件，而三维理论用于分析实际问题相当难以操作。为此，Liu 和 He[114]应用虚位移原理，创新地提出了一个计及高阶影响的复合材料层合板的非线性弯曲问题的简化理论。通过将板的总挠度分解为一个由弯曲而产生的挠度和一个由剪切而产生的挠度，进而建立了较为简单的控制方程。他们的理论首次说明了横向剪切应变在厚度上的抛物线变化的原因，并使得控制方程

的求解更简单，所得到的结果与 Reddy 的三维理论[115]所得的结果相当一致。沈惠申运用 Reddy 的高阶剪切理论研究了复合材料层合板热后屈曲[116]和复合材料层合剪切圆柱曲板在侧压作用下的后屈曲[117]。冯世宁和陈浩然[118]基于 Reddy 提出的层合板高阶剪切变形简化理论对复合材料层合板的非线性动力稳定性问题进行了研究。张雨和向锦武[119]对复合材料层合厚圆柱壳高阶理论进行改进，提出了一个改进的精化高阶理论。文献［120-127］给出了其他的二阶和三阶的剪切变形理论。

Matsunaga[124-129]在研究二阶和三阶理论的基础上提出了九阶理论，Reissner[130]给出了十二阶理论。在上述的各种高阶理论中，都是将位移函数展成厚度坐标的多项式系列。Ambartsumian 在文献［131］中综述了各种高阶剪切理论，并首次提出横向剪切应变形状函数。此外，还提出了一些新的高阶理论，在这些高阶理论中横向剪切应变形状函数不是取为多项式函数，而是取为其他的函数，如 Kaczkowski[132]，Pane[133]和 Reissner[49]，Levinson[134]，Reddy 和 Liu[101]等的理论。在 Ghugal 和 Sayyad[135]提出的一种高阶剪切理论里，其剪切应变形状函数为三角函数，El Meiche 等[136]引入一个双曲剪切变形函数作为剪切变形函数，Matsunaga[137]将位移场展成厚度坐标的幂级数，Karama 等[138-140]和 Aydogdu[141,142]运用了指数函数，Neves 等[143]采用了双曲函数和正弦函数。

4. 锯齿理论

现有的大多数层合板理论都建立在位移假设基础上，这要归因于基于位移的理论在发展层合板计算模型时表现出的简单性。按照位移假设方式的不同，这些理论可以归结为两类：一类是将非均匀的各向异性层合板等效为均匀各向异性单层板，并将板的位移沿整个厚度假设成厚度方向坐标的连续光滑函数。这方面的理论有基于面内位移沿厚度线性分布假设的一阶理论，以及基于面内位移沿厚度非线性分布假设的高阶理论。显然，由于这类理论中应变在层合板层间连续，因而横向剪应力在层间间断。另一类理论则采用逐层假设位移的方法，预先满足层间位移和横向剪切应力连续的条件，即锯齿理论。

锯齿理论最早由 Murakami[144]提出。为了模拟面内位移沿厚度方向呈之字形的分布情况，Murakami 在面内位移场上添加一个锯齿函数。之后，通过在勒让德多项式基础上添加一个锯齿函数，Murakami 和 Toledano[145]提出了一个改进的锯齿理论，并据此分析了层合板的圆柱弯曲问题。锯齿理论比单纯提高位移场阶数更为有效，不仅计算精度比高阶理论高，而且位移的未知量数目也比高阶理论少，因此被用来分析层合板壳的静动力性能。Carrera[146]将 Murakami 的锯齿函数（Murakami zig-zag function，MZZF）引入一阶和高阶剪切位移场，详细研究了简支层合板壳的静力、动力和热载荷响应，并分析了 MZZF 的有效性和局限性。Bhaskar 和 Varadan[147]运用 MZZF 分析了正交层合旋转壳的弯曲问题，并建立了基于 Murakami 的锯齿函数的高阶理论。Vidal 和 Polit[148]把 MZZF 添加到正弦位移场，并研究了复合材料层合梁弯曲问题。Ferreira 等[149]和 Rodrigues 等[150]提出了一种变异的 MZZF，并运用径向基函数配置法研究了层合板的静力问题、自由振动和屈曲问题。文献［151-155］介绍了基于 Murakami 的锯齿理论的各种有限元分析。

Murakami 的锯齿理论虽然满足层间位移连续的条件（C^0 连续），但不满足层间横向剪切应力的连续条件，因此不能用来确定层间的应力分布。为了克服这一问题，Di Sciuva 等[156-158]在假设层合板面内位移沿厚度呈折线分布，即面内位移沿厚向呈锯齿形分布的基础上，提出了预先满足层间应力连续的锯齿理论，并进一步推广此理论，用之分析复合材料板壳。然而，Di Sciuva 的理论中横向剪应力在厚度方向上是均匀的，不满足上下表面处的横向剪应力协调的条件。1992 年，Cho 等[159-161]在之字形的面内线性位移基础上叠加一个三次的位移场，提出了适于分析复合材料层合结构的锯齿理论。在此锯齿理论中，层合板壳结构中第 k 层任一点的位移场可以统一地表示为

$$\left.\begin{aligned} u^k(x,y,z) &= u_0(x,y) + z\psi_x(x,y) + z^2\varphi_x(x,y) + z^3\theta_x(x,y) \\ &\quad + \sum_{j=1}^{k-1} S_x^j(z-z_j)H(z-z_j) \\ v^k(x,y,z) &= v_0(x,y) + z\psi_y(x,y) + z^2\varphi_y(x,y) + z^3\theta_y(x,y) \\ &\quad + \sum_{j=1}^{k-1} S_y^j(z-z_j)H(z-z_j) \\ w^k(x,y,z) &= w_0(x,y) + zw_1(x,y) + z^2 w_2(x,y) \end{aligned}\right\} \tag{12}$$

式中，S_x^j（$z-z_j$）和 S_y^j（$z-z_i$）为第 j 层的位移增量，H（$z-z_i$）为阶跃函数。由层间横向剪应力的协调性和表面剪应力为零的条件，可以建立 S_x^j 与 S_x^{j-1}，S_y^j 与 S_y^{j-1} 的递推关系，并推导出 ψ_x，ψ_y 与 w_0 以及 ϕ_x，ϕ_y 与 w_1，w_2 关系表达式。

Cho 的锯齿理论适合于分析层合板壳由纯机械载荷引起的弯曲、屈曲及动力性能。对于受机械载荷与热载荷共同作用下的层合板/壳结构的静动力问题，横向的正应变就不能忽略，因为在此情况下，面外的热变形与面内热变形几乎相当[162]，为此，Cho 等[163,164]提出了考虑横向正应变的三阶锯齿理论。该理论被用来分析含多个分层的层合壳的弹性屈曲，以及黏弹性层合板的力学性能[165]和在不同边界情况下层合板的弯曲性能[166]。之后，Icardi[167,168]采用面内位移和面外位移沿厚度分别呈三次和四次分布，提出了适合厚壳的修正锯齿理论。

与此同时，锯齿理论在国内的研究也取得了令人瞩目的成果。1994 年，何陵辉和刘人怀等[169]发展了一种高阶锯齿理论，开创了锯齿理论在国内研究的新局面。国内外曾有过许多层合复合材料板壳的力学模型，但这些模型或基于分层的运动学假设，使未知量数目随层数急剧增加，因而计算极为复杂；或引入整个厚度上的光滑位移假设，虽使未知量数目与层数无关却不能满足层间应力协调条件。他们提出了逐层光滑的连续位移表达式，建立了任意形状层合壳体的线性和非线性一般理论，不但使表面和层间应力协调条件自动满足，而且未知量数目与层数无关（5 个）。通过一些特例的计算表明，他们提出模型的预言结果与相关精确解无论在局部还是在整体力学响应上都十分吻合。因此，在兼顾计算的简单性和结果的精确性方面，他们提出的模型优于国内外同类工作。之后，舒小平[170]提出了一种简单高阶剪切变形层板理论，其位移场满足层间位移和横向剪应力的连续性，并将该理论推广用于分析层合壳的热弹性动力响应[171]。白瑞祥和陈浩然[172]运用 zig-zag 模型和 Mindlin 一阶剪切变形板理论，考虑面

板的横向剪切变形和芯体的面内刚度对夹层板力学性能的影响，建立了一种分层模型和多标量损伤模型。任晓辉和陈万吉[173]发展了 C^0 型锯齿理论，通过虚位移原理推导出在热载荷作用下复合材料梁的平衡方程，并给出了简支复合材料层合梁解析解。

锯齿理论虽然能使层间的横向剪应力连续，但依据该理论的有限元模型需要满足位移的 C^1 连续条件。而要在任意四边形和三角形单元的接口处施加 C^1 连续条件极难实现，且其单元刚度矩阵的计算极其耗时。为了获得准确的层间应力还需要借助三维的平衡方程[174-178]。其次锯齿理论中没有考虑横向正应变的影响，不能精确地确定在材料间断处，如孔边、开孔处、上下界面、脱层尖端等处的应力情况。锯齿理论的另一个限制是横向剪切应力在层合板的边缘处不能满足自由边界条件。

5. 广义分层理论

对于较厚的层合梁、板和壳等结构，其层间剪切和拉、压应力呈三维应力状态，采用等效单层板的板壳理论分析，已不能满足精度要求。为此，Reddy[179]提出了广义分层理论。该理论采取在板的厚度方向取线性插值函数来描述每个数值层内位移沿厚度方向的变化规律。在线性理论基础上，Reddy 等[180,181]又考虑 von Kármán 几何关系，建立了非线性的分层理论，并提出了考虑脱层的层合板分层理论[182]，其位移场取为

$$\left.\begin{aligned} u(x,y,z) &= u_0(x,y) + \sum_{i=1}^{N} U_i(x,y)\phi_x^i(z) \\ v(x,y,z) &= v_0(x,y) + \sum_{i=1}^{N} V_i(x,y)\phi_y^i(z) \\ w(x,y,z) &= w_0(x,y) + \sum_{i=1}^{N} W_i(x,y)\phi_z^i(z) \end{aligned}\right\} \tag{13}$$

其中，u_0，v_0，w_0 为板中面的变形，N 为沿厚度方向层合板的子板层数，U_i，V_i，W_i 为第 i 个子层的位移分量，ϕ_x^i，ϕ_y^i，ϕ_z^i 是厚度坐标的分段连续插值函数，由相邻的两个子板决定。分层理论将层合板沿厚度方向分成若干子层，每个子层视为一个由相同材料构成的单层板，可由多个铺设层构成，总的子层板数可以小于层合板的铺设层数。位移在厚度方向是 C^0 连续的，但其导数在层间不连续，通过运用三维的本构方程保证了横向剪切应力在层间连续，并且上下表面的协调性也得到满足。

由于 Reddy 的分层理论中，面内位移为沿厚度线性分布的，因此所得的剪应力和剪应变在各层内为均匀分布，而非抛物线分布。为了克服这一问题，出现了各种高阶分层理论，即是在线性的分层位移基础上再叠加高阶位移场。Di Sciuva 和 Gherlone[183]基于 Hermite 分层位移场，构造了一种满足 C^1 连续的，含贯穿厚度损伤的有限元梁模型，并将该模型推广到含损伤的层合板[184]，建立了满足 C^0 连续的八节点板单元[185]。Pai 和 Palazotto[186]提出了一种高阶分层理论，在他们的理论中，每一层面内位移沿厚度呈二次和三次分布，而面外位移沿厚度呈二次分布 Carrera[187-189]发展了一种预先满足层间位移，横向剪应力和正应力连续条件的混合分层理论，并建立了有关的有限元模型。在其混合分层理论中，每一层的位移场和应力场被展成 N 阶的勒让德多项式。而 Carrera 和 Demasi[190]以及 Icardi[191]则运用混合分层理论分析了夹层板的力学性能。

由于分层理论能够描述层合板壳内每一层的独自变形情况，因此被广泛用于分析层合结构的静动力问题。Marjanovic 和 Vuksanovic[192]进一步改进了分层模型，以考虑层合板壳内脱层的影响，并运用改进的模型计算含脱层层合板的自然频率、振型和屈曲载荷。Alnefaie[193]运用分层有限元模型，对含层间损伤层合板的基本动力特性进行了分析。Basar 等[194,195]开发了一系列的多层壳单元模型来计算层间应力，并用分层理论的壳模型来研究层合结构的自由振动响应。为了降低计算成本，Botello 等[196]采用三角形单元导出基于分层理论的有限元模型，并提出了一种子结构技术以消除刚度矩阵装配过程中的面内自由度。Ghoshal 等[197,198]运用一种精细的分层理论分析了脱层层合板的线性和非线性振动，在他们的分析中考虑了脱层处的接触效应。更多关于分层理论的成果，可参阅文献［199-204］。

分层理论是一种准三维理论，其特点是未知变量个数依赖于层合板层数，因此，当层合板层数增加时，其计算量将大大增加。其次，分层理论不能预先满足层间应力连续条件，需要采用三维平衡后处理方法来计算层间应力。

总的来说，层合板壳层间界面的材料不连续性会导致层合板壳内部位移和应力分布的复杂性。文献［205-211］综述了最近几十年剪切变形理论的发展情况。

三、复合材料层合板壳非线性力学性能

对纤维增强复合材料层合板壳结构在弹性变形条件下的非线性力学分析，现有的研究已经由对单一结构、简单加载等问题的研究转为了对复合结构、在复杂加载环境下的性能研究，由线性材料推广到考虑断裂、蠕变等非线性材料性质，由完善结构到含缺陷的不完善结构。

1. 典型复合材料板壳结构的失效机理及优化设计

刘人怀与合作者[5,6,40-42,75-91,169]最早在我国开展复合材料层合板壳结构的一系列非线性力学问题研究。他们的研究涉及工程结构中最重要的 5 类板壳：层合圆板、层合矩形板、层合椭圆板、层合扁球壳和层合扁锥壳，同时还研究了复合材料夹层板壳，针对这些结构在典型载荷作用下的非线性弯曲、稳定和振动问题，进行了系统的分析和求解，获得了精确的解析解，他们的工作为开展复合材料层合板壳非线性理论问题的继续研究奠定了坚实的理论基础。

刘人怀等[212-214]研究了更复杂的尚未有人处理的问题，即考虑横向剪切的对称层合圆柱正交异性复合材料截头扁锥壳和线布载荷作用下的扁球壳的非线性稳定问题，用修正迭代法获得了十分精确的解析解。同时，还对复合材料浅球壳进行了优化设计。杨增涛等[215]进一步研究了三角脉冲作用下对称铺设正交异性复合材料层合扁锥壳的轴对称冲击屈曲，深入讨论了壳体几何参数、物理参数和边界条件对复合材料层合扁锥壳冲击屈曲的影响，并首次将 B-R 准则应用到集中冲击载荷和爆炸冲击载荷作用下的层合扁球壳和层合扁锥壳的非线性动力屈曲问题的研究中[216,217]。他们指出，当冲击载荷幅值达到某一值时，壳体的位移明显急剧增大，显示壳体发生了屈曲，此时对应的载荷即为临界屈曲载荷。他们的研究结果表明应用 B-R 准则确定壳体冲击动力屈曲临界载荷是非常有效的。王璠[218]考虑横向剪切、阻尼、强迫载荷对幅频特性曲线的影

响，研究了对称层合圆柱正交异性开孔扁球壳的非线性动态响应。研究表明，振幅随着阻尼的增大而减小，随着强迫载荷和剪切刚度的增大而增大。王志伟等[219-221]应用试验和有限元模拟方法分析了蜂窝纸板的动态冲击压缩过程，研究了层数、湿度和厚跨比对蜂窝纸板的冲击承载和能量吸收能力的影响，并首次开展了多次低强度冲击对蜂窝纸板缓冲性能影响的分析。研究结果表明，随着低强度冲击次数的增加，蜂窝纸板吸能呈上升趋势，蜂窝纸板剩余结构仍能承受冲击作用吸收能量。Song 等[222]对机织碳/环氧复合材料在动态压缩载荷作用下的力学行为和失效模式进行了试验分析。Fan 等[223,224]提出了整体夹芯连续纤维复合材料结构动态压缩失效模式和机理。

复合材料不仅给设计者提供了高比弹性模量和高比强度的材料，还可以让设计人员在一定范围内对这些性能参数进行改变，以使结构设计和材料使用最优化。何陵辉等[225]对含非均匀界面相碳/碳纤维复合材料的微结构特性和相应的物理模型进行了详细描述，导出了其等效热传导性质的一些基本关系，得到了等效热传导系数的解析公式，解决了复合材料一个经典难题。同时，又建立了球面各向同性球粒复合材料的基体均匀场，给出了复合材料等效膨胀系数与体积模量的精确公式[226]。而且，还建立了空心球增强复合材料的基体均匀场，得到了复合材料局部热弹性场之间以及等效热弹性性质之间的一些精确关系[227]。王璠等[228]通过制备不同铺层角度和不同引发模式的玻璃纤维/聚酯树脂基圆柱壳，开展了铺层对纤维增强复合材料圆柱壳吸能特性影响的冲击试验研究。郑伟玲等[229,230]分别介绍了碳纳米管增强复合材料中碳纳米管的制备方法和石墨烯与多壁碳纳米管增强环氧树脂复合材料的制备方法，分析了石墨烯与多壁碳纳米管的协同作用、两者的含量以及多壁碳纳米管功能化方法对复合材料力学和热学性能的影响。通过将改性碳纳米管按不同质量分数分别加入聚氨酯中，他们研究了碳纳米管/聚氨酯复合材料的制备，讨论了碳纳米管对复合材料性能的影响[231]。姜久红等[232,233]对蜂窝层合纸板的力学性能进行了研究。通过静态压缩试验，他们获得了蜂窝层合纸板平压性能，以及芯纸的克重与平压力学性能的关系，讨论了非线性系统中的阻尼比和阻尼系数对产品加速度及位移响应的影响；针对蜂窝纸板在中低应变率下的缓冲性能，他们的研究表明中应变率下蜂窝纸板平台应力和吸能特性随着应变率增大而提高。应变率对蜂窝纸板的力学性能具有重要影响。

2. 复合材料板壳结构在复杂环境下的破坏机理

当复合材料处于复杂环境下时，如高温、高压、磁电、湿热、氧化腐蚀和应力耦合，这些环境因素通过不同的机制作用于复合材料，使复合材料的树脂基体、增强纤维以及树脂/纤维粘接界面破坏而引起性能下降，状态改变，直至损坏变质。

在湿度和温度的协同作用下，复合材料会产生湿热老化现象，使其形态、质量、力学性能等指标发生改变。宁志华等[234,235]首次将复变函数的分区全纯函数理论、Cauchy 型积分与 Riemann 边值问题结合，研究了含非均匀界面层纤维增强复合材料受径向约束的热弹性问题。基于广义平面应变的假设，考虑材料参数呈线性变化的非均匀界面层，利用迭代法求解了均匀温变及径向约束下界面层、纤维及基体的位移及应力场。他们揭示了径向应力约束下各相的位移及应力分布规律，以及界面层的不同体积分数对位移及应力分布的影响，解决了线性温变下含界面层圆形夹杂的温度场难题，所发展的分析方法为求解复杂多连通域的平面热弹性问题提供了一条全新的途径。除了非均匀界面层纤维增强复合材料的热弹性分析，刘述伦等[236,237]又开展了正交各向异性复合材料层合板的力学性能受温度和湿度影响的研究，这是一个十分困难的理论问题，而且又有实际意义价值，他们考虑湿-热-力的多场效应，运用宏微观力学模型，分析了考虑含损伤层合板在湿热环境下的分层屈曲问题。他们的研究展示了当前复合材料非线性问题研究的新特点——多场、多尺度、跨学科，指出了该领域前沿问题的研究新趋势。同时，朱永安等[238,239]运用修正迭代法首次研究了计及横向剪切的对称层合扁球壳的非线性热屈曲问题。E 等[240,241]对不同湿度下蜂窝纸板的缓冲性能进行了评估。他们发现当相对湿度为 40%～75%时，对蜂窝纸板的缓冲性能无明显影响；当相对湿度超过 75%时，蜂窝纸板的缓冲性能和平台应力快速下降。

复合材料板壳结构的压电、磁电性能也是当前受到广泛关注的问题之一。陈炎等[242-245]率先研究了压电矩形薄板的非线性自由振动、强迫振动和动态屈曲问题，又建立了位移场和电场高阶分布的压电板的非线性理论。Wang 等[246,247]从压电线性理论出发，研究了两种结构在共振频率附近的非线性动力学特征以及在共振频率附近的电-力-电多场非线性耦合作用。傅衣铭等[248,249]研究了具界面损伤压电智能层合板的非线性自由振动。文献［250～254］对受磁-电-机械载荷作用的复合材料层合结构进行了研究。

除了湿热环境和电磁场的影响，高压腐蚀的环境会导致复合材料性能的急剧下降，引起复合材料的结构失稳。Xue 等[255]研究了深海保温夹层输油管道的结构失稳现象，以及腐蚀对管道结构承载能力的影响[256-258]。

3. 复合材料板壳结构的物理非线性特性

一般而言，在工程问题中有两种非线性——几何的和物理的。由于新材料的应用及结构向轻型化发展，先进复合材料板壳结构往往同时具有几何非线性和材料非线性特性。

纤维增强复合材料已被广泛应用于混凝土结构的加固中，此时复合材料和混凝土材料通过黏结构成层合结构，界面性能对结构破坏往往起到决定性作用[259-263]。然而，现有的工作主要涉及最终载荷和有效黏结长度的预测，极少注意到整个剥离过程的研究，因而在很大程度上限制了对界面剥离传播规律和破坏机理的认识。袁鸿等[264-267]基于刚性软化、双线性和三线性的内聚力模型，研究了 FRP 和混凝土界面的剥离破坏。他们采用线弹性断裂力学和非线性断裂力学理论，提出了一种简便有效的界面能量释放率的计算方法，首次给出了能够预言脱层传播全过程的解析解，建立了与 4 种非线性界面局部剪应力位移关系对应的应力、有效黏结长度、载荷位移关系的解析表达式，阐述了复合材料承载力、载荷位移关系和界面剥离传播之间的关系，提出了界面断裂力学破坏准则。在此基础上，他们又进一步开展了单向/双向粘贴纤维复合材料抵抗剥落的直梁/曲梁/平板试件的理论和实验研究[268-270]，分析了不同黏结长度对载荷位移曲线、极限载荷的影响，首次从理论上搞清楚了两条相邻裂纹之间的纤维复合材料补强的混凝土结构的界面性质。他们的研究为分析全过程载荷位移行为打下了严格和完备的理论基础，对于深入理解复合材料增强结构开裂后的界面黏结层破坏，更好地理解黏结接头的力学特性进而发展有效的复合材料锚固体系十分有益。

由于高温工作环境以及基体材料具有黏性性质，材料力学性能表现出与时间相依的特征。因此，在对由纤维增强复合材料构成的结构进行动力分析时，需要考虑材料的蠕变特征。向红等[271,272]率先分析了垂直与倾斜壁厚不等的非正六边形不规则蜂窝材料在面内单向应力作用下的蠕变屈曲行为。Wang 等[273]研究了层合圆柱曲板的蠕变失稳，得到了确定蠕变失稳临界载荷的解析方法。

4. 含脱层纤维增强复合材料板壳结构的破坏机理

脱层是复合材料层合板壳结构的一种常见的缺陷或损伤形式。根据不完全统计，应用于航空航天器的复合材料结构发生破坏时，有接近 60%的直接原因是脱层。所以，探究脱层产生的机理、含脱层板壳结构的剩余承载能力，对当前复合材料的科学应用有非凡的意义。

对复合材料板壳结构分层失效破坏机理的研究，是当前非线性力学领域的一个研究热点，受到研究人员的广泛关注。Xue 等[274]将脱层板分成 4 个区域，在冯·卡门非线性理论基础上，建立了复合材料脱层板的屈曲控制方程。通过各子板连接处位移和力的连续性建立了连续条件。求得满足边界条件和连续性条件的控制方程的解，并讨论了脱层长度和脱层深度等因素对临界屈曲载荷和屈曲模态的影响。由于含脱层的纤维增强复合材料板承受面内和面外载荷作用时，会在脱层界面上产生接触现象，引起接触力。如果忽略接触效应可能导致层合子板相互之间的互相贯穿。因此，Xue 等[275]又开展了含脱层层合板在面外载荷作用下的接触问题的研究。通过分析脱层界面处的变形机制，运用复合材料微观力学，他们建立了脱层界面处上下子板变形挠度与接触力之间的关系，并采用摄动法求解了受面外载荷作用的含脱层层合板的大挠度弯曲问题，首次获得了接触效应的解析解，解决了脱层板子板之间的接触效应，包括接触力的大小、接触区域的大小和位置等难以确定的难题。他们首次发现，位于分层界面的上下两个子板的挠度由两部分组成：完全一致的整体弯曲挠度和相反的局部微小屈曲模态；接触区域的位置和大小，由脱层所在的位置和大小决定，而与层合板的长宽比无关。据此，他们进一步研究了考虑接触影响的含脱层层合板受面内载荷作用的屈曲问题[276]，并与不考虑接触影响的情况进行对比[274]。他们的研究结果表明，在分析含脱层的复合材料层合板的屈曲问题时，是否考虑接触效应会对结果产生显著的区别。考虑接触效应的层合板的分层屈曲压力远高于不考虑接触效应的层合板的分层屈曲压力。最近，刘述伦等[237]又研究了含脱层复合材料层合矩形板在湿热环境下的屈曲问题。此外，Zhu 等[277]提出了一种压电层合圆柱壳的非线性接触效应分析模型，考虑了非线性接触效应对能量释放率和分层增长率的影响，同时他们[248,278-280]研究脱层的层合圆板的非线性动力响应、脱层层合圆柱壳的屈曲和振动等问题。Chen 等[281,282]用一阶剪切变形理论，Schwarts-Givli 等[283]和 Oh 等[174]用高阶剪切变形理论，Wang 等[284]用非线性反贯穿模型，以及 Kwon 等[285]用考虑接触冲击条件的有限元模拟等方法分别研究了接触对分层复合材料结构的动力性能的影响。文献［286-292］给出了更多关于含脱层复合材料层合板壳的非线性力学研究成果。

四、研究展望

在详细介绍过去复合材料层合板壳非线性力学研究进展的基础上，本文对今后的研究方向提出了如下的展望。

1. 等效单层板壳理论与多层板壳理论相结合的理论分析法

目前，对复合材料层合板壳的分析理论有两个方向：等效的单层板壳理论和多层板壳理论。等效的单层板壳理论将层合板壳视为一个具有复杂的本构关系的单层板壳，从最基本的 Kirchhoff 理论开始，在位移场上叠加沿厚度的高阶展开项。虽然位移场的阶数越高，得到的解的精度也越高，但各高阶项所对应的力学属性却难以确定。事实上，目前尚不清楚三阶以上各项的力学描述。而大量的研究结果表明，用等效单层板壳理论得到的层合板壳的整体性能分析与用三维连续介质理论得到的结果及实验结果有极高的吻合性。多层板壳理论则将层合板壳化为多个子层，对每个子层独立进行研究分析。研究表明，多层板壳理论能准确描述层合板壳的局部力学性能，但子层间位移和应力连续的必要条件导致随着子层数的增加计算量大大增加。

在层合板壳理论未来的研究发展中，上述两个方向将共存，而一个具体方法的使用取决于微观或宏观层面的研究，甚而两个方向的结合。

2. 含损伤复合材料板壳结构在复杂环境下的失效分析

对含损伤复合材料层合板壳结构在外载作用下的失效机理研究将导致多学科之间的交叉渗透。如分层复合材料在湿、热、机械力共同作用下的板壳结构整体稳定性问题，需要同时考虑弹性稳定性问题、接触问题和断裂问题。而湿热环境对复合材料层合板壳结构力学性能的影响又具有多场耦合的效果。深入分析在复杂环境中不完善层合板壳结构的失效机理，具有多尺度、多场耦合、跨学科的特点，不仅可在复合材料层合板壳结构力学的分析理论上取得突破进展，而且对复合材料在实际工程中的应用具有重要的指导意义。

3. 不完善复合材料层合板壳结构的可靠性与优化设计

复合材料具有良好的材料性能可设计性、制备的灵活性和易加工性。设计者可以通过改变组成材料的种类、含量和铺设方向，来达到在结构设计中对材料刚度、强度、稳定性和方向性的要求。但复合材料的力学指标受组分材料、界面性质、层合结构、载荷、环境等多种因素的影响，并含有许多不确定因素，而且含缺陷的不完善复合材料的失效具有多种模式，如基体开裂、脱胶、分层、分层裂纹偏转、多分层以及分层传播等，当存在多种失效模式时，需要建立针对各种破坏机理的失效模型，确定其主要失效模式以及建立相应的可靠性指标对其进行评估。由于问题的复杂性，目前的研究报道有限，尚需进一步研究。

参 考 文 献

[1]　刘人怀. 板壳力学. 北京：机械工业出版社，1990.

[2]　徐芝纶. 弹性力学（下册）. 4 版. 北京：高等教育出版社，2006.

[3]　沈观林，胡更开. 复合材料力学. 北京：清华大学出版社，2006.

[4]　von Kármán T V. Festigkeits Problem in Maschinenban，Enzyk-lopädie der Mathematjschen Wissenschaften，Bd. IV.，art. 27，1910.

[5]　Liu RH. Study on Nonlinear Mechanics of Plates and Shells. New York，Beijing，Guangzhou：Science Press and Jinan University Press，1998.

[6]　刘人怀. 复合材料层合板壳理论探索. 广州：暨南大学出版社，2006.

[7]　Stavsky Y. On the general theory of heterogeneous aeolotropic plates. Aeronautical Quarterly，1964，15：29.

[8]　Whitney JM，Leissa AW. Analysis of heterogeneous anisotropic plates. ASME Journal of Applied Mechanics，1969，36 (2)：261-266.

[9]　Chia CY. Nonlinear Analysis of Plates. New York：McGraw-Hill，1980.

[10]　Chia CY. Large deflections of heterogeneous anisotropic rectangular plates. International Journal of Solids and Structures，1974，10 (9)：965-976.

[11]　Chia CY，Prabhakara MK. Large deflection of unsymmetric crossply and angle-ply plates. ARCHIVE Journal of Mechanical Engineering Science，1976，18 (4)：179-183.

[12]　Prabhakafa MK，Chia CY. Finite deflections of unsymmetrically layered anisotropic rectangular plates subjected to the combined action of transverse and in-plane loads. ASME Journal of Applied Mechanics，1975，42 (2)：517-518.

[13]　Prabhakara MK，Chia CY. Nonlinear analysis of laminated crossply plates. Journal of the Engineering Mechanics Division，1977，103 (4)：749-753.

[14]　Prabhakara MK，Chia CY. Non-linear flexural vibrations of orthotropic rectangular plates. Journal of Sound and Vibration，1977，52 (4)：511-518.

[15]　Prabhakara MK. Finite deflection of unsymmetric angle-ply anisotropic rectangular plates under edge moments. ASME Journal of Applied Mechanics，1977，44 (1)：171-172.

[16]　Chia CY. Nonlinear bending of unsymmetric angleply plates with edges elastically restrained against rotation. Acta Mechanica，1984，53 (3-4)：201-212.

[17]　Chia CY. Large deflection of unsymmetric laminates with mixed boundary conditions. International Journal of Non-Linear Mechanics，1985，20 (4)：273-282.

[18]　叶开沅，刘人怀，平庆元，等. 在对称线布载荷作用下的圆底扁薄球壳的非线性稳定问题. 科学通报，1965，16 (2)：142-145.

[19]　叶开沅，刘人怀，张传智，等. 圆底扁薄球壳在边缘力矩作用下的非线性稳定问题. 科学通报，1965，16 (2)：145-147.

[20]　刘人怀. 在内边缘均布力矩作用下中心开孔圆底扁球壳的非线性稳定问题. 科学通报，1965，16 (3)：253-255.

[21]　刘人怀. 在内边缘均布载荷作用下中心开孔圆底扁薄球壳的轴对称稳定性. 力学学报，1977，13 (3)：206-212.

[22]　刘人怀. 波纹圆板的特征关系式. 力学学报，1978，14 (1)：47-52.

[23]　叶开沅，刘人怀，张传智，等. 弹性圆底扁球壳在边缘均布力矩作用下的非线性稳定问题. 应用数学和力学，1980，1 (1)：71-87.

[24]　刘人怀. 夹层圆板的非线性弯曲. 应用数学和力学，1981，2 (2)：173-190.

[25]　Liu RH. Non-linear thermal stability of bimetallic shallow shells of revolution. International Journal of Non-Linear Mechanics，1983，18 (5)：409-429.

[26]　Liu RH. Large deflection of corrugated circular plate with a plane central region under the action

of concentrated loads at the center. International Journal of Non-Linear Mechanics，1984，19 (5)：409-419.

[27] 刘人怀. 波纹环形板的非线性弯曲. 中国科学（A 辑），1984，(2)：247-253.

[28] Liu RH. Large deflection of corrugated circular plate with plane boundary region. Solid Mechanics Archives，1984，9 (3)：383-406.

[29] 刘人怀. 在复杂载荷作用下波纹环形板的非线性弯曲. 中国科学（A 辑），1985 (6)：537-545.

[30] Liu RH. Nonlinear bending of circular Sandwich plates under the action of axisymmetric uniformly distributed line loads. Progress in Applied Mechanics，Dordrecht：Martinns Nijhoff Publishers，1987，293-321.

[31] 刘人怀，成振强. 集中载荷作用下开顶扁球壳的非线性稳定问题. 应用数学和力学，1988，9 (2)：95-106.

[32] 刘人怀，李东. 均布载荷作用下开顶扁球壳的非线性稳定问题. 应用数学和力学，1988，9 (3)：205-217.

[33] 刘人怀. 复合载荷作用下波纹圆板的非线性分析. 应用数学和力学，1988，9 (6)：661-674.

[34] Liu RH，He LH. On the nonlinear stability of a truncated shallow spherical shell under axisymmetrically distributed load. Solid Mechanics Archives，1989，14 (2)：81-102.

[35] Liu RH，Li D. On the non-linear bending and vibration of corrugated circular plates. International Journal of Non-Linear Mechanics，1989，24 (3)：165-176.

[36] 刘人怀，朱金福. 夹层壳非线性理论. 北京：机械工业出版社，1993.

[37] 刘人怀. 精密仪器仪表弹性元件的设计原理. 广州：暨南大学出版社，2006.

[38] 刘人怀. 夹层板壳非线性理论分析. 广州：暨南大学出版社，2007.

[39] 刘人怀. 网壳结构的非线性弯曲、稳定和振动. 北京：科学出版社，2011.

[40] 刘人怀，何陵辉. 四边简支对称正交层合矩形板的非线性弯曲问题. 应用数学和力学，1990，11 (9)：753-759.

[41] 刘人怀. 对称圆柱正交异性层合扁球壳的非线性稳定问题. 应用数学和力学，1991，12 (3)：251-258.

[42] 刘人怀. 复合材料层合扁锥壳的非线性稳定问题. 第九届全国复合材料学术会议论文集. 北京：世界图书出版社公司，1996：249-255.

[43] Turvey GJ，Wittrick WH. The large deflection and post-buckling behaviour of some laminated plates. Aeronaut Quarterly，1973，24：77-86.

[44] Zaghloul SA，Kennedy JB. Nonlinear behavior of symmetrically laminated plates. Journal of Applied Mechanics，1975，42 (1)：234-236.

[45] Chandra R. Non-linear bending of antisymmetric angle-ply laminated plates. Fibre Science and Technology，1977，10 (2)：123-137.

[46] 周次青. 对称正交铺设矩形叠层板的非线性弯曲. 应用数学和力学，1985，6 (9)：819-832.

[47] 周次青. 不对称的各向异性叠层矩形板的非线性弯曲. 应用数学和力学，1986，7 (11)：1003-1020.

[48] 周次青. 对称铺设各向异性叠层矩形板的非线性弯曲. 应用数学和力学，1988，9 (3)：267-280.

[49] Reissner E. The effect of transverse shear deformation on the bending of elastic plates. Journal of Applied Mechanics，1945，12：A69-A77.

[50] Mindlin RD. Influence of rotatory inertia and shear on flexural motions of isotropic, elastic plates. Journal of Applied Mechanics, 1951, 18 (1): 31-38.

[51] Stavsky Y. On the theory of heterogeneous anisotropic plates. [PhD Thesis], MIT, Cambridge, MA, 1959.

[52] Yang P, Norris CH, Stavsky Y. Elastic wave propagation in heterogeneous plates. International Journal of Solids and Structures, 1966, 2 (4): 665-684.

[53] Ambartsumyan SA. Theory of Anisotropic Plates. Stanford: Technomic Publishing Co., Inc, 1970.

[54] Bert CW. Nonlinear vibration of a rectangular plate arbitrarily laminated of anisotropic material. Journal of Applied Mechanics, 1973, 40 (2): 452-458.

[55] Whitney JM. Structural analysis of laminated anisotropic plates. Stanford: Technomic Publishing Co., Inc, 1987.

[56] Dong SB, Tso FKW. On a laminated orthotropic shell theory including transverse shear deformation. Journal of Applied Mechanics, 1972, 39 (4): 1091-1097.

[57] Reddy JN. A small strain and moderate rotation theory of elastric anisotropic plates. Journal of Applied Mechanics, 1987, 54 (3): 623-626.

[58] Schmidt R, Reddy JN. A refined small strain and moderate rotation theory of elastic anisotropic shells. Journal of Applied Mechanics, 1988, 55 (3): 611-617.

[59] Palmerio AF, Reddy JN, Schmidt R. On a moderate rotation theory of elastic anisotropic shells. Part 1. Theory. International Journal of Non-Linear Mechanics, 1990, 25 (6): 687-700.

[60] Palmerio AF, Reddy JN, Schmidt R. On a moderate rotation theory of elastic anisotropic shells. Part 2. FE analysis. International Journal of Non-Linear Mechanics, 1990, 25 (6): 701-714.

[61] Kreja I, Schmidt R, Reddy JN. Finite elements based on a first-order shear deformation moderate rotation shell theory with applications to the analysis of composite structures. International Journal of Non-Linear Mechanics, 1997, 32 (6): 1123-1142.

[62] Iu VP, Chia CY. Effect of transverse shear on nonlinear vibration and postbuckling of anti-symmetric cross-ply imperfect cylindrical shells. International Journal of Mechanical Sciences, 1988, 30 (10): 705-718.

[63] Fu YM, Chia CY. Non-linear vibration and postbuckling of generally laminated circular cylindrical thick shells with non-uniform boundary conditions. International Journal of Non-Linear Mechanics, 1993, 28 (3): 313-327.

[64] Fu YM, Chia CY. Multi-mode non-linear vibration and postbuckling of anti-symmetric imperfect angle-ply cylindrical thick panels. International Journal of Non-Linear Mechanics, 1989, 24 (5): 365-381.

[65] Sivakumaran KS, Chia CY. Large-amplitude oscillations of unsymmetrically laminated anisotropic rectangular plates including shear, rotatory inertia and transverse normal stress. ASME Journal of Applied Mechanics, 1985, 52 (3): 536-542.

[66] Xu CS, Xia ZQ, Chia CY. Non-linear theory and vibration analysis of laminated truncated, thick, conical shells. International Journal of Non-Linear Mechanics, 1996, 31 (2): 139-154.

[67] Fares ME. Non-linear bending analysis of composite laminated plates using a refined first-order

theory. Composite Structures，1999，46 (3)：257-266.

[68] Fares ME，Zenkour AM. Mixed variational formula for the thermal bending of laminated plates. Journal of Thermal Stresses，1999，22 (3)：347-365.

[69] Fares ME. Mixed variational formulation in geometrically non-linear elasticity and a generalized nth-order beam theory. International Journal of Non-Linear Mechanics，1999，34 (2)：685-691.

[70] Fares ME. Generalized non-linear thermoelasticity for composite laminated structures using a mixed variational approach. International Journal of Non-Linear Mechanics，2000，35 (3)：439-446.

[71] Whitney JM，Pagano NJ. Shear deformation in heterogeneous anisotropic plates. Journal of Applied Mechanics，1970，37 (4)：1031-1036.

[72] Fortier RC，Rossettos JN. On the vibration of shear deformable curved anisotropic composite plates. Journal of Applied Mechanics，1973，40 (1)：299-301.

[73] Sinha PK，Rath AK. Vibration and buckling of cross-ply laminated circular cylindrical panels. Aeronautical Quarterly，1975，26：211-218.

[74] Bert CW，Chen TL. Effect of shear deformation on vibration of antisymmetric angle-ply laminated rectangular plates. International Journal of Solids and Structures，1978，14 (6)：465-473.

[75] 刘人怀. 考虑横向剪切的对称层合圆柱正交异性扁球壳的非线性稳定问题. 中国科学（A辑)，1991 (7)：742-751.

[76] 刘人怀，王璠. 复合材料层合扁球壳的非线性强迫振动. 力学学报，1997，29 (2)：236-241.

[77] Liu RH，Wang F. Nonlinear dynamic buckling of symmetrically laminated cylindrically orthotropic shallow spherical shells. Archive of Applied Mechanics，1998，68 (6)：375-384.

[78] 刘人怀，徐加初. 直线型正交异性层合圆板的大挠度问题. 暨南大学学报（自然科学版)，1998，19 (4)：1-10.

[79] 刘人怀，徐加初. 剪切变形对直线型正交异性层合圆板大幅度受迫振动的影响. 应用数学和力学，1998，19 (2)：105-112.

[80] Liu RH，Xu JC，Zhai SZ. Large-deflection bending of symmetrically laminated rectilinearly orthotropic elliptical plates including transverse shear. Archive of Applied Mechanics，1997，67 (7)：507-520.

[81] Liu RH. On non-linear buckling of symmetrically laminated，cylindrically orthotropic，truncated，shallow，spherical shells under uniform pressure including shear effects. International Journal of Non-Linear Mechanics，1996，31 (1)：101-115.

[82] 刘人怀，钟诚. 考虑横向剪切的对称层合圆柱正交异性中心开孔扁球壳的非线性屈曲. 暨南大学学报（自然科学版)，1994，15 (1)：1-12.

[83] 刘人怀，王璠，肖谭. 阻尼对层合复合材料中心开孔扁球壳非线性振动的影响. 刘人怀等主编. 应用力学研究与实践，广州：暨南大学出版社，2000：35-43.

[84] 王璠，刘人怀. 复合材料层合开顶扁球壳的非线性动态屈曲，固体力学学报，2001，22 (3)：309-314.

[85] Liu RH. Non-linear buckling of symmetrically laminated，cylindrically orthotropic，shallow，conical shells considering shear. International Journal of Non-Linear Mechanics，1996，31 (1)：89-99.

[86] 刘人怀，王志伟. 复合材料面层夹层板中转动一致有效理论. 上海力学，1996，17 (3)：222-228.

[87] 刘人怀，王志伟. 考虑横向剪切应力连续的复合材料面层夹层壳非线性一致有效理论. 刘人怀等主编. 应用力学研究与实践，广州：暨南大学出版社，2000：1-12.

[88] 王志伟，刘人怀. 正交复合材料面层夹层板非线性理论及应用. 应用力学学报，1998，15 (2)：18-24.

[89] Wang ZW，Liu RH. Nonlinear buckling for sandwich shallow shells accounting for continuity conditions of stresses at interfaces//Proceedings of the 3rd Int. Conf. on Nonlinear Mechcanics，Shanghai：Shanghai University Press，1988：381-386.

[90] 王志伟，刘人怀. 复合材料面层夹层扁壳非线性精化理论及应用. 应用力学学报，2000，17 (4)：86-91，184-185.

[91] 徐加初，王乘，刘人怀. 具有夹层的正交异性复合材料圆锥壳体的非线性轴对称屈曲分析. 工程力学，2001，18 (3)：81-87.

[92] Chow TS. On the propagation of flexural waves in an orthotropic laminated plate and its response to an impulsive load. Journal of Composite Materials，1971，5 (3)：306-319.

[93] Whitney JM. Shear correction factors for orthotropic laminates under static load. Journal of Applied Mechanics，1973，40 (1)：302-304.

[94] Noor AK，Burton WS. Assessment of shear deformation theories for multilayered composite plates. Applied Mechanics Reviews，1989，42 (1)：1-13.

[95] Noor AK，Burton WS. Stress and free vibration analyses of multilayered composite plates. Composite Structures，1989，11 (3)：183-204.

[96] Vlachoutsis S. Shear correction factors for plates and shells. International Journal for Numerical Methods in Engineering，1992，33 (7)：1537-1552.

[97] Laitinen M，Lahtinen H，Sjölind SG. Transverse shear correction factors for laminates in cylindrical bending. Communications in Numerical Methods in Engineering，1995，11 (1)：41-47.

[98] Reddy JN. A simple higher-order theory for laminated composite plates. Journal of Applied Mechanics，1984，51 (4)：745-752.

[99] Reddy JN. Energy and Variational Methods in Applied Mechanics with an Introduction to the Finite-Element Method. New York：Wiley，1984：364-371.

[100] Reddy JN. A refined nonlinear theory of plates with transverse shear deformation. International Journal of Solids and Structures，1984，20 (9-10)：881-896.

[101] Reddy JN，Liu CF. A higher-order shear deformation theory of laminated elastic shells. International Journal of Engineering Science，1985，23 (3)：319-330.

[102] Simitses GJ，Anastasiadis JS. Shear deformable theories for cylindrical laminates— equilibrium and buckling with applications. AIAA Journal，1992，30 (3)：826-834.

[103] Huang NN. Influence of shear correction factors in the higher order shear deformation laminated shell theory. International Journal of Solids and Structures，1994，31 (9)：1263-1277.

[104] Sheikh AH，Chakrabarti A. A new plate bending element based on higher-order shear deformation theory for the analysis of composite plates. Finite Elements in Analysis and Design，2003，39 (9)：883-903.

[105] Nayak AK，Moy SSJ，Shenoi RA. Free vibration analysis of composite sandwich plates based on Reddy's higher-order theory. Composites：Part B，2002，33 (7)：505-519.

[106] Nayak AK，Moy SSJ，Shenoi RA. A higher order finite element theory for buckling and vibration analysis of initially stressed composite sandwich plates. Journal of Sound and Vibration，

2005，286 (4-5)：763-780.
[107] Shahrokh HH，Fadaee M，Taher HRD. Exact solutions for free flexural vibration of Levy-type rectangular thick plates via third-order shear deformation plate theory. Applied Mathematical Modelling，2011，35 (2)：708-727.
[108] Di S，Rothert H. Solution of laminated cylindrical shells using an unconstrained third-order theory. Composite Structures，1995，32 (4)：667-680.
[109] Nayak AK，Moy SJ，Shenoi RA. Free vibration analysis of composite sandwich plates based on Reddy's higher-order theory. Composites：Part B—Engineering，2002，33 (7)：505-519.
[110] Kant T，Swaminathan K. Free vibration of isotropic，orthotropic，and multilayer plates based on higher order refined theories. Journal of Sound and Vibration，2001，241 (2)：319-327.
[111] Kant T，Swaminathan K. Analytical solutions for free vibration of laminated composite and sandwich plates based on a higher-order refined theory. Composite Structures，2001，53 (1)：73-85.
[112] Kant T，Swaminathan K. Analytical solutions for the static analysis of laminated composite and sandwich plates based on a higher order refined theory. Composite Structures，2002，56 (4)：329-344.
[113] Barut A，Madenci E，Heinrich J，et al. Analysis of thick sandwich construction by a {3，2} -order theory. International Journal of Solids and Structures，2001，38 (34-35)：6063-6077.
[114] Liu RH，He LH. A simple theory for non-linear bending of laminated composite rectangular plates including higher-order effects. International Journal of Non-Linear Mechanics，1991，26 (5)：537-545.
[115] Kuppusamy T，Reddy AN. A three-dimensional nonlinear analysis of cross-ply rectangular composite plates. Computers & Structures，1984，18 (2)：263-272.
[116] 沈惠申. 高阶剪切变形板理论 Kármán 型方程及在热后屈曲分析中的应用. 应用数学和力学，1997，18 (12)：1059-1073.
[117] 沈惠申. 复合材料层合剪切圆柱曲板在侧压作用下的后屈曲. 应用数学和力学，2003，24 (4)：357-366.
[118] 冯世宁，陈浩然. 基于高阶剪切理论的层合板非线性动力稳定性. 大连理工大学学报，2006，6 (1)：1-6.
[119] 张雨，向锦武. 复合材料层合厚圆柱壳高阶理论的改进及其应用. 复合材料学报，2002，19 (4)：86-91.
[120] Whitney JM，Sun CT. A higher order theory for extensional motion of laminated composites. Journal of Sound and Vibration，1973，30 (1)：85-97.
[121] 傅晓华，陈浩然，王震鸣. 复合材料多层厚板的精化高阶理论及其有限元法. 复合材料学报，1992，9 (2)：39-47.
[122] 范业立，曾加雄. 筒形弯曲和球壳弯曲的一个新的高阶理论解. 华南理工大学学报（自然科学版），1991，19 (4)：44-53.
[123] 陈浩然，温玄玲，左庄太. 含分层损伤轴对称复合材料层合壳的高阶理论及有限元分析. 大连理工大学学报，1999，39 (5)：601-606.
[124] Matsunaga H. Assessment of a global higher-order deformation theory for laminated composite and sandwich plates. Composite Structures，2002，56 (3)：279-291.
[125] Matsunaga H. A comparison between 2-D single-layer and 3-D layerwise theories for computing

interlaminar stresses of laminated composite and sandwich plates subjected to thermal loadings. Composite Structures, 2004, 64 (2): 161-177.

[126] Matsunaga H. Interlaminar stress analysis of laminated composite and sandwich circular arches subjected to thermal/mechanical loading. Composite Structures, 2003, 60 (3): 345-358.

[127] Matsunaga H. Stress analysis of functionally graded plates subjected to thermal and mechanical loadings. Composite Structures, 2009, 87 (4): 344-357.

[128] Matsunaga H. Effects of higher-order deformations on in-plane vibration and stability of thick circular rings. Acta Mechanica, 1997, 124 (1): 47-61.

[129] Matsunaga H. Vibration and buckling of multilayered composite beams according to higher order deformation theories. Journal of Sound and Vibration, 2001, 246 (1): 47-62.

[130] Reissner E. A twelfth-order theory of transverse bending of transversely isotropic plates. Zeitschriftfiir Angewandte Mathematic und Mechanic, 1983, 63 (7): 285-289.

[131] Ambartsumian SA. On theory of bending plates. Isz Otd Tech Nauk ANSSSR, 1958, 5: 69-77.

[132] Kaczkowski Z. Plates-Statical Calculations. Arkady: Warsaw, 1968.

[133] Pane V. Theories of Elastic Plates. Prague: Academia, 1975.

[134] Levinson M. An accurate simple theory of the statics and dynamics of elastic plates. Mechanics Research Communications, 1980, 7: 343-350.

[135] Ghugal YM, Sayyad AS. Free vibration of thick orthotropic plates using trigonometric shear deformation theory. Latin American Journal of Solids and Structures, 2011, 8 (3): 229-243.

[136] El Meiche N, Tounsi A, Ziane N, et al. A new hyperbolic shear deformation theory for buckling and vibration of functionally graded sandwich plate. International Journal of Mechanical Sciences, 2011, 53 (4): 237-247.

[137] Matsunaga H. Vibration and stability of cross-ply laminated composite plates according to a global higher-order plate theory. Composite Structures, 2000, 48 (4): 231-244.

[138] Karama M, Afaq KS, Mistou S. Mechanical behaviour of laminated composite beam by the new multi-layered laminated composite structures model with transverse shear stress continuity. International Journal of Solids and Structures, 2003, 40 (6): 1525-1546.

[139] Karama M, Afaq KS, Mistou S. A new theory for laminated composite Plates//Proceedings of the IMechE Conference 2009: 223.

[140] Karama M, Abou Harb B, Mistou S, et al. Bending, buckling and free vibration of laminated composite with transverse shear stress continuity model. Composites: Part B-Engineering, 1998, 29 (3): 223-234.

[141] Aydogdu M. Comparison of various shear deformation theories for bending, buckling and vibration of rectangular symmetric cross-ply plate with simply supported edges. Journal of Composite Materials, 2006, 40 (23): 2143-2155.

[142] Aydogdu M. A new shear deformation theory for laminated composite plates. Composite Structures, 2009, 89 (1): 94-101.

[143] Neves AMA, Ferreira AJM, Carrera E, et al. A quasi-3D sinusoidal shear deformation theory for the static and free vibration analysis of functionally graded plates. Composites Part B-Engineering, 2012, 43 (2): 711-725.

[144] Murakami H. Laminated composite plate theory with improved in-plane responses. Journal of Applied Mechanics, 1986, 53 (3): 661-666.

[145] Toledano A, Murakami H. A higher-order laminated plate theory with improved in-plane response. International Journal of Solids and Structures, 1987, 23 (1): 111-131.

[146] Carrera E. On the use of the Murakami's zig-zag function in the modeling of layered plates and shells. Computers & Structures, 2004, 82 (7-8): 541-554.

[147] Bhaskar B, Varadan TK. A higher-order theory for bending analysis of laminated shells of revolution. Computers and Structures, 1991, 40 (4): 815-819.

[148] Vidal P, Polit O. A sine finite element using a zig-zag function for the analysis of laminated composite beams. Composites: Part B-Engineering, 2011, 42 (6): 1671-1682.

[149] Ferreira AJM, Roque CMC, Carrera E, et al. Radial basis functions collocation and a unified formulation for bending, vibration and buckling analysis of laminated plates, according to a variation of Murakami's zig-zag theory. European Journal of Mechanics -A/Solids, 2011, 30 (4): 559-570.

[150] Rodrigues JD, Roque CMC, Ferreira AJM, et al. Radial basis functions-finite differences collocation and a unified formulation for bending, vibration and buckling analysis of laminated plates, according to Murakami's zig-zag theory. Composite Structures, 2011, 93 (7): 1613-1620.

[151] Rao KM, Meyer-Piening HR. Analysis of thick laminated anisotropic composite plates by the finite element method. Composite Structures, 1990, 15 (3): 185-213.

[152] Brank B, Carrera E. Multilayered shell finite element with interlaminar continuous shear stresses: a refinement of the Reissner-Mindlin formulation. International Journal for Numerical Methods in Engineering, 2000, 48 (6): 843-874.

[153] Carrera E, Parisch H. An evaluation of geometrical nonlinear effects of thin and moderately thick multilayered composite shells. Composite Structures, 1997, 40 (1): 11-24.

[154] Auricchio F, Sacco E. Refined first-order shear deformation theory models for composite laminates. Journal of Applied Mechanics, 2003, 70 (3): 381-390.

[155] Carrera E, Demasi L. Multilayered finite plate element based on Reissner mixed variational theorem, part I: theory and part Ⅱ: numerical analysis. International Journal for Numerical Methods in Engineering, 2002, 55 (2): 191-231.

[156] Di Sciuva M. An improved shear-deformation theory for moderately thick multilayered anisotropic shells and plates. Journal of Applied Mechanics, 1987, 54 (3): 589-596.

[157] Di Sciuva M. Bending, vibration and buckling of simply supported thick multilayered orthotropic plates: an evaluation of a new displacement model. Journal of Sound and Vibration, 1986, 105 (3): 425-442.

[158] Tessler A, Di Sciuva M, Gherlone M. A refined zigzag beam theory for composite and sandwich beams. Journal of Composite Materials, 2009, 43 (9): 1051-1081.

[159] Cho M, Parmerter PR. An efficient higher-order plate theory for laminated composites. Composite Structures, 1993, 20: 113-123.

[160] Cho M, Parmerter PR. Efficient higher order composite plate theory for general lamination configurations. AIAA Journal, 1993, 31 (7): 1299-1306.

[161] Cho M, Kim KO, Kim MH. Efficient higher-order shell theory for laminated composites. Composite Structures, 1996, 34 (34): 197-212.

[162] Ali JSM, Bhaskar K, Varadan TK. A new theory for accurate thermal/ mechanical flexural

analysis of symmetric laminated plates. Composite Structures, 1999, 45 (3): 227-232.

[163] Cho M, Oh J. Higher order zig-zag plate theory under thermoelectric-mechanical loads combined. Composites: Part B-Engineering, 2003, 34 (1): 67-82.

[164] Oh J, Cho M, Kim JS. Buckling analysis of a composite shell with multiple delaminations based on a higher order zig-zag theory. Finite Elements in Analysis and Design, 2008, 44 (11): 675-685.

[165] Nguyen SN, Lee J, Cho M. Efficient higher-order zig-zag theory for viscoelastic laminated composite plates. International Journal of Solids and Structures, 2015, 62: 174-185.

[166] Azhari F, Boroomand B, Shahbazi M. Exponential basis functions in the solution of laminated plates using a higher-order Zig-Zag theory. Composite Structures, 2013, 105 (8): 398-407.

[167] Icardi U. A three-dimensional zig-zag theory for analysis of thick laminated beams. Composite Structures, 2001, 52 (1): 123-135.

[168] Icardi U. Higher-order zig-zag model for analysis of thick composite beams with inclusion of transverse normal stress and sublaminates approximations. Composites: Part B-Engineering, 2001, 32 (4): 343-354.

[169] 何陵辉，刘人怀. 一种考虑层间位移和横向剪应力连续条件件的层合板理论. 固体力学学报，1994，15 (3): 319-326.

[170] 舒小平. 满足板面载荷和横向剪应力连续的层板理论. 复合材料学报，1994，11 (3): 61-68.

[171] 舒小平. 复合材料层合板壳热弹性动力响应. 复合材料学报，1997，14 (2): 103-107.

[172] 白瑞祥，陈浩然. 含界面脱粘及表板基体开裂损伤的复合材料夹层板非线性稳定性的研究. 复合材料学报，2002，19 (2): 80-84.

[173] 任晓辉，陈万吉. C^0 型高阶锯齿理论及复合材料厚梁热应力分析. 计算力学学报，2015，32 (1): 77-82.

[174] Oh J, Cho M, Kim JS. Dynamic analysis of composite plate with multiple delaminations based on higher-order zigzag theory. International Journal of Solids and Structures, 2005, 42 (23): 6122-6140.

[175] Cho M, Choi YJ. A new postprocessing method for laminated composites of general lamination configurations. Composite Structures, 2001, 54 (4): 397-406.

[176] Chalak HD, Chakrabarti A, Iqbal MA, et al. Free vibration analysis of laminated soft core sandwich plates. Journal of Vibration and Acoustics-Transactions of the ASME, 2013, 135 (1): 011013.

[177] Wu Z, Lo SH, Sze KY, et al. A higher order finite element including transverse normal strain for linear elastic composite plates with general lamination configurations. Finite Elements in Analysis and Design, 2012, 48 (1): 1346-1357.

[178] Xiaohui R, Wanji C, Zhen W. A C0-type zig-zag theory and finite element for laminated composite and sandwich plates with general configurations. Archive of Applied Mechanics, 2012, 82 (3): 391-406.

[179] Reddy JN. A generalization of two-dimensional theories of laminated composite plates. Communications in Applied Numerical Methods, 1987, 3 (3): 173-180.

[180] Reddy JN, Barbero EJ, Teply JL. A plate bending element based on a generalized laminated plate theory. International Journal for Numerical Methods in Engineering, 1989, 28 (10):

2275-2292.

[181] Barbero EJ, Reddy JN, Teply JL. General two-dimensional theory of laminated cylindrical shells. AIAA Journal, 1990, 28 (3): 544-553.

[182] Barbero EJ, Reddy JN. Modeling of delamination in composite laminates using a layer-wise plate theory. International Journal of Solids and Structures, 1991, 28 (2): 373-388.

[183] Di Sciuva M, Gherlone M. A global/local third-order Hermitian displacement field with damaged interfaces and transverse extensibility: FEM formulation. Composite Structures, 2003, 59 (4): 433-444.

[184] Gherlone M, Di Sciuva M. Thermo-mechanics of undamaged and damaged multilayered composite plates: a sub-laminates finite element approach. Composite Structures, 2007, 81 (1): 125-136.

[185] Gherlone M, Di Sciuva M. Thermo-mechanics of undamaged and damaged multilayered composite plates: assessment of the FEM sub-laminates approach. Composite Structures, 2007, 81 (1): 137-155.

[186] Pai PF, Palazotto AN. A higher-order sandwich plate theory accounting for 3-D stresses. International Journal of Solids and Structures, 2001, 38 (30-31): 5045-5062.

[187] Carrera E. Mixed layer-wise models for multilayered plates analysis. Composite Structures, 1998, 43 (1): 57-70.

[188] Carrera E. Evaluation of layerwise mixed theories for laminated plates analysis. AIAA Journal, 1998, 36 (4): 830-839.

[189] Carrera E. Theories and finite elements for multilayered plates and shells. Archives of Computational Methods in Engineering, 2003, 10 (2): 215-296.

[190] Carrera E, Demasi L. Two benchmarks to assess two-dimensional theories of sandwich composite plates. AIAA Journal, 2003, 41 (7): 1356-1362.

[191] Icardi U. Layerwise mixed element with sublaminates approximation and 3D zig-zag field, for analysis of local effects in laminated and sandwich composites. International Journal for Numerical Methods in Engineering, 2007, 70 (1): 94-125.

[192] Marjanovic M, Vuksanovic DJ. Layerwise solution of free vibrations and buckling of laminated composite and sandwich plates with embedded delaminations. Composite Structures, 2014, 108 (1): 9-20.

[193] Alnefaie K. Finite element modeling of composite plates with internal delamination. Composite Structures, 2009, 90 (1): 21-27.

[194] Basar Y, Itskov M, Eckstein A. Composite laminates: nonlinear interlaminar stress analysis by multi-layer shell elements. Computer Methods in Applied Mechanics and Engineering, 2000, 185 (2-4): 367-397.

[195] Basar Y, Omurtag MH. Free-vibration analysis of thin/thick laminated structures by layer-wise shell models. Computers & Structures, 2000, 74 (4): 409-427.

[196] Botello S, Onate E, Canet JM. A layer-wise triangle for analysis of laminated composite plates and shells. Computers and Structures, 1999, 70 (6): 635-646.

[197] Ghoshal A, Kim HS, Chattopadhyay A, et al. Effect of delamination on transient history of smart composite plates. Finite Elements in Analysis and Design, 2005, 41 (9-10): 850-874.

[198] Chattopadhyay A, Kim HS, Ghoshal A. Non-linear vibration analysis of smart composite

structures with discrete delamination using a refined layerwise theory. Journal of Sound and Vibration, 2004, 273 (1-2): 387-407.

[199] 胡明勇，王安稳，姜伟，等. 复合材料层合板的动力响应和横向应力分析. 船舶力学，2008，12 (4): 778-784.

[200] 胡明勇，王安稳. 纤维增强粘弹性复合材料层合板的自由振动和应力分析. 工程力学，2010，27 (8): 10-14, 20.

[201] 赵龙胜，吴锦武，赵飞，等. 复合材料简支板固有频率与振型分析. 南昌航空大学学报（自然科学版），2013，27 (2): 10-17.

[202] 杨少红，王安稳. 粘弹层合板的稳态振动和层间应力. 应用力学学报，2003，20 (4): 112-117, 168.

[203] Pandey S, Pradyumna S. A new C0 higher-order layerwise finite element formulation for the analysis of laminated and sandwich plates. Composite Structures, 2015, 131: 1-16.

[204] Plagianakos TS, Saravanos DA. Higher-order layerwise laminate theory for the prediction of interlaminar shear stresses in thick composite and sandwich composite plates. Composite Structures, 2009, 87 (1): 23-35.

[205] Reddy JN, Robbins DH. Theories and computational models of composite laminates. Applied Mechanics Reviews, 1994, 47 (6): 21-35.

[206] Noor AK, Burton WS, Bert CW. Computational models for sandwich panels and shells. Applied Mechanics Reviews, 1996, 49 (3): 155-199.

[207] Kant MT. A critical review and some results of recently developed refined theories of fiber-reinforced laminated composites and sandwiches. Composite Structures, 1993, 23 (4): 293-312.

[208] Altenbach H. Theories for laminated and sandwich plates. Mechanics of Composite Materials, 1998, 34 (2): 243-252.

[209] Zhang YX, Yang CH. Recent developments in finite element analysis for laminated composite plates. Composite Structures, 2009, 88 (1): 147-157.

[210] Kreja I. A literature review on computational models for laminated composite and sandwich panels. Central European Journal of Engineering, 2011, 1 (1): 59-80.

[211] Khandan R, Noroozi S, Sewell P, et al. The development of laminated composite plate theories: a review. Journal of Materials Science, 2012, 47 (16): 5901-5910.

[212] Liu RH, Su W. Nonlinear stability of symmetrically laminated cylindrically orthotropic truncated shallow conical shells including transverse shear. International Journal of Applied Mechanics and Engineering, 2009, 14 (2): 769-790.

[213] 安娜，王璠，刘人怀. 复合材料层合圆柱正交异性扁球壳在线布载荷作用下的非线性稳定性. 暨南大学学报（自然科学版），2006，27 (3): 398-405.

[214] 丁磊，刘人怀，王璠. 复合材料浅球壳模态分析与结构优化设计. 材料科学与工程学报，2007，25 (3): 439-442.

[215] 杨增涛，徐加初，王璠. 复合材料层合扁锥壳的冲击屈曲. 振动工程学报，2007，20 (1): 19-23.

[216] 徐加初，杨增涛. 复合材料层合开顶扁球壳在集中冲击下的非线性动力屈曲. 振动与冲击，2007，26 (1): 1-3, 7.

[217] 徐加初，杨增涛. 爆炸冲击下复合材料层合扁球壳的动力屈曲. 爆炸与冲击，2007，2 (2): 116-120.

[218] 王璠. 开孔复合材料扁球壳的非线性动态响应. 哈尔滨工业大学学报，2000，32 (1)：37-40.

[219] 王志伟，姚著. 蜂窝纸板冲击压缩的试验研究和有限元分析. 机械工程学报，2012，48 (12)：49-55.

[220] Wang DM，Wang ZW. Experimental investigation into the cushioning properties of honeycomb paperboard. Packaging Technology and Science，2008，21 (6)：309-316.

[221] 王志伟，王立军，徐晨翼. 多次低强度冲击对蜂窝纸板缓冲性能的影响. 应用力学学报，2015，32 (2)：441-445，8.

[222] Song ZH，Wang ZH，Ma HW，et al. Mechanical behavior and failure mode of woven carbon/epoxy laminate composites under dynamic compressive loading. Composites：Part B-Engineering，2014，60 (3)：531-536.

[223] Fan H，Zhao L，Chen H，et al. Ductile deformation mechanisms and designing instructions for integrated woven textile sandwich composites. Composites Science and Technology，2012，72 (12)：1338-1343.

[224] Fan H，Zhao L，Chen H，et al. Dynamic compression failure mechanisms and dynamic effects of integrated woven sandwich composites. Journal of Composite Materials，2014，48 (3)：427-437.

[225] 何陵辉，成振强，刘人怀. 含非均匀界面相碳海纤维复合材料等效热传导性质的研究. 应用数学和力学，1997，18 (3)：193-201.

[226] 何陵辉，刘人怀. 球面各向同性颗粒复合材料膨胀系数的界限. 应用数学和力学，1997，18 (4)：317-323.

[227] 何陵辉，成振强，刘人怀. 空心球复合材料热弹性性质的一些精确结果. 应用数学和力学，1998，19 (5)：399-406.

[228] 王璠，何一帆，宋毅，等. 引发方式、铺层对纤维增强复合材料圆柱壳吸能特性影响的冲击试验研究. 振动工程学报，2013，26 (1)：33-40.

[229] 郑伟玲，肖潭，朱朦琪，等. 聚苯乙炔包覆多壁碳纳米管的制备及其分散性. 物理化学学报，2009，25 (11)：2373-2379.

[230] 卫保娟，肖潭，李雄俊，等. 石墨烯与多壁碳纳米管增强环氧树脂复合材料的制备及性能. 复合材料学报，2012，29 (5)：53-60.

[231] 朱朦琪，肖潭，郑伟玲，等. 混杂功能化碳纳米管/聚氨酯复合材料的制备及性能. 材料研究学报，2012，26 (2)：191-198.

[232] 姜久红，王志伟. 蜂窝型缓冲包装系统的振动特性分析. 湖北工业大学学报，2006，21 (3)：18-20.

[233] Wang ZW，E YP. Energy-absorbing properties of paper honeycombs under low and intermediate strain rates. Packaging Technology and Science，2012，25 (3)：173-185.

[234] 宁志华，刘人怀，刘稀南. 径向约束下含非均匀界面层纤维增强复合材料的热弹性分析//第十三届现代数学和力学学术会议暨钱伟长诞辰 100 周年纪念大会论文集，上海，2012-10-6-2012-10-8.

[235] Ning ZH，Liu XN. Thermal analysis for fiber-reinforced composites containing inhomogeneous interphase subject to a radial constraint. Advanced Materials Research，2013，652-654：77-83.

[236] 刘述伦，薛江红，王璠. 纤维增强正交各向异性复合材料层合板的湿热屈曲. 暨南大学学报

（自然科学版），2013，34（5）：489-494.

［237］ 刘述伦，薛江红，王璠，等. 含脱层纤维增强复合材料层合板在湿热环境下的屈曲分析. 工程力学，2015，32（3）：1-8.

［238］ 朱永安，王璠. 均布载荷和温度场联合作用下的扁球壳的热屈曲. 南昌大学学报，2007，31（增刊）：205-207.

［239］ 朱永安，王璠，刘人怀. 考虑横向剪切的对称圆柱正交异性层合扁球壳的热屈曲. 应用数学和力学，2008，29（3）：263-271.

［240］ E YP，Wang ZW. Effect of relative humidity on energy absorption properties of honeycomb paperboards. Packaging Technology and Science，2010，23（6）：471-483.

［241］ E YP，Wang ZW. Plateau stress of paper honeycomb as response to various relative humidities. Packaging Technology and Science，2010，23（4）：203-216.

［242］ 陈炎，韩景龙，刘人怀. 横观各向同性压电矩形薄板的非线性振动. 南京航空航天大学学报，2003，35（1）：18-24.

［243］ 陈炎，刘人怀. 压电矩形薄板的非线性强迫振动. 华南理工大学学报（自然科学版），2003，31（S1）：63-66.

［244］ 陈炎，刘人怀. 单向轴压下压电矩形薄板的动态后屈曲问题. 振动工程学报，2008，21（3）：335-342.

［245］ 陈炎，刘人怀. 考虑位移场和电场高阶分布的压电板耦合非线性理论. 刘人怀等主编. 工程力学科学与实践，广州：暨南大学出版社，2010：183-214.

［246］ Wang H，Hu Y，Wang J. On the nonlinear behavior of a multilayer circular piezoelectric plate-like transformer operating near resonance. IEEE Transactions on Ultrasonics，Ferroelectrics and Frequency Control，2013，60（4）：752-757.

［247］ Wang H，Xie X，Hu Y，Wang J. Nonlinear analysis of a 5-layer beam-like piezoelectric transformer near resonance. Acta Mechanica Solida Sinica，2014，27（2）：195-201.

［248］ 傅衣铭，李升，姜叶洁. 具界面损伤压电智能层合板的非线性自由振动分析. 应用数学和力学，2009，30（2）：127-141.

［249］ 郑玉芳，傅衣铭，王锋. 具损伤压电层合板的非线性动力稳定性分析. 力学学报，2006，38（4）：570-576.

［250］ Rao MN，Schmidt R，Schröder KU. Geometrically nonlinear static FE-simulation of multilayered magneto-electro-elastic composite structures. Composite Structures，2015，127：120-131.

［251］ Xiao Y，Zhou HM，Cui XL. Nonlinear resonant magnetoelectric coupling effect with thermal，stress and magnetic loadings in laminated composites. Composite Structures，2015，128：35-41.

［252］ Yu GL，Zhang HW，Li YX，et al. Theoretical study of nonlinear magnetoelectric response in laminated magnetoelectric composites. Composite Structures，2014，108：287-294.

［253］ Shi Y，Gao YW. Theoretical study on nonlinear magnetoelectric effect and harmonic distortion behavior in laminated composite. Journal of Alloys and Compounds，2015，646：351-359.

［254］ 舒小平. 压电复合材料层板柱面弯曲的精化理论. 应用力学学报，2005，22（1）：44-48.

［255］ Xue JH，Wang YO，Yuan D. A shear deformation theory for bending and buckling of undersea sandwich pipes. Composite Structures，2015，132：633-643.

［256］ Xue JH. Asymptotic analysis for buckling of undersea corroded pipelines. ASME Transaction—Journal of Pressure Vessel Technology, 2008, 130 (2): 021705.

[257] Xue JH. A non-linear finite-element analysis of buckle propagation in subsea corroded pipelines. Finite Elements in Analysis and Design, 2006, 42 (14-15): 1211-1219.

[258] Xue J, Hoo Fatt MS. Symmetric and anti-symmetric buckle propagation modes in subsea corroded pipelines. Marine Structures, 2005, 18 (1): 43-61.

[259] Williams JG, Hadavinia H. Analytical solutions for cohesive zone models. Journal of the Mechanics and Physics of Solids, 2002, 50 (4): 809-825.

[260] Blackman BRK, Hadavinia H, Kinloch AJ, et al. The use of a cohesive zone model to study the fracture of fibre composites and adhesively bonded joints. International Journal of Fracture, 2003, 119 (1): 25-46.

[261] Pike MG, Oskay C. XFEM modeling of short microfiber reinforced composites with cohesive interfaces. Finite Elements in Analysis and Design, 2015, 106: 16-31.

[262] Ouyang ZY, Li GQ. Cohesive zone model based analytical solutions for adhesively bonded pipe joints under torsional loading. International Journal of Solids and Structures, 2009, 46 (5): 1205-1217.

[263] De Lorenzis L, Zavarise G. Cohesive zone modeling of interfacial stresses in plated beams. International Journal of Solids and Structures, 2009, 46 (24): 4181-4191.

[264] Yuan H, Teng JG, Seracino R, et al. Full-range behavior of FRP-to-concrete bonded joints. Engineering Structures, 2004, 26 (5): 553-565.

[265] Yuan H, Lu XS, Hui D, et al. 'Studies on FRP-concrete interface with hardening and softening bond-slip law. Composite Structures, 2012, 94 (12): 3781-3792.

[266] Chen JF, Yuan H, Teng JG. Debonding failure along a softening FRP-to-concrete interface between two adjacent cracks in concrete members. Engineering Structures, 2007, 29: 259-270.

[267] 赵玉萍，袁鸿，韩军. 基于弹性-塑性内聚力模型的纤维拔出界面宏观行为分析. 力学学报，2015, 47 (1): 127-134.

[268] 刘三星，袁鸿. FRP抗剪加固混凝土梁中间裂缝的扩展过程. 暨南大学学报（自然科学版），2014, 35 (1): 104-112.

[269] 关天发，袁鸿. CFRP加固钢筋混凝土曲梁的抗弯性能试验研究. 中山大学学报（自然科学版），2008, 47 (增刊2): 109-113.

[270] 袁鸿，李庆，萧澎伟，等. 侧面黏贴纤维复合材料的抗剪加固梁破坏机理. 暨南大学学报（自然科学版），2012, 33 (5): 463-467.

[271] 向红，傅衣铭，洪力. 受面内双向荷载蜂窝材料的弹性屈曲模式及其影响因素. 应用力学学报，2009, 26 (2): 321-324.

[272] 向红，傅衣铭，洪力. 不规则蜂窝材料在单向荷载作用下的蠕变屈曲分析. 固体力学学报，2008, 29 (2): 181-186.

[273] Wang YJ, Wang ZM. Creep buckling of cross-ply symmetric laminated cylindrical panels. Applied Mathematics and Mechanics, 1993, 14 (4): 313-318.

[274] Xue JH, Luo QZ, Han F, et al. Two-dimensional analyses of delamination buckling of symmetrically cross-ply, rectangular laminates. Applied Mathematics and Mechanics, 2013, 34 (5): 597-612.

[275] Xue JH, Zi L, Yuan H, et al. Contact analysis for fiber-reinforced, delaminated laminates

with kinematic nonlinearity. Acta Mechanica Solida Sinica，2013，26 (4)：388-402.

[276] Tang MQ，Xue JH，Yuan H，et al. Macro-micro analysis for anti-penetrating postbuckling of T300/QY8911 carbon fiber reinforced，deboned laminates with contact effects. International Journal of Applied Mechanics，2014，6 (4)：1450044.

[277] Zhu F，Fu Y，Chen D. Analysis of fatigue delamination growth for piezoelectric laminated cylindrical shell considering nonlinear contact effect. International Journal of Solids and Structures，2008，45 (20)：5381-5396.

[278] 杨金花，傅衣铭. 具脱层复合材料层合圆柱壳的屈曲分析. 湖南大学学报（自然科学版），2005，32 (5)：25-30.

[279] 杨金花，傅衣铭. 具脱层轴对称层合圆柱壳的振动模态分析. 应用力学学报，2007，24 (3)：509-513，682.

[280] 陈得良，傅衣铭. 考虑接触效应的具脱层轴对称层合圆板的非线性动力响应. 固体力学学报，2007，28 (4)：406-410.

[281] Chen HR，Wang M，Bai RX. The effect of nonlinear contact upon natural frequency of delaminated stiffened composite plate. Composite Structures，2006，76 (1-2)：28-33.

[282] Chen HR，Hong M，Liu YD. Dynamic behavior of delaminated plates considering progressive failure process. Composite Structures，2004，66 (1-4)：459-466.

[283] Schwarts-Givli H，Rabinovitch O，Frostig Y. High-order nonlinear contact effects in the dynamic behavior of delaminated sandwich panels with a flexible core. International Journal of Solids and Structures，2007，44 (1)：77-99.

[284] Wang J，Tong L. A study of the vibration of delaminated beams using a nonlinear anti-interpenetration constraint model. Composite Structures，2002，57 (1-4)：483-488.

[285] Kwon YW，Aygunes H. Dynamic finite element analysis of laminated beams with delamination cracks using contact-impact conditions. Computers & Structures，1996，58 (6)：1161-1169.

[286] Whitcomb JD. Analysis of a laminate with a postbuckled embedded delamination，including contact effects. Journal of Composite Materials，1992，26 (10)：1523-1535.

[287] Yeh MK，Fang LB，Kao MH. Bending behavior of delaminated composite plates with contact effects. Composite Structures，1997，39 (3-4)：347-356.

[288] Yeh MK，Fang LB. Contact analysis and experiment of delaminated cantilever composite beam. Composites：Part B-Engineering，1999，30 (4)：407-414.

[289] Luo H，Hanagud S. Dynamics of delaminated beams. International Journal of Solids and Structures，2000，37 (8)：1501-1519.

[290] Chen HR，Bai RX. Postbuckling behavior of face/core debonded composite sandwich plate considering matrix crack and contact effect. Composite Structures，2002，57 (1-4)：305-313.

[291] Hu N，Fukunaga H，Sekine H，et al. Compressive buckling of laminates with an embedded delamination. Composites Science and Technology，1999，59 (8)：1247-1260.

[292] Bruno D，Greco F，Lonetti P. A coupled interface-multilayer approach for mixed mode delamination and contact analysis in laminated composites. International Journal of Solids and Structures，2003，40 (26)：7245-7268.

第七章　网格扁壳非线性力学

网格扁壳的非线性弯曲理论①

由杆系组成的空间网格结构，由于空间刚度大、受力合理等力学特点，低成本、轻质省材、易于包装、运输、安装等经济特点，以及造型新型、美观等结构特点，多年来尤其是近年来在机械、建筑、航空航天等领域得到了越来越广泛的应用。

对网络结构进行分析和研究基本上有两大途径：一是离散化方法，二是等效连续化方法。前一方法中，有直接使用的有限元法，有对周期性结构进行分析时所采用的有限元与转换矩阵相结合的方法或精确刚度表示法，还有一种离散场分析方法，即先对网络结构建立特征节点的差分方程，然后直接求解或者采用级数展开将差分方程转化为微分方程来分析，详细的阐述参见文献［1］。由于网格结构本身所呈现的离散化格局，运用离散化分析方法更直接、更直观。并且计算精度较高。然而，当网格结构比较复杂，尤其是在网格分布稠密、节点数目大的情况下，运用有限元、差分等离散化方法直接求解，将是一件耗时费力的大工程，并且对于分析结构的稳定（弯曲、后弯曲）、振动或进行非线性探讨，其繁复的程度将是不言而喻的。针对这一点，等效连续化方法的运用将使问题的处理变得相对简单。

本节通过直接分析矩形网格扁壳的变形与内力，建立等效模型，利用虚功原理推导得到了用于该类网格结构非线性分析的大挠度基本控制方程。

一、基本假设

考虑一双向布置的矩形网格扁壳，壳的中面由两向肋条（离散梁元）中线相交的点组成，其底面如图 1 所示。x_1，x_2 方向的肋条长度分别为 L_2，L_1；弹性模量、截面积、惯性矩、曲率分别为 E_1，A_1，I_1，I_{10}，k_1 和 E_2，A_2，I_2，I_{20}，k_2。

现在作如下假设：

（1）网格分布很稠密，即保证肋条长度 L_1，L_2 大大小于底面尺寸；

（2）肋条截面尺寸远小于 L_1（L_2），每根肋条可当作细长曲梁来处理；

（3）肋条之间的连接为刚性固结；

（4）忽略材料的泊松比 ν 对肋条横向弯曲、扭转和轴向变形的影响；

（5）由于结构表面形状为扁壳，分析时可运用已有的扁壳性质，并采用 Kirchhoff-Love 假定。

---

① 本文原载《江西工业大学学报》，1991，13（2）：186-192. 作者：刘人怀，聂国华.

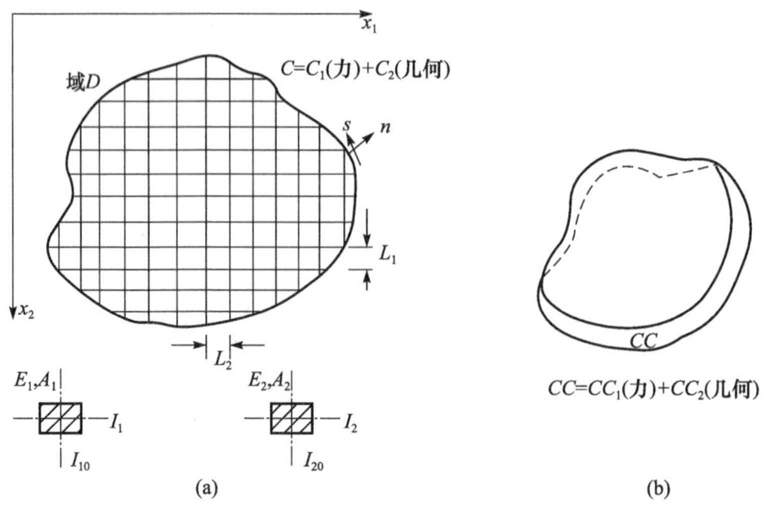

图 1　双向布置的矩形网格扁壳

二、网格扁壳结构内力、变形与等效模型

从网格扁壳中取出典型单元框架（如图 2 所示），肋条截面上的轴向力、剪力、横向剪力、弯矩和扭矩分别为 N_1（N_2），N_{12}（N_{21}），Q_1（Q_2），M_1（M_2），M_{12}（M_{21}）.

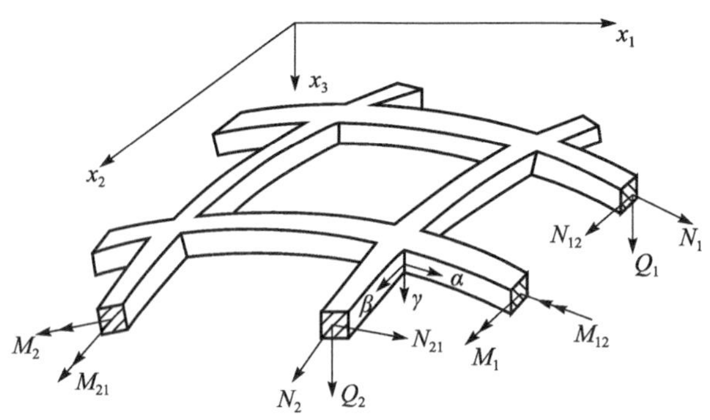

图 2　典型单元框架

根据肋条正交布置，在网壳中面上正交曲线坐标系（α，β，γ）下，结构变形的位移设为 u_1，u_2，w，其中 w 即为壳中面挠度。由于变形，扁壳中面各点的主曲率和扭率的改变可表达为

$$\chi_\alpha = -w_{,11}, \quad \chi_\beta = -w_{,22}, \quad \chi_{\alpha\beta} = -w_{,12}, \tag{2.1}$$

其中

$$w_{,1} = \frac{\partial w}{\partial x_1}, \quad w_{,2} = \frac{\partial w}{\partial x_2}.$$

基于假设（4），弯矩、扭矩和轴向力可表示如下：

$$\begin{aligned} &M_1 = E_1 I_1 \chi_\alpha, \quad M_2 = E_2 I_2 \chi_\beta, \quad M_{12} = G_1 J_1 \chi_{\alpha\beta}, \quad M_{21} = G_2 J_2 \chi_{\alpha\beta}, \\ &N_1 = E_1 A_1 \varepsilon_\alpha, \quad N_2 = E_2 A_2 \varepsilon_\beta, \end{aligned} \tag{2.2}$$

其中 ε_α 和 ε_β 为轴向应变，G_iJ_i（$i=1$，2）为扭转刚度。

考虑肋条自身的平衡可知，剪切力 N_{12}（N_{21}）实际上由三部分构成：其一为网壳切面内的纯剪力，其二为弯矩 M_1（M_2）的影响，其三为扭矩 M_{12}（M_{21}）的影响，即

$$N_{12}=N_{12}^{(1)}+N_{12}^{(2)}+N_{12}^{(3)},$$
$$N_{21}=N_{21}^{(1)}+N_{21}^{(2)}+N_{21}^{(3)},\tag{2.3}$$

这里

$$N_{12}^{(2)}=M_{12}(k_1-\chi_\alpha),\quad N_{12}^{(3)}=M_1\chi_{\alpha\beta},$$
$$N_{21}^{(2)}=M_{21}(k_2-\chi_\beta),\quad N_{21}^{(3)}=M_2\chi_{\alpha\beta},\tag{2.4}$$

两纯剪力 $N_{12}^{(1)}$ 和 $N_{21}^{(1)}$ 可由面内弯曲平衡（如图 3（a）所示）得到关系

$$N_{12}^{(1)}L_2=N_{21}^{(1)}L_1\tag{2.5}$$

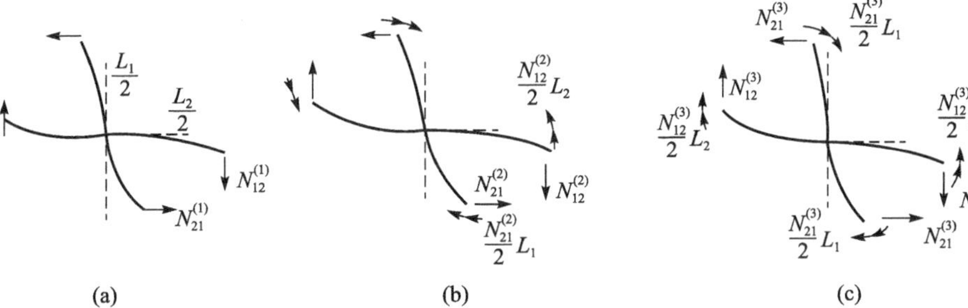

图 3　面内弯曲平衡

现在考虑网壳单元面内的剪切变形，如图 3 所示。剪切变形由纯剪切力（$N_{12}^{(1)}$，$N_{21}^{(1)}$），两个平衡力系

$$\left(N_{12}^{(2)},\frac{N_{12}^{(2)}}{2}L_2,N_{21}^{(2)},\frac{N_{21}^{(2)}}{2}L_1\right)$$

和

$$\left(N_{12}^{(3)},\frac{N_{12}^{(3)}}{2}L_2,N_{21}^{(3)},\frac{N_{21}^{(3)}}{2}L_1\right)$$

三部分面内弯曲变形所引起，亦即

$$\gamma_{\alpha\beta}=\gamma_{\alpha\beta}^{(1)}+\gamma_{\alpha\beta}^{(2)}+\gamma_{\alpha\beta}^{(3)}$$

利用假设（2.2）、（2.3），变形后节点处两条切线仍成直角，节点为反弯点，由此得到剪应变

$$\gamma_{\alpha\beta}^{(1)}=\frac{N_{12}^{(1)}L_2^2}{12E_1I_{10}}+\frac{N_{21}^{(1)}L_1^2}{12E_2I_{20}},\quad \gamma_{\alpha\beta}^{(2)}=-\frac{N_{12}^{(2)}L_2^2}{24E_1I_{10}}-\frac{N_{21}^{(2)}L_1^2}{24E_2I_{20}},$$
$$\gamma_{\alpha\beta}^{(3)}=-\frac{N_{12}^{(3)}L_2^2}{24E_1I_{10}}-\frac{N_{21}^{(3)}L_1^2}{24E_2I_{20}}$$

运用式（2.5），总剪应变为

$$\gamma_{\alpha\beta}=\frac{N_{12}^{(1)}L_2}{12}\left(\frac{L_2}{E_1I_{10}}+\frac{L_1}{E_2I_{20}}\right)-\frac{L_2^2}{24E_1I_{10}}(N_{12}^{(2)}+N_{12}^{(3)})$$
$$-\frac{L_1^2}{24E_2I_{20}}(N_{21}^{(2)}+N_{21}^{(3)})\tag{2.6}$$

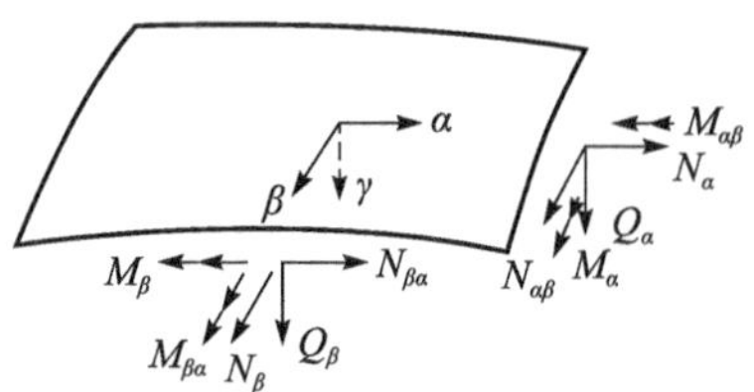

图 4　微元体内的内力

至此，我们已得到了网格扁壳的内力与应变的关系式（2.2）和（2.6）。

接下来讨论网壳单元的等效连续体微元的情况。微元体中的内力（单位长度）显示在图 4 中。

由于壳的扁平性，我们近似认为

$$N_{\alpha\beta}=N_{\beta\alpha},\quad M_{\alpha\beta}=M_{\beta\alpha}$$

根据稠密性假设（2.1），建立如下内力关系[2]：

$$\begin{aligned}
&N_\alpha=\frac{N_1}{L_1},\quad N_\beta=\frac{N_2}{L_2},\quad N_{\alpha\beta}=\frac{1}{2}\left(\frac{N_{12}}{L_1}+\frac{N_{21}}{L_2}\right),\\
&Q_\alpha=\frac{Q_1}{L_1},\quad Q_\beta=\frac{Q_2}{L_2},\\
&M_\alpha=\frac{M_1}{L_1},\quad M_\beta=\frac{M_2}{L_2},\quad M_{\alpha\beta}=\frac{1}{2}\left(\frac{M_{12}}{L_1}+\frac{M_{21}}{L_2}\right)
\end{aligned}\tag{2.7}$$

从式（2.7）中的平错力 $N_{\alpha\beta}$ 的关系，可表示出纯剪力

$$N_{12}^{(1)}=N_{\alpha\beta}L_1-\left(\frac{1}{2}N_{12}^{(2)}+N_{12}^{(3)}\right)-\frac{1}{2}\left(N_{21}^{(2)}+N_{21}^{(3)}\right)\frac{L_1}{L_2}$$

将上式代入式（2.6）中，并且把式（2.7）代入式（2.2），再利用式（2.1）和（2.4），最后得到连续体模型的内力与应变（位移）的物理关系：

$$\begin{aligned}
&\varepsilon_\alpha=\frac{L_1}{E_1A_1}N_\alpha,\quad \varepsilon_\beta=\frac{L_2}{E_2A_2}N_\beta,\\
&\gamma_{\alpha\beta}=\frac{L_1L_2}{12}\left(\frac{L_2}{E_1I_{10}}+\frac{L_1}{E_2I_{20}}\right)N_{\alpha\beta}+\frac{L_1L_2}{24}\left[\left(\frac{2}{E_1I_{10}}\frac{E_2}{L_1}+\frac{1}{E_2I_{20}}\right)G_1J_1K_1\right.\\
&\qquad\left.+\left(\frac{1}{E_1I_{10}}+\frac{2}{E_2I_{20}}\frac{L_1}{L_2}\right)Q_2J_2k_2\right]w_{,12}\\
&\qquad+\frac{L_1L_2}{24}\left(\frac{2}{E_1I_{10}}\frac{L_2}{L_1}+\frac{1}{E_2I_{20}}\right)(G_1J_1-E_1I_1)w_{,11}w_{,12}\\
&\qquad+\frac{L_1L_2}{24}\left(\frac{1}{E_1I_{10}}+\frac{2}{E_2I_{20}}\frac{L_1}{L_2}\right)(G_2J_2-E_2I_2)w_{,22}w_{,12},\\
&w_{,11}=-\frac{L_1}{E_1I_1}M_\alpha,\quad w_{,22}=-\frac{L_2}{E_2I_2}M_\beta,\\
&w_{,12}=-\frac{2}{\dfrac{G_1J_1}{L_1}+\dfrac{G_2J_2}{L_2}}M_{\alpha\beta}
\end{aligned}\tag{2.8}$$

至于位移与应变 ε_α，ε_β，$\gamma_{\alpha\beta}$ 的几何关系，从大变形的角度分析可以得到[3,4]

$$\begin{aligned}
&\varepsilon_\alpha=u_{1,1}-k_1w+\frac{1}{2}(w_{,1})^2,\\
&\varepsilon_\beta=u_{2,2}-k_2w+\frac{1}{2}(w_{,2})^2,\\
&\gamma_{\alpha\beta}=u_{1,2}+u_{2,1}+w_{,1}w_{,2}
\end{aligned}\tag{2.9}$$

考虑到与横向挠度相比，面内变形较小，故上式中忽略了 $u_{1,1}$，$u_{1,2}$，$u_{2,1}$，$u_{2,2}$ 的二次项的影响。

三、网格扁壳的非线性基本控制方程与边界条件

设扁壳所受外力（单位面积上的）在 α，β，γ 方向的分量分别为 f_α，f_β，f_γ，边界 CC_1 上给定单位面积的外力 $\overline{F}_\alpha$，$\overline{F}_\beta$，$\overline{F}_\gamma$，边界 CC_2 上给定几何（位移）边界条件。

运用虚功原理[5]，写出表达式如下（$\mathrm{d}\alpha\mathrm{d}\beta \cong \mathrm{d}x_1x_2$）：

$$\iint_D (N_\alpha\delta\varepsilon_\alpha + N_\beta\delta\varepsilon_\beta + N_{\alpha\beta}\delta\gamma_{\alpha\beta} - M_\alpha\delta w_{,11} - M_\beta\delta w_{,22} - 2M_{\alpha\beta}\delta w_{,12})\mathrm{d}x_1\mathrm{d}x_2$$

$$-\int_{C_1}(\overline{N}_{\alpha n}\delta u_1 + \overline{N}_{\beta n}\delta u_2 + \overline{V}_\gamma\delta w)\mathrm{d}s - \int_{C_1}(\overline{M}_{\alpha n}\delta w_{,1} + \overline{M}_{\beta n}\delta w_{,2})\mathrm{d}s$$

$$-\iint_D (f_\alpha\delta u_1 + f_\beta\delta u_2 + f_\gamma\delta w)\mathrm{d}x_1\mathrm{d}x_2 = 0$$

将几何关系（2.9）代入上式，运用分部积分和线面积分之间的格林关系，最终可以得到

$$\iint_D (N_{\alpha,1} + N_{\alpha\beta,2} + f_\alpha)\delta u_1\mathrm{d}x_1\mathrm{d}x_2 + \iint_D (N_{\alpha\beta,1} + N_{\beta,2} + f_\beta)\delta u_2\mathrm{d}x_1\mathrm{d}x_2$$

$$+\iint_D [M_{\alpha,11} + 2M_{\alpha\beta,12} + M_{\beta,22} + (w_{,1}N_\alpha + w_{,2}N_{\alpha\beta})_{,1} + (w_{,1}N_{\alpha\beta} + w_{,2}N_\beta)_{,2}$$

$$+ N_\alpha k_1 + N_\beta k_2]\delta w\mathrm{d}x_1\mathrm{d}x_2 - \int_{C_1}(N_{\alpha n} - \overline{N}_{\alpha n})\delta u_1\mathrm{d}s - \int_{C_1}(N_{\beta n} - \overline{N}_{\beta n})\delta u_2\mathrm{d}s$$

$$-\int_{C_1}[(M_{\alpha,1} + M_{\alpha\beta,2})l + (M_{\alpha\beta,1} + M_{\beta,2})m + w_{,1}N_{\alpha n} + w_{,2}N_{\beta n}$$

$$+ M_{ns,s} - (\overline{V}_\gamma + \overline{M}_{ns,s})]\delta w\mathrm{d}s + \int_{C_1}(M_n - \overline{M}_n)\delta w_{,n}\mathrm{d}s = 0,$$

其中

$$(\quad)_{,s} = \frac{\partial(\quad)}{\partial s}, \quad w_{,n} = \frac{\partial w}{\partial n}$$

另外，还有

$$N_{\alpha n} = N_\alpha l + N_{\alpha\beta}m, \quad N_{\beta n} = N_{\alpha\beta}l + N_\beta m,$$

$$M_n = M_\alpha l^2 + 2M_{\alpha\beta}lm + M_\beta m^2, \quad M_{ns} = -(M_\alpha - M_\beta)lm + M_{\alpha\beta}(l^2 - m^2),$$

$$(\overline{N}_{\alpha n}, \overline{N}_{\beta n}, \overline{V}_\gamma) = \int(\overline{F}_\alpha, \overline{F}_\beta, \overline{F}_\gamma)\mathrm{d}\gamma,$$

$$(\overline{M}_{\alpha n}, \overline{M}_{\beta n}) = \int(\overline{F}_{\alpha\gamma}, \overline{F}_{\beta\gamma})\mathrm{d}\gamma,$$

$$\overline{M}_n = \overline{M}_{\alpha n}l + \overline{M}_{\beta n}m, \quad \overline{M}_{ns} = -\overline{M}_{\alpha n}m + \overline{M}_{\beta n}l$$

这样，便得到平衡方程：

$$N_{\alpha,1} + N_{\alpha\beta,2} + f_\alpha = 0, \tag{2.10a}$$

$$N_{\alpha\beta,1} + N_{\beta,2} + f_\beta = 0, \tag{2.10b}$$

$$M_{\alpha,11} + 2M_{\alpha\beta,12} + M_{\beta,22} + (w_{,1}N_\alpha + w_{,2}N_{\alpha\beta})_{,1}$$

$$+ (w_{,1}N_{\alpha\beta} + w_{,2}N_\beta)_{,2} + N_\alpha k_1 + N_\beta k_2 + f_\gamma = 0 \tag{2.10c}$$

C_1 上力的边界条件为

$$N_{\alpha n}=\overline{N}_{\alpha n},$$
$$N_{\beta n}=\overline{N}_{\beta n},$$
$$(M_{\alpha,1}+M_{\alpha\beta,2})l+(M_{\alpha\beta,1}+M_{\beta,2})m+w_{,1}N_{\alpha n}+w_{,2}N_{\beta n}+M_{ns,s}=\overline{V}_\gamma+\overline{M}_{ns,s},$$
$$M_n=\overline{M}_n$$

C_2 上几何（位移）边界条件可写成

$$u_1=\bar{u}_1,\quad u_2=\bar{u}_2,\quad w=\bar{w},\quad w_{,n}=\bar{w}_{,n}$$

接下来采用应力函数 Φ 来表示薄膜内力：

$$N_\alpha=\Phi_{,22}-f_1,\quad N_\beta=\Phi_{,11}-f_2,\quad N_{\alpha\beta}=-\Phi_{,12},\tag{2.11}$$

其中

$$f_1(x_1,x_2)=\int f_\alpha \mathrm{d}x_1,\quad f_2(x_1,x_2)=\int f_\beta \mathrm{d}x_2$$

这样，平衡方程（2.10a，b）将自动满足，式（2.10c）变成

$$\begin{aligned}&\frac{E_1I_1}{L_1}w_{,1111}+\left(\frac{G_1J_1}{L_1}+\frac{G_2J_2}{L_2}\right)w_{,1122}+\frac{E_2I_2}{L_2}w_{,2222}\\&=w_{,11}\Phi_{,22}-2w_{,12}\Phi_{,12}+w_{,22}\Phi_{,11}+k_1\Phi_{,22}+k_2\Phi_{,11}-f_1(w_{,11}+k_1)\\&\quad-f_2(w_{,22}+k_2)-f_\alpha w_{,1}-f_\beta w_{,2}+f_\gamma\end{aligned}\tag{2.12}$$

上式中已将物理关系（2.8）和式（2.11）代入。

对于用 w，Φ 表示的协调方程，可以从 $u_{1,1}$，$u_{2,2}$ 和（$u_{1,2}+u_{2,1}$）之间的协调来考虑。从式（2.9）容易得到

$$\varepsilon_{\alpha,22}+\varepsilon_{\beta,11}-\gamma_{\alpha\beta,12}=\left(\frac{1}{2}(w,1)^2-k_1w\right)_{,22}+\left(\frac{1}{2}w_{,2}-k_2w\right)_{,11}-(w_{,1},w_{,2})_{,12}$$

将式（2.8）和（2.11）代入上式，运算后可得到协调方程

$$\begin{aligned}&\frac{L_1}{E_1A_1}\Phi_{,2222}+C\Phi_{,1122}+\frac{L_2}{E_2A_2}\Phi_{,1111}\\&=(w_{,12})^2-w_{,11}w_{,22}-k_1w_{,22}-k_2w_{,11}+C_1w_{,1122}+C_2(w_{,11}w_{,12})_{,12}\\&\quad+C_3(w_{,22}w_{,12})_{,12}+\frac{L_1}{E_1A_1}f_{1,22}+\frac{L_2}{E_2A_2}f_{2,11},\end{aligned}\tag{2.13}$$

上式中，材料常数 C，C_1，C_2，C_3 表达式为

$$\begin{aligned}&C=\frac{L_1L_2}{12}\left(\frac{L_2}{E_1I_{10}}+\frac{L_1}{E_2I_{20}}\right),\\&C_1=\frac{L_1L_2}{24}\left[\left(\frac{2}{E_1I_{10}}\frac{L_2}{L_1}+\frac{1}{E_2I_{20}}\right)G_1J_1k_1+\left(\frac{1}{E_1I_{10}}+\frac{2}{E_2I_{20}}\frac{L_1}{L_2}\right)G_2J_2k_2\right],\\&C_2=\frac{L_1L_2}{24}\left(\frac{2}{E_1I_{10}}\frac{L_2}{L_1}+\frac{1}{E_2I_{20}}\right)(G_1J_1-E_1I_1),\\&G_3=\frac{L_1L_2}{24}\left(\frac{1}{E_1I_{10}}+\frac{2}{E_2I_{20}}\frac{L_1}{L_2}\right)(G_2J_2-E_2I_2)\end{aligned}\tag{2.14}$$

如果将物理关系（2.8）和几何关系（2.9）直接代入式（2.10a，b，c），就可以得到用位移分量 u_1，u_2 和 w 表示的基本方程[1]。

四、总结

本节采用连续模型，建立了矩形网格扁壳的非线性弯曲理论。网格扁壳结构的特性，尤其是宏观力学特性可通过数学表达式（2.12）、（2.13）等来表征。主曲率 k_1，k_2 决定了网格结构的表面曲面形状。特别地，当 $k_1=k_2=0$ 时，网格结构即为一网格平板。

从式（2.12）、（2.13）可以看出，与网格扁壳等效的连续模型实际上是一个各向异性扁壳。通过求解偏微分方程的边值问题可具体分析网格扁壳的非线性性质。

参 考 文 献

[1] 聂国华．网格扁壳的非线性理论．上海工业大学/上海市应用数学和力学研究所博士学位论文，1990.

[2] Kollar L. Continuum equations of timber lattice shells. Acta Technica Academiae Scientiarum Hungaricat，1982，94（3/4）：133-141.

[3] 诺沃日洛夫　B B. 非线性弹性力学基础．朱兆祥，译．北京：科学出版社，1958.

[4] 沃耳密尔 A C. 柔韧板与柔韧壳．卢文达，等，译．北京：科学出版社，1959.

[5] 鹫津久一郎．弹性和塑性力学中的变分法．老亮，郝松林，译．北京：科学出版社，1984.

网格扁壳的非线性自由振动分析①

一、网格扁壳非线性自由振动的初边值问题

考虑到网格结构的横向振动，而忽略壳面内的振动因素，由横向挠度 w 和应力函数 Φ 表达的控制方程可表达成如下形式[1]：

$$\frac{EI}{L_1}w_{,1111}+GJ\left(\frac{1}{L_1}+\frac{1}{L_2}\right)w_{,1122}+\frac{EI}{L_2}w_{,2222}$$
$$=w_{,11}\Phi_{,22}-2w_{,12}\Phi_{,12}+w_{,22}\Phi_{,11}+k_1\Phi_{,22}+k_2\Phi_{,11}-\rho\ddot{w}, \tag{1}$$
$$\frac{L_1}{EA}\Phi_{,2222}+C\Phi_{,1122}+\frac{L_2}{EA}\Phi_{,1111}$$
$$=(w_{,12})^2-w_{,11}w_{,22}-k_1w_{,22}-k_2w_{,22}$$
$$+C_1w_{,1122}+C_2(w_{,12}w_{,11})_{,12}+C_3(w_{,12}w_{,22})_{,12}, \tag{2}$$

式中 ρ 是网格扁壳单位面积质量，材料常数 C，C_1，C_2，C_3 见文献［2］中的式（14）（也见文献［1］中的式（3.1.3）），这里已考虑了两向布置的肋条材料相同、截面一样的情形，

$$w=w(x_1,x_2,t),\quad \Phi=\Phi(x_1,x_2,t),\quad \ddot{w}=\frac{\partial^2 w}{\partial t^2}$$

矩形底面边界可移的简支条件可写成：

当 $x_1=0$，a 时，

$$w=0,\quad w_{,11}=0,\quad N_\alpha=0,\quad N_{\alpha\beta}=0;$$

当 $x_2=0$，b 时，

$$w=0,\quad w_{,22}=0,\quad N_\beta=0,\quad N_{\alpha\beta}=0 \tag{3}$$

初始条件为

$$w(x_1,x_2,0)=0,\quad \Phi(x_1,x_2,0)=0 \tag{4}$$

二、初边值问题的求解与幅频关系

为满足式（3）表示的边界条件，设定 w，Φ 为如下形式：

$$w(x_1,x_2,t)=\sum_{m=1}^{\infty}\sum_{n=1}^{\infty}A_{mn}f(t)\sin m\alpha x_1\sin n\beta x_2,$$
$$\Phi(x_1,x_2,t)=f(t)\sum_{p=1}^{\infty}\sum_{q=1}^{\infty}B_{pq}X_p(x_1)Y_q(x_2)$$
$$+f^2(t)\sum_{p=1}^{\infty}\sum_{q=1}^{\infty}C_{pq}X_p(x_1)Y_q(x_2), \tag{5}$$

① 本文原载《江西工业大学学报》，1991，13（2）：193-198. 作者：刘人怀，聂国华.

式中 A_{mn}，B_{pq}，C_{pq} 为待定常数，$f(t)$ 为未知时间函数，$X_p(x_1)$，$Y_q(x_2)$ 为梁振动特征函数，

$$\alpha=\frac{\pi}{a},\quad \beta=\frac{\pi}{b},$$

$$X_p(x_1)=\cosh\alpha_p\frac{x_1}{a}-\cos\alpha_p\frac{x_1}{a}-\nu_p\left(\sinh\alpha_p\frac{x_1}{a}-\sin\alpha_p\frac{x_1}{a}\right),$$

$$Y_q(x_2)=\cosh\alpha_q\frac{x_2}{b}-\cos\alpha_q\frac{x_2}{b}-\nu_q\left(\sinh\alpha_p\frac{x_2}{b}-\sin\alpha_q\frac{x_2}{b}\right),$$

这里，系数 ν_p，α_p，ν_q，α_q 有下列关系：

$$\nu_p=\frac{\cosh\alpha_p-\cos\alpha_p}{\sinh\alpha_p-\sin\alpha_p},\quad \cos\alpha_p\cdot\cosh\alpha_p=1,$$

$$\nu_q=\frac{\cosh\alpha_q-\cos\alpha_q}{\sinh\alpha_q-\sin\alpha_q},\quad \cos\alpha_q\cdot\cosh\alpha_q=1$$

将式（5）代入协调方程（2），两边同乘 $X_i(L_1)$ $Y_j(x_2)$，并在底面的区域上积分，合并关于 $f(t)$ 和 $f^2(t)$ 的同类项，最后可得到系数之间的关系：

$$\begin{aligned}&\frac{L_1}{EA}\left(\frac{\alpha_j}{b}\right)^4abB_{ij}+C\sum_p\sum_q B_{pq}k_1^{pi}L_1^{qj}+\frac{L_2}{EA}\left(\frac{\alpha_i}{a}\right)^4abB_{ij}\\&=\alpha^2\beta^2\sum_m\sum_n\left(\frac{k_1}{\alpha^2}n^2+\frac{k_2}{\beta^2}m^2+C_1m^2n^2\right)M_4^{im}N_4^{jn}A_{mn},\end{aligned}\tag{6}$$

$$\begin{aligned}&\frac{L_1}{EA}\left(\frac{\alpha_i}{b}\right)^4abC_{ij}+C\sum_p\sum_q C_{ij}+C\sum_p\sum_q c_{ij}k_1^{pi}L_1^{qj}+\frac{L_2}{EA}\left(\frac{\alpha_i}{a}\right)^4abC_{ij}\\&=\alpha^2\beta^2\sum_m\sum_n\sum_r\sum_s(mnrsM_6^{imr}N_6^{jns}-m^2s^sM_1^{imr}N_1^{jns})A_{mn}A_{rs}\\&\quad-\frac{\alpha^2\beta^2}{4}\sum_m\sum_n\sum_r\sum_s(C_2\alpha^2m^2+C_3\beta^2n^2)rs[(m+r)M_7^{i(m+r)}\\&\quad+(m-r)M_7^{i(m-r)}]\cdot[(n+s)N_7^{i(n+s)}+(n-s)N_7^{i(n-s)}]A_{mn}A_{rs},\quad i,j=1,2,\cdots\end{aligned}\tag{7}$$

式中

$$M_1^{pmi}=\int_0^a X_p(x_1)\sin m\alpha x_1\sin i\alpha x_1\,dx_1,$$

$$M_2^{pmi}=\int_0^a X_{p,11}\sin m\alpha x_1\sin i\alpha x_1\,dx_1,$$

$$M_3^{pmi}=\int_0^a X_{p,1}\cos m\alpha x_1\sin i\alpha x_1\,dx_1,$$

$$M_4^{pi}=\int_0^a X_p\sin i\alpha x_1\,dx_1,$$

$$M_5^{pi}=\int_0^a X_{p,11}\sin i\alpha x_1\,dx_1,$$

$$M_6^{imr}=\int_0^a X_i\cos m\alpha x_1\cos r\alpha x_1\,dx_1,$$

$$M_7^{im}=\int_0^a X_i\cos m\alpha x_1\,dx_1,$$

$$K_1^{pi} = \int_0^a X_{p,11} X_i \mathrm{d}x_1,$$

$$L_1^{qj} = \int_0^b Y_{q,22} Y_j \mathrm{d}x_2, \tag{8}$$

用 N，Y，x_2，b，n，q，j 分别替代 M，X，x_1，a，m，p，i 即可得 $N_1 \sim N_7$。

同样，将式（1）两边同乘 $\sin i\alpha x_1 \sin j\beta x_2$，积分后可得到决定未知时间函数 $f(t)$ 的振动方程

$$\ddot{f} + \omega_0^2 f + \alpha_1 f^2 + \alpha_2 f^3 = 0, \tag{9}$$

式中

$$\omega_0^2 = \frac{1}{\rho}\left[\frac{EI}{L_1} i^4 \alpha^4 + GJ\left(\frac{1}{L_1} + \frac{1}{L_2}\right)\alpha^2\beta^2 i^2 j^2 + \frac{EI}{L_2} j^4 \beta^4\right]$$
$$- \frac{4}{pabA_{ij}} \sum_p \sum_q (k_1 M_4^{pi} N_5^{qj} + k_2 M_5^{pi} N_4^{qj}) B_{pq},$$
$$\alpha_1 = \frac{4}{pabA_{ij}} \sum_m \sum_n \sum_p \sum_q (m^2\alpha^2 M_1^{pmi} N_2^{qnj} + 2mn\alpha\beta M_3^{pmi} N_3^{qnj}$$
$$+ n^2\beta^2 M_2^{pmi} N_1^{qnj}) A_{mn} B_{pq} - \frac{4}{pabA_{ij}} \sum_p \sum_q (k_1 M_4^{pi} N_5^{qj} + k_2 M_5^{pi} N_4^{qj}) C_{pq},$$
$$\alpha_2 = \frac{4}{pabA_{ij}} \sum_m \sum_n \sum_p \sum_q (m^2\alpha^2 M_1^{pmi} N_2^{qnj} + 2mn\alpha\beta M_3^{pmi} N_3^{qnj}$$
$$+ n^2\beta^2 M_2^{pmi} N_1^{qnj}) A_{mn} C_{pq}, \tag{10}$$

式中 $M(N)$ 表达式见式（8）。

下面用摄动法求解方程（9）。选取 $w(x_1, x_2, 0)$ 无量纲化的幅值 ε 作为摄动参数，设

$$f(\tau) = \sum_{m=0}^{\infty} \varepsilon^m f_m(\tau),$$

$$\omega = \sum_{m=0}^{\infty} \varepsilon^m \omega_m, \tag{11}$$

式中 $\tau = \omega t$，$f(0)A = h\varepsilon$，A 为 $\sum_m \sum_n A_{mn} \sin m\alpha x_1 \sin n\beta x_2$ 的幅值，h 为厚度。

初始条件由式（4）可改写成

$$f_1(0) = \frac{h}{A},\ f_0(0) = f_2(0) = f_3(0) = \cdots = 0,$$

$$f_0'(0) = f_1'(0) = f_2'(0) = \cdots = 0 \tag{12}$$

此时式（9）变成

$$\omega^2 f'' + \omega_0^2 f + \alpha_1 f^2 + \alpha_2 f^3 = 0, \tag{13}$$

这里

$$f'' = \frac{\partial^2 f}{\partial \tau^2}$$

将式（11）代入式（13），比较 ε^m 的系数可得到相应各阶的基本方程

$$\varepsilon^0: \omega_0^2 f_0'' + \omega_0^2 f_0 + \alpha_1 f_0^2 + \alpha_2 f_0^3 = 0,$$

$$\varepsilon:\omega_0^2 f_1'' + \omega_0^2 f_1 + 2\omega_0\omega_1 f_0'' + 2\alpha_1 f_0 f_1 + 3\alpha_2 f_0^2 f_1 = 0,$$
$$\varepsilon^2:\omega_0^2 f_2'' + 2\omega_0\omega_1 f_1'' + \omega_0 f_2 + \alpha_1 f_1^2 = 0,$$
$$\varepsilon^3:\omega_0^2 f_3'' + 2\omega_0\omega_1 f_2'' + \omega_1^2 f_1'' + \omega_0^2 f_1'' + 2\omega_0\omega_2 f_1'' + 2\alpha_1 f_1 f_2 + \alpha_2 f_1^3 = 0, \tag{14}$$
$$\cdots\cdots$$

结合式（12）可逐阶求解方程（14），根据消除长期项的原则，计算可得到

$$\omega_1 = 0,$$
$$\omega_2 = \frac{1}{4}\left(\frac{h}{A}\right)^2\left(\frac{3\alpha_2}{2\omega_0} - \frac{5\alpha_1^2}{3\omega_0^3}\right)$$

这样，频率 ω 由式（11）第二式可写成

$$\omega = \omega_0 + \varepsilon^2\omega_2 + o(\varepsilon^3)$$

或

$$\frac{\omega}{\omega_0} = 1 + \frac{1}{4}\left(\frac{h}{A}\varepsilon\right)^2\left(\frac{3\alpha_2}{2\omega_0^2} - \frac{5\alpha_1^2}{3\omega_0^4}\right) + o(\varepsilon^3)$$
$$= \left[1 + \left(\frac{h}{A}\varepsilon\right)^2\left(\frac{3\alpha_2}{4\omega_0^2} - \frac{5\alpha_1^2}{6\omega_0^4}\right)\right]^{\frac{1}{2}} \tag{15}$$

如果讨论（m，n）振型，并取 $A_{mn}=h$，即

$$w(x_1, x_2, t) = hf(t)\sin m\alpha x_1 \sin n\beta x_2$$

此时，式（15）改写成

$$\frac{\omega_{mn}}{\omega_{mn}(0)} = \left[1 + \varepsilon^2\left(\frac{3\alpha_2}{4\omega_0^2} - \frac{5\alpha_1^2}{6\omega_0^4}\right)\right]^{\frac{1}{2}} \tag{16}$$

上式即为对应于（m，n）振型的幅频特征关系式。

三、数值计算结果与讨论

为便于计算，通过引入如下的无量纲量和比例常数可对式（16）中的 α_1，α_2，ω_0^2 进行无量纲化[1]：

$$K_1^* = \frac{k_1 a^2}{h},\quad K_2^* = \frac{k_2 b^2}{h},$$
$$m_1 = \frac{a}{b},\quad m_2 = \frac{L_1}{L_2},\quad m_3 = \frac{h}{L_2},$$
$$m_4 = \frac{a}{L_2} \tag{17}$$

肋条的截面取定为圆截面。

图 1～图 3 显示了各种振型的幅频关系。无论对网格扁壳（球和柱形）还是对网格板来说，随着幅值的增加，频率将越来越高。对于振型 $m=2$，$n=2$（x_1，x_2 方向都为一个全波）和 $m=3$，$n=3$（x_1，x_2 方向都为三个半波）情况，非线性振动频

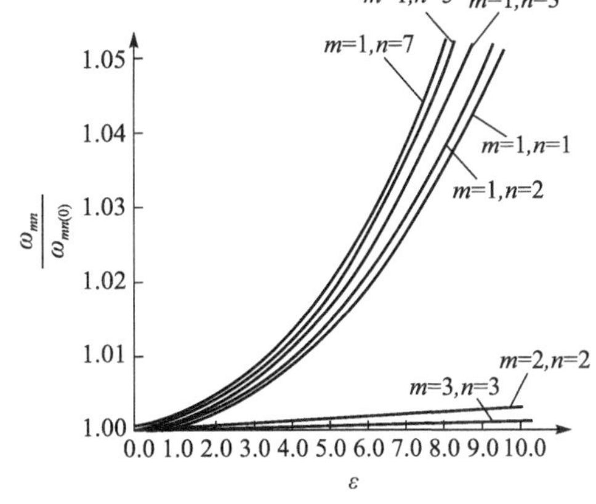

图 1　网格扁球壳自由振动幅频关系

（$K_1^* = K_2^* = 25$，$m_1 = m_2 = 1$，$m_3 = 0.05$，$m_4 = 40$）

率随幅值增加缓慢增大。

比较振型 $m=1$，$n=1$；$m=1$，$n=2$；$m=1$，$n=3$ 所对应的幅频关系可知，网格扁圆柱壳的振动特征较网格扁球壳更接近网格平板的情况。

从图 3 也可看出，网格平板自由振动特性与各向异性板的非线性振动特性相一致[3]。

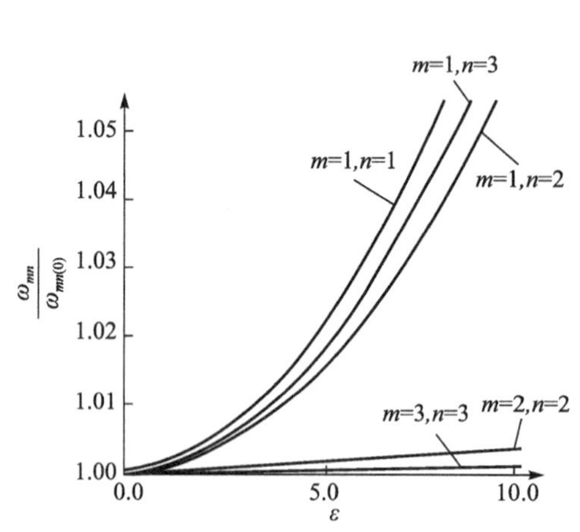

图 2　网格扁圆柱壳自由振动幅频关系

($K_1^*=0$，$K_2^*=25$，$m_1=m_2=1$，$m_3=0.05$，$m_4=40$)

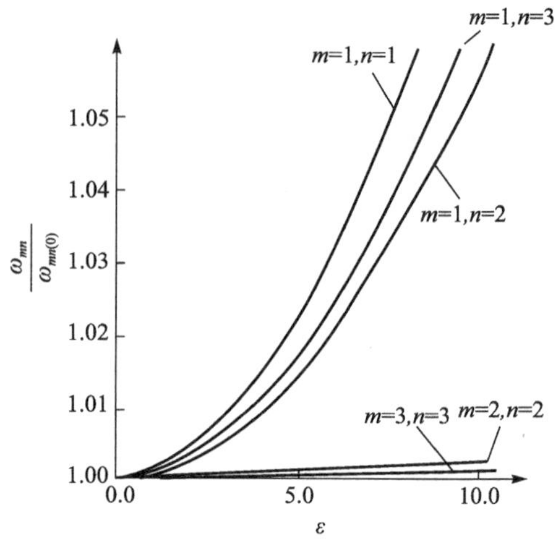

图 3　网格平板自由振动幅频关系

($m_1=m_2=1$，$m_3=0.05$，$m_4=40$)

参 考 文 献

[1]　聂国华．网格扁壳的非线性理论．上海工业大学/上海市应用数学和力学研究所博士学位论文，1990.

[2]　刘人怀，聂国华．网格扁壳的非线性弯曲理论．江西工业大学学报，1991，13 (2)：186-192.

[3]　Chia C Y. Nonlinear Analysis of Plates. New York：McGraw-Hill，1980.

Non-linear buckling of squarely-latticed shallow spherical shells①

中 文 摘 要

本文使用连续化方法创立了方形网格扁球壳大挠度分析的理论，推出了卡门型非线性微分方程，用变分原理得到了承受均匀分布载荷的网络扁球壳的近似边值问题，求解了在竖向均布载荷作用下一个石油储罐的方形网格顶盖的实际问题，给出了在夹紧固定和铰支两种边界条件下屈曲载荷的数值。此问题尚无人研究过。

1. Introduction

Because of the increasing utilization of latticed structures in practical engineering, the investigation of such structures has received constant attention during the last three decades. Generally, research in this area has been performed in two ways. In the first way, the latticed structures were studied with discrete analysis, such as finite difference analysis[1]. In the second way, the equivalent continuum method[2-4] was applied to predict the deformations or critical buckling loads of the structures. The studies of the nonlinear buckling of these latticed structures had not become possible until the increasing use of nonlinear finite element analyses were realized[5]. However, when the number of the lattices of the structures is very great, numerical analysis is complex, expensive, and time-consuming. As for predicting overall buckling loads or vibration frequencies of latticed structures, we can utilize a reasonable continuum model of the structures to simplify the problem and obtain useful results applicable to engineering design.

Latticed shallow spherical shells are often used as the cover of large spatial structures such as domes of gymnasia, and the caps of large petroleum vessels in chemical engineering, due to their light self-weight and vast capacity to carry external loads such as rain, snow, wind or gas pressure. The importance of such shells has led to various studies, theoretical or experimental[6], but most of the previous studies were mainly concerned with the triangularly-latticed shallow shells, and were carried on by complex numerical analyses. So far, there is no analysis on the study of the nonlinear buckling of squarely-latticed shallow spherical shells with an effective equivalent continuum method.

In this paper, we have constructed a theory for the analysis of large deflections of squarely-latticed shallow spherical shells by adopting the continuum method. The von Kármán type nonlinear differential equations are derived. An approximate boundary

① Reprinted from *International Journal of Non-Linear Mechanics*, 1991, 26 (5): 547-565. Authors: Liu Renhuai, Li Dong, Nie Guohua and Cheng Zhenqiang.

value problem for a uniformly-loaded latticed shallow spherical shell is formulated with variational principles. Using this theory, we have analysed a practical case of a squarely-latticed cap of a petroleum vessel under vertical uniform loads. The predicted buckling loads are numerically presented for clamped/hinged boundary conditions.

2. Basic Assumptions

We now consider a squarely-latticed shallow spherical shell as is shown in Fig. 1 (a). All the members are of the same cross-section and placed on the same spherical surface, the middle surface of the shell is defined as the surface interwoven by the centroids of all the cross-section of the members. Let us first construct a fixed coordinate system $Oxyz$ and a moving coordinate system $o\xi\eta\zeta$, as shown in Fig. 1 (b), where O is the center of the base circle of the shell, the x-axis and y-axis run parallel with the two directions of the placed members respectively, o moves along the centroid of the cross-section of each member, the ξ-axis and η-axis run in the tangent directions of the centroids of the cross-sections of the members, and the ζ-axis coincides with the internal normal of the middle surface of the shell.

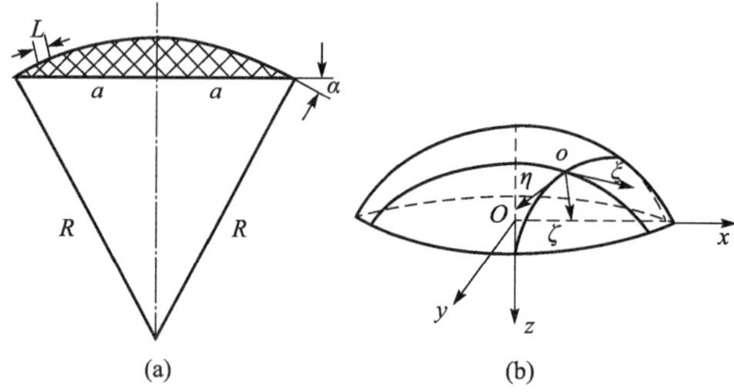

Fig. 1　Geometry and the coordinate of a latticed shell

Next, we introduce the following basic assumptions:

(a) The members are placed on the same spherical surface with a radius of curvature of R and a span of $2a$. All the members are of identical and regular cross-section of area A.

(b) All lattices of the shell, except those connected with the boundary ring, are regarded as square ones of the same side-length L.

(c) The members are placed densely enough so that $L \ll \alpha$.

(d) The characteristic size of the cross-section of each member is much smaller than L, the distance between the two nearest parallel members, so each curved member between the two nearest joints is regarded as a long and thin curved beam.

(e) Each member is made of the same linear elastic material, the Young's modulus is E and shear modulus is G.

(f) The effect of Poisson's ratio of the material on the transverse bending, twist and axial deformation of each member is neglected.

(g) The rotation of each joint is negligible.

(h) The initial curvature of each member is very small, so the corresponding latticed shells are shallow shells, so α, the maximum slope angle of the tangent to a meridian does not exceed 30°.

3. Internal Forces and Deformation

Let us separate a typical latticed element from the shell, as shown in Fig. 2. The normal stress σ and shearing stress τ on the cross-section of each member arising from the deformation is now reduced towards o, the centroid, which yields the six pairs of internal forces as follows.

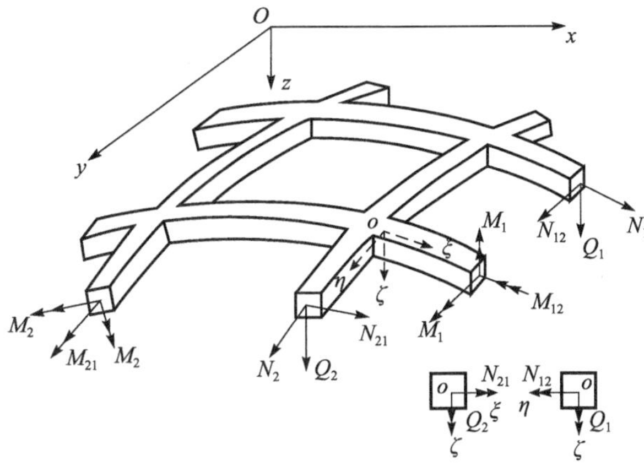

Fig. 2　Internal forces and sign convention

Normal forces:

$$N_1 = \iint_A \sigma_\xi \mathrm{d}\zeta \mathrm{d}\eta$$

$$N_2 = \iint_A \sigma_\eta \mathrm{d}\xi \mathrm{d}\zeta$$

$$N_{12} = \iint_A \tau_{\xi\eta} \mathrm{d}\eta \mathrm{d}\zeta$$

Shearing forces:

$$N_{21} = \iint_A \tau_{\eta\xi} \mathrm{d}\xi \mathrm{d}\zeta$$

$$Q_1 = \iint_A \tau_{\xi\zeta} \mathrm{d}\eta \mathrm{d}\zeta$$

Transverse shearing forces:

$$Q_2 = \iint_A \tau_{\eta\zeta} \mathrm{d}\xi \mathrm{d}\zeta$$

$$M_1 = \iint_A \sigma_\xi \zeta \mathrm{d}\eta \mathrm{d}\zeta$$

Transverse bending moments:

$$M_2 = \iint_A \sigma_\eta \zeta \mathrm{d}\xi \mathrm{d}\zeta$$

$$M_{12} = \iint_A (\zeta \tau_{\xi\eta} - \eta \tau_{\xi\zeta}) \mathrm{d}\eta \mathrm{d}\zeta$$

Torsional moments:

$$M_{21} = \iint_A (\zeta \tau_{\eta\xi} - \xi \tau_{\eta\zeta}) \mathrm{d}\xi \mathrm{d}\zeta$$

$$M_1^* = \iint_A \sigma_\xi \eta \mathrm{d}\eta \mathrm{d}\zeta$$

Lateral bending moments:

$$M_2^* = \iint_A \sigma_\eta \xi \mathrm{d}\xi \mathrm{d}\zeta$$

Here, lateral bending indicates the bending of each member about the ζ-axis.

The components of the displacement of the middle surface of the shell are (u, v, w) in coordinates (ξ, η, ζ). The overall deformation of the shell can be decomposed into transverse bending, twist, shearing and stretching of the middle surface of the shell. Because the transverse bending w is much larger than the stretching deformation u or v, the lateral moments M_1^* and M_2^*, which are associated with the second derivatives of u or v, are neglected when the equilibrium of the shell is considered.

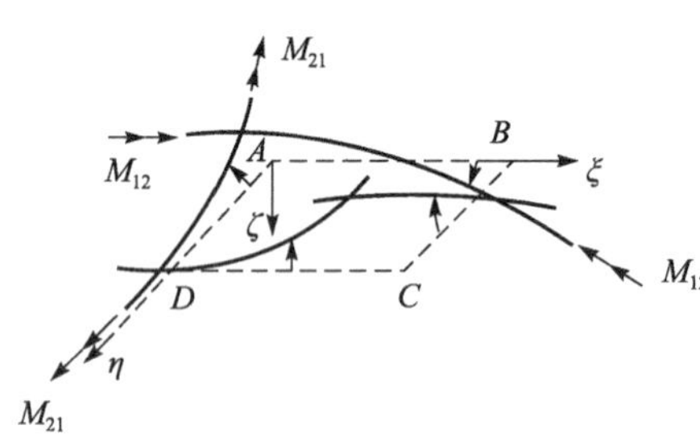

Fig. 3 Twist of the members and sign convention

According to assumption (h), we can approximately obtain the variation of the curvature and torsion of the middle surface arising from the deflection w as

$$k_\xi = -\frac{\partial^2 w}{\partial x^2}, \quad k_\eta = -\frac{\partial^2 w}{\partial y^2}, \quad k_{\xi\eta} = -\frac{\partial^2 w}{\partial x \partial y}.$$

The non-uniform deflection of lattices results in the twist of the members. As is shown in Fig. 3, the relative rotation of member BC to member AD is:

$$\theta_{AB} = \frac{w_B - w_C}{\overline{BC}} - \frac{w_A - w_D}{\overline{AD}} = -\left(\frac{\partial w}{\partial y}\right)_B + \left(\frac{\partial w}{\partial y}\right)_A$$

therefore the twist per unit length of member AB is

$$\bar{\theta}_{AB} = \frac{\theta_{AB}}{\overline{AB}} = \frac{1}{L}\left[\left(\frac{\partial w}{\partial y}\right)_A - \left(\frac{\partial w}{\partial y}\right)_B\right] \cong -\left(\frac{\partial^2 w}{\partial x \partial y}\right)_A = \bar{\theta}.$$

Similarly, the twist per unit length of member AD is

$$\bar{\theta}_{AD} = -\left(\frac{\partial^2 w}{\partial y \partial x}\right)_A = \bar{\theta}.$$

The positive directions of unit twist $\bar{\theta}_{AB}$ and $\bar{\theta}_{AD}$ are the same as those of M_{12} and M_{21}, and obviously we have $M_{12} = M_{21} = GJ\theta$, where GJ indicates the torsional stiffness of each member.

Because of the large deformation of the shell, the shearing forces N_{12} and N_{21} cannot be equivalent, but we can represent N_{12} and N_{21} in the following forms

$$N_{12} = N_{12}^0 + \overline{N}_{12}$$

$$N_{21} = N_{21}^0 + \overline{N}_{21} \tag{3.1a,b}$$

where N_{12}^0 or N_{21}^0 are the shearing forces induced by pure shearing deformation of the lattices. They are equivalent, i.e., $N_{12}^0 = N_{21}^0$. Also, $\overline{N}_{12}$ and $\overline{N}_{21}$ are effective shearing forces associated with the deflection and twist of the curved members, which can be determined by considering the equilibrium of a bent and twisted member placed along the ξ-axis or η-axis. As shown in Fig. 4, summing up all the moments projected to the ζ-axis will yield

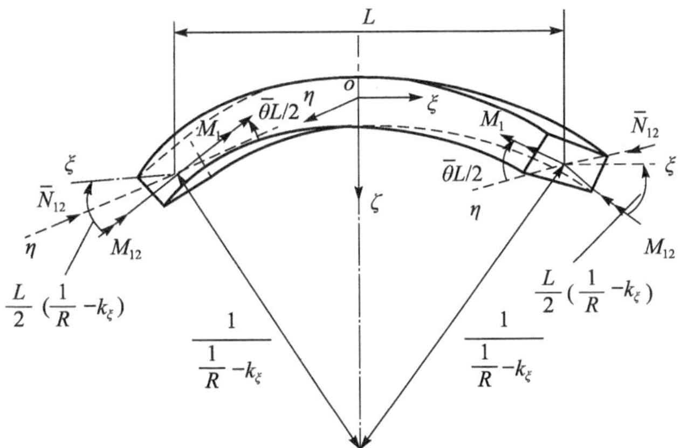

Fig. 4　Equilibrium of a deformed member placed along the ξ-axis

$$\overline{N}_{12}L - \left[L\left(\frac{1}{R} - k_\xi\right)M_{12} + M_1\bar{\theta}L\right] = 0 \tag{3.2}$$

in which $\frac{1}{2}\left[M_1\bar{\theta}L + L\ (1/R - k_\xi)\ M_{12}\right] = \overline{M}_1^*/2$ can be interpreted as the effective lateral moments exerted on both ends of the member placed along the ξ-axis.

Substituting (3.2) into (3.1a) we have

$$N_{12} = \overline{N}_{12} + N_{12}^0 = \left(\frac{1}{R} - k_\xi\right)M_{12} + M_1\bar{\theta} + N_{12}^0. \tag{3.3a}$$

Similarly, we have

$$\begin{aligned} N_{21} &= N_{21}^0 + \overline{N}_{21} = N_{21}^0 + \frac{\overline{M}_2^*}{L} \\ &= N_{21}^0 + \left(\frac{1}{R} - k_\eta\right)M_{21} + M_2\bar{\theta} \end{aligned} \tag{3.3b}$$

where $\overline{M}_2^*/2$ is the effective lateral moment exerted on both ends of the member placed along the η-axis.

Let us assume the latticed element in Fig. 2 to be a continuum element of the same area as shown in Fig. 5. The displacement components of such a continuum element are the same as those of the latticed element, namely, (u,v,w) in coordinate (ξ,η,ζ), but the internal forces of the equivalent shell element are $N_\xi, N_\eta, N_{\xi\eta}, N_{\eta\xi}, Q_\xi, Q_\eta, M_\xi, M_\eta$, $M_{\xi\eta}$ and $M_{\eta\xi}$. Taking into account the basic assumption (c) and using the continuum method [2,3], we have the following relations

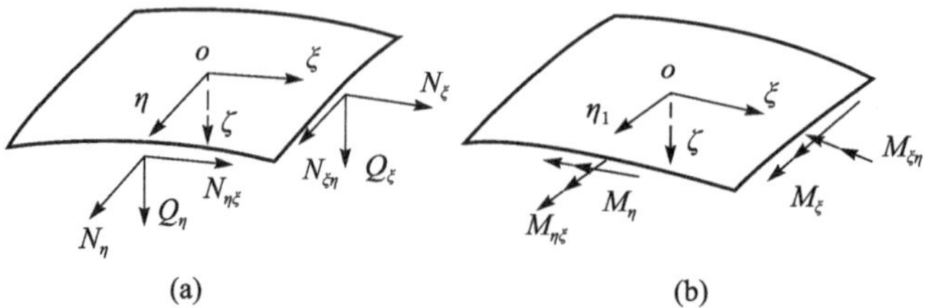

Fig. 5 Equivalent continuum element of the latticed shell

$$N_\xi = \frac{N_1}{L},\ N_\eta = \frac{N_2}{L},\ N_{\xi\eta} = N_{\eta\xi} = \frac{N_{12}+N_{21}}{2L},$$

$$Q_\xi = \frac{Q_1}{L}, Q_\eta = \frac{Q_2}{L},\ M_\xi = \frac{M_1}{L},\ M_\eta = \frac{M_2}{L},$$

$$M_{\xi\eta} = M_{\eta\xi} = \frac{M_{12}+M_{21}}{2L}. \tag{3.4a-h}$$

4. Elastic Relations and Strain-Displacement Equations

To establish the elastic relations of the latticed shell, we take out an element $ABCD$ shown in Fig. 6 and regard it as a continuum element of area L^2. Using assumptions (a), (d), (e), (f), (h), we have the following expression for the complementary energy of the latticed element subject to actions of extension, bending, and twisting forces

$$V^{(\mathrm{I})} = \frac{N_1^2}{2EA}L + \frac{N_2^2}{2EA}L + \frac{M_1^2}{2EI_1}L + \frac{M_2^2}{2EI_1}L + \frac{M_{12}^2}{2GJ}L + \frac{M_{21}^2}{2GJ}L$$

in which EI_1 is the transverse bending stiffness of each member.

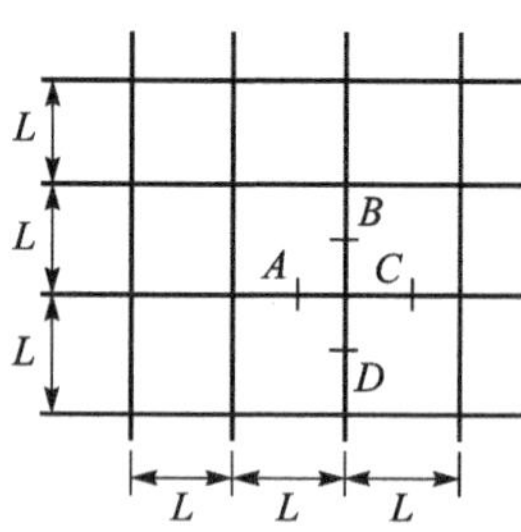

Fig. 6 A latticed element

Using (3.4) and the relation $M_{12}=M_{21}$ leads to

$$V^{(\mathrm{I})} = \frac{L^3}{2EA}(N_\xi^2 + N_\eta^2) + \frac{L^3}{2EI_1}(M_\xi^2 + M_\eta^2) + \frac{L^3}{GJ}M_{\xi\eta}^2. \tag{4.1}$$

The shear deformation of the latticed element is more complex. We must take into account the unified actions of equivalent shearing forces $\overline{N}_{12}, \overline{N}_{21}$ and the effective moments $(1/2)\ \overline{M}_1^*$, $(1/2)\ \overline{M}_2^*$ exerted on both ends of AC and BD due to transverse bending and twist of the element, where

$$\overline{M}_1^* = \overline{N}_{12}L,\quad \overline{M}_2^* = \overline{N}_{21}L. \tag{4.2}$$

As shown in Fig. 7, this unified deformation can be regarded as the deformation induced by pure shearing forces N_{12}^0, N_{21}^0 superposed on/or that caused by the self-equilibrium forces $\overline{N}_{12}$, $\overline{N}_{21}$ and $\overline{M}_1^*$, $\overline{M}_2^*$. It is also noted that in pure shearing deformation, the middle point of each member between two nearest joints is an inflexion point where the lateral bending moments vanish because the deformation pattern of the element must be antisymmetrical, thus only N_{12}^0 and N_{21}^0 remain on the ends of AC and BD in the case of pure shearing deformation of the element. It is clearly noted that the shearing stiffness of the lattices except those connected with the boundary ring is contributed by EI_2, the lateral bending stiffness of the members, which is in no way different from the shearing stiffness of the boundary lattice elements.

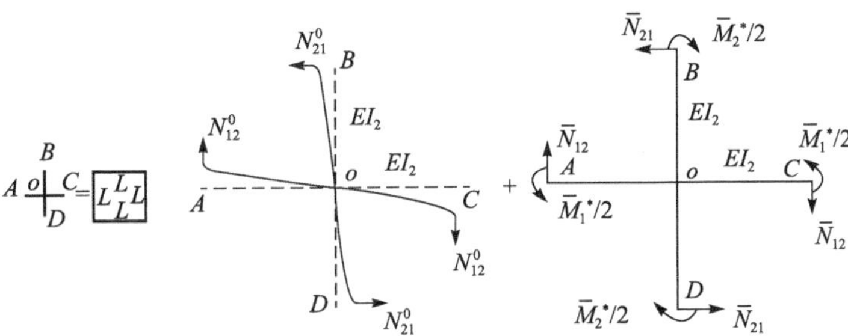

Fig. 7 Unified actions of shearing, twist and bending

The total complementary energy of such a lattice due to the unified actions can be represented as

$$V^{(\mathrm{II})}=\frac{L^3}{EI_2}\Big[\frac{1}{24}N_{12}^2+\frac{1}{24}N_{21}^2+\frac{1}{8}(\overline{N}_{12}^2+\overline{N}_{21}^2)-\frac{1}{8}N_{12}\overline{N}_{12}-\frac{1}{8}N_{21}\overline{N}_{21}\Big] \tag{4.3}$$

where $N_{12}=N_{12}^0+\overline{N}_{12}$, $N_{21}=N_{21}^0+\overline{N}_{21}$.

From Eqs. (3.1) and (3.4c) we have

$$N_{12}=LN_{\xi\eta}+\frac{1}{2}(\overline{N}_{12}-\overline{N}_{21})$$

$$N_{21}=LN_{\xi\eta}+\frac{1}{2}(\overline{N}_{21}-\overline{N}_{12})$$

substituting the above relation into (4.3) yields

$$V^{(\mathrm{II})}=\frac{L^3}{EI_2}\Big[\frac{L^2}{12}N_{\xi\eta}^2-\frac{L}{8}N_{\xi\eta}(\overline{N}_{12}+\overline{N}_{21})+\frac{1}{4}\overline{N}_{12}\overline{N}_{21}+\frac{1}{12}(\overline{N}_{12}-\overline{N}_{21})^2\Big]. \tag{4.4}$$

Therefore, the total complementary energy per unit area of the equivalent continuum element is

$$U=\frac{V^{(\mathrm{I})}+V^{(\mathrm{II})}}{L^2}$$

which is converted into

$$U=\frac{L}{2EA}(N_\xi^2+N_\eta^2)+\frac{L}{EI_2}\Big[\frac{L^2}{12}N_{\xi\eta}^2-\frac{L}{8}N_{\xi\eta}(\overline{N}_{12}+\overline{N}_{21})$$
$$+\frac{1}{4}\overline{N}_{12}\,\overline{N}_{21}+\frac{1}{12}(\overline{N}_{12}-\overline{N}_{21})^2\Big]$$
$$+\frac{L}{2EI_1}(M_\xi^2+M_\eta^2)+\frac{1}{GJ}M_{\xi\eta}^2.$$

Then, the elastic relations for the equivalent shell can be derived from Green's relations,

$$\varepsilon_\xi=\frac{\partial U}{\partial N_\xi}=\frac{L}{EA}N_\xi$$
$$\varepsilon_\eta=\frac{\partial U}{\partial N_\eta}=\frac{L}{EA}N_\eta$$
$$k_\xi=\frac{\partial^2 w}{\partial x^2}=\frac{\partial U}{\partial M_\xi}=\frac{L}{EI_1}M_\xi$$
$$k_\eta=-\frac{\partial^2 w}{\partial y^2}=\frac{\partial U}{\partial M_\eta}=\frac{L}{EI_1}M_\eta$$
$$2k_{\xi\eta}=-2\frac{\partial^2 w}{\partial x\partial y}=\frac{\partial U}{\partial M_{\xi\eta}}=\frac{2L}{GJ}M_{\xi\eta}$$
$$\gamma_{\xi\eta}=\frac{\partial U}{\partial N_{\xi\eta}}=\frac{L^3}{6EI_2}N_{\xi\eta}-\frac{L^2}{8EI_2}(\overline{N}_{12}+\overline{N}_{21}). \qquad (4.5\text{a-f})$$

Using (3. 3a,b) and (4. 5c-e) in (4. 5f) yields

$$\gamma_{\xi\eta}=\frac{L^3}{6EI_2}N_{\xi\eta}+\frac{L^3}{8EI_2}\Big[\frac{GJ}{L}\Big(\frac{2}{R}+\nabla^2 w\Big)\frac{\partial^2 w}{\partial x\partial y}-\frac{EI_1}{L}\frac{\partial^2 w}{\partial x\partial y}\nabla^2 w\Big]. \qquad (4.6)$$

Thus relations (4. 5a-e) and (4. 6) constitute the elastic equations of the equivalent shell, in which ε_ξ, ε_η and $\gamma_{\xi\eta}$ are equivalent strains of the middle surface of the shell.

Using the well-known large deflection geometrical equations for shallow shells, we have the following strain-displacement equations for the equivalent shell

$$\varepsilon_\xi=\frac{\partial u}{\partial x}+\frac{1}{2}\Big(\frac{\partial w}{\partial x}\Big)^2-\frac{w}{R}$$
$$\varepsilon_\eta=\frac{\partial v}{\partial y}+\frac{1}{2}\Big(\frac{\partial w}{\partial y}\Big)^2-\frac{w}{R}$$
$$\gamma_{\xi\eta}=\frac{\partial u}{\partial y}+\frac{\partial v}{\partial x}+\frac{\partial w}{\partial x}\frac{\partial w}{\partial y}. \qquad (4.7)$$

5. Equilibrium and Compatability Equations

The external loads exerted on the equivalent shell surface can be decomposed as (P_ξ, P_η, P_ζ) in the coordinate (ξ,η,ζ). We here define the boundary of the shell is Γ, which lies in-plane Oxy and encircles an area Σ of the same plane. Let $\hat{n}$ and s indicate the unit normal and tangent vector of Γ in the plane Oxy, and onsζ be the boundary local coordinate where o is on Γ and n is a unit vector tangent to a meridian of the shell but normal to Γ. It should be noted that n differs from $\hat{n}$ because n does not lie in plane Oxy, as shown in Fig. 8.

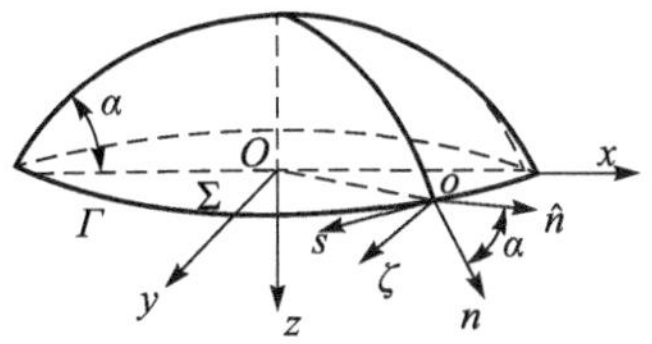

Fig. 8　Boundary local coordinate

The displacement of the shell on Γ is decomposed into (u_n, u_s, w) in coordinates (n, s, ζ). Supposing the outer edge of the shell is immovably clamped on hinged, we have

$$\delta u_n = \delta u_s = \delta w = \delta\left(\frac{\partial w}{\partial s}\right) = 0 \quad \text{on } \Gamma \tag{5.1}$$

where δ denotes variation. Since the shell is a shallow one, we have

$$\mathrm{d}\xi\mathrm{d}\eta = \mathrm{d}x\mathrm{d}y.$$

From the virtual work principle we have

$$\begin{aligned}
&\iint_{\Sigma}\Big[\delta\zeta_{\xi}N_{\xi} + \delta\gamma_{\xi\eta}N_{\xi\eta} + \delta\varepsilon_{\eta}N_{\eta} + M_{\xi}\delta\left(-\frac{\partial^2 w}{\partial x^2}\right) \\
&+ 2M_{\xi\eta}\delta\left(-\frac{\partial^2 w}{\partial x\partial y}\right) + M_{\eta}\delta\left(-\frac{\partial^2 w}{\partial y^2}\right)\Big]\mathrm{d}x\mathrm{d}y \\
&-\iint_{\Sigma}(P_{\xi}\delta u + P_{\eta}\delta v + P_{\zeta}\delta w)\mathrm{d}x\mathrm{d}y \\
&+\oint_{\Gamma}\overline{M}_n\delta\left(\frac{\partial w}{\partial n}\right)\mathrm{d}\Gamma = 0
\end{aligned}$$

where $\overline{M}_n$ is a given boundary normal bending moment.

Substituting Eqs. (4.7) into the above equations, using integration by parts, condition (5.1) and Green's theorem yield

$$\begin{aligned}
&\iint_{\Sigma}\left(\frac{\partial N_{\xi}}{\partial x} + \frac{\partial N_{\xi\eta}}{\partial y} + P_{\xi}\right)\delta u\,\mathrm{d}x\mathrm{d}y + \iint_{\Sigma}\left(\frac{\partial N_{\xi\eta}}{\partial x} + \frac{\partial N_{\eta}}{\partial y} + P_{\eta}\right)\delta v\,\mathrm{d}x\mathrm{d}y \\
&+\iint_{\Sigma}\Big[\frac{\partial^2 M_{\xi}}{\partial x^2} + 2\frac{\partial^2 M_{\xi\eta}}{\partial x\partial y} + \frac{\partial^2 M_{\eta}}{\partial y^2} + \frac{\partial}{\partial x}\left(N_{\xi}\frac{\partial w}{\partial x}\right) + \frac{\partial}{\partial y}\left(N_{\eta}\frac{\partial w}{\partial y}\right) + \frac{1}{R}(N_{\xi} + N_{\eta}) \\
&+\frac{\partial}{\partial x}\left(N_{\xi\eta}\frac{\partial w}{\partial y}\right) + \frac{\partial}{\partial y}\left(N_{\xi\eta}\frac{\partial w}{\partial x}\right) + P_{\zeta}\Big]\delta w\,\mathrm{d}x\mathrm{d}y + \oint_{\Gamma}(M_n - \overline{M}_n)\delta\left(\frac{\partial w}{\partial n}\right)\mathrm{d}\Gamma = 0
\end{aligned} \tag{5.2}$$

in which M_n is the normal bending moment.

The first two integrals of (5.2) yield

$$\begin{aligned}
&\frac{\partial N_{\xi}}{\partial x} + \frac{\partial N_{\xi\eta}}{\partial y} + P_{\xi} = 0 \\
&\frac{\partial N_{\xi\eta}}{\partial x} + \frac{\partial N_{\eta}}{\partial y} + P_{\eta} = 0
\end{aligned} \tag{5.3}$$

Using (5.3) in the third integral of (5.2) we obtain

$$\begin{aligned}
&\iint_{\Sigma}\Big[\frac{\partial^2 M_{\xi}}{\partial x^2} + 2\frac{\partial^2 M_{\xi\eta}}{\partial x\partial y} + \frac{\partial^2 M_{\eta}}{\partial y^2} + N_{\xi}\frac{\partial^2 w}{\partial x^2} + 2N_{\xi\eta}\frac{\partial^2 w}{\partial x\partial y} + N_{\eta}\frac{\partial^2 w}{\partial y^2} \\
&- P_{\xi}\frac{\partial w}{\partial x} - P_{\eta}\frac{\partial w}{\partial y} + \frac{1}{R}(N_{\xi} + N_{\eta}) + P_{\zeta}\Big]\delta w\,\mathrm{d}x\mathrm{d}y = 0.
\end{aligned} \tag{5.4}$$

The last integral of (5. 2) yields

$$\iint_{\Gamma}(M_n-\overline{M}_n)\delta\left(\frac{\partial w}{\partial n}\right)\mathrm{d}\Gamma=0. \tag{5. 5}$$

Equations (5. 3) and (5. 4) are the basic equilibrium equations for the equivalent shell. (5. 1) and (5. 5) are corresponding boundary conditions.

If the extermal loads are axisymmetric, we have

$$P_\xi = xf(r)$$
$$P_\eta = yf(r)$$
$$P_\zeta = P_\zeta(r) \tag{5. 6}$$

where r is the distance from a point on the middle surface to the central symmetric axis of the shell, i. e.

$$r=\sqrt{x^2+y^2}$$

and $f(r)$ is a continuous function.

Obviously the following relation holds true

$$\frac{\partial P_\xi}{\partial y}=\frac{\partial P_\eta}{\partial x}$$

therefore P_ξ and P_η can be expressed by a potential $F(r)$ as

$$P_\xi=\frac{\partial F}{\partial x}$$
$$P_\eta=\frac{\partial F}{\partial y} \tag{5. 7}$$

where

$$F(r)=\int f(r)r\mathrm{d}r.$$

Invoking the stress function $\phi(x,y)$ such that

$$N_\xi=\frac{\partial^2\phi}{\partial y^2}-F$$
$$N_\eta=\frac{\partial^2\phi}{\partial x^2}-F$$
$$N_{\xi\eta}=-\frac{\partial^2\phi}{\partial x\partial y}. \tag{5. 8}$$

Thus Eq. (5. 3) is automatically satisfied. Using (4. 5c-e) and (5. 7), (5. 8) in Eq. (5. 4) leads to

$$\iint_{\Sigma}\left[L(w,\phi)-P_\zeta+\frac{2}{R}F\right]\delta w\mathrm{d}x\mathrm{d}y=0 \tag{5. 9}$$

in which

$$L(w,\phi)=\frac{EI_1}{L}\nabla^4 w+\frac{2(GJ-EI_1)}{L}\frac{\partial^4 w}{\partial x^2\partial y^2}$$
$$-\frac{\partial^2\phi}{\partial x^2}\frac{\partial^2 w}{\partial y^2}-\frac{\partial^2\phi}{\partial y^2}\frac{\partial^2 w}{\partial x^2}+2\frac{\partial^2\phi}{\partial x\partial y}\frac{\partial^2 w}{\partial x\partial y}$$

$$-\frac{1}{R}\nabla^2\phi+F\nabla^2 w+\frac{\partial F}{\partial x}\frac{\partial w}{\partial x}+\frac{\partial F}{\partial y}\frac{\partial w}{\partial y}.$$

The variational Eq. (5.9) can be used in approximate analysis.

To obtain the compatability equation, first let some components of the displacement of Γ be given as

$$u_n=\bar{u}_n, u_s=0, w=0, \frac{\partial w}{\partial s}=0, \frac{\partial w}{\partial n}=0 \quad \text{on } \Gamma. \tag{5.10}$$

Applying the stationary complementary energy principle[7] to a shallow shell, we have

$$\iint_\Sigma\left[\varepsilon_\xi\delta N_\xi+\varepsilon_\eta\delta N_\eta+\gamma_{\xi\eta}\delta N_{\xi\eta}+\left(-\frac{\partial^2 w}{\partial x^2}\right)\delta M_\xi+\left(-2\frac{\partial^2 w}{\partial x\partial y}\right)\delta M_{\xi\eta}+\left(-\frac{\partial^2 w}{\partial y^2}\right)\delta M_\eta\right]\mathrm{d}x\mathrm{d}y$$
$$+\iint_\Sigma\left[\frac{1}{2}\left(\frac{\partial w}{\partial x}\right)^2\delta N_\xi+\left(\frac{\partial w}{\partial x}\right)\left(\frac{\partial w}{\partial y}\right)\delta N_{\xi}n+\frac{1}{2}\left(\frac{\partial w}{\partial y}\right)^2\delta N_\eta\right]\mathrm{d}x\mathrm{d}y-\oint_\Gamma\bar{u}_n\delta N_n\mathrm{d}\Gamma=0. \tag{5.11}$$

Equations (5.3) and (5.4) require that

$$\frac{\partial\delta N_\xi}{\partial x}+\frac{\partial\delta N_{\xi\eta}}{\partial y}=0$$
$$\frac{\partial\delta N_{\xi\eta}}{\partial x}+\frac{\partial\delta N_\eta}{\partial y}=0$$
$$\frac{\partial^2\delta M_\xi}{\partial x^2}+2\frac{\partial^2\delta M_{\xi\eta}}{\partial x\partial y}+\frac{\partial^2\delta M_\eta}{\partial y^2}+\delta N_\xi\frac{\partial^2 w}{\partial x^2}+2\delta N_{\xi\eta}\frac{\partial^2 w}{\partial x\partial y}+\delta N_\eta\frac{\partial^2 w}{\partial y^2}+\frac{1}{R}(\delta N_\xi+\delta N_\eta)=0.$$

Using the above equations and conditions (5.10) in (5.11) yields

$$\iint_\Sigma\left\{\left[\varepsilon_\xi-\frac{\partial u}{\partial x}-\frac{1}{2}\left(\frac{\partial w}{\partial x}\right)^2+\frac{w}{R}\right]\delta N_\xi+\left[\gamma_{\xi\eta}-\frac{\partial u}{\partial y}-\frac{\partial v}{\partial x}-\frac{\partial w}{\partial x}\frac{\partial w}{\partial y}\right]\delta N_{\xi\eta}\right.$$
$$\left.+\left[\varepsilon_\eta-\frac{\partial v}{\partial y}-\frac{1}{2}\left(\frac{\partial w}{\partial y}\right)^2+\frac{w}{R}\right]\delta N_\eta\right\}\mathrm{d}x\mathrm{d}y+\oint_\Gamma(u_n-\bar{u}_n)\delta N_n\mathrm{d}\Gamma=0.$$

Substituting (4.5a,b),(4.6) and (5.8) into the above equation and taking

$$\delta\phi=\delta\left(\frac{\partial\phi}{\partial s}\right)=\delta\left(\frac{\partial\phi}{\partial n}\right)=0 \quad \text{on } \Gamma$$

we finally arrive at

$$\iint_\Sigma\left[L^*(\phi,w)-\frac{L}{EA}\nabla^2 F\right]\delta\phi\mathrm{d}x\mathrm{d}y+\oint_\Gamma(u_n-\bar{u}_n)\delta N_n\mathrm{d}\Gamma=0 \tag{5.12}$$

where

$$L^*(\phi,w)=\frac{L}{EA}\nabla^4\phi+\left(\frac{L^3}{6EI_2}-\frac{2L}{EA}\right)\frac{\partial^4\phi}{\partial x^2\partial y^2}$$
$$-\frac{GJL^2}{4EI_2R}\frac{\partial^4 w}{\partial x^2\partial y^2}+\frac{1}{R}\nabla^2 w$$
$$-\frac{L^2(GJ-EI_1)}{8EI_2}\frac{\partial^2}{\partial x\partial y}\left(\frac{\partial^2 w}{\partial x\partial y}\nabla^2 w\right)$$
$$-\left(\frac{\partial^2 w}{\partial x\partial y}\right)^2+\frac{\partial^2 w}{\partial x^2}\frac{\partial^2 w}{\partial y^2}.$$

Equation (5.12) can be separated as follows:

$$\iint_{\Sigma}\left[L^{*}(\phi,w)-\frac{L}{EA}\nabla^{2}F\right]\delta\phi\,\mathrm{d}x\mathrm{d}y=0 \tag{5.13}$$

and

$$\oint_{\Gamma}(u_n-\bar{u}_n)\delta N_n\,\mathrm{d}\Gamma=0. \tag{5.14}$$

Equation (5.13) is the compatible equation in variational form.

Thus, the analysis of the deformation of latticed shallow shells is reduced to solving equations (5.9) and (5.13) with corresponding boundary conditions.

6. Approximate Governing Equations in a Polar Coordinate System

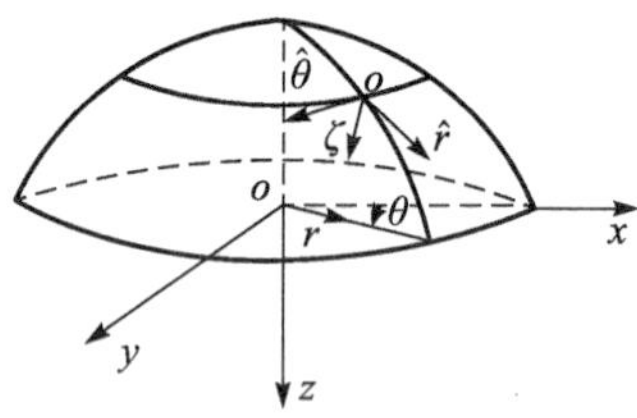

Fig. 9　Polar coordinate systems of the shell

Now we set up a cylindrical polar coordinate system (r,θ,z) and a moving coordinate system $(\hat{r},\hat{\theta},\zeta)$ as shown in Fig. 9. The displacement of the shell is decomposed into (u_r,u_θ,w) where u_r is tangent to a meridian and u_θ is tangent to a latitude of the middle surface of the spherical shell, and w is normal to the middle surface. Then, the basic equations (5.9) and (5.13) can be transformed into the following forms

$$\begin{aligned}&\int_0^{2\pi}\int_0^{a}\Big[\frac{EI_1}{L}\nabla^4 w+\frac{2(GJ-EI_1)}{L}K_1(w)-\frac{1}{R}\nabla^2\phi-K_2(w,\phi)\\&\quad+F\nabla^2 w+\frac{\partial F}{\partial r}\frac{\partial w}{\partial r}-P_\zeta+\frac{2}{R}F\Big]\delta w r\,\mathrm{d}r\mathrm{d}\theta=0\\&\int_0^{2\pi}\int_0^{a}\Big[\frac{L}{EA}\nabla^4\phi+\left(\frac{L^3}{6EI_2}-\frac{2L}{EA}\right)K_1(\phi)+\frac{1}{R}\nabla^2 w\\&\quad-\frac{GJL^2}{4EI_2R}K_1(w)+\frac{1}{2}K_2(w,w)-\frac{L}{EA}\nabla^2 F\\&\quad-\frac{L^2(GJ-EI_1)}{8EI_2}K_0(K_0(w)\nabla^2 w)\Big]\delta\phi r\,\mathrm{d}r\mathrm{d}\theta=0\end{aligned} \tag{6.1a,b}$$

and (5.5), (5.14) are converted into

$$\int_0^{2\pi}(M_r-\overline{M}_r)\delta\left(\frac{\partial w}{\partial r}\right)2\pi a\,\mathrm{d}\theta=0$$

and

$$\int_0^{2\pi}(u_r-\bar{u}_r)\delta N_r 2\pi a\,\mathrm{d}\theta=0 \tag{6.2a,b}$$

where M_r is the radial bending moment, and N_r is the radial membrane stress. Also,

$$\begin{aligned}M_r&=M_\xi\cos^2\theta+M_\eta\sin^2\theta+2M_{\xi\eta}\sin\theta\cos\theta\\&=-\left(\frac{3EI_1+GJ}{4L}-\frac{EI_1-GJ}{4L}\cos 4\theta\right)\frac{\partial^2 w}{\partial r^2}\\&\quad+\frac{EI_1-GJ}{2L}\sin 4\theta\left(\frac{1}{r}\frac{\partial^2 w}{\partial r\partial\theta}-\frac{1}{r^2}\frac{\partial w}{\partial\theta}\right)\end{aligned}$$

$$-\frac{EI_1-GJ}{4L}(1-\cos 4\theta)\left(\frac{1}{r}\frac{\partial w}{\partial r}+\frac{1}{r^2}\frac{\partial^2 w}{\partial \theta^2}\right) \tag{6.3}$$

The differential operators ∇^2, ∇^4, K_0, K_1 and K_2 are as follows

$$\nabla^2=\frac{\partial^2}{\partial r^2}+\frac{1}{r}\frac{\partial}{\partial r}+\frac{1}{r^2}\frac{\partial^2}{\partial \theta^2},\ \nabla^4=\nabla^2\nabla^2$$

$$K_0()=\sin\theta\cos\theta\frac{\partial^2()}{\partial r^2}+\frac{\cos^2\theta-\sin^2\theta}{r}\frac{\partial^2()}{\partial r\partial\theta}-\frac{\sin\theta\cos\theta}{r}\frac{\partial()}{\partial r}$$
$$-\frac{\cos^2\theta-\sin^2\theta}{r^2}\frac{\partial()}{\partial\theta}-\frac{\sin\theta\cos\theta}{r^2}\frac{\partial^2()}{\partial\theta^2}$$

$$K_1()=\frac{1-\cos 4\theta}{8}\frac{\partial^4()}{\partial r^4}+\frac{1+3\cos 4\theta}{4r}\frac{\partial^3()}{\partial r^3}-\frac{1+15\cos 4\theta}{8r^2}\frac{\partial^2()}{\partial r^2}+\frac{1+15\cos 4\theta}{8r^3}\frac{\partial()}{\partial r}$$
$$+\frac{\sin 4\theta}{2r}\frac{\partial^4()}{\partial r^3\partial\theta}-\frac{3\sin 4\theta}{r^2}\frac{\partial^3()}{\partial r^2\partial\theta}+\frac{7\sin 4\theta}{r^3}\frac{\partial^2()}{\partial r\partial\theta}-\frac{6\sin 4\theta}{r^4}\frac{\partial()}{\partial\theta}$$
$$+\frac{1+3\cos 4\theta}{4r^2}\frac{\partial^4()}{\partial r^2\partial\theta^2}-\frac{1+15\cos 4\theta}{4r^3}\frac{\partial^3()}{\partial r\partial\theta^2}+\frac{1+11\cos 4\theta}{2r^4}\frac{\partial^2()}{\partial\theta^2}$$
$$-\frac{\sin 4\theta}{2r^3}\frac{\partial^4()}{\partial r\partial\theta^3}+\frac{3\sin 4\theta}{2r^4}\frac{\partial^3()}{\partial\theta^3}+\frac{1-\cos 4\theta}{8r^4}\frac{\partial^4()}{\partial\theta^4}$$

$$K_2(w,\phi)=\frac{\partial^2 w}{\partial r^2}\left(\frac{1}{r}\frac{\partial\phi}{\partial r}+\frac{1}{r^2}\frac{\partial^2\phi}{\partial\theta^2}\right)+\left(\frac{1}{r}\frac{\partial w}{\partial r}+\frac{1}{r^2}\frac{\partial^2 w}{\partial\theta^2}\right)\frac{\partial^2\phi}{\partial r^2}-2\frac{\partial}{\partial r}\left(\frac{1}{r}\frac{\partial w}{\partial\theta}\right)\frac{\partial}{\partial r}\left(\frac{1}{r}\frac{\partial\phi}{\partial\theta}\right).$$

Now we consider the case of the latticed shell subject to uniform vertical load q. We have

$$F(r)=\frac{r^2}{2R}q. \tag{6.4}$$

The relation between P_ζ and q can be expressed as

$$q=P_\zeta+\frac{r}{R}\frac{\mathrm{d}F}{\mathrm{d}r} \tag{6.5}$$

and the membrane force-stress function relations are

$$N_r=\frac{1}{r}\frac{\partial\phi}{\partial r}+\frac{1}{r^2}\frac{\partial^2\phi}{\partial\theta^2}-F$$
$$N_\theta=\frac{\partial^2\phi}{\partial r^2}-F$$
$$N_{r\theta}=-\frac{\partial}{\partial r}\frac{1}{r}\frac{\partial\phi}{\partial\theta}. \tag{6.6a-c}$$

Because of the geometrical symmetry of the latticed shell about the x-axis or y-axis and axisymmetry of external loads and boundary restraint, it is reasonable to approximate the displacement and stress state of the shell as

$$u_r(r,\theta)=u^*(r)+\sum_{j=1}^{N}(A_j(r)\cos 4j\theta+B_j(r)\sin 4j\theta)$$
$$u_\theta(r,\theta)=\sum_{j=1}^{N}C_j(r)\sin 8j\theta$$

$$w(r,\theta)=w^{*}(r)+\sum_{j=1}^{N}(D_j(r)\cos 4j\theta+E_j(r)\sin 4j\theta)$$

$$\phi(r,\theta)=\phi^{*}(r)+\sum_{j=1}^{N}(F_j(r)\cos 4j\theta+G_j(r)\sin 4j\theta) \tag{6. 7a-d}$$

where $u^{*}(r), w^{*}(r), \phi^{*}(r), A_j(r), B_j(r), C_j(r), D_j(r), E_j(r), F_j(r)$ and $G_j(r)$ are unspecified continuous functions.

Substituting the above expressions into (6. 1a,b), (6. 2a,b) and other appropriate boundary conditions, using integration of θ over the interval $[0,2\pi]$ yields a set of nonlinear coupled ordinary differential equations and corresponding boundary conditions. The analysis is very complex. We however can reasonably simplify the problem by considering that the governing terms in (6. 7) are those indicating the symmetrical part, i. e., $u^{*}(r), w^{*}(r)$ and $\phi^{*}(r)$.

Thus, we substitute $u^{*}(r), w^{*}(r)$ and $\phi^{*}(r)$ for $u_r(r,\theta), w(r,\theta)$ and $\phi(r,\theta)$, and set $u_\theta=0$, then put them into (6. 1a,b). After using (6. 4)-(6. 6) and integrating the equation, we finally have

when $0<r<a$

$$\frac{3\,EI_1+GJ}{4L}r\frac{\mathrm{d}}{\mathrm{d}r}\frac{1}{r}\frac{\mathrm{d}}{\mathrm{d}r}(r\psi_2)-rN_r(\psi_1+\psi_2)=\frac{1}{2}qr^2 \tag{6. 8}$$

$$\left(\frac{L^3}{48EI_2}+\frac{3L}{4EA}\right)r\frac{\mathrm{d}}{\mathrm{d}r}\frac{1}{r}\frac{\mathrm{d}}{\mathrm{d}r}(r^2N_r)-\frac{GJL^2}{32EI_2R}r\frac{\mathrm{d}}{\mathrm{d}r}\frac{1}{r}\frac{\mathrm{d}}{\mathrm{d}r}(r\psi_2)$$
$$-\frac{(GJ-EI_1)}{64EI_2}L^2\left(r\frac{\mathrm{d}f}{\mathrm{d}r}+2f\right)+\psi_2\left(\psi_1+\frac{1}{2}\psi_2\right)=-\frac{1}{R}\left(\frac{L^3}{12EI_2}+\frac{2L}{EA}\right)qr^2 \tag{6. 9}$$

in which

$$f=\left(\frac{\mathrm{d}\psi_2}{\mathrm{d}r}\right)^2-\frac{1}{r^2}\psi_2^2$$

$$\psi_1=\frac{r}{R},\quad \psi_2=\frac{\mathrm{d}w^{*}}{\mathrm{d}r},$$

$$N_r=\frac{1}{r}\frac{\mathrm{d}\phi^{*}}{\mathrm{d}r}-F$$

$$N_{r\theta}=0$$

$$N_\theta=\frac{\mathrm{d}^2\phi^{*}}{\mathrm{d}r^2}-F$$

7. Approximate Boundary Conditions

The outer edge of the shell is immovably clamped or hinged. Using the assumned axisymmetric displacement and stress state, we can express the exact boundary conditions as

$$w^{*}=0\quad \text{at } r=a$$

$$\frac{\mathrm{d}w^{*}}{\mathrm{d}r}=0,\quad rN_r=0 \text{ at } r=0.$$

According to (6.2a, b), the approximate boundary conditions are

$$\int_0^{2\pi}(u^* - \bar{u}^*)\delta N_r \mathrm{d}\theta = 0$$

$$\text{or} \int_0^{2\pi}(M_r - \overline{M}_r)\delta\left(\frac{\mathrm{d}w^*}{\mathrm{d}r}\right)\mathrm{d}\theta = 0 \quad \text{at } r = a \tag{7.1a,b}$$

in our case $N_r = N_r(r)$, $w^* = w^*(r)$ and $\bar{u}^* = 0$, $\mathrm{d}w^*/\mathrm{d}r = 0$ (immovably clamped) or $\bar{u}^* = 0$, $\overline{M}_r = 0$ (hinged), therefore, the boundary conditions are finally reduced to

(a) Immovably clamped:

$$\begin{aligned} &w^* = 0, \quad \frac{\mathrm{d}w^*}{\mathrm{d}r} = 0, \quad \int_0^{2\pi} u^* \mathrm{d}\theta = 0 \text{ at } r = a \\ &\frac{\mathrm{d}w^*}{\mathrm{d}r} = 0, \quad rN_r = 0 \qquad\qquad \text{at } r = 0. \end{aligned} \tag{7.2a-e}$$

(b) Hinged:

$$\begin{aligned} &w^* = 0, \quad \int_0^{2\pi} M_r \mathrm{d}\theta = 0, \quad \int_0^{2\pi} u^* \mathrm{d}\theta = 0 \quad \text{at } r = a \\ &\frac{\mathrm{d}w^*}{\mathrm{d}r} = 0, \quad rN_r = 0 \qquad\qquad \text{at } r = 0. \end{aligned} \tag{7.3a-e}$$

In practical case, the latticed shell is bounded by a circular ring, as shown in Fig. 10, so the lattices connected with the ring are irregular curved triangular or quadrilateral ones. As the members are placed densely enough, most of these boundary lattices are triangles (see Fig. 10 (a), the shaded triangles), the shearing stiffness of the boundary lattices is created mainly by EA, the axial extension or compression stiffness of the ring. A simple and useful model is that a right isosceles, as shown in Fig. 10 (b), in which the hypotenuse is a part of the ring, is substituted for a boundary lattice. The shearing forces in the boundary equivalent continuum element are $N_{\xi\eta}$ and $N_{\eta\xi}$, resulted from N_3, the axial force of the ring.

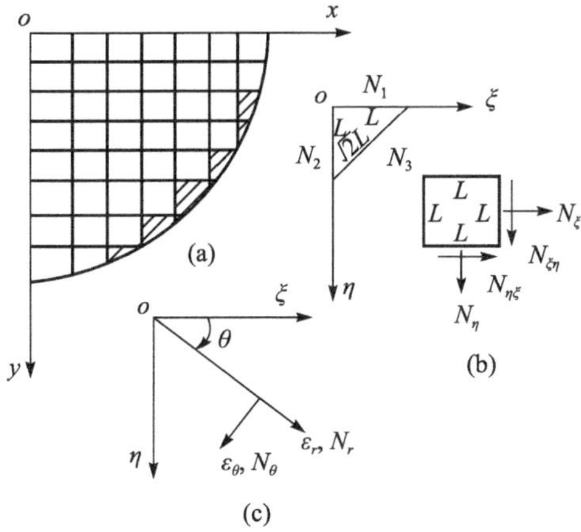

Fig. 10 Boundary lattice element

From Fig. 10 (b), (c), using coordinate transformation leads to

$$N_\xi = \frac{N_1}{L} + \frac{N_3}{\sqrt{2}L}, \quad \varepsilon_1 = \varepsilon_\xi$$

$$N_\eta = \frac{N_2}{L} + \frac{N_3}{\sqrt{2}L}, \quad \varepsilon_2 = \varepsilon_\eta$$

$$N_{\xi\eta} = -\frac{N_3}{\sqrt{2}L}, \quad \varepsilon_3 = \frac{1}{2}\varepsilon_\xi + \frac{1}{2}\varepsilon_\eta - \frac{1}{2}\gamma_{\xi\eta} \tag{7.4a-f}$$

and Hooke's law requires that

$$N_1 = EA\varepsilon_1, \quad N_2 = EA\varepsilon_2, \quad N_3 = EA\varepsilon_3. \tag{7.5a-c}$$

Using (7.4) and (7.5) yields the following relations

$$\varepsilon_\xi = \frac{L}{EA}(N_\xi + N_{\xi\eta})$$

$$\varepsilon_\eta = \frac{L}{EA}(N_\eta + N_{\xi\eta})$$

$$\gamma_{\xi\eta} = \frac{L}{EA}(N_\xi + N_\eta + 2(1+\sqrt{2})N_{\xi\eta}). \tag{7.6a-c}$$

We have assumed $N_{r\theta}=0$, hence we have (Fig. 10 (c))

$$\begin{Bmatrix} N_\xi \\ N_\eta \\ N_{\xi\eta} \end{Bmatrix} = \begin{bmatrix} \cos^2\theta & \sin^2\theta & -2\sin\theta\cos\theta \\ \sin^2\theta & \cos^2\theta & 2\sin\theta\cos\theta \\ \sin\theta\cos\theta & -\sin\theta\cos\theta & \cos^2\theta - \sin^2\theta \end{bmatrix} \begin{Bmatrix} N_r \\ N_\theta \\ 0 \end{Bmatrix} \tag{7.7}$$

and the circumferential membrane strain is

$$\varepsilon_\theta = \varepsilon_\xi \sin^2\theta + \varepsilon_\eta \cos^2\theta - \gamma_{\xi\eta}\sin\theta\cos\theta. \tag{7.8}$$

Using (7.6) and (7.7) in (7.8) we finally have

$$\varepsilon_\theta = \frac{L}{EA}[(1 - \sin 2\theta + 2\sqrt{2}\sin^2\theta\cos^2\theta)N_\theta - 2\sqrt{2}N_r\sin^2\theta\cos^2\theta].$$

In the polar coordinate system, using the assumed displacement and stress field, we have

$$\varepsilon_\theta = \frac{u^*}{r} - \frac{w^*}{R}$$

$$N_\theta = r\frac{\mathrm{d}F}{\mathrm{d}r} + \frac{\mathrm{d}}{\mathrm{d}r}(rN_r) = \frac{r^2}{R}q + \frac{\mathrm{d}}{\mathrm{d}r}(rN_r).$$

Noticing that w^* vanishes at $r=a$, so we have $u^* = a\varepsilon_\theta$ at $r=a$, or

$$u^* = \frac{La}{EA}\left\{[(1 - \sin 2\theta + 2\sqrt{2}\sin^2\theta\cos^2\theta)\left[\frac{\mathrm{d}}{\mathrm{d}r}(rN_r) + \frac{a^2}{R}q\right]\right.$$

$$\left. - 2\sqrt{2}N_r\sin^2\theta\cos^2\theta\right\} \quad \text{at } r = a. \tag{7.9}$$

Putting (6.3) and (7.9) into (7.2c), (7.3b,c) and integrating the equations on θ yields

(a) Immovably clamped:

$$\frac{L}{EA}\left(1 - \frac{2}{\pi} + \frac{\sqrt{2}}{4}\right)\left[a^2\frac{\mathrm{d}N_r}{\mathrm{d}r} + (1-n)aN_r + \frac{a^3}{R}q\right] = 0 \text{ at } r = a \tag{7.10}$$

(b) Hinged:

$$\frac{L}{EA}\left(1-\frac{2}{\pi}+\frac{\sqrt{2}}{4}\right)\left[a^2\frac{\mathrm{d}N_r}{\mathrm{d}r}+(1-n)aN_r+\frac{a^3}{R}q\right]=0$$

$$-\frac{3EI_1+GJ}{4L}\left(\frac{\mathrm{d}^2w^*}{\mathrm{d}r^2}+\frac{m}{r}\frac{\mathrm{d}w^*}{\mathrm{d}r}\right)=0\quad \text{at } r=a \tag{7.11a,b}$$

where

$$m=\frac{EI_1-GJ}{3EI_1+GJ},\quad n=\frac{\sqrt{2}}{\left(4+\sqrt{2}-\frac{8}{\pi}\right)}.$$

8. Non-Dimensional Boundary Value Problem and Its Solution

To make the subsequent analysis simpler, we introduce the following non-dimensional quantities:

$$\rho=\frac{r}{a},\quad W=\frac{4w^*}{\sqrt{(3EI_1+GJ)\left(\frac{3}{EA}+\frac{L^2}{12EI_2}\right)}},\quad \varphi=\frac{\mathrm{d}W}{\mathrm{d}\rho}$$

$$S=\frac{4La^2\rho}{3EI_1+GJ}N_r,\quad k_1=\frac{4a^2}{R\sqrt{(3EI_1+GJ)\left(\frac{3}{EA}+\frac{L^2}{12EI_2}\right)}}$$

$$k_2=\frac{GJL^2}{8EI_2R\sqrt{(3EI_1+GJ)\left(\frac{3}{EA}+\frac{L^2}{12EI_2}\right)}}$$

$$k_3=\frac{1}{R}\left(\frac{4}{EA}+\frac{L^2}{6EI_2}\right)\sqrt{\frac{3EI_1+GJ}{\frac{3}{EA}+\frac{L^2}{12EI_2}}}$$

$$k_4=\frac{1}{2R}\sqrt{(3EI_1+GJ)\left(\frac{3}{EA}+\frac{L^2}{12EI_2}\right)}$$

$$k_5=\frac{(GJ-EI_1)L^2}{32EI_2a^2},$$

$$Q=\frac{8La^4}{(3EI_1+GJ)\sqrt{(3EI_1+GJ)\left(\frac{3}{EA}+\frac{L^2}{12EI_2}\right)}}q.$$

Using the above quantities, we can convert Eqs. (6.8), (6.9) and boundary conditions (7.2a,b,d,e), (7.10) or (7.3a,d,e), (7.11a,b) into the following non-dimensional forms

$$\varphi=\frac{\mathrm{d}W}{\mathrm{d}\rho}$$

$$\mathscr{L}(\rho\varphi)=S(k_1\rho+\varphi)+Q\rho^2$$

$$\mathscr{L}(\rho S)-k_2\mathscr{L}(\rho\varphi)=-\varphi\left(k_1\rho+\frac{1}{2}\varphi\right)+k_5\left[\left(\frac{\mathrm{d}\varphi}{\mathrm{d}\rho}\right)^2-\left(\frac{\varphi}{\rho}\right)^2\right]$$

$$+\frac{k_5}{2}\rho\frac{\mathrm{d}}{\mathrm{d}\rho}\left[\left(\frac{\mathrm{d}\varphi}{\mathrm{d}\rho}\right)^2-\left(\frac{\varphi}{\rho}\right)^2\right]-k_3Q\rho^2$$

and for the cases of:

(a) Immovably clamped:

$$W=0,\quad \varphi=0,\quad \frac{\mathrm{d}S}{\mathrm{d}\rho}-nS+k_4Q=0\quad \text{at } \rho=1$$

$$\varphi=0,\quad S=0\qquad \text{at } \rho=0$$

(b) Hinged:

$$W=0,\quad \frac{\mathrm{d}\varphi}{\mathrm{d}\rho}+m\varphi=0,\quad \frac{\mathrm{d}S}{\mathrm{d}\rho}-nS+k_4Q=0\quad \text{at } \rho=1$$

$$\varphi=0,\quad S=0\qquad \text{at } \rho=0$$

where $\mathscr{L}$ is an operator

$$\mathscr{L}()=\rho\frac{\mathrm{d}}{\mathrm{d}\rho}\frac{1}{\rho}\frac{\mathrm{d}}{\mathrm{d}\rho}(\ ).$$

It is also noted that putting $m\to\infty$ in (b) yields (a), therefore we only need to consider case (b) in the subsequent analysis.

We now solve the above boundary value problem by means of the modified iteration method[8-10]. We begin with the following linear boundary value problem

$$\varphi_1=\frac{\mathrm{d}W_1}{\mathrm{d}\rho}$$

$$\mathscr{L}(\rho\varphi_1)=Q\rho^2$$

$$\mathscr{L}(\rho S_1)=k_2\mathscr{L}(\rho\varphi_1)-\varphi_1\left(k\rho+\frac{1}{2}\varphi_1\right)+k_5\left[\left(\frac{\mathrm{d}\varphi_1}{\mathrm{d}\rho}\right)^2-\left(\frac{\varphi_1}{\rho}\right)^2\right]$$

$$+\frac{k_5}{2}\rho\frac{\mathrm{d}}{\mathrm{d}\rho}\left[\left(\frac{\mathrm{d}\varphi_1}{\mathrm{d}\rho}\right)-\left(\frac{\varphi_1}{\rho}\right)^2\right]-k_3Q\rho^2.\tag{8.1a-c}$$

$$W_1=0,\quad \frac{\mathrm{d}\varphi_1}{\mathrm{d}\rho}+m\varphi_1=0,\quad \frac{\mathrm{d}S_1}{\mathrm{d}\rho}-nS_1+k_4Q=0\quad \text{at } \rho=1\tag{8.2a-c}$$

$$\varphi_1=0,\quad S_1=0\quad \text{at } \rho=0.\tag{8.3a-b}$$

Integrating Eq. (8.1b) and using appropriate boundary conditions yields

$$\varphi_1=\frac{Q}{8}(\rho^3-\beta_1\rho)\tag{8.4}$$

where

$$\beta_1=\frac{3+m}{1+m}.$$

Since the deflection of the outer edge of the shell vanishes, we have the nondimensional central deflection of the shell as

$$W_m=W\mid_{\rho=0}=-\int_0^1\varphi\mathrm{d}\rho.\tag{8.5}$$

Substituting (8.4) into (8.5) we obtain

$$Q=32\beta_2W_m\tag{8.6}$$

where

$$\beta_2 = \frac{1+m}{5+m}.$$

Inserting (8.6) into (8.4) leads to

$$\varphi_1 = \beta_4(\rho^3 - \beta_1\rho) \tag{8.7}$$

in which

$$\beta_4 = 4\beta_2 W_m.$$

Upon placing solution (8.7) into Eq. (8.1c), integrating the equation and using boundary conditions (8.2c), (8.3b), we have

$$S_1 = \left[-\frac{1}{96}\rho^7 + \frac{1}{24}\beta_1\rho^5 - \frac{1}{16}\beta_1^2\rho^3 + k_5(\rho^5 - \beta_1\rho^3) + A_1\rho\right]\beta_4^2 + \left[-\frac{k_1}{24}\rho^5 + (k_2 - k_3 + \frac{1}{8}k_1\beta_1)\rho^3 + A_2\rho\right]\beta_4 \tag{8.8}$$

in which

$$A_1 = \frac{6(3-n)\beta_1^2 - 4(5-n)\beta_1 + 7 - n}{96(1-n)}$$

$$A_2 = \frac{(5-n)k_1 - 3(3-n)\beta_1 k_1 - 24(3-n)(k_2 - k_3) - 192k_4}{24(1-n)}.$$

Next, we have to solve the following modified boundary value problem

$$\varphi_2 = \frac{dW_2}{d\rho}$$

$$\mathscr{L}(\rho\varphi_2) = S_1(k_1\rho + \varphi_1) + Q\rho^2 \tag{8.9a,b}$$

$$W_2 = 0, \quad \frac{d\varphi_2}{d\rho} + m\varphi_2 = 0 \text{ at } \rho = 1$$

$$\varphi_2 = 0 \qquad \text{at } \rho = 0. \tag{8.10a-c}$$

After integrating Eq. (8.9b) and using (8.10b,c) we can obtain the expression of φ_2, then substituting this expression into (8.5) again we finally arrive at

$$Q = \alpha_1 W_m + \alpha_2 W_m^2 + \alpha_3 W_m^3 \tag{8.11}$$

in which

$$\alpha_1 = -\beta_2^2\left\{-\frac{9+m}{24(1+m)}k_1^2 + \frac{16(7+m)}{9(1+m)}\left[\frac{1}{8}k_1^2\beta_1 + (k_2 - k_3)k_1\right] + \frac{4(5+m)}{1+m}A_2 k_1 + 32\beta_2\right\}$$

$$\alpha_2 = -\beta_2^3\left\{-\frac{2k_1(11+m)}{15(1+m)} + \frac{4(9+m)}{1+m}\left[k_2 - k_3 + \frac{5k_1}{24}\beta_1\right] + \frac{16(5+m)}{1+m}[A_1 k_1 - A_2\beta_1] + \frac{64(7+m)}{9(1+m)}[A_2 - (k_2 - k_3)]\beta_1 - \frac{3}{16}k_1\beta_1^2] + k_1 k_5\left[\frac{4(9+m)}{1+m} - \frac{64(7+m)\beta_1}{9(1+m)}\right]\right\}$$

$$\alpha_3 = \beta_2^4\left\{\frac{2(13+m)}{27(1+m)} - \frac{8\beta_1(11+m)}{15(1+m)} + \frac{5\beta_1^2(9+m)}{3(1+m)}\right.$$

$$-\left(A_1+\frac{1}{16}\beta_1^3\right)\frac{256(7+m)}{9(1+m)}+\frac{64A_1\beta_1(5+m)}{1+m}$$

$$+k_5\left[-\frac{256(11+m)}{25(1+m)}+\frac{32\beta_1(9+m)}{1+m}-\frac{256\beta_1^2(7+m)}{9(1+m)}\right]\Bigg\}.$$

Thus we have derived the load-deflection relation of the shell.

The extremum condition

$$\left.\frac{\mathrm{d}Q}{\mathrm{d}W_m}\right|_{W_m=W_m^*}=0$$

yields two values of W_m^* as

$$W_m^*=\frac{-\alpha_2\pm\sqrt{\alpha_2^2-3\alpha_1\alpha_3}}{3\alpha_3}. \tag{8.12}$$

Putting these values of W_m^* into (8.11) we have

$$Q^*=\alpha_1W_m^*+\alpha_2W_m^{*2}+\alpha_3W_m^{*3} \tag{8.13}$$

where Q^* with respect to the positive and negative signs of (8.12) are the non-dimensional lower and upper critical buckling loads, respectively. In practical engineering, only the upper buckling loads must be predicted.

If we put $m\to\infty$ in the foregoing analysis, we can obtain the corresponding results for the same shell but with an immovably clamped outer edge.

9. Calculation Examples and Discussions

Let us now consider a squarely-latticed shallow cap of a petroleum vessel. The span of the cap and the radius of the curvature of the cap are the same, i.e. $R=2a=28$m. Each member has the same rectangular tube cross-section with the size of $100\times60\times5$mm, and is so placed that the angle of the two closest parallel members with respect to the center of the sphere is 2.5°. Besides, the shell is covered with a 4 mm-thick steel skin. The material quantities are given as $E=2.10\times10^6\,\mathrm{kg/cm^2}$ and $\nu=0.3$. The self-weight of the shell per square meter is

$$q_1=50.85\,\mathrm{kg/m^2}$$

and the initial pressure is

$$q_0=125.0\,\mathrm{kg/m^2}.$$

Lastly, the cap carries an external uniform vertical load q_2.

The calculated upper buckling loads from formula (8.13) are q^*, so the actual buckling loads q_2^* are

Case (a) Immovably clamped:

$$q_2^*=q^*-q_0-q_1=501.70\,\mathrm{kg/m^2}.$$

Case (b) Hinged:

$$q_2^*=q^*-q_0-q_1=744.99\,\mathrm{kg/m^2}.$$

In an actual case, the latticed shell is elastically restrained against rotation, therefore the actual buckling loads vary between the predicted buckling loads via formula

(8.13) for case (a) and case (b). It then can be suggested that in the practical estimation of buckling loads, the two results of case (a) and (b) must be compared, then the smaller value of the two results can be chosen as the buckling load for the sake of safety.

References

[1] S. E. Forman and J. W. Hutchinson, Buckling of reticulated shell structures. *Int. J. Solids Struct.* 6, 909 (1970).

[2] D. T. Wright, Membrane forces and buckling in reticulated shells. *Proc. ASCE, Struct. Div.* 91, 173 (1965).

[3] L. Kollar, Continuum equations of timber lattice shells. *Acta Technica Academiae Scientiarum Hungaricae* 94, 133 (1982).

[4] V. V. Ponomarev and G. I. Belikov, Analysis of latticed shells of revolution. *Soviet Appl. Mech.* 17, 645 (1981).

[5] I. M. Kani and R. E. McConnel, Collapse of shallow lattice domes. *J. Struct. Eng.* 113, 1806 (1987).

[6] K. Heki, *Shells, membranes and space frames // Proceedings of the IASS symposium on Membrane Structures and Space Frames*, Osaka, Japan (September 1986).

[7] W. Z. Chien, Variational principles and generalized variational principles for nonlinear elasticity with finite displacement. *Appl. Math. Mech.* 9, 1 (1988).

[8] K. Y. Yeh, Liu Renhuai et al. Nonlinear stability of thin circular shallow spherical shell under action of axisymmetric uniform distribution line loads. *Bull. Sci.* 2, 142 (1965). (in Chinese).

[9] K. Y. Yeh, Liu Renhuai et al. Nonlinear stability of thin circular shallow spherical shell under the actions of uniform edge moment. *Bull. Sci.* 2, 145 (1965). (in Chinese).

[10] Liu Renhuai, Nonlinear stability of circular shallow spherical shell with a hole in the center under the action of uniform moment at the inner edge. *Bull. Sci.* 3, 253 (1965). (in Chinese).

双层正交正放网格扁壳结构的非线性弯曲理论①

一、引言

双层网格扁壳结构具有刚度大、自重轻、造型优美等许多优点，是大跨度、大空间结构的主要结构形式之一。由于结构复杂和非线性数学上的困难，故尚未有非线性弯曲分析方面的文献。刘人怀[1,2]曾运用等效连续化的思想，建立了单层网格扁壳结构的非线性弯曲理论。本文在上述工作基础上进一步对双层网格扁壳结构进行了分析。

二、基本假定与等效模型

如图 1 所示，直角坐标 x 和 y 轴在网壳底面内且与网格杆件同向，z 轴垂直底面。x，y 方向杆件（图 2）长度是 L_2 和 L_1，截面积是 A_1 和 A_2 并对通过质心 z 方向轴线的惯性矩是 I_{10} 和 I_{20}，结构腹杆中（图 3），x，y 方向斜杆长度是 L_{c2} 和 L_{c1}，截面积是 A_{c1} 和 A_{c2}，竖杆的截面积是 A_h；上下表层中面距离 h，整个网格扁壳结构的中面沿 x，y 方向曲率是 k_1 和 k_2，材料弹性模量是 E，剪切模量是 G。我们假定：①双层网格扁壳结构上、下表层相应位置的结构形式和几何尺寸相同，表层之间距离远小于网壳结构的曲率半径，网格分布稠密，网格划分均匀规则；②网壳结构中各杆件截面尺寸远小于网格单元尺寸和表层间距离，杆件固结；③杆件材料完全弹性；④仅考虑结构的反对称变形。

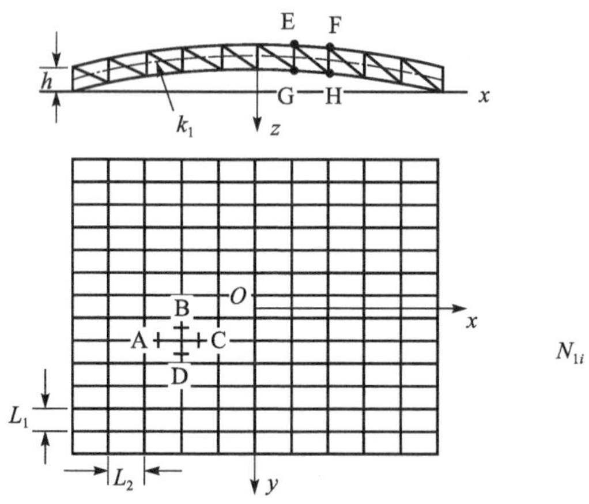

图 1　双层正交正放网格扁壳结构

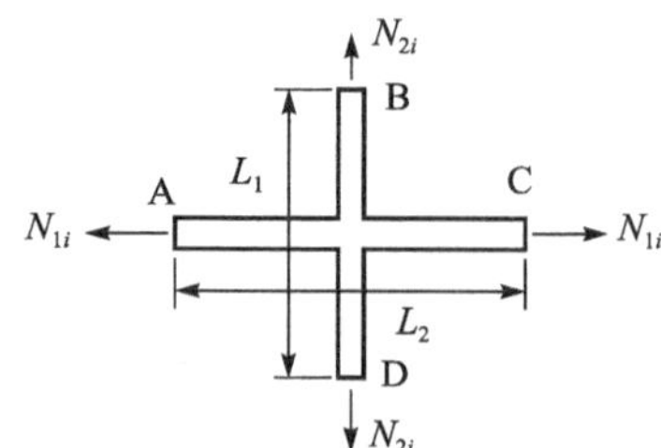

图 2　网格扁壳结构表层单元

把双层正交正放网格扁壳结构等效连续化成几何外形相同的夹层扁壳，即把结构

① 本文原载《现代力学与科技进步（庆祝中国力学学会成立 40 周年论文集）》，第 3 卷，北京：清华大学出版社，1997：1212-1215. 作者：刘人怀，肖潭．

的上、下弦等效为夹层扁壳的上、下表层，把它的腹杆折算成夹层扁壳中厚度为 h 的夹心层，并假设夹层扁壳表层处于薄膜应力状态，夹心中面法线在变形后保持直线，夹心沿板面方向不能承受载荷，且夹心横向不可压缩。结构表层单元、腹杆单元内力和夹层扁壳微元体内力标示在图 2～图 4 中。由于壳的扁平性，可有 $N_{xyi}=N_{yxi}$。于是，结构和夹层壳的内力等效关系如下

$$N_{xi}=\frac{N_{1i}}{L_1},\quad N_{yi}=\frac{N_{2i}}{L_2},\quad N_{xyi}=\frac{1}{2}\left(\frac{N_{12i}}{L_1}+\frac{N_{21i}}{L_2}\right),$$
$$Q_x=\frac{Q_1}{L_1},\quad Q_y=\frac{Q_2}{L_2},\quad i=1,3 \tag{1}$$

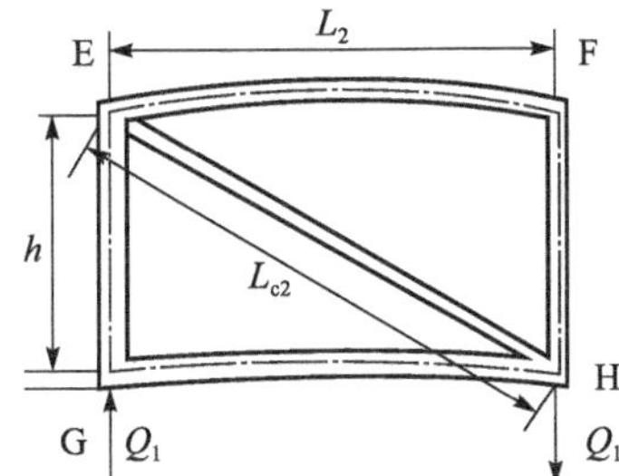

图 3　网格扁壳结构腹杆单元

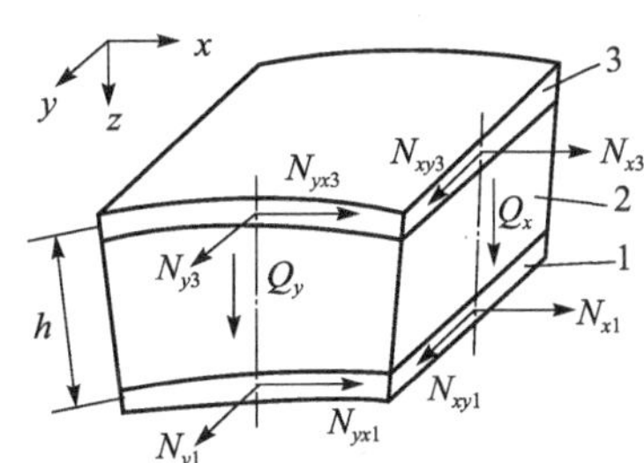

图 4　等效夹层扁壳微元体

记 u_i，v_i，w_i 和 ε_{xi}，ε_{yi}，ε_{zi}，γ_{xyi}，γ_{yzi}，γ_{xzi}（i=1，2，3）为夹层扁壳的上表层、夹心、下表层的位移分量和应变分量，u，v 和 w 为夹层扁壳中面的位移分量，ψ_x 和 ψ_y 为夹心中面法线在 xz 和 yz 平面内的转角，则有 $u_i=u\pm h\psi_x/2$，$v_i=v\pm h\psi_y/2$，$w_i=w$（当 i=1，3 时分别取正和负）和 $u_2=u+z\psi_x$，$v_2=v+z\psi_y$，$w_2=w$，代入几何方程得到

$$\begin{aligned}
\varepsilon_{xi}&=u_{i,x}+\frac{1}{2}w_{i,x}^2-k_1w_i\\
&=u_{,x}\pm\frac{h}{2}\psi_{x,x}+\frac{1}{2}w_{,x}^2-k_1w,\\
\varepsilon_{yi}&=v_{i,y}+\frac{1}{2}w_{i,y}^2-k_2w_i\\
&=v_{,y}\pm\frac{h}{2}\psi_{y,y}+\frac{1}{2}w_{,y}^2-k_2w,\\
\gamma_{xyi}&=v_{i,x}+u_{i,y}+w_{i,x}w_{i,y}\\
&=v_{,x}+u_{,y}\pm\frac{h}{2}(\psi_{y,x}+\psi_{x,y})+w_{,x}y_{,y},\quad i=1,3
\end{aligned}\tag{2}$$

$$\begin{aligned}
\gamma_{xz2}&=u_{2,z}+w_{2,x}\\
&=\psi_x+w_{,x},\\
\gamma_{yz2}&=v_{2,z}+w_{2,y}\\
&=\psi_y+w_{,y}
\end{aligned}\tag{3}$$

三、本构方程

网格扁壳结构表层单元如图 2 所示。受载变形后，表层杆件的轴力为 $N_{1i}=EA_1\varepsilon_{xi}$，

$N_{2i}=EA_2\varepsilon_{yi}(i=1,3)$，考虑表层肋条自身的平衡有 $N_{12i}L_2=N_{21i}L_1(i=1,3)$，根据假设(2)，表层肋条变形后节点处两条切线仍成直角，节点为反弯点，由此得剪应变

$$\begin{aligned}\gamma_{xyi}&=\frac{N_{12i}L_2^2}{12EI_{10}}+\frac{N_{21i}L_1^2}{12EI_{20}}\\&=\frac{N_{12i}L_2}{12E}\left(\frac{L_2}{I_{10}}+\frac{L_1}{I_{20}}\right)\\&=\frac{N_{21i}L_1}{12E}\left(\frac{L_2}{I_{10}}+\frac{L_1}{I_{20}}\right),\quad i=1,3\end{aligned}\tag{4}$$

利用式（1），得到夹层扁壳模型表层内力和变形的关系是

$$\begin{aligned}N_{xi}&=\frac{EA_1}{L_1}\varepsilon_{xi},\\N_{yi}&=\frac{EA_2}{L_2}\varepsilon_{yi},\\N_{xyi}&=\frac{12E}{L_1L_2\left(\frac{L_2}{I_{10}}+\frac{L_1}{I_{20}}\right)}\gamma_{xyi},\quad i=1,3\end{aligned}\tag{5}$$

网壳结构中 x 方向一腹杆单元如图 3 所示。根据假定②、③和腹杆受力特点，我们仅考虑腹杆单元整体的剪切刚度而忽略它整体的抗拉刚度和弯曲刚度，把它近似等效成节点铰接的桁架结构。当它在一对大小相等、方向相反的横向力 Q_1 作用下，腹杆单元中竖杆内力为 Q_1，斜杆内力 $N_{c2}=Q_1L_{c2}/h$，那么腹杆单元的剪切变形余能为

$$V_c=\frac{Q_1^2h}{2EA_h}+\frac{N_{c2}^2L_{c2}}{2EA_{c1}}=\left(\frac{h}{2EA_h}+\frac{L_{c2}^3}{2EA_{c1}h^2}\right)Q_1^2\tag{6}$$

对应的等效夹心单位面积的变形余能为 $U_c=V_c/L_1L_2$，将式（1）代入，然后运用卡氏第二定理 $\gamma_{xz2}=\partial U_c/\partial Q_x$ 求得

$$Q_x=G_x\gamma_{xz2}\tag{7}$$

同理

$$Q_y=G_y\gamma_{yz2}\tag{8}$$

其中

$$G_x=\frac{EL_2}{L_1\left(\frac{h}{A_h}+\frac{L_{c2}^3}{A_{c1}h^2}\right)},\quad G_y=\frac{EL_1}{L_2\left(\frac{h}{A_h}+\frac{L_{c2}^3}{A_{c2}h^2}\right)}\tag{9}$$

四、非线性基本方程

假定夹层扁壳区域 D 上单位面积所受的外力分量分别为 f_x、f_y、f_z，边界 C 上给定力或位移边界条件，根据虚功原理有

$$\begin{aligned}&\iint_D(N_{x1}\delta\varepsilon_{x1}+N_{y1}\delta\varepsilon_{y1}+N_{xy1}\delta\gamma_{xy1}+N_{x3}\delta\varepsilon_{x3}+N_{y3}\delta\varepsilon_{y3}+N_{xy3}\delta\gamma_{xy3})\mathrm{d}x\mathrm{d}y\\&+\iint_D(Q_x\delta\gamma_{xz2}+Q_y\delta\gamma_{yz2})\mathrm{d}x\mathrm{d}y-\oint_c(\overline{N}_{xn}\delta u+\overline{N}_{yn}\delta v+\overline{N}_z\delta w)\mathrm{d}s\\&-\oint_c(\overline{M}_{xn}\delta\psi_x+\overline{M}_{yn}\delta\psi_y)\mathrm{d}s-\iint_D(f_x\delta u+f_y\delta v+f_z\delta w)\mathrm{d}x\mathrm{d}y=0\end{aligned}\tag{10}$$

其中$\overline{N}_{xn}$，$\overline{N}_{yn}$，$\overline{N}_z$ 和 $m\,\overline{M}_{xn}$，$\overline{M}_{yn}$是夹层扁壳边界C上给定的外力和外力矩。

将式（2）、式（3）代入式（10），分部积分后得到基本方程

$$
\begin{aligned}
&N_{x,x}+N_{xy,y}+f_x=0,\\
&N_{yx,x}+N_{y,y}+f_y=0,\\
&M_{x,xx}+2M_{xy,xy}+M_{y,yy}+N_x(w_{,xx}+k_1)+2N_{xy}w_{,xy}+N_y(w_{,yy}+k_2)\\
&-w_{,x}f_x-w_{,y}f_y+f_z=0,\\
&M_{x,x}+M_{xy,y}-Q_x=0,\\
&M_{xy,x}+M_{y,y}-Q_y=0
\end{aligned}
\tag{11}
$$

和边界条件

$$
\begin{aligned}
&lN_x+mN_{xy}=\overline{N}_x \text{ 或 } u=\bar{u},\\
&lN_{xy}+mN_y=\overline{N}_y \text{ 或 } v=\bar{v},\\
&l(M_{x,x}+M_{xy,y})+m(M_{xy,x}+M_{y,y})+w_{,x}N_{xn}+w_{,y}N_{yn}=\overline{N}_z \text{ 或 } w=\bar{w},\\
&lM_x+mM_{xy}=\overline{M}_{xn} \text{ 或 } \psi_x=\bar{\psi}_x,\\
&lM_{xy}+mM_y=\overline{M}_{yn} \text{ 或 } \psi_y=\bar{\psi}_y
\end{aligned}
\tag{12}
$$

其中，l，m 为边界C外法线方向余弦，

$$
\begin{aligned}
&N_x=N_{x1}+N_{x3},\\
&N_y=N_{y1}+N_{y3},\\
&N_{xy}=N_{xy1}+N_{xy3},\\
&M_x=\frac{h}{2}(N_{x1}-N_{x3}),\\
&M_y=\frac{h}{2}(N_{y1}-N_{y3}),\\
&M_{xy}=\frac{h}{2}(N_{xy1}-N_{xy3})
\end{aligned}
\tag{13}
$$

将式（5）代入式（13），并利用式（2），同时又将式（3）代入式（7），求得

$$
\begin{aligned}
&N_x=E_x\varepsilon_{x0},\\
&N_y=E_y\varepsilon_{y0},\\
&N_{xy}=G_{xy}\gamma_{xy0},\\
&M_x=D_x\psi_{x,x},\\
&M_y=D_y\psi_{y,y},\\
&M_{xy}=D_{xy}(\psi_{y,x}+\psi_{x,y}),\\
&Q_x=G_x(\psi_x+w_{,x}),\\
&Q_y=G_y(\psi_y+w_{,y})
\end{aligned}
\tag{14}
$$

其中

$$
E_x=\frac{2EA_1}{L_1},\quad E_y=\frac{2EA_2}{L_2},\quad G_{xy}=\frac{24E}{L_1L_2\left(\dfrac{L_2}{I_{10}}+\dfrac{L_1}{I_{20}}\right)},
$$

$$D_x=\frac{EA_1h^2}{2L_1},\quad D_y=\frac{EA_2h^2}{2L_2},\quad D_{xy}=\frac{6Eh^2}{L_1L_2\left(\frac{L_2}{I_{10}}+\frac{L_1}{I_{20}}\right)},\tag{15}$$

$$\varepsilon_{x0}=u_{,x}+\frac{1}{2}w_{,x}^2-k_1w,\quad \varepsilon_{y0}=v_{,y}+\frac{1}{2}w_{,y}^2-k_2w,$$
$$\gamma_{xy0}=v_{,x}+u_{,y}+w_{,x}w_{,y}\tag{16}$$

引入以下力函数

$$N_x=\varphi_{,yy}-F_x,\quad N_y=\varphi_{,xx}-F_y,\quad N_{xy}=-\varphi_{,xy}\tag{17}$$

其中

$$F_x=\int f_x\mathrm{d}x,\quad F_y=\int f_y\mathrm{d}y\tag{18}$$

则在利用式（14）和（16）后，方程组（11）简化为

$$\begin{aligned}&D_x\psi_{x,xxx}+2D_{xy}(\psi_{y,xxy}+\psi_{x,xyy})+D_y\psi_{y,yyy}+(\varphi_{,yy}-F_x)(w_{,xx}+k_1)\\&-2\varphi_{,xy}w_{,xy}+(\varphi_{,xx}-F_y)(w_{,yy}+k_2)-w_{,x}f_x-w_{,y}f_y+f_z=0,\\&D_x\psi_{x,xx}+D_{xy}(\psi_{y,xy}+\psi_{x,yy})=G_x(\psi_x+w_{,x}),\\&D_{xy}(\psi_{y,xx}+\psi_{x,xy})+D_y\psi_{y,yy}=G_y(\psi_y+w_{,y})\end{aligned}\tag{19}$$

最后，由式（16）导出协调方程

$$\frac{\varphi_{,yyyy}}{E_x}+\frac{\varphi_{,xxxx}}{E_y}+\frac{\varphi_{,xxyy}}{G_{xy}}=(w_{,xy})^2-w_{,xx}w_{,yy}-k_1w_{,yy}-k_2w_{,xx}+\frac{F_{x,yy}}{E_x}+\frac{F_{y,xx}}{E_y}\tag{20}$$

于是，式（19）和（20）便成为双层正交正放网格扁壳结构的非线性基本方程组。

参 考 文 献

[1] Liu Ren-huai, Li, D. and Nie, G. H. et al. Non-linear buckling of squarely-latticed shallow spherical shells. Int. J. Non-Linear Mechanics, 1991, 26 (5): 547-565.

[2] 刘人怀，聂国华. 网格扁壳的非线性弯曲理论. 江西工业大学学报，1991，13 (2)：186-192.

矩形底双层网格扁壳的非线性弯曲[①]

一、引言

双层网格扁壳结构具有刚度大、自重轻、造型丰富美观、综合技术经济指标好的特点，是大跨度、大空间结构的主要结构形式之一，在大型建筑屋盖结构中有着重要的应用[1]。目前国内外对此类结构主要采用离散方法进行非线性分析。由于双层网格扁壳结构构造复杂、网格分布稠密、节点数目大，离散方法显得耗时费力，在分析结构的稳定、振动时尤为繁复。针对这一问题，刘人怀和聂国华等[2-6]运用等效连续化方法建立了单层网格扁壳结构的非线性弹性理论，并求解了一些具体问题，而后，刘人怀和肖潭[7]又进一步提出了双层网格扁壳的非线性弯曲理论。本文根据这一理论，对横向载荷作用下具有边缘不可移简支边界条件的矩形底双层网格扁壳进行了非线性弯曲分析。

二、基本方程

如图 1 所示，双层网格扁壳矩形底面的长和宽是 a 和 b，整个网格扁壳的中面沿 x、y 方向的曲率是 k_1 和 k_2，结构承受分布横向载荷 $q(x, y)$，壳的四边皆是不可移简支，并且无底面面内位移和横向位移。根据文献［7］并使用相同的符号，这一双层网格扁壳中面的位移分量 u，v，w 和中面法线转角 ψ_x，ψ_y 所满足的基本方程组为

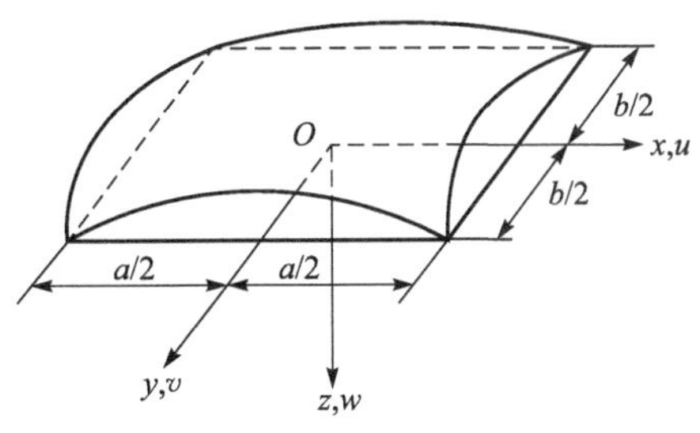

图 1 矩形底双层网格扁壳

$$
\begin{aligned}
&E_x u_{,xx} + G_{xy} u_{,yy} + G_{xy} v_{,xy} + E_x (w_{,x} w_{,xx} - k_1 w_{,x}) + G_{xy} (w_{,y} w_{,xy} + w_{,x} w_{yy}) = 0,\\
&E_y v_{,yy} + G_{xy} v_{,xx} + G_{xy} u_{,xy} + E_y (w_{,y} w_{,yy} - k_2 w_{,y}) + G_{xy} (w_{,y} w_{,xx} + w_{,x} w_{,xy}) = 0,\\
&D_x \psi_{x,xxx} + 2D_{xy} (\psi_{y,xxy} + \psi_{x,xyy}) + D_y \psi_{y,yyy} + E_x \left(u_{,x} + \frac{1}{2} w_{,x}^2 - k_1 w\right)(w_{,xx} + k_1)\\
&+ E_y \left(v_{,y} + \frac{1}{2} w_{,y}^2 - k_2 w\right)(w_{,yy} + k_2) + 2G_{xy} (v_{,x} + u_{,y} + w_{,x} w_{,y}) w_{,xy} + q = 0,\\
&D_x \psi_{x,xx} + D_{xy} (\psi_{y,xy} + \psi_{x,yy}) = G_x (\psi_x + w_{,x}),\\
&D_{xy} (\psi_{y,xx} + \psi_{x,xy}) + D_y \psi_{y,yy} = G_y (\psi_y + w_{,y})
\end{aligned}
\qquad (1a\text{-}e)
$$

不可移简支边界条件是

当 $x = \pm \dfrac{a}{2}$ 时，$u = v = w = \psi_y = 0, M_x = 0$；

当 $y = \pm \dfrac{b}{2}$ 时，$u = v = w = \psi_x = 0, M_y = 0$；

① 本文原载《暨南大学学报》，1998，19（3）：1-5. 作者：刘人怀，肖潭.

即

$$\text{当 } y=\pm\frac{a}{2}\text{ 时}, u=v=w=\psi_y=\psi_{x,x}=0,$$

$$\text{当 } y=\pm\frac{b}{2}\text{ 时}, u=v=w=\psi_x=\psi_{y,y}=0 \tag{2}$$

三、边值问题求解

如果取位移函数为以下双重傅里叶级数形式

$$u=\sum_{i=1}^{\infty}\sum_{j=1,3,\cdots}^{\infty}U_{ij}\sin\frac{2i\pi x}{a}\cos\frac{j\pi y}{b},$$

$$v=\sum_{i=1,3,\cdots}^{\infty}\sum_{j=1}^{\infty}V_{ij}\cos\frac{i\pi x}{a}\sin\frac{2j\pi y}{b},$$

$$w=\sum_{i=1,3,\cdots}^{\infty}\sum_{j=1,3,\cdots}^{\infty}W_{ij}\cos\frac{i\pi x}{a}\sin\frac{j\pi y}{b},$$

$$\psi_x=\sum_{i=1,3,\cdots}^{\infty}\sum_{j=1,3,\cdots}^{\infty}G_{ij}\sin\frac{i\pi x}{a}\cos\frac{j\pi y}{b},$$

$$\psi_y=\sum_{i=1,3,\cdots}^{\infty}\sum_{j=1,3,\cdots}^{\infty}H_{ij}\cos\frac{i\pi x}{a}\sin\frac{j\pi y}{b} \tag{3a-e}$$

则边界条件（2）已被满足。这里 U_{ij}，V_{ij}，W_{ij}，G_{ij} 和 H_{ij} 是待定常数，双重傅里叶级数满足如下正交关系

$$\int_{-a/2}^{a/2}\sin\frac{2i\pi x}{a}\sin\frac{2m\pi x}{a}\mathrm{d}x=\begin{cases}0, & i\neq m\\ \frac{a}{2}, & i=m\end{cases}\quad(i,m=1,2,3,\cdots)$$

$$\int_{-a/2}^{a/2}\cos\frac{i\pi x}{a}\cos\frac{m\pi x}{a}\mathrm{d}x=\begin{cases}0, & i\neq m\\ \frac{a}{2}, & i=m\end{cases}\quad(i,m=1,2,3,\cdots) \tag{4}$$

再将横向载荷 $q(x,y)$ 展开成以下双重傅里叶级数

$$q(x,y)=\sum_{i=1,3,\cdots}^{\infty}\sum_{j=1,3,\cdots}^{\infty}q_{ij}\cos\frac{i\pi x}{a}\cos\frac{j\pi y}{b} \tag{5}$$

其中

$$q_{ij}=\frac{4}{ab}\int_{-a/2}^{a/2}\int_{-b/2}^{b/2}q(x,y)\cos\frac{i\pi x}{a}\cos\frac{j\pi y}{b}\mathrm{d}x\mathrm{d}y \tag{6}$$

将式（3）和（5）代入方程（1a-c），用 $\sin\frac{2m\pi x}{a}\cos\frac{n\pi y}{b}$ 乘方程（1a），用 $\cos\frac{m\pi x}{a}\sin\frac{2n\pi y}{b}$ 乘方程（1b），用 $\cos\frac{m\pi x}{a}\cos\frac{n\pi y}{b}$ 乘方程（1c），在它们各自的区间对 x 和 y 进行积分，并利用式（4），便得非线性代数方程组如下

$$-\frac{ab\pi^2}{4}\left(\frac{4m^2E_x}{a^2}+\frac{n^2G_{xy}}{b^2}\right)U_{mn}-\sum_{i=1,3,\cdots}^{\infty}\sum_{j=1,3,\cdots}^{\infty}\frac{2ij\pi^2G_{xy}}{ab}K_1^{im}J_2^{jn}V_{ij}$$

$$+\sum_{i=1,3,\cdots}^{\infty}\frac{i\pi bk_1E_x}{2a}K_1^{im}W_{in}+\sum_{i=1,3,\cdots}^{\infty}\sum_{j=1,3,\cdots}^{\infty}\sum_{p=1,3,\cdots}^{\infty}\sum_{q=1,3,\cdots}^{\infty}\pi^2\Big[\Big(\frac{ip^2E_x}{a^3}$$

$$+\frac{iq^2G_{xy}}{ab^2}\Big)M_1^{ipm}Q_1^{jqn}-\frac{jpqG_{xy}}{ab^2}M_1^{pim}Q_1^{jqn}\Big]W_{ij}W_{pq}=0,$$

$$m=1,2,3,\cdots;\quad n=1,3,5,\cdots$$

$$-\frac{ab\pi^2}{4}\Big(\frac{4n^2E_y}{b^2}+\frac{m^2G_{xy}}{a^2}\Big)V_{mn}-\sum_{i=1,3,\cdots}^{\infty}\sum_{j=1,3,\cdots}^{\infty}\frac{2ij\pi^2G_{xy}}{ab}K_2^{im}J_1^{jn}U_{ij}$$

$$+\sum_{j=1,3,\cdots}^{\infty}\frac{j\pi ak_2E_y}{2b}J_1^{jn}W_{mj}+\sum_{i=1,3,\cdots}^{\infty}\sum_{j=1,3,\cdots}^{\infty}\sum_{p=1,3,\cdots}^{\infty}\sum_{q=1,3,\cdots}^{\infty}\pi^3\Big[\Big(\frac{iq^2E_y}{b^3}$$

$$+\frac{jp^2G_{xy}}{a^2b}\Big)P_1^{ipm}N_1^{jqn}-\frac{ipqG_{xy}}{a^2b}P_2^{ipm}N_1^{qjn}\Big]W_{ij}W_{pq}=0,$$

$$m=1,3,5,\cdots;\quad n=1,2,3,\cdots$$

$$q_{mn}-(k_1^2E_x+k_2^2E_y)W_{mn}-\pi^3\Big(\frac{m^3D_x}{a^3}+\frac{3mn^2D_{xy}}{ab^2}\Big)G_{mn}$$

$$-\pi^3\Big(\frac{2m^2nD_{xy}}{a^2b}+\frac{n^3D_y}{b^3}\Big)H_{mn}+\sum_{i=1}^{\infty}\frac{4i\pi k_1E_x}{a^2}K_2^{im}U_{in}$$

$$+\sum_{j=1}^{\infty}\frac{4j\pi k_2E_y}{b^2}J_2^{jn}V_{mj}+\sum_{i=1,3,\cdots}^{\infty}\sum_{j=1,3,\cdots}^{\infty}\sum_{p=1,3,\cdots}^{\infty}\sum_{q=1,3,\cdots}^{\infty}2\pi^2\Big[\frac{k_1E_x}{a^3b}(ipP_2^{ipm}$$

$$+2p^2P_1^{ipm})Q_1^{jqn}+\frac{k_2E_y}{ab^3}(jqQ_2^{jqn}+2q^2Q_1^{jqn}P_1^{ipm})\Big]W_{ij}W_{pq}$$

$$+\sum_{i=1,3,\cdots}^{\infty}\sum_{j=1,3,\cdots}^{\infty}\sum_{p=1,3,\cdots}^{\infty}\sum_{q=1,3,\cdots}^{\infty}\sum_{r=1,3,\cdots}^{\infty}\sum_{s=1,3,\cdots}^{\infty}2\pi^4\Big(\frac{4iqrsG_{xy}}{a^3b^3}P_4^{irpm}Q_4^{qsjn}$$

$$-\frac{ipr^2E_x}{a^5b}P_4^{iprm}Q_3^{jqsn}-\frac{jqs^2E_y}{ab^5}P_3^{iprm}Q_4^{jqsn}\Big)W_{ij}W_{pq}W_{rs}$$

$$-\sum_{i=1,3,\cdots}^{\infty}\sum_{j=1,3,\cdots}^{\infty}\sum_{p=1,3,\cdots}^{\infty}\sum_{q=1,3,\cdots}^{\infty}8\pi^3\Big(\frac{ip^2E_x}{a^4b}M_2^{pmi}Q_1^{jqn}+\frac{jpqG_{xy}}{a^2b^3}M_1^{pmi}Q_2^{jqn}\Big)U_{ij}W_{pq}$$

$$-\sum_{j=1,3,\cdots}^{\infty}\sum_{i=1,3,\cdots}^{\infty}\sum_{p=1,3,\cdots}^{\infty}\sum_{q=1,3,\cdots}^{\infty}8\pi^3\Big(\frac{jq^2E_y}{ab^4}P_1^{ipm}N_2^{qnj}+\frac{ipqG_{xy}}{a^3b^2}P_2^{ipm}N_1^{qnj}\Big)V_{ij}W_{pq}=0$$

$$m,n=1,3,5,\cdots \tag{7a-c}$$

其中

$$K_1^{im}=\int_{-a/2}^{a/2}\sin\frac{i\pi x}{a}\sin\frac{2m\pi x}{a}\mathrm{d}x,$$

$$K_2^{im}=\int_{-a/2}^{a/2}\cos\frac{2i\pi x}{a}\cos\frac{m\pi x}{a}\mathrm{d}x,$$

$$P_1^{ipm}=\int_{-a/2}^{a/2}\cos\frac{i\pi x}{a}\cos\frac{p\pi x}{a}\cos\frac{m\pi x}{a}\mathrm{d}x, \tag{8}$$

$$P_2^{ipm}=\int_{-a/2}^{a/2}\sin\frac{i\pi x}{a}\sin\frac{p\pi x}{a}\cos\frac{m\pi x}{a}\mathrm{d}x,$$

$$P_3^{ipm}=\int_{-a/2}^{a/2}\cos\frac{i\pi x}{a}\cos\frac{p\pi x}{a}\cos\frac{r\pi x}{a}\cos\frac{m\pi x}{a}\mathrm{d}x,$$

$$P_4^{iprm} = \int_{-a/2}^{a/2} \sin\frac{i\pi x}{a}\sin\frac{p\pi x}{a}\cos\frac{r\pi x}{a}\cos\frac{m\pi x}{a}\mathrm{d}x,$$

$$M_1^{pmi} = \int_{-a/2}^{a/2} \sin\frac{p\pi x}{a}\cos\frac{m\pi x}{a}\sin\frac{2i\pi x}{a}\mathrm{d}x,$$

$$M_2^{pmi} = \int_{-a/2}^{a/2} \cos\frac{p\pi x}{a}\cos\frac{m\pi x}{a}\cos\frac{2i\pi x}{a}\mathrm{d}x$$

用 J，Q，N，y，b，j，n，q 和 s 代替式（8）中的 K，P，M，x，a，i，m，p 和 r，可以得到式（7a-c）中的常数 J_1，J_2，$Q_1 \sim Q_4$，N_1 和 N_2。

再将式（3c-e）代入方程（1d，e），比较方程等号两边的系数，又可得到如下线性方程组

$$\left(\frac{m^2D_x}{a^2}+\frac{n^2D_{xy}}{b^2}+\frac{G_x}{\pi^2}\right)G_{mn}+\frac{mnD_{xy}}{ab}H_{mn}=\frac{mG_x}{\pi a}W_{mn},\quad m,n=1,3,5,\cdots$$

$$\frac{mnD_{xy}}{ab}G_{mn}+\left(\frac{m^2D_{xy}}{a^2}+\frac{n^2D_y}{b^2}+\frac{G_y}{\pi^2}\right)H_{mn}=\frac{nG_y}{\pi b}W_{mn}\quad m,n=1,3,5,\cdots \tag{9a,b}$$

四、数值结果

我们以均布横向载荷 q_0 作用下双层正交正放网格扁壳为例进行计算，壳体的有关数据是：底面长 $a=35\mathrm{m}$，宽 $b=35\mathrm{m}$，网格长 $L_2=0.625\mathrm{m}$，宽 $L_1=0.625\mathrm{m}$，厚 $h=0.625\mathrm{m}$，壳的曲率 $k_1=k_2=0.001$，网格表层杆件的截面积 $0.03\times0.03\mathrm{m}^2$，夹心和腹杆截面积 $0.021\times0.021\mathrm{m}^2$，弹性模量 $E=2.1\times10^{10}\mathrm{kg/m}^2$，剪切模量 $G=0.81\times10^{10}\mathrm{kg/m}^2$。

将上述数据代入方程组式（7）和式（9），解方程组后，所得计算结果在图 2 中，这里 w_0 是网格扁壳中心挠度。特别地，当 $k_1=k_2=0$ 时，网格扁壳退化成网格平板，计算结果如图 3 所示。在图 2 和图 3 中，(a) 是本文线性解，× 是文献［8］的线性解，(b) 是 u，v，w，ψ_x 和 ψ_y 每个级数取一项得出的结果，(c) 是 u，v，w，ψ_x 和 ψ_y 每

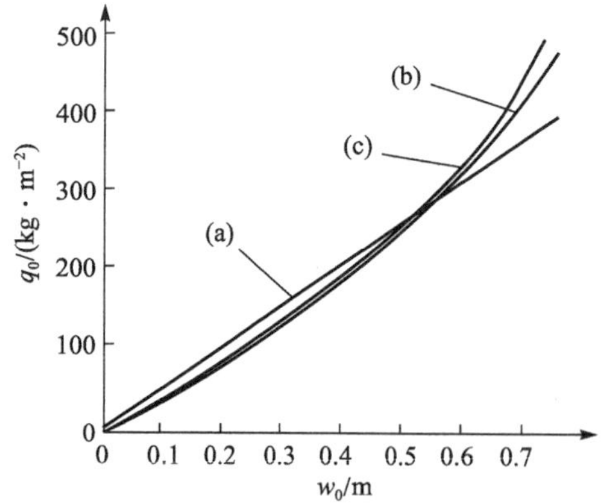

图 2　双层网格扁壳的载荷-挠度曲线

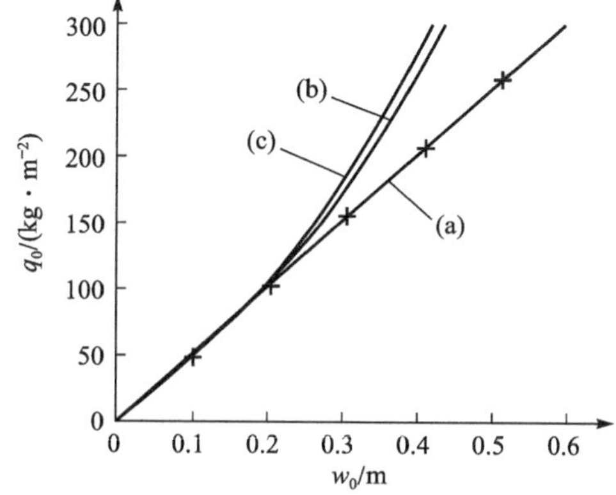

图 3　双层网格板的载荷-挠度曲线

个级数取四项得出的结果。(b) 和 (c) 之间的差值都在 4%以内。由此可见，级数收敛很快，取级数较少几项就能得到很精确的结果。

参考文献

[1] 董石麟，姚谏．中国网壳结构的发展与应用．第六届空间学术会议论文，广州，1992.

[2] Liu Ren-huai, Li, D., Nie, G. H. et al. Nonlinear buckling of quarely-latticed shallow spherical shells. Int. J. Non-Linear Mech., 1991, 26 (5): 547-565.

[3] 刘人怀，聂国华．矩形网格扁壳结构的非线性弯曲理论．江西工业大学学报，1991，13 (2)：186-194.

[4] 刘人怀，聂国华．网格扁壳的非线性自由振动分析．江西工业大学学报，1991，13 (2)：193-198.

[5] 聂国华，刘人怀，翁智远．网格扁壳结构的非线性弯曲与稳定问题研究．上海力学，1994，15 (2)：17-27.

[6] 聂国华，刘人怀．矩形网格扁壳结构的非线性弹性理论．应用数学和力学，1994，15 (5)：389-397.

[7] 刘人怀，肖潭．双层正交正放网格扁壳结构的非线性弯曲理论//庄逢甘．现代力学与科技进步．3 卷．北京：清华大学出版社，1997：1212-1215.

[8] 董石麟，夏亨熹．正交正放网架结构的拟扳（夹层板）分析法．建筑结构学报，1982，3 (2)：14-25.

矩形底双层网格扁壳的非线性屈曲[①]

一、引言

网格扁壳是曲面型的网格结构，兼有杆系结构和薄壳结构的固有特性，受力合理，覆盖跨度大，是一种有着广阔发展前途的空间结构。

自从1961年罗马尼亚布加勒斯特国家经济展览馆93.5m直径的网状穹顶失稳坍塌以后，国外对大跨度网壳结构稳定性的研究非常重视。由于网格扁壳的厚度和跨度相比是一个小量，因此它是一个大型的空间柔性结构。在工程设计中，人们除了关心它的位移和应力外，也很关心它的稳定性。因为这类结构往往不是因为强度不足而失效，而是在足够强度下发生屈曲而坍塌。因此，对网格扁壳的非线性稳定性进行分析是十分必要的。

目前国内外对网格结构进行分析主要有两种方法：离散方法和等效连续化方法。在网壳非线性稳定分析中，如果采用离散方法来计算网壳屈曲后的载荷反应，在临界点附近，由于网壳结构刚度矩阵接近奇异，迭代不易收敛，为了使计算顺利通过临界点，只有凭实际工作经验来选择计算参数并反复进行试算。因此，在网壳非线性稳定性分析中，离散方法的适应性、稳定性和计算效率不能满足工程设计单位的需要。连续化方法则有涉及参数少、计算工作量小、物理意义明确等优点，在进行复杂网格结构的稳定、振动分析时和离散方法相比有较大的优越性。

前不久，刘人怀和聂国华等[1-6]采用等效连续化方法，开创性地建立了单层网络扁壳的非线性理论，成功地求解了几个有实际意义的非线性弯曲、屈曲和振动问题。接着，在上述工作的基础上，本文作者[7,8]又将双层网格扁壳等效为夹层扁壳，首次提出了双层网格扁壳的非线性弯曲理论。本文按照上述的双层网格扁壳的非线性弯曲理论，应用双重傅里叶级数法，对横向载荷作用下，具有固定简支边界条件的矩形底双层网格扁壳进行了非线性屈曲分析，求得临界载荷。据作者所知，这一问题尚未被人研究过。

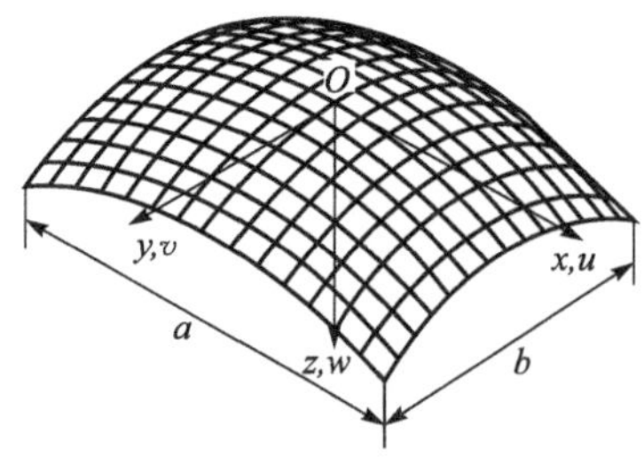

图1　矩形底双层网格扁壳示意图

二、基本方程

考虑如图1所示的矩形底双层网格扁壳，其中底面的长和宽是a和b，网格扁壳的中面沿x，y方向的曲率是k_1和k_2，结构承受均匀分布横向载荷q_0，壳的四边皆是固定简支。根据文献［7］并使用相同的符号，此双层网格扁壳中面的位移分量u，v，w和中面法线转角ψ_x，ψ_y所满足的非线性弯曲的控制方程组为

① 本文原载《暨南大学学报》，1999，20（1）：1-6. 作者：刘人怀，肖潭.

$$E_x u_{,xx} + G_{xy} u_{,xy} + G_{xy} v_{,xy} + E_x (w_{,x} w_{,xx} - k_1 w_{,x}) + G_{xy} (w_{,y} w_{,xy} + w_{,x} w_{,yy}) = 0,$$

$$E_y v_{,yy} + G_{xy} v_{,xx} + G_{xy} u_{,xy} + E_y (w_{,y} w_{,yy} - k_2 w_{,y}) + G_{xy} (w_{,y} w_{,xx} + w_{,x} w_{,xy}) = 0,$$

$$D_x \psi_{x,xxx} + 2D_{xy} (\psi_{y,xxy} + \psi_{x,xyy}) + D_y \psi_{y,yyy} + E_x \Big(u_{,x} + \frac{1}{2} w_{,x}^2 - k_1 w\Big)(w_{,xx} + k_1) + E_y \Big(v_{,y} + \frac{1}{2} w_{,y}^2 - k_2 w\Big)(w_{,yy} + k_2) + 2G_{xy}(v_{,x} + u_{,y} + w_{,x} w_{,y}) w_{,xy} + q_0 = 0,$$

$$D_x \psi_{x,xx} + D_{xy} (\psi_{y,xy} + \psi_{x,yy}) = G_x (\psi_x + w_{,x}),$$

$$D_{xy} (\psi_{y,xx} + \psi_{x,xy}) + D_y \psi_{y,yy} = G_y (\psi_y + w_{,y}) \tag{1a-e}$$

相应的固定简支边界条件是

$$x = \pm \frac{a}{2} \text{ 时}, u = v = w = \psi_y = \psi_{x,x} = 0,$$

$$y = \pm \frac{b}{2} \text{ 时}, u = v = w = \psi_x = \psi_{y,y} = 0 \tag{2}$$

三、求解非线性边值问题

假定网格扁壳的屈曲模态是

$$w = w_0 \cos \frac{\pi x}{a} \cos \frac{\pi y}{b} \tag{3}$$

其中，w_0 是壳中心的挠度值。

同时，取位移分量 u 和 v 为以下双重傅里叶级数形式

$$u = \sum_{i=1}^{\infty} \sum_{j=1,3,\cdots}^{\infty} U_{ij} \sin \frac{2i\pi x}{a} \cos \frac{j\pi y}{b},$$

$$v = \sum_{i=1,3,\cdots}^{\infty} \sum_{j=1}^{\infty} V_{ij} \cos \frac{i\pi x}{a} \sin \frac{2j\pi y}{b} \tag{4}$$

将式（3）代入方程（1d，e），我们可求得满足边界条件（2）的解

$$\psi_x = \frac{D_1}{D} w_0 \sin \frac{\pi x}{a} \cos \frac{\pi y}{b},$$

$$\psi_y = \frac{D_2}{D} w_0 \cos \frac{\pi x}{a} \sin \frac{\pi y}{b} \tag{5}$$

其中

$$D = \Big(\frac{D_x}{a^2} + \frac{D_{xy}}{b^2} + \frac{G_x}{\pi^2}\Big)\Big(\frac{D_{xy}}{a^2} + \frac{D_y}{b^2} + \frac{G_y}{\pi^2}\Big) - \Big(\frac{D_{xy}}{ab}\Big),$$

$$D_1 = \frac{G_x}{\pi a}\Big(\frac{D_{xy}}{a^2} + \frac{D_y}{b^2} + \frac{G_y}{\pi^2}\Big) - \frac{G_{xy} G_y}{\pi a b^2},$$

$$D_2 = \frac{G_y}{\pi b}\Big(\frac{D_x}{a^2} + \frac{D_{xy}}{b^2} + \frac{G_x}{\pi^2}\Big) - \frac{G_{xy} G_x}{\pi a^2 b} \tag{6}$$

将式（3）和（4）代入方程（1a，b），使用伽辽金方法并利用三角级数的正交性

质，得到

$$\frac{ab\pi^2}{4}\left(\frac{4m^2E_x}{a^2}+\frac{n^2G_{xy}}{b^2}\right)U_{mn}-\sum_{i=1,3,\cdots}^{\infty}\sum_{j=1}^{\infty}\frac{2ij\pi^2G_{xy}}{ab}K_1^{im}L_2^{jn}V_{ij}$$

$$=\frac{\pi bk_1E_x}{2a}K_1^{1m}\delta_{n1}w_0+\pi^3\left[\left(\frac{E_x}{a^3}+\frac{G_{xy}}{ab^2}\right)M_1^{11m}Q_1^{11n}-\frac{G_{xy}}{ab^2}M_1^{11m}Q_2^{11n}\right]w_0^2,$$

$$m=1,2,3,\cdots;n=1,3,5,\cdots$$

$$\frac{ab\pi^2}{4}\left(\frac{4m^2E_y}{b^2}+\frac{n^2G_{xy}}{a^2}\right)V_{mn}+\sum_{i=1}^{\infty}\sum_{j=1,3,\cdots}^{\infty}\frac{2ij\pi^2G_{xy}}{ab}K_2^{im}L_1^{jn}U_{ij} \tag{7}$$

$$=\frac{\pi ak_2E_y}{2b}L_1^{1n}\delta_{m1}w_0+\pi^3\left[\left(\frac{E_y}{b^3}+\frac{G_{xy}}{a^2b}\right)P_1^{11m}N_1^{11n}-\frac{G_{xy}}{a^2b}P_2^{11m}N_1^{11n}\right]w_0^2$$

$$m=1,3,5,\cdots;n=1,2,3,\cdots$$

其中

$$\delta_{m1}\begin{cases}1, & 当\ m=1\ 时,\\0, & 当\ m\neq 1\ 时\end{cases} \tag{8}$$

积分常数的表达式如下

$$K_1^{im}=\int_{-\frac{a}{2}}^{\frac{a}{2}}\sin\frac{i\pi x}{a}\sin\frac{2m\pi x}{a}\mathrm{d}x,$$

$$K_2^{im}=\int_{-\frac{a}{2}}^{\frac{a}{2}}\cos\frac{2i\pi x}{a}\cos\frac{m\pi x}{a}\mathrm{d}x,$$

$$P_1^{ipm}=\int_{-\frac{a}{2}}^{\frac{a}{2}}\cos\frac{i\pi x}{a}\cos\frac{p\pi x}{a}\cos\frac{m\pi x}{a}\mathrm{d}x, \tag{9}$$

$$P_2^{ipm}=\int_{-\frac{a}{2}}^{\frac{a}{2}}\sin\frac{i\pi x}{a}\sin\frac{p\pi x}{a}\sin\frac{m\pi x}{a}\mathrm{d}x,$$

$$M_1^{pmi}=\int_{-\frac{a}{2}}^{\frac{a}{2}}\sin\frac{p\pi x}{a}\cos\frac{m\pi x}{a}\sin\frac{2i\pi x}{a}\mathrm{d}x$$

用 L，Q，N，y，b，j，n，q 和 s 代替式（9）中的 K，P，M，x，a，i，m，p 和 r，可以得到积分常数 L_1，L_2，Q_1，Q_2 和 N_1。

线性方程组（7）的解有如下形式

$$\begin{aligned}U_{mn}&=A_{mn}w_0+B_{mn}w_0^2, \quad m=1,2,3,\cdots; \quad n=1,3,5,\cdots\\V_{mn}&=C_{mn}w_0+D_{mn}w_0^2, \quad m=1,3,5,\cdots; \quad n=1,2,3,\cdots\end{aligned} \tag{10}$$

其中，A_{mn}，B_{mn}，C_{mn} 和 D_{mn} 是方程组（7）的系数有关的常数。

横向均布载荷 q_0 展开成双重傅里叶级数

$$q_0=\sum_{i=1,3,\cdots}^{\infty}\sum_{j=1,3,\cdots}^{\infty}q_{ij}\cos\frac{i\pi x}{a}\cos\frac{j\pi y}{b} \tag{11}$$

把式（3）～(5）和（11）代入基本方程（1c），并利用式（10），使用伽辽金方法后得到

$$q_{mn}=I_1'(m,n)w_0+I_2'(m,n)w_0^2+I_3'(m,n)w_0^3, \quad m,n=1,3,5,\cdots \tag{12}$$

其中

$$I_1'(m,n)=\pi^3\left(\frac{m^3D_x}{a^3}+\frac{2mn^2D_{xy}}{ab^2}\right)\frac{D_1}{D}\delta_{m1}\delta_{n1}+\pi^3\left(\frac{2m^2nD_{xy}}{a^2b}+\frac{n^3D_y}{b^3}\right)\frac{D_2}{D}\delta_{m1}\delta_{n1}$$

$$+(k_1^2E_x+k_2^2E_y)\delta_{m1}\delta_{n1}-\sum_{i=1}^{\infty}\frac{4i\pi k_1E_x}{a^2}K_2^{im}A_{in}-\sum_{j=1}^{\infty}\frac{4j\pi k_2E_y}{b^2}L_2^{jn}C_{mj},$$

$$I_2'(m,n)=-2\pi^2\left[\frac{k_1E_x}{a^3b}(p_2^{11m}+2p_1^{11m})Q_1^{11n}+\frac{k_2E_y}{ab^3}(Q_2^{11n}+2Q_1^{11n})P_1^{11m}\right]$$

$$-\sum_{i=1}^{\infty}\frac{4i\pi k_1E_x}{a^2}K_2^{im}B_{in}-\sum_{j=1}^{\infty}\frac{4j\pi k_2E_y}{b^2}L_2^{jn}D_{mj}$$

$$+8\pi^3\sum_{i=1}^{\infty}\sum_{j=1,3,\cdots}^{\infty}\left(\frac{iE_x}{a^4b}M_2^{1mi}Q_1^{j1n}+\frac{jG_{xy}}{a^2b^3}M_1^{1mi}Q_2^{j1n}\right)A_{ij}$$

$$+8\pi^3\sum_{j=1}^{\infty}\sum_{i=1,3,\cdots}^{\infty}\left(\frac{jE_y}{ab^4}P_1^{i1m}N_2^{1nj}+\frac{jG_{xy}}{a^3b^2}P_2^{i1m}N_1^{1nj}\right)G_{ij},$$

$$I_3'(m,n)=-2\pi^4\left(\frac{4G_{xy}}{a^3b^3}P_4^{111m}Q_4^{111n}-\frac{E_x}{a^5b}P_4^{111m}Q_3^{111n}-\frac{E_y}{ab^5}P_3^{111m}Q_4^{111n}\right)$$

$$+8\pi^3\sum_{i=1}^{\infty}\sum_{j=1,3,\cdots}^{\infty}\left(\frac{iE_x}{a^4b}M_2^{1mi}Q_1^{j1n}+\frac{jG_{xy}}{a^2b^3}M_1^{1mi}Q_2^{j1n}\right)D_{ij}$$

$$+8\pi^3\sum_{j=1}^{\infty}\sum_{i=1,3,\cdots}^{\infty}\left(\frac{jE_y}{ab^4}P_1^{i1m}N_2^{1nj}+\frac{iG_{xy}}{a^3b^2}P_2^{i1m}N_1^{1nj}\right)D_{ij}\tag{13}$$

新的积分常数的表达式如下

$$P_3^{iprm}=\int_{-\frac{a}{2}}^{\frac{a}{2}}\cos\frac{i\pi x}{a}\cos\frac{p\pi x}{a}\cos\frac{r\pi x}{a}\cos\frac{m\pi x}{a}\mathrm{d}x,$$

$$P_4^{iprm}=\int_{-\frac{a}{2}}^{\frac{a}{2}}\sin\frac{i\pi x}{a}\sin\frac{p\pi x}{a}\cos\frac{r\pi x}{a}\cos\frac{m\pi x}{a}\mathrm{d}x,$$

$$M_2^{pmi}=\int_{-\frac{a}{2}}^{\frac{a}{2}}\cos\frac{p\pi x}{a}\cos\frac{m\pi x}{a}\cos\frac{2i\pi x}{a}\mathrm{d}x\tag{14}$$

用 Q，N，y，b，j，n，q 和 s 代替式（14）中的 P，M，x，a，i，m，p 和 r，可以得到积分常数 Q_3，Q_4 和 N_2。

当 $x=y=0$ 时，将式（12）代入式（11）得到

$$q_0=\sum_{i=1,3,\cdots}^{\infty}\sum_{j=1,3,\cdots}^{\infty}q_{ij}=I_1w_0+I_2w_0^2+I_3w_0^3\tag{15}$$

其中

$$I_1=\sum_{m=1,3,\cdots}^{\infty}\sum_{n=1,3,\cdots}^{\infty}I_1'(m,n),$$

$$I_2=\sum_{m=1,3,\cdots}^{\infty}\sum_{n=1,3,\cdots}^{\infty}I_2'(m,n),$$

$$I_3=\sum_{m=1,3,\cdots}^{\infty}\sum_{n=1,3,\cdots}^{\infty}I_3'(m,n)\tag{16}$$

根据驻值条件

$$\frac{\mathrm{d}q_0}{\mathrm{d}w_0}=0 \tag{17}$$

由式（15）得到网壳失稳时的临界挠度

$$w_{\sigma}=\frac{-I_2\pm\sqrt{I_2^2-3I_1I_3}}{3I_3} \tag{18}$$

将值 w_{σ} 代入式（15），便得到结构失稳时的临界载荷

$$q_{\sigma}=I_1w_{\sigma}+I_2w_{\sigma}^2+I_3w_{\sigma}^3 \tag{19}$$

令式（18）中的根式为零

$$I_2^2-3I_1I_3=0 \tag{20}$$

可以从中解出网壳产生失稳的最小曲率值。

四、数值结果

以均布横向载荷作用下矩形底双层正交正放网格扁壳为例：首先求解线性方程组(7)，求得常数 A_{mn}，B_{mn}，C_{mn} 和 D_{mn} 后代入式（13），再根据式（16）计算方程式（15）的系数 I_1，I_2 和 I_3，最后由式（18）和式（19）求出固定简支情况下双层网壳的临界挠度和临界载荷值。网壳的固定参数是：底面长 a=35m，宽 b=35m，厚 h=0.6m，网格表层杆件的截面积 0.03×0.03m^2，夹心和腹杆截面积 0.021×0.021m^2，弹性模量 E=2.1×10^{5}MPa，剪切模量 G=0.81×10^{5}MPa。网壳的网格尺寸 L_1 和 L_2 以及网壳的曲率 k_1 和 k_2 是变化参数。这里，取 $L_1=L_2$，$k_1=k_2$，位移分量 u 和 v 级数取 324 项，数值结果如图 2 和图 3 所示。

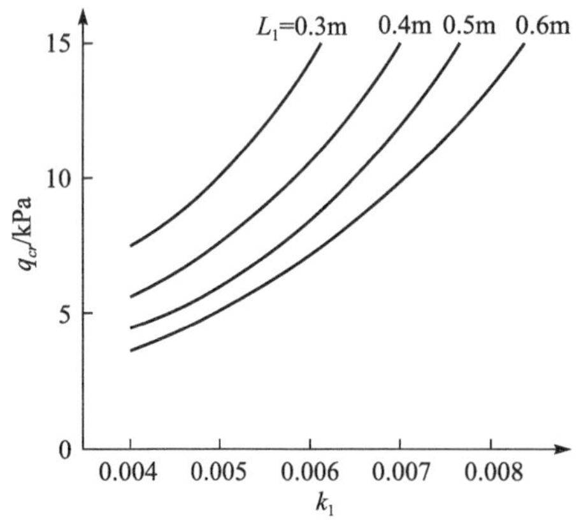

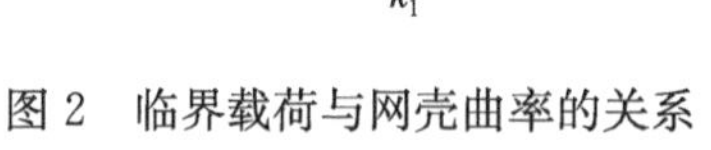
图 2　临界载荷与网壳曲率的关系

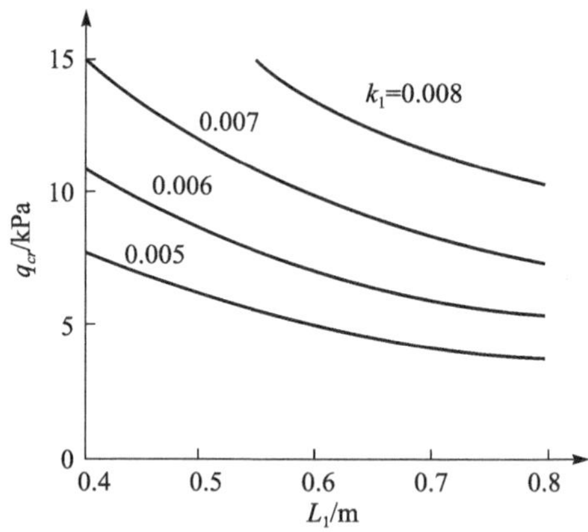

图 3　临界载荷与网壳网格尺寸的关系

实际上，在工程实际中，网壳只出现上临界失稳，故仅计算上临界载荷。

在图 2 中，随着网壳曲率的增大，网壳的上临界载荷不断增加。当然，对于非常扁平的网格扁壳，将不会出现失稳现象。另外，不同网格尺寸的同一形式网壳开始失稳时的曲率值相同。图 3 所表示的是网壳网格尺寸的变化对其临界载荷的影响。随着网壳网格尺寸 L_1 的增大（网格由密变稀），网壳的上临界载荷下降。

参考文献

[1] Liu Ren-huai, Li Dong, Nie Guohua, et al. Non-linear buckling of squarely-latticed shallow spherical shells. Int. J. Non-Linear Mech., 1991, 26 (5): 547-565.

[2] 刘人怀，聂国华. 网格扁壳的非线性弯曲理论. 江西工业大学学报，1991，13 (2)：186-192.

[3] 刘人怀，聂国华. 网格扁壳的非线性自由振动分析. 江西工业大学学报，1991，13 (2)：193-198.

[4] 聂国华，刘人怀，翁智远. 网格扁壳结构的非线性弯曲与稳定问题研究. 上海力学，1994，15 (2)：17-27.

[5] 聂国华，刘人怀. 矩形网格扁壳结构的非线性弹性理论. 应用数学和力学，1994，15 (5)：389-397.

[6] 聂国华，刘人怀. 矩形网格扁壳的非线性特征关系. 土木工程学报，1995，28 (1)：12-21.

[7] 刘人怀，肖潭. 双层正交正放网格扁壳结构的非线性弯曲理论//庄逢甘. 现代力学与科技进步. 3卷. 北京：清华大学出版社，1997：1212-1215.

[8] 刘人怀，肖潭. 矩形底双层网格扁壳的非线性弯曲. 暨南大学学报，1998，19 (3)：1-5.

第八章 厚、薄板壳弯曲分析

高压聚乙烯反应器厚壁筒体径向开孔的应力计算①

以前国内生产聚乙烯，大多采用管式反应器生产，产量较小，无法满足国内需求。因此，高压聚乙烯反应器的试制具有重大意义。该反应器是超高压容器，其设计压力高达2300个大气压，试制成功后，将是我国自行制造的最高压力容器。但是，在反应器的厚壁筒体上需要开几个径向孔洞，这会产生应力集中问题，对研制产生了极大困难，就我们所知，国内外还没有这方面的理论研究。为此，我们进行了初步研究尝试，考虑了厚壳的因素，使用混合法，给出了厚壁圆柱壳径向开孔处最大环向应力的计算公式。最后，进行了数值计算，其理论值与实验值符合。所得结果已供原国家第一机械工业部和化工部联合设计小组使用。

一、问题的简化

高压聚乙烯反应器的筒体上开了一些径向孔洞，其中以两个安装爆破帽的孔洞的直径为最大，该孔截面形状和尺寸如图1所示。

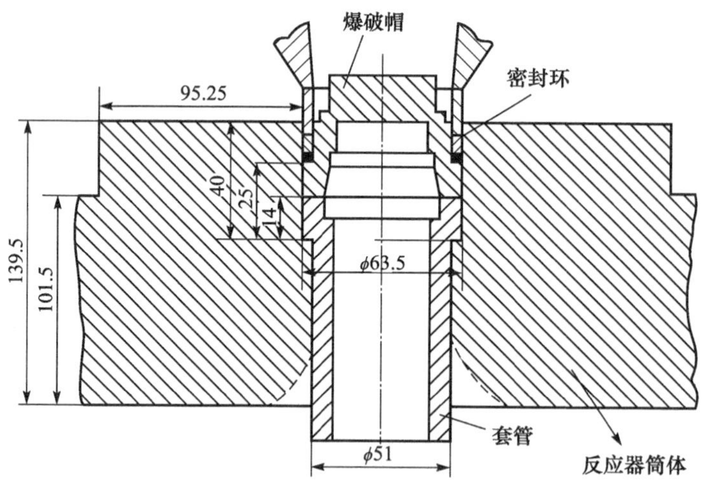

图1 ϕ51孔的截面形状和尺寸（mm）

① 本文是提交给国家第一机械工业部和化工部联合设计小组的研究报告，1970年7月27日. 原载《压力容器和压力管道的分析与计算》，北京：科学出版社，2014，80-94.

由于高压聚乙烯反应器筒体是厚壁圆柱壳，它在径向开孔后形成为三维弹塑性力学问题。为了计算简单、易于获得结果，我们采用了以下几项简化措施：

(1) 孔洞本来是变直径的，现忽略直径的变化，将孔洞视为等直径孔洞，并以中间段直径，即 51mm 作为计算直径。

(2) 孔洞本来是处在变厚度圆柱壳上，且在一个横截面内有 4 个孔洞。由于孔洞直径小，我们忽略壳厚的变化和孔洞间的彼此影响，认为在一个等厚度圆柱壳上仅开了一个 $\phi 51$ 孔洞。

(3) 由于孔洞中设有保护孔壁的套管，除下缘 25mm 高度外，它代替孔洞壁直接承担了内压力 P_0，而且孔洞上部有一小区域 40mm 高度部分不承受压力，这就造成孔边缘法向压力分布的不均匀性。为易于求解，并尽可能使解答接近实际情况，我们在直孔的假定下，设孔内壁仍承受 P_0 的均匀内压力。

(4) 反应器筒体是一相当厚的壳体。由于厚壳甚至是中等厚度壳体理论都未达到成熟的阶段，所以我们仍采用薄壁圆柱壳理论来研究孔洞对应力的干扰。处理时，计及了壳壁过厚的影响。

(5) 由于孔洞边缘温度分布情况不清楚，故不计及温度应力。

二、简化问题的解

在上述简化措施下，我们的问题就成为在孔洞边缘承受线布载荷（由爆破帽安装所引起）和均布内压力的情况下，计算等厚度圆柱壳孔洞区域的应力集中。简化孔洞截面形状和圆柱壳的内力、内力矩的正方向分别如图 2、图 3 所示，其中 ρ_0 为孔洞半径，R 为圆柱壳中曲面半径，a、b 分别为圆柱壳内、外壁半径，h 为圆柱壳厚度，q_0 为孔洞边缘的线布载荷。

仿效文献［1，2］处理薄壁壳体孔洞应力集中问题办法，认为厚壁圆柱壳径向孔洞的应力状态区分为无孔壳体的基本状态和由于集中因素所产生的干扰状态。于是，在极坐标（ρ，λ）下，孔洞附近区域的内力和内力矩为

$$
\begin{aligned}
&T_\rho = T_\rho^0 + T_\rho^*, \quad T_\lambda = T_\lambda^0 + T_\lambda^*, \\
&T_{\rho\lambda} = T_{\rho\lambda}^0 + T_{\rho\lambda}^*, \quad M_\rho = M_\rho^0 + M_\rho^*, \\
&M_\lambda = M_\lambda^0 + M_\lambda^* \quad \bar{N}_\rho = \bar{N}_\rho^0 + \bar{N}_\rho^*,
\end{aligned}
\tag{1}
$$

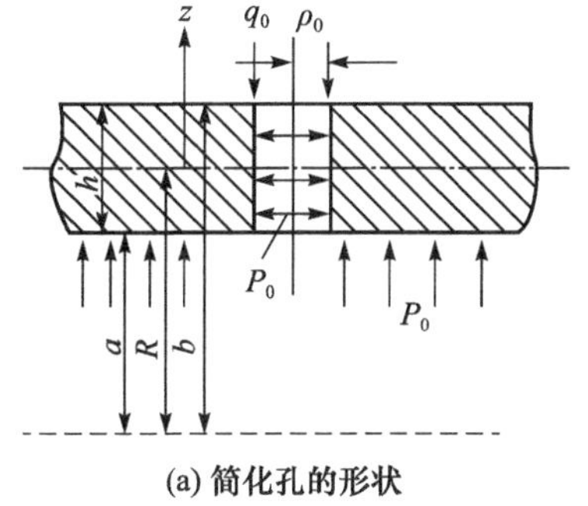

(a) 简化孔的形状

x
O
λ
ρ
y

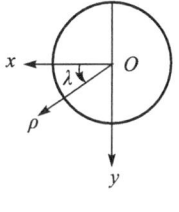

(b) 孔的极坐标

图 2　简化孔洞

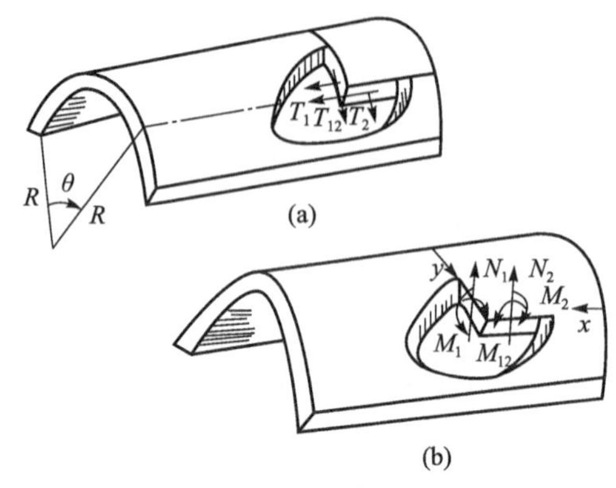

图 3　圆柱壳的内力和内力矩

其中 T_ρ、T_λ、$T_{\rho\lambda}$ 为内力，M_ρ、M_λ 为内力矩，$\overline{N}_\rho$ 为广义横向力，带“0”指标表示厚壁圆柱壳无孔时的解，带“ * ”指标表示孔洞存在时的干扰解。

由文献［3］，在圆柱坐标系（R，θ，x）下，封闭厚壁圆柱壳在承受均匀内压力 P_0 的情况下，其应力的精确解为

$$\sigma_x=\frac{a^2}{b^2-a^2}P_0,$$
$$\sigma_\theta=\frac{a^2}{b^2-a^2}\left[1+\frac{b^2}{(R+z)^2}\right]P_0, \tag{2}$$
$$\sigma_z=\frac{a^2}{b^2-a^2}\left[1-\frac{b^2}{(R+z)^2}\right]P_0$$

利用式（2），求得厚壁圆柱壳无孔时的内力和内力矩的精确解：

$$\begin{aligned}
T_1^0&=\int_{-h/2}^{h/2}\left(1+\frac{z}{R}\right)\sigma_x\mathrm{d}z=\frac{a^2}{2R}P_0,\\
T_2^0&=\int_{-h/2}^{h/2}\sigma_\theta\mathrm{d}z=aP_0,\\
M_1^0&=T_{12}^0=0,\\
M_2^0&=\int_{-h/2}^{h/2}\sigma_\theta z\,\mathrm{d}z=-\frac{a^2b^2}{2Rh}\left[\ln\left(\frac{b}{a}\right)-\frac{Rh}{ab}\right]P_0,\\
\overline{N}_1^0&=\overline{N}_2^0=0
\end{aligned} \tag{3}$$

转化到（ρ，λ）的极坐标系统，厚壁圆柱壳无孔时的内力和内力矩的精确解可写为

$$\begin{aligned}
T_\rho^0&=\frac{1}{2}(T_1^0+T_2^0)+\frac{1}{2}(T_1^0-T_2^0)\cos(2\lambda),\\
T_\lambda^0&=\frac{1}{2}(T_1^0+T_2^0)-\frac{1}{2}(T_1^0-T_2^0)\cos(2\lambda),\\
T_{\rho\lambda}^0&=-\frac{1}{2}(T_1^0-T_2^0)\sin(2\lambda),\\
M_\rho^0&=\frac{1}{2}[1-\cos(2\lambda)]M_2^0,\\
M_\lambda^0&=\frac{1}{2}[1+\cos(2\lambda)]M_2^0,\\
\overline{N}_\rho^0&=0
\end{aligned} \tag{4}$$

下面，要计算式（1）中带“ * ”的项，即求有孔存在时的干扰解。为书写简单起见，先下标注“ * ”号。

在无表面载荷作用下，薄圆柱壳位移函数 φ 的方程[3]为

$$\nabla^4\varphi+\frac{12\ (1-\nu^2)}{R^2h^2}\frac{\partial^4\varphi}{\partial x^4}=0, \tag{5}$$

其中 x 是圆柱壳母线坐标，y 是环向坐标，θ 是子午面的方位角，ν 是泊松比，∇ 是拉普拉斯算符，

$$y=R\theta,$$

$$\nabla=\frac{\partial^2}{\partial x^2}+\frac{\partial^2}{\partial y^2} \tag{6}$$

从方程（5）求得 φ 后，便可由下列公式求得轴向位移 u、环向位移 v 和径向位移 w，轴向内力 T_1、环向内力 T_2 和切向内力 T_{12}，轴向弯矩 M_1、环向弯矩 M_2 和扭矩 M_{12} 以及广义横向力 $\overline{N}_1$、$\overline{N}_2$：

$$
\begin{aligned}
&u=\frac{1}{R}\frac{\partial}{\partial x}\left[\nabla-(1+\nu)\frac{\partial^2}{\partial x^2}\right]\varphi,\\
&v=-\frac{1}{R}\frac{\partial}{\partial y}\left[\nabla+(1+\nu)\frac{\partial^2}{\partial x^2}\right]\varphi,\\
&w=\nabla^2\varphi,\\
&T_1=\frac{Eh}{R}\frac{\partial^4\varphi}{\partial x^2\partial y^2},\\
&T_2=\frac{Eh}{R}\frac{\partial^4\varphi}{\partial x^4},\\
&T_{12}=-\frac{Eh}{R}\frac{\partial^4\varphi}{\partial x^3\partial y},\\
&M_1=-D\left(\frac{\partial^2}{\partial x^2}+\nu\frac{\partial^2}{\partial y^2}\right)\nabla^2\varphi,\\
&M_2=-D\left(\frac{\partial^2}{\partial y^2}+\nu\frac{\partial^2}{\partial x^2}\right)\nabla^2\varphi,\\
&M_{12}=-D(1-\nu)\frac{\partial^2}{\partial x\partial y}(\nabla^2\varphi),\\
&\overline{N}_1=N_1+\frac{\partial M_{12}}{\partial y}=-D\frac{\partial}{\partial x}\left[\frac{\partial^2}{\partial x^2}+(2-\nu)\frac{\partial^2}{\partial y^2}\right]\nabla^2\varphi,\\
&\overline{N}_2=N_2+\frac{\partial M_{12}}{\partial x}=-D\frac{\partial}{\partial y}\left[\frac{\partial^2}{\partial y^2}+(2-\nu)\frac{\partial^2}{\partial x^2}\right]\nabla^2\varphi,
\end{aligned} \tag{7}
$$

其中 E 是弹性模量，D 是抗弯刚度，

$$D=\frac{Eh^3}{12(1-\nu^2)} \tag{8}$$

方程（5）可以由下面方程组来代替

$$
\begin{aligned}
&\nabla^2\varphi-\sqrt{\frac{12(1-\nu^2)}{R^2h^2}}\frac{\partial^2\varphi_1}{\partial x^2}=0,\\
&\nabla^2\varphi_1+\sqrt{\frac{12(1-\nu^2)}{R^2h^2}}\frac{\partial^2\varphi}{\partial x^2}=0
\end{aligned} \tag{9}
$$

引入辅助复函数

$$\Phi^* = \varphi + \mathrm{i}\varphi_1, \tag{10}$$

方程组（9）简化为

$$\nabla^2\Phi^* + \frac{2\mathrm{i}\sqrt{3(1-\nu^2)}}{Rh}\frac{\partial^2\Phi^*}{\partial x^2} = 0 \tag{11}$$

为了计算方便，再引入一个函数

$$\Phi = \nabla^2\Phi^* \tag{12}$$

显然 Φ 也满足 Φ^* 所满足的方程（11）：

$$\nabla^2\Phi + \frac{2\mathrm{i}\sqrt{3(1-\nu^2)}}{Rh}\frac{\partial^2\Phi}{\partial x^2} = 0, \tag{13}$$

于是，式（7）中的内力和内力矩公式可用 Φ 的实部和虚部来表示，为

$$\begin{aligned}
T_1 &= -\frac{D}{h}\sqrt{12(1-\nu^2)}\frac{\partial^2 \mathrm{Im}\Phi}{\partial y^2},\\
T_2 &= -\frac{D}{h}\sqrt{12(1-\nu^2)}\frac{\partial^2 \mathrm{Im}\Phi}{\partial x^2},\\
T_{12} &= \frac{D}{h}\sqrt{12(1-\nu^2)}\frac{\partial^2 \mathrm{Im}\Phi}{\partial x\partial y},\\
M_1 &= -D\left(\frac{\partial^2}{\partial x^2} + \nu\frac{\partial^2}{\partial y^2}\right)\mathrm{Re}\Phi,\\
M_2 &= -D\left(\frac{\partial^2}{\partial y^2} + \nu\frac{\partial^2}{\partial x^2}\right)\mathrm{Re}\Phi,\\
M_{12} &= -D(1-\nu)\frac{\partial^2 \mathrm{Re}\Phi}{\partial x\partial y},\\
\overline{N}_1 &= -D\frac{\partial}{\partial x}\left[(2-\nu)\frac{\partial^2}{\partial y^2} + \frac{\partial^2}{\partial x^2}\right]\mathrm{Re}\Phi,\\
\overline{N}_2 &= -D\frac{\partial}{\partial y}\left[(2-\nu)\frac{\partial^2}{\partial x^2} + \frac{\partial^2}{\partial y^2}\right]\mathrm{Re}\Phi
\end{aligned} \tag{14}$$

现在求解方程（13），先将它记为

$$L_1L_2\Phi = 0, \tag{15}$$

其中

$$\begin{aligned}
L_1 &= \nabla + 2(1-\mathrm{i})\beta\frac{\partial}{\partial x},\\
L_2 &= \nabla - 2(1-\mathrm{i})\beta\frac{\partial}{\partial x},\\
\beta &= \frac{\sqrt[4]{3(1-\nu^2)}}{2\sqrt{Rh}}
\end{aligned} \tag{16}$$

由于算子 L_1、L_2 是连着使用的，所以两个方程式

$$L_1\Phi = 0, \qquad L_2\Phi = 0 \tag{17a, b}$$

的解都是方程（15）的解。为此，在方程（4.17a，b）中分别令

$$\begin{aligned}
\Phi &= \mathrm{e}^{-(1-\mathrm{i})\beta x}F(x, y),\\
\Phi &= \mathrm{e}^{(1-\mathrm{i})\beta x}F(x, y),
\end{aligned} \tag{18}$$

则 $F(x, y)$ 由下面方程确定

$$\nabla F+2\mathrm{i}\beta^2 F=0 \tag{19}$$

因 $\mathrm{e}^{-(1-\mathrm{i})\beta x}$ 和 $\mathrm{e}^{(1-\mathrm{i})\beta x}$ 都可用克雷洛夫函数来表达，其偶函数和奇函数解分别为

$$\Omega_1(\beta x)-2\mathrm{i}\Omega_3(\beta x);\quad \Omega_2(\beta x)-2\mathrm{i}\Omega_4(\beta x)$$

那么，解（18）改写为

$$\begin{aligned}\Phi&=[\Omega_1(\beta x)-2\mathrm{i}\Omega_3(\beta x)]F(x,y),\\ \Phi&=[\Omega_2(\beta x)-2\mathrm{i}\Omega_4(\beta x)]F(x,y)\end{aligned} \tag{20}$$

为了求解孔洞区域的应力，引入极坐标是有益的，故设

$$x=\rho\cos\lambda,\qquad y=\rho\sin\lambda \tag{21}$$

这里 $\lambda=\text{const}$ 是一螺旋线。如果在平面上把柱面展开，那么这些螺旋线就变成一族从原点出发的径向直线，而曲线 $\rho=\text{const}$ 将变成一族同心圆，特别是孔洞将变成为一半径为 ρ_0 的圆形切口。

引入式（21）后，式（14）成为

$$\begin{aligned}
T_\rho&=-\frac{D}{h}\sqrt{12(1-\nu^2)}\left(\frac{1}{\rho}\frac{\partial}{\partial\rho}+\frac{1}{\rho^2}\frac{\partial^2}{\partial\lambda^2}\right)\mathrm{Im}\Phi,\\
T_\lambda&=-\frac{D}{h}\sqrt{12(1-\nu^2)}\frac{\partial^2\,\mathrm{Im}\Phi}{\partial\rho^2},\\
T_{\rho\lambda}&=\frac{D}{h}\sqrt{12(1-\nu^2)}\frac{\partial}{\partial\rho}\frac{1}{\rho}\frac{\partial\,\mathrm{Im}\Phi}{\partial\lambda},\\
M_\rho&=-D\left(\frac{\partial^2}{\partial\rho^2}+\frac{\nu}{\rho}\frac{\partial}{\partial\rho}+\frac{\nu}{\rho^2}\frac{\partial^2}{\partial\lambda^2}\right)\mathrm{Re}\Phi,\\
M_\lambda&=-D\left(\nu\frac{\partial^2}{\partial\rho^2}+\frac{1}{\rho}\frac{\partial}{\partial\rho}+\frac{1}{\rho^2}\frac{\partial^2}{\partial\lambda^2}\right)\mathrm{Re}\Phi,\\
M_{\rho\lambda}&=-D(1-\nu)\frac{\partial}{\partial\rho}\frac{1}{\rho}\frac{\partial\,\mathrm{Re}\Phi}{\partial\lambda},\\
\overline{N}_\rho&=-D\left[\frac{\partial}{\partial\rho}\Delta+(1-\nu)\frac{1}{\rho}\frac{\partial}{\partial\rho}\frac{1}{\rho}\frac{\partial^2}{\partial\lambda^2}\right]\mathrm{Re}\Phi,\\
\overline{N}_\lambda&=-D\left[\frac{1}{\rho}\frac{\partial}{\partial\lambda}\Delta+(1-\nu)\frac{\partial^2}{\partial\rho^2}\frac{1}{\rho}\frac{\partial}{\partial\lambda}\right]\mathrm{Re}\Phi,
\end{aligned} \tag{22}$$

其中 Δ 是极坐标中的拉普拉斯算子，

$$\Delta=\frac{\partial^2}{\partial\rho^2}+\frac{1}{\rho}\frac{\partial}{\partial\rho}+\frac{1}{\rho^2}\frac{\partial^2}{\partial\lambda^2} \tag{23}$$

考虑到坐标 λ 的周期性，可令方程（19）的解为三角函数：

$$F(\rho,\lambda)=\sum_{n=0}^{\infty}[F_n(\rho)\cos(n\lambda)+\overline{F}_n(\rho)\sin(n\lambda)] \tag{24}$$

再考虑到孔洞受载荷的对称性质，可令

$$\overline{F}_n(\rho)=0 \tag{25}$$

将式（24）代入方程（19），便得关于 $F_n(\rho)$ 的贝塞尔方程

$$\frac{\mathrm{d}^2F_n}{\mathrm{d}\rho^2}+\frac{1}{\rho}\frac{\mathrm{d}F_n}{\mathrm{d}\rho}+\left(2\mathrm{i}\beta^2-\frac{n^2}{\rho^2}\right)F_n=0 \tag{26}$$

方程（26）的通解为

$$F_n = c_1^* J_n(\beta\rho\sqrt{2\mathrm{i}}) + c_2^* H_n^{(1)}(\beta\rho\sqrt{2\mathrm{i}}), \tag{27}$$

其中 c_1^*、c_2^* 为待定常数，J_n 是贝塞尔函数，$H_n^{(1)}$ 是第一类汉克尔函数。

因为孔洞的影响是局部的，而 $J_n(\beta\rho\sqrt{2\mathrm{i}})$ 是正则函数，故应令

$$c_1^* = 0, \tag{28}$$

而

$$H_n^{(1)}(\beta\rho\sqrt{2\mathrm{i}}) = \psi_n(\beta\rho) + \mathrm{i}\chi_n(\beta\rho) \tag{29}$$

是按负指数规律变化的函数。将 ψ_n 和 χ_n 在坐标原点附近分解为关于 ρ 的幂级数，同样也把克雷洛夫函数在坐标原点附近分解为 ρ 的幂级数，在计及解的偶函数性质后，方程（13）在 $\beta\rho$ 很小时的解就成为

$$\Phi_n = (\alpha_n + \mathrm{i}\beta_n)\cos(n\lambda), \tag{30}$$

其中 γ 是欧拉-马斯克隆常数，$\ln\gamma = 0.57722$，

$$\begin{aligned}
\alpha_0 &= \frac{1}{2} + \frac{1}{\pi}\beta^2\rho^2\left\{[2+\cos(2\lambda)]\ln\left(\frac{\gamma\beta\rho}{\sqrt{2}}\right) - 1\right\} + \cdots,\\
\beta_0 &= \frac{2}{\pi}\ln\frac{\gamma\beta\rho}{\sqrt{2}} - \frac{1}{4}\beta^2\rho^2[2+\cos(2\lambda)] + \cdots,\\
\alpha_1 &= \cos\lambda\left[\frac{2}{3\pi}\beta^2\rho^2 - \frac{1}{3\pi}\beta^2\rho^2\cos(2\lambda) - \frac{2}{\pi}\beta^2\rho^2\ln\left(\frac{\gamma\beta\rho}{\sqrt{2}}\right)\right] + \cdots,\\
\beta_1 &= \cos\lambda\left(\frac{1}{2}\beta^2\rho^2 - \frac{2}{\pi} + \cdots\right),\\
\alpha_2 &= -\frac{2}{\pi}\frac{1}{\beta^2\rho^2} - \frac{1}{2\pi}\beta^2\rho^2\left[\frac{1}{4} + \cos(2\lambda) - \frac{2}{3}\cos(4\lambda) + \ln\left(\frac{\gamma\beta\rho}{\sqrt{2}}\right)\right] + \cdots,\\
\beta_2 &= \frac{1}{\pi}\cos(2\lambda) + \frac{1}{8}\beta^2\rho^2,\\
\alpha_3 &= \cos\lambda\left\{-\frac{8}{\pi\beta^2\rho^2} + \frac{2}{3\pi}\beta^2\rho^2\left[\frac{1}{4} - \frac{1}{2}\cos(2\lambda) + \frac{2}{5}\cos(4\lambda)\right]\right\} + \cdots,\\
\beta_3 &= -\frac{2}{3\pi}\cos\lambda[1 - 2\cos(2\lambda)] + \cdots,\\
\alpha_4 &= -\frac{2}{\pi\beta^2\rho^2}[5 + 3\cos(2\lambda)] + \cdots,\\
\beta_4 &= \frac{24}{\pi}\frac{1}{\beta^4\rho^4} + \cdots,\\
&\cdots\cdots
\end{aligned} \tag{31}$$

可以看出，解（30）中的 Φ 及其各阶导函数在 $\rho\to\infty$ 时是趋于零的，这符合孔洞区域的远端边界条件，现在需要去满足孔洞的边界条件。

由于孔半径较小，亦即

$$\frac{\rho_0}{R} \ll \sqrt{\frac{h}{R}}, \tag{32}$$

所以仅研究忽略$\dfrac{\rho_0^2}{Rh}$二次以上项的近似解。为此，先将复变量应力函数Φ记为

$$\begin{aligned}\Phi=&(\overline{A}+\mathrm{i}\overline{B})\Phi_0+(\overline{C}+\mathrm{i}\overline{D})\Phi_1+(\overline{E}+\mathrm{i}\overline{F})\Phi_2+(\overline{H}+\mathrm{i}\overline{K})\Phi_3\\&+(\overline{L}+\mathrm{i}\overline{S})\Phi_4+\cdots,\end{aligned}\tag{33}$$

这里，为满足边界条件，将按如下方式配置常数：

$$\begin{aligned}&\overline{A}=A_0+A_1\beta^2+\cdots,\\&\overline{B}=B_1\beta^2+\cdots,\\&\overline{C}=C_0+C_1\beta^2+\cdots,\\&\overline{D}=D_1\beta^2+\cdots,\\&\overline{E}=E_2\beta^4+\cdots,\\&\overline{F}=F_1\beta^2+F_2\beta^4+\cdots,\\&\overline{H}=H_2\beta^4+\cdots,\\&\overline{K}=K_3\beta^6+\cdots,\\&\overline{L}=L_4\beta^8+\cdots,\\&\overline{S}=S_3\beta^6+\cdots,\\&\cdots\cdots\end{aligned}\tag{34}$$

将式（30）、式（34）代入式（33），在略去高于β^2级的项后，便得

$$\begin{aligned}&\mathrm{Im}\Phi=f_0+\beta^2f_1,\\&\mathrm{Re}\Phi=w_0+\beta^2w_1,\end{aligned}\tag{35}$$

其中

$$\begin{aligned}f_0=&\frac{2A_0}{\pi}\left[\ln\left(\frac{\rho}{\rho_0}\right)+\gamma_1\right]-\frac{C_0}{\pi}[1+\cos(2\lambda)]-\frac{2F_1}{\pi}\frac{\cos(2\lambda)}{\rho^2},\\f_1=&-\left\{\frac{A_0}{4}\rho^2[2+\cos(2\lambda)]-\frac{C_0}{4}\rho^2[1+\cos(2\lambda)]-\frac{2A_1}{\pi}\left[\ln\left(\frac{\rho}{\rho_0}\right)+\gamma_1\right]\right.\\&\left.-\frac{B_1}{2}+\frac{2F_2}{\pi}\frac{\cos(2\lambda)}{\rho^2}+\frac{C_1}{\pi}[1+\cos(2\lambda)]\right\},\\w_0=&\frac{A_0}{2},\\w_1=&\frac{1}{\pi}\left\{\frac{\pi A_1}{2}-2B_1\gamma_1+D_1-\frac{F_1}{\pi}+A_0\rho^2\left[2\ln\left(\frac{\rho}{\rho_0}\right)+2\gamma_1-1\right]\right.\\&-2B_1\ln\left(\frac{\rho}{\rho_0}\right)+c_0\rho^2\left[\frac{1}{4}-\ln\left(\frac{\rho}{\rho_0}\right)-\gamma_1\right]+\cos(2\lambda)\left\{A_0\rho^2\left[\ln\left(\frac{\rho}{\rho_0}\right)+\gamma_1\right]\right.\\&\left.+C_0\rho^2\left[\frac{1}{6}-\ln\left(\frac{\rho}{\rho_0}\right)-\gamma_1\right]+D_1-\frac{2E_2}{\rho^2}-\frac{4H_2}{\rho_2}\right\}-\cos(4\lambda)\left(\frac{C_0}{12}\rho_2\right.\\&\left.\left.+\frac{F_1}{2}+\frac{4H_2}{\rho^2}+\frac{24S_3}{\rho^4}\right)\right\},\\\gamma_1=&\ln\left(\frac{\gamma\rho_0\beta}{\sqrt{2}}\right)\end{aligned}\tag{36}$$

在式（35）中令 $\beta=0$，便得有孔的平板情形。此时，应有径向位移

$$w=\mathrm{Re}\Phi=0 \tag{37}$$

所以，应在式（33）的右端加入常数$\left(-\dfrac{A_0}{2}\right)$，才能满足上面条件。显然，$\Phi=\mathrm{const}$ 也是方程（13）的解。

引用极坐标变换（21）后，圆柱壳孔洞区域解就相当于平面应力状态中的有孔板应力分布状态的克尔希问题。根据式（1）、式（4），孔洞边缘的法向力和切向力条件为

在 $\rho=\rho_0$ 时，

$$\begin{aligned}
T_\rho &=T_\rho^0+T_\rho^*\Big|_{\beta=0}\\
&=\frac{1}{2}(T_1^0+T_2^0)+\frac{1}{2}(T_1^0-T_2^0)\cos(2\lambda)-\frac{D}{h}\sqrt{12(1-\nu^2)}\left(\frac{1}{\rho}\frac{\partial}{\partial\rho}+\frac{1}{\rho^2}\frac{\partial^2}{\partial\lambda^2}\right)f_0\\
&=-hP_0,\\
T_{\rho\lambda}&=T_{\rho\lambda}^0+T_{\rho\lambda}^*\Big|_{\beta=0}\\
&=-\frac{1}{2}(T_1^0-T_2^0)\sin(2\lambda)+\frac{D}{h}\sqrt{12(1-\nu^2)}\frac{\partial}{\partial\rho}\frac{1}{\rho}\frac{\partial f_0}{\partial\lambda}\\
&=0
\end{aligned} \tag{38}$$

将式（36）中的 f_0 值代入式（38），便得

$$\begin{aligned}
A_0&=\frac{\pi\rho_0^2}{2Eh^2}\sqrt{3(1-\nu^2)}(T_1^0+T_2^0+2hP_0),\\
C_0&=\frac{\pi\rho_0^2}{Eh^2}\sqrt{3(1-\nu^2)}(T_1^0-T_2^0),\\
F_1&=-\frac{\pi\rho_0^4}{4Eh^2}\sqrt{3(1-\nu^2)}(T_1^0-T_2^0)
\end{aligned} \tag{39}$$

要达到 $\beta^2\rho_0^2$ 级精确度，还应在 $\rho=\rho_0$ 时，使

$$\begin{aligned}
&\left(\frac{1}{\rho}\frac{\partial}{\partial\rho}+\frac{1}{\rho^2}\frac{\partial^2}{\partial\lambda^2}\right)f_1=0,\\
&\frac{\partial}{\partial\rho}\frac{1}{\rho}\frac{\partial f_1}{\partial\lambda}=0
\end{aligned} \tag{40}$$

由此得

$$\begin{aligned}
A_1&=\frac{\pi\rho_0^2}{2}\left(A_0-\frac{C_0}{2}\right),\\
C_1&=\frac{\pi\rho_0^2}{2}(C_0-A_0),\\
F_2&=-\frac{\pi\rho_0^4}{8}(C_0-A_0)
\end{aligned} \tag{41}$$

于是，在 $\beta^2\rho_0^2$ 级精确度，已完全满足了孔洞边缘的法向力和切向力边界条件。

至于孔洞边缘上的力矩和横向力边界条件，还有：

在 $\rho=\rho_0$ 时，

$$
\begin{aligned}
&M_\rho = M_\rho^0 + M_\rho^* = 0, \\
&\overline{N}_\rho = \overline{N}_\rho^0 + \overline{N}_\rho^* = q_0,
\end{aligned} \tag{42}
$$

这里，q_0 是孔洞边缘线布载荷，其值由孔洞的力平衡条件求得。

因为

$$2\pi\rho_0 q_0 + \pi\rho_0^2 P_0 = 0,$$

故

$$q_0 = -\frac{\rho_0}{2}P_0 \tag{43}$$

而 M_ρ^*、$\overline{N}_\rho^*$ 可按照式（22）、（35）求得

$$
\begin{aligned}
M_\rho^* = -\frac{D\beta^2}{\pi}\Bigg\{ & 4A_0\left[(1+\nu)\ln\left(\frac{\rho}{\rho_0}\right)+(1+\nu)\gamma_1+1\right]+\frac{2B_1}{\rho^2}(1-\nu) \\
& -C_0\left[\frac{5+\nu}{2}+2(1+\nu)\ln\left(\frac{\rho}{\rho_0}\right)+2(1+\nu)\gamma_1\right] \\
& +\cos(2\lambda)\bigg\{A_0\left[2(1-\nu)\ln\left(\frac{\rho}{\rho_0}\right)+2(1-\nu)\gamma_1+3+\nu\right] \\
& -C_0\left[2(1-\nu)\ln\left(\frac{\rho}{\rho_0}\right)+2(1-\nu)\gamma_1+\frac{4(2+\nu)}{3}\right] \\
& -\frac{4\nu D_1}{\rho^2}-\frac{12E_2}{\rho^4}(1-\nu)-\frac{24H_2}{\rho_4}(1-\nu)\bigg\} \\
& -\cos(4\lambda)\left[\frac{1-7\nu}{6}C_0-\frac{8\nu F_1}{\rho^2}+\frac{24H_2}{\rho^4}(1-3\nu)+\frac{480S_3}{\rho^6}(1-\nu)\right]\Bigg\},
\end{aligned} \tag{44}
$$

$$
\begin{aligned}
\overline{N}_\rho^* = -\frac{D\beta^2}{\pi}\Bigg\{ & 8\left(A_0-\frac{C_0}{2}\right)\frac{1}{\rho}-\cos(2\lambda)\bigg\{\frac{4(1-\nu)}{\rho}A_0\left[1+\gamma_1+\ln\left(\frac{\rho}{\rho_0}\right)\right] \\
& -\frac{4(1-\nu)}{\rho}C_0\left[\frac{5}{6}+\ln\left(\frac{\rho}{\rho_0}\right)+\gamma_1\right]-\frac{4(3-\nu)}{\rho^3}D_1+\frac{24(1-\nu)}{\rho^5}E_2 \\
& +\frac{48(1-\nu)}{\rho^5}H_2\bigg\}+\cos(4\lambda)\left[\frac{4(1-\nu)}{4}\frac{C_0}{\rho}-\frac{8F_1}{\rho^3}(3-\nu)\right. \\
& \left.-\frac{192}{\rho^5}(2-\nu)H_2-\frac{1920S_3}{\rho^7}(1-\nu)\right]\Bigg\}
\end{aligned}
$$

应用式（42）、(4)、(44)，再分别令常数项和 $\cos(2\lambda)$、$\cos(4\lambda)$ 的系数为 0，便得下面六个方程：

$$4A_0[(1+\nu)\gamma_1+1]-C_0\left[\frac{5+\nu}{2}+2(1+\nu)\gamma_1\right]+\frac{2(1-\nu)\beta_1}{\rho_0^2}=\frac{\pi M_2^0}{2D\beta^2}, \tag{45a}$$

$$
\begin{aligned}
& A_0[2(1-\nu)\gamma_1+3+\nu]-C_0\left[2(1-\nu)\gamma_1+\frac{4(2+\nu)}{3}\right]-\frac{4\nu D_1}{\rho_0^2} \\
& -\frac{12(1-\nu)E_2}{\rho_0^4}-\frac{24(1-\nu)H_2}{\rho_0^4}=-\frac{\pi M_2^0}{2D\beta^2},
\end{aligned} \tag{45b}
$$

$$\frac{1-7\nu}{6}C_0-\frac{8\nu F_1}{\rho_0^2}+\frac{24(1-3\nu)}{\rho_0^4}H_2+\frac{480(1-\nu)}{\rho_0^6}S_3=0, \tag{45c}$$

$$-\frac{8D\beta^2}{\pi\rho_0}\left(A_0-\frac{1}{2}C_0\right)=-\frac{\rho_0}{2}P_0, \tag{45d}$$

$$A_0(1-\nu)(1+\gamma_1)-C_0(1-\nu)\left(\frac{5}{6}+\gamma_1\right)-\frac{3-\nu}{\rho_0^2}D_1+\frac{6(1-\nu)}{\rho_0^4}E_2+\frac{12(1-\nu)}{\rho_0^4}H_2=0, \tag{45e}$$

$$\frac{1-\nu}{3}C_0-\frac{2(3-\nu)}{\rho_0^2}F_1-\frac{48(2-\nu)}{\rho_0^4}H_2-\frac{480(1-\nu)}{\rho_0^6}S_3=0 \tag{45f}$$

将式（39）代入方程（45d），显然已经满足。其余五个方程（44a～c，e，f）包含五个常数，求解这组方程，可得

$$\begin{aligned}
&B_1=\frac{\rho_0^2}{2(1-\nu)}\left\{\frac{\pi M_2^0}{2D\beta^2}-4A_0[(1+\nu)\gamma_1+1]+C_0\left[\frac{5+\nu}{2}+2(1+\nu)\gamma_1\right]\right\},\\
&H_2=\frac{\rho_0^4}{48(3+\nu)}\left[(1-3\nu)C_0-12(1+\nu)\frac{F_1}{\rho_0^2}\right],\\
&S_3=-\frac{\rho_0^6}{1440(1-\nu)(3+\nu)}\left[(3-19\nu+10\nu^2)C_0-6(3+6\nu-5\nu^2)\frac{F_1}{\rho_0^2}\right],\\
&D_1=\frac{\rho_0^2}{12(3+\nu)}\left[\frac{3\pi M_2^0}{D\beta^2}+24(1-\nu)(A_0-C_0)\gamma_1+6(5-\nu)A_0-2(13-\nu)C_0\right],\\
&E_2=\frac{\rho_0^2}{12(1-\nu)(3+\nu)}\left\{\frac{\pi(3-\nu)M_2^0}{2D\beta^2}+[6(1-\nu)^2\gamma_1+9-4\nu+3\nu^2]A_0\right.\\
&\qquad\left.-[6(1-\nu)^2\gamma_1+2(4-\nu+\nu^2)]C_0-24(1-\nu)(3+\nu)\frac{H_2}{\rho_0^4}\right\}
\end{aligned} \tag{46}$$

因为应力集中将出现在孔洞边缘处，所以我们仅给出孔洞边缘的内力和内力矩公式。由式（1）、（4）、（22）、（35）得到：

在 $\rho=\rho_0$ 时，

$$\begin{aligned}
&T_\rho=-h\rho_0,\\
&T_\lambda=\left[\frac{1}{2}(T_1^0+T_2^0)-\frac{1}{2}(T_1^0-T_2^0)\cos(2\lambda)-\frac{D}{h}\sqrt{12(1-\nu^2)}\frac{\partial^2\operatorname{Im}\Phi}{\partial\rho^2}\right]\\
&\quad=hP_0+T_1^0+T_2^0-2(T_1^0-T_2^0)\cos(2\lambda)+\frac{\pi\rho_0^2\sqrt{3(1-\nu^2)}}{4Rh}\left\{h[1\right.\\
&\qquad\left.+\cos(2\lambda)]P_0+T_0^2+\frac{1}{2}(3T_2^0-T_1^0)\cos(2\lambda)\right\},\\
&M_\rho=0,\\
&M_\lambda=\left\{\frac{1}{2}[1+\cos(2\lambda)]M_2^0-D\left(\nu\frac{\partial^2}{\partial\rho^2}+\frac{1}{\rho}\frac{\partial}{\partial\rho}+\frac{1}{\rho^2}\frac{\partial^2}{\partial\lambda^2}\right)\operatorname{Re}\Phi\right\}\\
&\quad=\left[1+\frac{1+\nu}{3+\nu}\cos(2\lambda)\right]M_2^0\\
&\qquad-\frac{h\rho_0^2(1+\nu)}{8R}\left\{2\left[1+2\gamma_1+\frac{1+3\nu-4(1-\nu)\gamma_1}{2(3+\nu)}\cos(2\lambda)\right]P_0\right.
\end{aligned} \tag{47}$$

$$-\frac{1}{2h}T_1^0+\frac{5+8\gamma_1}{2h}T_2^0-\frac{1}{6(3+\nu)}\left[\frac{7+5\nu-12(1+\nu)\gamma_1}{h}T_1^0\right.$$

$$\left.\left.-\frac{13+23\nu-36(1-\nu)\gamma_1}{h}T_2^0\right]\cos(2\lambda)+\frac{1-\nu}{2(3+\nu)h}(T_1^0-T_2^0)\cos(4\lambda)\right\}$$

在孔洞边缘上，环向应力将是最大应力值，故先按文献［4］给出厚壁圆柱壳的环向应力公式：

$$\sigma_2=\frac{T_2}{h}-\frac{\nu\left(\frac{z}{R}\right)^2}{h(1-\nu^2)\left(1+\frac{z}{R}\right)}T_1+\frac{1}{h^2}\left(\frac{h}{R}+\frac{12\,\frac{z}{h}}{1+\frac{z}{R}}\right)M_2-\frac{\nu}{h^2}\left(\frac{h}{R}-\frac{12\,\frac{z^2}{Rh}}{1+\frac{z}{R}}\right)M_1 \tag{48}$$

同时，在孔洞边缘 $\lambda=0$，π 处，还有下面公式成立：

$$\sigma_2=\sigma_\lambda,\quad T_1=T_\rho,\quad T_2=T_\lambda,\quad M_1=0,\quad M_2=M_\lambda \tag{49}$$

由上面公式看出，厚壁圆柱壳最大的环向应力 $\sigma_{\lambda\max}$ 将发生在壳体内表面 $z=-\frac{h}{2}$ 的孔洞边缘 $\lambda=0$，π 处。由式（48）和（49），可得

$$\sigma_{\lambda\max}=\frac{T_{\lambda0}}{h}=\frac{\nu h}{2R(1-\nu^2)(2R-h)}T_{\rho0}+\frac{1}{h^2}\left(\frac{h}{R}-\frac{12R}{2R-h}\right)M_{\lambda0}, \tag{50}$$

其中 $T_{\lambda0}$、$T_{\rho0}$ 和 $M_{\lambda0}$ 分别为孔洞边缘 $\lambda=0$，π 处的环向薄膜力。径向膜膜力和环向弯矩，

$$\begin{aligned}&T_{\lambda0}=hP_0-T_1^0+3T_2^0+\frac{\pi\rho_0^2\sqrt{3(1-\nu^2)}}{4Rh}\left[2hP_0+\frac{1}{2}(5T_2^0-T_1^0)\right],\\&T_{\rho0}=-hP_0,\\&M_{\lambda0}=\frac{1}{3+\nu}\left\{2(2+\nu)M_2^0-\frac{h\rho_0^2(1+\nu)}{8R}\left[(7+5V+8(1+\nu)\gamma_1)P_0\right.\right.\\&\qquad\left.\left.-\frac{13+11\nu-12(1-\nu)\gamma_1}{6h}T_1^0+\frac{55+41\nu+12(3+5\nu)\gamma_1}{6h}T_2^0\right]\right\}\end{aligned} \tag{51}$$

三、实例与讨论

目前正在研制的高压聚乙烯反应器数据为 $R=222.25\text{mm}$，$h=139.5\text{mm}$，$a=152.5\text{mm}$，$b=292\text{mm}$，$\rho_0=25.5\text{mm}$，$\nu=0.3$，最大设计压力 $P_0=23\text{kg/mm}^2$，材料：34CrNi3MoA，材料屈服极限 $\sigma_s=75\text{kg/mm}^2$。

将上面数据代入式（50），便得反应器最大应力值为

$$\sigma_{\lambda\max}=4.735P_0 \tag{52}$$

由式（4），可求出此处不开孔时的环高压力值为

$$\sigma_\lambda^0=1.81P_0 \tag{53}$$

则开孔后的最大应力与不开孔时应力的比值为

$$\omega=\frac{\sigma_{\lambda\max}}{\sigma_\lambda^0}=2.62 \tag{54}$$

由文献［5］，该处的实验值为

$$\sigma_{\lambda\max}^{*}=4.956P_0 \tag{55}$$

本节理论值与实验值的相对误差为 4.459%，显然，相当符合。这说明本节的理论研究是正确的。

在该处，三个主应力为

$$\sigma_1=4.735P_0,\quad \sigma_2=-P_0,\quad \sigma_3=-P_0 \tag{56}$$

将这些数据代入第四强度理论公式

$$(\sigma_1-\sigma_2)^2+(\sigma_2-\sigma_3)^2+(\sigma_1-\sigma_3)^2=2\sigma_3^2 \tag{57}$$

便求得最大应力点处的屈服内压力为 13.08kg/mm²，它远小于最大设计压力。这说明联合设计小组研制的聚乙烯反应器的开孔强度设计是安全的。

参考文献

[1] Лурbе А И. Концентрация напряжений в области отверстия на поверхности крутового цилиндра. ПММ，1946，10 (3).

[2] Гузb О М. Про наближений метод визначения концентрації напруженb bіля криволінійних отворів в оболонках. Прикладна Механика，1962，8 (6).

[3] 杜庆华，等. 材料力学. 北京：高等教育出版社，1958.

[4] 符拉索夫 В З. 壳体的一般理论. 薛振东，朱世靖，译. 北京：人民教育出版社，1960.

[5] 兰州石油机械研究所强度组. 高压聚乙烯反应器筒体径向开孔应力集中的光弹性试验. 化工炼油机械通讯，1972，(1)：53-61.

尿素合成塔底部球形封头开孔的应力计算①

大型尿素合成塔是生产优质化肥尿素的关键设备，是我国农业方面的急需设备之一。1970 年，某厂将尿素合成塔这个外国“王牌”产品结构进行了大胆革新，试制成功了具有先进水平的大型尿素合成塔。但是，在新的尿素合成塔底部球形封头上，采用的是直接开孔加补强的方法。即在一厚达 97mm 的球壳孔洞边缘，需要焊接一块厚达 80mm 的补强圈。这样做的结果，造成焊接工作量大，操作条件恶劣，而且产品质量得不到保证。若预热温度不够，焊接稍不注意，封头就会产生裂纹。针对这种情况，提出了是否能取掉补强圈的问题。此问题属于厚壳理论范围，研究极少，难度极大。为此，我们采用了较简捷的办法，改进了薄球壳理论公式，算出了较厚球壳的理论值，并且，做了现场试验。理论和试验结果的一致性，证实了取掉补强圈的意见是可取的。两年来，取掉补强圈后的尿素合成塔仍然正常运转，这说明了计算和试验的可靠性。

一、基本方程的建立

尿素合成塔底部球形封头如图 1 所示。在封头顶部一同心圆上，等距地开了三个孔洞，其中 B、C 两个孔洞尺寸相同，A 孔较 B、C 孔小。在孔洞中套有管子，管与壳中间是松动的，管在壳内侧与封头焊接。为了易于计算孔洞附近的应力，我们采用了下面几项简化措施。

(1) 由于孔彼此间的距离较孔径大得多，故不计及孔彼此间的影响，而只对应力较大的大孔进行应力计算。

(2) 以与直径面一致的孔代替实际的变直径孔，并以中间阶梯处的直径作为计算直径。

(3) 尿素合成塔是高压容器，其球形封头壳壁较厚（$h/R=0.13$），已属厚壳理论研究范围（$h/R>0.05$）。但是，我们以简捷的办法进行处理，即以薄球壳理论公式为基础，考虑了壳壁过厚的因素，改进了薄球壳理论的部分方程和公式。

(4) 忽略温度和自重的影响。

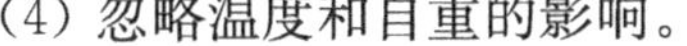

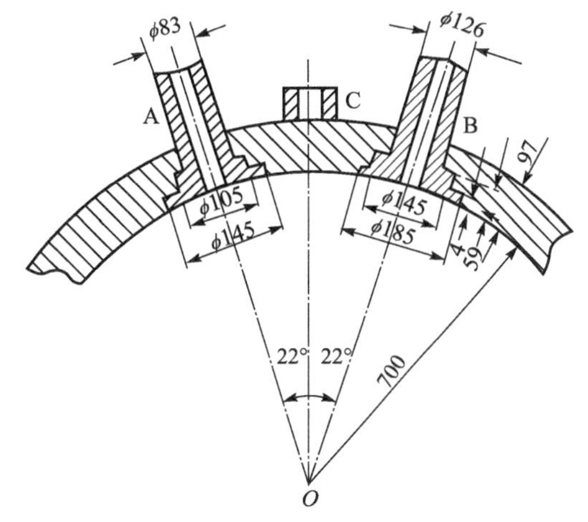

图 1　尿素合成塔底部球形封头

在以上简化措施情况下，我们的任务便成为求解中心开孔球形壳在内压力及内边缘线布载荷作用下的轴对称应力问题。球壳的截面形状和内力、内矩、载荷的正方向分别如图 2、图 3 所示，其中 R 为球壳中曲面半径，b 为中心孔半径，h 为球壳

① 本文原载《科技专利（兰州大学）》，1973，(1)：1-13. 作者：刘人怀，陈山林.

厚度，q 为均匀内压力，p 为孔边缘线布载荷，N_φ 和 N_θ 为法向力，Q_φ 为横向力，M_φ 和 M_θ 为弯矩。而 p 值可按内压力对管截面的作用由孔洞中间台阶承担的事实来推出：

$$\pi b^2 q=2\pi b p,$$

即

$$p=\frac{1}{2}bq \tag{1}$$

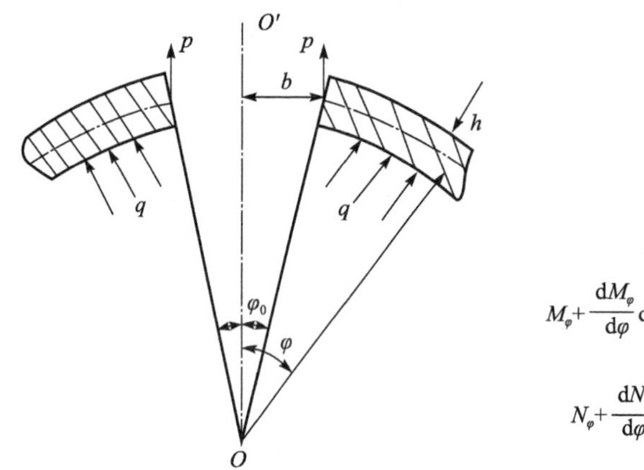

图 2　中心开孔球壳的截面形状

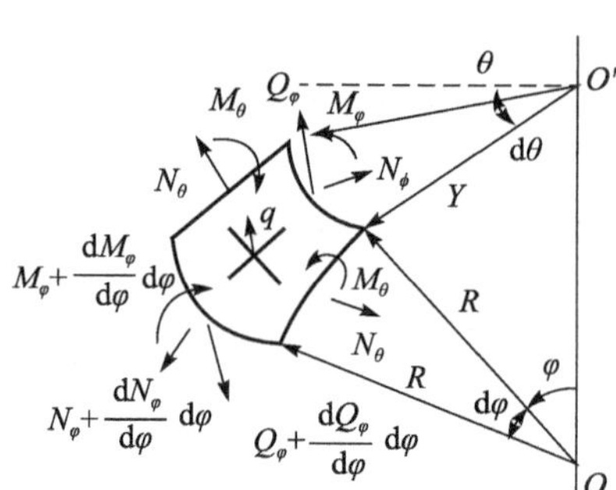

图 3　内力、内矩、载荷的正方向

首先，我们将受内压的薄球壳轴对称应力问题的基本方程及应力公式列在下面[1]：

(1) 平衡方程：

$$\frac{\mathrm{d}(N_\varphi\sin\varphi)}{\mathrm{d}\varphi}-N_\theta\cos\varphi-Q_\varphi\sin\varphi=0, \tag{2a}$$

$$\frac{\mathrm{d}(Q_\varphi\sin\varphi)}{\mathrm{d}\varphi}+N_\theta\sin\varphi+N_\varphi\sin\varphi=Rq\sin\varphi, \tag{2b}$$

$$\frac{\mathrm{d}(M_\varphi\sin\varphi)}{\mathrm{d}\varphi}-M_\theta\cos\varphi=RQ_\varphi\sin\varphi \tag{2c}$$

(2) 弹性定律：

$$N_\varphi=\frac{D}{R}\left[\left(\frac{\mathrm{d}v}{\mathrm{d}\varphi}+w\right)+\nu(v\cot\varphi+w)\right], \tag{3a}$$

$$N_\theta=\frac{D}{R}\left[(v\cot\varphi+w)+\nu\left(\frac{\mathrm{d}v}{\mathrm{d}\varphi}+w\right)\right], \tag{3b}$$

$$M_\varphi=\frac{K}{R^2}\left[\frac{\mathrm{d}}{\mathrm{d}\varphi}\left(\frac{\mathrm{d}w}{\mathrm{d}\varphi}-v\right)+\nu\left(\frac{\mathrm{d}w}{\mathrm{d}\varphi}-v\right)\cot\varphi\right], \tag{3c}$$

$$M_\theta=\frac{K}{R^2}\left[\left(\frac{\mathrm{d}w}{\mathrm{d}\varphi}-v\right)\cot\varphi+\nu\frac{\mathrm{d}}{\mathrm{d}\varphi}\left(\frac{\mathrm{d}w}{\mathrm{d}\varphi}-v\right)\right] \tag{3d}$$

其中 w 和 v 分别为球壳中曲面的法线方向与径线的切线方向上点的位移；v 在坐标 φ 的增长方向为正，而 w，若它从径线曲率的中心向外指向，则视为正。E 为弹性模量，ν 为泊松比。D、K 分别为抗拉、抗弯刚度：

$$D=\frac{Eh}{1-\nu^2},\quad K=\frac{Eh^3}{12(1-\nu^2)} \tag{4}$$

(3) 应力公式（由法向力和弯矩占主要地方的矩形截面梁而来）：

$$\sigma_\varphi=\frac{N_\varphi}{h}-\frac{12M_\varphi z}{h^3},\quad \sigma_\theta=\frac{N_\theta}{h}-\frac{12M_\theta z}{h^3}, \tag{5}$$

其中 z 为球壳中曲面法线方向坐标，向外为正。σ_φ 和 σ_θ 分别为经向、纬向正应力，这两式中的 N 项称为直接应力，M 项称为弯曲应力。

方程（2）和（3）是含有七个未知量的七个独立方程，因此是可解的。通过式（2）、（3）、（5）诸方程，便可求得薄球壳的轴对称应力解。为了较简单地求得不太厚的厚壁球壳轴对称应力问题的解答，且使解答的误差在工程上的允许范围以内，我们认为，基本借助上述薄球壳方程的办法是可取的。下面，我们就试图改进薄球壳理论的方程。

大家知道，平衡方程（2）是通过两对相邻坐标线所截的球壳微元的平衡得到的。方程（2b）是 z 轴方向力的平衡方程，当时是在中曲面上建立的。现在，考虑壳壁较厚，有必要重视法向应力的影响。鉴于此，我们对方程（2b）进行改进。我们认为，由于内压力 q 未直接作用在中曲面上（其微元面积为 $rR\mathrm{d}\theta\mathrm{d}\varphi$），所以考虑 q 的实际作用面积（其微元面积为 $\left\langle R-\frac{h}{2}\right\rangle^2\sin\varphi\mathrm{d}\theta\mathrm{d}\varphi$）是可取的。当然，在薄球壳时，此差别很小，可以不用考虑。这样，在较厚球壳微元体的 z 轴方向，就有下面四个力存在：

$$-N_\theta R\sin\varphi\mathrm{d}\theta\mathrm{d}\varphi,\quad -N_\varphi r\mathrm{d}\theta\mathrm{d}\varphi,\quad -\frac{\mathrm{d}(rQ_\varphi)}{\mathrm{d}\varphi}\mathrm{d}\theta\mathrm{d}\varphi,\quad q\left(R-\frac{h}{2}\right)^2\sin\varphi\mathrm{d}\theta\mathrm{d}\varphi$$

它们处于平衡状态，在舍弃了公因子 $\mathrm{d}\theta\mathrm{d}\varphi$ 后，得平衡方程：

$$RN_\theta\sin\varphi+rN_\varphi+\frac{\mathrm{d}(rQ_\varphi)}{\mathrm{d}\varphi}=q\left(R-\frac{h}{2}\right)^2\sin\varphi \tag{6}$$

而

$$r=R\sin\varphi \tag{7}$$

将式（7）代入方程（6），便有

$$\frac{\mathrm{d}(Q_\varphi\sin\varphi)}{\mathrm{d}\varphi}+N_\theta\sin\varphi+N_\varphi\sin\varphi=\frac{q}{R}\left(R-\frac{h}{2}\right)^2\sin\varphi \tag{8}$$

于是，适用于较厚球壳的改进的平衡方程为（8）和（2a，c），对于弹性定律仍使用式（3）。

对于应力公式（5），我们做下述改进。因为壳壁厚度与曲率半径相比不十分小，则横截面的梯形形状是值得考虑的。这时，应力沿壳厚是非线性分布情况。为此，我们直接利用文献［2］的以下公式来替代公式（5）：

$$\sigma_\varphi=\frac{R}{R+z}\left(\frac{N_\varphi}{h}-\frac{12zM_\varphi}{h^3}\right),$$
$$\sigma_\theta=\frac{R}{R+z}\left(\frac{N_\theta}{h}-\frac{12zM_\theta}{h^3}\right) \tag{9}$$

经过上述推导，我们就得到了用于不太厚的厚壁球壳的基本方程（2a，c）、（8）、（3）以及应力公式（9）。下面，我们致力于求解这组基本方程。

二、基本方程的求解

为了求解简单起见，我们先引入一个辅助变量：

$$\psi=\frac{1}{R}\left(\frac{\mathrm{d}w}{\mathrm{d}\varphi}-v\right), \tag{10}$$

它表示当变形时经线微元 $R\mathrm{d}\varphi$ 旋转的角度。利用式（10），可将弯矩公式（3c，d）改写为下列形式：

$$M_\varphi=\frac{K}{R}\left(\frac{\mathrm{d}\psi}{\mathrm{d}\varphi}+\nu\cot\varphi\right),$$
$$M_\theta=\frac{K}{R}\left(\psi\cot\varphi+\nu\frac{\mathrm{d}\psi}{\mathrm{d}\varphi}\right) \tag{11}$$

平衡方程（2a，c）、（8），弹性定律（3a，b）、（11），以及定义（10）合在一起成为关于八个未知量（N_φ，N_θ，Q_φ，M_φ，M_θ，v，w，ψ）的八个方程组。这组方程可以缩减为关于 Q_φ 和 ψ 的一对方程。其中之一易于求得，只须将方程（11）代入式（2c）就可得到

$$\frac{\mathrm{d}^2\psi}{\mathrm{d}\varphi^2}+\cot\varphi\frac{\mathrm{d}\psi}{\mathrm{d}\varphi}-(\cot^2\varphi+\nu)\psi=\frac{R^2}{K}Q_\varphi \tag{12}$$

另一个方程必须包含方程（2a）、（8）和（3a，b）。由方程（3a，b）可得

$$\frac{\mathrm{d}v}{\mathrm{d}\varphi}+w=\frac{R}{D(1-\nu^2)}(N_\varphi-\nu N_\theta), \tag{13a}$$

$$v\cot\varphi+w=\frac{R}{D(1-\nu^2)}(N_\theta-\nu N_\varphi) \tag{13b}$$

我们微分方程（13b），得

$$\cot\varphi\frac{\mathrm{d}v}{\mathrm{d}\varphi}-\frac{v}{\sin^2\varphi}+\frac{\mathrm{d}w}{\mathrm{d}\varphi}=\frac{R}{D(1-\nu^2)}\left(\frac{\mathrm{d}N_\theta}{\mathrm{d}\varphi}-\nu\frac{\mathrm{d}N_\varphi}{\mathrm{d}\varphi}\right) \tag{14}$$

由（13a，b）、（14）三个方程消去 $\frac{\mathrm{d}v}{\mathrm{d}\varphi}$ 和 w 后，我们得到关于 $\left(\frac{\mathrm{d}w}{\mathrm{d}\varphi}-v\right)$ 的表达式，根据定义（10），它等于 $R\psi$，即

$$R\psi=\frac{R}{D(1-\nu^2)}\left[\frac{\mathrm{d}N_\theta}{\mathrm{d}\varphi}-\nu\frac{\mathrm{d}N_\varphi}{\mathrm{d}\varphi}+(1+\nu)(N_\theta-N_\varphi)\cot\varphi\right] \tag{15}$$

现在利用平衡方程（2a）、（8），设法以 Q_φ 来表示 N_φ 和 N_θ。当我们消去 N_θ 后，就得到

$$\frac{\mathrm{d}(N_\varphi\sin^2\varphi)}{\mathrm{d}\varphi}+\frac{\mathrm{d}(Q_\varphi\sin\varphi\cos\varphi)}{\mathrm{d}\varphi}-\frac{q}{R}\left(R-\frac{h}{2}\right)^2\sin\varphi\cos\varphi=0 \tag{16}$$

这个方程表示平行圆 φ=const 以上壳体区域的平衡条件。对方程（16）积分一次，就有

$$N_\varphi \sin\varphi + Q_\varphi \cos\varphi - \frac{q}{2R}\left(R-\frac{h}{2}\right)^2 \sin\varphi = -\frac{P}{2\pi R\sin\varphi} \tag{17}$$

其中积分常数 P 是整个竖直合力。

由方程（17）可解得

$$N_\varphi = -Q_\varphi \cot\varphi + \frac{q}{2R}\left(R-\frac{h}{2}\right)^2 - \frac{P}{2\pi R\sin^2\varphi} \tag{18}$$

再由方程（8）得

$$N_\theta = -\frac{dQ_\varphi}{d\varphi} + \frac{q}{2R}\left(R-\frac{h}{2}\right)^2 + \frac{P}{2\pi R\sin^2\varphi} \tag{19}$$

最后，将表达式（18）和（19）代入方程（15）的右端，当按照 Q_φ 的导数排列这些项时，就得到所需的微分方程：

$$\frac{d^2Q_\varphi}{d\varphi^2} + \cot\varphi\frac{dQ_\varphi}{d\varphi} - (\cot^2\varphi - \nu)Q = -D(1-\nu^2)\psi \tag{20}$$

值得注意的是含有载荷 q 和积分常数 P 的项都已被消去。

我们所求得的方程（12）和（20）是一对变量为 ψ 和 Q_φ 的变系数二阶齐次微分方程。很容易看出，这两个方程的左端相互之间是十分相似的。这点相似性使得我们可以定义一个线性微分算子：

$$L(\cdots) = \frac{d^2}{d\varphi^2}(\cdots) + \cot\varphi\frac{d}{d\varphi}(\cdots) - \cot^2\varphi(\cdots) \tag{21}$$

应用这个算子，方程（12）和（20）便写为下列形式

$$L(\psi) - \nu\psi = \frac{R^2}{K}Q_\varphi, \tag{22a}$$

$$L(Q_\varphi) + \nu Q_\varphi = -D(1-\nu^2)\psi \tag{22b}$$

现在，使用代入法，可以将方程（22a，b）中的未知量分开：

$$LL(\psi) - \nu^2\psi = -\frac{D(1-\nu^2)R^2}{K}\psi, \tag{23a}$$

$$LL(Q_\varphi) - \nu^2 Q_\varphi = -\frac{D(1-\nu^2)R^2}{K}Q_\varphi \tag{23b}$$

只要使用（23a,b）的任一方程，就可以解决我们的问题。譬如，由方程（23b）求得了 Q_φ 后，便可从方程（22b）解得 ψ，而其他各未知量都能从前面公式求出。

既然这样，我们就从方程（23b）着手处理问题。先将方程（23b）改写为下面形式

$$LL(Q_\varphi) + 4\beta^4 Q_\varphi = 0 \tag{24}$$

其中

$$\begin{aligned}\beta^4 &= 3(1-\nu^2)\frac{R^2}{h^2} - \frac{\nu^2}{4}\\ &\approx 3(1-\nu^2)\frac{R^2}{h^2}\end{aligned} \tag{25}$$

我们可以将方程（24）写为下列任一形式：

$$
\begin{aligned}
&L[L(Q_\varphi)+2\mathrm{i}\beta^2 Q_\varphi]-2\mathrm{i}\beta^2[L(Q_\varphi)+2\mathrm{i}\beta^2 Q_\varphi]=0,\\
&L[L(Q_\varphi)-2\mathrm{i}\beta^2 Q_\varphi]+2\mathrm{i}\beta^2[L(Q_\varphi)-2\mathrm{i}\beta^2 Q_\varphi]=0
\end{aligned}
\tag{26}
$$

由此看出，两个二阶方程

$$L(Q_\varphi)\pm 2\mathrm{i}\beta^2 Q_\varphi=0 \tag{27a,b}$$

的解就能满足方程（23b）。方程（27）的解具有复数值，而且两个方程的解互为共轭。这两对解是线性无关的，因而共同构成方程（23b）的一组四个解。鉴于此，我们只需要解（27）中的任一方程就行了。

由于本节关心的是求距下边缘较远的球壳中心孔附近的应力解，故可做一些简化。在 $\varphi=0$ 的邻近，可将 $\cot\varphi$ 展成罗朗级数：

$$\cot\varphi=\frac{1}{\varphi}-\frac{\varphi}{3}-\frac{\varphi^3}{45}-\cdots$$

若孔径较小，就可只取级数第一项 φ^{-1} 来逼近 $\cot\varphi$。我们将这一近似用于方程（27a）：

$$\frac{\mathrm{d}^2 Q_\varphi}{\mathrm{d}\varphi^2}+\frac{1}{\varphi}\frac{\mathrm{d}Q_\varphi}{\mathrm{d}\varphi}-\frac{1}{\varphi^2}Q_\varphi+2\mathrm{i}\beta^2 Q_\varphi=0 \tag{28}$$

这个方程近乎贝塞尔方程，只要引入新自变量

$$\xi=y\sqrt{\mathrm{i}}=\beta\sqrt{2\mathrm{i}}\varphi \tag{29}$$

方程（28）便成为贝塞尔方程的标准形式：

$$\frac{\mathrm{d}^2 Q_\varphi}{\mathrm{d}\xi^2}+\frac{1}{\xi}\frac{\mathrm{d}Q_\varphi}{\mathrm{d}\xi}+\left(1-\frac{1}{\xi^2}\right)Q_\varphi=0 \tag{30}$$

此方程的解是复变量 ξ 的一阶圆柱函数：

$$Q_\varphi=AJ_1(\xi)+BH_1^{(1)}(\xi), \tag{31}$$

其中 A、B 为待定常数。

为了得到实数解，我们来给出 $J_1(\xi)$、$H_1^{(1)}(\xi)$ 与汤姆生函数的关系式。因为一阶与零阶圆柱函数的关系式为

$$J_1(\xi)=-\frac{\mathrm{d}J_0(\xi)}{\mathrm{d}\xi},\quad H_1^{(1)}(\xi)=-\frac{\mathrm{d}H_0^{(1)}(\xi)}{\mathrm{d}\xi}, \tag{32}$$

而

$$
\begin{aligned}
&J_0(\xi)=\mathrm{ber}y-\mathrm{i}\,\mathrm{bei}y,\\
&H_0^{(1)}(\xi)=-\frac{1}{\pi}(\mathrm{kei}y+\mathrm{i}\,\mathrm{ker}y)
\end{aligned}
\tag{33}
$$

其中 $\mathrm{ber}y$、$\mathrm{bei}y$、$\mathrm{ker}y$、$\mathrm{kei}y$ 是汤姆生函数。

将式（33）代入式（32），便有

$$
\begin{aligned}
&J_1(\xi)=-\frac{1}{\sqrt{2}}[(\mathrm{bei}'y-\mathrm{ber}'y)+\mathrm{i}(\mathrm{bei}'y+\mathrm{ber}'y)],\\
&H_1^{(1)}(\xi)=\frac{\sqrt{2}}{\pi}[(\mathrm{ker}'y+\mathrm{kei}'y)+\mathrm{i}(\mathrm{ker}'y-\mathrm{kei}'y)]
\end{aligned}
\tag{34}
$$

这里的撇号表示汤姆生函数对其变量 y 的导数。因为 $J_1(\xi)$ 和 $H_1^{(1)}(\xi)$ 的实部及虚部，以及它们的线性组合都是方程（23b）的解，所以我们可取汤姆生函数的导数作

为基本解。这样，我们就得到方程（30）的实数解：

$$Q_\varphi = A_1 \operatorname{ber}' y + A_2 \operatorname{bei}' y + B_1 \operatorname{ker}' y + B_2 \operatorname{kei}' y \tag{35}$$

其中 A_1、A_2、B_1、B_2 是待定常数，需要由边界条件来确定。

因为讨论的是小孔情况，离下边缘甚远，而含 A 各项是 y 因而也是 φ 的正则函数，它随 φ 增加而增加，在本公式不适用的下边缘取极大值，在上边缘却很小，这与含 B 的项恰恰相反，因此为了简单起见，我们可舍弃含 A 的各项，于是解（35）成为

$$Q_\varphi = B_1 \operatorname{ker}' y + B_2 \operatorname{kei}' y \tag{36}$$

为了从式（36）求出关于倾角和内力的公式，我们给出汤姆生函数的高阶导数公式：

$$\begin{aligned} \frac{d^2}{dy^2}\operatorname{ker} y &= -\operatorname{kei} y - \frac{1}{y}\operatorname{ker}' y, \\ \frac{d^2}{dy^2}\operatorname{kei} y &= \operatorname{ker} y - \frac{1}{y}\operatorname{kei}' y \end{aligned} \tag{37}$$

将解（36）代入方程（22b），并利用公式（37），便得

$$D(1-\nu^2)\psi = B_1(2\beta^2 \operatorname{kei}' y - \nu \operatorname{ker}' y) - B_2(2\beta^2 \operatorname{ker}' y + \nu \operatorname{kei}' y) \tag{38}$$

由式（18）、（19）和（11），还可得到法向力与弯矩的公式：

$$N_\varphi = -\frac{1}{\varphi}Q_\varphi + \frac{q}{2R}\left(R - \frac{h}{2}\right)^2 - \frac{P}{2\pi R\varphi^2}, \tag{39a}$$

$$N_\theta = \beta\sqrt{2}\left[B_1\left(\operatorname{kei} y + \frac{1}{y}\operatorname{ker}' y\right) - B_2\left(\operatorname{ker} y - \frac{1}{y}\operatorname{kei}' y\right)\right] + \frac{q}{2R}\left(R - \frac{h}{2}\right)^2 + \frac{P}{2\pi R\varphi^2}, \tag{39b}$$

$$\begin{aligned} M_\varphi = \frac{K\beta\sqrt{2}}{DR(1-\nu^2)}\Bigg\{ & B_1\left[2\beta^2\left(\operatorname{ker} y - \frac{1-\nu}{y}\operatorname{kei}' y\right) + \nu\left(\operatorname{kei} y + \frac{1-\nu}{y}\operatorname{ker}' y\right)\right] \\ & + B_2\left[2\beta^2\left(\operatorname{kei} y + \frac{1-\nu}{y}\operatorname{ker}' y\right) - \nu\left(\operatorname{ker} y - \frac{1-\nu}{y}\operatorname{kei}' y\right)\right]\Bigg\}, \end{aligned} \tag{39c}$$

$$\begin{aligned} M_\theta = \frac{K\beta\sqrt{2}}{DR(1-\nu^2)}\Bigg\{ & B_1\left[2\beta^2\left(\nu \operatorname{ker} y + \frac{1-\nu}{y}\operatorname{kei}' y\right) + \nu\left(\nu \operatorname{kei} y - \frac{1-\nu}{y}\operatorname{ker}' y\right)\right] \\ & + B_2\left[2\beta^2\left(\nu \operatorname{kei} y - \frac{1-\nu}{y}\operatorname{ker}' y\right) - \nu\left(\nu \operatorname{ker} y + \frac{1-\nu}{y}\operatorname{kei}' y\right)\right]\Bigg\} \end{aligned} \tag{39d}$$

下面，我们来决定解（39）中的积分常数 P 值。根据其物理意义，并注意壳体较厚的因素，我们考虑 $\varphi=\text{const}$ 至 φ_0 间的壳体部分竖直方向力的平衡条件：

$$2\pi RN_\varphi \sin^2\varphi + 2\pi RQ_\varphi \sin\varphi\cos\varphi - 2\pi\int_{\varphi_0}^{\varphi}\left(R - \frac{h}{2}\right)^2 q\sin\varphi\cos\varphi d\varphi - 2\pi bp = 0 \tag{40}$$

利用式（1），化简式（40），最后得到

$$N_\varphi = -Q_\varphi \cot\varphi + \frac{q}{2R}\left(R - \frac{h}{2}\right)^2 + \frac{1}{2}\left[1 - \left(1 - \frac{h}{2R}\right)^2\right]\frac{R\sin^2\varphi_0}{\sin^2\varphi}q \tag{41}$$

将式（41）与式（18）比较，便得关于 P 的公式：

$$P = -\pi R^2 \sin^2\varphi_0\left[1 - \left(1 - \frac{h}{2R}\right)^2\right]q \tag{42}$$

现在，我们来确定解中的待定常数 B_1 和 B_2。首先需要给出封头孔洞的边界条件。因为套管在孔洞中是松动的，且在内侧与铝衬焊接，而台阶处承担了横向力，故在孔

洞边缘处有条件：

当 $\varphi=\varphi_0$ 时，　　$Q_\varphi=p\cos\varphi_0$，　$M_\varphi=0$　　(43)

将解（36）和（39c）代入边界条件（43）中，便得关于 B_1、B_2 的公式：

$$B_1=\frac{\alpha_2\cos\varphi_0}{\alpha_2\mathrm{ker}'y_0-\alpha_1\mathrm{kei}'y_0}p,\quad B_2=-\frac{\alpha_1\cos\varphi_0}{\alpha_2\mathrm{ker}'y_0-\alpha_1\mathrm{kei}'y_0}p,\tag{44}$$

其中 y_0 是 y 在孔洞边缘的值，

$$\alpha_1=2\beta^2\left(\mathrm{ker}y_0-\frac{1-\nu}{y_0}\mathrm{kei}'y_0\right)+\nu\left(\mathrm{kei}y_0+\frac{1-\nu}{y_0}\mathrm{ker}'y_0\right),\quad \alpha_2=2\beta^2\left(\mathrm{kei}y_0+\frac{1-\nu}{y_0}\mathrm{ker}'y_0\right)-\nu\left(\mathrm{ker}y_0-\frac{1-\nu}{y_0}\mathrm{kei}'y_0\right)\tag{45}$$

三、数值计算

已知尿素合成塔底部球形封头的有关数据为

$R=748.5\mathrm{mm}$，　$h=97\mathrm{mm}$，　$b=72.5\mathrm{mm}$，　$\varphi_0=0.0970$

封头材料：18MnMoNb 钢，在常温下的 E、ν 值为

$E=2.1\times10^4\mathrm{kg/mm^2}$，　$\nu=0.3$

材料的屈服极限 $\sigma_\mathrm{S}=45\mathrm{kg/mm^2}$。

设计压力：$q=2.20\mathrm{kg/mm^2}$。

试验压力：$q=3.00\mathrm{kg/mm^2}$。

将以上数据代入式（39）、（42）、（44）、（45）和（9）中，便可得到尿素合成塔底部球形封头大孔，即图 1 中的 B、C 孔附近的应力值。表 1～表 3 分别列出了内力、内矩以及在设计压力和试验压力下球壳内、外表面上应力的数值结果，其中 $z=-h/2$ 和 $z=h/2$ 表示球壳的内、外表面。

表 1　孔洞附近的内力、内矩值　　（单位：kg/mm 和 kg·mm/mm）

φ	N_θ/q	N_φ/q	M_θ/q	M_φ/q
0.0970	729	2.31	−3020	0
0.102	698	32.2	−2850	−28.0
0.124	580	144	−2190	−345
0.140	529	191	−1850	36.8
0.190	442	266	−1170	203

表 2　在设计压力下的孔洞附近的应力值　　（单位：$\mathrm{kg/mm^2}$）

φ	σ_θ		σ_φ	
	$z=h/2$	$z=-h/2$	$z=h/2$	$z=-h/2$
0.0970	19.5	13.2	0.0492	0.0560
0.102	18.6	12.7	0.723	0.739
0.124	15.2	10.8	3.52	2.97

续表

φ	σ_θ		σ_φ	
	$z=h/2$	$z=-h/2$	$z=h/2$	$z=-h/2$
0.140	13.7	10.1	4.02	4.69
0.190	11.0	8.96	5.40	6.76

表 3　在试验压力下的孔洞附近的应力值　　（单位：kg/mm^2）

φ	σ_θ		σ_φ	
	$z=h/2$	$z=-h/2$	$z=h/2$	$z=-h/2$
0.0970	26.6	17.9	0.0671	0.0764
0.102	25.4	17.3	0.985	1.01
0.124	20.8	14.7	4.80	4.06
0.140	18.7	13.7	5.48	6.39
0.190	14.9	12.2	7.36	9.21

由应力公式（9）和表 1～表 3 看出，最大应力是纬向应力 σ_θ。σ_θ 的最大值发生在球壳外表面的孔洞边缘处，它是拉伸应力，在设计压力情况下，其值为

$$\sigma_{\theta\max}=19.5\text{kg/mm}^2 \tag{46}$$

关于屈服限的安全系数有下面公式成立：

$$n_S=\left|\frac{\sigma_S}{\sigma_{\theta\max}}\right| \tag{47}$$

将式（46）代入式（47）中，便得

$$n_S=2.31 \tag{48}$$

故孔洞边缘的强度是足够的，因而不需要设护强板。另外，由表 2 和表 3 看到，对于纬向应力来说，壳外表应力大于壳内表应力，特别在孔边上大很多，故从强度考虑，孔中的中间台阶以靠内侧近一些为好。这两点意见均被厂里采用，起到了积极作用。

为了验证本节改进理论公式的正确性，我们在实际产品上做了一次常温下的应力测定。测点布置在从封头顶点至二大孔圆心的经线上，共测得了两组结果。但未测得孔边缘应力值，这是实验的不足之处。考虑到工艺加工原因，实际的封头并不是一个等厚球壳，而是顶部厚、下边缘薄的变厚度壳体，由于它的中曲面近似于球面，所以仍用前面的球壳公式来处理。因为我们研究的是顶部区域应力问题，故以顶部厚度（还包括了铝内衬厚度）作为球壳厚度。这样，实际的数据为

$$R=752\text{mm},\quad h=116\text{mm},\quad b=72.5\text{mm},\quad \varphi=0.0966$$

实验压力为 $q=2.20\text{kg/mm}^2$，$E=2.18\times10^4\text{kg/mm}^2$，$\nu=0.292$。

应用这些数据于公式（39）、式（42）、式（44）、式（45）、式（9）中，便得在实验压力（等于设计压力）下的本节公式的理论结果。表 4 和表 5 给出了这些结果以及实验值，为了便于比较，还在图 4 上绘出了应力分布曲线。

为了说明本节改进公式的优越性，我们还使用了薄球壳理论公式[1]（2）、（3）、（5），将其结果一起给在表 4 和表 5 以及图 4 中。实验和本节理论结果的较好符合，说明了前面建立的改进公式是正确的，结论是可靠的。

表 4　孔洞附近壳外表面的 σ_θ 的理论值与实验值比较（q=2.20kg/mm²）

（单位：kg/mm²）

φ		0.0966	0.112	0.137	0.174
本节结果		13.7	12.1	10.1	8.52
文献［1］结果		16.9	14.9	12.4	10.6
实验值	第一组	/	/	9.50	7.68
	第二组	/	/	/	9.39
理论与实验结果误差（%）	本节	/	/	6.32	10.9～9.27
	文献［1］	/	/	30.5	38.0～12.9

表 5　孔洞附近壳外表面的 σ_φ 的理论值与实验值比较（q=2.20kg/mm²）

（单位：kg/mm²）

φ		0.0966	0.112	0.137	0.174
本节结果		0.0742	1.53	3.15	4.20
文献［1］结果		0.0799	1.90	3.92	5.26
实验值	第一组	/	/	3.43	3.94
	第二组	/	/	/	3.10
理论与实验结果误差（%）	本节	/	/	−8.16	6.60～35.5
	文献［1］	/	/	14.3	33.5～69.7

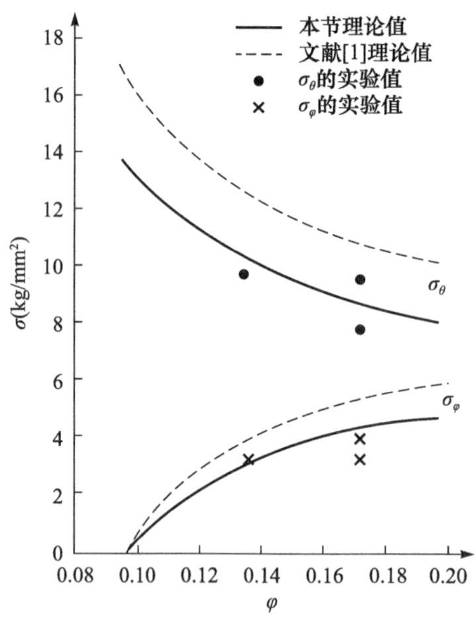

图 4　孔洞附近壳外表面应力的理论值与实验值比较

（q=2.20kg/mm²）

四、小结

从前面的计算和实验结果看出，将本节改进公式与薄球壳理论公式相比，本节与实验结果误差小，精确度高，能满足工程需要。这一事实说明，使用本节改进公式来计算厚壁球形容器开孔附近的应力分布，是可取的。

致谢：兰州大学数学力学系王廷栋同志负责本节的试验工作。

参考文献

[1]　弗留盖 W. 壳体中的应力. 薛振东，龙驭球，叶耀先，等译. 北京：中国工业出版社，1965.

[2]　拉包德诺夫 Ю. Н. 材料力学. 北京：高等教育出版社，1956.

铂重整装置反应器椭球封头中心开孔接管的强度问题①

在工业容器制造中，将常常碰到在椭球封头中心孔上连接管子的情况。接管选用厚壁好还是薄壁好？这是搞工程的同志很关心的一个问题。我们将通过一具体例子的强度分析来得出比较正确的结论。

铂重整装置反应器便有此一问题存在，这种装置是石油炼制工业中的重要装置，产品先进，用途很广，只要改变工艺条件，就可生产不同的石油制品。铂重整装置反应器是高温（设计温度为300℃）、中压（设计压力为 $80\mathrm{kg/cm^2}$）容器，因此，对其设计和制造说来，都提出了比较严格的要求。1969年下半年，某厂要我们对其制造的这种反应器的封头中心开孔接管的强度进行分析。由于当时尚无厚壁圆柱壳弯曲理论，我们进行了理论上的一点探讨，提出了实用的厚壁圆柱壳弯曲理论，并做了现场产品试验，初步解决了这个产品上的有关问题，同时对椭球封头中心开孔处采用厚壁和薄壁接管的优劣提供了初步看法。

一、椭球封头的基本方程和公式

我们仅研究承受内压的、只开了一个中心小孔的薄椭球壳。于是，问题便成为轴对称的，而且可略去下边缘条件对小孔附近的影响。薄椭球壳的内力、内矩和位移的正方向示于图1中，其中 φ 为决定纬线圆上点位置的角度，θ 为中面法线和壳体轴线所夹的角，θ_0 为 θ 在中心孔的边缘值，h 为壳厚，q 为均匀内压力，M_1、M_2 为经向和纬向弯矩，T_1、T_2 为经向和纬向法向力，Q_x 为垂直于轴线方向的力，Δ_x 为垂直于轴线方向的位移分量，ψ 为经线的切线转角。在上述情况下，薄椭球壳轴对称应力问题中的复内力 T 所满足的基本方程为[1]

$$
\begin{aligned}
&\frac{\mathrm{d}^2T}{\mathrm{d}\theta^2}+\left[\left(\frac{2R_1}{R_2}-1\right)\cot\theta-\frac{1}{R_1}\frac{\mathrm{d}R_1}{\mathrm{d}\theta}\right]\frac{\mathrm{d}T}{\mathrm{d}\theta}+\mathrm{i}\,\frac{R_1^2}{R_2C}T\\
&=\mathrm{i}\,\frac{R_1^2}{R_2C}\left[qR_2-\frac{1}{\sin^2\theta}\left(\frac{1}{R_1}-\frac{1}{R_2}\right)\left(\overline{C}-\int_{\theta_0}^{\theta}qR_1R_2\sin\theta\cos\theta\mathrm{d}\theta\right)\right],
\end{aligned}
\tag{1}
$$

其中 a、b 为椭球中曲面的短、长半轴，R_1、R_2 为中曲面的两个主曲率半径，$\overline{C}$ 为积分常数，

$$T=T_1+T_2-\mathrm{i}\,\frac{M_1+M_2}{(1+\nu)\;C},$$

$$R_1=b\sqrt{\frac{1+\gamma}{(1+\gamma\sin^2\theta)^3}},$$

$$R_2=b\sqrt{\frac{1+\gamma}{1+\gamma\sin^2\theta}},$$

① 本文原载《科技专利（兰州大学）》，1973，(1)：14-28. 作者：刘人怀，陈山林.

$$\gamma=\frac{b^2}{a^2}-1,$$
$$C=\frac{h}{\sqrt{12\ (1-\nu)^2}}, \tag{2}$$

ν 为泊松比。

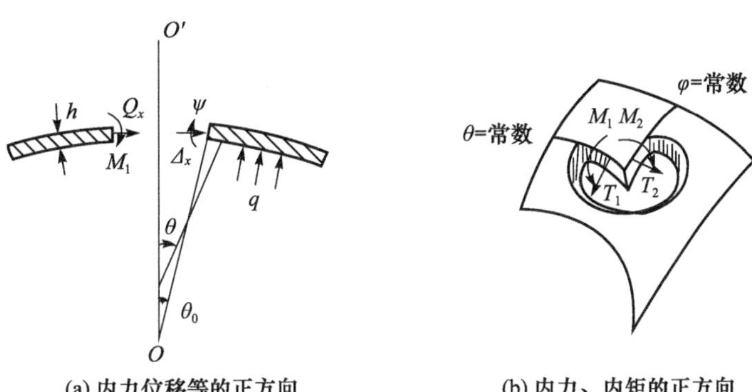

(a) 内力位移等的正方向　　(b) 内力、内矩的正方向

图 1　薄椭球壳的内力和位移

由变系数的常微分方程（1）解得复内力 T 后，便可求出内力、内矩和位移的公式。为避免文章繁杂起见，我们略去推导，直接从文献［1］中引出误差相对于 1 为 $\sqrt{\frac{h}{R}}$ 级的有关结果：

$$\begin{aligned}
\Delta_x&=\frac{\sqrt{2}R_2\sin\theta}{Eh}\left[-\sqrt{\frac{R_{20}}{C}}(Q_0-Q_0^*)\sin\theta_0\cos\beta-\frac{M_0}{C}\cos\left(\beta+\frac{\pi}{4}\right)\right]e^{-\beta}\\
&\quad+\frac{qR_2^2\sin\theta}{2Eh}\left(2-\nu-\frac{R_2}{R_1}\right),\\
\psi&=-\frac{1}{Eh}\sqrt{\frac{2R_2}{C}}\left[\sqrt{\frac{R_{20}}{C}}\sin\theta_0(Q_0-Q_0^*)\cos\left(\beta-\frac{\pi}{4}\right)+\frac{1}{C}M_0\cos\beta\right]e^{-\beta},\\
T_1&=\sqrt{\frac{2C}{R_2}}\cot\theta\left[\sqrt{\frac{R_{20}}{C}}\sin\theta_0(Q_0-Q_0^*)\cos\left(\beta+\frac{\pi}{4}\right)-\frac{1}{C}M_0\sin\beta\right]e^{-\beta}+\frac{1}{2}qR_2,\\
T_2&=\left[-\sqrt{\frac{2R_{20}}{C}}\sin\theta_0(Q_0-Q_0^*)\cos\beta-\frac{\sqrt{2}}{C}M_0\cos\left(\beta+\frac{\pi}{4}\right)\right]e^{-\beta}+\frac{1}{2}qR_2\left(2-\frac{R_2}{R_1}\right),\\
M_1&=\left[\sqrt{2}M_0\cos\left(\beta-\frac{\pi}{4}\right)+\sqrt{2R_{20}C}\sin\theta_0(Q_0-Q_0^*)\sin\beta\right]e^{-\beta},\\
M_2&=\nu M_1,
\end{aligned} \tag{3}$$

其中 E 为弹性模量，S 为弧长，凡带下角标“0”的表示该量在 $\theta=\theta_0$ 时的值（以后类似如此时，不再另述），

$$Q_0^*=\frac{1}{2}qR_{20}\cos\theta_0,$$
$$\beta=\frac{1}{\sqrt{2}}\int_{\theta_0}^{\theta}\frac{R_1\mathrm{d}\theta}{\sqrt{R_2C}}\approx\frac{S}{\sqrt{2R_{20}C}} \tag{4}$$

需要指出，公式（3）已略去了下边缘影响，并满足了孔洞边缘条件：

$$\text{当 } \theta=\theta_0 \text{ 时，} \quad M_1=M_0, \quad Q_x=Q_0 \tag{5}$$

为了确定 M_0、Q_0 值，我们需要位移 Δ_x 和转角 ψ 在孔洞边缘处的表达式，因此，设

$$\text{当 } \theta=\theta_0 \text{ 时，} \quad \Delta_x=\Delta_0', \quad \psi=\psi_0' \tag{6}$$

将式（3）中的第一、二式代入式（6）中，我们记为

$$\begin{aligned} \Delta_0' &= \alpha_{11}' Q_0+\alpha_{12}' M_0+\alpha_1' q, \\ \psi_0' &= \alpha_{21}' Q_0+\alpha_{22}' M_0+\alpha_2' q, \end{aligned} \tag{7}$$

其中

$$\begin{aligned} \alpha_{11}' &= -\frac{2\sqrt[4]{3(1-\nu^2)}}{E}\sqrt{\frac{R_{20}^3}{R^3}}\sin^2\theta_0, \\ \alpha_{12}' &= \alpha_{21}'=-\frac{\sqrt{12(1-\nu^2)}}{Eh^2}R_{20}\sin\theta_0, \\ \alpha_{22}' &= -\frac{4\sqrt[4]{27(1-\nu^2)^3}}{Eh^2}\sqrt{\frac{R_{20}}{h}}, \\ \alpha_1' &= -\frac{1}{2}R_{20}\alpha_{11}'\cos\theta_0+\frac{R_{20}^2\sin\theta_0}{2Eh}\left(2-\nu-\frac{R_{20}}{R_{10}}\right), \\ \alpha_2' &= -\frac{1}{2}R_{20}\alpha_{21}'\cos\theta_0 \end{aligned} \tag{8}$$

这里的单撇号表示与封头有关的量。

由公式（3），我们给出直接计算孔边内力、内矩的公式。

当 $\theta=\theta_0$ 时，

$$\begin{aligned} T_{10} &= (Q_0-Q_0^*)\cos\theta_0+\frac{1}{2}qR_{20}, \\ T_{20} &= -\sqrt{\frac{2R_{20}}{C}}(Q_0-Q_0^*)\sin\theta_0-\frac{M_0}{C}+\frac{1}{2}qR_{20}\left(2-\frac{R_{20}}{R_{10}}\right), \\ M_{10} &= M_0, \\ M_{20} &= \nu M_0 \end{aligned} \tag{9}$$

在求得了内力、内矩后，便可由下面公式计算经向和纬向应力：

$$\begin{aligned} \sigma_1 &= \frac{T_1}{h}+\frac{12M_1 z}{h^3}, \\ \sigma_2 &= \frac{T_2}{h}+\frac{12M_2 z}{h^3}, \end{aligned} \tag{10}$$

其中 z 为中曲面法线方向坐标，向外为正，T 为薄膜应力，M 为弯曲应力。

二、厚壁圆柱接管的基本方程和公式

我们仅研究承受内压的较长厚壁圆柱接管下端的情况。这样，这们涉及的是半无限长厚壁圆柱壳体的轴对称应力问题。由于目前尚无厚壁圆柱壳弯曲理论，为简单起见，我们借助较精确薄壁圆柱壳体的理论[2]，将它们简单改进，建立实用的厚壁圆柱壳弯曲理论。注意壳厚过大的因素，此时，内压力 q 作用在壳内表面，于是，给出了

经线方向位移 u 和中曲面法线方向位移 w 所满足的基本方程（其内力、内矩、位移、坐标的正方向示于图 2 中）：

$$\frac{d^2u}{d\alpha^2}+\nu\frac{dw}{d\alpha}-d^2\frac{d^3w}{d\alpha^3}=0, \tag{11a}$$

$$\nu\frac{du}{d\alpha}-d^2\frac{d^3u}{d\alpha^3}+d^2\frac{d^4w}{d\alpha^4}+(1+d^2)w=\frac{1-\nu^2}{Eh}R^2\left(1-\frac{h}{2R}\right)q, \tag{11b}$$

其中 R 为中曲面半径，α 是以 R 作比例系数的经向无量纲坐标，N_1 为经向横向力，其余符号意义与椭球壳符号相同，

$$d^2=\frac{h^2}{12R^2} \tag{12}$$

因为 d^2 与 1 相较甚小，故在（$1\pm d^2$）项出现时，我们可以略去 d^2，于是方程（11b）成为

$$\nu\frac{du}{d\alpha}-d^2\frac{d^3u}{d\alpha^3}+d^2\frac{d^4w}{d\alpha^4}+w=\frac{1-\nu^2}{Eh}R^2\left(1-\frac{h}{2R}\right)q \tag{13}$$

由方程（11a）和（13）解得 u 和 w 后，便可按下面公式求内力、内矩：

$$T_1=\frac{Eh}{(1-\nu^2)R}\left(\frac{du}{d\alpha}+\nu w-d^2\frac{d^2w}{d\alpha^2}\right), \tag{14a}$$

$$N_1=\frac{1}{R}\frac{dM_1}{d\alpha}, \tag{14b}$$

$$M_1=\frac{D}{R^2}\left(\frac{du}{d\alpha}-\frac{d^2w}{d\alpha^2}\right), \tag{14c}$$

其中

$$\text{抗弯刚度 } D=\frac{Eh^3}{12(1-\nu^2)} \tag{15}$$

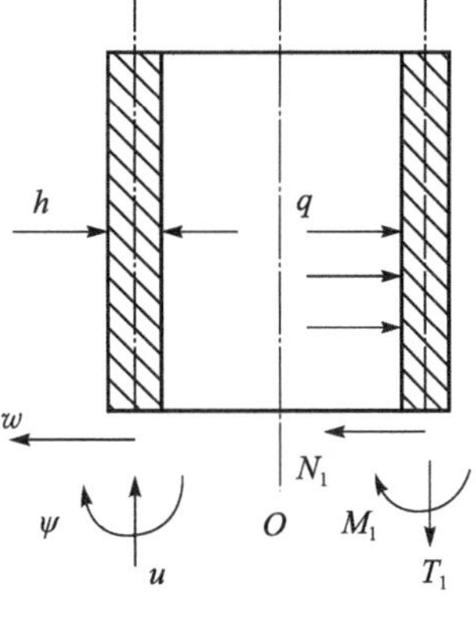

图 2　圆柱壳各量的正方向

实际上，由方程（11a）和（14a）看出，T_1 是一常数。我们可以通过圆柱壳某一横截面的平衡单独解出 T_1。因为

$$2\pi RT_1=\pi r^2q,$$

其中 r 为圆柱壳内半径。

则有

$$T_1=\frac{1}{2R}r^2q \tag{16}$$

现在，我们来求解方程式（11a）和式（13）。首先，需要引入一个新的函数 $\phi(\alpha)$，它与位移的关系为

$$u=d^2\frac{d^2\phi}{d\alpha^2}-\nu\phi+m\alpha, \tag{17a}$$

$$w=\frac{d\phi}{d\alpha}, \tag{17b}$$

其中，m 是一待定常数，它将通过关于 T_1 的平衡式来决定。

将式（17）代入式（14a）中，便有

$$T_1=\frac{Eh}{(1-\nu^2)R}m \tag{18}$$

由式（16）和（18）的相等，便决定了 m 值：

$$m=\frac{1-\nu^2}{2Eh}r^2q \tag{19}$$

用代入法可以证明，$\phi(\alpha)$ 已经自动满足了方程（11a）。将式（17）代入方程（13），并利用结果（19），就得关于 $\phi(\alpha)$ 的五阶微分方程：

$$\frac{\mathrm{d}^5\phi}{\mathrm{d}\alpha^5}+2\nu\frac{\mathrm{d}^3\phi}{\mathrm{d}\alpha^3}+k^2\frac{\mathrm{d}\phi}{\mathrm{d}\alpha}=\frac{R^2rq}{D}\left(R-\frac{1}{2}\nu r\right), \tag{20}$$

其中

$$k^2=\frac{1-\nu^2}{d^2} \tag{21}$$

先研究式（20）的齐次方程：

$$\frac{\mathrm{d}^5\phi}{\mathrm{d}\alpha^5}+2\nu\frac{\mathrm{d}^3\phi}{\mathrm{d}\alpha^3}+k^2\frac{\mathrm{d}\phi}{\mathrm{d}\alpha}=0 \tag{22}$$

我们可以引入一个新的函数 $P(\alpha)$，以使方程（22）降低一阶。令

$$P(\alpha)=\frac{\mathrm{d}\phi}{\mathrm{d}\alpha} \tag{23}$$

将式（23）代入方程（22）中，得关于 $P(\alpha)$ 的四阶方程：

$$\frac{\mathrm{d}^4P}{\mathrm{d}\alpha^4}+2\nu\frac{\mathrm{d}^2P}{\mathrm{d}\alpha^2}+k^2P=0 \tag{24}$$

此方程的四个特征根为

$$\pm\sqrt{-\nu\pm\mathrm{i}k}$$

于是，方程（24）的通解写为

$$P=\mathrm{e}^{-\lambda_1\alpha}[C_1\cos(\lambda_2\alpha)+C_2\sin(\lambda_2\alpha)]+\mathrm{e}^{\lambda_1\alpha}[C_3\cos(\lambda_3\alpha)+C_4\sin(\lambda_2\alpha)], \tag{25}$$

其中 C_1、C_2、C_3、C_4 为待定常数，

$$\begin{aligned}\lambda_1&=\sqrt{k}\cos\left[\frac{1}{2}\arctan\left(-\frac{k}{\nu}\right)\right],\\ \lambda_2&=\sqrt{k}\sin\left[\frac{1}{2}\arctan\left(-\frac{k}{\nu}\right)\right]\end{aligned} \tag{26}$$

利用式（25）和（23），并加上方程（20）的特解，就得到方程（20）的通解：

$$\begin{aligned}\phi(\alpha)=&\frac{r^2q}{Eh}\left(\frac{R}{r}-\frac{1}{2}\nu\right)\alpha+\int\mathrm{e}^{-\lambda_1\alpha}[C_1\cos(\lambda_2\alpha)+C_2\sin(\lambda_2\alpha)]\mathrm{d}\alpha\\ &+\int\mathrm{e}^{\lambda_1\alpha}[C_3\cos(\lambda_2\alpha)+C_4\sin(\lambda_2\alpha)]\mathrm{d}\alpha+C_0,\end{aligned} \tag{27}$$

其中，C_0 为待定常数。

显然，C_0 表示刚性位移，加之我们只讨论半无限长圆柱壳的下端情况，故可令

$$C_0=C_3=C_4=0 \tag{28}$$

于是，解（27）简化为

$$\phi(\alpha)=\frac{r^2q}{Eh}\left(\frac{R}{r}-\frac{1}{2}\nu\right)\alpha+\int\mathrm{e}^{-\lambda_1\alpha}[C_1\cos(\lambda_2\alpha)+C_2\sin(\lambda_2\alpha)]\mathrm{d}\alpha \tag{29}$$

现在，我们来决定解中的待定常数 C_1 和 C_2。先利用解（29），从式（14b，c）和

(17b) 求出

$$w=\frac{r^2}{Eh}\left(\frac{R}{r}-\frac{1}{2}\nu\right)q+\mathrm{e}^{-\lambda_1\alpha}[C_1\cos(\lambda_2\alpha)+C_2\sin(\lambda_2\alpha)], \tag{30a}$$

$$\begin{aligned}M_1=&\frac{r^2}{k^2}\left(\frac{1}{2}-\frac{\nu R}{r}\right)q-\frac{D}{R^2}\{(\lambda_1^2-\lambda_2^2+\nu)[C_3\cos(\lambda_2\alpha)+C_2\sin(\lambda_2\alpha)]\\&+2\lambda_1\lambda_2[C_1\sin(\lambda_2\alpha)-C_2\cos(\lambda_2\alpha)]\}\mathrm{e}^{-\lambda_1\alpha},\end{aligned} \tag{30b}$$

$$\begin{aligned}N_1=&\frac{D}{R^3}\{\lambda_1(\lambda_1^2-3\lambda_2^2+\nu)[C_1\cos(\lambda_2\alpha)+C_2\sin(\lambda_2\alpha)]\\&+\lambda_2(3\lambda_1^2-\lambda_2^2+\nu)[C_1\sin(\lambda_2\alpha)-C_2\cos(\lambda_2\alpha)]\}\mathrm{e}^{-\lambda_1\alpha}\end{aligned} \tag{30c}$$

设圆柱壳下边缘处，有

$$\text{当}\ \alpha=0\ \text{时},\quad M_1=M_0,\quad N_1=N_0 \tag{31}$$

将式（30b，c）代入条件（31）中，便得决定 C_1 和 C_2 的公式：

$$\begin{aligned}C_1&=\eta\left\{2\lambda_1RN_0+(3\lambda_1^2-\lambda_2^2+\nu)\left[\frac{r^2}{k^2}\left(\frac{\nu R}{r}-\frac{1}{2}\right)q+M_0\right]\right\},\\C_2&=\frac{\eta}{\lambda_2}\left\{R(\lambda_1^2-\lambda_2^2+\nu)N_0+\lambda_1(\lambda_1^2-3\lambda_2^2+\nu)\left[\frac{r^2}{k^2}\left(\frac{\nu R}{r}-\frac{1}{2}\right)q+M_0\right]\right\},\end{aligned} \tag{32}$$

其中

$$\eta=-\frac{R^2}{D[(\lambda_1^2+\lambda_2^2)^2+2\nu(\lambda_1^2-\lambda_2^2)+\nu^2]} \tag{33}$$

另外，由式（30a）还可得到经线转角 ψ 的公式：

$$\begin{aligned}\psi&=-\frac{1}{R}\frac{\mathrm{d}w}{\mathrm{d}\alpha}\\&=\frac{1}{R}\{\lambda_1[C_1\cos(\lambda_2\alpha)+C_2\sin(\lambda_2\alpha)]\\&\quad+\lambda_2[C_1\sin(\lambda_2\alpha)-C_2\cos(\lambda_2\alpha)]\}\mathrm{e}^{-\lambda_1\alpha}\end{aligned} \tag{34}$$

令

$$\text{当}\ \alpha=0\ \text{时},\quad w=\Delta_0'',\quad \psi=\psi_0'' \tag{35}$$

再将式（30a）和（34）代入条件（35）中，我们记为

$$\begin{aligned}\Delta_0''&=\alpha_{11}''N_0+\alpha_{12}''M_0+\alpha_1''q,\\\psi_0''&=\alpha_{21}''N_0+\alpha_{22}''M_0+\alpha_2''q,\end{aligned} \tag{36}$$

其中

$$\begin{aligned}\alpha_{11}''&=2\lambda_1R\eta,\\\alpha_{12}''&=(3\lambda_1^2-\lambda_2^2+\nu)\eta,\\\alpha_1''&=\frac{D\alpha_{12}''}{R^2}\left(\frac{\nu R}{r}-\frac{1}{2}\right)+\frac{r^2}{Eh}\left(\frac{R}{r}-\frac{1}{2}\nu\right),\\\alpha_{21}''&=(\lambda_1^2+\lambda_2^2-\nu)\eta,\\\alpha_{22}''&=2\lambda_1(\lambda_1^2+\lambda_2^2)\frac{\eta}{R},\\\alpha_2''&=\frac{r^2\alpha_{22}''}{k^2}\left(\frac{\nu R}{r}-\frac{1}{2}\right)\end{aligned} \tag{37}$$

这里的双撇号表示与接管有关的量，薄壁接管情况也同此。

三、薄壁圆柱接管的公式

因有以上现成的公式，我们就不再单独求解薄壁接管的基本方程。从第二部分公式中略去厚壁因素项，便可得薄壁圆柱壳方程和公式。不过，考虑到所推的薄接管公式需要与薄椭球壳公式误差一致（相对于 1 的$\sqrt{\frac{h}{R}}$级误差），我们宁愿从第一部分结果来推出所需要的公式。这样，只需令

$$R_1\to\infty,\quad R_2=R=\text{const},\quad \theta=\frac{\pi}{2},$$

便从式（6）～(8) 推得薄圆柱壳（符号同厚壁圆柱壳情况）的有关结果：

$$\text{当 }\alpha=0\text{ 时},\quad w=\Delta_0'',\quad \psi=\psi_0'',\tag{38}$$

其中

$$\begin{aligned}\Delta_0''&=\alpha_{11}''N_0+\alpha_{12}''M_0+\alpha_1''q,\\ \psi_0''&=\alpha_{21}''N_0+\alpha_{22}''M_0+\alpha_2''q;\end{aligned}\tag{39}$$

$$\begin{aligned}\alpha_{11}''&=-\frac{2\sqrt[4]{3(1-\nu^2)}}{E}\sqrt{\frac{R^3}{h^3}},\\ \alpha_{12}''&=\alpha_{21}''=-\frac{\sqrt{12(1-\nu^2)}}{Eh^2}R,\\ \alpha_{22}''&=-\frac{4\sqrt[4]{27(1-\nu^2)^3}}{Eh^2}\sqrt{\frac{R}{h}},\\ \alpha_1''&=\left(1-\frac{\nu}{2}\right)\frac{R^2}{E},\\ \alpha_2''&=0\end{aligned}\tag{40}$$

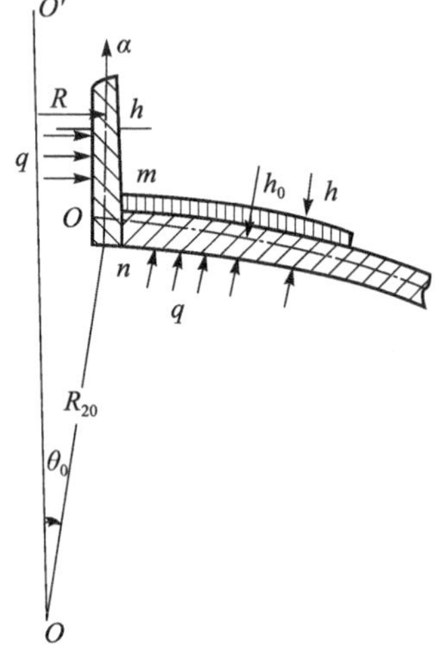

图 3 薄椭球壳中心开孔接管截面

四、椭球封头与接管的连接条件

在椭球封头中心开孔处，接管与封头是用焊缝连接起来的（位于图 3 中的 mn 截面处）。有时，在孔边缘会增置一块护强板，造成孔附近封头为双层壳情况。因护强板的尺寸较 mn 截面尺寸大得多，材质又与封头相同，而且双层壳总厚度又不太大，所以只要选取离封头下表面距离为 h_0（其值为双层壳总厚度之半）的平行面作为坐标参考面，我们就可以照常应用第一部分单层壳的公式来计算这种双层壳情况下的焊缝应力。

现在需要决定前面公式中的 M_0、Q_0 未知值。考虑到焊接部分的变形情况，我们将接管与封头连接部分的连续条件写为

$$\Delta_0'=\Delta_0'',$$

$$\begin{aligned}&\psi_0'=-\psi_0'',\\&Q_0=-N_0,\\&M_{10}'=M_{10}''=M_0\end{aligned}\tag{41}$$

将式（7）、（36）或式（7）、（39）代入式（41）中，便得关于 M_0 和 Q_0 的二元联立方程组：

$$\begin{aligned}&l_{11}Q_0+l_{12}M_0=l_1q,\\&l_{21}Q_0+l_{22}M_0=l_2q,\end{aligned}\tag{42}$$

其中

$$\begin{aligned}&l_{11}=\alpha_{11}'+\alpha_{11}'',\\&l_{12}=\alpha_{12}'-\alpha_{12}'',\\&l_{21}=\alpha_{21}'-\alpha_{21}'',\\&l_{22}=\alpha_{22}'+\alpha_{22}'',\\&l_1=\alpha_1''-\alpha_1',\\&l_2=-(\alpha_2'+\alpha_2'')\end{aligned}\tag{43}$$

解方程组（42），得决定 M_0、Q_0 值之公式：

$$\begin{aligned}&Q_0=\frac{l_1l_{22}-l_2l_{12}}{l_{11}l_{22}-l_{12}l_{21}}q,\\&M_0=\frac{l_2l_{11}-l_1l_{21}}{l_{11}l_{22}-l_{12}l_{21}}q\end{aligned}\tag{44}$$

于是，我们就能由式（3）、（9）、（10）计算封头孔洞附近和焊缝处的应力，从而达到本章分析椭球封头中心开孔接管强度的目的。下面，我们用实际例子来说明这个问题。

五、实例

铂重整装置反应器椭球封头上的开孔情况示于图 4 中，中心孔最大，旁侧的催化剂卸出口次之，其余三个孔很小，且离中心孔较远。为了避免计算这复杂的复连通区域的强度问题，我们假定封头仅有一个中心孔（这一孔洞的边缘应力是本文要计算的）。另外，中心孔的接管是由圆柱壳和锥壳组合而成的，由于感兴趣的是开孔接管连接处的强度，故我们以一半无限长圆柱壳来代替。这样，便可以应用前面的公式进行计算。可以预料，图 4 中 A 点的理论值将在误差允许范围以内；B 点的理论值将较实际值低，这是旁侧孔干扰的结果。我们将通过试验值来估计这干扰引起的误差，以便解决铂重整装置反应器产品上的强度问题。下面，我们分四种情况来进行计算。

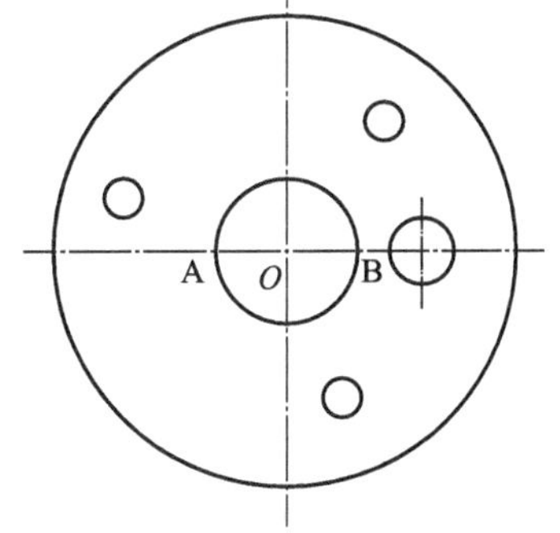

图 4　反应器封头开孔情况

例 1　对现有产品的计算。厂内已生产的几台产品是修改了原设计方案后制造的，此时，接管是厚壁的，孔边未置护强板，有关数据为

对于薄椭球壳而言，有

$$a=480\text{mm},\quad b=930\text{mm},\quad h=60\text{mm}$$

对于厚壁接管而言，有

$$R=211.5\text{mm},\quad h=90\text{mm}$$

制造产品的钢材为20CrMo9，其泊松比、弹性模量和屈服极限为

$$\nu=0.3,\quad E=2.1\times10^4\text{kg/mm}^2,\quad \sigma_S=35.0\text{kg/mm}^2$$

容器的设计压力：$q=0.80\text{kg/mm}^2$

将以上数据代入式（8）、式（37）和式（43）中，便有

$$\begin{aligned}&\alpha'_{11}=-2.80\times10^{-4},\quad \alpha'_{12}=\alpha'_{21}=-9.25\times10^{-6},\quad \alpha'_1=3.44\times10^{-1},\\&\alpha'_{22}=-6.10\times10^{-7},\quad \alpha'_2=8.10\times10^{-3};\end{aligned}\tag{45}$$

$$\begin{aligned}&\alpha''_{11}=-4.47\times10^{-4},\quad \alpha''_{12}=-4.56\times10^{-6},\quad \alpha''_1=2.27\times10^{-2},\\&\alpha''_{21}=-3.93\times10^{-6},\quad \alpha''_{22}=-7.77\times10^{-8},\quad \alpha''_2=-1.73\times10^{-5};\end{aligned}\tag{46}$$

$$\begin{aligned}&l_{11}=-7.28\times10^{-4},\quad l_{12}=-4.68\times10^{-6},\quad l_1=-3.21\times10^{-1},\\&l_{21}=-5.31\times10^{-6},\quad l_{22}=-6.88\times10^{-7},\quad l_2=-8.10\times10^{-3}\end{aligned}\tag{47}$$

再将这些值代入公式（44）中，得

$$M_0=7.04\times10^3\text{kg/mm}^2,\quad Q_0=308\text{kg/mm}\tag{48}$$

由于焊接条件关系，在接管与封头开孔的连接处，难以焊透，造成 mn 截面中间空隙，致使 mn 截面承担应力的厚度减少（我们把这一厚度称作有效厚度，记为 h^*），这将成为容器强度最薄弱之处。严格来讲，由于 mn 截面中部含有许多小空洞，造成复杂的应力分布，故精确计算这样焊缝的应力是很困难的，甚至不可能得出解。我们从简出发，认为这有效厚度所承担的内力和内矩与按等厚壳推出的值相同，应用这些值于式（3）、(10）中，就可计算焊缝 mn 的应力。值得指出，封头开孔附近区域的最大应力就发生在壳外表面的焊缝处。由于工程上感兴趣的关系，我们在下面只给出这些值。

表 1　在设计压力下的封头外表面孔边缘应力值　　（单位：kg/mm²）

h^*/mm	48	51	54	57	60
σ_1	22.3	20.3	18.6	17.1	15.8
σ_2	23.5	21.9	20.5	19.3	18.2

另外，我们还有关于屈服极限 σ_S 的安全系数公式

$$n_S=\left|\frac{\sigma_S}{\sigma_{max}}\right|,\tag{49}$$

其中 σ_{max} 是指封头开孔焊缝处的最大应力值。

由表1看出，在 mn 截面处，都是纬向应力 σ_2 取最大值。将表1的最大应力值代入公式（49），便得安全系数 n_S 值，结果给在表2中。

表 2　在设计压力下的封头开孔处的 n_S 值

h^*/mm	48	51	54	57	60
n_S	1.49	1.60	1.71	1.81	1.92

经过超声波探伤检验，在产品的 nm 截面处，一般都有 12mm 未焊透部分，这时，焊缝有效厚度仅有 48mm。由表 2 看到，开孔边缘的强度是不符合设计要求的（设计要求 $n_S \geqslant 1.6$）。如果考虑实际产品中催化剂卸出口的影响，安全系数 n_S 将较表 2 所给的值更小，强度将更不够。

下面，我们将用试验结果来说明上述理论结果的精确度。在设计压力下的产品的常温水压试验获得了封头中心孔边缘附近的应力值。由于某些原因，未能测得焊缝处的应力，只测得了距 A 点 20mm 和距 B 点 34mm 处的壳外表应力值。为叙述方便起见，这两处的试验应力符号的角标记以“$*$”“A”“B”字样。我们以壳外表面应力试验值来做比较。试验值为

$$\begin{aligned}
&\text{当 } \theta=\theta_A, \quad z=\frac{h}{2}\text{时}, \quad \sigma_{1A}^*=14.6\text{kg/mm}^2, \quad \sigma_{2A}^*=19.8\text{kg/mm}^2,\\
&\text{当 } \theta=\theta_B, \quad z=\frac{h}{2}\text{时}, \quad \sigma_{1B}^*=19.2\text{kg/mm}^2, \quad \sigma_{2B}^*=27.5\text{kg/mm}^2;
\end{aligned} \tag{50}$$

而由本章公式所得的理论值为

$$\begin{aligned}
&\text{当 } \theta=\theta_A, \quad z=\frac{h}{2}\text{时}, \quad \sigma_{1A}=15.4\text{kg/mm}^2, \quad \sigma_{2A}=17.6\text{kg/mm}^2,\\
&\text{当 } \theta=\theta_B, \quad z=\frac{h}{2}\text{时}, \quad \sigma_{1B}=15.0\text{kg/mm}^2, \quad \sigma_{2B}=17.2\text{kg/mm}^2
\end{aligned} \tag{51}$$

试验值（50）与理论值（51）之误差为 5.48%、11.1%和 21.9%、37.4%，由于感兴趣的 A 点应力误差较小，这就说明本章计算结果是可靠的。B 点附近应力误差大，主要是催化剂卸出口距中心孔很近所造成的。

例 2　对改进后的产品计算。

例 1 理论与试验结果的一致性，说明了原来生产的几台铂重整装置反应器的封头中心开孔接管处的质量是不符合设计压力要求的。从易于弥补的途径着手，我们建议增置一块厚度为 32mm 的与封头同质的护强板，以使开孔接管处的强度提高，符合设计要求，这样，有关的数据对于双层薄椭球壳而言，有

$$a=496\text{mm}, \quad b=946\text{mm}, \quad h=92\text{mm}, \quad h_0=46\text{mm}$$

而厚壁接管尺寸、设计压力及材料有关数据仍与例 1 相同。

将这些数据应用于式（8）、(37) 和（43）中，得

$$\begin{aligned}
&\alpha_{11}'=-1.47\times10^{-4}, \quad \alpha_{12}'=\alpha_{21}'=-3.93\times10^{-6}, \quad \alpha_{22}'=-2.10\times10^{-7},\\
&\alpha_1'=1.94\times10^{-1}, \quad \alpha_2'=3.50\times10^{-3};
\end{aligned} \tag{52}$$

$$\begin{aligned}
&l_{11}=5.95\times10^{-4}, \quad l_{12}=6.32\times10^{-7}, \quad l_{21}=3.75\times10^{-10},\\
&l_{22}=-2.87\times10^{-7}, \quad l_1=-1.71\times10^{-1}, \quad l_2=-3.40\times10^{-3};
\end{aligned} \tag{53}$$

而 α_{11}''，α_{12}''，α_{21}''，α_{22}''，α_1''，α_2'' 与式（46）中的值相同。

将这些值代入公式（44）中，得

$$M_0=9.58\times10^3\text{kg/mm}^2, \quad Q_0=240\text{kg/mm} \tag{54}$$

我们要求将护强板全焊透，这样 mn 截面焊缝的有效厚度为

$$h^*=80\text{mm}$$

那么，按照公式（10），便得到了最大应力所在的焊缝处壳外表面的应力值：

$$\sigma_1 = 11.5\text{kg/mm}^2, \quad \sigma_2 = 13.9\text{kg/mm}^2 \tag{55}$$

此时，纬向应力 σ_2 取最大值。将此值代入式（49）中，得

$$n_S = 2.52 \tag{56}$$

当然，距催化剂卸出口最近的 B 点处的应力将较式（55）所给值大，但由前面的试验结果估计，现在的改进情况是符合强度要求的。

例 3　对原设计方案的计算。此时，接管是薄壁的，孔边设置了一块厚达 50mm 的护强板，有关数据如下。

对于双层薄椭球壳而言，有

$$a = 505\text{mm}, \quad b = 955\text{mm}, \quad h = 110\text{mm}, \quad h_0 = 55\text{mm}$$

对于薄壁接管而言，有

$$R = 177.5\text{mm}, \quad h = 22\text{mm}$$

制造反应器的钢材及设计压力均与前面情况相同。

将以上数据代入式（8）、（40）和（43）中，便有

$$\begin{aligned} &\alpha'_{11} = 7.92\times10^{-5}, \quad \alpha'_{12} = \alpha'_{21} = -2.31\times10^{-6}, \quad \alpha'_{22} = -1.35\times10^{-7}, \\ &\alpha'_1 = 1.17\times10^{-1}, \quad \alpha'_2 = 2.00\times10^{-3}; \end{aligned} \tag{57}$$

$$\begin{aligned} &\alpha'_{11} = -2.81\times10^{-3}, \quad \alpha''_{12} = \alpha''_{21} = -5.77\times10^{-5}, \quad \alpha''_{22} = -2.37\times10^{-6}, \\ &\alpha''_1 = 5.80\times10^{-2}, \quad \alpha''_2 = 0; \end{aligned} \tag{58}$$

$$\begin{aligned} &l_{11} = -2.90\times10^{-3}, \quad l_{12} = l_{21} = 5.54\times10^{-5}, \quad l_{22} = -2.51\times10^{-6}, \\ &l_1 = -5.85\times10^{-2}, \quad l_2 = -2.00\times10^{-3} \end{aligned} \tag{59}$$

再将这些值代入公式（44）中，得

$$M_0 = 1.76\times10^3\text{kg}, \quad Q_0 = 49.9\text{kg/mm} \tag{60}$$

于是，通过式（9），我们得到 mn 截面的内力和内矩值：

$$\begin{aligned} &T_1 = 124\text{kg/mm}, \quad T_2 = 1.30\times10^3\text{kg/mm}, \\ &M_1 = 1.05\times10^3\text{kg}, \quad M_2 = 316\text{kg} \end{aligned} \tag{61}$$

按照设计要求，mn 截面的焊缝的有效厚度（包括焊缝的突出部分）为

$$h^* = 88\text{mm}$$

将以上有关数据代入公式（10）中，便得到了最大应力所在的焊缝处壳外表面的应力值：

$$\sigma_1 = 2.22\text{kg/mm}^2, \quad \sigma_2 = 15.0\text{kg/mm}^2 \tag{62}$$

这时，纬向应力较经向应力大得多。将 σ_2 值代入式（49）中，得

$$n_S = 2.33 \tag{63}$$

故 mn 处的强度是符合要求的。

例 4　对比方案的计算。

为了对厚、薄接管情况进行比较，我们在保持接管内径不变的情况下，使封头厚度与例 2 相同，使接管厚度与例 3 相同，关于钢材和设计压力亦同前面数据。

将上面数据代入式（8）、（40）和（43）中，便有

$$\begin{aligned} &\alpha'_{11} = -1.04\times10^{-4}, \quad \alpha'_{12} = \alpha'_{21} = -3.30\times10^{-6}, \quad \alpha'_{22} = -2.10\times10^{-7}, \\ &\alpha'_1 = 1.47\times10^{-1}, \quad \alpha'_2 = 2.90\times10^{-3}; \end{aligned} \tag{64}$$

$$l_{11}=-2.92\times10^{-3},\quad l_{12}=l_{21}=5.44\times10^{-5},\quad l_{22}=-2.58\times10^{-6},$$
$$l_1=-8.89\times10^{-2},\quad l_2=-2.90\times10^{-3} \tag{65}$$

至于 α''_{11}，α''_{12}，α''_{21}，α''_{22}，α''_1，α''_2 的值，与式（60）相同。

将这些数据代入式（44）中，得

$$M_0=2.34\times10^3\text{kg},\quad Q_0=68.3\text{kg/mm} \tag{66}$$

为便于比较，我们使 mn 截面焊缝的有效厚度与例 2 相同，即

$$h^*=80\text{mm}$$

于是，由式（10），便得到最大应力所在的焊缝处壳外表面的应力值：

$$\sigma_1=3.29\text{kg/mm}^2,\quad \sigma_2=16.8\text{kg/mm}^2 \tag{67}$$

此时，纬向应力 σ_2 取最大值。将 σ_2 值代入公式（49）中，得

$$n_S=2.08 \tag{68}$$

这说明 mn 截面强度基本合乎要求。

六、讨论

根据上节的实际计算，我们有下面几点看法。

(1) 由表 2 看到，对于薄椭球封头仅开了一个中心孔的情况而言，只要焊接情况良好，焊缝有效厚度在 51mm 以上，强度就符合要求。这一事实说明，使用厚壁管来代替封头开孔补强的办法是可行的。但由于焊接工艺较难达到这种要求，故影响了此种措施的效果。

(2) 将采用厚壁接管的例 2 和薄壁接管的例 4 对比，两种情况的接管内径、封头厚度、焊缝有效厚度完全相同，只是接管壁厚相差 4 倍，这时，例 2 的安全系数值提高了 17%，不过例 4 要省一些钢材。

(3) 从四种情况的计算结果看到，孔边缘应力很大，而 mn 截面处不能焊透，致使多费钢材，增加焊接难度。为避免这种情况，最好采用锻压而成厚壁接管环结构，即在接管壁下端有凸缘存在，将凸缘与封头焊接，这样可不设置护强板，而且焊缝又离接管根部一定距离，焊接质量也易保证。

致谢：兰州大学数学力学系王廷栋同志负责本章的试验工作。

参考文献

[1]　诺沃日洛夫 B B. 薄壳理论. 北京石油学院材料力学教研组，译. 北京：科学出版社，1959.
[2]　符拉索夫 B 3. 壳体的一般理论. 薛振东，朱世靖，译. 北京：人民教育出版社，1960.

500 万吨/年常减压装置减压塔下端部分壳体的应力分析①

我国的石油炼制工业，正在蓬蓬勃勃地向前发展。在我国当时最大的 500 万吨/年炼厂的设计中，有一些亟待解决的力学课题，常减压装置减压塔下端部分壳体（包括大筒体、锥体和小筒体）在常温下直立水压试验时的强度问题便是其中之一。现把我们所做的这方面的工作汇总如下。

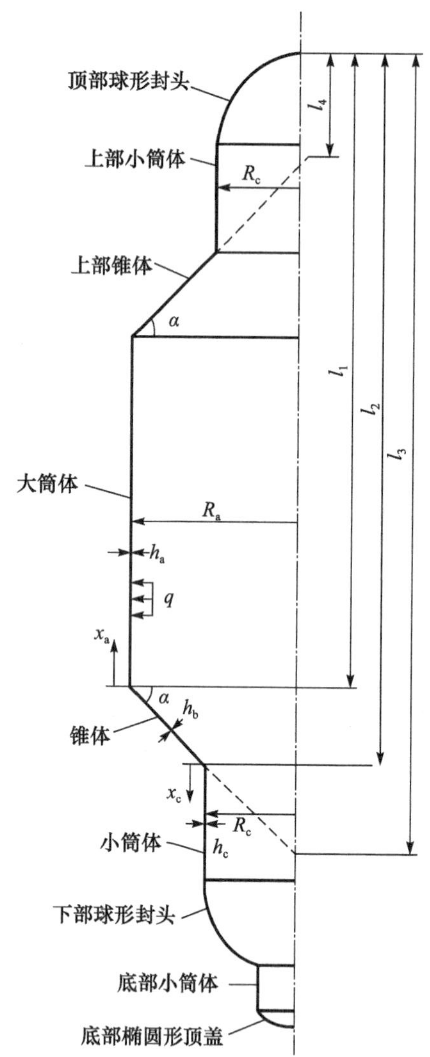

图 1　减压塔的简化计算模型

一、简化原则

由于任务急，我们从简着手，采用下面三个办法，以节省计算量。

（1）常减压装置减压塔是由多个壳体组合而成的容器，在两个不同壳体的连接处，都是圆弧过渡。我们忽略圆弧过渡的影响，以壳体直接连接为计算模型；

（2）忽略减压塔本身重量的影响；

（3）在常减压装置减压塔下部壳体中，大筒体的轴向长度很长，小筒体的轴向长度要短些，我们把它们都作为半无限长圆柱壳来处理。于是，常减压装置减压塔的实际计算模型就如图 1 所示。显而易见，这是薄壁容器的轴对称应力问题。这里，q 为内压力，R 为半径，h 为壳体厚度，x 为筒体轴向坐标，α 为锥角，l 为高度。右下角标 a、b、c 分别表示减压塔下部壳体中大筒体、锥体和小筒体的量。

下面，分六个部分进行分析（本节位移、法向力和力矩的正方向与文献［1］相同，不再赘述）。

二、大筒体的基本方程及其解

大筒体是薄壁圆柱壳。为书写方便起见，除易于混淆者外，都先不标右下角标，下面第 3 小节和 4 小节也是如此。众所周知，轴对称变形问题圆柱壳的法向位移 w 所满足的微分方程为[1]

① 本文原载《压力容器》，1975，(3)：1-28. 作者：刘人怀，王凯.

$$K\frac{d^4w}{d\rho^4}+DR^2(1-\nu^2)w=q_aR^4-\nu R^3N_x, \tag{1}$$

其中

K 为抗弯刚度，$K=\dfrac{Eh^3}{12(1-\nu^2)}$， (2)

D 为抗拉刚度，$D=\dfrac{Eh}{1-\nu^2}$， (3)

E 为弹性模量，

ν 为泊松比，

ρ 为无量纲轴向坐标，$\rho=\dfrac{x}{R}$， (4)

q_a 为大筒体承受的内压力，由均匀内压力 q_0 和静水压力所组成，

$$q_a=q_0+\gamma(l_1-R_\rho), \tag{5}$$

γ 为水的比重，

N_x 为筒体轴向力。

由方程 (1) 求得 w 后，便可按下面公式计算转角 ψ，环向力 N_θ，轴向和环向力矩 M_x、M_θ，横向力 Q_x：

$$\begin{aligned}&\psi=\frac{1}{R}\frac{dw}{d\rho},\\&N_\theta=\frac{D}{R}\left(w+\nu\frac{du}{d\rho}\right),\\&M_x=\frac{K}{R^2}\frac{d^2w}{d\rho^2},\\&M_\theta=\nu M_x,\\&Q_x=\frac{1}{R}\frac{dM_x}{d\rho}\end{aligned} \tag{6}$$

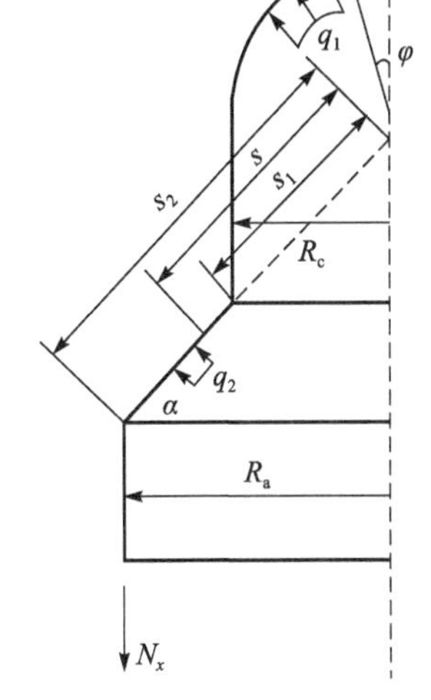

图 2 减压塔上部结构的竖向力平衡

下面，我们先确定轴向力 N_x 的值。为此，从减压塔中截出大筒体以上结构，由图 2 可知，它的竖向力平衡条件为

$$2\pi R_aN_x-\int_0^{\pi/2}2\pi R_c^2q_1\cos\varphi\sin\varphi d\varphi-\int_{s_1}^{s_2}2\pi sq_2\cos^2\alpha ds=0, \tag{7}$$

其中

s 为上部锥体母线方向的坐标，

s_1、s_2 为上部截锥体上、下端的位置，

φ 为余纬度，

q_1 为顶部球形封头承受的内压力，

$$q_1=q_0+\gamma R_c(1-\cos\varphi), \tag{8}$$

q_2 为上部锥体承受的内压力

$$q_2=q_0+\gamma(l_4+s\sin\alpha) \tag{9}$$

将式 (7) 化简，得

$$N_x = \frac{1}{2} R_a q_0 + \frac{\beta_a}{R_a} \gamma, \tag{10}$$

其中

$$\beta_a = \frac{1}{6} R_c^3 + \frac{1}{2} (R_a^2 - R_c^2) l_4 + \frac{1}{3} (R_a^3 - R_c^3) \tan\alpha \tag{11}$$

将式（10）代入方程（1）中，应用通常的方法，可得通解：

$$\begin{aligned} w = & \mathrm{e}^{-\eta\rho} [c_1 \cos(\eta\rho) + c_2 \sin(\eta\rho)] + \mathrm{e}^{\eta\rho} [\bar{c}_1 \cos(\eta\rho) + \bar{c}_2 \sin(\eta\rho)] \\ & + \frac{R^2}{Eh} \left[\left(1 - \frac{\nu}{2}\right) q_0 + \gamma \left(l_1 - R_\rho - \nu \frac{\beta_a}{R^2} \right) \right], \end{aligned} \tag{12}$$

其中 c_1、c_2、$\bar{c}_1$、$\bar{c}_2$ 为待定常数，

$$\eta = \sqrt[4]{3(1-\nu^2) \frac{R^2}{h^2}} \tag{13}$$

因为大筒体是半无限长壳体，故可取

$$\bar{c}_1 = \bar{c}_2 = 0 \tag{14}$$

再由式（6），可得

$$\begin{aligned} & \psi = -\frac{\eta}{R} \mathrm{e}^{-\eta\rho} [(c_1 - c_2)\cos(\eta\rho) + (c_1 + c_2)\sin(\eta\rho)] - \frac{\gamma R^2}{Eh}, \\ & M_x = \frac{2K\eta^2}{R^2} \mathrm{e}^{-\eta\rho} [c_1 \sin(\eta\rho) - c_2 \cos(\eta\rho)], \\ & M_\theta = \nu M_x, \\ & Q_x = \frac{2K\eta^3}{R^3} \mathrm{e}^{-\eta\rho} [(c_1 + c_2)\cos(\eta\rho) - (c_1 - c_2)\sin(\eta\rho)], \\ & N_\theta = \frac{4K\eta^4}{R^3} \mathrm{e}^{-\eta\rho} [c_1 \cos(\eta\rho) + c_2 \sin(\eta\rho)] + R[q_0 + \gamma(l_1 - R\rho)] \end{aligned} \tag{15}$$

求得了法向力和力矩后，就可按下式计算轴向和环向应力：

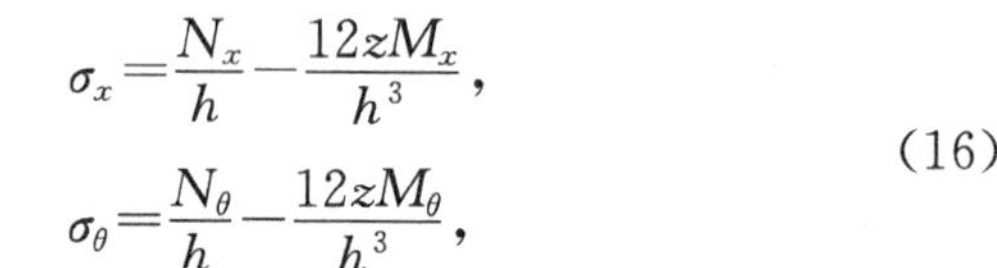

$$\begin{aligned} & \sigma_x = \frac{N_x}{h} - \frac{12 z M_x}{h^3}, \\ & \sigma_\theta = \frac{N_\theta}{h} - \frac{12 z M_\theta}{h^3}, \end{aligned} \tag{16}$$

其中 z 为中曲面外法线方向的坐标。

三、小筒体的基本方程及其解

小筒体也是薄壁圆柱壳，其基本方程与式（1）相同，仅 N_x 和载荷 q 的值不一样。小筒体承受的内压力为

$$q_c = q_0 + \gamma(l_2 + R_c \rho) \tag{17}$$

为计算 N_x，我们从减压塔中截取部分壳体（图 3），显然，此部分壳体的竖向力平衡条件为

$$2\pi R_c N_{xc} + \int_{s_1}^{s_2} 2\pi s q_b \cos^2\alpha \mathrm{d}s - 2\pi R_a N_{xa} = 0, \tag{18}$$

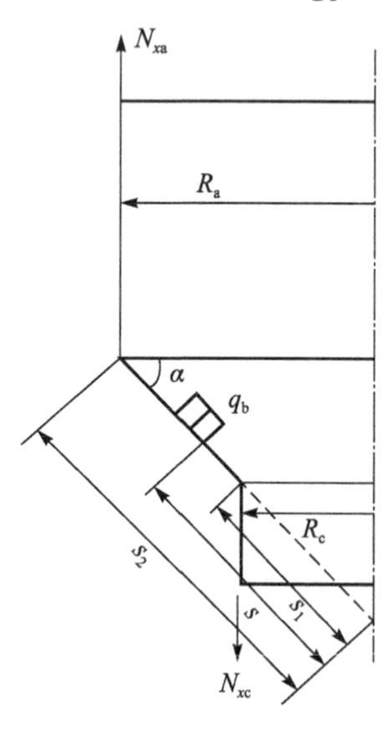

图 3　减压塔下端部分壳体的竖向力平衡

其中 q_b 为锥体承受的载荷，

$$q_b = q_0 + \gamma(l_3 - s\sin\alpha) \tag{19}$$

应用式（10），式（18）成为

$$N_{xc} = \frac{1}{2}R_c q_0 + \frac{\beta_c}{R_c}\gamma, \tag{20}$$

其中

$$\beta_c = \beta_a - \frac{1}{2}(R_a^2 - R_c^2)l_3 + \frac{1}{3}(R_a^3 - R_c^3)\tan\alpha \tag{21}$$

因为视小筒体为半无限长壳体，故解与大筒体相似，为

$$\begin{aligned}
&w = e^{-\eta\rho}[c_7\cos(\eta\rho) + c_8\sin(\eta\rho)] + \frac{R^2}{Eh}\left[\left(1-\frac{\nu}{2}\right)q_0 + \gamma\left(l_2 + R\rho - \nu\frac{\beta_c}{R^2}\right)\right],\\
&\psi = -\frac{\eta}{R}e^{-\eta\rho}[(c_7 - c_8)\cos(\eta\rho) + (c_7 + c_8)\sin(\eta\rho)] + \frac{\gamma R^2}{Eh},\\
&M_x = \frac{2K\eta^2}{R^2}e^{-\eta\rho}[c_7\sin(\eta\rho) - c_8\cos(\eta\rho)],\\
&M_\theta = \nu M_x,\\
&Q_x = \frac{2K\eta^3}{R^3}e^{-\eta\rho}[(c_7 + c_8)\cos(\eta\rho) - (c_7 - c_8)\sin(\eta\rho)],\\
&N_\theta = \frac{4K\eta^4}{R^3}e^{-\eta\rho}[c_7\cos(\eta\rho) + c_8\sin(\eta\rho)] + R[q_0 + \gamma(l_2 + R\rho)],
\end{aligned} \tag{22}$$

其中 c_7、c_8 为待定常数。

应力公式与式（16）相同。

四、锥体的基本方程及其解

锥体也是薄壁壳体。它的母线方向坐标与上部锥体相同。由文献［2］和［3］，可推得在轴对称变形时，锥壳横向力 Q_s 所涉及的微分方程为

$$LL(sQ_s) + \mu^4 sQ_s = F(q_b, s), \tag{23}$$

其中

$$F(q_b, s) = \cot\alpha L\left[\frac{d(s^2 q_b)}{ds} - \frac{1}{s}\int s q_b ds\right], \tag{24}$$

$$\mu^4 = \frac{12(1-\nu^2)}{h^2}, \tag{25}$$

$$L(\cdots) = \cot\alpha\left[s\frac{d^2(\cdots)}{ds^2} + \frac{d(\cdots)}{ds} - \frac{1}{s}(\cdots)\right] \tag{26}$$

应用式（19），式（24）成为

$$F(q_b, s) = -8\gamma s\cot\alpha\cos\alpha \tag{27}$$

由方程（23）解得了 Q_s 后，就可按下列公式计算转角 ψ，经向和环向力 N_s、N_θ，经向和环向力矩 M_s、M_θ：

$$\psi = -\frac{1}{D(1-\nu^2)}\left\{\cot\alpha L(sQ_s) - \frac{P}{2\pi s\sin^2\alpha} - \cot^2\alpha\left[\frac{d(s^2 q_b)}{ds} - \frac{1}{s}\int s q_b ds\right]\right\},$$

$$N_s = -Q_s\cot\alpha + \frac{\cot\alpha}{s}\int q_b s\mathrm{d}s - \frac{P}{2\pi s\sin\alpha\cos\alpha},$$

$$N_\theta = \cot\alpha\left[q_b s - \frac{\mathrm{d}(sQ_s)}{\mathrm{d}s}\right],$$

$$M_s = K\left(\frac{\mathrm{d}\psi}{\mathrm{d}s} + \nu\frac{\psi}{s}\right),$$

$$M_\theta = K\left(\frac{\psi}{s} + \nu\frac{\mathrm{d}\psi}{\mathrm{d}s}\right), \tag{28}$$

其中 P 为待定常数。

方程（23）的齐次式可化为两个二阶贝塞尔方程，其解已在文献［2］和［3］中列出。应用观察法，方程的特解也易于求得。于是，有方程（23）的通解：

$$Q_s = \frac{1}{s}[c_3(\mathrm{ber}y - 2y^{-1}\mathrm{bei}'y) + c_4(\mathrm{bei}y + 2y^{-1}\mathrm{ber}'y) + c_5(\mathrm{ker}y - 2y^{-1}\mathrm{kei}'y) + c_6(\mathrm{kei}y + 2y^{-1}\mathrm{ker}'y)] - 8\gamma\mu^{-4}\cot\alpha\cos\alpha, \tag{29}$$

其中 c_3、c_4、c_5、c_6 为待定常数，ber$y$、bei$y$、ker$y$、kei$y$ 是汤姆生函数，带撇号表示各函数对其变量 y 的导数。

$$y = 2\mu\sqrt{s\tan\alpha} \tag{30}$$

下面，我们决定式（28）中的 P 值。为此，从锥体中截出一部分壳体（图 4），它的竖向力的平衡条件为

$$2\pi sN_s\cos\alpha\sin\alpha + 2\pi sQ_s\cos^2\alpha - \int_{s_1}^{s} 2\pi sq_b\cos^2\alpha\mathrm{d}s - 2\pi R_c N_{xc} = 0 \tag{31}$$

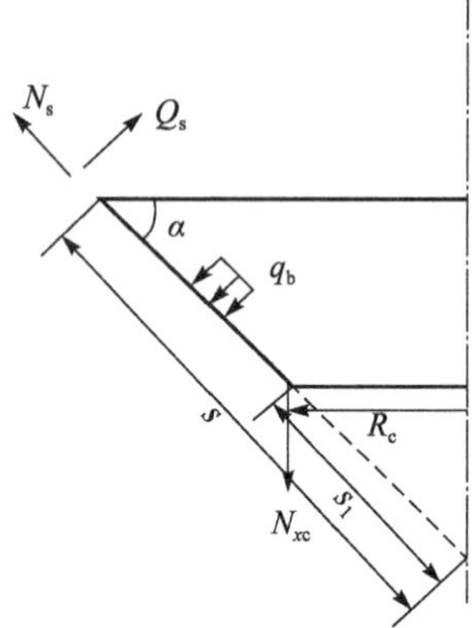

图 4　减压塔下部锥体部分壳体的竖向力平衡

将式（19）代入式（31），得

$$N_s = -Q_s\cot\alpha + \cot\alpha\left[\frac{1}{2}(q_0 + \gamma l_3)s - \frac{1}{3}\gamma s^2\sin\alpha\right] - \frac{\cot\alpha}{s}\left[\frac{1}{2}(q_0 + \gamma l_3)s_1^2 - \frac{1}{3}\gamma s_1^3\sin\alpha\right] + \frac{R_c N_{xc}}{s\cos\alpha\sin\alpha} \tag{32}$$

而按式（28）的第二式，还有

$$N_s = -Q_s\cot\alpha + \cot\alpha\left[\frac{1}{2}(q_0 + \gamma l_3)s - \frac{1}{3}\gamma s^2\sin\alpha\right] - \frac{P}{2\pi s\sin\alpha\cos\alpha} \tag{33}$$

比较式（32）和（33），并利用式（20），便确定了 P 值：

$$P = \gamma\pi\left(l_3R_c^2 - \frac{2}{3}R_c^3\tan\alpha - 2\beta_c\right) \tag{34}$$

应用式（29）和（34），由式（28）得

$$\psi = \frac{1}{Eh}\left\{\frac{2\sqrt{3(1-\nu^2)}\cot\alpha}{h}[c_3(\mathrm{bei}y + 2y^{-1}\mathrm{ber}'y) - c_4(\mathrm{ber}y\right.$$

$$-2y^{-1}\mathrm{bei}'y)+c_5(\mathrm{kei}y+2y^{-1}\mathrm{ker}'y)-c_6(\mathrm{ker}y-2y^{-1}\mathrm{kei}'y)]$$

$$+\cot^2\alpha\left[\frac{3}{2}(q_0+\gamma l_3)s-\frac{8}{3}\gamma s^2\sin\alpha\right]+\frac{P}{2\pi s\sin^2\alpha}\Bigg\},$$

$$N_\theta=-\cot\alpha\Bigg\{\frac{1}{2s}[c_3(y\mathrm{ber}'y-2\mathrm{ber}y+4y^{-1}\mathrm{bei}'y)+c_4(y\mathrm{bei}'y-2\mathrm{bei}y$$

$$-4y^{-1}\mathrm{bei}'y)+c_5(y\mathrm{ker}'y-2\mathrm{ker}y+4y^{-1}\mathrm{kei}'y)+c_6(y\mathrm{kei}'y-2\mathrm{kei}y$$

$$-4y^{-1}\mathrm{ker}'y)]-8\gamma\mu^{-4}\cot\alpha\cos\alpha-s(q_0+\lambda l_3)+\gamma s^2\sin\alpha\Bigg\},$$

$$M_s=2y^{-2}\{c_3[y\mathrm{bei}'y-2(1-\nu)(\mathrm{bei}y+2y^{-1}\mathrm{ber}'y)]-c_4[y\mathrm{ber}'y$$
$$-2(1-\nu)(\mathrm{ber}y-2y^{-1}\mathrm{ber}'y)]+c_5[y\mathrm{kei}'y-2(1-\nu)(\mathrm{kei}y$$
$$+2y^{-1}\mathrm{ker}'y)]-c_6[y\mathrm{ker}'y-2(1-\nu)(\mathrm{ker}y-2y^{-1}\mathrm{kei}'y)]\}$$
$$+\frac{K}{Eh}\left\{\cot^2\alpha\left[\frac{3}{2}(1+\nu)(q_0+\gamma l_3)-\frac{8}{3}(2+\nu)\gamma s\sin\alpha\right]-\frac{(1-\nu)P}{2\pi s^2\sin^2\alpha}\right\},\tag{35}$$

$$M_\theta=2y^{-2}\{c_3[\nu y\mathrm{bei}'y+2(1-\nu)(\mathrm{bei}y+2y^{-1}\mathrm{ber}'y)]-c_4[\nu y\mathrm{ber}'y$$
$$+2(1-\nu)(\mathrm{ber}y-2y^{-1}\mathrm{bei}'y)]+c_5[\nu y\mathrm{kei}'y+2(1-\nu)(\mathrm{kei}y$$
$$+2y^{-1}\mathrm{ker}'y)]-c_6[\nu y\mathrm{ker}'y+2(1-\nu)(\mathrm{ker}y-2y^{-1}\mathrm{kei}'y)]\}$$
$$+\frac{K}{Eh}\left\{\cot^2\alpha\left[\frac{3}{2}(1+\nu)(q_0+\gamma l_3)-\frac{8}{3}(1+2\nu)\gamma s\sin\alpha\right]+\frac{(1-\nu)P}{2\pi s^2\sin^2\alpha}\right\}$$

求得了法向力和力矩后，就可按下式计算经向应力和环向应力：

$$\sigma_s=\frac{N_s}{h}-\frac{12zM_s}{h^3},\qquad \sigma_\theta=\frac{N_\theta}{h}-\frac{12zM_\theta}{h^3}\tag{36}$$

对于边缘上的连续条件，还需要建立锥壳水平位移 Δ 和水平力 H 的公式。由图 5 知，水平位移 Δ 为

$$\Delta=v\cos\alpha+w\sin\alpha\tag{37}$$

由锥壳的弹性定律能够容易得到：

$$v\cos\alpha+w\sin\alpha=\frac{s\cos\alpha}{Eh}(N_\theta-\nu N_s)\tag{38}$$

于是，在利用式（33）和（35）的第二式后，得

$$\Delta=-\frac{\cot\alpha\cos\alpha}{Eh}\Bigg\{c_3\left[\frac{1}{2}y\mathrm{ber}'y-(1+\nu)(\mathrm{ber}y-2y^{-1}\mathrm{ber}'y)\right]+c_4\left[\frac{1}{2}y\mathrm{bei}'y\right.$$
$$\left.-(1+\nu)(\mathrm{bei}y+2y^{-1}\mathrm{ber}'y)\right]+c_5\left[\frac{1}{2}y\mathrm{ker}'y-(1+\nu)(\mathrm{ker}y-2y^{-1}\mathrm{kei}'y)\right]$$
$$+c_6\left[\frac{1}{2}y\mathrm{ker}'y-(1+\nu)(\mathrm{kei}y+2y^{-1}\mathrm{ker}'y)\right]-8(1-\nu)\gamma\mu^{-4}s\cos\alpha\cot\alpha$$
$$-\left(1-\frac{\nu}{2}\right)(q_0+\gamma l_3)s^2+\left(1-\frac{\nu}{3}\right)\gamma s^3\sin\alpha-\frac{\nu P}{2\pi\cos^2\alpha}\Bigg\}\tag{39}$$

另外，由图 6 知，水平力 H 为

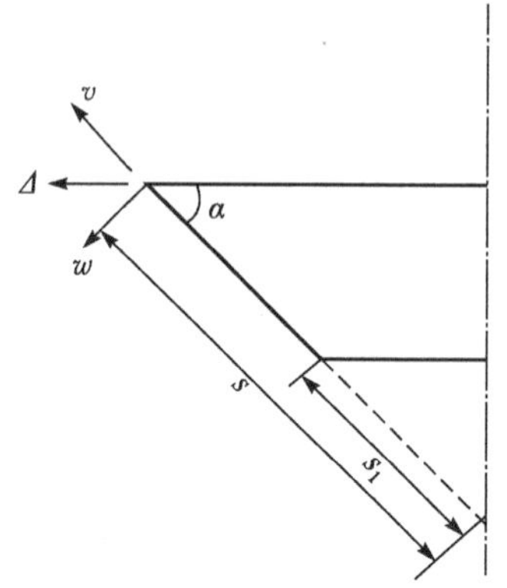

图 5　锥壳的水平位移

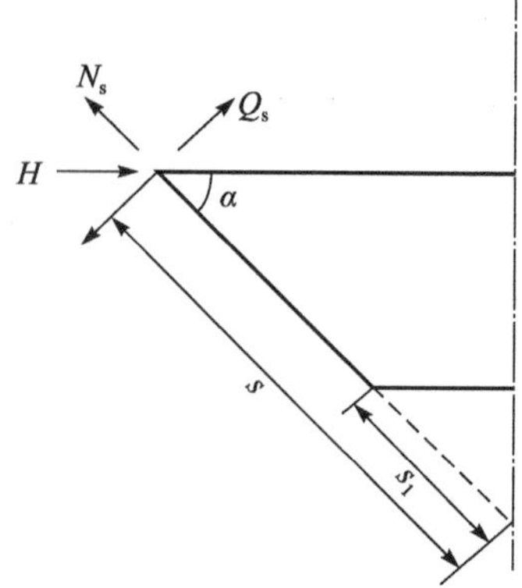

图 6　锥壳的水平力 H

$$H=Q_s\sin\alpha-N_s\cos\alpha \tag{40}$$

将式（33）代入上式中，得

$$H=\frac{Q_s}{\sin\alpha}-\cot\alpha\cos\alpha\left[\frac{1}{2}(q_0+\gamma l_3)s-\frac{1}{3}\gamma s^2\sin\alpha\right]+\frac{P}{2\pi s\sin\alpha} \tag{41}$$

五、连续条件

在锥体和大筒体以及锥体和小筒体交接的地方，是用电焊连接起来的，因此，它们应该满足下面的连续条件。

在 $\rho_a=0$，$s=s_2$ 处，

$$w'_a=\Delta'',\quad \psi'_a=\psi''_b,\quad M'_{xa}=M''_s,\quad Q'_{xa}=H''; \tag{42}$$

在 $s=s_1$，$\rho_c=0$ 处，

$$\Delta'=w'_c,\quad \psi'_b=-\psi'_c,\quad M'_s=M'_{xc},\quad H'=-Q'_{xc} \tag{43}$$

这里的右上角标"′"表示壳体上边缘的量，"″"号表示壳体下边缘的量。交界处各量的正方向示于图 7 中。

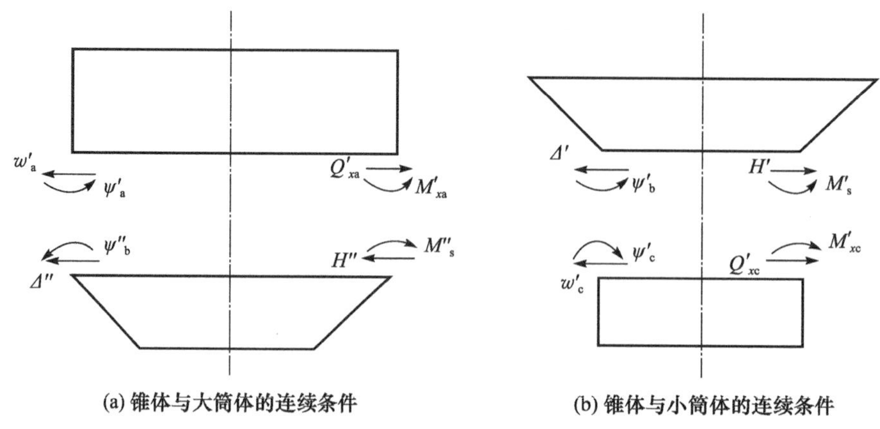

图 7　交界处各量的正方向

将第 2 小节～第 4 小节中的有关公式代入连续条件（42）和（43）中，便得关于

八个待定常数 c_1、c_2、…、c_8 的八元线性代数方程组：

$$\begin{aligned}&\alpha_{11}c_1+\alpha_{12}c_2+\cdots+\alpha_{18}c_8=\omega_{11}q_0+\omega_{12}\gamma,\\&\alpha_{21}c_1+\alpha_{22}c_2+\cdots+\alpha_{28}c_8=\omega_{21}q_0+\omega_{22}\gamma,\\&\cdots\\&\alpha_{81}c_1+\alpha_{82}c_2+\cdots+\alpha_{88}c_8=\omega_{81}q_0+\omega_{82}\gamma,\end{aligned}\tag{44}$$

其中

$\alpha_{11}=-1,$

$\alpha_{12}=\alpha_{17}=\alpha_{18}=0,$

$$\alpha_{13}=-\frac{\cot\alpha\cos\alpha}{Eh_b}\left[\frac{1}{2}y_2\,\mathrm{ber}'y_2-(1+\nu)(\mathrm{ber}y_2-2y_2^{-1}\mathrm{bei}'y_2)\right],$$

$$\alpha_{14}=-\frac{\cot\alpha\cos\alpha}{Eh_b}\left[\frac{1}{2}y_2\,\mathrm{bei}'y_2-(1+\nu)(\mathrm{bei}y_2+2y_2^{-1}\mathrm{ber}'y_2)\right],$$

$$\alpha_{15}=-\frac{\cot\alpha\cos\alpha}{Eh_b}\left[\frac{1}{2}y_2\,\mathrm{ker}'y_2-(1+\nu)(\mathrm{ker}y_2-2y_2^{-1}\mathrm{kei}'y_2)\right],$$

$$\alpha_{16}=-\frac{\cot\alpha\cos\alpha}{Eh_b}\left[\frac{1}{2}y_2\,\mathrm{kei}'y_2-(1+\nu)(\mathrm{kei}y_2+2y_2^{-1}\mathrm{ker}'y_2)\right];$$

$$\alpha_{21}=-\alpha_{22}=\frac{\eta_a}{R_a},$$

$$\alpha_{23}=\frac{2\sqrt{3(1-\nu^2)}\cot\alpha}{Eh_b^2}(\mathrm{bei}y_2+2y_2^{-1}\mathrm{ber}'y_2),$$

$$\alpha_{24}=-\frac{2\sqrt{3(1-\nu^2)}\cot\alpha}{Eh_b^2}(\mathrm{ber}y_2-2y_2^{-1}\mathrm{bei}'y_2),$$

$$\alpha_{25}=\frac{2\sqrt{3(1-\nu^2)}\cot\alpha}{Eh_b^2}(\mathrm{kei}y_2+2y_2^{-1}\mathrm{ker}'y_2),$$

$$\alpha_{26}=-\frac{2\sqrt{3(1-\nu^2)}\cot\alpha}{Eh_b^2}(\mathrm{ker}y_2-2y_2^{-1}\mathrm{kei}'y_2),$$

$\alpha_{27}=\alpha_{28}=0;$

$\alpha_{31}=\alpha_{37}=\alpha_{38}=0,$

$$\alpha_{32}=\frac{2K_a\eta_a^2}{R_a^2},$$

$\alpha_{33}=2y_2^{-2}[y_2\,\mathrm{bei}'y_2-2(1-\nu)(\mathrm{bei}y_2+2y_2^{-1}\mathrm{ber}'y_2)],$

$\alpha_{34}=-2y_2^{-2}[y_2\,\mathrm{ber}'y_2-2(1-\nu)(\mathrm{ber}y_2-2y_2^{-1}\mathrm{bei}'y_2)],$

$\alpha_{35}=2y_2^{-2}[y_2\,\mathrm{kei}'y_2-2(1-\nu)(\mathrm{kei}y_2+2y_2^{-1}\mathrm{ker}'y_2)],$

$\alpha_{36}=-2y_2^{-2}[y_2\,\mathrm{ker}'y_2-2(1-\nu)(\mathrm{ker}y_2-2y_2^{-1}\mathrm{kei}'y_2)];$

$$\alpha_{41}=\alpha_{42}=-\frac{2K_a\eta_a^3}{R_a^3},$$

$$\alpha_{43}=\frac{\cot\alpha}{R_a}(\mathrm{ber}y_2-2y_2^{-1}\mathrm{bei}'y_2),$$

$$\alpha_{44}=\frac{\cot\alpha}{R_a}(\mathrm{bei}y_2+2y_2^{-1}\mathrm{ber}'y_2),$$

$$\alpha_{45}=\frac{\cot\alpha}{R_{\mathrm{a}}}(\mathrm{ker}y_2-2y_2^{-1}\mathrm{kei}'y_2),$$

$$\alpha_{46}=\frac{\cot\alpha}{R_{\mathrm{a}}}(\mathrm{kei}y_2+2y_2^{-1}\mathrm{ker}'y_2),$$

$$\alpha_{47}=\alpha_{48}=0;$$

$$\alpha_{51}=\alpha_{52}=\alpha_{58}=0,$$

$$\alpha_{53}=\frac{\cot\alpha\cos\alpha}{Eh_{\mathrm{b}}}\left[\frac{1}{2}y_1\mathrm{ber}'y_1-(1+\nu)(\mathrm{ber}y_1-2y_1^{-1}\mathrm{bei}'y_1)\right],$$

$$\alpha_{54}=\frac{\cot\alpha\cos\alpha}{Eh_{\mathrm{b}}}\left[\frac{1}{2}y_1\mathrm{bei}'y_1-(1+\nu)(\mathrm{bei}y_1+2y_1^{-1}\mathrm{ber}'y_1)\right],$$

$$\alpha_{55}=\frac{\cot\alpha\cos\alpha}{Eh_{\mathrm{b}}}\left[\frac{1}{2}y_1\mathrm{ker}'y_1-(1+\nu)(\mathrm{ker}y_1-2y_1^{-1}\mathrm{kei}'y_1)\right],$$

$$\alpha_{56}=\frac{\cot\alpha\cos\alpha}{Eh_{\mathrm{b}}}\left[\frac{1}{2}y_1\mathrm{kei}'y_1-(1+\nu)(\mathrm{kei}y_1+2y_1^{-1}\mathrm{ker}'y_1)\right],$$

$$\alpha_{57}=1;$$

$$\alpha_{61}=\alpha_{62}=0,$$

$$\alpha_{63}=\frac{2\sqrt{3(1-\nu^2)}\cot\alpha}{Eh_{\mathrm{b}}^2}(\mathrm{bei}y_1+2y_1^{-1}\mathrm{ber}'y_1),$$

$$\alpha_{64}=-\frac{2\sqrt{3(1-\nu^2)}\cot\alpha}{Eh_{\mathrm{b}}^2}(\mathrm{ber}y_1-2y_1^{-1}\mathrm{bei}'y_1),$$

$$\alpha_{65}=\frac{2\sqrt{3(1-\nu^2)}\cot\alpha}{Eh_{\mathrm{b}}^2}(\mathrm{kei}y_1+2y_1^{-1}\mathrm{ker}'y_1),$$

$$\alpha_{66}=-\frac{2\sqrt{3(1-\nu^2)}\cot\alpha}{Eh_{\mathrm{b}}^2}(\mathrm{ker}y_1-2y_1^{-1}\mathrm{kei}'y_1),$$

$$\alpha_{67}=-\alpha_{68}=-\frac{\eta_{\mathrm{c}}}{R_{\mathrm{c}}};$$

$$\alpha_{71}=\alpha_{72}=\alpha_{77}=0,$$

$$\alpha_{73}=-2y_1^{-2}[y_1\mathrm{bei}'y_1-2(1-\nu)(\mathrm{ber}y_1+2y_1^{-1}\mathrm{ber}'y_1)],$$

$$\alpha_{74}=2y_1^{-2}[y_1\mathrm{ber}'y_1-2(1-\nu)(\mathrm{ber}y_1-2y_1^{-1}\mathrm{bei}'y_1)],$$

$$\alpha_{75}=-2y_1^{-2}[y_1\mathrm{kei}'y_1-2(1-\nu)(\mathrm{kei}y_1+2y_1^{-1}\mathrm{ker}'y_1)],$$

$$\alpha_{76}=2y_1^{-2}[y_1\mathrm{ker}'y_1-2(1-\nu)(\mathrm{ker}y_1-2y_1^{-1}\mathrm{kei}'y_1)],$$

$$\alpha_{78}=-\frac{2K_{\mathrm{c}}\eta_{\mathrm{c}}^2}{R_{\mathrm{c}}^2};$$

$$\alpha_{81}=\alpha_{82}=0,$$

$$\alpha_{83}=\frac{\cot\alpha}{R_{\mathrm{c}}}(\mathrm{ber}y_1-2y_1^{-1}\mathrm{bei}'y_1),$$

$$\alpha_{84}=\frac{\cot\alpha}{R_{\mathrm{c}}}(\mathrm{bei}y_1+2y_1^{-1}\mathrm{ber}'y_1),$$

$$\alpha_{85}=\frac{\cot\alpha}{R_{\mathrm{c}}}(\mathrm{ker}y_1-2y_1^{-1}\mathrm{kei}'y_1),$$

$$\alpha_{86}=\frac{\cot\alpha}{R_c}(\mathrm{kei}y_1+2y_1^{-1}\mathrm{ker}'y_1),$$

$$\alpha_{87}=\alpha_{88}=\frac{2K_c\eta_c^3}{R_c^3};$$

$$\omega_{11}=\frac{2-\nu}{2E}\left(\frac{R_a^2}{h_a}-\frac{s_2^2\cot\alpha\cos\alpha}{h_b}\right),$$

$$\omega_{12}=\frac{R_a^2}{Eh_a}\left(l_1-\frac{\nu\beta_a}{R_a^2}\right)-\frac{\cot\alpha\cos\alpha}{Eh_b}\left[8(1-\nu)\mu^{-4}R_a\cot\alpha+l_3s_2^2\left(1-\frac{\nu}{2}\right)\right.$$
$$\left.-s_2^3\sin\alpha\left(1-\frac{\nu}{3}\right)+\frac{\nu P}{2\gamma\pi\cos^2\alpha}\right],$$

$$\omega_{21}=-\frac{3s_2\cot^2\alpha}{2Eh_b},$$

$$\omega_{22}=-\frac{1}{E}\left[\frac{s_2\cot^2\alpha}{h_b}\left(\frac{3}{2}l_3-\frac{8}{3}s_2\sin\alpha\right)+\frac{R_a^2}{h_a}+\frac{P}{2\gamma\pi s_2h_b\sin^2\alpha}\right],$$

$$\omega_{31}=-\frac{3(1+\nu)K_b\cot^2\alpha}{2Eh_b},$$

$$\omega_{32}=-\frac{K_b}{Eh_b}\left\{\cot^2\alpha\left[\frac{3}{2}l_3(1+\nu)-\frac{8}{3}s_2(2+\nu)\sin\alpha\right]-\frac{(1-\nu)P}{2\gamma\pi s_2^2\sin^2\alpha}\right\},$$

$$\omega_{41}=\frac{1}{2}R_a\cot\alpha,$$

$$\omega_{42}=\frac{1}{\sin\alpha}\left(8\mu^{-4}\cot\alpha\cos\alpha-\frac{P}{2r\pi s^2}\right)+R_a\left(\frac{1}{2}l_3\cot\alpha-\frac{1}{3}R_a\right),$$

$$\omega_{51}=\frac{1}{E}\left(1-\frac{\nu}{2}\right)\left(\frac{s_1^2\cot\alpha\cos\alpha}{h_b}-\frac{R_c^2}{h_c}\right),$$

$$\omega_{52}=\frac{1}{E}\left\{\frac{\cot\alpha\cos\alpha}{h_b}\left[8(1-\nu)\mu^{-4}R_c\cot\alpha+l_3s_1^2\left(1-\frac{\nu}{2}\right)-s_1^3\sin\alpha\left(1-\frac{\nu}{3}\right)\right.\right.$$
$$\left.\left.+\frac{\nu P}{2\gamma\pi\cos^2\alpha}\right]-\frac{R_c^2}{h_c}\left(l_2-\nu\frac{\beta_c}{R_c^2}\right)\right\},$$

$$\omega_{61}=-\frac{3s_1\cot^2\alpha}{2Eh_b},$$

$$\omega_{62}=-\frac{1}{E}\left[\frac{s_1\cot^2\alpha}{h_b}\left(\frac{3}{2}l_3-\frac{8}{3}s_1\sin\alpha\right)+\frac{P}{2\gamma\pi h_bs_1\sin^2\alpha}+\frac{R_c^2}{h_c}\right],$$

$$\omega_{71}=\frac{3k_b\cot^2\alpha(1+\nu)}{2Eh_b},$$

$$\omega_{72}=\frac{K_2}{Eh_b}\left\{\cot^2\alpha\left[\frac{3}{2}l_3(1+\nu)-\frac{8}{3}s_1(2+\nu)\sin\alpha\right]-\frac{(1-\nu)P}{2\gamma\pi s_1^2\sin^2\alpha}\right\},$$

$$\omega_{81}=\frac{1}{2}R_c\cot\alpha,$$

$$\omega_{82}=\frac{1}{\sin\alpha}\left(8\mu^{-4}\cot\alpha\cos\alpha-\frac{P}{2\gamma\pi s_1}\right)+R_c\left(\frac{1}{2}l_3\cot\alpha-\frac{1}{3}R_c\right),$$

$$y_1=y(s_1),$$

$$y_2=y(s_2) \tag{45}$$

当减压塔的数据为已知时，我们可遵循下列顺序求解锥体、大筒体和小筒体的应力：首先从方程组（44）解出 $c_1 \cdots c_8$，然后根据前 3 小节中的有关公式求法向力和力矩值，最后，应用公式（16）和（36），便得锥体、大筒体和小筒体的应力值。

六、数值计算

已知原设计方案中减压塔的数据如下：

$$R_a=501.0\text{cm},\quad R_c=320.8\text{cm},\quad h_a=2.0\text{cm},$$

$$h_b=2.6\text{cm},\quad h_c=1.6\text{cm},\quad l_1=2184.07\text{cm},$$

$$l_2=2363.52\text{cm},\quad l_3=2684.32\text{cm},\quad l_4=289.28\text{cm},$$

$$s_1=453.680\text{cm},\quad s_2=708.521\text{cm},\quad \alpha=\pi/4$$

在立试水压试验时，

$$q_0=2\text{kg/cm}^2,\quad \gamma=0.001\text{kg/cm}^3$$

减压塔使用的材料为 A_3 钢，它在常温下的 E、ν 值，以及屈服限、强度限为

$$E=2.1\times10^6\text{kg/cm}^2,\quad \nu=0.26,$$

$$\sigma_s=2300\text{kg/cm}^2,\quad \sigma_b=3800\text{kg/cm}^2$$

将以上数据代入前面的公式中，便可求得大筒体、锥体和小筒体内、外表面（$z=\mp h/2$）的应力值和位移值，其数值结果见表 1～表 3。为了醒目起见，还在图 8～图 10 上绘制了应力曲线。由图表可见，大筒体的最大应力为轴向应力，发生在与锥体交界处的内表面上，其值为

$$\sigma_{x\max}=5581\text{kg/cm}^2 \tag{46}$$

锥体的最大应力为经向应力，发生在与大筒体交界处的内表面上，其值为

$$\sigma_{s\max}=3358\text{kg/cm}^2 \tag{47}$$

小筒体与锥体连接处的最大应力为环向应力，发生在壳体内表面上，其值为

$$\sigma_{\theta\max}=822.9\text{kg/cm}^2 \tag{48}$$

另外，还发现它们的位移值都是很小的，最大的仅有 3mm，说明刚度是足够的。

由国家规范知，安全系数为

$$n_s=1.6,\quad n_b=2.7 \tag{49}$$

则许用应力为

$$[\sigma]=1407\text{kg/cm}^2 \tag{50}$$

由此看出，小筒体与锥体交界区域的强度是足够的，而大筒体与锥体交界区域的强度是远远不够的。强度不够的区域为，大筒体下端轴向 15cm 内，锥体大头径向 10.5cm 内。为此，应将大筒体和锥体的厚度增加为

$$h_a=3.5\text{cm},\quad h_b=4.0\text{cm} \tag{51}$$

这时，大筒体、锥体和小筒体的应力、位移分布见表 1～表 3 中的建议方案（1），由表可见，强度基本满足。

实际上，式（49）中的值仅适用于薄膜应力计算情况。鉴于我国规范对壳体边缘的局部应力无具体规定，我们再按 ASME 锅炉及压力容器规范（第Ⅷ篇第二分册附录），取许用应力为

表 1　常减压装置减压塔中大筒体下端部的应力和法向位移值　　（应力单位：kg/cm^2，位移单位：mm）

ρ_a	原设计方案					建议方案（1）					建议方案（2）			
	σ_x		σ_θ		w	σ_x		σ_θ		w	σ_x		σ_θ	
	$z=\frac{h}{2}$	$z=-\frac{h}{2}$	$z=\frac{h}{2}$	$z=-\frac{h}{2}$		$z=\frac{h}{2}$	$z=-\frac{h}{2}$	$z=\frac{h}{2}$	$z=-\frac{h}{2}$		$z=\frac{h}{2}$	$z=-\frac{h}{2}$	$z=\frac{h}{2}$	$z=-\frac{h}{2}$
0	−4965	5581	−2643	98.42	−3.227	−1105	1457	−792.1	−126.0	−1.204	−1687	2098	−1117	−133.4
0.005	−4043	4659	−2439	−176.3	−3.311	−928.3	1281	−754.2	−179.9	−1.224	−1408	1819	−1056	−216.7
0.010	−3210	3826	−2206	−377.1	−3.273	−764.9	1117	−706.9	−217.6	−1.212	−1151	1562	−980.5	−275.3
0.015	−2464	3080	−1955	−513.9	−3.137	−614.5	966.8	−652.0	−240.8	−1.174	−915.6	1327	−895.4	−312.4
0.020	−1804	2420	−1694	−596.0	−2.923	−476.9	829.1	−591.1	−251.5	−1.114	−702.5	1113	−802.8	−330.7
0.025	−1226	1842	−1430	−632.0	−2.651	−351.8	704.0	−525.8	−251.3	−1.036	−510.4	921.4	−705.1	−332.8
0.030	−726.0	1342	−1168	−629.7	−2.335	−238.7	591.0	−457.5	−241.8	−0.9434	−338.8	749.7	−604.3	−321.3
0.035	−299.1	915.5	−912.2	−596.4	−1.991	−137.3	489.6	−387.4	−224.4	−0.8390	−186.5	597.5	−502.2	−298.3
0.040	60.28	556.2	−667.2	−538.3	−1.629	−46.96	399.2	−316.4	−200.4	−0.7258	−52.66	463.6	−400.3	−266.0
0.050	599.4	17.07	−218.5	−369.9	−0.8932	102.9	249.3	−175.7	−137.6	−0.4830	164.3	246.6	−201.9	−180.5
0.060	937.5	−321.0	165.0	−162.2	−0.1878	215.9	136.3	−40.96	−61.66	−0.2317	321.6	89.32	−16.81	−77.22
0.070	1119	−502.6	479.1	57.48	0.4489	297.2	55.02	83.79	20.82	0.01554	428.8	−17.80	149.9	33.84
0.080	1184	−567.8	725.5	267.0	0.9963	351.9	0.3359	196.1	104.7	0.2495	494.9	−83.92	295.8	145.3
0.090	1168	−551.3	910.1	463.1	1.447	384.8	−32.59	294.6	186.0	0.4640	528.5	−117.6	419.8	251.8
0.100	1098	−481.6	1041	630.1	1.802	400.4	−48.19	378.9	262.2	0.6554	537.4	−126.4	522.3	349.7
0.120	884.3	−267.8	1177	877.8	2.260	395.0	−42.79	506.6	392.7	0.9635				
0.140	661.7	−45.31	1204	1020	2.462	361.4	−9.114	586.5	490.1	1.175				
0.160	484.9	131.5	1177	1084	2.504	317.2	35.06	629.2	555.8	1.304				
0.180	367.6	248.8	1129	1098	2.465	273.6	78.67	645.6	594.9	1.371				
0.200	302.7	313.8	1084	1087	2.398	236.4	115.9	645.5	614.1	1.393				
0.300	295.7	320.7	1003	1010	2.210	166.6	185.7	587.7	592.6	1.299				
0.400	309.8	306.6	997.8	996.9	2.188	173.3	179.0	567.6	569.0	1.247				
0.500	308.2	308.2	985.5	985.5	2.160	176.5	175.7	562.6	562.4	1.233				

表 2　常减压装置减压塔中锥体的应力和水平位移值　（应力单位：kg/cm^2，位移单位：cm）

s	原设计方案					建议方案（1）					建议方案（2）			
	σ_s		σ_θ		Δ	σ_s		σ_θ		Δ	σ_s		σ_θ	
	$z=\frac{h}{2}$	$z=-\frac{h}{2}$	$z=\frac{h}{2}$	$z=-\frac{h}{2}$		$z=\frac{h}{2}$	$z=-\frac{h}{2}$	$z=\frac{h}{2}$	$z=-\frac{h}{2}$		$z=\frac{h}{2}$	$z=-\frac{h}{2}$	$z=\frac{h}{2}$	$z=-\frac{h}{2}$
453.680	−125.9	106.7	712.0	776.3	0.1141	−62.72	45.49	545.2	570.0	0.08552	−76.72	58.85	597.3	630.1
460	−82.02	83.50	721.4	770.7	0.1155	−57.07	55.21	530.0	557.4	0.08425				
470	−24.75	58.16	743.9	773.9	0.1194	−42.08	63.06	515.8	543.5	0.08339				
480	17.52	47.18	769.2	786.2	0.1243	−24.04	66.51	511.3	537.2	0.08384				
490	47.15	48.39	793.4	803.0	0.1297	−6.059	69.06	514.2	537.5	0.08541				
500	67.12	58.84	814.8	821.6	0.1350	10.33	72.57	522.3	543.5	0.08790				
510	80.05	75.90	833.3	840.8	0.1403	24.73	77.65	534.3	554.0	0.09116				
520	87.88	97.62	849.6	860.6	0.1455	37.56	84.04	549.4	568.1	0.09507				
530	92.12	122.5	865.0	881.5	0.1509	49.82	90.89	567.0	585.0	0.09953				
540	94.30	149.0	881.1	904.5	0.1566	62.90	96.89	587.1	603.6	0.1045				
550	96.49	175.3	900.1	930.6	0.1630	78.50	100.4	609.4	623.1	0.1098				
560	101.9	198.2	924.1	960.4	0.1703	98.42	99.68	633.6	641.9	0.1154				
570	115.3	213.1	955.1	993.2	0.1788	124.5	92.90	658.8	658.3	0.1210				
580	143.2	213.8	994.8	1027	0.1884	158.1	78.48	683.7	669.6	0.1261				
590	193.4	192.4	1044	1057	0.1987	200.4	55.35	705.8	672.7	0.1303				
600	274.3	140.7	1099	1077	0.2090	251.2	23.35	721.5	664.0	0.1327				
610	393.1	51.26	1157	1077	0.2175	308.8	−16.27	726.3	639.2	0.1324				
620	552.6	−79.19	1206	1043	0.2219	369.7	−60.30	714.4	594.5	0.1282				
630	747.4	−246.2	1230	961.0	0.2185	427.4	−102.9	679.3	526.4	0.1189				
640	959.1	−432.7	1207	817.3	0.2033	472.3	−135.3	614.0	433.2	0.1034				
650	1150	−603.5	1108	602.3	0.1716	491.1	−144.8	512.2	315.6	0.08074				

续表

s	原设计方案					建议方案（1）					建议方案（2）			
	σ_s		σ_θ		Δ	σ_s		σ_θ		Δ	σ_s		σ_θ	
	$z=\frac{h}{2}$	$z=-\frac{h}{2}$	$z=\frac{h}{2}$	$z=-\frac{h}{2}$		$z=\frac{h}{2}$	$z=-\frac{h}{2}$	$z=\frac{h}{2}$	$z=-\frac{h}{2}$		$z=\frac{h}{2}$	$z=-\frac{h}{2}$	$z=\frac{h}{2}$	$z=-\frac{h}{2}$
660	1261	−699.8	901.3	316.9	0.1191	466.9	−115.3	369.3	178.6	0.05073				
670	1202	−636.2	557.1	−20.61	0.04391	379.2	−27.05	183.6	32.66	0.01407	533.1	−125.3	252.0	22.67
680	859.6	−299.8	54.46	−365.7	−0.05229	205.3	142.3	−40.73	−104.8	−0.02701	308.2	94.42	−38.86	−157.5
684	613.0	−59.33	−191.8	−489.3	−0.09502	106.3	238.1	−139.1	−152.0	−0.04383	173.7	224.9	−170.0	−219.6
688	290.2	255.3	−462.3	−594.9	−0.1389	−11.65	352.0	−240.8	−191.6	−0.06033	9.897	383.6	−307.7	−271.8
692	−117.9	653.2	−754.4	−674.4	−0.1827	−150.0	485.6	−344.4	−221.4	−0.07608	−185.6	573.1	−450.0	−310.5
694	−356.8	886.2	−907.4	−701.5	−0.2041	−227.2	560.2	−396.4	−231.8	−0.08351	−296.0	680.1	−522.2	−323.6
696	−620.2	1143	−1064	−718.4	−0.2248	−309.9	640.2	−448.1	−238.8	−0.09056	−415.1	795.6	−594.7	−331.9
698	−909.2	1426	−1224	−723.9	−0.2447	−398.4	725.6	−499.5	−242.0	−0.09713	−543.1	919.9	−667.1	−334.8
700	−1225	1734	−1386	−716.4	−0.2634	−492.5	816.8	−550.0	−240.9	−0.1031	−680.4	1053	−738.9	−331.8
702	−1567	2069	−1549	−694.5	−0.2807	−592.6	913.7	−599.5	−235.3	−0.1085	−827.1	1196	−809.7	−322.3
704	−1938	2432	−1714	−656.6	−0.2962	−698.6	1016	−647.5	−224.7	−0.1132	−983.3	1348	−879.0	−305.5
706	−2338	2823	−1877	−601.0	−0.3095	−810.6	1125	−693.8	−208.7	−0.1170	−1149	1509	−946.4	−281.0
708	−2766	3244	−2037	−526.0	−0.3203	−928.6	1240	−737.9	−186.8	−0.1199	−1325	1680	−1011	−248.1
708.521	−2883	3358	−2079	−503.4	−0.3227	−960.4	1271	−748.9	−180.0	−0.1204	−1372	1727	−1027	−238.0

表 3　常减压装置减压塔中小筒体上端的应力和法向位移值　　(应力单位：kg/cm^2，位移单位：cm)

ρ_c	原设计方案					建议方案 (1)				
	σ_x		σ_θ		w	σ_x		σ_θ		w
	$z=\frac{h}{2}$	$z=-\frac{h}{2}$	$z=\frac{h}{2}$	$z=-\frac{h}{2}$		$z=\frac{h}{2}$	$z=-\frac{h}{2}$	$z=\frac{h}{2}$	$z=-\frac{h}{2}$	
0	−321.0	293.3	663.2	822.9	0.1141	−534.1	506.4	421.0	691.5	0.08552
0.01	−229.5	201.8	685.6	797.7	0.1138	−353.1	325.5	482.4	658.9	0.08772
0.02	−154.0	126.3	711.9	784.7	0.1149	−208.7	181.0	547.2	648.5	0.09188
0.03	−93.79	66.12	739.5	781.1	0.1167	−97.68	70.02	610.7	654.3	0.09717
0.04	−47.60	19.94	766.5	784.1	0.1190	−16.03	−11.63	669.8	670.9	0.1029
0.05	−13.69	−13.97	791.6	791.6	0.1215	40.71	−68.37	722.5	694.1	0.1088
0.06	9.812	−37.47	814.1	801.8	0.1240	77.00	−104.7	768.0	720.7	0.1143
0.07	24.79	−52.45	833.4	813.4	0.1263	97.09	−124.7	805.9	748.2	0.1193
0.08	33.01	−60.67	849.6	825.3	0.1285	104.8	−132.5	836.5	774.8	0.1236
0.09	36.10	−63.76	862.8	836.8	0.1304	103.6	−131.2	860.3	799.2	0.1273
0.10	35.45	−63.11	873.1	847.4	0.1320	96.13	−123.8	878.1	821.0	0.1303
0.12	27.46	−55.12	886.5	865.0	0.1343	71.49	−99.15	899.5	855.1	0.1346
0.14	15.93	−43.60	892.9	877.5	0.1358	43.78	−71.44	907.2	877.3	0.1369
0.16	4.859	−32.52	895.0	885.3	0.1365	19.74	−47.40	907.3	889.8	0.1378
0.18	−3.942	−23.72	894.8	889.7	0.1369	1.943	−29.60	903.9	895.7	0.1380
0.20	−10.01	−17.65	893.9	891.9	0.1369	−9.548	−18.11	899.7	897.4	0.1378
0.30	−15.41	−12.25	893.9	894.7	0.1372	−17.01	−10.65	893.3	894.9	0.1371
0.40	−13.80	−13.86	900.5	900.5	0.1381	−13.68	−13.98	900.4	900.3	0.1381

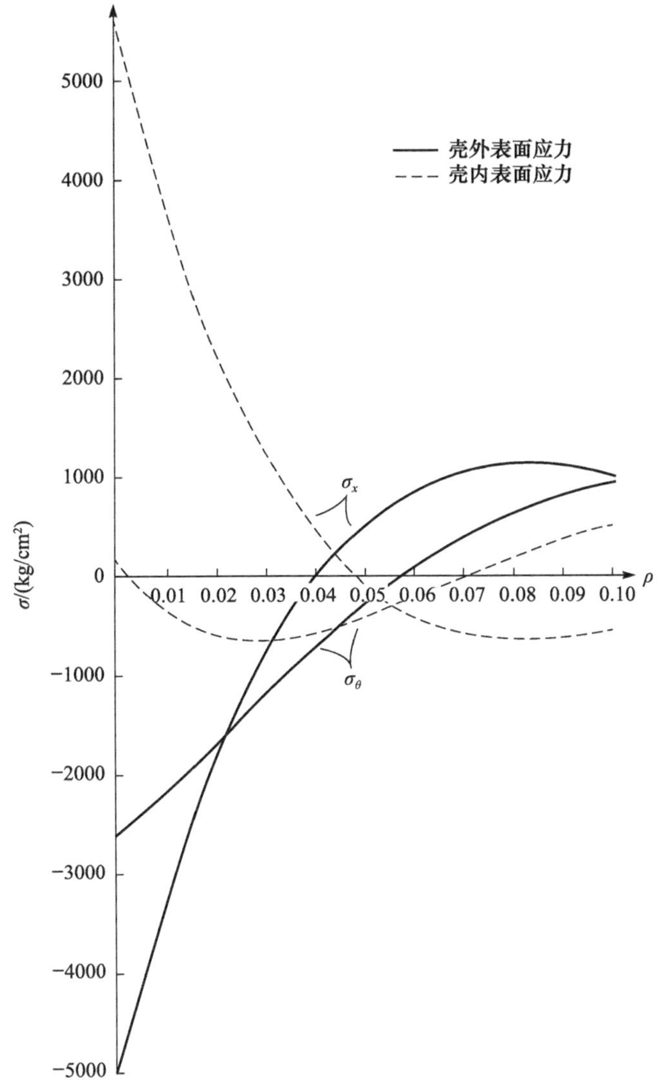

图 8　原设计方案中大筒体的应力

$$[\sigma]=\sigma_s, \tag{52}$$

这时，原设计方案中的大筒体和锥体交界区域的应力仍大于此许用应力，为此，将大筒体和锥体厚度增加为

$$h_a=3.0\text{cm}, \qquad h_b=3.5\text{cm} \tag{53}$$

此时的应力值见表 1 和表 2 中的建议方案（2），由表可知，强度能满足要求。

七、结束语

由上述计算结果看到，减压塔下端锥体及其邻近的大筒体有一局部区域的应力过高，超过了许用应力值，故我们给出了两个近似方案，供设计组参考。值得指出，由于未考虑交界区域的圆弧过渡，故我们所得的应力值与真实情况有一定误差。

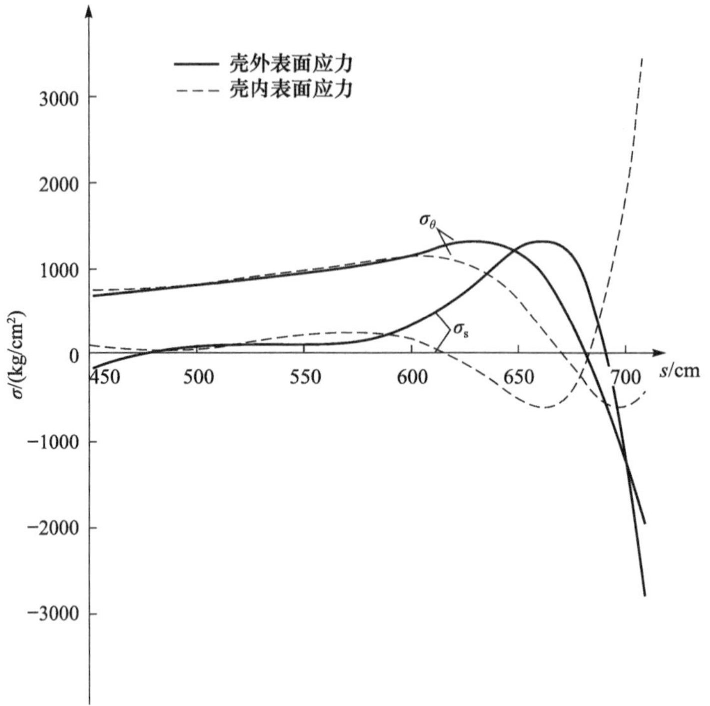

图 9　原设计方案中锥体的应力分布

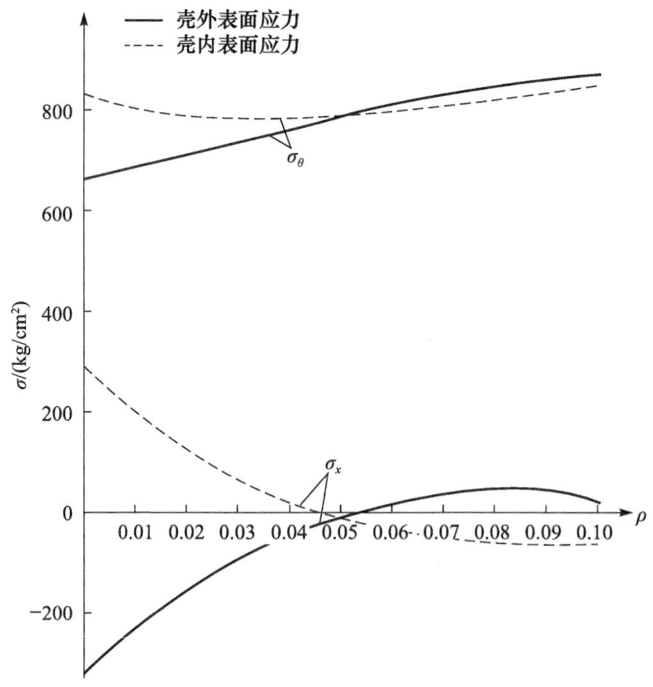

图 10　原设计方案中小筒体的应力分布

一般说来，我们的结果是偏大的。实际上，只要局部加强应力最大区域就行了，由于本节采用了三条简化措施，故进行进一步的实验和理论计算是必要的。

本节的数值计算工作是由兰州地震大队黄龙秋和邵世勤在 DJS-6 型电子计算机上完成的，特此致谢。

参 考 文 献

[1] 刘人怀. 薄壳理论讲义. 兰州大学数学力学系力学教研组，1974.

[2] 弗留盖 W. 壳体中的应力. 薛振东，龙驭球，叶耀先，等译. 北京：中国工业出版社，1965.

[3] Timoshenko S，Woinowsky-Krieger S，Theory of Plates and Shells. 2nd ed. New York：McGraw-Hill，1959.

在轴向压力与均匀外压力共同作用下薄壁截头圆锥形壳的稳定性①

圆锥形壳体在石油化学、航空、造船等工业部门中应用较广，因此，研究它的稳定性具有很大的实际意义。但由于它是一个很复杂的数学问题，所以许多人都只研究简单的承受单一载荷的锥壳稳定性。至于较复杂的，如在轴向压力与均匀外压力共同作用下的截锥壳稳定性问题，则研究较少。这种复合载荷作用可以有两种不同的受力方式，一种是沿壳体母线方向受压，外压力遍及壳外侧和上、下底面；另一种是沿壳体轴线方向受压，外压力仅作用于壳体外侧。现有的文献［1～3］针对前一种情况进行了研究。后一种情况虽然在实际情况中常常遇到，但却很少人探讨，可能是求解时较为困难的缘故。

1974 年底，我们在兰州石油化工机器厂二分厂结合生产实际，对一炼油塔器中的截锥壳进行了稳定性研究，它的受载情况与上面后一种情况相同。为了求解简单起见，我们将轴向压力进行分解，一个分量沿着壳体的母线方向，另一个分量垂直于壳体的轴线方向。由薄膜理论所知，壳体仅能承受前一个力，而后一个力需要一个加强环来承担。于是，我们的问题便转化为在壳体母线方向受压和在壳外侧承受均匀外压力的截锥壳稳定性问题。

与文献［1］一样，我们使用布勃诺夫-迦辽金方法求解，得到了可供工程应用的计算临界载荷的解析公式。借助电子计算机，对这一公式进行了数值计算，得出了那个炼油塔器中截锥壳的临界载荷值，可供该厂设计参考，由于缺乏实验资料验证，我们只好用单一载荷的实验数据和其他理论结果[4]来进行比较。数值比较表明，本节结果与实验值吻合较好。

附带指出，文献［5］所推导的稳定方程有商榷之处。本节所得的关于法向位移函数 W 的八阶常微分方程以及中面位移函数 U、V 与 W 的关系式并不比文献［5］采用简化运算措施后所得的复杂。

一、在轴向压力与均匀外压力共同作用下薄壁截头圆锥形壳的稳定方程

我们所采用的坐标系为 s、θ、z，它们分别表示圆锥壳的母线方向、圆周方向和壳体中面的外法线方向。与它们相应的位移为 u、v、w，其正方向示于图 1 中。

由文献［6］知，在轴向压力（在小端处等于 p）与均匀外压力 q 作用下薄壁截头圆锥形壳的稳定方程（亦称作 Donnell 型方程），在以法向位移 w 和应力函数 f 作未知函数的情况下，为

$$\nabla^4 f+\frac{Eh\cot\alpha}{s}\frac{\partial^2 w}{\partial s^2}=0,$$

① 本文原载《兰州大学学报》，1975，(2)：16-25.

$$D\,\nabla^4 w-N_{s_0}\frac{\partial^2 w}{\partial s^2}-\frac{N_{\theta_0}}{s}\left(\frac{1}{s}\frac{\partial^2 w}{\partial\theta_1^2}+\frac{\partial w}{\partial s}\right)-\frac{\cot\alpha}{s}\frac{\partial^2 f}{\partial s^2}=0, \tag{1}$$

其中 α 为锥壳的半顶角，h 为壳厚，E 为弹性模量，ν 为泊松比，D 为抗弯刚度，N_{s_0} 和 N_{θ_0} 分别为屈曲前的纵向和周向薄膜内力，

$$N_{s_0}=\frac{s_1}{s}\left(\frac{1}{2}qs_1\tan\alpha-\frac{p}{\cos\alpha}\right)-\frac{1}{2}qs\tan\alpha, \tag{2}$$

$$N_{\theta_0}=-qs\tan\alpha;$$

$$N_s=\frac{1}{s}\frac{\partial f}{\partial s}+\frac{1}{s^2}\frac{\partial^2 f}{\partial\theta_1^2},$$

$$N_\theta=\frac{\partial^2 f}{\partial s^2}, \tag{3}$$

$$N_{s\theta}=-\frac{\partial}{\partial s}\left(\frac{1}{s}\frac{\partial f}{\partial\theta_1}\right);$$

$$D=\frac{Eh^3}{12(1-\nu^2)},$$

$$\theta_1=\theta\sin\alpha,$$

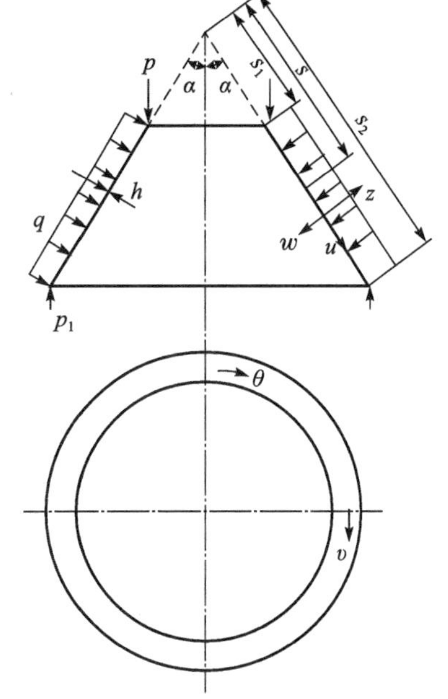

图 1　圆锥壳方向示意图

$$\nabla^4(\cdots)=\left(\frac{\partial^2}{\partial s^2}+\frac{1}{s}\frac{\partial}{\partial_s}+\frac{1}{s^2}\frac{\partial^2}{\partial\theta_1^2}\right)\left[\frac{\partial^2(\cdots)}{\partial s^2}+\frac{1}{s}\frac{\partial(\cdots)}{\partial s}+\frac{1}{s^2}\frac{\partial^2(\cdots)}{\partial\theta_1^2}\right] \tag{4}$$

附带指出，在采用一项简化运算措施后，文献［5］给出了稳定方程的另一种形式。这似无必要，为此，将本节所得结果列在下面，以供比较。

若以 u，v，w 为未知函数，并取成下面形式：

$$\begin{aligned}u&=U(s)\cos(n\theta),\\ v&=V(s)\sin(n\theta),\\ w&=W(s)\cos(n\theta),\qquad n=1,2,\cdots\end{aligned} \tag{5}$$

则经过一系列变换后，可将稳定方程化为关于 $W(s)$ 的八阶常微分方程：

$$(\Delta_s^4-\Delta_s^2)W+\frac{h^2\tan^2\alpha}{12(1-\nu^2)}L(s^2\Delta^4 W)=\frac{\tan^2\alpha}{Eh}L[N_{s_0}(\Delta_s^2-\Delta_s)W+N_{\theta_0}(\Delta_s-n_1^2)W], \tag{6}$$

其中

$$\begin{aligned}&\Delta_s(\cdots)=s\,\frac{\mathrm{d}(\cdots)}{\mathrm{d}s},\\ &\Delta_s^2(\cdots)=\Delta_s[\Delta_s(\cdots)],\\ &\Delta_s^4(\cdots)=\Delta_s^2[\Delta_s^2(\cdots)],\\ &\Delta^4(\cdots)=\left(\frac{\mathrm{d}^2}{\mathrm{d}s^2}+\frac{1}{s}\frac{\mathrm{d}}{\mathrm{d}s}-\frac{n_1^2}{s^2}\right)\left[\frac{\mathrm{d}^2(\cdots)}{\mathrm{d}s^2}+\frac{1}{s}\frac{\mathrm{d}(\cdots)}{\mathrm{d}s}-\frac{n_1^2}{s^2}(\cdots)\right],\\ &L(\cdots)=\Delta_s^4(\cdots)-2(1+n_1^2)\Delta_s^2(\cdots)+(1-n_1^2)^2(\cdots),\\ &n_1=\frac{n}{\sin\alpha}\end{aligned} \tag{7}$$

中面内的位移函数$U(s)$、$V(s)$ 与 $W(s)$ 之间的关系为

$$
\begin{aligned}
&L(U)=\cot\alpha[\nu\Delta_s^3W-\Delta_s^2W+(n_1^2-\nu)\Delta_sW+(1-n_1^2)W],\\
&L(V)=-n_1\cot\alpha[(2+\nu)\Delta_s^2W+(1-\nu)\Delta_sW+(1-n_1^2)W]
\end{aligned}
\tag{8}
$$

由文献［7］可知，方程（6）、(8）是正确的。

我们所讨论的边界条件如下：对于法向位移 w 而言是固定的，对于母线方向位移 u 而言是可自由滑动的，而且无周向应变，即

在 $s=s_1$ 和 $s=s_2$ 上，

$$
w=0,\quad \frac{\partial w}{\partial s}=0; \tag{9}
$$

$$
N_s=0,\quad \varepsilon_\theta=0 \tag{10}
$$

将式（10）应用应力函数表示后，成为

$$
\frac{\partial f}{\partial s}+\frac{1}{s}\frac{\partial^2 f}{\partial\theta_1^2}=0,\quad \frac{\partial^2 f}{\partial s^2}=0 \tag{11}
$$

二、稳定方程的求解

我们应用方程（1）和边界条件式（9）、式（11）来进行计算。首先，引进下列变换：

$$
\begin{aligned}
&f=F(s)\cos(n_1\theta),\\
&w=W(s)\cos(n_1\theta_1)
\end{aligned}
\tag{12}
$$

将式（12）代入式（1）和（9）、(11)，便有

$$
\begin{aligned}
&\Delta^4F+\frac{Eh\cot\alpha}{s}\frac{\mathrm{d}^2W}{\mathrm{d}s^2}=0,\\
&\Delta^4W-\frac{1}{D}\left\{\left[\frac{s_1}{s}\left(\frac{1}{2}qs_1\tan\alpha-\frac{p}{\cos\alpha}\right)-\frac{1}{2}qs\tan\alpha\right]\frac{\mathrm{d}^2W}{\mathrm{d}s^2}\right.\\
&\left.-q\tan\alpha\left(\frac{\mathrm{d}W}{\mathrm{d}s}-\frac{n_1^2}{s}W\right)+\frac{\cot\alpha}{s}\frac{\mathrm{d}^2F}{\mathrm{d}s^2}\right\}=0;
\end{aligned}
\tag{13}
$$

在 $s=s_1$ 和 $s=s_2$ 上，

$$
W=0,\quad \frac{\mathrm{d}W}{\mathrm{d}s}=0; \tag{14}
$$

$$
\frac{\mathrm{d}F}{\mathrm{d}s}-\frac{n_1^2}{s}F=0,\quad \frac{\mathrm{d}^2F}{\mathrm{d}s^2}=0 \tag{15}
$$

我们再令

$$
x=\ln\frac{s}{s_1}, \tag{16}
$$

将式（16）代入（13）和（14）、(15)，便有

$$
\begin{aligned}
&\frac{\mathrm{d}^4F}{\mathrm{d}x^4}-4\frac{\mathrm{d}^3F}{\mathrm{d}x^3}+2(2-n_1^2)\frac{\mathrm{d}^2F}{\mathrm{d}x^2}+4n_1^2\frac{\mathrm{d}F}{\mathrm{d}x}+n_1^2(n_1^2-4)F\\
&+Ehs_1\mathrm{e}^x\cot\alpha\left(\frac{\mathrm{d}^2W}{\mathrm{d}x^2}-\frac{\mathrm{d}W}{\mathrm{d}x}\right)=0,
\end{aligned}
\tag{17a}
$$

$$
\mathrm{e}^{-4x}\left\{\frac{\mathrm{d}^4W}{\mathrm{d}x^4}-4\frac{\mathrm{d}^3W}{\mathrm{d}x^3}+2(2-n_1^2)\frac{\mathrm{d}^2W}{\mathrm{d}x^2}+4n_1^2\frac{\mathrm{d}W}{\mathrm{d}x}\right.
$$

$$
\begin{aligned}
&+n_1^2(n_1^2-4)W-\frac{s_1\cot\alpha}{D}e^x\left(\frac{d^2F}{dx^2}-\frac{dF}{dx}\right)\\
&+\frac{ps_1^2}{D\cos\alpha}e^x\left(\frac{d^2W}{dx^2}-\frac{dW}{dx}\right)+\frac{qs_1^3\tan\alpha}{D}\left[\frac{1}{2}(e^{3x}\right.\\
&\left.\left.-e^x)\frac{d^2W}{dx^2}+\frac{1}{2}(e^{3x}+e^x)\frac{dW}{dx}-n_1^2e^{3x}W\right]\right\}=0;
\end{aligned}
\tag{17b}
$$

在 $x=0$ 和 $x=\ln\frac{s_2}{s_1}=x_2$ 上，

$$
W=0,\quad \frac{dW}{dx}=0;\tag{18}
$$

$$
\frac{dF}{dx}-n_1^2F=0,\quad \frac{d^2F}{dx^2}-\frac{dF}{dx}=0;\tag{19}
$$

为了解此边值问题，设

$$
W=A_0\sin^2(m_1x),\tag{20}
$$

其中 A_0 为待定常数，

$$
m_1=\frac{m\pi}{x_2},\quad m=1,2,\cdots\tag{21}
$$

显然，已预先满足了边界条件（18）。

将式（20）代入方程（17）中，有

$$
\begin{aligned}
&\frac{d^4F}{dx^4}-4\frac{d^3F}{dx^3}+2(2-n_1^2)\frac{d^2F}{dx^2}+4n_1^2\frac{dF}{dx}+n_1^2(n_1^2-4)F\\
&=EhA_0m_1s_1e^x\cot\alpha[\sin(2m_1x)-2m_1\cos(2m_1x)]
\end{aligned}
\tag{22}
$$

此方程是四阶常系数的非齐次常微分方程，其通解为

$$
\begin{aligned}
F=&C_1e^{(2+n_1)x}+C_2e^{(2-n_1)x}+C_3e^{n_1x}+C_4e^{-n_1x}\\
&+A_0\omega_{mn}e^x[2m_1\cos(2m_1x)-\sin(2m_1x)],
\end{aligned}
\tag{23}
$$

其中 C_1、C_2、C_3、C_4 为待定常数，

$$
\omega_{mn}=-\frac{Ehs_1m_1\cot\alpha}{(1+4m_1^2)^2-2n_1^2(1-4m_1^2)+n_1^4}\tag{24}
$$

将式（23）代入边界条件（19），便确定了 C_1、C_2、C_3、C_4：

$$
\begin{aligned}
&C_1=\frac{m_1A_0\omega_{mn}}{1+n_1}\left(l_{mn}+\frac{1}{2}N_{mn}\right),\\
&C_2=-\frac{m_1A_0\omega_{mn}}{1-n_1}l_{mn},\\
&C_3=-\frac{m_1A_0\omega_{mn}}{n_1(1-n_1)}[H_{mn}+K_{mn}+G_{mn}e^{-(1+n_1)x_2}+M_{mn}],\\
&C_4=-\frac{m_1A_0\omega_{mn}}{n_1(1+n_1)}[H_{mn}+G_{mn}e^{-(1-n_1)x_2}+M_{mn}],
\end{aligned}
\tag{25}
$$

其中

$$
\begin{aligned}
&l_{mn}=\frac{\lambda_1}{2\beta_1}N_{mn}e^{-x_2},\\
&N_{mn}=4m_1^2+n_1^2+1,
\end{aligned}
$$

$$
\begin{aligned}
&H_{mn}=\frac{\lambda_1}{2\beta_1^2}N_{mn}\left[(2-n_1)e^{(1+n_1)x_2}-(2+n_1)e^{(1-n_1)x_2}\right],\\
&K_{mn}=2(2-n_1)m_1^2+\frac{1}{2}(2-n_1-2n_1^2-n_1^3),\\
&G_{mn}=\frac{n_1\lambda_1}{\beta_1^2}N_{mn},\\
&M_{mn}=\frac{1}{\beta_1}\left[\left(1-\frac{n_1}{2}\right)N_{mn}e^{(2+n_1)x_2}-K_{mn}e^{n_1x_2}-2n_1^2e^{x_2}\right],\\
&\lambda_1=1-e^{(1+n_1)x_2},\\
&\beta_1=e^{n_1x_2}-e^{-n_1x_2}
\end{aligned}
\tag{26}
$$

将式（20）和（23）代入方程（17），按照布勃诺夫-迦辽金方法，将它乘以 $e^{2x}\sin^2(m_1x)\ \mathrm{d}x$，然后在 $0\sim x_2$ 区间积分所得方程，便求得决定临界载荷的特征方程：

$$
\phi_1=a_{mn}\phi_2+kb_{mn}+c_{mn},
\tag{27}
$$

其中

$$
\begin{aligned}
&\phi_1=\frac{p}{Eh},\\
&\phi_2=\frac{s_1q}{Eh};
\end{aligned}
\tag{28}
$$

$$
k=\frac{h^2}{12(1-\nu^2)s_1^2};
\tag{29}
$$

$$
\begin{aligned}
a_{mn}=&\sin\alpha\left[\frac{\beta_2}{64m_1^4}(1-e^{x_2})\left(2m_1^2+3n_1^2-\frac{4n_1^2}{1+4m_1^2}+\frac{n_1^2-2m_1^2}{1+16m_1^2}\right)+\frac{1}{2}\right],\\
b_{mn}=&\frac{\beta_2\cos\alpha}{32(1+m_1^2)(1+4m_1^2)}(1-e^{-2x_2})(16m_1^4+24m_1^2+8m_1^2n_1^2+3n_1^4-4n_1^2+8),\\
c_{mn}=&-\frac{\beta_2\cos\alpha\cot^2\alpha}{4[(1+4m_1^2)^2-2n_1^2(1-4m_1^2)+n_1^4]}\left\{\frac{1}{4m_1^2+(1+n_1)^2}[(2+n_1)\lambda_1c_{10}\right.\\
&\left.-n_1\lambda_4c_{40}]+\frac{1}{4m_1^2+(1-n_1)^2}[(2-n_1)\lambda_2c_{20}+n_1\lambda_3c_{30}]-\left(1+\frac{1}{4m_1^2}\right)x_2\right\};
\end{aligned}
\tag{30}
$$

$$
\begin{aligned}
&\lambda_2=1-e^{(1-n_1)x_2},\\
&\lambda_3=1-e^{-(1-n_1)x_2},\\
&\lambda_4=1-e^{-(1+n_1)x_2},\\
&\beta_2=\frac{1+16m_1^2}{1-e^{-x_2}},\\
&c_{10}=\frac{C_1}{m_1A_0\omega_{mn}},\\
&c_{20}=\frac{C_2}{m_1A_0\omega_{mn}},\\
&c_{30}=\frac{C_3}{m_1A_0\omega_{mn}},\\
&c_{40}=\frac{C_4}{m_1A_0\omega_{mn}}
\end{aligned}
\tag{31}
$$

在已知壳体数据的情况下，对于 m 与 n 的任意一组值，方程（27）就表示了无量纲载荷 ϕ_1 与 ϕ_2 两量间的某一线性关系。将 ϕ_1 与 ϕ_2 分别作为直角坐标系的纵、横坐标，则方程（27）能定出许多直线。对于一既定的横坐标，具有最小纵坐标的诸线段便形成了一折线，这折线可用来决定载荷的临界值。

三、实例计算并比较

实例 1　某炼油塔器中的截锥壳的简化模型便属于本节研究情况。已知数据如下：

$$\alpha=45°,\quad h=2.6\text{cm},\quad s_1=453.7\text{cm},$$
$$s_2=708.5\text{cm},\quad E=1.64\times10^8\text{kg/cm}^2,\quad \nu=0.26$$

将上述数据代入方程（27）中，在 DJS-6 型电子计算机上进行了数值计算。由计算可知，在塔器处于大气压力，即 $q=1\text{kg/cm}^2$ 的操作情况下，临界轴向力为

$$p^*=8113\text{kg/cm} \tag{32}$$

但是，塔器中的截锥壳实际承受的轴向力仅为

$$p=218.6\text{kg/cm} \tag{33}$$

故塔器中的截锥壳的稳定性是很好的。

实例 2　文献［4］给出了 $\alpha=14°30'$的截锥壳在轴向压力作用下的临界载荷实验值和理论计算值，它的边界条件与本文基本相同。对于这种单一载荷作用情况，它只是我们所研究问题的特殊情形。令

$$q=0\quad（或者\ \phi_2=0）$$

并应用下列数据

$$E=2.1\times10^6\text{kg/cm}^2,\quad \nu=0.3,$$

我们便可由方程（27）求得临界轴向载荷。为便于比较，我们将所获结果连同文献［4］的实验和理论值一起一并绘制在图 2 中。图中的 ρ_{cp}是相当圆柱壳体的半径，σ_s^*是临界纵向应力。由图可见，本节结果与实验值吻合较好。

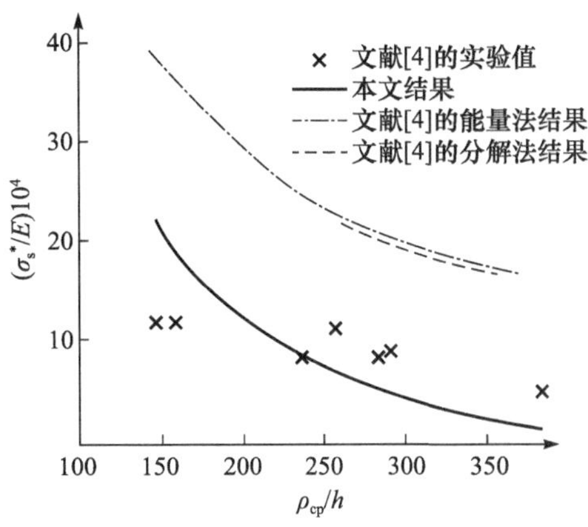

图 2　理论与实验之临界值的比较

本节的数值计算工作是由兰州地震大队邵世勤和黄龙秋同志完成的，特此致谢。

参 考 文 献

［1］　Муштари Х М，Саченков А В. Об устойчивости цилиндрнческих и конических оболочек кругового сечения при совместном действии осевого сжатия и внещнего пормалbного давления. ПММ，1954，18（6）：667-674.

［2］　Seide P. Calculations for the stability of thin conical frustums subjected to external uniform hydrostatic pressure and axial load. Jour. Aerospace Sci，1962，29（8）：951-955.

［3］　Weingarten V Ⅰ，Seide P. Elastic stability of thin-walled cylindrical and conical shells under combined external pressure and axial compression. AIAA Journal，1965，3（5）：913-920.

[4] 唐照千，沈亚鹏. 圆锥形薄壳在轴向压力作用下的稳定性. 力学学报，1966，(1)：70-78.

[5] 唐照千，沈亚鹏. 圆锥形壳体在侧向均压和均布液压作用下的稳定计算. 西安交通大学学报，1963，(4)：1-17.

[6] Seide P. A donnell type theory for asymmetrical bending and buckling of thin conical shells. Jour Appl Mech，1957，24 (4)：547-552.

[7] Seide P. Note on "stability equations for conical shells". Jour Aero Sci，1958，25 (5)：342.

加氢反应器顶部厚壁壳体的应力分析①

符号说明

x　圆柱壳的轴向坐标

φ　球壳的余纬度

z　壳体中曲面外法线方向的坐标

l_a　大法兰的轴向长度

N_φ　球壳的径向内力

N_θ　壳体以及双锥密封垫的环向内力

M_x　圆柱壳的轴向弯矩

Q_φ　球壳的横向力

H　球壳的水平力

M_0　大法兰上端的径向弯矩

w　壳体的法向位移，双锥密封垫的径向位移

v　球壳的径向位移

ψ　壳体经线微元的转角

σ_φ　球壳的经向应力

σ_θ　壳体的环向应力

n_s　屈服限安全系数

D　抗拉刚度

$C_1\cdots C_4$，$B_1\cdots B_4$，P，c　待定常数

$y_1\cdots y_4$，$y_1'\cdots y_4'$　超几何级数及其一阶导数

η，k，$a_{11}\cdots a_{88}$，$\omega_{11}\cdots\omega_{82}$，$\lambda_1$，$\lambda_2$，$\delta_1\cdots\delta_4$，$\mu_1\cdots\mu_4$，$\alpha_m$，$\beta_m$，$\overline{\alpha}_m$，$\overline{\beta}_m$　辅助量

右下角标 max 极大值

m，n　自然数列

r_δ　主螺栓中间细长部分的截面半径

s　主螺栓在单纯加载状态下的轴向力

b　双锥密封垫的高度

ρ　圆柱壳的无量纲轴向坐标

R　壳体的中曲面半径

h　壳体厚度

φ_0　球形封头孔边缘的 φ 值

N_x　圆柱壳的轴向内力

M_φ　球体的经向弯矩

M_θ　壳体的环向弯矩

Q_x　圆柱壳的横向力

N_0　大法兰上端的轴向力

Q_0　大法兰上端的横向力

u　圆柱壳的轴向位移

Δ　球壳的水平位移

q　均匀分布的内压力

σ_x　圆柱壳的轴向应力

σ_s、σ_b　壳体材料的屈服限和强度限

K　抗弯刚度

ν　泊松比

E、E_δ、E_k　壳体、主螺栓、双锥密封垫材料的弹性模量

y_α、y_β、$\overline{y}_\alpha$、$\overline{y}_\beta$　超几何方程的复数解

右下角标 a，b，c　分别代表大法兰、球形封头和筒体的量

R_δ　主螺栓至对称轴的距离

l_δ　主螺栓的有效工作长度

m_δ　主螺栓数目

s_0　主螺栓在预紧状态下的轴向力

d　双锥密封垫的外缘高度

α　双锥密封垫的锥角

ϵ_θ　圆柱壳的环向应变

T　在单纯加载状态下密封面上单位长度的反作用力

F　大法兰内横截面积

σ　密封比压

---

① 本文原载《化工炼油机械通讯》，1975，(6)：40-55.

ε_{x0}，$\varepsilon_{\theta0}$，k_x，k_θ　圆柱壳中曲面的伸长应变和曲率变化

r　双锥密封垫的半径

f，β　双锥密封垫与顶盖、筒体之间的摩擦系数和摩擦角

T_0　在预紧状态下密封面上单位长度的反作用力

G　自紧系数

ε_x　圆柱壳的轴向应变

20世纪70年代以来，在石油炼制工业中，由于加氢裂化的新工艺优点甚多，故在国内外受到极大重视。实现加氢裂化的关键设备是加氢反应器，因为它属于大型高温高压厚壁容器，操作条件相当恶劣，所以在设计时特别慎重。最近，某厂设计了一台我国最大直径的加氢反应器，在顶部 ϕ2m 的球形封头中央开了一个 ϕ800mm 的大孔。欲研究此大孔边缘以及顶部球形封头与筒体交界区域的应力状态，仍使用以往常用的薄壳理论，显然是极不合理的。鉴于此，有必要建立精确度较高的公式来进行分析计算。

作为弹性力学空间问题之一的厚壳理论，由于数学上的困难，还很少有人触及。在我国社会主义建设中，随着高压容器工业的飞跃发展，急需建立既简单又精确的计算厚壁壳体的应力公式。在文献［1］中，我们在薄壳理论使用的基本假定基础上，对厚壁球壳开小孔所引起的应力集中进行过初步探讨，注意了壳体厚度因素，改进了平衡方程，同时使用了文献［3］的应力公式，这样，所获的理论值与实验值符合很好。在文献［2］中，我们继续了文献［1］的工作，进行了全面研究，所得的理论值较由文献［3］以及文献［4］的精确薄球壳理论公式所算的值更接近实验数据。在本节中，除应用文献［2］的公式外，还仿照文献［2］的方法，建立了工程实用的承受内压力作用的厚壁圆柱壳的近似计算公式。应用这些公式，我们得到了加氢反应器顶部球形封头大孔边缘以及球形封头和筒体交界区域的应力值，可供工程设计时参考。此外，我们还用文献［3］的厚壁球壳理论和相当于文献［3］精确度的厚壁圆柱壳理论与以文献［4］的精确薄球壳理论和相当于文献［4］精确度的薄圆柱壳理论进行了比较计算，所获结果与本节结果相比，误差是比较大的。实际上，文献［3］仅给出了扁厚壁球壳的内力公式，不适用于本节问题的计算。但为了比较，我们按其精神建立了适合于本节问题的计算公式，仍称作文献［3］的理论。最后，我们以承受内压力的厚壁圆筒为特例，进行了对比计算，本节值极接近精确值。

一、加氢反应器顶部壳体的计算模型

设计中的加氢反应器顶部结构比较复杂，它包括顶盖、双锥密封垫、主螺栓、大法兰、球形封头和筒体等构件。欲计算球形封头孔边缘以及球形封头和筒体交界区域的应力状态，我们作如下简化：①虽然球形封头和筒体是双层结构，但由于材质相同，故仍用单层壳体理论进行研究；②由于筒体很长，故作半无限长圆柱壳处理；③由于顶盖等构件离孔缘较远，故粗略计及它们的影响，建立近似的大法兰上端的边界条件；④忽略反应器的自重作用。在上述简化后，加氢反应器顶部壳体的简化计算模型便如

图 1 所示。

下面，我们给出 M_0、Q_0 和 N_0 的公式。按照力的等效原理，我们将密封面上和主螺栓的作用力（$T+T_0$）、$f(T-T_0)$、$m_\delta(S+S_0)$ 转移到大法兰的上端中曲面上（图 2）。利用文献 [5] 所得的公式，便有

$$M_0=\mu_1 q+\mu_2 T_0,\qquad Q_0=\mu_3 q+\mu_4 T_0,\qquad N_0=\frac{Fq}{2\pi R},\tag{1}$$

其中

$$\begin{aligned}
\mu_1&=\frac{F}{2\pi R}\left\{G(R-R_\delta)-(G-1)\left[R-r+\frac{b-d}{4}\tan(\alpha+\beta)\right]\right\},\\
\mu_2&=\frac{r\cos(\alpha-\beta)}{R\cos\beta}\left[r-R_\delta-\frac{b-d}{4}\tan(\alpha-\beta)\right],\\
\mu_3&=\frac{F}{2\pi R}(G-1)\tan(\alpha+\beta),\\
\mu_4&=\frac{r\sin(\alpha-\beta)}{R\cos\beta}
\end{aligned}\tag{2}$$

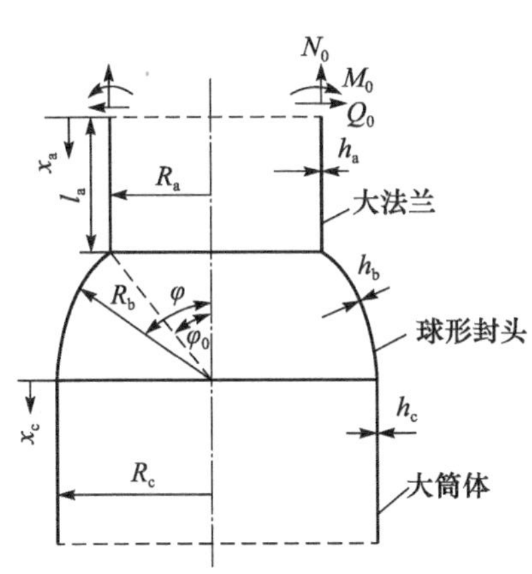

图 1　加氢反应器顶部壳体的简化计算模型

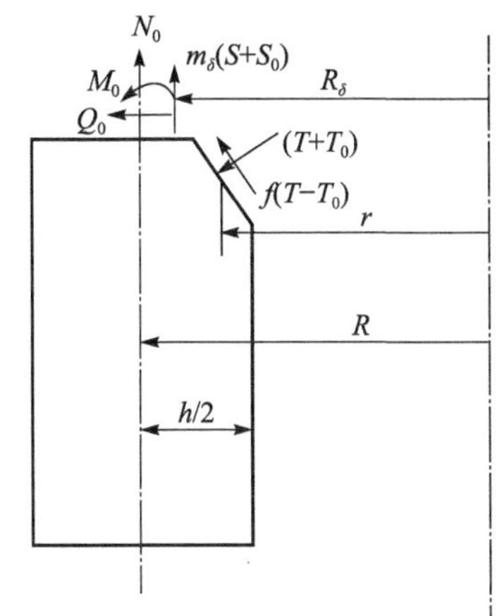

图 2　大法兰上端边界条件的建立

二、厚壁球壳的基本方程及其解

加氢反应器球形封头的厚度与中曲面半径的比值高达 0.148，属于厚壁球壳理论的研究范围。在厚壁球壳轴对称变形的近似理论中，以横向力 Q_φ 作未知数的基本方程为[1,2]

$$\frac{d^3 Q_\varphi}{d\varphi^2}+\cot\varphi\frac{dQ_\varphi}{d\varphi}-Q_\varphi\cot^2\varphi\pm 2i\eta^2 Q_\varphi=0,\tag{3}$$

其中

$$\eta^2=\sqrt{3(1-\nu^2)\frac{R^2}{h^2}-\frac{\nu^2}{4}},\tag{4}$$

求得 Q_φ 后，便可按以下公式求内力和转角

$$N_\varphi=-Q_\varphi\cot\varphi+\frac{1}{2R}\left(R-\frac{h}{2}\right)^2 q-\frac{P}{2\pi\sin^2\varphi},$$

$$
\begin{aligned}
&N_\theta=-\frac{\mathrm{d}Q_\varphi}{\mathrm{d}\varphi}+\frac{1}{2R}\left(R-\frac{h}{2}\right)^2 q+\frac{P}{2\pi R\sin^2\varphi},\\
&M_\varphi=\frac{K}{R}\left(\frac{\mathrm{d}\psi}{\mathrm{d}\varphi}+\nu\psi\cot\varphi\right),\quad M_\theta=\frac{K}{R}\left(\psi\cot\varphi+\nu\frac{\mathrm{d}\psi}{\mathrm{d}\varphi}\right),\\
&\psi=-\frac{1}{Eh}\left[\frac{\mathrm{d}^2Q_\varphi}{\mathrm{d}\varphi^2}+\cot\varphi\frac{\mathrm{d}Q_\varphi}{\mathrm{d}\varphi}-(\cot^2\varphi-\nu)Q_\varphi\right],
\end{aligned}
\tag{5}
$$

其中 P 为积分常数，

$$
K=\frac{Eh^3}{12(1-\nu^2)} \tag{6}
$$

众所周知，变系数二阶常微分方程（3）的精确解法是化为超几何方程进行的，由于计算量过大，过去人们都不愿用它，而宁愿简化方程（3）来寻求解。但在厚壁球壳范围内，一般是行不通的。我们借助电子计算机，克服了计算量过大的困难，获得了方程（3）的精确解。

将方程（3）化为超几何方程后，得通解[2,4]：

$$
Q_\varphi=(c_1y_1+c_2y_2+c_3y_3+c_4y_4)\sin\varphi, \tag{7}
$$

其中 $c_1\cdots c_4$ 为待定常数，$y_1\cdots y_4$ 为超几何级数，

$$
\begin{aligned}
&y_1=\frac{1}{2}(y_\alpha+\overline{y}_\alpha),\\
&y_2=\frac{1}{2}(y_\beta+\overline{y}_\beta),\\
&y_3=\frac{\mathrm{i}}{2}(y_\alpha-\overline{y}_\alpha),\\
&y_4=\frac{\mathrm{i}}{2}(y_\beta-\overline{y}_\beta),\\
&y_\alpha=1+\sum_{n=1}^{\infty}\frac{1}{(2n)!}\left(\prod_{m=1}^{n}\alpha_m\right)\cos^{2n}\varphi,\\
&y_\beta=\cos\alpha\left[1+\sum_{n=1}^{\infty}\frac{1}{(2n+1)!}\left(\prod_{m=1}^{n}\beta_m\right)\cos^{2n}\varphi\right],\\
&\overline{y}_\alpha=1+\sum_{n=1}^{\infty}\frac{1}{(2n)!}\left(\prod_{m=1}^{n}\overline{\alpha}_m\right)\cos^{2n}\varphi,\\
&\overline{y}_\beta=\cos\varphi\left[1+\sum_{n=1}^{\infty}\frac{1}{(2n+1)!}\left(\prod_{m=1}^{n}\overline{\beta}_m\right)\cos^{2n}\varphi\right],\\
&\alpha_m=\frac{(4m-1)^2-5}{4}-2\mathrm{i}\eta^2,\\
&\beta_m=\frac{(4m+1)^2-5}{4}-2\mathrm{i}\eta^2,\\
&\overline{\alpha}_m=\frac{(4m-1)^2-5}{4}+2\mathrm{i}\eta^2,\\
&\overline{\beta}_m=\frac{(4m+1)^2-5}{4}+2\mathrm{i}\eta^2
\end{aligned}
\tag{8}
$$

应用解（7），公式（5）成为

$$
\begin{aligned}
N_\varphi &= -(c_1y_1+c_2y_2+c_3y_3+c_4y_4)\cos\varphi+\frac{1}{2R}\left(R-\frac{h}{2}\right)^2 q-\frac{P}{2\pi R\sin^2\varphi},\\
N_\theta &= N_\varphi-(c_1\dot{y}_1+c_2\dot{y}_2+c_3\dot{y}_3+c_4\dot{y}_4)\sin\varphi+\frac{P}{\pi R\sin^2\varphi},\\
M_\varphi &= \frac{R}{\delta_1^4}[c_1(2\eta^2\dot{y}_3-\nu\dot{y}_1)+c_2(2\eta^2\dot{y}_4-\nu\dot{y}_2)-c_3(2\eta^2\dot{y}_1+\nu\dot{y}_3)\\
&\quad -c_4(2\eta^2\dot{y}_2+\nu\dot{y}_4)]\sin\varphi+\frac{Rk}{1-\nu}[c_1(2\eta^2y_3-\nu y_1)+c_2(2\eta^2y_4-\nu y_2)\\
&\quad -c_3(2\eta^2y_1+\nu y_3)-c_4(2\eta^2y_2+\nu y_4)]\cos\varphi,\\
M_\theta &= \frac{\nu R}{\delta_1^4}[c_1(2\eta^2\dot{y}_3-\nu\dot{y}_1)+c_2(2\eta^2\dot{y}_4-\nu\dot{y}_2)-c_3(2\eta^2\dot{y}_1+\nu\dot{y}_3)\\
&\quad -c_4(2\eta^2\dot{y}_2+\nu\dot{y}_4)]\sin\varphi+\frac{Rk}{1-\nu}[c_1(2\eta^2y_3-\nu y_1)+c_2(2\eta^2y_4-\nu y_2)\\
&\quad -c_3(2\eta^2y_1+\nu y_3)-c_4(2\eta^2y_2+\nu y_4)]\cos\varphi,\\
\psi &= \frac{1}{Eh}[c_1(2\eta^2y_3-\nu y_1)+c_2(2\eta^2y_4-\nu y_2)-c_3(2\eta^2y_1+\nu y_3)\\
&\quad -c_4(2\eta^2y_2+\nu y_4)]\sin\varphi,
\end{aligned}
\tag{9}
$$

其中 $\dot{y}_1\cdots\dot{y}_4$ 分别是由 $y_1\cdots y_4$ 对 φ 求导数得到的，

$$k=\frac{h^2}{12R^2},\quad \delta_1^4=\frac{1-\nu^2}{k} \tag{10}$$

为了建立球形封头与大法兰、球形封头与筒体之间的连续条件，还需要给出球壳水平位移 Δ 和水平力 H 的公式。由图 3，球壳中曲面上任一点的水平位移为

$$\Delta=w\sin\varphi+v\cos\varphi \tag{11}$$

按照文献［1］的式（13b），还有

$$w\sin\varphi+v\cos\varphi=\frac{R\sin\varphi}{Eh}(N_\theta-\nu N_\varphi) \tag{12}$$

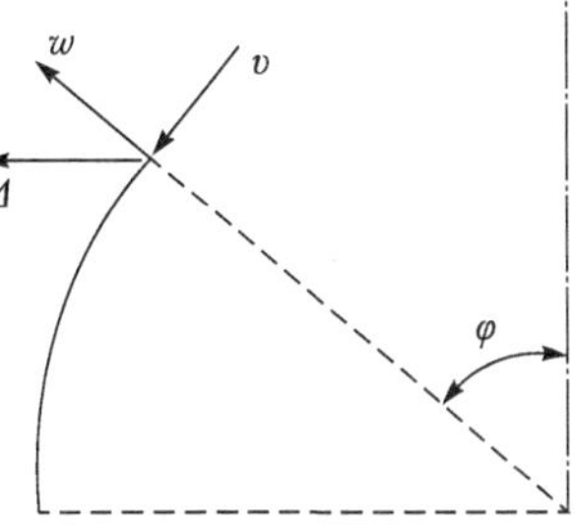

图 3　球壳的水平位移

应用式（9）中的前二式，得到

$$
\begin{aligned}
\Delta=-\frac{R\sin\varphi}{Eh}\Big[&(1-\nu)(c_1y_1+c_2y_2+c_3y_3+c_4y_4)\cos\varphi\\
&+(c_1\dot{y}_1+c_2\dot{y}_2+c_3\dot{y}_3+c_4\dot{y}_4)\sin\varphi\\
&-\frac{1-\nu}{2R}\left(R-\frac{h}{2}\right)^2 q-\frac{(2+\nu)P}{2\pi R\sin^2\varphi}\Big]
\end{aligned}
\tag{13}
$$

由图 4，球壳中曲面上任一点的水平力为

$$H=Q_\varphi\sin\varphi-N_\varphi\cos\varphi \tag{14}$$

将式（7）和（9）的第一式代入，得

$$H=c_1y_1+c_2y_2+c_3y_3+c_4y_4-\frac{\cos\varphi}{2R}\left(R-\frac{h}{2}\right)^2 q+\frac{P\cos\varphi}{2\pi R\sin^2\varphi} \tag{15}$$

求得内力后，我们通过下面公式计算厚壁球壳的径向和环向应力[3]：

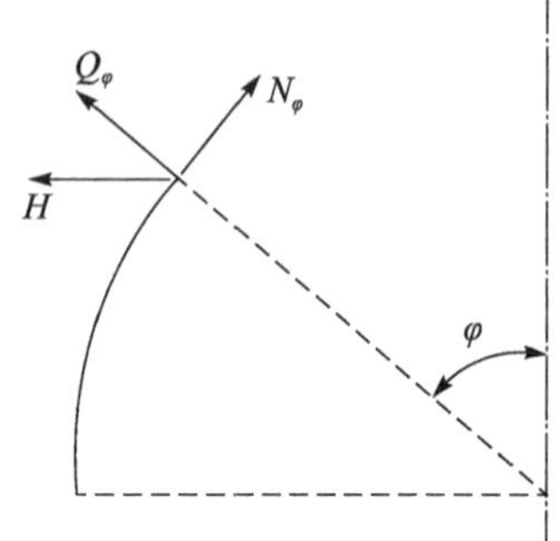

图 4　球壳的水平力

$$\sigma_\varphi=\frac{1}{1+z/R}\left(\frac{N_\varphi}{h}-\frac{12zM_\varphi}{h^3}\right),\quad \sigma_\theta=\frac{1}{1+z/R}\left(\frac{N_\theta}{h}-\frac{12zM_\theta}{h^3}\right) \tag{16}$$

三、厚壁圆柱壳的基本方程及其解

加氢反应器的大法兰和筒体的厚度与中曲面半径的比值分别高达 0.40、0.148。故都属于厚壁圆柱壳理论的研究范围。下面，我们来建立承受内压力作用的厚壁圆柱壳近似理论。

由文献［4］可知，在内压力 q 的作用下，薄壁圆柱壳轴对称变形理论的平衡方程和弹性定律为

$$\frac{\mathrm{d}N_x}{\mathrm{d}x}=0, \tag{17a}$$

$$\frac{\mathrm{d}M_x}{\mathrm{d}x}-Q_x=0, \tag{17b}$$

$$R\frac{\mathrm{d}Q_x}{\mathrm{d}x}+N_\theta-qR=0; \tag{17c}$$

$$\begin{aligned} N_\theta&=D\left(\frac{w}{R}+\nu\frac{\mathrm{d}u}{\mathrm{d}x}\right)+\frac{K}{R^3}w,\\ N_x&=D\left(\frac{\mathrm{d}u}{\mathrm{d}x}+\frac{\nu}{R}w\right)-\frac{K}{R}\frac{\mathrm{d}^2w}{\mathrm{d}x^2},\\ M_\theta&=K\left(\frac{w}{R^2}+\nu\frac{\mathrm{d}^2w}{\mathrm{d}x^2}\right),\\ M_x&=K\left(\frac{\mathrm{d}^2w}{\mathrm{d}x^2}-\frac{1}{R}\frac{\mathrm{d}u}{\mathrm{d}x}\right), \end{aligned} \tag{18}$$

其中内力和位移的正方向如图 5 所示，

$$D=\frac{Eh}{1-\nu^2} \tag{19}$$

采用文献［1］和［2］中处理厚壁球壳相类似的办法，在建立微元体 z 方向力的平衡方程时，注意 q 的作用面是壳体内表面的事实，这样，方程（17c）改写为

$$R\frac{\mathrm{d}Q_x}{\mathrm{d}x}+N_\theta-q\left(R-\frac{h}{2}\right)=0 \tag{20}$$

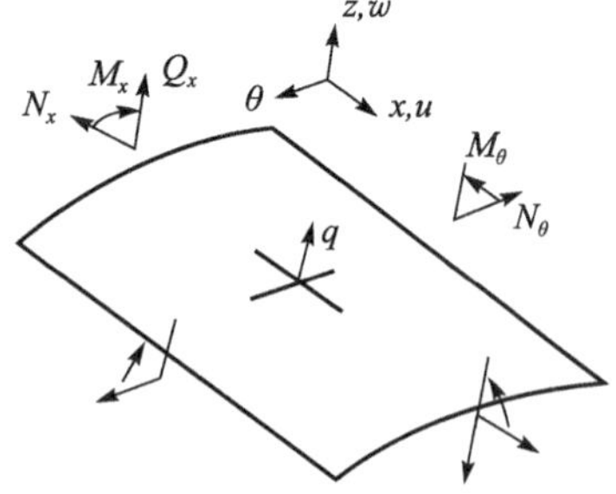

图 5　内力和位移的正方向

于是，平衡方程（17a，b）、（20）和弹性定律（18）组成了在内压力 q 作用下厚壁圆柱壳轴对称变形近似理论的方程组。七个方程包含了七个未知量，故是可解的。下面，我们来化简这方程组。

首先，积分方程（17a），有

$$N_x=C, \tag{21}$$

其中 C 为积分常数。

通过圆柱壳某一横截面轴向力的平衡，可得

$$N_x=\frac{1}{2R}\left(R-\frac{h}{2}\right)^2 q \tag{22}$$

则

$$C=\frac{1}{2R}\left(R-\frac{h}{2}\right)^2 q \tag{23}$$

其次，将方程（17b）、（20）和（18）的首末二式综合为一个方程。将方程（17b）对 x 求导数，得

$$\frac{\mathrm{d}Q_x}{\mathrm{d}x}=\frac{\mathrm{d}^2 M}{\mathrm{d}x^2} \tag{24}$$

然后，将此式代入方程（20）中，则有

$$R\frac{\mathrm{d}^2 M_x}{\mathrm{d}x^2}+N_\theta-q\left(R-\frac{h}{2}\right)=0 \tag{25}$$

最后，将式（18）的首末二式代入上式中，便得

$$R^2 k\frac{\mathrm{d}^4 w}{\mathrm{d}x^4}+\frac{1+k}{R^2}w-Rk\frac{\mathrm{d}^3 u}{\mathrm{d}x^3}+\frac{\nu}{R}\frac{\mathrm{d}u}{\mathrm{d}x}=\frac{1}{RD}\left(R-\frac{h}{2}\right)q \tag{26}$$

另外，将式（18）中的第二式代入式（21）中，得

$$\frac{\mathrm{d}u}{\mathrm{d}x}=\frac{1}{D}\left[\frac{K}{R}\frac{\mathrm{d}^2 w}{\mathrm{d}x^2}+\frac{1}{2R}\left(R-\frac{h}{2}\right)^2 q\right]-\frac{\nu}{R}w \tag{27}$$

将此式代入方程（26）中，在略去（$1+k$）中的小量 k 后（以下仍用此近似，不再叙述），我们得到关于法向位移 w 的方程：

$$R^2 k\frac{\mathrm{d}^4 w}{\mathrm{d}x^4}+2\nu k\frac{\mathrm{d}^2 w}{\mathrm{d}x^2}+\frac{1-\nu^2}{R^2}w=\frac{1}{DR}\left(R-\frac{h}{2}\right)\left[1-\frac{\nu}{2R}\left(R-\frac{h}{2}\right)\right]q \tag{28}$$

引入新自变量

$$\rho=\frac{x}{R}, \tag{29}$$

方程（28）成为

$$\frac{\mathrm{d}^4 w}{\mathrm{d}\rho^4}+2\nu\frac{\mathrm{d}^2 w}{\mathrm{d}\rho^2}+\delta_1^4 w=\delta_2 q, \tag{30}$$

其中

$$\delta_2=\frac{R}{kD}\left(R-\frac{h}{2}\right)\left[1-\frac{\nu}{2R}\left(R-\frac{h}{2}\right)\right] \tag{31}$$

方程（30）是常系数四阶非齐次的常微分方程，其通解易于求得，为

$$w=\mathrm{e}^{-\lambda_1\rho}[B_1\cos(\lambda_2\rho)+B_2\sin(\lambda_2\rho)]+\mathrm{e}^{\lambda_1\rho}[B_3\cos(\lambda_2\rho)+B_4\sin(\lambda_2\rho)]+\frac{\delta_2}{\delta_1^4}q, \tag{32}$$

其中 $B_1\cdots B_4$ 为待定常数，

$$\lambda_1=\delta_1\cos\left[\frac{1}{2}\arctan\left(\frac{\delta_1^2}{\nu}\right)\right],\qquad \lambda_2=\delta_1\sin\left[\frac{1}{2}\arctan\left(\frac{\delta_1^2}{\nu}\right)\right] \tag{33}$$

应用解（32），由式（18）和（17）得

$$
\begin{aligned}
N_\theta=&\frac{D}{R}\left(\mathrm{e}^{-\lambda_1\rho}\{[1-\nu^2+\nu k(\lambda_1^2-\lambda_2^2)][B_1\cos(\lambda_2\rho)+B_2\sin(\lambda_2\rho)]\right.\\
&+2\nu\lambda_1\lambda_2 k[B_1\sin(\lambda_2\rho)-B_2\cos(\lambda_2\rho)]\}+\mathrm{e}^{\lambda_1\rho}\{[1-\nu^2+\nu k(\lambda_1^2-\lambda_2^2)][B_3\cos(\lambda_2\rho)\\
&\left.+B_4\sin(\lambda_2\rho)]-2\nu\lambda_1\lambda_2 k[B_3\sin(\lambda_2\rho)-B_4\cos(\lambda_2\rho)]\}\right)\\
&+\left(R-\frac{h}{2}\right)q,\\
M_x=&\frac{K}{R^2}\left(\mathrm{e}^{-\lambda_1\rho}\{(\lambda_1^2-\lambda_2^2+\nu)[B_1\cos(\lambda_2\rho)+B_2\sin(\lambda_2\rho)]\right.\\
&+2\lambda_1\lambda_1[B_1\sin(\lambda_2\rho)-B_2\cos(\lambda_2\rho)]\}+\mathrm{e}^{\lambda_1\rho}\{(\lambda_1^2-\lambda_2^2+\nu)[B_3\cos(\lambda_2\rho)\\
&\left.+B_4\sin(\lambda_2\rho)]-2\lambda_1\lambda_2[B_3\sin(\lambda_2\rho)-B_4\cos(\lambda_2\rho)]\}\right)+\delta_3 q,\\
M_\theta=&\frac{K}{R^2}\left(\mathrm{e}^{-\lambda_1\rho}\{[1+\nu(\lambda_1^2-\lambda_2^2)][B_1\cos(\lambda_2\rho)+B_2\sin(\lambda_2\rho)]\right.\\
&+2\nu\lambda_1\lambda_2[B_1\sin(\lambda_2\rho)-B_2\cos(\lambda_2\rho)]\}+\mathrm{e}^{\lambda_1\rho}\{[1+\nu(\lambda_1^2-\lambda_2^2)][B_3\cos(\lambda_2\rho)\\
&\left.+B_4\sin(\lambda_2\rho)]-2\nu\lambda_1\lambda_2[B_3\sin(\lambda_2\rho)-B_4\cos(\lambda_2\rho)]\}\right)+\delta_4 q,\\
Q_x=&-\frac{K}{R^3}\left(\mathrm{e}^{-\lambda_1\rho}\{\lambda_1(\lambda_1^2-3\lambda_2^2+\nu)[B_1\cos(\lambda_2\rho)+B_2\sin(\lambda_2\rho)]\right.\\
&+\lambda_2(3\lambda_1^2-\lambda_2^2+\nu)[B_1\sin(\lambda_2\rho)-B_2\cos(\lambda_2\rho)]\}\\
&-\mathrm{e}^{-\lambda_1\rho}\{\lambda_1(\lambda_1^2-3\lambda_2^2+\nu)[B_3\cos(\lambda_2\rho)+B_4\sin(\lambda_2\rho)]\\
&\left.-\lambda_2(3\lambda_1^2-\lambda_2^2+\nu)[B_3\sin(\lambda_2\rho)-B_4\cos(\lambda_2\rho)]\}\right),
\end{aligned}
\tag{34}
$$

其中

$$
\begin{aligned}
\delta_3&=\frac{R}{\delta_1^4}\left(R-\frac{h}{2}\right)\left[\nu-\frac{1}{2R}\left(R-\frac{h}{2}\right)\right],\\
\delta_4&=\frac{R}{\delta_1^4}\left(R-\frac{h}{2}\right)\left[1-\frac{\nu}{2R}\left(R-\frac{h}{2}\right)\right]
\end{aligned}
\tag{35}
$$

另外，对于转角，还有公式

$$
\begin{aligned}
\psi=&\frac{\mathrm{d}w}{\mathrm{d}x}\\
=&-\frac{1}{R}\left(\mathrm{e}^{-\lambda_1\rho}\{\lambda_1[B_1\cos(\lambda_2\rho)+B_2\sin(\lambda_2\rho)]+\lambda_2[B_1\sin(\lambda_2\rho)-B_2\cos(\lambda_2\rho)]\}\right.\\
&\left.-\mathrm{e}^{-\lambda_1\rho}\{\lambda_1[B_3\cos(\lambda_2\rho)+B_4\sin(\lambda_2\rho)]-\lambda_2[B_3\sin(\lambda_2\rho)-B_4\cos(\lambda_2\rho)]\}\right)
\end{aligned}
\tag{36}
$$

求得了内力后，再按文献［6］中所得的公式计算轴向和环向应力：

$$
\begin{aligned}
\sigma_z=&\frac{1}{h}\left[1-\frac{\frac{z}{R}}{1-\nu^2}\left(1-\frac{\nu^2}{1+\frac{z}{R}}\right)\right]N_x+\frac{\nu\frac{z}{R}}{h(1-\nu^2)}\left(1-\frac{\nu^2}{1+\frac{z}{R}}\right)N_\theta\\
&+\frac{1}{h^2}\left[\frac{h}{R}-\frac{12z}{h}\left(1-\frac{\nu^2}{1+\frac{z}{R}}\right)\right]M_x-\frac{12\nu z}{h^3\left(1+\frac{z}{R}\right)}M_\theta,
\end{aligned}
$$

$$\sigma_\theta=\frac{N_\theta}{h}-\frac{\nu\left(\frac{z}{R}\right)^2}{h(1-\nu^2)\left(1+\frac{z}{R}\right)}N_x-\frac{1}{h^2}\left(\frac{h}{R}+\frac{12\frac{z}{h}}{1+\frac{z}{R}}\right)M_\theta+\frac{\nu}{h^2}\left(\frac{h}{R}-\frac{\frac{12z^2}{Rh}}{1+\frac{z}{R}}\right)M_x \tag{37}$$

四、加氢反应器顶部壳体待定常数的确定

由式（1），大法兰上端的边界条件为

在 $\rho_a=0$ 处，

$$M'_{xa}=M_0,\quad Q'_{xa}=Q_0, \tag{38}$$

这里的撇号表示壳体上边缘的量。

由第二条简化措施，在筒体的公式（32）中应令

$$B_{3c}=B_{4c}=0 \tag{39}$$

因为反应器是一组合壳体，在大法兰与球形封头、球形封头与筒体之间，都是用电焊固结起来的，所以在交界处，它们的内力与变形必是连续的，则有下列连续条件成立：

在 $\rho_a=\frac{l_a}{R_a}$，$\varphi=\varphi_0$ 处，

$$\begin{aligned}&w''_a=\Delta',\quad \psi''_a=\psi'_b,\\&M''_{xa}=M'_\varphi,\quad Q''_{xa}=H';\end{aligned} \tag{40}$$

在 $\varphi=\frac{\pi}{2}$，$\rho_c=0$ 处，

$$\begin{aligned}&\Delta''=w'_c,\quad \psi''_b=\psi'_c,\\&M''_\varphi=M'_{xc},\quad H''=Q'_{xc},\end{aligned} \tag{41}$$

这里，各量的正方向表示在图 6 中，带单撇号的表示上边缘的量，带双撇号的表示下边缘的量。

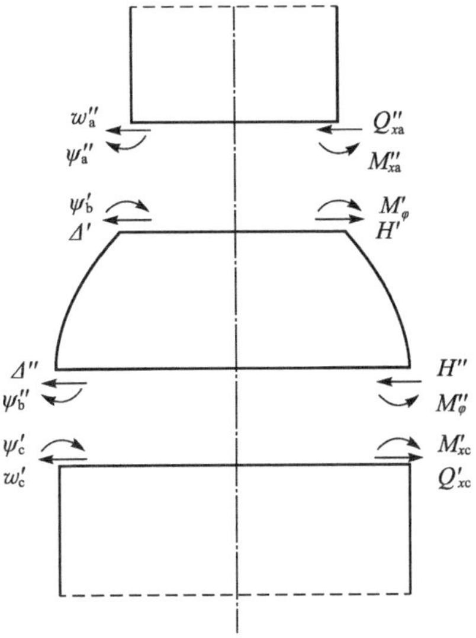

图 6 壳体边缘内力和位移的正方向

至于公式（9）中的积分常量 P，将由球形封头任一平行圆以上部分的竖直方向力的平衡条件来确定，显然，在本问题情况，

$$P\equiv 0 \tag{42}$$

将第 2、3 节中的式（9）、（13）、（15）、（32）、（34）和（36）代入式（38）、（40）和（41）中，并应用式（39）和（42），便得关于待定常数 $B_{1a}\cdots B_{4a}$、$C_1\cdots C_4$、B_{1c}、B_{2c}的十元联立代数方程组：

$$\begin{aligned}&a_{1,1}B_{1a}+a_{1,2}B_{2a}+a_{1,3}B_{3a}+a_{1,4}B_{4a}+a_{1,5}C_1+a_{1,6}C_2+a_{1,7}C_3\\&+a_{1,8}C_4+a_{1,9}B_{1c}+a_{1,10}B_{2c}=\omega_{1,1}q+\omega_{1,2}T_0,\\&a_{2,1}B_{1a}+a_{2,2}B_{2a}+a_{2,3}B_{3a}+a_{2,4}B_{4a}+a_{2,5}C_1+a_{2,6}C_2+a_{2,7}C_3\\&+a_{2,8}C_4+a_{2,9}B_{1c}+a_{2,10}B_{2c}=\omega_{2,1}q+\omega_{2,2}T_0,\end{aligned}$$

$$\cdots\cdots$$

$$a_{10,1}B_{1a}+a_{10,2}B_{2a}+a_{10,3}B_{3a}+a_{10,4}B_{4a}+a_{10,5}C_1+a_{10,6}C_2+a_{10,7}C_3 \tag{43}$$
$$+a_{10,8}C_4+a_{10,9}B_{1c}+a_{10,10}B_{2c}=\omega_{10,1}q+\omega_{10,2}T_0,$$

其中

$$a_{1,1}=a_{1,3}=\frac{K_a}{R_a^2}(\lambda_{1a}^2-\lambda_{2a}^2+\nu),$$

$$a_{1,2}=-a_{1,4}=-\frac{2K_a\lambda_{1a}\lambda_{2a}}{R_a^2},$$

$$a_{1,5}=a_{1,6}=a_{1,7}=a_{1,8}=a_{1,9}=a_{1,10}=0;$$

$$a_{2,1}=-a_{2,3}=-\frac{K_a\lambda_{1a}}{R_a^3}(\lambda_{1a}^2-3\lambda_{1a}^2+\nu),$$

$$a_{2,2}=a_{2,4}=\frac{K_a\lambda_{2a}}{R_a^3}(3\lambda_{1a}^2-\lambda_{2a}^2+\nu),$$

$$a_{2,5}=a_{2,6}=a_{2,7}=a_{2,8}=a_{2,9}=a_{2,10}=0;$$

$$a_{3,1}=e^{-\lambda_{1a}\rho_0}\cos(\lambda_{2a}\rho_0),$$

$$a_{3,2}=e^{-\lambda_{1a}\rho_0}\sin(\lambda_{2a}\rho_0),$$

$$a_{3,3}=e^{\lambda_{1a}\rho_0}\cos(\lambda_{2a}\rho_0),$$

$$a_{3,4}=e^{\lambda_{1a}\rho_0}\sin(\lambda_{2a}\rho_0),$$

$$a_{3,5}=\frac{R_a}{Eh_b}[(1-\nu)y_1(\varphi_0)\cos\varphi_0+y_1^{\cdot}(\varphi_0)\sin\varphi_0],$$

$$a_{3,6}=\frac{R_a}{Eh_b}[(1-\nu)y_2(\varphi_0)\cos\varphi_0+y_2^{\cdot}(\varphi_0)\sin\varphi_0],$$

$$a_{3,7}=\frac{R_a}{Eh_b}[(1-\nu)y_3(\varphi_0)\cos\varphi_0+y_3^{\cdot}(\varphi_0)\sin\varphi_0],$$

$$a_{3,8}=\frac{R_a}{Eh_b}[(1-\nu)y_4(\varphi_0)\cos\varphi_0+y_4^{\cdot}(\varphi_0)\sin\varphi_0],$$

$$a_{3,9}=a_{3,10}=0;$$

$$a_{4,1}=-\frac{1}{R_a}[\lambda_{1a}e^{-\lambda_{1a}\rho_0}\cos(\lambda_{2a}\rho_0)+\lambda_{2a}e^{-\lambda_{1a}\rho_0}\sin(\lambda_{2a}\rho_0)],$$

$$a_{4,2}=-\frac{1}{R_a}[\lambda_{1a}e^{-\lambda_{1a}\rho_0}\sin(\lambda_{2a}\rho_0)-\lambda_{2a}e^{-\lambda_{1a}\rho_0}\cos(\lambda_{2a}\rho_0)],$$

$$a_{4,3}=\frac{1}{R_a}[\lambda_{1a}e^{\lambda_{1a}\rho_0}\cos(\lambda_{1a}\rho_0)-\lambda_{2a}e^{\lambda_{1a}\rho_0}\sin(\lambda_{2a}\rho_0)],$$

$$a_{4,4}=\frac{1}{R_a}[\lambda_{1a}e^{\lambda_{1a}\rho_0}\sin(\lambda_{2a}\rho_0)+\lambda_{2a}e^{\lambda_{1a}\rho_0}\cos(\lambda_{2a}\rho_0)],$$

$$a_{4,5}=-\frac{\sin\varphi_0}{Eh_b}[2\eta^2y_3(\varphi_0)-\nu y_1(\varphi_0)],$$

$$a_{4,6}=-\frac{\sin\varphi_0}{Eh_b}[2\eta^2y_4(\varphi_0)-\nu y_2(\varphi_0)],$$

$$a_{4,7}=-\frac{\sin\varphi_0}{Eh_b}[2\eta^2y_1(\varphi_0)+\nu y_3(\varphi_0)],$$

$$a_{4,8}=\frac{\sin\varphi_0}{Eh_b}[2\eta^2 y_2(\varphi_0)+\nu y_4(\varphi_0)],$$

$$a_{4,9}=a_{4,10}=0;$$

$$a_{5,1}=\frac{K_a}{R_a^2}[(\lambda_{1a}^2-\lambda_{2a}^2+\nu)e^{-\lambda_{1a}\rho_0}\cos(\lambda_{2a}\rho_0)+2\lambda_{1a}\lambda_{2a}e^{-\lambda_{1a}\rho_0}\sin(\lambda_{2a}\rho_0)],$$

$$a_{5,2}=\frac{K_a}{R_a^2}[(\lambda_{1a}^2-\lambda_{2a}^2+\nu)e^{-\lambda_{1a}\rho_0}\sin(\lambda_{2a}\rho_0)-2\lambda_{1a}\lambda_{2a}e^{-\lambda_{1a}\rho_0}\cos(\lambda_{2a}\rho_0)],$$

$$a_{5,3}=\frac{K_a}{R_a^2}[(\lambda_{1a}^2-\lambda_{2a}^2+\nu)e^{\lambda_{1a}\rho_0}\cos(\lambda_{2a}\rho_0)-2\lambda_{1a}\lambda_{2a}e^{\lambda_{1a}\rho_0}\sin(\lambda_{2a}\rho_0)],$$

$$a_{5,4}=\frac{K_a}{R_a^2}[(\lambda_{1a}^2-\lambda_{2a}^2+\nu)e^{\lambda_{1a}\rho_0}\sin(\lambda_{2a}\rho_0)+2\lambda_{1a}\lambda_{2a}e^{\lambda_{1a}\rho_0}\cos(\lambda_{2a}\rho_0)],$$

$$a_{5,5}=-\frac{R_b}{\delta_{1b}^4}\{\sin\varphi_0[2\eta^2\dot{y}_3(\varphi_0)-\nu\dot{y}_1(\varphi_0)]+\cos\varphi_0(1+\nu)[2\eta^2 y_3(\varphi_0)-\nu y_1(\varphi_0)]\},$$

$$a_{5,6}=-\frac{R_b}{\delta_{1b}^4}\{\sin\varphi_0[2\eta^2\dot{y}_4(\varphi_0)-\nu\dot{y}_2(\varphi_0)]+\cos\varphi_0(1+\nu)[2\eta^2 y_4(\varphi_0)-\nu y_2(\varphi_0)]\},$$

$$a_{5,7}=\frac{R_b}{\delta_{1b}^4}\{\sin\varphi_0[2\eta^2\dot{y}_1(\varphi_0)+\nu\dot{y}_3(\varphi_0)]+\cos\varphi_0(1+\nu)[2\eta^2 y_1(\varphi_0)+\nu y_3(\varphi_0)]\},$$

$$a_{5,8}=\frac{R_b}{\delta_{1b}^4}\{\sin\varphi_0[2\eta^2\dot{y}_2(\varphi_0)+\nu\dot{y}_4(\varphi_0)]+\cos\varphi_0(1+\nu)[2\eta^2 y_2(\varphi_0)+\nu y_4(\varphi_0)]\},$$

$$a_{5,9}=a_{5,10}=0;$$

$$a_{6,1}=-\frac{K_a}{R_a^3}[\lambda_{1a}(\lambda_{1a}^2-3\lambda_{2a}^2+\nu)e^{-\lambda_{1a}\rho_0}\cos(\lambda_{2a}\rho_0)$$
$$+\lambda_{2a}(3\lambda_{1a}^2-\lambda_{2a}^2+\nu)e^{-\lambda_{1a}\rho_0}\sin(\lambda_{2a}\rho_0)],$$

$$a_{6,2}=-\frac{K_a}{R_a^3}[\lambda_{1a}(\lambda_{1a}^2-3\lambda_{2a}^2+\nu)e^{-\lambda_{1a}\rho_0}\sin(\lambda_{2a}\rho_0)$$
$$-\lambda_{2a}(3\lambda_{1a}^2-\lambda_{2a}^2+\nu)e^{-\lambda_{1a}\rho_0}\cos(\lambda_{2a}\rho_0)],$$

$$a_{6,3}=\frac{K_a}{R_a^3}[\lambda_{1a}(\lambda_{1a}^2-3\lambda_{2a}^2+\nu)e^{\lambda_{1a}\rho_0}\cos(\lambda_{2a}\rho_0)$$
$$-\lambda_{2a}(3\lambda_{1a}^2-\lambda_{2a}^2+\nu)e^{\lambda_{1a}\rho_0}\sin(\lambda_{2a}\rho_0)],$$

$$a_{6,4}=\frac{K_a}{R_a^3}[\lambda_{1a}(\lambda_{1a}^2-3\lambda_{2a}^2+\nu)e^{\lambda_{1a}\rho_0}\sin(\lambda_{2a}\rho_0)$$
$$+\lambda_{2a}(3\lambda_{1a}^2-\lambda_{2a}^2+\nu)e^{\lambda_{1a}\rho_0}\cos(\lambda_{2a}\rho_0)],$$

$$a_{6,5}=-y_1(\varphi_0),$$

$$a_{6,6}=-y_2(\varphi_0),$$

$$a_{6,7}=-y_3(\varphi_0),$$

$$a_{6,8}=-y_4(\varphi_0),$$

$$a_{6,9}=a_{6,10}=0;$$

$$a_{7,1}=a_{7,2}=a_{7,3}=a_{7,4}=a_{7,5}=a_{7,7}=a_{7,8}=a_{7,10}=0,$$

$$a_{7,6}=\frac{R_b}{Eh_b},$$

$a_{7,9}=-1$;

$a_{8,1}=a_{8,2}=a_{8,3}=a_{8,4}=a_{8,6}=a_{8,8}=0$,

$a_{8,5}=-\frac{\nu}{Eh_b}$,

$a_{8,7}=-\frac{2\eta^2}{Eh_b}$,

$a_{8,9}=\frac{\lambda_{1c}}{R_c}$,

$a_{8,10}=-\frac{\lambda_{2c}}{R_c}$;

$a_{9,1}=a_{9,2}=a_{9,3}=a_{9,4}=a_{9,5}=a_{9,7}=0$,

$a_{9,6}=\frac{\nu R_b}{\delta_{1b}^4}$,

$a_{9,8}=\frac{2R_b\eta^2}{\delta_{1b}^4}$,

$a_{9,9}=-\frac{K_c}{R_c^2}(\lambda_{1c}^2-\lambda_{2c}^2+\nu)$,

$a_{9,10}=-\frac{2K_c\lambda_{1c}\lambda_{2c}}{R_c^2}$;

$a_{10,1}=a_{10,2}=a_{10,3}=a_{10,4}=a_{10,6}=a_{10,7}=a_{10,8}=0$,

$a_{10,5}=1$,

$a_{10,9}=\frac{K_c\lambda_{1c}}{R_c^3}(\lambda_{1c}^2-3\lambda_{2c}^2+\nu)$,

$a_{10,10}=-\frac{K_c\lambda_{2c}}{R_c^3}(3\lambda_{1c}^2-\lambda_{2c}^2+\nu)$;

$\omega_{1,1}=-\frac{R_a}{\delta_{1a}^4}\left(R_a-\frac{h_a}{2}\right)\left[\nu-\frac{1}{2R_a}\left(R_a-\frac{h_a}{2}\right)\right]+\mu_1$,

$\omega_{1,2}=\mu_2$,

$\omega_{2,1}=\mu_3$,

$\omega_{2,2}=\mu_4$,

$\omega_{3,1}=\frac{R_a}{E}\left\{\frac{1-\nu}{2R_bh_b}\left(R_b-\frac{h_b}{2}\right)^2-\frac{1}{h_a}\left(R_a-\frac{h_a}{2}\right)\left[1-\frac{\nu}{2R_a}\left(R_a-\frac{h_a}{2}\right)\right]\right\}$,

$\omega_{5,1}=-\frac{R_a}{\delta_{1a}^4}\left(R_a-\frac{h_a}{2}\right)\left[\nu-\frac{1}{2R_a}\left(R_a-\frac{h_a}{2}\right)\right]$,

$\omega_{6,1}=-\frac{\cos\varphi_0}{2R_b}\left(R_b-\frac{h_b}{2}\right)^2$,

$\omega_{7,1}=\frac{1}{E}\left\{\frac{R_c}{h_c}\left(R_c-\frac{h_c}{2}\right)\left[1-\frac{\nu}{2R_c}\left(R_c-\frac{h_c}{2}\right)\right]-\frac{1-\nu}{2h_b}\left(R_b-\frac{h_b}{2}\right)^2\right\}$,

$\omega_{9,1}=\frac{R_c}{\delta_{1c}^4}\left(R_c-\frac{h_c}{2}\right)\left[\nu-\frac{1}{2R_c}\left(R_c-\frac{h_c}{2}\right)\right]$,

$$\omega_{3,2}=\omega_{4,1}=\omega_{4,2}=\omega_{5,2}=\omega_{6,2}=\omega_{7,2}=\omega_{8,1}=\omega_{8,2}=\omega_{9,2}=\omega_{10,1}=\omega_{10,2}=0 \quad (44)$$

由方程组（43）解得十个待定常数后，就可应用前面的式（9）、（22）和（34）计算内力，最后由式（16）和（37）计算球形封头和筒体的应力值。

五、加氢反应器顶部壳体的应力分布

某厂设计的加氢反应器的有关数据如下：

$R_a=50\text{cm}$，　$R_b=R_c=108\text{cm}$，　$l_a=32.76\text{cm}$，

$h_a=20\text{cm}$，　$h_b=h_c=16\text{cm}$，　$\varphi_0=27°35'$，

$\alpha=60°$，　$\beta=8°30'$，　$r_0=42.1\text{cm}$，

$b=10\text{cm}$，　$d=3\text{cm}$，　$R_\delta=48.5\text{cm}$，

$l_\delta=32.2\text{cm}$，　$r_\delta=3.7\text{cm}$，　$m_\delta=16$.

在设计压力下，$q=210\text{kg/cm}^2$；在试验压力下，$q=305\text{kg/cm}^2$；密封比压 $\sigma=500\text{kg/cm}^2$。

大法兰、球形封头和筒体的材料为 20CrMo9 钢，在常温下，其物理常数及屈服限、强度限为

$$E=2.18\times10^6\text{kg/cm}^2,\quad \nu=0.3^{①},$$
$$\sigma_s=4500\text{kg/cm}^2,\quad \sigma_b=6500\text{kg/cm}^2$$

主螺栓的材料为 $25Cr2MoV_A$ 钢，双锥密封垫的材料为 1Cr18Ni9Ti 钢，在常温下，它们的弹性模量为

$$E_\delta=2.19\times10^6\text{kg/cm}^2,\quad E_k=2.03\times10^6\text{kg/cm}^2$$

将上述数据代入到前面的有关公式中，便求得加氢反应器顶部球形封头及邻近的筒体在设计压力和试验压力下的内、外表面（$z=\mp h/2$）的应力值。全部的数值运算都是在 DJS-6 型电子数学计算机上进行的，结果给在表 1～表 4 中。由表 1 可知，球形封头的最大应力为环向应力，发生在孔边缘的壳外表面上。在设计压力下，其值为

$$\delta_{\theta\max}=1219\text{kg/cm}^2$$

此与屈服限的比值为

$$n_s=0.2709$$

它远小于规范值 $n_s\leqslant2$，故球形封头孔边缘的强度非常好。

至于在球形封头下边缘应力中，最大应力仍是环向应力，发生在与筒体交界的内表面上。在设计压力下，其值为

$$\sigma_{\theta\max}=1021\text{kg/cm}^2$$

此与屈服限的比值为

$$n_s=0.2269$$

它满足 ASME 锅炉及压力容器规范值 $n_s\leqslant1$，故球形封头下边缘强度也是很好的。

在筒体上边缘应力中，最大应力为环向应力，发生在 $\rho_c=0.90$ 处的内表面上。在设计压力下，其值为

---

① 是近似值。

表 1　加氢反应器顶部球形封头在设计压力下的应力值　　（单位：kg/cm²）

φ	σ_θ						σ_φ					
	$z=h/2$			$z=-h/2$			$z=h/2$			$z=-h/2$		
	本节公式	文献［3］公式	文献［4］公式	本节公式	文献［3］公式	文献［4］公式	本节公式	文献［3］公式	文献［4］公式	本节公式	文献［3］公式	文献［4］公式
27°35′	1219	1420	1525	944.6	1105	1022	1166	1361	1462	−567.0	−660.6	−626.9
28°	1211	1410	1515	938.8	1098	1016	1134	1323	1421	−505.7	−589.2	−545.8
30°	1162	1354	1454	914.8	1070	990.9	993.4	1159	1245	−243.1	−283.2	−262.2
32°	1105	1289	1385	895.4	1048	970.2	876.2	1023	1099	−25.88	−30.08	−27.85
34°	1045	1220	1310	878.5	1028	952.1	778.8	909.6	977.0	153.4	178.7	165.1
36°	985.0	1150	1235	862.9	1010	935.4	697.9	815.8	876.3	300.8	350.2	324.3
38°	925.8	1082	1162	847.9	992.8	919.2	631.1	738.5	793.2	421.6	490.5	454.1
40°	869.4	1017	1092	833.4	975.8	903.5	576.1	675.1	725.1	520.0	604.6	559.8
50°	649.0	763.0	819.5	772.1	902.1	835.3	419.8	498.7	535.6	790.6	914.0	846.3
60°	538.9	636.4	683.6	753.7	873.5	808.8	359.0	437.4	469.8	884.4	1012	937.4
70°	531.9	624.1	670.4	801.5	914.9	847.2	331.0	414.8	445.5	925.7	1048	970.0
80°	639.0	731.6	785.7	909.6	1020	944.4	362.3	452.2	485.7	897.2	1012	936.8
82°	676.4	769.3	826.3	934.6	1045	967.4	384.4	475.4	510.6	872.4	985.5	912.5
84°	719.5	812.8	873.0	959.3	1069	989.9	414.6	506.8	544.3	837.5	949.2	878.9
86°	768.3	862.1	926.0	982.7	1092	1011	454.5	547.8	588.4	790.5	900.8	834.1
88°	822.8	917.1	985.0	1004	1113	1030	505.7	600.1	644.6	729.2	838.1	776.1
90°	882.6	977.4	1050	1021	1130	1046	569.9	665.5	714.8	651.4	759.0	702.7

表 2 加氢反应器筒体在设计压力下的应力值 (单位：kg/cm^2)

ρ_c	σ_θ						σ_x					
	$z=h/2$			$z=-h/2$			$z=h/2$			$z=-h/2$		
	本节公式	文献 [3] 公式	文献 [4] 公式	本节公式	文献 [3] 公式	文献 [4] 公式	本节公式	文献 [3] 公式	文献 [4] 公式	本节公式	文献 [3] 公式	文献 [4] 公式
0.00	880.1	974.6	972.4	1018	1127	1124	568.3	663.5	714.8	649.6	756.8	702.7
0.02	914.2	1009	1007	1031	1139	1136	609.4	705.2	757.5	606.5	713.0	660.0
0.04	946.7	1042	1039	1045	1153	1150	645.0	741.6	794.7	569.1	674.9	622.8
0.06	977.3	1073	1070	1060	1168	1165	675.7	772.8	826.7	536.9	642.2	590.8
0.08	1006	1102	1099	1076	1185	1181	701.7	799.3	853.9	509.7	614.4	563.6
0.10	1033	1129	1127	1093	1202	1198	723.5	821.5	876.6	486.9	591.2	540.9
0.20	1142	1239	1236	1182	1290	1287	780.1	879.7	936.1	427.8	530.3	481.4
0.30	1210	1308	1305	1263	1372	1368	777.3	877.7	934.0	431.0	532.7	483.5
0.40	1247	1346	1343	1328	1438	1433	744.7	845.3	900.6	465.5	567.0	516.9
0.50	1263	1362	1358	1374	1484	1480	702.0	802.5	856.7	510.5	612.1	560.8
0.60	1264	1363	1360	1403	1514	1509	661.0	761.3	814.4	553.6	655.5	603.1
0.70	1259	1358	1354	1419	1530	1525	627.4	727.4	779.6	588.9	691.1	637.9
0.80	1250	1349	1346	1425	1536	1532	602.9	702.6	754.2	614.6	717.1	663.3
0.90	1242	1340	1337	1426	1537	1533	587.0	686.5	737.7	631.3	734.0	679.8
1.00	1234	1334	1330	1423	1535	1530	578.0	677.4	728.3	640.8	743.6	689.2
1.50	1222	1321	1318	1409	1521	1516	577.4	676.6	727.6	641.3	744.3	689.9
2.00	1223	1323	1319	1408	1519	1515	581.7	681.0	732.1	636.8	739.8	685.4

$$\sigma_{\theta\max}=1426\mathrm{kg/cm^2}$$

此与屈服限的比值为

$$n_s=0.3169$$

它也满足规范值 $n_s\leqslant1$，故筒体的强度也是足够的。

表 3　加氢反应器顶部球形封头在试验压力下的应力值　　（单位：$\mathrm{kg/cm^2}$）

φ	σ_θ		σ_φ	
	$z=h/2$	$z=-h/2$	$z=h/2$	$z=-h/2$
27°35′	1759	1364	1727	−865.1
28°	1748	1356	1678	−775.3
30°	1682	1323	1470	−390.1
32°	1603	1296	1297	−70.74
34°	1519	1273	1152	193.4
36°	1433	1251	1032	411.2
38°	1348	1230	932.2	590.0
40°	1267	1210	849.9	736.2
50°	947.2	1123	614.4	1141
60°	785.3	1097	522.1	1283
70°	773.6	1165	480.0	1345
80°	928.2	1322	525.3	1304
82°	982.4	1358	557.4	1268
84°	1045	1394	601.4	1217
86°	1116	1428	659.4	1149
88°	1195	1458	733.8	1060
90°	1282	1483	827.1	946.8

表 4　加氢反应器筒体在试验压力下的应力值　　（单位：$\mathrm{kg/cm^2}$）

ρ_c	σ_θ		σ_x	
	$z=h/2$	$z=-h/2$	$z=h/2$	$z=-h/2$
0.00	1278	1479	824.7	944.1
0.02	1328	1497	884.4	881.5
0.04	1375	1517	936.3	827.1
0.06	1419	1539	980.9	780.4
0.08	1461	1563	1019	740.8
0.10	1500	1588	1050	707.7
0.20	1658	1716	1133	621.7
0.30	1758	1834	1129	626.2
0.40	1812	1928	1081	676.1
0.50	1834	1995	1020	741.4
0.60	1836	2037	960.0	804.0
0.70	1828	2060	911.2	855.3
0.80	1816	2070	875.7	892.6
0.90	1803	2071	852.6	916.9
1.00	1793	2067	839.5	930.6
1.50	1775	2047	838.6	931.5
2.00	1777	2045	844.9	924.9

此外，我们还得到双锥密封中的自紧系数值：

$$G=1.081$$

这说明，所设计的双锥密封垫的自紧密封作用很小。

最后，我们还使用了文献［3］的厚壁球壳近似理论公式和我们所推导的相当于文献［3］精确度的厚壁圆柱壳近似理论公式，以及文献［4］的薄壁球壳精确理论公式和我们所推导的相当于文献［4］精确度薄壁圆柱壳理论公式，进行了比较计算，其结果也列在表 1 和表 2 中（为简单起见，称为文献［3］或文献［4］公式结果）。为醒目起见，还以球形封头外表面的环向应力和筒体外表面的经向应力为例，把三种理论公式的结果绘制在图 7 和图 8 中，显而易见，相对误差是很可观的，以图 7 中的最大应力值为例，文献［3］、文献［4］公式的结果与本节结果的相对误差分别为 16.49%和 25.10%；以图 8 中的最大应力值为例，文献［3］、文献［4］公式的结果与本节结果的相对误差分别为 12.77%和 20.00%。

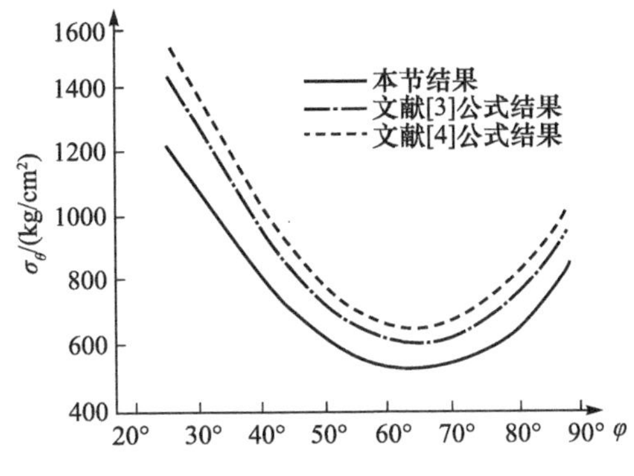

图 7　在设计压力下顶部球形封头外表面的 σ_θ 值

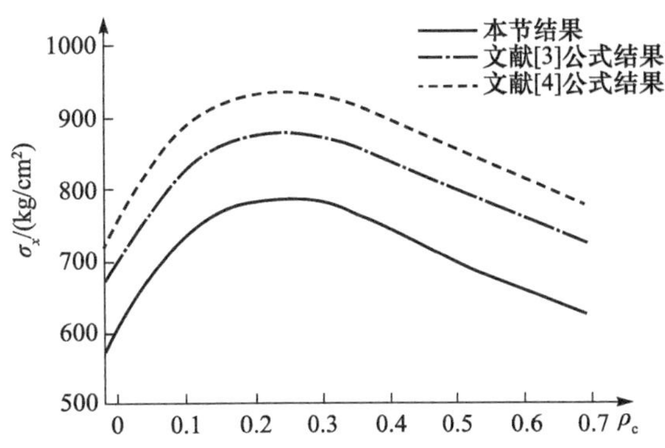

图 8　在设计压力下筒体外表面的 σ_x 值

对于本节所建立的厚壁圆柱壳轴对称问题的近似计算公式，我们用一特殊例子进行评价。对于承受均匀分布内压力的厚壁圆筒，众所周知，有拉梅公式的精确解（参见文献［5］）。我们使用此公式和前面三种近似理论的公式，对在设计压力下的加氢反应器筒体远离边缘的区域，进行了内、外表面环向应力的计算，所得结果载于表 5 中。由表 5 可见，本节结果与精确解的误差最小，仅在 1%左右。自然，把近似公式与边缘应力区域的实验结果做比较，那将是更完善的评价。

表 5　厚壁圆筒 σ_θ 值的比较　　（单位：kg/cm^2）

$z=-h/2$							$z=h/2$						
拉梅公式	本节公式		文献［3］公式		文献［4］公式		拉梅公式	本节公式		文献［3］公式		文献［4］公式	
	值	误差/%	值	误差/%	值	误差/%		值	误差/%	值	误差/%	值	误差/%
1425	1408	−1.193	1519	6.596	1515	6.316	1215	1223	0.6541	1323	8.889	1319	8.560

六、结论

通过上述计算结果，我们有下面几点粗浅看法。

（1）本节的数值结果具有一定的可靠性，能做加氢反应器设计时的参考。从强度角度看，还可减薄加氢反应器顶部球形封头和筒体的厚度；

（2）由文献［2］和本节建立的厚壁球壳及厚壁圆柱壳轴对称问题的近似计算公式比较简单，其计算量与精确薄壳公式差不多，因而便于工程应用；

（3）为了确定文献［2］和本节公式的精确度及使用范围，做一些实验研究是大有必要的；

（4）在厚壁壳体的边缘应力中，与薄壳不同，而是平缓变化。

参 考 文 献

［1］ 刘人怀，陈山林. 尿素合成塔底部球形封头开孔的应力计算. 科技专刊（兰州大学），1973，(1)：1-13.

［2］ 刘人怀. 厚壁球壳的弯曲理论及其在高压容器上的应用. 化工炼油机械通讯，1980，(3)：1-10.

［3］ Рабинович А Л. Приближенный метоя расчета толстостенной сферической оболочки. Тр. Моск. физ.-техн-ин-та，вып. з，1959.

［4］ 弗留盖 W. 壳体中的应力. 薛振东，龙驭球，叶耀先，等译. 北京：中国工业出版社，1965.

［5］ 刘人怀. 双锥密封中的内力分析. 化工炼油机械通讯，1975，(6)：55-57.

［6］ 刘人怀. 厚壁圆柱壳轴对称变形近似理论的应力公式. 化工炼油机械通讯，1975，(6)：58-59.

双锥密封中的内力分析①

在研究我国最大直径加氢反应器应力分析问题中，遇到了双锥密封结构的力学分析问题。鉴于原有规范设计公式[1]以及大家常使用的捷克法公式比较粗糙，无法满足本问题的需要，为此我们进行了初步尝试，给出了较为精确的计算公式。

我们在弹性范围内研究，分两步进行分析：第一步是预紧阶段，第二步是加载阶段。我们先分析第二步，并且暂不计及预紧力，此时称作单纯加载阶段。

一、单纯加载阶段

在容器内压力 q 的作用下，每一个主螺栓伸长了 Δl_δ，产生了轴向力 S。按照胡克定律，有

$$\Delta l_\delta=\frac{l_\delta S}{E_\delta U_\delta},\tag{1}$$

其中 l_δ 是主螺栓的有效工作长度，U_δ 是主螺栓中间细长部分的截面积，E_δ 是材料弹性模量，r_δ 是主螺栓中间细长部分的截面半径，

$$U_\delta=\pi r_\delta^2\tag{2}$$

容器的顶盖和大法兰的密封面与双锥密封垫的密封面是相互紧贴在一起的。我们视双锥密封垫为薄圆环，则在内压力 q 的作用下，它的径向伸长 w 与 Δl_δ 间存在关系（图 1）：

$$w=\frac{1}{2}\Delta l_\delta\cot\alpha,\tag{3}$$

其中 α 是双锥密封垫的锥角。

将式（1）代入式（3），得

$$w=\frac{l_\delta S}{2E_\delta U_\delta\tan\alpha}\tag{4}$$

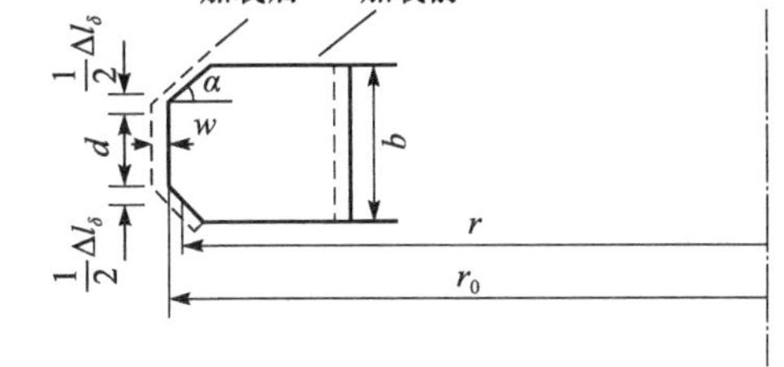

图 1　双锥密封垫直径截面的径向变形

其次，我们研究在内压力 q 作用下双锥密封垫的平衡情况。显然，可认为双锥密封垫处于轴对称变形情况，于是，在其直径截面上无剪力和横向力作用。设它的上、下两个密封面上的力大小相等，而且摩擦力满足最大静摩擦力公式，则其微元体（图 2）在 y 方向上的力平衡方程为

$$2N_\theta\sin\frac{\mathrm{d}\theta}{2}+2T(r+w)\mathrm{d}\theta\sin\alpha+2fT(r+w)\mathrm{d}\theta\cos\alpha-bq(r+w)\mathrm{d}\theta=0,\tag{5}$$

其中，N_θ 是双锥密封垫的环向内力；T 是单纯加载状态下密封面上单位长度的反作用力；f 和 β 是双锥密封垫与顶盖、筒体之间的摩擦系数和摩擦角；r 和 r_0 是双锥密封垫的半径和外缘半径；b 和 d 是双锥密封垫的高度和外缘高度；$\mathrm{d}\theta$ 是双锥密封垫微元体

① 本文原载《化工炼油机械通讯》，1975，(6)：55-57.

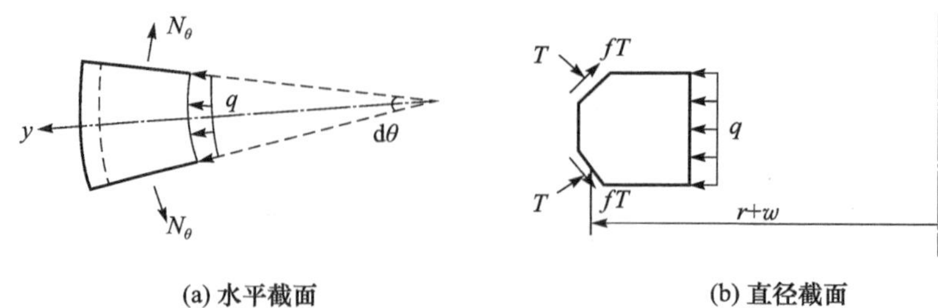

(a) 水平截面　　(b) 直径截面

图 2　变形后的双锥密封垫微元体

的两个直径截面间的夹角，

$$r=r_0-\frac{b-d}{4}\cot\alpha,\qquad f=\tan\beta \tag{6}$$

由于 w 和 $\mathrm{d}\theta$ 都是很小的，故可使用近似公式：

$$\sin\frac{\mathrm{d}\theta}{2}\approx\frac{\mathrm{d}\theta}{2},\quad (r+w)\mathrm{d}\theta\approx r\mathrm{d}\theta \tag{7}$$

在此情况下，式（5）化简成为

$$N_\theta=r\left[bq-\frac{2\sin(\alpha+\beta)}{\cos\beta}T\right] \tag{8}$$

下面，我们推导双锥密封垫的径向位移和环向内力的关系式。首先，对于双锥密封垫的环向应变，有

$$\varepsilon_\theta=\frac{2\pi(r+w)-2\pi r}{2\pi r}=\frac{w}{r} \tag{9}$$

其次，我们假定，双锥密封垫在变形时，纵向纤维互不挤压、材料服从胡克定律，故有

$$\varepsilon_\theta=\frac{N_\theta}{E_\mathrm{k}U_\mathrm{k}} \tag{10}$$

其中，E_k 是双锥密封垫材料的弹性模量，U_k 是双锥密封垫直径截面的面积。

综合式（9）、（10）两个公式，则得

$$w=\frac{rN_\theta}{E_\mathrm{k}U_\mathrm{k}} \tag{11}$$

另外，将式（8）代入式（11），还有

$$w=\frac{r^2}{E_\mathrm{k}U_\mathrm{k}}\left[bq-\frac{2\sin(\alpha+\beta)}{\cos\beta}T\right] \tag{12}$$

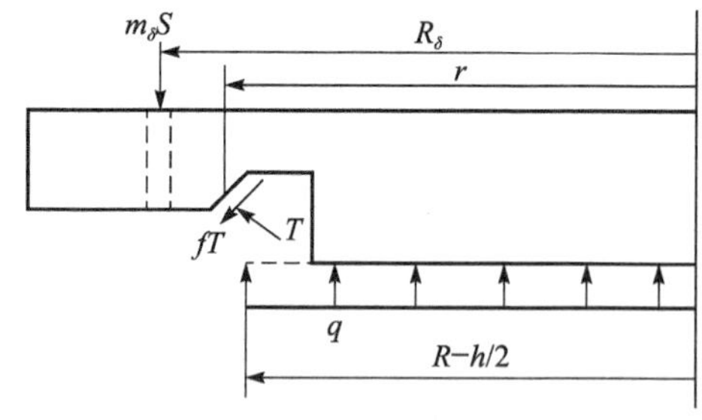

图 3　顶盖受力情况

按照图 3，可建立顶盖在竖直方向力的平衡方程：

$$2\pi rT(\cos\alpha-f\sin\alpha)+Fq-m_\delta S=0, \tag{13}$$

其中 F 是大法兰内横截面积，R 是大法兰的中曲面半径，h 是大法兰的厚度，m_δ 是主螺栓数目，

$$F=\pi\left(R-\frac{h}{2}\right)^2 \tag{14}$$

从式（13）解出 T：

$$T=\frac{(m_\delta S-Fq)\cos\beta}{2\pi r\cos(\alpha+\beta)} \tag{15}$$

将式（15）代入式（12），在利用式（4）后，得计算主螺栓轴向力的公式：

$$S=\frac{1}{m_\delta}GFq, \tag{16}$$

其中 G 是自紧系数，

$$G=\frac{\tan(\alpha+\beta)+\dfrac{\pi br}{F}}{\tan(\alpha+\beta)+\dfrac{\pi E_k U_k l_\delta}{2rm_\delta E_\delta U_\delta\tan\alpha}} \tag{17}$$

由式（16）看出，在单纯加载阶段中，主螺栓的作用力与自紧系数和内压力的乘积成正比。

将式（16）代入式（15），还得到密封面反作用力 T 的公式：

$$T=\frac{(G-1)Fq\cos\beta}{2\pi r\cos(\alpha+\beta)} \tag{18}$$

二、预紧阶段

在预紧阶段，容器内无内压力作用。为分析简单起见，仍假定摩擦力满足最大静摩擦力公式。按照图 4，可建立顶盖竖直方向力的平衡方程：

$$2\pi rT_0(\cos\alpha+f\sin\alpha)-m_\delta S_0=0, \tag{19}$$

其中 T_0 是预紧状态下密封面上单位长度的反作用力。

由式（19），得主螺栓预紧轴向力 S_0 的公式：

$$S_0=\frac{2\pi r\cos(\alpha-\beta)}{m_\delta\cos\beta}T_0 \tag{20}$$

由于密封面之间置有极薄的纯铝垫片，在预紧时将完全屈服，故可设

$$T_0=\frac{b-d}{2\sin\alpha}\sigma \tag{21}$$

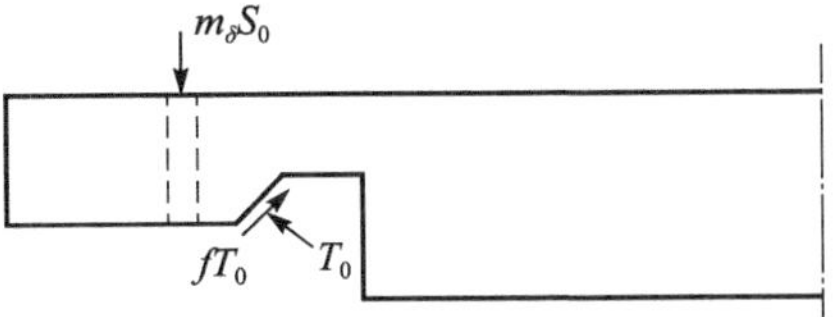

图 4　预紧状态下顶盖受力情况

这里密封比压 σ 的值介于纯铝的屈服限和强度限之间。

归结起来，在容器操作情况下，主螺栓将受力（$S+S_0$），密封面上受力（$T+T_0$）和 $f(T-T_0)$。由公式（18）看出，在 $G\leqslant1$ 时，双锥密封垫起不到自紧密封作用。实际上，只能使 G 略大于 1，故双锥密封垫的自紧密封作用很小，主要靠预紧力来密封。

参考文献

[1]　化工设备设计手册编写组. 高压设备. 上海：上海人民出版社，1973：38-40.

厚壁圆柱壳轴对称变形近似理论的应力公式①

我们讨论厚壁圆柱壳轴对称变形情况，其坐标系和内力、内力矩的正方向示于图 1 中。

基于文献［1］讨论厚壁球壳的办法，本文将推导厚壁圆柱壳的应力公式。设 ε_x、ε_θ 分别为厚壁圆柱壳轴对称变形时的轴向和环向应变，按文献［2］，有

$$\varepsilon_x=\frac{\mathrm{d}u}{\mathrm{d}x}-z\frac{\mathrm{d}^2w}{\mathrm{d}x^2},\qquad \varepsilon_\theta=\frac{w}{R+z},\tag{1}$$

其中 u、w 分别为轴向和法向位移，R 为圆柱壳半径。

引入圆柱壳中曲面的伸长应变 ε_{x0}、$\varepsilon_{\theta0}$ 和曲率变化 k_x、k_θ：

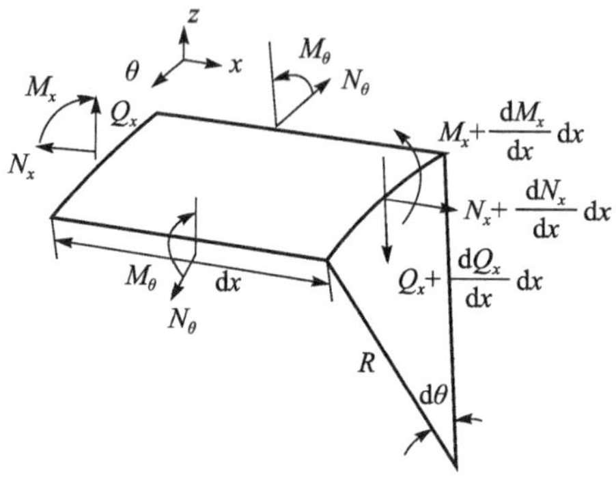

$$\varepsilon_{x0}=\frac{\mathrm{d}u}{\mathrm{d}x},\qquad \varepsilon_{\theta0}=\frac{w}{R},\qquad k_x=\frac{\mathrm{d}^2w}{\mathrm{d}x^2},\qquad k_\theta=\frac{w}{R^2}\tag{2}$$

图 1　圆柱壳微元

则式（1）改写为

$$\varepsilon_x=\varepsilon_{x0}-zk_x,\qquad \varepsilon_\theta=\varepsilon_{\theta0}-\frac{z}{1+z/R}k_\theta\tag{3}$$

已知胡克定律

$$\sigma_x=\frac{E}{1-\nu^2}(\varepsilon_x+\nu\varepsilon_\theta),\qquad \sigma_\theta=\frac{E}{1-\nu^2}(\varepsilon_\theta+\nu\varepsilon_x),\tag{4}$$

其中 σ_x、σ_θ 是轴向和环向应力，E 是弹性模量，ν 是泊松比。

将式（3）代入式（4），得

$$\sigma_x=\frac{E}{1-\nu^2}\left[(\varepsilon_{x0}+\nu\varepsilon_{\theta0})-z\left(k_x+\frac{\nu}{1+\frac{z}{R}}k_\theta\right)\right],$$

$$\sigma_\theta=\frac{E}{1-\nu^2}\left[(\varepsilon_{\theta0}+\nu\varepsilon_{x0})-z\left(\frac{\nu}{1+\frac{z}{R}}k_\theta+\nu k_x\right)\right]\tag{5}$$

① 本文原载《化工炼油机械通讯》，1975，(6)：58-59.

由内力定义，还有

$$
\begin{aligned}
&N_x=D(\varepsilon_{x0}+\nu\varepsilon_{\theta0})-\frac{K}{R}k_x,\\
&N_\theta=D(\varepsilon_{\theta0}+\nu\varepsilon_{x0})+\frac{K}{R}k_\theta,\\
&M_x=K\left(k_x-\frac{1}{R}\varepsilon_{x0}\right),\\
&M_\theta=K(k_\theta+\nu k_x),
\end{aligned}
\tag{6}
$$

其中 D 是指抗拉刚度，K 是抗弯刚度，

$$
\begin{aligned}
&D=\frac{Eh}{1-\nu^2},\\
&K=\frac{Eh^3}{12(1-\nu^2)}
\end{aligned}
\tag{7}
$$

解方程组（6），得

$$
\begin{aligned}
&\varepsilon_{x0}=\frac{1}{Eh}\left(N_x-\nu N_\theta+\frac{1-\nu^2}{R}M_x+\frac{\nu}{R}M_\theta\right),\\
&\varepsilon_{\theta0}=\frac{1}{Eh}\left(N_\theta-\nu N_x-\frac{1}{R}M_\theta\right),\\
&k_x=\frac{1}{Eh}\left(\frac{Eh}{K}M_x+\frac{\nu}{R^2}M_\theta+\frac{1}{R}N_x-\frac{\nu}{R}N_\theta\right),\\
&k_\theta=\frac{\nu}{Eh}\left(\frac{Eh}{\nu K}M_\theta-\frac{Eh}{K}M_x+\frac{\nu}{R}N_\theta-\frac{1}{R}N_x\right)
\end{aligned}
\tag{8}
$$

将式（8）代入式（5），便得厚壁圆柱壳轴对称变形近似理论的应力公式：

$$
\begin{aligned}
\sigma_x=&\frac{1}{h}\left[1-\frac{z/R}{1-\nu^2}\left(1-\frac{\nu^2}{1+z/R}\right)\right]N_x+\frac{\nu z/R}{h(1-\nu^2)}\left(1-\frac{\nu^2}{1+z/R}\right)N_\theta\\
&+\frac{1}{h^2}\left[\frac{h}{R}-\frac{12z}{h}\left(1-\frac{\nu^2}{1+z/R}\right)\right]M_x-\frac{12\nu z}{h^3(1+z/R)}M_\theta,\\
\sigma_\theta=&\frac{N_\theta}{h}-\frac{\nu(z/R)^2}{h(1-\nu^2)(1+z/R)}N_x-\frac{1}{h^2}\left(\frac{h}{R}+\frac{12z/h}{1+z/R}\right)M_\theta\\
&+\frac{\nu}{h^2}\left(\frac{h}{R}-\frac{12z^2/Rh}{1+z/R}\right)M_x
\end{aligned}
\tag{9}
$$

略去上式中的小量，可得熟知的薄壁圆柱壳的应力公式：

$$
\begin{aligned}
&\sigma_x=\frac{N_x}{h}-\frac{12zM_x}{h^3},\\
&\sigma_\theta=\frac{N_\theta}{h}-\frac{12zM_\theta}{h^3}
\end{aligned}
\tag{10}
$$

参考文献

[1] Рабинович А Л. Приближенный метод расчета толстостенной сферической оболочки. Тр. Моск. физ-техн. ин-та，вып. 3，1959.

[2] 弗留盖 W. 壳体中的应力. 薛振东，龙驭球，叶耀先，译. 北京：中国工业出版社，1965.

双层套箍式厚壁压力容器环沟部位的应力状态①

符号说明

x　圆柱壳的轴向坐标

ρ_1　环沟处内筒的无量纲轴向坐标的边缘值

R　圆柱壳的中曲面半径

$2l$　环沟宽度

M_x，M_θ　圆柱壳的轴向和环向弯矩

σ_x，σ_θ　圆柱壳的轴向和环向应力

σ_θ^+，σ_x^-　圆柱壳外、内壁的环向应力

$\sigma_{\theta,0}^+$，$\sigma_{\theta,1}^-$　环沟处内筒外壁中央和内壁边缘处的环向应力

ρ　圆柱壳的无量纲轴向坐标

z　圆柱壳中曲面外法线方向的坐标

h　圆柱壳的厚度

N_x，N_θ　圆柱壳的轴向和环向内力

Q_x　圆柱壳的横向力

σ_x^+，σ_x^-　圆柱壳外、内壁的轴向应力

u，w　圆柱壳的轴向和法向位移

ψ　圆柱壳母线微元的转角

E，ν　壳体材料的弹性模量和泊松比

k　弹性系数

δ，α_1，α_2，a_0，a_1，a_2，a_{11}，a_{12}，a_{21}，a_{22}，λ_1，λ_2，λ，ω，μ，β，φ　辅助量

右下角标 max　极大值

$A_1\cdots A_4$，$B_1\cdots B_4$，$C_1\cdots C_4$，$\overline{A}_1\cdots\overline{A}_4$，$N$　积分常数

D，K　抗拉和抗弯刚度

q，P^-，P^+　壳体的均匀分布内压力和内、外壁法向载荷

右下角标 a，b，c　分别表示环沟处内筒、不包含环沟的内筒和套箍的量

双层套箍式容器是一种新型的结构形式，它的筒体部分采用双层套箍式结构，即内筒用焊接连成整体，外筒是以一节节环箍形式套合在内筒上，且外筒间环缝不加连接，留有小的间隙。此种结构形式对于需要用厚钢板制造的高压容器来说，开辟了一条用较薄钢板代替厚钢板的新途径。并且，还简化了焊接工艺，节省了外筒环缝自动焊工作量。但是，对于这种新型结构形式压力容器强度的理论计算和试验研究，却缺乏完整的参考资料。1961 年，Palmer[1] 曾对套箍式压力容器做过初步研究，但他所研究的外箍是由多层薄板组成的，且未提出关键部分即环沟附近的应力计算公式。20 世纪 70 年代，兰州石油化工机器厂和兰州石油机械研究所结合加氢反应器的试制，对双层套箍式厚壁压力容器环沟部位的强度进行了模型试验研究和理论计算[2]，但试验数据较分散，理论计算又是错误的。因此，有必要进一步对环沟部位的应力状态进行分析。由于环沟结构特殊，加之此种容器的内筒和套箍都属于厚壁圆柱壳（$h/R>0.05$），故给分析研究带来了很大的困难。在文献［3］中，作者曾建立了关于厚壁圆柱壳轴对称弯曲的近似理论。本节应用这一理论，研究了双层套箍式厚壁压力容器环沟部位的

① 本文原载《兰州大学学报》，1977，(4)：9-25.

应力状态，给出了计算应力的解析公式。通过实例计算，说明由本节公式所获的数值结果与现有的试验数据是一致的。这说明，本节的公式是可靠的，可供工程设计部门参考使用。最后，我们还附带推得了套箍或内筒是薄壳情况下的计算公式。

一、环沟部位的计算模型

为计算简单起见，我们采用如下 4 条假定：

(1) 假定套箍不承担轴向应力，全部轴向应力由内筒承担；

(2) 鉴于内筒和套箍的长度较环沟宽度大得多，且我们仅研究环沟附近的应力状态，故视内筒和套箍为半无限长圆柱壳；

(3) 不计及套箍端部导角的影响，以两相邻套箍与内筒实际接触间距为环沟计算宽度；

(4) 不计及热套残余应力的影响。

在上述假定下，我们将容器环沟部位(图 1) 的套箍分离开，注意到套箍对内筒的作用犹如一个弹性支承，因而可设介面压力为 kw_b，又由于两相邻套箍对于环沟的对称关系，故我们只需研究下述三个厚壁圆柱壳：属于短圆柱壳的承受均匀内压力 q 的环沟处内筒 (图 2(a))、属于半无限长圆柱壳的承受均匀内压力 q 和介面压力 kw_b 的不包含环沟的内筒 (图 2(b)) 以及属于半无限长圆柱壳的承受介面压力 kw_b 的套箍 (图 2(c))。这里，对于环沟处内筒，将轴向坐标原点选在中央；对于不包含环沟的内筒和套箍，将轴向坐标原点选在靠环沟的边缘处。

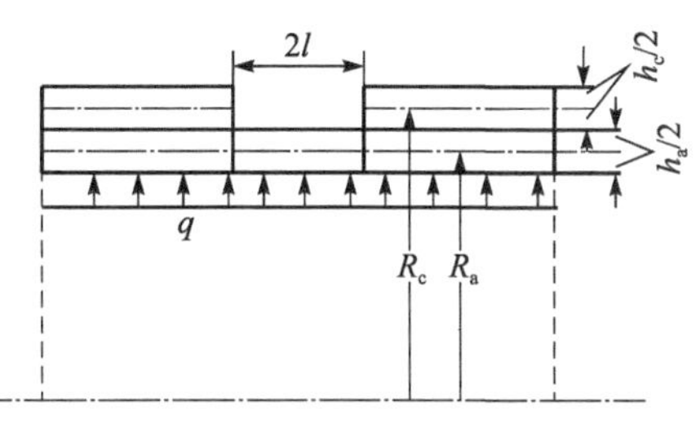

图 1　环沟部位示意图

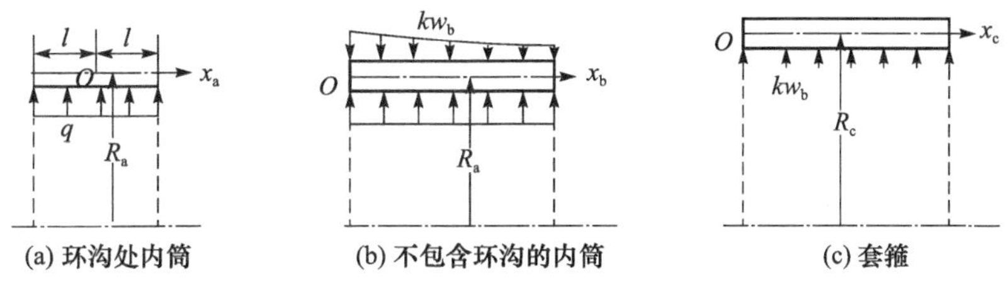

(a) 环沟处内筒　(b) 不包含环沟的内筒　(c) 套箍

图 2　分离后的环沟部位壳体

于是，我们的问题化为研究三个承受法向载荷的厚壁圆柱壳的轴对称应力分布。

二、在内、外法向载荷共同作用下厚壁圆柱壳轴对称弯曲的近似理论

在文献 [3] 中，我们已经推导出在均匀内压力作用下厚壁圆柱壳的基本方程。借助这一方法，我们容易推得厚壁圆柱壳在仅与 x 有关的内、外法向载荷 $P^{\mp}$ 共同作用下的轴对称弯曲近似理论的平衡方程和弹性定理：

$$\frac{\mathrm{d}N_x}{\mathrm{d}x}=0, \tag{1a}$$

$$\frac{\mathrm{d}M_x}{\mathrm{d}x}-Q_x=0, \tag{1b}$$

$$R\frac{\mathrm{d}Q_x}{\mathrm{d}x}+N_\theta-\left(R-\frac{h}{2}\right)P^- -\left(R+\frac{h}{2}\right)P^+ =0 \tag{1c}$$

$$N_\theta=D\left(\frac{w}{R}+\nu\frac{\mathrm{d}u}{\mathrm{d}x}\right)+\frac{K}{R^3}w, \tag{2a}$$

$$N_x=D\left(\frac{\mathrm{d}u}{\mathrm{d}x}+\frac{\nu}{R}w\right)-\frac{K}{R}\frac{\mathrm{d}^2w}{\mathrm{d}x^2}, \tag{2b}$$

$$M_\theta=K\left(\frac{w}{R^2}+\nu\frac{\mathrm{d}^2w}{\mathrm{d}x^2}\right), \tag{2c}$$

$$M_x=K\left(\frac{\mathrm{d}w^2}{\mathrm{d}x^2}-\frac{1}{R}\frac{\mathrm{d}u}{\mathrm{d}x}\right), \tag{2d}$$

其中

$$D=\frac{Eh}{1-\nu^2},\qquad K=\frac{Eh^3}{12(1-\nu^2)} \tag{3}$$

方程（1）和（2）一共七个方程组，表达了五个内力 N_θ、N_x、M_θ、M_x、Q_x 和两个位移 u、w 的关系。我们化简这组方程，以求出一个仅含 w 的微分方程式来进行求解。

微分方程（1b），有

$$\frac{\mathrm{d}Q_x}{\mathrm{d}x}=\frac{\mathrm{d}^2M_x}{\mathrm{d}x^2}$$

将此式代入方程（1c），得

$$\frac{\mathrm{d}^2M_x}{\mathrm{d}x^2}+\frac{N_\theta}{R}-\left(1-\frac{h}{2R}\right)P^- -\left(1+\frac{h}{2R}\right)P^+ =0 \tag{4}$$

对于方程（2a），可改写为

$$N_\theta=D\left(\frac{1+\delta}{R}w+\nu\frac{\mathrm{d}u}{\mathrm{d}x}\right), \tag{5}$$

其中

$$\delta=\frac{h^2}{12R^2} \tag{6}$$

由于 δ 与 1 相比是一个很小的量，故可略去小量 δ（以后仍要应用此近似方法，不再说明），于是，式（5）简化为

$$N_\theta=D\left(\frac{w}{R}+\nu\frac{\mathrm{d}u}{\mathrm{d}x}\right) \tag{7}$$

将式（7）和（2d）代入方程（4），得

$$R^2\delta\frac{\mathrm{d}^4w}{\mathrm{d}x^4}+\frac{w}{R^2}-R\delta\frac{\mathrm{d}^3u}{\mathrm{d}x^3}+\frac{\nu}{R}\frac{\mathrm{d}u}{\mathrm{d}x}=\frac{1}{D}\left[\left(1-\frac{h}{2R}\right)P^- +\left(1+\frac{h}{2R}\right)P^+\right] \tag{8}$$

积分方程（1a），可知 N_x 是常数：

$$N_x=N \tag{9}$$

将此值代入方程（2b），便可解得

$$\frac{\mathrm{d}u}{\mathrm{d}x}=\frac{1}{D}\left(\frac{K}{R}\frac{\mathrm{d}^2w}{\mathrm{d}x^2}+N\right)-\frac{\nu}{R}w \tag{10}$$

最后，将式（10）代入方程（8）中，便得到只含法向位移 w 的一个微分方程：

$$\frac{\mathrm{d}^4w}{\mathrm{d}x^4}+\frac{2\nu}{R^2}\frac{\mathrm{d}^2w}{\mathrm{d}x^2}+\frac{1-\nu^2}{R^4\delta}w=\frac{1}{R^2\delta D}\left[\left(1-\frac{h}{2R}\right)P^-+\left(1+\frac{h}{2R}\right)P^+-\frac{\nu}{R}N\right] \tag{11}$$

引入新自变量：

$$\rho=\frac{x}{R}, \tag{12}$$

方程（11）成为

$$\frac{\mathrm{d}^4w}{\mathrm{d}\rho^4}+2\nu\frac{\mathrm{d}^2w}{\mathrm{d}\rho^2}+\frac{1-\nu^2}{\delta}w=\frac{R^2}{\delta D}\left[\left(1-\frac{h}{2R}\right)P^-+\left(1+\frac{h}{2R}\right)P^+-\frac{\nu}{R}N\right] \tag{13}$$

弹性定律（2）成为

$$N_\theta=\frac{D}{R}\left(w+\nu\frac{\mathrm{d}u}{\mathrm{d}\rho}\right), \tag{14a}$$

$$M_\theta=\frac{K}{R^2}\left(w+\nu\frac{\mathrm{d}^2w}{\mathrm{d}\rho^2}\right), \tag{14b}$$

$$M_x=\frac{K}{R^2}\left(\frac{\mathrm{d}^2w}{\mathrm{d}\rho^2}-\frac{\mathrm{d}u}{\mathrm{d}\rho}\right) \tag{14c}$$

方程（13）是常系数四阶非齐次的常微分方程，其通解易于用标准的方法求得。求得通解后，便可按式（14）确定内力。最后，按文献［3］给出的公式计算环向和轴向应力：

$$\begin{aligned}
\sigma_\theta&=\frac{N_\theta}{h}-\frac{\nu\left(\frac{z}{R}\right)^2}{h(1-\nu^2)\left(1+\frac{z}{R}\right)}N_x-\frac{1}{h^2}\left(\frac{h}{R}+\frac{12\frac{z}{h}}{1+\frac{z}{R}}\right)M_\theta+\frac{\nu}{h^2}\left(\frac{h}{R}-\frac{12\frac{z^2}{Rh}}{1+\frac{z}{R}}\right)M_x,\\
\sigma_x&=\frac{1}{h}\left[1-\frac{\frac{z}{R}}{1-\nu^2}\left(1-\frac{\nu^2}{1+\frac{z}{R}}\right)\right]N_x+\frac{\nu\frac{z}{R}}{h(1-\nu^2)}\left(1-\frac{\nu^2}{1+\frac{z}{R}}\right)N_\theta\\
&\quad+\frac{1}{h^2}\left[\frac{h}{R}-\frac{12z}{h}\left(1-\frac{\nu^2}{1+\frac{z}{R}}\right)\right]M_x-\frac{12\nu z}{h^3\left(1+\frac{z}{R}\right)}M_\theta
\end{aligned} \tag{15}$$

三、环沟处内筒的解

为书写简单，在本小节和第 4，5，8 小节中，除特殊情况外，不在符号右下角标使用 a，b，c 字母。

对于环沟处的内筒，仅有均匀内压力 q 作用，则

$$P^-=q,\qquad P^+=0 \tag{16}$$

由环沟处内筒某一横截面轴向力的平衡得

$$N_x = N = \frac{R}{2}\left(1-\frac{h}{2R}\right)^2 q \tag{17}$$

将式（16）、（17）代入方程（13），便得环沟处内筒的基本方程：

$$\frac{\mathrm{d}^4 w}{\mathrm{d}\rho^4} + 2\nu \frac{\mathrm{d}^2 w}{\mathrm{d}\rho^4} + \alpha_1^4 w = \alpha_2 q, \tag{18}$$

其中

$$\alpha_1^4 = \frac{1-\nu^2}{\delta}, \quad \alpha_2 = \frac{R^2}{\delta D}\left(1-\frac{h}{2R}\right)\left[1-\frac{\nu}{2}\left(1-\frac{h}{2R}\right)\right] \tag{19}$$

方程（18）的通解为

$$w = \mathrm{e}^{-\lambda_1\rho}[\overline{A}_1\cos(\lambda_2\rho)+\overline{A}_2\sin(\lambda_2\rho)] + \mathrm{e}^{\lambda_1\rho}[\overline{A}_3\cos(\lambda_2\rho)+\overline{A}_4\sin(\lambda_2\rho)] + \frac{\alpha_2}{\alpha_1^4}q, \tag{20}$$

其中

$$\lambda_1 = \alpha_1\cos\left[\frac{1}{2}\arctan\left(\frac{\alpha_1^2}{\nu}\right)\right], \quad \lambda_2 = \alpha_1\sin\left[\frac{1}{2}\arctan\left(\frac{\alpha_1^2}{\nu}\right)\right] \tag{21}$$

用双曲函数代替式（20）的指数函数，即

$$\mathrm{e}^{-\lambda_1\rho} = \cosh(\lambda_1\rho) - \sinh(\lambda_1\rho), \quad \mathrm{e}^{\lambda_1\rho} = \cosh(\lambda_1\rho) + \sinh(\lambda_1\rho) \tag{22}$$

经过某些简化后，式（20）成为

$$\begin{aligned} w = {} & A_1\sin(\lambda_2\rho)\sinh(\lambda_1\rho) + A_2\cos(\lambda_2\rho)\cosh(\lambda_1\rho) + A_3\sin(\lambda_2\rho)\cosh(\lambda_1\rho) \\ & + A_4\cos(\lambda_2\rho)\sinh(\lambda_1\rho) + \frac{\alpha_2}{\alpha_1^4}q \end{aligned} \tag{23}$$

由于问题的对称性，法向位移 w 必定对称于原点。为了保证解中只有对称的项出现，A_3 和 A_4 一定要等于零，于是解（23）简化为

$$w = A_1\sin(\lambda_2\rho)\sinh(\lambda_1\rho) + A_2\cos(\lambda_2\rho)\cosh(\lambda_1\rho) + \frac{\alpha_2}{\alpha_1^4}q \tag{24}$$

由此，还可得转角

$$\begin{aligned} \psi = \frac{\mathrm{d}w}{\mathrm{d}x} = \frac{1}{R}[& (\lambda_2 A_1 + \lambda_1 A_2)\cos(\lambda_2\rho)\sinh(\lambda_1\rho) \\ & + (\lambda_1 A_1 - \lambda_2 A_2)\sin(\lambda_2\rho)\cosh(\lambda_1\rho)] \end{aligned} \tag{25}$$

将式（24）代入式（14），得内力

$$\begin{aligned} N_\theta = {} & \left\{\frac{D}{R}A_1[(1-\nu^2)\sin(\lambda_2\rho)\sinh(\lambda_1\rho) + 2\nu\lambda_1\lambda_2\delta\cos(\lambda_2\rho)\cosh(\lambda_1\rho)]\right. \\ & \left. + A_2[(1-\nu^2)\cos(\lambda_2\rho)\cosh(\lambda_1\rho) - 2\nu\lambda_1\lambda_2\delta\sin(\lambda_2\rho)\sinh(\lambda_1\rho)]\right\} \\ & + R\left(1-\frac{h}{2R}\right)q, \end{aligned} \tag{26a}$$

$$\begin{aligned} M_\theta = {} & \frac{K}{R^2}\{A_1[(1+\nu(\lambda_1^2-\lambda_2^2))\sin(\lambda_2\rho)\sinh(\lambda_1\rho) + 2\nu\lambda_1\lambda_2\cos(\lambda_2\rho)\cosh(\lambda_1\rho)] \\ & + A_2[(1+\nu(\lambda_1^2-\lambda_2^2))\cos(\lambda_2\rho)\cosh(\lambda_1\rho) - 2\nu\lambda_1\lambda_2\sin(\lambda_2\rho)\sinh(\lambda_1\rho)]\} \\ & + \frac{\alpha_2 K}{\alpha_1^4 R^2}q, \end{aligned} \tag{26b}$$

$$M_x=\frac{K}{R^2}\{A_1[(\lambda_1^2-\lambda_2^2+\nu)\sin(\lambda_2\rho)\sinh(\lambda_1\rho)+2\lambda_1\lambda_2\cos(\lambda_2\rho)\cosh(\lambda_1\rho)]$$
$$+A_2[(\lambda_1^2-\lambda_2^2+\nu)\cos(\lambda_2\rho)\cosh(\lambda_1\rho)-2\lambda_1\lambda_2\sin(\lambda_2\rho)\sinh(\lambda_1\rho)]\} \tag{26c}$$
$$+\frac{R^2}{\alpha_1^4}\left(1-\frac{h}{2R}\right)\left[\nu-\frac{1}{2}\left(1-\frac{h}{2R}\right)\right]q$$

对于横向力 Q_x，由式（1b）、(26c) 得

$$Q_x=\frac{K}{R^3}\{A_1[\lambda_2(3\lambda_1^2-\lambda_2^2+\nu)\cos(\lambda_2\rho)\sinh(\lambda_1\rho)+\lambda_1(\lambda_1^2-3\lambda_2^2$$
$$+\nu)\sin(\lambda_2\rho)\cosh(\lambda_1\rho)]+A_2[\lambda_1(\lambda_1^2-3\lambda_2^2+\nu)\cos(\lambda_2\rho)\sinh(\lambda_1\rho) \tag{27}$$
$$-\lambda_2(3\lambda_1^2-\lambda_2^2+\nu)\sin(\lambda_2\rho)\cosh(\lambda_1\rho)]\}$$

四、不包含环沟的内筒的解

对于不包含环沟的内筒，除承受均匀内压力 q 外，尚有介面压力 kw 作用，即

$$P^-=q,\quad P^+=-kw \tag{28}$$

关于 N 值，显而易见，与环沟外内筒的公式相同。将这些值代入方程（13），便得不包含环沟的内筒的基本方程：

$$\frac{\mathrm{d}^4w}{\mathrm{d}\rho^4}+2\nu\frac{\mathrm{d}^2w}{\mathrm{d}\rho^2}+\alpha_1^4w=\alpha_2q, \tag{29}$$

其中

$$\alpha_1^4=\frac{1-\nu^2}{\delta}\left[1+\frac{kR^2}{Eh}\left(1+\frac{h}{2R}\right)\right],$$
$$\alpha_2=\frac{R^2}{\delta D}\left(1-\frac{h}{2R}\right)\left[1-\frac{\nu}{2}\left(1+\frac{h}{2R}\right)\right] \tag{30}$$

方程（29）的通解为

$$w=\mathrm{e}^{-\lambda_1\rho}[B_1\cos(\lambda_2\rho)+B_2\sin(\lambda_2\rho)]+\mathrm{e}^{\lambda_1\rho}[B_3\cos(\lambda_2\rho)+B_4\sin(\lambda_2\rho)]+\frac{\alpha_2}{\alpha_1^4}q, \tag{31}$$

其中 λ_1 和 λ_2 的公式形状与式（21）相同。

由于不包含环沟的内筒为半无限长圆柱壳，故应有

$$B_3=B_4=0, \tag{32}$$

则解（31）简化为

$$W=\mathrm{e}^{-\lambda_1\rho}[B_1\cos(\lambda_2\rho)+B_2\sin(\lambda_2\rho)]+\frac{\alpha_2}{\alpha_1^4}q \tag{33}$$

由此，得转角和内力

$$\psi=\frac{1}{R}\mathrm{e}^{-\lambda_1\rho}[(\lambda_2B_2-\lambda_1B_1)\cos(\lambda_2\rho)-(\lambda_2B_1+\lambda_1B_2)\sin(\lambda_2\rho)], \tag{34a}$$

$$N_\theta=\frac{D}{R}\mathrm{e}^{-\lambda_1\rho}\{B_1[(1-\nu^2)\cos(\lambda_2\rho)+2\nu\lambda_1\lambda_2\delta\sin(\lambda_2\rho)]$$
$$+B_2[(1-\nu^2)\sin(\lambda_2\rho)-2\nu\lambda_1\lambda_2\delta\cos(\lambda_2\rho)]\} \tag{34b}$$
$$+\left[\frac{\alpha_2Eh}{\alpha_1^4R}+\frac{\nu R}{2}\left(1-\frac{h}{2R}\right)^2\right]q,$$

$$M_\theta=\frac{K}{R^2}\mathrm{e}^{-\lambda_1\rho}\left(B_1\{[1+\nu(\lambda_1^2-\lambda_2^2)]\cos(\lambda_2\rho)+2\nu\lambda_1\lambda_2\sin(\lambda_2\rho)\}\right.$$
$$\left.+B_2\{[1+\nu(\lambda_1^2-\lambda_2^2)]\sin(\lambda_2\rho)-2\nu\lambda_1\lambda_2\cos(\lambda_2\rho)\}\right)+\frac{\alpha_2K}{\alpha_1^4R^2}q, \tag{34c}$$

$$M_x=\frac{K}{R^2}\mathrm{e}^{-\lambda_1\rho}\{B_1[(\lambda_1^2-\lambda_2^2+\nu)\cos(\lambda_2\rho)+2\lambda_1\lambda_2\sin(\lambda_2\rho)]$$
$$+B_2[(\lambda_1^2-\lambda_2^2+\nu)\sin(\lambda_2\rho)-2\lambda_1\lambda_2\cos(\lambda_2\rho)]\}$$
$$+\delta\left[\frac{\nu\alpha_2D}{\alpha_1^4}-\frac{R^2}{2}\left(1-\frac{h}{2R}\right)^2\right]q, \tag{34d}$$

$$Q_x=-\frac{K}{R^3}\mathrm{e}^{-\lambda_1\rho}\{B_1[\lambda_1(\lambda_1^2-3\lambda_2^2+\nu)\cos(\lambda_2\rho)+\lambda_2(3\lambda_1^2-\lambda_2^2+\nu)\sin(\lambda_2\rho)]$$
$$+B_2[\lambda_1(\lambda_1^2-3\lambda_2^2+\nu)\sin(\lambda_2\rho)-\lambda_2(3\lambda_1^2-\lambda_2^2+\nu)\cos(\lambda_2\rho)]\} \tag{34e}$$

五、弹性系数 k 的确定

对于套箍，仅在内壁作用有界面压力 kw_b。因为内筒和套箍在界面上要满足法向位移连续条件，故可将 kw_b 改写为 kw_c。又由假定知，轴向力 N_x 为零。则有

$$P^-=kw,\quad P^+=0,\quad N=0 \tag{35}$$

将这些值代入方程（13），便得套箍的基本方程：

$$\frac{\mathrm{d}^4w}{\mathrm{d}\rho^4}+2\nu\frac{\mathrm{d}^2w}{\mathrm{d}\rho^2}+\alpha_1^4w=0, \tag{36}$$

其中

$$\alpha_1^4=\frac{1-\nu^2}{\delta}\left[1-\frac{kR^2}{Eh}\left(1-\frac{h}{2R}\right)\right] \tag{37}$$

方程（36）的通解为

$$w=\mathrm{e}^{-\lambda_1\rho}[c_1\cos(\lambda_2\rho)+c_2\sin(\lambda_2\rho)]+\mathrm{e}^{\lambda_1\rho}[c_3\cos(\lambda_2\rho)+c_4\sin(\lambda_2\rho)], \tag{38}$$

其中 λ_1 和 λ_2 的公式形状与式（21）相同。

由于套箍是半无限长圆柱壳，故应有

$$c_3=c_4=0 \tag{39}$$

则解（38）简化为

$$w=\mathrm{e}^{-\lambda_1\rho}[c_1\cos(\lambda_2\rho)+c_2\sin(\lambda_2\rho)] \tag{40}$$

由此得轴向弯矩和横内力：

$$M_x=\frac{K}{R^2}\mathrm{e}^{-\lambda_1\rho}\left\{c_1[(\lambda_1^2-\lambda_2^2+\nu)\cos(\lambda_2\rho)+2\lambda_1\lambda_2\sin(\lambda_2\rho)]\right.$$
$$\left.+c_2[(\lambda_1^2-\lambda_2^2+\nu)\sin(\lambda_2\rho)-2\lambda_1\lambda_2\cos(\lambda_2\rho)]\right\},$$
$$Q_x=-\frac{K}{R^3}\mathrm{e}^{-\lambda_1\rho}\left\{[\lambda_1(\lambda_1^2-3\lambda_2^2+\nu)\cos(\lambda_2\rho)+\lambda_2(3\lambda_1^2-\lambda_2^2+\nu)\sin(\lambda_2\rho)]\right.$$
$$\left.+c_2[\lambda_1(\lambda_1^2-3\lambda_2^2+\nu)\sin(\lambda_2\rho)-\lambda_2(3\lambda_1^2-\lambda_2^2+\nu)\cos(\lambda_2\rho)]\right\} \tag{41}$$

在套箍端部，需要满足自由边界条件：

$$\text{当}\ \rho=0\ \text{时}, \quad M_x=0, \quad Q_x=0 \tag{42}$$

将式（41）代入边界条件（42），得

$$\begin{aligned}&(\lambda_1^2-\lambda_2^2+\nu)c_1-2\lambda_1\lambda_2c_2=0,\\&\lambda_1(\lambda_1^2-3\lambda_2^2+\nu)c_1-\lambda_2(3\lambda_1^2-\lambda_2^2+\nu)c_2=0\end{aligned} \tag{43}$$

这是关于 c_1 和 c_2 的线性齐次方程组，欲使 c_1 和 c_2 有一个不恒为零的解组，就须使方程组（43）的系数行列式为零，即

$$\lambda_2[(\lambda_1^2+\lambda_2^2)^2+2\nu(\lambda_1^2-\lambda_2^2)]=0 \tag{44}$$

这是一个关于弹性系数 k 的超越方程，解此方程，就能确定 k 值。

要使式（44）成立，就须

$$(\lambda_1^2+\lambda_2^2)^2+2\nu(\lambda_1^2-\lambda_2^2)=0 \tag{45a}$$

或

$$\lambda_2=0 \tag{45b}$$

先研究方程（45a）。将 λ_1 和 λ_2 的公式代入，得

$$\alpha_1^2\left\{\alpha_1^2+2\nu\cos\left[\arctan\left(\frac{\alpha_1^2}{\nu}\right)\right]\right\}=0 \tag{46}$$

要使此式成立，就须

$$\alpha_1^2+2\nu\cos\left[\arctan\left(\frac{\alpha_1^2}{\nu}\right)\right]=0 \tag{47a}$$

或

$$\alpha_1^2=0 \tag{47b}$$

对于方程（47a），可改写为

$$\cos^2\left[\arctan\left(\frac{\alpha_1^2}{\nu}\right)\right]=\frac{\alpha_1^4}{4\nu^2} \tag{48}$$

应用熟知的三角公式：

$$\cos^2\varphi=\frac{1}{1+\tan^2\varphi}$$

方程（48）成为

$$\left(\frac{\alpha_1^2}{\nu}\right)^4+\left(\frac{\alpha_1^2}{\nu}\right)^2-4=0 \tag{49}$$

这是一个关于 $\left(\frac{\alpha_1^2}{\nu}\right)^2$ 的二次方程，其根为

$$\left(\frac{\alpha_1^2}{\nu}\right)^2=-\frac{1}{2}\ (1\mp\sqrt{17}) \tag{50}$$

将式（37）代入式（50），有

$$\frac{1-\nu^2}{\delta}\left[1-\frac{kR^2}{Eh}\left(1-\frac{h}{2R}\right)\right]+\frac{\nu^2}{2}(1\mp\sqrt{17})=0 \tag{51}$$

略去第二项小量，上式简化为

$$\alpha_1^4=\frac{1-\nu^2}{\delta}\left[1-\frac{kR^2}{Eh}\left(1-\frac{h}{2R}\right)\right]=0 \tag{52}$$

由此方程，弹性系数 k 得以确定：

$$k=\frac{Eh}{R^2\left(1-\frac{h}{2R}\right)} \tag{53}$$

显而易见，对于尚未研究的方程式（45b）、式（47b），将获得与此相同的公式。

六、积分常数的确定

环沟处的内筒和不包含环沟的内筒本是一个整体，故在其交界处，应满足下列连续条件：

当 $\rho_a=\frac{l}{R_a}=\rho_1$，$\rho_b=0$ 时，

$$w_a=w_b,\quad \psi_a=\psi_b,\quad M_{x,a}=M_{x,b},\quad Q_{x,a}=Q_{x,b} \tag{54}$$

将式（24）～(27)、(33)、(34a，d，e) 代入连续条件（54），便得关于 A_1、A_2、B_1、B_2 的四元联立代数方程组：

$$\begin{aligned}&A_1\sin(\lambda_{2,a}\rho_1)\sinh(\lambda_{1,a}\rho_1)+A_2\cos(\lambda_{2,a}\rho_1)\cosh(\lambda_{1,a}\rho_1)-B_1\\&=\alpha_{2,a}\left(\frac{1}{\alpha_{1,b}^4}-\frac{1}{\alpha_{1,a}^4}\right)q,\end{aligned} \tag{55a}$$

$$\begin{aligned}&[\lambda_{2,a}\cos(\lambda_{2,a}\rho_1)\sinh(\lambda_{1,a}\rho_1)+\lambda_{1,a}\sin(\lambda_{2,a}\rho_1)\cosh(\lambda_{1,a}\rho_1)]A_1\\&+[\lambda_{1,a}\cos(\lambda_{2,a}\rho_1)\sinh(\lambda_{1,a}\rho_1)-\lambda_{2,a}\sin(\lambda_{2,a}\rho_1)\cosh(\lambda_{1,a}\rho_1)]A_2\\&+\lambda_{1,b}B_1-\lambda_{2,b}B_2=0,\end{aligned} \tag{55b}$$

$$\begin{aligned}&[(\lambda_{1,a}^2-\lambda_{2,a}^2+\nu_a)\sin(\lambda_{2,a}\rho_1)\sinh(\lambda_{1,a}\rho_1)+2\lambda_{1,a}\lambda_{2,a}\cos(\lambda_{2,a}\rho_1)\cosh(\lambda_{1,a}\rho_1)]A_1\\&+[(\lambda_{1,a}^2-\lambda_{2,a}^2+\nu_a)\cos(\lambda_{2,a}\rho_1)\cosh(\lambda_{1,a}\rho_1)-2\lambda_{1,a}\lambda_{2,a}\sin(\lambda_{2,a}\rho_1)\sinh(\lambda_{1,a}\rho_1)]A_2\\&-(\lambda_{1,b}^2-\lambda_{2,b}^2+\nu_a)B_1+2\lambda_{1,b}\lambda_{2,b}B_2\\&=\frac{1}{D_a}\left\{\frac{\nu_a\alpha_{2,a}D_a}{\alpha_{1,b}^4}-\frac{R_a^2}{2}\left(1-\frac{h_a}{2R_a}\right)^2-\frac{R_a^2}{1-\nu_a^2}\left(1-\frac{h_a}{2R_a}\right)\left[\nu_a-\frac{1}{2}\left(1-\frac{h_a}{2R_a}\right)\right]\right\}q,\end{aligned} \tag{55c}$$

$$\begin{aligned}&[\lambda_{2,a}(3\lambda_{1,a}^2-\lambda_{2,a}^2+\nu_a)\cos(\lambda_{2,a}\rho_1)\sinh(\lambda_{1,a}\rho_1)+\lambda_{1,a}(\lambda_{1,a}^2-3\lambda_{2,a}^2\\&+\nu_a)\sin(\lambda_{2,a}\rho_1)\cosh(\lambda_{1,a}\rho_1)]A_1+\lambda_{1,a}[(\lambda_{1,a}^2-3\lambda_{2,a}^2+\nu_a)\cos(\lambda_{2,a}\rho_1)\sinh(\lambda_{1,a}\rho_1)\\&-\lambda_{2,a}(3\lambda_{1,a}^2-\lambda_{2,a}^2+\nu_a)\sin(\lambda_{2,a}\rho_1)\cosh(\lambda_{1,a}\rho_1)]A_2+\lambda_{1,b}(\lambda_{1,b}^2-3\lambda_{2,b}^2+\nu_a)B_1\\&-\lambda_{2,b}(3\lambda_{1,b}^2-\lambda_{2,b}^2+\nu_a)B_2=0\end{aligned} \tag{55d}$$

解此方程组。先由方程（55a）得

$$B_1=A_1\sin(\lambda_{2,a}\rho_1)\sinh(\lambda_{1,a}\rho_1)+A_2\cos(\lambda_{2,a}\rho_1)\cosh(\lambda_{1,a}\rho_1)+\alpha_{2,a}\left(\frac{1}{\alpha_{1,a}^4}-\frac{1}{\alpha_{1,b}^4}\right)q \tag{56}$$

将此式代入方程（55b），得

$$\begin{aligned}B_2=\frac{1}{\lambda_{2,b}}\Big\{&A_1[\lambda_{2,a}\cos(\lambda_{2,a}\rho_1)\sinh(\lambda_{1,a}\rho_1)+\lambda_{1,a}\sin(\lambda_{2,a}\rho_1)\cosh(\lambda_{1,a}\rho_1)\\&+\lambda_{1,b}\sin(\lambda_{2,a}\rho_1)\sinh(\lambda_{1,a}\rho_1)]+A_2[\lambda_{1,a}\cos(\lambda_{2,a}\rho_1)\sinh(\lambda_{1,a}\rho_1)\\&-\lambda_{2,a}\sin(\lambda_{2,a}\rho_1)\cosh(\lambda_{1,a}\rho_1)+\lambda_{1,b}\cos(\lambda_{2,a}\rho_1)\cosh(\lambda_{1,a}\rho_1)]\\&+\lambda_{1,b}\alpha_{2,a}\left(\frac{1}{\alpha_{1,a}^4}-\frac{1}{\alpha_{1,b}^4}\right)q\Big\}\end{aligned} \tag{57}$$

再将式（56）、（57）代入方程（55c，d），便得关于 A_1 和 A_2 的二元联立代数方程组：

$$\begin{aligned}a_{11}A_1+a_{12}A_2&=a_1q,\\ a_{21}A_1+a_{22}A_2&=a_2q,\end{aligned}\tag{58}$$

其中

$$\begin{aligned}
a_{11}=&(\lambda_{1,a}^2-\lambda_{2,a}^2+\lambda_{1,b}^2+\lambda_{2,a}^2)\sin(\lambda_{2,a}\rho_1)\sinh(\lambda_{1,a}\rho_1)\\
&+2\lambda_{1,a}\lambda_{2,a}\cos(\lambda_{2,a}\rho_1)\cosh(\lambda_{1,a}\rho_1)+2\lambda_{1,b}[\lambda_{2,a}\cos(\lambda_{2,a}\rho_1)\sinh(\lambda_{1,a}\rho_1)\\
&+\lambda_{1,a}\sin(\lambda_{2,a}\rho_1)\cosh(\lambda_{1,a}\rho_1)],\\
a_{12}=&-2\lambda_{1,a}\lambda_{2,a}\sin(\lambda_{2,a}\rho_1)\sinh(\lambda_{1,a}\rho_1)+(\lambda_{1,a}^2-\lambda_{2,a}^2+\lambda_{1,b}^2\\
&+\lambda_{2,b}^2)\cos(\lambda_{2,a}\rho_1)\cosh(\lambda_{1,a}\rho_1)+2\lambda_{1,b}[\lambda_{1,a}\cos(\lambda_{2,a}\rho_1)\sinh(\lambda_{1,a}\rho_1)\\
&-\lambda_{2,a}\sin(\lambda_{2,a}\rho_1)\cosh(\lambda_{1,a}\rho_1)],\\
a_{21}=&-2\lambda_{1,b}(\lambda_{1,b}^2+\lambda_{2,b}^2)\sin(\lambda_{2,a}\rho_1)\sinh(\lambda_{1,a}\rho_1)\\
&+\lambda_{2,a}[3(\lambda_{1,a}^2-\lambda_{1,b}^2)-\lambda_{2,a}^2+\lambda_{2,b}^2]\cos(\lambda_{2,a}\rho_1)\sinh(\lambda_{1,a}\rho_1)\\
&+\lambda_{1,a}[\lambda_{1,a}^2+\lambda_{2,b}^2-3(\lambda_{1,b}^2+\lambda_{2,a}^2)]\sin(\lambda_{2,a}\rho_1)\cosh(\lambda_{1,a}\rho_1),\\
a_{22}=&-2\lambda_{1,b}(\lambda_{1,b}^2+\lambda_{2,b}^2)\cos(\lambda_{2,a}\rho_1)\cosh(\lambda_{1,a}\rho_1)\\
&+\lambda_{1,a}[\lambda_{1,a}^2+\lambda_{2,b}^2-3(\lambda_{1,b}^2+\lambda_{2,a}^2)]\cos(\lambda_{2,a}\rho_1)\sinh(\lambda_{1,a}\rho_1)\\
&+\lambda_{2,a}[\lambda_{2,a}^2-\lambda_{2,b}^2-3(\lambda_{1,a}^2-\lambda_{2,b}^2)]\sin(\lambda_{2,a}\rho_1)\cosh(\lambda_{1,a}\rho_1),\\
a_1=&\alpha_{2,a}(\nu_a-\lambda_{1,b}^2-\lambda_{2,b}^2)\left(\frac{1}{\alpha_{1,a}^4}-\frac{1}{\alpha_{1,b}^4}\right)\\
&+\frac{\nu_aR_a^2}{E_ah_a}\left(1-\frac{h_a}{2R_a}\right)\left[\frac{\nu_a}{2}\left(1-\frac{h_a}{2R_a}\right)-1\right]+\frac{\nu\alpha_{2,a}}{\alpha_{1,b}^4},\\
a_2=&2\lambda_{1,b}\alpha_{2,a}(\lambda_{1,b}^2+\lambda_{2,b}^2)\left(\frac{1}{\alpha_{1,a}^4}-\frac{1}{\alpha_{1,b}^4}\right)
\end{aligned}\tag{59}$$

解方程组（58），得

$$\begin{aligned}A_1&=a_0(a_1a_{22}-a_2a_{12})q,\\ A_2&=a_0(a_2a_{11}-a_1a_{21})q,\end{aligned}\tag{60}$$

其中

$$a_0=\frac{1}{a_{11}a_{22}-a_{12}a_{21}}\tag{61}$$

七、算例

计算内径 144mm 套箍式小容器环沟附近的应力。已知：

$$h_a=0.6\text{cm},\quad h_c=0.5\text{cm},\quad R_a=7.5\text{cm},$$
$$R_c=8.05\text{cm},\quad l=0.375\text{cm},\quad q=100\text{kg/cm}^2$$

容器内筒和套箍材料为 $20^{\#}$ 钢，在常温下，其弹性模量和泊松比为

$$E=2.02\times10^6\text{kg/cm}^2,\quad \nu=0.26$$

在此例中，

$$\frac{h_a}{R_a}=0.080,\quad \frac{h_c}{R_c}=0.062,$$

因此，此小容器内筒和套箍均属于厚壁圆柱壳，故应使用前面公式计算。

将上述数据代入式（15）、（17）、（26）、（34b～d）、（53）、（56）、（57）、（60），我们得到环沟附近内筒内、外壁（$z=\mp h/2$）的应力值，结果给在表1、表2和图3、图4中。为了比较，我们给出了内筒内壁环向应力 σ_θ^- 的试验值[2]，并绘制了图5和图6。试验有两组，一组用“×”表示，我们称为第一组试验；一组用“•”表示，我们称为第二组试验。由图表看出，理论值与试验值是一致的。本节的理论值非常接近第一组试验值，相对误差在7.39%以内。

表1　环沟处内筒的应力值　　（单位：kg/cm^2）

x/mm	σ_θ^+ 理论值	σ_θ^-			误差/%	σ_x^+ 理论值	σ_x^- 理论值
		理论值	试验值				
			第一组	第二组			
0	823.6	808.6				705.6	442.9
0.25	823.5	808.7	838		−3.50	705.5	443.0
0.75	823.1	808.8		1078	−25.0	704.2	444.3
1.25	822.1	809.2	827	1113	−2.15～27.3	701.7	446.9
1.75	820.7	809.8	857		−5.51	697.9	450.8
2.75	816.6	811.5				686.6	462.4
3.75	810.6	814.1				670.2	479.3

表2　不包含环沟的内筒的应力值　　（单位：kg/cm^2）

x/cm	σ_θ^+ 理论值	σ_θ^-			误差/%	σ_x^+ 理论值	σ_x^- 理论值
		理论值	试验值				
			第一组	第二组			
0	810.7	814.1	879		−7.39	670.2	479.2
0.25	793.9	819.0		791	3.54	627.9	522.7
0.50	778.0	819.8		908	−9.71	595.2	556.3
0.80	761.8	816.9		960	−14.9	566.9	585.4
1.50	731.7	802.2				535.9	617.2
2.10	717.7	788.4		710	11.0	533.5	619.6
3.00	710.1	773.4				545.8	607.0
4.00	710.8	766.1				560.5	591.9
5.10	713.8	764.7	791		−3.32	568.9	583.2

由计算结果看到，在环沟宽度很小的情况下，环沟处内筒的应力分布有两个特点：①应力沿着环沟宽度变化不剧烈；②环向应力沿着壳厚的变化很小。

由图表可知，环沟部位的最大应力是环向应力，发生在环沟处内筒外壁的中央处，其值为

$$\sigma_{max}=\sigma_{\theta,0}^+=823.6kg/cm^2$$

按照拉梅公式计算，内筒和套箍结构筒体的最大应力分别为 $1252kg/cm^2$ 和 $708.1kg/cm^2$。

这说明，当环沟宽度较小时，一方面，套箍对环沟有加强作用，最大应力仅为不加强情况的 65.8%；另一方面，与没有环沟的套箍结构情况相比，应力增加 16.3%，亦即环沟削弱了容器强度。

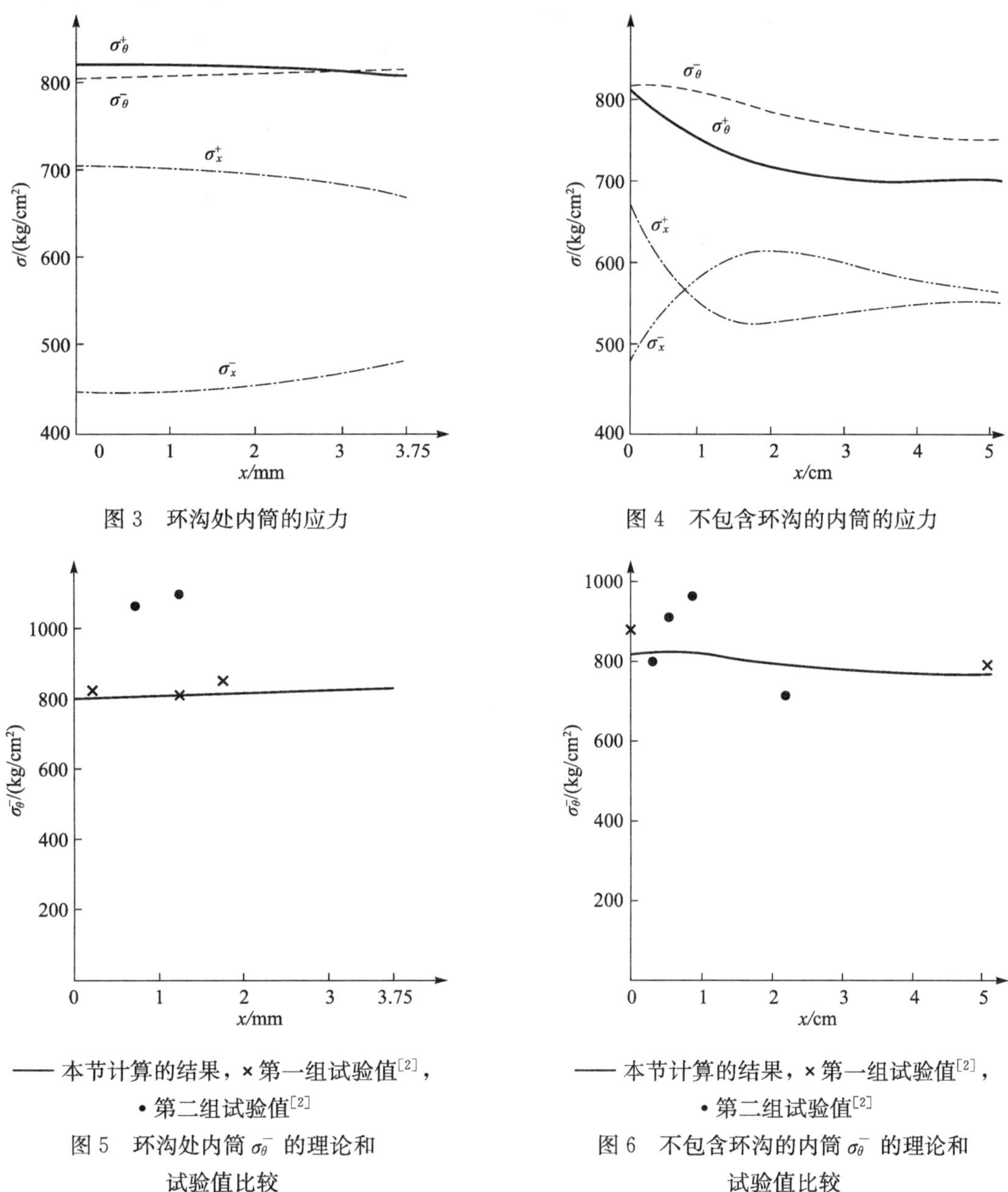

图 3　环沟处内筒的应力

图 4　不包含环沟的内筒的应力

—— 本节计算的结果，× 第一组试验值[2]，• 第二组试验值[2]

图 5　环沟处内筒 σ_θ^- 的理论和试验值比较

—— 本节计算的结果，× 第一组试验值[2]，• 第二组试验值[2]

图 6　不包含环沟的内筒 σ_θ^- 的理论和试验值比较

另外，我们还研究了环沟宽度的变化对最大应力的影响。由计算可知，在环沟宽度不太大的范围内，环沟部位的最大应力一般都是环沟处内筒的环向应力，发生在外壁的中央或内壁的边缘处，且随环沟宽度的增大而增大，图 7 示出了环沟处内筒外壁中央和内壁边缘的环向应力曲线。从图 7 容易看出，在环沟宽度小于 0.6 倍套箍结构筒体厚度的情况下，最大应力是内壁边缘的环向应力，反之，是外壁中央的环向应力。一方面，在环沟宽度小于四倍筒体厚度时，环沟处内筒的最大应力比由拉梅公式所计算的内筒最大应力小，这说明套箍对环沟处内筒有加强作用；另一方面，环沟处内筒的最大应力又比由拉梅公式所计算的套箍结构筒体的最大应力大，这说明环沟的存在又削弱了容器的强度。为了醒目起见，我们给出了环沟处内筒的最大应力和由拉梅公式所计算的套箍结构筒体的最大应力的相对差（图 8），以表示环沟宽度对环沟处内筒强度的削弱情况。

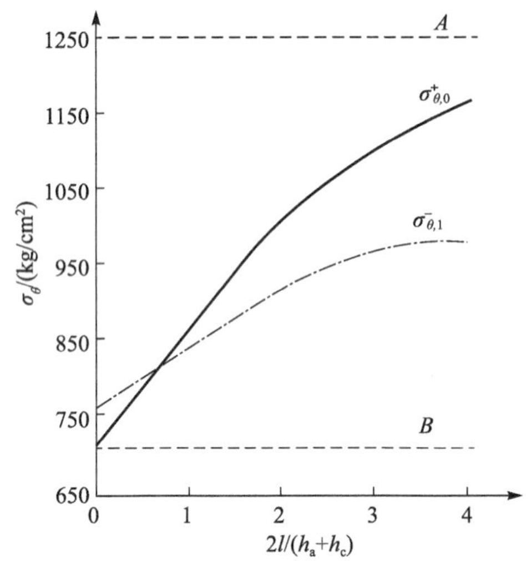

图 7　环沟处内筒 $\sigma^+_{\theta,0}$ 和 $\sigma^-_{\theta,1}$ 曲线
A 为内筒 σ_θ^- 的拉梅公式值；B 为套箍结构筒体 σ_θ^- 的拉梅公式值

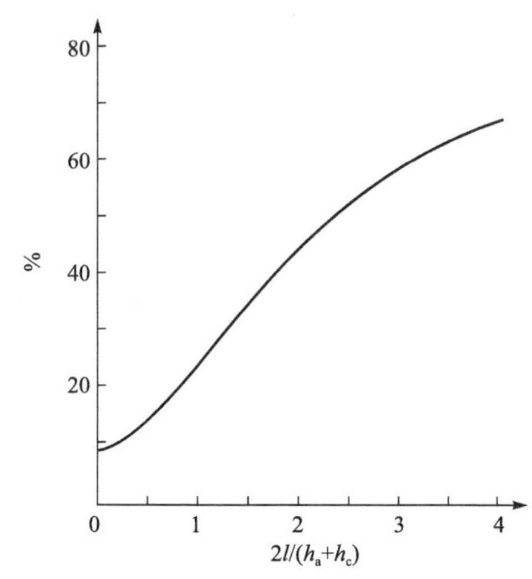

图 8　环沟宽度对环沟处内筒强度的削弱情况

八、小结

由前面实例的数值计算和分析，我们有下面几点粗浅看法。

(1) 本节理论值和试验值的一致性，说明本节所建立的公式是可靠的，可供工程设计部门参考应用。

(2) 套箍式容器的环沟削弱了容器的强度，而且环沟宽度越大，削弱越厉害，故制造时，应尽量减小环沟宽度。

(3) 工程设计中，仅计算环沟处内筒的最大应力即可。由于最大应力位置不固定，为计算简单起见，以环沟处内筒外壁中央环向应力作为最大应力。

(4) 对于套箍是薄壳，或内筒是薄壳，或套箍和内筒都是薄壳的情况，用前面公式计算过于复杂了。此时，可在前面公式中略去小量即能得到所需公式。下面，我们给出上述三种情况下环沟处内筒的向应力公式：

(a) 在内筒是厚壳而套箍是薄壳的情况下，仍使用前面的公式，仅将弹性系数 k 的公式改为下式：

$$k=\frac{E_c h_c}{R_c^2} \tag{62}$$

(b) 在内筒和套箍均是薄壳的情况下

$$\sigma_\theta=\frac{N_\theta}{h}-\frac{12M_\theta z}{h^3}, \tag{63}$$

其中

$$N_\theta=\frac{Eh}{R}[A_1\sin(\lambda\rho)\sinh(\lambda\rho)+A_2\cos(\lambda\rho)\cosh(\lambda\rho)]+qR,$$

$$M_\theta=\frac{2\nu k\lambda^2}{R^2}[A_1\cos(\lambda\rho)\cosh(\lambda\rho)-A_2\sin(\lambda\rho)\sinh(\lambda\rho)],$$

$$A_1=\mu q[2\beta\sin\omega\sinh\omega-(1-\beta^2)\cos\omega\sinh\omega+(1+\beta^2)\sin\omega\cosh\omega],$$

$$A_2=\mu q[2\beta\sin\omega\cosh\omega+(1+\beta^2)\cos\omega\sinh\omega+(1-\beta^2)\sin\omega\cosh\omega],$$

$$\mu=\frac{R^2(2-\nu)(\beta^4-1)}{Eh[(1-\beta^2)^2\sin(2\omega)+2\beta(1-\beta^2)\cos(2\omega)+(1+\beta^2)^2\sinh(2\omega)+2\beta(1+\beta^2)\cosh(2\omega)]},$$

$$\beta=\sqrt[4]{\frac{Eh}{Eh+kR^2}},$$

$$\lambda=\sqrt[4]{3(1-\nu^2)R^2/h^2},$$

$$\omega=\lambda\rho_1; \tag{64}$$

$$k=\frac{E_c h}{R_c^2} \tag{65}$$

(c) 在内筒是薄壳而套箍是厚壳的情况下，仍使用式（63）、（64），仅将式（65）改为式（53）。

参考文献

[1] Palmer P J. Band reinforced and layer built pressure vessels. British Welding Journal，1961，8 (2)：51-57.

[2] 兰州石油化工机器厂，兰州石油机械研究所．套箍式容器试验．化工炼油机械通讯，1975，(6)：1-21.

[3] 刘人怀．加氢反应器顶部厚壁壳体的应力分析．化工炼油机械通讯，1975，(6)：40-59.

厚壁球壳的弯曲理论及其在高压容器上的应用①

在石油、化学、原子能和动力工业中，高压容器是关键设备之一。随着这些工业部门的飞跃发展，对高压容器的设计和制造，提出了更高的要求。在高压容器中，用球形壳体制造容器和封头的情况是很多的。一般说来，它们是厚壁的壳体。然而，现有的文献大多是以薄壁球壳为研究对象。若仍依据这些文献来进行厚壁球壳的设计，则显然是不合理的。有的设计中即使采用了厚壁壳体理论的拉梅公式，但仍没有解决极其重要的弯曲应力的计算问题。因此，为高压容器设计建立厚壁球壳弯曲应力的计算公式是必要的。

厚壁球壳弯曲理论，属于弹性力学的空间问题，由于数学上的困难，过去只有少数人进行了研究。Рабинович[1]采用了通常的薄壳理论假定，在给出沿球壳厚度遵循双曲线规律分布的法向应力公式的基础上，进行了扁厚壁球壳轴对称问题的近似计算。但是，该文无实验结果做比较，其精确度也不够高，因此，适用范围较窄。

本节注意了上述问题，进行了改进，获得了精确度较高的公式。此公式简单，适用于研究厚壁球壳在内压力作用下的各种实际情况中的弯曲应力分布，对于开孔（大孔或小孔）以及边缘应力的计算有理论和实用意义。

一、基本方程

1. 平衡方程

在高压容器设计中，常常要进行常温下的强度计算。通常，有两种载荷：均匀内压力 q 和壳体自重。由于后一载荷与前者相比甚小，因而可以忽略不计。这样，本节所涉及的仅为厚壁球壳在均匀内压力作用下的强度计算。

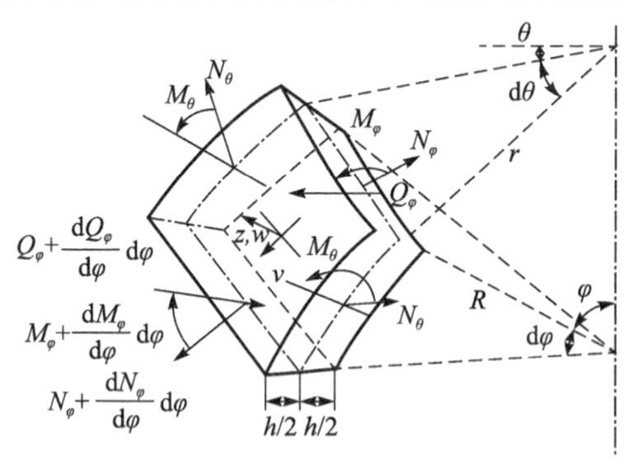

图 1　球壳微元

为了建立这一轴对称问题的平衡方程，我们以两对相邻坐标线从厚壁球壳中截出一个微元（图 1），并在其上示出作用于各截面上的内力及位移的正方向。图中，φ 是余纬角，θ 是表征经线位置的角，z 是壳体中曲面外法线方向的坐标，r 是壳体中曲面上点到旋转轴的距离，R 是壳体中曲面的半径，h 是壳体厚度，N_φ 是经向力，N_θ 是纬向力，M_φ 是经向弯矩，M_θ 是纬向弯矩，Q_φ 是横向力，w 是法向位移，v 是经向位移。

在球壳微元中曲面外法线 z 方向，有下面几个内力作用：两个纬向力 $N_\theta R\,d\varphi$ 构成

① 本文原载《化工炼油机械通讯》，1980，(3)：1-10.

水平合力 $N_\theta R\,\mathrm{d}\varphi\mathrm{d}\theta$，此合力在 z 的负方向有分量

$$N_\theta R\,\mathrm{d}\varphi\mathrm{d}\theta\sin\varphi,$$

两个经向力 $N_\varphi r\,\mathrm{d}\theta$ 也在 z 的负方向有合力

$$N_\varphi r\,\mathrm{d}\theta\mathrm{d}\varphi,$$

两个横向力大小不等，其差是 z 的负方向上的力

$$\frac{\mathrm{d}(Q_\varphi r\,\mathrm{d}\theta)}{\mathrm{d}\varphi}\mathrm{d}\varphi=\frac{\mathrm{d}(rQ_\varphi)}{\mathrm{d}\varphi}\mathrm{d}\varphi\mathrm{d}\theta$$

另外，还需引进外力。由于壳体较厚，故应该用中曲面上静力等效的力来代替作用在壳体内表面上的均匀压力 q。显然，此静力等效的力为

$$q\left(R-\frac{h}{2}\right)\mathrm{d}\varphi\cdot\left(r-\frac{h}{2}\sin\varphi\right)\mathrm{d}\theta,$$

它指向 z 的正方向。

将上述力加在一起，得 z 方向力的平衡方程

$$\begin{aligned}&\frac{\mathrm{d}(rQ_\varphi)}{\mathrm{d}\varphi}\mathrm{d}\varphi\mathrm{d}\theta+N_\theta R\,\mathrm{d}\varphi\mathrm{d}\theta\sin\varphi+N_\varphi r\,\mathrm{d}\theta\mathrm{d}\varphi\\&=q\left(R-\frac{h}{2}\right)\left(r-\frac{h}{2}\sin\varphi\right)\mathrm{d}\varphi\mathrm{d}\theta,\end{aligned}\tag{1}$$

注意到

$$r=R\sin\varphi,\tag{2}$$

并约去公因子 $\mathrm{d}\varphi\mathrm{d}\theta$，式（1）简化为

$$\frac{\mathrm{d}(Q_\varphi\sin\varphi)}{\mathrm{d}\varphi}+N_\theta\sin\varphi+N_\varphi\sin\varphi=qR\left(1-\frac{h}{2R}\right)^2\sin\varphi\tag{3a}$$

至于经线切线方向力的平衡方程和对于纬线切线的力矩平衡方程，仍与原来相同：

$$\frac{\mathrm{d}(N_\varphi\sin\varphi)}{\mathrm{d}\varphi}-N_\theta\cos\varphi-Q_\varphi\sin\varphi=0,\tag{3b}$$

$$\frac{\mathrm{d}(M_\varphi\sin\varphi)}{\mathrm{d}\varphi}-M_\theta\cos\varphi-RQ_\varphi\sin\varphi=0\tag{3c}$$

2. 弹性定律

$$N_\varphi=\frac{D}{R}\left[\left(\frac{\mathrm{d}v}{\mathrm{d}\varphi}+w\right)+\nu(v\cot\varphi+w)\right],\tag{4a}$$

$$N_\theta=\frac{D}{R}\left[(v\cot\varphi+w)+\nu\left(\frac{\mathrm{d}v}{\mathrm{d}\varphi}+w\right)\right],\tag{4b}$$

$$M_\varphi=\frac{K}{R}\left(\frac{\mathrm{d}\psi}{\mathrm{d}\varphi}+\nu\psi\cot\varphi\right),\tag{4c}$$

$$M_\theta=\frac{K}{R}\left(\psi\cot\varphi+\nu\frac{\mathrm{d}\psi}{\mathrm{d}\varphi}\right),\tag{4d}$$

其中，ν 是泊松比，E 是弹性模量，抗拉刚度 $D=\dfrac{Eh}{1-\nu^2}$，抗弯刚度 $K=\dfrac{Eh^3}{12(1-\nu^2)}$，经线微元旋转角 $\psi=\dfrac{1}{R}\left(\dfrac{\mathrm{d}w}{\mathrm{d}\varphi}-v\right)$。　(5)

3. 应力公式

求得内力和内矩后，按下列公式[1]计算经向应力和纬向应力：

$$\sigma_\varphi=\frac{1}{1+z/R}\left(\frac{N_\varphi}{h}-\frac{12zM_\varphi}{h^3}\right),\quad \sigma_\theta=\frac{1}{1+z/R}\left(\frac{N_\theta}{h}-\frac{12zM_\theta}{h^3}\right) \tag{6}$$

方程式（3）～（5）合在一起，是包括八个未知量，即 N_φ，N_θ，M_φ，M_θ，Q_φ，v，w，ψ 的八个独立方程，因此是可解的。经过不太复杂的数学运算，这组方程可以缩减为关于 Q_φ 的两个变系数二阶线性微分方程

$$\frac{\mathrm{d}^2Q_\varphi}{\mathrm{d}\varphi^2}+\cot\varphi\frac{\mathrm{d}Q_\varphi}{\mathrm{d}\varphi}-Q_\varphi\cot^2\varphi\pm 2\mathrm{i}\eta^2Q_\varphi=0, \tag{7a,b}$$

其中

$$\eta^2=\sqrt{3(1-\nu^2)\frac{R^2}{h^2}-\frac{\nu^2}{4}}\approx\sqrt{3(1-\nu^2)}\frac{R}{h}$$

由方程（7）解得 Q_φ 后，按下列公式计算转角和内力

$$\begin{aligned}\psi&=-\frac{1}{D(1-\nu^2)}\left[\frac{\mathrm{d}^2Q_\varphi}{\mathrm{d}\varphi^2}+\cot\varphi\frac{\mathrm{d}Q_\varphi}{\mathrm{d}\varphi}+(\nu-\cot^2\varphi)Q_\varphi\right],\\ N_\varphi&=-Q_\varphi\cot\varphi+\frac{qR}{2}\left(1-\frac{h}{2R}\right)^2-\frac{P}{2\pi R\sin^2\varphi},\\ N_\theta&=-\frac{\mathrm{d}Q_\varphi}{\mathrm{d}\varphi}+\frac{qR}{2}\left(1-\frac{h}{2R}\right)^2+\frac{P}{2\pi R\sin^2\varphi},\end{aligned} \tag{8}$$

这里，P 是积分常数，由球域的平衡条件决定。

这样，厚壁球壳的弯曲问题就归结为求解方程（7）。由于式（7）中两个方程仅有最后含虚数 i 的项的符号不同，因而方程（7a）的解与方程（7b）的解是共轭复数。于是，我们只需求解它们中的任何一个方程即可。下面，我们来寻找 Q_φ 的解答。

二、基本方程的解

分下面两种情况求解二阶微分方程（7）。

（1）余纬角 φ 很小，亦即属于扁壳的情况（$\varphi\leqslant 21°30'$）。

因为 φ 很小，则可用 $\cot\varphi$ 的罗朗级数的首项 φ^{-1} 来代替函数本身，即

$$\cot\varphi\approx\frac{1}{\varphi} \tag{9}$$

将此近似引进方程（7a），便转化为熟知的贝塞尔方程

$$\frac{\mathrm{d}^2Q_\varphi}{\mathrm{d}\xi^2}+\frac{1}{\xi}\frac{\mathrm{d}Q_\varphi}{\mathrm{d}\xi}+\left(1-\frac{1}{\xi^2}\right)Q_\varphi=0, \tag{10}$$

其中

$$\xi=\sqrt{2\mathrm{i}}\eta\varphi=\sqrt{\mathrm{i}}y$$

于是，得到用汤姆生函数的一阶导数所表示的关于方程（10）的通解

$$Q_\varphi = C_1 \operatorname{ber}'y + C_2 \operatorname{bei}'y + C_3 \operatorname{ker}'y + C_4 \operatorname{kei}'y, \tag{11}$$

这里 C_1、C_2、C_3、C_4 是待定常数，由边界条件决定。

将解（11）代入式（8）和（4c，d）后，得

$$\begin{aligned}
\psi &= \frac{1}{D(1-\nu^2)}\left[C_1(2\eta^2 \operatorname{bei}'y - \nu \operatorname{ber}'y) - C_2(2\eta^2 \operatorname{ber}'y + \nu \operatorname{bei}'y)\right. \\
&\quad \left. + C_3(2\eta^2 \operatorname{kei}'y - \nu \operatorname{ker}'y) - C_4(2\eta^2 \operatorname{ker}'y + \nu \operatorname{kei}'y)\right], \\
N_\varphi &= -\frac{1}{\varphi} Q_\varphi + \frac{qR}{2}\left(1 - \frac{h}{2R}\right)^2 - \frac{P}{2\pi R \sin^2\varphi}, \\
N_\theta &= \sqrt{2}\eta\left[C_1\left(\operatorname{bei}y + \frac{\operatorname{ber}'y}{y}\right) - C_2\left(\operatorname{ber}y - \frac{\operatorname{bei}'y}{y}\right) + C_3\left(\operatorname{kei}y + \frac{\operatorname{ker}'y}{y}\right)\right. \\
&\quad \left. - C_4\left(\operatorname{ker}y - \frac{\operatorname{kei}'y}{y}\right)\right] + \frac{qR}{2}\left(1 - \frac{h}{2R}\right)^2 + \frac{P}{2\pi R \sin^2\varphi}, \\
M_\varphi &= \frac{\sqrt{2}\eta K}{RD(1-\nu^2)}\left\{C_1\left[2\eta^2\left(\operatorname{ber}y - \frac{1-\nu}{y}\operatorname{bei}'y\right) + \nu\left(\operatorname{bei}y + \frac{1-\nu}{y}\operatorname{ber}'y\right)\right]\right. \\
&\quad + C_2\left[2\eta^2\left(\operatorname{bei}y + \frac{1-\nu}{y}\operatorname{ber}'y\right) - \nu\left(\operatorname{ber}y - \frac{1-\nu}{y}\operatorname{bei}'y\right)\right] \\
&\quad + C_3\left[2\eta^2\left(\operatorname{ker}y - \frac{1-\nu}{y}\operatorname{kei}'y\right) + \nu\left(\operatorname{kei}y + \frac{1-\nu}{y}\operatorname{kei}'y\right)\right] \\
&\quad \left. + C_4\left[2\eta^2\left(\operatorname{kei}y + \frac{1-\nu}{y}\operatorname{ker}'y\right) - \nu\left(\operatorname{ker}y - \frac{1-\nu}{y}\operatorname{kei}'y\right)\right]\right\}, \\
M_\theta &= \frac{\sqrt{2}\eta K}{RD(1-\nu^2)}\left\{C_1\left[2\eta^2\left(\nu\operatorname{ber}y + \frac{1-\nu}{y}\operatorname{bei}'y\right) + \nu\left(\nu\operatorname{bei}y - \frac{1-\nu}{y}\operatorname{ber}'y\right)\right]\right. \\
&\quad + C_2\left[2\eta^2\left(\nu\operatorname{bei}y - \frac{1-\nu}{y}\operatorname{ber}'y\right) - \nu\left(\nu\operatorname{ber}y + \frac{1-\nu}{y}\operatorname{bei}'y\right)\right] \\
&\quad + C_3\left[2\eta^2\left(\nu\operatorname{ker}y + \frac{1-\nu}{y}\operatorname{kei}'y\right) + \nu\left(\nu\operatorname{kei}y - \frac{1-\nu}{y}\operatorname{ker}'y\right)\right] \\
&\quad \left. + C_4\left[2\eta^2\left(\nu\operatorname{kei}y - \frac{1-\nu}{y}\operatorname{ker}'y\right) - \nu\left(\nu\operatorname{ker}y + \frac{1-\nu}{y}\operatorname{kei}'y\right)\right]\right\}
\end{aligned} \tag{12}$$

（2）一般情况（$0<\varphi<180°$）

在厚壳情况下，η 值较小，这就使得我们不能采用薄壳中的方法来简化方程（7）。在这种情况下，我们只有采用收敛较慢的超几何级数，求取方程（7）的精确解。

引入新的变量：

$$x = \frac{Q_\varphi}{\sin\varphi}, \qquad t = \cos^2\varphi \tag{13}$$

方程（7a）成为超几何方程

$$\frac{\mathrm{d}^2 x}{\mathrm{d}t^2} + \frac{1-5t}{2t(1-t)}\frac{\mathrm{d}x}{\mathrm{d}t} - \frac{1-2\mathrm{i}\eta^2}{4t(1-t)}x = 0 \tag{14}$$

按照微分方程理论，立即得到方程（14）的通解。将此通解转化为实数解，得

$$Q_\varphi=\sin\varphi(C_1x_1+C_2x_2+C_3x_3+C_4x_4) \tag{15}$$

其中 C_1、C_2、C_3、C_4 是待定常数，由边界条件决定。x_1、x_2、x_3、x_4 是超几何级数，为

$$x_1=\frac{1}{2}(x_a+\overline{x}_a),$$

$$x_2=\frac{1}{2}(x_b+\overline{x}_b),$$

$$x_3=\frac{i}{2}(x_a-\overline{x}_a),$$

$$x_4=\frac{i}{2}(x_b-\overline{x}_b),$$

$$x_a=1+\sum_{n=1}^{\infty}\frac{1}{(2n)!}\left(\prod_{m=1}^{n}a_m\right)t^n,$$

$$x_b=\sqrt{t}\left[1+\sum_{n=1}^{\infty}\frac{1}{(2n+1)!}\left(\prod_{m=1}^{n}b_m\right)t^n\right],$$

$$\overline{x}_a=1+\sum_{n=1}^{\infty}\frac{1}{(2n)!}\left(\prod_{m=1}^{n}\overline{a}_m\right)t^n,$$

$$\overline{x}_b=\sqrt{t}\left[t+\sum_{n=1}^{\infty}\frac{1}{(2n+1)!}\left(\prod_{m=1}^{n}\overline{b}_m\right)t^n\right],$$

$$a_m=\frac{(4m-1)^2-5}{4}-2i\eta^2,$$

$$b_m=\frac{(4m+1)^2-5}{4}-2i\eta^2,$$

$$\overline{a}_m=\frac{(4m-1)^2-5}{4}+2i\eta^2,$$

$$\overline{b}_m=\frac{(4m+1)^2-5}{4}+2i\eta^2$$

这里的超几何级数对于 $0\leqslant t<1$ 收敛。

将式（15）代入式（8）和（4c，d），得

$$\psi=\frac{\sin\varphi}{D(1-\nu^2)}[C_1(2\eta^2x_3-\nu x_1)+C_2(2\eta^2x_4-\nu x_2)-C_3(2\eta^2x_1+\nu x_3)$$
$$-C_4(2\eta^2x_2+\nu x_4)],$$

$$N_\varphi=-\cos\varphi(C_1x_1+C_2x_2+C_3x_3+C_4x_4)+\frac{qR}{2}\left(1-\frac{h}{2R}\right)^2-\frac{P}{2\pi R\sin^2\varphi},$$

$$N_\theta=N_\varphi-\sin\varphi(C_1\dot{x}_1+C_2\dot{x}_2+C_3\dot{x}_3+C_4\dot{x}_4)+\frac{P}{\pi R\sin^2\varphi},$$

$$M_\varphi=\frac{K}{RD(1-\nu)}\left\{\frac{\sin\varphi}{1+\nu}[C_1(2\eta^2\dot{x}_3-\nu\dot{x}_1)+C_2(2\eta^2\dot{x}_4-\nu\dot{x}_2)\right.$$
$$-C_3(2\eta^2\dot{x}_1+\nu\dot{x}_3)-C_4(2\eta^2\dot{x}_2+\nu\dot{x}_4)]$$
$$+\cos\varphi[C_1(2\eta^2x_3-\nu x_1)+C_2(2\eta^2x_4-\nu x_2)-C_3(2\eta^2x_1+\nu x_3)$$
$$\left.-C_4(2\eta^2x_2+\nu x_4)]\right\},$$

$$M_\theta=\frac{K}{RD(1-\nu)}\left\{\frac{\nu\sin\varphi}{1+\nu}\left[C_1(2\eta^2x_3'-\nu x_1')+C_2(2\eta^2x_4'-\nu x_2')\right.\right.$$
$$-C_3(2\eta^2x_1'+\nu x_3')-C_4(2\eta^2x_2'+\nu x_4')]$$
$$+\cos\varphi[C_1(2\eta^2x_3-\nu x_1)+C_2(2\eta^2x_4-\nu x_2)-C_3(2\eta^2x_1+\nu x_3)$$
$$\left.-C_4(2\eta^2x_2+\nu x_4)]\right\},\tag{16}$$

其中 x_1'、x_2'、x_3'、x_4' 分别是由 x_1、x_2、x_3、x_4 对 φ 求导数得到的四个超几何级数。

三、算例与比较

例 1　在均匀内压力 q 作用下的封闭厚壁球壳（图 2）。

由半球的平衡条件，可知

$$2\pi RN_\varphi=\pi\left(R-\frac{h}{2}\right)^2q\tag{17}$$

由此式解得 N_φ，并代入应力公式（6），注意到 M_φ 为零，便得封闭厚壁球壳切向应力的计算公式

$$\sigma_\varphi=\frac{q}{2\left(1+\dfrac{z}{R}\right)\dfrac{h}{R}}\left(1-\frac{h}{2R}\right)^2\tag{18}$$

显然，最大应力发生在壳体内表面$\left(z=-\dfrac{h}{2}\right)$，并为

$$(\sigma_\varphi)_{\max}=\frac{q}{2h/R}\left(1-\frac{h}{2R}\right)\tag{19}$$

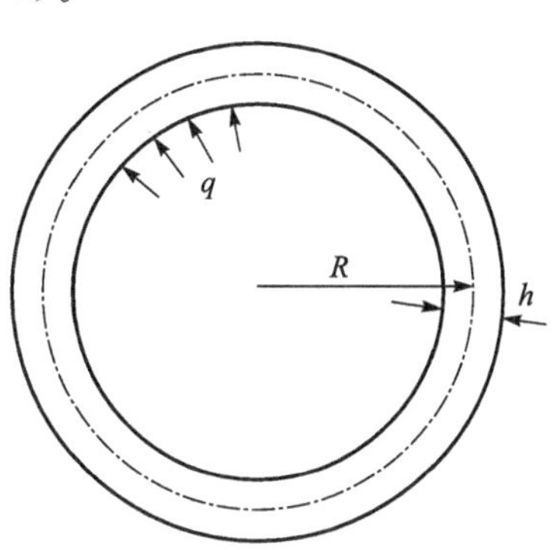

图 2　均匀内压力作用下的封闭球壳

众所周知，此问题有拉梅公式[2]的精确解。我们使用拉梅公式、薄壳理论公式[3]、本节公式（18），对壳内、外表面（$z=\mp h/2$）的应力进行了比较计算，其数值结果如表 1 所列。

表 1　球壳内、外表面应力 σ_φ/q 值的比较

h/R	壳内表面应力					壳外表面应力				
	拉梅公式	本文公式		薄壳理论公式		拉梅公式	本文公式		薄壳理论公式	
		值	误差/%	值	误差/%		值	误差/%	值	误差/%
0.05	9.768	9.750	−0.18	10.00	2.4	9.268	9.274	0.069	10.00	7.9
0.10	4.783	4.750	−0.70	5.000	4.5	4.283	4.298	0.33	5.000	17
0.15	3.134	3.083	−1.6	3.333	6.4	2.634	2.653	0.74	3.333	27
0.20	2.136	2.250	−2.9	2.500	7.9	1.816	1.841	1.4	2.500	38
0.25	1.833	1.750	−4.5	2.000	9.1	1.333	1.361	2.1	2.000	50
0.30	1.516	1.417	−6.6	1.667	9.9	1.016	1.047	3.1	1.667	64

由结果看出，在 $h/R<0.05$ 的薄壳情况下，才能应用薄壳公式；在 $h/R>0.05$ 的情况下，本文公式能满足工程的要求。即使对于壳体相当厚的情况，如 $h/R=0.30$，本文公式与精确公式之间的相对误差仅为 6.6%，而薄壳公式的误差竟高达 64%。

例 2　开小孔的某高压容器球形封头。

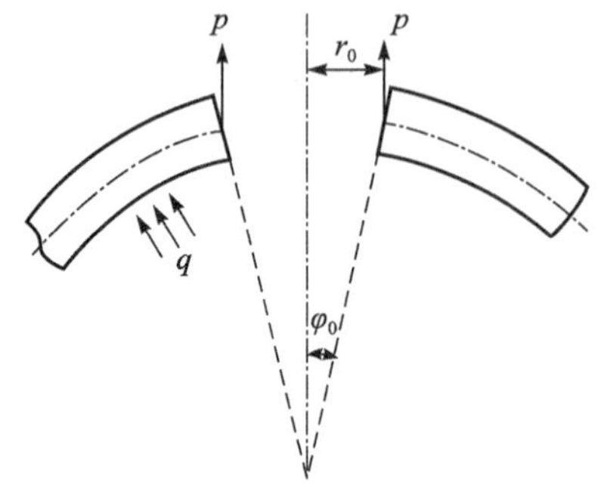

图 3 中心开小孔的球壳

某高压容器球形封头上开有等直径为 $2r_0$ 的小孔，在小孔边缘上承受了均布竖直载荷 p，它与均匀内压力 q 的关系为

$$p=\frac{\pi r_0^2 q}{2\pi r_0}=\frac{1}{2}r_0 q \tag{20}$$

为了计算简单，首先进行以下简化。将等直径的小孔视为变直径的小孔，孔的边界与直径面一致，孔在壳中曲面处的直径为 $2r_0$，p 作用在壳中曲面上（图 3）。

已知数据：

$$R=752\text{mm},\quad h=116\text{mm},\quad \varphi_0=0.0966=5°32',$$

$$q=2.20\text{kg/mm}^2,\quad E=2.18\times10^4\text{kg/mm}^2,\quad \nu=0.292$$

此时

$$h/R=0.154,$$

故此封头属于厚壁壳体理论研究范围。

由于仅研究小孔附近的应力状态，所以我们采用前面第一种情况的公式进行计算。

首先，应决定式（12）中的 P 值。为此，从某一余纬角 φ 处，截出部分球壳（图 4）。此部分壳体的竖直力平衡条件为

$$\begin{aligned}&2\pi RN_\varphi\sin^2\varphi+2\pi RQ_\varphi\sin\varphi\cos\varphi\\&-2\pi\int_{\varphi_0}^{\varphi}\left(R-\frac{h}{2}\right)^2 q\sin\varphi\cos\varphi\mathrm{d}\varphi-2\pi r_0 p=0\end{aligned} \tag{21}$$

化简此式，得

$$N_\varphi\approx-\frac{\cos\varphi}{\varphi}Q_\varphi+\frac{qR}{2}\left(1-\frac{h}{2R}\right)^2+\frac{qR\varphi_0^2}{2\varphi^2}\left[1-\left(1-\frac{h}{2R}\right)^2\right] \tag{22}$$

将此式与（12）的第二式相比较，便确定了 P：

$$P=-\pi R^2\varphi_0^2\left[1-\left(1-\frac{h}{2R}\right)^2\right]q \tag{23}$$

然后，我们来决定式（11）中的 C_1、C_2、C_3、C_4 值。因为涉及的仅是小孔附近区域，故可认为下边缘离孔很远。函数 $\mathrm{ber}'y$ 和 $\mathrm{bei}'y$ 是 y 因而也是 φ 的正则函数，并随 φ 的增加而增加。在离孔远处将取极大值，在孔边缘却很小，这与函数 $\mathrm{ker}'y$ 和 $\mathrm{kei}'y$ 恰恰相反。因此，在研究小孔边缘附近的应力状态时，可在式（11）中令

$$C_1=C_2=0 \tag{24}$$

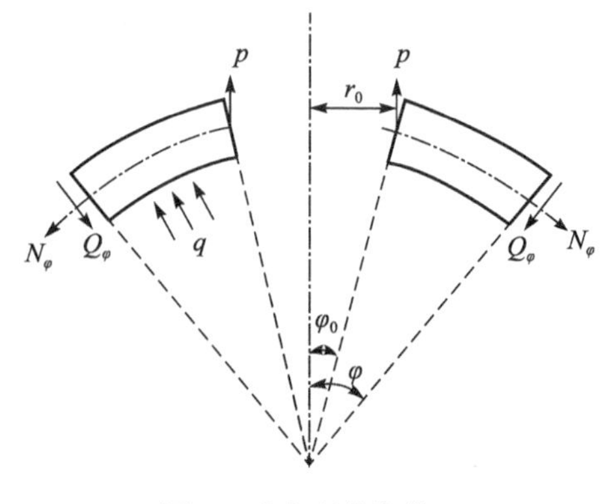

图 4 确定积分常数 P

至于 C_3、C_4 将由小孔边缘条件确定：

$$\text{在 }\varphi=\varphi_0\text{ 处},\quad Q=p\cos\varphi_0,\quad M_\varphi=0 \tag{25}$$

将解（11）和（12）中的第四式代入边界条件（25），便确定了 C_3 和 C_4：

$$C_3=\frac{\alpha_2\cos\varphi_0}{\alpha_2\mathrm{ker}'y_0-\alpha_1\mathrm{kei}'y_0}p,\qquad C_4=-\frac{\alpha_1\cos\varphi_0}{\alpha_2\mathrm{ker}'y_0-\alpha_1\mathrm{kei}'y_0}p, \tag{26}$$

其中

$$\alpha_1=2\eta^2\left(\mathrm{ker}y_0-\frac{1-\nu}{y_0}\mathrm{kei}'y_0\right)+\nu\left(\mathrm{kei}y_0+\frac{1-\nu}{y_0}\mathrm{ker}'y_0\right),$$

$$\alpha_2=2\eta^2\left(\mathrm{kei}y_0+\frac{1-\nu}{y_0}\mathrm{ker}'y_0\right)-\nu\left(\mathrm{ker}y_0-\frac{1-\nu}{y_0}\mathrm{kei}'y_0\right),$$

$$y_0=\sqrt{2}\eta\varphi_0$$

将已知数据代入式（12）、（6）、（23）、（26）中，就得到了小孔边缘附近壳内、外表面的应力值。应力分布曲线已绘制在图 5 和图 6 中。

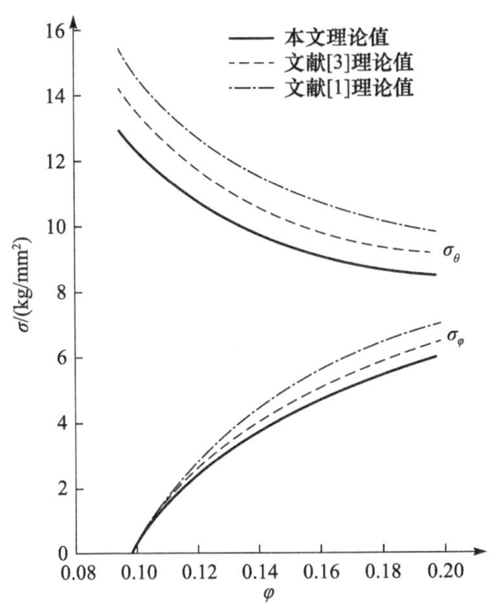

图 5　壳内表面应力的理论值比较

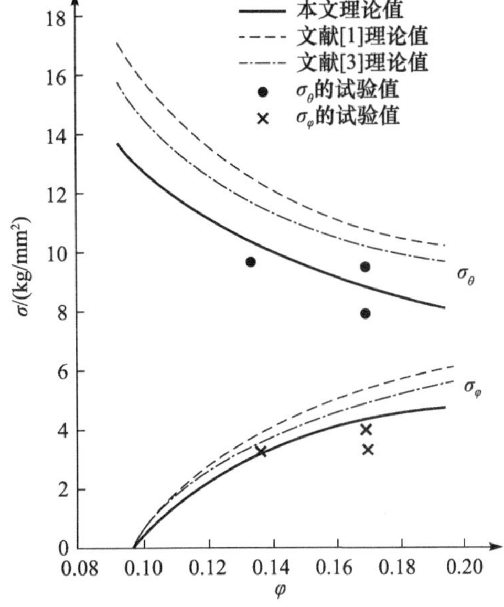

图 6　壳外表面应力的理论值与试验值的比较

为了验证和比较本文公式的精确度，我们除了在产品的外表面进行了常温下的应力测定试验，并将其所得的试验值绘于图 6 中之外，还采用文献［1］的厚壁球壳理论公式以及文献［3］的精确薄壁球壳理论公式进行了数值计算，也将其结果一并绘入图 6 中。

由该图看出，三种理论公式所得结果之间的差异是可观的，本文公式结果与试验值的相对误差最小。以最大应力（孔边缘壳外表面的纬向应力）为例。文献［3］，［1］公式与本文公式结果的相对误差分别为 23.1％和 14.3％。以所测的试验值中最大应力（φ=0.137 处的壳外表面的纬向应力）为例，本文公式和文献［3］，［1］公式结果与试验值的相对误差分别为 6.32％、30.5％和 22.1％。

由于按本文公式计算所得的最大应力较小，因而设计的球形封头就比较薄，这就节省了钢材，简化了工艺。

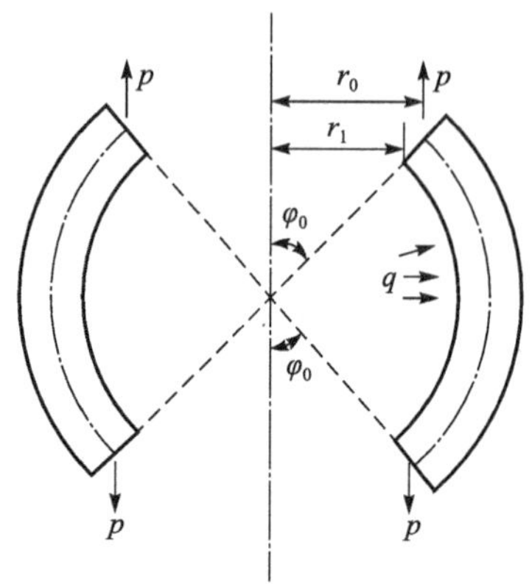

图 7　开两个大孔的球壳

例 3　开大孔的某球形高压容器（图 7）。

某球形高压容器上开有两个半径为 r_0 的相同大孔，在孔边缘上承受均布竖直载荷 p，它与均匀内压力 q 的关系为

$$p=\frac{\pi r_1^2 q}{2\pi r_0}=\frac{qR\sin\varphi_0}{2}\left(1-\frac{h}{2R}\right)^2 \tag{27}$$

此球壳的边界条件如下。

在 $\varphi=\varphi_0$ 处，

$$Q_\varphi=p\cos\varphi_0,\quad M_\varphi=0; \tag{28}$$

在 $\varphi=180^\circ-\varphi_0$ 处，

$$Q_\varphi=-p\cos\varphi_0,\quad M_\varphi=0 \tag{29}$$

已知数据：

$$R=1080\text{mm},\quad h=160\text{mm},\quad \varphi_0=27°35',\quad q=2.10\text{kg/mm}^2,$$

$$E=2.12\times10^4\text{kg/mm}^2,\quad \nu=0.292$$

此时

$$h/R=0.148$$

故此容器属于厚壁壳体理论研究范围。因为球域的余纬角很大，所以我们采用前面第二种情况的公式进行计算。

首先，应决定式（16）中的 P 值。为此，从某一余纬角 φ 处，截出部分壳体，考虑此部分壳体的竖直力平衡，有

$$2\pi RN_\varphi\sin^2\varphi+2\pi RQ_\varphi\sin\varphi\cos\varphi-2\pi\int_{\varphi_0}^{\varphi}\left(R-\frac{h}{2}\right)^2 q\sin\varphi\cos\varphi\mathrm{d}\varphi-2\pi r_0 p=0 \tag{30}$$

化简此式，得

$$N_\varphi=-Q_\varphi\cot\varphi+\frac{qR}{2}\left(1-\frac{h}{2R}\right)^2 \tag{31}$$

将此式和式（8）的第二式相比较，便确定了 P：

$$P=0 \tag{32}$$

然后，我们来决定式（15）中的待定常数 C_1、C_2、C_3、C_4。由于容器本身的对称性和载荷对于平行圆 $\varphi=90°$的对称性，所以解中只能保留含有相同对称性的函数 x_2 和 x_4，于是应令

$$C_1=C_3=0 \tag{33}$$

至于 C_2，C_4，将由容器任一孔的边界条件确定。将式（15）和式（16）的第四式代入边界条件（28），便确定了 C_2 和 C_4：

$$\begin{aligned}C_2&=\frac{\beta_2\cot\varphi_0}{\beta_2 x_{20}-\beta_1 x_{40}}p,\\C_4&=-\frac{\beta_1\cot\varphi_0}{\beta_2 x_{20}-\beta_1 x_{40}}p,\end{aligned} \tag{34}$$

其中

$$\beta_1=\frac{\sin\varphi_0}{1+\nu}(2\eta^2\dot{x}_{40}-\nu\dot{x}_{20})+\cos\varphi_0(2\eta^2x_{40}-\nu x_{20}),$$

$$\beta_2=-\left[\frac{\sin\varphi_0}{1+\nu}(2\eta^2\dot{x}_{20}+\nu\dot{x}_{40})+\cos\varphi_0(2\eta^2x_{20}+\nu x_{40})\right],$$

$$x_{20}=x_2(\varphi_0),$$

$$x_{40}=x_4(\varphi_0),$$

$$\dot{x}_{20}=\dot{x}_2(\varphi_0),$$

$$\dot{x}_{40}=\dot{x}_4(\varphi_0)$$

将已知数据代入式（16）、式（6）、式（34），则得到球形高压容器内、外表面的应力值，并把应力分布曲线绘入图 8 和图 9 中。

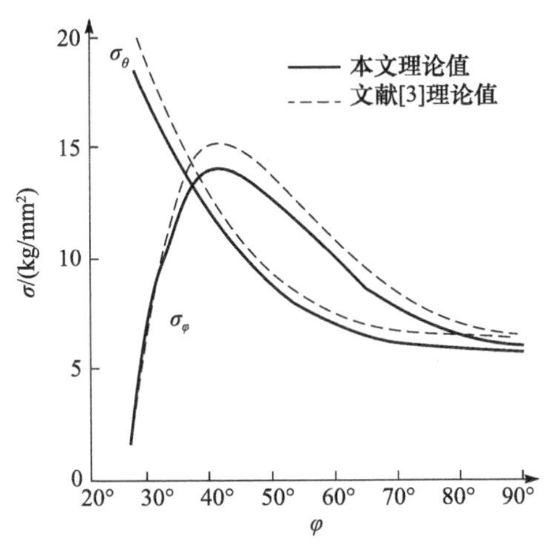

图 8　壳内表面应力的理论值比较

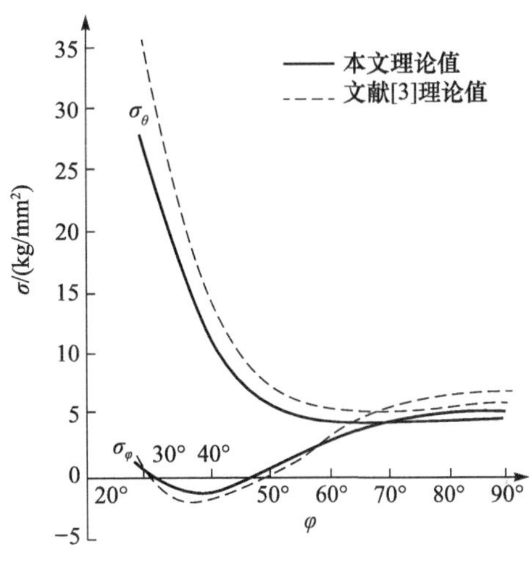

图 9　壳外表面应力的理论值比较

为比较起见，我们还在图中给出了应用文献［3］的精确薄球壳理论公式所计算的结果。

由图看出，两种理论公式结果之间的差异是可观的。以最大应力（孔边缘壳外表面的纬向应力）为例，文献［3］公式与本文公式结果的相对误差高达 25.3%。按本文公式计算所得的最大应力较小，因而设计时所求得的容器就比较薄，于是节省了钢材，简化了工艺。

四、结束语

（1）本文在处理表面载荷时，注意了厚度较大的因素，获得了计算厚壁球壳弯曲应力的公式。与精确值和试验值相比，本文公式较已有的理论公式精确。加之，本文公式简单，适用范围广，所以适于工程应用。

（2）对于承受均匀外压力和其他轴对称载荷情况，读者不难按本文方法推出公式。

（3）对于厚壁球形壳体与圆柱形壳体连接情况弯曲应力的计算，读者可参看作者过去的论文[4]。

(4) 虽然超几何级数收敛较慢，计算量较大，但可借助电子计算机计算，困难易于克服。

致谢：兰州大学数学力学系王廷栋同志负责例 2 的试验，75 届李树勋同学在 DJS-21 电子计算机上进行了例 3 的超几何级数的数值计算。

参考文献

[1] Рабинович А Л. Приближенный метод расчета толстостенной сферической оболочки. Тр. Моск. физ.-техн. ин-та，вып. 3，1959.

[2] 铁摩辛柯，古地尔．弹性理论．徐芝纶，吴永禛，译．北京：人民教育出版社，1964.

[3] 弗留盖 W. 壳体中的应力．薛振东，龙驭球，叶耀先，等译．北京：中国工业出版社，1965.

[4] 刘人怀．加氢反应器顶部厚壁壳体的应力分析．化工炼油机械通讯，1975，(6)：40-59.

厚管板的设计①

符号说明

r、θ、z 圆柱坐标

x、x_1 无量纲坐标及其边缘值

l 管子和筒体的计算长度，即两管板中面距离的一半

l_1 管板与第一块折流板中面距离

l_2 相邻两折流板中面距离

h、h_p 管板和筒体的厚度

d 管板的计算直径，即筒体的内直径

d_1、d_2 管子的内、外直径

n、s 管子和折流板数

q、q_1、q_2 有效、管程、壳程压力

f_1、f_2 管板受 q_1、q_2 作用的面积与整个面积之比

D 管板的抗弯刚度

ψ 开孔对管板抗弯刚度的影响系数

E、E_r、E_p 管板、管子、筒体材料的弹性模量

ν 管板材料的泊松比

w 管板的挠度

φ_r 管板的径向转角

M_r、M_θ 管板的径向、切向弯矩

Q_r、Q 管板的剪力及其边缘值

σ_r、σ_θ、σ_z 管板的径向、切向、法向正应力

$\tau_{r\theta}$、τ_{rz}、$\tau_{\theta z}$ 管板的剪应力

η 管板的应力减弱系数

m 由管束弯曲所产生的单位面积径向反力矩

R 基础反力

k_1 管束的纵向弹性系数

k_2 管束的弯曲刚度系数

K_s 管子的弯曲系数

J_r 管子的截面惯性矩

δ 管子和筒体的温度变形差的一半

δ_p 由 Q 引起的筒体纵向伸长之半

t_r、t_p 管子、筒体的温度

t_1、t_2 管子、筒体沿 z 轴的始末处温度

t_0 装配温度

α_r、α_p 管子、筒体材料的热膨胀系数

U_0、V_0 金尼克函数

U_0'、V_0' U_0、V_0 的一阶导数

C_1、C_2 待定常数

σ 管板的无量纲最大应力

σ_{max}、w_{max} 管板的最大应力、最大挠度

μ_1、μ_2 管板的无量纲特征参数

β_1、β_2、$\gamma_1 \cdots \gamma_9$、$\lambda_1 \cdots \lambda_8$、$\omega_0 \cdots \omega_3$、$\varphi$、$a_s$、$b_s$、$c_s$、$d_s$ 辅助量

∇^2、∇^4 单、双调和算子

近代的许多化工过程，大都要求在高压下进行。特别是在石油和化工领域里，由于工艺技术的进展，高压的要求显得更为迫切。因此，常常采用固定式换热器来进行热交换。厚管板是这种换热器比较关键的部件，一般采用大型锻件，不仅金属消耗大，

① 本文原载《化工炼油机械通讯》，1980，(4)：1-12. 作者：刘人怀，魏俊.

而且机械加工的工作量也相当大。过去，国内外所发表的大量文献都是基于薄板理论方面的工作，这对于中低压换热器管板的设计来说是有实际意义的。但若将这些公式用于高压换热器管板的设计则过于粗略。因此迫切需要为厚管板的设计提供一个比较精确的理论公式。

1970 年以来，上海锅炉厂[1]、原国家一机部机械科学研究院和化工部第一设计院等单位先后两次对厚管板进行试验，但没有进行理论分析。这年下半年，作者[2]接受了某厂自行设计的我国新产品——换热、分离氨组合设备中水冷却器厚管板的应力分析任务，应用厚板理论对这一问题进行过初步探讨，使该厂厚管板设计制造得以顺利进行。1974 年，Мельников[3]也从理论上研究了这一问题，但考虑管子对管板的影响不够。1979 年，作者[4]应用 Reissner[5]的厚板理论，再一次研究了这一问题，获得了较精确的理论公式，并编制了 DJS-8 电子计算机的计算程序，绘制了图表，以便于工程设计的应用。

一、厚管板的理论公式

众所周知，管板计算涉及的因素较多，要精确计算中低压换热器中的薄管板就十分困难了。而高压换热器管板比较厚，即为一块多孔的厚圆板，超出了薄圆板理论所研究的范围（薄板理论要求板厚与直径的比值小于 0.1），是一个三维问题，显然，这比属于二维问题的薄管板的计算要困难得多。为了使厚管板的分析计算易于进行，有必要将问题进行一些简化。在薄管板理论研究中，乌班诺夫斯基[6]等采用了一些很好的处理方法，这里，结合厚管板的具体情况，我们将适当利用这些方法。

图 1 所示的厚管板标明了坐标、载荷、挠度、弯矩和剪力的正方向。对管板进行力学分析，可看到有下面几个载荷作用在管板上。

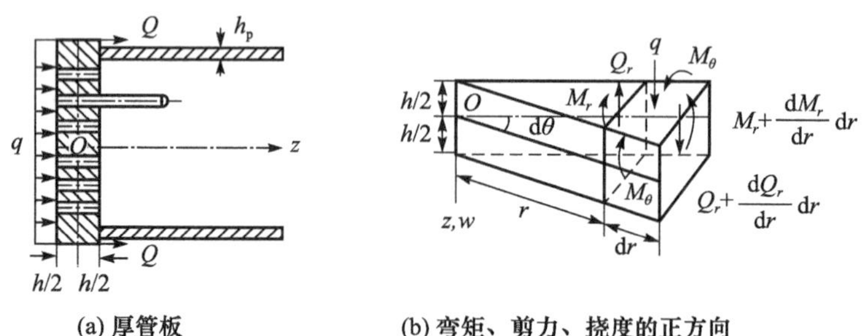

图 1　厚管板

(1) 作用在管板上、下表面的管程和壳程压力 q_1、q_2，这两个载荷属于横向载荷。

(2) 由于换热器两端管板的相对弯曲，引起管子的压缩和弯曲，这就使得管板在其下表面上受到管子所给予的横向反力和径向反力矩的作用。

(3) 由于筒体和管束存在温度变形差，这就使得管板的下表面上又受到管子所给予的一个横向反力的作用。

我们先假定用一个无孔的厚圆板来代替有孔的厚圆板，并将管束视为连续弹性基础，然后考虑开孔对管板的强度和刚度的削弱，则管板所承受的上述载荷可写成以下形式：

(1) 横向载荷（或称有效压力）

$$q=f_1q_1-f_2q_2, \tag{1}$$

式中

$$f_1=1-n\left(\frac{d_1}{d}\right)^2,$$

$$f_2=1-n\left(\frac{d_2}{d}\right)^2$$

(2) 基础反力

$$R=k_1(w+\delta-\delta_p), \tag{2}$$

式中

$$k_1=\frac{nE_r(d_2^2-d_1^2)}{d^2l},$$

$$\delta=\frac{1}{2}\int_0^{2l}[\alpha_r(t_r-t_0)-\alpha_p(t_p-t_0)]dz,$$

$$\delta_p=\frac{Ql}{E_ph_p}$$

为了便于计算温度变形差 δ，可设管子和筒体的温度沿 z 轴方向呈线性变化，它们的某一截面的温度以其内外壁温度的平均值来代替，于是便有

$$t_{r(p)}=\frac{t_2-t_1}{2l}z+t_1 \tag{3}$$

(3) 基础径向反力矩

$$m=k_2\varphi_r, \tag{4}$$

式中

$$k_2=\frac{4nE_rJ_rK_s}{\pi d^2l_1}$$

$$J_r=\frac{\pi}{64}(d_2^4-d_1^4)$$

考虑折流板的影响，由材料力学连续梁公式可导出管子弯曲系数 K_s 的计算公式：

$$K_0=2,$$

$$K_1=4,$$

$$K_2=\frac{4(1+3l_2/2l_1)}{1+2l_2/l_1},$$

$$K_3=\frac{4(1+3l_2/4l_1)}{1+l_2/l_1},$$

……

这些公式可以写成下面的递推形式：

$$\begin{aligned}K_s&=\frac{4(a_s+b_sl_2l_1)}{c_s+d_sl_2/l_1},\\K_{s+2}&=\frac{4[(a_s+b_s)+3(c_s+d_s)l_2/4l_1]}{(a_s+b_s)+(c_s+d_s)l_2/l_1},\end{aligned} \tag{5}$$

式中下角标 s 为折流板数，即 0，1，2，…。

当 $s=0$ 时，$a_0=b_0=0.5$，$c_0=d_0=1$；

当 $s=1$ 时，$a_1=b_1=c_1=d_1=1$。

由式（5）可看出，由 K_0 开始，可推得角标为偶数的 K_s 公式；而由 K_1 开始，则可推得角标为奇数的 K_s 公式。

在上述载荷作用下，按照 Reissner 的厚板理论，我们推得厚管板的基本方程组：

$$
\begin{aligned}
&\frac{\mathrm{d}M_r}{\mathrm{d}r}+\frac{M_r-M_\theta}{r}=Q_r+m,\\
&\frac{\mathrm{d}(rQ_r)}{\mathrm{d}r}=r[k_1(w+\delta-\delta_{\mathrm{p}})-q],\\
&M_r=D\left\{\frac{\mathrm{d}\varphi_r}{\mathrm{d}r}+\frac{\nu}{r}\varphi_r+\frac{6\nu(1+\nu)}{5Eh}[q-k_1(w+\delta-\delta_{\mathrm{p}})]\right\},\\
&M_\theta=D\left\{\frac{\varphi_r}{r}+\nu\frac{\mathrm{d}\varphi_r}{\mathrm{d}r}+\frac{6\nu(1+\nu)}{5Eh}[q-k_1(w+\delta-\delta_{\mathrm{p}})]\right\},\\
&\varphi_r=\frac{12(1+\nu)}{5Eh}Q_r-\frac{\mathrm{d}w}{\mathrm{d}r},
\end{aligned}
\tag{6}
$$

式中

$$D=\psi\frac{Eh^3}{12(1-\nu^2)},\quad \psi=f_1^{7/3},$$

这里，开孔对管板弯曲刚度的影响系数 ψ 的公式是按照文献［7］的实验公式确定的。

方程组（6）包含五个常微分方程，共涉及五个未知量：挠度 w、径向弯矩 M_r、切向弯矩 M_θ、剪力 Q 和径向转角 φ_r，显然是可解的。但是，若要直接求解这组方程，又很不方便，为此，施用消元法，我们将这组方程化为一个含挠度 w 的四阶变系数常微分方程：

$$
\begin{aligned}
&D\,\nabla^4 w-k_2\,\nabla^2 w-\beta_1\psi h^2\,\nabla^2[k_1(w+\delta-\delta_{\mathrm{p}})-q]\\
&+\left[\frac{12k_2(1+\nu)}{5Eh}+1\right][k_1(w-\delta-\delta_{\mathrm{p}})-q]=0,
\end{aligned}
\tag{7}
$$

式中

$$\nabla^2=\frac{\mathrm{d}^2}{\mathrm{d}r^2}+\frac{1}{r}\frac{\mathrm{d}}{\mathrm{d}r},\quad \nabla^4=(\nabla^2)^2,\quad \beta_1=\frac{2-\nu}{10(1-\nu)}$$

在解方程（7）时，我们考虑下面两种边界条件：

（1）固定边界条件

当 $r=\dfrac{d}{2}$ 时，　$w=0$，　$\varphi_r=0$；

（2）简支边界条件

当 $r=\dfrac{d}{2}$ 时，　$w=0$，　$M_r=0$。

经过繁杂的数学运算，可得上述边值问题的精确解：

$$w=C_1U_0(x,\varphi)+C_2V_0(x,\varphi)+\frac{q}{k_1}-(\delta-\delta_{\mathrm{p}})\tag{8}$$

这里 C_1 和 C_2 在不同边界条件下有不同的公式：

（1）固定边界条件

$$C_1=\lambda_2\left(\frac{Q}{\lambda_7}-\frac{q}{k_1\lambda_1}\right),$$

$$C_2=\lambda_3\left(\frac{Q}{\lambda_7}-\frac{q}{k_1\lambda_1}\right),$$

$$Q=\frac{\lambda_7 E_p h_p \delta}{\lambda_7 l+\lambda_1 E_p h_p};$$

（2）简支边界条件

$$C_1=\lambda_5\left(\frac{Q}{\lambda_8}-\frac{q}{k_1\lambda_4}\right),$$

$$C_2=\lambda_8\left(\frac{Q}{\lambda_8}-\frac{q}{k_1\lambda_4}\right),$$

$$Q=\frac{\lambda_8 E_p h_p \delta}{\lambda_8 l+\lambda_4 E_p h_p};$$

式中

$$x=\omega_0 r\left(\text{边缘极值:}x_1=\frac{1}{2}\omega_0 d\right),$$

$$\varphi=\frac{1}{2}\arccos\left(\frac{\omega_2}{2D\omega_0^2}\right),$$

$$\omega_0=\sqrt[4]{\frac{k_1}{D}(\omega_1 k_2+1)},$$

$$\omega_1=\frac{12(1+\nu)}{5Eh},$$

$$\omega_2=-(\beta_1\omega_3+k_2),$$

$$\omega_3=\psi k_1 h^2,$$

$$\lambda_1=\lambda_2 U_0(x_1,\varphi)+\lambda_3 V_0(x_1,\varphi),$$

$$\lambda_2=\gamma_9 U_0'(x_1,\varphi)-\gamma_8 V_0'(x_1,\varphi),$$

$$\lambda_3=\gamma_8 U_0'(x_1,\varphi)+\gamma_9 V_0'(x_1,\varphi),$$

$$\lambda_4=\lambda_5 U_0(x_1,\varphi)+\lambda_6 V_0(x_1,\varphi),$$

$$\lambda_5=\lambda_2 U_0(x_1,\varphi)-\gamma_1 V_0(x_1,\varphi),+\frac{\gamma_4}{x_1}U_0'(x_1,\varphi)-\frac{\gamma_3}{x_1}V_0'(x_1,\varphi),$$

$$\lambda_6=\gamma_1 U_0(x_1,\varphi)+\gamma_2 V_0(x_1,\varphi),+\frac{\gamma_3}{x_1}U_0'(x_1,\varphi)+\frac{\gamma_4}{x_1}V_0'(x_1,\varphi),$$

$$\lambda_7=(\gamma_6\gamma_9-\gamma_7\gamma_8)[U_0'^2(x_1,\varphi)+V_0'^2(x_1,\varphi)],$$

$$\lambda_8=\lambda_5[\gamma_6 U_0'(x_1,\varphi)+\gamma_7 V_0'(x_1,\varphi)]-\lambda_6[\gamma_7 U_0'(x_1,\varphi)-\gamma_6 V_0'(x_1,\varphi)],$$

$$\gamma_1=\frac{1}{2}\omega_2+\omega_3\beta_1,\quad \gamma_2=-\frac{1}{2}\omega_2\tan(2\varphi),$$

$$\gamma_3=\frac{1}{5}\omega_3\cos(2\varphi)+\frac{\omega_2(1-\nu)}{2\cos(2\varphi)},$$

$$\gamma_4=\frac{1}{5}\omega_3\sin(2\varphi),\quad \gamma_6=-\frac{k_1}{\omega_0}\cos(2\varphi),$$

$$\gamma_7=-\frac{k_1}{\omega_0}\sin(2\varphi),\quad \gamma_8=\omega_1\gamma_6-\omega_0,\quad \gamma_9=\omega_1\gamma_7,$$

$$U_0(x,\varphi)=1+\sum_{i=1}^{\infty}\frac{(-1)^i}{\prod_{j=1}^{i}(2j)^2}x^{2i}\cos(2i\varphi),$$

$$V_0(x,\varphi)=\sum_{i=1}^{\infty}\frac{(-1)^i}{\prod_{j=1}^{i}(2j)^2}x^{2i}\sin(2i\varphi),$$

$$U_0'(x,\varphi)=\sum_{i=1}^{\infty}\frac{(-1)^i 2i}{\prod_{j=1}^{i}(2j)^2}x^{2i-1}\cos(2i\varphi),$$

$$V_0'(x,\varphi)=\sum_{i=1}^{\infty}\frac{(-1)^i 2i}{\prod_{j=1}^{i}(2j)^2}x^{2i-1}\sin(2i\varphi)$$

最后，应用平均应力概念，取开孔引起的应力减弱系数：

$$\eta=f_2,\tag{9}$$

则得厚管板的径向正应力 σ_r、切向正应力 σ_θ、法向正应力 σ_z，以及水平剪应力 $\tau_{r\theta}$、径向横向剪应力 τ_{rz} 和切向横向剪应力 $\tau_{\theta z}$（这些应力的正方向规定与一般弹性理论中规定是一样的）的计算式：

$$\begin{aligned}
&\sigma_r=\frac{12M_r z}{\eta h^3},\\
&\sigma_\theta=\frac{12M_\theta z}{\eta h^3},\\
&\sigma_z=\frac{1}{\eta}\left\{[q-k_1(w+\delta-\delta_p)]\left(\frac{3z}{2h}-\frac{2z^3}{h^3}\right)-\frac{1}{2}[f_1q_1+f_2q_2+k_1(w+\delta-\delta_p)]\right\},\\
&\tau_{rz}=\frac{3Q_r}{2\eta h}\left[1-\left(\frac{2z}{h}\right)^2\right],\\
&\tau_{r\theta}=\tau_{\theta z}=0,
\end{aligned}\tag{10}$$

式中

$$\begin{aligned}
M_r=&C_1\left[\gamma_1U_0(x,\varphi)+\gamma_2V_0(x,\varphi)+\frac{\gamma_3}{x}U_0'(x,\varphi)+\frac{\gamma_4}{x}V_0'(x,\varphi)\right]\\
&-C_2\left[\gamma_2U_0(x,\varphi)-\gamma_1V_0(x,\varphi)+\frac{\gamma_4}{x}U_0'(x,\varphi)-\frac{\gamma_3}{x}V_0'(x,\varphi)\right],\\
M_\theta=&C_1\left[\gamma_5U_0(x,\varphi)+\nu\gamma_2V_0(x,\varphi)-\frac{\gamma_3}{x}U_0'(x,\varphi)-\frac{\gamma_4}{x}V_0'(x,\varphi)\right]\\
&-C_2\left[\nu\gamma_2U_0(x,\varphi)-\gamma_5V_0(x,\varphi)-\frac{\gamma_4}{x}U_0'(x,\varphi)+\frac{\gamma_3}{x}V_0'(x,\varphi)\right],\\
Q_r=&C_1[\gamma_6U_0'(x,\varphi)+\gamma_7V_0'(x,\varphi)]-C_2[\gamma_7U_0'(x,\varphi)-\gamma_6V_0'(x,\varphi)],
\end{aligned}$$

$$\gamma_5=\frac{\nu}{2}\omega_2+\omega_3\beta_2,$$

$$\beta_2=\frac{\nu}{10(1-\nu)}$$

值得指出，中、低压固定式换热器的薄管板的理论公式仅仅是上述公式的特殊情况。我们只要令 $\omega_1=\omega_3=0$ 便可由上述公式推得适用于薄管板的理论公式。

二、验证与结论

我们通过三个实例应用试验结果来讨论本节厚管板理论公式的精确度，同时，也与本节薄管板理论公式进行比较，内容见文献［4］。

由三个实例的计算，有下面几点结论：

(1) 在固定和简支两种边界条件下，管板的最大挠度都发生在中心。对同一管板来说，按固定边界条件下所算得的最大挠度要比按简支边界条件下所算得的最大挠度小一些。但最大应力发生位置却不同，固定边界条件下发生在管板边缘的表面上；简支边界条件下则发生在管板中心的表面上。同一管板就绝对值而言，按固定边界条件下所算得的最大应力要比按简支边界条件下所算得的最大应力小一些。

(2) 厚管板理论计算值比薄管板理论计算值更接近试验值，这说明用厚管板理论公式进行设计计算更为准确。

(3) 厚、薄管板理论计算值有显著的差值，即按薄管板理论公式算得的最大应力要大于按厚管板理论公式算得的最大应力。但对于最大挠度则恰恰相反。

(4) 在设计厚管板时，选择何种边界条件进行计算是个很重要的问题。一般说来，当管板与较厚的法兰等构件连接时，可视管板的边界条件为固定，如文献［4］的例 1。当管板与较薄法兰等构件连接时，可视管板的边界条件为简支，如文献［4］的例 2，而且试验时又未设置筒体，所以管板边界条件偏于简支。即使如此，最靠近板边缘一点的可以预料，试验值还是接近固定边界条件的理论曲线。若试验时设置筒体则管板更接近于固定边界条件。因此，只要法兰厚度较大就可认为管板是固定边界条件。

(5) 当管内走热载体时，由压力和温度变化引起的应力和挠度的符号相反。因此，我们可以不管操作情况如何，只要把常温下由压力所产生的应力和挠度作为设计数据即可，这种设计既简单又较安全。

三、厚管板的设计公式和图表

按照上述分析，现在给出厚管板在常温下由压力 q 所产生的最大挠度和最大应力的计算公式。由式 (8) 和式 (10) 推得。

(1) 在固定边界条件下

$$\text{最大挠度 } w_{\max}=\frac{q}{k_1}\left(1-\frac{\lambda_2}{\lambda_1}\right), \tag{11a}$$

$$\text{最大应力 } \sigma_{\max}=\pm\frac{6q}{f_2k_1h^2\lambda_1}(\lambda_2\lambda_6-\lambda_3\lambda_5); \tag{11b}$$

（2）在简支边界条件下

$$\text{最大挠度 } w_{\max}=\frac{q}{k_1}\left(1-\frac{\lambda_5}{\lambda_4}\right), \tag{12a}$$

$$\text{最大应力 } \sigma_{\max}=\pm\frac{6q}{f_2k_1h^2\lambda_4}\left\{\frac{\omega_2(1+\nu)}{\varphi}[\lambda_5+\lambda_6\tan(2\varphi)]+\frac{\omega_3\lambda_5}{10(1-\nu)}\right\}, \tag{12b}$$

这里，式（11b）和（12b）中的“±”号分别对应于管板的上、下表面。

由于最大挠度值常常很小，在设计中处于次要地位，加之计算公式比较简单，所以我们仅将最大应力绘成图表，以供设计时使用。

设管板材料的泊松比 ν 为 0.3。并引入两个无量纲特征参数：

$$\begin{aligned}\mu_1&=\frac{0.195nEh_rK_s}{Ehd^2l_1}(d_2^4-d_1^4),\\ \mu_2&=\frac{10.92nE_rh\psi}{Ed^2l}(d_2^2-d_1^2),\end{aligned} \tag{13}$$

那么，公式（11b）和（12b）成为

$$\sigma_{\max}=\pm\frac{\psi q}{f_2}\sigma, \tag{14}$$

这里，σ 为管板的无量纲最大应力，它依赖于 μ_1、μ_2 和 x_1 三个变量在不同边界条件下取不同的值。

使用式（14）时，必须先求下列各式：

$$\begin{aligned}x_1&=\frac{\sqrt[4]{\mu_2\ (\mu_1+1)}}{2\sqrt{\psi}}\frac{d}{h},\\ q&=f_1q_1-f_2q_2,\\ f_1&=1-n\left(\frac{d_1}{d}\right)^2,\\ f_2&=1-n\left(\frac{d_2}{d}\right)^2,\\ \psi&=f_1^{7/3}\end{aligned} \tag{15}$$

应用 DJS-8 电子计算机，我们将管板的无量纲最大应力 σ 绘制于图 2～图 8 和表 1～表 6，同时也将管子的弯曲系数 K_s 制成表 7。由于篇幅所限，图表中的变量间距不可能取得很密，读者在使用时可应用通常的插值公式或按平均分配法读出结果。在使用时，若遇到特征参数 $\mu_1<0.0001$ 时，可用 $\mu_1=0.0001$ 的图表代替，其误差不大，当折流板数大于 10 时，可用 K_{10} 去代替真实的管子弯曲系数，也不会导致较大误差。

四、图表的应用

厚管板设计时，按下述步骤计算：

（1）计算 l_2/l_1 值，根据 l_2/l_1 和 s 值查表 7，确定管子弯曲系数 K_s 值；

（2）将管板的已知数据和 K_s 值代入公式（13），确定特征参数 μ_1、μ_2；

（3）将已知数据和 μ_1、μ_2 代入公式（15），得无量纲半径 x_1 的值；

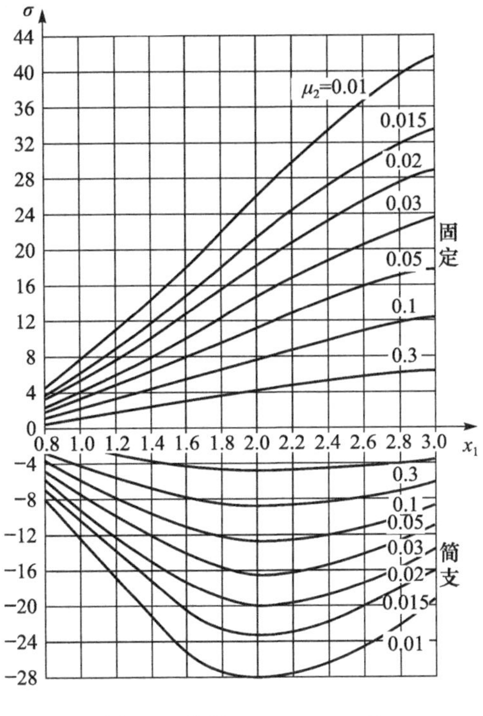

图 2　σ 曲线（μ_1=0.0001）

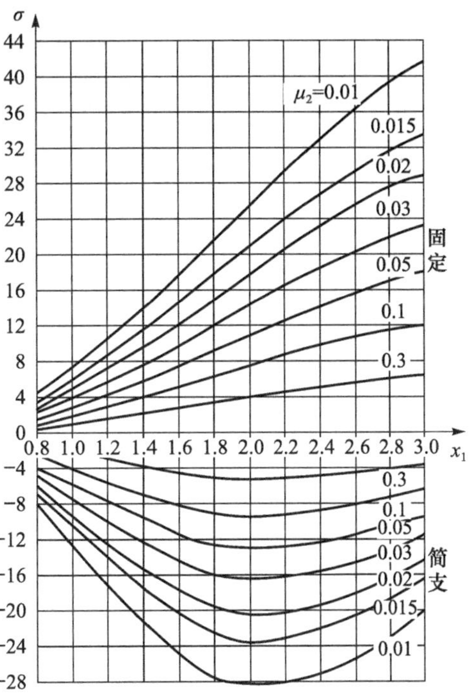

图 3　σ 曲线（μ_1=0.001）

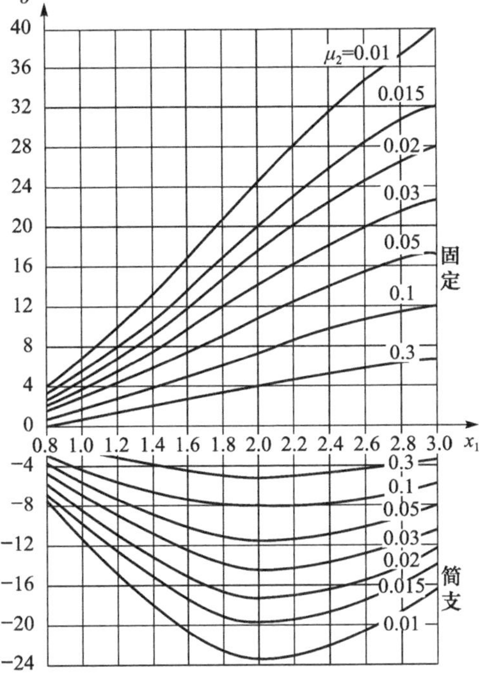

图 4　σ 曲线（μ_1=0.01）

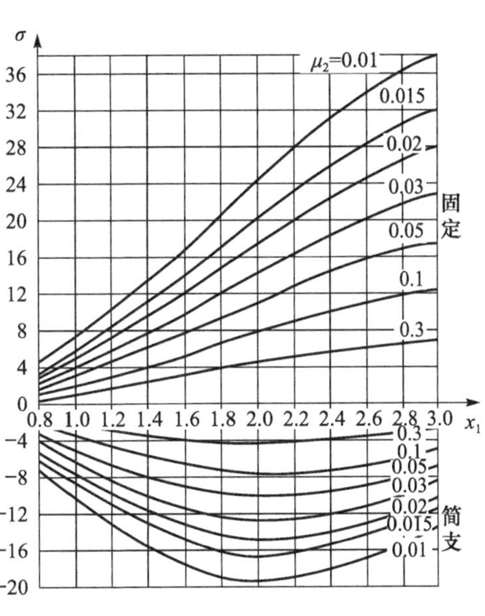

图 5　σ 曲线（μ_1=0.02）

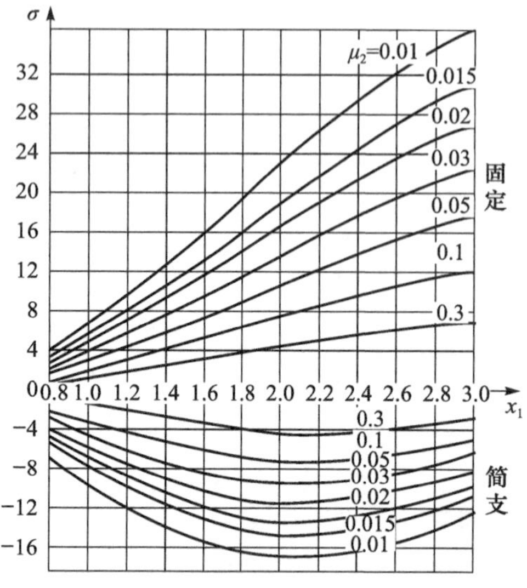

图 6　σ 曲线（$\mu_1=0.03$）

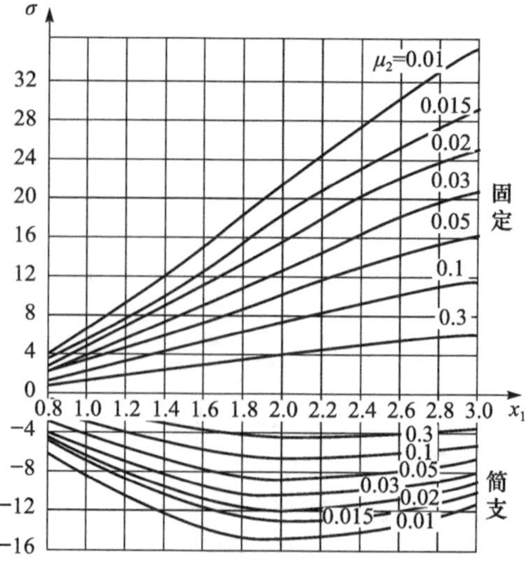

图 7　σ 曲线（$\mu_1=0.04$）

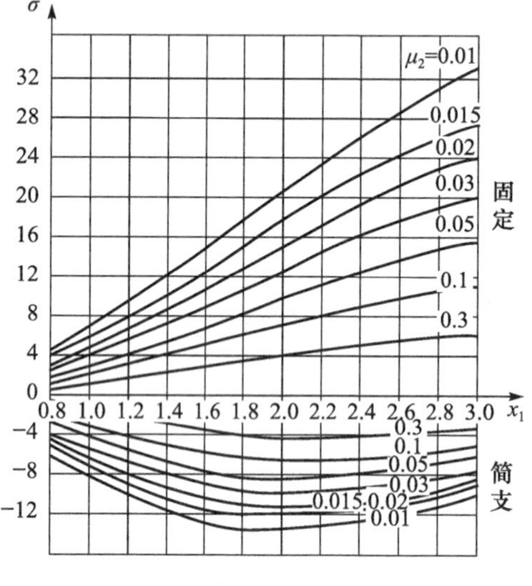

图 8　σ 曲线（$\mu_1=0.05$）

(4) 按照 x_1、μ_1、μ_2 三个值，查图 2～图 8 或表 1～表 6，确定无量纲量大应力 σ 值；

(5) 应用式（15）、（14），可计算出管板的最大应力。

反过来说，若预先不知道管板尺寸，而只知道操作压力、材料性能等数据，亦可按逆顺序定出管板尺寸。

下面举例说明计算过程。

例　计算一厚管板的最大应力。

已知：$l=4904\text{mm}$，$l_1=713\text{mm}$，$l_2=254\text{mm}$，$h=190\text{mm}$，$d=700\text{mm}$，$d_1=12\text{mm}$，$d_2=18\text{mm}$，$n=745$，$s=33$，$q_1=3.2\text{kg/mm}^2$，$q_2=0.05\text{kg/mm}^2$，常温下 $E=E_r=2.1\times10^4\text{kg/mm}^2$。

解：(1) 首先，算得 $l_2/l_1=0.356$，因 $s=33$，按 $s=10$ 查表 7，并使用插值法，得 $K_s=3.710$。

(2) 将已知数据代入公式（13）得

$$\mu_1=\frac{0.195\times745\times3.710}{190\times700^2\times713}\times(18^4-12^4)=0.0006840,$$

$$\mu_2=\frac{10.92\times745\times190\times0.5618}{700^2\times4904}\times(18^2-12^2)=0.06505$$

(3) 根据 μ_1、μ_2 和其他数据求 x_1：

表1 σ值（$\mu_1=0.001$）

x_1 \ μ_2	0.01		0.015		0.02		0.03		0.05		0.1		0.3	
	固定	简支	固定	简支	固定	简支	固定	简支	固定	简支	固定	简支	固定	简支
0.8	4.5074	−7.7128	3.6315	−6.3000	3.1092	−5.4564	2.4896	−4.4544	1.8681	−3.4480	1.2427	−2.4336	0.60526	−1.3969
0.9	5.5554	−9.6375	4.6494	−7.8729	3.9899	−6.8888	3.2075	−5.5665	2.4226	−4.3082	1.6328	−3.0393	0.82771	−1.7422
1.0	7.1378	−11.697	5.7767	−9.5565	4.9650	−8.2773	4.0019	−6.7568	3.0358	−5.2285	2.0636	−3.6869	1.0725	−2.1103
1.1	8.6480	−13.845	7.0078	−11.313	6.0295	−9.7986	4.8689	−7.9983	3.7045	−6.1881	2.5325	−4.3614	1.3380	−2.4925
1.2	10.278	−16.026	8.3358	−13.096	7.1776	−11.343	5.8033	−9.2589	4.4245	−7.1620	3.0368	−5.0452	1.6222	−2.8787
1.3	12.016	−18.175	9.7521	−14.854	8.4016	−12.867	6.7990	−10.502	5.1911	−8.1221	3.5726	−5.7185	1.9229	−3.2578
1.4	13.851	−20.228	11.247	−16.534	9.6927	−14.322	7.8486	−11.689	5.9984	−9.0384	4.1359	−6.3604	2.2376	−3.6179
1.5	15.768	−22.117	12.807	−18.080	11.040	−15.661	8.9435	−12.782	6.8396	−9.8816	4.7217	−6.9505	2.5633	−3.9478
1.6	17.750	−23.782	14.419	−19.443	12.432	−16.843	10.074	−13.745	7.7072	−10.625	5.3247	−7.4698	2.8969	−4.2371
1.7	19.778	−25.174	16.069	−20.582	13.856	−17.830	11.229	−14.550	8.5929	−11.245	5.9390	−7.9030	3.2352	−4.4774
1.8	21.832	−26.255	17.739	−21.467	15.297	−18.597	12.397	−15.176	9.4876	−11.727	6.5584	−8.2389	3.5746	−4.6630
1.9	23.891	−27.006	19.412	−22.083	16.740	−19.130	13.567	−15.610	10.382	−12.062	7.1766	−8.4714	3.9117	−4.7905
2.0	25.933	−27.422	21.072	−22.424	18.170	−19.425	14.725	−15.851	11.268	−12.247	7.7870	−8.5995	4.2431	−4.8597
2.1	27.938	−27.515	22.699	−22.500	19.572	−19.491	15.860	−15.904	12.134	−12.287	8.3836	−8.6265	4.5655	−4.8728
2.2	29.884	−27.306	24.279	−22.329	20.933	−19.343	16.961	−15.784	12.974	−12.193	8.9606	−8.5596	4.8762	−4.8337
2.3	31.754	−26.827	25.796	−21.937	22.239	−19.004	18.017	−15.506	13.779	−11.979	9.5129	−8.4087	5.1724	−4.7480
2.4	33.531	−26.115	27.237	−21.355	23.480	−18.499	19.019	−15.094	14.542	−11.660	10.036	−8.1852	5.4522	−4.6220
2.5	35.203	−25.208	28.592	−20.612	24.646	−17.856	19.961	−14.569	15.259	−11.255	10.527	−7.9011	5.7139	−4.4625
2.6	36.759	−24.145	29.854	−19.742	25.731	−17.101	20.837	−13.954	15.925	−10.780	10.982	−7.5683	5.9565	−4.2760
2.7	38.193	−22.960	31.016	−18.773	26.731	−16.261	21.644	−13.268	16.538	−10.251	11.401	−7.1980	6.1793	−4.0687
2.8	39.503	−21.688	32.076	−17.731	27.643	−15.359	22.380	−12.532	17.098	−9.6827	11.783	−6.8004	6.3823	−3.8463
2.9	40.689	−20.357	33.036	−16.642	28.468	−14.415	23.045	−11.762	17.603	−9.0884	12.129	−6.3845	6.5657	−3.6136
3.0	41.753	−18.991	33.897	−15.524	29.209	−13.447	23.642	−10.972	18.057	−8.4790	12.438	−5.9580	6.7304	−3.3751

表 2 σ 值（$\mu_1=0.01$）

x_1 \ μ_2	0.01		0.015		0.02		0.03		0.05		0.1		0.3	
	固定	简支	固定	简支	固定	简支	固定	简支	固定	简支	固定	简支	固定	简支
0.8	4.4475	−7.3340	3.5883	−6.0425	3.0748	−5.2606	2.4644	−4.3213	1.8509	−3.3660	1.2322	−2.3908	0.60019	−1.3813
0.9	5.6673	−9.0573	4.5866	−7.4785	3.9403	−6.5191	3.1716	−5.3629	2.3985	−4.1831	1.6183	−2.9745	0.82095	−1.7196
1.0	7.0127	−10.857	5.6882	−8.9847	4.8955	−7.8426	3.9522	−6.4617	3.0028	−5.0475	2.0442	−3.5936	1.0637	−2.0774
1.1	8.4756	−12.685	6.8867	−10.522	5.9350	−9.1975	4.8018	−7.5903	3.6604	−5.9381	2.5071	−4.2329	1.3267	−2.4477
1.2	10.046	−14.491	8.1744	−12.049	7.0521	−10.547	5.7148	−8.7138	4.3670	−6.8306	3.0040	−4.8763	1.6081	−2.8130
1.3	11.714	−16.223	9.5419	−13.520	8.2388	−11.851	6.6849	−9.8115	5.1175	−7.6992	3.5313	−5.5020	1.9056	−3.1834
1.4	13.465	−17.832	10.979	−14.893	9.4859	−13.071	7.7044	−10.839	5.9060	−8.5175	4.0847	−6.0941	2.2166	−3.5269
1.5	15.284	−19.271	12.473	−16.128	10.783	−14.172	8.7646	−11.768	6.7258	−9.2607	4.6593	−6.6333	2.5383	−3.8398
1.6	17.156	−20.505	14.010	−17.192	12.118	−15.124	9.8561	−12.574	7.5694	−9.9070	5.2498	−7.1034	2.8675	−4.1128
1.7	19.062	−21.505	15.577	−18.059	13.478	−15.902	10.968	−13.236	8.4288	−10.439	5.8506	−7.4917	3.2010	−4.3384
1.8	20.985	−22.256	17.158	−18.714	14.851	−16.492	12.091	−13.740	9.2954	−10.846	6.4555	−7.7895	3.5354	−4.5113
1.9	22.906	−22.752	18.737	−19.151	16.223	−16.887	13.212	−14.080	10.160	−11.123	7.0586	−7.9924	3.8673	−4.6292
2.0	24.806	−22.908	20.299	−19.373	17.579	−17.021	14.320	−14.258	11.015	−11.269	7.6537	−8.1006	4.1936	−4.6940
2.1	26.668	−23.008	21.830	−19.391	18.908	−17.112	15.406	−14.280	11.852	−11.290	8.2350	−8.1178	4.5110	−4.7018
2.2	28.476	−22.801	23.316	−19.221	20.198	−16.964	16.458	−14.159	12.662	−11.196	8.7972	−8.0509	4.8168	−4.6628
2.3	30.215	−22.400	24.743	−18.883	21.436	−16.666	17.468	−13.910	13.439	−10.998	9.3356	−7.9089	5.1085	−4.5801
2.4	31.872	−21.833	26.103	−18.401	22.615	−16.238	18.429	−13.550	14.177	−10.712	9.8460	−7.7019	5.3842	−4.4597
2.5	33.437	−21.126	27.386	−17.797	23.727	−15.701	19.334	−13.098	14.872	−10.352	10.326	−7.4408	5.6423	−4.3079
2.6	34.905	−20.306	28.587	−17.095	24.766	−15.076	20.179	−12.571	15.519	−9.9316	10.772	−7.1359	5.8818	−4.1306
2.7	36.269	−19.397	29.701	−16.317	25.729	−14.383	20.961	−11.987	16.118	−9.4644	11.183	−6.7971	6.1022	−3.9338
2.8	37.529	−18.421	30.728	−15.482	26.676	−13.639	21.680	−11.359	16.667	−8.9630	11.560	−6.4334	6.3034	−3.7226
2.9	38.684	−17.400	31.667	−14.607	27.425	−12.860	22.335	−10.702	17.166	−8.4381	11.902	−6.0528	6.4857	−3.5017
3.0	39.737	−16.350	32.521	−13.708	28.160	−12.060	22.928	−10.027	17.617	−7.8993	12.211	−5.6623	6.6499	−3.2752

表 3　σ 值（μ_1=0.02）

μ_2 / x_1	0.01		0.015		0.02		0.03		0.05		0.1		0.3	
	固定	简支	固定	简支	固定	简支	固定	简支	固定	简支	固定	简支	固定	简支
0.8	4.3831	−6.9548	3.5418	−5.7805	3.0376	−5.0594	2.4372	−4.1828	1.8323	−3.2796	1.2208	−2.3452	0.59465	−1.3645
0.9	5.5733	−8.4893	4.5192	−7.0844	3.8869	−6.2158	3.1328	−5.1539	2.3723	−4.0527	1.6026	−2.9060	0.81356	−1.6941
1.0	6.8800	−10.053	5.5938	−8.4244	4.8211	−7.4104	3.8986	−6.1631	2.9671	−4.8610	2.0231	−3.4956	1.0541	−2.0422
1.1	8.2941	−11.602	6.7583	−9.7634	5.8343	−8.6104	4.7298	−7.1834	3.6129	−5.6834	2.4795	−4.0991	1.3145	−2.3999
1.2	9.8050	−13.093	8.0042	−11.064	6.9191	−9.7821	5.6204	−8.1862	4.3062	−6.4990	2.9686	−4.6998	1.5928	−2.7596
1.3	11.401	−14.486	9.3218	−12.289	8.0672	−10.893	6.5636	−9.1429	5.0387	−7.2783	3.4867	−5.2804	1.8868	−3.1049
1.4	13.068	−15.745	10.700	−13.408	9.2692	−11.912	7.5518	−10.027	5.8075	−8.0047	4.0295	−5.8235	2.1938	−3.4313
1.5	14.791	−16.844	12.127	−14.392	10.515	−12.813	8.5765	−10.813	6.6049	−8.6557	4.5922	−6.3134	2.5111	−3.7269
1.6	16.556	−17.760	13.591	−15.220	11.792	−13.576	9.6282	−11.484	7.4236	−9.2143	5.1696	−6.7363	2.8355	−3.9833
1.7	18.347	−18.481	15.076	−15.879	13.090	−14.187	10.697	−12.024	8.2558	−9.6679	5.7561	−7.0821	3.1639	−4.1939
1.8	20.146	−19.003	16.570	−16.362	14.396	−14.638	11.773	−12.427	9.0934	−10.009	6.3459	−7.3441	3.4930	−4.3543
1.9	21.938	−19.329	18.059	−16.670	15.698	−14.929	12.846	−12.691	9.9283	−10.235	6.9332	−7.5196	3.8195	−4.4626
2.0	23.707	−19.466	19.530	−16.808	16.983	−15.064	13.905	−12.818	10.753	−10.347	7.5134	−7.6096	4.1492	−4.5189
2.1	25.439	−19.427	20.969	−16.788	18.241	−15.053	14.941	−12.816	11.559	−10.353	8.0779	−7.6181	4.4523	−4.5258
2.2	27.121	−19.231	22.366	−16.624	19.462	−14.909	15.946	−12.698	12.339	−10.260	8.6248	−7.5516	4.7529	−4.4869
2.3	28.741	−18.894	23.711	−16.332	20.636	−14.647	16.912	−12.474	13.088	−10.079	9.1488	−7.4184	5.0398	−4.4073
2.4	30.290	−18.437	24.994	−15.931	21.756	−14.284	17.832	−12.161	13.801	−9.8227	9.6462	−7.2275	5.3111	−4.2926
2.5	31.761	−17.878	26.210	−15.437	22.816	−13.835	18.701	−11.772	14.473	−9.5031	10.114	−6.9883	5.5654	−4.1485
2.6	33.148	−17.237	27.355	−14.868	23.811	−13.316	19.516	−11.322	15.102	−9.1321	10.551	−6.7100	5.8017	−3.9807
2.7	34.448	−16.531	28.424	−14.239	24.740	−12.743	20.274	−10.823	15.686	−8.7209	10.955	−6.4013	6.0195	−3.7944
2.8	35.659	−15.775	29.417	−13.567	25.600	−12.128	20.975	−10.289	16.223	−8.2799	11.326	−6.0702	6.2168	−3.5947
2.9	36.783	−14.984	30.335	−12.862	26.393	−11.485	21.618	−9.7293	16.716	−7.8182	11.664	−5.7236	6.3999	−3.3858
3.0	37.821	−14.171	31.178	−12.138	27.120	−10.824	22.206	−9.1542	17.164	−7.3439	11.971	−5.3676	6.5634	−3.1713

表 4　σ 值（μ_1=0.03）

x_1 \ μ_2	0.01		0.015		0.02		0.03		0.05		0.1		0.3	
	固定	简支	固定	简支	固定	简支	固定	简支	固定	简支	固定	简支	固定	简支
0.8	4.3209	−6.6132	3.4966	−5.5406	3.0015	−4.8734	2.4106	−4.0532	1.8140	−3.1978	1.2095	−2.3015	0.58921	−13482
0.9	5.4830	−7.9883	4.4542	−6.7301	3.8351	−5.9400	3.0951	−4.9610	2.3467	−3.9306	1.5872	−2.8407	0.80631	−1.6701
1.0	6.7535	−9.3591	5.5032	−7.9299	4.7494	−7.0236	3.8468	−5.8912	2.9323	−4.6882	2.0025	−3.4032	1.0446	−2.0083
1.1	8.1223	−10.687	6.6357	−9.1058	5.7376	−8.0935	4.6604	−6.8181	3.5668	−5.4500	2.4525	−3.9739	1.3025	−2.3542
1.2	9.5782	−11.936	7.8429	−10.226	6.7922	−9.1202	5.5297	−7.7156	4.2454	−6.1947	2.9341	−4.5370	1.5778	−2.6982
1.3	11.109	−13.078	9.1145	−11.261	7.9045	−10.076	6.4477	−8.5591	4.9628	−6.9012	3.4434	−5.0763	1.8684	−3.0304
1.4	12.701	−14.087	10.440	−12.188	9.0649	−10.938	7.4068	−9.3266	5.7130	−7.5501	3.9761	−5.5763	2.1716	−3.3410
1.5	14.340	−14.947	11.807	−12.987	10.263	−11.687	8.3985	−9.9997	6.4894	−8.1245	4.5275	−6.0232	2.4846	−3.6208
1.6	16.012	−15.648	13.204	−13.646	11.489	−12.310	9.4136	−10.565	7.2848	−8.6113	5.0923	−6.4056	2.8044	−3.8621
1.7	17.702	−16.185	14.618	−14.158	12.731	−12.799	10.443	−11.012	8.0918	−9.0014	5.6653	−6.7151	3.1279	−4.0592
1.8	19.395	−16.559	16.036	−14.523	13.977	−13.150	11.477	−11.339	8.9027	−9.2898	6.2409	−6.9469	3.4518	−4.2082
1.9	21.078	−16.778	17.447	−14.744	15.217	−13.368	12.506	−11.546	9.7099	−9.4761	6.8135	−7.0995	3.7731	−4.3078
2.0	22.738	−16.849	18.838	−14.828	16.441	−13.457	13.521	−11.637	10.506	−9.5631	7.3777	−7.1745	4.0885	−4.3585
2.1	24.362	−16.787	20.200	−14.786	17.638	−13.427	14.514	−11.620	11.284	−9.5570	7.9284	−7.1760	4.3954	−4.3627
2.2	25.941	−16.604	21.522	−14.631	18.799	−13.289	15.477	−11.504	12.038	−9.4658	8.4611	−7.1104	4.6911	−4.3241
2.3	27.464	−16.316	22.796	−14.376	19.917	−13.056	16.403	−11.302	12.762	−9.2993	8.9718	−6.9850	4.9734	−4.2474
2.4	28.924	−15.937	24.015	−14.034	20.986	−12.742	17.287	−11.026	13.452	−9.0678	9.4570	−6.8080	5.2405	−4.1379
2.5	30.316	−15.482	25.174	−13.621	22.001	−12.359	18.125	−10.686	14.104	−8.7817	9.9143	−6.5877	5.4911	−4.0008
2.6	31.636	−14.965	26.270	−13.148	22.959	−11.920	18.913	−10.296	14.716	−8.4512	10.342	−6.3324	5.7243	−3.8416
2.7	32.881	−14.398	27.300	−12.628	23.857	−11.437	19.649	−9.8650	15.285	−8.0859	10.739	−6.0497	5.9396	−3.6650
2.8	34.051	−13.793	28.264	−12.074	24.694	−10.920	20.334	−9.4038	15.813	−7.6945	11.104	−5.7466	6.1371	−3.4767
2.9	35.145	−13.162	29.161	−11.493	25.471	−10.379	20.967	−8.9210	16.298	−7.2848	11.439	−5.4293	6.3169	−3.2777
3.0	36.165	−12.512	29.993	−10.896	26.190	−9.8231	21.549	−8.4247	16.743	−6.8636	11.744	−5.1033	6.4798	−3.0744

表 5　σ 值（$\mu_1=0.04$）

x_1 \ μ_2	0.01		0.015		0.02		0.03		0.05		0.1		0.3	
	固定	简支	固定	简支	固定	简支	固定	简支	固定	简支	固定	简支	固定	简支
0.8	4.2608	−6.3039	3.4528	−5.3203	2.9663	−4.7010	2.3847	−3.9318	1.7961	−3.1203	1.1985	−2.2596	0.58385	−1.3324
0.9	5.3962	−7.5431	4.3913	−6.4098	3.7850	−5.6879	3.0585	−4.7824	2.3218	−3.8159	1.5721	−2.7786	0.79918	−1.6468
1.0	6.6326	−8.7543	5.4161	−7.4902	4.6802	−6.6753	3.7966	−5.6427	2.8985	−4.5276	1.9823	−3.3158	1.0354	−1.9757
1.1	7.9594	−9.9042	6.5186	−8.5308	5.6448	−7.6351	4.5934	−6.4884	3.5220	−5.2355	2.4263	−3.8564	1.2907	−2.3104
1.2	9.3647	−10.965	7.6896	−9.5046	6.6710	−8.5416	5.4425	−7.2962	4.1835	−5.9196	2.9006	−4.3854	1.5631	−2.6476
1.3	10.836	−11.915	8.9188	−10.389	7.7499	−9.3725	6.3369	−8.0451	4.8896	−6.5615	3.4014	−4.8878	1.8503	−2.9597
1.4	12.360	−12.738	10.195	−11.167	8.8721	−10.110	7.2687	−8.7172	5.6222	−7.1443	3.9244	−5.3496	2.1498	−3.2556
1.5	13.924	−13.426	11.508	−11.827	10.028	−10.741	8.2298	−9.2986	6.3788	−7.6546	4.4649	−5.7588	2.4587	−3.5208
1.6	15.514	−13.974	12.845	−12.361	11.206	−11.256	9.2113	−9.7796	7.1524	−8.0819	5.0179	−6.1059	2.7740	−3.7483
1.7	17.117	−14.384	14.196	−12.767	12.397	−11.653	10.204	−10.155	7.9360	−8.4200	5.5781	−6.3843	3.0928	−3.9331
1.8	18.719	−14.658	15.547	−13.047	13.591	−11.932	11.200	−10.423	8.7222	−8.6661	6.1402	−6.5904	3.4118	−4.0719
1.9	20.309	−14.807	16.890	−13.208	14.776	−12.096	12.190	−10.587	9.5039	−8.8208	6.6989	−6.7237	3.7280	−4.1637
2.0	21.876	−14.838	18.212	−13.256	15.945	−12.153	13.166	−10.651	10.274	−8.8879	7.2491	−6.7863	4.0394	−4.2094
2.1	23.409	−14.764	19.506	−13.202	17.087	−12.110	14.120	−10.623	11.027	−8.8773	7.7861	−6.7823	4.3403	−4.2112
2.2	24.900	−14.595	20.763	−13.056	18.197	−11.980	15.045	−10.513	11.757	−8.7846	8.3055	−6.7177	4.6312	−4.1730
2.3	26.341	−14.345	21.977	−12.830	19.267	−11.771	15.936	−10.329	12.458	−8.6303	8.8037	−6.5993	4.9091	−4.0990
2.4	27.727	−14.024	23.141	−12.535	20.292	−11.496	16.788	−10.082	13.127	−8.4196	9.2777	−6.4345	5.1722	−3.9942
2.5	29.053	−13.645	24.251	−12.182	21.267	−11.165	17.597	−9.7824	13.760	−8.1616	9.7250	−6.2308	5.4193	−3.8637
2.6	30.315	−13.218	25.305	−11.782	22.191	−10.787	18.361	−9.4397	14.356	−7.8648	10.144	−5.9954	5.6496	−3.7122
2.7	31.513	−12.752	26.301	−11.344	23.062	−10.373	19.078	−9.0629	14.913	−7.5377	10.534	−5.7353	5.8625	−3.5445
2.8	32.645	−12.257	27.237	−10.878	23.878	−9.9318	19.748	−8.6602	15.431	−7.1875	10.894	−5.4565	6.0582	−3.3648
2.9	33.711	−11.740	28.115	−10.391	24.640	−9.4703	20.370	−8.2390	15.910	−6.8211	11.226	−5.1648	6.2368	−3.1767
3.0	34.713	−11.209	28.935	−9.8907	25.350	−8.9958	20.946	−7.8059	16.351	−6.4444	11.529	−4.8650	6.3990	−2.9836

表 6　σ 值（$\mu_1=0.05$）

μ_2 / x_1	0.01		0.015		0.02		0.03		0.05		0.1		0.3	
	固定	简支	固定	简支	固定	简支	固定	简支	固定	简支	固定	简支	固定	简支
0.8	4.2027	−6.0226	3.4102	−5.1172	2.9321	−4.5407	2.3593	−3.8179	1.7786	−3.0467	1.1877	−2.2194	0.57859	−1.3171
0.9	5.3127	−7.1450	4.3306	−6.1190	3.7364	−5.4568	3.0228	−4.6166	2.2974	−3.7081	1.5573	−2.7194	0.79218	−1.6243
1.0	6.5171	−8.2226	5.3323	−7.0968	4.6134	−6.3603	3.7479	−5.4147	2.8655	−4.3781	1.9626	−3.2331	1.0263	−1.9443
1.1	7.8046	−9.2272	6.4066	−8.0237	5.5557	−7.2259	4.5287	−6.1895	3.4786	−5.0375	2.4006	−3.7460	1.2792	−2.2683
1.2	9.1632	−10.138	7.5438	−8.8776	6.5552	−8.0317	5.3587	−6.9202	4.1316	−5.6683	2.8679	−4.2440	1.5477	−2.5874
1.3	10.581	−10.938	8.7338	−9.6411	7.6029	−8.7598	6.2307	−7.5892	4.8190	−6.2538	3.3606	−4.7131	1.8327	−2.8925
1.4	12.044	−11.621	9.9658	−10.302	8.6896	−9.3967	7.1370	−8.1821	5.5349	−6.7800	3.8743	−5.1409	2.1285	−3.1748
1.5	13.540	−12.180	11.229	−10.853	9.8057	−9.9335	8.0696	−8.6885	6.2729	−7.2359	4.4044	−5.5170	2.4334	−3.4265
1.6	15.057	−12.617	12.513	−11.292	10.941	−10.366	9.0200	−9.1019	7.0261	−7.6135	4.9461	−5.8333	2.7444	−3.6414
1.7	16.583	−12.935	13.806	−11.619	12.086	−10.692	9.9799	−9.4195	7.7877	−7.9086	5.4941	−6.0848	3.0586	−3.8149
1.8	18.106	−13.140	15.098	−11.838	13.232	−10.915	10.941	−9.6420	8.5510	−8.1200	6.0435	−6.2688	3.3728	−3.9445
1.9	19.615	−13.241	16.380	−11.955	14.369	−11.041	11.895	−9.7725	9.3091	−8.2493	6.5891	−6.3857	3.6842	−4.0293
2.0	21.101	−13.246	17.642	−11.979	15.488	−11.074	12.835	−9.8168	10.056	−8.3007	7.1262	−6.4379	3.9827	−4.0704
2.1	22.556	−13.167	18.877	−11.918	16.583	−11.025	13.754	−9.7818	10.785	−8.2798	7.6503	−6.4295	4.2869	−4.0703
2.2	23.973	−13.012	20.078	−11.781	17.647	−10.901	14.646	−9.6758	11.492	−8.1938	8.1573	−6.3661	4.5733	−4.0323
2.3	25.344	−12.792	21.239	−11.579	18.674	−10.713	15.505	−9.5074	12.173	−8.0504	8.6439	−6.2540	4.8469	−3.9609
2.4	26.666	−12.516	22.354	−11.321	19.660	−10.469	16.328	−9.2855	12.823	−7.8575	9.1073	−6.1000	5.1062	−3.8605
2.5	27.935	−12.195	23.422	−11.016	20.601	−10.179	17.112	−9.0186	13.439	−7.6231	9.5452	−5.9108	5.3500	−3.7359
2.6	29.148	−11.835	24.439	−10.672	21.495	−9.8505	17.854	−8.7150	14.021	−7.3548	9.9564	−5.6929	5.5774	−3.5917
2.7	30.304	−11.444	25.403	−10.298	22.340	−9.4916	18.553	−8.3822	14.566	−7.0597	10.340	−5.4525	5.7880	−3.4321
2.8	31.402	−11.301	26.315	−9.9007	23.137	−9.1096	19.209	−8.0271	15.075	−6.7443	10.696	−5.1951	5.9819	−3.2611
2.9	32.442	−10.600	27.174	−9.4859	23.885	−8.7108	19.821	−7.6561	15.548	−6.4145	11.024	−4.9258	6.1593	−3.0822
3.0	33.425	−10.158	27.981	−9.0598	24.585	−8.3008	20.391	−7.2747	15.985	−6.0753	11.325	−4.6490	6.3209	−2.8985

表 7　K_s 值

l_2/l_1 \ s	0	1	2	3	4	5	6	7	8	9	10
0.20	2.0000	4.0000	3.7143	3.8333	3.8065	3.8140	3.8120	3.8125	3.8124	3.8124	3.8124
0.25	2.0000	4.0000	3.6667	3.8000	3.7692	3.7778	3.7755	3.7761	3.7760	3.7760	3.7760
0.30	2.0000	4.0000	3.6250	3.7692	3.7353	3.7447	3.7422	3.7429	3.7427	3.7427	3.7427
0.35	2.0000	4.0000	3.5882	3.7407	3.7042	3.7143	3.7116	3.7123	3.7121	3.7122	3.7122
0.40	2.0000	4.0000	3.5556	3.7143	3.6757	3.6863	3.6835	3.6842	3.6840	3.6841	3.6840
0.45	2.0000	4.0000	3.5263	3.6897	3.6494	3.6604	3.6574	3.6582	3.6580	3.6581	3.6581
0.50	2.0000	4.0000	3.5000	3.6667	3.6250	3.6364	3.6333	3.6341	3.6339	3.6340	3.6340
0.55	2.0000	4.0000	3.4762	3.6452	3.6024	3.6140	3.6109	3.6118	3.6115	3.6116	3.6116
0.60	2.0000	4.0000	3.4545	3.6250	3.5814	3.5932	3.5901	3.5909	3.5907	3.5907	3.5907
0.65	2.0000	4.0000	3.4348	3.6061	3.5618	3.5788	3.5706	3.5714	3.5712	3.5713	3.5712
0.70	2.0000	4.0000	3.4167	3.5882	3.5435	3.5556	3.5523	3.5532	3.5530	3.5530	3.5530
0.75	2.0000	4.0000	3.4000	3.5714	3.5263	3.5385	3.5352	3.5361	3.5358	3.5359	3.5359
0.80	2.0000	4.0000	3.3846	3.5556	3.5102	3.5224	3.5191	3.5200	3.5198	3.5198	3.5198
0.85	2.0000	4.0000	3.3704	3.5405	3.4950	3.5072	3.5040	3.5049	3.5046	3.5047	3.5047
0.90	2.0000	4.0000	3.3571	3.5263	3.4808	3.4930	3.4897	3.4906	3.4903	3.4904	3.4904
0.95	2.0000	4.0000	3.3448	3.5128	3.4673	3.4795	3.4762	3.4771	3.4768	3.4769	3.4769
1.00	2.0000	4.0000	3.3333	3.5000	3.4545	3.4667	3.4634	3.4643	3.4641	3.4641	3.4641

$$f_1=1-745\times\left(\frac{12}{700}\right)^2=0.7811,$$

$$\psi=0.7811^{7/3}=0.5618,$$

$$x_1=\frac{\sqrt[4]{0.06505\ (0.0006840+1)}}{2\sqrt{0.5618}}\times\frac{700}{190}=1.241$$

(4) 根据 μ_1、μ_2 和 x_1 查图 2～图 8，确定 σ：

为简便起见，我们仅按固定边界条件情况查图表。因为 $\mu_1=0.0006840\approx0.001$，故应查表 1。应用插值方法，首先得

对于 $x_1=1.24$，　$\mu_2=0.06$ 情况，　有 $\sigma=4.25$；

对于 $x_1=1.24$，　$\mu_2=0.07$ 情况，　有 $\sigma=3.95$。

再一次使用插值方法，便有

对于 $x_1=1.24$，　$\mu_2=0.065$ 情况，　得 $\sigma=4.10$。

最后，再用图 3 进行校核。

(5) 利用式 (15)、(14) 求最大应力：

$$f_2=1-745\times\left(\frac{18}{700}\right)^2=0.5074,$$

$$q=0.7811\times3.2-0.5074\times0.05=2.474,$$

$$|\sigma_{max}|=\frac{0.5618\times2.474}{0.5074}\times4.1=11.2(\text{kg/mm}^2)$$

由这一数例计算看到，应用本节所给的图表设计厚管板，计算将变得十分简单，

极适于工程应用。

参 考 文 献

[1] 上海锅炉厂. 水冷器管板应力的测定. 化工炼油机械通讯，1974，(3)：18-25.

[2] 刘人怀，程昌钧，陈庆益，等. 高压固定式热交换器管板的应力计算——复变元圆柱函数的应用. 数学的实践与认识，1973，(1)：52-64.

[3] Мельников Н П. Исследование толстых перфорированных плит. Нзбранные проблема прикладлой механики，Москова，1974，497.

[4] 刘人怀. 高压固定式热交换器管板的弯曲问题. 中国科学技术大学 30 周年校庆报告会宣读，1979 年 10 月 26 日.

[5] Reissner E. The effect of transverse shear deformation on the bending of elastic plates. J Appl Mech，1945，12：A69-A77.

[6] 乌班诺夫斯基 W. 热交换器管板的设计. 力学学报，1960，4 (2)：94-111.

[7] НИИХИММАШ. Сосуды и аппараты，методика и нормы расчета на прочность узлов и деталей. Руководящие Технические Материалы，Москва，1960.

固定式厚管板的弯曲问题①

在化学、石油、石油化学、原子能、动力及其他许多工业部门中，固定式热交换器占有很重要的地位，而管板是热交换器的关键部件，它一直为研究者所注意，迄今所发表的基于薄板理论方面的工作，对中、低压热交换器的管板设计来说，是有实际意义的。但近年来，工业部门大量采用高压固定式热交换器，这种热交换器的管板厚度变得很大，已超出薄板理论研究范围。可是，由于厚管板问题比较复杂，因此研究的人较少，使得至今尚未得到一个能供工程设计应用的公式。因此，工程上不得不借用薄管板的公式进行粗略设计。显然，这是不合理的。

1970 年以来，上海锅炉厂[1]等单位先后两次对厚管板进行试验研究，但未进行理论分析。1974 年，Мельников[2]从理论上研究了厚管板，由于考虑管子对管板的影响不够，所以解决问题不够全面。此外，在 1970 年前后，Slot[3,4]还对另一类型，即 U 型管热交换器厚管板的热应力问题作过理论和实验研究，但未接触更重要的本类型的厚管板问题。

本文采用 Reissner[5-8]的厚板理论，足够精确地求解了厚管板的轴对称弯曲问题，为工程设计提供了可靠的理论公式。

一、厚管板的基本方程

因为热交换器结构相当复杂，所以要使厚管板的分析计算易于进行，就必须采用一些简化措施。考虑到厚管板（其几何尺寸、坐标及弯矩、剪力、挠度的正方向示于图 1 中）的弯曲特性在某些方面与薄管板相似，所以，我们参照对薄管板的处理[9-11]，对厚管板采用下列假定。

(1) 用一无孔的厚圆板来替代开孔的厚管板，在替代时考虑了开孔对板的刚度和强度的削弱。于是，在厚圆板上作用的有效压力为

$$q=f_1q_1-f_2q_2, \tag{1}$$

其中 d 为管板的计算直径，其值为筒体的内直径；d_1、d_2 分别为管子的内、外直径，n 为管子数目，

$$f_1=1-n\left(\frac{d_1}{d}\right)^2,\qquad f_2=1-n\left(\frac{d_2}{d}\right)^2 \tag{2}$$

设无孔厚圆板的弯曲刚度为 D_0，径向和切向弯曲应力为 $\sigma_{r,0}$、$\sigma_{\theta,0}$，横向剪应力为 $\tau_{rz,0}$，则厚管板的弯曲刚度和应力为

$$D=\psi D_0=\psi\frac{Eh^3}{12(1-\nu^2)}, \tag{3a}$$

$$\sigma_r=\frac{\sigma_{r,0}}{\eta}, \tag{3b}$$

---

① 本文原载《力学学报》，1982，(2)：166-179.

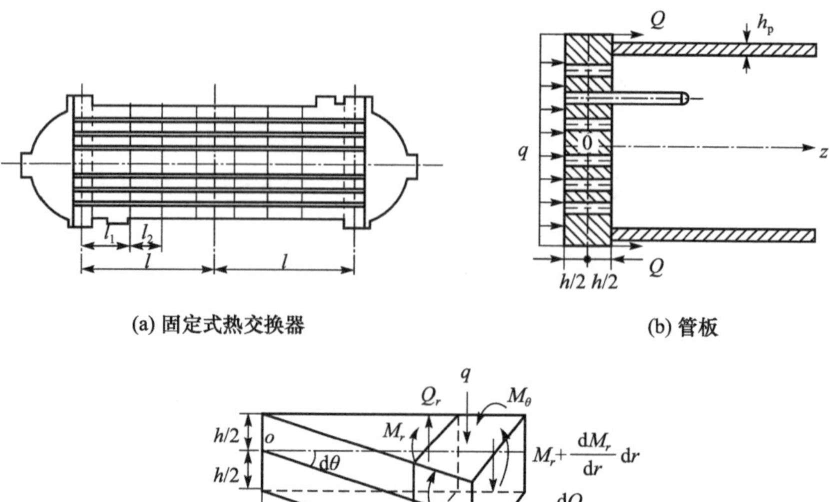

(a) 固定式热交换器　　(b) 管板

(c) 弯矩、剪力、挠度的正方向

图 1　厚管板特性

$$\sigma_\theta=\frac{\sigma_{\theta,0}}{\eta}, \tag{3c}$$

$$\tau_{rz}=\frac{\tau_{rz,0}}{\eta} \tag{3d}$$

这里，开孔对管板弯曲刚度的影响系数 ψ 可按文献［11］的实验公式定出，而由开孔引起的应力减弱系数 η 则可简单地按平均应力概念给出，亦即

$$\psi=f_1^{7/3},\qquad \eta=f_2 \tag{4}$$

（2）把密集、均匀分布在管板上的管束视为连续弹性基础，则管板成为弹性基础上的厚圆板。于是，由板的挠度以及筒体与管束存在温差的关系，在厚圆板上将有基础反作用力作用，其强度为

$$k_1(w+\delta-\delta_p) \tag{5}$$

其中

$$k_1=\frac{nE_r(d_2^2-d_1^2)}{d^2 l}, \tag{6a}$$

$$\delta=\frac{1}{2}\int_0^{2l}[\alpha_r(t_r-t_0)-\alpha_p(t_p-t_0)]dz, \tag{6b}$$

$$\delta_p=\frac{Ql}{E_p h_p} \tag{6c}$$

这里 E_r、E_p 分别为管子和筒体材料的弹性模量，α_r、α_p 分别为管子和筒体材料的热膨胀系数，t_r、t_p 分别为管子和筒体的温度，t_0 为装配温度。

（3）把管束弯曲而在管板上产生的反力矩视为连续分布，并且考虑折流板对管子弯曲的限制作用。于是，在厚圆板上将作用有管束的反力矩，其强度为

$$m=k_2\varphi_r, \tag{7}$$

其中

$$k_2=\frac{4nE_rJ_rK_s}{\pi d^2 l_1},$$

$$J_r=\frac{\pi}{64}(d_2^4-d_1^4) \tag{8}$$

至于管子弯曲系数 K_s（下角标符号 s 表示折流板的数目）的计算公式，则由材料力学连续梁的公式导出：

$$\begin{aligned}
&K_0=2,\\
&K_1=4,\\
&K_2=\frac{4(l_1+3l_2/2)}{l_1+2l_2},\\
&K_3=\frac{4(l_1+3l_2/4)}{l_1+l_2},\\
&\cdots\cdots\\
&K_s=\frac{4(a_sl_1+b_sl_2)}{c_sl_1+d_sl_2},\\
&K_{s+2}=\frac{4[(a_s+b_s)l_1+3(c_s+d_s)l_s/4]}{(a_s+b_s)l_1+(c_s+d_s)l_2},\quad s=0,1,2,\cdots
\end{aligned} \tag{9}$$

应该指出，在 $s>10$ 时，K_s 的值用 K_{10} 代替计算即可，此时能准确到五位有效数字以上。

(4) 认为管子、筒体的温度沿 z 轴方向呈线性变化，对于它们的某一截面的温度则以其内外壁温度的平均值近似代替。于是

$$t_{r(p)}=\frac{t_2-t_1}{2l}z+t_1 \tag{10}$$

其中 t_1、t_2 分别为管子和筒体沿 z 轴的始末处温度。

在上述假定下，我们按照 Reissner[5-8] 的厚板理论，经过简单的推导，便得到厚管板所满足的基本方程组：

$$\frac{\mathrm{d}M_r}{\mathrm{d}r}+\frac{M_r-M_\theta}{r}=Q_r+m, \tag{11a}$$

$$\frac{\mathrm{d}(rQ_r)}{\mathrm{d}r}=r[k_1(w+\delta-\delta_p)-q], \tag{11b}$$

$$M_r=D\left\{\frac{\mathrm{d}\varphi_r}{\mathrm{d}r}+\frac{\nu}{r}\varphi_r+\frac{6\nu(1+\nu)}{5Eh}[q-k_1(w+\delta-\delta_p)]\right\}, \tag{11c}$$

$$M_\theta=D\left\{\frac{\varphi_r}{r}+\nu\frac{\mathrm{d}\varphi_r}{\mathrm{d}r}+\frac{6\nu(1+\nu)}{5Eh}[q-k_1(w+\delta-\delta_p)]\right\}, \tag{11d}$$

$$\varphi_r=\frac{12(1+\nu)}{5Eh}Q_r-\frac{\mathrm{d}w}{\mathrm{d}r} \tag{11e}$$

现在，我们有方程（11），共有五个未知量：挠度 w，径向和切向弯矩 M_r、M_θ，剪力 Q_r 和转角 φ_r，故是可解的。

下面，我们化简方程组（11）。这组方程可以缩减为关于挠度 w 的一个高阶微分方程。先将方程（11e）代入方程（11c，d），并应用方程（11b），便得

$$
\left.\begin{aligned}
M_r&=-D\left\{\frac{\mathrm{d}^2w}{\mathrm{d}r^2}+\frac{\nu}{r}\frac{\mathrm{d}w}{\mathrm{d}r}-\frac{12(1-\nu^2)}{5Eh}\frac{\mathrm{d}Q_r}{\mathrm{d}r}-\frac{6\nu(1+\nu)}{5Eh}[k_1(w+\delta-\delta_\mathrm{p})-q]\right\},\\
M_\theta&=-D\left\{\frac{1}{r}\frac{\mathrm{d}w}{\mathrm{d}r}+\nu\frac{\mathrm{d}^2w}{\mathrm{d}r^2}-\frac{12(1-\nu^2)}{5Eh}\frac{Q_r}{r}-\frac{6\nu(1+\nu)}{5Eh}[k_1(w+\delta-\delta_\mathrm{p})-q]\right\}
\end{aligned}\right\}\tag{12}
$$

然后，将式（12）代入平衡方程（11a），并应用方程（11b），又得

$$
\begin{aligned}
&\frac{6(1-\nu^2)}{5Eh}\left(\nabla^2Q_r+\frac{2}{r}\frac{\mathrm{d}Q_r}{\mathrm{d}r}+\frac{Q_r}{r^2}\right)-\frac{Q_r}{D}\\
&=\frac{\mathrm{d}}{\mathrm{d}r}(\nabla^2w)-\frac{6(1+\nu)}{5Eh}\frac{\mathrm{d}}{\mathrm{d}r}[k_1(w+\delta-\delta_\mathrm{p})-q]\\
&\quad+\frac{12(1-\nu^2)}{5Eh}\frac{1}{r}[k_1(w+\delta-\delta_\mathrm{p})-q]+\frac{m}{D},
\end{aligned}\tag{13}
$$

其中

$$
\nabla^2=\frac{\mathrm{d}^2}{\mathrm{d}r^2}+\frac{1}{r}\frac{\mathrm{d}}{\mathrm{d}r},\tag{14}
$$

最后，应用平衡方程（11b）和式（7），我们便由方程（13）得到关于挠度 w 的四阶变系数常微分方程：

$$
\begin{aligned}
&D\,\nabla^4w-k_2\,\nabla^2w-\beta_1\psi h^2\,\nabla^2[k_1(w+\delta-\delta_\mathrm{p})-q]\\
&+\left[\frac{12k_2(1+\nu)}{5Eh}+1\right][k_1(w+\delta-\delta_\mathrm{p})-q]=0,
\end{aligned}\tag{15}
$$

其中

$$
\nabla^4=(\nabla^2)^2,\tag{16}
$$

$$
\beta_1=\frac{2-\nu}{10(1-\nu)}\tag{17}
$$

下面，我们致力于求解方程（15），在求解此方程时，将考虑两种边界条件：

（1）固定边界条件

$$
当\ r=\frac{d}{2}时,\qquad w=0,\qquad \varphi_r=0;\tag{18}
$$

（2）简支边界条件

$$
当\ r=\frac{d}{2}时,\qquad w=0,\qquad M_r=0\tag{19}
$$

为求解简单起见，引入新自变量

$$
x=\omega_0r,\tag{20}
$$

其中

$$
\omega_0=\sqrt[4]{\frac{k_1}{D}(\omega_1k_2+1)},\quad \omega_1=\frac{12(1+\nu)}{5Eh},\tag{21}
$$

则方程（15）简化成为下面形式：

$$
\frac{\mathrm{d}^2w}{\mathrm{d}x^4}+\frac{2}{x}\frac{\mathrm{d}^3w}{\mathrm{d}x^3}-\left(\frac{1}{x^2}-2\rho\right)\frac{\mathrm{d}^2w}{\mathrm{d}x^2}+\left(\frac{1}{x^3}+\frac{2\rho}{x}\right)\frac{\mathrm{d}w}{\mathrm{d}x}+w=\frac{q}{k_1}-(\delta-\delta_\mathrm{p}),\tag{22}
$$

其中

$$
2\rho=\frac{\omega_2}{D\omega_0^2},\quad \omega_2=-(\beta_1\omega_3+k_2),\quad \omega_3=\psi k_1h^2\tag{23}
$$

而边界条件（18）和（19）成为：

（1）固定边界条件

$$当\ x=\frac{1}{2}\omega_0 d=x_1\ 时，\quad w=0,\quad \varphi_r=0; \tag{24}$$

（2）简支边界条件

$$当\ x=x_1\ 时，\quad w=0,\quad M_r=0 \tag{25}$$

最后，由方程（11b）、（12）、（13），我们得到弯矩、剪力、转角和 x 的关系式：

$$\begin{aligned}
M_r&=\frac{\omega_3}{5x}\frac{\mathrm{d}^3 w}{\mathrm{d}x^3}+\left(\frac{\omega_3}{5x^2}-D\omega_0^2\right)\frac{\mathrm{d}^2 w}{\mathrm{d}x^2}-\frac{1}{x}\left[\frac{\omega_3}{5}\left(\frac{1}{x^2}-2\rho\right)+\nu D\omega_0^2\right]\frac{\mathrm{d}w}{\mathrm{d}x}\\
&\quad+\beta_1\omega_3\left(w+\delta-\delta_{\mathrm{p}}-\frac{q}{k_1}\right),\\
M_\theta&=-\left\{\frac{\omega_3}{5x}\frac{\mathrm{d}^3 w}{\mathrm{d}x^3}+\left(\frac{\omega_3}{5x^2}+\nu D\omega_0^2\right)\frac{\mathrm{d}^2 w}{\mathrm{d}x^2}-\frac{1}{x}\left[\frac{\omega_3}{5}\left(\frac{1}{x^2}-2\rho\right)-D\omega_0^2\right]\frac{\mathrm{d}w}{\mathrm{d}x}\right.\\
&\quad\left.-\beta_2\omega_3\left(w+\delta-\delta_{\mathrm{p}}-\frac{q}{k_1}\right)\right\},\\
Q_r&=-\frac{k_1}{\omega_0}\left[\frac{\mathrm{d}^3 w}{\mathrm{d}x^3}+\frac{1}{x}\frac{\mathrm{d}^2 w}{\mathrm{d}x^2}-\left(\frac{1}{x^2}-2\rho\right)\frac{\mathrm{d}w}{\mathrm{d}x}\right],\\
\varphi_r&=-\frac{\omega_1 k_1}{\omega_0}\left[\frac{\mathrm{d}^3 w}{\mathrm{d}x^3}+\frac{1}{x}\frac{\mathrm{d}^2 w}{\mathrm{d}x^2}-\left(\frac{1}{x^2}-2\rho-\frac{\omega_0^2}{\omega_1 k_1}\right)\frac{\mathrm{d}w}{\mathrm{d}x}\right],
\end{aligned} \tag{26}$$

其中

$$\beta_2=\frac{\nu}{10(1-\nu)} \tag{27}$$

于是，归结为求解常微分方程边值问题（22）、（24）和（22）、（25）。由这些边值问题求得挠度 w 后，再按式（26）求弯矩和剪力，最后，按照式（3b），我们得到计算管板的径向、切向弯曲应力 σ_r、σ_θ 以及横向剪应力 τ_{rz} 的公式：

$$\begin{aligned}
\sigma_r&=\frac{12M_r z}{\eta h^3},\\
\sigma_\theta&=\frac{12M_\theta z}{\eta h^3};
\end{aligned} \tag{28}$$

$$\tau_{rz}=\frac{3Q_r}{2\eta h}\left[1-\left(\frac{2z}{h}\right)^2\right] \tag{29}$$

值得指出，在前面的方程（22）、表达式（26）和边界条件（24）、（25）中，只要将计及横向剪切的项取为零，亦即令

$$\omega_1=\omega_3=0, \tag{30}$$

我们便得到通常的薄管板理论中的挠度方程，弯矩、剪力、转角表达式以及相应的边界条件。因此，只要我们求得了厚管板的解，同时也就得到了通常的薄管板的解。

二、边值问题的求解

现在，我们来研究方程（22）的齐次方程：

$$\frac{\mathrm{d}^4 w}{\mathrm{d}x^4}+\frac{2}{x}\frac{\mathrm{d}^3 w}{\mathrm{d}x^3}-\left(\frac{1}{x^2}-2\rho\right)\frac{\mathrm{d}^2 w}{\mathrm{d}x^2}+\left(\frac{1}{x^3}+\frac{2\rho}{x}\right)\frac{\mathrm{d}w}{\mathrm{d}x}+w=0 \tag{31}$$

很容易证明，方程（31）等价于下面两个二阶方程

$$\frac{\mathrm{d}^2 y}{\mathrm{d}x^2}+\frac{1}{x}\frac{\mathrm{d}y}{\mathrm{d}x}+g^2 y=0, \tag{32a}$$

$$\frac{\mathrm{d}^2 w}{\mathrm{d}x^2}+\frac{1}{x}\frac{\mathrm{d}w}{\mathrm{d}x}+\varepsilon^2 w=y, \tag{32b}$$

其中

$$\varepsilon^2=\rho+\mathrm{i}\sqrt{1-\rho^2},\qquad g^2=\rho-\mathrm{i}\sqrt{1-\rho^2} \tag{33}$$

先解方程（32a），它是零阶贝塞尔方程，其通解为

$$y=AJ_0(gx)+BY_0(gx), \tag{34}$$

其中 $J_0(gx)$、$Y_0(gx)$ 均是复变元圆柱函数，A、B 为待定常数。

将解（34）代入方程（32b），并求解方程，得通解：

$$w=\overline{C}_1 J_0(\varepsilon x)+\overline{C}_2 J_0(gx)+\overline{C}_3 Y_0(\varepsilon x)+\overline{C}_4 Y_0(gx), \tag{35}$$

其中 $\overline{C}_1$、$\overline{C}_2$、$\overline{C}_3$、$\overline{C}_4$ 为待定常数，而

$$\varepsilon=\sqrt{\frac{1+\rho}{2}}+\mathrm{i}\sqrt{\frac{1-\rho}{2}},\qquad g=\sqrt{\frac{1+\rho}{2}}-\mathrm{i}\sqrt{\frac{1-\rho}{2}}, \tag{36}$$

再令

$$\rho=\cos(2\varphi), \tag{37}$$

则式（36）成为

$$\varepsilon=\mathrm{e}^{\mathrm{i}\varphi},\qquad g=\mathrm{e}^{-\mathrm{i}\varphi} \tag{38}$$

于是，解（35）改写为

$$w=\overline{C}_1 J_0(x\mathrm{e}^{\mathrm{i}\varphi})+\overline{C}_2 J_0(x\mathrm{e}^{-\mathrm{i}\varphi})+\overline{C}_3 Y_0(x\mathrm{e}^{\mathrm{i}\varphi})+\overline{C}_4 Y_0(x\mathrm{e}^{-\mathrm{i}\varphi}) \tag{39}$$

为了得到实数解，我们用金尼克函数来替代复变元圆柱函数：

$$\begin{aligned}
U_0(x,\varphi)&=\frac{1}{2}[J_0(x\mathrm{e}^{\mathrm{i}\varphi})+J_0(x\mathrm{e}^{-\mathrm{i}\varphi})],\\
V_0(x,\varphi)&=\frac{1}{2\mathrm{i}}[J_0(x\mathrm{e}^{\mathrm{i}\varphi})-J_0(x\mathrm{e}^{-\mathrm{i}\varphi})],\\
\overline{U}_0(x,\varphi)&=\frac{1}{2}[Y_0(x\mathrm{e}^{\mathrm{i}\varphi})+Y_0(x\mathrm{e}^{-\mathrm{i}\varphi})],\\
\overline{V}_0(x,\varphi)&=\frac{1}{2\mathrm{i}}[Y_0(x\mathrm{e}^{\mathrm{i}\varphi})-Y_0(x\mathrm{e}^{-\mathrm{i}\varphi})]
\end{aligned} \tag{40}$$

由于挠度 w 在板中心有界，而 $\overline{U}_0(x,\varphi)$ 和 $\overline{V}_0(x,\varphi)$ 在中心处变为无限大，故应去掉解中的 $\overline{U}_0(x,\varphi)$ 和 $\overline{V}_0(x,\varphi)$。在考虑到方程（22）的特解后，我们就得到方程（22）的实数解：

$$w=C_1 U_0(x,\varphi)+C_2 V_0(x,\varphi)+\frac{q}{k_1}-(\delta-\delta_{\mathrm{p}}), \tag{41}$$

其中 C_1、C_2 为待定常数，而 $U_0(x,\varphi)$、$V_0(x,\varphi)$ 及其导数公式为

$$U_0(x,\varphi)=1+\sum_{i=1}^{\infty}\frac{(-1)^i}{\prod_{j=1}^{i}(2j)^2}x^{2i}\cos(2i\varphi),$$

$$V_0(x,\varphi)=\sum_{i=1}^{\infty}\frac{(-1)^i}{\prod_{j=1}^{i}(2j)^2}x^{2i}\sin(2i\varphi),$$

$$\frac{\mathrm{d}U_0(x,\varphi)}{\mathrm{d}x}=U_0'(x,\varphi)=\sum_{i=1}^{\infty}\frac{(-1)^i 2i}{\prod_{j=1}^{i}(2j)^2}x^{2i-1}\cos(2i\varphi),$$

$$\frac{\mathrm{d}V_0(x,\varphi)}{\mathrm{d}x}=V_0'(x,\varphi)=\sum_{i=1}^{\infty}\frac{(-1)^i 2i}{\prod_{j=1}^{i}(2j)^2}x^{2i-1}\sin(2i\varphi),$$

$$\frac{\mathrm{d}^2U_0(x,\varphi)}{\mathrm{d}x^2}=-U_0(x,\varphi)\cos(2\varphi)+V_0(x,\varphi)\sin(2\varphi)-\frac{1}{x}U_0'(x,\varphi),$$

$$\frac{\mathrm{d}^2V_0(x,\varphi)}{\mathrm{d}x^2}=-U_0(x,\varphi)\sin(2\varphi)-V_0(x,\varphi)\cos(2\varphi)-\frac{1}{x}V_0'(x,\varphi),$$

$$\begin{aligned}\frac{\mathrm{d}^3U_0(x,\varphi)}{\mathrm{d}x^3}=&\frac{1}{x}U_0(x,\varphi)\cos(2\varphi)-\frac{1}{x}V_0(x,\varphi)\sin(2\varphi)\\&+\left[\frac{2}{x^2}-\cos(2\varphi)\right]U_0'(x,\varphi)+V_0'(x,\varphi)\sin(2\varphi),\end{aligned}$$

$$\begin{aligned}\frac{\mathrm{d}^3V_0(x,\varphi)}{\mathrm{d}x^3}=&\frac{1}{x}U_0(x,\varphi)\sin(2\varphi)+\frac{1}{x}V_0(x,\varphi)\cos(2\varphi)\\&-U_0'(x,\varphi)\sin(2\varphi)+\left[\frac{2}{x^2}-\cos(2\varphi)\right]V_0'(x,\varphi)\end{aligned}\tag{42}$$

将解（41）代入式（26），得

$$\begin{aligned}M_r=&C_1\left[\gamma_1U_0(x,\varphi)+\gamma_2V_0(x,\varphi)+\frac{\gamma_3}{x}U_0'(x,\varphi)+\frac{\gamma_4}{x}V_0'(x,\varphi)\right]\\&-C_2\left[\gamma_2U_0(x,\varphi)-\gamma_1V_0(x,\varphi)+\frac{\gamma_4}{x}U_0'(x,\varphi)-\frac{\gamma_3}{x}V_0'(x,\varphi)\right],\end{aligned}\tag{43a}$$

$$\begin{aligned}M_\theta=&C_1\left[\gamma_5U_0(x,\varphi)+\nu\gamma_2V_0(x,\varphi)-\frac{\gamma_3}{x}U_0'(x,\varphi)-\frac{\gamma_4}{x}V_0'(x,\varphi)\right]\\&-C_2\left[\nu\gamma_2U_0(x,\varphi)-\gamma_5V_0(x,\varphi)-\frac{\gamma_4}{x}U_0'(x,\varphi)+\frac{\gamma_3}{x}V_0'(x,\varphi)\right],\end{aligned}\tag{43b}$$

$$Q_r=C_1[\gamma_6U_0'(x,\varphi)+\gamma_7V_0'(x,\varphi)]-C_2[\gamma_7U_0'(x,\varphi)-\gamma_6V_0'(x,\varphi)],\tag{43c}$$

$$\varphi_r=C_1[\gamma_8U_0'(x,\varphi)+\gamma_9V_0'(x,\varphi)]-C_2[\gamma_9U_0'(x,\varphi)-\gamma_8V_0'(x,\varphi)],\tag{43d}$$

其中

$$\begin{aligned}&\gamma_1=\frac{1}{2}\omega_2+\omega_3\beta_1,\quad \gamma_2=-\frac{1}{2}\omega_2\tan(2\varphi),\quad \gamma_3=\frac{1}{5}\omega_3\cos(2\varphi)+\frac{\omega_2(1-\nu)}{2\cos(2\varphi)},\\&\gamma_4=\frac{1}{5}\omega_3\sin(2\varphi),\quad \gamma_5=\frac{\nu}{2}\omega_2+\omega_3\beta_2,\quad \gamma_6=-\frac{k_1}{\omega_0}\cos(2\varphi),\\&\gamma_7=-\frac{k_1}{\omega_0}\sin(2\varphi),\quad \gamma_8=\omega_1\gamma_6-\omega_0,\quad \gamma_9=\omega_1\gamma_7\end{aligned}\tag{44}$$

为了今后使用方便，我们在式（41）和（43a-c）中令 x 为零，这样便得到直接计算厚管板中心处的挠度、弯曲和剪力的简单公式：

$$
\begin{aligned}
&w(0)=C_1+\frac{q}{k_1}-(\delta-\delta_p),\\
&M_r(0)=M_\theta(0)=\frac{1+\nu}{4}\omega_2[C_1+C_2\tan(2\varphi)]+\frac{1}{10(1-\nu)}C_1\omega_3,\\
&Q_r(0)=0
\end{aligned}
\tag{45}
$$

下面，我们分两类情况决定待定常数 C_1 和 C_2。

（1）仅有压力 q 作用，此时

$$\delta-\delta_p=0$$

（a）在固定边界条件下，将式（41）、（43d）代入边界条件（24），得决定待定常数 C_1 和 C_2 的公式：

$$C_1=-\frac{\lambda_2}{k_1\lambda_1}q,\qquad C_2=-\frac{\lambda_3}{k_1\lambda_1}q, \tag{46}$$

其中

$$
\begin{aligned}
&\lambda_1=\lambda_2U_0(x_1,\varphi)+\lambda_3V_0(x_1,\varphi),\\
&\lambda_2=\gamma_9U_0'(x_1,\varphi)-\gamma_8V_0'(x_1,\varphi),\\
&\lambda_3=\gamma_8U_0'(x_1,\varphi)+\gamma_9V_0'(x_1,\varphi)
\end{aligned}
\tag{47}
$$

（b）在简支边界条件下，将式（41）、（43a）代入边界条件（25），得决定待定常数 C_1 和 C_2 的公式：

$$C_1=-\frac{\lambda_5}{k_1\lambda_4}q,\qquad C_2=-\frac{\lambda_6}{k_1\lambda_4}q, \tag{48}$$

其中

$$
\begin{aligned}
&\lambda_4=\lambda_5U_0(x_1,\varphi)+\lambda_6V_0(x_1,\varphi),\\
&\lambda_5=\gamma_2U_0(x_1,\varphi)-\gamma_1V_0(x_1,\varphi)+\frac{\gamma_4}{x_1}U_0'(x_1,\varphi)-\frac{\gamma_3}{x_1}V_0'(x_1,\varphi),\\
&\lambda_6=\gamma_1U_0(x_1,\varphi)+\gamma_2V_0(x_1,\varphi)+\frac{\gamma_3}{x_1}U_0'(x_1,\varphi)+\frac{\gamma_4}{x_1}V_0'(x_1,\varphi)
\end{aligned}
\tag{49}
$$

（2）仅有筒体和管束温差的影响，此时

$$q=0$$

（a）在固定边界条件下，除边界条件（24）外，尚需引入边缘剪力连续的条件：

$$\text{当 } x=x_1 \text{ 时，}\qquad Q_r=Q, \tag{50}$$

将式（41）、（43c，d）代入条件（24）、（50），并应用式（6c），得决定待定常数 C_1 和 C_2 的公式：

$$C_1=\frac{\lambda_2}{\lambda_7}Q,\qquad C_2=\frac{\lambda_3}{\lambda_7}Q \tag{51}$$

其中

$$
\begin{aligned}
&\lambda_7=(\gamma_6\gamma_9-\gamma_7\gamma_8)[U_0'^2(x_1,\varphi)+V_0'^2(x_1,\varphi)],\\
&Q=\frac{\lambda_7E_ph_p\delta}{\lambda_7l+\lambda_1E_ph_p}
\end{aligned}
\tag{52}
$$

(b) 在简支边界条件下，将式 (41)、(43a，c) 代入条件 (25) 和 (50)，并应用式 (6c)，得决定待定常数 C_1 和 C_2 的公式：

$$C_1=\frac{\lambda_5}{\lambda_8}Q,\qquad C_2=\frac{\lambda_6}{\lambda_8}Q, \tag{53}$$

其中

$$\begin{aligned}&\lambda_8=\lambda_5[\gamma_6U_0'(x_1,\varphi)+\gamma_7V_0'(x_1,\varphi)]-\lambda_6[\gamma_7U_0'(x_1,\varphi)-\gamma_6V_0'(x_1,\varphi)],\\&Q=\frac{\lambda_8E_ph_p\delta}{\lambda_8l+\lambda_4E_ph_p}\end{aligned} \tag{54}$$

三、实例计算

例 1　绘制一常温下试验厚管板的应力曲线和挠度曲线。这一试验是在 1970 年由上海锅炉厂、一机部机械科学研究院、化工部第一设计院完成的。

已知数据：

$$\begin{aligned}&l=1532.5\text{mm},\quad l_1=532.5\text{mm},\quad l_2=500\text{mm},\\&h=65\text{mm},\quad d=350\text{mm},\quad d_1=6\text{mm},\\&d_2=14\text{mm},\quad n=232,\quad s=5,\\&q_1=3\text{kg/mm}^2,\quad q_2=0,\quad E=E_r=2.1\times10^4\text{kg/mm}^2,\\&\nu=0.3\end{aligned}$$

此管板的厚度与直径比为 $h/d=0.186$，显然，需用本节的厚管板理论公式进行讨论。

显而易见，此时 $\delta-\delta_p=0$。

故我们仅计算由压力 q 引起的解。将上述数据代入式 (28)、(41)、(43)、(46) 和 (48) 中，便得在固定和简支两种边界条件下的应力和挠度值。由理论计算结果和试验数据可知，按绝对值而言，管板表面上的径向应力 σ_r 取最大值。而工程设计感兴趣的也是最大应力，所以我们仅将管板上表面的径向应力沿半径的变化表示在图 2 上。关于管板挠度沿半径的变化曲线则绘在图 3 中。为了便于比较，我们还在图中给出了按薄管板理论公式（从本节公式略去含 ω_1 和 ω_3 的厚板因素项即可得到）所计算的值。由结果看出：

(1) 在固定边界条件下，最大应力发生在板的边缘，且在板的表面上；最大挠度发生在板中心。在简支边界条件下，最大应力发生在板中心，且在板的表面上；最大挠度发生在板中心。

(2) 应力试验值与厚管板在固定边界条件下的理论值比较一致，特别在靠近板边缘最大径向应力的区域内，更为吻合。以最大的试验值（在靠近板边缘的 $r=165$mm 处）为例，理论值与试验值的相对误差仅为 1.13%。而且，厚管板理论公式值比薄管板理论公式值更接近试验值。

(3) 将厚、薄管板理论公式值进行比较，则发现挠度之间的相对误差较应力大。以固定边界条件下的最大挠度和最大应力为例，它们的相对误差分别为 31.8% 和 −9.00%。而且还可看到，按薄管板理论公式计算所得的最大应力要大于按厚管板理论公式计算所得的最大应力。对于最大挠度，则恰恰相反，是前者小于后者。

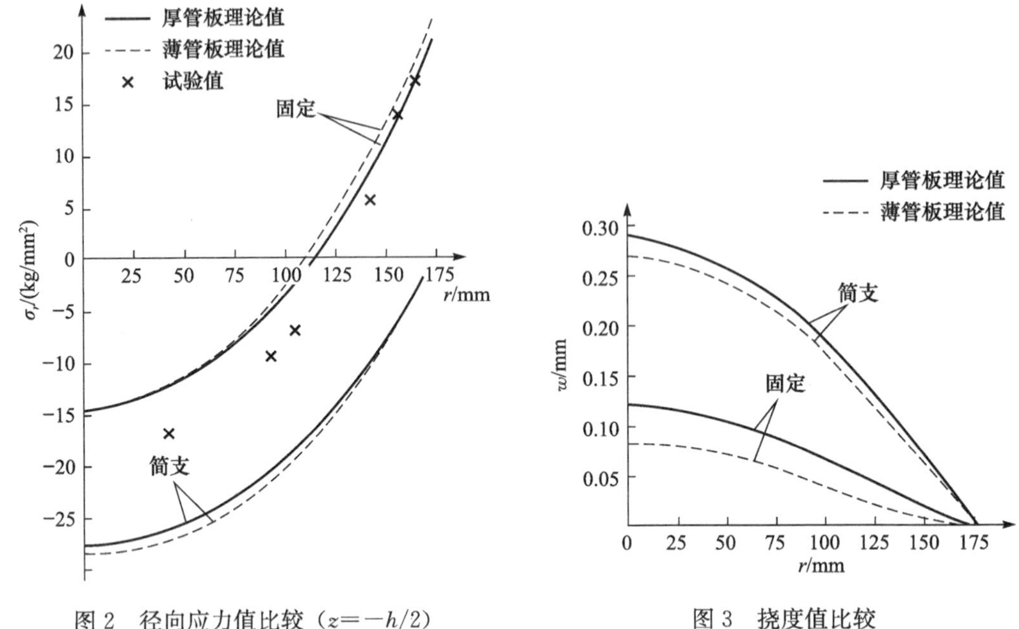

图 2　径向应力值比较（$z=-h/2$）　　图 3　挠度值比较

例 2　绘制一常温下试验厚管板[1]的应力曲线和挠度曲线。

已知数据：

$$l=2935.5\text{mm},\quad l_1=535.5\text{mm},\quad l_2=300\text{mm},$$
$$h=135\text{mm},\quad d=550\text{mm},\quad d_1=6\text{mm},$$
$$d_2=14\text{mm},\quad n=505,\quad s=17,$$
$$q_1=3.20\text{kg/mm}^2,\quad q_2=0,\quad E=2.02\times10^4\text{kg/mm}^2,$$
$$E_r=2.1\times10^4\text{kg/mm}^2,\quad \nu=0.3$$

此管板的厚度与直径比为

$$h/d=0.245,$$

显然，需用本节的厚管板理论公式进行讨论。

由于试验时，

$$\delta-\delta_p=0,$$

所以我们仅计算由压力 q 引起的解。与前例一样，将上述数据代入式（28）、（41）、（43）、（46）和（48）中，便得在固定和简支两种边界条件下的应力和挠度值。为了简明起见，我们仍然只给出最大应力即管板上表面径向应力以及挠度的曲线（图 4，图 5）。同时，我们也将薄管板理论值一起给在图中。由结果知，我们仍然有前例所得的第一点结论。此外，还有

（1）应力试验值与厚管板在简支边界条件下的理论值比较符合，特别在靠近板中心最大径向应力的区域内更趋一致。以最大的试验值（在靠近板中心的 $r=19$mm 处）为例，理论值与试验值的相对误差仅为 0.15%。而且，厚管板理论公式值比薄管板理论公式值更接近试验值。

（2）将厚、薄管板理论公式值进行比较，则再一次发现，挠度之间的相对误差比应

力大。以简支边界条件下的最大挠度和最大应力为例，它们的相对误差分别为 13.4% 和 −2.25%。与前例一样，还可看到，按薄管板理论公式计算所得的最大应力要大于按厚管板理论公式计算所得的最大应力。对于最大挠度，也是刚好相反，是前者小于后者。

例 3　计算一管板在工作情况下的最大应力和最大挠度。

已知数据：

$l=4924\text{mm}$，　$l_1=733\text{mm}$，　$l_2=254\text{mm}$，　$h=230\text{mm}$，

$h_p=12\text{mm}$，　$d=700\text{mm}$，　$d_1=12\text{mm}$，　$d_2=18\text{mm}$，　$n=745$，

$s=33$，　$q_1=3.2\text{kg/mm}^2$，　$q_2=0.05\text{kg/mm}^2$，　$t_0=20℃$

在高温下（100℃以上），$E=E_p=1.9\times10^4\text{kg/mm}^2$，$E_r=1.87\times10^4\text{kg/mm}^2$，$\nu=0.316$，$\alpha_r=18\times10^{-6}/℃$，$\alpha_p=11\times10^{-6}/℃$。

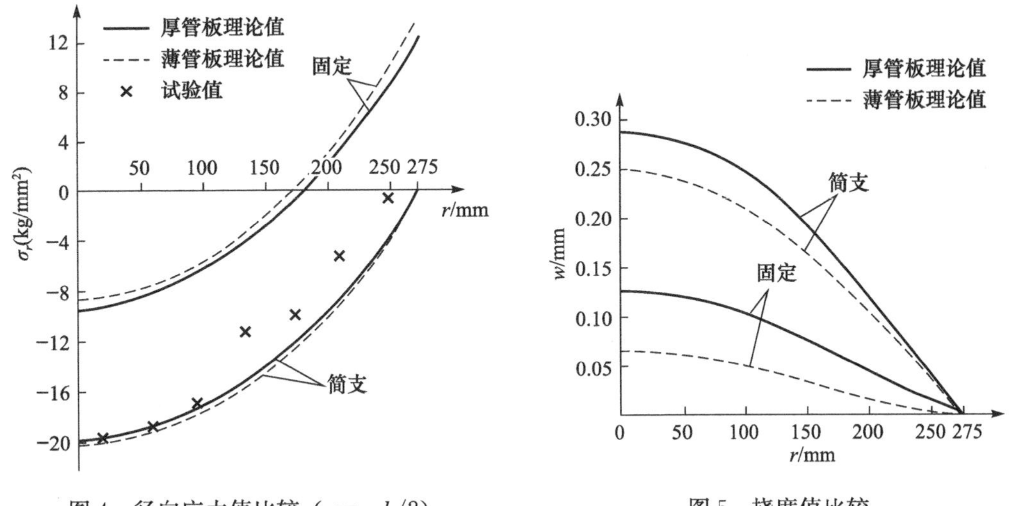

图 4　径向应力值比较（$z=-h/2$）　　图 5　挠度值比较

管子和筒体在工作情况下的温度分布情况见表 1。

此管板的厚度与直径比为

$$h/d=0.329,$$

显然它属于厚管板理论的研究范围。

表 1　管子、筒体沿 z 轴的温度分布情况　（单位：℃）

z	管内侧	管外侧	筒体
0	200	200	200
$h/2$	200	100	100
$2l-h/2$	20	20	20
$2l$	20	20	20

众所周知，材料的弹性常数在温度变化时并不是常数。但为了简单起见，我们不计及这种变化，而以其高温下的值近似地作为计算数据。

将以上数据代入式（28）、(41)、(43)、(46)、(48)、(51) 和 (53) 中，即可求得管板的应力和挠度。与上面两个实例一样，在固定和简支边界条件下，最大应力仍分别是管板边缘和中心的板表面上的径向应力；最大挠度仍是管板的中心挠度。为了便于比较，我们在表 2 中分别按照压力以及筒体与管束温差的影响来给出计算结果。由表 2 的数值结果看出，当管内走热载体时，由筒体和管束温差引起的热应力和挠度与由压力引起的应力和挠度反号。显然，这种情况对设计来说，是很有利的。

表 2　管板的最大应力 ($z=-h/2$) 和最大挠度值

承载情况	σ_{max}/ (kg/mm^2)		w_{max}/mm	
	固定	简支	固定	简支
$q\neq0$，$\delta-\delta_p=0$	7.51	−13.1	0.100	0.221
$q=0$，$\delta-\delta_p\neq0$	−2.33	4.15	−0.0310	−0.0702
合计	5.18	−8.95	0.0690	0.151

四、结论

(1) 由上面的实例计算看到，本节公式与试验值相当符合。这说明，采用厚板理论去研究高压管板的强度与刚度，才能给出比较精确的结果。

(2) 在固定和简支两种边界条件下，最大挠度都发生在板中心；最大应力都是径向应力，分别发生在板边缘和中心的表面上。对同一管板来说，按固定边界条件公式所计算的最大应力和最大挠度要比简支边界条件情况的小一些。

(3) 在具体情况中，我们如何决定边界条件呢？一般说来，对于与管板连接的法兰或其他结构较厚的情况，可视管板边界条件为固定。例 1 就是这种情况，法兰与管板的厚度比达 2.42。而在例 2 中，此比值仅为 1.74，加之未设置筒体，故管板边界条件偏于简支。即使如此，靠近板边缘一点的试验值却远离简支情况的理论曲线，而与固定情况的理论曲线吻合，误差仅为 0.09%。可以预料，如果设置了筒体，此管板边界条件趋于固定。因此，一般来说，由于法兰较厚，高压管板的边界条件大多属于固定情况。

(4) 当管内走热载体时，由筒体和管束温差所引起的应力和挠度将与由压力 q 所产生的应力和挠度反号，这对管板的强度和刚度来说是有益的。

参 考 文 献

[1] 上海锅炉厂. 水冷器管板应力的测定. 化工炼油机械通讯，1974 (3)：18-25.

[2] Мельников Н П. Исследование толстых лерфорированных плит. Избранные Проблема Прикладлой Механики，Москова，1974：497.

[3] Slot T. Stress Analysis of Thick Perforated Plates. New York：Technomic，1972.

[4] Slot T. Theoretical and experimental analysis of a thermal stress problem in tube sheet design//Proceedings of First International Conference on Pressure Vessel Technology. New York：ASME，1969.

[5] Reissner E. The effect of transverse shear deformation on the bending of elastic plates. J Appl

Mech，1945，12：A69-A77.

[6]　Timoshenko S，Krieger S W. Theory of Plates and Shells. New York：McGraw-Hill，1959：165.

[7]　Frederick D. On some problems in bending of thick circular plates on an elastic foundation. J Appl Mech，1956，23：195-200.

[8]　玉手統，伊藤金弥. 円形厚板の轴对称非線形たわみ. 日本机械学会論文集，1962，28（191）：812-818.

[9]　乌班诺夫斯基 W. 热交换器管板的设计. 力学学报，1960，4（2）：94-111.

[10]　Urbanowski W. Niektóre przypadki zginania plyty okraglei polaczonej z podložem sprezystym o wlanośćiach uogolnionyck. Zeszyty Politechniki Warszawskiej，Mechanika，1955：3.

[11]　НИИХИММАШ. Сосуды и аппараты，методика и нормы расчета на прочность узлов и деталей. Руководящие Технические Материалы，Москва，1960.

层合圆薄板的轴对称弯曲问题①

一、引言

多层层合薄板是在近代工业中广泛应用的一种承载构件。这种薄板结构复杂，材料特性沿厚度方向逐层变化，这就使其应力分析尤其是层间应力的分析变得十分困难。而层间剪应力是引起层合板破坏的最主要原因，因此对之作精确的计算是十分必要的。

按照经典的 Reissner-Mindlin 型薄板理论及后来的高阶理论，虽然可以较精确地计算板的整体响应（挠度、临界屈曲载荷和固有频率等），但所得局部响应误差非常大，例如所得横向剪应变在层间连续而剪应力在层间间断，这正与事实相反。为此，Sun 和 Whitney[1]、Srinivas[2] 以及 Whitney 和 Sun[3,4] 提出了一类对各层分别作位移或应力假设的分层理论。然而这类理论的控制方程数目随层合板层数的增多而骤增，对于层数较多的层合板计算极其复杂甚至根本不可能。近年，Ren[5] 将多层层合梁的结果推广到层合板，较好地计算了矩形层合板的弯曲问题。此外，Pagano[6,7] 曾应用三维线弹性理论给出了矩形层合板一些简单问题的精确解，显然，三维理论也只能研究层数极少的简单问题。

本文研究了各向同性多层层合圆板的轴对称小挠度问题，考虑了剪应变在各层的抛物线分布形式和剪应力在层间的连续条件，运用虚功原理导出了仅含两个未知量的控制方程，给出了控制方程通解的求法，并针对两层板的特例进行了数值计算。

本文理论很容易推广应用于各向异性层合板的弯曲问题。

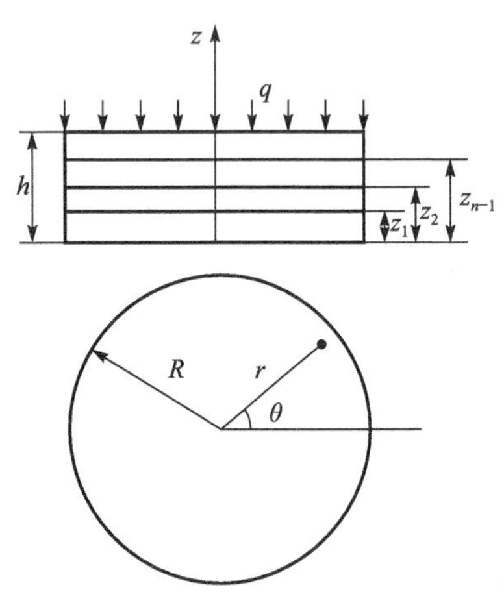

图 1　均布载荷作用下的层合圆形板

二、控制方程的推导

考虑一块均布载荷 q 作用下由 n 层不同材料的等厚圆薄层叠合而成的圆形薄板（如图 1），板的半径为 R，厚度为 h。由下至上将各薄层记为 $1,2,\cdots,n$，其中第 i 层薄层的厚度、弹性模量和泊松比分别为 h_i、E_i 和 ν_i（$i=1,2,\cdots,n$）。板的参考面取为下表面，板内各分界面方程为 $z=z_i$。

假设各薄层间没有相对滑动，略去厚度方向的应变和参考面内的位移，则柱坐标下板的应变可表示为[8]

① 本文原载《江西工业大学学报》，1991，13（2）：199-205. 作者：刘人怀，何陵辉.

$$
\begin{aligned}
\varepsilon_r &= -z\frac{\mathrm{d}^2 w(r)}{\mathrm{d}r^2} \\
&\quad + \left\{z^2 - \frac{2}{3h}z^3 - \sum_{i=1}^{n-1}\left[\frac{z^3}{3h^2} - (z-z_i)H(z-z_i)\right]\alpha_i\right\}\frac{\mathrm{d}\psi(r)}{\mathrm{d}r}, \\
\varepsilon_\theta &= -\frac{z}{r}\frac{\mathrm{d}w(r)}{\mathrm{d}r} \\
&\quad + \left\{z^2 - \frac{2}{3h}z^3 - \sum_{i=1}^{n-1}\left[\frac{z^3}{3h^2} - (z-z_i)H(z-z_i)\right]\alpha_i\right\}\frac{\psi(r)}{r}, \\
\gamma_{zr} &= \left\{2\left(z - \frac{z^2}{h}\right) - \sum_{i=1}^{n-1}\left[\frac{z^2}{h^2} - H(z-z_i)\right]\alpha_i\right\}\psi(r)
\end{aligned} \tag{1}
$$

式中 $w(r)$ 为参考面挠度，$\psi(r)$ 为未知函数，$H(z-z_i)$ 为 Haveside 函数，

$$
H(z-z_i) = \begin{cases} 1, & z \geqslant z_i \\ 0, & z < z_i \end{cases} \tag{2}
$$

α_i 为与材料有关的常数，由以下联立方程确定

$$
\begin{aligned}
&2(Q_3^{(2)} - Q_3^{(1)})\left(z - \frac{z_1^2}{h}\right) - (Q_3^{(2)} - Q_3^{(1)})\sum_{j=1}^{n-1}\frac{z_1^2}{h^2}\alpha_j = -Q_3^{(2)}\alpha_1, \\
&2Q_3^{(i+1)} - Q_3^{(i)}\left(z_i - \frac{z_1^2}{h}\right) - (Q_3^{(i+1)} - Q_3^{(i)})\sum_{j=1}^{n-1}\frac{z_i^2}{h^2}\alpha_j + (Q_3^{(i+1)} - Q_3^{(i)})\sum_{j=1}^{i-1}\alpha_j = -Q_3^{(i+1)}\alpha_i
\end{aligned} \tag{3}
$$

其中

$$
Q_3^{(i)} = \frac{E_i}{2(1+\nu_i)} \tag{4}
$$

板中第 i 层薄层的应力为

$$
\begin{aligned}
\sigma_r^{(i)} &= Q_1^{(i)}\varepsilon_r + Q_2^{(i)}\varepsilon_\theta, \\
\sigma_\theta^{(i)} &= Q_2^{(i)\varepsilon_r} + Q_1^{(i)}\varepsilon_\theta, \\
\tau_{zr}^{(i)} &= Q_3^{(i)}\gamma_{zr}
\end{aligned} \tag{5}
$$

其中

$$
Q_1^{(i)} = \frac{E_i(1-\nu_i)}{(1+\nu_i)(1-2\nu_i)}, \quad Q_2^{(i)} = \frac{E_i\nu_i}{(1+\nu_i)(1-2\nu_i)} \tag{6}
$$

由虚功原理可导出固支边界条件下层合圆板弯曲问题的控制方程及边界条件如下

$$
\begin{aligned}
&D_1\frac{\mathrm{d}^4 w}{\mathrm{d}r^4} - (D_1 + 2D_2)\frac{1}{r^2}\frac{\mathrm{d}^2 w}{\mathrm{d}r^2} + (D_1 + 2D_2)\frac{1}{r^3}\frac{\mathrm{d}w}{\mathrm{d}r} - J_1\frac{\mathrm{d}^3\psi}{\mathrm{d}r^3} \\
&+ (J_1 + 2J_2)\frac{1}{r^2}\frac{\mathrm{d}\psi}{\mathrm{d}r} - (J_1 + 2J_2)\frac{\psi}{r^3} = q, \\
&J_1\frac{\mathrm{d}^3 w}{\mathrm{d}r^3} - (J_1 + J_2)\frac{1}{r^2}\frac{\mathrm{d}w}{\mathrm{d}r} - K_1\frac{\mathrm{d}^2\psi}{\mathrm{d}r^2} + (K_1 + K_2)\frac{\psi}{r^2} = 0
\end{aligned} \tag{7a,b}
$$

在 $r=R$ 时，　$w=0$，　$\frac{\mathrm{d}w}{\mathrm{d}r}=0$，　$\psi=0$　　　(8a-c)

式中

$$D_j = \int_0^h Q_j^{(i)} z^2 \mathrm{d}z,$$

$$J_j = \int_0^h Q_j^{(i)} \left\{ z^2 - \frac{2}{3h} z^3 - \sum_{k=1}^{n-1} \left[\frac{z^3}{3h^2} - (z - z_k) H(z - z_k) \right] \alpha_k \right\} z \mathrm{d}z,$$

$$K_j = \int_0^h Q_j^{(i)} \left\{ z^2 - \frac{2}{3h} z^3 - \sum_{k=1}^{n-1} \left[\frac{z^3}{3h^2} - (z - z_k) H(z - z_k) \right] \alpha_k \right\}^2 \mathrm{d}z.$$

$$j = 1,2 \tag{9}$$

三、控制方程的通解

设

$$S = \frac{\mathrm{d}w}{\mathrm{d}r} \tag{10}$$

则方程组（7a，b）化为

$$D_1 \frac{\mathrm{d}^3 S}{\mathrm{d}r^3} - (D_1 + 2D_2) \frac{1}{r^2} \frac{\mathrm{d}S}{\mathrm{d}r} - (D_1 + 2D_2) \frac{S}{r^3} - J_1 \frac{\mathrm{d}^3 \psi}{\mathrm{d}r^3}$$

$$+ (J_1 + 2J_2) \frac{1}{r^2} \frac{\mathrm{d}\psi}{\mathrm{d}r} - (J_1 + 2J_2) \frac{\psi}{r^3} = q, \tag{11a,b}$$

$$J_1 \frac{\mathrm{d}^2 S}{\mathrm{d}r^2} - (J_1 + J_2) \frac{S}{r^2} - K_1 \frac{\mathrm{d}^2 \psi}{\mathrm{d}r^2} + (K_1 + K_2) \frac{\psi}{r^2} = 0$$

再作代换

$$r = \mathrm{e}^t \tag{12}$$

代入方程（11a，b）得

$$D_1 \frac{\mathrm{d}^3 S}{\mathrm{d}t^3} - 3D_1 \frac{\mathrm{d}^2 S}{\mathrm{d}t^2} + (D_1 - 2D_2) \frac{\mathrm{d}S}{\mathrm{d}t} + (D_1 + 2D_2) S - J_1 \frac{\mathrm{d}^3 \psi}{\mathrm{d}t^3}$$

$$+ 3J_1 \frac{\mathrm{d}^2 \psi}{\mathrm{d}t^2} - (J_1 - 2J_2) \frac{\mathrm{d}\psi}{\mathrm{d}t} - (J_1 + 2J_2)\psi = q\mathrm{e}^{3t}, \tag{13a,b}$$

$$J_1 \frac{\mathrm{d}^2 S}{\mathrm{d}t^2} - J_1 \frac{\mathrm{d}S}{\mathrm{d}t} - (J_1 + J_2) S - K_1 \frac{\mathrm{d}^2 \psi}{\mathrm{d}t^2} + k_1 \frac{\mathrm{d}\psi}{\mathrm{d}t} + (K_1 + K_2)\psi = 0$$

式（13b）可写为

$$\frac{\mathrm{d}^2}{\mathrm{d}t^2}(J_1 S - K_1 \psi) - \frac{\mathrm{d}}{\mathrm{d}t}(J_1 S - K_1 \psi) - (J_1 S - K_1 \psi) = J_2 S - K_2 \psi \tag{14}$$

令

$$J_1 S - K_1 \psi = u, \quad J_2 S - K_2 \psi = v \tag{15a,b}$$

则式（14）变为

$$\frac{\mathrm{d}^2 u}{\mathrm{d}t^2} - \frac{\mathrm{d}u}{\mathrm{d}t} - u = v \tag{16}$$

（1）如$\frac{J_1}{J_2} \neq \frac{K_1}{K_2}$，则可由式（15a，b）解得

$$S = a_1 u - b_1 v, \quad \psi = a_2 u - b_2 v \tag{17a,b}$$

其中

$$a_1 = \frac{K_2}{J_1K_2 - J_2K_1},\quad a_2 = \frac{J_2}{J_1K_2 - J_2K_1},$$
$$b_1 = \frac{K_1}{J_1K_2 - J_2K_1},\quad b_2 = \frac{J_1}{J_1K_2 - J_2K_1} \tag{18}$$

将式（17a，b）代入方程（13a），并结合式（16）消去 v，得到关于 u 的五阶常系数线性微分方程

$$\begin{aligned}&(D_1b_1 - J_1b_2)\frac{\mathrm{d}^5u}{\mathrm{d}t^5} - 4(D_1b_1 - J_1b_2)\frac{\mathrm{d}^4u}{\mathrm{d}t^4} - (D_1a_1 - J_1a_2 - 3D_1b_1\\&+ 3J_1b_2 + 2D_2b_1 - 2J_2b_2)\frac{\mathrm{d}^3u}{\mathrm{d}t^3} + (3D_1a_1 - 3J_1a_2 + 3D_1b_1 - 3J_1b_2\\&+ 4D_2b_1 - 4J_2b_2)\frac{\mathrm{d}^2u}{\mathrm{d}t^2} - (D_1a_1 - 2D_2a_1 - J_1a_2 + 2J_2a_2 + 2D_1b_1\\&- 2J_1b_2)\frac{\mathrm{d}u}{\mathrm{d}t} - (D_1a_1 + D_1b_1 + 2D_2a_1 + 2D_2b_1 - a_2J_1 - b_2J_1\\&- 2a_2J_2 - 2b_2J_2)u = - q\mathrm{e}^{3t}\end{aligned} \tag{19}$$

其特解为

$$u^* = \frac{q\mathrm{e}^{3t}}{4(D_1a_1 - 5D_1b_1 - D_2a_1 + 5D_2b_1 + 5J_1b_2 - J_1a_2 - 5J_2b_2 + J_2a_2)} \tag{20}$$

而相应齐次方程的特征方程可变形为

$$\begin{aligned}&(\lambda - 1)[(D_1b_1 - J_1b_2)\lambda^4 - 3(D_1b_1 - J_1b_2)\lambda^3 - (D_1a_1 - J_1a_2\\&+ 2D_2b_1 - 2J_2b_2)\lambda^2 + (3D_1b_1 - 3J_1b_2 + 2D_1a_1 - 2J_1a_2 + 2D_2b_1\\&- 2J_2b_2)\lambda + D_1b_1 - J_1b_2 + D_1a_1 - J_1a_2 + 2D_2b_1 - 2J_2b_2\\&+ 2D_2a_1 - 2J_2a_2] = 0\end{aligned} \tag{21}$$

求出上式中方括号内一元四次代数式的四个根，便可得方程（19）的通解，进而再由式（16）、（17a，b）、（12）和（10）求出 w 和 ψ 的通解。

（2）如 $\dfrac{J_1}{J_2} = \dfrac{K_1}{K_2} = \dfrac{1}{f}$，则由方程（15a，b）知

$$v = fu \tag{22}$$

于是方程（16）化为

$$\frac{\mathrm{d}^2u}{\mathrm{d}t^2} - \frac{\mathrm{d}u}{\mathrm{d}t} - (1 + f)u = 0 \tag{23}$$

解之得

$$u = C_1^*\,\mathrm{e}^{\left(\frac{1}{2}+\sqrt{\frac{5}{4}+f}\right)t} + C_2^*\,\mathrm{e}^{\left(\frac{1}{2}-\sqrt{\frac{5}{4}+f}\right)t} \tag{24}$$

将式（24）代入式（15a），再由式（15a）和（13a）消去 ψ，即得关于 S 的三阶常系数微分方程

$$\begin{aligned}&\left(D_1 - \frac{J_1^2}{K_1}\right)\frac{\mathrm{d}^3S}{\mathrm{d}t^3} - 3\left(D_1 - \frac{J_1^2}{K_1}\right)\frac{\mathrm{d}^2S}{\mathrm{d}t^2} + \left[(D_1 - 2D_2) - \frac{J_1^2(1 - 2f)}{K_1}\right]\frac{\mathrm{d}S}{\mathrm{d}t}\\&+ \left[(D_1 + 2D_2) - \frac{J_1^2(1 + 2f)}{K_1}\right]S\end{aligned}$$

$$
\begin{aligned}
&= q\mathrm{e}^{3t} - \frac{J_1}{K_1}\Bigg\{C_1^*\left(f\sqrt{\frac{5}{4}+f}+\sqrt{\frac{5}{4}+f}-2f^2-\frac{5}{2}f\right.\\
&\quad\left.-\frac{1}{2}\right)\exp\left[\left(\frac{1}{2}+\sqrt{\frac{5}{4}+f}\right)t\right]-C_2^*\left(f\sqrt{\frac{5}{4}+f}\right.\\
&\quad\left.+\sqrt{\frac{5}{4}+f}+2f^2+\frac{5}{2}f+\frac{1}{2}\right)\exp\left[\left(\frac{1}{2}-\sqrt{\frac{5}{4}+f}\right)t\right]\Bigg\}
\end{aligned}
\tag{25}
$$

其通解为

$$
\begin{aligned}
S &= C_1\mathrm{e}^t + C_2\exp\left[\left(1+\sqrt{2+2\frac{D_2K_1-fJ_1^2}{D_1K_1-J_1^2}}\right)t\right]\\
&\quad + C_3\exp\left[\left(1-\sqrt{2+2\frac{D_2K_1-fJ_1^2}{D_1K_1-J_1^2}}\right)t\right]\\
&\quad + \frac{K_1q\mathrm{e}^{3t}}{4[D_1K_1-D_2K_1-(1-f)J_1^2]}\\
&\quad - J_1\Bigg\{C_1^*\left(f\sqrt{\frac{5}{4}+f}+\sqrt{\frac{5}{4}+f}-2f^2-\frac{5}{2}f-\frac{1}{2}\right)\exp\left[\left(\frac{1}{2}\right.\right.\\
&\quad\left.\left.+\sqrt{\frac{5}{4}+f}\right)t\right]\div\left[D_1K_1\left(f\sqrt{\frac{5}{4}+f}-\frac{3}{2}f-1\right)\right.\\
&\quad\left.+D_2K_1\left(1-2\sqrt{\frac{5}{4}+f}\right)+J_1^2\left(f\sqrt{\frac{5}{4}+f}+\frac{f}{2}+1\right)\right]\\
&\quad + C_2^*\left(f\sqrt{\frac{5}{4}+f}+\sqrt{\frac{5}{4}+f}+2f^2+\frac{5}{2}f+\frac{1}{2}\right)\exp\left[\left(\frac{1}{2}\right.\right.\\
&\quad\left.\left.-\sqrt{\frac{5}{4}+f}\right)t\right]\div\left[D_1K_1\left(f\sqrt{\frac{5}{4}+f}+\frac{3}{2}f+1\right)\right.\\
&\quad\left.-D_2K_1\left(1+2\sqrt{\frac{5}{4}+f}\right)+J_1^2\left(f\sqrt{\frac{5}{4}+f}-\frac{f}{2}-1\right)\right]\Bigg\}
\end{aligned}
\tag{26}
$$

由式（15a）、(12）和（10）可得 w 和 ψ 的通解。

四、算例

作为特例，现考虑两层层合板变曲问题的解。为此假设

$$
z_1=\frac{h}{2},\quad E_1=\frac{E_2}{2}=E,\quad \nu_1=\nu_2=0.3 \tag{27}
$$

由式（6)、(5)、(3）和（9）得

$$
\begin{aligned}
&D_1=0.8413Eh^3,\quad D_2=0.3606Eh^3,\quad J_1=0.2690Eh^4,\\
&J_2=0.1153Eh^4,\quad K_1=0.1140Eh^5,\quad K_2=0.04887Eh^5,\\
&\frac{D_1}{D_2}=\frac{J_1}{J_2}=\frac{K_1}{K_2}=\frac{1}{f}=2.3333
\end{aligned}
\tag{28}
$$

将式（28）代入式（16)，由式（24)、(15a)、(12)、(10）和（8）并结合 $r=0$ 时 $\frac{\mathrm{d}^4w}{\mathrm{d}r^4}$ 有限的条件可得

$$w = 0.4323\frac{q(r^4 - R^4)}{Eh^3} - 0.8646\frac{qR^2(r^2 - R^2)}{Eh^3},$$
$$\psi = 4.0800\frac{qr(r^2 - R^2)}{Eh^4} \tag{29}$$

有了 w 和 ψ 的表达式。便可由式（1）和（5）计算出板中各点的应力。图 2 绘出了板中 $r=\frac{R}{2}$ 处横向剪切应力 τ_{zr} 沿厚度方向的分布情况。由图 2 可见，τ_{zr} 在板的上下表面为零，而在层间连续。

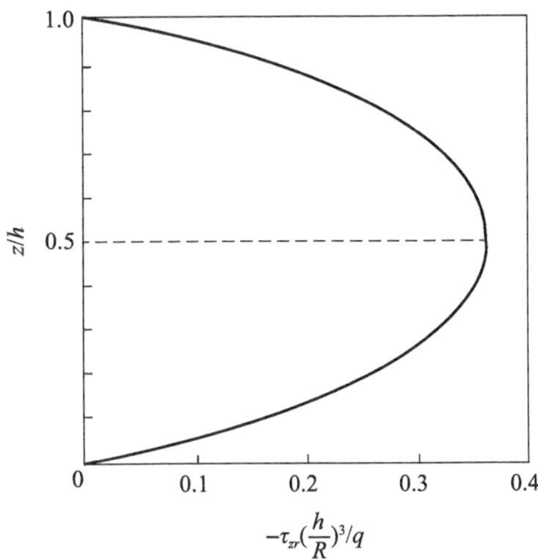

图 2　横向剪切应力沿厚度的分布（$r=\frac{R}{2}$）

参考文献

[1] Sun C T, Whitney J M. Theories for the dynamic response of laminated plates. AIAA J., 1973, 11 (2): 178.

[2] Srinivas S. A Refined analysis of composite laminates. J. of Sound and Vibration, 1973, 30 (4): 495.

[3] Whitney J M. The effect of transverse shear deformation on the bending of laminated plates. J. of Composite Mater., 1969, 3 (7): 534.

[4] Whitney J M, Sun C T. A higher order theory for extensional motion of laminated composite. J. of Sound and Vibration, 1973, 30 (1): 85.

[5] Ren J G. A new theory of laminated plates. Composites Science and Technology, 1986, 26: 225.

[6] Pagano N J. Exact solutions for composite laminates in cylindrical bending. J. of Composite Mater., 1969, 5 (3): 20.

[7] Pagano N J. Exact solutions for rectangular bidirectional composites and sandwich plates. J. of Composite Mater., 1970, 4 (1): 20.

[8] 何陵辉. 各向异性层合壳体的非线性理论. 上海：上海工业大学博士论文，1991.

焦炭塔鼓胀与开裂变形机理及疲劳断裂寿命预测的研究进展①

焦炭塔又称焦化塔或热裂化反应器，是炼油工业中延迟焦化的关键设备。它把价值低的劣质油转化为价值高的汽油和中馏分油，产生巨大的经济效益。世界上第一套延迟焦化装置于1930年8月在美国印第安纳州怀延炼油厂建成，到1998年全世界建有焦炭塔的国家为28个。中国从20世纪60年代开始使用延迟焦化装置，目前正在运行的焦化装置有24套，焦炭塔一般为板焊结构的壁塔式容器，如图1所示。由于延迟焦化工艺的特点，焦炭塔经历循环的升温、降温，16～48h内，在室温～480℃之间循环变化，同时容器承载的介质由气态到液态至固态，工作环境复杂、恶劣，致使焦炭塔在运行若干年后普遍存在以下两方面的问题：①筒体鼓胀；②筒体、焊缝和裙座开裂。探讨焦炭塔的鼓胀与开裂变形机理，合理预测焦炭塔的疲劳寿命，进而改进其设计与工作条件，使其更安全可靠运行是多年来人们致力研究的目标。

图1　焦炭塔

对焦炭塔的鼓胀变形与开裂问题，人们尝试从多学科、多角度展开研究与探讨，因此存在多种机理与观点，但普遍认同焦炭塔问题不是单一因素所造成的，且在剩余寿命预测方面也提出了多种方法。

1. 鼓胀变形机理

美国石油协会（API）对焦炭塔进行的三次（1968、1979和1996年）大规模的调查结果[1,2]表明，在焦炭塔寿命的早期阶段，鼓胀变形仅局限于底部，随着时间的推移，上部产生的增长变得更明显。由于环焊缝具有较高的屈服强度，而它们又比母材稍微厚些，因而显示出较小的增长，容器就产生一个强制的气球状鼓凸（见图2)。对于碳钢制造的焦炭塔，这个变形是最为明显的。在这些塔上的某些鼓凸，可以使直径增大20.3～25.4cm。碳-铝钢和铬-铝钢焦炭塔表现出较小的径向增长。

由于焦炭塔受到的内压、介质重量和塔的自重引起的应力数值很小，远未达到塔体材料的屈服极限，而鼓胀表明塔内出现了显著的塑性变形。这一现象长期困扰着人们，众多的研究者试图从各个角度进行研究，主要提出以下几种鼓胀变形机理。

① 本文原载《压力容器》，2007，24（2)：1-8. 作者：刘人怀，宁志华.

1）屈曲

Penso[3]将屈曲作为焦炭塔鼓胀的一种可能原因，但是没有给出进一步的论证说明。由于焦炭塔为承受轴压与内压的圆柱薄壳，很可能发生屈曲。影响薄壳结构极限承载力的因素包括：结构形状尺寸、荷载的不均匀性、材料的力学性能、结构存在的缺陷和残余应力等，其中结构缺陷的影响最明显。结构缺陷包括几何缺陷和物理缺陷。几何缺陷主要是指制造和运输过程中产生的加工误差、焊接残余变形等初始挠度；物理缺陷则指材料冶炼和结构制造过程中出现的裂纹、空穴、夹杂、焊接未熔合，以及服役中产生的划伤、疲劳裂纹和腐蚀裂纹等。Holst 等[4]综合考察了周向轴对称缺陷和焊接残余应力对圆柱薄壳屈曲的影响，结果表明残余应力能提高圆柱壳的屈曲临界荷载。但当壳体内同时承受很大的内压时，可能在局部首先屈服，进而加速壳体的屈曲，发生“象脚”破坏，如图 3 所示。

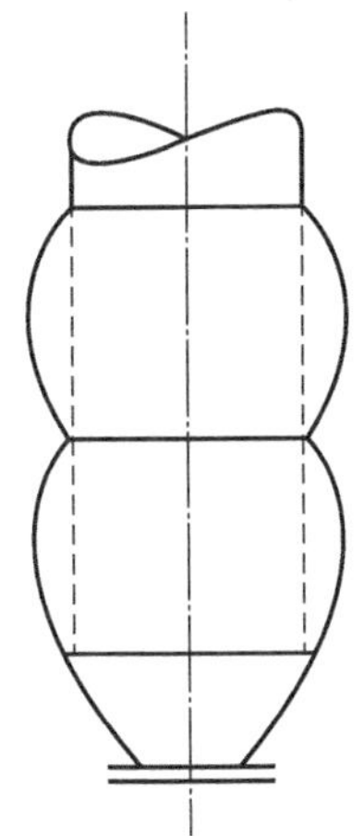

图 2　受约束的气球状的变形塔

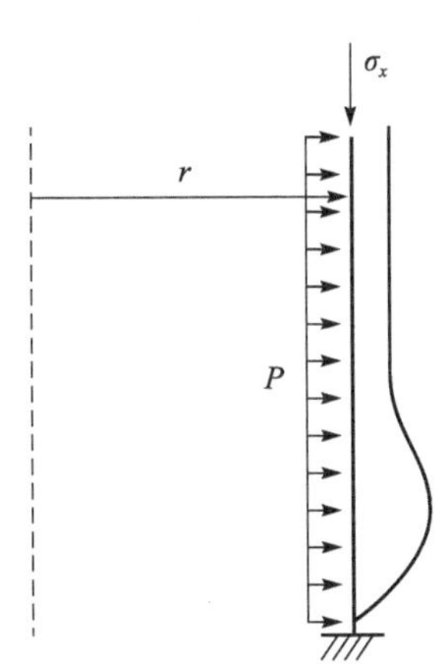

图 3　圆柱壳的“象脚”破坏示意

由于焦炭塔的环焊缝对筒体的变形具有一定的约束作用，若将环焊缝视作边界，则筒体的鼓胀变形与“象脚”破坏非常相似，Rotter[5,6]与 Teng[7]的研究表明，圆柱壳在边界及环肋处出现“象脚”破坏时强度较低。焦炭塔属于大型的圆柱壳体，在制造和装配过程中容易产生几何和物理初始缺陷，且由于工作环境恶劣，运行若干年后，材质发生变化，加上疲劳裂纹与腐蚀裂纹的产生，这些因素都可能导致塔的屈曲临界荷载降低。有研究结果[8]表明，在循环的轴向热梯度与轴压作用下，圆柱薄壳的径向位移随着循环次数增多而增大，直至发生屈曲。但至今仍未见有从屈曲的角度来探讨焦炭塔的鼓胀问题的相关成果，因此对这一观点的证实还有待于进一步研究。

2）蠕变-疲劳交互作用

蠕变是材料在高温工作条件下，即使承受低于屈服极限的恒定应力作用，也会发生永久变形的行为。对于金属材料或合金钢，通常在温度高于 $0.3T_m$（T_m 为材料的熔点）时，蠕变就已比较明显。焦炭塔的材料为碳钢、碳钼钢或铬钼钢，它们的熔点为 1300～1600℃，而焦炭塔的温度在室温～480℃之间变化，因此蠕变是可能发生的。但测量结果[9]显示，在一个循环中，塔内温度超过 400℃时，塔壁的应力并不高（约为

69MPa)，由此产生的蠕变变形不明显。从焦炭塔鼓胀变形最严重处取样，进行金相分析的结果也证实了这一点[10,11]。由此可见，蠕变并非形成鼓胀的主导因素。

3）局部应力

在焦炭塔的循环操作过程中，内压、介质重量和塔的自重引起的应力数值很小，远未达到屈服极限[12]。但在焦炭塔的应力测量中，发现局部区域出现了接近和超过材料屈服极限的应力[9,13]。Penso 等[13]认为形成局部应力主要有以下几个原因：

（1）局部的几何特性，例如环焊缝的焊趾处，会引起应力集中。

（2）在不同材料的界面上，由于热膨胀系数的不同，也会引起峰值应力。

（3）残余应力。在机械设备的制造和加工过程中，若某一局部区域出现不均匀温度场，就会导致残余应力的出现，这些残余应力会严重影响该部分的强度。在焦炭塔的焊接和整直过程中会引起残余应力[14-17]。

（4）焊缝与母体金属抗拉强度的不同。Boswell 等[18]认为，当环焊缝的屈服极限高于母体金属时，母体金属变形较大，受到焊缝的约束，因此筒体就在焊缝上方或下方向外鼓凸。与此相似的现象是，当焊缝强度低于母体金属时，在焊缝处或附近出现了鼓凸，但这些鼓凸通常比前一种情况提到的鼓凸小[18,19]。因此有研究认为，为降低这一因素引起的局部应力，应使焊接金属与母体金属的强度差异控制在 10%以内[3]。

（5）热冲击。在焦炭塔急冷阶段，在塔体内表面引起了严重的热梯度，进而产生热应力，在早期研究中，Weil 和 Rapasky[20]认为，当轴向温差大于 2.19℃/cm 时，由此产生的热应力就会超过屈服极限。Weil 还定义了急冷因子（UQF)，即水冷时间（分钟）与单塔焦炭产量（吨）的比值，来判别焦炭塔是否会出现鼓胀：若 UQF＞0.5，鼓胀可以忽略；若 UQF＞0.8，则根本不会出现鼓胀。

4）焦炭对鼓胀的影响

根据 Ellis 和 Hardin[21]的测量发现。生焦的热膨胀系数高于焦炭塔塔壁的热膨胀系数。他们认为，如果缓慢、均匀地冷却焦炭和塔壁，则在塔壁中不会出现向外的压力；但如果塔壁的冷却速度比焦炭快，则塔壁就会受到焦炭的压力，从而可能导致塔体的鼓胀。另有测试表明[22]，在焦炭塔中的焦炭实际上并非多孔性，仅在靠近塔壁处孔口稍微多些，因而当水冷却速率过高时，冷水就会流进焦床的外围去冷却塔壁。这一点也有助于解释急冷时塔体鼓胀的现象。

5）热棘轮效应

当结构承受的荷载或温度循环变化时，如果荷载或温度高于某一极限值，应力或应变会随着循环次数的增加而不断地积累，最终导致结构由于超过承载容限或发生过度变形而失效。这种现象称为棘轮效应。Boswell[18]认为，当焊缝强度高于母体金属的强度，并且存在轴向温差时，应力热棘轮才会发生，这正与焦炭塔的情况相符。

6）局部冷热斑点

当机械设备或结构的各部分温度不同时，如果某一部分的伸缩受到相邻部分的约束，就会引起热应力。当受约束的试件内存在温度梯度时，高温部分受压且由于温度升高屈服极限降低，容易出现塑性变形，使得该部分横截面面积增大；而当试件冷却时，原来的高温部分由于截面面积增大，此时的应力值比相邻部分低，并且由于温度降低，屈服极限恢复，与相邻部分相比，很难产生塑性变形。经过多次循环之后，塑性变形不断积累，最终在高温区出现鼓胀，而相邻区域出现瓶颈现象[23,24]。焦炭塔的操作过程中也出现同样的现象。由于急冷阶段的非均匀冷却，高温区域受到周围低温区域的压缩，产生了永久的变形[18]。

综上所述，可归纳出焦炭塔鼓胀变形机理为：在初始预热和急冷阶段，塔体产生了严重的热梯度，在局部区域引起了超过塔体材料屈服极限的热应力，并且由于环焊缝金属与塔体母材金属强度的差异，筒节变形受到约束，向外鼓凸，随着循环次数增多，塑性变形不断积累，最终出现了严重的鼓胀。

2. 开裂

根据美国石油协会对焦炭塔的第三次考察报告[2]（1996～1998 年），焦炭塔的开裂有：裙座与塔体的焊接处开裂、筒体环焊缝以及镀层开裂。

裙座的开裂主要出现在裙座与塔体焊接处的两侧，以及狭槽的扩大端和锁眼处，另外裙座本身也有开裂。被调查的焦炭塔中，裙座有狭槽扩大端的 89%都出现了裂纹，而没有扩大端的只有 22%出现了开裂；没有出现裙座开裂的塔中有 75%是经过裙座焊接打磨处理的[2]。因此，Boswell 等[18]认为，裙座的开裂是由于预热时，裙座的膨胀程度不如塔体充分，在连接处产生了热应力，从而在焊缝、狭槽和锁眼这些尺寸突变处出现应力集中，产生裂纹。

筒体的裂纹主要出现在焊缝附近，即使在非鼓胀区域裂纹也很常见，既有从内表面萌生，也有从外表面萌生。这些裂纹有环向、纵向，甚至不规则开裂，但主要是环向的[2]。研究表明[24]，当塔内存在严重温度梯度时，轴向应力大于环向应力，在轴向应力作用下，许多环向浅裂纹扩展并连接发展成深裂纹。Penso 的实测结果表明[13]，最危险的是热影响区裂纹，它们通常会扩展成深裂纹，并且裂纹与峰值应力方向垂直。同时，由于出现鼓胀后，引起局部的曲率增加，塔内的轴向温度梯度以及焦炭与塔壁的相互作用变得更为严重，进一步加速裂纹的扩展[13]。另外，如前所述，当环焊缝材料强度高于母体金属时，筒体出现鼓胀，这种情况下裂纹会从热影响区的外表面萌生，并向内扩展。此时裂纹容易从外部检测，在裂纹进一步扩展前可以迅速修复；反之，若环焊缝强度低于母体金属，则鼓胀会出现在环焊缝处，裂纹从内表面萌生，则裂纹很难被发现，直至穿透塔壁[18]。另外，在保温支持圈处以及镀层处也出现开裂[2]。裂纹大多出现在几何尺寸变化处，例如焊缝处的焊堆、焊趾，可见，应力集中起了重要作用。

由于现有焦炭塔中出现的裂纹主要在环焊缝附近。且大多是环向的，美国芝加哥桥梁钢铁公司（CB&I）提出了一种新的焦炭塔的制造方法[25]：垂直焊接焦炭塔，即焦炭塔的筒节是由若干钢板垂直焊接而成。这种设计方法旨在通过减少环焊缝的数量来减少环向裂纹。但后续的跟踪研究显示[3]，在同样的操作条件下，这种垂直焊接的焦炭塔中沿着纵向焊缝同样出现了开裂。

美国石油协会进行的三次焦炭塔考察[1,2]认为，穿透裂纹发生在水冷却、吹蒸汽冷却，或刚启动时。第二次考察对容器厚度与首次发生裂纹之前的使用年限的关系作了分析，发现对于碳钢塔，发生穿透性裂纹以前的使用时间与塔的厚度是敏感的，但对碳-钼钢塔、铬-钼钢塔而言，发生裂纹的时间与容器厚度之间没有明显的关系。1980年的考察表明，焦炭塔的破裂问题比1968年考察时少。这可归因于改进了材料的选择(20世纪60年代以前的塔主要是用碳钢制造的)，20世纪60年代安装的塔主要是碳-钼钢塔，而20世纪70年代安装的主要是铬-钼钢（含1%或1.25%的Cr）以及在已经改进急冷速度的情况下改善操作的缘故。考察结果表明急冷速率越低，则焦炭塔在首次出现穿透性裂纹前经历的工作循环次数明显增高。

Penso等将焦炭塔开裂的原因总结为以下几方面[13]：热冲击是浅裂纹萌生的直接原因；浅裂纹在塔体内表面随处可见，特别是几何尺寸变化处，在热影响区浅裂纹会进一步扩展为深裂纹，并最终发展成穿透塔体的裂纹；其次，环境和介质的腐蚀（硫化腐蚀、渗碳、蚀损斑等）被认为是导致开裂的另一诱因。由各种腐蚀产生的微缺陷成为应力集中点，从而引起裂纹萌生和加速裂纹扩展。另外，腐蚀与疲劳的交互作用也是一个可能的原因。由于环境和介质的腐蚀，金属的疲劳性能会降低（例如变脆），从而加速表面裂纹的扩展。

3. *剩余疲劳寿命预测*

由于焦炭塔产生着巨大的经济效益，且制造费用昂贵，故目前大量的焦炭塔都在超期服役。老龄的焦炭塔存在的问题尤为严重，维护和检查费用非常高，故对其进行疲劳剩余寿命预测有着重要的意义。目前普遍认可焦炭塔的失效模式为低周热机疲劳，其寿命是一定应力范围下经历的循环次数，而不是单纯时间的函数。焦炭塔的低周疲劳失效可划分为以下三个阶段[18]。

(1) 微观裂纹的萌生。这一阶段肉眼观察不到裂纹，也没有穿透塔壁；

(2) 裂纹的稳定扩展。这一阶段占了焦炭塔整个寿命的50%～95%，裂纹会发展到其临界尺寸的1/2；

(3) 裂纹迅速扩展直至最后断裂。

因此，对焦炭塔寿命的预测主要是对低周疲劳裂纹的扩展至断裂所经历循环次数的估计。目前关于焦炭塔疲劳寿命预测的各种方法可归纳为以下几类。

1) 缺陷容限法

这是一种基于断裂力学的分析方法，假设材料存在初始缺陷，将焦炭塔的寿命定义为裂纹从初始尺寸扩展到某个临界尺寸所经历的循环次数。而裂纹临界尺寸的计算基于断裂强度、最大荷载、容许应变。该方法对裂纹扩展规律的选择依赖于实际经验。最早应用的疲劳断裂参量是应力强度因子幅ΔK[26]。在小范围屈服条件下，ΔK是影响疲劳裂纹诸因素中最强烈的一个参量，它与裂纹长度的变化率$\frac{\mathrm{d}a}{\mathrm{d}N}$之间的关系为

$$\frac{\mathrm{d}a}{\mathrm{d}N}=C\Delta K^m, \tag{1}$$

式中，C，m为与材料有关的常数。

英国的ERA焦炭塔寿命评估法[27]采用的就是把ΔK作为疲劳断裂控制参量。

Dowling和Begley[28]提出用J积分参量来描述疲劳裂纹的扩展，认为裂纹长度的变化率$\frac{\mathrm{d}a}{\mathrm{d}N}$与$(\Delta J)^m$成正比，其中$m$的含义与式(1)中类似。但焦炭塔运行过程中的

非比例加载以及弹性卸载的出现，与 J 积分参量对断裂问题进行预测的应用条件相冲突。

裂纹嘴张开位移幅值△COD 也被用于疲劳裂纹扩展描述。Dover[29] 的试验表明裂纹扩展速率与△COD 呈现良好的线性关系。但实践表明[30]，△COD 容易受到表面变形与裂纹尖端不规则性的影响。

2）应变-寿命曲线

在高周疲劳条件下，材料的变形主要是弹性的，结构的寿命用应力幅来表征；而在低周疲劳下，应力较高，使得结构失效之前会出现明显的塑性应变，因此疲劳寿命用应变幅来表征。焦炭塔的情况属于后者，因此适合采用应变寿命法。

在所有应力-寿命曲线中，Manson-Coffin[31,32]公式应用最为广泛，即

$$\begin{aligned}\varepsilon_a &= \varepsilon_{ea} + \varepsilon_{pa} \\ &= \frac{\sigma_f'}{E}(2N_f)^b + \varepsilon_f'(2N_f)^c,\end{aligned} \tag{2}$$

式中，σ_f'为疲劳强度系数，ε_f'为疲劳延性系数，b 为疲劳强度指数，c 为疲劳延性指数。

在双对数坐标系中，疲劳寿命 N_f 与弹性应变分量 ε_{ea}、塑性应变分量 ε_{pa}之间呈线性关系。采用上式预测疲劳寿命的方法也称为当量应变法。

Manson-Coffin 公式是一个经验表达式，对于有些金属材料的疲劳寿命预测得很好，但对某些金属材料预测的精度较差。何雪宏[33]、黄宏发[34]的研究认为，用当量应变范围评估轴对称问题的热疲劳寿命，对于工程问题是可行的。张文孝等[35]、赵克勤等[36]应用 Manson-Coffin 公式来预测某焦炭塔的剩余寿命，得到了较好的结果。

3）Palmgren-Miner 法[37,38]

真实的工程结构所经历的不是常应力循环，每个循环的应力幅值、平均应力以及荷载的频率通常是变化的。Palmgren-Miner 法就是预测在应力幅值不同的一组常应力循环作用下结构的寿命预测。设有一组应力幅值分别为 σ_{ai}的应力循环，$i=1,2,\cdots,m$，n_i 为第 i 种循环作用的次数，N_{fi}为在第 i 种应力循环下构件的疲劳寿命，则当下式成立时：

$$\sum_{i=1}^{m}\frac{n_i}{N_{fi}} = 1 \tag{3}$$

结构发生疲劳失效。这一方法实际上是一个线性疲劳累计损失理论，能较好地预测疲劳寿命的均值。但它没有考虑载荷次序的影响，而实际上加载次序对疲劳寿命影响很大。该方法对于随机载荷的疲劳试验结果比较吻合[39]。

4）神经网络方法

神经网络已被广泛用于各个领域，研究结果显示了该方法的应用潜力和卓越性能[40,41]。神经网络具有很强的学习和映射能力，可以很方便地拟合出许多复杂的非线性关系，解决常规方法无法处理的一些难点。最近。我们[42]采用改进了的 BP 神经网络方法来预测焦炭塔的剩余寿命，得到了较好的结果。但神经网络在学习过程中，需要大量的训练样本以保证其结果的正确。对于焦炭塔而言，需要有连续若干年的现场实测的温度、应力、变形以及裂纹等数据。模糊神经网络是模糊理论和神经网络的学

习功能，实现模糊系统的自学习和自适应，在没有完全信息的时候更显示出优势，因此在焦炭塔剩余寿命的预测当中具有较大的应用潜力和优势。

5）寿命外推法

蠕变是高温设备和构件破坏的主要因素。由于蠕变和持久强度的规定时间一般为100kh或200kh，难以在实验室的条件下进行试验，所以一般采用提高应力和温度的方法得到短时区内的断裂时间与应力、温度之间的关系，进行寿命外推。寿命外推法中精度较高的是时间-温度参数外推法，如Larson-Miller公式[43]：

$$T(C+\lg\tau)=P(\sigma), \tag{4}$$

式中，C为材料常数，由试验测定；T为热力学温度；$P(\sigma)$为随应力变化的函数，称为热强参数，可通过试验用统计回归方法求得。

6）θ投影法

英国学者Evans和Wilshire[44]提出的θ投影法以蠕变过程的物理模型为基础，提出了描述整个蠕变曲线（应变ε和时间τ的关系曲线）的特征方法，进而预测高温构件的寿命，具体的θ函数公式为

$$\varepsilon=\theta_1(1-e^{\theta_2\tau})+\theta_3(1-e^{\theta_4\tau}), \tag{5}$$

式中θ_i（i=1，2，3）与材料本身，以及应力σ和温度T有关。

寿命外推法以持久强度为主要指标，而θ投影法则以蠕变变形为主要指标，但这两类方法都以蠕变是构件破坏的主要因素为理论基础，而目前的研究结果表明焦炭塔的蠕变变形并不明显，因此这些方法对焦炭塔寿命预测的可行性还有待进一步考察。

以上各种预测方法当中所涉及数据均需对焦炭塔进行现场实测得到。工程应用中已有比较成熟、有效的一套测量和评估的方法与措施，如英国的ERA公司已经为全球焦炭塔用户提供了十年的寿命预测服务[27]，美国的CiTGO Petroleum公司也摸索出一套适合自己的预测方法[18]。他们共同的做法是

（1）首先通过各种探测或扫描技术，并根据检修与维护记录，以及操作者的经验判断焦炭塔危险点的位置；

（2）然后在这些确定的危险点处布置耐高温应变仪、热电耦，测量并记录足够的循环数据；

（3）对采集得到的数据再进行统计分析，得到平均应力、应力范围等低周疲劳计算数据；

（4）从塔上取出材料样本进行金相分析与材料性能测试，得到材料性能的有关数据。根据得到的数据，采用上述方法进行剩余寿命预测，同时考虑检修与维护的因素，使预测值与事实寿命更接近。

4. 结语

目前认为焦炭塔的鼓胀变形以及开裂是由多种原因造成的，最主要的原因是在初始预热和急冷阶段产生严重的热梯度，引起了超过屈服极限的热应力，导致塑性变形的产生以及浅裂纹的萌生，随着循环次数的增加，塑性变形不断积累、裂纹持续扩展，最终出现鼓胀和开裂。

现有对焦炭塔存在问题的研究，大多停留在有限元分析与模拟、试验研究上。众

所周知，有限元分析的结果并不适合进行参数分析，而现场实测数据受到单个焦炭塔的结构特点以及操作工艺的影响，有限的实验数据很难给出普遍适用的结论。同时由于焦炭塔的寿命周期长达数十年，采取更换塔材、改进设计或操作工艺对焦炭塔问题所产生的影响在短时间内不便于进行对比。因此，美国耗费庞大的人力、物力和财力开展了大规模的三次焦炭塔调查后，即使许多用户公司都强烈要求美国石油协会（API）能够出台一个焦炭塔的设计以及操作的规范，但是鉴于数据的分散性，至今仍无法实现。

对于减少鼓胀与开裂，目前的认识仅限于通过降低急冷速率以及改善应力集中来减小应力峰值，其他因素诸如焦炭塔的结构参数等对二者的影响规律尚不清楚。因此很有必要建立合理的理论研究模型来探讨焦炭塔问题，以便进行参数分析，从本质上揭示焦炭塔的鼓胀变形机理、裂纹扩展规律，以及各种结构、操作参数对焦炭塔问题的影响规律。需要指出的是，由于焦炭塔工作条件恶劣，运行多年后，材料老化以及环境和介质腐蚀等因素将导致塔体材料性能参数发生变化，因此在建立理论研究模型和进行应力计算时，应充分考虑材质变化所产生的影响。

参考文献

[1] Thomas J W. API survey of coke drum cracking experience. Pressure Vessel and Tank Developments，1981：80-88.

[2] Bagdasarian A，Horwege J，Kirk S，et al. Integrity of coke drums（Summary of 1998 API Coke Drum Survey). American Society of Mechanical Engineers，Pressure Vessels and Piping Division（Publication）PVP，2000，411，Service Experience and Fitness-for-service in Power and Petroleum Processing：265-270.

[3] Penso J A. Fundamental study of failure mechanisms of pressure vessels under thermo-mechanical cycling in multiphase environments. The Ohio Stare University，2001.

[4] Holst J M F G，Rotter J M，Calladine C R. Imperfections and buckling in cylindrical shells with consistent residual stresses. Journal of Constructional Steel Research，2000，54：265-282.

[5] Rotter J M. Local inelastic collapse of pressurized thin cylindrical steel shells under axial compression. Joumal of Structural Engineering，2000，116（47）：1955-1970.

[6] Rotter J M. Calculated buckling strengths for the cylindrical wall of 10，000 tonnes silos at port Kembla. Investigation Report S663，School of Civil and Mining Engineering，University of Sydney，1988.

[7] Teng J G. Plastic collapse at lap-joints in pressurized cylinders under axial load. Journal of Engineering Mechanics，1994，120（1）：23-45.

[8] Ignaccolo S，Cousin M，Jullien J F，et al. Interaction of mechanical thermal stresses on the instability of cylindrical shells，Res Mechanica：International Journal of Structural Mechanics and Materials Science，1988，24（1）：25-33.

[9] Allevato C，Richard S，Boswell P E. Assessing the structural integrity and remaining life of coke drums with acoustic emission testing，strain gaging，and finite element analysis. Engineering Sources Technology Conference & Exhibition，1999，Houston，Texas.

[10] 李一玮. 延迟焦化装置焦碳塔的变形、开裂机理和安全分析. 压力容器，1989，6（4）：61-66.

[11] 赵莹，周鸿. 焦炭塔的鼓凸损伤分析. 西安石油学院学报（自然科学版），1998，13（6）：38-41.

[12] Capstone Engineering Service Inc，Houston，Texas. 1996 API coke drum survey final report. 1998：50.

[13] Penso J A，Lattarulo Y M，Seijas A J，et al. Understanding failure mechanisms to improve reliability of coke drums. American Society of Mechanical Engineers，1999，395：243-253.

[14] Jones D P，et al. Residual stresses in weld-deposited clad pressure vessels and nozzles. Journal of Pressure Vessel Technology，1999，121：423-429.

[15] Ohta Akihiko，et al. Effect of residual stress on fatigue of weldments//IIW International Conference on Performance of Dynamically Loaded Welded Structures. Welding Resarch Council. San Francisco，1997：108-122.

[16] Kim D S，McFarland D B，Reynolds J T，et al. Fitness for service assessment utilizing field residual stress measurements. PVP-ASME Conference Proceedings，1995，315：327-334.

[17] Kim S，Boswell R. Residual stress measurements of coke drum welding coupon. ASME-PVP Conference Proceedings，Fatigue，Fracture，and Residual Stresses，1998：405-409.

[18] Boswell R S，Ferraro T. Remain life evaluation of coke drums. Houston Plant engineering，operations，design and reliability symposium. Energy Engineering Conference，1997.

[19] Dunham C，Clark R. Laser mapping of coke drums-what has been learned. New Orleans：AIChE Spring National Meeting，1998.

[20] Weil N A，Rapasky F S. Experience with vessels of delayed coker units. 23rd Midyear Meeting of API Division of Refining，1958.

[21] Ellis P J，Hardin E E. How petroleum delayed coke forms in a drum. Light Metals 1993. The Minerals，Metals and Materials Society，1992.

[22] Ellis P J，Paul C A. Ddayed coking fundamentals. AIChE Spring National Meeting，New Orleans Louisiana，1998.

[23] Taler J，Weglowski B，Zima W，et al. Analysis of thermal stresses in a boiler drum during start-up. Journal of Pressure Vessel Technology. Transaction of the ASME，1999，121：84-93.

[24] Satapathy A K. Thermal analysis of an infinite slab during quenching. Communications in Numerical Methods in Engineering，Mechanics Engineering，2000，16：529-536.

[25] Antalffy L P. Innovations in delayed coking coke drum design. American Society of Mechanical Engineers，1999，388：207-217.

[26] Paris P C，Endogon F. Critical analysis of crack propagation laws. Trans. J Basic Eng，1963，85：528-534.

[27] Chruch J M，Lim L B，Jarvis P，et al. Crack growth modelling and probabilistic life assessment of coke drums operating under fatigue conditions. International Journal of Pressure and Piping，2001，78：1011-1020.

[28] Dowling N E，Begley J A. Fatigue crack growth during gross plasticity and the J-integral. ASTM STP，1976，590：82-103.

[29] Dover W D. Fatigue crack growth under COD cycling. Engineering Fracture Mechanics，1973，5：11-21.

[30] 雷月葆. 应变疲劳裂纹扩展与应力疲劳裂纹扩展的同一规律. 上海：华东理工大学博士学位论文，1992.

[31] Manson S S. Behavior of materials under conditions of thermal stress. NACA TN-2933，1953.
[32] Coffin L F，Jr. A study of the effects of cyclic thermal stresses on a ductile metal. Transactions of the American Society of Mechanical Engineers，1954，76：931-949.
[33] 何雪宏. 复杂应力状态低周疲劳寿命评价和裂纹扩展规律研究. 大连：大连理工大学博士学位论文，1992.
[34] 黄宏发. 多维应力状态 LD8 铝合金高温低周疲劳与热疲劳寿命评价方法的研究. 大连：大连理工大学博士学位论文，1994.
[35] 张文孝，郭成璧，张振华，等. 焦炭塔的热机械疲劳剩余寿命分析. 压力容器，1995：12 (1)：69-72.
[36] 赵克勤，胡海龙. 焦炭塔热疲劳寿命估算. 石油化工设备技术，1991，12 (1)：18-20.
[37] Miner M A. Cumulative damage in fatigue. Journal of Applied Mechanics，1945，12 (1)：159-164.
[38] Palmgren A G. Die Lebensdauer von Kugellagern. Zeitschrift des Vereins Deutscher Ingenieure，1924，14：339-341.
[39] 姚卫星. 结构疲劳寿命分析. 北京：国防工业出版社，2004.
[40] 蒋宗礼. 人工神经网络导论. 北京：高等教育出版社，2003.
[41] 李孝安，张晓绩. 神经网络与神经计算机导论. 西安：西北工业大学出版社，1995.
[42] 傅继阳，王璠，刘人怀，等. 基于改进 BP 神经网络的焦炭塔热机械疲劳剩余寿命预测. 压力容器，2005，22 (5)：4-7.
[43] Larson F R，Miller J. A time temperature relationship for rupture and creep stresses. Trans ASME 1952，74：765-771.
[44] Evans R W，Wilshire B. Creep of Metals and Alloys. London：The Institute of Metals，1985：60-65.

膜盒基体的理论与设计①

一、引言

膜盒基体是精密仪表的一个基本构件，不仅消耗金属材料较多，而且用量大，因而对它进行正确设计是有实际意义的。它是一块环形厚板，属于三维弹性理论的研究范畴，若要进行精确分析，将是十分困难的。过去，限于研究的困难，人们采用薄板弯曲理论进行分析，因而不能进行较精确的设计。为改变这一现状，适应精密仪表工业发展的形势，作者在以往研究精密仪表波纹膜片弹性元件工件[1-11]的基础上，应用Reissner厚板理论[12]的假定，建立了膜盒基体的轴对称弯曲理论。最后，本文给出了膜盒基体位移和应力的解析解。通过数值计算，与薄板理论结果进行了比较。所获设计公式可直接用于精密仪表膜盒基体的刚度和强度的设计。

二、基本方程与求解

在对膜盒基体进行研究时，我们忽略它的表面波纹刻槽的微小影响，而认为它是一个如图1所示的承受均布横向荷载

$$q = q_1 - q_2 \tag{1}$$

的内边缘自由、外边缘固定的环形厚板。这里，我们采用（r，θ，z）圆柱坐标系。q_1和q_2分别是作用在板上、下表面（$z=\mp\dfrac{h}{2}$）的均布压力，h为板的厚度，a为外半径，b为内半径。

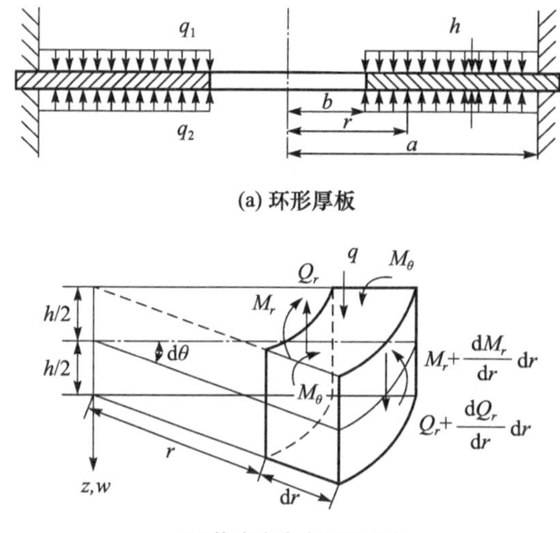

(a) 环形厚板

(b) 挠度和内力的正方向

图1　膜盒基体

---

① 本文原载《澳门科技大学学报》，2009，3（1）：111-116.

按照计及剪切变形影响的 Reissner 厚板理论的假定[12]，经过适当推导，可得膜盒基体的轴对称弯曲理论的基本方程如下

$$
\begin{aligned}
&\frac{\mathrm{d}M}{\mathrm{d}r}+\frac{M_r-M_\theta}{r}=Q_r, \\
&\frac{\mathrm{d}(rQ_r)}{\mathrm{d}r}=-qr, \\
&M_r=D\left[\frac{\mathrm{d}\varphi_r}{\mathrm{d}r}+\nu\frac{\mathrm{d}\varphi_r}{r}+\frac{6\nu(1+\nu)}{5Eh}q\right], \\
&M_\theta=D\left[\frac{\varphi_r}{r}+\nu\frac{\mathrm{d}\varphi_r}{r}+\frac{6\nu(1+\nu)}{5Eh}q\right], \\
&\varphi_r=\frac{12(1+\nu)}{5Eh}Q_r-\frac{\mathrm{d}w}{\mathrm{d}r}
\end{aligned}
\tag{2}
$$

其中，M_r 和 M_θ 为径向、切向弯矩，Q_r 为剪力，φ_r 为径向转角，w 为挠度，E 为弹性模量，ν 为泊松比，D 为抗弯刚度，

$$D=\frac{Eh^3}{12(1-\nu^2)} \tag{3}$$

为便于求解，我们用消元法，将基本方程组（2）缩减为如下一个关于挠度 w 的三阶变系数常微分方程

$$\frac{\mathrm{d}}{\mathrm{d}r}\frac{1}{r}\frac{\mathrm{d}}{\mathrm{d}r}r\frac{\mathrm{d}w}{\mathrm{d}r}=\frac{q}{2D}\left(r-\frac{b^2}{r}\right) \tag{4}$$

与此方程相应的膜盒基体的边界条件为

（1）内边缘自由

$$当\ r=b\ 时，\quad M_r=0 \tag{5}$$

（2）外边缘固定

$$当\ r=a\ 时，\quad w=0,\quad \varphi_r=0 \tag{6}$$

其中

$$
\begin{aligned}
&M_r=-D\left(\frac{\mathrm{d}^2w}{\mathrm{d}r^2}+\frac{\nu}{r}\frac{\mathrm{d}w}{\mathrm{d}r}\right)+\overline{\omega_1}\frac{\mathrm{d}Q_r}{\mathrm{d}r}-\overline{\omega_2}q, \\
&Q_r=-\frac{q}{2r}(r^2-b^2), \\
&\varphi_r=-\frac{\mathrm{d}w}{\mathrm{d}r}+\frac{6(1+\nu)}{5Ehr}(b^2-r^2), \\
&\overline{\omega_1}=\frac{h^5}{5}, \\
&\overline{\omega_2}=\frac{\nu h^2}{10(1-\nu)}
\end{aligned}
\tag{7}
$$

由边值问题（4）～（6）求得 w 以后，便可按以下公式计算膜盒基体任一点的径向和切向位移 u、v，径向、切向和法向应力 σ_r，σ_θ 和 σ_z，水平剪应力 $\tau_{r\theta}$，径向与切向的横向剪应力 τ_{rz}，$\tau_{\theta z}$：

$$u = z\varphi_r, \quad v = 0 \tag{8}$$

$$\begin{aligned}
&\sigma_r = \frac{12M_r z}{h^3}, \quad \sigma_\theta = \frac{12M_\theta z}{h^3}, \\
&\sigma_z = \frac{qz}{2h}\left[3 - \left(\frac{2z}{h}\right)^2\right] - \frac{1}{2}(q_1 + q_2), \\
&\tau_{rz} = \frac{3Q_r}{2h}\left[1 - \left(\frac{2z}{h}\right)^2\right], \\
&\tau_{r\theta} = \tau_{\theta z} = 0
\end{aligned} \tag{9}$$

其中

$$M_\theta = -D\left(\frac{1}{r}\frac{\mathrm{d}w}{\mathrm{d}r} + \nu\frac{\mathrm{d}^2 w}{\mathrm{d}r^2}\right) + \overline{\omega_1}\frac{Q_r}{r} - \overline{\omega_2}q \tag{10}$$

连续积分方程（4），并应用边界条件（5）和（6），我们得到挠度 w 的解析解

$$w = \frac{\overline{c_1}}{4}x^2 + \overline{c_2}\ln x + \overline{c_3} + \frac{qa^4}{8D}\left(\frac{1}{8}x^4 - \alpha^2 x^2 \ln x + \alpha^2 x^2\right) \tag{11}$$

其中

$$\begin{aligned}
\overline{c_1} =& \frac{qa^2\lambda_0}{D}\left\{a^2\left[(1+\nu)\alpha^2\ln\alpha - \frac{1+3\nu}{2}\alpha^2 - 1 + \nu\right.\right. \\
&\left.\left. - \frac{1-\nu}{2}\alpha^{-2}\right] - 4\,\overline{\omega_1}(1+\alpha^{-2}) - 8\,\overline{\omega_2}\right\} \\
\overline{c_2} =& \frac{qa^2\lambda_0}{D}\left\{\frac{a^2}{2}\left[2(1+\nu)\alpha^2\ln\alpha + \frac{1-\nu}{2}\alpha^2 + \frac{1+\nu}{2}\right]\right. \\
&\left. - \frac{2\,\overline{\omega_1}}{1-\nu}[(1+\nu)\alpha^2 + 1 - 3\nu] - 4\,\overline{\omega_2}\right\}, \\
\overline{c_3} =& -\frac{\overline{c_1}}{4} - \frac{qa^4}{8D}\left(\alpha^2 + \frac{1}{8}\right), \\
x =& \frac{r}{a}, \quad \alpha = \frac{b}{a}, \quad \lambda_0 = \frac{\alpha^2}{4[(1+\nu)\alpha^2 + 1 - \nu]}
\end{aligned} \tag{12}$$

将上面所得的解（11）代入式（7）、（8）和（10），便得

$$\begin{aligned}
u =& -\frac{z}{a}\left[\frac{\overline{c_1}}{2}x + \frac{\overline{c_2}}{x} + \frac{qa^4}{8D}\left(\frac{1}{2}x^3 - 2\alpha^2 x\ln x + \alpha^2 x\right) + \frac{\overline{\omega_1}qa^2}{2D(1-\nu)}\left(x - \frac{\alpha^2}{x}\right)\right], \\
M_r =& -\frac{D}{a^2}\left(\frac{1+\nu}{2}\overline{c_1} - \frac{1-\nu}{x^2}\overline{c_2}\right) - q\left\{\frac{a^2}{8}\left[\frac{3+\nu}{2}x^2 - 2(1+\nu)\alpha^2\ln x - (1-\nu)\alpha\right]\right. \\
&\left. + \frac{\overline{\omega_1}}{2}\left(1 + \frac{\alpha^2}{x^2}\right) + \overline{\omega_2}\right\}, \\
M_\theta =& -\frac{D}{a^2}\left(\frac{1+\nu}{2}\overline{c_1} + \frac{1-\nu}{x^2}\overline{c_2}\right) - q\left\{\frac{a^2}{8}\left[\frac{1+3\nu}{2}x^2 - 2(1+\nu)\alpha^2\ln x + (1-\nu)\alpha^2\right]\right. \\
&\left. + \frac{\overline{\omega_1}}{2}\left(1 - \frac{\alpha^2}{x^2}\right) + \overline{\omega_2}\right\}, \\
Q_r =& \frac{qa}{2}\left(x - \frac{\alpha^2}{x}\right)
\end{aligned} \tag{13}$$

三、设计公式与图表

在膜盒基体的刚度和强度设计中，挠度、径向应力、切向应力和径向的横向剪应力是最主要的量。为了应用方便，我们引入下列无量纲量

$$
\begin{aligned}
&S_r=\frac{\sigma_r}{q},S_\theta=\frac{\sigma_\theta}{q},T_{rz}=\frac{\tau_{rz}}{q},\\
&W=\frac{E}{12(1-\nu^2)q}\frac{w}{h},H=\frac{h}{a},c_1=\frac{D}{qa^4}\overline{c_1},\\
&c_2=\frac{D}{qa^4}\overline{c_2},c_3=\frac{D}{qa^4}\overline{c_3},\omega_1=\frac{H^2}{5},\\
&\omega_2=\frac{\nu H^2}{10(1-\nu)}
\end{aligned}
\tag{14}
$$

于是，由式（9）～(13)，得到

$$
\begin{aligned}
W&=\frac{1}{H^4}\left[\frac{1}{4}c_1x^2+c_2\ln x+c_3+\frac{1}{8}\left(\frac{1}{8}x^4-\alpha^2x^2\ln x+\alpha^2x^2\right)\right],\\
S_r&=-\frac{12z}{hH^2}\left\{\frac{1+\nu}{2}c_1-\frac{1-\nu}{x^2}c_2+\frac{1}{8}\left[\frac{3+\nu}{2}x^2-2(1+\nu)\alpha^2\ln x-(1-\nu)\alpha^2\right]\right.\\
&\quad\left.+\frac{\omega_1}{2}\left(1+\frac{\alpha^2}{x^2}\right)+\omega_2\right\},\\
S_\theta&=-\frac{12z}{hH^2}\left\{\frac{1+\nu}{2}c_1+\frac{1-\nu}{x^2}c_2+\frac{1}{8}\left[\frac{1+3\nu}{x^2}x^2-2(1+\nu)\alpha^2\ln x+(1-\nu)\alpha^2\right]\right.\\
&\quad\left.+\frac{\omega_1}{2}(1-\frac{\alpha^2}{x^2})+\omega_2\right\},\\
T_{rz}&=-\frac{3}{4H}\left[1-\left(\frac{2z}{h}\right)^2\right]\left(x-\frac{\alpha^2}{x}\right)
\end{aligned}
\tag{15}
$$

其中

$$
\begin{aligned}
c_1&=\lambda_0\left[2(1+\nu)\alpha^2\ln\alpha-\frac{1+3\nu}{2}\alpha^2-1+\nu-\frac{1-\nu}{2}\alpha^{-2}-4\omega_1(1+\alpha^{-2})-8\omega_2\right],\\
c_2&=-\lambda_0\{(1+\nu)\alpha^2\ln\alpha-\frac{1-\nu}{4}\alpha^2+\frac{1+\nu}{4}\alpha^2\\
&\quad-\frac{2\omega_1}{1-\nu}[(1+\nu)\alpha^2+1-3\nu]-4\omega_2\},\\
c_3&=-\frac{1}{4}c_1-\frac{1}{8}\left(\alpha^2+\frac{1}{8}\right)
\end{aligned}
\tag{16}
$$

至于法向应力 σ_z，仍使用式（9）中的公式进行计算。

以 $\alpha=0.1$ 和 $H=0.3$ 的膜盒基体为例，按式（15）进行数值计算，在 $\nu=0.3$ 的情况下，我们分别在图 2～图 4 上给出了无量纲挠度沿膜盒基体半径的分布曲线、无量纲径向和切向应力沿膜盒基体的下表面半径的分布曲线以及无量纲径向横向剪应力沿膜盒基体中面半径的分布曲线。

由上述公式和计算可知，挠度的最大值出现在膜盒基体的内边缘；最大径向应力是在膜盒基体的外边缘，且在上表面上；最大切向应力是在膜盒基体的内边缘，在下表面上，并且大于最大径向应力。

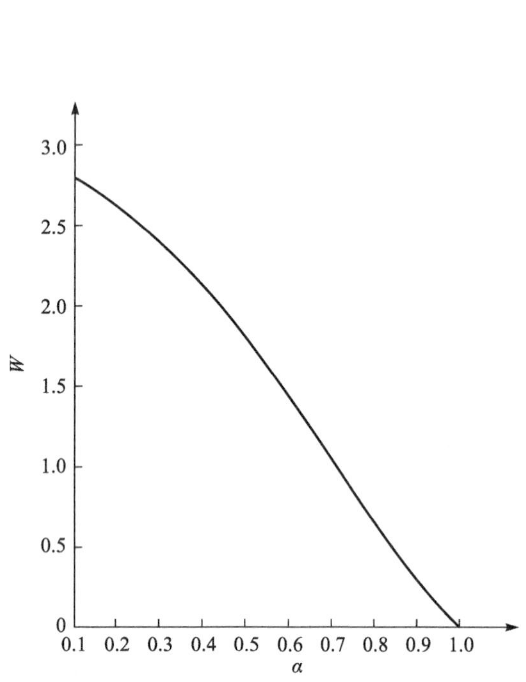

图 2　膜盒基体的挠度分布

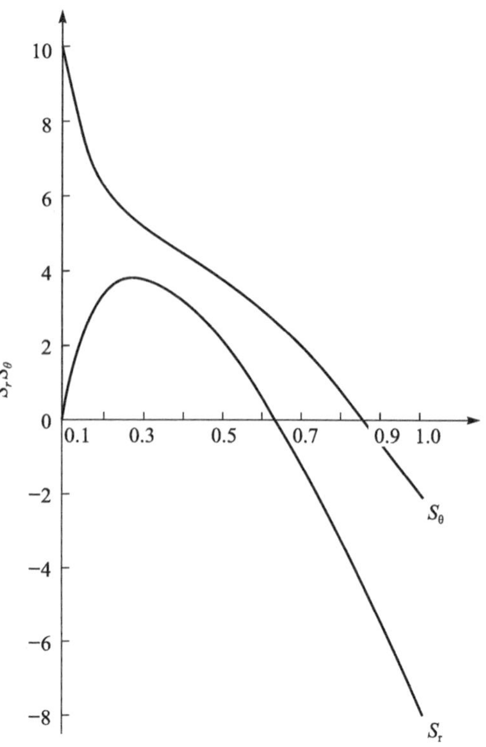

图 3　膜盒基体下表面的正应力分布

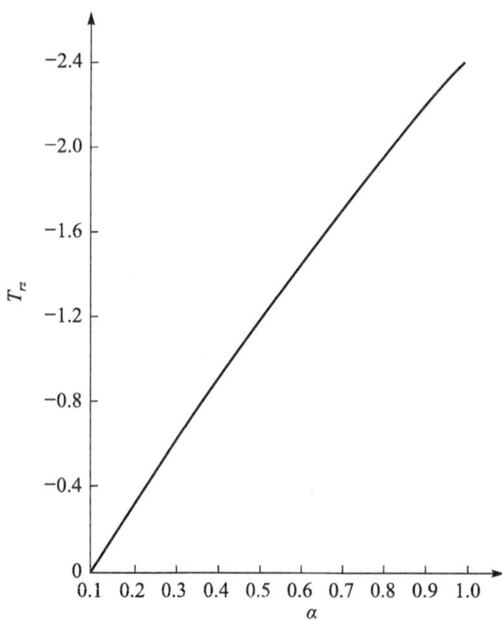

图 4　膜盒基体中面的径向横向剪应力分布

按照绝对值而言，板的上、下表面 $\left(z=\mp\dfrac{h}{2}\right)$ 有数值相等的正应力分布，数值的正、负将分别是拉应力的压应力。最大拉应力的所在位置将是膜盒基体最易破坏的地方，由式（15），我们得到如下膜盒基体的无量纲最大挠度和无量纲最大正应力的设计公式

$$
\begin{aligned}
W_{\max} = \frac{2\lambda_0}{H^4}\Bigg\{ & \frac{7+3\nu}{32}\alpha^4 - \frac{1+\nu}{2}\alpha^2\ln^2\alpha \\
& - \left[\frac{5-\nu}{8} - \frac{(1+\nu)H^2}{5(1-\nu)}\right]\alpha^2\ln\alpha \\
& - \left[\frac{1+7\nu}{32} + \frac{H^2}{10(1-\nu)}\right]\alpha^2 \\
& - \left[\frac{1+\nu}{8} - \frac{(1-2\nu)H^2}{5(1-\nu)}\right]\ln\alpha \\
& - \frac{7-5\nu}{32} + \frac{\nu H^2}{10(1-\nu)} \\
& + \frac{1}{2}\left(\frac{1-\nu}{16} + \frac{H^2}{5}\right)\alpha^{-2}\Bigg\},
\end{aligned}
$$

$$(S_\theta)_{\max} = -\frac{3(1-\nu^2)\lambda_0}{H^2}\left\{\alpha^2 - 4\ln\alpha - \left[1+\frac{8\nu H^2}{5(1-\nu^2)}\right]\alpha^{-2}\right\} \tag{17}$$

对于膜盒基体上表面的无量纲最大径向应力，则用如下公式计算：

$$(S_\theta)_{\max} = \frac{6\lambda_0}{H^2}\left[2(1+\nu)\alpha^2\ln\alpha - \frac{1+12\nu+3\nu^2}{2}\alpha^2 + 2\nu\left(1+\frac{H^2}{5}\right) + \left(\frac{1-\nu}{2} - \frac{2\nu H^2}{5}\right)\alpha^{-2}\right] \tag{18}$$

使用式（17）进行数值计算，在 $\nu=0.3$ 的情况下，分别在图 5 和图 6 上绘制了无量纲最大挠度和下表面无量纲最大切向应力随中心孔半径和厚度变化的曲线。由图 5 看到，膜盒基体的最大挠度 $W_{\max}$ 随相对厚度 H 的增大而减小；在无量纲内缘半径 α 比较小的情况下，随 α 的增大而缓慢增大，而在 α 较大的情况下，却随 α 的增大而缓慢减小。由图 6 看到，无量纲最大切向应力 $(S_\theta)_{\max}$ 随相对厚度 H 的增大而减小，随无量纲内缘半径 α 的增大而降低。

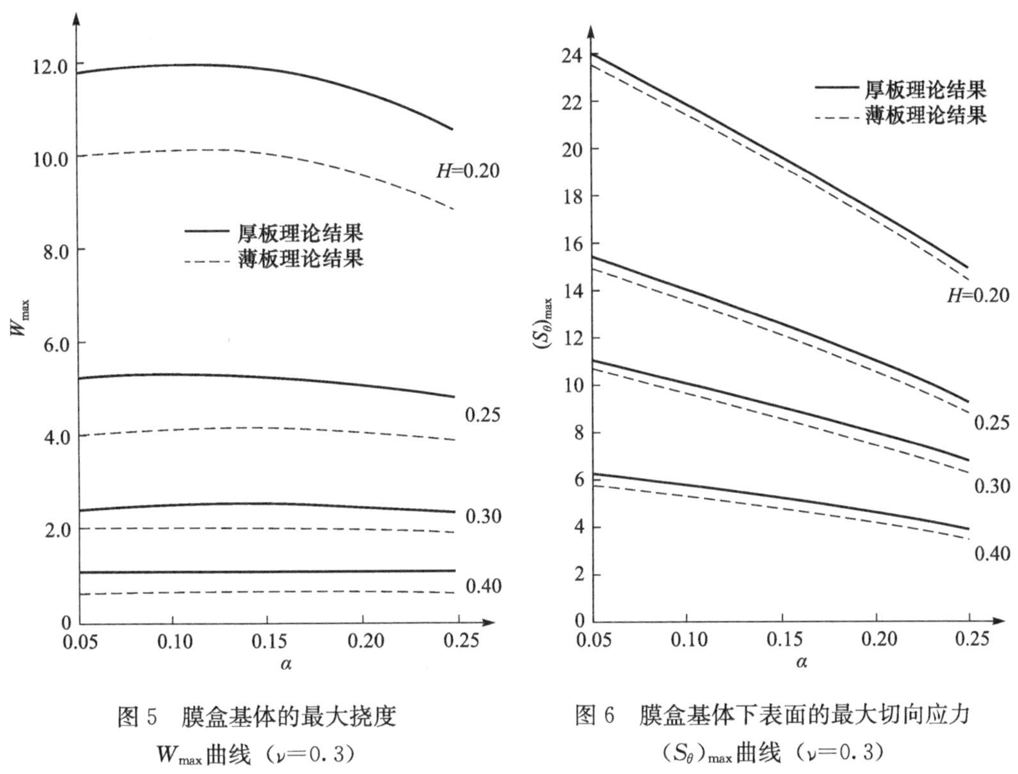

图 5　膜盒基体的最大挠度 $W_{\max}$ 曲线（$\nu=0.3$）

图 6　膜盒基体下表面的最大切向应力 $(S_\theta)_{\max}$ 曲线（$\nu=0.3$）

为使膜盒基体的设计更加合理，我们还把用薄板计算的数值结果给在图 5 和图 6 中，并在图 7 和图 8 中给出厚、薄板理论数值结果的相对误差。由图看到，薄板理论结果小于厚板理论结果，且相对误差数值一般较大。同时，对最大挠度而言，两种理论数值结果的相对误差既随相对厚度 H 的增大而增大，又随内缘半径 α 的增大而缓慢减小；对最大切向应力而言，两种理论数值结果的相对误差既随相对厚度 H 的增大而增大，又随内缘半径 α 的增大而增加。

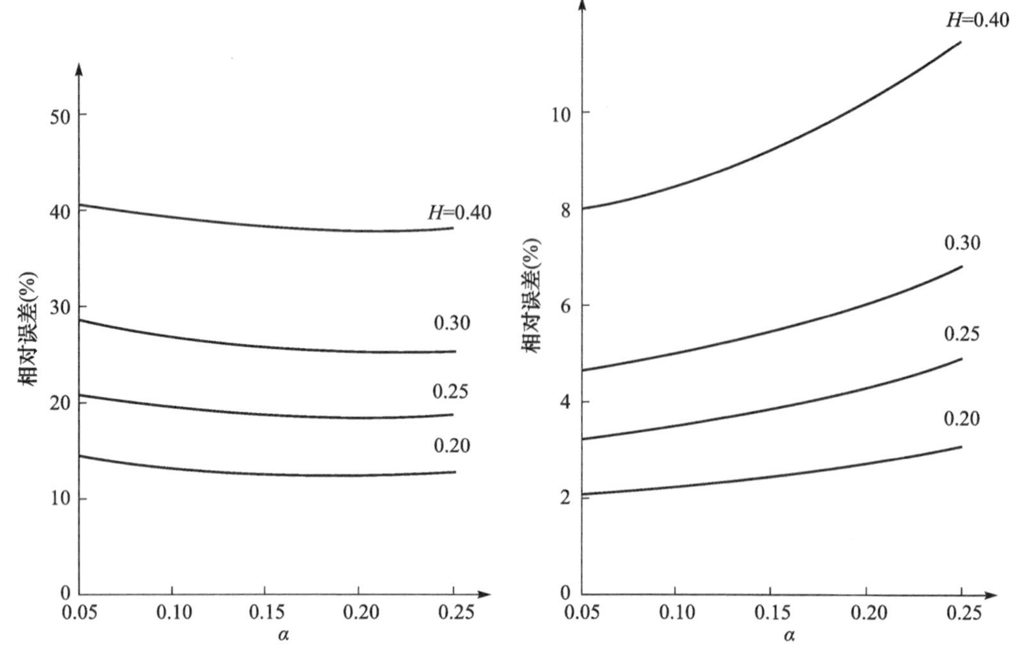

图 7　膜盒基体最大挠度的相对误差与 α 的关系曲线（ν=0.3）

图 8　膜盒基体最大切向应力的相对误差与 α 的关系曲线（ν=0.3）

四、结束语

通过厚板弯曲理论微分方程的求解，给出了膜盒基体的所有位移和应力的计算公式，并指出最大的位移是内边缘的挠度，最大的应力是内边缘处下表面的切向应力。按所提供的最大挠度和最大应力的公式，可进行膜盒基体的刚度的强度的设计。本文所给的设计公式简单、精确，便于运算，便于使用。

参考文献

[1]　刘人怀．精密仪器仪表弹性元件设计的力学原理．广州：暨南大学出版社，2006

[2]　Liu Ren-huai. Study on Nonlinear Mechanics of Plates and Shells. New York，Beijing，Guangzhou：Science Press and Jinan University Press，1998.

[3]　刘人怀．波纹圆板的特征关系式．力学学报，1978（1）：47-52.

[4]　刘人怀．波纹环形板的非线性弯曲．中国科学，A 辑，1984（3）：247-253.

[5]　Liu Ren-huai. Large deflection of corrugated circular plate under the action of concentrated loads at the center. International Journal of Non-Linear Mechanics；1984，19（5）：409-419.

[6]　Liu Ren-huai. Large deflection of corrugated circular plate with plane boundary region. Solid Mechanics Archives，1984，9（4）：383-406.

[7]　Liu Ren-huai and Li Dong. On the non-linear bending and vibration of corrugated circular plates. International Journal of Non-Linear Mechanics，1989，24（3）：165-176.

[8]　Liu Ren-huai and Zou Ren-po. Non-linear bending of a corrugated annular plate with a plane boundary region and a non-deformable rigid body at the center under compound load. Internation-

al Journal of Non-Linear Mechanics，1993，28 (3)：353-364.

[9] Liu Ren-huai and Yuan Hong. Nonlinear bending of corrugated annular plate with large boundary corrugation. Applied Mechanics and Engineering，1997，2 (3)：353-367.

[10] Liu Ren-huai and Wang Fan. Non-linear stability of corrugated shallow spherical shell. International Journal of Applied Mechanics and Engineering，2005，10 (2)：295-309.

[11] Yuan Hong and Liu Ren-huai. Nonlinear vibration of corrugated shallow shells under uniform load. Applied Mathematics and Mechanics，2007，28 (5)：573-580.

[12] Reissner E. The effect of transverse shear deformation on the bending of elastic plates. Journal of Applied Mechanics，1945，12 (2)：A69-A77.

刘人怀院士文集

第三卷

刘人怀◎著

科学出版社
北京

内 容 简 介

本文集由著者在力学和管理科学领域60多年中所发表的文章汇编而成，分为上、下篇。上篇是力学部分，结合工程需要，对单层板壳、波纹板壳、双层板壳、夹层板壳、复合材料层合板壳和网格扁壳等六类板壳的非线性弯曲、稳定和振动问题，以及厚、薄板壳弯曲问题进行理论探索。下篇是管理科学部分，联合实际，对工程管理、公共管理、工商管理、科技管理和教育管理的问题进行了研究。

本文集可作为高校力学和管理科学专业，以及相关专业老师、研究生、本科生的参考书，适合政府、高等学校、科研院所、科技社团、企业等领域的领导、管理人员和科研人员参考阅读，又可作为航天、航空、航海、机械、建筑和交通等工程设计师与工程师的设计制造指导书。

GS京（2025）0637号

图书在版编目（CIP）数据

刘人怀院士文集 / 刘人怀著. -- 北京 ：科学出版社，2025. 4. -- ISBN 978-7-03-080897-4

Ⅰ. O3-53；C93-53

中国国家版本馆CIP数据核字第2024JH1723号

责任编辑：陈会迎/ 责任校对：姜丽策

责任印制：张　伟/ 封面设计：有道设计

科学出版社出版

北京东黄城根北街16号

邮政编码：100717

http://www.sciencep.com

北京中科印刷有限公司印刷

科学出版社发行　各地新华书店经销

*

2025年4月第　一　版　开本：787×1092 1/16

2025年4月第一次印刷　印张：153 3/4

字数：3 646 000

定价：698.00元（全五卷）

（如有印装质量问题，我社负责调换）

目　　录

下篇　管理科学理论与应用

第九章　管理学科 …… 985

谈谈创建现代管理科学中国学派的若干问题 …… 985

再谈创建现代管理科学中国学派的若干问题 …… 995

三谈创建现代管理科学中国学派的若干问题：四条定义与三点建议 …… 1004

大平台、聚义厅及其他——四谈创建现代管理科学中国学派的若干问题 …… 1013

《学科目录》第 12 学科门类与管理科学话语体系——五谈创建现代管理科学中国学派的若干问题 …… 1021

当前管理科学研究中的若干问题——几个疑点的澄清和两种研究方法的评析 …… 1030

中国管理科学的现状和走向 …… 1040

传统文化基因与中国本土管理研究的对接：现有研究策略与未来探索思路 …… 1041

Research on Chinese school of modern GUANLI science …… 1056

第十章　工程管理 …… 1068

上海浦东新区建设工程 …… 1068

中国制造业的生存哲学 …… 1097

东水西调工程 …… 1100

一、关于改善我国北方水资源缺乏的建议 …… 1100

二、关于实施“东水西调”工程的建议 …… 1100

关于“发展中国家的工业化道路”论坛的讨论 …… 1133

绿色制造与学科会聚 …… 1140

工程管理是管理对国民经济的深度介入 …… 1144

城市矿产工程 …… 1147

一、关于治理垃圾的建议 …… 1147

二、推进“垃圾分类，从我做起”科普宣传活动 …… 1148

三、开展“垃圾分类，从我做起”科普资源包研发及宣讲活动 …… 1150

四、开展“垃圾分类”宣讲与培训服务 …… 1152

五、关于推进“餐厨废弃物变废为宝”的建议 …… 1155

六、开展美丽广东科普教育示范户工作 …… 1157

七、早日实现“美丽广东梦” …… 1158

八、促进城市矿产资源化利用，共建美丽中国 …… 1159

九、绿色再制造的探索 …… 1159

十、珠三角城乡生活垃圾统筹治理战略研究 …… 1168

十一、城市餐厨垃圾回收逆向物流系统构建的研究 …… 1184
十二、让人民过上好日子 …… 1192
十三、关于推进餐厨垃圾治理装备高技术制造业发展的建议 …… 1193
大型工程项目管理的中国特色及与美苏的比较 …… 1195
建设低碳社会关键在制造创新 …… 1207
工程管理信息化的内涵与外延探讨 …… 1209
转变经济发展方式关键在制造业创新 …… 1215
公共安全工程 …… 1217
一、公共安全相关工程科技研究的重要性 …… 1217
二、信息安全相关工程科技发展战略 …… 1219
三、生物安全相关工程科技发展战略 …… 1224
四、食品安全相关工程科技发展战略 …… 1232
五、土木工程安全相关工程科技发展战略 …… 1244
六、生产安全相关工程科技发展战略 …… 1256
七、社会安全相关工程科技发展战略 …… 1268
八、生态安全相关工程科技发展战略 …… 1279
工程管理信息化架构研究 …… 1292
大力促进科技与经济融合 …… 1300
勇担重任　敢攀高峰 …… 1301
有效推动轨道交通结构健康监测和整治修复技术的发展提升 …… 1303
第十一章　公共管理 …… 1304
征求意见　完善报告 …… 1304
领导用人标准 …… 1306
有效利用外资　扩大对外开放 …… 1309
如何防止公共关系庸俗化 …… 1314
领导科学与领导艺术 …… 1315
城市政府工作目标管理与治理整顿、深化改革 …… 1329
实施城市管理系统工程建设　开创广州可持续发展新格局 …… 1332
建议重视我省仪器仪表工业的发展，以迎接 21 世纪挑战 …… 1338
广东省发展高新技术的若干意见和建议 …… 1340
珠海前山镇（街）转型与社区建设研究 …… 1345
广东的治安状况与投资环境 …… 1368
教育彩票作用惊人 …… 1370
建设广州石牌大学城 …… 1372
办好院士之家 …… 1374
坦诚建言 …… 1376
光大华侨文化　建设文化大省 …… 1377
传承岭南文化　服务文化大省 …… 1380

建立粤港澳综合协调机构 …… 1381
泛珠三角：推进科技、教育和文化的区域合作 …… 1382
促进广东省职业教育发展 …… 1385
关于尽快制定国家统一法的建议 …… 1392
激励民办专科学校升为本科学校 …… 1393
让象牙塔成为顶梁柱 …… 1394
大力发展我省高中阶段教育 …… 1397
关于允许市民在节假日有条件燃放烟花的建议 …… 1404
在推进和谐社会建设中切实解决“农民工”身份问题 …… 1405
关于改善财政宏观调控深化分税制财政体制改革的调研报告 …… 1407
岁月留声 …… 1418
黄石应该在建设特色城市中凸出优势 …… 1420
关于将清明节设为国家法定假日的建议 …… 1421
爱心和匠心 …… 1423
发展“乡村旅游”促进广东新农村建设 …… 1426
关于城市基础设施建设投融资体制改革研究 …… 1428
关于将香港、澳门特别行政区的所有统计数据纳入全国性统计数据的建议 …… 1437
大规模引进和培训人才为广东产业结构优化升级服务 …… 1438
大力推动“政产学研金”合作创新为广东省经济社会发展做贡献 …… 1447
积极推进知识产权事业发展 …… 1449
关于完善我省应对台风灾害预防措施的建议 …… 1455
坚持重点　保持特色 …… 1456
中国的过去和现在 …… 1457
系统工程与领导科学 …… 1473
关于实行九年一贯制办校的建议 …… 1474
2012年中国工程院院士广州咨询活动中心工作总结 …… 1476
百年追梦　科技兴国 …… 1481
充分发挥桥头堡作用，推进孟中印缅经济走廊建设 …… 1488
建设低碳社会　托起美丽中国梦 …… 1490
研究传染病突发事件的危机管理十分重要 …… 1492
以智能制造促进产业转型升级 …… 1493
网络强国战略与实践 …… 1502
两化深度融合与质量管理 …… 1503
推动广东非开挖产业快速发展 …… 1512
谈谈标准化工作对当前的重要意义 …… 1517
为推动仪器仪表产业发展做出新贡献 …… 1522
加强数字产业化和产业数字化双轮驱动 …… 1526

下篇　管理科学理论与应用

第九章　管理学科

谈谈创建现代管理科学中国学派的若干问题[①]

一、我们面临的形势

在世界范围内，20世纪的管理科学曾经有两大强势的主流学派：计划经济体制下的管理学派，属于社会主义阵营，以苏联为代表（不妨称为苏联学派）；市场经济体制下的管理学派，属于资本主义阵营，以美国为代表（不妨称为美国学派）。两大阵营、两大学派曾经界限分明，严重对立。随着苏联解体和东欧剧变，苏联学派不复存在，美国学派似乎可以独步天下了，但是事情并没有那么简单。从1978年开始，中国实行改革开放，逐步建立健全社会主义市场经济体制，现在，一个新的学派——现代管理科学的中国学派出现了曙光。

我们应该积极推进现代管理科学中国学派的创建工作。

中国管理科学界现在形势很好。最近有两个很重要的好消息令人振奋。

(1) 党的十七大高度重视管理工作。据我们统计，胡锦涛同志在报告中说到“管理”一词有45处，说到“治理”一词有6处，说到“法治”一词有16处，说到“依法治国”有7处；而按一个“管”字（含“管理”）统计，则有57处。按一个“治”字（含“治理”“法治”，不含“政治”）统计，则有51处；报告的第6部分专门有一段“（五）加快行政管理体制改革，建设服务型政府”，报告的第8部分专门有一段“（六）完善社会管理，维护社会安定团结”，两段话共有720多字。

党的十七大把中国特色社会主义伟大旗帜的科学内涵界定为坚持中国特色社会主义道路和中国特色社会主义理论体系。毋庸置疑，现代管理科学的中国学派是这个理论体系的组成部分。

(2) 2008年1月19日，胡锦涛同志看望著名科学家钱学森院士。胡锦涛同志谈起系统工程理论时说：“20世纪80年代初，我在中央党校学习时，就读过您的有关报告。您这个理论强调，在处理复杂问题时一定要注意从整体上加以把握，统筹考虑各方面因素，这很有创见。现在我们强调科学发展，就是注重统筹兼顾，注重全面协调可持续发展。”

钱学森等[1]有一个著名的论断：组织管理的技术——系统工程。30年来，系统工程与改革开放共生共荣，与时俱进。

伟大的实践孕育伟大的理论，辉煌的成就铸造辉煌的模式。中国的革命和建设、

---

① 本文原载《管理学报》，2008，5（3）：323-329，作者：刘人怀，孙东川.

中国的改革开放是中国人民的伟大实践，已经取得了辉煌成就，其中必然孕育出伟大的理论，铸造出辉煌的模式。创建现代管理科学的中国学派，把它贡献给全人类，是中国管理科学界当仁不让、责无旁贷的重大历史使命。

国家自然科学基金委员会（简称“基金委”）管理科学部[2]在“十一五”期间的第一项战略目标就是“奠定在未来10～20年中逐步建立管理科学中国学派的学科基础”。近几年召开的“中国管理科学与工程论坛”“中国管理学年会”“中国系统工程学会学术年会”等重要学术会议，以及在《管理学报》等刊物上，都有这方面的文章[3-6]。

我们认为，创建现代管理科学的中国学派，不但具有必要性与可行性，而且具有紧迫性。

历史上有不少事例说明，中国人必须积极关注中国的事情，无论从学者的角度，还是从中国公民的角度，我们都应该热情关注中国的管理工作和管理科学。我们不关注，别人很关注；我们掉以轻心，别人抓住机会跑到前面去，我们就会落后，陷入被动局面。创建现代管理科学的中国学派是我们中国人的历史使命。

二、“管理科学”与“管理学”

“管理科学”与“管理学”2个术语，2个基本概念，谁包含谁？有2种理解。第一，把“管理科学”作狭义理解，把“管理学”作广义理解，其理由大概有2点：①依据美国人的理念；②文献［7］中，第12门类是“管理学”，一级学科1201是“管理科学与工程”。第二，把“管理科学”作广义的理解，把“管理学”作狭义的理解。笔者赞成后者，即广义的管理科学包含狭义的管理学。

不妨对比一下“社会科学”与“社会学”，“系统科学”与“系统学”，大概就没有多少争议了。钱学森院士不仅阐述了关于系统科学与系统学的关系，还提出把“地理学”扩大为“地理科学”，把“建筑学”扩大为“建筑科学”等[8]。

在文献［7］中“管理学”门类对应的英文是management science。该目录正在修订之中，其中第12门类存在若干不妥之处，1201一级学科是焦点之一，人们已经提出了不少批评意见和修改建议。

我们必须注意一个基本事实：英文单词management并不等同于中文的“管理”，“管理”一词的含义比management丰富得多，对应“管理”的还有administration、supervise等多个英文单词，如热门的专业学位MBA即master of business administration（工商管理硕士）。所以，中文的“管理科学”比英文的management science广义得多。

基金委管理科学部对管理科学的界定接近广义。教育部科学技术委员会管理科学部也是这样。

笔者作以上说明，只是解释一下自己的思考和选择，大家不必在名称上争来争去，要紧的是先把研究工作做起来。谁的工作做得好，卓有成效，说服力强，谁就有话语权。但是，没有必要过于迁就美国人的理念，目前风行世界的管理科学（广义的）其实是“美国学派”，而且大体上是美国版的企业管理学（偶尔有一点日本人的东西）。它的特点是五花八门，不成系统；在许多方面是见物不见人。美国人做学问喜欢标新立异——这是好的，不喜欢持之以恒——这就不大好了。例如，好端端的systems engineering（系统工程），在其发源地美国早就不大提起了，我国自1978年引进以来，30年坚持不懈，无论在理论上还是在应用上都取得了巨大的成功，以至于我们今天可以很自豪地说：中国的系统工程，是系统工程的中国学派，是高高飘扬的一面旗帜。美国人发明的PERT/CPM① (1958/1956)，是很好的计划管理与成本管理相结合的方法。但是在美国国内并没有广泛使用，甚至连美国运筹学家都知之甚少；在中国，由于华罗庚和钱学森等学者的推动，现在每一个建筑队施工都用它，而且用得很不错；中国的每一本运筹学教科书都讲述它（称为统筹法或计划协调技术等）。所以，美国人的这种作风是不值得效法的。再看，美国的管理工作搞得怎么样？实在不敢恭维：从宏观到微观，各个领域、各个层面，问题成堆。管理不是政治，但是，管理工作不能远离政治，尤其是宏观管理——国家治理与国际关系管理——与政治是密切相关的；管理伦理更是如此。大量事实证明：美国和一些西方国家经常用双重标准对中国说三道四，大言不惭地责难中国，似乎中国无论做什么事情都是错的，我们对此必须有足够清醒的头脑。美国的管理理论与方法，该学习的我们一定要虚心学习，但是，确有许多是不值得效法的，我们不能良莠不分、全盘照抄。中国绝对不可能全盘接受美国人的旨意和做法。我们应该扬弃美国学派，创建高于美国学派的现代管理科学的中国学派。

三、学习和借鉴钱学森院士的工作

有关系统科学与系统学的关系，是钱学森院士的观点。30年来，钱学森院士一直大力倡导和推动系统科学与系统工程[1,8-11]。他的科学探索精神，他的一系列观点和工作，是值得我们学习和借鉴的。

1. 对于系统工程与运筹学的重组

钱学森院士指出：从20世纪40年代以来，国外对定量化系统思想方法的实际应用相继取了许多名称，如operations research（运筹学）、management science（管理科学）、systems engineering（系统工程）、system analysis（系统分析）、systems research（系统研究）、cost-effectiveness analysis（费用-效益分析）等。运筹学是指目的在于增加现有系统效果的分析工作；管理科学是指大企业的经营管理技术；系统工程是指设计新系统的科学方法；系统分析是指对若干可供选择的执行特定任务的系统方案进行比较和选择；如果系统分析着重在费用与效益方面，就是费用-效益分析；系统研究是指拟制新系统的实现程序。由于历史原因形成的这些不同名称，混淆了工程技术与其理论基础即技术科学的区别，用词不够妥当，认识也不够深刻，国外曾经有人试图给这些名词的含义以精确的区分，但是未见取得成功。其实，用定量化的系统方法处理大型复杂系统的问题，无论是系统的组织建立，还是系统的经营管理，都可以统一地看成工程实践。engineering（工程）这个词，18世纪在欧洲出现时，本来专指作战兵器的制造和执行服务于军事目的的工作。从后一种含义引申出一种更普遍的看法，把服务于特定目的的各种工作的总体称为工程，如水利工程、机械工程等。如果这个特定的目的是系统的组织建立或者是系统的经营管理，就可以统统看成系统工

① program evaluation and review technique/critical path method，即计划协调技术与关键路径法.

程。国外所称的运筹学、管理科学、系统分析、系统研究以及费用-效益分析的工程实践内容，均可以用系统的概念统一归入系统工程；而其数学理论和算法，可以统称为运筹学。

中国系统工程学会前理事长、中国工程院院士许国志教授评价说，钱学森院士关于系统工程的定义，以及上面这段话，把“人各一词，莫衷一是”的情况澄清为“分门别类，共居一体”。同时，他对于系统工程给出了一个确切的描绘，提出系统科学体系，进而提出现代科学技术体系和人类知识体系，并论述系统工程在其中的地位。

钱学森院士和许国志院士的论述，对于我们开展现代管理科学中国学派的创建工作具有指导意义。

2. 系统科学体系结构

钱学森院士非常重视研究系统科学体系、现代科学技术体系和人类知识体系。他指出，现代科学技术体系具有许多部门，如自然科学、社会科学、系统科学、地理科学、军事科学等。在每一部门中，直接与改造客观世界的实践活动相联系的是工程技术；稍微远离工程实践的是工程技术的理论基础——技术科学；再远一些的是这一部门科学技术的基础科学；基础科学再经过一座过渡的桥梁与马克思主义哲学相联系[8]。

1981 年和 1985 年，钱学森院士两次描绘了包含图 1 的系统科学体系。他指出，在系统科学这个部门中，系统工程属于工程技术，其理论基础是运筹学、控制论和信息论这三门技术科学；关于系统的一般理论——系统学，是系统科学这个部门中的基础科学；系统科学从各门系统工程到运筹学、控制论和信息论，再到系统学，形成现代科学技术体系中的一个分体系；这个分体系通向马克思主义哲学的桥梁，是大约一百年前启示的，后来经过现代科学技术大大丰富了的系统论（又称为“系统观”）；系统论将充实科学技术的方法论，并为马克思主义哲学的深化和发展提供素材[9]。

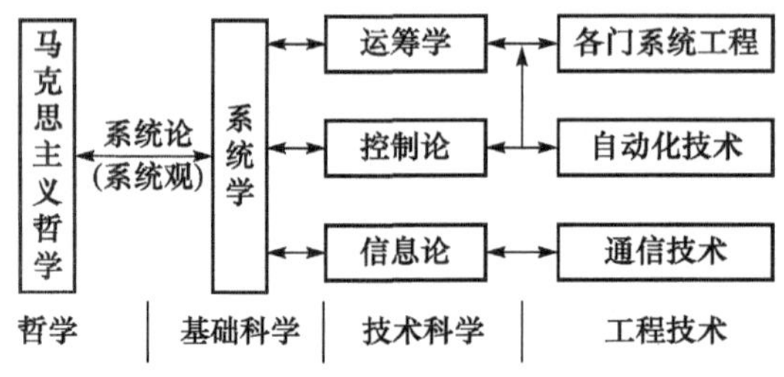

图 1　系统科学体系结构

四、中国管理落后与不落后之辨析

世界在关注中国。西方人士惊呼：中国的经济发展使得一切传统的经济理论失效。诚哉斯言！不但如此，新的经济理论——经济理论的中国学派——正在孕育，管理科学也是这样。中国的管理工作尽管目前还是问题较多，还要大力改进和改革，但就总体和主流而言是卓有成效的，在全世界独树一帜。

经济与管理是正相关的，不是不相关，更不是负相关。如果中国的管理工作是

"一团糟"，那就无论如何也不能解释长达30年（而且还将长期延续下去）中国经济的巨大发展和中国社会的安定团结。不妨把经济与管理比作一对双胞胎，一个长大了，另一个也长大了，不可能一个不断成长，另一个不成长；现在，中国的经济长大了，中国的管理也长大了。经济与管理又好比是"夫妻店"（这里不含贬义，世界各国经济与管理都可以比作"夫妻店"），两个人相辅相成，谁也离不开谁，恩恩爱爱，欢乐同享，患难与共；经济发展了，有他的一份功劳，也有她的一份功劳。实践之树常青，理论常常是灰色的。具有中国特色的管理模式正在形成，现在需要管理科学的理论研究紧紧跟上。

我们较早开展了与本项目直接相关的研究，例如，1998年就写了《中国管理落后与不落后之辨析》[11]一文；当时在MBA论坛和多种场合讲述，听到不少赞成的和不赞成的意见。现在看来，这篇文章的观点是经得起时间考验的。今天，很值得用同样的题目重新写一篇大文章。我们认为对于中国管理落后与不落后的问题，应该从理论与实践、宏观与微观、现代与古代等多个角度仔细辨析，而不能简单地一概而论。

（1）中国的宏观管理堪称世界一流：国家治理井井有条，各方面都在前进，综合国力蒸蒸日上，国际地位日益提高。在中观、微观层次，我们有不少落后面，但是并非一无是处，而且正在不断改进与完善。越来越多的中国企业走出国门，走向世界；"中国制造"风靡全球，而且正在转化为"中国设计""中国创造"——目前还不显著，随着时间的推移，将会迅速凸显。

（2）在中国古代，曾经有过卓越的管理思想和管理实践。这是任何其他国家无法比拟的，只要加强古为今用，就会转变为巨大的力量。即便在改革开放之前，中国的管理也不乏闪光点，如鞍钢宪法、大庆精神、统筹法与优选法的推广，以及"两弹一星"的管理工作等。今天，中国特色社会主义管理模式正在发展之中，我们应该本着一分为二、实事求是的原则，发展管理科学的中国学派，指导管理实践；同时，发展管理教育的中国学派，培养管理人才。

五、关于现代管理科学中国学派的思考

笔者认为，现代管理科学的中国学派将具有以下特点[3-5]。

（1）它是中国的：包含中国从古到今的管理科学成果，适合中国特色社会主义市场经济建设和社会发展，而且，它应该是中国人创建的，或者是以中国人为主导而创建的。

（2）它是现代的：充分应用现代计算机和互联网技术，反映世界各国最新的管理科学成就，适用于现代社会各方面的长期发展。

（3）它是创新的、先进的：博采众长，推陈出新，与时俱进。

（4）它是世界的：具有充分的包容性与普适性，是全人类的共同财富。

我们认为，创建现代管理科学中国学派的基本途径是：洋为中用，古为今用，近为今用，综合集成。其中"近为今用"是重点，是笔者提出的理念，其含义是：从我国近期的社会实践中总结经验和教训，上升到理论高度，指导当前和今后的社会实践。

"近"主要是指20世纪和21世纪以来的岁月，有"两大板块"：①中国共产党领

导革命、建设和改革开放事业的近 90 年历程；②20 世纪上半叶，我国民族资本主义工商业有所发展，20 世纪下半叶至今，台湾、香港、澳门市场经济繁荣，从管理科学角度都值得研究和总结。这个“近”，与泰罗制产生以来的西方管理科学发展期大致吻合。近为今用还可以上溯到清王朝后期的洋务运动。

大陆较早开展创建工作的著名学者有成思危、李京文、潘承烈、席酉民、苏东水、赵纯均等，台湾学者有钱穆、曾仕强、林国雄等，美籍华裔学者有成中英、杜维明等。文献［12-28］及其 10 多名作者，只是露出水面的“冰山之尖”而已。有些人采用不同的名称，如“有中国特色的管理科学”“中国特色的管理科学”“中国管理学”“中国管理科学”“中国现代管理科学”“有中国特色社会主义管理科学”，以及很有创意的“和谐管理”“和合管理”“中道管理”等。笔者查阅诸多参考文献之后，认为这些名称后面的主张乃大同小异，目标是高度一致的。

现代管理科学必然具有多种学派，因为现在世界上有多种经济模式、多种社会制度、多种文化形态。世界是多元的，管理科学的学派必然也是多元的。

笔者曾经用过“有中国特色的现代管理科学”[4,5]，后来看到基金委提出的“管理科学的中国学派”，认为很好，欣然接受，加上“现代”二字[3,29]，其理由见本小节叙述的 4 个特点之（2）、（3）。“现代”是“古代”的延续，“古为今用”是创建工作的基本途径之一。如果我们对于“科学”与“管理科学”做广义理解，则可以说：古代曾经有很出色的“管理科学的中国学派”，而且有许多“支派”，如儒家、法家、道家等，儒家还有“思孟学派”等支派。

学术界已经做了很多创建工作，概述如下。

（1）洋为中用。改革开放 30 年，就是西方管理科学“洋为中用”的 30 年，功不可没。毫无疑问，我们应该继续重视洋为中用，目光紧紧盯住国外管理科学的发展，第一时间看到外国的新发明、新创造，引进、消化、吸收、改造、创新。“外国”不光是“西方七国”，也包括所有其他各国，尤其是印度、巴西、俄罗斯和东方近邻韩国、新加坡等。洋为中用要防止炒作行为，坚持实事求是，坚持本土化。现在亟须做一项工作：从 20 世纪 80 年代初获得巨大成功的 TQC（total quality control，全面质量管理）到后来的 MRP（material requirement planning，物资需求计划）、MRPⅡ（manufacturing resource plan，制造资源计划）、BPR（business process reengineering，业务流程重组）、CRM（customer relationship management，客户关系管理）、SCM（supply chain management，供应链管理）和物流管理等，包括不怎么成功的 ERP（enterprise resource planning，企业资源计划）等，应该做一番全面的回顾、反思与总结，提高自觉性，减少盲目性。

（2）古为今用。我们已经出版了多部专著，如文献［20-28］等。还有人从《三国演义》《红楼梦》等古典文学作品中研究管理思想；日本、韩国、新加坡等国学者对此很热衷，而且受到这些国家企业界的高度重视，有些企业甚至将它们用作培训经营管理人员的教材。欧美学者对中国传统管理思想也很关注，例如，美国学者雷恩[30]的著作《管理思想的演变》，就论述了孔子在管理思想方面所做的贡献。1988 年 1 月，一批诺贝尔奖获得者，在法国巴黎召开的“人类如何面向 21 世纪”讨论会上，1970 年诺贝尔物理学奖获得者瑞士阿尔文博士说："人类要在 21 世纪生存下去。必须回到 25 个世纪以前，去汲取孔子的智慧"[30]①。我们认为，根据历史的演变和现实的存在，古为今用要以儒家为主、兼顾百家；要以近几年的"国学热"为契机，深入开展典籍研究，系统整理和发掘古代管理科学成就。

(3) 近为今用。虽然我们还没有看到其他学者使用这一术语，但是，许多学者已经做了这方面的实际工作，如文献［14，18］等。文献［31］则独辟蹊径开展近为今用的研究工作，该书的封面上写着："中国共产党是中国近 100 年来管理得最成功的一个现代组织；在我们耳熟能详的党史常识（1921～1949 年）中，蕴藏着博大精深的'管理之道'"。

(4) 综合集成。这是钱学森院士提出的系统工程方法论的关键词。许多学者虽然没有使用"综合集成"术语，但是他们实际上提出了类似的主张。例如，郭重庆院士[32]提出："从'照着讲'到'接着讲'"，"接着中国传统文化讲；接着西方管理学讲；接着中国管理实践讲"。苏东水教授等[14]说："中国管理学不仅研究中国传统管理思想与实践，更要研究中国近代、现代管理思想与实践；中国管理学不仅要研究中国管理实践中的成功经验，也要研究失败的教训；中国管理学不仅要聚焦研究中国管理实践，也要研究西方的管理思想与实践。"席酉民教授等[18]在创建和谐管理理论中也是这么做的[22]。

笔者认为，明确提出"近为今用"和"综合集成"比较好，有利于从必然王国走向自由王国。"三用"可以合并为"两用"：洋为中用，中为中用。"三用"或"两用"，哪个用得多一些，哪个用得少一些，在不同的层面上、不同的领域中是不一样的。"三用"是途径，其目的是一个：创建现代管理科学的中国学派。

我们提出一种"三室一厅"工作方案（图 2）。"三室"是指洋为中用、古为今用、近为今用等三个研讨室，"一厅"即钱学森院士倡导的"综合集成研讨厅"——把"三室"中的内容汇集起来，反复研讨，从定性到定量综合集成。图中左边以近为今用为重点，洋为中用和古为今用分别影响近为今用；三者都有向右的箭头，就是说：通过综合集成，创建现代管理科学的中国学派（the Chinese school of modern management science，CSMMS）。

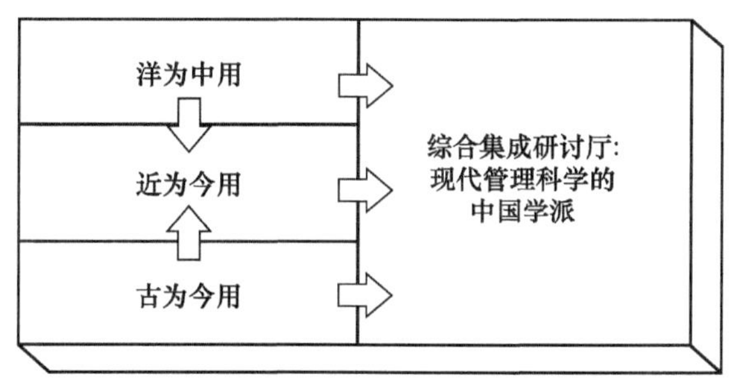

图 2　"三室一厅"工作方案

① 这段报道是由以色列记者帕特里特写的，1988 年 1 月 24 日发表在澳大利亚《堪培拉时报》上。在流传的过程中，阿尔文博士个人的学术见解被误认为讨论会的集体宣言，后来产生了"有"与"无"之争。

我们把管理工作比作一座多层结构的大厦，见图 3 右：基层管理者进行现场作业管理，高层管理者进行战略管理，中层管理者则进行承上启下的战术管理。每一层次，尤其是中层，还可以细分。基层比较庞大，越往上越小，大厦呈现金字塔形。

管理科学也是一座多层结构的大厦，见图 3 左：基层是各种职能管理的方法与技巧，包括泰罗制、TQC/TQM（total quality management，全面质量管理）、会计电算化、OA（office automation，办公自动化）、MIS（management information system，管理信息系统）、CRM、ERP、SCM 等；中间层是管理的一般理论、技术和方法，如管理学（狭义的，又称为“一般管理学”，general management）、行为科学、IT（information technology，信息技术）、O. R.（operations research，运筹学）、控制理论与技术等，为基层提供理论的、技术的支持；大厦的顶层是管理哲学与管理伦理。每一层次都可以细分。例如，可以把管理学分出来，作为紧靠管理哲学与管理伦理的一个层次。管理哲学与管理伦理具有丰富的内容，管理科学大厦并非金字塔形。现在我们所说的管理科学大厦，是一座 4 层楼。它与系统科学体系结构相类似：如果图 1 顺时针旋转 90°，也是一座 4 层楼。

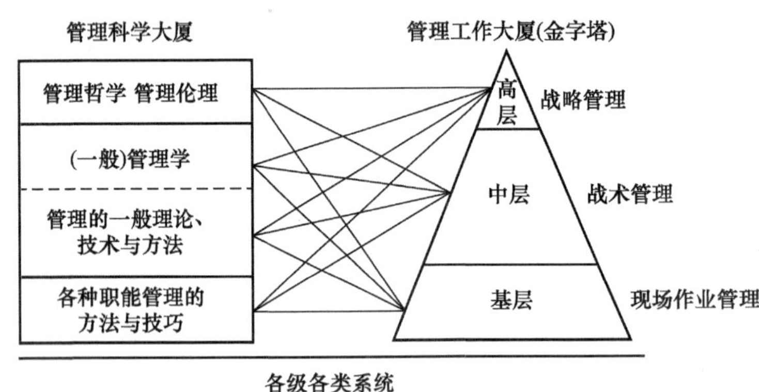

图 3　管理科学大厦与管理工作大厦示意图

管理科学大厦的每一层都有图 2 所示的“三室一厅”的“套房”。

两座大厦是连通的。从管理科学大厦的每一层次都通向管理工作大厦的任何层次，尤其是管理哲学与管理伦理对于所有的管理者来说，都是必不可少的思想指导。优秀的管理者必须具有哲学思维。另外，管理工作每一层次的实践经验加以总结都可以丰富管理科学大厦的内涵，所以，在图 3 中，两座大厦之间的连线都是双向作用的。在图面上，两座大厦是等高的，实际上，在管理层级扁平化的进程中，管理工作金字塔将会有所降低，而管理科学大厦随着管理科学的发展将会有所升高。

六、对于创建工作的建议

创建现代管理科学的中国学派，是一项艰巨而复杂的系统工程，需要千军万马长期作战。我们的建议是：求同存异，积极行动。

求同存异在这里有两重含义：第一，存异；第二，求同。目前学术界存在着一些不同意见，在有些问题上的分歧还比较大，相互之间应该尽量容忍，避免过多的争论，耗费不必要的时间与精力。有些问题需要展开讨论，各抒己见，有一点争论也是可以的，但是一定要心平气和，实事求是，一定要尊重对方，费厄泼赖（fair play）。求同不得，继续存异。“君子和而不同，小人同而不和。”目前创建工作方兴未艾，应该鼓励标新立异，提倡发散思维而不是收敛思维，尤其需要存异而不是求同。

积极行动也有两重含义：第一，自组织行为；第二，他组织行为。自组织行为就是有志者发挥各自的积极性，自由探索。他组织行为就是希望有关部门、有关机构加以适当的组织与引导。我们首先寄希望于基金委。基金委一直是关心和支持管理科学发展的，基金委提出的战略目标我们是非常赞成的。实现战略目标则需要赋予一定的资源，希望在主任基金项目、面上项目、青年基金项目、重点项目乃至重大项目等各方面都鼓励申请有关创建管理科学中国学派的项目。希望有关刊物设立专栏刊登有关文章。《管理学报》已经这样做了，而且发挥了很好的作用。

“求同存异，积极行动”是一个完整的概念。创建现代管理科学的中国学派，需要求真务实，真抓实干，加紧干，早出成果，快出成果。如果光是争论，放松实际的研究工作，不开花不结果，那有什么意义呢?

最后，我们郑重建议：“中国学派”与“中国特色”都是很好的褒义词。不要把它们弄成贬义词。引导中国革命夺取胜利的毛泽东思想，引导中国改革开放获得成功的邓小平理论等，就是马克思列宁主义的“中国学派”——有中国特色的马克思列宁主义。

参考文献

[1] 钱学森，许国志，王寿云. 组织管理的技术——系统工程. 文汇报，1978-09-27.

[2] 国家自然科学基金委员会管理科学部. 管理科学发展战略——暨管理科学“十一·五”优先资助领域. 北京：科学出版社，2006，25，26.

[3] 孙东川，张振刚，孙凯. 一项重大历史使命：创建现代管理科学的中国学派. 美中经济评论，2008，(1)：57-63.

[4] 孙东川，林福永，孙凯. 创建现代管理科学的中国学派及其基本途径研究. 管理学报，2006，3 (2)：127-131.

[5] 孙东川，林福永. 洋为中用，古为今用，近为今用——创建有中国特色的现代管理科学. 系统工程，2004，(增刊)：13-15.

[6] 王克强，孙东川. 管理科学与物理学、医学的分析比较——论创建有中国特色的现代管理科学. 生产力研究，2005，(11)：211-213.

[7] 国务院学位委员会. 授予博士、硕士学位和培养研究生的学科、专业目录（1997). 北京：高等教育出版社，1999，543.

[8] 北京大学现代科学与哲学研究中心. 钱学森与现代科学技术. 北京：人民出版社，2001：183-188，216-221.

[9] 钱学森. 论系统工程（新世纪版）. 上海：上海交通大学出版社，2007：141-143，293-298.

[10] 中国系统工程学会，上海交通大学. 钱学森系统科学思想研究. 上海：上海交通大学出版社，2007.

[11] 孙东川. 中国管理落后与不落后之辨析. 系统工程与管理科学研究（论文集）. 广州：暨南大学出版社，2004，72-77.
[12] 李京文. 创新发展有中国特色的管理科学——兼评“和合管理”. 管理学报，2007，4（2）：141-143.
[13] 成思危. 建立中国特色的管理科学. 煤炭企业管理，2003，（2）：8-11.
[14] 苏东水，彭贺. 中国管理学. 上海：复旦大学出版社，2006，3.
[15] 马庆国. 中国管理科学研究面临的几个关键问题. 管理世界，2002，（8）：105-115，140.
[16] 许广玺. 关于中图现代管理科学“本土化”建设的思考. 考试周刊，2007，（28）：154-155.
[17] 宁玉琼. 建构有中国特色社会主义管理科学的几个认识问题. 广西社会科学，2000，（1）：74-77.
[18] 席酉民，尚玉钒. 和谐管理理论. 北京：中国人民大学出版社，2002.
[19] 黄如金. 和合管理. 北京：经济管理出版社，2006.
[20] 曾仕强. 中道管理——M 理论及其应用. 北京：北京大学出版社，2006.
[21] 曾仕强. 中国式管理. 北京：中国社会科学出版社，2005.
[22] 潘承烈，虞祖尧. 中国古代管理思想之今用. 北京：中国人民大学出版社，2001.
[23] 胡祖光，朱明伟. 东方管理学十三篇. 北京：中国经济出版社，2002.
[24] 许倬云. 从历史看管理. 桂林：广西师范大学出版社，2005.
[25] 黎红雷. 儒家管理哲学. 广州：广东高等教育出版社，1997.
[26] 林国雄. 新儒学经济与管理. 台北：慈惠堂出版社，1997.
[27] 成中英. C 理论：中国管理哲学. 北京：中国人民大学版社，2006.
[28] 杜维明. 儒家传统与文明对话. 石家庄：河北人民出版社，2006.
[29] 雷恩 D. 管理思想的演变. 孔令济，译. 北京：中国社会科学出版社，2000.
[30] 贾顺先. 孔子智慧与 21 世纪人类和平，儒学与当代文明. 国际儒学联合会纪念孔子诞生 2555 周年国际学术研讨会论文集（卷一）. 北京：九州出版社，2005，209.
[31] 周大江. 党史商鉴. 北京：人民出版社，2006.
[32] 郭重庆. 中国管理学界的社会责任与历史使命. 中国科学院院刊，2007，（2）：132-136.

再谈创建现代管理科学中国学派的若干问题①

创建现代管理科学的中国学派，这是中国管理科学界光荣的历史使命。为此，几年来笔者在多种场合宣传和呼吁。2008 年 3 月在西安“管理学在中国”的研讨会上，做出题为“谈谈创建现代管理科学中国学派的若干问题”的大会报告[1]。

通过研究，发现学术界在一些基本概念和基本问题上还缺乏共识，尤其是受中西方文化差异影响较大，妨碍进一步地深入研究，因此有必要做一番正本清源的探讨。

一、现代管理科学中国学派的基本特点及创建工作的基本途径

现代管理科学中国学派的基本特点及创建工作的基本途径，笔者已经做过多次论述[1-6]。作为再谈，有必要复述前面的一些论述——并非简单重复，而是做出一些修改和完善。

1. 现代管理科学中国学派的基本特点

(1) 它是中国的，具有显著的中国特色，它是在中华大地上生长起来的，适合中国国情，能够有效地解决当代中国的经济发展与社会进步问题，而且，中国人应该是创建中国学派的主体。

(2) 它是现代的，运用现代计算机技术和互联网技术，体现世界上最新的管理科学成就。

(3) 它是先进的，博采众长，推陈出新，综合集成。

(4) 它是世界的，具有普适性，是全人类的共同财富，尤其是可以为发展中国家所借鉴。

(5) 它是开放而与时俱进的，所以有强大的生命力，将会不断完善与发展。

这里要特别说明中国人应该是创建中国学派的主体。第一，这是中国管理科学界义不容辞、当仁不让的历史使命，中国管理科学界应该为人类做出杰出的贡献；第二，如果掉以轻心，可能会出现一些不应有的尴尬。例如，下面的一些事情应该引以为鉴、引以为戒。

不要重复这样的悲剧：“敦煌是中国的，敦煌学是日本的。”这是 20 世纪 80 年代中期的一场风波，不管说话者究竟是日本人还是中国人，总之是一个悲剧，与中国在鸦片战争以后的积贫积弱密切相关。

也不要重复这样的“喜剧”：第一部《中国科学技术发展史》的鸿篇巨制是由英国友人李约瑟博士自发地领头编撰的，我们当然要感谢李约瑟博士“替天行道”，为中国、为世界做了一件大好事。但是，如果这件事情是由中国人自己做出来的，或者是以中国人为主导、联合国际朋友一起做出来的，岂不是更好吗？

“出口转内销”：1960 年，中国出现了以“两参一改三结合”（干部参加劳动，工人

---

① 本文原载《中国工程科学》，2008，10（12）：24-31，作者：刘人怀，孙东川.

参加管理，改革不合理的规章制度，工人群众、领导干部和技术员三结合）为主要内容的“鞍钢宪法”，作为政治口号喊了一阵子就销声匿迹了。但是东邻日本人却很在意，他们据此创造出了 Total Quality Control（TQC），20 世纪 80 年代初称为“全面质量管理”被引入中国，发挥了很大的作用，可谓是“出口转内销”。

近年来很无奈的一件事：好莱坞热销大片《功夫熊猫》全部是中国元素，但是片子却不是中国制造，可爱的熊猫为好莱坞的票房价值打工，很令人心酸。

2. 创建现代管理科学中国学派的基本途径是洋为中用，古为今用，近为今用，综合集成

1）洋为中用

改革开放的 30 年，就是西方管理科学洋为中用的 30 年，功不可没。我们应该继续重视洋为中用，目光紧紧盯住国外管理科学的进展，第一时间看到外国的新发明新创造，引进、消化、吸收、改造、创新。外国不仅仅是西方 7 国，也包括世界上其他各国，尤其是印度、巴西、俄罗斯和东方近邻韩国、新加坡等。洋为中用要防止盲目跟风和炒作行为，坚持实事求是，坚持本土化。

2）古为今用

中国古代曾经有过卓越的管理思想和管理实践。中国的文明史从来没有间断过，这在全世界是独一无二的。先秦诸子百家争鸣、百花齐放，留下了丰富的文化遗产。后世学者又持续不断地研究和阐述，其蕴含可以说比山高、比海深。中国大陆学者、中国香港与澳门学者、中国台湾学者以及美籍华裔学者和其他学者在古为今用方面已经做出不少研究，出版了多部专著。但是，还有许多工作要做。

孔夫子的影响是深远的，是全世界的。我国从 2004 年 11 月以来，在全球已经兴办了 260 多所孔子学院，计划到 2010 年建成 500 所。孔子热、汉语热在世界范围内方兴未艾。北京奥运会与残奥会的巨大成功，有力地向全世界宣扬了中国文化，包括现代文化和历史悠久的优秀传统文化。根据历史的演变和现实的存在，古为今用要以儒家为主、兼顾百家；要以近几年的国学热为契机，深入开展典籍研究，系统整理和发掘古代管理思想。

3）近为今用

近为今用是重点，这是笔者提出的理念，其含义是：从我国近期的社会实践中总结经验和教训，上升到理论高度，指导当前和今后的社会实践。

“近”，主要是指 20 世纪和 21 世纪以来这一段时期，历经 100 多年。这个“近”，正是从泰罗制开始至今的西方管理科学发展期。“近”还可以上溯到清王朝后期的洋务运动（始于 1860 年）。从管理科学角度，需要研究和总结三个方面的经验和教训，即中国共产党与中国的革命事业，新中国的建设与改革开放；我国香港、澳门和台湾的发展；海外华人华侨的奋斗与贡献。

近为今用需要消除民族虚无主义情绪，消除“文革”后遗症和逆反心理。近两年愈演愈烈的美国次贷危机、金融海啸以及可能接踵而来的经济衰退（不但是美国的，也拖累了世界上许多国家）给我们上了十分生动的一堂课。在全世界一片风声鹤唳之中，“风景这边独好”——我国的金融管理是独树一帜的，为我国筑起了一道有效的防

火墙。所以，现在事情很清楚：一方面我国的金融管理还要继续改革，另一方面不应该盲目否定我们自己，更不能盲目模仿美国。

近为今用面广量大、头绪多、难度大。近为今用是研究学术问题，不是研究政治问题。

4）综合集成（meta-synthesis）

综合集成是系统工程术语，著名科学家钱学森院士提出了综合集成方法论[6]。对于创建现代管理科学的中国学派可以打一个比喻，“洋为中用”“古为今用”“近为今用”不是三盘子菜，也不是一个大拼盘，而是要烹调为一道又一道美味的佳肴，办成一桌又一桌丰盛的筵席，让广大的中外“食客”普遍满意和赞赏。

综上所述，笔者提出“三室一厅”的工作方案，如9.1节图2所示。“三室”是指洋为中用、古为今用、近为今用等三个研讨室，“一厅”即钱学森院士倡导的综合集成研讨厅[7]，把“三室”中的内容汇集起来反复研讨，从定性到定量综合集成。

二、什么是管理　什么是管理科学——正本清源的探讨

什么是管理？什么是管理科学？这是两个最基本的问题，也是很有争议的问题，众说纷纭，仁者见仁，智者见智。通过对这两个问题进行专门研究，笔者的观点是：管理活动是人类的第二类活动，它为第一类活动（作业活动）服务；管理科学是研究管理活动规律与做好管理工作的知识体系。

目前我国的管理科学界具有很深的西方烙印，主要是美国的烙印。中西方之间、中英文之间对于“管理”的定义是有很大差别的。第一，汉语的“管理”一词，对应于英语的多个单词：management、administration、run、supervision、governance 等，这些英语单词不能互相覆盖，任何一个单词都不能完全替代别的单词。有人认为 management 的含义比较宽，可以作为其他单词的代表，与汉语“管理”相对应。这种说法恐怕站不住脚。例如，著名的专业学位工商管理硕士 MBA（master of business administration）与公共管理硕士 MPA（master of public administration），用的都不是 management。第二，汉语“科学”一词及其对应的英语单词 science 都有广义与狭义之分。science 的本义是知识、知识体系，是广义的，可以是自然科学，也可以是社会科学与人文科学。后来变得狭义了，成为自然科学的代名词，甚至成为“真理”“正确”的代名词。汉语的“科学”也是这样。

笔者主张从“管理”与“科学”的广义上来界定“管理科学”，即管理科学是关于人类第二类活动——管理活动（含管理工作）的知识体系，它研究管理活动规律和做好管理工作的理论与方法，包含所有一切有关管理活动和管理工作的理论知识与实践经验。管理科学是一个内容丰富的体系，其结构如9.1节图3所示，左边是管理科学体系的理论部分，右边是管理科学体系的实践部分——管理工作。

英语的 management science（MS）译为“管理科学”，似乎无可厚非，其实两者的口径是大不一样的，不能等同视之，来回翻译。MS 注重于建立数学模型、进行计算与定量研究，它大体上相当于运筹学（operations research，OR）。这是一种客观存在，无法否认。既然对“管理科学”作广义的界定，就不能把它翻译为 management science（MS）。

图 1 比较了中西方的管理与管理科学概念，分别以中文与英文表示，中西方文化差异很显著：管理>management>MS，管理科学>MS。

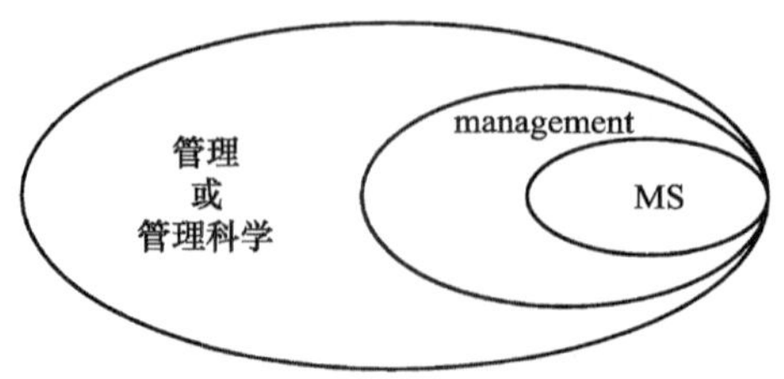

图 1　中西方文化差异

鉴于上述理由，笔者建议直接采用汉语拼音 guanli 作为“管理”对外语的翻译——首先是对英语的翻译，那么，“管理科学”则为 guanli science（GS）。GS>MS，GS 包含 MS。现代管理科学的中国学派可以翻译为 The Chinese School of Modern Guanli Science（CSMGS）。建议把 management science（MS）称为“狭义的管理科学”或“管理的数量方法”，避免与管理科学（GS）混淆。应该积极地把 guanli 推向世界，让外国人知道 guanli，熟悉 guanli，讲 guanli，写 guanli。这其实是把中国的管理理念和管理科学推向世界，不光是一个单词的事情。

现在，“关系”的拼音 guanxi 已经成为新的英语单词了，尽管英语有单词 relation，但是，英语界认为 relation 不足以表达汉语“关系”一词的含义，只有 guanxi 才能表达“关系”。好莱坞热销大片 KUNG FU Panda（《功夫熊猫》）也是一个有力的例证（严格地说，“功夫”的英文 KUNG FU 是有读音缺陷的，应为 Gongfu，但是，美国人采用他们习惯的拼音法，我们唯有“笑纳”）。而 guanli（管理），比 guanxi 或 gongfu 要重要得多。

9.1 节图 3 所界定的管理科学，消除了管理究竟是科学还是艺术之类的质疑。管理科学作为研究管理活动规律与做好管理工作的知识体系，它不但包含了狭义的管理科学，也包含了其之外的管理艺术以及其他用于管理工作的各种知识。

笔者还要郑重提出一条建议：中国学派与中国特色都是很好的褒义词，不要把它们弄成贬义词。引导中国革命获得胜利的毛泽东思想，引导中国改革开放获得成功的邓小平理论等，就是马克思列宁主义的中国学派——有中国特色的马克思列宁主义。

三、在工程管理领域开展中国学派创建工作的试验性研究

工程管理的中国学派（或者有中国特色的工程管理，或者工程管理的中国特色）是很值得梳理和研究的。国外有很多典型案例，我国古代和近现代也有很多典型案例，可以总结出很多成功的经验和失败的教训。

1. 洋为中用的若干典型案例

1）苏联的案例

（1）世界上第一颗人造地球卫星：1957 年 10 月 4 日成功发射，表明人类开始了航天时代。

(2) 世界上第一艘宇宙飞船：1961 年 4 月 12 日成功发射世界上第一艘绕地球轨道飞行的载人飞船“东方 1 号”，宇航员是尤里·加加林少校。

2) 美国的案例

(1) 曼哈顿工程：第二次世界大战后期研制原子弹的工程项目，从技术和工程的角度看，它是很成功的。

(2) 北极星导弹计划：1957～1960 年，美国海军研制导弹核潜艇系统的工程项目，为了加快进度，发明了 PERT (program evaluation and review technique)，即计划协调技术，在我国又称为统筹法。

(3) Apollo 登月计划：1961～1972 年，耗资 300 多亿美元的工程项目；1969 年 7 月 20 日，Apollo-11 号实现人类首次成功登月。有一个很中肯的评论：“两位宇航员在月球上迈出了一小步，人类在历史上迈出了一大步。”后来发射的 Apollo 飞船又多次载人登月，进行科学考察与试验。

2. 古为今用的若干典型案例

中国古代杰出的工程实践很多，如都江堰、万里长城、京杭大运河、丁谓工程、北京故宫等。下面简单说明都江堰和丁谓工程。

都江堰是公元前 256 年前后，秦太守李冰及其儿子李二郎率领当地群众修筑起来的。它有鱼嘴岷江分水工程、飞沙堰分洪排沙工程、宝瓶口引水工程三大主体部分，加上一系列灌溉渠道网巧妙地结合，形成一个完整的系统。并且李冰父子总结了“深淘滩，低作堰”六字口诀，指导人们进行养护维修工作。两千多年以来，都江堰一直造福于四川人民。都江堰主体工程在汶川大地震中安然无恙，只是附属建筑物，如二王庙有些损伤，不难修复。

丁谓工程不是很出名，但是很典型。宋真宗时期（公元 998～1022 年），有一次宫殿被毁于火。大臣丁谓受命限期修复皇宫。这项工程怎样进行？丁谓提出：首先把皇宫旧址前面的一条大街挖成沟渠，用挖沟的土烧制砖瓦；其次把附近的汴水引入沟内形成航道，从外地运输砂石木料；最后待皇宫修复后，把沟里的水排掉，用建筑垃圾填入沟中，恢复原来的大街。这是一个杰出的方案，它把皇宫修复全过程划分成几个阶段，统筹兼顾，综合考虑。

3. 近为今用的若干典型案例[8]

(1) 两弹一星。刚刚度过三年困难时期的中国，1964 年 10 月 16 日，第一颗原子弹成功试验；仅仅过了两年零八个月，尽管全国陷入“文革”浩劫，1967 年 6 月 17 日，第一颗氢弹成功试验；1970 年 4 月 24 日，第一颗人造地球卫星发射升空，全球响彻悦耳的《东方红》乐曲。先后 5 年半时间，中国在一穷二白的情况下，把这些惊天动地的大事办成了。

(2) 航天事业。1999 年 11 月 20～21 日，中国第一艘“神舟”无人试验飞船飞行试验获得了圆满成功；2001 年初至 2002 年底又相继研制并发射成功了神舟二号、神舟三号和神舟四号无人试验飞船，获得了宝贵的试验数据，为实施载人航天打下了坚实的基础；2003 年 10 月 15 日，神舟五号飞船成功发射升空，飞船乘有 1 名航天员杨利伟，在轨运行 1 天，环绕地球 14 圈后在预定地区着陆，实现了中华民族千年飞天的愿望；2005年10月12日上午9：00，神舟六号飞船成功发射升空，费俊龙和聂海胜两名中国航天员被送入太空，飞行5天，成功返回地面；2007年10月24日18时05分，嫦娥一号探测器成功发射，卫星发射后，经过8次变轨，于11月7日正式进入工作轨道，11月18日卫星转为对月定向姿态，11月20日开始传回探测数据，11月26日，中国国家航天局正式公布嫦娥一号卫星传回的第一幅月面图像；神舟七号今年9月成功发射，航天员翟志刚在浩瀚太空中成功出舱活动（舱内还有2名航天员刘伯明、景海鹏，详细情况大家记忆犹新，这里从简）；夸父计划——世界上唯一的日地空间探测计划于2012年开始……

（3）三峡工程。兴建三峡工程是近一个世纪以前孙中山先生的理想。1992年开工，2009年竣工，总工期17年。最终投资总额预计在2000亿元左右。三峡工程的综合效益主要有：防洪（可以有效阻挡百年一遇的大洪水）、发电（装机224×10^5千瓦，年发电846.8×10^8千瓦·时，可以照亮大半个中国）、航运（能够较为充分地改善重庆至武汉间通航条件）。另外，在养殖、旅游、保护生态、净化环境、开发性移民、南水北调、供水灌溉等方面均有巨大效益。

（4）青藏铁路。青藏铁路是我国在21世纪的战略决策，是西部大开发的标志性工程，对加快青藏两省区的经济、社会发展，增进民族团结，造福各族人民，具有重要意义。青藏铁路由西宁至拉萨，全长1956千米；2001年6月29日开工，2006年7月1日全线通车。青藏铁路是目前世界上海拔最高的铁路，沿线常年平均气温在0摄氏度以下，空气中的含氧量仅为平原地区的一半。铁路穿越海拔4000米以上的地段为960千米，其中翻越唐古拉山最高点海拔达到5072米。铁路建设克服了多年冻土、高寒缺氧、生态脆弱等三大难题。

汶川大地震之后的抗震救灾工作、北京奥运会与残奥会的组织工作，也可以作为工程管理的典型案例（或包含着工程管理的社会系统管理的典型案例）。

4. 钱学森院士与系统工程

美国的工程管理具有美国特色，中国的工程管理更加具有中国特色。中国工程院院士郭重庆教授说：一个奥本海默，一个邓稼先，分别把美国的原子弹和中国的原子弹搞成了，如果把他们两个人互换位置，恐怕是两个都搞不成的。

著名科学家钱学森院士为“两弹一星”做出了巨大贡献，他从中总结出“总体设计部”的思想和工作方式——工程管理或工程系统工程的工作模式。

钱学森院士指出：“我们把处理开放的复杂巨系统的方法定名为从定性到定量综合集成法，把应用这个方法的集体称为总体设计部。总体设计部由熟悉所研究系统的各个方面专家组成，并由知识面比较宽广的专家负责领导，应用综合集成法（或综合集成研讨厅体系）对系统进行总体研究。总体设计部设计的是系统的总体方案和实现途径。总体设计部把系统作为它所属的更大系统的组成部分来进行研究，对它们所有要求都首先从实现这个更大系统相协调的观点来考虑；总体设计部把系统作为若干分系统有机结合的整体来设计，对每个分系统的要求都首先从实现整个系统相协调的观点来考虑，对分系统之间、分系统与系统之间的关系，都首先从系统总体协调的需要来考虑，进行总体分析、总体论证、总体设计、总体协调、总体规划，提出具有科学性、可行性和可操作性的总体方案。”[9]

在总体设计部的思想和实践的基础上，在借鉴美国出现的新兴学科系统工程（systems engineering，SE）的基础上，1978年9月27日，钱学森、许国志、王寿云联名在上海《文汇报》发表重要文章“组织管理的技术——系统工程”。文章说：“系统工程是组织管理系统的规划、研究、设计、制造、试验和使用的科学方法，是一种对所有系统都具有普遍意义的科学方法。”30年来，系统工程的中国学派已经形成，在世界上别具一格。

钱学森院士非常重视研究系统科学体系和现代科学技术体系。他指出：现代科学技术体系具有许多部门，如自然科学、社会科学、系统科学、地理科学、军事科学等。在每一部门中都有4个台阶：直接与改造客观世界的实践活动相联系的是工程技术；稍微远离工程实践的是工程技术的理论基础技术科学；再远一些的是这一部门科学技术的基础科学；基础科学再经过一座过渡的桥梁与马克思主义哲学相联系[10]。

1981年和1985年，钱学森院士两次描绘系统科学体系——它是现代科学技术体系中的一个部门。它包含4个台阶——或称4个层次，如9.1节图1所示。9.1节图3左半部，也是类似的4个层次。

5. 北京的城市建设

北京故宫又称紫禁城，是世界上最大、最完整的古代宫殿建筑群。始建于明永乐四年（1406年），建成于永乐十八年（1420年）。明清两朝皇帝在这里君临天下。

北京故宫显著地区别于法国的凡尔赛宫、俄罗斯的克里姆林宫。故宫有一条贯穿宫城南北的中轴线，整个布局端庄大方，气势恢宏，十分壮美，具有显著的中国特色。

北京老城区为很规则的矩形，以故宫中轴线为严格对称：东单—西单，日坛—月坛，东直门—西直门，等等。新中国开辟了天安门广场，广场东西两侧建设了中国历史博物馆和人民大会堂，南面建设了人民英雄纪念碑、毛主席纪念堂，仍然保持中轴线对称；为了举办2008年奥运会和残奥会，中轴线往北延伸，两侧新建了奥运会场馆鸟巢和水立方。两个奥运会已经成功举办，高水平、有特色，举世称赞。

尽管加进了许多现代元素，北京城在整体上仍然保持着南北中轴线对称的中国特色，体现了一种中国模式。但是，现在看来，对于半个多世纪以来的北京城市建设应该进行一些反思。

（1）拆除城墙的利弊与得失如何？

（2）作为首都，也一度大力发展工业生产，近郊（现在已经成为市区，如石景山地区）发展重化工业，其利弊得失如何？

（3）缺水问题如何解决？南水北调对南方会不会有负面影响？

（4）大气污染和春天的风沙如何迅速根治？

（5）如何避免首都城市继续膨胀、消除交通堵塞……

6. 中国工程管理的若干特色

根据上面列举的案例，笔者总结出以下几点。

（1）集中力量办大事，唯有如此，大事才能办得成。

（2）办大事必须以政府行为和计划机制为主导，不但中国如此，美国也是如此。

(3) 必须要有正确的战略决策，战略决定方向、决定成败，要通过科学化、民主化的决策程序来避免错误的决策。

中国为什么能在一穷二白的条件下搞出“两弹一星”？首先是因为当时有毛主席和所有中央领导人的坚强决心和英明决策，同时，有一批顶尖科学家具有使命感的追求，有全体参加人员的忘我奉献，上上下下齐心协力。“两弹一星”的研制成功具有十分重大的意义。邓小平同志指出：“如果60年代以来中国没有原子弹、氢弹，没有发射卫星，中国就不能叫有影响的大国，就没有现在这样的国际地位。这些东西反映一个民族的能力，也是一个民族、一个国家兴旺发达的标志。”

据报道，美国航天局负责人不久前说：“Apollo登月之后，美国停止探索月球是一个重大失误。为了避免被中国超过，美国要重返月球。”媒体上还有一种说法：如果美国不是中止登月30年，今天的月球恐怕早已成为美国的“第51个州”了，别的国家没有插足的余地——姑妄言之，姑妄听之。

(4) 宏观着眼，微观着手。“战略上藐视敌人，战术上重视敌人”，毛主席的这句名言在工程管理上也是非常适用的。

(5) 精心设计，精心施工，精心运作，精心管理。每一个细节都可能影响成败，工程管理一环扣一环，不容许任何环节马虎大意，不能放过一个螺丝钉；火箭发射要实行倒计时，以便高度集中大家的注意力（也不排除在最后一秒钟叫停）。

(6) 统筹兼顾、综合集成。Apollo计划的总指挥韦伯说：“Apollo计划中没有一项新发明的自然科学理论和技术，而是现成技术的应用，关键在于综合，综合就是创造。”我们说，综合就是创造，综合也是创新。每一个大型工程项目，都是多种技术的综合集成，技术与管理的综合集成，技术、经济、环境、政治等多种因素的综合集成。

(7) 工程管理需要“总体设计部”或“工程指挥部”。

(8) 工程问题不光是技术问题，工程系统存在于社会经济系统之中，所以，工程管理是社会经济系统管理的组成部分。

(9) 人民子弟兵在工程建设中发挥了巨大的作用，中国人民解放军的战斗作风在工程管理中发挥了巨大的作用。从“两弹一星”到汶川大地震的抗震救灾工作，全世界都是有目共睹的。

(10) 工程建设在追求正面效应的同时，应该认真研究并尽量避免负面效应，对于可能出现的负面效应要有切实可行的防范措施。负面效应往往需要很长的时间才能显露出来。例如，围湖造田、三门峡水利工程，其负面影响现在尽人皆知，但当初是缺乏认真研究的。现在有不少人担心三峡工程、南水北调，有没有负面效应？现在还看不出来，有待于时间的考验。负面效应一旦出现，应该及时采取积极措施化解。

以上各点，与美国等国比较，有一些共性，但更多的是特性，即体现了中国特色——中国共产党领导的社会主义的特色。

四、创建现代管理科学是重大历史使命

创建现代管理科学的中国学派是当代中国人尤其是中国管理科学界的重大历史使命。现代管理科学的中国学派也可以称为有中国特色的现代管理科学。创建中国学派，

现在正当其时，应该不失时机地抓紧抓好。

管理科学是广义的，它是一个内容丰富的知识体系，具有类似著名科学家钱学森院士提出的系统科学体系结构，它包含管理学。西方的 management science（MS）是狭义的，不能等同于中国的管理科学；西方的 management 或者 administration 等单词的任何一个也不能等同于中国的“管理”一词。鉴于中西方文化差异及其造成的国内学术界的分歧，笔者建议采用“管理”的汉语拼音 guanli 作为对外翻译，则“管理科学”可以翻译为 guanli science（GS），GS 包含 MS；“管理学”可以翻译为 guanliology，“现代管理科学的中国学派”可以翻译为 the Chinese school of modern guanli science（CSMGS）。

创建现代管理科学中国学派是一项艰巨复杂的系统工程，需要千军万马长期作战。创建的基本途径是洋为中用、古为今用、近为今用、综合集成，其中近为今用是重点。

让我们大家共同努力，创建现代管理科学的中国学派。争取创建得好一些、快一些，出色地完成我们的历史使命，为中国和世界做出应有的贡献！

参考文献

[1] 刘人怀，孙东川．谈谈创建现代管理科学中国学派的若干问题．管理学报，2008，5（3）：323-329.

[2] 孙东川，张振刚，孙凯．一项重大历史使命：创建现代管理科学中国学派．美中经济评论，2008，(1)：57-63.

[3] 孙东川，林福永，孙凯．创建现代管理科学的中国学派及其基本途径研究．管理学报，2006，3（2）：127-131.

[4] 孙东川，林福永．洋为中用，古为今用，近为今用——创建有中国特色的现代管理科学．系统工程，2004，(增刊)：13-15.

[5] 王克强，孙东川．管理科学与物理学、医学的分析比较——论创建有中国特色的现代管理科学．生产力研究，2005，(11)：211-213.

[6] 国务院学位委员会．授予博士、硕士学位和培养研究生的学科、专业目录（1997）．北京：高等教育出版社，1999，543.

[7] 钱学森．创建系统学（新世纪版）．上海：上海交通大学出版社，2007.

[8] 常平，刘人怀，林玉树．20 世纪我国重大工程技术成就．广州：暨南大学出版社，2002.

[9] 钱学森．论系统工程（新世纪版）．上海：上海交通大学出版社，2007，141-143，293-298.

[10] 许国志．系统科学．上海：上海科技教育出版社，2000.

三谈创建现代管理科学中国学派的若干问题：四条定义与三点建议①

在文献［1］的基础上，旨在更深入、更系统地探讨什么是管理，什么是管理科学，以及与之相关的若干基本概念与基本问题。它们常常被忽略，并且有些概念现在大有争议，有些争议与中英文语义差异和中西方文化差异有关，有些争议与一些人的思想方法有关。所以，这些概念是值得探讨的，否则，管理科学大厦的基础是不扎实、不牢固的。现在，越来越多的人士谈论创建现代管理科学的中国学派[1-4]或中国管理学等[5-8]，如果对一些基本概念与基本问题不做一番深入的系统的研究，不作出合理的界定，后续工作是难以进行的。笔者将引用和延伸此前已经发表的一些观点，所以在文字上会有一些重复[1,4]，否则会造成不连贯、不系统，阅读上也会有所不便，这是希望读者能够理解的。

一、什么是管理

什么是管理？从不同的角度有各种各样的说法。例如，M1 管理就是通过别人把事情做好；M2 人人都是管理者，人人都是被管理者；M3 管理的职能有计划、组织、领导、控制；M4 管理就是决策；M5 管理就是优化配置资源；M6 三百六十行，行行有管理；M7 现代经济是一辆自行车，技术与管理是它的两个轮子；M8 三分技术，七分管理等。

M1 是很经典的一句话，到处都在引用。这句话说明管理的主体是人（管理者），客体是人（被管理者）和各种事情（其中含有物）。但是这句话把管理者与被管理者截然分开了，其实，他们是不能截然分开的，M2 可以弥补它。M3 说明了管理的职能，这是“四职能说”，还有“五职能说”“七职能说”等。诺贝尔经济学奖获得者 H. 西蒙特别强调管理的决策职能，M4 是他的高论，人们因此把他称为管理的决策学派。M5 当然也有道理，但是有见物不见人之虞。M6 说明了管理的普遍性。M7，M8 强调管理的重要性，在改革开放初期，这两句话发挥了巨大的作用，促使上上下下重视管理工作。

还可以列举很多说法。各种说法都具有一定道理，仁者见仁、智者见智。但是有一种说法笔者是不敢苟同的，即管理学是应用经济学的分支。经济固然需要管理，社会更需要管理，那么岂不是更加有理由说管理学是社会学的分支吗？教育也需要管理，医疗卫生事业也需要管理，依此类推可以说管理学是教育学的分支，是医学的分支等，这等于什么都没有说。或者只是说明了一个共同的现象，各行各业都有管理工作，都离不开管理，管理学有许多分支，这就自动否定了管理学是应用经济学的分支这个命题。笔者认为，M1～M8 的综合性都不够，也没有揭示管理活动的起源和本质，难以作为管理术语的基本界定，因此提出以下见解。

---

① 本文原载《中国工程科学》，2009，11（8）：18-23，63，作者：刘人怀，孙东川，孙凯.

二、管理活动与管理科学的四条定义

文献［1］提出，管理活动是人类的第二类活动，它为第一类活动（作业活动）服务；管理科学是研究管理活动规律与做好管理工作的知识体系。这两句话比较简略，笔者把它展开为四条定义，并做出必要的论述。

定义 1：管理活动是人类的第二类活动，它为第一类活动服务。

人类的全部活动可以分为两大类[3]，第一类是作业活动（记为活动Ⅰ），第二类是为作业活动服务的管理活动（记为活动Ⅱ）。第二类活动是基于第一类活动而产生的。

作业活动可以分为两类，生活的作业活动（记为Ⅰ-1），生产的作业活动（记为Ⅰ-2）。相应的管理活动也分为两类，对于生活作业的管理活动（记为Ⅱ-1）和对于生产作业的管理活动（记为Ⅱ-2）。人类的生活，包括物质生活和精神生活（其活动分别记为Ⅰ-11，Ⅰ-12）。物质生活的作业包括衣食住行、体育活动、婚姻、生儿育女等；精神生活更加丰富多彩，不但包括亲情、文化娱乐、休闲、旅游、探险，而且包括学习、科学研究、理想与信仰等。这些作业活动都需要加以管理。生产包括物质产品和精神产品的生产（其活动分别记为Ⅰ-21，Ⅰ-22），如农民种庄稼、工人开机床、科学家做试验、文学家写小说、音乐家演奏等，形成人类社会中的各行各业。各行各业的生产活动也需要加以管理。管理活动使得作业活动有条不紊地进行，提高效率和效益，促进和谐。相对于作业活动的细分，管理活动可以细分为Ⅱ-11，Ⅱ-12，Ⅱ-21，Ⅱ-22，两类活动都可以继续细分下去。有些活动是跨类别的，有些体育活动是物质活动特征明显（主要是体力活动），如田径运动和球类运动；有些体育活动是精神活动特征比较明显（主要是脑力活动），如棋类比赛。

在人类社会中，群体以组织的形式存在。有人类就有作业活动，有作业活动就有管理活动，管理活动的历史与人类一样久远。M1～M8 都可以作为定义 1 的演绎。作业活动有个人的与群体的，管理活动也有个人的与群体的。

定义 2：组织的管理活动称为管理工作，这是组织中的一大类工作。

定义 3：组织委派某些人员专门从事管理工作，这些人员就成为管理工作者，简称管理者。

管理工作的基本宗旨是服务，即第一类活动服务，为员工服务，为用户服务，承担社会责任。图 1 表现了两类活动的密切关系，即管理活动是为作业活动服务的。表 1 说明了人类全部的活动与分类。

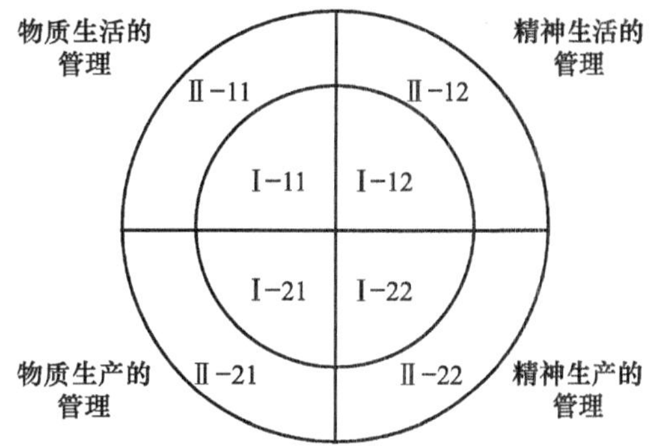

图 1 人类的两类活动及其相互关系

下面把人类的活动与哺乳动物的活动作一番比较。哺乳动物也具有作业活动Ⅰ-11（吃、喝、拉、撒、睡、运动、繁殖等），与人类相比一样不少，区别在于哺乳动物的作业活动只是简单的初级形态，人类的作业活动是复杂的高级形态。兔子吃草，有草就吃，否则就挨饿；老虎捕食其他动物，也是有东西就吃，否则就挨饿。它们都是简

表 1　人类的活动与分类

<table>
<tr><td colspan="4">第一类活动Ⅰ：作业（逐层次往下细分）</td></tr>
<tr><td colspan="4">Ⅰ-1：生活</td></tr>
<tr><td colspan="2">Ⅰ-11：物质生活</td><td colspan="2">Ⅰ-12：精神生活</td></tr>
<tr><td>Ⅰ-111：个人</td><td>Ⅰ-112：群体</td><td>Ⅰ-121：个人</td><td>Ⅰ-122：群体</td></tr>
<tr><td>Ⅱ-111</td><td>Ⅱ-112</td><td>Ⅱ-121</td><td>Ⅱ-122</td></tr>
<tr><td colspan="2">Ⅱ-11：物质生活管理</td><td colspan="2">Ⅱ-12：精神生活管理</td></tr>
<tr><td colspan="4">Ⅱ-1：生活管理</td></tr>
<tr><td colspan="4">Ⅰ-2：生产</td></tr>
<tr><td colspan="2">Ⅰ-21：物质生产</td><td colspan="2">Ⅰ-22：精神生产</td></tr>
<tr><td>Ⅰ-211：个人</td><td>Ⅰ-212：群体</td><td>Ⅰ-221：个人</td><td>Ⅰ-222：群体</td></tr>
<tr><td>Ⅱ-211</td><td>Ⅱ-212</td><td>Ⅱ-221</td><td>Ⅱ-222</td></tr>
<tr><td colspan="2">Ⅱ-21：物质生产管理</td><td colspan="2">Ⅱ-22：精神生产管理</td></tr>
<tr><td colspan="4">Ⅱ-2：生产管理</td></tr>
<tr><td colspan="4">第二类活动Ⅱ：管理，对第一类活动提供服务（逐层次往上细分）</td></tr>
</table>

注：Ⅰ-111 为个人物质生活的作业；Ⅱ-111 为个人物质生活的管理，其余依此类推

单地向大自然索取，不需要洗涤，更不需要烹饪。人类的吃则非常讲究，讲究卫生、烹饪，有生吃有熟吃，食不厌精，脍不厌细，形成一套又一套吃的文化或餐饮文化。设想在野外看到一棵硕果累累的桃树，你会选择一个又大又熟的桃子，摘下来，擦拭又擦拭，如果有条件还要洗涤，去皮，然后才把桃子吃下去，猴子可不是这样，它是摘下来就咬的。再设想一只猴子到了大富豪的宴会上，大概也只是胡乱抓一点水果吃吃，什么山珍海味，它是不理会的。哺乳动物只具有极其少量的精神生活Ⅰ-12，如母子情，发情期的异性追求与调情，人类则有家庭温暖与天伦之乐，有梁山伯与祝英台式的爱情故事，更有文化娱乐和理想信仰等。全部生产活动Ⅰ-21 与Ⅰ-22 都是人类所独有的，某些哺乳动物具有极其少量的、可以忽略不计的Ⅰ-21，如松鼠储藏过冬的食物（老虎似乎不会），Ⅰ-22 则是没有的。管理活动Ⅱ-1，不能说哺乳动物完全没有，如猴群有猴王，羊群有头羊，猴王和头羊也可以看成管理者，它们对于自己的群体进行很少的原始的管理工作，这种管理工作相对于人类而言只能勉强算作“小儿科”，可以忽略不计。管理活动Ⅱ-2，哺乳动物是无从谈起的。因为没有管理活动，所以，哺乳动物的作业活动谈不上什么秩序、效率、效益、和谐等。因此，可以说第二类活动是人类所独有的，哺乳动物基本没有。正因为人类具有丰富多彩的第一类活动Ⅰ-12 和复杂多样的第二类活动Ⅱ-21 与Ⅱ-22，人类才得以超脱动物界。但是人类不能过分骄傲，因为即便是一些低等动物也有“过人之处”，如蜂房结构和蚂蚁社会就令学者惊叹不已。仿生学的研究使人类从生物界学习到许多知识来制造器具。

这里要专门说一下科学研究活动。科学研究活动既可以是科学家的生活作业（研究、发明、创造成为执着的科学家生活的组成部分），也可以是科学家的生产作业；既可以是个人行为，也可以是群体行为。艺术创作活动也是这样。

定义 4：管理科学是研究管理活动规律与做好管理工作的全部知识的总和，是一个内容丰富的知识体系。

相对于狭义的管理科学（management science，MS），这里说的是广义的管理科学。文献［1］已经用文字和图示说明了管理科学的体系结构，它包括两大部分，即管理科学的理论部分和实践部分（即管理工作），都是具有层次结构的。

三、管理工作具有基本性与普遍性

管理工作具有基本性与普遍性。管理工作的基本性是由管理活动的基本性所决定的，因为管理活动是人类的第二类活动。管理工作的普遍性是由它的基本性决定的。前面的 M6 和 M2 已经描述了管理工作的普遍性，即事事有管理，人人搞管理。管理无处不在，无时不在。各种作业活动与管理活动都具有共性与个性。如果说，各种作业活动是个性比较突出而显得千差万别的话，那么，各种管理工作是共性比较突出而显得大同小异。例如，企业与军队是两种不同的组织，在结构上与功能上有很大的差别，两者的作业活动是大不一样的，但是企业管理与军事管理的差别则要小得多，企业管理从军事管理中吸取了很多营养。例如，统筹法、运筹学（又称计划协调技术、计划评审技术）和物流管理，都是直接由军事领域产生与发展起来，然后移植到经济领域和其他领域的管理工作中。

管理工作离不开信息，信息流驱动物质流和人员流，支配作业活动和管理活动本身，使得一个组织协调运行。信息也需要管理，称为信息管理；为了有效地开展信息管理，需要建立先进的管理信息系统，信息管理和管理信息系统是管理科学研究的重点领域。在所有一切管理工作中，信息管理更加具有基本性与普遍性，管理信息系统是最基本的管理工具。计算机与信息网络的出现和发展，与信息管理是互相推动、相得益彰的。

从系统工程的角度看，管理问题都是系统问题（系统之中的问题，或者具有系统性的问题），管理工作都是系统工程。必须用系统工程的理论与方法来研究与求解管理问题，才能做好管理工作。

四、中国的管理科学研究要走出中西方文化差异造成的困境

1. 困境是什么

管理学与管理科学两个概念究竟是什么关系，谁包含谁？现在有两种截然相反的说法，文献［1，4］已经作了介绍，这里再以图 2、图 3 加以说明。在图 2 中，(a) 表示管理科学包含管理学，(b) 表示管理学包含管理科学。主张 (b) 的主要依据是管理科学对应的英语是 management science（MS），MS 强调建立数学模型进行定量研究，是很狭义的，大体上相当于 OR（operations research）。

笔者主张图 2 (a)，有两条重要的理由。理由之一是对比社会科学与社会学，以及钱学森院士提出的系统科学与系统学[9]，如图 3 所示。钱学森院士等还主张把地理学扩展为地理科学，把建筑学扩展为建筑科学，都是符合大科学的理念的。军事学已经扩展为军事科学了。

笔者的理由之二是西方的 MS 与我们的管理科学是不对等的，管理科学的口径比 MS 大得多。

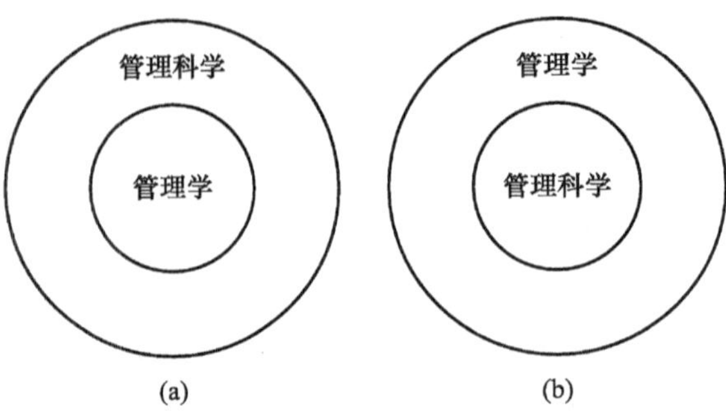

图 2　两种理解：管理科学与管理学的关系

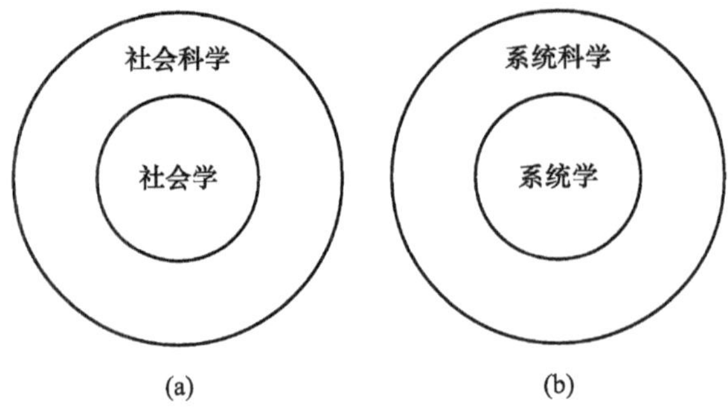

图 3　社会科学与社会学、系统科学与系统学的关系

首先，汉语管理一词的含义比 management 要大得多。除 management 之外，管理还对应其他若干英语单词，是“一对多”的情况。请看几种比较权威且当前比较通行的词典的解释（管理既是动词也是名词，英语相应的几个单词也是这样）。

精选英汉汉英词典（*CONCISE English-Chinese Chinese-English Dictionary*），第三版，商务印书馆，牛津大学出版社，2005：管理 guanli（动）manage；administer；run；～部门 administrative office；企业～business management。

新时代汉英大词典（*NEW AGE CHINESE ENGLISH DICTIONARY*），吴景荣，程镇球主编，商务印书馆，2006 年，北京：管理 guanli①manage；administer；run；supervise：～生产 manage production/～企业 run（or manage）an enterprises/～侨务 administer affairs concerning overseas Chinese/市场～supervise a market/～国家大事 administer（or be in charge of）state affairs/经济～economic administration/商业～business management（or administration）②take care of；look after：～宿舍 look after the dorm/～仓库 take care of the warehouse ③watch over；tend；look after：～犯人 watch over prisoner；…

在这些引述中，管理以一对多的情况很清楚。此外，control 一词也有管理的含义。

其次，management，administration，run，supervision，control 这些词是不能互相覆盖的，其中任何一个词都不能完全替代别的词。有人说 management 用得比较普遍，有代表性。其实不然，著名的专业学位 MBA（master of business administration）与 MPA（master of public administration），用的都不是 management。

牛津高阶英汉双解词典（*OXFORD ADVANCED LEARNER' S English-Chinese Dictionary*），第6版，商务印书馆，牛津大学出版社，2004：management 1 [U] the act of running and controlling a business or similar organization 经营；管理 2 [C+sing./pl. v.，U] the people who run and control a business or similar organization 经营者；管理部门；资方 3 [U] the act or skill of dealing with people or situation in a successful way（成功的）处理手段；（有效的）处理能力。

administration 1 [U] the activities that are done in order to plan，organize and run a business，school or other institution（企业、学校等的）管理，行政 2 [U] the process or act of organizing the way that sth is done 施行；执行 3 [C] the people who plan，organize and run a business，institution，etc.（企业、机构等的）管理部门，行政部门 4（often administration）[C] the government of a country，especially the US（尤指美国）政府 5 [U]（formal）the act of giving a drug to sb（药物的）施用。

两个英语单词 management 与 administration 各有其适用范围，看不到它们互相覆盖或互相替代的情况。单词 run，supervision 等从略。

最后，科学及其对应的 science 也有广义与狭义之分。从广义（也是本义）来说，science 来源于拉丁文 scientia，意为知识，任何系统化的知识都可以称为 science。science 一词在 narural science 出现之前早就出现了。汉语原来没有科学一词，是大约 100 年前引进的日语对于 science 的翻译。曾经一度出现科学与格致（格物＋致知）两词并存的局面。狭义的 science 或科学仅仅是指自然科学，并且蕴含着真理、正确等含义。

牛津高阶英语词典第 7 版（牛津大学出版社，2007）对 science 的解释是：1. Knowledge about the structure and behaviour of the natural and physical world，based on facts that you can prove，for example by experiments；2. The study of science；3. A Particular branch of science；4. A system for organizing the knowledge about a particular subject，especially one concerned with aspects of human behaviour or society：life sciences，natural science，political science，social science。

其中，解释 4 可以翻译为：对一个特定主题的组织化的知识体系，特别是针对人类行为或人类社会的，如生命科学、自然科学、政治科学、社会科学等；解释 1 可以翻译为：关于自然和物质世界结构和行为的知识，这些知识基于可被验证的事实（如实验）。4 是广义的，1 是狭义的，2 与 3 比较含糊。

现代汉语词典（商务印书馆，1994 年，北京）对科学的解释是：①反映自然、社会、思维等客观规律的分科的知识体系。②合乎科学的：这种工作方法不～；革命精神和～态度相结合。其中解释①显然是广义的，解释②则比较含糊，可以理解为广义的，也可以理解为狭义的。

笔者主张从管理与科学的广义上来界定管理科学，这就是前面提出的定义 4 与图 2 的描述。

2. 走出困境的三点建议

管理一词在汉语中历史悠久，早在《旧唐书》中已经出现，在《四库全书》中它出现了 17 400 多次，它的基本含义与今天汉语语境基本相当[10]。管理科学作为新词只有 30 年左右的历史，宜取广义，如果取狭义就是作茧自缚了。

在英语世界，MS 相当于 OR，已经习惯成自然了，我们无法改变外国人的习惯，所以，必须另辟蹊径，寻找解决的办法。如果说以前把 MS 翻译成管理科学是一种无奈（也许是“一不小心”造成的疏忽），情有可原，那么，现在弄清楚了管理与 management 的区别，弄清楚了 science 的狭义与广义之后，继续把管理科学认定为 MS 则是不明智的，起码有两点不妥：大材小用和削足适履。在英译汉时，把 management 翻译成管理，把 administration 翻译成管理，都是可以的，但是如果在汉译英时把管理唯一地翻译成 management 或者唯一地翻译成 administration，就不妥当了。双方不对等，不能来回翻译而把双方画上等号。必须走出中英文语义差异，在实际上是中西方文化差异造成的困境。把文献［1］提出的建议梳理为以下三条。

建议 1：把 MS 称为狭义的管理科学，或管理的数量方法，或西方的管理科学。

建议 2：在中文翻译为英文的时候，管理的翻译采用汉语拼音 guanli 来表达。

建议 3：在中文翻译为英文的时候，管理科学翻译为 guanli science（GS），而现代管理科学的中国学派则翻译为 The Chinese school of modern guanli science（CSMGS）。

建议 1 体现了实事求是，建议 2 可以克服词不达意、口径错位的尴尬。在行文时，如果有必要，可以参照前述《精选英汉汉英词典》等工具书加注一下。

Guanli：management；administration；run；supervise；governance；…

建议 3 是顺理成章的，管理科学翻译为 guanli science，则 guanli science 包含 management science 即 GS 包含 MS。现代管理科学的中国学派，以前译为 the Chinese school of modern management science（CSMMS），仍然受了 MS 的束缚，现在译为 the Chinese school of modem guanli science（CSMGS）可以摆脱其束缚。

澳门科技大学有一个管理学院，其英文名称是 The administration and Management School，但是正式的中文名称却是行政与管理学院。可见，主事者也感到困惑。如果中文名称为管理学院，英文名称为 The Guanli School，恐怕是可取的。

可能有人会问：外国人会不会接受我们的建议？这是不成问题的问题，因为行文如说话，我们按照我们的实际情况说 guanli，说 guanli science 以及 CSMGS，外国人在听的时候，理所当然地会想方设法理解我们所说的意思。大可不必替外国人担忧什么，我们言之有理，就有说服力。如果“屈尊”去俯就外国人，词不达意地跟着他们说 management 或 administration…，说 MS 和 CSMMS，他们是不会思考，也不会意识到我们与他们的差异，也就难以接受我们的理念了。关键在于我们自己要有自信心，要坚持不懈，把 guanli、GS 和 CSMGS 推向世界。

翻译的要旨是“信，达，雅”三个字。“信”就是准确，这是第一位的，不准确就什么都谈不上。在意译有困难的时候，采取音译是可取的。音译可以避免失真，可以给人留出思考的空间，可以为今后找到好办法保留希望。音译在中外文互译的历史和现实中是比较常见的现象。例如，英语已经有新词 guanxi（关系），Kung Fu（功夫，这是美国人的拼法，按照汉语拼音应为 gongfu），taikongren（太空人）等；俄语有 чай（茶叶，茶），доуфу（豆腐）等；汉语有雷达（radar），坦克（tank）以及布尔什维克（来自俄语），英特那雄耐尔（国际歌词，来自法语）等。激光（laser）一开始没有找到合适的意译，就采用音译“莱塞”。而乌托邦（utopia）、互联网（internet）、星巴克（starbucks）则采取音译+意译的办法，是一种很好的选择。在《和谐管理理论》的英文摘要中，“和谐”一词就译为 Hexie[5]

其实，外国人很重视中国的事情，很重视“与中国接轨”。他们很愿意使用一些中国元素，很愿意讲几句汉语，如 nihao（你好），xiexie（谢谢），ganbei（干杯），zaijian（再见）。在北京奥运会上，外国人也大喊 jiayou（加油）！这类现象势必对英语的发展产生影响，我们可以顺水推舟。不妨看看英语发展史，美式英语的影响早已超过了伦敦英语，印度式英语在世界上的影响现在也很大。一个国家、一个民族，如果人多势众，有自信心，实力强大，科技先进，国际交往多，它的语言的传播速度和影响力是很大的，它的语言对其他语言的影响力也是很大的。语言是很重要的软实力，汉语发展史及其在东亚的影响也证明了这一点。笔者相信，汉语和英语的互动会日益增强，汉语将会给英语很大的影响。汉语是联合国的 6 种官方语言之一，汉语在全世界的话语权正在不断扩大。

2008 年 12 月 18 日，胡锦涛总书记在纪念党的十一届三中全会召开 30 周年大会上发表重要讲话，讲了“不动摇、不懈怠、不折腾”①。折腾一词是什么含义？中国老百姓人人知晓，但是在英语里却找不到对应的单词，不得不采用音译 Zheteng，现在 Zheteng 一词在英语世界已经迅速流传开来。

五、积极推介中国人的管理理念

研究中国的管理，研究创建现代管理科学的中国学派，中西方应该双向互动，双向交流。我们应该向外国人积极推介中国人的管理理念，推介中国人在管理研究中的创意；要敢于打破外国人有意无意构筑的藩篱，避免跟着外国人亦步亦趋。我们应该积极地把 guanli 推向世界，让外国人知道 guanli，熟悉 guanli，讲 guanli，写 guanli、GS 与 CSMGS 更是这样。笔者相信：只要坚持，假以时日，guanli，GS 与 CSMGS 也会成为新的英语词汇。这其实是把中国的管理理念和管理科学推向世界，不光是一个单词的问题，也是中国人取得国际话语权的事情。

鲁迅先生说得好：“其实地上本没有路，走的人多了，也便成了路”②。话语权也是这样，其实世界上本没有话语权，说的人多了，也便成了话语权。

参考文献

[1] 刘人怀，孙东川. 再谈创建现代管理科学中国学派的若干问题. 中国工程科学，2008，10 (12)：24-31.

[2] 孙东川，张振刚，孙凯. 一项重大历史使命：创建现代管理科学的中国学派. 美中经济评论，2008，8 (1)：57-63.

---

① 胡锦涛在纪念党的十一届三中全会召开 30 周年大会上的讲话，https：//www. gov. cn/test/2009-10/13/content_1437699_6. htm.

② 鲁迅. 故乡//鲁迅. 呐喊. 北京：人民文学出版社. 1979.

[3] 孙东川，林福永，孙凯. 创建现代管理科学的中国学派及其基本途径研究. 管理学报，2006，3（2）：127-131.

[4] 刘人怀，孙东川. 谈谈创建现代管理科学中国学派的若干问题. 管理学报，2008，5（3）：323-329.

[5] 李京文. 创新发展有中国特色的管理科学：兼评《和合管理》. 管理学报，2007，4（2）：141-143.

[6] 苏东水，彭贺. 中国管理学. 上海：复旦大学出版社，2006：3.

[7] 马庆国. 中国管理科学研究面临的几个关键问题. 管理世界，2002，(8)：105-115，140.

[8] 曾仕强. 中国式管理. 北京：中国社会科学出版社，2005.

[9] 钱学森. 论系统工程（新世纪版）. 上海：上海交通大学出版社，2007：141-143，293-298.

[10] 赵树进. 管理活动的认识理论研究. 广州：华南理工大学，2005.

大平台、聚义厅及其他——四谈创建现代管理科学中国学派的若干问题①

一、研究背景

对于创建现代管理科学中国学派，笔者已经发表了多篇文章[1-9]，阐明了我们的基本观点。“嘤其鸣矣，求其友声！”这些观点不管是引起共鸣还是争鸣，都是我们所欢迎的友声，唯有这样，才能把研究工作引向广泛而深入的发展。我们还将继续研究，提出一些观点、思路、意见和建议。

在已经发表的文章中，我们提出并且初步回答了以下问题。

(1) 为什么要对“管理”“科学”和“管理科学”，以及 management、administration、science、management science 等基本概念做一番正本清源的探讨?

(2) 什么是管理? 包括什么是管理活动? 什么是管理工作? 什么是管理者?

(3) 什么是管理科学? 管理科学与管理学是什么关系?

(4) “管理”与 management、administration 等的差异是什么?

(5) “管理科学”有广义与狭义之分——笔者主张广义，而且不加说明的管理科学就是广义的管理科学——广义的管理科学是什么? 狭义的管理科学是什么?

(6) “管理科学”与 management science (MS) 的差异是什么? 这是可以忽略不计的差异，还是反映了中英文语义的差异、中西方文化的差异? 是必须高度重视的差异吗?

(7) 我们是跟着西方亦步亦趋、人云亦云呢，还是要有自己的思考与见解、要走自己的路? 我们相信，大多数人的回答一定是否定前者，肯定后者。但是，迄今为止，在客观上我们是不是还在跟着西方亦步亦趋、人云亦云呢?

(8) 什么是现代管理科学中国学派? 这个名称由何而来? 它的定义是什么? 它有哪些基本特点?

(9) 为什么要研究和创建现代管理科学中国学派?

(10) 创建现代管理科学中国学派的时机成熟了吗?

(11) 创建现代管理科学中国学派的基本途径是什么?

(12) 中国现在的管理怎么样? 很落后吗? 或者很先进吗?

(13) 管理与经济(或者经济与管理)，两者是不相关还是密切相关? 是负相关还是正相关?

(14) 蓬勃发展的经济伴随着一团糟的管理，可能吗?

(15) 为什么要把“管理”的汉语拼音 guanli 向世界(包括英语世界)推出?

(16) “管理科学”为什么要翻译为 guanli science (GS)? 为什么不采用司空见惯的、许多人习以为常的 management science (MS)?

---

① 本文原载《管理学报》，2009，6 (9)：1137-1142，作者：刘人怀，孙凯，孙东川.

(17) “现代管理科学中国学派”为什么要翻译为 the Chinese school of modern guanli science (CSMGS)？为什么不采用 the Chinese school of modern management science (CSMMS)？

(18) 创建现代管理科学中国学派，作为基本途径之一的“近为今用”是什么意思？为什么要以它为重点？

(19) 创建现代管理科学中国学派大概要用多长时间？“10～20 年奠定基础”，时间是否太长了？是否可以短一些？

以上问题，笔者均已做出了回答，可能不充分，可能有不足之处或者有谬误，欢迎朋友们批评指正。本节将着重讨论以下几个问题。

(1) 创建现代管理科学中国学派，需要哪些人来做？需要多少人来做？

(2) “现代管理科学中国学派”与其他提法，如“和谐管理”“和合管理”“中国式管理”“中国管理学”“中国管理学派”“有中国特色的现代管理科学”等，是什么关系？

(3) 现代管理科学中国学派有代表人物吗？谁是代表人物？

(4) 创建现代管理科学中国学派，还需要做哪些工作？我们准备做哪些工作？

为了便于开展讨论，写出前面提出的几个观点是必要的：

——管理活动是人类的第二类活动，它为第一类活动（作业活动）提供服务（定义 1[8]）；

——管理科学是研究管理活动规律与做好管理工作的全部知识的总和，是一个内容丰富的知识体系（定义 4[8]）；

——现代管理科学中国学派具有如下特点：它是中国的、现代的、先进的、世界的、开放的、与时俱进的[8]；

——创建现代管理科学中国学派的基本途径是洋为中用、古为今用、近为今用、综合集成。这是一项艰巨复杂的系统工程，需要千军万马长期作战[6]。

二、定义与建议

文献 [8] 已经提出四条定义、三条建议，下面按其顺序提出定义 5 与建议 4。

定义 5：现代管理科学中国学派，是现代中国人研究管理科学所形成的学派。

使用这个定义的时候，希望联想到现代管理科学中国学派的六个特点。事实上，之前我们一直是用这些特点来描述现代管理科学的中国学派的。

推论 1：现代管理科学中国学派，是现代中国人对管理科学研究成果之总和，具有中国特色和时代特征。

这个推论是根据定义 4 和定义 5 所做的直接演绎，是对定义 5 的补充。

推论 2：现代管理科学中国学派的核心，是现代中国人研究中国的管理所形成的学问，是现代中国人研究中国的管理所得到的研究成果之总和。

中国是世界的一部分，中国的事务是世界事务的一部分，中国的管理也是世界管理的一部分。中国人研究的管理科学包括全世界的管理，其中核心部分当然是中国的管理，那么，推论 2 就是很自然的了。

推论 3：外国人研究中国管理所获得的研究成果也可以纳入现代管理科学中国学派。

中国人历来强调自力更生，依靠自己的力量做自己的事情，但是并不排外。事实上，许多外国人对中国的事情是很感兴趣的，无论是朋友还是敌人，他们对中国的事情进行研究，也不乏真知灼见，可以为我们提供参考，正所谓“他山之石，可以攻玉”。

建议 4：“管理学”的英文名称建议为 guanliology。

建议 4 在文献 [7] 中已经提及，这里重提是为了加强论述。有人可能会问：建议 4 有没有必要？我们认为：有必要，因为有一个合适的名称比没有名称要好，可以避免一些歧义和尴尬。例如，研讨会“管理学在中国”，其英文是 Management in China[10,11]，似乎不够好。试问：如果有人不了解这个研讨会，尽管他的英文水平很高，他由英文直接回译将会如何？恐怕会译为“管理在中国”——即不一定会在“管理”之后加一个“学”字，或者会译为“管理工作在中国”——那就差得更远了。同理，文献 [11，12] 用 Chinese management 来翻译“中国管理学”和“中国管理学科”，其中的“学”或“学科”都是根据对内容的了解加上去的，不了解情况的人把英文回译成中文，可能是“中国的管理工作”或“中国式管理”“中国人的管理”等。不是有些不方便，或者有些尴尬吗？笔者查过不少词典，management 均没有解释为“管理学”和“管理学科”。把“管理”与“管理学”区别开来，也许是中国特色，这个特色是好的，是应该坚持的。那么，顺理成章的事情就是我们造一个单词出来——哪怕是面对英语，我们也是理直气壮的。

如今有了新词 guanli 和 guanliology，则“管理学在中国”可以翻译为 Guanliology in China；“中国管理学”可以翻译为 Chinese guanliology，“中国管理学科”可以翻译为 Chinese disciplines of guanli，或 disciplines of Chinese guanli，或 Chinese guanli's disciplines。如果接受定义 4。那么，“管理学在中国”也可以翻译为 Guanli Science in China，“中国管理学”也可以翻译为 Chinese guanli science，等等。在这里，管理科学（GS）相当于广义的管理学，或者说，广义的管理学相当于管理科学（GS）——两者可以视为等同（暂时不谋求谁代替谁）——是否妥当？仅供参考。

议题与我国的《授予博士硕士学位和培养研究生的学科专业目录》（简称《学科目录》）及《授予博士硕士学位和培养研究生的学科专业简介》有关。现行版的《学科目录》是 1997 年 6 月颁布实施的，10 多年了。它对于我国管理科学发展和管理科学人才培养发挥了很大作用；同时，也存在不少问题，引起了越来越多的争议，迫切需要修订。事实上，《学科目录》目前正在修订之中。

笔者认为，无论在《学科目录》修订之前还是修订之后，我们都不应该墨守成规。学术界应该引导《学科目录》的修订、再修订，使之不断完善，而不是以某一个版本的《学科目录》作为定论，作为检验真理的标准，从而约束学术探讨。尤其是现在，《学科目录》第 12 门类中暴露出很多问题，管理学与管理科学的矛盾明显存在，更不应该把它奉为圭臬。

三、现代管理科学中国学派：大平台、聚义厅、大家庭

研究中国的管理，在理论和方法上有所建树，建立一个体系或框架，并且为它起一个好的名称——国内许多学者都想到了，而且采取了行动。目前的名称是多种多样

的，例如：①中国管理科学[13]；②中国现代管理科学[14]；③中国特色的管理科学[15]，或有中国特色的管理科学[16]；④有中国特色的现代管理科学[3,4]；⑤中国特色社会主义管理科学[17]；⑥中国管理学[14]；⑦东方管理学[18,19]；⑧中国式管理[20]；⑨中道管理[21]；⑩和谐管理[22]；⑪和合管理[16,23]；⑫合作管理[24]；⑬秩序管理[25]等。

这些名称都是有道理的。本节笔者之一发表文章，曾经使用的名称是“有中国特色的现代管理科学”[3,4]。2005 年 8 月，笔者看到国家自然科学基金委员会管理科学部（简称“管理学部”）在《国家自然科学基金“十一五”发展规划》纲要中用的名称是“管理科学中国学派。”[26]，觉得很好，就拿过来用了。同时加上了“现代”二字，即“现代管理科学中国学派”。笔者认为，按照定义 4，管理科学是研究管理活动规律与做好管理工作的全部知识的总和，是一个内容丰富的知识体系[8]；先秦诸子及其后继者对于管理的论述、千百年来大量卓越的工程实践（如都江堰、郑国渠，以及皇宫和园林等）中闪耀的管理思想，可以称之为古代管理科学中国学派，而且包含有多种流派或分支；那么，现在所要研究和创建的就是现代管理科学中国学派了。几年来，笔者及其团队积极开展中国学派的研究，不但在《中国工程科学》《管理学报》和《美中经济评论》等刊物上发表了多篇文章，而且积极参与“中国工程管理论坛”“中国系统工程学术年会”“中国管理科学与工程论坛”“中国管理学年会”和“管理学在中国”等学术会议，还承担了以现代管理科学中国学派研究为主题的国家自然科学基金项目和广东省科技厅项目。

我们是自发研究＋响应号召。创建现代管理科学中国学派，我们认为是“到时候了”，义不容辞，责无旁贷，应该挺身而出。

《管理学报》扉页印了“办刊宗旨”4 句话，有 1 句话是“力助中国管理学派成长”。我们觉得“中国管理学派”与“现代管理科学中国学派”是很相近的。

我们认为研究中国的管理，创建一些理论与方法，有益于改进和提高中国的管理工作，并且为世界的管理做出中国人应有的贡献。所以，大家不必在名称上争来争去，要紧的是把研究工作做起来。邓小平当年在“（经济特区）姓社姓资”等问题上主张“不争论”，他说：“不搞争论，是我的一个发明。不争论，是为了争取时间干。一争论就复杂了，把时间都争掉了，什么也干不成。”实践证明，这是十分英明的，何况我们现在的问题并没有那样的“大是大非”呢。

我们建议：每个人都可以继续使用自己喜欢的名称，也可以改用别的名称；谁的工作做得好，卓有成效，说服力强，他就会不断加强话语权，大家就会采纳他的意见——我们相信这是学者的良心和良知。

从 2006 年以来，我们一直使用“现代管理科学中国学派[1,2,6-9]”（简称“中国学派”）这一名称。我们认为：现代管理科学中国学派具有最大的兼容性。可以打比喻：中国学派是个大平台（大舞台），是个聚义厅，是个大家庭。

“大平台”是说：任何中国人研究管理的学问，都可以到这个平台上来。中国学派这个平台足够大、充分大，要多大有多大，完全容纳得下前文列出的种种学术团队以及更多的其他学术团队到平台上来做学问；外国友人研究中国管理的学问，也可以到这个平台上来。“大舞台”的意思相类似：这个舞台足够大、充分大，要多大有多大，各种艺术团体和个人演员都可以到这个舞台上来尽情表演。

“聚义厅”——例如，水泊梁山108将的聚义厅——是说：中国学派可以聚集各路英雄好汉，共同做一番“替天行道”的事业，但是要警惕“投降主义”——满脑子想着接受洋大人的“招安”。

创建现代管理科学中国学派，是一项艰巨复杂的系统工程，需要千军万马长期作战[6]。以中国之大、学者之多，面对如此丰富的多样性的中国的管理实践，创建现代管理科学中国学派是人才济济、英雄辈出、前赴后继的，坚持洋为中用、古为今用、近为今用、综合集成，是一定可以成功的，而且时间上也不会太长。

“大家庭”好比是中华民族大家庭，这就更好了：有56个兄弟民族，有多种民族语言和文字；有多种风格的音乐、歌曲、舞蹈和戏剧等。在大家庭中，大家互相尊重，携手前进，共同创造美好的生活、美好的明天——创建现代管理科学中国学派。

我们作为“现代管理科学中国学派”这个名称的提出者、作为“管理学部”的战略目标的积极响应者，甘当“革命军中马前卒”（我国清末资产阶级革命者邹容写《革命军》一书所用的笔名），希望随着时间的推移，越来越多的学者采用“现代管理科学中国学派”这一名称。“管理学部”前副主任陈晓田是“管理科学中国学派”的积极倡导者，2009年4月在“管理学部”部署“十二五”发展战略研究的会议上，他说：“中国学派是迄今为止最好的提法。如果以后有更好的提法，可以考虑采用更好的提法，现在还没有，所以，我赞成继续采用中国学派的提法。”

中国学派可以有许多流派或分支，如南方学派、北方学派，具有特征A、B、C的学派，具有特征D、E、F的学派，等等，如同先秦有诸子百家一样。

有人问：“中国学派有没有代表人物？代表人物是谁？”我们说：已经有一批代表人物了——前面不完全列举的各种提法，都是有代表作的（有的是发表了文章，有的已经出版了专著），它们的作者都是现代管理科学中国学派的代表人物。可以说，这些学术团队之于现代管理科学中国学派而言相当于多军种、多兵种之于中国人民解放军一样。

“天下为公”，“振兴中华”。大平台是共用的、公有的，聚义厅也是共用的、公有的，不由谁家私有私用。大家庭更是和谐的，其乐融融，齐心协力。在创建现代管理科学中国学派的进程中，对于创建工作会有一些不同意见，需要大家求同存异。在学术观点方面必然会有种种不同主张，这是正常现象，尤其是在管理科学研究领域，因为我们都知道一个基本事实：管理问题不是只有唯一解，而是具有许多可行解、满意解，但是不见得有最优解；即便是同一个管理者在不同的情境之下处理同样一个管理问题，他所选择的解和得到的效果都可能是不一样的。学术领域，应该百花齐放，和而不同。

有人问：中国学派是不是可以细分为具有地方特色的若干学派？例如，西安学派、上海学派、广州学派和某省某市学派等。笔者认为这是完全可以的，因为各地的情境有所不同。比如，歌曲茉莉花、信天游、敖包相会和金珠玛米颂歌等，各具地方特色，都很优美，都是中国的。中国地方特色的总和，就是区别于外国的中国特色。正如出国旅游团到了欧洲，当地人不难辨认：“来了一群中国人!”就是说，这一群人自有他们的共性——中国特色。现代管理科学中国学派也是这样，可以有差异或大或小的若干支派，但是它们一定具有显著的共性——中国特色——因而它们都是中国学派。

有人说：美国的管理很先进，为什么没有美国学派？不，现在风行于全世界的管理理论与方法实际上就是美国学派（加上一些“日本元素”），这是独霸天下的学派，其他学派都比较弱势，所以美国学派不必冠以“美国”二字。如同美国的互联网的网站地址不必写美国的后缀一样。如果有了强大的中国学派和其他学派，那时候，美国学派大概就不得不标注“美国”二字了。所谓“美国的管理很先进”也要加以分析，不能一言以蔽之。美国的次贷危机引发了美国的金融危机，殃及了全世界，起码说明这个第一号的资本主义大国没有搞好金融管理。美国在其他方面的管理也是问题多多，这里不去细说。

中国人研究与创建中国学派是理所当然的，这是中国人的优势所在，责任所在。外国人研究中国学问也是可以的，我们非常欢迎。

中国人研究外国学问也是可以的，应该的。研究学问必须放眼世界。中国人研究外国的管理而得到的研究成果，也可以属于中国学派的一部分。

“中国学派”其实在许多领域都存在。毛泽东思想就是马克思列宁主义的中国学派。又如，中国的运筹学比较严谨，与美国的运筹学有比较大的区别，以至于美国人把中国的运筹学称为“运筹数学”——也可以称为运筹学的中国学派。中医中药，就是医学和药学的中国学派。中国的系统工程具有显著的中国特色，就是系统工程的中国学派。

四、中国学派与中国模式

与中国学派相关的术语较多，如中国模式、中国特色、中国风格，以及中国元素等。它们之间既有联系又有区别。

“中国模式”现在出现频率很高，外国人很喜欢说“中国模式”一词。改革开放30年来，中国经济蓬勃发展、经久不衰、势不可挡，不但使得西方预言家们的“预言”一次又一次失效，而且使得传统的经典的西方经济理论都无法解释，于是称为“中国模式”。中国社会安定团结令一些西方人诧异，也称为“中国模式”。

我们认为：“中国模式”上升到理论高度，提炼出新的理论，就会成为“中国学派”。换言之，“中国模式”的学术形式是“中国学派”，“中国学派”的实践形式是“中国模式”。实践先行，理论随后。有了实践形式，才能产生学术形式；有了“中国模式”，才能产生“中国学派”。这种产生不是自动的，不是从天上掉下来的，而是要有一大批有心人、有志者执着地开展研究工作，包括调查、归纳、提炼、总结和验证等——这就是管理科学界的责任，尤其是中国管理科学界的历史使命。我们不能等待像李约瑟博士那样的外国友人来创建现代管理科学中国学派，必须自己动手，主要依靠自己的力量开展创建工作。

高盛公司提出了区别于“华盛顿共识”的“北京共识”，我们认为它与中国模式、中国学派是相辅相成的。

中国模式与中国学派，必定具有鲜明的中国特色、中国风格，否则难以立足于世界学术之林。“中国特色”是褒义词而不是贬义词，中国特色意味着优势和长处、独到之处。

中国模式与中国学派，必定含有中国元素（不可能没有），但是仅有少量的中国元素还不够。量变引起质变，中国元素多了而且很强势，才能形成中国特色。外国元素放在“中国底色”上，才是中国特色。

“沧海横流，方显出英雄本色。”席卷全世界的金融危机，又一次雄辩地证明了中国模式的优越性，证明了“美国模式”并非尽善尽美、并不具有普世价值。我们要向西方、向美国学习一切优秀成果，但是绝不是“全盘西化”“全盘美国化”。一句话：洋为中用。

笔者认为：现在应该重新审视“中学为体，西学为用”这个口号（把它作为一项原则、一种途径）。洋务运动中提出的这 8 个字，在当时只是反映了一批仁人志士想要进行改革的美好愿望，并未圆满实现。因为当时清王朝这个“体”太腐朽了，当时的主流文化这个“体”太保守了，所以当时先进的西方文化无法运用，“虚不受补”。今天，我们建设中国特色社会主义，构建社会主义和谐社会，应该提倡和实行“中学为体，西学为用”。

“洋为中用”，“洋”的比例是多少？如果比例为 0，则为关门主义；比例为 100%，则为全盘西化。“中学为体，西学为用”，体者，主也；中学与西学之比例，起码应该是六四开或七三开。总之，以中（中学）为主，而不是以洋（西学）为主。在没有先例、没有经验的情况下，是摸着石头过河，而不是跟在洋人后面亦步亦趋。

笔者在此申明：我们说的“洋为中用”，是“中学为体，西学为用”的简称与缩写。

五、建立共识继续探讨

本节的目的，一是在于和朋友们进行学术交流与沟通，把我们的基本观点告诉大家。希望得到批评指正，建立共识，继续开展研究；二是循着我们已有的思路，又向前做了一点探索。

在文献 [8] 定义 4 的基础上提出了定义 5，把它们合起来就是说：管理科学是研究管理活动规律与做好管理工作的全部知识的总和，是一个内容丰富的知识体系；现代管理科学的中国学派，是现代中国人研究管理科学所形成的学派。这样界定的现代管理科学中国学派是个大平台（大舞台），是个聚义厅，是个大家庭，可以容纳所有一切研究中国管理的人士开展研究，提出种种关于中国管理的理论与方法。

在文献 [8] 建议 1 的基础上提出了建议 4，把它们合起来是说：把“管理”的汉语拼音 guanli 作为英译，替代与汉语口径不一的 management、administration 等单词，把“管理学”英译为 guanliology。这样，可以解决“管理学在中国”“中国管理学”和“中国管理学科”此前在翻译上遭遇的尴尬。

还分析了中国学派与中国模式、中国特色、中国元素的关系，认为：中国模式上升到理论高度，提炼出新的理论，就会成为中国学派；中国模式的学术形式是中国学派，中国学派的实践形式是中国模式。

参考文献

[1] 孙东川，张振刚，孙凯．一项重大历史使命：创建现代管理科学的中国学派．美中经济评论，2008，8（1）：57-63.

[2] 孙东川，林福永，孙凯．创建现代管理科学的中国学派及其基本途径研究．管理学报，2006，3（2）：127-131.

[3] 孙东川，林福永．洋为中用，古为今用，近为今用——创建有中国特色的现代管理科学．系统工程，2004，(增刊)：13-15.

[4] 王克强，孙东川．管理科学与物理学、医学的分析比较——论创建有中国特色的现代管理科学．生产力研究，2005，(11)：211-213.

[5] 孙东川．系统工程与管理科学研究．广州：暨南大学出版社，2004：72-77.

[6] 刘人怀，孙东川．谈谈创建现代管理科学中国学派的若干问题．管理学报，2008，5（3）：323-329.

[7] 刘人怀，孙东川．再谈创建现代管理科学中国学派的若干问题．中国工程科学，2008，10（12）：24-31.

[8] 刘人怀，孙东川，孙凯．三谈创建现代管理科学中国学派的若干问题．中国工程科学，2009，11（8）：18-23，63.

[9] 孙东川．三分法与管理工作．管理学报，2009，6（7）：861-866.

[10] 韩巍．管理学在中国—本土化学科建构几个关键问题的探讨．管理学报，2009，6（6）：711-717.

[11] 彭贺，苏勇．也从批判性和建设性的视角看“管理学在中国”——兼与韩巍商榷．管理学报，2009，6（2）：160-164，186.

[12] 罗珉．中国管理学反思与发展思路．管理学报，2008，5（4）：478-482.

[13] 马庆国．中国管理科学研究面临的几个关键问题．管理世界，2002，(8)：105-115，140.

[14] 苏东水，彭贺．中国管理学．上海：复旦大学出版社，2006：3.

[15] 成思危．建立中国特色的管理科学．煤炭企业管理，2003，(2)：8-11.

[16] 李京文．创新发展有中国特色的管理科学——兼评《和合管理》．管理学报，2007，4（2）：141-143.

[17] 宁玉琼．建构有中国特色社会主义管理科学的几个认识问题．广西社会科学，2000，(1)：74-77.

[18] 胡祖光，朱明伟．东方管理学十三篇．北京：中国经济出版社，2002.

[19] 苏东水．东方管理学．上海：复旦大学出版社，2005.

[20] 曾仕强．中国式管理．北京：中国社会科学出版社，2005.

[21] 曾仕强．中道管理——M理论及其应用．北京：北京大学出版社，2006.

[22] 席酉民，尚玉钒．和谐管理理论．北京：中国人民大学出版社，2002.

[23] 黄如金．和合管理．北京：经济管理出版社，2006.

[24] 李常洪，李铁，范建平，等．合作管理研究框架．管理学报，2008，5（2）：169-176.

[25] 谭人中．秩序管理概论．管理学报，2008，5（3）：345-357.

[26] 林国雄．新儒学经济与管理．台北：慈惠堂出版社，1997.

《学科目录》第12学科门类与管理科学话语体系——五谈创建现代管理科学中国学派的若干问题①

一、修订《学科目录》的建议

要谈论现代管理科学中国学派及其创建工作，必须有一个合适的话语体系。这个话语体系涉及管理工作与管理科学一些基本概念的界定：一是要准确与明朗：二是要合理与和谐，能够恰如其分地表达相关理念、观点、思路和研究成果，形成一个内容丰富的理论体系。管理科学话语体系涉及我国现行的《授予博士、硕士学位和培养研究生的学科、专业目录》（简称《学科目录》）第12学科门类的修订。

我国现行的《学科目录》是国务院学位委员会和国家教育委员会于1997年6月颁布的。它是1990年版本的修订版，其中增设了第12学科门类“管理学”，这是一大亮点。此前的《学科目录》只有11个学科门类，没有“管理学”这一门类，管理类专业是“寄人篱下”，分散在其他学科门类中。例如，管理工程专业在工学门类，企业管理专业在经济学门类。1998年颁布的培养本科生的学科目录也增设了“管理学”门类，也是第12个学科门类。这两个目录都增设了“管理学”门类之后，管理教育有了很大的发展，因此也就有了管理学的博士、硕士和学士。

现行版《学科目录》使用至今已经十多年了，由于形式不断发展变化，产生了不少问题，所以修订的呼声很高。国务院学位委员会办公室（简称“学位办”）几年前已经组织力量开展修订工作。这是全国教育界、科技界都十分关注、翘首以待的一件大事。本节将对现行版《学科目录》的第12学科门类的修订提出一些建议。

二、第12学科门类的问题

表1是“管理学”学科门类下设的一级学科和二级学科中英文名称[1]，在这里面有两个方面的问题：一是门类本身的问题；二是学科名称的英语译名问题。

表1　第12学科门类管理学及下设学科、专业的中英文名称

学科代码	中文学科、专业名称	英文学科、专业名称
12	管理学	management science
1201	管理科学与工程	management science and engineering
1202	工商管理学*	science of business administration
120201	会计学	accounting
120202	企业管理学*（含：财务管理、市场营销学、人力资源管理学）	corporate management (including financial management, marketing, and human resource management)

① 本文原载《学位与研究生教育》，2010，(8)：67-73，作者：刘人怀，孙东川.

续表

学科代码	中文学科、专业名称	英文学科、专业名称
120203	旅游管理学*	tourist management
120204	技术经济及管理学*	technology economy and management
1203	农林经济管理学*	agricultural and forestry economics&management
120301	农业经济管理学*	agricultural economics&management
120302	林业经济管理学*	forestry economics&management
1204	公共管理学*	science of public management
120401	行政管理学*	administration management
120402	社会医学与卫生事业管理学*	social medicine and health management
120403	教育经济与管理学*	educational economy and management
120404	社会保障学*	social security
120405	土地资源管理学*	land resource management
1205	图书馆、情报与档案学*	science of library，information and archival
120501	图书馆学	library science
120502	情报学	information science
120503	档案学	archival science

注：1. 表中学科代码是四元编号者为一级学科，六元编号者为二级学科

2. 标有“*”号的中文学科、专业名称，在国务院学位委员会、国家教育委员会 1997 年颁布的《学科目录》中，是没有“学”字的。但由于参考文献［1］的读者面广，影响大，所以本节还是就此展开讨论

1. “管理学”学科门类本身的问题

（1）没有厘清管理学与管理科学的关系。按照大科学大工程的理念来理解管理，管理就应该是大管理，管理科学则应该是管理研究领域最大的概念。我们曾撰文把管理科学定义为：管理科学是研究管理活动规律与做好管理工作的全部知识的总和，是一个内容丰富的知识体系[2,3]。这样定义的管理科学包含管理学。但是，在现行的《学科目录》中看到的不是这样的理念与关系，恰恰相反：“管理学”是第 12 门类最大的概念，它包含管理科学。

（2）一级学科“管理科学与工程”的名称太大，不易界定其内涵，不易区分它与其他学科的差异和分工。“管理科学与工程”是管理科学与管理工程的简称，管理工程也是一个很大的概念，两个大概念加起来就更大了。在一级学科“管理科学与工程”下没有明确设立二级学科，而是给出了 6 个“学科研究范围”，它们是很宽泛的。事实上，各高校自主设立的二级学科很多（2005 年笔者做过统计，有 100 多个），有一些是很不切题的。

（3）一级学科“工商管理”内部结构不合理。一级学科“工商管理”下设 4 个二级学科，其分类标准很不一样。二级学科“会计学”，是个十分专业化的学科（“条条”型的）；而“企业管理”与之不同，是“块块”型的，《学科目录》上就注明（含：财务管理、市场营销、人力资源管理）。其实，“旅游管理”也可以归入“企业管理”，因为旅游企业需要管理，但是，更有理由说它是行业管理；“技术经济及管理”的质疑更多：是两个专业还是一个专业？其中一个“及”字，显得“技术经济”为强势而“管理”为弱势，“技术经济”属于经济学，但是研究生毕业时授予的是管理学学位，颇为

矛盾。参考文献［1］第 512～513 页对于该二级学科有整整一页的解释，也不能自圆其说，反而令人产生更多的疑问。

2. 英语译名问题

英语单词 management、administration、management science（MS）都是比较狭义的概念，相对于大管理，它们表达的都是“小管理”。由于混淆了或者忽视了大管理与小管理之差异，第 12 学科门类及下设的各学科的英语译名存在很多问题[2-5]。请继续看表 1。

（1）学科门类名称“管理学”译为 management science（MS）是很不妥当的。MS 是很狭义的概念，它的内容是建立数学模型，进行计算分析，相当于运筹学（operations research，OR），而运筹学在我国属于应用数学。MS 应该实事求是地翻译为“管理的数量方法”或“西方的管理科学”，它与“管理学”有很大差异。而管理科学与 MS 的口径也大不一样，前者比后者大得多，后者是前者的一个局部、一个子集。

（2）学科门类的英文名称是 management science。下属的一级学科“管理科学与工程”的英文名称是 management science and engineering，这是把“大坛子装入小坛子”里面去了。

（3）一级学科“工商管理”的英文名称是 science of business administration，前后矛盾。把它联系学科门类名称看，是 management 包含 administration，而联系二级学科“企业管理”的英文名称 corporate management 看，则是 administration 包含 management；你包含我，我包含你，意味着 management＝administration，对吗？当然不对。

（4）一级学科“农林经济管理”的英语名称是 agricultural and forestry economics&management，其中的“&”很容易令人误解它是两个学科，即经济学科与管理学科的组合。该一级学科下的两个二级学科“农业经济管理”和“林业经济管理”的英文名称也是这样的问题。

从概念和逻辑上看，第 12 学科门类还有以下问题。

（5）一级学科“公共管理”译为 science of public management，合适吗？其下属的二级学科“行政管理”译为 administration management，又作何解释？

（6）第 12 学科门类共有 5 个一级学科、14 个二级学科，除一级学科“管理科学与工程”之外，每个学科名称最后都有个“学”字；它们的英文名称，少数带有 science 而多数没有，但是，根据词典，英文单词 management 和 administration 都没有“管理学”的含义。所以，学科代码分别为 120202、120203、120204、1203、120301、120302、120401、120402、120403、120404、120405 的学科、专业的中英文名称是不完全对应的。

上面所提出的问题中最主要的两大问题：一是颠倒了管理学与管理科学的关系，不符合大科学、大工程和大管理的理念；二是忽视了中英文语义差异，等同了管理科学、管理学与 MS 的关系，把 3 个差异较大的概念混为一谈。这些问题造成了许多人的困惑，对我国的管理科学研究产生了误导。所以，第 12 学科门类需要做出比较大的修订。

3. 什么是管理?

有几句话是广泛的共识:“三百六十行,行行有管理”;“人人都是管理者,人人都是被管理者”;人人都要学习管理,变不自觉为自觉,从必然王国到自由王国。

现在要回过头来再思考:什么是管理?说法很多,但是都有所不足。参考文献[2,3]对此做了分析,并且提出了别具一格的定义:管理是人类的第二类活动,它为第一类活动提供服务。人类的第一类活动是生活作业和生产作业(统称“作业活动”或“作业”),第二类活动使得第一类活动能够井井有条,提高效率,提高效益,促进和谐。“二分天下有其一”,管理乃是人类全部活动的“半壁江山”,其作用可谓大矣!

对管理的定义要做一些说明。人类乃万物之灵,哺乳动物的生理本能,如吃、喝、拉、撒、睡、运动、繁殖等,人类也都具有,这就是人类的生活作业——第一类活动的第一部分。人类的生活作业要高雅得多、复杂得多:演变为衣、食、住、行、恋爱、婚姻、家庭、生儿育女、教育培养,等等。

一般哺乳动物的生活资料都是直接取之于大自然,如兔子吃草,老虎吃兔子。人类也从大自然直接采集或猎取,但是更多的是自己生产,于是有了种植业、养殖业、加工业、制造业,等等。人类不但有物质生活和物质生活资料的生产,而且有精神生活和精神生活资料的生产。生产作业是人类的第一类活动的第二部分,是其他哺乳动物所没有的。生活作业和生产作业都需要管理——人类的第二类活动,更是其他哺乳动物所没有的。笔者在参考文献[2,3]中有更多的说明。

管理问题都是系统问题,各级各类管理工作都是系统工程——仅仅是规模大小和复杂性高低的区别而已。

三、构建中西合璧的管理科学话语体系

大管理与小管理之差异,实际上是中西方文化的差异。据考查,“管理”一词在《旧唐书》中已经出现,在《四库全书》中出现了17 400多次,其含义与今天基本一致[6]。“管理”对应的英语单词比较多:management,administration,control,run,supervise,governance,…。其中management用得多一些,但是它并不能覆盖或替代administration。例如,很热门的专业学位——工商管理硕士(master of business administration,MBA)、公共管理硕士(master of public administration,MPA),用的单词都是administration,全面质量管理(total quality control,TQC)则是用control,“三百六十行,行行有管理”,汉语的“管理”是一个词,英语就是多个单词了。笔者归纳出以下等式和不等式:

$$\text{管理} \neq \text{management,或 administration,或 control,}\cdots \tag{1}$$

$$\text{管理} = \text{management} + \text{administration} + \text{control} + \cdots \tag{2}$$

$$\text{管理科学} \neq \text{management science} \tag{3}$$

$$\text{管理科学} = \text{management science} + \cdots + \cdots \tag{4}$$

如果把不等号简单地改为等号,可谓是削足适履、作茧自缚,或者“自废武功”“被绑架”。为了克服中英文差异和中西方文化差异,我们必须正本清源,另辟蹊径,标新立异。

笔者查了很多词典，无论 management 还是 administration，都不具有"管理学"的含义，英语没有对应于"管理学"的单词。且看一个尴尬的例子："管理学在中国"于 2009 年 3 月在西安召开第一届学术会议，2009 年 11 月下旬在武汉召开第二届，两次会议都开得很好[7]。会议的汉语名称很明确，没有歧义，但是，它的英语名称 management in China 就很不明确了。第一，如果不知道汉语名称，直接看英语名称，很可能理解为一群企业家或政府官员在商谈管理事务，而想不到是一群学者在研讨管理学。我们设想，请 10 名不知情的英国汉学家或中国的英语教师来翻译 mamgement in China，恐怕很少有人能够翻译成"管理学在中国"。第二，会议内容不但有 management，也有 administration（讨论 MBA 的培养问题），以及 governance（研究公司治理问题），supervise，等等，但是，management in China 能够包含它们吗？第三，如果把单词 management 换成 administration，甚至换成 management&administration，情况也差不多。这样，进行中西方交流就有困难了，如同广东人所说的"鸡同鸭讲。"

再举一例。某高校管理学院的英文名称为"school of management"，另一所高校的管理学院的英文名称为"school of business administration"，还有一所高校的管理学院的英文名称为"school of administration and management"。它们的英文名称差别很大，似乎是性质迥异的学院，其实这三个学院几乎没有区别，都有"管理科学与工程"与"企业管理"博士点，都培养 MBA，也都研究公司治理（corporate governance）。如果你不知道它们的中文名称，凭借英文名称能够准确说出它们的中文名称吗？恐怕很难。英文名称，词不达意，以偏概全，又是中英文差异造成的尴尬。怎么办？

笔者提出的解决办法是：构建一个中西合璧的管理科学话语体系，具体包括以下几点。

（1）用汉语拼音 guanli 来表达"管理"一词。在国际交往中，词汇难以意译的情况是经常发生的，采用音译是很常见的，如巧克力、咖啡、坦克、雷达，都是音译；国际歌歌词"英特耐雄纳尔"、苏联的"布尔什维克（党)"，也都是音译；guanxi（关系)、taikongren（太空人）已经成为英语单词；好莱坞大片"Kung Fu Panda"（功夫熊猫）的 Kung Fu 则是美国人"发明创造"的音译来翻译他们无法准确意译的汉语单词"功夫"。

2008 年 12 月 18 日，胡锦涛总书记在纪念中国共产党十一届三中全会召开 30 周年大会上发表重要讲话，讲了"不动摇、不懈怠、不折腾"①。"不折腾"是什么意思？中国老百姓人人知晓，但是在英语里却找不到对应的词语，不得不采用音译 zheteng 来表示"折腾"，现在，zheteng 作为英语单词已经在世界上流传开来。

（2）在 guanli 一词的基础上，"管理科学"可以翻译为 guanli science（GS)。式（2）、式（4）现在可以改写为

$$\text{guanli} = \text{management} + \text{administration} + \cdots \tag{5}$$

$$\text{guanli science} = \text{management science(MS)} + \cdots + \cdots \tag{6}$$

"现代管理科学中国学派"则可以翻译为 the Chinese school of modern guanli science

---

① 胡锦涛在纪念党的十一届三中全会召开 30 周年大会上的讲话，www.gov.cn/test/2009-10/13/content_1437699_6.htm.

(CSMGS)，而不是 the Chinese school of modern management science（CSMMS)。

（3）根据英语的构词法，用 guanliology 来表示“管理学”，或者整个用音译 guanlixue 来表示“管理学”。于是，前面说的三个管理学院，它们的英文名称可以统一叫做“the guanli school”。学术会议“管理学在中国”的英语名称问题也就迎刃而解：guanliology in China，或 guanlixue in China。

英语世界的外国人一开始会看不懂 guanli、guanliology、guanlixue、guanliscience（GS)，这不要紧，相信他们有足够的学习精神和学习能力，他们是不难搞懂的——如同我们搞懂了 management、administration、management science（MS）一样。要交流就要互相学习，互相理解。中国人为洋人担忧，如同今人“看三国故事，为古人掉泪”一样，大可不必。

（4）英语单词 management、administration 等，在合用的时候仍然继续使用 MBA、MPA、TQC、engineering management（EM）等。

这样，音译的汉语单词，加上原有的英语单词，就构成了一个中西合璧的管理科学话语体系，如图 1 所示。凭借这个话语体系，可以更好地开展学术交流，有利于准确表达中国人的管理理念和内容丰富的管理科学体系，有利于现代管理科学中国学派的创建工作，有利于争取中国管理科学界在世界上的话语权。

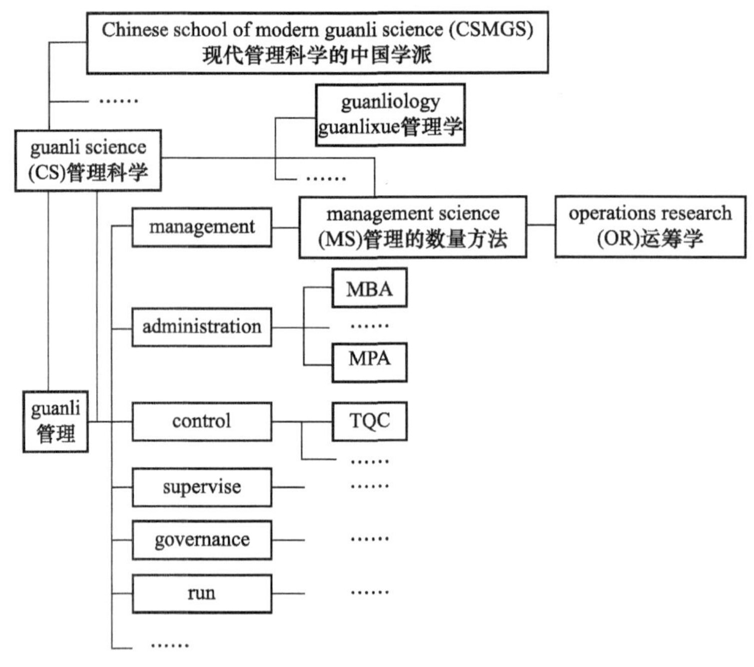

图 1　中西合璧的管理科学话语体系

四、第 12 学科门类的问题如何解决

1. 建议把第 12 学科门类称为“管理科学”

无论从中国人的传统理念来看，还是从当今世界的大科学大工程潮流来看，管理都应该是大管理，“管理科学”应该是管理研究领域最大的概念，第 12 学科门类应该称为“管理科学”而不是“管理学”。不妨类比一下：社会科学包含社会学，系统科学包含系统学，地理科学包含地理学，建筑科学包含建筑学，军事科学包含军事学，农业科学包含农学，等等。顺理成章，管理科学包含管理学。

现在，除中国科学院与中国工程院以外，还有众多的“××科学院”或“××科学研究院”，它们都体现了大科学大工程的理念，如中国社会科学院（全国各省市都有社会科学院）、中国农业科学院（全国各省市都有农业科学院）、中国热带农业科学院、中国医学科学院、中国中医科学院、中国人民解放军军事医学科学院、中国地质科学院、中国科学院数学与系统科学研究院、中国电力科学研究院、中国纺织科学研究院、中国环境科学研究院、中国计量科学研究院、中国建筑科学研究院、中国林业科学研究院、中国气象科学研究院、中国水产科学研究院、中国铁道科学研究院、中国原子能科学研究院、交通运输部公路科学研究院、中国管理科学研究院等。

不妨设想一下：如果把“××科学院”“××科学研究院”改为“××学研究院”行不行呢？例如，社会科学院更名为“社会学研究院”，农业科学院更名为“农学研究院”，建筑科学研究院更名为“建筑学研究院”，……合适吗？那就体现不出大科学大工程的理念了，它们的研究范围就会缩小许多、狭窄许多。管理科学研究院大概是不会更名为“管理学研究院”的。

2. 第 12 学科门类内部的学科名称修改建议

修改建议如表 2 所示，有以下几点。

（1）第 12 学科门类的名称建议改为“管理科学”，其英文名称为 guanli science。

（2）一级学科“管理科学与工程”名称和内涵都太大，建议分为两个一级学科，一个是“管理科学研究”，另一个是“管理工程研究”；分别设立若干二级学科，如表 2 所示。

（3）一级学科与二级学科的名称一般都加上一个后缀“研究”，既整齐划一，也突出了研究之意。

（4）一级学科和二级学科的英文名称均可采用 guanli 一词，这是方案一；方案二，如果 management 或 administration 比较合用的话，则继续使用，如代码分别是 *120104、*1202、*120208、*1203、*1205 等的学科、专业。

表 2　第 12 学科门类修改的建议

*12	管理科学	Guanli science
*1201	管理科学研究	research of guanli science
*120101	管理学研究（含管理哲学研究与管理伦理研究）	guanliology research 或 guanlixue research
*120102	发展战略与规划研究	
*120103	管理理论与方法研究	research of guanli theory and methodology
*120104	管理的数量方法研究	research of management science
⋮		
*1202	管理工程研究	research of management engineering

续表

*12	管理科学	Guanli science
*120201	信息管理与管理信息系统研究	
*120202	物流与供应链管理研究	
*120203	人力资源管理研究	
*120204	会计学与财务管理研究	
*120205	行业管理研究	
*120206	工业工程研究	
*120207	系统工程研究	
*120208	工程管理研究	engineering management research
*120209	科技管理研究	
⋮		
（以下暂略二级学科）		
*1203	企业管理研究	corporate management research
*1204	农林经济管理研究	
*1205	公共管理研究	public administration research
*1206	图书馆、情报与档案学研究	

注：加“*”号是为了区别于现行的《学科目录》中相应的编号

五、进一步加强管理科学的地位与作用

鉴于管理是人类的第二类活动，“三百六十行，行行有管理”，管理科学在《学科目录》中的地位与作用应该进一步加强。

现行版的《学科目录》在第 1 学科门类至第 11 学科门类中，许多学科的培养目标都包含管理人员的培养[1]。例如，一级学科“哲学”下设的二级学科“伦理学”，要求培养的人才“毕业后，可在有关部门和单位从事理论宣传、思想教育、新闻出版、行政管理等性质的工作”；二级学科“宗教学”的培养目标为“本学科培养……独立从事宗教教学、科研和宗教事务管理等工作的高级专门人才”；一级学科“经济学”下设的二级学科“世界经济”，要求培养的人才可“……在实际业务部门从事涉外经济的管理和调研等工作”；一级学科“中国语言文学”下设的二级学科“文艺学”，要求培养的人才可“在相关的部门从事专业性的管理工作”；一级学科“新闻传播学”下设的二级学科“新闻学”，要求培养的人才“能够胜任”……各新闻单位高层次的研究和管理工作”，等等，不胜枚举。

即便某学科在培养目标中没有写出类似的话语，该学科培养出来的博士、硕士在他们的业务工作中难免要带领一个学术团队开展研究工作吧，而且有可能担任所在研究所的所长或副所长职务，这些都是管理工作。切莫认为管理工作谁都能做，无师自通。实际上，许多人自以为“通”的，并不怎么通，常常办错事，闹笑话。学习一下与不学习大不一样。所以，人人都应该学习一点“管理科学 ABC”，然后，根据工作需要，再多学一些。

在现行版《学科目录》中，许多学科的“主要相关学科”都有“管理学”及相关

课程。我们建议：各学科均开设《管理科学概论》课程，其内容分为两部分：一是公共内容即管理科学一般理论与方法；二是本学科本行业专门的管理知识。

六、敢于构建中西合璧的管理科学话语体系

管理领域存在的中英文差异和中西方文化差异。中国人说的“管理”是大管理。符合当今世界大科学大工程的潮流，而 management、administration、management science（MS）等英语词汇均属于小管理，两者的口径相差很大。把中国人的大管理强行纳入英语世界的小管理中，是很别扭的，毫无必要，毫无益处。故此，我们提出了体现中国特色的词语 guanli、guanliology 或 guanlixue、guanli science（GS），加上现有的英语词汇，构成中西合璧的管理科学话语体系，这样有利于准确表达中国人的管理理念，有利于现代管理科学中国学派的创建工作，有利于争取中国管理科学界在世界上的话语权。

图 1 表示的话语体系还不够完善，需要继续研究，相信可以迅速趋于完善。首要的、关键的一点是，要敢于构建中西合璧的话语体系，要走出第一步！不妨借用鲁迅先生的话：“这正如地上的路；其实地上本没有路，走的人多了，也便成了路。”（鲁迅：《呐喊·故乡》）在管理科学领域是这样，在其他领域也是这样。

《学科目录》在我国具有很强的导向作用，它应该与时俱进，及时吸纳学术界已有的成果，引领学术研究的潮流。《学科目录》的修订工作是一项颇为复杂的系统工程，希望抓紧抓好，尽快完成。

笔者从大处着眼，对现行版《学科目录》第 12 学科门类的修订提出了一些建议。它们主要是修订的思路与观点，可操作的具体意见还有待于继续研究。

嘤其鸣矣，求其友声！同行朋友的共鸣或争鸣都是我们企求的友声。

参考文献

[1] 国务院学位委员会办公室，教育部研究生工作办公室. 授予博士、硕士学位和培养研究生的学科专业简介. 北京：高等教育出版社，1999.

[2] 刘人怀，孙东川. 再谈创建现代管理科学中国学派的若干问题. 中国工程科学，2008，10（12）：24-31.

[3] 刘人怀，孙东川，孙凯. 三谈创建现代管理科学中国学派的若干问题. 中国工程科学，2009，11（8）：18-23，63.

[4] 刘人怀，孙凯，孙东川. 大平台、聚义厅及其他——四谈创建现代管理科学中国学派的若干问题. 管理学报，2009，（9）：1137-1142.

[5] 霍恩比. 牛津高级英汉双解词典. 第四版增补本. 北京：商务印书馆，2002.

[6] 赵树进. 管理活动的认识理论研究. 广州：华南理工大学，2005.

[7] 长松. 争鸣强劲 青年崛起 任重道远——第 3 届“管理学在中国”学术研讨会纪要. 管理学报，2010，7（9）：1290.

当前管理科学研究中的若干问题——几个疑点的澄清和两种研究方法的评析①

一、几个疑点的澄清

1. “管理科学部”还是“管理学部”

最近友人来访，谈起一些文章似乎故意把国家自然科学基金委员会管理科学部写成“管理学部”，不知为什么？其中包括拙文《大平台、聚义厅及其他——四谈创建现代管理科学中国学派的若干问题》[1]。笔者表示：不能一概而论，就我们而言，一直都是称呼“管理科学部”的，在拙文[1]中确有 4 处“管理学部”的提法，其实在原稿上都是“管理科学部”；其中第 1139 页说“……国家自然科学基金委员会管理科学部（简称‘管理学部’）”，括弧里的注释把问题说明，当然很有必要，但是，括弧及里面的文字是原稿所没有的。

经询问《管理学报》编辑部，答复如下：关于国家自然科学基金委员会管理科学部的缩写，当时我们认为“管理学”与“管理科学”本身有无实质差别，或“管理学”是否更合适，学界存在争议，故简称“管理学部”；后来知道有教育部科技委员会管理学部。为区别起见，后来的文章将前者简称为“管理科学部”，后者简称为“管理学部”。

编辑部用心良苦，可叹可赞，而且“后来”已经分开称呼，这就好了。

机构名称如同人名，应该一字不差才好，差一个字就不是这个机构或这个人了。“国家自然科学基金委员会管理科学部”，这是国家规定的正式名称，不管谁喜欢不喜欢，都要使用这个名称而不是给它改名字。

国内的现状确如《管理学报》编辑部所说，学术界对于“管理学”与“管理科学”的差别是存在争议的。我们认为管理科学包含管理学，有些人认为管理学包含管理科学。为了不影响讨论，这里约定：大管理学等同于我们所说的管理科学，所以，如果主张大管理学，则可以把“管理科学中国学派”理解为“管理学中国学派”来开展讨论。

2. 正本清源探讨基本概念与“梳理名词术语”两者的区别

《管理学报》2010 年第 9 期刊文《争鸣强劲 青年崛起 任重道远——第 3 届“管理学在中国”学术研讨会纪要》[2]，其中说：“另一个焦点是构建‘管理的中国理论’起点何在？刘人怀认为应从梳理名词术语入手（现状是学者们对基本概念理解不一，使深入探讨没有可能）；席酉民认为应该‘轻概念辨析’，重从实践中挖掘、归纳真正具有中国要素的有关管理规律理论。”我们认为从字面上看，两种“认为”似乎有较大分歧而相互对立，其实不然。我们（笔者及其合作者）也很重视“从实践中挖掘、归纳真正具有中国要素的有关管理规律理论”，在研讨会结束前的发言中已经做出说明。实际上，我们在前几年已经开展了工程管理领域中的“挖掘”与“归纳”，写出了研究报

① 本文原载《管理学报》，2011，8（9）：1263-1268，1352，作者：刘人怀，孙凯，孙东川.

告，发表了学术论文[3]。

关于“构建管理的中国理论”的起点问题，笔者在这次研讨会上的发言，旨在阐明：中英文之间、中西方文化之间，管理理念差异很大，不可忽视；对于“什么是管理”“什么是管理科学”这些基本概念应该进行一番正本清源的探讨。作为“起点”，这种探讨是无可回避的一项重要工作。

几年来，我们学术团队做出较多的探讨（见文献［1，3-19］。本节后面对它们多有引述，不尽标注）。为了阐明问题，有必要概述我们所做的部分探讨：

——什么是管理？管理是人类的第二类活动，是为第一类活动（生活的与生产的作业）服务的。这个定义是我们比较了教科书上许多种定义之后，反复推敲才提出来的。

——什么是管理科学？管理科学是人类研究管理活动规律和做好管理工作的全部知识之总和。管理科学是一个内容丰富的知识体系，包含理论部分与应用部分，两部分都是多层次结构。管理科学横跨自然科学与社会科学两大领域，包含国家自然科学基金资助的管理科学研究以及国家社会科学基金资助的管理学研究。

——中英文之间、中西方文化之间在管理理念上的差别可以用这样的式子表示：

管理＝management＋administration＋……

如果认为管理＝management，或管理＝administration，都是不妥当的。

——中英文之间、中西方文化之间在管理科学理念上的差别可以用这样的式子表示：

管理科学＝MS＋AS＋……，其中，MS 即 management science，AS 即 administration science。

如果认为管理科学＝MS，或管理科学＝AS，都是不妥当的。

——我们建议：把“管理”的拼音 guanli 推向国外包括推向英语世界，把“管理科学”翻译为 guanli science（GS）。

——我们建议：把“现代管理科学中国学派”翻译为 the Chinese school of modern guanli science，缩写为 CSMGS，为了避免 M 被误认为 management，索性去掉 M，把 CSMGS 写成 CSGS。为了行文简洁，本文后面多采用 CSGS 来表达。

——正视中英文之间的差异、中西方文化之间的差异，构建中西合璧的管理科学话语体系，争取中国人在管理领域的话语权和影响力，等等。

显见，上面所列举的各点，并不仅仅是“梳理名词术语”而已。如果简单地认同英语世界和西方人的理念，人云亦云，放弃中国人丰富而优秀的固有理念，无异于削足适履，是不可取的。

希望中国的管理学界重视构建中西合璧的管理科学话语体系。

3. 有没有“管理科学美国学派”

作为对于 CSGS 的质疑，有人问：有没有“管理科学美国学派”？我们的回答是：有！

“管理科学美国学派”无疑是有的：现在风行于全世界的就是美国学派，而且主要是企业管理的美国学派。试看世界管理百年史，从泰罗（“科学管理”创始人）到德鲁克等一系列管理学界的著名人物，几乎都是美国人。他们对管理科学贡献甚大，许多理论与方法具有永恒的价值，应该学习和借鉴。同时，不能忽视：美国人所研究的对象毕竟只是美国的管理，而且大多是企业管理，他们的研究成果首先是适用于美国，别的国家可以移植；但是必须“本土化”，否则就会“水土不服”。美国学派也有不同流派，如被称为泰罗制的“科学管理（学派）”，Mayo（梅奥）等开创的行为科学学派，Simon（西蒙）开创的决策理论学派，等等。中国学派主要靠中国人自己来创建。中国人研究中国管理，获得一系列研究成果且具有中国特色，就能形成 CSGS。同时，外国人研究中国管理，有些成果也可以纳入 CSGS。

日本人在管理领域颇有建树，有可能形成日本学派。韩国人则有可能创建韩国学派，还可能有新加坡学派、印度学派，等等。CSGS 加上这些学派，就可以形成东方学派。

东方学派，加上美国学派，加上欧洲学派，等等，这就是世界性的现代管理科学。

也可以这样说：现代管理科学可以分为东方学派、西方学派等；西方学派又可以分为美国学派、德国学派、英国学派等，东方学派又可以分为中国学派（CSGS）、日本学派、韩国学派等；其中，CSGS 又可以分为若干学派。

不管大学派、小学派，都可以统一命名为“（现代）管理科学××学派”，其中“现代”二字省略亦可。

现在在中国占主导地位的管理理论与方法实际上是美国学派。改革开放 30 多年来，中国的管理界很重视“洋为中用”。大量引进美国学派的成果。本土的古为今用和近为今用有一些，不占主导地位。洋为中用无疑还要继续做，积极做，但是光有洋为中用是不够的。我们认为[1,3-19]：创建 CSGS 的基本途径是洋为中用，古为今用，近为今用，综合集成，开拓创新。

目前管理科学的美国学派没有标以“美国学派”几个字，不是因为它不存在，而是因为它很强大，独一无二，所以无须标注。笔者相信：当管理科学中国学派屹立于世界管理科学之林的时候，当其他国家的学派也成长起来的时候，“管理科学美国学派”也就成为常见的一般术语了，就是说，为了区别起见，“美国学派”这几个字就不得不标注了。

二、我们为什么赞成“管理科学中国学派”这个提法

1. “管理科学中国学派”是一个很好的科学术语

我们认为“管理科学中国学派”是一个很好的科学术语，例如，它比“中国管理科学学派”要好。管理科学是世界性的，属于全世界全人类。管理科学中国学派既是中国的，也是世界的。管理科学中国学派这个提法符合国际国内惯例，如奥运会中国队、世博会中国馆、联合国大会中国代表团，等等，很合理，如果说成“中国奥运会运动队”“中国世博会展览馆”“中国联合国大会代表团”则不好。

另以北京的人民大会堂为例。大家经常说，人民大会堂上海厅、人民大会堂江苏厅、人民大会堂广东厅，等等，很合理，如果说成“上海人民大会堂会议厅”“江苏人民大会堂会议厅”“广东人民大会堂会议厅”则不好。

研究管理、研究中国管理的学者很多，采用的名称多种多样。例如，和谐管理、和合管理、中国管理学、东方管理学、中国式管理、中道管理、中国管理科学、中国现代管理科学、中国特色管理科学、有中国特色的管理科学、中国特色社会主义管理科学，等等。2005 年及以前，我们用的名称是“有中国特色的现代管理科学”[5,6]，后

来看到基金委管理科学部的提法“管理科学中国学派”，认为很好，就一直采用至今。虽然前面列举的名称在字面上差异比较大，其实学者们的心愿都差不多：研究管理、研究中国管理，创建一些理论与方法，有益于改进和提高中国的管理工作，并且为管理科学的发展做出中国人应有的贡献。

不同的名称，只要有自己的研究成果，都可以称为一个“学派”。有些人把“学派”看得很神秘，高不可攀，这是不必要的。何谓学派？一个学派，就是一群做学问的人，他们有共同的观点或思路，在一起开展研究工作，出了一些成果而且具有某些特色，就成为一个学派。学派有大有小，一个大学派可以包容若干小学派。一个小学派，可以是一个项目组，一个学术团队，人数可以是几十个人或十几个人，也可以是几个人。中国春秋战国时期出现了诸子百家，如儒家、道家、法家、墨家，等等，就是一个个学派，而且都是中国学派。

学派可以有包容与被包容、嵌套与被嵌套的关系。CSGS具有最大的包容性，是个大平台（大舞台）、聚义厅，是个大家庭[1]。

2009年4月份，基金委管理科学部在北京召开的一次研讨会上，前常务副主任陈晓田明确表示：“管理科学中国学派是迄今为止最好的提法。如果以后有更好的提法，可以考虑采用更好的提法，现在还没有，所以，我赞成继续采用中国学派的提法。”2年多的时间过去了，依然没有更好的提法。

我们认为，相互有所区别的各个学派（以各种不同的名称为代表），都可以为实现基金委管理科学部提出的战略目标贡献力量。我们认为：这些学派实际上都是管理科学中国学派之一、之二等。

有人问：“CSGS的代表人物是谁？”我们说：上述各种提法的代表人物，实际上都是CSGS的代表人物。我们开展CSGS研究是“自发研究加上响应号召”，自命为“革命军中马前卒”[1]。

我们还要说明：尽管我们主张CSGS这个提法，但是我们认为各个学派或各个研究团队可以自愿选择，继续使用或者改用自己喜爱的名称，无须强求一致。

海纳百川，有容乃大。无论是经济领域还是管理领域或其他领域，都会产生中国学派。任何领域的中国学派都可以有不同的流派或支派。但是，相比外国学派而言，它们都是中国学派。和而不同，大同归一。

2. 我们为什么加上“现代”两个字

“管理科学中国学派”加上“现代”两个字，是为了强调我们和同行们所研究的管理科学具有很强的现代性，是在计算机和互联网时代的管理科学。

我们主张的“管理”“科学”是大管理、大科学，“管理科学”则是大管理科学——它是一个内容丰富的知识体系，包括管理科学的理论部分和管理科学的实践部分，两部分都是多层次结构的大厦或金字塔[9,20]。

从这样的意义上理解管理科学，则不但现代有管理科学，古代也有管理科学。我国古代的学者们提出了丰富多彩的管理思想；古代有许多杰出的工程实践，如举世闻名的都江堰、万里长城，古人建造它们的时候，必定伴随着杰出的管理实践（包括今天可以称之为管理办法或管理条例等初步的管理理论与方法）。古代的管理思想与管理实践汇总起来，就是古代管理科学，可以称之为“古代管理科学中国学派”。

我们认为，基金委管理科学部希望创建的管理科学中国学派，当然是现代管理科学中国学派。所以，我们说的“现代管理科学中国学派”（CSGS），与基金委管理科学部说的“管理科学中国学派”，并无什么区别。在不至于与古代管理科学中国学派混淆的情况下，“现代”两个字可以不写。“古代”的英文是 ancient，则“古代管理科学中国学派”可以译为 the Chinese school of ancient guanli science，也可以是 the ancient Chinese school of guanli science，依据后者，则简记为 ACSGS，CSGS 与它就显著地区别开了。创建 CSGS，要继承和发扬 ACSGS 和外国各种学派。CSGS 之研究方兴未艾，我们相信，将会形成新的诸子百家。

三、创建 CSGS 的路线图与研究方法

我们在多篇文章中指出，创建 CSGS 的基本途径是“三室一厅”：洋为中用，古为今用，近为今用，综合集成[3-10,12-24]。这里再加一句话：开拓创新。五句话 20 个字，可以简化为两句话 8 个字：“三室一厅”，开拓创新。

1. 创建 CSGS 的路线圈：从中国模式到中国学派

要说明路线图，还要先讨论一下：什么是中国模式？中国模式与中国学派是什么关系？

较早提出“中国模式”的是高盛公司研究中国问题的资深专家 Joshua Cooper Ramo（乔舒亚·库珀·雷默），他于 2004 年 11 月发表文章“The Beijing Consensus”，文中提出“中国模式”。此后研究“中国模式”在国内外成为一大热门，文献［15-18］列举了几本专著（可以找到几十本之多）。

何为“中国模式”？简言之，就是中国人办事的方式方法。有人说：每一个中国人办事的方式方法都有所不同，怎么可能有“中国模式”呢？辩证法告诉我们：共性存在于个性之中。人们常说：世界上没有两片相同的树叶。是的，但是梧桐树叶与白杨树叶不会被认错，它们各有自己的特色。同理，既有美国模式，也有中国模式和其他模式，中国模式与美国模式是不一样的，它们各有自己的特色。

现在，谈论中国模式的外国人比中国人还要多。这种形势令人想起马克思和恩格斯在 1848 年所著的《共产党宣言》开头那段精彩的话语。

一个幽灵，共产主义的幽灵，在欧洲游荡。为了对这个幽灵进行神圣的围剿，旧欧洲的一切势力，教皇和沙皇、梅特涅和基佐、法国的激进派和德国的警察，都联合起来了。

……

从这一事实中可以得出两个结论：

共产主义已经被欧洲的一切势力公认为一种势力；

现在是共产党人向全世界公开说明自己的观点、自己的目的、自己的意图并且拿党自己的宣言来反驳关于共产主义幽灵的神话的时候了。

仿之。我们可以说：中国模式已经被西方和全世界公认为一种模式；现在是中国人向全世界公开说明中国模式的特点、目的、意图的时候了。

“中国模式”今天面临的形势与 1848 年“共产主义幽灵”面临的形势相比较，有两点不同：①关注中国模式的人不限于欧洲，而是遍布于全世界，尤其是美国政界和学术界；②各国政要、学者和大众对于中国模式的态度大不一样，有的是敌视或围剿，有的是怀疑与迷惑，有的则是欢迎和期待。“兼听则明，偏听则暗”，我们对各种声音都要听，都要加以分析，不要简单地肯定或否定；要保持清醒的头脑，上下齐心，落实科学发展观，构建社会主义和谐社会，让中国模式发扬光大，为全世界全人类树立了榜样。

我们认为，中国模式与中国学派的关系是：中国模式是中国学派的实践形式，中国学派是中国模式的理论形式（学术形式）；有中国模式，就会有中国学派；先有中国模式，后有中国学派；中国模式已经是客观存在（当然，它还会不断发展和完善），中国学派的研究与创建工作还有许多事情要做；包含 CSGS 在内的中国学派不会自动生成，要靠当代中国人（尤其是学者们）积极创建——这是当代中国人的历史使命和历史机遇。

从中国模式到中国学派——这就是创建 CSGS 的基本路线图。而研究中国模式，也有一个路线图，即对于企事业单位开展案例研究，从个别的、具体的模式逐步上升到一般模式。

2. 案例研究与问卷调查式的实证研究（empirical research，ER）两种实证研究方法的评析

我们前几年的研究工作侧重于基本概念的正本清源的探讨，这是起始阶段（即文献［2］所说的“起点”）必不可少的工作。这一工作做到一定程度就应该“转型”，多角度、多层次、多领域开展紧密联系实际的研究。

研究与创建 CSGS，应该理论与实践并重，运用多种研究方法。研究方法包括理论研究＋应用研究，文献研究＋实证研究，定性研究＋定量研究（钱学森先生提出。“从定性到定量综合集成”，以及综合集成研讨厅体系），等等。其中符号“＋”表示两者相结合并且交互作用。实证研究方法包括案例研究、问卷调查、田野调查等。

本节不打算全面论述各种研究方法，仅对案例研究与问卷调查式的实证研究（ER）这两种实证研究方法作一些评析。

一是案例研究评析。

案例研究主要是对实践进行采样研究，是一种很重要的研究方法。“麻雀虽小，五脏俱全”，一个典型案例就是一个具体模式，从多个具体模式可以上升到一般模式[5-6]。例如，研究中国的钢铁企业宝钢、鞍钢、武钢、邯钢等，可以分别得到宝钢模式、鞍钢模式、武钢模式、邯钢模式等，从这些具体的钢铁企业模式可以总结出钢铁企业的中国模式和钢铁行业的中国模式。

研究中国的家电企业海尔、美的、春兰等，可以得到海尔模式、美的模式、春兰模式等，从众多具体的家电企业模式可以总结出家电企业的中国模式和家电行业的中国模式。

同理，研究中国的 IT 企业联想、华为、北大方正，研究中国的汽车制造企业一汽、二汽、吉利，可以分别得到 IT 行业的中国模式、汽车行业的中国模式。

从众多行业的中国模式，可以总结和提炼出“工业经济的中国模式”；

同理，通过研究，可以总结和提炼出“农业经济的中国模式”“第三产业的中国模式”等。更上一层楼，则可总结提炼出“经济发展的中国模式”。

上面所说，是从行业的角度（横向角度）研究中国模式，还可以从地区的角度（纵向角度）研究中国模式（案例研究也可以采用各种层次的地区作为案例）。改革开放初期，乡镇企业就有苏南模式、温州模式、珠江三角洲模式等。当前，自下而上，我们可以从乡镇—县—地市—省开展案例研究，总结、提炼出省级区域模式，如广东模式、江苏模式、上海模式、河南模式、甘肃模式、新疆模式等，然后上升到东部地区模式、中部地区模式、西部地区模式，再进一步可以得到“经济发展的中国模式”。

横向角度的研究＋纵向角度的研究，可以得到更完整的“经济发展的中国模式”。

由“经济发展的中国模式”上升到理论的或学术的高度，可以创建“现代经济学中国学派”或“现代经济科学中国学派”。

同理，通过研究可以总结提炼出“管理（工作）的中国模式”，并且上升为“现代管理科学中国学派”（CSGS）。

上面说的典型案例，既包括成功的案例——这是主要的，也包括失败的案例。两方面的案例研究汇集起来才比较完整，不失偏颇。

企业案例既包括国企案例，也包括三资企业案例和民企案例，这样才比较完整，不失偏颇。

案例研究不光是研究中国的当代案例——这属于“近为今用”范畴，还要研究外国案例，做到“洋为中用”。

“古为今用”主要通过历史文献进行研究，也可以开展一些案例研究，如对都江堰、大运河、故宫建筑等进行实际考察。

笔者和项目组的同事们，采用文献研究与案例研究相结合的方法，研究了大中型工程项目管理的中国特色[10]。其中，洋为中用的案例有苏联成功发射世界上第一颗人造地球卫星、第一艘宇宙飞船，美国研制原子弹的曼哈顿计划、研制导弹核潜艇系统的北极星计划及阿波罗登月计划等；古为今用的案例有都江堰和大运河等水利工程、宋代修皇宫的丁谓工程等；近为今用的案例有“两弹一星”（原子弹、导弹、人造地球卫星）、神舟飞船、嫦娥一号与奔月工程、三峡工程、青藏铁路等，还研究了负面案例1958年大炼钢铁。对于三方面的案例研究分别归纳，最后总结、提炼出工程管理的中国特色——包括五个方面：决策机制与宏观管理方面，运作机制与管理方面，监督与防范机制及其管理方面，评估机制与管理方面，管理伦理、道德及其他方面，共计30条。进一步开展研究可以提炼出大型工程项目管理的中国模式，并且上升为“工程管理中国学派”——它是CSGS的重要组成部分。

二是问卷调查式的实证研究（ER）评析。

ER是一种问卷调查式的实证研究方法，最近几年在国内很流行。有人把ER与实证研究画等号，认为ER是全部的、唯一的实证研究方法。其实不然，实证研究的方法还有案例研究、田野调查，等等。

最近几年，很多博士生做学位论文都采用ER方法，而且形成一种环境压力：似乎不做ER就是“落伍”，就是“没有水平”，就可能通不过。ER俨然成为研究方法之“王”。

我们认为，ER本身是一种很好的研究方法，但是，现在过于“时髦”，被滥用了，以至于走了样、变了味，出现许多“假冒伪劣产品”。一些人的ER是怎么做的呢？且看：

——作者根据自己的需要编造几个“假设”，选择一种“量表”，设计一种“问卷”，发放给一批“专家”填写，然后以很高的回收率收回，据说“有效率”也很高；

——于是，很漂亮地通过“统计检验”，全部“假设”都获得“支持”（有的“研究者”则故意让少量假设不获得“支持”，或者仅仅获得“部分支持”，这就显得更加“高明”了）；

——于是，万事大吉，学位论文完成，答辩会轻易通过，皆大欢喜。

不过，其中有一个“小小的秘密”：“专家”是谁？——远在天边近在眼前：“专家”大多是MBA班的学员、总裁培训班的学员，这是比较好的情况；差一些的情况是找一些本科生包括大一新生（他们有多少管理实践呢？天晓得）：甚至还有“闭门造车”、捉刀代笔的现象——这就无须评论了。

为了揭示这种变了味的ER之弊端，不妨在此做一个简单的“模拟”：用这种方法来“研究”哥白尼日心说。且看：

——提出两个假设，假设1，地球绕着太阳转，假设2，太阳绕着地球转；

——设计问卷，参照国外权威方法设计“量表”；

——选择足够数量的专家（几百名、几千名，应有尽有）：山区老农民和海边老渔民。入选资格：66岁以上（寓意“六六大顺”），文化程度无须高，文盲也可以，关键是一个“老”字，因为年老，他们对于太阳的日常运行有足够多的观察天数，堪称“资深专家”了；而且，如果他们不知道哥白尼则更好，免得受哥先生的“精神污染”；

——发放问卷请专家们填写（可能有些老人不会填写，那不要紧，可以现场举办“填写问卷培训班”，耐心地“教会”他们）；

——接着，回收问卷（回收率和“有效率”要多高就有多高）；

——然后，运用权威软件进行“统计检验”，可以轻轻松松得出结论：假设2获得支持。假设1没有获得支持。

这样的ER能够发现真理吗？这样的ER只可能模糊真理，毁坏真理。

回到积极的意义上说，ER本是一种很有用的研究方法，但是，并不是唯一的研究方法，它不能“包打天下”。“尺有所短，寸有所长”，各种方法和模型各有自己的优越性与局限性。研究一个复杂问题，应该采用多种研究方法（形成方法体系），建立多种模型（形成模型体系）。例如，打仗需要多兵种、多种武器配合作战，不是光靠一挺机关枪或者一门大炮就能取得胜利的。

如何得到真知灼见，发现真理？笔者认为，发现真理主要有三种途径：经验的总结、理性的思考、天才的直觉。谓予不信，试问：爱因斯坦提出相对论，是依靠ER吗？邓小平同志提出“一国两制”伟大构想，是依靠ER吗？或者反过来问：依靠ER能够提出相对论吗？依靠ER能够提出“一国两制”伟大构想吗？

ER以及其他实证研究方法，应该与发现真理的多种途径结合起来。

人类的思维可以分为逻辑思维、形象思维和创造性思维，尤其是创造性思维。

总之，创建CSGS既有必要性，又有可能性。我们相信：CSGS一定可以创建起来，而且时间不会太长，大概再有5～10年时间可以基本完成，只要仁人志士努力去做。

参考文献

[1] 刘人怀，孙凯，孙东川. 大平台、聚义厅及其他——四谈创建现代管理科学中国学派的若干问题. 管理学报，2009，6（9）：1137-1142.

[2] 长松. 争鸣强劲 青年崛起 任重道远——第3届“管理学在中国”学术研讨会纪要. 管理学报，2010，7（9）：1290.

[3] 刘人怀，孙凯，孙东川. 大型工程项目管理的中国特色及与美苏的比较. 科技进步与对策，2009，26（11）：5-12.

[4] 孙东川，林福永，孙凯. 创建现代管理科学的中国学派及其途径研究. 管理学报，2006，3（2）：127-131.

[5] 孙东川，林福永. 洋为中用，古为今用，近为今用——创建有中国特色的现代管理科学. 系统工程，2004，（增刊）：13-15.

[6] 王克强，孙东川. 管理科学与物理学、医学的分析比较——论创建有中国特色的现代管理科学. 生产力研究，2005，（11）：211-213.

[7] 国务院学位委员会. 授予博士、硕士学位和培养研究生的学科、专业目录（1997）. 北京：高等教育出版社，1999，543.

[8] 刘人怀，孙东川. 谈谈创建现代管理科学中国学派的若干问题. 管理学部，2008，5（3）：323-329.

[9] 刘人怀，孙东川. 再谈创建现代管理科学中国学派的若干问题. 中国工程科学，2008，10（12）：24-31.

[10] 刘人怀，孙东川，孙凯. 三谈创建现代管理科学中国学派的若干问题. 中国工程科学，2009，11（8）：18-23，63.

[11] 孙东川. 三分法与管理工作. 管理学报，2009，6（7）：861-866.

[12] 刘人怀，孙东川.《学科目录》第12门类与管理科学话语体系——五谈创建现代管理科学中国学派的若干问题. 学位与研究生教育，2010，（8）：67-73.

[13] 刘人怀，孙凯. 工程管理信息化的内涵与外延. 科技进步与对策，2010，27（19）：1-4.

[14] 孙东川，朱桂龙. 系统工程基本教程. 北京：科学出版社，2010：60-69.

[15] 郑永年. 中国模式—经验与困局. 杭州：浙江人民出版社，2010.

[16] 雅克 M. 当中国统治世界：中国的崛起和西方世界的衰落. 张莉，刘曲，译. 北京：中信出版社，2010.

[17] 张维为. 中国震撼：一个“文明型国家”的崛起. 上海：上海人民出版社，2011.

[18] 丁学良. 辩论“中国模式”. 北京：社会科学文献出版社，2011.

[19] 孙东川. 软科学研究的方法与模型体系. 华东工学院学报（哲学社会科学版），1990（1）：100-109.

[20] 孙东川，张振刚，孙凯. 一项重大历史使命：创建现代管理科学的中国学派. 美中经济评论，2008，8（1）：57-63.

[21] 韩巍. 管理学在中国——本土化学科建构几个关键问题的探讨. 管理学报，2009，6（6）：711-717.

[22] 刘人怀，杨东进，朱峰. 国际化视野与本土化关注——MBA战略管理案例精选集. 北京：科学出版社，2011.
[23] 王礼恒，刘人怀，郭重庆，等. 20世纪中国知名科学家学术成就概览管理学卷，第一分册，北京：科学出版社，2013.
[24] 王礼恒，刘人怀，郭重庆，等. 20世纪中国知名科学家学术成就概览管理学卷，第二分册，北京：科学出版社，2013.

中国管理科学的现状和走向[①]

今天，在景色美丽、气候温和的澳门，由国家教育部科学技术委员会管理学部和中国工程院院士广州咨询活动中心联合主办的“首届中国管理科学论坛”隆重开幕了！请让我代表国家教育部科技委员会管理学部、中国工程院院士广州咨询活动中心和中国（澳门）综合发展研究中心，向来自海峡两岸、香港、澳门的朋友们表示热烈的欢迎和真诚的敬意！

1840 年以来，中华文明的发展在历史合力的作用下，经过几代人的努力，发生了巨大的转折和变化，一度在西方资本主义冲击下衰落的中华文明，今天已经重新走上了伟大复兴之路！

面对实践，中国的管理科学研究，走到了哪里？该去向何方？为此，我们组织了本次论坛，请来自海峡两岸、香港、澳门的管理学者进行研讨，欢迎大家畅所欲言。

明天上午，还要举行 2012（澳门）经济合作交流峰会暨第二届文化传媒论坛和我们管理科学论坛的联合开幕式，请大家参加。明天下午再举行我们管理科学论坛的会议。

最后，要特别感谢澳门特别行政区政府和中央人民政府驻澳门特别行政区联合办公室对我们论坛的关心和支持，要特别感谢澳门商报和广州生产力促进中心对本次论坛的大力支持！

① 本文是首届中国管理科学论坛开幕词，澳门，2012 年 12 月 12 日；现代管理的中国实践，北京：科学出版社，2016，79.

传统文化基因与中国本土管理研究的对接：现有研究策略与未来探索思路①

2008 年韩巍[1]在《管理学报》发表了《从批判性和建设性的视角看“管理学在中国”》一文，对国内学者所创建的三类本土管理理论进行了深入评述。文章认为，除缺少清晰的认识论与本体论基础之外，本土学者所建立的理论，尤其是东方管理学与和合管理都存在着用文化认同来替代文化解剖、以意识形态来代替理论“实然性”分析等嫌疑。之后，彭贺等[2,3]分别于 2009 年、2011 年在《管理学报》上发表论文来回应韩巍的评价。他们认为，作为中国学者，基于文化认同来发展本土管理理论是中国学人必须经历的阶段[2]，而管理研究作为一种社会科学研究，必然会出现“价值涉入”，不可能完全脱离意识形态[3]，因此东方管理学并不是意识形态，而是科学[2]。

这场“火力十足”的学术争论，不仅得到了学术界的广泛关注，也引起人们对“如何选择研究方法？如何完善研究规范？”等问题的思考，更凸显出人们在如何借助中国传统文化的滋养来发展本土管理学的具体道路上存在的分歧。实际上，围绕着这一“古为今用”的重大问题，学术界一直存在着激烈的争论。一些学者认为，中国悠久独特的历史文化是中国式管理得以成立的关键以及成长的根基[4-7]。中国管理实践之所以与西方存在差异，主要是其本体价值——文化价值观的不同[8]，因此主张基于中国传统文化精髓来创建本土管理理论。还有一些学者则认为，尽管西方管理理论与中国文化的冲突，使得创建管理科学的中国流派成为必然[9]，但中国管理学的本体只能是本土管理实践，而不是本土管理思想[10]。由此，传统文化只能作为可供借鉴的一个重要的思想库，却不能成为发展中国式管理理论的根基。甚至有人认为中国传统文化自身的保守倾向与陈旧观念难以支撑中国本土管理研究的发展[10-14]。尤其是境外的学者，在承认中国传统文化独特性的同时，又推崇西方管理理论的普适性。他们主张以西方管理理论为基础，通过改良的方式来提高其在中国环境中的解释能力[15-20]。这两种观点的相互抵触，不仅说明中国本土管理学尚处在幼年时期，还告诉我们要想发展本土的管理理论，必须在深入认识中国传统文化独特基因的基础上，客观分析上述两种观点的优劣，由此发现传统文化与中国本土管理理论对接的途径。本文首先对中国传统文化的独特性进行简要探讨，其次全面评价当前有关本土管理研究，最后提出新的研究策略。

一、中国传统文化的独特基因分析：层次与关联

文化是个难以界定而又应用广泛的概念。广义上的文化分为物质与精神两个层面，而狭义的文化仅指精神层面。笔者借鉴人类学大师马凌诺斯基[21]的文化定义，并以文

① 本文原载《管理学报》，2013，10（2）：157-167，作者：刘人怀，姚作为.

化学者泰勒[22]所提出的狭义“文化”概念为基础，将中国传统文化定义为中国的先贤经过几千年的积累与进化，在特定的自然环境、经济社会环境、政治架构的作用下所创造和流传下来的精神财富总体，包括民族思维方式、价值观念、宗教信仰、风俗习惯、伦理道德、性格特征、知识结构、科学技术与各种制度等。传统文化不仅能显示出中国人的心理图像与文化品格，并且还积淀为一种文化遗传基因，深深融入中国人的社会、经济、法制、精神等层面，以巨大的力量影响着人们的思想意识和行为，决定着社会历史的发展进程。同样传统文化也是影响管理科学等社会科学发展与建构的内生变量。

尽管中国传统文化的众多特征早在商朝就已经基本定型[23]，但由于深受儒、道、法、释等不同学派的影响，中国传统文化成分繁杂且含义交叠，使得人们对其独特基因的概括还存在着分歧。例如，郭济兴[24]提出天人合一、以人为本和“民本”思想、和合文化是中国传统文化最为重要的 3 个特点。黄小军[25]则认为中国传统文化的最本质特征可概括为 3 点：天人合一的宇宙观、君权至上的政治思想和以孝为核心的伦理观念。笔者根据我国传统哲人的经典著作（如《论语》《大学》《道德经》《韩非子》等），结合众多研究中国传统文化的重点文献①，将中国传统文化的独有特征总结为核心概念与具体应用结构两个层面。天人合一、王权神授、祖先崇拜与阴阳思想（含金木水火土五行结构）[23]等，还有道[26,27]都是中国传统文化最为核心的概念或基因。而且这些结构相互影响，形成复杂的共生关系。由这些核心思想引申出的第二层核心概念包含：和合（和谐）思想②、君权至上、中庸和家族-血缘思想、以孝道为基础的仁义道德等。这些结构运用到实践中，可分为国家、家庭与个人三个层面来简要描述。国家层面的行为是“平天下”，以达到“国泰民安”的目标，其具体思想结构包含仁义道德、小康大同与民本思想，应用结构包含内儒外法、权威主义、和合、集体主义等。家庭层面的主要行为是“齐家”，目标追求“家和万事兴”，其思想结构包含血缘意识、仁义道德、和合，应用结构包含家族至上、长者权威、缘圈结构、孝道纲常（孝、顺、悌、慈等）。个人层面主要追求“正心修身”，以达到“人和”的目标，其思想结构包含和合、中庸与伦理道德等，应用结构包含内圣外王、道德规范（仁义礼智信）、忠孝与义利观等。有关中国传统文化各基因的层次与简要关联、这些特殊基因在组织内部所呈现出的具体结构见图 1，中国人的思维具有整体观、结果导向、直觉思维与伦理本质等特点[28]。

---

① 重点文献的选择与分析方法如下。首先，以“中国/文化＋传统文化”“中国文化＋特征”“中国文化＋精髓”等为题名在中国知网上搜索，按顺序选这 3 个搜索列表中引用率最高的前 60 篇文章、前 20 篇文章与全部 12 篇文章作为初步的分析群体。其次，在这 92 篇文章中，再按主题是否与“中国传统文化”或与企业管理紧密相关进行筛选，确定 27 篇论文作为备选分析论文集。最后，在厘清了最基本的传统文化基因及其含义之后，以这些词汇作为题名继续搜索并结合引用率又选择 27 篇文献纳入备选分析论文群。经由 2 位作者一道运用内容分析法对这 54 篇文献合并进行分析，初步确定了传统文化因子结构的早期版本，再由广东行政学院硕士研究生张崇根对以“传统文化＋企业管理”作为主题选择的引用率最高的 22 篇文献进行传统文化基因词汇的用词频率分析，最终确认了中国传统文化的独特基因图谱，见图 2.4。因篇幅所限，有关具体分析过程在这里不再赘述。

② 有研究表明，和合与和谐是类同的概念，笔者不做区别。具体见张立文．中国文化的精髓——和合学源流的考察．中国哲学史，1996，(1)：13-57.

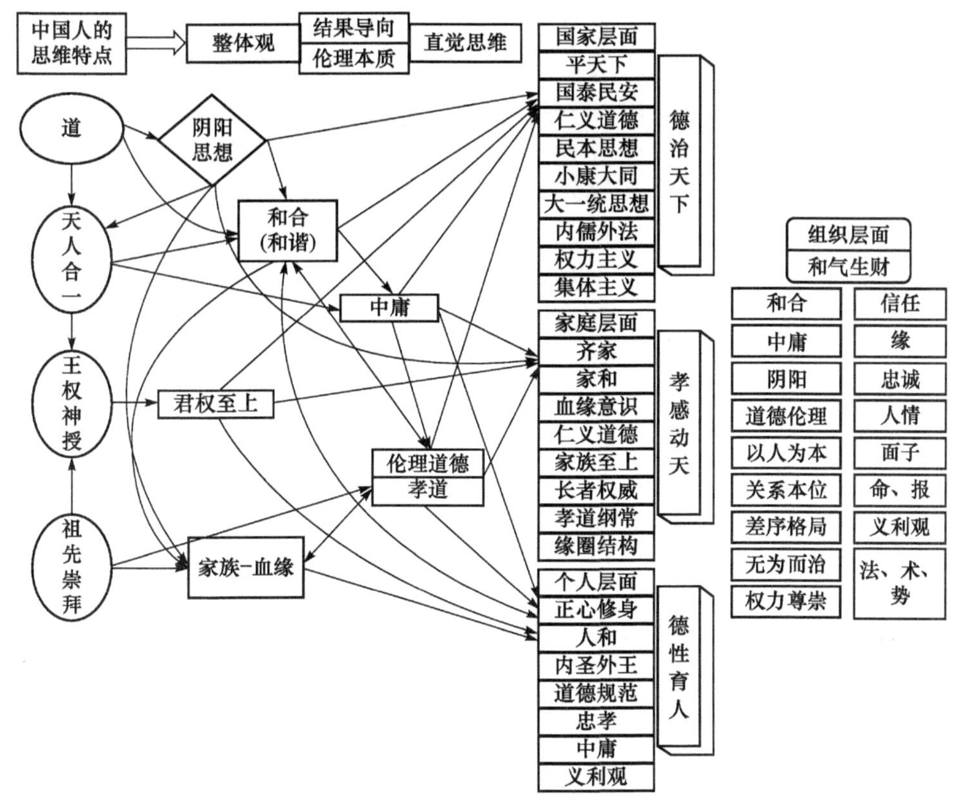

图 1 中国传统文化基因层次与关联简图

二、现有本土管理研究与中国传统文化基因的对接：两种策略的比较

从当前国内外文献来看，学术界在利用传统文化来创建中国本土管理学理论的过程中，主要采取了两类研究策略：文化核心植入式与文化情境嵌入式。

1. 文化核心植入式的研究策略

所谓文化核心植入式的研究策略是将中国传统文化的精髓作为内核放入西方管理理论的框架之中，来创建中国本土管理理论。实际上，在罗纪宁[4]所界定的全盘西化派、洋为中用派与中国式管理学派等中国管理研究学派中，后两者均采用了文化核心植入式的研究策略。采用此策略的学者认为，中国悠久灿烂的传统文化不仅是创建中国式管理科学理论的重要基础，更是构建有关理论的核心内容[4-7,29,30]。他们或者以西方成熟的管理理论为基础，重新诠释中国传统文化中的管理思想；或者将基于西方的某个理论与中国传统文化进行糅合提出一种新的理论；或者将中国传统哲学作为内核并与西方管理理论相结合创建一种新的管理理论。其代表性理论包含和合理论、东方管理学与 C 理论。

1）代表性理论简介

(1）和合管理理论。和合管理理论试图在中国文化基因——传统“和合”哲学思想的基础上，实现与现代经济学和管理学理论的融会贯通[29]。该理论认为管理是以人为主体的社会性合作活动。其主要内容包含：构建以人为本与和合相互叠生的价值观体系[31]，形成独特的和合管理原理、方法与理念[32]。其中，“和”，即和谐、和睦、和平、谐和、中和；“合”，即合作、联合、结合、融合、组合。“和”是“合”的基础和前提，“合”是“和”的选择和结果，“和合”表达了和气生财、合作制胜的哲学理念[29]。其采纳的研究方法是“以社会存在决定社会意识为基本研究纲领，将中庸原则与矛盾分析相结合的规则、历史和逻辑相结合的规则、借鉴与创新相结合的规则与社会科学研究一般方法（包括个体分析与整体分析相结合的规则、实证分析与规范分析相结合的规则）有机统一”[33]。和合管理在实践中提倡的是，立足于“和合”方式方法，通过对竞争的包容和矛盾的化解，来形成“和合发展力”，促成“中和”状态与稳定发展，即最终实现企业利益和消费者效用共赢的目标。和合发展力就是在和合理念及其战略思想指导下，通过合作伙伴之间各种相关因素的优势互补，包括生产、技术、价格、市场、管理等各个方面的有机整合，形成有利于共同发展和增加盈利的能力[29]。

(2) 东方管理学。东方管理学是由复旦大学苏东水教授首创并由其团队经过 30 多年的研究形成的，是具有中国特色、全球视野的当代管理新理论。其研究对象并不限于企业管理领域（即中国与华商管理实践中的现象与问题)，还包括政治、经济、科技等方面的管理，即涵盖宏观（国家)、中观（产业）与微观（企业、家庭与个人）三个层面的管理问题[30]。该理论以中国管理、西方管理以及华商管理的理论与实践为理论基础[34]，选择“道、变、人、威、实、和、器、法、信、筹、谋、术、效、勤、圆”15 个哲学要素，萃取出“以人为本”“以德为先”“人为为人”的“三为原理”，形成“治国”“治生”“治家”“治身”的四治体系，构建“人道行为”“人心行为”“人缘行为”“人谋行为”“人才行为”的五行管理理论，其管理目标是构建和谐社会的“和贵”“中和”“和合”[34]。东方管理学“应以问题为中心，根据问题来选择研究范式，寻找东方管理独特的研究方法，综合运用主位和客位研究策略将是未来东方管理方法论方面研究的重要方向”[35]；“以能有效解决问题作为判断、选择研究方法的标准，量化的方法与质化的方法均是研究问题的有效途径”[35]。很显然，东方管理学全方位地运用了诸如以人为本、阴阳五行、和合与道等传统文化基因，并试图融合儒、道、释、法等各派思想的精髓。

(3) C 理论。由美国学者成中英[36]所创立的 C 理论，“以《易经》为基础，以中国传统智慧与西方科学精神的融会贯通为目的，以‘中国管理科学化、科学管理中国化’为宗旨，以集科学、文化、艺术三位一体为特征，总结中外管理理论而建立起来的一套具有中国特色的现代管理哲学理论体系。”[36]该理论注重企业伦理，以“金、木、水、火、土”五行相生相克原理为基础，以人性、智慧的结合互动，形成一种阴阳互补、相辅相成的中国管理基本模式。该理论将管理最重要的环节归纳为决策、领导、权变、创新与统合人才等，提出“守成知变、穷化创新、定位断疑、简易即时”4 个管理原则和“知、行、体、用、主、客、内、外”8 个管理要素[36]。“以道、法、兵、墨、儒的精神分别代表管理中的决策、领导、生产、营销、人事 5 项活动，模仿五行原理建立管理要素运作与评估整合系统模型”[36]。C 理论主张采用系统分析与案例研究的研究方法，强调体用互动与主客兼容的具体研究策略，并赞成吸取一切有益的科学知识与中华管理哲学的精华来开展中国本土管理学的创新。

2）策略的优点

（1）有助于开发中国文化独特基因的管理学韵味。按照西方管理学的思路来对传统文化的精髓进行有针对性的梳理，有助于深度解读与开发中国文化独特基因的管理学韵味。例如，黄如金与东方管理学对“和合”的现代管理学意义的解读与开发，苏东水团队对“以人为本”“德治思想”的全方位解读与运用，成中英对“易经文化”的管理学价值的深度利用，这些都在一定程度上开拓了中国文化与管理学对接的途径。

（2）有助于发掘中国管理理论的独特之处。采用整体观的认识角度来组建中国本土管理理论，有助于发掘中国管理理论的独特之处。例如，苏东水及其团队在东方管理学的研究中提出需在宏观、中观到微观三个层次，从“和合”的角度组建中国管理理论，并且围绕着人与自然、人与社会、人与组织、人与人之间的关系提出了较全面的治理模式。这种观点实际上凸显出中国人对“管理”含义的认知与西方人对“management”的认知之间的区别[37]。这比较有利于整体地开发中国传统文化的价值，以实现文化的传承与发展，并对中国本土管理学的创建有所贡献。

3）策略的不足

（1）在本体设定上存在偏差。采用这一策略的学者无一例外地声称其理论研究的本体是需要“直面”的中国本土管理实践，但实际上在对包含“和合”“易经文化”在内的中国传统文化的管理意蕴的挖掘中，却有意无意地将文化本身当作了研究对象。正如吕力[10]一直强调的那样，中国本土管理学的本体只能是本土管理实践，而不是本土管理思想。这种偏差使得本土管理研究难以触摸到真正的研究对象。

（2）未能提供较为清晰的认识论。采用这种策略的学者多数采取的是传统上的整体认识论。这种认识论往往用笼统的规定性来反映复杂多样的管理实践中的关键命题[38]，用中国哲人的直觉感悟来代替科学的论证过程，甚至没能具体地描述整体认识论的分析路向。这样会引发两种结果。一是因认识论的缺位，而导致中国传统思想的“不可言喻本体论”[39]占主导作用，即学者往往从意会知识推理出需要经验支持的实践知识。例如，C 理论把《易经》所表达的晦涩难懂的五行理论推而广之为管理理论中的 5 个职能。二是无法弥合不同流派思想的差异。例如，C 理论试图用道、法、兵、墨、儒来代表与整合决策、领导、生产、营销、人事 5 项活动。但实际上这 5 个流派之间的差异，如没有科学的整体认知思路是难以弥合的。

（3）较少使用现代的研究手段。采用这一策略的学者多数采用传统中常见的偏重现象描述与机理分析等定性研究方法，很少采用西方流行的以实证研究为代表的定量研究方法。诚然“实证研究”的绝对主导地位不利于更全面地展开科学研究[8,40]，也不能深入地把握研究的实质，但仅偏重定性研究，也只能局限于哲理层面，无法深入地认知中国本土管理现象背后的原因与运作机制。此外，整体观本身也存在一定问题，因为没有必要的分解，在一定程度上根本无法展开管理研究[11]。就最具整体观的中医理论来看，其构建的基础必须是对人体构造与心理的分解与剖析。显然，仅靠模糊的整体定性的研究方法，难以达成发展中国本土管理理论的战略任务。

（4）缺少对传统文化的辩证性解读。在价值取向上，采纳这一策略的学者多数对中国传统文化精华持有强烈的认同感，并由此整体地将其纳入西方管理框架之中。作为中国学者，应该倡导对优秀民族文化的强烈认同，这也有利于从主体的角度来培育本土管理研究。但这并不意味着对传统文化的全盘接受，不恰当地拔高往往导致将本土文化基因神圣化与固态化[11]，使其变得难以理解，降低其实践价值。例如，“和合”作为具有至上价值的、中华文化的核心，被众多学者用作创建中国本土管理理论的基础。但实际上我国各传统思想流派在“和合”的达成途径上存在较大分歧。儒家着眼于德性化人的人为措施，道家主张无为自然，法家提倡“法、术、势”，佛家则强调因缘际会。由此看来，法、道与儒三家均认为：和合是可以达到的一种境况，但在佛家看来，那只是一种终极的目标，硬求则是痴妄。显然，要将“和合”作为管理的内核，需按照科学探索的本质要求，对“和合”的传统实现机制进行深入的现代化解读，以解决其存在的歧义。况且，中国人所追求的“和合”背后的等级制度、权威至上、血缘主义、孝道纲常等制约因素，对人的控制、对组织的管理发挥了至关重要的作用，这与西方人的“和谐”背后的个性解放是完全不能相提并论的。

（5）缺乏真实可靠的实践经验积累。从理论建构的标准来看，任何科学理论的诞生需要直面现实的管理经验作基础[38]。但就现有这些本土理论而言，多数学者将构建的重点集中在宏观层面，对于依赖实践经验支持的微观（企业）管理涉及较少。这种基于“应然性”的叙事性研究，既缺少满足“实然性”要求的实践指导，更缺乏真正具有科学内涵的理论建构。例如，和合管理理论只是讲“和合”管理如何重要，能够达到什么目的，却较少涉及具体操作的内容，也缺乏对管理的众多子理论（如人力资源管理理论与战略管理理论）的清晰构建。

2. 文化情境嵌入式的研究策略

文化情境嵌入式的研究策略是将文化作为研究本土管理学理论的情境（即环境）变量，实施西方管理理论的情境嵌入，即将管理理论嵌入到一定的情境之中来创建中国本土管理学派。郭毅[42]将其表达为寻找管理理论的地方性知识。应该说采取这个策略的研究目前在境外学术界与国内学术界都占主导地位[15,18,19,40,42-55]。

1）具体操作

运用此策略的学者认为，由于人类活动的社会嵌入属性[56]，所有的理论都因情境的约束而未必能在新的情境中适用[49]。受悠久历史文化传统影响的中国具体管理情境就是本土管理研究发展的基础[19]。这样，中国本土管理研究可被看成是西方管理理论在中国环境中的情境化问题。情境是指“与现象有关并且有助于解释现象的周边环境，也特指超越现有研究中分析层次的有关因素”[57]。由此可知，情境化研究也就是把组织所在国家的社会、文化、法律和经济因素作为情境变量，探讨这些因素对于组织特征这些因变量的影响[19,58]。围绕着西方理论与中国情境之间的不匹配状态，Tsui[19]提出了两个解决途径：理论应用与理论创新。前者是将现有的理论直接应用于中国的情境中，后者是指过通过发展新的理论来解释中国环境下的独特管理现象。Barney 和 Zhang[42]将其归纳为中国管理理论与管理的中国理论的区隔。Whetten[49]从理论与情境的关系出发，借用 Peterson（彼得森）的知识创造的整体观框架，指出理想情况下的学术探究是将现有的知识（文化普遍性）与新情境中的关键知识（文化特殊性）嫁接结合的迭代、循环的过程。由此他进一步强调：中国本土管理研究的焦点在于如何实现现有理论的情境敏感性与跨情境运用问题，提出要从情境化管理理论（情境嵌入理论）与理论化情境理论（情境效应理论）两个途径来实现理论创新的累进。他认为，前者是将西方理论嵌入到中国情境中，而后者则是考虑中国情境的变动对西方理论的影响。实质上，在情境被视为可变动因素的情况下，情境效应理论也可归入情境嵌入理论。在此基础上，Child[15]以 Weber（韦伯）的情境构成框架为基础，从物质体系（经济、技术）、理念体系（文化价值与理性、宗教价值、政治价值）与制度产出（政府、中介机构、国际规章与标准）三个方面勾画了情境的实际构成，并对具体的情境化研究方法进行了较全面的探讨。邹国庆等[55]、田恒[52]则从中国现实出发，将中国本土管理研究的情境从制度、文化与社会结构三个方面进行了详细分析，其中转型经济中的政府与市场共同主导资源分配、伦理价值导向、非正式制度的盛行与以差序结构为特征的关系性运作机制都成为表述中国情境特殊性的重要元素。Cheng 等[43]则以他们的亲身研究经历，为大家提供了一个开展中国本土情境化研究的具体操作路径。按照深度理解华人传统、发现值得关注的管理议题、实地观察、理论建构、理论实证检验与理论再修正六个阶段的研究步骤，就可以实现本土化研究的完整运作，由此来保证研究的切题性与严谨性的统一[18]。

采纳这一策略的研究日见增多，如境外不少学者运用社会资本理论对中国情境中的关系的研究[42]、Cheng 等[59]对中国组织中家长式领导所进行的研究、李新春和刘莉[60]有关嵌入式关系对家族企业创业成长的影响研究、蓝海林等[61]对中国战略行为的情境嵌入式研究等，但其数量还不足以完成有关理论的系统化建构。而运用情境式研究策略的本土研究多出现在社会学与心理学领域。有关社会学涉及中国文化的文献参见周雪光的综述，而中国本土心理学方面的研究见文献［16］。

2）策略的优点

（1）有助于解读、把握中国文化的构成因素。国内外学者可以利用这一策略，多层次地剖析中国文化环境的各种因素，从国际的视野来捕捉与解读具有中国特色的文化基因，运用各种西方成熟的研究方法来循序渐进地推进中国管理的本土化进程，这样更有利于全面把握中国文化的构成因素及其影响效应。

（2）有助于中外学者间的学术对话。由于这一策略天然地与西方管理理论的接轨性，有助于促进本土学者与国际学者之间的全面对话，推动中国本土管理研究成果的国际化，在一定程度上可提升中国本土管理学的国际地位。

3）策略的不足

就其想要达成的发展中国本土管理研究的目标来看，该策略也存在着以下的问题。

（1）错误地将西方管理理论奉为圭臬。不管是理性认可[15,18,19,42,43,48,52-55]还是否认[49]中国式管理理论存在的学者，都认为中国本土管理研究的发展必须以西方管理理论为核心内容。因此，中国本土管理研究或者是西方管理理论在中国情境下的“借鉴与改良”[42]，或者是现有理论经过多文化跨情境运用而不断改善与扩展的结果[49]。有学者甚至认为，除非现有的管理理论都不能解释中国独特的管理现象，中国本土管理研究才有存在的必要[49]。而瞄准提高中国企业绩效的本土管理研究很可能面临符合切题性标准，却难以达到严谨性要求的困境[18]。

（2）不适当地将情境化研究等同于本土化研究。采纳这个研究角度的学者或许无意将情境化研究等同于本土化研究，但从国内学者对此的呼应来看，西方理论的中国情境化研究俨然已经成为本土管理研究的主流。不过李平[17]认为，情境化研究与本土化研究存在两点差异。第一，目的不同。情境化可以瞄准多种因素，而非针对本土文化因素，而本土化研究则针对寻找本土独特和新颖的启示与结论。第二，路径依赖不同。情境化研究既可涉及历史又可涉及现实，但本土化研究必然对本土历史文化有路径依赖。李平还认为到本土研究的高级阶段，必须淡化情境化研究，暗示所谓的情境化研究只能算是低级的本土研究。吕力[62]认为，Tsui 的情境嵌入式研究所产生的只能是西方理论的扩展，而特定的情境未必能产生新的理论。郭毅[41]指出，情境化研究与本土化研究的不同在于：前者将重点放在研究的途径和研究方法的优劣比较上；后者则将情境作为前提，力图建立一个相对独立而特色鲜明的“地方性知识”系统。由此，笔者认为，即使不断优化情境构成、完善具体的操作方法，也未必能保证情境化研究产生期望中的中国本土化管理理论。

（3）对本土因素的把握出现了偏差。这一问题源于两个方面。一是因为操作上的问题。不管被操作为自变量还是调节变量，此策略都将文化看成是影响企业和企业中的人的外围环境因素。然而，作为能影响人的价值观与行为的重要因素之一，文化是内化在人的心理与行为之中，进而内化在企业的行为之中，因此将文化看成是环境变量，难以把握到文化的真正含义。von Glinow 和 Teagarden[18]指出，仅把关系与社会资本理论联系在一起，却不涉及关系背后深刻的传统背景，难以保证情境化研究的科学性。二是因为情境化设定上的问题。根据嵌入性理论，情境化研究实际上将文化情境（作为社会结构的一部分）看成是既定的、外生的因素[63]。这一假设忽视了经济行为和文化环境共同演化的特征[64]。实际上，人的行为会受文化环境的影响，但也同样反过来影响文化环境。从 Polanyi 的嵌入性理论出发，管理是整体被嵌入到文化情境之中[65]，需要使用整体主义方法来分析文化对管理的影响，但西方管理学的研究方法大多是解析式的，难以提供可靠的答案。此外，尽管有学者将嵌入性区分为四种形式：结构嵌入（行动者之间的物质特征与结构关系）、认知嵌入（引领经济逻辑的结构化的心智过程）、文化嵌入（形塑经济目标的共享信念与价值观）与政治嵌入（限制经济权力的国家角色与制变法规，很多情况下与制度嵌入同义）[65]，但 Granovetter 的将社会因素简化为关系网络的观点[57]，无疑大大降低了嵌入性在管理理论中的解释力。显然，文化不仅具有内生性，还具有整体性（文化与行为的不可分割性）、演变性等特点。西方在发展自己的管理理论时，并未涉及对情境的描述与析出，而是对蕴含着文化意味的管理行为与活动进行了研究。这说明，情境化研究并不是把握本土因素的唯一途径。

（4）理论的情境化假设无法被证实。Tsang[47]并不认为情境化研究是一个真正的议题。他运用迪昂-奎因论题指出：理论（或基本假设）不可能孤立地经受检验，必须与辅助假设组一起经受联合检验。由此，一般学者所认为的只有当一种或多种情境影响改变了理论的预测时，才有可能对理论有贡献的观点是错误的，因为即使基本假设被证实，也不表示辅助假设能被证实。但一个假设被改变时，要清楚地证明此改变是因情境因素的影响而发生的是非常困难的。吕力[62]将其表达为情境化假设无法被证实也无法被证伪。

三、中国本土管理研究的未来探索思路

作为5000年悠久历史的产物，传统文化早已经融入中国人的心理与行为之中。尽管近代以来，经过几代人前赴后继的“西化”努力，再加上当前全球化进程日益深入的影响，中国文化的某些特质经历了一定程度的改变或者具体操作的方式有所进化，但其基本内核却未发生根本性的变化。例如，有研究显示，中国人对于收入差距的容忍度有所提高，但对于分配公平的要求依旧占主导地位[66]。又如，中国人对于组织内冲突所扮演的作用开始有现代意义上的认识[67]，但这种认识上的变化依靠的是对外来文化的借鉴与融合，以及对其在中国文化框架内的创新性运用。因此，笔者认为要想实现利用传统文化基因来推进中国本土管理研究的目标，就必须关注文化的整体性、内含性、长期性、演化性与隐形约束的特点，采取聚焦于文化内生性的研究策略。这一策略将文化看成隐含在一个国家的人们心理与行为背后的内生因素，并承认其与企业的管理行为的整体纠结，由此来研究中国企业的管理实践活动，其具体的运作应注意以下几个方面。

1. 秉持辩证的认识论与本体论

基于文化内生性的本土管理研究应以现实的中国企业实践为自己的研究本体，并采取以“感性认识—理性认识—认知证伪”为主要演进阶段的辩证认识论。其要点可从四个方面来阐述。①要以提高对中国管理实践的指导价值为目的。具体来说，就是在有效借鉴本土管理思想的基础上，围绕着蕴含传统文化意味的、复杂多样的中国管理实践需要，并以此为研究的出发点，考虑组织内外部的各种因素的影响，借鉴社会学、心理学、经济学等多个学科的理论与方法，来有效推进中国本土管理研究。②建立有效的整体认识论的分析框架。至关重要的是对整体分析进行从认识论到具体操作方法等多个层面的研究，完成对其的现代化提升，力求实现整体分析与解析分析的相互借鉴、直觉思维与理性思维的实效结合。③要从历史演变与整体分析的角度来研究传统文化独特基因在中国组织中的运用方式及其效果，由此发掘其理论与实践价值，尤其要选择那些影响本土组织的生存理性[68]的文化基因进行深入研究。④要保持开放的心态来借鉴西方管理理论，既要继承其中的有益经验，又要辨识出那些不适合中国国情的成分，在注重双方互动协调的基础上进行中国本土管理的创新，并力求提升与世界管理知识库对话的能力。例如，由席酉民及其团队创建的“和谐管理”理论尽管其体系也因对中国传统文化基因——“和谐”的理想化规定而引起一定的争议[1]，但从其所持的演化主义认识论来看，尤其是以“不确定性—支配权”为特征的本土化领导理论的出现[70]及其后续研究[70]，可以说是运用科学认识论的成功案例。

2. 逐步完成对独特文化基因的现代化解读

在中国传统文化中，具有独特性的基因数量众多，历史悠久，但不少文化基因或因内涵陈旧，或因历代学者的不同阐述出现“理念”与“实践”之间的严重偏差，

已经无法适应新时代的要求，故而难以实现与西方管理理论的对接。为了完成这一对接，要辩证科学地对待中国文化的独特基因，以不卑不亢、庄敬自重的心态审视自我，实现对中国文化的现代化解读。具体来说，既要从文化认同的角度使用传统的识辨方法来找回传统文化经典的“本原”意义，又要从文化借鉴与融合的角度运用先进的手段来发掘独特文化基因的现代价值。下面以“和谐/和合”这一备受推崇的文化基因为例来阐述一下具体的做法。一方面，传统文化中对“和谐/和合”基本含义的界定是基本清晰的，但道、儒、法、释等流派对达成“和谐”的途径却存在差异性的理解。这就引出了一个传统文化经典理论中根本没有解决的问题——如何在出现各自利益矛盾的情况下做到“和而不同”？实际上，在中国封建社会的集权体制之下，“和谐”很容易在权力的运作下得以实现，但却会隐藏背后真正的“不和谐”，反过来可能直接威胁到“和谐”的可行性。这些问题的解答都需要我们走进历史，从概念构成的初级阶段去寻找它最初的含义，了解其演变的进程。这样才能确定我们所了解的“和谐”与老祖宗所指的“和谐”有什么区别？才能掌握“和谐”的具体成分，找到继承与扬弃的出发点。另一方面，为实现对“和谐”的现代化解读，还应分析中西方对“和谐”的认知存在着哪些区别？从现有研究看，西方人的和谐是一种清楚界定各自利益基础上的和谐，而中国人的和谐是一种暧昧意味下的和谐。那么，哪种“和谐”更有利于中国组织的发展呢？在何种机制下，中国人的“和谐”与西方人的“和谐”在一定程度上能够趋同？从阴阳的角度来看，和谐本身是否也存在着对组织的负面效应？这种负面效应该如何消解？这些问题的解决，都需要以开放的心态去借鉴西方文化的精华，推动与本土文化的有效对接，来提炼“和谐”概念的现代化价值。更为关键的是要寻找到达成“和谐”的具体方法。应该说席酉民等[71]在和谐管理理论中所创立的“和则”与“谐则”之间的互动机制对此问题有一定的贡献，但依旧需要对其具体的机制持续进行研究。需要特别强调的是，走进历史，并不是拒绝现代化进程，奉老祖宗的学说为不可更改的圭臬，而是要找回其不受历史迷雾遮蔽的“本来面貌”。唯有如此才算是真正地在承认文化演变的同时，实现对中华文化的认同与继承。

3. 审慎地选用多元的研究方法

要实现中国本土管理理论的审慎、科学与客观的发展，中国管理学界应该重视研究方法的选择。

(1) 根据问题来选择合适的研究方法。为了避免产生重复的研究成果以及不够成熟的理论创建，本土管理研究须直面中国管理现实[1,11,68]中出现的问题。尽管李平[17]认为并非西方管理已研究过的问题，中国学者就不能研究。但是问题的质量直接影响到研究的质量与未来的走向。有学者指出，对问题的选择需要分析其对中国管理实践影响的程度（是否重大）、其具有的新颖程度（是否具有独创性）[72]。在确定了问题之后，还需要根据问题来选择合适的研究方法。例如，想挖掘“和谐”的现代内涵，就需要采用中国传统上的历史文献解读方法来找回其在本土文化意蕴中的具体成分，同时分析西方文化视野下“和谐”的含义，并通过跨文化比较的方法来实施意义比较与整合，由此实现对组织中“和谐”意义的现代化解读。如果要寻找“和谐”在组织内运作的具体机制，就需要运用案例研究方法、深度访谈等定性研究方法，再辅以定量研究方法来检验构想的理论与假设。对于属于狭义管理科学领域中的数量化问题，则应采取定量研究方法。

（2）要根据不同的研究阶段来选择合适的研究方法。在研究初期，主要目标是围绕本土企业的管理实践来发现重大问题，可采用扎根理论、诠释学等方法，以确定问题的实质与基本轮廓。在研究中期，重点目的是在透视西方管理理论的基础上，提出中国本土的独有理论或者假设，可采取案例研究、比较分析等方法。在研究后期，主要焦点是检验中国本土的管理理论或假设，可采用实证研究等定量研究方法。当然大面积的样本调查与多文化样本对比也可成为方法的备选项。

（3）要注意研究具体操作的规范性。尤其应注意整体分析的规范性。建议运用郭毅[42]所强调的“特质-深描-理解”研究途径，也可参考 Cheng 等[43]的六阶段研究操作模型来不断规范本土化研究。

4. 择取独特的研究视角

要实现中国传统文化与中国本土管理研究的有效对接，就必须科学地展现中国文化的独特基因。为此，可采取具有中国文化特点的研究视角[62]。笔者认为，视角可以是宏观上的，如可从国家文化的角度来研究“和谐”管理在国家与地区层面的运作机制。如从他者与文化、政党间互动的角度来研究“中国共产党的成功之道”[73]。视角也可以是微观上的，如从关系的角度、人情与面子的角度，来分析中国文化在企业管理决策过程中所扮演的角色。还可从文化因素与管理活动之间的多层次互动来把握文化与管理活动的整体演变。更应该借鉴社会学[11]与心理学[16]有关中国本土的研究视角与方法。笔者认为视角的选择应该囊括情境化研究，尽管后者作为一种研究方式存在一定问题，但对其充分利用可帮助我们从一个侧面来分析文化因素（作为内部因素或环境因素）是如何参与到管理活动的运作过程中的。此外，情境化研究有利于通过跨文化比较，发现中国本土管理与西方管理理论之间的区别与相同点，帮助中国学者勾勒本土管理理论的边界。

四、中国管理科学理论发展的科学路径

研究结果表明：中华传统文化的独特基因历史悠久、结构繁多，形成一个从核心结构到应用结构的多层次复杂体系。现有研究中，将传统文化基因与中国本土管理研究进行对接的策略包含两个：文化核心植入式与文化情境嵌入式。文化核心植入式的研究策略是将中华传统文化的特有基因当成本土管理学的内核，与现有的西方管理学理论实现对接。文化情境嵌入式的研究策略则将西方管理理论嵌入中国特有的文化情境中，来创建中国本土管理理论。前一策略有助于深度开发文化独特基因的管理学韵味与从整体观的角度来创建中国本土管理理论，但由于本体设定与认识论存在的缺失，并且缺少对传统文化的辩证性解读，缺乏长期的、真实可靠的经验积累，无法解决传统文化思想与现实管理活动之间的契合问题。后一策略注重借鉴各种西方成熟的管理理论与研究方法，有利于全面剖析与把握中国文化环境的构成因素及其影响效应，能循序渐进地推进中国管理的本土化与国际化进程，但因先

验地将西方管理理论当成模仿的经典，用无法被证实的情境化研究来代替本土化研究，难以把握本土文化的内生性等特征。未来中国管理科学理论的发展应采取将文化作为内生变量的研究策略，秉持辩证的认识论与本体论，有步骤地开展对传统文化独特基因的现代化解读，择取中国文化特有的研究视角，依据问题来选择有效的研究方法。唯有在结合东西方交流所带来的文化演变与对传统文化精髓的批判性继承的基础上，直面中国企业的现实管理问题，关注研究的严谨性，才可能寻求到中国本土管理理论发展的科学路径。

参考文献

[1] 韩巍. 从批判性和建设性的视角看“管理学在中国”. 管理学报，2008，5（2）：161-168，176.

[2] 彭贺，苏勇. 也从批判性和建设性的视角看“管理学在中国”——兼与韩巍商榷. 管理学报，2009，6（2）：160-164，186.

[3] 彭贺. 管理学研究中的“价值无涉”与“价值涉入”. 管理学报，2011，8（7）：949-953.

[4] 罗纪宁. 创建中国特色管理学的基本问题之管见. 管理学报，2005，2（1）：11-17.

[5] 罗纪宁. 中国管理学研究的实践导向与理论框架——一个组织管理系统全息结构. 管理学报，2010，7（11）：1646-1651，1670.

[6] 王学秀. 文化传统与中国式管理价值观选择. 科学学与科学技术管理，2006，27（2）：156-160.

[7] 周建波. 中国管理学建构与演化——基于哲学四分法与管理文化结构的推演. 管理学报，2008，5（6）：781-791.

[8] 吕力. 管理学的元问题与管理哲学——也谈《出路与展望：直面中同管理实践》的逻辑瑕疵. 管理学报，2011，8（4）：517-523.

[9] 本刊特约评论员. 试问管理学——管理学在中国刍议. 管理学报，2007，4（5）：549-555.

[10] 吕力. 什么是“中国管理学”研究的本体. 管理观察，2009，29（16）：16-17.

[11] 韩巍. 学术探讨中的措辞及表达——谈《创建中国特色管理学的基本问题之管见》. 管理学报，2005，2（4）：386-391.

[12] 罗珉. 中国管理学反思与发展思路. 管理学报，2008，5（4）：478-482.

[13] 吕力. 论“中国管理学”的新文化保守主义立场. 知识经济，2009，11（8）：179-180.

[14] 谭力文. 中国管理学构建问题的再思考. 管理学报，2011，8（11）：1596-1603.

[15] Child J. Context，comparison，and methodology in Chinese management research. Management and Organization Review，2009，5（1）：57-73.

[16] 梁觉，李福荔. 中国本土管理研究的进路. 管理学报，2010，7（5）：642-648.

[17] 李平. 中国管理本土研究：理念定义及范式设计. 管理学报，2010，7（5）：633-641，648.

[18] von Glinow M A，Teagarden M B. The future of Chinese management research：rigour and relevance redux. Management and Organization Review，2009，5（1）：75-89.

[19] Tsui A S. Contextualization in Chinese management research. Management and Organization Review，2006，2（1）：1-13.

[20] Zhao S M，Jiang C Y. Learning by doing：emerging paths of Chinese management research. Management and Organization Review，2009，5（1）：107-119.

[21] 马凌诺斯基. 文化论. 费孝通，译. 北京：华夏出版社，2002.

[22]　泰勒 A B. 原始文化. 连树声，译. 上海：上海文艺出版社，1992.
[23]　艾恺. 中国文化形成的要素及其特征. 传统文化与现代化，1994，(3)：83-91.
[24]　郭济兴. 现代管理和中国传统文化. 经济经纬，2003，20 (4)：76-79.
[25]　黄小军. 中国文化的原始特征. 云南学术探索，1995，(4)：16-20.
[26]　邹顺康. 论中国传统文化的特征. 西南师范大学学报：人文社会科学版，2002，28 (2)：111-114.
[27]　成中英，郭桥. 儒家和道家的本体论. 人文杂志，2004，30 (6)：1-6.
[28]　连淑能. 论中西思维方式. 外语与外语教学，2002，24 (2)：40-46，63-64.
[29]　黄如金. 和合管理：探索具有中国特色的管理理论. 管理学报，2007，4 (2)：135-140，143.
[30]　苏东水. 21 世纪东西方管理融合与发展的趋势——当代中国东方管理科学的创新与实践. 上海管理科学，2008，(5)：4-10.
[31]　黄如金. 和合管理的价值观体系. 经济管理，2006，28 (12)：11-22.
[32]　黄如金. 和合管理的真谛：和气生财，合作制胜. 管理学报，2007，(3)：258-265.
[33]　黄如金. 中国式和合管理的方法论问题. 经济管理，2006，(18)：4-13.
[34]　彭贺，苏宗伟. 东方管理学的创建与发展：渊源、精髓与框架. 管理学报，2006，3 (1)：12-18.
[35]　彭贺，苏东水. 论东方管理的研究边界. 学术月刊，2007，39 (2)：71-79.
[36]　成中英. C 理论：中国管理哲学. 北京：中国人民大学出版社. 2006.
[37]　刘人怀，孙东川. 谈谈创建现代管理科学中国学派的若干问题. 管理学报，2008，5 (3)：323-329.
[38]　林福永. 一般系统结构理论//许国志，顾基发，车宏安. 系统科学与工程研究. 上海：上海科技教育出版社，2000：183-195.
[39]　成中英，郭桥. 儒家和道家的本体论. 人文杂志，2004，30 (6)：1-6.
[40]　韩巍. 论“实证研究神塔”的倒掉. 管理学报，2011. 8 (7)：980-989.
[41]　郭毅. 地方性知识：通往学术自主性的自由之路——“管理学在中国”之我见. 管理学报，2010，7 (4)：475-488.
[42]　Barney J B，Zhang S J. The future of Chinese management research：a theory of Chinese management versus a Chinese theory of management. Management and Organization Review，2009，5 (1)；15-28.
[43]　Cheng B S，Wang A C，Huang M P. The road more popular versus the road less travelled：an “insider's” perspective of advancing Chinese management research. Management and Organization Review，2009，5 (1)：91-105.
[44]　Leung K. Never the twain shall meet? Integrating Chinese and western management research. Management and Organization Review，2009，5 (1)：121-129.
[45]　Leung K. Indigenous Chinese management research：like it or not，we need it. Management and Organization Review，2012，8 (1)：1-5.
[46]　Li P P，Leung K，Chen C C，et al. Indigenous research on Chinese management：what and how. Management and Organization Review，2012，8 (1)：7-24.
[47]　Tsang E W K. Chinese management research at a crossroads：some philosophical considerations. Management and Organization Review，2009，5 (1)：131-143.
[48]　Tsui A S. Editor's introduction-autonomy of inquiry：shaping the future of emerging scientific communities. Management and Organization Review，2009，5 (1)：1-14.

[49] Whetten D A. An examination of the interface between context and theory applied to the study of Chinese organizations. Management and Organization Review，2009，5 (1)：29-55.

[50] 韩巍. “管理学在中国”——本土化学科建构几个关键问题的探讨. 管理学报，2009，6 (6)：711-717.

[51] 韩巍. 管理研究认识论的探索：基于“管理学在中国”专题论文的梳理及反思. 管理学报，2011，8 (12)：1772-1781.

[52] 田恒. 中国情境下的管理学研究探索——基于理论发展脉络的观角. 科技管理研究，2011，31 (1)；226-230，242.

[53] 韵江，陈丽，李青霞，等. 中国管理学的效用：基于四维矛盾的解析. 管理学报，2011，8 (4)：486-492.

[54] 周建波. 中国管理环境：暧昧文化因子、管理真实形态与情境嵌入机理. 管理学报，2012，9 (6)：785-791，817.

[55] 邹国庆，高向飞，胥家硕. 中国情境下的管理学理论构建与研究进路. 软科学，2009，23 (2)：135-139，144.

[56] Granovetter M. Economic action and social structure：the problem of embeddedness. American Journal of Sociology，1985，91 (3)：481-510.

[57] Cappelli P，Sherer P D. The missing role of context in OB：the need for a meso-level approach. Research in Organizational Behavior，1991，13 (1)：55-110.

[58] Cheng J L C. Notes：on the concept of universal knowledge in organizational science：implications for cross-national research. Management Science，1994，40 (1)：162-168.

[59] Cheng B S，Chou L F，Wu F Y，et al. Paternalistic leadership and subordinate responses：establishing a leadership model in Chinese organizations. Asian Journal of Social Psychology，2004，7 (1)：89-117.

[60] 李新春，刘莉. 嵌入性—市场性关系网络与家族企业创业成长. 中山大学学报（社会科学版），2009，19 (3)：190-202.

[61] 蓝海林，李铁瑛，王成. 中国企业战略管理行为的情景嵌入式研究. 管理学报，2009，5 (1)：78-83.

[62] 吕力. 中国本土管理学何以可能——对“独特性”的追问、确证与范式革命，管理学报，2011，(12)：1755-1761.

[63] 王凤彬，李奇会. 组织背景下的嵌入性研究. 经济理论与经济管理，2007，27 (3)：28-33.

[64] William F，Sewell. A theory of structure：duality，agency，and transformation. American Journal of Sociology，1992，98 (1)：1-29.

[65] 符平. “嵌入性”：两种取向及其分歧. 社会学研究，2009，24 (5)：141-164.

[66] 李强. 中国社会分层结构的新变化. http：//www. sociologyol. org/yanjiubankuai/xueshuredian/xueshurediansan/xues hurcdian3liebiao/2007-03-23/476. html [2012-08-28].

[67] Tjosvold D. Cooperative and competitive goal approach to conflict：accomplishments and challenges. Applied Psychology：An International Review，1998，47 (3)：285-342.

[68] 郭毅. 活在当下：极具本上特色的中国意识——一个有待开发的本土管理研究领域. 管理学报，2010，7 (10)：1426-1432.

[69] 韩巍，席酉民. 不确定性—支配权—本土化领导理论：和谐管理理论的视角. 西安交通大学学报（社会科学版），2009，29 (5)：7-17，27.

[70] 韩巍，席酉民. 两个中国本土领导研究的关键构念. 管理学报，2012，9 (12)：1725-1734.

[71] 席酉民，韩巍，尚玉钒. 面向复杂性：和谐管理理论的概念、原则及框架. 管理科学学报，2003，6（4）：1-8.

[72] 陈春花. 中国企业管理实践研究的内涵认知. 管理学报，2011，8（1）：1-5.

[73] 郭毅. 论本土研究中的他者和他者化——以对中国共产党成功之道的探讨为例. 管理学报，2010，7（11）：1675-1684.

Research on Chinese school of modern GUANLI science①

中文摘要

本文是期刊社论，研究了共建现代管理科学中国学派问题。特别提出，用汉语拼音词 GUANLI 作为一个世界学术语言基本单词，它包含多个英语单词：management，administration，governance 等的含义。

1. Introduction：GUANLI is our basic terminology

The development of the Chinese economy has achieved a great success over the past 38 years since China's reform and opening up. According to the statistic figures，based on the international exchange rate，provided by the World Bank，etc.-China's gross domestic product（GDP）reached US＄5.88tn in 2010，while the GDP of the USA was US＄14.58tn and that of Japan was US＄5.5tn. China's GDP accounted for 40.33 per cent of that of the USA and 106.91 per cent of Japan. China has overtaken Japan，being the world's second largest economy. In 2015，the GDP of China was US＄10.86tn，while that of the USA and Japan was US＄17.97tn and US＄4.12tn，respectively. China's GDP accounted for 60.43 per cent of that of the USA and 263.59 per cent of that of Japan. The international community referred to China's continuous rapid economic development as the "economic blowout" and expected that China will soon exceed the USA and become the world's biggest economy. According to the GDP calculation based on purchasing power parity provided by the International Monetary Fund，China's GDP was US＄17.6tn in 2014，while that of the USA was US＄17.4tn. China has already surpassed the USA as the world's largest economy. However，China is quietly looking at this thing and keeping a cool head. China's science and technology are also developing rapidly at the same time. "Technology blowout" has begun as well. All-rounded industries in China are thriving. The Chinese society is stable and united，without school shootings and much social violence and terror. Chinese people and foreign tourists have both felt a high level of sense of security. According to a survey conducted by the US Pew Research Center，the satisfaction level of Chinese citizens toward Chinese Government and personal life status ranked the highest in the word. That means：along with China's economic development，there is a progress in science and technology and social stability，and the management，the administration，and the governance of China has proved to be very successful. In Chinese，management，adminis-

① Reprinted from *Chinese Management Studies*，2017，11（1）：2-11. Authors：Liu Renhuai，Sun Kai and Sun Dongchuan.

tration, and governance can be used interchangeably as one Chinese character: GUANLI. The application of GUANLI is very broad, not only including the above-mentioned three English words but also having the meaning of control, run, rule, supervise, regulate and other words, which have synonyms meaning as the above three words that can be expressed as GUANLI. GUANLI can be used both as a verb and as a noun.

GUANLI is a word that embodies the characteristics of typical Chinese culture, for which no counterpart exists in English. GUANLI should be introduced solely to the world as a unique, significant scientific term on its own. To highlight it, we advocate that all letters should be typed in upper case. It can also be expressed in the following mathematical formula:

$$\text{GUANLI} = \text{management} + \text{administration} + \text{governance} + \cdots \quad (1)$$

In this article, and in all other articles within our series, GUANLI exists as a fundamental academic term. It is applied as a replacement for management, administration, governance and many other words that embody the similar meaning as these three words.

What needs to be particularly emphasized is that: we present Management Science (MS) and Administration Science (AS) in one term GUANLI Science (GS), formulating in a mathematical equation:

$$\text{GS} = \text{MS} + \text{AS} + \cdots \quad (2)$$

We study GUANLI and the GS to research and set up the Chinese School of Modern GUANLI Science (CSMGS). This is a great cause with massive work to do and needs to be done evolving many people's endeavor. After 5 to 10 years'efforts, this great cause could make a great success.

2. Derivatives of GUANLI and Other Related Words

We need to derive some English words and terms for more accurate description of Chinesestyle GUANLI concepts and GUANLI tasks on the basis of basic terminology GUANLI to precisely communicate with the English-speaking world.

If we use those English words like management, administration or governance directly to explain their Chinese meanings, but not GUANLI and its derivatives, then we will fail to make effective and efficient communication and cooperation.

Let us take the following two examples. The first one reads: "三百六十行，行行有管理", how to translate this Chinese idiom into English? It is impossible to translate it into English if not adopted the word GUANLI. The first half of the sentence will be easy to translate: *There are various trades and professions*. The second half of the sentence will be difficult to translate: how to translate "管理" (GUANLI)? The word management or administration used alone here is inappropriate, even the combination of management and administration fails to convey the exact meaning underneath, as there are several underlying meanings. Generally speaking, management refers to profitability, as the job responsibility is due to the head of a factory or a manager; administration refers to public welfare, as the job responsibility is bore by such as a school president, the chairman of a hospital or a government officer. Their job duties are highly different but are all included in the variety of trades. However, as we create the word GUANLI, this sentence can be easily translated as: *There are various trades and professions, which all involvee the GUANLI.*

The second example: "人人都是管理者", how to translate this sentence into English properly? There is a Chinese word "管理者" (GUANLIZHE). We can create a word GUANLIST (meaning the people of GUANLI) based on the word GUANLI. In this way, this sentence can be translated as: *Everyone is a GUANLIST*. It is simple and straightforward. GUANLIST can be a manager, administrator, president, executive, head, commander, conductor, etc. All of them fulfill the job of GUANLI. If we do not use GUANLIST and rather translate this phrase into *Everyone is a manager/president.* Then it is a big mistake.

Another word GUANLIER can be created based on the word GUANLI by reference to the word engineer, which has the same meaning of the word GUANLIST. Similarly, this sentence can be translated as: *Everyone is a GUANLIER.*

We explain why there is a saying that everyone is a GUANLIST/GUANLIER. A single violist can perform himself solely, but an orchestra needs an impresario in the group. Therefore, we can say that a single person can be a GUANLI himself, but a group of people requires a specialized GUANLIST. Even though it is just one person, he/she also needs to GUANLI his/her own business (including one's economic and non-economic affairs, and one's material and spiritual life).

The conductor of a band is actually a GUANLIST, engaging in the technical GUANLI job of the band. A band consists of financial manager as well as administrator (the band leader or boss). It is the same case for enterprises. A large enterprise consists of many GUANLISTs; they work together with different job specifications, and they are divided into different levels: top GUANLISTs (president, general manager, chief engineer, chief economist and chief accountant, engaged more in strategic management levels), middle GUANLISTs (branch director, workshop director, etc., engaged in production control level), grass-root GUANLISTs (squad leader and team leader, engaged in daily operations). The middle level can be further divided into several sublevels, and the GUANLIST team across the enterprise forms a "pyramid". The pyramid should not contain too many hierarchies. A large number of hierarchical levels will reduce the efficiency; therefore, the less hierarchical levels, the better.

The Chinese word "管理" (GUANLI) is composed of two one-syllable words: "管" (GUAN) and "理" (LI), which shows the typical characteristics of Chinese culture. "管" (GUAN) represents constraints for people, so that human beings abide by the norm; "理" (LI) represents categorization, categorizing to make things in good order. "管理" can also be simplified as a single-syllable word "治" (ZHI). There is an old Chinese saying that goes: "ZHI army when gets on the horse, ZHI citizens when gets down the horse." (It means that a competent general can GUANLI well both the army and the citizens.) "治国理政" (ZHIGUOLIZHENG) is to GUANLI national affairs. Lao Zi says: "Governing a Big Country is as easy as Cooking a Small Dish".

According to the research, the word "管理" (GUANLI) appeared in the history book *OLD BOOK of TANG* (旧唐书) in the 945 A.D. (Song Dynasty) for the first time, and appeared more than 14 700 times in the encyclopedia *SIKUQUANSHU* (四库全书) in 1778 A.D. (Qing Dynasty); their meanings are basically equivalent to that of the present day.

The Chinese term "管理学" (GUANLIXUE) does not have an appropriate counterpart in English that can correspond fully to the same meaning in Chinese. Management and administration are often used to represent "管理学". For example, a conference's title reads "管理学 in China", and the title is translated into "Management in China". However, this translation may result in some ambiguity. The meaning of "管理学 in China" in Chinese is accurate: a group of scholars hold a conference for the discussion of knowledge of GUANLI. But after being translated into English, it could be understood that a group of entrepreneurs or government officials hold a meeting for discussion of management affairs. Based on the word GUANLI, we can make up a word GUANLIOLOGY, the title of the conference can be exactly translated into "GUANLIOLOGY in China". Accuracy achieved, with no ambiguity.

The Chinese term "管理工作" (GUANLI GONGZUO) is hereby suggested to be translated into GUANLIWORK, GUANLIJOB or the work of GUANLI, the job of GUANLI.

Transliteration is a very common phenomenon when communication occurs between different languages. For example, Chinese terms like 雷达, 坦克 and 咖啡 are transliterated from radar, tank and coffee, respectively. 星巴克咖啡 is Starbucks Coffee, partial transliteration and partial paraphrase. The English word tea is the transliteration of China's Minnan dialect, Russian word÷aé is a transliteration of Mandarin.

The Chinese Pinyin "guanxi" (means special social relationships) has been recognized as an English word. The Hollywood movie "KUNG FU Panda", KUNG FU is the Wade-Giles of "功夫" (gongfu).

The words Confucius and Mencius are the transliteration of 孔夫子 (Confuci) and 孟子 (Menci) by the Wade-Giles and plus the suffix "-us". Presently, these words should adopt the Chinese Pinyin directly instead, and then they are Kongfuzi (or Kongzi) and Mengzi, respectively.

3. China Model and Chinese School

China's development has formed uniquely the "China Model". A number of people worldwide are investigating the "China Model"; this terminology was first proposed by foreigners. Chinese should advocate more actively in the study of the "China Model".

With this "China Model", there will be a corresponding "Chinese School". The China Model is a practical form of the Chinese School, and the Chinese School is the academic form of the China Model. China Model is the creation of numerous Chinese labors, and the Chinese School is the blank academic areas for future creation and contribution.

Western economic theories failed to explain China's economic development. There is an urgency to research and create a Chinese School of economics. Western theories of management, administration and governance cannot fully explain China's GUANLI-WORK, so to create the CSMGS has become the most urgent requirement for the time being in research field. Since 2004, we have been committed to the research and creation of CSMGS, and fruitful research achievements have been obtained.

In 2005, The Department of Management Sciences of National Natural Science Foundation Committee of China (NNSFC) suggested that 10 to 20 years in the future, the academic foundation of Chinese School of Management Sciences (CSMS) would be fully and well laid. Before then, our research was temporarily phrased as the "Modern GUANLI Science with Chinese Characteristics" (Sun and Lin, 2004; Wang and Sun, 2005). We are greatly satisfied with the term CSMS and changed it to CSGS: the Chinese School of GUANLI Science, and decided to apply this term as the theme for future research.

We propose the following definition:

Definition. All the Chinese research results of GS are called the Chinese School of GUANLI Science (CSGS).

GS and the Chinese School of GUANLI Science (CSGS) have existed since ancient times.

This definition has two corollaries:

Corollary 1. All research results of GS by ancient Chinese have been called the Chinese School of Ancient GUANLI Science (CSAGS).

Corollary 2. All research results of GS by contemporary Chinese have been called the CSMGS.

Prior to 1840 in contemporary Chinese history, the CSGS are referred as CSAGS. The Hundred Schools (HS) are outstanding during the period from pre-Qin Dynasty, which belonged to a great group of different schools of thoughts. Lao Zi, Kong Zi, Mo Zi, Meng Zi, Hanfei Zi, etc., who are representatives of different schools; these schools have coexisted peacefully with complementary or opposite thoughts. HS's thoughts and theories are still remarkable nowadays and getting highly recognized worldwide.

CSMGS is the CSGS of computer and the internet eras. In terms of time, CSMGS can also accommodate the relevant research results after 1840.

Now, many foreign researchers also, especially overseas Chinese scholars, investigate the Chinese GUANLI, and their research results can also be included into CSMGS (Zhu, 2007; Nonaka and Zhu, 2012; Li, 2013).

CSMGS is also a large school; it is a group of schools. Within the group, each school can make up their own name. For instance, "Modern GUANLI Science with Chinese Characteristics", "GUANLI Science of China", "GUANLIOLOGY of China", etc. Just as the same case as HS of pre-Qin Dynasty, CSMGS will become a place for the new HS.

For the sake of brevity, CSGS is used as short for CSMGS.

CSMGS has the following basic features:

- it is Chinese, with distinct Chinese characteristics, which is growing up in China, suitable for China's national conditions, effectively resolving contemporary problems in China's economic development and social progress, and the Chinese scholars are the main force of created CSMGS;
- it is modern, using modern computer and internet technology, and reflects the latest achievements of GS in the world;
- it is an advanced, be of universal significance absorbing, innovation and integration;
- it belongs to the world, which is universal, and it is a common wealth of all mankind, especially is the model for developing countries; and
- it is open, keeps pace with the time, has a very strong vitality and will continuously improve and develop.

At present, the GUANLI research in China is deeply influenced by the USA (US schools). Textbooks, management/administration concepts and terminologies are all introduced from the USA. Since China's reform and opening up, the introduction of the magnum opuses of the US schools has played a very important role in China's GUANLI research. However, their theory is not entirely suited to China, and many Chinese scholars have blind faith for US schools. We will continuously insist "make foreign things serve China", but at the same time, moving out from the shadow of US schools as soon as possible.

4. To Construct the Combined Chinese-Western Discourse System of GUANLI Science

For the purpose of communication between the Eastern and Western culture, the construction of a combined Chinese-Western discourse system is necessary. To study the GS, it is inevitably necessary to build the combined Chinese-Western discourse system of GUANLI science (CCWDSGS).

Earlier in this article, some words and terms have been constructed: GUANLI, GS, GULIOLOGY, GUANLIST/GUANLIER, GUANLIWORK/work of GUANLI, GUANLIJOB/job of GUANLI. They are the basic building bricks.

The basic thoughts are as follows:

- Above words and terms are all incorporated into the current English system, and new up-coming terms will continue be included into the system.
- The words GUANLI, etc., from the ancient Chinese are not found in the West; these can only be introduced into CCWDSGS by transliteration.
- The corresponding transliterations are all applied with the standard Hanyu Pinyin, but do not follow the Wade-Giles Romanization.
- The existing Wade-Giles Chinese words and names should all be converted to Hanyu Pinyin.

The China's Hanyu Pinyin System was officially promulgated in 1958. In 1979, the United Nations passed a resolution that all Wei-Giles should be replaced by Hanyu Pinyin System. Since 1982, the International Standard Organization (ISO) began using the Hanyu Pinyin System as the international standard of spelling Chinese. However, it is still a common practice that some foreign scholars apply Wei-Giles spelling Chinese names and place names while composing China-related historical documents.

Wade-Giles has been popular in China and worldwide more than 100 years. Its creditability is undeniable. Since the declaration of Hanyu Pinyin System in 1958, the historical mission of Wade-Giles was accomplished. It is time for it to be sent to the historical museum. The original Chinese word expressed in Wade-Giles can hardly be understood to people in modern days. They should be converted into Hanyu Pinyin. Otherwise, some ambiguity and jokes will be resulted as "常凯申" and "门修斯".

Here are some examples of conversion (the former Wade-Giles):

Lao Tze-Lao Zi (老子); Confucius-Kong Fuzi (孔夫子, i.e. 孔子, Kong Zi); Mencius-Meng Zi (孟子); CHUNGHUA-ZHONGHUA (中华); Chiang Kai-shek-Jiang Jieshi;

Tao-Dao (道); Tao Te Ching-Daode jing (《道德经》);

Peking-Beijing (北京); Nanking-Nanjing (南京); TsingTao-Qingdao (青岛); Canton-Guangzhou (广州); Amoy-Xiamen (厦门), etc.

- Some standard translations of the refined Chinese phrases can also be incorporated into CCWDSGS. For example, one country, two systems (一国两制), Belt and Road (一带一路), macro regulation and control, micro enliven (宏观调控, 微观搞活), top-level design, interactive up and down (顶层设计, 上下互动), from the masses, to the masses (从群众中来, 到群众中去) and so on.

The newspaper *China Daily* is an important reference, which can provide many standard translations, in particular the new reference.

We believe that as time passes, the proportion of Chinese will definitely increase in the CCWDSGS. We highly recommend that contemporary Chinese people should enhance their confidence in four aspects of self-confidence, especially cultural self-confidence, to improve humanity accomplishment and know more about Chinese history and traditional culture.

5. Summary of Our Other Studies

What we have studied in the CSMGS in addition to the content described above includes the following research works and results (Sun et al., 2006, 2008; Liu and Sun, 2008a, 2008b; Liu et al., 2009a; Sun, 2009; Liu et al., 2009b, 2009c; Liu and Sun, 2010; Liu et al., 2011):

- conducted research of study two fundamental questions: What is GUANLI? What is GS? We propose a totally different point of view (Liu et al., 2009a, 2009b, 2009c);
- proposed the basic route to establishing the CSMGS, named "Three Rooms and One Hall, exploitation and innovation" (Sun and Lin, 2004; Wang and Sun, 2005; Zhu, 2007; Nonaka and Zhu, 2012; Li, 2013; Sun et al., 2006, 2008; Liu and Sun, 2008a, 2008b);
- studied the GUANLI philosophy and ethics (Liu et al., 2009a, 2009b, 2009c); and
- studied the China Model of project management (Liu et al., 2009a, 2009b, 2009c) and so on.

Due to the limitation of space, above-mentioned items can only be introduced in future other articles separately.

We argue that the creation of the CSMGS requires combining the theories to practices. We should not only carry out the theoretical studies and literature research but also conduct case study and applied research at the same time. The establishment of the CSMGS plays a critical role in the process of improving the GUANLI of China.

We should pay special attention to the case study. A typical case illustrates a specific model. Several similar enterprises' development and management models can be summarized into a specific unique industrial development and management model. The development and management models of a number of similar enterprises in a specific region can sum up as the unique regional development and GUANLI model. The development and GUANLI models of a number of industries can sum up as the unique development and GUANLI model for a big region or even at a national degree. Based on this series of studies, together with literature study and theory study, China can extract and create the Chinese School of Modern Economics, the Chinese School of Modern Sociology and the CSMGS.

Something special to be explained is that the research and creation of the CSMGS can capture experience from the Chinese School of Systems Engineering (CSSE). Since China adopted its reform and opening up policy, with great efforts exerted by Academician Qian Xuesen, Systems Engineering (SE) discipline was successfully introduced to China. An important article titled "*The technology of Organization and GUANLI-Systems Engineering*" was published on September 27, 1978 in Shanghai's newspaper *Wenhui Bao* (Qian, 2007). This article was written by Qian Xuesen, Wang Shouyun and Xu Guozhi, three scholars of SE in China with Chinese characteristics as a definition, which influenced the development pattern of SE in China. SE gained strong support from Chinese top leaders, which is unique over the world and the only exception among more than 300 disciplines in China. SE at the macro level (country), middle levels (provinces, departments) and microlevels (businesses, schools, villages and so on) are widely applied and attained good outcomes. SE has reached great achievements both in theories and practical methods. Academician Qian Xuesen proposed the "open complex giant system", "Meta-Syntheses methodology from qualitative to quantitative" with the world advanced level of outcomes. In the 1990s of the last century, it has formed the CSSE, which was also known as the School of Qian Xuesen (the Dr Qian Xuesen's School) (Sun and Zhu, 2011; Sun et al., 2015; Sun and Liu, 2016). In China, SE is equal to GS. All the theories and methods of the CSSE could be applied in CSMGS.

6. The Postscript

G. W. F. Hegel said: Whatever existed is reasonable. We extend it with two more sentences: Whatever is developed is reasonable. Whatever succeeded is reasonable. China has never collapsed, although some Westerners advocate "the collapse of China" over and over again. However, China is developing rapidly, and China has achieved tremendous success. It is no doubt that China will continue to develop and continue to achieve much more success in the future. The great rejuvenation of the Chinese Nation's Dream will certainly come true.

GUANLI (includes management, administration, governance and so on.) and its service objects are positively correlated, neither unrelated nor negatively correlated. Along with China's development and success, China's GUANLI is also proved to be reasonable, well-developed and successful. Some Chinese scholars always comment China's GUANLI as "a complete mess", which is totally absurd. Western Economics failed to explain China's economic development, and Western theory of GUANLI cannot explain GUANLIWORK in China either. Who is unreasonable after all? The only possible answer is: Western theories cannot be applied fully for the development of China. Therefore, it is essential to create China's theory including the Chinese School of Economics, the Chinese School of Sociology, the CSMGS and so on.

The basic route to creating Chinese School is: making foreign things serve China, making ancient things serve the present, making former things serve the today and future, Meta-Syntheses. "Making former things serve the today and future" is mean: to summarize former experiences and lessons, and theorizing it to an upper level to guide current and future acts and practices. The so-called "former" refers to the nearly 40 years since 1978—nearly 70 years since 1949—nearly 100 years since 1921, also can be traced to the 1911s Revolution—the Reform Movement of 1898, until the Opium War of 1840. We name the basic route as "Three Rooms and One Hall, exploitation and innovation", on which we are prepared to write special articles for further explanation.

It is needless to say that there are still lots of problems in China, including in the GUANLI field. These problems will continue to be gradually solved in the continuous China's reform and development. New problems will come up while old problems are being solved and settled; problems producing in a loop, problems solving in a loop, this is the dialectics. We have full confidence in solving problems, as well as in China's development and future.

Practice comes first and then it is followed by theory. We first have "China Model", followed by the "Chinese School". The "China Model" has already been there, and the "Chinese School" relies on our proactive research and innovation.

It is just the right time for us to study and create the CSMGS. This is the very historical mission, and opportunity awaits contemporary Chinese.

References

Li, P. P. (2013), "Indigenous research on Chinese management and Chinese traditional philosophies", *Chinese Journal of Management*, Vol. 10 No. 9, pp. 1249-1261 (in Chinese).

Liu, R. and Sun, D. (2008a), "Discussion on some problems of establishing the Chinese school of modern GUANLI science", *Chinese Journal of Management*, Vol. 5 No. 3, pp. 323-329 (in Chinese).

Liu, R. and Sun, D. (2008b), "On some issues about setting up the Chinese school of modern GUANLI science", *Engineering Sciences*, Vol. 10 No. 12, pp. 24-31 (in Chinese).

Liu, R. and Sun, D. (2010), "Some issues about setting up the Chinese school of modern GUANLI science (Ⅴ): 12th department of disciplines catalogue and basic term system", *Academic Degrees&Graduate Education*, Vol. 8, pp. 67-73 (in Chinese).

Liu, R., Sun, D. and Sun, K. (2009a), "Some issues about setting up the Chinese school of modern GUANLI science (Ⅲ): four definitions and three suggests", *Engineering Sciences*, Vol. 11 No. 8, pp. 18-23, 63 (in Chinese).

Liu, R., Sun, K. and Sun, D. (2009b), "Some issues about setting up the Chinese school of modern GUANLI science (Ⅳ): the big platform, the big hall for justice, and others", *Chinese Journal of Management*, Vol. 6 No. 9, pp. 1137-1142 (in Chinese).

Liu, R., Sun, K. and Sun, D. (2009c), "The Chinese characteristics of large-scale projects management and its comparison with US and USSR", *Science and Technology Progress and Policy*, Vol. 26 No. 21, pp. 5-12 (in Chinese).

Liu, R., Sun, K. and Sun, D. (2011), "Some issues in current research in Guanli science: clarification of some doubts and comments on two research methods", *Chinese Journal of Management*, Vol. 8 No. 9, pp. 127-142 (in Chinese).

Nonaka, I. and Zhu, Z. (2012), *Pragmatic Strategy: Eastern Wisdom, Global Success*, Cambridge University Press, London.

Qian, X. (2007), *On The Systems Engineering* (*New Century version*), Shanghai Jiao Tong University Press, Shanghai, pp. 11-12.

Sun, D. (2009), "On the tripartite thinking method and management", *Chinese Journal of Management*, Vol. 6 No. 7, pp. 861-866 (in Chinese).

Sun, D. and Lin, F. (2004), "Set up modern management science with Chinese characteristics", in Sun, D., Lin, F. and Zhou, G. (Eds), *Systems Engineering and Management Science Research*, Jinan University Press, pp. 128-136.

Sun, D. and Liu, K. (2016). *Behold a High Mountain, Pattern Always Exist* (*Gaoshan Yangzhi, Fengfan Yongcun*, 高山仰止 风范永存), Chinese Communist Party History Publishing House, Science Press Co. Ltd. et al, Beijing, pp. 155-197 (in Chinese).

Sun, D. and Zhu, G. (2011), *Basic Course in Systems Engineering*, Science Press, Beijing (in Chinese).

Sun, D., Lin, F. and Sun, K. (2006), "Set up the Chinese school of modern management science", *Chinese Journal of Management*, Vol. 3 No. 2, pp. 127-142 (in Chinese).

Sun, D., Lin, F., Sun, K. and Zhong, Y. (2015), *Introduction to Systems Engineering*, Tsinghua University Press, Beijing (in Chinese).

Sun, D., Zhang, Z. and Sun, K. (2008), "An important historical mission: set up the Chinese school of modern management science", *China-USA Business Review*, Vol. 8 No. 1, pp. 57-63 (in Chinese).

Wang, K. and Sun, D. (2005), "The analysis and compare of GUANLI science with physics and medicine", *Research of Productivity*, Vol. 2005 No. 11, pp. 211-213 (in Chinese).

Zhu, Z. (2007), "Reform without a theory: why does it work in China?", *Organization Studies*, Vol. 2007 No. 28, pp. 1503-1522.

Further reading

GDP Data from Countries/Regions of the World Rankings (2015), available at: http://blog.sina.com.cn/s/blog_416ba4c90102w5oi.html(In Chinese).

Hanyu Pinyin (2017), available at: http://baike.so.com/doc/1970158-2085114.html.

IFM (2014), "In terms of purchasing power parity, China has become the world's largest economy", available at: www.chinanews.com/gj/(2014),/10-09/6657307.shtml(In Chinese).

Kai-shek, C., available at: http://baike.so.com/doc/2333451-2467987.html(In Chinese).

Mencius, available at: http://baike.so.com/doc/5620854-5833471.html(In Chinese).

Pew Opinion Survey (2015), "Global support for more than half", available at: http://news.sina.com.cn/w/(2015),-06-25/100231986804.shtml(In Chinese).

Ramo，J. C.，"Beijing consensus"，available at：www. doc88. com/p-9935409490191. html.

Wade-Giles Romanization，available at：http：//baike. so. com/doc/2995801-3159529. html(In Chinese).

World Bank Edition (2010)，"World countries and regions GDP rankings"，available at：www. 360doc. com/content/11/0924/16/5931940_150892824. shtml(In Chinese).

第十章　工程管理

上海浦东新区建设工程①

一、新技术应用与开发战略

开发浦东是上海经济发展战略的重要组成部分，也是加快发展以上海为中心的区域性外向型经济的重要步骤，党中央、国务院对开发、建设浦东新区十分重视，多次对浦东新区开发做出指示。根据这些指示的精神，经过各方面研究，提出大力开发浦东新区，使之成为对内对外开放的枢纽化、国际化、现代化新市区，要创造条件吸引外资，以老市区支持新区开发。同时以新区开发改善老市区，尽快地使上海有一个良好的投资、经营和生产的环境，逐步使其成为太平洋西岸最大的经济贸易中心之一。

浦东地处长江口，背靠老市区，面对太平洋，便于建成以出口创汇为导向的外向型经济基地，对外参与国际大循环，吸收外资发展产业，对内则加强横向联系，吸引国内企业来上海参与投资，扩大出口。

当前的国际形势有利于浦东新区开发。世界经济进入一个和平发展的阶段，工业发达国家和地区正在进行产业结构调整，以应对新的挑战。我们一定要抓住时机，参与国际分工，吸引外资，引进技术，调整我们的产业，壮大上海的经济[1-3]。目前浦东的潜在优势深深吸引着国外投资者，要不误时机地制订和实施开发浦东的计划。由于众所周知的原因，我们错过了几次机会，这次再也不能错过了。

显而易见，浦东开发主要是利用外资，要以高技术产业为开发重点，这是浦东开发建设必须认真贯彻的指导思想。

浦东新区开发要以出口创汇为导向改变产业结构，改善投资环境，要改造传统产业，创建新技术产业。除政策和体制外，上述一切发展一定要以科技进步为基础。目前上海市（含浦东）经济增长主要还是依靠劳动力和资金等外延因素的增长，而技术进步的贡献不明显，远远落后于工业发达国家。世界经济发展实践证明，科学技术进步将日益成为经济增长的决定因素。如果仍靠扩大外延投入来发展经济，不但无法达到“现代化”和“中心”的目标，还可能会拉大与发达国家的差距。中央提出的“翻两番，一半靠科学技术进步”“经济建设必须依靠科学技术，科学技术必须面向经济建设”等方针，明确指出了科学技术对经济增长的作用。

---

① 本文是 1989 年上海市科学技术委员会重大科研项目报告，收录在北京科学出版社 2015 年出版的《工程管理研究》第 73～108 页．作者：刘人怀、于应川、汤万方、薛沛丰、王伟、周民立、张跃、曹庆弘、屠寿康、王子来、李克明、刘浚、谢岚、戈云、陈笃平、陈瑛、龚日清、高法华．

综上所述，我们提出浦东新区建设中新技术应用与开发的目标：为浦东向外向型经济产业结构转轨服务。为改造传统产业，开创新产业，协调发展第三产业服务。为浦东改善投资环境，完善社会服务系统。促进社会、经济繁荣，达到现代化经济贸易中心的总目标。

本报告将从以下几个方面展开浦东新区建设中新技术的应用与开发。

(1) 应用与开发新技术，促进技术进步，改变产业结构，形成出口创汇导向的外向型产业结构。

(2) 应用与开发新技术，加快浦东传统工业的改造。

(3) 应用与开发新技术，发展完善浦东新区的社会服务系统，改善基础设施，改善投资环境，吸引外资。

(4) 应用与开发新技术，发展浦东现代化近郊农业，为城市提供多品种高质量副食品，并发展名优产品出口创汇。

二、促进浦东向外向型经济的产业结构转换

在我国地区发展战略格局中，沿海地区发展外向型经济，实行经济发展战略转轨。这是浦东经济发展、产业结构转换的指导方针。

外向型经济发展战略是国家或地区以国际市场的需求为依据，通过出口创汇和利用外资，引进先进设备、技术，引进原料和中间产品，建立和优化以出口产业为主导的产业结构，进一步扩大出口创汇和利用外资，这样周而复始，循环不已，推动本地区经济发展的战略。

我国沿海地区逐步走向外向型经济发展轨道是国际上的“示范效应”和我国经济体制改革对沿海产生“挤压效应”的结果。浦东新区开发只有一条路，走外向型经济发展的路子。由内向型经济转变为外向型经济，这是经济发展模式的根本转变。这个转变中最关键的是主动参与国际分工，建立适应国际市场变化的产业结构及其运行方式。

1. 外向型经济发展战略对产业结构转换的要求

产业结构有多个层次，最高层次是指第一产业、第二产业、第三产业间的结构，第二个层次是指各次产业内部行业结构。还可以第三、第四层次再细分，我们仅研究第一、第二两个层次结构的转换。

外向型经济是以“出口导向”和出口贸易占国民总收入的较大比重为特征的。它要求区域产业结构的建立基本上以国际市场需求和供给状况为依据，优势产业的选择以国际市场价值作为判断标准。而现在浦东地区的产业结构基本是内向的，缺乏高层次，又缺乏弹性，不符合外向型经济发展战略的要求。因此，浦东地区产业结构要实现以下转换。

1) 产业结构要向外向型转换

进口替代和出口导向是对外贸易战略中的两种战略。我国长期采取内向的进口替代发展战略，用进口数量“差别性”控制和“倾斜性”关税筑起重重壁垒。在此庇护下，产业结构具有明显的内向型，不利于产品出口。

采取进口替代发展战略的国家和地区为了取得外汇也会有出口贸易的增长，但由

于其产业结构是内向型的，出口实绩明显不如产业结构外向型的国家和地区。巴西、墨西哥、阿根廷、哥伦比亚是实行进口替代发展战略的国家，可以把它们作为产业结构内向型的样本；韩国、新加坡、中国台湾是实行出口导向发展战略的国家和地区，可把它们作为产业结构外向型的样本。现在来比较它们的出口差异。

从表1可见，实现了产业结构外向型转变后的新加坡、韩国和中国台湾无论是制成品出口与制成品产量之比的增长，还是出口与国内生产总值之比的增长都比产业结构仍然为内向型的巴西、墨西哥、哥伦比亚、阿根廷快。这就给我们一个启示：我们沿海地区包括浦东要发展外向型经济，只有打破原先封闭型的产业格局，形成面向世界市场、参与国际分工、力图在国际市场竞争中生存和发展的产业结构。

表1　七个发展中国家和地区的出口产量和国内生产总量的比较　　单位：%

项目	年份	巴西	墨西哥	阿根廷	哥伦比亚	新加坡	中国台湾	韩国
制成品出口占制成品产量的比重	1960	0.4	2.6	0.8	0.7	11.2	8.6	0.9
	1966	1.3	2.9	0.9	3.0	20.1	19.2	13.9
	1973	4.4	4.4	3.6	7.5	42.6	49.9	40.5
制成品出口与制成品产量之比的增长率	1960～1966	3.6	3.2	1.0	7.7	28.4	24.8	24.8
	1966～1973	5.6	5.5	6.5	11.4	47.5	56.4	45.7
出口占国内生产总值的比重	1960	6.1	6.4	8.9	11.3	9.9	9.5	1.5
	1966	7.1	5.4	7.3	9.5	26.6	17.1	6.5
	1973	9.8	4.3	8.1	11.8	44.6	47.8	26.1
出口与国内生产总值之比的增长率	1960～1966	12.3	4.3	5.3	3.4	52.0	24.7	13.0
	1966～1973	11.5	3.3	9.0	14.5	52.0	63.3	34.0

产业结构向外向型转换要注意以下四方面。

（1）以国际市场的需求为导向，从国际分工出发来调整产业结构。

（2）优势产业选择要以国际市场价值来衡量，要考虑其国际竞争力，而不是以国内价值来衡量。

（3）要使主导产业与出口产业相一致，资源的配置应有利于出口产业的扩张。

（4）要避免过分强调自成体系，要充分参与国际分工。

2）产业结构要向高度化转换

产业结构高度化有两个含义，产业结构的重型化和产业结构的“高加工度化”。产业结构高加工度化是指，不论何种产业都从以原材料为重心的浅度加工、低附加价值的结构向以加工组装为重心的深度加工、高附加价值的结构转换。出口产品构成不同，对经济增长的推动作用是不同的。如果出口的产品主要是以原材料为主的初级产品，对经济的推动作用是有限的。因此，浦东地区要使产业结构向高度化转换，要从以初级产品为主的轻型结构逐渐转向以制成品特别是重化产品为主的重型结构转换。而制成品出口反过来会进一步促使产业结构高度化。这是因为，制成品出口扩大，导致对机械、金属、化学部门的需求扩大，根据“后向联动效应”的要求，重化工业就可以作为第二项进口替代产业得到发展。某一产业的“后向联动效应”越大，这一产业规模的扩大就越能促进新的中间产品投入物产业的建立。“下游部门”的面向出口促进“上游部门”的进口替代、尔后进一步发展为面向出口，这样形成一个新的因果链，进一步促进产业结构高度化。

3）基础设施必须超前发展

良好的基础设施是发展外向型产业必备的条件。浦东产业结构中的电、热、水、气、港口、公路、机场等基础产业和设施“滞后”已成为加快外向型经济发展的“瓶颈”，应作为重点来发展。

4）第三产业协调发展

由于外向型经济与国际市场之间有广泛的联系，各国经济往来十分密切。交通运输、邮电通信、金融保险、信息、咨询等第三产业必须协调发展才能保证主导的出口产业不断发展和扩张。浦东的第三产业不发达，会影响外向型经济发展，应予以统筹安排，协调发展。结合上海市区第三产业滞后的现状，浦东相对来讲更需要超前发展。

2. 产品结构转移是一个长期的、多层次的过程

亚太地区的国际环境，20 世纪的 80 年代不如六七十年代。“四小龙”是靠六七十年代的产业转移发展起来的。80 年代情况已有不同，我国单以劳动密集型产品出口加入国际大循环已不可能。针对我国沿海地区工农业发展不平衡的特点，宜采用多层次发展战略。在发展劳动密集型产品出口的同时，在生产发展水平高的地区，实行更加开放的政策，推动资金技术密集型产品出口，让重加工工业直接面向国际市场。80 年代以来，劳动密集型的优势正在消失。看不到这一点，就会放松对高技术的发展。

产业结构多层次的特点是我们制定浦东新区战略的一个依据。

浦东地区工业是多层次的，有高桥石化公司、沪东造船厂、上钢三厂等现代化程度较高的大企业，以及浦西工业辐射影响，也有郊县地方工业，以及乡镇企业。因此在向外向型经济产业结构转换时也应是多层次的。

第一个层次是乡镇企业和县地方企业等。这些企业以劳动密集型为特点，开始以国内资源加工出口为主，逐渐扩大出口门类，提高加工深度，迅速增加创汇，积极发展进料加工、来料加工、来件装配，提高用国外资源加工产品的比重，最后发展为以国外资源加工产品出口的外向型企业。这些产品大概分为三类：一是劳动密集型的轻工产品，主要是纺织和服装、食品和饮料、工艺品和轻工杂品等；二是劳动密集型的机电产品；三是城郊创汇农业产品。

第二个层次是技术较先进的传统工业，如钢铁、石油加工、造船。这些企业在国内尚属中上游，有强烈的国内市场需求，在向外向型经济转轨过程中，引进开发新技术，提高产品质量和数量，逐步由以内销为主变为内销外销均衡，最后变为以外销为主，其原料来源也是由内供变为内供外供各半，最后变为外供为主，还要注意引进新技术使传统产业“脱胎换骨”成为新兴产业。例如，电子行业与机械行业结合，发展为机器人产业。造船工业发展为上海海洋工程、石化工业和钢铁工业发展为新型材料产业。

第三个层次是新兴产业，目前浦东几乎是空白。从长远看，可形成新材料、生物

工程、海洋石油开发、机器人等新产业。在新兴产业政策的支持下，尽快地“消化吸收”新兴产业基础产品，由加工型新兴产业转向装配型新兴产业，以二次开发的高附加值的高技术产品推向国际市场，参加国际竞争。

3. 加快技术进步，促进产业结构的转换

技术进步是产业结构变化的动力。技术进步产生了新兴产业，促进了传统产业的生产要素利用率的提高，使资源的分配更趋于合理，从而促进了产业结构变化。浦东产业结构要向外向型转换，向高度化转换，基础设施和第三产业要超前发展，这些全要靠加快技术进步来保证。考虑到浦东目前生产力水平较低，其主体是传统技术和传统产业，新兴技术和新兴工业比重相当小，新技术应用应重视传统产业的技术改造，有重点地开发新兴产业，加快“瓶颈”基础设施和服务产业的现代化。

浦东新技术发展战略目标可以这样设想：从实际出发，有重点地开发新技术，改造老企业，逐步建一些新兴产业、一些科学研究园区，使浦东的产业结构转向外向型，逐渐以新兴产业为主，促进经济、社会、科技协调发展。

政府通过宏观指导和干预使浦东技术开发与应用服从于上述目标，通过各种调节，使企业在竞争的条件下，加快技术进步的步伐，极大地发挥技术进步的微观机制，以促进产业结构的转轨。此外，还要注意下列问题。

(1) 建立浦东技术开发调节机构。协调和调节浦东技术改造、技术引进工作，避免盲目发展，保证重点项目的需要。

(2) 注意控制技术构成的合理比例。应根据浦东实际情况量力而行，使引进或开发的高技术和适用的先进技术之间有一个合理比例，以适应浦东生产力产业结构多层次的状况。

(3) 注意重点开发区与一般地区的协调发展关系。在外高桥、陆家嘴、花木、北蔡四个地区，分别开发劳动密集型出口加工产业、第三产业、科学园区和高技术新兴产业。在一般地区，要重点开发与推广“短、平、快”适用先进技术，以适应中小企业、乡镇企业和农业生产对新技术的需要。

(4) 正确处理技术开发与环境保护关系。

(5) 通过各种形式教育，提高人民的文化和科技水平，以消除对采用先进技术的阻碍，可以采用人口流动、引进人才来改变浦东新区的人口结构。

4. 促进产业结构转换的产业政策

浦东新区要实现产业结构的转换需要变产业政策的内向偏向为外向偏向。产业政策主要包括产业结构政策和产业组织政策。

(1) 制定产业导向政策。在外向型经济条件下，产业结构转换的主要杠杆之一是利用国际资源，特别是引进先进技术和利用外资，有选择地引进外资外技，发展出口产业。外资作为沿海地区投资结构的一个构成因素，随着其在投资中所占比重的上升，外资通过投资效应，对沿海地区的未来产业结构的影响会不断加强，因此有必要尽快建立和完善我国对外商直接投资的产业导向政策，制定各种外资法和投资优惠政策，有选择地将外资外技引向出口产业，加快产业结构高度化或扩大产品出口，正确的产业导向政策有利于把对发达国家来说已相对落后的，但对发展中国家仍适用的优势的产业“移植”进来，以此缩小同发达国家的经济技术水平差距，占领发达国家产业结构转换而出现的市场缺口。

（2）要重视扶植出口大中型企业、企业集团的形成和发展。这是因为，出口大中型企业、企业集团尤其是金融、生产、科技一体化的企业集团较之于小企业参与国际竞争更有利。第一，它们能集聚更多的资产和资金，扩大出口拳头产品的能力。第二，它们具有强大的新产品研究和开发、产品升级换代的能力，以适应国际市场的变化。第三，它们有消化、吸收、创新技术的能力，可以成为传递先进技术和先进管理的“二传手”。第四，它们有直接对外、及时掌握国际动态的能力。

5. 外高桥工业贸易区作为浦东向外向型经济转轨的突破口

浦东外高桥一带，地势平坦、开阔，住宅户少。外高桥地处长江下游，随着新港口建设，对内对外交通方便，又能依托高桥石化总公司，是一个发展出口加工、以外贸为主的工贸区的好地方，除配合港口建设发展仓储、运输、服务业等第三产业外，应把外高桥作为浦东外向型经济转轨的突破口，统一规划，给予政策优惠，尽快吸引外资，发展各种外向型产业。

外高桥地区的产业结构大致如上节分析，有三个层次。

第一个层次是劳动密集型的出口加工工业。由外国投资或合资经营。产品可以是机电、纺织。由于高桥石化总公司存在，可依托其增加原油炼制设备，组织一些油品、化工产品、塑料橡胶制品。原料可采取进口原油。

第二个层次是高桥石化工业扩充发展，扩大石油产品、化工产品生产，采用进口原油，产品输出国外（详情见下面三节）。

第三个层次是高桥石化工业引入新技术，向化工新材料新兴产业发展。

以下专门有两节研究传统产业改造和发展基础设施的问题，本节专门就利用外资发展新兴产业做进一步研究。

6. 利用外资引进新技术发展浦东新兴工业

浦东新区开发建设的指导思想是利用外资和以高技术产业（或称新兴产业）开发为重点。但是新兴工业利用外资有三个特殊困难。一是新兴工业投资大、见效慢、风险大。新技术、高技术产业化的过程往往是远期效益大而近期效益并不大，或是社会经济效益大而企业自身的效益并不大，由于缺少相应的特殊的扶助政策和鼓励措施，企业缺乏发展新兴工业的内在活力和动力。这样，新兴工业利用外资外技工作就难以大面积推开。目前，如要求企业外汇平衡，企业就更没有积极性。二是产品难以打进国际市场。三是新兴技术不少是国际上保密和禁止转让的，捕捉引进和合作的机会就比较困难。

只有采取切实措施，力排困难，才能实现调整产业结构，开发新兴产业的目标。

如何积极利用外资，把浦东新兴工业搞上去，以下几条思路可供参考。

1）利用外资的重点应放在建立中外合资、合作企业等外商直接投资形式上

这有四方面的好处：①有利于冲破封锁，获得我们科研开发和产业化所需的技术。②有利于利用外商的销售渠道和经营能力，早日和尽可能多地把产品打进国际市场。③有利于减少风险，合资、合作经营，把中外企业利益捆在一起，风险共担，利益均

沾，外商就会尽其全力经营企业，使中方企业的风险大大减少。④有利于培养和引进发展上海新兴工业所急需的中高级人才。

2）可采取先易后难、循序渐进的方针

吸引外商来经营新兴工业，不是一件轻而易举的事。外商有个观望和探路的过程，中方也有个了解和适应的阶段。为此，有些项目，不能急于求成。可先搞“三来一补”，单件和零配件合作，等有了一定的条件和基础，吸引某些低层次项目，再向合资、合作经营和整机合作过渡，这对双方都有利。

3）引进外资改造传统产业，使其孕育新兴产业

直接利用外贷，引进先进技术，不失为加快新兴工业发展的一个有效途径。但是，要求新兴工业自身“借外汇，还外汇”，尤其是每个项目做到这一点，是很难的。我们认为，应引进外资改造传统产业，使其孕育新兴产业，如高桥石化工业发展化工材料工业，造船工业发展海洋工业，钢铁工业发展新材料工业等，开始要“以老养新”，用传统工业创汇来补足部分新兴工业的外汇用款，逐渐发展为新兴工业自给。

4）外资工业、科研联合经营

由于新兴工业产品技术性强、规格复杂、变化快，并且还有个售后服务问题，所以，外贸部门不太熟悉，很难承担开拓国际市场的工作，工业部门对外贸的经营、信息和政策等更是生疏。因此，要使新兴工业产品能更多地、不断地打进国际市场，必须开辟出口渠道。建立“贸工研”联合体，外贸、工业、科研部门联合经营、通力合作，发挥各自优势，开拓国际市场，单独或与外商联合在国外设销售点，开展新兴工业的情报收集、市场开拓和售后服务工作，充分利用外商的经营经验和销售渠道，寻找可靠的经销代理商等。

5）在外高桥、北蔡等地建立新兴工业园区，采取特殊政策，吸引外资

台湾地区在新竹建设了科学公园，用政策引导从事研制、开发高技术的研究所、大学和企业汇聚该地，吸引高级专门人才流向科学公园。在此基础上，采取优惠政策，吸引海外高技术经营者和专门人才到新竹经营和工作，对促进台湾技术密集型工业的开发和发展起到了很大的推进作用。浦东可借鉴这种经验，在外高桥或北蔡建设新兴工业园区，吸引科研、生产单位和各类专门人才，到工业园区来搞开发和经营。对于本市一些能生产出口的新技术产品但加工能力不足或生产环境不佳的企业，可优先鼓励它们到新兴工业园区来发展生产。

6）采取扶植措施，调动新兴工业企业利用外资的积极性

目前，企业缺乏利用外资的内在动力，这是影响上海新兴工业利用外资向前发展的一个重要因素。因此，迫切需要制定一些配套的扶植政策鼓励企业在从事新技术产业化时积极利用外资。例如，对于那些利用外资（不论是以合作、合资，还是举借外贷的形式）来开发生产新技术、高技术产品的企业应免征调节税和产品税，若干年内免交所得税，外汇全部留用；对新兴工业发展所必须进口的零配件，关税应优惠。这样一方面可以增强这类企业自我发展的能力，另一方面有利于它们到外汇调剂市场上去调剂外汇余缺。

三、加快浦东传统工业改造

1. 浦东传统工业改造的意义

浦东现有工业中，冶金、造船、石化工业为三大支柱，另外还有建材、机械、家电、纺织等，一部分企业近年来引进了一些先进设备和工艺，开发了新的产品，但是总体上与浦西相比，设备陈旧、技术落后，经济效益普遍低下，新兴产业极少，浦东经济发展更要依靠科技进步。

新技术开发首先要从“传统企业改造”出发，因为传统工业在国民经济中占的比重绝对大，要使目前国民经济增长速度提高，只有依靠传统工业改造。世界范围的经验，包括美国、日本等发达国家的经验证明用新技术改造产业所取得的效益极为显著，新技术对促进管理技术、产品特性、质量、生产效益等的改进与提高起到了大幅度的“增值效应”。《中共中央关于制定国民经济和社会发展第七个五年计划的建议》明确指出，坚决把建设重点切实转到现有企业的技术改造和改建扩建上来，走内涵为主的扩大再生产的路子。对现有工业企业进行技术改造和设备更新，既能改变重工业任务不足的状况，使当前的经济发展保持一定的速度，并且增加新的生产能力，又能使我国工业生产技术达到一个新的水平，为今后整个国民经济的现代化创造条件，储备力量，这是使我国经济走向顺利发展的一个关键。

2. 开发与应用新技术，促进传统工业改造

世界范围的技术改造正向以下三个方向发展。

(1) 采用新技术，把劳动密集型企业改造成资本与技术密集型的企业，以提高产量，增加品种，节省人力，并把节省下来的人力用于发展第三产业。

(2) 采用新技术，在节约能源、节省材料和防止环境污染上下工夫。

(3) 大搞优化设计、优化生产和优化管理，加快产品更换的速度，降低生产成本，大大增强企业和产品的竞争能力。

浦东传统工业的改造应在向外向型经济转轨的总的目标下，向着提高生产技术水平、提高产品质量水平、增加新产品、节约能源、节约原材料、减少污染等方向努力。并创造条件，孕育新兴产业。

浦东各企业根据不同的特点，可以进行以下形式的技术改造。

(1) 以提高现有生产技术水平为主的改造，即用先进的设备代替陈旧的设备，或者使设备现代化。使原来已有的拳头产品质量提高，经济效益提高，如上钢三厂的钢材生产、高桥石化总公司的石油产品和化工产品基本是这种类型。

(2) 改变生产行业而进行的改造，即开发新技术、新工艺，提高原有产品质量的同时开发新行业的产品。浦东造船业基本就是这种类型，既提高造船的质量，又开发非船产品钢结构生产。在这种科技改造的基础上，造船工业可孕育海洋工程，石化与钢铁可孕育新材料工业。

在技术改造过程中新技术发展领域的应用强度大体如下排列：①机电一体化技术；②微电子与计算机技术（含计算机软件）；③信息技术；④新材料技术；⑤机器人技术；⑥传感技术；⑦激光技术。由此可以建立技术开发网络，以推动各企业的技术改

造工作。

根据我国实际情况，一些企业已摸索到一些有效的技术改造的技术措施，可以提出的有以下几方面。

(1) 原有设备加电脑控制。以武汉汽轮发电机厂为例，他们以此路线完成技术改造项目 120 多项。若按淘汰更换的办法需投资 3000 万～4000 万元，而采用技术改造办法，只花了 270 万元。与技术改造前相比，国家资产没有增加，总产值却从 2000 万元上升到 6000 万元，由亏损 240 万元跃为盈利 600 万元。因此，“原有设备加电脑控制”是一条投入少、产出大，真正靠内涵扩大再生产的老企业技术改造的新路。

(2) 抓“工艺突破口”，提高产品质量。

(3) 产品升级换代。

3. 对企业技术改造的建议

(1) 技术改造应讲求实效，不图形式。我国曾有过若干次重大投资上的失误，其最主要的原因是追求形式上的技术先进而忽视实效，或在实效估计上的失算误判。要时刻着眼于产品质量和企业的经济效益。

(2) 要有明确的技改重点，不能把摊子铺得过大。

(3) 逐步提高现有固定资产的折旧率。

(4) 改进技术改造资金的管理办法。对使用技术改造资金筹建的项目，要严格把关，切忌挪用于基本建设。

(5) 为企业的技术改造提供优惠信贷政策。对企业技术改造的贷款利率要较大幅度地低于基本建设贷款利率，对老企业或重点技术改造项目可考虑实行无息贷款。

(6) 严格控制固定资产的投资规模。

(7) 提高技术改造的投资质量，推动企业的技术进步。

(8) 引进外资，包括合资经营改造现有企业。

(9) 理顺管理体制，扩大企业自主权，鼓励企业实行技术改造的积极性。

(10) 加强科研机构、高校、企业进行技术改造的合作。

(11) 加强领导与组织协调，统一规划，集中力量，实行重点技术改造攻关。

以下分别就造船、石化和钢铁进行详细论述。

4. 造船业

上海是我国最大的港口和船舶工业的重要基地。上海现有大小修造船企业八九十家，其中上海船舶公司所属的六家船厂占了绝对优势。它们分布在浦东地区的有沪东造船厂、上海造船厂、中华造船厂的沪南分厂，在浦西地区的有江南造船厂、求新造船厂、东海造船厂和中华总厂，其产量约各占一半。中国船舶公司是我国船舶生产的主要力量，上海船舶公司又是中船总公司系统中的主要生产力量，其船舶的产量、产值一直占总公司一半左右。到 2000 年，中船公司规划的年造船量为 240 万～300 万 t，上海将承担 120 万～150 万 t。

1) 国内外造船业的发展形势

(1) 世界造船业起伏发展，我国造船业稳定增长。世界船舶需求量周期性变化。20 世纪 70 年代中期，造船业兴旺发达，80 年代初逐渐萧条，预计 90 年代回升。由于海洋工程的开发，到 90 年代后期造船业将出现新的较持续的兴旺时期。

随着我国国民经济的稳步发展，内河沿海和远洋水运需求量不断增加，预计每年需要新造载重 80 多万吨江船海轮和远洋船舶，才能满足国内运输和国际贸易发展的需求。我国对船舶工业采取了扶植政策，并重视引进先进的造船技术，促进了船舶生产。“六五”期间，中国船舶工业公司造船产量年均增长 13.6%，工业总产值年均增长 10.6%。1985 年，中船公司年造船能力为 120 万 t。预计将以 10%的速度增长，到 2000 年，年造船量达 240 万～300 万 t。现在我国是世界第六位造船国。

(2) 世界造船重心东移，我国在世界船舶市场中大有发展余地。第二次世界大战前，西欧造船量占世界造船总量的 4/5。20 世纪 50 年代，日本跃居世界第一造船大国，以后韩国发展为世界第二造船国，80%的世界造船量由西欧东移到亚洲。当前，由于新技术的兴起，一些传统造船国家，将人力、财力转移到新兴产业，造船力量开始削减，于是，劳动力廉价又有一定技术基础的国家，造船业趁机崛起。我国正是在这种形势下，从 1980 年起走向世界船舶市场的。1980～1987 年，我国出口民用船舶 180 万 t、军舰 34 艘和船用柴油机、柴油发电机等，共创汇 16 亿美元。目前，在每年世界 2000 万 t 船舶订货中，我国只承接到 2%，而日本承担 42%、韩国承担 19%，相比之下差距很大，我国需大力发展造船业，参加国际市场竞争，为实现我国成为世界第三造船大国而奋斗。

(3) 船舶市场竞争加剧，船舶新技术发展迅速。当前萧条的世界船舶市场，不仅没有影响船舶技术的发展，相反市场的压力加剧了市场的激烈竞争，实质上是促进了科学技术的激烈竞争，推动了船舶新技术的应用研究，加快了船舶的更新换代。据测算：今后 10 年中，通过应用新技术，造船生产率可提高 20%～25%，成本可降低 20%～30%，工时可减少 30%～40%。当代，各造船国比以往任何时候都注重新的技术应用与开发。日本十分重视新船型的创新研究，不断扩大现有技术优势。韩国在保持产量优势的同时，增强技术竞争力。第二次世界大战后造船力削弱的西欧，在原有技术基础上，已有重新起步的势头。在新的造船技术挑战面前，唯一的出路是加速开发以高科技为核心的新技术的应用研究，提高我国产品在市场的竞争力，即提高质量，降低船价，缩短周期，确保服务。

目前，我国船舶的竞争力，与世界先进造船国相比，差距较大：①质量方面，船东对船舶质量的要求不断提高，需要多品种、高质量、技术先进、能耗低、自动化程度高、安全可靠并配备有新技术设备的高档船。我国船舶技术还不先进，质量不稳定，造的大都是普通船。②船价方面，我国尚有竞争力。这几年船价一跌再跌，近五年已下降了 30%以上。我国造的船价格一般比日本还低 10%～15%，这主要因为我国劳动力成本低。但近几年，我国每工时成本上涨，而且我国劳动力生产率低，人均年造船量为 6t，仅相当于日本的 1/8～1/10，这就大大削弱了劳动力成本低的优势。随着新技术的应用，自动化、机械化程度不断提高，造船业劳动密集度将从现在的 30%下降到 10%～15%，劳动力成本低的优势将逐渐丧失。所以，我国劳动力成本低的优势是表面的、暂时的。要降低船价，根本办法只有采用先进的造船技术，提高劳动者素质，降低成本。③制造周期方面，我国投入的造船工时一般比日本高 5 倍，周期长 1 倍左右。除技术水平低外，还有管理落后的因素。所以必须改革不适应应用新技术的体制，调整工艺布局，提高科学管理水平，尽可能缩短周期。④售后服务方面，我国差距更大，连我国自己制造的远洋船也不愿采用国产化配套设备。

通过船舶需求量预测、船舶市场和我国现状分析可以发现，国内外造船业形势对我国造船业的发展比较有利，但要稳固地实现奋斗目标，必须抓紧世界航运业尚未复苏的时机，依靠科技进步，振兴造船业。

2）浦东造船业的发展方向

（1）应用新技术，改造造船业，适应外向型经济的发展。新技术的研究与应用，促进了船舶工业的更新与发展。一方面，应用新材料、新能源和现代化的造船新技术，建造现代化的新船舶。例如，采用计算机综合导航、卫星通信、光纤通信等先进技术，在增加造船吨位、增多船舶规格品种（如新型船、专用船）的同时，提高船舶的性能质量，提高船舶的可靠性和自动化程度，降低船舶的能耗和污染。应用新的设计理论与方法，采用计算机辅助设计、辅助制造、辅助检测方法，采用二氧化碳自动、半自动焊接技术，研究应用机器人切割、焊接和装配。另一方面，应用现代科学管理技术，较大幅度地提高劳动生产率、降低成本：增加经济效益。目前，我国管理技术约落后先进国家 20 年。因此我国造船业是有潜力、有发展前景的。通过技术改造，预计可将生产率提高 6 倍以上。

要吸引外资、引进先进技术，积极争取承接国外造船业务，快造船、造新型船，包括生产国外销路较好的玻璃钢游艇、豪华旅游船和轮船出口。同时，应发展修船创汇业务。上海造船业除六家主要船厂外，均是修船厂，这六家也可兼营修船业务。现在每年进入上海港的外轮不断增加，达 4000 艘以上。通常，每艘均需要航次修理。平均每检修 15 艘外轮的创汇值，相当于造一艘万吨级出口船的净创汇值。因此要重视修船业。

（2）应用新技术，开发海洋工程新产业。海洋工程是世界的新兴产业，海洋开发技术是当代的新技术之一，其中海洋石油的开采是发展的一个重点。勘探开采海洋石油所需要的海洋机械、海洋结构物是先进造船国研制的重点。美国、日本、韩国，新加坡海洋工程装备产值已相当于船舶产值的 1/3。我国也有了一定的研究生产基础，已建造 10 座海洋钻井平台. 其中出口 3 座，还出口了 2 座平台模块。上海是我国生产海洋工程装备的主要基地，为此，造船业要集中主要技术力量，进一步研制海洋钻探采油平台、辅助船舶和油气集输管道等海洋工程装备，以适应海洋石油的开发。

上海浦东面临的东海大陆架蕴藏着丰富的油气资源，上海市应组织力量，尽早开发东海油气田。如能把油气引上岸，浦东的能源就能得到补充。而且海洋工程能带动其他产业，如冶金、造船和石油化工等的发展，这对建设浦东有着特殊的意义。我国目前资金不足、技术水平还不高，要尽快赶上世界先进水平，使东海早日出气冒油，除调动本市及国内的资金技术外，还要积极吸引外资，引进新技术，把起点放在 20 世纪 70 年代世界开采技术水平上，加速东海油气田开采的步伐。

（3）应用新技术，发展非船产品生产与经营。国外主要造船国家利用造船业技术门类俱全、综合加工能力强的优势，重视发展大型机械装备，如建筑机械（挖掘机、铲运车、翻斗车等）、起重机（桥式、塔式、桁架式等起重机）、矿山采掘机、陆用柴油机等；并大量承接大型钢结构工程，如高层建筑、桥梁、隧道等建筑工程。日本1970年非船产品产值已占总产值的60%，1982年上升到80%。

浦东造船业可以利用过剩的生产能力和雄厚的技术力量来支持浦东的现代化市政工程建设。下述的项目是现代化市政工程的重头项目，浦东造船业已积累经验，可开发以下新的建筑技术以替代外国公司。① 建筑工程的钢结构技术，包括大型钢结构高层建筑、单层大跨度建筑和桥梁。②钻孔灌注桩技术。③顶管隧道技术。

3）落实措施建议

（1）制定船舶业科技发展战略，推动新技术应用研究，加速船舶业的技术改造。

（2）研究与完善新技术法律对策，制定我国造船工业法、海洋工程法、外资及引进技术管理法等，改造造船业的外部环境。

（3）通过对外招标，选定中外合资、国外独资项目。

美、日等先进造船国和海洋石油开发国都有技术输出的愿望，他们各有所长，我们采取对外公开招标，在各国的竞争中选择哪些项目与哪个国家签订合资或独资开发协议，甚至把同一项目分别与两个国家签约，以促进相互竞争。对中标的国家，优惠供应开发所得成果。

（4）加强造船业内部的组织协调和全市的横向联合，开拓多种经营。

选定钢结构和大型机电装备作为多种经营的支持产品，创造条件，增加产品的竞争力，制定新的具体政策，促进各行业相互渗透、竞争和联合。

（5）全面规划船厂在黄浦江外的新修造船基地。

5. 石油化工

1）浦东石化工业的现状

浦东石化工业仅有高桥石化公司。该公司是在新中国成立之后，以炼油厂为起点，逐步通过炼油能力的增加，带动了石油化工、有机化工和精细化工的发展。随着化工产品在市场的扩展，反过来又促进了炼油能力和技术水平的提高，通过油品和化工产品的出口创汇，积累外汇资金来引进国外先进技术和装备来发展自己。

目前高桥石化公司是一个包括生产、经营、科研、教育、建筑、服务等内容的综合性联合企业。具有一定先进水平的大型石油炼制、石油化工、合成塑料、合成橡胶、合成化纤、精细化工等综合性生产的能力。

但是，高桥石化公司的生产水平和技术与世界先进水平有不少差距。例如，生产的内燃机燃料质量未能达到优质要求；适合化工原料的优质轻油率尚停留在45%左右（发达国家超过70%）；原油的化工利用率仅4%（发达国家超过8%）；有机原料的生产布局分散，规模过小，20世纪五六十年代建成的装置，大部分还是因陋就简，逐步改造、扩建过来，技术水平的经济指标都落后于国际平均水平，70年代引进部分大装置后，初步缩短了差距。但是，由于消化吸收改造创新的步伐未能及时跟上，先进技术的推广尚未对整个石油化工行业起巨大推动作用，影响了整个石油化工行业水平的提高。

2）浦东石化工业的发展前景

浦东新区的发展对高桥石化有着非常重要的影响，因为浦东建设为世界一流的城区，将会吸引大量的外资和新技术，在这个过程中石化产业的发展一定会得到充分的发展。反之高桥石化的发展同样也会对浦东新区的发展产生不可估量的作用。由于高桥石化公司的长期建设，目前已具备相当的生产规模和技术水平，它是上海及华东地区的主要燃料供应点和原材料基地之一，每年提供相当数量的油品、合成塑料、橡胶、纤维、活性剂等化工产品，为本地区的工业发展和建设提供原材料。将来浦东的发展也会用到大量的石化产品。

因此，设想浦东石油化工业发展目标是充分利用浦东大规模开发的有利时机，积极吸引外资和先进技术，改造传统工艺和装置，大力发展深、精加工产品，中、下游产品和出口产品，使之成为高技术、高效益、高创汇、低消耗、少污染的高度现代化的石油化工基地，进一步发展为科研、生产、经营、出口紧密结合（以科研为先导，生产、经营、出口为主体）的石化科学园区。

A. 提高生产水平，发展新产品。

（1）石油产品。首先要调整汽油和柴油的产品结构。增加优质汽油比重，提高汽油辛烷值，满足汽车工业的需求。柴油产品必须加氢精制达到出口标准。增加节能产品和出口产品（如高速道路沥青）。发展急需的石油焦和针状焦（针状焦是生产高功率和超高功率电极的优质原料）。此外，还要增产高档次的润滑油和石蜡附加价值较高的产品。在满足国内基本需求后还要努力出口创汇。

（2）化工产品。以塑料为例，目前高桥石化公司虽有 30 多年的塑料生产经验，但无论是品种还是质量，都与世界水平相差甚远。因此，不仅要求生产苯乙烯塑料，还应扩大到生产其他聚烯燃塑料；不仅生产通用塑料，还要生产能填补国内空白的有关工程塑料，如 PPD 改性塑料，它是以每吨售价分别为 6000 元、10000 元的聚苯乙烯和聚苯醚各 50%聚合而成，在电器和化纤生产上有广泛的用途，售价可达每吨 1.5 万元。此外，如丙烯腈（acrylonitrile）、苯乙烯-丙烯腈共聚体（styrene-acrylonitrile copolymer，SAN）、聚苯乙烯（polystyrene，PS）、绝热用模塑聚苯乙烯泡沫塑料（moulded polystyrene foam for thermal insulation，EPS）、高抗冲聚苯乙烯（high impact polystyrene，HIPS）等也应酌情增产。发展塑料的原料除少量买进外，大部分可以从炼油重整芳构化和乙烯裂解的芳构化中得到。如果将化工二厂（原化纤二厂）的专业技术与化纤行业相结合，还可以生产多种增强工程塑料等。

以合成橡胶为例，目前世界上不仅顺丁橡胶有 100 多种牌号，乙丙橡胶也有 100 多个品种，而我们还不到 10 种。况且目前全国合成橡胶产量只能满足市场的 1/3，缺口很大。因此在这方面也大有可为，可以发展丙烯酸乙酯或丙烯丁酯为主体的丙烯酸橡胶（这是一种优质合成橡胶，具有良好的耐臭氧、耐大气老化、耐高温、耐挠曲性和气密性好等特点），还可以发展乙丙橡胶、丁腈橡胶、丁基橡胶等新品种，以填补国内空白。还要考虑各种性能不一的牌号，以形成完整的系列。

以精细化工产品和化工原料为例，可以在重点发展环氧乙烷系列非离子表面活性剂、烷基酚聚氧乙醚、工业清洗剂、石油破乳剂、甲基叔丁基醚（MTBE）、添加剂和助剂的基础上，进一步发展聚氨酯黏合剂、光敏胶、厌氧胶、增塑剂等新产品。此外，还可以酌情发展苯酐、丙酮、环乙酮、乙丙酰胺、双酚 A、环氧氯丙烷、甘油、丙烯酸、丙烯腈、甲基丙烯酸酯类等。

从长远看，可以在高桥石化公司发展新兴的化工材料工业，并带动整个高桥及外高桥地区产业，使之形成一个以高技术为重点的工业贸易区。

B. 依靠科学技术进步，开发应用新技术

(1) 加速科技进步。高桥石化公司的大部分企业设备老化，工艺比较落后，需要进行有计划、有重点的改造。炼油企业要提高两次加工能力和加工深度，提高油品质量，千方百计降低能耗。后加工企业要进一步采用新的催化裂化和聚合新工艺、新技术。如能改进重质油深度加工的重油催化裂解工艺，超临界溶剂脱沥青、生产针状焦的延迟焦化装置等，就能使轻油产率提高到70%左右，重油收率降低到10%左右。这是我国现在炼油技术水平能达到的，也是在不增加原油需要量的前提下为两次加工增加资源的关键。

为了增加石油化学工业的优质原料，炼油工业的工艺条件也必须有相应的调整，才能满足石油化工对轻质、优质的原料要求。例如，适合于裂解制乙烯的原料——汽油馏分油、石脑油、轻柴油的馏分，现在我国产率为45%，美国为69%，德国为62.6%，英国为57.5%，法国为58.3%。我国与美国相差24%左右（日本炼油工业的重点是兼顾燃料油的供应，因此轻质油比率只占43.8%，比我国还低），同时我国炼油工业以热加工裂解为主，所采用的工艺尚不完备，生产的内燃机燃料未能达到优质要求，亟待改革，争取把适合于化工原料的优质轻油的产率从现在的45%提高到70%左右，把内燃机用的石油产品提高到国际市场的要求水平。这是我国现在炼油技术经验和水平所能达到的。

(2) 要合理地利用油气资源，搞好深、精加工和综合利用。同时要尽可能利用国内外的各种资金，引进和制造新的工艺装置，使老的工艺装置得到更新换代。特别要密切注意和吸收国际石油化工的先进技术、先进工艺，引进先进设备，进行消化和推广。总之要依托全国最大开放城市和现代化新区开发的有利地位，尽快让高桥石化公司的生产技术赶上和超过国际先进水平，这是提高经济效益、扩大出口创汇、搞好环境保护的基础和保证。

尽快采用国内外现成易行的科研成果和新技术、高技术，改造炼厂和石油化工企业的传统技术和正在老化的“新装置”，实现现有企业挖潜改造和扩大产品的需求，如重质油加工的热裂解、减黏、焦化、抽提等工艺，代之以深度加氢裂解、缓和加氢裂解、两段催化裂解、新型超临界抽提，以达到提高石油产品质量和石油化工产品质量的要求，达到油、化、纤结合，深度开发，提高经济效益。国外也把旧装置的技术改造作为今后石油化工发展的主要方向之一。

(3) 要推动技术进步，更要重视科学研究。高桥石化公司的几个化工厂的发展，可以说都是采用了国内科研成果建立起来的。目前，公司本身已形成了一支科研力量，建立了研究院。并在催化、裂化方面有较好的研究基础，上海又是一个化工科研力量比较集中的地方。因此，要紧紧抓住科研这个重要环节，十分重视吸收国内外的科研成果来发展，充实自己，提高自己，这是一条投资省、见效快的发展途径。高桥石化公司积极依靠上海雄厚的科研力量，依靠化工、纺织、仪表、轻工、建材、医药、机电等各局的支持和合作，以石油化工研究院为骨干，形成浦东的石化科研基地，有计划地针对重点项目组织协同攻关，形成有自己特色的石化生产技术，建立各种形式的科研、生产、经营、出口这四个环节的联合体，将科研成果尽快地形成新的生产力，开拓新的国际市场，把高桥石化公司建设成一个现代化的科学园区。

(4) 注意控制石化产业的“三废”治理，对废水要尽量循环利用，减少不易降解的有毒有害物质；对废气要减少 COD（chemical oxygen demand，化学需氧量）、硫化物的含量；对废料要综合利用。

(5) 采用必要的和先进的节能技术，把石油化工的能耗损失节约出来，移做石油化工原料之需，既可大大降低成本，提高竞争能力，又能弥补原料不足。现在石油炼制和石油化工生产工艺，主要是热加工或催化热加工过程，耗能过大，而且这些能源都是珍贵的石油、天然气和裂解气。现在炼油厂及石油化工厂万元产值耗煤达 14t 以上，比全国工业平均万元产值能耗的 8.15t 高 70%。全国的炼油厂、石油化工厂每年能耗达千万吨煤之多，可见节能之重要。石油、石油化工企业对节能颇为重视，一般通用节能措施包括热量节约和能级节约。石化产业是能源消耗的大户，应该有针对地关注一些能耗过大的生产装置，通过挖潜改造和开发补用，使产品的单位能耗能有所降低，从而达到节约能源的目的，进而扩大再生产。

6. 钢铁工业

浦东钢铁工业就其现状来说主要就是上钢三厂，上钢三厂位于浦东南端，有隧道可通，交通方便。

1）上钢三厂技术改造前景

上钢三厂分为两个大的生产模块，即炼钢和轧钢。炼钢目前主要是提高合金比，不是提高产量。即使是普通钢产品，上钢三厂也通过增加低合金钢的比例而使质量提高，产值提高。轧钢部分的产量、产值的增长有相对的独立性，因为可以通过厂际的锭、坯协作来解决。

近年来，上钢三厂型材增产显著，这主要是由于技术力量足和近年来市场需求的增长。板材的生产量则受到设备限制不能增产，基本保持不变。

目前轧钢方面设备较陈旧，检测、控制技术落后，且二次能耗高，热处理能力不足。

总之，上钢三厂的客观条件是，设备老化，而目前还没有足够的资金进行大范围的更新。由于矿源问题，也很难进行大规模的自动化炼钢。

(1) 上钢三厂应走高技术的发展道路。发挥品种全和质量上的优势，尤其注意发挥目前合金钢、低合金钢、中板的优势，像桑塔纳国产化这样的重点工程，需要大量的优质、多品种钢材，而上钢三厂对于其中的某些钢材目前还不能生产，这个事例说明发展高质量的钢材是十分迫切的。

(2) 稳定产量吨数。钢铁工业主要配置的钢潜力有一定的限度，而新建炉除个别情况（如原已配套的）外，以另建新厂、选点外高桥等地为好，吨产量不变并不是产值不变。据上钢三厂的同志的观点，通过发展高质量钢材而使产值翻番是可以做到的。

(3) 搞好横向联系。这包括两个方面。一方面，外地的一些钢厂需要技术，有些小铁厂想自己炼钢，上钢三厂可以输出技术。由于上钢三厂拟转产质量较高的品种，这样做不但不矛盾，而且还可获得社会效益。另一方面，由于许多铁厂争取自炼钢，铁源受到威胁，上钢三厂应同时搞好铁的来路，以便生产得到保证。上钢三厂目前的位置，与宝山钢铁总厂的位置不同，不宜自己炼铁。因为常规炼铁设备相当大，污染（火焰、烟雾和吹尘）也要大得多。采用等离子法炼铁炼钢可以降低污染，但技术要求高，目前尚属研究项目，不能作常规工程考虑上马。上钢三厂目前已有一些关于铁源的协作关系，应该扩充，可以考虑联营、调价、技术支援等方针。只要上海炼钢工业的技术优势确实还在，加上正确的经营方针，即使不靠国家调拨，也能获得足够的生铁炼出好钢，并达到提高产值、利润的目的。

（4）正确处理好国家控制和市场调节的关系。从对世界各国情况的简介中可知，各国的钢铁企业都要有一定的控制权由国家掌握，因为钢铁工业的稳定如同石油等的稳定产、销一样关系到国家经济的稳定。另外，为了适应不断变化的市场的需要，企业又必须有一定的自主权。钢材的国家调拨和市场价差距较大，要严格做好审计工作，以免出现谋私现象。

（5）参加世界技术市场。可以抽出部分人力研究世界一流技术，以加强交流，提高企业素质。

2）主要炼钢和轧钢技术

上钢三厂的设备水平主要都停留在相当于国外五六十年代的水平，迫切需要结合浦东的炼钢业现状来决定先采用哪些新技术以及引进哪些新设备。

（1）连铸技术。连铸技术由于可以做到节能、节材，又较易上马，所以近年来在国内正在推广。以首钢为例，首钢有 3 座 30t、2 座 5t 和 2 座 210t 的转炉。除 30t 转炉正在配制连铸设备外，其他的各炉都已经实现连铸生产。今后的计划是使连铸比接近 100%，仅例外情况才用模铸。首钢公司使用连铸技术后，降低成本效果显著。

上钢三厂现在也正在努力提高连铸比。以对不锈钢板采用的连铸技术为例，可提高成材率 7%。今后要进一步提高连铸比以达到降低成本、缩短生产流程的目的。要采用顶底复合吹炼技术。向铁水中吹入氧，使氧与铁水中的碳结合成 CO 或 CO_2，即炼钢的主要化学过程。顶吹即从钢包的上部（空气）吹入氧气，底吹则是从钢包的底部吹入氧气。比较起来，底吹效率较高，顶吹则技术较简单。

顶底复合吹炼技术是从上下两个方向同时吹入氧气，因此效率更高。首都钢铁公司两座大转炉（210t）目前正在搞复合吹炼，其他的炉都已实现了复合吹炼。根据他们的经验，可以做到降低消耗（钢铁消耗少，吹尘也少），搅拌性好，质量提高。

上钢三厂也正在搞顶底复合吹炼，并运用了一些引进技术，目前发展的方向是全面实现顶底复合吹炼。

上钢三厂目前平炉的产量仍占总产量的近 1/3，目前不能一下子就取消。但平炉产钢质量较难提高，应作规划分期分批由电炉和转炉代替。

（2）连轧技术是国际上的先进技术，往往成为衡量一个国家钢铁业先进程度的指标。近年来各先进国的连轧率都在不断提高，某些欧洲国家已达 100%。这些都主要是在近十年内完成的。

但是就上钢三厂目前的情况说，则不宜搞连轧项目。因为连轧虽然工艺先进，生产流程短，节省能源，但上马时需要将厂房全部重建，造成原有设备的浪费。因此从目前的经济效益上来看是不可取的。从长远的角度考虑，如要上马连轧项目，从北京、上海各地的专家的意见来看，也以另建新厂效益较高。这样原有设备可以继续生产。

(3) 自动控制技术。自动化程度是现代工业的又一个指标，上钢三厂的轧钢部分可以通过运用自动制技术来提高产量质量，减轻强度。但是炼钢部分目前则不能实现全面的自动控制。主要原因是我国的铁矿往往含铁低，属贫铁矿。更有些矿含量较杂，因此我国一般对不同的铁源采用不同的冶炼技术，这样才能使钢质量提高、稳定。而自动化的炼钢一般要求铁矿来源稳定，因此上钢三厂为了今后能适应不同的铁源的炼钢，就不宜上马全面自动控制炼钢，这样可保持一定的灵活性，效益上讲是正确的。

轧钢部分由于原料来源一般波动不大，所以可以搞如在线检测、实时控制等自动化项目。

(4) 自动取样。现在已经具备了自动取样的能力。但是现在自动取样的成本相对较高，因此从效益的角度考虑不常用到。但自动取样对于生产高质量的钢材是完全必要的。所以从现在起就要积极学习国外先进经验，提高样品分析能力，降低自动取样分析的成本。这样做的目的是为以后生产优质钢、特种钢打好基础。至于目前是否用这项技术，取决于具体经济效益的投入产出分析。

3) 具体的炼钢和轧钢技术

转炉车间三座转炉缺乏自动化系统，炉外精炼不够完善，连铸机质量也参差不齐，这些都需要较多的投资才能改善。在开发顶底复合吹炼的同时，也要进行双流顶吹技术的研究，双流顶吹技术上较易实现，同时开展对喷氧枪等具体细节的研究，使质量有新的突破。连铸机要攻克 95×125 连铸机的生产技术，实现全面连铸。

平炉车间已采用双枪顶吹，但模铸生产收得率低，劳动强度高。平炉车间现在在研究钢包喷粉和大板坯连铸技术以提高钢的收得率。

电炉炼钢则可通过 VOD（vacuum oxygen decarburization，真空吹氧脱碳法）炉外精炼及不锈钢连铸工艺来提高不锈钢的内在质量。现在的主要问题是电炉容量低，装备落后不配套，可以进行采用金属球围冶炼、热装铁水等生产工艺的研究。

在轧钢的中小型材方面，在线检测控制低，二次能耗高。这些都是目前可以进行改善的。同时要扩大品种范围，目前以轻轨、槽钢、轮相钢材、圆钢、螺纹钢为主。

中厚板生产的热处理能力要进行提高。热处理可以改变板的力学性质，实现液压 AGC 以提高板的厚度精度、研制无损在线探伤。

薄板生产也存在能耗高、成材率低、装备落后的问题，同时劳动强度高。通过自动控制来提高成材率，提高精度，降低劳动强度，同时研究轧辊堆焊、工艺润滑技术，采用森吉米尔轧机等先进设备。中厚板在上钢三厂的生产受其他工业的影响。上钢三厂生产的中厚板主要用于造船、石化等工业。如果上海（尤其是浦东）进一步发展造船、石化工业，则上钢三厂要相应地配套增产中厚板。

在环境保护方面，各项指标与冶金部的标准还略有差距。主要体现在用水量、降尘量和钢渣利用率还略超出部标准。随着浦东的全面开发和繁荣，将对环保提出更高的要求。现在的第一步是要使各项指标达到部标准。

4）新兴技术

由于其他领域的科技成果，钢铁工业能采用一些投资少的改革以达到改进生产的目的。这比更新炼轧基本设备效益高，因此被各国广泛采用。

A. 信息技术

现代信息技术使钢铁生产技术有了新的突破。20 世纪 70 年代后，微电子技术、计算机技术、光纤通信和激光等新兴技术的发展，导致了信息技术时代的到来，使生产面貌大为改观。我们钢铁业也开始采用这一技术。下面简介有关技术的发展现况，作为上钢三厂技术改造的借鉴。

（1）氧气转炉的计算机动态（实时）控制系统的发明，大大减少了操作人员的负担，并可以使控制更为准确，是国外炼钢新技术不可缺少的一部分。虽然因我国矿源不稳定而不能照搬西方的现有成果，但应对适合我国矿源的自动化控制系统进行研究，搞出适合国情的自动系统。

（2）电视会议。日本的神户钢铁公司自从采用了电视会议系统，不但加快了节奏，提高了生产管理的实时性，同时节约了经费。扣除电视会议系统的费用后，每月还可净节约 400 万日元。虽然以我们目前的经济技术实力不能照搬，但也可考虑采用类似的信息技术来节约差旅费，并加强企业间的联系，如电话会议系统等。

（3）光纤通信技术用于管理系统，可以克服钢厂占地面积大，常规通信显得缓慢的毛病，已被世界上一些先进钢厂采用。光纤通信频带宽、容量大、抗干扰力强、体积小，但技术性强，需要专业人员研究开发。

计算机在钢厂内有两种用途，一是技术信息系统，二是实时控制（当然还有科学运算）。上钢三厂在这两方面都已开始起步。国内各大钢厂，如首钢、马钢等也都已列入专项计划。

计算机技术属于高技术范畴，运用时除添加机器外（上钢三厂已有不少微机），更要注意人员培训，才能达到预期的效果。具体措施是从经济效益出发对计算机运用的项目进行核算，减少资源的浪费。

B. 检测技术

检测技术包括温度、流量、压力、高度控制回路以及计算机和各计量设备，这些作为控制中心的感官，已成为现代钢厂的必备部分。先进国家一个产钢 2000 万 t 的钢铁公司约有温度控制回路 2000 个。

C. 等离子体炼钢

等离子态时物质可达很高的温度。用等离子法炼钢，原料可不经造块，不需用焦炭。瑞典 SKF 公司用等离子法炼铁，生铁成本下降 20%，能耗下降，污染减少。这种等离子技术在钢铁工业中用途很广，目前是世界上的热门课题。我国也有部门在研究，具体实现则需要有很强的技术、经济实力做后盾。

以上新兴技术代表了世界钢铁工业革新的一种趋势。有些技术能马上在浦东使用，有些则还不能。但我们都应对它们进行研究，以确定合适的发展战略。

四、发展和完善浦东新区社会服务系统

1. 形成良好投资环境

利用外资开发浦东是我们的战略方针。上海吸引外资的成功与否，在于能否形成一个良好的投资环境。投资环境是指投入资金循环增殖的外部条件。上海包括浦东的这种外部条件虽然已经大大改善，但是与国外甚至国内某地区相比，仍然有差距。要吸引更多的外资，最大限度地发挥资金的使用效益，首先应健全三大机制，创造良好的投资环境。这就是：①优惠政策机制，包括优惠税收政策和优惠价格政策等。②投资管理机制，包括建立管理机构，担当调剂、平衡、控制、监督的责任。③社会发育机制，包括对社会各服务子系统（图 1）的改造和完善，使整个社会系统向现代化开放型转变。

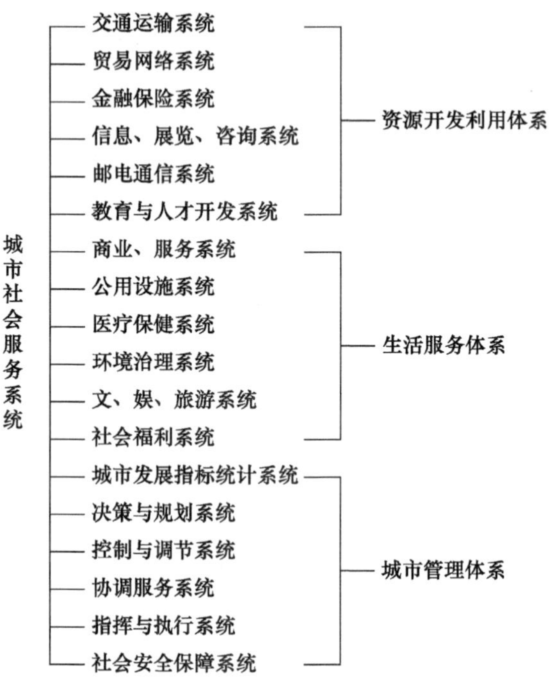

图 1　城市社会服务系统

上海市在上述几方面迈出了一大步，促进了投资大幅度增长。要增强社会发育机制，使资金在浦东循环增殖速度快、增殖高，就必须使浦东新区具备以下条件。

（1）现代科学标准的合理城市布局。

（2）高效能的基础设施，先进的地上与地下建筑等，这包括信息、交通、给排水、供电、供热能各项设施，以及商业、金融、服务业等。

（3）发达的第三产业和信息咨询事业。

（4）高水平的城市管理、高质量的生活环境，居民能在清洁、优美、安静、安全的环境中工作和生活。

(5) 先进、高效的生产和经营手段。

科学技术进步促进了城市服务系统的飞速发展，总的来讲表现在两个方向：一是沿着纵向渗透，服务系统的每个环节不断地提高，出现更先进的设施，更先进的服务。二是沿着横向扩展，新型的服务项目不断扩展，如电视数据、办公室自动化、新型交通工具等。在这样一个高效的城市服务系统条件中，资金循环增殖的速度很快，形成了良好的投资环境。

2. 浦东新区发展基础设施和社会服务系统的策略

由于黄浦江一水之隔，浦东发展受到局限，与浦西差距不小，与世界现代化城市差距更大得多，要赶上世界水平，必须以较高的效率、较高的速度来改善浦东的条件，但是目前我国处在社会主义初级阶段，生产力水平不高，国家不可能投入很多资金来改善浦东条件，吸引外资也要有一个过程。为解决这个矛盾，必须根据实际条件，选择一些优化的策略。

采取"有限目标、突出重点、分期实施"的方针。要避免"全面赶超""齐头并进"。要制定规划，选择目前对改善投资环境最关键的"瓶颈"项目，实施重点突破。对于浦东来讲，可选择电信、城市交通、港口建设作为改善的重点项目。(能源不属于本课题研究范围，我们在此较详细地研究这三项，在分报告中还将研究公用系统管线地下共同沟和办公室自动化等)。

在技术引进或开发某一个项目时，应根据上海市现有的科技基础，选择那些花钱少、见效快的技术，要注意阶段性。

要充分重视"软技术"的开发或引进，过去存在"轻软重硬"的倾向，只引进设备，不注意引进管理技术，使许多工作事倍功半。现在港口建设中要注意引进管理技术，在市内交通中要注意引进交通管理的各种技术。

信息化已是现代化城市服务系统的特征。因此，应用电脑和信息技术来改善服务系统的效率是今天的一个重要方向。

在完善社会服务系统方面，新技术应用与开发的目标是：有重点地、阶段性地改善电信、交通、港口等方面的服务设施和服务手段，使其成为现代化的高效率服务系统。

3. 电信

目前各经济发达国家和地区都已建成了以模拟为主、模数并存的通信网，并正向数字化过渡。市内电话和长途电话实现了自动化，市话普及率一般都在30%以上。除了普通电信业务的普及，一些适合现代社会需求的信息通信业务也在不断推出。在技术方面，正在积极采用程控交换机；大容量光纤和电缆传输，并且和微波、卫星通信构成多路由多手段的电信网；大量安装数字新设备。世界电信的发展正趋向于综合数字信息网。电信对现代工业社会的发展起着难以估量的巨大作用。

浦东地区的电信目前总交换容量19 000门，其中已有近8000门程控交换容量，但电话普及率和接通率都很低，与浦西沟通的中继电路不足，线路陈旧老化，使得浦东内部、浦东与浦西电话不易畅通，更谈不上发展直拨电话和开展新业务。按照常规，浦东难以超过市区的发展速度，近五六年上海电话增长率为14%，而待装用户年增长率超过30%，电话接通率降至50%以下。为解决浦东地区通信紧张，市政府和邮电部门已做了很大努力，如用程控交换机对老式电话局进行扩容改造，和去年新浦东话局第一期工程的4000门纵横制交换设备的安装等，但考虑到要将浦东建成一个现代化的新区，要求电信在一个不太长的时期内，越过一个在业务、技术和管理上的跨距，以较快的速度缩小与经济发达国家城市的差距。

浦东新区的电信应该充分满足现代化城区的需求，除开放常规电信业务外，还要为居民、企业和政府等提供更多的服务。新区的电信要考虑到开发重点的需求，特别是在陆家嘴和外高桥地块。要确保这些地块的国内外联系畅通，优先提供先进的通信手段。新区的电信应该有全面的长远规划和阶段目标。起步阶段应集中投资，开通最迫切又可迅速见效的电信业务，电话普及率应超过10%，并使1/3人口可享有住宅电话；展开阶段电话普及率将超过30%，力争接近目前香港水平，并使那时每户人家都有电话。并且开通数据通信等信息通信业务，到新区建成使电话普及率达到60%～80%，接近或超过东京现有水平，并能提供综合信息服务。

根据浦东新区电信发展的目标，必然要向综合业务数字网过渡，电信技术装备具有国际先进水平，电话密度到达发达国家城市的水平，可设想：①应在浦东新区建成统一的本地网，不再有市话网和郊县网之分。新区的电信网仍然是上海市网的一部分。具有汇接局、端局和用户线路三个网络层次，有陆家嘴等处多个汇接局可沟通与浦西的联系。②电话局所的设置在原有基础上全面考虑，将密切联系开发重点及其阶段的要求。可在闹市区高楼集中地块建设程控选端模块地下无人值导机务站。考虑到减少网络节点，原则上不宜再增加电话局数，而可扩大交换容量。最终话局控制在15～16个，平均容量在6万～8万门。③为确保通信畅通，所有中继线路都采用两种传输手段，过江中继采用水缆和微波并举（在新建陆家嘴电视塔、浦江大桥和越江隧道时应有所考虑），一般中继采用光纤和微波并举（在建高楼时要考虑空中微波通道）。选用先进的地下线路铺设方式，可参考日本东京等城市在街区道路建设时就建设地下通信线路管道与其他管线共用的“共同沟”，在浦东新区彻底消灭“路面开剖”的现象。所有楼房在建造时就考虑到电话线和电源线一样铺设。④建立浦东电信大楼，它既是新区电信管理机构和营业机构的所在地，又是浦东地区电信网络监控管理的中心。电信大楼又是一个主要汇接局，以建在陆家嘴附近为宜。

从目前发展趋势看，浦东新区在开发初期可大胆使用国外20世纪70年代末80年代初成熟的先进技术，如本地网技术、光纤通信、卫星通信、微波通信、移动通信和数字程控交换技术等。

电信是一项技术密集型和资金密集型产业。加速发展电信还和政府政策、投资能力与策略、城建规划、经营思想甚至与国际经济气候等诸多因素有关。脱离这些软环境的支持，单纯以技术观点去发展电信是难以获得成功的。然而，浦东新区的电信建设有可能获得较快的发展速度，这是由于已经得到肯定的政策开放度和有可能充分吸取国内外发展电信的正反两方面经验教训，从而确定一个合理的电信发展模式。

对浦东新区电信建设提出如下建议。

(1) 马上着手新区电信建设的规划。一个城区的电信规划是长远大计，要严格履行科学程序，不可草率。通常需要先进行可行性研究，然后筹划详尽的技术规划。由于浦东新区发展的特殊性，基本数据的缺乏，这项工作更不能掉以轻心。邮电部规划所曾对上海经济区电信发展做过规划，上海市邮电部门也有过 2000 年规划，新区的规划应该在这些工作基础上进行。

（2）加强领导部门的协调功能。电信牵涉问题太多，是非由政府出面协调不可的。目前上海市虽有市通信领导小组，但不是一个有权威的日常领导机构。可设想在新区管理委员会下建立一个主管信息产业的机构，除主要抓电信建设外，还可兼管局部计算机网、数据库和信息中心。必要时可就某一工程专门组建领导建设办公室，协调和监督工程的进程。

（3）倡导电信企业化经营。这涉及电信体制的敏感问题，但浦东新区有可能跨出一大步。这是由于它已获得中央肯定的政策开放度和有限面积的地理区域。可考虑：成立浦东电信局，技术体制仍受市电信局管辖，行政管理可由市电信局提供适当指导帮助，而经营管理完全由新区电信局承担，自负盈亏。这样既可保持上海电信网的一致性，浦东电信建设发展又可不受上海市管理方式的束缚，呈现极大的灵活性，适应新区的需要，更广泛的讨论将涉及这个电信企业是否可采取“三资”企业的形式。

（4）保证足够的电信投资。电信投资具有促进社会再生产的资本作用，事实上是一种效益最高的投资，但电信投资的主要效益是被全社会获取，电信部门仅受惠极小部分。苏联的研究表明，1 卢布电信投资可为国民经济节省 3 卢布。日本统计表明，1 日元的电信投资可为社会创造 2.5 日元价值。我国邮电研究部门的研究成果也表明，1 元人民币电话投资可带来 6.78 元社会效益。具体体现在：提高社会劳动生产率，节约人力资金，加快资金周转和节约能源。必须大力宣传这种观点。浦东新区电信的发展速度必须高于新区的经济发展速度。如果不能在电信投资上有一个超出常规的增长，新区建成现代化城区的战略是难以兑现的。

4. 交通运输

1）浦东交通现状问题严重

浦东地区交通难的问题主要表现在道路少、公交系统不发达，交通管理系统不健全。另外还有一个特殊问题是过江难。目前交通问题严重，常堵塞的道路有浦东南路等若干条。浦东开发后，若不重视交通，问题就更严重。

浦东要开发成一个具有世界一流水平的新区，不具备先进的道路结构、发达的公共交通系统和提高交通管理水平是不可能适应这种开发需要的。

2）浦东的交通建设要考虑的几个方面

（1）布置合理的道路结构。浦东的外高桥、陆家嘴、北蔡、花木四个开发区，每个区域内都必须具备与区域发展相协调的干道、次干道、支路系统，并配置一定规模的停车场、广场、绿化地。干道和次干道要形成快、慢车分流，机动车、非机动车分流，人行和车流分行，道路宽度和交叉口设计要根据交通流调查及发展来确定。干道宜建成六车道，次干道建成四车道，尽管开始可能占地多，但对今后城市的发展有好处。而区域与区域间应建立高速干道网，使客、货运能各行其道。在高速干道下面可铺设各种管道（电力、电信、供水、供气等）和公路专用通信网，提高道路信息的快速传递和通信，降低道路的造价。

(2) 建立必要的公交系统。随着汽车工业的发展，加上小汽车舒适性灵活性的优势，小汽车在各个大城市的保有量在近几十年飞速增长，但是在我国，大众化的、大运量的客运交通工具还得靠公交系统来实现，这在许多发达国家也得到了证明。关键是在浦东新区要建设能快速输送流量在100万人次/日的主要公交系统，如地铁、市郊铁路、有轨交通、磁力悬浮车等。

从目前浦东地区的地理情况来看，地铁的造价在2亿元/km，有轨交通系统造价在50万元/km左右，而运输量地铁是有轨交通的2.5倍左右。另外，磁力悬浮车由于车辆沿导轨飞驶，速度可达190km/h。从性能价格比来看，地铁方式具有快速、正时、干扰少等优点，但造价贵、要求高（通风、排水、铝合金不锈钢车厢、紧急出口等）、施工难度大。而有轨交通具有价低、易于施工的特点，缺点是噪声大（需隔离墙）、干扰多（要考虑交叉道口）。但在浦东开发建设中，这种公共交通运输系统是一定要搞的。光靠单一的公共汽车是难以解决实际问题的。要组成一个在小区内有小型（公共汽车、小汽车、公共车辆优先行驶、专用车道等）公共交通系统，区域间有大运输量的公共交通系统的格局，并与浦西的交通联系起来。

(3) 过江交通的安排。过江交通要采用轮渡、隧道、大桥一体化的运输方式，随着时间的推移和新区的发展，轮渡方式应逐步减少，而由隧道（包括自行车专用隧道）和过江桥梁来承担。从布局来看，从闵行至外高桥地区几十个主要道口越江工具都需要发展。从流量来看，客、货流最好是分流，特别是大宗货物的分流（如集装箱运输，30t的集装箱要从隧道过江相当困难），因此这一地区还要考虑建造第二座浦江大桥，选址可进一步分析。自行车的过江问题可考虑建设专用隧道。

(4) 管理系统的改善。交通管理的提高能明显改善交通条件，而所花费的资金却不多，是一条很好的途径。国外研制并日趋成熟的英国SCOOT（split，cycle and off-set optimization technique，绿信比、周期、相位差优化技术）、澳大利亚SCATS（Sydney coordinated adaptive traffic system，悉尼自适应交通控制系统）、日本omron（ohm dragon，欧姆龙）系统、交通诱导控制，这些系统从原理上都能解决交通紧张所带来的交通混乱问题，但有些参数不太适合中国国情，而这些系统的软件还没有完全公开，因此可以在借鉴这些系统优点的基础上研制适合我们自己道路情况的交通管制系统。当然也可以通过技术引进、消化，对其修改后再进行使用。

(5) 外高桥的建设。根据发展设想首先要开发外高桥地区，因为那里有优越的港口和高桥石化工业区，可以开发成一个外贸工业加工区，那样的话，该地区港口的货物运输量和周转量将有大幅度的提高，人口也会在短时期内有一个新的突破。形成一片新的出口导向型加工区，并随着发展产生一系列生活区、商业区和娱乐区，随着外高桥地区的发展，公路网建设将贯穿浦东的各个主要地区。

3) 建议和措施

(1) 制定合理的交通规划。交通规划必须从现在做起，使其保证和浦东新区的规划开发同步发展，具有整体性、合理性和持续性。

(2) 综合治理交通运输。要适当引进一些新技术、新设备和新材料，如城市道路施工设备、交通管制设备、公交管运工具、反光材料等，并充分利用我国现有基础和材料开发的仪器、设备来武装浦东交通运输。

(3) 注重城市环境。城市交通系统的水平提高既不能以破坏城市生活环境和质量为代价，也不能过分强调城市功能而忽视交通。城市的发展依靠交通运输的高效率，交通运输的发展需要与城市的环境保持一致，包括景观的保护、与建筑协调、与城市功能相吻合。

5. 港口建设

浦东有两处建港地——外高桥和卢潮港，其中外高桥地区岸线长 7km，万吨泊位可设百个，水深 10～15m，条件相当好。浦东发展战略中已确定外高桥作为对外交通和外贸工业区重点开发。

1) 浦东港口发展的设想

初步设想的浦东港口发展三阶段如下。

在 2000 年以前，作为第一阶段，首先开发建设外高桥新港区，以港口及其航运业带动浦东地区迅速发展起来。纵观中外经济发展史，最早发展起来的城市几乎都是依靠港口的。利物浦、纽约、神户、横滨、上海、天津都是明显的例证。

2000 年以后 15～20 年，作为第二阶段，随着大系统交通体制的完善和发达的资料信息系统的建立，应进一步建设发展外高桥新港区，发展全集装箱码头，使其作为华东地区的对外门户，能主要承担上海及长江流域的外贸运输的繁重任务。

在以后更长的一段时间内，也即第三阶段，要进一步开发建设金山嘴大型、深水港区，使包括浦东在内的整个上海港成为大型的、具有世界先进水平的国际枢纽港，从而为实现把上海港建成西太平洋最大的经济贸易和文化中心之一的战略目标打下坚实基础。

2) 浦东港口与世界上先进港口的差距及努力方向

(1) 泊位建设。每个高效率的现代化港口，为了能迅速接纳来港的大量船舶，都建有足够的泊位，外高桥目前仅几个运油专用码头，要在统一规划下，加快建设泊位的速度。

(2) 深水航道和深水港区的建设。浦东深水航道和深水港区的范围不大，长江口内侧的铜砂浅滩阻塞了航道，有待于疏浚和扩大。

(3) 导航设施的建设。20 世纪 70 年代末期出现了应用电子计算机技术的船舶交通管理系统（vessel traffic management system，VTS)。我国青岛港口在港在区设置 VTS，并向附近海域扩展。台湾与香港也在建设。其他地区，包括上海均未建设 VTS。上海市水上交通拥挤，水上事故多，1977～1981 年的五年间发生在上海的海损事故达 57 起，占全国港口海损事故总数的 51.8%，这说明上海亟须建立 VTS。

(4) 装卸设备专业化，高效率化。上海港口与世界现代港口相比，在泊位数量、深水航道和港区、导航设施、陆域和库场、集疏设施以及劳动生产率等方面差距是明显的。若仅就装卸技术及设备而言，从整体上说，专业化、机械化程度较高，水平不低，属国内先进水平，在世界上也处于较先进行列。但还有差距。主要表现在两个方面：成组化、集装化比例小，应用电子计算机对单机以至装卸系统进行自动控制与管理不广泛、水平低。

（5）库场容量扩大。由于采用大型船舶，货运集中，每次装卸量大，货物一时运不走，要设库场存放，逐步疏运。

外高桥港口附近可用土地面积大，应率先规划好，逐阶段建设现代化的货场。

（6）疏运设施的完备。如果忽略了港口的疏运能力，造成港口堵塞，造成的损失是巨大的。世界上先进港口均有畅通的疏运设施，尤其是铁路、高速公路、内河运输连接成网，外高桥区应考虑铁路建设，尽早使铁路过江，与中国广大腹地接通。

（7）扩大修船能力。现代化大港口一般都具备强大的修船能力，浦东造船业十分发达，近年来修船业务也有发展，应有计划地在外高桥新港区建立修船业务。

（8）加强港口的计算机管理。广泛引进电子计算管理技术，逐步实现管理操作自动化，是世界先进港口特点，现已有码头遥控、自控、装卸设备自动化。

烟台港率先开发了港口调度信息管理系统，自 1987 年初以来一直成功地运行着。他们的经验值得外高桥港口借鉴。

（9）要重视环境保护，绿化美化港区。可考虑在港口与浦东新区内设置绿化带，规划和建设游览区，严格控制污染。

3）分阶段引进、开发、应用新技术

2000 年以前的第一阶段，重点建设多用途泊位和扩建泊位。

（1）尽快地开发 VTS 技术，以高效管理船舶交通，减少交通堵塞和交通事故。

（2）开发港口管理信息系统。

（3）尽可能使大宗散货装卸专业化、高效化，使杂化装卸成组化、集装箱化。

（4）采用各种高效连续卸船机、多用途国产起重机、高架轮胎起重机等先进的装卸车、船设备。

（5）广泛应用电子计算机进行单机以至于系统的自动控制和管理、自感故障测量和管理。

（6）加强技术和人才引进，同时加强培训工作。

2000 年以后的阶段，则应在逐渐减少引进先进技术和设备、引进人才的同时，立足于自己开发和发展适合中国情况的新技术、新设备、并使整个装卸技术及设备处于当时的世界先进水平。

4）策略

（1）港口建设包括基础设施和经营性设施建设两部分。基础设施无利可图，应立足于国家投资，以内币为主；经营性设施则应积极引进外资，以外币为主。可以多渠道、多方式引进外资，可以利用世界银行、亚洲开发银行或其他政府或金融机构低息贷款，或采用外商独资、合资及其他形式。在第一阶段还可搞保税仓库、出口加工区等以提供多种经营服务方式活跃投资环境，更多地吸引外资。

（2）多渠道、多方式引进智力或技术及培训人才。可以搞联合咨询、合作设计、联合投标（施工或生产）或引进人才、送出国外培训等。

（3）承包施工或采购设备主要应采用国际招标方式。交通部利用世界银行贷款，三次通过国际招标，为上海、天津、黄埔三港共采购各类设备 525 台，合同总金额 6640.1 万美元。因日元、马克升值，实际付款约 6900 万美元。由于公开招标竞争激烈，对降低标价、节省投资、保证质量都十分有利。例如，第一次招标，总标价3104.5万美元，为原概算的53.3%；第二次总标价1697万美元，仅为原标价一半左右。

五、建立浦东新区现代化城郊农业

作为一个国际化、枢纽化和现代化的世界第一流的浦东新区的城郊农业，必定是外向型的、现代化的。新区农业的主要任务是：①生产多品种、高质量的农副产品，以保证现代化大城市，特别是对外开放的供应；②利用种植养殖方面的条件与优势，针对国际市场需求引进资金、技术和良种，发展出口农产品，发展创汇农业；③依据旅游、游乐休息的需求，发展“旅游农业”，开辟新的旅游资源。

浦东原农业结构较为单一，属于以种植业为主体的传统农业。传统农业作为农业发展的一个基本阶段，其特征是以人力、畜力和手工劳动为主进行自给自足的生产；其目的是维持劳动者及其家庭成员的生存而生产。现代农业的特征则是：产品商品化，生产专业化，服务社会化，经营集约化，生态良性化，工具机械化，技术科学化。显然，为完成浦东新区发展所赋予的任务，浦东农业应加快由传统农业向现代化农业转变的进程。在产品商品化、生产专业化、技术科学化等方面下功夫。

以下分四个部分来展开浦东农业的发展。

1. 发展设施园艺，生产多品种、高质量蔬菜果品

城郊农业要向城市提供大量的蔬菜果品。由于城市居民对蔬菜果品要求越来越高，不但要求多样化的品种，而且要求周年生产，四季供应。先进国家，如日本、美国及西欧各国，相继发展了设施园艺。从结构上分类，设施园艺包括：玻璃温室，钢筋塑料大棚，塑料管棚，塑料小棚，简易棚等。各种结构形式，又区分若干不同栽培方式，以适应不同的蔬菜果品的要求。

在设施园艺内，采用了先进的设备和技术。几乎所有温室和大棚装置了比较现代化的加温、灌水、通风设备，还有电子计算机控制和管理，普遍采用了多种薄膜，多层覆盖保温。由于设施园艺的发展，居民随时随地都能买到无季节限制的质量上乘的菜。

设施园艺的发展有两个方向：①向节能方向发展；②向养液栽培方向发展。养液栽培，又称无土栽培或水培。

设施园艺是一项复杂的工程，涉及农业、科研、工业等许多部门和学科。上海市具有全国最强的工业和科技基础，尤其在农业科技力量方面，遥居全国之首，上海已与日本建立中日合作设施园艺试验场协议。由日方无偿提供全套园艺设备和农膜、农药、化肥、测试仪器、小型农机具，并派出专家来华进行技术指导。上海试验场已投入试验生产，取得了一定的经济效益和社会效益。在此经验的基础上，会较迅速地发展设施园艺。

对于浦东新区设施园艺发展，可采取更灵活的政策。除政府补贴外，可争取外资。与某些饮食、旅游联营集团签订合同，由他们提供资金。菜农以高质量蔬菜、果品补偿，以取得较多资金发展设施园艺，尤其是在蔬菜采后保鲜和批发市场方面，极尽地吸收先进的科学技术和管理方法，以提高蔬菜品质和营养价值，方便消费者，减轻城市的垃圾污染，使新区以一个新的面貌出现。

2. 发展现代化养殖业，生产高质量的肉、蛋、乳食品

作为世界一流城区，浦东的肉、蛋、乳食品的供应量必须丰富多样。除实行体制改革措施外，还应注重吸收世界先进养殖业技术，以求较快发展养殖业。

养殖业工厂化、机械化、自动化、电脑化。品种健康高产，饲料全价组合，厂房环境、设备卫生，管理科学，效率高，料肉比符合标准，经济效益高。

养猪、奶牛（牛）、鸡（蛋鸡）、鱼等的现代化新技术可简要概括为①育种采用新技术，疾病防治技术先进。②喂养机械化、自动化、电脑化。③饲料制备科学化、机械化、自动化。④加工、包装、运送、销售专门化、现代化。

浦东地区，主要是川沙县①，发展养殖业的条件在上海郊县中属上乘。据估计可发展淡水养殖1万亩（1亩≈666.67平方米），可建立奶牛和乳制品生产基地以及瘦型猪和良种禽蛋基地。作为世界一流城市郊区，浦东地区肉、乳、禽蛋需求量增加很快，相对应的养殖业发展要走集约化的道路，以技术进步，即不以扩大外延而增加内涵来增加生产的道路。浦东养殖业发展策略可考虑如下几点。

（1）建立专业生产基地。根据地理环境、技术基础条件等选定各类养殖业生产基地。这样便于资金、科学技术引进，便于销售和管理。

（2）建立技贸农联合体，实行科研、生产、贸易三结合。由贸易部门引进优良品种及新设备、新技术，科研单位和高校担任技术攻关，而农民进行生产。可实行横向联营，由食品公司、旅馆餐厅投资，建立现代化饲养工厂，将产品优先供应它们。

（3）国家提供一定的补贴等政策，扶持养殖业迅速发展。

（4）瞄准市场，尤其是国际市场需求，发挥本地优势，创造一批名优品种，占领国际市场，出口创汇。

（5）建立养殖信息技术咨询产业，向农民提供国际新技术、国际市场信息，以发展养殖业生产。

（6）培养技术人才，引进生物工程等新技术，提高养殖工厂的经营管理水平，以及科学技术水平。

3. 发展花卉新产业，美化城市，增加创汇

花卉，作为雅致、整洁、文明、进步的象征，它可以反映出一个国家、一个地区、一个城市科学、文化和艺术水平的高低。随着世界科学技术的进步和人类生活水平的提高，不少国家已把花卉作为一项重要商品进行生产，并迅速形成一项新兴产业。其迅速发展的原因有以下几点。

（1）具有较高的经济效益。花卉业是一项经济效益较高的生产门类。在传统的农产品中，低产值的居多。而花卉则属于高产值的高档商品，如小麦、棉花与杜鹃花产值比为1∶3.5∶162。

（2）带动附属行业的发展。花卉生产是一种劳动密集型的生产，可以吸收大量的就业人口，广开就业门路，对于人多地少的地区有特殊意义。

（3）推动旅游事业的发展。

---

① 于1993年正式撤消，建立浦东新区。

(4) 促进食品、香料和药材工业的发展。

(5) 改善环境质量。

世界市场对花卉的急切需求大大刺激了花卉生产的发展，从育苗、栽培到管理逐渐趋向采用最先进的工业化生产体系，形成一个专业化、现代化、立体化的花卉生产体系。

花卉育苗的全过程几乎都在水泥地上进行，看不到真正的土地，比一般工厂还干净，从拌土、下种、催芽到育苗由计算机控制。

随着花卉工厂化育苗生产技术的发展，园艺事业发达的国家早已突破了自然条件的限制，在人工控制温度、湿度和光照强度等设施园艺条件下生产各种花卉植物。同时，花卉栽培技术也发生了重大变革，大量采用无土栽培，浇灌营养液的方法。

现代化、国际化的浦东新区是一个容量很大的花卉市场。发展花卉生产，供应新区，美化新区环境，同时逐渐出口创汇，有极重要意义。

浦东川沙县花木乡以培养花卉著名，有较好条件。现有 60 亩盆景场、50 亩园艺场及 265 亩果园、350 亩树木面积。但是由于品种少、质量差、种植技术落后，目前仍采用瓦盆、泥土、粪肥养花，检疫往往通不过，花卉出口停滞不前，创汇能力很差。为了迅速发展花卉生产，提出以下四条建议措施。

(1) 建立花卉基地，包括生物工程育苗基地、生产基地、出口基地，采用先进的技术管理，发展设施园艺，即温室、大棚、营养盆栽、营养液栽培等。逐渐建立可控温、光、水、气的现代化温室园艺。

(2) 抓特色、拳头产品。充分利用浦东及上海基础，培育特有的优秀花卉，尽快形成独有的花卉优势，占领国际市场。例如，可发展日本、中国香港、欧洲欢迎的月季、菊花和香石竹。

(3) 建立技贸农联合体，实行科研、生产、外贸三结合。例如，由外贸引进良种，科研单位负责繁殖和栽培技术攻关，然后组织花农批量生产，由外贸部门组织销售。这样可以发挥各方优势，以形成高效率的花卉生产体系。

(4) 培养花卉专业人才，引进与研究先进花卉生产理论及生产技术，将花卉生产提高到世界先进水平。

4. 引进与开发生物工程技术，发展新型农业

现代科学的发展，给农业带来了一场革命，世界上先进国家纷纷开发生物工程技术研究，将其应用于农业、医药卫生、轻工食品、化工、能源开发、冶金、环境保护等领域、取得了惊人的成就。

生物工程是近几年来发展的一项高技术产业，它是指综合运用生物学、化学和工程等的手段，直接或间接地利用生物体本身或某些特殊机能得到产品，为社会服务的科学体系。

生物工程产品在农业、医药卫生、环境保护、冶金等领域广泛应用，其优越性在于能以地球上丰富的再生性资源为原料，在缓和的条件下进行反应，过程简单，节约能源，降低成本，减少污染，可提供一种新途径解决传统或常规方法难以解决的问题，能提供新的动植物品种满足人的需要。

从城市总体功能看，发展生物工程有助于建设一个环境最清洁、最卫生、最优美的一流城市，有助于促进产业向技术密集型转化。

世界上先进国家在将生物工程应用于农业生产方面做了许多工作。下面结合浦东情况谈一些设想。

浦东农业可应用的生物工程技术有以下七点。

（1）植物组织培养技术。在这项植物遗传操作技术中包括三大课题：快速繁殖植物株系；无病植株的获得；植物无性系诱变，培育新品种。

（2）花药培养、单倍体育种。

（3）农用单克隆抗体的研究。

（4）家禽胚胎移植及胚胎分割技术。

（5）控制鱼类性别比例的技术。

（6）生物防治。

（7）基因工程的应用。（详细见分报告）

上海市有我国最强的生物工程技术力量，中国科学院在上海的各研究所及其他研究机关和高校聚集了雄厚的技术力量。外国生物工程公司早有意向在上海投资经营，浦东可选独资或合资生物工程公司引进外国生物工程产品或技术，利用本市充足的科技人才，优先生产进口替代或填补空白的生物工程产品，进而扩大规模，生产出口产品，占领国际市场，创取外汇。

值得一提的是在浦东建立植物组织培养中心条件基本成熟，可采用外商独资或合资的形式，生产无毒试管苗（花卉、蔬菜、水果等）供应国外和当地。我国在这方面的技术已达世界领先水平，国内也形成了产业，广西柳州以及广东顺德、新会均建立企业，经济效益和社会效益十分显著。例如，广东顺德办厂两年收回投资，生产香蕉苗供给果农，大大增加了产量也保证了质量。

参考文献

[1]　中国城市经济学会秘书处，兰州市经济研究中心．城市发展战略与管理．兰州：甘肃人民出版社，1988.

[2]　汪道涵．城市经济与区域经济．城市发展战略与管理．兰州：甘肃人民出版社，1988：218-228.

[3]　刘人怀，史乐毅．有效利用外资 扩大对外开放．国际商务研究，1987，(6)：1-5.

中国制造业的生存哲学[①]

一、中国制造业的两难境地

制造业，承载着科学家、科技工作者和企业家的责任和理想，承担着“纸上富贵”至物质财富的转化任务。如果说设计方案是上层建筑，那么科学严谨、一丝不苟的制造业则是实实在在的经济基础。从美国和日本的制造业在20世纪末的进步可以看出，制造业不断与新技术相结合是一个国家经济发展和综合实力提高的支柱。

重温人类制造业的历史：

制造业萌芽于人类进化之初，猿人学会制造和使用工具，人的概念由此而生；

170万年前，中国元谋人开始使用火；

5000年前，青铜时代开始；

公元1380年，欧洲建成最早的炼铁高炉；

1765年，瓦特发明蒸汽机，标志着第一次产业革命的开始；

19世纪后半叶至20世纪初，电气化发展拉动第二次产业革命。

时至今日，任何一国的国民总收入中都有80%以上与物质生产和消费密切相连，其工业化程度的高低直接决定社会的发展阶段和人民的生活水平，而在工业部门中，制造业占有中心地位。

在今日中国，制造业直接创造国民总收入的1/3，占整个工业生产的4/5，出口总额的九成，提供8000多万就业岗位，是1/3国家财政的来源。制造业是我国国民经济的主要组成部分，出口的主力军，就业的重要市场。

新中国成立以来，我们的制造业从无到有，逐步形成完整的工业体系。1978年以前。我们独立自主自力更生，取得了以“两弹一星”为代表的重大成就；最近这二十多年来，我国制造业获得巨大发展。制造出了秦山300兆瓦核电站、数字程控交换机、银河及神威高性能计算机等有形产品。

现在，我国制造业的工业增加值（工业企业在报告期内以货币形式表现的工业生产活动的最终成果）居世界第四位，仅次于美国、日本和德国。

世界制造业正一步步向中国转移：装配电脑整机所需零配件，95%以上可在东莞市采购；格兰仕微波炉的销售规模占全球市场的35%；江苏电脑鼠标的产量占全球总量的65%；早在1999年，全球彩电销售量的四成在中国生产，而复印机更占到六成；日本的五大汽车厂，丰田、本田、日产、三菱、马自达已全部在中国设厂；奥迪轿车已经实现国产化，中国产宝马轿车已下线，奔驰汽车也开始了和北汽福田的全面合作。

但是，事物总是有它的正反面，正如我们常说的，科学是一把双刃剑。

制造业在创造辉煌的同时，也在侵蚀、污染着我们赖以生存的环境。统计数字触

---

① 本文原载《科技中国》，2004，创刊号，56-57.

目惊心：我国制造业每年产生约 55 亿吨无害废物和 7 亿吨有害废物；所有造成环境污染的排放物中 70%来源于制造业；每年全国因大气污染损失 740 万个工作日。据世界银行估计，环境污染给中国带来相当于 3.5%～8%的 GDP 损失；中国科学院牛文元教授更认为，中国经济增长中 18%的 GDP 是依靠资源和生态环境的“透支”获得。

二、绿色化是中国制造业的生存哲学

中国制造业，面临着是否开展绿色改革的艰难选择。

我们有必要重温恩格斯的论述：“我们不能过分陶醉于对自然界的胜利，不管我们取得了多么辉煌的成就，我们连同我们的血、肉和头脑都属于和存在于自然界。”

正因为工业社会在创造巨大的物质财富的同时，也创造了巨大的环境灾难，人们越来越意识到：我们不只是继承了父辈的地球，而且是借用了儿孙的地球。因此，自 20 世纪以来，人类对环境的关注一直没有停止过：

1962 年，美国学者出版了《寂静的春天》，引起全球对人类生活环境的关注；

1972 年，波托马克协会、罗马俱乐部和麻省理工学院研究小组联合出版《增长的极限》，其中阐述的“合理的持久的均衡发展”，为孕育可持续发展的思想萌芽提供了土壤；

1978 年，德国率全球之先，实行绿色产品制度，产品经国家权威部门审评后贴绿色标志；

1985 年，德国的胡伯（Huber）教授提出生态现代化理论，认为应利用人类智慧去协调经济发展和生态进步；

1987 年，联合国世界环境和发展委员会发表《我们的共同未来》，阐述可持续发展的思想，它告诫人类：在满足当代人需求的同时，不要损害子孙后代的基本生存条件。

1992 年，联合国环境与发展大会通过全球《21 世纪议程》，要求各国就“环境标志产品认证”制定和实施相应政策；

1996 年，国际标准化组织正式颁布“绿色通行证”——ISO14000 系列国际环境标准。

2003 年 5 月，美国举办“绿色汽车”拉力赛，中国制造的“黑豹”电动汽车首次亮相；

2003 年 6 月，世界第二大汽车制造商福特宣称将建造绿色制造基地；

20 世纪末至今，“绿色消费”观念席卷全球。从绿色食品、绿色设计、到绿色制造，绿色浪潮正在改变人类的生活和生产方式。

我国是人口众多、资源相对不足的国家（人口占世界人口的 22%，耕地和草地的人均占有量仅是世界平均水平的 32.3%，淡水和森林更是仅为世界平均水平的 28.1%和 9%），中国制造业经历了 50 多年的发展，但依然没有走出资源型经济增长路线，传统的以“高投入、高消耗、高污染、低质量、低效益、低产出”为特征的发展方式不可能长期持续下去，环境污染和资源匮乏是悬在中国制造业头上的两把利剑。

可是，我们又不能放弃制造业。

十六大报告提出全面建设小康社会的目标之一是：可持续发展能力不断增强，生态环境得到改善，资源利用效率显著提高，促进人与自然的和谐，推动整个社会走上生产发展、生活富裕、生态良好的文明发展道路[①]。

既要持续发展，又不能放弃，中国制造业只能走绿色化的道路。绿色制造是可持续发展战略在制造业中的体现，是中国制造业的必由之路。

如果说全球化是中国制造业面临的经营环境，信息化是制造业运作的有力手段，那么，绿色化更是中国制造业的生存哲学、一种价值取向。

三、中国制造业绿色化任重道远

绿色制造的概念最早由美国制造工程师学会提出，该学会1996年发表的关于绿色制造的蓝皮书（*Green Manufacturing*）中阐述：绿色制造是一种综合考虑环境影响和资源效率的现代制造模式，它的目标是使产品从设计，制造、运输、使用到报废处理的整个产品生命周期中，对环境的负面影响最小、资源效率最高。简单地说，绿色制造要综合考虑制造、环境和资源这三大领域。

在技术层面上，绿色制造包含：绿色产品设计技术、绿色制造技术，产品的回收和循环再制造技术，它是精益生产、柔性生产、敏捷制造的延伸和发展；作为一种指导思想，绿色制造则充满诗意并洋溢着温暖的人文关怀；具体到企业，绿色制造体现了企业家的管理道德和社会责任。

可见绿色制造是一种大制造、大过程，它学科交叉、观念现代，对它的深入研究必将推动制造科学的发展。

由于中国在市场潜力，劳工价格等方面有比较优势，发达国家制造业已经正在向中国实行梯度转移，出于就业压力、经济发展和学习经验的考虑，现阶段，我国制造业承接此类转移不失为一个良策。但是，我们不能满足于被“锁定”在欠发达状态，只有通过绿色制造这一途径，使生态与经济协调发展，我们应该并且能够达到发达国家水平。

为此，我们必须软硬兼施。

在软环境方面．我们要加强绿色制造的技术开发和基础性研究，大学应该设立相关专业；我们要通过媒体，宣传环境保护意识，企业要具有环保的责任心。

在硬环境上，绿色制造企业需要相关政策予以指导，法律、法规对其进行约束，尤其应该把ISO14000系列标准和环境评价纳入企业评价体系。

绿色制造，从某种程度上讲，还是一个概念，对全世界漫长而又波澜壮阔的制造历史来说，它不过是“小荷才露尖尖角”，而中国制造业的绿色化道路更是刚刚起步，任重而道远。

---

① 江泽民在中国共产党第十六次全国代表大会上的报告，https：// www. gov. cn/test/2008-08/01/content_1061490_4. htm.

东水西调工程

一、关于改善我国北方水资源缺乏的建议①

众所周知，我国华北和西北地区水资源极其匮乏。据统计，黄河、淮河、海河三大流域的河川径流量不到全国的6%，而耕地面积却占全国的40%。就连首都北京附近的海河流域耕地亩均水量，也低于以干旱著称的以色列。一些城市已长期对居民实行水的限制供应，一些城市（包括首都北京）大量超采地下水，造成地面大幅沉降。长此下去，后果不堪设想。这必将制约我国经济社会的可持续发展，影响工农业生产，影响人民日常生活，影响我国现代化的实现，威胁中华民族的生存繁衍。

党中央、国务院已着手南水北调工程，以改善北方水资源严重缺乏状况。但是，就我国整体来说，人均水资源本来就少得可怜，不到世界平均水平的1/4。笔者认为，单靠南水北调一项工程，不能根本解决问题，故建议再增加一个解决办法，即实施东水西调工程。

地球上的海洋之水丰富，取之不尽。建议从渤海和黄海取海水，用管道将海水输向华北和西北，沿途建咸水湖。湖水的蒸发，会改善干旱的气候，使土地不致沙漠化，同时再沿途实施海水淡化，用于居民日常用水和耕地灌溉。

二、关于实施“东水西调”工程的建议②

（一）“东水西调”工程的目的和意义

2004年，笔者基于我国北方严重缺水的形势，向国家提出实施“东水西调”工程的建议[1]。随后便开始了本课题的研究。2011年的中央一号文件《中共中央 国务院关于加快水利改革发展的决定》，是一个特别重要的文件，是关系我国人民和子孙生活幸福的文件。为此，更加深了本课题的研究深度。

本项研究认为东水西调工程是解决我国北方缺水问题的战略性工程，是着眼于海水综合利用的系统工程。东水西调不仅是海水西调，也可以直接从长江口取水。不应局限于长距离调水和沿途蓄水，而应具有调整工业布局、优化用水结构及复原特定地区生态系统等一系列功能。

东水西调工程是增加我国水资源总量的新思路。在近期，可与南水北调及各地区的局部调水工程，以及各种节水及优化水资源配置的措施相配套；在远期，可逐渐减少北方地区对“南水”的过分依赖，保证南方用水丰富。东水西调工程是集长距离输水、海水综合利用、清洁能源替代及信息化建设等多种高新科技应用为一身的系统工程。

---

① 本文原载《参事建言（2004—2005年）》，广东省人民政府参事室编，香港：中国评论学术出版社，2006，229.

② 本文是中国工程院课题报告（2009—2012年），收录在北京科学出版社2015年出版的《工程管理研究》第203～242页，作者：刘人怀，孙东川，孙凯，朱丽，刘泽寰.

中国是一个严重缺水的国家，淡水资源总量为28 000亿m^3，占全球水资源的6%，人均水资源量只有2300m^3，仅为世界平均水平的1/4，是全球人均水资源最贫乏的国家之一。然而，中国又是世界上用水量最多的国家。2008年，全国淡水取用量达到5910亿m^3，大约占世界年取用量的13%。

中国不仅水资源匮乏，地区分布也很不平衡，长江流域及其以南地区，国土面积只占全国的36.5%、人口占53%、耕地占35%、GDP占55%，其水资源量占全国的81%；长江以北地区，国土面积占全国的63.5%、人口占47%、耕地占65%、GDP占45%，其水资源量仅占全国的19%。全国600多个城市中，有2/3城市处于缺水状态，缺水城市有400多个，其中严重缺水城市114个。严重缺水城市中，北方城市占71个，南方城市占43个[2]。

中国已成为世界少数几个最缺水的国家之一，目前有16个省（区、市）人均水资源量（不包括过境水）低于1000m^3严重缺水线，有6个省、区（宁夏、河北、山东、河南、山西、江苏）人均水资源量低于500m^3极度缺水线。而在现有的水资源中，有78%的淡水污染物超标，40%的水源已不能饮用，50%的地下水被污染。因此，治理污染，保护水资源，维护生态平衡已到了刻不容缓的地步。

我国的水资源不仅匮乏，而且分布不均匀。随着北方地区经济的发展，对水资源的需求日益强烈。例如，河北省近20年超采了1000亿m^3地下水，地下水位大幅沉降，现在只有打500m左右的深井才能出水；山西省也面临同样的问题，煤炭开采造成地下水位沉降，不仅造成了取水困难，还使得煤矿透水事故频繁发生，2010年3月的王家岭矿难就是例证。

我国的中西部地区，特别是北方中西部地区，由于多年大兴水利，开垦农田及过度放牧，造成了大量内陆河流断流，尾闾干涸，形成了大量的干盐湖，周围土地沙化和盐碱化，沙漠范围不断扩大，形成的沙尘暴每年对北方城市造成了严重的威胁。2010年4月的特大沙尘暴覆盖了西北大部及几乎整个华北地区。

水资源长期的超额使用，已打破了原有的生态平衡。尽管国家近年来实施了一系列大型的调水工程，但由于中国的水资源总量不多，有限的资源重新分配并不能从根本上解决问题。南方地区只是在丰水年才显得水量稍多，当遇到枯水年也同样面临无水可用的局面。如遇2010年春季西南大旱及2011年的长江中下游旱情，可向北方调取的水量就可能大为减少。

东水西调工程正是在这一背景之下提出的，将海水、淡化海水或长江口淡水调往内陆干旱地区以改善气候，恢复生态，改变中国水资源匮乏的现状所带来的益处是显而易见的。但是，对于这样一项庞大的系统工程，如何计划与实施、如何将对环境的负面影响降到最低、如何确保不对沿线经济造成新的负担，以及调水的手段、调水的总量、调水的方式等问题，都需要仔细研究、小心论证、谨慎思考。

（二）东水西调工程方案

1. 调水规模与线路

东水西调工程计划将东部海水和长江口淡水通过华北平原调入黄土高原及内蒙古高原，最终引入新疆地区。工程的目的一方面是用以补充当地水资源的不足，改善当地的生态结构；另一方面通过优化当地的用水结构，达到调整产业布局，提升地区竞争力的目的。

东水西调工程的实施应在确定各地用水需求的前提下确定规模，先行试点、逐步推进，以调整用水结构为主线，以工程可实现性为前提。

1）调水规模

东水西调工程的用水地区包括北京、天津、河北、内蒙古、山西、陕西、甘肃、宁夏及新疆9省（区、市），工程水源地位于河北、天津、山东、辽宁沿海以及江苏省长江入海口北岸地区。整个工程涉及13个省（区、市）。各受水地的用水现状如表1所示。

表1　东水西调工程沿线各地用水现状　　单位：亿 m^3

省（区、市）	年份	工业用水	农业用水	生活用水	生态用水	合计	南水北调
北京	2008	5.20	12.00	14.70	3.20	35.10	10
天津	2008	5.36	13.21	3.11	0.65	22.33	10+10
河北	2007	25.11	155.75	19.76	1.21	201.83	60+10
内蒙古	2004	10.40	149.40	10.90		170.70	
山西	2007	10.00	28.81	5.23	0.45	44.49	
陕西	2004	12.47	51.30	8.96	2.80	75.53	
甘肃	2005	14.46	97.48	7.95	3.08	122.97	
宁夏	2008	3.33	69.13	1.72		74.18	
新疆	2003	8.23	457.70	10.36	18.13	494.42	

表1根据各地当年水资源公报整理而成。因各份报告中用词不一致，表中数据进行了相应的处理。其中，工业用水包括第二产业用水、第三产业用水；农业用水包括第一产业用水、农田、灌溉、林牧渔畜；生活用水包括城镇公共用水、居民生活用水、农村居民（或人畜）供水。南水北调数据包括中线和东线两部分，加号之前为中线水量，之后为东线水量。

通过对相关省（市、区）近年用水现状的分析，以工业用水、农业用水、生活用水、生态用水等的使用现状作为基础数据，以确定未来用水需求量，并以此估算东水西调工程的调水规模。

对表1数据进行分析，我们可以将数据分为三组。

第一组包括北京、天津及河北3省（市）。其共同特点是人口密度大，大部分城市都在海拔200m以下，距海岸线较近，且地势平坦，没有大的山脉阻隔。工业用水合计在9个省区市中占37.7%，农业用水占17.5%，生活用水占37.6%。近期的用水缺口基本可以由南水北调补充，如河北省目前每年超采地下水约50亿 m^3，南水北调工程可以补足这一缺口。但是，南水北调工程调水量的多少会因南方地区当年的水情发生变化。如遇南方旱情，来水量势必减少；因而，还需要通过海水利用进行补充。此地区的海水利用采用海边淡化，再以适当口径管道定点输送到特定地区或企业，淡化海水的用途以工业用水和生活用水为主。在工业用水方面，也可以采用管道直接输送至工矿企业的方式，由于海拔较低，所需的能源是非常有限的。

第二组包括内蒙古南部地区、山西、陕西、甘肃及宁夏5省（区）。其共同特点是均为内陆省（区），地域广大，人口密度相对较低，且基本都是重要的产煤区。离海岸很远，海拔大都在 1200m 左右，需要长距离、高扬程输水。工业用水合计在 9 省区市中占 23%，农业用水占 59%，生活用水占 27%。这些省（区）是东水西调工程将海水淡化后调往的主要地区。在调水方式上，东水西调工程已采用大口径管道长距离输送之后再分水，送至当地使用。

第三组包括新疆及内蒙古北部地区。其特点是地域广阔，多为沙漠及干旱地区，人烟稀少，海拔大都在 1000m 以上。输水距离长，沿途有大量干涸或半干涸的湖盆可用于蓄水改善生态环境及发展海水养殖业，建议直接将海水调往这些地区，用湖盆蓄水，视需要再进行淡化处理。但由于海水远距离运输尚无先例，技术方案有待进一步研究，对生态环境可能造成的负面影响还需要进一步论证；此外，还需要对工程的运营模式做进一步的探讨，以使工程的成本与经济效益达到平衡，确保工程运营的长期稳定。

考虑工程规模、成本效益及技术条件等方面的因素，我们建议东水西调工程主要用于解决工程沿线各省（区、市）的工业及生活用水。一方面由于工程提调的水量有限，必须计划使用，以最经济的方式解决关键性的问题，特别是农业用水的水量巨大，工程由于技术条件及成本等方面的限制，尚不能全面满足沿线的用水要求。另一方面农业及生态用水的大部分可通过再生水来补充，这样就降低了总体用水成本。

虽然东水西调的调水成本较高，但海水淡化后的水质非常好，故应主要用于置换和增补现有的工业用水和生活用水。这也就相当于提高了农业用水及生态用水的可用量。根据各省（区、市）的用水状况，工程输水区里的距离远近及工程实施的难易程度等因素，将东水西调工程分为近期规划和远期规划：近期规划分为三期工程实施，采用“先淡化，后调水”的模式；远期规划拟采用“直接抽调海水”的模式。

一期工程供水范围：北京、天津、河北，调水规模 20 亿～35 亿 m^3。

二期工程供水范围：山西、内蒙古（鄂尔多斯高原为主），调水规模 20 亿～35 亿 m^3。

三期工程供水范围：陕西、甘肃、宁夏，调水规模 30 亿～50 亿 m^3。

远期规划供水范围：内蒙古（鄂尔多斯高原以外的大部分地区）、新疆，调水规模 25 亿～40 亿 m^3。

鉴于上述方案中海水淡化的成本较高，故提出第二套方案。即由长江口直接取水，以解决上述省（区、市）的缺水问题。

2）线路规划

东水西调工程的调水线路分为两类，A 类为近期规划的淡水线路，B 类为远期规划的海水线路。从分散布置海水淡化设施的角度考虑，A1 线路自渤海取水，A2 线路自黄海取水；S 线为长江口取水线路，自江苏启东境内、临近长江北入海口处取水，经净化后输送至 A1、A2 线，线路长度为 500～800km。东水西调工程的线路总体情况如图 1 所示。

图中各线路距离由地图测量估算得出，情况简述如下。

A1：唐山→北京→张家口→大同→鄂尔多斯，1000km。

该线在河北省境内滦河入海口附近取渤海水，在淡化后经过河北、山西两省，过黄河后进入内蒙古，在位于鄂尔多斯高原的库布齐沙漠与毛乌素沙地之间与 A2 线汇合。调水量为 20 亿～30 亿 m^3/a。

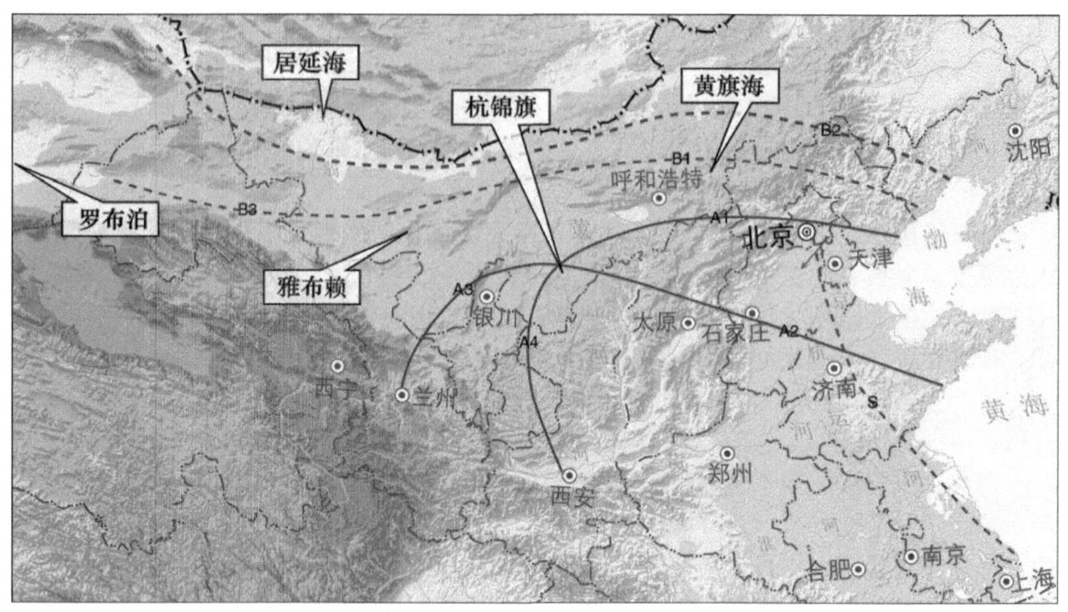

图 1　东水西调线路规划示意图

A2：日照→济南→石家庄→太原→鄂尔多斯，1000km。

该线在山东省境内取黄海水，在淡化后经过山东、河北及山西，过黄河后进入内蒙古，在位于鄂尔多斯高原的库布齐沙漠与毛乌素沙地之间与 A1 线汇合。调水量为 30 亿～55 亿 m^3/a。

A3：鄂尔多斯→银川→兰州，600km。

该线由鄂尔多斯高原起，向西过黄河，经银川到达兰州，满足宁夏及甘肃两省（区）的用水需要。调水量为 15 亿～25 亿 m^3/a。

A4：鄂尔多斯→西安，600km。

该线由鄂尔多斯高原起，向南至西安，经过宁夏、甘肃及陕西，满足 3 省（区）的用水需要。调水量为 15 亿～25 亿 m^3/a。

B1：秦皇岛→阴山以南→阿拉善高原南部→玉门→敦煌→罗布泊，3000km。

B2：辽东湾→阴山以北→居延海→哈密东北部盆地，3000km。

A 类线路工程为“先淡化，再调水”，主要用于置换和增补沿线地区的工业及生活用水，调整用水结构和产业结构。在 A1 与 A2 线汇合地的杭锦旗境内有大量的干湖盆，可选择适当地点建立大型水库，一方面供本地使用，另一方面作为 A3 与 A4 线的水源。鄂尔多斯高原是内蒙古、陕西、甘肃、宁夏、山西 5 省（区）的交界点，是我国主要的煤炭产区，可以建设大型坑口电站群，为东水西调工程提供充足的能源；更重要的是，东水西调工程引入的水资源也为煤电一体化基地的建设提供了新的契机，通过建设大型电站群，减少原煤外运，降低物流成本，实现煤电及化工联产，对于改善产业结构有着积极的意义。此外，A1 线路起于渤海，A2 线路起于黄海，这样可在一定程度上减少海水淡化设施过于集中对海洋产生的污染。

S类线路为长江口取水线路。其作用是与海水淡化方案类似的水源地工程建设，即A类线路的水源可选择海水淡化和江口取水两个方案。

B类线路为远期规划线路，直接调取海水，海水可直接用于沿线地区的生态用水并发展海水农业，用于改善沿线植被及气候。部分海水在当地淡化后，可用于工业及生活用水。线路沿线是我国主要的风场，调水所需能源中的相当一部分可通过风力发电提供。线路所经过的居延海等地区海拔较低，在注入海水的同时可利用高度差建设水电站，所得收益可弥补工程的运营成本。工程输水进入新疆前，海拔为1200m左右，而新疆境内的几处注水地点海拔较低，其中，罗布泊海拔约780m，吐哈盆地最低点艾丁湖海拔约为－155m，准噶尔盆地最低点艾比湖湖面海拔约为190m，在注入过程中可建立多个梯级水电站，在很大程度上补充工程前段提水至1200～1300m的能源消耗，同时也为新疆当地提供强劲的电力供应。

3）网络化调水

相比较而言，东水西调工程采用多线路组成的网络化调水模式是比较可行的方案，与单线路调水方案相比，其优点有以下几点。

（1）覆盖区域广阔。由于东水西调工程需要为沿线服务，因此考虑各线路沿能源基地及工业基地展开。每个途经省份均有两条以下的线路经过，使得工程可以充分惠及沿线，发挥最大效益，平衡各地的用水需求，并减少支线管道的埋设距离。

（2）降低施工难度，提高工程进度。调水网络各线路基本沿河谷、公路及铁路布置，降低施工难度，减少施工道路的修筑，便于安装提升设施。同时网络化的线路配置，可实现多线路多工程段同时开工，加快工程的进度，单条线路完成后就可以局部开始输水，使工程早日发挥效益。

（3）方便配套能源项目布局。由于东水西调工程所需要的电力巨大，采用风力发电需要大面积地布置风力发电机。而我国的地形特点是，海拔200m过渡到1200m的区域比较狭窄。如线路过于集中，风力发电设备规模布置难度较大。而采用多线路的网络化布置线路，这一问题就可得到妥善解决。

（4）多水源地建设。工程不仅考虑了调取海水，还考虑以长江口取水作为第二套方案。整个工程设计了多个取水点，一方面避免水源地的建设在一处占用较多土地；另一方面当一处地点出现污染或疫情导致水源中断时，整个输水网络仍能够运转。

（5）就地生产或购买管道。工程所需夹砂玻璃钢管的生产工艺并不复杂，全国各地的生产企业很多，多线路的布局可以方便在工程沿线就近购买或生产管道，降低运输成本。同时也可以提高工程沿线的生产制造水平。

（6）安全保障提高。当某条线路出现故障或所经地区发生自然灾害时，可及时关闭该条或其中一段线路，工程的其他部分仍可继续输水，使工程运作的安全性得到充分保障，为线路维修提供充足的时间。

（7）推广海水淡化。随着陆地水资源的日益匮乏，海水淡化是大势所趋。但是由于建设大规模海水淡化厂需要占用土地，消耗大量能源，如果所有的海水淡化厂都建在海边，势必造成土地匮乏。调水网络可以将海水输送到使用地，按需淡化，促进资源的合理利用。为降低成本和增加效益，在海水淡化的同时，还可以深度开发盐、镁、钾、锂等相关资源。

(8) 汛期调蓄洪水。调水网络沿黄河及其支流布置，一期工程基本覆盖黄河流域中上游流域。如遇汛期，可从河中抽取多余水量进入输水管道，降低洪水的威胁。如遇较大洪水的发生，甚至可关闭海边的提升设施，让抽取的洪水从抽水点向管道两边流动，则每条线路理论上可抽取 75.56m^3/s 的洪水，作用非常明显。

(9) 便于建设配套风电设施。除主体工程之外，工程还考虑建设与之配套的能源设施用于提水和输水。工程可在沿线配套风力发电设施，用于对管道进行加压。另有部分小型独立设施可采用太阳能发电。因为工程所经过的区域都是风能储量极高的地区，有采用风力发电的极好条件。采用配套的风力发电网络可以减少购置电力的成本，同时也不会对沿线造成庞大的电力负担。

(10) 全线通信配套。东水西调工程沿线将全部铺设通信光缆，并配套建设包括检测设施在内的自动控制系统，用于实现工程的实时信息传输和远程集中控制。同时，工程配套的通信工程实施实际上已构成了一个覆盖北方大部分地区的网络，加以充分利用，不仅可以满足工程运行的需要，还可以发挥更大的社会作用并带来相应的效益。

2. 工程技术特点分析

东水西调工程面临两个主要问题，一是如何利用现有技术长距离、高扬程的提升和输水，二是如何使工程对沿线生态环境的影响降到最低。

1) 海水淡化

综合多方面的因素分析，工程的 A 类线路宜采用“先淡化，再输水”的方式。主要有以下几个原因。

(1) 海水长距离运输尚无先例，相关技术有待完善。A 类线路所经大都为人口稠密地区及主要的工农业基地。输送海水需要特别防护，将大幅提高工程成本；且一旦发生事故，会对周边环境造成极大污染。

(2) 海水淡化的产出比为 30%～40%，高浓度的海水尾液在内地难以处理，对环境影响极大。目前全国盐产量为 6000 万 t，而 20 亿 m^3 海水的含盐量已经达到 7000 万 t 以上，故海水淡化的尾液目前尚无法在内地消化。一个出路是逐渐减少现有海盐产量，用尾液产盐替代，并开展相关化工业，提取镁、钾、锂有用原材料。

(3) 如果“先调水，再淡化”，则相当于有 60%～70%的运能浪费，大幅降低工程的效益。相对于长距离水利设施建设的投入而言，海水淡化设施投入为输水设施投入的几分之一。此外，随着技术的不断进步，海水淡化的各项成本还会逐步降低。

(4) 输送淡水可与现有的水利设施相结合，充分利用既有资源。

(5) 输送淡水的技术已经非常成熟，产业完善。而输送海水需要解决防腐等一系列问题，需要进行大规模的设备改造，大量的研发投入，且会推后工程的实施。

随着海水淡化技术的发展，生产 1m^3 淡水的成本已降至 5 元以下，且随着技术的不断更新及装置规模的扩大，最终的生产成本有望降至 3 元左右，与城市的水价相当。

由于淡化海水的水质远高于饮用水标准，无须再处理即可直接进入城市自来水或直接饮水管道；相比长距离调水后再处理而言，两者的综合成本相当。此外，与调水工程所不同的是，海水淡化的水源地建设不会淹没土地及产生移民问题，建设成本高度可控，建设周期比较短，且治水成本呈现不断下降的趋势，可以根据具体的需求分阶段建设。

目前成熟的海水淡化技术主要包括低温多效蒸馏和膜反渗透法两种方式，两者生产淡水的成本大致相当。通过参考有关案例[3-6]，对海水淡化的成本进行了分析，相关案例的数据，如表 2 所示。

表 2　两种海水淡化技术的成本比较　　单位：元

成本项目	低温多效蒸馏			膜反渗透法		
	案例 A	案例 B	取值	案例 C	案例 D	取值
药剂消耗	0.36	0.28	0.28	0.4	0.23	0.23
电力消耗	0.2	0.3	0.3	1.17	1.2	1.17
蒸汽消耗	1.6	2.2	1.6			
膜更换费				0.96	0.88	0.88
维修费用	0.41	0.08	0.08	0.25	0.22	0.22
运行成本	2.87	2.86	2.26	2.78	2.53	2.5
工资福利	0.45	0.05	0.05	0.45	0.07	0.07
资产折旧	1.18	1.49	1.18	0.79	1.87	0.79
管理费用	0.12		0.12	0.12	0.13	0.12
管理成本	1.63	1.54	1.16	1.24	1.55	0.98
总成本	4.5	4.4	3.61	4.02	4.57	3.48

表 2 中的数据取自参考文献，根据比较分析的需要对部分数据进行了变换、合并等处理。

案例 A 和 C 为山东黄岛电厂（山东青岛），日产淡水规模均为 3000t。

案例 B 为沧州国华电厂（河北沧州），日产淡水规模为 10 000t。

案例 D 为华能玉环电厂（浙江台州），日产淡水规模为 30 000t。

表 2 中所列 4 个案例的日产淡水规模从 3000t 到 30 000t 不等，考虑到大型海水淡化装置的单位成本将会低于小型装置，因而表中的“取值”一律采取小中取小的原则确定，分别得到低温多效蒸馏法的成本为 3.61 元/m^3，膜反渗透法为 3.48 元/m^3。

分析表 2，在运行成本中，低温多效蒸馏法的电力消耗及蒸汽消耗之和与膜反渗透法中的电力消耗及膜更换费之和相当，随着规模的扩大和工艺的改进，这部分的成本将会有所降低。此外，表中的用电以 0.3 元/度计算，如电厂能实现电水联产，以发电成本约 0.25 元/度计算，这一部分的费用约可降低 1/4。药剂消耗则视海水的污染程度而定，维修费用一般也会随着规模的增加而降低。

依托滨海电厂建立海水淡化项目，通过充分利用滨海电厂机组的产能实现电水联产是较好的海水淡化产业模式。其优势主要体现在以下几方面。

（1）利用电厂的成本电价，可以大幅降低海水淡化的能源成本。

（2）不增加当地输电网络负荷，同时可避免输电损耗。

（3）海水淡化装置可与发电机组共用海水抽取及排放设施，避免重复建设。

（4）以海水淡化装置作为调节手段，可使发电设备的利用效率维持在较高水平。

(5) 可以提高电厂锅炉的热效率，减少蒸汽损耗。

(6) 以电厂机组冷却后的高温海水作为冷却水可提高淡水产量，进一步降低能耗。

(7) 可充分利用电厂土地资源及运输设施。

(8) 滨海电厂都有海水淡化装置，在设施建设及维护方面已有成熟的经验。

海水淡化单套设备的产水规模相对不大，且建设周期较短，可视实际需要逐渐增加设备的数量，并视电厂多余电量动态调节产量。此外，独立运行的海水淡化厂只能采用膜反渗透方法，而依托电厂建立的海水淡化厂可以同时采用低温多效蒸馏和膜反渗透两种方法，其中低温多效蒸馏可以充分利用电厂锅炉的低压蒸汽，使锅炉效率由40%提高到60%。综合分析，大规模海水淡化的单位成本大约可降到3元。

初步估计，年产1亿m^3淡水的装置投资约为20亿元（以年运行6000h计算），则A线工程的海水淡化装置总投资为1400亿～2400亿元，其中：一期工程年输水规模为20亿～35亿m^3，海水淡化装置投资为400亿～700亿元；二期工程年输水规模为20亿～35亿m^3，海水淡化装置投资为400亿～700亿元；三期工程年输水规模为30亿～50亿m^3，海水淡化装置投资为600亿～1000亿元。

2) 江口取水

江口取水主要指从江苏启东境内、崇明岛北侧长江临近入海口处取水，经净化后输送山东省境内，汇入东水西调工程A1、A2线. 此项工程内容为可选性质，按实际用水考虑是否建设。在具体实现上，可考虑利用南水北调东线部分河道送水，或另行建设500km管道输送至山东省境内，再送至京津冀地区。

在长江口取水不会对流域的生态及用水产生影响，所取水量在长江年平均入海水量为9600多亿m^3中所占的比例极小。例如，取1%即可达96亿m^3/a，不会对长江入海口的生态环境产生明显的影响。

在长江近入海口处所取淡水只需进行简单的过滤，减少了海水淡化过程中的能源消耗；江水过滤后的污水排放也较少，避免了海水淡化对生态的影响。考虑到江水抽取、过滤及输送需要消耗一部分能源，按照目前的技术水平，将其单位成本设定在1元。

江口取水虽然因调水线路加长而增加了管道工程建设投资，但多水源地、多种取水方式的综合建设，可以最大限度地保证用水安全，增强应对突发事件的能力。此外，还可以起到免去海水淡化成本、减少能源消耗、净化海洋环境的作用。

以下计算长江口取水的管道工程投资。管道采用与A类线路相同的、直径4m夹砂玻璃钢管，双线铺设考虑，即每输送10亿m^3水，每公里投资约为6000万元（具体数据计算参考下面第3小节）。长江口取水的总规模为70亿～120亿m^3，管道总投资为5160亿～8790亿元。其中，一期工程年输水规模为20亿～35亿m^3，沿S线输水至京津冀地区，管道长度约为800km，线路投资为960亿～1680亿元，二期工程年输水规模20亿～35亿m^3，沿S线和A1线输水至山西、内蒙古境内，管道长度约为1100km，线路投资为1320亿～2310亿元，三期工程年输水规模30亿～50亿m^3，沿S线和A2、A3线输水至陕西、甘肃、宁夏境内，管道长度约1600km，线路投资为2880亿～4800亿元。

3) 管道输水

工程计划全线采用管道输送的方式输水，采用管道方式虽然工程成本较高，但可以把对周边环境的影响降到最低，还可以大幅提升工程运作的可靠性。其优点有如下几个方面。

（1）减少土地占用，并有利于环境保护。采用管道输水时，管道埋在地下管沟里，回填覆土的厚度不小于 1m（具体深度根据当地最大冻土层厚度和地面荷载确定）。除水泵站、调节阀门站等局部地段外，其余绝大部分地段仍可在覆土上种植作物，可大大减少占地面积，并有利于环境保护。

（2）土建施工难度低，工程量较小。渠道输水是依靠重力自流，所有渠段都必须沿等高线布置，并保持合理的纵向坡降。管道输水则不受此限制. 管道总是按最简捷（最短）的路线布置，可大大减少土建施工工程量。当输水线路与河流、道路以及其他管线等交叉时，采用管道时的处理方式要比渠道简单很多。部分管线可沿公路布置，无须修建专门的施工道路。

（3）不存在冬季和早春结冰问题，有利于保证长年正常供水。东水西调工程所处地区冬季和早春的气温低，处于结冰期。采用渠道输水方式，轻则减小输水流量，重则引起断流，并需要对调节闸门等各种活动结构件采取防冻措施。如不能很好地解决这两个时节输水过程中的结冰问题，则工程的功能将大打折扣。显而易见，如采用管道输水，管道埋在冻土层以下，该问题将迎刃而解。

（4）可避免输水过程中的渗漏和蒸发，输水损失小。水在输送过程中的损失形式主要有三种，即渗漏、蒸发和泄漏。输水距离越长，损失比例越大；土质越差（渗透能力越强），损失比例也越大。东水西调工程输送的水量毕竟有限，如沿途渗漏和蒸发过大，则将影响对预计的发挥，沿途渗漏也会对当地环境产生不可预知的影响。

（5）输水保证率高，并有利于分期施工投运。对于渠道输水方式，当因自然灾害或人为因素造成渠堤溃破时，势必造成停水。东水西调工程的各线路基本都采用双管并铺的方式，可以使风险分散，保证率提高。另外，修复管道所需的时间一般也比恢复渠堤所需的时间短，检修维护也非常方便。再者，采用管道非常有利于分期施工建设，以后再根据需水情况，采取增加管道条数的方式续建和完建。

（6）可在不停水或基本不停水的条件下对输水线路进行维修和更换。不管采用土渠、衬砌渠道或管道，当达到一定使用寿命后，都需要进行修复或更换。假如采用渠道，要想在达到寿命期后修复或更换衬砌层，则必须先停水才能进行作业；如采用管道输水，当需要修复或更换时，逐条管道进行，可保证持续输水。

（7）便于实施远程和集中控制。由于东水西调工程线路较长，且需穿越沙漠等人烟稀少的地带，工程运作的安全性是需要考虑的重要因素。采用管道方式输水，可以间隔一定距离设置电动调节阀门和监控设备，通过结合沿管道布置的通信网络远程调节输水的流量和方向，实时掌握水情和水质的变化，减少人工维护成本。如遇突发事件，可立即关闭阀门，将灾害影响降到最低。

（8）维护费用低。管道的寿命大都在 50 年以上。在此期间，管道基本不需要维护。而渠道（尤其是土梁）则不同，很容易出现淤积、冲蚀、杂草滋生、泄漏等问题，可以说每时每刻都需要维护。可以预见，假如采用渠道输水，当工程建成投入运行时，沿线将需要配备一支规模庞大的看管维护队伍，其费用可能也相当可观。

相关资料[7-10]显示，目前国内已经产品化的、可用于长距离输水的管道的最大直径为4m。如按照通常采用的1.5m/s的流速计算，则管道每秒的流量大约为18.84m^3，年流量大约为5.94亿m^3。考虑到实际应用的情况，同时也为了简化计算，本工程按单管5亿m^3/a的倍数设计各线路的输水规模。

按照南水北调中线应急供水工程等数据分析，以直径4m夹砂玻璃钢管双线铺设考虑，则每输送10亿m^3水，每公里投资约为6000万元。A线输水总规模为70亿～120亿m^3，线路工程总投资为4284亿～7257亿元，其中：

一期工程：年输水规模为20亿～35亿m^3，沿A1线输水至京津冀地区，管道长度约170km，线路投资为204亿～357亿元；

二期工程：年输水规模20亿～35亿m^3，沿A1（或沿A2线）输水至山西、内蒙古境内，管道长度约为1000km，线路投资为1200亿～2100亿元；

三期工程：年输水规模30亿～50亿m^3，沿A1和A3线（或沿A2、A3线）输水至陕西、甘肃、宁夏境内，管道长约1600km，线路投资为2880亿～4800亿元。

4）配套能源项目

东水西调工程是高能耗项目，能源需求主要是两个方面：一是海水淡化的消耗，每生产1m^3淡化海水大约需要4kW·h；二是提水的消耗，将1m^3水提高1200m大约需要6kW·h。两者合计为10kW·h。

如将二期及三期工程预估的50亿～85亿m^3水输送到目的地就需要500亿～850亿kW·h，相当于三峡工程年发电量约850亿kW·h的58.8%或全部。由此可见，工程所需要的电量是沿线现有设施无法承受的，必须配套相应的能源项目；此外，如采用向沿线购电的方式运作，即使按0.5元/(kW·h)计算，平均到每立方米的输水成本将为5元，将使工程的效益大打折扣。

就我国的地形特点而言，东水西调工程的主要目的是将水从地理第一阶梯提升至第二阶梯，如图2所示。

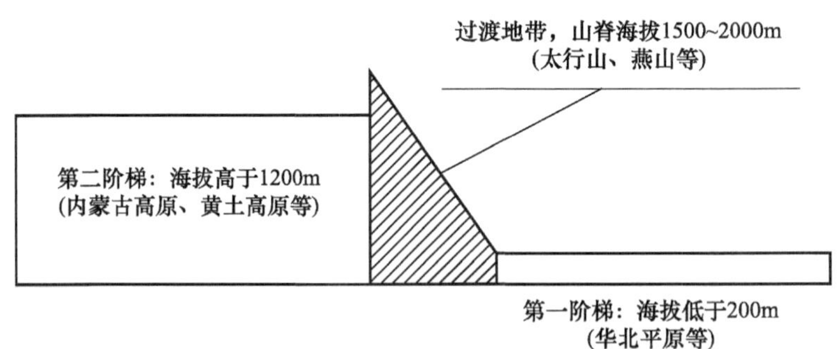

图2　提水阶段情况

我国地理结构的特点对配套能源方式的选择有着重要的影响，根据具体情况分析[10-16]，东水西调工程的能源配套项目宜采用“火电为主，风电为辅”的形式建设，通过采用大规模超超临界机组，将火力发电对环境的影响降低到最低程度，以下就此进行分析。

如采用火力发电作为能源，以目前最先进的超超临界 660MW 或 1000MW 发电机组为例，如以全年满负荷发电量（6000h）计算，每年生产 10 亿 m^3 淡化海水配套 2 台 660MW 机组，每年提升 10 亿 m^3 淡化海水配套 2 台 1000MW 机组即可满足需求。

按每千瓦造价 4500 元计算，则年产 10 亿 m^3 淡化海水的能源项目投资约为 30 亿元。年提升 10 亿 m^3 水的能源项目投资约为 45 亿元。在工程提水设施分布的地区也是主要的煤炭产地，也为火电配套项目建设提供了必要的条件。如按最低输水规模考虑，则一期工程需要 4 台 660MW 机组，二期工程需要 660MW 机组和 1000MW 机组各 4 台，三期工程需要 660MW 机组和 1000MW 机组各 6 台。

如以火电作为主要的能源配套项目，则年产 10 亿 m^3 淡化海水的能源项目投资约为 30 亿元，年提升 10 亿 m^3 水的能源项目投资约为 45 亿元。第一、二、三期工程的总投资为 435 亿～720 亿元，其中：

一期工程：年输水规模为 20 亿～35 亿 m^3，淡化海水的能源投资为 60 亿～105 亿元；

二期工程：年输水规模为 20 亿～35 亿 m^3，淡化海水的能源投资为 60 亿～105 亿元，提水的能源项目投资为 90 亿～135 亿元；

三期工程：年输水规模为 30 亿～50 亿 m^3，淡化海水的能源投资为 90 亿～150 亿元，提水的能源项目投资为 135 亿～225 亿元。

5）工程管理信息化

在国家“信息化与工业化融合”的整体战略中，工程管理信息化是其中的重要组成部分。随着现代计算技术、网络技术和通信技术的快速发展及其在工程管理领域中的广泛应用，与大型工程配套的信息系统建设是必不可少的。

工程管理信息化的实现涵盖了工程从论证、设计、施工、使用直至报废的全生命周期过程。信息化的有效实现降低了工程管理中的协作成本和重复投资，实现资源共享，对工程整个生命周期的设计、建设、运行和维护等各阶段实施的有效控制，以及信息化手段使工程的社会效益得以大幅度地提升。

东水西调工程尚处于论证的前期阶段，在整个论证过程中，需要收集大量的相关案例，以及工程沿线的气象、水文、地质、地震及自然灾害等方面的资料、建立完善的数据库、案例库及专家系统。需要建立协作平台，供参与工程论证的各单位共享和交互信息。这些数据的收集也是工程设计、施工、运行阶段所需要参考和不断扩充的。

此外，国内外尚没有如此大型的同类工程建设的先例，在工程的论证及设计期间还需要建立完善的仿真系统以验证工程方案的可行性。在工程的整个生命周期之中，不同阶段的参与方不同，所面对的主要矛盾也有所区别；在工程设计与建设期间，由于工程技术涉及面广，参与的协作单位多，信息沟通复杂，社会影响面广，因而需要多层次的系统综合集成，信息系统需要支持跨组织、跨地域、跨时间段的协作。在工程运行和维护期间，通过信息化的途径对能源及人力物力方面的投入进行合理控制，可以提高运行效率，降低使用成本、延长使用寿命，以使工程产生更大的经济效益；此外，还需要对环境及民生等的影响进行综合统筹，以使其发挥最大的社会效益。

参考“八横八纵”大容量光纤通信网建设案例：

呼和浩特—广西北海：工程总长4000km，投资8亿多元；

西宁—拉萨：工程总长2454km，投资6亿多元；

北京—兰州：工程总长2052km，投资4亿多元。

以每公里20万元，A类线路长度3200km计算，合计为6.4亿元；考虑到光缆除沿管道埋设之外，还需要在不同线路之间建立联络线路等，实际的线路长度将会大幅超过管道长度，因此将线路投资调整为10亿元。另外，考虑到还需要建立与之配套的通信基站等设施以及远程控制及监测设备才能构成完整的系统；一般在系统中，线路方面的投资只占较少部分，如按1/5考虑，则将信息化建设的投资调整为50亿元。

3. 工程投资与运营成本分析

东水西调工程经过中国广袤的北方地区，工程浩大，经济及社会影响广泛。工程建设涉及海水淡化、江口取水、管道运输、水库调蓄、清洁能源等多个领域。本节将对海水淡化和江口取水两套方案的投资规模、单位水价等进行分析比较。

工程分阶段进行，一期供水范围为北京、天津、河北，二期供水范围为山西、内蒙古（鄂尔多斯高原为主），三期工程供水范围为陕西、甘肃、宁夏。

采用海水淡化方案的工程总投资计6119亿元，第一、二、三期调整后的水价分别为5元/m^3、12元/m^3和15元/m^3。

采用江口取水方案的工程总投资计5385亿元，第一、二、三期调整后的水价分别为6元/m^3、11元/m^3和14元/m^3。

1）海水淡化方案

A. 工程投资

工程总投资计6119亿元（未计入工程征地费用、调蓄水库投资及银行贷款利息等），其中一期工程的投资为664亿元，二期工程投资为1750亿元，三期工程投资为3705亿元，如表3所示。

表3 工程投资情况 单位：亿元

项目	海水淡化		输水设施		合计
	淡化设施	配套火电	输水管道	配套火电	
一期工程	400	60	204		664
二期工程	400	60	1 200	90	1 750
三期工程	600	90	2 880	135	3 705
合计	1 400	210	4284	225	6 119

在各投资项目中，淡化设施合计1400亿元，输水管道合计4284亿元，配套能源投资为435亿元。

B. 单位水价

单位水价中的运营成本主要考虑海水淡化及输水能源消耗。综合前述分析，按海水淡化成本3元/m^3计；输水电力消耗按0.25元/kW·h计，二期及三期工程输水消耗电力为6（kW·h）/m^3，单位水价增加1.5元。

管理成本中，因淡化设施及输水设施的投资分摊已分别计入运营成本，则此处主要是管道设施投入按 20 年分摊的费用，以及年度产生的 10%维护及管理费用。如表 4 所示。

表 4　输水成本分析　　　　单位：元/m^3

项目	海水淡化	输水电力	运行费用	小计	管道折旧	合计
一期工程	3	0	0.05	3.05	0.5	3.55
二期工程	3	1.5	0.3	4.8	3	7.8
三期工程	3	1.5	0.48	4.98	4.8	9.78

上述仅为干线输水的成本水价，考虑支线供水、输水损耗等因素，以增加 50%左右成本计算，按四舍五入原则，第一、二、三期的成本水价分别调整为 5 元/m^3、12 元/m^3 和15 元/m^3。

2）江口取水方案

A. 工程投资

工程总投资计 5385 亿元（未计入工程征地费用、调蓄水库投资及银行贷款利息等），其中一期工程的投资为 960 亿元，二期工程投资为 1410 亿元，三期工程投资为 3015 亿元，如表 5 所示。

表 5　工程投资情况　　　　单位：亿元

项目	输水设施		合计
	输水管道	配套火电	
一期工程	960		960
二期工程	1 320	90	1 410
三期工程	2 880	135	3 015
合计	5 160	225	5 385

在各投资项目中，输水管道合计 5160 亿元，配套能源投资为 225 亿元。

B. 单位水价

单位水价中的运营成本主要考虑 A 类线路输水能源消耗，S 类线路的能源消耗已计入成本。综合前述分析，按江口取水成本 1 元/m^3 计，输水电力消耗按每度电 0.25 元计，二期及三期工程输水消耗电力为 6（kW·h)/m^3，单位水价增加 1.5 元。

管理成本中主要是管道设施（包括 A 类和 S 类线路）投入按 20 年分摊的费用，以及年度产生的 10%维护及管理费用，如表 6 所示。

表 6　输水成本分析　　　　单位：元/m^3

项目	江口取水	输水电力	运行费用	小计	管道折旧	合计
一期工程	1	0	0.24	1.24	2.4	3.64
二期工程	1	1.5	0.165	2.67	1.65	4.32
三期工程	1	1.5	0.48	2.98	4.8	7.78

上述仅为干线输水的成本水价，考虑支线供水、输水损耗等因素，以增加 50%左右

成本计算，按四舍五入原则，第一、二、三期的成本水价分别调整为 5 元/m^3、6 元/m^3 和 12 元/m^3。

3）两方案的数据比较

综合海水淡化方案和江口取水方案的分析结果，将工程投资及干线输水成本水价的有关数据归纳如表 7 所示。

表 7 两方案数据比较表

项目	工程投资/亿元		成本水价/(元/m^3)	
	海水淡化	江口取水	海水淡化	江口取水
一期工程	664	960	3.55	3.64
二期工程	1 750	1 410	7.8	4.32
三期工程	3 705	3 015	9.78	7.78
合计	6 119	5 385		

相比较而言，采用江口取水方案的投资总额减少约 12%。其中，一期工程增加约 45%，二期工程减少约 19%，三期工程减少约 19%。在干线输水的成本水价方面，一期工程增加约 2.5%，二期工程减少约 45%，三期工程减少约 20%。

综合上述分析，在工程总投资和成本水价方面，江口取水方案较海水淡化方案有优势，仅仅是在一期工程时，采用海水淡化方案有优势。

4. 生态影响及对策

东水西调工程的实施是解决长期困扰我国中西部地区干旱缺水问题的重要举措之一。工程实现可以改善相关地区的用水结构；调节气候，减少沙尘天气的危害。由于工程浩大，影响面广，对相关地区产业结构的调整，提升装备制造水平，新能源开发，盐化工业及海水养殖业的开拓等都有很大的促进作用，但其对生态环境的负面影响也是需要得到重视和有效应对的。

1）能源消耗与排放

海水淡化虽然可以解决北方地区的水资源匮乏问题，减少对地下水资源的过度依赖，但其对生态环境的影响也是不容忽视的。生态影响既有正面的也有负面的，正面影响体现在可以通过充分利用海水淡化厂排放的高浓度尾液制盐，提高盐产量，通过减少盐田面积置换出滨海土地，可从根本上解决露天晒盐带来的滩涂盐碱化问题；同时，大规模的抽排海水可加速渤海的水体循环。

负面影响还体现为海水淡化产业对海洋与大气的污染。按照目前的海水淡化方法计算，1m^3 海水可以产出 30%～40%的淡水。例如，年产 30 亿 m^3 淡水的同时，需要排放大约 60 亿 m^3 浓缩海水。若以 6%浓度计算，则其中所含的盐分大约为 3.6 亿 t，相当于全国年产原盐 6000 万 t 的 6 倍，年产海盐 1800 万 t 的 20 倍。

由于盐化工产业对原盐的需求有限，海水淡化厂的高浓度尾液中的大部分仍将直接排回海洋之中，这将造成近海海水浓度及温度增加，对海洋生物的生存环境产生较大的影响。此外，在生产过程中加入的处理药剂，也会对海洋造成污染。特别是对于渤海这样的半封闭内海，其影响尤其值得关注。

海水淡化产业对生态的负面影响是较大的，控制海水淡化产业对生态的负面影响主要包括：控制海水淡化产业的规模，做到按需生产，避免盲目重复投资；同时合理分布海水淡化设施，避免在一地过于集中；此外，应尽量减少渤海区域的海水淡化规模，如山东的海水淡化设施以布置在黄海岸边为宜。

由于海水淡化过程中需要消耗大量电能，其对大气造成的污染也是不容忽视的。按照消耗300g/(kW·h) 标准煤的较先进水平计算，生产70亿～120亿m^3淡水需要消耗210万～360万t标准煤，产生二氧化碳550万～943万t，二氧化硫1.79万～2.97万t，氮氧化物1.56万～2.69万t，以及大量的粉尘、废渣等。

远距离调水，特别是由低处向高处调水的过程也需要消耗大量的能源。由海平面提升至1200m高度消耗的能源相当于海水淡化的1.5倍。而采用风力发电整个工程需要5000km^2以上的有效风场才能满足需求，而且仍然面临着发电有效小时数较低、电力供应不稳定等诸多不确定因素。因此，如何合理规划线路、按需取用、高效利用是东水西调工程需要慎重考虑的首要问题。

2）采用风力发电的可行性分析

风力发电是新兴的清洁能源产业，但风力发电具有发电时效短（2000h/a）、电力不连续、占地面积大等缺点。火电项目虽然污染较高，但能保证工程运行的连续性，且占地面积较少，东水西调工程所经区域同时也是我国主要的煤炭产区。

在海水淡化方面如采用风力发电，每年生产10亿m^3淡化海水需要40亿kW·h，理论上需要的风场面积约为800km^2（每平方公里2.5MW，年有效发电时数2000h），能源项目投资约170亿元（以国电每千瓦造价8500元计）。东水西调一、二、三期工程合计的70亿～120亿m^3/a的输水量就需要5600～9600km^2的风场及1190亿～2040亿元投资。显然，在沿海地区为单一工程建设如此大规模的风电场是难以实现的，且由于发电时效过低，工程将在一年的大部分时间内无法运作。

在提水方面同样存在类似的问题，对于二期及三期工程而言，需要每年由海平面提水50亿～80亿m^3至1200m以上，需要300亿～510亿/(kW·h)。理论上需要的风场面积为6000～10 200km^2，投资为1200亿～2380亿元。如图2所示，图中的过渡地带是完成提水工作的主要区域，这一过渡地带很窄，要建设大规模的风电场同样存在较大的困难。

如以风电作为主要的能源配套项目，则年产10亿m^3淡化海水的能源项目投资约为170亿元，年提升10亿m^3水的能源项目投资约为255亿元。第一、二、三期工程的总投资为2465亿～4207.5亿元，其中：

一期工程：年输水规模为20亿～35亿m^3，淡化海水的能源投资为340亿～595亿元；

二期工程：年输水规模为20亿～35亿m^3，淡化海水的能源投资为340亿～595亿元，提水的能源项目投资为510亿～892.5亿元；

三期工程：年输水规模为30亿～50亿m^3，淡化海水的能源投资为510亿～850亿元，提水的能源项目投资为765亿～1275亿元。

综合上述分析，由于采用风力发电项目投资较大，需要占用大量土地，配套输电网络的建设也需要大量投入，并且难以找到大规模的风场，且风电场的年发电量较低，受自然因素的影响较大. 难以保证工程运行的连续性，同时风力发电投资过大，又将加大供水成本，提高单位水价。因此对于东水西调工程而言，以火力发电为主是比较现实的选择。

如配套能源全采用风电，配套能源的投资将增加 2030 亿元，工程总投资将增加 30%，达到 8145 亿元（未计入征地费用）。

如配套设施全部采用风力发电，则单位电价需调整为 0.5 元/(kW·h)，则一期工程单位成本水价需增加 1 元，二期及三期单位水价需增加 2.5 元。

需要说明的是，由于风力发电的年有效小时数为 2000，只有火力发电 6000h 的 1/3，因而如全部采用风力发电作为能源一方面会占用大量的土地，另一方面会造成整个工程的使用效率降低。东水西调工程的能源应该如何配套是工程在实际建设过程中需要面对的重要问题。

（三）东水西调一期工程的可行性分析

本部分通过对京津冀地区 1999～2008 年供用水情况的研究，在分析该地区用水变化趋势的基础上，通过数值分析探讨了在该地区推进海水淡化产业发展的可行性。

1. 工程的目的与意义

京津冀所在的海河流域水资源总量 372 亿 m^3，仅占全国的 1.3%，人均水资源占有量仅 $305m^3$，比 2000 年全国人均用水量还少 $125m^3$。人均和耕地亩均水资源量分别为全国平均水平的 1/5 和 1/6。由于长期干旱缺水，过度开发利用地表水、大量超采地下水、不合理占用农业和生态用水以及使用未经处理的污水，海河流域基本处于“有河皆干、有水皆污”和地下水严重超采的严峻局面。水资源严重短缺、环境污染和生态破坏已经成为今后环渤海地区可持续发展的主要制约因素。

自 20 世纪 80 年代以来，国家先后建设了包括引滦入津在内的多个大型水利工程，并多次实施了引晋水入京、引冀水入京、引黄济津、引黄济淀等应急性调水措施。虽然在一定程度上缓解了即时的需求，但仍无法改变当地水资源缺乏的困境。

海河流域平原河道长期干涸，被迫大量超采地下水，已累计超采 1000 亿 m^3 造成地下水位大面积持续下降。由于长期超采地下水，华北平原形成了大量的地下漏斗，严重威胁着当地的安全。根据《海河流域水资源公报（2008）》，海河流域各行政区 16 个地下漏斗总面积 2.85 万 km^2，其中深层漏斗 6 个。水资源的过量开发，导致河湖干涸、河口淤积、湿地减少、土地沙化、地面沉陷以及海水入侵等生态环境问题日趋恶化，严重制约经济社会的可持续发展。

随着南水北调中线及东线工程的建设，水资源匮乏的情况将会有很大的改观。但南水北调工程新增的水量只够用于替换目前超采的地下水量，而长年积累的近 1000 亿 m^3 的超采水量还难以得到有效回补，地下漏斗现象仍将长期持续。因此，要从根本上解决水资源匮乏的问题，开发海水淡化及再生水回用等新型水资源就显得十分必要。

对于京津冀环渤海城市而言，完善合理的供水体系应是多种水资源之间相互配合，系统使用、适时调整的模式。各类水资源按取用先后排序分别是：地表水、外调水、再生水、淡化海水、地下水。其中，地下水从主要的取水源转变为水资源的战略储备和调节供水的手段，用于当发生干旱时本地地表水减少、来自水源地的外调水减少以及海洋污染等特殊情况下的应急水源。

本报告通过分析京津冀地区的历年供用水情况，估算海水淡化作为新型水资源供给的可行性，并对取水成本、供水水价及可能带来的生态影响进行分析。研究对于促进水资源的综合利用，探讨提升海水淡化产业的规模和技术应用的途径都是有益的。

就地理分布而言，京津冀处于同一地区，拥有共同的水资源。但就行政划分而言，三地又分别属于不同的省级区域，具有各自独立的供用水体系。建立由多种水资源构成的完善供水体系，以及地区之间的有效用水协调机制是解决京津冀地区水资源匮乏的根本途径。

随着技术应用的日趋成熟，海水淡化作为新兴的水资源在供水体系中的比例将会不断扩大。与传统水资源的多少受自然因素的制约所不同的是，海水淡化是人类以工程技术方法创造的水资源，其供给规模及生产方式可以完全由人为因素控制，因而可以做到按需取用，成为调节供水结构的重要手段。与远距离调水相比，海水淡化虽然在供水成本方面并不占优势，但因其不会带来取水地环境先期治理、大规模的土地淹没及移民等一系列问题，而且不会发生因上游来水减少而产生的供水不足问题，因而其综合成本效益具有较大的优势。

海水淡化不仅可以作为开发水资源的新途径，同时也可以作为充分挖掘电力行业生产能力的手段。依托滨海电厂建设的海水淡化设施，可以通过调整特定时段的淡水产量，起到调节电网负荷的作用。

在开发海水资源的同时，淡化产业对生态的负面影响也是不容忽视的。海水淡化是高耗能及高污染的行业，在上马相应工程的同时应充分估计其对大气及海洋的影响，并配套相应的环保措施。如何在增加水资源的同时减少对生态环境的破坏，是决策者所需要面对的主要问题。

2. 需求及海水淡化潜力分析

1）供用水情况分析

以下通过表 8～表 10 对北京、天津及河北三地的供用水情况进行分析。

表 8　北京 1999～2008 年的供用水情况　　单位：亿 m^3/a

年份	供水量				用水量				
	地表水	地下水	其他	合计	农业	工业	生活	生态	合计
1999	14.95	26.76	0.10	41.81	18.50	10.60	12.70	—	41.80
2000	13.30	27.20	0	40.50	16.50	10.50	13.40	—	40.40
2001	11.70	27.20	0	38.90	17.40	9.20	12.40	—	39.00
2002	9.65	24.24	0.73	34.62	15.45	7.54	11.63	—	34.62
2003	8.34	25.42	1.25	35.01	12.92	7.65	13.49	0.95	35.01
2004	5.71	26.80	2.04	34.55	13.50	7.66	12.78	0.61	34.55

续表

年份	供水量				用水量				
	地表水	地下水	其他	合计	农业	工业	生活	生态	合计
2005	7.00	24.90	2.60	34.50	12.67	6.80	13.93	1.10	34.50
2006	6.36	24.34	3.60	34.30	12.05	6.20	14.43	1.62	34.30
2007	5.67	24.19	4.95	34.81	11.73	5.75	14.60	2.72	34.80
2008	5.81	22.97	6.30	35.08	11.35	5.20	15.33	3.20	35.08
平均	8.85	25.40	2.16	36.41	14.21	7.71	13.47	1.02	36.41

表 9　天津 1999～2008 年的供用水情况　　单位：亿 m^3/a

年份	供水量				用水量				
	地表水	地下水	其他	合计	农业	工业	生活	生态	合计
1999	18.44	7.07	0	25.51	13.00	7.00	5.60	—	25.60
2000	14.40	8.20	0	22.60	12.10	5.30	5.20	—	22.60
2001	11.20	8.00	0	19.20	10.00	4.50	4.70	—	19.20
2002	11.74	8.20	0	19.94	10.71	4.50	4.75	—	19.96
2003	13.37	7.14	0.02	20.53	11.17	4.86	4.20	0.30	20.53
2004	14.89	7.07	0.10	22.06	12.18	5.20	4.20	0.48	22.06
2005	16.02	6.98	0.10	23.10	13.59	4.51	4.54	0.45	23.09
2006	16.10	6.76	0.10	22.96	13.43	4.43	4.61	0.49	22.96
2007	16.49	6.81	0.07	23.37	13.84	4.20	4.82	0.52	23.38
2008	15.96	6.25	0.12	22.33	12.99	3.81	4.88	0.65	22.33
平均	14.86	7.25	0.05	22.16	12.30	4.83	4.75	0.29	22.17

表 10　河北 1999～2008 年的供用水情况　　单位：亿 m^3/a

年份	供水量				用水量				
	地表水	地下水	其他	合计	农业	工业	生活	生态	合计
1999	50.16	172.05	0.80	223.01	173.30	27.30	22.40	—	223.00
2000	44.80	164.90	0.90	210.60	160.30	27.30	22.80	—	210.40
2001	39.00	169.80	0.80	209.60	159.90	27.10	22.60	—	209.60
2002	38.83	170.04	0.88	209.75	160.02	26.75	22.97	—	209.74
2003	33.35	164.32	0.49	198.16	148.21	26.21	23.41	0.32	198.15
2004	36.06	157.80	0.50	194.36	151.04	25.17	17.70	0.45	194.36
2005	36.96	160.68	0.51	198.15	148.28	25.61	23.32	0.95	198.16
2006	38.45	162.30	0.66	201.41	150.36	26.09	23.79	1.17	201.41
2007	38.97	159.54	0.54	199.05	149.27	24.99	23.59	1.21	199.06
2008	37.53	153.56	1.08	192.17	140.85	25.07	23.07	3.19	192.18
合计	39.41	163.50	0.72	203.63	154.15	26.16	22.57	0.73	203.61

三地 1999～2008 年的供用水量变化情况如图 3 和图 4 所示。

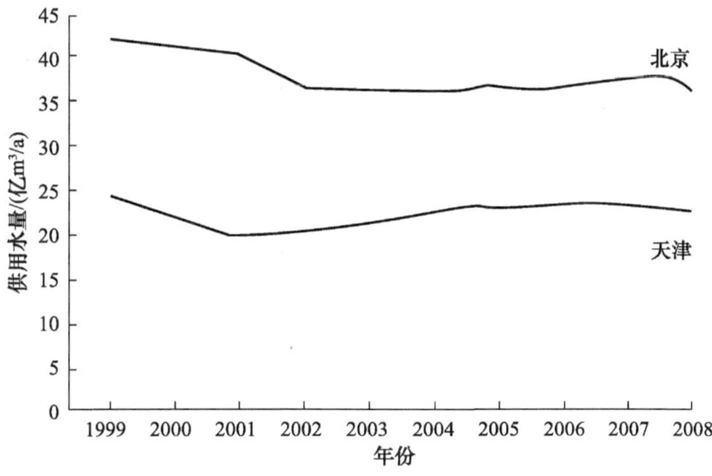

图 3　京津年度供用水量变化（1999～2008 年）

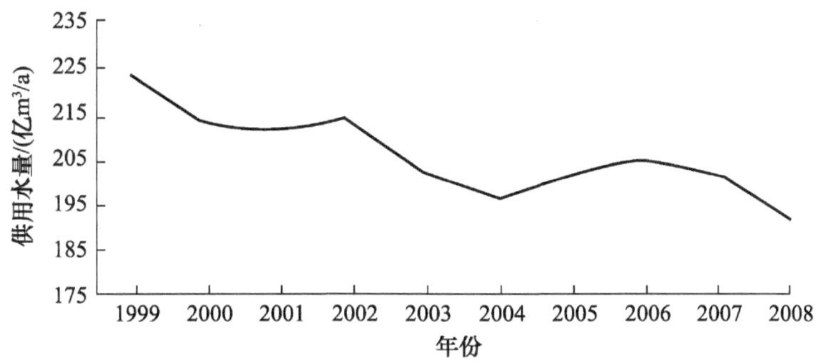

图 4　河北省年度供用水量变化（1999～2008 年）

由图表可以看出，在 1999～2008 年的 10 年间，三地的供用水总量呈逐年下降趋势，其中京津的供用水量已逐渐趋于平稳，河北省还有一定的下降空间。这一趋势的产生主要得益于近年产业结构的调整及节水措施的推行。但是，地下水供用过多应引起关注。

三地在 2003～2008 年万元 GDP 用水量及变化趋势如表 11 和图 5、图 6 所示。

由图 5 和图 6 可知，三地的万元 GDP 用水量呈逐年下降趋势。2008 年，北京市及天津市的数据分别为 30m^3/万元和 34m^3/万元，已接近万元 GDP 用水量 28m^3 的先进

表 11　京津冀万元 GDP 用水量（2003～2008 年）　　单位：m^3/万元

年份	北京	天津	河北
2003	96	86	279
2004	76	75	234
2005	51	63	199
2006	44	53	165
2007	34	45	144
2008	30	34	119

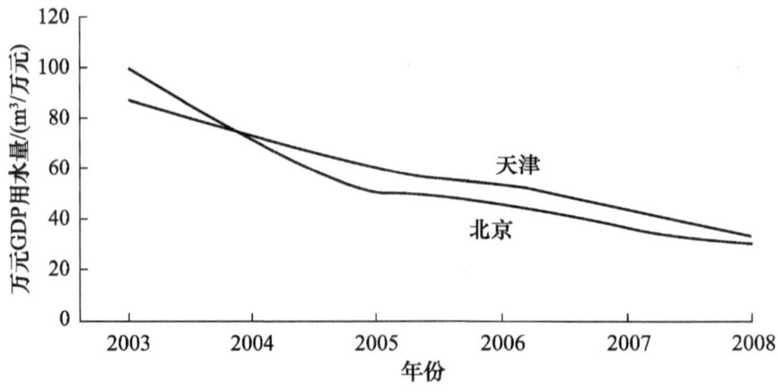

图 5　京津万元 GDP 用水量（2003～2008 年）

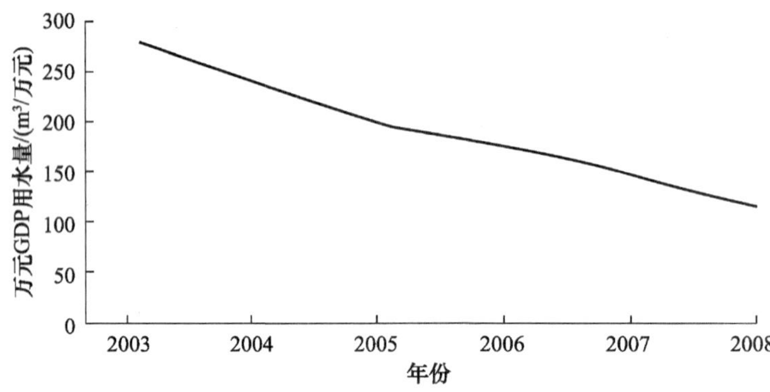

图 6　河北省万元 GDP 用水量（2003～2008 年）

水平。河北省由于农业用水占比为 75%以上，万元 GDP 用水量远高于京津，以其 2008 年的数据 119m³/万元为例，在各省份中仍处于极高的水平。随着地区产业结构调整的深化，三地的工业及农业用水呈逐年下降趋势，其减少量远高于生活及生态用水的增加量，其用水结构变化情况如图 7 所示。

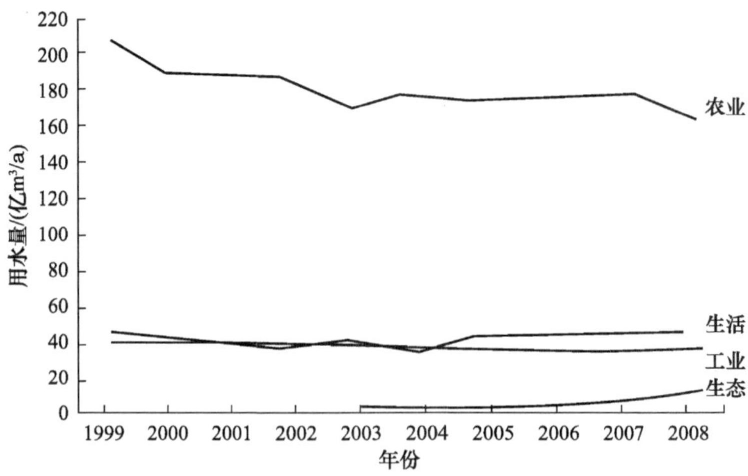

图 7　京津冀用水结构变化情况（1999～2008 年）

由上述分析可知，京津冀三地近 10 年来的供用水总量呈逐年下降趋势，近年的总用水情况已趋于平衡。随着产业结构的调整及节水措施的推行，各地 GDP 增长并未增加新的用水量。据此，本报告假设在未来 10 年乃至较长的时期，上述地区总的用水量将保持在相对稳定的水平，不会有较大的增长；因而，可以以 1999～2008 年的平均值作为确定供水需求的标准。

相比于传统的地表和地下水资源而言，新兴的水资源包括外地调水、海水淡化及再生水回用等。由于其处理成本较高，供水量相对有限，并不适用于农业大规模、低成本用水的需要；此外，由于农村地区缺乏管网设施，也无法有效使用此类水源；因而在整个供用水体系中，此类水源主要应用于增补当前的工业及生活用水。通过调整用水结构，解决工业及生活用水挤占农业用水的问题，同时提高地表水在农业用水中的比重，使其降低对地下水资源的依赖，从而使得地下水资源逐渐得到自然回补。

2）南水北调工程增补供水情况分析

南水北调中线及东线工程是有效缓解京津冀地区水资源匮乏的重要举措。随着南水北调东线及中线工程分别于 2013 年及 2014 年正式通水，受水地区水资源短缺的现象将会有所缓解。根据有关资料的说明，在南水北调供水前，受水区地下水开采量基本维持可开采量，但河北省、天津市仍需要继续超采；南水北调通水后，地下水总开采量将有所减少，地下水超采现象将逐步被禁止。

南水北调工程将成为京津冀地区未来的主要水源之一，在供水需求方面，《南水北调城市水资源规划》中的有关数据如下。

北京市根据供需分析各水平年的缺水量和远期以南水北调为主要水源的水厂建设规划，确定多年平均需调入境水量为 12 亿 m^3。遇枯水年需加大调水量。

天津市在采取开源和节水措施后，2010 年和 2030 年分别缺水 12 亿 m^3 和 18 亿 m^3 左右，相应需调净水量分别为 12 亿 m^3 和 18 亿 m^3。

河北省提出 2010 年和 2030 年的需调水量均为 45 亿 m^3，在调水初期，除满足城市用水外，尚有少量余水用于缓解地下水超采，改善地下水环境。随着城市用水量的增加，南水北调水量将逐步全都用于城市。

……

在可供水量方面《南水北调中线工程规划（2001 年修订）》及《南水北调东线工程规划（2001 年修订）》中的有关数据如下。

汉江流域地表水资源总量为 566 亿 m^3，现状总耗水量为 39 亿 m^3，其中丹江口水库大坝以上，地表水资源量为 388 亿 m^3，预计 2010 年上游耗水量约为 23 亿 m^3，中下游需水库下泄补充 162 亿 m^3，在剩余的 203 亿 m^3 水中规划可调走 97 亿 m^3。

……

将丹江口水库、汉江中下游及受水区作为一个整体进行供水调度及调节计算，在近期可调水量 97 亿 m^3 中，有效调水量为 95 亿 m^3。

……

近期有效调水量分配如下：北京 12 亿 m^3，天津 10 亿 m^3，河北 35 亿 m^3，河南 38 亿 m^3（含刁河灌区现状引水）。

……

根据东线工程供水范围内江苏省、山东省、河北省、天津市城市水资源规划成果和《海河流域水资源规划》、淮河流域有关规划，在考虑各项节水措施后，预测 2010 年水平，供水范围需调水量为 45.57 亿 m^3，其中江苏 25.01 亿 m^3，安徽 3.57 亿 m^3，山东 16.99 亿 m^3；2030 年水平需调水量 93.18 亿 m^3，其中江苏 30.42 亿 m^3，安徽 5.42 亿 m^3，山东 37.34 亿 m^3，河北 10.00 亿 m^3，天津 10.00 亿 m^3。

……

根据上述资料，归纳南水北调中线及东线工程分配水量，如表 12 所示。

表 12　南水北调中线及东线工程分配水量　　单位：亿 m^3/a

地区	中线		东线		
	一期（在建）	二期	一期（在建）	二期	三期
北京	12				
天津	10	8		5	5
河北	35	13		7	3
河南	38	14			
山东			16.81	0.05	20.41
安徽			3.29	0.14	1.82
江苏			19.22	2.9	6.08

上述资料显示，南水北调中线一期工程总调水量 95 亿 m^3，即使在枯水年份也是可以得到保证的，在汉江水量较丰时，还可以增加调水量。二期工程供水总量将增加到 130 亿 m^3，但由于汉江水量的限制、移民搬迁及工程投资等问题，实施的难度较大。此外，目前已建成的中线京石段及天津干线工程是按照年输水规模 10 亿 m^3 和 8 亿 m^3 建设的，即使调水总量增加，也无法再增加京津两地的供水量。

东线工程自长江下游提水，终点是天津市。目前在建的东线一期工程只向江苏和山东供水，规划的二期及三期工程将分别向天津和河北各供水 10 亿 m^3。相比于中线而言，东线的水源较为丰富，供水线路利用现成河湖，工程规划的输水规模是有保障的，但是由于东线工程确定的是先治污后供水的原则，正式向津冀两地供水还需要数年时间。

综合分析，在扣除实际供水过程中 20%的输水损失后，南水北调工程对京津冀三地的实际供水规模为：北京 9.6 亿 m^3/a、天津 16 亿 m^3/a（中线 8 亿 m^3/a、东线 8 亿 m^3/a）、河北 35 亿 m^3/a（中线 27 亿 m^3/a、东线 8 亿 m^3/a）。

3）海水淡化需求分析

此处的供水需求是指在各地水资源需求总体变化较小的情况下，用于调节不合理供水结构的替代性水资源需求。新增的水资源主要用于工业及居民生活用水，置换出相应数量的农业用水，以解决地下水超采的问题，并在水资源富余的情况下，逐年回补地下水资源。

在前述各项分析的基础上，将各地用水情况及南水北调可供水情况的各项数据综合，如表 13 所示。

表 13 京津冀用水和南水北调增补情况 单位：亿 m^3/a

项目		北京	天津	河北	合计
1999～2008 年供用水量平均数	工业	7.71	4.83	26.16	38.70
	生活	13.47	4.75	22.57	40.79
	生态	1.02	0.29	0.73	2.04
减：南水北调中线（扣除输水损失 20%）	中线供水量	9.60	8	27	44.60
	需求值（高）	12.60	1.87	22.46	36.93
减：南水北调东线（扣除输水损失 20%）	东线供水量		8	8	16
	需求值（低）	12.60	−6.13	14.46	20.93

根据表 13，考虑各地再生水等利用水平的提高及节水措施的进一步推广等因素，京津冀三地潜在的海水淡化量估算为每年 20 亿～35 亿 m^3。其中，北京市为 10 亿～15 亿 m^3/a。在南水北调工程所调水量中，北京从中线得到约 10 亿 m^3/a 水量分配，东线则不向北京供水。从多年平均的供水数据中可以看出，北京市对地下水的依赖程度达到 70%，每年超过 20 亿 m^3。综合分析，北京市对淡化海水的潜在需求约为 15 亿 m^3/a。考虑管道建设的成本效益，先期建设以 10 亿 m^3/a 规模为宜。

天津市为 5 亿～10 亿 m^3/a。从表 13 数据分析，如南水北调中线及东线工程全部完成，天津市的缺水情况将大为改善。但考虑到天津市本地水资源缺乏，目前的水源主要依赖引滦入津工程及不定期的引黄济津调水的实现，多年平均值约为 6 亿 m^3/a，如表 14 所示。天津市有滨海的地域优势，上马海水淡化工程可增加本地的水资源供给。此外，由于南水北调工程并不对河北省的滦河流域供水，而滦河流域由于首钢迁入及曹妃甸新区的建设，用水需求扩大。若天津市能减少引滦水量，将对整个地区的水资源平衡有很大的益处。以引滦多年平均水量作为参考，天津市对淡化海水的潜在需求为 5 亿～10 亿 m^3/a。

表 14 天津市外调水量（2004～2008 年） 单位：亿 m^3/a

年份	引滦入津	引黄济津	合计
2004	3.89	3.87	7.76
2005	4.21	0.61	4.82
2006	7.01		7.01
2007	6.13		6.13
2008	4.41		4.41
平均			6.03

河北省为 10 亿～15 亿 m^3/a。南水北调一期工程分配给河北省的水量与该省每年超采地下水的水量相当。除南水北调中线及东线工程分配给河北省的水量以外，京津地区通过南水北调工程及海水淡化产业得到的水量增加可以使其减少对河北省水资源的依赖，进而也就增加了河北省的水资源总量。综合分析表 14 数据，河北省滨海地区对淡化海水的需求为 10 亿～15 亿 m^3/a。

4）滨海电厂的海水淡化潜力分析

除新建机组专门用于海水淡化的方式之外，还可以通过充分挖掘现有机组的潜力来实现海水淡化的能源供应。由于近年火电建设的大规模投入，大型机组往往处于开工不足的情况，如表 15 所示。

表 15　全国电力行业年发电小时数（2002～2009 年）　　单位：h

年份	水电	火电	核电	风电	行业平均
2002	3 289	5 272			4 860
2003	3 229	5 767			5 245
2004	3 462	5 911			5 455
2005	3 642	5 876			5 411
2006	3 434	5 633	7 774		5 221
2007	3 532	5 316	7 737		5 011
2008	3 621	4 911	7 731	2 046	4 677
2009	3 264	4 839	7 914	1 861	4 527
多年平均	3 434	5 441	7 789	1 954	5 051

根据中电联的统计数据，2002～2009 年，全国火电机组（600MW 以上）利用小时数多年平均约为 5400h，且波动较大，其中 2004 年接近 6000h，而 2008 年及 2009 年均低于 5000h，相差 20%，近年的下降趋势明显，如图 8 所示。

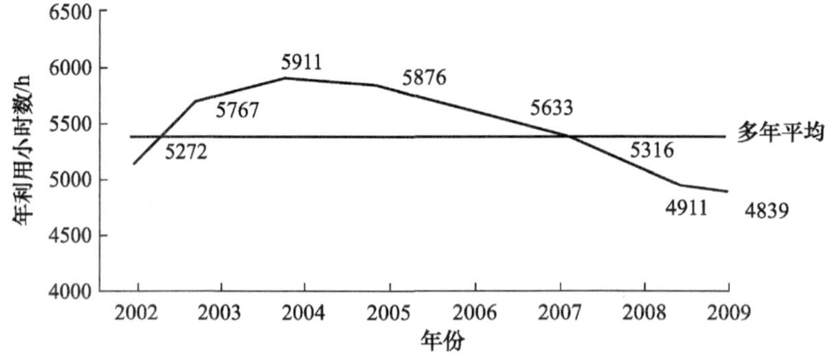

图 8　全国火电机组年利用小时数（2002～2009 年）

滨海电厂可以将未上网的多余电量用于海水淡化。通过增加配套的海水淡化设施，根据用电需求和用电负荷，动态调节上网电量，可实现错峰供电。将用电负荷较低时段的发电量用于海水淡化，高峰用电时段则向电网供电，这样可以使发电机组处于较高的使用水平，同时也可有效减少电网负荷。根据相关资料收集整理，河北及天津两地滨海电厂的装机情况如表 16 所示。

表 16　河北省和天津市滨海电场装机情况　　单位：MW

电厂	已建	在建	拟建	合计
国华沧东电厂（沧州）	600×2	660×2	1 000×4	6 520
大唐王滩电厂（大连）	600×2	1 000×2	1 000×4	7 200
河北曹妃甸电厂（唐山）	300×2	1 000×2	1 000×4	6 600
天津北疆电厂	1 000×2	1 000×2		4 000

续表

电厂	已建	在建	拟建	合计
天津大港发电厂	1 314			
秦皇岛热电厂	1 000			
合计	7 314	7 320	12 000	24 520

按目前各电厂已建及在建的总装机容量 14 634MW 计算，如淡化海水每立方米耗电 4kW·h，则增加机组年利用小时数分别为 300h、600h、900h，可分别生产 10.89 亿 m^3、21.96 亿 m^3、32.85 亿 m^3 淡水。随着海水淡化设施规模的增加和技术的改进，海水淡化耗电量将会有所降低；当每立方米耗电减少到 3kW·h 时，可分别生产淡水 14.63 亿 m^3、29.28 亿 m^3、43.91 亿 m^3。此外，如上述电厂规划中的机组全部建成，还将增加一倍左右的生产能力。

此外，还可以通过电网的合理调控，挖掘周边地区各电厂的生产能力。通过降低滨海电厂的上网电量，提高其他电厂的上网电量，还可以进一步提高海水淡化的能力。由此可见，充分利用目前河北、天津两地滨海电厂的产能即可满足京津冀三地对海水淡化的需求。

3. 工程投资及水价分析

1）工程投资

工程投资主要分为两个部分，一是海水淡化装置的投资，为 400 亿～700 亿元；二是远程输水管道的投资，为 204 亿～357 亿元。其中，海水淡化装置的投资可视需要分步建成；远程输水管道的建设则需要一次性完成主要线路。

A. 淡化装置投资

淡化装置的投资相当于南水北调的水源地建设投资，但与后者不同的是，建设海水淡化项目不存在水源地环境先期治理、淹没土地及移民搬迁等问题。在资金投入方面，不需国家直接投入，只需要辅以合理的市场调节及产业扶持政策，海水淡化项目的建设都可以通过企业行为实现。

根据预测，到 2015 年，我国反渗透海水淡化装置的吨水投资可下降到 4200 元/(t·d)，低温多效蒸馏下降到 5500 元/(t·d)。以目前各大滨海电厂的装机规模计算，每厂都可建设不小于年产 10 亿 m^3 的海水淡化基地。按工程综合造价 5000 元/(t·d) 估算，则年产 1 亿 m^3 淡水的装置投资约为 20 亿元（以年运行 6000h 计算），以总规模年产 20 亿～35 亿 m^3 考虑，总造价为 400 亿～700 亿元。

B. 输水管道投资

由于海水淡化后的水质已达到直接饮用标准，大大高于自来水的供水标准，因而可以直接进入城市自来水管网或直饮水管网。对于天津市及河北省滨海城市而言，可利用现有管网而无须对管道建设进行大规模投资。对于北京市而言，则需要建立主干管道接入城市管网。

考虑到海水淡化尾液排放及渤海污染等问题，海水淡化基地的建设不宜过于集中。由于天津市自身也需要建立大规模的海水淡化设施，因而将对北京供水的海水淡化设施建在河北省境内相对较为合适，就地点而言，唐山或沧州沿海距离北京市均为

250km 左右，且都有新建的大型滨海电厂，是比较合适的备选地点。

输水管道工程的单位工程量及投资规模可参考南水北调中线京石段工程及天津干线工程的案例，有关数据如表 17 所示。

表 17 输水管道工程投资计算分析

名称	总投资/亿元	公里数	每公里造价/万元
南水北调中线京石段工程	200	306	6 536
南水北调中线天津干线工程	85	155	5 484
每公里造价平均（4m 双管线路）：			6 000
双管线路	150	250	6 000
单管线路（单位造价增加 20%）	90	250	3 600

以上述两项工程采用的 4m 管道为例，其年输水能力约为 5 亿 m^3，当输水规模为 10 亿 m^3 时，宜采用单线路双管铺设方式，工程投资约为 150 亿元。但输水规模增大为 15 亿 m^3 时，可另增加一条线路，投资约为 90 亿元。

2）水价分析

对于电水联产的滨海电厂而言，增加淡水产量将减少发电收入。从企业经营的角度考虑，需在发电收入减少与投资海水淡化设施收益之间取得平衡。以生产每立方米淡水减少上网电量 4kW·h，上网电价 0.5 元/(kW·h)，成本电价 0.25 元/(kW·h) 计算，其中的差价为 1 元，因而淡化海水的出厂价应在 4 元/m^3 左右。

对于供水企业而言，由于淡化海水不需再处理可直接进入管网，因而只需考虑输水成本及相应的效益。以增加 1 元/m^3 计算，则淡化海水的水价应在 5 元/m^3 左右。如现有水价为 4 元/m^3，结合 50%淡化海水，则综合水价应为 4.5 元/m^3 左右。

对于北京市而言，如果管道投资按 15 年折旧及考虑维护费用，则每立方米淡化海水成本增加 1.5 元左右，为 6.5 元。如可改变财务核算方法，延长投资回收期，以管道寿命 30～50 年考虑，则淡化海水水价可降至 5.5～6 元/m^3。如现有水价为 4 元/m^3，结合 50%淡化海水，则综合水价应为 5 元/m^3 左右。

（四）东水西调工程远期规划探讨

东水西调工程远期规划应着眼于海水的综合利用。工程的作用并不局限于长距离调水和沙漠蓄水，同时具有调整工业布局、优化用水结构及复原特定地区生态系统等一系列功能。

远期工程是指图 1 中所表示的 B 类线路。与 A 类线路采取“先淡化、后输送”的调水方式不同，B 类线路采用直接输送海水的方式，按照当地需求直接取用或按需淡化后取用。

作为本项研究的成果，我们建议远期计划分两期进行，每期各调水 50 亿 m^3/a，总计调水 100 亿 m^3/a，其中，一期工程调水至内蒙古、山西、陕西、甘肃、宁夏 5 省区及河北、河南两省部分地区，二期工程再调水至新疆。

在一期工程的总水量中，其中 35 亿 m^3 用于工程沿线的工业用水，主要以电厂冷取水为主；并结合当地盛产煤炭的优势，以海水发展煤盐化工。另外 15 亿 m^3 用于注入干盐湖，以复原当地的生态，恢复种植业及养殖业，并可进一步发展旅游业。

1. 远期调水工业用途

考虑工程规模、成本效益及技术条件等方面的因素，我们建议东水西调工程主要用于解决工程沿线各省区市的工业及生态用水。一方面由于工程提调的水量有限，必须计划使用，以最经济的方式解决关键性的问题，特别是农业用水的水量巨大，所以工程由于技术条件及用水成本等方面的限制，尚不能全面满足沿线的用水要求。另一方面由于海水未经淡化不能直接用于农业及生活用水，但工业用淡水的减少，就相当于增加了农业及生活用淡水，这样减少了海水淡化的必要环节，也就降低了用水成本。更进一步，海水可以作为盐化工的原料，东水西调工程可以视为盐化工企业原料的管道运输方式，可大幅降低沿线相关企业的用盐成本。海水用于工矿企业，是东水西调工程沿线主要的用水方式。主要有两大领域的利用：一是工业冷却水；二是盐化工原料，特别是煤盐化工。

1）工业用水

在工业用水中，冷却水占比重最大。我国工业用水量中，冷却水占75%以上，化工、冶金部门冷却水占工业用水80%，电力工业中占90%以上。因为冷却水不参加生产过程本身，因而只需对冷却设施稍加改造，就可以将淡水冷却装置改用海水冷却。利用海水直接冷却遇到的主要技术问题是设备防腐。目前，海水冷却的防腐技术已经成熟，投资并不高。主要的防腐措施有涂保护层法、电化学保护和改进管网材质。此外，还可采用二级冷却系统，即以海水冷却装置循环冷却淡水，代替原来以冷却塔空冷循环水的方式。

根据有关资料，仅火电企业的冷却水一项就占到了整个工业用水的45%左右。工业用水用于发电厂冷却水已经有非常成熟的技术方案及实际案例，以2008年为例，用于电厂冷却的海水，广东省为203.8亿m^3，江苏省为156.1亿m^3，浙江省为119亿m^3。这样不仅解决了工业用水紧张的问题，也避免了大规模海水淡化的技术瓶颈及高额成本。经过使用的海水一部分蒸发进入大气，有助于调节当地气候，而余下的海水经过循环使用已经浓缩，可以输送至盐化工企业，作为生产原料，一举多得。

北方地区由于盛产能源，因而近年建设了大量的火电厂，一座1000MW装机容量的大型火力发电厂的耗水量相当于一个中等城市的用水量。由于冷却水的不足，发电机组往往不能满负荷运作。而如果采用空冷机组，虽然可以大幅减少冷却水的消耗，但是其煤耗要远高于湿冷机组，同时需向周边环境释放大量热量，既造成了对周边环境的影响，也使得这部分热量无法被有效利用。有关资料显示，截止到2008年7月，我国已投运及已核准的空冷机组总容量已经达到7355.7万kW。假设将其全部改为湿冷机组，则每年将需要约10.81亿m^3的水量，显然，在没有足够水源的情况下是难以实现的。

在确定工业用水总量时以考虑内蒙古、山西、陕西、宁夏、甘肃5省区为主，并兼顾河北、河南两省与之邻近的地区。5省区的工业用水合计为每年50.66亿m^3，如按平均75%的冷却水比例计算，则每年需要约38亿m^3。

如单独考虑火电厂冷却水一项，按45%的比例，则每年需要约22.8亿m^3，加上空改湿机组所需的水量则需要 33.6 亿 m^3。

综合上述分析，东水西调工程用于工业用水的水量约为每年 35 亿 m^3。这样各省区就可减少此部分淡水的供应，而将其用于其他方面。扣除海水中 3%左右的盐分及损耗，则相当于每年为沿线增加了至少 30 亿 m^3 的淡水供应，相当于间接增加了生活用水及农业用水的供应量，也可以视为减少地下水的抽取数量。

将海水用于工业冷却，减少了海水淡化的环节。如果按 5 元/m^3 的淡化成本计算，将这 35 亿 m^3 用于海水淡化，一年至少需要 150 亿元。同时，海水用于工业冷却后，其含盐浓度进一步提高，可再将其通过管道输送至盐化工企业作为其生产原料。

2）盐化工

除水之外，海水中的另一主要成分就是 3%左右的盐。按照前面确定的每年 35 亿 m^3 工业用水的总量计算，一年就可以提取 1.05 亿 t 的原盐，即使以渤海最低含盐量 2.3%计算也超过 7000 万 t，高于目前全国原盐年产量约 6000 万 t 的水平。这对大力发展盐化工业，特别是与当地煤炭结合的煤盐化工有着积极的促进作用。

对于制碱、化肥等用盐企业来说，所面临的问题一方面是原料不足，另一方面是运费高居不下。本工程都可以有效解决这两方面的困境，如前所述，原盐供应已可以充分满足需求；而对于运输问题，分析如下。

将调水网络视为原盐管道运输来计算，海拔提升 1200m，每立方米需 6kW·h，电费为 0.5 元/kW·h，则需 3 元。海水含盐 3%，则 1t 原盐需要 33.33m^3，不考虑海水浓缩为卤水的费用。则东水西调工程 1t 盐的运费为 100 元。

而如果采用陆路运输，公路为 0.6 元/(km·t)，铁路为 0.4 元/(km·t)，不考虑运力不足及仓储等方面的成本，按平均 0.5 元/(km·t)，运输 500km 计算，则每吨盐的运费就需要 250 元。

以上分析可解释为，如果某用盐企业处于 1200m 海拔地区，需从 1000km 以外购盐，则其成本每吨可以节省 150 元。

东水西调工程所带动的盐化工产业发展可以有多种模式。一是工业冷却水浓缩后制盐，同时可以得到很大数量的淡水。二是将海水注入沿线的干盐湖，进一步提高海水的含盐量，再抽取高含盐量的卤水制盐。再与当地盛产的煤炭进一步合成多种高附加值的化工制品。

目前全国每年生产 6000 万 t 左右的原盐，海盐只有 1800 万 t 左右。海盐的生产占用了沿海地区大量的土地资源，且会在一定程度上造成土地的盐碱化。随着沿海地区经济的发展，用地紧张的局面凸显，各地都在大力进行盐田置换，关闭中小规模的盐场。东水西调工程的建设为盐田置换提供了新的方式，为制盐业的产业布局调整创造了条件。

此外，北方是湖盐的主要产地，湖盐开采具有工艺简单、成本低、开采量大的优势。但由于气候干燥，大规模开采盐湖资源会出现卤水水位下降，周边环境日趋沙漠化等问题，因而需要对盐湖进行补水。东水西调工程所输送的海水可以有效解决这一难题，在盐湖综合开发、合理利用，提高盐资源生长量、质量和利用率，改善地区生态环境等方面效果显著。

2. 远期调水生态用途

东水西调远期工程可分两期进行，第一期将海水调入内蒙古等 5 省区，第二期则进一步调入新疆。第一期工程的 50 亿 m^3 水量中有 15 亿 m^3 作为生态用水。第二期工程的 50 亿 m^3 中有 40 亿 m^3 作为生态用水。

将一部分海水注入盐湖或咸水湖的目的是通过长期的适量补水，恢复干涸湖泊原有的生态系统和自然景观，减少风沙的侵蚀，进而可以在一定程度上减少沙尘暴对北方地区的威胁。恢复干涸湖泊是解决问题的最好办法也是最简便的方法，所需做的工作只是注水，有足够的水量，当地就会自行恢复，比任何其他的治沙措施都有效，而且事半功倍。海水的注入一方面有效改善了当地的生态环境，一方面也可在此基础上发展特色农业和进行旅游开发。

1）海水注入地点

东水西调工程的生态用水主要用于注入工程沿线的干盐湖或咸水湖。与淡水湖相比，盐湖的湖底构造大都十分坚硬，不宜向地底渗漏。而且由于经过了相当长的历史时期，周围的生态已经适应了高盐分的环境，且海水的含盐量并不高于盐湖的含盐量，因而不会对周边的生态产生影响。

海水注入盐湖一方面用于改善当地的环境，一方面也起着调节整个工程水量的作用。根据对相关资料的分析，我们选择了 4 处主要地点作为东水西调远期工程一期调水注入的备选地点，每个地点每年注入 3 亿～5 亿 m^3 海水，加上沿线一些小的地点，合计每年注入 $15m^3$，分别说明如下。

A. 黄旗海

黄旗海位于乌兰察布市察哈尔右翼前旗境内，因清代在正黄旗辖境内而得名。北魏时称南池、乞伏袁池，金代称白水泊，明代称集宁海子，清代始称黄旗海。20 世纪 50 年代末黄旗海湖面扩展到 130 多平方公里，60 年代～70 年代黄旗海已收缩到只有 70 多平方公里，但水质较好。湖水补给主要来源于霸王河、泉玉林河、磨子山河等 19 条河沟。湖盆封闭，无泄水路，湖水消耗于蒸发，水呈碱性。水面四周盛产芦苇。近 30 年，由于上游建库蓄水等原因，进湖水量大减，加之乌兰察布市集宁区的污水排放，水质变差，鱼类骤减，目前湖面已大为缩减，满眼尽是白色的碱滩，抑或零星的水域。

B. 鄂尔多斯盐湖群

鄂尔多斯高原的库布齐沙地和毛乌素沙地是内蒙古现代盐湖的主要分布区，盐湖成群成组出现，境内有大小不同规模的盐碱硝湖 100 多处：多数盐湖盐、碱、硝在一个湖内共生。鄂尔多斯地区近代产盐池就有十几处，较著名者有杭锦旗盐海子（蒙古名哈口芒乃淖尔），即汉时所称青盐泽；昌汉淖（杭锦旗霍洛柴登苏木境内），即汉时所称金莲盐泽，现仍在生产芒硝。但鄂尔多斯高原的盐湖面积都比较小，可选择多个小盐湖分别注入适量海水。

C. 民勤以北盐湖群

位于甘肃省民勤县北、内蒙古阿拉善左旗及阿拉善右旗交界处是巴丹吉林沙漠与腾格里沙漠交界处，分布有一系列的盐湖，可选择合适地点注入适量海水，对避免民勤成为第二个罗布泊，以及防止两大沙漠合拢都有较大的作用。可选的地点如雅布赖盐湖、和屯池，以及稍北的吉兰泰盐湖等。

D. 居延海

居延海也是东水西调一期工程的终点，位于内蒙古自治区阿拉善盟额济纳旗北部，形状狭长弯曲，犹如新月，额济纳河汇入湖中，是居延海最主要的补给水源。西居延海原有水面近 3000km²，自 1961 年干涸以来，长期被白茫茫的碱漠和荒沙覆盖着，成为飞扬沙尘的发源地。东居延海共干涸过 6 次，1992 年更是遭遇到长达 10 年的彻底干涸。西居延海的干涸是由额济纳河水量逐年减少所致，由此引发居延海绿洲萎缩、地区生态环境急剧恶化。

2）生态改善

“风起西伯利亚，沙起额济纳”，每年覆盖北方地区的沙尘暴的来源即内蒙古高原的各大沙漠。东水西调工程的四个主要注水点中有三个都位于这些沙漠之中。分别是：位于巴丹吉林沙漠之中的居延海，即“沙起额济纳”的源头。位于民勤县与阿拉善盟交界处的盐湖群处于巴丹吉林沙漠与腾格里沙漠交汇处，对于防止民勤成为“第二个罗布泊”、防止两大沙漠的合拢具有重要的作用。位于鄂尔多斯高原、库布齐沙漠与毛乌素沙地的盐湖群对于减少沙尘暴的威胁、防止沙漠的扩大也具有积极意义。

注入盐湖对当地的生态影响是正面的，因为盐湖周围的生态是千百年形成的，已经适应了高盐的环境，如西居延海在历史上就被称为“苦海”，其矿化度为 88.0～103.0g/L，是海水的 30 倍。黄旗海含盐量从 1972 年的 7.79‰上升到 1973 年的 9.04‰、1974 年的 11.68‰、1977 年的 18‰，即使按 1977 年的数据也是海水的 6 倍。注入海水可将湖底沉积的盐分稀释，可以降低盐碱化的程度。

东水西调工程注入的水量可以使得湖面得以长期保持，可以形成湿地及沼泽，促进植物的生长，稳固沙地，形成小块绿洲。同时水汽蒸发量的增加也会提高大气的湿度，减少风沙的形成。当某一处湖面注入一定面积时，可考虑将水引到周围的其他盐湖。

湖泊周围的生态是极易恢复的。1996 年，黄旗海上游水库冲垮，大发洪水，大量的水和鱼类流入黄旗海，湖面陡升，出现了大量的鱼；但这只是昙花一现，此后的黄旗海仍然摆脱不了枯竭的命运，但如果补水充足，黄旗海将有可能恢复原状。东居延海也是类似的例子，由于国家统一调度，黑河连续 6 年来第 15 次调水进入东居延海，累计达 2.48 亿 m³，湖面面积最大达 38.6km²，如今湖区周边的植被呈现良性演替趋势，还能在此见到天鹅、灰雁、黄鸭等珍稀鸟类出没，一度消失的黑河特有鱼类大头鱼也重新畅游东居延海。

3）特色农业

20 世纪 60 年代初，东居延海曾盛产各种鱼类，并栖居有天鹅、鹳、鸥和野鸭。旗政府曾经在这里设有渔场和野鸭饲养场，每年盛产鲜鱼 5 万多公斤；捕捉野鸭 3000～5000 只。每年还能收割芦苇牧草 150 万～200 万 kg，能满足驻地牧民牲畜过冬补饲需要。同样，黄旗海在干涸之前曾以盛产体肥肉美的“官村鲫鱼”闻名京津一带，并运往北京作为国宴佳肴。

20世纪40年代，东居延海被描述为："水色碧绿鲜明，水中富鱼族，大者及斤。鸟类亦多，千百成群，飞鸣戏水，堪称奇观。湖滨密生芦苇，粗如笔杆，高者及丈，能没驼上之人。时见马饮水边，鹅翔空际，鸭浮绿波，碧水青天，马嘶雁鸣，缀以芦苇风声，真不知为天上人间。"[17] 20世纪60年代以来，下游生态环境严重恶化，黑河尾闾西居延海、东居延海水面面积50年代分别为267km^2 和35km^2，于1961年和1992年相继干涸。

只要干涸的湖泊能得到有效恢复，长期保持足够水量注入，原有的生态系统就可以得以复原，本地区原来的农业就可以得到很好的恢复。在此基础上，可适时引入新的、适合当地发展的项目。利用湖面可以发展海水养殖业，如南美对虾已经经过试验，在内蒙古地区养殖成功；利用滩涂和沼泽，可以发展经济作物的种植等。

此外，高含盐量的盐湖大都盛产卤虫，可以作为优质饲料，大力发展养殖业。卤虫进一步深加工后，可以成为食品或药品，卤虫含有丰富的蛋白质，氨基酸组成齐全，粗脂肪含量比较高，其中不饱和脂肪酸高于饱和脂肪酸，带壳卵的不饱和脂肪酸为48.15％，脱壳卵的为54.82％。带壳卵的饱和脂肪酸是脱壳卵的1.5倍。从营养分析上可以确定卤虫含有丰富的蛋白质、氨基酸、不饱和脂肪酸和无机元素，可明显提高肝中Fe^{2+}的含量，Fe^{2+}是血红素的主要成分，因此卤虫可作为一种补血剂，食卤虫卵还可提高脑蛋白含量。

4）旅游开发

东水西调工程所达到的沙漠地区存在着大量的古代文明遗迹，以一期工程的终点居延海为例，居延海相传是老子成仙处，《水经注》中将"居延"二字译为弱水流沙，《尚书·禹贡》："导弱水至于合黎，馀波入于流沙。""东渐于海，西被于流沙，朔南暨声教，讫于四海"，其中"流沙"即是此地。《红楼梦》："任凭弱水三千，我只取一瓢饮。""弱水"即为注入居延海的额济纳河，其孕育的居延文化和黑城文化遗迹具有很高的旅游开发价值。

居延地区在汉代称"西海郡"，西夏称"黑水镇燕军司"，元灭西夏后，置"亦集乃路总管府"。古弱水三角洲上的文化遗产，底蕴厚重，遗址遗存颇多。与敦煌藏经卷齐名的开创了简牍学的居延汉简，奠定了西夏学基础的黑城西夏经卷，还有汉唐遗址、元代纸币等，则充分展现了辽、宋、夏、金，尤其是西夏时期的文化资源。

鄂尔多斯历史悠久，是人类文明的发祥地之一，萨拉乌苏文化、青铜文化源远流长。35 000年前，古"河套人"就在这块广袤的土地上繁衍生息，并创造了著名的"河套文化"十二世纪，一代天骄成吉思汗亲征西夏，途经鄂尔多斯，被这里的美丽景色打动，选为长眠之所。历史上的鄂尔多斯，曾经是一个水草丰美、"风吹草低见牛羊"的富庶之地。

类似这样的地点，具有优美的自然景观及文化底蕴，具有极高的旅游开发价值。如果通过东水西调工程的实施能基本或部分恢复其湖面，将成为人所向往的游览之地，可以使得当地目前零散开发的旅游设施得以整合，使得旅游资源的开发更具深度，更加系统化。

参考文献

［1］ 刘人怀，关于改善我国北方水资源缺乏的一个建议（2004年12月2日）．参事建言（2004—2005）．香港：中国评论学术出版社，2006：229.

［2］ 国家水利部，北京、天津、河北、内蒙古、山西、陕西、甘肃、宁夏及新疆各省区市水利厅（局）历年水资源公报.

［3］ 潘焰平，李青．海水淡化技术及其应用．华北电力技术，2003，(10)：49-52.

［4］ 刘金生，庞胜林．华能玉环电厂海水淡化工程设计概述．水处理技术，2005，(11)：73-75，83.

［5］ 王仁雷，刘克成，孙小军．滨海电厂万吨级低温多效蒸馏海水淡化工程．水处理技术，2009，35 (10)：111-114.

［6］ 阮国岭．海水淡化及其在电厂中的应用．电力设备，2006，(9)：1-5.

［7］ 陈玉春，欧阳越，徐忠辉，等．PCCP管道在南水北调中线京石段应急供水工程（北京段）中的应用．水利水电技术，2008，(5)：51-55.

［8］ 熊和金．南水北调的管道输送方案研究．综合运输，2000，(12)：22-24.

［9］ 兰才有，仪修堂．南水北调工程部分线段应采用管道输水．南水北调与水利科技，2004，(1)：22-24.

［10］ 陈湧城．长距离输水工程的管材选用．中国给水排水，1999，(8)：41-42.

［11］ 王建彦，贾义．超超临界机组选型分析．锅炉技术，2011，42 (1)：14-15，35.

［12］ 龙辉，严舒，王盾．超超临界机组设计技术集成化发展探讨．电力建设，2011，32 (2)：71-75.

［13］ 张静媛，刘明福．关于超临界超超临界发电机组的发展．山西科技，2006，(4)：7-8，10.

［14］ 李建锋，吕俊复，郝继江，等．海水西调用于火力发电的研究．现代化工，2009，29 (11)：85-88.

［15］ 于建辉，周浩．我国风电开发的现状及展望．风机技术，2006，(6)：46-50.

［16］ 刘春鸽，陈戈．我国风力发电的现状与发展思考．农业工程技术（新能源产业），2009，(2)：10-12.

［17］ 阿拉善盟政府．古居延——黑城绿洲文化史浅译．http://www.als.gov.cn/main/tour/whals/alsws/cb60d2e4-3e09-4943-9569-d2a3fe19cd13/［来自139邮箱］.

关于“发展中国家的工业化道路”论坛的讨论[①]

受世界工程师大会和中国工程院的委托，由我主持在暨南大学开办了“发展中国家的工业化道路”论坛，现在，我就论坛的情况向大家作一简要报告。

“发展中国家的工业化道路”论坛分为中文论坛和英文论坛两大板块，设立了8个议题，于今年5月8日正式开通，9月30日结束，为期145天。在广大工程师及相关人员的积极参加和大力支持下，本次论坛取得了令人满意的成果：有970个注册会员，论坛点击43 634人次，发表主题290个，回帖706篇，共计503 549字（具体情况见表1）。

表1　论坛的基本数据

	点击人次	发表主题数	回帖数	论坛字数
中文论坛	38 035	266	657	494 233
英文论坛	5 599	24	49	9 316

表2的数据表明，大家对“面对当前资源（能源）、环境等问题，发展中国家的工业化道路如何走”“信息化如何有效地带动工业化”两个专题的讨论比较热烈，点击人次分别占总量的36%和24.5%，发帖数分别占总量的40.7%和19.6%；其他几个议题的讨论则较为一般。这也反映出当前工程师们的一种思维取向，即大多数人比较注重效益性明显的工业化问题；对于工业化带来的负面影响，以及工程师个人的自身建设，如品德修养、成长道路等问题，则关注度相对较少。

表2　中文论坛中8个专题的讨论情况

专题名称	点击人次	发帖数	字数
面对当前资源（能源）、环境等问题，发展中国家的工业化道路如何走	13 679	376	270 567
信息化如何有效地带动工业化	9 325	181	84 804
如何克服温室气体带来的问题	1 205	55	21 181
工程师的责任和工程师的社会地位	1 567	45	20 790
如何有效提高大学中工科学生的动手能力；工程教育的实践环节如何建设	2 764	108	49 419
工程技术人员的择业问题（现在中国的普遍现象学理工科的人中有许多人选择与专业无关的工作）	3 251	99	27 107
工程技术人员的伦理道德和工程技术人员成长的道路	1 724	33	10 819
中国的三个产业结构中，服务业的比重仍然偏低，而工业的比重仍在持续增长，怎样使我国三个产业更合理地协调发展	4 520	26	9 546
总计	38 035	923	494 233

① 本文是世界工程组织联合会和联合国教科文组织发起的“2004年世界工程师大会”分组会上的主题演讲，上海，2004年11月5日；一个大学校长的探索，北京：高等教育出版社，2011，341-348.

下面，我就讨论情况向大家做概括性的介绍。

一、面对当前资源（能源）、环境等问题，发展中国家的工业化道路如何走

（一）发展中国家的工业化道路

大家的观点颇为一致：绝对不能走一些发达国家“先污染后治理”的路子，需要寻找一条适合本国国情、科技含量高、经济效益好、资源消耗低、环境污染少、人力资源优势得到充分发挥、人类与环境和谐共生、可持续发展的新型工业化道路。

（二）对于可持续发展的工业化道路，工程师们提出的措施

1. 增强自身科技力量，营造工程创新环境

“科学技术是第一生产力”。国家的工业化程度，终将取决于本国的科技发展水平和工程师的创新能力。政府需要对科技的发展方向进行宏观引导，加大绿色科技的创新力度和政策支持，建立工程应用研发基地，大力培养工程科技精英，重视科技研究与工程制造技术的衔接，鼓励有利于环境保护的工程技术的生成，并尽快转化成生产力。

2. 摒弃传统的工业化发展途径，开拓可持续发展的经济增长模式

（1）推行清洁生产，倡导建立生态工业，将综合预防的环境策略持续应用于生产全过程，构建经济效益、社会效益、环境效益同步规划、同步实施和同步发展的循环经济增长模式。

（2）大力扶持和加快发展信息技术、新材料技术、新能源技术、生态技术等高新技术产业，丰富工业化的结构。并运用信息技术改造传统产业，实现跨越式发展，提高工业化水平和国际竞争力，走复合的新型工业化模式。

（3）强调“开源节流”，大力开发推广可再生资源，并提高综合利用和优化配置水平，走“节约而不吝啬、集约而非粗放”的经济增长模式。

3. 妥善解决制约发展中国家工业化道路的突出问题

妥善解决在国际社会分工格局下，进行经济结构的调整、优化和升级问题；地区之间的均衡发展和资源禀赋优势的充分发挥及其功能互补与竞争的关系问题；第二、三产业的迅速发展与农业的关系，以及新型工业化与农民合法权益的关系问题；新型工业化与发展服务业的关系，以及提高劳动生产率和扩大劳动就业的关系问题；工业化过程中因贫富分化加剧而产生的社会问题，等等。

二、信息化如何有效地带动工业化

各位工程师的主要观点有以下几点。

（1）正确理解信息化的内涵及其和工业化的相互作用。

信息化涉及国民经济和社会生活众多领域，是社会发展过程的特定阶段。以人为本的信息化过程，反映经济发展形态由粗放型向集约型方式过渡。对业界而言，其主要内容表现在产业信息化、企业信息化和产品信息化三个方面。

信息化产生于工业化，反过来又极大地促进工业化的发展。两者相互作用，共同前进。

(2) 要充分利用信息化加速工业向“技术经济模式”的转移，使新型工业的科技含量高，并实现传统产业结构的优化升级。形成以高新技术产业为先导、基础产业和制造业为支撑、服务业全面发展的产业格局，带动工业化发展。

(3) 大力发展信息产业，特别是加速发展具有自主知识产权的信息产业，形成发展中国家的核心竞争力，以及国民经济的先导产业和新的成长链，实现工业化的跨越式发展。

(4) 加强信息基础设施的建设，大力开发不同层次、系统和种类的信息资源，加快建设“信息高速公路”，实现共享和利用。重视网络环境下企业信息系统软硬件产品的开发和企业信息化咨询服务体系的建设，为带动工业化发展创造条件。

(5) 在全社会广泛推广和使用信息技术，特别是推进电子商务、电子政务建设，重视培养信息人才，营造有序运行的信息环境，形成信息时代经济发展和管理的新机制。

三、如何克服温室气体带来的问题

大家讨论的焦点：第一，如何减少温室气体的产生；第二，减少温室气体排放的主要障碍。

(一) 减少温室气体的产生

(1) 减少石化燃料的燃烧，从根本上控制二氧化碳、甲烷等温室气体的产生。采用节能技术，取消高能耗的加工行业；开发利用新能源，并转向低碳和可再生物质燃料，改善农村家庭燃料和能源管理，提高热能利用率；采用零排放技术，减少工艺过程中有毒物质的排放，提倡清洁生产等。

(2) 以生物或工程措施将二氧化碳固定储存，如进行碳分离和储存；或者在利用排出气体的余热时，同时回收 CO_2 作为化工原料。

(3) 增加温室气体的吸收。大力开展植树造林，利用处理后的生活污水作为绿化用水，加大城市立体空间绿化力度。

(4) 由政府采取一定的经济和管制政策，经济政策手段包括征收排放税、碳税或能源税，以及可排放许可等；强制性政策手段涉及技术或性能标准、产品禁令等。

(二) 减少温室气体排放的主要障碍

(1) 由于技术、经济、政治、文化、社会、行为和体制上的种种壁垒，阻碍了各地区、各部门温室气体减排技术潜力的实现。

(2) 国际政策协调对于降低减排成本、避免碳泄漏等有着十分积极的意义。《京都议定书》中规定的排放贸易、联合履行（joint implementation，JI)、清洁发展机制（clean development mechanism，CDM）等措施，主要目的在于降低减排成本。但由于没有一个有效的国际政府，这些措施的实施比较困难。

四、工程师的责任和工程师的社会地位

讨论主要围绕对工程师责任、社会地位的界定，以及两者的相互关系进行。

(一) 工程师责任和社会地位的界定

工程师的责任就是按时按质地完成自身负责的项目。工程师不但要积极发挥自己

的主观能动性，而且要富于创新精神，促进团队合作，引领技术潮流，指引人类前进。

工程师的社会地位不仅仅是指工程师的福利待遇，还应包括工程师对其从事领域技术问题的权威性，从而引导经营者或领导者用科学的手段去管理或施政。

（二）工程师的责任和社会地位的相互关系

（1）由于特殊国情和教育体制，我国的工程师总量不少，总体质量却不是很高，这无疑影响工程师的整体素质，也影响到他们在社会中的地位。

（2）由于当前的企业主要依靠技术引进，对工程技术的消化吸收不重视，对科技成果转化不积极，因此工程师的责任降低，在一定程度上影响到工程师的社会地位。

（3）由于工程技术专家的工作在相当大程度上是接受经营者管理，自主权存在局限性，因此工程师的责任非常有限，影响创新能力的发挥，社会作用也有所减弱。

（4）工程师要更好地完成社会赋予自身的责任，提高社会地位，必须能综合运用现代科学理论和技术手段，精通经济、善于管理，兼备人文精神和科学精神（而不仅是科学知识），努力使自己成为复合型人才，才能在奉献中获得社会的认可。

五、如何有效提高大学中工科学生的动手能力；工程教育的实践环节如何建设

大家讨论的焦点在于：第一，为什么工科学生动手能力差；第二，如何改变这种现象。

（一）工科学生动手能力差的原因

1. 教育体制方面

（1）现代科技的发展使大学生在校期间需要学习的知识越来越多，造成一些学校理论课程所占比重加大，而教学方法和考试方法未能与时俱进，不利于学生把所学知识用于实践。

（2）传统的实践性教学体系存在工业社会特征，它既建造了实践教育的“流水生产线”，又塑造了人才培养的“单向模式”，妨碍学生综合性实践能力和个性化创新能力的提高，无法适应信息化发展要求。

（3）大部分学校的实验室追求“小而全”，资源不能共享，出现高投入、低效能的现象，相关学科无法融会贯通与相互渗透，抑制了学生的发散思维和动手能力的培养。

2. 社会方面

（1）当前社会上用人单位往往只重文凭，不太注重考察毕业生的实践能力，使工科学生过于追求分数，而忽视动手能力的培养。

（2）社会媒介的评价存在着潜在的因素，媒体上宣传的工程技术方面的学者屈指可数，也在一定程度上挫伤了工科学生学习的积极性。

3. 学生方面

当前学生的优越感太强，没有足够的学习压力和自制力，浪费了许多宝贵的可以动手实践的时间。

（二）对提高大学中工科学生的动手能力及加强工程教育实践环节的建议

（1）建立实践教学和理论课程紧密衔接，工程应用和研究能力相互补充，综合素

质和创新精神逐层深化的工程教育实践环节的新型培养模式。

（2）构建教师、设备和环境等实践教学资源可以共享，管理运作机制合理高效，实践教学体系和教育技术创新先进的开放式实践基地，为学生营造一片开拓思维，发挥科学想象力，提高工程能力的广阔天地。

（3）加大实践课时的比重，在实践教学内容中增加综合性和设计性实验类型，设置“基本实践”和“扩展实践”教学模块，提供学科交叉和技术组合的实践环境，为学生发展个性，鼓励创新预留空间。

（4）强化生产实习和毕业设计环节的训练，从中注入工程科技发展动态及学科前沿知识，系统地培养学生科技探索意识，铸造工程创新能力，让学生置身于运用已知，发现不知，探索未知的创新环境下，实现知识深化、能力突破和素质升华。

六、工程技术人员的择业问题

现在，中国学理工科的人中有许多人选择与专业无关的工作。

大家深入剖析了出现这种现象的原因。

（一）教育存在的问题

（1）一方面是社会发展使技术更新加快，学生在校学习的知识有的已被淘汰；另一方面是学生必须将较多时间放在英语、计算机等公共课程的学习，导致学科知识和实际需要不配套，解决不了具体的工程问题。

（2）大学教育与职业联系不够紧密。一些专业课程设置陈旧，特色不明显，与社会需求脱节，学生觉得“读了几年的专业知识派不上用场”，无法以最快速度在毕业后适应并进入高科技信息社会的发展浪潮中，成为有作为的弄潮儿。

（二）社会就业形势的影响

在市场经济条件下，社会上对经管类人才的需求有时也大于对理工科人才的需求。

（三）学生自身方面的原因

（1）目前就读理工科的部分学生，理论知识虽然较为丰富，但由于动手能力相对较差，显示不出优势，不得不另寻他业。

（2）当前不少选择理工科专业的学生是迫于升学压力，并不是自己真心喜欢，在就业时跳槽也就不足为奇了。

（3）在市场经济大潮中，“不以专业限制择业”是绝大部分毕业生的心态和观点，很多人为了追求高薪而找了不对口的工作，这也是无可厚非的现象。

七、工程技术人员的伦理道德和工程技术人员成长的道路

讨论的焦点：第一，是否要制定工程技术人员的伦理道德标准；第二，工程技术人员的伦理道德与工程技术人员成长的关系。

（一）工程技术人员的伦理道德标准的制定问题

（1）第一种观点认为，应该制定一部道德标准，用以规范工程人员的伦理道德和责任心，在此基础上，再充分发挥行业协会的自律监督作用。

（2）第二种观点认为，与其制定道德标准，不如制定完善的工程管理机制。因为

工程技术人员的伦理道德是个人的行为，而重要的是当工程技术人员做出违背伦理道德行为时，必须将其对工程造成的损失降到最低。因此，应该建立一套完善的工程管理机制，使工程技术人员的行为成为可控。

(3) 第三种观点认为，制定一部道德标准以及完善的工程管理机制都是非常必要的，但更重要的是工程技术人员道德修养的提高。因为工程管理机制对工程技术人员所起的只是一种保险作用，是治标不治本的做法。如果工程技术人员能提高自身的道德修养，违背伦理道德的行为自然会减少。

(二) 工程技术人员的道德修养与其自身成长道路的密切关系

一方面，学校的教育是其形成良好的基本道德观的重要因素；另一方面，工程技术界的氛围及老一辈工程技术人员对年轻一代的影响非常重要。所以，要提高工程技术人员自身的道德修养必须把好这两关。

八、中国的三个产业结构中，服务业的比重仍然偏低，而工业的比重仍在持续增长，怎样使我国三个产业更合理地协调发展

讨论的内容集中在两方面：第一，服务业比重偏低的原因；第二，三个产业合理协调发展的有效措施。

(一) 服务业比重偏低的原因

(1) 我国服务业对外开放程度不高，城市化的总体水平也不高，制约了服务业的发展。

(2) 居民收入差距加大，高收入阶层的收入更多地转化为金融资产和投资行为，造成消费结构的脱节。

(3) 外资倾向于工业，对其投资比重大于服务业，也影响了服务业的发展。

(4) 尽管工业的比重不断加大，但由于自动化程度加深，可提供的就业机会并不是很多，而且相当部分的工资水平仍然较低，对普遍提高消费水平作用不大，对服务业的发展无法起到推动作用。

(5) 服务业的支撑体系不完善，人力资本和工程技术方面的投资不够，服务基础设施跟不上需要，影响了现代服务业的发展。

(6) 当前我国服务业主要集中在商贸、餐饮、仓储等传统服务业上，金融、电信、房地产、信息服务、物流等现代服务业发展不足，导致服务业仍处于低层次的结构水平。

(二) 三个产业合理协调发展的措施

(1) 采取有效措施，大力发展服务业，进而提高第三产业的比重。①国家应该在经济管理政策方面对服务业给予优惠，鼓励银行对服务产业的资金投入，搞活服务业的经营机制。②加快旅游、会展、饮食业等行业的发展，并建立有效的预警应急机制，以防范一些紧急情况带来的风险。③服务业发展的当务之急是优化结构，提高水平。重点发展现代服务业；积极开拓新兴服务业，形成新的经济增长点；改造传统服务业，运用现代经营方式和服务技术，提高经营效率。

(2) 根据我国人力资源丰富的特点，在大力发展资金技术密集型产业的同时，既

要继续发展为生活服务的第三产业，也要大力发展为生产服务的第三产业，如金融、保险、法律等各类中介服务，使我国的人力资源优势得到充分发挥。

（3）改变就业结构，降低第一产业所占比重，通过工业化和与此相联系的城市化进程，把大批农村剩余劳动力转移到第二、第三产业中去，变成城市和乡镇的居民。

绿色制造与学科会聚[①]

制造业是我国国民经济的主要组成部分，但是环境的污染和资源的匮乏是我们国家制造业必须面对的两大课题，因此绿色制造就成了中国制造业的必由之路。学科会聚是一种新的研究范式，在绿色制造中有着不可替代的作用，是制造业绿色化的必要手段。我想先讲绿色制造，然后再讲学科会聚。

一、坚定走绿色制造之路

近年来，当人们把浪漫主义应用到科学工程领域，大唱知识经济和网络经济的赞歌，并扬言传统的制造业必然被新技术革命淘汰的时候，我们的制造业却仍然像一个任劳任怨的母亲，在默默无闻地把一碟碟美味佳肴端上饭桌，供这些夸夸其谈的孩子享用。

因为这个调子太高，就好像制造业没用了，但是实际上制造业还在干主要的活儿。我们要看清楚这一现实。如果说完美的设计、漂亮的方案和伟大的设想是上层建筑的话，那么一向科学严谨、兢兢业业、一丝不苟的制造业，则是实实在在的经济基础，它在任何国家还是占有主导地位的，各国国民总收入 80%以上还是与物质生产和消费紧密相连的。

在今天的中国，制造业工业产值占了国内生产总值（GDP）的 1/3，占整个工业生产的 4/5。1/3 的国家财政来自制造业，8000 多万就业岗位在制造业。制造业是我们国家国民经济的主要的组成部门，是出口的主力军，是就业的重要市场。

但是，事物总是有它的正反两面，正如我们常说的科学是一把双刃剑，制造业在制造它的辉煌的同时，也在侵蚀污染着我们赖以生存的环境。统计的数字非常触目惊心：我们国家的制造业每年产生约 55 亿吨无害废物和 7 亿吨有害废物，造成环境污染的排放物有 70%归咎于制造业；每年全国大城市有 17.8 万人死于污染；每年全国因为大气污染损失 740 万个工作日；世界银行估计，环境污染给中国带来相当于 3.5%～8%GDP 的损失。

我国是人口众多、资源相对不足、生产率相对低下的国家。我们国家的人均水资源是世界人均水量的 1/4，单位面积的污水负荷量是世界平均数的 16 倍多；我国制造业的劳动生产值为年均每人 3.82 万元，少得可怜，是美国的 4.38%、日本的 4.07%、德国的 5.56%。这些数据显示了我们跟发达国家的差距，并且我们国家制造业单位产品能耗平均比国际先进水平高出 20%～30%。

我们国家制造业虽然 50 多年来有了巨大的发展，但仍然没有走出资源型的经济增长模式。传统的以“高投入、高消耗、高污染、低效益、低产出，追求数量增长而忽略质量”为特征的发展方式不可能长期维持下去。环境污染和资源匮乏是悬在我国制

① 本文原载《学科会聚与创新平台——高新技术高峰论坛》，杭州：浙江大学出版社，2006，15-18.

造业头上的两把利剑。由于中国的市场潜力、劳工价格等方面有着比较大的优势，所以发达国家制造业已经或正在向中国实行梯度转移。出于就业压力、经济发展和学习经验的考虑，目前我国制造业承接转移不失为一个良策．但是我们不能因为长期落后，就被永远锁定在欠发达状态。相信通过绿色制造这个路径，使生态和经济协调发展，我们应该并且能够达到发达国家水平。

如果说全球化是中国制造业面临的经营环境，那么绿色化似乎是一种生存哲学、一种价值取向，它充满着温暖的人文关怀。

绿色制造的这个概念，最先是由美国制造工程师学会在 1996 年所发表的关于绿色制造的蓝皮书（*Green Manufacturing*）里提出来的。绿色制造的内涵是：它是一种综合考虑环境影响和资源效率的现代制造模式，它的目标是使产品从设计、制造、运输、使用到报废处理的整个产品生命周期中，对环境的副作用最小，资源效率最高。简单地说，绿色制造要综合考虑制造、环境和资源三大问题。在技术层面上，绿色制造包括：绿色产品设计技术、绿色制造技术、产品的回收和循环再制造技术，它是智能生产、精益生产、柔性生产、敏捷制造的延伸和发展。绿色制造是一种大制造、大过程，它学科交叉、观念现代，它不断吸收机械、材料、物理、化学、信息、生物和现代管理技术等方面的最新成果，并不断推动制造科学的发展。

二、学科会聚至关重要

其实，人类科学技术发展史就是一个各种学科技术互动融合、交叉发展的历史过程。科学发展的历史更是表明科学发展经历了综合、分化、再综合的过程。在科学萌芽时期，人类只能直观地认识自然界，这种直观的对自然界的认识是综合性的。

18 世纪开始，科学发展沿着分科治学的途径迈进，科学分裂为众多学科，它们又不断地产生亚学科，呈现出一种日趋精细化和逻辑严谨的形态。然而，科学在继续分化的同时，也呈现交叉和融合的趋势。科学的整合始于 20 世纪初的维也纳学派，20 世纪五六十年代后发展加快了。正如伟大的物理学家、量子论的创始人 M·普朗克所说：“科学是内在的整体，被分解为单独的部门不是取决于事物的本质，而是取决于人类认识能力的局限性。实际上存在着由物理学到化学、通过生物学和人类学到社会科学的链条，这是一个任何一处都不能被打断的链条。”这一句话也代表很多学者的态度。人们越来越认识到，解决人类问题的很多重大的复杂的科学问题、社会问题和全球性的问题，不是一门学科就能够单独解决的，而是需要会同相关学科的学者共同努力。相关学科间共同的目标、共同的工作假设、共同的理论模型、共同的研究方法和共同的语言构成科学整合的基础。它是科学技术飞跃发展的产物，是人类文明发展的必然要求。至于现代科学技术，就更是既高度分化又高度交叉。分化向着更精细、更深入方向前进，而交叉又集分化与综合于一体，实现了科学的整体化。学科会聚在分化和综合的方面已经发挥着不可替代的作用。

三、绿色制造和学科会聚必须紧密结合

绿色制造要求从产品的初期设计开始就必须考虑学科会聚的特点，从材料的选择、

产品的结构功能、生产加工的过程设计、包装和运输的方式都必须考虑资源优化和环境影响。这个过程涉及经济学、计算机科学、机械制造学、材料学、管理学、社会学和环境学等诸多学科内容。简单地说，就是要进行 3P 和 3E 的分析。3P 分析是指生产率（productivity）、可制造性（producibility）、可预测性（predictability）的英文的 3 个字头；3E 分析是指环境（environment）、能源（energy）、经济性（economy）的英文的 3 个字头。通过 3P 和 3E 的分析达到 4R 目标，也就是减量化（reduce）、再利用（reuse）、再循环（recycle）、再制造（reproduce），将这 4 个 R 作为绿色制造的目标。

至于减量化和再利用，大家都很清楚，因为我们要减少污染，减少资源的消耗。绿色制造要求在产品的生产制造过程中，采用的生产工艺势必最大限度地减少资源的消耗和环境的污染，这对我们中国特别地重要。去年我们国家统计，我们消耗了世界上总量 30%多的煤炭、10%的石油、40%多的水泥，我们国家的 GDP 高速增长，但是我们国家的资源的消耗过大。我们应该通过绿色制造来减少资源的消耗，要减少环境的污染，要提高材料和能源的循环利用。

这种大制造应该包括：光机电产品的制造、工业流程的制造、材料的制备等。从制造的方法来看，它包括了机械加工方法、高能束加工方法、硅微加工方法、电化学加工方法等。学科和亚学科的会聚，在绿色制造的过程中表现更为具体和精细。这种大制造涉及：管理学科关于可重组企业和可重组制造系统理论、企业管理方法和工业工程理论；计算机学科、半导体学科中的微电子器件和计算机器件的设计与制造；自动化学科的制造过程和制造系统控制理论和方法；光学和光电子学科的器件和仪器的设计与制造、光电测试理论和方法；物理学科中的纳米科学；力学学科中的机电系统动力学问题；机械工程学科中的零件和机器的设计制造理论与方法、机械构件及机电系统性能的模拟仿真；材料学科中的新材料制备科学；冶金学科中的材料成型科学；化学工程中的化工流程科学和化工产品的制造科学；与生物科学交叉的仿生科学和仿生机械学以及会聚技术中的纳米科技、生物技术、信息技术和认知科学。

关于会聚技术，这里再多说一句话。2001 年 12 月美国商务部技术管理局、美国国家科学基金会、美国国家科学技术委员会几个部门组织一个圆桌会议，由著名的科学家、工程师、政府官员参加，提出了会聚技术概念。从 2001 年开始，美国每年政府都在组织这个会议。美国现在把这个会聚技术，称作美国要在 21 世纪要继续在科技经济领域领先的着力点和基础。美国强调的会聚技术是讲四个方面，即：纳米科学技术、生物技术、信息技术和认知科学，也就是把以上四大科学技术集中会聚在一起。美国把这个作为整个科学技术最核心的东西，和我们今天讲的学科会聚是很一致的。会聚技术的目标是，如果认知科学家能够想到它，纳米科学家就能够制造它，生物科学家就能够使用它，信息科学家就能够监控它。从这个循环，我们看到了美国 21 世纪的重点，我想我们也该重视这个东西。

四、坚持走中国制造业的绿色化道路

对绿色制造而言，学科会聚固然需要多学科的参与，但绝不只是强调各学科对制造业面临的困境以及未来发展目标从不同学科的视野的简单介入。绿色制造的学科会聚是一种新的研究范式，在方法论上必须有所发展。不同学科、不同技术领域以及不同专业背景的科学家、学者、工程技术人员、伦理专家、政策分析家、政策制定者要突破学科的壁垒和打破科研机构的条块分割、各自为政的体制上的限制，要根据自己的学术背景、研究方法提出对问题的特定看法，彼此间广泛交流和密切合作，使参与的学科之间没有固定的边界，使研究摆脱单一视野的限制，构成一个多维视野、深层次的体系。

学科会聚是个抽象的概念。它在绿色制造的具体表现是科研机构、高等学校、产业界和政府间的合作。

绿色制造对全世界漫长而又波澜壮阔的制造历史来说，还是一个概念，而中国制造业的绿色化道路更是刚刚起步，任重道远。即便如此，学科会聚对中国制造业来说，并不是一个遥不可及的梦想。正如制造业的特长便是把图纸变成实物一样，在 21 世纪之初，如何发挥学科会聚的巨大潜力，如何把绿色制造这一崭新的概念变为现实，如何化解环境的危机，如何坚持走可持续发展的道路，中国制造业面临着前所未有的挑战，承载着责无旁贷的责任，更孕育着无限的生机。

工程管理是管理对国民经济的深度介入①

长城称得上是中国历史上最宏伟的工程，它延绵万里，横亘千古，让后人惊叹那个时代意志力的博大。长城虽然宏伟，却不是民心工程，孟姜女的眼泪让它千百年来经受着人文和道德的双重拷问。中国历史上还有一个工程，它没有显赫的名声，也没有突出的形象，却至今仍在发挥着原有的作用，它“润物细无声”，至今仍在灌溉着天府之国。它就是都江堰水利工程。

不但历史上的重大工程有优劣高低之分，在新中国的国民经济建设中，不同工程的作用更是有着天壤之别。我们经历过、看到过太多的豆腐渣工程、政绩工程、面子工程。

《中国固定资产投资统计年鉴》[1]显示，1958～2001 年我国投资项目失误率接近投资项目的 50%！我国的建设投资最近几年增长迅速，据测算，“十一五”期间，平均每年将超过 10 万亿元。如果按照这种失误率，我国今后平均每年要在建设投资上损失 5 万亿元，人均损失约 3800 元，而 2005 年农村居民人均年纯收入为 3255 元。

造成这种浪费和损失的主要原因是，管理工作跟不上形势的要求。因此，为了提高管理水平，工程管理领域迫切需要大量既精通管理业务又具有战略眼光的工程管理人才，让管理深度介入国民经济，成为中国经济建设的迫切需要。

历史上，管理对国民经济的介入，曾经产生过许多激动人心的成果：古典管理理论造就了英国、法国，这是第一代经济强国；科学管理和行为科学管理造就了美国，这是第二代经济强国；而企业文化理论造就了日本，这是第三代经济强国。

有时，甚至一种新的管理方法或技术、一种新的生产管理制度，也能造就一次飞跃。例如，甘特发明的甘特图，即生产计划进度图，以及在它的基础上发展起来的计划评审法、关键线路法等，是当时管理思想的一次革命，极大地促进了工程建设的科学性和时效性；再如，亨利·福特的标准化制造方式、流水式装配线在汽车行业的应用，成为日后大规模生产的里程碑。

尽管管理世界日新月异，各种流派精彩纷呈，但知识和信息呈几何级数的激增，现代经济建设的一日千里，令人目不暇接，更使管理学科的主流有从行为科学演变为管理理论丛林、使管理学科逐步沦为学问家的学问的危险。我们不少硕士研究生、博士研究生甚至教授所从事的研究，不乏从文献到文献的研究，所构建的数学模型往往是纯理论的模型，既未能从实践中来，更不能到实践中去。

美国学者威尔·杜兰特[2]在评价哲学时说过这么一段精辟的话：哲学这门一度把所有学科汇集在它的大旗之下、为这个世界树立一个井井有条的形象并绘出一幅诱人的图景的学问，曾几何时已经没有勇气再去承担协调如此艰巨的任务，而是从所有的

① 本文原载《中国工程管理环顾与展望——首届工程管理论坛论文集锦》，北京：中国建筑工业出版社，2007，260-262.

这些为真理而战的沙场上退避三舍，躲进深奥、狭窄的小胡同里，小心翼翼地回避起人世的问题和责任来了。

管理学科如果也这样“躲进小楼成一统”，势必会沦为和哲学一样的下场。因此，威尔·杜兰特继续说：哲学都躲起来，如果管理学再躲起来，这意味着人类的活动都成了没有思想的活动，这也许是人类正在面临或将要面临的最恐怖最可怕的问题。

现实虽然不容乐观，但也还没有悲观到威尔·杜兰特所担心的地步。

第二次世界大战以后，人类在科学技术和经济建设的突飞猛进，尤其是核能、计算机、新材料、空间技术以及生物工程的开发、应用和发展，给人类的生产和生活方式带来了革命性变化，现实使人们认识到，良好的管理在国民经济中起着日益重要的作用。

工程管理就是那个时代科学技术发展的产物，也是社会需要的结果。这是因为，系统学为人们提供了认识工程内在复杂性以及内外环境相关性的可能性，运筹学使人们能够进行多方案的优化，给人们提供科学决策的工具，此外，工程规模大型化、工程技术复杂化迫切需要一个专业化的组织或群体来完成这些任务，所有这些都催生了工程管理技术及其相应教育的诞生。

美国工业工程学会的调查发现，70％的工程师在 40 岁之后都要承担工程管理的工作。美国的工程管理教育已有 50 多年的历史，它的工程管理专业硕士研究生的数量，近 20 年来在逐年增长，到 20 世纪末，工程管理学科已在美国取得了成熟的发展。

进入 21 世纪，工程管理更是被世界范围内的工业界、企业界广泛认可、推广，时代在呼唤管理对国民经济做深度介入，尤其是对我们中国而言，管理比在许多发达国家中显示出更大的重要性。在这个时候大力开展工程管理教育，显得十分及时和必要。

我国工程管理作为一个专业，出现在 1998 年教育部颁布的普通高等学校本科专业目录中，工程管理目前是我国管理科学与工程学科下设的 6 个学科方向之一。

由中国工程院院士朱高峰、王众托担任负责人的《中国新型工业化进程中工程管理教育问题研究》[3]咨询课题，把工程管理定义为一门关于计划、组织、资源分配以及指导和控制带有技术成分经济活动的科学和艺术。可见，工程管理教育培养的是既掌握工程技术又具有管理知识和技能的复合型人才，工程管理专业是管理学、经济学的基本理论与工程技术的有机结合而形成的，具有交叉学科的特点。

其实，人类科学技术发展史就是一个各种学科和技术互动融合、交叉发展的历史过程。科学发展的历史更是表明，科学经历了综合、分化、再综合的过程。

在科学萌芽时期，人类只能直观地认识自然界，这种从直观上对自然界的认识是综合性的。

18 世纪开始，科学发展沿着分科治学的途径迈进，科学分化为众多学科，它们又不断产生新的亚学科，呈现出一种日趋精细化和逻辑严谨的形态。

然而，科学在继续分化的同时，也呈现交叉和综合的趋势。科学的整合始于 20 世纪初的维也纳学派，20 世纪五六十年代后发展加快。

“科学是内在的整体，被分解为单独的部门不是取决于事物的本质，而是取决于人类认识能力的局限性。”物理学家、量子论的创始人 M. 普朗克的这段话代表了很多学者的态度。

人们越来越认识到，人类面临的很多重大的、复杂的科学问题、社会问题以及全球性问题，不是一门学科的学者所能单独解决的，而是需要会同相关学科的学者共同努力。相关学科间共同的奋斗目标、共同的工作假设、共同的理论模型、共同的研究方法和共同的技术语言，构成科学整合的基础，它是科学技术飞跃发展的产物，是人类文明发展的必然要求。

管理和其他学科的整合，在人才教育实践中已经有了成功的先例。像MBA（工商管理硕士）教育、MPA（公共管理硕士）教育，为从事工商管理、公共管理并有实践经验的人员提供在职或脱产学习机会，满足了国家建设发展中对工商和公共管理人才的迫切需要。为此，大力提倡MEA（工程管理硕士）的培养，让MBA、MPA、MEA这三驾马车齐头并进，共同为国民经济建设提供合格人才，应该提上我国教育事业的议事日程，并且需要在近期大力开展。

如果说MBA是管理对工商业的深度介入，MPA是管理对公共事务的深度介入，那么，MEA则是管理对国民经济的一种深度介入，它介入的强度和深度与MBA、MPA相比，可以说是有过之而无不及。

为此，工程管理专业培养出来的毕业生首先应该是科学严谨的工程师，具有技术和专业知识的底蕴；其次是精打细算的经济师，懂得如何实现工程的成本最小化和效益最大化；再次是具有人文情怀的管理者，我们强调管理者的人文情怀，因为工程往往是百年大计，要经得起人民和历史的考验，缺乏人文关怀的决策，往往经不起时间的考验。像一座高架桥的架设绕过候鸟栖息的一片红树林、一条铁路的铺设为藏羚羊留出迁徙的后路，都是人文情怀的温暖表现。

我们今天所处的时代是激动人心的时代，改革开放以来，我国的经济发展举世瞩目，我们所从事的事业是前所未有的事业，建设中国特色社会主义的伟大实践，正在摸着石头过河，我们的理论界、哲学界没有躲起来。在这样的国民经济建设浪潮中，社会和百姓期待千千万万的工程是民心工程、科学工程、合理工程。管理学界在这样的时代也没有理由躲起来。寻常一样窗前月，才有梅花便不同。工程管理在这个时候介入国民经济，不是锦上添花，而是雪中送炭！

参考文献

[1]　国家统计局固定资产投资统计司. 中国固定资产投资统计年鉴. 北京：中国统计出版社，2001.

[2]　威尔·杜兰特. 哲学的故事. 梁春，译. 北京：中国档案出版社，2001.

[3]　朱高峰，王众托. 中国新型工业化进程中的工程管理教育问题研究. 中国工程院咨询课题总报告，2006.

城市矿产工程

一、关于治理垃圾的建议①

世界经济的迅速发展，加之人口的迅速增加，使资源消耗和环境污染日益加剧，因此可再生资源开发利用与环境保护就成为21世纪各国政府及科学家和经济界共同关注的焦点问题。

我们居住的地球已诞生了45.67亿年，多年以来，仅仅是地球科学家、天文学家将它作为行星关注。20世纪后期，地球日益成为热门话题。这是由于人类活动对自然系统的影响迅速增大，经济增长，人口膨胀，需求扩大，使人类与其赖以生存的自然环境之间的矛盾日益尖锐。1804年，全球人口为10亿。1927年，全球人口增加到20亿。1960年，全球人口为30亿。1975年，全球人口为40亿。1987年7月11日，全球人口达到50亿。1999年，全球人口为60亿。2007年，全球人口达到67亿。1910年，我国人口已达4亿。2007年，我国人口已增至13.8亿，占世界1/5。

在地球上，我们赖以生存的生物圈有多大？薄薄几米厚的土壤，几千米厚的大气层，几千米深的海洋。目前，无论土壤、大气，还是水资源、能源资源（石油、煤），以及其他许多资源，都在恶化中。

目前，石油、天然气仅有50年用量，煤炭仅有200年用量。

地球表面只有11%的土地适于耕种，其中大部分已使用。而且土壤流失严重。世界沙漠化的扩展速度，每年在5万～7万km^2。每年全球土壤损失量高达254亿t，其中中国43亿t，占世界1/6。研究表明，在自然力作用下，每生成1cm^3原土壤，需100～400年的漫长岁月。因此，每年全球的土壤损失量已超过新土壤的形成量。尤其我国近10多年的城市化和经济发展，引起耕地大幅度减少，我们更应爱惜耕地。

清洁的大气已成为城市居民的奢侈品。目前世界约1/4的人口居住在空气烟尘超标的地区。过量的城市烟尘及其夹带的氮、硫氧化物，造成呼吸疾患、癌症和男女不育。冰箱、空调机中的氯氟烃类制冷剂的大量使用和排放，造成大气臭氧层减薄，特别是出现了南极上空的臭氧空洞，地球的保护罩没有了，紫外线辐射加大，更威胁了人类和生物的生命！工业发展、燃煤增加、森林和植被减少，使大气中二氧化碳浓度增大，造成温室效应，使全球气候变暖。100年来，全球气温已升高1℃，且升温速度在加剧。全球的降雨量分布发生变化，从而引发水、旱灾情的概率加大。地球水资源总量为15亿km^3，其中97%是海洋，淡水仅有3%。人类淡水用量激增和水资源污染对工农业发展、生活质量改善造成严重制约。

我国人均淡水占有量仅为世界平均值的1/4，其中西北、华北等地区尤其严峻，人

① 本文是在广东省产学研结合高端论坛暨院士专家云浮行活动开幕式上的专题报告摘要，云浮，2008年8月19日；《工程管理研究》，北京：科学出版社，2015，243-244.

均水量只有世界人口的1/20，北方已有80%城市供水不足。我国是世界升温最为严重的国家之一，100年来，升温1.1℃，比世界平均值高。

滥用资源、破坏资源、扩大生态赤字，无异于自我毁灭。“生态文明”“绿色科技”“绿色制造”“清洁能源”“低碳社会”正为社会接受[1-6]。

在资源、能源、环境方面，世界科技发展很快。

在开发可再生资源方面，工业垃圾和餐厨垃圾处理和开发可再生资源问题，刻不容缓，应引起政府和民众注意。这是一项系统工程，对人类社会至关重要，值得大家去认真解决。

二、推进“垃圾分类，从我做起”科普宣传活动①

近年来，我市城市生活垃圾收运网络日趋完善，垃圾处理能力不断提高，城市环境有了较大改善。但由于城镇化快速发展，城市生活垃圾激增，垃圾处理能力相对不足，城市面临着“垃圾围城”的困境，严重影响城市环境和社会稳定。为贯彻落实《国务院批转住房城乡建设部等部门关于进一步加强城市生活垃圾处理工作意见的通知》（国发〔2011〕9号）及《广州市城市生活垃圾分类管理暂行规定》（广州市人民政府令第53号）的精神，切实提高城市生活垃圾处理减量化、资源化和无害化水平，改善城市人居环境，倡导绿色健康生活方式，促进城市可持续发展，广东省科学技术普及志愿者协会（以下简称“省科普志愿者协会”）拟发挥协会组织及人才优势，集中一批科普专家及广大科普志愿者，围绕“垃圾减量，垃圾分类，低碳生活”主题，开发集成一批适合青少年及社区居民喜爱的科普资源包，主要包括：“垃圾分类，从我做起”科普漫画宣传挂图、科普漫画宣传折页、科普漫画拼图、科普漫画扑克牌、科普动漫Flash游戏、科普漫画活动手册、垃圾桶分类模型及日常生活垃圾模型等科普资源；同时发起成立垃圾分类科普志愿者宣讲队，推进“垃圾分类，从我做起”科普宣传公益活动进校园，进社区。

实施开展“垃圾分类，从我做起”科普宣传公益活动具有十分的重要意义。它是全面建设小康社会和构建社会主义和谐社会的总体要求，是维护好群众利益的重要工作和城市管理的重要内容，是全民动员、科学引导、倡导节约、低碳消费模式的迫切需求，是综合利用、变废为宝、推进城市可持续发展的迫切需求，是城市文明程度的重要标志，关系着人民群众的切身利益。

本活动旨在倡导绿色健康的生活方式，促进垃圾源头减量和回收利用，将生活垃圾处理知识纳入青少年及社区居民学习知识，引导全民树立“垃圾减量和垃圾管理从我做起、人人有责”的社会意识和责任感。

因此，省科普志愿者协会计划利用两年时间，推进“垃圾分类，从我做起”科普宣传公益活动范围辐射到我市主要社区楼盘、小学及幼儿园等。

2011年主要任务是面向100所小学、幼儿园的儿童、家长及老师，宣讲垃圾分类知

① 本文是以广东省科学技术普及志愿者协会会长身份递交给广东省委常委、广州市委张广宁书记的报告，2011年7月1日；工程管理研究，北京：科学出版社，2015，249-251.

识，培养从小做起、从我做起的良好行为习惯。2012年主要任务是在全市各区街道建立100个垃圾分类科普宣传示范点，引导社区居民自觉做好垃圾分类，从源头减少垃圾。

为做好“垃圾分类，从我做起”科普宣传公益活动，结合“两进”（进校园、进社区）载体，协会一方面成立了活动组委会，另一方面组织开展“六个一”活动。

协会还设立了组委会办公室，具体负责“垃圾分类，从我做起”科普宣传公益活动的策划、执行与宣传。

1）整合资源，组织开展“六个一”活动

（1）编印“垃圾分类，从我做起”科普拼图、科普折页、科普扑克牌及科普宣传海报。为推动幼儿及儿童认知学习“垃圾分类”知识，结合幼儿儿童喜爱的漫画表现方式，围绕“垃圾减量，垃圾分类，低碳生活”主题，协会组织科普专家开发编印2万套科普拼图，拼图内容主要包括垃圾分类及垃圾箱（4张）、垃圾回收处理与利用（2张）；20万张“垃圾分类，从我做起”科普折页，2000套“垃圾分类，从我做起”科普挂图（8张/套）及5万副“垃圾分类，从我做起”科普扑克牌。拟全部发给我市主要幼儿园、小学及100个街道科普宣传示范点。

（2）开发制作“垃圾分类，从我做起”科普资源包互动器材。为确保“垃圾分类，从我做起”科普宣传公益活动具有持续影响力，结合科普志愿者宣讲队校园行、社区行活动，以及少年儿童喜爱兼可操作性的表现形式，围绕“垃圾减量，垃圾分类，低碳生活”主题，组织开发“垃圾分类，从我做起”科普活动手册5万册，设计制作垃圾回收箱模型及垃圾模型500套，以及“垃圾分类，从我做起”Flash互动游戏，在科普志愿者宣讲队校园行、社区行科普活动中配套使用，同时赠送给学校、幼儿园一套器材，供学校（或幼儿园）开展科普宣传使用。

（3）举办“垃圾分类，从我做起”科普拼图及绘画大赛。为培养少年儿童认知垃圾知识，从小养成垃圾分类与低碳生活的行为习惯，围绕“垃圾减量，垃圾分类，低碳生活”主题，引导广大少年儿童、老师及家长积极参与科普拼图大赛、绘画大赛及Flash绿色游戏，倡导绿色健康生活方式。同时，建立“垃圾减量，垃圾分类，从我做起”专题宣传网站，宣传垃圾分类宣讲及大赛活动成果。

（4）举办“垃圾分类，从我做起”科普宣讲“两进”活动。一是发起组建垃圾分类科普志愿者宣讲服务队，二是组织宣讲队成员到100所幼儿园及小学、100个街道科普宣传示范点开展“垃圾分类，从我做起”科普宣讲，宣讲队成员携带垃圾分类桶模型、日常生活垃圾模型、有奖抢答等道具配合宣传，进一步培养少年儿童及家长认知垃圾知识，积极参与“垃圾分类，低碳生活”体验活动。

（5）举办“垃圾减量，垃圾分类，低碳生活”科普创意大赛。为发动广州地区高等院校师生参加体验“垃圾分类，低碳生活”绿色健康生活模式，发挥大学生的想象力及创作灵感，组织“垃圾减量，垃圾分类，低碳生活，环境保护”等科普作品创作大赛。进一步提高市民自觉参与“垃圾减量，垃圾分类”行动。

（6）组织千栏万幅“垃圾分类，从我做起”科普宣传挂图巡展活动。为充分利用好分布在市社区、街道、校园等地2000多座科普宣传栏，围绕“垃圾分类，从我做起”主题，组织2000套科普宣传挂图在全市科普宣传栏巡展。

2）时间计划安排

本次活动计划分两年执行，2011 年主要任务一是组织资料开发编印，举办启动仪式；二是宣讲队进幼儿园、小学开展“垃圾分类，从我做起”科学知识宣讲，协助幼儿园、小学开展拼图及绘画大赛。2012 年主要任务是在全市建立 100 个科普宣传示范点，促进“垃圾分类，从我做起”活动取得实效。

2011 年主要时间计划安排如下。

第一阶段时间安排：2011 年 7 月 1 日～9 月 17 日

(1) 2011 年 7 月 1 日～8 月 31 日，活动宣传资料制作及发行，专题网站建议；

(2) 2011 年 9 月 1 日～9 月 15 日，向全市幼儿园、小学邮寄宣传学习资料；

(3) 2011 年 9 月 17 日，在全国科普日期间，举办“垃圾分类，从我做起”校园行科普宣传启动仪式，启动少年儿童绘画及大学生科普创意大赛。

第二阶段时间安排：2011 年 9 月 18 日～12 月 31 日

(4) 2011 年 9 月 18 日～12 月 31 日，组织科普志愿者宣讲队在百所幼儿园、小学宣讲垃圾分类与低碳生活，引导幼儿、家长及老师积极参与；同时，协助幼儿园及小学开展拼图大赛及绘画大赛。

三、开展“垃圾分类，从我做起”科普资源包研发及宣讲活动①

近年来，由于城镇化快速发展，城市生活垃圾激增，垃圾处理能力相对不足，城市已面临着“垃圾围城”的困境，严重影响城市环境和社会稳定。为此，国务院批转住房城乡建设部等部门《关于进一步加强城市生活垃圾处理工作意见的通知》，省政府及时转发了该通知（粤府〔2011〕63 号）。积极推进垃圾分类工作，已成为当前加强城市建设、构建和谐社会的重要内容。

为配合做好我省推行垃圾分类各项任务，提高城市生活垃圾减量、分类和再资源化，我们协会整合科普专家力量，拟以广州为试点开展“垃圾分类，从我做起”科普资源包的研发，推进“垃圾分类，从我做起”两进（进学校、进社区）百场科普宣讲活动。力争在五年内，在全省开展千场“垃圾分类，从我做起”科普宣讲活动。

前不久，胡锦涛总书记到深圳、广州等地视察，深入企业和社区时，特别提出“鼓励大家养成良好的垃圾投放习惯，减轻城市垃圾处理压力，用实际行动支持城市建设”。总书记的亲切关怀和殷切期望，说明做好垃圾分类工作的极端重要性，也为我们共同建设美好家园提供了强劲动力。汪洋书记在学习贯彻胡锦涛总书记重要讲话精神干部大会上提出“要以国家低碳省试点建设为契机，加快建设资源节约型、环境友好型社会步伐，构筑以珠江水系、沿海重要绿化带和北部连绵山体为主要框架的区域生态安全体系”。

但是，目前广大居民依然缺乏对垃圾分类的科学认识。据南方日报记者刘可英在广州部分街道采访调查，不少居民对垃圾分类依然一头雾水，近六成受访者认为垃圾

---

① 本文是以广东省科学技术普及志愿者协会会长名义递交给广东省黄华华省长的报告，2011 年 8 月 31 日；工程管理研究，北京：科学出版社，2015，251-253.

分类成效不大。这与我会科普志愿者在街道的调研结果基本一致。因此，开展“垃圾减量，垃圾分类”的科普宣讲工作显得十分重要。

我们协会成立以来，认真贯彻实施《全民科学素质行动计划纲要》，围绕省委省政府中心工作，开拓创新，整合资源，稳步推进我省科普志愿服务事业深入发展，推进我省公民科学素质的提高。目前，协会已拥有包括钟南山等德高望重的院士专家、李楚源等热心公益的企业领导，以及各类专业科普志愿者服务队等 15 万名会员，他们志愿为社会发展，普及低碳环保、绿色健康、安全文明等科学知识贡献自己的力量。

围绕“垃圾减量，垃圾分类，低碳生活”等主题，协会已组织科普专家开展“垃圾分类，从我做起”科普资源包研发工作。该资源包主要包括“垃圾分类，从我做起”科普宣传挂图、科普宣传折页、科普扑克牌、科普益智拼图、Flash 动漫科普教育游戏，以及“垃圾分类，从我做起”科普活动手册及配套多媒体光盘（2 种，适合幼儿园及小学生使用）、四种垃圾桶模型及各种垃圾模型（科普活动现场互动使用）等科普资源。科普资源包研发后，将在广州试点的基础上，免费赠送给我省各市幼儿园、小学及社区居民使用，普及垃圾分类知识，为协会扎实推进“垃圾分类，从我做起”进校园、进社区千场科普宣讲活动奠定基础。

由于协会是公益性社团，研发及印制“垃圾分类，从我做起”科普资源包费用较大，预计需 145 万元，见表 1。协会拟通过政府和会员单位支持，进行筹款，以更好地开展垃圾分类宣讲活动，推进城市居民自觉开展垃圾减量与分类，为我省建设幸福广东做出新贡献。

表 1　“垃圾分类，从我做起”科普资源包研制经费预算表

<table>
<tr><th>序号</th><th>项目</th><th>项目具体内容</th><th>金额/万元</th></tr>
<tr><td rowspan="9">1</td><td rowspan="9">“垃圾分类，从我做起”科普资源包研发</td><td>科普宣传挂图</td><td>3</td></tr>
<tr><td>科普宣传折页</td><td>1.5</td></tr>
<tr><td>科普益智拼图</td><td>3</td></tr>
<tr><td>科普扑克牌</td><td>2</td></tr>
<tr><td>Flash 动漫科普教育游戏</td><td>5</td></tr>
<tr><td>科普活动手册及配套多媒体光盘（2 种，适合幼儿园及小学生使用）</td><td>10</td></tr>
<tr><td>四种垃圾桶模型</td><td>1.5</td></tr>
<tr><td>各种生活垃圾模型</td><td>2</td></tr>
<tr><td>专题网站建设费</td><td>5</td></tr>
<tr><td rowspan="8">2</td><td rowspan="8">“垃圾分类，从我做起”科普资源包印制</td><td>科普宣传挂图（5000 套）</td><td>15</td></tr>
<tr><td>科普宣传折页（20 万张）</td><td>8</td></tr>
<tr><td>科普益智拼图（1 万套）</td><td>25</td></tr>
<tr><td>科普扑克牌（5 万副）</td><td>7</td></tr>
<tr><td>Flash 动漫科普教育游戏光盘（10000 张）</td><td>4</td></tr>
<tr><td>科普活动手册及配套多媒体光盘（2 种，各 2 万册）</td><td>12</td></tr>
<tr><td>四种垃圾桶模型（1000 套）</td><td>15</td></tr>
<tr><td>各种生活垃圾模型（1000 套）</td><td>6</td></tr>
</table>

续表

序号	项目	项目具体内容	金额/万元
3	2011 科普宣讲活动	百场“垃圾分类，从我做起”科普志愿者服务队宣讲活动及拼图、绘画大赛	20
合计（人民币）			145

四、开展“垃圾分类”宣讲与培训服务①

2012 年 1 月，广东省政府办公厅印发了《关于进一步加强我省城乡生活垃圾处理工作实施意见的通知》（粤府办〔2012〕2 号），标志着城乡生活垃圾处理工作正式列入省政府的重大工作部署，为全省城乡生活垃圾的分类处理及再资源化提出了明确的目标和要求。

解决“垃圾围城困局”有赖于各级政府的高度重视和强有力的政策措施，同时，也离不开广大群众的积极参与。2011 年 5 月，省科普志愿者协会（以下简称“协会”）针对广州城市生活垃圾分类的情况进行专门调查，发现近 6 成的市民对垃圾分类的相关知识和方法一无所知。为此，2011 年 7 月，笔者以广东省科普志愿者协会会长的名义向时任广东省委常委广州市委书记张广宁同志提交“推进‘垃圾分类，从我做起’科普宣传活动”的建议，得到张广宁同志赞同。2011 年 8 月，我向时任省长黄华华同志提交了“开展‘垃圾分类，从我做起“科普资源包研发及科普宣讲活动”的建议，得到了黄华华同志和时任副省长宋海同志的高度重视，批示省财政安排专项资金 80 万元支持省科普志愿者协会开展科普资源包研发及 100 场公益宣讲活动。

省领导的批示和支持，使全省 15 万名科普志愿者深受鼓舞。协会专门成立了活动组委会，成立了四支科普志愿者队伍，包括垃圾分类科普资源包研发队、垃圾分类科普志愿者宣讲队、科普艺术团志愿者服务队及绿色家园科普志愿者服务队，研发了“垃圾分类，从我做起”有声图书等 25 种科普资源包，制订了“垃圾分类，从我做起”进社区、进学校的活动方案，先后在广州空军机关幼儿园、荔湾康有为纪念小学、越秀登峰街等地开展了 100 多场公益科普宣讲活动，宣传生活垃圾分类的重要意义，普及相关科学知识，积极引导市民自觉开展生活垃圾分类减量和再资源化，促进市民参与家园绿化及低碳环保生活。取得了良好效果，得到广大市民的积极响应，省内外新闻媒体进行广泛报道。同时，协会还为华南理工大学、华南师范大学等 35 家广州地区的学生环保社团、NGO（non-governmental organization，非政府组织）组织提供“垃圾分类”科普资源包及宣讲培训服务，指导他们开展生活垃圾分类宣讲活动。

粤府办〔2012〕2 号文件的颁布，为协会进一步推进垃圾分类的科普宣传指明了方向。今年 5 月，协会召开一届四次理事会议，提出要积极响应省政府的号召，在“十二五”期间，继续在全省城乡广泛开展城市生活垃圾分类的宣讲及推广工作，在全省培训 1000 个高校社团及 NGO 组织，免费为他们提供“垃圾分类”科普资源包，引导

① 本文是以广东省科学技术普及志愿者协会会长名义递交给广东省朱小丹省长的报告，2012 年 6 月 28 日；工程管理研究，北京：科学出版社，2015，253-257.

广大社团及 NGO 组织积极推动生活垃圾分类宣讲工作。

作为老科技工作者，笔者非常高兴担任省科普志愿者协会会长，愿意继续发挥作用，为全省的科普工作和社会公益事业做出新贡献。协会在各级领导的重视支持下，在全省广大科普志愿者的共同努力下，近几年做了许多科普公益宣传工作，先后被广东省科学技术协会评为学会创新发展试点单位、全省学会改革工作先进集体。为更好地发挥协会在推进生活垃圾处理工作中的独特作用，恳请政府继续关心和支持协会工作：

一是在“十二五”期间，支持协会在全省各市开展生活垃圾分类宣讲服务工作；

二是继续加大公共财政的支持力度，用于开发“垃圾分类”科普资源包、开展群众性宣讲及培训活动，详细内容如下。

1）科普资源包开发成本预算：383.00 万元，如表 2 所示。

表 2　“垃圾分类，从我做起”科普资源包开发成本预算

序号	适合群体	具体内容	载体形式	数量	单价/元	总价/万元	备注
1	幼儿 2～6 岁	幼儿实践有声图书	图书	2 万册	6.00	12.00	
2		环状双面有声益智拼图	拼图	2 万盒	24.00	48.00	每盒 12 张
3		有声互动宣讲挂图	海报	1 万套	36.00	36.00	每套 11 种
4		点读笔	电子笔	3000 支	200.00	60.00	
5	小学生	小学生实践手册（1～3 年级）	图书	2 万册	6.00	12.00	内含垃圾分类贴纸及活动规则
6		小学生实践手册（4～6 年级）	图书	2 万册	6.00	12.00	
7		科普益智拼图（1～3 年级）	拼图	2 万册	15.00	30.00	每盒 6 张
8		爱心笔筒拼装玩具（立体）	玩具	1 万盒	35.00	35.00	用积木拼装出含有四种生活垃圾的笔筒或储钱柜
9		Flash 动漫科普教育游戏（2 种）	软件	免费下载			“垃圾分类，从我做起”及“垃圾再利用，变废为宝”游戏
10		动漫科普宣传教育片（12 分钟）	光盘	免费下载			介绍“垃圾分类，从点滴做起”，“垃圾再利用，变废为宝”及“善待地球，人人有责”
11	社区居民	垃圾分类，从点滴做起宣传挂图（4 张/套）	海报	1 万套	8.00	8.00	介绍四类生活垃圾如何分类，共 4 张，对开，适合楼盘大堂及宣传栏张贴宣传，以下三种海报同
12		垃圾再利用，变废为宝宣传挂图（4 张/套）	海报	1 万套	8.00	8.00	介绍各种生活垃圾经工业化后再利用情况
13		垃圾巧利用，变废为宝宣传挂图（4 张/套）	海报	1 万套	8.00	8.00	介绍各种生活垃圾经不同手艺后再利用情况

续表

序号	适合群体	具体内容	载体形式	数量	单价/元	总价/万元	备注
14	社区居民	善待地球，保护环境宣传挂图（4张/套）	海报	1万套	8.00	8.00	介绍在日常生活中如何践行低碳环保，促进垃圾减量
15		垃圾分类，从点滴做起宣传折页	折页	20万张	0.30	6.00	活动现场免费派发给居民
16		垃圾再利用，变废为宝宣传折页	折页	20万张	0.30	6.00	
17		垃圾巧利用，变废为宝宣传折页	折页	20万张	0.30	6.00	
18		漫画扑克牌	扑克	10万	2.00	20.00	以漫画方式展示各种生活垃圾，寓教于乐
19		宣讲展板（40张/套）	展板	50套	6000.0	30.00	介绍“垃圾分类，从我做起”（10张），“垃圾再利用及巧利用，变废为宝”（20张）“善待地球，低碳环保”（10张）三部分，规格1.2m×0.9m
20		厨余堆肥与家园绿化实践手册	图书	2万册	6.00	12.00	介绍如何厨余垃圾堆肥与家园绿化
21	机关（企事业）单位	垃圾分类，从点滴做起宣传挂图（4张/套）	海报	1万套	8.00	8.00	主要介绍四类生活垃圾如何分类
22		垃圾分类，从点滴做起宣传折页	折页	20万张	0.30	6.00	
23	科普宣传栏	垃圾分类，从点滴做起宣传挂图（10张/套）	挂图	2000套	20.00	4.00	
24		垃圾再利用，变废为宝宣传挂图（10张/套）	挂图	2000套	20.00	4.00	
25		垃圾巧利用，变废为宝宣传挂图（10张/套）	挂图	2000套	20.00	4.00	适合社区宣传栏张贴宣传
费用合计383.00万元							

2）垃圾分类公益宣讲培训计划经费预算：100.00万元。

为动员更多高校环保社团及NGO组织参与城市生活垃圾分类宣讲活动，促进城市生活垃圾分类减量与再资源化，为我省各市破解“垃圾围城”困局尽微薄之力，协会拟与南方日报合作，为环保社团及NGO组织免费提供“垃圾分类”资源包及宣讲培训服务。具体计划如下。

（一）培训时间

星期一～星期五：晚上7：30～9：00

星期六～星期日：上午9：30～12：00，下午2：30～5：00

（二）课程安排

（1）生活垃圾分类与减量；

（2）厨余垃圾堆肥与家园绿化；

（3）垃圾再利用（或巧利用），变废为宝；

（4）环保低碳，善待地球。

同时考虑针对不同的用户群，如何结合科普资源包进行宣讲，以达到最佳宣传效果。

（三）培训地点

广东科学馆。

（四）培训老师

每门课程由5位科普专家及10位助理负责，每天轮流担任培训宣讲老师。

（五）公益培训及宣讲经费预算

主要用于支付专家培训费、场地费及NGO组织公益宣讲培训费补助，每年预计上课35周，动员培训200家NGO组织开展公益宣讲活动，项目预计需经费：100万元。公益培训及宣讲经费明细如下：

（1）35周×5位/周×300元/位＝52 500元（星期一～星期五专家培训劳务费）；

（2）35周×10位/周×100元/位＝35 000元（星期一～星期五助手劳务费）；

（3）35周×2位/周×1000元/位＝70 000元（星期六～星期日专家培训劳务费）；

（4）35周×4位/周×300元/位＝42 000元（星期六～星期日助手劳务费）；

（5）场地费：按100平方米计算，10万元/年；

（6）支持NGO社会组织开展1000场公益宣讲活动补助费：1000场×600元/场＝60万元；

（7）媒体宣传服务费：10万元。

五、关于推进“餐厨废弃物变废为宝”的建议①

近年来，由于城镇化快速发展，城市已面临“生活垃圾围城”的困境。为此，2012年1月，广东省政府办公厅印发了《关于进一步加强我省城乡生活垃圾处理工作实施意见的通知》（粤府办〔2012〕2号），明确了生活垃圾的分类处理及再资源化的目标和要求。

解决“垃圾围城”的困境，有利于各级政府的高度重视和强有力的政策措施实施，同时，也离不开再资源化的科技创新手段和广大群众的积极参与。事实上，在日常生活中，大部分垃圾都是可回收再利用的，特别是社会和家庭生活垃圾中，除报纸等可回收物及电池等少量有害垃圾外，最常见的就是餐厨废弃物，约占生活垃圾比例的40％～60％，但由于餐厨废弃物具有高含水量、高有机质、高油盐含量、低热值等特点，简易的填埋和焚烧存在一定的安全隐患，一直都遭到群众的强烈反对，而合理地实现餐厨废弃物减量化和再资源化是解决“垃圾围城”最有效的手段。

早在2011年5月起，省科普志愿者协会就开始了城乡生活垃圾分类调研及公益科普宣讲活动。2011年8月，笔者以省科普志愿者协会会长的名义向时任省长黄华华同志提交了“开展‘垃圾分类，从我做起’科普资源包研发及科普宣讲活动”的建议，得到了黄华华同志、时任常务副省长朱小丹同志和时任副省长宋海同志的高度重视，批示省财政安排专项资金80万元支持省科普志愿者协会开展科普资源包研发及100场

---

① 本文是以广东省科学技术普及志愿者协会会长名义递交给中共中央政治局委员、广东省委书记汪洋的报告，2012年9月1日；工程管理研究，北京：科学出版社，2015，258-260.

公益宣讲活动。

省领导的批示和支持，使全省 15 万名科普志愿者深受鼓舞。协会积极开展工作，普及相关科学知识，引导市民自觉开展生活垃圾分类减量和再资源化，促进市民参与餐厨废弃物变废为宝试点及家园绿化工程。目前，协会已在广州、深圳等地发展了 100 名家庭科普志愿者利用厌氧发酵桶开展餐厨废弃物厌氧发酵堆肥及家园绿化工程试点，试点情况证明，家庭仅需较少的支出（平均每个厌氧发酵桶及酵母约 120 元，家庭备两个发酵桶即可），餐厨废弃物从源头上减量率就达 90%左右，转换成的有机肥料，成为阳台绿化、种盆菜的重要肥料来源，多余的有机肥料还可为社区绿化提供服务，既美化了家园，又节约了肥料，也减少了污染。

同时，协会还与暨南大学分子生物研究中心联合开展餐厨废弃物转换燃料乙醇的科技攻关工作，取得了可喜的自主技术突破，即开发出了能直接降解利用餐厨废弃物成为乙醇的转基因"嗜污酵母"，并在此"超级"酵母的基础上建立了"餐厨废弃物一燃料乙醇一步转化法"工艺技术，可一步到位地将餐厨废弃物变为燃料乙醇、酒糟营养饲料等一系列高附加值产品，真正实现餐厨废弃物的无害化、资源化、能源化、低碳化利用。据最新的试验结果统计，若利用该套新技术建立年处理 1500t 餐厨废弃物的中试示范线，则能产燃料乙醇 30 多 t、高蛋白酒糟营养饲料 70 多 t、废油脂 100t 左右，价值 100 多万元，并且排污极少、转化率高；若使用国内传统的处理方法如填埋、焚烧等，非但产生不了什么经济效益，反而带来很大的环境污染；若使用近年来引进和发展起来的一些制造"简单加工饲料"或制造"沼气＋堆肥"的工艺来处理，也只能产生具有同源性污染等潜在威胁的"简单加工饲料"或占地广、转化率低的沼气和堆肥等产品，同样经济效益不高，仅能达到该项新技术的 1/5 左右。由此可见，该项新技术远优于当前国内主流工艺技术，也丝毫不逊色于国际前沿技术，更完全符合我国新能源政策。在世界发达国家如美国等正在大力推动以玉米等粮食生产燃料乙醇作为石油替代品的"新能源"时代，该项新技术充分利用餐厨废弃物替代"粮食"生产燃料乙醇，既能大为减少粮食损失和石油消耗，也将从源头有效阻断地沟油、潲水猪等威胁广大人民群众健康的生产链，具有极高的推广价值。

为更好地发挥协会在推进生活垃圾处理工作的独特作用，恳请政府在百忙之中关心和支持协会工作：

一是在"十二五"期间，支持协会在全省各市开展生活垃圾分类宣讲及培训服务工作；

二是继续加大财政的支持力度，支持协会在广州萝岗区（或增城区）、惠州市、东莞市、中山市等地发展 1 万户家庭试点餐厨废弃物转换有机肥料工作，并初步建立家庭有机肥料与堆肥厂、有机蔬菜基地对接模式，预计需财政扶持 350 万元（详见表 3）；

三是支持协会与暨南大学分子生物研究中心在珠三角地区择地建立餐厨废弃物转化燃料乙醇的新技术中试示范线，定点收集处理当地餐馆、酒店产生的餐厨废弃物，建立起餐厨废弃物无害化、资源化、能源化、低碳化的示范性处理模式，预计需财政扶持 1000 万元（详见表 4）。

表 3　家庭餐厨废弃物转化成有机肥料项目经费预算

序号	项目主要内容	数量	经费/万元
1	厌氧发酵桶	2 万个	240
2	厨余堆肥与家园绿化实践手册	2 万册	12
3	场地、培训及指导服务	200 场	20
4	办公及通信服务费		8
5	科普志愿者服务补贴	3000 人次	15
6	公益宣讲活动物料费	200 场	15
7	科普志愿者住宿餐饮费	100 天	20
8	车辆租赁服务费	100 天	20
	合计		350

表 4　餐厨废弃物转化为燃料乙醇中试示范线项目经费预算

序号	项目主要内容	经费/万元
1	搭建和调试中试生产线费用	500
2	场地租赁、建设、装修等费用	100
3	员工工资等	200
4	水、电、蒸汽、交通等费用	40
5	知识产权、委托检测、信息检索等费用	20
6	调研、差旅、会议费	20
7	接待专家、领导参观访问等费用	10
8	日常办公设备、耗材费用	10
9	需要的原料、辅料、实验试剂、耗材等费用	100
	合计	1000

六、开展美丽广东科普教育示范户工作①

2013 年 1 月，省长朱小丹同志在广东省第十二届人民代表大会第一次会议上作政府工作报告时指出“积极推进生态文明建设。建设生态文明关系人民福祉、关乎广东未来。要坚持节约优先，加强节能减排，推进绿色、循环、低碳发展，建设美丽广东”。

近两年来，广东省科普志愿者协会紧紧围绕省政府有关“环保低碳、生活垃圾分类处理及再资源化”等文件内容，在省财政专项资金支持下，研制了包括垃圾分类、低碳生活等 30 多种科普资源，制订了“环保低碳，垃圾分类”进社区、进学校的活动方案，先后在广州荔湾康有为纪念小学、越秀登峰街社区等地开展了 100 多场公益科普宣讲活动，宣传低碳环保及生活垃圾分类的重要意义，普及相关科学知识，积极引导市民自觉开展生活垃圾分类减量和再资源化，促进市民参与绿色家园及低碳环保生活，为加快建设美丽广东做出了积极贡献。

① 本文是以广东省科学技术普及志愿者协会会长名义递交给广东省常务副省长徐少华的报告，2013 年 2 月 4 日；工程管理研究，北京：科学出版社，2015，260-261.

省政府工作报告为协会进一步开展美丽广东科普教育示范户的培训与推广指明了方向。2013 年 2 月 1 日，协会召开一届五次常务理事会议，提出要积极响应省政府的号召，在“十二五”期间，在全省城乡广泛开展美丽广东科普教育示范户的培训宣讲及推广工作，免费为城乡居民提供“环保低碳、垃圾分类、绿色家园”等科普资源包，培养他们成为争当“环保低碳达人、绿色家园达人、垃圾分类达人”等公益科普教育示范户。

为更好地发挥协会在开展美丽广东科普教育的独特作用，恳请政府在百忙之中关心和支持协会工作：

一是在“十二五”期间，支持协会在全省各市开展美丽广东科普示范户的培训宣讲及推广工作；

二是继续加大公共财政的支持力度。期盼省财政拨款支持 300 万元，用于开发“环保低碳、垃圾分类、绿色家园”等科普资源包、开展示范户培训宣讲及推广活动。

七、早日实现“美丽广东梦”①

今天，我们在这里举办“美丽城市，从垃圾分类做起”公益科普宣传活动启动仪式，首先请允许我代表广东省科普志愿者协会对支持、参与本次活动的广东省住房和城乡建设厅、惠州市政府、惠州学院、惠州市环卫局、惠州市科学技术协会、惠城区人民政府及广大科普志愿者表示衷心的感谢和崇高的敬意。

党的十八大提出“生态文明”“美丽中国”② 战略部署，就是希望我们的生活环境更加美丽。举办“美丽城市，从垃圾分类做起”公益科普宣传活动，既是省委省政府落实党的十八大精神的一项举措，也是一项功德无量的民生工程。

据《2012 年广东统计年鉴》数据，全省城市人口约 5238 万人，若按人日均排放生活垃圾 1kg 计算，则全省城市每天产生生活垃圾约 52 万吨，每年产生近 1.89 亿吨。这么多的生活垃圾对我们的生活环境造成相当大的影响，“垃圾围城”困境及垃圾导致的环境污染已相当紧迫，需要我们增强“节约资源，保护环境”的忧患意识。近两年来，我通过院士直通车制度，先后向中共中央政治局委员、时任广东省委书记汪洋同志、时任省长黄华华同志、省长朱小丹同志等中央及省级领导建议，要从源头上抓好生活垃圾分类、回收及再资源化等公益性科普宣传工作，并得到各位领导的大力支持。

其实，垃圾是放错地方的资源，实现垃圾分类与再资源化是人类可持续发展的必经之路，也是建设美丽家园的重要内容之一。例如，回收 1 吨废纸，可造好纸 850 千克，相当于少砍 17 棵大树，节约 100 吨水，节电 600 千瓦时，也相当于节约木材 4 立方米，节省化工原料 300 千克，减少 75％空气污染，35％的水污染。

因此，“建设美丽城市，从垃圾分类做起”公益科普宣传活动，实现生活垃圾减量化、无害化及资源化，是利国利民的大好事，也是今天举办宣讲活动的主要目的。

---

① 本文是在广东省“美丽城市，从垃圾分类做起”公益科普活动启动仪式上的讲话，惠州市，2013 年 5 月 18 日；工程管理研究，北京：科学出版社，2015，261-262.

② 胡锦涛在中国共产党第十八次全国代表大会上的报告，npc. gov. cn/zgrdw/npc/zggcddsbcqgdbdh/2012-11/19/content_1743312_8. htm.

我相信，只要大家一起来，践行垃圾分类，低碳生活，从点滴做起，从身边做起，我们一定能早日实现省委省政府提出的“美丽广东梦”。

八、促进城市矿产资源化利用，共建美丽中国①

在这秋风送爽的时节，欢迎各位来到“山城”重庆参加中国工程院工程管理学部城市矿产工程前沿技术论坛。在此，我谨代表论坛组委会，对远道而来的各位院士专家、各位领导和企业家表示诚挚欢迎，对长期投身于环保低碳和城市矿产领域发展的各界人士致以崇高敬意！

环保是中国当前的头等大事之一。习近平总书记指出：要自觉把经济社会发展同生态文明建设统筹起来，充分发挥党的领导和我国社会主义制度能够集中力量办大事的政治优势，充分利用改革开放40年来积累的坚实物质基础，加大力度推进生态文明建设、解决生态环境问题，坚决打好污染防治攻坚战，推动我国生态文明建设迈上新台阶。

城市是一国经济增长的发动机，创新的产生、技能的培育，基本都在城市进行。城镇化当然也有负面效应，人口密集会造成垃圾的快速产生，“垃圾围城”已成为城市治理最重要的问题之一。

本届论坛的主题是“促进城市矿产资源化利用，共建美丽中国”。“城市矿产”是为了应对矿产资源枯竭和资源价格日益上涨而提出来的概念，其主要思想是从城市垃圾中开发人类所需的各类矿产资源。

为落实中央提出的“美丽中国、健康中国”的战略部署，我们有幸邀请到各位院士专家、企业家共聚一堂，围绕“国家低碳环保政策”“城市矿产资源化利用”以及“中外企业低碳环保领域的最佳实践”三个议题，为我国生态文明建设、可持续发展献言献策，我相信，通过各界的共同努力，中国一定可以实现既有金山银山，又有绿水青山的宏大目标。

谢谢！

九、绿色再制造的探索②

1. 世界环境形势

世界经济近百年来的迅速发展，加之人口的迅速增加，使资源消耗和环境恶化日益加剧[1-3]。

1）人口膨胀

200多年前的1804年，全球人口才10亿人。过了123年到1927年，全球人口增加到20亿人。又过了60年到1987年，全球人口达50亿人。又过了26年到2013年，全球人口增加到70亿人。人口增速越来越快，200年来人口增长了7倍。

我国人口增加也一样。1910年，我国人口约为4亿人。1949年，我国人口约为

① 本文是中国工程院工程管理学部城市矿产工程前沿技术论坛开幕词，重庆，2018-11-10.

② 本文原载《中国管理科学的研究与实践》（第四届中国管理科学论坛论文集），北京：科学出版社，2019，1-10.

4.75 亿人。2016 年，我国人口已增加到 13.83 亿人，约为世界人口的五分之一，100 多年来约增长了 3.5 倍。

2）土地状况恶化

地球表面只有 11％的土地适于耕种，大部分已被使用。

土地流失严重，全球每年土地损失最高达 254 亿吨，我国土地损失 43 亿吨，约占世界土地损失的六分之一。

我国耕地不仅在减少，而且许多城市周边土地重金属含量显著增高。

3）大气变差

清洁的大气已成为奢侈品。目前全世界约有四分之一的人口生活在空气烟尘超标的地区。

工业的发展，使大气中二氧化碳浓度增大，造成温室效应。100 年来，全球升温高达 1 摄氏度以上，我国还高于世界平均值，达到 1.1 摄氏度，越来越影响人类的生存。

4）淡水缺乏且污染严重

随着经济的发展，人类淡水用量激增且水资源污染对生活质量改善造成严重制约。我国人均淡水资源占有量仅为世界平均值的四分之一，其中西北、华北，包括首都地区，人均水量仅为世界人口的二十分之一，而且水质污染也很严重。

5）资源紧张

许多资源都在恶化中。

特别是能源资源紧张，石油、天然气仅有 50 年用量，煤炭仅有 200 年用量。我国石油尤其缺乏，60％靠进口。

2012 年，我国经济总量约占全球的 11.5％，却消耗了全球 21.3％的能源、45％的钢、43％的铜、54％的水泥，排放的二氧化碳、氮氧化物总量居世界第一。

鉴于上述情况，环境保护与可再生资源开发利用就成为近年来全球关注的焦点问题。

2. 中国餐厨垃圾现状

2003 年后，笔者在工作中接触到垃圾处理之事，并开始关注固体废弃物处理问题。随后，又在组织广东省垃圾分类活动中，深刻地认识到其中餐厨垃圾问题的严重性，故牵头组织了一个团队，包括工程管理、生化工程、机械工程、环境工程、企业管理等学科领域的科技人员，集中力量研究餐厨垃圾治理问题。这是绿色再制造中的一个重要问题，与全球每个人有关。从有人类以来，每个人每天都在生产餐厨垃圾，6000 年来，一直未解决好，是一个大难题。解决这一难题，与人类健康有关，也与可再生资源有关。

1）中国餐厨垃圾情况

2016 年，全国年产餐厨垃圾 9700 万吨，约占生活垃圾的 50％，而且每年以 10％的增量不断增长。同时，全国城市还存在大量的有机余料，分布在农贸水果市场、海鲜肉类市场、大型超市、食品粮油企业和物流冷链企业。

国内的一线城市，如北京、上海、广州、深圳等，餐厨垃圾的日产量要达数千吨，已处于“垃圾围城”状态。

餐厨垃圾成分复杂，不仅包括剩饭、剩菜，还包括厨房的下脚料、报废旧餐具、破碎器皿，其特点有以下七个：①含水率高，可达75%～85%，不适合焚烧和填埋；②富含油脂、淀粉、蛋白质、纤维素等有机成分物质；③盐分含量偏高，不适合直接生产肥料；④偏酸性；⑤富含氮、磷、钾、钙及各种微量元素，营养充分；⑥存在细菌、病毒、寄生虫等病原微生物；⑦易腐烂、变质、发臭，孳生蚊蝇。

2）餐厨垃圾的危害

A. 地沟油的危害

在餐厨垃圾的成分中，约有6.5%的油脂可分离出来，俗称地沟油。全国每年共生产约631万吨地沟油。这种油脂是过氧化值、酸价、水分严重超标的非食用油，含有砷、铅、黄曲霉素、苯并芘等毒素，尤其是其中的黄曲霉素是国际卫生组织划定的一级致癌物，毒性比砒霜大68倍。一旦食用地沟油，就会破坏人体的白细胞和消化道黏膜，使人中毒，甚至致癌，包括胃癌、肠癌、肾癌、乳腺癌、卵巢癌等。因此，只能把这种油脂用作生物柴油，或者用于制造精细化工产品，如肥皂、洗涤剂、环保增塑剂、高碳脂肪醇等。这是可再生资源，特别是能源的主要原料。但是，每吨油价值约5000元，由于利益驱使，虽然国家和地方政府都对地沟油进行严格的监督，每年仍有60%的地沟油，约378万吨流回到人们的餐桌上。全国平均每人每年要食用2.73千克地沟油，即每人每月大约要吃228克，这就是许多人生病甚至患上癌症的缘由之一。

B. 餐厨垃圾用作饲料的危害

按照我国的传统习惯，人们常把餐厨垃圾直接用于饲养动物，特别是用来喂猪，这种猪现在又称为垃圾猪。经目前研究，这种饲养方法是错误的。

餐厨垃圾含有大量可以引起人畜共患传染病的原病体、病毒、寄生虫和虫卵。同时，在收集和运输中，容易孳生许多有害微生物，如黄曲霉素等。动物直接食用后，容易感染和诱发各种疾病，治病后会加大抗生素类药物的残留。人食用这些动物的肉，积累到一定程度后，就会导致肝脏、肾脏、脾脏等病变，严重时可致癌症。

餐厨垃圾富含动物蛋白，若未经无害化处理，直接给同类动物食用，可能导致同源性污染，容易形成一种特殊的蛋白病毒——朊病毒，造成类似疯牛病、疯羊病等动物疫情。人若吃了这种未煮熟的患病动物，可出现痴呆或神经错乱症状，甚至死亡。1986年英国发生疯牛病，波及十几个国家，恐慌震全球。

C. 渗滤液的危害

餐厨垃圾堆放场和填埋中将会产生大量垃圾渗滤液，它们将直接渗入土地直至地下水中。垃圾渗滤液含有多种有毒有害的污染物质，其中第一类是苯系物质，是有强烈芳香的气体，易燃，有血液毒性，对造血系统、神经系统、皮肤、黏膜和肝肾有破坏作用，联合国已将其中的苯和二甲苯列为强烈致癌物质。第二类是酚类，这里面以苯酚毒性最大。它们对皮肤和黏膜有强烈的腐蚀作用，也可影响中枢神经系统或损害肝肾功能。特别是五氯苯酚已被美国国家环境保护局列为“优先控制污染物”。第三类是多环芳烃类，共有200多种，其中已有许多被证明具有致癌毒性，美国已将其中16种列为环境“优先控制污染物”，特别是苯并芘致癌性最强。此外，还有邻苯二甲酸酯类，对内分泌有干扰作用，构成健康隐患。垃圾渗滤液中的重铬酸盐指数（dichromate oxidizability，CODcr）和五日生化需氧量（biochemical oxygen demand，BOD5）的数值可高达数千至几万，远高于城市污水的浓度。渗滤液将导致地下水严重污染，破坏土地微生物的正常生存环境，还造成重金属污染。

D. 二噁英的危害

将餐厨垃圾混在生活垃圾中焚烧，若燃烧不充分，极易产生二噁英，使大气受到污染，二噁英是一种无色无味、毒性严重的脂溶性物质，毒性极大，是砒霜的900倍，有“世纪之毒”之称，万分之一甚至亿分之一的二噁英就会给人类健康带来严重的危害。除致癌外，还具有生殖毒性和遗传毒性，直接危害子孙后代的健康和生活。它已被国际上列为人类一级致癌物。

3）餐厨垃圾常规处理技术简介

目前国内外处理餐厨垃圾的技术可归纳为以下五种方法。

A. 填埋

填埋处理是一种简单而且普通的垃圾处理方法，但浪费资源，且占用大量土地，污染环境，易产生渗滤液污染地下水。

B. 焚烧

使用餐厨垃圾焚烧产生的热量进行发电。由于垃圾含水多而影响炉温，既增加成本又易产生二噁英等剧毒致癌物质。炉温达到1300℃时，二噁英才能被除去99%。而许多城市焚烧炉为节省能源，炉温才800℃，产生的二噁英较多。

C. 简单加工为饲料

我国传统习惯是将餐厨垃圾直接喂猪，由于有害微生物的毒性物质和同源性污染，容易产生“垃圾猪”。

D. 好氧堆肥

依靠自然界广泛分布的细菌等微生物，在人工控制的条件下，把餐厨垃圾转变为有机复合肥，其缺点是盐分高而易盐化土地，耗时又长，臭味污染严重。

E. 厌氧发酵

厌氧工艺是指利用垃圾生产沼气并将其转化为电能与燃气，对厌氧消化罐中产出的残渣进行二次发酵堆肥处理。其缺点是占用大量土地，耗时长，而且沼泥含盐高，污染土地。

上述五种方法都未能将餐厨垃圾处理完善，弊端很多，经济效益也很差，每吨垃圾收入都未超过200元。

3. 治理餐厨垃圾的新技术

1）项目内容

经过六年多的艰辛努力，我们依靠科技创新，通过超前的战略理念和多学科的技术整合，设计搭建了一套高度资源化、无害化和减量化的餐厨垃圾处理新工艺，即联合生物加工技术以及配套的产业化装备线。核心的创新技术是通过遗传育种的方法筛选到一株多功能的酵母菌，现把它命名为“噬污酵母”，能直接对餐厨垃圾的复杂成分进行降解和转化。同时，还在广州开发区成功完成了日处理5吨餐厨垃圾的工业示范线建设工作，项目的工艺流程简图，如图1所示。

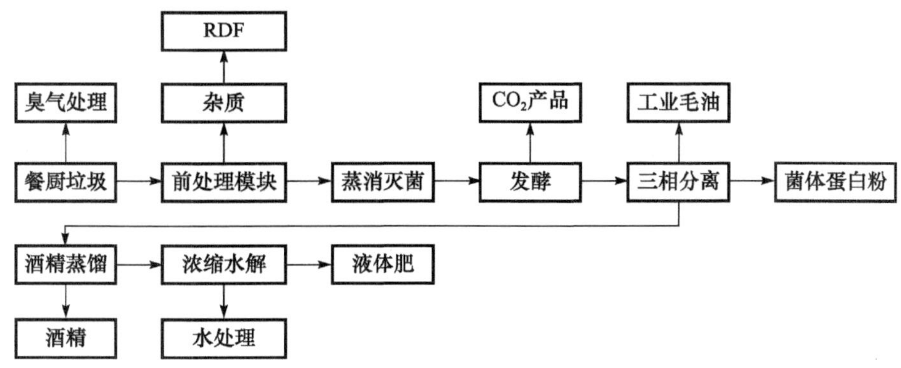

图1　项目的工艺流程简图

RDF：refuse derived fuel，垃圾衍生燃料

2015年8月18日，刘人怀院士团队的餐厨垃圾资源化项目通过了广州开发区“院士创新创业项目”的专业评审，成为广州开发区自2015年引进的唯一院士创新创业项目，已开始对外推广。

这一创新工作提出了一种原创性的颠覆性技术，彻底解决了餐厨垃圾的危害，这是人类6000年来第一次巧妙地解决了餐厨垃圾治理难题。我们拥有七项国际国内专利，在国际生物质领域权威期刊 *Bioresource Technology*（《生物资源技术》）上发表了学术论文。

2）项目的产品

（1）非粮乙醇（95%），约2.5%。用途：燃料乙醇、消毒。市场价格：4.5元/千克，收入：112.5元/吨。

（2）工业油脂，约5%。用途：生物柴油、洗涤。市场价格：4元/千克，收入：200元/吨。

（3）酒糟饲用蛋白粉，约2.5%。用途：养殖、宠物饲料。市场价格：3元/千克，收入：75元/吨。

（4）二氧化碳干冰，约2.5%。用途：消防、制冷。市场价格：5元/千克，收入：125元/吨。

（5）液体肥，约10%。用途：园艺植物肥料。市场价格：4.77元/千克，收入：477元/吨。

（6）无机渣滓燃料，约2.5%。用途：垃圾焚烧材料。市场价格：0.6元/千克，收入：15元/吨。

不计政府补贴，一吨餐厨垃圾收入为1004.5元，能耗成本为300元，净利润为704.5元。

3）项目的经济效益

一个垃圾处理能力为50吨/日（一年处理18 250吨）的餐厨垃圾厂，不计政府补贴，一年纯收入1833万元；一个垃圾处理能力为400吨/日（一年处理146 000吨）的餐厨垃圾厂，一年的纯收入14 666万元。不计政府补贴，扣除其全部成本（包括运营成本和折旧，其中折旧按 15 年分摊），净利润分别为 918 万元和 7750 万元。而投资则分别为 3000 万元和 18 000 万元。预计两年多时间就收回成本。

若全国城市的餐厨垃圾都得到处理，一年纯收入高达 974.4 亿元，再加上其他有机资源废料和农村餐厨垃圾的处理，全国总收入可高达 2000 亿元。若再加上制造产业装备的收入，经济效益将更可观。另外，此项目还可向国际推广。

4）项目的社会效益

A. 无害化

① 地沟油从餐桌上消失。

② 垃圾猪变成有机猪。

③ 减少有机湿垃圾的焚烧，节能的同时减少排放二噁英。

④ 减少有机湿垃圾填埋，减少渗滤液。

B. 资源化

① 地沟油变工业油脂。

② 垃圾代替粮食变乙醇，燃料乙醇是当今国际最公认和流行的清洁能源。

③ 垃圾变高蛋白饲用蛋白粉。

④ 二氧化碳得到回收利用。

⑤ 垃圾生产液体肥。

⑥ 垃圾中的无机渣滓变可再生燃料。

C. 减量化

① 随着资源化利用而实现减量。

② 最终实现“吃干榨尽”，不产生污水和二氧化碳排放，实现零排放。

③ 处理周期 48 小时左右，快速减量。

4. 与国内外其他餐厨垃圾处理技术的比较

该项目技术与国内外常规处理技术相比，具有显著优势，具体见表 5。

表 5　该项目技术与国内外常规处理技术的比较

比较项	国内外常规处理技术					本项目技术
	填埋	焚烧	饲料	好氧堆肥	厌氧发酵	
资源化	×	△	△	△	√	√
无害化	×	×	×	△	△	√
简易化	×	√	√	×	×	√
耗时	×	√	√	×	×	√
占地	×	√	√	×	×	√

注：“√”表示“较好”；“×”表示“较差”；“△”表示介于“较好”与“较差”

5. 项目的实施

该项目已完成创新课题的中试，每日可处理 5 吨餐厨垃圾，现在正进行推广实施，包括以下四个任务。

1）创建有机资源再生中心

创建有机资源再生中心，将配有以下装备：有机垃圾自动化分选设备、高效油脂回收设备、乙醇生产设备、菌体蛋白粉加工设备、液体肥加工设备。有机资源再生中心采用闭环式生产，不让臭气散发到该中心之外。

将城市的餐厨垃圾以及农贸水果市场、海鲜肉类市场、大型市场、食品粮油企业、物流冷链企业的有机余料收运在一起，由有机资源再生中心按该项目联合生物加工技术的方法进行处理，日处理不同规模的有机资源再生中心的经济指标，如表 6 所示。

表 6　日处理不同规模的有机资源再生中心的经济指标

项目	日处理规模			
	50 吨	100 吨	200 吨	400 吨
投资/亿元	0.3	0.58	1	1.8
土地需求/亩	10	20	40	80
年产品销售收入/万元	1 833	3 666	7 338	14 666
年运营成本/万元	715	1 429	2 858	5 716
折旧分摊（15 年）/万元	200	339	667	1 200
净利润/万元	918	1 898	3 813	7 750

注：1 亩≈666.67 平方米

由表 6 可见，一座日处理 200 吨的有机资源再生中心，不计政府补贴，只需投资 1 亿元，即可每年收入 7338 万元，净利润高达 3813 万元，两年多就收回成本。

2）创建工程菌种生产基地

有机资源再生中心将成为全国可复制的产业板块，预计在未来三年，可在全国各地复制 100 个日处理 200 吨的有机资源再生中心。为此，急需建设工程菌种生产基地，为全国各地有机资源再生中心提供工程菌种。

未来三年处理规模：每年 730 万吨有机资源。

菌种厂房的土地需求：10 亩。

总投资：1 亿元。

预期年产值：3.65 亿元。

3）创建产业装备生产基地

在有机资源再生中心内，已搭建了一套高度资源化、无害化、简易化的餐厨垃圾联合生物加工技术以及配套的产业化装备线。这些装备需在创建的产业装备生产基地制造。按上文预计，3 年内建 100 个日处理 200 吨的有机资源再生中心，则需建一座相应的设备制造厂。

预计年产值：20 亿元。

土地需求：20 亩。

4）创建有机资源转化技术研究院

在该项目实现产业化的基础上，我们团队拟继续围绕有机资源的转化和利用进行产业链的深入研发工作，包括专用自动化环保装备开发、工程菌种开发、转化利用新工艺开发等研究课题，为此，将建立一个研究院。

场地需求：1 万平方米。

初次资金投入：5000 万元。

预计研究成果转化年收入：1000 万～2000 万元。

6. 科技创新成果的产业化

1）采用“政产学研金”模式创立科技创新型企业

要顺利完成上述餐厨垃圾治理的艰巨任务，就必须采用“政产学研金”模式。

“政产学研金”是一个跨行业、跨部门、跨地区合作的系统工程，符合中国特色，有助于创新科技成果的转化。

政府是关键。政府应强化政策的激励引导作用，为企业、高校和研究所科研人员营造有利于科技创新的良好环境，引导创新要素向企业集聚。同时加大扶持力度，强化促进保障作用。

企业应成为研发投入的主体、科技创新的主体、创新成果的主体。

高校和研究院所是企业的坚强后盾，既培养科技创新人才，又提供科技创新成果。金融机构可为产学研提供经济支持，这是强大的支柱。

现在面对全社会都关注的餐厨垃圾治理工程，更应该利用“政产学研金”的模式进行科学治理。

2）坚决执行国家政策

2016 年 12 月，国家发展和改革委员会、住房和城乡建设部印发《“十三五”全国城镇生活垃圾无害化处理设施建设规划》，要求到“十三五”末，力争新增餐厨垃圾处理能力 3.44 万吨/日，城市基本建立餐厨垃圾回收和再生利用体系。接着，2017 年，国家发展和改革委员会发布《战略性新兴产业重点产品和服务指导目录（2016 版）》，将“餐厨废弃物资源化无害化利用”列入。

2017 年 3 月，《国务院办公厅关于转发国家发展改革委住房城乡建设部生活垃圾分类制度实施方案的通知》发布，鼓励利用易腐垃圾生产工业油脂、饲料添加剂等，严厉打击和防范“地沟油”的生产流通。

2017 年 4 月，《国务院办公厅关于进一步加强“地沟油”治理工作的意见》发布，提出“推动培育与城市规模相适应的废弃物无害化处理和资源化利用企业”和“引导废弃物无害化处理和资源化利用企业适度规模经营，符合条件的按规定享受税收优惠政策”。

“垃圾围城”已成为城市治理不可忽视的问题，习近平同志早在 2012 年 12 月就指出：“我们在生态环境方面欠账太多了，如果不从现在起就把这项工作紧紧抓起来，将来付出的代价会更大。”① 2013 年 4 月 25 日，习近平同志又说：“经济上去了，老百姓的幸福感大打折扣，甚至强烈的不满情绪上来了，那是什么形势？”② 习近平同志还指出：

---

① 曹滢．习近平为何如此看重“美丽”，http://www.chinanews.com/gn/2018/01-29/8435559.shtml，2018-01-29.

② 林孔仕．习近平擘画美丽中国 绿水青山换得人民幸福感，http://www.chinanews.com/gn/2017/08-26/8314638.shtml，2017-08-26.

“走向生态文明新时代，建设美丽中国，是实现中华民族伟大复兴中国梦的重要内容。”①

必须按习近平同志指示办事，按中央指示办事，建议用“政产学研金”模式把“垃圾围城”之事尽快妥善解决。

3）治理城市餐厨垃圾的建议

为使上文提出的治理餐厨垃圾技术尽快落地实施，为实现美丽中国、美丽城市服务，特提出以下几点建议。

（1）政府重视并直接管理餐厨垃圾治理工作。截至 2017 年底，我国有 12 个千万人口特大城市和 142 个百万人口大城市，每个城市餐厨垃圾量都很大，每日达数千吨，已经“垃圾围城”。餐厨垃圾治理是民生工程，其治理困难已达 6000 年之久，与所有人都有关系，且与国家的发展紧密相关，与广州的环境美好相关，故希望市、区、街道各级领导直接抓餐厨垃圾治理工程的组织实施工作，必要时，还需立法，进行制度建设，此事才能解决。

（2）希望政府推动建立和完善收运处理体系。生活垃圾源头分类处理是发达国家垃圾管理的主要经验，在广州和其他许多城市，已实施多年，只是还不理想，需要继续积极推广。建议采用简单易行的干湿分类模式，即将生活垃圾按干、湿两类进行分类收集，其中的湿垃圾直接运往将来建成的有机资源再生中心采用新技术进行处理，干垃圾部分待人工进一步分选后将剩余物运往垃圾使用、填埋和焚烧发电厂处理。

对收运的车辆和桶要统一大小和形状，利于运输和装卸。同时，每日定时收集，确保收运处无蚊蝇孳生、无臭味扑鼻。

（3）大力提高市民的责任心。做好餐厨垃圾治理工作，既是政府的职能，也是公众的责任。必须加强宣传和教育工作，让公众了解治理的重要性和迫切性，自觉地、积极地参与，每个公民都自觉地不随地乱扔垃圾，都按制度设计进行垃圾分类和投放。同时，需要用激励和惩罚相结合的方法来保证市民做好自己应尽的责任。

参 考 文 献

[1] 段显明. 2015. 控制和减少污染物排放的机制与政策研究. 北京：经济科学出版社.

[2] 范体军，楼高翔，王晨岚，等. 2011. 基于绿色再制造的废旧产品回收外包决策分析. 管理科学学报，14（8）：8-16.

[3] 方行明，刘天伦. 2011. 中国经济增长与环境污染关系新探. 经济学家，（2）：76-82. “广州城市生活垃圾”课题组. 2014，论广州城市生活垃圾治理的战略取向. 广州大学学报（社会科学版），（8）：44-50.

[4] 金嵩川，宋文芳. 2017. 创新不止变废为宝——记刘人怀院士及其餐厨垃圾能源化处理项目. 科技创新与品牌，（3）：12-17.

[5] 赖铭辉. 1998. 台北市垃圾清理工作. 环境教育季刊，87（5）：37-43.

[6] 李金惠，王伟，王洪涛. 2007. 城市生活垃圾规划与管理. 北京：中国环境科学出版社.

① 中国共产党新闻网：生态文明是中国梦重要内容，http://env.people.com.cn/n/2014/0708/c386617-25253450.html.

十、珠三角城乡生活垃圾统筹治理战略研究①

1. 引言

本文主要针对珠三角地区的广州、深圳、佛山、江门、惠州、东莞、珠海、中山、肇庆（以下简称珠三角地区）。根据广东省统计局数据，2015 年广东省总常住人口 10 849 万人，其中城镇常住人口 7454.35 万人，农村常住人口 3394.65 万人。珠三角地区每平方千米 1073 人。广州、佛山、东莞、中山四市人口密度高于北京和天津，而深圳则已超过了上海，成为全国人口密度最高的超大城市。珠三角地区的平均人口密度达到全国平均人口密度的 8 倍，广东省人口密度仅是珠三角地区的二分之一。珠三角地区常住人口约 6000 万人，因此，生活垃圾治理的压力极大。作为国内较发达的地区，在生活垃圾处理方面走在前列，有效处理率高于 90%。与中国其他地区一样，珠三角地区在生活垃圾分类、餐厨垃圾处理、垃圾减量化等方面尚未形成良好的治理模式。广州、深圳、佛山、东莞仅初步建立餐厨垃圾收运处理系统，未达到“十二五”规划任务所要求的“餐厨垃圾收运处理体系成熟完善”的预期目标。具体来讲，“十二五”定下的指标是：①广州、深圳能够初步建立行之有效的生活垃圾分类收运处理运行机制和实施保障体系，餐厨垃圾收运处理系统成熟完善；②初步建成餐厨垃圾收运处理系统，并规划建设 17 座餐厨垃圾处理设施，处理规模达到 3990 吨/日。但到 2015 年底，餐厨垃圾收运处理系统仍处于初步建设阶段。广州、深圳、东莞、佛山成为国家餐厨废弃物资源化利用和无害化处理试点城市。建成后处理规模达 1742 吨/日，离预期还具有一定的差距。因此，餐厨垃圾的治理是当务之急。

2. 珠三角生活垃圾治理现状

1）研究背景

当前，作为重要的战略部署，国务院正在积极推进粤港澳大湾区城市群建设的相关规划，如发布了《推动共建丝绸之路经济带和 21 世纪海上丝绸之路的愿景与行动》《国家发展改革委办公厅关于加快城市群规划编制工作的通知》；2017 年 3 月 5 日，粤港澳大湾区城市群发展规划出现在国务院总理李克强的《政府工作报告》中，标志着该战略正式得到国家层面的确认。

珠三角地区以庞大的经济体量、宜人的环境、包容的文化氛围、高效的社会治理能力成为国家和地区的经济中心，以强大的辐射能力带动周边区域经济的发展。可见，珠三角地区经济、社会全面发展对国家和区域发展有着重要的带动作用。而生活垃圾治理水平反映了城市管理、社会治理、经济发展和政府服务的综合水平。由于珠三角地区城镇化水平高、经济发展迅速、消费水平较高，生活垃圾的治理压力非常大，生活垃圾的综合治理水平与其战略地位不相匹配。

2）珠三角生活垃圾治理现状分析

A. 生活垃圾处理基础设施建设

① 本文原载《中国管理科学的研究与实践》（第四届中国管理科学论坛论文集），北京：科学出版社，2019，11-29，作者：刘人怀，江峰.

到 2015 年末，为了对生活垃圾进行无害化处理，广东省共建成 110 座相关设施，总处理规模为 77 996 吨/日，其中，能够进行卫生填埋的设备 83 座，处理规模 49 196 吨/日；焚烧发电厂 26 座，处理规模 27 800 吨/日；水泥窑协同处置设施 1 座，处理规模 1000 吨/日。珠三角地区生活垃圾处理量统计及预测，见表 7。

表 7　珠三角地区生活垃圾处理量统计及预测　　单位：吨/日

地区名称	处理方式	处理规模						2020 年城市生活垃圾产生量	2020 年城市生活垃圾无害化处理率
		2015 年处理规模	续建处理规模	新改扩建处理规模	封场/停用/转用处理规模	填埋场增加处理规模	2020 年处理规模		
珠三角地区	填埋	33 050	8 530	1 840	18 000	1 415	26 835	71 537	100%
	焚烧	24 250	28 140	14 900	1 200		66 090		
	水泥窑协同	1 000					1 000		
	合计	58 300	36 670	16 740	19 200	1 415	93 925		

资料来源：《广东省城乡生活垃圾处理“十三五”规划》

截至 2015 年末，广东省规划的 105 个项目中有 75 个建成，按项目数量计算，完成率为 71%，按规模计算，完成率为 54%。其建设进度缓慢的主要原因为“邻避效应”问题导致项目征地阻力大，难以落地。

由于项目推进受阻，生活垃圾焚烧处理能力建设未能达到规划目标，规划的生活垃圾焚烧处理能力占总能力比例为 65%以上，实际占比约 36%，指标完成率为 55%。

为妥善处理垃圾，节约集约利用土地资源，提高垃圾处理的效率与效益，珠三角地区的广州、深圳、珠海、佛山、惠州、中山、江门等多地积极筹建生活垃圾综合处理环境园区。生活垃圾综合处理环境园区日益显现出较好的集聚效应及其在资源、环境、社会方面的综合效益，有利于化解垃圾处理的“邻避效应”，优化配置要素资源，深入进行垃圾多元、综合、全程和依法处理。但在环境园区规划、设计与建设中，一些地区存在垃圾处理方式简单集中、园区主业不清和区域布局不合理等问题。

广东省强化“村村保洁”制度。试点探索农村相关垃圾区别减量和市场化运营，着力出台相关条文，基本形成了县域统筹的村收集、镇转运、县处理的模式。

B. 餐厨垃圾收运及治理

截至 2015 年末，珠三角地区的广州、深圳、佛山、东莞均已开展餐厨垃圾管理的相关工作，并制定出台了管理办法。广州、深圳、佛山、东莞仅初步建立餐厨垃圾收运处理系统，未达到“十二五”规划任务所要求的“餐厨垃圾收运处理体系成熟完善”的预期目标。截至 2016 年，广东省除珠三角地区外其他城市均未建立餐厨垃圾收运处理基础设施。因此，餐厨垃圾的治理是当务之急。

C. 生活垃圾分类及管理的要求

“十三五”规划中着重说明发展理念，表明党中央对城乡餐厨垃圾处理工作的深入关注。不仅对相关设备建设提出要求，更对工作实施中的细节问题给予了关注。经济发展与绿色建设相统一，城市与乡村绿色建设相统一。对于垃圾处理要加大力度，减少污染。对于生活垃圾的处理问题，广东省立项规划，推动建立长久高效的工作机制，响应国家政策，进一步提高垃圾分类的力度，制定落实条文。广州、深圳已于 2015 年 4 月创建并成为全国第一批生活垃圾分类示范城市，在部分社区建立垃圾分类试点，在中小学和社区开展垃圾分类教育等。珠三角地区其他城市均相应开展了生活垃圾分类方面的教育宣传工作，并出台了一系列政策，但缺乏配套的制度安排，政策落实不到位，基础设施建设跟不上，导致垃圾分类工作迟迟无法开展。

3）生活垃圾治理模式对比及社会总成本

本文将四种处理模式的用地需求和经济成本进行估算，具体见表 8。

表 8　生活垃圾处理模式用地需求及经济成本估算

模式	用地需求/千米^2	分类补贴/亿元	收运转运/亿元	设施投资/亿元	当期成本/（元/吨）	长期成本/（元/吨）
焚烧＋填埋（不分类）	15	0	16.5	75	170	180
全焚烧（不分类）	2.4～4	0	16.5	0	220	280
干湿分类（源头分类）	3～5	30	15	0	300	160
干湿分类（末端分类）	3.5～6	0	13	0	240	270

注：垃圾填埋场设计设备寿命按 20 年计，垃圾焚烧场设备寿命是 30 年，长期成本指 2020 年后的成本

生活垃圾治理的社会总成本还涉及诸多风险，如有害物质的分布、附近人员的生命财产、意外问题、源头治理等。对珠三角地区来说，还有垃圾焚烧发电的上网电价补贴、环境监测成本、税收优惠等。

按照《北京市城市生活垃圾焚烧社会成本评估报告》的估算，对北京居民来说，北京市 11 座垃圾焚烧场近 70 年来二噁英致癌风险均超过百万分之一，较全社会的所有成本，焚烧约占 70%，直接处理只占 15%，其余 15%为税收优惠、电价补贴等成本。考虑生活垃圾焚烧全过程，社会成本将超过 6000 元/吨。

从用地需求出发，全焚烧（不分类）模式为首选模式，但是考虑全过程的社会成本，全焚烧的社会成本最高；从短期经济成本角度出发，焚烧＋填埋（不分类）模式为首选模式，但是珠三角地区城市化程度高，多数城市已经没有土地进行垃圾处理设施建设；从经济成本和资源化角度出发，干湿分类（源头分类）模式是首选模式，其短期经济成本较高是因为推广源头分类需要政府补贴和宣传教育，如果居民主动进行干湿分类，那么经济成本和社会总成本都会降低。

珠三角地区生活垃圾成分与北京市相似，垃圾处理技术水平相当，此外，珠三角地区“十三五”期间人口将超过 6000 万人，垃圾焚烧引致的患癌人数每年约为 8500 人，考虑居民健康成本，垃圾源头分类刻不容缓。

3. 珠三角生活垃圾治理诊断

作为经济和社会发展较快的地区，珠三角地区各地市政府高度重视生活垃圾管理工作，陆续出台实施了一系列政策和法规，进行有益的生活垃圾治理实践探索，广东

省政府从2013年起连续将生活垃圾治理工作纳入省政府十件民生实事。在各级政府及社会各界的共同努力下，珠三角地区生活垃圾统筹治理工作取得一定成绩。不过，也还存在诸多问题，部分问题的解决尚有待深入，如顶层机制、分类、餐厨方面、产业链建设等存在不足。

1）生活垃圾处理基础设施建设滞后

垃圾治理设施建设难免会遇到“邻避效应”。“邻避效应”虽然延缓了垃圾治理项目的建设和运营，但是在一定程度上推动了垃圾治理的法治化、公开化水平，促进了政府管理和公民参与水平的提高。若能应对得当，可以将其负面作用降至最低。但事实上一些地方从观念到行为，还未能跟上这种变化。特别是在政府公信力不足、法治规范不完善、民意表达机制和舆论监督机制相对缺失的情况下，如何推进垃圾处理基础设施建设，需要各方反思，而不是简单的“立项—抗议—停滞”，只求息事宁人，不解决根本问题。

截至2017年底珠三角地区垃圾填埋库容快速消耗，情况乐观地区剩余库容仅够使用5～8年，广州等地区已经面临垃圾无处倾倒的困境，如不及时转变垃圾处理模式，垃圾围城的问题将更加突出。

垃圾填埋场属于“厌恶型”市政设施，却是无法避免的。因此，填埋场是现代城市必须具备的市政基础设施之一，填埋库容资源也相应是城市可持续发展必须拥有的战略资源之一。

新建垃圾焚烧设施难以落地有三个方面原因。其一，建成运行的垃圾焚烧厂建设与运行标准低，监管不到位导致公信力缺失；一些项目运营不善、污染超标，污染物处理不到位甚至偷排，经常冒黑烟，臭气扰民，影响整个产业形象；对投诉和信访处理缺乏力度，给公众造成不良印象并使他们产生抵触心理。因此，市民格外反对在自己的居住地周边建设焚烧厂，这导致了立项的垃圾焚烧厂无法建设运营。其二，部分地方管理部门未能有效担负起生活垃圾处理的主体责任，存在畏难情绪，导致处理设施建设推进较慢，或者垃圾收运工作不落实，建成设施不能正常运转。其三，公众对环境利益的维权意识加强，但政府及企业缺乏透明公开的项目信息、平等有效的沟通机制和公平合理的补偿机制。

2）生活垃圾清收、运输和分类过程粗放

出于历史遗留问题或者资金投入不足等原因，部分地区仍存在敞开式的生活垃圾清收、运输设施，存在污水横流、恶臭弥漫、蚊蝇孳生等问题，严重影响环境卫生和城市形象，使生活垃圾收运设施成为群众厌恶排斥的对象，急需进行升级改造，配套封闭式设施设备，实行密闭化作业。

大部分城市垃圾箱区分了可回收和不可回收，但是清收的过程并无区别，并未达到应有的效果。

3）不重视源头分类和减量，只关注末端处理

受土地资源的限制和市民维权意识的提高的影响，在城市化程度较高的珠三角地区，新建垃圾填埋场的难度非常大，特别是深圳地区，由于无地可用，即使在满足技术规范和排放标准的情况下仍然难以实施。因此，珠三角地区应当在生活垃圾源头分类和减量方面下功夫。

早在 1998 年，中央政府指定八个经济发展较好的城市进行试点，不过没有取得很好的结果。其主要原因就在于垃圾治理战略选择上“轻源头减量、重末端处理”。没有根据公平原则让污染者付费、没有将垃圾治理看作关系民生的重大工程，从而偏离了“无害化、减量化、资源化”的总体战略目标。

分类收集迟迟未能成功推广，珠三角地区生活垃圾仍然采用混合收集、混合处理的方式，加之厨余垃圾比例高、气候炎热潮湿，项目组调研的所有垃圾填埋场都恶臭扑鼻，蚊蝇孳生，下雨天气则污水横流，对周边污染严重。周边居民和企业叫苦不迭，同时还导致周边土地和物业贬值，群众反映强烈。

产业链的脱节也是垃圾分类试点效果不明显的重要原因。初始环节并未重视降低垃圾数量，也未要求所有人员严格进行垃圾分类。相反，当下依旧是实行“源头分类、中间混运，末端填埋”的固有做法。此举的后果便是产业链的脱节，居民的有效分类并没有成效，在很大程度上降低了居民的分类意愿。各城市的关注点还是在垃圾桶的款式、颜色、摆放的位置上下功夫，大搞形象工程，使垃圾治理止步不前，近年来造成了严重的环境负担。

长期来看，大城市的生活垃圾卫生填埋处置方式恐怕难以持续。珠三角地区其他城市也由于无害化处理能力不足，形成了各地未经处理的存量垃圾场。此类垃圾极难处理，危害程度深且容易引发大批垃圾堆放滑坡等。部分存量垃圾场使用年限长、垃圾体量大、堆放高，安全隐患很多。

各地对存量垃圾治理工作推进慢，工作积极性不够。当下，对垃圾进行统计检查的工作有待开展，这项工作将有利于收集相关信息，了解内在情况。从研究机理到有效处理道阻且长，所面临的挑战和困难都急需我们努力解决，对于已完成治理的存量垃圾场，亦未对其治理效果进行后续的监督评价，不能排除治理措施简单、治理效果不达标的情况。

4）垃圾处理场园设置缺乏整体规划且超负荷运行

珠三角地区城市主流的垃圾处理场园是“一区一场”模式。填埋仍是当前情况下主要依赖的途径，由于项目推进进程和处理能力存在不足，珠三角地区一半以上的处理途径是填埋，剩下的则是处理了再填埋。

5）餐厨垃圾治理推进缓慢

但餐厨垃圾分离处理工作有众多优点，其有利于减少含水量，进一步减少辅助热燃量。另外，餐厨垃圾分离可以使生活垃圾渗滤液产量率减少，二噁英浓度降低。这些变化一方面使渗滤液处理补贴降低；另一方面使焚烧厂处理垃圾的内部成本降低。餐厨垃圾的分离处理将降低公共财政在垃圾焚烧上的支出。

珠三角地区城市餐厨垃圾收运和处理都面临诸多难题。由于餐饮企业产生的餐厨垃圾出售给养殖户及非正规处理企业能产生一定的经济效益，故餐厨垃圾的清运工作较难进行。尤其是对于还需单独缴纳餐厨垃圾处理费用，经营者更是有抵触情绪。因此，虽然珠三角地区的餐厨垃圾处理能力缺口很大，但这些正规企业却常有收运的垃圾不够的情况。随着垃圾分类的进行与对餐饮业餐厨垃圾管理的正规化，对餐厨垃圾处理的需求将更大，故需加快立法进程，保证餐厨垃圾不直接进入食物链，加快相关处理设施的建设。

6）农村生活垃圾治理资金投入不足

虽然各级财政逐年加大对农村生活垃圾处理设施建设及运营的补贴力度，但农村地区经济基础相对薄弱，各级财政配套资金投入不足，部分农村还缺乏治理的监督运行流程，设施建设、清运、治理处于经费短缺状况，影响其成果和效率。部分县（市、区）尤其是粤东和粤西地区农村的村边、路边、田边、河边、环卫设施周边仍然存在垃圾乱堆乱放、露天焚烧等问题。

7）政府职能缺位、产业链尚未形成

垃圾治理的关键是源头的分类和减量。珠三角地区垃圾分类已经倡导多年，但是一直无法落地。原因是缺乏相关的激励和约束机制，各方都没有垃圾分类的积极性，甚至居民进行了垃圾分类，但是在收运过程中由于硬件设备缺失，只能再混合收运，因此，垃圾治理的制度体系需要衔接，垃圾分类各环节基础设施需要尽快配套。

以中国台北、日本等先进地区为标杆，广州、深圳等市在推行垃圾分类方面进行了很多探索和努力，但出于垃圾分类相关法制政策不健全、公众垃圾分类意识不足、分类后垃圾处理技术不明确、分类垃圾缺乏后续出路、对垃圾分类工作投入不足，以及监督执法力度不足等多方面原因，分类效果不尽如人意。

生活垃圾治理既是公众的责任，也是政府的职能。政府的职能是进行顶层制度设计，激励公众进行源头分类和减量，并推动垃圾处理的基础设施建设和产业链完善。正是由于政府职能的缺位，珠三角地区的生活垃圾治理工作压力重重、举步维艰。

垃圾分类处理是一个系统工程，需要广泛的公众参与，软件和硬件建设，还需要相关激励和约束制度衔接。因此要提高公众参与分类的主体责任意识，并按需配套垃圾分类各环节的基础设施。

珠三角地区有大量的企业参与生活垃圾处理，但尚未形成垃圾治理的完整产业链，存在垃圾处理过程中环境卫生缺乏相应的资金链支持、地区效益分配不平等、运行处理流程模糊、相应监督未落实等问题。设施工作水平依靠地方的支持力度，部分有待深入。大多垃圾填埋场存在场内填埋机械设备没有进行分层压实和每日覆盖等规范作业，填埋堆体没有进行严格意义上的雨污分流措施，部分垃圾填埋场渗滤液处理设备运转不正常，容易出现渗漏严重、臭气超标等问题。对焚烧飞灰的无害化处置问题认识不足，缺乏必要的政策引导和规范，飞灰处理处置过程监管缺乏力度，存在一定的安全隐患。部分地方生活垃圾收运设施存在清洗保洁不及时、作业安排不合理的情况，影响周边环境卫生。运营管理信息公开不足，公众难以参与监督，导致公众对政府和企业管理水平持怀疑态度，进而影响相关设施建设工作的推进和环卫管理工作的开展。

由于缺乏相关科学技术，在运行工作中存在一系列问题，如渠道受阻、基本信息不全、信息有效性存疑、监管不足、成效缓慢等。为切实提高珠三角地区环卫工作管理水平和工作效率，应结合云计算、物联网、大数据技术，将互联网与环卫相结合，建立数字化管理平台。通过技术支撑，各方位、多角度对治理进程进行监督与控制。

8）社会参与度不够

随着时代的发展和社会人口受教育程度的提高，公众的环保意识逐渐增强，对于生活垃圾可能带来的环境污染问题已有一定的认识。但是，作为生活垃圾的产生者，很多人仍缺乏环卫主体责任意识，认为环境卫生只是环卫工人或者环卫管理部门的事情。乱扔垃圾的行为仍然普遍存在，垃圾投放环节实施分类更是举步维艰，要进一步做好生活垃圾管理工作，公众环卫主体责任意识亟待提高。珠三角地区城市没有将全民垃圾分类的责任和权益落实到户，没有按照公平原则按量征收垃圾费、没有乱扔垃圾的处罚机制，公众参与的积极性不高。

我国的可回收垃圾主要依靠拾荒者和清洁工人进入循环利用。这也导致了其他不可回收垃圾的价值不高，导致垃圾处理的承包企业无利可图，只能依靠财政补贴来维持经营，其改进技术和管理的积极性不高。有的甚至将承包项目分包给没有资质的企业，导致垃圾的乱排乱放。

4. 珠三角生活垃圾治理战略选择

中国台湾省和台北市生活垃圾进厂量对比如表 9 所示，其在生活垃圾治理方面的成就受到全世界的赞誉。欧美等发达国家在餐厨垃圾治理方面的诸多做法也值得我们学习。因此，我们在学习先进经验的基础上，提出符合珠三角地区实际的生活垃圾治理战略。

表 9　中国台湾省和台北市生活垃圾进厂量对比

项目	台北市生活垃圾进厂量			
	总进厂量	一般废弃物	一般事业废弃物	一般废弃物占比
2005 年	621 851.55 吨	498 824.13 吨	123 027.42 吨	80%
2016 年	189 833.48 吨	77 198.46 吨	112 635.02 吨	41%
变化量	−432 018.07 吨	−421 625.67 吨	−10 392.40 吨	
变化率	−69.473%	−84.524%	−8.447%	
项目	中国台湾省生活垃圾进厂量			
	总进厂量	一般废弃物	一般事业废弃物	一般废弃物占比
2005 年	5 767 652.87 吨	4 523 993.70 吨	1 243 659.17 吨	78%
2016 年	6 441 999.22 吨	4 271 178.81 吨	2 170 820.41 吨	66%
变换量	674 346.35 吨	−252 814.89 吨	927 161.24 吨	
变化率	11.692%	−5.588%	74.551%	

经过十多年的推动，台北市的资源回收取得了巨大的成效。2011 年，台北市每人每日垃圾清运量降至 0.39 千克。根据相关数据，台北市 2015 年每人每日垃圾清运量进一步降至 0.29 千克。

1）台北市生活垃圾治理经验分析

20 世纪末，中国台湾省出现垃圾增量居高不下，生活垃圾处理能力严重不足的状况，于是决定启动“一县市，一焚化炉”庞大垃圾焚烧计划，该计划提出后遭到台湾民众的强烈抗议，最终计划搁浅。1998 年马英九出任台北市市长，面对民间“反焚烧”运动，马英九团队决定从“源头分类和减量”入手，把垃圾分类作为台北市政府施政的民生工程。为达到宣传垃圾分类的目的，马英九呼吁参加垃圾分类的公益电视节目，深入社区做动员、宣传。正是因为市长的重视和以身作则，政府的政策法规推行有力，社会各界的积极参与，台北市用了十多年时间形成了生活垃圾治理的完整产业链。从源头分类、中游清运、下游回收处理，产业上、下游无缝连接，环环相扣，实现了生活垃圾治理的产业化、规模化、集约化。分析发现台北市的经验主要有以下几个方面。

A. 源头减量、强制分类

台北市在民众的压力之下，着力破解垃圾围城的难题，办法是“源头分类和减量”。为推进生活垃圾分类，台北市政府尝试通过购买服务的方式，委托民间环保组织配合负责实施垃圾分类的宣传和教育，先在两个居民区约5000户进行试点计划，试点结果发现垃圾源头减量明显，回收效果显著。

在试点源头减量取得成功的基础上，台北市政府马上采取强制垃圾分类，要求台北市所有居民必须按照规定，对生活垃圾按“可回收”“厨余”“一般”三大类进行源头分类，职能部门定时、定点集中收集，否则拒收或处罚。强制分类使台北市生活垃圾由迅速增长转变为负增长。台北市政府将垃圾费从水费中剥离出来，运用市场化手段，对居民垃圾处理费由原来的“定额制”改为“从量制”，推出“按量计费”的政策，即“随袋征收垃圾费”政策。这一政策的实施，促进居民进行垃圾分类，效果显著。

B. 集中收集、全程监控

实施“按量计费”政策以来，台北市所有的生活垃圾出处都可查到。通过量化实现了垃圾治理的集中化和集约化，为管理部门进行全程监控提供了信息支持。台北市政府规定，生活垃圾必须有政府授权才可进行相应处理。自2000年以来，台北市逐渐降低对生活垃圾处理的费用，同时将变卖垃圾资源所得用于社会福利事业，实现了生活垃圾治理的可持续发展。因为该政策的实施，台北市生活垃圾产量大幅降低，原有的两座垃圾焚烧厂已经关了一家。

C. 污染者付费、严格执法

污染者付费是基于公平原则的生活垃圾治理理念，台北市实行强制性垃圾分类、“按量计费”的政策后，居民会逐步改变自己的生活习惯和消费习惯，降低生活垃圾产出，必然使自己所缴纳的垃圾费逐年降低。对生活垃圾的处理费用，政府财政支出不升反降，并有结余，政府也实现了生活垃圾治理的可持续发展目标。生活垃圾处理产业链上的企业和非营利组织均能实现持续发展。

经济激励只是一个方面，台北市政府也严格执法，对不遵守规矩和搭便车者坚决惩处。对不进行垃圾分类者，拒收并处罚；对随便乱扔生活垃圾者，严格处罚；对不定时定点扔垃圾者，处罚。处罚额度高达1200～6000新台币。举报市民可获得两成罚款作为奖金。

D. 产业链完整、社会参与

台北市按照循环经济的要求，居民先对塑料、电池、玻璃、纸张、金属、衣物等生活垃圾按其属性进行“大类粗分”，为了增加其附加值，对同类物质属性的垃圾，再次进行“同类细分”，做到“物尽其用”。设立专门的人工修复厂，对大件垃圾，如汽车、家具、家用电器等进行修复或拆解。对有害垃圾集中处理。以市场化的手段、公平的理念、长期的政策扶持和完整的产业链，最终使台北市成为生活垃圾治理的典范。

2）国内外主要地区生活垃圾治理经验分析

珠三角地区城市的生活垃圾物理成分相似，其比率随年份变化不大。本节以深圳为例，分析国内外主要城市生活垃圾的组成及其原因。

A. 国内主要城市生活垃圾特性

无机物含量较高、生物质比例偏高，但可回收物比例偏低、不可燃成分较多，这些是我国城市生活垃圾的总体特征。国内部分主要城市生活垃圾物理组分，如表 10 所示。

表 10　国内部分主要城市生活垃圾物理组分

城市	厨余	纸类	塑料	纺织	木竹	灰土	砖瓦	玻璃	金属	其他
北京	63.79%	9.75%	11.76%	1.69%	1.26%	—	—	1.70%	0.33%	0.20%
上海	61.11%	9.46%	19.95%	2.80%	1.48%	—	—	2.98%	0.28%	0.37%
武汉	52.41%	9.17%	17.06%	3.38%	2.47%	9.29%	2.37%	2.34%	1.50%	0
深圳	58.12%	13.30%	15.40%	7.12%	0.95%	0.17%	1.43%	2.33%	0.77%	0.42%

注：北京、上海的灰土、砖瓦没有数据，因此合计不为 100%

从表 10 中数据可以看出：厨余类垃圾是各个城市生活垃圾中最主要的部分，这些城市的厨余含量都在 50%以上，各地饮食习惯和经济发展水平在一定程度上会影响厨余含量。

根据香港环境保护署的年报，2010～2013 年香港的城市生活垃圾组分如图 2 所示。

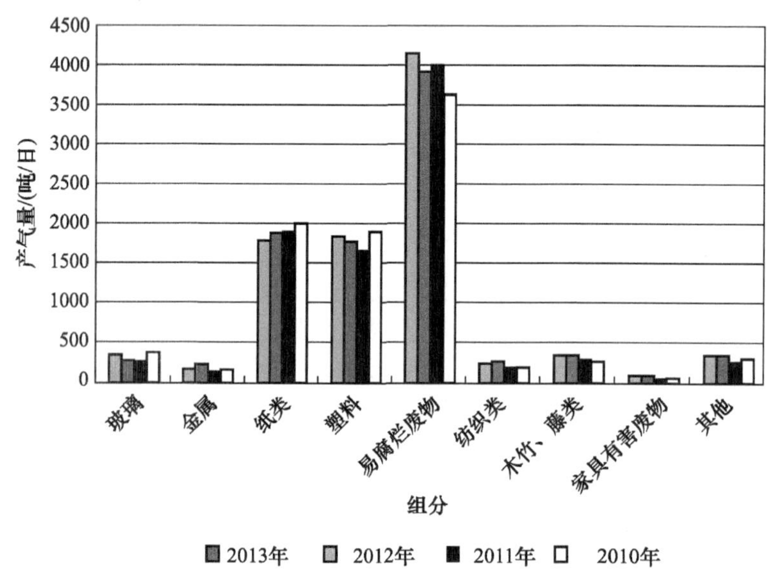

图 2　2010～2013 年香港的城市生活垃圾组分

香港城市生活垃圾中，易腐烂废物比例最高，达到 43.9%。对比深圳市生活垃圾数据，可以发现，两者生活垃圾组分相似，都是易腐烂废物（厨余）、塑料和纸类，而其他组分的含量都不太高，这说明深圳与香港的生活垃圾组成大体相同。

B. 国外主要国家生活垃圾特性

国外生活垃圾组分不仅受环境、气候、生活习性，以及经济发展水平等因素影响，还受其法治化管理的影响。

由于政策、生活方式及生活垃圾进入清运处理的过程不同，国内与国外典型地区的垃圾组分含量有较大差异。在国外，特别是美国，庭院垃圾成为其生活垃圾的典型组分，但在国内大部分居民并没有庭院式住宅，而道路小区中的枝叶等归入园林绿化垃圾，不属于居民生活垃圾。国外几个国家垃圾组分对比，如表 11 所示。

表 11 国外几个国家垃圾组分对比

国家	厨余	纸类	塑料	纺织	木竹	玻璃	金属	庭院垃圾	灰土	其他
美国	14.60%	27.00%	15.80%	6.00%	6.20%	4.50%	9.10%	13.50%	1.50%	1.80%
英国	17.84%	22.70%	10.00%	2.80%	3.70%	6.60%	4.30%	14.10%	4.50%	13.40%
日本	19.10%	36.00%	18.30%	9.50%	4.50%	0.30%	0	0	6.10%	6.20%
法国	25.50%	21.50%	11.50%	2.60%	2.50%	13.00%	3.00%	0	3.00%	17.40%
新加坡	20.16%	26.20%	25.40%	3.22%	3.20%	2.01%	2.40%	5.24%	0	12.10%

注：由于舍入修约，数据存在误差

从表 11 中可以看出，珠三角地区与国外生活垃圾组分一个较大不同就是厨余类垃圾的含量，还可以发现，可回收物，如纸类和塑料的含量有较大不同，这可能与国外较成功的生活垃圾分类回收有关，可回收物能较好地、系统地进行回收处理。主要影响因素应该是我国生活垃圾中一部分可回收物在被清运前，已被拾荒者或清洁工收走，并没有进入垃圾末端处理设施。这也可以解释表 10 中深圳市生活垃圾中玻璃类和金属类的含量要比欧美国家低得多。

总体来说，珠三角地区生活垃圾的组分相对于欧美发达国家及新加坡、日本等国来说，厨余类垃圾所占比例很大，总含水率高，生活垃圾中可燃成分含量少，生活垃圾的热值低。因此，如果珠三角地区生活垃圾不做前端分类，特别是不对厨余垃圾进行分类回收和处理，则垃圾焚烧和填埋的经济成本和社会成本远大于国外垃圾的处理成本。

3）珠三角城乡生活垃圾治理的预测

珠三角地区城市生活垃圾清运已基本实现全覆盖，随着垃圾分类工作的推进，预计“十三五”期间珠三角地区城市生活垃圾产生量的增长率相对于“十二五”时期将有所下降。

根据广东省统计局统计数据，2015 年广东省常住人口 10 849 万人，其中城镇常住人口 7454.35 万人，农村常住人口 3394.65 万人。《广东省国民经济和社会发展第十三个五年规划纲要》预计到 2020 年，广东省总常住人口控制在 11 400 万人，常住人口城镇化率达到 71.7%，即城镇常住人口 8173.8 万人，农村常住人口 3226.2 万人。珠三角地区人口约 6000 万人。

2015 年末，广东省城市生活垃圾人均产生约 1.09 千克/天，预计到 2020 年末，广东省城市生活垃圾人均产生约 1.20 千克/天。2020 年末，预计广东省城市生活垃圾产生量为 9.84 万吨/天，珠三角地区生活垃圾产生量为 7.2 万吨/天。

综合考虑生活垃圾的年均增长率，预计 2020 年广东省城乡生活垃圾产生量约为 12.90 万吨/天，其中，城市生活垃圾产生量约为 9.64 万吨/天，农村生活垃圾产生量约为 3.26 万吨/天。

截至 2015 年末，广东省生活垃圾无害化处理基本实现“十二五”规划预期目标，但生活垃圾无害化、资源化处理比例偏低，需进一步加大力度提高生活垃圾资源化处理的比例。争取到 2020 年，广东省城市生活垃圾无害化处理率达到 98%以上。

城市餐厨垃圾人均产生量按 0.35 千克/天计，预测 2020 年广东省城市餐厨垃圾产生量为 2.86 万吨/天。珠三角地区餐厨垃圾产生量为 1.9 万吨/天。

4）珠三角城乡生活垃圾治理战略目标

A. 战略总目标

加强顶层制度设计和立法，通过推进城乡生活垃圾强制分类，探索生活垃圾按量计费制度；通过教育和经济激励，明确生活垃圾治理中各方的主体责任，提高城乡生活垃圾减量化、资源化、无害化处理水平；加快提升餐厨垃圾的分类收集和资源化处理水平；运用物联网技术和信息技术，加大力度监管垃圾投放、收运和处理的全过程；通过税收、土地、财政等政策扶持，主要运用市场化手段，保障城乡生活垃圾收运设施与垃圾分类环节的衔接，促进生活垃圾治理形成完整的产业链。

B. 具体目标

按照《广东省城乡生活垃圾处理“十三五”规划》的要求，“十三五”时期珠三角地区生活垃圾的治理仍然以焚烧为主，同时提出加大对餐厨垃圾的分类回收力度，平均回收率达到 25%，生活垃圾总资源回收率达到 35%。该规划尚未纳入餐厨垃圾治理的相关指标，这是规划的一个重要局限。“十三五”期间，餐厨垃圾分类要达到 50%以上才能缓解生活垃圾带来的压力。“十三五”时期城乡生活垃圾处理发展目标指标如表 12 所示。

表 12 “十三五”时期城乡生活垃圾处理发展目标指标

序号	指标名称	现状值（2015 年）	目标值（2020 年）	属性
1	城市生活垃圾无害化处理率	90%	98%	约束性
2	城市生活垃圾焚烧处理能力占无害化处理总能力的比例	36%	60%	预期性
3	城市生活垃圾回收利用率	—	35%	预期性
4	餐厨垃圾有效处理率	—	50%	预期性
5	村庄保洁覆盖面	94%	100%	预期性
6	农村生活垃圾分类减量比例	24%	50%	预期性

5）珠三角生活垃圾治理的战略任务

A. 加快立法、推进生活垃圾强制分类，重点推进干湿分类

破解生活垃圾治理难题的关键是推行生活垃圾分类制度，特别是干湿分类。珠三角地区要尽快形成并完善生活垃圾分类制度，加快建立生活垃圾分类收运处理系统，努力提高生活垃圾分类制度覆盖范围，有效减少生活垃圾清运量和最终处理量，减轻末端压力，推动生活垃圾的资源化和有害垃圾分类立法工作。生活垃圾治理顶层制度体系如表 13 所示。

表 13 生活垃圾治理顶层制度体系

项目	生活垃圾	餐厨垃圾	有害垃圾
地方政府法规	垃圾分类管理条例	餐厨垃圾管理条例	有害垃圾管理条例
地方政府规章	管理办法及目标	管理办法及目标	管理办法及目标
技术规定	技术要求及标准	实施细则及标准	技术要求及标准

珠三角地区应根据《广东省城乡生活垃圾处理条例》进一步完善生活垃圾管理法规制度体系，明确各级政府、各相关部门、企事业单位和社会公众的义务及责任，明确相关的管理制度和追责制度，使生活垃圾管理得到更充分的法治保障。必须加快生活垃圾强制分类、餐厨垃圾处理和有害垃圾处理的相关立法工作。重点推进干湿分类，鼓励居民将生活垃圾分开放置并单独投放。

B. 探索按量计费制度，实施定时定点投放，促进源头减量

从世界经验来看，促进垃圾减量有两种措施：其一，按照公平原则，谁污染谁付费，也就是生活垃圾按量计费；其二，生活垃圾不得乱扔，必须定时定点投放。对珠三角地区来说，应当先完善垃圾收费制度，推动按行业平均成本、企业合理利润和居民承受能力来确定单位生活垃圾的收费标准。

当前普遍实施的生活垃圾"定额收费制"，将垃圾收费与水费绑定，难以形成垃圾减量的经济激励，中国台湾省台北市对生活垃圾"按袋计费"，很快就实现了生活垃圾的减量。只有"按量计费"方法才能从根本上鼓励居民改变消费习惯，倡导绿色低碳生活。所以我们建议，珠三角地区要在城市成熟社区和企事业单位尝试推出"按量计费"试点，探索制定垃圾按量计费制度。

C. 提升餐厨垃圾资源化处理水平

餐厨垃圾占生活垃圾的一半以上，其含水率高、热值低的特性会导致焚烧和填埋过程中产生二噁英和渗滤液。餐厨垃圾进入食物链，导致"地沟油"和"垃圾猪"问题难以从根本上解决，严重损害了人们的健康。本文建议，在推进生活垃圾干湿分类的基础上，加快制定《广东省餐厨垃圾管理办法》，珠三角地区各地市政府结合当地特点和实际需求，制定相应的管理办法，规范餐厨垃圾的处理过程。

在有条件的农贸市场、大型餐饮单位等场所，首先进行餐厨垃圾集中收运和处理。建议推广刘人怀院士团队研发的餐厨垃圾联合生物加工技术，鼓励其他餐厨垃圾收运企业与其对接，从而构建完整的餐厨垃圾资源化产业链。从土地、财政和税收方面支持厨垃圾处置企业产业化、可持续健康发展。

D. 运用物联网技术和信息技术，加强生活垃圾全过程监管

做好生活垃圾无害化处理设施建设和运营信息统计工作，实时监测焚烧厂和卫生填埋场主要设施运行状况，加强对焚烧设施烟气排放、渗滤液和填埋气体的监测。建立生活垃圾清扫、分类、收集、运输、处置信息管理系统，并与省城乡生活垃圾管理信息系统实时联网。

生活垃圾管理系统应包括环卫工作的全过程监管，生活垃圾收集、运输、处置设施运营在线监测，以及设施污染物排放处置监控等内容。

运用信息技术手段［如餐厨垃圾产生点设置信息监控设备、收运车辆安装全球定位系统（global positioning system，GPS）、处置场所安装视频监控装置等］，加强执法协助，建立并完善餐厨垃圾产生与收运处置各环节紧密联系的全程监控机制，防止"地沟油"回到餐桌。加强餐厨垃圾分类处置、日产日清和追踪制度，确保餐厨垃圾及其加工物流向的可查可控。严禁餐饮服务企业未经油水分离即将含废食油的潲水排入地沟或管网。清理规范收运队伍，加强对收运、处置企业的监管。收运企业应使用餐

厨垃圾收运专用车。禁止将餐厨垃圾及其加工品作为生产加工食品的原料，禁止使用未经无害化处理的餐厨垃圾饲养畜禽。防止餐厨垃圾未经处理进入食物链。

E. 保障分类、清运和处理环节的衔接，形成完整的产业链

按照区域统筹要求，在珠三角地区构建与生活垃圾处理配套的分类、清运、处理体系，有效推进生活垃圾分类、收运系统与再生资源回收系统衔接。建立健全物流中转设施，全面推进末端处理设施建设，最终建立有效衔接的生活垃圾分类体系和产业链。

珠三角地区应加快生活垃圾处理产业化发展和社会化运作，建立多元化投入机制。完善以公共财政为导向的城市垃圾处理设施建设投资体制，逐步形成“政府引导、社会参与、市场运作”的多元化投资机制。

F. 统筹垃圾处理基础设施建设，降低垃圾处理场园污染

对珠三角地区来说，要打破“一区一场”的垃圾处理设施布局，应当从市域统筹、系统规划的角度对各类生活垃圾处理设施进行协调整合，通过垃圾分类回收、焚烧发电、卫生填埋、生化处理等多种技术组合，实现各类垃圾处理工艺的优势互补，将环境厌恶型的生活垃圾处理设施整合并打造为垃圾综合处理环境园，减少不必要或低水平的重复建设。

G. 实现农村保洁常态化，加快存量垃圾治理

在珠三角地区的农村地区建立稳定的村庄保洁队伍，明确保洁员在垃圾收集、村庄保洁、资源回收、宣传监督等方面的职责。可再生资源应尽可能回收，对废弃的农资包装物进行回收利用或者无害化处理。加快农村存量垃圾治理，应全面调查了解陈年垃圾存量、分布和污染现状，集中精力尽早完成陈年垃圾清理工作，全面提升农村环境卫生综合水平。禁止城市向农村转移垃圾，防止村庄周围产生新的垃圾污染。

5. 结论

当前我国解决生活垃圾问题的主要方式是“焚烧＋填埋”的末端处理模式，并且在大部分地区，未来生活垃圾治理的战略方向仍然是加大焚烧的比例。这种处理模式很难从根本上解决垃圾围城的难题，而且会导致非常高的社会成本。首先，垃圾焚烧设备投资大、费用高，建设周期长；其次，末端治理会产生二次污染，二噁英、渗滤液和飞灰等给人民健康带来危害；最后，没有做干湿分类的生活垃圾，含水率高、热值低，焚烧效果不好。

在对珠三角地区垃圾治理现状进行诊断的基础上，参照国内外地区生活垃圾治理的经验，特别是中国台湾省的经验，项目组认为生活垃圾治理战略的关键是源头分类和减量。珠三角地区要想实现源头治理，要扎实做好三个方面的基础工作。

其一，管理部门做好立法和制度建设，果断推进生活垃圾强制分类，重点是干湿分类，探索垃圾“按量计费”制度。台北市人均产生生活垃圾的数量为 0.28 千克/天，而珠三角地区为人均 1.1 千克/天，说明台北市生活垃圾治理制度有效促进了生活垃圾的减量。台北市的制度建设经验包括三个方面，分别是强制分类、按袋计费和定时定点投放。珠三角地区生活垃圾治理制度建设包括三个层次：第一层次是强制性的生活垃圾分类条例、餐厨垃圾管理条例、有害垃圾管理条例等地方法规；第二层次是地方规章制度，包括生活垃圾分类的管理办法及实施细则、生活垃圾治理的年度目标任务

等；第三层次是技术规定，包括垃圾处理的实施办法及标准等。制度必须要配套，同时要严格执法，特别是要尽快推进强制性生活垃圾干湿分类，探索适合地区的垃圾“按量计费”模式。

其二，加快推进餐厨垃圾分类收集和处理，并对餐厨垃圾的产生、收运和处理进行全过程监管。餐厨垃圾在我国的生活垃圾中占比超过一半，由于对餐厨垃圾缺乏集中监管，餐厨垃圾在没有经过处理的情况下，以地沟油、饲料等方式进入食物链，带来诸多食品安全问题，严重危害人民健康，餐厨垃圾焚烧和填埋过程中会产生二噁英和渗滤液，对环境造成二次污染。珠三角地区气候炎热，未进行干湿分类的生活垃圾在存放和运输过程中会腐烂变质，导致蚊蝇孳生，影响城市形象。因此，餐厨垃圾的有效治理是生活垃圾处理的关键。

其三，改革生活垃圾管理体制，鼓励社会各界参与生活垃圾分类、清运、处理和回收工作，运用市场化手段，构建生活垃圾治理的完整产业链。生活垃圾治理需要“政企分开”，政府部门转变职能，社会各主体各司其职，政府主要进行政策制定和监督管理，垃圾的清运可以通过购买服务的方式外包给有资质的企业。对生活垃圾处理企业，要考虑行业发展和企业成本，通过一定的政策扶持、税收减免等促进企业可持续发展，同时引入了市场竞争机制，鼓励创新，提高垃圾处理效率。垃圾分类的主体是居民，根据中国台湾省台北市的经验，针对老人和小孩的宣传教育可以有效促进垃圾分类工作。垃圾分类工作需要坚持长期进行，切忌浅尝辄止，台北市也是经过了十多年的努力才取得良好的效果。

在全面建设小康社会的新时期，国家提出了建设粤港澳大湾区的战略部署，对珠三角经济的发展寄予厚望。生活垃圾治理工作是珠三角地区环境治理能力和社会治理能力的重要内容，需要政府和社会各界互相协作、脚踏实地，有耐心、有步骤、有重点地推进。

参考文献

[1] 昌道励，成广伟. 2015-07-10. 重点推广垃圾分类处理“广州经验”. 南方日报，(1).

[2] 陈琼. 2013. 广州市城市生活垃圾分类管理政策研究. 电子科技大学硕士学位论文.

[3] 陈秀珍. 2012. 德国城市生活垃圾管理经验及借鉴. 特区实践与理论，(4)：69-72.

[4] 邓鹏敏. 2015. 深圳城市垃圾管理的 SWOT 分析及对策研究. 华中师范大学硕士学位论文.

[5] 杜铭州. 2003. 环境政策执行过程之比较研究——以台北市及台中市垃圾清运为例. 东海大学学位论文.

[6] 付敏言，田秀娜，乔宇彤. 2015. 广州市城市生活垃圾处理的 PPP 模式研究. 中国集体经济，(33)：18-19.

[7] 广东省人民代表大会常务委员会. 2015.《广东省城乡生活垃圾处理条例》2016 年 1 月 1 日起施行. http://www.rd.gd.cn/xwdt/201509/t20150930_147658.html [2015-09-30].

[8] 广东省人民政府. 2015. 广东省人民政府关于加快推进城市基础设施建设的实施意见. http://zwgk.gd.gov.cn/758336165/201506/t20150612_585489.html [2015-06-11].

[9] 广东省人民政府. 2016. 广东省人民政府关于印发《广东省国民经济和社会发展第十三个五年规划纲要》的通知，http://zwgk.gd.gov.cn/006939748/201801/t20180113_748467.html [2016-04-20].

[10] 广东省人民政府办公厅. 2012. 广东省人民政府办公厅关于印发广东省生活垃圾无害化处理设施建设“十二五”规划的通知. http://zwgk.gd.gov.cn/006939748/201211/t20121122_355668.html [2012-11-13].

[11] 广东省住房和城乡建设厅. 2016. 广东省住房和城乡建设厅关于印发《广东省住房城乡建设事业“十三五”规划纲要》的通知. http://zwgk.gd.gov.cn/006939799/201605/t20160522_655981.html [2016-05-18].

[12] 广东省住房和城乡建设厅. 2017. 广东省住房和城乡建设厅广东省发展和改革委员会关于印发《广东省城乡生活垃圾处理“十三五”规划》的通知. http://www.gdcic.gov.cn/HTMLFile/shownews_messageid=145587.html [2017-05-11].

[13] “广州城市生活垃圾治理”课题组. 2014. 论广州城市生活垃圾治理的战略取向. 广州大学学报（社会科学版），(8)：44-50.

[14] 国务院办公厅. 2012. 国务院办公厅关于印发“十二五”全国城镇生活垃圾无害化处理设施建设规划的通知. http://www.gov.cn/zwgk/2012-05/04/content_2129302.html [2012-04-19].

[15] 洪池，吕永. 2016. 广州市生活垃圾分类人工分拣成本探讨. 广东科技，(6)：61-63.

[16] 黄大佑. 2013. 垃圾的伦理化？台北都会区垃圾治理与资源回收体制的转型. 台湾大学硕士学位论文.

[17] 赖铭辉. 1998. 台北市垃圾清理工作. 环境教育季刊，(36)：37.

[18] 郎国华，孙国英，谢苗枫，等. 2015-05-13. “盐田模式”如何成为创新范例？南方日报，(5).

[19] 林玲舟. 2015. 104 年污染防治支出统计调查报告，EPA-104-M104-02-101.

[20] 林清蓉. 2015-08-11. 填埋处理难以持续垃圾分类难解“近渴”，深圳特区报，(2).

[21] 刘怀宇，黄祖建. 2016. 垃圾分类“广州经验”领跑全国. 广西节能，(1)：37.

[22] 卢文刚，黎舒菡. 2016. 基于利益相关者理论的邻避型群体性事件治理研究：以广州市花都区垃圾焚烧项目为例. 新视野，(4)：90-97.

[23] 马小康，温丽琪，谢锦松，等. 2009. 垃圾费随袋征收与随水征收制度垃圾减量及回收成效评估专案工作计划.

[24] 内湖垃圾焚化厂. 2015. 台北市政府环境保护局内湖垃圾焚化厂营运管理及环境品质检测报告.

[25] 潘二波. 2012. 深圳生活垃圾分类的对策. 绿色科技，(8)：173-175.

[26] 钱玉兰. 1997. 一般废弃物（垃圾）收费技术研究，EPA-86-E3Hl-09-01.

[27] 申卉，穗外事. 2017. 广州处理餐厨垃圾创新技术世界领先 垃圾处理后变废为宝. http://news.sina.com.en/c/2017-04-12/doc-ifyecfnu8185764.shtml [2017-04-12].

[28] 沈大明. 2014. 广州城市生活垃圾分类处理的系统设计与实践. 广东技术师范学院硕士学位论文.

[29] 台北市政府环境保护局. 1999. 1998 年度台北市资源回收比率统计表. 台北市政府环境保护局年报.

[30] 台北市政府环境保护局. 2015. 台北市政府环境保护局木栅垃圾焚化厂营运管理及环境品质检测报告.

[31] 台北市政府主计处. 2002. 台北市垃圾处理概况（1999—2002). 台北市政统计周报.

[32] 台北市政府主计处. 2011. 台北市垃圾处理工作. 台北市统计年报.

[33] 台北市政府主计处. 2012. 台北市环境负荷指标（2001、2011). 台北市政统计周报.

[34] 王权典. 2015. 城市生活垃圾终端处理设施生态补偿机制——以广州相应的创制规范为例//生态文明法制建设——2014 年全国环境资源法学研讨会（年会）论文集（第二册）. 广州：中

国法学会环境资源法学研究会.
[35] 卫国栋. 1999. 废铝箔包及废纸盒包回收清除处理之成本收益评估，TMP-88-002.
[36] 温丽琪. 2010. 台北市垃圾费随袋征收之探讨与分析.
[37] 文灿. 2015-12-18. 深圳探索用购买服务打通垃圾回收产业链. 深圳商报，(2).
[38] 吴浩，王艳宜，吴燕琦，等. 2016. 深圳市生活垃圾分类对垃圾焚烧影响的研究. 环境卫生工程，24 (1)：40-43.
[39] 吴氏如霞. 2014. 深圳市生活垃圾处理监管体系研究. 华南理工大学硕士学位论文.
[40] 萧江碧，黄荣尧. 2003. 废弃混凝土再生利用成本效益分析之研究.
[41] 新华网. 2016. 中华人民共和国国民经济和社会发展第十三个五年规划纲要，http://www.xinhuanet.com/politics/20161h/2016-03/17/c_1118366322.htm [2016-03-17].
[42] 新浪网. 2016. 广环投集团：如何正确引导市民处理生活垃圾广环投有奇招，http://news.sina.com.cn/o/2016-10-12/doc-ifxwvpqh7269183.shtml [2016-10-12].
[43] 徐菊荣. 2013. 深圳城市生活垃圾产生量预测及南山区垃圾处理路线设计. 华中师范大学硕士学位论文.
[44] 杨朋，王芙蓉，陈红忠，等. 2016. 深圳市生活垃圾分类收运系统调研分析. 环境卫生工程，24 (4)：16-18.
[45] 杨清显，刘敏信，刘俊佑，等. 2006. 电子电器资源回收之经济分析.
[46] 叶国顺，张梅英. 2006. “一县市一焚化炉”政策终止后之垃圾处理方式研究：环境政策、规划管理与永续发展. 台北：台湾环境资源永续发展协会.
[47] 詹尚文. 2009. 国内工业废弃物处理成本效益分析探讨——以印刷电路板业含铜污泥为例. 台北大学学位论文.
[48] 张紧跟. 2014. 从抗争性冲突到参与式治理：广州垃圾处理的新趋向. 中山大学学报（社会科学版），(4)：160-168.
[49] 张钦华. 2016. 生活垃圾焚烧处理技术在广州的应用. 广东化工，(11)：194-195.
[50] 郑希黎，杜任俊. 2016. 走出邻避困局，探索垃圾综合处理设施新模式——以深圳国际低碳城节能环保产业园规划为例. 沈阳：中国城市规划学会.
[51] 中共中央. 2015. 中共中央关于制定国民经济和社会发展第十三个五年规划的建议. www.gov.cn/xinwen/2015-11/03/content_5004093.htm [2015-11-03].
[52] 中国发展网. 2017. 住房城乡建设事业“十三五”规划纲要. http://www.chinadevelopment.com.cn/ztbd/2017zt/2017/02/1123720.shtml[2017-02-20].
[53] 中华人民共和国国务院新闻办公室. 2015. 国务院关于加强城市基础设施建设的意见. http://www.scio.gov.cn/32344/32345/32347/33173/xgzc33179/Documen/1442976/1442976.htm[2015-07-31].
[54] 中兴工程顾问股份有限公司. 2008. 废食用油回收、PVC 管制及一般废弃物清除处理成本估算专案工作计划（第二年).
[55] Agency C E P. 2016. User manual for the hotspots analysis and reporting program air dispersion modeling and risk assessment tool version2.
[56] Kulkami P S，Crespo J G，Afonso C A M. 2008. Dioxins sources and current remediation technologies—a review. Environment International，34 (1)：139-153.
[57] Pesatori A C，Zocchetti C，Guercilena S，et al. 2017. Dioxin exposure and non-malignant health effects：a mortality study. Occupational and Environmental Medicine，55 (2)：126-131.
[58] Software L E. 2016. US EPA models—air quality models，documentation & guidelines. https://www.weblakes.com/download/us_epa.html [2016-11-28].

十一、城市餐厨垃圾回收逆向物流系统构建的研究①

1. 引言

截至2016年底，全国餐厅总数已经超过600万家，由餐饮行业和居民家庭产生的餐厨垃圾数量大，并且在持续上升中。在“十二五”期间，全国餐厨垃圾的产量约为10万吨/日，全国每年餐厨垃圾总产量约为4000万吨；到2016年，全国餐厨垃圾日产出量已达到25万吨，餐厨垃圾总产生量达9700多万吨。面对数量庞大的餐厨垃圾引起的环境污染和资源浪费，我国采取了许多垃圾回收措施和手段，但是成效比较小，由我国的餐厨垃圾带来的问题依然很严重。因此为提高餐厨垃圾回收逆向物流效率，规范餐厨垃圾回收逆向物流企业的物流活动，迫切需要构建循环经济下的餐厨垃圾回收逆向物流体系，进而实现餐厨垃圾的合理资源化、无害化利用。

2. 餐厨垃圾及其特征

1）餐厨垃圾定义

城市餐厨垃圾，俗称泪水、潲水，是城市固体废弃物垃圾的一种，是在食品加工与流通、餐饮服务及居民的日常生活过程中形成的生活废物，也是城市生活垃圾中的主要组成部分之一。其来源主要分为两种，一种是餐饮垃圾，主要来自家庭、学校、食堂，以及餐饮行业的残羹剩饭；另一种称为厨余垃圾，主要是在餐饮加工制造过程中产生的废料及过期商品等（林敏晖，2013）。

2）特性

显著的危害性与资源性二重性并存是餐厨垃圾区别于城市垃圾的主要特性。

A. 危害性

餐厨垃圾作为污染物的一种，与城市其他类型的垃圾一样，如果处理不当，会对城市的环境造成严重的污染与危害。主要表现在以下几个方面。首先，餐厨垃圾在性状上主要表现为湿淋淋、油漉漉，流动性强，堆放布局会影响城市市容。其次，餐厨垃圾成分复杂，含水率和油脂成分高，餐厨垃圾的集中堆放会造成污水渗滤到地表和地下，从而引起地表水和地下水的污染（肖汛和王杰，2007）。最后，餐厨垃圾容易腐烂发臭，特别是未经过及时处理的餐厨垃圾，是蚊蝇滋生地，会滋生大量细菌和病毒，同时餐厨垃圾释放的有毒有害气体，如硫化物、氢化物、甲烷等会造成大气污染，影响甚至严重危害人类健康。因此，餐厨垃圾具有很大的危害性。

B. 资源性

餐厨垃圾中含有大量营养物质，主要成分是淀粉、脂肪、无机盐、蛋白、纤维素等物质，以及氮、磷、钾、钙、铁等微量元素（王丹阳等，2010；王莉和刘应宗，2009a；邢汝明等，2006），营养丰富并且有毒有害化学物质含量少，回收价值高，资源丰富。据统计，餐饮行业中餐厨垃圾水分含量约占80%，纯干物料中的粗蛋白含量是玉米等农作物蛋白含量的2倍，粗脂肪含量是大豆等农作物脂肪含量的1.5倍。同

① 本文原载《中国管理科学的研究与实践》（第四届中国管理科学论坛论文集），北京：科学出版社，2019，30-39，作者：刘人怀，杨静，段显明，程翠云.

时，餐厨垃圾中的油脂成分经过处理可以制作生物柴油、厌氧发酵后产生清洁能源沼气等，随着人们生活水平的不断提高，餐饮中垃圾的营养成分也逐步增加，因此餐厨垃圾的资源性比较强，是具有较高价值的生物质资源。

3. 城市餐厨垃圾回收物流问题现状

1）缺乏相应的餐厨垃圾回收的政策法规

在法律法规方面，我国餐厨垃圾回收的法律法规不健全，缺乏专业性，由于餐厨垃圾具有一定的资源可利用性，因此我国餐厨垃圾的回收都是在一种自发利益驱动下进行的。近年来，随着人们对餐厨垃圾认识的不断提高及我国法律的不断发展与完善，在餐厨垃圾方面也制定了一系列的餐厨垃圾管理法规等，如北京、上海、广州、深圳、乌鲁木齐等地都制定了餐厨垃圾管理办法。但是这些法律法规分散性大，整合困难，难以形成一整套系统的餐厨垃圾处理及回收法律体系（徐福华和黄利华，2004；佚名，2015；王雪和刘丹，2016）。一些地区法律法规监管不力，使许多不法商贩偷钻法律的空隙与漏洞，餐厨垃圾不合理利用的状况仍然存在，因此我国餐厨垃圾回收工作进展比较缓慢。

2）居民回收意识淡薄，餐厨垃圾回收率低

长期以来，我国的城市餐厨垃圾一直被认为是废物垃圾，其外表不美观，容易腐烂，易滋生细菌，同时又散发着恶臭气味等，其具有的资源性为外表所掩盖，因此我国居民的餐厨垃圾的资源性回收利用意识不强。在我国，城市餐厨垃圾主要有以家庭为单位形成的餐厨垃圾和单位、公司、餐饮行业等产生的公共餐厨垃圾两个主要来源。来自家庭的餐厨垃圾由于其产量小，分布比较分散，大部分和其他城市生活垃圾混装在一起，丢弃在垃圾桶中，又由于其可溶于其他垃圾和具有流动性等特性，和城市垃圾分拣难度大，大部分餐厨垃圾被焚烧或者填埋，利用率低。而对于公司、单位、餐饮行业等的垃圾，由于其产量大，有些直接运往郊区，不经过处理就作为牲畜饲料，给人民生活带来了隐患；有些卖给制作“地沟油”的不法商贩及私营者，获得一些收入；只有少量单位或者企业会缴纳回收处理费将垃圾运往环卫处及回收处进行回收处理。

3）回收处理技术不成熟

在处理技术方面，我国的餐厨垃圾缺乏统一的技术标准，我国餐厨垃圾的主要处理方法包括填埋、焚烧、肥料化、饲料化、生物发酵化等方法（张庆芳等，2012）。

A. 填埋、焚烧技术

填埋技术和焚烧技术是我国传统的餐厨垃圾的主要处理方法，是针对和城市垃圾混合在一起的餐厨垃圾的处理所采用的技术（秦学等，2015）。餐厨垃圾中有许多成分可以通过填埋或焚烧降解与处理，并且这两种技术操作简单、需要的处理费用与其他餐厨垃圾处理技术相比较低、适用于大量的垃圾处理，但是垃圾中的化学成分比较复杂，餐厨垃圾的含水率高，容易和其他垃圾中的化学成分发生作用，长期填埋堆积会造成二次地下水污染、土壤污染及大气污染等，同时占地面积大。而且，焚烧又易造成大气污染，也不是一个完美方法。

B. 肥料化技术

由于餐厨垃圾中的有机物成分含量大，富含大量的微生物和微量元素，营养价值比较高，与传统的堆肥材料相比而言，餐厨垃圾堆肥更具有优势。所谓肥料化技术就是将餐厨垃圾作为一种肥料进行处理的技术，在我国，肥料化主要表现在将餐厨垃圾进行好氧堆肥和蚯蚓堆肥两方面。

好氧堆肥技术操作比较简单，运行成本也比较低，其主要操作原理是利用好氧菌对一部分餐厨垃圾中的易溶解有机物质进行溶解，对另一部分不易溶解的进行高温溶解堆肥，在溶解过程中实现微生物的新陈代谢，以实现餐厨垃圾向肥料的转换（农传江等，2014）。但是餐厨垃圾盐分和油脂含量比较大，如果处理不当，好氧堆肥会造成土壤污染，使土壤盐碱化程度增高，而高蛋白物质的溶解也会造成一定的空气污染。

与好氧堆肥不同，蚯蚓堆肥是利用蚯蚓自身在新陈代谢的过程中，运用自己身体中的酶和蛋白质相互作用的消化系统，对有机质进行分解，将餐厨垃圾中的有机物质转换为肥料中需要的营养物质，进而形成有机肥料。蚯蚓堆肥中的有机肥料的形成主要是靠蚯蚓的新陈代谢，因此可以降低土壤中的重金属含量。但是蚯蚓堆肥也有其缺点：蚯蚓生活的环境比较潮湿，对温度也有一定的要求，所以蚯蚓堆肥容易受到环境的限制。

C. 饲料化技术

由于餐厨垃圾中含有的油脂、蛋白质、有机物质等成分可以替代小麦、玉米、大豆等中的高蛋白物质，将这些成分进行加工就可以形成饲料供牲畜食用，其主要做法是将餐厨垃圾进行粉碎、脱水、杀菌、氧化处理等，进行合成，形成饲料，供牲畜食用。但是我国饲料化技术不成熟，耗费成本较高，许多企业直接将未经过处理的餐厨垃圾喂食牲畜。据统计，我国约 80％的餐厨垃圾未经处理，直接用来喂养牲畜，这种未经处理的餐厨垃圾可能会引起病毒传播进而影响人们的健康（王永会等，2014）。

D. 生物发酵化技术

生物发酵化技术主要是对餐厨垃圾进行微生物发酵处理以实现资源化，主要是通过厌氧发酵技术产生一些再生能源和清洁能源，如沼气。也可以用餐厨垃圾替代大豆、玉米、大米等作物生产乳酸等有机酸，用于工业生产。刘人怀院士团队采用联合生物加工技术，发明“噬污酵母”，对垃圾成分进行提炼制造乙醇（魏自民等，2013；段妮娜等，2013）。这种处理餐厨垃圾的方法成本低，经济和社会效益高，可以真正实现餐厨垃圾的无害化、减量化和资源化，已被积极推广。

虽然餐厨垃圾处理的方法比较多，但是由于受技术、资金等限制，以及各个企业的利益追求不同，还没有形成统一的回收处理技术和方法，回收技术相对落后，处理不彻底，容易引发二次污染。

4）缺乏健全的回收网络

到 2017 年为止，我国许多城市虽然建立了城市垃圾的回收体系，但是由于餐厨垃圾处置设施落后和政府职能部门的监管不当等因素，我国还没有形成一套完整的餐厨垃圾回收体系。据统计，从事餐厨废物回收工作的主要是城市郊区的农民、个体私营者等。由于对餐厨垃圾处理技术的缺乏和设备的落后，并且人们在餐厨垃圾回收方面意识淡薄，许多居民将餐厨垃圾和其他生活垃圾混在一起扔掉，使餐厨垃圾分拣困难，回收难度比较大，而对于收集来的大部分餐厨垃圾还是以一种传统落后的处理方式进行的回收，还有一部分的回收是不法商贩在经济利益的驱使下进行的回收，如由于餐厨垃圾回收方面疏于监管，许多企业为了逃避餐厨垃圾处理的费用将餐厨垃圾贩卖给不法商贩以增加收入。种种现象表明我国的餐厨垃圾回收一直处在一种无序混乱的状态，餐厨垃圾处理不当的现象突出，难以形成一个完整的体系，我国需要构建一个健全的餐厨垃圾回收网络（刘彭，2013）。

4. 餐厨垃圾回收中逆向物流体系的构建探讨

1）构建餐厨垃圾回收逆向物流体系的必要性

随着人们对环境保护和资源利用意识的不断增强，逆向物流逐渐引起人们的关注，越来越多的学者开始围绕逆向物流进行系统的研究。20 世纪 90 年代，欧美学者开始对逆向物流进行深入研究，逆向物流开始与人们的生活日渐紧密。当今时代，随着循环经济、绿色经济及可持续经济的发展，餐厨垃圾回收逆向物流也逐步进入人们的生活。

与传统物品的正向流向模式相反，餐厨垃圾物流方式运动是从餐厨垃圾制造者开始，通过对餐厨垃圾的回收、运输、处理而实现循环的一个过程（Kinobe et al.，2015）。具体地说就是对餐厨垃圾的回收、无公害处理的逆向物流运作模式，它是以“3R”［即减量化（reduce）、再利用（reuse）和再生资源化（recycle）］为原则，来强化和规范餐厨垃圾的源头回收分类、集中运输、分拣物流体系的建设与管理，使餐厨垃圾回收利用制度化、技术化、清洁化、网络化，以此来构建餐厨垃圾的产生—消费—回收—再利用—资源化的逆向物流循环经济体系，进而实现资源投入最小化、环境污染最小化、废物利用最大化（Liu J E and Liu P，2013；Li et al.，2013）。通过建设餐厨垃圾回收的逆向物流体系，可以有效地控制对环境的污染、减少餐厨垃圾的浪费、提高资源的利用效率、促进循环经济的发展，同时还可以避免不法商贩通过非法销售渠道处理餐厨垃圾，在一定程度上保证了食品安全，也逐渐培养了人们的环保意识，能够为整个社会带来良好的经济效益、社会效益和环境效益。因此构建餐厨垃圾回收逆向物流体系是非常必要的。

在提出“餐厨资源化”概念的初期，我国餐厨垃圾产量比较少，人们对餐厨垃圾认识不足，因此传统的餐厨垃圾处理模式主要是直接排放和非法利用，主要表现在居民和餐饮企业等两个方面。由于居民的餐厨垃圾产量比较少，而且分散，早期的餐厨垃圾大部分和城市生活垃圾混合在一起被运往垃圾场进行填埋或者焚烧；餐饮企业等的餐厨垃圾则被个体私营者购买，运往郊区作为养殖户饲料，或者被提炼制成“地沟油”返回市场中，形成一条具有危害性的循环链，影响着人类的身体健康。传统餐厨垃圾回收模式如图 1 所示。

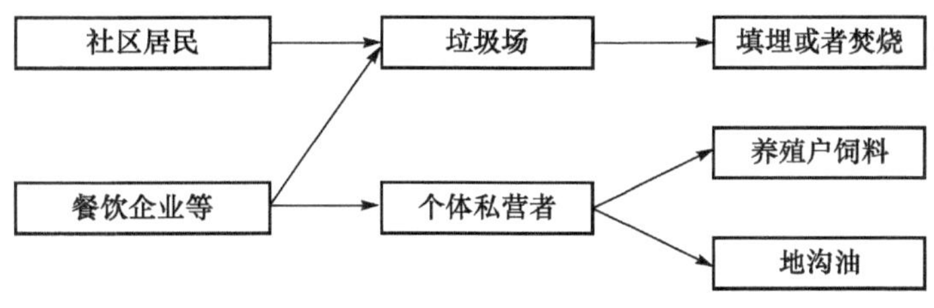

图 1　传统餐厨垃圾回收模式

2）探析构建餐厨垃圾逆向物流体系

餐厨垃圾逆向物流目前还没有引起我国广大消费者的足够重视，因此还有很多家庭存在肆意丢弃餐厨垃圾的现象，并且在一些地区很难建立餐厨垃圾回收渠道，尤其是在一些距离城市比较远的地方，餐厨垃圾分布比较分散，建立餐厨垃圾回收渠道的成本有些高，许多地区和企业可能资金不足或者不愿意投资。因此，建立餐厨垃圾回收逆向物流体系任重而道远。

餐厨垃圾逆向物流模式（图 2）是在政府的统一监督下，从单位、餐饮企业或者消费者出发，经由垃圾回收站、处理中心、无害化处理厂的一种逆向流动，主要通过“餐厨垃圾—回收—运输—处理—加工—销售”这几个方面实现废物利用；通过回收点，回收站，处理中心，无害化处理厂，再生产品生产、制造商，批发商、零售商之间的相互协调来实现餐厨的回收、运输、处理，实现产品的循环。

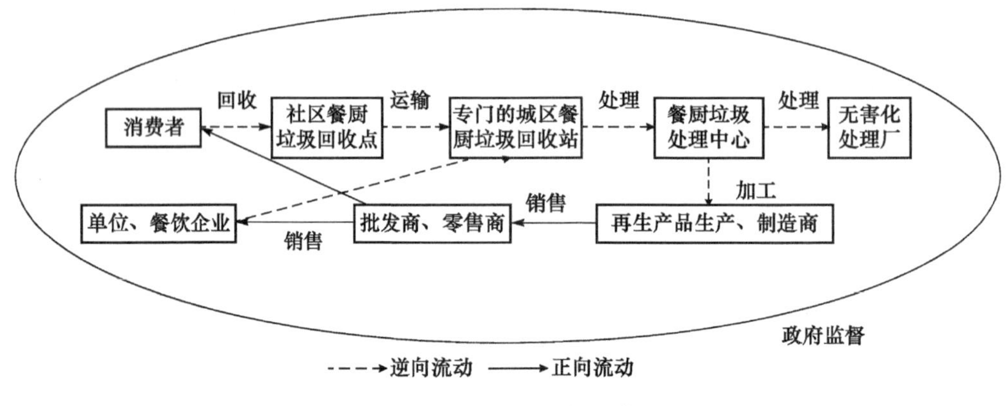

图 2　餐厨垃圾逆向物流模式

A. 回收环节——社区回收点、城区回收站

回收环节主要是运用无偿或者有偿的手段将所需要的东西从消费者或者其他单位那里进行收集集中，简单地说就是将许多零散分散的东西收集到一个区域的过程（米宁，2003）。在餐厨垃圾逆向物流中，对餐厨垃圾进行回收是实现餐厨垃圾逆向物流的第一步，因此我们首先要考虑回收点的设置。家庭、社区等产生的垃圾和餐饮业等部门产生的餐厨垃圾回收存在一些不同，因为家庭、社区产生的餐厨垃圾总体数量虽然比较大，但是比较分散，不易收集，而餐饮行业、学校、机关等公共部门由于餐饮垃圾产量大而且比较集中，因此回收比家庭、社区方便，可以对社区和餐饮行业分别进行回收。

对于回收点的设置，对家庭用户产生的垃圾设立专门的回收点，对于城市居民，可以在居民生活的社区设置一个社区餐厨垃圾回收点。具体来说，可以在居民居住小区内或者附近（包括超市、菜市场等）的每个垃圾桶旁边设置一个专门的餐厨垃圾桶，这样可以有效地避免城市餐厨垃圾和城市生活垃圾混合在一起，再由垃圾收集人员集中处理，运送到社区的专门回收点，这种做法可以将家庭中的零散、小批量餐厨垃圾进行集中汇总，实现餐厨垃圾的集中化存储，更加方便和简化居民餐厨垃圾的收集，减轻餐厨垃圾的分拣工作。城区回收站是社区垃圾回收点的餐厨垃圾和公众部门的餐厨垃圾的集中地区，当社区餐厨垃圾回收点和公众部门餐厨垃圾达到一定量时，可以将餐厨垃圾批量运输到城区回收站，同时回收站在进行回收时可以根据餐厨垃圾资源化的标准对回收垃圾成分进行测试，根据价值的高低将垃圾进行分类，制定科学的回收价格。

B. 运输环节——第三方物流运输

运输环节是餐厨垃圾逆向物流模式的第二步，也是餐厨垃圾回收逆向物流体系的核心环节。这一环节是将社区餐厨垃圾回收点的餐厨垃圾运送到回收站、处理中心、无害化处理厂及企业等，运输的主要目的是实现餐厨垃圾的中转，通过中转，将各个网点的餐厨垃圾运送出去，以保证整个餐厨垃圾逆向物流系统中各个环节的正常运作。餐厨垃圾的运输成本也是整个系统中的主要成本支出之一，因此，如果运输方式和路径搭配得当可以大幅度降低餐厨垃圾的回收成本。

传统的餐厨垃圾运输主要有两种：一种是个体工商户进行上门回收，运至城郊的小型饲养场，用以喂养家畜；另一种是餐饮企业自己运送到垃圾站，或者市政环卫上门运输。但是这种运输有时会因为信息沟通不当、运输信息传达错误等因素，造成运输路线、运载量、运输时间都不能合理搭配，因此，我国传统的餐厨垃圾在运输上比较分散，有时会产生信息不对称问题，缺乏一个系统性的体系，因此运输成本较高、效率较低。

餐厨垃圾在运输方面可以采用第三方运输，建立一个有效的餐厨垃圾物流运输系统。餐厨垃圾的第三方运输就是通过再生产品生产、制造商或与之相似的企业与政府进行合作，共同出资建立运输体系。第三方物流公司可以通过与互联网相结合，运用运筹学原理和模型软件对各个回收点、回收站、处理中心进行定位，同时结合人流、物流、交通状况等实时信息综合分析，合理安排运输班次、运输时间和运输路径，实现效益最大化。这样的运输方式既可以节省餐厨垃圾供应商和供应网点的时间，又可以在最短的时间内实现餐厨垃圾的回收与运送，同时也节约了餐厨垃圾供应网点和回收站的空间，达到共赢的局面（图 3）。

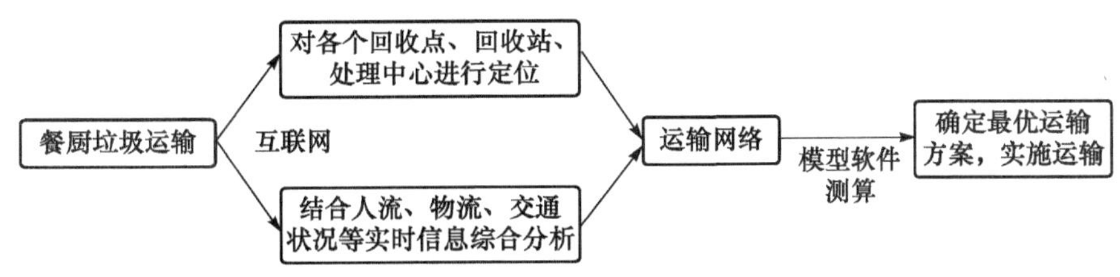

图 3 第三方运输设计思路

C. 处理环节

a）设置专业化餐厨垃圾处理中心

对于专业回收处理中心的设置，要尽量避开人群聚集区，防止处理过程中影响居民生活，同时又要考虑各个回收点的位置，以保证实现运输距离最短化、效益最大化。

在餐厨垃圾处理中心主要运用一些技术，如消毒杀菌、提纯除杂、垃圾净化等，实现餐厨垃圾的资源性转换，转为动物饲料、肥料或者可以利用其制造产生资源、能源等。这种做法可以实现餐厨垃圾的集中统一处理，实现规模效应，在一定程度上降低了处理成本，是餐厨垃圾逆向物流处理系统中的重要一环。

b）无害化处理厂的设置

无害化处理厂的设置主要是对餐厨垃圾处理中心中资源价值比较低或者没有任何价值的餐厨垃圾及残留的废弃物进行处理，一般采用填埋和焚烧两种方式，但是不管是焚烧还是填埋都需要按照环境保护规定的处理方式进行处理，防止由于处理不当产生二次污染。

D. 市场销售环节

再生产品生产、制造商及批发商、零售商是实现将餐厨垃圾资源化后的材料进行加工后运送到市场的主要载体，其主要方法是通过餐厨垃圾处理中心将餐厨垃圾进行资源化转换后，餐厨垃圾处理中心可以根据再生产品生产、制造商的产品要求，直接将需求产品配送给再生产品生产、制造商，与相关原料企业实现合作交易，再通过批发商、零售商的相互作用，实现营销网络的构建，完成餐厨垃圾逆向物流回收网络体系的完整运作。

3）建设餐厨垃圾逆向物流系统的建议

A. 完善法律法规以及政策，明确部门职责

目前我国在餐厨垃圾回收处理中的法律法规不健全，许多都是以章程制度、法律条例出现的，其权威性不高，大部分的法律及政策只具有指导性作用，而缺乏可操作性，约束力不强（王莉和刘应宗，2009b）。为了提高我国餐厨垃圾的回收管理效率，政府还需要制定更加完善的法律法规及政策，以法律手段来规范餐厨垃圾回收中的条件、范围、处置准则等，同时制定餐饮企业或者单位在餐厨垃圾回收过程中的一些准则和标准，明确政府各个部门的职责，防止多头管理的弊端出现，逐步实现分级管理，使得政府各个部门职权清晰，防止不法商贩有可乘之机，同时也要进一步强化地方各部门的管理职能，实现餐厨垃圾回收过程中各个部分相互紧密配合的局面。

B. 加强监管力度，实现信息公开

“地沟油”事件的发生，说明我国在餐厨垃圾回收过程中的监管力度不够，有些地区还是我国监管部分的盲区，餐厨垃圾的回收方面还缺乏一套系统的、严密的监管机制（陈建平和王文华，2011）。因此在监督监管方面可以实行信息公开制度，让群众可以借助一些社交软件和网络平台，如微博、微信、QQ等，实行公共监督，对非法餐厨垃圾交易的行为进行举报，经查询情况属实可以实行一些奖励激励措施，以这种手段来实现政府监管和群众监督的统一，更好地提升监管力度。

C. 提高居民环境意识，进行教育宣传

定期进行环保知识宣传教育，提高居民的环境保护意识，倡导绿色环保的生活方式。消费者的环境保护意识对餐厨垃圾的回收起着重要作用，通过宣传教育可以让人们深刻认识到餐厨垃圾的不当处理产生的危害，对居民的环境宣传教育主要有以下几

个方面：第一，在社区、工作及学习场所定期进行环保知识教育；第二，利用互联网，借助社交工具和网络信息平台，如微信、微博、人人网等，进行环保话题探讨等；第三，政府部门可以定期举办环保优秀人物评选和环保知识比赛等活动，给予一定的奖励，让环境问题紧紧围绕在他们的生活、工作、学习中，潜移默化地让居民自觉形成保护环境的意识，自觉参与环境保护，养成节约资源的好习惯。

D. 加大投入，培养专业人才

加大餐厨垃圾处理的投入，学习外国，如德国、日本等餐厨垃圾处理效率高的国家的先进餐厨垃圾处理经验，努力提高技术含量，研发和采用更加符合环境处理要求和资源化的新技术；同时也要加大对专业人才的培养力度，提高人才素质，为餐厨垃圾的处理提供技术和人才方面的支持。

5. 结束语

建立完善的餐厨垃圾逆向物流体系可以使餐厨垃圾“变废为宝”，给社会带来巨大的经济效益和社会效益，符合当代建立资源节约型、环境友好型社会，以及实现可持续发展理念的要求。逆向物流和餐厨垃圾回收相结合的发展方式能更好地实现垃圾废物的资源化，但是由于我国餐厨垃圾逆向物流体系不太健全，还需要政府、企业、消费者的共同努力，才能更好地实现餐厨垃圾的资源化利用，实现逆向回收体系，推动经济的发展。

参考文献

陈建平，王文华. 2011. 我国餐厨垃圾管理现状、问题及对策研究//中国环境科学学会. 2011中国环境科学学会学术年会论文集（第二卷）. 北京：中国环境科学学会：508-510.

段妮娜，董滨，李江华，等. 2013. 污泥和餐厨垃圾联合干法中温厌氧消化性能研究. 环境科学，34（1）：321-327.

胡新军，张敏，余俊锋，等. 2012. 中国餐厨垃圾处理的现状、问题和对策. 生态学报，32（14）：4575-4584.

林敏晖. 2013. 城市餐厨垃圾逆向物流回收系统设计探析. 物流工程与管理，35（12）：92-93.

刘彭. 2013. 区域性餐厨垃圾逆向物流系统重构与管理研究. 河北工程大学博士学位论文.

米宁. 2003. 基于逆向物流管理的产品回收网络规划研究. 大连海事大学硕士学位论文.

农传江，徐智，汤利，等. 2014. 餐厨垃圾特性及处理技术分析. 环境工程，（S1）：626-629，692.

秦学，李宁，李贵霞，等. 2015. 餐厨垃圾资源化处理技术现状及进展. 煤炭与化工，（7）：35-40.

王丹阳，弓爱君，张振星，等. 2010. 北京市餐厨垃圾的处理现状及发展趋势. 环境卫生工程，（1）：24-26.

王莉，刘应宗. 2009a. 城市餐厨垃圾分级回收处理模式探索. 西北农林科技大学学报（社会科学版），（3）：110-114.

王莉，刘应宗. 2009b. 基于逆向物流的餐厨垃圾回收体系构建研究. 西安电子科技大学学报（社会科学版），（2）：62-67.

王雪，刘丹. 2016. 基于循环经济的乌鲁木齐市餐厨垃圾逆向物流系统探析. 牡丹江大学学报，（3）：36-38.

王永会，赵明星，阮文权. 2014. 餐厨垃圾与剩余污泥混合消化产沼气协同效应. 环境工程学报，（6）：2536-2542.

魏自民，夏天明，李鸣晓，等．2013．不同湿热预处理条件对餐厨垃圾厌氧发酵产氢的影响．环境科学研究，26（11）：1239-1245．
肖汛，王杰．2007．循环经济模式在餐饮业中的实现及其逆向物流管理探讨．大连轻工业学院学报，（3）：284-288．
邢汝明，吴文伟，王建民，等．2006．北京市餐厨垃圾管理对策探讨．环境卫生工程，（6）：58-61．
徐福华，黄利华．2004．上海餐厨垃圾的资源化利用．中国环保产业，（4）：42-43．
佚名．2015．南京餐厨废弃物管理办法将实施．中国资源综合利用，33（5）：60．
张庆芳，杨林海，周丹丹，等．2012．餐厨垃圾废弃物处理技术概述．中国沼气，30（1）：22-26．
Kinobe J R，Gebresenbet G，Niwagaba C B，et al. 2015. Reverse logistics system and recycling potential at a landfill：a case study from Kampala City. Waste Management，42：82-92.
Li S S，Wu H P，Wu A L. 2013. Location-routing problem in food waste reverse logistics system. Journal of Shijiazhuang Institute of Railway Technology，1：89-94.
Liu J E，Liu P. 2013. Study on design and management of regional food waste reverse logistics systems. Logistics Technology，（8）：19-21.

十二、让人民过上好日子①

今天，万分高兴参加成都香城投资集团有限公司（以下简称香投集团）与广东利世康低碳科技有限公司合作建设的利世康低碳产业园的奠基典礼。

今天是一个新的开始，我的心情非常激动。因为我是成都人，今天能为我的故乡带来革命性、颠覆性的创新技术，能长远改善家乡人民的环境条件，我感到非常高兴。作为一个从成都走出去的科学家，我的一生，都在为科技创新奋斗。十年前，我的团队开始关注垃圾分类、垃圾处理这个人人头疼、天天遇到的大难题。如何让垃圾不再危害环境，能够变废为宝，如何让老百姓过上青山绿水的好日子，成了我们魂牵梦绕的大课题。十年来，我们一直围绕着餐厨垃圾资源化、无害化、减量化的目标，研发相关处理技术，进而延伸至“有机资源与新能源开发”领域。

终于，我们成功了。利世康国际领先的核心技术在于将生物工程领域的发酵工程、基因工程、蛋白质工程与工程管理和机械工程的顶尖式结合，克服了传统有机垃圾在产业上“效益难以覆盖减量成本”“二次污染”等问题，同时，在生物质能源领域实现了重大突破。我们设计并完成了一套资源化充分、无害化彻底、减量化明显的餐厨垃圾处理新工艺。

本项新技术可以成功将餐厨垃圾以及农副市场、食品企业和肉类海鲜市场余料转化为燃料乙醇、工业毛油、酒糟蛋白饲料、液体肥料以及干冰等多种高附加值产品，使得餐厨垃圾变废为宝，实现良好的经济效益；同时餐厨垃圾的无害化和减量化效果十分突出，将为妥善解决“地沟油”“二噁英”“渗滤液”“垃圾猪”等危害人民健康的环境污染问题提供解决出路，实现重要的社会效益。今天产业园的奠基，正是政产学研金的完美结合，希望在大家的共同努力下，利世康的技术从这里腾飞，从这里升华，从这里走向全国、走向全世界，服务全人类。

① 本文是采用“联合生物加工技术”治理餐厨垃圾示范性工程动工典礼开幕词，成都，2019年2月17日．

“生态文明”和“美丽中国”的概念已成为国家的发展方向，实现中华民族伟大复兴，比任何时候都更需要“青山绿水”。今天，我们大家在这里见证利世康低碳产业园的奠基，也一同见证了成都、新都美好未来的加速实现。最后，我衷心感谢新都区委、区政府的领导，感谢香投集团的同志们，对利世康事业的支持和付出，你们辛苦了！我们在一起为之奋斗的事业，必将功垂千秋，流芳万代。

我祝福我们伟大祖国、美丽四川、天府成都、创新新都，繁荣昌盛，人民健康幸福！

谢谢大家！

十三、关于推进餐厨垃圾治理装备高技术制造业发展的建议[①]

制造业是国民经济的支柱产业，是经济增长的主导部分。而且，由于新一轮科技革命的大力推动，制造业正在发生颠覆性变化。但在我们广东省，工业大体仍处于 2.0 向 3.0 过渡阶段，中低端产业仍占较大比例，高技术制造业增加值占规模以上工业的比重仅为 28.8%，远低于美国、德国、日本等发达国家 50%以上的水平。因此，高技术制造业的提升是将我省建设成为制造强省的当务之急。

2009 年，我在担任省科普志愿者协会会长之时，恰逢贯彻落实《国务院批转住房城乡建设部等部门关于进一步加强城市生活垃圾处理工作意见的通知》(国发〔2011〕9 号）及《广州市城市生活垃圾分类管理暂行规定》(广州市人民政府令，第 53 号）精神的时期，我带领协会会员，推进“垃圾分类，从我做起”科普宣传活动进校园，进社区。由此，我才发现生活垃圾处理中的问题。特别是，在生活垃圾中，最麻烦的是占其中一半的餐厨垃圾的处理问题。现有的焚烧、填埋、饲料化处理、好氧堆肥、厌氧发酵等处理技术既不能解决危害人类健康的污染问题，经济效益亦很差。这是我们五千多年来都未解决的问题，也是国际上未解决的问题，是世界难题。于是，我组建一个综合型跨界专项研发团队，就餐厨垃圾治理技术创新进行摸索，终于在 2013 年利用联合生物加工技术筛选出一种新型酵母，即噬污酵母，它能在 36 个小时内，将复杂的餐厨垃圾转化成乙醇、工业油脂、饲用蛋白粉等。我们拥有中国、美国、欧盟专利近 70 项，还在国际著名学术期刊上发表了论文。这是一种绿色生物处理原始创新技术，使得餐厨垃圾处理过程达到零排放、零污染，杜绝二次环境污染。同时变废为宝，再创价值，每吨餐厨垃圾处理成本仅 200 多元，但可产生 600 元以上收益、30 公斤乙醇燃料、50 公斤工业毛油、30 公斤高蛋白饲料添加剂、50 公斤 RDF（refuse derived fuel，垃圾衍生燃料)。

在广州市陈建华市长的支持下，我们创立了广东利世康低碳科技有限公司，于 2013 年作为院士项目进入广州经济技术开发区进行中间试验，并于 2016 年顺利完成。接着进行产业化推广，经历无数困难，终于今年 7 月在四川省成都市新都区，由广东利世康低碳科技有限公司和新都区香投集团兴城公司合作成立的四川利兴龙环保科技有限公司将我们的技术建成世界上第一个示范工厂，日处理餐厨垃圾 50 吨。

① 本文是提交给广东省人民政府的建议，广州，2020 年 12 月 29 日.

全国每一座城市都应建立这种垃圾处理厂，而且还可向全世界推广。这样一来，既完成了对餐厨垃圾的治理，又增加了一门新的高技术制造业。厂内的工艺设备设计有粉碎机、发酵容器、蒸馏装置等机械设备。例如，一座每日处理 400 吨餐厨垃圾工厂的设备价值达 1 亿元，如果广州每日 1 万 5 千吨餐厨垃圾都能得到处理，则需设备费 37.5 亿元。同时，还需要建造专用的餐厨垃圾运输车。所以，全国，全世界都需要这一行业装备高技术制造业，这是创业的大好机会，有助于我省高技术制造业的发展。建议我省迅速规划组织，做全国和全世界这一领域的领头制造强省。

大型工程项目管理的中国特色及与美苏的比较①

笔者正在开展以现代管理科学的中国学派[1-3]为主题的两个管理科学研究项目，一个是国家自然科学基金项目，一个是广东省科学技术厅项目。两个项目互相结合，一方面进行理论探讨，旨在提出现代管理科学的中国学派的框架体与研究规划，另一方面选择若干领域开展中国学派的试验性研究，大型工程项目管理（以下简称工程管理）就是其中的首选。

笔者虽然提出了工程管理的中国学派，但需要循序渐进，即先研究工程管理的中国模式。笔者认为：中国模式是中国学派的实践形式，中国学派是中国模式的学术形式或理论形式。实践在前，理论随后。理论不会自动生成，不会从天上掉下来，需要有心人、有志者去研究和创建。而研究中国模式，又需要先研究许多典型案例，同时还要与其他国家，如苏联和美国进行比较研究，从众多案例中加以归纳、总结、提炼、提高，以得到中国模式，再进一步上升到中国学派。

研究工程管理的中国模式和中国学派，其基本途径是：洋为中用，古为今用，近为今用，综合集成。近为今用是重点，即重点研究我国近期的大型工程项目及其管理的典型案例。所谓典型案例，既包括成功案例，也包括失败案例，研究前者主要是总结经验，研究后者主要是汲取教训。

一、洋为中用的典型案例

1. 苏联的案例及评述

（1）世界上第一颗人造地球卫星。1957 年 10 月 4 日，苏联成功发射了世界上第一颗人造地球卫星（Спутник-1），表明人类跨出了地球，开始了航天时代。第一颗人造地球卫星呈球形，直径 58cm，重 83.6kg。它沿着椭圆轨道飞行，每 96min 环绕地球一圈。1 个月后，1957 年 11 月 3 日，苏联又发射了第二颗人造地球卫星，它的重量增加了 5 倍多，达到 508kg。这颗卫星呈锥形，卫星上有 1 个密封生物舱，舱内有 1 只进行实验的名叫“莱卡依”的小狗。

（2）世界上第一艘宇宙飞船。1961 年 4 月 12 日，苏联成功发射了世界上第一艘绕地球轨道飞行的载人飞船“东方 1 号”（Восток-1），宇航员尤里·加加林少校成为飞出地球大气层进入外层空间的第一人，绕地球飞行 1 周。

同年 8 月 6 日至 7 日，苏联宇航员格尔曼·季托夫在跨越两天的 25h 飞行中，环绕地球 17 圈，飞行中他首次启动了宇宙飞船的人工控制系统。

（3）对苏联案例的评述。“国际悲歌歌一曲，狂飙为我从天落。”苏联在人类航天事业上建立了开创性的不可磨灭的功勋，尽管苏联已经于 1991 年解体，但是，它作为开创者的业绩永远令人缅怀。但没有安定团结的社会局面，就会影响科学研究。

---

① 本文原载《科技进步与对策》，2009，26（21）：5-12，作者：刘人怀，孙凯，孙东川.

苏联解体前后，曾经发生这样的故事：航天员上天的时候还是苏联人，返回时苏联已经不存在了，“返回不了苏联”。事实上，苏联解体之后，俄罗斯的航天事业受到了很大的挫折，它的经费不足，一些发射场地不在俄罗斯境内，苏联的加盟国都成为独立国家，相互之间的关系很复杂，经常产生一些纠纷。有人认为：苏联是被美国的星球大战计划拖垮的。是否如此，可以存疑。但有一点是可以肯定的：美苏两霸的军备竞赛和航天领域的激烈竞争，消耗了大量的人力财力物力，如果说美苏两霸的人口和人力（包括科技人才）相差无几，那么苏联的财力比美国就逊色多了。苏联还有东欧华约诸国的沉重负担，所以在经济上两霸是不对等的。还有人认为，美国并没有真正启动星球大战计划，只是虚张声势引诱苏联“中招”，把苏联拖垮。我们可以从中得到启示：不要争霸，要量力而行，有所不为而有所为，才能使自己立于不败之地。

2. 美国的案例及评述

(1) 曼哈顿工程。第二次世界大战后期，在著名科学家爱因斯坦的推动下，美国总统罗斯福决定研制原子弹，取名为曼哈顿工程。为了先于纳粹德国制造出原子弹，曼哈顿工程集中了当时西方国家最优秀的核科学家，动员了 10 万多人参加这一工程，其中技术人员 1.5 万人，历时 3 年，耗资 20 亿美元，于 1945 年 7 月 16 日成功地进行了世界上第一次核爆炸，接着造出两颗实用的原子弹投放于日本的广岛与长崎，促使日本宣布投降。

(2) 北极星计划。1957～1960 年，美国海军开展研制导弹核潜艇系统的工程项目，称为北极星计划。它由 8 家总承包公司、250 家分包公司、3000 家公司承担，其协调事务非常繁杂。为了加快进度，在顾问公司的协助下，发明了一种控制工程进度的先进的管理方法——计划协调技术（program evaluation and review technique，PERT），使得整个计划提前两年完成。

PERT 在我国又称为网络计划技术和统筹法。著名数学家华罗庚生前带领“双法”小分队在全国积极推广优选法、统筹法，取得很大成效。

(3) 阿波罗计划（project Apollo)。1961 年 5 月 25 日，美国总统肯尼迪提出要在 10 年内将美国人送上月球。他声称：苏联人把我们击败在地球上，我们要把苏联人击败在月球上。于是，美国国家航空航天局制订了阿波罗计划。

从 1961 年 5 月至 1972 年 12 月，阿波罗计划历时 11 年半，耗资 300 多亿美元，参加研制的美国与外国企业达 2 万多家，大学与研究机构 120 所，使用大型计算机 600 多台，参加研制工作的人员达 400 多万人，其中高级技术人员 42 万人。

1969 年 7 月 20 日（美国时间 20 日，北京时间 21 日)，乘坐阿波罗 11 号，人类第一次登上了月球。宇航员尼尔·阿姆斯特朗和巴兹·奥尔德林踏上了月面，第三名宇航员迈克尔·科林斯留在指挥舱内，继续沿着环月轨道飞行。两名登月宇航员在月面上展开了太阳电池阵，安设了月震仪和激光反射器，采集月球岩石和土壤样品 22kg。

阿姆斯特朗说：“（踏上月面）对一个人来说，这是小小的一步，但对人类来说，这是一个巨大的飞跃。”表 1 是 1969 年 7 月阿波罗 11 号在登月的主要环节上，实际时间与计划时间的比较。

表1 阿波罗11号宇宙飞船登月的主要环节上，实际时间与计划时间的比较

登月主要环节	计划			实际			相差	
	日	时	分	日	时	分	小时	分钟
飞船发射	16	20	32	16	20	32		0
进入飞向月球轨道		23	16		23	16		0
进入绕月球的椭圆轨道	20	0	26	20	0	22		4
登月舱进入接近月面轨道	21	2	10	21	2	08		2
登月舱在月面登陆		3	19		3	17		2
宇航员走出登月舱踏上月面		13	19		9	56	3	23
宇航员回到登月舱		15	42		12	11	3	31
登月舱离开月面开始上升	22	0	55	22	0	55		0
宇航员进入返回地球的轨道		11	56		11	55		1
在太平洋中部溅落	24	23	51	24	23	50		1

注：表1时间是北京时间。另据资料，"宇航员走出登月舱踏上月面"是美国东部夏令时1969年7月20日22时56分20秒，据此可以进行时间换算

后来发射的阿波罗飞船又多次载人登月，共计有18名宇航员登上月面，进行科学考察与实验。

阿波罗计划也采用了PERT，并且发展成为图示评审技术（graphical evaluation and review technique，GERT），GERT可以处理多种复杂的随机因素。

阿波罗计划是举世公认的工程运用系统工程方法的范例。阿波罗计划的成功，当然是很高的科学技术成就，但是它的总指挥小韦伯说："阿波罗计划中没有一项新发明的自然科学理论和技术，而是现成技术的应用，关键在于综合。"

"关键在于综合"，综合就是创造，这是系统工程的基本观点之一。每一个大型工程项目，都是多种技术的综合集成，技术与管理的综合集成，技术、经济、环境、政治等多种因素的综合集成。

（4）对美国案例的评述。曼哈顿工程和阿波罗计划都是总统亲自决策的，是计划机制而不是市场机制在发挥主要作用。如果不是这样，工程是推不动的。

竞争可以促进科技进步。美国的阿波罗计划在于抢先登上月球，"把苏联人击败在月球上"，美国在实现了竞争领先之后就改弦易辙，转向研制航天飞机和探测火星。今天，美国人提出"要重返月球"，为什么？因为中国启动了登月工程，日本、印度等国也争先恐后，美国有可能被超越，于是美国人着急了，要重返月球。

几个美国案例，都是系统工程案例，案例的成功是系统工程的胜利。我国大力宣传、推广和研究系统工程是很有必要的。事实上，系统工程对我国的"两弹一星"研制也发挥了巨大作用。

二、我国古为今用的典型案例[4]

（1）都江堰。都江堰位于四川省成都平原西部的岷江上，始建于公元前256年，是秦太守李冰及其儿子李二郎率领当地老百姓修筑起来的。它由鱼嘴岷江分水工程、飞沙堰分洪排沙工程、宝瓶口引水工程三大主体工程和百丈堤、人字堤等附属工程构

成，加上一系列灌溉渠道网，形成一个完整的水利系统。它科学地解决了江水自动分流、自动排沙、控制进水流量等问题，消除了水患，使川西平原成为“水旱从人”的“天府之国”。李冰父子还总结了“深淘滩，低作堰”六字口诀，指导人们进行养护维修工作。它是全世界迄今为止年代最久、唯一留存、以无坝引水为特征的宏大水利工程。两千多年以来，都江堰一直发挥着防洪灌溉作用，造福于四川人民。

都江堰不仅是举世闻名的中国古代水利工程，也是著名的风景名胜区。1982 年，都江堰被国务院批准列入第一批国家级风景名胜区。2007 年 5 月 8 日，青城山—都江堰旅游景区经国家旅游局正式批准为国家 5A 级旅游景区。2000 年，在联合国世界遗产委员会第 24 届大会上，都江堰被确定为世界文化遗产。都江堰主体工程在 2008 年汶川大地震中安然无恙，只是附属建筑物，如二王庙有些损伤，不难修复。

(2) 丁谓工程。丁谓工程不是很出名，但是很典型。宋真宗时期（公元 998～1022 年），有一次宫殿被毁于火。大臣丁谓受命限期修复皇宫。这项工程怎样进行？丁谓提出：首先把皇宫旧址前面的一条大街挖成沟渠，用挖沟的土烧制砖瓦；其次把附近的汴水引入沟内形成航道，从外地运输砂石木料；最后，皇宫修复好了，把沟里的水排掉，用建筑垃圾填入沟中，恢复原来的大街。这是一个杰出的工程方案，它把皇宫修复全过程划分成几个阶段，统筹兼顾，综合考虑。

中国古代的城市建设和宫殿、园林、庙宇建筑等都是很出色的，还可以找到许多案例，这里不再赘述。

(3) 对于中国古代案例的评述。都江堰作为水利工程，至今仍然发挥着当初设计建造时被赋予的作用，这在全世界是绝无仅有的。相类似的水利工程还有：郑国渠（今天陕西境内）、灵渠（今天广西境内）。京杭大运河也是杰出的水利工程，今天仍然发挥着运输、灌溉、排洪等作用，造福于运河两岸的人民群众。中国还有大禹治水的传说，那是更加遥远的年代里的大型水利工程，它强调了在大自然的灾难面前不悲观、不退缩、不消极等待，事在人为，顺应客观规律改造山河，劳动创造世界。

“不到长城非好汉”。万里长城，工程浩大，两千年中多次修建。在历史上，它是汉族防御北方民族进攻的最重要的军事设施。今天它虽然不再具有军事功能，但却充分发挥着旅游功能。

今天，古迹和文物保护工作得到了空前的重视，但也有一些似是而非的现象，有些提议不见得可取。例如，北京圆明园遗址的断垣残壁见证了英法联军的罪行，记录了中国人民的灾难和屈辱，留着它是个不可多得的教材，可以进行爱国主义教育，比在教室里上政治课要生动有力得多。但是，有人提出重建圆明园，谋取商业利益，笔者认为不可取。重建就把爱国主义教育基地毁掉了。即便重建，要不要“原汁原味”？不搞“原汁原味”没有意思，即便“原汁原味”也有难处，如大水法，在当初是世界顶级喷泉，但是，它比今天很多地方的音乐喷泉已经逊色多了。

有学者提出“管理没有新问题”，其理由是中国作为几千年的文明古国，历朝历代从朝廷到基层，大大小小的管理问题都经历和处理过了，有过多次太平盛世，所以，到今天，管理已经没有新问题了。这种说法不完全正确，但也有一定的道理。我们应该以史为鉴，古为今用，但是这还不够。事实上，管理既有老问题，也有新问题，新问题要研究和解决，老问题也要在新形势下用新观点、新技术、新办法加以研究和解决。更全面的提法是：洋为中用、古为今用、近为今用、综合集成，创建现代管理科学的中国学派，把中国的管理搞得更好。

三、近为今用的典型案例[5]

(1)“两弹一星”(原子弹，氢弹，人造卫星)。早在1955年冬天，时任中国人民解放军军事工程学院院长的陈赓大将移樽就教，会见刚刚回国不久的著名科学家钱学森，问他：“中国人能不能搞导弹?”钱学森说：“为什么不能搞?外国人能搞，我们中国人就不能搞?难道中国人比外国人矮一截?”陈赓大将说：“好!”在周总理、聂荣臻元帅的支持下，组建了国防部第五研究院等研究机构。

1964年10月16日，我国第一颗原子弹爆炸成功；1967年6月17日，我国第一颗氢弹爆炸成功；1970年4月24日，我国第一颗人造地球卫星发射升空，全球响彻悦耳的《东方红》乐曲。卫星直径约1m，重173kg，沿着近地点439km、远地点2384km的椭圆轨道绕地球运行，运行周期114min。

从原子弹到氢弹，我们仅用了两年零八个月的时间，比美国、苏联、法国所用的时间要短得多。在导弹和卫星的研制中所采用的新技术、新材料、新工艺、新方案，在许多方面跨越了传统的技术阶段。“两弹一星”是中国人民艰苦奋斗和创造活力的产物。

值得一提的是日本的竞争。日本于1970年2月11日抢先发射了第一颗人造地球卫星“大隅”号。卫星重约9.4kg，近地点339km，远地点5138km，运行周期144.2min。

(2)神舟飞船。1999年11月20日至21日，中国第一艘“神舟”号无人试验飞船飞行试验获得了圆满成功。2001年年初至2002年年底又相继研制并成功发射了神舟二号、神舟三号和神舟四号无人试验飞船，获得了宝贵的试验数据，为实施载人航天打下了坚实的基础。

2003年10月15日上午9：00，神舟五号飞船从我国酒泉卫星发射中心成功发射升空，飞船载有中国首飞航天员杨利伟，在轨运行1天，环绕地球14圈后在预定地区成功着陆。

2005年10月12日上午9：00，神舟六号飞船从我国酒泉卫星发射中心成功发射升空，费俊龙和聂海胜两名中国航天员进入太空，飞行5天后成功返回地面。

2008年9月25日21点10分04秒，搭载3名航天员的神舟七号从我国酒泉卫星发射中心载人航天发射场发射升空，飞行2天20小时27分，于2008年9月28日17点37分成功着陆于中国内蒙古四子王旗主着陆场，实现了中国历史上宇航员第一次的太空漫步，中国成为第三个有能力把航天员送上太空并进行太空行走的国家。

此外，还有嫦娥一号与探月工程、三峡工程、青藏铁路等典型案例，也有1958年的“大炼钢铁”等失败的典型案例。

四、中国工程管理的特色

美国的工程管理具有美国特色，苏联的工程管理也很有特色，中国的工程管理则具有显著的中国特色。中国工程院院士、国家自然科学基金委员会管理科学部主任郭重庆教授说："中美两个核弹之父，一个邓稼先，一个奥本海默，两种管理思维，两个都成功了。也有人说，若两人互换位置，可能都不成功。"[6]据此，笔者尝试总结中国工程管理的特色如下。

1. 决策机制与宏观管理方面

(1) 高瞻远瞩，战略决定成败。中国为什么能在一穷二白的条件下搞出"两弹一星"？首先是因为当时有毛泽东主席和中央领导集体的坚强决心与英明决策，同时，有一批具有使命感的顶尖科学家的追求，还有全体参加人员的忘我奉献，全国上下齐心协力。

我国要实现到 21 世纪中叶的"新三步走"战略，还需奋斗 40 年。现在世界上既有人叫喊"中国威胁论"，也有人在吹捧中国，"鼓励"中国做这做那，指望中国做"冤大头"，承担过多的责任，到处花钱，到处树敌。对于这些，中国要保持清醒的头脑，永远不做超级大国，不搞霸权主义。我们仍然需要韬光养晦，埋头苦干，实实在在做工作，一步一个脚印前进，才能比较顺利地实现强国梦。

(2) 集中力量办大事是显著的中国特色。在 20 世纪 60～70 年代，如果不是集中力量办大事，"两弹一星"是搞不出来的。今天，中国比以前富裕了许多、强大了许多，能够做的事情也比较多了。但是，仍然需要集中力量办大事，要有所不为才能有所为。所以，选择重点、确保重点是非常重要的。

美国一直号称是"世界上最富有的国家"，就阿波罗计划而论，在 40 多年前也只有美国才有财力搞得起。但是，美国的富有也是有限的，不可能想花多少钱就花多少钱。美国既然已经登上了月球，"把苏联人击败在月球上"了，已经耀武扬威了，与其继续在月球上逞能，不如换一个角度、换一个场所更加荣耀。于是，美国转向探测火星，转向研制航天飞机，前者可以使它继续大幅度领先世界各国，后者可能带来商业价值。所以，美国改变战略是有道理的，是精打细算的。即便"最富有"它也量入为出，讲究资源优化配置、最佳配置，这是管理工作的一项基本原则。

(3) 高起点、跨越式发展。在"两弹一星"项目中，在神舟飞船项目中，在嫦娥一号项目中，我国都是高起点、跨越式发展的，这是很值得赞扬的。跟在人家后面爬行"克隆"一个东西来"填补空白"，没有什么意思。只有高起点、跨越式发展，才能较快地赶上世界先进水平，争取领先地位。

(4) 坚持科学发展观，按照科学规律办事。1958 年"大炼钢铁"的闹剧，再也不能重复上演了。它违反客观规律——炼钢的规律、经济发展的规律，违背科学的决策机制。

科学技术不是政治，不能以政治口号代替科学技术。"人有多大胆，地有多高产"之类的口号，纯属主观唯心主义。工程项目不能搞成"政治任务"。苏联的航天事业多次发生可怕的灾难，近百人丧生，大多是不按照科学规律办事。例如，1960 年 10 月 24 日，为了执行长官意志，苏联宇航局高层官员、太空科学家们对"东方 1 号"飞船搞突击调试，飞船火箭突然发生爆炸，当场炸死 54 人，大多是精英人物。

在香港回归之前，曾经有一些人上书中央领导，提议把探月工程的火箭作为"礼炮”提前发射。中央领导认为，当时准备工作尚未做好，发射毫无科学意义，因此，予以否定。这是按照科学态度办事，不把科学事业混同于“政治任务”。

(5) 民主集中制是行之有效的决策机制。办大事必须以政府行为和计划机制为主导。不但中国如此，美国也是如此。曼哈顿工程、北极星计划、阿波罗计划，都是政府行为和计划机制主导的。美国应对金融危机的一系列措施也是政府主导的，以至于有人说“奥巴马搞社会主义”。

“民主是个好东西”[7]，但是民主也有其不足之处，如拖延时间、错失良机；消耗许多社会资源，特别是容易形成不同的利益集团，互相否定、制造对立，使得好事办不成。民主集中制的真正好处是它行之有效的决策机制，它是我们的传家宝。首先必须发扬民主，然后在民主的基础上集中。不发扬民主不行，光是发扬民主不集中也不行。

(6) 领导者尤其是决策者要有哲学思维。领导人和管理者，尤其是决策者，要有较高的文化素养，要有较好的哲学思维能力。

钱学森院士指出：“应用马克思主义哲学指导我们的工作，这在我国是得天独厚的……马克思主义哲学确实是一件宝贝，是一件锐利的武器。我们搞科学研究时（当然包括搞交叉科学研究），如若丢掉这件宝贝不用，实在是太傻了。”[8]他在给朋友的信中说：“我近30年来一直在学习马克思主义哲学，并总是试图用马克思主义哲学指导我的工作。马克思主义哲学是智慧的源泉!”[8]

《矛盾论》不仅是一般哲学（general philosophy）的经典著作，而且是重要的管理哲学，值得从事大型工程管理的管理人员和一切人员认真学、积极用。

近几年有两本书很畅销，一是《战略决定成败》，二是《细节决定成败》，后者比前者更畅销。笔者认为，战略是最重要的，战略错了，大局失败了，细节再好又有多少用处呢？但是，在战略是正确的、好的情况下，要把它落到实处，还必须有一整套规划、计划、政策、措施来加以保障，计划和措施之中就要解决细节问题，有可能“一着不慎，满盘皆输”。一颗螺丝钉的松动、脱落，可能引起发动机故障，那就影响大局了，可能决定成败了。

这两本书的矛盾，只有用《矛盾论》来分析和调解。《矛盾论》说：当有诸多矛盾的时候，其中必有一种矛盾是主要矛盾，它的存在和发展，影响其他所有矛盾的存在和发展，所以必须用全力抓住主要矛盾，解决主要矛盾，如同牵牛要牵牛鼻子一样；主要矛盾解决了，其他矛盾就迎刃而解了；当主要矛盾解决之后，原来的某种次要矛盾可能上升为主要矛盾，这时还是要抓主要矛盾。一颗螺丝钉之所以影响大局，是因为它成了主要矛盾，否则，在这颗螺丝钉影响大局之前，大局已经被其他矛盾影响了，“轮不着”这颗小小的螺丝钉来表演。

(7) 全国一盘棋，实现可持续发展。中国科学院院士、中国探月工程首席科学家欧阳自远2006年7月在第36届世界空间科学大会上说，包括建造、发射、运行，嫦娥一号的总费用是14亿元人民币，只相当于在北京修建约2公里地铁所需的费用；探月工程第二、第三阶段可能会多花一些钱，但是绝不会拖中国人民全面建设小康社会的后腿，只能是促进和推动。

全国一盘棋，各行各业各地区的发展要统筹兼顾。不搞重复建设，不搞政绩工程、形象工程、面子工程。要确保工程质量，杜绝豆腐渣工程。

有些大型项目，可以一步到位；有些大型项目，则要分步实施。具体问题具体分析，不搞一刀切。不要好大喜功、贪大求洋，动不动就要成为“世界第一”。有些事情，当个世界第二、第三也可以。例如，万里长江就是世界第三大河。我国 960 万 km^2 的陆地面积，也是世界第三。有些“世界第一”，当了不如不当。例如，中国要不要建造“世界第一高楼”？恐怕不一定。至于近几年一些地方刻意“创造”的吉尼斯纪录，如万人弹古筝、万人吃火锅、万人齐洗脚、万人齐唱歌，以及 2008m 长的婚纱、22.5km 长的鞭炮……有什么意义？大国要有大国风范，不要小家子气、搞哗众取宠的事情。

(8) 保持安定团结的社会局面。工程问题不光是技术问题，工程系统存在于社会经济系统之中，工程管理是社会经济系统管理的组成部分。

前面说过，苏联解体对俄罗斯的航天事业造成了很大的挫折。

我国在“文化大革命”中的教训十分惨重。除了“两弹一星”以外，几乎所有的科研项目、科研活动都停止了。“文化大革命”拉大了我国科学技术与世界先进水平本来已经在不断缩小的差距。

三峡工程大移民百万人，如果没有全国许多地方热情接受、妥善安置，三峡工程就难以开展了。现在常常看到一些建设项目遇到“钉子户”，搞得很尴尬。无理取闹的“钉子户”应该绳之以法，但是确有一些问题是因为建设项目方面（开发商）做得不合理、不周到而造成的，弱势群体利益受损，于是成为“钉子户”。简单粗暴地对待“钉子户”，会酿成社会问题，影响安定团结的社会局面。

2. 运作机制与管理方面

(1) 把大型工程项目作为系统工程来搞。《组织管理的技术：系统工程》，这是 1978 年 9 月 27 日钱学森、许国志、王寿云三位学者联名发表在上海《文汇报》上的一篇重要文章的题目，被誉为系统工程在中国第一声嘹亮的进军号。这是 30 年来系统工程在中国发展的基本定位和主要流派。系统工程在中国得到了很大的发展，可谓是家喻户晓、人人皆知，这是因为系统工程在中国得到了两个方面人士的大力支持和推动：一是以钱学森院士、关肇直院士、张钟俊院士、许国志院士等为代表的学术界的德高望重人士；二是改革开放以来的历任中央领导人以及各部门各地区领导人。

系统工程有一系列理论与方法，重视定性研究与定量研究相结合，实现从定性到定量的综合集成。西方有霍尔系统工程方法论、切克兰德软系统方法论，中国有钱学森院士等提出的综合集成方法论、顾基发研究员等提出的“物理-事理-人理系统方法论”。

系统工程强调系统观点，如宏观着眼，微观着手；统筹兼顾，综合集成。“战略上藐视敌人，战术上重视敌人”，这句名言在工程管理上也是非常适用的。战略着眼系统的宏观、大局和长远发展；战术在战略的指导下，研究解决各种微观问题。

(2) 建立总体设计部是行之有效的工作方式。建立总体设计部的工作方式是在研制“两弹一星”中建立和完善起来的。钱学森院士指出：“我们把处理开放的复杂巨系统的方法定名为从定性到定量综合集成法，把应用这个方法的集体称为总体设计部。

总体设计部由熟悉所研究系统的各个方面专家组成，并由知识面比较宽广的专家负责领导，应用综合集成法（或综合集成研讨厅体系）对系统进行总体研究。总体设计部设计的是系统的总体方案和实现途径。总体设计部把系统作为它所属的更大系统的组成部分来进行研究，对它们的所有要求都首先从实现这个更大系统相协调的观点来考虑；总体设计部把系统作为若干分系统有机结合的整体来设计，对每个分系统的要求都首先从实现整个系统相协调的观点来考虑，对分系统之间、分系统与系统之间的关系，都首先从系统总体协调的需要来考虑，进行总体分析、总体论证、总体设计、总体协调、总体规划，提出具有科学性、可行性和可操作性的总体方案。”[9]

研制“两弹一星”的经验、研制神舟飞船和嫦娥一号的经验，用于造大飞机、造航空母舰，相信也是行之有效的。

(3) 推行现代化管理体制。大型工程项目都要推行现代化管理体制。神舟飞船、奔月工程属于国家高科技项目，涉及国防与军事，要有特别的现代化管理体制。三峡工程、青藏铁路等工程项目则可以按照市场化运作，实行现代公司制度。

经国务院批准，中国长江三峡工程开发总公司（以下简称中国三峡总公司）于1993年9月27日成立。中国三峡总公司是国有独资企业，注册资本金39.36亿元；截至2008年12月31日，总资产达到2241亿元，2008年利润总额为113亿元，它的战略定位是以大型水电开发与运营为主的清洁能源集团，主要经营范围是水利工程建设与管理、电力生产、相关专业技术服务。中国三峡总公司全面负责三峡工程的建设与运营，实行总经理负责制，设立总工程师、总经济师、总会计师，协助总经理工作；设有科学技术委员会、投资委员会、预算委员会3个专业委员会，作为公司技术、经济决策咨询机构；设有总经理工作部、资产财务部、计划发展部、人力资源部、信息中心、科技环保部、党群工作部、新闻宣传中心等职能部门。

关于青藏铁路的建设与运作，国家成立了青藏铁路公司，网上可以找到有关资料。

(4) 精心设计、精心施工。工程管理一环扣一环，不容许任何环节马虎大意。精心设计、精心施工是我国大型工程项目运作与管理的成功经验。发现不合格的环节，坚决推倒重来，不能“带病运作”。若埋下隐患，可能酿成大祸。

(5) 奖优罚劣，奖勤罚懒。管理学基本原理之一就是要建立激励机制，对员工实行奖优罚劣，奖勤罚懒。

改革开放以来，“取消大锅饭”，“端走铁饭碗”，实行按劳付酬，收入与直接业绩挂钩。这些措施是很有效的，但收入差距不能拉得太大。现在有一些既得利益者，片面强调“与国际接轨”来为自己谋利益。他们向消费者收钱是高水平的，服务却是低水平的，群众意见很大。企业家是社会精英，具有不可推卸的社会责任。共产党员企业家是党的干部，要全心全意为人民服务，“先天下之忧而忧，后天下之乐而乐”。2009年5月，王岐山副总理在伦敦当着数百位银行家的面直言金融危机的起因是金融人士的贪婪，全场顿时鸦雀无声。这说明击中了他们的要害。社会上贪婪的不只是银行家，任何缺乏高尚道德的人员都可能贪婪，人类要警惕贪婪、拒绝贪婪。陈毅元帅的诗句“手莫伸，伸手必被捉”应该作为官员们的座右铭。

(6)“两弹一星”精神。1985年，诺贝尔物理学奖得主杨振宁在看望身患癌症的两弹元勋邓稼先时，问起国家为两弹研发有功人员颁发奖金的事情。邓稼先说："奖金 20 元，原子弹 10 元，氢弹 10 元。"原来，由于国家经济困难，发给整个"两弹"科研队伍的奖金总数仅 1 万元，受奖机构自身又拿出一部分钱，按照 10 元、5 元、3 元的级别分下去。邓稼先当时拿到了最高的奖励级别，但每一个"弹"只有 10 元钱。

"两弹一星"精神就是虽欠缺良好条件仍忘我地从事科学技术开发研究的精神，是爱国主义、集体主义、社会主义精神和科学精神的集中体现，是中国人民在 20 世纪为中华民族创造的宝贵的精神财富。要发扬光大这一伟大精神，使之成为全国各族人民在现代化建设道路上奋勇开拓的巨大推动力量。"人是要有一点精神的"，在金钱至上、物欲横流的时候，要有抵抗力，能够抵御"挡不住的诱惑"。

(7) 学习人民解放军。中国人民解放军是人民子弟兵，在我国大型工程建设中发挥了巨大的作用。解放军的战斗作风在工程建设与管理中发挥了巨大的作用，建立了不朽的功勋。从"两弹一星"的研制到汶川大地震的抗震救灾工作，全世界有目共睹，交口称赞。

解放军的优良传统和战斗作风，内容丰富，形象生动。例如，全心全意为人民服务，一不怕苦、二不怕死，雷厉风行、不折不扣，召之即来、来之能战，攻无不克、战无不胜，等等，这些对于大型工程项目及其管理都是十分有用的。

3. 监督与防范机制及其管理方面

(1) 警惕大型工程项目的负面效应。大型工程项目在追求正面效应的同时，应该尽量避免负面效应。未雨绸缪，对于可能出现的负面效应要有切实可行的防范措施。负面效应往往需要很长的时间才能暴露出来。

(2) 建立阳光机制，严厉打击贪官污吏，严格禁止挥霍浪费。事实表明，大型工程项目滋生了不少贪官污吏，如在高速公路建设中，倒下了一批交通运输厅厅长，以及处长、科长等。

全社会都呼吁建立阳光机制，落实信息公开制度。阳光机制和信息公开制度可以杜绝暗箱操作，杜绝少数人以权谋私、权钱交易、贪污腐败。

阳光机制和信息公开制度得到了广大老百姓的拥护，却受到了贪污腐败涉嫌者的顽固抵制。应该尽快在全国推行各级官员财产申报制度，尤其是领导干部。

4. 评估机制与管理方面

(1) 大型工程项目上马之前一定要认真开展可行性研究。可行性研究一定要具有客观性、公正性、科学性，要实事求是地找出项目的优缺点、上马或不上马的理由、上马以后的利弊得失，要计算工程项目全寿命周期总费用。可行性研究否定的项目一定不能上马，或者在作重大修改之后重新开展可行性研究。

可行性研究不能变成"可批性研究"。为了使得某项目能够上马，有关部门和人员隐瞒项目的缺陷与弊端，甚至昧着良心编造虚假情况和"理由"，使项目骗取上级部门审批通过，这种现象应该杜绝。

按照国际惯例，一个大型工程项目的可行性研究，至少要由两家具有资格的可行性研究机构来承担，其中至少一家机构与项目承担单位及其上级部门没有任何利害关系。

(2) 建立必要的评估制度和综合评估指标体系。大型工程项目要定期检查，消除隐患。评估是需要的，评估泛滥则不好。评估泛滥使得评估走了样、变了味，但是“倒脏水不能把孩子也倒掉”。

综合评估需要科学合理的评估指标体系。评估者与被评估者不能有利害关系。“拿了人家的手短，吃了人家的嘴软”，评估者要拒绝被评估者的威逼利诱。

5. 管理伦理、道德及其他方面

(1) 发扬革命传统，争取更大光荣。前面说了“两弹一星”精神，这里重点说说勤俭节约、艰苦奋斗，这是共产党人的革命传统。要发扬延安作风，杜绝贪污与浪费。奢靡之风不可长，办公楼、办公室搞豪华装修能够提高工作效率、保证工作质量吗？恰恰相反，豪华装修可能引起群众公愤。

(2) 全心全意为人民服务。管理者、领导者应该牢牢树立公仆意识，全心全意为人民服务；应该“关心群众生活，注意工作方法”，自觉接受群众监督。

(3) “中学为体，西学为用”。洋务运动提出的“中学为体，西学为用”思想，从理论上说是合理的，洋务运动的失败是因为清王朝那个“体”太腐败了，当时的传统文化太保守了，中国社会与科学技术和民主制度隔阂太大了。今天的“体”已经截然不同。改革开放30年后的中国，是朝气蓬勃的社会主义市场经济，正在构建社会主义和谐社会。中国已经而且正在继续阔步前进和迅速发展，中国的国际地位已经而且还要大幅度提高。总之，中国今天的“体”是健康之“体”、强壮之“体”、生机盎然之“体”，“中学为体，西学为用”这个原则完全适用于今天的中国。

(4) 学习“洋办法”不要鄙视“土办法”。“土办法”大有可为，如走群众路线，群策群力做事情，这是一条行之有效的经验。现在美国奥巴马政府宣布要采取全民反恐战略，搞群众反恐，这与我国的做法相类似。共产党历来相信群众，发动群众，依靠群众。美国的金融危机给人以启示：在金融管理方面美国“闯了大祸”；中国尽管受到美国的牵连，但是并无大碍。“已是悬崖百丈冰，犹有花枝俏。俏也不争春，只把春来报。待到山花烂漫时，她在丛中笑。”这正是中国的写照。

“洋办法”不一定先进，“土办法”不一定落后。以其“出生地”在中国还是在外国来判定落后与先进是不妥当的。“洋办法”其实是西方的“土办法”，有的也很落后。中国的“土办法”到了外国也是“洋办法”，有的也很先进。西方开始掀起“中国热”，孔子学院在全世界受到热烈欢迎，就是有力的证明。

(5) “两参一改三结合”。“两参一改三结合”是“鞍钢宪法”的核心内容：干部参加劳动，工人参加管理；改革不合理的规章制度；实行干部、技术人员和工人三结合。今天看来，哪一条没有现实意义呢？不能轻易废弃，而要充分珍惜。

(6) 学好用好《矛盾论》，正确处理矛盾，做好管理工作。根据《矛盾论》的观点，工作中时时有矛盾，处处有矛盾，旧的矛盾解决了，新的矛盾又会产生。作为管理者、领导者，不是回避矛盾，而是要正确处理矛盾，积极解决矛盾，推动工作前进。

(7) 重视思想政治工作。思想政治工作是革命的传家宝，一定要重视。领导要与群众心连心，深入群众，与群众打成一片，同甘共苦。

计算机网络发挥着越来越大的作用，要充分利用网络传递信息，宣传和发动群众，正确引导群众。不要惧怕网络监督，不要试图愚弄群众。

（8）共产党员要发挥模范带头作用。大型工程项目的领导者、管理者大都是共产党员，要发挥共产党员的模范带头作用。干部带头了，党员和群众就会跟上来。

（9）与时俱进。领导者、管理者要与时俱进，包括在思想观念上、业务技术上，以及在业余的娱乐活动中，否则就会和群众格格不入，脱离群众。

（10）积极反思与总结经验。大型工程项目及其管理工作应该提倡积极的反思，包括在项目开展过程中进行阶段性回顾、小结与总结，以及项目完工以后的回头看。

反思有个人的、自发的反思，有集体的、组织的回顾总结。从创建现代管理科学的中国学派而言，进行反思与总结，才有可能做到近为今用，提炼中国模式，打造中国学派。

五、案例研究的意义

本文研究了洋为中用、古为今用、近为今用3个方面的若干案例，分别进行了简单的评述，然后重点研究了中国工程管理的特色，归纳成5个主要方面：决策机制与宏观管理方面；运作机制与管理方面；监督与防范机制及其管理方面；评估机制与管理方面；管理伦理、道德及其他方面。相信进一步的研究可以提炼出大型工程管理的中国模式，并上升为工程管理的中国学派，它是现代管理科学的中国学派的重要组成部分。

参考文献

[1] 刘人怀，孙东川，孙凯. 三谈创建现代管理科学中国学派的若干问题. 中国工程科学，2009，11（8）：18-23，63.

[2] 刘人怀，孙东川. 二谈创建现代管理科学中国学派的若干问题. 中国工程科学，2008，10（12）：24-31.

[3] 刘人怀，孙东川. 谈谈创建现代管理科学中国学派的若干问题. 管理学报，2008，5（3）：323-329.

[4] 孙东川，林福永，孙凯. 系统工程引论. 2版. 北京：清华大学出版社，2009.

[5] 常平，刘人怀，林玉树. 20世纪我国重大工程技术成就. 广州：暨南大学出版社，2002.

[6] 郭重庆. 中国管理学界的社会责任与历史使命. 管理学报，2008，5（3）：320-322.

[7] 俞可平. 民主是个好东西. 学习时报，2006，（12）：28.

[8] 卢嘉锡. 院士思维. 合肥：安徽教育出版社，1998.

[9] 钱学森. 论系统工程（新世纪版）. 上海：上海交通大学出版社，2007.

建设低碳社会关键在制造创新①

2009年12月，哥本哈根气候大会向全世界七十亿人民宣布：要保护地球！全球随即进入低碳经济时代。

冰冻三尺非一日之寒！

地球在茫茫宇宙中，是人类赖以生存的唯一的一个得天独厚的乐园，早期人类与自然一直和谐相处。

但是，18世纪以来，人类进入工业时代以后，就进入了一个新时期，人类从大自然的手中夺得了权力，成为地球的主人，雄心勃勃地宣告：向大自然宣战，要战胜自然，要征服自然！

200年后的今天，全球能源告急！资源告急！环境告急！全球七十亿人的需求几乎要把自然拖向崩溃的边缘。

再看看我们国家，从1978年年底党的十一届三中全会以来，在改革开放的旗帜下，我国取得了举世瞩目的成就，经济持续高速发展，经济总量跃升为世界第三，即将登上世界第二的宝座，中华民族终于再次站起来了。

但是我国经济的高速发展，主要靠物质投入的传统发展方式，在一定程度上，是以环境资源的巨大牺牲以及经济发展的扭曲与不合理为代价的。以推动国家经济增长的消费、投资、出口这三驾马车为例，长期以来，我国经济越来越严重依赖出口这一驾马车的拉动。外贸依存度高达60%，珠三角尤为严重。特别是我国的外贸出口，主要依赖大量廉价劳动力，大量资源、能源的消耗以及随之而来的环境污染的巨大代价。我国的资源十分有限：我国的石油储量仅占世界的1.8%，天然气占0.7%，铁矿石占9%，铜矿石占5%，铝土矿占2%。在人均资源方面，我国人均45种主要矿产资源为世界平均水平的1/2，人均耕地、草地资源为1/3，人均淡水资源为1/4，人均森林资源为1/5，人均石油资源仅为1/10。以污染造成全球变暖为例，我国面临减排二氧化碳温室气体的严峻形势。2003～2006年，我国四年能耗增量超过了以前25年能耗增量的总和。2003年以前，我国发电能力不到2万亿kW，现在已经增加到7万亿kW，2003年钢产量为2亿t，现在高达7亿t。这些都要大量耗费碳资源。随之，我国二氧化碳排放量激增，2007年超过美国成为世界第一排放国，排放量为59.6亿t，占全球排放量的21%，与1990年相比，几乎翻了两番。我国的排放量激增，最大的推动力是发电、交通运输和制造业等。

大量的研究表明，大气中的二氧化碳是地球升温的祸首，世界著名气候科学家、美国国家航空航天局戈达德空间研究所詹姆斯·汉森认为："我们的排放水平早在20年前就已超标。"

---

① 本文是第二届深港澳节能减排论坛的开幕式主题演讲，深圳，2010年4月24日；工程管理研究，北京：科学出版社，2015，16-19.

超过临界点并不意味着会马上发生重大的灾难，但意味着一种从“可能”到“必然”的转变。

如果全球气温上升 5℃，风暴和干旱等极端天气出现的概率会急剧增加，海平面将上升 10m，海中的一些岛屿与大陆沿海地带特别是一些大城市会被淹没，全球一半物种可能面临灭绝。

如果气温上升 6℃，地球上的生命将会遭到毁灭性打击。

我国是升温最为严重的国家之一，近 100 年来平均升高气温 1.1℃，略高于同期全球平均升温幅度，近 50 年来变暖尤为明显。发生极端天气与气候事件的频率和强度出现了明显变化。因此我们应尽快行动起来，节能减排，建设低碳社会。

建设低碳社会，必须全国每个人都要参加。我们的首要任务是要建设以低碳排放为特征的产业体系和消费模式。要改变高能耗、高污染的粗放型增长方式，形成有利于我国可持续发展的经济发展模式。

要建设低碳社会，关键是将制造业改造为先进制造业，成为创新型企业。因此应在以下制造创新上做文章。

企业可以通过节能、提高能效实现低碳生产过程。

企业可以通过可再生能源开发利用进行能源结构调整、参与低碳能源发展。

企业可以通过提供低碳产品，引导全社会消费者进行低碳消费。

然而我国原有的企业与创新型企业差得很远，大多没有自主知识产权产品。在世界市场中，我国商品量大面广，其占有率年年增加，表面现象十分喜人，但是成绩背后的隐忧却不能忽视。因为这些产品大多不是低碳产品，同时产生的利润中仅 10%属于中国企业，其余 90%的利润都被国外企业拿走。

因此，企业必须建立自己的科研机构，坚持产学研合作，才有望成为创新型企业，这应该是国家层面上的一项战略性举措。

产学研结合是一个跨行业、跨部门、跨地区合作的系统工程。实际上，还应在产学研结合三方面的两头，分别加上政府和金融机构，即应改为政产学研金的合作。政府应负责营造科技创新的良好环境，强化政策的激励引导作用，为企业、高等院校、研究所人员营造有利于科技创新的政策环境，引导创新要素向企业集聚。在科技投入、成果奖励、创新平台建设等方面加大扶持力度，同时还应强化考核的促进保障作用。企业是主体地位，应提升自主研发和技术利用的创新能力，使企业成为研发投入的主体，成为科技创新主体，成为创新成果主体。这将大幅提升企业的核心竞争力，使企业健康发展。高等教育既培养科技创新人才，又提供科技创新成果。研究院所主要是提供科研创新成果，它们是企业的坚强后盾，它们将为企业源源不断地输入人才和成果，是创新型企业的生命源泉和基础。金融机构可为产学研结果投入提供风险管理服务，这是产学研的强大支柱。

目前，政产学研金合作的核心问题是要完善联盟的组织模式和运作方式。政府起引领作用，金融机构起支柱作用，产学研三个方面组成一个平台，五个方面均要受益才算成功。由于五个方面紧密联系，一损俱损，因此每个方面都要尽力，都要注意协调，在这里，“和为贵”是至理名言。

工程管理信息化的内涵与外延探讨①

在国家的“信息化与工业化融合”战略中，工程管理信息化是其中的重要组成部分。随着现代计算技术、网络技术和通信技术的快速发展及其在工程管理领域中的广泛应用，对工程管理信息化进行全面与深入的研究显得日益迫切和重要。

工程是人类为了生存和发展，实现特定的目的，运用科学和技术，有组织地利用资源所进行的造物或改变事物性状的集成性活动[1]。工程管理信息化是指为了更好更有效地实施工程管理，利用信息技术，构建信息系统，并在工程管理实践中加以应用的过程[2]。

大中型工程的投资较大，对社会及环境的影响广泛。通过工程管理信息化有效降低工程管理中的协作成本和重复投资，实现资源共享，对工程的设计、建设、运行和维护等各阶段实施有效控制，将使得整个工程全生命周期的总投入降低，工程质量及运作效率提高，同时也可以使得工程的环保效益及民生满意度得到改善，使得工程的经济效益与社会效益得以大幅度提升。

在特定工程从论证、设计、施工、运行直至报废的全生命周期过程中，不同阶段的参与方不同，工程管理信息化所面对的主要矛盾也有所区别。在工程设计与建设期间，由于技术涉及面广，参与的协作单位多，信息沟通复杂，社会影响面广，信息系统需要支持跨组织、跨地域、跨时间段的协作，需要实现多层次的综合系统集成，建立合作伙伴之间有效沟通的信息化模式。在工程运行和维护期间，通过信息化的途径对人力、物力、能源等的投入进行合理控制、有序安排，可以提高运行效率，降低运行成本，延长运行寿命，使工程产生更大的综合效益。

由于特定工程生命周期的各阶段是相对独立的，不同阶段的组织管理形式及所采用的信息技术平台可能存在着较大的差异。如何保证工程不同阶段的平稳过渡，确保信息资源共享的延续性是工程管理信息化所要研究的重要问题。更进一步讲，相关及同类工程之间的信息资源共享与知识创新平台的建立也是工程管理信息化所面对的重大课题。

一、工程管理信息化的概念框架

信息化建设不仅是信息技术系统建立的问题，同时也是与之相适应的组织架构与沟通机制、信息共享与知识创新模式不断调整、不断完善的过程，涉及不同组织内部、相关组织之间、不同工程之间以及工程与政府和社会公众之间的信息沟通问题。对工程的全生命周期进行分析，工程管理信息化的内涵可以归结为 4 个方面：运营管理、伙伴协作、公众服务与集成创新。其概念框架见图 1。

在工程的整个生命周期之中，工程管理信息化的实现涵盖了业主，施工、监理、

---

① 本文原载《科技进步与对策》，2010，27（19）：1-4，作者：刘人怀，孙凯.

设计、运营方，供应商、客户等经营伙伴，以及工程相关企业与社会公众和政府部门之间的信息采集、信息处理、信息存储和信息交互等一系列相关问题。在信息技术应用过程中，涉及各组织内外架构的调整、沟通机制的形成与完善等组织管理问题。

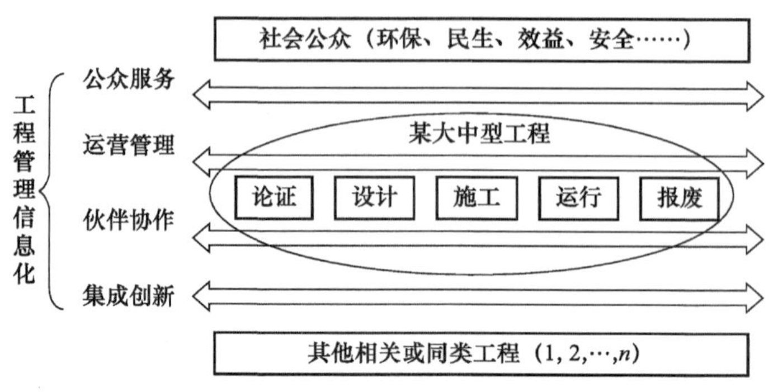

图 1　工程管理信息化概念框架

1）运营管理

运营管理是工程管理信息化的核心。以工程运行期的运营管理为主线，可将其简单划分为运营系统和管理系统两部分。前者主要用于支持工程的日常运营，实现计划调度、自动控制、远程监控、设备维护等，以确保工程日常运行。后者则主要用于应对工程运营企业的财务融资、人力资源、市场开拓等管理活动。

对于管理多个大中型工程的企业来说，一般存在多个工程运营系统并存的情况。这些系统的实现方式、技术水平、设备状况等可能存在较大的差别。实现多个运营系统的相互集成和资源共享，并与管理系统之间实现有效的数据交互是工程整体运营效率提高的重要途径。

2）伙伴协作

对伙伴协作的支持是工程管理信息化的重要特征。鉴于特定工程的阶段性和局限性，其所涉及的人力资源、设备资源及数据资源等大都是与其他工程共享的，在工程的不同阶段对这些资源的需求和具体使用方式也会有所不同，工程资源的调配需要有高效率的信息系统平台予以支持。

工程的所有者或管理者为确保工程的顺利运营而建立的沟通协作平台，通过及时的信息沟通和有效的信息共享，可以确保工程的设计方、施工方、供应商、运营商等一系列与工程直接相关的组织之间有效协作，提升工程的运作效益，实现对工程的全方位的管理。

3）公众服务

无论是公益性工程还是经营性工程，其目的都是为社会公众服务。建立工程的公众服务平台可以使得工程的所有者或管理者与社会公众之间的沟通更加有效，使得受工程影响的社会公众群体，如水利工程的移民、交通工程的乘客、房地产工程的拆迁户等，以及环境影响等的受益者或受害者都能及时准确地获取信息，形成信息公开与透明的机制，接受社会公众的建议与监督。在这一点上，电子政务方面的相关理论是可以充分借鉴的。

对于经营性的项目而言，如交通、能源、商业服务等设施在建成之后，此平台还可以与电子商务功能相结合，满足工程日常运营的需要，并借此带来可观的经营效益。作为工程管理信息化的重要组成部分，建立工程的公众服务平台，一方面为公众提供功能性的服务，另一方面，通过对社会公众及时发布信息，实现信息的公开透明，接受社会监督。

4）集成创新

集成创新是把已有的知识、技术创造性地以系统集成的方式创造出前所未有的新产品、新工艺、新的服务方式或新的经营管理模式，其新颖性表现在系统的集成思想和方式上[3]。任何一个工程都不是独立存在的，在全生命周期的工程管理中，需要不断吸收类似及相关工程的方法、经验和教训；同样，该工程的信息和知识资源也可以为其他工程所借鉴。通过不同工程之间的信息共享可以形成新的知识财富，使今后的工程项目受益。

更进一步讲，建立国家或行业层面的信息交换标准和信息共享机制，完善相应的信息技术基础设施的建设，建立为众多工程所共享的信息资源平台，以实现与其他相关或同类工程之间的资源共享，提升工程的知识管理水平。通过建立知识库、案例库、专家系统等，并在此基础上结合人工智能方法实现集成创新的途径，是工程管理信息化研究的重要内容之一。

二、工程管理信息化与企业管理信息化的区别与联系

工程管理信息化与企业管理信息化之间既有区别也有联系。区别在于，其所针对的组织形式、任务目标及管理对象等有着很大的差异，两者的研究对象不同。企业是假设永续经营的，而工程则是有生命周期的；企业规模是不断变化的，而工程规模是在一定时期内保持不变的；企业的经营状况是波浪式的，而工程的运行状况是相对稳定的；企业的组织机构和管理体制是持续变革的，而工程的管理机构和管理体制是相对稳定的。

从管理理论的角度而言，工程管理与企业管理的概念范畴和研究对象有着较大的区别，两者在信息化的实现方式上会有所不同。企业管理信息化更注重于组织自身的信息化，侧重于组织内部的信息基础设施建设与业务流程的优化，是以运营与控制为主的组织内部平台。而工程管理信息化则更侧重于不同组织之间的信息资源共享与相关业务流程的整合，在信息基础设施建设上更偏重于对公共基础设施的有效利用，是以协作和共享为主的跨组织平台，是典型的跨组织信息系统。

工程管理信息化与企业管理信息化具有很强的联系。从信息化应用的角度进行分析，两者在信息化实现上基于同样的技术方式和理论基础，在研究上可以相互借鉴、取长补短，在共同的规划框架基础上实现各自的应用模式。从信息技术实现的角度而言，信息化问题尽管面对的对象不同，实现的方式有差异，但归根结底都是如何更好地实现信息沟通与共享的问题，两者具有同样的理论基础和技术基础。

一个工程涉及多个企业，广义的工程管理信息化还包括工程施工企业及运营企业

在内的各类相关组织的信息化，这也是两类信息化的联系所在。工程管理信息化实现的重点之一，就是在工程生命周期的不同阶段、不同企业之间的信息系统集成、数据交换和信息共享。例如，在工程设计与施工期间，工程业主与施工、设计、监理等相关企业之间的集成；在工程运行期间，运营商和供应商、客户之间的集成。工程管理信息化整体水平是与工程相关各方自身信息化水平的高低密切相关的，没有相关企业信息化的实现就没有工程管理信息化的实现。

三、工程管理信息化的知识体系

工程管理信息化知识体系的建立有两条途径：一是对企业管理信息化等相关领域的理论借鉴；二是对各类工程管理案例的分析研究与归纳整理。

目前与信息化相关的研究大都集中在企业管理信息化、电子商务及电子政务等领域。对相关领域的研究成果进行借鉴，是研究工程管理信息化的有效途径。以工程交付使用为分界点，各类相关理论可大致分为两个部分。

在工程交付之前的设计、招标、建设等环节中，包括项目管理、成本控制、风险管理、招标管理、楼宇信息建模等理论已大量应用于工程管理的实践，在当前的工程管理信息化相关研究中也得到了较多的关注。对这些理论及相关应用案例进行归纳整理，结合信息化的基础理论和方法，是建立此阶段工程管理信息化整体应用模式的有效途径。

在工程交付之后的运行环节中，很多相关领域的研究如设备管理、资产管理、公用设施管理、基础设施管理、物业管理等都有着很完善的理论体系和丰富的应用案例，其研究对象及应用范畴有着很大的相似之处，但这些理论都专注于各自所应用的领域，将其有效借鉴和整合，以形成工程管理信息化在工程运行直至报废后期处理过程的理论体系。

在此基础上，将两部分理论进行合理的整合，可形成工程管理信息化的知识体系框架。在知识体系建立的过程中，应对工程交付前后的信息系统转换问题，以及工程生命周期结束之后信息系统的后期处理问题予以关注；还应对信息资源处理、存储、转换、查询等的组织机制和技术方案进行研究，以保障信息资源的延续性和长期可用性。

除了理论上的整合与借鉴之外，对以往及现有各类工程的实际工作成果进行分析、整理和挖掘，建立相应的工程管理信息化案例库，可以为建立工程管理的知识化体系作出贡献。通过将理论研究成果与实际应用案例相结合，探讨适合的实施方式和实现途径。工程管理信息化应用模式的研究将为其提供有效的理论支持和方法选择。

从信息系统规划的战略视角对工程管理中的信息化问题进行系统研究，使得工程管理中的信息技术从单一环节的控制、分散及单体运行的模式发展为全过程的管控，多流程、多环节的并行应用模式，将分散的“信息孤岛”纳入一个整合、统一的集成信息平台之上，形成跨组织的信息共享、业务数据的高度集成、运作流程的互联互通，以实现对工程全生命周期管理中的任务协同、资源协同、组织协同、地域协同和流程优化，促进工程管理信息系统向集成化发展。

工程管理信息化的具体实现是将特定工程及与之相关的各类组织纳入整体的信息系统架构之中，通过制订统一规划并付诸实施的过程。一个有效的战略规划可以使信息系统与用户有良好的关系，可以做到信息资源的合理分配和使用，从而可以节省信息系统的投资，还可以促进信息系统应用的深化[4]。通过对诸如企业资源规划、客户关系管理、供应链管理、全生命周期评价等信息系统理论，以及协同管理、知识管理等组织管理理论进行归纳和整理，在有所选择和甄别的基础上，将其全部或部分应用于工程管理信息化的研究，可以充实和完善工程管理信息化的理论体系。

四、对未来研究的展望

未来的研究趋势将主要体现在以下几个方面。

（1）从对工程单一阶段的研究拓展到全生命周期的工程管理信息化研究。有针对性地将信息化基础理论与工程管理的实际结合，形成具有通用性的工程管理信息化理论框架，对包括工程论证、设计、施工、运行直至报废的工程管理信息化问题进行系统研究，强调针对工程的整个生命周期运作制订统一的、分层次的信息系统规划。建立包括基础设施、信息资源、应用系统及服务交互等的分层次信息化架构，在整体规划的指导下实现对特定工程的信息化管理，保持信息系统的延续性、动态适用性和可扩展性。

（2）从对单一管理平台的研究拓展到包括协作平台、创新平台、服务平台集成的信息系统研究。从工程的本质出发，通过分析其参与者和使用者的需求，将工程管理信息化的单一运营管理平台应用模式拓展到包括伙伴协作、公众服务及集成创新功能在内的整体信息化应用模式。在信息化的研究中引入知识创新及服务创新的相关理论，以期通过工程管理信息化的实现，提高工程参与者之间的协同与交互水平，提升系统的整体运行效益；提高工程使用者及利益相关者的满意度，提升工程的社会经济效益。

（3）从基于特定技术平台的研究拓展到对通用技术平台与组织业务相匹配的综合集成研究。在特定工程的应用中，首先从分析工程业务需求出发，结合工程参与者的综合信息系统能力，以及项目所拥有的信息资源来制订相应的信息系统规划。摆脱从特定技术方案出发研究信息化的传统模式。以信息技术与组织业务的匹配关系为重点，建立工程管理信息化应用模式的原型，以实现理论的技术无关性和业务通用性，进而在统一的规划指导下，结合当前信息技术发展的最新成果，有针对性地选择技术平台和确定信息化解决方案。

（4）基于“现代管理科学中国学派”框架的中国特色工程管理信息化研究工程管理（信息化）的中国模式和中国学派，其基本途径是：洋为中用，古为今用，近为今用，综合集成。近为今用是重点，即重点研究我国近期的大型工程项目及其管理的典型案例。所谓典型案例，既包括成功案例，也包括失败的案例，研究前者主要是总结经验，研究后者主要是吸取教训。研究中国模式，又需要先研究较多的典型案例，同时还要与其他国家，如苏联和美国的案例进行比较研究，从众多案例中加以归纳、总结、提炼、提高，以得到中国模式，再进一步上升到中国学派[5]。

参考文献

[1] 何继善，陈晓红，洪开荣. 论工程管理. 中国工程科学，2005，7（10）：5-10.

[2] 朱高峰. 对工程管理信息化的几点认识. 中国工程科学，2008，10（12）：32-35.

[3] 王众托. 系统集成创新与知识的集成和生成. 管理学报，2007，4（5）：542-548.

[4] 薛华成. 管理信息系统. 5版. 北京：清华大学出版社，2007.

[5] 刘人怀，孙凯，孙东川. 大型工程项目管理的中国特色及与美苏的比较. 科技进步与对策，2009，26（21）：5-12.

转变经济发展方式关键在制造业创新①

我国自1978年改革开放以来，社会经济迅猛发展，取得了巨大的成就，成为世界第二经济大国，给世界带来了相当的震撼！中华民族终于再次站起来了！

但是，我国经济的高速发展，主要靠物质投入的传统发展方式，在一定程度上是以环境资源的巨大牺牲以及经济发展的扭曲与不合理为代价所取得的。以推动国家经济增长的消费、投资、出口这三驾马车为例，长期以来，我国经济已经越来越严重依赖出口这一驾马车的拉动。外贸依存度高达60%。特别是我国的外贸出口主要依赖的是大量廉价劳动力，大量资源、能源的消耗以及所付出的环境污染破坏的巨大代价。

我国资源十分有限：我国的石油储量仅占世界的1.8%，天然气占0.7%，铁矿石占9%，铜矿石占5%，铝土矿占2%。在人均资源方面，我国人均工业化所需要的45种主要矿产资源为世界平均水平的1/2，人均耕地、草地资源为1/3，人均淡水资源为1/4，人均森林资源为1/5，人均石油资源仅为1/10。

以污染造成全球气候变暖为例，我国面临减排二氧化碳温室气体的严峻形势。以2003～2006年为例，我国四年能耗增量超过了以前25年能耗增量的总和。2003年以前，我国发电能力不到2万亿kW，现在已经达到7万亿kW。2003年钢产量只有2亿t，现在有7亿t。这些都要大量耗费碳资源。随后，我国二氧化碳排放量大增，2007年，我国二氧化碳的排放量已达59.6亿t，超过美国，占全球排放总量的21%，成为世界第一排放大国。1990～2007年，我国的排放量几乎翻了两番，因此，我国在减排方面遭遇到的国际政治压力越来越大。我国的排放量激增，最大的推动力是发电、制造业和交通运输。

从长期来看，二氧化碳浓度的变化是一个逐渐累积的过程，超过了临界点标准，并不意味着马上会发生重大的灾难，但意味着一种从“可能”到“必然”的转变。

如果全球气温上升5摄氏度，风暴及干旱等极端天气的出现概率会急剧增加，海平面将上升10米，全球一半以上物种可能灭绝。

如果全球气温上升6摄氏度，则全球的物种将遭受毁灭性打击。

20世纪，全球气温上升了1.1摄氏度，近年来还在持续增长，值得我们重视、警惕！

显然，我国目前的继续采用大量廉价劳动力，依靠大量资源投入，造成环境污染的经济发展方式是难以为继的，是行不通的。

同时，我们还应注意到，我国的制造企业中，百分之九十多无自主知识产权产品。在国际市场出售的商品中，有很多是“中国制造”，而且其占有率每年都在增长，这个趋势非常令人高兴。但成绩背后的隐忧却不应被忽视，因为这些商品所产生的利润率，

① 本文是第七届沈阳科学学术年会的报告，沈阳，2010年10月20日；现代管理的中国实践，北京：科学出版社，2016，114-116.

中国所能获得的仅有 10%，其余 90%都被国外的创业公司或设计公司拿走。比如，DVD 制品，前年全国一年利润达 60 亿美元，我们一分钱都得不到，还要欠 10 亿美元。

因此，转变我国经济发展方式势在必行，其中的关键是要使我国的制造业企业，不仅是国有企业，而且也包括民营企业，不仅是大企业，而且也包括中小型企业，都成为创新型企业。

为此，制造业企业必须建立自己的科研机构，坚持产学研合作，才有望成为创新型企业，这应该是国家层面上的一项战略性举措。

产学研结合是一个跨行业、跨部门、跨地区合作的系统工程。实际上，还应在产学研结合三方面的两头，分别加上政府和金融机构，才全面，才符合中国特色，才能成功，即应改为“政产学研金”的合作。

政府的地位十分重要，是关键。政府应负责营造科技创新的良好环境，强化政策的激励引导作用，为企业、高校和研究所、科研人员营造有利于科技创新的良好环境，引导创新要素向企业集聚。在科技投入、成果奖励、创新平台建设等方面加大扶持力度，同时还应强化考核的促进保障作用。

企业是主体地位，应提升自主研发和技术利用的创新能力，使企业成为研发投入主体，成为科技创新主体，成为创新成果主体。这将大幅提升企业的核心竞争力，使企业健康发展。

高等学校既培养科技创新人才，又提供科技创新成果。研究院所主要是要提供科技创新成果，它们是企业的坚强后盾，是创新型企业的生命源泉和基础。

金融机构可为产学研提供经济支持，提供风险管理，这是强大的支柱。

目前，“政产学研金”合作的核心问题是要完善联盟的组织模式和运作方式。政府起引领作用，金融机构起支柱作用，产学研三个方面组成一个平台。五个方面均要受益才算成功。由于五个方面紧密联系，一损俱损，因此每个方面都要尽力，都要注意协调，在这里“和为贵”是至理名言。

公共安全工程

一、公共安全相关工程科技研究的重要性①

1. 中国公共安全形势严峻

新中国成立60多年来，各族人民同心同德、艰苦奋斗，战胜各种艰难曲折和风险考验，取得了举世瞩目的成就[1]。特别是改革开放30多年来，中国保持GDP年均9.8%的增长率，政治和社会大局稳定，有效应对了一系列危机和灾难，公共安全形势保持了总体稳定、趋向好转的态势。但是，中国公共安全形势依然严峻并面临着新挑战[2]。

1）自然灾害严重

由于特有的地质构造和地理环境，中国是世界上遭受自然灾害最严重的国家之一，其基本特征是灾害种类多、分布地域广。

2）事故灾难频发

由于经济快速发展，粗放型的经济增长方式、一些企业安全管理水平和技术落后、非法开采、违规操作等导致的煤矿、交通、化学品等事故频发。

3）公共卫生事件威胁着人民群众的生命和健康

全球新发现的30余种传染病已有半数在中国被发现，多种传染病尚未得到有效遏制；职业病危害严重；农村卫生发展滞后，传染病、慢性病和意外伤害并存；重大食物中毒事件时有发生。重大疫情的不时出现和公共卫生事件的不确定性及严重性已经成为一个新的重大问题。

4）影响国家稳定和社会安全的因素依然存在

经济社会快速发展中的结构性矛盾和不平衡问题以及各类经济纠纷等，危及我国安宁。

国际经验表明，当一个国家人均GDP处于1000～3000美元这一区间时，公共安全事件发生频率处于上升期，3000～5000美元时公共安全事件处于高发期。而我国在2008年人均GDP就已达到3260美元，正处于公共安全事件高发期。

据国家统计局近五年的数据估算，我国每年因自然灾害、事故灾难、公共卫生和社会安全等突发公共安全事件造成的经济损失相当于GDP的3.5%，远高于中等发达国家1%～2%的同期水平。目前我国正处于经济和社会转型期，公共安全面临的严峻形势越来越凸显，但我们还没有建立起一套完备的公共安全保障体系。在谋求经济与社会发展的过程中，人的生命始终是最宝贵的，应当像对待人口问题、资源问题、环境问题一样，把公共安全问题提升到国家战略层面。目前，不少发达国家已经将公共安全上升到国家安全高度，进行战略研究与产业化发展，在核心技术方面已经占领了

---

① 本文是中国工程院科研项目报告，北京，2010；工程管理研究，北京：科学出版社，2015，109-111.

制高点。如果我们不加快步伐，不仅将在这一领域始终受制于人，也难以保障我国的国家安全利益。

2. 中国公共安全管理工作面临新挑战

1）城镇化和城市现代化的挑战

改革开放以来，中国以年均增加 18 个城市和 1.4%城镇人口的速度在发展。2008 年，城镇人口为 6.07 亿，城镇化率为 45.7%，百万人口以上大城市有 118 座。高风险的城市和不设防的农村并存。城市灾害的突发性、复杂性、多样性、连锁性、集中性、严重性、放大性等，使城市应急管理工作任务繁重。

2）工业化、信息化的挑战

自然灾害、事故灾难、公共卫生事件和社会安全事件等各类突发事件的关联性越来越强，互相影响、互相转化，导致次生、衍生事件的产生或各种事件的耦合。水、电、油、气、通信等生命线工程和信息网络一旦被破坏，轻则导致经济损失和生活不便，重则会使整个国家的政治、经济或军事陷入局部或暂时瘫痪，社会秩序失控。

3）市场化的挑战

中国经济社会发展进入了一个关键时期，经济体制深刻变革，社会结构深刻变动，利益格局深刻调整，人们的思想观念深刻变化，再加上民族宗教问题的影响，不稳定、不确定、不安全因素增加。人民群众的法律意识、权利意识明显增强，舆论监督、社会监督力度空前加大，对公共安全的要求越来越高。

4）国际化的挑战

和平与发展是时代的主题，但世界并不太平，并不安宁，各种矛盾交织，错综复杂。恐怖袭击，局部战争，金融危机，对水资源、石油资源的争夺以及跨国性的重大疫情的传播等不时出现，境外涉我和境内涉外的突发事件增多。

随着经济发展水平的日益提高，安全问题越来越成为社会关注的焦点。促进与公共安全相关的工程科技建设，对我国建设更安全、更和谐的社会将起到举足轻重的作用，是我国面向未来必须重视和关注的重大工程科技问题。为此，我们将从以下几个公共安全领域开展研究工作：①信息安全；②生物安全；③食品安全；④土木工程安全；⑤生产安全；⑥社会安全；⑦生态安全。

3. 课题研究的指导思想

瞄准国家经济社会发展的需求和目标，力争做到具有前瞻性、战略性和创新性，做到既有现实性又有长远性。着力研究我国必须高度重视和关注的公共安全相关工程科技问题，促进中国工程科技更好地坚持中国特色、服务于国家经济社会发展、支撑创新型国家建设以及保障国家安全。

研究重点和亮点主要体现在监测、预警、应对和管理四个方面。

研究工作突出战略性、前瞻性和创新性，以最大限度地保护人民生命财产安全为出发点，重点研究公共安全相关工程科技中的共性问题，充分调研国内外研究现状和发展趋势，提出公共安全对相关工程科技领域的战略需求、发展思路和目标，并就重点发展领域进行展望，提出相应的政策建议。

二、信息安全相关工程科技发展战略①

1. 当前我国信息安全的研究现状

信息化作为推动经济社会变革的重要力量，是当今世界发展的大趋势。随着信息技术，特别是互联网的日益普及和广泛深入的应用，信息安全的重要性与日俱增。中共中央办公厅和国务院办公厅于 2006 年 3 月印发了《2006—2020 年国家信息化发展战略》，确立了我国信息化发展的指导思想和战略目标，其中特别强调全面加强国家信息安全保障体系建设，大力增强国家信息安全保障能力。

信息安全涉及信息的制作处理、存储访问、通信传播、使用呈现等多个环节，包括信息本身的安全保密以及相关系统和网络环境的安全运行，在工程技术上分属通信安全、网络与系统安全、信息安全保密以及电子商务/电子政务安全等领域。

由于政府部门的大力推进和商业市场规模的形成共同作用，我国在若干信息安全关键技术的研究开发和产业化方面，取得了一批较好的工程技术成果[3]。特别是“十一五”期间，在“863”计划等国家科技计划的支持下，已经在 PKI/CA（public key infrastructure/certificate authority，公钥基础设施/证书授权中心）技术、密码标准和芯片、网络积极防御、网络入侵检测与快速响应、网络不良内容的监控与处理、数字水印等方面取得了较大的进展。

纵观我国信息安全领域的发展历史，不难看出，虽然取得了不少优秀成果，但存在的问题依然很突出，主要体现在以下几个方面。

(1) 信息与网络安全的防护能力依然很弱，不少应用系统处于不设防状态，具有很高的风险性和危险性。

(2) 对引进的信息技术和设备，缺乏保护信息安全必不可少的有效管理和技术改造措施。

(3) 信息安全自主能力不强，缺乏控制能力；信息安全关键技术整体上比较落后，关键技术和产品受制于人，信息安全产业缺乏核心竞争力，尚未形成应有的规模。

(4) 利用网络信息的犯罪在我国有快速发展和蔓延的趋势。

(5) 网络与信息系统环境的整体防护水平不高，应急处理能力不强，网络泄密事件时有发生。

(6) 基础信息安全管理机构缺乏必要的权威，协调能力不够。

(7) 信息安全法律法规和标准不够完善，信息安全保障制度不够健全；现行法规不能完全适应信息安全的新形势。

(8) 信息安全服务机构专业化程度不高，行为不规范，服务类型有待拓展。

(9) 信息安全管理和技术人才缺乏，不能适应信息安全保障工作和信息化发展的需求。

(10) 全社会的信息安全意识及应急处理能力亟须提高。

---

① 本文是中国工程院科研项目报告，北京，2010；工程管理研究，北京：科学出版社，2015，111-118. 作者：刘人怀，方滨兴，蔡吉人，何明昕.

2. 当前信息安全研究的国际先进水平与前沿问题综述

信息安全研究主要以提高安全防护、隐患发现、应急响应及信息对抗能力为目标。针对现代信息安全的内涵，目前国内外在该领域的研究热点主要包括以下几个方面。

1）信息安全基础设施关键技术

信息安全基础设施关键技术涉及密码技术、安全协议、安全操作系统、安全数据库、安全服务器、安全路由器等技术领域。

2）信息攻防技术

在广泛使用的互联网上，由于黑客、病毒等有害程序的入侵破坏事件不断发生，不良信息大量传播，以及国家和组织出于政治、军事、经济目的而日益兴起的信息战，信息攻防技术已成为当前国内外的重要研究热点。信息攻防技术涉及信息安全防御和信息安全攻击两个方面，主要技术研究内容包括黑客攻防技术、病毒攻防技术、信息分析与监控技术、入侵监测技术、防火墙、信息隐藏及发现技术、数据挖掘技术、安全资源管理技术、预警、网络隔离等。信息攻防领域目前正处于发展阶段，研究还比较零散，缺乏系统性。

3）信息安全服务技术

信息安全服务技术包括系统风险分析和评估、信息安全检测和监测技术、应急响应和灾难恢复技术等。系统风险评估和信息安全检测、监测技术是有效保证信息安全的前提条件，也是制定信息安全策略和措施的重要依据，其中包含网络基本情况及异变的检测与监测、信息系统基本安全状况调查分析、网络安全技术实施使用情况分析、动态安全管理状态监视与分析、数据及应用加密情况分析、网络系统访问控制状况分析等。信息和信息系统的安全可靠是一个相对的概念，绝对安全可靠的信息系统并不存在，除非没有人的参与；而无人参与的信息系统是没有现实意义的。信息安全服务使入侵者的代价增高，同时可及时发现入侵，减少系统拥有者的损失。研究系统风险分析和评估方法，可为信息安全策略的制定提供依据；信息安全检测、监测服务可以较小的代价，获得系统安全可靠性的提升，同时大大降低因信息安全问题带来的损失。

4）信息安全体系

信息安全涉及技术和管理等诸多方面，是一个人、系统与网络、环境相结合的复杂大系统。针对这样的复杂环境，研究设计安全体系相当困难，然而十分必要。安全体系对规划具体安全措施具有指导意义和检验作用。信息安全体系研究涉及安全体系结构和系统模型研究、安全策略和机制研究、检验和评估系统安全性的方法和规则的建立，有助于信息安全检测、监测和防护服务的实施运营。目前，国内外都很重视这方面的研究，我国“863”计划已将部分内容作为重要研究课题。

3. 信息安全研究中的重大工程科技问题

1）网络攻击与防范问题

网络安全保护的核心是如何在网络环境下保证数据本身的秘密性、完整性与操作的正确性、合法性与不可否认性。而网络攻击的目的正相反，会以各种方式通过网络破坏数据的秘密性和完整性或进行某些非法操作。因此可将网络安全划分为网络攻击和网络安全防护两大类。

2）安全漏洞与安全对策问题

受限制的计算机、组件、应用程序或其他联机资源无意中留下的不受保护的入口点将是安全漏洞，是硬件软件或使用策略上的缺陷，它们会使计算机遭受病毒和黑客的攻击。

3）信息安全保密问题

信息安全保密技术诠释了信息安全保密的概念，构建了信息安全保密体系，从物理、平台、数据、通信、网络等层面全面、系统地介绍了信息安全保密的各项技术，给出了开展信息安全保密检查、保密工程和安全风险管理的规范和方法，以及典型的信息安全保密实施方案，具有较强的针对性和可操作性。

4）系统内部安全防范问题

随着网络的急剧扩展和上网用户的迅速增加，内部网络的应用也随之不断发展。内部网络的应用要远复杂于传统外部网络的应用，而且内部网络大多数的应用，对内部网用户来说是很重要的，甚至是严格保密的，这也使得内部网的安全越来越受到重视。

5）防范病毒问题

在网络环境下，防范病毒问题显得尤其重要。这有两方面的原因：一方面是网络病毒具有更大破坏力，另一方面是遭到病毒破坏的网络进行恢复非常麻烦，而且有时几乎不可能恢复。因此采用高效的网络防病毒方法和技术是一件非常重要的事情。网络大都采用“client-server”（客户端-服务器）的工作模式，需要结合服务器和工作站两个方面解决防范病毒的问题。

6）数据备份与恢复问题

计算机里面重要的数据、档案或历史纪录，不论是对企业用户还是对个人用户，都是至关重要的，一旦不慎丢失，会造成不可估量的损失，轻则辛苦积累起来的心血付之东流，严重的会影响企业的正常运作，给科研、生产造成巨大的损失。为了保障生产、销售、开发的正常运行，企业用户应当采取先进、有效的措施，对数据进行备份、恢复，防患于未然。

4. 信息安全中的关键技术

1）安全测试评估技术

安全是网络正常运行的前提。网络安全不是单点的安全，而是整个信息网的安全，需要从多角度进行立体防护。要知道如何防护，就要清楚安全风险来源于何处，这就需要对网络安全进行风险分析。网络安全的风险分析，重点应该放在安全测试评估技术方面，其战略目标是：掌握网络与信息系统安全测试及风险评估技术，建立完整的面向等级保护的测评流程及风险评估体系。

首先，安全测评要建立适应等级保护和分级测评机制的通用信息系统与信息技术产品测评模型、方法和流程，即不同的级别采取不同的测评方法，分级应符合等级保护机制；重点放在通用产品方面，即建成一个标准的流程，完全限于一事一议，以使相互之间有比较；建立统一的测评信息库和知识库，即测评要有统一的背景；制定相关的国家技术标准。其次，建立面向大规模网络与复杂信息系统安全风险分析的模型和方法；建立基于管理和技术的风险评估流程；制定定性和定量的测度指标体系。如果没有这个指标体系，只能抽象地表述，对指导意见来讲并没有太多的实际意义。

2）安全存储系统技术

在安全策略方面，重点需要放在安全存储系统技术上。其战略目标一是要掌握海量数据的加密存储和检索技术，保障存储数据的机密性和安全访问能力；二是要掌握高可靠海量存储技术，保障海量存储系统中数据的可靠性。

关于安全存储，首先采用海量，即太字节（terabyte，TB）级的分布式数据存储设备的高性能加密与存储访问方法，并建立数据自毁机制。为海量信息进行高性能加密，虽然有加密解密的过程，但对访问的影响并不明显，由此对算法的效率提出了很高的要求．一旦数据被非授权访问，应该执行数据自毁。

其次，采用海量存储器的高性能密文数据检索手段。加密的基本思路是使内容无规则化（杂化），使其根本看不到规则；而检索就是要尽量有规律。这就提出了一个折中的要求，如何加密对检索能够尽可能地支持，同时又具备一定的安全强度。这对密码算法和检索都提出了严峻的挑战。

再次，构建基于冗余的高可靠存储系统的故障监测、透明切换与处理、数据一致性保护等方面的模型与实现手段。高可靠的关键要依赖冗余，一旦系统崩溃，还有冗余信息。

最后，制定安全的数据组织方法，采用基于主动防御的存储安全技术。

3）主动实时防护模型与技术

在现有网络环境下，安全大战愈演愈烈，防火墙、杀毒、入侵检测“老三样”等片面的安全防护应对方式已经越来越显得力不从心，目前需要的不仅是片面的被动防护，而且要在防护的过程中强调主动实时防护模型与技术。

主动防护的战略目标应该是：掌握通过态势感知、风险评估、安全检测等手段对当前网络安全态势进行判断，并依据判断结果实施主动安全防护体系的实现方法与技术。传统的防护一般都是入侵检测，发现问题后有所响应，但是现在越来越多的人更加关注主动防护，通过态势判断，进行系统的及时调整，提高自身的安全强度。通过感知，主动作出决策，而不是事后亡羊补牢，事后做决策。

关于主动防护：一是建立网络与信息系统安全主动防护的新模型、新技术和新方法，建立基于态势感知模型、风险模型的主动实时协同防护机制和方法；二是建立网络与信息系统的安全运行特征和恶意行为特征的自动分析与提取方法，采用可组合与可变安全等级的安全防护技术。不同的系统会有不同的需求，应该具备一定的提取能力，进而监控其特征，通过监控判断所出现的各种情况。另外，如果通过检测发现恶意行为，应该对其特征进行提取，提取的目的就是为了进一步监测，或在其他区域进行监测，检查同样的情况是否存在，如果存在，就要对这个态势进行明确的分析，而这些都需要有自动的特征提取。

4）网络安全事件监控技术

监测是实现网络安全不可或缺的重要一环，其中重点强调的是实施网络安全事件监控技术。

实时监控的战略目标是：掌握保障基础信息网络与重要信息系统安全运行的能力，支持多网融合下的大规模安全事件的监控与分析技术，提高网络安全危机处置的能力。三网融合势在必行，不同网络状态实现融合，对网络安全监测提出了更高的要求，监测广度和监测水平都需要提升。

5. 保障信息安全的重要措施与政策建议

信息安全事关我国的经济安全、社会安全和国家安全，是我国信息化建设进程中具有重大战略意义的课题。近年来，信息安全已受到我国政府的高度重视。国家自然科学基金，“973”、“863”等计划已将信息安全列为重要研究方向，许多科研机构及企业也开展了相关研究，并取得一定成果。然而与发达国家相比，我国尚存在不小差距，相关产业的发展也比较落后。目前存在的主要问题是：低水平分散、重复研究、整体技术创新能力不足、信息安全产业规模相对较小且缺乏联合和集中、一些关键软硬件系统还不能自主研制生产等。为了推动我国信息安全工作的发展，应采取如下几项政策措施。

1）加强信息安全管理

我国信息安全领域所表现出的不足，在很大程度上归因于管理体制不健全。为此，应在中央和地方建立高效的信息安全领导机构，制定和实施重大决策并协调各方关系，振兴信息安全产业。此外，应建立完善的信息安全标准、法律法规体系，以规范信息安全领域的行为。

2）加强信息安全基础设施建设

信息安全基础设施是信息安全得以实现的基本条件，包含 PKI、密码等关键技术及产品，以及包括信息安全检测评估中心、应急响应中心等在内的信息安全服务体系。目前，我国信息安全基础设施尚不健全，需要加强建设。

3）开展信息安全关键技术研发

在信息安全技术和产品研发方面，应坚持“有所为、有所不为、重点突破、技术创新”的方针，开展关键安全芯片、安全操作系统、密码技术等方面的重点自主研发；并密切关注国外相关领域研究状况，进行关键技术创新研究，力求我国信息安全不受制于人。

4）发展信息安全产业

信息安全产业的强劲发展是提高国家信息安全水平的必然途径。我国需要大力发展信息安全产业，以振兴国家信息安全综合实力。除了信息安全技术和产品之外，应大力促进信息安全服务，特别是信息安全检测和监测服务的健康发展与产业化进程。

5）加强信息安全专业人才培养

目前，国内信息安全人才紧缺，不适应国家信息安全建设的需要和未来发展的需要，亟须加大安全技术、安全管理、安全教育及复合型人才的培养，造就一批能够解决信息安全重大系统工程技术难题的高级专家，以及一批适应市场竞争的科技创新人才和工程技术人员。

三、生物安全相关工程科技发展战略①

随着现代科技的发展和人们生活质量的提高，人们的安全意识和要求明显增强，转基因生物、人畜共患传染病、外来入侵生物以及药品（医药、农兽药及生物制品）对人们健康和环境安全的影响引起了公众广泛关注，尤其是近年来连续发生的“禽流感”、“转基因稻米市场化争议”、“植物杀手薇甘菊”以及“维C银翘片”等安全事件更加凸显了当今我国生物安全面临的严峻形势。为了缓解当前科技发展中存在的安全问题，提前应对未来可能发生的生物安全问题，开展以转基因生物、人畜共患传染病、外来生物入侵以及药品为主的生物安全工程科技发展战略研究意义重大。

1. 当前我国生物安全的研究现状

1）转基因生物安全

转基因生物安全属于高新技术领域，目前在我国已基本形成了以自主创新为主的系统布局，初步建立了一定的农业转基因生物安全评价体系，取得了一系列突破性进展，包括：①初步建立了转基因生物及其产品对人体健康影响的评价体系和动物模型，并以潮霉素抗性标记基因为代表建立了外源蛋白的食用安全性评价技术平台；②建立了转基因作物分子特征分析的基因组学和蛋白质组学分析方法，建立了一套快速、高通量分析外源基因插入和筛选稳定遗传的转基因植物的DNA检测方法；③初步建立了转基因作物环境安全评价体系，完成了转cry1Ab/Ac抗虫水稻“华恢1号”和“Bt汕优63”的环境安全评价，为国家给两个转基因水稻品种颁发安全证书提供了科学依据；④阐明了转基因棉花农田生态系统中生物群落结构的变化规律及有害生物优势种类的演化机理；⑤以转基因水稻为代表，基本探明了转抗虫基因作物和转抗除草剂基因作物的基因漂移规律，以及影响转基因漂移的主要环境因子，提出了转基因作物基因漂移的安全控制措施等。

2）人畜共患传染病控制

当前我国人畜共患传染病的研究现状主要表现在：①多种人畜共患传染病流行广泛，危害严重，而且不断面临外来与新发人畜共患传染病的威胁；②缺乏连续、系统的流行病学监测与研究，技术储备薄弱，空白太多，应对新发传染病的能力较弱；③缺乏完整有效的防控技术体系，高水平专业人才严重不足，研究项目低水平重复，缺乏系统性和连续性；④高等级生物安全实验室匮乏，严重制约相关研究的开展；⑤国民关于人畜共患传染病的公共卫生意识淡薄，国家相关法律法规不健全，执法监督诸多环节缺位。

3）外来生物入侵

进入21世纪，我国加大了对生物入侵的基础与应用研究的投资，形成了从外来入

① 本文是中国工程院科研项目报告，北京，2010；工程管理研究，北京：科学出版社，2015，118-128. 作者：刘人怀，方智远，彭于发，李云河，张建中，步志高，万方浩，王瑞，杜冠华，冯忠武，马志勇，戴小枫，刘泽寰.

侵物种普查、入侵机理的基础理论研究到入侵物种的预防与控制技术三大层面的研究体系，在基础与应用研究中均取得了一些突破性进展，初步形成了国家防控技术体系。在理论基础研究方面：发现了入侵物种B型烟粉虱与土著烟粉虱之间的“非对称交配互作”机制；B型烟粉虱与其传播的植物双生病毒存在互利共生关系；入侵物种红脂大小蠹与土著种黑根小蠹的协同入侵机制；明确了紫茎泽兰的化感作用机制与偏利效应。在应用研究方面：改进了入侵物种适生性风险评估的技术与方法，评估了70余种入侵物种的适生性风险，制定了控制预案；建立了30余种入侵物种的快速分子检测方法，并开发了多种快速检测与野外监测的试剂盒；构建了早期预警、应急控制、阻断与扑灭、可持续综合防御与控制四大技术体系；建立了重大入侵物种生态修复与区域控制的野外实验与观测基地。在基础性工作方面：已收集400余种我国的入侵物种的信息，初步建立了风险评估的指标筛选方法及风险程度定量分析方法；初步建立了外来物种入侵对生态环境的影响评估技术方法。在学科建设方面：在国内外学科发展的基础上，结合我国科研成果，提出并构建了入侵生物学学科的框架，为生物入侵研究提供了模式。总体而言，我国生物入侵的科学研究已步入国际先进水平，并具备了与国际一流研究机构平等对话与交流的能力。

4）药品安全（医药、农药、兽药、生物制品）

人类健康的可持续发展是经济、社会和环境可持续发展的前提，而药品是人类维持生命健康的物质基础，是公共卫生安全的重要保证。然而，药品存在着发展风险，我国现行药品安全监管体系还存在着许多不足之处，主要表现在以下几个方面。①我国药品安全法律法规体系还不够完善。首先，作为我国药品管理法律体系核心的《中华人民共和国药品管理法》存在不足和空白。例如，将新药定义为“未曾在中国境内上市销售的药品”，造成审批注册的新药90%以上均为仿制药，使原本用于监管新药安全的标准体系形同虚设。其次，法律制度规定过于粗糙。例如，药品召回、药品广告、药品标准等制度仅有原则上的阐述，而缺乏相应的实施细则，造成遵守难、执行难的局面。最后，惩罚力度较轻，对违法者威慑力不足。例如，《中华人民共和国药品管理法》规定，制售假药者一般要处以货值金额2～5倍的罚款，而制售劣药者只需处以1～3倍的罚款。药品属于批次生产和销售的商品，能够发现的违法行为可能只是其中的一两批，很难准确计算出全部违法所得，因此有时甚至会产生罚金小于利润的情况，导致制售假劣药者前赴后继。②药品安全标准化体系存在缺陷，导致药品质量堪忧。1998年，我国开始实行药品生产管理规范（good manufacturing practice，GMP）认证制度，目前已延伸到药品流通、药品管理、药品临床科研和药品广告管理等领域，初步形成了药品安全监管的标准化体系雏形。然而，包括GMP在内的许多标准体系认证，都比较强调事前审批，后续跟踪复核则非常薄弱，导致药品生产经营企业把应付审批作为主要工作，审批过后对药品质量的重视程度明显下降。③组织机构体系不健全，药品管制人员缺乏。据统计，2007年全国药监系统人员不足4万人，而国内的药品生产企业有接近7000家，销售企业最少也有1.5万家，有限的监管人员根本无法满足规模巨大的药品行业，造成药品管理过程漏洞不断。并且，国家食品药品监督管理局作为药品安全管理的唯一部门，由谁来进行第三方监管也是一个亟须完善的问题。

④药品安全信息宣传系统存在问题。我国近年来虽然不断提高药品安全信息和文档的公开化、透明化程度，但在不合理用药、药品不良反应等宣传方面仍存在不足，同时对药品广告的夸大宣传、误导患者的情况经常监管不力。

当前，农药仍是保证人类粮食安全的最重要生产资料。我国农药技术研究发展迅速，目前已经成为世界农药生产和使用大国。然而，我国农药发展过程中的安全问题不容忽视，主要表现在：①我国自主研发的农药品种少，创制研究能力相对薄弱，核心专利技术被国外控制，使得为我国粮食安全提供新农药受阻；②相对于发达国家，我国农药产品质量较低、剂型少、产品稳定性差，给农药使用安全埋下隐患；③农药器械种类少，自动化程度低，药械研究力量薄弱，基础技术储备少，直接导致农药利用率低，造成农药安全性问题；④农药种类多，利用传统农药安全评价方法不能全面评价其安全性，需要完善新农药安全评价体系研究，保证人类健康和环境安全。

兽药安全研究是多学科交叉领域。我国兽药安全研究基础薄弱，与国际先进水平差距较大，安全性问题突出。我国兽药安全问题主要涉及兽药评价体系、兽药生产体系和兽药应用体系等多个层面，兽药残留的安全、由兽药滥用导致的次生安全问题（如耐药性扩散、环境安全等）以及兽药安全评价标准和体系亟须完善与发展。

生物制品的安全性问题也比较突出，主要表现在：①种类繁多，安全质量参差不齐；②生物原材料来源复杂，潜藏着外源性微生物和有害物质污染的生物安全隐患；③基因工程等新技术产品的安全性评价体系尚不完善；④生物制品的生物安全质量标准不高；⑤对生物制品的生物安全问题的研究投入和重视程度不够。

2. 当前生物安全研究的国际先进水平与前沿问题综述

1）转基因生物安全

当前国际农业转基因生物安全研究主要集中在对食品和环境的安全性评价方面。①对食品安全性的评价。主要研究内容包括毒性抗生素标记基因 *aphA2* 产物及转基因作物中外源基因表达产物对人和动物的毒性影响及过敏反应，提出了利用标记基因删除技术研制不含标记基因的转基因作物来控制 *aphA2* 产物对人及高等动物可能带来的风险。②对环境安全性的评价。主要研究内容包括转基因植物生存竞争性、基因漂移及其生态后果、对非靶标生物的毒性影响和对生物多样性的影响、对靶标生物的抗性影响和非靶标生物对转基因作物的抗性进化等。研究发现转基因作物与常规作物在生长势、种子活力及越冬、抗病、抗逆能力方面没有差异；对转基因作物外源基因漂移的研究主要关注的问题是转基因植物花粉的数量、生命力及扩散能力、杂交亲和性、杂交后代可育性和繁殖能力等；在非靶标生物影响方面主要研究转基因作物外源基因表达物对非靶标生物的毒性及外源表达蛋白在食物链及农田环境中的传递、转移和降解等；在抗虫作物靶标昆虫的抗性进化方面的主要研究内容是监测昆虫的抗性进化规律、以生化和分子手段研究抗性进化机制、昆虫抗性进化与昆虫适合度的关系以及昆虫抗性进化的治理措施等。

2）人畜共患传染病控制

当前人畜共患传染病研究的国际前沿问题主要有以下几个方面。①人类新发传染病的病原主要来自动物，包括野生动物和养殖动物。②先进国家高度重视人畜共患传染病的基础研究和防控技术的创新研究，已经形成完善的流行病学监测网络与数据分析中心，基本能够及时、准确掌握疫情动态；注重病原的高通量快速筛选、鉴定技术研发与应用，尤其重视探知未知病原的技术平台建设。③先进国家普遍拥有健全的法律法规和严格的执法监督队伍，国民普遍拥有较强的公共卫生意识；已经形成高效、快速应对新发人畜共患传染病的防控体系。④在人畜共患传染病的防控上，注重多部门间协作以及资源整合，高度重视国际合作。

3）外来生物入侵

外来生物入侵的研究热点主要集中在入侵基础理论研究和入侵防治技术研究两方面。①外来生物入侵基础理论研究：国际上生物入侵基础理论研究产生了一些解析外来物种入侵机制的新概念、新假设与新推论，如最小种群维持理论、十数定则、新武器假说、物种多样性阻抗假说、自然平衡假说、空余生态位假说、生态系统干扰假说、干扰产生空隙假说、土著种适应性差假说、环境发生化学变化假说、天敌逃避假说、入侵进化假说、资源机遇假说、繁殖体压力假说、生态位与遗传多样性正比相关理论。②外来生物入侵防治技术研究：以分子生物学和计算机技术为代表的现代科学广泛应用于生物入侵灾变机制的研究，大力推动了防控技术的发展，如入侵物种的遗传分化和快速进化，入侵物种的快速分子检测与诊断，基于地理信息系统（geographic information system，GIS）、全球定位系统（global positioning system，GPS）、图像识别和分析技术、人工智能决策支持系统技术和计算机网络管理技术的入侵物种预警系统，基本实现了入侵物种防控决策和咨询的网络化。生物防治和生态调控技术广泛应用于外来入侵物种的防控。

4）药品安全（医药、农药、兽药、生物制品）

欧洲药品管理局/人用药品委员会与医务人员专业组织联合工作组定期举行会议，为建立医药合作框架提供建设性意见；欧洲药品管理局/人用药品委员会和患者与消费者工作小组则针对消费者提供建议和信息，提高公众警觉性，促进消费者合理用药，发展并培训患者组织网络。美国食品与药物管理局每年公布一次药品评价和研发指南，依据发现药品企业采取的新方法和新技术的效果不断改善现行药品 GMP 标准，保证其 GMP 理念的先进性和制度的时效性。针对药物安全性的评价分为上市前的评价和药物上市后的监测两个阶段。在药物被批准上市前，管理当局将作出药物风险评价；上市后继续监测其安全性，并完善药品召回机制。

在农业病虫草害防治药物研究中，农药已经从强调“杀死”转向“控制”，注重“绿色”分子靶标的研究。当前农药研究主要集中于在分子设计、新合成技术及新药效方法的基础上探索发现先导结构及作用靶标，即分子靶标导向的绿色化学农药创新研究，该研究方向已成为现阶段国际农药研究与开发的主流。开展了农药的靶标比较生物学研究；根据靶标与作用小分子的选择性，设计合成了一系列高效、安全的“绿色农药”；采用控制释放技术、对靶喷雾技术、合理混用轮用等措施，大幅度提高了药物的剂量传递效率。

兽药安全研究的国际先进水平与前沿问题主要集中在以下几个方面：①设计和开发更加安全、高效、环境友好的新兽药；②开发兽药监管（兽药残留监测）的新技术；

③重视细菌耐药性的发展；④关注兽药生产、使用过程中的环境安全；⑤将分子生物学、基因芯片、药物流行病学等技术应用于兽药的安全评价。

生物制品安全的国际前沿问题主要集中在以下三个方面：①生物安全是生物制品研究的新热点；②传统生物制品的生物安全控制技术相对成熟，但仍需要改进；③基因工程等新技术产品的安全评价体系尚不完善。

3. 生物安全研究方面的重大工程科技问题

1）转基因生物安全

现阶段我国转基因生物安全研究主要面临的科技问题有以下几点。①安全技术体系尚不完善，可用于风险评价的基础数据较少，风险预警和管理体系缺乏。对于新基因、新性状、新用途和多基因等转基因生物的安全性评价及检测能力不足。②自主知识产权技术较少，安全评价理论、方法和技术主要靠借鉴国外，对国内自主创新和国外新研制的部分转基因生物难以及时准确地进行安全性评价与检测。③关于转基因作物对作物起源中心和基因多样性的长期生态效应的影响，以及针对第二代（质量性状、医药和工业用途等）转基因作物新材料的安全研究技术还比较缺乏。④转基因生物安全评价技术体系和研究平台需要进一步完善。⑤我国的转基因生物风险监测、预警及应急处理能力，需进一步提高，需建立起覆盖全国政府管理部门、科研机构和地方环境监测机构一体化的转基因生物安全监测预警与快速反应体系。

2）人畜共患传染病控制

人畜共患传染病研究中的重大工程科技问题主要有：①重大人畜共患传染病的病原学、流行病学与生态学研究；②有关病原跨种传播、致病机制及免疫机制等基础研究；③病原学与血清学快速、敏感、特异的高通量检测诊断技术创新及其产业化；④新型治疗药物及疫苗的创新研制与产业化；⑤未知病原的快速筛查和鉴定技术平台与能力建设；⑥病原生物信息学与流行病学信息的数据集成和分析；⑦加强新兴生物技术，如基因组学、蛋白质组学技术在人畜共患传染病的快速筛查和鉴别诊断技术平台建设中的应用。

3）外来生物入侵

针对入侵生物的生态学过程、成灾机制及其预防与控制特点，分别从基础性工作、基础研究到应用技术研究及推广的国家主体科技计划，列出未来应优先发展的方向，具体如下所示。①基础性工作——科学技术部（以下简称科技部）基础性平台建设：农林外来入侵物种和安全性普查及其数据库的完善。针对我国外来入侵生物频发而基础数据信息严重匮乏的现状，在前期基础上，重点围绕生物入侵灾害严重的区域（主要为华东和西南地区）和主要生态系统，全面系统地开展农林外来入侵物种（节肢动物、无脊椎动物、杂草、植物病害病原物）的种类、分布、危害的调查和考察并进行数据的收集与整合；同时构建外来生物入侵信息数据库系统及共享技术平台和远程实时诊断与服务平台。外来入侵物种成灾机制的野外长期定点监测：针对局部分布的重要入侵物种，在其对应的入侵区和前沿扩散阵地，建立长期的定点野外观测试验站和观测点，以研究外来入侵物种种群适应、扩散扩张成灾的生态过程和机制。②基础研究——国家“973”计划项目：重要农业入侵物种的灾变机制与控制基础。以重大农林危险性外来入侵物种为主要研究对象，从宏观尺度（景观、区域以及全球尺度）、生态系统生态学核心尺度（群落/生态系统尺度）以及微观尺度（分子、个体以及种群尺度）三个尺度，着重研究生物入侵的机理、过程以及生态系统功能变化与反馈这三个前沿热点问题。全球气候变化对生物入侵的影响与生态学效应：综合当前全球气候变化以及外来生物入侵现状及其效应评估的热点问题，在不同层面系统研究气候变化对外来入侵生物的入侵特性、本地生态系统的可入侵性、外来与本地共生物种关系的影响，以及基于气候变化的外来入侵生物预警与控制基础的理论与方法，发展有效的监控技术体系。重要外来入侵物种对本地物种的竞争效应及控制的基础研究：以重要农林外来入侵物种为研究对象，围绕外来入侵物种入侵后至成灾的生态过程中的外来物种对本地物种的竞争排斥/替代/置换的这一核心科学问题，重点研究外来物种排斥本地物种的种间竞争类型及其相应的竞争机制，并创新研究外来物种防控策略与基础技术。③高技术发展——“863”计划项目：重要农业入侵物种辐射不育技术。以苹果蠹蛾/橘小实蝇/橘大实蝇为研究对象，通过实验室培育靶害虫的雌雄不同蛹色品系，进行大量饲养，分离雌雄虫，选择雌虫进行辐射促使其不育，并在域区大量释放不育雌虫，最终达到防控/根除靶害虫的目的。重要农业入侵物种野外、实时、远程监控的数字化平台：以重要的局部分布并具有快速扩张和危害的入侵物种，如紫茎泽兰、苹果蠹蛾、橘小实蝇等为对象，建立基于图像识别及野外实时调查的野外简便数据传输系统和实时远程监控的信息采集、信息处理与信息发布系统，以明确入侵的扩张趋势、路线及成灾机制以及分析环境或气候保护与入侵物种种群扩张的关系。

4）药品安全（医药、农药、兽药、生物制品）

目前我国医药安全方面的重大工程科技问题有以下几个方面：①常见与重要药品的安全标准研究；②药品不良反应信息采集、评价、分析利用和预警系统研究；③常见药品安全事故的主要原因和对策研究；④传统中药毒性成分和有效成分的提取分离、分析方法研究；⑤毒性中药材质量评价方法、定性与定量检测方法和限量标准研究；⑥建立一批具有自主创新能力、符合国际标准的药品研究开发示范基地；⑦全国范围内的药品生产管理数据库和安全监管信息化工程建设。

现阶段我国农药安全研究主要面临的科技问题有以下几点：①绿色农药创制问题以及农药毒理学机制研究欠缺；②农药产品质量有待提高，农药加工和农药应用技术储备不足；③农药职业健康风险评估研究尚处于起步阶段；④新的农药安全评价体系正处于补充和完善阶段。

我国兽药安全研究中的关键工程科技问题是：①兽药残留标准和兽药使用规范研究；②兽药残留检测技术研究；③耐药性监测和耐药机制研究；④兽药（兽药残留）安全性的评价及毒性机制研究；⑤兽药环境影响监测研究；⑥药物与机体相互作用规律以及比较药理学/毒理学研究；⑦新型安全高效兽药开发专项；⑧兽药良好实验室规范（good laboratory practice，GLP）安全评价认证和体系建设；⑨兽药安全评价新技术和新领域研究。

为了保证生物制品在防控动物疫病中的关键作用和最大限度地降低生物安全隐患，必须围绕生物制品的生物安全质量和生物安全隐患的关键科学问题开展研究：①弄清生物原材料对生物制品质量安全的影响；②探究生物制品的生物安全问题发生的机制；③开发生物制品原材料外源性微生物和有害物质的检测技术；④研究和完善传统生物制品和基因工程生物制品的安全评估技术体系；⑤制定更高标准的生物制品安全标准。

4. 保障生物安全的重要措施与政策建议

1）转基因生物安全

转基因生物安全管理是风险评估、风险管理和风险交流三位一体的。当前，我国在转基因生物风险评估方面的技术已经有了很大进步，逐步形成了既与国际接轨又有自己特色的技术体系，但在风险管理和风险交流方面还需要大力改进和完善。具体提出以下措施和政策建议。①继续加强转基因生物安全评估能力建设。加强专家队伍建设、稳定的机构建设、现代化的实验室设施建设、规范的野外基地建设等。②建立完善的全国转基因生物综合管理体制和协调机制，在国家层面上建立一个综合监督管理和专业管理相结合的管理体系。③建立完善的转基因生物环境风险监测和预警体系，提高应对有关转基因作物生物安全突发事件的能力。例如，建立地方环境监测机构、科研机构和政府相关部门间相互支持的一体化监测、预警和快速反应体系来应对转基因生物安全方面的突发事件。④加强对我国公众的生物技术科普教育，减少对转基因食品的恐慌和误解。政府建立并运行转基因生物安全信息交换机制，保证生物安全信息的畅通和公众对相关信息的知情权，使转基因技术方面的信息公开，让公众参与，让普通民众对转基因生物安全有基本的了解。

2）人畜共患传染病控制

首先，在流行病学监测和研究方面保持稳定持续的科研支出，确保基础研究的连续性和系统性，不断创新防控技术，提高防控能力；在国家层面统筹安排，合理布局，逐步建设网络完整的高等级生物安全设施［BSL-3/4（biosafety level-3/4，生物安全水平-第 3 类和第 4 类）实验室］；建立人畜共患传染病原体及其相关基因材料及诊断标准物质的保藏中心、病原生物信息与流行病学信息数据中心。其次，不断完善卫生防疫和兽医防疫方面的法律法规，加强执法队伍及其能力建设，严格执法；制定完善的人畜共患传染病突发事件应急预案；加强人畜共患传染病相关公共卫生教育，引导国民形成正确的公共卫生观念、健康的生活方式和社会活动方式：不断改进畜牧养殖业生产、加工、流通和消费环节，建立与生产消费规模相适应的先进的环境卫生管理系统。最后，强化卫生系统与兽医系统的联动和合作机制；建立多渠道的国际合作，实现我国人畜共患传染病基础研究和防控技术与能力的跨越式发展。

3）外来生物入侵

（1）保障措施：建立稳定的人、财、物保障措施。以稳定的项目支持来保持人才队伍、基础设施的稳定和合理可持续利用。扩展科技平台的基础和作用：在现有的相关国家实验室，国家重大科学工程，各级植保植检站、森保站等的基础上，稳定和拓宽这些平台的功能，并加强这些平台的合作和协作。改善科研和技术应用推广的创新环境：制定新型鼓励和支持产学研相结合的模式与考核机制，从而营造科研和技术应用推广的积极氛围，使实验室的科研成果真正转化为服务于“三农”的现实生产力。

（2）政策建议：加大科研经费的公共投入，完善基础设施。制定外来入侵物种防控策略的国家行动的长期规划和近期目标，加大政府对生物入侵预防与控制的科研和技术应用推广的投入，同时吸引相关的企业参与，并形成稳定持续支持的长效机制，以稳定人才队伍和防控措施的网络结构。培养和稳定人才队伍，不仅要培养生物入侵领域前沿研究的顶尖人才，以抢占科技制高点；而且要利用科研教学推广三农大协作，真正使外来入侵物种的防控策略、技术和公众认知深入人心。建立由农业部牵头、相关部门协调统一行动的合作机制，提升预警和快速反应能力；成立生物入侵的国家专业咨询委员会，负责统一发布相关的入侵物种信息。构建有效的组织管理机制，以项目合作为纽带，以原有的农技系统网络为平台，产学研结合。完善和健全相关的法律法规，确立国家统一的外来生物入侵风险评估指标体系。同时，通过宣传、培训和教育来强化公众意识，充分发动公众参与，使外来入侵物种防控策略和技术的实施变为公众的自觉行动。大力加强和促进国际合作，基于外来生物入侵学科本身的特点，加强国际合作，特别是与入侵物种的原产地国家之间的合作。

4）药品安全（医药、农药、兽药、生物制品）

（1）医药方面：①构建科学完善的药品和生物技术产品评价体系，加强药品注册管理法规建设，严格药品注册审批程序，加强对药物临床研究及临床前的过程监督检查；②加强药品生产质量监管，提高我国药品生产管理规范的实施水平，并逐步与发达国家接轨；③完善药品不良反应监测网络，对已上市药品分期分批开展再评价研究，建立并完善上市后药品监测、预警、应急、撤市、淘汰的风险管理长效机制；④加强药品检验检测方法研究，搭建药检系统技术平台，普及快速检测技术，建立与完善全国药品技术检验信息管理和数据交换系统；⑤建立完善中药标准规范和技术评价体系，研究构建具有中国特色、符合中医药规律的中药标准规范和技术评价体系基本框架，加大对民族药品的扶持和监管力度，积极倡导建立传统药物国际协调机制；⑥加强国际合作交流，促进药品安全监管水平的提升；⑦广泛建立药品安全使用知识培训体系，普及药品安全知识，正确引导公众对药品安全的认知。

（2）农药方面：①农药安全问题重在防范，对公共安全领域的重大工程科技问题研究要转变研究理念，在预防、监测、预报方面加大力度；②广泛建立农药安全使用知识培训体系和技术推广平台，加强农药使用技术标准化和使用技术培训；③建立农业病虫草害急性发生的预警与农药应对防控体系；④加强农药化学与应用基础研究；⑤完善农药安全评价体系；⑥建立农药废弃物回收站并加大宣传力度；⑦建立农药等污染物快速检测技术平台。

（3）兽药方面：①加大兽药安全研究的资金投入，尤其是加强兽药安全基础性研究，为兽药安全标准制定和监控提供技术支持；②加快兽药 GLP 安全评价体系认证和基础设施建设，消化吸收改进 VICH（Veterinary International Conference on Harmonization，兽药注册技术要求国际协调会）等国际组织的评价标准体系，提高兽药安全评价标准；③加快安全、高效新兽药的研制开发与相关科技成果转化；④加快兽药安全人才培养与引进，普及兽药安全知识，引导公众建立对兽药安全的信心。

（4）生物制品：①建立生物制品安全评价专用的无特定病原体动物实验中心；

②提高生物制品的生物安全质量标准。

四、食品安全相关工程科技发展战略①

1. 当前我国食品安全的研究现状

食品安全作为重大民生问题，影响公众健康乃至国家安全、社会稳定，在发达国家位列公共安全之首[4]。我国先后颁布了《中华人民共和国食品安全法》、《中华人民共和国农产品质量安全法》和《中华人民共和国食品安全法实施条例》，明确将实施风险评估和风险监测制度作为国家需要建立的制度，鼓励加强食品安全和农产品质量安全科技的投入。《突发公共卫生事件应急条例》明确将食物中毒等食品安全事故列入突发公共卫生事件。《国家中长期科学和技术发展规划纲要（2006—2020 年）》也将食品安全列入公共安全重点领域。

科技部组织、实施了“十五”国家重大科技专项、“十一五”国家科技支撑计划重大项目“食品安全关键技术”和“水体污染控制与治理”国家科技重大专项“饮用水安全保障技术研究与综合示范主题”，缩短了我国安全科技方面与发达国家之间的差距，在检测技术装备研制、风险评估、应急管理等方面的能力建设取得了跨越式发展；在关键检测技术、风险评估技术、监控预警技术等方面的自主创新能力有了较大的提高；示范区的示范和科技引领作用得到了充分发挥，安全状况逐年好转，强力支撑了我国食品产业的发展，保障了消费者的健康。

1）食品安全风险评估技术建设迈出可喜步伐

按照《中华人民共和国食品安全法》和《中华人民共和国农产品质量安全法》，国家食品安全风险评估专家委员会和国家农产品质量安全风险评估专家委员会已分别由卫生部和农业部批准成立。在科技支撑方面，风险评估是世界贸易组织（World Trade Organization，WTO）和国际食品法典委员会（Codex Alimentarius Commission，CAC）强调的成员国用于制定食品安全法规、标准和政策等控制措施的科学基础，而风险评估依赖食品污染、食源性疾病和食品中有害因素的监测和国家食物消费数据库的建立，同时也需要危害识别技术的支撑。针对我国风险评估基础数据缺乏、风险评估技术还未能全面应用等问题，重点开展了食品毒理学安全性评估技术、食品污染物暴露评估技术研究，建立和完善了我国食品病原微生物、农兽药残留、化学污染物（含生物毒素）、新资源食品等风险评估技术，初步开发具有自主知识产权的食品污染物评估模型和具有自主知识产权的风险评估计算机软件，培育了多个风险评估研究基地。例如，针对中国人膳食消费习惯的重金属、有机磷农药、氯丙醇、二噁英、丙烯酰胺、霉菌毒素等典型污染物进行了模型评估，逐步建立了我国农兽药残留、化学污染物等风险评估技术体系、模型和指南，开展了原产地溯源、污染物溯源、大型动物个体溯源和电子标签溯源等方面的研究。2008 年“三聚氰胺问题奶粉事件”中，我国在风险评估的基础上制定了三聚氰胺临时限量标准，并获得世界卫生组织（World

① 本文是中国工程院科研项目报告，北京，2010；工程管理研究，北京：科学出版社，2015，128-143. 作者：刘人怀，陈君石，吴永宁，吴希阳.

Health Organization，WHO）等国际组织的认可。但我国食品安全风险评估技术体系刚刚起步，符合国际规范的评估参数体系尚待建立。首先，我国评估工作仅局限于污染物的平均暴露量的比较，这对慢性长期暴露十分有用；但对于评估敏感人群，也就是高暴露和高风险人群，有些评估参数（如中国婴幼儿食物消费量分布数据库等）则尚未建立，农药最大残留限量标准中建立在急性毒性数据基础上的急性参考剂量还没有实施，而且即使近期实施也是采用欧、美、日、澳等发达国家和地区的数据，与我国实际膳食结构差别很大，不适用于保护我国公众。其次，我国食品中的不少污染情况仍然本底不清。生物性与化学性危害是目前中国食品安全的主要因素，而我国当前缺乏食源性危害的系统监测与评价资料。最后，目前食品污染并不是单一污染，而是多种污染物的联合暴露所产生的复合污染效应。我国农药开发过程中创新能力还有待提高，目前以仿制农药的复配制剂为主，联合暴露问题更加突出。鉴于食品中两种或两种以上的化学污染物存在潜在的相互作用，其复合效应及预测和评估技术是各国目前与今后进行化学污染物风险评估的重难点，我国在这方面的探索性研究才刚刚开始。

2）食品安全风险监测与预警技术已具雏形

《中华人民共和国食品安全法》规定国家建立食品安全风险监测制度和食品安全风险评估制度。为了解决食品安全情况不明、本底不清的问题，在“十五”国家重大科技专项“食品安全关键技术”的支持下，卫生部着手建立了全国食品污染监测网，化学污染监测覆盖 13 个省（区、市）食源性疾病，病原菌监测覆盖 16 个省（区、市），各监测点通过“国家食品安全监测信息系统”报告监测数据；农业部建立了农产品中农药和重金属污染与动物源产品的残留监控计划。通过监测，基本“摸清”了我国在污染物方面的家底，掌握了我国食品中的重要污染物及其污染状况，积累了大量的基础数据，向世界卫生组织食品污染监测和评估规划和 CAC 提交了我国的监测数据，并参与国际标准制定。卫生部发布了丙烯酰胺等风险预警通报，建立了重要食源性致病菌的分子分型和溯源分析数据库，并在四川、河北、河南、广东、福建等监测点的数起食品污染调查，食物中毒病因学诊断、溯源中应用。基于 SQL Server 的动物源病原菌耐药性数据库，可对区域耐药性趋势进行预测，并对耐药安全等级进行监控。这些工作成为国务院要求的“建立覆盖全国省、市、县并延伸到农村的食品污染和食源性疾病监测网”的基础。但目前食品污染和食源性疾病的监测数据资料还很有限，只有静态的数据而缺少动态数据，终端产品监测数据多，而覆盖整个食物链，特别是产品生命周期的前期危害物监测缺失，食品安全信息渠道不畅通，各方面的数据不能共享共用，还没有建立起人群食源性疾病症状监测网络，远远不能达到科学预警的要求。由于实验室能力参差不齐，只在有能力开展工作的省份建立了监测网络。在致病微生物造成的食源性危害方面，我国目前尚缺乏对于基于哨点实验室和城乡社区居民开展食源性疾病的主动监测，以及引起食物中毒的常见重要致病菌进行风险评估的背景资料。在化学污染方面，无论是监测覆盖面、监测项目、监测技术、数据库建设和应用等方面均与国家要求还有很大差距。对食品中农药和兽药残留以及生物毒素等的污染状况尚缺乏长期、系统的监测资料；一些对健康危害大而在贸易中又十分敏感的污染物，如二噁英及其类似物、氯丙醇酯和某些真菌毒素污染状况及其对我国居民健康的影响尚不清楚。与发达国家相比，我国食品污染物监测网的差距在于监测网点的代表性、监测目标物特别是农药品种的全面性和针对性较低，从而限制了监测结果对我国食品安全的警示作用。这些基础数据的缺乏使食品安全预警更多地停留在经验阶段。由于技术支撑薄弱，我国尚未建立食品安全预警系统。

3）食品安全检测与危害识别技术基本与国际接轨

“十五”和“十一五”期间，科技部重点部署食品安全检测技术能力建设。针对我国食品安全检测技术主要集中在化学分析方法、痕量和超痕量确证技术、残留检测技术上，且大多为单一成分分析的状况，建立了粮谷、茶叶、果蔬、果汁、肉类食品等农产品中农药多残留系列检测和确证方法，可同时检测 500 多种农药；建立了 20 余大类（激素、β-兴奋剂等）300 余种兽药残留确证检测技术，最多能同时检测 70 多种兽药。尤其是代表一个国家分析水平的二噁英和多氯联苯等超痕量检测技术，促进成为二噁英和多氯联苯国际标准物质的定值实验室，显著提升了我国食品安全科技领域的国际地位，使中国成为 CAC 二噁英、氯丙醇、丙烯酰胺等热点污染物的国际标准的起草国之一。在微生物检测技术方面，开发了适合肉制品和水产品中 19 种致病菌复合增菌和 PRC（polymerase chain reaction，聚合酶链反应）快速检测方法。建立了能与国际 PulseNet（病原菌分子分型实验室监测网络）接轨的沙门氏菌、大肠杆菌 O157∶H7、空肠弯曲菌、副溶血性弧菌、单增李斯特菌、阪崎肠杆菌等重要食源性细菌的脉冲场凝胶电泳技术和核糖体分型溯源技术平台。针对传统的食品毒理学安全性评价程序与方法，我国已经建立了国家标准，但对国际新技术的追踪投入较少。一是在投入比例上过分强调化学分析检测，使得食品毒理学检测技术更加薄弱，在相当大的程度上限制了对食品的危害识别与溯源能力；二是特殊毒性测试及其食品新资源评价技术（如动物代替试验、免疫毒性和致敏试验）尚未建立；三是毒理组学技术刚刚起步，生物标志物在人群生物监测中仅是个别应用；四是我国毒理学检测方法的标准化程度低，与国际良好 GLP 要求有相当差距。

另外，对于潜在污染物和非法添加物的识别等方面手段不多、技术储备不足，由于受不同类化合物结构性质以及基质等的限制，没有形成系统的样品前处理和筛选监测技术，难以应对我国复杂的食品安全局面。基于以现代多维色谱-高分辨质谱技术为基础的食品危害物高通量筛选识别技术目前正成为国际食品安全研究领域的热点，是食品安全危害识别的强有力技术手段。

4）已开发一批具有自主知识产权的快速检测试剂和装备

已建立小分子半抗原抗体制备技术平台，研制出针对盐酸克仑特罗、氯霉素、利血平等农药、兽药残留的食品安全检测试剂盒 170 余种，用于食品中农药、兽药、生物毒素、食品添加剂、饲料添加剂和违禁化学品及动植物病原体等的检测，研制了 50 余种相关检测设备，实现了“瘦肉精”“氯霉素”等主要违禁药物残留免疫快速检测技术产品的国产化，基本满足了我国食品安全领域快速检测的市场需求，提升了我国食品安全监测和突发事件快速应对的能力。食品安全检测装备开发取得重大突破，研制了拥有多项发明专利、集多项检测技术于一体的食品安全检测车；“系列食品安全快速检测装备”在军队和地方食物中毒原因的调查中多次得到应用；特别是研制的 H5、H7、H9 等不同亚型禽流感病毒的荧光 RT-PCR（reverse transcription-polymerase chain reaction，逆转录聚合酶链反应）检测试剂盒，使检测时间从 21 天缩短为 4 小时。致病性微生物引起的食源性疾病仍然是我国目前食品安全的主要问题，占食物中毒每年发病人数的 50%以上，但食品中致病微生物的免疫检测技术缺少；农兽药残留、各种污染物和非法使用添加剂在实际样品中是多种类多组分共存，目前缺少能够应对突发食品安全事故和针对国家整顿的食品“非法添加物”的储备抗体。我国在这方面的研究刚起步，还没有形成具有实用价值的抗体资源库。

5）应对食品安全突发事件的能力日益增强

已逐步形成了国际认可度高、高效快捷的食品安全检验检测体系，应急能力大幅度提升，在应对 2004 年的苏丹红，2005 年的孔雀石绿、啤酒甲醛、禽流感，2007 年的毒饺子，2008 年三聚氰胺等突发事件检测中发挥了重要作用。例如，专项课题率先研制出快速、准确的禽流感病诊断方法，解决了禽肉中微量禽流感病毒快速检测技术难题，在应对禽流感突发事件、促进我国的禽肉出口等方面都起到了重要作用。建立的基于症状和物质特征查询食物中毒数据库和基于电喷雾质谱的毒理学质谱库，已经成功应用于北京“山吧可乐定中毒”等多起食物中毒事件侦破和违法添加筛查。苏丹红事件和三聚氰胺事件中，疾病预防控制、检验检疫、农业等食品安全关键技术项目参加实验室迅速应对，在很短的时间内拿出与国际接轨的检测方法，为保障消费者的身体健康、维护社会稳定和保护国际国内贸易提供了技术支持。但我国目前尚缺乏一个系统的覆盖食品安全甄别、处置和响应的“金”标准和规范，特别是对未知物的筛查识别技术需要突破。

6）创建了各具特色的食品安全控制模式

食品安全共性技术的应用和综合示范，支撑了区域重点产业的发展，增强了核心竞争力，培育了一批放心企业品牌。针对部分农兽药等有毒有害物质的吸收、转化、代谢和消解规律以及食用农产品生产、加工包装、储藏和流通等过程的安全控制，研制开发了一系列处理技术和方法；以危害分析与关键控制点、良好农业规范等为基础，建立了主要食用农产品全程质量控制技术。形成了依托超市把关的江苏苏果超市、突出投入品控制的陕西洛川苹果、着力全程检疫的福建银祥猪肉、严把出口质量的广东恒兴水产、兼顾种植和加工安全的河南鹤壁小麦、政府主导与区域经济社会化的山东寿光蔬菜、保障重大活动安全的北京奥运和青岛奥帆餐饮、突出基地选择的浙江龙井茶叶和强化配送安全控制的北京物流等 9 个各具特色的食品安全示范模式。但缺乏对从种养殖业到最终消费环节整个产品生命周期的全程相对完整、配套、可操作的食品生产过程安全控制技术体系、标准和具体措施。

7）初步形成饮用水安全保障技术系统

在饮用水安全保障方面，我国已初步建立了城市饮用水安全保障体系。在水源开发保护、供水设施保障、水质监督监测、管理能力建设等方面做了大量工作。全国新增城市供水量 160 亿立方米，供水普及率达 96.8%，出厂水质合格率接近 90%，初步具备了饮用水安全的监督监测和管理能力，在一定程度上保障了城市饮用水的需求。

例如，国家实施了“水体污染控制与治理”科技重大专项“饮用水安全保障技术研究与综合示范主题”，选择深圳、天津、上海等城市作为试点，突破了饮用水源水水质改善、常规处理工艺强化、安全消毒与安全性评价等关键技术，建设了3个日产水量20万吨以上的饮用水安全保障示范工程。但随着经济社会发展、人口增加和城市化进程的加快，我国城市饮用水安全面临的形势依然严峻，水资源短缺、水环境污染、供水设施不配套、饮用水安全管理体制和机制不健全等问题仍然突出。目前，我国尚缺乏城市饮用水安全评价的标准和规范。全国城市饮用水不安全人口为9900万人（扣除水质、水量重复计算人口2171万人），占规划范围内现状城市总人口41 829万人的23.7%，其中水质不安全（末梢水合格率低于93%视为不安全）人口为7196万人，水量供给不足（供水保证率低于90%或地下水超采率大于115%）的人口为4875万人。废污水排放及面源污染严重、净水处理工艺技术陈旧、管网老化破损、自建供水和二次供水技术设施落后以及管理不善等成为影响饮用水水质的主要因素，水源地水量供给不足和部分城镇公共供水能力不足使该问题加剧。更加严重的是，许多农村地区，由于工农业生产和高自然本底（如高砷、高氟）造成水源污染，得不到清洁卫生的饮用水。此外，饮用水安全管理及应急供水能力薄弱也是影响饮用水安全的因素之一。在水源水质变化与水厂管理（预警）和操作（实时控制）相结合的技术领域基本上还是空白，在预处理和深度处理技术集成和应用方面，我国与发达国家仍存在较大的差距。

2. 当前食品安全研究的国际先进水平与前沿问题综述

1）实施从水源到龙头的全程保护是饮用水安全保障的核心

目前，国际上解决饮用水问题的战略，一种称为“退缩战略”，另一种是以大系统解决问题的“扩张战略”。饮用水的“退缩战略”可以表示为：河流→水库→自来水厂→饮水机→瓶装水。从国家保障河流水生态系统步步退缩，直至每个人以瓶装水保证饮用水质量。另一种“扩张战略”指导思想则恰恰相反，从修复水生态系统入手，保证河流水源地供水质量，从而保证自来水直饮，保证全民的饮用水质量。世界上很多国家对饮用水源水质评价开展大量的研究，提出了比较完善的水质指标体系、标准和评价方法。美国《安全饮用水法》要求各州和供水单位对所属饮用水水源水质进行调查、评价，确定水体遭受污染的脆弱性。新西兰地表和地下水水源地水质评价包括风险源等级评价和水源地水质等级评价、突发水污染事件的概率预测评价、污染物质在水体中的扩散行为、水体健康风险评估等。在系统评价基础上，可以有针对性地开展包括有机污染、重金属污染、持久性有机物污染、富营养化等污染控制技术研究和工程实施。由于水源地水质得不到有效的保证，给饮用水净化带来极大的难题。美国等发达国家对水源水质保护—预处理—水厂处理—安全消毒—输配过程的水质保证进行了系统的技术研究和应用实施。随着膜技术的发展及其在水处理方面应用水平的提高，国外已研究开发出适于饮用水深度净化的膜技术及其集成应用工艺，不少水厂在常规过滤和活性炭过滤的基础上，采取超滤、纳滤等工艺，使饮用水水质大大提高。国外一直将安全消毒作为饮用水技术研究的重点，将强化致病微生物控制、减少消毒副产物等作为关键技术。

2）世界各国加大食品安全科技投入

世界卫生组织（WHO）和联合国粮食及农业组织（Food and Agriculture Organization of the United Nations，FAO）等国际食品安全主管组织及国际食品法典委员会（CAC）近年来都大大加强了对食品安全工作的投入与重视。2000 年第 53 届世界卫生大会（world health assembly，WHA）通过的 WHA 53.15 决议将食品安全列为优先领域。2002 年，WHO 发布了《全球食品安全战略》，随之 WHO 与 FAO 合作建立了国际食品安全当局网络。2010 年 WHO 执行委员会于第 126 次和第 127 次 WHO 会议上再次将食品安全作为重要议题，分别提出了《推进食品安全行动》（EB 126.R7）和《人类饮用水安全管理战略》（EB 127/6）。要求加强食品安全并针对影响健康的生物、化学与核危害建立全球性快速响应机制；通过全面实施全球食品安全战略，加强国际食品安全当局网络合作，共享食品安全应急数据、信息、知识；与 FAO/国际兽疫局（Office International des Epizooties，OIE）合作建立食源性疾病和人畜共患传染病疾病负担（微生物和化学性）监测体系；像化学污染物一样，将食源性致病菌纳入全球环境监测系统中的食品污染监测和评估规划，提高暴露评估能力。

目前世界发达国家和地区，如美国、欧盟、澳大利亚、新西兰和日本均制定了详尽的食品安全保护战略并启动了相应的科技规划。美国在《健康公民 2010》和《健康公民 2020》计划中均将食品安全列为优先主题，以降低食源性疾病发病率和提高人民生活质量为目标。相关部门启动了相应的食品安全研究专项，仅美国农业部实施的 108 食品安全计划在 5 年中就投入了 5 亿美元。美国政府 2007 年针对食品安全面临的新挑战，如高危人群的增加、消费模式的改变、食品供应全球化、食物恐怖和新的食源性病原体启动了食品保护计划，由美国食品与药物管理局和美国农业部联合实施；2009 年美国国会批准了 1250 亿美元的预算用于实施《美国 2009 年食品安全加强法案》。俄罗斯政府 2010 年 2 月发布了《关于批准俄罗斯食品安全战略的法令》。欧盟第六框架计划中涉及食品安全的科研经费达到 6.85 亿欧元，第七框架计划则在食品安全、农业和生物技术方面的科研经费达到了 19 亿欧元。

3）食品安全风险评估是国际社会和各国政府共同遵循的准则

WTO 认可的以科学为基础的风险分析框架是建设有效食品安全控制体系的基础。强化本国食品控制体系、履行基于风险分析的食品安全控制策略是当前全球食品贸易的需要和各国在进行食品贸易中必须遵守的原则。FAO 和 WHO 联合建立了 CAC 作为风险管理机构，制定国际标准和规范，同时建立专门的联合专家委员会，如联合国粮农组织/世界卫生组织食品添加剂联合专家委员会（The Joint FAO/WHO Expert Committee on Food Additives，JECFA）和农药残留联席会议等，开展食品安全风险评估，实现了风险评估与风险管理的分离，进而使得食品安全法规、标准、准则和规范的制定更具科学性与透明度。欧洲食品安全局、日本内阁府食品安全委员会、德国联邦风险评估研究所等食品安全风险评估专门机构的建立，以化学、毒理学、营养学、微生物学和分子生物学等学科为基础建立起来的多学科机制的定量风险评估技术在食品安全保障工作中飞速发展。风险评估报告的发布增强了消费者对政府保障食品安全的信心。

WHO 组织实施的全球环境监测系统中的食品污染监测和评估规划为开展国际暴露评估已经建立了全球 13 个地区性的膳食数据库。欧盟和美国更加成熟，分别开发了只供本地区/本国机构使用的相关软件，建立了比较成熟的从点评估、分布评估到概率评估的一套解决不确定度的方案，以收集各国食品消费量和污染水平建立数据库为重点，在个体数据基础上的概率分布模型获得了高百分位数的精确估计。这些技术已逐步被 JECFA 采纳，如在镉限量的国际标准制定中采用。基准剂量模型及以生理学为基础的药动/药效模型、暴露边界法相关技术和软件也在不断推进，有关基准剂量模型的软件也由美国环境保护局和荷兰国家公共卫生与环境研究院开发成功。美国毒物和疾病登记署在各种方法研究的基础上，提出了对化学污染物的联合效应评估，以及从致癌效应和非致癌效应两个方面进行系统研究的模式指南。

4）在风险分析框架下建立食品安全标准

食品安全标准体系的有效实施，可以使食品生产和安全控制的全过程标准化、规范化；实现对食品安全各个关键环节和关键因素的有效监控，满足食品安全标准的规定，可全面保证和提升食品质量安全水平。同时，系统完善、科学合理的食品安全标准体系是国家食品安全监管部门规范市场行为的重要依据，为建立和完善食品市场规则体系、法律法规体系和市场管理体系奠定基础，提供技术依据和支撑，从而为消费者营造放心消费环境，提供安全卫生食品，切实保护消费者的健康安全和权益。欧盟、美国和日本等一直将很多精力和时间放在 CAC、OIE 等的国际标准化活动上，并依赖其风险评估研究起步早的优势主导食品安全国际标准的制定，不遗余力地试图将具有限制发展中国家食品出口作用的本国标准变成国际标准。为此，发达国家投入巨大，如美国国家标准与技术研究院，每年从政府得到的标准研究经费多达 7 亿美元。1999 年 6 月至 2001 年 9 月日本投资数亿日元，历时两年三个月完成了日本标准化发展战略的制定任务。

5）实施从农田到餐桌的食品安全全程控制

安全的食品是生产出来的，全程质量控制已成为世界各国公认的食品安全最佳防控模式。全程质量控制通过全程监管，对可能出现问题的环节预先加以评估和防范，世界各国纷纷实施良好农业规范以及各种生产规范，以求从源头上控制食品安全。发达国家和 CAC 等国际组织都已制定了大量的农兽药最大残留限量标准，成为农产品市场准入和质量安全监管的重要依据。近年来，欧美日等国家（地区）对食用农产品设定的最大残留限量值正在快速增长，限量指标也越来越严格。食用农产品的重要特性之一就是其“鲜活性”，因此要求检测速度要快，多目标、精准的检测技术也成为重要的发展趋势。发达国家食用农产品农兽药、违禁添加物快速、高通量检测方法和产品得到了很大的发展，确证检测方面也趋向于高精度、多残留检测方法的研发，检测水平达到了超痕量水平。以此为基础，实行问题食用农产品的追溯和召回，既可明显节约人力物力财力，又能最大限度地保证产品安全。良好农业规范的对象是大田作物、果蔬、牛羊、奶牛、猪、家禽、茶叶、水产的农业生产经营者或是由独立的农业生产经营者组成的联合组织，是对食品链源头实施管理的有效形式，良好农业规范要求初级农产品种植、养殖过程实施科学、系统、标准化的管理。所有这些工作的实施对于初级农产品生产组织提供安全产品给予了良好的保障。

6）食品加工和流通污染问题日益受到重视

食源性疾病是食品安全防范的重要目标，而安全加工可以有效降低致病微生物带来的风险。食品加工过程中有害物质的形成机理、变化规律及有害物的控制技术研究是目前加工过程安全研究的重点和热点。国际上目前对一些有害物的产生机制和控制技术已经有了较为深入的研究。随着食品加工工艺和食品原料供应的日益复杂化，加工过程的安全风险也在逐步增加，国际上对加工过程中致病微生物和有害化学物的监控、食品添加剂的合理使用等方面的研究非常重视，JECFA 等每年都对添加剂的安全性进行大量的风险评估研究。通过一些物理、化学、生物学方法，确定食品加工过程中影响食品安全的因素并建立系统的监测技术体系和控制技术体系是目前研究的重点领域。国际上食品的流通安全一直受到高度重视，低温冷链物流是目前主要的发展趋势，在低温冷链物流研究中深低温急冻技术、低温保鲜技术和常规低温微生物污染监控技术、食品安全包装是研究的重点与前沿领域，特别是近年来双酚 A 等一系列问题的出现使包装材料的安全问题受到广泛关注。而我国，由于技术和人力资源方面的问题，还没有涉足这些领域。

7）食品供应综合保护战略的核心要素

在食品安全水平最高的美国，经历 2007 年宠物食品大规模召回、肉毒杆菌污染红番椒事件后，2007 年 11 月美国政府启动了食品保护计划。该计划首次提出了美国国家食品供应综合保护战略的核心要素和工作重点：核心要素包括更好预防、更强干预和更快响应；工作重点放在先期介入并预防问题的发生上，然后应用基于风险分析的干预和控制措施确保预防措施得到有效实施，最大限度地降低风险，一旦发现问题，则快速反应并运用预警系统进行快速预警。食品保护计划不仅明确针对食品污染和天然存在的危害物的监管（传统意义上的食品安全），也针对包括违法掺假和以食物为载体进行恐怖活动的危害物监管（食品防卫）。

在食品安全溯源与预警方面，WHO 与 FAO 合作建立的国际食品安全网络正在向各成员国扩展，以建立全球性的食品安全预警应急对策机制。欧盟建立的欧盟食品和饲料快速预警系统是一个系统连接各成员国食品与饲料安全主管机构、欧盟委员会以及欧洲食品安全局等的网络。所有参与其中的机构都建有各自的联系点，并通过联系点彼此联系，形成沟通渠道顺畅的网络系统，系统及时收集源自所有成员的相关信息，以便各监控机构就食品安全保障措施进行信息交流并快速反应以保护消费者免受不安全食品和饲料危害。在此基础上，欧盟建立了基于网络、媒体信息搜集的食品安全预警系统。英国食品标准机构通过监控预警系统随时加强传染病控制。德国联邦消费者保护和食品安全局建立的农产品风险预警信息预报系统，提供食品和饲料安全监察工作的查询与服务。美国多家联邦政府机构于 1995 年联合推出基于核酸的脉冲场凝胶电泳技术食源性疾病监测网络，该网络由美国疾病控制和预防中心具体负责，监测对象包括 7 种细菌和 2 种寄生虫，通过人群监测（涵盖美国人口 14%）构建食品安全主动保障体系。该系统目前覆盖了美国，德国等 11 个欧洲国家，南美洲的巴西等国家和亚洲的日本、菲律宾、韩国等共计 23 个国家和地区，在全球食源性疾病的监测方面发挥了巨大的作用。目前，美国正在加强食品污染危害识别、危害消减、危害控制、食品化学危害物和食源性病原体筛选技术研究，以加强现代信息技术的应用集成来快速提升美国食品与药物管理局的信息收集、识别能力，形成早期预警监管和信息系统来降低损失。荷兰瓦格宁根大学研究中心基于整体方法策略正在开发一套预警系统用于潜在的食品安全危害的预警。

3. 食品安全研究中的重大工程科技问题

根据未来国家经济社会发展对食品安全（含饮用水安全和食用农产品安全）研究的战略需求，应设立“食品安全工程”、“饮用水安全保障工程”和“食用农产品质量安全科技工程”。

（1）建立国家食品与饮用水污染监测网络和食源性疾病监测网络体系，网络将覆盖全国所有的省、市、县的城镇并逐步延伸到农村地区。建立一批食品安全监测基准实验室，重点解决获得国际可比对性实验数据所需要的标准质量控制准则、标准物质、标准菌株，并参与实验室间国际比对，为获得准确可靠的污染本底和制定国家控制目标提供技术保障。开展重要食源性疾病和人畜共患传染病负担调查，为评估疾病负担提供基础数据。

（2）重点加强我国食品安全风险评估中亟须与国际接轨的基础数据库（如中国人群膳食消费数据库、污染物数据库、耐药性数据库、毒理学数据库）建设及其基准剂量、急性参考剂量、风险-获益平衡模型等参数建立。

（3）充分利用现代信息技术（垂直搜索、语义搜索、对等点搜索等）、计算机技术、数据挖掘以及数据仓库技术和数学建模，实现基于国家地理信息、产地环境污染（包括环境安全事故、全球食品安全事故）的食品危害物风险自动预测预警。

（4）突破从农田到餐桌全过程的食品和饮用水中危害识别关键技术，实现关键试剂（抗体）和产品（免疫试剂盒、胶体金试纸条、固相萃取柱填料、固相萃取柱、免疫亲和柱、分子印迹柱、饮用水和食品卫生指标生物传感器等在线监测设备、现场监测便携式仪器）的国产化，提高危害识别能力。继续制定和坚持可持续性预防措施，包括食品安全教育规划，以通过涵盖从农场到消费者的完整食品生产链的系统方式，减轻食源性疾病负担。

（5）突破不明原因食物中毒甄别、处理和处置技术，重点解决基于症状和毒物结构特征查询的大容量数据库建设与查询准确度问题，形成主动捕获食品安全隐患的网络体系和诊断体系。

（6）以合理使用农业投入品、降低产地环境污染传递为目标，加强食用农产品质量安全风险评估、源头治理、过程控制、安全限量及检验检测等技术研究，控制源头污染。

（7）注重食品安全从农田到餐桌的全过程控制，研究食品从生产到消费过程危害物的形成机制，通过优化工艺，突破关键工程技术，解决过程污染问题。

（8）积极开展饮用水水源保护及修复、净化处理、安全输配、水质监测、风险评估、应急处置技术开发及集成应用。

4. 食品安全研究中的关键技术

1）农产品质量安全监测控制技术

强化产地环境控制及源头治理技术研究，重点研究食用农产品产地安全评价指标体系与评估技术，建立产地环境质量安全监测数据库、信息网络以及数据挖掘技术研究。加强农业种养殖过程安全控制技术研究，重点研发鲜食产品中农药残留、畜禽产品中抗生素、抗菌药和促生长素等药物残留污染控制技术，开发抗菌促生长类植物提取物并进行安全评价与高效利用关键技术研究，研究畜禽产品初加工过程消毒保鲜剂安全性评价和降解规律。开展食用农产品中重要化学污染物的剂量-反应评估关键技术、暴露评估方法优化与风险关键因子、食用农产品安全性风险指数以及主要食用农产品中危害因素风险评估模式研究。建立我国农兽药残留准许列表体系，开展基于风险评估的农兽药残留等污染物限量标准制定、特色小宗作物农药合理使用准则与残留限量标准、畜禽产品农药残留消解转化行为及其限量标准、种植业产品产地主要污染物限量标准与安全评价分类技术研究。开展粮油、畜禽、果蔬产品重点危害因子控制技术集成与示范。

2）食源性病原微生物监控技术

建立基于哨点实验室的食源性疾病监测网络，重点研发重要食源性致病菌特异分子标志库及高通量快速检测技术、常见食源性病毒分布规律与快速检测及控制技术、食源性致病微生物快速检测关键技术及产品、食品中致病菌危害控制与生物防腐剂新产品等重点技术，通过启动微生物定量风险评估解决海产品中副溶血性弧菌限量标准，突破基于特异性生化反应和特征性代谢产物的新型食源性致病菌定量检测与鉴定技术、常见食源性病毒分布规律与快速检测及控制技术研究、重要食源性致病菌特异分子标志库的建立及高通量快速检测技术、生物防腐剂作用机理研究及新型产品研发等核心技术，形成较为成熟的食源性致病微生物监控技术体系。开展食源性危害多因素的风险评估技术和评估模型研究、建立食源性危害多因素的风险预测模型、构建食源性病原微生物高通量快速分型和溯源技术及相关数据库，全面实现食源性疾病由事后处置向主动防御转变。

3）食品污染监测与暴露评估技术

建立国家食品污染监测网络，重点针对我国食品安全风险评估急需的基准参考数据和基本评估方法，从化学污染物（含环境污染与农药残留）和真菌毒素的膳食暴露评估、抗生素暴露与细菌耐药性评估、食品和包装材料中添加剂的毒理学阈值等风险评估基础参数问题入手，建立我国化学污染物暴露评估的基本技术体系。近期优先解决食品安全风险评估中基准剂量、急性参考剂量和风险-获益平衡模型，镉限量标准和全民食盐加碘再评估及其膳食暴露污染物对甲状腺功能的影响等国家亟待解决的科学问题。重点开展食源性危害多因素的风险评估技术和评估模型、食源性危害多因素的风险预测技术和预测模型、新的不明原因危害因素风险评估和预测技术、基于多因素的食源性疾病的风险预测和预警技术研究；充分利用现代化学分析技术、现代分子细胞生物学、基因组学、代谢组学和毒理组学技术，研究有毒有害物质的代谢与转化途径、健康危害分子机制及信号转导通路。建立食源性病原微生物高通量快速分型、溯源技术和相关数据库，为食品安全风险评估提供技术数据和技术支持。完成基准剂量与暴露评估结合的随机模型及其软件研发，建立药物与个人护理产品污染的膳食暴露评估技术；开展复合污染的食品毒理学技术研究和膳食暴露评估模型新参数研究。在前期已有技术基础上，进一步研究建立抗生素耐药性的风险评估模型与技术体系，针对水产品消费量增加与污染增加的双重压力，开发食用安全性的风险-获益平衡模型与软件。

4）食品（含农产品和饮用水）中危害物识别技术与装备

研究食品（含农产品和饮用水）中危害物的检测技术，逐步实现快速检测试剂和装备的国产化。优先发展有害因素的免疫识别系列技术，解决现场在线检测技术和仪器研制问题；形成食品过敏原类群特征特异表位精准定位成分检测技术并研制标准物质；开发针对农兽药残留、真菌毒素、贝类毒素和藻毒素等的高通量检测技术平台和食品污染及添加剂系列集成检测技术，集成相关的检测技术产品及样品前处理设备，建立食品和饮用水（含瓶装水）包装材料中有害物质迁移评估技术。重点突破农兽药残留抗体库建立及产品研发、人畜共患病病原检测技术。针对目前食品安全风险物质和潜在的风险物质，研究食品安全基础标准中的关键技术，为完善我国的食品安全标准体系提供技术支持。在完善食品安全国家标准查询数据库基础上，加强数据库的应用开发，构建重要化学污染物标准物质的制备及标准物质数据库，建立食品安全检测实验室与参考实验室技术规范和食品安全检测质量控制技术规范。开发食品中有害物质分析全程质量控制技术并研制相关产品；研制污染监测质控样品；研究食品中生物毒素与代谢应答检测技术、食物过敏原分子鉴定技术。

5）食物中毒与群体性不明原因疾病病因因子溯源控制

利用现代信息技术，引入风险预测技术，在食源性病原微生物分子分型标准化技术和食品中有害残留物高通量表征关键技术基础上，建立我国食源性微生物污染分子分型溯源数据库。开展不明原因食物中毒甄别、处理和控制技术研究，基本形成以临床生命体征识别、毒理学质谱筛查技术和代谢组学技术为支撑的食物中毒诊断数据库。研究动物性食物中毒病因因子权重判别与风险预警技术、微生物性食物中毒分子分型溯源技术和化学性食物中毒成因解析技术。在医院-疾控部门联动机制基础上建立症状主动监测技术，完善并集成开发基于现代信息技术的临床症状病因搜索解析技术、多维色谱-高分辨质谱筛查技术、基于疾病-生物标志物的代谢组学病因解析技术、现场流行病学技术等进行危害物群的解析技术体系；利用脉冲场凝胶电泳技术建立我国食源性致病微生物的分子分型国家网络数据库，加入食源性致病微生物分子分型国际网络。充分利用建立的不明原因食源性疾病病因解析溯源技术，结合现代信息技术、计算机技术、数据挖掘技术和数学建模，建立群体性不明原因疾病诊断监测预警技术平台。研究建立食物中毒现场干预处置技术。

6）食品加工与流通过程有害物质的安全控制技术

重点突破食品加工过程中消毒控制技术，研发一批智能化的监控设备，开发高效安全消毒剂；针对食品加工与流通过程安全控制关键技术，优先研究传统发酵食品制造过程中有害物质的安全控制技术、食品热加工过程中有害物质的风险控制技术、食品加工过程中有害物质的动态监控技术、食品安全流通实时监控技术等，形成可集成示范的控制技术体系。研究建立食品流通安全技术标准体系。针对我国大宗食品在流通中的主要安全问题，以流通过程中病原生物和重金属、农药残留、霉菌毒素等污染监控为目标，重点开展食品中主要污染物残留动态监控技术，食品储藏、包装、运输过程的安全控制技术，安全流通包装技术和设备的研发；研究食品在流通过程中有害物质的动态变化规律和控制技术，着重加强流通过程中以食物为载体的人畜共患传染病的传播、发生规律和控制技术的研究；同时要注重对食源性致病微生物在流通过程中的增殖和食源性疾病诱发规律的研究，在检测技术、致病微生物的控制技术等方面开展技术创新和重大技术集成研究。重点突破食品流通安全信息化技术、食品流通安全溯源技术、食品流通安全技术标准体系、食品安全流通处理关键技术、食品流通安全包装、运输关键技术等方面的研究和重大装备开发。集成以上的技术并进行有效的领域示范。

7）饮用水安全控制与预警技术

基于我国水体普遍遭受污染的现实状况和不同水源类型、不同水质特征和不同供水系统存在的安全隐患，开展供水管网和输水设施改造，减少供水漏损、提高供水管网水质化学稳定性和供水管网水质生物稳定性、提高供水设施和水资源利用效率。强化城市供水水质监督管理，建设国家和省域的城市供水水质实验室平台，完善城市供水水质检测网络，形成完善的城市供水水质监测体系；提升城市供水水质监测网检测能力建设，检测项目与国际接轨；按照《生活饮用水卫生标准》要求，配置国家监测网中心站的检测设备，检测能力可以支持评价重大水质问题；建设供水水质安全信息管理系统，实现城市供水水质数据的规范化和信息化管理及社会共享，提高对突发事件的反应和决策能力。通过技术研发、集成和综合示范，构建集水源保护、净化处理、安全输配、水质监测、风险评估、应急处置于一体的饮用水安全保障技术和监管体系，持续提升我国饮用水安全保障能力。

8）非法添加物筛查与食品防恐

针对食品反恐的需要和在经济转型和企业诚信制度建设期间出现的非法添加问题，引进脆弱点评估与预警技术，发展新型监控技术和生物恐怖防御技术。利用现代质谱技术、免疫分析技术、生物传感器、基因芯片等现代生物技术、替代毒理学技术开发非法添加物和恐怖生物及毒素检测技术；利用垂直搜索等现代信息技术建立食品安全风险隐患收集体系。重点进行食品违禁化学添加物筛查技术、食品违禁工业微生物筛查技术、食品恐怖生物毒素检测技术、动物源食品中激素非法使用与内源性鉴别技术、重要食品掺伪识别技术、食品生产脆弱性评估与风险预警技术研究，初步建立起我国非法添加物快速筛查与食品防恐技术体系。利用现代信息技术建立我国非法使用、超范围使用有毒物质的科学预警和信息收集平台，集成建立基于科学的风险防御技术体系。

5. 保障食品安全的重要措施与政策建议

第一，进一步贯彻落实《中华人民共和国食品安全法》、《中华人民共和国突发事件应对法》、《中华人民共和国农产品质量安全法》和《中华人民共和国食品安全法实施条例》与《突发公共卫生事件应急条例》，理顺监管体制。

第二，加大食品安全（含农产品质量安全和饮用水安全）的科技投入，提高科技创新能力。结合当前监管执法过程中面临的科技瓶颈，争取中央及地方财政支持，继续加大科技研究的支持力度，逐步形成多元化科技投入格局。加强产学研结合，提升科技自主创新能力，加强科技创新和成果转化。

第三，加强食品安全（含农产品质量安全和饮用水安全）的基地、人才和学科队伍建设。建设一批食品安全与中毒诊断控制国家重点实验室、工程中心、基准实验室和区域网络中心。围绕国家重大技术需求，整合资源，形成跨单位、跨区域、跨系统的安全科技协作网络，构建全国性科技创新共享平台，促进农科教、产学研紧密结合。建立人才的引进、培养机制，通过项目带动等形式，造就一批科技领军人才、战略科学家和创新团队，形成一支具有世界前沿水平的创新人才队伍。鼓励高校和科研机构开展食品安全人才（含农产品质量安全和饮用水安全）基地建设，增强我国在食品安全领域的科技人才储备。

第四，加强食品安全（含农产品质量安全和饮用水安全）知识普及和培训。针对我国人民群众安全卫生基本知识缺乏、科普工作极为薄弱、一些虚假宣传误导群众等问题，建议有关部门采取群众喜闻乐见的形式，宣传食品安全科普知识，培养安全意识，增强抵制不安全、不卫生食品（含农产品和饮用水）的能力。建立多渠道、多形式的培训体系。重点培训食品从业人员、管理者，提高安全生产技术、意识和诚信水平等。

第五，要积极推进节水型社会建设，定期发布饮用水水质情况，使全体公民掌握科学的饮用水知识，采用卫生安全的饮用水方式。大力宣传和推广科学用水、节约用水经验，加强饮用水安全保障工作，在全社会形成节约用水、合理用水、防治水污染、保护水资源的良好生产和生活方式。

五、土木工程安全相关工程科技发展战略①

土木工程是人类建造的并固定于地面或地下的各类工程设施的科学技术的统称。土木工程按狭义理解是指建造的对象，如房屋建筑物、路桥、隧道、堤坝、码头、管道及地下防护工程和航天发射塔等；但按广义理解，土木工程还指建造这类工程设施所用的建材、设备和勘察、规划、设计、施工、维修、检测、鉴定、加固、改造、拆除等种种技术活动及与其相关的管理活动。

对于危及多数人生命、健康和财产安全的公共安全问题，土木工程应做到在各种自然灾害、人为灾害以及人因差错发生时，能够不因这些工程设施的破坏，保护公众的生命财产安全，防止或减轻其损失。地震和洪水等自然灾害对土木工程造成的损失最大；人为灾害与人因差错的区别，在于行为的故意或无意、主观与客观。工程设施的破坏程度往往取决于上述灾害和行为的组合；人因差错能放大灾害的破坏程度，个别的人因差错一般不至于造成很大损失，但多种人因差错若在同一工程设施中偶然组合，即使没有天灾，也可能造成严重后果。

---

① 本文是中国工程院科研项目报告，北京，2010；工程管理研究，北京：科学出版社，2015，143-157. 作者：刘人怀，陈肇元，陈厚群，袁鸿，王璠，徐加初.

1. 当前我国土木工程安全问题的现状与挑战

1）城镇化建设中的安全问题

城镇化是人类社会走向现代化文明的重要标志，城镇化水平已成为公认的现代化衡量标准之一。目前，发达国家城镇化程度普遍超过 70%，有些甚至达到 80%以上。在中国，城镇化是现代化的必然要求，是经济社会迈入新一轮发展的重要突破口。截至 2009 年，中国城镇人口已经达到 6.2 亿，城镇化率达到 46.6%，预计到 2015 年达到 52%左右，到 2030 年达到 65%左右。由于体制和政策不完善，当前中国城市发展尚存在诸多问题，如城市土地扩张与人口增长不匹配，城乡、区域间发展严重不平衡，收入差距扩大与居住分异加剧，各种城市社会问题日益凸显，城市空间开发无序现象严重，大城市膨胀等，以及由此而引发一系列公共安全问题。城镇化建设的公共安全问题，包括两个方面，一方面是所有的城市灾害都会影响到城镇化建设；另一方面是城镇化建设过程中的公共安全问题。城市灾害引起的安全问题包括：城市工业危险源带来的风险、城市人口密集的公共场所存在的风险、城市公共设施脆弱引起的风险、流行性疾病引发的城市公共卫生灾害、恐怖袭击与破坏、城市生态环境恶化引发的社会安全问题。城镇化建设过程中的公共安全问题包括：城镇化建设过程的防灾与减灾、建设施工安全和灾害、城市宏观空间模式与城市防灾、气候变化下大中城市内涝灾害、土建结构安全事故、安全设置水准偏低。

2）西南多震地区的高坝抗震安全问题

随着我国国民经济和社会的稳步发展，能源紧张和作为二次能源的电力短缺现象已日益突出。为缓解我国能源紧张状况、优化能源结构，大力发展水电是唯一有效途径。我国水力资源理论蕴藏量为 6.94 亿千瓦，技术可开发装机容量 5.2 亿千瓦。截至 2009 年年底，我国水电装机容量为 1.97 亿千瓦，水电开发率为 36%。水电是目前最有可能大规模开发利用且技术最为成熟的可再生清洁优质能源，是集国土整治、河流开发、防洪抗旱、地区经济振兴、扶贫、生态改善等综合效益于一体的可持续工程。高坝大库在调节性能好、装机容量大、综合效益高的水电工程建设中，具有无可替代的重大作用。随着全球环境意识和可持续发展要求的日益增强，国际社会对高坝大库功能和作用的认识正不断深化。因此，在充分重视移民安置、生态和环境影响的前提下，积极有序地进行水库大坝建设切合我国国情和社会经济发展的需要，已成为我国基础设施建设中不可或缺的重要组成部分。西部地区是我国水能资源最丰富的地区，主要集中在岷江、大渡河、雅砻江、金沙江、澜沧江、怒江、黄河上游等流域，地形地质条件适宜修建移民淹地相对较少、发电效率高和调节性能好的高坝大库。但西南地区处于我国地势第一阶梯和第二阶梯过渡地带，这里地质断裂发育，属地震多发区和地质灾害高发区。目前在建和拟建的一系列 200～300 米级的世界级高坝，工程规模巨大且少有先例、坝址地震烈度高且缺乏工程震害实例，面临一系列世界级技术难题。高坝大库一旦受震溃决，将导致大坝下游遭受严重次生灾害，给人民生命及财产造成巨大损失。

3）梯级电站群的高效安全运行问题

我国西部地区已建、在建和拟建的大型水电站有几十座，岷江、大渡河、雅砻江、金沙江、澜沧江、怒江和黄河上游将建一批梯级大型电站。截至 2009 年，西部地区水库群总库容达到 700 多亿立方米，预计到 2020 年，西部地区水库群总库容将达到 1500 多亿立方米，2030 年，西部地区水库群总库容将超过 3000 多亿立方米。

但是，这些梯级大型电站群都位于我国西部高山峡谷地区，其运行将面临许多前所未有的科技问题。第一，我国的水资源时空分布极不均衡，西部梯级电站群的运行要考虑防洪与水资源的高效利用问题，以及各梯级电站的联合调度问题。第二，西部建坝地区的地形地质条件十分复杂，山高谷深，水力落差大，因此需要解决复杂条件下与运行相关的各种关键技术问题。第三，水电建设的生态环境保护压力空前加大，西部梯级电站群的运行管理中需要优先考虑生态环境问题，实现资源开发利用与生态环境保护双赢。

4）桥梁安全问题

我国自改革开放以来，公路建设事业迅猛发展，尤其是高速公路建设，作为公路建设重要组成部分的桥梁建设也得到相应发展，跨越大江（河）、海峡（湾）的长大桥梁也相继修建，一般公路和高等级公路上的中、小桥，立交桥，形式多样，工程质量不断提高，为公路运输提供了安全、舒适的服务。中国的大桥跨度已名列前茅，与先进国家技术上的差距也大大缩小。据统计，2006 年年末我国有公路 345.7 万公里，有公路桥 53.36 万座、203.99 万米，我国已经成了世界桥梁大国。

虽然我国跨大江大河、跨海大桥质量和安全性好，但是从总体上看，我国桥梁安全状况不容乐观，施工或使用中，时有桥梁坍塌。总体来说，我国桥梁面临以下安全问题：第一，我国桥梁规范标准太低，而规定的材料设计强度又较高；第二，设计往往重美观而忽视了整体牢固性，我国设计规范强调单个构件的承载能力设计计算，较少涉及整体牢固性的要求；第三，目前“目标使用年限”模糊，在我国建筑法律法规中，未对不同类型建筑物的合理使用寿命（或年限）规定具体的量值；第四，轻维护、轻建设质量问题突出；第五，我国土建设计人员创新意识薄弱，过分依赖规范，不善于根据工程的具体特点去解决问题，施工一线工人的素质较低，难以及时发现和有效消除人为差错。

2. 当前土木工程安全的前沿问题研究综述

1）城镇化建设中安全问题的前沿研究综述

随着城市化程度的提高，城市建设在单位土地上的聚集程度比以往任何时候都高，灾害发生时的放大效应也就更加明显。单体建筑物越来越高、体量越来越大，城市建筑向高层、超高层、大跨度、大型空间建筑、城市大型地下建筑发展。因此，针对城镇化建设中的安全问题，应保证千米级大桥、五百米级超高层建筑、三百米级高坝等的安全。前沿研究旨在通过对重大工程在强地震动场和强/台风场动力作用下的损伤破坏演化过程的研究，揭示重大工程的损伤机理和破坏倒塌机制，建立重大工程动力灾变模拟系统，发展与经济和社会相适应的重大工程防灾减灾科学和技术，为保障重大工程的安全建设和运营提供科学支撑。主要研究内容包括以下几个方面。

（1）强地震动场和强/台风场的建模与预测研究：强地震动场的破坏特性、理论模型与预测方法；强/台风场的分布特性、时空模型与预测方法。

(2) 重大工程动力灾变的关键效应研究：材料、构件和结构的非线性动力效应研究；结构与环境介质的动力耦合效应；结构的空间动力作用效应；结构内部及与环境介质之间的能量转换和耗散效应。

(3) 重大工程动力灾变的全过程分析：建立考虑强非线性、多介质耦合、能量转换和耗散等效应影响的复杂重大工程系统的快速建模理论以及强地震动场和强/台风场动力作用下重大工程动力损伤演化的高效数值计算方法，发展基于高维动态数据场特征分析和提取的高效可视化技术；研究强地震动场和强/台风场作用下重大工程的损伤累积效应及其演化规律，揭示重大工程的动力灾变失效破坏机理，建立重大工程的动力灾变失效破坏准则；研究强地震动场和强/台风场作用下重大工程的构件破坏、局部结构破坏以及整体结构破坏之间的关系，揭示重大工程的动力灾变破坏与倒塌机制。

(4) 动力灾变过程控制研究：重大工程结构的失效模式及其高效分析方法；重大工程结构失效模式优化与结构整体抗震抗风能力提高的理论与方法；重大工程动力灾变过程的损伤与倒塌控制原理。

(5) 重大工程动力灾变模拟系统的集成与验证：包括系统集成方法研究；研制和开发强地震动场作用下重大工程动力灾变模拟的软硬件技术平台，并采用模型试验进行有效性验证或采用原型监测案例分析进行可靠性验证，再现强地震动场作用下重大工程的动力灾变过程；研制和开发强/台风场作用下重大工程动力灾变模拟的软硬件技术平台，并采用模型试验进行有效性验证或采用原型监测案例分析进行可靠性验证，再现强/台风场作用下重大工程的动力灾变过程。

(6) 重大工程结构的抗爆、耐火、耐腐研究：城市复杂环境中爆炸冲击波的传播、作用机理与荷载模型；工程结构遭受撞击的作用机理与防护措施；爆炸与冲击荷载作用下工程结构的损伤累积、破坏机理与倒塌机制；满足性能化目标的结构抗火设计准则；结构抗火设计中“抗力”和“荷载效应”的概率模型，以及模型参数随温度的定量变化规律；结构抗火的目标可靠指标；导致结构材料性能劣化（如钢材锈蚀、混凝土腐蚀）的环境作用等。

(7) 土建结构工程的安全性与耐久性研究：分析我国土建结构工程的安全性与耐久性现状，寻求存在的问题及其根源，探讨解决的途径、方法与对策，并为政府部门制定相关的技术政策提供建议，以期土建结构工程能够更好地满足我国现代化建设的需要，并适应我国经济转型后面向市场经济的需求。

2) 高坝大库抗震安全研究中的前沿问题研究综述

世界不少多地震国家，在地震区都修建有众多大坝。已有一些遭受震害实例，但迄今因地震溃决的极少，仅限于设计和施工不良的低坝。我国 2008 年的汶川大地震中，各类大坝总体上经受住了特大地震的考验。震区的众多中小型水坝虽有不同程度的震害，但无一垮坝。近震区 4 个不同类型百米以上高坝，虽经受超设计水准的强震，但均保持了结构整体稳定和挡水功能。

此外，迄今全球水库蓄水引发水库地震震例占水库总数的比例是极小的，并非修建高坝大库就一定会引发水库地震。在已有的被较普遍承认的水库地震震例中，绝大多数最大震级不超过 3～4 级，对工程和库区未造成危害。只有极少数震例，属于社会和工程界所关心的构造型水库触发地震，这类构造型水库触发地震，只在特定的地震地质和水文地质条件下才会发生，其最大震级不可能超过被其触发断层本身的最大震级。迄今全球仅有 4 个最大震级超过 6 级的水库地震震例，其最大震级不超过 6.5 级。高坝大库的库区如有发震断层，必须在抗震设计中经充分论证，并考虑其对工程抗震安全的影响。

因此，可以认为，在强震区修建高坝大库，只要按规范要求精心进行抗震设计、确保施工质量和运行管理到位，抗震问题并非颠覆性的制约因素。近年来，在我国大坝工程建设的推动下，我国在大坝抗震研究中取得了一定进展，增强了我国在西部强震区修建高坝大库的信心。

但也应当看到，高坝大库工程的安全问题极端复杂，涉及多学科的交叉。工程设计中的技术问题，不少都缺乏较深入的理论支撑，因而在相当程度上仍有赖于工程实践经验。特别是抗震设计，由于地震动的不确定性和结构动态响应的复杂性，困难更大。高坝大库地震安全性的评价，必须从坝址地震动输入、坝体-地基-库水体系地震响应分析、材料动态抗力这三个相互影响和配套的方面综合分析。目前大坝抗震研究的现状是，对结构地震响应的研究日益精细，而对地震动输入和结构抗力的研究相当粗放，形成“两头小、中间大”的局面，实际由粗放的两头决定了其抗震安全性评价的水平。因此，需加强对地震动输入和结构抗力这两方面的研究，以力求对工程抗震安全作出更系统全面合理的综合评价。此外，大坝抗震设计规范中，对大坝地震响应的分析，大多仍基本沿用基于已有工程实践经验和类比的传统理念与方法，其中有些并不能完全反映高坝在强震作用下的实际状态和很好解释震害实例。加上遭受过强震的大坝震例，远少于房屋建筑和道桥等工程，在汶川大地震中经受强震检验的少量高坝，其坝高也都在 160 米以下。而 160 米以下的高坝与 300 米级的高坝相比，可能存在从量变到质变的本质差异。此外，近期我国处于水电开发高潮，包括规划、设计和科研在内的前期工作不够充分，往往仍只能依据已有的规范和方法。因此，面对超大型世界一流工程建设中提出的前所未遇的前沿性工程技术难题，当前科研进展所提供的科技支撑赶不上工程建设发展的规模和速度，存在着很大的风险。

同时也要认识到，在当前科学技术发展和知识更新十分迅速的信息时代，众多跨学科的高新技术，为以往工程设计中许多难以解决的工程技术难题提供了前所未有的新的研究思路、方法和手段，为大坝工程抗震安全性的评估更接近实际创造了条件。诸如生成更符合坝区具体地震、地质和地形条件的场地相关地震动输入；建立更切合工程实际的大坝体系地震响应分析模型；确定接近坝体和地基材料真实性能的本构关系和损伤演化过程，采用更精确和有效的非线性方程求解方法和计算工具等。

因此，必须要从前瞻性和战略性的高度，在立足当前工程需要的同时，放眼长期战略目标，敢于突破某些已难以适应发展需求的传统理念和方法，在加强交叉学科的借鉴和协作、大力学习引用高新技术的基础上，为切实防止高坝大库地震灾变的研究开拓创新的理念、思路、技术途径和方法。当然，这是一个科学求实、谨慎认真，并需要在实践中不断检验改进的长期积累的发展过程。

3）梯级电站群高效安全运行中的前沿问题研究综述

（1）防洪与水资源高效利用。洪水具有高度的不确定性和风险性。防洪一直是西部梯级水电站群的重要任务之一。防洪与发电既有矛盾，又密切联系。采用先进的科学技术，针对梯级电站群各种实际情况，研究相应的最佳防洪规划方案和调度方案，不仅能够保障工程安全，满足防洪需要，同时也能够显著提高梯级电站群的水能资源利用效率。

（2）梯级流域水文预报技术。可靠的水文预报是保证水资源系统安全调度、充分发挥系统运行效益的前提条件。流域水文预报涉及的技术问题包括：①用水预报技术研究；②二元分布式水文预报技术研究；③梯级流域内长短嵌套水文预报方式研究；④汛末判断决策技术研究。

近 20 年来，随着自动化观测手段、信息传输和分析处理技术的高速发展，在短期入库径流预报方面取得了很大的进步，定量预报精度已基本满足工程需要。目前，水库汛限水位动态控制研究主要集中在复核水库功能任务、设计洪水复核、水库洪水预报、水库汛限水位分期控制，利用预报技术对水库汛限水位动态控制进行经济评价与风险分析、水库调度运用方式研究，取得了初步研究成果。但在全球气候变化背景下，梯级流域的水文预报还面临很多亟待解决的问题。

（3）梯级水库泥沙冲淤问题。河道上修建水电站后，必然改变原有河道的水沙关系，造成水库泥沙淤积、下游河道冲刷甚至河口侵蚀等一系列泥沙问题，伴随着规划、设计、施工和调度运行的全过程。与单个水库相比，梯级水库泥沙问题更加复杂。不仅要考虑上游的来水来沙和水库本身运用方式对本库泥沙冲淤影响，还应考虑到梯级开发次序、上库运用方式和下库回水对本库泥沙冲淤与出力的影响。系统研究梯级水库泥沙冲淤规律，为工程规划设计、施工安全、优化梯级联合调度、合理调控泥沙淤积、实现水库群长期有效运行提供科学依据。涉及的主要科技问题包括：①梯级水库排沙技术与电站进沙研究；②水库干流及支流变动回水区泥沙冲淤对有效库容和发电效益影响研究；③水库泥沙淤积调控与长期使用措施研究。

（4）大型水电工程的生态环境保护问题。发展具有防洪、发电、供水、航运等综合效益的水电工程是我国经济社会可持续发展的必然要求。然而，水电工程造成的河流阻隔效应和水文情势改变可能带来一系列深远的生态与环境问题。如何协调水电开发与生态环境保护之间的关系，保障经济-社会-环境-良性发展，成为我国未来水能资源开发中的一项重大挑战。西部梯级电站建设与运行过程中面临的主要生态问题包括：①流域梯级开发中的生态与环境累积效应；②流域梯级电站的生态调度；③中低坝过鱼设施；④泄水水温与溶解气体变化及其对鱼类的影响和对策；⑤库区水体富营养化防治以及水库消落区生态恢复和保护等。

（5）梯级水电站群联合优化调度问题。梯级水电站群系统是一个规模庞大、单元众多、结构复杂、关系错综的动态多目标复杂系统。联合优化调度的目标就是在确保安全的前提下，最大限度地发挥梯级枢纽群的效益，建立流域电站群之间的联合调度机制，实现优化调度。梯级水电站群联合优化调度是具有多目标、多阶段和多约束条件的非线性优化问题，涉及的主要问题有：①多目标优化问题求解的理论算法；②水库径流中长期预报方法；③梯级水电站中“龙头水库”的作用和补偿效益；④水库的动态汛限水位风险效益；⑤梯级水电站短期优化调度数学模型和求解方法。

4）桥梁安全中的前沿问题研究综述

（1）预应力混凝土桥梁的裂缝问题。一些预应力混凝土桥梁，由于梁体裂缝严重、挠度大、危及使用安全而实施加固，预应力混凝土主梁也有裂缝发生。从根本上讲，应从设计和施工工艺方面采取有效措施。经检查发现，采用传统的压浆工艺，钢束管道内浆体不饱满，钢束严重锈蚀导致有效预应力降低。因此，对于预应力混凝土桥梁，为保证钢束管道压浆质量，塑料波纹管及真空辅助压浆工艺的推广应用不容置疑。有专家提出，预应力混凝土连续钢构桥的跨径不宜过大，跨径为100～200米，矮塔斜拉桥、梁拱组合体系等桥型具有可比性。大跨径斜拉桥主梁的结构形式，应总结已建桥梁的经验，经充分论证比较确定。

（2）斜拉桥的拉索。平行镀锌钢丝拉索在我国已应用多年，近几年，无黏结镀锌钢绞线拉索和环氧涂层钢绞线拉索也先后被采用。三种形式的拉索，其构造、防腐、制作安装和实施换索的方式不同。在现有的技术条件下，斜拉桥在百年使用期内，拉索的更换不可避免，但应尽量做到在保证拉索的安全耐久性前提下，换索的次数最少、最方便。目前，我国桥梁界对三种形式拉索的认识还不尽一致，有必要从拉索的性能、安全耐久性、应用效果，以及建设、养护维修费的综合经济指标等方面，进行技术、经济的进一步研究论证，尽快取得共识。

（3）钢桥的桥面铺装和钢结构的防腐。我国以钢箱梁为主梁的悬索桥、斜拉桥，采用的桥面铺装形式较多，有的比较成功，但有的在通车后不久就因损坏而改建。其铺装的设计、材料、工艺问题，应通过研究和试验，尽快解决。对钢箱梁的防腐虽然比较重视，但尽管采用了较先进的防腐技术，严格的养护维修是不可或缺的。钢管混凝土拱桥拱肋构件的防腐，拱肋内混凝土脱空问题以及中承式、下承式拱桥吊杆易腐蚀、疲劳问题，也应认真对待。

3. 土木工程安全研究中的关键科技问题

1）城镇化建设安全研究的关键科技问题

（1）强地震动场的理论预测模型与数值预测方法。研究地震震源过程对近场强地震动场的影响规律，建立考虑精细震源过程、传播路径、复杂地形和局部场地条件的强地震动场全尺度理论预测模型与数值计算方法，揭示强地震动场的形成机理与分布规律。

（2）重大工程结构倒塌模式与防御。研究强地震动场和强/台风场作用下重大工程结构的倒塌模式及其高效分析方法，揭示重大工程结构的动力灾变和倒塌破坏的演变机理及共性规律，建立重大工程结构的动力灾变与破坏倒塌准则及防御方法。

（3）沿海台风的近地特性及结构致灾作用。研究我国东南沿海台风的近地特性及其对重大工程结构的致灾作用，建立模拟沿海台风近地特性的风洞试验方法和数值计算方法。

（4）内陆强风及其作用的现场实测与模拟。现场实测研究内陆强风的平均风特性和脉动风特性，建立内陆强风的平均风、脉动风普适性数值分析模型及模拟的风洞试验方法和数值计算方法。

(5) 现代钢结构的关键基础理论与设计方法。超高度和超跨度的现代钢结构的整体稳定性、地震荷载下的动态响应、风荷载下的动态响应研究；大型钢结构的防火研究、吸声、隔音研究、防腐研究。

(6) 城市地下工程结构安全关键科学问题。以城市地下工程的安全性为研究对象，围绕地层变形、破坏特点及演化规律、多体作用及灾变形成机制、安全性控制原理等关键科学问题，通过对施工扰动地层的破坏机理、地层变形传播及其与结构的相互作用特征、灾害演化过程、结构劣化评价及灾害控制的研究，揭示地层变形机理及灾害形成机制，构建我国城市地下工程安全性评估和控制的系统科学理论，为我国城市地下工程的发展提供科学的理论依据。

2) 西南多震地区高坝抗震安全评价中的关键科技问题

(1) 高坝抗震设防标准。研究西南多震地区的高坝地震损伤和破坏机理，确定高坝设防的地震强度与频谱特性，提出高坝安全检测内容与安全标准。

(2)“溃坝”极限状态的定量准则。研究高拱坝两岸坝肩岩体的超常变形，提出其失稳条件的定量指标和控制标准；开展大坝混凝土和大体积岩体的复杂应力状态下的拉、压损伤演化规律与循环加载下的残余应变问题研究；研究混凝土高坝在强震作用下的动态损伤破坏过程；发展高坝损伤破坏过程的数值模拟方法；提出表征大坝整体失稳的控制标准。

(3) 坝址地震动参数确定和地震动输入方法。开展西南地区坝址地震动衰减关系、水库地震机理和有限断层法的研究，生成更符合坝区具体地震、地质和地形条件的场地相关地震动输入；研究强震时坝体能量向远域扩散的辐射阻尼效应，提出更合理的坝址地震动参数确定方法。

(4) 筑坝材料的动态特性和抗力研究。研究大体积混凝土材料的动态特性与非线性动力本构关系；加载速率、地震变幅循环作用、双轴和多轴加载作用对混凝土材料的应力应变关系、吸能能力、强度和变形模量的影响。开展土石料材料级配及力学性能的原位测试技术研究；高围压下大颗粒堆石料的力学特性与渗透特性研究；开展已建工程的反演分析，提出更合理的土石料强度、变形指标的取值规定以及更为普遍适用的本构关系。

(5) 混凝土高拱坝的地震动响应分析。研究坝与无限地基的动力相互作用。坝与无限地基的动力相互作用对拱坝与重力坝的地震响应产生重要影响，相互作用改变了坝的固有频率与振动模态，使地震动输入发生变化，同时使振动能量向无限地基发生散逸。开展坝与无限地基的动力相互作用研究，揭示地基相互作用对拱坝地震响应影响。

开展高坝的非线性地震响应分析与地震损伤发展的数值模拟研究。由于目前大坝的抗震设计主要基于弹性动力分析，计算出的局部超强高应力难以对大坝的整体安全性作出可靠的估计。开展高坝的非线性地震响应分析，为高坝抗震安全的科学评价提供必要的技术基础。

(6) 土石坝的地震动响应分析。建立土石坝的地震响应与安全评价的计算模型，对土石坝进行静力仿真和地震作用全过程进行分析，研究土石坝的抗震性能、震害形态以及各种因素对土石坝抗震性能的影响。

3）梯级电站群高效安全运行的关键科技问题

（1）流域二元水循环模拟技术。从关注气候变化和人类影响的角度出发，以“自然-人工”二元驱动理论为指导，研究长江上游和其他西部河流域的大流域“大气—坡面—地下—河道”自然循环及“取水—输水—用水—排水”人工侧支循环过程。研究基于数字流域技术的流域二元水循环模拟平台，对水循环陆面全过程进行系统模拟，揭示“自然-人工”二元驱动下的水资源演化规律。

（2）全球气候变化背景下的流域水文预报模式。研究流域不同预见期长短嵌套的水文预报系统；根据实时作业预报中自动预报和人机交互预报的作业流程，研究大流域多节点耦合实时校正系统；针对西南地区气象水文特征，研究基于物理概念和统计方法的大流域实用型中长期预报模型。

（3）梯级水库群泥沙冲淤规律与调度。建立大型梯级水库泥沙淤积分析数学模型，考虑上游的来水来沙及水库运用方式对本库冲淤影响，研究梯级水库群的泥沙冲淤规律，合理配置泥沙淤积，为优化梯级调度、实现水库群长期高效运行提供科学依据。

（4）基于生态安全的水电站库群梯级调度。基于梯级水电工程对河流生态系统演变和水力调控机制进行研究，对生态安全进行综合评价，定量研究改善水库水质和富营养化状况、满足重要生物生命周期所需的生态水文过程、提供重要生态恢复与重建的水力条件以及保障河口生态流量等的生态补偿目标；研究梯级水电工程的生态安全补偿目标、梯级水库生态补偿的技术准则和技术方案。

（5）梯级枢纽联合调度综合优化技术。从整个梯级系统的角度出发，综合考虑防洪、发电、供水、生态、泥沙等方面因素，研究整个梯级系统综合效益最大的运行方式，研究多目标群决策的梯级水库联合调度优化模型；研究梯级各电站间水力联系与电力联系、相互间的水文补偿、库容补偿、电力补偿等，研究基于补偿调节技术的联合调度方案。

4）桥梁安全中的关键科技问题

（1）预应力混凝土桥梁的裂缝问题。重点研究高性能预应力梁混凝土配合比设计，桥梁工程预应力箱梁开裂分析及防治对策，预应力混凝土桥梁结构加固与裂缝处理，预应力混凝土连续箱梁桥腹板裂缝问题，混凝土桥梁结构形式施工方法与裂缝控制关系。

（2）斜拉桥的拉索问题。研究斜拉桥拉索线密度对拉索索力测量的影响，斜拉桥施工控制方法，索力对斜拉索动力特性的影响，钢拱塔斜拉桥拉索锚固区局部应力与敏感性，斜拉桥索-塔-梁耦合参数振动，碳纤维索斜拉桥的抗震性能分析，混凝土斜拉桥索力优化与合理施工状态，碳纤维与钢组合斜拉索设计方案及理论，混凝土斜拉桥换索工程施工控制。

（3）钢桥的桥面铺装和钢结构的防腐问题。研究钢桥面铺装脱层破坏的原因及对策，桥面铺装对钢桥面板疲劳应力幅的影响，环氧沥青钢桥面铺装设计理论与方法，大跨径钢桥铺装组合结构疲劳性能，正交异性钢桥面新型复合铺装结构，钢结构防腐涂料的制备及其性能，钢结构腐蚀机理及影响因素，钢结构材料在沿海地区的应用，全寿命周期的桥梁结构耐久性关键技术。

（4）超大跨度深水基础桥梁强震灾变过程及控制分析。探索大跨度桥梁深水基础的地震动输入模型以及多维多点非一致性地震动激励作用下大跨度桥梁的动力非线性模型，研究上部结构-桥墩-基础-水/土多介质动力相互作用效应与损伤破坏演化过程及其控制，建立考虑地震动空间效应、结构非线性效应和多介质动力相互作用效应的大跨度深水基础桥梁强震灾变行为的精细化模拟模型与分析方法以及安全控制体系，揭示强震作用下大跨度深水基础桥梁动力灾变的形成机理与控制原理。

4. 土木工程安全研究中的关键技术

1）城镇化建设安全的关键技术

（1）城市地震安全关键技术。研究大尺度地震动随机场分析中的数值稳定性；研究钢筋混凝土结构的地震倒塌机理及其地震倒塌过程的离散元模拟方法；复杂高层结构的抗震设防标准、抗震性态评价；重大工程抗震数字减灾系统。

（2）城市建筑结构抗风关键技术。近地风特性和建筑结构响应的实测。大中城市尺度风特性和小区尺度风环境的风洞模拟理论与方法；高雷诺数效应的模拟方法及模拟效应；大型复杂结构风荷载和响应机理、结构抗风性能和设计方法；群体建筑结构的风荷载和响应的干扰效应及抗风设计；强风作用下典型低矮建筑的破坏机理；强风暴作用下重大工程结构的控制。

（3）城市建筑结构抗火关键技术。高温下与高温后建筑材料特性；高温下与高温后高强混凝土爆裂规律和力学性能；高强混凝土结构、预应力混凝土结构、异型柱混凝土结构、已加固混凝土结构等现代混凝土结构的火灾行为与灾后性能；结构抗火设计中“抗力”和“荷载效应”的概率模型，以及模型参数随温度的定量变化规律；大空间建筑、大跨度结构、多高层结构火灾升温模型。

（4）城市建筑抗爆关键技术。城市复杂环境中爆炸冲击波的传播、作用机理与荷载模型；爆炸飞片对结构的破坏作用；爆炸冲击荷载与其他灾害荷载对结构的综合作用；地下爆炸波冲击下土与地面结构基础的动力相互作用及地面结构的动力响应：空爆作用下考虑场地效应的地面结构物灾害响应的数值模拟分析；爆炸波冲击下城市生命线工程的灾害响应；爆炸波冲击下工程结构的减灾控制措施与加固技术；民用建筑爆炸防护设计。

（5）城市建筑结构灾变健康监测关键技术。城市建筑结构的损伤评定与健康诊断方法；大型结构动力模态指纹分析技术；城市建筑结构的非线性损伤变量及其识别；工程结构的损伤尺度谱与损伤定位；重大工程结构和系统健康监测的信号转换接口、海量数据的远距离传输技术和智能处理方法；工程结构安全评定的灾害风险分析、确定性的体系安全评定方法和体系可靠度评定方法。

（6）城市生命线工程安全性监测与重大灾害应急处置关键技术。研究内容包括：先进传感技术、结构与系统智能监测技术、重大突发灾害预警技术、系统可靠性决策技术、重大突发灾害应急处置系统。

（7）城市道路交通安全管理决策支持系统关键技术。针对道路交通安全问题的定性多、定量少的状况，应用城市道路交通安全管理决策支持系统的理论和方法，对系统管理中的事故成因机理分析与预测、安全评价方法、安全对策与决策等关键技术进行深入研究，建立决策系统中相应各子系统的数据库和模型库。

(8) 城市区域性重大事故风险评价技术、城市安全功能区划分方法及其规划技术、城市整体（综合）安全规划技术、城市重大危险源综合整治安全规划技术和城市其他公共安全规划编制要点等。

2）西南多震地区高坝抗震安全评价中的关键技术

（1）高坝整体稳定性的实时监控与预警关键技术。包括高坝稳定性宏观效应现代监控理论与技术、高坝整体稳定性在线动态诊断理论与技术、高坝失稳风险识别和评估理论与方法、高坝深层抗滑稳定分析和预警软件技术。

（2）新技术在大坝质量控制、健康监测和抗震评价中的应用研究。利用各种新技术发展快速、轻型、简便、精确的实时质量控制和结构损伤诊断技术。例如，利用声波、地震波、雷达电磁波技术开发快速测定材料的质量、含水量的方法与设备；利用纳米级的硅酸作黏结剂具有高断裂与抗压强度的性能，发展对结构进行改性或对结构的裂缝、损伤进行快速修补的技术与方法；利用 GPS 测量技术、三维激光测量技术、摄影测量技术、中子和伽马量子与介质的相互作用原理发展快速、轻便地进行结构隐蔽处的损伤诊断方法。

3）梯级电站群高效安全运行中的关键技术

（1）气象水情自动测报系统。研究梯级流域气象水情自动测报系统，开发新型水情测报装置，进一步研究通信新技术在水情测报系统中的应用。

（2）梯级枢纽联合调度决策支持系统。搭建基于网络和 GIS 的面向对象、功能强大、反应快捷、使用方便灵活、界面美观的梯级枢纽联合调度决策支持平台，综合利用模拟技术、预报技术和优化调度技术，指导整个梯级系统的运行管理，并适合各电站建设运行不同时期，为充分发挥工程的综合效益提供技术支撑和管理工具。

（3）水库汛限水位动态控制关键技术。研究内容包括：梯级流域内暴雨洪水分布规律；水库分期设计洪水；汛限水位动态控制运用的可行性和具体的汛限水位变化范围；防洪调度与决策的风险因子和风险特征；汛限水位不同动态控制方案及其调度运用方式的风险。

（4）联合调度评估分析关键技术。研究梯级枢纽调度运行评估理论，建立评估指标体系，构建评估分析的理论框架和方法体系，建立梯级枢纽联合调度后评估模型，评估代表洪水年份调度方案的综合效益，分析影响效益发挥的主导因素及影响度，提出改进手段及相应的提升空间。

（5）梯级电站库群的泥沙处理技术。针对西部水电开发运行管理中存在的工程泥沙技术问题，研究低水头闸坝枢纽水沙联合优化调度运行方式，针对冲沙和发电的矛盾，研究电站进口、渠道、电站前池的输沙排沙特性，提出流域水工程水沙联合管理及与电站水沙优化调度运行管理有关的工程泥沙处理技术。

4）桥梁安全中的关键技术

（1）研发桥梁出现危险征兆初期作出预警的小型传感器。

（2）自然灾害对公路、桥梁影响的损害评价分析与思考。

（3）应对自然灾害的桥梁新技术和方法研究。

（4）千米以上级斜拉桥结构体系、设计及施工控制关键技术。

(5) 基于结构健康监测实测数据的伸缩缝评价系统方法。

(6) 考虑梯度、温度等因素的悬索桥主梁挠度控制阈值指标。

5. 保障土木工程安全的重要措施与政策建议

1) 低碳可持续发展约束下的城镇化建设

对正处于空前绝后的快速城镇化浪潮之中的我国，正确选择适合于国情和当地资源环境条件的城镇化模式已成为当务之急。倡导科学的建筑节能理念的今天，能源问题已成为人们关注的焦点。建筑节能是保证国家安全，建设资源节约型、环境友好型社会的重要举措。加强建筑节能是转变建筑经济增长方式的重要手段，是保障能源安全的迫切需要。针对新建建筑执行新的建筑节能标准，要严格执行强制性条文；对既有居住建筑进行节能改造，对既有的公共建筑包括政府办公建筑和大型公共建筑进行节能改造。大力推广可再生能源在建筑中的应用及解决可再生能源技术和建筑结合的问题，即可再生能源与建筑一体化问题。因此需要从材料、设计、施工以及各个层面积极关注建筑节能，在实际工程中采用新的结构体系。

2) 提高城镇建筑结构、桥梁和大坝工程的安全性与耐久性

在重大工程结构的安全性设计上，除了要提高工程的使用荷载及相应的安全储备外，更为重要的是必须加强工程的抗灾、减灾能力。

建议有关部门明确规定各类土建工程的设计使用年限（设计寿命），并规定在重要工程和可能遭受冰冻以及接触海水、除冰盐和腐蚀性气体、水体或土体的工程设计文件中，必须有耐久性设计的独立章节及论证。对于桥梁、大坝等基础设施工程，应该在设计中进行工程投资的全寿命投资成本分析，包括初始的建造投资和后期的维修投资之间的合理性评估。提高城市多、高层房屋，特别是大型建筑物和高层建筑的设计使用年限和桥梁等基础设施工程的设计使用年限。

3) 建立城市宏观空间模式与城市防灾体系

间隙式城市空间结构将在城市的整体形态上建立一个战略性的有利于城市防灾减灾的空间格局。

4) 合理规划城市地下空间

地下空间的规划应注意保护和改善城市的生态环境，科学预测城市发展的需要。坚持因地制宜，远近兼顾，全面规划，分步实施，使城市地下空间的开发利用同国家和地方的经济技术发展水平相适应。城市地下空间规划应实行竖向分层立体综合开发，横向相关空间互相连通，地面建筑与地下工程协调配合。

5) 真正树立和始终坚持质量安全第一的理念

建设单位、设计单位、施工单位、工程监理单位及其他与建设工程安全生产有关的单位，切实执行 2003 年 11 月国务院出台的《建设工程安全生产管理条例》，依法承担建设工程安全责任。政府主管部门要严格地监督检查，出台实施细则，坚持安全第一的认识，除国防、抢险工程外，在安全与投资、工期发生矛盾时仍坚持安全第一的原则。

6) 加强对建筑设计规范的研究

开展立足于更高安全标准的建筑设计规范研究，及时修订土木工程设计规范，满足新形势和新条件下的重大工程结构安全的设计需求。

六、生产安全相关工程科技发展战略①

1. 当前我国生产安全的研究现状

生产安全是人类为其生存与发展向大自然索取和创造物质财富的生产经营活动中一个最重要的基本前提。在生产经营活动中生产安全问题无所不在，无时不有。生产安全工作就是对生产经营活动中的事故风险进行识别、评价和控制过程的监测、预警、应对、管理与相关科技活动。

党的十六大报告明确提出“高度重视安全生产，保护国家财产和人民生命的安全”。生产安全是社会文明和进步的重要标志，是国民经济稳定运行的重要保障，是坚持以人为本的发展观的必然要求，是坚持人与自然和谐发展的前提条件，是全面建设小康社会宏伟目标的重要内容，是实践“三个代表”重要思想的具体体现。

我国正处于并将长期处于社会主义初级阶段。生产力、科技和教育还比较落后，实现工业化和现代化还有很长的路要走。现阶段我国的生产安全形势表现为总体稳定，趋于好转的发展趋势与依然严峻的现状并存[2,5]。如果不采取强有力的措施，生产安全形势在未来相当长的时间内，仍将十分严峻，事故发生率仍将在高位徘徊。随着我国社会经济和科技水平的发展，生产经营手段、设备和工艺愈加复杂，运行（或使用）条件愈加苛刻，更易出现各种重大、特大的灾难性事故，所造成的损失和社会影响也更大，因此，生产安全问题不仅不会自行消亡，反而会更加突出。同时，人们的生产安全理念发生了深刻变化，对生产安全的关注已经上升到前所未有的高度，追求人-社会-经济-环境的可持续发展成为全社会首要的共同目标。近年来，我国生产安全形势依然严峻，因此，高度重视生产安全问题仍是我国的主要任务之一。

2004 年国务院出台《关于进一步加强安全生产工作的决定》，明确了我国生产安全的中长期奋斗目标：到 2020 年实现全国生产安全状况的根本性好转，亿元 GDP 死亡率、10 万人死亡率等指标达到或接近世界中等发达国家水平。要把安全发展作为一个重要的理念纳入我国现代化建设的总体战略。要坚持把实现安全发展、保障人民群众的生命财产安全和健康作为关系全局的重大责任，与经济社会发展各项工作同步规划、同步部署、同步推进，促使生产安全与经济社会发展相协调。要经常分析生产安全形势，深入把握生产安全的规律和特点，抓紧解决生产安全中的突出矛盾和问题，有针对性地提出加强生产安全的政策举措。

生产安全的长期性、复杂性、艰巨性和紧迫性，决定了必须从我国经济和社会发展全局出发，依靠安全科技，防止和减少各类事故，保障人民群众的生命和财产安全。

我国生产安全科技发展严重滞后于经济和社会的发展，在科学技术整体中属于发展落后领域，但还没有得到科技界和全社会的广泛重视，其中存在的主要问题表现在以下几个方面。

---

① 本文是中国工程院科研项目报告，北京，2010；工程管理研究，北京：科学出版社，2015，157-173. 作者：刘人怀，钟群鹏，马宏伟，黄世清.

1）生产安全专业人才缺乏

生产安全事业的发展离不开专业人才。我国工矿商贸领域从业人员已达 4 亿多人，现有政府安全监管人员 4 万～8 万人，2020 年全国城镇就业人员将达 5 亿，按工业化国家万名职工配 2～4 名安全监察员的比例，2020 年我国政府安全监管人员总量需达到 10 万～20 万人。按目前 300 多万个企业的数量和规模，企业生产安全管理和安全技术人员数量应在 30 万人以上。目前，全国仅有 80 余所高校设立了安全工程本科专业，每年毕业生不足 5000 人。加强安全科学与工程学科建设，培养和造就百万安全技术和安全管理人才是未来 20 年的当务之急。

2）生产安全科技和装备落后

科学技术是生产安全的重要基础和技术保障。目前，我国安全科学理论研究滞后于实践，企业安全技术与装备落后，安全科技投入严重不足。必须研究开发适合我国国情的安全管理理论，提高安全技术、工艺与装备质量，通过安全科技创新引领生产安全发展，提高企业生产安全和政府安全监管的科技含量。重大事故风险辨识、评价，重大危险源监测、预警、控制与事故应急救援技术仍然是未来安全科技研究的中心任务。将生产安全工作转变到预防为主和依靠科技进步是长期的战略任务。《国家中长期科学和技术发展规划纲要（2006—2020 年）》已将公共安全列为重点发展领域。国家安全生产监督管理总局提出了科技兴安战略，印发了《“十一五”安全生产科技发展规划》。

3）生产安全标准滞后

生产安全标准的水平在很大程度上代表着生产安全水平，我国生产安全形势严峻的重要原因之一是生产安全标准的落后、滞后和不落实。我国现有生产安全相关国家标准近 1500 项，行业标准 3000 多项，大多数标准滞后、内容过时、技术水平低，与国际标准接轨程度低等问题严重，难以适应生产安全形势的需要。安全技术和产品安全标准已成为我国加入 WTO 后，国际贸易中的技术性贸易壁垒。调查显示，近年来技术性贸易壁垒已取代反倾销成为制约我国出口的最大障碍，71％的企业、39％的产品出口因此受挫。生产安全标准若不能与国际接轨，一方面将削弱中国出口产品的竞争力，另一方面将影响安全健康，导致污染环境的生产活动和产品可能向中国转移。建立健全与国际接轨的生产安全标准体系是未来的迫切任务之一。

4）生产安全教育培训严重不足

事故预防离不开生产操作岗位，离不开数以亿计的劳动者。2020 年我国就业总量将达到 8.4 亿人，城镇就业人员将达 5 亿人。目前工矿商贸企业 4 亿职工中的 2 亿农民工的职业技能培训严重不足，国务院研究室发布的《中国农民工调研报告》显示，目前 76.4％的农村劳动力没接受过技术培训；建筑行业 3200 万农民工，参加过培训的仅占 10％；矿山、建筑等高危行业事故中农民工伤亡人数占 75％以上。现阶段我国高等教育毛入学率仅为 17％，美国等工业化国家超过 80％，工业化国家农民工受过职业培训的比例都在 70％以上。如何提高几亿劳动者，尤其是只具有小学、初中学历的高危行业的劳动者的安全技术素质是实现生产安全形势好转的根本性任务之一。当前亟须解决的问题是要尽快扭转“高风险岗位、低安全素质、低收入报酬”的现象。

5）安全文化落后

安全文化是个人和集体的价值观、安全态度、能力和行为方式的综合产物，它决定着安全管理层的承诺、工作作风和效能。近20年来国内外生产安全的实践和理论研究表明，文化的伦理功能、社会定向和规范功能及先进的安全文化产生的凝聚力与约束力既是保障生产安全的有效力量，也是落实以人为本的科学发展观的必然要求。要提高全民的安全意识，树立“安全是相对的，风险是永存的，事故伤害是可以预防的”科学的安全理念，营造“关爱生命，关注安全”的文化氛围，培养遵章守纪的安全行为习惯，必须加强安全文化建设。

6）中小企业生产安全监管不力

中小企业生产安全问题是国内外生产安全领域长期面临的重要问题之一。中小企业工伤事故较多，由于安全技术、专业人才、安全管理和政府监管等多方面的原因，未来中小企业安全问题仍将是全社会关注的重点。工业化国家的经验表明，中小企业生产安全离不开生产安全准入制度，安全培训、安全技术中介服务网络等安全措施。目前，我国职业危害严重，预防和减少职业危害应作为我国现代化进程中生产安全工作的重点任务。

7）厂房选址与土地使用缺乏科学规划

当代著名的社会学家、风险社会学派的代表人物之一——乌尔里希·贝克认为，当今社会由于工业化带来的事故灾难风险不同于工业化以前人类所遭遇的各种自然灾害，事故灾难风险源于人们的决策，已不可避免地成为一个政治问题，社会成员、企业、国家机构、政治家都应该对工业化所造成的事故灾难风险负责。我国工业化和城市化进程中，由于缺乏土地使用安全规划法规和标准，留下的土地开发利用和工厂选址等布局性隐患已成为今后长时间内生产安全的重要风险，也是生产安全引发环境保护和公共安全问题的根源之一。吉林石化“11·13”爆炸及松花江水域污染事件后开展的全国石化化工项目环境风险排查结果显示，7555个石化化工建设项目中81%布设在江河水域、人口密集区等环境敏感区域，45%为重大风险源。据估计，现阶段全国范围内因安全距离不足，布局不合理需搬迁的危险化学品生产企业达2000多个。中国的工业化、城市化不能走入“盲目建设→搬迁→再盲目建设→再搬迁”的恶性循环，应尽快出台土地使用安全规划、厂房选址布局安全法规和政策，切实落实“安全第一，预防为主，综合治理”的方针。

8）农村生产安全未引起足够重视

随着全面小康社会和新农村建设目标的逐步实现，中国21世纪的现代化进程中不能不重视农业生产安全。2020年我国农村劳动力仍将保持在3亿左右。在农业生产过程中，目前每年因农机、农电、农药中毒和其他职业危害造成的伤亡人数较多。如何保障3亿农业生产者的安全是我们面临的一个新课题。

9）生产安全应急救援保障体系不完善

生产安全是公共安全的重要内容，生产安全问题极易引发环境污染、社会不稳定等公共安全问题。随着城市化和工业化进程的加速，重特大事故更具灾难性和社会危害性。与发达国家完善的公共安全保障体系相比，我国的差距仍然很大：与公共安全面临的巨大挑战相比，迄今为止的应对措施仍仅局限于部门或地区性、行业性的缺乏全局战略意义的宏观设计和考虑；应对的方式仍然是传统的行政式的陈旧办法，科学技术的支撑显得软弱无力；应对时间仍然是救火式的事后诸葛，缺乏预见性的事前监督措施；应对意识仍然是被动挨打式的心理，缺乏主动的科学的防范策略。社区安全基础差，还没有形成以社区为中心的矩阵式的、综合的事故灾害预防、管理与监督模式，部门分割、条块分割的现象严重；重大危险源分布分类不清，还没有建立起以社区为中心的重大危险源监控、事故预警和应急救援体系相结合的公共安全保障体系。建立健全多元化的、严密有效的公共安全保障体系，是建立生产安全长效机制、构建和谐社会和我国现代化进程中重要的战略任务。

回顾我国生产安全的历史经验和教训，综合分析和总结我国生产安全工作的发展规律和变化特点。在我国生产安全出现的三个较好历史时期，各级政府和企业对生产安全十分重视，注重解决实际问题和取得实效；国家与企业都加大对生产安全的投入；国家生产安全监察机制适时调整，各级政府的管理力度大；处理重、特大事故严肃、及时，执法严格。而在三次事故高峰期（仅就事故发生环境和条件而言），第一次事故高峰是人为地破坏了在计划经济条件下的经济规律，即大跃进时代；第二次事故高峰（两段）是由于“文化大革命”中破坏了经济工作（包括当时的生产安全管理架构）正常运行和不按经济规律办事，并在“文化大革命”结束不久，急于求成，生产安全管理架构未及时到位所造成的；第三次事故高峰主要是在我国由计划经济向市场经济转轨初期发生的。

总之，要做好生产安全工作必须做到坚持“安全第一，预防为主”方针，树立“以人为本”思想，不断提高生产安全素质；加强生产安全法治建设，有法可依，执法必严，违法必究，落实生产安全责任制；建立完善生产安全管理体制，强化执法监察力度；突出重点，专项整治，遏制重特大事故。归根结底，需要加大生产安全投入，依靠科技进步，标本兼治，全面改善生产安全基础设施和提高管理水平，做好生产安全工作。

2. 当前生产安全研究的国际先进水平与前沿问题综述

一个国家的生产安全状况与该国的社会经济发展水平有着密切的联系，发达国家的生产安全状况普遍好于发展中国家。通过对国外一些国家生产安全发展趋势的分析，人均 GDP 为 1000～3000 美元时，生产事故发生频率基本呈上升趋势，人均 GDP 达到 5000 美元之后事故发生频率才开始下降。但是，及时地制定生产安全法律法规，认真执法，加强监管，建设安全文化等做法，可以有效减少伤亡事故的发生。发达国家的生产安全也经历了一个由乱到治的过程。美国、日本、德国和俄罗斯等生产安全状况较好的国家，在以下几个方面有很好的经验。

1）立法先行确保生产安全

生产安全立法是为了保护劳动者的安全与健康，保障生产经营人员的利益和应享受的法定权利，保障社会生产资料和国家及人民财富安全。英国、德国、美国等工业发达国家是劳动安全立法最早和最为完善的国度。现在英、美等国已根据职业安全卫生的要求制定了不少法律法规，形成了较为完整的安全卫生法律体系，并具有以下几个特点。

(1) 立法层次高，权威性强。美国、俄罗斯、南非、韩国等国家对煤矿（矿山）的立法非常重视，煤矿（矿山）的安全立法都由国会审议通过，并由总统颁布。

(2) 法律体系严密，完整性强。美国、苏联、印度、韩国除了制定煤矿（矿山）安全法律外，还制定了一系列相配套的安全规程或实施细则，建立了完善的法律、法规体系。例如，美国矿山安全与健康管理局制定的矿山安全与健康标准，包括了煤矿和非煤矿的详细标准。由于有了全面严格的法规，煤矿安全工作走上了正规化、法治化的轨道，煤矿安全状况明显改善。进入20世纪90年代，煤矿事故持续减少，保持了世界最好水平。苏联制定了《煤矿和油页岩矿安全规程》，内容十分具体，为了更有效地实施该规程，还制定了实施细则。

(3) 法律条款明确，操作性强。美国《联邦矿山安全与健康法》明确规定，颁布与有毒和对身体有害药剂有关的法定标准时，劳工部部长必须在最可靠的基础上制定安全标准，以确保矿工在该安全标准规定下的有害环境中长期工作而健康或工作能力不会受到实质性的损害。

(4) 技术规程、标准上升到法律层次。美国、苏联、印度、澳大利亚等主要产煤国家都把技术规程、标准写入法律之中，以法律的形式颁布实施。美国《联邦法规》“矿产资源”卷由美国国会审议通过，是与1977年《联邦矿山安全与健康法》配套的全套法规标准，并每年进行修订出版。

预计随着社会、经济、生产的不断发展，生产安全立法一定会有新的发展，21世纪，人类的生产安全立法将体现出如下趋势和特点。

(1) 生产安全法规由专门化走向综合化，从分散逐步发展为体系。

(2) 生产安全法规的任务更突出预防性，更强调超前和本质安全化的特点。

(3) 生产安全立法的目标体系更趋明确。立法的目标不但包含防止生产过程的人员伤亡，还包括避免生产过程的危害（职业病）以及生产资料的安全保障和社会财产损失的控制等方面。

(4) 生产安全立法的层次体系更为全面。国际通用的生产安全法规（建议书、国际劳工公约等）、各国的国家生产安全法规、世界范围及本国的行业生产安全法规（石油、核工业等）、地区生产安全法规（欧盟、亚太）等，得到全面发展。

(5) 生产安全立法的功能体系更为合理。建议性法规（如《职业安全健康管理体系导则（2001年）》）、强制性法规（一般各国制定的国内生产安全法规都属此类）、承担不同法律功能的法规（如法律、技术标准、行政法规、管理规章等），各守其责，发挥各自的功能和作用。

2) 实施职业安全与健康计划

世界各国的生产安全理念已经从以控制伤亡事故为主向全面做好职业安全健康工作转变，把职工安全健康放在第一位。所以，很多国家在制定生产安全的相关立法之后，为了进一步提高职业安全状况和生产安全率，都制订了更加详细的职业安全与健康计划。综合各国近年提出的一些职业安全与健康计划，可以发现以下几个特点。

(1) 目标明确。2002年，欧盟提出了2002～2006年关于提高安全与健康标准的共同体战略，该战略讨论了在制定实施战略的政策和行动计划时要考虑的因素，包括必须与国际劳工组织、世界卫生组织及其他国际机构正在开展的工作联系起来。

美国职业安全与健康管理局于2003年制订了一项战略管理计划，支持劳工战略计划司的关于建立一支有准备的劳动力队伍的建议，提出实现筹建劳动力队伍和高质量工作场所目标的手段。战略目标反映了以下几个主题：加强职业安全与健康管理局的战略监督能力，将其资源集中在能得到最高回报的领域；采用直接干预和合作方式，在提高人们的自尊和建立安全与健康的工作场所文化方面取得更大的进展；确保职业安全与健康管理局具有国家层面上的领导职责，在履行对工作场所安全与健康职责方面具备专门知识和技术能力。

(2) 措施有力。欧盟提高安全与健康标准的共同体战略考虑到劳动世界的变化和新的危险，特别是心理素质变化的现实，对工作中的健康问题采取了综合解决办法，并以提高工作的质量为目的，将安全和健康工作环境作为实现的根本手段之一。该战略基于加强危险预防文化建设，基于将包括立法、社会对话、进步措施和最佳实践、公司法人责任和经济奖励等多种政治手段结合起来，以及在包括工人参加的安全与健康平台上建立伙伴关系。

3) 工伤保险制度预防在先

职业伤害是大工业生产的必然产物。工人在生产过程中，可能遭受伤害，也可能受到物理的、化学的、生物等因素的影响而导致职业病。工伤保险可以为受害者及其家庭提供得以在未来体面地生存下去的保障。近些年，世界各国工伤保险制度经历了一次新的变革，扩大了工伤保险的覆盖范围，调整和完善了工伤保险待遇标准。

德国工伤保险制度规定工伤保险由同业公会负责。为了保障企业的生产安全，减少工伤事故，各工伤保险同业公会在全国自上而下设立了安全技术监察部门，配备专职安全监督员，监督员在工作中发现企业存在生产安全问题时，一是能够及时提出指导性意见，督促企业整改，二是可提请国家生产安全监督管理部门监督企业整改。此外，同业公会内还设有技术支援机构、医院和研究室。技术支援机构可帮助企业培训和检测分析，指导企业改进工作，医院可医治一些较轻的伤员和职业病患者，研究机构可对一些影响职工安全与健康的危害因素做一些前瞻性的专题研究。由于工伤保险在事故预防方面发挥了积极作用，德国全行业工伤死亡人数年年下降，2001年已降低到1975年总量的75%。

工伤预防能大大降低事故的发生率，这是国外工伤保险经过100多年的实践得出的结论。各国推行工伤保险管理制度，其目的是促进生产安全，维护社会公平和稳定，保障企业可持续发展。以下是各国在实施工伤保险制度方面的一些共同特点。

(1) 职业伤害税制。其目的是通过收税方式筹集资金，用于工伤补偿或职业安全卫生研究和立法，它克服了工伤保险的弊端，被认为是建立工伤赔偿基金最直接的方式。

(2) 制定工伤保险费率的处罚机制。处罚机制的指导思想是，如果雇主欲避免高费率处罚，他们就必须搞好职业安全工作。

(3) 工伤保险与事故预防相结合的一体化体系。德国是工伤保险与事故预防一体化的样板国家。职业安全卫生专家直接面向生产安全记录不良的企业，并帮助这些企业改善职业安全卫生状况。同时，如果企业的投资或技术革新是将工作风险降低到法规或标准规定之下，工伤事故保险联合会通过返还部分保险费帮助企业实施这些安全技术，包括采用新工艺、增加设备的防护装置以及建立生产安全小组等。在该政策的实施期间，德国的职业伤害和职业病记录逐年大幅度下降。

4）严密的生产安全监察监管体制

世界上大多数国家实行“国家立法，政府监察，业主负责，员工守章”的生产安全管理体制。国家为了保障公民的生命安全与职业健康，颁布生产安全法律强制业主执行，健全完善国家生产安全法律体系；政府依照国家生产安全法律制定生产安全监督（察）法规，督导业主（企业）依法做好职业安全健康工作。一些非政府组织和行业协会密切配合政府的生产监察工作，加强同业监管。很多企业在内部也设立管理机构，建设职业健康安全管理体系，提升自身的社会形象，减少工伤事故给企业带来的损失。

（1）政府部门依法行政，职责清晰。美国联邦政府负责生产安全执法的部门众多，根据所涉及的具体行业领域由不同的部门参与。但是，为了保证执法的强制性和统一性，美国联邦政府成立了专门的机构，以加强生产安全的立法和执法。职业安全与健康管理局是美国负责生产安全管理的主要机构，其宗旨是通过建立并实施职业安全卫生标准、提供教育培训等服务、与其他联邦部门和地方政府建立伙伴关系、促进工作场所安全与健康水平的持续提高，保障美国所有从业人员的安全与健康。日本的劳动安全事故与职业病统一由日本厚生劳动省劳动基准局管理。《劳动基准法》明确了厚生劳动省负责一切与工人安全、健康有关的事务，包括制定标准与管理规章、行政监察、工伤保险和中介机构的管理等。

（2）非政府组织协助监督。国外的非政府组织非常发达，在生产安全管理方面，很多国家也采取小政府、大社会的模式，给予非政府组织明确的定位和职责，协助政府工作。在日本，除了厚生劳动省垂直领导的政府生产安全监管体系以外．还有一些与劳动安全卫生有关的非政府组织，在政府授权的条件下代行一些检查与监督职能。例如，日本工业安全卫生协会、日本建筑安全卫生协会、日本道路运输安全卫生协会、日本港口工伤事故预防协会、森林和木材加工事故预防协会、日本矿山安全卫生协会、日本锅炉协会、日本起重机协会、锅炉和起重机安全协会、劳动安全卫生综合研究所等都是被授权的检验和监察职能机构。除此以外，还有一些基金会、促进会等组织。德国的行业协会也具有公共管理部门的性质，可以依法强制企业缴纳工伤保险和采取安全防范措施，并由政府进行监督。行业协会既负责调查事故、办理赔付，也负责所辖范围内企业生产安全日常性监督检查，提供生产安全政策、技术等方面的咨询和指导。

（3）监察人员权力大、执法力度强。德国对于劳动保护的监察力量相当强大，全国范围内有 3000 多人的监察力量。监察人员可在不事先通知的情况下，在任何时间对企业进行抽查，对不符合安全规定的企业提出限期整改通知，如果企业在规定的时间内没有达到要求，监察人员有权力令企业停产。正是在这种严密的生产安全监管、监察体系之下，很多国家的生产安全已经从被动防范走向了源头治理，形成了生产安全的长效管理机制。

5）安全科学快速发展，减少事故发生

安全科学是人类生产、生活、生存过程中，避免与控制人为技术、自然因素或人为-自然因素所带来的危险、危害、意外事故和灾害的学问。它以技术风险作为研究对象，通过对事故与灾害的避免、控制和减轻损害及损失，达到人类生产、生活和生存的安全。随着安全科学学科的全面确立，21 世纪，人们更会深刻地认识安全的本质及其变化规律，用安全科学的理论指导人们的劳动与生产实践活动，保护劳动者与社会大众的安全与健康，发展生产，增长经济，创造物质和精神文明，推动社会进步。世界各国都高度重视发展安全技术，实现生产过程的安全系统工程，使技术系统的本质安全化提高到理想的水平。

发达国家普遍重视生产安全的科技与投入，很多国家政府都把生产安全科技的研发作为国家的研发重点。美国矿山安全与健康管理局财务预算也逐年增加，2000 年预算资金额度为 2.28 亿美元。这些预算资金主要用于建立煤矿安全监察信息系统和更新安全与健康监察仪器设备。英国健康与安全执行局每年发布几百项科研项目，科研费用大约有 3000 万英镑。法国国家科学研究中心以人机工程研究为基础，进行动作研究、工作空间（环境）研究、人与设备关系研究、新型机械性能研究等，并开发先进的劳动防护用品。

目前，欧、美等发达国家和地区已经普遍采用了多种现代化生产安全技术，以煤矿业为例，有以下几种非常先进的手段来确保煤矿的生产安全。

（1）瓦斯检测、吸收技术。瓦斯是煤矿生产安全的最大隐患。德国的煤矿都装有瓦斯检测、吸收装置，矿山的“自动断电系统”，会随时与检测装置相连，当瓦斯浓度超过 1.5%的警戒线、温度超过 25℃（上下 5℃）及新鲜空气输入量不够时，所有采矿设备用电就会跳闸，自动停止采掘。

（2）机器人替代装置。在危险性极高的矿山作业中，用机器代替人的工作是生产安全技术的重要分支。澳大利亚科学家研制出世界上第一台矿井用遥控紧急营救车。这种车的主要用途是在地下深处作业。它能够攀登和爬过瓦砾堆，能在被毁坏的矿井或被水淹没的坑道中涉水而过。发生事故时，这种车的传感器能够收集有关矿井里空气和自然环境的信息，为营救工作引路。

（3）井下环境多功能计算机监测系统。德国石煤股份公司、德国矿冶技术有限公司及多家科研机构共同研制出井下无线局域网系统，这种技术利用安装在矿工头盔上的摄像头传送地下煤矿实时图像，并通过手机、耳麦等移动通信设备，借助微型电脑进行数据传输等。

21 世纪，生产安全仍然是各国经济稳定发展的重要保障，安全科学技术也会迎来更快速的发展和应用。充分利用好安全科学技术这一强有力的武器，世界定会更加稳定、和谐，社会经济定会更加发达，人民定会更加平安、康乐。21 世纪人类的安全技术将在如下领域获得重大突破。

（1）深入开展安全人机学研究，建立不同国家和民族的人体要素尺寸数据库和人的可靠性数据库，为制定安全人机学标准提供依据。

（2）研究实现技术系统的安全自组织功能，如消防系统的高度阻燃材料、可靠的防爆电气、灵敏的自灭（喷淋系统）装置、高性能的个体防护用品和设备。

（3）实现可能的替代技术。能量的替代技术——用安全能源代替危险能源；用机器人代替人进入危险作业场所；在易发生火灾、泄漏等的危险生产过程中，使用安全卫生的高性能材料等。

（4）研究重大工业事故预防与控制技术，特别是火灾、爆炸、毒物泄漏监测、控制和防护技术，建立各国的重大危险源数据库和监控系统网络，建立地区重大事故应急系统。在不久的将来，人类重大工业事故将得到极大的消除和抑制。

（5）研究重大危险源监控技术。在对重大危险源进行识别、监测与风险分析评价的基础上，实施适时的监控，将其可能的危险与危害控制在许可的状态和水平上。

（6）研究矿山、建筑等事故多发行业的事故控制措施与对策，以及中小企业、外资企业、合资企业安全监察与管理措施。

（7）安全工程技术和劳动保护产业得到极大的发展。形成安全工程设施、设备，安全监测仪器，个体防护用品以及安全信息咨询与工程设计，安全教育与仿真培训系统等安全产业发展支柱。

6）建设安全文化，重视安全培训

安全文化指人类安全活动创造的生产安全、生活安全的观念、行为、环境、条件的总和。安全文化的目的是提高人的安全素质，建设本质安全的环境和氛围。其意义在于为预防事故构筑基础工程，具有长远的安全战略性意义。英国健康与安全委员会核设施安全咨询委员会把安全文化定义为“个人和集体的价值观、态度、能力和行为方式的综合产物，它决定于健康安全管理上的承诺、工作作风和精通程度”。

在生产中，引起事故的直接原因一般可分为两大类，即物的不安全状态和人的不安全行为。生产安全技术解决的只是物的不安全状态。不得不承认，科学技术和工程技术是有局限性的，并不能解决所有的问题，其原因一方面可能是科技水平发展不够，另一方面可能是经济上不合算。因此控制、改善人的不安全行为也是十分重要的，安全文化可以补充安全管理的不足，是和企业的生产安全实践活动紧密结合在一起的。

从观念上体现本质论的倾向，从行为上实现预防性的趋势，是21世纪安全文化突出的特征。各国各行业安全文化发展已成燎原之势。据国际劳工组织提供的资料，许多国家已经开始重视并促进安全文化的发展。例如，欧盟在其共同体战略中直接提到了加强危险预防文化建设；日本的五年计划的一个基本政策是加强安全文化建设，使公司和个人重视安全并保障职业安全，建立自我保护机制；英国健康与安全委员会的一个主要目标是在日益变化的经济中为建立并维持一种有效的安全文化寻求新的方法；美国在其五年计划中寻求在朝着创造一种根深蒂固的文化方面取得更大的进展，使企业更自觉地参与自我管理计划。

生产安全状况的好转与强化全民安全意识是分不开的。安全文化的作用在于产生一种内约束，即人们在作业行为方面主动、长效的自我约束。各类安全周和展览活动是各国政府向社会普及安全文化的绝佳形式。欧洲安全与合作组织时常开展以预防职业危害为主题的欧洲安全卫生年活动。英国健康与安全执行局 1992 年开始举办工作场所安全卫生周活动，以后每年下半年都开展一次安全卫生周活动。美国在每年的 10 月由美国国家安全委员会组织，开展全美安全大会及展览会。美国安全工程师学会把每年的 6 月 20～27 日作为全国作业车间安全周，通过活动达到减少工作场所事故的目的。加拿大安全工程协会每年 6 月都发起加拿大职业安全卫生周活动。日本每年 7 月 1 日至 7 日是全国安全周，10 月 1 日至 7 日是全国劳动卫生周，推进企业事故预防和劳动卫生管理活动，提高全民的安全意识。

安全培训是塑造员工的安全行为文化最为有效的途径。欧盟把培训教育作为搞好安全与健康工作的重要手段。在欧盟，职业安全与健康培训教育大多是免费的。培训内容比较广泛，从安全心理学到操作技能都培训。德国把企业的安全教育状况作为重点监管内容，促进企业将其变成自我行动，以不断提高工人的防护意识和防护能力；同时加强职业资格管理，对安全管理人员分层次进行培训，提高其安全管理能力。

3. 生产安全研究中的重大工程科技问题

生产安全是以人为本，是坚持全面、协调、可持续发展观的直接体现，是全面建设小康社会的重要内容。安全科学技术是保障生产安全的基础，人才、投入、科研设备、条件以及政策、法规、环境等都是重要保障措施。

立足于“高、新、深、实”（“高”就是把握生产安全全局，高屋建瓴；“新”就是开拓创新，耳目一新；“深”就是深谋远虑，体现战略思维；“实”就是增强可操作性，解决实际问题）思想，坚持“安全第一、预防为主、综合治理”的方针和“安全发展”理念，充分发挥生产安全科技在有效遏制重特大事故、促进生产安全形势根本性好转、推动安全保障型社会建设进程中的重大作用。

围绕生产安全科技的新任务和新要求，需要进一步深入开展以下几个方面研究工作。

第一，创新生产安全理论，在下述工作中寻求突破：①安全科学基本理论；②矿山重大灾害事故致因机理及动力学演化过程；③典型工业事故发生机理及动力学演化过程；④生产事故应急救援理论；⑤生产安全长效机制理论。

第二，针对生产环境和生产工艺过程中的灾变因素与危险源特性，争取在以下方面取得创新性研究成果：①煤矿重大灾害防治关键技术；②煤矿灾害连续监测、预警及防控技术；③煤矿重大灾害的救灾技术与装备；④海上油气勘探安全保障技术；⑤大型油轮泄漏油围堵清除设备；⑥深海资源开采生产安全关键技术；⑦特种设备失效模式、失效准则、风险评价、剩余寿命预测等的关键技术；⑧重大危险源辨识指标体系、监测与监控网络化技术；⑨重大事故模拟仿真与虚拟现实技术；⑩重大事故应急救援预案及指挥决策系统；⑪构筑物（地基、大坝、高陡边坡、尾矿库等）失稳监测预警技术；⑫埋地压力（气、水、油）管线检测、报警、关断及维护技术；⑬事故隐患辨识与评价、监测与控制及治理等关键技术研发。

第三，建设社会化、网络化科技服务力量，加强先进适用技术推广应用，加强国内外合作交流，及时对新技术、新成果进行消化、吸收、集成、再创新，逐步形成具有自主知识产权的技术成果；不断发挥专家队伍作用，持续培养生产安全战线的高水平科研队伍，为生产安全科技工作的明天积蓄力量。

4. 保障生产安全的重要措施与政策建议

新中国成立以来，我国生产安全工作取得了很大进步。特别是近几年来，我国在生产安全的专项治理和整顿方面，成绩显著，生产安全状况明显好转，但形势仍然十分严峻，任重道远。随着我国经济的发展、人民生活水平的不断提高，特别是社会主义市场经济体制的建立，生产安全工作亟须不断加强，以保障人民生命财产安全和经济可持续发展。

为使我国的生产安全工作尽快走上良性、可持续发展的轨道，我们应以邓小平理论重要思想为指导；继续坚持“安全第一，预防为主”的方针；大力加强生产安全法治建设；健全国家生产安全监察体制；大力推进安全培训教育和科技进步；以企业为主体，以人为本，标本兼治，综合治理；努力实现“四个转变”（即生产安全工作由事后查处向事前预防转变；生产安全监察重点从国有企业向多种所有制经济成分转变；生产安全管理方式逐步从计划经济下的传统方式向依法、依靠科技进步和运用市场经济手段的方式转变；生产经营单位的负责人和广大职工从“要我安全”向“我要安全、我会安全”转变)，按照十六大提出的全面建设小康社会的目标，力争通过十五到二十年的扎实工作，建立起适应社会主义市场经济的生产安全工作体制和生产安全的长效机制，使生产安全水平整体提高，实现生产安全状况明显好转，以满足新世纪全面建成小康社会的需要。具体措施建议有以下几个方面。

1）依靠科技进步，促进我国生产安全水平的提高

组织对生产安全领域重大的工程技术、管理问题开展科研攻关，并将其列入国家科技重大项目攻关计划。建立国家和省级安全专家组与专家库，充分发挥其在重特大事故调查、隐患评估和整改、重大危险源监控等方面的重要作用；在严肃事故责任追究的同时，重在分析事故原因和落实防范措施。

跟踪国际生产安全领域科技发展动向，引进、推广国内外先进的生产安全技术和管理方法，促进生产安全领域科技成果的推广应用和产业化，培育和发展生产安全产业，使保护人类生命安全与健康为目的的职业安全卫生仪器、设备、防护用品等产品的研制、生产和应用得到快速健康的发展。国家要定期公布明令淘汰、禁止使用严重危及生产安全的工艺、设备，禁止使用超期服役的运载工具，为建立新型工业化起保障作用。

2）促进企业建立“预防为主，持续改进”的自我约束和激励机制

企业是搞好生产安全工作的主体和基础。必须通过法律、行政、经济和社会舆论等各种手段，促使企业遵守生产安全法律法规，建立生产安全的长效机制，积极倡导企业像质量、环保管理那样建立职业安全健康管理体系，建立健全生产安全责任制，完善生产安全条件，确保生产安全。严格生产安全条件的市场准入，运用工伤保险费率和银行贷款等经济手段，按照违法必究和严明奖惩的原则，强化生产安全监督检查，强化工会、广大群众和社会舆论监督，形成全社会对企业的外部强大制约机制，使企业真正建立起“预防为主，持续改进”的生产安全自我约束和激励机制。

3）尽快建立适应社会主义市场经济的工伤保险机制

针对我国现行工伤保险存在的赔付水平低、覆盖面较窄、保险费率差异小、与事

故预防相脱节等问题，必须改革并完善工伤社会保险制度。根据不同行业、企业的危险程度、事故的概率、生产安全管理水平与业绩，实行差别费率和浮动费率制度；根据国内有关试点城市的经验，学习和借鉴国外做法，可以省（区、市）或地级市为单位，把不低于8%的工伤社会保险资金，由本地区生产安全主管部门负责，用于生产安全宣传教育和培训等工作。真正建立起强制性的，覆盖全社会的，赔偿、康复和事故预防相结合的社会主义市场经济的工伤保险机制。

4）建立健全生产安全的六大支撑体系

第一，建立健全适应社会主义市场经济体制的生产安全法律体系。围绕贯彻落实《中华人民共和国安全生产法》，力争在“十五”期间，与《中华人民共和国安全生产法》配套的主要法规、规章、标准陆续出台。

第二，建立健全生产安全信息体系。大力改进和完善伤亡事故统计方法并与国际接轨，积极推进生产安全信息网络建设，建立健全重特大事故隐患警告、事故统计分析、政策法规信息等生产安全信息发布制度，建成上下贯通、反应快速、信息准确的生产安全信息体系。

第三，建立健全生产安全的宣传教育体系。建立必要的生产安全宣传教育机构，并充分依靠和发挥各种宣传媒体的作用，组织开展多种形式的宣教活动，把以“关爱生命、关注安全”为主要内容的安全文化作为先进文化的重要内容，加强小学、中学、大学和各种岗位的全民安全素质教育，不断提高人们的安全意识和安全文化水平。

第四，建立健全生产安全培训体系。建成多层次、多渠道的生产安全培训网络，逐步使培训机构、考核标准、证书管理、师资和教材建设等管理工作规范化、制度化。

第五，建立健全生产安全技术保障体系。加强生产安全科技研究，大力发展并规范安全评价、评估、认证、咨询、检测检验、技术服务、技术培训等社会中介组织，积极培育具有执业资格的注册安全工程师队伍，为生产安全工作服务。

第六，建立矿山事故和危险化学品事故等综合性的统一应急救援体系，对现有应急救援资源进行整合和优化，增强对各类重、特大事故的应急救援能力。

5）加大国家和企业对生产安全的投入

针对长期以来国家和相当一部分企业对生产安全投入严重不足、历史欠账较多、生产安全基础薄弱的实际，国家和企业都必须加大对安全的投入。根据国际劳工组织统计和我国专家研究分析的保守估计，伤亡事故和职业危害造成的经济损失占年GDP的2%～4%，而我国20世纪90年代的生产安全投入与产出比为1∶4.505，生产安全的经济贡献率（安全产出/国内生产总值）为3.15%。因此国家和企业都应加大生产安全投入力度。首先是企业必须依照《中华人民共和国安全生产法》，确保生产安全条件的投入，并对安全投入不足导致的后果负责。

与此同时，国家对生产安全监管部门和生产安全科研工作的直接投入也应加大力度。美国2002年职业安全与健康管理局经费预算总额为4.26亿美元、矿山安全与健康管理局为2.46亿美元，合计6.72亿美元，而我国在这方面的经费预算与之无法相比，同年我国国家煤矿安全监察局的经费预算仅是美国的1/60，煤矿数量约是美国的300倍。2002年美国联邦政府预算中用于职业安全研究和事故伤害控制的经费为4.1

亿美元，而我国“十五”期间直接用于安全科研和事故预防的经费只有 8200 万元人民币，年平均不足 2000 万元，约为美国的 1/20。借鉴发达国家经验，我国在生产安全方面（监督管理、科技研究、事故预防、事故救援及调查处理、关系公共利益的重大事故隐患治理和建设项目等）的直接经费投入应该大幅增加，并像我国环保治理那样列为中央预算。

6）设立直属国务院的权威、高效的国家生产安全监察机构

要做好以上工作，需要进一步深化生产安全监察体制的改革，落实《中华人民共和国安全生产法》的执法主体，加强生产安全监察机构建设，改变当前生产安全监督管理工作上存在的职能交叉、职责重叠、多重多头执法、权威性差的状况，解决安全管理与安全监察不分和机制不顺的问题，做到既不脱离我国的实际，又能使生产安全监察工作有效开展，建议设立直属国务院的权威、高效的国家生产安全监察机构。实行中央与省级地方政府两级监察，以省（区、市）为责任区，省以下垂直监察的生产安全监察体制。

充实加强基层安全监察力量。目前，我国生产安全监察人员与每万名职工的比例小于 0.83，而英国为 4.5，德国为 3.5，美国为 2.1。建议我国可按每万名职工 1.5 人的比例配备专职生产安全监察员，使全国专职生产安全监察员总数至少为 3 万人。

七、社会安全相关工程科技发展战略①

1. 中国社会安全相关工程科技研究的意义

改革开放以来，在党和政府的领导下，我国经济、政治、文化、社会建设都取得了举世瞩目的伟大成就，中国的国内生产总值保持持续快速增长，与此同时，人均收入持续快速提高，人民生活实现了由解决温饱到总体上达到小康的历史性跨越。

世界发展进程的规律告诉我们，当一个转型国家发展到人均 GDP 处于 1000～3000 美元的阶段，人口、资源、环境、效率、公平等社会矛盾的瓶颈约束最为严重，也往往是经济容易失调、社会容易失序、心理容易失衡、社会伦理需要调整重建的关键时期。2008 年我国人均 GDP 达到 3260 美元，一举突破了 3000 美元大关。这些数据在一定程度上预示着中国正处在一个至关重要的社会转型时期。中国面临的最大挑战是要在一个相对较短的时间内完成两个重要的转型：第一要完成由传统经济形态向现代经济形态、由传统社会结构向现代社会结构过渡等一系列重大任务；第二是平稳实现由计划经济体制向市场经济体制转变的独特的艰巨任务。不管哪一项任务，其他国家都是用了很长一段时间才完成的。因此，无论从哪个角度来看，中国正在与即将经历的都是前所未有的、艰难和复杂的社会转型，这必然使中国面临着更大的社会压力和更多的不确定性因素，并已经形成或者正在形成更为广泛、多层次的和突出的种种社会风险[6]。

2010～2030 年，对于中国来说是个攸关发展的重要时期。一方面，2020 年前后，

① 本文是中国工程院科研项目报告，北京，2010；工程管理研究，北京：科学出版社，2015，173-187. 作者：刘人怀，王礼恒，孙永福，姚作为.

我国力求实现全面建成小康社会的目标。2030 年，我国要基本完成城市化目标，50%以上的人口将居住在城市。这一波澜壮阔的发展进程会将占全球人口 1/5 的大国带入繁荣昌盛、文明富强的现代化轨道。另一方面，2020 年前后，中国环境污染的势头才会出现真正的拐点；2024 年左右中国也将进入老龄社会。未来 20 年，中国众多国内供应资源都将面临严重短缺，尤其是水资源可能出现严重危机。这些问题的解决，不仅需要长足的经济发展提供支撑，更需要一个稳定与和谐的社会作为保障。因此研究社会安全相关工程科技发展战略问题不仅对推进中国的现代化进程、实现我国社会的长治久安有深远的战略性意义；而且对构建我国公共服务型政府、提高我国政府和群众的社会安全意识、建立政府主导的全民参与的社会安全管理新局面有重要的现实意义；对于弥补我国社会安全管理学科领域的空白，拓宽社会安全研究领域更具有非同一般的理论意义。

由于社会安全的含义比较广泛，为有效集中精力研究其中的热点、重点问题，本节并不涉及对社会安全问题的全方位探讨，只将重点放在社会安全（含金融安全）的工程技术发展战略问题上，重点研究领域包括危机的早期预警、危机的预警系统构建、危机的处置技术、金融安全的早期预警与预警系统构建等，其他问题则不作探讨。

2. 当前我国社会安全管理的现状分析与评价

社会安全是一个较为复杂的概念。由于其内容的广泛性，因此产生了广义和狭义之分。广义的社会安全是指整个社会系统能够保持良性运行和协调发展、最小化不安全因素和其影响度的社会运动状态；从能力建设的角度则是指全社会各个群体避免伤害的能力和机制。显而易见，广义的社会安全包括了国家安全、政治安全、军事安全、经济安全、文化安全、科技安全、社会生活安全等诸多方面，也包含有关机制与体系等操作性的内容。而狭义的社会安全是指除经济、军事、文化和政治等系统以外其他领域的安全，直接地说，主要是局限在人们的日常生活领域及其环境空间。基于上述两个方面的分析，本节采用狭义的社会安全概念，将社会安全定义为人群公共生活环境空间不受侵害并相对稳定的状态，它包括公民生命、财产、社会生活秩序和生态环境的安全，它直接体现了与公民密切相关的公共安全利益的需要。社会安全代表所有社会群体、利益集团在共同的社会生活中的公共安全利益。但要注意的是，影响社会安全的因素包括经济、技术、自然与社会等诸多方面[7]。

1）有关社会安全危机的早期预警系统尚未建立

自 2003 年“非典”事件以来，我国政府高度重视社会安全管理工作，不仅要求各级政府尽快建立公共危机预案，而且按传统的条块分割模式建立起政府的危机管理框架，各级政府、各个专管部门基本上建立起了相应的应急应对体系。此外，随着 2007 年 11 月《中华人民共和国突发事件应对法》的正式实施与 2008 年 5 月《中华人民共和国政府信息公开条例》的正式施行，与危机管理相配套的应急法律体系也趋于完备。目前，我国政府将危机管理的重点工作放在安全危机发生时的紧急应对与危机之后的简单反思，而对于社会安全危机发生前的预测与预防没有足够的重视。综观国内学术界相关方面的文献，可以发现有关研究对至关重要的社会安全危机的早期预警系统没有进行全面的分析，尤其是在社会安全危机的早期预测与预警方法、早期预警的智能决策与咨询系统等方面的研究基本上属于空白。此外现有研究对社会危机管理的分析比较偏重政府的主导地位，忽视了第三方组织的作用；而对危机处置技术的研究也没有什么进展。实际上，即使在国外学术界，有关危机的早期预警系统建设依旧是个前沿的课题，有待人们去探索。

2）全国社会安全危机早期预警指标体系尚未研制出来

早期的研究只是将社会安全问题等同于社会保障问题。随着我国近十年来社会安全事故的频发，人们开始重视对社会安全事件产生的背景与演变规律的研究。但从现有文献来看，学术界对社会安全的研究重点主要包括社会安全应急机制与社会（安全）预警等主题。我国学者在西方学者的研究成果基础上对我国社会预警指标的研究有了较为成熟的成果。这些研究可以分为以下四类。

（1）单一的社会安全指标体系，如文献［8］探讨了社会保障指标体系。

（2）基于社会-经济系统交叉影响的社会安全指标体系，如较具代表性的中国社会风险综合指标体系[9]和社会预警指标体系[10]等。这些指标体系的特点是基本上涵盖了经济与社会领域的众多方面，但一些指标体系缺少必要的实践检验支持。

（3）基于巨系统思维的社会安全指标体系，较具代表性的是运用社会物理学提出的社会稳定预警指标体系[11]和运用“自然-经济-社会”复杂巨系统理论所建立的社会稳定与安全预警指标体系[12]。这些指标体系的特点是结构相当复杂，几乎考虑到了影响社会安全稳定方面的所有因素。

（4）基于利益群体心理的社会安全指标体系，其中较具代表性的是文献［13］所提出的基于公众心理的社会预警指标体系。但令人遗憾的是尚没有一个能够经得起时间检验的、被公认的国家级社会安全预警指标体系被研究出来，这直接影响了我国社会安全早期预警指标体系的应用效果，也影响到我国社会安全危机管理的有效运作。

3）社会安全事件的风险评估能力和关键技术有待加强

对社会安全事件演变成危机的风险进行评估是我国开展有效的社会安全管理最为基础的工作。强化对社会安全事件或危机的风险评估能力本是社会危机管理的重中之重，风险评估就是通过对各种社会安全状态的监控，发现可能影响社会安全的风险因素，主动搜集与风险有关的各种信息，通过对信息的整合、处理、判断和相关数据的分析，掌握风险的各种变化和最新信息，监测风险发生的概率和趋势，运用一定的科学方法对风险趋势进行科学的评估，预测可能出现的社会危机，以便为不同类别或级别的社会安全危机制定相应的危机应对策略。及时发现风险因素、准确进行风险评估，有利于及时采取有效的处置行动，将危机消灭在萌芽之中，这是危机管理的最高境界。

但目前不管是从政府的运作还是从学术界的研究来看，有关社会安全的风险评估至今尚不属于重要的操作主题或者研究课题。政府多将风险评估看作是专家的任务，而不少文献只不过强调风险评估的重要性，却缺少对风险评估全面的理论探讨，尤其是缺少对风险评估方法与技术的研究。这直接导致我国政府在面对社会安全危机时对可能出现的危机后果常常预备不足，有关的危机物资储备时常短缺，危机预案内容空洞，应对技术手段落后，危机应对捉襟见肘。而国外对于各种风险的评估非常重视，不仅设置专门的机构来开展社会风险评估研究，还特别重视运用跨学科的方法对本国乃至国际上众多冲突热点问题进行风险评估，以利于及早把握应对的时机与策略。例如，美国兰德公司、日本野村综合研究所都拥有世界上最新的风险评估技术与方法，这对于帮助政府预测社会安全危机、及时采取应对措施具有重要的作用。

4）社会安全危机的处置能力还不够强

当危机发生之后，社会安全危机管理的重点就转向了危机的处置，危机的处置步骤、速度、技术与方法会直接影响危机处置的效果，更会极大地影响到政府的形象与信誉。而其中的关键环节是分门别类地区分社会安全危机，并根据不同类型社会安全危机的特点与演变规律详细确定危机处置的具体技术，这本是危机预案中非常重要的内容之一。应该说我国政界与学术界对于社会安全危机的处置相当重视，不仅在研究中非常关注这些问题，而且在公共危机管理机制中将社会安全危机的处置放在了非常重要的地位。但从对 2005 年哈尔滨水危机、2010 年大连沿海石油污染、2010 年富士康员工连续跳楼等重大事件的处置情况可以看出，有关部门依旧缺乏有效的危机处置方法与技术选择，这直接导致政府在社会安全危机应对过程中依旧采取传统的人海战术与运动应对的方法。这种状况说明我国目前的社会安全危机的处置能力还不够强，尤其是缺乏对关键环节危机处置的技术与方法的把握。

5）社会安全事件的预测方法还不够科学

从国际上研究来看，有关危机或者突发事件的预测方法最早是用于经济危机或者金融危机的预测或者预警，后来有关方法被运用于社会危机的预测。最经典的预测方法有：①KLD（Kullback-Leibler distance，库尔贝克-莱布勒距离）信号法；②FR（Frankel-Rose，弗兰克尔-罗斯）概率模型；③STV（Sachs-Tornell-Velasco，萨克斯-托内尔-贝拉斯科）横截面回归模型；④冯芸和吴冲锋[14]的多时标货币危机预警模型。这些方法的共同特点是具有类似的架构与内容：①搜集一个危机的样本；②选定一个指标序列；③选定一个观察期；④考察样本个体的各项指标在各自观察期的运动趋势；⑤总结所有样本个体各项指标的运动规律；⑥运用这些运动规律作为预警机制从而达到预测危机的目的。

我国在这方面的研究进展多数是沿用国外的研究成果，采取以上方法运用中国的数据进行危机预测。但这些方法本身就具有一定的缺陷，如方法的运用需要特定的条件；在实际运用时，常常面临难以准确预测危机的出现等问题。尽管人们已经认识到这些问题，开始运用计量经济模型、案例推理、神经网络、遗传算法、灰色系统理论等方法或理论来预测金融危机，但有关预测金融危机的方法如何能够被合理地移植到中国社会安全危机的预测当中依旧是个有待解决的课题。因此，我们无可避免地需要面对有关社会安全危机预测方法的本土化问题，尤其需要考虑如何借助更为先进的理论与方法来创新本土的研究思路。

6）缺少适用于中国实际的金融安全预警研究

金融安全一直是我国金融界与学术界关注的问题之一[15,16]。随着经济全球化的进程不断加快，人们的关注点转向金融全球化背景下的金融安全问题。基于西方学者的研究成果，中国的学者对金融安全的概念与内涵、金融安全与金融风险的关系、金融危机与金融安全的关系、金融安全与金融创新的关系、开放经济下的金融安全、金融安全与法治建设等问题进行了深入的研究。尤其是重点探讨了金融安全预警体系、机制、预警指标与方法。但现有研究在金融危机的早期预警方法上还有进展的空间。此外，涉及中国金融安全的特有因素，如金融业的效率低下、金融市场的开放、金融的虚拟化所引起的安全问题都给后人留下了研究的空间。

3. 当前社会安全研究的国际先进水平与前沿问题综述

目前国内外学术界有关社会安全领域的最新研究课题与前沿问题主要包括以下三个主题：社会安全的早期预警、社会安全危机的有效应对与金融安全的早期预警等。

1）社会安全的早期预警

（1）社会安全的早期预警系统与网络。目前国际上有关社会安全的早期预警系统与网络的研究尚处于起步阶段，前沿研究主题围绕着理论基础的整合与探索、早期预警决策模型的研制与社会危机的仿真模拟模型的建立。有学者在总结金融领域早期预警系统研究经验的基础，提出要运用一整套新的理论和技术，用来建立一种全面的、综合的、自动化的、可推广和有效的预警系统，以实现对社会危机的有效监测、评估和预报。有学者[17]认为早期预警系统应该建立在多学科交叉影响的理论基础上，并将社会看作是基于“自然-经济-社会”相互影响的复杂巨系统，由此出发提出有关社会安全早期预警系统的构成、完善与网络构建。有关早期预警决策模型研究则集中在研制基于知识积累、案例推理与逻辑推理的危机预警模型，基于情景模拟的决策模型和基于学习与认知的风险评估模型上。

（2）社会安全的早期预警指标体系。目前有关社会安全的早期预警指标体系研究主要集中在影响因素的选择及早期预警指标体系的系统设计与运行上。对应于解决类型不同的社会问题，中外学者分别提出了社会稳定预警指标体系、民族团结预警指标体系、社会风险预警指标体系、社会安全预警指标体系、心理角度的社会安全预警指标体系等多种既有相同点、又有不同点的指标体系。但令人遗憾的是，目前并没有一个被国际学术界所公认的社会安全早期预警指标体系。值得关注的是，西方学者多将社会预警指标体系的设计、运用与政策制定相结合，注重其实用性；而中国学者则更关注全方位地考虑影响社会安全的因素来设计早期预警指标体系。

（3）社会安全的早期预警方法。学术界对社会安全的早期预警方法的研究主要集中在以下几个方面：①基于案例推演的社会安全危机预测的混合软推理方法；②基于迁移学习的社会安全危机预测方法；③基于学习与案例推理的社会安全风险模糊评估方法；④基于案例仿真的社会安全危机的智能模拟技术。

2）社会安全危机的有效应对

（1）社会安全危机的整体应对。目前对社会安全危机应对的前沿研究主要集中在有关社会安全危机的整体应对上，主要包括社会安全危机的整合应对机制、应对组织的整合以及处置技术的整合运用等三个方面。文献［18］提出面对日益复杂的公共危机，需要建立一种全面整合的危机应对模式，其中的关键是建立一个资源整合、机制整合、组织整合与信息整合的高效的整合机制。这方面不管是国内还是国际都还处在摸索阶段。尤其是政府组织与第三方组织之间的整合与有关处置技术的整合更具有一定的迫切性。在这里需要指出的是，有关社会安全的整体应对是一项艰巨的系统工程，非常需要政府细化行政工作的各个过程。

（2）社会安全危机的处置技术。有关社会安全危机的处置技术探索基本上也处在一种不断完善的进程中。国际上相关的前沿研究主要集中在以下几个方面：①恐怖事件的现场处置技术与早期干预技术；②群体突发事件的现场处置技术与早期干预技术；③劳资纠纷事件的现场处置技术与早期干预技术；④民族团结事件的现场处置技术与早期干预技术；⑤非传统社会事件的现场处置技术与早期干预技术。

（3）社会安全危机应对的科学管理。有关社会安全危机应对的科学管理是老课题，又是新课题。西方国家在这方面的前沿研究，主要集中在以下几个主题：①有关安全应对物资的调配、储存与发放的规划管理；②安全应对物资仓库选点的规划；③社会安全危机中的人员使用规划与管理；④社会安全危机中的资金使用规划与管理；⑤社会安全危机应对中的后勤保障规划与管理。

3）金融安全的早期预警

（1）金融安全的早期预警系统与网络。目前国际上有关金融安全的早期预警系统与网络的研究有比较丰富的研究成果，前沿研究主题包括早期预警决策模型的完善与金融危机的仿真模拟模型的建立。目前学者对有关金融危机的形成机制的构思已经历了四个阶段，新一代金融危机理论所提出的造成金融危机的因素越来越复杂，各国迫切需要在总结金融领域早期预警系统研究经验的基础上，提出一套新的理论和技术，用来建立一种更加全面的、综合的、自动化的、可推广和有效的预警系统，以实现对金融危机的有效监测、评估和预报。有关金融危机早期预警决策模型研究则集中在研制基于知识积累、案例推理与逻辑推理的金融危机预警模型，基于情景模拟的金融危机应对决策模型和基于学习与认知的金融风险评估模型等方面。

（2）金融安全的早期预警指标体系。有关金融安全的早期预警指标体系研究已经形成了比较丰富的研究成果，目前的前沿研究主要集中在有关指标体系构成的完善及其效果的检验上。目前中外学者所提出的金融危机早期预警指标体系主要包括信用危机（货币安全）预警指标体系、银行危机预警指标体系、金融危机早期预警体系、资本市场预警指标体系等。但目前没有一个指标体系得到了国际学术界的公认，并具有可接受的预测准确性。

（3）金融安全的早期预警方法。有关金融安全（危机）的预警方法的研究有较长的历史，但至今为止依旧没有形成验证准确性高的研究成果。现有的金融安全（危机）预警方法包括 KLD 信号法、FR 概率模型、STV 横截面回归模型与多时标货币危机预警模型等。目前有关前沿研究多集中在运用新的理论与方法（如模糊神经网络理论、灰色系统理论、层次分析模型、计量经济学、熵权法、投影寻踪技术、遗传算法等）来研究新的金融安全的早期预警方法上，如基于案例推演的金融危机预测的混合软推理方法、基于迁移学习的金融危机预测方法、基于学习与案例推理的金融风险模糊评估方法、基于情景仿真的金融危机的智能模拟技术等。

4. 社会安全研究中的重大工程科技问题

1）社会安全的早期预警工程

（1）工程目标与内容。跟踪国内外相关领域的进展，运用最先进的技术手段、创新理论基础，将最新的理论与最新的技术融入社会安全早期预警的各个环节，重点开发一个面向认知和基于学习的早期预警系统框架及其模块。框架应能够描述如何指导基于学习的知识推理、基于案例软推理的预测和模糊风险评估、基于虚拟技术的社会安全危机事件模拟，以应对社会危机的综合预警和认知决策。这将成为一个系统级的使用案例库、知识容器、推理技术和预测模型的指南。按需要适时开发以下技术：社会安全早期预警的智能决策与咨询技术、社会安全事件的虚拟模拟技术、社会安全风险的早期预测方法、社会安全的风险评估技术、社会安全事件的群体心理干预技术、社会安全的舆情管理与干预技术、社会安全早期预警系统软件与维护技术等。

（2）关键技术[17]。①社会安全事件智能全程模拟技术。运用虚拟技术等多种技术，研制能够真实模拟社会安全事件全貌与演变趋势的智能全程模拟系统，以便为科学决策提供有力的依据。②社会安全危机的专家辅助决策技术。运用案例推理、专家决策、情景仿真等知识模块与数据库，总结与提升专家的知识和经验，形成社会安全危机的专家辅助决策技术。③社会安全事件的早期预测方法。总结前人研究的成果，运用多种理论与方法来开发研究适应中国社会发展状况的社会安全危机的早期预测方法。例如，可以开发基于案例的混合软推理方法。该方法将整合案例推理、软计算、规则推理、模型推理、约束满足问题求解和优化、机器自学习等多种算法，提出一个注重预警生成效率和预警生成的可靠性、准确性的早期风险预测技术；也可开发一个基于迁移学习的知识推理方法，这一方法与基于案例的软推理方法相结合，建立一个新的危机预测模型。该方法能通过机器学习将积累的知识转移到新的预测任务，并帮助实现问题的快速求解。④社会安全风险评估技术。利用各种先进的理论与技术，开发适应本国文化心理与经济发展态势的社会风险评估技术，如可以开发一个基于学习的模糊风险评估方法，该方法与基于案例的推理方法相关联，用于确定预警级别和辅助危机战略决策。⑤社会安全危机智能决策系统。在2020～2030年，应以研究社会安全危机的智能决策系统为主要研究方向。以前面的专家辅助决策系统为基础，利用预警系统提供的警情信息，基于知识库、案例库、指标体系、法律库等数据库，运用图像仿真、智能模拟等技术，形成社会安全危机智能决策系统。例如，可以开发一个基于认知驱动的决策过程模型，以支持在社会危机早期预警系统中的认知驱动决策。这个模型的关键元素是知识检索、心智模型和情景推演，其基础是足够深度的知识库（包含相关领域的经验知识与理论知识）与丰富的案例库。⑥社会安全预警系统软件与维护技术。应运用以上提及的各种方法与技术，开发一个用户友好的面向认知和基于学习的早期预警系统软件原型，并验证其合法性和有效性，之后提供给决策者，以便能系统地预测和响应未来可能发生的危机。应提出一个有效的应用方法与技术，用于指导、设计、开发和维护实际的早期预警系统，使之适用于一些专门的社会危机问题，并找到相应的解决方案。

2）社会安全信息的数字化工程与知识共享工程

（1）工程目标与内容。运用最新信息技术，以国家、省级社会安全早期预警研究中心为主体，开发社会安全危机管理领域的知识数字信息化工程。主要内容包括：警情信息数字化、案例库数字化、社会安全管理知识的数字化、对策库的数字化。建立国家与省级的知识共享平台，并通过一定技术手段满足人们查询、复制、运行、研究、咨询、演示等的需要，通过一定技术手段形成一个完整的社会安全管理的数字信息网络与知识共享平台。

（2）关键技术。①建立有关社会安全管理知识的数字信息网络。运用最新技术，开发将预警警情库、案例库、知识库、对策库数字化的先进技术，并以国家和省级社会安全早期预警研究中心为网络基点，形成国家的社会安全管理知识的数字信息网络。②基于互联网的社会安全知识库共享技术与平台。利用最新技术，在社会安全知识数字信息网络的基础上，通过开发基于互联网的社会安全知识库共享技术与平台，来有序集成和整合利用各种分处中央和地方的数字化的社会安全知识资源，以虚拟社会安全知识库为入口，以专题知识为线索整合理论资源、三维资源、视频资源、学术成果、虚拟书架、案例库、对策库等各类资源，来实现社会安全危机管理知识的全国覆盖与瞬时共享。③基于网络共享的重要文献门户系统。按照社会安全危机的种类与涉及的行业或者专业领域，专门定制一套用于构筑自主分类的互联网社会安全信息情报中心的软件工具，为局域网用户进行社会安全学术研究提供参考信息，同时适合物理隔离环境，通过网络开关实现定时信息更新和系统内外网切换。④基于用户友好的多媒体演示技术与平台。利用虚拟技术、多媒体、数据库技术，集声音、图像、文字、视频为一体，开发基于用户友好的多媒体演示技术与平台，通过多媒体导览、三维辅助陈列展示、视频点播等方式，为全国网络用户提供社会安全案例演示、事件真实回放、模拟仿真等个性化服务。

3）社会安全危机的整合管理科学工程

（1）工程目标与内容。基于管理学、系统工程、数学、政治学、公共管理与项目管理等多学科理论，通过理论与技术创新，开展社会安全危机的整合管理科学工程研究。主要内容包括：①社会安全危机管理的整体系统规划，如在不同社会安全危机事件中人、财、物等各种资源之间的整体管理与协调规划，各参与行政主体之间的协调运作管理规划，政府机构与第三方组织之间的协调运作管理等；②社会安全危机管理中的物资运筹规划与管理，如物资的前期储备点与运输线路选择规划；危机爆发时物资的调配、购置与运输、发放规划与管理；捐赠物资的登记、运输与发放规划与管理；在危机后期物资储备规划再评估；③社会安全危机管理中的人员规划与管理，如针对不同类型社会安全事件，在危机爆发时所需要的救援人员、医护防疫人员、工程科技人员、后勤保障人员、志愿人员与部队人员的协调合作机制；危机应对人员的紧急选拔与调配机制；危机后期人员调配机制的再评估；④社会安全危机管理中的资金策划规划与管理，如政府监管下的由第三方实施的募捐资金的登记、使用与审计机制；募捐资金在不同危机处置项目的分配规划等；⑤社会安全危机管理中受难人员救助规划，如对受难人员的资金与物资发放、就业救济、保险救济等综合规划；⑥社会安全危机情景中的政策规划与实施；⑦社会安全危机管理中的后勤保障规划；⑧危机中的人员撤离规划与管理；⑨应急处置中的在线决策支持系统。

（2）关键技术。①应急物资的资源储备与调度技术。运用排队论与选址理论、随机规划理论，研究涉及有关应急物资的前期储备种类、质量、布局与调度、运输问题。②社会安全危机状态下的物资调配与分配技术。运用项目管理理论、运筹学等理论研究危机状态下的物资调配技术，如危机处置中物资运用与管理的动态博弈网络技术、不确定信息背景中的多阶段随机规划技术、不确定信息背景中的多目标约束下的资源调度模型等。③社会危机状态下的人员调配技术。运用排队论与随机规划理论研究危机状态下的人员调配技术，如在危机应急处置阶段的人员调配技术等。④社会危机状态下的救援与后勤保障服务系统。基于排队论与选址理论、随机规划理论等，研究有优先权的多服务台的应急服务的排队系统（包括医疗救护、人员转移、警力、部队、道路疏通等），主要涉及应急服务设施的布点选址与服务区域的划分等。⑤不同社会安全危机事件中人、财、物等各种资源之间的整体管理与协调规划技术；社会安全危机管理中受难人员救助规划技术；危机中的人员撤离规划与管理技术等。

4）金融安全的早期预警技术工程

（1）工程目标与内容。跟踪国内外相关领域的进展，运用最先进的技术手段，创新理论基础，将最新的理论与最新的技术融入金融安全早期预警的各个环节。重点发展金融安全的早期预警系统及其重点板块，按需要适时开发以下技术：金融安全早期预警的智能化决策技术、金融安全事件的虚拟模拟技术、金融安全事件的早期预测方法、金融安全的风险评估技术等。

（2）关键技术。①金融安全危机智能全程模拟技术。运用虚拟现实技术等多种技术，研制能够真实模拟金融安全事件全貌与演变趋势的智能全程模拟系统，以便为科学决策提供有力的依据。②金融危机的专家辅助决策技术。运用案例推理、专家决策、情景仿真等知识模块与数据库，总结与提升专家的知识与经验，形成金融危机的专家辅助决策技术。③金融事件的早期预测方法。总结前人研究的成果，运用多种理论与方法来开发研究应对中国金融行业发展中出现的金融事件的早期预测方法。④金融危机智能决策系统。应以研究金融危机的智能决策系统为主要研究方向。以前面的专家辅助决策系统为基础，利用国家与省级预警中心的警情库，基于知识库、案例库、指标体系等数据库，运用图像仿真、智能模拟等技术，形成金融危机智能决策系统。

5. 社会安全研究的重要措施与政策建议

1）社会安全研究的重要措施

（1）加强领导，统筹兼顾，有序推进。深入开展社会安全（含金融安全）工程研究，开展有关社会安全（金融）危机早期预警的研究是建设和谐社会的战略需求，是实现社会平稳转型的重要保障。未来的社会安全（含金融安全）研究要满足社会经济飞速发展的需要，必须坚持党的领导，突出政府主导的作用，发挥社会主义制度的优势，按科学规律办事，使社会安全管理机制与社会转型的需要相协调，统筹兼顾中央与地方、政府与民间组织、行政机构与研究团体在社会安全（含金融安全）危机预警与应对、社会安全（含金融安全）警情监测与分析、社会安全（含金融安全）管理研究与人员培训等各方面的分工合作，充分调动社会各阶层的积极性，使我国社会安全（含金融安全）管理工作能够追上发达国家的步伐，实现我国加强社会安全（含金融安全）管理、建设和谐社会的伟大目标。

（2）结合实际，建设全国、区域与省级的社会安全早期预警研究中心。要以实效推动我国本土的社会安全研究，根据国家行政区划与社会发展态势，建立中国的各级社会安全（含金融安全）早期预警研究中心。首先，可以先在中国较发达的地区，如珠三角、长三角、京津冀、环渤海等地区，设立区域社会安全（含金融安全）早期预警研究中心；其次，适时建立省级社会安全（含金融安全）早期预警研究中心，可以先在重点区域与经济发达省份同步设立，再到非重点区域与经济不发达省份依次建立；再次，在区域与省级社会安全（含金融安全）早期预警研究中心建立之后，考虑建立国家社会安全（含金融安全）早期预警研究中心。在建立各级研究中心的进程中要注意几个问题：①要高标准、严要求地进行研究中心的规划与建设；②要预先留下省级中心、区域中心与国家中心之间的各种接口：包括数据库连接接口、与国家以及相应级别社会安全（含金融安全）早期预警网络与警情采集中心的连接接口；③要预留与其他不同领域的安全预警中心的接口，以利于建立覆盖全国的总体安全预警中心系统。

（3）创造条件，完善法治，有力推动社会安全早期预警体制的建立。功能完备、法治完善的社会安全（含金融安全）早期预警体制是我国监控社会安全态势、应对社会安全危机、建设和谐社会的重要体制，也是我国加强社会安全管理的战略基础。要建立社会安全（含金融安全）早期预警体制，应采取以下措施：①及早完善现有的危机预警组织结构与功能，使其成为社会安全（含金融安全）的早期预警管理组织；②在调研的基础上，适时推出我国社会安全（含金融安全）早期预警法律法规，为我国社会安全早期预警系统的有效运行提供法律支持；③引入第三方评价机制，对我国现有的危机预警机制进行评估，并提出整改意见；④在汲取国外经验与第三方所提整改意见的基础上，修正与完善现有危机预警体制的不足。建议做好以下工作：完善早期预警体制中有关不同利益群体的诉求机制与预警机制；研究并制定非政府组织参与社会安全（含金融安全）预警与应对的具体组织结构对接、协作机制，以实现各方在信息传递、物资调运与发放、志愿人员的组织与调配、心理干预与救助等多个方面的统一；研究并制定在危机状态下，危机处置的核心组织架构的建立、邻近地区的参与机制和协作机制、不同救灾部门的合作与分工；研究并制定政府机构与部队之间在社会安全危机状态下的分工和协作机制；根据不同时期的危机管理要求，研究社会安全（含金融安全）早期预警体制的运行与工作重点。

2）社会安全研究的保障条件

（1）加强社会安全工程技术研究，提升社会安全管理的技术支撑。在社会安全工程研究中，要特别重视加强社会安全技术研究。社会安全事件的应对与管理同其他领域的安全事件一样，需要运用科学的技术与管理手段。人们对诸如卫生领域的安全事件，或者食品安全事件的应对与管理技术与方法有一定程度的了解，而对于越来越容易诱发社会不稳定的社会安全危机的应对与管理技术却不是非常清楚。要解决这个困局，就应加强社会安全技术研究：①要对社会安全事件早期预测技术与方法进行深入研究，并在有关危机应对中有效实施；②结合未来社会安全形势的需要，组织各级预警研究中心与相关高校专业研究机构对社会安全管理中存在的关键技术问题进行技术攻关，尤其是对重点社会安全事件的风险评估技术与危机应对处置的具体方法进行技术攻关；③结合我国国情，对国家与省级的社会安全危机早期预警系统的构建与完善技术进行深入研究；④跟随我国社会发展的速度，积极开展专项社会安全事件的危机管理科学工程项目的研究，如研究第三方组织与政府在危机管理人员与物资调配方面的协作方法、群体冲突事件的具体监管技术问题等；⑤建立健全各级科技支撑体系，发挥各自功能，做好社会安全危机管理的科技指导与技术服务工作。用类似的方法与步骤，可同步加强金融安全技术研究。

（2）强化社会安全研究人员培养，提升社会安全管理研究能力。为了适应未来日益复杂的社会安全态势，需要有一支素质过硬、能力超群、有责任心、有奉献精神的研究人员来从事社会安全研究。为此，建议：①在部分重点高校设立社会安全管理与技术工程专业，培养能够从事社会安全管理与安全技术管理的基础管理人员；②在部分中心城市早期预警研究中心和部分重点高校设立社会安全管理与技术工程研究生学位培养点，培养高级的社会安全管理人员与研究人员；③制订计划，抽调师资，利用现有的资源，对现有社会安全管理人员进行培训，提升这些管理者探索问题和实际操作的素质与能力。与此同时，也要加强对金融安全研究人员的培育与金融安全管理人员的培训。

（3）加快社会安全管理的基础设施建设。基础设施建设是社会安全管理得以有效进行的基石和保障。要遵循适应需要、突出重点、量力而行、分步实施的原则，加快包括社会安全早期预警网络与警情采集点建设、社会安全早期预警中心建设、社会安全案例库与知识库的建设、社会安全早期预警阈值库等社会安全管理的基础设施建设；加强对社会安全状况的监测与预警，努力降低社会安全事件可能造成的损失与影响；利用信息、计算机等现代化技术和手段，建立快捷、有效的监测预警与信息管理体系，为政府的决策提供准确、及时、科学的依据。用类似的方法与步骤，可同步建设金融安全管理的基础设施。

（4）有效规划，瞄准热点，持续进行地方社会安全热点管理细化工作。应根据社会转型不同时期的特点，在深入研究的基础上，做好地方社会安全热点的管理细化工作：①在不同的时期关注不同的热点，通过规划来研究其演变规律，并据此采取有效的管理对策化解矛盾、增进和谐；②修改与细化地方社会安全危机应对预案，使之更具可操作性；③强本固基，加强对现有村一级党组织与行政机构的建设，转变工作作风，继承光荣传统，密切党、政府与群众的血肉联系，使其具备原有的传声筒、稳压器、降压阀的作用；④做好省级研究中心、早期预警系统同各地行政部门危机应对管理机构之间的互动，强化对各类安全隐患的监控。

3）关于社会安全研究的政策建议

（1）借助现有国家科研项目遴选机制，集中开展社会安全（含金融安全）危机项目研究。为了早日掌握社会安全（含金融安全）事件演变的规律以及对应的社会安全（含金融安全）管理对策、方法与技术，需要利用现有的国家科研项目遴选制度——国家、省、市三级社会科学基金与自然科学基金项目申请制度，结合我国社会发展进程，在不同时期围绕多个研究领域的研究主题提出项目申请目录，向社会发布，吸引社会人才参与社会安全（含金融安全）管理的各种研究。尤其要注意对社会安全（含金融安全）形势的新动向的持续研究、对社会安全（含金融安全）事件的模拟技术的持续研究、对社会安全（含金融安全）事件的处置技术的持续研究、对政府提高社会安全（含金融安全）应对能力的持续研究、对社会安全（含金融安全）热点问题的演变规律的研究、对社会大众的社会公平感心理感知的研究、对金融风险动态监控技术的研究。

（2）启动早期预警网络示范项目，推进国家社会安全早期预警系统与网络建设。要稳步推进国家社会安全早期预警系统与网络建设，应采取树立标杆、建立典型的方法。为此，要适时启动早期预警网络示范项目，在全国发达地区选择多个典型城市，试点建设早期预警网络示范项目，争取在预警网点布局与地点选取、警情采集布点、项目人员配备、组织架构、系统运作机制、设备采购与安装等方面形成一套可行的操作程序，并在专家审核的基础上通过总结提高，形成可供复制与仿效的操作手册，这样可以加快国家社会安全早期预警系统与网络的建设。可以根据金融业的特点同步建设金融安全早期预警系统与网络，也可仿照以上做法，推动国家金融安全早期预警系统与网络的建设。

（3）推进应对社会安全危机工作的法治化完善工作。要力求在法治的轨道上推进社会安全危机管理工作。为此，建议做好以下工作：①适时对《中华人民共和国突发事件应对法》进行修订，增加与完善有关社会安全早期预警法规的有关条文；②鼓励各地通过地方立法来推进社会安全早期预警管理条例的起草，力促各地社会安全危机预警预案内容的完善与修正，并要做到常修常新；③鼓励地方立法清晰界定赋予第三方组织参与危机应对的权利与义务；④鼓励地方政府通过立法来完善有关救助物资发放、募捐物资使用、责任的追究等不明晰的地方；⑤鼓励地方政府通过立法或者出台管理条例来处理有关社会安全危机中志愿人员出现伤亡事故的保险与赔付问题；⑥要特别关注关于金融运作监管的法治化完善工作。

八、生态安全相关工程科技发展战略①

1. 当前我国生态安全的研究现状

我国生态安全问题的提出始于 20 世纪 90 年代后期，主要背景是国内生态环境恶化，生态赤字膨胀，自然灾害加剧。特别是连续出现的特大洪灾和急剧扩大的荒漠化，引起全国上下的极大震动。我国西部大开发的生态环境保护和建设问题引起人们的普遍关注，我国西部地区生态环境脆弱，而西部地区又是全国生态环境的源头地区，事关全国的生态安全[19-22]；截至 2008 年 8 月 14 日 12 时，四川汶川地震确认 69 225 人遇难，374 640 人受伤，失踪 17 924 人，公路受损里程累计 53 295 公里；截至 2010 年 8 月 13 日 16 时，舟曲特大山洪泥石流灾害造成 1156 人死亡、588 人失踪。此外，还有

① 本文是中国工程院科研项目报告，北京，2010；工程管理研究，北京：科学出版社，2015，187-202. 作者：刘人怀，金鉴明，庞国芳，尹华.

俄罗斯和西方国家关于生态环境安全的理论与实践在我国产生的反响。

当前我国生态安全的研究内容主要集中于区域生态安全状况分析、生态系统健康状况评价、生态安全预报与预警、生态系统功能的可持续性分析和生态安全管理。其中区域生态安全状况分析包括了自然生态系统（水域、湿地、森林、草地）和人工生态系统（农田）的变化、生态演替、景观斑块动态、系统对外界干扰的恢复与阻抗能力等，重点是优势生态系统的稳定性和完整性。此外，区域生态安全状况分析的内容之一是研究重要生态过程的连续性，包括对过程强度、速率和方向的判断研究。生态系统健康状况评价主要是通过生态价位以及生态系统成熟度来实现，然后确定生态系统的安全系数。

在生态保育、修复与重建方面，研究了保护重要生态功能区生态功能、遏制区域生态恶化趋势的科学和技术问题，开展了草原退化、水土流失、矿区生态环境状况评价方法和理论研究；完善了环境标志、环境认证和政府绿色采购制度，研究制定了发展循环经济与建设生态补偿机制的政策、标准和评价体系。“十一五”期间，重点解决了重要生态功能区的系统保护与建设理论和支撑技术、生态脆弱区保护与建设研究、生态保护的技术支撑体系、生态环境监控技术、生态承载力与区域可持续发展理论和方法等环境科技问题。

在重大流域水污染和区域大气污染控制方面，重点解决了流域或跨流域水环境容量、生态环境容量测算技术方法和实施技术路线，重点流域水环境承载力和生态需水量阈值，饮用水安全保障技术、面源控制技术、水污染控制生物与物化技术，中小城镇污水处理厂成套技术与设备、污泥处理利用等环境科技问题。

在城市化快速发展进程中面临的突出环境问题方面，重点研究了城市大气环境复合污染、水环境复合污染、固体废物污染及其优化控制技术，大气细颗粒物和超细颗粒物的控制技术，城市生态综合调控系统，城市重污染水体修复、饮用水源地保护及饮用水安全保障技术，城市连绵带和城市群复合污染综合调控技术，环境安全、健康安全和经济合理的城市垃圾与危险废物处理技术，城市臭氧、大气有毒有害污染物、有毒化学品和持久性有机污染物污染控制方法，城市物理污染控制对策和方法，城市环境污染对公众健康的影响（包括污染暴露评估技术、污染-健康剂量反应关系评估以及室内污染防治技术等），城市物流、能流优化控制和管理技术等问题。

2. 当前生态安全研究的国际先进水平与前沿问题综述

1987 年，第四十二届联合国大会通过的 169 号决议确定 20 世纪后十年为“国际减轻自然灾害十年”，第四十四届联合国大会又通过了《国际减轻自然灾害十年国际行动纲领》。1992 年联合国召开环境与发展会议，专题商讨危害全球生态安全的环境问题，并通过了会议宣言和相关的公约。鉴于环境问题的严重恶化，联合国决定于 2002 年 9 月在约翰内斯堡再次召开环境与发展会议，进一步商讨生态安全大计。2000 年 2 月 21 日联合国环境规划署执行主任托普费尔在“环境安全、稳定的社会秩序和文化”会议上指出：“环境保护是国家或国际安全的重要组成部分，生态退化则对当今国际和国家安全构成严重威胁。”他还指出，有清晰的迹象表明，环境资源短缺在世界上许多地方可能促成暴力冲突。在未来几十年，日益加剧的环境压力，可能改变全球政治体系的基础。

美国国家安全部门早在20世纪70年代末就资助科学家进行全球环境变化的系列研究计划，研究成果成为美国在国际事务中处理全球环境问题的依据。1991年8月，美国首次将环境视为国家安全问题而写入新的国家安全战略。美国国防部1993年成立了"环境安全办公室"，并自1995年起每年向总统和国会提交关于环境安全的年度报告。美国前国防部长佩里1996年11月20日指出："一个强有力的环境规划，是一个强有力的国防的有机组成部分。"美国白宫1996年发表的《国家安全科学和技术战略》指出："环境压力加剧所造成的地区性冲突或者国家内部冲突，都可能使美国卷入代价高昂而且危险的军事干预、维护和平或者人道主义活动。"1997年，美国中央情报局成立"环境研究中心"，以维护国际生态安全、国家安全。美国国家环境保护局1999年9月提交了题为《环境安全：通过环境保护加强国家安全》的报告。

20世纪80年代末期，美国人就提出了生态安全的概念。冷战结束后，美国人认为国家安全政策的目标已开始从单纯的军事安全逐渐演变为包括环境安全、经济安全和军事安全在内的多重目标，具体包括以下四方面的内容：资源安全、能源安全、环境安全和生物安全。美国的国家生态安全所依据的是，外国的环境行为可以影响到本国的环境系统，引起所谓的域外环境损害，而对本国的环境造成威胁。美国的国家环境安全主要目标并不是针对本国的环境问题，其逻辑是任何发生在他国的、他国之间的、地区性的乃至全球性的事件，只要对美国的环境安全造成损害、威胁或者有潜在的威胁，美国就可以进行干预，以解除、减少对美国的损害或者威胁，以保护美国的环境安全。

为应对气候变化，欧盟在温室气体排放交易机制、能源政策、交通运输政策、适应能力建设等方面制定了相应的政策。欧盟近年来大力发展风能、太阳能、生物质能等清洁低碳的可再生能源，部分地取代石油、煤炭和天然气等化石能源的使用，减少温室气体排放（2008年CO_2排放总量前十名国家见表1）。可再生能源主要用于发电、制热、制冷、交通燃料等。2005年12月，欧盟委员会提出了一项专门针对生物质能发展的立法建议，即"生物质能行动计划"，其主要目的是：建立相应的市场激励机制，扫除开发生物质能的障碍，最大限度挖掘生物质能潜力，扩大生物质能在供热、发电和交通运输业中的应用。

表1 2008年CO_2排放总量前十名国家

位次	国家	CO_2排放总量/亿吨
1	美国	28
2	中国	27
3	俄罗斯	6.61
4	印度	5.83
5	日本	4
6	德国	3.56
7	澳大利亚	2.26
8	南非	2.22
9	英国	2.12
10	韩国	1.85

在温室气体控制方面，英国诺丁汉大学碳捕获和存储技术创新中心研制了二氧化碳存储新技术，利用了一种含硅酸盐矿物质捕获二氧化碳，使二氧化碳永久储藏在岩石中。

2001 年 3 月，英国外交部（环境政策司）和国际发展部（冲突事务司）在伦敦召开了“环境安全与冲突预防”国际研讨会。78 名与会代表主要来自欧美、非洲、亚洲国家，以及北约、欧盟、欧洲安全与合作组织等政府间组织和世界自然保护联盟等非政府组织，我国也应邀派代表参加了会议。伦敦研讨会的议题主要包括：环境与安全的相互关系；环境和暴力冲突的联系；环境压力的根源；环境与冲突预防；防务系统对促进环境安全的作用；种族问题和环境安全挑战等。英国外交国务大臣巴特尔在研讨会上提出：“我们需要严肃地对待环境问题，不论它是目前正在影响国家安全，还是将来可能影响国家安全。这种努力是值得的，也是符合成本效益原则的：因为投资于负责任的环境管理，不仅有助于预防可能产生的环境冲突，也有助于避免代价高昂的军事介入。”

2009 年 11 月，法国率先制定“碳税”法案，规定从 2010 年 1 月 1 日起对化石能源的使用按照每排放一吨二氧化碳付费 17 欧元的标准征税。

日本也较早提出“环境安全关系国家安全”的观点，并认为“只有在地球环境问题上发挥主导作用，才是日本为国际社会作贡献的主要内容”。俄罗斯、欧盟等也把生态安全列入安全战略目标。俄罗斯的环境资源法学界将生态安全作为环境资源法调整对象大致也始于 20 世纪 80 年代后期，《俄罗斯联邦宪法》将保障生态安全规定为俄罗斯及其各主体共同管辖的事项，《俄罗斯苏维埃联邦社会主义共和国自然环境保护法》将保障生态安全作为保证人的生态权利得以实现的保障措施，1995 年 11 月 17 日还通过了《俄罗斯联邦生态安全法》，作为保障生态安全方面的专门性联邦法律。

3. 环境安全研究中的重大工程科技问题

1）生态安全综合监测体系建设

建成环境、地质、气候和灾害的综合监测体系，为中央和各级政府提供及时、可靠的决策依据，为全社会的参与和监督提供丰富翔实的信息。主要内容包括：以水和大气质量为主的环境质量监测体系，特别是跨省河流水质的自动化定时监测和重点污染源的在线自动监测；以遥感和地面观测站相结合的生态与资源监测体系；重大自然灾害的监测、预报和应急系统。综合监测体系的建设要依靠高新技术改造现有的信息获取、加工、传输网络，并与传统方法相结合，提高系统的总体可靠性。

2）重点流域综合整治工程

在实现国家和地方节能减排目标的基础上，建立基于生态健康、足以支撑我国未来可持续发展的高功能河流水质评价新体系、水污染系统控制工程新体系和水环境综合管理新体系；实现从常规水质指标、痕量毒害物指标、水体生物毒性指标与水生态完整性指标等四方面系统评价河流水环境的安全性；在当前主要控制常规污染物达标排放的前提下，建立有毒有害物减排、废水脱毒减害深度处理及资源化、受纳排水河道水质净化与生态修复以及河流生态功能恢复等组合技术构成的高功能河流水污染控制工程体系；以发展布局优化、产业结构升级、工程减缓、综合调控等四大措施系统保证重点江河高动能目标的实现，并分阶段、分区域、有侧重地示范推广。

3）城市垃圾处理及资源化技术

实行城市垃圾减量化、资源化、无害化，加快城市垃圾无害化处理设施的建设。

4）水体石油污染强化生物修复工程

该工程将分别针对中低浓度石油污染土壤和高浓度石油污染土壤，开展关键技术研发、系统集成与应用示范，建立适合我国国情的石油污染土壤修复技术体系。开发高效生物修复技术，治理水体和土壤中的石油污染；研制可生物降解、高效安全的表面活性剂。

5）尾矿安全处置、利用和矿山污染区域生物修复

在矿产开采和选冶过程中所排放的重金属等有毒有害物质部分进入人们赖以生存的土壤系统，给矿山周边及下游民众的生活质量和农产品质量带来严重威胁。应针对我国矿区及其周边土壤存在大量砷、铅、镉、铜等重金属复合污染问题，开展关键技术研发、系统集成与应用示范，为保障矿区及其周边地区的农产品质量、生态安全和提高综合环境质量提供技术支撑。通过物理化学和多种生物联合修复技术稳定土壤、控制污染、改善景观、减轻污染对人类的健康威胁，在矿区土地建立一个可自我维持的良性生态系统。

开发酸性矿山废水回用浸出技术回收尾矿贵重金属；开发尾矿综合回收与利用技术、矿山采空区尾矿回填技术、尾矿高效整体利用技术，生产微晶玻璃原料、矿物肥料、土壤改良剂、尾矿砖、混凝土骨料、砂浆、铁路道砟、筑路碎石、井下回填料、复垦料，使矿山向无尾矿山目标迈进；开发矿区宜耕宜居复垦技术，对复垦土地的利用要因地制宜，采取不同的利用途径，可将废弃地恢复为农业用地（种植业、水产养殖业、林业用地），或改造为休养和娱乐场所，或为工业、建筑业所用。

6）燃煤污染控制工程

我国90％二氧化硫、67％氮氧化物、70％烟尘排放量来自煤炭的燃烧。其中，燃煤电站、燃煤工业锅炉、燃煤炉窑等烟气排放污染问题最为突出。应围绕我国大气污染控制方面的重大需求和国际技术前沿，通过关键技术研发和系统集成，开发具有自主知识产权的燃煤电站、燃煤工业锅炉、燃煤炉窑等烟气排放污染物控制技术与设备，推动我国大气环境质量改善。

7）生物质能源

以生物质发电、沼气、生物质固体成型燃料和液体燃料为重点，大力推进生物质能源的开发和利用。在粮食主产区等生物质能源资源较丰富地区，建设和改造以秸秆为燃料的发电厂和中小型锅炉。在经济发达、土地资源稀缺地区建设垃圾焚烧发电厂。在规模化畜禽养殖场、城市生活垃圾处理场等场所建设沼气工程，合理配套安装沼气发电设施。大力推广沼气和农林废弃物气化技术，提高农村地区生活用能的燃气比例，把生物质气化技术作为解决农村和工业生产废弃物环境问题的重要措施。努力发展生物质固体成型燃料和液体燃料，制定有利于以生物燃料乙醇为代表的生物质能源开发利用的经济政策和激励措施，促进生物质能源的规模化生产和使用。

8）气候安全

推进中国气候变化重点领域的科学研究与技术开发工作。加强气候变化的科学事实与不确定性、气候变化对经济社会的影响、应对气候变化的经济社会成本效益分析和应对气候变化的技术选择与效果评价等重大问题的研究。加强中国气候观测系统建设，开发全球气候变化监测技术、温室气体减排技术和气候变化适应技术等，提高中国应对气候变化和履行国际公约的能力。重点研究开发大尺度气候变化准确监测技术，提高能效和清洁能源技术，主要行业二氧化碳、甲烷等温室气体的排放控制与处置利用技术，生物固碳技术及固碳工程技术等。

9）城市地质灾害监测预报与应急救灾技术

城市地质灾害种类繁多，对人类的危害广泛而严重。因此，应开展城市地质灾害监测预报与应急救灾技术研究。重点对各类地质灾害的地质环境背景进行评价；研究各类地质灾害发生和发展的规律、强度、形成机制及作用速率；从工程建设和地质环境的相互作用出发，评价和预测各类工程建设可能产生的灾害性地质作用及危害程度；对区域性和严重危害人类生产、生活的地质灾害开展长期监测，进行时间、空间、强度的预测、预报；开展地质灾害的防治研究和指导防治工作。

4. 生态安全研究中的关键技术

1）生态安全综合监测体系建设

A. 水和大气环境质量监测体系

建立以水和大气质量为主的环境质量监测体系，特别是流域和跨国河流水质的自动化定时监测和重点污染源在线自动监测。

研究沉积物中毒害污染物的含量水平与化学组成特征，掌握沉积物中毒害污染物的分布特性及空间格局；研究沉积物的物理、化学组成特征，评估毒害污染物对水体的“释放”形成“二次污染”的程度、与微界面交换过程及在这一过程中的化学/生物作用与主导影响因素。

建立以海量数据库与网络技术为基础的水和大气环境质量监测体系信息系统平台。实现流域水环境毒害污染物信息、毒害污染物基准的查询、统计和分析。建立优先控制毒害污染物清单与动态信息管理系统，以应对由毒害污染物带来的永久性或突发性环境灾害。

关键技术：沉积物中毒害污染物的含量水平、化学组成特征、分布特性及空间格局。

B. 生态与资源监测体系

建立以遥感和地面观测站相结合，以海量数据库和网络技术为基础的生态与资源监测体系，形成具有现代化装备、技术优化、科学适用的生态与资源监测技术集成。

关键技术：遥感长期自动定标与数据校正技术；快速高效的遥感解译技术。

C. 重大自然灾害的监测、预报和应急系统

建立并优化自然灾害数据库，依托现场监测以及 GIS 的空间分析能力和图像功能，以发生地质灾害地段的地形、地物、地质情况、自然降水及地下水变化情况等空间图形数据和工程属性数据为计算分析基础，依据地质灾害类型及发生机理，确定地质灾

害的发生因子及各因子对地质灾害的影响程度，建立正确可靠的数学模型，科学预测、预报各类地质灾害发生的条件、影响的范围和危害的程度。

关键技术：地质灾害发生因子的筛选与赋值。

2）重点流域综合整治工程

A. 微污染饮用水源地水体净化工程

强化传统水处理工艺的处理效果，开发合适的预处理工艺和深度处理工艺，寻求新型微污染水源水处理工艺，从而有效地去除微污染水源水中的有机物、氨氮等污染物，避免消毒剂与原水中的有机物反应产生有毒的消毒副产物。

关键技术：高毒持久性污染物去除技术。

B. 典型毒害污染物/典型行业事故排放应急处理工程

重点关注含毒害有机物和重金属等典型毒害污染物废水的事故排放应急处理。

关键技术：毒害污染物快速吸附和去除技术。

C. 流域湖泊和湿地水生态修复、拓建和防洪工程

对蓄滞洪区进行合理分类，优先安排使用频率较高、在流域防洪体系中发挥重要作用的蓄滞洪区安全建设，进一步重视在蓄滞洪区内生活和生产的群众的安居问题，采取综合措施，为群众建设安全住房创造条件。加强国家确定的重点防洪城市和重要经济区的防洪工程建设，通过制定城市河湖治导控制线，避免城市向洪水高风险区发展，优化城市布局，加强城市水系综合整治，构建城市综合防洪减灾体系。

努力维护河流健康，通过生态补水、面源治理等综合措施，逐步修复部分生态脆弱河流、湖泊、湿地的水生态系统和部分城市水生态系统。充分利用水生生物的净化作用，改善水域生态环境。

关键技术：流域湖泊和湿地生态恢复技术。

3）城市垃圾处理处置及资源化技术

A. 城市生物质垃圾高温快速发酵制肥的关键技术

通过高温好氧堆肥技术将生物质垃圾堆积在发酵装置中，添加适量的调理剂，利用高效微生物将垃圾中易降解有机物逐步降解，最终形成稳定的腐殖质，制备肥料。

关键技术：垃圾高温快速发酵，以及发酵过程臭气的控制。

B. 城市生物质垃圾发酵制备清洁能源的关键技术

利用产氢、产甲烷菌在厌氧的条件下，发酵生物质垃圾，制备氢和甲烷，用作燃料，进行供热和发电，实现资源和能源的回收，将发酵后的固体残留物制备为高质量的有机肥料和土壤改良剂。

关键技术：垃圾厌氧发酵分为产酸和产甲烷两个阶段，产酸阶段将复杂有机物水解和发酵形成脂肪酸、醇类、二氧化碳和氢气等，产甲烷阶段将产酸阶段的产物进一步转化为甲烷和二氧化碳。由于产甲烷菌对 pH 变化敏感，最适 pH 为 6.8～7.2，而产酸阶段会导致系统 pH 降低，因此，需要解决产甲烷菌对 pH、DO 变化敏感这一关键技术问题。

C. 城市垃圾无害化焚烧高效发电技术

城市垃圾焚烧发电的环保性及发电的效率受焚烧温度、空气过剩系数及蒸汽温度

等关键因素影响。焚烧温度低于 850℃时，焚烧过程可能会产生二噁英等有毒污染物；蒸汽温度越低，发电效率越低。在规模化畜禽养殖场、城市生活垃圾处理场等建设沼气工程，合理配套安装沼气发电设施，解决垃圾焚烧发电关键技术。

关键技术：二噁英的控制与发电功效的优化。

4）水体石油污染强化生物修复工程

A. 亲脂无毒型表面活性剂开发与应用

石油烃污染物的乳化和分散是增加石油烃比表面积，促进石油烃和微生物相互接触的关键步骤。表面活性剂可以达到该目的。本技术的目的是开发出无生物毒性，可以高效地促进石油烃乳化和分散的亲脂型表面活性剂。

关键技术：表面活性剂的亲脂性和无毒性。

B. 亲脂无毒型石油烃降解菌营养物质开发与应用

开发亲脂无毒型营养物质，投放到受石油烃污染的水体，该类营养物质可以缓慢释放到含油水层中，加速降解菌对石油烃的降解，避免营养物质污染净水层。

关键技术：营养物质的亲脂性和缓慢释放效果。

C. 水体石油污染强化生物修复技术

建立亲脂无毒型表面活性剂和营养物质添加量与石油烃生物降解的数学模型，确定表面活性剂和营养物质的添加模式与添加量；开展水体石油烃微生物降解实验室和现场试验，建立土著微生物和外源微生物与石油烃生物降解的数学模型；筛选石油烃生物降解标记物，建立水体石油污染强化生物修复技术的评估方法。

关键技术：表面活性剂和营养物质等添加物对石油烃降解菌降解石油烃的促进作用。

5）尾矿安全处置、利用和矿山污染区域生物修复

A. 矿区重金属与持久性有机污染物污染土壤联合修复与生态恢复技术

针对我国矿区及其周边土壤存在大量铅、镉、铜、镍、砷、锌等重金属复合污染问题，采用物理、化学、根际微生物-土壤动物-超富集植物的联合技术进行修复，开展关键技术研发、系统集成与应用示范，为保障矿区及其周边地区的农产品质量、生态安全和提高综合环境质量提供技术支撑。

关键技术：超富集植物快速生长的培养技术；根际微生物和超富集植物处理重金属污染物的复合效应。

B. 酸性矿山废水回用淋滤尾矿耦合微生物采矿技术

开发酸性矿山废水回用浸出技术回收尾矿贵重金属，加快硫氧化菌氧化还原硫和铁的反应速度，提高硫酸及氧化态铁对尾矿的淋滤速率，实现贵重金属的回收。

关键技术：解决酸性矿山废水和浸出液对设备的腐蚀问题；硫氧化菌快速生长的控制技术。

C. 尾矿综合回收与利用技术

通过尾矿再选，矿山采空区尾矿回填，生产微晶玻璃原料、矿物肥料、土壤改良剂、尾矿砖、混凝土骨料、砂浆、铁路道砟、筑路碎石等，实现尾矿高效整体利用，使矿山向无尾矿山目标迈进。

关键技术：尾矿低成本综合利用技术。

D. 矿区宜耕宜居复垦技术

开发矿区宜耕宜居复垦技术，实现复垦后土地的重新利用，减少雨水和地表径流对尾矿废矿层的渗透，避免复垦后矿区对地下水的污染。

关键技术：外源植物快速生长技术。

6）燃煤污染控制工程

A. 脱硫副产物资源化利用技术

关键技术：脱硫副产物形成、加工、应用技术。

B. 燃煤过程 SO_2、NO_x 同步控制与治理技术

针对我国湿法脱硫工艺能耗较高的问题，以节能和降低脱硝成本为目的，开发节能低成本型脱硫脱硝技术，为我国大型燃煤烟气脱硫脱硝提供有力的技术支撑。基于现有的石灰石/石膏湿法脱硫系统，开发经济高效、可工业化使用的脱硝剂或添加剂，优化脱硝工艺及脱硫脱硝匹配技术，实现同一装置内同时脱硫脱硝；开发石灰石/石膏湿式脱硫系统降低能耗技术。

针对我国烟气半干法脱硫存在的运行可靠性差，对机组负荷、煤种、脱硫剂适应性差的问题，开发大型燃煤火电机组烟气半干法脱硫技术。

关键技术：脱硫脱硝技术的稳定性。

7）生物质能源

发展生物质固体成型燃料和液体燃料，利用能源型生物发酵，制备以乙醇为代表的生物质能源，促进生物质能源的规模化生产和使用。

关键技术：能源型生物的低成本预处理技术；发酵废水的再利用和高效处理技术。

8）气候安全

A. CO_2 捕集与利用技术

捕集和分离天然气井、煤化工、煤气化联产和燃煤电厂等行业及排放点所排放的 CO_2，开发油气储层和不可开采的甲烷煤层 CO_2 注入技术；制成高纯度干冰，用于食品、消防等行业领域，实现 CO_2 的利用。

关键技术：CO_2 与共存气体成分的分离技术。

B. CO_2 捕集与封存技术

对捕集点污染源的 CO_2，进行地质封存。

关键技术：CO_2 与共存气体成分的分离技术。

9）城市地质灾害监测预报与应急救灾技术

A. 城市地质灾害监测与预警预报新技术

主要包括耦合遥感、地面调查、地球物理勘探、钻探、原位及室内实验分析、地球化学勘探和动态监测技术，以及区域崩（滑）塌和泥石流预警技术、地质灾害气象预警预报技术。重点开展泥石流易发区地质灾害监测及预警预报技术方法研究。完善突发性地质灾害实时监测系统，加快突发性地质灾害应急救灾技术研究；针对主要突发性地质灾害，如泥石流，建立应急调查体系、治理技术体系、灾情评估技术体系。

关键技术：多种监测技术和方法的耦合技术。

B. 城市地质灾害形成机理与应急救灾技术体系建设工程

利用耦合遥感、地面调查、地球物理勘探、钻探、原位及室内实验分析、地球化学勘探和动态监测技术，开展突发性地质灾害早期识别技术研究，阐明灾害发生机理；根据城市主要地质灾害发育分布的规律，建立地质灾害数据库，建立城市地质灾害信息系统；逐步建立标准化的地质灾害危险性和风险性评价的技术方法体系，进一步完善地质灾害综合防治技术体系。

关键技术：遥感长期自动定标与数据校正技术。

5. 保障生态安全的重要措施和政策建议

1）加强研发资源配置

A. 面向全国，广聚人才

生态安全研究是多学科交叉的、长期的系统工程，要面向全国研究机构，广招人才，建立吸引和广聚人才的机制，构建结构合理、专业齐全、适度竞争、精干高效的研发队伍。

B. 建立协调、长效的筹资和投入机制

建立多元化的筹资机制，以保障长期研究开发和工程建设的资金需求。研究制定经费的筹集、管理、使用、监督和审计的法律、规章制度，确保研发活动投入的持续稳定供应，促进基础与能力的协调发展，实现研发投入产出的高效率和高效益。

C. 加大科研经费投入，加强科技基础条件建设

加大政府对生态安全相关科技工作的资金支持力度，建立相对稳定的政府资金投入渠道，确保资金落实到位、使用高效，发挥政府作为投入主渠道的作用。多渠道筹措资金，吸引社会各界资金投入生态安全的科技研发工作，将科技风险投资引入生态安全领域。充分发挥企业作为技术创新主体的作用，引导中国企业加大对生态安全领域技术研发的投入。积极利用外国政府、国际组织等双边和多边基金，支持中国开展生态安全域的科学研究与技术开发。

2）加强科技管理体制和机制建设

A. 构建科技决策与协调机制

依据政府相关部门的管理职责，在国家层面建立高放废物地质处置决策、规划、协调机制，完善重大决策议事程序，加强跨部门的协调。

B. 创新管理体制和运行机制，加大组织管理工作的力度

建立创新的组织管理体制和运行机制，适应投资多元化、全国多学科长期联合攻关任务的需要；形成勇于探索、科学求实的科技创新环境；树立“尊重知识，尊重人才，尊重首创”的良好风尚，发挥科技人才在研究开展中的主体地位；建立和完善竞争、评价、监督与激励机制，完善科技规划计划、执行评估及监督管理制度。

C. 建立和完善信息交流与公众参与机制

逐步建立健全信息交流与公众参与的机制；执行机构应当以适当的方式使公众了解有关计划、时间表、活动以及进展情况；认真听取公众意见，接受公众监督，促进高放废物地质处置研究开发工作的顺利开展。

3）加强部门合作

生态安全研究开发和建设本身是一项探索性工作，涉及立法、决策、审管、执行、实施等层面的众多政府行政部门或企事业单位，需要各部门加强沟通、协调和合作。

4）加强国际合作

A. 走出去，请进来，加强科技交流

通过聘请外国专家来华讲学，鼓励和支持研究人员参加国际会议、考察访问，充分借鉴国外的经验和教训。通过国际合作，开展对一些重大问题的研究和评估，促进国内相关工作开展。

B. 积极参与国际合作研究

积极开拓多种国际合作渠道，参加国际合作研究计划，利用国外的设施和装置，发挥共同研究的智力互补及融合的优势，提高研究起点，加快研究进度。

5）强化和细化产业扶持政策及激励措施

政府需在投资、税收、价格、财政等方面出台针对性比较强的激励扶持政策，有序有效引导和扶持生态安全产业的发展。

进一步推动生态安全产业的机制建设。按照政府引导、政策支持和市场推动相结合的原则，建立稳定的财政资金投入机制，通过政府投资、政府特许等措施，培育持续稳定增长的生态安全产业市场；改善生态安全产业的市场环境。

6）培育核心企业，完善产业链条

在重点扶植龙头企业的同时，积极培育目前规模不大，但具有一定产品优势和发展基础的公司。引导企业不断完善公司组织结构，形成长效的公司治理机制和适应大规模企业管理的组织结构，鼓励企业在既有价值链的基础上进一步进行分工与协作，拉长产业链条。

7）推进科技创新，掌握核心技术

大力推进科技创新，提高自主创新能力，建设创新型企业。一是抓企业创新能力建设，促核心竞争力提升。引导企业加强和新建一批国家、省、市级工程技术研究中心、企业技术中心，支持企业通过原始创新、集成创新和引进消化吸收再创新，用高新技术和先进适用技术改造提升传统产业，促进产业转型升级。二是推动企业技术创新项目实施，促进新产品开发。围绕战略性新兴产业和传统优势产业，组织实施重大科技项目攻关，形成一批拥有自主知识产权的核心技术和高新技术产品。三是加强企业品牌和信息化建设，促进生产经营规模扩大。推进企业信息化建设，提升企业技术装备水平，以信息化和先进装备扩大产能，提高效益。四是加强公共技术服务平台建设，改善创新服务环境。深化企业与高校、科研机构的合作，加强产学研结合，突破一批关键和共性技术瓶颈，加快转化一批科研成果，促进形成现实生产力。培育发展各类科技中介服务机构，并引导其向专业化、规模化和规范化方向发展，建设社会化、网络化的科技服务体系，提高科技服务水平。集中力量，多渠道筹集研发资金，多渠道、多形式引进高水平的研发人才，在整合的基础上成立研发中心，提升研发能力。搭建人才培训及流动平台、物流支持平台、信息平台、售后服务平台、资金流优化配置平台。通过委托培养或补贴的形式培养产业集群发展所需要的技术人才和经营管理人才。

8）制定关键扶持政策，支持重点工程建设

通过合理的税收制度和科学的价格形成机制，促进新能源和节能环保产业的健康快速发展。安排政府预算内投资和财政专项资金，采取补助、贴息、奖励等方式，支持重点工程建设。建立增强自主创新能力的体制机制，支持和引导科研机构围绕新能源与节能环保产业发展中的共性技术、关键技术进行研究开发。

采取财政补贴、税收减免、价格优惠等措施，重点开拓使用市场，显著提高新能源在能源消费中的比重；对采用合同能源管理方式实施的节能项目给予投资补助或财政奖励、税收减免；深入推进循环经济试点，依法设立循环经济发展专项基金，在投资、价格、财税等方面健全相关配套政策，细化完善有关方案，形成发展循环经济的激励与约束机制，对污水、垃圾处理企业和再生资源回收利用企业免征城镇土地使用税与房产税，出台节能环保和循环经济鼓励类产品、工艺和技术目录；进一步推进资源性产品价格改革，完善脱硫电价，尽快出台脱硝电价政策，加大财政建设性资金投入力度，提高污水垃圾处理收费标准，采取"以奖促治""以奖代补"等办法，进一步推动环保产业发展。

9）完善制度法规，强化监督管理

完善产业标准化制度建设。逐步提高重点用能产品能效标准，扩大终端用能产品能效标示范围，提高重点行业能耗限额强制性标准和污染物排放标准。建立健全节能环保产品认证体系、再制造产品标志管理制度。加强市场监督、产品质量监督，落实招投标各项规定。加强宣传教育，增强全社会节约环保意识，倡导绿色消费，引导消费者更多地购买节能环保产品。

建立严格的节约资源能源管理体制，制订节能、节水、节材等专项规划，严格控制资源消耗强度，创建节水型社会，实行区域总量控制和定额管理相结合的用水管理制度；建立资源性产品价格形成机制，完善差别化能源资源价格制度，建立绿色电价机制，实施太阳能光伏并网发电收购制度；优化政绩考核体系和干部考核制度，加大生态环保、节能降耗、开发新能源产品等的指标考核权重，引导各级政府及部门把工作重点转移到为新能源和节能环保产业主体营造环境与改善服务上来。

参考文献

[1] 常平，刘人怀，林玉树. 20世纪我国重大工程技术成就. 广州：暨南大学出版社，2002.

[2] 范维唐. 我国安全生产形势、差距和对策. 北京：煤炭工业出版社，2003.

[3] 吕新奎. 中国信息化. 北京：电子工业出版社，2002.

[4] 许世已. 新时期中国食物安全发展战略研究. 济南：山东科学技术出版社，2003.

[5] 刘铁民. 迈向新世纪的中国劳动安全卫生：二十一世纪安全生产宏观战略研究. 北京：中国社会出版社，2000.

[6] 吴忠民. 渐进模式与有效发展：中国现代化研究. 北京：东方出版社，1999.

[7] 魏永忠，员绍忠. 论城市社会安全与稳定预警等级指标体系的建立. 中国人民公安大学学报，2005，(4)：150-155.

[8] 王林. 社会安全体系的发展及其启示. 管理世界，1992，(6)：149-153.

[9]　宋林飞. 中国社会风险预警系统的设计与运行．东南大学学报（社科版），1999，(1)：69-76.

[10]　阎耀军. 城市社会预警基本原理刍议：从城市社会学视角对城市社会问题爆发的预警机理探索. 天津社会科学，2003，(3)：70-73.

[11]　牛文元，叶文虎. 全面构建中国社会稳定预警系统. 中国发展，2003，(4)：1-4.

[12]　杨多贵，周志田，陈邵锋，等. 中国社会稳定与安全预警系统的理论设计. 系统辩证学学报，2003，11 (4)：82-87.

[13]　白新文，王二平，周莹，等. 团队作业与团队互动两类共享心智模型的发展特征. 心理学报，2006，38 (4)：598-606.

[14]　冯芸，吴冲锋. 货币危机早期预警系统. 系统工程理论方法应用，2002，11 (1)：8-11.

[15]　王岚，刘人怀. 证券公司风险的实证分析及风险券商处置模型研究. 管理工程学报，2006，20 (1)：118-123.

[16]　何问陶，刘人怀．新形势下金融风险及防范对策的思考. 参事建言（2008). 香港：中国评论学术出版社，2008：223-228.

[17]　路节. 加强社会危机早期预警系统的研究. http://www.wrsa.net/11/10/25@3201.htm，2009-11-10.

[18]　张成福. 公共危机管理：全面整合的模式与中国的战略选择. 中国行政管理，2003，(7)：6-11.

[19]　刘人怀. 关于改善我国北方水资源缺乏的一个建议. 参事建言（2004～2005 年). 香港：中国评论学术出版社，2006：538.

[20]　刘人怀. 绿色制造与学科会聚//2006 年“学科会聚与创新平台”高新技术高峰论坛. 杭州：浙江大学出版社，2006：15-18.

[21]　刘人怀. 爱低碳生活创绿色校园. 澳门：绿色澳门建设研讨会. 2010.

[22]　徐乾清. 中国防洪减灾对策研究. 北京：中国水利水电出版社，2002.

工程管理信息化架构研究①

工程管理信息化是指为了更好更有效地实施工程管理，利用信息技术，构建信息系统，并在工程管理实践中加以应用的过程[1]。由于大中型工程的投资较大，所以对社会及环境的影响广泛。通过信息化实现资源共享，能够有效降低工程管理中的协作成本和重复投资，有效监控工程的设计、建设、运行和维护等各阶段，这将有助于降低工程全生命周期内的总投入，提高工程质量及运作效率，同时促进工程环保效益的实现和改善民生满意度，使得工程的经济效益与社会效益得以大幅度提升。

信息化建设不仅是信息技术系统的建立问题，同时也是与之相适应的组织架构与沟通机制、信息共享与知识创新模式不断调整、不断完善的过程。它涉及不同组织内部或组织之间、不同工程之间，以及工程与政府和社会公众之间的信息沟通等一系列问题。基于对工程的全生命周期分析，工程管理信息化的内涵可以归结为 4 个方面，分别是运营管理、伙伴协作、公众服务与集成创新[2]。

本文在以前研究的基础上，将对工程管理信息化的架构进行探讨。工程管理的信息化架构是用于描述在工程管理实现信息化过程中，其涉及的所有要素及要素间关系的一般框架。其中，要素包括与工程建设及运营相关的各类组织、业务流程、信息系统及人员。

一、工程管理信息化的利益相关者分析

工程管理信息化的实现涉及与特定工程相关的多个组织和社会公众，是典型的跨组织信息系统应用。一项工程，特别是大中型工程，往往具有实体设施投资巨大、技术标准复杂、建设及运营周期长等特点，且大都属于高耗能项目，对环境、社会民生及国民经济等都有着较大的影响。通过对特定工程利益相关者的分析，可以将工程管理信息化架构所涉及的各类组织及人群划分为 5 类角色，分别是运营组织、建设组织、政府部门、公共资源和社会公众。

（1）运营组织。工程运营组织是在工程的全生命周期之中，对不可移动的实体设施进行经营、维护和升级的组织。这类组织有明显的地域特征，大都拥有或管理着庞大的固定资产。运营、管理实体设施是该组织的主要业务，贷款偿还、资产折旧和能源消耗在其运营成本中占有较大比重。

针对特定工程而言，工程运营组织是工程管理信息化的主体。工程的所有者或管理者为确保工程的顺利运营，建立沟通协作平台。及时的信息沟通和有效的信息共享，可以确保工程设计方、施工方、供应商、运营商等一系列与工程直接相关的组织之间有效协作，以提升工程的运作效益，实现对工程全方位的管理[3-5]。

对于管理多个大中型工程的企业来说，一般存在多个工程运营系统并存的情况。

① 本文原载《中国工程科学》，2011，13（8）：4-9. 作者：刘人怀，孙凯.

这些系统的实现方式、技术水平、设备状况等可能存在较大的差异。实现多个运营系统的相互集成、资源共享，并与管理系统之间实现有效的数据交互，是提高工程整体运营效率的重要途径。

(2) 建设组织。建设组织主要指在工程交付之前，受业主委托完成工程的设计、建设及监理等一系列工作的组织，它也可以包括相关的物流企业等。

工程建设组织自身的信息化建设，是工程管理信息化建设的重要组成。没有相关企业信息化的实现，就没有工程管理信息化的全面实现。在工程设计与施工期间，工程业主、施工方、设计方、监理方等相关企业之间的信息系统集成、数据交换和信息共享水平是与工程相关各方自身的信息化水平密切相关的。

(3) 政府部门。在实现工程管理信息化的过程中，政府部门扮演的是政策引导、规划、监管、标准制定及应急指挥的角色。政府进行工程规划和监管的不同模式，对工程建设的进度和质量有着深远的影响[6]，也间接地决定了工程管理信息化的实现模式。

此外，建立国家或行业层面的信息交换标准和信息共享机制，完善相应的公共信息技术基础设施建设，建立为众多工程所共享的信息资源平台，通过建立知识库、案例库、专家系统等，提升各类工程的知识管理水平，也是研究工程管理信息化的重要内容之一。

(4) 公共资源。与工程管理信息化建设相关的公共资源包括：金融保险、能源交通、教育科研、设备租赁、电子商务及交易平台等相关组织。有效利用公共资源，可以降低工程的重复投资率和风险强度，实现不同组织、不同工程项目之间的资源共享，有利于更加高效地完成财务融资、人力资源培养、市场开拓等管理活动。

任何一个工程都不是独立存在的，在全生命周期的工程管理中，需要不断吸收相关及类似的工程方法、经验和教训；同样，该工程的信息和知识资源也可以为其他工程所借鉴。知识在今后的生产和社会活动中的重要性是与日俱增的，通过不同工程之间的信息共享，可以形成新的知识财富，使今后的工程项目受益。

(5) 社会公众。社会公众是工程的受益群体，是工程的服务对象和工程质量的最终评判者。在国民经济建设浪潮中，社会和百姓期待千千万万的工程是民心工程、科学工程、合理工程[7]。无论是公益性工程还是经营性工程，其目的都是为社会公众服务。建立公众服务平台，可以使工程的开发者或管理者与社会公众之间形成有效的沟通，并使得受工程影响的社会公众群体，如水利工程的移民、交通工程的乘客、房地产工程的拆迁户等，以及环境影响的受益者或受害者都能及时准确地获取信息，形成公开与透明的信息机制。

二、传统的工程管理信息化架构

工程管理信息化的实现涉及多个组织信息系统的有效集成。各组织现有的信息系统是在其不断发展的过程中逐步引入的，且大都是为适应特定需要而建立的专用系统。早期的信息技术在安全措施及传输速率等方面具有一定的局限性且大都造价昂贵，加之系统只能根据即时的需求，在遵循成本与效益对等的原则下逐步引入和更新，这些在客观上造成了技术应用的局限性。

由于没有统一的信息系统规划来进行指导，各组织的信息系统建设存在很大的无序性、重复性和局部性，工程整体的信息化架构在事实上是不存在的，所有的系统大都以孤立的方式运作，即通常所说的“信息烟囱”（information silo）或“信息孤岛”（information island），如图 1 所示。

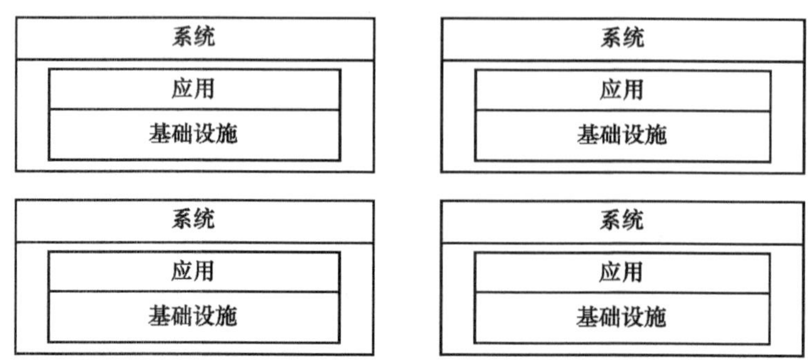

图 1　传统的信息化架构

图 1 也可视作孤立运作状态下的工程管理信息化架构。这种孤立运作的信息系统状态已无法适应现代工程建设的需要，主要源于以下几个方面的问题。

（1）系统目标不一致。各类系统都是根据特定时期的需求，由不同组织或部门根据自身需求而建立的。系统建立与运作的目标大都从属于业务单元的局部利益，而非工程管理的总体目标。

（2）技术体系缺乏标准化。由于缺乏统一的整体规划，各系统的网络、硬件、软件都是专用的，具有各自不同的技术标准、运行流程及维护团队，因此系统的互联互通性较差。

（3）信息交互渠道不畅。不同组织之间、不同信息系统之间的信息交互无法实时进行，很多环节依赖人工操作。数据的重复录入不仅会导致运营成本的增加，而且会导致错误概率的提高。

（4）风险隐患难以消除。没有统一的网络管理和监控措施，具有较大的风险隐患。因为独立的网络大都只能实现设备级的安全管理和较低层次的容错机制，所以安全性完全由设备决定，如硬件发生设备故障，可能导致所属网络发生瘫痪，并可能影响到其他系统的运作。

（5）潜在的安全问题。缺乏集中的安全管理，不能对网络、安全、数据、业务等进行统一管理，不能识别网络业务和用户行为。由于网络边缘安全设施的不足，各类终端技术标准不一，安全事件不可见、不可控，也无法预防。

（6）数据存储问题。各系统的数据存储方式和存储标准不一致，数据可靠性不足，资源共享性差，存储系统的可用性欠缺，不利于建立统一的备份和灾难恢复机制。

以上问题制约着工程管理信息化的实现，并与现代工程的建设与运营要求不相适应，因此应逐步过渡到具有一体化的系统应用、集中的数据存储、基于网际协议（internet protocol，IP）的统一网络，以及全面深入的安全机制和高效智能管理的模式，从而建立设施共用、数据共享、应用集成及服务统一的完善体系。

三、工程管理信息化的二层架构

在工程的全生命周期中，工程各利益相关者之间存在广泛的联系，任何一方的变化都会对其他组织产生一定的影响；重大的业务流程、运营模式及技术架构的变革只能在一定时期内，以相对平稳的方式发生。对旧系统的改造需要一个渐进的过程，二层架构的实现具有重要的现实意义，如图 2 所示。

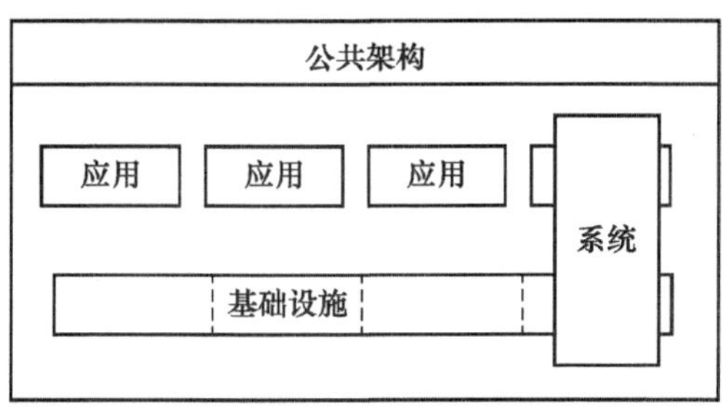

图 2　信息化的二层架构

二层架构可视为由孤立到全面整合的信息化过渡架构，其突出特点是实现了信息技术基础设施的互联互通。二层架构能对各组织的信息技术基础进行集成。在技术实现上，二层架构是在不改变网络拓扑结构及应用范围的情况下，对各个孤立的网络进行连接，对各网络的外部出口进行整合和规范。相比于系统孤立运作的方式，其特点十分突出，分别叙述如下。

(1) 统一的信息系统规划。基础设施的互联互通使得整体工程的管理信息化架构得到了初步建立，为实现统一的信息系统规划创造了物质条件；在此基础上，相关业务流程的优化及运作模式的变革成为可能。

(2) 单一的技术标准和维护体系。各组织间的网络优化、单一标准体系的建立，以及统一的系统维护等都可以在网络互联互通的过程中实现，这是技术架构进一步完善的基础步骤。

(3) 充分利用社会公共资源。不同工程建设和运营阶段的信息化需求是不同的，很多系统是临时建立的，且是短期使用的。基础设施层的互联互通，为充分利用社会公共网络及信息资源提供了便利，减少了工程建设中的信息化投入成本，加快了信息化实现的进度。

(4) 安全措施得到了极大的完善。基础设施整合的两个重要指标分别是内部网络的连通和外部接口的统一。内部网络的连通，使得安全标准得以统一、网络安全监控得以实现、各组织的网络安全水平得以提高。外部接口的统一将跨组织的数据交换纳入统一的管理之下，可以有效地杜绝各类信息安全隐患。

(5) 实现的难度相对较低。相比较而言，基础设施的技术标准更加通用，比直接进行应用系统层面的集成更容易实现；此外，基础设施与具体业务的关联度也小于应用系统，集成过程对整体运作的影响程度较小。

(6) 为各类系统的迁移和集成创造了条件。在系统孤立运作状态下，信息系统的升级往往需要更新大量硬件设备及延伸各自的网络。基础设施的整合则为网络和设备共用创造了条件，降低了系统更新方面的投资强度。

（7）降低了新系统引入的投资门槛和技术难度。在系统孤立运作状态下，新系统的建立必须考虑网络建设、设备采购、维护机制等一系列相关问题。基础设施的整合为降低新系统的投资规模和技术难度、缩短系统实施周期创造了条件。

（8）使用效率提高，产生环保效益。信息系统大都是按照最高承载标准和快速应对突发事件的原则建立的。在日常运作中，很多设备都处于较低的使用率状态。基础设施的整合可以实现跨组织的资源共享，使得整体系统冗余有所降低，而能源方面损耗的降低可以带来直接的环保收益。

就不足而言，二层架构只是整合了基础设施层面，系统在应用层面仍是独立运作的。因为相比于基础设施集成的单纯技术实现而言，应用系统直接与业务运作相关，由其变化产生的影响远大于基础设施。此外，应用系统的种类繁多，各类接口标准的变化很大。相比于基础设施，应用系统的整合难度较大。鉴于此，二层架构还无法实现各组织内部及外部资源的充分整合，架构的整体性、灵活性或可扩展性都还有待加强。

四、工程管理信息化的四层架构

工程管理信息化实现涉及不同组织的多个业务系统。由于各类应用系统种类繁多，建设年代先后不一，技术标准有异，供应商也各不相同，因此直接对应用系统进行集成的难度较大，耗费的成本较高；而且应用系统间的两两集成，会造成系统的灵活性下降，对于某些外部购买的系统，可能还会因为供应商的软件升级而变更集成方案。此外，不同业务流程之间的数据交换有时是临时性的，为此设置专门的集成方案则过于低效。因此，在充分保持现有应用系统状态的情况下，通过建立集成式的公共平台完成系统是比较合理的选择。

四层架构在二层架构的基础上增加了资源层和服务层：通过资源层，实现数据的集中传递和备份，借此实现数据的专门存储和建立灾难恢复机制；通过服务层，进行不同系统之间的数据交换和信息传递，形成跨组织沟通和远程应用。架构的四层分别是：服务（service）、应用（application）、资源（resource）和基础设施（infrastructure），取各层的第一个字母，将其称为SARI[8]，如图3所示。

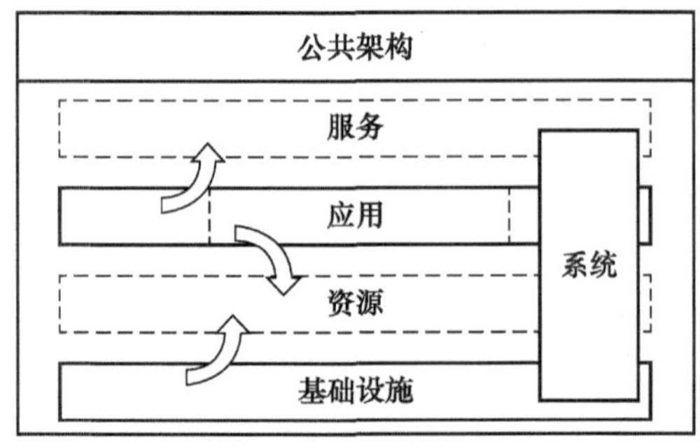

图3 信息化的四层架构（SARI）

1. 四层架构概述

SARI 可以适应工程全生命周期中多组织进行有效协作及持续变革的需求。相比较而言，二层架构只是基础设施层面的整合，而 SARI 则是各个层面的全面集成和共享，因而架构的整体性、灵活性、可扩展性和可操作性有了较大的提升。

（1）基础设施层。基础设施层为技术架构的其他层面提供统一的支持。作为供多个流程及系统共享的物理平台，基础设施层主要包括网络及共用设备两部分。基础设施的网络部分包括由有线、无线及其他特殊网络构成的园区网络，它涉及各类线路及相关的软件及硬件设备。园区网络与外部网络的整合能够采用虚拟专用网络（virtual private network，VPN）和虚拟局域网（virtual local area network，VLAN）等技术，通过构建虚拟网络来支持各类系统的运作，即在互联互通的基础上，通过建立单一的技术标准和规范，按照统一的规划对相关的网络进行分期分步合并，将各类外部网络接口纳入统一管理，进行物理上的集中，从而形成真正意义上的基础设施共享平台。随着技术的发展，诸如自动识别、楼宇控制、媒体传输、通信设备等都可以采用基于 IP 的技术进行改造。采用基于 IP 的技术，意味着这些设备可以运行于公共平台之上，相互之间的数据传输可以不再依赖于专门的网络。同时，这也意味着这类设备不再被特定的系统所独占，并可以为多个信息系统或业务流程提供支持。

（2）资源层。资源层可以实现对工程相关数据资源的集中管理，在具体实现上，它包括支持日常运作的运营数据库（operation data base，ODB）、支持管理事务的管理数据仓库（management data warehouse，MDW）及与之相关的存储、备份及灾难恢复的硬件、软件及专用网络，以及数据格式、数据接口标准等内容。ODB 与 MDW 有着很强的联系，前者是后者主要的数据源。两者基于相同的数据库技术，都需要对数据存储、传输、备份等相关的设施进行物理上的集中。两者的数据都来自多个不同的信息系统，因此，需在数据结构等方面进行逻辑上的集中，才能够有效共享。无论在物理上还是逻辑上，ODB 与 MDW 都不是一个单一的数据库，而是由一系列主题数据库构成的数据库体系。ODB 与 MDW 所不同的是，ODB 的主题数据库一旦确定将长期存在，不会轻易修改，而 MDW 中的主题数据库则可以根据不同的需要而动态设置；ODB 只存储与运作有关的内部数据，而 MDW 不仅包括内部数据，还包括经过整理汇总的外部数据。从数据备份及灾难恢复的角度而言，ODB 的重要程度更高一些。

（3）应用层。与其他 3 个层次都是作为公共平台所不同的是，应用层可看作一系列应用系统的集合。应用层是一种分散的结构，各系统可根据业务的需要灵活配置、增减或升级，保持了系统组合的灵活性、动态性和可升级性。应用系统的划分与业务流程划分相对应；在业务流程内部，各类相关系统应逐步整合为单一的系统；业务流程之间的数据交互和信息传递分别通过资源层和服务层实现。

（4）服务层。服务层在整个信息架构中起着重要的作用，既是应用系统界面集成及信息集成的技术平台，又是组织与合作伙伴之间的信息交流渠道。运营系统的远程应用、呼叫中心、办公自动化及知识管理都可以基于此平台实现。服务层主要包括两部分：企业服务总线（enterprise service bus，ESB）和公共服务平台。前者主要完成对应用层系统的信息集成及界面集成，并作为后者的技术支持平台。公共服务平台结合电子商务及远程协作办公等相关技术手段，通过整合应用层的各项远程服务及信息发布功能，为各组织提供包括在线服务、信息共享、远程应用、协同办公、电子商务等功能的在线基础应用平台。

2. 四层架构的优势

集成与共享是SARI建立的核心原则。SARI面向工程的全生命周期，提供对各类业务流程的整体支持，实现全面的信息资源共享。与二层架构相比，四层架构有以下四方面变化。

(1) 设施共用。在基础设施层面，通过统一规划基础设施，由简单的互联互通过渡到基于IP协议的整合网络，降低了网络的复杂程度，使得二层架构的各项优点得到了进一步的提升。

(2) 数据共享。将从属于基础设施的数据库硬件及网络资源进行物理集中，将从属于应用系统的数据库资源进行逻辑集中，两者结合形成独立的数据资源层。在此基础上建立统一的数据存储、备份与灾难恢复机制。

(3) 应用集成。对各类应用系统按照业务流程的优化和调整进行合并，通过资源层实现数据集成，通过服务层实现信息集成。这一方面实现了系统之间的有效交互，另一方面又保持了系统的灵活性和独立性。

(4) 服务统一。对各类应用系统的操作界面进行统一，形成基于浏览器的操作界面，提高了组织间的交互能力；对各类信息沟通渠道进行整合，形成统一的信息沟通渠道，避免了跨组织交互过程中可能产生的信息混乱和无序状态。

SARI是一种通用架构，可以为工程管理信息化建设提供借鉴[9]。在应用过程中，需要对各组织自身的信息化模式作出较大的调整，由完全分散的模式调整为集中与分散相结合的模式。在业务流程方面，SARI的实现要求将与信息技术应用有关的各类事项整合为一个独立的业务流程，以便为其他的业务流程提供统一的支持和服务。

工程管理者需要制定适用于各组织协作的技术标准与规范、日常维护机制，以确保各组织之间的有效联系，并建立统一的系统集成标准。它具体包括两方面：一是自有标准的确定，包括硬件设备的选择标准、软件接口标准、软件存储标准、网络协议等；二是通用标准的采用，对有关国际标准、地区标准、国家标准、行业标准、组织标准的选取。两者结合构成组织范围内适用的单一标准体系。

统一的技术保障体系是工程管理信息化架构的重要组成，具体包括整体的技术保障体系和支持团队建设。整体的技术保障体系包括对所有硬件、软件、网络及与之相关的弱电系统等其他设施的维护、更新等，以及用户的定期培训机制。支持团队建设则包括与信息技术有关的人力资源方面的工作，以及形成信息技术人员间相互沟通、资源共享、协作创新的氛围。

五、集成与共享模式

工程管理信息化架构的实现不仅包括单纯的技术方面，而且还包括与之相关的运作模式变革及业务流程优化等相关内容。整合与协调运作的信息系统集合，能够为工程项目的实施提供有效的技术支持和资源保障。

SARI架构的核心是集成与共享，涉及网络、数据、流程、界面等各个方面；集成包括物理上的集成和逻辑上的集成，具体为由物理集成形成共用平台，逻辑集成形成各类系统；各类系统在不同的层面通过共用平台进行交互，共享平台的各类资源。

从纯技术的角度而言，工程管理信息化的实现并不需要特定的专有技术，所有用到的技术都是现有的成熟与通用的技术。SARI架构所关注的并不是追求先进的技术设施，而是关注如何选择合适的技术，如何确定适用的技术组合，以及如何更有效地使用相关技术的问题；更重要的是，信息化架构应该关注与技术应用相关的组织运作模式变革及业务流程优化所带来的一系列问题。工程管理信息化的实现过程是各利益相关团体有效协作与资源充分共享的过程，这也是信息化架构逐步由专用与独占模式演变到集成与共享模式的过程。

参考文献

[1] 朱高峰. 对工程管理信息化的几点认识. 中国工程科学，2008，10（12）：32-35.

[2] 刘人怀，孙凯. 工程管理信息化的内涵与外延探讨. 科技进步与对策，2010，27（19）：1-4.

[3] 孙凯. 关于澳门国际机场实施虚拟化机场战略的初步设想. 亚洲战略管理学会：亚澳论坛文集，2008.

[4] Sun K，Guo W. Airport business model with integrated services platforms for transportation，commerce & information. Proceedings of the 2nd International Conference on Engineering and Business Management，Wuhan，2011.

[5] Sun K，Guo W. Analysis on partners & customers of airport management informatization. Proceedings of the 2nd International Conference on Engineering and Business Management，Wuhan，2011.

[6] 刘人怀，孙凯，孙东川. 大型工程项目管理的中国特色及与美英的比较. 科技进步与对策，2009，26（21）：5-12.

[7] 刘人怀. 工程管理：管理对国民经济的深度介入//中国工程管理环顾与展望编委会. 中国工程管理环顾与展望. 北京：中国建筑工业出版社，2007：260-262.

[8] Sun K，Lai W C. SARI：a common architecture for information systems planning. Energy Procedia，2011，13：8104-8111.

[9] Sun K，Lai W C. Integrated passenger service system for airport based on SARI. Energy Procedia，2011，13：8112-8119.

大力促进科技与经济融合[①]

首先，热烈祝贺深圳市先进制造业联合会成立！感谢深圳工业总会对我的信任，邀请我担任深圳先进制造业院士指导委员会主任委员。同时也感谢深圳市政府发展研究中心和深圳中科院院士活动基地积极推动协助“院指委”筹建工作，感谢各位院士委员的鼎力支持。

我认为，成立先进制造业院士指导委员会是一件非常有意义和价值的事，是从更高层次促进科技与经济融合的探索，又是促进院士工作与企业紧密结合的机制创新。

下面我就“院指委”的工作谈几点意见，供大家参考：

一是通过利用深圳中科院院士活动基地与中科院、工程院的学部工作局建立畅通的联系机制，借助广东院士联谊会的工作布局和院士智力资源，从更高层次促进深圳科技与经济融合；创建政产学研金紧密合作的新模式。

二是根据粤港澳大湾区城市群工业创新的实际需求，邀请院士进行区域工业化进程的战略性研究，为深圳市委市政府提供决策依据。

三是根据政府战略性新兴产业和未来产业发展导向，邀请院士建立协同创新机制，在先进制造业领域选择科研成果转化的试点企业以及有转型升级需求的企业，率先开展科研成果产业化的工作，并以设立院士工作站的形式引进相关学科的院士进行指导；对于已建立院士工作站的企业，根据产业发展需要，协调引进关联学科院士协同攻关，帮助企业提质增效。

四是组织开展形式多样的“院士讲坛”系列活动，包括“院士开讲”“院士报告会”“对话院士”请教会等。

我希望深圳的企业，尤其是深圳市先进制造业联合会的会员企业，能够利用好“院指委”平台，组合资源、加速发展，在粤港澳大湾区经济社会发展中发挥更大的作用。

① 本文是在深圳先进制造业院士指导委员会揭牌仪式上的讲话，深圳，2017 年 9 月 24 日.

勇担重任 敢攀高峰①

非常高兴来到具有悠久历史和灿烂文化的中原大地的郑州，参加由铁路 BIM 联盟主办，中国铁道科学研究院、翰阳国际、中铁第四勘察设计院承办的“基于 BIM 技术的装配式建造技术应用交流会”。首先，向多年来为祖国的铁道事业发展作出重要贡献的 BIM 联盟成员、企业、专家表示诚挚的敬意！

上午我们一同参观了郑州四环工程，现场介绍的工程师对我说，这是目前全球最大的桥梁工业化节段预制工程。无论是工程体量、技术含量、标准化建造工艺还是绿色建筑理念都是国际先进水准，而这一切都是由国内的工程师和技术人员完成的。作为基础设施建设事业的一员，我由衷地觉得自豪和骄傲，我们在一直向前奔跑。

回顾改革开放 40 年，我们共同耳闻目睹了伟大祖国发生的翻天覆地的变化：科技发展取得了举世瞩目的伟大成就，在重要领域跻身世界先进行列。特别是近几年来，随着高速铁路建设推进和“一带一路”倡议的引导，我们取得了一系列具有世界先进水平的创新成果，同时我们也更清晰地认识到中国铁路战略空间的迅速扩张对中国铁路的设计、施工、建设、运维能力提出了更高的挑战，加速推进铁路事业的科技创新和产业发展就具有特别重要的战略意义。我们需要跑得更稳、更好、更远。

铁路 BIM 联盟以“联合、共享、引领、创新”为理念，凝聚行业智慧，以服务国家战略、推动行业技术进步为宗旨，积极推进 BIM 技术在铁路工程设计、建设、运营全生命周期的应用，会员单位涵盖了国内铁路及相关行业顶尖的科研、勘察、设计、建管、施工、运维、制造企业和相关重点大学，可谓群英荟萃，具有全方位的政产学研金能力和全产业链技术集成创新优势。历史必将赋予其重任！

“科技是国之利器，国家赖之以强，企业赖之以赢，人民生活赖之以好”。习总书记在改革开放 40 周年大会的讲话上强调：“在我国发展新的历史起点上，把科技创新摆在更加重要位置，吹响建设世界科技强国的号角。”这次的交流会云集了中国铁路科技研发、技术应用的专家，汇聚了国家铁路高端科技智库。

我们手握资源、技术和人才，这是时代赋予我们的光荣使命。如何贯彻“创新、协调、绿色、开放、共享”的发展理念？如何有效提升绿色建设水平，让装配式建造更好地运用到全国各地的基础设施建设中？如何搭建信息资源共享平台，实现组织团队多方联动，高效协作？如何进一步加强 BIM 技术的研究、应用和推广，为新时代中国铁路现代化提供强有力的科技支撑？这些都是我们的共同课题和努力方向。

让我们更加紧密团结，不忘初心，紧紧围绕事关科技创新发展全局和长远问题，把握世界科技发展大势，敏锐抓住科技革命新方向，勇担重任，敢攀高峰，当好建设

① 本文是在基于 BIM 技术的装配式建造技术应用交流会上的讲话，郑州，2018 年 12 月 24 日.

铁道科技强国的先锋，为中华民族伟大复兴、为中国铁路事业发展作出新的更大的贡献！

最后，祝大会圆满成功！

谢谢大家！

有效推动轨道交通结构健康监测和整治修复技术的发展提升①

今天，我受到主办方的邀请，再一次来到东莞松山湖参加第二届粤港澳大湾区轨道交通结构健康监测与整治修复技术学术研讨会。2017 年第一届会议在东莞理工学院召开的时候，我就曾受邀来参会；这一次能再次与各位专家学者和行业一线的工程师共同交流、探讨轨道交通结构健康监测和整治修复领域的前沿技术，我感到非常高兴！同时，看到本届研讨会的规模更加扩大，参会的专家学者和工程一线人员更多，我对此表示由衷的欣慰和热烈的祝贺！

众所周知，结构的健康监测和整治修复技术近年来发展迅速，为不同类型建筑结构安全、稳定、高效地运行、服役提供了重要保障。我国轨道交通的快速发展，对结构健康监测和整治修复技术的应用需求也日益旺盛。目前，中国铁路运营里程超过 12 万公里，其中高铁运营里程超过 2 万公里，位居世界首位，同时，中国 20 多个城市建成了 3000 多公里的轨道交通线路，预计 2020 年将达到 6000 公里。“十三五”以来，珠三角地区共规划了 15 条城际轨道交通线路，合计里程约 1430 公里。随着港珠澳大桥的开通和未来深中通道、虎门二桥的投入使用，珠三角将形成“三环八射”城际轨道交通网络构架，粤港澳大湾区一小时生活圈也将在未来成为现实。因此，结构健康监测和整治修复技术在轨道交通领域大有用武之地。把握粤港澳大湾区建设这一重大国家战略的发展契机，着力研究夯实结构健康监测与整治修复技术的理论基础，提升其在轨道交通和城市生命线工程中的应用价值，是当前我们国家在该领域发展提升并立足于国际前沿的重要机遇，这也正是我们连续举办第二届学术研讨会的根本意义所在。

我注意到，参加本届研讨会的专家学者和工程技术人员来自国内外不同的高校、不同的单位，大家的专业学科背景和研究、从业方向都有所差异，这也更有利于不同专业方向之间的交流互补。我希望大家能利用好本届研讨会的契机，充分结合自身专业和技术优势，积极参与研讨，有效推动轨道交通结构健康监测和整治修复技术的发展提升。

最后，我预祝本届学术研讨会取得圆满成功！同时，我也衷心地希望这个高水平的研讨会能一直办下去！山再高，往上攀，总能登顶；路再长，走下去，定能到达。望与会同仁共同努力，为粤港澳大湾区助发展，为祖国谋复兴！

① 本文是在第二届粤港澳大湾区轨道交通结构健康监测与整治修复技术学术研讨会的讲话，东莞，2019 年 1 月 5 日.

第十一章　公 共 管 理

征求意见　完善报告[①]

去年 3 月，我们上海工业大学经济管理学院接受了市科委下达的“崇明县 2000 年经济、科技、社会长远发展规划战略研究”重大科研课题。经过一年多来的努力，在钱伟长先生的直接领导下，在崇明县委和县人民政府的直接参与下，在各协作单位——上海交通大学、上海财经大学、上海社科院情报所、上海农科院土壤肥料所、上海能源所、上海建设银行投资所等的大力帮助下，在许多专家的帮助与支持下（他们提了许多极其宝贵的意见和建议），我们才将战略报告的初稿完成。这份报告约 30 万字。为此，请允许我以上海工业大学的名义，向各协作单位和各位专家致以崇高的敬意，表示衷心的感谢！同时，提请给各位领导和专家审查。

崇明在祖国的千千万万岛屿中，大小位居第三，约有 1100 平方公里，占上海市总面积的六分之一。它具有许多显著的特点。

一、优越的地理位置

位于亚洲，在东太平洋海岸的中部，是洲际交通的好港口。

位于世界第三大河长江的出海口，面向大海。有广大的我国经济重点区域——长江流域腹地，因而是繁荣内河与海洋间的交通枢纽。

位于世界特大城市——上海市郊，有强硬的依靠。

二、优越的自然条件

在咸淡水交界处，环江环海岸线长达 204.5 公里，岛内河网密布，各种稀有名贵的鱼类在此繁衍生息，水产资源十分丰富。

土地日长夜大，每年增加国土 2 万亩。解放以来，已新增土地 68 万亩，该岛的面积新增 65%。而且土质肥沃，水量充沛，气候温和，动植物资源十分丰富。

可以毫不夸张地说，崇明是祖国的一颗明珠，极富前途，值得精雕细刻。

党的十一届三中全会以来，中央和上海市对崇明岛的开发与振兴一直高度重视，高瞻远瞩。1979 年，国务院的 202 号文件，就曾准备把它辟为经济特区。最近，中央

---

① 本文是在上海市科技委员会“崇明县 2000 年经济、科技、社会长远发展规划战略研究”重大课题评审会上的讲话，上海，1987 年 7 月 29 日。

领导和上海市委也多次指示，要很好考虑崇明岛的规划与开发。

根据以上精神，我们规划组（其中上海工业大学31人，崇明县24人）借鉴国内外岛屿经济开发的经验，运用系统工程的理论与方法，初步提出了崇明县2000年经济、科技、社会发展战略研究的报告。今天，由陈笃平和严志渊两位同志向大家汇报。我们衷心希望各位领导和各位专家多提意见，完善这份规划报告。

领导用人标准①

要认真读书，这是社会主义现代化事业对每一位领导干部的要求。作为一名合格的领导干部，必须有广博而精深的理论知识和丰富的实践经验。这里所说的理论知识是指社会科学和自然科学、管理科学等，而其中最重要的且必须掌握的知识有两个方面，第一是马克思主义理论，第二是领导科学和领导艺术。下面，着重谈谈领导干部必须掌握的“德才兼备”用人标准。

一、我国历史上对“德才兼备”的论述

“德才兼备”的用人标准不是我们社会主义时代才提出的，在我国古代就早已有之。在我国历史上，关于“德才兼备”的文字记载，可算是丰富。一方面，在漫长的我国五千年历史中，由于常常是大动荡、大变革，因而思想活跃、人才辈出。另一方面，历代统治者为创立或维持政权，总要煞费苦心地选才用人，这就形成了我国历史上丰富的用人之道。我们不应割断历史，应该了解历史，继承我们祖先的关于用人之道的珍贵的历史遗产，去其糟粕，取其精华，这对建立我们社会主义时代的用人标准将是不无裨益的。

在我国漫长的原始公有制、奴隶制、封建制和半殖民地半封建制历史上，对领导者，即官吏的选拔有以下十种制度。

1. 原始贤能制

在原始公有制下，以贤能作为酋长等部落人才的选拔标准。正如孔子在其《礼记·礼运篇》中所说：“大道之行也，天下为公，选贤与能，讲信修睦，故人不独亲其亲，不独子其子。”

2. 禅让制

随着生产力的发展，原始部落联盟不断扩大，在原始社会晚期出现了“禅让制”，即以贤能为准绳，通过让贤方法来选择首领。在《尚书·尧典》篇中有这样的记载：尧晚年选择继承人时，曾要四岳出仕，四岳认为自己不能胜任，便推荐隐居在民间的贤人舜。经过尧对舜三年的考核，尧便让位于舜。这种选拔理论是朴素的，方法是原始的，内容是简单的，但对中华民族的用人史产生了极其深刻的影响。

3. 传子制

由于畜牧业的发展和私有制的产生，便产生了废除禅让制、实行世袭传子制的社会经济基础。自禹的儿子启夺取王位建立“夏”朝后，中国社会进入奴隶制时代，开始世袭制的时期。可以说，任人唯亲的用人标准就是从这儿起源的。根据《尚书·盘庚》的记载，王位和大臣都是世袭的。但是，对官吏的任免，并非完全是世袭的，往

① 本文是在上海市委党校学习时的读书报告，上海，1987年11月13日；现代管理的中国实践，北京：科学出版社，2016，100-104.

往还是以贤者能者作为官吏的选拔标准。因此，人才选拔中的“唯贤”与“唯亲”的斗争，将作为一种特殊的矛盾存在并发展，从而推动社会前进。

4. 军功制

秦始皇统一中国后，废除了官吏的传子制，建立了与郡县制相适应的官吏任命制。此时，主要是按军事战争中功劳大小，同时也注意德行来安排职务。由于偏重了军功，忽视了官吏管理才能的鉴别，因而使一些缺乏治理才干的人担任了官吏。由于秦朝的统治时期过短，考选人才的军功制使用时间不长。

5. 察举制

察举制方法始于西汉武帝时期，主要是由中央或地方长官推荐人选，经过考核，任以官职。举荐人才的标准为德行、才能、知识、功绩四个方面。这是地主阶级在上升时期广开才路所采取的措施，是千方百计维护地主阶级利益的制度。

6. 九品制

九品制是魏晋南北朝时期确保世族特权的官吏考选制。官吏的选拔标准为：①家世，即门第；②状，即对道德才能的评述；③品，即品级。由于这种考法受主评者个人好恶因素影响很大，且又重视门第，因而限制和扼杀了许多人才，所以当隋朝建立后，就被科举制代替了。

7. 科举制

科举制起源于隋朝，历经唐、宋、元、明、清，是我国古代历时最长的考选人才制度。它比九品制进步之处，是以才能（文才为主）取人，代替以门阀取人，从而巩固地主阶级的统治。由于这种制度最后蜕变为贻误人才和束缚人才的桎梏，终于在清朝末年的维新运动中被政府废止。

8. 考绩制

考绩制是古代人才考核与选拔制度的重要组成部分。大凡一个朝代的兴盛时期，统治者总是重视官吏实绩的考核。例如，唐代以“四善”作为官吏的通用标准，又以“二十七最”作为不同职事官吏的特殊标准。

9. 保举制

保举制是太平天国时使用的官吏选拔制度。它吸取了历代考选制的长处，以德才为取人标准，如原始贤能制的贤能原则、禅让制的让贤风气、军功制的集中选拔、察举制的推举方法、考绩制的注重绩效。同时，摒弃了历代考选制的糟粕，如传子制的世袭、九品制的门第、科举制的八股文。尽管这种制度不是很完善，保留着一些封建专制的色彩，但比起封建王朝的任何一种考选制，无疑是有了进步。

10. 尚实制

尚实制是我国近代史上，地主阶级开明派和资产阶级早期维新派为吸取西方物质文明和科技进步经验，以使中国富强，而提出的“务归实用，不尚虚文”的选拔官吏的标准。

综合上述十种制度，可以概括出如下特点。

归根结底，历代统治者的用人标准就是“德才兼备”，只是根据统治阶级需要和不同历史条件而在德才关系上有所偏重。春秋初期的管仲明确指出，德、功、能，治乱之原也。三国时，曹操主张“唯才是举”和“治平尚德行，有事赏功能”。唐太宗也是“唯才是与”，说：“朕任官必以才……若才，虽仇如魏征，不弃也。”北宋司马光进一步说：“才者，德之资也；德者，才之帅也。”然而，由于阶级的局限和历史科技条件的限制，对官吏的选拔尚存在许多不可克服的弊病。他们的“德与贤”只是封建地主阶级的政治需要和伦理道德，带有很大的虚伪性和欺骗性，这是造成我国近几百年来经济发展缓慢的重要原因。

二、正确领会和执行社会主义初级阶段对干部的“德才兼备”用人标准

党的十一届三中全会以来的九年，是马克思主义路线高奏凯歌的九年。我国的社会主义建设进入了一个新的发展时期。目前，我们的一个非常艰巨的任务就是要在20世纪内实现四个现代化，把我国建设成为一个社会主义强国。为实现这一光荣而艰巨的任务，关键就是要有一大批热心改革并善于领导和组织实施改革的干部。我们有句俗话：事在人为。同样的事，不同的人去做，得出的结果可能会有很大差别。再好的改革方案计划，如果没有积极热情并善于动脑动手的人去做，最后也只能成为空中楼阁、纸上谈兵。毛泽东同志说，领导者的责任，归结起来，主要地是出主意、用干部两件事。领导干部选择得是否恰当，对领导活动的效能有着举足轻重的影响。

在广东北部山区，有一个出了名的穷县，即南雄县。在1984年时，全县的人均产值在全省倒数第一，连县委干部工资也发不出，只好伸手向上级要。1984年5月，改选了县委领导班子。新的县委领导一上台，就提出了“佼佼者上，平庸者让，不称职者下，以权谋私者撤”的用人原则，大刀阔斧地调整了中层领导班子。仅仅两年时间，南雄县就面貌巨变。全县乡镇企业由原来的40多家发展到1000多家，财政增长率居全省之冠。这一事实使人们认识到一个真理：在建设社会主义事业中，不仅要求有正确的路线方针政策，而且还要有执行这些路线方针政策的优秀领导干部。这里的优秀领导干部就是“德才兼备”的干部。

我国现正处于社会主义现代化建设的新的历史时期，现代化的关键就是科学技术的现代化。正因为如此，现代化事业的竞争归根结底就是科学技术的竞争，科学技术的竞争就是人才的竞争，而人才的竞争的关键就在于选用能尊重知识、尊重人才的知人善任的“德才兼备”的领导干部。

“德才兼备”是把“德”和“才”联结为一个整体，既不能割裂，也不能偏废。离开“德”，“才”就失去了正确的方向；没有“才”，“德”就成为空的东西。所谓“德”，主要是要考察干部是否坚定不移地贯彻执行十一届三中全会以来党的路线方针政策，是否坚持四项基本原则和改革开放的方针；所谓“才”，主要是看干部在改革当中的实际贡献和才干，看他们是否能够将两个基本点很好地统一起来。

综上所述，在社会主义现代化事业中，我们必须坚持“德才兼备”的用人标准。

有效利用外资 扩大对外开放①

利用国外各种资金和吸引外商直接投资，是实行对外开放的重要内容，也是补充国内资金不足、增强出口创汇能力、提高我国技术水平和管理水平的重要途径[1-4]。

去年10月，国务院和许多省市先后颁布了关于鼓励外商投资的规定。接着，各有关部门又陆续公布了十个实施细则。这些规定和实施细则为外商投资企业在实现外汇平衡、简化各种手续、降低各种费用、保证水电原材料供应、保障企业经营自主权等方面提供了法律依据。

总结我们前几年的实践经验，并借鉴国外有关经验，我们认为，有效利用外资，应该遵循以下三条原则。

第一，外债的总额要有控制、结构应该合理，并应与自己的偿还能力和消化能力相适应。

第二，应该用在生产建设上，重点是出口创汇企业、进口替代企业和技术先进的企业。

第三，应该讲求经济效益，并注意保持国际声誉。

下面，以上海市为例证，在调查研究外商投资企业现状的基础上，对如何有效利用外资、扩大对外开放问题，提出一些初步的看法。

一、外商在沪直接投资现状分析

1. 发展速度

截至1983年年底，上海的三资企业仅20家。1984年，上海被批准列为我国14个沿海开放城市之一以后，发展很快，三资企业当年累计64家；1985年，累计160家。1986年，由于对大楼、宾馆等项目有所控制，发展速度略为降低，当年累计220家。截至今年6月底，累计265家。其中，合资企业181家；合作企业81家；外商独资企业3家。

2. 投资金额

截至1987年6月底，外商在沪三资企业协议投资总额达到16.34亿美元，占上海吸收外资总额的81%，其中181家合资企业投资金额达6.9亿美元，占三资企业投资总额的42.2%；81家合作企业投资金额9.4亿美元，占三资企业投资总额的57.5%；3家独资企业投资金额406万美元，仅占总额的0.2%。

3. 投资结构

截至今年6月底，大楼、宾馆项目共34项，总计10.9亿美元，占协议投资额的66.7%。生产性项目98项，总计3.5亿美元，占协议投资额的21.4%。其中，大众汽车、福克斯波罗仪表、耀华皮尔金顿浮法玻璃、施贵宝制药、迅达电梯、贝尔电话设

① 本文原载《国际商务研究》，1987，(6)：1-5。作者：刘人怀，史乐毅.

备属于进口替代型项目，其余大多数属于出口创汇型项目。

4. 经营状况

根据上海市经济和信息化委员会系统已经生产营业的 18 家工业性合资企业统计，经营状况大致可分为以下三类。

(1) 办得较好的，约占四分之一。其主要特点是引进了先进技术和现代化管理，产品质量稳定，经济效益明显，出口创汇增加，外汇支出减少，达到了利用外资的基本要求。这类企业有联合毛纺、迅达电梯、福克斯波罗仪表、高仕香精等。

(2) 办得一般的，约占二分之一。其主要特点是引进了先进技术和设备，但产品的出口竞争能力薄弱，外汇难以平衡，经济效益尚未得到落实。这类企业有施贵宝制药、贝尔电话设备、大众汽车、易初摩托等。例如，桑塔纳轿车，各项技术指标比较先进，但由于德国马克大幅度升值，进口全散件装的成本大幅度提高，国产化一时又难以实现（目前国产化率仅 4%），外汇无法平衡。

(3) 办得较差的，约占四分之一。这类企业既未引进先进技术，经济效益又不理想。例如，上海兴中皮鞋有限公司，进口原料，生产皮鞋、运动鞋和旅游鞋，由于管理不善，质量不好，产品全部内销，外汇无法弥补。

5. 投资国别（地区）

目前在沪直接投资的共有 16 个国家和地区的外商。按投资比重排队，中国香港地区居首位，占 35%；美国第二，占 33%；日本第三，占 15%；接着依次为联邦德国、英国、比利时、泰国等。按各国（地区）在沪投资与该国（地区）在华全部投资的比值排队，美国居首位；联邦德国第二；中国香港地区第三。

6. 投资动向

(1) 投资金额多半集中在投资回收期较短的项目上。大楼、宾馆的比重较大；中小型工业项目较多。

(2) 投资于劳动密集型项目较多。根据 70 项工业性项目统计，技术密集型项目仅占 25.7%。

(3) 部分外商由于担心我方对外资的偿还能力，要求采取相对贸易的方式与我方合作，即由外商提供技术和设备，换取我方出口物资，实行间接补偿。

(4) 近年来，有些外商由于在直接投资中得到了利益，已经从原来的信心不足，打算撤走投资，逐渐转变为信心倍增，准备扩大再投资。

(5) 国务院和上海市颁布关于鼓励外商投资的规定以后，外商反响很大；外方经理十分重视，纷纷表示要进一步推动和促进双方的合作。

(6) “22 条”一公布，香港新鸿基有限公司驻沪代表立即打电话找市长，要求洽谈项目；美国领事馆汪德先生当天就访问财政局，了解上海对外资企业的税收政策，并表示美商与上海企业界进一步合作的兴趣很大；美国总统里根的经济顾问彼德·格雷斯当月就在上海开办了一家独资企业——格雷斯中国有限公司，专门生产密封胶产品，全部在上海销售。上海将用出口罐头所获得的外汇支付格雷斯中国有限公司。格雷斯先生说：“我对上海市场进行了六年的观察和研究，我认为上海工商业集中，市场繁荣，海陆空交通方便，进出口贸易发达，是外商最理想、最合适的投资场所。”

二、有效利用外资，扩大对外开放

资源短缺、资金不足、技术装备落后、基础设施薄弱、传统工业改造任务繁重，这是上海经济发展所面临的主要矛盾。要解决这些矛盾，根本出路在于按照上海经济发展战略的总目标，进一步实行对内对外两个开放，把改革搞好搞活。

在对外开放方面，应该抓紧当前有利时机，把利用国外资金、国外资源和积极开拓国外市场有机地结合起来，逐步形成外向型经济结构；要贯彻执行并充分运用国务院批准的有关优惠政策，进一步扩大利用外资。根据上海的实际情况，我们认为，当前应着重做好以下三方面的工作。

1. 继续吸引外商直接投资

要进一步明确投资方向，把重点放在生产性项目方面，特别是要鼓励技术先进型和产品出口型的投资项目；争取把国际资源和上海加工工业的优势结合起来，积极发展进料加工出口和来料加工出口，并按照国际市场变化和外向型经济发展的要求，不断调整产品结构，尽量增加工业制成品，特别是轻纺、电子和机械等深加工、精加工产品的出口，大力开拓国际市场，提高出口创汇能力。

2. 积极发行国际债券

利用外资的方式很多，各有其适用性和优缺点，我们应当择优选用。通过发行国际债券筹措资金是当前值得考虑的一种利用外资方式。

根据 1986 年年底统计数据，我国已有 18 次发行国际债券的经验。筹措的资金包括日元 2650 亿、德国马克 3 亿、港币 3 亿和美元 2.5 亿。

通过发行国际债券筹措资金，具有如下明显的优点。

（1）可以灵活地用来弥补财政预算的赤字和国际收支的逆差，而且发行手续简单方便。

（2）偿还的期限较长（通常为 10 年以上），可以满足地方政府和金融机构对长期资金的需要。

（3）各主要国家（美国、日本、英国、联邦德国等）都设有国际债券市场，债券发行人可以方便地获得以各种主要货币（美元、日元、英镑、德国马克等）为计值单位的国外资金。

发行国际债券是以某种外国货币作为计值单位的，有时，也以欧洲货币单位或特别提款权等一揽子货币作为计值单位，目的是谋求减少国际债券发行中的汇率风险。汇率风险是发行国际债券必须慎重考虑的问题，有时虽然资金成本较低，但偿还债务所用货币的汇率升值，往往使债券发行人得不偿失。

（4）大多数国际债券的利率以供求关系为基础，并且具有相当的竞争性，这对债券发行人来说，是比较有利的。

（5）属于国际债券市场的欧洲债券市场，由于不受任何国家税收条例的约束，很受债券投资人的欢迎。因之，其吸收能力很强，而发行成本较其他债券市场为低。

（6）债券发行成功，将会提高发行国家或发行企业的国际声誉，从而使有关发行人在选用其他利用外资方式时处于较为有利的地位。

鉴于上海是我国最重要的工业基地之一，也是全国最大的港口和沿海开放城市，素为世界所重视，对外颇有吸引力；此外，近一年来，日元和西欧货币升值，国际金融市场利率下降，香港及东南亚地区经济繁荣，国际游资大量增加，投资人的目光频频转向中国。我们认为：当前是利用国外资金的大好机会；充分发挥上海作为重要的金融中心和信息中心的优势，适当运用上海在国际上的声誉，积极主动地通过发行国际债券的方式，为加速改造和振兴上海筹措必要的资金，是大有可为的。

为了有效地利用外资，建议进一步扩大并完善金融市场，诸如进一步建立长期资金市场，必要时开放黄金市场；增加信用工具，放宽金融管制，健全金融体系，对外资银行扩大开放（如同意其升格为分行等）；在条件成熟时，可考虑设立“国际银行自由区”，进一步密切沪港金融合作，加强与国际金融市场的联系等。如果我们能把握世界经济重心东移和国际金融变化的有利时机，到 20 世纪末，上海争取成为亚洲金融中心的可能性是存在的。

3. 充分利用国外商业贷款

目前，上海大多数工业企业，利用国外商业贷款的偿还能力不强，积极性不高。

对本市 70 家工业企业所做的调查表明：愿意利用国外商业贷款进行技术改造的只有 10 家，仅占 14.3%。分析其原因，主要有以下几点。

（1）国外商业贷款的利率高，风险大，而企业得益甚少。许多厂长认为：借用国外商业贷款，风险由企业承担，收益归政府支配，得不偿失。

（2）企业的偿还能力薄弱。调查结果表明：有 40 家（占 57.1%）企业自认为对借用国外商业贷款缺乏偿还能力。

（3）企业的创汇能力很低。调查结果表明：有 47 家（占 67.1%）企业认为借用外资改造企业以后，虽然有利于提高产品质量，但仍然缺乏出口创汇能力。

近年来，美元的国际市场利率已从 20 世纪 80 年代初期的 20%逐步下降到 9%左右，这对我国利用国外商业贷款提供了一个十分有利的客观条件。我们应当抓住有利时机，加快改革步伐！

去年，国务院批准上海市可直接向国外举借外债，自借自还。截至今年五月底，已批准项目建议书或可行性报告 223 项，投资合计占第一批额度的 79.3%。

利用国外商业贷款，对工业部门的某些行业来说，要比吸引外商直接投资更为有利。例如，某些基础比较好、管理上轨道的项目，只要引进关键性技术或设备，就能大幅度提高产品质量和生产能力。这些项目如果与外商合营，让肥水滚滚外流，岂非十分可惜！

另外，有些大楼、宾馆等项目，是否可利用国外商业贷款由我们自己来干，何必要用合资或合作经营的方式，让外商来分成呢？

为了调动企业和职工利用国外商业贷款、引进先进技术和增加出口创汇的积极性，制定一套科学、合理的鼓励政策是十分必要的。在现行体制下，是否可以采取如下政策和措施？

（1）鉴于当前企业经营环境和国内市场条件的现状，利用国外商业贷款的企业只能与主管部门共同承担经营性风险，而借用外资的汇率风险，可由全市的外资统筹机构或社会保险公司，实行汇率保险，借以消除企业和职工借用外资的后顾之忧。

（2）由于利用国外商业贷款进行技术改造的项目，与三资企业一样，都具有引进外资和技术、推动上海经济发展的作用，所以，当其利用外资金额达到企业全部固定资产的三分之一时，是否应给予享受中外合资经营企业的优惠待遇？

（3）对利用国外商业贷款引进先进技术，而且创汇多、效益高的企业，应当给予奖励，并提高奖金税的起征点；在经济效益较低的初期，建议给予免除产品税、增值税、城市维护建设税等税收，以保证不降低职工的奖金水平和福利水平。

（4）在鼓励企业借用外资、积极创汇的同时，要求企业优先采用国产零部件和原材料，大力节约用汇。目前有些企业把大笔外汇花费在零部件和原材料的进口上，实为近期内“得过且过”之举，从长远着眼，既非可行，也不合理，应该从改造或引进零部件和原材料的生产线入手，逐步形成国产化配套生产体系，并具体规定零部件国产化的期限和指标。例如，桑塔纳轿车，目前的国产化率仅为4%，到1991年，要求国产化率达到33.3%。

（5）应该抓紧落实“以产顶进”项目的扶植政策，实行价格浮动、适当减免调节税，以提高其竞争能力，并及时制定“横向配套”的优惠措施，鼓励企业发展全国范围的联合和协作，争取全国性配套，加快国产化进程。

（6）国务院规定，地方借用外资要由地方财政作担保。所以，借用外资不但要受地方外汇收入的制约，而且还要受外汇分成比例的制约。鉴于上海的外汇收入大部分来源于具有附加价值的加工产品的出口销售，所以，其外汇收入的分成比例应该略高于以出口资源性产品或矿产品为主的省（区、市）。现行的外汇分成比例，对上海的经济发展和城市改造，都是不利的。为了使上海能跻身于国际市场，必须有较多的外汇使用额度，是否可以把上海的外汇分成比例从目前的25%提高到50%左右？

（7）建议企业主管部门切实下放人事权、财权和物权，并加强服务和指导。在当前条件下，特别需要加强计划指导和经营管理指导，及时为企业提供国际市场信息和技术咨询服务，帮助企业发展工贸、农贸之间的横向联合和协作。

总之，企业主管部门在加强宏观管理的同时，应该想方设法为企业和职工创造各种有利条件，把企业和广大职工利用外资、发展生产的积极性充分调动起来，同心协力地为进一步实行对内对外两个开放贡献智慧和力量。

参考文献

［1］　国务院. 关于大力发展对外贸易增加外汇收入若干问题的规定通知. ［1979-10-15］. http://www.reformdata.org/1979/0410/3696.shtml.

［2］　中共中央关于经济体制改革的决定. 北京：人民出版社，1984.

［3］　邓小平. 建设有中国特色的社会主义. 北京：人民出版社，1984.

［4］　史景星. 工业经济管理. 上海：上海人民出版社，1986.

如何防止公共关系庸俗化[①]

也不知道从哪一天开始，公共关系突然之间成了人们所追求的一种时髦，时下公共关系这个词频频出现在报纸上、招生广告上、名片上、职业讨论会上以及读书活动中，有关公共关系的书籍也越来越多。这种现象的出现自然有可喜的一面，说明我国的公共关系经过全体同仁的几年努力，其作用在日益有效地发挥出来。然而，问题的另一面是这种状态也很容易使公共关系庸俗化，使我国根底本来就不深的公共关系变为一种商品经济模式下的装饰品，一种缺少良好气质的时髦，而无法真正体现出公共关系在社会主义商品经济建设中的功能价值，那么，如何防止与改变这种现象呢？我们认为关键在于加强公共关系的基础理论与实验研究，建立与完善公共关系的评价指标体系。如果离开了这一点，在前几年问题还不大，但到了现阶段，即使出现再多的貌合神离的公共关系教材与专著，充其量也只是一个内容上的循环反复，这就像社会上流行的一种比喻，说简单原型思维痕迹很深的那些书是“蛋炒饭，饭炒蛋，炒来炒去蛋炒饭”，其理论水平实在难以符合社会实际工作的需要，引起的连锁反应则是导致大量的从事公共关系实际工作的同志在操作公共关系实务时没有一个相关的科学参考系，自我评价工作成绩与工作失误，心中也没有个谱，换句时髦的话则是摸着石头过河，而不管水深水浅，摸得着石头摸不着石头。事实上，世界上大多数公共关系发达国家在公共关系方面成功的经验，也充分说明了他们公共关系的有效性，是与加强公共关系基础理论研究与实验研究，以及具有一套行之有效的评价指标体系同步向前发展的。

在美国众多的公共关系硕士、博士论文就是以公共关系基础理论与实验为研究方向的，这批人很受企业与政府部门的欢迎，欢迎的理由正如一位公共关系博士所说的，是由于美国似乎所有的大企业都设有公共关系部，遍及全美的公共关系公司也有数千家，这些部门具有大量的实践经验，但他们更需要用理论来说明他们的实践，使他们在做一桩公共关系工作的时候，做到知其然，也知其所以然。美国每年众多的理论与实践研究方面的成果，保证了他们可以制定出比较科学的公共关系行为准则、公共关系工作岗位规范指标系统，以及公共关系顾问工作准则等一套完整的理论与工作评价系统。这一套完整的理论与工作评价系统又可以确保只有合适的人员才能够胜任公共关系工作，滥竽肯定不能在乐队中充数并有一席之地。在当代发达国家中，各大公司公共关系部经理、咨询顾问，以及比较正规的公共关系专业人员一般都经得起公共关系工作评价系统的检验，他们的工作绩效也是可以做到相对量化的。为此，我们在宣传普及公共关系的同时，现阶段应该是建立我国公共关系的理论与科学的工作评价系统的时候了，只有让这些内容与我们的公共关系工作结合为一体，才有可能使我们的公共关系工作的质量发生根本性的变化，做到工作中有理论指导，评价工作结果时做到客观公正，从而防止公共关系庸俗化。

---

① 本文原载《公共关系》，1989，(4)：10。作者：刘人怀，斯晓夫.

领导科学与领导艺术①

一、领导科学的概况

1. 此学科尚在建设中

领导科学是应运而生的新兴学科，尚在发展完善中。

何谓领导，何谓领导者，这是个难以准确回答的问题。

在《辞海》中，“领袖”一词被定义为“为人表率的人”和“国家、政治团体、群众组织等的最高领导人。”[1]这一定义显然没有包括数以千万计的我国上下各层次的领导者。

领导、领导者、领导行为和领导活动，有史以来就一直存在，人们并不乏对其的研究，只不过长期以来没有把它视为一门专门的学科来加以研究。

2. 建立领导科学的必要性

早在原始社会时期，人们处于蒙昧中。那时，生产力极为低下，人们运用棍棒、石器等简陋粗糙的工具，依靠民族集体的力量与大自然开展斗争。此时，民族的酋长就是领导者。

不过，那时人们的知识十分贫乏，生活简单，征服自然的能力有限。人们对生老病死、吉凶祸福都没有科学的认识，误以为人们的生活完全是由一种神力在进行控制和支配。在这种背景下，神权政治和迷信政治应运而生。

进入奴隶社会和封建社会之后，人与人之间关系发生了质的变化。奴隶主与奴隶、地主与农民，成了奴役与被奴役、剥削与被剥削、压迫与被压迫的关系。此时，行政官僚只对君主负责，横征暴敛，完全不顾劳动人民的死活，阶级矛盾十分尖锐。神权政治和迷信政治逐步为统治者利用。历代统治者出于维护本阶级利益的需要，必然研究“统治学”“奴役学”“压迫学”，拼命地维护其统治。

到了16、17世纪，欧洲产业革命促进资本主义兴起，诞生了工人阶级队伍。在工人阶级的持续冲击下，资产阶级的强权政治逐渐有所改变，使资本主义民主有了一定发展。为了适应资产阶级统治的需要，资产阶级社会学家研究了“权力学”。他们把人类社会中必不可少的“领导活动”简单地归结为“权力”，并渲染一些冒险家怎样玩弄权术和政治手段，从而登上权力宝座，以致把领导科学的研究引向了歧途。

19世纪马克思主义诞生，标志着工人阶级在人类历史上开始登上舞台。科学社会主义和共产主义理论为人类的未来描绘了最为绚丽、壮观的美景；同时，也为人类历史上第一次科学地确认领导与被领导、领袖与群众的关系奠定了基础。

1949年中华人民共和国的成立，标志着我国的国体、政体的根本改变。宪法规定：

---

① 本文是上海工业大学经济管理学院讲稿，1990年5月28日；现代管理的中国实践，北京：科学出版社，2016，85-100.

“中华人民共和国是工人阶级领导的、以工农联盟为基础的人民民主专政的社会主义国家。”“中华人民共和国的一切权力属于人民。”这些都标志着人类社会最进步的原则，是建立社会主义领导科学的最基本原则。

“文化大革命”中，林彪、江青反革命集团乘虚而入，继承历代封建王朝和资产阶级统治的衣钵，把“专制”和“权力学”一齐应用，给中国人民带来了十年内乱的深重灾难。

新中国成立以来，既取得了伟大成就，也发生了许多失误。这之中，也包含着没有把长期积累起来的领导经验进一步升华，没有建立一门领导科学。

积历史之经验，建立一门以马克思主义为指导的、具有中国特色的领导科学，这乃是时代的需要，是社会主义建设的需要。

3. 怎样建立中国特色的领导科学

任何一门科学理论的建立，绝不会取决于人们的主观想象或一时灵感，它必然是一个历史的科学研究，是一个承前启后的发展过程。

20 世纪以来，全世界对经济、社会的管理活动研究已经逐步从科学管理发展到管理科学，并建立了管理科学的许多分支学科，如行政管理、企业管理、工程管理、科技管理、教育管理等。

相应地，发展了管理的基础学科，如管理心理学、行为科学等。

同时，建立了一系列管理技术和方法论，如控制论、信息论、系统论、数学规划、管理信息系统、目标规划、策略规划、网络分析等。

以上内容差不多都涉及领导者和领导行为的研究，是现代领导者手中的重要工具，对建立领导科学起到积极的奠定基础的作用。

时代需要，为我们建立领导科学提供了必要条件。世界管理科学研究的科学成就，为我们建立这门学科提供了充分条件。

领导科学应运而生，实属历史的必然产物。我国从 20 世纪 80 年代初开始研究[2-5]，而国外至今还仅把领导作为管理工作中的一种职能、一个环节进行研究。

二、领导科学的定义、研究范围和对象

领导工作自古有之。领导科学是一门新兴的、年轻的科学，又是一门发展中的科学，也是一门非常重要的科学。

领导科学的定义：领导科学是研究领导活动的特点和规律的科学。

领导科学的研究范围：领导与其所在系统的关系；领导的群体活动；领导的个人行为。

领导科学的研究对象：领导活动的特点和规律。

三、领导科学与原有的管理科学的区别

1. 研究对象不同

原有的管理科学研究的对象主要是事业层的具体业务规律。例如，企业管理主要是研究在生产过程中如何对所能支配和影响的资源进行整合，从而提高效率的方法和

规律。

领导科学研究的对象是比事业层更高一层次的领导层的活动规律。换句话说，管理科学研究的是“将才”活动层，领导科学研究的是“帅才”层次。

2. 存在形式不同

大家都知道，一门科学理论的成熟程度取决于它应用数学的熟练程度。那些能使用数学工具的理论，通常称为“精密科学”，或者叫“硬科学”。反之，则称为“描述性科学”，或者叫“软科学”。

管理科学是一门综合性的学科，既有自然科学的成分，又有社会科学的成分。在应用数学工具方面，它有“硬科学”的特征；在定性的描述上，又有“软科学”的特征。

然而，领导科学却不是这样。它所涉及的问题，现在还很难用数学工具来描述。即使未来，可能也无法描述那些变化莫测的“领导艺术”和“兵诈之术”。所以，它只是一门“软科学”。

3. 应用范围不同

管理科学是非常专业化的理论，其知识面和应用面均较窄，例如，企业管理不能用于科研管理。

领导科学却不是这样。虽然也是一种专业理论，但其面宽，具有综合性的特征。

因此，一个“帅才”领导人对于更高一级领导人而言，他便成为“将才”。所以，现代“将才”管理者和现代“帅才”领导人的绝对界限会变得越来越模糊。他们常是二重性人物，既要懂得原有的管理科学，又要掌握现有的领导科学。

总之，领导科学是现有管理科学的发展，它的成熟和发展将会大力推进人类社会继续向前。

四、领导活动的要素

1. 领导活动的三个基本要素

领导活动有三个基本要素，即领导者、被领导者和他们共同的群体目标。

凡是需要两个人以上去实施完成一个群体目标时，就存在着领导和被领导的行为。领导者就会表现出一种领导行为，被领导者必须接受领导者的指令。

2. 群体目标是三要素中起支配作用的要素

领导者与被领导者之间所体现的只是一种体制形式，而实施完成群体目标才是领导者的活动舞台。因此，群体目标便成为领导活动中最为活跃且起支配作用的要素。

3. 领导活动三要素是相互作用和相互制约的有机体

领导活动的三要素不是各自独立存在的要素，而是一个相互作用、相互制约的有机体。

过去，许多人解决问题时，只注意其中一个要素，常收不到好的效果。例如，许多人解决问题时，只注意解决领导者的问题，忽视了群众，殊不知，这是一个相互作用、相互制约的完整系统，是极为重要的指导原则。

五、领导科学的规律

1. 领导活动的主体-主导律

1）被领导者是主体

大家常说，"人民创造历史""时势造英雄"，这实际上就是人类社会规律在领导活动中的具体体现：被领导者是主体，领导者是主导。

在原始社会，氏族酋长是氏族成员公推的，证明氏族成员的主体地位明显。

在奴隶社会和封建社会，君主专制是世袭制，被领导者的主体地位被颠倒了，被领导者面对统治阶级的暴力统治。

在资本主义兴起之后，产生了资产阶级民主的选举制度，出现了政党。领导者走向政治舞台就必须通过政党。政党又必须有相当大的社会基础才能取得执政党的地位。被领导者的主体地位得到了一定的体现。

在社会主义社会，被领导者的主体地位才得到全面的体现。

领导者，特别是高层次领导者的挑选，既要看到领导者必备的某些个体素质的重要性，又要看到符合担任这一职务个体素质的人选是很多的，究竟选谁？这就取决于他在被领导者中的影响力和特殊的客观规律。例如，有的国家当左、右两大党势均力敌之时，国家元首往往只能是一个温和的中间人物。在关键时刻，起关键作用的不一定是领导者的个体素质，可能是被领导层的选择和客观环境的政治风向。

人们以往较多地重视领导者的个体素质研究，反而忽视了对大环境的探索，这是一种很危险的倾向。

2）领导者是领导活动的主导

领导者是主导。当人类到资本主义社会以后，领导者在领导活动中的主导作用，再也不可能只凭个人的个体素质和才干来得到。特别是对于那些政党领袖和政府首脑，必须要借助于阶级和政党的群体力量来实现其主导地位和作用。同时，他的主导活动舞台也必然受到阶级、政党的主体制约。

主导作用永远不可越过主体因素的制约，这就是主体-主导律的核心和实质。违反主体-主导律必然受到客观规律惩罚。

2. 领导活动的反馈调节律

在控制论中有一个非常重要的概念，即反馈调节。反馈是指一个系统输出信息，作用到被控制对象之后，会产生一个结果。然后，将这一结果输送回来，并对信息的再输出发生影响。这一过程就称为反馈调节。在领导活动中，第一次输出的信息是指领导者的原始决策，而被领导者产生的结果回到领导者再经过调节后才输出。这就是客观存在的反馈调节的普遍规律。

领导活动是由领导者、被领导者和群体目标组成的系统。一般说来，系统有两种调节方法：①一次调节；②反馈调节，也称二次调节。

通常，领导活动过程中，一次调节甚多，常常是第一次决策就正确。但是不正确的决策也很多，"长官意志""官僚主义""主观主义"等均属此例。领导者总是习惯于把自己最初的决定当成“圣旨”，并希望下属“理解的要执行，不理解的也要执行”。其实，一次调节的成功常常是一种特例。有些人总是希望设计出十全十美的决策，管它几十年，这显然是不科学的。真正的办法，不仅有二次调节，还有继续调节，要使领导者建立起反馈调节的领导制度，具有自动反馈调节的能力，并不断地适应和促进社会的进步。

现代的领导活动的反馈调节机制如图 1 所示。其中，A 是领导活动群体目标；B 是现代智囊机构，即政策研究机构；C 是决策机构，即领导层；D 是执行机构，即中层、基层领导和执行人员。

领导活动由上面四个部分组成一个封闭的反馈调节系统，只有这样，才能可靠地运转。在这个封闭系统中，现代智囊机构实际上是一个信息鉴别机构，而且，鉴别、决策、执行三者必须要各自独立地完成自己的任务，不能互相干扰，更不能互相取代，这样才能使受不断变化着的客观情况影响的群体目标得以实现。

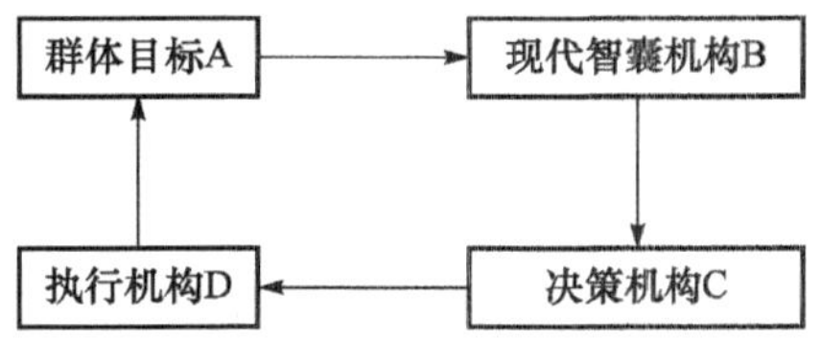

图 1 领导活动的反馈调节机制

3. 领导活动的统御力

1）统御力的定义

领导者能让被领导者发自内心的折服的能力就叫统御力。这是领导者拥有的正式权力以外的一种客观存在的影响力，即领导者不依赖领导特权，不依赖组织赋予的正式权力，也不依赖外在势力，而能说服并指导他人行动的能力。

统御力的存在是被领导者的一种心理反应，是一种客观存在，它不仅不依赖于领导者的在位与不在位，而且也不依赖于领导者的在世与不在世。杰出领袖的威望和成就被世代相传与歌颂，就反映了领导者的统御力的客观存在。

2）领导者使用两种力量

领导者在履行职责时，使用如下两种力量。

（1）领导权。任何人只要组织赋予其一个法定职务成为领导者后，就拥有相应的领导权力，这叫领导权。它由三部分组成。一是强制权。这是让被领导者感到惧怕的权力，因为不服从领导者的命令会遭到惩罚。二是法定权。这个权力来自组织授予领导者的法定地位，如省长比厅长有更大的权力。三是奖励权。这是强制权的对应物。它使被领导者认识到，完成工作职责和任务后能获得奖励。任何人无论是能干的或不能干的，只要组织赋予他成了领导者，他都会拥有上述三种权力。如果仅靠这三种权力来开展工作，一般说来，是不能搞好领导工作的。我们常常看到，只有领导权的领导者，当他在位时，下属慑于压力，只会表面上服他，一旦他离开岗位，失去领导权，就会出现“人走茶凉”现象。

（2）统御力。在领导活动中，大多数领导者不会只依赖其领导权，或多或少要依赖统御力。领导者的这种特殊才能和素质，是通过他平常的一言一行、一举一动十分自然地表现出来，而且很自然赢得人心，受人敬佩，深孚众望。这种统御力的社会价值不可估量。

3）统御力的永恒作用

统御力是领导者在领导活动中的一种客观存在的影响力。特别是对于一位杰出领导者，其统御力更是一种特别创造。经大众公认后，更成为一种精神存在和宝贵财富，它不依领导者的去向消失，甚至会随着时间的消逝，而被人们神化，永远歌颂，世代流传。

4）统御力的特殊作用

一个有统御力的人，不管他是否在领导岗位上，他对别人行为的影响一定存在。

统御力既是取得领导权的重要客观条件，又是加强领导权的重要基础。一个人若想成为领导者，其最稳妥的办法就是先干出一些成绩，以在他的周围环境中产生较强的影响力。此后，一旦他走上领导者岗位，统御力与领导权相结合，他作出的贡献就大了。

六、领导者的工作任务

领导者的工作任务包括两部分，即根本性工作和常规性工作，下面分别叙述。

1. 根本性工作

根本性工作包括以下四项。

1）规划目标

所谓目标，即是指在一定条件下，希望达到的结果。

目标包含短期目标、中期目标和长期目标。制定目标必须科学规划，必须正确和明确，这是领导者的首要职责。

规划的目标要集中体现人民群众的愿望和要求。首先，目标的水平高低要适当。规划的目标太高，可望而不可即，会使人民群众失去奋斗的积极性。规划的目标太低，则无法调动人民群众的积极性。其次，目标的范围宽窄要适当。规划的目标太窄，则可能失去环境变化带来的机会。规划的目标太宽，则因人、财、物力条件的限制而难以实现。

按性质分类，可将规划的目标分为以下六种。

（1）升位目标。即本单位在本系统在新的一年里由前一年的第几位上升到第几位，或保持在第几位。制定目标时，需要注意本单位的历史状况，注意全国同行同业同类先进单位以及国际同类单位的先进水平。

（2）实事目标。

（3）效率目标。即对工作的数量、质量、完成时间提出具体要求。

（4）成果目标。

（5）经济效益目标。

（6）规范目标。有些事情不容易提出具体目标，如企业素质，则可从程度上提出规范要求。

设立目标时，应从大系统开始，然后向子系统展开，以形成一个目标网络。

现实中，有些领导者尚未提出自己工作岗位的目标，都先要求下级部门提出奋斗目标，美其名曰走群众路线，这其实是领导的严重失职行为。

目标规划必须由领导者亲自抓。规划后，要交群众讨论，为大多数群众所接受，这样的目标成为群众意愿的集中表现，所以又可称为群体目标。这样，群众和领导者就会一起为实现群体目标而努力。

2）建立组织

提出目标后，就要根据目标建立合理的组织机构，以保证目标的实现。

组织机构包含一个中心（决策中心）和三个系统（执行系统、监督系统、反馈系统）。

3）制定规章制度

仅有组织机构而无合理有效的管理章法，即规章制度，工作就很难做好，规划的目标也难以实现。所以，一定要搞依法管理，而且要严格执行规章制度。

制定规章制度必须面向绝大多数群众，而且让他们乐于接受，这样才能真正发挥规章制度的作用。

一个单位的规章制度很多，绝大多数由执行部门制定。

领导者必须亲自动手制定的规章制度有如下两个。

(1) 岗位责任制。从领导者到每一个员工都要明确自己的责任。每个岗位都有明确的责任，才能使管理系统有效地运转起来。

(2) 赏罚法。要使下属按各自岗位要求积极工作，就必须有赏罚法。赏罚分明才能产生巨大的管理动力。必须严格执行赏罚法，才能推动管理前进。

管理的赏罚必须重视一时一事，不能让下属等待承袭积累才有结果。处理人之过错时要特别谨慎，一定要有实据时才能批评和处理，要实行“疑罪从无”原则。

评价一个人不可仅看一时一事，要看全面，要注意全部历史。否则，一过毁终生之风一长，自暴自弃的人就会越来越多。

4）选用人才

这是一项根本性的工作，在随后将专门叙述。

2. 常规性工作

常规性工作包含以下五项。

1）决策和决断

决策和决断是领导者的一种职能。大政方针一定要按科学决策进行。较小问题则要及时和明确决断，不能决断就不能领导，优柔寡断不可能是好的领导者。

现代领导者并不需要包揽一切进行决断，仅对以下两种情况进行决断。

(1) 发生非规范性事件。现实生活十分复杂。偶然的、例外的情况常常会发生。例如，上级下达一项临时任务，到底是让办公室管理，还是让人事科负责，这就需要领导者决断，而不能下放，以免扯皮。

(2) 下级请示的重大问题。重大问题一般是指大系统全局性的问题。下级需向上级领导请示，上级领导需要及时作出决断。

需要注意，一个只会上交问题而不提出建议的下属，那是不称职的下属，应考虑撤换。

下属请示的建议方案不能仅是一个，应该至少两个，并要列出各个方案的利弊，或至少附有反对者的意见。

决断必须及时、明确。不及时，不仅会耽误了处理问题的最佳时间，而且会使处理难度、成本加大，还会使问题越积越多，更趋复杂化。越是基层领导者，碰到的问题越直接，决断越要及时。凡是能当场决断的，绝不能推到以后。决断错了怎么办?关键是要及时反馈，根据反馈，及时纠正。不及时处理，就会酿成错误。

决断不明确，不仅会使下属无所适从，而且会使下属误解，更使问题复杂化。

有些问题比较复杂，决断及其实施将会遇到各种阻力。此时，领导者不可固执己见，要允许妥协。妥协只是暂时的，目标才是根本。没有目标的妥协，那是随波逐流；只有目标坚定，才有最大的灵活性。

2）指挥

确定了目标后，领导者就要推动下属和群众执行自己的决定，努力完成为实现目标而分配的任务，这就是指挥。指挥是领导者更具体、更直接的职能。

带有强制性，这是指挥的一般特性。在阶级利益根本对立的社会中，指挥是强迫，是压服。在社会主义制度下，是靠必要的强制和细微的思想工作，是靠党和国家的力量，是靠法律力量，以及领导手中的赏罚等权力。此外，还要靠群众的自觉。

一般说来，指挥有三种方式。

(1）命令的方式。在军事和现代社会中，在一切事关全局时间紧迫的关键时刻，指挥表现为命令。此时，有令必行，有禁必止，必须服从，如飞机起飞、火车行进等。

(2）说服的方式。向下属布置任务时，不但要讲清道理，善于听取下属意见，而且在某些问题上要有弹性，允许下属根据实际情况相机处理。

(3）示范的方式。领导者表扬先进，推广正确的做法，或者自己作出表率，带头工作。这是无声的命令，实际的指挥。

3）协调

协调是指领导者通过及时地调整，使各项工作、各个部门和谐地配合，以便完成任务、达到目标。

领导者在于统筹全局，使用协调方式使各个部门工作步调一致，不要互相矛盾。

4）调查研究

规划目标、制定规范、选用人才，实行指挥等，都需要进行调查研究，这是做好领导工作的基础。

5）联系群众

深入基层，深入群众，与群众建立广泛的联系。只有这样，才能了解下情和实情，才能施展领导才能，实现目标。

七、领导者的决策

1. 科学决策的含义

决策是领导者的基本职能。决策作为一门科学，是从20世纪50年代开始发展起来的，原意为“作出决定”“决定政策”。凡有领导，就有决策。

从前的决策一般是凭借领导者的知识、智力、阅历、经验进行。例如，一些正确的决策与客观实际能很好符合，诸葛亮提出的“隆中对”战略思想便与以后魏、蜀、吴三分天下的实际相符合。

我们新中国前 30 年出现的几次战略性决策的重大失误也证明了科学决策的重要性。20 世纪 50 年代，错误地批判了马寅初教授的新人口论，其后果是造成我国人口成倍快速增长，造成现在人口众多负担沉重的代价。1958 年“大炼钢铁”，损失严重。1966～1976 年，“文化大革命”，又损失严重，让国家经济濒于崩溃。

因此，对重大问题单凭经验来决策，不用科学方法和科学程序，那是不行的。决策科学化，就是为了使领导者决策与客观实际相符合。现实社会迫切需要从经验决策发展到科学决策。

科学决策是指领导者凭借科学的决策机构，运用科学的思维方法和决策技术，遵循科学的决策程序，在充分和自由讨论的基础上作出科学的决定。

2. 衡量标准

具备以下四个条件，就属于科学决策。

1）明确目标

作为一个正确的、科学的决策，应该有一个明确的目标。没有一个目标，就无从决策。

实例：从现在到 2000 年，我国国民经济总产值翻两番，达到小康社会，人均收入 800 美元，这就是一个目标。

2）目标是优化的目标

目标要求低，那不是科学决策；目标要求太高，实现不了，那也不是科学决策。目标是要经过相当的努力，且能够实现的，即称为可行的，并有如何实现的优化途径的方案，才能称为优化的目标。

3）目标要实施

一定要将目标付诸实施。仅仅纸上谈兵不能算科学决策。

4）多方案决策

决策时必须要有至少两个方案，也就是要进行多方案的决策。没有比较，就没有选择，单方案无法进行选择。如果下级只交来一个方案，则退回去重新做。

3. 决策程序

科学决策要经过下列八个步骤。

1）发现问题

决策通常是在发现问题前提下才提出来的，如宝山钢铁厂的筹建问题。国家和上海地区建设需要优质钢材，但钢厂的固定资产投资需要 207 亿元，而上海的整个固定资产才 200 亿元，这么大的企业要建吗?

2）确定目标

通常目标能用数量表示，不应是口号式的，如宝山钢铁厂，投资 207 亿元，钢的年产量 600 万吨，以及各种类型钢多少吨。

3）确定价值准则

确定价值准则是为了落实目标，以便作为以后评价和选择方案的基本依据。例如，一位青年要选择对象，马上就会建立一个价值准则，会用此准则去衡量对象。

价值准则的作用首先就是把目标分解为若干层次的、确定的价值指标，即细化目标。总目标很粗，是粗线条，如国民经济目标。然后，再考虑这一目标能否做到？怎样算才算做到？

价值准则一般说来包括三类：学术价值、经济价值、社会价值。

4）研制方案

提出的方案要 2 个以上。方案多便于选择，便于优化，当然也有限度。

实例：20 世纪 60 年代初，苏联赫鲁晓夫主席将导弹运到古巴，对准美国。智囊团拿出的解决问题的决策方案一共有 6 个：①不理不睬；②派飞机炸掉导弹基地；③派军队占领古巴；④通过联合国，用舆论迫使苏联撤走导弹；⑤与古巴卡斯特罗谈判；⑥封锁古巴。

5）分析评估

领导有了多个方案后，要对这些方案进行分析评估。

实例：仍是前一例子。肯尼迪得到上述 6 个方案后，便请国务卿、中央情报局局长以及美国的一些智囊团体（兰德公司、对外关系委员会等机构的研究专家）进行分析评估。根据各方案的优缺点，分析如下。①方案显得太软弱，美国是一个大国，导弹放在门口，不能不管。②和③方案有效，但冒险太大，有可能引起第三次世界大战，故不宜实施。④和⑤方案是制造舆论，进行谈判，有可行性，但时间可能拖得太久。夜长梦多，万一古巴发射一枚导弹就麻烦了，故这两个方案也不宜用。⑥方案，是一个中性方案，利多弊少，既不太激烈，又不太软弱，是一个给出路的方案。只要古巴把导弹撤走，就撤销封锁，这是给出路的，不是把门堵死，不是把人家推到绝路上，这是一个比较好的方案。于是，各方案的优缺点都评价出来了。

分析评估各方案时，如果能建立数学模型，就尽量这样做，这就是所谓的定量分析，并可用计算机算出数值结果。如无法做定量分析，则进行定性分析。

6）方案选优

对各种方案评价后，就要由领导人进行总的决策，这就是方案选优。

领导者是决策的人。专家智囊只是辅助决策，但不能代替决策。

实例：还是上一个例子。美国中央情报局局长把 6 个方案评价交给肯尼迪，肯尼迪与大家意见一致，选了⑥方案，于是，马上调动第六舰队，包围古巴。古巴遭到封锁，对外贸易立即被掐断，弄得焦头烂额。只用了五天时间，赫鲁晓夫就把导弹撤走了。国际上反响很大，认为把导弹弄去是冒险，而撤走又是逃跑。结果，赫鲁晓夫在决策失误后不久便在苏联下台了。

领导人决策后，用不了多少时间，便有成功与失败的结果，真是肩负千斤。决策过程中，领导者的素质很重要。

7）试验验证

试验验证通常就是指试点。

有一些决策，特别是一些比较大的政策性问题的决策，就必须先在局部地区进行试点。

通过试点，总结经验。一方面，看一看这样做，行不行。如果行，总结一些经验再推广。如果不行，便反馈回去，看一看哪个步骤需要修正，这就是决策修正。如果反馈回去，发现原来目标都搞错了，就要从目标开始，再重新进行决策，这就叫追踪决策。

实例：1981 年，国家财力不够，提出了“调整、改革、整顿、提高”的八字方针，要进行调整。当时全国 23 个大项目要调整，其中最大的项目就是上海宝山钢铁厂。因为我们一向都搞一刀切，如果上海宝钢不下马，其他都下不来。宝钢第一期工程要用 126 亿元，此时已用 90 多亿元。将近 100 亿元的设备已经搞好了，高炉砌了三分之一，厂房也盖了不少，发电厂、水厂、道路都已建好。如下马不搞了，这 90 多亿元就付之东流了。这样大的损失是世界上任何一个企业、任何一个工程都无先例的。所以专家们从实际出发，从国家财力考虑，提出了“缓中求活”方案，就是国家再花 26 亿元，按五年“4，5，7，7，3”投资，到 1985 年结束第一期工程。中央批准了这一方案，最后上海宝山钢铁厂建设成功了。

8）普遍实施

经过试验实证，可行的话，决策就进入了普遍实施阶段。在这一阶段，要尽可能把有些东西规范化，建立一些条例、法规，形成法治概念。

经过上面 8 个过程，决策就做到科学化了。

实际上，不是所有决策都要进行 8 个步骤。有时，可以合并一些，简化几个步骤。

决策时，还应选用科学方法进行分析研究。目前，已有以下 6 种方法为决策服务：调查研究法、预测研究、环境分析法、智囊技术、决策技术和可行性研究、效用理论法等。

八、领导者的选人和用人

领导者的根本职责就是要正确地选用人才[6]。

1. 领导者首先要树立科学人才观

1）人才是最宝贵的财富

按《新华字典》定义，人才是指有品德有才能的人[7]。换句话说，人才是具有优良品德、具有一定知识、具有一定技能，并能进行创造性工作的人。他们比一般人对社会创造的价值大。

实例：1923 年，美国福特公司一台大型电机出现故障。公司总经理请本公司的工程师们进行修理，四个月都无结果。随后请到一位留居美国的德国科学家斯特曼斯，他来工厂后，即在电机旁搭起帐篷，连续两天，进行诊断测量。最后，在电机上画了一条线，并要求打开电机，把记号处的线圈去掉 17 圈，结果电机故障排障，正常运转。斯特曼斯向总经理要报酬 1 万美元。有人说，这么一条线就值 1 万美元，是勒索。总经理回答：“用粉笔画一条线值 1 美元，知道在哪里画线值 9999 美元。”

这就是知识和人才的价值。

2）人才是事业成功的关键

事业要成功，关键在人才。从古至今，莫不如此。例如，我国春秋战国时代，诸侯争霸，要成功，首先就需要争取人才，得人才者得天下，事例很多。

3）我国人才资源有待开发

心理学家研究，在全体人口中，智力非常优秀者占1%。中华民族的智力在世界各民族中有突出地位。从美国科技界看，全美第一流科学家、工程师约有12万～13万名，其中中国人约有3万名，约占25%。而在全美人口中，中国人仅占0.5%。

我国人才资源是十分丰富的，有待开发。

2. 选人原则

领导者选人时需要按照下面三个原则：德才兼备、以德为主；专业内行；具有开拓创新能力。

3. 用人原则

领导者用人时需要应用下面三个原则。

1）扬长避短

每个人都有优缺点，要注意用其所长，避开其短处。特别是，一个人的才能越大，特长就越突出，其短处越明显，各方面的议论也越多，所以更要用好，发挥其长处。

实例：美国南北战争时，林肯总统先按照没有大缺点的用人标准，任命了几位将领。虽然有较强的人力和物力，但是却被打败了。他反观南方用将，将领们虽然各有大大小小的缺点，但都有特长，又用其所长。于是，他毅然起用格兰特为总司令，命令一出，舆论哗然。说格兰特好酒贪杯、难当大任。林肯笑着回答："如果我知道他喜欢喝什么酒，我还准备送他几桶酒。"林肯知道他好喝酒，但更知道他具有运筹帷幄、决胜千里的能力。果然，他打了胜仗。

所以，作为领导者，宁可用有缺点的人才，决不能用"无缺点"的庸才。

2）量才任职

领导者要按人的才能高低，安排相应职务。量才任职，才能使下属充分发挥才能，使事业成功。

实例：古代楚汉战争时，韩信投奔刘邦，而刘邦只封了一个管粮草的小官给他，把他气跑了。萧何识人才，向刘邦力荐。于是刘邦拜韩信为大将，最终打败了项羽。"萧何月下追韩信"讲的就是这个故事。

3）科学匹配

领导者组织领导班子时，一定要注意人才的科学匹配，要把不同学识、不同智能、不同专业、不同年龄、不同气质的个人，按互补原理组成一个具有合理结构、能合力做事的集体。这就是说，领导班子的结构要科学化。

（1）学识结构。班子内高、中、初级知识人数的比例以6∶3∶1为宜。

（2）智能结构。班子内成员不能都是帅才型的人，应由帅才型、将才型、智囊型的成员按一定比例组成。

（3）专业结构。班子成员不能只由一种专业人才组成，应由各种专业，如理科、工科、财务、文科等专业的人组成。

（4）年龄结构。班子成员应有老年人，他们经验丰富。还应有中年人，他们是中流砥柱。也应有青年人，他们勇于创新。

（5）气质结构。气质就是人的脾气、性格。一个领导班子成员的气质结构应能使领导班子成为朝气蓬勃、奋发向上的团结和谐的集体。

人的气质分为四种类型：胆汁质（急躁型，性情急躁、动作迅猛）、多血质（活泼型，性情活跃、动作灵敏）、黏液质（胶滞型，性情沉静、动作迟缓）、抑郁质（稳重型，性格脆弱、动作迟钝）。因此选人时，要注意成员气质的搭配，不可使班子成员是一种气质。作为班长，更应具有广阔的胸怀，善于团结不同性格的人，以便共同工作。俗话说，“宰相肚里好撑船”，领导班子的班长一定要有“宰相”的这种气质。

九、领导艺术

在一定知识、经验同创造性思维相结合的基础上的领导技能就称为领导艺术。三国时，诸葛亮的空城计，就是领导艺术的体现。

领导科学是对领导经验的理论概括，是对领导艺术的规范化。领导艺术是领导科学的基础。

领导艺术分为以下五种。

1. 处理人际关系的艺术

1）表扬的艺术

表扬的形式有当众表扬、个别表扬、背后表扬、集体表扬等，不同的形式具有不同效果，要善于使用。例如，背后表扬往往比当面表扬更好。

表扬的时机有及时表扬、集中表扬、年终表扬、长时间后的表扬等，时机抓得好，表扬就有效果。

表扬的方式要因人而异。对长者要尊重；对年轻人，语言可夸张；对性格机敏者，两三句话便心领神会；对疑虑心情者，则可用轻松的谈话方式。

表扬的内容要选好，要有表扬的侧重点，可表扬人，也可表扬具体事。

2）批评的艺术

领导者对下属要善于用好批评的武器，应使用“激励为主，处理为辅”的原则，即是说，表扬为主，批评为辅。

批评包括当面单独批评、当众点名或匿名批评、转达批评，不同批评有不同的效果。

3）谈话的艺术

领导者与下属谈话的艺术十分重要，效果可迥然不同。谈话的方式包括讲演、小型座谈会、几个人和单独两人面对面等形式，也包括热情洋溢、诙谐、严肃等方式，不同场合应有不同的谈话方式。

2. 提高工作效率的艺术

1）领导者必须干领导的事

一个好的领导者，一定要懂得领导者应该做哪些事情。决不能去代替下属和群众做事，决不能搞“事必躬亲”。

2）授权办事的分身术

一个优秀的领导者，要善于授权让下属去办事，决不要包揽一切，而且还要做到“大权不旁落”，让授权分身办事成功。

3. 提高会议效率的艺术

需要用会议来解决问题时，领导者要注意开会的会场形式，大会或小会形式、讲演式或座谈式等。不同的开会方式会有不同的效果。

4. 运筹时间的艺术

领导者要使所领导的事业成功，一定要精心安排自己的每时、每天、每周、每月、每季、每年、五年、十年，直至几十年的工作时间，要用好时间，才能使事业成功。有些事情要早办，而有些事要晚办；有些事情要大办，而有些事情要小办；有些事情要急办，而有些事情要慢慢办。千万不能事事一样办。

5. 领导方法的艺术

领导者应根据工作任务的情况，按轻重缓急、时间远近、大小不等进行分类，再选用方法，各个处理。切不可用一种方法解决所有问题。我们社会，喜欢用“一刀切”方法处理问题，这往往造成诸多后患。

参考文献

[1] 辞海编辑委员会. 辞海. 上海：上海辞书出版社，1980.
[2] 徐联仓. 关于“领导”的科学研究. 自然辩证法通讯，1981，3（3）：8-10.
[3] 夏禹龙，刘吉，冯之浚，等. 领导科学基础. 南宁：广西人民出版社，1983.
[4] 钱学森，李宝恒，杨沛霆，等. 现代领导科学与艺术. 北京：军事译文出版社，1985.
[5] 中共上海市委党校《领导科学》教学组. 领导活动案例选. 1987-03.
[6] 刘人怀，斯晓夫. 工业企业管理岗位要素设计. 北京：机械工业出版社，1990.
[7] 新华词典编纂组. 新华词典. 北京：商务印书馆，1989.

城市政府工作目标管理与治理整顿、深化改革[①]

治理整顿、深化改革是我国的重要任务，实行城市政府工作目标管理有助于这个任务的完成。城市政府工作目标管理是在我国经济政治体制改革中应运而生的，而它的产生和发展又促进了经济政治体制改革。在治理整顿、深化改革中，城市政府目标管理进一步充实和完善。

一、城市政府工作目标管理是改革的必然要求和重要内容

经济体制改革在农村取得了巨大成就后，转入以城市为重点。城市是我国经济、政治、科学技术、文化教育的中心，是现代工业集中的地方，在社会主义现代化建设中起着主导作用。只有以城市为重点进行经济体制改革，才能使城市经济兴旺繁荣，才能适应对内搞活、对外开放的需要，真正起到主导作用，推动整个国民经济更好更快地发展。城市经济是国民经济的重要组成部分，城市改革在全国经济政治体制改革中具有特殊地位。城市政府作为国家的一级政权组织，具有承上启下的特殊作用。以城市为重点的经济体制改革必然要求城市政府管理工作实行改革。

城市工作内容是极为丰富的，从行政管理来看，有市、区（县）、街的管理；从经济管理来看，有对工业、农业、商业、财税、土地、物质等的管理；从社会发展来看，有对城市建设、市容、科技文教事业等的管理，而它们之间又交叉、交流、交融。城市是一个复杂的开放系统，城市改革包含商业物资改革、企业管理改革、街区工作改革等。对于一个城市来讲，城市政府管理工作改革尤为重要，它是其他各项改革的基础和前提。经济政治体制的各单项改革汇聚在城市，城市政府管理工作改革将对各单项改革有着明显的综合、协调和推动作用。因此，实行城市政府目标管理是城市改革的重要内容。

二、在实行城市政府工作目标管理中治理整顿、深化改革

1. 实行目标管理是对无人负责和官僚主义的打击

由于种种原因，在城市各级政府工作中，往往存在无人负责的弊病，名之曰“集体负责”，实际上无人负责。由于责任不明确，工作推诿，该拍板的无人拍板，出了问题找不到人承担责任。官僚主义作风普遍存在，而又无制度、法规加以有效约束。结果是工作受损失，群众不满意。实行城市政府工作目标管理，科学构造目标体系，确定各级各类各项目标，实行目标分解，民主决策，反复研究，责任到人，定期考核评审，奖惩兑现。这就使政府工作制度化，强调了岗位责任，改变了无人负责的状况，打击了官僚主义作风。

---

① 本文原载《学习与实践》，1990，(11)：35-36.

2. 实行目标管理增加了城市经济和社会管理工作的透明度，有利于加强廉政建设，密切党和群众的联系

过去在城市经济社会管理工作中，搞的计划本子，城市各级管理机构中的许多干部不清楚，普通市民就更难知道了，不仅不能有效调动各级干部和广大市民的积极性，而且造成一些不必要的麻烦、误解和不满。实行目标管理以后，许多城市都立足于调动全市人民的积极性，每年的目标年初见报，公布于众；季度检查，一一对照；半年通报，年终总结评比。这样，既增加了城市经济社会管理工作的透明度，又调动了全市干部和市民的积极性，密切了党群关系。

3. 实行目标管理，注意中期和短期目标衔接，克服短期行为

治理整顿、深化改革需要克服“过热”现象和短期行为。历来考核政绩，主要看产值等数量指标。市长又是有限任期制。因此，普遍存在着急于求成、显示政绩的短期行为。这种“过热”现象和短期行为往往影响国民经济与城市经济的持续、稳定、协调发展，造成大起大落和后劲不足。在实行城市政府工作目标管理中，强调年度目标和中期目标结合，主张年度计划与中期计划（五年计划）衔接，着眼于中期目标，据以制定每年的城市发展目标，使中期目标年度化，注意目标的连续性，这就较好地避免了政府的短期行为，使政府工作目标不因市长人选的变易而发生突然的变化，保持了政府工作目标的延续，从而使城市的经济社会事业得到稳步发展。

4. 实行目标管理有助于政企分开

过去城市政府机构臃肿，职能庞杂，且又政企不分，政府往往直接干预企业工作，成了“政府办企业”。企业有许多需要由社会解决的问题又得不到解决，不得不“企业办社会”。政府和企业不能各司其职、各负其责、工作做不好，大家都苦恼。实行目标管理后，政府工作的目标是政府职能范围内的管理工作，即使是一些与企业有关的目标，也都从直接管理转向了以间接管理为主的轨道。这就使政府能“站得高，看得远”。政府从一些烦琐的干预企业的杂务中解放出来，而把注意力集中在发挥和完善市场的整体功能、研究和制定城市的经济社会发展战略上。企业也能在政府的帮助下，自如地处理自己的工作，把抓生产和抓改革结合起来、抓产值和抓效益结合起来，使企业能在改革中得到壮大，从而有助于整个国民经济的发展。

5. 实行目标管理有助于国民经济持续、稳定、协调发展

推进经济政治体制改革，完善社会主义体制，充分调动人民群众的积极性和创造性，促进国民经济走上持续、稳定、协调发展的轨道，这是我们面临的任务。城市政府工作在完成这项任务中起着重要作用。实行城市政府工作目标管理，城市在确定目标体系时，既有经济管理的目标，又有社会发展的目标。一般说来，有综合经济指标（如国内生产总值、国民收入、社会总产值等指标），工业生产指标（如工业总产值、企业技术改造、新产品开发等指标），农村经济指标（如主要农副产品产量、副食品生产基地建设、农田水利建设等指标），城市建设指标（如城市总体和分区的专项的规划、旧城改造和新区开发、园林绿化等指标），商业流通指标（如社会商品零售总额、重要商业设施建设等指标），社会事业指标（如教育、卫生、文化、体育事业发展等指标）等。这些指标可以综合反映经济增长、科技发展、城市功能、

社会事业和人民生活等方面的内容和要求。城市政府工作的目标，既有速度目标，又有效益目标；既有年度目标，又有中期目标；既有微观（城市）目标，又有宏观（国家）目标。它们的结合，有助于提高城市的整体功能和促进国民经济持续、稳定、协调发展。

由此可见，实行城市政府工作目标管理，是治理整顿、深化改革的重要内容。同时，也只有在治理整顿、深化改革中才能进一步充实和完善城市政府工作目标管理的目标体系、目标决策与组织实施，推进城市政府工作目标管理。

实施城市管理系统工程建设　开创广州可持续发展新格局①

一、城市科学管理的重要性

1993年，广州市领导决策机构明确提出，要用15年左右的时间，在21世纪初叶将广州建设成为现代化大都市[1]。改革开放以来，广州由于其政策、地理和人缘优势而获得可喜的发展。然而，随着改革开放的深入以及上海浦东的开发，这些相对优势已经明显减弱，发展呈现疲态，后劲乏力[2]。因此，如何利用已经取得的成果，培育新的突破口，推动广州第二次发展，已经成为一个突出的问题。

一个城市，尤其是现代化大都市是一个复杂系统，它的生存与发展遵循系统原理及规律。因此，对城市的科学管理是一项复杂的系统工程，应当自觉运用系统工程原理和方法逐步建成一个完善的城市管理系统。系统工程理论已经证明[3,4]，一个系统的演化与发展主要有两种途径：一是系统环境的推动；二是系统内部良性正反馈结构的运动和推动。这时系统具有决定和支配自身演化与发展的能力。广州过去十多年的发展主要是城市系统环境推动的结果。然而，由于广州过去所拥有的政策、地理和人缘优势已经明显衰减，城市系统环境在推动广州持续发展方面已显得力不从心。本文拟运用系统工程的原理，通过对广州城市系统分析，揭示广州可持续发展的突破口；在此基础上，探讨城市系统内部能够形成的良性正反馈结构，提出广州可持续发展的新格局；最后，研究如何通过实施城市管理系统工程建设，开创这一新格局。笔者在此谨以一个广州市民的名义，将这些建议提供给市领导参考。

二、广州城市系统分析

1. 广州城市系统目标分析

广州市已经提出在15年内实现现代化的目标。通过目标分解，我们得到若干子目标，如图1所示，其中，城市管理现代化、城市经济现代化和城市人口素质现代化是三个主要子目标。

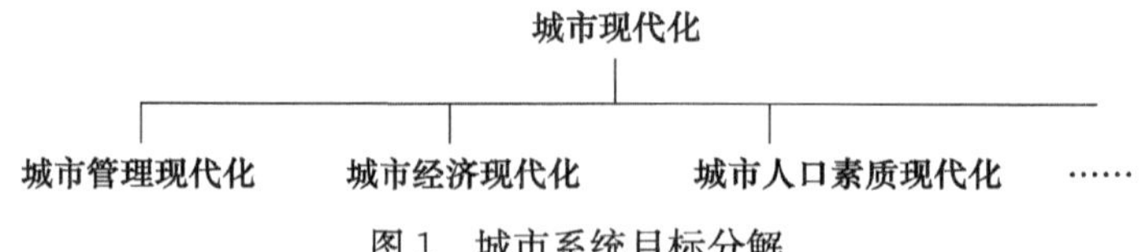

图1　城市系统目标分解

根据子目标与目标的关系，我们知道，只有所有的子目标都实现了，才能保证目标的实现。因此，我们面临这样一个问题，即如何开创一种优化的城市发展格局，保

① 本文是穗港澳地区十二所高校校长与广州市领导恳谈会上的主题报告，广州，1997年12月3日；羊城科技报，1998年2月8日，第4版.

证这些子目标，特别是三个主要子目标同步、协调发展。广州过去十多年的发展主要集中在城市经济现代化建设上，在城市现代化建设的初级阶段，这样做是必要和合理的，而且，没有城市的经济现代化，要实现整个城市现代化也是不可能的。然而，如果今天广州仍然片面地强调城市经济现代化建设而忽视了其他子目标，特别是城市管理现代化的建设，必然产生城市资源配置不合理、城市资源严重浪费以及其他问题，导致城市畸形发展，而畸形发展往往是打乱了系统的内部平衡，是不可持续的，其结果是我们最终要为矫畸付出沉重的代价，使城市现代化建设遭受波折。现在，我们已经到了必须高度重视和认真研究这一问题的时候了。

2. 广州城市系统环境分析

系统工程理论已经证明，一个系统的目标能否实现与系统环境密切相关，如图 2 所示。

图 2　系统与系统环境

系统环境是指与系统存在关联的系统外部部分的集合。有利的系统输入将推动系统的进化和系统目标的实现，而合理的系统结构调整将引起有利的系统输入变化。对于一个城市系统，其系统输入主要包括国家政策，国家和周边地区人、财、物及信息的输入。过去十多年，广州城市系统输入具有明显的优势，如国家的优惠政策、大量的资金流入，以及源源不断的大批高素质人才的流入。广州过去十多年的发展主要是这些有利的城市系统输入推动的结果。然而，随着改革开放的深入，其他城市和地区的开放都获得与广州相同的优惠政策，同时，港澳地区资本在流入广州的同时也流向其他开放地区及内陆城市，甚至以数倍于流入广州的速度流向了珠江三角洲的其他城市，并且，人才流入数量也在减少。进入 20 世纪 90 年代，广州城市系统输入优势已经明显减弱，发展呈现疲态，后劲乏力。面对这种态势，根据系统工程原理，广州可持续发展的唯一途径是合理调整城市系统结构，形成良性正反馈结构。这样，一方面使城市系统具有决定和支配自身可持续发展的能力，另一方面使城市系统输入发生变化，再造城市系统输入优势，这对于广州能否在 15 年内实现现代化的目标具有决定性的影响。

3. 广州城市系统结构分析

我们知道，一个城市是一个复杂系统，它由许多部分，如各种城市资源和设施、各种产业，以及交通、治安、环保、文教、科技、卫生、人才与人的素质等组成，这些组成部分之间存在复杂的相互联系、作用并相互制约，构成复杂的城市系统结构。根据系统工程原理，在一定的环境中，一个系统的目标能否实现完全由系统结构决定。因此，在目前外部条件下，广州能否在 15 年内实现现代化的目标完全由城市系统结构决定。这意味着，寻求合理的城市系统结构是保证广州可持续发展并在 15 年内实现现代化目标的唯一途径。

广州市已经把旅游、商业、房地产、金融保险、交通运输和高科技制造业列为支柱产业[5]。无疑，它们是城市系统的重要组成部分，直接承担着经济现代化建设的任务，它们的发展对于城市系统的发展具有重要的影响。广州在支柱产业方面有很大的优势和潜力。广州作为中国的南大门和华南地区的政治、经济、文化中心，并且毗邻港澳，它在发展旅游、商业、房地产、金融保险、交通运输业方面具有明显的优势，特别是珠江三角洲的快速发展以及港澳的回归，为广州的旅游、商业、房地产、金融保险、交通运输业的发展开拓了广阔的前景；同时，广州今天已经拥有的经济实力也为高科技制造业的发展创造了良好的条件。因此，广州在发展支柱产业方面无疑具有明显的优势和很大的潜力。但是，作为城市系统的组成部分，支柱产业与城市系统的其他组成部分之间存在密切的联系、作用与制约，它们的发展受到城市系统其他组成部分，如城市管理、城市社会环境与秩序、城市建设、人才与人口素质等因素的影响和制约。因此，必须高度重视和认真研究这些因素对六大支柱产业的影响和制约，改善有关因素，创造有利条件，发挥支柱产业的优势和潜力，促进和保证支柱产业的可持续发展。

城市管理是城市政府的主要职能，也是城市系统的重要组成部分，它指挥和控制着城市系统各个组成部分的演化与发展[6]。在过去的城市管理实践中，我们缺乏自觉运用系统工程原理和方法管理城市系统的观念，忽略了局部最优不等于全局最优，而局部不优也不等于全局不优的系统工程原理，片面强调支柱产业的发展，结果由于影响和制约支柱产业发展的城市系统的其他组成部分没有得到协调的发展而反过来制约着支柱产业的持续发展；同时，由于没有应用现代化的管理手段，如计算机化的管理信息系统，难以及时、准确、完整地掌握相关的管理信息，导致城市管理有效性和效率低下，不能适应城市可持续发展和现代化建设的需要。特别是，广州的支柱产业是那种紧紧依赖城市管理的产业。例如，旅游、商业和房地产业的发展在很大程度上依赖城市环境和秩序，而城市环境和秩序的状况则取决于城市管理。因此，实施城市管理系统工程建设，自觉运用系统工程原理和方法逐步建成一个完善的城市管理系统，是推动广州第二次发展（可持续发展）的突破口。

人才和人口素质同样是城市系统的一个重要组成部分，它影响和制约着城市系统其他组成部分的进化和发展，特别是影响和制约着城市管理和支柱产业的发展。现代发展经济学理论认为，由人口素质决定的劳动力素质才是经济增长的决定性因素。目前，广州人才和人口素质与它在全国大城市中居第三位的经济实力极不相称。据1991年人口普查统计，广州630万人口中不识字和半识字者有94万人，占全市总人口的14.9%；大学文化程度的人口所占百分比低于国内许多城市，只有5.46%；就业人口受教育平均年限是8年，平均文化程度只有初中二年级；同时，由于广州缺少一流的高校，所以，尽管有一些人才流入，高素质人才依然紧缺。广州人才紧缺和人口素质不高对城市发展的制约作用在其粗放型发展阶段尚未明显表现出来，但是当城市发展从粗放型转向集约型时——这是城市现代化建设的必由之路，它无疑将成为城市发展的主要制约因素之一。特别是，高科技制造业被列为广州市的六大支柱产业之一，它的发展取决于高精尖人才。同时，城市管理的有效性和效率也与人才及人口素质紧密相关，良好的人口素质将极大地提高城市管理的有效性和效率，极大地改善城市环境和秩序，推动支柱产业的发展。因此，造就和吸引一大批人才，特别是拔尖人才，以及普遍提高市民的素质是广州可持续发展的重要条件。

三、广州可持续发展新格局

广州城市系统分析表明，广州要在15年内实现现代化，其若干子目标，特别是城市管理现代化、城市经济现代化和城市人口素质现代化三个主要子目标必须同步、协调发展。这意味着承担实现这些子目标的城市系统组成部分必须可持续发展。目前，广州城市系统环境优势已经明显减弱，难以推动这些组成部分的可持续发展，因此，只能依靠城市系统内部形成良性正反馈结构，开创广州可持续发展新格局。

广州在支柱产业方面有明显的优势和很大的潜力，但目前这些优势与潜力由于明显地受到城市社会环境和秩序、城市形象、支柱产业管理的有效性、效率和一体化水平（由于支柱产业之间存在紧密的相互联系和影响），以及城市人口素质的限制而难以发挥，而所有这些方面的改善都有赖于建设一个完善的城市管理系统；同时，广州要在15年内实现现代化，实现城市管理现代化是主要子目标之一，没有城市管理的现代化就不可能有城市的现代化。因此，广州应当以实施城市管理系统工程建设为突破口，极大地改善城市社会环境和秩序，改善城市形象，提高支柱产业管理的有效性、效率和一体化水平，由此带动支柱产业的良性正反馈发展，如图3和图4所示，开创广州可持续发展新格局，保证广州可持续发展，并在15年内实现现代化。

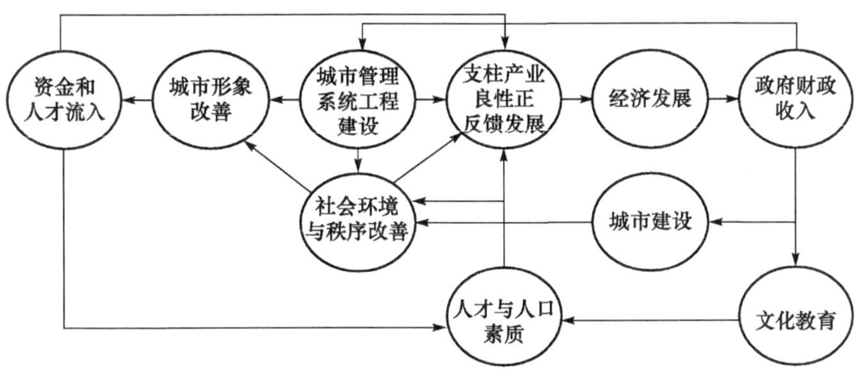

图3 广州可持续发展新格局

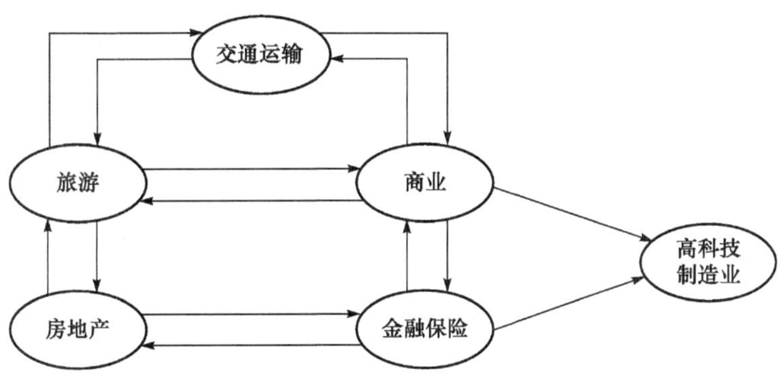

图4 广州支柱产业良性正反馈发展

四、广州城市管理系统工程建设

一个城市是一个复杂系统，开创广州可持续发展新格局是一项复杂系统工程，因此，应当自觉运用系统工程原理和方法以及现代管理手段。实施广州城市管理系统工程建设就是要运用系统工程原理和方法，以及现代管理手段，逐步建成一个一体化、标准化、程序化、半自动化、决策科学化的城市管理系统。它是一个由城市管理的系统工程观念、系统工程原理、系统工程工作方法和现代管理手段构成的系统。它的实体部分是计算机化和电台化的城市管理信息系统，具有城市交通、治安和环保监控、信息咨询、导游、导购、税收，以及城市管理数据和信息共享等功能，并且，以开发城市管理信息系统为动力，推动其建设。世界上几乎所有的现代化城市都建立了这样的城市管理系统，尽管其功能由于其不同的城市管理重点而有所不同。广州已经到了必须认真研究和建设这样的城市管理系统的时候了。

五、结论

建设这样的城市管理系统是一项复杂的系统工程，需要应用系统分解方法将它分解成若干相对独立的子系统，如城市管理宣教子系统、城市管理监控子系统和城市管理服务子系统。其中，城市管理宣教子系统具有城市管理的宣传教育、立法与执法及监督的功能，城市管理监控子系统具有城市交通、治安、卫生和环保监控的功能，而城市管理服务子系统具有信息咨询、导游、导购、税收以及城市管理数据和信息共享等功能，分别建设；然后根据子系统之间的相互影响和关系，对子系统进行综合协调，构成一个整体优化的城市管理系统。目前，广州迫切需要进行城市管理监控子系统和城市管理宣教子系统的建设，极大地改善城市社会环境和秩序，改善城市形象，提高支柱产业管理的有效性、效率和一体化水平，由此带动支柱产业的发展、外部资金和人才的流入，开创广州可持续发展新格局。

当然，要管理好城市，还要大力加强法规建设，不搞人治，要依法治市，依靠日臻完善的法规规范政府的行为，并以此管理好城市。

应用系统工程原理和方法研究广州可持续发展问题，取得如下主要结论。

（1）广州要在15年内实现现代化，必须开创一种优化的城市发展格局，保证其若干子目标，特别是城市管理现代化、城市经济现代化和城市人口素质现代化三个主要子目标同步、协调发展。这意味着承担实现这些子目标的城市系统组成部分必须可持续发展。

（2）广州城市系统输入优势已经明显减弱，发展呈现疲态，后劲乏力。面对这种态势，广州可持续发展的唯一途径是合理调整城市系统结构，形成良性正反馈结构。这样，一方面使城市系统具有决定和支配自身可持续发展的能力，另一方面使城市系统输入发生变化，再造城市系统输入优势，这对于广州能否在15年内实现现代化的目标具有决定性的影响。

（3）广州应当以实施城市管理系统工程建设为突破口，极大地改善城市社会环境和秩序，改善城市形象，提高支柱产业管理的有效性、效率和一体化水平，由此带动支柱产业的发展，并形成城市系统内部良性正反馈结构，如图 3 和图 4 所示，开创广州可持续发展新格局，保证广州可持续发展，在 15 年内实现现代化。

(4) 实施广州城市管理系统工程建设就是要运用系统工程原理和方法以及现代管理手段，逐步建成一个一体化、标准化、程序化、半自动化、决策科学化的城市管理系统。它是一个由城市管理的系统工程观念、系统工程原理、系统工程工作方法和现代管理手段构成的系统。它的实体部分是计算机化和电台化的城市管理信息系统，具有城市交通、治安和环保监控、信息咨询、导游、导购、税收，以及城市管理数据和信息共享等功能，并且，以开发城市管理信息系统为动力，推动其建设。

目前，广州迫切需要进行城市管理监控子系统和城市管理宣教子系统的建设，极大地改善城市社会环境和秩序，改善城市形象，提高支柱产业管理的有效性、效率和一体化水平，由此带动支柱产业的发展、外部资金和人才的流入，开创广州可持续发展新格局。

参考文献

[1]　伍亮. 1997 广州经济形势分析与预测（经济蓝皮书）. 广州：广州出版社，1997.

[2]　夏泉. 国际大都市与广州高等教育. 开放时代，1995，(5)：72-75.

[3]　林福永，吴健中. 一般系统结构理论及其应用（Ⅰ）. 系统工程学报，1997，12（3）：1-10.

[4]　林福永，吴健中. 一般系统结构理论及其应用（Ⅱ）. 系统工程学报，1997，12（4）：11-20.

[5]　刘人怀. 旅游工程原理与实践. 上海：百家出版社，1991.

[6]　刘人怀. 城市政府工作目标管理与治理整顿、深代改革. 学习与实践，1990，(11)：35-36.

建议重视我省仪器仪表工业的发展，以迎接21世纪挑战[①]

从100年世界科学技术和工业发展的历程可以看出，前半个世纪的许多重大发现、发明为后半个世纪的高新技术产业的兴起奠定了坚实的基础，并使当今世界发生了翻天覆地的变化。未来的21世纪，人类将由工业化时代全面进入信息化时代。这个时代的特征是以计算机为核心延伸人的大脑功能，扩展人的脑力劳动。

仪器是认识世界的工具，而机器是改造世界的工具。仪器和机器同等重要。仪器起着重要的信息源作用。

信息技术由测量技术、计算机技术和通信技术三部分组成。而测量技术则是关键和基础。现在人们一提信息技术，就是指计算机技术和通信技术，而关键的基础性的测量技术却往往被人们忽略了。它是重要的信息的源头技术。它代表着科技的前沿、工业的前沿。它涉及所有的产业，从天上到地上的，从地上到海上的，它的发展代表着国家科技和工业的水平。它相当于人的感官系统，起着眼、耳、口、鼻、舌的功能，它是科学研究的基础，它是工业的基础。如将科研和工业比作龙的话，它就是科研和工业的龙头。

因此在发展我省信息产业时，千万不要忽略现代仪器仪表产业的发展。我省在这一方面，已有一定基础，但这些年来受市场冲击，发展缓慢，十分困难。

分析诺贝尔物理和化学奖，四分之一是因为测试方法和仪器创新得到的。也就是说，科学技术的重大成就和科研新领域的开辟，往往是以检测仪器和技术方法上的突破为先导。

因此，能不能创造新式科学仪器仪表，体现了一个民族、一个国家的创新能力。所以，发展仪器仪表工业应当作为国家的战略。

当今仪器仪表技术最引人注目的发展是在生物、医学、材料、航天、环保、国防等直接关系到人类生存和发展的领域之中。对于我省而言，着力于生物、医学、材料、环保领域将是正确的选择。

预计我国仪器仪表行业的增长速度在10%左右，全国机械系统的工业总产值到2000年达到450亿元，2001年达到500亿元。

目前，仪器仪表行业在发展中存在以下问题。

(1) 企业内部结构不合理，经济效益不高。

(2) 产品结构不合理，市场占有率不高。

(3) 科研开发力量弱，成果转化困难，企业无发展后劲。

因此，加强国有企业改革，加大投入，包括科技投入，使技术创新，进行产品结构调整，十分重要。

---

① 本文是在广东省政府主办的2000年广东经济发展国际咨询会上的讲话，广州，2000年11月16日.

加入WTO（入世）在即，必须对我省仪器仪表行业的产品结构进行调整，这要根据国际竞争力的变化，根据我省在国际市场分工中的变化，根据我省在国际市场分工中的位置变化来调整目标和方向。在世界经济日益一体化的情况下，只有具有国际竞争力的产业和产品，才能在国内外两个市场中取胜。

广东省发展高新技术的若干意见和建议[①]

中国加入WTO后，我国高新技术产业将在一个开放的世界舞台上参与竞争，接受挑战。面对世界上发达国家在技术、资金、管理和人才等方面的竞争优势，加入WTO，对中国高新技术发展从长远看是“利好”的，但从短期来看，“阵痛”是不可避免的。面临的主要挑战包括国内市场的国际化压缩了我国高新技术产业的发展空间，国外高科技产品技术优势和价格优势将使一大批创新能力弱、核心冲力不强的高科技企业难以为继；全球跨国公司大举进入中国，推行研发力量本土化的战略，中外企业对人才的争夺将从“背靠背”变成“面对面”，优秀人才的流失也将对成长中的中国高新技术企业带来巨大的冲击。

广东毗邻港澳，处于改革开放的最前沿，对外贸易占全国40%以上。广东省外向型企业多，应积极采取应对策略，争取时间和主动，在这场汹涌的竞争潮流中站稳脚跟，并发展壮大。为此提出以下几点看法。

一、推行人才战略，构筑国际化人才高地

迈入21世纪，全球经济一体化明显加速。中国加入WTO后，国际化人才竞争加剧，广东实施外向带动的战略以及率先基本实现现代化、建设国际化大都市的目标等均使广东对国际化人才的需求日益迫切。广东高新技术发展要在科技全球化的竞争中脱颖而出，快速发展，拥有国际化人才是关键。

目前，广东高新技术企业要走向世界，面临许多不适应，包括人才素质不适应、技能不适应、人才结构不适应。具体表现在缺乏一批具有战略思维、世界眼光，通晓国际经济的“游戏规则”，具备跨文化操作能力的企业家和掌握现代化知识的技术人员，此外各行各业均缺乏适应WTO的各类人才。

加速构筑国际化人才高地主要应从如下几个方面着手进行。

1. 加速广东本土人才国际化

人才培养，教育先行。建议在省内若干大学成立培养适应国际化人才的学院或培训中心。例如，暨南大学去年率先在国内成立了采用全英语教学的国际学院，聘请国内外专家授课，专业设置文理兼容，与国际接轨，为广东本土人才国际化做了有益的尝试。

此外，有组织地选派干部和企业领导人去国外深造，鼓励科技人员赴国外参加学术交流，并予以专项资金资助。鼓励企业走出国门，在国外创办研究所，直接在海外环境中工作和成长。

在科技园或大学内采取种种措施吸收跨国公司设立研究机构，也是国际化人才培

① 本文是广东省政协八届五次会议大会发言，广州，2002年1月31日；政协广东省第八届委员会建言选编，广州：政协广东省委员会办公厅，2003，289-297.

养的路子。例如，中关村科技园活跃的技术创新氛围，吸引了像微软、英特尔、IBM等跨国大公司聚集中关村，不仅为中关村培养了大批国际型人才，也为中关村在全球科技创新链中谋求了一席之地。

2. 制订人才吸引计划，充分利用海外智力资源

美国“9·11”事件后，加上欧美近年来经济不景气，广东各类留学生交流会吸收了大批海外留学人员。政府应及时成立专门的机构和中介机构，负责招揽海外人才，制订优厚的人才吸收计划及一系列规范配套的政策，做好载体建设，优化创业环境，为归国海外学子创造优良的工作条件和解决生活后顾之忧。建立充分体现海外人才价值、灵活有效的薪酬机制，为国际化人才来粤工作提供动力保障。

引进海外智力资源时，要重视定向引进。在重视基础领域的高水平人才外，尤其要重视高技术领域和工程领域的高水平人才，在重视海外科研院所和高等学校人才的同时，还必须特别吸引在著名跨国公司工作的各类人才。

3. 充分利用国内的顶尖人才

充分利用区域外科技资源，满足本区域的技术创新需求和提升区域科技创新能力是弥补本地区顶尖人才不足的一种有效方法。广东两院院士少，顶尖人才缺乏，且高度集中在广州。广东应采取措施，吸收并充分利用省外优秀人才，建议推广深圳和宁波的做法，靠优越的创业环境和创新的文化氛围、先进的管理方式吸引国内的优秀人才，甚至不惜重金，对国内顶尖急需人才实行“候鸟”式或“飞人”计划，以短期工作方式，帮助解决重大技术和关键技术难题，培养高级专门人才。

二、重视基础研究和源头创新工作，增强国际竞争力

广东2000年研究与开发（research and development，R&D）支出为人民币107.1亿元，仅次于北京，位居全国第二。R&D是科技活动的核心部分，涵盖了基础研究、应用研究和试验发展三个阶段。

广东R&D活动高度集中在试验发展阶段，支出比重高达92.7%，而在基础研究和应用研究的支出仅占7.3%，远低于全国22.2%的相应指标，反映出广东目前许多高水平的基础研究和应用研究成果还需要从国内或国外引入。这种局面如不能迅速改变，源头创新能力将会进一步削弱，将来随着长江三角洲经济发展及内地其他省份经济发展和全球经济一体化，这种主要靠引进高新技术过日子，而没有源源不断创新能力的产业前景是十分令人担忧的。

广东2000年经济总量全国第一，约占全国1/10，R&D支出全国第二，创新能力则排在全国第三，且与北京、上海有较大差距，科技人力资源和科技意识在全国并无明显优势。据瑞士洛桑国际管理学院发布的关于世界竞争力的报告，我国已连续三年国际科技竞争力排名下降，与近年来基础研究未被充分重视、投入不足有明显关系。实际上，在21世纪技术创新速度明显加快，基础研究与商品应用之间的周期缩短，从商品竞争转化为基础研究的竞争在高科技领域已形成潮流，谁抢占了高新技术的制高点、谁拥有自主知识产权的专利，谁就将在未来的竞争中处于有利的地位。创新基础

薄弱、水平低已成为我省高科技产业未来发展的瓶颈问题。

基础研究与应用研究是高新技术产业链中不可或缺的一环，否则高新技术产业就会成为无源之水、无本之木。重视源头创新，必须从基础抓起。

1. 激活高校和科研院所存量产业资源

发挥广东高校和科研院所在科技进步与创新中的强大生力军作用对广东高科技发展十分关键。省政府应挑选若干所科研基础较强的高校，加大经费投入，支持其向研究型大学发展，逐步形成若干个科技创新能力强、特色鲜明、与国际接轨的研究型大学。省政府对省内为数不多的“211 工程”大学应加大扶持力度，对重点学科建设一视同仁，使之形成各自特色和优势，形成竞争、开放、滚动、优胜劣汰的局面，要充分发挥高校在高技术人才培养及关键技术和共性技术攻关上的优势。

2. 改革投入机制，加大对基础开发和应用研究的投入

要大力增加对基础研究和应用研究的投入，单靠政府主管部门的投入是远远不够的，应动员全社会的力量参与，多层次、多渠道、多形式筹集研究经费，实行政府资助和社会联动相结合。

3. 重视科技创新基地条件建设

至 2000 年，广东建有国家级和省级工程研究开发中心 157 家，其中国家级 46 家，省级 111 家，这 157 家工程中心所依托的企业总产值约占全省工业总产值 20%，说明这些研发机构对推动广东经济发展具有举足轻重的作用。全省大专院校 62 所，国家和省级实验室有 55 个。这是一支科技源头创新的载体和支撑力量，对提高广东省科技竞争力至关重要。

政府应关心和重视这些载体的条件建设，进一步激活这类积存的创业资源。建议政府在考虑特色和行业特点布局的基础上，为避免重复建设，应打破常规，与国家和部级实验室对接。广东省内的高校和研究机构，只要取得了部级以上重点实验室或工程中心，只要建设内容与广东省拟建的相似，广东省应直接挂牌并予以联动，给予等额的资助，使这些重点实验室或工程中心条件建设能锦上添花，尽快与国际接轨，形成强大的研发力量。

三、完善广东省各级风险投资机构，为高新技术发展提供动力支持和资金保障

我国科技成果商品化程度为 10%，产业化率为 5%左右，科技成果转化率相当于美国的 30%左右。发达国家在科技成果的研究开发、中试、商品化三个阶段的资金比例为 1∶10∶100，我国相应比例仅为 1∶1.1∶1.5，资金投入严重不足已成为严重阻碍我国科技成果产业化的关键因素。

广东省高校科技成果转化率在全国排名很后，2000 年全国高校科技型企业收入总额超亿元的 35 家高校中，广东无 1 家；全国 20 多家高校科技类上市公司，广东无 1 家；这与广东经济大省的地位极不协调，与广东省科技实力也不协调。究其深层原因，与广东省企业家和风险投资公司追求短期利益、缺乏战略眼光、不愿早期介入种子期或胚胎期科研成果有关。由于种子基金严重缺乏，大量的科技成果无法完成中试或无法成熟而被束之高阁。以广东省生物医药为例，目前有不少基因工程一类新药正在研

究与开发之中。

国家一类新药研制完成实验室阶段后，还要经过中试和一系列安全性试验及若干期临床试验，才能获得药证许可。这一过程在顺利的情况下费时 3～5 年，投入至少 600 万～1000 万元，这是高校及科研院所无法承担的。建议政府参照外省一些先进的做法，成立不同类型的风险公司，早期介入科研项目。例如，黑龙江省通过私募方式，加上政府引导资金成立了两家风险投资公司，一家专注于成长期企业的投资，一家专注于种子期企业的投资。“种子基金”投入将使大批新科技成果进入第二阶段。完成中试后，高校科技成果与企业的对接才能像接力棒似的不断进行下去，进入良性循环。政府的这种有限投入往往可以起到“四两拨千斤”的作用，大大提高我省科技成果的转化率，进而孵化一批上市公司。政府要采取措施，鼓励民间企业家对幼苗期，甚至胚胎期科技成果采取“提前介入，联合攻关，成果共享，风险共担”的投资模式。此外要对风险公司提供一系列政策支持，允许风险公司多渠道筹措资金，完善风险投资公司市场退出机制，使风险资金不断成长壮大，为高科技成果转化提供不竭动力。

四、重视高新技术发展规划，形成有鲜明特色的空间布局和产业发展格局

广东省发展高新技术产业要加强规划和布局研究，不要一拥而上，重复建设，形成恶性竞争。应制订统一、科学的高技术创新规划，结合国际高技术潮流和我省发展实际，有所为，有所不为，抓住高新技术发展的关键。各高新区要结合自己的优势和资源，在功能定位上要有分工，相互合作，优势互补，形成自己的特色。在产业创新链上要注意上中下游紧密结合，通过接力棒形式，在每个阶段都有相应的政策、法规、支撑体系及专门资金支持，形成上、中、下游密切结合的良性循环。

在高新技术发展过程中尤其要重视装备制造工业的发展，制造技术是高新技术产业化的重要载体，是科技成果转化为现实生产力的物化手段和物质基础。制造技术不仅是衡量一个国家工业发展水平的重要标志，也是国际科技竞争的重点，离开先进的制造技术和强大的装备制造工业作为支撑，高新技术的重要性和巨大作用等于空谈。

现代机械制造已不是传统的机械制造，它是高新技术与现代管理技术在高层次上的有机结合的综合技术。

目前世界装备制造工业发展趋势表现为以下几点。

第一，地位“基础化”。

信息化高度发达的工业化国家均重视装备制造业的发展，其在工业中所占比例和贡献均占前列。

第二，产品高技术化。

各种高新技术融入装备制造业，使之成为市场竞争取胜的关键。

第三，服务个性化。

第四，经营规模化。

我国装备制造业总产值在世界上位列第五，产品在 2000 年外贸出口中占 42.3%，表明装备制造业对我国发展外向型经济具有举足轻重的作用。我国是装备制造业大国，但不是强国，突出的问题主要有以下几点。①经济效益率低；②市场占有率低；③技

术创新力量薄弱，缺乏后劲；④部分产品能耗大，污染严重；⑤大企业不大不强，小企业不精不专，分散重复严重。

广东装备制造业底子弱，随着经济的发展，家用电器、电视机、冰箱、汽车、电梯、信息产品等机械制造发展较快，但与经济强省的要求尚有较大的差距。中国加入WTO后，广东的装备制造业要用信息技术武装，重视经济信息化、网络化、知识化，加快相关产品的开发。要抓住世界制造中心转移到中国的历史性机遇，在专业上要做深做透，做大做强。因此，广东企业要注意全套掌握外来技术设备，并充分利用好长期经济效益好、投资规模大的核心技术价值链，避免为尽快赢利而盲目引进的短期行为。

珠海前山镇（街）转型与社区建设研究①

一、转型中的前山

1. 基本情况

前山古为一城寨，具有悠久的历史和独特的文化底蕴，1987年建镇，2001年4月改设为珠海市香洲区前山街道。前山位于珠海市区的西南部，临江靠海，交通便利，属珠海市中心的一部分，辖区面积55.93平方千米，户籍人口4万人，常住人口12.8万人。

改革开放特别是建镇以来，前山的各项事业取得了长足发展，有各类企业300多家，其中外商投资企业250多家，涉及电器、医药、服装、建材、五金等诸多产业。2001年实现工业总产值41.9亿元，实现国内生产总值10.6亿元。

前山街道办事处现设有兰埔、翠微、前山、翠景、圆明、中山亭、莲塘、春晖、白石、岱山、造贝、夏村、逸仙、福石、梅溪、南溪、南沙湾、金钟、长沙、东坑、界冲共21个社区居民委员会（以下简称居委会），还有驻街单位上百个，如圆明新园、珠海百货等。全国知名高等学府——暨南大学珠海学院就坐落在前山。前山街道设党工委、街道办事处两套机构。街道党工委作为香洲区委的派出机构，全面领导街道的各项工作，并对街道工作负总责，是街道工作的领导核心。街道办事处是香洲区政府的派出机构，在街道党工委的领导下，依据政策、法规，对街道各项行政管理工作负责，并起主导作用。

2. 问题、挑战与机遇

本节中的转型是指前山镇政府（一级政府）转变为街道办事处（政府的派出机构）。前山撤镇改街，不仅是名称的变化，而且是全方位的转变，涉及观念、体制、机制、管理手段等多方面的变化。转型既是珠海市深化改革、转变职能、实现城市管理、全面推进城市现代化的迫切需要，更是前山改变基层社会管理薄弱和城市管理与服务不相配套的现状、大力推进社区建设、提高居民素质和文明程度、适应城市现代化建设的现实选择。转型中的前山街道办事处，面临着加强城市管理和加强社区建设双重任务，如何才能走好这一步？这正是研究的基本出发点和目标。我们认为，转型中的困难是存在的，挑战是现实的，机遇却是潜在的。

1）街道办事处的性质和职能

在新形势下，进一步明确街道办事处的性质和职能，对于实际工作有重要意义。

街道办事处是在城市以主要街道为标志的自然区域的基础上，由上级政府根据其行使职能的需要设立的派出机构，在城市管理中处于基础执行层的地位。根据1954年

---

① 本文是珠海市香洲区前山街道党工委和街道办事处科研项目报告，2002年4月5日；现代管理的中国实践，北京：科学出版社，2016，139-166. 作者：刘人怀，张永安，傅汉章，林福永，杨英，梁明珠，刘治江.

颁布的《城市街道办事处组织条例》的规定，街道办事处的主要工作有三项，即办理市、市辖区人民委员会有关居民工作的交办事项；指导居委会的工作；反映居民的意见和要求。其工作性质综合体现在以下几个方面。

（1）行政执行性。街道办事处作为政府的派出机构，主要职责是在街道居住区内贯彻执行政府的有关决定、指示、命令。

（2）派出代表性。街道办事处在街道居住区直接代表城市派出政府，具体办理派出政府所委办的行政事务。

（3）区域综合性。街道办事处是各种条条块块的结合点，城市管理的各项任务只有通过街道办事处才能落实。

在实践中，街道办事处的工作特点表现为：一是行使权力的局限性；二是行政事务的繁杂性；三是管理服务的直接性；四是所做工作的被动性；五是工作人员的边缘性。这说明，旧的街道管理体制是一种“辅助体制”，会影响街道办事处职能的发挥。

随着改革开放的不断深入，特别是城市管理体制的改革，城市管理的重心逐步下移，街道办事处正按照“小政府、大社会”的要求，全面负责辖区内的地区性、社会性、群众性工作，强化了街道办事处的综合管理职能，其地位和作用日益提高。在这种情况下，街道办事处的职能也有了重大变化，主要职能如下。

（1）管理。随着城市现代化建设步伐的加快，街道办事处将在城市管理、社区建设、社区服务、精神文明建设、社会治安综合治理等方面承担更多的任务，负有组织管理的职能。

（2）服务。街道要逐步确立其在城市管理、社区建设、社区管理、社区服务中的主体地位，增强服务意识，通过强化社区功能，完善社区服务体系，寓管理于服务之中，提高为人民服务水平。

（3）沟通。街道办事处处于“二级政府、三级管理、四级网络”运行模式的中介部位，起着承上启下、上令下达、下情上达的作用，在政府与居民群众之间建起一座沟通的桥梁，协调辖区内各行各业各方面的工作关系。

（4）指导。街道办事处要发挥接近基层的优势，有意识地发展社会组织，指导社区居委会和协调驻街单位的工作，增强政府与市民的亲和力，从而更好地推动城市建设和管理。

具体来说，街道办事处的主要职责如下。

（1）在区委、区政府的领导下，贯彻党和国家的各项方针、政策、法律法规，负责街辖内的地区性、社会性、群众性工作。

（2）加强基层党组织建设，充分发挥党组织的战斗堡垒作用和党员的先锋模范作用。

（3）加强精神文明建设，积极组织以提高市民素质为目的的活动，树文明新风。

（4）发展以税源为主的街道经济，管理好国有和集体资产，为各类经济组织提供信息和服务。

（5）负责街道执法队伍的建设和管理，做好辖区内市容环境卫生，园林绿化，市政维护，城建监察等监督、管理和服务工作。

(6) 负责辖区内社会治安综合治理，依照有关规定管理外来人口，做好民事调解和法律服务工作，维护居民的合法权益。

(7) 做好拥军优属、优抚安置、社会救济、社会福利、社区文化、科普教育等工作，兴办社会福利事业，发动和组织社区成员开展各类公益事业。

(8) 做好计划生育、劳动就业、安全生产、卫生保健、婚姻管理、民兵和兵役、侨务以及民族等工作。

(9) 做好社区建设、社区管理和社区服务工作，指导和帮助社区居委会搞好组织建设和制度建设，发挥社区居委会群众自治组织作用。

(10) 配合有关部门做好防汛、防风、防火、防震、防灾等工作。

(11) 及时向市、区政府反映居民群众的意见和要求，办理人民群众来信、来访事项。

(12) 承办区委、区政府交办的其他工作。

2) 转型中的主要问题和困难

前山在转型和社区建设中存在的主要问题和困难有如下几点。

(1) 城市管理基础薄弱。具体表现在：居民在思想观念上农村“烙印”深刻，城市意识淡薄；行政区划不够规范，街道辖区过大，居委会规模差别大；城中村遍布整个街道，改造任务重、难度大；城市基础设施建设相对滞后。

(2) 街道层面的综合管理体系尚未形成。由于“镇改街”时间较短，街道管理的任务大量增加，如市场管理、市容保洁、园林绿化、民政福利等，街道办事处尚未形成精简、统一、有效的城市综合管理架构和体系，领导体制、管理体制、工作机制、机构编制、管理队伍以及经济条件等方面，都还不能全面适应现代化城市管理的要求。

(3) 社区管理体系未能得到确立。街道辖区内没有形成系统的社区管理组织体系，街道办事处在社区建设与管理中的主体地位还没有得到确立。社区建设是我国城市经济和社会发展到一定阶段的必然选择，是面向21世纪我国城市现代化建设的重要途径。前山街道办事处仍然停留在传统城市管理阶段，社区建设与管理的要求非常迫切。

(4) 合法性和权力不到位。这是街道办事处和居委会在城市管理、社区建设中遇到的突出障碍。具体表现在：首先，街道办事处和居委会的多数管理工作缺乏法律、法规和政策依据，因而综合管理功能得不到很好的发挥；其次，上级政府的权力下放不到位，削弱了街道办事处管理的权威性，形成了“看得见的管不着，管得着的看不见”的现象；最后，尚未形成条块结合的“监督机制”，降低了管理效能。

(5) 社区居委会任务重、力量配备不足、公建配套设施严重滞后。在新形势下，居委会作为基层社区，面对流动人口、下岗职工、老龄工作、社会治安、计划生育等各种问题，工作任务重，在管理和服务上力不从心，存在着责权利不统一、职责任务不明确、管辖范围过大、人员过少、设施不配套、工作条件差、经费不足等问题，严重影响了居委会在社区建设、管理服务中作用的发挥。

3) 转型中面临的挑战与机遇

我们认为，前山在转型过程中既存在问题和困难，同时也面临着挑战与机遇，这些挑战是现实的，机遇却是潜在的。如何面对挑战、抓住机遇提升前山的城市管理水平和现代化水平，是研究的基本目标。

（1）“镇改街”首先面临观念的转变。观念必须先行，以观念的转变来带动职能转变、体制转变和机制转变。因此，前山街道办事处要丢弃镇（政府）的情结，强化城市意识，加强城市管理和社区建设，特别要突破只管市政维护、绿化养护、环卫保洁、城建监察等传统城市管理的旧模式，更不能沿用管理农村的方法来管理城市，树立城市管理、社区建设本身就是经济建设和在管理中经营城市、在经营中发展城市的观念。

（2）“镇改街”意味着管理职能的转变。要从作为一级政府所行使的行政管理职能转变到街道办事处管理、服务、沟通、指导等职能上来。因此，街道办事处要正确认识和处理城市管理与经济工作之间的关系，从直接抓经济工作中解脱出来，全力以赴进行城市管理和社区建设，强化街道办事处在城市管理和社区建设中的主体地位。通过启动“社区建设工程”，更好地实现镇向街、传统城市管理向现代城市管理的模式跨越，走出一条提升街道城市管理水平的新路子。

（3）“镇改街”意味着管理方式和方法的转变。由于管理职能、工作重心的转变，前山街道办事处应尽快实现由以“条条”管理为主向以“块块”管理为主的转变，由行政化向民主化、目标化的转变，以街道党工委为核心，建立覆盖街道辖区的横向联系网络，以人为本，通过资源共享、优势互补、联手共建等途径，形成城市管理和社区建设齐抓共管的新格局。积极培育“税源经济”“环境经济”“服务经济”。整章建制，使街道办事处的各项工作制度化、规范化和科学化。

（4）“镇改街”意味着获取资源方式的转变。“镇改街”后，大部分财源和财权要上交，经济实力会受到影响，人力、财力、物力等资源的获取方式以及责权利的关系将发生重大变化，“自给自足”的财政体系将成为过去。因此，要调动社会各方力量，整合社会资源，通过全面推进城市管理和社区建设来开创“税源经济”的新路子，以发展社区服务为先导，促进社区服务社会化、产业化和市场化。

（5）“镇改街”意味着管理体制的转变。在“二级政府、三级管理、四级网络”的城市管理架构下，要建立起新型的责权利相统一、条块相结合、管理与服务相协调的管理体制，由街道党工委、办事处构成街道城市管理和社区建设的领导系统，对城市管理和社区建设行使管理、服务、沟通、指导等职能；由街道各职能部门构成街道城市管理和社区建设的执行系统，行使规划、实施、综合管理等职能；由辖区内企事业单位、社会团体、居民及自治组织构成街道城市管理和社区建设支持系统，形成“小政府、大社会、大服务”的运行机制。

（6）单位体制主导型向社区体制主导型转变。这是我国城市发展的必然趋势，社区在城市管理和基层政权建设中的地位和作用更加重要。前山街道办事处应从实际出发，全面推进社区建设，把社区建设当成一项系统工程来抓，以发展社区服务和加强社区管理为龙头，以提高居民生活质量和文明程度为宗旨，以街道和社区居委会为依托，建设管理有序、环境优美、服务完善、资源共享、生活便利、治安良好、人际关系和谐、文化生活丰富多彩的现代文明社区，进而实现城市街道社区的可持续发展。

二、指导思想与发展目标

1. 指导思想与基本原则

1）指导思想

认真贯彻落实中央、省、市关于撤镇改街指示的基本精神以及加强社区建设的要求，围绕珠海市香洲区 2005 年率先基本实现现代化的奋斗目标，从前山街的实际出发，以推进具有鲜明的前山特色及较高水平的社区建设为基本方向，以发展社区服务和加强管理为龙头，以提高居民生活质量和文明程度为宗旨，以街道、社区居委会为依托，充分发挥党组织的核心作用，深化基层体制改革、强化社区功能、维护社会政治稳定，调动一切积极因素，促进本街道城市经济与社会协调发展，并成为香洲区中极具特色及较高发展水平的区域经济社会体系。

2）基本原则

（1）以人为本，服务居民。社区建设必须以“人”为中心和以实现人的全面发展为最终目标，并将服务社区居民及辖区机构作为社区建设的根本出发点和归宿。

（2）资源共享，共驻共建。运用市场机制合理配置本社区内分属不同主体（包括社区居民和辖区机构）的各种人力、物力、财力及其他社会资源，以支持及投资社区建设，并充分地调动全社区成员最广泛地参与社区建设及社区管理的责任感和积极性。

（3）责权统一，管理有序。按建设“两级政府、三级管理、四级网络”的城市管理体制的要求，建立健全社区组织，并相应地明确街道与社区的职责分工及权利划分，以形成有序的社区管理体制。同时，寓管理于服务之中，以增强社区的凝聚力。

（4）扩大民主，居民自治。坚持按地域性、认同感等社区构成要素科学合理地划分社区，在社区内实行民主选举、民主决策、民主管理、民主监督，逐步实现社区居民自我管理、自我教育、自我服务和自我监督。

（5）因地制宜，循序渐进。社区建设必须趋利避害，依托本地优势资源，以彰显本街道的基本特色及实施可持续发展战略；科学制订符合本街道实际的社区建设发展规划，从居民群众关注的热点、难点问题入手，有计划、有步骤地实现社区建设的战略目标。

2. 目标定位

1）确立前山街街区建设定位的依据

（1）珠海市建市以来，一直按现代化的基本要求，高起点进行城市规划建设，因而进入 20 世纪 90 年代以后连获全国环保模范城市、全国园林城市、国家卫生城市等荣誉称号，既被国家环境保护总局确定为“全国生态示范区建设试点”，也获得联合国“改善人居环境最佳范例奖”，最近又被评为“全国环境与文化竞争力最强的城市”，现已成为国内外闻名的环境优美、布局合理、生态有序的滨海花园城市。这使前山街笼罩在一座现代生态城市的大格局中，既为前山街的社区建设创造了良好的基本环境，也要求前山街街区建设要符合珠海滨海花园城市规划的大方向。

（2）占地面积近 1000 亩，已有 1000 名大学本科生就读的暨南大学珠海学院坐落于本街道。按其发展规划，该学院拟于五年内将办学规模发展至5000人。名牌大学孕育着先进的大学文化。同时，前山街辖区内有较为完善的中小幼教育体系及文化设施体系（其中有市一级学校、省特级文化站、省示范性成年教育学校及珠海唯一的专业电影院），又有在较深厚的传统文化根基上孕育起来的具有较高水平的群众文化。前山街的文化氛围颇为浓郁。

（3）位于前山街东南部的圆明新园、梦幻水城、前山城寨遗址和中山纪念亭等著名旅游景区，是珠海旅游城市的重头戏。游客云集，促进前山街的酒店、餐饮及商业服务业发展，凸显旅游经济功能。

（4）前山街地处城郊接合部，不仅环境优美，而且城市建设用地相对充裕（人口密度为3000人/千米^2），有建造可区别于中心城区高密度建筑物的郊野生态景观的空间条件。而其所临近的拱北口岸区及南屏科技工业园，前者人口密度（9524人/千米^2）及建设密度都很高，进一步建设宜居环境的空间资源有限；而后者以研发、生产及经营为主要功能的厂区布局基本确定，故此，前山街有建设优美的居住环境以吸纳拱北及南屏两镇人口的潜力。

（5）前山街具有较坚实的商贸基础，其中尤以建筑装饰材料贸易最为突出，已成为珠海市镇街特色经济的龙头。该专业贸易市场包括砂石土重型材料、水泥钢筋竹木材料、石雕夹板不锈钢材料、家具家饰材料、陶瓷洁具地板材料等五大市场，年交易额20亿元以上。为使该产业能提高市场竞争力，适应市场需求的发展趋势，宜进一步提升商业文化品位，造就良好市场环境，突出前山建筑装饰材料市场的品种齐全、环保生态的特色，达到上规模、上档次、以文化拉动商贸的目的。

2）定位设想

根据上述分析，我们认为，为凸显当地区位特色，满足珠海市滨海花园城市建设与发展的整体要求，前山街在城市建设上应体现“花园”及“生态”的基本特征：倚靠“特色经济”及“文化底蕴”的优势，前山街城区建设以城区商贸特色经济为依托，以构建浓郁的文化氛围及发达的文化产业为方向，努力将前山街建设成为“宜商宜住宜学宜旅的文化生态街区”。这是我们对前山城建定位的基本设想。

3. 总体目标及其实施安排

前山街街区建设的总体目标是：依前山街的定位要求，在2005年将前山街逐步建设成为一个文化气息浓郁、社区环境优美、社区服务功能完善、社区管理有序的“宜商宜住宜学宜旅的文化生态街区”。具体的实施安排如下。

（1）“两级政府、三级管理、四级网络”的城市管理体制构建，是一种由党委和政府领导、民政部门牵头、有关部门配合、社区居委会主办、社会力量支持、群众广泛参与的社区管理体制。具体的安排如下。

2002年及2003年，按国家、省、市关于大力推进城市社区建设的要求，全面完成城市基层管理体制改革工作；开展街道办事处管理体制改革，调整街道办事处设置；合理调整及划分社区范围；将居委会及村民委员会改造为社区居委会，并成立社区居民代表大会和社区议事协商委员会，以及社区居民小组，探索并逐步推行由居民直接选举社区居委会的领导机构，为社区建设与社区管理走向规范化打下扎实的基础，这一工作大部分将在 2002 年完成。

2004 年及 2005 年，进一步深化社区建设与社区管理体制改革：理顺区、街及社区相互关系；全面履行社区居委会作为居民自治组织所应承担的职能任务，推动社区建设与社区管理走上规范有序的轨道。

(2) 加强社区基础设施和社会信息网络建设。具体的安排如下。

2002 年及 2003 年，利用仍享受原一级政府财政的资源条件，积极按前山街区建设总体目标的要求，规划并加快建设辖区内主要的基础设施和社会信息网络，以有效引导各类经济要素进入前山，加快街区经济的发展和各类服务设施的建设。

2004 年及 2005 年，建立能及时不断地向市、区两级政府反映本地发展新趋势及社区设施建设要求的基本机制，以争取市、区有关部门在规划与建设方面的支持。

(3) 整合全街的社区资源，促进社区服务社会化、产业化和信息化，并形成完善的社区服务体系。具体的安排如下。

2002 年及 2003 年，摸清及规划好利用本街道的各类社区资源及引导外来资源发展社区服务业的基本方案，初步建成现代化社区服务体系的雏形。

2004 年及 2005 年，建立依本地“文化生态街区”发展要求建设社区服务业的动态机制，逐步完善本街区的现代化社区服务体系。

(4) 营造文体活动活跃、人际关系和谐及社会治安良好的现代文明社区环境。具体的安排如下。

2002 年及 2003 年，较系统地组织辖区内的各类社区文体活动及其他精神文明建设活动，探索新型的社区治安管理模式，以构建现代文明社区环境的雏形。

2004 年及 2005 年，实现社区文体活动及其他的精神文明建设活动制度化，并逐步完善新型的社区治安管理体制，不断优化现代文明社区环境。

三、基层组织建设与管理体系

街道办事处、居委会在城市管理中具有重要地位和作用，是社区建设的组织者和主办者。

1. 加强前山街道办事处在转型中的自身建设

通过进一步深化街道管理体制改革，按照党领导政权，“小政府，大社会”、管理与服务相统一、责权利相结合的原则，调整原有组织结构体系，建立和完善以街道党工委为领导核心，以街道办事处的城市管理、经济建设、社区管理、社区服务、社会治安综合治理、计划生育等职能部门为主体，以社区居委会、驻街单位和社会团体为依托的新型街道组织体制架构，充分发挥街道办事处的职能作用，调动社区居委会的积极性，做到精干、统一、高效，逐步形成街道依法为社区成员进行服务和管理与社区成员自我服务、自我约束、自我管理相结合的运行机制，从而提高街道办事处管理城市和社区建设的有效性。

在改革转型中，街道办事处和社区居委会可考虑采取以下组织结构，形式如图 1 和图 2 所示。

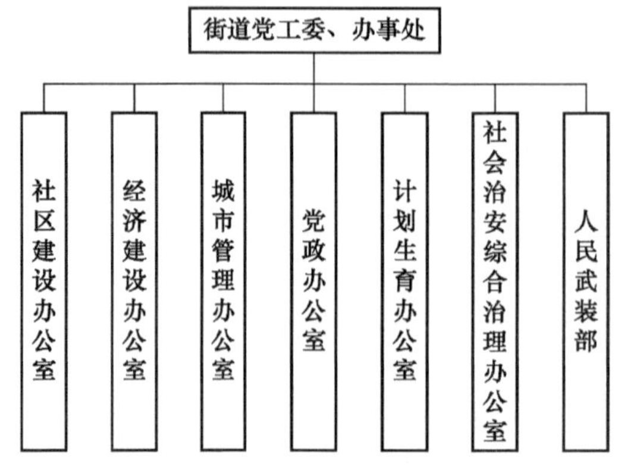

图 1 街道党工委、办事处组织结构

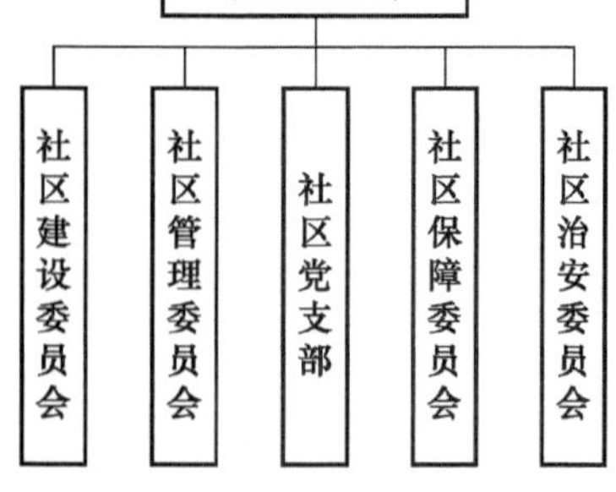

图 2 社区居委会组织结构

2. 加强社区党组织建设和领导核心作用

(1) 街道党工委对街道下属单位，以及辖区内无主管企业、社会组织中的党组织和党员实行全面的政治、思想、组织领导。

(2) 原则上以社区为单位建立党支部，实现“一社区一支部”，充分发挥社区党组织在社区建设中的战斗堡垒作用和党员的先锋模范作用。

(3) 街道党工委牵头建立“社区党建联席会议”制度，实现社区政治资源的整合，使党在社区形成整体上的政治优势和组织优势。

(4) 根据“属地管理”原则，街道党工委要对辖区内各类党组织和党建工作，流动党员等实行属地管理。

(5) 通过创办街道或社区党校，加强对党员的教育管理工作，通过建立“社区党风监督站”和“社区党员联络站”，加强社区党的作风建设。

(6) 加强党对群众团体工作的领导，充分发挥工会、团组织、妇联在社区建设中的作用。

3. 加强社区居委会建设，建立新型社区自治组织体系

(1) 街道办事处要以城市管理和社区建设为工作重点，组织发动和协调社会各界参与社区建设，形成社区建设的合力，优化和整合社区资源，制订辖区内社区建设发展规划，并认真组织实施。

(2) 按照“居民自治、议行分设”的原则，建立和健全社区居民代表大会、社区居委会、社区议事协商委员会“三位一体”的基层社区组织体系，分别行使决策、执行、议事和监督等职能，在街道办事处的指导下，全面领导和组织社区的建设、管理和服务工作，形成结构合理、功能完备、互动畅通的社区运转机制。社区自治组织体系的构成及其职能一般包括如下三个方面。

决策层——社区居民代表大会。其由社区居民代表和社区单位代表组成，定期讨论社区重大事项，选举产生社区居委会和社区议事协商委员会。

执行层——社区居委会。这是社区居民代表大会的办事机构，代表社区成员管理社区的公共事务，定期向社区居民代表大会和议事协商委员会报告工作。

议事层——社区议事协商委员会。这是社区居民代表大会的议事机构，由社区内

人大代表、政协委员、知名人士、居民代表组成，进行民主议事和民主监督。

通过社区民主政治建设，逐步形成以社区党支部为领导核心、居民代表大会决策、居委会办事、议事协商委员会议事、社区单位及居民群众广泛参与的社区管理新格局，进而实现社区居民自我管理、自我教育、自我服务、自我监督。

(3) 积极创造条件解决社区居委会办公用房问题，努力建设一支专业化、高素质的社区工作者队伍，积极发展志愿者队伍，培育和发展社会中介组织，广泛动员社会力量参与社区建设。

(4) 社区建设中，服务是宗旨，管理是保证，建设是基础。坚持服务先行，寓管理于服务中，通过社区服务来激发体制变革需求，进而配套进行社区建设和职能转换，从而建立起由社区党组织系统、社区居民自治组织系统和社会中介组织系统有机结合的社区组织体系。

四、社区服务体系与功能

城市管理的重心在社区，完善城市管理应以社区服务为中心。社区服务是社区建设的龙头，也是社区建设的基本任务。前山街道办事处在推进社区建设的进程中应坚持服务先行，建立和健全社区服务体系，发挥社区服务功能，走社会化、产业化、专业化和网络化的发展道路，进而提高居民生活质量，满足人民群众日益增长的物质文化需要。

1. 开展社区服务的意义

社区服务是以街道、居委会等社区组织为依托，以社区居民的自助互助为基础，面向社区全体居民和辖区内各企业、事业单位，动员社会各界和居民群众广泛参与，以解决本社区居民实际生活困难和不便、不断提高社区居民生活质量为目的的社会性服务。社区服务具有如下基本特征：一是福利性，二是社会性，三是互助性，四是区域性。

完善社区服务体系，发挥社区服务功能，可以增强居民的社区归属感、认同感和凝聚力，提高居民参与社区活动的积极性，强化社区整合和稳定机制，为社区建设奠定基础，推动社区建设其他方面的发展；有利于树立良好的社区形象，提升城市管理水平，推进社会保障事业的发展和社会的文明进步，为扩大对外开放和招商引资创造更好的环境；发展社区服务也是适应社会主义市场经济体制、建立健全社会化服务体系，不断提高人民生活质量的迫切需要。

2. 社区服务的内容

社区服务的领域广阔，内容丰富，根据社区居民的需求，主要开展以下几个方面的服务工作。

(1) 面向特殊群体的社会救助和福利性服务，主要有：老年人服务；残疾人服务；优抚对象服务；社会困难群体服务。

(2) 面向全体社区居民的便民利民服务，主要有：家政服务；幼少儿服务；生活服务；文教服务；文体娱乐服务；劳动就业服务；医疗卫生服务；社区中介服务；环保服务；安全保卫服务。

(3) 面向社区单位的社会化服务。此种服务的特点是“后勤”性的服务。社区服务组织与社区内的企事业单位、机关团体之间开展双向服务活动，既可以拓宽社区服务领域，提高社区服务的效益；同时，也有利于密切街道办事处、居委会与社区单位的关系，以达到资源共享、共驻共建的目的。

(4) 面向下岗失业职工的再就业服务和社会保障服务。在我国转型期，职工下岗、失业、再就业以及社会保险等不仅是政府的事情，社区应及时为下岗失业职工再就业创造条件和提供信息，特别是有针对性地开办一些再就业培训班，增强其再就业能力，缓解社会矛盾，维护和保障人民群众基本生活权益。

3. 社区服务体系的建立和完善

由于社区服务对象和服务项目的多样性，以及各社区的情况千差万别，因而社区服务的形式也是多种多样的，为了有效地开展社区服务，可根据各自不同特点，建立和完善社区服务体系。

(1) 建立、健全社区社会保障体系。要形成街、居二级社会劳动和保障网络，为老、弱、病、残、孤、抚、困等群体提供服务，为失业人员提供失业登记、职业介绍、再就业培训、社会保险登记等服务，逐步实现社区老有所养、孤有所抚、残有所助、贫有所济、难有所帮。

(2) 建立、健全社区卫生服务体系。以居民健康为中心、社区为范围、家庭为单位、需求为导向，积极开展以疾病预防、医疗、保健、康复、健康教育和计划生育技术等为主要内容的服务，充分发挥现有基层医疗卫生机构的作用，引入竞争机制，整合现有资源，尽快建立一批布局合理、规模适度、功能完善、便民利民、运转良好的社区卫生服务站，逐步向“小病在社区，大病上医院”的医疗服务模式转变。广泛开展群众性的爱国卫生运动，为社区创造优美、干净、整洁的生活环境。

(3) 建立、健全社区教育体系。以暨南大学珠海学院等教育机构为依托，以提高居民整体素质为目的，举办社区业余教育和老、中、青、幼教育，普及科学文化知识，进行公民道德教育和各种技能的培训，把学习和教育引入家庭，引导居民自我教育、自我管理、自我提高，逐步在社区建立、完善全方位、系列化的教育体系。

(4) 建立、健全社区文化体系。加强社区文化阵地、队伍、活动、服务 4 个网络建设，优化活动设施，充分利用文化站活动室、社区广场等文化活动设施，开展丰富多彩的社区群众性文化体育活动，形成社区文化特色，丰富社区居民文化生活，增强社区居民的认同感和凝聚力，促进居民的身心健康，共建、共创文明社区。

(5) 建立、健全社区中介服务体系。通过发展各种中介组织在社区范围内开展各式各样的服务活动，如信息服务、法律咨询与援助服务、家政服务、保险与理财服务、房屋租售服务、婚介服务以及会计审计服务等，建立外来人员服务机构，做好对外来人员的服务工作。

(6) 建立、健全社区治安体系。从增强社区居民法律意识和自觉遵纪守法意识入手，建立健全以责任区民警为龙头，以居委会治保主任为骨干，以区内保安联防队员和群众义务治安联防为依托，以化解社区居民内部矛盾纠纷、保证社区居民安全和社会稳定为目标的社区治安综合治理体系，创建“安全文明小区”。

(7) 建立、健全社区环境体系。社区环境是社区服务体系的重要内容，不仅影响社区的形象，而且直接关系社区的服务质量。因此，要从培养社区居民的环境意识、社区意识、城市意识出发，通过综合执法和“门前三包”等措施，大力整治脏、乱、堵，积极推进“绿色社区建设”，净化、绿化、美化社区，实现“一年一小变，两年一中变，2005年一大变”的目标。

4. 加强服务设施和服务队伍建设

(1) 社区服务设施（如社区服务中心、老年人活动中心、文化站、卫生站等）建设要纳入城市建设和城中村改造的公共配套设施建设规划，形成以社会投入为主，政府投入为辅的多层次、多渠道、多元化的社区服务业投资体制与运营机制。走服务设施共建、社区资源共享的路子，鼓励社区自筹资金、自办项目、自治管理、自我服务，建设完善的社区服务设施。

(2) 社区服务队伍是开展社区服务的首要条件。根据精干、高效的原则，尽快配备一支由专职人员、兼职人员和志愿者人员所组成的社区服务队伍。采取正规教育与短期培训相结合的方法加强专职服务队伍建设；采取岗位培训和职业道德教育加强兼职服务队伍建设；通过精神文明建设不断扩大志愿者队伍。

5. 逐步推进社区服务的社会化、专业化、产业化和网络化

(1) 实行投资主体多元化，加速发展社区服务业。社会救助和福利性的服务要发挥政府投入的主渠道作用，同时，也要多方面动员社区单位和居民捐助；有偿性服务要走社会化和市场化的路子，鼓励集体、私人等社会力量投资开展社区服务，开拓更多服务项目，扩大社区服务需求。

(2) 社区服务应朝专业化方向发展，即由专业组织来从事社区服务的组织与管理，配备专业知识和技能的服务人员来从事社区服务工作，提高社区服务的质量和水平，更好地满足社区居民各种服务需求。

(3) 随着市场经济体制的确立和社区服务社会化的发展，社区服务再也不能停留在过去的传统福利角色之中，必须走向市场，按照市场运行机制，拓宽社区服务领域，逐步实现社区服务的实体化、规模化、产业化、品牌化。

(4) 为了更好地利用和配置社区资源，社区服务应走网络化的道路。建立和健全社区服务网络，街道办事处要落实“五个一”建设：①开通一个“热线连万家，服务你我他”的社区服务热线电话；②设立一个社区服务信息收集和传递的网站；③建设一个综合性、多功能的社区服务中心；④在社区居委会抓好一个有特色、高效率的社区服务站；⑤按照“一街一所、一区一室”的模式建立一个社区警务室。

五、政府职能转型对街道经济管理的新要求

1. 街道经营城市与发展街道经济的指导思想

政府职能的转变，如从经济建设层面看，就是要实现政企分开，从过去的单纯对公有企业的管理转为对属地经济的管理，政府不再直接从事经济活动；从过去只从微观上对部分企业的管理转变为从宏观上规划、指导、引导属地经济，规范市场秩序，强化服务意识，树立“所在即所有”的“税源经济”观念。为此，珠海市区政府对街

道办事处提出过“三不”原则，即街道办事处领导不直接担任经济实体领导职务；不直接参与经济实体的经营决策；原则上不再投资兴办企业。这“三不”原则，体现了政府经济管理运作模式转型的基本要求，也符合建立和完善社会主义市场经济体制的必然趋势。

但是，政府职能的转型，并不意味着政府在经济领域的简单退出和政府功能的弱化，而是在“有所为有所不为”中，政府可以通过社会管理、经济调控、优质服务和行政执法手段来优化地区经济发展环境，从而促进本地区经济持续、快速、健康发展。

2. 摆脱传统城建困境，走经营城市之路

从目前全国大多数城市街道办事处的情况看，其管理功能还不是十分明确。1954年颁发的《城市街道办事处组织条例》规定，街道办事处的任务只有三条，但目前城市街道办事处所担负的任务已远远超过这三条规定，其工作范围涉及辖区内党、政、经、警、文化、劳动就业、计生、卫生、司法调解、社会治安、优抚安置等，而大部分精力则放在发展街道经济上。

城市街道办事处作为区政府的派出机构和“三级管理”的组成部分，在其辖区范围内同样肩负着如何把城市建设好、管理好的问题。城市街道管理是城市管理这个庞大系统中的一个子系统，就这个子系统来说，它又可以自成一个系统。

过去传统的城建弊端主要表现在：政府包办城建的一切，单一投资，投入大，周期长，公益性强，由于政府财力不足，往往导致“工程马拉松、投资无底洞”的局面，如要单靠财政去搞城建，势必举步维艰。“经营城市”的全新概念，就是把城建作为一种产业看待，产业的发展是要有回报的。城市是一个永续发展的空间，其中人是主体，生态环境是条件，经济是支柱。城建的好坏，不仅是看财政的投入，更重要的是看城市元素的科学利用。城市元素就是资本，土地是资本，道路是资本，环境是资本，人文资源是资本，简单说，一切可经营的城市元素都是资本。街道办事处要实施“富民强街”的目标，就必须强化经营城市意识，着力培育自身的优势。

3. 培育前山自身优势，在经营城市中发展城市

1）继续深化经济体制改革，彻底实行政企分开

为使原街（镇）企业成为真正的市场竞争主体和法人实体，享有充分的经营自主权，必须彻底实行政企分开。为此，珠海市前山街道资产经营中心已于2000年2月组建成，为镇属独立企业法人。公司的经营宗旨是：在授权范围内，通过审慎投资、有效管理和提供优质服务的方式，对所持有的镇属企业和物业进行积极运作、产权经营和实业开发，实现国有资产的保值与增值。现该中心拥有珠海市前山工业集团有限公司等七家控股企业，发展势头良好。

改革后的街道经济管理职能，应以对辖区内的宏观管理和间接管理为主，面向全社会的经济管理实施调控与监督。具体的管理职能主要包括四个方面：一是从本地资源出发，按照市场需要制订经济发展战略规划；二是组织协调职能，单个企业的自身发展计划，由各企业自主独立完成，街道办事处不参与经济实体的经营决策，主要是协调各经济单位的活动，用经济办法管理经济；三是控制监督职能，其目的是保证各经济单位有效地按发展总体规划有序运作；四是信息沟通职能，为了保证街道经济能

协调发展，街道经济管理机构必须发挥信息沟通作用，及时了解本辖区内外经济状况、市场动态、需求变化以及各经济单位的情况等。

2）发展第三产业，办出“特色经济”

街道经济的生命力，在于“因地制宜”和“具有特色”，而发展第三产业，又是街道经济的优势所在。前山镇在体制改革前已经把发展第三产业作为经济建设的重点，并确定以建筑材料销售业、社区生活配套服务业、旅游业和文体娱乐业为开发重点，2001年全街道第三产业营业总收入七亿四千多万元。

在发展“特色经济”方面，根据前山的地理环境优势和发展基础，通过近年的努力，前山现已形成全珠海市最大的建筑装饰材料专业市场。经营该类产品并具有一定规模的企业有27家，个体工商户400余家，经营品种一万多个，总投资额20多亿元，从业人员近两万人。建材业已成为第三产业的龙头，营业额占全街道第三产业的二分之一。

建材业是广东省传统支柱产业，随着房地产业的发展，建材需求总量在不断增加。但也应看到，从全省情况看，建材供大于求的局面短时期内不可能改变，国外建材业抢占市场又是不可避免，市场竞争将日趋激烈，而我们目前的建材市场仍多是传统的经营模式，即“建场—招租—收租”式的低业态市场，这种模式带来的弊病是：容易造成布局不合理，设施、功能单一，缺乏经营个性与先进管理手段，商品质量参差不齐。对此，前山已提出了建材专业市场升级问题，由粗放型向集约化、精品化、规模化方向发展，引进开发的世邦国际装饰广场项目以及建立分类专业市场，实现规模经济与竞争活力的统一，以新的经营理念，推动建材市场的发展。

3）调整经济结构，扶持民营企业发展

放手发展民营经济，是“富民强国”之路。同理，支持与鼓励民营企业的发展，也就成为发展街道经济的重要任务。目前前山街道有各类企业300多家，主要生产经营手提电视机、收录机、汽车音响、防盗器材、服装、鞋帽等20多类产品。

进一步发展民营经济，还应注意处理好以下关系：一是进一步破除在所有制方面的传统意识形态观念，全面认识发展民营经济对调整经济结构、发展街道经济的意义；二是在政策上，为民营经济的发展提供支持与保护，在行政职能上，强化服务意识；三是建立合理的制度，加强对民营企业的监督、管理与引导，扶持其上水平、上档次、上规模，逐步实现资本的社会化，走股份制和股份合作制的路子。

4）强化社区商业建设，改善和优化居民生活质量

强化社区商业建设，主要是为了提高、改善和优化社区居民生活服务。目前全国许多城区对此已有新的部署，上海把推进社区商业建设任务列入经济和社会发展规划，通过发展社区商业，满足社区居民多样化、就近化的消费需求，同时也利于凝聚人气，发挥商业带动效应。

强化社区商业建设，对推进传统商业的调整有积极作用，逐步把传统的百货店、各种小商店和各类市场改造成专卖店、专业店或转为服务性企业，开发家政服务、保安服务、保洁服务、保绿服务、票务服务、维修与回收服务等新型服务项目。前山在这方面已有一定的基础，目前亟须解决的是要有一个完整的发展规划，提高社区商业的组织化程度和商业层次。

5）以环境引人，以服务感人，做好招商引资工作

招商引资是发展街道经济的一种手段，其行为主体应是各类企业，但在当前的体制情况下，政府的主体作用还是不可忽视的，因为不论是物流、商流、资金流，都与市场环境密切相关，而市场环境就不是一个独立的企业所能影响的。例如，从街道办事处层面看，招商引资成功与否，除投资政策的落实外，市场环境建设以及一系列的跟踪配套服务，就显得特别重要，它决定了投资吸引力的强弱。

前山的招商引资工作，除港、澳、台地区外，已逐步向西欧、美、日、加等国和地区扩展，2001 年直接利用外资达 2529.4 万美元，引进区外资金达 6.3 亿元人民币，近年来连续保持了较好的增长势头。今后在持续发展招商引资工作时，以下几个问题还应引起注意。一是提高引资整体水平和质量，企业自身质量的提高是重要基础，如果自身经营水平低、负债率高、资金利润率低、投资成本高，此种“二低二高”情况，招商引资竞争力是不会强的。二是丰富和创新招商引资的方式与手段。随着国内外市场的变化，资金流动不断出现新的情况和特点，因而招商引资的方式和手段也应不断创新与丰富。例如，前山提出的招商代理制、网上招商、以商引商、招商与招才结合、引资与引智结合等，都是一些新的思路。三是把招商引资作为一项长期和经常性工作来抓，在人员安排和经费投入上要给予保证，对有关人员的奖励也要形成制度化。

六、社区文化建设与环境管理

1. 社区文化建设的主要内容

街道社区文化是城市文明程度的体现。社区文化建设对提高社区成员的综合素质，培养高尚的道德情操，提高文化品位，创建整洁、文明向上的社区风尚有着极其重要的意义。

社区文化是一种特有的文化现象，既包括社区意识、社区心理、社区风尚、社区公德、社区教育、社区艺术活动、社区生活方式等精神层面的要素，也包括社区文艺活动场所、宣传橱窗、公益广告、艺术雕塑、标志性建筑以及环境绿化等物质层面的要素。它是社区地域、人口、组织、经济、科学技术等历史与现实的综合反映，也是社区发展的动力源泉。社区文化就活动地域而言，主要包括广场文化、家庭文化、楼群文化、企业文化等。

社区文化建设主要包括：丰富社区群众的业余文化生活、搞好社区教育、进行科学普及、提高卫生保健水平、开展群众性体育活动、倡导文明健康的生活方式等。

2. 文化发展总体目标

依据前山街的实际情况及经济实力，街区文化发展应力争达到以下总体目标：以传统文化和现代文化为基础，以旅游文化和高教文化为载体，以增强社区认同感、归属感为原则，建立文化设施完善的街区文化体系，形成亲和度很高的新型文化氛围，以较高的生态环境品位，尽早建成珠海市的文明示范街区和广东省先进文明街区。

3. 文化建设发展规划

虽然前山街经济实力较强，但考虑到目前尚处在“镇改街”的转型期，许多方面尚未具备城市成熟街区的文化发展条件，因此我们认为，前山街的文化建设在硬件、

软件两手抓的同时，目前应以硬件建设为主，软件建设为辅，待硬件建设基本完善和形成网络之后，再将文化发展的重点转为软件建设。

1）文化硬件建设

（1）建立必备的街区公益性文化设施。首先以目前较为成形的街区和较为繁华的街道为主，建立起点、线、面结合，布局合理，大小得当，高低相宜的各种文化设施，如宣传栏、阅报栏等休闲活动设施等。可考虑重点在长沙、金钟、翠景、春晖等社区予以实施。

（2）建设群众性文化活动网络。建设大型文化广场，形成以文化馆为龙头、文化站为中心、文化广场为舞台、文化室为基础的群众性文化活动网络。争取在街区所辖范围内建设大型文化广场两个、文化活动中心一座；并按省有关规定建设省级特级文化站、一级文化站、二级文化站等；全部社区居委会都建有文化活动室和图书室。可考虑在长沙、新苑和街道办事处后面的二手市场临街的开敞空间各建立 3 万平方米左右的大型文化广场两座，并在前山路与逸仙路交会的繁华地点建立文化活动中心一座。对于文化广场的建设，可考虑与珠海市团委、暨南大学珠海学院等单位共建，并以青年文化广场、科幻广场为主题（图 3）

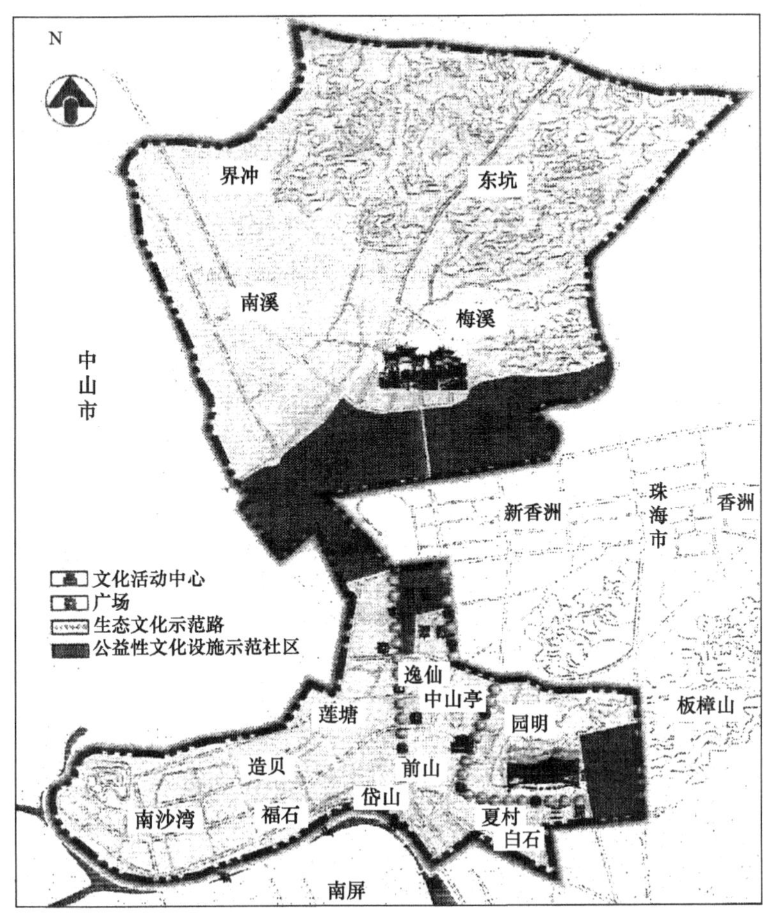

图 3　前山文化建设示意图

(3) 建设多元化的文化休闲场所和体育健身场馆。在各企事业单位、物业小区、各居民区楼与楼之间和建筑物之间的空地上增加休闲场所及休闲活动设施，合理规划与布局各类娱乐、体育健身场馆。可考虑在适当地点配套大型体育活动场馆，并要求每一个社区居委会都建有一个居民健身活动场地。

(4) 整合区内各类文化资源，构建特色文化社区。辖区内各类文化资源丰富。其中包括历史文化资源、商贸旅游文化资源、教育资源等，可依据各类资源的特色，构建不同类型的文化社区，如历史文化社区、商旅互动特色社区、智能型社区、工业文化型社区等。

前山历史悠久，人杰地灵。自南宋绍兴二十二年开埠，明代天启元年建设城寨以来，名人辈出，有前清举人郭以治、文学家苏曼殊、曾任广东省省长的陈席虞等，孙中山先生曾到前山宣传三民主义并亲自为中山纪念亭奠基。丰富的历史文物古迹，众多的历史名人为前山留下了可供挖掘历史文化资源，因此，可考虑在适当地点建立历史名人博物馆，以展示前山的历史文化底蕴。

辖区内拥有圆明新园、陈芳花园、梅溪牌坊、中山纪念亭、前山城寨遗址等旅游景点和初具规模的繁华商业街，可考虑以圆明新园、世邦国际装饰广场等为基础，构筑历史文化型和商旅互动型的特色街区（图 4）。

教育是文化之基，文明之本。辖区内拥有丰富的教育资源，暨南大学珠海学院、珠海教育学院、前山中学等多家学校落户前山，宜选择教育资源相对集中的区域，积极探索利用教育资源开展智能型社区的建设，建立校、区各类资源的共享机制和共建机制，建立校、区文化活动制度，实施“教育强街”战略。

(5) 调整街区产业发展方向，引导文化产业向暨南大学珠海学院外围集中，如电脑城、购书中心、电影城、文具城等。

(6) 建立覆盖全前山的社会信息网络、社区服务网络，开办社区服务热线和街道、居委会的图书馆（室）的网上阅览业务；建立社会共享网络，实施“光纤入户”。

(7) 依据国家规定的各项社区文化指标——人均公共图书馆藏书拥有量、公共文化活动场所数量、人均年文化消费支出额、户均电视机拥有量、居民对互联网的利用度和依赖度、居民对社区文化体育活动的参与率、居民对社区文化体育的满意度等，抓好文化硬件建设，力争达到国家先进指标。

2) 文化软件建设

(1) 开展文明社区意识教育，营造良好人文环境。在各个社区、楼院开展主题鲜明、富有特色的文明社区意识教育，倡导“文明礼貌、助人为乐、尊老爱幼、邻里团结”的社会风气，弘扬社会主义精神文明；制定街区文明公约，规范居民行为。

在公共场所和社区内设立永久性宣传栏与标语牌，大力宣传“共居一地，共建文明，共创繁荣”等社区理念，使社区意识深入千家万户。

引导广大居民群众和辖区内单位树立良好的环境意识，养成为了保护环境而不断调整自身经济活动和日常生活行为的自觉性。

广泛开展具有地域特色、健康向上的社区文化、庭院文化、楼道文化活动和全民健身的体育活动，活跃文化生活，增强人民体质。

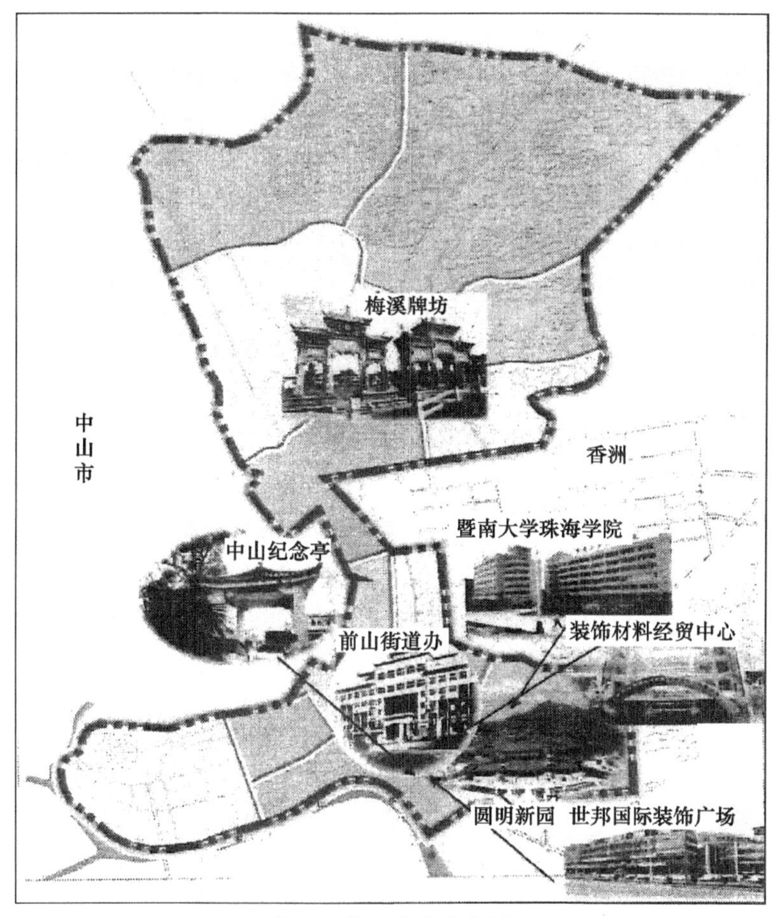

图 4 前山现状示意图

(2) 整合区内文化资源，形成共建联办的良好氛围。发挥区内各文化、企事业单位的潜能，做到“互补”“互促”“共建”“共享”，争取社区共建参与率达 60%以上，一半以上的单位和学校文体活动场地能向社会居民开放，形成资源共享、优势互补的格局，并合办各类社区学校、培训班等。例如，文化补习班、书画学习班、计算机辅导班、艺术学习班等，尤其是可以利用暨南大学珠海学院的优势，合办高层次的各类管理人员培训班。

(3) 发展休闲文化活动，丰富群众文化生活，促进文化生活社会化。主要是通过开展多形式、多层次的群众性文化活动，营造有利于提高人的文化素养和精神境界的氛围。一是要抓好以广场文化和家庭文化为重点的文化娱乐活动；二是要抓好以群众体育和竞技体育协调发展为基础的全民健身活动；三是要抓好节庆文化活动。例如，组建民间艺术、曲艺、体育团体（如社区体育代表队等），构筑民间艺术之乡、体育之乡的形象；举办大众化的文艺晚会、歌舞表演、书法、绘画、摄影、戏曲表演等文娱健身活动，形成都市社区一道独特的风景线。

3) 推进生态文化建设，改善社区生态环境

依据世界城市发展的趋势和可持续发展的需要，各地对人居环境的重视程度越来

越高。目前前山尚处于现代化城市建设的初级阶段，就整体而言，全区域的城市建设尚处于继续推进、不断完善的时期，因此，宜将发展远景目标定在较高水平，以便在推进城市建设的过程中，描绘出现代化街区的美好蓝图。因此，建议在文化建设中，注重生态环境的建设，在不断改善街区生态环境的基础上，创建“生态文化型”街区。

(1) 建设生态文化型示范道路。为提高城市环境文化品位，宜选择几条主要马路，沿道路两旁修建绿化带、街头建筑和环境小品，并沿道路两侧建设一批小公园、小绿地和休闲活动场所，部分建筑物的墙壁配以反映城市特色的图画并配置灯饰，构筑区内突出的文化景观带，并以“长藤结瓜”的形式建设珠海生态文化示范路。可考虑在明珠路、前山路、翠前路等几条主要干道予以实施。

(2) 突出生态文化亮点，建设“L”形绿色文化长廊。考虑到区内拥有生态环境很好的圆明园生态、旅游文化亮点和暨南大学珠海学院等智能型文化优势，同时，位于两点一线的“L”形地带亦是前山人口较为密集、商贸旅游活动较为频繁的地带，因此，建议建设这一“L”形绿色文化长廊，以凸显前山的生态文化品位。

(3) 制定区内生态、环保规范，创建“生态环保达标区”活动。积极改善基础设施，大力兴建治污工程，减少污染源。严格控制区内的大气、水体、声音的环保达标指标，区内各小区、企事业单位的生产、生活活动均应严格依据各项环保指标来进行；大力宣传生态环保意识和理念，使区内的公共区域和各种公共设施均体现强烈的环保设计理念，其中包括区内的垃圾桶、公厕和各种公共标识牌等。

(4) 实施“绿色工程”，展示“最佳人居示范区”的形象。力争全部道路绿树成荫，住宅小区凝芳拥翠，各建筑物屋顶空中花园争芳斗艳。抓好庭院绿化，开展创建“绿化先进单位”“花园式单位”活动。大力开展户外环境和绿化美化环境活动；动员社区成员积极投入花园式社区建设之中，共同创造“绿、静、美、安”的美好家园。依据国家规定的各项社区生态环境指标——人均绿地面积、绿化覆盖率、空气质量指数、水污染及其治理程度、噪声污染及其控制程度等，抓好生态环境建设，力争达到国家先进指标。

4) 建立和完善推进街区文化发展激励机制，实施“生态文化型社区”发展战略

(1) 开展“社区建设示范区”“示范街道”评选和创建“高标准社区”活动。考核标准主要为六个方面：组织网络化、制度规范化、干部专业化、功能全面化、设施配套化、工作一流化。

(2) 建立奖优罚劣的评估机制。将社区文化建设的实绩纳入工作目标考核范围，统一考核，实施奖励。建立考核、评估、表彰、实施奖励制度，激励社区单位和个人参与社区文化建设。

七、社区建设中的物业管理

1. 物业管理的基本机制

物业管理是社区建设与社区管理中的重要组成部分。物业管理所涉及的住宅小区治安防范、环境建设与管理、生活服务体系的发展均是社区建设的重要内容，为此，物业管理水平的高低及其质量的好坏，将直接影响社区建设与管理的水平和质量。搞

好社区建设，必须以提高社区中各住宅小区的物业管理水平为基本前提。

按社区物业管理主体的工作性质，我们可将物业管理的主体分为宏观管理主体及微观管理主体两类。

1）物业管理的宏观主体及其职能定位

物业管理的宏观主体有街道办事处和社区居委会两者。从社会管理职能的基本要求看，它们除履行《中华人民共和国城市居民委员会组织法》赋予的工作职能外，还要进一步强化与完善对社区各类组织的综合管理、服务、协调和监督职能，统管各自社区的公共事务和公益事业。因而，街道办事处和社区居委会在社区物业管理中主要的职能应作如下的定位：一是遵循方便管理及规模经济原则，开展住宅小区区划工作，将相对集中连片且整体性强的住宅区域划为一个个的区域“单元”，即规模化的住宅小区，指导并协助同一规模住宅小区内的业主成立“业主委员会”；二是协助“业主委员会”进行物业管理的招投标并遴选合适的物业管理公司管理小区；三是引导物业管理公司按照社区建设整体要求进行小区管理；四是在一般层面上协调物业管理公司与业主委员会以及社区居委会的关系，并监督物业管理公司的履约行为。

2）物业管理的微观主体及其职能定位

物业管理公司是物业管理的微观主体。它是根据其与业主委员会签订的物业管理委托合同，对受托的物业行使管理权的服务型经营企业。其基本职能是对小区的治安防范、园林绿化、环境卫生、设施的维护保养和生活服务等方面进行专业化管理。

3）两类物业管理主体的基本关系

虽然，物业管理的宏观主体及微观主体的工作性质不尽相同，它们相应的管理范围及职能也有较大的差别，但两者在社区建设方面却有着一个共同的管理领域：为住户提供整洁、文明、安全、舒适、方便的居住环境，使居民安居乐业，促进小区居民的精神文明建设，在小区良好的自然环境和文明环境中陶冶居民的高尚情操，提高文明水平，形成互助、互谅的社会风气，为社会稳定和经济增长创造必备的前提条件。受职能角色的规定，在进行社区建设时，社区居委会主要履行其社会公共职能，从全社区的整体建设出发，通过调动及整合全社区的资源来开展社区建设；而物业管理公司则将托管业主的利益放于首位，一般只按管理合同要求进行社区建设。为此，有序的社区建设体制应该表现为：街道办事处及社区居委会对物业管理公司行使指导、协调和监督职能；而物业管理公司则应配合和协助街道办事处及社区居委会共同搞好社区公共事务、公益事业管理。物业管理公司从事的是企业行为，街道办事处和社区居委会担当的是行政职能，其共同利益要求共同配合，为社区建设做好服务工作。

2. 前山物业管理现状分析

由于土地及管理政策等历史原因，前山住宅小区的物业管理水平处于明显较低的状况，具体表现在如下两个层面上。

1）物业管理普遍存在的问题

主要表现：一是辖区内除金钟花园等几个小区初具规模外，其他绝大多数住宅小区的规模均明显偏小，造成物业管理难以达到规模经济效益，且小区配套服务不足；

二是全街道只有三个住宅小区成立业主委员会，其他绝大多数住宅小区均未成立业主委员会，致使业主权益难以保证；三是物业管理公司的资质偏低，导致物业管理不规范及服务水平不高；四是物业管理公司的“界外负担”重，如小区规划设计不合理、小区建设与管理脱节、小区配套设施不完善、工程质量遗留等问题，此外，其还额外承担了小区出租屋管理、人口普查及社区文明建设等政府行政工作，致使物业管理营运成本居高。

2）城郊接合部物业管理的特有问题

前山刚由镇转为街道，在城市建成区中分布着 10 个城中旧村（全珠海市仅有 26 个城中旧村）。这些城中旧村的居民尽管在 1988 年土地统征后已转为非农业人口，但仍沿袭传统的农村管理模式和居住习惯（多建分散的单家独户小楼房），管理及监督水平落后，导致城中旧村在城市化进程中及物业管理方面存在着一系列与老城区不同的问题：一是建筑密度大且土地利用率低，建筑物布局混乱，以致其区域“物质外貌”与城市的整体形象不相称，物业管理的起点明显偏低；二是区域环境脏乱差，抗击洪涝、台风等自然灾害能力薄弱，火灾隐患极大，居住环境差，物业管理的物质基础差；三是多外来人口，治安管理难度大，物业管理更显复杂性；四是没有建立起现代物业管理的基本体系。

3. 前山提高物业管理水平的基本思路

住宅小区是社区活动及社区服务的基本载体，因而物业管理便是社区建设与管理的基本单元及着力点。搞好住宅小区的物业管理，将是提高城市居民居住与生活水平及社区管理水平的重要途径。

从规范管理及提高管理效益的角度出发，我们认为，前山提高物业管理水平应遵循如下的基本思路。

1）构建规范有序的物业宏观管理环境

提请市有关部门尽快依珠海市实际情况和市场经济发展的要求，修改完善《珠海市住宅小区物业管理条例》和实施细则，以便在市场准入、行为规范、权利义务、纠纷处理、法律责任等方面对物业管理营运主体进行约束和规范，为本街道物业管理的有序进行创造良好的法律环境。

成立前山街道办事处物业管理办公室，各社区居委会也指定专人负责处理社区内有关物业管理方面的事务。对下，协助辖区内业主委员会开展住宅小区物业管理的招投标工作、接受群众对物业管理的投诉及调停物业管理机构与业主委员会之间的一般性纠纷；对上，向香洲区物业管理主管部门提供有关住宅小区物业评级和相应的物业管理机构资质审核的咨询意见，以构建规范的物业管理运行机制。

指导各住宅小区依法成立业主委员会，并全面推行住宅小区物业管理的招投标工作，以推动物业管理按市场原则高效运作。

开展辖区内住宅小区物业评级工作，激发住宅小区相关利益主体争创“优质住宅小区”的热情，有效提高本街道物业管理水平。

2）着力提升物业管理水平

倡导以人为本，按“文化生态街区”的要求引导住宅小区开展建设先进文明社区活动，提高物业管理水平。

通过招投标方式引进具有较高素质及经营管理实力的物业管理公司，管理本街道的住宅小区。

因地制宜调整住宅小区范围，尽可能使住宅小区规模达到建筑面积在5万平方米以上，以促使住宅小区物业管理体现规模经济效益。具体做法：一是对新建小区应作统一规划（最好能进行成片整体开发）；二是对现有在地域上连成一片的若干规模偏小的住宅小区进行合并，以形成较具规模的小区。例如，位于街道办事处对面的农机站住宅区、自来水公司住宅区及其周围的若干住宅便可合并成一个较具规模的住宅小区，而位于南屏大桥旁、金鸡路南面的银苑新村、中臣花园和康宏花园也可圈在一个住宅小区之中。

对金钟花园这一上规模的住宅小区应促使其物业管理向上档次的方向努力，促进物业管理走社会化、专业化和市场化之路。

3）加快城中旧村改造步伐，为提高物业管理水平创造物质环境

为提高物业管理水平及社区建设与管理水平，必须尽快完成前山街城中旧村的改造工程。珠海于2000年提出将用3年的时间改造城中旧村并将其建设成文明社区。前山街将长沙村、新涌村和兰埔村当成全市城中旧村改造的先行村与示范村，其改建工程已基本完成。截至2001年3月，全街共丈量各类房屋面积107万平方米，完成7个村（含自然村）约72.9万平方米的确权工作，其他城中旧村的确权工作仍在进行中；共有5个村确定了开发商，4个村已动工兴建，其中长沙文明社区有100多栋楼已建成，村民已入住，新涌文明社区也已实现完工入住。

尽管前山街城中旧村改造的进程较快，但却存在着市政及公用配套设施建设严重滞后；市所出台的支持旧村改造的优惠政策未落实；因市有关涉及旧村改造审批的部门关系未理顺，造成审批效率低，影响改建进度。

八、对策与建议

做好前山街社区建设与管理工作，除加强党组织建设，充分发挥党在社区建设与管理中的领导核心作用外，还应该着重解决如下的几个问题。

1. 树立社区管理新观念

前山街社区的建设是处于国家推行全新的城市基层管理体制改革的大背景中、本辖区建制刚完成“镇改街”及珠海市在城市管理上从传统的城市管理模式向现代城市管理模式转变的过渡时期，这使社区管理的环境、机制及运作将进入一个全新的时期。因而，只有从如下几个方面树立起新的社区观念，才能适应新的形势，高效地进行社区建设。

1）适应“镇改街”的要求

前山辖区的管理机构原为镇一级政府，在进行社区建设与管理方面具有较为全面的职责及相应的权力，其一级财政的体制也为自身提供了相应的资源条件。在经过“镇改街”后，本级管理机构已由一级政府转化为香洲区政府的派出机构，在进行社区建设与管理的职责及权力上已发生了改变：辖区内有关城市设施的规划、投资建设、

经营及其相应的宏观管理等事项已交由上级政府进行统一管理，前山街道办事处将以城市管理和社区建设为工作重点，组织发动、协调社会各界参与社区建设，并只按照上级政府的授权，在辖区内行使一定的检查监督权、审批管理权、人事考核建议权、公共配套设施知情及参与验收权等职权。这样的建制转型，要求新成立的前山街道办事处要将上级政府授权管理的事项及辖区内的社区建设与管理做细致、做具体。为此，搞好社区建设与管理工作，前山街道办事处应该做好如下工作：建立系统观念，以将涉及社区建设的方方面面的工作做细做足做实；树立主导观念，扮演辖区社区建设组织者及领导者角色；有效地争取上级政府从财力、物力等多方面支持及吸引社会资源参与本街道的社区建设。

2）适应新管理体制的要求

传统上，各级政府及其派出机构对所辖社区基本上是进行直接的行政管理，在构建起新型的社区管理体制之后，各社区居委会将被塑造成真正的社区居民自治组织。街道办事处与社区居委会相互之间已不为简单的行政关系所维系了；随着社会经济体制改革的进展，居住在社区中的为数众多的“单位人”（在传统体制中，因单位办社会、单位包福利，居住于社区中的居民与其所服务的单位有着特殊的千丝万缕的关系）逐渐转变成真正意义上的“社区人”，对社区的依赖性会大幅增强；同时，由于互为依托的关系，设立于社区中的各类机构与社区的联系也将变得极为密切。在这种新的环境中，街道办事处及社区居委会必定要将现行的行政式的领导及管理模式改变为以沟通、服务、协调和指导为基本内容的管理模式。顺应管理模式的转变，应该建立整体观念，体现社区内所有成员（包括全体居民及机构）的意愿，从辖区整体发展的要求出发，进行整体定位，整合全街道所有资源进行整体建设，并建立相应的社区运作机制；树立服务观念，寓社区管理工作于服务之中，既服务社区居民，又服务设立于社区内的机构，以提高社区服务质量并通过高效的服务提升社区凝聚力。

3）适应推行现代城市管理模式的要求

珠海市及香洲区政府在城市管理上提出要超越“小城管”（传统的城市管理模式）而实施“大城管”（现代城市管理模式）的基本策略，按照这一策略，前山街辖区内的城市基础设施的规划、建设与维护均将主要落实在社区管理这一层面上。为此，前山街必须逐步强化社区观念。

2. 将社区建设与管理工作落到实处

前山街道的社区建设与管理工作是一个复杂的系统工程。要做好这方面的工作，街道办事处还得通过如下措施来具体落实。

1）建立组织体系

成立前山街道办事处社区建设与管理规划领导机构，具体领导并组织力量开展本街道社区建设与管理规划工作。

2）具体安排实施工作

在本街道社区建设与规划报告完成后，尽快组织有关力量，对规划安排进行分解及系统安排，制订出具体的实施方案，并将其具体地量化及有效地落实到各个部门及有关社区居委会。

3）进行广泛宣传推介

运用系统设计的方法，采取多种手段和不同方式，通过多种渠道和媒介大力宣传及推介本报告所提出的社区建设思路，树立前山街的主体形象，以使街道上下能对前山街社区发展形成一个明确的认识，动员和引导本街道内外各种社会力量推动及促进本报告所提及方案的实施。

3. 确保社区建设与管理的投入

社区建设与管理工作，必须要有相应的资源投入。资金筹措的基本途径有两种。

1）政府的财政投入

根据国家、省、市所确定的“财随事转”“费随事转”的原则，对前山街社区建设与管理的所有工作进行系统测算，匡算其静态投资及动态投资的基本规模。其中，所确定的常态维持经费，由辖区政府设立专项财政预算项目（今后按每年新增财政收入的一定比例，增拨社区建设与管理经费）解决（2003 年前由本街道自行支付，2004 年后开始由香洲区政府支付）；对于有关动态建设项目的投资则可争取市或区政府的有关专项资金解决。

2）多形式筹集社会资金

这主要包括动员社会各界力量积极参与社区建设，以及接受社会捐献以筹集资金等方式。

4. 营造育人才和集人才的良好环境

进行社区建设与管理，必然要求各类专业技术、管理及复合型人才队伍要有新的发展。为此，前山街应从以下几个方面营造育人才和集人才的良好环境，以满足这一需要。

1）营造育人才环境

制定并健全各种学习制度，采取如下多种方式培育人才。一是选送有关人员到高等院校进修、学习。二是通过与各高等院校开展横向联合，或邀请区外专家学者到本街道进行讲学或开展人才培训。特别是可利用暨南大学珠海学院坐落于辖区内的优势，提请市政府与暨南大学联合设立“社区建设与管理培训基地”，以结合本街道需要，有效地培育及向辖区以至华南地区输送社区建设与管理方面的人才。三是鼓励街道及社区居委会的干部和职工自学相关专业技术与知识，以尽快造就一支高素质的人才队伍。

2）营造集人才环境

坚持有计划、有步骤地从社会引进本街道发展所需的各类专业人才。主要通过如下措施营造聚集人才的良好环境，以有效引进人才：一是更新用人观念，改革本街道人事制度，建立健全公平竞争、能上能下的用人机制，使真正的人才能有用武之地；二是对有突出贡献的人才要在精神上、经济上给予相应的奖励，以调动人才的积极性；三是对人才要在思想上、政治上、生活上给予充分的关心和优先照顾，为他们解决后顾之忧，使人才能来得了、用得上、留得住。在引进各类专业人才的同时，还要特别重视对“外脑”的借用。

广东的治安状况与投资环境[①]

一、安全是社会发展的基本保障

安全，是一个古老的话题，也是一个世界性话题。自"9·11"事件以来，情况更是如此。"9·11"事件对美国经济乃至于全球经济的打击之大不亚于一场战争。如果美国人仍没有安全感，他们的经济可能会持续低迷。安全关系到民众的信心，决定着经济发展的环境。就中国和其他国家而言，情况也是如此。当然，我这里所讲的安全没有那么复杂，不牵涉国际政治或国际军事。这里不是谈国家安全的问题。我这里只想谈谈社会治安问题。具体到广东，所谓安全问题更是纯粹的社会治安问题。社会治安要解决的就是社会内部的稳定与安全，要打交道的对象不是某些敌对国家和国际恐怖组织，而是潜伏于社会内部的有害个人和团伙。这些力量与外部的威胁国家安全的力量相比可谓是微不足道，但他们对社会安全造成的危害却绝对不容小觑，社会治安的好坏直接影响社会的发展。社会邪恶势力一旦成了气候，对社会发展造成的破坏是难以估量的，而后还要花大力气、大本钱才能根治。社会治安与文明进步是联系在一起的。广东光有经济发展是不够的，老百姓也不仅满足于生活富裕，没有安全感的社会是不完满的，因为，生活在恐惧中的人们是没有安居乐业的心情的。

二、社会治安状况与投资环境

广东是中国的经济强省，堪称中国经济发展的火车头，不过，广东经济也面临着重大挑战，也存在着再上一个新台阶的问题。广东经济需要某种重大突破，也应该有重大突破，同时也有能力、有条件取得重大突破。广东经济不能满足于相对于内地经济的优势。广东经济应该瞄着"亚洲四小龙"，应该瞄着欧美经济强国，要与它们一争高低。广东要做经济强省或经济大省的目标应该定位在这里。如果这一突破能够实现，对整个中国经济的带动将是巨大的。当然，要实现这一目标，广东还需要更多的资金和技术。没有这个保障，广东经济的发展将受到很大的限制。广东需要外地和外国资金源源不断地注入，而不能眼看自己的资金流向外地或外国。

在吸引外资与内资方面，广东当然有得天独厚的优势。广东堪称中国经济发展的热土，而且经济发展的条条框框相对较少，资金在这里能产生较大的经济效益。现在，中国已经加入 WTO。加入 WTO 会改变政府职能，过去制约经济发展的许多体制上的原因或人为的因素可能会减少乃至消失。这无疑会对引进外资与内资创造更加有利的条件。加入 WTO 有利有弊，但至少对引进外资和改善投资环境有着明显的积极的影响。现在大家都在谈加入 WTO 给我们带来机遇，希望外资蜂拥而至。可是，我们不

① 本文是在广东省政协各界别委员代表座谈会上的发言，广州，2003 年 1 月 12 日；现代管理的中国实践，北京：科学出版社，2016，137-139.

能让治安问题成为阻碍广东经济发展的重大障碍。就广东而言，总的来说投资环境和经济发展环境很好，加入 WTO 之后可能会更好。但治安方面的负面影响以后会不会进一步增加呢？应该说，广东的社会治安还是需要花大力气整治的。

三、社会治安状况关乎政府威信

苏联解体之后，俄罗斯政府对西方采取一边倒的政策，欢迎西方在俄罗斯投资，希望西方国家帮助俄罗斯一步到位实现市场经济。然而，俄罗斯的经济转型非常缓慢，经历了较大、较长的痛苦。外国资金并没有蜂拥而至，外国投资者大多持观望的态度。这里面原因比较复杂。其中一个非常重要的原因就是俄罗斯治安状况较差，许多外国投资者都落得个血本无归的下场。所以，现在西方国家对在俄投资并不积极。这大大延缓了俄罗斯的经济发展。说到底，外国投资者对俄罗斯没有信心，对俄罗斯政府没信心。

经济转型国家或经济过热地区社会秩序有一定程度的不稳定乃至混乱都是不可避免的，甚至是正常现象。但这种不稳定或混乱应该尽快被控制，否则，经济发展就要受到极大的拖累。就广东的情况而言，社会是稳定的。不过，广东存在的影响社会治安的因素要比别的省份多。这是广东的特殊情况决定的。因为广东有全国最多的流动人口，这是广东社会治安问题面临的最大挑战。因此，广东维护社会稳定的任务更艰巨，困难更多。

经济过热地区必然吸引大量的流动人口。经济发展和经济市场化必然需要劳动力市场的高度开放。市场化首先应该是人才和人力资源的市场化。人口流动是必要的，不过，外来人口的涌入必然对当地的社会治安造成较大冲击。但我们不能因噎废食，不能为了保障治安而杜绝外来人口。这是不明智的做法，也是不负责任的做法或偷懒的做法。当然，我们也没有这么做。任何时候，维护社会治安都应该是头等大事。这是我们的首要任务。完成这一任务是无条件的。我们决不能认为社会治安的恶化属于正常情况。社会治安状况直接关系到政府的威信，也关系到民众对经济发展的信心问题。

政法部门当然是解决好治安问题的主力军。不过，社会治安是一个系统工程，需要全社会的投入和关心。要唤起全民的责任心，调动起全社会的力量。让犯罪分子感觉到整个社会强大的威慑力。这方面要有一个系统的筹划，同时也需要全社会献计献策。相信广东会成为人人向往的生活乐土。这才是投资环境得到彻底改变的根本标志，希望广东经济的发展与良好的治安环境同步实现。

教育彩票作用惊人①

彩票被国外经济学家称为第三次分配的“神奇之杖”，购买彩票则被称为“微笑纳税”。作为一种吸纳社会资金、调节社会收入和增进消费者福利的有效手段，彩票具有一定的积极意义。如果操控得当，教育彩票的收入便能真正做到“取之于民，用之于民”。

一、良好效益

首先，发行教育彩票可以带来良好的社会效益和巨大的经济效益。以广东省发行福利彩票为例，自 1987 年发行福利彩票以来，广东省已发行销售福利彩票约 93 亿元，筹集社会福利资金 23 亿元，累计投放使用近 22 亿元，资助各种社会福利和社会公益项目超过 1 万个，收到了良好的社会效益。仅 2002 年度“南粤风采”计算机福利彩票，中奖者所缴的税便达 1.2 亿元，增加了政府财政收入。如果发行教育彩票，可以说是利国利民利教育。

二、发行符合彩民心理

发行教育彩票符合彩民的社会游戏心理。彩票的无穷魅力在于微小的投入可能带来巨大的收益，对未来事物的猜测和对中奖的期望，可以在短时间内调动人们的积极性。而且，中国的多数彩民在购买彩票时比较理智，极少有彩民抱着孤注一掷的赌博心态去购买彩票，正因为如此，彩票发行以来社会各方面的反应平静。

三、具有深厚社会基础

发行教育彩票在中国具有深厚的社会道德基础。中华民族千百年来形成了扶贫济困、扶弱助残的优良传统，具有乐善好施的传统美德，教育彩票可以为中国这个深厚的社会基础赋予新的形式和载体。调查表明，51.5%的彩票购买者抱着“中奖幸运，无奖奉献”的心态：中彩得奖，可以使生活多点惊喜；空手而返，算是为社会做点公益。

四、可以激活巨额闲资

发行教育彩票可以将巨额闲散资金由“休眠态”转化为“市场态”。现在，中国城乡居民储蓄额达 10 万亿元，其中广东省城乡居民储蓄额为 1 万亿元，发展教育彩票具有广阔的空间。如果发行教育彩票，相信百姓也会踊跃购买，其中将不乏善举，即使将这笔庞大资金的一部分用于教育事业，也会产生惊人的推动作用。

---

① 本文原载《南方日报》A4 版，2003 年 1 月 24 日.

五、可以避免捐赠弊端

发行教育彩票，可以避免捐赠的一些弊端。从广义上讲，人们购买彩票是个体的捐助行为，但购买彩票的善行已经完全避免了捐助者和被捐助者之间面对面的接触。社会成员更关注彩票发行销售之后所募集的资金是否真正用于社会公益，是否被集中起来科学、合理地使用，是否得到有效的善用。这便会自觉形成一个资金使用监督机制，以达到发行彩票的初衷。

六、彩金四成用于教育

基于以上几点原因，建议在全国发行教育彩票，可以以广东省为试点，并由省政府直接管理。每两日为一期，彩票收入的50%用于奖金，10%用于发行成本，40%用于支持教育。投入教育的经费中，50%用于基础教育，50%用于高等教育。假使广东省教育彩票每年发行额12亿元，则用于基础教育的资金可达2.4亿元，如以贫困地区中小学生读书每人每年需1000元计算，便可资助24万名失学儿童。

七、其副作用微不足道

发行教育彩票给我们带来的好处是全局性的，副作用与之相比则是局部的、微不足道的。“两利相权取其重，两弊相权取其轻”，发行教育彩票这件有利于国计民生的大事，我们为什么不试一试呢?

建设广州石牌大学城[①]

广东的经济发展在改革开放20多年来已取得世人瞩目的成绩。当前，广东要进一步增强发展后劲，就必须大力发展高等教育，推进高等教育大众化。连续几年的高校扩招，已使广东不少学校的教学资源捉襟见肘。解决大学基础设施供求矛盾的最好办法便是兴建大学城，卢瑞华省长在广东省十届人大一次会议上所作的《政府工作报告》中特别指出要“建设广州大学城，实施高校‘强校建设工程’”，建设大学城的工作已经提上日程。

当前，不少人提出在番禺小谷围岛建设广州大学城，这当然是可行之道，但大学城建设的指导方针应是在充分利用现有条件的前提下，积极开辟新的教育资源。所以，我们更应该高起点地规划和建设好现有的广州石牌大学区来建设石牌大学城。

(1) 石牌建有多所大学，已初具大学城雏形。在广东进入国家“211工程”建设的四所大学中，石牌地区就有三所——暨南大学、华南师范大学、华南理工大学，连同附近的华南农业大学、广东技术师范学院、广东工业大学（五山校区），设在石牌的广东乃至全国知名大学已达六所。这些大学涵盖文、理、经、管、法、工、农、医、生命科学等多个学科，使石牌成为广州著名的大学园区，形成了良好的学术氛围。美国华裔著名科学家杨振宁先生曾经说过，一所名牌大学形成学术氛围，没有五十年的时间积累是不可能的。所以，在石牌建设大学城具有其他地区无法比拟的环境优势。

(2) 在石牌建设大学城，具有优良的资源优势。经过国家和地方政府的持续投入，石牌各高校均已建成完善的校园基础设施、教学设施和公共服务设施。石牌六校作为国家级文化素质教育基地，早已实现了图书馆藏书互借和跨校选课，具有较长时间的友好合作关系。而且各高校之间的距离适中，实验室、仪器设备、体育馆等教育设施和学生公寓、食堂等公共设施及师资方面均易实行共建共享，学生也可以共同开展课外讨论及专业学术活动，具有建设大学城所需的资源共享条件。

(3) 在石牌建设大学城，具有深厚的产业优势。大学城是一个集教育、科技和现代服务业为一体的宏观产业环境，在石牌高校附近已自发形成一个高科技产业区和文化产业区，最具代表性的便是以太平洋电脑城和广州科贸园数码城为代表的电子产业，以天河购书中心和天河娱乐广场为代表的文化产业，以天河城、天河购物中心、总统大酒店、华威达酒店等为代表的第三产业，效益良好，声名远播。据统计，太平洋电脑城年交易总额已超过北京中关村的交易额。一个以教育、科技和第三产业相结合的宏观文化环境在石牌一带已经形成，必然能进一步促进大学城的发展。

(4) 在石牌建设大学城，具有完善的服务体系。经过几十年的建设，石牌附近形成了完善的公共服务设施体系，东有天河公园、西有动物园、南有珠江公园、北有广

---

① 本文是在广东省做大做强高等教育座谈会上的发言，东莞，2003年4月2日；教育与科技管理研究，北京：科学出版社，2016，121-123.

东植物园，乘车 10 分钟可到天河体育中心，天河娱乐广场、天河购物中心则与高校比邻而居，银行、电信等服务设施齐全。广东省教育厅计划在华南农业大学附近建设的大型教师住宅和学生公寓，更解决了在石牌建设大学城的后顾之忧。一个以教育产业为龙头，高素质、高层次、多功能的知识园区已经在石牌一带呈现。

（5）在石牌建设大学城，有利于“泛教育”事业的进行。“泛教育”的核心思想是“跳出教育做教育”，从教育的产业价值和实用角度延伸教育的多种功能，以教育为形，用教育产业作龙头，充分利用各种产业的教育功能，培养综合型高素质人才。“泛教育”要求建设一个以教育为龙头，复合教育、科技、图书、文化、商业、休闲、娱乐、体育、旅游、培训等多种产业和行业的巨型产业群，在石牌建设大学城已初步达到“泛教育”的要求。

可见，石牌一带历史性地形成了高教园区，具备了建设大学城的许多必需条件。我们为什么要舍弃这么好的教育基础呢？一个全新的大学城固然可以在全面规划后快速建成，但浓厚学术氛围的形成和公共服务设施的完善绝非一日之功，而且重复建设将造成资源的极大浪费。所以，我们的当务之急应是尽快对石牌进行合理规划，以最少的投入获得最大的收益，早日把石牌一带建设成统一管理、分工负责、资源共享、优势互补的大学城。

办好院士之家[①]

今天，由中国工程院与广州市政府合作建立的中国工程院院士广州咨询活动中心在这里亮牌成立，这将成为院士之家。我代表中国工程院院士广州咨询活动中心，对前来参加仪式的各位领导、嘉宾表示衷心的感谢！

为加强中国工程院与广州市人民政府的合作，促进广州市经济和科技快速发展，2003 年 2 月 25 日，中国工程院与广州市政府商定，在广州市建立中国工程院院士广州咨询活动中心。中国工程院是我国工程科技界的最高荣誉性、咨询性学术机构，拥有一批在工程科技方面作出重大成就和贡献的院士，掌握一批高水平的科技成果。广州是华南地区的中心城市，一直走在全国改革开放和经济发展的前列。进一步加强工程院与广州市的紧密合作，有利于发挥中国工程院的人才优势和科技资源优势，促进广州市经济社会持久快速健康发展。

中心的成立是中国工程院与广州市政府开展合作的良好开端，也是工程院对如何与地方经济结合、为地方经济服务的一种有益探索和尝试。中心将依托中国工程院强大的技术力量，依靠广州市雄厚的经济实力，为双方的合作找准切入点，力争成为双方开展合作和交流的服务平台。

一是对广州市重大项目、重大工程开展决策咨询。围绕广州市经济和社会发展的重点与热点问题，组织专家进行战略研究，为广州市政府的一些重大决策提供咨询、论证服务，为科学决策提供指导。同时，对广州市拟建的重大项目、重大工程进行调查研究，开展可行性论证。

二是引荐适合转化和产业化的科技成果。中心将根据广州市经济社会发展的需求和实际，一方面，把工程院拥有的大量的、高水平的科技成果推荐给广州市进行转化和产业化；另一方面，利用工程院的科技资源加强信息交流，向广州市介绍、引进一批适合转化和产业化的国内外先进科技成果，为成果转化提供市场调研、技术评估等服务，加速科技成果转化，支持广州科技和经济的发展。

三是对技术难题进行攻关。围绕加快完善广州区域创新体系建设，推进关键技术创新和系统集成，充分利用工程院和广州现有的科技资源，进行攻关，突破一批对广州市经济社会发展至关重要的技术瓶颈；研制、开发一批科技含量高、市场容量大、竞争力强的重大产品；攻克一批对产业升级改造和持续发展有重大带动作用的共性关键技术，不断提高广州区域创新能力。

四是开展学术交流与研讨。中心将根据当前世界科学技术发展的趋势和我国科技发展战略，组织国内外有关专家，开展专题学术交流，跟踪了解国际科技发展动态和趋势，开展技术发展预测和战略研究；根据科技与经济结合的内在要求，组织专家，

---

① 本文是在中国工程院院士广州咨询活动中心亮牌仪式上的讲话，广州，2003 年 6 月 18 日；教育与科技管理研究，北京：科学出版社，2016，286-287.

针对经济与社会发展中亟须解决的问题进行专题研讨。

五是加强与在穗工程院、科学院院士的联系，做好为在穗两院院士服务的工作。

我相信，通过双方的共同努力，中国工程院院士广州咨询活动中心将成为中国工程院与广州市合作的一个重要桥梁和纽带，为推动科技与经济的紧密结合作出应有的贡献。

坦诚建言[①]

在新春佳节即将来临之际，省委、省政府以及省委统战部的领导与我们欢聚一堂，同贺新春，我们感到非常高兴。在这喜庆的时刻，我代表新聘参事和馆员祝在座各位身体健康，新春快乐！

参事室（文史馆）是在老一辈无产阶级革命家毛泽东主席和周恩来总理亲自关怀下成立的，参事工作是党的统一战线工作的重要方面，也是政府工作的组成部分，是我国民主政治建设的具体体现。特别是去年，中共中央、国务院为了加强新时期参事工作，专门颁发了中办发〔2003〕10号文件，省委、省政府也颁发了粤办发〔2003〕20号文件，对新时期参事工作提出了新的更高的要求，进一步明确了参事工作是政府了解民情、反映民意、集中民智、密切联系群众的重要渠道，加强参事工作也是发展社会主义民主政治、建设社会主义政治文明的有益措施。

今天，我们能够非常荣幸地在这里接受省长的聘任，加入参事、馆员队伍，成为其中的一员，感到非常高兴和光荣。我们十分感谢省委统战部、省政府参事室（文史馆）党组对我们的关注和信任，我们决不辜负省委、省政府对我们的期望，坚持以邓小平理论和“三个代表”重要思想为指导，围绕全面建设小康社会的奋斗目标，认真履行职责；并按照恪尽职守、坦诚建言、各展所长的要求，发挥我们自身的优势，为广东省率先基本实现社会主义现代化和发展社会主义民主政治而努力奋斗。

① 本文是在广东省人民政府参事、馆员迎春茶话会上的讲话，广州，2004年1月14日.

光大华侨文化　建设文化大省①

一、对建设文化大省的认识

自广东省委九届二次全会提出了建设文化大省的战略部署后，我省各条战线都积极围绕这一重大决定展开了热烈讨论和慎密布置，并根据各自的实际情况针对建设文化大省的工作提出了具体工作思路。今天，政协广东省委员会九届二次会议组织我们与省领导们一起以“弘扬岭南文化，建设文化大省”为主题进行座谈，这是对省委关于建设文化大省这一重要决策的贯彻和落实，是推动我省实现物质文明、政治文明和精神文明全面、协调发展的重要举措。

改革开放初期，广东作为中国改革开放的前沿阵地，在我国经济建设中发挥了桥头堡的重要作用。当我国改革开放的车轮驶入新的历史时期，广东又承担起了率先实现社会主义现代化的光辉使命。《中共广东省委关于认真学习贯彻党的十六大精神的决定》指出：全面建设小康社会，必须与建设经济强省相适应，大力发展社会主义文化，建设文化大省。改革开放以来，广东经济高速发展，一跃成为全国的经济强省，但在此过程中，其文化的发展与其经济强省的地位并不相称，与经济强省应有的文化地位还有一段距离。正是在这种背景下省委作出了建设文化大省的决定，这为我省今后的文化发展，为我省物质文明、政治文明和精神文明的全面、协调发展指明了方向，提出了新的要求。经济的健康、稳定、持续发展与文化建设密不可分，与参与经济建设的人的素质密不可分，与发展环境的文化内涵密切相关。广东要率先实现社会主义现代化，要全面建设小康社会，要保持强劲的发展态势，必须有相应的文化环境和文化内涵来支撑。因此，我们要把建设文化大省作为一种理念渗入全省每个人的心中，贯彻到我们的各项事业中去。我们每个公民、每个单位都有义务、有责任为建设文化大省承担力所能及的工作。我们应该认识到，建设文化大省，既是全面建设小康社会、率先实现社会主义现代化的需要，也是满足我省人民群众随着物质文化快速发展而日益增长的精神文化需求的需要。

二、华侨文化是建设文化大省的重要内容

去年5月，省侨办、暨南大学华侨华人研究所等4家单位为贯彻落实省委建设文化大省的决议精神，联合召开了“华侨文化建设与文化大省”研讨会。会议指出，研究华侨华人文化在建设广东省文化大省的工作中有着重要的意义。老一辈华侨华人的大量出现是中国历史发展的产物，其作为特殊时期产生的特殊群体，依然是中华民族不可分割的一部分。广东作为全国最大的侨乡，华侨渊源历史悠久，华侨文化积淀深厚。华侨文化既是我国文化的重要组成部分，更是广东文化的重要内容。

① 本文是在广东省政协“弘扬岭南文化，建设文化大省”专题座谈会上的发言，广州，2004年2月11日.

特别是在中国加入世界贸易组织以后，广东省的对外文化交流和贸易进一步扩大了，而华侨华人文化作为其中的一个重要桥梁，发挥了其独特的作用。因此，结合省委建设文化大省的战略部署，进一步加强对华侨文化的研究、宣传，并结合时代的特点进行重新塑造，是推进广东省文化大省建设的一项重要举措。在新的历史时期，建设好华侨文化，对增强民族凝聚力、促进文化开放和经济发展既有重要的现实意义，又有深远的历史意义。

三、暨南大学在光大华侨文化中的优势和作用

暨南大学作为中国第一所由国家创办的华侨学府，在我国高等教育界担负着特殊的使命，享有特殊的地位，发挥着特殊的作用。它是国家联系海外华侨华人和港澳台同胞的一座文化桥梁，负有培养海外华侨华人和港澳台青年的历史使命。我们要通过这座文化桥梁延续和传承中华优秀传统文化，努力团结海外华侨华人和港澳台同胞，不断增强民族凝聚力。

文化交流是文化发展和进步的重要内容。暨南大学这座文化桥梁通过传道授业解惑及其得天独厚的地理条件为中西文化交流提供了平台。学校坐落在祖国改革开放前沿的广东，华侨华人带来的西方文化与中华优秀传统文化以及广东本土文化在这里融合交汇，广东文化也在与西方文化的交流发展中吸纳了许多西方文明的有益成果，不断发展和进步。在建设文化大省、弘扬华侨文化的过程中，我们要通过文化交流，充分利用华侨华人既熟悉中国文化又熟悉居住地文化的优势，积极发挥华侨华人文化大使的联系作用，吸收异域文化的精华，丰富和发展中华优秀传统文化。

对于海外华侨华人和港澳台同胞来说，民族凝聚力来自他们对中华民族优秀传统文化的认同和延续。作为传播中华优秀传统文化和联系海内外华侨华人亲情的重要桥梁与纽带的暨南大学，以其 98 年的办学历史见证了这一事实。我们常说，有海水的地方就有华侨存在。在新生代华侨华人比重日益增长的今天，相对于老一辈华侨来说，新生代对中国历史文化和发展现状知之甚少。但随着中国经济的快速发展以及综合国力的迅速提高，这些新生代的华侨华人了解中国的迫切感和对中国的认同感正与日俱增。在这种特殊时期，暨南大学充分发挥自身优势，积极实践国家赋予的特殊使命，努力创造条件，不断加强基础设施建设，大力提升办学水平和综合实力，吸引更多的海外华侨华人和港澳台同胞青年前来学习。尤其是近两年来，暨南大学在海外及港澳台的招生人数连创新高，首次达到并超过了海外及港澳台学生与内地学生人数 1∶1 的比例。这种突破是基于我国综合国力的稳步增强，基于我国高等教育水平的不断提高。自 1978 年以来，先后有 99 个国家和地区的华侨华人青年在暨南大学学习和生活过，暨南大学共为上述国家和地区培养各类人才万余人，他们当中的许多人在实践“一国两制”、传播中华优秀传统文化和推动我国社会主义现代化建设以及居住地建设中发挥了重要作用。当前，暨南大学共有来自海外 52 个国家和中国港澳台 3 个地区的学生 7480 余人，这些人都是文化传播和文化交流的重要载体，他们在暨南大学这一文化摇篮中相互接纳、传播、交流和发展不同民族的文化，共同享受人类的文明成果。也正是由于他们的到来，为我们提供了更多的了解华侨文化、研究华侨文化、发展华侨文

化的机会，同时也为我们进一步宣传、弘扬华侨文化创造了条件。

在新的历史时期，暨南大学确立了“侨校＋名校”的发展战略。我们正在努力为海外及中国港澳台学生创造一个更好的学习生活环境，进一步为传播中华优秀传统文化，研究和宣传华侨华人文化做好工作，使华侨文化在这里得到交流、弘扬和发展，为国家精神文明建设和广东省建设文化大省作出新的贡献。

四、对弘扬华侨文化的几点建议

1. 建立广东省华侨华人博物馆

长期以来，广大华侨华人及港澳台同胞为祖国的建设和发展出谋献策，捐资捐物，殚精竭虑。尤其是改革开放后，他们为祖国的改革开放事业，为改变家乡贫穷落后的面貌作出了不可磨灭的伟大贡献，在中国发展史和华侨史上谱写了光辉的篇章。作为中国最大的侨乡，广东是最大的受益者。要将广东建成文化大省，大力弘扬华侨文化，就应该让人们更多地了解华侨华人在海外奋斗发展的创业足迹，展示华侨华人心系祖国、矢志报国的爱国情怀，铭记华侨华人参与祖国建设和发展的丰功伟绩，激励后辈进一步发扬中华民族伟大的优良传统。因此，我认为省委、省政府如果能在广州建设一座华侨华人博物馆，将其作为征集、收藏、展览、宣传与华侨华人有关史实和资料的重要基地，它将能够生动地展示华侨华人的历史，进一步树立侨乡的形象和品牌。

2. 加大对华侨华人文化建设的支持力度

首先，暨南大学是广东省最重要、最有优势的华侨华人文化教育和研究基地，在广东省华侨华人文化建设中起着不可替代的作用。省委、省政府如能进一步支持暨南大学，为学校创造更好的硬件设施和办学环境，暨南大学作为华侨华人进一步了解中国、了解广东的窗口，必将吸引更多的华侨华人来广东学习、工作和投资。暨南大学近几年的跨越式发展已经印证了这一事实。

其次，充分发挥我省高校华侨华人研究机构的学术优势，建立专门基金，有计划地设立有关华侨华人研究的课题，加大对华侨华人历史、现状以及未来发展的研究。

最后，组织海内外华侨历史专家、学者就华侨历史文化进行学术研讨，共同研讨广东华侨历史文化。

传承岭南文化　服务文化大省[①]

文化，是社会文明进步的先导和重要标杆。在文化因素日益取代自然资源和物质资本并成为决定社会发展最重要因素的今天，文化的地位和作用更受关注。建设“文化大省”，是广东省委、省政府在新时期做出的一个很有远见、很有气魄的重大决策，是广东可持续发展的又一重大战略选择，也是广东再次腾飞的一次重大历史机遇。

建设“文化大省”，必须大力弘扬优秀的传统民族文化。作为中华民族传统文化中最具特色和活力的地域文化之一，岭南文化不但最先接受了欧风美雨的冲击，而且作为中国改革开放的试验区开风气之先河，逐步形成了底蕴深厚而又富有地域和时代性的文化特色。正是有赖于这种生机勃勃的文化力量的强劲推动，广东创造了举世瞩目的经济辉煌。在建设“文化大省”的进程中，岭南文化这种与时俱进的生命力的重要性更是不言而喻。

时代的需要，将对岭南文化的研究推向深入。《岭南文史》为此进行了长期不懈的努力，他们积极挖掘岭南文史中最具特色和生命力的东西，冀以为当今的文化建设和社会进步提供有益的借鉴和启示。作为广东文化建设的一支生力军，《岭南文史》在当今广东的“文化大省”的建设过程中，在促进经济社会的协调发展中，正在发挥着独特而巨大的作用。

鉴往知来，探古以兴今。《岭南文史》可以当之矣。

① 本文原载《岭南文史》，2004，(2)：扉页.

建立粤港澳综合协调机构[①]

广东、香港、澳门三地毗邻，唇齿相依，在经济、文化、教育等方面的合作与交往历来频繁。自1978年改革开放以来，广东省借助于毗邻港澳的得天独厚的优势，实现了跨越式发展。近年来，尤其是港澳回归祖国和我国加入世界贸易组织后，粤港澳合作迈向新的历史时期，三地发展遇到了许多新情况、新问题，也遇到了前所未有的机遇，粤港澳的发展都有很大的空间。对此，省委、省政府已采取了许多行之有效的有力措施，效果显著。但是，目前的合作多是单位之间的合作，协调不够，影响了广东省经济的发展。当前，长江三角洲和渤海湾地区经济发展很快。为了使广东省具有更强的竞争力，继续保持在国内经济发展中排龙头的地位，建议在进一步健全粤港澳联席会议制度等工作的基础上，在广东省建立一个推动三地在各方面开展富有成效的协调与合作，促进粤港澳积极要素互动的综合协调机构，在粤港澳三地开展的经济、文化、教育、科技、人才、旅游等方面的交往与合作中担负组织协调职责，当是促进三方互动交流，提升广东持续发展竞争力，推动区域经济发展，实现共同繁荣的一大重要举措。

该机构直属省政府管理，其职能应不同于省港澳台事务办公室、外经贸委等单一职责机构所承担的职能。机构的建立，对于落实张德江书记提出的关于粤港澳合作要以“全局、前瞻、务实、互利”为指导思想的指示，深化粤港澳合作，进一步全方位、宽领域、多层次加强粤港澳合作，增强粤港澳三地互补互动互利的合力，推进广东现代化建设和港澳地区繁荣稳定将起到积极作用。

① 本文是政协广东省第九届委员会提案，广州，2004，731；《现代管理的中国实践》，北京：科学出版社，2016，112.

泛珠三角：推进科技、教育和文化的区域合作①

一、"泛珠三角"经济区发展应提升为国家级发展战略

2003年7月，中共中央政治局委员、广东省委书记张德江同志首次提出"泛珠三角"区域协作这个概念。他指出，广东要"积极推动与周边省区和珠江流域各省区的经济合作，构筑一个优势互补、资源共享、市场广阔、充满活力的区域经济体系"。"泛珠三角"经济区概念一经提出，立即获得周边省区的热烈回应，并迅速进入了实质性启动阶段。

从科学发展观来看，"泛珠三角"经济区的倡导，符合区域经济发展的客观规律和要求。20世纪90年代以来，经济全球化和区域化已成为国际经济发展的主要趋势，世界经济正趋向于形成欧盟、北美和亚太区特别是东亚区三大板块，区域间的竞争正成为时代的主要特征。就亚太区而言，亚洲金融危机后，东亚各国明显加快了区域经济合作的步伐，《清迈协议》的签署、东盟自由贸易区的启动以及"中国-东盟自由贸易区"（"10+1"）的倡导等，都是亚太区经济走向一体化的具体表现。

在经济全球化、区域化的国际背景下，CEPA（《内地与香港关于建立更紧密经贸关系的安排》）的签署加快促成了大珠三角经济区的崛起，并为"泛珠三角"经济区的提出和启动提供了客观基础。与长江三角洲地区相比，大珠三角经济区的主要弱点是经济腹地不足。"泛珠三角"经济圈的提出，既体现了粤港澳经济能量集聚亟须扩大释放经济腹地的内在要求，也反映了周边省区接受粤港澳经济辐射的强烈愿望。"泛珠三角"面积199.45万平方公里，人口4.46亿，分别占全国的20.78%和34.3%，GDP占全国的1/3。由于它的鲜明特点，即自然资源和人力资源丰富，同时拥有资金、金融、技术、人才、交通的优势，因此，"泛珠三角"经济区的提出，对加强粤港澳经济合作，统筹我国东、中、西部地区的协调发展，沟通我国大西南和东盟的经济联系，并提高整个区域的国际竞争力，无疑都具有极其重要的战略意义。

从这一视角看，应将"泛珠三角"经济区发展，提升为国家级发展战略，以更有效地统筹区域内各省区的合作与发展。

二、区域经济合作是一个系统工程

区域经济合作是一个包含众多领域的系统工程。就"泛珠三角"经济区而言，区域经济合作毫无疑问将首先在经济层面展开，现阶段可启动的合作领域包括：基础设施建设和协调发展、自然资源开发及合理配置、产业的合理分工和配套、旅游资源的互补、物流运输服务系统的形成、统一共同市场的筹建及区域内政策的协调等；但是，

① 本文原载《泛珠三角区域合作与发展论坛演讲录》，澳门，2004年6月2日，32-37；《现代管理的中国实践》，北京：科学出版社，2016，106-109.

考虑到经济与科技、教育、文化的密切互动关系，积极推进科技、教育、文化等方面的区域合作，也是其中极为重要、不可忽视的领域。

纵观人类社会的发展历史，经济发展固然推动了科技、教育、文化的发展，但实践证明，科技进步已经越来越成为生产要素中最重要的因素。研究表明，在西方发达国家，科技进步对经济增长的贡献率已经从 20 世纪 70 年代初的 50%提高到现在的 80%。近年来频繁出现的知识经济这一概念，实际上就是突出科技进步在经济发展中的重要作用。

目前，经济全球化、区域化这一发展趋势，实际上也是由世界科技进步所决定的。随着科学技术特别是信息技术的发展，高新技术的使用范围越来越广。以计算机及其他信息技术为基础的高新技术在企业中的广泛应用已成为 20 世纪最后 20 年的主要特色之一。在技术全球化的背景下，产品生命周期进一步缩短，产品的研制开发难度越来越大，面对同一机会可以参与竞争的企业越来越多，这就大大加剧了国际竞争的激烈程度。

在技术全球化背景下，各国的生产诸要素出现大规模的跨国界流动，跨国公司大量涌现并在全球经营，形成分散性生产和世界性生产网络，供应链被拉长并在全球范围内配置各节点，全球供应链管理亦应运而起。经济全球化、区域化有利于世界各国发挥比较优势，节约社会劳动；使生产要素得以在更广区域甚至在全球范围内得到合理配置，从而最大限度地节约成本，提高边际利润。因此，经济发展的背后，是科技、教育、文化等因素的有力支撑和配合。

可以预料，随着“泛珠三角”区域经济合作的逐步展开，区域内九省两区之间的科技、教育、文化等领域的合作与交流也必将提到议事日程上，并且将成为越来越重要的合作领域。

三、三个重要问题

当前，在“泛珠三角”经济区内，积极推进科技、教育、文化的区域合作，需要特别强调以下几方面。

第一，积极推进科技合作，构建区域科技交流与合作平台，创建区域科技创新体系，以提高整个区域的综合科技实力。

区域经济整合与发展，必然带动科技的交流与合作，并且需要后者的强有力支撑。去年 10 月，广东省与“泛珠三角”各省区签署的《泛珠三角区域科技创新合作框架协议》，在科技合作方面迈出重要的一步。根据《泛珠三角区域科技创新合作框架协议》，“泛珠三角”区域将实行科技资源的开放和共享，加快推进科技资源的联网共享，相互认可经科技行政管理部门认定的有关资质，建立科技项目合作机制，鼓励和支持区域内高校、科研院所、企业联合承担国家重大科技项目，并围绕“泛珠三角”的特色资源和共性技术开展联合攻关。不过，该协议仅属框架性协议，尚有许多原则性、技术性问题和细节需要跟进、落实，以构建区域科技交流与合作平台。

更进一步的合作是建设区域性的科技创新体系，按照“优势互补、互惠互利、合作共生”的基本原则，建立“泛珠三角”区域科技创新体系，以营造科技创新的宏观大环境，提高区域自主创新能力和区域科技综合实力，以适应经济的发展和产业结构的升级转型。

第二，推进区域内的高等教育合作，加强高中级人才的培训与交流，建立人才交流的统一平台。

科技进步需要教育支撑。“泛珠三角”区域内各省区的高等院校都各有其独特的优势和传统，加强区域内高等院校的交流与合作，实行优势互补，有利于提高区域内教育发展的整体水平，为区域经济发展提供科技发展动力和源源不断的人才资源。建议“9+2”省区借鉴科技合作的经验，签署《“泛珠三角”区域教育合作框架协议》，加强区域内各省区教育信息和经济社会发展情况的交流，定期举行教育合作交流活动及教育发展的学术研讨；推进区域内优质教育资源共享，扩大优质教育资源的辐射力；鼓励区域内高等院校开展校际教学合作，推动师资互聘、联合办学、联合攻关等。

目前，制约珠江三角洲经济发展的诸因素中，人力资源质量已成为经济发展的重要“瓶颈”。珠三角人才储备不足已成为未来经济发展的主要隐患。因此，广东应该加强与香港、澳门以及周边省区的人才交流与合作，积极吸引港澳人才参与广东经济发展。近年来，随着珠江三角洲经济的增长，已经有越来越多的港澳人才愿意北上工作、创业或是合作。据调查，香港已有接近 10 万个不同层次的技术与管理人员应聘到珠三角工作。从欧盟的经验看，在区域经济合作朝共同市场方向发展时，人才的交流及劳动市场的流动是难以阻挡的，需要一个合理的机制进行疏导。因此，应在加强“泛珠三角”人才交流的基础上，着手构建区域人才交流和培训的统一平台，从而更有效地整合区域人力资源，以适应经济合作与发展的需要。

第三，积极推进区域文化交流，培育及发展“珠江文化”。

“泛珠三角”的概念起源于珠江流域，因此，它不仅是个经济的概念，也是一个以珠江为核心的文化概念。有学者早已指出，珠江文化的地域范围包括下游的香港、澳门和中上游的广东等 8 个省区，与“泛珠三角”所涉及的区域基本一致。这种地域范围的一致性，有利于通过珠江文化的培养支撑“泛珠三角”区经济的发展，通过珠江文化可以将它们联系起来，在同一水域范围内进行合作。因此，有必要推进区域内文化的交流和发展，培养区域内珠江文化的广泛认同感，培育和发挥珠江文化的开放性、领潮性、务实性、兼容性的特点，以利于在区域经济合作中磨平保守的地方观念，消除观念上的隔膜和差异。

四、结语

需要指出的是，在科技、教育和文化的区域合作中，高等院校是重要的载体之一。暨南大学是全国著名侨校，也是进入“211 工程”的国家重点大学，创办至今已有近 100 年的悠久历史，与香港、澳门的高等院校以及海外众多教育机构有着长期、密切的联系，在海外享有良好声誉。目前在 2 万名研究生和本科生中，有来自世界五大洲 56 个国家和中国港澳台 3 个地区的学生 7772 名，在 2003 年全国 1533 所高校综合实力排名已提升为 36 名。在区域内科技、教育和文化的合作中，暨南大学将致力于加强与区域内兄弟院校的合作，为“泛珠三角”区域经济发展做出应有贡献。

促进广东省职业教育发展[①]

为促进我省职业教育发展，广东省政协教科文卫体委员会与民盟广东省委会联合组成了调研组，于2004年6月至9月间，走访了广东省教育厅、广东省劳动和社保厅，深入广州、深圳、佛山和韶关等地进行了调研。

一、广东经济与社会的发展，要求大力推进职业教育改革与发展

中国特别是广东正面临着成为新一轮“世界工厂”的重大机遇，我们能否抓住这次机遇，真正成为我国乃至世界最大的制造业基地之一，关键在于我们能否不失时机地大力发展职业教育。为适应广东经济与社会发展要求，必须大力推进我省职业教育改革与发展。

近年来，广东省下达了不少旨在推动我省职业教育发展的文件，显示出党政领导对发展职业教育的重视，但是，我省目前职业教育发展的状况依然大大落后于客观要求。其重要原因之一就在于没有真正认识广东经济与社会发展对职教发展的客观要求。

（1）大力发展职业教育是我省经济结构调整的必然要求。我省经济结构调整，一方面将产生大量的新兴产业，如信息技术、生物技术、光机电一体化技术和新材料等高新技术产业，形成对新型技能型人才的巨大需求；另一方面将淘汰一批不适应经济与社会发展的企业，导致大量的结构性转岗及失业人员。据调查，广州市制造业的24个行业中，在引进高新技术情况下，只有19.5%的技术工人能完全胜任新技术工作，其余80%需要在职培训提高。技术工人缺乏是制约广东高新技术产业发展的瓶颈之一。因而，广东必须不失时机地真正地大力发展职业教育。

（2）外来工和下岗职工的就业再就业需要大力发展职业教育。2003年9月，我省流动人口已达到2100万人，他们普遍文化偏低，技能缺乏。例如，初中及以下学历的超过75%，对他们的培训不仅有利于广东经济发展，也有助于广东社会治安的稳定。

广东下岗失业人员、新成长劳动力和农村富余劳动力增多，失业问题日趋严峻。1998～2001年，城镇失业人员总数逐年上升，分别为68万人、87万人、93万人、104万人，2002年120万人，到2003年失业人口达到125万人。广东省发展和改革委员会在2004年报告中，将就业再就业问题列为我省经济运行的首要问题。

流动人口或下岗职工的各种岗位的培训，要靠职业技术教育来实现。

（3）广东城市化进程对职业技术教育的需求。据有关部门测算，2005年广东城市

① 本文是广东省第九届政协教科文卫体委员会和民盟广东省委员会向政协提交的调研报告，广州，2004年9月；《教育与科技管理研究》，北京：科学出版社，2016，124-131。作者：刘人怀，韩大建，王则楚，钟韶，罗小平，王绵宁，梁仁，林维明，潘炬，张晓丹，谢可瑶，常会友，夏伟，陈朝填，乐军，林鸿伟，汪国强，李盛兵，董标，郑国强，高宏的，梅霭，周敏，王光飞.

化水平将达到40%，届时将有近600万人从农村向城镇转移；而到2010年，将达到800万人，这部分人当中的绝大部分由于文化程度低、缺乏技能，需要经过技术培训才能上岗工作。

(4) 我省技能型人才在数量及结构上与实际需求的严重不适应，要求大力发展职业教育。广东省现有城镇从业人员1260万人，技能人才400多万人，离实际需求的530万人相比，缺口达130万人，其中，仅操作数控机床的高级技工就缺口10万人。实际上技能人才的缺乏由来已久，如深圳市2001年下半年各类钳工需求职位数约5000个，而真正合格的应聘者不到千人。据预测，到2005年，全省城镇从业人员将达到1400万人，届时仅中级工以上技能人才就缺少180万人，缺口相当大。

在人才结构方面，发达国家的技术人才结构比例为，高级工35%，中级工50%，初级工15%。据调查统计，我省现有各技能人才478.15万人，其中，高级工17.31万人，中级工208.95万人，初级工248.83万人，分别占总数的3.6%、43.6%、52%，结构比例严重失调。广州市企业调查队对80家企业的专项调查结果表明，到2001年年底，被调查企业共拥有高级工以上技术等级证书的技术工人分别为高级工19.2%、初级工47.4%、中级工33.4%。

因此，无论从现有技能人才的数量上还是结构上看，我省必须大力发展职业教育。

二、广东职业教育发展的成就

改革开放以来，特别是近年来，我省党政领导及有关机关下达了大量推进职业教育改革与发展的文件，取得了很大的成就，表现在：

(1) 职业教育的数量、质量及结构方面都有较大的提升。截至2003年底，我省的中等、高等职业院校已发展为964所，在校生达118万人，其中，中等职业学校（含技校）903所，在校生达87万人；高等职业院校61所，在校生达31万人。在这些院校中，省级及以上重点中等职业学校228所，其中国家级重点中等职业学校93所。2000～2003年，我省的中等、高等职业院校共培养了108.7万名技能型、实用型人才。特别是我省的技工学校在2003年，取得了招生数、在校生数、校均在校生规模等五项指标全国第一的骄人成绩。

在结构方面，进行合理的布局和资源整合，将中等职业学校由2002年的1005所调整为804所，校均在校生规模由908人扩大到1082人，高等职业院校由2002年的40所增加到61所。初步形成了初、中、高三级职业教育体系。

(2) 我省职教不断深化改革，专业结构进一步调整优化。为了适应市场对相关专业人才的需求，我省大力改造传统专业，重点建设数控技术、环境保护等一大批重点骨干专业和示范性专业；启动省级技能型紧缺人才培养培训工程，建立一批省级技能型紧缺人才培养培训基地；开展示范性软件学院工程建设项目，扶持两所院校成为国家示范性软件职业技术学院建设项目，选择20所院校建立软件人才培养基地，以适应我省对软件技术应用型人才的需求；共有33所高职、中职学校承担国家职业院校制造业和现代服务业技能型紧缺人才培养培训工程项目；8所职校的8个专业被确定为全国重点建设的示范性专业点；建立全国高技能（机电）人才培养基地4个、全国技校实习指导教师基地1个（全国共10个）。改革调整优化产生积极影响，职业技术院校毕业生就业形势看好，不少专业的毕业生供不应求，各类职技院校毕业就业率普遍提高。

（3）鼓励社会力量参与职业技术教育的发展，多元化办学体制初步确立。目前我省民办职业技术教育发展很快，已经成为我省职业技术教育的重要力量，相继涌现了以白云职业技术学院、白云工商高级技工学校、广州华立学院、肇庆科技职业技术学院等为代表的一大批办得好、有一定社会影响力的民办教育机构，为各行各业培养输送了一大批技能型、实用型人才。据统计，2003年，全省民办高等职业技术学院有12所，民办中等职业学校111所，民办职业培训机构2438所，民办专修学院53所。民办职业技术教育的发展，为职业技术教育注入了生机和活力，有力推动了全省职业技术教育事业繁荣和发展。

（4）办学特色逐步凸显。实行弹性入学。中等职业技术学校招生，学生不受年龄、户口等限制，凭初、高中毕业证书报考或免试入读，允许学校春秋两季招生。

积极开展“大专业、宽基础、活模块”课程改革试验，组织编写了中等职业技术教育8大专业共120多种专业课程及教辅资料。

突出技能培养。全省建起了70个专业特色明显、布局基本合理、设施设备先进、各项功能齐全、服务辐射周边的中等职业教育实训中心，强化并培养了学生的专业技能和动手能力。

（5）各种形式的继续教育和岗前、在岗、转岗培训并存共进。

（6）逐步建立劳动准入制度。全社会推行职业资格证书制度是国家的重要方针，广东积极推行职业资格证书制度，2002年全省共有68万人参加职业技能鉴定，其中56万人获得相应的职业资格证书，已经连续5年在全国位居第一。

（7）各级各类职教院校初步形成了有一定数量、质量的师资队伍。至2003年底，广东省中等、高等职业院校（未含技工学校）的专任教师为45 478人，其中，中等职业学校专任教师为33 700人；专任教师中，本科以上毕业24 126人，占71.59%，双师型教师4970人，占14.75%；高等职业院校专任教师为11 778人，其中，正高职称383人，副高职称2632人，高职称占25.6%，具有硕士以上学位者2686人，占22.8%。

三、广东省职教发展中存在的困难和问题

与先进国家地区职业技术教育发展相比较，广东省职业技术教育发展中仍存在不少困难和问题。具体表现在以下几个方面。

（1）社会上甚至少数党政机关领导对广东省大力发展职业教育的必要性和紧迫性缺乏认识，不同程度上存在着有认识无共识的问题。对于发展职业技术教育，一些党政机关甚至少数领导不同程度上存在着有认识无共识的问题。这一问题对广东省职业技术教育发展造成不利影响，职教发展依然是说得多做得少，虚功多实功少，蜻蜓点水多深入实质少。表现在，职业技术教育至今被看成一个低层次的教育，政府远未将它纳入主流教育体系，普教与职教畸轻畸重的现象非常严重。例如，省重点扶持九所高校没有一所职教院校，普教的资金投入远远大于职教，招生制度大大有利于普教，

不利于职教等。其结果造成了人才培养上的严重畸形，一方面，不少“社会精英”——大学生为找不到工作而苦恼，另一方面，大量的企业却为找不到合适的“非社会精英”——高级技工而头痛，甚至出现了以年薪30万招高级技工无人应聘的情况。

传统观念、社会偏见也对广东省职业教育的发展造成不利影响。例如，“学而优则仕”“金榜题名”等传统观念会影响学生及家长对普教职教的选择，影响职业教育的发展，但关键的影响还是在党政机关特别是领导们的认识。

我们的领导应当纠正对职业教育的偏见，树立正确认识。要知道，德国、日本的腾飞不是靠德、日的“创造”——这方面他们比不过美国，而是靠“制造”，靠大力发展职业教育。广东应当从中获得深刻的启示，我们正在打造“广东制造”。

(2) 职业教育投入较低，导致办学经费严重不足。近年来，广东省积极实施“科教兴粤”和“教育强省”战略，教育经费投入逐年上升，有力推动了教育事业的发展。有关统计数据显示，1998～2002年，广东省预算内教育经费年均递增19.89%，但是，根据《中国教育统计年鉴》(表1) 对广东地方教育经费支出的结构分析，可以看出我省地方财政对中等职业教育经费投入不是逐年上升，而是呈下降趋势。

表1　广东地方教育经费结构支出情况　　单位：亿元

项目	2000年	2001年	2002年
广东地方教育经费支出	315.9	358.8	447.49
技工教育	3.65	2.8	1.75
职业中学	10.47	10.47	6.55

例如，揭阳市1996年本级财政预算内职业教育经费占预算内教育经费的5%，2001年降到3%；潮州2001年为2%，比2000年全国平均水平8.4%还要差。近年来，省财政每年对技工教育的专项投入只有2600万元，除广州、深圳外，绝大多数市未将技工教育专项经费纳入预算。在德国，培养一个一线技术工人要花去六七万马克，约合人民币23万～27万元，而我国技工教育的经费一般都是由办学企业的职工培训经费列支，尽管规定用于企业培训经费必须占到职工工资总额的1.5%，但这捉襟见肘的经费还经常不到位，甚至被克扣。致使现在许多技校的实训设备、仪器仪表以及实训用的原材料等普遍缺口较大，严重地制约了对学生技能的培养。

高等职业院校经费同样严重不足。在调查中，白云职业技术学院、广东轻工职业技术学院、番禺职业技术学院等职业学院都反映了经费严重不足的问题。广东轻工职业技术学院的省投经费多年一贯制，为2600万元，随着该院学生教师数量的大幅度增加，该投资占学院经费已由2000年的51%逐年下降到21%。番禺职业技术学院被定为广州职业院校的龙头，但一年的财政拨款仅有1800万元，为了维持正常运转，不得不去借款，甚至到教师口袋拿钱。白云职业技术学院反映了高职院校的共同问题：稳定与发展的基本矛盾是有钱维持，无钱发展。职业教育经费严重不足导致我省职业技术教育发展十分缓慢。

(3) 职教师资数量、质量、来源、结构缺陷已成为我省职教发展的瓶颈之一。从数量上看，全省现有各类初、中、高等职技院校教师约 5 万人。按照我省规划，到 2005 年，将有 250 万人接受职业教育、岗位培训，按 1∶16 的师生比计，届时，全省需要 15.6 万名职教教师，缺口将达 10 多万人。从质量上看，全省现有中等职教（未含技工学校）师资中，本科及以上毕业者 24 216 人，达标率仅有 71.6%，其中有 7 个市达标率低于 60%，最低的河源市竟然只有 34.7%，如果再把技工学校算上，达标率将更低。

从来源及结构上看，现有师资大多由原来的普教师资转行，渠道单一，真正懂得职教，来源于企事业单位及生产服务第一线的教师少而又少。据调查，全省中等职业学校专任教师 33 700 人中，双师型教师仅有 4970 人，占 14.75%；潮州、汕尾、阳江三市的双师型教师比例仅为 5%，师资本身已经实践经验不足，如何能培养出满足社会市场需求的高素质实用技能型人才?

广东省的职教师资队伍状况，显然完不成培养打造"广东制造"、打造"世界工厂"的生力军的目标；假如，广东要成为"9+2"的龙头，和"长三角"相抗衡，那么，广东省这样的职教状况只会对这一战略目标造成拖累。

职教师资缺陷问题无疑已成为我省职教发展的瓶颈之一。

（4）职业教育基础设施较为薄弱。① 在广东省职技院校中，不少校园校舍面积比较紧张，有的学校仅有一两幢教学楼和学生宿舍，如国家级师资培训基地、省定位为广东职教龙头的广东技术师范学院，其全日制普通生达到 8100 多人，但校园面积仅 23 万平方米，生均仅为 28.4 平方米，远远低于 54 平方米的最低要求。② 实习、实训基地不足，平均每 10 所中等职业学校才有一座实训中心，一些中等职业学校根本没有实习、实训基地。③ 教学仪器设备相当匮乏。教学仪器设备是培养动手型人才的最基本条件，其先进程度直接制约着所培养学生的质量。广东省中等职业学校的生均仪器设备值很低，如 2003 年均值仅为 3660 元，江门、清远、河源等市在 2750 元以下，其中，河源市仅为 1320 元。广东省技校的实训设备、仪器仪表以及实训用的原材料等普遍缺口较大，严重地制约了对学生技能的培养。西方经验表明，在教学仪器设备的投入上，职教应为普教的 2 倍。我省显然达不到这一标准。

（5）改革滞后。广东省职业技术教育的改革力度不大，步伐缓慢。在体制、机制上，职业学校条块分割，分属行业、部门、企业、地方政府办学，政府统筹不力；没有充分发挥部门、行业、企业和社会力量的作用，民办职业教育的发展不充分。办学观念上，用普教的理念办职教，片面追求升学率，职教特色不鲜明。有的名为职教，但专业及课程还是老一套，被人戏称为"打职教牌子，作普教压缩饼干"。改革力度上，在全国"3+2"和五年一贯制培养模式已普遍推广，而广东却还没有推开，五年一贯制也仅有少数院校中的英语等 5 个专业试点。一些有条件的学校，如深圳华强职校曾强烈呼吁举办"3+2"或五年一贯制的教育模式，却未能引起省里重视和支持。

（6）职教专业及课程设置的硬性规定与市场需求脱节的情况严重。调查中发现，绝大多数职业院校（未含技校）专业及课程基本是按照教育行政主管部门规定统一制定，体现不出各自的特色，而且僵化教条，与市场需求严重脱节，结果成为普教的"压缩饼干"。教育部要求高职院校学制三年改为两年，使得这一矛盾更突出。

(7) 职业教育及劳动就业中的资格证准入制度存在不适应职教发展的问题。对于资格证书，我国尚无完善的法律体系调整，资格证书行政收费过高，不利于职教发展，同时资格证书在企业用工上强制作用不够，一些企业宁愿用无资格证书的人员，而不用有资格证书的人员。劳动就业中的资格证准入制度不严格。

(8) 人事、教育等体制不完善，有关职教法规政策存在不配套不落实的情况，严重制约职教发展。例如，关于20%的教育附加投入职教的规定，大多未能兑现；从今年起财政按学生人头给职业院校拨款的政策未能兑现；在职称评定上，体制性束缚严重，动手能力强的工程师、技师、烹调师在评定职称上不被认可；公办教师进入民办院校其教龄工龄不能连续计算；人事编制大多是80年代制定的，远远不能适应客观需求。番禺介绍经验说，番禺在广州电工考试中数一数二，关键是让有经验的技师任教，但这又不符合师资必须本科以上的规定。职教院校评估体系照搬普教不合理。2001年我省第98号文件规定高级技工学校与高职院层次一致，但无配套措施很难落实，结果在毕业生考公务员时不承认其大专学历，干部不承认其厅级级别。对此问题，山东省专门下达配套措施予以确认。

四、大力推进广东职业教育改革与发展的建议

(1) 更新职业技术教育观念。突破传统的职业技术教育旧观念，按照联合国教科文组织《21世纪展望：技术和职业教育与培训》文件，确认职业教育也是终身教育的一部分，职教应由传统的技术能力的“需求驱动”观念转变为“需求驱动”和“发展驱动”相结合的观念。白云、番禺职业技术学院等很多院校意识到，职业教育绝不仅仅是就业教育，素质教育更为重要。

(2) 制定配套法规政策。以《中华人民共和国职业教育法》、国务院《关于大力发展职业技术教育的决定》《大力推进职业教育改革与发展》、教育部《2003—2007年教育振兴行动计划》法规政策为依据，结合我省实际情况制定相应的地方配套法规政策，加大对职业教育的支持保护力度。

(3) 政府应当承担起我省职业技术教育发展的统筹领导角色。在我国公有制为主体的体制下，行业协会及民营经济发展还不成熟且资源有限，各级政府在我省职业技术教育发展中具有不可替代的作用，政府应充分发挥这一作用。仿效国务院成立的六部委职业教育联席会议制度，我省也可建立相应的职业教育联席会议制度。

(4) 以政府投入为主，动员企业及社会力量加大对职业技术教育发展的资金投入力度。

(5) 合理设计中等及高等职业技术教育结构比例，构建职业教育与普通教育的立交桥，为职业技术教育观念转变创造条件。废除职教是层次不是类别的错误观念。明确高等中等职业院校及中高级技校的培养目标和比例。

(6) 着力培育我省职业技术教育师资培养的龙头院校，将之列入我省重点扶持的院校，以解决我省职业技术教育发展的瓶颈问题。发达国家对职教师资的培养要求严格，主要由高职师范学院负责培养。例如，德国要获得职教教师资格，须参加两次国家的职教教师资格考试，还要进入教师实习学院实习两年。

（7）给予职教院校以较大的专业及课程设置自主权，加强公共的及院校内的实训基地建设相结合。放开职教院校专业及课程设置的自主权，各院校可根据市场的变化调整自己的专业设置，优化课程结构。例如，白云职业技术学院的汽车营销专业，加强技能课的比例达到50%。

（8）政府应当统一协调教育、劳动、行业协会及企业，统一制定各行业岗位职业技术标准，强化资格证书及就业市场准入制度。

关于尽快制定国家统一法的建议[①]

国家统一是中华民族的最高愿望。在党中央、国务院的领导下，新中国的 55 年历史中，一直在为祖国的统一大业而奋斗；前几年，港澳已顺利回归，但台湾现状却越来越糟糕，陈水扁不断采用卑鄙的伎俩，一步一步向前，企图使台湾独立出去。形势已万分紧急，切不可疏忽。

为了祖国统一大业安全顺利，建议国家尽快制定统一法。既是国家法律，凡是中国人，就都得遵守，这从根本上防止了台湾的分离，完全杜绝陈水扁之流的幻想。法律条文写明，搞分裂者，就是中华民族的千古罪人！统一法将保护祖国的完整和尊严，定将给陈水扁之流的千古罪人以巨大的威慑力。

① 本文是广东省政府参事室建议，广州，2004 年 12 月 2 日；《参事建言（2004—2005 年）》，香港：中国评论学术出版社，2006，538.

激励民办专科学校升为本科学校①

我国的民办教育尤其是民办高等教育经过20年的风风雨雨，现已开始迈入一个新的发展阶段，其主要标志之一便是民办专科学校积极努力升格为本科学校。

“专升本热”不仅反映了专科层次学校追求向更高层次发展的一种期盼和要求，更反映了我国本科教育中仍旧突出的供需矛盾。当前，我国高等教育资源特别是优质资源仍很紧缺，与日益增长的高等教育需求形成了强烈反差，高等教育供给水平仅能满足大约1/6适龄人口的需求，而且在其中也仅有45%左右的青年能够获得接受本科教育的机会。因此，允许有条件的民办专科院校升格为本科，无论是在促进学校的发展、教师的成长、人才的培养方面，还是在优化国家教育资源、提高国民素质方面，均具有很重要的意义。

由于条件所限，且发展时间短，目前我国民办学校的教学水平和教育质量良莠不齐，但这并不能否定民办学校的美好发展，西方发达国家私立学校的发展就能很好地说明这一点。长期以来，利用社会力量办学是西方发达国家的一贯做法，公众对私立学校的认可程度明显高于公立学校。经过多年发展，西方发达国家的私立学校办学质量明显高于公办学校，且大多为世界著名高等学府，如英国的牛津大学、剑桥大学，美国的哈佛大学、麻省理工学院、斯坦福大学等，均为世界著名的私立大学。

所以，为了使民办学校有较好的发展，我们应当采取“区别对待、积极引导”的措施，激励民办专科学校升格为本科。

首先，在宏观上对民办专科院校升本科要加强引导，合理规划，分步实施，掌握好专科院校升格为本科的发展速度和节奏，防止出现一哄而起的情况。对于教学质量优异、办学水平较高的民办专科院校，应当允许其升为本科。

其次，要制定专科院校升格为本科的标准，并严格审核程序，保证专科院校升格为本科的必要办学条件和质量。这些标准包括学校的硬件设施、管理水平、经费保障、招生条件、学生毕业的标准、教师的标准等多个方面。

最后，政府要建立一套完整、系统、尽可能具有操作性的制度、法令，诸如经费的筹集、支出，人员的聘用、晋级，生源的录取、管理、分配，课程、专业的设置等，都应有一套相应的制度与之对应，以明确、可行的法律和法规去规范民办学校，以保证民办院校的教学质量。

① 本文是政协广东省第九届委员会提案，广州，2005年6月；《一个大学校长的探索》，北京：高等教育出版社，2011，382.

让象牙塔成为顶梁柱[①]

英国学者 R. G. 坦普尔认为，“现代世界赖以建立的基本创造发明和发现可能有一半多来自中国”[②]。

从古到今，中国人一直都是很有创意的民族，我们可以看到古代中国一系列至今仍令人引以为傲的发明：丝绸、茶叶、瓷器、指南针、纸张、印刷术、火药、纸币、热气球、枪炮、水雷、水力纺麻机、十进位数学、现代农业、现代海运、现代石油工业、现代天文观测、现代音乐甚至蒸汽机的基本设计等。

尽管古代中国拥有如此光辉灿烂的发明创造，工业革命却没有在中国如期而至。世界上的第一次工业革命，是在欧洲发生的。1765 年瓦特蒸汽机的发明，标志着第一次工业革命的开始；19 世纪后半叶和 20 世纪初，电气化的产生被称作第二次工业革命。两次工业革命极大地提高了社会劳动生产率和社会财富的积累速度和规模，形成了至今我们仍在享用的社会文明的主体。

几乎所有历史学家认定的启动西欧工业革命的因素，在中国都存在，但是，中国为何没能实现飞跃？1840 年鸦片战争之前，特别是 17 世纪前，中国一直领先其他文明，为什么中国现在不再领先？这就是著名的李约瑟之谜。

尽管对这个问题各有各的回答，比如，李约瑟博士的答案是中国的官僚体制阻碍了重商主义价值观的形成，阻碍了现代科学在中国的成长；诺思将之归咎于中国当时没有建立起一套有效地保护创新、调动人的积极性的产权制度；大卫·蓝迪斯认为是令人窒息的政府控制抹杀了科技进步；林毅夫教授则认为是中国的激励结构（如科举制度）使知识分子无心从事科学事业。但是，有一个不争的事实是中国漫长的历史中充满了科技被淹没、消失或退化的例子，除了农业以外，中国在大部分的科技领域中，从来没有人尝试对发明进行改良、进行商业应用[③]。这与欧洲人对新发明、新机械、新动力运用于商业生产的狂热形成了强烈的对比。

其实，人类社会发展进程的每一步跨越，都凝结着技术进步的贡献。综观整个人类社会的生产发展过程，其实质是一个科学技术不断突破、不断积累、不断飞跃、不断应用于生产过程、不断对经济增长和经济发展产生革命性影响的过程。

古代中国甚至是当代中国对科技的追求，往往是浅尝辄止、点到为止，从科技创新的严格意义上来讲，这种对科技的追求过程，是残缺不全的。

让我们重温一下技术进步的完整过程。

科技，即科学技术，包含着科学和技术两个方面的内容。科学属于认识世界的范畴，科学发展表明人类认识世界能力的提高，科学知识为技术进步奠定了必要的理论

① 本文是中国中部科技创新与风险投资国际论坛的主题报告，合肥，2005 年 9 月 21 日.

② R. G. 坦普尔.《西方欠中国的“债”》.《哲学译丛》，1992，(3)：71.

③ 大卫·蓝迪斯. 新国富论. 兰州：敦煌文艺出版社，2000：61.

基础，它构成潜在的生产力；技术是改造世界的手段，技术进步把科学知识转化为直接的、现实的生产能力，技术进步意味着人类改造世界能力的增强。

经济合作与发展组织（Organisation for Economic Co-operation and Development，OECD）在1988年的《科技政策概要》中指出，技术进步包括三种要素：第一个要素是发明，即有关新的或改进的技术设想，它的重要来源是科学研究；第二个要素是创新，即发明首次被商业应用；第三个要素是扩散，即创新出现后被许多使用者应用。

与此对应，技术进步过程包括研究与开发（R&D，简称研发）、技术创新、技术扩散这三个环节，这三个环节环环相扣，相互重叠，相互作用，缺一不可。

在研发环节中，虽然知识总量增加了，但是技术与经济仍处于分离状态。

在技术创新环节中，人们依据社会需求把技术发明应用于生产，把新知识转化为物质产品，把潜在的生产力转变为现实的生产力，实现了技术、知识与经济的结合。技术创新实现了技术进步过程中质的飞跃。

在技术扩散环节中，新的技术和知识通过一定的渠道向潜在的采用者转移，并在生产中取得广泛应用。我们知道，质的突破只有经过量的进一步扩张，才能显示出它的经济意义，科技创新这种量的进一步扩张是在技术扩散中实现的。

由此可见，科技创新的力量，不仅取决于研发、取决于科技创新的数量与质量，更取决于技术扩散的速度与范围。

之所以要把技术进步的完整过程阐述得如此详细，是为了说明古代中国和近代中国虽然拥有众多举世瞩目的发明创造，但由于缺乏实际应用，缺乏商业运作，从经济意义上讲，这些发明在一定程度上是纸上谈兵，它们带给后人的，也只能是纸上宝贵了。

如果把近代中国对科技追求过程的残缺不全归结为当时人们的历史局限性，尚且情有可原的话，那么，如果现代中国的大学仍对如火如荼的经济建设隔岸观火，高校仍像个密不透风的“象牙塔”的话，则是对社会的漠视，是其自身的一种退化。

一直以来，中国教育和研究很大程度上蜷缩于“象牙塔”中，理论和实际脱离，科技与经济脱节，无法根据国家国防、民生的特点提出问题，组织有效攻关。高等院校和科研院所靠吃“皇粮”搞科研，游离于企业和市场之外。一些人关起门来搞科研，一门心思想着出论文、单纯追求学术水平、得奖，其结果是闭门造车。至今仍存在这样的现象：一方面是高校大量的科技成果束之高阁，另一方面是企业技术匮缺，国家原本有限的科技投入，没有发挥应有的效益。

大学应当服务于整个人类，为国家的进步作贡献，大学精神也应该是与时俱进的。“传道、授业、解惑”已不能满足时代对中国高校的要求。全面建设小康社会、加快推进社会主义现代化的经济社会发展目标是宏伟而艰巨的。这也意味着促进科技创新与产业化面临着更现实、更艰巨的历史使命。

在现代高等教育传授知识、创造知识、应用知识的三大职能中，科技创新是贯穿其中的最核心的驱动力，在加强基础性和前沿性研究的同时，必须重视技术创新，必须适应国家与社会发展的需求，实现知识产业化和产业知识化，促进经济增长，提高综合国力。丰富的人才资源、门类齐全的学科设置、活跃的学术研究氛围以及人才和知识的快速流动等，形成了高校在科技创新活动中科研院所和企业所无法替代的种种优势，奠定了其在国家创新体系中不可或缺的重要地位。

我们追求这样一种良性循环：产业依靠知识来提升和发展，即产业知识化，而在经济增长和社会进步过程中，产业的发展又会不断地提出新的要求，成为高校创新的外部动力。

我们知道，技术进步的动因有两种：一种是自发的，是人类对自然规律的探寻精神；另一种是引致的，是追逐经济利益的动机。自发的技术成果可能具有更为深远的意义和价值，但引致的技术发明往往具有直接的应用价值，它的进一步发展和推广将取决于需求的强度。

技术进步能否有效地推动经济发展，既取决于技术进步的动因，更决定于是否存在着广泛的企业家精神。科技创新、技术扩散过程的高创新性、高难度性、知识密集性和低成功率，需要倡导者、实行者具有独特的素质和精神：首创性、成功欲、冒险和以苦为乐、精明与敏锐，以及强烈的事业心。正是这种企业家精神促使人们去探索新的生产技术和方法，以取得利润最大化和实现自我价值。我们的风险投资家正是这样的一个群体，我们尚不算成熟的股票市场亦为科技创新投资提供了场所，高校的科技创新有了走向市场的媒介，而高校在中国的经济建设中，同样有了用武之地。

人不能两次踏进同一条河流，国家最好也不要犯同样的错误，如果头发花白的历史老人站出来说话，他一定会这样告诫我们。在我国的社会发展史中，有着太多的知识分子归隐、科技发明沉没的教训。知识分子、科技工作者独善其身、徘徊在“象牙塔”之中，是社会财富的一种巨大浪费，也不符合中国读书人己达达人、兼济天下的初衷。今天我们生活在一个开明的时代，在计划经济向市场经济转轨的进程中，追求经济增长的动因让高校必须更加强调其运用知识的职能，这是历史的潮流。科技的“象牙塔”必须成为区域发展的顶梁柱，这是历史赋予的机遇，更是时代提出的挑战。

大力发展我省高中阶段教育①

近年我省高等教育实现了跨越式的发展，但是，作为义务教育和高等教育的衔接阶段，我省高中阶段教育无论是办学规模还是教学质量，都与高校扩招要求和经济社会发展需要存在较大差距，成为教育事业快速、均衡发展的“瓶颈”。为促进这一问题的解决，省政协将“大力发展我省高中阶段教育”作为今年专题协商内容，派出调研组，于4月15日至5月25日，赴广州、湛江、揭阳、河源、中山市进行专题调研。调研组认真听取了地方政府及有关职能部门的情况介绍，考察了部分公办和民办高中及中等职业技术学校，召开有主管部门、学校和教师参加的座谈会。在此基础上进行了研究和讨论，提出了大力发展我省高中阶段教育的意见和建议。

一、我省高中阶段教育取得了显著的成绩

近年来，我省高中阶段教育以高校扩招为契机，加快了发展的步伐，主要表现在以下五个方面。

（1）投入逐年增加，办学规模持续扩大。2000年以来，全省高中教育投入超过50亿元，新建普通高中学校50多所，扩建200多所，普通高中学校由947所增至998所。与此同时，校均规模人数由765人增至1320人。2000～2004年，高中阶段教育招生数由55.62万人增至86.85万人，年均增长11.79%；在校生从152.56万人增至224.85万人。其中普通高中在校生从72.5万人增至131.3万人，每万人口普通高中在校生由99.76人增至165.1人；中等职业学校（含技工学校）招生数由26.9万人增至36万人，在校生（含技工学校）由80万人增至93.5万人。目前省级以上重点中等职业学校237所，其中国家级重点中等职业学校126所。全省各类职业技术学校为生产第一线输送了120多万名技能型、实用型人才。

（2）办学条件有所改善，优质学位数量增加较快。随着创建国家级、省级示范性普通高中步伐的加快，我省高中学校在校舍、办学场地、设备等方面有较大的变化，优质学位比例增加较快。普通高中生均预算内教育事业费支出由2000年的1892.32元增至2003年的2803.76元。2000～2004年，生均校舍面积由7.94m^2增至9.09m^2；生均图书由36.49册增至39.48册；省一级普通高中由134所增至262所，增长95.5%。目前全省普通高中优质学位（省、市一级）占70%以上。

（3）教师数量增长较快，整体素质有所提高。2000～2004年，我省普通高中教师由43 941人增至75 330人；具有本科毕业及以上学历的从67.51%提高到79.41%。教

① 本文是广东省第九届政协教科文卫体委员会向政协常委会提交的专题议政材料，广州，2005年10月；《教育与科技管理研究》，北京：科学出版社，2016，247-254。作者：刘人怀，陈万鹏，韩大建，周明理，林维明，吴明兴，陈年强，彭文晋，王绵宁，翁宗奕，柳柏濂，潘史扬，张焜，周国贤，林伟健，李宗桂，刘晖，李定安，祁海，陈晓薇，丁义，余国慧.

师队伍结构和质量均有所改善和提高。中等职业教师 3.38 万人。三年来共培训专业课骨干教师近 3000 人，考取职业技能证书的 2250 多人，考取高级技能证书的 1083 人、中级技能证书 748 人。“双师型”专业教师（教师+工程师或技师）占 28.7%。

（4）中等职业教育布局结构调整初见成效，多元办学体制初步确立。近年来，我省对中等职业学校布局结构进行了调整和优化。学校数由 2002 年的 1005 所，调整到 2004 年的 820 余所，省级以上重点中等职业学校 237 所，国家级、省级专业点 131 个；校均规模达 1135 人，比 2002 年增加 241 人。民办职业教育发展较快，成为我省职业教育的重要力量。目前全省民办中等职业学校 103 所，在校生 5.9 万人；民办职业培训机构 2438 个。

（5）教育教学改革全面推进，办学特色逐步显现。从 2004 年秋季开始，作为首批全国高中新课程实验省区（广东、山东、宁夏和海南）之一，我省在普通高中一年级全面开展以实施素质教育为目的的新课程改革实验，有 3000 多名教育教学管理人员和 2.4 万名学科教师接受了国家级和省级培训。组织编写的高中新课程实验教材共有 7 个科目通过教育部评审，并在全国发行，居全国各省首位。

中职教育以服务为宗旨，以就业为导向，强化专业技能和动手能力的培养。目前我省有职业教育实训基地（中心）84 个，市场紧缺的技能型人才培训学校 32 所。职业教育实行弹性入学，招生不限年龄、地域，学生凭初、高中毕业证书报考或免试入读，春秋两次或多次招生。组织编写了 120 多种专业课教材及资料。职业学校与劳动力市场联系紧密，“订单式”的培养及培训取得一定的成效。珠三角不少中职学校与东西两翼和山区中职学校建立了合作办学关系。2004 年，我省中职学校毕业生就业率平均达 95%。

目前启动的“广东省百万农村青年技能培训工程”，将培训与就业、培训转移就业与扶贫助康、培训转移就业与缓解企业技工短缺相结合，由各级财政资助全免费对农村百万青年进行岗前技能培训。这一举措，对于加快推进我省工业化、城市化和现代化进程，促进产业升级，增强发展后劲，增加农民收入，推动城乡和区域协调发展，构建和谐广东将产生重大的影响。

二、我省高中阶段教育发展面临的问题

虽然我省高中阶段教育发展较快，取得了明显的成绩，但整体发展水平仍不高，“瓶颈”的制约仍未突破，存在着不少亟须解决的问题。

（1）发展水平在全国处于中下游，与高等教育跨越式的发展不相适应，制约了发展的后劲。2004 年，广东每万人口普通高中在校生为 165.1 人，低于全国平均水平 14 人，在全国排第 21 位；每万人口中等专业学校在校生为 82.4 人，低于全国平均水平 13 人，在全国排第 17 位；普通高中教师达标率为 79.4%，低于全国平均数 0.2%，在全国排第 17 位；初中毕业生升学率我省只有 65.13%，而同期的江苏、浙江两省均超过 80%。从这些方面看，我省高中阶段教育发展水平与先进省相比差距仍较大，与我省经济大省地位不相称。

从教育发展一般规律来看，各级学校的录取率应与教育层次成反比，即从小学、

初中、高中到高等教育，录取率应呈金字塔形，越往上录取率越低，淘汰率越高，这符合人才培养规律和需求结构特点。推进高等教育大众化的前提是高中阶段教育的充分发展和质量保证。由于我省高等教育大众化是在高中阶段教育发展并不充分的情况下启动的，以致 2004 年我省初中毕业生升学率低于普通高等学校录取率。这一年，普通高中和职业高中录取率为 65.13%，但高校本、专科录取率为 83.14%（中职升大学人数很少，省略未计）。这些情况表明，加强高中阶段教育十分紧迫。

由于高中阶段教育总体发展滞后，加上有些地区初中生辍学率高，因此山区和东西两翼不少地方高中阶段升学率低下，大批适龄青少年得不到较好的教育。揭阳市高中（含职中）录取率仅为 31%，不及全省平均水平的一半，每年考不上高中和辍学的初中学生约 10 万人。该市建市 10 多年来，累计已超过 100 万人。这种现象的直接后果，就是人力资源质量和层次的下降，严重制约了发展的后劲。因此，在抓高校扩招的同时，抓好高中阶段教育，巩固九年义务教育成果刻不容缓。

（2）地区发展不均衡，优质资源主要集中在珠三角地区。经济发展不平衡，直接影响教育的均衡发展。珠三角地区与东西两翼和山区、城市与农村的差距，主要体现在发展规模、师资、校舍、设备等方面。以初中毕业生升学率为例，2004 年，我省为 65.13%，其中珠三角地区的广州、深圳、佛山等 6 市的平均升学率为 90.23%，东西两翼和山区的汕头、韶关、湛江等 9 市的平均升学率只有 49.08%，相差 41.15 个百分点。人均预算内教育经费投入差距更为悬殊，2003 年，排在前三位的深圳市（3975 元）、东莞市（1258 元）、珠海市（951 元）的平均数，是排在后三位的汕尾市（172 元）、湛江市（160 元）、揭阳市（155 元）的 12.7 倍。从中等职业教育来看，珠三角在校生占全省总数的 73%，而占全省人口 72%的东西两翼和山区在校生仅占全省总数的 27%。从每万人口中等职业学校在校生数来看，珠三角地区为 268 人，东西两翼和山区仅为 39 人，差距十分明显。

（3）中职教育体制不顺、发展迟缓，技能型人才的培养滞后于经济社会发展的需求。我省中等职业教育分属不同的部门、行业。职业高中由教育部门管理，而技工学校由劳动部门管理。条块分割，政出多门，造成宏观统筹调控不力，发展规模、力量整合、资源合理配置等还存在不少亟待解决的问题，严重影响了中职教育的发展。目前我省获得职业资格证书的技术技能劳动者共有 296.5 万人，其中高级以上技工（包括技师、高级技师）11.7 万人，仅占 4%，与发达国家 30%～40%的比例相去甚远。技能型人才的严重缺乏，制约了广东打造世界制造业基地的发展后劲。

受传统观念影响，社会上普遍存在重普高轻职高思想，职业教育被视为非正规、低层次的“二等教育”，很多学生不是自愿而是迫不得已才接受职业教育。一些地方未将其纳入主流教育体系，致使职业教育投入严重不足，人才培养结构不合理。一方面，不少大学生为找不到工作而苦恼；另一方面，大量的企业却为找不到合适的技师和高级技工而头痛。广东省地方教育经费结构支出（表 1）显示，全省地方教育经费支出普通高中占 32%以上，而中等职业教育（含技工学校）仅占 3%左右。由于投入不足，相当部分职业学校办学条件差，教学设备简陋，难以满足专业教学的需要，如清远市中等职业学校从 2000 年至今，5 年市财政投入仅 80 万元。

表 1　广东省地方教育经费结构支出表　　单位：亿元

项目	2000 年	2001 年	2002 年	2003 年	2004 年
教育经费支出	315.95	358.85	447.49	524.58	605.97
普通高中	101.82	116.06	149.31	181.24	211.47
中等职业教育（含技工学校）	14.12	13.28	15.3	18.71	21.57

（4）投入不足与加快人才培养矛盾较大。经过连年快速的发展，我省高中教育规模不断扩大，按照生均成本测算，每增加一个高中学位约需投入 1 万元。近年来我省新增加高中学位 60 万个，各级财政共投入 60 亿元。按照基本普及高中阶段教育要求，到 2010 年我省还需增加 80 多万个高中学位，增加投入 80 亿元。从 2001～2004 年，省财政每年拨出 5000 万元专项经费扶持经济欠发达地区发展普通高中教育，2003 年一次性拨出 5700 万元用于扶持 16 个贫困县发展普通高中。这对于欠债较多的全省高中教育来说，只能起到一定的引导作用。按目前的体制，发展高中阶段教育主要靠地方政府投入，但对经济欠发达地区来说，政府保九年义务教育已经很困难，每年还要拿出几千万元甚至更多的资金来发展高中阶段教育，实在难以支撑。为了加快高中教育发展步伐，各地不得不通过财政以外的渠道来增加高中办学经费。例如，创建示范性高中，各地均不同程度地依靠银行贷款，通过公办高中和依靠名校发展国有民办初中收取择校费的办法来偿还贷款。这种做法虽加快了高中发展，但也带来了一些突出问题。茂名每县（市）建一所示范性普通高中，全市共建五所示范性普通高中，每所都向银行贷款 1 亿多元。对欠发达地区来说，这是一笔不小的数目。通过择校费来偿还贷款，必然加重家长的负担，增加学生的心理压力，而且在缺少政策支持的情况下，其潜在的金融风险也令人担忧。

（5）贫困地区优秀教师外流现象严重，高中阶段教师明显不足。粤东、粤西的领导及学校反映最集中的问题之一是难以留住人才。近几年湛江市每年外流的优秀教师 200～250 名。随着高中课程改革和高中教育规模的扩大，需要的教师不断增加，造成高中教师严重缺乏。不少地方不得不用初中教师上高中课程，致使高中教育质量难以保证。根据目前教师缺编情况和今后三年教育事业发展需求计算，2005～2007 年，揭阳市需增配高中阶段教师 3630 人，而每年返回本市任教的师范类本科毕业生仅 200 名。2000～2004 年，我省普通高中师生比由 1∶16.5 增加到 1∶17.4。教师数量不足，使高中阶段教育扩大规模和提高质量受到限制。

从职业教育方面看，在数量上，2005 年，我省职教教师缺口达 10 万人；在质量上，全省现有中等职业学校（未含技工学校）师资中，本科及以上毕业生 24 216 人，达标率仅为 71.6%，有 7 个市达标率低于 60%，最低的河源市只有 34.7%；在来源及结构上，现有教师大多是由普通教育转行而来，渠道单一，来自企事业单位及生产服务第一线，既有理论又有实践经验的教师少而又少，这是影响职业教育质量的一个重要原因。

三、大力发展我省高中阶段教育的对策建议

根据《广东省教育现代化建设纲要（2004—2020 年）》，到 2010 年我省高中阶段毛

入学率要达到 80%，这是建设教育强省的重要标志之一。围绕这一目标，必须加快高中阶段教育发展步伐，巩固和提高基础教育已取得的成果。

(1) 抓好落实，分类指导，加大投入。提高认识，协调发展。由于高中阶段教育的法律地位不明确（义务教育和高等教育有相应的《中华人民共和国义务教育法》和《中华人民共和国高等教育法》保障），对政府缺乏法律和制度上的刚性约束；欠发达地区政府重点在保九年义务教育，致使对发展高中阶段教育的投入不足。因此，要大力发展高中阶段教育，必须克服“奉命办学，奉命发展”的思想，提高对其基础性和战略性的认识①，以科学发展观为指导，坚持协调发展方针，致力于各个阶段教育、普教与职教、发达地区与欠发达地区的协调发展。近期出台的《广东省教育现代化建设纲要实施意见（2004—2010 年）》，已对未来一个时期高中阶段教育的发展提出了具体要求和保证措施，建议认真组织实施，以在不太长的时间内突破这一瓶颈。

统筹兼顾，分类指导。全省经济发展不平衡，差距较大的状况将长期存在。省里统筹兼顾、资金支持的重点，应是山区和两翼地区。在珠三角发达地区，资金投入应以县（市、区）级政府为主，镇级政府大力配合。高中教育应加大示范性高中建设的力度，扩大优质高中办学规模，全面推进这一地区普通高中学校向省一级优质学校发展。职业教育要扩大规模，面向市场，增创职教特色，创办一批现代化示范性职业学校和一批名牌专业，增强珠三角地区职业教育的辐射力和影响力。在东西两翼和山区，县（市、区）级政府应加大投入的力度，同时，省、地级市政府要大力扶持，促进这些地区高中阶段学校规模进一步扩大，普及程度不断提高。要通过整合教育资源，努力使每县新建、扩建 1 所以上优质普通高中。在市、县两级建设一批省级以上重点职业学校、实训中心和重点专业（点）。

调整政策，增加投入。基础教育的政策是“分级管理、以县为主”。因此，发展高中阶段教育，县级政府责无旁贷，要采取多种形式，加大对高中阶段教育的投入。近年来，省委、省政府狠抓了高等教育，使我省高等教育实现了跨越式发展；同时，也着力抓九年义务教育，如省财政投入 9.5 亿元完成了 3000 所老区学校的改造；对人均 1500 元以下的困难家庭，年投入 3.85 亿元，免除其子女的书杂费等。这些措施影响巨大。对高中阶段教育发展（不包括技工学校），省财政每年投入 1.3 亿元，2005～2007 年，省财政每年再安排 1 亿元资金用于支持经济欠发达地区高中阶段教育。相比对九年义务教育的投入，力度小些。我们认为，为突破瓶颈的制约，加大政策支持力度和投入力度是十分必要的。建议省政府近期内专题研究加快高中阶段教育发展问题。重点研究支持欠发达地区发展示范性高中的支持政策，如贴息贷款等。通过省、市、县共同努力，逐步解决高中阶段教育的瓶颈问题。

(2) 积极推进办学体制、投资机制改革，形成多元投资与多元办学的格局。高中阶段教育应坚持以政府办学体制为主的原则，同时鼓励学校由公办向学校公办和民办并存转变，由政府包揽办学向政府体制内办学与社会、企业、境外机构等体制外办学并存转变。要加快建立高中阶段教育成本分担机制，适当提高高中阶段教育收费标准；

① 刘佳炎. 中小学教育与人才培养. 百名专家谈人才. 北京：党建读物出版社，2012：138-139.

鼓励以“国有民办”“国有联办”等形式发展高中阶段教育；引进国外优质教育资源，鼓励符合办学资格要求的国外组织和个人，与省内教育机构合办高水平的高中阶段学校及职业培训机构；支持民办学校的发展，给予民办高中阶段学校与公办学校在招生、教师进修和职称评定等方面一视同仁的政策。近几年，惠州市采取学校管理、企业投资的办法，吸纳社会资金5亿～6亿元，缓解了高中教育资金不足的问题。茂名市采取财政拨一点，社会捐一点，银行贷一点，择校费收一点，行政规费减一点的五个一点措施筹措教育经费。这些做法，都可借鉴。

(3) 调整结构，加强衔接，促进各类中等职业学校协调发展。统筹规划各级、各类及各地区职业教育结构比例，使职业教育与普通教育比例大体相当。加强教育行政部门对职业教育的协调管理，理顺各部门所属的技工学校与教育、劳动部门统管院校的关系。国家已经建立了职业教育工作部际联席会议制度。建议我省也相应建立由教育、劳动等部门及企业、学校参加的联席会议制度，共同研究解决职业教育发展问题。要加强中等职业教育与高等职业教育在招生、专业设置等方面的衔接，加大高职院校对中等职业学校对口专业招生的力度，改变中等职业教育作为终结性教育的不利地位。把高等职业教育纳入职业教育系列，统一管理，避免把高等职业教育办成翻版的大专。

(4) 采取有效措施，加强教师队伍建设。抓好定编工作，解决师资紧张问题。目前采取的教师每三年一核定编的做法，已不能满足高中阶段教育快速发展和新课程改革对教师迅速增加的需求。对此，建议及时修订教师编制标准，根据当年学生数，每年进行教师一核定编并入编，以缓解教师数量的不足问题，确保教育质量稳步提升。

制定稳定教师队伍的相应政策。建立针对农村地区高中教师的培养及学习交流制度。推进“青年志愿者扶贫接力计划”，鼓励大学生志愿者从教于条件艰苦的农村中学。设立“大学生到农村从教奖学金”，对已经考取大学，但因家庭经济困难不能继续深造的学生，可采取“委托代培”办法，定向培养。改善欠发达地区农村教师工作和生活条件，研究贫困地区农村教师工资补助政策。鼓励和组织退休教师到农村薄弱学校任教。推行中心城镇及经济发达地区优质学校教师对口支援欠发达地区薄弱学校的制度。

加强职教师资队伍建设。重点培育1～2所省级职教师资培养龙头院校，将之列入省重点扶持院校，以培养更多的高素质职教教师。多渠道引进师资，除通过正规的职业技术师范学院培养外，还可广泛吸引和鼓励企事业单位工程技术人员、管理人员和特殊技能人员担任专兼职教师，到2010年，将“双师型”教师比例提高到60%以上。改革职教职称评定制度，确认工程师、技师、烹调师等在评定职称上的同等资格。

(5) 巩固九年义务教育成果，夯实高中教育基础。巩固和发展九年义务教育是确保高中阶段教育高质量发展的前提，要不断完善中小学的办学条件，实现办学条件基本均衡化。严格控制初中阶段辍学率，建立“控辍”机制，落实校长、教师责任制，将控制学生辍学情况作为一项重要的考核内容，巩固和提高毕业率和升学率，要特别重视农村基础教育。我省农村人口占全省的72.23%，其教育发展落后状况不改变，全省教育现代化就不可能实现。通过不断改善农村学校办学条件，增加对农村学校投入等措施，巩固农村普教成果。使教育体系中各个环节相互适应，相互促进。

要重视创建示范性高中带来的新问题。创建示范性普通高中应量力而行，克服盲目攀比、求全的思想，充分考虑我省各地的实际情况，评估标准不宜全省一刀切，不能以牺牲优质初中资源为代价。要致力于高中教育与义务教育的协调发展。对于目前群众反映强烈的“名校办民校”“股份分校”等问题要给予关注，出台相应的政策予以规范。使九年义务教育与高中阶段教育相互衔接，相得益彰，实现我省基础教育的协调发展。

做好计划生育工作是发展教育、提高人力资源素质的前提和基础。东西两翼和山区由于人口增长过快，带来了学生入学难、大班额及教学质量不高等问题。2003 年，全国平均每万人口小学在校生 910 人，广东却达 1304.8 人，湛江、揭阳两市超过 1500 人，其中湛江的雷州市达到 1700 人；2004 年，全国平均每万人口小学在校生为 909.7 人，而我省却攀升到 1319.6 人，两年均居全国第一位。因此，在认真解决高中阶段教育发展问题的同时，要坚定不移地抓紧、抓实计划生育工作，有效控制人口增长速度，促进经济与社会协调发展。

关于允许市民在节假日有条件燃放烟花的建议①

国家要求对烟花爆竹市场加强管理、规范整顿，出于公众安全的考虑，当然是无可非议。但是任何事情都应该辩证地看待，对于市民燃放烟花爆竹而言，我们应该尽力将其负面影响降到最低，而不是全面禁止，一堵了之。如果因其负面影响的存在而因噎废食，不允许市民涉足，一方面体现了市民对政府行为的理解，但另一方面也反映了市民对政府行为的无奈。每逢传统节日珠江沿岸市民观看燃放烟花爆竹的盛大场面，以及居民区稀稀落落的爆竹声是市民对重开烟花之禁的心态的真实写照。

当前，我们正在全力构建和谐社会，在构建过程中增强百姓对生活的满意度、对社会的满意度，融会成一个群体性的认同，这是构建和谐社会之源。近年来，伴随国民生活水平的提高，追求心情的欢畅、追求精神的愉悦，已成为百姓日益关心的问题，在节假日的表现尤为突出。燃放烟花爆竹这种有中华文化特色的民俗文化活动，正是百姓表达满意心情的一种载体，也在燃放烟花中寄托着对于和谐未来的一份希望。因此，我认为允许市民在节假日有条件地燃放烟花爆竹既可以满足多数市民的愿望，烘托传统节日氛围，同时还可以推动经济的发展，西安、青岛、四川等地的做法就值得借鉴。

根据 2003 年 12 月 9 日西安公布施行的《西安市销售燃放烟花爆竹管理条例》，在西安市禁放 9 年多的烟花爆竹，将在今年新春佳节之际，重新绽放和响彻古城西安。

2003 年 12 月 18 日，青岛市通过了关于修改《青岛市禁止制作销售燃放烟花爆竹的规定》的决定。明确规定在春节期间的六天可以燃放烟花爆竹。

四川和宁夏等地方实行烟花爆竹连锁配送经营，实现经营行为规范化和标准化，从源头上堵住假冒伪劣商品进入市场。

据了解，青岛市有关部门在调查中发现，赞成继续禁放烟花爆竹的占 24%，而赞成限制燃放的占 65%，赞成全面开禁的占 11%左右。

新浪网站做了一个“你是否赞成节日期间燃放烟花爆竹”的新闻调查，调查结果显示：持“赞成”态度，并认为“应该保留”者占 65.33%，持“反对”态度者占 27.6%，持“无所谓”态度者占 6.87%。在该网站的另外一个调查中，67.24%的人认为最能体现“年味”的行为是燃放烟花。

烟花爆竹是国家严格管理的易燃易爆物品，必须坚持“安全第一、预防为主”的方针，这一点毋庸置疑，但整顿和规范烟花爆竹经营秩序要标本兼治，重在治本。我们应该在加强管理、规范经营、安全教育、有条件燃放等方面下功夫，而不是一味去堵。省人大可通过在新闻媒体上发布消息、举行立法听证会、开展问卷调查等方式，进行调研、论证、征求群众意见，真实了解市民心中的想法。我们期待在不远的一天，广州市民在节日来临之际在指定的范围内能够重新点燃象征喜悦和欢乐的绚丽烟花。

---

① 本文原载《参事建言（2004—2005 年）》，香港：中国评论学术出版社，2006，518-519.

在推进和谐社会建设中切实解决“农民工”身份问题[①]

伴随中国改革开放和工业化、城镇化过程的深化，我国产业队伍中涌现出一支新型的由农村到城镇工作的劳动大军，他们为城市繁荣、农村发展和国家现代化建设做出了重大贡献。国家统计局 2004 年调查显示，全国进城务工和在乡镇企业就业的“农民工”总数已超过 2 亿人，其中进城务工人员 1.2 亿人左右，并且广泛分布在各个行业。可以说，过去 20 多年，如果没有农民工，我国的工业化、城镇化进程就不会那么快，新兴产业和开放型经济就不可能迅猛发展。

对于“农民工”问题，党中央、国务院始终给予高度重视，已经制定一系列保障农民工权益和改善农民工就业环境的政策措施，明确认定“进城就业的农民工已经成为产业工人的重要组成部分”。2006 年 3 月 27 日，国务院又制定并公布了一个全面研究解决农民工问题的重要指导性文件——《国务院关于解决农民工问题的若干意见》。各地区、各部门也做了大量工作，取得了明显成效。可是，从现实情况看，农民工面临的问题仍然十分突出，主要是：工资偏低，拖欠工资现象严重；劳动时间长，安全条件差；缺乏社会保障，职业病和工伤事故多；培训就业、子女上学、生活居住等方面也存在诸多困难，经济、政治、社会权益得不到有效保障。这些问题引发不少社会矛盾和纠纷，在一定程度上影响到改革发展的稳定和建设社会主义新农村的进程。

有人认为，从长远工作效果和建设社会主义新农村的角度考虑，农民工权益保障亟须解决的是农民工身份问题。

古人有言，“名不正，则言不顺；言不顺，则事不成”。只有科学地称谓“农民工”，明确其社会身份，才能更好地确定其权益保障归属地区或部门。其实，“农民工”本身就是个不伦不类的称谓，其目的是把进城务工的农民与城镇居民区别开来，这是一个带有歧视性的称谓。农民、工人的称谓都是就社会分工而言的，农民进城务工，其本身已不再从事农业生产，从一定意义上讲，他们已经是现代工人。农民当了工人，就是工人，不应再冠以“农民”二字。而且，绝大多数进城就业的农民工与土地之间已经没有劳动和收入的关系，他们长期在城市就业，取得工资收入，不少优秀分子还成为现代企业的管理者和高级技工。国家也已经把进城就业的农民工定性为产业工人的重要组成部分，我们也应该转变传统观念，视农民工为城镇居民一员，取消“农民工”的称谓，从政策层面自上而下改变目前城镇并未真正接纳这些在当地长期工作并且做出积极贡献的工作者的身份问题，将他们的经济、政治及社会权益纳入工作所在地的管理范围，使相关权益得到切实保障。

因此，广东省在解决农民工问题上应该先行一步，作为一种过渡性办法，可以采取以下措施来推进体制创新，逐步建立有利于改变城乡分割二元结构的体制。

---

① 本文原载《参事建言（2006 年）》，香港：中国评论学术出版社，2007，312-315.

一、建立农民工身份转换制度

以配额制的方式，在广东省建立“农民工身份转换制度”。该制度可以以在城市工作达到一定工作年限为主要条件，给予农民工城市户籍，从而逐步转换农民工身份。

具体实施措施为：由省主管部门每年给各地市配以一定的转换名额，并统一为农民工身份转换设定一些必要条件，如在城市连续工作达到一定年限，未达到一定年限的可以根据其对城市社会发展有重大贡献表现、获得过特别嘉奖、在工作中获得的重大奖励等予以考虑。可在广州、深圳等珠三角等经济发达地区率先实施，在东西两翼以及粤北山区逐步推行这一制度，如在珠三角等发达地区连续工作满五年以上的农民工可转换户口，在东西两翼以及粤北山区一些城市则可放宽至七年。

通过对农民工身份的转换，可在基本维持现有的以城市户籍为基准的城市人口管理制度基础上，逐步解决农民工在薪酬、劳动保障、社会保障、福利待遇、子女就学等方面应享受的与城市劳动者同等待遇的问题。

二、实行居民证制度

对户籍人口以及在城市有房产、有正当职业或有其他合法居住资格的外来人员，将统一发放居民证，持有者享受“市民待遇”。此举目的在于淡化“户籍”概念，强调“居民”身份。

随着国家对于户籍制度改革的进一步深入，户籍制度的本来功能，即证明身份和满足社会管理需要的功能将凸现，同时将逐渐剥离依托在户籍制度上的其他功能。这样，所有持有“居民证”的市民，无论其有无“户籍”，皆应当在就业、居住、教育、社保等各方面同等对待，享受同等社会待遇。在广东省如率先实行居民证制度，可以从制度设计层面和社会管理方式上为农民工在城市生存创造条件，进一步体现公民的平等权。

三、用新称谓取代“农民工”一词

词语是一种文化现象，带有相应的文化含义。含有某种成见、偏见或歧视意义的用语不是先进文化理念用词的体现，也不利于构建和谐社会目标的实现。目前，我们一些文件、媒体常会出现“农民意识”“小农思想”“农民食品”“打工者”等带有明显歧视性用语。这些歧视性用语在生活中是一种身份地位的象征。歧视性称谓用语不改，政策法规再多，也难以从根本上解决农民工问题。

国内已有一些地区的政府部门为了根本解决“农民工”问题，出台了一些相关政策和法规为“农民工”一词正名，如西安市雁塔区委、区政府要求对“外来人口、外来务工人员、打工者、农民工”等称谓统一规范为“新市民”，该区还出台了具体办法。在北京、山东、江苏等省市下发的有关文件中，也抛弃了“农民工”的称谓，统称“外来务工人员”等。因此，广东省也可考虑在一些文件、法规和正式场合用“务工新市民”等新称谓取代“农民工”一词。

关于改善财政宏观调控深化分税制财政体制改革的调研报告①

今年以来，我们经济组在4月、5月、9月、10月先后前往广东省地税局、广东省财政厅、江门市（包括台山市）、汕尾市（包括陆丰市）和潮州市（包括潮安县）政府及财税单位，就“改善财政宏观调控，深化分税制财政体制改革”课题进行调研，重点就“1994年实行分税制财政体制以来，市、县级地方政府的事权、财权与财力三者是否统一？有何矛盾？矛盾如何解决”等若干相关问题进行座谈、研讨，被调研的有关部门和市、县政府针对分税制财政体制改革实施十多年以来我国经济形势的发展变化，结合本地区的现实情况，在某些重大的理论问题与实践的结合方面，取得了很多共识，提出了很多真知灼见与建议，并对分税制财政体制的利弊关系作了比较客观的分析，现将此次调研的情况作如下汇报。

一、实行分税制财政体制的成果

分税制财政体制是社会主义市场经济体制的重要组成部分，1994年实行分税制财政体制改革是我国改革开放以来财政体制最深刻的变革。这次改革基本理顺了中央与省的财政分配关系，调动了中央与省的理财积极性，大大提高了全国财政收入占全国GDP的比重和中央财政收入占全国财政收入的比重，初步建立起与市场经济相适应的财政体制，基本确立了我国分级财政管理体制的总体框架，逐步建立了财政收入稳定增长机制，确立了中央财政的主导地位，提高了中央财政收入的集中程度，有利于增强财政宏观调控能力[1-3]。

“十五”时期全国财政收入为11.5万亿元，比“九五”时期增加6.4万亿元，增长125.5%；全国财政支出为12.8万亿元，比“九五”时期增加了7.1万亿元，增长124.6%，国家经济建设和各项事业发展都取得了新的成就。

我省从改革开放以来，特别是自1994年分税制财政体制改革以来，全省财政收入进入了一个健康快速发展的新时期。1994年我省地方一般预算收入为299亿元，2005年达到1806.01亿元；全省一般预算支出，1994年为416.8亿元，2005年达到2287.54亿元。全省一般预算收入自1991年以来连续15年位居全国第一。2005年来源于广东的财政收入完成4431.91亿元，剔除652.5亿元的出口退税后完成5084.47亿元，比上年增长17.07%，有力地推进了全省经济和社会事业的全面发展。

二、现行分税制财政体制存在的问题

分税制财政体制改革以来，由于种种因素的制约，当年的许多改革内容并未能一

① 本文原载《参事建言（2006年）》。香港：中国评论学术出版社，2007，75-92. 作者：刘人怀，周裕新，何问陶，吴厚德，毛蕴诗，于正林，陈婉玲，彭赟，刘凤英，黄莹莹.

步到位，而是采取了很多变通性、随意性的不规范的做法，仅仅是初步搭建了中央与省之间的财政关系的初级框架，这个框架只是勾勒了大体的轮廓，其中许多涉及分税制的关键之处，如事权、财权和财力模糊不清，基本不对称，特别是省以下的地方政府财力吃紧，财政处境艰难。为了“吃饭”和机构运转，到处抓收入，引发了各种各样的违法分配关系，留下了诸多的“先天不足”的后遗症。总体而言，可以说，我国省以下财政体制并没有真正进入分税制轨道，即使在发达地区的县、乡（镇）层级政府也没能真正实行分税制。

现行分税制财政体制存在的主要问题表现如下。

1. 政府职能缺乏科学界定，财权层层上收，事权层层下放，造成省以下政府财权与事权严重分离

主要表现在以下几方面。

（1）“共享税”越来越多，比重越来越大，严重地削弱了地方政府财力。被调研的市、县财政税务部门同志反映，实行分税制后，财权与财力上收，特别是中央集中的力度过大。1994 年中央与地方的共享税只有 3 个，而目前则增加到 12 个，由 1994 年占税种总数的 10％，上升到 31％；共享税税收收入占全国税收的比重由 1994 年的 55％增至 2003 年的 70％左右，而中央共享的比例最大。加上省与县的分成，县的实际所得很少。共享税比例的提高和比重的加大，很自然地弱化了本来就脆弱的地方税体系，再加上一些原属于地方的税种，如农业税、农业特产税和屠宰税等的取消，县、乡（镇）财政就更加困难了。比如，地处珠三角腹地的江门市是一个农业大市，农业税在县、乡（镇）级财政占相当大的比重，尤其是那些经济不发达、收入来源主要是农业税、农业特产税的乡（镇）财政运转能力相当紧张，而分税制确立的转移支付资金又不到位，农村社会保障工作实施难度加大。虽然农村合作医疗在江门市的参保率超过 70％，但由于省级要求进一步提高农村合作医疗标准，即从原来的 30 元提高到 50 元，这将进一步加大地方财政负担。由于财力不足，该市至今无法启动农村养老保险工作。

（2）分税制财政体制在明确收入范围的同时，没有对市、县政府的事权和支出范围做出相应的界定和调整，常有交叉和错位的现象。比如在潮州市，一些属于中央财政管理的事业费——武警、边防、消防、气象经费等，都要求地方财政每年安排一定的资金；市、县政府一些执法、执罚部门上划省级管理，如工商、技监等，一方面削弱了市、县政府职能，使市、县政府事权责任大，另一方面带来了财政职能弱化和财政税收管理权、执法权分散。

（3）现阶段我国各级政府的职能分配是，越是基层政府，所承受的社会公共服务职能就越沉重。世界银行在 2002 年的一份报告中指出，“中国县乡两级政府承担了 70％的预算内教育支出和 55％～60％的医疗支出，地级市和县级市负责所有的失业、养老保险和救济”。中国工程师协会会长殷大奎透露，“我国人口占世界总人口的 22％，而卫生总费用仅占世界卫生总费用的 2％，财政投入占医院当年支出的比重，省级以上医院约占 5％，市县医院一般约占 1％，乡镇卫生院在 1％至 5％之间”。他提供的数据还显示，2003 年中国医疗总费用 6600 亿元，政府只负担 17％，而欧共体为 80％～

90%，美国为45.6%（老人、穷人、残疾人全免费）、泰国为56%。卫生部一位官员提供的数据也表明，政府预算支出在中国卫生总费用中的比例，已从1978年的33%左右下降到2003年的17%。

（4）法律法规规定县级财政各种法定支出比例太多，事权范围不断扩大，使县级财政难以招架。例如，《中华人民共和国教育法》规定教育支出增长要高于经常性财政收入增长；“科技三项费用”支出要达到总支出的1%；《中华人民共和国农业法》规定农业支出要高于经常性财政收入的增长等。

2. 现行分税制财政体制是一种权力主导型的分税制，缺乏公共财政民主化与公共参与的机制

县级政府财权在上级政府的财政管制下，常常处于被动的状态，其核心就是存在制度性的缺陷，突出表现在以下方面。

（1）省以下地方政府财权缺乏自主权。一是目前地方税收立法权全部集中在中央，不利于地方政府根据本地的实际开征与停征税种。中央集中全国范围内普遍征收的地方税的立法仍然必要，但一些具有典型的地方税特点的税种的立法权应下放给地方。二是地方政府缺少税收的调整权。税收调整权包括税目、税率、征税方式方法的调整。地方税的调整权应由地方掌握，使地方在规定的范围内，依据需要与可能对地方税的征收作因地制宜的调整，以保证税收制度更具高效性和实用性。三是地方政府缺乏税收减免权。税收减免权的运用，最能体现税收政策，地方政府不具有这方面的权限，将大大限制地方政府运用地方税收政策的权力。

（2）上级财政对下级财政的各种专项补助，强制性地要求下级财政要按上级政府要求，进行配套资金全额配足，作为检查地方政府或主要领导是否执行政策的条件和标准。例如，教育、科技、农业支出等这些实行经常性财政支出增长比例挂钩的做法，从实际和长远来看，将使县级地方政府财政无力承担。法定支出在本级财政支出中的占比随着政府级次的下降而上升，越是到基层政府法定支出对本级财政支出的重要性越大，而地方支出自主越难以保证。因此，从现行的分税制财政体制来看，作为下级地方政府很难有相应的税权的独立性、完整性和主动性，在财力相应减少，财政缺乏自主性的情况下，地方财政尤其是县、乡财政陷入困境就成为必然。

（3）权力主导型的分税制无疑是造成政府行为异化的深层次原因之一。政府行为异化所造成的负面影响很大，表现在很多地方政府官员为了“吃饭”，到处抓钱，找钱下锅，没有多少时间去抓经济、公共服务、解决民生问题，为此，不少地方政府为了确保机构的正常运转，除了借债和巧立名目乱收费，别无他路可走。

3. 分税制改革未能从深层次上解决社会收入分配不公的差距问题，甚至还扩大了社会收入分配差距

当前，我国在分配领域出现了较为严重的收入分配不公的问题，主要表现为居民之间、城乡之间以及地区之间收入差距越来越大。据世界银行的统计数据，我国的基尼系数在改革开放前为0.16，2003年达到0.458，超过了国际公认的警戒线0.4，2005年逼近0.47，而这个“系数”不包括非正常收入，如果把非正常收入也算进去，基尼系数可能要更大一些。“系数”越大，贫富悬殊越大，社会越不稳定。据有关方面统

计，城市居民收入最低的1/5人口只拥有全部收入的2.75%，仅为收入最高的1/5人口拥有收入的4.6%。国家统计局等单位2006年发布的《农村经济绿皮书》中指出，我国城乡收入差距继续扩大，2005年城乡人均收入比例高达3.22∶1，比2002年的2.8∶1扩大了。2004年农村居民人均生活性消费支出为2185元，而城镇则为7182元。地区之间居民收入差距也呈现扩大趋势，2003年西部地区人均GDP仅为东部地区的40%。从2000年到2003年，西部与东部地区人均GDP的差距已由7548元扩大到9250元。2003年东、中、西部地区城镇居民人均可支配收入分别为10 366元、7036元和7096元，比2002年分别增长10.8%、10.5%和8.4%。收入分配不公已成为经济和社会发展的障碍。

4. 现行的转移支付制度存在不少问题

主要表现在以下方面。

(1) 一般转移支付不能满足地方政府最低公共服务水平的提供，一些贫困乡镇出现了“抓工资保稳定”的反常现象。在财力向中央集中的过程中，中央一般转移支付没有作相应的调整，而是通过增加专项补助的方式来平衡地区间的财力差异。专项补助在设置方面已经出现项目设置过多过滥、覆盖面过宽、重点不突出、零星分散的现象，在实际操作中被挤占和挪用的现象十分严重。

(2) 大多数专项转移支付下达后，不仅不能增加地方政府可支配财力，地方政府还需要额外增加配套资金，加重了地方财政的困难。

(3) 由于缺乏有效的约束和效益评估，转移支付资金的使用效率不高，有的地方甚至将转移支付资金用于形象工程建设。

在现实实践中，转移支付补助是以1993年基数给予地方税收返还，没有触及不同地区长期以来形成的财力分布不均衡的格局。随着经济的发展，1993年确定的税收返还和一般性转移支付补助等基数已发生了巨大变化，但各地所得的各项补助仍按当年的基数计算，上级财政至今仍未调整税收返还基数。如汕尾市，2005年的各项税收返还基数为19 546万元，而各项上划收入达51 819万元，为各项税收返还基数的2.65倍。返还基数未随着上划收入的增长而提高。再如潮州市财政，1994年上划中央“两税”收入26 478万元，2005年达到88 586万元，而税收返还基数占地方财政收入的比重逐年下降，“两税”由1994年66.4%下降至2005年27.16%。1996年上划省“四税”收入7025万元，2005年上划省“四税”收入17 526万元，“四税”返还基数占地方财政收入由1996年的20.54%下降至2005年的7.93%。

5. 市县政府历史债务沉重，金融风险、企业债务风险及社会债务已经转嫁为财政风险

我省的陆丰县是一个以农为主，经济基础十分薄弱、债务包袱十分沉重的贫困县(2005年全市生产总值人均仅为4869元，而2004年广东省人均GDP已达到19 683元，其中珠三角地区达到43 054元)，其财政运行的基本经费支出中，三分之二以上来源于税收返还和转移支付。但2001年以来，由于归还化解金融风险再贷款和财政临时借款等，该市2.6亿元的税收返还及转移支付基本用于抵偿债务，财政供需矛盾十分突出，财政实际可支配能力很小。为确保工资和离退休费发放，财政不得不向上级借款、调

剂部分专款和压缩单位预算资金供给，这又直接影响了各单位的正常运转，形成了现实的财政支付风险。

江门市由于化解金融风险、借款等历史债务，欠款数额巨大，加上以前年度政府债务扣款未能全部消化，造成财政负担越来越重，截至 2005 年底，共有政府债务 141 亿元，其中市（区）级共有政府债务约 70 亿元，镇级政府债务约 71 亿元。因化解债务难度加大，又受债务扣款影响，该市各级财政资金调度十分困难，严重影响社会经济的均衡发展。

被调研的其他市县级政府财政，都有不同程度的受债务负担过重而深感支出缺口加大，地方财政面临困境的情况。

各市、县财税单位的同志们一致认为：实行分税制，富裕地区越富，贫穷的地区就越穷，因为富的地区有较好的工业和服务业基础，税基大，地方政府按“基数法”测算出来所取得的收入就越多，相反，贫穷的地区，税基小，地方政府所取得的收入就越少。因此，分税制财政体制必须改革，才能更好地调节地区之间分配不公的差距，调动地方政府理财的积极性。

三、改革取向与建议

1994 年分税制改革的大方向和基本制度成果，必须肯定和维护。但实际情况是，1994 年以后，省以下财政体制在分税制改革方向上却几乎没有取得实质性的进展，分税制在实施过程中的一些负面因素却在积累和增强。“共享税”越来越多，越靠近基层，越倾向于采用“讨价还价”的各种包干制和分成制。

我们认为，财政体制如要适应市场经济的客观要求，必然要处理好“两个基本经济关系”——处理好政府与企业、中央与地方的分配关系。除了分税制之外，再不能走回头路搞“分成制”或“包干制”，这是由我国的国情决定的。

如果以“跳出财政看财政”的全局思维和前瞻思维看问题，应该看到，在我国省以下推进分税制财政体制的大方向是正确的，问题的关键和当务之急是应努力解决分税制如何过渡的问题，以及如何抑制其负面作用、发挥其正面作用，这才是我们的问题。为此，我们提出以下建议。

1. 必须按照市场经济的客观要求，加快政府职能的转变

（1）明确政府职能范围。党的十六届三中全会对我国各级政府职能进行了明确界定，即“经济调节、市场监管、社会管理、公共服务”。随着构建和谐社会这一任务的确立以及经济形势的不断变化，政府职能应该逐渐向支持和谐社会的构建转变，并据此决定财政职能的转换。

从战略目标来看，即以 2020 年基本完成经济体制改革，建立全面小康社会与和谐社会为目标，界定与经济体制、政治体制、社会经济发展水平相适应的政府职能，从而确定与财政职能相适应的财政体制改革目标。

总的说来，政府职能包括社会管理职能和经济管理职能两个方面。随着向市场经济体制的转变，政府应该有进有退，逐渐从竞争性、营利性的领域退出，避免对微观经济主体的干预，培养企业的自主创新能力和竞争能力；同时，政府应该通过财政手段对经济加强宏观调节，改善宏观调控，对新出现的问题，如环境保护、生态平衡、教育与医疗、社会保障、突发性事件、国家安全等问题加强管理。因此，市场经济要求政府收缩经济管理职能，强化社会管理职能。有什么样的政府职能，就有什么样的财政职能。

按国际惯例，社会保障和救济通常由中央政府提供，教育和医疗是中央和省级政府的责任。在公共财政制度较完善的国家，中央政府提供教育、卫生、社会保障与福利、住房与社区环境等社会服务的支出占绝对主导地位，这是中央财政的主要职能。

(2) 转换政府职能重点。在构建社会主义和谐社会中，我们应该打破体制性障碍，在界定各级政府职能的基础上，即在明确各级政府事权的条件下，正确地划分各级政府的支出责任，并将政府职能转换的重点放在实现由经济建设型政府向公共服务型政府转变；实现由单纯支持经济增长到与社会发展保持和谐的全面转变。可以考虑将地方政府能有效提供的公共服务，作为地方政府的事权，中央只承担地方政府难以有效行使或不宜由地方政府行使的事权。对于涉及社会保障和公共卫生体系建设等共同事务，应明确规定各级政府应承担或分担的比例，并使之稳定和规范化。

各级政府财权的划分，是以各级政府职能的划分为基础，因此，科学界定政府职能并由此确定各级政府财政收支范围，加快财政公共化的趋势，是长期必须解决的基础性和体制性障碍，而优化各级政府的财权与事权，是提高财政职能的体制实现能力的关键。

2. 推进依法治税，对税收立法权限做出统一规定

在这次调研中，市、县政府财政部门普遍认为地方税制十分混乱，地方税概念不明确，地方税收法规制定权、解释权、税目税率调整权以及减免税收权等都集中在中央，地方税管理体制变成了一种变相的收入分成，存在税收征管主体权限交叉和税收执法权保障不足。此外，地方税划分标准杂乱无序，有按税种划分的，有按行业划分的，也有按企业隶属关系划分的。比如，将营业税划分为地方固定收入，但却又不包括铁道部门、各银行总行、各保险总公司集中缴纳的营业税（这部分划入中央固定收入）；将企业所得税分为中央企业所得税和地方所得税，这种交叉重叠的划分方法，违反了分税制的规范要求。此外，地方税制不尽合理，主要表现在税种虽多，但缺乏对地方财力具有影响且长期稳定的主体税种。目前占地方税收比重比较大的税种只有营业税和企业所得税，但营业税和企业所得税又都不是实际意义上的地方税，因有中央分享的部分，而且随着增值税的改革，如果将建筑安装、交通运输企业的营业税改为增值税，营业税的税基将会随之减少，收入占地方税收的比重也会下降，更难成为地方主体税种。

为此，必须推进依法治税，对税收立法权限做出统一规定，我们建议：

(1) 制定《税收通则》(或《税收基本法》)，对税收立法权限做出统一的规定。根据国际惯例，为了充分发挥税收立法权的作用，必须由法律对税收立法权问题做出统一、明确的规定。税收立法权是全部税收问题的逻辑起点和核心内容，它明确国家对哪些行为征税，不同层级间如何分配等基本问题。由于宪法是国家根本法，指望其对全部税收立法权都予以明确、细化是不现实的。在这种情况下，《税收通则》是合理划

分税权、健全税法体系的必然选择，它是规范国家税收问题的宪法性文件，而《中华人民共和国税收征收管理法》的法律级次偏低。

(2) 加强税收立法，建立健全税收法律法规体系。税收立法应以立法机关立法为主，提升税收立法级次，逐步形成以税收法律为主体，并以税收行政法规为配合、税收部门规章和规范性文件为补充的税法梯级结构，尽量避免税收立法的行政化和低层次化。主要内容及做法大致是：①有关税收法律的基本原则、通用条款、税务机关及纳税人的权利、义务、税收立法的权限、税收管理权限等基本内容，需要通过《税收通则》加以规范，由全国人大立法。②凡全国性税种的立法权，包括全部的中央税、共享税和部分在全国范围内普遍征收的地方税，其税法的规定、颁布、实施权和修改、补充、废止权，属于全国人大及其常委会。③某些全国性税种经全国人大及其常委会授权可以先由国务院以条例或暂行条例的形式发布，但要规定试行期限，并尽快由全国人大上升为正式法律。④国务院有制定税法实施细则，增减个别税目，调整个别税率的权力及对税法的解释权。⑤省级人大及其常委会，负责部分地方性税收法规的制定、颁布、实施、解释和调整，并可依据法律规定，由省级人民政府制定地方性税收法规实施细则和征管办法。⑥国务院财税主管部门为了贯彻执行税法和税收行政法规，有权制定规章并拥有一定的行政解释权。

(3) 赋予地方对某些地方税种一定的税收立法权。广东省地区经济发展不平衡，需要进行适度的分权，逐步建立完善的地方税体系。其思路是：①对地方影响较大的税种，如营业税、企业所得税等在不违背中央统一规定的前提下，可赋予地方在一定范围、一定幅度内的政策调整权。②全国统一开征但区域性特征比较明显的地方税种，如资源税、房产税、城市维护建设税等，具体的实施办法、税目、税率的调整、税收减免以及征收管理等权限，赋予地方。③征收成本较高的税种，如车船使用税、契税、印花税、城镇土地使用税等，其立法权、征税权、管理权全部划归地方。④省级政府在报请中央批准后，可以开征某些具有地方特色的税种。⑤中央赋予地方税收立法权是一个渐进过程，在当前只能下放到省一级地方政府。

3. 配套改革：压缩政府规模，减少政府层次，节约行政成本

(1) 压缩政府机构，缩小政府规模和财政供养人员。有什么样的政府职能，就有什么样的政府规模，政府规模随着政府职能的变化而变化。

改革开放以来，市场已取代政府成为资源配置的基础，但我国政府职能并没有按照市场经济的要求发生实质性转移，突出表现在从中央到地方仍然保留着庞大的行政管理费用。据统计，1990～2001 年，我国行政管理费用年均增速为 29.2%，12 年间增长了 7.3 倍，是各项财政支出增长中最快的一项。比如，在 20 世纪 90 年代，全国有 350 万辆公务车，一年耗费 3000 亿元，而在 2001 年，用于低保的费用只有 100 亿元，失业保险额只有 190 亿元，国家财政用于公共产品的民生供给严重不足，而用于行政开支的供给却大大超前，这令人深思。

在我们的调研过程中，有的市（地市）财政部门反映："由于财政供给人员过多，财政用于养人的支出越大，政府用于发展性的支出就越小，财政支出压力大，必然导致政府要抓收入。过度的抓收入必然增加企业和农民负担，加剧社会不安与矛盾。即

使每年财政收入增加了，增加部分也被新增人员吃掉，这样政府无法拿出更多的资金去发展经济和公共服务事业。”虽然中央经过几次大的机构改革，但改革的分流人员大多进了事业单位，而事业单位改革至今尚未全面推开。

政府职能转变滞后，形成如此庞大的政府规模和如此之多的财政供养人员（2004年底公务员的统计数字为640万人），必然使各级政府不堪重负，特别是县乡政府尤其是贫困地区更为突出。因此，要实现政府事权与财权相对称，必须大力压缩政府机构，缩小政府规模和财政供给人员规模，使其拿出更多的财力，加大对县乡财政的转移支付，为完善省以下的分税制财政体制打下良好基础。

（2）减少政府层次，使五级政府变为三级政府。在现行的省以下的各级政府分税中，省、地（市）拿去中央与地方共享税地方分成的绝大部分和地方大税种，到县乡基本无税可分。这种局面的形成与我国实行中央、省、地（市）、县、乡（镇）五级政府有很大的关系。在取消农业税和农业特产税的基础上，即将启动的新一轮税制改革将要统一内外资企业所得税，合并房产税、土地税、车船使用税、车辆购置税等税种；取消屠宰税、筵席税、固定资产投资方向调节税等税种，这将使我国税种进一步减少，使五级政府之间的税种划分更为困难，县、乡政府基本无税可分。因此，要完善省以下分税制财政体制，确保各级政府都有自己的稳定的主体税种，必须对现行的政府体制从纵向上进行改革。为此，建议取消地（市）级政府，将乡（镇）政府变成县级政府的派出机构，实行中央、省、县三级和乡（镇）派出机构的政府体制，这是符合中央十六届五中全会“减少行政层次”的要求的，可以使省以下的分税制，由原来五级架构下的“无税收政策”变为三级架构下的柳暗花明。

我们对政府体制提出这样的改革建议，除了确保各级政府能够确立与自己事权相对应的主体税种外，主要还有以下两方面的原因：①在税务、工商、公安、司法实行条条垂直管理和教师工资上划县级管理后，乡（镇）政府已不是一级完整的政府，但完全取消乡（镇）政府由于县辖区域大，县级政府难以直接管理农村事务，乡（镇）政府可作为县级政府派出机构。②地（市）级政府本来就是省级政府派出机构演化而来，至今也尚未有宪法地位，并且地（市）级政府为了本级利益往往上收县级财力，截留中央和省级政府对县级政府的转移支付，而且由于省内区域经济发展不平衡，地（市）级政府对县级财力的调控能力十分有限。在这种情况下，取消地（市）级政府，有利于省级政府对县级财力进行直接调控，可以最大限度地实现全省范围内公共产品供给的均等化。这种改革有力地促使事权的划分更加清晰化、合理化和构建与事权相匹配的分税制财政体制，降低行政体系的运行成本，县级财政财力更大、更好地促进县域经济发展，再配之以中央、省两级自上而下转移支付制度的加强与完善，必将会从根本上缓解基层财政困难，从而形成有利于欠发达地区进入“长治久安”的机制。

如果从我国的历史和发达国家的情况来看，也大多是实行三级政府。我国自秦朝到民国2100多年中，除了疆域横跨欧亚的元朝实行五级政府体制外，其他各朝基本上都实行三级和四级政府体制。在国外发达国家大多实行三级到四级政府体制，美国实行联邦、州、市三级体制；英国实行中央、郡、区三级体制；日本实行中央、都道府县、市町村三级体制；德国实行联邦、州、区三级体制；法国实行中央、大区、省、市镇四级体制。

4. 完善政府转移支付制度

针对我国现行转移支付制度存在的问题，我们建议：中央对地方转移支付制度的改革应突出公平，加大对省以下地方政府财政之间财力差距的调节，使省以下各级地方政府享有其与事权相对应的财力支持。

改革的思路是：

(1) 科学界定转移支付目标。转移支付目标是为实现社会分配公平目标而为各地提供相对均等化的公共产品和服务，必须将转移支付的资金用于最需要的地方，而不是用于获得最大经济回报的地方。

(2) 合理确定转移支付计算依据。在这次调研过程中，陆丰市财政局提出“一般性转移支付的计算应根据地方总人口、财政供养人口、地方一般预算收入、税收收入、经济发展程度、全年平均支出及每年增支因素等设定，充分体现公平和均等化原则、分析制定统一的计算公式。我们认为，一级政府正常运转基本支出应以全省平均工资为基数，结合人口及各地国内生产总值，分不同地区设置不同系数，但差别不应该过大，总体保证正常运转基本需要和保障水平应一致”。陆丰市财政局提出的转移支付这一计算依据是从实践中总结出来的，比较实事求是地反映客观要求，可以作为参考。

(3) 应简化转移支付方式。将现行多种转移支付方式统一为均等化转移支付和专项转移支付，不要搞其他花样很多、又不现实的不规范的转移支付方式。

(4) 优化转移支付结构。重点是严把专项拨款立项关，取消配套的专项资金补助，防止因地方政府缺乏专项补助配套资金而出现的“半截子工程”。为此，必须建立严格的专项拨款项目准入机制和审批制度，减少立项的随意性和盲目性，防止资金的浪费。

(5) 清理整合专项转移支付项目。根据政府支出责任范围内的事务，应取消现有专款中名不符实和过时的项目，归并重复交叉的项目，严格控制专项转移支付规模；对年度之间变化不大，且将永久存在的项目列入体制补助，冲减地方上缴。

5. 尽快地解决地方政府历史债务，特别对贫困县地区的历史债务应根据有关政策，重新进行一次彻底的清理

对地方政府的历史债务应根据区别对待，认真扶贫的原则，区别几种不同情况，采取不同做法。①对偿还能力较强的地区，争取在三年内还清，每年按一定比例还债；②对偿还能力不强，又需要利用债务支付最需要的经常性预算支出，如工资、离退休费、义务教育等，应给予适当的减债（如台山市）；③对暂时无力还债，又急需用债务支付经常性预算支出的，在今后三五年内有能力支付债务的地区（如陆丰市），可以考虑给予延期三五年才还债，以便让地方政府财政渡过这个偿债难关，保证社会稳定；④对于经济基础十分脆弱、又无其他门路“找米下锅”的贫困地区，应给予全免一切债务的优惠政策，并在免除一切债务的基础上，上级政府还应给予更加优惠的税收政策和专项补助资金，发展以“造血功能”为主的工业和服务产业，以带动该地区的经济起飞和发展。

6. 国税、地税机构合并的可行性分析与建议

从 1994 年分税制改革开始，我国设立了国税、地税两套税务机构，分别履行对中央收入与地方收入的征管职能。十多年来，国税、地税机构分设对于组织税收收入，进行宏观调控，提高中央税收收入比重，增强中央财政宏观调控能力，促进市场经济体系培育和经济的发展等，发挥了重要的作用。但是国税、地税机构分设出现的各种弊端也暴露无遗，难以适应新的形势要求。

(1) 国税、地税机构分设的弊端。征管效率低。主要体现在对改制、合并、改组企业的企业所得税管理上，双方互不相让，争抢税源现象经常发生；在对违法案件的查处上，相互扯皮推诿，谁都不愿意行使管辖权，致使案件久拖不决，客观上放纵了违法行为，造成了很坏的影响；存在大量的重复劳动，如税务登记、征管资料、信息处理、税收票证、税收宣传、政策咨询、税收科研等事项性质相同，但国税、地税两机构各行其是，互不通气，造成大量的资源浪费。

征税成本高。两套机构的分设无疑是造成税收成本过高的主要原因。1995 年我国仅预算内支出的征税成本就在 5%以上，而同期美国为 0.6%，日本 0.8%，法国 1.9%，加拿大 1.6%。我国征税成本高的主要原因有：人员费用增长过快，1995 年的 60 万财税人员增至 1998 年的 100 万人（不包括临时工），以年人均 4 万元估算，就需要 400 亿元经费；基建投入大，两套机构，两套办公场所，国税、地税的各种征、管、查机构和办税服务厅设置雷同。此外，国税、地税各干各的，分别开发两套网络，各自为政，互不兼容，不能对接共享，造成人、财、物的巨大浪费。

分头管理、分庭抗衡现象突出。将税收分为国家税收和地方税收，极易引起地方政府重视地税，而不重视国税，造成很多矛盾与冲突。

隶属关系不同，利益分配不同。国税属中央垂直领导，从业人员具有很强的优越感，待遇也比地税好，造成利益分配不公，地税人员反感情绪大，不利于国税、地税团结协作，影响税收征纳。

(2) 两套税务机构合并的有利条件。我国已实行分税制 10 多年了，中央财政收入得到了切实的保证，打下了比较坚实的经济基础。目前中央税收收入占税收总收入的 68%左右，中央本级财政收入约占全国财政总收入的 56%左右，比改革前的 22.02%提高了一倍多，从而确立了中央财政在分配格局中的主导地位，有利于中央宏观调控的加强。

我国实行分税制是一项正确的决策。实行分税制是世界各国的通行财政体制，但分税制绝不是分机构，即使在实行分税制的西方国家中，也不是一定要分设两套税务机构。如法国，地方政府不单设税务机构，而由国家税务机构的派出机构代为征收。其实，分税制的归宿集中反映在税金在不同政府级次间的分配，国税、地税两套机构都共同地承担着这一国民收入分配和再分配的任务，这为国税、地税两税机构合并提供了理论支持与基础。

信息化建设取得了突破性进展，充分发挥税收监控功能，监管方式实现了根本性变革。上级政府和税务部门可以通过信息网络即时监督下级政府和税务部门的执法情况，对出现的违规违法现象及时通报和纠正，消除了传统层层上报和下达导致信息容

易失真和整改滞后的弊端，大大提高了工作效率，降低税务成本。

（3）国税、地税两套税务机构合并的建议。要建设廉价高效的税务机构。“税收是政府的奶娘”，一个真正的民主社会是无法容忍昂贵低效的政府的。一个政府机构的设、存、废必须考虑其运行成本。我国税收成本居高不下是毋庸置疑的事实，国税、地税机构的合并本身是“税收成本最低化”的必然要求，应在全国组建新的、单一的国家税务机构，实行从中央到地方的垂直管理体制。在机构合并之后，更要树立税收成本理念，加强成本控制，转变税收管理上粗放型管理观念，加大税务系统内部机构改革、人事制度改革和从中央到地方的精简力度，消除富余人员，精简机关，充实征管向基层倾斜的政策；大力减少各种会议、文件报表等，简化办事程序，提高办事效率。

构建税收监督制约机制。机构合并以后，各级政府的利益倾向不会自然消失，各种违规操作问题仍可能存在。约束乏力，是导致权力滥用的重要原因，因此，构建税收监督制约机制至关重要，应做到决策、执行、监督相分离又协调的要求，在税务系统内部实现征收权、管理权、稽查权的相互制约和彼此协调与平衡。在这基础上，逐步做到财政、国库、银行、税务部门之间实现微机一体化联网，做好信息联通共享，互相监督制约。此外，还要从纳税人监督、社会新闻舆论监督角度共同建立起独立的税务监察审计机构，从而形成全社会监督制约机制。

税务机构的设置要打破行政区划的限制，其密度和人员应视税源状况而定。

整合信息化资源，全国统一联网开发，降低运行维护费用，提高行政管理效率。

加强干部队伍建设。提高税务人员素质。现代管理理论突出体现了以人为本、以人为中心的理念，机构合并以后税务人员更要努力加强自身建设，加强学习与修养，增强依法行政的业务素质；要分期分批、定期和不定期地进行培训学习，并把它作为考核干部业绩的要求，所有这些都要从制度上规定下来。

参考文献

[1] 楼继伟. 分步实施税收制度改革. //本书编写组.《中共中央关于完善社会主义市场经济体制若干问题的决定》辅导读本. 北京：人民出版社，2003：206-220.

[2] 张元元，李金亮. 社会主义市场经济学. 广州：暨南大学出版社，1993：255.

[3] 郭熙保. 发展经济学理论与应用问题研究. 太原：山西经济出版社，2003：284-296.

岁月留声[①]

2006年，是不平常的一年，是丰收的一年。在这一年里，我国科技界取得了丰硕的成果：家蚕基因芯片与表达图谱的诞生、新粒子在正负电子对撞机上被发现、口蹄疫基因工程疫苗的研制成功、青藏铁路全线开通、龙芯2E通过验收等等。每一项成果都离不开两院院士的源源智慧和辛勤劳动。新年的钟声就要敲响，中国工程院院士广州咨询活动中心衷心祝愿各位院士新年快乐、身体健康、合家幸福！

过去一年，我们一如既往，在各位院士、领导的大力支持下，开拓创新，做了一些工作，创造了一些亮点，取得了一定的成绩：

首先是借用外脑，开展院士沙龙活动，为广州发展献计献策。在去年开展工作的基础上，按照市科技局领导的指示，我们加强了与外地院士机构的合作，邀请外地院士专家和外地机构共同在广州开展学术交流等活动，为广州经济和科技活动提供咨询和建议。2006年，中心分别邀请了外地的江欢成、项海帆、杨胜利等院士到广州，结合本地院士专家，分别举办了以“可持续发展和结构优化”和“中国桥梁科技发展的战略思考”、“光纤通信材料与器件”、“如何持续快速发展生物医药技术产业”为主题的三期院士沙龙活动。院士、专家、企业与政府代表在活动中都积极阐述自己的观点，展开深入讨论，提出了不少新思路和好建议。会后与会代表都一致认为能借用“外脑”为广州科技经济社会持续发展出谋划策非常必要而且非常有效，对促进广州贯彻科学发展观和提高自主创新能力将起到积极的作用。“借用外脑，为广州科技经济服务”也成为今年院士中心工作的一个亮点。

其次是举办院士专家讲座，为科普工作作出贡献。为了营造崇尚科学、相信科学、依靠科学的社会氛围，提高干部和市民的科技意识，广州市科普工作领导小组办公室、广州市科技局制定了科普工作计划，向政府部门、企事业单位的领导干部、管理人员、工程技术人员及广州市民宣传科普知识，举办广州院士专家科普系列讲座。中国工程院院士广州咨询活动中心受其委托，负责讲座的组织工作，2006年共举办了三期院士专家讲座，包括钟世镇院士的“数字人与数字医学”；杨胜利院士的“医药生物技术的现状与发展趋势”；梁桂主任的“科技创新与服务”。参加人数达1300多人次。在中国工程院和市政府有关部门的支持下，院士活动中心积极开展工作，充分利用本地及外地院士、专家人才的资源优势，精心策划和组织，把讲座活动办得卓有成效，受到各方的肯定。通过组织举办讲座和院士中心的影响，树立了品牌。

再次是加强对外联络交流，提高服务能力。为了不断提高我们的工作水平，完善管理，丰富院士活动内容和形式，更好地做好服务工作，中心今年非常重视与工程院及外地院士机构的联络和交流学习。在中心领导的带动下，我们主动采取走出去，请进来的方式，进行交流学习，利用出席“两院院士大会”、《院士通讯》年会等机会，

① 本文原载《岁月留声——院士活动剪影（二）》，中国工程院院士广州咨询活动中心编，2007，1.

向全国各地的优秀院士活动中心取经，开展交流。与会期间，我们及时向工程院领导汇报了工作情况并听取了他们的意见和建议。交流学习，使我们的视野更加开阔、思路更加清晰，为提高我们的服务水平和工作能力起到了促进作用，并且为院士活动异地合作创造了有利条件。

最后是增强服务意识，主动做好服务保障工作。加强与在穗院士的联系，做好为院士服务的工作，建设真正的“院士之家”是我中心主要任务和职责。我们在春节和中秋佳节分别组织了院士新春联谊会和中秋联谊会，市委原副书记方旋、市政府甘新副市长等领导及有关部门领导、代表出席了联谊会，中秋、春节共有 40 多名院士出席了联谊会。在了解到有院士患病在医院疗养的消息，院士活动中心也派出人员前往探望问候。根据中国工程院及上级部门的安排，我们还协助做好外地院士专家来穗交流访问的接待工作，如承办了“中国工程院土木、水利与建筑工程学部常委扩大会议”的会务组织工作，以及协助广州市政府联络有关院士出席各界人士新春招待会等活动。各级领导、在穗院士及到访的外地院士对我们的服务工作都给予了高度的肯定。

这里特意收集了过去一年里开展各项活动的部分剪影，制作成《岁月留声》画册第二期，仅供各位指导。

黄石应该在建设特色城市中凸出优势[①]

黄石市矿产资源丰富、人文景观优美，工业经济和社会发展有一定的基础，综合经济实力连续两年进入全国百强城市，先后获得全国先进科技城市、双拥模范城市、文明城市等称号，桂冠很多，充分说明市委、市政府的正确领导和在改革开放中做出了显著成绩。

但我有一个感觉，黄石想全面发展，既要发展重工业，还要发展其他的方方面面，当然理想是好的，但我觉得不要把发展目标定得太宽泛了。比如说湖北省是很好的教育科技省份，重点大学多，科技成果多，湖北人文化素质比较高，这是湖北的优势。在这个前提下，黄石不可能再发展高等教育。怎么把自己的布局搞好，我建议你们要集中发挥优势，建设特色城市，不要像其他城市一样搞全面发展。你们提出举全市之力争创全国优秀园林城市、卫生城市、旅游城市、生态宜居城市等，目标定得太多。

我们在发展经济中把环境搞坏了。环境恶化的结果，会反过来影响经济的发展，花钱去治理环境污染。所以我建议黄石市委、市政府首先要给老百姓一个好的环境。黄石的环境保护得还不算太差，应该花大力气把工业发展与环境治理结合起来。否则，人家不会来的，污染重了，把人都吓跑了。

工业发展既已形成了黑色金属、有色金属、水泥等重工业，特别是汉冶萍一百多年传承下来，首先要保护这个品牌。我建议你们打汉冶萍这个世界牌子。因为19世纪的中国诞生了汉冶萍的牌子，我们不要丢掉。中国城市都搞全面发展，这是不行的。你看美国、欧洲几乎每个城市都有自己的特色，如文化城市、旅游城市、教育城市、重工业城市、汽车工业城市等。黄石可以发挥自身钢铁优势，特种钢、有色金属、黄金、银搞得好，人家自然会找上门来。搞生物、医药行业，没有高校、科研单位的支持，要想搞好，谈何容易啊。在原有重工业的优势上，发展产业链。中国人喜欢黄金，把黄金加工成首饰，并保证是真的。到黄石买黄金首饰，货真价实，我相信黄石的旅游一定会上去。黄石很漂亮，有几千年的铜绿山，有一百多年的汉冶萍，文化旅游与风光旅游结合，我相信一定会迎来旅游业的发展。而重工业要发展产业链，这样才能更好地发展。

我建议广为宣传黄石重工业的优势和特点，让全国人民了解。同时狠下力气对老企业进行技术改造，大幅度地提升技术管理水平，突出重点，把重工业做大做强。我国还有相当多的钢种不能生产，你们应该在这方面多下功夫，做到不进口，到黄石来买。在这方面，企业是主体，政府只能引导。为此，我建议你们搞一个企业论坛，让企业为黄石发展建言献策。企业家们不行动，我们讲得再好也没用。在广州等沿海地区，每年都要举办这样的论坛。比如新冶钢可邀请宝钢、鞍钢的专家来黄石传经送宝，把国内的先进经验学一学，再把日本等国外的专家请进来，找出我们的差距在哪里。现在的关键是形成合力，在主产品上做文章，就可把黄石建成一个好的特种钢基地。

---

① 本文原载《同舟行》，2007，专刊：5.

关于将清明节设为国家法定假日的建议[①]

“清明时节雨纷纷，路上行人欲断魂。借问酒家何处有？牧童遥指杏花村。”

唐代诗人杜牧所作的这首广为流传的《清明》，不仅写出了清明节的特殊气氛，也揭示出清明拜祭的习俗由来已久。清明节，在中国诸多传统节日中有着鲜明的自身特色，它既是二十四节气中唯一具有人文历史的节日，同时，也是整合了寒食节的一个大节日，其习俗活动之丰富足以和春节一比高下。将清明节设为国家法定假日，应该提上议事日程。

探究清明扫墓风俗的形成，应从古代墓祭制度及清明节的由来综合考察。墓祭之风早在西周时期就已形成，但对于祭墓时间未有确切规定。祭祖之风俗融入清明节是在后来历史进程中形成的，民间流传甚广的起源说是春秋时期介子推的故事。晋国晋献公死后，诸子争位，公子重耳逃难，途中一日饥饿难耐，介子推从自己大腿上割掉一块肉，煮熟后让重耳充饥。后来，重耳复国，就是著名的春秋五霸之一——晋文公。晋文公赏赐有功之臣，单单遗忘了介子推，“有功不言禄”的介子推又不愿前去邀功，于是携同老母亲前往绵山隐居。介子推的邻居将此事报告给了晋文公，晋文公马上派人到山上搜寻介子推，搜寻未果后，于是听从侍臣意见，放火烧山，想逼迫侍母至孝的介子推下山。但是，耿直孤傲的介子推坚决不肯下山，最后与其母亲一起被活活烧死在一棵柳树下。晋文公对自己的行为十分后悔，于是将绵山改为介山，下令规定自介子推被烧之日全国禁火三日，不吃烟火食，因此便有了“寒食节”。第二年，晋文公又在介子推殉难之日上山拜祭，发现枯死老柳树又发新枝，便将这棵柳树封为“清明柳”，将当日定为“清明节”。

因为寒食节和清明节日子相近，而古人在寒食节的活动又往往延续至清明节，久而久之，拜祭介子推的日子转化为扫墓祭祖的时间，寒食节逐渐淡化，直至隐没至清明节，在唐代已完成转换。这点在白居易《寒食野望吟》诗中可以得到验证：“乌啼鹊噪昏乔木，清明寒食谁家哭。风吹旷野纸钱飞，古墓垒垒春草绿。棠梨花映白杨树，尽是死生别离处。冥冥重泉哭不闻，萧萧暮雨人归去。”白居易将清明寒食与祭祀扫墓相提并论，清楚地说明了这一问题。至宋代，朝廷又规定。“寒食至清明三日，各地均须祭扫陵墓”。后来更沿世成为风俗。随着社会的发展，古代的清明节风俗在祭祀扫墓之外，又增加了踏青、宴游、插柳、蹴鞠、拔河、植树、荡秋千、放风筝等习俗，成为中华民俗的一个重要载体。

清明节是如此重要的民族传统节日，建议国家将其设为法定假日，使大家有机会扫墓祭祖。除上述原因外，还有以下几个原因。

1. 通过清明祭祖，在一定程度上有利于和谐社会的构建

每逢清明节，不少海外华人、华侨不远万里，回到祖先的坟前扫墓，祭祖敬宗；

① 本文原载《参事建言（2007年）》，香港：中国评论学术出版社，2008，327-329；《广东政协》，2007，(6)：10-11.

远亲近邻则因扫墓而相聚，拜祭先人亡魂，庄重地对始祖、对先人呈上自己的思念和敬意。“慎终追远，民德归厚矣”（《论语·学而》），通过清明祭祖这种敦亲睦族的具体仪式，更容易让大家感念到同祖同根，抒发人们尊祖敬宗、继志述事的道德情怀，缔结浓浓的亲情友谊，加强伦理观念，调和思想情感，进而激发强大的团结力量，使家庭生活、人际关系趋于和谐。

2. 追忆先辈，可以激励青年一代进一步发扬艰苦奋斗精神

清明节不论以何种形式祭祀，其基本内涵都是缅怀先人、祭祀祖先，这是一种有形的祭奠方式和内在的心理感受相结合的礼仪形式，是表达情感的礼仪之举。在对先辈进行真切哀思的同时，今人不免会追忆先人的一生行状，追念他的事业功勋，回忆祖先的奋斗历史，进而审慎自己的灵魂，清扫心境中的灰尘与污垢，继承先人遗志，发扬光大其进取精神，将祖辈留给青年一代的美好记忆作为继续前行的起点。

3. 将清明节列为法定假日，有利于更好地保护并弘扬民族文化

民俗是民族文化中极具民族个性的组成部分，是一个民族带有文化胎记的身份确证，是一个国家民族精神的重要载体，是一个民族共同创造、共同参与的一种文化，这种文化自然包含着深深的民族认同感和民族凝聚力，它的精神影响力是巨大的。近年来，随着中国对外交流的加深，越来越多的优秀民俗传统远离了人们的日常生活，保护优秀传统民俗已成为急需进行的一项重要工作。2004 年，端午节被韩国向联合国教科文组织申报为本国文化遗产，已经为我们的文化遗产保护工作敲响警钟。清明节作为中华民族传统节日文化的一个重要组成部分，不但有着悠久的传承历史和深厚的文化积淀，而且是一个仍受现代人重视的节日，将清明节设为法定假日，也不失为保护、弘扬优秀文化传统的一个良策。

爱心和匠心[①]

今天，应贵市卫生局领导邀请，来做医学人文精神讲座，实在不敢当。本人不是研究人文精神学科的专家，也不是医务工作者，只是一名教师。好在我自已热爱医学，年轻时，在 1965 年和 1968 年，曾先后两次当过业余的赤脚医生，加之我的家属又是医生，因此对医学只是热爱，略懂一点皮毛。盛情难却，接受邀请，今天就讲“爱心和匠心”这个题目。

一、医务工作是伟大和崇高的职业

医师、护士、药剂师、麻醉师等，所有医务工作者，是全人类群体中最特殊的群体之一，最受人爱戴，最受人尊敬。毫不夸张地说，医务工作是伟大和崇高的职业！为什么这样说呢？这是因为，古今中外，任何一个社会，其核心任务是要让人健康地活着。社会上的人都有求于医务工作者，概莫能外。这一任务的完成者，正是光荣的医务工作者！

举一例：本人生在四川省成都平原上一个小县城，名为新繁县，地处成都市北面 30 公里处。新中国成立时，我才 9 岁。那时，我们县里有两位名医，我至今仍记得他们两人的名字：周子豪和雷益周。但是，我的许多小学同学的名字，以及当时县长等的名字，却记不得了。这在某种程度上说明医师在社会上的地位是很高很高的。

二、医疗工作现状

经过改革开放 30 年的奋斗，我们国家的 GDP 已居世界第四位，仅次于美、日、德三个国家。这是我们伟大的祖国几百年来第一次进入盛世时代。这一了不起的成就，也包含着医疗事业所做的贡献！我国人均寿命已从新中国成立前的不到 50 岁提高到现在的 73 岁。

但是，在伟大的成就面前，却出现了“上学难，看病难，购房难”的局面，为此，大家议论纷纷。

以“看病难”来说吧，老百姓中一些人反映，甚至媒体也在炒作：

（1）看病贵！有时，住一天医院，费用高达 1 万多元！

（2）看病效果差。

（3）医疗事故多。

（4）医生收红包才认真看病，甚至在手术室门口收红包。

（5）就诊排长队。

（6）医患关系紧张。

（7）报考大学学习医科的积极性降低。

而医务工作者反映，看病难不是医务人员造成的。他们处在高难职业，责任大，

① 本文是在珠海市卫生系统医学人文精神论坛的报告，珠海，2008 年 2 月 26 日；《现代管理的中国实践》，北京：科学出版社，2016，169-172.

负担重，工作时间长，假期也要上班，太辛苦；学习时间长，收入低，又是高风险职业，常被投诉，有时还挨病人和家属追打。

现在医疗系统处于困难阶段，从最受重视的光荣职业降落到受人非议的职业。究其原因，个人认为：

(1) 与发达国家相比，政府投入医疗经费偏少，人家高达80%，而我国才10%。有的医院还要少，如我校华侨医院是国家三甲医院，一年国家投入经费仅占1.5%。真是巧妇难为无米之炊！医院设施落后，医务工作者的收入偏低。同时，我国医务工作者数目也偏少。

(2) 医疗管理存在问题，特别是我们国家医疗保险做得不太成功。同时，许多医院的管理还不是科学管理。

(3) 医务工作者自身也存在一些问题。

要解决上述问题，使医疗事业的工作搞好，从加强人文精神建设上着手，无疑是有很大意义的。

三、医务工作者人文精神的基本点

医务工作者的人文精神内涵十分丰富，见仁见智。但本人认为，其核心有两点：

(1) 对患者要有关爱的精神，简称为爱心。

(2) 对医疗技术要有精益求精、追求完美、追求卓越的精神，简称为匠心。

四、爱心

医者爱心，亦即医者父母心，本着仁爱之心为病人的健康利益据理力争，并在第一时间安慰和支持病人，缓解患者恐惧、彷徨的心情，这正是医者父母心的体现。相信每一位有医德、有仁爱之心的医务工作者都会抱着医者父母心对待每一位病人，都会为病人的健康利益而坚持。

2400多年前，古希腊著名医生希波克拉底（公元前460～前377年），他被誉为世界医学之父，有一句名言，“医生有三件法宝：语言、药物、手术刀”。语言恰是爱心最关键的体现之处，医生一句鼓励的话，可以使病人转忧为喜、精神倍增、病情开始好转。相反，一句泄气的话，可以使病人抑郁焦虑、卧床不起，甚至不治而亡。

我在报纸上曾看到过一个故事，三句话说死了一个人。一天上午十一时，一位病人到医院看病。医生第一句话对他说：“你的病呀，来晚啦!”病人听完便马上求医生：“大夫呀，我大老远慕名而来，求您想想办法。”医生回他第二句话：“你这个病呀，没法了。”病人马上又求医生，医生说了第三句话：“你早干吗去了?”病人听完，好像一盆冷水兜头浇下，心想完了，回到家中，当天夜里就死了。

所以，医生语言十分重要。一定要用微笑的面孔对待病人，切忌冷面孔。

对病人具有爱心，我想医患关系肯定会和谐，病人的健康状况会更好。

五、匠心

前面讲的古希腊医生名言中，还有药物和手术刀，这是治病的物品，其深刻含义就是医生的治病能力。治病能力越精湛，则治病效果会越好。优秀的能力来源于医务工作者在技术上孜孜不倦的、精益求精的追求，且独具匠心。这一能力水准的检验，通常用医疗质量来说明。过去，医院属于社会公益性事业单位，是一个特殊行业。从现代意义上讲，在国民经济中，医疗行业属于第三产业，应该用企业化的管理模式。

医疗质量，是医院的生命！

质量，就相当于射箭、打靶，必须命中靶心！一次就把事情做对。实际上，在医疗中，往往不是一次就把事情做对，事故和差错并不罕见，而是常见。这说明，医疗质量有待大大加强。

以自己为例，在前半生中，我亲身遭遇过三次医疗事故。①1971 年 10 月，在兰州市一家最有名的医院住院治疗。尽管领导和医生工作热情，但治疗质量不行，不仅把病完全搞错，将良性肿瘤诊断为结肠癌，而且手术中又误伤其他的器官，造成我血尿一年多，肠梗阻时有发生，吃了几十年的苦。②1971 年 6 月，在兰州一家有名的医院住院治疗，我是青霉素过敏病人，而一位护士粗心地用了一支打过青霉素的针管给我注射，几乎弄得我丢掉性命。幸亏我机敏，才幸免于难。③1996 年 9 月，在广州一家有名的医院，我因吃鱼肉喉头被鱼刺卡住，尽管医生热情，但发生事故，造成咽喉部和食管损伤，甚至气胸，让我又遭受一次痛苦。

我是千千万万病人中的一人，都遇到过三次大医疗事故。可见，医疗质量亟待重视。这说明，一位医务工作者，不仅要有爱心，而且必须要有匠心！要求做事高质量，就要一次就把事情做对，才能使病人满意。对生命健康问题，不能因为一次病痛，就一定要多次反复治疗，病人既痛苦，病的治疗效果差，且要多花钱，还要影响工作。

六、质量管理的基本原则

笔者从事企业管理研究，认为医疗质量管理的基本原则可以借用其思想[1]，其要点如下。

（1）质量的定义必须是符合要求。

（2）质量系统的核心在于预防，而不是检验和评估。

（3）工作标准必须是“零缺陷”，是第一次就把事情完全做对，而不是“差不多就好”。

（4）质量是用不符合要求的代价衡量的，所有做错的事情导致的花费是它的代价。

笔者建议用上面原则来管理医院的医疗质量，将“零缺陷”“第一次就把事情做对”作为每个医务工作者做人做事的准则，成为医务工作者人文精神“匠心”的最高境界标准。

我认为，如果每位医务工作者都具有“爱心”和“匠心”，都有精湛医术和高尚医德，那我们珠海的卫生事业肯定会受到珠海人民的赞赏，推而广之，我们中华民族的健康状况会更好，我们的祖国会更加伟大、光辉！

参考文献

[1]　刘人怀. 质量是企业生存的根本. 中国工程院和深圳市人民政府主办，中国青年科技企业家管理论坛论主题报告，深圳，2002；工商管理研究，北京：科学出版社，2015，172-177.

发展“乡村旅游”促进广东新农村建设[①]

广东气候宜人，山川河流、沙滩海滨、温泉岛屿，旅游资源丰富，加上交通便捷、信息畅通，因而，通过发展“乡村旅游”可增加农民收入，缩小城乡差别，是建设社会主义新农村的可选择途径之一。

一、“乡村旅游”已成为创新型旅游

随着工业化、城市化进程的加速，休闲、旅游、度假已成为城市人生活不可或缺的组成部分。旅游消费也成了经济增长的重要助推器之一。传统主流旅游模式大多以旅游中介组团为组织形式，以景点旅游为主要方式，旅游购物两兼顾。这种旅游模式省去了旅客找交通、寻旅店、迷路的麻烦，很受欢迎。但这种跑景点式的旅游方式花在路途上的时间太多，游客疲于奔波，缺乏休闲和深度，而且一家人出游费用也不菲。近年来，一种休闲度假、联谊交友、合家团聚、体验采风、回归自然、丰俭由人、时间自由的“乡村旅游”正在兴起。旅游中的“农家乐”已作为一种品牌在全国取得共识。

这里所指的“乡村旅游”并不是泛指在乡村办餐厅、酒楼接待食客，而是以村镇为依托，以农户为载体（农户家庭式旅游既可以一家一户为主体经营，也可以联合经营），以家庭日常生活方式安排的一种休闲度假方式。深度游、自然游、生态游、体验游，近年来这种独特的旅游方式越来越受到城市人的欢迎，甚至供不应求。

优质“乡村旅游”之所以供不应求，是因为这种旅游并不是任何农户、任何村镇想办就可以办起来的，它对整个农村大环境和农户家庭小环境都有较高的质量要求，现在很多地方还难以达到。例如，在优美的生态环境和古朴的民俗的大前提下，要求旅游地交通通信便利、水电畅通、道路及建筑整齐清洁、基本家电齐备、环境卫生、生活文明、服务周到等。要达到这些要求，村镇和农户要自觉改善居住环境，并以旅游业带动农、林、牧、副、渔业发展，改善就业，学会经营，促进消费，提升整体素质。中国人口众多，旅游需求正在向个性化、多元化转变。“乡村旅游”有着广阔的发展前景。

二、“乡村旅游”在广东发展相对滞后

“乡村旅游”在发达国家早已兴起并日趋成熟，随着我国人民生活水平的提高和农村环境的改善，这种旅游模式大约在20世纪90年代中后期才发展起来，最早是依托村、镇的自然环境办起了“农家乐”餐厅酒楼。由于其价格低廉，风味特色突出，许多城市人自发到这些地方去消费，或吃一顿饭，或自娱自乐玩一天，基本上不住宿。随着农村居住环境的改善，“农家乐”也从酒楼饭店拓展到农户家庭，融吃、喝、玩、乐、住、游于一体的“乡村旅游”模式。游客根据自身的偏好，住一两天可以，住十天半月也行。每个人大约花400～600元就可以享受一个月的休闲。如果要加菜式或提

---

① 本文原载《广东政协》，2008，(3)：21-22；参事建言（2008年），香港：中国评论学术出版社，2008，197-200. 作者：刘人怀，何问陶.

供特色风味饮食项目，游客另外加钱，大家都方便。目前，“乡村旅游”在四川成都平原、贵州遵义地区、云南丽江地区、长三角地区发展势头很好，吸引了全国各地游客，暑假期间还要提前预订才能成行。

广东有发展“乡村旅游”的自然资源，但这种创新方式并未很好地铺开，大多数“农家乐”还仅限于提供餐饮服务。其原因大致有：第一，广东农村，特别是珠三角虽然住房条件都比较好，不是小别墅就是小洋房，比起城市宽敞明亮多了，但村镇规划落后，环境脏、乱、差，缺乏吸引力。第二，农村住房设施不配套，特别是浴室和卫生间还达不到旅客要求。第三，农村的文化、娱乐、体育活动、现代服务欠缺，其他配套的特色或生态旅游项目跟不上，难以久留旅客。第四，广东人待人接物观念比较内向，不善于与陌生人交往，更不愿意外人住在家里，农户主动发展“乡村旅游”的意愿不强。第五，虽然广东也有许多地方具备开展这种旅游的条件，但缺乏宣传、引导和推广，“乡村旅游”还没有作为一个产业和增加就业的手段来策划，更没有把它作为建设社会主义新农村的重要途径之一来重视，仅停留于自发阶段，发展迟缓。

三、“乡村旅游”发展对策及注意事项

发展“乡村旅游”既是经济主体的市场行为，又是涉及包括政府和农户在内的方方面面配套的系统工程，需要认真调研、精心策划。第一，建议由省旅游局或旅游协会牵头对省内外“乡村旅游”状况进行充分的调查研究，找出广东发展“乡村旅游”的困难和差距，并在此基础上提出广东发展“乡村旅游”的初步规划或指导性意见。同时，在珠三角、粤东、粤西、粤北各选一个条件比较成熟的村镇作为“乡村旅游”试点，总结经验，加以推广。第二，各级财政要加大对村镇的基础设施建设和公共服务的投入，营造适宜“乡村旅游”发展的大环境。例如，村村通公路、通水电、通宽带，村村有医疗卫生站或社区医院、文化站或文化馆、体育健身场所等。第三，村、镇政府要组织和引导农户自觉搞好清洁卫生，保护生态环境，提倡健康文明的生活方式，同时科学有序地规划村镇建设，维护社会治安，而农户自身则要改造住房和周边环境，既方便自身居住，又利于发展旅游。第四，引导农户根据所在地的人文历史，发展传统民间工艺和旅游产品，既满足了旅客要求又增加了农民的就业和收入。充分利用现有旅游中介机构为“乡村旅游”做好宣传推荐、组织客源、推销产品和业务培训的工作。第五，严格审查，加强管理。旅游主管部门应对“乡村旅游”的准入资格、经营范围、卫生条件、服务质量等制定管理规则，条件成熟后纳入立法管理。第六，发展“乡村银行”为农户提供“乡村旅游”小额贷款，用于住房改造、家电申购、环境卫生改善以及其他副业，并监管资金合理使用和有效循环。第七，“乡村旅游”本身收费较低，房产属农民自己，劳动力价格又低，加上经营时间不稳定，成本收益难以合理计算，因此建议对这一旅游项目免收承办农户的税收，放水养鱼。

“乡村旅游”是一种创新业务，没有现成的制度规范和统一的标准模式，需要我们在实践中摸索。但应该注意“乡村旅游”就是“乡村旅游”，要保持自然环境、古朴民风，各具特色，千万不要相互攀比和复制城市样本，同时特别注意不要因为游客增多、管理不善而加剧农村的环境污染，应坚持可持续发展。

关于城市基础设施建设投融资体制改革研究①

广州市政府出台的《关于推进广州市城市基础设施投融资体制改革的工作方案》指出，广州将组建七个专业性基础设施建设投资集团，打破过去城市建设由政府单一投入的模式。这一重大举措，表明了政府将破解当前城建资金主要依靠政府单一投入、责任制不够完善等难题的决心，走出一条政府主导、市场运作、资金筹措多样化、投资主体多元化的新路子。受广州市政府委托，笔者分别就“城市基础设施投融资模式”“城市基础设施投融资体制改革”“城市基础设施经营管理机制创新”等问题结合广州城市基础设施建设投融资实际进行了广泛深入的研究。笔者查阅了国内外相关文献[1-3]，邀请市建委、市政园林局、市交委、市建设投资发展有限公司、市政公司等相关政府部门和企业召开了座谈会，并对近年来广州具有代表性的重大城市基础设施项目进行了实地考察与调研。在此基础上，对城市基础设施建设投融资体制的内涵、国内外城市基础设施建设投融资模式、广州城市基础设施建设投融资体制现状、存在的问题以及对策建议进行了认真的讨论，撰写了如下建议。

一、城市基础设施建设投融资体制内涵

1. 城市基础设施建设投融资体制定义

城市基础设施建设投融资体制，是城市基础设施建设投融资活动运行机制和管理制度的总称，由投资主体、投融资方式和政府用以扶持、引导、限制和管理的投融资调控体系三项要素组成。投融资体制解决由谁投资、如何投资以及由谁决策、由谁承担风险和责任的问题。

2. 城市基础设施建设投融资体制构建的内容

以政府为主导，以产业化、市场化为特征的新型城市基础设施投融资体制主要由城市基础设施项目的工程建设机制、定价机制和资产经营机制三大环节构成。

（1）城市基础设施工程建设机制。在市场经济条件下，基础设施的建设单位（即招标方），建设工程承包商（即投标方）和监理方围绕着建设商品所构成的关系，决定了基础设施工程建设的运作机制。具体来说，招标制、投标制、监理制、项目法人制等构成我国城市基础设施建设的现行体制。

（2）城市基础设施定价机制。价格管制是基础设施运作中最为广泛的一种管制方式，适用于供水、煤气、污水和垃圾处理、通信等城市基础设施。不管是在高收入国家还是在低收入国家，政府对这一类基础设施的收费价格都采取不同程度的控制。促进社会的公平分配，保证广大城市居民的消费水平往往是政府控制城市基础设施价格的主要原因。

（3）城市基础设施资产经营机制。在基础设施运作中引入资产经营是必要的，它

① 本文原载《决策与咨询》，广州市人民政府研究室编，2009，(5)：1-6.

是民间投资力量引入城市基础设施运作的桥梁，是形成政府主导下的产业化、市场化运作机制不可或缺的重要组成部分。当前城市基础设施资产经营包括实物资产经营、资本经营与产权经营三方面。

二、广州城市基础设施建设投融资体制现状分析

1. 广州城市基础设施建设投融资主要成就

广州市政府于 1996 年 10 月成立了广州市建设投资发展有限公司，主要职能是广州城市建设投融资以及资产管理和运营。

广州市建设投资发展有限公司的成立，突破了城市建设单纯依靠财政的传统观念，标志着城建资金由“投”向“融”的跨越，城市建设不再是政府统包统揽，而是在政府不断投入的基础上，由政府领导下的专业公司以企业举债的形式，多元化筹措城市建设资金，广州 1995～2006 年社会固定资产投资情况详见表 1～表 3。

表 1　1995～2006 年广州全社会固定资产投资资金来源　　单位：万元

项目	年份					
	1995	1996	1997	1998	1999	2000
本年资金来源合计	7 978 652	7 944 084	8 185 958	9 955 334	10 983 747	11 679 071
上年末结余资金	961 511	1 513 379	1 394 838	1 433 763	1 548 943	1 552 005
本年资金来源小计	7 017 141	6 430 705	6 791 120	8 521 571	9 434 804	10 127 066
国家预算内资金	67 608	51 257	92 550	200 069	233 267	196 772
国内贷款	1 058 442	867 596	906 033	1 279 909	1 998 101	2 174 224
债券	27 815	6 388	3 112	1 029	23 600	62 353
利用外资	1 256 012	1 656 655	1 657 279	1 185 066	909 234	883 291
自筹资金	2 413 074	2 003 513	2 429 018	3 196 317	3 330 138	3 356 781
其他资金	2 194 190	1 845 296	1 703 128	2 659 181	2 940 457	3 453 645
项目	年份					
	2001	2002	2003	2004	2005	2006
本年资金来源合计	11 649 579	12 856 766	14 649 881	16 786 548	19 247 238	22 251 012
上年末结余资金	1 546 436	1 649 716	1 674 756	1 874 335	2 186 460	2 776 720
本年资金来源小计	10 103 143	11 207 050	129 175 125	14 912 213	17 060 778	19 474 292
国家预算内资金	123 638	211 676	353 310	311 727	3 411	33 390
国内贷款	1 912 787	2 696 256	2 653 393	2 922 647	4 172 521	5 172 610
债券	22 888					130 000

续表

项目	年份					
	2001	2002	2003	2004	2005	2006
利用外资	706 179	910 991	957 866	1 212 633	1 156 940	1 508 470
自筹资金	4 230 371	3 938 859	5 241 604	6 321 305	7 016 193	7 321 261
其他资金	3 107 280	3 449 268	3 768 952	4 143 901	4 711 713	5 308 561

资料来源：广州统计信息网

表 2　1995～2006 年广州全社会固定资产投资资金来源构成　　单位：%

年份	本年资金来源合计	上年末结余资金	本年资金来源小计	国家预算内资金	国内贷款	债券	利用外资	自筹资金	其他资金
1995	100.00	12.05	87.95	0.85	13.27	0.35	15.74	30.24	27.50
1996	100.00	19.05	80.95	0.65	10.92	0.08	20.85	25.22	23.23
1997	100.00	17.04	82.96	1.13	11.07	0.04	20.25	29.67	20.81
1998	100.00	14.40	85.60	2.01	12.86	0.01	11.90	32.11	26.71
1999	100.00	14.10	85.90	2.12	18.19	0.21	8.28	30.32	26.77
2000	100.00	13.29	86.71	1.68	18.62	0.53	7.56	28.74	29.57
2001	100.00	13.27	86.73	1.06	16.42	0.20	6.06	36.31	26.67
2002	100.00	12.83	87.17	1.65	20.97		7.09	30.64	26.83
2003	100.00	11.43	88.57	2.41	18.11		6.54	35.78	25.73
2004	100.00	11.17	88.83	1.86	17.41		7.22	37.66	24.69
2005	100.00	11.36	88.64	0.02	21.68		6.01	36.45	24.48
2006	100.00	12.48	87.52	0.15	23.25	0.58	6.78	32.90	23.86

资料来源：广州统计信息网

表 3　2002～2005 年广州城市建设固定资产投资额　　单位：万元

项目	2002 年	2003 年	2004 年	2005 年
本年完成投资	1 377 443	1 603 774	1 996 322	2 325 685
按构成分				
建筑工程	814 275	952 315	1 407 650	1 451 637
安装工程	46 959	34 761	33 875	78 714
设备及工器具购置	79 634	86 178	104 840	209 690
其他费用	436 575	530 520	449 957	585 644
按系统分				
供水	18 790	48 237	132 218	24 610
燃气	6 761	5 764	13 413	63 053
道路桥梁	586 615	875 621	936 656	891 640
排水	87 572	252 743	155 117	248 962
园林绿化	16 797	22 002	65 472	82 823
市容环境卫生	33 377	34 714	50 861	40 932
公共交通			524 195	702 380
其他	627 531	364 693	118 390	271 285

资料来源：《广州建设年鉴》

截至 2007 年 12 月，广州市建设投资发展有限公司资产总额约 600 亿元，银行贷款约 360 亿元。2003 年城建财政资金约 30 亿元，自 2004 年起，投资公司平均每年在城市基础设施项目建设方面实际使用新增贷款 35 亿元左右，到 2008 年已超过 120 亿元。

20 世纪 90 年代以来，广州城市基础设施建设投资虽然增长较快，但与高速增长的经济发展水平相比仍需进一步提高。根据世界银行 1994 年发展报告的研究和建议，发展中国家的城市基础设施建设投资比例，一般应维持在同期国内生产总值的 3%～5% 的水平上，或固定资产投资的 10%～15%，这样才能确保城市基础设施与经济增长的需求相协调。广州城市基础设施建设投资所占国内生产总值和全社会固定资产投资比例见表 4。

表 4　广州城市建设投资与国内生产总值和全社会固定资产投资的比例

年份	城市建设投资额/万元	全社会固定资产投资额/万元	国内生产总值/万元	城市建设投资占全社会固定资产投资比例/%	城市建设投资占国内生产总值比例/%
2001	1 178 300	11 649 579	28 416 511	10.1	4.1
2002	1 377 443	12 856 766	32 039 616	10.7	4.3
2003	1 603 774	14 649 881	37 586 166	10.9	4.3
2004	1 996 322	16 786 548	44 505 503	11.9	4.5
2005	2 325 685	19 247 238	51 542 283	12.1	4.5

2. 广州城市基础设施建设投融资体制存在的问题

（1）政府职能不明确，政企不分，企业无法发挥投融资主体的地位。一是由行政审批替代应有的投资决策。目前项目决策制度仍由各级政府的审批权限构成，审批的方式和内容还没有随着投资的多元化做出相应的调整。二是政府投资管理不到位。一方面，政府对市政、公用基础设施的投资管理分属计划、建设、财政等多个部门，政出多门，影响了管理效果。另一方面，政府部门的投资管理方式还没有形成一套有效的间接调控市政公用基础设施投融资的办法和手段。

城市建设投资公司作为政府投资建设经营城市的载体，就实际运营情况来看，其主要职能是作为政府向金融机构贷款特别是向国家开发银行贷款的工具，并没有真正成为城建投融资的主体。具体表现：工程项目前期准备阶段，城投公司没有投融资决策权，实质上是“政府决策，公司筹资”，即公司作为政府的“统借统还”的法人实体向银行贷款，同时贷款多少并不是根据公司承担债务能力的强弱来确定，而是根据当年地方政府城建资金需求量和财政能力而定，由政府财政作担保；项目实施和运营阶段，实施“收支两条线”的财务管理制度，由政府部门统筹安排资金的使用和偿还，公司对自己借入的资金没有处置权，仅充当“出纳”角色；公司的领导由政府官员兼任，政企不分，没有按照现代企业制度的要求建立法人治理机构。这种体制上的弊端造成城投公司无法发挥投融资主体的地位。

（2）按照现行管理制度，资金的计划、融资、使用、监管分属不同的职能部门和单位，投资者主体地位未形成，项目投资缺乏风险约束机制。在项目建设管理体制方面，由于通过行政手段将投融资建设责任主体分离，融资单位只负责融资和拨付项目建设资金，不负责项目建设和管理，难以对资金使用进行有效监管；建设单位只管建设，不管资金来源，整个工程缺乏科学有效的资金管理，这种分割管理导致投资的具体主体不明确，缺乏明确的业主负责，导致项目不能及时结算和竣工决算，未形成有效的固定资产，投资效益不能得到有效的控制，资源的综合效益也难以发挥。目前，城市基础设施投资的法人在具体运作过程中存在着项目法人实体不明的问题，即基础设施项目在建设中和在建设后的运营管理的主体是“两张皮”。基础设施项目在决策、建设和运营三个环节中相互脱节，项目由政府有关部门决策投资，政府通过组建一个项目公司来向银行借款，项目建设完成后，则将基础设施的资产所有权交予另一个法人负责具体经营。这样就造成借款人不负还款责任，而还款人以未参与项目的具体筹建和融资过程为由拖延还款。最终的结果是银行债券虚置。一些地方政府在经济资源短缺的情况下，利用这种方法无偿获得了用于经营城市的资金，但对未来的还款却缺乏系统的整体思考。只管借，不管还，导致金融风险的累积。所以，要充分发挥项目法人制的作用必须使其在建设和经营中实现责、权、利的紧密结合。

(3) 投资主体单一，融资渠道狭窄。一是城市基础设施建设项目存在资金需求量大与筹资不足的矛盾。传统的投融资体制投资渠道单一，不利于民间资金、企业资金和外资进入城市基础设施建设领域。各大企业集团期望出资参与城市基础设施的建设，但缺少准入规则，投资管理呈封闭式运行，使外界参与投资出现瓶颈效应，抑制了企业参与经营性项目公平竞争的渠道，阻碍了政府财政之外的资金投入城市基础设施建设项目。

融资的资金结构与资金需求不相适应，融资方式和渠道单一（表 5），融资成本偏高且政策性风险、金融风险大。根据城建项目投资大，回收期长的特点，银行间接融资和资本市场直接融资相结合的融资资金结构最为科学合理。目前城建融资的资金结构与城建资金的需求不相适应，较为单一的银行贷款融资方式受金融货币政策的刚性制约和影响较大，导致融资成本偏高且基础设施建设的资金链面临较大的政策性风险、金融性风险，融资系统缺乏风险预警和化解机制。

表 5　2001～2005 年广州城市建设资金来源构成　　单位：亿元

年份	合计	城建税费收入	银行贷款	地方财政拨款	土地出入金	企业自筹	业主专项部分筹措资金
2001	145.97						
2002	133.42						67.01
2003	163.35	28.30	48.07	0.12	1.78		
2004	219.32	31.92	60.53	1.8	8.01		
2005	260.14	40.54	61.57	3	10	5.95	39.08

注：不含电力设施建设

二是城市基础设施举债建设的现实性与公益项目还贷可能性的矛盾。城市市政公用基础设施产业所提供的产品和服务具有公共产品或准公共产品的性质，它为许多企业和个人服务或享用，具有较明显的共享性。公益性项目以获取社会效益和环境效益为主，不以谋取经济效益为目的，这些项目理应由政府出资建设。政府若按照传统投资渠道、方式筹资，筹资的数量不能满足需求，而举债建设必然遇到还贷的效益性问题。如果公益性项目长期通过举债方式，必然导致重复进行从举债到建设或从举债到还贷的单向运行，不仅巨额举债有困难，而且又无法进入城市基础设施建设的良性循环，最终导致企业资不抵债、城市建设瘫痪。

（4）现行城建资产建设管理模式难以实现城建资产的保值增值。城市供水、道路、排水、燃气、供热、公共交通等由不同的政府部门进行建设和管理，具有很强的垄断性，长期以来形成了“自我封闭，自我循环”的体制，即垄断设计、垄断施工、垄断经营、排斥竞争，造成城市基础设施建设投入高成本、质量管理低水平、经营管理低效益的局面。同时，城市基础设施缺乏资产经营机制和机构，对于大量的城市基础设施存量资产未能由政府授权和明确权益代表，以企业化的运作模式加以经营和管理，不仅造成城建资产产权主体不清晰，经营主体模糊，政府投资“只进不出”或“多进少出”，而且使得城建项目在投资权、经营权、收益权方面比较分散，无法利用资本市场，开展资本运营，盘活存量资产，实现城建资产的保值增值。

（5）当前城市基础设施的政策性收费机制和价格体系尚未形成以市场为导向的价格形成机制，导致难以吸引外部资金的投入。由于城市基础设施行业的价格体系和收费机制具有社会福利性，从而忽视了它们的商业属性，导致城市基础设施行业提供的产品和服务价格不合理，严重影响了城市资源的合理配置。主要表现：大部分城建项目仅靠财政补贴方式实现，这就导致政府建得越多，包袱就越重；价格偏低，导致社会的过度消费需求，造成资源浪费，进一步加剧了城市基础设施紧张的矛盾；政策性补贴原则不明确，方法不规范，缺乏激发企业加强管理、降低消耗、减少亏损的内在动力；由于缺少投资收益的刺激，难以吸引外部资金的投入。虽然近年来城市基础设施产品和服务价格调整步伐逐步加快，但由于价格形成机制不合理，难有标本兼治之效。

（6）财政资金目前操作模式不能充分发挥财政资金的杠杆作用，投资公司融资平台难以适应社会主义市场经济的发展及金融政策的形势变化，持续融资能力减弱。一是未建立保障性的还本付息机制，而五部委下文禁止政府部门对银行以承诺函的形式为贷款提供信用支持，以往采用的政府信用担保贷款的融资模式无法继续实行。二是由于财政收支两条线，投资公司投资的污水、年票制项目等收入纳入财政体系，项目配套资本金由财政直接集中支付给项目建设单位，导致投资公司缺乏经营性收入，无现金流，也难以利用财政资金结合现有的资产进行企业债券、项目融资等多元化融资。多年来，广州城建投融资建设一直是政府主导的行政行为，没有按照现代化企业进行管理，政府已成为法律事实的债务人，直接面对债务风险。随着投资公司贷款余额的增加和金融环境的变化，投资公司可持续融资能力减弱。

三、城市基础设施投融资体制改革的对策研究

城市基础设施建设项目，尤其是市政公用设施建设项目，就其整体而言主要属于公益性投资项目，这一特征决定了该领域投融资体制改革的取向既要面向市场，又不

能完全走市场化道路，而是要政府调控与市场配置、财政补偿与市场补偿、社会效益与经济利益三结合。其改革的方向应是投融资管理法制化、投资主体法人化、投融资渠道多元化。

1. 城市基础设施建设投融资体制改革的目标

（1）改政企不分为政企分开，将政府的投资、建设、运营职能分离出来，组建新的市场主体。政府与新的市场主体，在资产和项目管理经营上是一种授权委托关系；在市场角色上是一种监管和被监管的关系。

（2）改投、建、管三位一体为投、建、管三分离，将政府部门的投资、建设、运营职能分开，组建和培育投资主体、建设主体和运营管理主体；对分离投资、建设和运营职能的部门和单位进行撤并或改制为公司。

（3）改投融资的单一性为投融资的多元化。改革后的投融资市场面向社会开放，利用民间资本、信贷市场和资本市场，拓宽投融资渠道，发挥政府财政资金的杠杆和发酵作用。

（4）改分散管理城市资产为城市一体化经营，按“城市是最大的国有资产”理念，实现存量资产由部门分散管理转变为政府集中统一管理，由部门资源转换为政府资源，使城市资产部门分割的低效率配置转换为市场化的高效配置。

2. 关于城市基础设施投融资体制改革的建议

（1）转变政府职能，政企分开，实现企业投资主体归位。为了使城市基础设施投融资体制更加有效，首先应当转变政府职能。基础设施的投资建设与经营管理必须通过引入市场机制，探索产业资本与金融资本的互相融合，开拓社会化、市场化和国际化的筹融资渠道，从而建立起政府主导下多元化的基础设施投融资体系，在这一体系下进行基础设施的建设与运营。市场经济条件下的城市基础设施投资运营中，政府的主要职能是做好宏观调控、社会管理和公共服务。具体包括：政府应制定基础设施领域的中长期发展规划，确定重大的基础设施建设项目；统筹预算内、外等政府专项资金，使其在引导社会各类资金中起基础性、决定性作用；制定市场运作所需要的各项配套政策与法律法规，管好基础设施产品价格与服务质量；确定社会效益显著的非经营性项目的政府投资。从宏观上引导、促进基础设施建设的发展。政府应当把有限的财政资金从竞争性领域撤出，更多地投向基础设施领域，同时在保持基础设施领域控制地位的同时，也应该适当引入竞争机制，特别是在经营性基础设施领域。同时要改革政府机构，压缩政府规模使之与市场经济条件下政府职能的结构相适应。其次，应建立公司法人治理结构，实现政企分开。一方面，政府对国家出资兴办和拥有股份的基础设施企业，要通过出资人代表，按出资额行使所有者职能，不具体干预企业日常经营活动；另一方面，对企业经营者的选择、激励和约束可转移给企业，加强和改善基础设施国有企业的管理。这样，政府就可以专注于行政管理权，而由专职部门或企业来代管、代行投资经营权，从而提高基础设施的投资效益。

（2）正确定位城市基础设施项目的经营属性，打破垄断，实现投资主体多元化。在正确定位城建项目经营属性的前提下，积极打破垄断，实现投资主体多元化。第一类非经营性项目的投资主体自然是政府部门，资金来源主要是财政拨款，除此之外，要更多地采用政策性融资和经营性融资、政府和企业相结合的方式，来缓解单一依赖财政资金、运营效率低下等矛盾。第二类准经营性项目建议采用项目贴息、税收返还等补偿形式来吸引社会资金的投入，而非全额财政资金的投入。第三类可经营性项目，这类项目政府不必包揽，建设资金完全可以通过市场化运作来筹集。对于这类项目，允许各种所有制的企业参与投资，确立企业的投资主体地位。对于经营性和准经营性基础设施项目，可以允许私人资本的完全或部分介入，采取衍生工具进行资本补偿。同时根据设施的不同性质，安排不同的运行机制，以充分激励城市基础设施建设的发展。

(3) 建立区域性基础设施的共享体制、举债和还贷并存机制以及投资风险机制和投资约束机制。一是建立区域性基础设施的共享体制。通过建立并实施科学合理且全局性的区域基础设施建设规划，形成区域性基础设施的合理共享，来科学配置资源，获得尽可能高的投资效益。二是建立举债和还贷并存的机制。政府可根据经济实力，在每年的财政预算中安排一笔资金，作为融资的还本付息基金，用于支付贷款利息和归还贷款本金。投融资机构要充分争取、利用政府大力度的支持，发挥投资建设优势，积极建立发展基地，进行开发建设，增加收入，作为还贷基金。推行同步土地储备机制。三是建立责、权、利相结合的投资风险机制和投资约束机制。首先要公开基础设施项目的建设与借贷计划，由有资质的评估机构对政府的偿债能力进行评估后再决定建设与否和建设的最佳时机。其次是确立谁投资、谁受益、谁决策、谁承担风险责任的原则。最后是建立投资风险约束与监督机制，加强资金规划和管理，提高资金使用效益。具体措施如下：第一，建立贷款信息披露制度和资金使用的审核、评价制度，加强贷款信息披露的透明度，让社会参与监督贷款的使用；第二，积极研究建立起一套完善的监管体制，包括事先审查、事中检查、事后评价；第三，设立专门的“偿债基金”账户，保证贷款资金的偿还。

(4) 扩大融资渠道，增加融资工具。利用城市经营和基础设施市场化经营，把基础设施的建设运营纳入良性循环的轨道，使公共设施价格趋于合理化，真正实现以城养城，以设施养设施。也可以利用资本市场来扩展城市建设资金来源，如利用债券、股票、信托、投资资金、银行贷款、民间资本以及吸引外资等方式，来广泛地、多元化地吸引资金，加快城市建设的步伐，还可以通过发行市政债券这种方式融资。此外，应加强对广州市建设投资发展有限公司的改革，改革运作模式，建立现代企业制度，使之由“贷款的机器”“政府的出纳”成为真正意义上的自主经营、自负盈亏的法人实体和市场主体企业实体，从而更好地为社会主义市场经济服务。我们可以利用建设-运营-转让模式（build-operate-transfer model，BOT）、移交-经营-移交模式（transfer-operate-transfer model，TOT）等项目融资方式，积极推进政府和私人投资者的合作，并开展以资产证券化形式筹措城建资金的模式，借鉴西方发达国家的先进经验，推进基础设施的建设与发展。

(5) 加快体制创新。转变城市基础设施经营管理机制，推行特许经营、委托经营、租赁经营和承包经营，盘活存量资产。一是实行所有权和经营权的分离，有偿出让经营权，实行市场准入“特许经营权”招标拍卖制度，逐步向特许经营权转让过渡。二是重视价值形态的保值增值作用，将一些无形资产作为经营城市的资产推向市场，有偿出让城市基础设施的“冠名权”“广告权”和有限经营权等，缓解城建资金紧缺的矛盾，拓宽了城建资金的筹措渠道。三是扩大城建项目收益权质押贷款范围，以城建国有资产、土地资源、专项收费、基金收益作为质押、抵押担保，申请银行贷款。四是建立土地储备制度，成立土地储备中心，有效经营城市土地资产，最大限度地提高城市土地收益。五是积极推行资本运营，盘活现有基础设施的存量资产。除城市道路、市政管线等一些可转换性差的资产外，对相当一部分可转换性较强的资产，可以从原有经营体系中分离出来作为独立的经营实体，实现资产经营；又如城市公交线路、户外广告经营等在基础设施领域中的政府特许经营权、市场准入限制作为无形资产，在一定条件下也可以转化为资产经营；再如污水处理、垃圾无害化处理等基础设施在实现收费突破以后，也可望实现资产化经营。应改变这部分可转换性强的良性资产现存的零散、封闭、分割状态，使之在更高层次上以资产为纽带将其重组整合起来，形成规模效益的资本产业。对一些条件较为成熟的企业可推行资本运作、资产经营，转让现有的一部分国有资产，吸引社会各类投资主体进入这一领域。通过盘活存量资产，政府就可将相当部分的营运资金或补贴转而投入城市其他基础设施的建设。

(6) 尽快制定城市基础设施建设相关的投融资法律法规，为城市基础设施投融资体制建立良好的政策环境。尽快制定基础设施投融资方面的法律法规，规范基础设施投资，促进基础设施的建设与发展。在政府制定城建投融资优惠政策的方面，我们应不断地调整吸引外商投资的政策，逐步给予外商投资企业国民待遇，优惠待遇的调整应结合产业政策来进行，应有取舍，有升有降。对于为国家所限制的外商投资项目，就应逐步取消其各种优惠待遇，对基础设施项目不但要保留现有的优惠待遇，而且还要扩大这些优惠的使用范围，并增加一些新的优惠规定。除外汇保证、配套资金供应外，提出如下建议：一是将基础设施项目15%的所得税率及五免五减扩大到内陆地区，即凡举办基础设施项目所得税率均为15%，减免期均为五免五减，不受地区限制。二是放宽对基础设施项目注册资本与总投资比率的限制。三是放宽出资期限的规定，基础设施项目出资期限应视具体情况而定。四是对于建设经营公路的外商投资企业给予一些灵活的进出口政策。建议对这些承包商进口专门用于该项目的筑路机械设备予以免税，但要海关进行监管，待工程完工后由海关予以核销，如转卖他人或挪作他用，必须重新征税。另外，对于进口国内无法生产或技术性能达不到标准的筑路原材料，海关应予免税。

参考文献

[1] 庄俊鸿. 投资学. 北京：中国财政经济出版社，1997.
[2] 里格斯 J L. 工程经济学. 吕薇，等译. 北京：中国财政经济出版社，1989.
[3] 郭仲伟. 风险分析与决策. 北京：机械工业出版社，1987.

关于将香港、澳门特别行政区的所有统计数据纳入全国性统计数据的建议[①]

最近，我翻阅了近几年的《中国统计年鉴》，发现全国性统计数据中未包含香港、澳门特别行政区（简称“港澳特区”）的全部统计数据，深感不妥。因为香港和澳门早已回归祖国了，而《中华人民共和国统计法》第一条就明确指出：要“科学、有效地组织统计工作，保障统计资料的真实性、准确性、完整性和及时性，发挥统计在了解国情国力、服务经济社会发展中的重要作用，促进社会主义现代化建设事业发展”。为此，我特建议将港澳特区统计数据纳入国家统计总数，其原因如下。

从统计角度来讲，《中国统计年鉴》是一部全面反映中华人民共和国经济和社会发展情况的资料性年刊，如果全国性统计数据中未计入港澳特区的各项统计数据，就不能说全面地反映了国家整体的经济和社会发展情况。根据《世界概况》（由美国中央情报局出版）最权威的报道，2008 年底世界各国 GDP 前五名是：美国 14.33 万亿美元；日本 4.84 万亿美元；中国 4.22 万亿美元；德国 3.82 万亿美元；法国 2.98 万亿美元。排在第三位的中国与第二位的日本相差 0.62 万亿美元。而根据同样的统计，2008 年中国香港特别行政区的 GDP 是 0.22 万亿美元，中国澳门特别行政区的 GDP 是 0.02 万亿美元，如果把港澳特区的 GDP 加起来，中国的 GDP 应为 4.46 万亿美元，与日本只差 0.38 万亿美元。我们可以这样设想：只要今年中国的经济增长保 8%，人民币对美元的汇率相对稳定，中国的 GDP 将很快超过日本跃居世界第二位。

从政治影响来讲，香港、澳门虽然是单独关税区，但港澳特区的统计是构成国家统计总体的一部分，《中国统计年鉴》已经将其行政区划、国土面积和森林资源纳入全国性统计数据，那么港澳特区的其他相关统计数据，也应纳入全国性统计数据。

从法律规定来讲，中华人民共和国“香港特别行政区基本法”和“澳门特别行政区基本法”的有关原则规定：香港、澳门与内地作为相对独立的统计区域，根据各自不同的统计制度和法律规定，独立进行统计工作。而在全国性统计数据中纳入港澳特区统计数据，与此工作原则并不违背，《中国统计年鉴》中的港澳特区统计资料仍然由特区分别统计、提供，再由国家统计局进行汇总编辑。

① 本文原载《广东省政府参事建议》，广州：广东省人民政府参事室，2009，(82)：1-3.

大规模引进和培训人才
为广东产业结构优化升级服务①

改革开放初期，较为宽松的政策环境、相对丰厚的待遇使得广东这片南国改革热土成为当时的人才聚集高地，吸引了众多国内外英才。可以说，国内充裕的劳动力配合广东依靠“三来一补”政策所形成的以劳动密集型产业为龙头的产业发展格局，为广东的经济发展插上了腾飞的翅膀。斗转星移，时空变幻。30年后的今天，广东在经历了多年的快速发展之后，正处在一个前所未有的关键时期。一方面，广东在完成基本工业化进程之后，经济结构开始出现工业化中后期与以知识经济为特征的第二次现代化初期的表征，广东未来如何走的问题迫切需要寻找创新的思路。另一方面，发展成本的上升，资源环境压力的增大，产业结构的趋同，整体效率和竞争力减弱等众多问题的累积性爆发，预示着广东传统的以劳动密集型产业为龙头的经济发展格局已经走到了重要的转折点。更为棘手的问题是由于多年来在科技开发与教育投人方面欠账过多，使得我省的人力资源开发能力与速度不仅远远落后于国外先进国家，而且还落后于国内部分先进省市，难以满足广东未来经济发展的需要。

因此，为了从根本上解决我省人才储备匮乏的局面，我们必须站在战略的高度，透彻分析我们拥有的优势与面临的机遇，存在的困难和问题，准确把握我省人才开发的全局与发展思路，全方位实施“人才强省”战略。其关键点在于：一要通过多方位创新，实施广东“人才储备”新战略；二要重点搞好人才的培养工程，开创“人才培养”新模式。

一、目前我省人力资源开发工作存在的问题

1. 科技投入与本省的经济规模相比相对较低

从2007年的数据来看，尽管该年广东的科技人员数量排全国第一，但其科技经费内部支出总额为684.22亿元，与其他省份相比处在中间行列，远低于江苏（900.15亿元）和北京（825.42亿元）。再从2007年的人均科技经费支出额来看，广东省科技人员人均科技经费内部支出额为15.24万元，略低于全国平均水平（15.62万元/人），远低于北京（20.55万元/人）、上海（23.20万元/人）、江苏（20.55万元/人）与山东（18.21万元/人）的指标。而从2007年科技经费内部支出总额占GDP的比例来看，广东的指标为2.20%，也低于全国平均水平（2.84%），不仅低于山东（2.31%）、辽宁（2.61%）、浙江（2.71%）等省的指标，更远低于北京（8.82%）、上海（4.33%）与江苏（3.49%）等省市的指标，详见表1。

---

① 本文原载《民主与决策》，香港：中国评论学术出版社，2010，126-140.

表 1　2007 年广东与其他省份科技人员数量、科技经费内部支出总额与教育经费占 GDP 比例

项目	科技人员数量/人	科技经费内部支出总额/万元	人均科技经费支出额/（万元/人）	科技经费内部支出总额占 GDP 比例/%	财政性（预算内）教育经费占 GDP 比例/%
全国	4 543 868	70 988 733	15.62	2.84	3.28
北京	401 595	8 254 203	20.55	8.82	3.51
上海	227 867	5 287 128	23.20	4.33	2.65
江苏	437 923	9 001 516	20.55	3.49	1.99
浙江	347 787	5 093 635	14.55	2.71	2.20
广东	448 946	6 842 205	15.24	2.20	1.97
山东	330 500	6 021 574	18.21	2.31	1.77
河南	192 165	2 208 101	11.49	1.47	2.70
辽宁	188 663	2 888 715	15.31	2.61	2.48

注：资料来源于 2008 年《中国统计年鉴》，中国统计出版社，2008

2. 广东的教育投入也相对较低

根据 2008 年《中国统计年鉴》的数据，广东的财政性（预算内）教育经费（亿元）占 GDP 的比例（1.97%），仅与江苏（1.99%）、山东（1.77%）的指标相当，不仅远低于北京（3.51%）、上海（2.65%）两个直辖市的指标，更低于河南（2.70%）、辽宁（2.48%）两个省份的对应指标。虽然广东近年来迎头赶上，在教育经费总投入方面已达全国第一，但其占财政支出的比重和占 GDP 的比重仍然较低。如 2003～2006 年，广东教育经费支出年平均增长 14%，排在全国第 21 位；教育经费支出占地区生产总值的比例，排在全国第 22 位。另据国家有关部门联合公布的 2006 年与 2007 年《全国教育经费执行情况统计公告》的数据可知，2006 年全国有 12 个省区市预算内教育拨款增长低于财政经常性收入增长，广东名列其中。2007 年全国有 12 个省区市预算内教育拨款增长仅比财政经常性收入增长高出不超过 5%，广东也名列在内。

3. 广东的人力资源区域分布存在较大差距

广东省人才相对集中在经济发达的珠江三角洲。2005 年该地区拥有全省近 2/3 的中专以上学历人才和 3/4 中级技工以上人才，而经济发展相对滞后的山区和东西两翼，人才比重和人才密度较低。从各类专业技术人才看，珠江三角洲地区占全省的 86.97%，东翼占全省的 9.77%，西翼占全省的 3.19%，山区人才占全省的 1.32%。珠三角、粤东、粤西的人才密度分别是粤北的 7.8、2.66、1.6 倍，可见人才分布的区域差异之悬殊。从各类专业技术人才数量的增长情况可知，这几个地区的差异也较大。从 2001 年到 2005 年，珠三角和粤东地区 4 年间增长了 1 倍多，粤西基本没有变化，而粤北却减少了 1/2 多；从每万人拥有的人才分布情况看，珠三角 4 年增长了 36.05%，而其他 3 个地区不但没有增长，反而都不同程度地降低了，特别是粤北降低了 164.37%。这显示人力资源区域分布差异在不断扩大。

4. 广东的人力资源总量与质量均处于较低水平

2007 年广东高等学校的在校学生数为 111.97 万人，每万人口普通高校在校学生数增加到 120.34 人。与北京、江苏、上海等地相比还有不小差距，如 2007 年，江苏省

的普通高校在校学生人数为156.88万人。而从高级人才的质量来看，2007年，广东、江苏、上海、北京四省市在校的研究生总数依次为19 751人、96 546人、91 763人、187 414人，广东不仅与江苏、上海有较大差距，与北京的差距更大，两者相差8倍多。从院士的数量来看，广东在全国各省区市中排名第8位，84名院士的数量与江苏省的315人、浙江省的220人等相比差距不小。

5. 广东人力资源开发的一体化程度比较低

虽同属一个省份，但广东的众多城市在人力资源引进、人才培养专业设置、高校建设等方面均存在着结构趋同、低水平重复建设等问题。比如广东的大学城数量不少，但缺乏明确的定位与区隔，尚未形成独特的专业培养能力与优势。各大学城内不同高校之间的合作还处在初级阶段，难以对广东的技术创新能力的提高产生积极的影响。再加上各城市在产业结构、经济政策上相互参照，相互模仿，这就使得广州、深圳、珠海、东莞等各城市之间出现了比较严重的零和竞争，而这些城市在人才开发的不少方面基本上各自为政，如人才的准入制度各不相同，人才的激励制度相互攀比，人才的征信制度各自独立，这些都严重阻碍了我省人力资源开发迈向新阶段的步伐。

6. 广东对人力资源的吸引力不够

第一，30年基本不变的劳动报酬已经难以吸引蓝领阶层人才的到来，而广东现在的产业结构原本就对高科技创新型人才存在挤出效应，以目前的报酬水平很难吸引到合适的高素质人才。第二，国内其他省市经过多年的发展，经济有了长足的进步，愿意并能够为引进人才付出较高的工资与福利，成为与广东争夺高级人才的重要对手。这在一定程度上加重了广东人才流失的态势。第三，广东虽然已建立了较全面的法律体系，但出于对招商引资工作的重视，司法系统在办案或执行过程中或多或少地会出现难以有效保护劳动者利益等问题，形成广东的劳资纠纷比较多的局面。这也会影响人才来广东的积极性。第四，广东的企业主出于“找快钱”的思想，简单地将人才当作“高级打工仔”，而不是看作一项长期投资。很少有企业主能够从长远战略出发，创造各种条件，鼓励员工通过创新来长期参与企业的运作。

二、开展多方位创新，实施广东“人才储备”新战略

应该看到的是，尽管面临不少困难，但由于广东处在承接国际产业转移的第一线，而自身又需要顺应逐步深化工业化、信息化、城镇化、市场化、国际化的要求，加之粤港澳三地的经济快速融合，必定在人才开发方面为我们提供更加广阔的空间。特别是广东经过30年发展所积累的雄厚物质基础与经济实力，成为广东实施战略性人才引进与培养计划的坚强后盾。

1. 首先要转变我省传统的人力开发观念，提倡战略性人才开发观念

过去，我省在引进人才的战略观念上一直采取“人才能用钱买来”的观念，这一观念导致了很多企业，甚至包含政府机构的不当行为——只注重人才引进，却缺乏对人才的有效管理与激励；只看重短期效益，缺乏长远眼光。为此，我们应树立战略性人才开发观念。这一观念的核心是将人才看作通过培养与开发可以增值的人力资本，并同等关注人的经济需要与精神需求。首先，政府要从人才长远培养开发的角度出发，

制定相关的广东人力资源开发政策，持之以恒，一以贯之，以保证政策的长期导向性。其次，要树立人才平等观念，摒弃“外来的和尚好念经”与“学历至上”的错误思想，对外来人才与本地人才、高学历人才与低学历人才、高技能人才与高创新人才、理工类人才与社科类人才都一视同仁，唯贤是用，唯能是用。最后，选择紧跟经济需要、超前筹划的人才资源持续开发战略。在人才引进与培养问题上要树立人才可持续培养的观念，按照人才成长的规律来确定人才产生效益的时间。不能简单地按照产业发展的现实需要来选择人才引进目标，而应根据产业结构演变的规律，超前3～10年来引进或培养未来所需要的人才。对于属于基础研究领域的人才，更应采取比较特殊的引进与储备办法。

2. 改善与创新我省人才引进的相关机制

（1）要坚持政府主导，由政府负责制定人才引进的具体标准、评价体系、奖惩制度与相关产业的人才引进目录，确定人才引进工作的大方向与战略问题。但在制定人才引进战略的时候，要注意广泛听取各界意见，引入专家咨询环节。要鼓励企业根据自己的需要适时从国外引进高精尖人才，建立政府补一部分、企业拿一部分的方法来支撑人才能力发展的培养与维持机制所需的经费。

（2）创新政府人才引进工作机制。要开创我省人才引进的新局面，就必须破除一切束缚创新型人才发展的禁锢，简化人才引进工作程序，推进政府职能由管理为主向服务为主转变。充分发挥人才市场在人才资源配置中的主渠道作用，采用环境引才、项目引才、事业引才、感情引才、高薪引才等多种方式，拓宽高层次创业创新人才引进的绿色通道。尤其应加快建立全省海外留学人员统一信息网络，进一步完善和落实留学归国人员项目资助、创业创新、子女入学等各项扶持政策，使广东成为海外留学人员归国创业创新的热土。

（3）完善我省引进高层次创业创新人才的产业平台。建议政府以高新技术产业园区、科技企业孵化器、大学科技园、留学生创业园等各级各类创业创新载体为依托来建设人才引进平台，加强政策配套和集成支持，完善这些产业平台的企业孵化、技术转化、人才培养、团队建设与风险投资筹措等职能，形成面向海外高层次创新人才、国内与本身高层次创业创新人才等多层面高效的人才引进平台。

（4）完善面向企业的人才引进与培养机制。充分发挥企业在高层次创业创新人才过程中的主体作用。实施“广东省高层次创业创新人才引进与培育计划”，建议由省财政每年安排专项经费，重点面向高成长性的科技型中小企业选择引进与培养对象。引导企业加大技术创新和人才开发投入，把高层次创业创新人才的引进和培养纳入企业发展的战略目标，结合企业产品创新和重大项目的实施，吸纳和培养高层次创业创新人才。通过落实多种培养和支持措施，充分发挥企业的主体作用，到2012年左右，初步建成一支具有较强技术创新和经营管理能力，能够带领企业在科技自主创新、发展速度和规模效益以及核心竞争力等方面居于国内同行前列的创业创新领军人才队伍。

（5）要建立既符合国际潮流，又具有广东特色的人才评价机制。目前的评价机制比较偏重对人才的经济效用的评价，而缺乏对人才的全面评估。为建立更加公正的评价体系，广东省应学习国外先进经验，率先在国内建立第三方人才评估体系，制定相应的地方性法规规范人力资源评价机构的发展，在政府、企业之外，建立一个更加公平、更加高效的人才评估体系。评估标准应考虑涉及人才能力与未来发展的众多指标，如人才当前创造的经济效益与未来潜在经济效益，人才的发展前景，人才所拥有的技能与创新能力，能力的适用范围大小，人才的领导能力与培养能力，人才所在专业的衍生影响领域，人才的可替代程度等。

3. 要稳步推进广东省人才开发的一体化进程

人才开发的一体化是未来一段时间内广东面临的重要工作。为了加快全省人才开发一体化进程，广东省应在遵循市场主导原则、开放自主原则、互惠共享原则、优势互补原则的基础上，着手建立广东省人才开发共享机制、促进机制、互认机制和协调机制。为此要做好以下工作。

（1）逐步统一我省各地人才市场在准入标准、设立程序、营运规则等方面的规定，推进和实行区域内人才流动政策、吸引政策、培训政策和社会保障制度等方面一体化的政策框架，降低区域内人才流动和开发成本，逐步搭建我省区域一体化的人才交流互动平台，促进人才自主、自由地流动。

（2）共同构建公平竞争的人才法治环境和人才生态环境，防止过度竞争和无序竞争，并联手进行区域内人才市场监管和人力资源保护工作，从而真正实现人才的跨市资源配置和资源共享。

（3）充分运用网络技术，构建广东省人才征信系统，建立和健全全省人才信息交换和发布机制，逐步实现全省范围内的人才信息联网，构筑畅通、快捷的人才信息平台。

（4）推进各种资格证书的互认或衔接，实现教育、培训、考试的资源互通、共享及在服务标准上的统一。以组建跨地区或跨市一体化分支机构、技术转让、技术嫁接、科研专家交流等多种形式，共同培养各地的紧缺、急需人才，逐步形成人才共育的全新格局。

（5）适应我省各地区企事业单位和各类人才的不同需要，拓展人事人才服务领域和内容，通过异地人事代理等人才服务项目，搭建区域内共通的人才服务框架，形成区域内统一的公共人事服务体系。

4. 逐步建立人才引进的地区平衡机制

针对广东地域经济发展与人才分布地域不平衡的现状，结合《珠江三角洲地区改革发展规划纲要（2008—2020年）》对我省各地区的具体定位，根据不同产业发展重点，来确定珠江三角洲、粤东、粤西与粤北地区各自的人才引进目录。为了减少人才倒流现象，除了在政策上要有所偏重之外，我省对粤东、粤西与粤北地区的人才引进工作还应更多地在报酬与用人机制上想办法。在报酬政策上不仅可以沿袭珠江三角洲所采取的成功经验，还可以寻找与尝试新的办法，如技术入股、项目入股、利润分成等报酬分配办法。在用人制度上，为鼓励高层次创新人才到边远山区去创业与创新，可以采取以下措施，如户口留在广州、深圳等城市，工作在边远城市，享受中心城市的工资福利待遇；工作满一定年限，就可以享受相应的职称与福利待遇等，也可以采取定期挂职、技术嫁接、技术指导等方式。

5. 不断改善人才的使用环境

(1) 加强政策配套和集成支持。各级政府和相关部门要制定高层次创业创新人才队伍建设的专项政策措施，从财税、金融、政府采购、知识产权保护等方面给予支持，加强各个部门之间政策的配套衔接，整合资源，形成合力。在省级财政支持的基础上，各地也应给予相应的配套支持，政府主导的创业风险投资基金、重大科技成果转化资金、中小企业发展资金以及省市有关部门的产业化项目要向创业创新人才倾斜。把高层次创业创新人才的培养、引进工作与重大工程、重大项目的实施结合起来，以项目为载体，以公共财政为引导，加强对创业创新人才的培育。

(2) 加强人才服务体系建设。着力加强以科技中介服务机构为主的人才服务体系建设，注重对创业创新人才的辅导，做好提升创业创新能力培训工作，加强创业创新人才之间的交流与沟通，组织境内外的科技合作与交流等活动，为培育创业创新人才提供专业化、特色化、个性化的服务。

(3) 建立和健全科学、合理的人才激励机制。进一步完善与社会主义市场经济体制相适应、与工作业绩紧密联系、鼓励人才创业创新的分配制度和激励机制。坚持按劳分配和按生产要素分配相结合，将人才的收入与岗位职责、工作绩效、实际贡献及成果转化产生的效益直接挂钩，鼓励一流人才做出一流贡献、获得一流报酬，鼓励技术入股、专利入股，允许高层次创业创新人才兼职兼薪。广东应提高给予合格人才的经济报酬，建立平衡划一的报酬等级机制，使同等级人才不管在何地都能得到基本上一致的经济收入。

6. 建立合理的退出机制

过去我们比较重视人才的引进，而忽视对人才的退出管理。实际上这也是一个需要重视的问题。如果处理不当，就会给未来的人才引进工作带来影响。应采取合理的评估制度，根据人才在不同阶段的具体贡献来评估退出人才的贡献价值，按最终贡献价值给予应有的待遇，不能采取硬性一刀切的政策。对于个别有特殊才能的人才与稀缺人才，可采取灵活的方法安排决定其在广东的具体工作时间，尽力维系其与广东的联系。

三、推进产学研合作，开创广东“人才培养”新模式

要改善我省人才储备目前的困境，不仅要重视人才的引进工作，更应该重视人才的培养工作。为此我省必须审时度势，深入贯彻“科学发展，先行先试”的重要思想，敢于创新与突破旧有的制度与方法，以高校、企业、政府、研究机构为合作主体，从多层面来促进产学研合作，不断提高人才培养的质量与针对性，开创岭南“人才培养”新模式。

1. 建立战略层面的人才培养咨询与管理机构

广东应该在学习先进的国际或地区人才培养经验的基础上，结合我们广东的特点，建立以下人才培养战略机构。①成立广东省人才培养战略委员会，在省政府的直接领导下，全面负责全省人才培养的战略规划、紧缺专业目录的编制与修改，普通高等教育重点发展专业与核心专业目录的制定与修改、高等职业教育重点发展专业与核心专

业目录的制定与修改。②成立广东省产学研合作发展委员会，主要负责有关我省推进产学研合作政策的制定与修订，产学研合作平台的建立，鼓励地方政府、企业与高校协作建立多种形式的产学研开发基地，依托我省已有的国家级、省部级实验室和技术开发中心来搭建平台，并出台建立相应的支撑与保障体系，形成集研究开发、中试生产、企业孵化与人才培养于一体的格局，努力支撑广东现代产业体系的发展。

2. 以高校为主体，全面建立广东人才培养综合体系

人才培养工作是必须持续进行的战略工作，不仅需要冲天的干劲，还需要脚踏实地的措施；不仅需要短时间内见效的临时性举措，更需要放眼世界、长期坚守的战略决断。这一切均需要审时度势，站在广东发展全局的高度，以现有的广东高校教育系统为基础，通过多方位创新，建立一套适合我国国情、反映广东特点的人才培养综合体系。

(1) 要继续加大工作力度，办好现有的大学城。对现有的广州、深圳、珠海、东莞等地的大学城进行全面评估，尤其要分析这些大学城对广东经济发展、产业结构调整与培养高新技术人才和创业创新人才等方面工作的影响，把握其现状与问题。在此基础上，需通过重新明确各个大学城的定位，通过政策扶持与功能完善，以更好地发挥其作用。目前，建议将广州大学城定位于我省最重要的综合性人才教育培养基地。其主要任务是为广东全方位培养高级行政管理人员、商务管理人才、教育类人才、服务类人才、高科技创业创新人才与高级职业技术人才，应瞄准重点产业升级换代与经济发展所需要的基础研究专业与科技创新专业培养。深圳大学城应定位为我省珠三角东岸地区的科技创新型人才教育培养中心，其主要任务是为我省珠江三角洲东岸地区与港澳地区培养合格的高科技创新型人才、先进制造业人才、港口服务类人才、金融服务类人才与高级职业技术人才。珠海大学城应定位为我省珠江三角洲西部地区的服务型人才教育培养中心，其主要任务是为我省珠江三角洲西部地区与港澳地区培养合格的旅游休闲类人才、高科技创业创新人才、现代服务类人才、医疗制药等制造业人才与高级职业技术人才。东莞大学城可以定位为我省东部地区的高中级职业技术、技能型人才的培养基地，作为对深圳大学城的有效补充，其主要任务是为深惠莞产业带的飞速发展培养合格的高级制造业人才，尤其是高技能专才。

(2) 筹划建设汕头大学城与湛江大学城。作为所在地区的中心城市，汕头与湛江分别在粤东、粤西地区的发展中起着不可替代的作用。但目前来说人才缺乏是限制这两个地区经济社会发展的重要因素。因此，建议省委、省政府总结我省发展大学城的经验教训，在整合当地现有大学教学资源的基础上，根据我省粤东、粤西地区的产业发展规划，适时筹划建设汕头大学城与湛江大学城，明确它们的定位与重点发展专业，引入我国优秀高校，开拓办学思路，出台鼓励政策，善用海外与华侨资源，逐步完善功能，使其成为我省粤东、粤西地区两地的高级人才培养中心。

(3) 深化国际合作，提升我省高等教育办学水平。我省应以新的思维和机制推动高等教育发展上水平，继续深化广东省内各高校与国际不同层次高水平大学的合作，强化与香港、澳门各高校的合作与交流，尤其应鼓励省内重点大学率先建成国内一流、国际先进的高水平大学。与国外及港澳台学校开展多层次的科技合作和联

合办学，引进和利用国外及港澳台优质教育资源，组建联合学院和联合研究生院，以及具有独立法人资格和独立校区的中外合作学院，开展科学研究和人才培养领域的实质性合作，为广东现代产业的发展培养一大批具有全球视野、具备国际竞争力的复合型高层次人才。

3. 围绕现代产业体系建设的需求，推进产学研紧密合作，创新人才培养模式，着重培养引领行业发展的高素质人才

（1）广东省应尽快出台政策鼓励高校根据广东经济社会发展需求，围绕广东现代产业体系的建设，积极发展和调整优化学科专业，使学校学科门类覆盖广东支柱产业、高新技术产业和重点发展产业，加快培养产业发展紧缺人才。目前应采取优先发展服务类专业（如金融类专业、科技服务专业、商务服务专业、会展管理、信息服务类、旅游类等）、加快发展高精尖的制造类专业（如石油、钢铁、汽车工程、船舶制造工程、核工程、数控机床制造等专业）、大力发展高技术专业（如电子信息工程、生物工程、材料工程、环保工程、新能源工程、海洋工程等专业）。

（2）推进产学研紧密合作，拓宽人才培养渠道。进一步深化高校教育科研体制改革，优化研究生培养模式，强化创业创新意识和能力的培养，提升广东高校培养高层次创业创新人才的能力，使之成为高层次创业创新人才培养中心。适应企业提高经营管理水平的迫切需要，有计划地选拔一批高校优秀研究生和青年教师到国外高校、企业学习进修，为我省高新技术企业培养高素质的经营管理人才。加快建立以企业为主体、产学研紧密合作的有效机制，促进高层次人才培养与产学研合作互动融合。进一步深化人才管理体制改革，打通高校、科研院所和企业之间高层次人才流动通道，为产学研联合培养高层次创业创新人才提供制度保障。目前可制定有关不同系列职称之间的转换政策。

（3）加强创业创新载体建设，建立高层次创业创新人才培育基地。认真借鉴国内外成功经验，进一步加强高新技术产业园区、大学科技园、科技企业孵化器等载体建设，在风险投资、科技基础设施、科技公共服务平台等方面提升水平。吸引、支持省外国家级科研院所、工程技术研究中心、重点实验室等研发机构到广东设立分支机构。探索建立各级各类创业创新载体资源共享、合作研发、联合培养人才的融合互动机制。在继续加大硬件建设的基础上，着力加强为创业创新服务的软件建设，吸引国内外各类创业创新服务资源在载体内的集聚，把创业创新载体建设成高层次创业创新人才的培育基地。

（4）着力培养能够引领行业发展的高素质人才，将培养掌握产业关键技术并能解决工程实际问题的现代工程师和集专业技术、市场经验和管理才能于一身的科技型企业家作为人才培养的重点；坚持走“产学研合作”人才培养道路，加强与珠三角地区龙头企业的战略合作，以企业需求为导向，探索“订单式”教育培养模式，满足企业对人才个性化的要求。探索研究生培养的新机制，促进复合型创新人才的成长，实行“工作学习交替制”培养模式，鼓励企业与学校联合培养研究生。

4. 建立地区产业人才培养基地

为努力消除我省经济发展的不均衡状态，必须逐步减少或消除珠江三角洲地区与

粤东、粤北、粤西等地区的人力资源差距，降低人才倒流比例。建议省委、省政府利用现有的大学城，以及未来拟建的湛江、汕头两地的大学城，并在我省不同地区选择重点城市，设立地区产业人才培养基地。需结合不同地区的产业发展需要，鼓励各高校与企业采取多种形式来共建产业人才培养重点基地，在资金上给予支持，在培养方式上可以采取传统形式与创新形式相结合的办法，如委托培养、定向培养等传统方法，也可采取新的方式——与企业共同在生产一线设置培训现场与考场，采取现场培训或现场考试等方法。

人力资源的引进与培养是关系到我省未来经济社会发展的战略大计，只要遵循“科学发展，先行先试”的原则，抓住我省产业结构调整的战略契机，转变人才开发观念，全方位开拓创新，改善用人环境与相关机制，提高我省高校的人才培养能力与质量，我们就能迎难而上，重新占据人才储备的战略高地。

大力推动“政产学研金”合作创新为广东省经济社会发展做贡献①

自我国1978年改革开放以来，广东社会经济迅猛发展，成为国家经济发展的排头兵，其GDP总量已占全国1/9，源自广东的财政收入占全国财政收入的1/7，省外贸进出口总额占全国的1/3。广东从原来资源贫乏的边远省份一跃而为全国省区第一，实现了经济社会发展的历史性跨越，为国家改革开放和社会主义现代化建设做出了重大贡献。

在经历了30多年的飞速发展之后，又碰上了美国次贷引发的金融危机，这使我们广东处在一个前所未有的关键时期。特别是2008年12月，中央通过《珠江三角洲地区改革发展规划纲要》赋予广东“科学发展，先行先试”的重大使命，从国家战略的角度提出加快改革，增创新优势，更上一层楼，实现经济社会又好又快发展。

中央的殷切期望，人民也要求快速发展，我感到要实现伟大目标，其中的一个关键问题是要培育许多创新型企业。

过去，我们主要靠的是劳动密集型企业，靠来料加工，靠仿制，这在新的时期是不行的。

只有大力推动产学研合作，才能有大批创新型企业出现，才能增强自主创新能力，才能建设现代化产业体系，才能加快转型升级，才能加快经济发展方式转变，才能使广东继续担当全国的排头兵！这应该成为广东省的一项战略性举措！

产学研合作是一个跨行业、跨部门、跨地区合作的系统工程。实际上，还应在产学研三方面的两头，分别再加上政府和金融机构，即应写为政、产、学、研、金的合作。没有政、金两方面，产学研合作是搞不好的。

政府的地位十分重要，是关键。政府应负责营造科技创新的良好环境，强化政策的激励引导作用，为企业、高等学校、研究所人员营造有利于科技创新的政策环境，引导创新要素向企业集聚，在科技投入、成果奖励、创新平台建设等方面加大扶持力度，同时还应强化考核的促进保障作用。

企业是主体地位，应提升自主研发和技术创新能力，使企业成为研发投入主体，成为技术创新主体，成为创新成果应用主体，这将大力提升企业的核心竞争力，使企业健康发展。

高等学校是重要支撑体系，是坚强后盾，既要培养科技创新人才，又要提供科技创新成果（主要是应用基础理论成果）。

研究院（所）也是重要支撑体，要与高等学校合作，为企业提供能应用的科技创新成果。

① 本文是在广东省委、省政府领导接见院士专家会上的发言，广州，2010年2月8日；《一个大学校长的探索》，北京：高等教育出版社，2011，358-359.

所以，高等学校和研究院（所）是创新型企业的生命源泉和基础。

金融机构可为产学研提供投人，提供风险管理服务，这是产学研的强大支柱。

目前，“政产学研金”合作的核心问题是要完善联盟的组织模式和运作模式，政府起引领作用，金融机构起支柱作用，产学研三个方面组成一个平台，五个方面均要受益才算成功。由于五个方面紧密联系，一损俱损，因此每个方面都要尽力，都要注意协调，在这里，“和为贵”是至理名言。

为此，建议成立“广东省政产学研金合作发展委员会”，由省发展和改革委员会牵头，有关部门由省教育厅、省科技厅、省科协、省风险投资银行、省信息产业厅、省环保局、省农业厅等部门组成，主要负责有关我省推进“政产学研金”合作政策的制定与修订，产学研合作平台的建立，鼓励地方政府、企业与高校协作建立多种形式的产学研开发基地，依托我省已有的国家级、省部级实验室、工程中心来搭建平台，并建立相应的支撑与保障体系，形成集研究开发、中试生产、企业孵化与人才培养于一体的格局，努力培育许多创新型企业，为广东经济社会创新一轮发展做出贡献！

积极推进知识产权事业发展[①]

一、发挥知识产权促进科技创新的重要作用

（一）知识产权的内容

17 世纪中叶，法国卡尔普佐夫（Carpzov）首先提出知识产权的概念，至今已有 300 多年了。

知识产权是指法律规定人们对于自己脑力劳动创造的精神财富所享有的权利。

有以下七种知识产权：著作权（版权）、专利权、商标权、商业秘密权、地理标志权、植物新品种权、集成电路布图设计权。

前三种最为重要。

知识产权的特点是：它是一种无形财产权，它保护知识财产，与传统所有制度所保护的动产和不动产等有所不同，它保护的是知识技术、信息等知识财产。该权利必须经立法才能被直接确认；它与有形财产权相比，具有专有性、地域性和时间性。

应该指出，知识就是财富，更确切地说有产权的知识才能成为财富。脱离法律保障的知识可能被任何人独占，因此只有具备知识产权形式的知识，才可能转化为个人财富。比尔·盖茨正是凭借 Windows 操作系统的软件版权和“微软”这一著名商标，连续 13 次跻身世界首富。

1967 年，联合国国际局提出建立世界知识产权组织，于是召开了有 51 个国家参加的斯德哥尔摩会议，签订了《建立世界知识产权组织公约》，成立了世界知识产权组织。

1974 年 12 月，世界知识产权组织成为联合国组织的一个专门机构：World Intellectual Property Organization（简称 WIPO），总部设在日内瓦，我国于 1980 年申请加入该组织。

从 2001 年起，将每年 4 月 26 日定为世界知识产权日，今年 4 月 26 日已是第十个世界知识产权日，世界知识产权组织将今年的主题定为“创新——将世界联系在一起”。

（二）知识产权制度是一个国家走向现代化的必然选择

是否保护知识产权？

对哪些知识赋予知识产权？

如何保护知识产权？

对于上述三个问题，一个国家要根据现实发展状况和未来发展需要来进行制度选择和安排。

知识产权制度是西方国家 300 多年来不断发展的制度文明，对促进经济发展、推动科技进步、繁荣文化和教育起到了重要作用。知识产权制度已经成为每个国家走向

---

① 本文是兰州大学管理学院讲稿，兰州，2010 年 8 月 14 日.

现代化之路的必然选择，没有例外。无论是发达国家，如美国，还是发展中国家，如印度，都是以知识产权为战略武器，去占领国际竞争的制高点，去提升自己的国际竞争力。

美国专利商标局发布了《21 世纪战略计划》，形成了对知识产权的快速反应机制；

日本政府制定了《知识产权战略大纲》《知识产权基本法》；

澳大利亚政府提出了促进本国知识产权战略的“创新行动计划”；

韩国制定“知识产权制度”，提升科技竞争力，希望在 2025 年进入科技领先国家的行列；

印度政府制定“知识大国的社会转型战略”，确定了未来发展目标。

2008 年 4 月 9 日，国务院审议并通过《国家知识产权战略纲要》，6 月印发了纲要，标志着中国知识产权战略正式启动实施，这是我国知识产权事业发展的纲领性文件。文件有近 5 年的阶段目标，以及到 2020 年把我国建设成为知识产权创造、运用、保护和管理水平较高的国家的战略目标，凝聚了中央和国务院对知识产权的最新认识，是对把我国建设成为创新型国家内涵的进一步丰富和深化。有效利用知识产权这种先进的法律制度，以便缩小我国与发达国家差距、实现跨越式发展、把我国建设成为创新型国家。

（三）知识产权是国际贸易体制的基本规则

知识产权既是我国自身发展的需要，同时也是遵守国际条约的需要。在世界贸易组织框架内，知识产权保护已经成为国际贸易规则的重要组成部分。世界贸易组织主要有三大主体制度，即《货物贸易协议》《服务贸易协议》《与贸易有关的知识产权协定》。

上面的第三个制度是以国际法的形式表达了知识产权保护与国际贸易的合法关系。知识产权与国际贸易的融合主要因为下面两个因素。

(1) 现代国际贸易中技术因素的不断增长。特别是高科技产品的增长速度最快，如计算机、半导体芯片、通信设备等。它们能够为产品所有人带来高附加值。这高附加值来源于相应的技术优势，体现了知识产权所凝结的价值，与货物贸易相对应的服务贸易（相应上面第二个制度），归根结底就是一种思想、技术、信息的商业变换，与知识产权也是息息相关的。

(2) 现代国际贸易中技术优势与成本优势的激烈较量。在国际贸易中，发达国家占据技术优势，是一种头脑经济，高科技产品在发达国家出口中占很大比重，因此，发达国家不遗余力地倡导在全球贸易中保护知识产权，以维护他们的贸易利益，巩固其技术优势。相对而言，发展中国家在国际贸易中，具有的是一种成本优势，主要是劳动资源优势，这是一种“肢体经济”，以低工资、低福利、低劳保的劳动力价格来维持竞争优势，而且以污染环境、耗费原材料为代价。这两种优势，构成了当代国际贸易冲突的现状，经过长期斗争后才达成乌拉圭谈判的一揽子协议，即上述三个主体制度。

（四）知识产权对科技创新的重要作用

21 世纪，是知识经济时代。经济增长更加依赖于知识的积累与创新，知识已经成为经济活动中最重要的生产要素，知识产权制度作为制度创新的结果，保障着知识生产、传播和利用的智力劳动过程，服务于知识社会化、产业化、产权化的发展目标。这时所指的知识，主要是以自然科学为基础的技术知识。建设创新型国家，首先就是要在科技上创新，要形成一批具有自主知识产权的核心技术，并形成国际核心竞争力，而知识产权制度正是鼓励和保护科技创新的制度，在促进科技创新中越来越重要。

专利权是科技领域最重要的知识产权，也是知识产权类型的核心内容，科技创新建立在专利制度之上。专利权包括三种类型：发明专利、实用新型专利和外观设计专利，其中发明专利是指对产品、方法或其改进提出的新的技术方案，是最具经济价值、最有战略意义、最富有科技创新内涵的专利。故在知识产权中，要高度重视专利权，合理运用专利战略，这对国家的科技创新具有极为重要的意义。

二、大力推进知识产权的事业发展

（一）总结过去，吸取教训

1. 自然科学为何未在我国产生

在过去6000多年人类文明历史中，显而易见，是科学技术推动了人类文明的发展。

世界人类文明的4个发源地：中华文明、古埃及文明、古巴比伦文明、古印度文明，仅有中华文明一直延续至今。

从汉朝到明朝，16个世纪，中华文明领先于世界各国的文明，中国的科学技术领先于世界，中国的经济GDP甚至占世界一半以上。

17世纪至19世纪下半叶：

英国剑桥大学和牛津大学是世界科学中心，英国成为世界第一强国。

19世纪末至20世纪初：

德国哥廷根大学是世界科学中心，有世界闻名的哥廷根学派，德国成为世界第一强国。

20世纪上半叶至今：

美国哈佛大学、麻省理工学院是世界科学中心，美国成为世界第一强国。

中国从明朝以后，落后了，其根本原因是科技落后了。现代自然科学没有在中国产生，其根本原因是中国当时的政府管理落后了，不尊重科技创新。

2. 新中国前30年科技创新状况与落后原因

1）科技成就

A. 两弹一星

从20世纪50年代开始，我国开始独立自主地研制“两弹一星”，取得长足进步。

导弹：

1960年11月，近程地地导弹；

1964年6月，中近程导弹；

1966年10月，导弹核武器；

1966年12月，中程地地导弹；

1970 年 1 月，中远程地地导弹。

原子弹、氢弹：

1954 年 10 月，在广西发现铀矿并采集了第一块铀矿石；

1964 年 10 月，第一颗原子弹；

1967 年 6 月，第一颗氢弹。

卫星：

1970 年 4 月 24 日，东方红一号卫星，173kg；

1975 年 11 月，返回式卫星（发射与回收）。

B. 石油

1949 年，全国加工能力 11.6 万吨/年，自给率不到 10%；

1958 年，兰州炼油厂建成（100 万吨）；

1959～1963 年，发现和开发大庆油田；

1976 年，大庆油田年产量达 5000 万吨；

1978 年，全国原油产量达到 1.04 亿吨，进入世界石油大国行列。

C. 电气化

1949 年，全国发电装机容量 184.86 万千瓦，世界第 21 位；

全国发电量 43.1 亿千瓦，世界第 25 位；

1955 年，1 万千瓦水电机组投入运行（北京官厅水电站）；

1956 年，6000 千瓦火电机组投入运行；

1969 年，22.5 万千瓦水电机组投入运行（刘家峡水电站）（装机容量 122.5 万千瓦水电站）（坝高 147m）；

1974 年，30 万千瓦火电机组投入运行；

1972 年，330 千伏刘天关输变电工程投入运行。

D. 计算机

1958 年 8 月 1 日，第一台电子管计算机（103 型）诞生；

1965 年，第一台大型晶体管通用数字计算机（109 乙机）；

1973 年，百万次集成电路电子计算机（150 机）（第一台中小规模集成电路计算机）；

显而易见，我国科技成果较新中国成立前有飞跃，但与发达国家相比，差距却很大。（对比实例：1969 年 7 月 20 日至 1972 年 12 月，美国阿波罗飞船 6 次登月，已把 18 人先后送上了月球，累计在月球表面考察了 300 多个小时）。

2）落后原因

（1）旧中国底子薄；

（2）新中国的管理落后，主要是“尊重人才”“尊重知识”做得不好，知识分子积极性未能发挥，特别是在“文化大革命”中，知识分子成了“臭老九”。

知识分子列为第九等来源于元朝，人分为十等：官、吏、僧、道、医、工、猎、娼、儒、丐。

儒生排在娼妓和乞丐之间。

抗日战争时期，士兵被称为“丘八”，大学生被称为“丘九”。

在“文化大革命”中，更在“老九”之前加个“臭”字。

(3) 封闭的社会，群众不知道世界科技创新情况。

(4) 知识产权未列入我国法律制度中。

(5) 观念落后，仅有“奉献”的观念，没有“知识是财富”的观念。

(6) 科技创新的氛围差。

“木秀于林，风必摧之”。

“枪打出头鸟”。

(7) 教育、科技落后。大学时没有设置管理专业，教育、科技人才缺乏，国家投入少。

(8) 做事不认真，应树立质量第一原则思想。

(二) 加强知识产权工作

我国于1980年参加世界知识产权组织，才开始有知识产权事业。

1. 牢记小平同志的著名论断

1977年，邓小平同志就提出“尊重知识，尊重人才”。

1978年3月18日，全国第一次科学大会，郭沫若讲“科学的春天”。邓小平提出两个著名论断:“知识分子是工人阶级的一部分”“科学技术是生产力”。

我国科技创新事业在科学的春天迈开了新的步伐。

2. 改革开放30年来我国科技和经济的飞跃

科技创新成果丰富。

经济实现了跨越式发展。

解决了温饱问题，进入了小康社会，是几百年来出现的一次盛世。

成了制造大国。

目前，全世界70%的鞋、70%的玩具、50%的PC微机、50%的手机、50%的彩电、50%的空调、40%的纺织品产自中国。

美国，萨拉所写的《离开中国制造的第一年》，结论是“你无法拒绝中国出售的产品”。

但我国还不是“创造大国”。

3. 实施知识产权战略，建设创新型国家

1) 贯彻《国家知识产权战略纲要》(2008年4月)

国家知识产权战略的核心在于制度变革，重点是完善知识产权制度，其内涵是在履行国际义务的前提下，进一步完善符合我国国情的知识产权法律法规；建立健全知识产权执法和管理体制机制；在制定经济、社会、文化政策时强化知识产权导向作用。

完善知识产权制度是激励自主创新的制度保障。

自主创新需要知识产权制度，知识产权保护激励自主创新。自主创新以知识创新为先导，以知识产权保护为后盾。知识产权法律制度的完善程度、知识产权的实际保护水平，是一个国家科技水平、经济实力的法律体现，可以客观地反映出一个国家的科技水平和经济实力。

2）拥有自主知识产权是提高国家核心竞争力的战略重点

世界未来的竞争是自主知识产权的竞争。

宏观上，国家核心竞争力主要表现在科技实力和经济实力，而评价一国科技实力和经济实力的标准主要是自主知识产权的数量和质量。

微观上，国家之间的竞争最终细化为各国企业之间的竞争，而考量企业综合竞争力的尺度主要是科技竞争力和品牌竞争力，即知识产权的数量与质量所表现出的竞争力。

例 1：洛桑国际管理学院 2007 年度世界经济竞争力排行榜，中国由上年的第 28 名上升为第 15 名，升 13 位：而科学基础设施由上年的第 17 名升为第 15 名，升 2 位；技术基础设施由上年的第 33 名升为第 27 名，升 6 位。

结论：关于科技竞争力，它的明显提升对我国核心竞争力的提高起到了显著的推动作用。

例 2：世界品牌实验室编制的 2009 年世界品牌 500 强中，部分国家拥有的 500 强品牌数如表 1 所示。

表 1　2009 年主要国家拥有世界品牌数

排名	国家	数量
第 1	美国	241 个
第 2	法国	46 个
第 3	日本	40 个
第 4	英国	39 个
第 5	德国	24 个
第 6	瑞士	22 个
第 7	中国	18 个

结论：关于品牌竞争力，著名品牌拥有量的排名高低反映了一个国家核心竞争力的强弱。

3）加强知识产权保护是促进市场经济健康发展的重要举措

市场经济的健康发展，有赖于明晰产权归属，规范产权市场，加强产权保护与管理。在全球化的今天，无论在国内市场还是国际市场，中国经济的发展都离不开对知识产权的有效保护。

我国经济对外贸易的依存度已经达到了 60%以上，因此不能脱离对外贸易来谈经济发展；而知识产权已经成为国际贸易体制的基本规则，所以也不能离开知识产权保护来谈中国的对外贸易。完善的知识产权制度既是技术引进的先决条件，也是外商投资合作的环境要素。

有效利用知识产权这种先进的法律制度，是我们缩小与发达国家差距、实现跨越式发展、建设创新型国家的战略抉择。

所谓创新型国家，指的是以知识创新（包括科技创新和文化创新）为基本政策，提高自主创新能力和形成国际核心竞争力的先进国家。

知识产权制度对建设创新型国家具有重要的推动作用和保护作用。

关于完善我省应对台风灾害预防措施的建议[①]

由于我国疆域辽阔，山脉多，海岸线长，地质构造和地理环境复杂，自然灾害频繁，是世界上遭受自然灾害最严重的国家之一。灾害种类多，分布地域广，发生频率高，造成的损失严重，1990～2008 年，年均约 3 亿人次受灾，倒塌房屋 300 多万间，经济损失超过 2000 亿元。近期，如四川汶川地震有 69 225 人遇难，374 640 人受伤，17 939 人失踪，公路受损里程达 53 295 公里，受损供水管道达 48 275.5 公里；甘肃舟曲特大山洪泥石流灾害造成 1156 人死亡，588 人失踪。在我省台风年年有，每次到来均有很大损失。面对这些常常发生的灾害，我们国家已积累了许多应对方法，并及时进行了许多处理，减少了许多损失。但是，主要的方法还是应急处理，尚缺乏预防措施。如果事前预防措施得力，必将大幅度减少人的死伤和财物的损失。

为此，建议先将各种自然灾害分门类，根据历史上的积累数据，以严重、中等和一般三个等级，划分可能发生灾害地区，然后针对每一地区制定该地区进行工程建设的规范和设计标准。同时，在可能发生严重自然灾害地区，编写相应防御和应对灾害来临对策的教材，列入初中或小学高年级教学必读课程，使每一位公民在少年时代就掌握应对灾害的知识和能力，这样年复一年，我国所有人就都具备了防灾的本领。那么，当灾害来临时，就必将使人的生命和财物的损失减少到最低程度。显而易见，这种办法与现有临时措施相比较，人的死伤最少，经济损失也最少。

先在我省沿海的城乡初中或小学高年级开设应对台风课程，应是明智之举。在粤东地震带内的城乡中小学，还应加开预防地震灾害的课程。

① 本文原载《广东省政府参事建议》，第 81 期，2010 年 12 月 20 日.

坚持重点　保持特色[①]

端州“十一五”时期有很大飞跃，市委徐书记用“五个最”概括端州发展，非常赞同。上海闸北区、广州天河区，原来是比较落后的地区，现在已经发展为市中心了。现在的端州区相当于 2000 年时期的天河区。作为顾问，看到端州的发展感到非常高兴。政府工作报告写得很细，包括方方面面。建议端州谋划未来发展要坚持重点，保持特色。每一个城市以它的特色存在于这个世界上，让人们所认识。日内瓦，是一个美丽的风景城市、旅游城市、会议城市，是国际上开会的地方，非常漂亮。中国现在每一个城市都有一个品牌，都在追求 GDP。其实可以有很多的分类，旅游城市、教育城市、科技城市、会议城市、会展城市。端州一直以来就是个历史名城、风景名城。七星岩是鼎鼎有名的风景区，西江从你们这里经过，环境优美，端砚是你们的特色品牌，所以一定要保护这个核心东西。你们的文化、历史积淀不要被别的东西冲掉了，冲掉了就可惜了，再也回不来了。去年我国 GDP 已超越了日本，成为世界第二大经济体了。珠三角、长三角、京津冀是中国三大地区，珠三角最大。肇庆市作为珠三角九个城市中的一员，是光荣的责任，也是机遇。端州是肇庆市的核心区，一定要发挥好自己的优势，围绕打造文化名城、历史名城、旅游名城做文章。端砚是中国四大名砚（端砚、歙砚、洮河砚和澄泥砚）之首，中国笔、墨、纸、砚里面，搞得比较好的是端州。端州在保护和发扬光大中华文化方面做了积极的贡献。我认为要进一步做大做强端砚文化产业，建设规模较大的端砚博物馆，做解说，做故事，做电影，做宣传；开设全国性的端砚文化论坛，让全国老百姓对砚文化有更深的认识；鼓励企业把端砚作为产业，做大产业和保护中华文化；扩大文化产业范围，除端砚外，其他三大名砚以及纸、笔、墨等文房四宝相关产业也可开发；在中小学里面提倡用毛笔写字，提高中小学生的艺术修养，并形成一种特色；制作砚文化电视专题片，充分利用媒体在全国范围内加大砚文化宣传力度，进一步弘扬端砚文化，传承中华文化传统，把端州打造成为南方著名、全国知名的文化名城、历史名城和旅游名城。通过砚文化和旅游提高端州知名度，城市的名气越大，来旅游的人就更多了，从而带动端州区跨越发展。也可考虑发展光电产业、信息产业等战略性新兴产业。端州的地块不是很规则，未来五年还要做好规划，把城市建设搞好，带动经济发展，建设幸福端州。

① 本文原载《肇庆市端州区人民政府工作简报》，2011 年，(1)：6-8.

中国的过去和现在[①]

1949 年，中华人民共和国成立，到现在，中华人民共和国为代表全中国的唯一合法政府。世界上只有一个中国，其中文全名为中华人民共和国，中文简称为中国。英文全名为 The People's Republic of China，英文缩写为 P. R. C。

一、地理形势

1. 地理概况

中国位于欧亚大陆东部，太平洋西岸。中国约有 4257 年（含“夏朝”）的已知文明史（以出现国家体制为始），是亚非大河流域四大文明古国之一。它的版图被形象地比作雄鸡。中国是中华民族的主要聚居地，领土总面积约为陆地面积 960 万平方千米，内海和边海的水域面积约 470 万平方千米，总面积约 1430 万平方千米。

中国地势西高东低，山地、高原和丘陵约占陆地面积的 67%，盆地和平原约占陆地面积的 33%。山脉多呈东西和东北—西南走向。全国共有三十四个省级行政单位，二十三个省（河北省、河南省、湖北省、湖南省、江苏省、江西省、辽宁省、吉林省、黑龙江省、陕西省、山东省、山西省、四川省、青海省、安徽省、海南省、广东省、贵州省、浙江省、福建省、台湾省、甘肃省、云南省），五个自治区（内蒙古自治区、宁夏回族自治区、新疆维吾尔自治区、西藏自治区、广西壮族自治区），四个直辖市（北京、上海、天津、重庆）和香港、澳门两个特别行政区。

中国人口占据世界人口数目的第一位，2009 年 7 月统计数据约 13 亿 3185 万人。由 56 个民族构成，他们分别是：汉族、蒙古族、回族、藏族、维吾尔族、苗族、彝族、壮族、布依族、朝鲜族、满族、侗族、瑶族、白族、土家族、哈尼族、哈萨克族、傣族、黎族、傈僳族、佤族、畲族、高山族、拉祜族、水族、东乡族、纳西族、景颇族、柯尔克孜族、土族、达斡尔族、仫佬族、羌族、布朗族、撒拉族、毛南族、仡佬族、锡伯族、阿昌族、普米族、塔吉克族、怒族、乌孜别克族、俄罗斯族、鄂温克族、德昂族、保安族、裕固族、京族、塔塔尔族、独龙族、鄂伦春族、赫哲族、门巴族、珞巴族、基诺族。

2. 名山大川

中国的名山众多，一向以其俊秀的英姿、绚丽的风采吸引着全世界的游客流连忘返，其中以“黄山归来不看岳”闻名的安徽黄山、“匡庐奇秀甲天下”著称的江西庐山的名气最大，享有“世界级名山”的声誉。

中国名山中的五岳尤为著名，五岳是道教崇奉的中国五大名山的总称。道教认为每岳都有岳神，东岳泰山岳神“齐天王”，南岳衡山岳神“司天王”，西岳华山岳神“金天王”，北岳恒山岳神“安天王”，中岳嵩山岳神“中天王”，各领仙宫玉女几万人

---

① 本文是为德国访华代表团“中国业务开发专题讲座”所作的报告，北京，2011 年 4 月 24 日.

治理其地。

中国名山中还有著名的四大佛教名山，相传山西五台山曾是文殊菩萨的道场，四川峨眉山曾是普贤菩萨的道场，浙江普陀山曾是观音菩萨的道场，安徽九华山曾是地藏菩萨的道场，故称之为“佛教四大名山”，民间就有“金五台、银普陀、铜峨眉、铁九华”之说。另外中国还有著名的四大道教名山，它们分别是湖北武当山、江西龙虎山（道教发祥地）、安徽齐云山、四川青城山。

我国主要河流有长江、黄河、黑龙江、辽河、海河、淮河、钱塘江、闽江、瓯江、珠江以及塔里木河等。

其中长江（The Changjiang River；The Yangtze River）是中国第一长河，是世界第三长河。干流全长6380公里（以沱沱河为源），一般称6300公里。年平均入海水量约9600余亿立方米。流域介于北纬24°30′～35°45′，东经90°33′～112°25′，面积180余万平方公里（不包括淮河流域），约占全国土地总面积的1/5。

长江的北源沱沱河出自青海省西南边境唐古拉山脉格拉丹冬雪山，与长江南源当曲汇合后称通天河；南流到玉树县巴塘河口以下至四川省宜宾市间称金沙江；宜宾以下始称长江，扬州以下旧称扬子江。长江流经西藏、四川、重庆、云南、湖北、湖南、江西、安徽、江苏等省区市，在上海市注入东海。有雅砻江、岷江、沱江、嘉陵江、乌江、湘江、汉江、赣江、青弋江、黄浦江等支流，在江苏省镇江市同京杭大运河相交。

长江在湖北省宜昌市以上为上游，水急滩多；宜昌至江西省湖口间为中游，曲流发达，多湖泊（鄱阳、洞庭两湖最大）；湖口以下为下游，江宽，江口有冲积而成的崇明岛。长江水量和水利资源丰富，盛水期，万吨轮可通武汉，小轮可上溯宜昌。

长江流域是中国人口密集、经济繁荣的地区，沿江重要城市有重庆、武汉、南京、上海。

长江可供开发的水能总量达二亿千瓦，是中国水能最富集的河流。长江干流通航里程达两千八百多公里，素有“黄金水道”之称。

长江在重庆奉节以下至湖北宜昌为雄伟险峻的三峡江段（瞿塘峡、巫峡、西陵峡），世界最大的水利枢纽工程三峡工程就位于西陵峡中段的三斗坪。

而黄河（The Yellow River）则是世界上含沙量最多的河流。黄河，被誉为中国的母亲河。若把中国比作昂首挺立的雄鸡，黄河便是雄鸡的动脉。黄河流程达5464千米，流域面积达到79.5万平方公里，上千条支流与溪川相连，犹如无数毛细血管，源源不断地为祖国大地输送着活力与生机。

黄河是中国第二长河，世界第五长河，源于青海巴颜喀拉山，干流贯穿九个省、自治区：青海、四川、甘肃、宁夏、内蒙古、陕西、山西、河南、山东，年径流量574亿立方米。但水量不及珠江大，沿途汇集有35条主要支流，较大的支流在上游，有湟水、洮河，在中游有清水河、汾河、渭河、沁河，下游有伊河、洛河。两岸缺乏湖泊，黄河下游流域面积很小，流入黄河的河流很少。

黄河中游河段流经黄土高原地区，因水土流失，支流带入大量泥沙，使黄河成为世界上含沙量最多的河流。最大年输沙量达39.1亿吨（1933年），最高含沙量920千克/立方米（1977年）。

3. 风景名胜

中国的风景名胜众多，如表1所示：

表1 中国的风景名胜

地区	风景名胜
北京	八达岭、十三陵、颐和园、石花洞、天坛、故宫、圆明园、北海、公主坟、恭王府
天津	盘山
上海	外滩、东方明珠塔、上海环球金融中心、豫园
重庆	大足石刻、朝天门、解放碑、巫峡
河北	承德避暑山庄及外庙、秦皇岛北戴河
山西	大同云冈石窟、平遥古城、五台山、介休绵山、北岳恒山
辽宁	沈阳故宫、清东陵、清北陵、世界园艺博览会、大连发现王国、张氏帅府、千山、鸭绿江
吉林	松花湖、长白山
黑龙江	镜泊湖
江苏	南京明城墙、明孝陵、总统府、中山陵、秦淮河、夫子庙、玄武湖、金牛湖、雨花台、太湖、苏州古典园林、扬州瘦西湖
浙江	杭州西湖、杭州千岛湖、普陀山、雁荡山
安徽	黄山、九华山、安徽古村落、巢湖、紫微洞
福建	永定土楼、武夷山、鼓浪屿-万石山、老君岩
江西	庐山、井冈山、三清山、龟峰、铜钹山、婺源
山东	泰山、曲阜、蒙山、崂山、蓬莱、趵突泉
河南	龙门石窟、少林寺、白马寺、关林、龙亭、清明上河园、殷墟、红旗渠、嵩山、云台山、伏牛山、云梦山、白云山、木扎岭、重渡沟、鸡冠洞、六龙山、石鸭子山
湖北	武汉东湖、武当山、黄鹤楼、三峡、神农架、木兰山
湖南	衡山、岳阳楼、洞庭湖、马王堆、岳麓书院、张家界
广西	桂平西山、桂林漓江、宁明花山
广东	丹霞山、七星岩、河源万绿湖、新丰江水库、东源赖东亨、黄崇岩
海南	三亚热带海滨、天涯海角
四川	峨眉山、乐山大佛、黄龙池、九寨沟、青城山、都江堰
陕西	秦始皇兵马俑、华山、华清池、大雁塔、法门寺、壶口瀑布、半坡、陕西历史博物馆、黄帝陵、乾陵、茂陵、蓝田猿人遗址、西安钟鼓楼、明城墙、碑林、大唐芙蓉园、太白山、西安钟馗故里遗址、延安革命纪念馆、耀州窑、宝塔山旅游区、大兴善寺、水陆庵、金丝大峡谷国家森林公园、炎帝陵、贵妃墓、诸葛亮庙、张骞纪念馆、辋川溶洞、鸿门宴遗址、普照寺、灵岩寺、秦陵地宫展览馆、武侯墓、张良庙、司马迁祠、周公庙、秦岭野生动物园
贵州	黄果树瀑布
云南	三江并流、丽江古城、西双版纳、香格里拉、大理古城、苍山洱海、梅里雪山、玉龙雪山、泸沽湖、滇池、龙门
甘肃	敦煌莫高窟、天水麦积山、嘉峪关
新疆	天山天池、博格达峰
青海	青海湖
宁夏	西夏王陵
西藏	布达拉宫、大昭寺

续表

地区	风景名胜
内蒙古	呼伦贝尔草原、昭君墓、古长城遗址、锡林郭勒草原
香港	海洋公园、太平山、维多利亚港、迪士尼乐园

其中中国十大名胜如下。

1）天安门

天安门坐落在广场北端，天安门城楼虽饱经沧桑却依旧朝气蓬勃、威严四仪。天安门是北京城的眼睛，是中华人民共和国的象征。

天安门城楼始建于明永乐十五年（公元 1417 年），是明代皇城的正门，当时叫“承天门”，有承天启运之意。刚开始修建的时候，可不是现在看到的这个模样，而是五座木牌坊，后来改建成九开间门楼。清顺治八年（公元 1651 年），清世祖福临重新修建这座城楼。新中国成立后，人民政府又重建了城楼上的木建筑、加厚城墙，才成了现在的样子。明、清两朝，这儿是禁地，到公元 1911 年清王朝覆灭为止，除了皇亲贵族，老百姓不准过往。它最大的用途，是国家有大庆典（如皇帝登基、册立皇后）时在此举行“颁诏”仪式。而如今只要有机会，你随时都能登上这座城楼，去眺望目前世界上最大的广场——天安门广场。

在晴朗的天宇下，城楼上有黄色琉璃瓦闪耀着灿烂的光辉，朱红的柱子和城台，白色的华表、石栏杆、石狮子、金水桥一一浮现。碧水青天，丹墙绿树，石栏黄瓦，画梁朱柱，色彩丰富，轮廓美丽，宏伟端庄。

站在城楼上，放眼望去，人民大会堂、人民英雄纪念碑、毛主席纪念堂、中国国家博物馆，这些气势轩昂的现代大建筑竖立在宽阔的广场上，使广场呈现出前所未有的新气象，庄严的布局、磅礴的气势，会使每一个中国人油然而生自豪之情。

历史的回声仿佛还在回响，历史的步伐依然在前行，雄伟壮观的天安门同共和国一道迎接着一个又一个新的黎明。

2）八达岭

八达岭长城位于北京延庆县，是开放最早的一段长城，城墙全长 3741 米，八达岭长城的战略地位非常重要，所以此段长城修筑工程非常宏大，城墙高大坚固，箭楼密集。城墙随着山峰的走势，蜿蜒起伏，如巨龙盘绕。

八达岭长城驰名中外，誉满全球。它是万里长城向游人开放最早的地段。“不到长城非好汉”。迄今。八达岭已接待中外游人一亿三千万，先后有尼克松、里根、撒切尔、戈尔巴乔夫、伊丽莎白、希思等 372 位外国首脑和众多的世界风云人物，登上八达岭观光游览。这种情况，在世界风景名胜景点中，实属罕见。八达岭长城给来访者，留下了深刻的印象和无穷的回味。

八达岭长城，作为万里长城的精华，正以古老而年轻的雄姿，迎接慕名而至、纷至沓来的天下游人。登过长城的人，莫不叹为观止。

3）九寨沟

没有一个到过九寨沟的人，能否认它超凡的魅力。人们说，如果世界上真有仙境，那肯定就是九寨沟。这是一个佳景荟萃、神奇莫测的旷世胜地；是一个不见纤尘、自然纯净的“童话世界”。“黄山归来不看山，九寨沟归来不看水”，九寨沟的精灵是水、湖、泉、瀑、滩连缀一体，飞动与静谧结合，刚烈与温柔相济，不愧为“中华水景之王”。

九寨沟位于在四川阿坝藏族羌族自治州境内，巍峨的岷山山脉深处，距成都400多公里，纵深40多公里，总面积6万多公顷，三条主沟形成Y形分布，总长度达60余公里。原来这里由于交通不便，几乎成了一个与世隔绝的地方。仅有九个藏族村寨坐落在这片崇山峻岭之中，九寨沟因此得名。这里保存着具有原始风貌的自然景色，因而有着自己的特殊景观。

九寨沟，因沟内有九个寨子而得名。这九个寨子又称为“何药九寨”。这里藏胞的语言、服饰和习俗，与四邻的藏胞有着明显的差异。据考证，他们的祖先原来生活在甘肃省的玛曲，属阿尼卿山脚下的一个强悍的部落，随松赞干布东征松州时留在了白水江畔。《唐书·吐蕃传》中记载了唐初叶蕃东征时，松赞干布以勇悍善战的河曲部为先锋，一举占领松州，后部分人马被留在了弓杠岭下。他们将原河曲的俄洛女神山的传说及部落出生传说均带到了九寨沟内。九寨沟的色嫫山名及蛾洛色莫的传说都源于河曲。

4）桂林山水

桂林市是世界著名的风景游览城市和历史文化名城，享有桂林山水甲天下之美誉。它地处广西壮族自治区东北部，东经109°36′～111°29′，北纬24°15′～26°23′，市辖秀峰、象山、七星、叠彩、雁山五城区和灵川、兴安、全州、临桂、阳朔、平乐、荔浦、龙胜、永福、恭城、资源、灌阳十二县，行政区域总面积27 809平方公里，其中市区面积565平方公里。

5）张家界

张家界市（原名大庸市）位于湖南省的西北部，总面积为9516平方公里，武陵山脉横贯市境南部，其气候属中亚热带山原型季风湿润气候，年平均气温16.8℃，四季宜人。早在五千多年前就已有人类活动。在古代，张家界被称为九洲以外，圣人听其自然的南蛮荒芜之地。土家族、苗族、白族等少数民族占全市总人口的百分之六十。在长期的历史发展中，形成了他们独特的、绚丽多彩的风情习俗。

张家界国家森林公园位于武陵源风景名胜区南部，公园管理处驻锣鼓塌，距张家界市城区32公里，距武陵源区人民政府驻地约28公里，均有公路通达。张家界地域古属朝天山，因明崇祯邑人张再弘“蒙恩赐团官”设衙署于此而得名。也曾称张家界为马鬃岭。1958年建立国营张家界林场。1982年经国务院批准为国家森林公园。

1994年由林业部命名为“国家示范森林公园”。境内峰密岩险，谷深涧幽，水秀林碧，云缭雾绕。集雄、奇、幽、野、秀于一体，汇峰、谷、壑、林、水为一色。有金鞭溪、黄石寨、琵琶溪、腰子寨、畲刀沟、袁家界等6个小景区游览线，已命名景点90多个。有标准石板游道6条，总长21.8公里；车行游道总长29.8公里。

张家界国家森林公园，有三千多座奇峰异石，似人似物，神形兼备，或粗犷，或细密，或奇绝，或诡秘，浑朴中略带狂狷，威猛中又带妖媚，危岩绝壁，雍容大气。张家界集山奇、水奇、石奇、云奇、动物奇与植物奇六奇于一体，汇秀丽、原始、幽静、齐全、清新五绝于一身，纳南北风光，兼诸山之美，是大自然的迷宫，也是中国画的原本！

6）黄山

位于安徽省南部的黄山，为中国十大风景名胜之一，1990年被联合国教科文组织列入《世界遗产名录》，蜚声中外，令世人难忘。明代旅行家、地理学家徐霞客两游黄山，赞叹说，“登黄山，天下无山，观止矣!”，又留下“五岳归来不看山，黄山归来不看岳”的美誉。黄山将以它雄奇的容貌迎接四海宾客。

黄山，古称黟山，唐天宝六年（公元747年）因轩辕黄帝曾在黄山炼丹羽化升天的传说，唐玄宗敕改黟山为黄山。它地跨市内歙县、休宁、黟县和黄山区、徽州区，面积1200平方公里，现划入黄山风景区的154平方公里，是号称“五百里黄山”的精华部分。

黄山是以自然景观为特色的山岳旅游风景区，奇松、怪石、云海、温泉素称黄山“四绝”，令海内外游人叹为观止。黄山有名可数的72峰，或崔嵬雄浑，或峻峭秀丽，布局错落有致，天然巧成。天都峰、莲花峰、光明顶是黄山的三大主峰，海拔高度皆在一千八百米以上，并以三大主峰为中心向四周铺展，跌落为深壑幽谷，隆起成峰峦峭壁，呈现出典型的峰林地貌。

黄山延绵数百里，千峰万壑，比比皆松。黄山松，它分布于海拔800米以上高山，以石为母，顽强地扎根于巨岩裂隙。黄山松针叶粗短，苍翠浓密，干曲枝虬，千姿百态。或倚岸挺拔，或独立峰巅，或倒悬绝壁，或冠平如盖，或尖削似剑。有的循崖度壑，绕石而过；有的穿罅穴缝，破石而出。忽悬、忽横、忽卧、忽起，“无树非松，无石不松，无松不奇”。

7）故宫

故宫位于北京市中心，旧称紫禁城，曾经是明清两代的皇宫。始建于明永乐四年（1406年），建成于永乐十八年（1420年），距今约有600年的历史。它占地面积72万平方米，共有各式宫室九千余间，是世界上规模最大、保存最完整的宫殿建筑群。明清两代先后有24位皇帝居住在这里。1924年北京政变后，被废黜的清末代皇帝溥仪出宫，1925年成立了故宫博物院。新中国成立后，故宫进行了大规模的修缮，成为我国最大的国家博物馆。

紫禁城宫殿保持着明代的布局，宫殿按前朝后寝的制度分外朝内廷两部分。外朝为皇帝和大臣们举行大典、朝贺、筵宴和行使权力的地方，建筑高大堂皇，以太和殿、中和殿、保和殿三大殿为中心，文华、武英两殿为两翼。内廷是皇帝处理日常朝政和后妃、皇子们居住、游玩和奉神的地方，以乾清宫、交泰殿、坤宁宫为中心，东西六宫为两翼，布局严谨有序。紫禁城中有御花园、慈宁宫花园，另有宁寿宫花园（也称乾隆花园）。四个城角都有，建造精巧美观。

紫禁城北有天然屏障万岁山（清改名景山），南有金水河，构成前水后山的格局。紫禁城宫城周围环绕着高10米，长3440米的宫墙，共有四个宫门，南为午门，东为东华门，西为西华门，北为玄武门（清改为神武门）。宫墙四角各有一设计独特、精巧玲珑的角楼——9梁18柱72条脊。宫墙外有52米宽的护城河，独成体系。

明清故宫（英文名称：Imperial Palace of the Ming and Qing Dynasties）于1987年根据文化遗产遴选标准C（Ⅲ）（Ⅳ）被列入《世界遗产名录》（编号：200-003）。世

界遗产委员会评价：紫禁城是中国五个多世纪以来的最高权力中心，它以园林景观和容纳了家具及工艺品的9000个房间的庞大建筑群，成为明清时代中国文明无价的历史见证。

8）西湖

位于杭州市的西湖，可谓一首诗，一幅天然图画，一个美丽动人的故事，不论是多年居住在这里的人还是匆匆而过的旅人，无不为这天下无双的美景所倾倒。阳春三月，莺飞草长。苏白两堤，桃柳夹岸。两边是水波潋滟，游船点点，远处是山色空蒙，青黛含翠。此时走在堤上，你会为眼前的景色所惊叹，甚至心醉神驰，怀疑自己是否进入了世外仙境。

而西湖的美景不仅春天独有，夏日里接天莲碧的荷花，秋夜中浸透月光的三潭，冬雪后疏影横斜的红梅，更有那烟柳笼纱中的莺啼，细雨迷蒙中的楼台……无论你在何时来，都会领略到不同寻常的风采。

西湖十景形成于南宋时期，基本围绕西湖分布，有的就位于湖上。苏堤春晓、曲苑风荷、平湖秋月、断桥残雪、柳浪闻莺、花港观鱼、雷峰夕照、双峰插云、南屏晚钟、三潭印月，西湖十景各擅其胜，组合在一起又能代表古代西湖胜景精华，所以无论杭州本地人还是外地山水客都津津乐道，先游为快。

新西湖十景是一九八五年经过杭州市民及各地群众积极参与评选，并由专家评选委员会反复斟酌后确定的，它们是：云栖竹径、满陇桂雨、虎跑梦泉、龙井问茶、九溪烟树、吴山天风、阮墩环碧、黄龙吐翠、玉皇飞云、宝石流霞。

9）东方明珠

东方明珠广播电视塔高468米，位居亚洲第一、世界第三，是新上海的标志性建筑，2001年5月开放的上海城市历史发展陈列馆成为新的亮点，形成了“上塔观新上海、下塔看老城厢、登船游母亲河”三位一体的都市旅游格局；控股的上海国际会议中心包括1家5星级宾馆、大小各异的多功能会议厅和一个目前上海规模最大的大宴会厅，2001年10月成功承办了上海APEC部分会议；东方明珠传输公司承担着整个上海地区无线广播和电视发射以及数据传输等任务，24小时不间断地把几十套广播电视节目传输到上海乃至华东部分地区的千家万户，并利用其独特优势，开拓传输增值业务；上海东方绿舟管理中心是集团管理品牌输出的成功范例，中心管理着占地面积达5600亩的上海青少年校外活动基地“东方绿舟”，已经成为上海又一个重要的青少年校外活动素质教育、旅游观光、休闲度假相结合的综合性旅游景观；集团投资教育产业，建设和管理了松江大学城学生公寓项目。

10）泰山

山东省的泰山在五岳中名声最著，其原因亦与秦汉之际开始的封禅活动有关。封建帝王的封禅活动是政治和迷信的混合物。封禅是一种祭祀性的礼仪活动，“封”是在泰山上堆土为坛，在坛上祭祀天神，报答上苍的功绩；“禅”是在泰山下扫除一片净土，在净土上祭祀土神，报答后土的功绩。古人认为“天以高为尊，地以厚为德”“天高不可及于泰山”，祈愿“天地交泰”。于是，凡所谓“受命于天”的帝王，为答谢天帝的“受命”之恩，便到接近天神的泰山之巅，积土为坛，增泰山之高以祭天，表示功归于天；然后，再到泰山之前近地祇的梁父、社首、云云等小山丘设坛祭地，表示厚上加厚，福广恩厚以报地。这就是历代帝王狂热追求的泰山封禅大典。一代帝王登封泰山，被视为国家鼎盛、天下太平的象征，皇帝本人也俨然成为“奉天承运”的“真龙天子”了。

历代帝王封禅泰山活动，有其开始和发展的过程。封禅泰山，大约可溯源于原始人群对自然山川的崇拜。司马迁在《史记·封禅书》中引《管子·封禅》说，“古者封泰山禅梁父者七十二家”。东汉哲学家王充在他的《论衡·书虚》中也说，“百王太平，升封太山，太山之上，封可见者七十有二，纷纶湮灭者，不可胜数”。司马迁从《管子》中找到名字的是十二位，他们是：无怀氏、伏羲氏、神农氏、炎帝、黄帝、颛顼氏、帝喾、尧、舜、禹、汤、周成王。这些人大都是古代比较强大的部落首领，是传说中的人物。《尚书·舜典》记载：“岁二月，东巡狩，至于岱宗。柴，望秩于山川。遂觐东后。”所谓“巡狩”是指游牧民的巡行狩猎。“柴”，烧柴火。“望”，是一种祭祀形式。这是一种原始的自然崇拜，与后世有目的的神道设教，告成功于天，以强调帝王统治权力的礼仪活动，有明显区别。

二、历史

中国历史悠久，考古证据显示224万年至25万年前，中国就有直立人居住，目前考古发现的有元谋人、巫山人、蓝田人、南京直立人、北京直立人等，这些都是目前所知较早的原始人类踪迹。而中国的华夏文明形成于黄河流域中原地区。早期的历史，口口相传。神话中有盘古开天地、女娲造人的说法。

传说中的三皇五帝，是夏朝以前数千年杰出首领的代表，具体而言有不同的说法。一般认为，三皇是燧人、伏羲、神农以及女娲、祝融中的三人，五帝一般指黄帝、颛顼、帝喾、尧、舜。自三皇至五帝，历年无确数，最少不下数千年。

最早的世袭朝代夏朝约在前21世纪到前16世纪，由于这段历史目前没有发现文字性文物做印证，所以只能靠后世的记录和出土文物互相对照考证。

之后的商朝是目前所发现的最早有文字性文物的历史时期，存在于前16世纪到约前1046年。

到了大约前1046年，周武王伐纣，在牧野之战中取得决定性胜利，商朝灭亡，西周正式建立。

前770年，由于遭到北方游牧部落犬戎的侵袭，周平王东迁黄河中游的雒邑（今河南洛阳），东周开始。此后，周王朝的影响力逐渐减弱，取而代之的是大大小小一百多个小国（诸侯国和附属国），史称春秋时期，春秋时期的大国共有十几个，其中包括了晋、秦、郑、齐及楚等。先后有五个国家称霸，即齐、宋、晋、楚、秦，合称春秋五霸。春秋战国时期学术思想比较自由，史称百家争鸣。出现了多位对之后中国有深远影响的思想家（诸子百家），例如老子、孔子、墨子、庄子、孟子、荀子、韩非等人。出现了很多学术流派，较出名的有十大家，即道家（自然）、儒家（伦理）、阴阳家（星象占卜）、法家（法治）、名家（修辞辩论）、墨家（兼爱非攻）、杂家（合各家所长）、农家（君民同耕）、小说家（道听途说）和纵横家（纵横捭阖）等。文化上则出现了第一个以个人名字出现在中国文学史上的诗人屈原，他著有《楚辞》《离骚》等文学作品。孔子编成了《诗经》。战争史上出现了杰出的兵法家孙武、孙膑、吴起等。科技史上出现了墨子，建筑史上有鲁班，同时此时中国的冶金也十分发达，能制造精良的铁器，在农业上出现了各种灌溉机械，大大提高了生产率，从而为以后人口大大膨胀奠定了基础。历史上出现了《春秋左氏传》《国语》《战国策》。中华文化的源头基本上都可以在这一时期找到。

前 221 年，秦并其他六国后统一了中国主体部分，成为中国历史上第一个统一的中央集权君主统治国家，定都咸阳（今西安附近）。由于秦王政自认“功盖三皇，德过五帝”，于是改用皇帝称号，自称始皇帝，人称秦始皇（图 1）。前 206 年刘邦围攻咸阳，秦王子婴自缚出城投降，秦亡。

图 1 秦始皇像

前 202 年 12 月，项羽被汉军围困于垓下（今安徽灵璧），四面楚歌。项羽在乌江自刎而死。楚汉之争至此结束。汉高帝刘邦登基，定都长安（今陕西西安），西汉开始。

25 年刘秀复辟了汉朝，定都洛阳，史称东汉，而他就是汉光武帝。

220 年，曹操逝世，长子曹丕废汉献帝自立，建立魏国，定都洛阳，同时还有刘备的蜀汉定都成都和孙权的吴定都建业（今南京），历史进入了三国时期。

266 年，司马昭之子司马炎称帝，建立晋朝，定都洛阳，史称西晋。

从 304 年到 409 年，北部中国陆陆续续有多个国家建立，包括了成汉、前赵、后赵、前燕、前凉、前秦、后秦、后燕、西秦、后凉、北凉、南凉、南燕、西凉、夏和北燕，史称十六国。

581 年，杨坚取代北周建立了隋朝，定都长安。

618 年，唐高祖李渊推翻隋朝建立了唐朝，它是中国历史上延续时间最长的朝代之一。626 年，唐太宗李世民即位，唐朝开始进入鼎盛时期，史称贞观之治。

从公元 907 年朱温灭唐到 960 年北宋建立，短短的五十四年间，中原相继出现了梁、唐、晋、汉、周五个朝代，史称后梁、后唐、后晋、后汉、后周。同时，在这五朝之外，还相继出现了前蜀、后蜀、吴、南唐、吴越、闽、楚、南汉、南平（即荆南）和北汉等割据政权，这些政权统称“十国”。这就是中国历史上的“五代十国”。

960 年北宋建立后控制了中国大部分地区，到 1127 年（靖康二年）金国攻破北宋首都汴京（今河南开封），俘虏三千多皇族，其中包括了当时的皇帝宋钦宗和太上皇宋徽宗，因为钦宗其时的年号为靖康，史称靖康之难，北宋灭亡。同年宋钦宗的弟弟赵构在南京应天府（今河南商丘）即皇位，定都临安（今浙江杭州），史称南宋，偏安江南。1271 年忽必烈建立元朝，定都大都（今北京）。元军于 1279 年与南宋进行了崖山海战，陆秀夫背着 8 岁的小皇帝赵昺惨烈地跳海而死。崖山海战以元朝的胜利告终，南宋随之灭亡。

1368 年，农民起义军领袖朱元璋推翻元朝并建立了明朝。

明朝晚期，居住在东北地区的满族开始兴盛起来，终于在 1644 年攻克北京后不久，驱逐李自成，进入北京，建立了清朝。

中华民国，1912—1949 年，定都南京。

中华人民共和国 1949 年 10 月 1 日成立，首都北京。

中国各朝代的时间划分表如表 2 所示。

表 2　中国各朝代的时间划分表

朝代		起讫年代	都城	今地	开国皇帝	备注
夏朝		约前 2146—前 1675 年	阳城	河南登封	禹	亡于商
商朝		约前 1675—前 1046 年	亳	河南商丘	汤	亡于周
周	西周	约前 1046—前 771 年	镐	陕西西安	周武王	（亡于犬戎）东迁
	东周	前 770—前 256 年	雒邑	河南洛阳	周平王	亡于秦
	春秋	前 770—前 476 年				
	战国	前 475—前 221 年				
秦朝		前 221—前 207 年	咸阳	陕西咸阳	始皇帝赵政	亡于汉朝
汉	西汉	前 206—公元 8 年	长安	陕西西安	汉高帝刘邦	
	新朝	9—23 年			王莽	
	东汉	25—220 年	雒阳	河南洛阳	汉光武帝刘秀	亡于魏
三国	魏	220—265 年	洛阳	河南洛阳	魏文帝曹丕	亡于晋
	蜀汉	221—263 年	成都	四川成都	汉昭烈帝刘备	亡于魏
	吴	222—280 年	建业	江苏南京	吴大帝孙权	亡于晋
晋	西晋	265—316	洛阳	河南洛阳	晋武帝司马炎	亡于赵
	东晋	317—420	建康	江苏南京	晋元帝司马睿	亡于刘宋
十六国		304—439				
南朝	宋	420—479	建康	江苏南京	宋武帝刘裕	亡于齐
	齐	479—502	建康	江苏南京	齐高帝萧道成	亡于梁
	梁	502—557	建康	江苏南京	梁武帝萧衍	亡于陈
	陈	557—589	建康	江苏南京	陈武帝陈霸先	亡于隋

续表

朝代		起讫年代	都城	今地	开国皇帝	备注
北朝	北魏	386—534	平城	山西大同	魏道武帝拓跋珪	分裂成东西魏
	东魏	534—550	邺	河北临漳	魏孝静帝元善见	亡于齐
	西魏	535—556	长安	陕西西安	魏文帝元宝炬	亡于周
	北齐	550—577	邺	河北临漳	齐文宣帝高洋	亡于周
	北周	557—581	长安	陕西西安	周孝闵帝宇文觉	亡于隋
隋朝		581—618	大兴	陕西西安	隋文帝杨坚	亡于唐
唐朝		618—907	长安	陕西西安	唐高祖李渊	亡于后梁
五代十国	后梁	907—923	汴京	河南开封	梁太祖朱晃	亡于后梁
	后唐	923—936	洛阳	河南洛阳	唐庄宗李存勖	亡于后晋
	后晋	936—947	汴京	河南开封	晋高祖石敬瑭	亡于契丹
	后汉	947—951	汴京	河南开封	汉高祖刘暠	亡于后周
	后周	951—960	汴京	河南开封	周太祖郭威	亡于宋
宋	北宋	960—1127	开封	河南开封	宋太祖赵匡胤	亡于金
	南宋	1127—1279	临安	浙江杭州	宋高宗赵构	亡于元
辽国		907—1125	皇都	辽宁	辽国耶律阿保机	亡于金
大理		937—1254	太和城	云南大理		
西夏		1032—1227	兴庆府	宁夏银川		亡于大蒙古国
金国		1115—1234	会宁	黑龙江阿城	金太祖完颜旻	亡于大蒙古国
元朝		1206—1368	大都	北京	元世祖忽必烈	亡于明
明朝		1368—1644	北京	北京	明太祖朱元璋	（前期定都南京，朱棣迁往北京）亡于清
清朝		1616—1911	北京	北京	皇太极	（皇太极迁盛京辽宁沈阳、顺治迁往北京）亡于民国
中华民国		1912—1949	南京	南京		

中华人民共和国 1949 年 10 月 1 日成立。
首都北京。

三、改革开放与经济发展

1. 概况

1949 年新中国的成立使中国彻底摆脱了被压迫的境地。东方睡狮开始慢慢觉醒，但却步履维艰。直到 1978 年，国家全面实行改革开放的新决策，从此改革开放的春风使中华大地再次焕发了活力，中华民族终于踏上了民族复兴的伟大征程！33 年的征程，中华民族以崭新的姿态重新屹立于世界民族之林；33 年的沧桑巨变，33 年的光辉历程，成就了一个民族近百年的梦想：中国强盛了！

改革开放的33年，是中国农业完美突破的33年！中国恢复了以前的包产到户，以人性为本，广大的农民开始为自己劳动，农民们更加充满干劲，生产积极性高涨，加上杂交水稻技术的发明和推广，农民们富裕了起来，粮食产量也不断提高，每亩高产800多公斤已不再是不可能！

改革开放的33年，是中国经济迅速蓬勃的33年！幢幢高楼拔地而起，人民生活水平不断提高，1978年到2010年间，中国经济总量迅速扩张，国内生产总值从3645亿元（0.243万亿美元）增长至397 983亿元（6.04万亿美元），增长近110倍！中国的经济成就不仅写在了中国历史之上，也在世界历史上刻下了辉煌的一页！

改革开放的33年，是中国社会和谐稳定的33年！自粉碎“四人帮”以后，中华民族犹如钢铁长城一般坚不可摧！1997年香港回归，1999年澳门回归；1998年面对南方历史罕见的特大洪水，2003年面对让人闻风丧胆的非典疫情，2008年面对十几个省份百年不遇的冰雪灾害，四川汶川大地震，中华儿女众志成城，手挽手将一个个磨难阻击在脚下！

改革开放的33年，是教育事业稳步发展的33年！1983年，邓小平同志提出，教育要面向现代化，面向世界，面向未来！高考制度恢复之初，全国有570万人参加高考，却仅录取27万名；而到2010年，全国普通高校招生报名人数达到946万！伴随着教育规模的发展，更有越来越多的中华儿女在世界高精尖人才中占据着日益重要的位置！

改革开放的33年，是中国航天事业不断创新的33年！从1979年远程火箭发射试验成功，到2003年“神五”升天，首次载人航天飞行成功，再到2005年神舟六号载人航天卫星顺利返回，中国航天人在摸索中让祖国一跃成为航天科技强国！2010年，我国第二颗探月卫星“嫦娥二号”发射升空，中华民族的千年奔月梦成为现实！“神七”的成功发射，“神七”在太空漫步，让中国人第一次在太空留下了自己的足迹！

改革开放的33年，也是我国体育事业蒸蒸日上的33年！1984年许海峰摘得中国奥运首枚金牌，自此之后，中华体育健儿奋勇争先：2000年悉尼奥运，中国代表团收获28枚金牌，取得了金牌榜和奖牌榜均名列第三的佳绩；2004年雅典奥运，中国军团更是将金牌总数扩增到32枚，位列金牌榜第2位！而2008年，奥运大幕在中华大地上拉开，我们成为奥运的主人！51枚金牌，稳居金牌榜第一位。

2. 改革开放33年的十大成就

1）建立全面物质生产体系

中国改革开放33年，保证了和平稳定，建立起全面的物质生产体系，经济建设取得显著成就。中国已经由初级工业经济转变为高级工业经济，包括钢铁、家用电器在内的许多工业产品生产居世界第一位。与此同时，中国经济规模和经济总量也不断扩大。

2）国际地位持续不断提高

中国改革开放33年，中国的国际地位不断提高。快速经济增长使中国在世界经济中的地位不断上升。

3）全面融入世界经济体系

以加入 WTO 为标志，中国经济已经完成市场化和国际化进程，融入世界经济体系和经济全球化浪潮之中。

4）社会经济取得全面进步

5）经济增长变得更加稳健

中国的改革开放释放出巨大的生产力，政府主导、大力投资和不断强化的工业经济使中国经济增长一直高于世界经济增长水平。国家统计局发布的 2010 年国民经济和社会发展统计公报（下称统计公报）显示：2010 年国内生产总值（GDP）为 39.798 3 万亿元（相当于 6.04 万亿美元），为 1978 年 GDP 总值 0.365 万亿元的 109.036 倍。

6）经济发展水平不断提高

中国改革开放不断深入的同时，经济发展水平大幅度提高。1978 年中国人均 GDP 为 381 元，按照 1980 年 1 美元兑换 1.53 元人民币汇率计算，约合 249 美元。2010 年中国人均 GDP 上升到 4361 美元，比改革开放前增长了 16.5 倍。

7）人民生活水平显著改善

8）教育发展取得长足进步

教育发展是衡量一个国家发展水平和发展潜力的重要指标。改革开放 33 年，中国教育发展取得长足进步。1978 年中国普通高等学校毕业生数只有 16.5 万，占当时中国人口总量 96 259 万的 0.0171%。2010 年中国普通高等学校毕业生数达到 531.1 万，占当年中国人口总量 13.6 亿的 0.3905%，33 年间增长了 21.83 倍。

9）国民预期寿命明显提高

预期寿命是衡量一个国家社会经济发展的综合指标，预期寿命提高不但意味着经济发展水平提高，也意味着社会保障能力的提升。根据中国人口普查数据，1982 年中国人口平均预期寿命是 67.77 岁，2010 年中国人口平均预期寿命上升到 72.50 岁，增加了 4.73 岁。

10）人民生活更加丰富多彩

中国人民的物质生活和文化生活已经进入一个与世界同步的时代。

3. 改革开放取得成就的原因

改革开放以来我们取得一切成绩和进步的根本原因，归结起来就是：形成了中国特色社会主义理论体系，开辟了中国特色社会主义道路

改革开放 33 年来，中国共产党在建设中国特色社会主义的实践中，不断深化对执政规律、社会主义建设规律和人类社会发展规律的认识，从“发展是硬道理”到“发展是第一要务”，再到科学发展观，执政理念不断丰富和发展，中国特色社会主义建设取得了巨大成就，社会生产力得到了空前发展，人们的物质生活得到了极大丰富，战胜自然灾害的能力也越来越强。

中国 33 年的改革开放取得了巨大成功，基本经验在于中国遵循了经济增长“四色定理”。中国和世界发展历史充分证明，经济增长“四色定理”——和平稳定、开放结构、人力资本、结构增长是经济增长的充分必要条件，也是中国改革开放的基本经验。遵循“四色定理”，经济就发展，社会就进步。

4. 改革开放33年之思想转变

中国改革开放33年在经济发展取得巨大进步的同时，社会经济思想理论也发生了积极转变。

第一，国际战争理论转向国际和平理论，中国进入和平发展的年代。对于世界形势的判断决定一个国家的基本战略，20世纪前中期是美苏争霸的冷战时代，国际战争危险依然存在。进入20世纪80年代，世界政治发生重要变化。邓小平同志高瞻远瞩，审时度势，正确提出和平是世界主流的判断。从而为中国和平发展、改革开放提供了重要理论依据。

第二，计划经济理论转向市场经济理论，中国开始全面的市场经济改革。新中国建立在社会主义计划经济基础之上，计划经济缺乏市场竞争，也缺乏市场灵活性，使得中国经济发展面临许多困难。计划经济理论转向市场经济理论使中国经济变得更有活力，也为中国的改革开放奠定了理论基础。

第三，自力更生理论转向对外开放理论，中国经济全面走向世界。自力更生是中国社会主义建设和经济发展的重要理论，也使中国建立起相对独立的经济体系。然而，一个国家的资源和市场总是有限的，转向开放无疑会增加发展的资源与空间，对外开放理论使中国走向世界。随着改革开放，涌现出许多国际合作理论，许多跨国公司开始大量进入中国。

第四，社会公平理论转向社会竞争理论，社会分配差距不断扩大。中国的社会主义理论重要基础之一是公平理论，然而在生产力水平较低的时候，强调分配公平在一定程度上影响了生产者的劳动积极性。由公平理论转向竞争理论是提高经济效率的需要，也是由计划经济转向市场经济的逻辑基础。

第五，农村化理论转向城市化理论，城市人口越来越多。中国曾一度出现农村化理论，大量城市人口随着上山下乡从城市流向农村。随着改革开放，城市化理论成为主流，越来越多的农民进入城市。

5. 改革开放33周年感想

33年，对于历史长河就那么短短的一瞬间，然而，对于我们这样一个从贫穷落后一步步走向发达富裕文明和谐的国家来说，又是一个丰富而值得铭记的过程。

33年前的那个春天，一位颇具设计天才的老人邓小平先生，为中国社会主义现代化建设设计出了一条宽广坦途——实行改革开放，实事求是，解放思想，转移工作重心，放到经济上来。于是，1978年，我们成功召开了党的十一届三中全会。1979年，设立了经济特区。1982年，实行了家庭联产承包责任制。1986年，启动了全民所有制企业改革。1987年，提出了“一个中心、两个基本点”基本路线。1992年，确立了社会主义市场经济体制改革目标。1992年、1994年，施行了医疗及住房市场化改革。1997年，提出了党在社会主义初级阶段的基本纲领。1999年，吹响了西部大开发战略的冲锋号。2001年，中国正式成为WTO成员。2002年，确定了全面建设小康社会的奋斗目标。2005年，废止了农业税条例，提出了建设社会主义新农村的重大历史任务。2006年，做出了构建社会主义和谐社会的重大决定。2007年，科学发展观被写入党章……

33年改革开放，33年功勋卓著。33年辉煌成就有力地证明：以改革开放为动力，走有中国特色的社会主义道路的决策，是最科学的、最正确的、最有利于广大人民群众的决策。33年来，从农村到城市，从农民到市民，从农田到工厂，从森林到牧场……先进代替了落后，机械取代了人工，旧貌换新颜。我们体验了收复港澳的快慰，“一国两制”的成功，三峡工程建设的壮观，神舟号飞船升天的壮举，西部开发的激情，奥运成功举办……

我们一次又一次将社会主义经济建设推上新的高潮，一个又一个宏伟计划交替闪烁在电脑屏幕上，一扇又一扇窗口争先恐后向大海开放，一条又一条高速公路马不停蹄地追逐着车轮的速度。

33年后的今天，我们伟大的祖国阔步前进在建设中国特色社会主义道路上，经济发展、政治稳定、社会进步、民族团结、各项事业兴旺发达蒸蒸日上。

“雄关漫道真如铁，而今迈步从头越”，我们欣喜地看到，在新的历史条件下，继续把改革开放伟大事业推向前进，让我们的中国更美好！

6.33年中国外交理念之演变

33年来中国在外交领域非常之活跃，通过不断提出新主张、新倡议，来达到优化环境、扩大影响、树立形象和趋利避害的目的。

可以说，如今中国的外交理念已形成一个体系，其基石是和平共处五项原则，杠杆是独立自主的和平外交政策，旗帜是和平、发展、合作，远景是推动建设持久和平、共同繁荣的和谐世界，道路是和平发展，办法是互利共赢。

继承传统，与时俱进。

中国外交理念变化的第一特征，是在继承传统的基础上与时俱进。

中国率先于1953年酿成，随后于1954年与印度和缅甸共同提出和平共处五项原则。这些原则反映了二战结束之后广大新独立国家维护主权和建立国际新秩序的共同愿望。在过去33年里，中国领导人在传承与弘扬五项原则的基础上，力图开拓创新。

其一，奉行独立自主的和平外交政策。“对内和谐发展，对外和平合作”，这是把和平共处五项原则政策化，其中包含睦邻友好、互利合作的传统主张。随后的发展是：“睦邻”派生出“以邻为伴、与邻为善”；“友好”派生出各种说法的“伙伴关系”如“战略协作伙伴关系”“全面合作伙伴关系”；“互利”派生出“双赢、共赢”；“合作”被提升到与时代主题“和平与发展”并列的高度。中国外交中还注入了富有感染力的文化传统因素如“和为贵”“和而不同”“以人为本”，这些反映了中国人民与世界各国人民“同声相应，同气相求”。

其二，从不同角度确立对象国在中国外交中的地位，叫作周边国家关系是首要，大国关系是关键，发展中国家关系是基础。中国奉行全方位外交，提出要积极营造和平稳定的国际环境、睦邻友好的周边环境、平等互利的合作环境、互信协作的安全环境和客观友善的舆论环境。将处理国家关系的原则完整化，包括国家利益原则、平等互利原则、不结盟原则、反对霸权主义原则、不以意识形态划线原则、共同发展原则、面向未来原则、多样性原则和联合国核心作用原则。中国从原先重视发展双边关系，到后来越来越积极参与多边活动。

其三，坚持走和平发展的道路。和平发展的意思，是要营造一个良好的周边环境与国际环境，在安全、稳定的条件下加速社会经济建设；是要本着以人为本的目的，造福于中华民族，同时造福于全人类；是要采取和平的方式并通过和平的途径，全面建设小康社会，避免来自内外的冲突和对抗；是要维护世界和平，促进人类进步，亦即中国几代领导人所倡导的“中国应当对于人类有较大的贡献”。中国之所以选择和平发展的道路，是由中华民族的根本利益、中国的社会主义性质和时代发展潮流几方面因素决定的。

其四，提出推动建设以“持久和平、共同繁荣”为内涵的和谐地区、和谐亚洲、和谐世界的主张。

谢谢！

系统工程与领导科学①

编写《系统工程干部读本》很有必要。系统工程与干部和干部工作的关系非常密切，从理论上说，干部工作就是系统工程，干部应该就是系统工程工作者。

现在很多人讲“领导科学”。其实，系统工程是领导科学的重要组成部分。干部工作，尤其是领导干部的工作，一是复杂，有时候难以找到突破口；二是变化快，经常出现“计划跟不上变化”的情况。系统工程基本原理与方法论能够助他们一臂之力。

编写者认为：系统工程中国学派已经形成，这就是钱学森学派。该书力求准确地按照钱学森学派来阐述系统工程基本原理与方法论。

系统工程的基本原理与方法论并不深奥难懂，反而一学就会。20 世纪 80 年代，系统工程培训班在全国普遍举办，发挥了巨大的作用。今天，如果能够“梅开二度”，再次普遍举办系统工程培训班，必将再次发挥巨大作用。关键是要有好的教材，能够准确而深入浅出地阐述系统工程基本原理与方法论。该书的作者们具有自觉的使命感。

《系统工程干部读本》的主干“原理篇”包含 8 章，抓住了系统工程的基本内容、核心内容，而且有一些特色内容。例如，“发展战略与规划研究”（第 5 章）对于任何社会经济系统都是首要任务，是各级干部尤其是领导干部需要重视的大事；“信息系统工程”“农业系统工程”两章介绍了两个重要的系统工程专业；第 8 章“系统工程的发展前景”，提出“系统工程将永葆青春”，在第 8.4 节“系统工程概要 ABC”中，归纳了 60 多个基本命题，提纲挈领地涵盖了系统工程的基本内容。作者说：“这样归纳的目的，是为了‘简明扼要，记得住，用得上’。”可谓是“用心良苦”，值得嘉许。

系统工程理论与方法还要继续深入研究，尤其是复杂巨系统的运作机制及其研究方法，包括数学模型与求解方法，可能要用到比较高深的数学知识。这是一方面。另一方面，已有的系统工程基本原理与方法论已经可以解决许多问题，关键在于联系实际，加以宣传、普及与应用。就后者而言，这些知识应该像加减乘除四则运算一样，像九九表一样，为干部与群众所熟知，可以信手拈来，运用自如。

该书的编写者是三位老教授，他们的年纪都在“奔七”或“奔八”了。他们是“系统工程老战士”——资深系统工程工作者，对于系统工程一往情深，在系统工程领域从事教学、研究和应用工作 30 多年，指导了较多的硕士研究生和博士研究生。他们有较好的系统工程素养，在系统工程学术界具有较高的声誉。他们编写系统工程教材具有丰富的经验，先后编写和出版了多种系统工程教材。他们不但对大学生和研究生开展过大量教学工作，也从事过较多的系统工程和管理的干部培训工作。他们还开展过系统工程应用项目研究，取得了出色的成果。

① 本文是《系统工程干部读本》序，广州：华南理工大学出版社，2012，1-2.

关于实行九年一贯制办校的建议①

教育是中华民族振兴和国家富强的基石。百年大计，教育为本。在教育中，基础教育承担着培养国民素质的重任，是建设创新型国家的基础[1,2]。2008 年，我国城乡小学和初中义务教育已全面实现，覆盖全国城乡 1.6 亿名义务教育阶段学生，成绩显著。现在规定的义务教育是九年，即小学六年、初中三年。现有的办学模式继承着过去的传统方法，基本上仍是初中与小学独立办学。据我所见，这种办学模式存在如下弊端。

一、不利于社会的稳定

现在全国的小学升初中，“择校热”高烧不退。虽然教育行政部门要求就近升学，免试入学，划片入学，但在实际升学过程中，许多家长为了把孩子送到有名的学校，上各种辅导班、特长班（如有名的奥数班）来提高孩子知识的含金量，并动用各种关系上有名的学校，美其名曰，不要输在起跑线上。而学校为了招收好的学生搞择优录取，更助长了这一风气，最终形成了新的教育不公平现象，显然不利于社会稳定。

二、不利于社会和谐

在“择校热”的背后，下列问题应运而生：大量的辅导机构、教师的家教市场、不透明的择校费用、托关系的请客送礼等，给学生增加了不应有的学习负担，给家长增加了经济负担，给学校增加了危险负担，给中间环节增加了利益，甚至导致腐败现象发生，这些都对社会和谐造成了影响。

三、不利于人才的健康成长

小学和初中基础教育是未来人才的摇篮，其目的是培养孩子的基本素质，包括品行道德和基础知识两个方面的内容，十分关键，十分重要。由于“择校热”，小学教学应试行为严重，不利于学生品行道德的培养，不利于学生学习兴趣的培养，不利于学生学习能力的培养。小学教师为了追求考试成绩，教育管理机构为了评价小学的教学成绩，组织小学毕业考试。中学看重小学毕业考试成绩，用毕业成绩对学生进行筛选，造成小学过于注重考试成绩，因而加大作业量，增加考试频率，延长学生每天在校时间，使小学生苦不堪言。这样培养的人才，是模仿型人才，是听话型人才，是考试型人才，是对学习失去兴趣的人才。造成孩子们从小没有创新意识，没有独立思考意识，因而使素质教育落空，仅停留在口头上。显然，不利于人才的健康成长。

四、不利于教育教学模式的延续

教育的属性决定了其长期性和延续性，小学和初中分段管理和实施，容易造成各

① 本文原载《广东参事馆员建议》，2012，(52)：1-3；《教育与科技管理研究》，北京：科学出版社，2016，254-256.

办各的学，课程的设置和安排不尽合理，不利于教育循序渐进地发展。

五、不利于政府办学

学校布局不科学筹划，不尽合理，不能按地域、按人口居住区域合理分布学校，造成学生上学路程不均衡。各学校投入不均衡，硬件设施差距过大。教师队伍不稳定，教师想到好的学校去，不安心在薄弱学校。名校老师的工作量明显加大，不利于教师身心健康。教育行政部门对学校的评价不好实施，管理职能发挥不理想。因而造成政府办学费力不讨好，群众意见大。

六、不利于学校的发展

名校压力越来越大，招生人数越来越多，班容量越来越大。有的初中学校班容量竟达 80 多人，严重不符合国家 50 人以下的标准。教学硬件、软件跟不上快速增加的学生人数，教育质量下滑，安全隐患增多，名校最终也要被拖垮。薄弱学校办学的积极性受到影响，招生人数锐减，老师多，学生少，生源差，老师工作没有积极性，教育资源无端闲置，造成浪费。有的学校最终因学生人数少，不得不撤并，造成新的教育布局不合理。

鉴于上述原因，为了保证义务教育顺利实施，保证教育的公平性，保证人才的健康成长、快乐成长，建议广东省率先将小学和初中合并成一个学校，实行九年一贯制教育，进而推广到全国各省、市。这是发展之需和兴国之要。

小学与初中合并后，就不再有小学升初中的问题，这就从根本上解决了“择校热”的问题，而且更能提高人才培养的质量。在合并时，要先将学校布局合理，特别要注意山区和农村地区的学校设置，要注意课程设置的合理性，要注意硬件的保证，合理安排师资，保证老师的工作和生活条件，保证人人有学上，确保起点公平。

参 考 文 献

[1]　刘佳炎．中小学教育与人才培养//中共中央组织部人才工作局．百名专家谈人才．北京：党建读物出版社，2012：138-139.

[2]　刘人怀，郭广生，徐明稚，等．试答“钱学森之问”．中国高校科技，2011，(10)：4-7，14.

2012年中国工程院院士广州咨询活动中心工作总结①

2012年，在中国工程院和广州市政府的领导下，在广州市科技和信息化局和有关部门的大力支持下，中国工程院院士广州咨询活动中心（以下简称“院士中心”）坚持科学发展，以服务广州地区院士为宗旨，以促进院地合作为目标，开展了一系列院士工作，完成了计划任务，积累了经验，进一步扩大了影响，取得良好的社会效果。全年共计邀请到本地及外地的院士专家共计71人（次）来广州地区开展系列服务。现将本年度工作情况总结如下：

一、百名院士广州行的情况及成效

“2012年百名院士专家广州行”（以下简称“院士专家广州行”）自5月启动以来，院士中心已成功举办了8场大型院士活动，邀请到45人（次）院士和全国知名专家，院士沙龙5人（次），院士企业行5人（次），高端论坛2场，科学讲堂4期，参加总人数达约3000人，包括广州电视台、广州日报、羊城晚报、信息时报、南方网等在内的共计9家主流媒体进行全程报道。

（一）院士行启动仪式

5月科技活动周期间，经多次策划和部署，在中国工程院及在穗两院院士的大力支持下，院士中心先后邀请北京大学周其仁教授、清华大学于永达教授、刘人怀院士、钱清泉院士、罗锡文院士、刘颂豪院士等6位院士、专家分别参加了“广州民营科技企业发展高端论坛”、“集聚优势与广州科技创新专题报告会”和“2012年百名院士专家广州行自启动仪式暨科技进步与创新高端讲堂”3场活动，分别围绕民营经济面临的挑战和机遇、如何充分发挥好科技创新的支撑引领作用以及广州产业转型升级等热点和难点问题，共同探讨在新形势下民营经济面临的挑战、机遇和对策，并为广州的新型城市化建设、产业结构调整转型、战略性新兴产业发展等方面出谋划策、指导建言。全市各界代表共有1300余位听众参加了以上活动。

（二）院士沙龙

为配合广州市将生物与健康产业列入“十二五”时期重点发展的战略性新兴产业，加强广州生物与健康产业的产学研金合作，院士中心受市科信局委托，邀请姚新生院士参加第九届广州生物医药沙龙并作专题演讲，为打造千亿级生物与健康产业集群出谋献策。

11月杨胜利院士应院士中心邀请，赴穗参加万孚院士沙龙，围绕POCT（快速现场检验）产业发展战略方向和在未来发展规划中企业如何占先机等主题与企业领导、技术骨干进行深入讨论和交流，杨胜利院士还为企业提出战略性发展思路，极大地帮助了企业进一步在行业竞争中提高竞争力。

① 本文原载《岁月留声——院士活动剪影（八）》，中国工程院院士广州咨询活动中心编，2012-12-31，1-2.

11月上旬，院士中心邀请刘人怀院士参观考察中新知识城，同时与中新知识城共同举办中新院士沙龙，刘人怀院士与中新知识城的有关领导共同探讨园区创新管理问题。

11月下旬，高分子化工专家毛炳权院士来广东参加系列学术讲座，院士中心邀请毛炳权院士与华工、中大的高分子材料专家和广州高分子材料企业一同座谈，毛炳权院士将多年心得经验与大家分享，并对广州的化工企业治污提出了中肯的意见和宝贵的建议。

12月中旬，院士中心邀请到我国空间物理学家，中国科学院院士魏奉思以及国家空间天气监测预警中心主任王劲松教授来穗，参加由市科信局、中国地球物理学会空间天气专业委员主办、院士中心和广州气象卫星地面站承办的"第二十九期广州院士沙龙"。魏奉思院士作关于"空间天气与广东经济建设的关系"的专题报告，王劲松教授结合广东地区情况介绍了空间天气对地方经济的影响实例，他们从专业角度介绍了中国空间天气技术发展方向和广阔的应用前景。

院士中心为策划组织每期院士沙龙活动，均做了大量前期调研和准备，包括积极主动地与企业和院士多方面沟通，努力搭建良好的院地交流平台，力求充分利用国家级的高端科技智力资源，为广州区域经济发展提供新的创新思路，推动相关专业的产学研合作，尽可能在产业发展方面选准方向、优先方法和路径，并落到实处。

（三）院士咨询和院士座谈会

2012年，市科信局召开《广州市科学技术普及"十二五"发展规划》（征求意见稿）征求意见会，院士中心根据市科信局要求，邀请刘人怀院士参加意见会，发挥院士专家智慧库作用，为广州新型城市化建设助力。

中秋时节，院士中心利用广州地区院士中秋联谊会的时机，举办广州新型城市化发展和建设智慧广州院士座谈会，邀请17位广州地区院士与市委、市政府领导和市科信局领导以及相关部门领导共同畅谈对广州智慧城市建设的意见和建议，同时增进在穗院士与政府之间及院士之间的交流，了解院士的实际需求。

（四）科学大讲堂

至2012年11月，院士中心协助"珠江科学大讲堂"顺利举办5期，邀请邬贺铨、杨胜利、钟世镇、李立涅、魏奉思等五位院士参加，分别以"智慧城市与广州新型城市化"、"转化医学的现状与发展"、"智能电网是智慧城市的物理基础"、"空间天气与广东经济建设的关系"四个主题作报告，通过"珠江科学大讲堂"有效推动科学技术普及，加强科普宣传，促进建设国家创新型城市。讲堂通过院士中心的努力，邀请国内外一流的专家、学者来穗演讲，面向公众开展多学科领域的科普讲座，为公众提供更多了解社会热点、科技时事和科技发展前沿的机会，搭建一个高端的、面向公众的科学讲坛，全力提升广州科普工作的影响力，为广州新型城市化发展营造良好的创新氛围。

（五）院士企业行

院士企业行关注广州重点产业需求，院士中心与企业和院士经多次磋商，确定具体的院士行方案，邀请对口院士参加。由于院士中心组织举办的院士企业行针对性强，产学研合作成效大，对企业创新发展帮助力度大，深受企业青睐，下半年已有多家企业提交院士行需求征询表，希望通过院士中心邀请到对口院士为企业创新发展助力献策。院士中心积极主动联络院士及相关地区兄弟单位，并协助广州企业与院士之间互动沟通，为进一步合作创造机会和搭建平台。

9 月 19 日，院士中心邀请杨胜利院士到广州万孚生物技术有限公司参观考察，并在技术创新、可持续发展产学研合作方面开展咨询活动。

12 月，院士中心受广州毅昌科技股份有限公司委托，邀请到中国工程院院士、上海交通大学阮雪榆教授来穗开展“模具及三维智能设计专题”院士企业行。阮雪榆院士实地参观考察了毅昌科技创新大楼和生产流水线，并与集团领导和技术团队进行了深入充分的座谈。在参观座谈过程中阮雪榆院士多次赞扬毅昌科技在中国工业设计领域的积极进取和突出贡献，鼓励广州民营企业努力创新，向智能化、集团化发展，真正做到“中国创造”。阮院士还与企业达成进一步深入合作的意向。

百名院士行中每场院士活动均邀请多家广州主流媒体和门户网络同期报道，进行全程报道的有包括广州电视台、广州日报、羊城晚报、信息时报、科技日报、新快报、南方网、网易新闻、广州电台等共计 9 家主流媒体，不仅发挥主流媒体的宣传引导作用，还充分利用网络媒体巨大的传播力量，进一步扩大院士活动的影响范围和影响力度，使院士中心的工作价值得到进一步提升。

二、配合和开展联谊活动，促进营造“尊重劳动、尊重知识、尊重人才、尊重创造”的良好社会氛围

2012 年 1 月 19 日，王东副市长在市科信局领导和院士办领导的陪同下，看望了部分在穗工作的中国科学院、中国工程院院士，为在穗工作的两院院士送去了美好的新春祝福。

院士中心根据广州市委市政府要求，邀请六位在穗院士代表参加 2012 年广州各界人士迎春招待会。

9 月 20 日，院士中心成功举办了市政府中秋院士慰问联谊活动，共邀请 17 位院士及其夫人到场共庆中秋，会上院士们与市领导自由交流，畅谈工作、展望未来，共庆佳节。市委常委、副市长张骥出席会议并讲话，代表市委、市政府送上节日的祝福和问候。市委组织部副部长、市人才办主任李志昌，市科信局领导、市人社局领导等出席了联谊会。院士中心还专程登门拜访和慰问 20 余位未能到场的在穗院士，将市委、市政府的中秋慰问传达给各位院士。院士们纷纷表示将一如既往地配合和支持广州的科技发展和社会建设。

三、参加两院院士大会及院地合作座谈会，及时了解最新政策

院士中心办公室主任薛峰同志受中国工程院邀请，于 6 月 11 日赴京参加中国科学院第十六次院士大会、中国工程院第十一次院士大会。此次盛会在人民大会堂隆重开幕，中共中央总书记、国家主席胡锦涛出席会议并发表重要讲话。会议期间，还与全国各地的院士服务机构代表进行了广泛沟通和交流。

2012 年中国工程院院地合作会议在山东济南召开，周济院长、干勇副院长等院领

导出席会议并作了重要报告。会议期间，院士中心参会人员还咨询了分管院地合作的学部工作局李仁涵副局长关于广州与工程院开展院地合作的相关事项。

四、拓展功能，创造更广阔的学术交流氛围

院士中心积极主动地配合相关单位及相关部门的活动，参与筹备、承办了多种形式的院士活动，邀请各地院士紧紧围绕经济发展方式转变，以科学发展为主题、以转型升级为主线，为广州科技活动周、首届中国管理科学论坛及留交会相关院士活动等大型活动助力。

院士中心配合教育部管理学部于 2012 年 12 月 12 日在澳门举办了首届中国管理科学论坛，在中国科学院、中国工程院和教育部科技委等各级领导的大力支持下，邀请到 10 多位院士专家和 40 多位各地学者共赴澳门共同探讨中国管理科学的现状、发展和动向。院士中心通过此次活动，积累了经验，拓展了视野，加强了与各相关机构和单位之间的沟通与交流。

紧随首届中国管理科学论坛之后，院士中心受留交会委托，成功承办了“珠江创新论坛”和“大使科技论坛”两项活动，从“搜索科技革命新方向、寻找新型产业新举措”和“国际化与人才引进”两个主题为广州创新发展和科技水平得以国际化提升两个方面助力。

五、策划和正在进行中的活动

（一）院士行

2012 年底至 2013 年初，院士中心正在筹备举办多场院士企业行活动，在积极邀请中有张伯礼、王永炎、侯惠民、陈志南、姚新生、曾益新等数位生物医药方向的院士，准备在 2013 年初来穗举办三到四场院士行活动。院士中心根据众多企业反馈的“院士需求征询表”，挑选有创新成果的项目向有需求的企业进行推介，并开展院士行活动。为更好地落实企业的需求，院士中心还将逐个落实院士企业行的时间和对口业务及项目，希望通过院士这一高端智慧人才，引领广州高新科技领域在创新中不断发展壮大。

（二）科技讲坛

院士中心正在积极准备第六期科学大讲堂，根据策划方案的各类主题与全国多位院士联络洽谈，预约时间及演讲内容。

（三）其他院士活动

为配合院士行活动，院士中心另有 2～3 场院士沙龙和院士讲座正在策划组织中。

六、积极主动，做好院士之家的日常工作

院士中心在年内一一走访了广州地区各个院士服务机构，进行交流和沟通，为更好更细致地全面掌握院士情况打好了基础，并在各兄弟单位的协助下完成了在穗院士统计工作，建立更为完善的院士档案资料。

为不断提高院士中心的工作水平，完善管理，丰富院士活动内容和形式，宣传报道在穗院士的学习生活，更好地做好服务工作，院士中心非常重视相关的宣传报道工

作。在每次院士活动结束后，都及时地编写各种新闻、简讯，并积极收集其他主流媒体对在穗院士的相关报道，通过网站、《院士通讯》杂志等媒体报道活动的信息与成效。

在做好以上工作的同时，院士中心还完成了院士办公室日常管理工作，编撰印制“院士画册”和“院士之家”的宣传画册等。

百年追梦　科技兴国[①]

一、灿烂的古代中华文明

科学技术是人类在认识和改造自然的伟大实践中获得的丰厚知识财富。科学技术推动人类文明的发展。

世界人类文明的四个发源地：中国、古埃及、古巴比伦、古印度。

中国是世界文明古国，中华民族是四个世界人类文明中唯一延续至今的民族。

从汉朝（公元前206～220年）到明朝（1368～1644年），14个世纪，中国的科学技术领先于世界，中国的经济GDP甚至占世界一半以上。据统计，迄今为止，全人类百分之五十以上的科技发明创新属于我们中国人，我们为此而自豪!

二、屈辱的中国近代史

（1）第一次鸦片战争（1840年）

英国派来4000官兵，48艘船只，装有540门大炮，侵略我国，清朝战败，一个4亿人口的大国战败，被迫签订屈辱的第一个不平等条约《南京条约》，割让香港岛，赔2100万银元，我国开始沦为半殖民地半封建社会。

（2）第二次鸦片战争（1856年）

英法联军发动第二次鸦片战争，2.5万人攻占北京，放火烧毁了圆明园，签订屈辱的《北京条约》，再次割让土地，丧失大量主权。

（3）中日甲午战争（1894年）

清朝再败给日本，签订屈辱的《马关条约》，又割地（台湾等）又赔巨款（2亿两白银）。

（4）八国联军侵占北京（1900年）

1900年，英、美、德、法、俄、日、意、奥八国组成联军3万人，攻占北京，又签订屈辱的《辛丑条约》，赔巨款4.5亿两白银。

（5）抗日战争（1931年）

① 九一八事变（1931年9月18日）日本在1931年9月18日发动“九一八事变”，蒋介石实行不抵抗政策，日本轻而易举占领了我国东北三省。

② 七七事变（1937年7月7日）

1937年7月7日，日本在北京卢沟桥发动“七七事变”，再次发动对我国的侵略战争。直至1945年8月15日，日本无条件投降。在这期间，日本占领了我国大部分国土，我国伤亡3500多万人。

---

① 本文是中国科学技术协会2013年度弘扬科学道德 践行“三个倡导”奋力实现“中国梦”报告会的主题报告，贵阳，2013年5月24日；《现代管理的中国实践》，北京：科学出版社，2016，116-123.

1840～1949年，旧中国共与外国签署了1182个不平等条约，仅在清朝就赔给外国16亿两白银。割让土地面积达数百万平方千米。旧中国成为半殖民地半封建的落后国家，被称为“东亚病夫”。

三、现代自然科学推动人类社会的迅速发展

现代自然科学是从15世纪下半叶开始，在欧洲摆脱了神学的统治之后，伴随着资本主义工业大生产的兴起而发展起来的。现代自然科学的产生和发展，促进了技术科学和工程科学的发展，使得英国、德国、美国先后成为强大国家。

现代自然科学包括七个学科：数学、力学、物理、化学、生物、地理、天文。

初期的代表人物，如：

达·芬奇（1452～1519），意大利的艺术家和科学家（绘画、数学、力学），名言：“数学是科学的皇后，力学是数学的天堂”。

中期的代表人物，如：

牛顿（1643～1727），英国的物理学家和数学家。贡献如下。

光学：发现白色光的组成；

力学：提出万有引力定律；

数学：首创微积分学。

此时，英国成为世界科学中心，拥有世界顶级大学：剑桥大学、牛津大学。于是英国成为世界强国，号称“日不落帝国”。

接着，19世纪末，德国哥廷根大学成为世界顶级大学（1737年，由乔治二世创立），出现著名的哥廷根学派，德国成为世界科学中心，代表人物有：

克莱茵（1849～1925），德国数学家；

普朗特（1875～1953），德国物理学家，享有“空气动力学之父”称号。

于是，德国成为世界强国。

哥廷根学派的杰出传人：

爱因斯坦（1879～1955），生于德国，美国物理学家，提出相对论，被誉为20世纪世界最伟大科学家；

冯·卡门（1881～1963），美籍匈牙利人，美国力学家，被誉为20世纪世界第二大科学家。

德国希特勒排挤犹太人，爱因斯坦和冯·卡门于20世纪30年代移居美国。

美国成为世界科学中心，拥有世界顶级大学；哈佛大学、普林斯顿大学、加利福尼亚州理工学院、麻省理工学院、斯坦福大学……美国成为世界头号强国。

由于现代自然科学的发展，技术与工程科学也飞速发展，20世纪和21世纪与以前的人类社会相比，发生翻天覆地的变化。

四、李约瑟之谜

中国有五千多年文明史，中华民族聪明勤劳，中国人民在科技上有许多伟大发明创造，在世界人类文明发展史上留下无数灿烂的记录。但是，从15世纪开始，中国迅

速衰落了！

英国科学家李约瑟（1900～1995）的名著《中国科学技术史》展示了中国古代科技的辉煌历史，但留下了一个末尾结论未写，即著名的“李约瑟之谜”：“中国为什么没有发展出现代自然科学”。

实际上，是旧中国管理上出了问题，使得中国科技落后，从而国家衰落了。当欧洲产生自然科学理论时，中国却处在不尊重知识、不尊重人才的时期。

知识分子被贬为“老九”的历史：

在元朝（1206～1368）时，元世祖忽必烈将中国人分为十等：

官、吏、僧、道、医、工、匠、娼、儒、丐。

知识分子被列为第九等，被打入底层，仅比乞丐高一等，连妓女都不如，这就是知识分子被称为“老九”的来历。

从此，知识分子和知识不受尊重，科技发展严重受挫！

民国时期，士兵被称为“丘八”，知识分子被称为“丘九”，仍是“老九”。

“文化大革命”时期，知识分子更被称为“臭老九”。

科技创新的主体被贬为最底层，何谈受尊重，何谈工作环境和条件。因此，现代自然科学不会产生在我国。科技的落后，以致国弱民穷，导致 1840 年后的屈辱历史，一个拥有 4 亿人口的文明大国成了“东亚病夫”。

五、中华民族在中国近代史时期科技领域的奋斗和探索

旧中国在第一次和第二次鸦片战争的惨败，深深刺激了聪敏勤劳的中国人。19 世纪 60 年代至 90 年代，以曾国藩、李鸿章、左宗棠、张之洞为代表的清政府官僚主张学习西方资本主义的科学技术，以挽救中国的落后，历史上称为“洋务运动”。

开办江南制造局、福州船务局、各省机器局，成立北洋海军，形成近代军事和民用工业。

同时，第一次派遣留学生去国外，学习西方科技。

中国第一代留学生代表人物：

容闳（1828～1912），1854 年毕业于美国耶鲁学院（今耶鲁大学）；

严复（1854～1921），1879 年毕业于英国海军学校，我国第一位系统介绍西方哲学的人；

詹天佑（1861～1919），1881 年毕业于耶鲁大学，中国第一条铁路（京张铁路）的工程师，火车自动挂钩发明者。

典型实例：容闳，广东省珠海市人。1847 年，随美国传教士布朗去美国留学，考入耶鲁大学，成为中国最早的留美学生。刻苦攻读，立下志向“以西方之学术灌输于中国，使中国日趋文明富强之境”。毕业后，谢绝挽留，回到祖国，投资办学，在 1872～1875 年筹资组织四批幼童共 120 名赴美留学，涌现出詹天佑等杰出人才。

甲午战争的失败，让一切归零，宣告了洋务运动的破产。日本打断了中国第一次科技兴国的梦想。

接着，中国先进的知识分子痛定思痛，毫不气馁，转向经过明治维新（1868 年）

后强大的日本留学，代表人物：

黄兴（1874～1916），1902 年赴日本留学；

宋教仁（1882～1913），1904 年赴日本留学；

鲁迅（1881～1936），1902 年赴日本留学（仙台医学院）；

郭沫若（1892～1978），1914 年赴日本留学（九州帝国大学医科）。

我的祖父刘良也是在 1903 年赴日本留学，就读于东京振武学校，参加了同盟会，他的救国梦想影响了我一生。

孙中山先生领导的辛亥革命的成功，推翻了腐朽的清王朝。在北伐战争后，国家终于在 1928 年重新统一，又开始新一轮现代化建设。但是，1937 年七七事变，日本再一次打破了中国人第二次科技兴国的梦想。最近，美国在第二次世界大战结束时缴获的日本绝密档案解密，从中获知，早在 1935 年，日本内阁会议就决定，计划在完全征服中国之后，将首都从东京迁到北京，并移民 100 万日本人到北京，期望永远统治中国。实际上，早在 1584 年，日本丰臣秀吉就计划在 1594 年将日本首都迁到北京。日本的狼子野心昭然若揭！这种灭我思路为日本后来的多个统治者所继承。我们应该警惕！

第三代留学生，总结教训，继续出国留学，向西方学习科技，代表人物：

钱学森（1911～2009），1935 年先到美国麻省理工学院留学，1936 年成为加利福尼亚州理工学院冯·卡门第一位中国博士生，1955 年回国；

钱伟长（1913～2010），1940 年赴加拿大多伦多大学攻读应用数学博士学位，1942 年成为冯·卡门的助手；

钱三强（1913～1992），哥廷根学派传人居里夫人的女儿的博士生，1937 年赴法国留学，1948 年回国；

郭永怀（1909～1968），1941 年冯·卡门的中国博士生，1956 年回国；

李四光（1889～1971），1931 年获英国伯明翰大学地质博士学位；

茅以升（1896～1989），1919 年获美国卡内基理工学院土木工程博士学位；

周培源（1902～1993），1927 年获美国加利福尼亚州理工学院博士学位。

他们寻求科学救国之路，成为新中国成立之后科技事业的中坚力量。

典型实例：我的恩师钱伟长先生自幼对文史感兴趣，1931 年考入清华大学文学院，文科成绩满分，数学不及格。进校后第三天，发生九一八事变，他弃文学理，要走科学救国之路，死缠系主任整整一周，坚持要求转入吴有训任系主任的物理系读书，刻苦读书，终于成功。接着，留学加拿大、美国，并在冯·卡门身边做科学研究。1941 年，冯·卡门出版 60 岁祝寿文集，他是唯一发表论文的青年学者和中国学者。1946 年，他谢绝美国的优厚环境和待遇，返回中国，报效祖国，创建中国现代力学第一个研究机构。

六、中华民族在新中国时期科技领域的奋斗和探索

1）新中国前 30 年（1949～1978）

1949 年 10 月 1 日，新中国成立了，中国人民站起来了！

但是，国家十分衰弱，经济是一个烂摊子：

1949 年，全国经济 GDP179.56 亿美元，仅为 2012 年的 0.21%；全国钢产量 16 万吨，仅为 2012 年的 0.022%。制造业十分差，当时许多物品都带“洋”字——洋火、洋钉、洋蜡、洋车等，机器、汽车、飞机均不会制造。

向苏联和东欧派出留学生学习科技，代表人物：

谷超豪（1926～2012），中国科学院院士，著名数学家，留苏第一位博士；

杜祥琬（1938～），中国工程院院士，应用物理学家；

傅依备（1929～），中国工程院院士，放射化学专家；

……

从 20 世纪 50 年代开始，我国开始独立自主地研制“两弹一星”，这是新中国第一项最伟大的成就，震惊世界。

第一颗东方红人造地球卫星，1970 年 4 月 24 日升空，星重 173 千克。

第一颗原子弹，1964 年 10 月 16 日成功爆炸。

第一颗氢弹，1967 年 6 月 17 日成功爆炸。

第一枚近程导弹，1960 年 11 月 5 日试射成功，射程 590 千米。

第一枚地空导弹，1964 年 4 月 29 日试射成功。

第一枚中高空地空导弹，1965 年 6 月 29 日试射成功。

1978 年，我国国内生产总值已增长为 3645 亿元，为 1949 年的 12 倍。

典型实例：

(1) 郭永怀先生于 1956 年抛弃美国优厚的生活和工作条件（时任康奈尔大学教授），冲破阻力，毅然回到阔别 16 年的祖国。

他出生在山东省荣成县，经历过“九一八”和卢沟桥事变，目睹了日本飞机的狂轰滥炸，因此立志改学航空工程，为原子弹和氢弹试制作出贡献，1968 年他在参加一次原子弹试验回到北京时，因飞机失事牺牲。人们事后发现，在他和警卫员紧紧抱住的身躯间，竟然保藏着一份完好无损的重要绝密试验资料文件。钱学森先生在纪念他牺牲 20 周年的纪念会上深情地说，“他把力学理论和火热的改造客观世界的革命运动结合起来了。这里没有胆小鬼的藏身处，也没有私心重的活动地。这里需要的是真才实学和献身精神。郭永怀同志的崇高精神就在这里”。

(2) 我于 1958 年在兰州参加了我国第一颗东方红人造地球卫星的试制工作，担任一个三人小组的组长，因是绝密级的任务，无名无利，又无老师指导，但我们艰苦努力，做成一支纸火箭，升到几千米高空，让我经受了锻炼。但是，我们与美、苏差距仍然很大。苏联已于 1957 年发射了人造卫星。美国从 1969 年 7 月 20 日至 1972 年 12 月，用阿波罗飞船 6 次登上月球，累计 18 人在月球上考察了 600 多小时。

2) 新中国改革开放后的 34 年（1978～2013）

1978 年 3 月 18 日，全国第一次科学大会在北京举行。中国科学院院长郭沫若先生万分高兴地说“科学的春天来到了!”

就是在这次全国科学大会上，邓小平同志发表讲话，提出两个著名论断：“知识分子是工人阶级的一部分”“科学技术是生产力。”后来，1988 年，他又讲：“科学技术是第一生产力。”

这样，“尊重知识”和“尊重人才”成为国策，使“臭老九”帽子被扔掉，使知识分子由被改造对象升为领导阶级，使“知识”受到重视。

于是34年来，我国科技大发展，推动经济高速发展，国强民富，出现几百年来第一次太平盛世。

派出大量知识分子出国留学，学习科技。

从1872年到1978年，我国先后有13万人出国留学。而从1979年至今，先后有250万人出国留学，大大超过了以往一百年。海归人数也很多，仅2011年，就达到18.62万人。

实例：1981年3月1日，我作为中国第一批（8人）由德国挑选的洪堡研究会员前往联邦德国留学，受到联邦德国卡斯滕斯总统接见，得到深造。其间，我抵制了台湾特务的“策反”。1983年4月18日离开联邦德国，回国报效。出发前，得到钱三强先生接见、指教。我是从中国科技大学去联邦德国的。我在中国科技大学工作时，校长是郭沫若先生，系主任是钱学森先生，我深受他们的教诲。当年我被全校师生誉为中国科技大学的一颗明星。

2012年，我国GDP：519 322亿元，为1949年的467倍，为1978年的142倍。2008年，超过德国，居世界第三位。2010年超过日本，居世界第二位。人均GDP已从1949年的35美元增加到2012年的6000美元，增长171倍。人均寿命已从1949年的35岁延长到2012年的74岁。

标志性成就：

(1) 2003年10月15日，神舟五号飞船发射，我国第一位宇航员杨利伟进入太空。2005年10月12日，神舟六号飞船发射，我国两位宇航员费俊龙、聂海胜进入太空。2012年6月16日，神舟九号飞船发射，我国三位宇航员（包括一位女同志刘洋）进入太空。

上面后两次发射，我都在酒泉发射中心现场观看，真是惊心动魄。

(2) 2012年11月24日，我国第一艘航空母舰辽宁号完成歼击机15成功起降试验，举国欢腾！

建造中，我曾登上航空母舰参观，而且也参加了歼击机15研制的科研任务，深深体会到航空母舰制造之艰难。

(3) 2012年6月24日，我国的第一艘载人潜水器“蛟龙”号在太平洋马里亚纳海沟深潜到海平面下7020米，在那里，潜水器表面一平方厘米的面积上要承受702千克的压力。

中国人民的两弹一星梦、飞天梦、潜海梦、航空梦一一实现，科技工作者奋力拼搏，为实现中华民族伟大复兴的中国梦铸就坚强基石。

典型实例：

(1) 2012年11月24日，航空母舰舰载机试飞成功，举国欢腾。十几个小时后，传来了不幸的消息，担任歼15研制的现场总指挥的中航工业沈阳飞机工业（集团）有限公司董事长、总经理罗阳却因突发心肌梗死，心源性猝死，经抢救无效殉职，年仅51岁。人们说：“罗阳用生命撑起了舰载机起飞。”

他为了工作，已连续在外出差 17 天，其间路过家门而未入，为了航母，他太累了！在舰上坚持工作，下舰几小时后的庆功宴都未能参加，就走了！这就是科技工作者为了航母梦付出的辛劳。

我因参加歼 15 的科研任务与罗阳见过面，留给我深刻记忆！

(2) 我的恩师钱伟长先生在 1957 年被错打成右派后，1981 年才平反。但他仍然坚持进行科学研究，取得杰出成就，而且从 1983 年担任上海工业大学校长起，任校长整整 28 年，从未拿过学校一分钱津贴，未享受福利分房，虽然已经高龄，还为搞科学研究经常熬夜工作到深夜。这就是科学家的高尚道德情操。

(3) 我的成名作《波纹圆板的特征关系式》，是为仿制美国 P2V 低空侦察机的测高度仪表研制服务，不计名利，坚守爱国信念，历经千辛万苦和磨难，用 14 年时间（1964～1978 年）才得以应用。

七、“中国梦”鼓舞中华民族扬帆奋进

2001 年中国的 GDP 总量是美国的 1/8，十年之后的 2011 年就已达到其近 1/2。

权威国际机构纷纷预测，到 21 世纪 20 年代结束时，中国的 GDP 总量可能超过美国，国际货币基金组织甚至预测，2016 年中国将取代美国成为全世界最大经济体。

习近平总书记指出，我坚信，到中国共产党成立 100 年时全面建成小康社会的目标一定能实现，到新中国成立 100 年时建成富强民主文明和谐的社会主义现代化国家的目标一定能实现，中华民族伟大复兴的梦想一定能实现[①]。

作为科技工作者，要牢牢记住习近平总书记的指示。要沿着前辈科学家们的道路，记住历史，坚守科学道德，坚持创新，不畏艰险，扎扎实实，脚踏实地，用勤劳的双手托起伟大的“中国梦”！

---

① 习近平谈“全面建成小康社会”. http://news.cnr.cn/native/gd/20150304/t20150304_517875157.shtml [2024-08-27].

充分发挥桥头堡作用，推进孟中印缅经济走廊建设[①]

5月21日，习近平总书记在亚信第四次峰会上再次提出，“中国将同各国一道，加快推进丝绸之路经济带和21世纪海上丝绸之路建设”[②]。充分发挥云南桥头堡作用、推进孟中印缅经济走廊建设对于实施国家“一带一路”大战略、构建周边命运共同体，意义重大。由于国家战略地方化等原因，云南作为我国面向西南开放重要门户的综合优势没有得到充分发挥，表现在五个方面。

第一，对外协调层级偏低，权威性不够。

第二，对外合作以沿边为主，缺乏面向印度洋的大战略格局。

第三，偏重经济贸易合作，基于项目的合作多，机制化合作较少，形式单一且水平层次较低。

第四，对外投资以资源开发为主，短期化倾向明显，关注民生不够。缅甸2012年严控翡翠原石出口；2014年4月禁止原木出口；密松水电站也因民意反应强烈被叫停。

第五，国际大通道建设通而不畅，缺乏通向印度洋的战略主干线和战略支点。

针对上述情况，提出六点建议，供参考。

第一，提升云南在区域协商对话中的地位和层次，化对外开放前沿为区域合作中心。争取在以下方面获得中央支持：在昆明定期召开孟中印缅领导人会议，举办四国部长级会议和工作组会议，建立定期磋商机制；参照上海合作组织模式，建立昆明合作组织，由四国共同签订合作协议，将昆明作为永久会址；为南亚国家在昆明设立领事机构和商贸、金融、旅游等办事机构提供便利；建立昆明银行，力争把亚洲基础设施投资银行总部设在昆明；在人员往来、对外投资等涉外事务方面，争取中央赋予云南一定的审批自主权。

第二，实施面向印度洋的蛙跳式对外开放战略。在继续重视和加强沿边合作的同时，坚决走出去，深入缅甸、孟加拉国腹地，以密支那、曼德勒、皎漂、吉大港、达卡等城市为重点，实施资源就地转化工程，加大农业、教育、医疗卫生、食品安全、科技等民生领域投资力度，加强公共福利设施和基础设施建设，打好政治、经济、科技、文化“组合拳”，构建利益共同体。

第三，高起点推动大通道建设。推动国家尽快签订孟中印缅经济走廊交通运输合作协定，依托先进的高铁技术和高速公路设计建筑技术，加快国际大通道境外段建设。争取国家优先支持昆明—瑞丽—皎漂高速铁路规划设计建设，加快推进昆明—皎漂高速公路建设，加快伊洛瓦底江陆水联运通道建设。争取把芒市、腾冲机场建成国家口岸机场。

---

① 本文原载《云南省党政领导与院士专家座谈会院士专家发言提纲汇编》，昆明，2014年5月25日，1-3.

② 习近平在亚信第四次峰会作主旨讲话（全文）. https：//www.cs.com.cn/xwzx/hg/201405/t20140521_4397057.html[2024-08-27].

第四，努力把云南打造成为孟中印缅科技教育交流中心。借鉴湄公学院模式，在昆明建设“印太学院”；依托云南沿边州市现有高等教育机构建设“国门大学”，加强人力资源开发合作；启动中国—南亚科技伙伴计划，设立区域专项基金，支持在孟、缅、印建设农业科技示范园、高新技术产业园，推动在老挝万象、缅甸皎漂和密支那建设境外经贸合作区。

第五，打造“孟中印缅信息高地”。支持云南 IT 企业走出去，在信息基础设施建设、网络安全保障、应急响应、互联网管理体制机制建设、政策法规等领域，向周边国家提供技术援助和人员培训。争取在云南建设中国—东盟大数据、云计算、宽带网络中心，打造“孟中印缅信息化高地”。

第六，加大对外文化交流与合作。实施孟中印缅文化工程，建立“中国—南亚文化交流基金”，建设南亚民族民间艺术博览园、“中国—东盟”数字文化产业基地和文化产业公共服务平台，鼓励云南主流媒体以独资、合资或合作方式到境外发展，推进广播电视等重点媒体境外落地。利用周边国家丰富的旅游文化资源，大力发展旅游观光产业，加强文化交流与合作，惠及周边国家广大民众。

建设低碳社会　托起美丽中国梦[①]

2个月前，即2014年9月19日，世界银行和国际货币基金组织按货物购买力平价宣布中国GDP已超过美国，是世界第一位。当然，我们国家谦虚，目前尚未承认这一事实。总之，中国人民终于站起来了，再不是“东亚病夫”了！

但是，我国经济的高速发展，主要是靠物质投入的传统发展方式，在一定程度上，是以资源环境的巨大牺牲以及经济发展的一些不合理为代价所达到的。以推动国家经济增长的消费、投资、出口这三驾马车为例，长期以来，我国经济已经越来越严重地依赖出口这一驾马车的拉动，外贸依存度高达60%。特别是我国的外贸出口，主要依赖于大量廉价劳动力、大量资源能源的消耗，以及随之而来的环境污染的巨大代价。农民是廉价的劳动力，目前用度已到顶。我国的资源十分有限：石油储量仅占世界的1.8%，天然气占0.7%。在人均资源方面，我国人均45种主要矿产资源为世界平均水平的1/2，人均耕地、草地资源为1/3，人均淡水资源为1/4，人均森林资源为1/5，人均石油资源为1/10。以污染造成的全球变暖为例，我国面临减排二氧化碳温室气体严峻的形势。

世界卫生组织发布的空气质量建议标准为：PM2.5年均值不得高于每立方米10微克。我国的国家健康空气质量标准是每立方米35微克。按我国自己的标准，在2012年，全国超过70%的中国人即约10亿人已生活在超过这一标准的空气中，而且主要是在我国东部发达地区，特别是以首都北京为中心最为严重。有1.57亿中国人生活在PM2.5年均值超过每立方米100微克（高于国家3倍多）的环境中。

PM2.5多了，将引起四大疾病：中风、肺癌、冠心病和慢性阻塞性肺疾病。雾霾一年，导致一年67万中国人过早死亡。

2014年11月3日，联合国政府间气候变化专门委员会发布的综合报告宣布，为避免全球变暖的严重负面影响，有必要在21世纪末使温室气体排放量趋近于零。

完成这一目标的一条路径是，到2050年，使温室气体排放量相比2010年减少40%～70%，到21世纪末达到零排放。

如果不减排，则到21世纪末，平均气温将比20世纪末升高4.8℃，人类将达到危险时刻！

2007年，我国已超过美国，成为世界第一排放大国，排放量达到59.6亿吨，占全球总排放的21%，与1990年相比，几乎翻了两番。而且，更为严重的是，碳排放量继续增长，大幅度上升，去年（2013年），我国排放量又翻了一倍，达到119亿吨，占全球排放量的27.7%。我国现在的排放量已超过美国和欧盟的总和。最大的推动力是发电、交通运输和制造业。旅游业也占有较大份额。

大量的研究表明，大气中的二氧化碳是地球升温的祸首。超过临界点，并不意味

① 本文是在第三十四次中国科技论坛——绿色建设美丽中国论坛上的致辞，广州，2014年11月19日.

着马上会发生重大的灾难。但意味着一种从“可能”到“必然”的转变。

如果全球气温上升 5℃，风暴及干旱等极端天气的出现概率会急剧增加，海平面将上升 10 米，海中的一些岛屿与大陆沿海地带，特别是一些大城市将被淹没，全球一半物种可能面临灭绝。

如果全球气温上升 6℃，地球上的生命将遭遇到毁灭性的打击。

我国是升温最为严重的国家之一，近 100 年来，平均气温升高 1.1℃，略高于同期全球平均升温幅度，近 50 年来变暖尤其明显，而且极端天气与气候事件的频率和强度出现了明显变化。

2009 年 12 月，哥本哈根气候大会向全世界七十亿人民宣布：要保护地球！全球随即进入低碳经济时代。

地球在茫茫宇宙中，是人类赖以生存的唯一的一个得天独厚的乐园。一万年来，人类与自然一直和谐相处。现在，全球能源告急！资源告急！环境告急！七十亿人的需求几乎要把自然拖向崩溃的边缘。

我们要和全世界人民一起，尽早行动起来，坚持走节能减排、保护环境之路，坚持走可持续发展的绿色经济增长之路，托起美丽中国梦！

研究传染病突发事件的危机管理十分重要[①]

传染病等公共卫生事件是公共危机的重点范畴。随着社会经济发展、交通便捷和人员流动频繁，某一地区发生传染病如果管控不当，都有发生突发性公共卫生事件进而影响全球的可能，澳门地区也无法置身事外。本书作者围绕澳门传染病突发事件的危机管理能力问题进行了系统的研究、分析。

本书作者李炜毕业于中山大学生物化学专业，任职于澳门特区政府卫生局多年，曾多次参与澳门的传染病危机应对工作；其后攻读战略管理方向博士学位，从卫生防疫社会管控的角度展开专题研究。

本书的研究是从澳门特别行政区的实际出发，以传染病突发事件的危机管理为研究对象，运用公共危机管理相关理论成果，对传染病突发事件的危机管理体系进行研究。目的在于研究和评价我国实施“一国两制”的澳门地区，在其特殊的社会政治和经济环境下，传染病突发事件的危机管理问题，填补了该方面研究的空白，并为加强地区公共卫生危机管理体系建设提供决策参考。

作者围绕传染病及其公共卫生危机的管理，阐述了传染病及其特点，运用危机管理理论分析了传染病突发公共卫生危机的成因和危害，论述了加强公共卫生建设对于传染病等公共卫生危机管理的意义，综述了传染病公共卫生危机管理的要点。通过对澳门特别是回归以来改革医疗卫生体制的分析，总结了澳门公共卫生建设的成绩与不足。通过对澳门在公共卫生危机管理方面建立法律规章、健全机构和协调机制、加强初级卫生服务建设、应对多次传染病突发事件的分析，详细阐述了特区政府的医疗卫生方针和应对公共卫生危机的管理能力；通过实证分析进一步论证了澳门特区公共卫生危机管理体系的实际效能，同时也指出了尚待加强和改进之处，并有针对性地提出了今后体系改进的构思，具有借鉴和指导意义。

① 本文是《澳门传染病突发事件的危机管理能力研究》序，北京：中国文史出版社，2016，1-2.

以智能制造促进产业转型升级①

以大数据、移动互联网和云计算为代表的新技术，加速了智能制造技术的发展，正在润物无声地改变着我们的生产和生活方式。以《美国先进制造业国家战略计划》、德国“工业4.0”和《中国制造2025》为代表的国家发展战略，将深刻影响国际产业竞争的格局。

一、智能制造成为世界主要国家的国家战略

近几年来，国际上掀起了新一轮科技革命和产业变革的热潮，世界主要国家顺应潮流，纷纷制定刺激实体经济增长的国家战略发展计划，希望通过技术进步和产业政策调整重获在制造业上的竞争优势。

（1）美国实施“再工业化”计划。2011年6月，正式启动包括工业机器人在内的“先进制造伙伴计划”；2012年2月又出台《美国先进制造业国家战略计划》；2013年，美国将先进制造业国家战略聚焦到先进制造的感知控制、智能制造技术平台与先进材料制造三大技术的优先突破。美国的目的就是要保持美国制造业对全球的控制者地位。

（2）德国发起的“工业4.0”国家发展战略计划，非常有代表性，下面进行重点介绍。

先回忆一下德国的强国战略。19世纪末期，哥廷根学派声名鹊起，为一百年来世界科学技术发展做出杰出贡献。20世纪世界顶尖科学家爱因斯坦和冯·卡门都是哥廷根学派的传人。德国的强国战略就是抓科技创新。

第一次工业革命（1760～1840年），蒸汽时代。英国的瓦特（1736～1819年），经过26年的研究，发明了高效率蒸汽机，使英国成为强大国家，号称“日不落国”。本身国土仅24.4万平方公里，却曾有殖民地3000万平方公里。

第二次工业革命（1840～1950年），电气时代。法国的库仑（1736～1806年）于1785年提出电的库仑定律。美国的爱迪生（1847～1931年）于1879年发明白炽灯，使纽约在1881年建立了世界上第一个发电站和配电系统。此后，电力开始迅速地在世界推广。英国的汤姆孙（1856～1940年）在1897年发现电子。随后，二极管（1904年）、三极管（1907年）相继问世，标志电学向电子学过渡。电气化和汽车、飞机是标志。德国和美国强大起来。

第三次工业革命（1950～2000年），信息时代。1946年，第一台电子数值积分计算机（Electronic Numerical Integrator and Calculator，ENIAC）在美国投入运行，每秒可进行5000次加减运算。1981年，美国IBM公司推出个人计算机（微机）。在20世纪八十年代，美国的超级计算机每秒可进行100万亿次计算。如果用过去的计算器计算，则要花1000万年才行。1962年，美国麻省理工学院一位研究生克兰罗克发明了

① 本文是应广东省经济和信息化委员会邀请在中共广东省委党校所作的报告，广州，2016年6月21日.

互联网技术。1972年，美国的汤姆林森推出了电子邮件。1999年，互联网已有1.5亿用户。计算机和互联网深刻地改变了人们的生活和工作方式，它们是20世纪技术科学的最伟大成就，是20世纪的重要象征，带动了世界性的技术革命，使信息时代来临！

第四次工业革命（2000年～），带来颠覆性变化。以互联网产业化、工业智能化、工业一体化为代表，以人工智能、清洁能源、无人控制技术、量子信息技术为主的新技术革命，在国际上掀起了新一轮科技革命和产业变革的浪潮。我们国家第一次与美国、欧盟、日本等国家和地区站在同一起跑线上正式发动和创新绿色工业革命！

2006年，德国提出《德国高技术战略》，由德国工程院、西门子公司等提出。

2010年，德国政府制定了《高技术战略2020》。

2013年4月，三家德国协会（德国信息技术、通信与新媒体协会，德国电子电气制造商协会，德国机械及制造商协会）共同组建了“工业4.0”平台。

显然，德国是要通过技术和制度创新，推动工业的智能化、数字化。

“工业4.0”基本理念包含以下七个方面：①建设智能工厂；②搭建信息物理系统（又称智能系统CPS[①]）；③发展智能设备和智能决策系统；④物联网；⑤产品和服务开发的新流程；⑥柔性制造；⑦绿色、可持续发展。

“工业4.0”基本框架是通过传感器、计算机和互联网将虚拟世界和真实世界相连接，实现智能制造和智能生活。

德国的“工业4.0”计划，就是要组建多组合、智能化的工业制造系统，应对以智能制造为主导的第四次工业革命。

（3）日本提出《创新25战略》计划，通过科技和服务创造新价值，以“智能制造系统”为核心理念，促进日本经济持续增长，应对全球大竞争时代。

日本是全世界最大工业机器人生产和运行的国家，又是存在严重问题的国家——低出生率、老龄化和基础建设老化，期待通过机器人来解决这些问题。

2015年1月23日，日本公布“机器人新战略”，两大目的：①扩大机器人应用领域；②加快新一代机器人技术研发。

2015年5月，日本机器人革命促进会成立，主要任务是实现物联网技术对日本制造业的变革。

日本通过机器人战略，要成为世界机器人创新中枢，要成为世界领先的利用机器人的社会。

（4）英国实施“高价值制造战略”，应用智能化技术和专业知识，要重整英国制造业。

（5）韩国提出“新增长动力规划及发展战略”，确定3大领域17个专业为发展重点，加强对智能制造的支持。

（6）印度实施“印度制造”计划，要用智能制造技术将印度打造成“全球制造中心”。

---

① 信息物理系统，cyber physical system，CPS。

二、推动智能制造的代表性新技术

1. 大数据

智能制造有三个关键的基本元素，即大数据、智能机器和知识工人。大数据是指工业大数据，它源源不断地产生，基于云计算进行集中、传输以及分析处理，从而释放价值。智能机器就是将世界上各种机器、设备组、设施和系统网络与用于控制的传感器和计算机软件连接。知识工人是指工作中的人，在任何时候，无论是在办公室还是在室外，都能联系，以支持更加智能的设计、运营、维护以及更高质量的服务和安全性。

大数据与传统工业数据相比，有十个不同，即十个 V：①数量（volume），大量；②种类（variety），多种形式；③速率（velocity），不断地产生；④价值（value），提升效率更有用；⑤真实（veracity），一致且可行；⑥愿景（vision），来自有意义的流程；⑦挥发性（volatility），生命周期有限；⑧验证（verification），数据的产生应遵循规范，保证工程的测量是正确的；⑨效度（validation），数据应遵循其愿景，保证过程背后的假设和联系的透明度；⑩变化性（variability），具有一定程度的不确定性和不精确性。

大数据通过大数据平台，主要应用于以下几个方面：①数据可视化，包括文本可视化、网络（图）可视化、时空数据可视化、多维数据可视化，等等；②商业智能，包括大数据分析和智能化决策；③数据挖掘；④安全与隐私保护等。

2. 物联网

物联网是第四次工业革命的主要推动力之一，也是工业 4.0 的核心。物联网应用范围非常广，包含九大领域应用：①智能工业；②智能农业；③智能物流；④智能交通；⑤智能电网；⑥智能环保；⑦智能安全防护；⑧智能医疗；⑨智能家居。

举例：

1）每一片药都被全程控制

提供联网的智能药瓶，确保病人谨遵医嘱。

2）每一个人都在线上工作

可在办公环境使用可穿戴 Wi-Fi 对讲机，将人员接入物联网。

3）人脸将成为万能钥匙

以后大家都不用钥匙了，将人脸识别、计算机图像处理以及机器学习技术结合起来，人走到哪儿，门会自动打开，机器打开开始运作，等等。

3. 信息物理系统

智能系统应用于智慧城市、智能电网、智能“任何东西”。

4. 智能制造技术

智能制造技术来源于人工智能，兴于 20 世纪末、21 世纪初期，包括数控技术、Agent 技术（代理技术）、云技术等。最近，更加关注应用。

1）人工智能的含义

人工智能是对人的意识、思维的信息过程的模拟，即人类使用计算机对人类智能的模仿，让机器学会人类在某一领域的专业技能。这一学科诞生于20世纪50年代中期，当时由于计算机的产生与发展，人们开始了人工智能的研究。人工智能的应用范围很广，目前包括机器人、语言识别、图像识别、自然语言处理（翻译）、专家系统等。

2）人工智能实例

今年3月，机器人AlphaGo采用大数据（利用现存于人类数据库中的围棋棋谱）、人工智能系统和自我学习系统（自我对局三千万盘的方式训练），得到了完整的围棋程序，4∶1，战胜韩国围棋大师李世石。

3）移动智能终端的发展

让数据收集更加容易，更加多样化。通过一系列过程，从MP3→彩色屏幕→照片→手写输入→录音→TV功能→GPS→3G和Wi-Fi→触摸屏、二维码→嵌入传感器功能→多核处理器、图形处理器、遥控器，重力感应、光线感应、高清（4K）显示、手势输入、语音搜索和语音翻译在移动终端成为物联网节点。预测，未来10年，地球上将存在4万亿传感器。

4）移动通信的换代演进

从1G→2G→3G→4G→5G，十年一代，峰值速率十年千倍！目前是4G。1G时，我国是空白。2G时，我国是跟随。3G时，我国是突破。4G时，我国与国际同步。5G时，预计2020年实行商用，我国的目标是要引领。

5）互联网的演进

从20世纪70年代底的Web1.0到20世纪90年代的Web2.0，再到21世纪初的Web3.0，再到20世纪10年代的Web4.0；亦即从信息分发技术到信息交互技术，再到语义技术，再到智能应用技术。

6）水下机器人液压作业与推进系统

我国自主研制的首台4500米级深海遥控无人潜水器“海马”号首次使用，2014年就在珠江口盆地发现可燃冰矿藏，且易开采。这是2014年度中国海洋十大科技进展之一。

接着，又有无人驾驶汽车、超高速列车、波音747、空客的未来工厂的出现。

5. 云计算/云制造

1）云计算

云是网络、互联网的一种比喻说法。用户通过电脑、笔记本、手机等方式接入网络上的数据中心，按自己的需求进行运算。因此，可以让你体验到每秒10万亿次的运算能力。可以模拟核爆炸、预测气候变化、街道的人流量和市场发展趋势等计算。

2）云制造

云制造是一种新型网络化制造模式。各种网络化的制造资源被有序地组织在一个平台上，用户按需选择制造服务。

3）概况

（1）2015年8月5日，美国通用电气公司（General Electric，GE），这是一家百年老店，宣布了进入云服务市场的计划。它采用云计算和云制造，使得机器产量提升20％。

(2) 数字化身份。10年前，进入数字化时代还只意味着一个手机号码、一个邮箱地址，或许再加上一个个人主页。现在，人们在数字世界中的存在则包括进行互动，以及他们在各种线上平台和媒体上留下的痕迹。很多人拥有不止一个数字身份。在未来，建立并维护好自己的网络形象会如同我们在现实生活中通过打扮、言语及行为来展示自己一样。

(3) 视觉成为连接互联网的新媒介。眼镜式、头戴式及眼球追踪设备都会变得越来越智能，谷歌眼镜只是第一个成功尝试。在未来，人眼与视觉将成为连接互联网及数字设备的新媒介。通过视觉与互联网中各种应用与数据直接连接，从而更具身临其境之感。通过提供指令、信息可视化及交互作用，可使视觉成为一个即时、直接的交互界面，改变人们的学习、导航、指引和反馈方式，从而有助于人们更充分地与世界互动。

(4) 可穿戴设备联网。科技越发个人化了。早期的电脑需要几个大房间才放得下，后来能放在桌子上，现在则能放在膝盖上。当前的科技水平从我们口袋中的智能手机上可见一斑，而未来的科技将会被直接嵌入衣物和配饰中。

(5) 可穿戴设备与物联网及云计算结合。带有智能芯片的鞋子、手环、功能强大的智能手表、虚拟现实（virtual reality，VR）技术都可广泛地应用于城市规划、室内设计、工业仿真、古迹复原、桥梁道路设计、地质灾害、教育培训等众多领域。

(6) 可植入技术。设备不再仅是可穿戴的，还能够被植入体内，发挥通信、定位、行为监控及健康管理等功能。起搏器及人工耳蜗的发明只是一个开始。这些设备将能感知疾病参数，提醒用户采取措施，发送数据至监测中心，甚至有可能自动投放药物进行治疗。

(7) 普适计算。当前全球有43%的人能连接到互联网。据估计，过不了几年，全球3/4的人都能经常使用互联网。由于无线技术所需的基础设施少于其他许多公共服务（如电力、道路与水），任何国家的任何人都可以与地球上任一角落的人进行交流与互动。信息的产生和传播会比以往更加便利。今后原始部落分配打猎任务都有可能使用智能系统。

(8) 便携式超级计算机。现代智能手机和平板电脑所具备的运算能力比以前要占据一整个房间的许多“巨型电脑”还要强大。

(9) 全民无线存储。存储空间商品化已是大势所趋，其中一个原因就是存储价格飞速递减。世界的发展趋势是存储空间实现全面商业化，可以让用户免费享用无限量的存储空间。

(10) 万物互联。由于运算能力不断上升，硬件价格持续下降，几乎将任何东西连接上互联网都十分划算可行。智能传感器的价格也已相当合理。所有物品都会智能化并能联网，从而促进更广泛的交流和数据驱动的新型服务。未来世界的每一样（实体）产品都可以与无处不在的通信基础设施相连，同时无处不在的传感器也能使人们充分感知周围环境。

(11) 数字化家庭。随着家庭自动化快速发展，人们可以通过互联网控制照明、厨房家电、空调、音视频播放设备、安防系统和其他家用电器。可提供各种服务的联网机器人也可以帮上忙，如进行真空吸尘等。

(12) 智慧城市。许多城市会将服务、公共设施以及道路接入互联网。这些智慧城市将能够对能源、物料流、物流运输及交通等领域进行智能管控。目前率先践行这一理念的地区，如新加坡和巴塞罗那，已经开始实施许多数据驱动的新服务，如智能停车方案、智能垃圾回收以及智能照明等。智慧城市正不断拓展其传感技术网络，在此基础上依靠数据分析和预测模型拓展出新的公共服务。

(13) 运用大数据进行决策。大数据的运用能够让诸多行业及应用领域的决策过程变得更快更好。让政府和企业能够提供全方位的实时服务和支持，从与顾客的互动到税务的自动申报与缴纳，无所不包。

(14) 无人驾驶汽车。奥迪、谷歌、百度等大公司已经开始致力于无人驾驶汽车的开发试验。无人驾驶汽车有望在能效和安全性能上超越需要有人掌握方向盘的普通汽车。到时，它们也可以缓解交通压力，降低排放，并对现有的交通及物流模式产生颠覆性的影响。

(15) 人工智能与决策工作。除了驾驶汽车，人工智能还能从以往的情境中获取经验，提供建议，自动完成一些复杂的决策过程，更简单快捷地制订出具体方案。人工智能擅长模式匹配及自动化处理，使其可担任大型组织中的很多工作。智能机器具有自我学习能力，没有创新意识的领导可能被智能机器代替。

(16) 机器人与服务。汽车制造过程中有80%的工作都由机器人完成。机器人的应用可以显著提高生产和供应链效率。

6. 3D 打印

3D打印技术（又称“增材制造技术”）作为工业4.0中实现“智能生产”和“智能工厂”的重要路径，实现了信息流到物理实物的转变。

1) 3D打印与人类健康

在未来，3D打印机不仅能够打印物品，还能够制造人体器官——这一过程被称为“生物打印”。3D打印在满足顾客的定制化需求方面具有巨大潜力，人体器官最需要定制。当前，骨骼打印、皮肤打印、肾脏和心脏打印在我国都有突破性进展。生物器官3D打印技术是多学科交叉的一项前沿技术：①组织器官再生的划时代技术，解决器官短缺世界性难题新途径；②生物疾病机理模型构建新技术，破解疑难疾病新方法；③生物芯片制造新技术，新药物筛选新方法。

2) 生物器官3D打印全球竞争形势

美国的七大计划都包含生物3D打印计划，在《2020年制造技术的挑战》中被列为11项重要发展技术之一。欧盟《制造业的未来：2015—2020》将3D打印列为重要内容方向。日本“机械学会技术路线图”将支持组织再生作为10个重点之一。

由此可见，生物3D打印技术是国际前沿与热点，引领产业革命；生物3D打印是21世纪人类生命健康领域最重要支撑技术之一；坚持源头创新、突破生物打印关键技术与装备，支撑产业发展。

3) 国家目标

为此，《中国制造2025》是实现生物3D打印等新技术重点突破方向。《国家增材制造产业发展推进计划 2015—2016 年》将它作为重点发展方向之一。《国家中长期科学和技术发展规划纲要 2006—2020 年》把发展生物 3D 打印技术、研制组织工程和再生医学治疗产品列为重点。中国工程院将新材料产业体系国家重大专项生物医用材料第 9 项列为重点支持。中国科学院，中国工程院国家战略性新兴产业重大专项：生物医用材料专题给予重点支持。中国工程院生物医药“十三五”专家组规划和生物材料“十三五”专家组规划都列为重点。

上面介绍的六大智能技术和“互联网+”与交通、物流、金融、制造、安全、医疗、能源等行业的实体系统相融合，就会提升资源利用效率，提升社会经济效益，提升技术附加值，提升服务水平和安全水平。

智能制造是智能技术与制造技术不断融合、发展和应用的结果。智能制造有三个不同的层面：①制造对象或产品的智能化；②制造过程的智能化；③制造工具的智能化。

智能制造的领域：装备制造、轻工、纺织、化工、医药、电力、建材、冶金、飞机、火车、汽车、船舶、物流。

智能制造的发展程度与下面五方面紧密相关：①技术融合创新的进展；②高端智能设备制造能力；③新型传感器网络研发进展；④经济竞争力；⑤政策创新和体制创新的进展。

三、我国应对第四次工业革命的国家战略部署

我国不能再缺席第四次工业革命，还要在重点领域引领技术发展和产业升级。

1.《中国制造 2025》的伟大意义

我们国家在改革开放后，中国制造是以代加工、贴牌闻名，跟在人家后面大搞工业制造，由于正确的管理，我国高速前进，已成为世界第二大经济体。

我们成为世界第二，对我们中华民族来说，我们终于摆脱了落后挨打受欺负的位置，可以扬眉吐气了！我们的快速崛起是天大的好事！但对于其他国家说来，特别是对于欧美发达国家说来，他们不是由衷高兴。我们必须改变这个难受的位置。我们必须要继续向前！今年 5 月 30 日，习近平总书记号召，我们要建设世界科技强国，要实现中华民族伟大复兴的中国梦![①] 因此，就要按照国家 2015 年制定的《中国制造 2025》发展战略前进。这个战略不再是过去第三次世界工业革命那样，我们只是跟随者。这次第四次工业革命，我们要起引领作用，走在国际先进行列。这是一个十年纲领，要由一个制造大国向制造强国转变。智能制造战略的核心是创新，创新是灵魂。创新包括技术创新和制度创新。

2. 智能制造的核心含义

对于智能制造而言，需要大搞科技创新，特别要重视三个融合：知识融合、学科融合、技术融合。由技术集成到技术融合，由集成创新到融合创新，即由 1+1=2

---

① 习近平：为建设世界科技强国而奋斗.（2016-05-31）[2023-11-20]. http://www.xinhuanet.com/politics/2016-05/31/c_1118965169.htm.

到 1＋1＞2。协同创新的结果，将是功能、效益和技术附加值的大幅度提升。

3. 融合的实例

实例 1：乔布斯的苹果手机是 IT 与人文艺术的融合，开创了高附加值的新兴产业。

实例 2：我国的高铁工程就是集成创新的结果。现在已有 19 000 公里，居世界第一。速度可达 300 公里/小时至 350 公里/小时。

4. 制定政策

“十三五”期间，智能制造装备将成为我国国民经济重点产业的转型升级和战略性新兴产业培育发展的重大需求！《国务院关于加快培育和发展战略性新兴产业的决定》进行政策导向，未来 10 年，我们将形成完整的智能制造装备产业体系，总体技术水平迈入国际先进行列，部分产品成为创新第一。

2015 年，国家的《中国制造 2025》正式发布。这一强国战略引领全国各省市出台了地方政策，如《广东省智能制造发展规划（2015—2025 年）》《福建省实施〈中国制造 2025〉行动计划》等。

为做好“中国制造”试点示范工作，国家制造强国建设战略咨询委员会定于今日至 24 日在广东珠江西岸六市一区（珠海、中山、佛山、肇庆、阳江、江门、顺德）开展现场考察并做好示范评估工作。由中国工程院周济院长担任组长，我也是成员，要参加这项工作。

由上看出，中央对此工作高度重视。

5. 战略部署

《中国制造 2025》战略部署如下：

（1）制造强国战略方针为创新驱动，质量为先，绿色发展，结构优化，人才为本。

（2）制造强国战略原则为市场主导，政府引导，立足当前，着眼长远，整体推进，重点突破，自由发展，开放合作。

（3）分三步走：到 2025 年，迈入制造强国行列；到 2035 年，我国制造业整体达到世界制造强国阵营中等水平；到 2049 年（新中国成立 100 年时），综合实力进入世界制造强国前列！

6. 九项战略

《中国制造 2025》的九项战略为：①提高国家制造业创新能力；②推进信息化与工业化深度融合；③强化工业基础能力；④加强质量品牌建设；⑤全面推行绿色制造；⑥大力推动重点领域突破发展；⑦深入推进制造业结构调整；⑧积极发展服务型制造和生产性服务业；⑨提高制造业国际化发展水平。

7. 重点领域

《中国制造 2025》的十大重点领域为：①新一代信息技术产业；②高档数控机床和机器人；③航空航天装备；④海洋工程装备及高技术船舶；⑤先进轨道交通装备；⑥节能与新能源汽车；⑦电力装备；⑧农机装备；⑨新材料；⑩生物医药及高性能医疗器械。

8. 重点领域实例

例如，对于航空航天领域，重点是大飞机研制、发动机研制以及航天火箭、卫星、飞船的发展。

9. 智能制造政策导向

(1) 要实现制造技术、过程和产品的智能化。

(2) 要采用九大关键智能基础共性技术：①新型传感技术；②模块化、嵌入式控制系统设计技术；③先进控制与优化技术；④系统协调技术；⑤故障诊断与健康维护技术；⑥高可靠实时通信网络技术；⑦功能安全技术；⑧特种工艺与精密制造技术；⑨识别技术。

(3) 要采用八项核心智能测控装置与部件：①新型传感器及其系统；②智能控制系统；③智能仪表；④精密仪器；⑤工业机器人与专用机器人；⑥精密传动装置；⑦伺服控制机构；⑧液气密元件及系统。

(4) 要使用八类重大智能制造成套设备：①石油化工智能成套设备；②冶金智能成套设备；③智能化成型和加工成套设备；④自动化物流成套设备；⑤建材制造成套设备；⑥智能化食品制造生产线；⑦智能化纺织成套设备；⑧智能化印刷设备。

10. 企业积极布局智能制造

(1) 随着互联网技术及理念渗透，制造企业要着手推动生产方式、商业模式和组织方式转型。

案例：海尔、美的等传统家电制造企业积极改变传统的以出售硬件产品赚取成本差价的商业模式，通过构建智能云平台，发挥数据、交互、服务等优势而获得收益。

(2) 面对智能制造发展的需求，国内企业纷纷走智能化、柔性化、定制化的道路，如阿里巴巴企业。

(3) 智能制造行业热点为家电、汽车等领域，具有产品直接面向消费者的先天优势，将提升智能化和互联网化，如百度实现无人驾驶汽车，海尔、美的构建智能家居生态系统。

(4) 行业发展趋势。智能制造发展是一项复杂的系统性工程，企业间兼并与合作将更为频繁。

11. 政府职能：体制与制度创新

(1) 面对新一轮工业革命，政府部门要加大体制与机制创新的力度。要做好顶层设计，加强统筹规划，完善相关领域标准，积极推进各行业、各领域规划政策的出台。

(2) 扩宽资金渠道，促进核心技术研发及应用，要扶持智能制造关键核心技术研发和产业化。

(3) 依托相关科研机构、学会、协会和产业园区，分行业打造智能制造试点示范工程。

结束语：大数据、移动互联网、云计算等新一代信息技术加速了智能制造技术的发展，深刻地影响我们的生产和生活。实现伟大的中国梦，要靠智能制造。这是时代赋予科学家和在座领导同志的光荣任务，让我们携手共进，为顺利实现中国制造的战略部署共同努力。

网络强国战略与实践[①]

习近平总书记就建设网络强国发表了一系列重要讲话，是治国理政新理念新思想新战略的重要组成部分。浙江是习近平总书记多年工作过的地方，时任浙江省委书记的习近平同志大力推进“数字浙江”建设，为浙江以信息化驱动引领现代化奠定了坚实基础。党的十八大以来，浙江贯彻落实习近平总书记系列重要讲话精神，以信息经济创新发展、电子政务深化应用、网络文化滋养社会、信息惠民造福百姓的实践行动，树立了实施网络强国战略的区域样板。

《网络强国战略与浙江实践》一书，研究阐述了习近平总书记关于网络强国建设重要论述的要义和浙江贯彻实施网络强国战略的创新实践。作者有多年研究国家特别是浙江信息化发展战略的积累，本书是在深入学习研究习近平总书记关于网络强国战略的系列重要讲话精神，系统分析我国特别是浙江以信息化驱动现代化实践的基础上完成的。具有以下特点。

一是思想性和针对性强。本书把习近平总书记关于“没有网络安全就没有国家安全，没有信息化就没有现代化”等网络强国战略思想贯穿于我国和浙江网信工作实际进行研究阐述，既有对网络强国战略的科学分析和理论解读，又有对浙江实施网络强国战略实践的深度剖析和经验提炼。

二是系统性和广泛性强。网络强国战略是一个内涵丰富、内容充实的体系，是目标与任务的统一体。本书研究涉及互联网对经济、政治、文化、社会、生态建设的广泛深刻影响，尤其对浙江信息经济、电子政务、网络文化、网络安全与网络治理等诸多领域都有机理分析与实践总结。

三是时代性和创新性强。本书紧密结合当今世界网络信息技术日新月异，全面融入社会生产生活，深刻改变着全球经济格局、利益格局、安全格局的时代特征，客观分析我国从“网络大国”走向“网络强国”取得的成效与存在的问题，以浙江区域样板为代表，创新性地论述网络强国战略的推进机制、实施途径及经验做法。

本书理论联系实际，研究内容丰富，对我国实施网络强国战略的研究具有良好的理论价值，总结的路径模式和行动经验对推进网络强国建设实践也有现实指导作用。

---

① 本文是《网络强国战略与浙江实践》序，北京：科学出版社，2016，3-4.

两化深度融合与质量管理①

2008 年金融危机以后，主要发达国家为了寻找促进经济增长的新出路，纷纷发布新型工业化发展的国家级战略，我国也在 2015 年 5 月发布《中国制造 2025》战略。党的十六大报告指出，“以信息化带动工业化，以工业化促进信息化”，即“两化融合”的论断促使“中国制造”向“中国智造”转变。

一、强国战略的近代历史

1. 第一次工业革命（1760～1840 年），蒸汽时代，以英国为代表。

瓦特（1736～1819 年），用 26 年时间，发明高效率瓦特蒸汽机，使英国成为强大国家。

标志性：英国被称为“日不落帝国”，殖民地曾达 3000 万平方公里，有 3 个中国大，英国本土面积仅 24.4 万平方公里。

2. 第二次工业革命（1840～1950 年），电气时代。

库仑（1736～1806 年），法国物理学家，电的库仑定律提出者。

爱迪生（1847～1931 年），美国发明家，发明白炽灯。在 1881 年，在纽约市建立第一个发电站和配电系统。

汤姆孙（1856～1940 年），英国物理学家。发现电子，随后是二极管、三极管相继问世。电气化和汽车、飞机是标志。德国和美国成为强大国家。

3. 第三次工业革命（1950～2000 年），信息时代。

1946 年，第一台电子计算机 ENIAC 在美国投入运行，每秒可进行 5000 次加减运算。影响巨大！

我在 1958 年，在中国科学院兰州分院第一次见到计算机，是从苏联买来的 M3 型，全国有 6 台，是我国最早的计算机，一台计算机占整整一层楼面。

1981 年，美国 IBM 公司推出了个人计算机，即微机（PC）。

在 20 世纪 80 年代，美国的超级计算机，每秒可进行 100 万亿次计算。如果用以往的计算器计算，则要花 1000 万年才行。我在 20 世纪 60 年代则使用这种计算器，比我们的算盘好得多。

1962 年，美国麻省理工学院的一名研究生克兰罗克发明了互联网技术。

1972 年，美国的汤姆林森推出了电子邮件。

1999 年，全世界互联网已有 1.5 亿用户。2015 年有 30 亿用户。

计算机和互联网深刻地改变全世界人们的生活和工作方式。我虽然不是计算机专业的学者，但我对计算机和互联网还是比较敏感的。1991 年底来到暨南大学工作，就将学校财务处的财务管理从算盘改为计算机。将校办打印文件从键盘打字机改用计算

① 本文是为中华人民共和国人力资源和社会保障部举办的高级研修班所作的报告，杭州，2016 年 11 月 29 日.

机，并提出建立互联网站，于 1994 年建成，是广东高校第一家，离现在 22 年，发展多快啊！

互联网技术不由任何公司、企业或国家占有或控制，它通过计算机、光纤、卫星、电话线把全世界的人即时连接起来，正在改变文化模式、商业活动、工业方式，以及研究和教育工作、网上办公、网上读书、网上查阅资料、网上购物、网上聊天交朋友等，正在深刻地改变世界。

计算机和互联网的发明是 20 世纪技术科学的最伟大成就，是 20 世纪的重要象征，它带动了一次世界性的技术革命，使信息时代来临！

4. 第四次工业革命（2000 年～），带来颠覆性变化。

以互联网产业化、工业智能化、工业一体化为代表，以人工智能、清洁能源、无人控制技术、量子信息技术为主的新技术革命！国际上掀起了新一轮科技革命和产业变革的浪潮。

我国是第一次与美国、欧盟、日本等发达国家站在同一起跑线上，正式发动和创新绿色工业革命。

1）美国“先进制造”计划

2011 年 6 月，美国正式启动包括工业机器人在内的“先进制造伙伴计划”。

2012 年 2 月，又启动《美国先进制造业国家战略计划》。

2013 年，美国将先进制造业国家战略聚焦到先进制造的感知控制、智能制造技术平台与先进材料制造三大技术的优先突破。美国希望信息技术优势带动制造业，重整制造业实体经济，保持在全球的领导地位。

2）德国“工业 4.0”计划

2006 年，德国提出《德国高技术战略》。

2010 年，德国政府制定《高技术战略 2020》。

2013 年 4 月，三大德国协会，德国信息技术、通信与新媒体协会，德国电子电气制造商协会，德国机械及制造商协会共同组建“工业 4.0”平台。德国发起的“工业 4.0”概念非常有代表性，引领了第四次工业革命浪潮。

“工业 4.0”基本概念包含以下七个方面：①建设智能工厂。②搭建信息物理系统。③发展智能设备和智能决策系统。④物联网。⑤产品和服务开发的新流程。⑥柔性制造。⑦绿色、可持续发展。

“工业 4.0”基本框架：通过传感器、计算机、互联网将真实世界和网络世界相连接，实现智能制造和智能生活。

“工业 4.0”的目的：提升制造业的自动化、数字化与智能化是德国的国家战略。

3）法国“新工业”计划

2013 年 9 月，法国提出《新工业法国》，希望充分发挥法国高科技工业领域优势，研发了 3D 打印、智能机器人等高科技智能化技术，推动了传统制造业向未来高端制造业转型，重振法国制造业。

4）英国“高价值制造”战略

2013 年 10 月英国推出《英国工业 2050 战略》，提出的新型制造业模式“服务＋再制造”，复苏英国制造业。

5）日本“机器人新战略”

2015年1月，日本公布了“机器人新战略”，主要有两大目的，即“扩大机器人应用领域”与“加快新一代机器人技术研发”，主要任务为实现物联网技术对日本制造业的变革，使日本成为世界的机器人创新中枢。

6）韩国“新增长动力规划及发展战略”

确定3大领域17个产业为发展重点，推进数字化工业设计和制造业数字化协作建设，加强对智能制造基础的开发。

7）印度“印度制造”计划

以基础设施建设、制造业和智慧城市为经济改革战略的三根支柱，通过智能制造技术的广泛应用将印度打造成新的“全球制造中心”。

8）中国“中国制造2025”战略

2015年5月，中国发布《中国制造2025》，这是中国企业智能制造战略，就是以信息化带动工业化，以工业化促进信息化，使两化深度融合，促进中国制造业转型，使“中国制造”向“中国智造”转变，由制造大国向制造强国转变！

我国不再是前三次世界工业革命的跟随者，我国在第四次世界工业革命中要起引领作用，要走在国际先进行列！

二、“中国制造”的概述

1. 战略的伟大意义

在第一次工业革命时，我们中国未参加，而且在1840年，即第一次工业革命的末尾时，英国成为强国，对我们发动了第一次鸦片战争，来了4000个英国兵，48艘船只，装有540门大炮，侵略我国，清朝战败，一个4亿人口的大国被打败了！签了不平等的《南京条约》，割让香港岛。赔款2100万银元，是中国的奇耻大辱，我国开始沦为半殖民地半封建社会！

直到1949年10月1日，旧中国一直被动挨打，与外国签了1182个不平等条约，平均每个月签一个，割让土地数百万平方公里，仅清朝就赔给外国16亿两白银，中国成了半殖民地半封建的落后国家，被称为“东亚病夫”！

深刻教训：经过一代又一代人的努力，直到新中国成立后，我们才步入正轨。在第三次工业革命时，我们主动参加，但仅是跟随者，特别是在1978年改革开放后，我们的步伐大大加快了，但是，中国制造是以代加工、贴牌生产闻名。不过，经过30余年高速度发展，我国成为世界上第二大经济体，对我们中华民族来说，终于摆脱了落后挨打受欺辱的位置，丢掉了“东亚病夫”的帽子，扬眉吐气了！

今年5月30日，习近平总书记号召，我们要建设世界科技强国，要实现中华民族伟大复兴的中国梦！因此，我们一定要按国家2015年制定的《中国制造2025》发展战略前进！就是说，在第四次工业革命中，这个战略的伟大意义，就是我们中华民族在世界近代史上，第一次成为世界先进行列一员，是引领者！要由制造大国向制造强国转变！

《中国制造 2025》开宗明义提出，“制造业是国民经济的主体，是立国之本、兴国之器、强国之基”。我国制造业经过 67 年发展，建成了门类齐全、独立完整的产业体系，有力地推动着工业化和现代化进程，显著增强了综合国力。根据世界银行统计，2012 年，我国制造业增加值为 23 306.8 亿美元，超过美国的 18 532.7 亿美元，位居世界第一位。中美两国分别占全世界制造业增加值的 20.3%和 16.2%，显然，我国已是制造大国。然而，与世界先进水平相比，我国制造业仍然大而不强，在自主创新能力、资源利用效率、产业结构水平、信息化程度、质量效益等方面差距明显。

例如，在自主创新能力方面，我国创新能力不强，核心技术薄弱，共性技术缺位。根据我国社会科学文献出版社公布的《世界创新竞争力发展报告（2001～2012）》，2010 年我国的创新竞争力排名在全世界的第 15 位。我国自行研制的大型客机 C919 的全部发动机均靠进口。正在研制的重型燃气轮机的关键材料和关键零部件，如高温合金定向及单晶工作叶片仍依靠进口。近年来，我国加大了对数控机床研发的投入。但是，80%的高端数控机床至今仍靠进口。2013 年，我国已是机器人消费大国，位居世界第一位，但 80%的机器人是进口的。在产品质量方面，国际上畅销的“中国制造”产品，大量的还是中低端产品，国内消费品的质量，比起出口产品，在不少方面还有差距。

因此，我们只是制造大国，而不是制造强国。

制造强国的具体定义，国际上没有统一的标准。一般认为应该有以下五大特征：①拥有一定数量世界知名的企业；②具备高创新能力及竞争力；③掌握尖端技术和核心技术；④效率提升与质量安全兼具；⑤具备可持续发展的潜力。

2. 智能制造是建设制造强国的根本路径

由“中国制造”到“中国智造”。这是走向制造强国的制造业发展的重要方向，必须大力发展。智能制造是制造技术与数字技术、智能技术及新一代信息技术的融合，是面向产品全生命周期的具有信息感知、优化决策、执行控制功能的制造系统，旨在高效、优质、柔性、清洁、安全、敏捷地制造产品，服务用户。

智能制造源于人工智能，兴于 20 世纪末 21 世纪初期。

（1）它包括以下五个方面内容：①制造装备的智能化；②设计过程的智能化；③加工工艺的优化；④管理的信息化；⑤服务的敏捷化、远程化。

（2）它有以下三个主要特征。①信息感知。智能制造需要大量的数据支持，通过利用高效、标准的方法实时进行信息采集、自动识别，并将信息传输到分析决策系统。②优化决策。通过面向产品全生命周期的信息挖掘提炼、计算分析、推理预测，形成优化制造过程的决策指令。③执行控制。根据决策指令，通过执行系统控制制造过程的状态，实现稳定、安全的运行。

（3）它需要三个融合：①知识融合；②学科融合；③技术融合。

于是便由技术集成到技术融合，由集成创新到融合创新。即从 1+1=2 到 1+1>2，这样协同创新的结果便是功能效益和技术附加值的大幅度提升。

融合的实例：①乔布斯的苹果手机，是 IT 与人文艺术的融合，开创了商业附加值的新兴产业；②我国的高铁工程，就是集成创新的结果，现有 19 000 公里，世界第一，速度 300km/h～350km/h。

（4）制定政策。2015 年，国家的《中国制造 2025》制造强国战略出台后，中央和全国各省据此文件也出台了具体政策，如：《国务院关于加快培育和发展战略性新兴产业的决定》《广东省智能制造发展规划（2015—2025 年）》《福建省实施〈中国制造 2025〉行动计划》《浙江省国民经济和社会发展第十三个五年规划纲要》……

5 月 30 日，在全国科技创新大会上，习近平总书记、李克强总理发表了重要讲话，我也在会场聆听。我作为《中国制造 2025》发展战略计划项目第三课题组成员，从 2013 年起，参加了项目的研究工作。今年 6 月，为做好《中国制造 2025》试点示范工作，国家制造强国建设战略咨询委员会在广东珠江西岸六市一区（珠海、中山、佛山、肇庆、阳江、江门、顺德）开展现场考察并做好示范评估工作，中国工程院周济院长亲自担任组长，我也作为成员参加了这项工作，我亲身体会到中央对此工作的高度重视。

（5）《中国制造 2025》战略部署。

制造强国战略方针：创新驱动、质量为先、绿色发展、结构优化、人才为本。

制造强国战略原则：市场主导、政府引导、立足当前、着眼长远、整体推进、重点突破、自由发展、开放合作。

分三步走：到 2025 年，迈入制造强国行列；到 2035 年，我国制造业整体达到世界制造强国阵营中等水平；到 2049 年（新中国成立 100 年时），我国综合实力进入世界制造强国前列！

（6）九项战略。①提高国家制造业创新能力。②推进信息化与工业化深度融合。③强化工业基础能力。④加强质量品牌建设。⑤全面推行绿色制造。⑥大力推动重点领域突破发展。⑦深入推进制造业结构调整。⑧积极发展服务型制造和生产性服务业。⑨提高制造业国际化发展水平。

（7）十大重点领域。①新一代信息技术产业。②高档数控机床和机器人。③航空航天装备。④海洋工程装备及高技术船舶。⑤先进轨道交通装备。⑥节能与新能源汽车。⑦电力装备。⑧农机装备。⑨新材料。⑩生物医药及高性能医疗器械。

例如，航空航天领域的重点是大飞机研制、发动机研制、航天火箭、卫星、飞船的发展。

我从 1958 年开始，参加中国第一颗东方红卫星研制项目，即“581”工程项目，担任小组长，回想起来，至今仍使人激动不已。此后，又参加了神舟飞船的科研项目。以及第一艘航空母舰辽宁舰上的歼- 15B 飞机的研制项目，深深感到这一领域的重要性。

在信息技术产业方面，重点是精密仪器仪表和传感器。我参加了一些精密仪器仪表和传感器的研制（如 P2V 军事侦察飞机的高度表），涉及核心弹性元件的研制（包括波纹膜片），此外，也研究了跳跃膜片、波纹管等。目前我国急需在这一领域有所突破。

（8）智能制造政策导向。

①实现制造技术、过程和产品智能化。②九大关键智能基础共性技术：新型传感技术，模块化、嵌入式控制系统设计技术，先进控制与优化技术，系统协调技术，故

障诊断与健康维护技术，高可靠实时通信网络技术，功能安全技术，特种工艺与精密制造技术，识别技术。③八项核心智能测控装置与部件：新型传感器及其系统、智能控制系统、智能仪表、精密仪器、工业机器人与专用机器人、精密传动装置、伺服控制机构、液气密元件及系统。④八类重大智能制造成套设备：石油化工智能成套设备、冶金智能成套设备、智能化成型和加工成套设备、自动化物流成套设备、建材制造成套设备、智能化食品制造生产线、智能化纺织成套设备、智能化印刷设备。

（9）企业积极布局智能制造。随着互联网技术及理念渗透，制造企业要着手推动生产方式、商业模式、组织方式转型。实例：海尔、美的等传统家电制造企业积极改变传统的以出售硬件产品赚取成本差价的商业模式，通过构建智能云平台，发挥数据、交互、服务等优势而获得收益。

面向智能制造发展的需求，国内企业纷纷走智能化、柔性化、定制化的道路，如阿里巴巴企业。

智能制造行业热点：家电、汽车等领域，具有产品直接面向消费者的先天优势，将提升智能化和互联网化，如百度要实现无人驾驶汽车，海尔、美的构建智能家居生态系统。

行业发展趋势：智能制造发展是一项复杂的系统性工程，企业间兼并与合作将更为频繁。

（10）政府职能：体制与制度创新。①面向新一轮工业革命，政府部门要加大体制与机制创新的力度。首先要做好顶层设计，加强统筹规划，完善相关领域标准，积极推进各行业、各领域规划政策的出台。②拓宽资金渠道，促进核心技术研发及应用，要扶持智能制造关键核心技术研发和产业化。③依托相关科研机构、学会、协会和产业园区，分行业打造智能制造试点示范工程。

结语：智能制造是四次工业革命以来，最先进的工业生产方式，是未来企业生产发展的趋势，是制造业发展的重要方向，是国际制造业科技竞争的制高点，是我国建设制造强国的根本路径。

3. “质量为先”是实现制造强国的根本保障

《中国制造 2025》战略，制造业必须走“质量为先”的发展道路，真正靠质量取胜，质量就是建设制造强国的生命线。

1）质量的定义

国际标准化组织对质量所给的定义是“一组固有特性满足要求的程度”。这个定义既强调质量特性的形成，同时强调满足要求，质量特性是固有特性，其重要性在于质量特性的形成存在于产品的全寿命周期，质量是设计出来、制造出来的，而不是检测出来的。

2）质量的重要性

如今美国有哪些世界级的巨型企业？美国哪几家企业名列世界前茅的时间能持续100年？答案恐怕就只有一家：美国通用电气公司。成立于1892年的通用电气，历经百年，风采依旧，至今仍位居《财富》“全美最受推崇公司”排行榜最前列，总市值一直在5000亿美元以上，已成为世界上最具盈利能力和增长能力的公司。商海浮沉，风云变幻，是什么使英雄不倒？是质量！

海尔集团是一家土生土长的中国企业，原本是一个生产电动葫芦的集体企业。自张瑞敏 1984 年出任厂长以来，步步为营，保持了年平均 80%以上的高速增长，发展成年销售额收入达 897 亿元的跨行业、跨国界的大型企业集团。海尔集团成功崛起的主要原因是什么？是质量！

质量是企业生存的根本，没有质量的企业，是没有生命力的企业。新近一项统计数据显示，中国大多数企业一般的寿命为 2～3 年，成功的企业寿命为 8～10 年，而能生存 40 年以上的企业仅为约 2%，这也就说明了为什么像同仁堂这样的“百年老店”贵乎稀有！

水能载舟亦能覆舟，成也萧何败也萧何，酿成这种局面的原因是什么呢？还是质量！

3）质量管理的历史

第一次工业革命前（1760 年）：

产品质量由工匠或手艺人自己控制。

1875 年：

美国的泰勒制产生，这是科学管理的开端。最初的质量管理——检验活动与其他职能分离，出现了专职的检验员和独立的检验部门。

1925 年：

美国的休哈特提出统计过程控制（statistical process control，SPC）理论，应用统计技术对生产过程进行监控，以减少对检验的依赖。

1930 年：

美国的道奇和罗明提出统计抽样检验方法。

20 世纪 40 年代：

美国贝尔电话公司应用统计质量控制技术取得成效；美国军方在军需物资供应商中推进统计质量控制技术的应用，制定了最初的质量管理标准。

20 世纪 50 年代：

美国的戴明指出：在生产过程中，造成质量问题的原因只有 10%～15%来自工人，而 85%～90%是企业内部在管理系统上有问题。他用统计学的方法进行质量和生产力的持续改进；强调大多数质量问题是生产和经营系统的问题；强调最高管理层对质量管理的责任。最终形成对质量管理产生重大影响的“戴明十四法”。

20 世纪 60 年代初：

美国朱兰、费根堡姆提出全面质量管理（total quality management，TQM）的概念。他们提出，为了生产具有合理成本和较高质量的产品，需要对覆盖所有职能部门的质量活动进行策划。这一全面质量管理理论在日本备受推崇。日本企业创造了全面质量控制（total quality control，TQC）的质量管理方法。

20 世纪 60 年代中：

北大西洋公约组织（North Atlantic Treaty Organization，NATO）制定了 AQAP 质量管理系列标准，引入了设计质量控制的要求。

20 世纪 70 年代：

TQC使日本企业竞争力极大提高，轿车、家用电器、手表、电子产品等占领了大批国际市场，促进了日本经济的极大发展。日本企业的成功，使全面质量管理的理论在世界范围内产生巨大影响。

20世纪80年代：

美国克劳士比提出“零缺陷”概念。他指出，高质量将给企业带来高经济回报。质量运动在许多国家展开，包括中国、美国、欧洲等许多国家设立了国家质量管理奖，以激励企业通过质量管理提高生产力和竞争力。质量管理不仅被引入生产企业，而且被引入服务业，甚至医院、机关和学校。

1987年：

由国际标准化组织所制定的质量管理和质量保证的一系列国际标准，即ISO 9000面世。质量管理与质量保证开始在世界范围内对经济和贸易活动产生影响。

1994年：

ISO 9000系列改正版面世。新的标准更加完善，为世界绝大多数国家所采用。

20世纪90年代末：

全面质量管理成为许多“世界级”企业的成功经验，证明这是一种使企业获得核心竞争力的管理战略。

在围绕提高质量、降低成本，缩短开发和生产周期方面，新的管理方法层出不穷，其中包括：并行工程（concurrent engineering，CE）、企业流程再造（business process reengineering，BPR）等。

21世纪初：

随着知识经济的到来，知识创新与管理创新必将极大地促进质量的迅速提高。质量管理的理念和方法更加丰富，不断突破旧的范畴而获得极大的发展。

4）质量管理的基本原则

①质量的定义必须是符合要求，清楚地了解要求，然后消除一切阻碍。②质量系统的核心在于预防，而不是检验和评估。③工作标准必须是“零缺陷”，是第一次就把事情完全做对，而不是“差不多就好”。④质量是用不符合要求的代价来衡量的，所有做错的事情引致的花费是它的代价，而不是用指数或图表。

5）质量管理的具体内涵

（1）全面质量的概念。质量不光是产品的技术性能，还包括服务质量和成本质量（价格要低廉）；质量由设计质量、制造质量、使用质量、维护质量等多种因素构成。质量是设计和制造出来的，而不是检验出来的。

（2）全过程质量管理。其范围是产品质量产生形成和实现的全过程，包括市场调查、研究、开发、设计、制造、检验、运输、储存、销售、安装、使用和维护等多个环节和整个过程的质量管理。

（3）全员参与的质量管理。调动企业所有人员的积极性和创造性，使每一个人都参与到质量管理工作中来。

（4）全企业质量管理。企业各个管理层次都有明确的质量管理活动内容，产品质量职能分散在企业各有关部门，形成一个有机体系。

6）六西格玛质量理论

六西格玛质量理论是当下风行的引领人们进行质量革命的有效方法之一，它使美国通用电气公司和海尔集团受益。

六西格玛就是阿拉伯数字 6 加上希腊字母 σ，即“六个标准差”。

在 1 西格玛的情况下，所做的工作只有约 30％是正确的，即 1σ＝690 000 次失误/百万次操作。

在 2 西格玛的情况下，每 100 万次机会将出现 30 多万个缺陷，即 2σ＝308 000 次失误/百万次操作。

在 3 西格玛的情况下，出现 66 800 个缺陷，即 3σ＝66 800 次失误/百万次操作。

在 4 西格玛的情况下，大约存在 6000 个缺陷，即 4σ＝6210 次失误/百万次操作。

在 5 西格玛的情况下，出现 230 个缺陷，即 5σ＝230 次失误/百万次操作。

在 6 西格玛的情况下，出现 3.4 个缺陷，即 6σ＝3.4 次失误/百万次操作。

大部分公司是在 3 至 4 西格玛状态下工作，在 3.8 个西格玛的时候意味着达到了 99％的工作是处于正确的状态。但是，即使出现 1％的缺陷，也将导致大量的差错，相当于美国全国每小时丢失 2 万件邮件，或每周做错 5000 例外科手术，更有甚者类同于繁忙的大机场每天发生 4 起事故！就我们熟悉的工作来估算，1％的缺陷将导致怎样的损失和灾难？令人心惊！

数字既能让你清晰地表述问题，也能使你达到期望的准确的目标，六西格玛的最终目标是改善质量。它是一套管理方法，与其他质量理论的根本差别是强调不仅要挑出次品，更要寻根溯源，解决产生次品的根本性问题。

“百年大计，质量为先”就是我们永恒的主题！

为了实现制造强国的战略，《中国制造 2025》已为我们未来十年进行了顶层设计和规划了路线图，通过两化深度融合与质量管理，努力实现中国制造向中国创造、中国速度向中国质量、中国产品向中国品牌三大转变，推动中国到 2025 年基本实现工业化，迈入制造强国行列，为中华民族伟大复兴做出贡献！

推动广东非开挖产业快速发展①

2002 年 3 月 10 日，广东省非开挖技术协会在广州成立，至今已经成立了有 15 个年头，经历了 2 届的成长。协会的成立反映了广东的综合经济实力和改革创新精神，也反映了广东非开挖业界团结协作，奋发进取的精神。

协会成立以来，在省科协的指导及会员的大力支持下，有序开展各项工作。协会推动了广东非开挖产业的快速发展，也带动了华南地区甚至国内非开挖产业的发展。我们经历了由国外引进现代非开挖技术，然后本土化，进而再国际化的过程。

协会全体会员，为行业的发展，为协会工作的顺利开展，做出了不懈的努力，取得了一定的成绩。

现将本届理事会的主要工作总结如下。

一、协会各项工作成果

目前，国内地区性非开挖协会有广东、上海、北京和福建四个地方协会及中国地质学会非开挖技术专业委员会（China Society for Trenchless Technology，CSTT）。广东省非开挖技术协会现有会员单位 77 家，个人会员 5 位。协会全体理事为行业的发展，为协会工作的顺利开展，做出了不懈的努力，取得了一定的成绩，较好地推动了华南地区的非开挖事业的发展。

1. 认真抓好协会建设

协会自成立以来，就致力于协会的内部建设工作。随着各项工作的顺利开展，各项规章制度日趋完善，协会的机构设置也更加完善。采用了理事长、副理事长对部门的分管负责制度，指导监督和协调各部门开展相关工作，秘书处负责日常工作。

在协会的正规化建设方面，我们注重并要求协会的每个工作活动程序均要按照相关规范的流程进行。对协会工作认真严肃地对待，保证协会各项工作的顺利开展和进行。

协会的经费主要来源于会费、培训及外联赞助。协会严格按照规定收取会员会费，在经费管理方面，主要由协会秘书处负责，协会所有会费、赞助费用皆设立了账册及财务清单，由协会办公室相关负责成员详细列载经费收支，并定期向协会负责人进行汇报，接受协会全体成员的监督。协会经费全部用于协会的相关工作和活动，经费使用以高效和节约为原则。每次活动前由承办部门详列预算，活动过后提交正规、清楚的收据或发票等凭据，做到财务收支协调合理。

2. 积极开展学术交流及技术培训

本届理事会积极开展专题研讨活动和新技术推广活动，接待各地学者讲学，举办非开挖技术研讨会。

---

① 本文是广东省非开挖技术协会第二届理事会工作报告，广州，2017 年 12 月 15 日.

2012 年 12 月 9 日上午，在暨南大学召开了 2012 年技术交流会，参会代表达 100 多人，涉及省内近 20 家相关单位。会上广州市市政集团安关峰教授级高级工程师、广州地铁廖鸿雁高级工程师、广东水电二局林世友高级工程师、暨南大学吴起星副教授等围绕管道修复、顶管、盾构等典型非开挖技术作主题报告。

2013 年 5 月 8～9 日，协会与住房和城乡建设部市政给水排水标准化技术委员会及广州市市政集团有限公司在暨南大学共同举办《城镇排水管道检测与评估技术规程》宣贯暨非开挖技术研讨会。广州市水务局吴学伟副局长、住房和城乡建设部市政给水排水标准化技术委员会吕士健秘书长等嘉宾出席会议并致辞，同时吸引了来自全国各地的管道检测、修复单位，相关高校，广东省各水务部门、排水公司等共百余人参与此次会议。

2014 年 4 月 18～20 日，协会理事安关峰总工代表协会参加了在东莞市广东现代国际展览中心举办的第十八届中国国际非开挖技术研讨会暨展览会，并作了广东非开挖技术现状及发展趋势的主题报告。

2014 年 11 月 6～7 日，由广东省非开挖技术协会、广东省市政行业协会、广州市市政集团有限公司主办，暨南大学广东省城市生命线工程结构力学应急技术研究中心、广州市微资环境治理有限公司承办了“2014 广东省首届内涝防治研讨会”，研讨会在暨南大学学术报告厅举行，共举办 10 个专题报告会。

2015 年 3 月 13～14 日，协会在广州欧亚酒店主办了第一期“城镇排水管道检测与评估及非开挖修复技术岗位培训班”，来自广州、深圳、福建等地的 36 家单位 100 多人参加了培训。培训班的主要内容有：宣讲国家的排水政策、住建部行业标准《城镇排水管道检测与评估技术规程》（CJJ 181-2012）、《城镇排水管道非开挖修复更新工程技术规程》（CJJ/T 210-2014）的内容及主要条文解析，同时对参加培训人员进行结业考试，并为合格者颁发了技术岗位证书。

2015 年 8 月 6～7 日，由广东省非开挖技术协会主办，佛山鼎汉文化传播有限公司承办，佛山市钻龙市政工程有限公司、徐工集团工程机械股份有限公司等单位协办的 2015 年广东省非开挖高峰研讨会在广东省佛山市凯利莱国际酒店召开，来自全国各地的近百位非开挖行业代表出席了此次会议。

2016 年 4 月 15～16 日，协会在暨南大学举办 2016 年城镇排水管道检测与评估技术及城镇排水管道非开挖修复技术岗位培训班，参加学员有 39 位。

2015～2016 年，为推动我省城镇排水管道非开挖修复技术的推广应用，协会在会员的大力支持下组织编制了《广东省排水管道非开挖修复工程预算定额》。2016 年 4 月 28 日在广州市暨南大学蒙民伟理工楼 516 会议室召开了该定额专家评审会，会后定额编制组结合评审会专家组及各专家的意见对其进行了认真修改完善。2016 年 7 月 20 日发布公告实施《广东省排水管道非开挖修复工程预算定额》。

2016 年 11 月 26 日，由协会主办、暨南大学力学与建筑工程学院承办了“广东省非开挖技术协会 2016 年年会暨技术交流会”，会议在学校蒙民伟理工楼 103 室召开，来自协会的近 40 位代表出席了此会议。

2017 年 4 月 15～16 日，协会在暨南大学举办 2017 年城镇排水管道检测与评估技术及城镇排水管道非开挖修复技术岗位培训班，参加学员数为 57 人。

2017 年 11 月 16 日，由广东省非开挖技术协会主办、暨南大学力学与建筑工程学院承办的 2017 年广东省科协学术活动周学术讲座之“非开挖技术的应用及发展”在暨南大学学术报告厅胜利召开，非开挖协会会员、非开挖行业代表、暨南大学力学与建筑工程学院师生代表等近百人出席了本次讲座。广州市市政集团有限公司安关峰总工、广州金土岩土工程技术有限公司董事长陈雪华博士分别作了题为“非开挖技术的应用及发展”“大断面长距离矩形顶管在地下空间开发中的应用研讨”的讲座。

2017 年 11 月 18～19 日，协会将技术岗位培训申请纳入 2017 年广东省科协学术活动周活动，在暨南大学举办 2017 年第二期城镇排水管道检测与评估技术及非开挖修复技术岗位培训班，有 25 位学员参加了培训。

3. 联合企业，做好行业服务工作

2015 年 6 月 4 日，与佛山市钻龙市政工程有限公司合作，组织水平定向钻技术讲座，与佛山市钻龙市政工程有限公司、廊坊钻王科技深远穿越有限公司、武汉地网非开挖科技有限公司等作了技术交流。

2016 年 9 月，协会王和平副教授和吴起星副教授到佛山调研，与佛山市三丰市政工程有限公司、武汉中仪物联技术股份有限公司相关负责人等探讨非开挖交流合作事宜。

2017 年 10 月 27 日，会员广州万顺建筑工程有限公司在广州市广东迎宾馆碧海楼隆重举办了“广州万顺建筑—2017 水平定向钻行业交流会”。协会理事安关峰总工、秘书处吴起星副教授及王和平副教授参加了此次会议，并与行业代表进行有益交流。

非开挖产业是一个高新技术、高竞争的行业，特别是设备制造业，国外入粤强手如林，面对设备、技术上的差距和国际主流厂商在粤登陆的市场态势，协会工作向本土企业、本地企业倾斜，特别是向有实力和活力的民企设备制造厂商倾注了精力，协会就企业定位、海外市场、技术创新等向本土企业提出建议，提供服务。

4. 建立和加强协会间的联系

协会与国内的上海、北京、福建及香港非开挖技术协会建立了稳定的联系，建立了定期互通信息的机制。

多年来，协会与上海、北京、福建非开挖技术协会保持紧密联系，定期沟通信息，并作为地区间会员单位的沟通平台。通过香港非开挖技术协会，我协会保持与国际协会的联络，保持与世界各国非开挖协会的联系。

2012 年 8 月副理事长兼秘书长马宏伟教授与北京非开挖技术协会代表会谈，随后北京、上海以及广州三家地方协会已达成共识，同意联合启动中国非开挖技术协会的筹备成立工作，建议以北京、上海以及广州三家协会为主体组织全国协会，组建筹备小组（由每个协会派 2 人）。加强了各个地区非开挖技术协会的联系。

2015 年 4 月 10～12 日，中国地质学会非开挖技术专业委员会 2015 年 CSTT 会展/学术年会（ITTC①2015）在青岛国际会展中心举行，协会秘书处吴起星副教授前往参加交流学习。

---

① ITTC，International Trenchless Technology Conference，国际非开挖技术研讨会暨展览会。

2015年11月24日，我作为协会理事长，代表广东省非开挖技术协会，应邀参加在中山大学学人馆举办的“巨型城市内涝检测预警关键技术研究与示范”项目启动会。

2016年7月6～9日，由中国地质大学（武汉）中美联合非开挖工程研究中心主办、广州市市政集团有限公司、广东省非开挖技术协会及圣戈班穆松桥协办的“第七届管道工程与非开挖技术国际研讨会（ICPTT①2016）”在广东省广州市广州珠江宾馆成功召开，协会田管凤博士作了大数据技术在盾构施工智能预测的应用研究的专题报告。

2016年10月10～12日，协会派代表参加北京召开的第34届国际非开挖技术协会（International Society for Trenchless Technology）年会暨展览会，协会为该会议的支持单位。

2017年3月24日，在广东省住房和城乡建设厅、广州市住房和城乡建设委员会、广州市科学技术协会的指导下，由广州市市政工程协会主办的“2017中国（广州）城市地下综合管廊高峰论坛”在广州越秀宾馆隆重举行。协会副理事长及秘书长、东莞理工学院副校长马宏伟教授做专题演讲。

5. 促进行业自律，提高自我管理水平

协会的宗旨之一，在于提高行业的自我管理水平。通过建立施工企业安全事故通报机制、工程材料质量事故通报机制、工程设备质量问题通报机制，提高工程的质量水平。

对施工企业的主要负责人和技术骨干进行登记，对技术人员的流动进行跟踪，加强技术人员的管理。

协助政府的相关部门，制定合适的管理条例，有效加强施工安全的管理。

二、协会工作的不足

1. 自身建设有待加强

几年来，协会为行业发展做了大量的工作，但是，协会的内部存在重技术、轻管理的问题，对自身的建设力度不够。

2. 行业管理的功能有待加强

几年来，在行业的规范管理方面，做得不够。未能制定清晰的行业管理思路，尚未形成有力的管理规定。

三、工作展望

非开挖产业是新兴的科技含量高的行业，过去多年，已为广东的基础设施建设发挥了独特的作用；广东的非开挖装备制造、施工力量和市场备受国内、国际同行的关注。我们衷心希望，广东的非开挖产业在这场深刻的公用事业改革中得到新的整合，在已形成的规模中得以提高和发展，并打造出品牌。

继续向本土的企业倾斜服务；积极配合公用事业的深入改革，推介管道修复技术，推动城市综合管廊建设中非开挖技术的应用。

① ICPTT，International Conference on Pipelines and Trenchless Technology，管道工程与非开挖技术国际研讨会。

继续开展非开挖技术培训，尽快实施非开挖作业的单位资质认证工作，推动非开挖行业市场的规范化。

继续开展市场调查、经济分析；主动争取省科协、省建设厅对我们的业务指导；沟通与市政、燃气等协会的交流与联系。

以解决实际问题为切入点，逐步凝聚科研力量，开展非开挖技术方面的科研工作。

相信新一届理事会，在省科协的指导下，在政府相关部门的协助下，在各会员单位的支持下，会取得更好的成绩。

谈谈标准化工作对当前的重要意义[①]

这几年，国家标准化管理委员会授予我标准化科学家的身份，我经常在各地走访调研。今天，又正逢庆祝第49届世界标准日，借此机会，我想谈谈开展标准化工作的意义和重要性。

一、标准化水平是一个国家、一个区域、一个产业、一个企业的综合竞争力和核心竞争力的重要体现

标准通常是产业发展的秩序和规则，一项关键标准往往发挥着某个领域、一定时期内底层设计的作用。一项标准的出台和真正实施，在某种程度上决定了这个地区的产业质量水平，关系到一个企业、产业甚至一个区域的兴衰成败。事实上，标准的产生过程既是产业竞争的动态博弈过程，也是区域间综合竞争力和核心竞争力的博弈过程。

产业的发展实践告诉我们，谁掌握了标准话语权，谁就掌握了产业的主导权。国际上，大部分产业国际标准制定长期由英、美、德、日等工业发达国家主导，我国在国际上的标准制定声音十分微弱，我国产业在经济微笑曲线中基本上还没有笑的勇气和地位。

每个区域都有自己的主导产业，对主导产业的掌控能力和引领能力是这个地区综合竞争力和核心竞争力的重要标志，而标准在其中扮演着关键作用。从这种意义上讲，标准的竞争就是产业领导权的竞争，通过标准竞争获得的产业领先能力，通常可以转化为产业持久的“比较优势”，从而形成垄断力量和先发优势。

毫不夸张地说，标准在某种程度上决定了一个产业或区域的兴衰。另外，关联度较强的产业，标准竞争还会进而影响到上下游产业的竞争绩效。例如，家具产业—家具涂料产业—层压板产业—家具机械产业就是一个产业链条，家具甲醛有害物质限量标准出台后，不仅影响的是家具产业本身，直接受影响的还有涂料和家具层压板等相关产业。

在新形势下，国家强调创新驱动，是要以企业为主，正是由于我国许多产业在产业链中还处于低端地位，在国际环境中受到方方面面制约，并希望通过实施标准化战略，建立创新型国家，逐步取得产业领导权，提升综合竞争力和核心竞争力。

二、标准通常代表着先进生产力的发展方向，引领产业发展

生产力主要有三个要素：劳动者、劳动工具和劳动对象。显然，标准被劳动者掌握，便成为劳动生产力，成为提高质量、降低风险、获得最佳效益的工具；而劳动者在长期的实践中，结合新知识、新技术，不断地总结经验又形成新的标准，这当中标

① 本文是广东省2018年世界标准日宣传庆祝活动主旨演讲，广州，2018年10月14日。

准也可以物化为劳动工具和劳动对象，这时标准就成为物质生产力。随着科学技术的发展，劳动工具的形式会不断提升，从传统工具、自动化工具，再到信息化系统等，随着劳动工具的不断改进，标准会不断创新，标准化水平也就愈来愈高。

由于标准体现了先进的技术，使生产力诸要素更有效地组成一个整体，并最大限度地发挥作用，从而形成最佳秩序和效益。因此，标准代表着速度、代表着质量、代表着安全、代表着健康、代表着效益、代表着最低的风险。尤其在当前创新驱动和加快产业转型升级的过程中，标准体现更多的是产业在社会发展中的超前引领功能。

另外，标准是推动科学技术实践化的重要载体，是促进科学技术不断创新的重要平台，标准的每次制订和修订，既是科学技术在实践基础上的进一步总结，又融入了最新的技术成果，因此标准代表着先进生产力的发展方向和潮流，是发达国家借助环保、健康、质量等合理原则惯用的市场调节手段。

当然，标准也有双重性。一方面，在发达国家，新修订标准代表着先进技术的应用和推广，是技术的发展方向和潮流，也是标准超前引领功能的体现；另一方面，对发展中国家生产企业而言，虽然凭技术手段在一定时间内能够达到其要求，但这必然会增加设计成本、元件成本、管理成本及产品测试成本等，要马上采标就有一定的代价，但这对技术进步的推动是毋庸置疑的。因此说，从某种意义上讲，市场促进了新技术，新技术促进了标准的制订和修订，标准是实实在在引领着产业发展。

三、标准是完善社会主义市场经济的重要市场手段

标准化工作主要是通过制定和实施标准来实现的，标准是调整和规范人们在生产、服务、贸易、消费和创造等活动中的行为和利益关系，是市场经济发达国家用来管理经济和社会的重要手段和市场工具，是有形之手。目前，从我国对标准的分类性质看，国家标准分强制性标准和推荐性标准，企业一旦使用，则承担民事担保责任。

由于标准的应用十分广泛，涵盖了工业、农业及服务业，又分布在公共安全、建筑安全、产品质量、节约能源、社会管理、交通、航空、航天等各领域，加强对标准实施监督本身就是市场经济的有形之手。特别是在产业发展过程中，如何维护公平竞争，如何提高产品质量，如何不断提升产业竞争力，积极运用“标准”手段探索规范经济秩序，提升产品质量的路径和方法，是对社会主义市场经济有形之手的重大补充。

以中山红木家具为例，改革开放后，红木家具逐步发展成为中山市大涌镇的支柱产业。但企业数量多、生产规模小、技术含量低、行业自律性差和恶性竞争激烈，大大影响了产业健康发展。

此时，在当地政府主管部门的指引推动下，按照“镇政府＋行业协会＋企业”的产业集群标准模式，红木家具联盟标准应运而生，相关企业纷纷入盟，制定了红木家具行业标准和国家标准，产品质量大幅度提高，大涌镇区域品牌声名远播，在全国的市场份额跃升至60％以上，年产值提高到20多亿元。

四、标准是加快产业转型升级的好手段

过去，我们常常使用行政的手段发号施令，在一定时间内也能取得很好的效果，

但弊端也不少，不符合市场经济的运行原则。例如，审批权过大，出现权力寻租的问题；无条件招商，造成的无条件关停并转的政府诚信问题；与国际惯例不符而受到国际质疑等问题。按照国际通行的做法，总结各产业、各行业的技术与管理规律，将标准手段作为调节工具，标准确实也能起到“风向标”“加快器”“调节器”的作用。

实体经济是国民经济的命脉，调整产业结构和加快产业转型升级，事关科学发展大局，也是国家长治久安的必由之路。“工欲善其事，必先利其器”。当前的“事”即“积极稳妥地加快转型升级，实现创新驱动，促进科学发展，转变发展方式”。当前的“器”就是充分依靠法律规定，“在一些环境污染大、能源消耗高的领域或行业，积极探索制定和实施严于国家标准、行业标准及地方标准的途径和方法”。在一些发达地区，如制定产品质量高标准、限额能耗标准、环境质量指标标准、厂界噪声或粉尘限额标准等。当然，使用这种方法，必须有配套的抽查检测、政府监督以及强有力的后处理措施、政府激励约束政策等。这种措施也可以以标准化的形式公布于众，做到公开、公正、公平，并有法可依，依法办事。通过这种有形之手，促进了企业自身转型升级。标准在其中能够发挥重大的调节作用，用标准促进转型和创新驱动应是科学发展的客观规律，也是发达国家的一条经验。

标准的竞争与标准的制定关乎一个地区和一个行业的产业结构调整和转型升级的质量，标准背后的争夺隐含着对经济领导权的争夺和区域品牌的维护。实践证明，科学运用标准工具，是保证产品质量、环境质量、医疗卫生质量、建筑质量、服务质量以及人民幸福指数不断提升的方法论，也是实现节能减排、促进自主创新和产业结构调整的好手段。

五、先进标准是引领创新成果的好方法

从目前情况看，按照标准趋向划分，可分为规范性标准和引领性标准。规范性标准就是对出现的问题，制定标准，加以规范，多用于传统成熟的产业，而引领性标准是将标准制定时机提前到设计阶段之前，是提前设置更高的指标要求，主动防止或防范未来恶性竞争或风险的一种先进方法，多用于设定创新目标。超前标准化通常也表现在某个产业刚刚发展起来，仅由为数不多的创新型大企业、大集团提供产品或服务，他们为了保证质量，主动提出制定标准。这种方法还可以表现在由政府有关部门从地区发展角度，借鉴更发达地区的经验，预设更高要求的指标，用于引领发展。总之，是一种主动处理潜在问题或预防某类问题发生的超前性约束文件，因此，从某种意义上说，引领性标准通常与科技创新密切联系在一起。

引领性标准与创新是密不可分的，创新成果与标准的深度融合将是未来科技竞争的焦点。标准是科技创新体系的高级表现形式；标准包含着创新成果，并将创新成果的风险降至最低，标准体现了经验积累及创新成果应用；标准是创新的重要结果；创新是标准的重要过程。

创新的最终目的是应用与产业化，但任何创新都有可能存在着未知的风险，也有可能存在着无法实现的创新风险，而标准是创新成果迅速转化为生产力的重要载体，是创新成果减少风险的最好方式，是创新成果产业化的重要接口。从这个意义上看，

标准是引领创新的。没有了标准这个接口，创新成果往往会失去意义，发展就不完整。

另外，标准的每次修订反映了创新成果的成熟，并基于对已知风险的认识，也反映了标准本身在不断创新，标准只有不断创新才能促进创新成果转化，才符合科学发展规律。标准创新是标准走向成熟的动力源泉，是创新驱动的结果。

六、标准已成为保障社会管理、促进公平正义的重要支撑

近几年来，在规范和推动产业发展的同时，标准作为获得最佳秩序的工具，在保障社会管理和公共服务质量、服务治理能力现代化、维护社会公平正义方面，发挥着越来越重要的作用，甚至发挥了引导社会管理的“方向盘”作用。尤其是在当前我国相关法律法规不健全的情况下，标准作为产品质量、环境质量、建筑工程质量、优化公共服务、健全城市管理、完善社区管理和居家养老服务、提升教育文化管理水平等主要评判依据，其科学性与合理性，直接影响着企业之间的公平竞争，直接影响着消费者人身健康与安全，直接影响着政府的公信力，已成为社会管理和公共服务领域法律法规的重要补充和支撑。

在社会管理和公共服务方面，广东行政审批标准化工作让老百姓办事就方便了。民政推行的养老标准化，让老年人更有保障了，取得了明显的成效。政府及公共服务组织按照“管理规范，程序公开，服务高效，群众满意”的要求构建标准体系，标准在微观层面具有无可替代的支撑功能。

七、标准化是社会全面协调可持续发展的软实力

现在各方面都在谈文化，将标准化形成文化是重要的方面，是制度文化的组成部分。尤其是在过去我国法律法规不完善的情况下，标准化不仅可以完善机制体制，同时还可以传播文化，不仅是智慧型服务业的重要组成部分，还是推动区域和企业全面、协调、可持续发展的软实力，其完善程度是一个区域或企业整体档次的象征，体现着这个区域或企业的领导者的综合素质和软实力。

八、标准已构成我国法律、法规体系的重要组成部分

按照我国《中华人民共和国标准化法》等法律、法规规定，从范围上划分，标准主要包括：国家标准、行业标准、地方标准、团体标准、企业标准。国家标准包括强制性和推荐性标准。按照法律规定，强制性标准本身就是法律、法规体系的重要组成部分，是必须实施的，而推荐性标准一经选用，也必须执行。团体标准和企业标准一旦企业制定或采用，也必须要声明并且实施。

在国民经济和社会发展的各个领域，由于重复性事物均可以通过标准加以规范，《中华人民共和国标准化法》等法律、法规均赋予了标准支撑保障功能，实际上，即使是推荐性标准，由于受《中华人民共和国合同法》民事责任约定的约束，也已成为事实的法律法规体系的有益补充。

我们国家将长期处于社会主义市场经济的初级阶段，党的十八大和党的十九大，中央对标准的定位更加清晰，在众多微观层面，标准已成为重要的法律、法规补充，

已构成了维护社会主义市场经济秩序的主要制度支撑体系。从标准制定的属性看，又是共同治理的工具和手段，已成为国家与地方治理体系与治理能力现代化的底层设计，尤其是当产品质量出现问题、环境出现恶化、生态文明遭到破坏、国家利益受到危害、人民健康与安全受到威胁时，用标准加以规范，已成为相关部门进行有效管理的工具和手段，广泛应用于技术、经济、文化、社会等领域。另外，当法律还不完善或法律缺位时，标准往往由于其制定速度快，程序相对简单，在某个时期又发挥着技术法规的作用。因此说，标准已成为法律、法规的重要组成部分。

这些年，标准化工作发生了很多新变化，希望广大领导干部和企业家更加关注标准，更加关注标准的实施，更加关注标准化战略的普及推广，这样，我们广东的经济就大有希望，会永远在全国前列！

为推动仪器仪表产业发展做出新贡献[①]

我受第二届理事会委托，向会员大会作工作报告，请各位代表审议，并请其他与会同志提出意见。

一、第二届理事会工作回顾

广东省仪器仪表学会第二届理事会换届以来，在省科协的指导下，以锐意创新的精神和奋发进取的意识，励精图治，努力开拓，携手全体理事和广大会员，稳健有序地开展各项工作。学会全力推动广东仪器仪表产业快速发展，积极搭建产学研有机结合的平台，有效促进了华南地区乃至全国仪器仪表产业的迅猛发展。

总结本届理事会的主要工作，体现在以下“三个注重”。

1. 注重抓好学会内部建设，管理水平实现新提升

学会新一届理事会对标“开放型、枢纽型、平台型”学会建设的高度，十分重视并持之以恒抓好学会自身建设。2012 年 12 月，在学会新一届理事会履职伊始，便迅速召开第二届理事会第一次工作会议，我明确提出了本届学会理事会的总体目标：进一步发挥广东省仪器仪表学会在学术、技术与应用、咨询等方面的优势，团结广东省仪器仪表行业的企业、研究机构、高校从事世界先进仪器仪表技术的引进、研究、推广，推动先进技术的研究与应用，为国家和地方经济建设服务、为政府提供行业咨询与建议，将广东省仪器仪表行业推到新的高度。2015 年 11 月，召开了学会第二届常务理事会，我在工作报告中强调学会要完善自身、努力发展，加强内部建设，增强社会影响力，为推动广东仪器仪表产业的快速发展做出更大的贡献。

为使本届理事会总体目标能够顺利达成，我们从以下“三个强化”入手，不断加大内部建设力度。

一是强化理事长、副理事长对部门的分管负责制度，指导、监督和协调各部门开展相关工作。

二是强化学会规范化建设。我们严格要求学会的每个工作程序均须依照相关规范流程进行。以敬畏之心和敬业之情严肃认真对待学会各项工作，切实保障学会各项工作顺利开展和有序推进。

三是强化学会经费管理。学会的经费来源于会费收取及外联赞助，以外联赞助为主，会员缴纳会费为辅。学会严格按照规定收取会员会费。在经费管理方面，主要由学会办公室负责，学会所有会费、赞助费均设立了账册及财务清单，由学会办公室相关负责成员详细列载经费收支，并定期向学会负责人进行汇报，接受学会全体成员的监督。学会经费全部用于学会的相关工作和活动，经费使用以高效和节约为原则。每次活动前由承办部门详列预算，活动过后提交正规、清晰的发票或收据等凭据，做到

① 本文是广东省仪器仪表学会第二届理事会工作报告，广州，2019 年 3 月 22 日.

财务收支合理。

2. 注重开展学术交流暨产学研对接，助力仪器仪表行业新发展

本届理事会积极开展仪器仪表行业前沿技术交流活功、专题研讨活动以及产学研对接活动，多次主办高质量、高水平、高层次的学术交流论坛以及产学研对接峰会，热情邀请各大高校和科研机构权威学者讲学，为广东、华南乃至全国仪器仪表行业创新驱动助力。

2012 年 12 月，学会成功主办“传感技术暨第八届国际仪器仪表与测控自动化高峰论坛”。论坛汇聚业界专家、媒体、科研人员、厂商于一堂，包括仪器仪表、传感器与应用企业人员及政府职能部门领导、国内测试测量类科研院校专家代表 200 多人参会，现场气氛热烈，互动性强，嘉宾精彩的演讲内容亦引发参会者共鸣，纷纷表达对传感器产业发展的看法并提出建议。其作为中国仪器仪表与自动化领域备受瞩目的高规格学术会议已经持续举办八届，为广东乃至华南仪器仪表与自动化行业的繁荣发展，发挥了不可替代的桥梁、纽带作用，受到产业界、学术界、科技界的一致点赞；学会骨干单位《仪器仪表商情》作为论坛的组织机构，继续发挥专业媒体优势，为仪器仪表与自动化产业发展搭建起了一个高端、高质、高效的沟通、交流舞台。

此后，学会牢牢把握国际仪器仪表与自动化发展新趋势，每年坚持不懈主办“国际仪器仪表与测控自动化高峰论坛”，至 2018 年，“国际仪器仪表与测控自动化高峰论坛”薪火相传，已经连续举办十五届，成为中国仪器仪表与自动化行业不可或缺的知名高峰论坛。

2014 年 12 月，学会继续主办“LTE① 测试测量核心技术大会暨第十届国际（深圳）仪器仪表与测控自动化高峰论坛”，来自通信及物联行业国际知名企业、运营商、权威专家、资深工程师等 200 多位嘉宾参会，就 LTE、100G、物联网测试测量应用等业界热门话题进行深度剖析。本次大会采用多种交流形式，多场精彩报告，最大限度地为参会者提供高效解决方案，解决了有关“LTE 关联产业与电子测试测量技术”的疑难问题，创造了与权威专家现场交流学习的难得机会。参会者表示，参加这一“LTE 上下游产业链的聚会”，与行业精英及有志之士欢聚一堂，结识了更多的商业伙伴，为企业的发展积累了宝贵的人脉资源。

2015 年 12 月，学会顺利主办“2015 中国 5G 测试测量核心技术大会暨 5G 应用与产品展示会——第十一届国际（深圳）仪器仪表与测控自动化高峰论坛”。来自通信测试测量领域的行业专家、行业领袖、有关微波射频、毫米波、太赫兹的上中下游企业及检测认证机构、科研机构、高校实验室等 300 余位代表齐聚一堂，共同探讨 5G 测试测量技术现状、未来发展趋势以及 5G 测试测量的应用之路，为通信行业测试测量企业、供应商、用户打造了一个深层交流、经验分享、共享资源的平台。

2016 年 8 月，学会圆满举办“2016 智能汽车与测试测量技术大会暨第十三届国际仪器仪表与测控自动化高峰论坛”。政府职能部门领导、有关学会领导、业界研究机构专家、知名企业负责人及汽车产业的上中下游产业链的从业者近 200 人参加了

① LTE，long time evolution，长期演进技术。

本次高峰论坛。为更好地满足业界之间的技术交流与市场深入探讨的需要，本次大会采用“高峰论坛+产品展示”的形式，大会现场设置了汽车测试仪器产品、汽车测检仪器产品和汽车通用仪器产品展区，业内知名企业在现场展示了最新的汽车测试产品和解决方案，为观众提供了更全面的交流形式和展示平台，在资源对接和技术交流的同时针对性地满足受众的需要。

2017 年 11 月，学会再次主办“2017 中国 5G 核心技术大会”暨“第十四届国际仪器仪表与测控自动化高峰论坛”。政府职能部门领导、有关学会领导、国内外科技界专家学者嘉宾、产业链科技人员、专业观众、媒体记者约 300 位代表参加会议。重量级专家学者、科研精英等嘉宾发表精彩的专题主讲，以中国 5G 总体规划与技术研发试验为热点课题，在技术研发测试、参与国际 5G 标准、5G 频谱等方面，就如何实现创新与垂直行业融合，增强全方位布局 5G 能力进行深层交流与探讨。来自国内外运营商、国内检测认证中心及科研机构的嘉宾，仪器仪表产业链厂商等专业观众，与主讲嘉宾积极互动，共同探讨仪器仪表产业如何运筹帷幄 5G 技术前沿及解决方案。

2018 年 5 月，学会成功主办“第十五届国际仪器仪表与测控自动化高峰论坛”暨“2018 中国智能汽车测试测量技术大会”和“EMC① 国际电磁兼容及测试技术峰会”。集结行业领军企业、权威机构、学会协会组织、汽车产业链企业技术人员 250 位行业精英共同探讨汽车产业与电磁兼容测试热门话题、热点技术；峰会邀请科技、产业界专家学者发表主题演讲，深入研讨汽车电磁兼容测试与设计技术，解决测试及设计当中遇到的诸多问题，助力企业把握技术前沿及解决方案，推动汽车产业与电磁兼容测试测量技术创新发展。

早在 2012 年 12 月，在学会第二届理事会换届当日，迅即召开了学会第三次学术会议，邀请仪器仪表学界翘楚做主题报告。工业和信息化部电子第五研究所所长谢少锋、华南师范大学梁瑞生教授、华南理工大学刘桂雄教授、暨南大学王璠教授、深圳大学李学金教授、广东工业大学吴黎明教授、广州大学朱萍玉教授等七位专家学者分别作了题为《科学仪器的可靠性管理》《基于 MIM② 结构的等离子体共振波导光子器件的研究》《仪器科学与新兴战略产业》《波纹壳的非线性动态屈曲》《光纤温度传感器在电力系统中的应用》《机器视觉在工业过程精密检测的应用实践》《分布式光纤传感技术在堤防隐患监测上的应用》的报告。学术报告精彩纷呈，与会代表踊跃发言，与主讲人深入探讨相关技术问题。

2015 年 11 月，学会与中国仪器仪表学会仪表元件分会、广州市仪器仪表学会圆满举行了联合学术会议。沈阳仪表科学研究院原院长徐开先教授，杭州电子科技大学磁电子中心主任钱正洪教授等来自国内各个科研院校和公司部门的 80 多位专家学者济济一堂，共同探讨我国仪器仪表行业的前端科技以及在新形势下的发展方向。徐开先教授在会上作了题为《仪器仪表制造业四基战略研究》的报告，徐教授在报告中结合当今的发展趋势，立足我国仪器仪表行业的发展现状和“中国制造 2025”产业升级转型

---

① EMC，electromagnetic compatibility，电磁兼容性。

② MIM，metal-insulator-metal，金属-绝缘体-金属。

战略目标，指出我国仪器仪表行业具有远大的发展前景。钱正洪教授、谢少锋所长、刘桂雄教授、郑党儿高级工程师、钟金钢教授以及其他专家学者在此次会议上也纷纷发表其研究成果，内容涉及仪器仪表元件加工、机器视觉应用、光纤传感、无损检测技术、可靠性研究等前沿研究成果，报告内容多样而富有深度。

3. 注重加强会际联系，学会影响力喜获新增强

学会与中国仪器仪表学会、深圳市仪器仪表学会、广州市仪器仪表学会构建了联络沟通机制，建立了长期稳定关系，定期互通信息；与此同时，密切交流合作，推进资源整合，联合举办大型学术交流会。

在 2013 年 9 月召开的中国仪器仪表学会第八次全国会员代表大会选举产生的第八届理事会中，我被推举为名誉理事，我会副理事长兼秘书长袁鸿当选为常务理事，我会副理事长刘桂雄、我会常务理事邵火当选为理事，四人顺利入选中国仪器仪表学会理事会，彰显了我会在中国仪器仪表行业的强大影响力。

2015 年 11 月，学会与中国仪器仪表学会仪表元件分会、广州市仪器仪表学会共同举行联合学术会议，并取得圆满成功。我在会上致辞指出，此次联合举办学术会议是一个创举，极大地促进了各个学会会员之间的交流沟通，实现了资源的有效整合，提升了学会在仪器仪表学术领域的影响力和推动力。

2018 年 11 月，学会在广东省科协的倡导与指导下，与广东省机械工程学会、广东省自动化学会等 11 家制造领域与智能制造有关的广东省科协所属省级学会共同发起成立广东省科协智能制造学会联合体，该联合体广泛联合省内 16 所高等院校、20 家相关科研机构、10 家行业组织和产业联盟、36 个省内智能制造与特色龙头企业参与组建，我被聘为联合体顾问，袁鸿副理事长兼秘书长被聘为联合体主席团副主席，邵时副秘书长被聘为联合体智能制造研究所副所长、秘书处副秘书长。

在看到学会取得可喜成绩的同时，我们也清醒地意识到学会存在一些不足之处，主要体现在以下“两个尚需”。一是自身建设尚需进一步加强。几年来，学会自身建设虽然可圈可点，然而打造“开放型、枢纽型、平台型”学会依然任重道远。二是助力行业创新驱动能力尚需进一步提升。几年来，学会在产学研对接方面虽然喜创佳绩，但要实现“把广东省仪器仪表行业推向新高度”的目标仍然“在路上”。

二、对新一届理事会工作的希望

各位代表、同志们：当前，广东仪器仪表产业呈现出迅猛发展的喜人态势，为推动广东制造业转型升级发挥着日益重要的作用，也为广东省仪器仪表学会健康发展提供了良好机遇和广阔舞台。我们衷心希望新一届理事会不忘初心，砥砺前行，切实担负起改革创新、持续发展的历史重任，全力推进“开放型、枢纽型、平台型”学会建设，努力把学会建设成更具影响力和美誉度的现代学术团体，携手产业界、学术界、科技界的同仁再接再厉、再铸辉煌，为推动广东、华南乃至全国仪器仪表产业繁荣发展做出新贡献！

加强数字产业化和产业数字化双轮驱动①

非常高兴能够参加在海宁举行的2020数字化转型与高质量发展论坛暨浙江省首席信息官峰会。

自2018年数字经济发展战略上升为国家战略，国家层面、省级层面、各地区层面积极推进，已经形成了上下一致的发展思维，并加速向经济社会全领域贯彻，数字产业化和产业数字化双轮驱动的模式形成了很好的发展效果。产业数字化对我们现阶段国民经济特别是制造业的改造提升尤为关键，加快制造业的数字化转型的步伐，能够为当前经济发展注入全新的动能，创造全新的发展模式。

浙江省委、省政府将数字经济作为“一号工程”，将产业数字化定位为促进制造业高质量发展的新引擎，旗帜非常鲜明；提出利用数字化技术对传统产业进行全方位、全角度、全链条的改造，建立政策体系、目标体系、工作体系、评价体系“四位一体”的促进体系，考虑非常全面；提出梯度推进，从机器换人、智能化改造、企业上云到工业互联网梯度推进、层层深化，浙江各地区各企业探索出各具特色的产业数字化转型之路，路径非常清晰；根据2019年国家网信办发布的《数字中国建设发展报告》，浙江省“产业数字化”发展水平位列全国第一，成果非常突出。

来之前我特地了解了一下我们海宁的情况，海宁是浙江省产业数字化转型20强地区，提出了以产业数字化助力规上工业迈向2000亿元的目标，非常震撼！海宁作为数字化转型示范地区大格局、大手笔地将数字化赋能传统产业作为经济发展的第一要务。同时海宁的工作还非常细致、非常落地、非常专业，提出“一业一策一案”，不同行业推不同方案，为200家企业免费开展咨询诊断，这说明我们海宁的领导充分认识到，不同行业、不同企业的数字化方式方法和推动路径是不一样的。这和我们工信智库提出的关键点完全一致，并且已经落地实践，起到了很好的作用。

今天众多智库专家、全省的骨干企业首席信息官（chief information officer，CIO）、很多地方经信部门的产业数字化负责人、专业服务机构共聚于此，探讨数字化如何赋能制造业的高质量发展，具有非常重要的意义。借此机会我想提出几点建议。

一是，希望地方政府能够多关注一下中小微企业的数字化，帮助他们解决“不会转、不能转、不敢转”的问题，骨干企业特别是我们的CIO群体主动贡献自己的经验和力量，帮助小企业更快、更好地掌握数字化能力。

二是，我们的智库力量能够更好地融入地方，真正发挥智库的知识赋能、资源赋能、人才赋能，我们政产学研金能够形成合力、打通通道、构建模式，加快对地方产业集群的数字化赋能，推动制造业高质量发展。工信部工信智库联盟拥有100多个高层级智库机构，杭州电子科技大学创新与发展研究院也是浙江的重要机构，今明两天

① 本文是2020备战“十四五”数字化转型与高质量发展论坛暨浙江省首席信息官峰会开幕词，海宁，2020年12月19日.

的会上大家可以多研究一下智库怎么与像海宁这样的地区深度融合发挥作用。

三是，专业人才的组织机构应该发挥更大的作用。首席信息官协会拥有全省骨干企业的优秀 CIO 力量，CIO 是技术、知识、产业实践综合能力人群，是稀缺力量，是产业数字化转型的主力军。CIO 为所在企业贡献能力是个人的本分，为更多企业、政府部门贡献力量是时代的使命。希望各级政府能够多支持 CIO 组织发展，积极发挥组织的力量培养出更多、更优秀的产业数字化专业人才。

最后，我预祝本届大会圆满成功！

刘人怀院士文集

刘人怀◎著

第四卷

科学出版社
北京

内 容 简 介

本文集由著者在力学和管理科学领域 60 多年中所发表的文章汇编而成，分为上、下篇。上篇是力学部分，结合工程需要，对单层板壳、波纹板壳、双层板壳、夹层板壳、复合材料层合板壳和网格扁壳等六类板壳的非线性弯曲、稳定和振动问题，以及厚、薄板壳弯曲问题进行理论探索。下篇是管理科学部分，联合实际，对工程管理、公共管理、工商管理、科技管理和教育管理的问题进行了研究。

本文集可作为高校力学和管理科学专业，以及相关专业老师、研究生、本科生的参考书，适合政府、高等学校、科研院所、科技社团、企业等领域的领导、管理人员和科研人员参考阅读，又可作为航天、航空、航海、机械、建筑和交通等工程设计师与工程师的设计制造指导书。

GS 京（2025）0637 号

图书在版编目（CIP）数据

刘人怀院士文集 / 刘人怀著. -- 北京 ：科学出版社，2025. 4. -- ISBN 978-7-03-080897-4

Ⅰ. O3-53；C93-53

中国国家版本馆 CIP 数据核字第 2024JH1723 号

责任编辑：陈会迎/ 责任校对：姜丽策

责任印制：张　伟/ 封面设计：有道设计

科 学 出 版 社 出版

北京东黄城根北街16号

邮政编码：100717

http://www.sciencep.com

北京中科印刷有限公司印刷

科学出版社发行　　各地新华书店经销

*

2025 年 4 月第　一　版　开本：787×1092 1/16

2025 年 4 月第一次印刷　印张：153 3/4

字数：3 646 000

定价：698. 00 元（全五卷）

目　　录

第十二章　工商管理 …… 1529
上海旅游交通的症结与对策研究 …… 1529
上海华亭集团旅游宾馆摆脱当前困境的对策研究 …… 1535
人才开发是搞好企业管理的关键 …… 1544
上海——旅游业的春光与希望 …… 1545
旅游工程理论及其在浦东开发中的应用 …… 1548
上海旅游交通中的若干问题 …… 1552
旅游工程学的提出 …… 1561
开启思想的眼睛 …… 1564
质量是企业生存的根本 …… 1566
对某跨国公司绩效考评系统的评价 …… 1571
公司治理：理论演进与实践发展的分析框架 …… 1575
以价值工程方法全面提升荔湾商旅核心竞争力 …… 1581
绿色化是中国制造业的必由之路 …… 1633
旅游教材为旅游教学之本 …… 1638
关系质量研究述评 …… 1640
消费者行为研究是营销理论的基石 …… 1648
从 CSSCI 旅游研究文献看旅游学学科发展 …… 1649
旅游中间产品转移价格的确定 …… 1657
我国旅游价值链管理探讨 …… 1665
旅游业零负团费的运行机制及危害性探析 …… 1669
文化生产力：管理的视角 …… 1673
旅游交通与航空运输规划 …… 1680
办好旅游教育至关重要 …… 1682
旅游标志景区研究有意义 …… 1683
Preface …… 1685
产业工人的中国梦：从低技能劳工到专业技术工人的人资转型升级战略 …… 1687
企业战略管理概要 …… 1698
“技工荒”困扰中国高尔夫球具代工产业 …… 1715
系统论视角下的旅游发展与旅游研究——中国工程院工程管理学部
刘人怀院士访谈 …… 1720
旅游城市品牌的塑造 …… 1725
保护遗产至关重要 …… 1726

城市遗产旅游景观的研究 …… 1727
应用旅游工程理论探讨阳江旅游发展规划 …… 1728
创业拼凑对创业学习的影响研究——基于创业导向的调节作用 …… 1733
澳门会展业的经济效应与发展策略研究 …… 1747
创业拼凑、创业学习与新企业突破性创新的关系研究 …… 1755
数字经济　创新未来 …… 1768
互补性资产对双元创新的影响及平台开放度的调节作用 …… 1770
系统论视角下旅游学科提升发展的思考 …… 1782
Guest editorial …… 1785
发挥社会创业的重要作用 …… 1789
第十三章　科技管理 …… 1790
谈谈科研中的几个问题 …… 1790
青年人的奋斗方向 …… 1796
“科学技术是第一生产力”的趋势与我国高新技术发展的战略 …… 1798
人才是振兴国家的关键 …… 1804
传播科学思想　提高国民素质 …… 1805
大力重视非开挖工程技术 …… 1806
认真评选科技奖 …… 1807
与时俱进　开拓创新 …… 1808
以科技创新加快推进全面建设小康社会步伐 …… 1809
加强基础研究　实现科技强省 …… 1815
成立广东省仪器仪表学会是紧迫的历史使命 …… 1819
培养青少年的创新精神 …… 1822
大力推动科技事业的发展 …… 1824
迎接新挑战 …… 1826
同心同德　开拓前进 …… 1827
创新路上的感想 …… 1829
为提高全民族科学素质做出新贡献 …… 1844
稳步推进科普志愿服务事业 …… 1847
努力开创仪表元件分会工作的新局面 …… 1851
关注世界科技创新态势 …… 1861
推进振动工程学科发展 …… 1874
积极投身科技创新活动 …… 1876
认真做好“2011 计划”重大战略部署工作 …… 1877
组建“政产学研金”合作平台　推动协同创新迈上新台阶 …… 1878
开拓创新　做好科普工作 …… 1881
复合材料创新成果丰硕 …… 1884
肩负使命与责任　继续向前 …… 1886

去除浮躁现象　建立公正科技评价机制 …… 1889
切实做好科技评价工作 …… 1891
专家学者不妨多点科普和传播意识 …… 1892
推动力学学科发展 …… 1896
汇聚人才为广东发展效力 …… 1897
扎实开展科普志愿服务工作，为全面提高我省公民科学素质而努力 …… 1899
应该高度关注青少年的视觉健康问题 …… 1904
为中华民族伟大复兴做出更多更大的贡献 …… 1905
全省科技工作者应履行科普社会责任 …… 1907
坚守初心　坚持创新 …… 1909
追求真理　奉献国家 …… 1911
让未来祖国的科技天地群英荟萃 …… 1912
众人拾柴火焰高 …… 1913
努力推动压力容器技术进步 …… 1914
大力加强高校科技创新 …… 1917
创新争先　自立自强 …… 1918

第十二章　工 商 管 理

上海旅游交通的症结与对策研究①

现代化交通是产生旅游需求的“三要素”之一，搞好旅游交通是发展旅游业的必要条件。党的十一届三中全会以来，上海市的旅游业在改革开放方针指引下，取得了迅猛发展。但是近年来，上海旅游基础服务设施不足，特别是旅游交通设施不足与旅游业迅猛发展之间的矛盾也变得越来越尖锐，旅游交通问题已成为发展上海旅游业的“瓶颈”。为此，本文力求从上海旅游业的历史和现状中找出上海旅游交通的症结，提出改善上海旅游交通的对策建议。

一、上海旅游交通的症结

随着上海旅游业的发展，大批宾馆已经完工迎客，因此前几年上海旅游业中住宿难的矛盾已经转化为旅游交通不畅的矛盾。上海旅游交通不畅的原因主要可归结为下面四点。

1. 上海与西安、桂林之间航线运力严重不足

在上海接待的入境旅游者中，从西安、桂林飞抵上海，或是从上海飞向桂林、西安的人数都占乘坐国内航班进出上海的入境旅游者总人数的50%以上，而桂林、西安二市机场受跑道、导航设备等因素的限制，不能停降大型客机，空运能力无法大幅度增长，这就使得上海与桂林、西安之间的客运极度紧张。以1988年10月上海旅游旺季为例，据统计，该月通过民航由上海发送到桂林的有组织入境旅游者多达12 000人次左右，由西安飞抵上海的有组织入境旅游者多达10 000人次左右。而同期上海到桂林的正常航班运力约为7010人次，西安到上海的正常航班运力约为10 000人次。因此，即使在上海到桂林和西安到上海这两条航线上的全部运力都来接待有组织入境旅游者也满足不了一、二类旅行社的需要，况且一、二类旅行社所接待的入境旅行者只占入境旅游者总数的57.2%，而入境旅游者又只占民航客运总量的50%左右。面对如此悬殊的供需矛盾，在这两条旅游交通线路上必然会出现混乱局面。据统计，仅1988年10月10日至31日，上海的国际旅行社、中国旅行社、青年旅行社三家旅行社在通过旅游、民航协调会协商后，仍缺上海到桂林机票3361张，其中最多一天缺上海到桂林的机票520张。据上海某旅行社统计，在1988年10月10日至21日中10多天中，就有190人因没有去桂林的机票而被迫取消桂林行程而直飞广州。同样，西安飞抵上

---

① 本文原载《旅游学刊》，1990，5（2）：10-14，47. 作者：刘人怀，于英川，王怡然，王荣瑀，汪正元.

海的航班也几乎到了失控的程度。据上海某旅行社在1988年7月对64批由西安抵沪的旅行团队的抽样统计，准时抵沪的仅3批，占4.7%，有95%以上的航班脱班。其中延误半天的共28批，占44%；延误一天的共16批，占25%。有些旅行团队甚至在西安、桂林延误五六天，使得旅游计划全部被打乱。在旅游交通链中，每个环节互相依存，相互支持，形成一个完整的整体。如果有一环节受到损坏，就会互相影响，甚至引起整个旅游交通链的断裂。桂林、西安这两个城市对民航客运量的承受力远远不如北京、广州，而它们的客运实际承受量又大大高于北京、广州。可以说，桂林、西安这两个城市是上海旅游交通链中压力最重、承受力很差的环节。现在桂林、西安这两个上海旅游交通链中的重要环节无法承受巨大压力而产生旅客大量积压滞留等混乱现象，这必然会产生连锁反应，影响整个旅游交通链，进而引起上海以至全国旅游交通的混乱。

2. 上海与苏州、杭州之间外宾专用车厢不足

来自苏州、杭州或从上海发送到苏州、杭州的入境旅游者要占上海接待的乘坐火车的入境旅游者总数的80%以上。仍按1988年10月统计，上海各旅行社发送到苏州的入境旅游者约13 000人次，发送到杭州的入境旅游者约12 000人次，而同期上海到苏州的火车2列，再加上到无锡的4列，共6列；开往杭州的火车4列。按当时火车编组，每次火车挂一节软席车厢，68个座位。那么每月由上海到苏州的软席运力是12 240人次，上海到杭州的软席运力是8160人次。从中可以看到，即使把上海到苏州、杭州二地的全部软席座位都供给上海一、二类旅行社也满足不了各旅行社的需要，况且乘坐火车软席到苏州、杭州两地的乘客中，有组织入境旅游者仅占60%～70%。据上海某旅行社统计，1988年10月，该社有25.6%的入境旅游者不得不乘坐硬席到苏州、杭州旅游，而散客临时要票更是难上加难。正因为上海与苏州、杭州之间外宾专用车厢严重不足，中外旅客常常混坐一节车厢，在列车严重超载情况下就更容易产生混乱不堪的现象。例如，中外旅客争抢一个座位；外宾一直站到苏州、杭州；同一团队一半游客坐软席，一半游客坐硬席等，混乱现象不胜枚举。特别是由于上海与苏州、杭州之间火车运力紧张，许多旅游团队不得不更改车次，打乱了原定的旅游计划，进一步加剧了旅游交通的混乱现象。据上海某主要旅行社统计，1988年该旅行社有34.3%的团队因无票而被迫更改车次，给整个旅游线路的安排带来了很大影响。又如上海另一主要旅行社1988年9月对该社组织的“苏州一日游”统计表明，火车更改率竟高达94%，使该社的“苏州一日游”面临失控状态。

3. 交通运输部门的配套设施不足，服务质量不高

上海的交通运输部门为了适应当前改革开放的形势，采取了一系列改革措施，增加了上海交通运输部门的活力。但是，由于为旅游服务的措施不配套，一些改革措施和政策的效果不太明显，加重了旅游交通的压力。例如，中国民航成立地方航空公司，从加强旅游交通服务方面来看，效果就不如以前。以前在旅游旺季，通过上海民航局就可以调动华东各地的客机来抢运游客，但是现在东方航空公司无权调动属于其他省市地方航空公司的飞机，这使旅游旺季的民航运力不能在较大范围内统一协调。又如以前上海民航局统一管理机场、航站、飞行大队等飞行业务，现在成立了企业性的东方航空公司。由于东方航空公司内部关系一时没有理顺，对外口子

过多，机场、飞行大队、售票处、机修厂分属不同的上级单位，效率大为降低。过去旅游包机只需要和民航公司一家打交道，现在必须和四五家单位打交道，而且每一家都有否决权。再如目前飞国际航线利润大、津贴高，因此航空公司和飞行员都抢着飞国际航线。还有现在从国外引进的都是大飞机，而国内许多旅游城市的机场只能停中小型飞机，引进的飞机大小不配套，就不能有效地调配运力。铁路部门也存在同样的矛盾。以前由铁路分局统一调度分局范围内的客运业务，1988 年开始铁路分局把权力下放到车站，各车站都只管自己这一段的业务，这样就丧失了铁路整体协调的优势。据旅游部门反映，铁路从 1988 年上半年开始，供需矛盾突然变得尖锐，这也许和铁路分局的权力下放措施有一定联系。上级部门对交通运输部门的某些规定和政策也挫伤了交通运输部门职工的积极性，进一步加剧了旅游交通的紧张程度。例如，规定民航飞行员每月只能飞行 100 小时。飞行员加班飞一次桂林的津贴不如汽车驾驶员加班开车到苏州。又如铁路部门曾给一、二类旅行社设立一个专门订票点，适当收取一些手续费，对此各旅行社反映都很好，但物价局却认为这是额外收费，责令取消这个专门订票点。再如铁路上海站规定，火车进站后一律不准打扫车厢，因此即使有人想在外宾上车前对外宾用的硬席车厢进行打扫也没有办法，厕所卫生简直使人无法容忍。

交通运输部门服务质量不高也是引起入境旅游者不满的重要原因。特别是民航任意改变航班几乎到了使人无法容忍的地步。航班临时更改，旅客在机场少则白等几小时，多则要白等 30 多小时。例如，1988 年 7 月某日由西安抵沪的某旅游团队，原定该日乘 2501 航班下午 4 时飞沪，旅客从下午 3 时到达机场，一直等到深夜 11：30，机场才宣布该航班取消，改为次日 8：30 起飞，次日清晨 7：30 该团队旅客再次匆匆赶到机场，结果一直等到下午 1：30 才起飞，前后一共在机场白等近 15 小时，气得该旅游团队的客人取消下站去桂林的行程而提前回国。1988 年 7 月对上海某主要旅行社的抽样调查表明，飞机和火车的班次更改率已达 61%，如表 1 所示。

表 1　某旅行社旅游交通班次变更统计

时间	批数	班次变更数	变更百分比/%	抵沪变更数	抵沪飞机变更数	抵沪火车变更数	离沪变更数	离沪飞机变更数	离沪火车变更数
1987 年 7 月	303	135	45	46	34	12	89	56	33
1987 年 10 月	571	287	50	132	82	50	155	95	60
1988 年 7 月	270	164	61	67	29	38	97	53	44

从表 1 可以看出，1988 年 7 月旅游交通的变更率比上年同期要高出 16%，比上年旺季（1987 年 10 月）要高 11%。特别是 1987 年火车的班次变更率远远低于飞机的变更率，而 1988 年 7 月火车班次变更率已接近飞机航班变更率，这说明火车的运输矛盾也正在急剧上升。

除了任意更改火车、飞机的班次时间外，行李搞错、行李迟到、行李划破等现象也时有发生。有的入境旅游者的行李甚至在其出境后仍然无法找到。这些都引起了外国旅游者极大的不满，也进一步增加了上海旅游交通的混乱程度。

4. 各旅游经营单位“以我为主”，缺乏统一计划调度

上海各旅行社组团已涉及全国 61 家一类社和部分二类社。上海一、二类旅行社接待的总社外联团、外地地联团和自联团大约各占 45%、35%和 20%。总社外联团、外地地联团和自联团各自为政，组团时不顾国内交通状况及各条旅游线路的承受能力，盲目争取客源。结果往往使游客在同一季节、同一线路特别集中，旅游交通链中的某些环节出现断裂。旅游季节剪刀差近年来越来越大。据统计，1979 年旅游淡、旺季接待量的剪刀差为 1∶5.5，而 1988 年旅游季节剪刀差增长到 1∶10。旅游季节剪刀差的增大，给旅游交通的平衡增加了困难，给旅游旺季的旅游交通更是增加了巨大压力。面对越来越大的旅游季节剪刀差，旅游经营部门不是想方设法尽量缩小剪刀差，反而认为这是旅游业的特殊规律，要求由交通运输部门来适应这个规律，这就使得旅游旺季的旅游交通矛盾变得更加尖锐。

此外，由于各地旅行社缺少协调，特别是在旅游旺季各地来沪的旅游团队经常改变计划，有的则根本既无计划又无预报，直到旅游团队即将抵沪时才通知，搞得上海方面措手不及。这样就进一步加剧了上海旅游交通的拥挤。据上海某旅行社 1988 年 5 月统计，该社在一个月内接到这种改变计划或毫无计划的旅游团队达 100 余批。这就使得本来已经十分紧张的机票、车票更加紧张。同时由于上海方面临时无法妥善安排这些旅游团队，上海的旅游交通显得更加混乱。例如，某旅游团因外地走不掉临时转入上海，上海当日下午到厦门的飞机正好有座，就安排该团去厦门游览，当日晚上 8 时，该团飞抵厦门，谁知次日上午 8 时厦门方面又把该团送回上海。这种安排当然引起了外国旅游者的强烈不满。

二、缓解上海旅游交通的对策建议

1988 年初，中国的旅游交通陷入了前所未有的困境。“冰冻三尺，非一日之寒”，造成民航和铁路运力全面紧张的原因是多方面的。主要原因是近年来国民生产总值增长迅速，对交通运输的需求急剧增长和国家对交通运输业的投入不足。1985 年以来，国民生产总值急剧上升，而对交通运输的相对投资却由占国民生产总值的 2.9%下降到 1986 年的 2%和 1987 年的 1.87%。对交通运输的投入减少，影响设备增加、更新和人员的培训，造成了交通运输紧张、状况恶化。虽然采取了一些应急措施，仍无济于事，旅游交通状况自然也无法改善。因此，国家应该增加对交通运输的投入。但是，我们也应该看到，在当时情况下靠国家短期内增拨巨额投资发展交通是不可能的。即使国家能拿出资金来购买飞机、车厢，但是机场、铁路、人员等基础设施也无法在一两年内完全解决。只要桂林、西安的现代化机场不建成，沪宁、沪杭铁路双轨工程不完工，上海乃至全国的旅游交通问题就难以根本解决。当前我们一方面要依靠国家增加对交通运输业的投入，力争迅速改变我国交通运输基础设施的落后面貌，另一方面应该根据现有交通运输的最大能力，采取一些切实可行的措施，尽可能地改善目前旅游交通的混乱状况。旅游交通不仅与旅游部门和交通运输部门有关，而且与社会各部门都紧密相关。不能把解决旅游交通问题看成旅游部门和交通运输部门的事，要解决旅游交通问题，必须得到全社会的谅解和支持。根据上海旅游交通状况，本文拟提出下列缓

解上海旅游交通的对策建议。

1. 重提“外宾优先服务”的口号，保证外宾得到优先服务

（1）适当压缩国内客运。

（2）使“外宾优先服务”成为行动准则。

2. 采取经济措施，限制国内客流

（1）统一旅游交通票价。

（2）开辟旅游专线，旅游专线发票不作出差报销。

3. 旅游部门加强宣传，加强计划协调，主动调节，引导客流

（1）加强淡季旅游推销，减少旅游季节剪刀差。综合安排旅游线路，努力平衡南北流量。

（2）大力宣传一地游和区域游，减轻长线旅游交通压力。

（3）开辟上海旅游新景点，延长外宾在沪逗留时间。

4. 建立现代化计划调度中心，加强综合平衡

据调查，国际旅行社上海分社和上海中国旅行社等主要旅行社以及民航、铁路部门都已引入微型计算机参加部分计划管理工作，可根据现有条件，设想计算机旅游交通信息系统分为两个阶段来实现。第一阶段以现有设备为基础，组织技术力量编制统一的计算机软件交各单位使用。各旅行社每天把计算机处理结果存入软盘后由专人把软盘送民航和铁路部门的计算中心，再由民航和铁路部门的计算中心对各旅行社送来的软盘进行统一处理。处理结果也以软盘方式分送各旅行社。第二阶段把旅行社、民航、铁路、海运等计算机系统都纳入计算机旅游交通信息系统网络，通过计算机网络直接传输、处理和获取有关信息，进一步加速信息的控制和反馈，加强旅游交通的综合计划平衡。

5. 采用多种方式，发展旅游交通

（1）发展短途公路客运。在上海的旅游客流中有30%左右流向苏州、杭州。因此除加速沪宁、沪杭高速公路建设外，短期内可改造现有的几条公路，使其成为质量高一点的公路，使旅客能通过公路到达邻近的旅游城市，以减轻铁路客运的压力。例如，锦江旅游公司把40%去苏州的旅客用汽车输送，取得了较好的效果。

（2）增加民航机动运力，发展包机业务。世界上无论是发达国家还是发展中国家的国际旅游业务，在旅游旺季都有运力不足的问题。它们主要采用旅游包机这种特殊形式来缓解各主要旅游城市之间的交通困难。日本赴夏威夷的大批游客，基本上是包机运送的。马来西亚、朝鲜、印度、墨西哥、罗马尼亚等国家都允许外国包机运送游客。因此，在旅游旺季，一方面中国民航应增加机动运力，扩大旅游包机业务，另一方面也可以放宽政策，让国际游客乘坐外国包机来中国，并在规定的几个旅游城市之间飞行。中外结合，可解决旅游交通的燃眉之急。

6. 加强建设旅游交通设施

（1）修建上海民航候机楼。1984 年虹桥机场扩建的 1.8 万平方米候机楼现已拥挤不堪。目前上海民航客运量每年递增 25%～30%，拟建中的 2.8 万平方米候机楼显然不够。我们应考虑修建较大的候机楼，还要修建相应的指挥塔和第二跑道。

(2) 扩建上海国际客运码头。上海国际客运码头只有 1 个泊位，旅游船争泊位现象十分严重；高阳路码头现有 6 个泊位，国际客运只使用了 1 个泊位。建议改建高阳路码头泊位，使其成为一个与上海这个国际大都市相适应的国际客运码头。

(3) 加速桂林机场、西安咸阳机场和沪杭、沪宁铁路复线工程。桂林、西安、苏州、杭州这 4 个城市是上海旅游交通的“瓶颈”，国家对这 4 个城市的交通设施已有全面规划，现在需要尽快地把规划付诸实践。

7. 加快改革开放步伐，增加交通旅游部门的自主权

随着改革开放的不断深入，来我国旅游的外国游客不断增多，但是我国的主要出入境口岸仍然只有北京、上海、广州三个，建议增加西安、桂林、南京、杭州等城市作为新的出入境口岸，加快改革开放的步伐，合理平衡客流，吸引更多的外宾来华旅游。

随着改革开放的不断深入，给交通旅游部门足够的自主权已显得越来越重要。交通旅游部门应有权在国家规定的幅度内自行决定票价，如上海到西安，上海到桂林航线可采用倾斜性票价（即来回票价不等），有的航线可采用季节性票价（即季节不同，票价不等）。这样，交通旅游部门一方面可以通过经济手段来主动调节客流，另一方面也有利于在国家统一指导下，开展各种合理竞争。

本文的主要工作是在 1989 年 6 月以前进行的，因此本文中提到的规划和预测数据都是按 1989 年 6 月以前的情况而设定的。1989 年 6 月以后，中国的旅游业陷入了低谷。面对暂时萧条的旅游市场，有人认为现在旅游交通问题已经不会再阻碍上海旅游业的发展，当前只要一心一意抓客源就可以了。我们认为这种想法是不全面的、缺乏长远考虑的。近年来，旅游宾馆、旅游客源和旅游交通之间的发展是不平衡的，上海旅游业整体结构不合理的状况并没有因为游客的暂时减少而解决，客源达到一定数量，旅游交通问题就会再次突出。从整体和长远着眼，旅游交通问题仍然是阻碍上海旅游业发展的主要矛盾。旅游市场的萧条是暂时的，也许用不了两年，上海的旅游市场又会出现欣欣向荣的局面，如果到那时再考虑旅游交通问题就又会措手不及，陷入被动。有关部门应抓紧当前游客较少的时机，进行整顿治理和结构调整，加快机场、铁路等交通基础设施的建设和人员培训；加快旅游交通运输部门体制的改革，完善各种规章制度；促进旅游部门和交通运输部门的紧密合作；合理调整旅游交通布局，正确调节引导客流，深入开展旅游交通的研究，为即将来临的又一次旅游高潮做好必要的物质和精神准备。

上海华亭集团旅游宾馆摆脱当前困境的对策研究①

一、上海华亭集团旅游宾馆的困境

1987 年成立的华亭集团[1]，是上海旅游局所属的饭店集团。该集团现有所属旅游涉外饭店 10 家，即华亭宾馆一号楼、华亭宾馆二号楼、上海宾馆、新苑宾馆、程桥宾馆、金沙江大酒店、虹桥宾馆、白玉兰宾馆、樱花度假村及上海旅游专科学校实验饭店——天马大酒店。除了上海宾馆位于市中心之外，其他饭店都分散于虹桥区和市区边缘。这 10 家饭店都为内资国有企业，其中高级宾馆 1 家，中级 7 家，中级偏下 2 家。目前华亭集团已拥有客房 3920 间，资产 68 810 万元，职工 6037 人。我们在下面的分析中，因考虑到天马大酒店是教学实验宾馆，故在华亭集团经济效益分析中不计在内。

(1) 华亭集团从 1988 年开始，经济效益急剧下降。1988 年，华亭集团所属 9 家饭店（其中两家为 1988 年开业）的总利润不及 1987 年 7 家饭店总利润的 1/3，即由 5100 多万元降为 1800 多万元；税金除华亭宾馆外其他饭店普遍减少，而工资总额却有 68.7%的高幅增长，如表 1 所示。

表 1 华亭集团宾馆效益表

项目	1987 年（7 家饭店合计）	1988 年（9 家饭店合计）	1988 年（与 1987 年比较）
客房收入/万元	9 424	9 067	−3.57
占总收入比重/%	54.56	46.66	−7.9
利润/万元	5110.3	1 804.9	−3305.4
利润率/%	29.59	9.29	−20.3
出租率/%	78.09	61.72	−16.37
创汇率/%	85.2	78.3	−6.9
平均房价/元	138	127.19	−10.81
工资总额/万元	993.8	1 677	683.2

(2) 由于各方面影响，集团效益进一步恶化，1989 年 1～10 月共计亏损近 1500 万元，如表 2 所示。

表 2 1989 年 1～10 月华亭集团效益表

项目	1989 年 1～5 月	1989 年 6～10 月	合计
营业收入/万元	7 849.3	3 800.6	11 649.9
利润/万元	262.7	−1 741.55	−1 478.85
利润率/%	3.3	—	—

① 本文是上海市对外经济贸易委员会重大科研项目报告，上海，1990；《旅游工程管理研究》，北京：科学出版社，2014，110-119。作者：刘人怀，厉无畏，范家驹，钱幼森，姜文豹，郑琦.

续表

项目	1989 年 1～5 月	1989 年 6～10 月	合计
平均出租率/%	48	31	39.5
平均房价/元	122.5	116.8	119.65

(3) 地理位置差，出租率在全市平均水平之下，亏损局面无法扭转。华亭集团除了上海宾馆位于市中心之外，其他饭店都分散于虹桥区和市区边缘，很难吸引国外散客，其客源的 80%主要靠团体游客。所以，华亭集团的客房出租率在全市平均水平以下，随着新建客房的大量开业，全市客房将达 1.8 万间，但客源却不会大量增加，这样全市的客房出租率将不足 30%，加上地理位置的劣势，华亭集团很难走出亏损的低谷，如表 3 所示。

表 3　华亭集团与全市平均出租率比较表　　单位:%

项目	1988 年出租率	1989 年 1～10 月出租率	保本出租率	1990～1995 年预计出租率
全市	67.00	55.34	—	29.9
华亭集团	61.72	39.50	40～55	20.0

(4) 每年还本息近 1 亿，集团部分企业濒临绝境。除程桥宾馆、樱花度假村投资已回收之外，华亭集团的总贷款额高达 5.9 亿元，按 10 年还贷计划，每年还本付息近 1 亿元，按目前经营水平，集团付息都艰难，还本更难实现，如表 4 所示。

表 4　华亭集团 10 年还款分析表　　单位：万元

单位	贷款额	10 年总利息	每年利息	每年本息
华亭宾馆	27 986	17 631.18	1 763.12	4 561.72
上海宾馆	4 117	2 593.71	259.37	671.07
白玉兰宾馆	3 742	2 358.09	235.80	610.10
金沙江大酒店	4 382	2 760.66	276.10	714.30
新苑宾馆	2 412	1 519.56	151.96	393.20
虹桥宾馆	16 146	10 171.98	1 017.20	2 631.80
合计	58 785	37 035.18	3 703.55	9 582.19

部分饭店还款年限已由 10 年延长至 25 年左右，而 10 年后的追加利息将接近贷款额，在债务、客源的重压之下，华亭集团在“八五”期间亏损已成为定局。

二、上海华亭集团旅游宾馆摆脱困境的对策研究

造成当前上海宾馆业经济效益严重滑坡的一个重要的客观原因是，上海客房数量的严重过剩。到 1990 年上海立项开业的涉外饭店达 54 家，客房 1.8 万间，其接待容量可达到 208.5 万人次。但据预测，到 1991 年能恢复到 1988 年的接待水平是很困难的，即使以后几年能够达到 1988 年的 91.6 万人次的水平，其客房利用率也只有 30%左右，若加上立项以外的 6000 间客房，则客房利用率更低。如果按 10%的客源增长率推算，到 1995 年，客房出租率也达不到 50%。在这种情况下，要迅速扭转严重亏损局面，必

须果断采取应急措施，适当压缩宾馆数量，尽快实现市场供求关系的平衡，提高宾馆的经济效益和宏观效益，以解决目前困境，暂渡难关。应急措施可采取保留、经营转向、承包、出卖、合资、合作 6 种方式。

（一）保留

保留，即保留现有的营业饭店。

1. 实行保留的优点

（1）保留大部分中方宾馆，能使中方宾馆在本市宾馆业占主体地位，有利于国家的宏观调控，有利于宾馆行业性管理，更有利于维护民族旅游业的进一步发展。

（2）保留那些地理位置好、竞争能力强、经济效益较好的宾馆，仍可为国家创收创汇。即使某些宾馆目前经济效益不理想，但若其他经营条件尚可，一旦宾馆在总体上得到压缩，其客房出租率即可迅速回升，带来一定的经济效益。

2. 实行保留的缺点

（1）由于中方宾馆与“三资”宾馆在政策、信息网络、客源渠道、管理水平及服务质量等方面的差异，中方宾馆相对缺乏竞争力。

（2）在本市宾馆尚不能适当压缩、客源不乐观的情况下，加上“三资”宾馆的竞争压力，保留的中方宾馆仍将面临不可避免的困难，特别是还贷任务重的宾馆则压力更大。

3. 注意点

（1）对保留的中方宾馆，国家和市政府必须在政策上进一步放宽，给予一定的优惠，并采取相应措施，给予必要扶持，以共渡难关。

（2）保留下来的中方宾馆，必须深化改革，加强管理，努力提高经营水平，提高人员素质和服务质量，积极开拓客源渠道，迅速提高经济效益。

（3）保留的宾馆，要加强宏观平衡，做到高、中、低三个档次的比例协调。按国际惯例，旅游城市高、中、低档宾馆的最佳比例为 1∶4∶5，而目前本市的高档宾馆过多（占 35%），应注意适当压缩，使本市的宾馆比例趋向合理，以提高宏观经济效益。

（二）经营转向

经营转向，即把接待国际旅游者的涉外宾馆转成接待国内旅客的高、中档旅馆，或转成外省市的商务办事机构，贸易中心以及商务、饭店、商场等相结合的综合性企业。

1. 实行经营转向的优点

（1）能使企业扭亏为盈，增加收入。上海在今后相当长时间内国际客源不容乐观，全市平均客房出租率只有 30%，还达不到保本出租率 45%～55%的水平，大部分宾馆将亏损，但把涉外中低档宾馆相应转为接待国内旅客的高、中档旅馆，出租率就可以从 30%上升到 80%以上，从而扭亏为盈，如表 5 所示。

表 5　宾馆转向后效益比较表

项目	涉外中档转国内高档		涉外低档转国内中档	
	涉外中档宾馆	国内高档旅馆	涉外低档宾馆	国内中档旅馆
每间房价/元	120	80	70	40
出租率/%	30	80	30	80

续表

项目	涉外中档转国内高档		涉外低档转国内中档	
	涉外中档宾馆	国内高档旅馆	涉外低档宾馆	国内中档旅馆
每间客房年营业额/元	13 140	23 360	7 665	11 680
收入增加量/元	—	10 220	—	4 015
收入增加比/%	—	77.8	—	52.4

目前国内旅馆严重短缺，特别是40元一间的双人房（即20元一个床位），长年供不应求。若涉外低档宾馆转成国内中档旅馆，估计出租率可达80%以上，则国内中档旅馆每间客房的年营业收入要比涉外低档宾馆高出4000多元，200间客房规模的旅馆一年就能增加80多万元。同样，涉外中档宾馆转为国内高档旅馆，每间客房年收入可增加1万多元，400间客房规模的旅馆一年就能增加400多万元。这样就可扭亏为盈。

（2）经营转向，宾馆设施破坏不大。一旦上海的国际客源大大增加，还可马上转回来继续作宾馆，它是目前减少客房过剩的较理想的过渡措施，具有较强的适应性和灵活性。

（3）收入全归本市，无外流可能。

（4）外省市对在上海设办事机构或商务贸易中心很有积极性，这样通过经营转向可缓和上海客房过剩矛盾。

（5）转为商务、饭店、商场等相结合的综合性企业最为理想，能使宾馆扩大经营业务，增加收入，减少风险。尤其是虹桥区宾馆林立，商业、娱乐设施极不发达，若有部分宾馆搞综合经营，就能弥补商业、娱乐设施的不足，改善国际游客的住宿环境。

2. 实行经营转向的缺点

（1）没有外汇收入。那些有外汇贷款的宾馆，外汇将难以平衡。

（2）只能靠薄利多销，经济效益要比宾馆预计效益差。

（3）由于国内部分客人的不文明行为，饭店的设施如电视机、地毯、床单等损耗较大。

3. 注意点

（1）需经营转向的宾馆最好无外汇贷款，债务全部还清的则更佳。

（2）优先考虑转给能支付外汇的单位。

（3）优先考虑把现有宾馆转为经营其他业务的项目。在转为他用中，则优先考虑转成综合经营、经济效益高的项目。

（三）承包

承包，即在一定年限内把宾馆的经营权出让给外商或国内企业，实行基数承包，超额分成。承包基数，就是承包方定期给我方固定的收入，它是营业额扣除成本、费用、税金、职工工资奖金、福利等之后的部分纯收入。超额分成，指按合同规定，承包方收入超过规定水准的超额部分，由双方分成。

1. 实行承包的优点

（1）给外商承包，有利于扩大客源渠道，提高出租率。

（2）确定承包基数，超额分成，能给企业和国家带来稳定的收入。例如，程桥宾馆，地段差、档次低、规模小，效益不理想，在虹桥区饭店林立的不利条件下，已缺乏竞争力。最近有港商有意承包程桥宾馆，他们把程桥宾馆作为沪港贸易的联络地和办事中心，改变程桥宾馆以接待团体旅游客为主的单纯经营业务，而以安置港商的贸易客户、吸收商务客、方便港方贸易活动为经营宗旨。港方 1991 年将给程桥宾馆 100 万元纯利润作为承包基数，在此基础上，若承包方纯收入超过宾馆 100 万元，则超额部分实行中外双方各得 40%，余下的 20%则作为职工奖励基金的分成方式，承包期限为 5 年。根据历年经验，100 万元利润，出租率必须保持在 65%～70%的水平，这对程桥宾馆来说，如果今后几年单靠自我复苏是很难达到的，如表 6 所示。

表 6　程桥宾馆历年经营情况及承包效益分析表

项目	1986 年	1987 年	1988 年	1989 年（1～10 月）	承包预计	保本点
利润/万元	113.99	192	10.8	−55.7	100	0
出租率/%	66	76	56.89	30.48	65～70	40
房价/元	65	91	83	70	70	70

这 5 年承包期间，正是新饭店大量开业，客源缓慢恢复的困难时期，它能使程桥宾馆走出困境，渡过难关。

（3）与出卖相比，承包不存在产权和土地使用权的出让问题。

2. 实行承包的缺点

（1）外方承包，必有相当部分的外汇收入流入国外。

（2）承包者为在承包期获取更多利益，容易产生短期行为，造成设备设施超负荷运转和消耗，甚至服务质量下降，影响本市旅游业声誉。

（3）由于客源不可能在近期内大规模增加，承包者必然会争抢国内市场的部分客源，加剧市场竞争。

3. 注意点

（1）承包基数不得低于宾馆原计划投资回收期限的年平均还本付息额。

（2）承包一般以 5 年为期限，过短则承包方无利可图，不愿承包，或加剧短期行为。过长则不利于我方对承包方案的调整。

（3）对外汇尚未还清的饭店搞承包，必须明确双方还贷责任。

（4）承包者若改变经营业务，不作宾馆而做他用，则应优先考虑。

（四）出卖

出卖，即出卖宾馆所有权与土地使用权。可以“外卖”，实行外商独资形式，也可“内卖”，即卖给外省市或本市其他行业。

1. 实行“外卖”的优点

（1）能一次还清全部债务，包括外汇债务，收回成本。

以白玉兰宾馆为例，由于其地理位置差，经营管理不善，客房出租率 1988 年只有 27.1%，1989 年只有 20%左右，最低月份只有 4.4%。1988 年创汇率只有 25.9%，1989 年创汇率只有 15%左右，经济效益极其低下，已完全丧失竞争能力，处于投资高、还贷艰难的困境，如表 7 所示。

表 7　白玉兰宾馆 10 年还贷分析表

投资贷款/万元	10 年还款每年应付利息/万元	10 年还款每年应付本息/万元	原预计投资回收期/年	10 年共计还本付息/万元
3743	235.8	610.1	10 年	7358

原预计投资回收期为 10 年，这一测算是以年平均客房出租率 75％～85％而定，而在以后几年中，其客房出租率最多只能维持在 20％～30％，因此，毫无偿还能力，而且每年还要利上滚利，以复利计算，包袱将越背越重，不堪负担。如果能尽快出售给外商，则可彻底甩掉包袱、摆脱困境，不仅还清全部债务，收回成本，如果出售价高，还可有盈余，如表 8 所示。

表 8　白玉兰宾馆还贷和出卖费用表

投资贷款/万元	折算成美元/万美元（1 美元＝3.7 元人民币）	其中人民币贷款/万元	其中外汇贷款/万美元	出售最低保本价/万美元
3743	1000	1422	664	1300

如表 8 所示，1300 万美元的出售最低保本价包括投资额 1000 万美元，历年来的利息以及谈判等其他各项费用。按照白玉兰现有的规模与设施，这一价格是不算高的。这样，除成本收回外，还可比原来的外汇贷款 664 万美元多获得 636 万美元的外汇。若能以高出 1300 万美元的价格出售给外商，则回收全部成本外，尚可有盈余。

（2）国家可稳定地获得一部分营业所得税。

（3）可一次还清人民币贷款。

2. 实行“外卖”的缺点

（1）出卖给外商后，如果仍作宾馆，则宾馆过剩的问题仍未能得到解决，并将对中方宾馆增加市场竞争的压力。

（2）出卖后，中方将完全失去宾馆的控制权。

（3）外汇还贷能力无法得到保证。

3. 注意点

“内卖”与“外卖”，应以“外卖”为主。因为：第一，由于目前国内财政紧缩，“内卖”的可能性比“外卖”要小；第二，“外卖”相比“内卖”，不仅收益要高，而且可一次解决外汇还贷问题。

（五）合资

合资，即出卖部分股权给外商，双方共同经营，利益共享。

1. 实行合资的优点

（1）出卖部分股权，可迅速偿还部分贷款，缩短投资回收年限，减轻利息支付压力。以虹桥宾馆为例，虹桥宾馆的全部投资为 4550 万美元（向银行贷款）。若按现在的造价，可值 6500 万美元。如果外商能入股 50％，支付给中方 3250 万美元，则相当于在合营期间为虹桥宾馆减轻了 71.4％的还贷负担。那么还剩 1300 万美元债务，折合人民币 4810 万元。按最低保本经营测算，虹桥宾馆的保本营业额为每年 2400 万元，它包括每年600万元的折旧费（按25%营业额提取），其中90%即540万元可用于还贷，10年共计5400万元。只要虹桥宾馆能在前5年维持保本经营，以后逐年增加盈利，则在10年内就能还清全部本息。但实际上外商合资后总能带来一些稳定客源，将比预计效益要高，则还本付息时间还能缩短2～3年。

当然，当合营期满时还需偿还外方入股的本金，但由于这笔本金是要在期满时还款，因此，这实际上是推迟了原来一部分贷款的还贷时间。当度过这几年困难时期，企业已有盈余，合营期满时（假定15年）再偿还这笔贷款便比较容易了。

作为有意经营宾馆的外商来说，花费3250万美元能拥有700多间客房的中档偏上宾馆的使用权，可省去建造时间，直接经营，是很经济合算的。

（2）共同经营、分担责任、分担风险、分享利益，有利于调动中外双方的积极性。

（3）能带来国外先进管理经验与技术，提高宾馆的服务质量与经营管理水平。

2. 实行合资的缺点

（1）合资方式虽然可为实行合资的宾馆带来经济效益，但本市旅游业的总体经济效益将相对减少。特别是一旦本市旅游业形势好转，易造成肥水外流。

（2）合资后的宾馆仍将参与或部分参与本市的客源竞争，给中方宾馆增加竞争压力，不能缓和当前尖锐的供求矛盾。

3. 注意点

（1）合资对象应以外商为主，这样可保证外汇收入与外汇贷款。

（2）凡转业经营的合资对象应优先考虑，这样不仅能提高转业经营企业的经济效益，还有利于缓和旅游市场尖锐的供求矛盾。

（3）合资年限以10～15年为妥。因为，根据市场预测，在5年内客源不可能大幅度增长，所以合营的前5年不会有多大的利润，故合资年限如果太短，外商则无利可图，而且还因合资年限太短，期满后偿还外商的入股投资仍然是很困难的。

（4）利用外资应遵循平等互利和保障效益的原则。国家应致力于不断改善投资环境，各宾馆则应努力取得应有的经济效益和社会效益，必须自行求得外汇收支平衡，偿还贷款，既为国家创汇，又使外商得利。

（六）合作

合作，即由中方出土地，外方出资金，共同经营，分享盈利，共担风险。目前在中方宾馆已经建成的情况下实行合作方式，是指中方提供土地、劳动力，而外商则全部或部分地提供还贷资金和经营费用。

1. 实行合作的优点

（1）合作双方均以各自的法人身份进行合作，因此更有利于合营的协议或合同的达成。

（2）合作各方可以用不同的方式投资，不一定以货币计算股权比例，现金部分基本上由外方提供，易于解决宾馆的还贷困难。同时各方责任和权益都在合同中具体规定，使得利润分配和分配比例更加灵活具体。

（3）合作经营允许外方投资者用提取设备折旧费摊入成本的方法收回投资，合作结束后，所有财产归中方所有，这样合作就减少了外商投资风险，对中方也有益。

2. 实行合作的缺点

实行合作的缺点是中方受益相对要小。

3. 注意点

(1) 尽量选择财力雄厚，信誉好，与我方有长期真诚合作意向的企业集团为合作对象，与世界著名的各种饭店集团，携手合作，必定能通过其在世界各国联号饭店的客源网络，大大开拓国际客源，解决宾馆的客源问题。这样，也有利于加快技术改进和设备更新，培养管理人才。

(2) 合作期限不宜过长。因为考虑到不需要建造期，投资回收期可相对缩短。另外，日后上海旅游业形势好转，盈利增加，外商会得到更多利润，故合作期限以 10～15 年为妥。

(3) 必须为外商在饭店合营中创造良好的投资环境。

综观以上六种方式，各有利弊得失。

保留大部分宾馆，使中方宾馆在本市旅游业中占主体，虽然目前经济效益不高，但是这对于国家掌握我国旅游业的经济命脉，实行宏观调控，是至关重要的。

实行合作、合资、出卖即“三资”形式，对于解决目前宾馆严重亏损、客源不足以及还贷困难，是一种有效的应急措施。同时也有利于引进国外先进管理经验与技术，改善服务质量，提高上海旅游业的声誉，但也存在着增加中方宾馆竞争压力以及日后肥水外流的问题。为尽量减少“三资”形式带来的弊端，在三种形式中，应广泛采用合作经营方式。因为合作方式双方利益紧密联系，经营方式灵活，形式多样，手段简便，容易达成协议，合作到期后，宾馆的所有权与经营权即可全部收回。

在承包方式中，洋承包的外商因为不出投资，只搞经营，所以应在政策上尽量鼓励其转业经营。内承包如果要继续用作宾馆经营的，要选择国内有实力并在国际上有一定声誉的饭店集团来经营，否则应尽量鼓励其转业经营或搞综合经营，以缓和当前旅游市场尖锐的供求矛盾。

转向经营是解决当前上海宾馆严重过剩的最有效的应急措施。根据预测，到 1995 年前国际客源不可能大幅度增长，客房出租率达不到 50%，因此在 1995 年之前，上海的涉外饭店将过剩 1/3 左右。所以，转向经营的宾馆，包括“三资”形式中的转业经营、承包形式中的转业经营以及现有宾馆本身的业务转向或作他用，应达到上海现有宾馆总数的 20%～25%为妥，这样才可能在近期内尽快地实现市场供求的基本平衡，以保证上海旅游业的宏观经济效益。还应特别指出的是，转向经营应优先考虑那些少破坏或不破坏原有宾馆设施的经营项目，以便若干年后本市客源大幅度增长，客房不足时，能及时将它们转为宾馆经营使用。

以上六种方式，应根据各个宾馆的不同情况，分门别类地加以处置。

保留的宾馆，主要是地理位置好、档次结构合理、经济效益相对较好，具有较强竞争能力的宾馆。即使有些宾馆目前的经济效益不理想，但在其他方面若具备旅游经营的良好条件，从长远考虑也应保留，如华亭宾馆一号楼、华亭宾馆二号楼、上海宾馆、新苑宾馆。

实行“三资”形式的宾馆，应以地理位置差、档次较高、还贷任务艰难、经济效益低、丧失客源竞争能力的宾馆为主，如白玉兰宾馆、虹桥宾馆、金沙江大酒店。

实行承包或转向经营的宾馆，最好是投资还清或只剩小额债务、经济效益差、地理位置差、缺乏良好的旅游经营环境的中档与低档宾馆，如程桥宾馆、樱花度假村。

在采取上述几种应急措施的过程中，必须强调，一定要有专家参加可行性研究，对经济效益与宏观效益进行定量定性的分析，根据实际情况，做出最佳选择。同时，政府应加强统一领导与宏观调控，避免宏观失控[2]。

参考文献

[1] 上海市旅游事业管理局，上海交通大学管理学院，上海市经济学会. 上海经济区旅游概貌（上海市卷）. 上海：学林出版社，1990：46-47.

[2] 刘人怀. 城市政府工作目标管理与治理整顿、深化改革. 学习与实践，1990，(11)：35-36.

人才开发是搞好企业管理的关键①

自贯彻治理整顿方针以来，我国经济形势正在逐步好转，未来十年是我国走向现代化的关键时期，将给我国企业带来勃勃生机。面对这样的经济态势，企业首先要眼睛向内，强化企业管理，向管理要效益。一般来说，我国的企业已具有一定的技术设备和资金，并有合格的管理制度，但人的因素导致管理水平徘徊在原有水准线上，使管理效益难以大幅度地提高。在人的因素中，又以企业各级领导岗位上的管理者为核心因素。因此，摆在企业面前的首要任务[1]就是按照企业岗位对其任职人员客观上存在的指标要求，挑选合适的管理者，以使企业管理岗位与相应的管理者之间建立一种科学的、客观的、和谐的关系，从而大大提高企业管理效益。与此同时，要注意应用先进的科技成果和先进的技术代替落后的产品和技术，即通过技术改造，实现技术进步，以带动生产过程各组成要素质量的提高，促进产品更新换代，节约能源，降低物耗，提高劳动生产率。这是企业一项带有战略意义的长期性和根本性的任务。实践证明，一个企业的兴衰成败，就其企业内部来说，归根结底是人才开发和技术改造，因此企业应该高度重视。

参考文献

[1]　刘人怀，斯晓夫. 工业企业管理岗位要素设计. 北京：机械工业出版社，1990.

① 本文原载《上海管理科学》，1991，(1)：6.

上海——旅游业的春光与希望①

一日之计在于晨，一年之计在于春。早晨，东方升起鲜红的太阳；春天，使人充满信心和希望。上海，这个具有国际知名度的特大城市，是中国共产党的诞生地，商品经济的发展源远流长。如今它又带来了旅游业的春光与希望。

长期以来，我们主要从政治、外交角度来对待国际旅游，它的经济性、商品性被掩盖、被淹没。改革开放的洪流，社会主义商品经济与社会主义初级阶段的理论，使旅游作为一种产业进入商品市场，参与竞争，同时，又保留了它为政治服务的属性。旅游可以促进改革开放，促进社会主义经济建设的发展。旅游业，具有综合性的特点，它不仅涉及地理位置、交通条件，而且涉及工业产品水平、文化素养和艺术造诣；它不仅与国内经济形势有关，而且与世界经济形势有关；它不仅由旅游接待、服务部门组成，而且由旅游资源、宣传部门组成。总而言之，旅游业具有综合性。增强旅游吸引力，发展旅游业，是一项系统工程，需要理论部门与实际部门、旅游部门与非旅游部门共同努力才有成效，而上海旅游业在许多方面显示了它的优势与生命力[1]。

上海的最大优势在于它的国际知名度。它是中国最大的城市，也是世界五大城市之一。1988 年末市区非农业人口达 732.65 万，而每天的流动人口已超过 200 万。许多人生活在一个什么样的环境之中？他们是怎样生活的？“上海一瞥”本身就具有吸引力。到中国不到上海，是很大的遗憾。这种遗憾之感就是吸引游客再来中国的重要因素之一。上海的知名度是由它的若干优势综合而成的。这些优势主要是：①上海是我国最大的工业城市，经济发达、实力雄厚，工业总产值、财政收入和社会商品零售总额大约均占全国的 1/6。②上海是我国最大的港口城市、口岸城市。1988 年上海的进口总额为 26.4 亿美元，出口总额为 46.05 亿美元，总计为 72.45 亿美元，比上年增长 20.8%，约占全国进出口总额的 1/10。③上海是我国的贸易中心，汇聚了全国的名特优产品，有发展成为世界第一流旅游购物中心的基础和条件。④上海地理位置好，交通方便。地处我国东海岸中部，海陆空运发达，是交通枢纽，利于游客出入境和到内地游览。⑤上海附近省区旅游资源丰富，杭州西湖、南京钟山、江苏太湖、安徽黄山等都是国家重点风景名胜区。⑥上海的旅游设施基础较好，管理服务人员素质较高。1988 年上海接待外宾的饭店有 33 家（其中 4 星级以上饭店 5 家，中外合资饭店 9 家），客房房间数为 9649 间。主要旅游购物商店 18 家，不仅有综合性商店，而且有工艺品展销公司、文物商店等专业性商店。因此，上海接待旅游者人数较多，增长较快，特别是外国游客。以中国国际旅行社上海分社为例，该社接待的外国旅游者，1977 年为 2.46 万人，1980 年为 12.1 万人，三年内接待人数增加近 4 倍，平均每年递增 70%。1981 年接待外国旅游者人数为 15.76 万人，1985 年增长到 26.3 万人，平均每年递增 16.7%。

党中央和国务院关于沿海地区实行外向型经济发展战略，上海旅游业的优势得到

① 本文原载《旅游工程原理与实践》，上海：百家出版社，1991，1-6.

进一步发挥和显示。1988 年上海共接待入境旅游者 91.6 万人，较 1987 年增长 19.4%。其中，外国人 66.4 万人，华侨 1.1 万人，港澳台同胞 24.1 万人。旅游外汇收入 10.56 亿元（外汇券），较 1987 年增加 19.1%。可见，上海旅游业十分兴旺发达。当前整顿经济秩序，治理经济环境，深化改革的形势，对上海旅游业的发展也是有利的。正如吴学谦副总理在 1989 年全国旅游局长会议上所说，“当前在治理整顿中，按照国家的产业政策，一些产业要适当降低发展速度，一些产业则要支持发展，旅游业作为创汇行业，应当继续保持一定的增长速度”“旅游业是无烟工业，基本上不存在与其他行业争夺原材料的问题。旅游业的进一步发展，会对经济调整产生一定的支持促进作用”。上海旅游业应抓住时机，大力发展。

进一步发展上海旅游业，需要处理好以下几个关系。

(1) 大旅游和小旅游的关系。传统观念把旅游局限在小旅游范围内，即旅游仅被看成旅游部门的事。事实上，旅游业是包含政治、经济、科技、文化、教育、卫生在内的综合性事业。既然如此，大力发展旅游业光靠旅游部门是不行的。要有大旅游的观念，发展旅游业必须发展与旅游有关的各种行业，从多方面为旅游发展创造条件；不仅是旅游部门关心旅游，而且与旅游有关的非旅游部门也要关心旅游。有的省、市成立了旅游事业管理委员会，由有关方面的负责人组成，这是大旅游观念的体现。实践证明，旅游事业管理委员会的成立有助于领导和协调省、市有关部门之间的关系，以便共同努力发展旅游业。

(2) 国际旅游和国内旅游的关系。传统观念把旅游局限在国际旅游范围内，即旅游被认为仅是外国人、华侨等来华旅游，而经济社会的发展已使国内旅游业兴起。国内旅游业的发展不仅有利于满足人民物质文化生活的需要，而且有利于国家货币回笼。在改革开放的今天，国内旅游业迅速发展，使交通紧张、购物不便、景点拥挤等矛盾突出。上海作为国际大城市尤其需要处理好国际旅游与国内旅游的关系。由于上海是华东的门户，出入境口岸，应重点发展国际旅游，开拓国际客源市场，重视涉外饭店基础设施建设，整顿旅游市场秩序，加强对购物商店、机场、车站、港口服务人员的文明礼貌教育。对国内旅游业的发展要注意抓饭店的规划，不能一哄而上。抓市场物价管理，住房、客运不能漫天要价，损害上海的形象。

(3) 硬件和软件的关系。传统观念重旅游基础设施、设备等“硬件”，轻管理与服务等“软件”。发展旅游业，饭店设施、交通通信条件十分重要，始终要摆在重要位置，但饭店管理与服务、翻译导游服务也不能忽视。在旅游市场上，在旅游设施基本相同的条件下，服务质量的高低就成为竞争中获胜的关键。许多外宾反映，饭店不豪华可以理解，服务态度不好则无法忍受。良好的服务可以在一定程度上弥补设施的不完善。管理人员、服务人员的外语水平、文化素质都急需提高。

进一步发展上海旅游业，还要开展下列工作。

(1) 抓交通。上海地理位置特殊，处于对内对外“两个扇面”的枢纽。对外，要吸引世界各国来华旅游的客人，对内，要向内地旅游胜地转送客人，使旅客由海陆空“进得来，出得去”。空运和陆运对上海尤为重要。出入上海的国际游客约 2/3 乘飞机，1/3 乘火车。空运、陆运要增加航班、车次，完善设备和提高服务质量。市内交通、道路也急需改善，道路不畅、交通堵塞仍是上海市旅游业令人头疼的老问题，交通安全也是扩大客源的关键之一。

（2）抓外联内联。发展旅游外联十分重要，要及时了解、分析世界旅游市场行情，有针对性地加强旅游对外宣传招待工作。市内各外联单位在对外推销旅游路线时，要互相配合，顾全大局，不低于保护价竞销或相互贬低，最好是自愿组织起来，开展对外联合推销。为了全方位、多层次、多样化发展上海旅游，需要加强内联，上海发挥出入境口岸和人文资源、购物优势，内地发挥风景资源优势，协调配合开辟旅游新路线和专题旅游项目，增加吸引力。如果把上海旅游业局限在上海市内，是没有前途的。

（3）抓多样化旅游。综合旅游、团体旅游是传统的形式，已不能完全适应新的要求。散客旅游、家庭旅游、背包旅游的兴起，就要求开展多样化旅游，以适应各种旅游者的需要，进一步扩大客源。开展专题旅游值得下功夫。上海具有开展多形式旅游的条件，如召开国际会议、交通、通信设施设备的条件和人才优势，因此，可以开展国际学术会议旅游、公务旅游等。

（4）抓旅游管理体制改革。机构改革实行党政分开、政企分开、精简统一。企业实行承包经营责任制，旅行社向建立企业联合体和企业集团方向发展，将在上海旅游业进入一个新的发展阶段。

沐浴着改革的雨露阳光，上海旅游业茁壮成长，而上海旅游业的发展必将带动内地旅游业的发展。让我们携起手来为促进上海旅游业的兴旺发达而共同努力吧！

参 考 文 献

[1]　刘人怀. 旅游工程原理与实践. 上海：百家出版社，1991：1-6.

旅游工程理论及其在浦东开发中的应用①

近年来，旅游业的发展极为迅速，以至成为许多国家的支柱产业。由于旅游业对各国的经济、文化的推动作用越来越大，发达国家竞相研究旅游业发展战略，制订发展规划，大力发展旅游事业。我国过去也制订了一些发展规划，对推动我国旅游业的发展起到了积极的作用。但是，以往的决策，大多采用传统的经验方法，未能从系统观点去分析，水平不高，可操作性不强。因此，往往容易造成决策失误，造成发展不平衡，导致大的经济损失。

旅游工程学为旅游开发与旅游发展规划的制订提供了一种科学的理论、技术和方法。现在，上海浦东开发的宏伟蓝图为上海的发展注入了强大的动力，也激发了我们开发浦东旅游的信心，如何利用旅游工程原理，探讨浦东旅游发展规划在今天也显得更为重要。

一、旅游开发中的规划决策与导向

制订地区旅游发展规划是地区旅游领导机构战略级决策的重要方面，也是一项典型的非结构问题的决策，而科学的、严密的规划又是地区旅游领导机构对其战略级问题进行决策的重要依据。一般前者称为规划决策，后者称为规划导向。

地区旅游发展规划同旅游资源分配密切相关，是一种合理调配、组织旅游资源的决策活动。其目的是：①满足旅游者了解自然和社会，完善自我的需要；②给旅游企业带来一定的经济效益，社会也能通过旅游业积累建设资金。因此，规划制订在很大程度上依赖于自然旅游资源、人文旅游资源、旅游交通、旅游饭店建设和社会经济的发达程度等自然和社会因素。

为了促进地区旅游发展规划真正具有科学预见性，进行规划决策时应当重视下面几个方面的内容。

（1）规划的内容、进度和措施要符合客观规律，要达到技术上的先进性、经济上的合理性和实践上的可行性。

（2）规划不仅是具体计划的制订，还必须顾及资金筹集、人才结构、设备更新、管理现代化等多方面的相关内容。

（3）规划导向作用主要表现在预见性上，其基础是预测技术。预测是使决策避免盲目性，增强自觉性的重要手段。在制订地区旅游规划时事先进行科学的预测，就会使决策者做到心中有数，增加主动性和预见性，起到良好的导向作用。

地区旅游发展规划的模式，如图 1 所示。

由图 1 可以看到，正确的规划来自充分而可靠的信息以及对这些信息的深入研究，特别是基于经济结构、市场销售与人才开发这三大主要方面的研究。

---

① 本文原载《旅游工程原理与实践》，上海：百家出版社，1991，43-54. 作者：刘人怀，钱幼森.

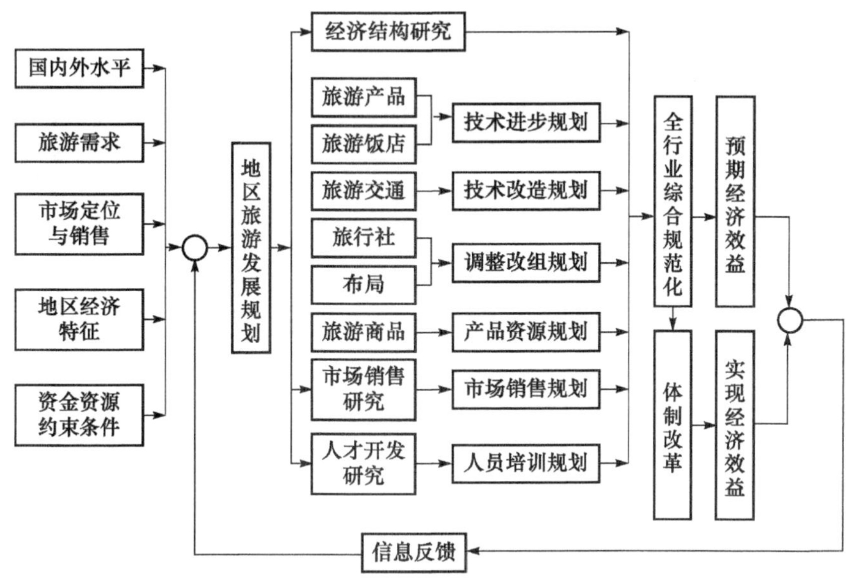

图 1　地区旅游发展规划模式

总之，根据旅游工程学的原理，地区旅游发展规划的制订，实质上是研究地区旅游部门从输入到输出“处理”过程中的管理与组织技术，并通过系统分析、统筹安排和综合平衡，充分发挥地区旅游部门的人力、物力与财力的潜力，实现地区旅游事业的最优化。

二、科学制订浦东旅游发展规划的几点看法

“开发浦东，开放浦东”像一声洪亮的春雷，从太平洋西岸响起，这是中国进入 20 世纪 90 年代的一大动作。开发和开放浦东不仅会带来上海本身的经济振兴和城市发展，还会进一步扩大其对外开放的广度和深度，并推进对外经济技术交流与合作，而且它还将起着关键和龙头的作用，启动上海经济区和长江流域的经济全面发展，使其在国内外的辐射力和吸引力十分巨大和深远。因此，开发浦东应把浦东建设成为高科技、高文化、具有世界第一流水平的城市，使它既是旅游的观赏主体，也是旅游网络的中心枢纽。可以相信，浦东的开发必然会促进上海旅游业的发展。

在制订浦东旅游发展规划时，应用旅游工程原理[1]，系统地、全面地去分析解决旅游开发中的具体问题，从开发浦东、发展浦东旅游业的整体出发，全面规划、统筹安排，使各项工作处于最优状态，力求以有限的资源取得最好效益。

在浦东发展旅游业的潜力很大，尤其是在开发起步阶段，为了配合开发浦东的整体战略，我们谈两点看法。

1. 结合浦东开发，把浦东建设成现代城市

再过十年就是 21 世纪了，在这世纪交替的时刻，在浦东——我国的东大门，建立一些反映现代科技成果的纪念性艺术群雕以庆贺新中国成立 50 周年，祝愿下一世纪人民幸福，并可成为我国物质和精神文明发展到一定高度的象征，理所当然也将成为具

有永久观瞻价值的人文景观和旅游胜地，并且还将为开发浦东、振兴上海、造福人民做出积极的贡献。

景胜人聚，建设富有特色的旅游景观，无疑将大大促进当地的经济繁荣和旅游事业的发展，国际上的一些著名城市都十分重视城市的美化和旅游设施的建设。有些城市，自然景观不多，但旅游业非常发达，就是靠城市优美的环境、具有特色的建筑、民族文化艺术和科学技术发展起来的。旅游景观（包括旅游设施）现代化是科学技术发展的必然结果，也是旅游业赖以发展的必由之路。现代高科技成果总会迅速地应用到旅游业中来，如美国的“迪士尼乐园”和“迪士尼世界”不仅以它们美丽的自然景观著称，而且以它们的现代化游乐设施取胜。

在旅游业中可应用的高科技成果十分丰富，如电子、激光设备；生态模拟设施，如建造人工景观、宇宙空间、海底世界、未来世界等；机器人，如日本筑波的机器人乐队等；现代建筑设施；现代管理设施，如计算机管理、无人售货等自动服务。通过这些成果可以创造具有时代水准的高科技形象。

用先进的科学技术来建造旅游设施装备旅游业，使现代科学技术成果为人们的旅游消费服务，将有利于不断满足人们精神文化消费的数量增长和质量提高的需要，也有利于旅游业的发展壮大。

在艺术群雕具体系统的开发中，可应用旅游工程学原理按艺术群雕的实际情况细分为若干步骤加以探讨。图 2 展示了群雕建设规划途径的框架图。

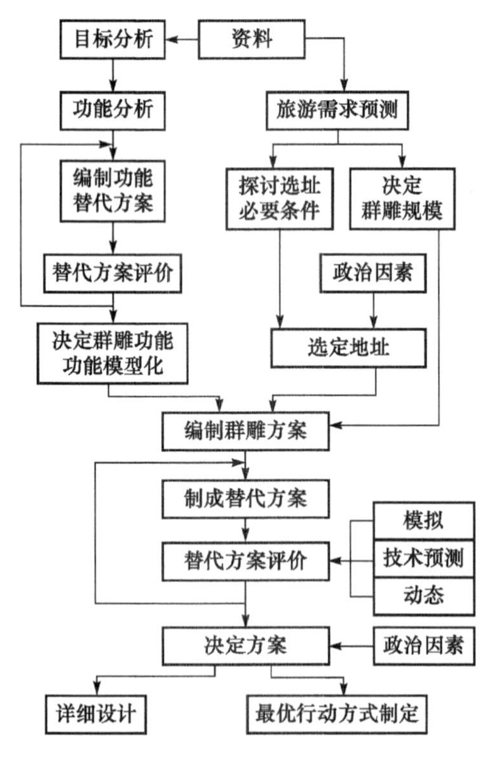

图 2　群雕建设规划途径

图 2 中关键步骤的简单说明如下。①目标分析是根据资料进行分析，明确目的、范围、环境及使命并掌握关键问题所在。②旅游需求预测是做出按年度变化的客流量预测表并预测建成后的客流量。③选定地址是根据地理位置、地貌、环境和交通等提出的要求来定。④功能分析是明确建立艺术群雕的必要特性、功能和目标，并且明确实现这些特性所必需的技术和研究课题。⑤合成的步骤是把子系统具体地综合起来提出几个具有不同特点的代替方案并使之最优化。⑥评价是从费用和效果、利益方面做出定量的评价。

目前浦东地区的开发已经起步，陆家嘴 450 米高的电视塔及附属旅游设施已在筹建，从发展现代旅游业角度出发，依赖科技，深入进行科学、全面、综合的研究，制订浦东旅游发展规划必须引起重视。

2. 结合浦东开发，举办大型国际博览会

国外的经验告诉我们，差不多所有的国

际大城市都举办过博览会。通过举办博览会，都取得了较大的经济效益，也提高了城市的知名度。从发展旅游事业角度来看，举办国际博览会至少有下列三点好处。

(1) 促进旅游事业的发展。大型国际博览会既是一个对外经济贸易活动，也是一个旅游活动。它涉及的面十分广泛，要对它做出正确的规划决策必须要有充分而可靠的信息以及对这些信息的深入研究，特别是对经济结构、市场销售与智力开发这三大主要方面的研究。我们认为这与上文所述地区旅游发展规划模式颇为相似，因此国际博览会的举办必然会促进旅游饭店、旅游交通、旅行社等的发展。

(2) 给旅游业带来直接的经济效益。因为参加国际博览会的人与一般旅游者不同，他们投宿时间长，消费水平高，所以旅游业的经济效益将会直线上升，同时参加国际博览会的人还可能去其他地方旅游，可同时促进其他省市旅游业的发展。

(3) 为上海真正成为旅游购物天堂创造良好的条件。国际博览会可以将全国名特产品集中展出，同时也可包括外国的东西，使国际旅游者能方便地在上海购到中国和世界各国的名特产品。

当然，举办一个国际博览会，不仅涉及实物的展出，而且要把组织海内外技术、经济、文化交流活动纳入计划，同时，还必须预计其社会效果。筹建国际博览会既要考虑外部大环境，又要考虑内部进程的种种变化，它是一个开放的动态大系统工程。举办国际博览会至少需要考虑：①计划的实施；②决策与实施的科学性和高效率；③某种国际合作方式；④组织机构与指挥的连续性；⑤监督、评价机制的介入。

三、结论

(1) 旅游工程是一门新兴学科，它突破了传统观念，消除了科学分隔的弱点，引入系统论、信息论和控制论[2,3]指导旅游事业，把自然科学、社会科学、技术科学和工程科学有机结合起来，对它的研究和发展，必将推进我国旅游事业的协调发展，并为促进四个现代化建设做出贡献。

(2) 上海拥有较强的工业基础，科技实力强大，各类人才众多，又处于环太平洋地区，具有战略性的优越地位，腹地纵深，实现浦东新区的目标有很多优势条件，在市委、市政府的领导下，上海旅游事业必将有一个大的发展。

参考文献

[1] 刘人怀. 旅游工程原理与实践. 上海：百家出版社，1991.

[2] 钱学森. 工程控制论. 戴汝为，何善堉，译. 北京：科学出版社，1958.

[3] 钱学森，许国志，王寿云. 组织管理的技术：系统工程. 文汇报，1978-09-27 (1).

上海旅游交通中的若干问题[①]

旅游交通是指为旅游者完成旅游目的而提供的交通服务，它是旅游和交通服务的交汇，包括水上运输、陆地运输、航空运输和相应的各项服务。现代化交通是产生旅游需求的“三要素”之一，搞好旅游交通是发展旅游业的必要条件。对上海来说，旅游交通主要是陆地运输和航空运输。上海是全国七个重点旅游城市之一，又是国务院公布的国家历史文化名城，它在我国经济、地理、交通、文化、科技等方面所处的地位，使它在我国旅游事业的发展中占有举足轻重的地位。党的十一届三中全会以来，上海市的旅游业在改革开放方针指引下，取得了迅猛的发展。1979～1988 年，上海共接待入境旅游者 518.37 万人次，年均增长率 18%，旅游创汇累计达 40.66 亿元，年均增长率 34%，1988 年旅游创汇 10.6 亿元，为 1977 年 8077 万元的约 13 倍。但是近年来，上海旅游基础服务设施不足，特别是旅游交通设施不足与旅游业迅猛发展之间的矛盾也变得越来越尖锐。“乘车难”“乘机难”，已成为上海旅游业的主要矛盾，旅游交通问题已成为发展上海旅游业的“瓶颈”。要迅速发展上海的旅游事业，首先必须治理、整顿、发展上海旅游交通业，解决入境旅游者进出上海的交通问题。如果不改善上海旅游交通现状，上海旅游业的进一步发展，将会受到严重阻碍。为此，本文力求从上海旅游业的历史和现状中找出上海旅游交通的症结，并提出改善上海旅游交通的对策建议。

一、上海旅游业概况

我们用图表的方式来描述上海旅游业的概况。从表 1 可以看出以下内容。

表 1　上海接待的入境旅游者人数

年份	旅游者人数/万人	年增长率/%	旅游部门接待比例/%
1983	44.0	18.6	67.4
1984	53.2	20.9	58.7
1985	60.2	13.2	58.7
1986	65.9	9.5	65.3
1987	76.8	16.5	54.5
1988	91.6	19.3	57.2
1989	105.4	15.1	54.6
1990	120.6	14.5	53.0

注：1989 年和 1990 年数据均为预测数据

(1) 1985 年和 1986 年两年上海接待的入境旅游者的年增长率明显放慢，从 1987 年开始年增长率又呈上升趋势。预计到 1990 年上海的入境旅游者人数将达到 120 万左

① 本文原载《旅游工程原理与实践》，上海：百家出版社，1991，96-126. 作者：刘人怀，王荣谲，汪正元，于英川，王怡然.

右，这说明上海对境外旅游者来说是有吸引力的，只要我们做好各项接待工作，上海的旅游客源是有一定保证的。

(2) 相对于1984年旅游部门接待入境旅游者的比例在减少，这一方面说明随着改革开放的不断深化，上海的国际交往范围正在不断扩大，另一方面也说明上海旅游部门接待入境旅游者的客源潜力很大。

从表2可以看出以下内容。

表2　1988年上海入境旅游者客源构成

国家和地区	接待人数/万人	占有率/%	年增长率/%
日本	32.5	35.5	13.0
中国港澳台	24.1	26.3	98.3
美国	10.0	10.9	−13.4
法国	2.7	2.9	7.5
联邦德国	2.7	2.9	6.0
英国	2.0	2.2	−0.1
加拿大	1.7	1.9	10.7
菲律宾	1.5	1.6	13.2
华侨	1.1	1.2	−1.9
新加坡	1.0	1.2	−19.8
澳大利亚	0.9	1.0	−11.9
印度尼西亚	0.7	0.76	−8.7
苏联	0.7	0.76	8.7
泰国	0.44	0.48	6.2
合计	91.6	100.0	19.4

(1) 上海前三位来客的国家和地区是日本、中国港澳台和美国。其中，日本来客占上海入境旅游者总数的35.5%，中国港澳台来客占26.3%，美国来客占10.9%，三者相加共占上海入境旅游者总数的72.7%，形成了上海旅游业的接待重点。

(2) 1988年来上海的台胞比上年增加3倍，达13.5万人。从另一方面讲，如果没有大量台胞来沪，上海接待的入境旅游者人数将维持在1987年水平。

(3) 1988年来上海的外国旅游者的年增长率只有4.4%左右，大大低于19.4%的平均水平。特别是美国和西欧市场，年增长率很慢（美国市场出现了滑坡），客源占有率很低。这反映了这些国家旅游市场开拓潜力很大。

(4) 随着中苏关系的改善和泰国、菲律宾的经济复兴，预计这三国的来沪游客会进一步增加。对此上海旅游业也要有所准备。

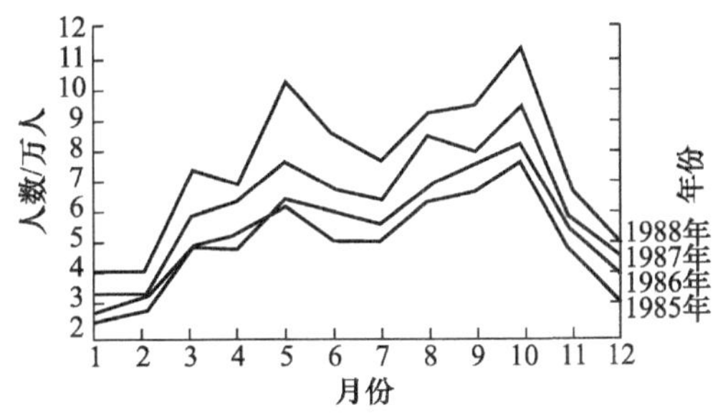

图1　上海1985～1988年各月接待入境有组织旅游者人数曲线

从图1可以看出以下内容。

(1) 淡季和旺季的接待量差距不断拉大，现已接近1∶10。

（2）旺季越来越长，淡季越来越短。除了 1 月、2 月和 12 月三个月作为淡季，10 月作为特旺季，其他各月几乎都可看成旺季。

（3）上海的各月接待人数呈锯齿形，峰点为 10 月。

表 3　1988 年上海部分旅行社接待的入境旅游者人数

旅行社名称	接待人数/万人	各旅行社接待比例/%
国际旅行社上海分社	27.5	52.5
上海中国旅行社	11.1	21.2
青年旅行社上海分社	5.0	9.6
锦江国际旅行社	2.5	4.8
春秋国际旅行社	1.0	1.9
上海铁路旅行总社	1.0	1.9
东湖旅游公司	0.8	1.5
扬子旅游公司	0.5	0.9
天马国际旅行社	0.5	0.9

从表 3 可以看出以下内容。

（1）上海现有 24 家二类旅行社。其中年接待量超过 1 万人次的有 6 家；年接待量在 5000 人次以下的有近 10 家。

（2）国际旅行社上海分社、上海中国旅行社、青年旅行社上海分社、锦江国际旅行社四家旅行社的接待量占全市旅游部门接待量的 88.1%，而其他 20 家旅行社的接待量只占全市旅游部门接待总量的 11.9%，因此应重点抓好这 4 家旅行社的服务质量。

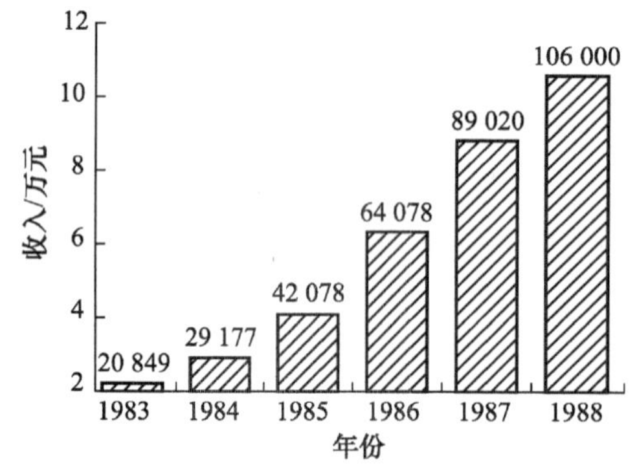

图 2　上海 1983～1988 年旅游外汇收入

从图 2 可以看出以下内容。

（1）1988 年上海旅游创汇已超过 10 亿元，上海旅游业在上海经济中的作用越来越大。

（2）从 1985 年开始，上海旅游创汇的年增长率达 40%，大大高于入境旅游者人数的年增长率。这说明只要有关方面共同努力，旅游创汇的潜力是很大的。据当时预测，到 1990 年上海的旅游创汇将达到 15 亿元。

表 4　上海 1983～1988 年旅游外汇收入构成　　单位：%

年份	商品	交通	住宿	饮食	邮电及其他	旅游服务
1983	43.9	20.0	13.9	8.2	5.3	8.7
1984	47.2	20.1	14.5	7.2	6.8	4.2
1985	42.2	22.6	15.4	8.4	6.4	5.0
1986	38.3	27.7	17.3	6.0	5.9	4.8
1987	37.0	22.5	22.1	5.9	7.9	4.6
1988	34.9	29.4	20.4	5.2	5.6	4.5

从表 4 可以看出以下内容。

(1) 旅游交通的收入在外汇收入中已接近 30%，超过住宿收入而居第二位，这充分说明了旅游交通在旅游业中的重要性。

(2) 近年来，旅游交通的收入在旅游创汇中的比例逐渐提高，这说明入境旅游者对旅游交通的需求越来越大，对旅游交通的要求越来越高。

从表 5 可以看出：上海 1983～1990 年客房数增长迅速，而客房出租率却逐渐下降，特别是今后一两年内，客房出租率将会降到 60%以下。如果不迅速采取有效措施，将会造成整个旅游业经济效益下降、投入产出严重失调的被动局面。

表 5 上海 1983～1990 年旅游饭店客房数和出租率

年份	饭店/家	客房数/间	出租率/%
1983	11	2 395	91.2
1984	12	3 048	92.5
1985	17	4 772	89.0
1986	20	5 524	75.8
1987	24	6 780	73.4
1988	37	10 987	72.0
1989	49	16 551	58.0
1990	55	19 317	41.4

注：1989 年和 1990 年数据是预测数据

从表 6 可以看出，外国旅游者在上海的人均逗留天数逐年减少，尽管在 1987 年有所回升，但仍低于 1983 年水平。这对提高客房出租率和缓解旅游交通拥挤都会产生许多不利影响。特别需要指出的是，根据国际旅行社上海分社和上海中国旅行社的统计，近几年来入境旅游者的平均逗留时间已下降到 1.7 天和 1.2 天。

表 6 1983～1988 年外国旅游者在沪人均逗留天数

年份	人均逗留天数/天
1983	3.43
1984	3.23
1985	3.49
1986	2.83
1987	2.96
1988	3.05

综上所述，可以看出，近几年来上海旅游业获得了迅速发展，取得了很大的经济效益，但是也面临客源增长缓慢、旅游交通不畅、客房出租率过低、旅游者人均逗留时间减少等诸多矛盾。其中旅游交通不畅将是今后一段时间内上海旅游业发展的主要矛盾。

二、上海旅游交通状况

上海是中国最大的港口城市，是我国经济、贸易中心，也是我国重要的出入境口岸，因此，对国内外游客都有很大的吸引力。上海地处东海之滨，内联广博腹地，位于江海之交，是中国对外开放、对内联合两个辐射扇面的枢纽和接合部，是各国旅游者出入我国的“中转站”，是中国旅游交通线路中的枢纽之一。上海内外交通四通八达，海、陆、空交通在国内都具有一定优势。

（1）民航。国内航线 36 条，可联结国内主要城市。国际和地区航线 6 条，可与日本的东京、大阪、长崎、福冈，美国的旧金山、纽约、洛杉矶、西雅图，加拿大的多伦多、温哥华，意大利的罗马，以及新加坡和中国香港直接通航。1988 年上海民航客运发送量达 175 万人次。上海民航注册的包括 5 架 A-310、7 架 MD-80、3 架 BAE-146 在内的数十架飞机，形成了一支有一定数量的现代化客机队。另外，在未来两年内，上海民航准备引进或租赁 3 架 MD-82、5 架 DC-11、3 架 A300-600、3 架 B-757。这将进一步大幅度提高上海民航的运输能力。

（2）铁路。通过沪宁、沪杭两条干线与全国铁路结成网络，通往全国各地。上海新客站每天接发旅客列车能力为 72 对。华东旅游线上现有列车：上海到南京 5 对，上海到无锡 4 对，上海到苏州 2 对，上海到杭州 4 对。1988 年上海铁路客运发送量达 2838 万人次。

（3）水路。国际和地区航线 2 条，可通过水运渠道直达日本的大阪、神户，以及我国香港。国内沿海航线 13 条，长江航线 6 条，可通往国内各主要沿海和沿江城市。1988 年上海水运发送量达 679 万人次。上海国际海运有“锦江”“上海”“海兴”三艘客货轮船航行我国香港，有“鉴真”一艘客货轮船航行日本，有外虹桥国际客运码头一个泊位供国际客货轮、外国旅游船以及友好访问的船只停靠。

1988 年是中国旅游交通矛盾大爆发的一年，大批旅游者滞留北京、西安、桂林等城市进不去，出不来。特别是旅游旺季，一些旅游热线的交通更是一片混乱，给中国旅游事业造成极坏影响，使中国旅游业在国际旅游市场的声誉受到损害。1988 年上海基本上没有发生大批境外旅游者滞留积压现象，这与交通运输部门的大力支持是分不开的。上海交通运输部门为了支持旅游事业，采取了许多有效措施，取得了明显效果。例如，中国东方航空公司上海民航与国际旅行社上海分社、上海中国旅行社、青年旅行社上海分社三家主要旅行社在旅游旺季定期召开协调会，想方设法、多方协调、统一安排，用包机和加班机及时输送游客，有效地解决了游客的积压滞留。又如铁路部门为一、二类旅行社单独设立订票点，尽量保证入境旅游者的票源，给一、二类旅行社带来很大方便。1988 年上海共接待入境旅游者 91.6 万人次。据统计，上海发送的有组织入境旅游者中，外国旅游者占 72.5%，华侨和港澳台胞占 27.5%。在外国旅游者中，约有 30%乘国际和地区航班直接由上海出境，约有 35%乘国内航班由上海飞向桂林、北京、西安、广州等其他城市，另有近 30%乘火车由上海到苏州、杭州、南京、无锡等城市，只有 5%左右乘其他交通工具继续旅游。华侨和港澳台胞旅游者，约有 10%乘国际和地区航班直接由上海出境，约有近 30%乘国内航班由上海飞向北京、桂林、福州、厦门等城市，另有 60%左右乘火车由上海到杭州、苏州、南京、无锡等城市，只有极少数人乘坐其他交通工具继续旅游。

在进入上海的有组织入境旅游者中，外国旅游者约有 30%从境外乘国际和地区航班直接由上海入境，约有 36%从西安、北京、桂林、广州等城市乘国内航班飞抵上海，另有 30%从苏州、杭州、南京、无锡等城市乘火车抵达上海。华侨和港澳台胞旅游者约有 8%从境外乘国际和地区航班直接由上海入境，约有 38%从北京、广州、桂林、西安乘国内航班飞抵上海，另有 53%从杭州、苏州、南京、无锡乘火车抵达上海。

乘坐国内民航抵达上海的有组织入境旅游者中，约有 32%来自北京，约 30%来自西安，约 20%来自桂林，另有 10%来自广州。也就是说，乘国内民航抵沪的有组织入境旅游者中，有 90%以上来自北京、西安、桂林、广州 4 个城市。从表 7 可以看出，乘坐国内民航从上海飞向外地的有组织入境旅游者中，约 35%去桂林，约 26%去北京，约 20%去西安，另有 10%去广州。也就是说，乘国内民航离沪的有组织入境旅游者中有 90%以上去桂林、北京、西安、广州 4 个城市。

表 7　1988 年有组织入境旅游者国内主要航线客流（从上海出发）　单位：人次

月份	桂林	北京	西安	广州	福州	厦门
1	1 570	1 360	680	490	95	10
2	1 820	1 370	640	670	66	41
3	3 720	3 860	3 780	760	88	28
4	6 080	4 080	3 210	960	376	207
5	7 540	6 160	3 290	1 500	769	1 164
6	4 240	4 470	1 110	1 520	1 690	1 185
7	5 470	4 190	3 570	2 250	751	585
8	7 920	5 670	5 530	1 610	492	385
9	7 830	6 330	4 820	1 890	678	704
10	10 500	7 630	5 500	2 790	1 347	846
11	3 510	8 560	1 380	1 513	946	514
12	2 660	2 010	1 350	1 070	231	710

乘火车从外地抵达上海的有组织入境旅游者中，约 52%来自苏州，约 28%来自杭州，约 7%来自南京，另有 5%来自无锡。也就是说，乘坐火车来上海的有组织入境旅游者中，有 90%以上来自苏州、杭州、南京、无锡 4 个城市。

从表 8 可以看出，乘火车从上海去外地的有组织入境旅游者中，约 55%去苏州，约 30%去杭州，约 7%去南京，另有 5%去无锡。也就是说，乘火车离开上海的有组织入境旅游者中，有 90%以上去苏州、杭州、南京、无锡 4 个城市。

表 8　1988 年有组织入境旅游者火车客流（从上海出发）　单位：人次

月份	苏州	杭州	南京	无锡
1	2 110	880	250	310
2	2 180	1 760	770	620
3	7 300	3 180	1 370	1 490
4	7 270	5 210	1 360	1 510

续表

月份	苏州	杭州	南京	无锡
5	8 610	5 970	1 940	2 290
6	6 520	5 960	1 560	2 960
7	5 880	6 090	1 670	2 850
8	7 830	7 080	1 790	2 210
9	9 600	7 980	1 610	3 430
10	11 360	9 280	2 370	4 300
11	5 200	4 430	1 820	2 240
12	4 170	2 010	620	860

综上所述，从上海出发的旅游交通线路中，航空线路主要是从上海到桂林、北京、西安、广州、福州、厦门 6 条，火车线路主要是从上海到苏州、杭州、南京、无锡 4 条。这 10 条旅游交通线路占了上海旅游部门接待入境旅游者国内旅游交通客流量的 90%以上，上海旅游交通运力不足的主要矛盾也暴露在这 10 条线路中。目前这 10 条旅游交通线的运力详见表 9 和表 10。

表 9　民航六条主要旅游线运力（从上海出发）

方向	周航班数/个	周运力/人次	月运力/人次
桂林	12	1 630	7 010
北京	22	4 140	17 800
西安	20	2 650	11 390
广州	32	5 470	23 550
福州	8	850	3 660
厦门	12	1 630	7 020

表 10　铁路四条主要旅游线运力（从上海出发）

方向	日车班数/个	日运力/人次	月运力/人次	月软席运力/人次
苏州	2	4 200	126 000	3 840
杭州	4	8 400	252 000	7 680
南京	5	10 500	315 000	10 200
无锡	4	8 400	252 000	7 680

1988 年有组织入境旅游者占入境旅游者总数的 57.2%。据估计，国内航班，上海到桂林航线的总客流中，入境旅游者大约占 60%；上海到西安航线，入境旅游者大约占 50%；上海到福州航线，入境旅游者大约占 30%；上海到广州航线，入境旅游者大约占 10%；上海到北京航线，入境旅游者大约占 40%。在铁路上海站接发的软席旅客中，上海到苏州、杭州、南京、无锡线路入境旅游者约占 80%。再根据表 9 和表 10 可以大致推算出从上海到 10 个重点旅游城市各月所需运力总量及运力缺口。

需要特别指出的是，表 11 和表 12 所列出的数据仅是按计算得到的结果，其中有些火车和飞机的班次不适合外国旅游者乘坐，再加上旅客实际上不可能每天平均分布，因此实际情况将比表 11 和表 12 列出的更严重。

表 11 民航五条主要旅游线所需运力及缺口（从上海出发） 单位：人次

月份	桂林		北京		西安		广州		福州	
	需	缺	需	缺	需	缺	需	缺	需	缺
1	4 580	−2 430	5 440	−12 360	2 280	−9 110	8 570	−14 980	560	−3 100
2	5 300	−1 710	5 990	−1 180	2 240	−9 150	11 720	−11 840	390	−3 270
3	10 830	+3 820	16 870	−930	13 220	+1 830	13 290	−10 260	510	−3 140
4	17 730	+10 120	17 850	+50	11 230	−160	16 780	−6 760	2 190	−1 460
5	21 970	+14 960	26 920	+9 120	11 500	+110	26 230	+2 673	4 480	+820
6	12 370	+5 360	19 540	+1 740	3 880	−7 510	26 573	+3 070	9 880	+6 220
7	15 940	+8 930	18 310	+510	12 480	+1 090	39 340	+15 780	4 390	+730
8	23 070	+16 060	24 860	+7 060	19 340	+7 950	28 150	+4 600	2 870	−780
9	22 800	+15 790	27 670	+9 870	16 850	+5 460	33 040	+9 490	3 970	+310
10	30 600	+23 590	33 340	+15 540	19 230	+7 840	48 780	+25 230	7 870	+4 210
11	10 240	+3 230	37 420	+19 620	4 830	−6 560	26 450	+2 901	5 510	+1 860
12	7 760	+750	8 810	−8 990	4 720	−6 670	18 710	−4 840	1 350	−2 310

注：+表示缺，−表示余

表 12 铁路四条主要旅游线所需软席运力及缺口（从上海出发） 单位：人次

月份	苏州、无锡、南京		杭州	
	需	缺	需	缺
1	5 840	−15 880	1 930	−5 750
2	7 800	−13 920	3 850	−3 830
3	21 900	+180	6 940	−740
4	22 160	+440	11 380	+3 700
5	28 040	+6 320	13 050	+5 370
6	24 120	+2 400	13 030	+5 350
7	22 740	+1 020	13 320	+5 640
8	25 870	+4 150	15 480	+7 800
9	31 990	+10 270	17 430	+9 750
10	39 390	+17 670	20 380	+12 700
11	20 230	−1 490	9 690	+2 010
12	12 340	−9 380	1 350	−6 330

注：+表示缺，−表示余

据有关部门反映，目前国内航线正常航班只能满足需求的 50%左右，火车也只能满足需求的 70%左右。因此，尽管 1988 年上海采取了许多应急措施，基本上没有发生大批外国旅游者滞留积压现象，但是 1988 年上海旅游交通也是“险象丛生”“危机四伏”。

这充分暴露了上海旅游交通不能适应旅游业进一步发展的各种矛盾。1988 年上海旅游交通中出现了许多“危险的第一次”。第一次出现一天有 300 多人不能按时离沪；第一次出现受飞机航班更改影响的团队达 60%以上；第一次出现乘火车的矛盾与乘飞机的矛盾一样尖锐；第一次出现有这么多的旅行社经理在旅游旺季为了搞机票而天天忙到深夜十一二点；第一次出现外国旅游团因交通问题而中止旅游提前出境。这么多的第一次充分说明如果不迅速改变上海旅游交通的现状，上海的旅游业就不能得到进一步发展，需要指出的是这许多“危险的第一次”是在上海一年接待 91 万人境旅游者的情况下发生的。根据国际旅游市场的变化发展以及来上海国际游客逐年增长，上海旅游业规划到 1990 年接待国际游客 120 万人次左右，这必将进一步增加上海旅游交通的压力。

旅游工程学的提出①

1986 年，我怀着促进中国旅游事业发展的强烈愿望，在综合分析国内外旅游科学研究工作的基础上，概括性地提出了一门新兴的综合性学科——旅游工程学[1]。实践证明，这一学科不但对研究旅游业的发展规划和计划有十分重要的作用，而且对促进旅游事业的发展也十分重要，特别是在旅游宏观和微观科学管理中，更具有广泛的适用性。

一、旅游工程学的基本概念

旅游工程学是一门新兴的边缘学科。它运用自然科学、技术科学、工程科学和社会科学的有关思想、理论、方法和手段，根据系统论的原理[2,3]，应用系统工程学的方法，对旅游系统的构成要素、组织结构、信息交换和反馈控制等问题进行综合分析、设计、试验、实施，以便充分地发挥人力、物力和财力的作用，最合理、最有效、最经济地实现旅游系统的整体效益优化。所以，这是一门由自然科学、技术科学、工程科学和社会科学相互交叉、相互渗透而形成的边缘性和综合性的学科。它是一棵幼苗，有待今后同行、专家和应用者通过实践加以培养，使之茁壮成长，以利于旅游科学的研究工作和旅游事业的发展。

旅游工程学作为一种崭新的思维观点，它将冲破过去传统的保守思维方式，冲破自然科学、社会科学、技术科学和工程科学之间的鸿沟，促进大学科和多学科之间相互渗透、交叉和联系，从而产生新的思想和理论观点，以指导旅游实践。这不仅是繁荣学术之必需，更是发展旅游事业的要求。

旅游工程学作为高层次的决策分析方法应用于旅游业，对发展旅游业的宏观规划、计划、未来预测、旅游理论和学术研究、旅游开发的科学论证等将起到十分重要的咨询作用，从而求得所涉问题的整体合理和可行性。

旅游工程学作为旅游管理的最先进的具体方法论，它还将在不同层次的旅游实践和旅游决策中参与多方案、对策性、优选和复杂的定量分析。

总之，旅游工程学促进了我们对旅游未来前景的认识。我们应当借助于这一理论来增强我们对旅游这一事物的观察能力、分析能力和运筹能力。

二、旅游工程学的研究方法

旅游工程学从旅游系统的总体高度出发，把系统当成一个相互作用、相互依存的超巨系统来进行研究，其研究方法如下。

（1）全面搜集资料，弄清问题所在。

① 本文是在日本神户商科大学的演讲稿，神户，1993 年 4 月 20 日；《旅游工程管理研究》，北京：科学出版社，2014，1-3.

(2) 确定和分析系统的评价目标。首先要了解建立这个系统的目的和要求，进而确定系统的目标。确定目标时要做到技术上先进，经济上合理，并考虑到与其他系统的兼容性和对客观环境的适应性。其次要分析实现系统目标所涉及的限制条件有哪些，达到总目标应完成哪些具体的项目，这些项目又应由哪些子系统来承担，进而确定各子系统的目标。

(3) 系统的模型化。明确了系统的目标以后，可根据系统的性质与功能的要求，提出几个可能的方案，找出相应的约束条件和各要素之间的逻辑关系，用数学方式把它们表达出来。应该指出，旅游工程学的研究对象不仅局限于某一个实物，它涉及的面常常十分广泛，而且都是很复杂的系统，因而不能用形象的实体模型来代替抽象的数学模型。

(4) 系统的最优化。系统的最优化是通过系统的模型进行的。不同的模型有不同的优化理论和优化方法。例如，对一些确定性的问题，可采用线性规划、整数规划、非线性规划、动态规划等理论和方法进行优化；对于那些非确定性的问题，则可用对策、决策理论和方法进行最优化。

(5) 综合评价。对于多目标的系统来说，不同的目标可以建立各种不同的模型，而不同的模型又可以求出几个不同的最优解，对这些最优解就要根据具体条件进行综合分析和评价，找出各方面的所得与所失，把不同性质、不同度量单位的几个方面的评价，用同一评价指标来比较，从中选出技术上先进、经济上合理、综合效益最优的方案。

图 1 描述了这些研究过程。

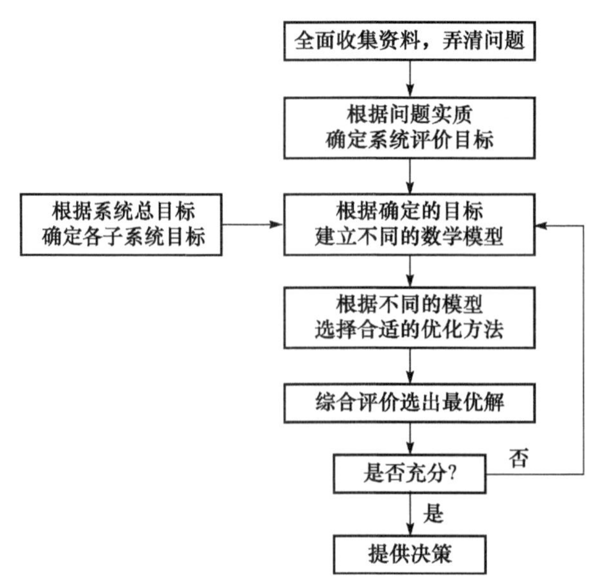

图 1　旅游工程学研究过程

应当指出，系统的模型化是系统最优化和综合评价的基础。要想准确而方便地求得系统的最优方案，就必须使所建立的模型既能真实地反映现实，而又不过于复杂。

如果建立的模型不确切，系统的最优化和综合评价也就失去了意义。因此，从事旅游工程学的研究，不仅要对所研究的旅游系统的全貌有深入的了解，而且还要熟练地掌握现代数学工具和各种模型的优化技术。

综上所述，我们已对旅游工程学进行了简要的介绍。它是一门新兴学科，突破了传统观念，消除了科学分隔的弱点，引入系统论、信息论和控制论指导旅游事业，并将自然科学、社会科学、技术科学和工程科学有机结合起来。对旅游工程学的研究和发展，必将促进中国和世界旅游事业的协调发展。

参 考 文 献

[1]　刘人怀. 旅游工程原理与实践. 上海：百家出版社，1991.

[2]　钱学森. 工程控制论. 戴汝为，何善堉，译. 北京：科学出版社，1958.

[3]　钱学森，许国志，王寿云. 组织管理的技术：系统工程. 文汇报，1978-09-27 (1).

开启思想的眼睛①

生活是由思想造就的。一个人活得愈久，就愈深信思想的力量。

新一届工商管理硕士（MBA）毕业了，看到年轻人学业精进、事业有成，我深感欣慰。业界精英能利用繁忙公务的间隙，亲身引领学习的革命并将所学所用积淀成书，值得称道；作为教学成果及满足市场渴求，遴选优秀毕业生学位论文中上乘之作结集出版，值得尝试。

当教育将理论与实践分离时，通常就会没有效率。

案例教学法是由哈佛商学院首创的管理教学方法。这种方法贴近实践，将管理理论与经验有效融合，大大拉近了管理教学与实践商战之间的距离。案例教学也是我国MBA研究生培养的重要环节。但目前国内尚无专门的案例教材，所用案例多选自国外的样本。为切合中国MBA教育的发展规划，造就跨世纪"实务型"人才，急需大量从国情出发，真实描述我国社会主义市场经济环境下企业状态、危机与挑战的优秀案例，积累大批反映中国特色的企业运行规则和操作实务的资源储备。此书作为暨南大学在MBA教材和案例建设、MBA论文学以致用等方面所作尝试之一，旨在为实践"管理教育、兴国之道"的方略探路寻踪。

时代已经不同，我们对才华的定义应该扩大。教育最大的帮助，是引导人进入适应性的领域，使其因潜能得以发挥而获得最大的成就感。

本书作者多为企业金字塔顶端的高级主管，实际上是EMBA（Executive Master of Business Administration，直译为高级管理人员MBA）。他们超越同侪志存高远，将企业远景与个人潜能熔于一炉，渴望淋漓发挥铸就工商领袖。书中涉及企业发展战略、IT运用等一个个生动的分析与规划个案，是企业高级经理们所面对的众多真实管理问题的记录，分述于案例背景、案例分析两部分，具有多样性、综合性、全面性的特点。可作工商管理专业人士论文写作的参考范本，也可作MBA教学中的备用案例，更把读者引入企业实际，置于一个实际经营者的立场上，从实践的环境出发学习什么是经营和如何经营。本书的特点在于不仅提出问题而且提供解决方案。这些观点来源于实践经验。"春江水暖鸭先知"，企业管理阶层对市场态势、企业困境、因应之道体认最深。但正如案例教学的规律所示：不存在绝对正确的答案，而在于培养个人对经营状况理解和判断的情境感。面对瞬息万变的企业环境，立于高层次来把握和分析问题，运筹帷幄追求成功经营的思路和方法。

同时，我们深知，没有一种经验可以完全翻版。行胜于言。

一个人的真正动力来自对自己所向往的方向的憧憬。如果我们知道要去哪，我们会很稳定地朝目标前进。事实是，一个人从来不是完全清楚自己的人生目标。所以，开启思想的眼睛，持续地聆听逐渐浮现的目的感，探明阶段性目标，十分重要；拭亮

① 本文是《MBA案例》序，广州：暨南大学出版社，2000，1-2.

思想的眼睛，看到目标，并保持达到目标的决心不变，培养适应生命顺逆的能力，更为关键。只有了解内心的欲求，找到个人的存在理由，才能激发抱负和承诺的独特力量。

正如作者在论文谢辞中所述：MBA 教育中心的老师们在理论和实践两方面都给予了悉心指导，他们知识渊博，诲人不倦的精神令人感受到跨世纪的中国高等院校学者的典型风范。在此，我也想对我的同事们的辛勤工作表示感谢。一所学府的美誉传承于桃李芬芳、卓尔不群的莘莘学子，更依托于渊博笃实、追求杰出的校风师德。让我们开启思想的眼睛，让憧憬飞扬，共创新世纪的激情与梦想。

是为序。

质量是企业生存的根本①

没有人能够不瞄准便命中“成功”的靶心。即使我们会有一点偏差，但这样至少比我们闭上眼睛盲目射击更接近靶心吧。

人生如此，企业也如此。

如果有人问，如今美国有哪些世界级的巨型企业？可以说不用动脑子人们就能说出许多；但如果再问，美国哪几家企业名列世界前茅的时间能持续 100 年？道琼斯 30 种工业股价指数自 1896 年设立以来，首批入选而至今未名落孙山者又有几家？答案恐怕就只有美国通用电气公司（GE）[1]。

美国《财富》杂志 1900 年列举了 12 家最著名的公司，成立于 1892 年的通用电气公司位列其中，历经百年，风采依旧。1998～2002 年通用电气公司连续五年位居《财富》“全美最受推崇公司”排行榜首位；通用电气公司的总市值超过 5000 亿美元，已成为世界上最具盈利能力和增长能力的公司。

商海浮沉，风云变幻，是什么使英雄不倒？

是质量。

海尔集团[2]是一家土生土长的中国企业，原本是一个生产电动葫芦的小企业，通过争取才获得我国最后一个生产冰箱的定点资格，资不抵债、濒临倒闭，销售收入只有 300 多万元。自张瑞敏 1984 年出任厂长以来，步步为营，保持了年平均 80%以上的高速增长，发展成为拥有 99 个紧密层企业，年销售收入达 160 亿元的跨行业、跨国界的大型企业集团。

海尔集团成功崛起的主要原因是什么？

是质量。

企业是社会经济的细胞，质量是企业生存的根本，没有质量的企业，既是没有信誉的企业，也是没有生命力的企业。新近一项统计数据显示，中国大多数企业一般的寿命为 2～3 年；成功企业的寿命为 8～10 年；而能生存 40 年以上的企业仅为约 2%。这也就说明为什么像同仁堂这样的“百年老店”贵乎稀有了。

不向消费者提供劣质品本应是企业生产的道德底线，但一些质量曝光却让人触目惊心：“国家质检总局检查酸奶合格率仅为 67.4%”“国家质检总局检查发现部分安全防护产品不安全”“国家质检总局抽查显示电水壶合格率为 66.7%”“全国 111 种食用盐抽查产品逾三成不合格”“质监部门检查捂伤口的药用棉纱竟然半数不合格”等。

水能载舟亦能覆舟，几多企业成也萧何败也萧何，酿成这种局面的成因之一是什么呢？

还是质量！

---

① 本文是中国工程院和深圳市人民政府主办的中国青年科技企业家管理论坛主题报告，深圳，2002 年 6 月 18 日；《中国青年科技企业家管理论坛论文集》，深圳，2002，30-35.

《财富》杂志自从 1955 年以来，每年都统计出美国 500 家最大企业的排行榜。能够进入《财富》500 强，是企业经营取得成功的一种标志。一项以"《财富》500 家企业中的 TQM"为题的调查结果，说明了实施全面质量管理（TQM）战略与企业经营获得成功之间的关系。

（1）91.5%的企业制订了正式的质量计划。

（2）81.3%的企业采用了戴明、朱兰、费根堡姆、克罗斯比的 TQM 理论。

（3）96.7%的企业成立了质量团队，并利用质量团队来发现组织中的问题和提出改进建议。

竞争就像一场战役，加入 WTO 以后，企业的竞争将不再是国有制企业与非国有制企业之间的竞争，而是中国企业与国外优秀企业的竞争。地方保护主义将在一定程度上折翅断戟，企业很快就可能体验到竞争无疆界的切肤之痛。改革开放 30 多年的成功实践证明，企业的健康发展，经济的持续繁荣，得益于产品质量水平的不断提高。在市场经济条件下，产品质量是市场竞争的重要内容。质量的高低，直接影响产品的市场竞争力，关系着我国产品在国际市场的信誉。我们只埋怨那些诸如底子薄、基础弱等无法解决的因素将于事无补，而应该把精力放在能够解决的问题上。

企业竞争力就像一个靶环，人力资源、组织架构、资本结构、创新机制、产品质量等是它的层层圆环，它们的整合凝聚成企业的综合竞争力。但在重重内涵与外延中，剥茧抽丝，产品质量是企业生存的关键，是经济发展的一个具有根本性的基础问题，是企业龙争虎斗的众矢之的，是一箭定乾坤的靶心！

你把问题确定得越准确，你的目标就越清晰，你命中靶心的机会就越多。如果说"我们的质量很不稳定"，这意味着什么呢？这样的说法能让你保证质量始终如一？不能！如果你要改善某件事情，首先你必须清楚目前的情况和你要达到的目标，否则事情就不可能得到改善。因此我们要做的正是阐明问题、阐明问题、再阐明问题！

第一，我们来回顾"质量管理"[3,4]的百年历程。

工业革命前，产品质量由各个工匠或手艺人自己控制。

1875 年，泰勒制诞生——科学管理的开端。最初的质量管理——检验活动与其他职能分离，出现了专职的检验员和独立的检验部门。

1925 年，休哈特提出统计过程控制理论——应用统计技术对生产过程进行监控，以减少对检验的依赖。

1930 年，道奇和罗明提出统计抽样检验方法。

20 世纪 40 年代，美国贝尔电话公司应用统计质量控制技术取得成效；美国军方在军需物资供应商中推进统计质量控制技术的应用，制定了以休哈特、道奇、罗明的理论为基础的战时标准 Z1.1、Z1.2、Z1.3——最初的质量管理标准。

20 世纪 50 年代，戴明提出质量改进的观点——在休哈特之后系统和科学地提出用统计学的方法进行质量和生产力的持续改进，强调大多数质量问题是生产和经营系统的问题，强调最高管理层对质量管理的责任，最终形成了对质量管理产生重大影响的"戴明十四法"。

20 世纪 60 年代初，朱兰、费根堡姆提出全面质量管理的概念。他们提出，为了生产具有合理成本和较高质量的产品，以适应市场的要求，只注意个别部门的活动是不够的，需要对覆盖所有职能部门的质量活动进行策划。朱兰、费根堡姆的全面质量管理理论在日本备受推崇，日本企业创造了全面质量控制（TQC）的质量管理方法。统计技术，特别是“因果图”“流程图”“直方图”“检查单”“散点图”“排列图”“控制图”等被称为“老七种”工具的方法，被普遍应用于质量改进中。

20 世纪 60 年代中期，北大西洋公约组织制定了 AQAP 质量管理系列标准，AQAP 标准以 MIL-Q-9858A 等质量管理标准为蓝本。所不同的是，AQAP 引入了设计质量控制的要求。

20 世纪 70 年代。TQC 使日本企业的竞争力极大地提高，轿车、家用电器、手表、电子产品等占领了大批国际市场，极大地促进了日本经济的发展。日本企业的成功，使全面质量管理的理论在世界范围内产生巨大影响。这一时期产生了石川馨、田口玄一等世界著名质量管理专家。产生的管理方法和技术包括 JIT（just in time，准时化生产）、Kanban（看板生产）、Kaizen（质量改进）、QFD（quality function deployment，质量功能展开）、田口方法、新七种工具。田口博士的努力和贡献使质量工程学开始形成并得到巨大发展。

20 世纪 80 年代。菲利浦·克劳士比[5]提出“零缺陷”的概念。他指出，高质量将给企业带来高的经济回报，突破了传统上认为高质量是以低成本为代价的观念。质量运动在许多国家展开。中国、美国、欧洲的许多国家设立了国家质量管理奖，以激励企业通过质量管理提高生产力和竞争力。质量管理不仅被引入生产企业，而且被引入服务业，甚至医院、机关和学校。许多企业的高层领导开始关注质量管理。全面质量管理作为一种战略管理模式进入企业。

1987 年，ISO 9000 系列国际质量管理标准问世，质量管理与质量保证开始在世界范围内对经济和贸易活动产生影响。

1994 年，ISO 9000 系列标准改版，新的 ISO 9000 标准更加完善，为世界绝大多数国家所采用。第三方质量认证普遍开展，有力地促进了质量管理的普及和管理水平的提高。

20 世纪 90 年代末，全面质量管理成为许多“世界级”企业的成功经验，被证明是一种使企业获得核心竞争力的管理战略。质量的概念也从狭义的符合规范发展到以“顾客满意”为目标。全面质量管理不仅提高了产品与服务的质量，而且在企业文化改造与重组的层面上，对企业产生深刻的影响，使企业获得持久的竞争能力。

在围绕提高质量、降低成本、缩短开发和生产周期方面，新的管理方法层出不穷。其中包括并行工程（concurrent engineering，CE）、企业流程再造（business process reengineering，BPR）等。

21 世纪初，随着知识经济时代的到来，知识创新与管理创新必将极大地促进质量的迅速提高，包括生产和服务的质量、工作质量、学习质量直至人们的生活质量。质量管理的理论和方法更加丰富，不断突破旧的范畴而获得极大的发展。

第二，我们要领悟“质量管理”的基本原则。

原则之一，质量的定义必须是符合要求，清楚地了解要求，然后消除一切阻碍，而不是“好”。

原则之二，质量系统的核心在于预防，而不是检验和评估。

原则之三，工作标准必须是“零缺陷”，是第一次就把事情完全做对，而不是“差不多就好”。

原则之四，质量是用不符合要求的代价来衡量的，所有做错的事情导致的花费是它的代价，而不是用指数或图表。

第三，我们来研究让通用电气公司和海尔集团受益的六西格玛质量理论。

六西格玛[6]正是当下风行的引领人们进行质量革命的有效方法之一，并且也是最具实践性的一种。六西格玛就是阿拉伯数字 6 加上希腊字母 σ，即“六个标准差”。它在商业和工业领域含义很广，是一组统计测量法和管理理念。它运用统计数据测算一件产品接近其质量目标的程度。“西格玛”用来确定一个公司到底出了多少差错，无论这个公司从事何种行业。而“6”则是努力追求的西格玛的完美程度。

在 1 西格玛的情况下，所做的工作只有约 30%是正确的，即 $1\sigma=690\ 000$ 次失误/百万次操作。

在 2 西格玛的情况下，每 100 万次机会将出现 30 多万个缺陷，即 $2\sigma=308\ 000$ 次失误/百万次操作。

在 3 西格玛的情况下，出现 66 800 个缺陷，即 $3\sigma=66\ 800$ 次失误/百万次操作。

在 4 西格玛的情况下，大约存在 6000 个缺陷，即 $4\sigma=6210$ 次失误/百万次操作；

在 5 西格玛情况下，$5\sigma=230$ 次失误/百万次操作；

在 6 西格玛情况下，$6\sigma=3.4$ 次失误/百万次操作。

大部分公司是在 3～4 西格玛的状况下工作的，在 3.8 个西格玛的时候意味着达到了 99%的工作是处于正确的状态。但实践证明，即使出现 1%的缺陷也将导致大量的差错。相当于美国全国每小时丢失 2 万个邮件，或每周做错 5000 例外科手术，更有甚者，类同于繁忙的大机场每天发生 4 起事故！从我们熟悉的工作来估算，1%的缺陷将导致怎样的损失和灾难？着实令人心惊！

这正是推行六西格玛的动因所在。数字既能让你清晰地表述问题，也能使你达到期望的准确目标。六西格玛的最终目标是改善质量，但在六西格玛中，改善质量只是达到结果的一种手段，并不是结果本身；它是一套管理方法，与其他质量理论的根本差别是强调不仅要挑出次品，更要寻根溯源，解决产生次品的根本性问题。

改进质量在于转变观念。我们要转变大众购买次等品、削价处理品、出口转内销产品的热情；改变员工认为“虽然犯了错误，但只要最后改正了，就还算是成功的”陈旧观念。购买次品类同于纵容企业的粗制滥造，而不符合要求的代价是第一次没做对而产生的所有额外费用。

六西格玛是一种“持续改进”的观念，对人们所要完成的任务给出明确的定义和清楚的结构，要求每一个工作程序、每一个产品、每一项服务都达到一种“近乎完美”的质量境界。它通过防止失误，强调消除错误、减少浪费以及避免重复劳动来使企业赢得先机，提高顾客满意度。倡导“零缺陷”，树立“第一次就把事情做对”的思维是六西格玛质量理论的精华。

质量始终是一个永恒的主题，20 世纪 80 年代，我们曾经喊出“质量就是生命”的口号，浙江民营企业高层质量论坛上发出过“质量是我们的根”倡议书。从“全面质量管理”到“质量奖”，从 ISO 9000 到六西格玛的方法，我们对质量的认识有很大的提高，“质量”在高层的舌尖与基层的实务之间翻滚起伏，艰难起舞。

但事情常常是这样，一件事最引人注目的阶段往往是它不尽如人意的时期，是整个社会感到匮乏和危机的时期，也是人们觉醒和自救，重整旧山河的时期。朱镕基总理曾指出，质量问题是经济发展的一个战略问题，产品质量代表了一个国家的形象，一个民族的精神，要发展经济，要提高经济增长的质量和效益，产品质量是关键。

新世纪是一个质量的世纪。企业的竞争，归根结底是文化的竞争。我们应致力于创建一种“质量文化”，用“零缺陷”的思想使企业摆脱长期的管理困惑，用“第一次就把事情做对”的思维指导我们做人做事的方式。

命中靶心，才能获得成功；第一次就把事情做对，才有高质量的商品、高质量的生活。这将使我们的国家、我们的企业、我们自身受益无穷。

参考文献

[1] 王如君. 通用电器，一直走在最前面. 环球时报，2002-05-30.
[2] 波顿 M D. 通用筹码与海尔策略. 北京：民主与建设出版社，2002.
[3] 伍爱. 质量管理学. 广州：暨南大学出版社，1996.
[4] 黄瑞荣，伍爱，王燕. 现代企业管理学. 广州：暨南大学出版社，1995.
[5] 克劳士比 P. 质量无泪. 北京：中国财政经济出版社，2002.
[6] 乔杜里 S. 六西格玛的力量. 郭仁松，朱健，译. 北京：电子工业出版社，2002.

对某跨国公司绩效考评系统的评价①

随着信息革命、知识经济时代进程的加快，企业面临着前所未有的竞争环境的变化，传统的组织模式和管理理念已越来越不适应环境，我国国有企业要在快速变化的市场中生存与发展，就要建立现代化的体制。本文对一著名跨国公司人力资源考评系统进行评价与分析，希望对我国国有大中型企业有一定的借鉴作用[1]。

一、绩效考评系统评价的标准

绩效考评是一种正式的员工评估制度，它是通过系统的方法、原理来评定和测量员工在职务上的工作行为和工作效果。绩效考评是企业管理者与员工之间的一项管理沟通活动。绩效考评的结果可以直接影响到薪酬调整、奖金发放及职务升降等诸多员工的切身利益。

绩效评估要同员工的生涯规划、企业的培训计划有机地结合起来，而不仅局限于员工的薪资、奖金、升免。评估过程中要体现公正、公平、公开的原则，既能真实地反映员工的工作实绩，同时又应尽量地避免绩效评估的负面影响。评估之后，对被评估人进行评估意见的反馈是很重要的，因为进行绩效评估的一个主要目的就是改进绩效。所以，主管和员工应合力安排绩效，改进计划。

二、某跨国公司绩效考评系统

该跨国公司是目前全世界最大的感光材料生产厂商，创办于 1880 年，至今占据世界市场的 42%左右。事实上，从 1993 年到现在，它在全中国 250 个城市建成 5500 余个快速彩扩店，其产品在中国的市场占有率由 26%提高到 53%，把竞争对手踢下了中国市场盟主位置。同时，中国在全球市场的排名由第 17 位上升到仅次于美国的第2 位。

在市场份额扩大的同时，该企业的组织架构也发生巨大变化。事实上，从 1993 年到现在，它在中国厦门、汕头、上海、无锡等地先后设立了工厂，办事处也由 4 个变为 23 个，公司为典型的事业部结构。

该公司绩效考评系统采用人事部制定考核标准，直接上司评估个人在工作上的行为表现的方法。考评时间为每年的 4 月，伴随着每年加薪与提升决定，考评每年一次。

人事部制定考核标准为目标管理，采用等级评估与关键事件相结合的方法。考评内容分为结果评估（result assessment）和价值观评估（value assessment）。

1. 结果评估

在结果评估中，直接主管根据工作分析，将被考评岗位的工作内容划分为相互独立的几个工作目标，在每个目标中用明确的语言描述完成该工作需要达到的工作标准。同时，将标准分为如下几个等级选项。

① 本文原载《暨南学报（哲学社会科学版）》，2002，24（1）：29-33. 作者：刘人怀，王学工.

（1）结果远超过所定目标（分值为 A）；

（2）结果部分超过所定目标（分值为 B）；

（3）结果全部达到所定目标（分值为 C）；

（4）结果部分达到所定目标（分值为 D）；

（5）结果未达到所定目标（分值为 E）；

（6）未在现工作岗位——请假等（分值为 I）；

（7）接受培训（分值为 T）。

直接上司根据被考评人的实际工作表现，利用关键事件对每个目标的完成情况进行评估。在写出考评结果的同时，直接主管被要求列出该员工所取得的工作成绩作为附录，并写出具体评语。总成绩便为该员工的考评成绩。

2. 价值观评估

该公司具有以下六个价值观。

（1）尊重个人（respect for individual dignity）——尊重个人的选择及每个人的差异性，避免在公开场合表露出愤怒与指责。

（2）正直不阿（uncompromising integrity）——诚实、道德的行为，对待他人以高度的诚信。

（3）相互信任（trust）——积极寻求他人的支持与帮助，自由地共享信息，相信他人能完成他们的工作。

（4）诚信可靠（credibility）——对行为承担个人责任，敢于承认错误及作出现实的承诺。

（5）不断进取（continuous improvement）——不断学习、提高，显示出新的能力；接受有建设性的反馈意见，不断取得事业的成功。

（6）认可及庆祝（recognition and celebration）——乐于恭贺他人、团队或供应商等所取得的成就。

直接主管根据该公司的六个价值观对员工进行考评。考评结果分为以下三个等级。

（1）为该价值观的表率（分值为第 1 等）。

（2）持续体现该价值观（分值为第 2 等）。

（3）非持续体现该价值观（分值为第 3 等）。

直接主管对每个价值观分别打分，最后打出总评分（总评分并非各项平均分）。

考评步骤完全通过该公司内部的 Lontus Notes 网络系统进行，具有严格的程序控制及安全性。首先由员工自我考评，考评结果不计入考评成绩，其次由直接主管在计算机中填写考评表，考评表填好后通过网络递交上级主管批准，上级主管只有批准与否的权限，而无修改的权限。上级主管批准后发回给直接主管，直接主管与员工面谈后再发给员工签名（员工只有浏览的权限），员工签名后考评表会自动传回给直接主管，这时，考评表的修改功能已被自动关闭，没人能再修改考评表。最后直接主管完成考评表并报给人事部等相关部门。

图 1 显示考评流程图。

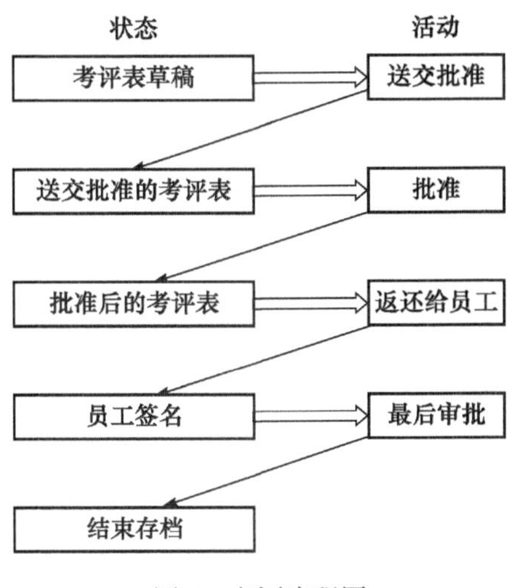

图1 考评流程图

三、考核系统的评价与改进建议

1. 优点

1）适用性与针对性

该考评系统适用于各个工作岗位，以工作业绩与工作态度两方面为基础对员工进行评价。该系统既照顾了共性（对该公司六个价值观的认同），也照顾了个性，各部门可结合自己的工作说明书与具体特点制定自己的考核目标。因此，它既有适用性也有针对性。

2）公正性与公平性

公平、公正可分为过程的公正和结果的公平。这二者中往往前者是因，后者是果，员工往往会很注重过程的公正。该系统全过程在计算机网络中自动运行，各环节清楚，权限与控制明确，具有很强的安全性及保密特征，并且，由于公司的事业部制结构，各部门间相对独立，考评成绩也不进行各部门间的横向比较，因此，给人很强的过程公平感。

3）简便性和有效性

该系统采用计算机管理。一旦设置完毕，操作非常简便，各资料保存于中央数据库，便于人事部门和其他相关部门的资料汇总与管理。对于考评结果要求与员工面对面地讨论，并将讨论与员工的未来发展计划（EDP）结合在一起，这样，既对员工绩效进行了及时的反馈，也将绩效评估与绩效改进联系起来。

2. 缺点

1）目标确定与衡量的困难

由于人力资源管理的复杂性，对各岗位的考评往往需要专门的知识。这套系统对各种不同部门考评目标的制定缺乏一般的指引，而需要直接主管自己制定。对于有些部门，工作成绩相对难以量化，因此目标的确定与衡量带有较强的主观性，容易引起员工对其考评结果的异议。对有些部门如销售部，制定的目标与现实往往有一定距离，这样导致目标需要经常修改，使最初制定的目标缺乏严肃性和稳定性。

2）易产生趋中性错误

对于工作态度的考评，直接主管往往缺乏考评依据，对于内容本身，就会有不同的看法，因而造成最难在执行上，直接主管往往应付了事，到考评期限截止时，匆匆交卷，而给予员工一个中性的分数，这样，未发挥其考评应有的作用。

3）评估的及时性

由于评估仅选择一年一次来测定，容易忽略被评估对象的一贯表现。这时评估者容易在未彻底了解事实的情况下评估员工，常会给员工造成一种错觉，只需在最后几个月努力工作，也可加薪提职。

4）未见可量化的绩效改进成果

该系统的投入花费了大量的成本，但通过系统投入前后相比较，难以发现明显的、可量化的员工绩效改进，因此难以用成本收益原则对该系统进行评价。

3. 改进建议

（1）不仅组织经理们有效地使用绩效管理系统，而且训练一般员工有效使用该绩效管理系统，也就是说，训练组织内所有人员明白如何使用绩效管理系统。这样，可以使绩效管理更具有效果。

（2）对有些部门使用职能（competency）作为绩效管理系统的主要工具，以衡量相关人员的行为而不是目标管理。

（3）对有些部门可采用同事评分、客户反馈、上司考核等多角度评量方式，并且将团队相关目标应用在个人的绩效计划上，而不单是采用上司评估这一项来评价个人在工作上的行为表现。

（4）适当提高评估频率。可考虑改为每季一次或每半年一次，并对被考评人提出改进意见，在年末进行正式的考评时，可参考季度考评记录得出正确考评结果。这样有助于消除近期误差并使考评更具前瞻性。

参考文献

[1]　克雷曼 L S. 人力资源管理. 孙非，等译. 北京：机械工业出版社，1999.

公司治理：理论演进与实践发展的分析框架①

公司治理问题是当前一个世界性的理论研究与实践课题，也是一个世界性的难题，并已形成全球性的公司治理浪潮。公司治理问题之所以如此重要，在于良好的公司治理是现代市场经济健康运作的微观基础，不仅影响到企业和个人，也关系到国家经济的稳定和增长。在我国，公司治理问题自 20 世纪 90 年代以来已引起了广泛的关注和讨论。

一、公司治理的内涵之争

公司治理问题的本质被认为是所有者与经营者的委托代理问题，公司治理问题的产生过程实际上就是企业产权关系的演变过程和相应的代理风险产生过程。从古典企业历经“法人革命”和“经理革命”，再到具有相对完善的公司治理结构的现代企业制度，其实是一个伴随着企业经济环境变化的公司治理问题的产生和演化进程。

公司治理概念构成公司治理问题研究的起点和基础。对公司治理的概念，国内外学者的研究并未取得一致的认识。

Hart[1]归纳了公司治理理论的分析框架，认为只要两个条件存在，公司治理问题就必然在组织中产生：一是代理问题，确切地说是组织成员（如所有者、经理人、工人和消费者）之间存在利益冲突；二是交易费用之大，致使代理问题不可能通过契约解决。如果存在代理问题并且合约不完全，则公司治理结构至关重要。Cochran 和 Wartick[2]对公司治理基本问题作了深入解释；Mayer[3]把公司治理定义为一种制度安排。钱颖一[4]认为公司治理结构是一套制度安排，用来支配若干在企业中有重大利害关系的团体的关系，各利益团体从这种制度中实现各自的经济利益；Shleifer 和 Vishny[5]认为，公司治理的问题是向公司提供资金的供给者如何保证他们能够从他们的投资中获得收益；Laffont 和 Tirole[6]认为，公司治理的标准定义为对股东利益的保护。他们的观点体现了以委托代理为核心的公司治理理论。

经济合作与发展组织（OECD）在总结各学者的理论学说和世界公司实践基础上，将公司治理定义为“一种据以对工商业公司进行管理和控制的体系，公司治理结构明确规定了公司的各个参与者的责任和权利分布，如董事会、经理层、股东等其他利益相关者，并且清楚地说明了决策公司事务所应遵循的规则和程序，同时，它还提供了一种结构，使之用以设置公司目标，提供了达到这些目标和监控运营的手段”。

结合中国的经济实践，国内学者对公司治理问题进行了深入探索，具有代表性的公司治理的定义有如下几种。①强调公司治理是一种制衡关系。吴敬琏[7]认为所谓公司治理是指由所有者、董事会和高级经理三者组成的一种组织结构，在这种结构中，上述三者之间形成一定的制衡关系。②强调所有权在公司治理中的作用。张维迎[8]认为，公司治理结构狭义地讲，是指有关公司董事会的功能、结构和股东权利等方面的

① 本文原载《经济体制改革》，2003，(4)：5-8. 作者：刘人怀，叶向阳.

制度安排，广义地讲，是指公司控制权和剩余索取权分配的一套法律、文化和制度安排，公司治理的目的是解决两个基本问题，一是激励问题，二是经营者选择问题。③强调利益相关者在公司治理中的利益保护。杨瑞龙[9]认为，国企改革应实现治理结构的创新，其核心是扬弃“股东至上主义”，遵循“共同治理”，强调利益相关者的权益。李维安[10]认为公司治理应该从更广泛的利益相关者的角度，从权力制衡和决策科学两个方面去理解。公司治理不是为了制衡而制衡，而是应如何使公司最有效地运行，如何保证各方面的参与人的利益得到维护，因此，科学的公司决策不仅是公司管理的核心，而且是公司治理的核心。公司治理的目标不是相互制衡，只是保证公司科学决策的方式和途径。

对公司治理理解的差异反映了不同学者处于特定的经济环境对公司中各利益主体关系所强调重点不同这一事实，但是，有一点是基本一致的，即认为公司治理问题产生于公司各个利益主体利益取向的不相融，公司治理的宗旨在于协调利益关系，促进利益相融程度和降低企业运作成本。

二、公司治理的外延拓展

公司治理是一个多层次的概念，从公司治理这一问题的产生与发展来看，可以从广义和狭义两个层次来理解，或称为内部公司治理和外部公司治理[3]。狭义的公司治理是指股东与经营者之间的权利与责任的关系，主要是通过股东大会、董事会、监事会及管理层所构成的公司治理结构的内部治理，治理目标是股东利益最大化，要解决的主要问题是所有权和经营权分离条件下的代理问题，即通过建立一套既分权又能达到相互制衡的制度来降低代理成本和代理风险，防止经营者对股东利益的背离。

然而，在现代市场经济中，企业是一个通过一系列合约联结起来的经济组织，企业的运行目标函数不应仅局限于股东的利益参数，而应该是全体签约人共同剩余的最大化。在现代社会中，公司治理的主体除了股东以外，还包括债权人、内部职工、供应商、消费者、政府和社区居民等。因此广义的公司治理则不再局限于股东对经营者的制衡，而是涉及广泛的利益相关者，包括股东、债权人、供应商、雇员、政府、社区等与公司有利害关系的利益集团。公司治理就是通过一套包括正式的或非正式的、内部的或外部的制度或机制来协调公司与利益相关者之间的利益关系，保证公司决策的科学化，从而最终维护各方面的利益。这一观点被称为利益相关者共同治理理论。因此，从广义上说，公司不再仅是股东的公司，而是一个利益共同体，公司治理机制不仅限于以治理结构为基础的内部治理，而是利益相关者通过一系列的内部、外部机制来实施共同治理，治理目标不仅是股东利益的最大化，而且是保证公司决策的科学性，保证各方面利益相关者的利益最大化。因此，企业不仅要重视股东的利益，也要重视其他利益相关者的利益，公司的治理需要全体利益相关者的共同参与，即利益相关者共同治理。

Bradford 和 Shapiro[11]曾深刻揭示了公司治理结构发展的这一趋势。他们认为公司治理在传统意义上主要集中在股东所有权和管理者控制权的福利问题，但现在人们开始认识到公司治理应该有一个更宽泛、更动态的概念，即企业必须对更多的利益相关者的预期做出反应，必须对多元的利益加以协调，以实现长期的价值最大化。公司治理不仅是一套静态的组织机构和制度安排，更是一个实际运行及监督指导的动态过程。公司治理应是委托人和代理人、所有者与经营者、债权人与债务人、管理者与被管理者之间的互动和博弈。

公司的有效运行和决策科学不仅需要通过股东大会、董事会和监事会发挥作用的内部监控机制，而且需要一系列通过资本市场、产品市场、经理人市场等来发挥作用的外部治理机制。在 OECD 制定的《公司治理原则》中，已不单纯强调公司治理结构的概念和内容，还涉及许多具体的治理机制，该原则主要包括以下五个方面：①股东的权利；②对股东的平等待遇；③利益相关者的作用；④信息披露和透明度；⑤董事会的责任。治理机制是一个比治理结构更广泛、更深层次的公司治理观念。除股东外，来自其他利益相关者的制衡对公司经营者正在产生越来越大的影响。

三、公司治理的典型模式

伴随着公司治理实践的进展，已形成了一些具有代表性的公司治理模式，这是公司治理在不同经济环境中应对特定问题的产物，但是，近年来，随着经济全球化和公司制度的扩散效应，世界各国的公司治理结构和机制出现了趋同化的趋势。公司治理理论的发展也正是遵循此线索演进的。

不同国家的经济文化历史各不相同，其公司治理模式也不相同，目前，世界各国具有代表性的公司治理模式大致可以划分为如下几种：英美市场导向模式、日德银行导向模式、东南亚家族控制模式以及转轨经济模式。在实践中，各国的公司治理模式不可能是单一的治理模式，往往是多种模式并行，区别只是在于侧重点的差异。

1. 英美市场导向模式

这一模式的特点是公司治理作用的发挥以发达的股票市场和积极的公司外部控制权市场的存在为前提。公司股权极为分散，股东很难直接影响经营者，出现“弱所有者、强管理者”现象，但同时证券市场发达，外部市场对公司控制起着重要作用，股东主要通过“用脚投票”来实现对公司的控制。该模式流行于美国、英国、加拿大和澳大利亚。极其发达的股票市场是公司融资的重要工具，成为有效配置市场资源、实现企业治理的重要手段之一。这一模式也存在弊端，一是“内部人控制问题”，由于股权的高度分散化，职业经理人成为公司的实际控制者，股东大会形同虚设，公司经理可能违背股东的利益寻求个人利益；二是这种模式使得公司经营行为具有明显的短期化倾向。敌意收购的频频发生对公司经理层构成了严重威胁，经理阶层往往以维持自身眼前利益代替公司长远的战略打算，公司长期发展问题被搁置。

2. 日德银行导向模式

该模式的特点是公司股权较为集中，银行在融资和公司治理中发挥巨大的作用，银行和企业之间形成典型的关系治理[12]。公司股票由机构股东包括银行和非银行的金融中介集中持有，机构股东掌握了足以控制股东会和董事会的股权，可以直接任免经理层。公司之间交叉持股，银行既有对公司的债权，还持有公司大量股票，从而使银行及利益相关者直接监控公司成为可能。公司外部控制权市场不发达，股票市场弱小，外部治理者与企业经济关系稳定。这种模式的弊端是明显的，进入 20 世纪 90 年代以来，日本泡沫经济破灭，经济陷入长期衰退，人们开始对日本的公司治理进行反思，许多学者认为企业的资产负债率过高、银行和企业关系过分密切是日本泡沫经济形成的一个重要原因。日本一些银行作为公司的主银行，对经营不善的公司，没有进行严格监管，反而向其大量注资，产生大量呆账、坏账。

3. 东南亚家族控制模式

这种模式以家族资本主义或裙带资本主义为特征，公司股权由家族控制，表现为具有绝对控股权的单一大股东和交叉持股的普遍存在，企业主要经营权掌握在家族成员手中，经营权与所有权很少分离。这种模式盛行的国家和地区，儒家文化思想影响深远，导致在用人制度上强调家族观念，企业员工管理家庭化，企业的所有权与控制权大都采用继承制。很显然，家族治理模式的优点正是其先天缺陷的天然生长点，家族内的关系治理具有灵活性和环境的适应性的优点，但是由于公司内部以决策独断、家长式约束为管理基调，而这恰恰是正式高效的激励监督机制的天敌，因此，一旦公司中的家族“权威”受到冲击或出现漏洞，整个治理体系将轰然倒塌，不像其他公司治理模式那样，可以通过“机制修复”来实现治理结构的恢复。

4. 转轨经济模式

这种模式主要存在于苏联、中东欧和中国等转轨经济国家。在这些国家，经济体制、法律制度都处于转轨时期，经济中存在数量众多、规模庞大的需要通过资本运营实现所有制形式改变和活力再现的国有企业，公司治理的矛盾问题表现为内部人控制问题。

5. 简要评述

各种公司治理模式（表 1）都是与各国不同的历史时期的经济条件相适应的结果，各有利弊。日本、德国的国家银行在金融体制中占主导地位，资本市场地位相对较弱，英美国家的情况则相反，这正是产生两种不同模式的基本原因。日德模式的优点在于采用低成本，由银行和大股东对公司进行直接监督，英美模式则通过证券市场采用高成本的接管方式来达到对企业经营进行监督的目的。另外，各国公司治理模式都受到本国相关法律法规的深刻影响。在英美国家，对机构投资组合、内部交易和反托拉斯的规定条例等，使得公司治理主要依靠证券市场而非股东和银行的直接监督。在日、德，对非银行融资的限制，则意味着通过证券市场的机构投资者的交易成本大为提高。一般认为，英美模式更适合在具有较成熟资本市场的国家中的高技术和高风险特征的企业，而在资本市场不成熟、市场合约范围相对小且约束力不够强的国家，具有标准化生产过程和广泛运用成熟技术的企业则更适用日德模式[12]。

表 1　不同公司治理模式及其关键问题

治理模式	关键问题
英美市场导向模式	强管理者、弱所有者
日德银行导向模式	利益相关者的利益组合
东南亚家族控制模式	强家族股东、管理层，弱中小股东
转轨经济模式	内部人控制

四、公司治理的发展趋势

随着世界公司经营国际化，资本市场全球化，经济全球一体化和信息技术的发展，公司治理模式出现了一些新的发展趋势，归纳起来主要有以下几个方面。

1. 公司治理模式的趋同化

自20世纪90年代以来，全球一体化加速发展，公司经营国际化，资本市场全球化，各国公司治理模式出现了趋同的现象。

英美国家的公司开始重视“用手投票”。占绝对控股地位的单一股东在高科技公司中出现，机构投资者持股比例不断上升并成为企业的大股东。美国董事会原来被比喻为橡皮图章，然而，自20世纪90年代以来，随着机构投资者逐步成为股票市场的重要力量之后，由于控股比例的提高，他们不再采用“用脚投票”的方式来保护其资本价值，转而采取积极干预的方法，对经营不善的公司采取积极的方法，如撤换经理人、改变公司战略和关键人事安排等。

美国经理层以股权为基础的激励机制大量出现，根据Hall和Liebman[13]的估计，基于股权激励性质的报酬比例增加导致CEO的收入与其工作表现之间的敏感度从1980年到1998年上升了10倍。经理人的利益与股东的利益更加紧密地联系起来，有助于解决企业的委托代理问题，完善公司治理。

在欧洲，公司的外部控制权市场发展迅速，收购大量出现，公司透明度增加；在日本、德国企业中，直接融资比重增加，内部融资成为企业资本的重要来源之一，金融危机的阵痛使得企业和银行之间更加相互警惕，企业对银行资本的依赖明显减弱，主办银行的作用日益降低，交叉持股比例不断减小，持股结构不断被调整。Wójcik[14]对上市的德国公司进行的研究发现，德国大型银行和保险公司出售了他们的部分权益，这些股权有一部分转让给投资基金，大型银行和保险公司在公司治理中的角色正在发生变化，从原来的控制投票权改为控制投票权和管理投资基金相结合。所有权集中度的降低、交叉持股现象的减少、银行机构的角色变化，使得德国的公司治理更加接近于英美模式。

在东南亚，随着竞争的加剧、东西方文化的交流、信息技术和经济的日益发展，传统文化中的成分逐渐丧失，家族观念日趋淡薄，公司治理朝民主化、规范化方向发展。

2. 机构投资者日益发挥出积极作用

随着养老基金、保险基金以及投资基金等机构投资者的发展，机构投资股东的持股数量越来越多，到20世纪90年代中期，这些集中的机构投资者大约持有美国所有公开交易公司股票的50%以上，这些机构投资者不再满足于在投票中跟随公司的安排，不再满足于“用脚投票”来表达他们的不满，而是不断向公司施加压力，要求公司提供更详细真实的信息，积极参与公司治理的重大决策，体现他们的意志。一些机构投资者，如CalPERS（California Public Employee’s Retirement System，美国加利福尼亚州公共雇员退休基金组织）已成为积极的投资活动家，向绩效较差的公司施加压力，要求他们改变公司的战略，提出具体的政策建议，游说公司董事会解雇公司CEO。股东行动主义和关系投资者出现，股东和公司管理人员之间的对话和沟通加强，股东行动与管理人员之间的关系系统化、正式化。公司也更加重视与投资者的联系沟通，以保持公司良好的资本市场形象。

3. 银行在公司治理中的作用发生变化

在日本，由于泡沫经济的破灭，银行本身产生的大量的呆账、坏账，使银行重新评价其在公司治理中的作用，更加关注良好的财务控制，减少外围业务。企业也开始理顺与银行过分密切的关系，自主进入资本市场融资。在英美国家，银行参与公司治理的作用被重新认识，自 20 世纪 80 年代以来，英美国家开始重视银行在公司治理中的监督作用，逐渐放松对银行的限制，银行间接融资比重不断上升。美国银行可以通过其成立或控股投资银行、信托公司等非银行金融机构间接持有公司股份，通过这些子公司间接进入公司董事会和监事会，参与公司决策。

4. 董事会的独立性和权威性开始增强

自 20 世纪 90 年代以来，大量具有专业背景的独立董事进入董事会，使董事会的独立性、专业性大为提高。根据科恩/费瑞国际公司的研究报告，《财富》美国公司 1000 强中，董事会的平均规模为 11 人，其中董事 2 人，独立董事 9 人，占 81.8%。2002 年 6 月，纽约交易所（The New York Stock Exchange，NYSE）用白皮书的形式提出了更为严格的独立董事定义，以及独立董事要占大多数的建议，独立董事制度作为公司治理的一项重要内容正在全球迅速推广。

参考文献

［1］ Hart O. Firms：Contracts and Financial Structure. Oxford：Oxford University Press，1955.

［2］ Cochran P L，Wartick S L. Corporate Governance：A Review of the Literature. Financial Executives Research Foundation，1988.

［3］ Mayer C. Corporate governance in market and transition economics. International Conference on Chinese Corporate Governance. Shanghai，October，1995.

［4］ 钱颖一. 中国的公司治理结构改革//青木昌彦，钱颖一. 转轨经济中的公司治理结构：内部人控制和银行的作用. 北京：中国经济出版社，1995：133.

［5］ Shleifer A，Vishny R W. A survey of corporate governance. The Journal of Finance，1997，52 (2)：737-783.

［6］ Laffont J J，Tirole J. Competition in Telecommunications. Cambridge：MIT Press，2001.

［7］ 吴敬琏. 市场经济的培育和运作. 北京：中国发展出版社，1993.

［8］ 张维迎. 所有制、治理结构及委托：代理关系：兼评崔之元和周其仁的一些观点. 经济研究，1996，(9)：3-15，53.

［9］ 杨瑞龙. 国有企业治理结构创新的经济学分析. 北京：中国人民大学出版社，2001.

［10］ 李维安. 现代公司治理研究：资本结构、公司治理和国有企业股份制改造. 北京：中国人民大学出版社，2001.

［11］ Bradford C，Shapiro A C. Corporate Stakeholders and Corporate Finance. Tucson：Financial Management Association，1987.

［12］ 青木昌彦. 比较制度分析. 周黎安译. 上海：上海远东出版社，2001.

［13］ Hall B J，Liebman J B. Are CEOs really paid like bureaucrats? The Quarterly Journal of Economics，1998，113 (3)：653-691.

［14］ Wójcik D. Change in the German model of corporate governance：evidence from blockholdings 1997—2001. Social Science Research，Network，2001：1-36.

以价值工程方法全面提升荔湾商旅核心竞争力①

随着经济全球化和我国加入WTO，中央要求广东省首先实现与国际接轨并走向世界，因此，构筑珠三角整体竞争优势、保持经济持续高速增长势头的问题备受关注。在此背景之下，广州市需要进一步提高核心城市的作用，增强辐射力和影响力，建成更具国际竞争力的商贸旅游中心、科技研发中心和现代服务中心，成为珠三角大区域腾飞的领头羊。

广州是我国华南地区的中心城市，是广东省的省会和政治、经济、文化中心。广州又是一座具有2200多年悠久历史的文化古城，是我国历史上最早对外开放的商业都市。现代的广州是我国改革开放的前沿，是对外通商的重要口岸城市，是我国商贸旅游业发展最快的城市之一。

荔湾区作为广州的老城区，有着良好的商贸业发展基础和颇具特色的历史街区，具有深厚的历史文化底蕴。因此，着眼于挖掘老城区的潜力、实现荔湾商贸旅游资源的有效整合与商贸旅游的互动互促，是实施城市形象工程、提升区域竞争力的必然选择。

荔湾区是广州市历史最为悠久的商业带，是广州市文化品位积淀最深厚的地方之一。经过长期的发展，特别是近年来的商业环境整治，荔湾区已成为知名度较高、影响力较大、辐射力较强的商业中心之一。我国加入WTO后所面临的机遇与挑战增多，运用“大流通、大市场、大商业、大旅游”的理念来塑造荔湾区商贸旅游业的整体平台，具有尤为重要的现实意义。

荔湾区在广州旅游业发展中也扮演着重要的角色。陈家祠是广州旅游景点中艺术价值最高、知名度最大的旅游精品；上下九步行街、西关风情、沙面欧陆风情等人文资源是广州风情的代表资源；此外，十三行、詹天佑故居等众多尚未开发的各类旅游资源，构筑了荔湾在广州旅游业发展中不可替代的优势。但目前荔湾的旅游资源开发尚处于起步阶段，缺乏系统整合的有效步骤与方法。

本文通过价值工程[1]及商贸旅游市场营销理论，以提高荔湾商贸旅游资源价值为目的，以分析商贸旅游产品功能为核心，以创新思维为指导，以系统整合为主要手段，探讨整合商贸旅游资源与构筑区域竞争力的方法，促进荔湾城区经济的发展。

一、荔湾区商贸旅游业状况分析

1. 区域概况

荔湾区位于广州市西侧，东与越秀区接壤，南与芳村、海珠两区隔江相望，西临珠江并可经大坦沙至南海、佛山，北与白云区罗涌围、同德围、矿泉街等地区相接，

---

① 本文是广州市荔湾区人民政府科研项目报告，广州，2003；旅游工程管理研究，北京：科学出版社，2014，37-96. 作者：刘人怀，张永安，傅汉章，谭浩邦，梁明珠，诸凤鸣，王桂林，管宇，雷方.

全区面积共16.2平方千米。

荔湾区在广州市“南拓北优，东进西联”规划中占有重要的战略位置，区内处于“广佛都市圈”承东启西的关键节点上，它是广州西联的要道和窗口。荔湾区常住总人口为474 830人。

荔湾区近年经济增长平均速度均超过10%。根据《广州市十五年基本实现现代化总体发展方案》及《广州市国民经济和社会发展第九个五年计划及2010年远景目标纲要》，预测广州市在未来15年内国民生产总值年均递增12%，预计在规划期内荔湾区大致可达到这个经济增长速度。

随着广州城市建设“三年一中变”的完成和“十年一大变”的展开，广州市的经济和社会发展不断提速，带动荔湾区的城市建设、交通运输、环境绿化等方面都有较大的变化，为今后的经济发展打下了良好的基础。按广州市规定的行政街道面积和人口标准，全区调整为13条行政街，包括金花街、西村街、南源街、逢源街、多宝街、龙津街、昌华街、岭南街、华林街、沙面街、站前街、彩虹街、桥中街。经过街道体制改革，各街确立了“两级政府、三级管理、四级网络”的城市管理体制以实现管理重心下移，把城市管理具体工作落实到基层。荔湾区辖下各行政街具有一定的商业基础，交通方便，环境优美，为商贸旅游业发展提供了良好的基础。

2. 商贸旅游业的发展状况

近几年，广州市交通环境的整治以及荔湾区西北部污染大厂的搬迁使沿江岸线的环境逐渐美化，大大地促进了荔湾区商贸旅游业的发展。同时，广州市大力发展以高新技术为重点的产业结构调整，推动了荔湾区产业结构的优化升级。

广州在强化商贸业的主导产业地位的同时，将旅游业发展摆在突出位置。荔湾是广州历史文化名城的重要组成部分，历史文化积淀较深，人们归纳荔湾区的特点有六多：文物古迹多、特色建筑多、历史名流多、土特产品多、风味食品多、专业街市多。为了适应现代化旅游城市发展的需要，需要建立富有荔湾特色的旅游项目，塑造荔湾旅游形象，在不久的将来，荔湾区一定会成为广州旅游的必到景点。

近五年来，荔湾区经济增长均保持两位数快速发展，其中，第三产业占主导地位，保持较快速度增长。2002年，荔湾区第三产业增长值为34.44亿元，占区国内生产总值的83.38%，五年年均递增13.04%。2002年，荔湾区社会消费品零售总额为83.44亿元，比上年增长10.5%，五年年均递增11.00%。2002年全地区商品销售总额为646.70亿元，绝对额在广州市10区中居首位。

从以上数据来看，商贸业是荔湾区最为重要的社会经济支柱，它在广州市商业体系中占有重要的地位。

随着广州市“一年一小变”“三年一中变”工程的完成，旅游基础设施和配套服务设施不断完善，有力地促进了旅游生产力的发展。目前，荔湾区旅游要素比较齐全，接待规模不断壮大，接待游客人数和营业收入不断增加，2002年全区接待海内外旅客总人数为988万人次，旅游业总收入62.96亿元。

旅游业方面，全区共有住宿设施209家，客房13 560间。其中，涉外旅游星级饭店18家，客房4723间，白天鹅宾馆是我国首家五星级宾馆，坐落于区内沙面。

餐饮业方面，全区有旅游团体推荐接待餐馆 9 家，其中，广州酒家、泮溪酒家、莲香楼、陶陶居等一批“中华老字号”单位是饮誉中外的国家特级酒家。

从商品市场结构来看：荔湾区的商贸业已突破了只局限于小百货、传统小食、“菜篮子”的小圈子，而逐步形成了多层次、多元化、多业态的商业群体。重点培育了一批具有荔湾特色、发展基础良好、单个经营场地面积 3000 平方米以上、营业额达到或超过 2500 万元、缴纳税金 100 万元以上的专业市场，如清平中药材市场、黄沙水产交易市场、站西路鞋城、留香针织袜业市场、谊园文具玩具精品市场、广东电器市场、广东童装妇婴用品市场等。此外，还大力培育发展潜力大、附加值高、创税能力强的批发行业，如通信设备、电器、机电、五金、装饰材料等，建成结构完善的商品市场体系，引导市场与现代营销方式相结合，大力发展连锁经营，积极培育发展名牌产品和名牌企业。

从旅游业结构来看：树立大旅游经营理念，发挥好地区系统功能和整体环境的效应，把荔湾旅游纳入广东省、广州市的旅游一体化当中，根据省、市的要求，创造出具有岭南特色的旅游景点、美食、文化项目来打造名牌产品，吸引珠江三角洲、全国乃至海外游客。同时，建立区域协作机制，由于地域相邻，区内与周边地区的旅游具有相连性，加强与周边的南海、佛山等地区域联合和协作，积极探索发展旅游的新路子。

3. 商贸旅游业的优势与劣势、机遇与挑战

1）商贸业的优势

近几年来，荔湾区商贸市场发展旺盛，消费优势引人注目。以商贸餐饮业为主体的服务业增长态势良好。上下九、第十甫、宝华路、文昌路等，它们是广州最有传统气息的工商业马路，路段上曾经汇集众多响当当的老字号、名牌店，虽然有些店铺已经易手经营，但依然给上下九、第十甫留下了浓浓的西关韵味，再加上各式各样的传统美食，商业和风情使这里成为游客必到之地，成为旅游者最喜欢游览购物的好去处。

长期以来，兼有现代和传统风情的上下九步行街是广州购物最旺的商业地带之一，特别是近两年来，上下九步行街的人流量不断增加，日平均人流量达到 35 万人次，而节假日时可达到 70 万人次。

荔湾区商贸业迅猛发展还体现在商品交易批发市场快速递增，成交额大幅度增长，市场规模从小到大，市场供给能力由弱到强上。专业市场成为荔湾区商贸市场的主要支柱，据统计，2001 年专业市场 101 个，占全区市场的 67.79%，商品成交额 41.16 亿元，占全区市场的商品成交额的 78.51%。专业市场中，以水产品、中药材、建筑及装饰材料、电子电器及通信器材、鞋类、服装等经营为主体。

2）旅游业的优势

荔湾区有深厚的文化底蕴、文物古迹、商业特色、饮食文化、民俗风情等丰富的旅游资源，而且广州发展的历史赋予了荔湾区“西关岭南文化”这一巨大的无形资产，这些都是其他市区所无法比拟的，给荔湾区旅游业带来了巨大的发展空间。近几年，旅游业已发展成为荔湾区经济支柱产业之一，区内旅游总收入和创汇呈持续增长的势头。

现代科学技术特别是信息技术在旅游业中的应用对荔湾旅游业的发展起了巨大的推动作用，现代科学技术在旅游业中的多方渗透给旅游业带来了新景象，如在上下九步行街的旅游建筑物中，应用新材料将大大增加观赏效果、降低能耗、减轻环境污染。

3）商贸业的劣势

（1）网点业态结构不合理。具体表现为：经营品种门类比较单调，千店一面，经营特色不明显；营业面积小，购物环境差；经营商品档次多为中低档，经营手法雷同；新型的经营方式和先进的经营手段不多，大型专业店的新业态数量太少。

（2）零售网点过于零散，规模太小，难以形成高效的规模效应。这种传统的小型、分散、落后的商业设施，对外影响力弱，难以发挥区域的辐射力及聚集效应。

（3）商业街发展方向比较盲目，趋同发展现象较为突出，市场定位不明确，特色不明显，这在一定程度上制约了荔湾区商贸业潜力的发挥。

（4）商业中心区域建设缺乏配套的环境设施。荔湾商业区位多集中于老城区核心地段，路面狭窄，人流、车流密度大，横街窄巷多，停车困难，限制了本区商业对区外顾客的吸引力。同时，在一定程度上制约了商业街档次的提高。

（5）批发市场受空间布局和交通环境的制约。由于荔湾区批发市场大部分分布在内圈层的南岸与黄沙老市区，一方面市场面积受用地限制，市场规模扩大受到制约；另一方面受内圈层的交通、环保、卫生等因素制约，从长远来看，批发市场难以在内圈层的老城区发展。

4）旅游业的劣势

（1）荔湾区的旅游资源丰富，但缺乏科学的、系统的、总体的开发与整合。荔湾区有十分宝贵的旅游文化资源，知名的“老字号”和文化古迹众多，但缺乏系统的优势整合，因此总体形象不够鲜明。

（2）文化资源挖掘欠缺深度，品牌效应有待进一步发挥。荔湾区文化资源相当丰富，应该将文化产业做大做强，进一步加大文化与旅游结合的力度，推出具有荔湾特色的文化旅游精品和品牌。

（3）配套设施不完善，制约了旅游业的发展。荔湾区面积较小，道路狭窄，路网建设不完善，食、住、购物、娱乐等环境设施不配套，这些因素成为制约旅游业发展的重要原因之一。

（4）缺乏高素质的人才队伍。荔湾区要实施商贸旅游业发展战略，必须要有一支高素质的人才队伍作保障，以加快推进商贸旅游业的规范化、标准化、制度化建设，为顾客提供优质服务，为旅游业向信息化、网络化、国际化方向发展积极创造条件。

5）商贸旅游业发展的机遇

从全国来看，经济全球化，我国加入 WTO，国家全面实施科教兴国、可持续发展、西部大开发和城镇化战略，这些有利政策都为荔湾区加快经济发展提供了较为宽松的外部环境。

从广州市来看，广州现正进入率先基本实现现代化，建设现代化中心城市的重要阶段，其发展目标是充分发挥中心城市的政治、经济、文化、商贸、旅游、信息中心和交通枢纽等功能，使广州发展成为一个繁荣、高效、舒适、适宜创业发展和适宜居住生活的国际大都市。广州市要成为两个适宜的国际性中心城市，就需要市内各区以功能强大的现代商贸旅游业作支撑。同时，根据广州发展规划所确定的"西联"发展战略，广州与南海、佛山将共建"广佛都市圈"，而荔湾区处于未来"广佛都市圈"承东启西的战略要点上。因此，荔湾要充分发挥商贸文化旅游区的地位与作用，认真做好"西联"工作，使其为"广佛都市圈"的发展发挥不可替代的作用。

从荔湾区来看，随着广州市实行"一年一小变、三年一中变、十年一大变"，荔湾区环境有了很大改观，交通投资环境进一步改善，经济结构调整取得明显成效，经济发展质量和效益显著提高，科技创新迈出实质性步伐，信息技术的普及与应用有较大的发展，以上这些均有力地促进了全区经济的持续发展。

承前启后，继往开来，与时俱进，开拓创新。在新的形势下，荔湾区抓好一批新的经济增长点，在新的起点上谋求更大的发展，把现代化的商贸文化旅游区建设推向新的阶段。

6）商贸旅游业面临的挑战

荔湾区商贸旅游业面临以下三大挑战。

（1）来自竞争对手的挑战。目前，商贸旅游业的竞争范围空前扩大，不仅要与国内、省内、市内同行业进行竞争，还要参与国际同行业的竞争。当前，周边城区和珠三角城市正在筹建或即将建成的新的商业街、广州西部商贸中心、大型现代化商业广场以及超大型购物中心等，均来势甚猛。据报道，未来 3 年珠三角将诞生 15 个面积达 10 万～60 万平方米的超大型购物中心，它们均瞄准整个珠三角消费圈，而广州近一半客源均来自珠三角，这将会对荔湾商业构成较大的威胁。业内人士已指出，广州商业应有危机感，荔湾处于边沿位置，其冲击将会更明显。此外，国际旅游市场、国内尤其是港澳台旅游市场的开拓，将会给荔湾区旅游业发展带来一定的压力。

（2）来自顾客的挑战。随着经济的发展，顾客的期望值越来越高，对服务质量的要求也越来越高。例如，在购物方面，顾客要求商品质量要好、价格要便宜、品种要齐全、服务要周到、购买要方便；在旅游方面，顾客要求诚信、服务周到，品牌效应更加明显。

（3）来自环境变化的挑战。由于经济全球化和高新技术的迅猛发展，人们的思维方式、生活方式、消费方式都在发生着变化。荔湾区是广州市的老城区，辖区面积小，人口密度大，人流、物流、车流过度集中，环境负荷重，给商贸旅游业发展带来一定的负面影响。

随着新城区的建设，家庭电气化的普及，零售业态的创新，大型综合超级市场迅速取代小百货店、小副食品店和小菜店。广州市传统的商业黄金地段向天河区、东山区推移，无疑对荔湾区的商贸旅游业发展提出了新的挑战。

二、商业步行街的价值提升与功能整合

当前国内许多城市都在急剧发展商业街或商业步行街，也都急需整合经营业态与经营资源。整合，就是重新对业态定位和经营方法进行反思与修正，通过对各种要素及资源进行调配与界定，从而提高整体经营的核心竞争力。整合的重点是经营定位、

商品构成、卖场布局、环境的整饰以及与其相匹配的后方支撑系统，从而筛选并锁定自己的主攻业态和经营战略。

商业步行街是商业街的一种特殊形式。要研究商业街或商业步行街价值的提升与功能的整合，首先必须对其发展的趋势与特点有所认识，其次再对特定城区的商业街进行分析。

1. 城区商圈的变化特点及其走向

随着城市规模不断扩大，原城区商圈所受到的制约更为凸显，再加上城镇的综合实力不断有所发展，就地消费的欲望加强，因而商业、服务业的商圈也在变化，并呈现出三大变化特点。

(1) 城镇型商圈高速成长。过去，由于社会经济发展的不平衡，城市与乡镇的差别较大，一般所指的商圈均为都市商圈，乡镇零售商业多为满足当地居民的日常生活小商品买卖，谈不上商圈。20 世纪 80 年代后，随着城镇经济的发展，次级城市消费力量已明显增长，大型零售商业和各类专业市场纷纷转往次级都会发展，预料今后这一趋势将会更为明显，城镇商圈将明显成形且高速成长，对都市商圈的影响也日益明显。城镇型商圈已形成不可忽视的聚散市场。

(2) 都市型商圈从线状形向块状形变化。过去都市商圈多是沿街建立，呈街道形的直线扩散。这种扩散模式已不适应现代都市的发展要求，其原因在于都会商业街内楼价高、铺租高，迫使一些中小经营者向四边扩散，而一些大卖场在商业街内又难以找到合适的地方，只能在周边开设，而此种布局又使消费者购买方便，人群流动从直线形向网状形转变，消费者对商品选购、人流特点更为适应。

(3) 超级都市型商圈向偏远副都市发展。由于大型都会区存在众多严重问题，如交通问题、停车问题、环境污染问题、空间布局拥挤问题、房租及地价高等问题，这就使得副都市型商圈在都会干道偏远地方得以发展。特别是大型商业企业向都会进入，包括国际商业巨头的加入，这就使得在都会与副都市的交通转接枢纽地，形成新的大型商圈。

以上三种变化趋势，并不否定大都会中商圈、商业街、商业带的存在，但更值得注意的是，市场流通是一个整体概念，各种商业形态、网络设置、商品流向都有内在关联，我们在考虑商业街的经营策划时，对城区整体商圈的变化趋势不能不加以考虑。

2. 商业步行街的功能分析

1) 商业步行街的功能及其发展阶段

商业步行街是商业街的一种形态，有的是在原传统步行街基础上加以改造而成，也有的是对近代商业街实施以步行化为主改造而成。步行街不仅具有一般的商业购物功能，还在一定程度上成为当地城市的旅游吸引物。成功的商业步行街不仅是一条购物街，还是一个旅游景点，更是一个文化、经济、历史含量甚高的都市品牌。因此，我国许多大中城市均纷纷成立商业步行街。具体说，商业步行街的功能主要有以下几个方面。

(1) 为消费者打造一个舒适的、安全的、方便的购物环境。步行街既能保证购物者在街道上穿梭流动畅通无阻，又能为消费者提供驻足休息、享用服务的设施；既是购物好天地，又是精神上美的享受。

(2) 促使交通秩序改善，减少汽车交通空间，增加步行人数，避免交通事故产生，增强消费者选购欲望，增加其逗留时间。一般估计，步行街至少可以增加50%的人流量。

(3) 有利于发掘当地的历史内涵，大打文化牌，将传统文化与现代商业结合起来。同时，各步行街也可在特色上做文章，打造各具特色的步行街，吸引各类个性消费者。

(4) 有利于商贸与旅游的有机结合，达到商旅互动的目的，形成购物、娱乐、观光、休闲的综合发展带。

(5) 形成当地经济的新亮点，加强人们的地域感与认同感，步行街也就成为该城市的象征或是该城区的标志性地带，起代表整个城市社会形象的作用。

总的来说，成功的商业步行街，能起到经济效益、社会效益和环境效益三个方面有机结合的作用。

从国内外有关商业步行街的发展史来看，现代步行街区的发展往往有三个不同的阶段。这些不同阶段反映出步行街的成熟程度和社会的影响度。这三个阶段大致情况如下。

第一阶段，为了留住和吸引顾客。这一阶段的步行街产生的根本原因，主要有两个方面：一是为了加强城市中心区的交通管理，在一些主要商业街道实行人车分离，保护并刺激中心商业区零售业的发展；二是由于郊区购物中心和新的商业带的兴起，城市中心区的商业功能日益衰退，为适应市场竞争新的形势和维持中心区的吸引力，通过设立步行街，加强竞争环境优势，这是纯粹商业性步行街的典型特征，也是步行街的起源及存在的基础。

第二阶段，文化的提升与环境的改善。“留人要留心”，步行街要吸引顾客，同样是这个道理。这一阶段的主要特点是加大步行街环境的整治，特别是文化内涵的提升，这是具有长久影响力的要素。在环境改善方面，要体现出对步行购物者的关心，处处为顾客着想，通过广场与绿地的建设，房屋立面与路面的整治，街头雕塑的建立，座椅休憩处的安排等，以人性为本，增强情调与气氛，当然，这一切还是以商业利润为根本目标。

第三阶段，社区活动中心的形成，社会效益更为明显。随着步行街的发展，可以预见其未来的方向是多功能和多要素的聚集，商业街的联网、地下购物中心的出现、空中天桥系统的建设，使步行街朝着网络化、立体化发展。步行街不仅是购物、旅游之地，也是社区活动中心，人际交往的好去处，其社会效益更为凸显。

2) 商业步行街存在问题分析

据媒体报道，至2002年中，全国已形成相当规模的商业街有120多条，其中，步行街也在不断发展，规模也日益扩大，其处境及经营情况，《中国商报》在2002年11月5日一篇专题报道中进行了概括：“前些年，中国各大城市几乎同时兴起‘步行街热’。数年过去，这些步行街上的风景各不相同。财源滚滚者有之，惨淡经营者也不在少数。”另一些媒体在综述步行街现状时，也提到现在有些步行街处于进退两难的尴尬境地，即“进也难，退也难”，而此种情况虽不能说是普遍现象，但也不是“仅此一处，别无其他”。因此，探究步行街存在的问题，研究其前因后果，对进一步健康地发展步行街是很有必要的。

在我国步行街发展过程中，有“全国八大步行街”及“五朵金花”之说。其中一些近年处境艰难，距离预期目标相差很大，比较典型的案例有以下三个。

武汉江汉路步行街，号称全国步行街“五朵金花”之一，也称武汉“金街”。在2002年“十一”黄金周前三天，该街三大零售业的“世界之窗”购物中心突然宣布清仓甩卖，关门歇业。据媒体引述专家的研究，这绝不是一店经营失策，而是带有普遍性的问题。

北京王府井大街，其美誉甚多，包括“中国商业金街”“中国‘银座’商业街”“北京商业第一街”等，尤其是经过一期、二期（2000年9月11日竣工）改造后，新王府井大街的确是今非昔比。整体商业风格是古今结合，以“现代、新潮”为主：在街头保留了象征性的铜匾，也有“拉洋车”“剃胎头”“弹三弦”等人工摆设；在810米长略呈波浪形的大街两侧集聚了京城最多、装修最现代化的12个大商场，其中，营业面积1万平方米以上的就有七八家之多，东方广场更在10万平方米以上。但这一“金街”目前在经营中也存在一些突出问题需要解决，首先是经济效益问题，据统计，该商业街日客流量为30万人次，节假日达50万人次，2001年全街纯商业零售总额为36亿元。但要知道，这一营业额是在地头旺、硬件优、营业面积大的条件下取得的。王府井大街的营业面积超过50万平方米，按此推算，每平方米营业面积年营业额仅为7200元，如何做到“人财两旺”，是一个重大挑战。其次王府井大街是一条曾经诞生过205个老字号的地方，如何“扶老”创新，又是另一个重要问题。目前有一些老字号已湮没在茫茫商海之中，一些正为明天生存而发愁，一些耳熟能详的名字，今后也许只能在电视上与它见面。

大连欧式商业步行街，是我国首家室内欧式商业步行街，是由某房地产经营集团投资2.5亿元兴建的，2001年9月29日面世，其广告促销口号是“不出国门逛欧洲”“我在‘欧洲’等你”。其设计与布局也很有特色，欧式商业文化在商业街内或零售商场内，都能有所体验，这种以景衬商、以商带景的经营模式令众多旅游者大开眼界。社会传闻，如此气魄的景物商场，只有美国拉斯维加斯、日本东京和俄罗斯莫斯科这三大城市才有。但是，该商业街（城）开业半年后就已进入“困境”，而且难以自拔，只得关张大吉，留给人们的只是一连串的思考与参观者的摄影留念。

以上三个案例，具有典型性和代表性。问题出自何处？存在的主要问题是什么？为什么各处所出现的困境，有着惊人的相似之处？这是值得我们借鉴与思考的。

当前，商业步行街存在的主要问题有以下五个方面。

（1）购物街变为游览街，人流不息，只逛不买。这是商业步行街商家普遍反映的问题。经过打扮的步行街确有引人之处，但购物环境与观光环境毕竟有许多差异，除了一些随意性和纪念性商品外，难以想象人们会在人头涌动、摩肩擦背中去购买商品，特别是高档、耐用、大件消费品。武汉江汉路“金街”在2000年“十一”黄金周开街当日创下了百万市民游的“辉煌历史”。但在两个月后，情况大变，商业利润迅速下滑，“逛街人多，购物人少”的怪现象，使商家伤透了心，只能以击掌、吆喝方式招徕生意。

大连欧式商业步行街更为典型。进入该商城的基本上都是观光游览者，目的是开眼界，饱眼福，他们对商街的赞誉是“名字美、街景美、风格美”，特别是对其中30多条异国风情步行街，更是感到新颖，结论是“真有特色，只缺欧元”“感而不动，看而不买”。

(2) 费用高昂，经营严重亏损，商家频繁更迭。租金高，成本大，项目多，不少企业处于惨淡经营状态，一些商家也只能勉强支撑。其中一些装修豪华、内部设施高档的企业更感困难重重。武汉江汉路步行街的“世界之窗”购物中心更具代表性，该企业2000年投建，营业后每天的水电、劳务、设备折旧、物业管理费等支出高达数万元，而每天的利润收入往往比不上支出。开店后的前两年全店亏损已高达1800万元，2002年情况似乎好一点，但亏损额也在200万元，结果只能关门歇业。此外，商家频繁更迭是步行街的另一特点，尤其是分摊租赁者为多，每月都有商家黯然退出，当然，也有不甘服输者又继续租赁拼搏。

(3) 经营定位出偏差，服务管理跟不上。一种较为普遍的定位偏差是步行街特色不明显，雷同化突出，而特色又恰恰是步行街的灵魂，没有自己的个性和与众不同的风格，步行街是难以持续发展和有吸引力的，即使是一时兴旺，也只是昙花一现而已。另一种定位偏差是，步行街虽有特色，但脱离市场，脱离消费者，大连欧式商业步行街就是这一类案例。其经营定位是中高档欧洲品牌服装，价位奇高，就是一般品牌的价格也在几千元左右，单价商品最贵的是40万元，这与普通消费者的购买力相差太远。同时，当地的服装市场供大于求的现象已十分突出，已出现几年负增长趋势。定位偏差还有一种表现是，宜商环境的宏观定位不明，商业和旅游资源的整合缺乏共享系统，较多地注重店外形象而忽视店内服务，一些步行街由于服务内容与服务质量不能令顾客满意，留给顾客的只是一种高级集贸市场管理印象，在顾客心目中的定位是“到此只是逛街而已”。

(4) 交通难始终是一个心病。步行街一大优点就是在一条路上避免了人车抢道，步行者具有安全感，更可以随意游览观光。但这也带来了另一个方面的问题，由于步行街周边停车不便，以车代步的消费群体对此感到十分不便，常因此而转到车流畅通的地方购物或消费，而这些人又常常是购买力较强的购买者，因而对大店、名店和中高档商店甚为不利。对此，一些老字号名店深有体会。为什么目前在国内不少步行街出现档次偏低、名优品牌不愿进场、老字号名店难以立足的现象，这有其内在原因。

(5) 体制改革不到位，管理制度不规范。步行街如何管理，各地各施各法，均在探索之中。《中国商报》在2002年6月25日对此有一篇专论，其中一段说：“目前全国120条商业街中，几乎都存在着政府行政管理的影子，充满强烈行政色彩，甚至成为对外交流的‘面子工程’，这就必然影响商业街的商业特色。政府对商业街的管理应侧重抓立法与规划，不宜直接参与商业运作，更不宜设置机构去左右商业经营。”这是一种主流观点，其要求与当今国际上一些商业街或步行街的管理模式类似。例如，巴黎香榭丽舍大街的管理机构，从1860年起就是由175名商人和团体组成的。该机构实际上是商业协会，其成员代表的95%来自临街店铺代表及街道的居民代表，它是一个不依附于政府的纯粹民间组织。再如，丹麦的哥本哈根步行街，这是一条从1962年11月17日起建立的供民众步行休闲购物的商业步行街，历史悠久，全街拥有275间店铺，每天接待顾客5万～10万人次。其管理模式是设立“步行街管理有限公司”，这是一个职能独立的管理公司，其费用由所有商铺的经营者共同承担，连步行街的外观、橱窗摆设都由该公司提出指导意见。

与上述不同的另一种管理模式是以行政管理为主的模式。北京王府井商业街具有代表性，该商业街成立了“王府井地区管理委员会”，直接代表政府实施管理，该委员会下辖一个行政色彩非常浓的“王府井地区建设管理办公室”，由东城区发展计划委员会领导。为什么目前许多步行街或商业街采用类似管理模式，其理由是：政府行政管理具有权威性，有利于统筹规划，协调各方；有利于改造、建设资金的筹划与运用；有利于招商引资工作；当然，也有利于城区形象工程的实施。但是，由于管理主体所处的地位不同，往往考虑的重点不同，采取的策略、措施也不同，常常出现管理者与商店经营者之间的许多矛盾，包括经营定位上的矛盾、土地资源利用上的矛盾、“政绩”与“业绩”的矛盾等。正如一些步行街管理工作经验中所归纳的：“行政管理越凸显，商人心理压力越大。”这些矛盾有待于今后在实践中解决。

除上述的管理体制所存在的问题外，步行街或商业街还存在“四益”的矛盾，四益即投资者利益、部门利益、区域利益、商家利益，这“四益”矛盾如何协调处理，也是要慎重研究的。

3. 广州商业街的今天与未来走向

荔湾是广州市城区之一，荔湾城区的发展和商业布局要适应全市的整体规划。分析或规划荔湾的商业街和步行街，必须对广州市的整体规划及发展趋势有所认识，这样局部才能明确有所为、有所不为、有所难为。

1）商业街的概念、类型及广州商业街的现状

商业街通常是指具有一定的长度（长度＞100 米），具备商品零售、批发、餐饮服务和文化旅游等功能，各种商业网点沿街道两旁分布且相对密集（商业网点的密度≥70%）的商业功能街道。

商业街，从商业街的功能定位及经营范围区分，可分为综合性商业街和专业性商业街两大类；从交通特点区分，可分为普通商业街（非步行商业街）和步行商业街两大类。

据统计，至 2000 年底，广州市共有商业街 36 条，其中，综合街 22 条，专业街 14 条，分别占商业街总数的 61.1%和 38.9%；在综合街中，步行街 2 条，其余均为非步行街。专业街的种类较多，以服装及其配料、饰物精品、建筑装饰材料、鞋类及相关产品、电子信息类产品等为主。

广州的商业街基本上是自然形成的。从空间分布来看，内圈层的商业街共有 24 条，占全市商业街总数的 66.7%。其中，越秀区有 8 条，荔湾区有 4 条（上下九步行街、十三行服装街、西关玉器古玩工艺街、南岸路装饰材料街），东山区有 7 条，海珠区北部有 2 条，天河区西部有 3 条。在中圈层和外圈层，商业街数量较少，只有 12 条，分布较为分散，其中，白云区有 4 条，主要分布在靠近老城区的南部地区，东村区、花都区各有 2 条，番禺区、黄埔区、从化市、增城市各有 1 条。

2）广州近中期商业街发展趋势分析

关于商业街发展设想，《广州市商业发展规划主报告（2000—2010）》明确提出其

发展方向是：以“以人为本”为出发点，以特色化为导向，围绕现代化中心城市的定位，控制数量，突出重点，将商业街的发展与城市建设和改造相结合，全面提升商业街的品牌形象和营销水平，使商业街成为反映城市历史风貌特色，适宜居民购物、休闲娱乐和社会交往的重要商业景观和场所，形成有较高起点、较高档次，在国内乃至国外有较高知名度的重点商业街，进一步强化广州商都形象，提升商贸中心的地位，促进全市商业繁荣。近期目标主要是重点建设9条特色商业街，包括北京路商业步行街、上下九步行街、珠江滨水风情休闲街、农林下路综合商业街、西关玉器古玩工艺街、状元坊饰物精品街、文德路古玩字画文化街、一德路商业街和黄石路汽车汽配街。

中远期重点改造和建设的商业街共有11条，包括中山路商业街、江南路商业街、东圃商业街、新华商业街、荔城商业街、康王路商业街、市桥商业街、大沙商业街、街口商业街、新城商业街和南沙商业街。

商业街的改造与建设的要求是什么呢？根据广州市商业街发展规划所提出的要求，未来广州商业街的发展方向，在经营服务、功能划分、环境设计和景观风貌等方面均应向特色化方向发展。以人为本，以消费旅游者的需要作为根本出发点，合理设置人行道、厕所、座椅、停车位、治安岗亭、公共交通体系等，营造一个舒适的购物环境。商业街还将按现代社会的消费需求和生活方式的特点，设立将购物、饮食、娱乐、游憩、文化交流、休闲服务等多种功能融为一体的公共活动社区。

4. 全面提升荔湾区上下九步行街竞争力

1）商业街竞争力，客源是基础，销售是目的

近几年，广州商业结构调整加速，商业业态变化也大，新的业态不断涌现，商业街与各类市场也如雨后春笋般出现。一句话，商业竞争加剧了，竞争的核心是上形象、争客源、创利润。广州的情况如何？消费者需求的特点什么？顾客到哪里去旅游和购物消费？这是营销决策必须掌握的基本点。为此，广州市旅游局对2002年游客的基本情况进行了调查分析，在38 726位境内外游客的调查中，重要的参考数据见表1～表5。

表1 广州游客来源构成及人均消费

项目	广州（含郊区）游客	省内游客	省外游客	境外游客	平均消费
游客来源构成/%	44.70	23.26	21.62	10.42	—
人均消费/（元/人）	185.36	394.25	517.77	1281.40	429.94

资料来源：根据《广州日报》2003年1月9日报道与《南方日报》2003年4月17日报道综合整理

表2 游客选择景区景点排序表

序号	综合	省内游客	单位组织	个人结伴	旅行社组织
1	白云山	白云山	白云山	白云山	白云山
2	北京路	北京路	上下九路	北京路	陈家祠
3	上下九路	上下九路	陈家祠	上下九路	天河城
4	天河城	珠江游	动物园	天河城	越秀公园
5	越秀公园	越秀公园	珠江游	越秀公园	珠江游

续表

序号	综合	省内游客	单位组织	个人结伴	旅行社组织
6	香江/长隆	动物园	天河城	香江/长隆	上下九路
7	珠江游	陈家祠	北京路	陈家祠	北京路
8	中山纪念堂	香江/长隆	越秀公园	珠江游	中山纪念堂

资料来源：根据《广州日报》2003年1月9日报道与《南方日报》2003年4月17日报道综合整理

表3　广州游客分类　　单位:%

项目	观光/休闲	探亲	公务经商	其他
比例	41.11	7.31	43.89	7.69

资料来源：根据《广州日报》2003年1月9日报道与《南方日报》2003年4月17日报道综合整理

表4　三大购物广场游客结构　　单位:%

项目	上下九	北京路	天河城
广州居民	48.08	33.82	29.11
省内游客	27.18	44.49	36.04
省外游客	20.26	19.55	30.69
境外游客	4.49	2.13	4.16

资料来源：根据《广州日报》2003年1月9日报道与《南方日报》2003年4月17日报道综合整理
注：数据之和不为100%是数据修约所致

表5　三大购物广场人均消费　　单位：元/人

项目	境内游客	境外游客
上下九	130.88	582.96
北京路	136.90	227.55
天河城	128.29	593.23

资料来源：根据《广州日报》2003年1月9日报道与《南方日报》2003年4月17日报道综合整理

从表1～表5中可以看出如下问题。

(1) 旅游目的构成以公务经商和观光/休闲为两个大头，两者比例非常接近，前者占43.89%，后者占41.11%，但应注意，观光/休闲比例在逐年提升，这说明广州市作为现代化旅游中心和旅游目的地的特征已经非常明显。

(2) 游客来源构成中，广州（含郊区）游客占44.70%，省内游客占23.26%，两者合计约近七成，这说明在城市旅游中，当前广州还是以区域性旅游为主。但应注意的是，省外与境外游客虽约占三成，但这部分客源潜力大，消费能力强，人均消费水平大大高于本地游客。调查资料显示，省外游客人均消费水平为广州（含郊区）游客消费水平的2.8倍；境外客源消费为广州（含郊区）游客消费的6.9倍。当然，这与住宿、餐饮、交通等费用支出有关，不完全是购物的问题。

(3) 在三大商圈（北京路、上下九、天河城）游客构成及消费比较中，各商圈均有自己的特色，吸引的游客各有侧重，而游客的消费支出也有所不同。

从游客结构上看如下。

上下九商圈——以广州居民为主（48.08%），其他依次为省内游客（27.18%）、省外游客（20.26%）、境外游客（4.49%）。

北京路商圈——以省内游客为主（44.49%），其他依次为广州居民（33.82%）、省外游客（19.55%）、境外游客（2.13%）。

天河城商圈——以省内游客为主（36.04%），其他方面省外游客与广州居民比例甚为接近，都在30%上下，境外游客（4.16%）。

从购物心理上看如下。

广州居民——首选上下九商圈，次为北京路商圈和天河城商圈，主要购物心理是：上下九路商业街商品种类多，价格实惠，具有岭南特色，符合一般市民的购物心理。

省内游客——首选北京路商圈，次为天河城商圈和上下九商圈，主要购物心理是：北京路商圈交通方便、名店名牌多、中高档商品多、宣传促销知名度大、观光与购物能较好结合。

省外游客——首选天河城商圈，次为上下九商圈与北京路商圈，主要购物心理是：天河城是广州市现代化大都市的标志性商圈，现代化气息浓，购物环境好，购物、购书、娱乐、饮食结合得好。

境外游客——首选上下九商圈与天河城商圈，次为北京路商圈，主要购物心理是：以观光为主，满足好奇心理。上下九商圈具有岭南特色，可以了解中国传统岭南文化；天河城具有现代商业和改革开放新貌特色，可以了解中国改革开放以来的新面貌。这两方面是境外游客，特别是西方游客的心理需求。

从上述分析看出，上下九步行街的客源组织可分为三个商圈：第一商圈（基本商圈）是广州市居民及广佛经济圈居民；第二商圈（发展商圈）是珠三角与港澳台地区；第三商圈（外围商圈）是省外游客及国外游客。

2）准确定位，特色经营，是上下九商业街生命所在

商业步行街具有生命力、能持续发展的关键在于"三有"，即有文化意蕴，有历史传承，有个性特点。文化意蕴是指文化氛围，寓商于文化之中；历史传承是指商业流通中的承前启后，继承与创新的结合，现代与传统的融合；个性特点是指在众多的商业街中，要有自己的特点，唯我独有，别人难以模仿。

一个城区、一条商业街在人们的心目中都会自觉或不自觉地形成一定的形象，并形成一定的引力。这种现象不是简单的"造势"所能达到的，它必须有相应的内涵来支撑，其中包括城区定位、产业定位、市场定位等因素，而任何定位都是由文化、环境、区位及历史等许多因素综合形成的，按照唯一性、排他性和权威性的原则，找出自身的个性、灵性和理念。

A. 广州市民对上下九步行街的认识与评价

在市民和游客的心目中，对各类的商业街都存在一个心理定位问题，即"哪里更适合我去"的问题，这是"买方市场"存在的经济现象。广州市民对本市两大步行街，特别是对上下九步行街消费环境是如何评价的呢？2002年，广州市统计局对一万户常住居民进行了抽样访问调查，而后荔湾区统计局又进行了细化调查分析，两次调查的结果综合表述见表6。

表 6　广州市民对商业步行街消费环境满意度调查　　单位：%

调查单位	步行街名称	总体满意度	文化特色	商业结构	公共设施	建筑风格	医疗急救	休息场所	交通	环境卫生	市场秩序	治安状况
市统计局	北京路步行街	50.5	45.9	48.8	34.4	57.9	32.2	23.1	47.8	44.2	46.2	50.4
市统计局	上下九步行街	57.2	64.4	52.2	30.5	76.2	32.5	22.1	43.9	45.5	50.9	56.0
区统计局	上下九步行街	56.2	64.2	52.9	—	75.5	—	25.5	—	—	54.9	60.6

资料来源：根据荔湾区统计局《荔湾统计》2002 年第 9 期和第 12 期及《广州日报》2002 年 6 月 24 日报道整理

从表 6 看出，市、区统计局的调查结果基本一致，即对上下九步行街的总体满意度接近六成，其中，文化特色和建筑风格两项比值较高，而公共设施和休息场所两项比值最低。再综合其他调查，可得知市民对两大步行街的反映存在“三难”问题，即“方便之处”难找，“休息场所”难见，“特色商品”难买。“特色商品”难买是指两街所经营的商品雷同，主要以百货、服装、鞋类为主，地方性特色产品难买。这些调查反映，对今后步行街的整治、提升，提出了许多值得注意的问题。

B. 历史的凝聚，消费者的认同，是商业街定位不可忽视的因素

国内外一些著名商业街都有其发生、发展的历史条件，当然也有一些因新兴城市、新兴产业的出现而形成新的商业聚散地。一般来说，出名的商业街都有其历史沿革，都有自己的特色，雷同性太多的商业街，不是好经营的地方。广州目前相对稳定的五个商业圈带，也都有各自的发展“家史”，这五个商业圈带是：以天河城为中心的现代化大都市商业圈带；以北京路步行街为中心的旅游时尚商业圈带；以上下九步行街为中心的具有岭南文化的商业圈带；以农林下路、中华广场为中心的城区消费商业圈带；以江南大道为中心的河南区消费商业圈带。从以上五大商业圈带的基本状况看出，由于各自所处城区的地域不同，经济发展的历史不同，当前业已形成的规模不同，其商圈定位已逐渐呈现出不同特点。当然，任何“定位”，都不是固定不变的，近年来，随着改革开放及经济发展，出现了不少“重新定位”的问题。“变”是绝对的，但“变”又是客观的、科学的，而不是人们主观随意的。

C. 上下九步行街定位设想

(1) 功能定位。以商业为主体，以岭南文化西关风情为底蕴，以观光休憩为引力，集购物、饮食、旅游、观光、休憩于一体的商业步行街，可以概括为“岭南民俗西关风情商旅文化步行街”。

在功能定位上要明确的是：是“商街”还是“景点”？谁为主谁为辅？上下九步行街的基本功能是“商街”，这是名副其实的。但也不是一般“商街”，它具有明显的“景点”特征，对客源起重要的引力作用，两者互为带动，相辅相成，在处理各种关系时，必须兼顾各方的特殊需求。

(2) 形象定位。根据上下九步行街的功能定位及商旅文化特征，其形象定位是：体验式消费理想地。近期在国外盛行“体验式消费模式”，这一概念在空间上可在商场

内、商业街内或周边环境内；在内容上包括多个方面。一般来说，体验式消费是指在购物过程中注重消费者的心理感受，在旅游过程中注重参与性活动，使客人在愉悦的精神状态下，从容购物和观光休闲，把购物与观光、休闲，消费与文化结合起来，强调人性化、个性化消费。在上下九步行街消费不是一般的“逛街购物”，更为重要的是得到精神体验，包括岭南民俗风情的体验、西关文化传统的体验、西关“雕—彩—绣”精湛艺术的体验、西关时尚变化的体验，给人一种既旧又新，传统与现代融合的感受。

(3) 客源定位与经营定位。客源定位与经营定位密切相关，客人需求的就是商店要经营的。目前人们普遍反映广州市两大步行街，甚至所有综合性的商业街所经营的商品基本上是雷同的、缺乏特色的，这是一个难以解决而又不能避开的问题。

客源定位：上下九步行街的目标客源，以中青年女性的需求及旅游购物为主。提出这一定位设想的理由是：首先，有历史成因，几百年来，上下九步行街就为女性所青睐，“要扮靓到上下九”，“脚上穿的、身上换的、手上戴的、家里用的，上下九可以找到你喜欢的”，这一特点让人印象深刻；其次，目前上下九步行街所经营的主打商品，包括服装、妇儿用品、鞋类、金银饰品、佩戴商品以及日用百货，这正是女性购物的主要商品；再次，国内外市场营销专家认为，21 世纪是女性购物年代，谁能掌握这一特点谁就能占领市场，国际市场调查公司——AC 尼尔森于 2001 年对北京、上海、广州等七大城市 6000 个样本（家庭）（代表了 1200 万家庭）进行了“购物习惯”调查，其结果显示，七成以上家庭已由 25～44 岁的女性掌握了“购物大权”。这不仅包括零星日用小商品，也包括耐用大宗消费品；最后，为了与广州其他城区商圈，特别是与天河城、北京路商圈拉开差别，上下九步行街实行差别化市场策略，树立了鲜明的个性特点。

经营定位：通过对全街商业业态进行调整，建立整体业态布局。为凸显上下九步行街的文化、名牌的商业品牌意境，商业业态应以品牌连锁店，名牌旗舰店、专营店、特色店等为主，大型百货店、大型广场应尽可能实行主题化经营，以适应部分购买者需求。一般超级市场及分租零散摊档可引导到主街道的周边经营，不占用步行街的黄金地段，避免冲淡经营特色。

3）网状布局，功能区分，创造顾客价值

A. 商圈发展的一般形态

点状、线状、网状是商圈发展的基本形态。点，一般是指一个大卖场、大市场，它虽具有一定规模，百货齐全，但周边商业气氛不浓，难成气候；线，是指各类商店呈线状聚集，起聚集效应，人气旺，商机多，如各种商业一条街；网，也有说是“块”状，是指呈网状的商带，它比线状商带更进一步，也是线状商带发展的一种趋势。网状商圈的优点有以下几个方面：①网状商圈较之线状商圈便于消费者走动和选购；②便于设立不同类型的专业购物区或功能区，有利于客源的分流；③多几个交通进出口，进出方便，也有利于疏导人流；④扩大商业街商圈，带动周边商业的发展，而周边的商旅优势又加强主街道的资源整合，提高知名度。

B. 上下九步行街布局的历史借鉴

上下九步行街现由上九路、下九路和第十甫路组成，全长 1218 米。清末民初，该路段已有许多商人在此经商，但最繁华的不是该处，而是靠近十三行和十八甫路一带。

20 世纪 40 年代中后期，该地段逐步形成西关繁华商业一条街，号称“衣食住行、购物娱乐一应俱全”。当时，该商业街的布局与行业分布，是经过长期的反复变迁而自发地形成的，其分布状况大致如下。

上九路北侧多是金铺，南侧多是花纱棉布庄，与光复南路、杨巷路的布庄经营相呼应。下九路北侧大部分是鞋店，包括著名老店鹤鸣皮鞋店、吴志记小圆头礼绒鞋店、潘常兴胶鞋店等共 22 家。此外，还有大吉、福生等床上用品店。南侧以绸缎店、服装店和百货店为主，其中，绸缎店就有 16 家之多，著名的有纶章、仁章、上海、唯一、同章、天生、广公诚等。此外，还有多家摄影店，为迎合女士照相心理，多设在楼上。市民常说，下九路是一条地道的“扮靓街”。

第十甫路以食肆居多，当时有“百步必有小食”之说，各类小食店星罗棋布，广州饮食业名店也多出于该处。新中国成立初期，第十甫西关食街共有大小饮食店 29 家，“文化大革命”期间停业 12 家，并店 4 家，迁店 2 家，只剩 11 家，整个市场显得萧条冷落。

从历史发展角度考察上下九与第十甫的商业情况，虽经起落兴衰，但在各街各段都已形成不同经营特色，各有功能分工，主街与周边相互配合，形成颇有影响的西关商业带。

C. 一轴两翼布局的设想

为使上下九步行街从线状向网（块）状发展，商圈辐射形成一轴两翼态势，主轴即现已形成的步行街，全长 1200 多米，不再延长；两翼是在上下九路南北向形成的。以主轴带两翼，两翼润滑主轴，加速主轴运行。

一轴两翼地理位置示意图如图 1 所示。

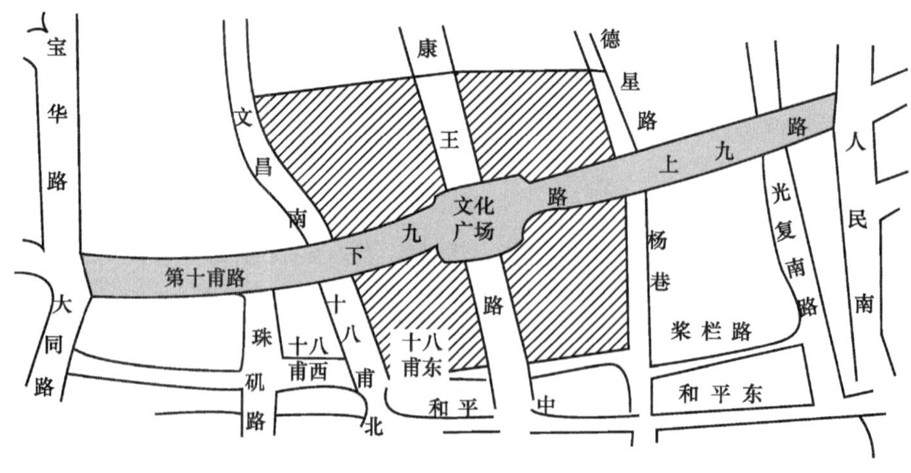

图 1　上下九步行街网状发展示意图

主轴的经营业态应有概念性分工，有利于发挥聚集效应，吸引消费者注意和方便选购。

上九路——以经营服装、金银首饰为主，包括穿的、戴的，并与光复南路、杨巷路经营相呼应。

下九路——以经营鞋帽、百货类为主，包括女性用品、儿童用品、装饰用品等，与文昌路、十八甫路经营相呼应。

第十甫路——以经营百货及饮食为主，与文昌路、宝华路经营相呼应。

两翼的大致范围是：北翼包括文昌南路和德星路；南翼包括十八甫北路、杨巷路，还可考虑延伸至桨栏路、十三行路。

对原有的专业街，因其已有一定的知名度，要加强利用并善待它们，特别是以下专业街。

（1）华林玉器街。该街在清朝同治年间已在长寿路一带摆卖玉器，清光绪年间建成“玉器圩”，以经营玉器、钻石、宝石、戒指、珍珠项链等饰物为主，成为我国南方最大的珠宝玉器商场。1988 年后，该处经营由长寿西路转入新胜街、华林寺前、西来正街一带，全长近 1000 米，成为名副其实的“玉器珠宝一条街”。

（2）源胜陶瓷玉石工艺街。该专业街开设于 1990 年，现有固定点档 170 多个，经营品种不断扩大。除玉石、玉器及各种玉石工艺品外，还有紫砂茶壶、字画、邮集、木雕、古玩等。

（3）德星路服装配件专业街。该处早在明清年代就已形成“苏杭杂货”店铺，逐步形成广州市有名的小百货批发一条街，经营品种十分繁杂，1984 年正式被确认为服装配件专业街，此后经营不断发展，现经营的品种多达 2000 余种。

（4）西来初地酸枝街。该地从明代起就以生产红木家具闻名，新中国成立初期成立了两个生产合作社，继后合并为一个木雕家具厂。该处的复兴是在 20 世纪 80 年代，重新成立了红木酸枝家具专业街。不少驻广州的外国领事馆官员也常到此街选购。

总的来说，关于上下九步行街的一轴两翼网状布局设想，只是一种初步的概念性设想，具体的安排、规划，还有待进一步研究。

4）加强管理，用诚信提高核心竞争力

商业的核心竞争力是什么，可以有各种提法，我们认为商业的核心竞争力是诚信。社会诚信制度和规则体系，以及社会成员和经济主体的诚信品质，是市场经济健康发展的重要基础。诚信是价值工程价值观的核心，是一种社会制度，是一种行为方式，它能降低交易成本，能为一个城市、一个企业创造最大的效益。良好的市场信用，已成为吸引资本和商贸往来的核心引力。

上下九商业在消费者心目中，已逐渐形成一个品牌，这与上下九的市场特点有关。人们在北京路或天河城购物，总会说出在××商场购买，而在上下九路购买往往只说在上下九购买，很少会说是在××商场购买，原因是人们已经把上下九作为一个市场主体。因此，在上下九商街，企业的诚信问题已不是一个企业的问题，而是一个特定商圈的问题。

上下九商业打诚信牌是有一定基础的，其在几年前就曾先后获得“不经销假冒伪劣商品一条街”“全国百城万店无假货示范街”“打假维权满意街”等称号。2003 年 3 月 30 日广州市物价局、质量技术监督局还授予上下九步行街“价格、计量信得过一条街”荣誉称号，并正式授牌。此前，一般获取“双信”称号的是单个商铺或企业，而首先以一条街获得这个荣誉的还是上下九步行街，这就为给广大的旅游者和购物者树立“诚实、守信”的商业形象奠下良好基础。

但从总体看，上下九步行街虽无不良声誉，但“诚信”之风还不浓，影响力也不够强，以此为“卖点”还不够。为此还需要加强以下四个方面。

（1）街、店、品三诚信。广泛宣传上下九步行街是诚信之街，使其成为诚信社会的典范；街内所有商店也都是诚信之店，严防街内龙蛇混杂，使顾客消除“走错门”的顾虑；店内商品也是诚信之品，严防假冒伪劣。其实，街、店、品三者密不可分，产品是商店经营的物质要素，店是街的组成部分，虽然树大难免有枯枝，但枯枝过多，此树也就难活了。

（2）实施名牌战略，有助于扩大诚信影响。一般而言，名牌就是信誉，谁的名牌响，谁就能在竞争中占有优势。在目前的情况下，价格战虽然是常规战的武器，也有其用武的空间，但名牌战越来越凸显其重要性。上下九步行街实施名牌战略的方法，主要是大力引进名牌企业与名牌产品，引进一批名牌企业群或名牌企业的连锁店，改造那种集市贸易式的吆喝经营方式，为塑造信用上下九步行街打下基础。

（3）建立市场信用评估核查体系，根治信用缺失。当前商品市场上信用缺失主要集中在假冒伪劣和价格欺骗问题上，而要有效地避免和防止信用缺失，必须建立社会化的市场信用评估核查系统，使信用缺失企业的失信行为能被市场及时、准确地“记录在案”，使失信企业增大“成本”，得不偿失，从而强化企业自律意识，同时也发挥群众的监督作用。

（4）建立对企业信用的有效监管机制，加强信用保障。企业信用靠企业的自律，同时也靠政府的监督，发挥有效的监管机制的作用，特别是加强对各类小企业、小摊档的监管，严防害群之马。建立企业信用考评系统，实行定期考评；建立个人信用体系，对企业法人代表进行信用考核记录；对在上下九步行街内销售的重点产品建立信用等级考核，凡有信用劣迹记录的，在一定时期内不准在商业街内销售。总的来说，要规范信用秩序，惩治失信行为，务使旅游购物者在上下九“玩得开心，购物放心”。

总之，强调商业诚信，似乎是“老生常谈”，但却是提升核心竞争力的重要因素。一个城市、一条商业街、一个企业在群众中失去诚信，要想恢复，那就要付出很长时间和很大的代价。

三、整合旅游资源，提升荔湾竞争优势

1. *荔湾旅游资源评析*

荔湾区是广州市的老城区之一，旧称“西关”，因区内有“一湾青水绿，两岸荔枝红”美誉的“荔枝湾”而得名。荔湾作为有百年底蕴的老城区，商贸旅游资源丰富，名胜古迹众多。区内有全国重点保护单位两处——陈氏书院和沙面古建筑群，以及市级文物保护单位多处；有曾独享外贸特权近百年的十三行；有创下岭南园林之冠和为发展岭南文化做出过贡献的海山仙馆；有广州佛教“五大丛林”之一的华林寺、道教真武帝庙；有近代广州工人反帝斗争丰碑“六二三”沙基惨案纪念碑、蒋光鼐先生故居、李文田的泰华楼、西关大屋等。另外，有孙中山、詹天佑、陈少白、汤廷光等名

人生活、工作过的足迹多处，还有清末洋务派官员张之洞创办的广雅书院和民国陈济棠主粤时期开发的西村工业区遗址等。西关古迹和传统文化成为荔湾的旅游优势，凸显了老城区厚重的历史文化底蕴，构筑了荔湾旅游业发展的基础，展示了广州老风情的精华。

为了正确评析荔湾的旅游资源，认清优势，整合资源，提高荔湾旅游竞争力，我们依据价值工程理论与方法，对荔湾的旅游资源和在旅游发展中有重要地位的核心资源进行价值和价值链分析，试图通过价值链的分析，找到旅游价值链的优势环与劣势环，并从构筑整体竞争优势的角度，提出发挥荔湾旅游资源优势的价值指向及战略取向。

1）荔湾旅游资源整体评析

从广义上说，凡对旅游者有吸引力的事物都是旅游资源，既包括有形的物质形态的东西，又包括精神文化的无形内容。

荔湾现在的主要旅游景点和资源共有 23 处（不含美食类），包括重点文物保护单位，国家级 2 处，省级 1 处，市级 16 处，共计 19 处；城市公园 2 处；博物馆 2 处。另外还有旅游购物场所 4 处。其中，上下九步行街在全国享有盛名；陈家祠被评为“新世纪羊城八景”后，旅游知名度、接待游客量均大幅增加，从而使荔湾的旅游业更受关注。此外，荔湾还有许多著名的老店和精彩的西关风情等无形的旅游资源。

经过一定的筛选，荔湾区内有旅游开发价值的旅游景点及旅游资源见表 7。

表 7　荔湾主要旅游景点及旅游资源简表

序号	名称	类型	级别	规模	位置	特征
1	陈家祠	历史文物	国家	1.3 万平方米	中山七路	①岭南古建筑精品；②宗祠文化；③岭南民间艺术博览；④羊城新八景
2	上下九步行街	购物商街		1218 米	上下九路	①旅游购物商街；②骑楼文化；③荔湾风情
3	沙面欧陆风情区	历史街区	国家	2.2 平方千米	沙面岛	①欧陆风情；②酒吧、咖啡厅休闲；③西洋建筑；④珠江景观
4	华林寺	宗教文化			下九路西来初地	佛教文化
5	荔湾博物馆	博物馆		2000 平方米	逢源北街	①西关文化博览；②西洋建筑
6	华林玉器街	购物商街			华林街	玉器专业街
7	荔湾湖公园	城市公园		27.8 万平方米	荔湾湖	园林
8	仁威庙	宗教文化	市级		昌华街	①宗教文化；②岭南古建
9	小画舫斋	历史文物	市级	2100 平方米	逢源大街	西关大屋精品
10	十三行	历史遗址			杉木栏路	①历史遗址；②洋务文化
11	八和会馆	历史文物			恩宁路	粤剧会馆
12	广州文化公园	城市公园		8.7 万平方米		休闲娱乐场所
13	沙基惨案纪念碑	历史文物	省级		西堤	①纪念碑；②珠江景观；③城市街景
14	西关大屋建筑保护区	历史文物	市级		昌华街	古民居

续表

序号	名称	类型	级别	规模	位置	特征
15	南方大厦	历史文物	市级		西堤	①古建筑；②大型购物场所
16	蒋光鼐故居	名人故居	市级	650 平方米	逢源北街	故居
17	广雅书院	历史文物			钟村镇北部	①书院文化；②岭南古建
18	海山仙馆	仿古建筑			荔湾湖内	园林别墅
19	詹天佑故居	名人故居			十二甫	故居
20	文塔	历史文物			龙津西路	古塔
21	清平市场	购物场所		7000 平方米	清平路	中药材市场
22	黄沙水产批发市场	购物场所			南岸路	①全国三大水产批发市场之一；②设有海鲜食肆
23	邮电博物馆	博物馆			西堤	①古建筑；②邮电博览
24	广州酒家	著名餐饮店			文昌南路 2 号	老字号
25	泮溪酒家	著名餐饮店			龙津西路 151 号	老字号
26	陶陶居	著名餐饮店			第十甫路 20 号	老字号
27	莲香楼	著名餐饮店			第十甫路 67 号	①老字号；②著名月饼店

A. 荔湾旅游资源类别分析

荔湾区十分重视旅游业的发展，长期以来进行了大量的工作，尤其是在宣传造势方面取得了很大成效，因此具有较大的旅游知名度。但为了客观地审视荔湾的旅游资源状况，为今后发展提出针对性较强的改进意见，我们选取了新的资源分析方法，力求对荔湾旅游资源能有新视角。

为了说明问题，便于和广州市的相关数据进行对应分析，现将表 7 进一步归类形成表 8。

表 8　荔湾区主要景点及资源分类情况

大类	景点类别	数量/个	比例/%
文博类	历史古迹	13	48.2
	宗教寺庙	2	7.4
	博物馆	2	7.4
现代城市景观类	购物美食	8	29.6
	城市公园	2	7.4
合计		27	100

从表 8 和表 9 可知，荔湾区的旅游资源结构与广州市有一定相似性，但也有其特殊性。为了便于说明问题，我们将广州市纳入旅游统计的各类景区点，分为观光休闲、文博、现代城市景观、主题公园与游乐场四大类，文博类所占比重最高，这与广州市历史文化名城的背景是相符的。荔湾区文博类资源所占比重超过了资源总数的一半，比广州的比重还大；历史古迹与购物美食两大类型所占比重高达 77.8%。这说明荔湾的历史文化底蕴十分深厚，旅游资源的类型高度集中，有利于利用以上两类资源，凸显荔湾旅游业发展的核心竞争力。但从广州市拥有的四大类型来看，荔湾区仅有两大类型，而观光休闲类、主题公园与游乐场类均缺失。这又说明荔湾旅游资源类型的丰

富度不够，资源结构较为单一，旅游拓展空间受到一定限制。

表 9 广州市主要景点及资源分类情况（纳入旅游统计的景点）

大类	景点类别	数量/个	比例/%
观光休闲类	风景名胜区 森林公园 休闲度假 观光农业 其他观光休闲	25	33.8
文博类	历史古迹 宗教寺庙 近现代史迹 博物馆	22	29.7
现代城市景观类	城市公园 文化艺术 购物美食 珠江游	18	24.3
主题公园与游乐场类	主题公园 游乐场	9	12.2
合计		74	100

资料来源：暨南大学旅游研究所《2003 年广州市旅游景区点调研》

B. 荔湾区旅游景区（点）评 A 情况分析

旅游景区（点）评 A 工作，是国家旅游局为了规范旅游景区（点）建设而推行的重要举措。截止到 2003 年 7 月，广州市现有的旅游景区（点）中，共有 8 个景区（点）被国家旅游局评定审核为 2A 级以上旅游区，占被调查旅游景区（点）总数的 10%左右，其中，4A 级旅游区 6 个，主要分布在番禺区和市内。荔湾区目前尚无参与评 A 的旅游景区点，两大王牌景区点——陈家祠和上下九步行街也未参与评 A。由此可见，荔湾区要实施“旅游带动”发展战略，尚缺乏品牌带动的龙头。

C. 荔湾旅游资源总体评价

旅游资源的评价工作，是旅游业发展的基础性工作。一个地区旅游资源的基底，决定了其旅游发展规模和旅游拓展空间，同时也是制定该地区旅游业发展战略的重要依据。荔湾在制订旅游发展规划时，已多次进行过旅游资源评价工作，可以说，荔湾旅游业发展的基底已基本摸清。本次研究，我们将要素分析法和层次分析法结合，对构成荔湾旅游业发展的核心资源以及竞争优势进行进一步探析，并由此得出一些新的建议。

我们通过筛选，确定了 12 个与旅游业发展相关性较大的评价因子，并赋予不同的权重，以此计算出每一资源点在地区旅游业中的地位和级别，进而应用价值工程的思想方法，对地区旅游资源的开发、利用和创新提出建议。

经过筛选的 12 个评价因子如下：①区位，以到主要中心城市（广州）的时间距离为标准；②可达性，包括交通条件、交通工具和到客源地转运条件；③容量规模，以用地面积和可游时间衡量景区发展潜力和效益潜力；④美学价值，综合美感度评价；⑤独特性，以有无替代和同类资源中的地位为标准；⑥资源结构，资源的密集程度和不同类型资源的组合状况；⑦用地条件，旅游开发、设施建设的用地类型和开发条件；⑧整体环境，包括旅游区及相邻地区的综合环境；⑨基础设施，区域和城市的基础设施配套；⑩旅游设施，旅游区和旅游接待区的服务设施配套；⑪客源基础，以现已达到的游客量为标准；⑫综合效益，以现已达到的综合经济效益为标准。

同时，给出了如表 10 所示的评价指标。

表 10 评价指标表

因子（权重）	一	二	三	四	五
可达性（1）	到中心城高速或一级路直达	1～2 级路直达，接转时间＜1 小时	有 1～2 级公路转 1～2 小时	3～4 级公路行车时间＞2 小时	2 小时以上砂石路、土路
容量规模（0.5）	＞10 万公顷，二日以上	2 万～10 万公顷，二日	0.5 万～2 万公顷，一日	0.1 万～0.5 万公顷，半日	＜0.1 万公顷，2 小时
美学价值（1.5）	很美	美	一般	较差	差
独特性（2）	独特不可替代	地位高可替代	地位较高可替代	一般可替代	较差可替代
资源结构（1.5）	集中结构良好	集中结构较好	较集中但单调	结构分散且单调	孤立且单一
用地条件（0.5）	良好	较好	一般	较差	差
整体环境（0.5）	内外均优美	外一般，内较好	内外均一般	内一般，外较差	内外均较差
基础设施（0.5）	良好	较好	一般	较差	差
旅游设施（0.5）	配套齐全设备良好	配套较全设施较好	配套不齐全设备一般	不够配套设施较差	不配套设施差
客源基础（0.25）	＞50 万人	20 万～50 万人	10 万～20 万人	5 万～10 万人	＜5 万人

依据表 7，选择荔湾区较具代表性、有旅游发展潜力的旅游资源及景区（点）进行综合评价，并依据其开发利用价值及可能贡献度划定其级别，见表 11。

表 11 旅游资源加权综合评价表

名称＼项目	区位	可达性	容量规模	美学价值	独特性	资源结构	用地条件	整体环境	基础设施	旅游设施	客源基础	综合效益	综合	位次	档次
陈家祠	10.0	10.0	2.0	15.0	20.0	12.0	3.0	4.0	4.0	5.0	2.5	2.5	90.0	1.0	一
沙面欧陆风情区	10.0	10.0	3.0	15.0	20.0	12.0	4.0	5.0	3.0	4.0	2.0	1.5	89.5	2.0	一
上下九步行街	10.0	9.0	3.0	14.0	20.0	12.0	4.0	4.0	4.0	4.0	2.5	2.5	89.0	3.0	一
荔湾湖公园	10.0	10.0	3.0	9.0	10.0	9.0	4.0	5.0	4.0	4.0	2.5	2.5	73.0	4.0	二
华林寺	10.0	8.0	1.0	13.0	12.0	9.0	3.0	3.0	4.0	3.0	2.0	2.5	70.5	5.0	二
华林玉器街	10.0	8.0	2.0	10.0	12.0	9.0	3.0	3.0	3.0	3.0	2.0	2.5	67.5	6.0	三
西关大屋建筑保护区	10.0	10.0	2.0	10.0	12.0	6.0	4.0	3.0	4.0	3.0	1.0	1.0	66.0	7.0	三

续表

项目 名称	区位	可达性	容量规模	美学价值	独特性	资源结构	用地条件	整体环境	基础设施	旅游设施	客源基础	综合效益	综合	位次	档次
广州文化公园	10.0	10.0	3.0	9.0	12.0	3.0	4.0	3.0	4.0	4.0	1.0	2.5	65.5	8.0	三
荔湾博物馆	10.0	8.0	1.0	12.0	12.0	9.0	2.0	3.0	3.0	3.0	1.0	1.0	65.0	9.0	三
清平市场	10.0	10.0	3.0	9.0	10.0	6.0	3.0	3.0	3.0	3.0	2.5	2.5	65.0	10.0	三
黄沙水产批发市场	10.0	10.0	2.0	7.0	12.0	5.0	3.0	2.0	2.0	2.0	2.5	2.5	60.0	11.0	三
仁威庙	10.0	10.0	1.0	7.0	11.0	9.0	2.0	2.0	2.0	3.0	0.5	2.5	60.0	12.0	三
邮电博物馆	10.0	10.0	1.0	10.0	13.0	6.0	2.0	2.0	2.0	3.0	0.5	0.5	60.0	13.0	三
沙基惨案纪念碑	10.0	10.0	1.0	6.0	12.0	3.0	2.0	3.0	4.0	2.0	0.5	0.5	54.0	14.0	四
十三行	10.0	10.0	1.0	0.0	15.0	6.0	3.0	2.0	2.0	2.0	0.0	0.0	51.0	15.0	四
八和会馆	8.0	8.0	1.0	3.0	10.0	3.0	2.0	1.0	2.0	2.0	1.0	0.5	41.5	16.0	四

根据各景点综合评价，其可开发性及对荔湾旅游业的可能贡献度排列如下。

一类：陈家祠、沙面欧陆风情区、上下九步行街。

二类：荔湾湖公园、华林寺。

三类：华林玉器街、西关大屋建筑保护区、广州文化公园、荔湾博物馆、清平市场、黄沙水产批发市场、仁威庙、邮电博物馆。

四类：沙基惨案纪念碑、十三行、八和会馆。

其中，一类资源的潜在吸引范围可波及省际或大区；二类、三类资源的吸引范围以省内或珠三角地区为主；其他孤立景点一般仅具有地方意义。以上旅游资源的分级，通常可作为旅游开发和投资的指引，以及政府制定旅游业相关政策时的参考依据。

需要说明的是，采用以上分析方法进行评价，虽有一定的科学性，但也有一定的局限性。因为在选取评价因子时，采用了基础设施、旅游设施、客源基础、综合效益四项指标，这四项指标对已开业的旅游景区（点）较为有利，而对潜在的旅游资源不利（在对潜在的资源进行评价时，可能会造成某项指标分值为0），会造成潜在资源评价的分值过低，因而不能很好地反映其旅游开发价值。为解决这一偏差，在对某些旅游资源进行评价时，常对选取的评价因子进行一定的调整或采用其他方法。本文对十三行的评析就是采用另一专门分析方法。采用以上方法得出的旅游资源分级，只是一种相对性判定，其等级划分均可能因为旅游开发、管理经营和市场策略的不同而发生改变。

通过上述分析，我们发现荔湾的旅游资源数量虽多，但具有较大影响力的资源不多，一类资源仅有三个，其余均为二类、三类或三类以下的资源，这些资源具有点小、级别低、市场辐射力不强的特点。从荔湾旅游业发展的优势看，荔湾的旅游资源高度集中在文博类、购物美食类和人文风情类；从地域分布上看，这些资源又具有相对集中在一定的地域范围内的特点。因此，这为围绕一定主题，或者以某一知名度较大的旅游资源为核心，整合周边具有互补性或烘托性的资源，规划建设亮点突出的蛛网状、长藤结瓜状的旅游片区提供了便利条件和可能性。

2）荔湾旅游资源对比分析

本部分所研究的旅游资源对比分析，包括周边地区、广州市区及荔湾区内各主要旅游资源的对比分析，其目的是更好地利用优势资源，提高其旅游价值，分析荔湾旅游资源未来的开发策略和与周边协调发展、优势互补的策略。

A. 旅游景区（点）数量对比分析

虽然荔湾区旅游资源不少，但从纳入广州旅游统计范围的数量来看，荔湾区的情况并不乐观。广州市纳入旅游统计范围的旅游景区（点）共有 78 处。从地理（地区）分布来看，目前主要集中在东山、越秀和番禺三个区，共有 36 个，占总景点的 50%左右，而荔湾区被纳入广州旅游统计范围的旅游景区（点）仅有 4 个，在全市属于中等偏下的水平。其原因在于，荔湾的大多数旅游景区（点）规模较小，级别不高，如图 2 所示。

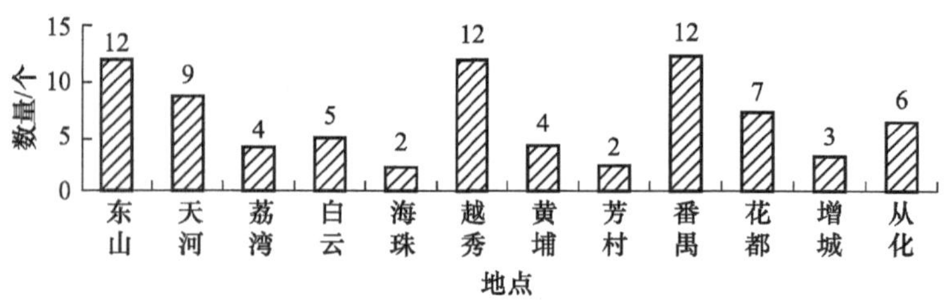

图 2　广州市主要景点地理位置分布

资料来源：暨南大学旅游研究所《2003 年广州旅游景区点调研》

B. 旅游景区（点）经营绩效对比分析

近年来，全市各旅游景点在广州市大力发展旅游业的政策指引下，根据市场消费需求和景点的发展目标，在增加旅游景区开发深度、挖掘旅游产品内涵和完善旅游景区功能配套等方面，都做出了积极的努力，并取得了良好的成效，其营业收入和接待游客人次不断地增长。

荔湾旅游业也有明显的发展，特别是拥有弘扬西关古迹和传统文化这一天然优势，更使西关风情旅游品牌不断增添活力。2001 年，全区接待海内外旅游者总人数为 819.10 万人次，其中过夜旅游者 318.38 万人次，一日游旅游者 500.63 万人次，旅游业总收入为 56.39 亿元。在此基础上，2002 年，全区接待海内外旅游者近 1000 万人次，旅游业总收入 62.96 亿元，达到历史最高水平。上下九步行街和陈家祠等主要旅游景点的知名度和吸引力日益扩大，经营绩效不断提高。

但是，随着旅游业竞争日趋激烈，不足之处也应“警钟长鸣”。例如，2000～2002 年三年广州市旅游业收入最高的前 10 名景点是：香江野生动物世界、长隆夜间动物世界、白云山风景名胜区、花都芙蓉度假村、广州市花卉博览园、广州动物园、莲花山风景名胜区、宝墨园、东方乐园和增城百花山庄。显然，荔湾区没有一处景点进入广州市前 10 名。

这一情况说明，荔湾区旅游业虽名声在外，旅游资源不少，却缺乏经营绩效突出的旅游景区（点）支撑；同时也说明，荔湾的旅游资源虽多，但多数规模小，级别低，缺乏品牌带动。这些问题必须引起高度重视，并实施资源整合，重组旅游空间格局，以构筑“以点联网，以网成片”的旅游发展优势，打造荔湾新的旅游竞争力。

C. 资源的整合利用分析

荔湾的旅游资源具有点多、规模小和高度集中在文博类与美食购物类的特点。根据这一特点，对区内及区外周边资源进行整合利用，是提升资源价值，形成旅游优势的一个重要的战略考虑。

就文博类旅游资源而言，可考虑以下几个方面的整合利用：①荔湾的西堤及其众多的各类博物馆资源，与越秀区的长堤、沿江西路等是广州市文博类资源高度集中的区域，不仅地域上连为一体，而且形成了竞争与互补，同时也是近代广州标志性建筑的集中区域和珠江夜游最具特色的风景河段，因此，应考虑围绕一定主题整合旅游资源，使之形成一个跨区域的文博游览带；②对于宗教类资源，荔湾的华林寺与越秀区的光孝寺有一定的历史联系，可借助达摩东渡的历史故事将相关的宗教资源整合，形成广州老城区宗教史迹游览带；③沙面欧陆风情区，由于水道的阻隔，其西南两面的珠江水域过宽，跨区的资源整合不太现实，但是与区内西堤、十三行等地区在地域上进行一体化开发考虑是有必要的。

就美食购物类资源而言，荔湾的美食购物具有突出优势，广州许多名店老字号集中在荔湾，西关美食知名度很高，但老的品牌名店存在着如何与时俱进，实现菜式的创新的问题。从旅游业的角度看，整体旅游竞争力的打造需要旅游六大要素的合理配套及价值链的协调运作。例如，上下九步行街如能采用“一轴两翼”的空间重组战略，实现购物与美食优势的叠加，实现旅游要素的协调运作，将使旅游功能大大提高，旅游价值进一步提升。此外，如能利用省里筹建“广东旅游之窗”的机遇，争取该项目落户大坦沙岛，也是凸显荔湾优势资源的绝好机会。

2. 荔湾旅游竞争力评析

旅游业是荔湾经济的支柱产业之一，要提升荔湾的旅游业就必须对其竞争力做出比较全面的评价分析，探索一套较为科学的评价指标体系，进而用于对比研究，明确提升、改造、整治方向。

城区旅游竞争力不属于企业、产业、区域竞争力的任何一种，与城市旅游竞争力虽属同一基本范畴，但仍有所不同。城区旅游毕竟是都市旅游的一个部分，食、住、游、行、购、娱六要素的要求均有差别，城区旅游在许多方面还要依靠所在地的都市旅游，如荔湾旅游业的发展在许多方面依赖于广州市旅游业的发展。

1）旅游竞争力评价框架与方法

评价与测度旅游竞争力的基本目的，是培育和提高旅游竞争能力与竞争优势，并制定出更好的竞争战略。因此，评估旅游的竞争力，除进行目前的业绩评估外，还要对影响旅游竞争力的主要决定因素进行评估，从而测度该城区旅游的可持续发展能力。荔湾城区旅游的特点，主要从三个方面进行测评，即旅游竞争业绩、旅游核心竞争力和旅游竞争环境。

（1）旅游竞争业绩。此部分虽只反映城区旅游业过去与现在竞争行为的结果，但能以此测度城区旅游业未来的竞争趋势。具体要素包括：①旅游业总收入（含外汇收入）。②旅游接待量（含广州地区及区外接待量）。③旅游企业经济效益（含景点及住宿饮食业企业）。

（2）旅游核心竞争力。此部分主要是对城区旅游的核心竞争力的测评，特别是对其后续竞争力的测评，此为发展旅游业的基础，也是旅游竞争力的主要源泉。具体要素包括：①旅游资源条件——资源类型、资源空间分布、资源特色与垄断度；②旅游市场——旅游市场的规模、客源的分布与潜力、旅游形象的知名度与信誉度；③旅游产品开发与管理——产品品位与吸引力、开发资金与技术、产品的管理与利用；④旅游服务要素综合——旅游六要素的配置。

（3）旅游竞争环境。此部分主要反映影响旅游发展的各种环境要素，它对旅游发展起重要的支持作用。如果环境不好，即使有丰富的旅游资源和众多的客源潜力，旅游业也难以持续发展。具体要素包括：①旅游区位——离主要客源地的距离、交通的通达性、区域的进入途径；②城区景观——城区建设风格、城区治安与环保、城区居民的经济与文化；③基础设施——城区基础设施的完善程度；④旅游接待人力资源——接待人员的素质和接待人员的数量。

测评的程序分为四步：第一步确定评价要素及评价指标；第二步经专家商讨，确定评价指标的权重；第三步邀请若干专家为各项指标评定分数，即基本分值；第四步依据下列公式计算出荔湾旅游竞争力的加权总分，以此判断荔湾旅游竞争力在广州市中的水平。荔湾旅游综合竞争力评价表，见表 12，计算公式为

$$M = \sum K_i X_i, \quad i = 1, 2, 3, \cdots, 22$$

其中，M 为加权总分，表示旅游综合竞争力水平；K_i 为第 i 个因子的权重，即各项指标的权重；X_i 为第 i 个因子赋予的平均分数。

表 12　荔湾区旅游综合竞争力评价表

评价系统	评价要素	评价指标	指标权重	基本分值
旅游竞争业绩（3.0）	旅游业总收入	（含外汇收入）	1.0	10
	旅游接待量	（含广州地区及区外接待量）	1.0	10
	旅游企业经济效益	（含景点及住宿饮食业企业）	1.0	10
旅游核心竞争力（4.0）	旅游资源条件（1.2）	资源类型	0.4	10
		资源空间分布	0.4	10
		资源特色与垄断度	0.4	10
	旅游市场（1.2）	旅游市场的规模	0.4	10
		客源的分布与潜力	0.4	10
		旅游形象的知名度与信誉度	0.4	10
	旅游产品开发与管理（0.8）	产品品位与吸引力	0.4	10
		开发资金与技术	0.2	10
		产品的管理与利用	0.2	10
	旅游服务要素综合（0.8）	旅游六要素的配置	0.8	10
旅游竞争环境（3.0）	旅游区位（0.8）	离主要客源地的距离	0.3	10
		交通的通达性	0.3	10
		区域的进入途径	0.2	10

续表

评价系统	评价要素	评价指标	指标权重	基本分值
旅游竞争环境（3.0）	城区景观（1.2）	城区建设风格	0.4	10
		城区治安与环保	0.4	10
		城区居民的经济与文化	0.4	10
	基础设施（0.5）	城区基础设施的完善程度	0.5	10
	旅游接待人力资源（0.5）	接待人员的素质	0.3	10
		接待人员的数量	0.2	10

总分：

2）荔湾旅游竞争力评价结果

依据表 12，我们邀请了广州市内各高校、科研院所、旅游局和旅游企业的专家、学者和企事业负责人等 21 人，针对荔湾区旅游业竞争业绩、旅游核心竞争力、旅游竞争环境三个层面的因素进行测评，每项指标基本分值为 10 分，旅游综合竞争力满分为 100 分，分数越高表明相应竞争力水平越高。荔湾区旅游综合竞争力评价结果见表 13。

表 13 荔湾区旅游综合竞争力评价结果

评价系统	评价要素	评价指标	指标权重	平均分数
旅游竞争业绩（3.0）	旅游业总收入	（含外汇收入）	1.0	5.765
	旅游接待量	（含广州地区及区外接待量）	1.0	6.735
	旅游企业经济效益	（含景点及住宿饮食业企业）	1.0	6.294
旅游核心竞争力（4.0）	旅游资源条件（1.2）	资源类型	0.4	6.200
		资源空间分布	0.4	6.313
		资源特色与垄断度	0.4	7.240
	旅游市场（1.2）	旅游市场的规模	0.4	6.613
		客源的分布与潜力	0.4	7.350
		旅游形象的知名度与信誉度	0.4	7.438
	旅游产品开发与管理（0.8）	产品品位与吸引力	0.4	6.400
		开发资金与技术	0.2	5.800
		产品的管理与利用	0.2	6.225
	旅游服务要素综合（0.8）	旅游六要素的配置	0.8	6.700
旅游竞争环境（3.0）	旅游区位（0.8）	离主要客源地的距离	0.3	8.025
		交通的通达性	0.3	7.708
		区域的进入途径	0.2	7.625
	城区景观（1.2）	城区建设风格	0.4	6.963
		城区治安与环保	0.4	7.213
		城区居民的经济与文化	0.4	6.753
	基础设施（0.5）	城区基础设施的完善程度	0.5	7.065
	旅游接待人力资源（0.5）	接待人员的素质	0.3	6.767
		接待人员的数量	0.2	6.400

加权总分：67.040

从表 13 的计算结果来看，荔湾区旅游综合竞争力分值为 67.040 分，这说明荔湾区旅游业目前在全广州市所属各区市中位于中等水平，旅游综合竞争力水平还有待进一步提高。

同时，根据高等院校、科研院所、旅游局、旅游企业四组人士的评价结果，分别计算其加权平均分，如图 3 所示。

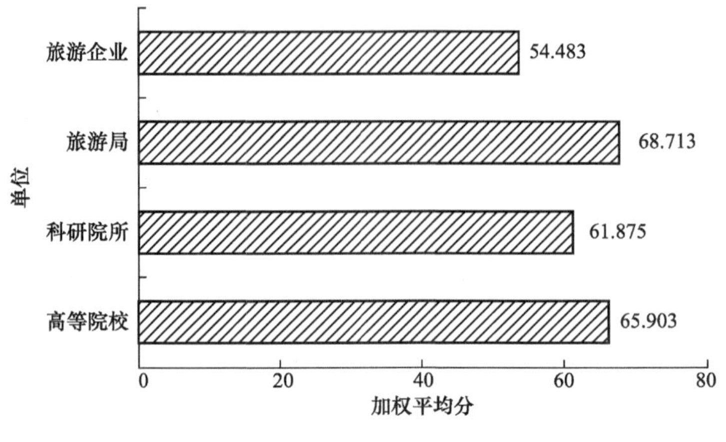

图 3　各类人员评价结果

统计说明，各级旅游局对荔湾区旅游综合竞争力评价反应较好，给出的分值为 68.713 分，而来自旅游企业的评价较差，只给出了 54.483 分。

若从三个不同层面的评价结果看，旅游竞争环境得分相对较高，旅游竞争业绩得分最低。从分项指标评分结果看，旅游区位要素和旅游市场要素得分相对较高；而旅游产品开发与管理要素得分最低，说明要提升荔湾旅游竞争力应加强对旅游产品的开发与管理。

从总体来看，荔湾旅游竞争力评价的各项指标分值均表现为中等水平，没有特别突出的强弱项。旅游产品开发与管理要素这一相对薄弱环节，与旅游资源的丰度不高和旅游资源点的规模不大有直接关联，也与西关风情的开发利用缺乏有效载体和无形资源转化为旅游产品难度较大有关。除了旅游区位要素的三项评价指标得分最高外，作为分项指标，旅游形象的知名度与信誉度得分较高，这给荔湾大力发展旅游业、实施“旅游带动”战略提供了很好的发展基础。

通过对荔湾区旅游竞争力进行测评，我们对荔湾区旅游业的发展与竞争战略提出如下建议。

（1）重组空间格局，重构价值链。21 世纪的旅游竞争不再是单个企业之间的竞争，而是一条价值链与另一条或几条价值链的竞争。空间格局的重组，能使商旅价值链得到优化，实现商旅价值链整体竞争能力的提升和加强。价值链的重构，比零散旅游景点的改善，更能从根本上改变价值链的效率，同时，还有利于改变原有空间活动的内容，实现功能与优势的叠加，加强基础，消除或减弱对不利因素的敏感性，提高价值链的整体效益。据此，我们建议，荔湾应以一级资源点为核心，实施“以点联网，以网成片”的发展战略，以陈家祠、上下九步行街、沙面岛为核心，整合周边各类旅游资源，形成“网”“片”的优势，争取在城市区域旅游竞争中重新洗牌。

（2）整合资源，拓展旅游功能。在空间格局重组和价值链重构的基础上，依据荔湾文博类、购物美食类和人文风情类资源丰富、特色突出的特点，从以下方面考虑资源的整合和旅游功能的拓展：①历史古迹点与购物功能的整合，如陈家祠；②购物商街与美食功能、西关风情的整合，如上下九步行街；③人文风情与观光休闲功能的整合，如沙面岛。

（3）从单一资源品牌向荔湾城区品牌发展。在价值链与价值链竞争的市场上，单一资源品牌在竞争中势单力薄，不利于凸显区域旅游优势，结合城区竞争优势，能借助商旅价值链使整体竞争能力提升。

品牌是一种无形资产，具有市场开拓力、形象扩张力、资本内蓄力，品牌作为市场竞争的武器，常常会带来意想不到的效果。以往研究旅游问题时，通常只注重旅游形象及其营销推广，而忽略了品牌的文化、质量、管理与服务等其他要素，更谈不上将品牌提升到价值链的高度，分析品牌与竞争力的关系。因此，各地花费大量的人财物塑造旅游形象，而持久的旅游竞争力始终未能建立，以品牌为核心的价值链始终未能形成，品牌的价值始终未能彰显。

利用品牌的聚合效应、磁场效应、带动效应能使其他相关产品顺利进入市场，并为消费者所接受。以上下九步行街为例，以历史文化底蕴作为构建品牌的基础，能凸显步行商街的个性，形成难以替代的旅游竞争力。同时，以上下九为核心，“以点联网”的周边区域，也会共享旅游品牌的三大效应。

荔湾的旅游发展优势，在于底蕴深厚的历史文化内涵和传统的人文风情。从荔湾三个一级资源点的特点和文化内涵来看，陈家祠的岭南建筑文化、中国的宗祠文化具有不可模仿、难以替代的特性，因而具有超常的生命周期，沙面的西洋建筑和欧陆风情也具有以上特点。但购物的特性及传统商业中心的特点会随着人们消费习惯的改变、高消费群体的转移而兴衰，因此，上下九步行街在打造商旅品牌时，应实施商旅互动，凸显西关风情特色和骑楼建筑特点，体现人本主义，增强休闲功能，招引名店汇聚，限制连锁超市进入和高层楼宇的建设，避免商街风格和业态的冲突，保持持久竞争优势。

3. 陈家祠文化休闲区的价值提升

区域内的核心资源对提高区域竞争力有着重要作用。为了进一步探讨提升荔湾旅游竞争力的方式，在对荔湾旅游资源进行全面分析的基础上，通过对荔湾区的核心旅游资源进行较为全面的价值分析，研究其价值与功能，探求整治、利用与提升的途径。

1）陈家祠的旅游功能分析

陈家祠又名陈氏书院，建于清光绪年间（1890～1894年）。整座建筑坐北朝南，由三进五间九堂六院大小十九座建筑组成。陈家祠以其巧夺天工的装饰艺术著称，荟萃了岭南民间建筑装饰艺术之大成，以其“三雕、三塑、一铸铁”著称，号称“百粤冠祠”。

陈家祠是广州典型的民间宗祠建筑，既体现了我国古代建筑的传统风格，又具有我国南方建筑的鲜明特色。其布局严谨对称、空间宽敞、主次分明。在建筑的处理上，以中轴为主线，两边以低矮偏间、廊庑围合，衬托出主殿堂的雄伟气概，形成纵横规整而又突出主体的格局。建筑外围有青砖围墙，形成外封闭内开放的建筑群体。

陈家祠以其精湛的装饰工艺著称于世，在它的建筑中广泛采用木雕、石雕、砖雕、陶塑、灰塑、壁画和铜铁铸等不同风格的工艺进行装饰。雕刻技法既简练粗放，又精雕细琢、相互映托，使书院在庄重淡雅中透出富丽堂皇。

陈家祠的木雕，数量最多，规模亦大，内容丰富。砖雕艺术十分高超，大门左右两侧的代表作运用了砖雕中的圆雕、浮雕、镂空、挂线砖雕等方法，再现了许多历史故事。屋脊装饰整体采用“三塑”工艺，上层是陶塑，基座是灰塑。在广东建筑工艺中，陈家祠的花脊是规模最大、最精美的一部分，陈家祠共有 11 条这样的陶塑花脊，有 1800 米的灰塑装饰，堪称广州之最。

祠堂落成后，一直作为陈姓子弟读书办学的地方，光绪三十一年（1905 年）废科举后，书院改为陈氏实业学堂。新中国成立后，其 1959 年被辟为广东民间工艺馆，1960 年经省人民政府批准被列为广东省文物保护单位，1988 年由国务院颁布为全国重点文物保护单位。

2）陈家祠旅游利用价值的局限性

作为“百粤冠祠”，陈家祠建筑艺术之精美众人皆知。但因其建筑规模与我国著名的古建筑群相比差之甚远，直至 1988 年才被评为全国重点文物保护单位。而获此殊荣后，又因为其周边环境杂乱，宣传力度不够等原因，这一岭南建筑的瑰宝一直处于“藏在深巷人未识”的境地。直至 2002 年被评为“新世纪羊城八景”，才引起游客的极大兴趣，继而出现黄金周爆满的情形。但是，目前的陈家祠仍有许多不尽如人意之处，如若不引起重视并加以改善，势必影响其品牌作用的发挥和可持续发展。

A. 规模是制约陈家祠进入高级别旅游景区（点）的限制性因素之一

在旅游景区（点）评级制度中，规模是重要指标之一，而规模的大小与景区（点）的游客容量密切相关。同时，游客容量又对景区（点）的游客总量和经营绩效产生直接影响，如图 4 所示。

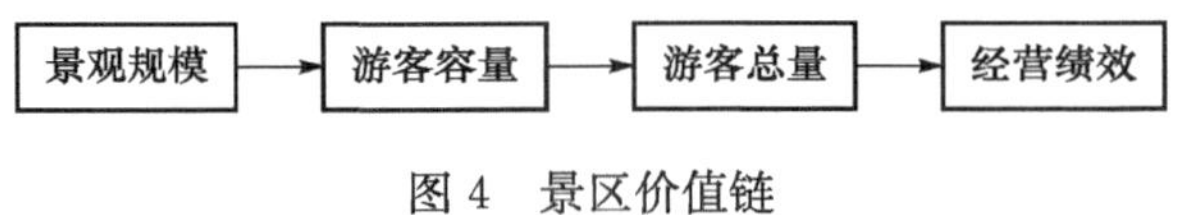

图 4　景区价值链

从图 4 可以看出，景观规模、游客容量、游客总量和经营绩效形成了紧密相关的价值链，而在这一价值链中，景区规模是关键的一环。

对于陈家祠的这一不足，有关部门曾采取一定的措施。例如，利用修建地铁站的机会，配套建设了陈家祠绿化广场。但遗憾的是，广场未能与陈家祠的文化内涵很好地协调，无法起到辅助陈家祠旅游空间拓展、游赏规模扩大、停车场地扩展的效果。因此，陈家祠景区规模过小的问题，仍然是值得关注并应予以改善的问题。

我国旅游景区（点）评 A 标准是围绕如下价值链制定的，如图 5 所示。

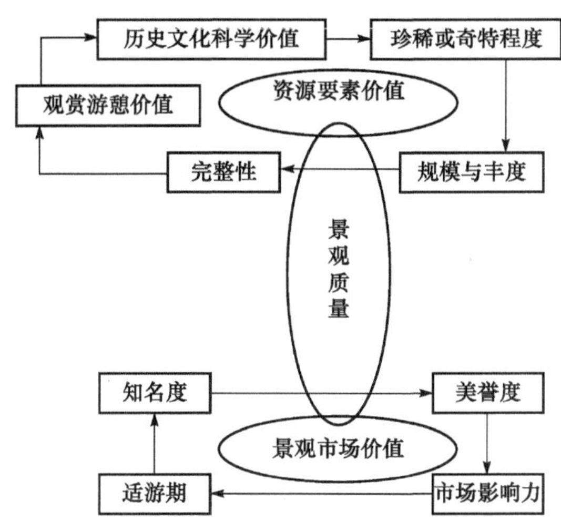

图 5　我国旅游景区（点）评 A 标准的依据因素

下面我们借助我国旅游景区（点）评 A 标准，尝试对陈家祠进行总体评价。我国旅游景区（点）评 A 的景观质量评分标准：AAAA 级旅游景区（点）需达到 85 分，AAA 级旅游景区（点）需达到 75 分，AA 级旅游景区（点）需达到 60 分，A 级旅游景区（点）需达到 50 分。可以看到，陈家祠的得分总和已初步达到了我国 AAAA 级旅游景区（点）景观质量评分的标准（表 14）。但在资源要素价值评价中，规模与丰度指标是整个价值链中得分较低的一环；在景观市场价值链中，市场影响力又是相对弱项。

表 14　陈家祠景观质量评分表

评价项目	评价因子	等级赋值				本项得分
		1	2	3	4	
资源要素价值（65）	观赏游憩价值（25）	20～25	13～19	6～12	0～5	24
	历史文化科学价值（15）	13～15	9～12	4～8	0～3	14
	珍稀或奇特程度（10）	8～10	5～7	4～6	0～3	9
	规模与丰度（10）	8～10	5～7	4～6	0～3	6
	完整性（5）	4～5	3	2	0～1	5
景观市场价值（35）	知名度（10）	10～8	5～7	4～6	0～3	6
	美誉度（10）	8～10	5～7	4～6	0～3	9
	市场影响力（10）	8～10	5～7	4～6	0～3	7
	适游期（5）	4～5	3	2	0～1	5
总分	—	—	—	—	—	85

注：此表为我国旅游景区（点）评 A 标准系列表格之一，引用时略作简化处理

从上述评价可见，陈家祠已初步具备参评 AAAA 级旅游景区（点）的基本条件，因此应围绕创建 AAAA 级旅游景区（点）的目标，切实推进陈家祠旅游区的建设，实现荔湾旅游景区（点）评 A 工作零的突破。

B. 周边环境协调性较差是陈家祠旅游发展限制性因素之二

广州市的旅游景区（点）其周边建筑物与景观格调基本协调，但也有部分景点（11家）的周边环境较差（图6），主要表现为景观混杂在周边的建筑群中、出入口绿地面积小、缺乏停车场所、周边的环境与景观的内涵不相协调等。

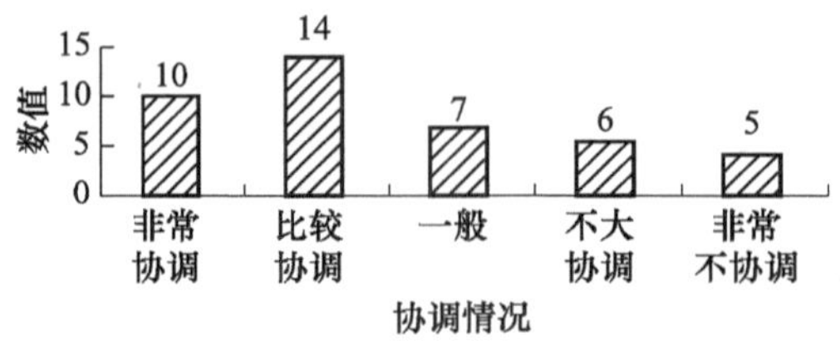

图6　广州旅游景点与周围环境的协调情况

资料来源：暨南大学旅游研究所《2003年广州旅游景区点调查》

陈家祠是广州旅游景点与周边环境不相协调的景点之一。陈家祠正面被32中学挡住，其西北面残旧民居影响观瞻，而东面新建的绿化广场虽对周边环境有所改善，但广场的文化内涵与陈家祠本身协调不够，留下了许多遗憾。

C. 旅游六大要素存在弱环是陈家祠旅游发展限制性因素之三

吃、住、行、娱、游、购是旅游业发展的六大要素，由此构成了一条旅游产业链和互为依存的系统。系统内部的价值活动是构筑竞争优势的基石，最优化和协调一致的运作能带来竞争优势。一般而言，由六大要素形成的产业链不宜出现断环和弱环，否则会影响整条产业链的运作。当然，对于具体的景点，我们虽不强调“五脏俱全”，但旅游六大要素的明显断环和弱环会构成突出的限制性因素，最终影响经营绩效。

就陈家祠而言，可以应用价值链分析的方法，将一些能够利用城市共享体系的非优势环节分解出去。例如，除吃、住等环节外，陈家祠自身的旅游购物环节也明显偏弱，行的环节由于停车场太小也会造成制约，参与性的娱乐环节也因场地限制而无法安排，从而使其市场影响力和经营绩效大受影响。

旅游购物与旅游收入之间的弹性远大于吃、住、行、游等方面与旅游收入之间的弹性。对景区（点）而言，旅游购物是改变“门票经济”的最佳途径。

3）提升陈家祠旅游价值的思考

在提升价值时，要充分运用价值工程方法。在设计和评审升级、改造方案时，要全面考虑提升价值与增加综合效益的相互关系。

（1）将争创“AAAA级”旅游区作为近期目标，并采用如下举措：①通过整体规划，在考虑较大手笔地改善陈家祠周边环境的前提下，创建环境幽雅的文化休闲游憩区；②力争32中学搬迁，调整陈家祠的正门并改善停车场；③争取拆迁陈家祠后面、西侧杂乱民居，并用于强化陈家祠的民间艺术博览功能。

（2）文化内涵的挖掘与体现使顾客价值得以提高。陈家祠的艺术魅力、文化品位具有很好的口碑。除陈家祠作为岭南建筑的精品大受赞赏外，其内部展览的广东民间艺术精品、广州百年风情陶塑展等也让游客大饱眼福。但目前的游览内容局限于静态的观赏，缺乏动态的参与性项目将其文化内涵衍射、提升。虽然场地较小，一些民间工艺的现场制作、游客参与动手的旅游开发手法受到制约，但可以利用其外围的绿化广场，在黄金周或传统节庆时，将一些经过精心编排的广东民间艺术表演放在绿化广场中进行。这样，既能使游览内容动静结合，又能使陈家祠的开发利用做精、做深。从价值工程的角度看，这些举措能提高旅游产品的功能，使旅游产品的价值得以提升。

(3) 通过对广东民间艺术工艺品的挖掘，强化旅游购物功能。陈家祠发展旅游购物有得天独厚的优势。首先，作为广东民间艺术博物馆的陈家祠，具有挖掘民间工艺精华，制作独特、精美、文化品位高的旅游纪念品的先决条件；其次，许多民间工艺的制作过程便是很好的旅游观赏项目，若采用先参观加工过程后购物的模式，会大大提升游客的购物兴趣。考虑到陈家祠西边杂乱的民居必须拆除，可借此配套旅游购物商街，或者建设以陈家祠为核心，整体协调的工艺博览购物区。

此外，还可将荔湾众多的民间艺术、民间工艺纳入陈家祠工艺博览购物区统一开发，即以陈家祠为核心，将旅游购物的资源优势在此凸显。

4. 沙面欧陆风情区的功能提升

1) 沙面的文化特征与旅游价值

沙面是广州北城区西南珠江边上的一个小岛，南临白鹅潭，西面隔河湾与黄沙相望，北、东两面有濠涌与市区隔开。全岛呈椭圆形，总面积 0.3 平方千米。沙面较完整地保留了 19 世纪英法租界欧陆风情风貌。沙面的建筑集中了欧洲各国的建筑风格：新古典主义、浪漫主义、折中主义和现代主义，因此，这里被称为欧陆建筑的大观园。其清代西式建筑群在 1996 年底被国务院定为全国重点文物保护单位，沙面岛上 169 幢建筑中，被列为文物建筑的有 53 幢。此外还有 100 多株古树名木被市政府挂牌保护。全岛绿树婆娑，异国情调建筑掩映其间。

100 多年前，沙面是与沙基相连的沙洲，叫拾翠洲。在广州十三行“一口通商”时期是行商人的仓库区。1858 年《天津条约》签订后，侵略者要恢复在广州被烧毁的商馆，英国人柏克选沙面为址。1861 年，英法殖民主义者“租借”了沙面。沙面当局挖了一条宽 40 米，长 1200 多米的小涌（即现沙基涌），用花岗石在沙面周围筑起高出水面 1.5～1.8 米的堤围，然后用河沙对泥土进行了平垫，使沙面成为一个小岛。同时将沙面西边 4/5 的地划为英租界，东面 1/5 的地划为法租界，不许中国船只靠近沙面的河涌停泊。1865 年，英领事馆迁入沙面，并出卖了部分租界地给外国人。法租界也于 1889 年拍卖了部分土地。

沙面从 1859 年成为英法租界，至 1945 年回归祖国，其间主要用作领事馆区、外商公司及侨民商住用地、来穗外国人文体和宗教活动场地等。共有 19 个国家相继在沙面设领事馆，在沙面租界里先后设有 9 家外国银行，40 多家洋行和企业的公司、分行、支行、办事处及代理处。建筑形式有新巴洛克式、新古典式及所谓的殖民式，沙面的建筑及其布局体现了欧洲乡土田园风貌，其建筑体现出 19 世纪末 20 世纪初欧美流行的各种艺术风格。

现存沙面的建筑都是 19 世纪以后建设的，有领事馆、教堂、学校、银行、洋行、俱乐部、旅馆、小住宅等。1959 年周恩来总理视察沙面时，指示要保护沙面的原貌和环境。国家曾拨款 1000 万元用于测绘和规划保护，沙面现已确定为历史文化保护区。

2）沙面旅游利用的优劣势分析

位于广州市白鹅潭畔的沙面岛，古树和西洋建筑群形成了一个静谧、优雅、协调的整体。沿街众多咖啡馆、酒吧向游人提供纯正的法国或意大利咖啡、英国的红茶等，充满欧陆风情的特色，没有矫饰，清纯自然，与广州古城的遗迹、羊城古老的传说形成了鲜明的对比。沙面的建筑、古树、流动的“风情画”让游人耳目一新。沙面的魅力所在是滨江夜色、欧陆风情、典雅环境与开放气息。它是广州珠江风情游览带的一个重要的组成部分，是广州近代史中一个保留完好的历史文化区域。相对封闭的空间，保存了其完整性，避免了经济建设的干扰，使其旅游利用价值大增。

但众多的西洋建筑为中央驻穗机构所占，为旅游的有效利用设置了障碍；岛内较多的常住居民使可为旅游利用的空间与建筑面积大为减少；岛外周边用地紧张，使得为进岛游览的车辆配备的大型停车场选址无法落实；白天鹅宾馆临江的引桥破坏了沙面优美滨江景观的完整性，等等，均增加了沙面旅游利用的难度。

3）沙面旅游景点的价值提升

沙面旅游价值的提升，可通过功能成本分析，运用价值工程方法进行优化、评价与决策。

（1）采取有效措施保护好沙面历史文化区及其建筑群。由于历史的原因，沙面部分建筑的使用功能已经被改变，有些则经过多次的改建，周边还建起了许多新的建筑，新的办公楼和住宅楼穿插其中，受保护的文物建筑群和非文物保护建筑混杂在一起，对建设沙面欧陆风情区造成一定影响。

沙面的西洋建筑群、街巷、环境等是欧陆风情的载体，也是一个和谐的整体。因此，要建设沙面欧陆风情区就必须将这些建筑群进行分类并区别对待：对文物建筑应保证其外观、结构、内部特有装饰不改变；对景观风貌建筑要确保其外观能继续构成沙面建筑群的风貌；对不协调的建筑和违规建筑应在适当的时候予以整治或拆除。这样才能确保欧陆风情的特色，并让游人通过这些精心保护的建筑群，体验沙面的魅力。

（2）按高档旅游区要求，全面整治岛内环境，确保沙面的优雅、静谧。沙面岛占据了珠江优美的河段，滨江、畔潭、珠水环绕，构筑了优雅的封闭空间，西洋建筑群与参天古树相映，一直是广州的高雅商住区。但因为历史的原因，岛内有较多的常住居民，旅游可利用空间受限，所以应合理地迁移部分常住居民，并在全面整治岛内环境的基础上，建设高档旅游区。

（3）鼓励发展共享场所，限制私人独占空间的利用。保护沙面的目的，是能让更多的人能享受这一优雅环境。因此，提升其旅游利用价值最好的方式是，将整饰过的景观建筑改为大多数人能够享受的公共空间，如建立博物馆，开设酒吧、咖啡厅、文化沙龙等。这样既能使欧陆文化得以传承、展示，又能让更多的人有机会进入这些建筑，从外到里近距离地欣赏。加上岛内部分外国领事馆及在广州工作的外国人喜欢到此区活动和消费，随处可听到的外国语言和见到的外国文化习俗，为岛内的欧陆风情加上了浓浓的色彩，形成了游动的欧陆风情画，使之拥有别处无法模仿的独特魅力。

（4）进一步推进光亮工程的实施，配合珠江夜游凸显欧陆特色。珠江夜游有效地整合了广州的资源，形成了广州著名的旅游品牌。但目前的开发利用仍有许多不足之处。例如，两岸的建筑物缺乏韵律，灯光的变化缺乏动态效果，特别是一些具有特殊意义的地段文化内涵未能展现等。因此，珠江夜游有必要做深、做精。沙面是整个珠水链中最亮丽的一段，除光亮工程要继续推进之外，更重要的是通过光亮工程凸显沙面文化内涵，凸显沙面欧陆文化特色。

(5) 考虑与西堤、十三行的有效沟通，形成以广州近代史为主线的洋务景观区。西堤是广州近代代表性建筑高度集中的区域之一，如粤海关大楼、邮电大楼等，现有些已改为博物馆供人参观，十三行的原样重建方案也已提出，若能将沙面、西堤、十三行的旅游开发围绕一定主题，整合成片或带状开发，将会使其旅游开发效果凸显。

5. 十三行商埠文化区建设

1) 十三行的历史及其影响

十三行在历史上曾一度是中国唯一、世界第四大的对外通商口岸，在近代中西方贸易交流史上扮演过极其重要的角色。清康熙二十四年 (1685 年)，清政府为发展对外贸易，便“开南洋之禁”“置粤海税务使”。官方指定商人为对外贸易经纪，凡外商购买或销售茶叶、烟、丝等国货或进入内地，都必须经过“官商”。

广州十三行“一口通商”是在清乾隆二十二年 (1757 年) 正式开通的，此后洋务制度发展起来。同时，广州城外西南处，也建起了丹麦、西班牙、法国、章官、美国、宝顺、帝国、瑞典、旧英国、炒炒、新英国、荷兰、小溪 13 间“夷馆”，供外国人居住。今西堤和文化公园一带，便成为外货登陆之地，也是洋行集中地，自此人们称之为十三行。

随着中外商贸交往的繁盛，中外文化交流也日渐增多，来华的西洋人中，既有传教士，也有手工艺人、画匠或精通医学、天文、数学的专门人才。

在英国侵华的鸦片战争期间，十三行被愤怒的群众烧毁。英国强迫清政府签订《南京条约》后，十三行洋行制度便宣告废除。清咸丰九年 (1859 年)，清政府批准英法租界设在沙面，十三行在外贸方面的地位逐步被沙面所取代。

当时十三行的覆盖范围大约是今天的东至仁济路，西至杉木栏路，南至珠江岸边，北至十三行路的一片地区。现在的十三行就只剩下一个“十三行路”地名，以及由当时的洋行名改的街道名，如怡和大街、宝顺大街、普源街、同文路等，人们只能从这些名称中寻觅到一段历史痕迹。

2) 十三行遗址建设设想

十三行遗址见证了广州在历史上中国对外贸易执牛耳达数百年之久的历史，又是海上“丝绸之路”的最早起点，清乾隆二十二年 (1757 年)，清政府关闭江、浙、闽三海关，广州便成为中国唯一对外贸易大港，直至鸦片战争前夕。这说明十三行是难得的历史文物旅游景点，开发十三行遗址具有重大的历史商贸文化价值，并为目前广州的经济发展起衬托作用，开发十三行遗址是毋庸置疑的。

由于十三行古迹已荡然无存，如何开发，虽经专家论证，也向社会公开征集开发方案，但是意见分歧较大，主要构想有如下几种。

(1) 原样重建。这是最大胆的设想。其主要根据是，清代十三行面临珠江，建筑华丽，气势雄伟，长时间为我国对外贸易、发展国际商务的唯一口岸，具有重要的历史文物意义，而且该区内还有多家具有浓厚岭南建筑风格的洋行，构成了独立的风景。要打造“十三行商埠文化区”只有下大决心，花大力气，重现十三行原貌，其才能真正成为荔湾旅游亮点。

(2) 文化遗产展馆。这一构想认为，十三行确是一项具有重要历史意义的文化遗产，但鉴于十三行古迹现已荡然无存，要全面重现历史十三行，不论从资金投入还是从社区建筑、民房拆建等方面看，都是不现实的，难度也很大，如再从经济投入与产出以及人工重现的旅游吸引力看，也不宜搞简单的复制重建。因此，建议筹建十三行展馆，地点设在文化公园内，可以利用现有场地和环境，待条件成熟时，扩大范围，建设新馆。这一建议，已被广州市文化局列入广州市文化事业发展“十五”规划，为建设十三行历史陈列馆，许多当年的图片、实物等已得到多方支持，不断有新的充实。

(3) 艺术演示。此类设想主张在不变化现有建筑群的基础上，对现今建筑进行改造和修复，在街区道路上用城市雕塑反映当时的商贸风情，或者在十三行路出入口处建设牌坊、牌楼等标志性建筑，也可以采用修建广场的方式在适当地址建设具有十三行标志的雕塑、石刻文字等，介绍曾经辉煌的港口历史。总之，此类设想是通过部分环境的整饰、城市雕塑的建立，重现当年十三行的商贸风情，使游人在艺术演示中去想象和感受当年的历史气息。

还有许多其他建议，但较为集中的是以上三种。我们认为，从可行性和“只争朝夕”的时间看，艺术演示是“最优方案”，圆明园的开发、保护思路可供我们借鉴，全国许多人造景点失败的教训也值得我们吸取。当然，如条件许可，城市雕塑、开设展馆和建标志性建筑物等可同时或分步进行。过于简陋、只设立几座雕塑、用意模糊、意向不清，不如再候时机。须知“夹生饭再煮”是很困难的。但目前必须马上着手实施的是：合理划定十三行遗址保护区范围、控制周边建筑物高度、防止再次出现类似新中国大厦的高层建筑在十三行遗址上崛起的事件发生，为今后实施选定方案留有余地。

无论采用哪一种方案，都要运用价值工程方法进行价值分析，谋求以“最低”（相对）的寿命周期成本，实现使用者需要的必要功能。

四、挖掘文化内涵，增强发展后劲

1. 文化与经济发展的相互关系

近年来，有关文化与经济发展的关系的文章不少，议论颇多，已基本上形成一个共识。从本源上看，经济是文化的基础；从发展上看，文化是经济发展的根本动力与源泉；从整体上看，决定一个国家或地区经济的发展和水平的是这个国家和地区的文化内容。

长期以来，人们对文化与经济建设的关系虽有一定的认识，但往往停留在表面层次上，对文化的内涵理解不深。例如，人们往往把文化只看成一种事业，是文化事业、公益事业，可以投入但不必计较成本，更不会把文化建设与经济建设挂钩，似乎经济建设只有技术竞争与规模竞争，只要加大技术含量和扩大规模就能在竞争中立于不败之地。殊不知在当代的生产力发展水平上，对于技术与规模在企业所占的比重，其可挖掘的潜力是有一定限度的，但文化则不同，由于人们对文化消费需求越来越大，文化资源可以形成巨大商机，文化的介入，会提升企业和产品的档次和文化品位。在文化与旅游的经济关系上，一般都能体会到文化可以提升旅游的品位，但也多从增加旅游品位的附属品角度去认识，似乎文化仅是旅游吸引物的添加剂而已。其实，文化在当代已不仅是一种精神需求和精神生活，它已成为重要的生产力，特别是在激烈的社会经济竞争中，文化具有不可忽视的战略意义，任何战略都必须立足于文化战略，当经济与文化对接，就能以文化产业作支点发力，其爆发出来的能量会是惊人的。商贸经济如此，旅游经济也如此，商贸的持续发展的后劲看文化的底蕴，特别是当经济发展到一定阶段，决定发展后劲的就是文化。旅游归根到底都是为了满足人们一种精神上和心理上的享受，文化是旅游经济的灵魂。

2. 文化的传承对荔湾有着更为特殊的意义

岭南文化是一个由多个部分组成的文化圈，是一个多样化的整体。广州的广府文化是岭南文化的组成部分，广州是一个具有 2200 多年历史的文化名城，文化资源厚实，现代的文化特色，人们将其概括为“四地”文化——岭南文化中心地、海上丝绸之路发祥地、近现代中国革命策源地、改革开放前沿地。这是一笔丰富的文化资源财富。

荔湾，可以说是广州四大老城区中最具历史底蕴的一个区，也是最能代表广州商贸发展史的区域。广州市民或广州周边城镇居民在谈论到荔湾——西关时，往往自觉或不自觉地联想到西关文化、西关风情，外地人被这种文化氛围所吸引，西关人有自己的生活方式和生活习惯，对西关有着浓厚的感情。

西关指广州古城西门关外地区。清乾隆二十二年（1757 年），清政府实行“海禁”政策，只批准广州作为唯一对外贸易口岸，西关也就变成清朝对外开放的“特区”。

西关文化具有深厚的历史渊源，在中西文化交融中，具有兼收、并蓄的特点；在经济贸易形成上，具有开放、重商的特质；在民俗文化上，具有正宗广府文化的特点。因此，西关文化在岭南诸文化中独领风骚，内容纷繁而典雅，精致而博约。

西关文化具有多元化的丰厚底蕴，表现形式丰富多彩，反映在语言、戏曲、建筑、饮食、民俗风情、民间艺术等多个方面，它代表一个城区的形象，也反映出一个城区的风格。西关文化是荔湾区的一张名片，特别是西关文化建设有雄厚的经济实力作后盾，随着政府投入的不断增加和多元资金的引入，文化基础设施必将得到快速发展。

当然，我们在提倡城市文化建设的同时，一方面既要看到岭南文化（包含西关文化）的优秀品质，另一方面也要看到由于文化精神的传统与惯性，其内在的不足也是显而易见的。因此，在强化城区文化建设时，必须注意与时代精神的结合，实现文化的现代提升。

3. 加大文化内涵，强化荔湾商旅核心竞争力

美国学者在研究了近年市场经济的变化发展状况后，提出了“文化经济学”的概念，其基本观点认为，一个国家或地区的经济情况应同其所在国民特性联结起来，全球各地所有市场都与它们在历史、社会结构、心理、宗教和政治状况方面的独特性相适应。历史学家戴维·兰德斯在《国富国穷》一书中断言：“如果说经济发展给了我们

什么启示，那就是文化乃举足轻重之因素。”

上下九商圈的核心竞争力靠什么？上下九、北京路、天河城广州三大商圈日均人流量为30万～35万人，地理位置在广州东、西、中部，各自拥有基本消费群，也各自拥有龙头企业。北京路商圈的广州百货大厦的营业额约占北京路总营业额的1/7，近期又在建设另一特大型综合商场。上下九商圈的两家龙头企业——妇儿公司和名汇新大新合计营业额与广州百货大厦相差很远，更不用说与天河城的特大企业相比较了。

上下九商圈的核心竞争力是什么？核心竞争力是指企业或企业群的特殊能力，它具有价值的优越性、异质性、难模仿性、不可替代性以及心理感受的差异性。核心竞争力的形成是一个过程，能把众多企业的多种创新构成一个新的有机整体。上下九商圈的核心竞争力就是西关文化，这是西关企业持续竞争的源泉和基础，也是西关企业文化承载和创新的切入点。

同样，荔湾旅游业的发展，应以资源创新为主线，以西关文化为底色，把文化资源作为重要的旅游资源来开发。从十三行古商业街到上下九现代商业街，从古海上丝绸之路到沙面现代欧陆风情，都是西关内外贸易发展的历史见证。西关的风土人情、古旧建筑、西学东渐，是广州的岭南人文风采的缩影，这也是荔湾旅游经济的核心竞争力的载体。

4. 西关骑楼——使人留恋的历史情结

1）骑楼建筑的历史沿革

骑楼是岭南特殊的建筑形式，在广东、广西及湘西许多城市都有。据历史记载，骑楼建筑结构原型起源于1800年前南方百越人的“干栏”房屋。干栏也称干阑或麻栏，魏晋南北朝时，中国南方的僚人（百越民族）已有这样的房屋，其结构分为两层，一般用木料、竹料制作桩柱、楼板和上层的墙壁，下层四面皆空，无遮拦。上层住人，下层圈养牲畜和放置农具等物。近现代，在壮族、傣族、布依族等百越后代聚居的村落中，都可以看到这种形式的建筑物。

任何建筑形式，都与一定的自然条件和人文条件有关。骑楼的结构，是因为南方河流纵横，多雨潮湿，夏季、秋季酷热，地面不利于居住之故，住上层有利于健康并防止猛兽毒蛇袭击。这类房子在湘西一带叫吊脚楼，两广叫骑楼，因为在城市之中，它们都“骑”在人行道上，故称骑楼。

东南亚一些国家也有骑楼，如在越南河内等城市就有骑楼建筑，因他们祖先也是越人，早期也住在“干栏”。至于新加坡、马来西亚等国家城市的骑楼，那是我国华侨照搬过去的建筑形式。

广州现存的商业骑楼建筑，多是在20世纪初广州开辟马路时，在原传统形式基础上，吸取了西方古典建筑中的券廊等形式，如在上层建筑柱顶，加浮雕、圆拱等装饰，故现存商业骑楼实为中外文化交融的结果。

2）骑楼的文化内涵及对生活的影响

骑楼文化在西关几代人的心目中已烙下深刻的印象，并且已经融入市民的生活中，在历史的长河中，西关与其他城区一样，经历了翻天覆地的变化，但人们在那长长的骑楼走廊下仍能找到自己亲切而熟悉的元素。

骑楼文化在人们的生活上、经商贸易上都有着独特的情趣。骑楼既是私人空间的外延部分，又是公共空间的内收敛部分。居民走出家门进到骑楼廊下，就意味着离开私人居室而进入公共空间；经商者的居室就是商品仓库，从仓库里搬出商品，就像从家里拿出来一份礼物；购物者走进骑楼，就有一种走进别人家的感觉，由购物者变成客人。这些也可以说是骑楼商店与现代超市的区别。如果说，骑楼商业能够吸引购买者，其中这种现代社会中家庭生活的传统因素起着作用。

广东省作家协会的张柠同志在2003年2月2日的《羊城晚报》上发表了一篇关于骑楼文化的感怀，对其文化影响有一段生动的描述："广州上下九的骑楼，现在要让你住肯定不舒服，但当某个灯光昏暗的晚上，你在骑楼下走，想象着过去的老广州生活，你会拉近和广州这个城市的距离，你会觉得你曾经参与了广州的历史。如果那个地方已经是中信大厦那样的现代化高楼，钢铁闪闪发光，肯定只会令你敬畏，你会觉得跟你没有关系。"

3）骑楼文化的传承与运用

骑楼文化是岭南文化的一部分，当人们看到骑楼，走在骑楼下，就会想象到广州的过去——是整整一代人对城市的过去与现状的情感关联，也是一个城区的独特所在，是别的城区难以培育的、真正的人文气息，这种人文气息是要很长时间培育的。所以，骑楼街是荔湾旅游的"亮点"之一，特别是荔湾骑楼街长达2600多米，投资近3000多万元加以整饰改造，全街楼宇均不超过三层，以淡黄、淡绿和淡蓝色外立面居多，偶尔也有紫红色立面，显得别具一格。将骑楼文化作为西关人真实生活的旅游资源进行开发，有利于将荔湾打造成一个"城中有景，景中有城"的西关文化人文风光的城市旅游区。

如何全面规划荔湾骑楼街，需要我们慎重研究。从骑楼街整体看，其在宜旅、宜商、宜居三大功能上的价值作用是依次递减的，在宜商、宜居方面都有许多不足之处，不是现代商业和现代家居理想的建筑结构。骑楼街的主要亮点是作为休闲观光之用，附加功能是购物。目前的基本情况是，上下九路、第十甫路的骑楼，商业氛围较深，凸显出其宜商的一面，十甫名都的气派就在于它是骑楼式商业建筑；而恩宁路、龙津西路则相对宁静，居住性的价值较大。

骑楼街在多个城市都有，在同一城市中也分布在多个城区，不具唯一性与排他性。况且目前的骑楼与早年的骑楼（木栏）已有很大不同，原始味道也较淡，现时可以争世界之最的主要是"长度"，所以将现在的上下九骑楼街延伸到恩宁路、龙津西路的泮塘处，总长达2600米，这是有必要的，具有其历史价值。

从商贸旅游的价值工程上看，将骑楼功能定位加以区分，则更有利于商贸旅游文化城区的发展。整个荔湾区有3条总长约为1300米的街道，可将其区分为3个不同的功能街区。

（1）上下九路、第十甫路商业步行街，总长度约为1300米，主要功能是商业步行街，以购物、观光为主。

（2）恩宁路、龙津西路（至荔湾湖泮塘处），总长度约为1300米，主要功能是民居民俗风情街，以荔枝湾风貌、西关小秦淮、西关大屋、西关骑楼、西关名食等为引领，打造西关民居民俗风情区。

(3) 荔湾路（北至东风西路，西接龙津西路），总长度约为 1300 米，是社区健康休闲之路。在该路段约 5 万平方米的大型综合体育场馆于 2003 年下半年落成；投资 1.5 亿元，占地 3 万平方米的“西关文化广场”也即将兴建，其中的水疗馆占地 3000 多平方米，是一个集游泳、玩乐、健身于一体的综合性场馆，是广州市目前最大的水疗中心。正如媒体报道所说，荔湾路有望成为广州“健康之路”。这也是荔湾旅游新的景点。

以上 3 个都是长度约为 1300 米的旅游购物商业步行街，具有不同的、鲜明的功能，但也可以整合成一个，其功能明确、印象深刻、吸引力强。

5. 西关大屋——老广州的沧桑繁梦

西关大屋的概念来自清代同治、光绪年间开始兴起的位于西关的大屋建筑群，是旧西关的名门望族、商贾富绅和洋行买办阶层等新兴富豪的高檐深宅，可以说是清末时期广州传统民居中的代表作。西关大屋建筑风格独特，布局巧妙细致，与广州古城历史一样，堪称历史弥珍。

西关大屋以门庭高大、装饰讲究而闻名。还未进门，只在门前一站，西关大屋就给人一种自然、纯朴的亲切感。屋前的麻石街是一条长长的、被磨得滑滑的青石板，外墙是岭南石脚水磨青砖高墙，正门是矮脚吊扇门、趟栊门和硬木大门，其功能不亚于现代的双层防盗门，但艺术视野更开阔。典型的屋内设计平面为三开间，正间以厅堂为主，从临街门廊、门官厅、轿厅、正厅、头房、二厅（饭厅）到尾房，形成一条纵深很长的中轴线。两旁偏间左为书房及小院，右为偏厅和客房，其中后部为卧室、楼梯间和厨房等。传闻西关大屋的兴建很是夸张，青砖墙铺砌用的是糯米饭拌灰浆，砌好砖墙后还要在外面再贴一层水磨青砖，所以西关大屋的青砖墙永远是平滑的，没有一丝缝隙。

随着岁月的流逝，现有西关大屋的相当部分房屋立面的青砖已破旧，屋檐花窗损坏，山花女儿墙脱落，但其古风古韵犹存。它记录着昔日西关豪门富户的风花雪月和广州西关的繁华岁月，也记录着众多的西关人的生活痕迹。这一切虽已是历史，但如能在旧城改造中，采取修旧如旧的手法，力显“古陈”风格，就能重现百年西关民居建筑历史。西关大屋是西关民俗民居的活见证，不失为荔湾游的一处亮丽景点。

西关大屋经过多年的整饰，现已姿容初露。西关民居民俗风情区的建设规划已纳入广州市旅游整体规划之内，西关大屋社区的整治、修复工作也在进行中，部分旧民居正进行外墙修饰，部分街道重新铺设了具有西关特色的麻石路面，还新设了具有西关风情的童谣、线画浮雕，西关民俗馆已于 2000 年 9 月 26 日正式对外开放，全方位恢复昔日西关大屋的建筑风格，包括“西关大屋建筑意境”“西关民俗风情”“婚嫁风俗”和“节令习俗”等展览。

当前，西关大屋社区的整饰要处理好如下一些问题。

(1) 在旧城危房改造中，如何做到既保留原始的西关特色建筑，同时又使整个城区面貌焕然一新，做到危房维修、改造和保护历史街区、传统建筑相结合。

典型西关大屋的特点是有 20～30 米大进深的民宅，采光差，建筑用地利用率低，不符合现代小家庭的居住模式，因而在改造旧房中只能通过增加横向街道、缩小建筑进深的办法来改善采光和环境，并将目前1～2层的建筑加高变成3～4层来提高单位用地的使用率，从而增加绿地建设。这一工程不论是在设计施工上还是在资金筹集上，难度都是很大的。

(2) 旧城区房屋基本上存在三类不同情况，应进行不同的处理：一是清末民初建筑，这是典型的西关大屋；二是民国后期、新中国成立初期的小洋楼式建筑；三是20世纪80年代初建的现代建筑。此三类建筑风格各异，很不协调，处理难度很大。

(3) 对现已保留已粉饰过的西关大屋，如何加强宣传力度，使其真正作为西关民居民俗风情推介的代表，还有很多工作要做。

6. 西关饮食——食在广州味在西关

目前我国荣获“中国优秀旅游城市”称号的共有138个城市，广州是其中之一。国家旅游局决定在2005年评审出全国首届9个“中国最佳旅游城市”，广州市已正式向国家旅游局提出申报。其中，“中国最佳旅游城市”将设9个专项最佳称号，广州市根据自身优势，已确定申报两个专项称号，分别是“中国最佳餐饮旅游城市”和“中国最佳购物旅游城市”。广州市这一决定是有其根据的，据2003年6月的一次民意调查，66%的被访者认为广州申报“中国最佳餐饮旅游城市”最有优势。这一新形势对荔湾饮食、购物提出了新的要求。

1) 西关饮食的特色与行业结构特点

饮食就餐是广州人的生活特色之一，也是岭南文化的一大特色。“食在广州”之所以闻名中外，其特色主要有以下三个方面。

(1) 食肆林立，摊店众多，日夜营业，方便消费。广州市内茶楼、酒家、餐馆、小食店遍布大街小巷，星罗棋布。营业时间长，“三茶两饭一夜宵”全天候服务，大小食档交错开市，适应多层次消费需求。

(2) 用料广而精，配料多而巧，模仿与创新兼有。广州饮食特点之一是食品丰富，款式多，取料范围广，飞禽走兽、山珍海味、中外食品，无所不有，可称全国之冠。

(3) 别具风味，特色菜点，特殊技巧。食味着重“清、鲜、爽、滑、嫩”，讲究镬气香；调味遍及“酸、甜、苦、辣、咸、鲜”；菜肴有“香、酥、脆、肥、浓”之别，即“五滋六味”俱全。西关饮食是广州饮食的代表，更以名小食出名。

从行业结构与布局看，西关饮食业有两大特点。

一是老店、名店高度集中，多在第十甫路及其周边地区。其中，历史悠久、颇有声望的饮食、食品企业如下。

陶陶居，位于第十甫路，创建于清光绪六年（1880年），“陶陶居”三字为康有为题写，其寓意为来此品茗，乐也陶陶，是广州最具特色的茶楼之一。鲁迅、许广平、巴金等都曾是陶陶居的座上客。

莲香楼，位于第十甫路，创建于清光绪十五年（1889年），以气派与众不同著称，采用华丽的科林斯柱廊装饰。

多如茶楼，位于珠玑路，创于清光绪十七年（1891年）。

广州酒家，位于第十甫与十八甫交界处，创建于1928年。

南信牛奶甜品店，位于第十甫路，创建于 1934 年。

银龙酒家，位于宝华路，创建于 1937 年。

趣香饼家，位于第十甫路，创建于 1938 年。

皇上皇腊味店，其分店在下九路，始创于 1940 年。

欧成记面食店，位于第十甫，创建于 1940 年。

清平饭店，位于清平路和上下九路，创办于 1964 年。

二是西关饮食中各类名小食店星罗棋布，广州市民形容西关为“百步之内必有名小食”。这些名小食中影响颇大的有莲香楼的莲蓉食品、南信的双皮奶、欧成记的鲜虾云吞面、老坑公的桑枝粥、广茂香的花生、欧成记的艇仔粥、南泰的肠粉白粥等。新中国成立初期，第十甫路的西关食街共有大小饮食店 29 家。

2）西关饮食期待振兴

荔湾饮食业的振兴，不仅是发展荔湾商贸旅游业之需，也是广州市申报“中国最佳餐饮旅游城市”之需。但是，目前的状况告诉我们，振兴的任务是艰巨的。

荔湾区餐馆数量不断增加，从 1998 年的 233 户发展到 2001 年的 312 户，增长 33.91%，其中，限额以上企业（指年营业额超 200 万元以上企业）从 14 户增加到 27 户，增长 92.86%，但总营业额则有所下降，2001 年完成销售额约为 1.67 亿元，比 1998 年下降 12.32%，年平均下降 4.29%（表 15）。

表 15　荔湾区餐饮业限额以上企业户数及营业额

年份	营业额/万元	户数/户
1998	19 075	14
1999	29 691	21
2000	13 731	14
2001	16 725	27

资料来源：荔湾统计局，《荔湾统计》2002 年第 18 期

荔湾区餐饮业限额以上企业营业额总收入比东山区和越秀区都低，2001 年按区属口径统计，东山区比荔湾区高 39.76%，比越秀区高 38.55%。其突出原因是，荔湾区餐饮业大户比重小，2001 年在限额以上的 27 户中，营业额超 1000 万元的企业只有 2 家（荣华饮食服务公司与新牛车水酒家）。同时，还有部分限额以上的企业，因经营不足以维持简单的扩大再生产而被迫关闭，如愉园酒家、孖宝酒家、焕发风味海鲜酒家等，其中，清平饭店营业额下滑速度更为突出，1998 年营业收入为 10 058 万元，到 2001 年下降为 97 万元，下降了 99.04%。

以上存在的问题，其原因是复杂的。反映比较突出的是，市场竞争十分激烈，有的地段餐馆数量过多，全行业处于微利状态，不少企业可持续发展能力较差。不少名店、老店之所以败下阵来，其主要原因除了体制与管理方面外，在经营方式、经营观念和经营内容上的陈旧落后，也是重要原因。总之，从目前状况看，要振兴西关饮食业，或是要打造新荔湾饮食一条街，如果没有创新独特之处，没有较强的吸引力，没有可观的区外客源，是很难实现的。

7. 西关曲艺——源远流长，基础广泛

西关人爱粤剧、爱曲艺，粤剧、曲艺可以说是西关文化特色之一。百余年前西关就是粤剧曲艺兴盛之地，粤剧曲艺界的名伶老馆长期聚居此地，为西关留下了深厚的文化传统，奠定了广泛的群众基础。至今，众多旧时的粤剧曲艺艺术形式，艺人的文物遗迹、历史传说还在传颂。据史料记载，早在 1884 年，广州第一个粤剧演出营业机构在黄沙承祥坊建立。1889 年，粤剧界的第一个行业组织——八和会馆也在这里建立，之后的曲艺歌坛“初一楼”和曲艺茶座“红荔音乐厅”也建于此。这代表了粤曲一代的鼎盛时期。

西关曲艺文化有如下一些特点。

(1) 对海外，特别是东南亚、美国等地华侨有很强的吸引力和影响力。八和会馆历史沧桑，几经劫难，最终在 1986 年正式更名为“广东省粤剧八和联谊会”，世界各地的“八和”都尊广州“八和”为祖，八和会馆成了粤剧的“祠堂”，“八和”子弟探祖、兴祖，这对广州的旅游商贸发展有着较大的推动作用。

(2) 有广泛的群众基础，私伙局成“群众运动”。私伙局，是荔湾区的一种群众性的曲艺社团，经区文化局登记并纳入文化部门管理的私伙局就有 40 多个，参加人数达 1500 多人，其中有 80 多岁的老人，也有三四岁的儿童。至于邀上三五知己一起玩乐器、唱粤曲的，那就无法统计了。

(3) 名声在外，品位不断提高。随着时代的发展，西关的“曲艺之乡”也在不断发展，品位高了，形式多了，普及也更广泛了。1997 年，荔湾已有 4 条街道被授予“广东省民族民间粤曲艺术之乡”称号，全区 12 个街道文化站，都开放给群众，供他们吹拉弹唱。2003 年初，荔湾区被中国文学艺术界联合会和中国曲艺家协会授予“中国曲艺之乡”称号。这一称号全国仅有 13 家，荔湾是广东唯一一个荣获该称号的城区。

(4) 后继有人，童星闪烁，前景辉煌。粤曲未来前景如何，是人们普遍关心的问题。西关“曲艺之乡”的特点之一，就是从来都注意新苗的培育。早在 1907 年，西关就成立了全省第一个粤剧童子班、名新少年班，继后又办起了超群乐童子班，新中国成立后开办了“小红兵”粤曲表演培训班，1995 年成立了荔湾区少年儿童艺术学校，培养了一批又一批少年粤剧爱好者和优秀小演员。更值得书写一笔的是，广州市唯一一所省级民乐教育示范学校就在荔湾三元坊小学，该校没有不会民乐的孩子，有超过 60%的小学生会两种以上乐器。

总之，荔湾的曲艺文化虽不能为荔湾旅游业带来人数众多的旅游者和旅游收入，但其潜在的、间接的影响是不可漠视的。

8. 西关国医——深情关注您的健康

西关文化不仅反映在民风、民俗、民居等方面，还反映在医疗保健方面，西关国医同样存在历史的文化特点。1899 年 12 月 12 日，美国女西医富马利在广州西关成立中国第一所女医校“广东女医学堂”，后改名为“夏葛女医学院”。其旧址就在现广州市第二人民医院的大院内。在大院的两侧，有一栋古老建筑“林护堂”，建于 1937 年，是当时办医的历史见证。

1935 年 11 月，“博济医院”建于长堤珠江河畔，孙中山先生早年学医就在此处，也是我国建立最早的西医院。随着医学的发展，西关的中药材、中医学也在不断发展，至今荔湾中医文化已形成一道独特的风景线，逐步树立寻医问药的荔湾商贸旅游文化形象。

为发扬西关医药文化的特点，有关部门计划对广州市第二人民医院按照百年多历史的背景重新设计布局，并把医院定位为一所具有西关风情的资深医院。为发扬医药国粹，在西关建立别具特色和特效的、连锁形式的国医馆，将治疗、饮食、娱乐结合起来，形成一体。2002 年 5 月西关国医馆开业，地点在长寿路，受到当地居民的热情欢迎；沙面国医馆是西关国医馆的连锁店，主要对象是健康和亚健康人群，为人们提供针灸、推拿、足疗、艾疗传统中医项目。此外，还拟议在清平中药材市场附近再设一家以食疗药膳为主的国医馆。

西关国医馆虽为连锁形式，但其经营特点各自不同，其中，药膳国医馆更具特色，市民可以吃到各种药材炖品、药食膏羹、药酒、专病药膳，甚至还有各种不同特色的药材火锅。更能吸引食客和旅游者的是，在医馆内有中医师驻扎，可为进馆者号脉，针对其经脉虚实以及不同体质开出不同的食疗处方，这使客人产生浓厚兴趣。

不同的国医馆各有特色，它们将互相补充，交互辉映，联结成荔湾中医文化一道独特的风景线，是荔湾旅游的另一文化特色。当前的关键是如何与旅游部门加强沟通与合作，扩大社会宣传，让游客在游览参观的同时，还可以亲身试试中医针灸、推拿，尝尝药膳，买买药材，增加荔湾旅游项目，诱发重游欲求，强化西关医疗保健旅游品牌。

五、拓宽外源增创新的经济增长点

一个都市或城区，要把经济搞上去，内源固然重要，但外源也不能忽视，特别是开放城市或资源缺乏的地方，外源更是一个重要因素。从荔湾的具体情况看，地域狭小，工业基础薄弱，商贸发展空间不大，如何拓宽外源，是一个重要战略考虑。本部分只从实施“西联”战略和发展“总部经济”两个方面，提出一些方向性思考的问题。

1. 荔湾是广州市“西联”战略的黄金结合点

1）“西联”战略的提出

广州城市经济发展规划中，有一个战略性的考虑，就是“北优、南拓、东进、西联”。这一战略思想的要求是，城市空间结构从单中心向多中心转变，“组合城市”是一个发展大趋势，加强区域性的分工与合作，以共生共荣为前提，整体协调为手段，达到创新优势的目的。

广州、佛山两市不论是在历史渊源和地理环境上，还是在经济社会关系上都有密切关系。从历史渊源上看，广州与佛山同为一家，广州是一个具有 2200 多年历史的文化名城，曾是南越王的都城，从古至今一直是华南地区政治、经济和文化中心；佛山是我国古代四大名城之一，历来受广州的影响至深，并依靠广州发展壮大。从地理环境上看，广州、佛山两市同属珠江下游的珠三角河网区，自然地域连为一体，广州的白云区、荔湾区、芳村区与南海区仅是一河之隔或是直接相联，行政区相互接壤。在经济社会中，许多基础设施已日渐融为一体，两地的交通路网通过 105 国道、325 国道、324 国道、广佛高速公路、广三高速公路、广佛铁路等路网，已紧密地连接起来。此外，信息宽带网、供水网等城市大型基础设施，已将两地紧密联系在一起。改革开放以来，两地在经济社会发展过程中，已自觉或不自觉地形成互动关系，企业间的联系越来越紧密，互补性越来越明显，产业链也逐步形成。广佛都市圈的经济社会发展基本状况见表16。

表16　广佛都市圈2001年经济社会发展状况

类别	广佛都市圈	占珠三角比重/%	占广东比重/%
面积	11 181平方千米	26.8	6.2
人口	1 528.1万人	37.5	20.2
城市人口	1 233.7万人	42.0	26.0
国内生产总值	3 754.1亿元	44.9	35.6
工业总产值	5 754.2亿元	47.9	31.0
社会消费品零售总额	1 623.1亿元	52.0	36.0
城乡集贸市场成交额	692.7亿元	—	32.9
全社会固定资产投资额	1 216.4亿元	47.4	34.3
外贸进出口总额	341.0亿美元	—	19.3
外贸出口总额	179.7亿美元	19.8	18.8
实际利用外资	43.4亿美元	30.6	27.6
城乡居民储蓄存款	3 731.3亿元	51.3	39.6
财政收入	329.5亿元	44.2	28.5
邮电业务量	251.5亿元	40.6	25.1
货运量	36 494万吨	—	27.7
城市化水平	80.7%	—	—

资料来源：2003年3月28日《羊城晚报》杨再高所写的《广佛都市圈时代已来临》一文

注：①珠三角指包括广州市、佛山市、东莞市、江门市、深圳市、珠海市、中山市和惠州市的市区、惠阳市、惠东县、博罗县、肇庆市的市区、高要市、四会市在内的珠三角经济区；②城市人口包括市区人口和镇区人口

广佛都市圈概念的提出，是与“西联”战略的设想相关的，而都市圈的概念早已存在，如大伦敦、大巴黎都市圈等，这是发达国家在城市化进程中所形成的相互协调的发展区域。其实，在珠三角经济圈中，已逐步形成了三个都市圈，即广佛都市圈、港（香港）深（深圳）都市圈、珠（珠海）澳（澳门）都市圈。究其原因，这是城市经济发展的必然，城市的本质是集中而不是分散，城市达不到必要的规模，集聚不起市场，集聚不起服务，集聚不起人气，也就不能称其为城市，城市中建起的各类市场和服务也不可能达到规模经济。

2）广佛都市圈，荔湾处在黄金结合点

在广佛都市圈内，广州、佛山两市共辖12个区、6个县级市、116个镇和109个街；广佛都市圈面积为11 181平方千米，总人口1528.1万人（其中居住在市区镇区的城市人口为1233.7万人）。经济发达，经济外向度高，广佛都市圈2001年GDP为3754.1亿元，占全省的35.6%；人均GDP为3.3万元，投资的国外跨国公司超过110户。这些就是广佛都市圈的基础条件。

广佛都市圈的成长，目前主要表现在广州、佛山、南海核心区的融合。广州的公路、轨道交通与佛山、南海的主要干道连成一体，构成了物流、商流、信息流的网络。从城区空间布局看，广州荔湾区的芳村与佛山的南海之间将会成为广佛都市圈的中轴线，它将促进两地发展物流业和房地产业，从而形成新的广佛新城，前景广阔。从商

贸交往、产业互补的角度看，荔湾是广州与佛山市场对接的黄金结合点。荔湾是广州商贸旅游文化区，内源经济深厚，外源经济潜力大。荔湾与南海的大沥镇、黄岐镇、盐步镇近在咫尺，两地商贸互动频繁，佛山大量居民常到广州购物、消费与旅游观光休闲，而广州也有不少人常到佛山、南海等地购物、消费与旅游。可以说，荔湾是广州“西联”战略中东承西拓的“窗口”。

3）“联”什么？如何“联”？

“联”是联合、协调，当然其中也有竞争，甚至可能还会出现彼长此短的现象。因而在“西联”中，要激励正因素，减少负因素。在市场经济运行中，要会竞争但更要会竞合。竞合作为一种新的战略，正越来越被广泛运用并取得成功。荔湾处于中心城区，在商品、人才、科技、信息和服务等方面具有集聚优势，应积极主动参与广佛都市圈的产业分工和资源组合，加强与佛山等周边地区的经济合作和产业协调互动，形成资源共享和优势互补，从而提升综合服务功能和经贸合作水平。

“西联”的重点可抓以下六个方面。

（1）产业链的衔接。广州、佛山两市产业互补性强，产业联系紧密，产业链早已初步形成。广州的汽车及其配件、电子工业及其配套产品，以及荔湾发展的三大工业园区产业的轻工产品都能优势互补，佛山许多企业直接为广州企业提供配套服务，不少企业产品供应广州企业和居民消费，如家电、陶瓷、家具、农副产品等。规划中的黄岐汽车城，其面积达 20 多万平方米，将集贸易、消费、会展、购物、商务、信息、旅游、娱乐、休闲等于一体，以经营整体与配套用品为主，这与广州汽车工业的发展紧密相关。

在“西联”的产业链衔接中，荔湾的大坦沙岛起着桥头堡的作用，通过对大坦沙岛开发，对岛东岸珠江大桥南北沿江一带、桥中中路沿线一带以及岛内民营科技产业园周边环境进行整治，不论大坦沙岛的定位最后如何确定，都将会成为广佛都市圈产业链的重要节点，强化荔湾在“西联”中的地位。

（2）商流的互动。广州与佛山的商贸往来，自古以来就有密切关系，广州许多工业品直接供应佛山居民消费，佛山也有大量居民常到广州购物消费，尤其是上下九步行街、北京路步行街等繁华商业街及专业性市场等是商流、人流旺地。当然，也有不少广州人到佛山著名购物广场消费，如黄岐广客隆、乐从家具城、平洲食街等，尤其是荔湾几个专业批发市场对佛山的互补性很大，同时通过佛山直接辐射珠三角地带。当然，南海大沥步行街建成后，由于其规模投资大、档次高、特色明显，将是上下九步行街的强有力竞争对手。

（3）通勤互动日益频繁。随着城市化发展和广州城区的扩展，广州至佛山之间形成了高度城市化的地区，尤其是南海东部的黄岐、平洲等地区与广州的荔湾、芳村连成一片。一些人居住在广州，就业在佛山；另一些人居住在佛山，就业在广州。这种“城市通勤”现象已日渐扩大。据统计，广佛间每天最大客流量超过 46 万人次，并且每年以 5%的速度增长，这就促进了通勤地域中房地产业的发展，尤其是促进了黄岐、平洲等地房地产业的日益繁荣。

（4）价值链的延长与提升。广州、佛山两地的融合，必然要求进一步整合资源，调整发展布局，促进城市功能合理分工，加快城市化进程。佛山与广州发展格局接轨，主动接受广州辐射，分流广州城市功能，构建核心城市边缘聚合功能，通过区域协调，就可降低摩擦成本，提高珠三角城市群的价值链；广州也可通过佛山的辐射带动，使珠三角城市群的价值链得以向粤西、西南等腹地纵深挺进，延长价值链。

（5）整合旅游资源，发展“广佛旅游圈”。广州，特别是荔湾与佛山在旅游资源上有许多共同点，它们都具有许多岭南文化特色，如名城的文物、岭南风貌的古建筑、地方特色的饮食文化和老字号品牌名店等，都可以进行互补，形成具有特色的“广佛旅游圈”。佛山不但地理条件优越，河网纵横，基塘相间，生态环境优异，而且佛山自唐代以来，就是岭南综合性生产基地。宋朝时佛山已设有对外贸易的市舶务，成为商业市镇，有“佛山成聚，肇于汴宋”之说，与北京、苏州、武汉三处一起成为我国商业繁盛的天下“四聚”。清代中期，佛山与湖北的汉口镇、江西的景德镇、河南的朱仙镇，并称为我国“四大名镇”。只要将广佛两地的旅游资源加以巧妙整合，“广佛旅游圈”是大有文章可做的，其引力不仅可以吸引珠三角的旅游观光休闲客，还可以吸引港、澳、台同胞及东南亚华侨。此外，还因为广佛两地商贸日渐发达，专业市场类型多、特色明显，区域性商业物流中心逐渐形成，也能吸引大量商务旅游客。

（6）加快路网的衔接。国内外经验证明，都市圈的建设与路网的建设密切相关，要加快都市圈内外的物流、商流、人流的周转，就要有快速路网。为适应广佛都市圈发展的形势要求，佛山市政府投资 120 亿元兴建 12 项交通项目，于 2003 年 6 月 18 日同时动工。这些项目中，“一环”“二环”是重中之重，“一环”城市快速干线跨 3 个区、14 个镇，全长 98 千米，途经佛山主要发达镇区，辐射范围超过 1000 平方千米，是广佛都市圈南北向的重要通道；“二环”与“一环”相连，建成后，将使佛山范围内的广三、佛开、广珠以及正在兴建的广明高速公路全部连通。这一工程也将成为广州、佛山与珠三角西部的重要交通走廊，这一工程将在三年内完成。对广州市来说，如何衔接配合，需要做出周密规划。

2. 发展“总部经济”，拓展荔湾城区经济价值

1）什么是“总部经济”

如何发展“总部经济”，已是当代国内外许多大都会进行研究探讨的重要课题之一，它关系到一个城市的形象和经济发展的影响力。在经济全球化的大趋势下，“总部经济”已成为许多国家（地区）中心城市所追求的新的经济形态。

“总部经济”，是指中心城市通过创造各种有利条件并利用其影响力，吸引跨国公司和外埠大型企业集团总部入驻。企业总部虽在某一中心城市布局，但其所属的加工生产基地、营销网络及所属的各区域营销总部，则安排在营运成本较低的周边地区或外地，从而形成合理的价值链分工。“总部经济”对发展区域经济具有集聚性、扩延性、示范性、辐射性、吸引性的重要意义，从而带动总部所在地的旅游业、饮食业、零售百货业、房地产企业、会展业等相关行业的发展，创造巨大的经济效益。

2）广州市发展“总部经济”的一些情况分析

在我国，“总部经济”发展最快的是上海，给其带来的各方效益也很明显。例如，一些大型跨国公司将自己的业务总部或研发中心迁往上海，这就加大了上海对于相关

人才的需求。据 2002 年资料，上海职位增幅达到 74.5%。与此相关，许多行业的经济效益快速增长，薪资行情也普遍上涨，据中华英才网 2003 年发布的薪资指数报告，上海年薪均值为 49 180 元，首次跃居全国第一，比第二位的深圳高出 1237 元，比第三位的北京高出 2569 元，比第四位的广州高出 7803 元。

广州情况如何？我国首部城市发展白皮书——《（2001—2002）中国城市发展报告》的首席科学家、第三世界科学院院士牛文元教授在接受记者采访，谈到有关我国国际化大都市时说，广州的定位已十分明确，即面向港澳地区及东南亚各国。几年后穗港经济必将走向高度一体化，10 年后东南亚经济联合体也将成立，这将进一步确立广州在此区域的国际化大都市的地位。他还指出，作为珠三角的中心、明星城市以及国际化大都市，广州有责任在新一轮财富集聚过程中，不只是成为单项冠军，还要引导整个珠三角城市群成为全能冠军。

在确定未来发展规划时，逐步明确广州已具备发展“总部经济”的条件与基础，大力发展“总部经济”有利于推动区域经济的发展，提升城市形象。在《广州市 2001—2010 年商业网点发展规划》中，已经把环市东路和天河北、珠江新城一起列为广州市中央商务区重点发展区域，还具体提出，通过环市东路和广州大道的有机连接和周边环境、配套设施的改造提升，打造一个国际化的中央商务区。经过近年的努力，这一设想已有相当进展。据报道，环市东路现有的 20 多万平方米写字楼已趋于饱和，花园酒店附近写字楼的出租率已超过 95%。

广州市“总部经济”的发展，以东山、天河两区最快，其中，东山区更为突出。据报道，2001 年总部设在东山区的公司、企业共有 609 家（2003 年上半年又有 8 家知名企业落户东山）。这些公司、企业创造的利润达 101 亿元，占东山区企业利润的七成多。东山全区 2001 年地税总收入达 84.1 亿元，占全市地税收入的 1/3，是粤东、粤北、粤西 10 市地税的总和。其中，缴纳地税额超过 300 万元的大公司有 232 户，约占东山区地税总收入的八成，全区前 20 位的地税大户都是大企业、集团等总部企业，这正是“总部经济”对区域经济发展起重要作用的体现。

为把“总部经济”做大，东山区委、区政府于 2003 年年初制定了《关于进一步加快总部经济发展的若干意见》，并计划在 5 年内再造 20 万平方米的优质写字楼，相当于再造一个环市东路。

3）荔湾打造“总部经济”的战略思考

荔湾是广州市众所周知的具有商贸传统的老城区，其明显特点是地域狭小，人口密度大，空间有限。2001 年 4 月广州市统计局发布的广州市第五次全国人口普查数据显示，荔湾区辖属区域仅为 12 平方千米，与广州 7000 多平方千米相比，是微不足道的。全区总人口数为 47.48 万人，比第四次全国人口普查时减少 7.98 万人，减幅为 14.38%，但人口密度为 4.02 万人/千米^2，人口密度从第四次全国人口普查时的全市第二位上升为第一位，比广州市人口密度 1337 人/千米^2 大得多。这一切都成为荔湾老城区进一步发展的瓶颈。从扬长避短的战略考虑，要扩大地方税源基础和带动城区经济的发展，大力发展“总部经济”是很有必要的。

荔湾具有发展“总部经济”的可行性。

首先，荔湾符合广州市发展经济的总体规划要求。虽然在市场规划中提出中央商务区是在环市东路、广州大道和珠江新城一带，但并不是说其他城区不能发展，而且“总部经济”形态多样，行业复杂，各公司、企业特点不同，战略各异，不可能在一地一处包揽一切。

“总部经济”一般可分为四种类型：一是大型企业集团总部，包括境内外大型企业集团总部；二是跨国公司研发中心，多与当地重点行业或重点环节合作；三是营销类公司总部，特点是机制灵活，写字楼面积不大，但经营额相当可观；四是发展贸易总部，其特点是主要从事进出口贸易或大型成套设备贸易等。由于总部类型不同，要求也有差异，从全国来说，一些企业集团的业务总部可能选择上海，但另一些企业集团出于其他原因可能选择广州。同理，在广州选择的空间也会有所不同，从目前条件来看，“东山”优于“西关”，但前者写字楼趋于饱和且价格奇高，不同的企业会有不同的价值比较选择。从荔湾的具体情况看，第三、第四类型的公司总部选址荔湾的机会是较多的。

其次，荔湾的区位优势突出。荔湾位于广州城区西部，广州西部进出口，在广佛都市圈中，是两大都市的黄金结合点，有利于吸纳珠三角的企业集团进驻。此外，近年来荔湾的市内交通已有很大改善。中山七路和中山八路房地产业日益兴旺、康王路两侧的商业地块和现代商务大楼建设，都是吸引外地“总部”的亮点工程。

最后，利用荔湾现有商贸行业优势，建立“行业商务区”。荔湾商贸行业有其传统特色，包括上下九步行街和各类专业市场都有不同的经营特色，其中包括零售服务业、水产业、中药材、服装针织、玉器精品、通信电器等，吸引不同行业的品牌厂家、商家到区内设总部或销售总部，发展区域性总经销、总代理，形成“行业商务区”，发挥行业的集聚效应。

总的来说，“总部经济”是一个多元化、多层次的概念，打造“广州西部商务区”（把广州中心商务区分为“东部商务区”与“西部商务区”）、“广州次中心商务区”或“广州行业商务区”等，规划必须先行，配套设施要跟上，写字楼要有一定规模，市场氛围要形成。

六、措施与建议

1. 依靠科技兴商旅

1）加大科技兴商旅力度

20 世纪 90 年代以来，江泽民同志根据邓小平同志关于“科学技术是第一生产力”的理论和坚持教育为本，把教育摆在优先发展的战略地位的思想，确立“科教兴国”的跨世纪国策。

（1）科技创新是生产力现代化的核心。中国现代化进程，始终是生产关系的现代化与生产力的现代化不断向前推进的过程。在这个过程中，制度创新是生产关系现代化的核心，而科技创新则是生产力现代化的核心。

（2）信息生产力是生产力现代化的标志。信息技术改变了经济增长的方式，经济增长方式从粗放型向集约型转变，这是经济发展过程中的必由之路。

（3）走科技与商旅相结合的创新路子。经济增长依靠科技的进步，反过来，科技进步也需要经济强有力的支持。在改革创新当中，如果没有先进的科学技术手段解决“大市场、大流通、大旅游、大服务”的规范、有序、高效的运作流程问题，则将贻误进程，影响科技兴商旅的全局。

具体措施与建议包括以下三个方面。

(1) 以政府为主导、以商旅企业为主体、以高等院校或科研机构为依托，走一条商贸旅游与科技紧密结合的产学研合作道路。

(2) 由政府或行业协会牵头，建立以市场需求为导向的荔湾区商旅经济研究所，承担战略性研究课题，对荔湾商旅的可持续发展进行深层次、前瞻性的研究。

(3) 加大科技对商旅的投入与运用，提高商旅经营的科技含量。一是加大科技基金对商旅经营管理研究的投入；二是加强对区内外商旅信息的收集、整理、分析、交流和服务工作；三是改变随机的、无序的投入运作模式，建立以提升商旅经营价值为目标的运行机制，把科技投入与运用贯穿到整条商旅价值链中去。据了解，区科技局2002年商旅科研经费投入为130万元，2003年科研经费投入为200万元，2010年前，每年科研基金经费的投入速度，以不低于20%递增为宜。

2) 构筑信息化、网络化平台，推进电子商务

高新技术是一个动态的概念，过去的高新技术发展到今天已成为传统技术，今天的高新技术也必将成为明天的传统技术。一个市区发展高新技术，除了实现自身产业化以外，还有一个重要任务就是要对传统产业进行改造，将高新技术注入传统产业，促使传统产业不断更新换代，向高质量、高效能、高竞争力的方向发展。商贸旅游业引进高新技术（如信息化、网络化、电子商务、物流、数码、网上购物等），使商贸旅游业结构得到优化，产业结构、产品结构得到升级，经营素质、经济效益也不断提高。加快建设统一的区信息化网络大平台，将政务、商务、社区和教育网络资源结合起来，实现全区网络资源共享。同时，充分发挥网络大平台的优势，大力推进商旅业网点电子商务的应用，利用信息技术提升传统产业的档次，带动全区经济的发展。

从电子计算机角度来看，文本、图像、多媒体等只不过是不同的数字文本而已。在信息数字化的基础上，更易于实现信息网络化，从而可有利、方便地实现资源共享。网络技术的发展带动了整个网络业的发展，而各类电子信息设备同计算机技术结合，正向智能化方向发展。可以预见，不久的将来在商贸旅游业经济中，数字化、智能化的应用有着广阔的前景。

具体措施与建议包括以下四个方面。

(1) 帮助几个大中型专业批发市场建设网络平台，有计划、有步骤地推进电子商务。

(2) 以上下九步行街为龙头，逐步构筑零售业信息网络。

(3) 运用互联网等信息技术，广泛宣传和招商引资。

(4) 对传统的商旅业进行电子化、数字化改造，以提高经营效率，减少人力、物力耗费，降低经营成本。

2. 深化改革，理顺关系

1) 企业生态与商贸旅游资源整合

自然界有个生态问题，企业界也有个生态问题，从管理哲学的角度来看，企业之

间有竞争的一面，但同时又有合作、互补的另一面。企业之间过度竞争，必然会带来互相“残杀”和恶性竞争，给企业带来内耗，其结果是企业失去市场的竞争力。企业之间如果只有竞争，而没有很好地分工协作，就必然会两败俱伤。企业生态就是指企业之间不仅要有竞争，更要相互支持、相互合作、优势互补，利润大家赚，风险大家担，企业才有真正的活力。只有这样，荔湾区商贸旅游资源才能真正地整合，这就是我们要提倡的企业生态。

2）树立“一盘棋”的全局观念

全市（区）商贸旅游业发展是一个系统工程，必须要有“一盘棋”的全局观念，“一哄而起、盲目上马、重复建设”的现象，给国家经济带来了极大的浪费和损失。同时，还应树立商贸旅游业开发与城市建设一体化的理念。将荔湾区整体经济作为一个大的商贸文化旅游区来看待，区内各产业之间既相对独立，又相互联系，难以分割。区内一切经济工作都要以强化商贸旅游业开发为中心，推进经济和社会进步。

3）加大扶持民营企业发展的力度

民营商贸旅游企业是第三产业发展的主力军。政府应加大力度促进民营企业的发展，民营商贸旅游企业对荔湾区经济发展发挥着越来越重要的作用。

4）建立现代企业制度

现代企业制度是符合社会化大生产特点，适应社会主义市场经济体制需要，真正体现企业是法人实体和市场竞争主体要求的一种企业体制。建立现代企业制度是社会主义市场经济对企业的客观要求，是规范和完善企业法人制度所必需的，是企业制度的一种创新。

具体措施与建议包括以下三个方面。

(1) 制订荔湾区商贸旅游一体化发展战略和发展规划，以及相应的实施计划和措施。同时建议跨区域商贸旅游携手合作，共同建设好长堤休闲带、广佛经济圈、西联价值链等新的经济增长点。

(2) 引导上下游的商旅企业以自愿为原则构建价值链并进行优化管理，增强整体竞争力，共同增创经营价值。

(3) 通过现代企业制度的建立和实施，用现代经营管理技术改造目前普遍存在的传统经营管理方法，推动商贸旅游业向现代化、集约化经营管理演进。

3. 政府转变职能，加强服务意识

(1) 切实转变政府职能，加强宏观管理和提供优质服务，以政企分开为切入点，以政社、政事分开为重点，加快社区建设，实现政府经济管理职能向社会管理职能的转变，建立高效廉洁、运转协调、优质服务、行为规范的行政管理体制。

(2) 实现政府职能由微观管理向宏观调控转变。政府逐步减少对经济事务的行政性审批，把精力集中到制定政策、营造环境、搞好服务上来。

(3) 鼓励和支持行业协会的发展。

具体措施与建议包括以下五个方面。

(1) 对商贸旅游业举办的一些大型促销活动，从审批制逐步过渡为登记制和备案制，从而实现以法律和经济手段为主的间接协调管理机制。

（2）理顺税费缴纳体系，实行“一站式”服务，既避免税收流失和乱收费现象，又便利经营者简化手续并减轻其负担。

（3）为加强法治建设和行业自律，营造良好的经营环境，应对历史上遗留下来的有关政策法规进行一次清理，逐步清除计划经济体制遗留下来的政府管理痕迹，改为运用市场手段进行调节。

（4）通过全国工商业联合会和行业协会加强对会员行为的规范化管理，推进商贸旅游业内外的交流、合作与发展。

（5）逐步实施电子政务和政务公开。

4. 加快商贸旅游企业集团化、连锁化的发展进程

商贸旅游业不仅要做大，更要做强，要突出主营业务，通过主辅分离，精干主体，增强竞争优势。

连锁经营是现代商贸旅游业的主流。连锁经营有强大的生命力，连锁经营具有“三大优势”，即降低风险、减少宣传费用、货源有保证。

连锁经营还具有以下七大有利条件：①集中化经营管理，有利于提高效率与效能；②标准化的经营，有利于扩大销售范围，改善服务；③现代科学技术的广泛应用，有利于加速科技进步；④专业化分工，有利于提高经营管理水平；⑤有利于提高市场竞争能力；⑥有利于减少商业投资风险；⑦有利于商贸旅游体系产业化、规模化、社会化。

具体措施与建议包括以下内容。

（1）通过价值链的构建与管理，推进商贸旅游业向集团化、连锁化发展。

（2）在有条件的专业批发市场，逐步构建商务中心，为商铺提供“一条龙”（结算、运输、质保等）和“一卡式”服务。

5. 建立产学研一体化的人才培养体系

（1）实施“引才工程”，加大人才引进力度。

（2）实施“育才工程”，加大人才培养力度，提高人才资源整体素质。

（3）实施“选才工程”，造就一批优秀人才，形成一支高水平、高素质、年富力强的科学技术带头人队伍。

（4）实施“人才创业”工程，不仅要留住人才，而且要使人才能够充分发挥其聪明才智。

具体措施与建议包括以下内容。

（1）组建一所商贸旅游业人才培训中心，持续培养商贸旅游经营管理者及业务骨干，提高他们的业务、法律及经营水平和管理能力。

（2）加强与开设有商贸、旅游专业高校的联系，建立健全区府与高校科研单位的联谊交流体制，充分利用高校的科技资源，推进产学研的紧密合作。

（3）在大力引进人才的同时，更要十分注意用好内部现有的各类人才，着力为他们提供施展才能的平台，构建相应的用人机制，使他们在商旅升级中发挥更大的作用。

参考文献

[1] 谭浩邦. 价值工程方法研究. 广州：广东科技出版社，1992.

绿色化是中国制造业的必由之路①

在成长的过程中，我们总是会遇到两难境地。

当一个人废寝忘食、拼命工作并取得辉煌成绩时，一纸医生诊断书告诉他，他的身体出现了问题；当经过几代人的奋斗，汽车成为普遍的代步工具时，我们发现城市的道路拥挤不堪，而郊外的鲜花、草地和蓝天变得渐行渐远。同样地，当中国制造业有了长足的发展，中国即将或正在成为世界制造中心时，环境问题和可持续发展这两大课题，似乎又让刚刚感到扬眉吐气的中国制造业陷入了迷惘。

制造业，承载着科学家、工程师和企业家的理想和希望，把一个个蓝图变成现实，把“纸上富贵”转化为看得见、摸得着的物质财富。如果说完美的设计、漂亮的方案和伟大的设想是上层建筑的话，那么一向科学严谨、兢兢业业、一丝不苟的制造业，则是实实在在的经济基础。纵观美国和日本的制造业在 20 世纪末的变化，我们知道，制造业不断与新技术结合，一如既往是一个国家经济发展和综合实力提高的支柱。

近年来，当人们把浪漫主义应用到科学工程领域，大唱知识经济和网络经济的赞歌，并扬言传统制造业将被新技术革命所淘汰时，我们的制造业像一位任劳任怨的母亲，默默无闻地把一碟碟美味佳肴端上饭桌，供夸夸其谈的孩子享用。

一个社会的进步，需要物质文明和精神文明两条腿走路，制造业作为物质文明的支柱，怎么可能像有的人说的，是夕阳行业呢？除非我们得意忘形、飘飘欲仙，逐渐不食人间烟火，成为网络上虚拟的网虫，或者干脆我们喝西北风就能填饱肚子。

让我们重温制造业辉煌的历史。

人类社会制造业的萌芽在几百万年前，那时猿人学会了制造原始工具，可以说是直立行走和制造业创造了人类。

170 万年前中国的元谋人懂得用火。

青铜器时代开始于 5000 年前。

欧洲最早的炼铁高炉建于公元 1380 年。

1765 年瓦特蒸汽机的发明，标志着第一次产业革命的开始。

19 世纪后半叶和 20 世纪初，电气化的产生被称作第二次产业革命。

两次产业革命极大地提高了社会劳动生产率和社会财富的积累速度，扩大了其规模，形成了至今我们仍在享用的社会文明的主体。

时至今日，不管一个国家的先进程度如何，它的国民生产总值有 80%以上与物质生产和消费密切相连。在所有国家，工业化程度决定社会的发展阶段和人民的生活水平，而在工业部门中，制造业又占有中心地位。

在今天的中国，制造业[1,2]直接创造国民生产总值的 1/3，占整个工业生产的 4/5。

① 本文是 2003 年中国机械工程学会年会特邀主题报告，深圳，2003 年 11 月 29 日；工商管理研究，北京：科学出版社，2015，177-181.

1/3 的国家财政来自制造业，90%的出口总额来自制造业，有 8000 多万名就业人员奋斗在制造业，制造业是我国国民经济的主要组成部分，既是出口的主力军，也是就业的重要市场。

新中国成立以来，我国的制造业从无到有，从小到大，逐步形成完整的工业体系；改革开放前，我们独立自主，自力更生，取得了以“两弹一星”为代表的重大成就[3]；最近 20 多年来，我国制造业有了巨大的发展，制造出了像首次载人的神舟五号飞船、秦山 300 兆瓦核电站、数字程控交换机、银河及神威高性能计算机等有形产品。

我国制造业的工业增加值已居世界第四位，仅次于美国、日本和德国。世界制造业正进一步转向中国，例如，装配计算机整机所需的零配件，95%以上都可以在珠三角的东莞市场配齐；格兰仕微波炉的年销售规模占全球市场的近 35%；江苏的计算机鼠标占世界总量的 65%；早在 1999 年，全球彩电销售量有四成在中国生产，而复印机则占六成；日本的五大汽车厂，丰田、本田、日产、三菱和马自达全部在中国设厂；奥迪轿车已经国产化，国产宝马轿车已于去年底下线，奔驰已经和北京汽车厂全面合作。

但是，事物总是有它的正反两面性，正如我们常说的科学是一把双刃剑。

制造业在创造它的辉煌的同时，也在侵蚀、污染着我们赖以生存的环境。

统计数字触目惊心，我国制造业[4]每年产生约 55 亿吨无害废物和 7 亿吨有害废物，造成环境污染的排放物有 70%归咎于制造业；每年中国大城市有 17.8 万人死于污染，全国每年因为大气污染损失 740 万个工作日；世界银行估计，环境污染给中国带来相当于 3.5%～8%的 GDP 的损失。

由于工厂排污，许多城市用水告竭，沿河两岸臭气熏天，鱼不能生，水生植物也不能长，在广州居住过的人都知道，珠江一度成了排污渠。

由于大量使用含硫量高的原煤作为能源，我国已成为重要的酸雨源，被称为世界上三大酸雨的发源地。

由于氯化物（如广泛用作制冷剂的氟利昂）对大气的污染，臭氧层逐渐变薄，相当于两个半中国面积的臭氧洞出现在南极上空，地球的防护衣正在日益破旧、褴褛。

由于大量燃烧像石油、天然气这样的化石燃料，大气的二氧化碳浓度直线上升，温室效应会使冰山融化，城市将变成泽国，印度洋上的美丽明珠群岛——马尔代夫将被海水淹没。

制造业，究竟是怎么回事？居然有七成的污染是它造成的。

难道我们生产防毒面具，仅仅是为了过滤被生产过程污染的空气？

难道我们制造飞机、火箭，仅仅是为了载着我们飞离这个已经不适合人类生存的星球？

难道我们的空调、冰箱制造企业，有意无意地制造温室效应以增加它们的销售量？

难道工业文明所引发的机器吃人的历史，将在新世纪变本加厉地重演？

难道制造业对中国经济的作用，摆脱不了成也萧何败也萧何的宿命？

回答当然是否定的！

中国制造业，面临着是否开展绿色改革的艰难选择。

我们有必要重温恩格斯的科学论述：我们必须时时记住，我们不能过分陶醉于对自然界的胜利。不管我们取得了多么辉煌的成就，我们连同我们的血、肉和头脑都是属于和存在于自然界的。

正因为工业社会在创造巨大的物质财富的同时，也创造了巨大的环境灾难，人们越来越意识到：我们不是继承了父辈的地球，而是借用了儿孙的地球。因此，从 20 世纪以来，人类对环境的关注一直没有停止过。

1962 年，美国学者卡逊出版了《寂静的春天》一书，引起了全球对人类生活环境的关注。10 年后，波托马克协会、罗马俱乐部和麻省理工学院研究小组联合出版了《增长的极限》一书，引发人们对人类发展前途的争论，人们开始思考地球上的资源和环境能否支撑人类的长期发展。

1985 年，德国的胡伯（Huber）教授提出了生态现代化理论，提出利用人类智慧去协调经济发展和生态进步。

1987 年，联合国环境与发展委员会发表《我们的共同未来》，阐述可持续发展的思想，它告诫人类：在满足当代人的需求的同时，不要对子孙后代满足其需要的能力构成危害。用通俗一点的话说就是不要涸泽而渔、杀鸡取卵，要保持人与自然和谐共存，给子孙后代留下美好、完整的发展空间。

20 世纪末，绿色消费引发了全球的绿色浪潮，从绿色食品、绿色设计到绿色制造等，绿色浪潮正在改变人类的生活和生产方式。

我国是人口众多、资源相对不足的国家（我国人口占世界人口的 22%，但是，我国淡水的人均水平是世界平均水平的 28.1%、耕地为 32.3%、森林为 9%、草地为 32.3%），我国制造业虽然 50 多年来有巨大的发展，但仍然没有走出资源型的经济增长路数，传统的以“高投入、高消耗、高污染、低效益、低产出、追求数量增长而忽视质量”为特征的发展方式不可能长期维持下去，环境污染和资源匮乏是悬在我国制造业头上的两把利剑。

但是，如果放弃制造业，我们就不能屹立于世界民族之林。这是中国人民经过 100 多年的浴血抗争所明白的真理。对此，我们老一代的科技工作者更是有着切身的体会。

因此，在党的十六大报告中提出的全面建设小康社会的目标之一便是：可持续发展能力不断增强，生态环境得到改善，资源利用效率显著提高，促进人与自然的和谐，推动整个社会走上生产发展、生活富裕、生态良好的文明发展道路[4]。

既要持续发展，又不能放弃，中国制造业只能走绿色化的道路。绿色制造是可持续发展战略在制造业中的体现，是中国制造业的必由之路。

如果说全球化是中国制造业面临的经营环境，信息化是制造业运作的有力手段，那么，绿色化似乎是一种生存哲学、一种价值取向。

绿色制造的概念，最先是由美国制造工程师学会于 1996 年所发表的关于“绿色制造”（green manufacturing）的蓝皮书中提出来的，它的内涵是：绿色制造是一种综合考虑环境影响和资源效率的现代制造模式，它的目标是使产品从设计、制造、运输以及使用到报废处理的整个产品生命周期，对环境的副作用最小，资源效率最高。简单地说，绿色制造要综合考虑制造、环境和资源这三大领域。

在技术层面上，绿色制造包含绿色产品设计技术、绿色制造技术和产品的回收与循环再制造技术，它是精益生产、柔性生产、敏捷制造的延伸和发展；作为一种指导思想，绿色制造则充满诗意并洋溢着温暖的人文关怀；具体到企业，绿色制造体现了企业家的社会责任和管理道德。

让我们来看看绿色制造的一些重要活动。

1978 年，德国最早实行绿色产品制度，产品由国家权威部门审评合格后，贴上绿色标志。

1992 年，联合国环境与发展会议通过全球《21 世纪议程》，要求各国就“环境标志产品认证”制定和实施相应政策。

1996 年，国际标准化组织正式颁布 ISO 14000 系列国际环境标准，该标准成为通往国际市场的“绿色通行证”。

2002 年，中国工程院“绿色制造与钢铁工业——钢铁工业绿色化问题”咨询项目立项，并于 2003 年结题。

2003 年 5 月，新一届美国“绿色汽车”拉力赛落下帷幕，中国制造的“黑豹”电动汽车首次亮相，并获得好评。

2003 年 6 月，世界第二大汽车制造商福特宣称将建造绿色制造基地。

近几年，“绿色壁垒”成为对外国商品进行市场准入限制的技术壁垒。

此外，美国的国家重点实验室和著名高校，纷纷开展了绿色制造的研究。

可见，绿色制造方兴未艾。绿色制造是一种大制造、大过程，它学科交叉、观念现代，对它的研究必将推动制造科学的发展。

由于中国在市场潜力、劳工价格等方面有比较优势，发达国家制造业正在向中国实行梯度转移，出于就业压力、经济发展和学习经验的考虑，现阶段，我国制造业承接转移不失为一个良策。但是，我们不能因为长期落后，便被永远“锁定”在欠发达状态，相信通过绿色制造这一路径，生态与经济能够协调发展，我们应该并且能够达到发达国家水平[5-7]。

为此，我们必须软硬兼施。

在软环境方面，我们要加强绿色制造的技术开发和基础性研究，大学应该设立相关的专业。我们要通过媒体，宣传环境保护意识，企业要有环保的责任心。

在硬环境方面，绿色制造需要法律、法规对制造企业进行约束，需要相关政策予以指导。我们应该把 ISO 14000 系列国际环境标准和环境评价纳入企业评价体系。

绿色制造，从某种程度上讲还是一个概念，对全世界漫长而又波澜壮阔的制造历史来说，它不过是“小荷才露尖尖角”，而中国制造业的绿色化道路更是刚刚起步，任重道远。

雄关漫道真如铁，而今迈步从头越。正如制造业的特长便是把图纸变为实物一样，在 21 世纪初，如何把“绿色制造”这一崭新的概念变为现实，如何化解环境危机，如何坚持走可持续发展的道路，中国制造业面临着前所未有的挑战，承担着责无旁贷的重任，更孕育着无限的生机。

我们的制造业必将唱响“春天的故事”，实现绿遍大江南北的梦想！

参 考 文 献

[1]　十六大报告辅导读本编写组. 十六大报告辅导读本. 北京：人民出版社，2002.

[2]　朱高峰. 全球化时代的中国制造. 北京：社会科学文献出版社，2003.

[3]　常平，刘人怀，林玉树. 20 世纪我国重大工程技术成就. 广州：暨南大学出版社，2002.

[4]　范维唐. 我国安全生产形势、差距和对策. 北京：煤炭工业出版社，2003.

[5]　郑必坚，杨春贵. 中国面向 21 世纪的若干战略问题. 北京：中共中央党校出版社，2000.

[6]　中国现代化战略研究课题组，中国科学院中国现代化研究中心. 中国现代化报告 2003——现代化理论、进程与展望. 北京：北京大学出版社，2003.

[7]　中国工程院. 工程科技与发展战略咨询报告集. 北京：中国工程院，2002.

旅游教材为旅游教学之本[①]

旅游活动是当今世界参与人数最多、规模最大的社会活动之一，旅游业是世界最大的产业之一。20 世纪 70 年代末期以来，中国旅游业快速增长，在 20 多年的时间里，我国已实现了从旅游资源大国向亚洲旅游强国和世界旅游大国的历史性跨越，现在正向世界旅游强国的目标阔步迈进。近几年，中国旅游业发展的内外部条件发生了根本性的变化。2001 年我国加入世界贸易组织，为了进一步与国际接轨，规范旅游业的管理，我国在旅游政策法规上做出了许多调查：在旅游资源管理方面出台了《旅游资源分类、调查与评价》；在旅行社管理方面修订了《旅行社管理条例》，制定了《设立外商控股、外商独资旅行社暂行规定》；在旅游饭店管理方面制定了《中国旅游饭店行业规范》，修订了《旅游饭店星级的划分与评定》；在旅游区管理方面修订了《旅游区（点）质量等级的划分与评定》；此外还制定和修订了《旅游发展规划管理办法》《中国优秀旅游城市检查标准》等。党中央国务院十分重视旅游事业的发展。胡锦涛总书记在 2003 年底召开的中央经济工作会议上指示："把加快发展服务业作为扩大就业、优化产业结构、提高国民经济整体效益和促进经济社会协调发展的重大举措。"温家宝总理在世界旅游组织第 15 届全体大会上指出，"目前，中国入境旅游人数和旅游外汇收入跃居世界前列，出境旅游人数迅速增加，已经成为旅游大国""我们要把旅游业培育成为中国国民经济的重要产业"。

在旅游业快速增长的同时，我国旅游教育和旅游科研蓬勃开展。目前全国共有高中等旅游院校（系、专业、所）1100 余个，在校生达 30 多万人，各种形式（全日制和非全日制）和各种层次（研究生、本科、大专、中专等）的旅游教育方兴未艾，为旅游业输送着大批专门人才。过去 20 多年，我国学者从地理学、经济学等学科对旅游问题进行研究，并逐步走向多学科的综合研究。在学者的不懈努力下，关于旅游学基本理论、旅游开发、旅游营销、旅游经营管理等的理论研究和应用研究不断有新的成果涌现。旅游实践的发展，为我们提供了广阔的研究空间。而旅游现象十分复杂，涉及许多要素，对此需要进行系统研究。

暨南大学是一所侨校，拥有侨校优势，较早地和境外旅游行业进行了接触。1978 年，商学系教师就应邀为香港中国旅行社开展旅游市场营销讲座。为适应旅游业发展对人才的需求和开办旅游专业对教师的需要，暨南大学于 1986 年在企业管理专业招收了旅游企业管理方向的研究生。1987 年，暨南大学在商学系开办了旅游经济本科专业。20 世纪 90 年代，又在深圳开办了我国第一个旅游学院，在珠海创办了设有旅游管理专业的珠海学院。这些年来，暨南大学为世界数十个国家和地区培养了大量旅游管理专门人才。与此同时，暨南大学旅游专业也在学习兄弟院校的经验中不断地发展和完善自己。

---

① 本文是《现代饭店管理》序，广州：暨南大学出版社，2004，1-2.

多年的教育实践，使我们深刻地认识到旅游教育的发展壮大需要教材建设的支持。在过去的20年间，我国旅游教育界出版了多种专业教材，为旅游学科发展奠定了基础，为旅游教育做出了贡献。但旅游业快速发展，旅游科研不断有新的突破，作为旅游教材就必须吸收这些内容，以最新的知识奉献给读者。高校旅游专业教师需要不断更新教学内容，补充新知识，教材也要体现与时俱进的特点。鉴于教材建设紧跟形势发展，反映旅游最新实践动态和最新研究成果的需要，我们组织了一批长期从事旅游专业相关课程教学和科研的教师编写了“21世纪旅游专业系列规划教材”[1,2]。

编写教材既要合乎学科的特点和规律，又要有创意。这一系列规划教材选题涉及旅游学科的各分支领域，涵盖旅游管理专业的各相关课程，内容紧密结合国际、国内旅游活动，旅游业和旅游科研发展的实际，从较高的理论起点阐述现代旅游经营管理的一般规律，总结学科、行业的经验教训，以最新的实际材料和旅游研究成果展现旅游学科体系的理论知识和实践技巧。在编写风格上，借鉴国内外旅游学科及其他学科教材的经验，力图使该套教材呈现理论的全面性、知识的丰富性、结构的合理性、形式的活泼性、内容的科学性和文字的生动性等特色。该套教材主要供高校旅游管理专业教学使用，也可作为高等职业教育、自学考试，以及旅游行业中高级管理人员的培训教材。

旅游学科作为一门发展中的学科，具有丰富的理论内涵和综合的知识结构。旅游专业教材的写作需要为此进行深入研究和精心归纳，不断吸收国内外学者的相关研究成果，同时还要真诚地接受专家及读者的批评意见，使其不断完善。最后，我们期待着该套教材能够得到读者的肯定，为旅游教育事业做出贡献。

参考文献

[1] 张永安. 现代饭店管理. 广州：暨南大学出版社，2004.
[2] 傅云新. 旅游学概论. 广州：暨南大学出版社，2004.

关系质量研究述评①

目前探讨关系质量的文献主要出现在B2B（business to business，企业对企业电子商务）、服务营销、关系营销等众多领域。研究的主题除了关系质量的维度、领域等理论问题之外，还涵盖了企业比较关注的一些实践问题，如关系质量与企业绩效的关系、关系质量对顾客购买行为的影响等。

本文将在文献梳理的基础上，结合当前研究的最新趋势，对营销理论中的关系质量研究现状做评述，并在此基础上提出有关未来研究的建议。在这里要特别指出的是，本文中的顾客是一个广义上的概念，既可指消费者，也可指购买工业产品或其他专业服务的企业，或者合作伙伴关系中的买方企业。

一、营销活动中的关系与关系质量

1. 营销活动中的关系

传统上，人们认为营销关系一般是指企业与顾客之间的关系。但关系营销需要从更为宽广的视角来探讨营销活动中的关系[1]。科特勒（Kotler）认为，企业在构建自己的营销系统时，不仅需要与顾客、供应商、分销商、竞争对手建立关系，而且还要受到营销环境的制约[2]。之后，他在“大市场营销”概念中，进一步指出要重视政府与公众等环境因素在营销活动中的影响。1992年，科特勒深化了对企业与影响企业绩效的各主体相互关系的认识，提出了“整体营销”（total marketing）概念。他认为，企业为使顾客对自己的产品与服务产生满意感，并建立顾客忠诚，必须和企业内部与外部的十种关系主体建立合作关系。关系主体就是具备影响企业实现营销目标能力的利益相关者，基本上覆盖了企业可能与之发生联系的所有力量。

Payne[3]所提出的“六市场模型”是关系营销理论中清晰描述企业营销活动中各种关系的一个著名模型。与科特勒相同，Payne尽可能地将所有的利益相关者纳入自己的研究范围。但与科特勒不同的是，他不仅从关系营销的角度分析企业要面对的关系市场构成，而且将“潜在交换”的概念融入自己的理论之中。Payne认为，企业要想在复杂多变的环境中获得成功，就必须持续使自己的顾客满意。为此，企业不能仅考虑单纯的顾客市场，而应该全面关注包括顾客市场在内的六个市场[4]，即顾客市场（现有和潜在顾客）、相关市场（现有与潜在中介组织，如批发商、零售商、代理商等）、供应商市场（现有与潜在的原材料、零部件供应商）、内部市场（员工以及企业各部门）、就业市场（待聘人员）、影响者市场（政府部门、法律部门、社会团体等），并从互惠、合作的角度来建立、维持和加强与它们的关系。“六市场模型”与Porter[5]在研究竞争优势时所提出的“五力模型”有异曲同工之妙。只不过，Porter强调利益相关者之间的相互竞争，而Payne则更注重各方力量之间的协作。

---

① 本文原载《外国经济与管理》，2005，27（1）：27-33. 作者：刘人怀，姚作为.

显然，企业在营销活动中可能涉及的关系包含四个层面：企业与内部员工、部门的关系，企业与价值链上的供应商、经销商、合作者或竞争对手等的关系，企业与宏观环境中的政府、公众的关系，以及企业与顾客的关系等。因此，企业在营销活动中，不仅需要深入探讨企业与顾客的关系质量问题，也需要面对企业与其他利益相关者的关系质量问题。根据李蔚提出的大顾客概念，除了与顾客之间的核心交易关系之外，其他利益相关者也与企业之间存在着现实或隐性的交易关系，也就是说，都可以看作企业的（广义）顾客[3,4]。据此，营销学者大多将关系质量研究的重点集中在企业与（狭义）顾客的关系质量问题上。

顾客在整个购买过程中，出于了解信息、回避风险、认知利益、抒发情感等多重目的，与企业的各方面建立起各种关系。这些关系是多层次的，具体表现为：企业的员工、经销商或分店、有形设备（如服务企业的设施）、形象、其他顾客、网站、产品/服务、品牌与顾客之间的多层关系[6]。其中，员工、经销商或分店、产品/服务、品牌与顾客之间的关系比较直接，并会对双方关系的形态、延续时间以及顾客购买行为的约束力量产生重要影响。如果顾客对企业的产品或服务感到满意，就会增加其下次选择同一品牌的机会[7]，促进重复购买，进而使其试图与企业建立关系[8]。无论是理论分析还是实证研究均表明，仅满意并不能导致顾客忠诚。只有高度满意的顾客才能持续青睐企业的品牌，在寻求利益与心理满足的同时，形成对企业及其品牌的忠诚[9]，并引发关系型倾向与需求[6]。实际上，学者 Jacoby 和 Kyner[10] 早在 1973 年就已经指出，顾客忠诚本质上属于一种关系现象。这说明，在顾客忠诚的状态下，顾客与企业之间往往建立起一种可持续的高质量的关系。

2. 关系质量的定义

尽管有关关系质量的文献不少，但对关系质量的定义依旧是不清晰的。在现有文献中，多数学者基本上均参考了 Crosby 等[8] 在 1990 年给出的关系质量定义。从人际关系角度出发，Crosby 等认为，（销售人员与顾客之间的）关系质量就是顾客在过去满意的基础上，对销售人员未来行为的诚实与信任的依赖程度。深入探究其研究的推理过程，可以发现 Crosby 等已经将顾客与企业之间的关系做了情节（一次交易）与关系（多个连续情节）的区分[8]。基于以上研究，Liljander 和 Strandvik[11] 根据顾客感知的特点，详细界定了情节与关系的含义，进一步提出了情节质量与关系质量的区别，将服务行业中的关系质量定义为顾客在关系中所感知到的服务与某些内在或外在质量标准进行比较后的认知评价。应该说，上述研究框架实现了一般质量概念与关系质量概念的对接，为后来的关系质量理论演进奠定了基础[12]。

也有学者对其他背景下的关系质量进行了探索。Johnson[13] 将营销渠道成员之间的关系质量解释为成员关系的总体深度与气氛。而 Holmlund[14] 则在前人研究的基础上，提出了更具适应性的 B2B 状态下的关系质量定义。他指出："感知关系质量是指商业关系中合作双方的重要人士根据一定的标准对商业往来（效果）的综合评价和认知。"很明显，该定义是依照商业伙伴之间的人际关系感知来确定双方的关系质量，这不仅在方法上与 Crosby 等[8] 的定义类型相同，而且在研究思路上也沿用了 Liljander 和 Strandvik[11] 的顾客感知评价。此外，与上述作者的行业性定义不同，Hennig-Thurau

和 Klee[15]试图提出具有一般意义的关系质量概念，他们认为，关系质量一般被看作关系对顾客的关系型需求的满足程度，又可以归结为顾客对营销者及其产品的信任与承诺。这里的产品是指广义产品，既包括服务也包括传统产品。不难看出，Hennig-Thurau 和 Klee 的关系质量概念实际上继承了 Morgan 和 Hunt[16]有关成功关系特征的描述。

综合众多学者对关系质量内涵的认识，可以给出以下定义：作为感知总质量的一部分，关系质量是关系主体根据一定的标准对关系满足各自需求程度的共同认知评价。其实质就是指能够增加企业提供物的价值，加强关系双方的信任与承诺，维持长久关系的一组无形利益。

二、关系质量的基本维度与领域

弄清楚关系质量的基本维度、领域等重大问题，对于企业而言，其重要性与搭建通向成功关系的阶梯无异。在现有文献中，多数涉及关系质量维度的研究，基本上都是围绕着企业或顾客各自的特征、互动的界面和关系管理三个方面来展开的。

信任与满意无疑是多数文献所公认的关系质量结构的重要维度。这个二维结构首先是由 Crosby 等[8]在零售业的背景下提出的。他们认为信任是指对推销人员的信任（Swan 等[17]），而顾客满意是指对推销人员的满意（Crosby 和 Stephens[18]）。值得关注的是，Crosby 和 Stephens[18]将信任与满意等互动特征作为关系质量结构中的内生维度，而将相似服务领域的专长等个人特性与关系型交易行为作为关系质量的外生变量。这一划分无疑运用了人际关系理论的研究方法。

以后的学者所提出的维度模型，在 Crosby 及其同事的研究基础上，主要沿着两个方向演变。一种是从关系双方的互动角度出发，着眼于关系的管理，增加对承诺、沟通质量、冲突解决以及双方关系管理等因素的重视，以减少机会主义行为等现象。例如，Mohr 和 Spekman[19]在实证分析的基础上，提出任何成功的伙伴关系的基本特征是承诺、合作、信任、沟通质量、参与以及冲突的共同解决等。此模型是对 Dwyer 等[20]提出的由满意、信任以及减少机会主义行为因素等组成的模型的扩展。另一种是从关系赢利的角度考虑关系质量的维度。例如，Storbacka 等[21]运用新古典经济学与交易成本理论的研究方法，以提高企业盈利能力为目标，构建了一个包含服务质量、顾客满意、关系力量、关系长度与关系盈利能力等因素的关系质量动态模型，他们所指出的关系质量维度包含满意、承诺、沟通和联系等因素。此外，也有学者[22]特意将人际交往与关系相区别，把营销关系的变量分为两个层面——人际交往与关系，并将承诺、共同目标和关系利益三个因素作为关系质量的维度。更有学者从整体质量感知的角度将关系质量的维度确定为信任、承诺和总体质量感知三个因素。值得关注的是，学者研究问题的理论视野已经从早期的人际关系理论单一层面扩大到人际关系、交易成本和新古典经济学等多理论整合的层面。

Holmlund[14]的关系质量维度研究是与领域研究交织在一起的。他认为，作为价值创造过程中的无形提供物，商业关系不仅涉及社会因素，而且还牵涉合作双方的技术与经济因素。因此，有关研究不能仅从社会的角度，还应从更广阔的视野来进行分析[15]。Holmlund以早期的服务质量模型为基础，把服务质量的过程与结果维度扩展为关系质量的过程与结果领域。每一个领域均包含技术、社会和经济三个维度。他认为，作为价值创造过程中的两个竞争范围，过程与结果领域代表着商业伙伴对关系质量的感知过程与感知结果。而技术、社会和经济这三个维度又是指对关系质量竞争方向的具体划分，代表着对关系质量的过程与结果的不同评价指标。通过对领域与维度进行组合，Holmlund将技术维度进一步分为过程类型（设计、生产、库存控制、运输、维修与补救等)、过程特性（可靠、创新、能力的使用、速度、有形资源的使用、灵活与安全）和技术结果（可靠性、创新性、一致性、美观性与耐久性）三个子维度。在社会维度中，结果被分为个人（感染力、信任、相知、尊敬、亲和力与喜悦）与企业（内部凝聚力、吸引与信任）两个层次。过程的子维度则与个人结果的子维度完全相同。在经济维度中，结果按成本收益被区分为关系利益（具有竞争力的价格、规模、边际利润、生产率提高与隐性关系奖励）与关系成本（直接关系成本、间接关系成本与隐性关系成本）两个部分。而过程则被分为定价、成本计算和生产率三个子维度。

应该说，Holmlund的研究企图纠正关系质量研究过分重视社会因素的问题，力求综合运用多种理论的研究方法，从动态、广泛的角度建立一个更具适用性的模型。但由于其对有关研究领域的界定过于宽泛，导致其维度描述与其他主流文献的研究出现了较大偏差，如经济维度中的关系利益与关系成本，在其他学者[23]的研究中被当作关系质量的前置因素或关系价值的构成部分[24]，而技术维度的几乎所有子维度及其内含维度，基本上反映的是关系所涉及的产品或可服务本身的技术质量，此类质量的好坏会影响顾客对产品或服务的质量感知，进而影响顾客对关系质量的感知，但能否作为关系质量的维度还有待深入探讨。此外，Holmlund的维度结构过于复杂，很难用来进行实证分析。

综合以上分析可以看出，关系质量包含过程与结果这两个涉及关系价值创造活动的竞争领域，而关系质量的维度应该更多地从社会交往的角度，根据不同研究行业来具体选择。例如，在B2B行业与营销渠道成员关系中，沟通质量、参与、冲突的处理、关系投资等理应成为关系质量的维度备选因素。但毋庸置疑，不论在什么行业背景下，信任、满意和承诺均是主要的关系质量维度。

三、关系质量与顾客购买行为

在关系质量研究领域，有关顾客购买行为的研究基本上是围绕着顾客期望—顾客感知—顾客满意—购后行为这一链条进行的。而学者所提出的研究模式，尽管均认同关系质量可以作为一个影响顾客购买行为的重要干涉变量，但对于其在研究链中的位置，以及其与主要研究变量的关系有不同的看法。

营销文献显示，长期以来，学者都将服务质量作为顾客满意的前置因素[12]。在关系质量概念出现之后，其与服务质量的相互关系就成了亟待解决的问题。Liljander和Strandvik[11]的理论研究，通过引入情节质量（单次交易的服务质量）和关系质量等新概念，较好地解释了两者的关系，并且在传统的行为研究链中植入了情节价值与关系价值等评价顾客感知价值的新指标，以解释关系对顾客感知、顾客满意和顾客购后行为的影响。值得注意的是，Liljander 和 Strandvik 认为，关系质量是服务质量的结果因素，又是关系满意的前置因素，即把关系质量与服务质量当作两个不同的结构。这一重要的结论从实证角度看，至少在零售领域[25]以及专业服务背景中[26]已经得到验证，与 Liljander 和 Strandvik[11]等的观点不同，有些学者在研究中运用了广义的服务质量概念（包含产品质量在内），并指出关系质量与服务、产品质量至少是内涵相互重叠的概念。Hennig-Thurau 和 Klee[15]及 Hennig-Thurau[27]认为，服务、产品质量应该是关系质量的基础，可以用总质量概念来概括地表示这两个概念。

学者大多认同顾客购后行为（如顾客忠诚、顾客维持等）是关系质量的后果，但对于关系质量与顾客满意的关系也有不同的看法。一种意见认为[15,27]，顾客满意是关系质量的前置因素。而另一种意见则认为[23,28]，顾客满意应该成为关系质量的基本维度。不过，这两种观点均得到了实证支持。出现此类争论的原因就在于前者将满意剔除在关系质量结构之外，而后者将满意包含在同样的结构之中。

此外，尽管多数学者的实证结果均确认了关系营销活动、关系质量与购后行为之间的正向关系，但还应该看到，目前学术界对于关系质量究竟是不是独立的行为解释变量缺乏有力的实证支持。Roberts 等[26]的研究在此问题上有所突破，他们以消费者服务行业为研究对象，通过实证分析，证明了关系质量对顾客购后行为意图解释能力是一个比服务质量更具解释力的基本结构，这无疑是重要的学术贡献。然而，受选取对象的局限，对 Roberts 等的研究尚需进一步验证。图 1 描述了目前主要涉及顾客购买行为的关系质量文献所采用的结构框架与具体研究结果。

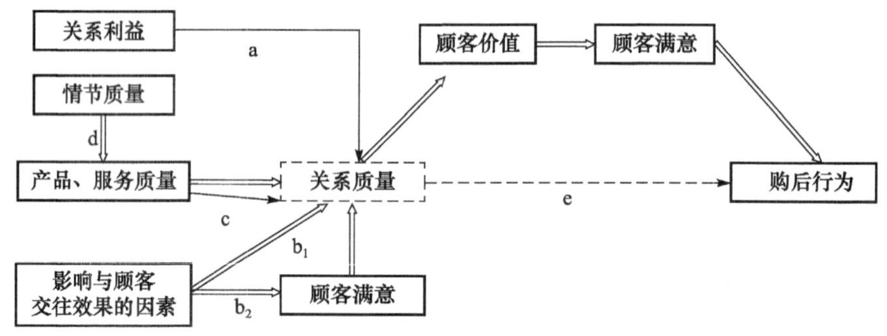

图 1　顾客购买行为研究结构中的关系质量

a 为多个服务行业[23]；b_1 为零售业[8]、美发业[28]；b_2 为电子消费品制造业[15,27]；c 为服务业[26]；d 为服务业[11]；e 为多数研究均公认的关系

四、关系质量与企业绩效

从企业的角度来看，关系质量与企业绩效的关系要比关系质量结构对顾客购买行为的解释力显得更加重要。因此，在关系质量文献中，此类问题也成为学者的重点研究课题。

尽管后期的学者也对关系质量、顾客维持行为与盈利能力的关系进行了理论或实证探讨，但他们的研究均无法同早期学者[21]的研究成果相比。Storbacka 等[21]考察了双方关系质量中的信任、满意与承诺等因素，并将顾客感知价值、关系力量、关系长度与关系盈利能力联系起来，以衡量企业与顾客关系的影响力以及关系所产生的绩效，从而实现交易与关系的对接。他们的研究思路显然为探讨关系质量与企业绩效的联系建立了一种动态的研究框架（图 2）。要特别指出的是，以上研究中所运用的关系力量与关系质量在内涵上有较大的重叠。Storbacka 等认为，关系力量的主要维度为联系与承诺等。这两个维度与关系质量的维度基本吻合，而企业与顾客关系的强势状态无疑是关系质量好的重要表现。

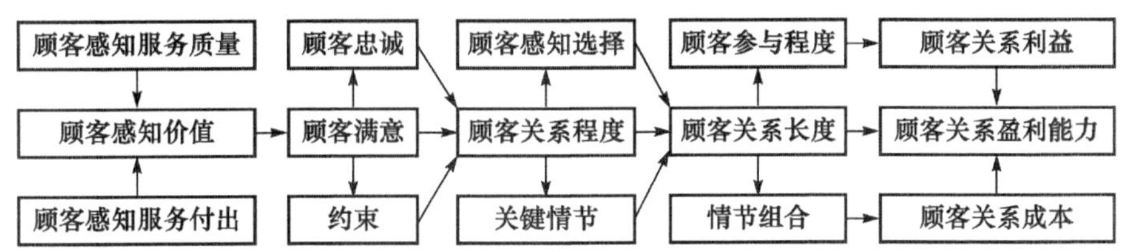

图 2　关系质量与企业绩效的联系[21]

五、小结与研究展望

关系质量理论是萌芽于关系营销理论的一种新理论[15]，它以人际关系研究范式为主，整合了交易成本、关系接触与新古典经济学等多方面的研究方法，运用经济学、社会学和心理学等多学科知识，研究顾客与企业之间的关系满足双方需求的程度，并对关系效果进行认知评价。关系质量的领域涉及过程与结果，都包含在关系的价值创造活动中，而关系的基本维度常因所研究的行业差异而不同，其中，信任、承诺和满意是公认的质量维度。实证研究证明，关系质量会受到顾客满意的影响，它本身却是顾客购后行为的前置因素。此外，关系质量还是预示企业绩效大小的重要指标。

显然，目前的关系质量研究虽然涉及众多的研究课题，也取得了一定成果，但在许多方面还需要进一步深化。未来的关系质量研究应该注重以下几个方面。

（1）选取多行业样本，通过实证分析来验证关系质量结构是一个具有一般性的独立结构，并且是比服务质量或产品质量更有解释力的顾客购买行为变量。要深入分析关系质量结构的前置、后果因素与基本维度内部各因素之间的关系，并扩大购后行为所涉及的变量（如口碑、抱怨等），继续验证关系质量对购后行为的影响效果。应注意综合运用多学科的知识，整合多个研究范式来改善关系质量的研究方法与研究思路，当前可考虑引入心理学、社会学等研究范式。

（2）在不同的行业背景下，研究关系质量在顾客购买行为中的具体位置。即使在同一个行业，也可利用细分标准将其分为不同的样本，以期通过对照比较来拓展研究。鉴于顾客价值在解释顾客购买行为过程中的重要作用，可考虑研究产品、服务质量、顾客价值、关系质量、顾客满意与顾客购后行为之间的复杂关系，尤其要探讨在关系质量的影响下不同的研究变量对于顾客购买行为的解释力有无变化。

（3）通过多个行业的实证分析来验证关系质量与企业绩效的关系，为 Storbacka 等[21]的理论提供支持，因此可考虑从关系资产的角度来分析问题。

（4）检验关系质量结构在不同文化背景下是否存在差异性。当务之急是，构建我国本土文化背景下的关系质量模型，对我国特有的混合型关系营销现象进行理论与实证分析。

（5）将关系质量理论运用于其他类型的营销关系，如企业与政府之间的关系、企业与公众之间的关系和企业与非营利机构之间的关系等。

参考文献

[1] 克里斯托弗 M，培恩 A，巴伦廷 D. 关系营销——如何将质量、服务和营销融为一体. 李宏明，李涌，译. 北京：中国经济出版社，1998：31.

[2] Kotler P. Marketing Management：Analysis，Planning，Implementation，and Control. 9th ed. Englewood Cliffs：Prentice-Hall，1997：15.

[3] Payne A. Relationship marketing：A broadened view of marketing//Payne A. Advances in Relationship Marketing. London：Kogan Page Ltd.，1995：29-40.

[4] Gummesson E. Making relationship marketing operational. International Journal of Service Industry Management，1994，5（5）：5-20.

[5] Porter M E. How competitive forces shape strategy. Harvard Business Review，1979，57（2）：137-145.

[6] Sheth J N，Parvatiyar A. Relationship marketing in consumer markets：antecedents and consequences. Journal of the Academy of Marketing Science，1995，23（4）：255-271.

[7] Oliver R L. Whence consumer loyalty?. Journal of Marketing，1999，63（4）：33-44.

[8] Crosby L A，Evans K R，Cowles D. Relationship quality in services selling：An interpersonal influence perspective. Journal of Marketing，1990，54（3）：68-81.

[9] Jones T O，Sasser Jr W E. Why satisfied customers defect. Harvard Business Review，1995，73（6）：88-101.

[10] Jacoby J，Kyner D B. Brand loyalty vs. repeat purchasing behavior. Journal of Marketing Research，1973，10（1）：1-9.

[11] Liljander V，Strandvik T. The nature of customer relationships//Swartz T A，Bowen D E，Brown S W. Advances in Services Marketing and Management. London：JAI Press，1995：141-167.

[12] Grönroos C. Marketing services：The case of a missing product. Journal of Business and Industrial Marketing，1998，13（4，5）：322-338.

[13] Johnson J L. Strategic integration in industrial distribution channels：Managing the interfirm relationship as a strategic asset. Journal of the Academy of Marketing Science，1999，27（1）：4-18.

[14] Holmlund M. The D & D model-dimensions and domains of relationship quality perceptions. Service Industries Journal，2001，21（3）：13-36.

[15] Hennig-Thurau T，Klee A. The impact of customer satisfaction and relationship quality on customer retention：A critical reassessment and model development. Psychology & Marketing，1997，14（8）：737-764.

[16] Morgan R M，Hunt S D. The commitment-trust theory of relationship marketing. Journal of Marketing，1994，58（3）：20-38.

[17] Swan J E, Trawick I F, Silva D W. How industrial salespeople gain customer trust. Industrial Marketing Management, 1985, 14 (3): 203-211.

[18] Crosby L A, Stephens N. Effects of relationship marketing on satisfaction, retention, and prices in the life insurance industry. Journal of Marketing Research, 1987, 24 (4): 404-411.

[19] Mohr J, Spekman R. Characteristics of partnership success: Partnership attributes, communication behavior, and conflict resolution techniques. Strategic Management Journal, 1994, 15 (2): 135-152.

[20] Dwyer F R, Oh S. Output sector munificence effects on the internal political economy of marketing channels. Journal of Marketing Research, 1987, 24 (4): 347-358.

[21] Storbacka K, Strandvik T, Grönroos C. Managing customer relationships for profit: The dynamics of relationship quality. International Journal of Service Industry Management, 1994, 5 (5): 21-38.

[22] Parsons A L. What determines buyer-seller relationship quality? An investigation from the buyer's perspective. Journal of Supply Chain Management, 2002, 38 (1): 4-12.

[23] Hennig-Thurau T, Gwinner K P, Gremler D D. Understanding relationship marketing outcomes: An integration of relational benefits and relationship quality. Journal of Service Research, 2002, 4 (3): 230-247.

[24] Ravald A. Grönroos C. The value concept and relationship marketing. European Journal of Marketing, 1996, 30 (2): 19-30.

[25] Wong A, Sohal A. Customers, perspectives on service quality and relationship quality in retail encounters. Managing Service Quality, 2002, 12 (6): 424-433.

[26] Roberts K, Varki S, Brodie R. Measuring the quality of relationships in consumer services: An empirical study. European Journal of Marketing, 2003, 37 (1-3): 169-196.

[27] Hennig-Thurau T. Relationship quality and customer retention through strategic communication of customer skills. Journal of Marketing Management, 2000, 16 (1-3): 55-79.

[28] Shamdasani P N, Balakrishnan A A. Determinants of relationship quality and loyalty in personalized services. Asia Pacific Journal of Management, 2000, 17 (3): 399-422.

消费者行为研究是营销理论的基石①

此书是我的学生姚作为博士在他的博士论文《基于品牌关系的服务消费决策行为研究》的基础上修改与丰富之后形成的作品。

消费者行为研究是营销理论的基石，任何营销理论的发展均需要结合时代特点而不断更新的消费者行为理论。无论从世界还是从中国来看，作为经济发展的主要动力，服务业对改善就业，促进产业结构优化，提高人民生活水平均有着不可忽视的作用。跟随当前经济改革的深化与产业结构的调整，中国经济的发展已经开始从依赖传统的制造业转向依靠服务业，从产品竞争走向基于知识创新的服务竞争。因此，运用新的理论来研究中国消费者的服务消费决策行为理论对于企业应对未来的新挑战具有重要的意义。

在服务经济时代，个性独特的品牌不仅可以帮助消费者简化购买决策过程，降低风险，而且可以通过提供卓越的服务体验与顾客价值，使消费者感受品质生活与品味不同的人生。从这个意义上讲，品牌已经成为消费者彰显自我、寻求认同的形象符号。而企业界则需要加强对品牌的培养来吸引消费者的目光，通过建立有关服务消费决策行为方面的品牌知识库，制订有效的营销方案来满足顾客不断提高的需求与欲望。因此，面对日益复杂的竞争环境与互动任务，品牌就成为企业与顾客之间进行情感沟通的工具与桥梁。作为西方服务营销理论的重要组成部分之一，服务消费决策理论虽然已经形成了相对成熟的理论体系，但要将有关理论运用于中国实践，必须通过本土化创新来探讨中国文化对服务消费行为的影响。姚作为博士在深挖中国文化内涵的基础上，运用国际上规范的研究方法，从品牌关系的前沿角度来研究服务品牌消费决策行为。这项研究不仅深化了人们对有关理论的认识，并且为有关理论与中国文化的接轨提出了自己的见解。

本书在大量阅读国内外相关文献的基础上，以关系营销理论与顾客价值理论为基础，将品牌关系视为品牌与顾客互动沟通所形成的隐性心理环境变量，运用品牌关系质量作为反映这一环境本质的重要结构，来研究品牌关系对顾客的服务消费决策行为变量——顾客价值、顾客满意与购后行为的影响，并据此提出了一个基于品牌关系的服务消费决策行为理论框架。为检验此理论，作者以位处广州、佛山与珠海等地的来自证券、银行与美容美发等行业的多个品牌的顾客作为调查总体，实施了两轮问卷调查，运用科学、规范的研究方法对此概念模型进行了实证检验。研究结果表明，标志着消费者与品牌之间的情感沟通程度的品牌关系质量，不仅影响消费者的服务消费决策行为，还是一个比传统的服务质量更为重要的、有着自己独立解释力的行为变量。本研究提出了一些创新的研究成果，在运用前沿的品牌关系理论解释实际消费行为研究方面获得了一定的进展。

姚作为博士学术思想活跃，学风务实。此书的出版有重要的学术价值，故乐为之序。

---

① 本文是《服务消费决策行为研究——基于品牌关系的角度》序，北京：中国标准出版社，2007，3-5.

从 CSSCI 旅游研究文献看旅游学学科发展①

由南京大学与香港科技大学联合研制成功的“中文社会科学引文索引”（Chinese Social Science Citation Index，CSSCI）是国家、教育部重点攻关项目，是一个文摘型数据库，是我国第一个大型人文社会科学科研成果检索和评价工具。作为我国社会科学主要文献信息统计、查询与评价的重要工具，CSSCI 提供多种信息查询、检索途径，可以为社会科学研究者提供国内社会科学研究前沿信息和学科发展的历史轨迹；为社会科学管理者提供地区、机构、学科、学者等多种类型的统计分析数据，从而为制订科学研究发展规划和科研政策提供科学合理的决策参考。利用 CSSCI 来对某一学科的发展状况和期刊影响力进行评价，已经在许多学科得到应用[1-8]，但利用 CSSCI 来研究旅游学和旅游期刊，据我们所知，本文还是第一个。

一个科学工作者在撰写论文时，为了给自己的论文提供佐证或背景材料，或者是为了指明自己的写作依据、尊重别人的劳动，一般都会在文后列出自己的“参考文献”，以便读者了解该文吸取或采用了何人、在何处提出的概念、理论、方法等。图书馆情报界把作者所撰写的论文称为“发文”“载文”或“来源文献”，文后的“参考文献”称为“引文”或“被引文献”（citation）。文献的引证关系，比较深刻地反映了科学文献之间的内在联系[9]。旅游发文，又称旅游论文，特指篇名中含“旅游”字段的论文；旅游引文，特指篇名中含“旅游”字段的被引文献。旅游发文和旅游引文排除了许多篇名中仅含“餐饮业”“饭店业”“旅行社”等字段而不含“旅游”字段的论文，这是为了统计旅游研究文献所采取的一种技术措施。旅游期刊，指登载旅游发文的期刊，即该种期刊登载篇名中含“旅游”字段的论文。

一、从 CSSCI 来源文献看旅游学学科发展

CSSCI 来源期刊的评定不是按主办单位的行政级别评定的，而是按引文量、影响因子、专家意见等评定的，只要具有正式刊号，又是学术性的期刊，则不论其行政级别都可参加评选[10]。南京大学中国社会科学研究评价中心咨询委员会于 2004 年 4 月在成都召开会议，对 CSSCI 来源期刊进行了讨论。经过一系列的工作和相关程序，对因停刊、合并及不符合 CSSCI 来源期刊收录原则的，以及根据 1998～2001 年的期刊影响因子平均值分学科排序且位于该学科应选期刊总数之外的，按照“末位淘汰”的原则，对原有的 18 种期刊进行了删除，并在来源期刊总量上增加了 42 种，最后按影响因子的位序，增补了位于相关学科前列的 60 种期刊，经调整后 CSSCI 来源期刊为 461 种。

1. 旅游发文篇数的年际分布

从表 1 可以看出，1998～2005 年，旅游发文年平均 435 篇，呈逐年上升趋势，年增长率平均为 17.14%，尤其 2004 年增幅最大，达 47.03%。近几年来，已有许多学者

---

① 本文原载《人文地理》，2007，22（4）：71-81. 作者：刘人怀，袁国宏.

对中国旅游研究状况进行了分析、评价[11-15]。但几乎都集中在对《旅游学刊》的研究，用对《旅游学刊》的文献研究来代替对中国旅游文献的研究，似有以偏概全之嫌。虽然《旅游学刊》是“全国中文核心期刊”，已经连续多年名列“旅游经济类核心期刊”第一名，“在一定程度上反映了中国旅游研究的水平和发展动向”“公认的比较权威的旅游专业学术刊物”。但在 1998～2005 年，《旅游学刊》共登载旅游发文 633 篇，CSSCI 来源期刊共登载旅游发文 3483 篇，《旅游学刊》登载的旅游发文篇数只占 CSSCI 旅游发文总数的 18.17%。

表 1　1998～2005 年 CSSCI 旅游发文年篇数及年增长率

项目	1998 年	1999 年	2000 年	2001 年	2002 年	2003 年	2004 年	2005 年	平均
旅游发文年篇数	248	335	336	384	393	438	644	705	435
旅游发文年增长率/%	—	35.08	0.30	14.29	2.34	11.45	47.03	9.47	17.14

资料来源：CSSCI 数据库（检索时间：2006-07-21）

2. 旅游发文篇数的学科门类分布

尽管我国旅游管理专业的本科生都要学习“旅游学概论”这门课程，但“旅游学”还没有获得正统的学科地位。在我国现行的学科分类体系中，只有二级学科“旅游经济学”和三级学科人文地理学下的“旅游地理学”研究方向；学位管理体系中，只有二级学科“旅游管理”；中国图书分类法中，只有经济类下的交通运输经济类下的“旅游经济类”[16]。因此，在 CSSCI 期刊学科门类中就没有“旅游学”这一类，旅游学的特定期刊《旅游学刊》《旅游科学》都划归到人文、经济地理这一门类。

2003～2005 年 CSSCI 旅游研究文献的期刊学科门类分布情况见表 2，表 2 将旅游期刊学科门类按旅游发文总数进行排序，旨在从期刊的视角了解旅游研究到底横跨了哪些学科门类，旅游研究的综合性程度到底有多大，各学科对旅游研究涉足有多深。从表 2 中可以看出：在 CSSCI 划分的 25 类期刊学科门类中，涉足旅游研究的期刊学科门类有 24 类，旅游学科综合广度为 96%。这与旅游现象的复杂性以及因基础研究薄弱致使旅游研究对象范围至今尚无定论有很大关系。旅游发文主要登载在人文、经济地理，经济学，综合性社科期刊，高校综合性社科学报，民族学，管理学，环境科学 7 类学科门类上，占旅游发文总数的 89.65%。因此旅游学科综合深度体现在人文、经济地理，经济学，民族学，管理学，环境科学 5 大类，这 5 类学科门类应该是旅游研究文献的 5 大载体。

表 2　2003～2005 年 CSSCI 旅游研究文献的期刊学科门类分布情况

期刊学科门类	2003 年篇数	2004 年篇数	2005 年篇数	三年篇数合计
人文、经济地理	175	273	289	737
经济学	84	76	103	263
综合性社科期刊	25	79	76	180
高校综合性社科学报	41	57	29	127
民族学	25	33	47	105
管理学	27	33	37	97

续表

期刊学科门类	2003 年篇数	2004 年篇数	2005 年篇数	三年篇数合计
环境科学	22	37	34	93
体育学	7	10	22	39
统计学	1	12	15	28
图书、情报与档案学	4	8	8	20
社会学	5	5	8	18
政治学	6	2	10	18
新闻与传播学	2	4	7	13
历史学	1	4	2	7
教育学	2	4	0	6
哲学	1	1	3	5
法学	0	1	3	4
宗教学	1	2	0	3
语言学	0	2	1	3
艺术学	1	0	2	3
心理学	0	1	1	2
马克思主义	1	0	1	2
考古学	0	0	2	2
外国文学	0	0	1	1
中国文学	0	0	0	0
总计	431	644	701	1776

资料来源：根据 2003～2005 年 CSSCI 来源期刊和 CSSCI 数据库统计而成

3. 旅游发文篇数的期刊分布

期刊文献计量学评价是采用数学、计算机科学、系统科学等方法对期刊文献的产生、发展、管理等各项过程以及各种技术动态特征进行统计、分析、研究的一门科学。早在 1943 年英国著名的文献学家塞缪尔·克莱门特·布拉德福（Samuel Clement Bradford）发现并提出定量描述文献序性结构的经验定律，认为对某一特定的主题，有关它的学术论文大量地集中在为数不多的刊物之上，其余的论文则分散在为数众多的大量刊物之中。美国《科学引文索引》（*Science Citation Index*）的创始人 E. 加菲尔德（E Garfield）通过研究文献被引用的情况，也得到了类似结果，即大量被引文献出自少数期刊，而其余少数被引文献则分散在大量的刊物上[9]。2003～2005 年 CSSCI 来源期刊中，旅游期刊 197 种，共登载旅游发文 1776 篇，其中，旅游发文数量前 40 位的期刊登载了旅游发文 1416 篇，即 20.30％的期刊登载了旅游发文总数的 79.73％，正好符合 20∶80 分布规律。三年来旅游发文量超过 100 篇的期刊有《旅游学刊》《经济地理》《旅游科学》。

4. 旅游期刊种数的学科门类分布

2003～2005 年旅游期刊在各学科门类中的分布广度排序见表 3，可以看出：2003～2005 年，CSSCI 的 463 种来源期刊中有 197 种登载了旅游论文，即 42.55％的期刊已经介入旅游研究，说明旅游科学作为综合性交叉科学，学科的发展非常活跃。民族学，人文、经济地理，环境科学，体育学，统计学类的期刊对旅游研究的介入程度最高，这 5 个学科类的期刊全部都登载了旅游发文。综合性社科期刊、高校综合性社科学报、管理学、经济学、新闻与传播学类的期刊有一半以上登载了旅游发文。历史学、中国文学类的期刊对旅游研究的介入程度最低。

表 3　2003～2005 年旅游期刊占各学科来源期刊种数的百分比排序

期刊学科门类	CSSCI 来源期刊种数	旅游期刊种数	旅游期刊占来源期刊种数的百分比/%
民族学	12	12	100
人文、经济地理	9	9	100
环境科学	8	8	100
体育学	7	7	100
统计学	4	4	100
综合性社科期刊	39	29	74.36
高校综合性社科学报	35	23	65.71
管理学	26	15	57.69
经济学	69	39	56.52
新闻与传播学	14	7	50
哲学	11	4	36.36
图书、情报与档案学	20	7	35
政治学	36	10	27.78
社会学	8	2	25
宗教学	4	1	25
法学	21	4	19.05
艺术学	17	3	17.65
外国文学	6	1	16.67
马克思主义	13	2	15.38
考古学	7	1	14.29
心理学	7	1	14.29
教育学	30	4	13.33
语言学	19	2	10.53
历史学	26	2	7.69
中国文学	15	0	0
总计	463	197	42.55

资料来源：CSSCI 数据库（检索时间：2006-07-21）

二、从 CSSCI 被引文献看旅游学学科发展

引文索引是指将来源文献（发文）和被引文献（引文）有机组织形成的二次文献，它既可作为一种新型的文献信息查询工具，又可作为对论文、学者、学科、机构、地区等科研情况的统计以及对期刊进行分析的有力工具[17]。通过引文的统计与分析，可以从一个重要侧面揭示学科研究与发展的基本走向，评价科学研究质量，为人文社会

科学事业发展与研究提供第一手资料[18]。

1. 旅游引文总被引频次的年际分布

总被引频次（total citations）指期刊在评价当年被其他期刊及期刊本身所引用的总次数。它可以反映学术期刊在学术界被注意的程度，是衡量期刊学术影响大小的重要指标之一，一般来说，其值越大，期刊作用就越重要[19]。从表 4 可以看到，1998～2005 年，旅游引文的年总被引频次平均为 1628 篇，呈逐年上升趋势，年增长率平均为 36.62%，尤其 2004 年增幅最大，达 86.60%。

表 4　1998～2005 年 CSSCI 旅游引文年总被引频次及年增长率

年份	1998 年	1999 年	2000 年	2001 年	2002 年	2003 年	2004 年	2005 年	平均
旅游引文的年总被引频次	447	650	739	1037	1395	1806	3370	3584	1628
旅游引文年增长率/%	—	45.41	13.69	40.32	34.52	29.46	86.60	6.35	36.62

资料来源：CSSCI 数据库（检索时间：2006-07-21）

2. 旅游发文篇均被引频次的年际分布

从表 5 可以看出，1998～2005 年，旅游发文篇均被引频次平均为 3.74 次，同样呈不断上升趋势，年增长率平均为 16.49%，尤其 2002 年度增幅最大，达 31.44%。一方面说明，旅游发文引起学术界重视程度提高快，旅游期刊学术影响力和学术成果扩散速度加快。这与旅游产业地位的快速提升是一致的。1998 年中央经济工作会议首次将旅游业列为国民经济新的增长点，2000 年 1 月国家旅游局提出“建设世界旅游强国，培育新兴支柱产业”的宏伟目标，2001 年 1 月国务院在北京召开全国旅游发展工作会议，出台了“关于进一步加快旅游发展的建议”，随后全国有 24 个省提出把旅游业作为“支柱产业/主导产业/先导产业”来发展。另一方面说明，旅游期刊质量稳步提高，学术影响力不断提升。

表 5　1998～2005 年 CSSCI 旅游发文篇均被引频次及年增长率

年份	1998 年	1999 年	2000 年	2001 年	2002 年	2003 年	2004 年	2005 年	平均
旅游发文总篇数	248	335	336	384	393	438	644	705	435
旅游引文的年总被引频次	447	650	739	1037	1395	1806	3370	3584	1629
旅游发文篇均被引频次	1.80	1.94	2.20	2.70	3.55	4.12	5.23	5.08	3.74
篇均被引频次年增长率/%	—	7.78	13.40	22.73	31.48	16.06	26.94	−2.87	16.50

资料来源：CSSCI 数据库（检索时间：2006-07-21）

三、结论与讨论

（1）随着网络的普及与广泛应用，旅游学科已从传统的应用型学科发展成为一个充满时代特色、极具现代化色彩的新兴学科。作为一门应用型学科，它与时代发展紧密相关，已从早期构筑在传统理论、技术、方法、工具的学科，发展成为以现代科学理论与方法为基础、借助计算机与网络等现代化技术手段实现其目标的崭新学科。1998～2005 年，CSSCI 共收录旅游发文 3483 篇，旅游引文 13 028 次，篇均引文数为 3.74 次，这些论文和引文为我们多角度、全方位地分析评价旅游学各类成果的学术影响提供了数据保证。8 年来旅游发文数年增长率 17.14%，旅游引文数年增长率 36.62%，有力地证明了旅游研究越来越成为学术热点，旅游发文的影响力越来越大，旅游学科是一门名副其实的热门学科、新兴学科、朝阳学科。

(2) 在国外旅游研究中，旅游期刊（tourism research）与接待业期刊（hospitality research）属于两种不同类型的期刊，因此旅游研究与接待业研究也应属于两种不同类型的研究领域。从这一视角看，本书根据“篇名”中是否含“旅游”二字来判断一篇论文是否为旅游论文具有一定的合理性。根据 Mckercher 等[20]对旅游与接待业杂志的排名，旅游领域前 10 名杂志与接待业领域前 10 名杂志见表 6。

表 6 旅游与接待业期刊排行前 10 名列表

排名	旅游期刊	接待业期刊
1	《旅游研究纪事》 *Annals of Tourism Research*（Annals）	《康奈尔饭店和餐饮管理季刊》 *Cornell Hotel and Restaurant Administration Quarterly*（CHRAQ）
2	《旅游管理》 *Tourism Management*（TM）	《国际接待业管理杂志》 *International Journal of Hospitality Management*（IJHM）
3	《旅行研究杂志》 *Journal of Travel Research*（JTR）	《接待业与旅游研究杂志》 *Journal of Hospitality & Tourism Research*（JHTR）
4	《可持续旅游杂志》 *Journal of Sustainable Tourism*（JST）	《国际现代接待业管理杂志》 *International Journal of Contemporary Hospitality Management*（IJCHM）
5	《旅行与旅游营销杂志》 *Journal of Travel & Tourism Marketing*（JTTM）	《接待业与旅游教育杂志》 *Journal of Hospitality & Tourism Education*（JHTE）
6	《国际旅游研究杂志》 *International Journal of Tourism Research*（IJTR）	《FIU 接待业评论》 *FIU Hospitality Review*
7	《旅游分析》 *Tourism Analysis*（TA）	《接待业与休闲营销杂志》 *Journal of Hospitality & Leisure Marketing*（JHLM）
8	《旅游研究的亚太杂志》 *Asia Pacific Journal of Tourism Research*（APJTR）	《国际接待业和旅游管理杂志》 *International Journal of Hospitality and Tourism Administration*（IJHTA）
9	《旅游研究杂志》 *Journal of Tourism Studies*（JTS）	《食品服务业研究杂志》 *Journal of Foodservice Business Research*（JFBR）
10	《旅游经济学》 *Tourism Economics*（TE）	《接待业与旅游人力资源杂志》 *Journal of Human Resources in Hospitality & Tourism*（JHRHT）

(3) 有效地分析评价旅游学成果的学术影响，对旅游学科的教育、人才培养、学科建设等有很大帮助，可以有效地指导青年学者学习、阅读和研究，帮助学者更准确地把握学科研究热点和发展趋势。在我国，旅游专业教育是在改革开放以后，伴随着旅游业的发展而快速发展起来的。随着旅游业的高速发展，旅游专业教育进一步完善，

逐渐形成了多层次、多学科视角的教育体系，已经形成研究生、本科、专科（含离职）和中等职业教育这样四个培养层次。截止到 2003 年底，全国共有高等、中等旅游院校（包括完全的旅游院校和只开设有旅游系或旅游专业的院校）1207 所，其中，高等院校 494 所，中等职业学校 713 所。2003 年旅游院校在校生总计为 45.90 万人，其中，旅游高等院校 19.97 万人，旅游中等职业学校 25.93 万人。2003 年全国旅游院校共有旅游专业教师 1.87 万人，其中，旅游高等院校 9298 人，旅游中等职业学校 9382 人[21]。

（4）CSSCI 的研制成功，为我们提供了旅游学科绩效评价的有力工具，随着今后的进一步完善，它必将推动我国的旅游教育部门和期刊编辑出版部门工作的规范化，也将为旅游科学研究提供强大的文献支撑，并为繁荣我国旅游科学研究事业做出更大贡献。CSSCI 提供了丰富的查询和统计功能，能从多角度、多方位对文献进行查询和统计，这些统计分析有助于我们对旅游学领域进行细致的对比和分析。首先，了解旅游学领域内重要的期刊、学者、论著和机构，从而了解旅游研究现状，以帮助我们更系统地从事旅游研究。其次，利用 CSSCI 从发文和被引两个方面考察了旅游学领域的期刊、作者、机构和地区的影响力，给出了旅游学领域的重要期刊、核心学者以及有影响力的论著等，以此为旅游研究提供参考。最后，CSSCI 不仅可以通过对个人发文量、机构发文量等指标的统计进行总体排序，来定量评价社会科学研究机构、高校、地区、作者的科研生产能力和学术影响，而且可以对期刊被引频次，期刊影响因子，期刊论文作者的地域分布、学科分布，期刊引文的年代分布等一系列定量指标数量的排序来实施对期刊的学术影响评价[22]。

参 考 文 献

[1]　程刚. 基于 CSSCI 对《心理科学》的定量分析. 心理科学，2001，24（4）：421-511.

[2]　姜春林. 从 CSSCI 透视我国管理学研究的地域和学科差距. 新世纪图书馆，2003，（5）：62-65.

[3]　苏新宁. 图书馆、情报与文献学学术影响力研究报告（2000—2004）：基于 CSSCI 的分析. 情报学报，2006，25（2）：131-153.

[4]　刘俊婉，苏新宁，邓三鸿. 经济学研究现状：基于 CSSCI 的评析. 经济学家，2004，（4）：73-80，91.

[5]　张晓阳. 基于 CSSCI 的长三角地区经济学研究评析. 中国市场，2005，（36）：124-125.

[6]　凌斌. 中国主流法学引证的统计分析：以 CSSCI 为数据基础的一个探索性研究. 中国社会科学，2004，（3）：97-107，207.

[7]　刘俊婉，苏新宁. 哲学研究现状：基于 CSSCI 的评析. 情报科学，2006，24（1）：52-59.

[8]　赵宪章，白云. 中国文学学者与论著影响力报告——2000～2004 年中国文学 CSSCI 描述. 文艺争鸣，2006，（2）：104-110.

[9]　邹志仁. 中文社会科学引文索引（CSSCI）之研制、意义与功能. 南京大学学报（哲学・人文科学・社会科学），2000，37（4）：145-154.

[10]　叶继元. CSSCI 的来龙去脉. www.acriticism.com [2006-03-01].

[11]　赵幼芳. 1990～1999 年《旅游学刊》文献库统计分析初步. 旅游学刊，2000，（4）：57-63.

[12]　吴必虎，宋治清，邓利华. 中国旅游研究 14 年——《旅游学刊》反映的学术态势. 旅游学刊，2001，16（1）：17-21.

[13] 张进福. 关于国内旅游研究文献的分析与思考. 华侨大学学报（哲学社会科学版），2002，(1)：52-58.

[14] 徐菊凤. 论旅游学术期刊与旅游人才培养及教学科研的关系——以《旅游学刊》为例. 旅游学刊，2003，(增刊)：167-172.

[15] 宋子千，吴巧红，廉月娟，等 . 回顾与展望：2000～2004 年的《旅游学刊》. 旅游学刊，2005，(4)：80-84.

[16] 陈德广. 从旅游研究博士论文看旅游学学科发展. 旅游学刊，2004，19 (6)：9-14.

[17] 苏新宁. 中国社会科学引文索引设计. 情报学报，2000，19 (4)：290-295.

[18] 邱均平，宋恩梅. 我国社会科学领域的重要工具——评《中文社会科学引文索引》(2000). 南京大学学报（哲学·人文科学·社会科学），2002，39 (5)：155-159.

[19] 姜春林. CSSCI 与科学学类源期刊学术影响力初步分析. 情报学报，2002，21 (4)：476-480.

[20] Mckercher B，Law R，Lam T，et al. Rating tourism and hospitality journals. Tourism Management，2006，27 (6)：1235-1252.

[21] 吴必虎，黎筱筱. 中国旅游专业教育发展报告. 旅游学刊，2005，(增刊)：9-15.

[22] 姜春林. 基于 CSSCI 的管理科学部分源期刊学术影响力分析. 情报科学，2002，20 (2)：145-147.

旅游中间产品转移价格的确定①

零负团费经营是指旅行社以低于行业平均成本的价格组接旅游团队，行程开始后，地方导游和旅游车司机依靠购物、自费项目等方式，巧妙地从游客手中不断收取其他费用，来填平“买”团所亏负的费用。2006 年 6～12 月，海南采取九大措施从根源上铲除零负团费经营方式，其中前两条措施涉及旅游价格的确定问题。第一，统一零负团费认定标准。根据有关法律、法规，综合海南旅游实际和行业惯例，由省物价、旅游部门牵头研究确定零负团费的认定标准，原则上确定为等于或低于旅游产品的行业平均成本价格。第二，规范旅游要素价格。要求制定旅行社价格行为规则，公布旅行社组接团成本信息的合理价格水平信息，制定旅游线路参考价格；完善旅游饭店、旅游车船政府指导定价标准并向社会公布；完善旅游景区、景点（含特种旅游项目）门票分级定价标准并向社会公布；制定旅游团队餐饮价格参考标准并向社会公布[1,2]。在旅游接待系统中，纵向联合是大势所趋，我国不少省份成立了旅游集团，如浙江省旅游集团。纵向联合是指旅游产品的生产过程包括不止一个阶段，如果中间产品的价格定得不正确会影响旅游集团总公司的总利润。特别是如果每个分部的决策者都企图使本单位利润最大，旅游集团总公司的总利润就有可能减少。因此，中间产品价格的确定应使旅游集团总公司的总利润最大，而不是分部的利润最大，这是很重要的。这一目标要求旅游集团高层管理当局参与旅游中间产品价格的确定。中间产品价格之所以重要还因为它要用来考核各旅游分公司的经营成果，如果价格是随意决定的，很可能会对分公司的工作做出不公正的评价。

一、旅游产品的概念界定

一个大学生通过四年的学习，获得的最终产品是“大学学历”和“学位品牌”，最终提高了自己的能力和素质，获得了知识、体验、转型及其物化（如文凭、校徽、教材、实用物等耐用品），但他必须由大学教育系统提供平台。旅游产品就是一个人需要的最终产品，然而，这个旅游产品必须由旅游接待系统提供平台，各旅游接待单位提供的产品是旅游中间产品，或者称旅游要素产品。从旅游者角度定义的旅游产品，是指旅游者花费一定的时间、费用和精力所获得的目的地阅历。一个人从到达、逗留到离开旅游目的地的全部过程中获得的最终产品是“目的地阅历”和“到此一游称号”。“目的地阅历”就是一个人在旅游目的地所寻求、购买的可记忆、可回味、可持续的经济价值提供物，如体验、转型；纪念品、照片等是体验、转型的物化。旅游产品的生产与消费表现为一个时间过程；在国际旅游中，以“旅游者人次数”表示推销出的旅游产品数量；旅游产品不是出自生产装配线，每个产品都具有独特性，因时间、地点和参与人群而异[3]。

---

① 本文原载《经济问题探索》，2007，(11)：107-111. 作者：刘人怀，袁国宏.

旅游产品量＝离开目的地时的旅游者状态－到达目的地时的旅游者状态

用公式表示为

$$Q = S_z(t = t) - S_z(t = 0)$$

海南省旅游产业组织结构相当于模拟分权制，如图 1 所示。模拟分权制的特点是模拟事业部相对独立经营、独立核算的性能，以达到改善经营管理的目的。按生产阶段把产业分成许多“组织单元”，这些“组织单元”拥有较大的自主权，有自己的管理机构。各个“组织单元”之间按内部的“转移价格”进行产品交换并计算利润，进行模拟性的独立核算，从而促进各“组织单元”改善经营管理。在旅游产业中，旅行代理商提供委托代办和问询服务（如火车、飞机、汽车、轮船票代售点），旅游区内旅游交通、旅游餐饮、旅游卫生（如高速公路两旁的休息站）、旅游住宿、旅游景区（点）、旅游娱乐、旅游购物等。旅游接待单位相当于“制造分公司”，生产旅游中间产品；旅行社相当于“销售分公司”，生产旅游最终产品；旅游政府主管部门相当于进行总体协调的“总公司”[4]。

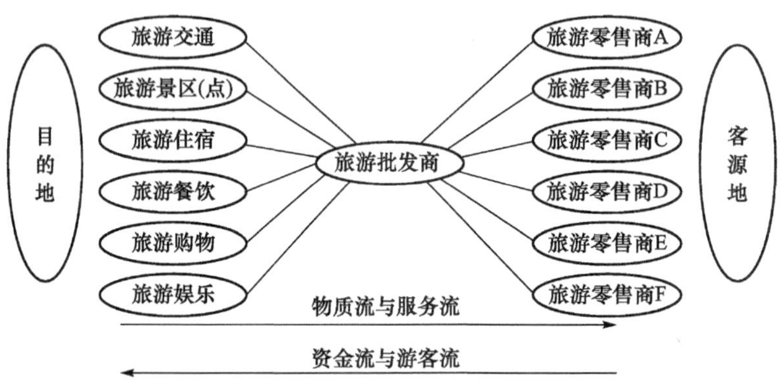

图 1　旅游产业组织结构模型

汽车公司、景点、饭店、餐馆、购物商店、演艺厅等旅游接待单位一般都自负盈亏，自主经营。转移价格是指旅游接待单位（前方）向旅行社（后方）转移中间产品时的价格。例如，一家大的旅游集团拥有旅游饭店、旅游购物商店和旅行社。旅游饭店、旅游购物商店除了向外部市场提供床位、商品外，还要向旅游集团内部销售。这里，旅游饭店、旅游购物商店向旅游集团内部销售床位，商品的价格就是转移价格。怎样确定转移价格？转移价格对于出售中间产品的部门来说，是它的收入；对于买入中间产品的部门来说，是它的成本。转移价格定得高，前方（出售方）分公司的利润就会增加，后方（购买方）分公司的利润就会减少。如果定得低，情况又会倒过来。所以，转移价格的高低直接影响利润在各分公司之间的分配，如果定得不当就会使利润分配不公。但尤其重要的是，如果转移价格定得不当，每个分公司都按利润最大化原则进行决策，分公司的产量决策就会与旅游集团总公司（或旅游目的地接待系统）的最优决策发生矛盾，导致旅游集团总公司总利润最大化目标无法实现。所以，中间产品的转移价格应当由旅游集团总公司来定。它制定得是否正确，对于正确处理各公司与旅游集团总公司之间在经济利润和决策上的矛盾，调动各分公司的生产积极性，共同为实现旅游集团总公司总利润最大化目标而努力，有重要意义[5]。

为了使分析简化，假定旅游集团总公司下属只有两个旅游接待单位：前方分公司 A（指旅游汽车公司、景点、饭店、餐馆、购物商店、演艺厅等旅游组织单元，相当于制造分公司）和后方分公司 B（指旅行社、旅游批发商等旅游组织单元，相当于销售分公司）。生产一件最终产品 F（即目的地阅历），正好需要一件中间产品。

二、无外部市场条件下旅游中间产品转移价格的确定

中间产品没有外部市场的情况，即前方分公司生产的中间产品，全部供应给后方分公司，后方分公司所需的中间产品，必须全部购自前方分公司。例如，有些中间产品，是前方分公司专为后方分公司定做的，后方分公司无法从外部购得，前方分公司也无法向外部出售，就属于这种情况。在旅游市场中，旅游购物商店就符合前方分公司必须具备的理想条件。旅游购物商店特指考虑到目标游客的时间-空间预算，根据可预测的旅行路线，“见缝插针”地创造、插入的购物机会，通常位于交通干线旁（如海南中线高速公路旁的京润珍珠馆、杭州赴上海途中的杭白菊商店）、风景区内（如杭州西湖风景区内的宝树堂制药有限公司、无锡太湖珍珠馆），有时甚至位于土特产品生产地（如苏州针织工厂、杭州梅家坞茶场）。这些地方通常布置有专为游客准备的产品加工流水线、工艺作坊、曲艺节目等“舞台”背景，完全依赖导游和旅游车司机为其提供顾客来源（散客很难找到），因其所处地理位置的独特性、销售商品的特色性和顾客群的异地性而享有“飞行购物”之称。

既然没有外部市场，也就不可能按市场价格来定价。那么，怎样定价呢？在这种情况下，首先是确定能使旅游集团总公司利润最大的产量和价格是多少。在图 2 中，D_F 和 MR_F 分别是最终产品的需求线和边际收入线，MC 是总公司生产最终产品的边际成本（$MC=MC_T+MC_F$）。使 $MR_F=MC$，就可得出总公司生产最终产品的最优价格 P_F 和最优产量 Q_F。这里 Q_F 当然也是分公司 B 生产最终产品的数量，由于前面假定一件中间产品生产一件最终产品，所以也应当是分公司 A 生产中间产品的数量。为了使分公司 A 根据利润最大化原则，愿意生产中间产品 Q_F，中间产品的价格就必须定在 Q_F 垂直线与 MC_T（中间产品的边际成本）线的交点上，即定在 P_T，此时 $P_T=MC_T$。由此可见，在无外部市场条件下，中间产品的最优产量应该按旅游集团总公司的最优产量来定，它的价格应该定在这个最优产量的边际成本上。

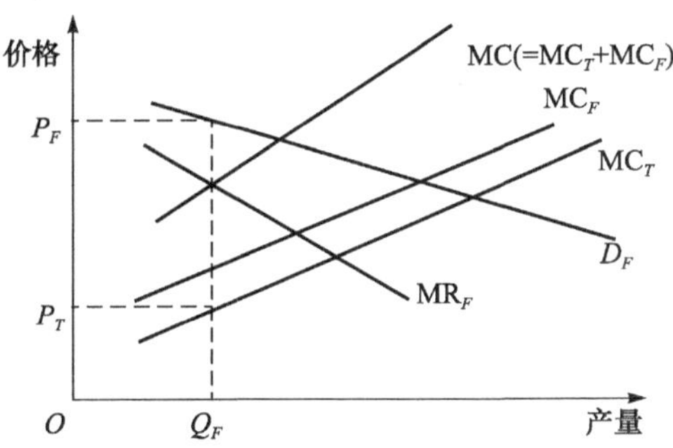

图 2　旅游中间产品无外部市场时的调拨价格

在实际生活中，由于计算边际成本有一定困难，中间产品的价格也可以按单位变动成本近似求出。需要指出的是，按成本加利润的方法来定是不正确的（参见例 1）。

例 1：假如某旅游集团总公司下面有两个自负盈亏的分公司：旅游购物商店和旅行社。旅游购物商店生产中间产品 T，卖给旅行社组织来的团队游客（为了分析方便，这里假定地方导游、旅游车司机、全程导游、地接社是一个“利润中心”，合称为“旅行社”，地方导游是该“旅行社”的代理人）。这种中间产品无外部市场。旅游购物商店生产产品 T 的标准成本[①]为单位变动成本 4 元；单位固定成本 3.5 元；全部成本 7.5 元（4＋3.5）。旅行社的标准成本为变动销售成本 2 元（不包括产品 T 的转移价格）；固定销售成本 0.5 元。旅行社出售的最终产品 F 的销售价格（本文中称为“挂牌价”）为 11 元。问：旅游集团总公司应如何确定中间产品 T 的价格（本文称为“协议价”）？

解：首先计算这一经营活动对旅游集团总公司是否有利。

销售价格	11.00
变动成本	6.00
其中：旅游购物商店	4.00
旅行社	2.00
旅游集团总公司贡献	5.00（元）

上面的计算表明，如果整个旅游集团总公司生产和销售产品 F，每件产品可得贡献 5 元，说明这一经营活动对整个公司是有利的。假如旅游购物商店的中间产品按成本加利润定价，即把转移价格定为 9.75 元（7.50＋7.50×30％），旅行社在这一经营活动中可得的贡献计算如下。

销售价格	11.00
变动成本	11.75
其中：旅行社变动成本	2.00
转移价格	9.75
旅行社贡献	－0.75（元）

上面的计算表明，如旅游购物商店的中间产品按成本加利润定价，旅行社的贡献将为负值，说明经营和销售产品 F 对旅行社不利。这样，旅行社的决策就会与旅游集团总公司的决策发生矛盾。但旅游购物商店的中间产品如按变动成本定价，这一矛盾就不再存在。

销售价格	11.00
变动成本	6.00
其中：旅行社变动成本	2.00
转移价格	4.00
旅行社的贡献	5.00（元）

上面的计算表明，旅游购物商店的中间产品按变动成本定价，旅行社的贡献（5 元）

① 在为中间产品定价时，应使用标准成本，而不是实际成本，因为这样有利于提高生产者降低成本的积极性。

就会和旅游集团总公司的贡献（5 元）相等，这说明两者会做出一致的决策，即对旅游集团总公司有利的活动，一定对旅行社也有利。这是因为按变动成本定价，会使旅行社的全部变动成本恰好与整个公司的全部变动成本相等。

但在实际生活中，按变动成本定价也有一个问题，即它会使旅游购物商店的收入不能弥补其全部成本支出①，这样，不仅由于固定成本得不到补偿，它的生产不可能持久，而且由于不能公平评价旅游购物商店的业绩，也会挫伤其生产积极性。

解决的办法是对中间产品实行双复位价的方法，即除了用变动成本定价外（目的是使旅行社的决策能与旅游集团总公司的决策保持一致），还要在各分公司之间合理分配利润的基础上来定价（目的是调动所有分公司的生产积极性）。

假定在例 1 中，旅游购物商店每年向旅行社提供中间产品 1 万件，旅行社每年销售最终产品也是 1 万件（由游客买走）。例如，按变动成本来确定中间产品价格，旅行社、旅游购物商店和旅游集团总公司的成本、利润数据见表 1。

表 1　按变动成本确定中间产品价格时的利润（一）

旅游购物商店		旅行社		旅游集团总公司	
转移价收入	40 000	销售收入	110 000	销售收入	110 000
自身的成本	75 000	自身的成本	25 000		
				旅游购物商店成本	75 000
		转移价支出	40 000	旅行社成本	25 000
亏损	35 000	利润	45 000	利润	10 000

为了使旅游购物商店有生产积极性，就应当把旅游集团总公司的利润在旅行社与旅游购物商店之间进行合理分配。假如双方协议按成本大小分配（如双方同意，也可以按其他标准来分配），即按 3∶1 来分配，也即旅游购物商店应得利润 7500 元，旅行社应得利润 2500 元。问：中间产品 T 应如何定价?

旅游购物商店为了得到利润 7500 元，总收入应为 82 500（75 000+7500），因此每件中间产品的价格应定为 8.25 元（82 500÷10 000）。把转移价格（俗称“协议价”）定在 8.25 元/件上，分公司和旅游集团总公司的成本利润数据的变化见表 2。

表 2　按变动成本确定中间产品价格时的利润（二）

旅游购物商店		旅行社		旅游集团总公司	
转移价收入	82 500	销售收入	110 000	销售收入	110 000
自身的成本	75 000	自身的成本	25 000		
				旅游购物商店成本	75 000
		转移价支出	82 500	旅行社成本	25 000
利润	7 500	利润	2 500	利润	10 000

由此可见，转移价格的变化（从每件 4 元改为每件 8.25 元），并不影响旅游集团

① 如果不是按变动成本，而是按边际成本定价，就不存在这个问题，因为在最优决策中，边际成本总是大于全部平均成本（MC 曲线位于 AC 曲线之上）。

总公司的总利润（仍为 10 000 元），只是利润在分公司之间的分配发生了变化。

这里，对中间产品 T 制定了双重价格。以变动成本（4 元）定价，是为了旅行社能正确地进行决策。以 8.25 元定价，是为了合理地分配利润，正确评价各分公司的绩效。这样就能较好地处理各分公司之间以及它们与旅游集团总公司之间的经济关系，建立利益共同体，调动旅游集团总公司内部所有成员的积极性，以实现旅游集团总公司的目标。

三、对旅游回扣现象的经济学解释

旅行社组接团"零负团费"经营模式具有自相似性（即部分与整体相似），旅游景点也出现了零负景点费，旅游购物商店也出现了零负购物点费，从而在海南旅游业中呈现出"一半是火焰（高挂牌价），一半是海水（低协议价）"的奇特现象。例如，海南岛在 2006 年 7 月旅游景点门票价格调整之前，定安黑熊园的价格曾涨到 148 元，而旅行社只需支付给景区 8 元的费用；兴隆热带植物园对外 45 元的门票，对旅游团队也只收 1 元费用，但通过在景区内设立咖啡厅等营业设施赚钱；海南东山岭景区对外 35 元的门票，对旅游团免费，甚至倒付 50～100 元的停车费，但靠缆车、拍照等经营收入来返还；黎村苗寨对外 65 元的门票，对旅游团免费，但同样靠景区内的商业一条街营利。2002 年新华社两位记者在海南采访时曝出惊人内幕：成本价 100 元左右的珠宝，在珠宝商场标价为 17 800～23 800 元；20 元左右的普通珠宝标价在 1880 元以上[6]。这是因为：一方面，旅游者对旅游产品价值的感知高，"消费者剩余"多，因而愿意以高价购买旅游中间产品，推动了旅游中间产品的"挂牌价"高涨；另一方面，旅游购物商店等接待单位没有完全竞争的外部市场，不是以"成本＋利润"，而是以边际成本来定价的，因而对中间产品的转移价格即"协议价"定得低。"挂牌价"与"协议价"之间的差价是作为回扣或佣金返还给旅行社的。本文中假定"地方导游、旅游车司机、全程导游、地接社是一个'利润中心'，合称为'旅行社'，地方导游是该'旅行社'的代理人"，而实际情况却要复杂得多。在普通团业务模式下，地接社得到的是"签单"，凭"签单"按月与旅游购物商店结算，一般每人 5～20 元。地接社导游拿到的回扣，先要扣除停车费缺口（约 500 元），旅游车司机还要求先扣除因购物和"加点"而增加的汽油费（约 150 元），然后才在地方导游、旅游车司机、全程导游（简称地陪、司机、全陪）三者间按一定的比例分配。按某旅行社的规定，购物回扣的分配比例为，地陪∶司机∶全陪＝4∶4∶2；"加点"看夜景收费的分配比例为，地陪∶司机∶全陪＝4∶3∶3。

所以，治理旅游回扣现象的手段也应分为两个方面。一方面，降低旅游接待单位的"挂牌价"。旅游产品价值是旅游者活动中所获取的一切经济价值物，包括从自然界和人类社会中直接提取的初级资源（如自然资源中的阳光、沙滩、海水、植物、动物、空气，人文资源中的历史古迹、纪念地、标志物、典故、民俗等），制造出的有形商品（如旅游设施的折旧部分、餐饮食品、旅游购物品），无形的服务（如导游服务、问询服务、交通运输服务、餐饮服务、卫生服务、住宿接待服务、游览娱乐服务、购物服务、邮件通信服务、护照签证服务），难忘的体验（如娱乐体验、教育体验、逃避现实体验、审美体验），引导的转型（如体能得到恢复和发展、智能得到恢复和发展、个人的精神世界获得一种“满足感”、激发出对物质世界的探索兴趣、改变对目的地社区居民的看法与态度）。这些价值提供物不只是由旅游接待单位生产出来的，任由旅游接待单位榨取消费者剩余是不公平的。另一方面，打破、削弱旅游接待单位的垄断地位。经济学家认为，完全垄断的企业有两大弊端。一是分配不公平。垄断企业能保持垄断利润，是以牺牲全社会消费者的利益为代价的，是对消费者的剥夺。二是生产不足。在垄断企业里，价格大于边际成本（$P>\mathrm{MC}$），即社会对所生产的产品价值的评价要高于把资源用于生产其他产品所能产生的价值，因而从社会合理分配资源的角度看，说明该企业生产的产品产量不足，应增加产量（同时减少其他产品的生产），才能更好地满足消费者的需要。由于垄断企业有以上弊端，就需要政府进行干预。政府对垄断企业的干预措施主要有两个。一是反对垄断行为，防止垄断企业的产生，主要办法是制定和执行《中华人民共和国反垄断法》。例如，将旅游购物商店聚集到一起，形成旅游购物市场；非少数民族聚居地禁止建造“民俗风情村”“野人谷”“黎村苗寨”等以少数民族为主题的公园；旅游娱乐场所应聚集到一起，形成演艺区；建造旅游风味餐厅一条街。二是政府对自然垄断企业进行管制。有些企业，如位于原产地的旅游购物商店、带有独特性的特种旅游项目、资源依托型的景区（点）、环境依赖型的民俗风情村、旅游干线两旁的休息站（通常含购物、餐饮设施），在时间、空间、人势方面拥有自然的垄断性。对于这类企业，政府不实行《中华人民共和国反垄断法》，而是准许其实行垄断，但由政府直接控制其“挂牌价”。其目的是，不让它们有超额利润，并促使它们增加产量。

四、我国旅游业持续快速发展的客观要求

（1）将例1中的“旅游购物商店”替换为“旅游风味餐厅”“白天自费景点”“夜晚娱乐场所”等游客自费活动项目，例1的结论仍然成立，因为这些旅游经营场所也无外部市场。“旅游风味餐厅”特指专为游客提供当地特色菜的餐厅，品尝“风味餐”的费用通常不包含在团费之中，如游客自费品尝海南的海鲜、云南的野生菌、杭州的十大名菜、北京的烤鸭等。“白天自费景点”特指地方导游、旅游车司机和全程导游为了向游客收取费用或从自费景点处拿取高额回扣，鼓励旅游者自费参观一些景点，如海南的“野人谷”“黑人谷”“民俗村”等。“夜间娱乐场所”指地方导游、旅游车司机和全程导游带领游客参观夜景或进自费娱乐点，如海南兴隆的“泰国人妖”表演、南京的夜游秦淮河、上海的夜游黄浦江、杭州的夜游西湖阮公墩岛等。

（2）将例1中的“旅游集团总公司”替换为“旅游目的地接待系统”，例1的结论仍然成立。欧美旅游经营商一般拥有航空公司、酒店、景点、交通工具、娱乐购物场所，良好的供应链条衔接和利益的一致性使得其价格优势无可比拟[7]。与欧美相比，我国旅游供应链建设滞后，旅游企业“小、散、弱、差”，因此许多地方的旅游政府主管部门就起到了“旅游集团总公司”的作用。在我国，各旅游接待单位生产“中间产品”，相当于旅游产品的制造分公司（前方分公司）；旅行社组合这些“中间产品”成为“最终产品”，相当于旅游产品的销售分公司（后方分公司）。要搞好这众多的“分

公司”之间的协调，必须要由“总公司”来“补位”，以完成国际联络与合作、政策与法规制定、旅游发展战略与计划制订等工作。因此，实行政府主导型旅游发展方针就成为我国旅游业持续快速健康发展的客观要求[4]。但是，政府主导型旅游管理体制并非永恒不变的，随着我国旅游企业走股份合作、兼并、重组之路，大型旅行社集团化、中型旅行社专业化、小型旅行社通过代理制成为大集团的网点，政府主导型旅游管理体制将逐渐向政府干预型、市场主导型旅游管理体制转变。

参考文献

[1] 高虹．转型升级．海南旅游的必然选择．海南日报，2006-07-06.

[2] 陈国华．海南要做旅游产业“试验基地”．http：//www. nstrip. com/Article-Show. asp? ArtieleID=587. 2006-10-22/ [2006-12-14].

[3] 袁国宏，张月芳．基于体验经济理论的旅游产品价值分析．旅游论坛，2006，(4)：399-403.

[4] 袁国宏，张月芳．旅游管理知识题解．北京：中国旅游出版社，2003：509-510.

[5] 吴德庆，马月才．管理经济学（修订本)．北京：中国人民大学出版社，1998：230-233，280-288.

[6] 卜云彤，周正平．破解旅游高额回扣怪圈．旅游管理，2002，(6)：23-26.

[7] 李万立，李平，张萍萍．欧洲旅行社供应链管理实践与启示．桂林旅游高等专科学校学报，2006，17 (1)：112-115.

我国旅游价值链管理探讨①

价值链就是从“原材料加工”到“产成品”并最终到达用户的过程中，所有增加价值的步骤所组成的，全部有组织的一系列活动。价值链管理的目标是建立价值链战略，这个战略的目的是满足和超越游客的需要和欲望，并使价值链上的相关节点企业达到无缝整合。在价值链管理中，最终客户定义什么是价值，以及怎样制造和提供价值。价值链在企业管理中被广泛用于分析企业生产经营活动各个环节中的价值转移是否出现价值流失或价值增值现象，以及如何对价值链进行优化与重构来提高产品的最终价值。对于旅游业来说，“原材料”指（出发前）没有目的地阅历的人，“产成品”指（返回后）拥有目的地阅历的人，这里的“阅历”是指体验、转型等可记忆、可回味、可持续的经济价值物。旅游价值链管理就是对涵盖旅游者完整经历的一系列活动和信息的全部过程进行管理，这些活动和信息是关于在旅游价值链上流动的产品的活动和信息，而且这些活动和信息是共享的和相互关联的。

一、旅游价值链的定义与模式

1. 定义

价值链管理以整合链上节点企业的内外部资源为基点，以信息技术为依托，以提高质量、服务水平，降低总成本为目标，不断自我改造以适应环境变化，是目前最为先进的作业管理模式。据调查[1]，进行价值链管理的益处包括提高客户服务水平（44%）、降低成本（40%）、加快运送速度（40%）、提高质量（39%）、减少设施闲置（35%）、提高后勤管理水平（27%）、提高销售量（26%）以及增加市场份额（20%）。

旅游价值链是围绕旅游批发商或旅游景区（点），通过对服务流、物质流、资金流、游客流、信息流、商务流和文化流的控制，经由问、行、食、卫、住、游、娱、购等一系列活动过程，将旅游供应商、旅游批发商、旅游零售商、旅游者连成一个整体的网链结构模式。互联网和电子商务手段的应用弥补了传统旅游价值链的不足，并带来了更多的灵活性。旅游价值链成员之间不再是固定的联系，可以交叉联系，也可以跨环节联系。例如，旅游供应商不仅可以与多个旅游批发商协作，还可以越过批发商和零售商而直接向旅游者销售，从而创造出多种价值链模式。价值链成员可以在广泛的选择机会中进行有效的资源优化整合，从而提高价值链的效率。旅游价值链的优化与重构将以“顾客让渡价值”最大化为导向，目的是通过最优的资源配置使价值链能发挥最大的效能和实现最大的增值。因为顾客让渡价值＝顾客价值－顾客成本，旅游价值链要实现顾客让渡价值最大化，一方面要增加顾客价值，包括产品价值、服务价值、人员价值和形象价值；另一方面要降低顾客成本，包括时间成本、金钱成本、精力成本和体力成本。

---

① 本文原载《生态经济》，2007，(12)：102-104. 作者：刘人怀，袁国宏.

2. 以旅游批发商为核心的适于观光旅游产品的“点线式”旅游价值链模式

不同出行目的的游客对行、游、住、食、购、娱的需求是不同的，但目前只有以“观光游览”为目的的团体包价旅游才可以较方便地获得成熟的产品。其他出游目的和出游方式的成熟产品相对较少，需要游客零星地向各行业的旅游供应商购买。目的地的旅游线路是指旅游批发商在研究旅游者的出游目的、消费特点、消费规模、消费水平等前提下，把目的地的各种旅游要素产品按时间顺序有机地组合起来而形成的产品链，如丝绸之路旅游线、华东五市旅游线。旅游批发商首先从单个的旅游供应商那里，分别购置组成产品的各个部件，即零散的旅行要素组件，包括交通、食宿接待、游览娱乐等。然后将它们包装，并命名为一个旅游线路品牌。旅游批发商的核心职能是生产，而不是中介。因为旅游批发商虽然批量购买接待行业的各种旅游服务项目，但它必须根据市场需求状况进行组装、加工，并融入旅游批发商的各种服务项目，进而形成线路产品，如图 1 所示。

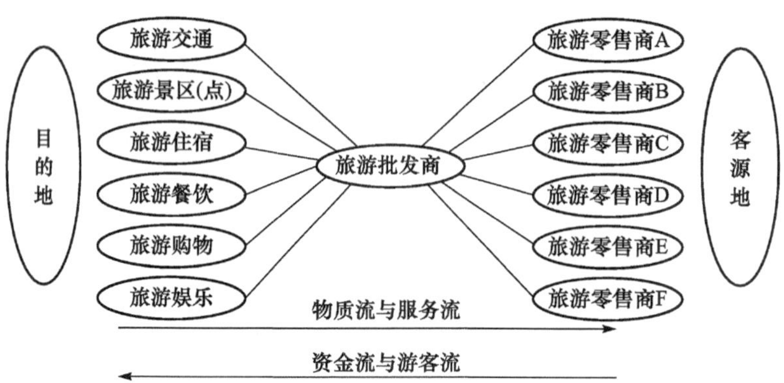

图 1　适于观光旅游产品的“点线式”旅游价值链模式

3. 以旅游景区（点）为核心的适于度假旅游产品的“板块式”旅游价值链模式

当前，我国旅游业正处于观光旅游占主流向度假旅游占主流转型的过渡时期：散客日渐增多，团体游客逐渐减少，而且传统的团体包价旅游正向“交通＋住宿”的“自由行”模式转变；旅游景区（点）的自我宣传力度和积极性不断提高，与游客的直接接触更加紧密；旅游者的需求种类日渐繁多，旅游产品组合层出不穷等。因此，路科[2]认为，新形势下的旅游价值链模式应以“重新确认链条核心节点”为目标而再造，确立以旅游景区（点）为核心的旅游价值链模式。设立旅游质量监督部门，对旅游供应商（即旅游接待单位）以维护“游客体验”为准则进行监控。对旅游景区（点），则以“资源保护”和可持续旅游发展为准则进行监控。对旅游价值链上的节点企业以执行规定或标准为主，对整条价值链则以收集旅游者反馈信息为主，进行综合评价，并提出改进意见。以旅游经营商为核心的价值链模式主要是由于经济发展的驱动，其出发点是“需求引致供给”；而以旅游景区（点）为核心的旅游价值链模式更注重法律的约束与公众的监督，其出发点是“供给导致需求”。

二、我国旅游价值链管理的问题

欧美旅游经营商一般拥有航空公司、酒店、景点、交通工具、娱乐购物场所，良好的价值链节点衔接和利益一致性使得其价格优势无可比拟。我国旅游企业“小、散、弱、差”“各自为战”，旅游价值链建设滞后。

第一，组织障碍方面。组织障碍是管理人员最难对付的障碍，它包括相互间不愿意分享信息，不愿意改变原有地位状态，以及对网络安全问题的顾虑。世界最大的旅游经营商美国运通公司在全球设有 1700 多个旅游办事处，拥有 84 000 多名员工，2000 年旅游业收入达 146 亿美元。欧洲最大的旅游经营商途易（TUI）集团占领了欧洲旅游市场的近九成，在全世界拥有 7 万名员工、81 家大旅游公司、3700 家旅游零售商、287 家饭店和 88 架飞机，2002 年总营业额为 200 亿欧元，其中 60%的收入来自旅游业。2000 年，我国的 8993 家旅行社总营业额为 56 亿美元，仅为美国运通公司的 1/3。在旅游价值链建设滞后，旅行社、饭店、交通、景区（点）没有整合的情况下，我国旅行社在应对国际竞争时显得势单力薄。

第二，文化和态度方面。价值链节点企业之间“缺乏信任”和“过于信任”都是不利的。当缺乏信任时，节点企业就不愿意分享信息、能力和生产过程；过多的信任容易导致窃取知识产权，以及担心与内外部成员达成合作以后无法掌握自己的命运。

第三，能力要求方面。价值链成员需要满足许多方面的能力要求，如良好的协调合作、整合满足供应商和客户的产品/服务、对于内外部成员的培训等，而这些能力都很难提高或达到。例如，景区（点）瓶颈和航空瓶颈。2001 年，我国国际旅游外汇收入中，长途交通支出占旅游支出的 28%，食宿支出占 27%，购物支出占 18.7%，三项共占总支出的 73.7%，而民航占长途交通费用的 70%。此外，旅游产业还存在着专吃门票的“偏食症”、购物与娱乐的“低消费”[3]。

第四，人员方面。没有组织成员坚定不移的承诺，旅游价值链管理是不可能成功的。如果员工不愿意或拒绝变得灵活，组织便无法在环境发生变化后做出灵活的反应。实施价值链管理，需要大量员工在时间和精力上的付出，而激励员工并非易事。此外，我国还缺乏有经验的管理人员带领组织实现价值链管理。

三、我国旅游价值链管理的对策

(1) 价值链上各节点企业的合作，包括分享信息资源以及灵活地决定“由谁来做什么”。旅游理论研究应该根据旅游产业的运行规律和产业关联理论，使用定量研究方法，确定旅游产业内各行业协调发展的比例关系，包括数量结构和层次结构。政府应根据本地区的实际情况和旅游产业的发展规律，制订本地的旅游产业发展规划，消除瓶颈环节（俗称“短腿”）；同时，对一些超出市场需求的环节予以限制，并制定相关政策，促使相应资源能够有效地退出。旅游企业应该根据旅游价值链的内在经济规律，积极参与价值链的分工与合作，尤其是要做好市场需求前景的预测，在动态中追求与价值链上其他环节的平衡，并在市场环境发生变化时有效地进入和退出。

(2) 恰当的组织流程，包括更好的需求预测、合作工作以及衡量在价值链上发生的各种活动所取得成效的标准。加强价值链中各节点企业之间的合作，需要建立合理的利益分配机制，限制上游企业向下游企业转嫁风险，避免出现“三角债”和“零负团费”现象；价格机制和利益分配机制的建立不仅需要运用经济学的一般理论，还需要用博弈论的思想分析利益主体的决策行为；旅游景区（点）不仅是一个企业，因此在涉及旅游景区（点）如何参与利益分配的问题时，要加入政府这个重要的利益主体，以强调游客教育和资源保护。

（3）强大而尽职的领导，包括强化旅游价值链核心企业的地位。一是要培育大型的旅游集团。旅游集团可以通过股权联结的方式组建，也可以通过特许经营、租赁经营、管理合同等契约方式组建，还可以通过企业战略联盟的方式组建。二是要强化目的地政府主导。各旅游接待单位生产“中间产品”，相当于旅游产品的制造分公司；旅行社组织这些“中间产品”成为“最终产品”，相当于旅游产品的销售分公司。要搞好这众多的“分公司”之间的协调，必须要有“总公司”来“补位”，以完成国际联络与合作、政策与法规制定、旅游发展战略与计划制订等工作。我国许多地方的政府旅游主管部门起到了“总公司”的作用。

（4）恰当的员工制度，包括灵活的职务设计、有效的招聘程序和持续的培训。旅游价值链管理中的职位设计，需要将主要精力放在职位设计是否能够更有利于员工为游客创造和提供价值上。这就需要员工的工作内容和工作形式具有灵活性。灵活的工作需要灵活的员工，组织招聘工作就是甄选那些有更强学习能力和适应能力的员工。员工不仅要认识到游客的重要性，而且要接受财务、信息技术和公司战略方面的培训。一个良好的价值链可以使各节点企业像团队般工作，每个节点企业都为全部过程增加相应的价值，如快速组装、更准确的信息、更快的游客反应速度和更好的服务等。

（5）适宜的、起支持作用的组织文化和理念，包括分享、合作、开放、灵活、相互尊重、信任。其强调打破区域的行政性界限，依照构建价值链的要求配置旅游资源和要素；强调政府在制定区域间旅游发展政策，统一协调区域旅游产业布局，统一规划区域重大旅游基础设施体系，以及生态环境保护、开展技术协作和信息交流等方面的作用。

（6）投资建立一个支持这种合作和分享的技术框架。传统的层级式信息流、资金流由于因特网的推动而被打断，所有传统经营者都成了信息中介。一是传统旅游企业的变革，有的成为网上旅行代理商，有的成为因特网推动下的批发商；二是旅游目的地营销组织的变革，其网站有的成为目的地网络门户，有的成为旅游市场项目管理机构，有的甚至成为网络平台；三是计算机预订系统和全球分销系统的变革，如建立新的商业模式和新的竞争策略[4]。

参考文献

[1] 斯蒂芬·罗宾斯，玛丽·库尔特. 管理学. 孙建敏，黄已伟，王凤彬，等译. 北京：中国人民大学出版社，2004：558-578.

[2] 路科. 旅游业供应链新模式初探. 旅游学刊，2006，(3)：30-33.

[3] 许韶立. 关于拉长我国旅游产业链的几点思考. 中州学刊，2005，(4)：38-42.

[4] 吴恒. 网络环境下旅游产业链的思考. 科技进步与对策，2005，(10)：161-163.

旅游业零负团费的运行机制及危害性探析①

零负团费是旅游商“发明”的一种经营方式，但在近年来却快速在我国旅游市场蔓延。1988年，我国内地首批开放公民自费前往新马泰和我国港澳地区旅游；而在1995年左右，泰国游便开始出现零负团费，并很快波及了新加坡和马来西亚两国。为杜绝“零团费”这一不正常的旅游市场现象，2000年中国、马来西亚、新加坡、泰国在昆明举行了四国联合的治理地区旅游市场部长级会议，就维护我国旅游者的合法权益达成了相关协议。

2004年9月，欧洲游向我国开放，首发团就有旅行社采取了零负团费的做法，如导游“自费”50%个人往返欧洲机票的费用作为“买团”的人头费。海南大量出现零负团费始于2003年下半年；到2004年底，云南游已发展到1700元可以“北京—昆明—大理—丽江6日双飞”的程度；2005年，华东五市游、九寨沟成都游也频繁出现了这一模式。

近几年，旅游业界也尝试推出“品质之旅”“诚信之旅”“纯玩团”来突破低价重围，但“品质团”却在夹缝中艰难生存。本文以系统理论和可持续发展理论为基础[1]，探讨零负团费的运行机制及其危害性，为当前我国整治零负团费行动提供理论支持。

一、旅游零负团费现象及其运行机制

1. 旅游零负团费现象

以我国旅游胜地海南为例[2]，长期低价运营的旅游业所产生的一组数字让人震惊：全岛众多旅行社年平均利润仅有1万余元。海南全岛的酒店有70%左右处于亏损状态，超过一半以上的旅游项目将80%的收入以回扣的方式返还给带游客前来的导游、旅行社和旅游车司机。这样的现状让海南的旅游企业、旅游从业者和旅游主管部门忧心忡忡，众多旅游车司机和导游人员也要求改变现状。其中所反映的海南旅游业发展中的深层次矛盾引起了更为广泛的关注和讨论。

国内其他旅游大省同样面临着低团费和零负团费的困扰，游客也期待零负团费的结束，因为零负团费可能意味着旅途难以得到保障甚至被欺诈。很多参团到海南旅游的人并不了解零负团费组接团的概念，仅凭借宣传知道，去海南旅游的价钱低得只相当于一顿大餐。到海南后却发现，导游和旅游车司机强行带着购物，在景点只能走马观花，根本没有真正领略到海南的热带风情，一些不愿购物的游客甚至被导游和旅游车司机甩在途中。

旅行社零负团费经营模式的资金流有两条：一条资金流根源于游客在客源地预付的初次旅游费（或刚性消费“打包”），资金流沿着“游客→组团社→中间社→地接社→司陪人员→旅游供应商”这样一条经营链条流动，运行不畅时会产生“三角债”现象。

---

① 本文原载《商业时代》，2007，(25)：87-88. 作者：刘人怀，袁国宏.

另一条资金流根源于游客在目的地的二次消费（或弹性消费），资金流沿着“游客→购物点和自费项目→司陪人员和地接社→旅游供应商（在“负团费”运作模式下包括组团社）”这样一条经营链条流动，购物点和自费场所的司陪人员、地接社的“回佣”现象就是其表面现象。“零团费”“负团费”就是旅行社以等于甚至低于行业平均成本的团费报价（第一条资金流）来吸引游客报名参团，到了旅游目的地再通过带游客参加自费旅游项目和购物等方式赚取“回佣”（第二条资金流），从游客身上把钱挣回来。

2. 组接团低价团费运行机制

组接团低价团费、零利润业务模式如图 1 所示，箭线的方向表示资金流动的方向，箭头的粗细表示资金流动的大小；实心线表示资金沿单一渠道流动。重叠线表示资金沿多个渠道流动。在这种模式中，客源地组团社付给目的地地接社的只是游客的基本旅游费用，如基本景点费、日常餐饮费（俗称“吃饭费用”）、住宿费、区内交通费；地接社的运营费用以组团社所付团费为主，购物、自费活动佣金和导游人头费为辅；地接社和导游的“回扣”项目少。“旅游初次费用”表示游客在出游前上交给客源地组团社的费用，“旅游二次费用”表示游客在旅途中所发生的、付给旅游服务供应商（即旅游接待单位）的费用，通常是购物和自费项目。

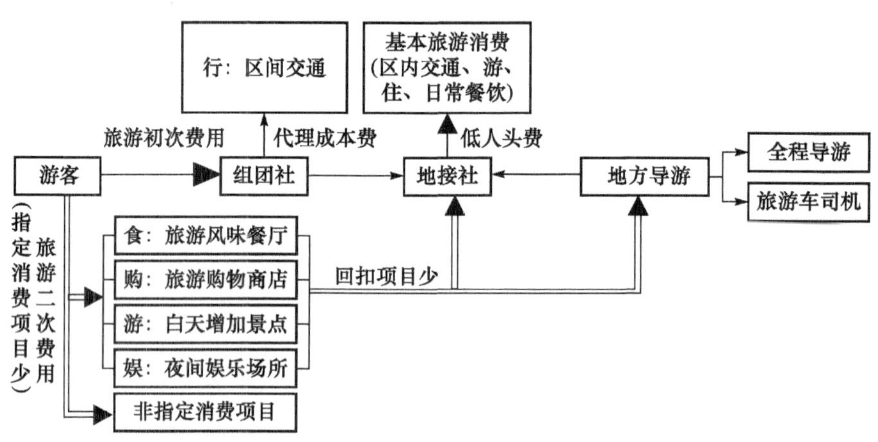

图 1 组接团低价团费、零利润业务模式

“旅游风味餐厅”是指为游客提供当地特色菜的餐厅，品尝“风味餐”的费用通常不包括在团费之中，如游客自费品尝海南的海鲜、云南的野生菌、杭州的十大名菜、北京的烤鸭等。“旅游购物商店”特指位于交通干线旁、风景区内、土特产品生产地的旅游指定商店，通常布置有专为游客准备的产品加工流水线、工艺作坊、曲艺节目等“舞台”背景，完全依赖导游和旅游车司机为其提供顾客来源（散客很难找到），因其所处地理位置的独特性和顾客群的异地性而拥有“飞行购物”之称。“白天增加景点”，指地方导游、旅游车司机和全程导游为了向游客收取费用或从自费景点处拿取高额回扣，鼓励旅游者自费参观一些景点；“夜间娱乐场所”指地方导游、旅游车司机和全程导游带领游客参观夜景或进自费娱乐点。“非指定消费项目”是指游客在非指定场所消费，这些场所通常不给导游回扣，有时甚至要向旅游车司机收停车费，如游客在上海的“八佰伴”商厦或南京路购物等。

3. 组接团“零团费”“负团费”运行机制

在“零团费”运行模式下，客源地组团社不付给目的地地接社任何资金，只输送客源。在这种模式中，游客的基本旅游消费，以及地接社的折旧费、利润、税金、导游服务费等都来自导游所交的“高人头费”和旅游服务供应商的“签单”；地接社的运营费用以购物和自费活动佣金为主，导游人头费为辅；地接社和导游的“回扣”项目多。

在“负团费”运行模式下，目的地地接社不仅不向组团社收取任何接待费用，反而向组团社“买团”。地接社的运营费用以导游人头费为主，购物和自费活动佣金为辅；游客必须在指定消费项目花费一定数额。风味餐馆、偏僻宾馆、观光缆车、上岛游船、出租车司机、低劣景点、购物商店、特种演艺厅等自费场所的旅游服务供应商为了保证客源，只得将大部分营业收入都“返还”给地接社和导游，陷入了低价经营循环中，而游客则陷入高价陷阱。例如，海南的地接社导游在带团前要先交6000元的押金给旅行社，地方导游真正成了“鱼鹰”“刀手”，陷入了高压力带团的境地，服务意识全无，整个人的心灵和行为都被扭曲。

二、旅游零负团费现象的危害性

1. 损害旅游目的地形象且损害目的地的整体利益

实行零负团费的“普通游”具有以下特征：“吃”，饮食简单，环境恶劣；“住”，经常安排住在偏僻的地方，出行不方便，如果游客没有完成购物指标，住得更差；“行”，低价旅游的车费经常不能给足，甚至零车费，所以司机会在旅游车上推销高价旅游纪念品弥补车费；“游”，行程中所列景点是一些不收费或低收费的，自费项目偏多；“购”，整个行程围绕着购物来进行，购物店里的商品基本上价高质次，个别商店销售假货，强迫消费；“娱”，必须参加自费项目，另外地方导游还推荐比实际价格高出很多的自费项目甚至不健康的自费项目；利润方面，购物和自费项目是旅行社和导游获得利润的来源，购物和自费项目就成了欺骗游客的陷阱，商家在欺骗中得到巨额的利润。

零负团费的机会成本巨大，甚至会导致“旅游摧毁旅游”。旅游者的口碑宣传和多种新闻媒体对“零团费”“负团费”负面影响的报道和炒作，严重损害了目的地的形象，不仅给旅游业下游企业带来严重灾难，而且损害了目的地的整体利益。

2. 损害旅游接待系统中各要素之间的代内公平

游客、组团社、地接社、地方导游、全程导游、旅游车司机、购物商店、景点、餐馆等各利益主体失去了共同的目标，也就失去了相互协调配合的基础，甚至互相伤害。在“与其自杀，不如自相残杀”理念的支配下，旅游团队接待系统中各要素之间形成了一条“食物链”，顶端的组团社将游客“卖”给地接社，地接社再将游客“卖”给导游，他们的风险较小，收益稳定。

但到了导游一层，风险骤增。导游预先支付了“人头费”和景点门票等费用，通过带游客购物和参加自费项目赚取回佣的行为，就被称为“填坑”。导游接团各凭运气，如果碰到“豪爽团”，回佣收入大于支出，则可获得利润，反之则亏钱，这在旅游业内被称为“赌团”。旅游经营行为一旦变成了“赌博”，势必陷入恶性循环。

3. 损害旅游接待系统中各要素的代际公平

（1）损害了游客的利益，影响回头客。零负团费虽然在报名时给游客短暂的惊喜，但在整个旅游过程中带给游客的是重重“陷阱”：降低餐标，吃不到当地特色餐；降低住宿标准，“能住郊区不住城里，能住乡下不住郊区”；更改行程、擅自增加自费景点；导游变成导购，听不到应有的旅游介绍；延长购物时间，缩短景点参观时间等。不仅如此，如果游客拒绝参加自费项目还可能被导游甩团。

（2）损害了地方导游的利益，优秀导游流失。有的旅行社为转移经营风险，把旅游团以倒卖的方式交给导游，从导游那里收取相应的“人头费”，使导游面临巨大的工作压力。当导游无法收回成本时，便将其不满发泄到游客身上，导致“甩团”等损害游客利益的行为。在这种现象成为常态以后，旅游行业的人才将会频繁“跳槽”，以寻求稳定的工作环境。

（3）损害了旅行社的自身利益，使旅行社无法成长。旅游企业的不公平竞争环境，使旅行社陷入低价的泥潭。最终导致旅行社行业的诚信尽失，新产品的研发严重滞后，企业无法做大做强。

（4）损害了饭店、景点、餐饮等旅游服务供应商的利益，难以为继。对散客来说，旅游要素产品的价格虚高，造成商品和服务质价不符，抑制了需求；对团队来说，旅游服务供应商陷入低价泥潭，难以保证服务高质量，企业发展后劲不足。

参考文献

[1] 高群. 东亚区域旅游、环境合作与经济可持续发展. 东北亚论坛，2001，(2)：38-42，96.

[2] 范莉娜. 背负恶名谁之过——浅议导游从民间大使到千夫所指之原由. 贵州民族学院学报（哲学社会科学版），2005，(6)：80-82.

文化生产力：管理的视角①

面对激烈的市场竞争，管理过程中的无形资源，如智力资本、组织资本、企业叙述结构、企业网络结构等所蕴含的力量由隐而显，成为获得持续竞争优势的强有力来源。在十多年前，程恩富[1]学者就开始著文论述其中的文化力和文化生产力的意义、结构及其功能。在国外文献中，近年来对文化与经济之间关系的讨论也一直不断。一般认为，文化在经济学中受到普遍的重视应该归功于Baumol和Bowen[2]的“行为艺术论”，而包亚明[3]的文化资本说却引发人们对蕴含在叙述中的规范力量的关注。具体而言，在生产管理过程中，应该更加关注文化叙述所蕴含的问题。

一、文化生产力

人们在生活中就可以感受到文化的生产力量，然而文化本身过于笼统，似乎难以找到它的真实所在。赞成者深切地感受到文化的巨大魅力，如直切主题的书名《不发达是一种心态》；反对者则恒反对。本文试从管理的角度，分析文化叙述结构所蕴含的文化生产力问题。关于文化生产力，李德顺[4]认为，它是创作和制造文化产品及提供文化服务的社会能力。一般来说，构成文化生产力的基本要素主要包括：一是主体的文化劳动者或生产者，主要是他们的素质、积极性和社会组织状况；二是文化资源，包括历史资源和现实资源，是指作为文化创造来源的对象和条件，如一切可依靠或有待开发的思想材料、风格样式、技术手段、人才环境等。金元浦[5]则强调文化生产力像科技一样，作为物质生产力的渗透性力量，作为一种产业而独立存在。李春华[6]认为，文化生产力指的是文化本身就是一种生产力形式，是生产满足人们精神需要的文化产品的水平和力量。文化生产力也表现为三个基本的层次或环节：一是以科学研究（尤其是人文社会科学研究）和文学艺术创作为代表的基础性或“原创型”精神生产；二是以文化教育、大众传播和推广普及为代表的应用性或“中介型”再创造活动，是精神文化的再生产；三是以社会和大众的文化消费和精神生活为代表的大众文化参与活动，是推动文化生产力发展的原动力[4]。

这里讨论的一个共同点就是强调文化生产的能力，不管是文化产品还是文化服务，这是一个积极的方面。对文化的基本界定，不论是泰勒[7]的描述性定义，“文化或文明，就其广泛的民族学意义来说，是包括全部的知识、信仰、艺术、道德、法律、风俗以及作为社会成员的人所掌握和接受的任何其他的才能和习惯的复合体”；还是马凌诺夫斯基[8]功能主义的定义，“文化是指那一群传统的器物、货品、技术、思想、习惯及价值而言的，这一概念包容着及调节着一切社会科学，我们亦将见，社会组织除非视作文化的一部分，实是无法了解的”；还是马克思主义的经典论述，“文化，从广义来说，指人类社会历史实践过程中所创造的物质财富和精神财富的总和。从狭义来说，

① 本文原载《生产力研究》，2008，(7)：54-56. 作者：刘人怀，龙先东.

指社会的意识形态，以及与之相适应的制度和组织机构”（《辞海》）。在这些界定中，隐而不张的是生产者，既在其中又在其外的是生产者。所不同的地方在于，功能主义提倡“文化为工具，生活乃主体”。

就生产的管理过程而言，外在的文化生产能力是相对容易的事情，而文化叙述结构所蕴含的释放功能，却是我们更加需要关注的因素。在生产过程中，文化对生产者的束缚，以及因此导致的巨大浪费也远不在我们的视野之内。在科技不发达的年代，我们苦于人类自身能力的不足，这限制了人类社会的生产能力。而文化却如影随形，人们被笼罩在这个巨大的阴影之中，这个阴影钝化人们的睿智，在庭院洒扫之间，举手投足之际，反身而诚。因此，就生产过程的管理而言，文化生产力就不应该仅局限于文化产品和文化服务的生产水平方面，而更应该强调在企业专用资源这个基础上文化的释放功能，释放生产者，解放企业的心灵。具体到企业组织，从巴雷特[9]的一个价值转换模型或许可以更加清楚地看到它们之间的关系。在这个模型中（图 1），巴雷特把个人的动机归纳为员工意识的七个层次，与企业意识的七个层次相匹配。这是一个诊断和转化工具，通过检测员工和企业的现实与理想的价值指标，确定其需要转变的行为和方向，通过双方的开放性期待，实现价值转换和价值创造。

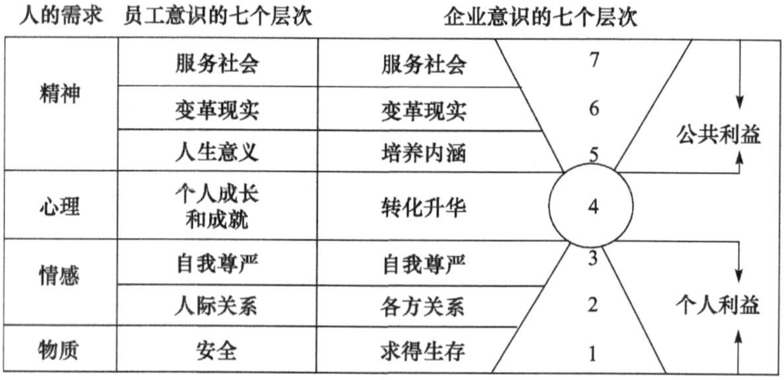

人的需求	员工意识的七个层次	企业意识的七个层次		
精神	服务社会	服务社会	7	
	变革现实	变革现实	6	公共利益
	人生意义	培养内涵	5	
心理	个人成长和成就	转化升华	4	
情感	自我尊严	自我尊严	3	
	人际关系	各方关系	2	个人利益
物质	安全	求得生存	1	

图 1　员工意识的七个层次与企业意识的七个层次之间的关系[9]

二、叙述结构所蕴含的故事

与新古典经济学把经济行为过分孤立于社会文化约束不同，Granovetter[10]强调经济行为的社会嵌入性，把生产、交换和消费过程还原到相应的社会文化环境当中来理解，触发人们对社会网络的关注。Lounsbury 和 Glynn[11]也从一个叙述的故事结构中（图 2）看到镶嵌在文化系统里面的认同与合法性的力量。在这个过程模型中，文化创业（cultural entrepreneurship）是一个调节现有创业资源存量及其后的资本获得和财富创造的说故事的过程，强调创业故事对组织成员认同的促进，以及由此获得的来自投资者、竞争者和专家认可的合法性。通过两个关键的创业资本，即企业专用资源资本和行业水平的制度资本所形成的核心内容，这样的故事有助于创造来自文化和组织认同的竞争优势。不同于故事策略观[12,13]把熟悉故事陌生化，文化创业的创业故事试图以启动可理解而具合法性的新业务使人们对不熟悉的世界熟悉起来。当然这样的文化创业故事并不仅局限于新启动企业，更适合成熟企业的整个成长过程，以不断的创业激励来调动组织及其成员的积极性。

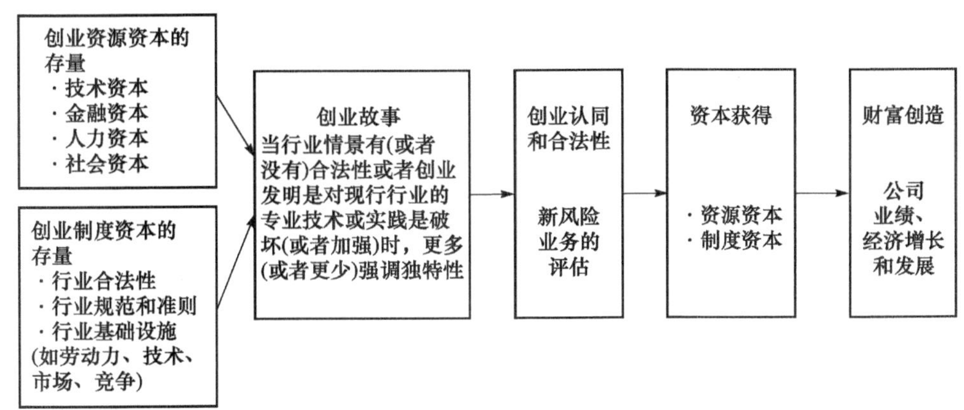

图 2　文化创业的过程模型[12]

文化生产力带来的企业间的竞争同时也蕴藏在对消费者需求的文化解读过程中。离开对消费者心理及其需求的理解和把握，离开独特的文化意义和价值，增强文化竞争力就是抽象的[4]，近年已经有不少学者注意到蕴含在商品文化中的巨大商机[14]。

另外，从贝克尔[15]的一个扩展效用函数，也可以分析叙述结构对消费者选择的影响。贝克尔和李杰将人生经历和社会力量引入到对偏好或者口味的影响。在这个扩展的效用函数（在某个时间 t）中

$$u = u(X_t, Y_t, Z_t, P_t, S_t)$$

商品 X，Y，Z 的消费受到 P 和 S 的影响。其中，P 表示个人资本，它包括有关影响当前和将来效用的过去消费和其他的个人经历；S 表示社会资本，它包括个人社交网络和控制体系中的同辈人和其他人以往活动的影响。P 和 S 构成了他全部人力资本存量的一部分，引入这个观念之后，在任何时点上，这个扩展的效用函数不仅取决于所消费的不同商品，还取决于该时点上个人和社会资本的存量。因此，对于文化生产力的认识不应该，也没有必要仅局限于对文化产业的分析，把消费者的消费放在历史文化环境中来理解，那么工厂生产的产品就不仅具备经济价值，消费过程也就不仅是经济消费。

三、文化生产力与企业文化资本积累

人类学所理解的文化是一个由社会-历史规定了的并从内部塑造而成的象征系统，因此，若将其简化为实践运用（praxis）社会结构或者“基本的人类冲动”就不能掌握文化概念——将其泛化也无法理解它[16]。

企业文化尽管没有人类学所理解的社会-历史规定那样坚硬，但也远不是可有可无的，不是功能主义者的方便法门。企业文化先是从企业中的每一个个体开始，这些个体都是在具体的文化环境中长成的，不是任意拿捏的标准件，却是可以塑造的、蕴含巨大潜能的企业主体。企业文化是由社会-历史规定了的并从内部塑造的一个象征系

统，这是一个依次积淀的过程，在建设企业文化的时候也应该从这个内部结构着手，释放其中的能量。文化生产力，除了它的文化生产功能之外，从企业文化资本的积聚的角度，还具备挖掘潜藏的企业内外资源的能力。

企业文化资本来自布迪厄的文化资本理论。他认为，资本是一种铭写在客体或主体结构中的力量，可以表现为三种基本的形态：经济资本、文化资本和社会资本。其中文化资本以三种形式存在：①具体的状态，以精神和身体的持久“性情”的形式；②客观的状态，以文化商品的形式（图书、书籍、词典、工具、机器等）；③体制的状态，以一种客观化的形式[3]。这样界定的文化资本，它的强大功能在于它的规范性力量。文化资本如果能对一个社会产生未来收益，并且创造和维持该项制度要付出昂贵的代价，那么这项制度可被视为文化资本的一种形式[17]。在经济领域，首次正式定义文化资本的是澳大利亚文化经济学家思罗斯比[18]，文化资本被定义为在经济学框架内除实物资本、人力资本和自然资本之外的第四种资本，是具有文化价值的财产，即文化资本是以财产的形式具体表现出来的文化价值的积累。这种积累紧接着可能会引起物品和服务的不断流动。有形的文化资本的积累存在于被赋予文化意义的人工品之中，无形的文化资本包括一系列与既定人群相符的想法、实践、信念、传统和价值。Klamer[19]进一步将文化资本定义为激发与被激发的能力，认为文化资本不仅是个人财产，也是企业、城市乃至国家的财产组成部分。这样，对于企业文化，从资本的角度进行整合，就将企业之间的竞争延伸到了企业日常生活的细节。企业文化资本的积累，包括对管理技巧的资本化，更加有助于企业获得持续的竞争优势[20]。因此，文化生产力对企业经营者而言，可以转化为更加易于管理的企业文化资本，从企业的文化智力培育到企业产品的文化价值生产，从被企业文化资本激活的生产者到在全球跨文化视野中的系统识别，锻造企业的核心能力，这是可能而必要的。

四、文化生产力与持续竞争优势

文化生产力与持续竞争优势的关系是一个争论激烈而又危险的话题，科特的实证研究是一个强有力的支持[21]。最近王辉等[22]对中国企业的实证研究更进一步证明了组织文化对企业绩效的影响，同时还证明了对于企业的所有制结构，组织文化的影响更大。然而，伯特[23]的研究却发现公司文化与公司绩效的关系远不是那么简单、直接，文化与绩效的关系主要取决于市场。在某些领域，它们之间没有什么关系。但是，在另外一些领域强有力的文化对绩效的影响随着市场竞争有效水平的提升而变得越来越大。因此，企业文化并不是一个不得不投的投资，也并不总是企业的一项竞争性资产。当面临竞争激烈的市场环境时，强有力的文化确实可以带来强大的优势。

有趣的是，20 世纪 80 年代中期，劳伦斯·哈里森通过南美洲一些国家一对一的对比研究之后撰写了《不发达是一种心态——拉丁美洲事例》，在每一组对比中劳伦斯·哈里森都能够以文化因素说明发展的差距。受到这个研究的启发，格龙纳多[24]构造了一组包含 20 种因素的“价值观体系”，这些因素包括宗教、对个人的信任、道德规范、两种财富观、两种竞争观、两种公平观、工作价值观、离经叛道的作用、教育不是洗脑、实用性的重要、轻轻的美德、时间上的重点、理性、权力、世界观、生命观、救己与救世、乌托邦、不同的乐观主义和两种民主观等。在这组因素中，格龙纳多总是能够找出发达经济体与不发达经济体选择上的一些差异，前者多是选择那些有利于经济发展的文化，而后者则相反。结论当然让人不胜感慨。归根结底，一个社会的经济发达与否，并不是外界所强加的，而是社会自身所选择的。

这对企业组织的文化建设有所启示。如果说企业文化是在一定的社会经济条件下通过社会实践所形成的并为全体成员遵循的共同意识、价值观念、职业道德、行为规范和准则的总和，是一个企业或一个组织由自身发展过程中形成的以价值为核心的独特的文化管理模式[25]，那么作为企业的一种生产能力，文化生产力一方面表现为文化的潜在能量和企业之间的文化竞争，另一方面，它还表现为在组织过程中能够做什么样的选择。

在动态、竞争的市场环境中培养企业的竞争优势，如果以 Barney[26] 的 VRIO 分析标准，即价值（value）问题、稀缺性（rarity）问题、难以模仿性（inimitability）问题和组织（organization）问题来选择能够给企业带来持续竞争优势的资源，组织文化是很容易符合这个标准的，而且这是在一个漫长的生产过程中逐步积淀下来的，有一定的历史依赖性和结构的互补性。但是在传统的企业内部行政管理架构中，组织文化也会随着企业的成熟而形成组织惰性和思维惯性，不利于发挥组织文化的主动性，甚至使组织僵化。这很容易陷入循环论证，但它是事实。首先需要做一个区分，传统的管理模式还是创业型管理模式；自由、全面发展的还是单向度的；是在守业还是创业。因为在传统的管理架构下形成的文化氛围中，是按照目标设置、计划、监控等，把组织成员重新束缚在组织计划之下。许多企业发现它们识别和创造性地利用机会的能力会随着企业的成熟而下降。在激烈的竞争、动态而不确定的环境中，企业有必要采纳创业战略，即通过持续创新以有效利用机会来获得竞争的优势。而要使得这种战略取得成功，一个关键的问题是在企业内部培育一个能动的经济和政治生态系统，这个系统不会阻碍以创新或创业为目的的各种资源的重新配置。为此，Sathe[27] 提出从表层到深层的公司创业问题。对于表层创业的公司，创业被认为是管理层试图推进和完成的一项重要的业务目标；深层创业意味着在公司内部创业不是一个管理目标而是一个共享的价值观，是一个在有意识和无意识中驱动管理行为的理所当然的假设。Ramachandran 等[28] 在这个基础上提出集中创业和组织全民创业的问题，这就像光谱的两个极端，前者是组织遇到困境之后的一种权宜之计，创业被“注入”到组织中，这通常很难让组织顺利协调各种资源，把创业贯穿到操作性的活动当中，更有甚者还会造成组织内部的不平衡而导致组织的效率下降。在组织全民创业的框架中，创业是组织共享的价值观，从组织创业伊始就有这个性格，贯穿始终并获得组织的全方位支持。集中和全民创业的对比贯穿在组织过程的每一个要素——使命、战略、结构、系统、流程、人事、技能和态度等，形成建构持续竞争优势的关键理念。

这是对比鲜明的两种管理模式、两种冲突的企业文化。更重要的是，除了带来经济上的业绩（这个方面的实证研究很多，如 Zahra 和 Covin[29] 等），创业的管理模式还带来整个组织的活力，培养了组织学习的能力，通过全民动员调动起整个价值创造链条上每一个敏感的环节。它所形成的组织系统还能够把组织成员的注意力自觉地导向重要的被组织期待的创新上。这些都是企业重要的文化生产力，能够为企业赢得不可替代且持续的竞争优势。

参考文献

[1] 程恩富．文化生产力与文化资源的开发．生产力研究，1994，(5)：14-16，21，81.

[2] Baumol W J，Bowen W G. On the performing arts：The anatomy of their economic problems. The American Economic Review，1965，55 (1/2)：495-502.

[3] 包亚明．文化资本与社会炼金术——布尔迪厄访谈录．包亚明译．上海：上海人民出版社，1997.

[4] 李德顺．形成强大的文化生产力．人民日报，2005-02-17.

[5] 金元浦．文化生产力与文化产业．求是，2002，(20)：38-41.

[6] 李春华．“文化生产力”初探——文化生产力研究之一．生产力研究，2005，(3)：85-86，186.

[7] 泰勒．原始文化：神话、哲学、宗教、语言、艺术和习俗发展之研究．连树声译．桂林：广西师范大学出版社，2005.

[8] 马凌诺夫斯基．文化论．费孝通译．北京：华夏出版社，2002.

[9] 理查德·巴雷特．解放企业的心灵——企业文化评估及价值转换工具．公茂虹，李汀，译．北京：新华出版社，2005.

[10] Granovetter M. Economic action and social structure：The problem of embeddedness. American Journal of Sociology，1985，91 (3)：481-510.

[11] Lounsbury M，Glynn M A. Cultural entrepreneurship：Stories，legitimacy，and the acquisition of resources. Strategic Management Journal. 2001，22 (17)：545-564.

[12] Barry D，Elmes M. Strategy retold：toward a narrative view of strategic discourse. The Academy of Management Review，1997，22：429-452.

[13] Ireland D R，Hitt M A. Strategy-as-story：Clarifications and enhancements to Barry and Elmes'arguments. Academy of Management Review，1997，22：844-847.

[14] 胡平．中国商业文化新论．山西财经学院学报，1996，(6)：1-6，18.

[15] 加里·贝克尔．口味的经济学分析．李杰，王晓刚译．北京：首都经济贸易大学出版社，2000.

[16] 爱德华·李普马，陈刚．实践理论中的文化与文化概念//薛晓源，曹荣湘．全球化与文化资本．北京：社会科学文献出版社，2005.

[17] 克利斯朵夫·克拉格，索珊娜·G. 斯哥茨曼．文化资本与经济发展导论//薛晓源，曹荣湘．全球化与文化资本．北京：社会科学文献出版社，2005.

[18] 戴维·思罗斯比，潘飞．什么是文化资本?．马克思主义与现实，2004，(1)：50-55.

[19] Klamer A. Accounting for social and cultural values. De Economist，2002，(4)：453-473.

[20] Harvey M G，Novicevic M M. The challenges associated with the capitalization of managerial skills and competencies. The International Journal of Human Resource Management，2005，16 (8)：1374-1398.

[21] 汤任海．约翰·科特．管理与财富，2004，(6)：66.

[22] 王辉，忻榕，徐淑英．影响企业绩效：组织文化比所有制更重要．商业评论，2006，(7)：14，32-35.

[23] 罗纳德·伯特. 公司文化什么时候是一种有竞争力的资产？//芝加哥大学商学院，欧洲管理学院，密歇根大学商学院，等. 把握战略——MBA战略精要. 王智慧译. 北京：北京大学出版社，2003.

[24] 马里亚诺·格龙纳多. 经济发展的文化分类//塞缪尔·亨廷顿，劳伦斯·哈里森. 文化的重要作用——价值观如何影响人类进步. 北京：新华出版社，2002.

[25] 沙因. 组织文化与领导. 台北：五南图书出版有限公司，1996.

[26] Barney J B. Organizational culture：Can it be a source of sustained competitive advantage? The Academy of Management Review，1986，11（3）：656-665.

[27] Sathe V. From surface to deep corporate entrepreneurship. Human Resource Management，1988，27（4）：389-411.

[28] Ramachandran K，Devarajan T P，Ray S. Corporate entrepreneurship：How?. Vikalpa：The Journal for Decision Makers，2006，31（1）：85-97.

[29] Zahra S A，Covin J G. Contextual influences on the corporate entrepreneurship-performace relationship：A longitudinal analysis. Journal of Business Venturing，1995，10（1）：43-58.

旅游交通与航空运输规划①

飞翔，人类最大的梦想之一。

没有翅膀，我们如何能像鸟一样翱翔蓝天？自从美国的莱特兄弟完成了世界上首次有动力飞行，成功试飞了第一架飞机后，人类飞翔的故事就不再是神话。虽然莱特兄弟的首次升空时间只有 12 秒，飞行距离不过 36.58 米，但是他们却开创了人类航空的新纪元。

1903 年，人类飞翔梦想由此实现。

100 多年来，航空技术迅猛发展，极大地改变了人们的生活。航空百年史，有人说是人类的飞翔史、人类的战争史、人类的文明史，更有人说是人类与天斗，不断挑战自我的历史。

如果说航空器对人们来说稍显遥远，民航飞机对大家来说就触手可及。有人感觉坐飞机不安全，一有空难就要伤亡上百人。实际上，民航发动机的空中事故率要求不能超过十万分之一，平均死亡率仅 0.04 人/亿千米，是汽车和轮船的几十分之一，航空运输已然成为最安全、最快捷的交通方式。

鉴于航空器的高速运行特性和昂贵的成本特性，以及航空运输的相对固定性和市场的不稳定性，航空运输规划对航空运输业的发展发挥着重要作用。航空运输规划不仅是一门重要的专业课程，而且掌握它所涉及的航空运输各领域的决策理论、方法和技术是成为航空运输业重要专业人才、高级管理人士的必由之路。

朱金福教授的《航空运输规划》[1]一书终于问世了。他忠厚严谨、默默耕耘，走过的是一条曲折的求学、成才之路。他曾三次无学可上，却又三次幸运地坐进教室，继而走出国门成为德国洪堡学者。他的成功，是永不倦怠、执着追求的最好注解。

该书作为交通运输专业的教材，是我国第一部关于航空运输规划的学术专著，涵盖了航空运输各领域的规划、计划等决策问题。其内容包括机场运行规划、空中交通规划、机队规划、飞机维修计划和航材库存计划、航线网络规划、航班计划、航空公司运行控制及不正常航班恢复优化等，深入讨论了从机场、空中交通管制等航空运输保障系统到航空公司管理的战略层面、战术层面和操作层面。它的问世将对交通运输特别是航空运输专业的建设起到推动作用，同时提高这一领域的学术研究水平。

该书系统性强、例题丰富、图文并茂，既独立成章、自成体系，又相互关联、相互支撑，构筑了一幅航空运输决策体系图。解题细致、脉络清晰，各章配有参考文献、思考题和练习题，充分体现了作为教材易于领会的特色。同时，它又是一本很好的专业参考书，囊括了该作者最新研究成果及独到见解，丰富了运筹学内容，经过分析提炼，形成浑然一体的知识体系。

该书适合从事航空运输各领域技术工作和管理工作的工程师、市场分析师、航线

---

① 本文是《航空运输规划》序，西安：西北工业大学出版社，2009，1-2.

航班分析师、航班舱位控制员研读，也适合航空运输决策支持信息系统开发人员。

过去 100 多年，航空技术已经取得了巨大成就，并推动了人类文明的进步。随着新的理论、方法持续发展和应用，航空技术必将取得更加辉煌的成就。在此衷心祝愿该书能获得成功，为把我国建设成为民航强国锦上添花。21 世纪，中国能够为世界的航空发展做出更大贡献。

参 考 文 献

[1]　朱金福. 航空运输规划. 西安：西北工业大学出版社，2009：1-2.

办好旅游教育至关重要①

欢迎大家莅临第三届亚太地区旅游会展教育与产业发展国际研讨会暨穗港澳会展业对接长三角论坛开幕式。本人代表澳门科技大学，对各位的到来表示最衷心的感谢!

本次研讨会由澳门科技大学与澳门展贸协会联合主办，澳门科技大学持续教育学院和国际旅游学院承办，汇聚一百余位来自亚太各地的旅游会展领域的知名学者专家、产业界精英，可谓是一次会展界的盛会。

在旅游产业迅猛发展的有力推动下，近十年亚太地区旅游教育事业获得了空前的发展。为了更有效地提高亚太地区旅游教育的办学水准，加强旅游及相关专业的品牌建设和课程改革，澳门科技大学与澳门展贸协会成功地连续举办了两届亚太地区旅游会展教育研讨会。在此基础上，研讨会逐步发展成为汇聚会展学界、业界精英的交流会议。本次研讨会亦将作为第十四届澳门国际贸易投资展览会的系列活动之一，获得各方机构如澳门旅游局、澳门成人教育学会等的鼎力支持。在此向他们致以诚挚的感谢。

本届研讨会除了邀请到业界七十多位专业团体代表，更有幸邀请到湖北大学旅游发展研究院院长、湖北大学中国会展研究中心主任和教育部工商管理（旅游）教学指导委员会委员马勇教授；华侨大学旅游学院院长郑向敏教授；重庆旅游学院院长罗兹柏教授；埃及亚历山大大学旅游学院系主任 Hanan Kattara（哈娜·凯塔尔）教授等共二十多位来自中国各地和亚太旅游教育领域内的专家学者参加，经由学术委员会评审，将有近二十篇论文于十月二十三日进行演说与宣读。欢迎各位届时光临澳门科技大学持续教育学院聆听并参与讨论。

最后，祝本次研讨会取得圆满成功!

① 本文是第三届亚太地区旅游会展教育与产业发展国际研讨会暨穗港澳会展业对接长三角论坛开幕词，澳门，2009 年 10 月 22 日.

旅游标志景区研究有意义[①]

旅游业是中国对外开放最早、发展速度最快的行业之一。回顾新中国成立 60 年的历史，中国旅游业经历了对外接待和市场经营的不同阶段，改革开放政策实施以后才真正作为一个经济产业开始了大踏步地前进，直至今日已经成为世界旅游大国，并且有希望在不远的将来成为世界旅游第一强国。产业的高速发展带来了专业学术研究的蓬勃兴旺，管理学、经济学、地理学、社会学、历史学、建筑学、生态学等多种学科的学者都在参与旅游产业的研究与开发，目前已经形成和积累了大量科研成果。

我本人与旅游的结缘开始于 20 世纪 80 年代初担任安徽省省长顾问时参与制定黄山旅游规划的工作。随着不断地接触和了解，我认识到旅游业是一个经济关联广泛的产业，有必要从系统管理科学的角度来对其进行深入研究，并且旅游产业的重要性将在中国经济快速增长和第三产业迅猛崛起的过程中体现得越来越明显。于是，之后我在上海工业大学（现上海大学）又主持建立了国内第一个旅游管理本科专业。2004 年，暨南大学管理学院旅游管理专业博士点建立，鉴于旅游产业发展对于高层次专业人才的迫切需要，我开始招收旅游管理专业的博士研究生。

文彤是我 2004 年招收的第一批旅游管理专业博士学生，他在之前的地理、规划专业学习中已经打下了扎实的理论基础，后来又通过在旅游系统的实际工作积累了丰富的实践经验。进入博士研究学习后他广泛阅读、勤于思考、勇于请教、乐于交流，主持和参加了多项课题研究，表现出较强的科研和管理能力。在博士论文研究中，小文针对实际问题大胆创新，提出标志景区的新概念，并对旅游目的地集群演化提出了独到的见解，他的博士论文得到了评审委员的一致好评。

《从极核到集群：旅游目的地标志景区发展研究》一书正是文彤在其博士论文基础上不断修改、充实、提高而成的，书中吸收了管理学、旅游学、地理学等多学科领域的新方法、新观点和新成果，将增长极核、产业集群、旅游标志景区持续发展、旅游目的地开发管理等内容有机地融合在一起，形成了自己的新认识、新观点和新理论。本书的主要创新之处在于：

（1）针对实际当中概念繁杂而不一致的问题，借鉴相关学科概念，大胆提出了标志景区这一新的概念，并建立了相应的评估标准体系，有助于后期研究在基本概念上的统一。

（2）构筑了标志景区形成基本模型，揭示了标志景区发展演变的规律特征，并从不同维度分析了具体因素对于标志景区培育的影响作用机制，对于旅游高端资源开发管理的实际工作提供了理论的借鉴和启示。

（3）从产业集群理论入手，建立了基于标志景区的旅游目的地“极核-集群”发展模式，深入探讨了旅游目的地在标志景区影响带动下所经历的不同阶段演化特征，并

① 本文是《从极核到集群：旅游目的地标志景区发展研究》序，北京：经济科学出版社，2010，1-2.

就旅游集群发展提出了理论解释和策略选择。

本书的出版是文彤在博士学习研究阶段的成果总结，作为他的博士生导师，我感到由衷的高兴！同时，这也是他专业学术研究的又一个新起点，希望文彤发挥管理学、旅游学、地理学、规划学的多学科背景优势，在今后的教学、研究、实践当中严格要求自己，不断进步，取得更加丰硕的成果，作出新的贡献！

《从极核到集群：旅游目的地标志景区发展研究》一书的出版，将为旅游管理研究、产业集群研究、经济地理与区域规划等领域的专业技术人员和管理人员提供有益的参考。

Preface①

中文摘要

第五届绿色旅游管理国际论坛在广州成功举行，90多位学者在绿色旅游管理领域，充分交换学术观点，这是对绿色旅游管理的伟大创新和奉献。

The Fifth International Symposium on Green Hospitality and Tourism Management was held successfully in Guangzhou city from June 25th to June 27th. The symposium was hosted by the Tourism Management Department, School of Management of Jinan University and jointly organized by the Management Department of Technology Committee of the Ministry of Education of China, the School of Management of Jinan University, the School of Business of Georgia State University in USA, The American Scholars Press, the Management School of Xiamen University, and the Management School of Shandong University. The theme of the symposium "Green Service and Tourism Management Innovation" has attracted scholars, university faculties, and graduate students from China and overseas, for example, Nankai University, Xiamen University, Zhongshan University, Sichuan University, Chi Nan University Taiwan, National Kaohsiung University of Hospitality and Tourism, DePaul University in Chicago, and China Tourism Academy of National Tourism Administration in Beijing.

During the opening and closing ceremony, several scholars delivered keynote addresses. To name a few, they were Professor Bin Dai, Director of China Tourism Academy of National Tourism Administration, Professor Mingzhu Liang, Director of the Tourism Management Department of Management School of Jinan University, Professor Debby Cannon, Director of the School of Management of Georgia State University in USA, and Professor Shiyan Lin from Department of Leisure Studies and Tourism Management of Chi Nan University Taiwan. They had respectively explained and inquired theoretically on green tourism management and innovation from the perspectives of demand on the development of the Chinese tourism industry, sustainable development of intangible cultural heritage, practice in the tourism hospitality industry, green tourism service and product standardization, and development of ecotourism. Delegates from Danxia Mountain, Kaiping Diaolou, and Villages which are both world heritages in China were also invited to present their practices on local community participation, designs and construction of tourist trails in the heritage sites. They addressed the importance of the green tourism management concept in the development and protection of

① Reprinted from Proceedings of The Fifth International Symposium on Green Hospitality and Tourism Management, Marietta: American Scholars Press, 2011, 56.

world heritage tourism and pointed out the concept will be the core of sustainable tourism development in the future.

In addition to keynote addresses, the symposium organized six concurrent sections for group presentations based on the theme of green tourism management and innovation. Participants discussed and shared ideas on the management of world heritage and resorts, green tourism and low-carbon tourism, hotel and tourism service and management, urban tourism and destination marketing, tourism market and tourist behavior, tourism education and tourism development.

The symposium has accepted over 150 academic papers. Only about 90 papers have been selected for publication. The topics included in those papers include the hot topics in China in terms of management of world heritage and resorts, and tourism development. For example, scholars' structured theory on the development and protection of world heritage sites using examples of some famous Chinese sites like Zhangjiajie, Jiuzhaigou and Sanxia of Yangzi River, etc. Some papers focused on the consumption demand of tourism development, management of carbon emissions, harmony of ecological environment under the idea of low-carbon and proposed corresponding tactics for management. The urban city was another key research subject in the symposiums. Sustainable development of urban tourism was studied and discussed from the aspects of management mechanism, public marketing and stakeholder analysis, and tourism planning. The tourist is a critical group in green tourism management. Some scholars conducted research on tourism market segmentation and tourist behavior. They talked about the impacts of service quality of ecotourism on tourist experience, tourist motivation, behavioral intentions and group characteristics. Some papers are concerned with the embodiment and practice of green tourism services and management in tourism education. They hope to guarantee the supply of human resources for the implementation of green concept throughout the tourism industry by talent training.

Relying on the Fifth International Symposium on Green Tourism Management, the symposium proceedings absorbed both domestic and abroad wisdom on green tourism management, which can definitely provide a platform for scholars to exchange academic viewpoints, strengthen communication of tourism research and broaden horizons. On the one hand, the conference proceedings are a theoretical summary of the symposium, on the other hand, it is a great innovation and contribution to the research of green tourism management.

产业工人的中国梦：从低技能劳工到专业技术工人的人资转型升级战略①

一、改革开放为中国梦的实现奠定雄厚基础

1978年以来，中国逐步走上了对外开放之路，改革开放使社会生产力得到极大解放，经济总量迅猛扩张，开创了中国经济发展史上前所未有的“高速”时代。国内生产总值（GDP）在1986年首度突破1万亿元人民币，2010年超过日本，达到40万亿元人民币，我国成为仅次于美国的世界第二大经济体，2012年上升到约52万亿元，不到30年就增长了50余倍（图1）。而中国的外汇储备也迅猛增长，从1978年的1.67亿美元迅速扩大到2012年底的3.31万亿美元，连续多年稳居世界第一位，在2013年第一个季度，达到了3.44万亿美元，相当于整个德国的经济总额。

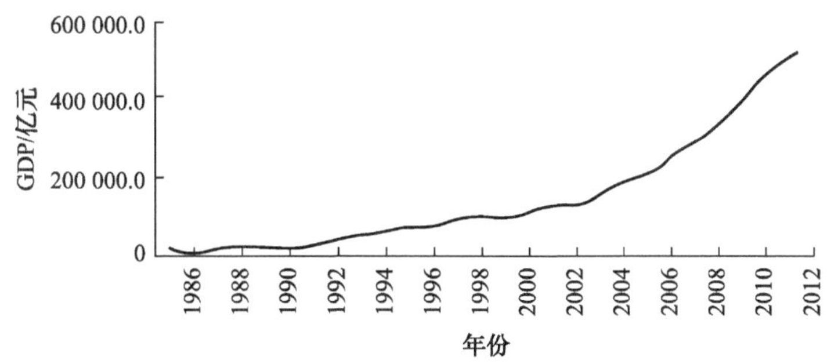

图1　中国GDP增加态势

资料来源：http://www.stats.gov.cn/tjsj/ndsj/2012/indexch.htm

近年来，除了谷物、肉类、花生、茶叶、水果等农产品产量稳居世界第一，中国工业生产能力也迅速扩张，在220多种工业品产量上居世界第一位，“中国制造”遍布全球，而原厂委托制造（original equipment manufacture，OEM）是其中最重要的组成部分，中国因此被称为“世界工厂”[1,2]。据美国经济咨询公司环球通视数据，2010年中国制造业产出占世界的比重为19.8%，已超过美国，成为全球制造业第一大国。

此外，中国从1993年起一直是吸收外商直接投资最多的发展中国家。2012年，中国货物进出口总额达到38 676.6亿美元，连续两年成为世界货物贸易第一出口大国和第二进口大国（仅次于美国）。而随着全球经济一体化进程的加快，中国已从大规模“引进来”进展到大踏步“走出去”的阶段，商务部估计，中国企业以及主权财富基金在海外的投资在2011～2015年将达到5600亿美元。

综上所述，改革开放使中国一跃成为世界经济大国，奠定了雄厚的物质基础，铺

① 本文原载《战略决策研究》，2013，4（5）：37-50. 作者：刘人怀，覃大嘉，梁育民，左晓安.

就了扎根中华、联通世界的联络渠道，为中华民族的伟大复兴，为中国各行业、各族人民实现中国梦提供了可靠保障。

二、产业工人中国梦的实现与中国经济转型升级紧密关联

1. 中国经济和产业发展必须转型升级

（1）过分依赖出口、投资拉动的经济增长模式不可持续。前述骄人的经济成就说明，中国已再次崛起，有能力创造21世纪的先进文明，可以与西方的任何国家平起平坐，但是，以出口与代工制造业务为主力军的发展方向，还是有很大缺陷的。2008年国际金融危机与2010年欧债危机接连发生之后，全球市场发生了巨大变化，发达国家的需求至今仍积弱不振、复苏缓慢。近几年来，美国失业率一直徘徊在9%左右，欧元区失业率也达到了近10年来的最高值，这就暴露出中国过去高度依靠出口的经济增长模式存在严重不足。简言之，中国在贸易、能源、资源等方面对西方成熟市场的依赖性过强。金融危机爆发前，中国进出口贸易额占GDP的比例高达67%。而金融危机后，欧美居民消费能力持续下滑，因此开始盛行对外贸易保护主义，这就严重地影响了所有依赖欧美市场的出口导向型经济体，包括中国。在这样的压力下，中国为了保持增长速度，出台了大规模的经济刺激计划，投资率快速升至50%左右的历史最高水平，最终导致中国地方债务风险、房价过高、铁路投资负债率过高等问题的凸显。中国投资驱动的经济增长已经走到尽头，必须向内需、投资、外需协调驱动的方式转型。换言之，中国产业结构已遇到瓶颈，需要大幅调整与改变，要从原本高度依赖全球价值链低端OEM业务的“制造业大国”转型升级成“服务业大国”，经济才能持续增长。

（2）利用劳动力比较优势参与国际竞争模式不可持续。如前所述，随着改革开放大幅加速了中国OEM产业的国际化进程，很多国外劳动密集型企业为了利用中国低廉的劳动力优势与出口优惠政策，纷纷进入中国建立制造基地[3]。因此，不少可以吸收外资的沿海城市，如东莞、天津、深圳等，在过去30多年享受了伴随外资投入而来的、社会经济高速发展的红利，但也吸引了大量的农民工前仆后继地涌入这些城市挣钱，让“外来工"成为沿海OEM产业劳动力的主要构成部分，这成了中国普遍的社会现象[4]。但随着基本工资不断增加，人民币持续升值，从2006年至今已升值超过28%，造成OEM企业沉重的压力，过去利用劳动力比较优势参与国际竞争的模式受到严峻考验。

此外，虽然2010年中国人均国民总收入达到了4270美元的水平，并且首次超过世界银行当年界定的中高收入国家3976美元的分界线，但是中国人均国民总收入仅仅相当于美国目前水平的10%。未来20年，中国将面临人口老龄化的压力。“人口红利”的渐渐逝去，也将对中国的劳动力低成本优势产生影响。利用低工资劳动力成本参与国际分工，使中国陷入“贫困的增长”之中。日本在经济快速增长时期，其工资的增长速度比美国快70%，到1980年就已经与美国持平，这一段路程大概用了30年时间，而1978～2004年，中国经济高速增长了近20多年，工资却只有美国的1/20、日本的1/24。这种以牺牲劳工利益为代价的经济增长模式所带来的负面后果，越来越以触目

惊心的方式表现出来，不断增加的社会矛盾警示了中国经济由劳动力密集型产业推动经济发展的粗放式增长向资本、技术密集型产业驱动经济发展转型的迫切性。

(3) 既面临“中等收入陷阱”又面临双重竞争压力。目前，我国一方面面临“中等收入陷阱”，另一方面，又面临低收入国家和发达国家的双重竞争压力。中国正处于由中等收入国家迈入高收入国家的关键时期。一个国家在达到中等收入水平后要想更加富裕，就必须适时采取新的发展战略，即专业分工＋规模经济战略。如果没有专业分工和规模经济，而是持续基于要素积累发展战略，将可能陷入“中等收入陷阱”。拉丁美洲和中东地区就是例证。

同时，我国还面临发达国家和低收入国家的双重产业竞争压力。一方面是作为竞争对手的低工资水平的贫困国家，它们的成熟产业将逐步占主导地位，将给中国发展带来严峻的挑战；另一方面，作为创新者的发达国家，它们在那些技术变化较快的产业领域中具有巨大优势，中国很难同它们竞争。关键是能够在全球市场上找到“市场空缺”。如果能够在全球市场上找到“市场空缺”，在全球化的今天，其生产规模就可以扩大，以充分利用规模经济获取竞争优势。

在全球化的今天，中国要维持自己的竞争优势，巩固自己在全球中的市场位置，避免陷入“中等收入陷阱”和有效应对发达国家与低收入国家的双重产业竞争压力，关键是在专业分工和规模经济的基础上向创新转型，促进产业在全球价值链中的攀升。

(4) 技术依赖导致核心竞争力缺乏。通过廉价劳动力参与国际产业价值链的低端分工，导致中国对发达国家的技术依赖。资方通过压低工资赚取巨额利润，降低了提高技术和改进管理水平的动力，使得中国的产业升级面临重重困难。就资方而言，由于没有较大的竞争压力，因而也就缺乏提升技术的动力。大量外商到中国投资，其首要动机就是看中了中国的廉价劳动力。中国沿海地区在早期开放过程中，使用了大量的内地农民工。但过去30年里，这些沿海企业没有明显的技术进步和产业升级，金融危机爆发使大量的企业倒闭。同时，大量使用廉价劳动力的工厂的存在，也使得劳动者没有动力来提升自己的技术和技能。打破技术依赖的怪圈，建立中国经济国际核心竞争力，必须坚决打破对粗放型产量和路径的依赖，转向核心技术、核心创新能力和创新产业。

2. 产业工人中国梦的实现要适应中国转型需要

在中国经济和产业转型的艰巨攻关中，在“中国制造”向“中国创造”跨越的征程中，产业工人必须扮演不可替代的重要角色。

目前，中国工业产品的平均合格率只有70%，每年因不良产品造成的经济损失高达2000亿元。另外，中国机械工业劳动生产率约相当于美国的1/12、日本的1/11，劳动生产率每增加或减少1%都会影响产值上百亿元，影响工资成本10多亿元[5]。

产业工人中的技能型人才不仅是企业产品的生产者和制造者，直接决定着企业的核心竞争力，更是国家软实力的重要组成部分。产业工人是制造业的重要生产资源，在目前越来越大的市场竞争压力下，谁有一支稳定的高素质员工队伍，谁就可能取得行业的主动权。

在中国经济和产业转型升级的过程中，高技能产业工人无论是知识、技能、对社会的贡献、社会对他们的需求度还是收入水平和消费水平，均不亚于白领。

在现代制造业最为典型的行业如电子、汽车、通信技术行业中，高级技师所做工作既动手又动脑的复合程度，远远超过了传统制造业中的相同角色。而在作为制造业大国的中国，最缺的是既具有高学历又具有高动手能力的技术工人。

一方面，中国经济转型升级，为中国产业工人发展提供了新的广阔发展空间；另一方面，中国产业工人的中国梦，必然也需要通过自身的发展转型，在中国经济和产业的成功转型中实现。

三、产业工人的梦想与困惑

1. 产业工人在多方面发生结构性变化

近年来，由于产业结构调整、所有制结构变化、城镇化提速，中国的产业工人情况出现了一些新特征。

(1) 职工所依存的经济组织的所有制形式日益多样化。产业工人开始由过去集中在全民和集体企业的单一性，发展到在流向全民、集体企业的同时，也向非公有制企业流动。在国有单位就业的工人有所减少，由 1996 年的 11 244 万人减少到 2011 年的 6704 万人。流向私营企业和个体经营企业的人员急剧增长，由 1996 年的 2329 万人迅速扩张到 2011 年的 12 139 万人，如图 2 所示。目前，约有 1 亿多名职工在私营企业、个体企业、“三资”企业等各类非公有制经济组织中就业。

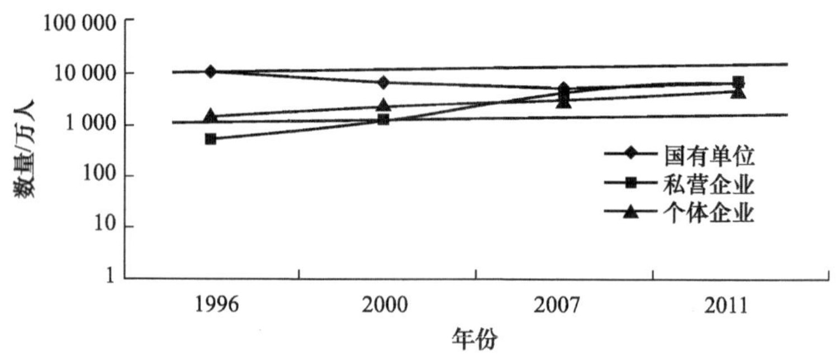

图 2　城镇就业人员所有制形式

资料来源：《中国统计年鉴》(1997～2012 年)

(2) 农民工正成为中国产业工人的主体。在中国改革开放初期的 1978 年，全国 99%的产业工人是城镇居民[6]。但在 20 世纪末 21 世纪初，这种情况迅速发生变化。根据 2004 年底劳动和社会保障部的调查，农民工已占企业员工总数的近 60%，农民工已经成为中国产业工人的主体。在第二产业中，农民工占从业总人数的 57.6%，其中，在加工制造业中占到 68%，在建筑业中占到近 80%。在第三产业的批发、零售、餐饮业中，农民工占从业人员人数的 52%以上[7]。总的趋势是，农民工数量快速增加，增速快过城镇就业人员增速。2006 年，中国农民工总数是 13 181 万人，与城镇就业人员的比值是 46.56%；2011 年，农民工数量增加到 25 278 万人，与城镇就业人员的比值增加到 70.38%（图 3）。

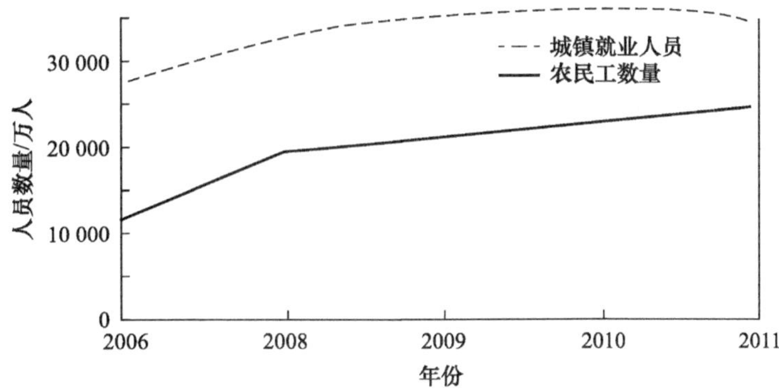

图 3　农民工与城镇就业人员数量比较图

资料来源：《2011 年我国农民工调查监测报告》，《第二次全国农业普查主要数据公报（第一号）》，《人力资源和社会保障事业发展公报》（2006～2011 年）

（3）从文化结构看，科技文化素质有所提高。工人的文化水平，主要指工人接受教育、培训的程度。新中国成立初期，中国工人阶级的文化水平基本上是文盲、半文盲，如今与全国教育水平的发展和就业人员文化程度提高相一致（图 4），中国产业工人的科技文化素质明显提高。改革开放初期，从事体力劳动的职工基本上是文盲、半文盲，而现在至少是初中学历，有些职工是大学文化。与老一代农民工相比，新一代

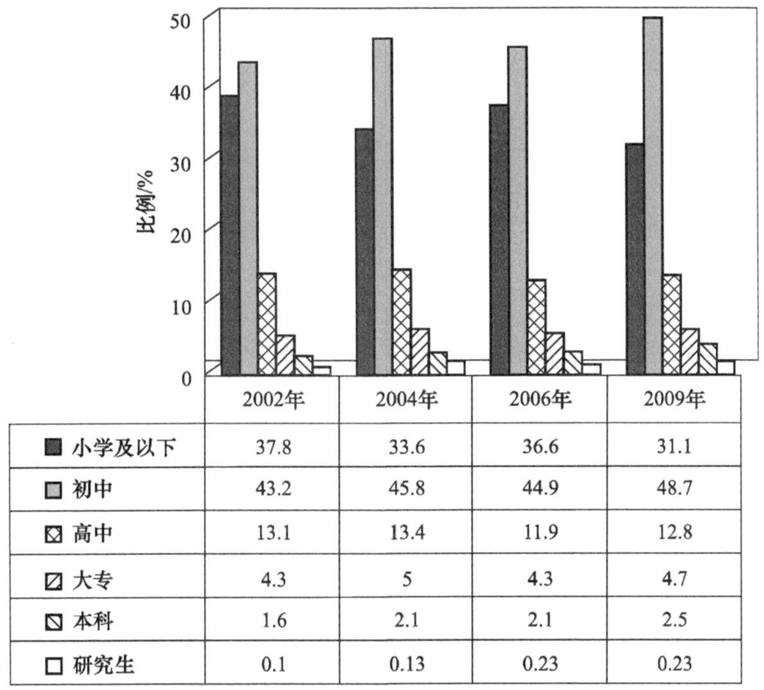

	2002年	2004年	2006年	2009年
小学及以下	37.8	33.6	36.6	31.1
初中	43.2	45.8	44.9	48.7
高中	13.1	13.4	11.9	12.8
大专	4.3	5	4.3	4.7
本科	1.6	2.1	2.1	2.5
研究生	0.1	0.13	0.23	0.23

图 4　全国就业人员文化结构

资料来源：《中国劳动统计年鉴》（2003 年、2005 年、2007 年、2010 年），国家统计局住户调查办公室《新生代农民工的数量、结构和特点》

数据之和不为 100%是数据修约所致

农民工的受教育年限、文化层次已经有所提高。小学文化层次以下农民工下降到总数的6.7%，中专文化层次的农民工由2.1%提高到9%，大专及以上文化层次的农民工由1.4%提高到6.4%（图5），农民工的劳动特点发生了由体力劳动型向智力劳动型的变化。传统意义上的工人，特指工业体力劳动者，现在正向脑力劳动、脑体相结合的复合型工人转变。

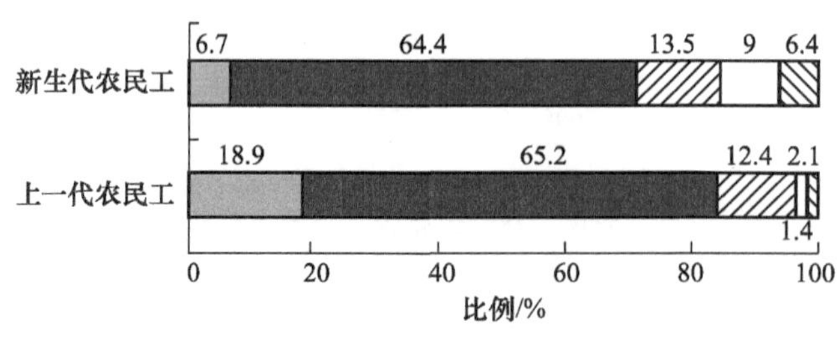

图5　新老两代农民工人力资本特征比较

（4）工人阶层的年龄结构发生变化。现在的很多劳动者都是“80后”甚至是“90后”。在青年职工中，艰苦条件下奋斗苦干早已不再时兴，勤劳创业、实业致富也远没有“挣快钱”受欢迎。“80后”“90后”职工认同“对社会应该无私奉献”的分别占37.4%和27.1%，与“50后”职工相比低了15～25个百分点[8]。

中国产业工人在年龄、受教育程度、家庭背景、就业载体等方面的变化导致了管理、培训等方面的一些变化和问题，当然，也导致了产业工人在理想和追求方面的变化。

2. 产业工人面临困惑

产业工人就业流动性增加，对企业的依附感和归属感下降。产业工人在不同产业、不同行业、不同所有制企业间的流动变得频繁起来。很多工人逐渐从被动转岗、下岗到主动竞争择业，就业观念开始自觉地与市场经济接轨。一些企业急功近利，片面追求经济效益，忽略了对工人技术、技能的培训，不重视工人民主参与、民主管理的权利，不注意培养和维护职工群众主人翁精神，不少产业工人认为自己就是靠打工吃饭，直接影响了工人对企业的归属感。

产业工人特别是农民工权益被侵害的现象增多，制度保护、政府保护、组织保护等缺失。涉及产业工人的劳动争议数量不断攀升，在劳动关系的矛盾中，工人合法权益受到侵害具有一定的普遍性，拖欠、克扣工资、强迫超时工作等现象屡有发生。由于在激烈的就业竞争中农民工处于相对弱势地位，加之劳动力市场管理不够完善，农民工特别是外来劳动者在经济地位、社会保障、民主权利等方面受到不平等待遇的问题尤其严重，主要表现为整体工资收入偏低、各项社会保险参保率低、民主参与渠道少，在有些地方还存在着农民工人身权利和自由受到侵犯等问题。

管理者与被管理者之间的亲和力有所减弱。越来越多的产业工人由过去留恋计划经济时代“铁饭碗”过渡到普遍接受劳动合同制，明确了工人与企业在权、责、利方面的关系。企业经营管理者与普通工人在经济利益、民主权利和社会保障等方面的差距拉得过大，矛盾有逐渐尖锐的趋势。管理者与被管理者同在一个屋檐下，心态各不同，以往的那种亲和力正逐渐淡化和消失。

3. 产业工人的梦想和追求多元化

（1）产业工人价值取向多元化。教育部人文社会科学重大课题的研究揭示，当代工人的价值取向呈现多元化趋势。他们的价值取向是品格追求、公共利益和工作成就，对法规、人伦前提和家庭价值的强调居中，而对金钱、权力和从众的价值取向与理论平均数相比则稍向负性侧面偏离[9]。

这一结果表明，在改革开放和经济的变革环境中，中国工人一方面依然重视中国传统价值对品格的看重；另一方面也重视对民主、平等、公正的社会秩序的追求，强调公共利益和工作成就，弱化金钱权力和从众的世俗观念。这体现出工人群体价值取向的相对独立性，也与社会主义市场经济强调的以实现人的自由和全面发展为本位的价值取向观念相符。但这并不表明工人对待遇问题的忽视。在回答“今后，您最期望的是什么”问题时，有 56％的工人选择“涨工资”，有 29.33％的工人选择“工作稳定”[10]。

（2）年轻一代产业工人的需求和定位由经济向经济、文化等多方面转变。现在的很多劳动者都是“80 后”甚至是“90 后”。以“80 后”“90 后”为主体的产业工人，他们的劳动已经上升到享受的层次，好多工人听着 MP3、MP4 去上班。如果让以前的工人加班，他们通常没什么意见，但现在的“80 后”“90 后”就不好说了，说不定直接辞职了。

1990 年后，没有从事过农业生产、直接从学校进城的务工者越来越多。他们有较高的文化水平，也就有较高的精神生活需求，渴望继续学习、重视技能培训，希望融入城市主流社会的理想特别强烈。他们要求和城里人一样平等就业、平等享受公共服务，甚至得到平等的政治权利。他们被称为“新生代农民工”。

（3）受过较高文化教育的产业工人渴望在帮助国家振兴中实现人生价值。他们希望立足岗位，树立“精准”意识，把产品做得精益求精，把“中国制造”的牌子在世界上打响，通过自己的技术和智慧在全球树立“中国创造”的品牌。他们希望企业的高管在一线工人中多交朋友，真实地了解工人的所思所想，给工人创造更好的工作条件和环境，激发出他们更多的智慧和力量，使诚实劳动、创新劳动、有尊严的劳动成为常态，企业实现高质量、高效益的生产。

4. 企业需要分析人力资源管理上的不足之处

中国产业工人面临的困惑和困难，一部分要归因于中国经济社会发展阶段的客观条件局限，另一部分也与中国企业在管理中存在的问题有关。协助中国产业工人实现中国梦，一方面要促进社会经济发展，在发展中解决问题；另一方面，广大企业经营管理层要剖析自身存在的问题和不足，寻求解决之道。

（1）过于以效率为核心，忽视当代产业工人的价值追求和多种需求。科学管理之父泰勒把提高工作效率等同于员工发挥最大限度的积极性，认为只要员工能够本着勤奋努力的道德品质，与他所设计的工作动作相配合，工人就能成为“一等工人”，就能减少不必要的时间损耗，就可以提高劳动生产率，实现企业的目标。中国富士康之类的代工企业，在对一线操作工的管理上沿用这样的管理逻辑，企业把员工当作机器的一个零部件，当作雇主实现利润的工具，忽视员工作为人的需求[11]。现代企业需要改变这样的管理价值，企业的目标和宗旨应该是实现雇主利益和雇员利益的共赢，企业实现利润的过程也是员工自我价值实现的过程。有的企业在日常管理工作中，过于重视对机器和工作中工人的管理，而轻视或忽略了员工的社会交往等需求和所存在问题的管理。当员工与员工、员工与基层管理人员之间由于日常人际关系或利益纠葛而产生摩擦和冲突时，企业往往采用粗暴的打压方式对付，而忽视了员工的心理需求和利益诉求的实现。

(2) 人力资源管理的机制、思想与方式过于陈旧。现阶段中国大多数企业的人力资源管理还处于传统行政性人事管理阶段，有的企业的人力资源管理机构只是发挥确定人事关系、审阅工资安排、管理员工档案等低效的职能，对于职能产生的作用来讲，与将员工作为一类资本进行开发、发挥作用与管理的目标还存在不小距离。而且，有许多企业在人力资源管理上还处于僵化地对员工实施考勤、奖励、处罚等层次的控制和管理阶段，远没有达到“以人为本”的水平，对员工主要是用制度来控制与强制，还处在“对立式管理”水平，基本没有达到使用人本管理方式来激励员工的潜能与热情的“情感管理”水平。

(3) 人力资本投入不足，培训机制不完善。人力资源作为企业一项非常重要的资源，其价值在于可以不断被开发，并持久使用。但许多企业没有做好人才的选拔和储备工作，对招聘工作不重视。企业管理者在人才认识方面存在不足：一是将人才当作商品，认为只要有钱就可以招到员工，不需要时就可以辞退；二是认为内部人都不行，外部人才又找不到，企业就是缺乏管理人才。问题的根源在于没有从根本上认识到人力资源的重要性，更未认识到人力资源还可以开发。当然，有些企业是认识到人力资源的重要性了，但不知如何开发，要么忽视培训，要么做大量没有效果的培训。

(4) 员工配置和薪酬机制、流动机制不健全。新时期很多企业人员流动频繁，人才流失严重，影响团队士气及整个组织气氛，影响到了企业的正常运行。有些企业缺乏动态的人才配置机制，忽视人员的素质和能力应与其所担负职责相匹配，更不用谈考虑员工未来工作的适应性和稳定性。同时，不能激发、挖掘员工的潜能，使人才的自我发展、提升空间受到限制，也是最终导致员工寻求外部更好的发展空间和机会的重要原因之一。

四、携手助推产业工人实现中国梦

1. 基本目标：开展产业工人凝聚力“5度”工程建设

企业的管理层是否真诚关怀基层员工，对员工忠诚度的影响十分关键。只有管理层愿意给员工提供事业上公平竞争的环境，关心员工生活，企业内部才会有适宜基层员工生存、发展机制的建立，以及对生活与工作环境的改善，企业也才能留住员工。从基层员工的角度来说，能否认同公司的企业文化，也是决定去留的重要因素。只有在价值观上认同甚至欣赏企业文化，他们才会愿意长期留在该企业工作。

(1) 提高企业与员工的沟通程度，包括企业与管理层对员工、员工对企业与管理层的态度。

(2) 提高员工职业发展平台的“三公”程度，主要包括基层员工职业发展平台与薪酬机制分析。

(3) 提高员工对企业文化的认同程度，从各个环节、各个层面，提高员工对企业文化的了解和认同程度。

(4) 提高员工对工作环境的满意度，包括安全、环境保护、劳动强度等方面。

(5) 提高员工对生活环境的满意度，包括食宿条件、文化娱乐等方面。

“5度”工程如图6所示。

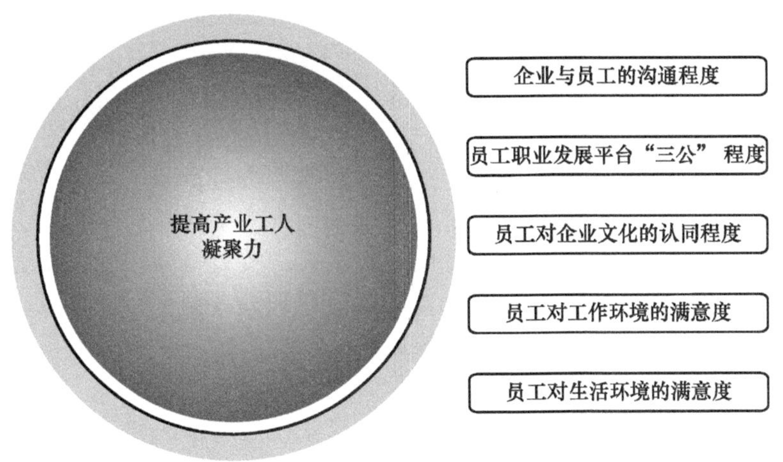

图6 产业工人凝聚力“5度”工程

2. 重点举措

(1) 以人为本，规范管理，增强产业工人融入感、归属感、文化认同感。全国总工会做“您认为企业文化最重要的是什么”问题调研时，有59%的工人选择“尊重、理解、信任工人，维护工人合法权益”[9]。这说明绝大多数工人希望企业能够营造一种尊重、提高工人地位和保障工人合法权益的文化，各级政府部门和相关企业需要携手把主要精力用在提高工人的地位和维护工人合法权益上。企业在人力资源管理过程中要充分体现人性化管理的要求，树立以人为本的思想，并深入贯彻到人力资源管理模式中，注重人的差异性、层次性，尊重员工个人追求的内在价值的自我实现，关心员工的自我进取。

把人文理念融入管理制度之中。要把人文理念作为管理制度的灵魂，作为制定管理制度的基点，有效渗透到改革发展、安全生产、经营考核、收入分配等管理制度当中，提高管理制度的人本性和可操作性。例如，在建立人事管理制度时，我们要大力倡导人文理念，建立和完善用人制度上的激励和约束机制，创造鼓励职工干事业的良好环境。在协调利益分配方面，我们要在人文理念的指导和检验下，建立和完善“限高托低”的收入分配机制，坚持按劳分配和效率优先的原则，合理协调各业务系统之间、干部与职工之间、机关与一线之间、新老职工之间的利益关系，既要有合理差别，又不能相差悬殊，既要考虑劳动强度，又要考虑技术含量，既要注重完成任务量的多少，又要强调劳动价值大小，进一步规范收入分配秩序，让全体干部职工共享改革发展成果。

加强对员工的精神关怀。以经济建设为中心的社会，人们往往更注意金钱和财富的作用，而常常忽略了人的精神需求，但实际上人的精神需求从来也没有降低过，现代社会中出现了不少抑郁症患者，许多时候人们感叹世态炎凉、人心冷漠，这其实恰恰反映了人们对精神需求的渴望。好的企业管理，绝不能只是把人当作流水线上的机器。现代社会中，金钱绝不是万能的，精神关怀绝对不可或缺。对员工的精神关怀不但可以改变员工的精神面貌，还可以让员工迸发出意想不到的创造能量，在无形中改变企业的人文环境和精神面貌，实现企业的发展目标。

(2) 建立和完善职业培训教育制度。为中国培养产业工人队伍，是提升中国制造业水平的关键。全国总工会的调研显示，近40%的工人认为技能对于工人是最重要的。在德国，有的蓝领收入比白领还高，有些工人几代人都是产业工人，甚至形成了技术的传承。普通工人阶层之所以在改革过程中呈现向下流动的趋势，其“文化资源”占有处于劣势是一个重要原因。因此，要加大对工人的关爱力度，把提高工人的技能作为当前一项十分重要的工作。要建立完善的职业培训教育制度，加强对农民工、体力工的职业培训，提升其文化资源和进入机会，不断提高其收入分配水平，使之逐步进入到新社会中产阶层。

(3) 做好职业生涯规划。企业为员工做好职业规划，强化员工的归属感，这是激励员工为企业发展而努力贡献的重要措施。每个企业在选拔或招用员工之初，就要根据员工的工作性质、岗位特点、人才性格、专业技能和个人特长等因素，为他们今后工作中的发展确定一个方向，也就是说，为不同文化层次、职业技能的员工量身定做，开展职业生涯规划设计，使他们的目标与企业的目标发展方向一致，从而使他们有较高的归属感，这样一来，就会为培养企业战略性的后备和骨干力量奠定一个坚实的基础。

(4) 提供宽松的职业发展空间。优秀的人才希望通过发展自己来实现自我价值，让这些人以最快的速度成长是非常重要的，不要担心当他们羽翼日益丰满就有可能离开企业，如果他们感觉自己有发展余地，而组织能给予这种空间，他们就会留下来。为了给予人才宽松的发展空间，企业一方面为其提供培训的机会，如派员工参加研讨会、专业会议等，以适应员工的需要；另一方面企业要给予其新的挑战，让其承担更重要、范围更广、更多责任的工作，以此发展和激励人才。

(5) 建立和完善社会保障制度。在中国社会阶层分化加剧的历史时期，加快建立和完善社会保障制度，能在全社会的范围内，特别是为社会弱势群体提供基本的生活保障，缓解数量巨大的产业工人的生活压力和顾虑，同时有利于消除和缓解不同社会阶层间的矛盾，起到维护社会稳定的“减震器”和“安全阀”的双重作用。

习近平同志提出中华民族要实现伟大复兴的中国梦，中华民族伟大复兴的中国梦的实现有赖于全国各族人民、各行业、各群体中国梦的实现。数量庞大的中国产业工人的力量与汗水曾经为改革开放后中国经济社会的快速发展开辟了通道，擦亮了“中国制造”的品牌。在未来，他们还要为“中国创造”贡献不可替代的技能和智慧。政府、企业和社会有责任发挥各自的作用，从制度、物质、文化和精神等层面携手合作，为他们铺平实现梦想之路。

参 考 文 献

[1] Chin T. An exploratory study on upgrading by FDI OEMs in China. International Business Research，2013，6（1）：199-210.

[2] 覃大嘉，吴东旭，毛蕴诗. 金融危机对中国天生国际化 OEM 企业的影响及其战略反应研究. 学术研究，2011，（9）：61-69.

[3] Gao C Y，Murray J Y，Kotabe M，et al. A "strategy tripod" perspective on export behaviors：Evidence from domestic and foreign firms based in an emerging economy. Journal of International Business Studies，2010，41（3）：377-396.

[4] BSR. A study on the labor shortage and employment guidelines for manufacturers in China. https：//www. bsr. org/en/reports/a-study-on-the-labor-shortage-and-employment-guidelines-for-manufacturers-i[2010-09-27].

[5] 郭奎涛. 中国产业工人需升级. 工程机械周刊，2010，（6）：25.

[6] 赵东辉，吴亮，江国成. 我国产业工人结构发生历史性变化. https：//news. sina. com. cn/c/2002-12-19/153318770s. shtml[2022-12-19].

[7] 阮煜琳. 农民工正成为中国产业工人的主体. https：//news. sina. com. cn/o/2004-02-22/09161864689s. shtml[2004-02-22].

[8] 郑莉，张锐. 壮大工人力量托起"中国梦". 工人时报，2012-03-05（第 2 版）.

[9] 金盛华，刘蓓. 当代中国工人价值取向：状况与特点. 心理科学，2005，（1）：244-247.

[10] 赵文祥，杨立华，纪德云. 当代中国工人文化研究. 中国工运，2008，（4）：48-50.

[11] 张劲松，李娟. 论中国现代企业管理的人文关怀——基于富士康"跳楼"事件的思考. 生产力研究，2011，（9）：6-7，35.

企业战略管理概要[①]

一、何谓战略？

记住六个字："做对事"而不是"做事对"。

简而言之，做正确的事情！首先"方向搞对"而不是"做事正确"。

1. 战略的由来

战略，原为军事用语，源于兵法。在我国古代，开始时"战"和"略"分开使用。战是指战斗、战争；"略"是指谋略、策略、谋划。后来，合并起来，称为"战略"。西方也是这样，"战略"来源于古代战争。

有关战略的举例如下。

(1)"王不如远交而近攻"(《战国策・秦策三》)(范雎向秦昭王献策，力主先打毗邻的韩国、魏国后，再打远的齐国)。

(2)"兵法贵在不战而屈人"(《三国志・魏书・陈泰传》)。

(3)"高筑墙，广积粮，缓称王"(朱升献给朱元璋的九字谋略)。

(4)"隆中对"以及后来的"联吴抗曹"(诸葛亮献给刘备的谋略)。

(5)"深挖洞，广积粮，不称霸"(毛泽东借鉴明朝谋略，演绎而来)。

上述都是长期性、全局性谋略，通称战略。

而刘备、孙权合谋赤壁之战火攻大败曹操，亦是谋略，但是短期性、局部性的谋略，则属于战术。

2. 企业战略

现在，"战略"一词已经从军事延伸开来，扩展到非军事领域。大约在 20 世纪 60 年代，战略进入企业领域，并从 80 年代以来，被全世界广泛使用。

各类非军事战略的举例如下。

(1) 国家：科教兴国战略、可持续发展战略。

(2) 地区：沿海开放战略、西部大开发战略。

(3) 城市：上海浦东开发战略。

(4) 学校：素质教育战略；暨大"侨校＋名校"战略。

(5) 个人：韬略——韬光养晦。

(6) 企业战略。

企业战略即企业的全局性、长远性谋划，内容包括企业的长期目标、总体目标以及达到该目标的基本方略。

战略的构成要素：全局＋长远/目标＋手段。

---

① 本文是西南交通大学经济管理学院博士生课程讲稿，成都，2014 年 6 月 23 日.

3. 企业战略的特点

(1) 具有全局性。事关大局、根本，事涉整个组织的生死存亡（如古代帝王将相的“江山社稷”）。

(2) 具有长远性（即未来性）。功效持久，影响深远，远见卓识，至少管 3 年，一般 5～10 年，多则 10 年以上。

(3) 具有目标性。确立企业要达到的长期、总体与核心目标。

(4) 具有谋略性。即一并提出达到长期、总体与核心目标的基本手段——方略。

(5) 具有纲领性。即对组织的核心目标及达到目标的基本手段的“粗线条”，框架性谋划，具有权威性、路线性和指导性。

(6) 具有抗争性（即竞争性）。制定企业战略就是为了企业能在激烈的市场竞争中发展壮大自己的实力，占有相对的优势。商场如战场，竞争如同没有硝烟的战争。“战略”一词正好反映了商场的这种抗争性。正因为如此，也有人认为“战略”二字的火药味太浓，而主张改称“策略”。

(7) 具有合作性。不只是与竞争对手展开竞争，还要在竞争中有条件地与竞争对手进行合作，以取得双赢的结果。

(8) 具有稳定性。企业战略必须在一定时期内保持稳定性，才能在企业经营实践中具有指导意义。

(9) 具有动态性。当企业外部环境与内部条件发生较大变化时，就必须对战略进行必要的调整修改，要对环境有动态适应特点。

(10) 具有系统性。大型企业的战略是一个庞大而又统一的完整系统，既有战略的层次问题，又要使各层战略不互相矛盾，要做到同步化和协调化，这就是系统性。

(11) 具有预应性。即对未来的预测和筹划，也是一种“预谋”。

(12) 具有风险性。正因为是对未来的预测和筹划，而未来又充满了种种变数和不确定性，因而战略不可避免地具有风险性。

案例：某企业决定实行亏本销售、食堂外包，算不算战略？

答案：否。

因为任何形式的亏本销售甚至倾销都只能是一个企业的临时措施或权宜之计，不可能是一种持久性方略，不具备长远性也就不能成为战略。

食堂外包可能是实施某一项战略的举措之一，但单就这项措施而言，不具有全局性影响，因此不足以称为战略。

4. 企业战略的层次

(1) 公司战略。公司战略是一个企业的整体战略，是最高管理层指导、影响和控制组织一切行为的战略总纲。

(2) 业务战略。业务战略又称“业务单位战略”“业务竞争战略”或“经营战略”，是在公司战略指导下，各个战略事业单位——子公司、分公司或事业部制定的分项业务（产品、服务）战略。

(3) 职能战略。职能战略是为辅助、支持公司战略与业务战略而在相关职能管理领域制定的部门战略，如市场营销战略、生产管理战略、研究与开发战略、财务管理战略、人力资源开发与管理战略等（图 1）。

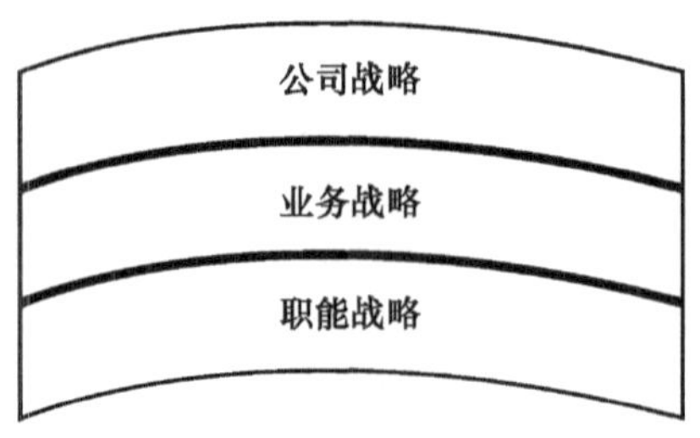

图 1　战略的层次示意图

注意：经营单项业务（产品、服务）的专业化公司，公司战略就是业务战略，业务战略就是公司战略。

5. 企业战略的分类

对前人已采行战略的总结如下（切不可死啃兵书，照搬硬套，受此局限）。

1）公司战略

A. 扩张战略（或进攻战略）

（1）走出去战略。

（2）联盟战略。

（3）一体化战略（输入端与输出端的企业联合或相同业务企业联合）：①横向一体化战略，②纵向一体化战略，③多向一体化战略。

（4）多元化战略，又称多角化战略或多种经营战略（同时在 2 个或 2 个以上行业从事经营活动）。

第一种分类：关联多元化战略和混合多元化战略。

第二种分类：一是单一业务型战略，95％以上销售收入来自一个业务单位。二是主导业务型战略，70％～95％的销售收入来自一个业务单位。三是相关多元化战略，70％以下销售收入来自一个业务单元。相关多元化战略又分为：①技术相关多元化战略，②生产相关多元化战略，③市场相关多元化战略。四是非相关多元化战略，业务单位之间没有技术或市场方面密切联系。非相关多元化战略又分为：①纵向多元化战略（上下游），②集成型多元化战略。

B. 稳定战略（或防御战略）

C. 收缩战略（或退却战略）

（1）转向战略（或转进战略）。

（2）剥离战略。

（3）清算战略。

2）业务战略（波特三大通用战略）

（1）成本领先战略（又称低成本战略）。

（2）差异化战略（差异化就是追求与众不同）。

（3）集中化战略（即专一经营，以在一个产业内狭窄的竞争范围里进行选择为基础）。

3）职能战略

（1）营销战略：①引领战略，②挑战战略，③追随战略，④补缺战略。

（2）财务战略。

（3）生产战略（或作业战略）。

（4）研发战略（R&D）战略：①贸-工-技战略，②代工-研发战略，③自主知识产权战略。

注意：从某种意义上讲，任何战略都是个性化战略，这就需要不断进行战略创新。

二、何谓战略管理？

企业战略管理是指对企业战略进行的分析制定、评价选择以及实施控制，使企业能够达到其战略目标的动态管理过程。

1. 战略管理的由来

从运用计划控制到实施战略管理，西方企业大致经历了以下四个发展阶段。

（1）计划控制阶段。计划控制阶段在 20 世纪初出现于美国，重点在控制偏差。其基本假定是：过去的情况，必须重视。

（2）长期计划阶段。长期计划阶段开始于 20 世纪 50 年代初，重点是预测企业的成长，在此基础上制订企业长期计划。其基本假定是：过去的情况，必定会延续到将来。

（3）战略计划阶段。战略计划阶段开始于 20 世纪 60 年代后期，核心是制定有效的经营战略，以适应经济、市场的变化和冲突。其基本假定是：长期计划的连续性预测已经不够。

（4）战略管理阶段。战略管理阶段兴起于 20 世纪 70 年代中后期，特点是将战略思维与规划贯穿于企业管理的全过程。其基本假定是：面对迅速变化的外部环境，单纯的周期性计划工作已不能完全满足变革的需要。

注意：上述各阶段的演进，并不是后者对前者的替代，而是补充。因此，最后形成的战略管理方式包含了前三者的内容。

2. 战略管理理论的发展

（1）以环境为基点的经典战略管理理论。20 世纪 60 年代初，美国管理学家钱德勒（Chandler）在《战略与结构》一书中，开创了企业战略问题研究之先河，提出企业经营战略应适应环境和满足市场需要，而组织结构需适应企业战略。

（2）以产业及市场结构为基础的竞争战略理论。其代表人物为迈克尔·波特（Michael Porter），代表作是 20 世纪 80 年代相继出版的竞争三部曲。其中提出了著名的三大通用战略：成本领先战略、差异化战略、集中化战略。

（3）以知识为基础的核心竞争力理论。1990 年，普拉哈拉德（美）和哈默尔（英）首先提出核心竞争力，后获得广泛认同，并对主流战略管理理论产生了深刻影响。此后，企业的核心能力被列为战略研究的起点和战略规划的要件。

通俗地讲，企业像一棵大树，树根就是企业的核心能力，树干是企业的核心产品，树叶、花、果就是企业的最终产品。

任何企业的优势都是暂时的，而不断创造企业优势的能力才是最宝贵的，这种不断创造优势的能力就是企业的核心能力，它是指企业在长期生产经营中逐渐积累的知识、技能，与其他资源相协调并有机结合而形成的经营体系。

3. 战略管理的内涵

战略管理要解决的核心问题如图 2 所示。

从性质上看，企业战略管理是一种：①高层次管理，②全局性管理，③前瞻性管理，④长期性管理，⑤主动性管理，⑥方向性管理（即导向性管理），⑦控制性管理，⑧动态性管理。

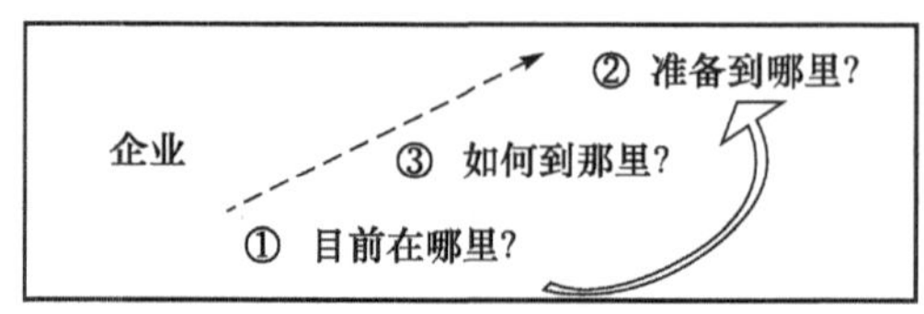

图 2　战略管理要解决的核心问题

4. 战略管理的过程

战略管理一般分为四个过程：①战略分析（或研究），②战略制定，③战略实施，④战略评价。

战略管理也可分为三个阶段：①战略制定（即将前面①、②阶段合并），②战略实施，③战略评价。

战略管理还可分为以下三个阶段（即将前面③、④阶段合并）：①战略研究，②战略制定，③战略实施。

根据战略管理四个过程，可建立起战略管理的通用模型（图 3）。

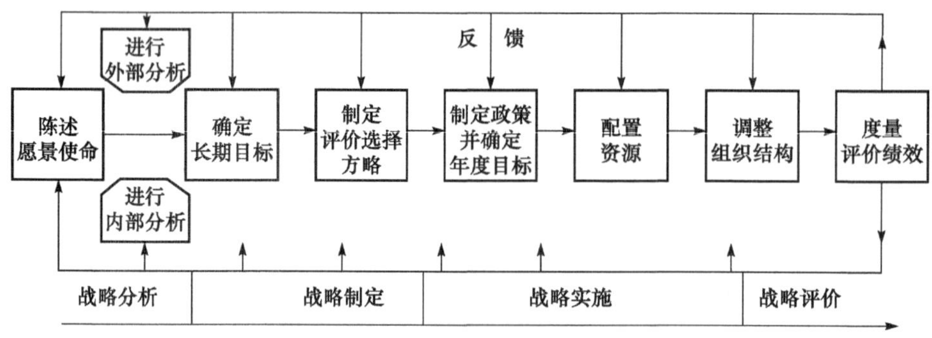

图 3　战略管理的通用模型

三、为何要有战略管理?

美国管理学家格林利（Greenley）从正面对战略管理意义做了较全面的归纳，给出了战略管理的 14 条益处。

（1）使人们识别、重视和利用机会。

（2）使人们客观地看待管理问题。

（3）加强对业务活动的协调与控制。

（4）将不利条件和变化的作用降至最小。

（5）使重要决策更好地支持已树立的目标。

（6）使时间和资源更有效地分配于已确定的目标。

（7）使企业将更少的资源和时间用于纠正错误或专项决策。

（8）建立了企业内部人员沟通的环境和条件。

（9）将个人的行为综合为整体的努力。

（10）为明确个人的责任提供了基础。

（11）鼓励向前式思维。

（12）提供了对待问题和机会的、合作的、综合的工作方法和积极的工作态度。

（13）鼓励对变化采取积极的态度。

（14）加强了企业管理的纪律和正规化。

通过负面例子，可以说明战略管理的必要性。

有专家统计，世界上每破产 100 家企业，就有 85 家是因为企业家的战略失误造成。而战略失误最多的又是盲目扩张。美国由盛而衰的国际数据、安然、世界通信等，我国昙花一现的巨人集团、飞龙集团、亚细亚集团、秦池集团、猴王集团、蓝田集团、亿安科技、广东国信等都是实例。

因此，现代企业不能没有战略管理！

现代企业家不能不关注战略问题！

再强调一下，对战略管理，最应记住的是以下六个重点。

1）增强主动

“凡事预则立，不预则废”。

本质上，战略管理是一种主动性、预应性管理，寻求对环境变化及组织发展作出预测、预期和计划，可最大限度地减少被动、盲动并避免遭遇“措手不及”。

2）着眼长远

“人无远虑，必有近忧”。

对企业来说，与人一样，何尝不是如此！

战略管理正是“从长计议”，即从长远考虑问题，为企业确立长期目标、制订长远规划和选择长远发展方向。

3）把握整体

“运筹帷幄，统揽全局，方能决胜于千里之外”。

战略管理就是要从全局出发考虑问题，为企业确立总体目标，制订总体规划和选择总体发展方向。

4）明确方向

先解决“做正确的事”，再来“正确地做事”。

这是西方学者对战略管理之精髓的精辟阐释！

若不首先解决好“做正确的事”，则愈是“正确地做事”可能错得愈远！如“南辕北辙”。例如，我国的“大跃进”“引洮上山”。

许多企业之所以失败并不是没有“做对事情”，主要是因为没有“做对的事情”。

战略管理的使命正是在于帮助决策者“做对的事情”!

5）坚定信念

“在战略上藐视敌人，在战术上重视敌人”。

大多必胜的信念来自战略判断!

许多中国企业要“跻身全球500强”等都是基于一种战略信念。此类战略信念是激励企业奋发图强和催人奋发的强大力量。

6）享受过程

战略管理最重要的意义不在结果而在于过程。

正如战略管理学家戴维（David）说：“战略制定过程是一种学习、帮助、教育和支持活动，而不是高级领导人之间的文书活动。战略家所做的最糟糕的事情便是自己制订战略计划，然后交给执行经理去实施。通过参与战略制定过程，基层管理者便成为战略的‘所有者’，战略的实施者成为战略的主人，这是成功的关键。”

四、如何进行战略管理?

（一）如何进行战略分析?

我国古代战略家说“知己知彼，百战不殆”。

这就是说，要先了解自己目前的处境，掌握敌我双方（商场上是竞争对手双方）的优劣势，打仗才有胜算。

商场如战场，企业在制定战略方案之前，必须先进行严密的战略环境分析。战略分析就是从企业的愿景（即企业的理想）和使命（即企业的责任）出发，通过分析外部环境，明确自身面临的机会与威胁，以判断自己该做什么和不该（回避）做什么；通过分析内部环境，认清自身的优势与劣势，以知道自己能够做什么和不能做什么。一般而言，企业的战略分析应遵循先外部再内部，先宏观再微观的逻辑（图4）。

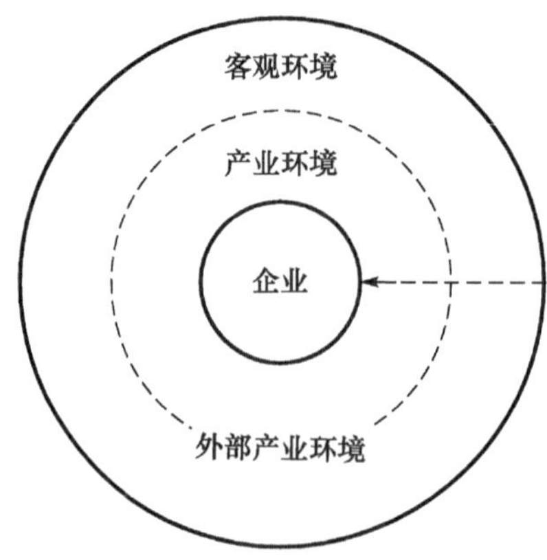

图4　战略分析顺序

1. 陈述愿景和使命

1）愿景

愿景即理想。

我们想要创造或成为什么？

愿景陈述举例：美国石油公司（Amoco Corporation）将成为一家全球性的企业。

它将被遍及世界的雇员、客户、竞争者、投资者和公众看作卓越的企业，将成为其他企业衡量其绩效的标杆。公司的标志将是创新、主动和团队合作。公司将预测变化并对其作出有效反应。公司将创造机会。

2）使命（或宗旨、信条、任务）

使命即责任。

我们的事业或业务是什么？

使命陈述举例：美国石油公司是一家在全球范围内经营的、综合性的石油企业。

企业勘探、开发石油资源并为客户提供高质量的产品和服务。企业高度负责地从事经营活动，以得到一流的资金回报。同时兼顾长期成长、股东收益并履行对社区和环境的责任。

3）并述

也可将愿景与使命合并陈述。

举例：华为技术有限公司指出：华为的追求是在电子信息领域实现顾客的梦想，并依靠点点滴滴而不舍的艰苦追求，成为世界级领先企业。

其中既有愿景——成为世界级领先企业，又有使命——在电子信息领域实现顾客的梦想。

大多数管理学家将明确的愿景与使命看作战略管理工作的起点，也是战略家的首要责任。这为战略管理确定了大方向，并赋予了我们为之奋斗的事业一份神圣。

2. 外部分析

外部分析也叫环境分析、市场分析或企业分析。

外部分析主要是分析影响企业生存和发展的各种外部环境因素，了解自身面临的机会与威胁。关键外部环境因素及其与企业的关系参看图 5。

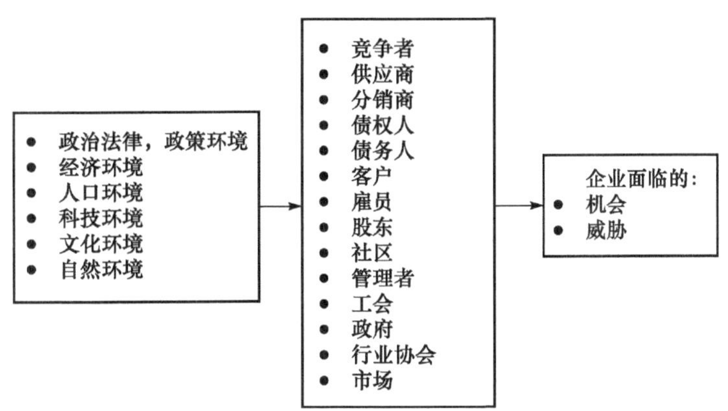

图 5　关键外部环境因素及其与企业的关系

外部环境分析包括以下几个方面。

1）企业宏观环境分析

(1）企业政治与法律环境分析。政治环境是指制约和影响企业的各种政治要素及其运行所形成的环境系统。内容主要包括五个方面：政治制度、政党制度、政治性团体、党和政府的方针政策、国家的政治气候。

法律环境是指与企业相关的社会法制系统及其运行所形成的环境系统。内容主要包括三个方面：国家的法律规范、国家的司法及执法机关、企业的法律意识。

(2) 企业经济环境分析。经济环境是指构成企业生存和发展的国际国内经济发展状况、国家经济政策及其运行所形成的环境系统。内容主要包括四个方面：企业所处的社会经济结构、经济发展水平、经济体制、宏观经济政策。

(3) 企业科技环境分析。科技环境是指企业所处的社会中的各种科技要素及其运行所形成的环境系统。主要包括四个方面：企业所处的社会科技水平、社会科技力量、国家的科技体制、国家的科技政策和科技立法。

(4) 企业的社会文化环境分析。社会文化环境主要包括企业所处国家的人口状况、社会生活方式、主导意识形态、信仰状况、教育状况、财富状况、社会环境状况、婚姻状况、文化状况、社会服务状况等。

2) 行业环境分析

(1) 行业竞争结构分析。一个行业中存在五种基本竞争力量（波特模型），见图 6。

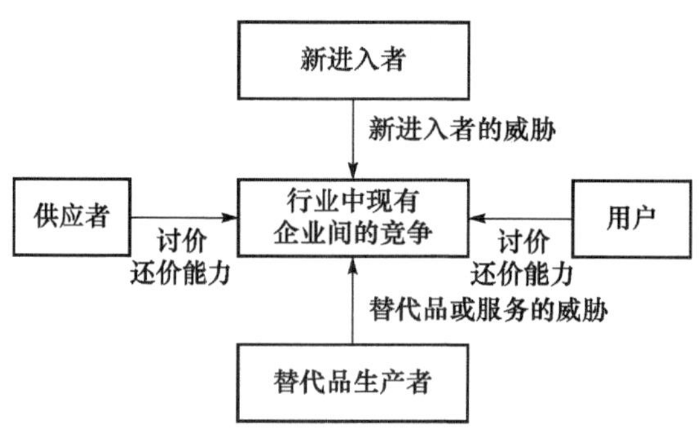

图 6　行业竞争结构

由图 6 知，这五种力量的现状、消费趋势及其综合强度决定了行业竞争的激烈程度和行业的获利能力。

在竞争激烈的行业中，一般不会出现某个企业具有非常高的收益率情况。在竞争相对缓和的行业中，有可能出现比较多的企业具有较高的收益率情况。

在制定战略过程中，要对上述五种竞争力量逐一加以分析，尤其要注意：要找到本企业在行业中的准确定位，从而找出防御这五种力量的办法，甚至要对这五种力量施加影响，使它们有利于本企业的发展而不利于竞争对手。

(2) 行业寿命周期分析。行业的寿命周期包括四个阶段：幼稚期、成长期、成熟期、衰退期（图 7）。

行业寿命周期曲线的形状是由社会对该行业的产品需求状况决定的。一般一个周期要一百年至几百年时间。其中成熟期分为成熟期前期及成熟期后期。成熟期后期包括两类，第一类型长期处于成熟期（如曲线 1），长期稳定型行业，第二类型（如曲线 2），行业迅速进入衰退期。

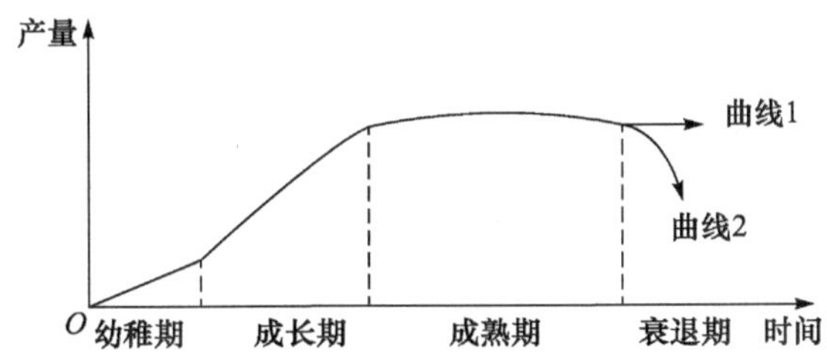

图 7　行业寿命的周期曲线

(3) 行业环境其他方面分析。行业环境其他方面分析包括：行业在社会经济中的地位分析，行业特征分析，行业中企业规模结构、企业数量结构、企业市场结构和企业组织结构分析，行业在发展过程中社会环境方向的限制等。

3) 企业竞争对手分析

在制定及实施企业战略过程中，要始终明确企业在国内、国外的竞争对手及其优劣势，竞争对手的战略及战略目标，竞争对手的管理特色、核心能力等。只有对竞争对手了解得比较清楚，才能更好地在本企业实行战略管理。

3. 内部分析

1) 资源分析

资源分析指分析企业的人力、财力、物力和技术等资源状况，以及土地储备（对房地产企业尤其重要）、无形资产（专利、品牌、企业文化等）情况。

2) 绩效分析

绩效分析指分析企业的经营绩效并做纵向、横向比较。

3) 价值链分析

遵循满足客户需求，即价值增值的路线来分析企业的作业流程，区分业务流程/直接增值活动（如进货后勤—生产作业—发货后勤—市场营销—售后服务）、支持流程/辅助价值活动（膳食—物管—研发—人资—公关—IT）、战略流程/导向和驱动活动。

4) 优势和弱势分析

各部门列出公司的优势和弱势，然后从中筛选出最重要的优势和弱势，需要深入分析和商讨并作出很多判断，这一过程有助于厘清我们的战略思路。

5) 核心能力分析

排在首位且对其他优势最具影响力者一般是核心优势，即公司的核心竞争力。明确自身的核心能力是制定战略的关键。

4. SWOT 分析①

(1) 内部分析的重点是明确自身的优势和弱势。

(2) 外部分析的重点是了解外在的机会和威胁。

(3) 战略分析中的内部外部分析并称为 SWOT 分析，它是战略分析的基本方法。

(4) 四者（优势、弱势、机会、威胁）是管理者制定战略的重要匹配工具，它们

① S 为 strengths，优势；W 为 weaknesses，弱势；O 为 opportunities，机会；T 为 threats，威胁。

相互之间的不同搭配构成不同的特色战略：SO战略、WO战略、ST战略、WT战略。

该匹配模型被称为TOWS矩阵，建立此矩阵的难点是列出关键内外部因素，参见表1。

表1 TOWS矩阵

	优势 优势1…… 优势2…… 优势3…… 优势4……	弱势 弱势1…… 弱势2…… 弱势3…… 弱势4……
机会 机会1…… 机会2…… 机会3…… 机会4……	SO战略 1）……利发 2）……用挥 3）……机优 4）……会势	WO战略 1）……克利 2）……服用 3）……弱机 4）……势会
威胁 威胁1 威胁2 威胁3 威胁4	ST战略 1）……回利 2）……避用 3）……威优 4）……胁势	WT战略 1）……回减 2）……避小 3）……威弱 4）……胁势

“威胁—机会—弱势—优势”四者之间不同匹配，形成不同的战略。不存在一种最佳匹配，采用哪种匹配还需要良好的判断。

（二）如何制定战略？

前面已讲，战略管理要解决的核心问题是：目前在哪里？准备到哪里？如何到那里？

前面战略分析已解决了“目前在哪里”的问题。

接下来，则需要通过“战略制定”来解决“准备到哪里”和“如何到那里”的问题。

1. 确立战略目标

1）战略目标

战略目标即组织（集团、公司、子公司、分公司、事业部、职能部门等）的总体、长远目标，也称“长期目标”。

2）战略目标内容

战略目标内容包括：规模指标（如资产额、销售额），效益指标（如成本、利润、资本收益），位势指标（如市场份额、行业排名、区域排名），扩张指标（如异地经营、海外投资、海外销售），技术创新，品牌知名度，文化建设，环境保护等。

3）战略目标与使命（责任）、愿景（理想）的区别

战略目标：主要回答“我们这一步要跨到哪里”，具有期限明确（如3～5年）、以定量描述为主、内容体系化等特征。

使命：主要回答“我们的事业是什么”，具有期限永久（假设）、描述定性、内容单一等特点。

愿景：主要回答“我们的理想是什么”，具有期限遥远（模糊）、描述定格（未来的生动图像）、内容简略等特点。

2. 提出达到目标的方略

1）方略，即如何达到战略目标的方法谋略，是组织的大政方针和行动指南

例如，孙中山先生提出的《建国方略》和民主建设“三部曲”（军政、训政、宪政），邓小平同志提出的中国现代化建设“三步走”（初级阶段、小康、中等发达）发展战略等，联想 1996～2020 年围绕其战略目标采用的五条战略路线为：坚持信息产业领域内的多元化发展（有限多元化战略），国际、国内市场同时发展，以国内为主，走贸、工、技道路，积极发展产品技术，以此为基础逐步逼近核心技术，充分运用股市筹资。

方略也被看作狭义的战略，战略的广义与狭义之分见图 8。

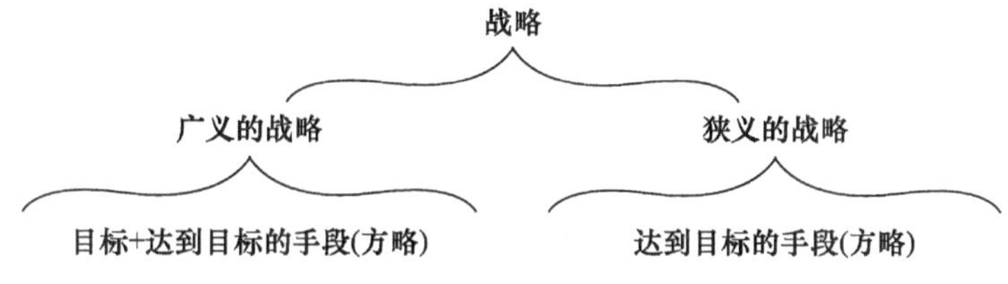

图 8　战略的广义与狭义之分

2）运用现代决策工具

西方现代战略学中有许多分析工具，如波士顿咨询集团（Boston Consulting Group，BCG）矩阵对多元化企业选择业务结构调整方略就颇有帮助，图 9 是全球著名咨询商麦肯锡 2000 年为 CMSK（China Merchants Skekou，招商蛇口）产业编制的 BCG 矩阵。

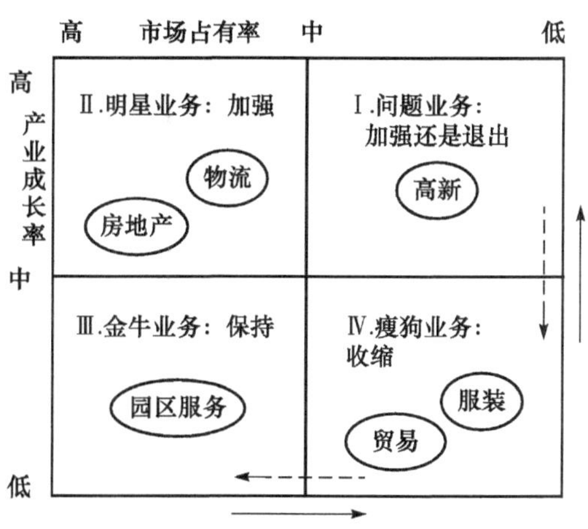

图 9　列入 BCG 矩阵的 CMSK 产业

图 9 矩阵所表达的咨询建议，CMSK 产业将居于“明星”象限的物流、房地产确立为优先发展的加强性产业；将居于“金牛”象限的供电、供水、视讯等园区服务业

列为维持性产业；将居于“问题”象限的高新技术产业作为探索性产业（去留尚待观察）；将居于“瘦狗”象限的服装、贸易等业务出售或清理。

3）借鉴经典战略

经典战略是对前人已采用战略的总结，多已成为经典案例或写进了教科书（如兵法里的三十六计），对我们制定战略具有一定启发、参考和借鉴意义。像波特提出的“成本领先”“差异化”“集中化”三大著名通用战略，像商家常常采用的市场引领、挑战、追随、补缺等营销战略，以及大家熟知的横向一体化、纵向一体化、关联多元化、混合多元化等扩张战略，都是成熟完善且很有指导意义的既有战略模式。

4）制定个性化战略

借鉴经典战略切不可受其局限，照搬照套。因为各个企业的情况千差万别，每个企业所面临的具体环境和形势也是千变万化的。从某种意义上讲，任何成功战略都是个性化战略，或者说战略方案的核心价值在于其独创性，这就需要我们创造性地制定战略并不断寻求战略创新。

3. 战略方案的选择

（1）中国古代谋略讲求上、中、下策之分，这里就包含战略方案的评估和选择的问题。

（2）不论面临什么内外部环境，不管采取何种工作方式（自拟、外包），公司都可能遇到多种战略方案的选择问题。一般工作程序是通过全员参与、企划人整理提出若干建议方案，交由最高决策班子评估选择，从中筛选出最佳方案或排出上下序列。

（3）西方战略学中的许多分析工具都只可能得出一张备选战略清单，如 BCG 矩阵、TOWS 矩阵、SPACE（strategic position and action evaluation，战略地位与行动评价）矩阵、IE（internal-external，内部-外部）矩阵。

例如，CM 地产的 TOWS 矩阵（表 2）。

表 2　CM 地产的 TOWS 矩阵：其中包含多个战略方案

	优势 （1）拥有较高品牌知名度 （2）拥有紧靠西部通道的土地资源储备 （3）拥有多种融资渠道 （4）擅长中档住宅区开发	弱势 （1）土地资源储备有限 （2）不良资产处理尚未了结 （3）体制不活 （4）人员流动过于频繁
机会 （1）城市化进程加快 （2）分房货币化 （3）《内地与香港关于建立更紧密经贸关系的安排》及深港一体化 （4）深港西部通道建设	SO 战略：发挥优势，利用机会 （1）循序开发高附加值口岸物业 （2）扩大中档住宅区开发规模	WO 战略：利用机会，克服弱势 （1）深度开发，提高资源利用率 （2）调整外地停建项目处理策略
威胁 （1）政府控制土地投放 （2）房地产投资持续升温，有“过热”之虞 （3）房贷收紧	ST 战略：利用优势，回避威胁 （1）扩大证券市场与境外融资 （2）并购项目公司	WT 战略：减小弱势，回避威胁 （1）深度开发，提高资源利用率 （2）采取灵活储地方式

(4) 对备选战略方案的评估结论往往是“各有所长，各有所短”。迄今开发出来的技术评价模型不多且缺乏可操作性，因此最后的拍板还离不开决策者的经验与直觉判断，并受到文化、政治等因素的影响。

4. 战略方案的确认

(1) 战略方案确认指由相当决策机构审议、通过、批准，获得必要的权威性的方案，由此成为组织的行动纲领。

(2) 国民经济与社会发展的战略规划（如五年计划）由国家最高权力机构——全国人大审议批准。

(3) 公司战略、战略方案或战略规划一般由公司最高决策机构——董事会审查批准，必要时提交股东大会审议通过。

(4) 董事会对战略审查的内容：对内外部环境是否作出了客观准确的分析和研判？建议的战略目标是否先进可行？提出的方略是否科学有效？战略实施是否有足够的资源保障？是否作出了足够的风险分析并备有预应措施？对未来作出多种假设，管理者如何应对？

（三）如何实施战略？

(1) 列宁：一个行动胜于一打纲领。

(2) 麦克斯韦尔：毛泽东是伟大的战略家，周恩来是出色的组织与实践家。

(3) 戴维：实际做一件事情（战略实施）总是比决定做这件事情（战略制定）要困难得多。

许多公司的成败证明，没有强大的执行力和有意义的实际行动，战略方案无论怎么完美都无济于事，无论其具有多高价值，若不能实现也等于零。因此战略实施是与战略制定同等重要的工作，也是战略管理过程不可或缺的内容。战略实施贯彻于企业工作的方方面面，体现在公司的每一件具体工作中，下面举出一些比较重要的环节。

1. 业务布局

围绕战略目标按“条条”和“块块”两条线部署全盘业务。“条条”指各业务种类，如柳传志所说的“核心业务（碗里的），成长业务（锅里的），种子业务（田里的）”；“块块”指业务的区域甚至全球布局。

2. 资源配置

资源配置指按战略目标及与之相关的年度目标的轻重缓急来预先安排资源，减少随意性和盲目性。企业的基本资源有：财力资源、物力资源、人力资源和技术资源。预算就是一种常用的资源配置手段。

3. 组织调整

组织调整指通过必要的调整和重组来使组织结构与战略相匹配。战略的变化将导致组织结构的变化。组织结构的重新设计应能够促进公司战略的实施。

离开了企业的使命或战略，组织结构将没有意义。钱德勒（Chandler）就战略-组织结构的关系编制了如下模型，参看图 10。

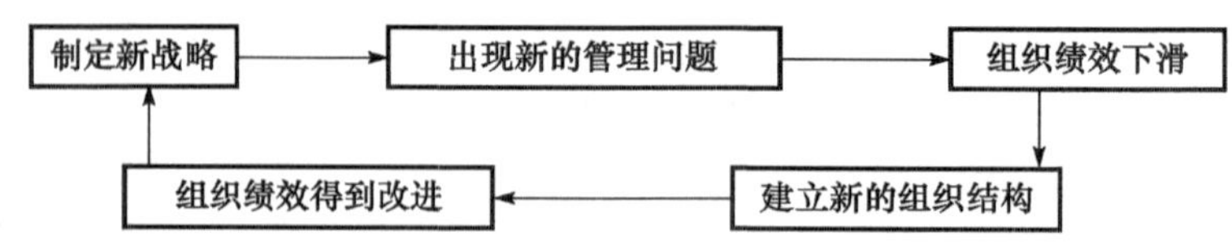

图 10　战略与组织结构的互动模型

4. 政策措施

实施企业战略，需要有具体的政策指导和措施保证，如质量管理办法、成本控制方案、品牌推广步骤、融资举措、促销新招、异地甚至跨国拓展与管控措施、激励政策与奖罚规则等。政策措施好比战术，没有好的战术再好的战略也是空谈。

5. 文化保障

培育支持企业新战略的文化，包括保留原有文化中与新战略相适应的方面，改变与之相矛盾的方面。改变的方法有：招募新员工，人员培训，组织结构变革，管理人员调整，榜样示范以及改变标识等。例如，美国电报电话公司在 20 世纪 90 年代放弃了人们熟悉的旧标识（圆圈内一个钟），采用了新标识（由电子通信标志所环绕的地球），象征将在电子通信领域及全球范围内展开的新战略。

6. 协调机制

在战略实施过程中需协调各种资源、力量、内外关系、意见分歧、矛盾冲突甚至危机事件等，因此，一个强有力的协调机制和协调人（或班子）是非常重要的。许多公司的战略失败正是因为缺少这样的协调力量。由此，也催生出了一门关于公司权力的学问——公司政治。

7. 年度计划

年度计划自古以来就与人类经济生活的节奏密切相关（源于狩猎和农耕）。同样，年度计划是实施战略规划的里程碑，是配置资源的基础，是财务核算的通用时段，是绩效考核的基本时间尺度。因此，安排并完成战略规划期中的每一年度计划是战略实施的重要步骤。

8. 跟踪监控

整个战略实施过程同时也是跟踪监控过程，目的在于监控环境变化、掌握实施进度、确保战略航向、纠正行动偏差并考核实施绩效。关键执行指南考核、季度检查、年度总结、中期检讨（回顾展望）等都是常用手段。

9. 领导作用

战略实施需要的领导素质包括组织力、行动力、协调力等以及战略实施艺术。这里就有两类领导才能的问题，即战略家与执行者、战略思想家与战略实践家、设计师与工程师、帅才与将才的区别问题。多数领导只在一方面表现出突出才能，少数同时具备二者。因此企业在战略实施中还得尽量避免两类领导错位的问题，确保由具备强大执行力的领导来负责战略实施工作。

（四）如何评价战略？

（1）中国名言：计划不如变化。

（2）西方名言：世界上唯一不变的是变化。

（3）戴维：当企业的外部及内部环境发生变化时，制定和实施得再好的战略也会过时。因此，对战略的实施进行系统化的检查、评价和控制就成为战略家的一项重要工作。

1. 战略评价过程

战略评价过程包括事前评价、事中评价和事后评价三个阶段。因此，广义的战略评价贯穿于战略管理的全过程，如图 11 所示。

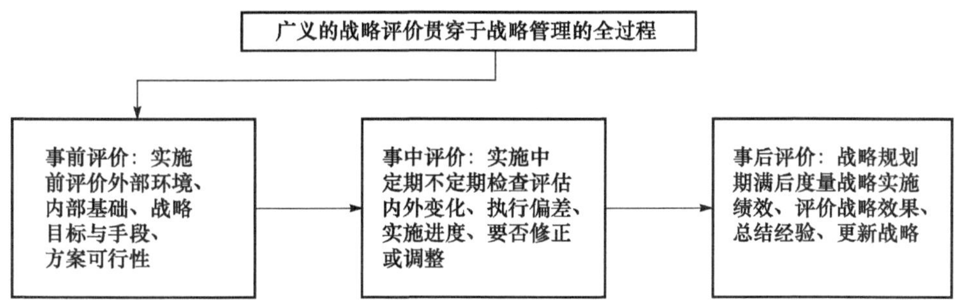

图 11　战略评价过程三个阶段及其内容

2. 战略评价框架

西方战略学根据战略评价的步骤和内容，制定了战略评价框架或模型，参见图 12。整个战略评价框架以审视战略基础、度量企业绩效、采取纠正措施为主干，可供我们的战略评价工作借鉴和参考。

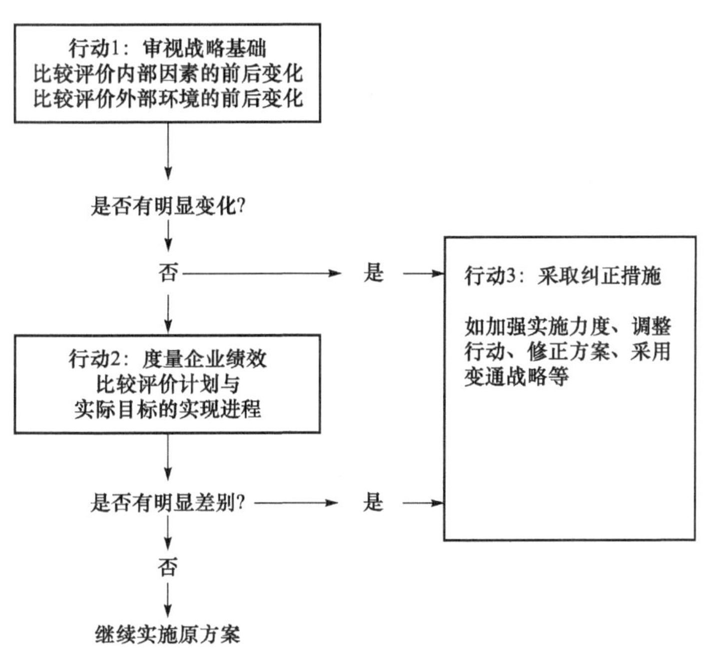

图 12　战略评价框架

3. 战略评价决策矩阵

西方战略学开发出来的战略评价决策矩阵，从应当考虑的关键问题、对这些问题的各种答案以及企业应当采取的适当行动诸方面概括了战略评价活动。除了最后一行“内外部环境均未发生重大变化”“战略实施进度令人满意”的情况外，纠正措施几乎总是需要的，参看表3。

表3　战略评价决策矩阵

内部环境是否已发生重大变化	外部环境是否已发生重大变化	战略实施进度是否令人满意	结果
否	否	否	采取纠正措施
是	是	是	采取纠正措施
是	是	否	采取纠正措施
是	否	是	采取纠正措施
是	否	否	采取纠正措施
否	是	是	采取纠正措施
否	是	否	采取纠正措施
否	否	是	继续战略进程

4. 战略评价结果的运用

(1) 对照战略评价结果改善战略实施进程。

(2) 运用战略评价结果提高经营绩效。

(3) 依据战略评价结果修正战略方案。

(4) 积累战略评价结果丰富战略管理案例。

(5) 参照战略评价结果制定新战略。

(6) 借鉴战略评价结果改善整个战略管理工作。

企业战略就是企业的全局性、长远性谋划，包括企业的长期、总体目标及达到该目标的基本方略。企业战略管理即企业为实现战略目标、研究战略环境、制定战略决策、实施战略方案、评价战略绩效的整个动态管理过程。或者说，战略分析、战略制定、战略实施和战略评价构成了战略管理的一般过程。

成功的战略管理除了有赖于掌握和运用上述一般知识外，还贵在或重在战略创新。

“技工荒”困扰中国高尔夫球具代工产业①

在中国强调经济结构转型、产业结构升级的今天，那些劳动力技术水平要求较高的制造业普遍面临中高等技术工人匮乏的瓶颈问题，让不少制造业者除了面对“民工荒”外，还要面对严重的“技工荒”[1]。不少学者认为，其根源在于中国制造业结构与劳动力素质结构仍然存在巨大偏差，职业教育无法培养足够的技能型劳动力，以满足制造业者从全球价值链低端向高端升级的需求。因此，培养并留住中高等技术工人甚至吸引受高等教育人才的加入，已成为中国制造业打破人力资源瓶颈的重要抓手。鉴于此，本文选择对技能型劳动力有大量需求的高尔夫球具代工产业作为典型案例，对相关议题进行深入的分析探讨。

一、球具代工产业发展概况

鉴于高尔夫球具市场主要由几个国际知名品牌所控制，而高尔夫球具代工制造又需要相当的专业技术水平，因此超过60%的市场份额，都是由3～4家知名台资企业设立于中国东南沿海的生产基地所完成的，这也让中国成为高尔夫球具代工产业的主要区域。

1. 高尔夫球运动刚起步，球具制造业有前途

和其他经济发达国家相比，中国的高尔夫球运动发展只有20多年，起步较晚。虽然中国人口总数早已超过13亿人，但高尔夫球场仍比较少。只有3.9亿人口的美国有2万多座高尔夫球场，等于中国的100倍，而只有1.3亿人口的日本则有约3000座高尔夫球场，相当于中国的15倍。而随着中国人均生活水平逐步提高，人们对体育运动(包括高尔夫球)、文娱活动及相关商品的需求也正不断提升。换言之，高尔夫球运动的普及与专业运动员人数的高速增长，自然而然会带动人们对高尔夫球具的需求。因此，高尔夫球具制造业在中国的发展方兴未艾，潜力巨大。

随着中国经济的不断增长，高尔夫球运动的发展速度也逐渐加快。有资料显示，中国高尔夫球行业从20世纪90年代以来快速增长，2011年的增长速度高达11.4%，高尔夫设施总数达到440家，共9769个球洞，折合约543个18洞高尔夫球场（图1），分布在全国30个省份。

2. 球场税率逐步降低，有效带动产业发展

过去，高尔夫球场的税收是按娱乐行业征收的，所有赋税超过营业收入的20%，高额税收在很大程度上限制了高尔夫行业的发展。然而，随着高尔夫球运动的不断发展，以及这项运动2009年重返奥运会，特别是自2011年以来，中国众多省市加入降低高尔夫球场税率的行列中。目前，全国各地高尔夫球场税率标准不尽相同，标准最低的是云南和广东的5%，其次是河南的7%，天津、浙江、上海及北京等地的10%。

① 本文原载香港《信报财经月刊》，2014，(4)：124-127. 作者：刘人怀，覃大嘉，梁育民.

税率降低促进了高尔夫球运动在中国的发展，也为中国高尔夫球具代工产业提供了一个更为广阔的市场前景。

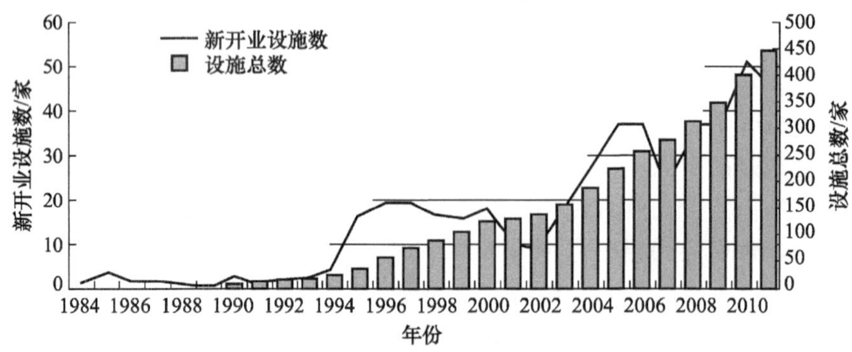

图 1　中国高尔夫设施数量及增长趋势

资料来源：朝向集团《朝向白皮书——中国高尔夫行业报告（2010）》

3. 工艺水准较高，“技工荒”是最大难题

虽然中国经济持续增长，廉价劳动力的竞争优势却逐渐缩小，劳动密集的代工制造产业普遍面临“民工荒”问题。和普通的劳动密集型代工产业有所不同，高尔夫球具代工产业要求劳动力具有一定的技术水准，方能满足其生产过程的工艺水准要求，因而除了面临“民工荒”问题，更面临“技工荒"问题。

所以说，高尔夫球具代工产业应当率先进行技术改造与革新，以创新技术创立自己的品牌，通过生产自动化来减少工人需求，进而解决招工难与员工离职问题；以创新管理提升自己的企业，争取形成比较完整的产业链，努力提高工人技术水准并解决“技工荒”问题，进而实现整个产业的结构升级和企业转型。

二、球具代工产业人力资源问题

一般来说，国家制定的政策法规等对高尔夫球具代工产业劳动力市场的供需情况会造成一定影响，比如，最低工资与最高工时标准会影响企业的成本和收益，会影响其对劳动力的需求情况，而职业教育的课程设计与财政支持缺失、户籍制度和技术工人认证等则会影响劳动力的素质与供给情况。

1. 企业用工成本不断提升

全球贸易壁垒的缩小，让中国以出口为主的高尔夫球具代工产业更容易进入其他新兴市场。但随着人民币持续升值、最低工资不断提高、医疗社保等社会保障要求上升，中国劳动密集型制造业者的用工与经营成本大幅上涨，部分企业只好转移到越南、柬埔寨等国家。虽说就高尔夫球具代工产业而言，生产自动化可节省用工数量与成本，但自动化设备需要较大的资金投入，而且工人也必须学习和掌握全新的技能，如机器的操控、管理和修缮等，因此在安顿老员工、招聘新员工与培养合格技工等一系列劳动力转型升级问题上，高尔夫球具代工业者面对相当大的挑战。

中国目前的高尔夫球具代工产业仍集中在东南沿海一带，大部分一线工人由外来务工人员组成。根据国际经验，当外来人口达到一定比例时，就能将外来文化融入本土文化中。但中国特殊的户籍制度，使本地居民产生对本土文化的优越感，外来文化只能得到外来工自身的认同。劳动力的频繁流动，也增加了企业的离职率和用工成本。

此外，高尔夫球具代工产业还有一个问题，就是工作环境不利于身体健康，噪声很大、粉尘很多，因而许多年轻人不愿意去工作。为解决这一问题，代工企业除了提高工资福利外，关键的是要在厂房设备、环保设施等方面增加投入，由此增加许多成本。

2. 高技能劳动力不足，年轻工人要求高

首先，传统教育理念对技术工人的供给产生不利影响。虽然中国推行学业文凭、职业资格证书并重制度，但人们长期以来对教育认知存在偏差（如“上大学才有好前途”）的想法，导致技术职业学校面临招生难问题，如报名率低、报到率低、生源素质较低等。同时，高校和普通高中扩招又使生源大量流失。另有大量学生在基础教育结束后，选择直接就业。

其次，劳动力群体需求层次的提高也会对劳动力供给施加影响。新生代技术工人主要以“80 后”“90 后”为主，其中超过 50%的工人受过高中或技校的教育，约有 11%的技工获得大专或本科学历，较高的学历使他们在满足生理与安全需求外，会追求更高的自我实现需求，因而期望公司能提供更多样化的专业培训课程，以便将来获得更高收入，从而发展事业。因此，企业培训是否有助于新生代工人的未来职业发展，已成为其就业时的考虑要素。

最后，代工企业的生产作业活动安排在淡旺季差别很大。旺季时，由于订单量很大，常需安排生产线工人加班加点地工作，造成工时过长，工作强度加重，工人劳累过度，容易导致员工反感。加上新生代工人在劳务报酬之外，往往也追求更充足的个人时间和业余生活，导致出现高离职率、劳动力流失现象。

3. 制度无法满足制造业转型升级需求

目前，中国沿海地区面临经济转型、产业结构升级的战略关口，对高技能劳动力存在巨大需求。但整个国家的职业教育体系没有跟上这种发展步伐，总体来说体能型（初中以下）与知识性（大学以上）劳动力所占比重比较大，但高技能劳动力不足，因而需要高技能劳动力的高尔夫球具代工产业，面临相当大的困难。其人力资源发展的核心问题，就是产业转型的劳动需求与目前中国劳动力素质的培养不匹配，由此造成“技工荒”。

首先，各地政府对职业技术教育的财政投入远小于其他基础教育的投入，导致职业技术学院发展缓慢、技工供给不足的现象。

其次，技工认证制度及对应的考试体制给技术水准高、文化水准低的技工实现工种升级造成一定障碍。在制造业中，取得初级技工认证的工人约为 40%，取得中级技工认证的工人约为 43%，而取得高级工认证的工人约为 15%，取得技师以及高级技师认证的工人仅 2%左右。再者，中国现有的户籍制度及建立在户籍制度上的就业、社会保障、医疗与教育等制度，造成对新生代技术工人融入务工城市的制度性排斥，间接导致企业招工难问题。在这种情况下，技术工人更倾向于在家乡附近找工作，由此降低了有关城市的技术工人市场供给。

三、高尔夫球具代工产业人力资源创新发展建议

怎样从技工荒、技术水准低等问题中突围，实现人力资源的创新发展，已成为中国高尔夫代工制造业者从全球价值链低端向高端迈进的战略要点[2]。

1. 生产逐步自动化，降低成本提高效率

生产自动化不仅能够提高技术水准，极大地释放产能，还能节省企业的人力成本、创造规模效益、提高经营效率，甚至还能改善工作环境，让员工远离嘈杂或危险的加工制造现场，并且降低工人因长时间工作而造成的职业病发病概率，满足企业员工的安全需求。

为了推动和加快企业的自动化进程，必须尽快引进和培养专业技术人员，以及大批掌握一定技术的一线工人；尽快让所有企业员工掌握自动化建设的有关知识和技术；尽快提升企业的自主研发与创新能力。企业在自动化研发方面的投入往往十分昂贵，因此可考虑以项目制的形式委托研究机构（实验室或高校），即企业提供一定的资金支持，优先拥有研究机构的成果及专利转让权。

同时，人力资源部门可到人才市场上招聘自动化设计与运营等方面的合适人才，也可多利用校园招聘，为大三和大四学生提供实习机会，强化高校人才对企业的接触与了解。另外，企业还可为高校设置奖学金，资助品学兼优的学生完成学业。当然，企业内部的培训、选拔机制也是发掘自动化人才的重要途径，比如，设立企业学院来培养合适且又急需的员工，普遍提高员工的文化素质和技能水准。

2. 用新能源新材料，满足员工较高需求

高尔夫球具代工产业除了引进自动化技术、实现生产流程的自动化及半自动化外，还应尽量使用新能源与新技术，尽量提高员工技能，以便满足新生代员工的较高需求。该产业比其他代工产业拥有较高的技术含量，可在使用新材料、复合材料等方面开展创新，比如，在碳纤维材质上做创新，才能减少劳动力的需求量和降低离职率。自动化生产、流程优化与材料创新，将是高尔夫球具代工产业未来发展的重要方向及突破口。

虽然降低一线工人离职率对代工企业很重要，但员工流动性过低却可能导致企业失去活力与创新动力。毕竟过于稳定的内部结构不利于公司充分挖掘员工潜能，提升优秀人才的工作积极性。因此需要引入公平的竞争机制，一方面，鼓励优秀人才发挥潜力；另一方面，激励老员工不断提升自身的工作效率。

新生代员工普遍带有明显的个人主义特色，其工作追求已从单纯的薪资待遇向自我实现、自我提升转变。对于优秀的新生代人才来说，他们更渴望个人表现机会，希望实现自我超越，因而企业应考虑给予他们更多关注，并在文化上给予更多包容。如采用“导师制”，即可帮助新晋员工了解、融入公司的企业文化。

3. 从各方面努力，提供转型升级人力资源保障

从用工企业的角度看，用工企业应设法提高员工对企业的归属感与认同感。除了提供文体设施与场所、组织和开展健康的文化娱乐体育活动外，还可举办各种讲座，

用来充实员工生活，提升员工的科技和文化知识水准。从社会文化的角度看，我们有必要借鉴国外先进企业的做法，设立不同特色群体的文化协会。

从国家政策的角度看，一方面要改革教育制度，增加技术工人的供应量；另一方面要改革户籍制度，尽量为高技能工人解决后顾之忧。例如，广州、中山、东莞和深圳等城市已采用外来工"积分入户"制度。以中山市为例，2009 年底，中山市在全国率先推行流动人员积分管理制度，以积分排名方式来解决外来工入户、子女入读公校等问题，进一步推动公共服务平等化。截至 2013 年底，4 年中共有 10 765 名流动人员通过这种方式取得中山市的入户资格[3]。笔者建议更多城市采用这种办法，以便吸引和留住所需人才或技术工人，为本地区企业和经济发展做出贡献。

参 考 文 献

[1] 刘人怀，覃大嘉，梁育民，等. 产业工人的中国梦：从低技能劳工到专业技术工人的人资转型升级战略. 战略决策研究，2013，(5)：37-50.

[2] 王志华，董存田. 我国制造业结构与劳动力素质结构吻合度分析：兼论"民工荒"、"技工荒"与大学生就业难问题. 人口与经济，2012，(5)：1-7.

[3] 王文傑，肖伟. "积分入户"四年 10765 名外地人成"新中山人". 南方都市报，2013-12-17.

系统论视角下的旅游发展与旅游研究
——中国工程院工程管理学部刘人怀院士访谈①

文彤、闫婷婷（以下简称“文”“闫”）：刘老师，您好！作为工程管理学部的院士，您在管理学科尤其是旅游管理领域做了大量的教学和科研工作，我们这次受《社会科学家》杂志的委托对您进行采访，想听听您对中国旅游学科建设和发展的一些想法。

刘人怀院士（以下简称“刘”）：好啊，欢迎欢迎！说起来我主要的研究工作集中在力学、工程学领域，但是回顾一下我的教学和科研历程，旅游与我的缘分一直贯穿其中，真的是很有意思的事情。由于出身于教师家庭的书香渊源，我从小对旅游就很感兴趣，对于名山大川、人文风光都很喜欢，在读书、工作时如果有机会我都尽可能到处看看。现在我还记得第一次去青城山、峨嵋山的旅游经历，那个时候没有什么旅游设施，整个过程还比较惊险。看看现在的青城山、峨嵋山已经建设得很好了，中国旅游业的发展真是快！后来我得到洪堡基金资助到德国访学以及到希腊等国讲学期间，也到巴黎、柏林、日内瓦、维也纳、威尼斯、罗马、梵蒂冈、雅典等欧洲有名的旅游地游览。这个阶段我对旅游只是爱好，还没有理论知识的系统化，但对于旅行社、导游、旅游景区、服务体系也有了自己的一点点体会。

1983 年 4 月我结束访学回国在中国科技大学担任教授，一个很巧的机会成为当时安徽省杨纪珂副省长的顾问，协助他处理一些事务，开始接触到旅游业。1984 年 8 月 7 日，安徽省及地市刚刚成立旅游局不久，安徽省计委（现发改委）会同省旅游局召开“皖南旅游区发展规划座谈会”，省计委主任兼省旅游局局长王胜才主持会议，讨论怎么发展安徽旅游，我到会代表杨副省长致辞。那个时候全国都刚刚开始发展旅游，旅游行业的干部都非常迫切地想了解相关的知识，旅游局局长就说：“刘教授，你从国外留学回来，见多识广，你跟我们讲一讲怎么搞旅游工作吧。”盛情难却之下，我就说：“我就讲讲欧洲见闻吧。”于是我就讲了出国期间我所见到的欧洲旅游发展的一些情况，比如说旅行社、导游、旅游景点设计等这些方面。讲完以后，这个旅游局长又问：“刘老师，安徽旅游应该从哪着手?”实际上我毫无准备，但我从小就知道“黄山归来不看山”这个说法，就说：“安徽旅游首先就应该从黄山开始，黄山很美丽，就从黄山开始，打造黄山旅游。”这次会议安徽省旅游局很满意，也接受了我的建议，开始着手黄山旅游开发，我随后也参与了制订黄山旅游规划的工作，也和安徽省旅游部门建立了良好的关系，他们经常邀请我参与商议安徽旅游发展的相关工作，后来又参加了九华山、天柱山等名山的开发规划工作，我和旅游的缘分开始从兴趣爱好转向实际工作了。

1986 年，我的老师钱伟长先生邀请我到上海工业大学（现上海大学）工作，主要任务之一是创建经济管理学院，我任副校长、经济管理学院首任院长和预测咨询研究所所长，开始涉足管理学科领域。20 世纪 80 年代之前全国大学几乎没有什么管理学科

---

① 本文原载《社会科学家》，2014，(11)：3-6. 作者：刘人怀，文彤，闫婷婷.

的专业基础，因此首要工作就是创建企业管理、会计、外贸、旅游等各个管理专业，当时我创建的旅游管理专业应该是全国最早的旅游管理本科专业了。在上海期间，我开始从事管理研究，并指导管理工程专业的硕士生，成立了上海旅游工程学会，倡导用系统论的原理和方法来研究旅游工程，慢慢开始积累教学科研成果。后来，复旦大学、上海交大、上海财大以及其他几个大学组织建立第一届上海管理学院院长联谊会，我担任主席，也作为课题组长开始针对上海进行酒店业发展战略、城市交通规划等管理研究工作，还包括崇明岛的战略规划、浦东发展研究、城市信息化等课题，不断在管理科学研究上、学院管理建设上、学生培养指导上往前走。

1991 年 11 月，我调到广州暨南大学工作，一直工作到现在，先后担任了副校长、校长职务。在我兼管经济学院期间，一直努力推动暨南大学经济学科、管理学科的建设发展，1992 年创立 MBA 中心，1998 年成立管理学院，那时就有旅游管理专业招生，1996 年在管理学院旅游管理专业基础上创建深圳旅游学院，这也是国内第一家旅游学院，在这个过程中我的科研教学工作不断地与管理学科加深关系。也正是在这一阶段，我于 2000 年 9 月当选为中国工程院工程管理学部首批院士，2004 年暨南大学管理学院旅游管理专业博士点建立，我作为暨南大学旅管专业的第一位博士学位指导教师，开始招收旅游管理专业的博士研究生，算是和旅游真正形成了全面深入的教学科研关系。

现在细想起来，我和旅游结缘是个很凑巧的过程，但也是历史的需要、国家的需要、社会的需要，使我一步步走进旅游这个领域。

文：相对于传统学科而言，旅游还是一个正在成长的年轻学科，不少人对旅游学科仍然存在一些质疑，您怎么看待这个问题？

刘：旅游业是世界第一大产业，对国家、经济，对所有人都有很重要的作用，旅游业不但有经济效用，而且可以净化人的思想，给人带来快乐。早期在欧洲旅游时我就发现，欧洲的穷人和富人都在旅游，而且都从旅游当中获得了快乐，真正实现了旅游的大众属性。人生需要快乐，旅游可以让人忘掉忧愁，多走出去看看别的国家、不同的地方风景可以让人在紧张的工作之后获得快乐，忘记烦恼，比如，巴西的狂欢节，在那里就可忘记一切忧愁、烦恼。

谈到旅游学科，我认为旅游是一门综合性学科，它是自然科学、技术科学、工程科学，再加上社会科学结合起来的学科，是非常特殊的学科。旅游业是中国对外开放最早、发展速度最快的行业之一。回顾新中国成立 65 年的历史，中国旅游业经历了对外接待和市场经营的不同阶段，改革开放政策实施以后才真正作为一个经济产业开始了大踏步地前进，直至今日中国已经成为世界旅游大国，并且有希望在不远的未来成为世界旅游第一强国。旅游产业在中国的发展带来了专业研究的蓬勃兴旺，现在管理学、经济学、地理学、社会学、历史学、生态学等不同学科都在参与旅游产业的研究与开发，恰恰说明了旅游学科的综合性。

具体到旅游产业，传统上大体归纳出“吃、住、行、游、娱、购”六要素，相关的研究也集中在这六个方面，也有部分研究关注了旅游发展的外部环境。其实这些还是不够的，一个原因是旅游消费涉及方方面面，甚至因为社会的发展、技术的进步而不断衍生创新，如互联网、手机等就让现在的旅游消费出现了很多新变化。人们对于旅游消费的动机、目的和期望也开始出现各种各样的变化，不同地区、不同人群、不同文化都存在这个现象，这就对旅游学科研究的领域范围提出新的要求和问题。另一个原因是全球经济一体化引发了不同产业之间的融合，你中有我，我中有你，旅游发展中这个现象也很突出，如旅游与医疗结合而成的医疗旅游，这个在欧洲很早就有了，还有旅游与工业结合形成的工业旅游，非常多的例子，有新的有旧的，但都反映出现代经济社会产业的边界开始模糊化，不同产业的融合可能成为新的研究热点，我们现在正在进行广东旅游产业融合发展方面的研究工作，就是一个尝试。当然，还存在其他的原因和影响因素，这些都使得旅游学科还表现为一个不断生长发展的学科，其他学科的人看起来觉得不成熟，有些质疑，这是正常的。

实际上，每一个学科的建立与发展，每一个学术问题的提出和解决，都是在质疑之中坚韧前进的。正是这种所谓的“不成熟”“质疑”给从事旅游研究的人提供了宽广的学术空间，大家要有忍耐的心态，耐得住寂寞，坚韧不拔地扎进问题之中进行探索。针对旅游学科的综合性，大家的研究视角应该拓宽，不要仅局限于某一个小系统，而应该从系统论的视角去重新认识旅游，旅游的发展绝不会局限于上面那六个要素，旅游引发的影响和问题也绝不仅在旅游范围内产生。旅游产业实际上是一个整体的外向系统，不同子系统互相合作产生的原动力，外部关联系统融合支撑产生的辅动力，共同组成旅游发展的动力，而内部不同子系统、外部不同关联系统之间往往又存在局部的关系和问题，这就使得旅游的问题既可能是大系统引发小系统产生的，也可能是小系统反馈大系统形成的，这方面的旅游现象与问题都可以通过系统动力学来进行分析解释。我觉得旅游这个学科很好，中国有很多人研究，但是还没有去好好把它系统化，值得从系统论的角度去深入探讨。具备了系统论的大视野和大思路，旅游研究者就有可能在细小的现象中发现贯穿旅游学科的学术问题，从而形成真正的理论贡献，推动我们旅游学科逐步走向成熟，去回答那些“质疑”。

文：刘老师，您刚刚出版了一部《旅游工程管理研究》著作，能否详细介绍一下这部著作的情况？

刘：由于我的力学、工程学背景，我对管理学科的认识与理解也集中在工程管理学方面，对旅游学科的介入也是由工程管理开始。1986 年我在上海工作时，结合旅游方面的研究和实践提出旅游工程学的概念，并成立了上海旅游工程学会。旅游工程学是一个新兴概念，就是将系统工程的原理运用到我们旅游专业，分析现有的旅游发展。它运用自然科学、技术科学和社会科学的有关思想、理论、方法和手段，根据系统论的原理，对旅游系统的构成要素、组成结构、信息交换和反馈控制等问题应用系统工程的方法，进行综合分析、设计、试验、实施，以便最充分地发挥人力、物力和财力，达到最合理、最有效、最经济地实现旅游系统的整体效益优化。我的这个想法也得到了老师钱伟长先生的认可，在我 1989 年主编的《旅游工程原理与实践》一书中，钱先生专门题词“以系统工程的科学理论指导旅游事业的运行”。

近代以来，工程科技就成为经济产业发展的主要驱动力，每一次技术革新与应用都引发了新的产业革命。在这个过程中，系统工程管理同样功不可没，正是在系统管理的指挥下，一项项新技术、新发明被纳入工艺流程系统，最终推动整个产业系统的整体革命。例如，现在国内正在推动的智慧城市、智慧旅游等经济新导向，就是要将各个细分的信息化技术通过系统工程管理的分工、统筹、构架、组织实现在同一个系统中的优化组合，从而最终实现系统整体的智慧化。在现代经济社会中，工程管理通过对技术、资金、信息、人力、物质等要素的合理布局组织，在各个方面给人们的生产生活带来便利和高效，不但成为推动人类社会发展的重要力量，而且正在给人类社会发展带来新的机遇。在旅游领域，对于信息技术的吸收、纳入、应用、推广，已经改变了人们的旅游消费和旅游活动方式，也改变了旅游企业经营管理和盈利模式，旅游工程学也因而有了更大的作用空间。

2014 年由科学出版社出版的《旅游工程管理研究》一书应该是对我近 30 年来从事旅游教学研究工作，尤其是旅游工程学理论探索的一个回顾和总结。这本书从旅游工程学的理论角度集中梳理了我在城乡旅游、旅游交通、旅游规划、旅游酒店、旅游市场营销等方面所进行的近 30 年的理论探索，并联系中国旅游实际开展理论方面的学术讨论，是在 1989 年提出旅游工程学概念的基础上对这一新兴方向更进一步的全面、深入阐释与探究。围绕着如何建立旅游研究的系统论视角，如何运用系统科学理论去分析与解释旅游发展、发现和解决旅游问题，进行了重点回答。举一个例子，这本书中有一篇《以价值工程方法全面提升荔湾商旅核心竞争力》文章，是 2002 年我主持的广州市荔湾区商业旅游发展研究课题的成果。荔湾区是广州市的中心城区，商业基础优势很明显，旅游方面的发展主要依靠底蕴深厚的历史文化内涵和传统的人文风情，两者关联发展虽然有基础但一直不显著。我就从系统工程的理论角度提出用价值链方法来指导荔湾区商业和旅游业的融合发展，结合城区竞争优势，通过价值链将单一资源品牌整合成荔湾城区品牌，构建商旅互动的价值链，推动荔湾区整体竞争能力提升。

以往分析旅游发展问题时，通常只注重旅游形象建设推广，而忽略了品牌的文化、质量、管理与服务等其他要素，更谈不上将品牌提升到价值链的高度，分析品牌与竞争力的关系。因此，出现了各地花费大量的人力、财物塑造旅游形象，而持久的旅游竞争力始终未能建立的问题。荔湾旅游业在价值工程方法的指导下，围绕品牌核心构建商旅产业价值链，利用品牌的聚合效应、磁场效应、带动效应促使相关产品被消费者认可与接受，在实际发展中已经取得了明显的效果。2010 年广州亚运会前期，荔湾区结合城市水系整治工作重新恢复了古时的广州荔枝湾河涌风光，现在已经成为广州城市旅游的名片之一，外地游客到了广州必定要沿着河涌散散步或者坐船游览，体会一下老广州河涌密布、水网纵横的岭南水乡特色。这就是从系统整体角度来认识旅游、发展旅游、管理旅游的良好效果。

闫：刘老师，您从事研究和教学工作 50 多年了，能否谈一下研究工作的经验以及对新一代研究人员的期望？

刘：回想几十年的工作，我觉得最主要的就是创新，这也是我个人获得成功的关键因素。那么何为创新呢？在原有的事物上添点东西不叫创新，跟着别人的步伐走也不叫创新。创新，是颠覆，是改造，是做人家没有做过的、没有想过的事情。要做到创新，首先必须要有好奇心。作为一个研究者，作为一个要有创新思想的人，好奇心很重要，必须要对事物好奇，要有兴趣。一个人对生活没有兴趣，根本就不可能提出问题。生活中的问题，工作中的问题，国家的问题，世界的问题，只要有兴趣，你就能在这种兴趣中提出一些新的思想。但是创新并非一件易事，创新的过程中可能会遇到种种困难，遭受各方面的反对，创新者还要承受失败的风险，这就要求我们要有耐心，要持之以恒。研究本来就是一条枯燥孤独的路，只有耐得住寂寞，坚持到最后，才能获得丰厚的成果。

只有创新的思维和灵感是远远不够的，要想做到创新还必须具备四个方面的条件。第一要有丰厚的知识积累，人类已经有几千年的知识积累了，形成了很多优秀的东西。你如果没有什么知识，不知道这些东西，你提出来，而别人早就提过了，早就有人做过了，这不是创新。牛顿就说过他是在巨人的肩膀上成长起来的。前人的东西已经非常丰富，所以你想要创新，就得有丰富的知识，不仅要有本专业的知识，还要有其他专业的知识。我在德国访学时遇到一个导游，精通德语、英语、法语、意大利语四种语言，他给游客提供服务时深受欢迎，这就是丰厚知识积累的好处。第二就是勤奋，要想拥有宽厚的知识积累就必须要勤奋，想到一个问题时，要通读关于这个问题的历史文献，看国内外的人对这个问题是怎么看的，怎么样解决这个问题的，还有什么没有解决或是解决得不够好的问题，我有没有办法解决它？如果有，那就继续往前走。第三要有发现的眼睛和联想的思维。有人能从苹果掉落发现万有引力，而有人想到的就是把苹果吃掉，这就是不懂得去联想才造成了天才与庸人的差别。生活中，要用心去观察周围的事物，将看到的现象与自己的知识结合起来，这样才能不断发现问题。第四要抓住机遇。法国科学家巴斯德说过“机遇只偏爱那些有准备、有头脑的人”。我们绝不能等到机遇来了才临时抱佛脚，那是不能成功的，要创新必须要善于抓住机遇。我的人生归纳起来，就是碰到机会，一个新的机会来到我面前时，我会用学工程、学科学的思想去对待这个问题，并且想办法尽力解决这个问题，跟旅游学科结缘也是这样。你们也一样，要抓住那一瞬间的灵感，一瞬间的感觉，抓住这一瞬间，你就可能触碰到了创新的边缘，当机会来临时，要思考应该怎么做，做好了就是创新。

中国人由于受到几千年文化体制的限制，缺乏创新，跳跃性思维较弱。我希望你们年轻人在这个纷繁复杂的社会中，要学会创新，并且敢于创新。相信自己，想到什么就付诸行动，并且不要太在乎周围人的眼光，只要你觉得是对的，你就去做，不要害怕失败。不然，随波逐流，只会被埋没在这片沙滩中，永无出头之日。所以一定要走创新的道路，中国缺乏创新，你们这一代更要注意这个问题。

文：刘老师，今天听了您的一席话，我们对旅游工程学以及系统工程科学在旅游领域的应用有了更深的理解。感谢您的精彩讲解！也感谢您在百忙之中接受我们的访谈！

刘：对于旅游我只做了一点点工作，这个学科还需要和你们一样的更多的有志者一起努力！谢谢你们对我的采访！也祝愿《社会科学家》杂志越办越好！

［责任编校：唐鑫］

旅游城市品牌的塑造[①]

当今中国旅游正以前所未有的速度向前发展，旅游因其综合性强的特征而不断拓展其概念边界，给旅游研究者带来越来越多新的挑战。在激烈的国内外旅游市场竞争中，塑造城市旅游品牌成为提升区域综合竞争力的重要手段。旅游城市品牌塑造是一项复杂的系统工程，涉及政治、经济、文化、社会、环境等方面。本书从品牌的灵魂要素“品牌个性”角度着眼，通过实证分析的方法，研究入境游客如何感知当代中国旅游城市的品牌个性，这是一个兼具理论和实践意义的研究项目，也是中国旅游品牌国际化道路上亟待研究的基础性课题。

国际旅游市场的竞争实质是旅游目的地之间的品牌竞争，而品牌竞争的关键是确立品牌的个性。旅游管理部门的主要职能是对外宣传和推广旅游目的地品牌形象，只有确立了旅游目的地的品牌个性，围绕品牌而开展的一系列营销活动才能做到有的放矢。旅游城市的品牌管理是对全域资源的统筹协调，以品牌为中心，整合分属不同利益相关者的、内部结构松散的资源，最终实现区域整体目标最优。本书最大的学术价值在于，综合运用了管理学、心理学、语言学、统计学等多学科理论和方法，通过长时间跨度、多语种的问卷抽样调查，构建了当代中国旅游城市品牌个性的多维框架，由活力、真诚、现代、文化、粗犷、随和六个因子构成，并验证了六个因子结构的信度和效度。书中既有理论性和逻辑性的思考，又有立足国际旅游营销最前线的实际应用探讨，对于旅游工程学是一本很有实用价值的书作。

本人是力学学者，在 20 世纪 80 年代改革开放初期又有幸从事经济管理理论与应用研究工作，从 1984 年开始，先后参加和主持开展了黄山旅游区规划、崇明岛发展战略、上海旅游交通和酒店管理等研究工作。1986 年，我担任上海工业大学经济管理学院首任院长，开始指导经济管理专业的研究生，并讲授“旅游工程学”等课程，提出中国旅游学研究应该建立“旅游工程学”。本书作者梁江川是我在暨南大学旅游工程研究方向上招收和指导的弟子。他本科就读于哈尔滨工业大学，硕士在韩国汉阳大学观光系学习，随后在暨南大学旅游管理系攻读博士学位，并先后在广东省旅游局旅游发展研究中心和国家旅游局驻首尔办事处工作，对国际旅游市场具有扎实的理论基础和丰富的实践经验。作为他的导师，我非常欣喜地看到此书的出版，中国旅游城市品牌个性是一个值得持续研究的课题，我相信，这本书将成为该领域研究的基石之一，也将为广大旅游实务管理者提供有益的借鉴。

---

① 本文是《中国旅游城市品牌个性感知研究——基于广东入境游客视角》序，广州：暨南大学出版社，2016，1-2.

保护遗产至关重要[①]

无论是自然遗产还是文化遗产，都是老天爷、老祖宗留给我们的宝贵财富。这些遗产积淀和凝聚着深厚的自然、文化内涵，成为反映地球演变、人类创造力以及人类与自然关系的有力物证，是全人类的共同财产。我国拥有辽阔而多样的国土以及几千年悠久的历史文化，我们有幸继承了丰富的自然、文化遗产，对这些宝贵遗产的保护与利用承载着我们民族历史、现在与未来的责任。

我国日益重视对遗产的保护。从 2006 年起，就将每年六月的第二个星期六定为“文化遗产日”，同时也采取多种措施保护遗产。然而，仅凭政府的力量，很难完全做好遗产保护工作。目前社会上也有很多人认为遗产保护是国家的责任、政府的事，与己无关。而事实上，我们每个人都是遗产的主人，都有责任保护它们，并将它们传承给我们的子孙后代。在这样的背景下，研究公众参与遗产保护有着重要的现实意义。

以往遗产保护的研究更多关注政府如何采取保护行动，往往忽略了公众在遗产保护中的主观能动性及应承担的责任与义务。本书则旗帜鲜明地提出应积极引导全民共同参与遗产保护，尤其关注遗产利益相关者在遗产保护中的利益诉求及其作用。在本书中作者能够很好地将质性研究与量性研究有机结合，深入探讨公众参与遗产保护的动力与阻力，提出并验证了公众的遗产认知、权利意识、责任意识以及参与能力等个人因素与人们参与行为之间的关系。本书所设计的公众参与激励机制以及提出的公众参与路径选择，为公众参与遗产保护提供了新的思路。

本书作者刘小蓓是我的博士生，她长期从事旅游规划、遗产旅游、文化景观保护等方面的教学、研究与实践活动，具备较为扎实的理论基础并积累了丰富的实践经验。攻读博士期间，她勤奋好学，广泛阅读，勤于思考，治学严谨，对自然、文化遗产的保护与利用问题一直抱有浓厚的兴趣。我曾经对我的博士生说，要和博士论文谈一辈子的恋爱。刘小蓓一直和她的遗产保护与利用问题相恋，自她博士毕业之后，相继主持的广东省自然科学基金、广东省教育厅青年人才培养计划课题和广州市“十三五”哲学社会科学项目均与遗产保护有关。本书就是在其博士论文的基础上不断修订、完善与拓展而成。她持续关注本书的案例开平碉楼遗产保护情况，读博士期间以及毕业之后多次到该地进行调研，与相关管理部门负责人交流，深入各村落与村民访谈，获取了大量宝贵的一手素材。作为她的导师，我很高兴看到她的研究成果出版，也希望她能够在这方面继续深入研究，因为公众参与遗产保护是一个值得持续研究的课题。我相信，这本书的研究成果将成为该领域相关理论的重要组成部分，也将为我国正在如火如荼兴起的遗产保护及遗产旅游开发实践提供有益的借鉴与启示。

是为序。

---

① 本文是《公众参与遗产保护的激励机制研究》序，广州：暨南大学出版社，2017，1-2.

城市遗产旅游景观的研究[①]

历史街区是一个城市的重要文化遗产构成，在当今中国现代化进程中面临着更新改造的社会事实。如何保护及利用遗产，则成为社会及学术的关注焦点，特别在当下城市消费社会转型背景下具有深刻的意义。廖卫华的《消费主义视角下城市遗产旅游景观的空间生产：成都宽窄巷子个案研究》一书则是对这一问题的深度探讨，以消费主义视角、运用空间生产理论解读宽窄巷子这一遗产旅游景观的空间生产问题。指出历史街区更新改造不仅是一个空间生产的实践过程，同时也是各利益相关群体博弈的结果，并导致了社会、经济、政治及文化空间的生产。因此，这一空间生产既是一个物质过程，同时也是一个社会关系的生产过程，并对其生产及消费逻辑进行了分析，在理论研究及实践应用方面具有创新意义及现实价值。本书运用管理学、旅游学、城市社会学和文化地理学等相关理论进行跨学科研究。此外，作者多次到研究样地——成都宽窄巷子进行实地调查，并对相关群体包括游客、本地居民、街区原住民、商家、政府人员、规划师、管理运营商、专家学者等进行深度访谈，获得大量的一手资料。可见，该书的研究结论建立在扎实的数据及理论基础之上。

廖卫华是我带的第一批旅游管理专业博士生。2004 年入学以来，廖卫华博士就对城市及遗产旅游产生浓厚兴趣，并致力于对该问题的研究。学习期间，她还参与多项旅游规划及科研项目，积累了丰富的实践经验，理论及实践水平得以提高。在学位论文的写作过程中，她虽然经历了一些挫折和困难，但最终都得以克服并完成了博士论文。在论文答辩时得到答辩委员会专家的一致好评，尤其是论文独特的理论视角、深度的田野调查，以及流畅且具有可读性的文字，均给答辩老师留下良好印象，并对其进一步的研究有着较高的期许。本书则是她在博士论文基础上进行持续研究并加以完善的成果。作为她的导师，我欣慰地看到她在学术道路上的成长。是为序。

① 本文是《消费主义视角下城市遗产旅游景观的空间生产：成都宽窄巷子个案研究》序，北京：科学出版社，2017，1.

应用旅游工程理论探讨阳江旅游发展规划[①]

旅游业是世界最大产业之一，一百年来发展极为迅速。2015 年，全世界旅游业对全球 GDP 的综合贡献为 7.8 万亿美元，占全球 GDP 高达 10%。再看我们国家，旅游业的发展几乎从零起步，一路突飞猛进。2015 年，我国国内旅游人数、出境旅游人数、国内旅游消费、境外旅游消费均列世界第一；国内游高达 40 亿人次，出境游 1.28 亿人次，入境游 1.38 亿人次，旅游总收入综合贡献为 7.34 万亿元，占 GDP 的 10.8%。而在 39 年前的 1978 年，我国国内游和出境游几乎为零，入境游 30 万人次，大多是外事活动人员。这种转变是我们国家遵循邓小平同志指引，采用改革开放政策，才使旅游业走上了高速发展的正确道路。阳江市有得天独厚的优势，位于南海之滨，拥有如“南海一号”这样独特的旅游资源，理所当然，阳江应大力发展旅游业，把它培育为战略性的支柱产业，使阳江成为国际旅游名城，为阳江市的发展规划实现，建成“富美阳江”、“海洋经济强市”和“海洋经济创新发展示范市”做出贡献！

为了使阳江旅游产业成为阳江市的战略支柱产业，就必须研究旅游业发展战略，制订好发展规划。为此，先将旅游工程学介绍给大家，以便为阳江旅游发展开发与旅游发展规划的制订提供一种科学的理论、技术和方法。同时，再谈几点建议。

一、旅游工程学的基本概念和研究方法

1984 年夏天，一个偶然的机会，我出席了安徽省皖南旅游区发展规划座谈会，因而被带进了旅游业领域，提出了安徽省应先制订黄山旅游规划、成立安徽航空公司、修建黄山机场的建议。事后这些建议得到了省委、省政府的批准，从而参与黄山旅游规划制订的工作，最终使黄山旅游规划和建设得以成功。遗憾的是，成立安徽航空公司和修建黄山机场的建议由于时机不对，而未得到上级批准，因而作罢。1986 年春，我由上级指令调往上海，钱伟长先生让我筹建经济管理学院，我才发现我们全国的高校中竟然没有设立旅游管理的本科专业，当时的上海旅游管理专科学校是我国旅游界的名校。为此我在全国最早设立旅游管理本科专业，同时又建立旅游学术团体——上海旅游工程学会，并承担上海旅游交通、上海酒店管理等旅游课题的研究。为促进中国旅游业发展，在 1986 年，我提出建立一所新兴的综合性学科——旅游工程学。1992 年，在暨南大学工作时，我又创办中国第一所旅游学院——暨南大学深圳中旅学院（现名：暨南大学深圳旅游学院）。2003 年，我成为旅游管理博士学位的指导老师，也是全国最早的几位旅游管理博士导师之一，开始直接指导旅游管理博士生。通过旅游管理的教学和科研工作，让我更感到旅游工程学的重要性。钱伟长先生于 1989 年曾为我们题词鼓励“以系统工程的科学理论指导旅游事业的运行，题赠《旅游工程原理与实践》”。《旅游工程原理与实践》于 1991 年在上海百家出版社出版，我任主编。现在，

① 本文是应邀为广东省阳江市党政领导干部大会所做的报告，阳江，2017 年 8 月 9 日.

请让我把旅游工程学的基本概念和研究方法介绍给大家。

旅游工程学是一个新兴的边缘学科。它运用自然科学、技术科学、工程科学和社会科学的有关思想、理论、方法和手段，根据系统总体协调的需要，对旅游系统的构成要素、组织结构、信息交换和反馈控制等功能，运用系统工程的方法，进行综合分析、设计、试验、实施，以便最充分地发挥人力、物力和财力的作用，达到最合理、最有效、最经济地实现旅游系统的整体效益的优化。所以，这是一个由自然科学、技术科学、工程科学和社会科学相互交叉、相互渗透而形成的综合性学科。

旅游工程学作为一种崭新的思维观点，它将冲破过去传统的保守思维方式，冲破自然科学、技术科学、工程科学和社会科学之间的鸿沟，从而产生新的思想和理论观点，以指导旅游实践。

旅游工程学作为高层次的决策分析方法应用于旅游业，对旅游业的发展战略、宏观规划、未来预测、旅游理论与学术研究、旅游开发的科学论证等将起到十分重要的咨询作用，从而求得所涉问题的整体合理和可行。

旅游工程学作为旅游管理的最先进的具体方法论，它还将在不同层次的旅游实践和旅游决策中参与多方案、对策性、优选和复杂的定量分析。

旅游工程学从旅游系统的总体高度出发，把系统当成一个相互作用、相互依存的超巨系统来进行研究，其研究方法如下。

（1）全面地搜集资料，弄清问题所在。

（2）确定和分析系统的评价目标。首先要了解建立这个系统的目的和要求，进而确定系统的目标。确定目标时首先要做到技术上先进、经济上合理，并考虑到与其他系统的兼容性和对客观环境的适应性。其次要分析实现系统目标所涉及的限制条件有哪些，达到总目标应完成哪些具体的项目，这些项目又应由哪些子系统来承担，进而确定各子系统的目标。

（3）系统的模型化。明确了系统的目标后，可根据系统的性质与功能的要求，提出几个可能的方案，找出相应的约束条件和各要素之间的逻辑关系，用数学方式把它们表达出来。应该指出，旅游工程学的研究对象不仅局限于某一个实物，它涉及的面常常十分广泛，而且都是很复杂的系统，因而不能用形象的实体模型来代替抽象的数学模型。

（4）系统的最大优化。系统的最大优化是通过系统的模型进行的。不同的模型有不同的优化理论和优化方法。例如，对一些确定性的问题，可采用线性规划、整体规划、非线性规划、动态规划等理论和方法进行优化；对于那些非确定性的问题，则可用对策、决策理论和方法进行最优化。

（5）综合评价。对于多目标的系统来说，不同的目标可以建立各种不同的模型。而不同的模型又可以求出几个不同的最优解，对这些最优解就要根据具体条件进行综合分析和评价，找出各方面的所得与所失，把不同性质、不同度量单位的几个方面的评价，用同一评价指标来比较，从中选出技术上先进、经济上合理、综合效益最优的方案。

图 1 描述了这些研究过程。

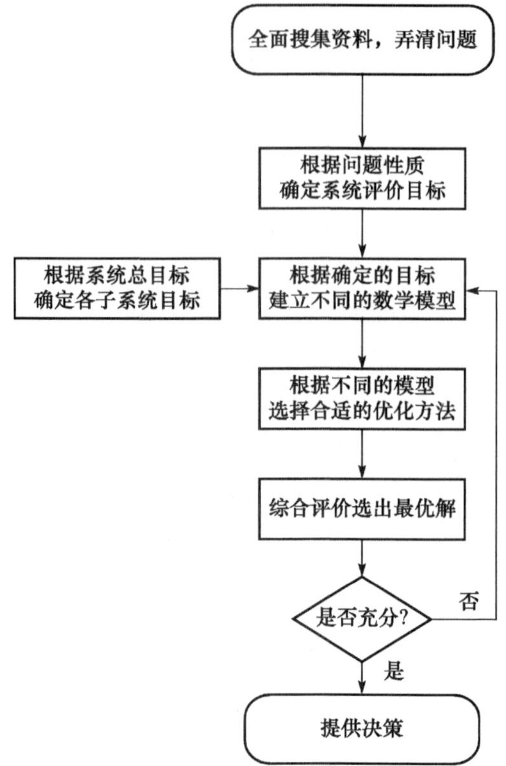

图 1　旅游工程学的研究过程

应当指出，系统的模型化是系统最优化和综合评价的基础。要想准确而方便地求得系统的最优方案，就必须使所建立的模型既能真实地反映现实，而又不过于复杂。如果建立的模型不确切，系统的最优化和综合评价也就失去了意义。因此，从事旅游工程学的研究，不仅要对所研究的旅游系统的全貌有深入的了解，而且还要熟练地掌握现代数学工具和各种模型的优化技术。

二、探讨阳江市滨海旅游发展规划

制订阳江市滨海旅游发展规划是阳江市旅游领导机构战略级决策的重要方面，也是一项典型的非结构问题的决策，而科学的、严密的规划又是阳江市旅游领导机构对其战略问题进行决策的重要依据。

滨海旅游发展规划同旅游资源分配密切相关，是一种合理调配、组织旅游资源的决策活动，其目的是：①满足旅游者了解自然和社会、完善自我的需要；②给旅游企业带来一定的经济效益，社会也能通过旅游业积累建设资金。因此，规划制订在很大程度上依赖于自然旅游资源、人文旅游资源、旅游交通、旅游饭店建设和社会经济的发达程度等自然和社会因素。恰好，阳江滨海已具备了这些条件：位于美丽的南海之滨，古代“一带一路”的重要海路名城，闻名于世的“南海一号”博物馆，美丽的海陵岛等，故值得高度重视，制订旅游发展规划。

为了使滨海旅游发展规划真正具有科学预见性，进行规划决策时应当重视下述内容。

(1) 规划的内容、进度和措施要符合客观规律，要达到技术上的先进性、经济上的合理性和实践上的可行性。

(2) 规划不仅是具体计划的制订，还必须顾及资金筹集、人才结构、设备更新、管理现代化等多方面的相关内容。

(3) 规划导向作用主要表现在预见性上，其基础是预测技术。预测是使决策避免盲目性、增强自觉性的重要手段。在制订滨海旅游规划时事先进行科学的预测，就会使决策者做到心中有数，增加主动性和预见性，起到良好的导向作用。

滨海旅游发展规划模式，如图 2 所示。

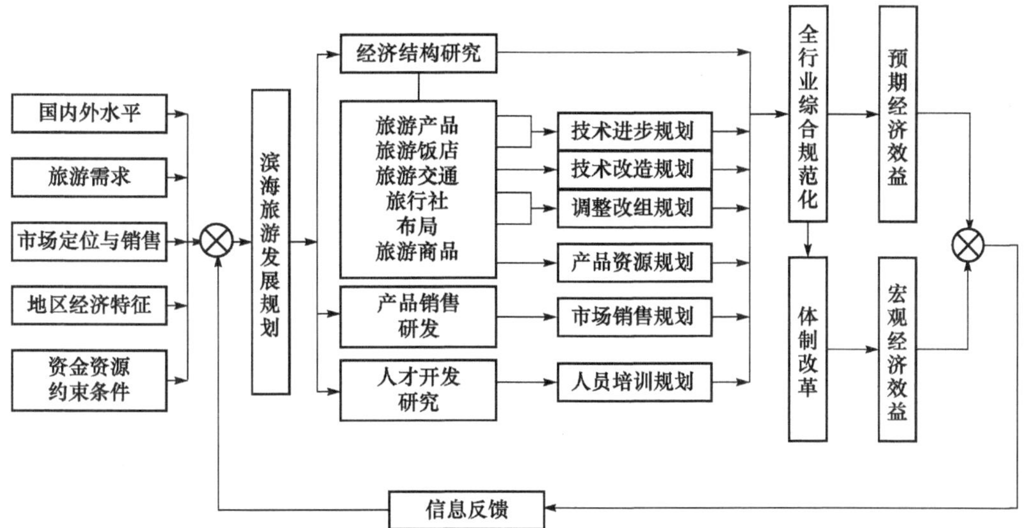

图 2　滨海旅游发展规划模式

由图 2 看到，正确的规划来自充分而可靠的信息以及对这些信息的深入研究，特别是基于经济结构、市场销量与人才开发这三大主要方面的研究。

总之，根据旅游工程学的原理，滨海旅游发展规划的制订，实质上是研究滨海旅游部分从输入“处理”过程中的管理与组织技术，并通过系统分析、统筹安排和综合平衡，充分发挥滨海旅游部门的人力、物力与财力的作用，实现滨海旅游事业的最优化。

三、科学制订滨海旅游发展规划的几点看法

要把阳江市建成“富美阳江”和“海洋经济强市”，不仅要着眼于先进装备制造业和水产品深加工，而且还要发挥滨海旅游优势。为了配合开发阳江的整体战略，在滨海旅游规划方面，我谈几点看法。

(1) 结合“一带一路”建设，把阳江建设为一座国际旅游名城。经过 39 年的努力，我们伟大的祖国经过高速发展，已经成为世界第二大经济的国家。今年 5 月 14 日

至 15 日，我国在北京主办“一带一路" 国际合作高峰论坛。习近平总书记在论坛开幕式上说：“‘2000 多年前，我们的先辈筚路蓝缕，穿越草原沙漠，开辟出联通亚欧非的陆上丝绸之路；我们的先辈扬帆远航，穿越惊涛骇浪，闯荡出连接东西方的海上丝绸之路。’”“古丝绸之路绵亘万里，延续千年。”“这是人类文明的宝贵遗产。”阳江市正处于海上丝绸之路的开端要道之处——南海之滨，1987 年在阳江海域发现的“南海一号”古沉船，这是南宋初期一艘在海上丝绸之路向外运送瓷器时失事沉没的木质船，距今 800 年，是国内发现的第一艘古沉船，也是迄今为止世界上发现的海上沉船中年代最早、船体最大、保存最完整的远洋贸易商船。船上载有文物 6 万至 8 万件，且有不少是价值连城的国宝级文物。这件事轰动世界，它将为复原海上丝绸之路的历史以及陶瓷史提供极为难得的实物资料，甚至可以获得文献和陆上考古无法提供的信息。“南海一号”的发现是阳江的幸运和特大优势，理所应当首先发挥它的作用。从旅游业来看，随着“一带一路”的建设，“南海一号”的发现为阳江建设国际化旅游名城带来了千载难逢的机会，这是难得的机遇，要紧紧抓住。首先要用旅游工程学的理论方法来规划此事。然后就要大力宣传，同时要好好包装，使之更有色彩，多多吸引全世界游客。并且还要把旅游交通搞好，既要有高铁和高速公路，而且还要建设机场，还要有停靠海上游船的港口。对于酒店，不仅要有高级的酒店还要多建设年轻人的酒店，使各类游客来往方便。

（2）围绕“南海一号”中心，再多建设几个旅游景观，使阳江旅游资源更加丰富多彩。首先，将现有的海陵岛旅游区建设得更有特色，吸引更多的国内外游客来参观。同时，阳江滨海海岸线长，多搞几个以海洋特色为题材的旅游项目完全可能。我相信，阳江旅游事业必将有一个大的发展。

创业拼凑对创业学习的影响研究
——基于创业导向的调节作用①

近些年，在“大众创业，万众创新”的号召下，我国的创业活动如火如荼地开展，大批新创企业如雨后春笋般地涌现出来。但同时，新创企业较高的失败率与死亡率也在创业者心头蒙上了一层阴霾。如何提高新创企业的存活率已然成为创业者与学者关注的热点话题。

众所周知，资源短缺是大多数新创企业的生存常态[1]，这在转型经济背景下的我国新创企业中尤为突出。资源是创业行为的载体，是新创企业生存的基础和必要条件[2]，资源缺乏势必会直接威胁企业生存。尽管一些创业者、学者诉诸外部资源获取以弥补企业的资源缺口，但新创企业严重的“新创劣势”（liability of newness）以及我国相关制度与市场机制的不完善使得这种应对方式举步维艰。在这种情况下，依靠自身主动性的拼凑行为便受到越来越多的学者的推崇[3-5]，即通过创造性地使用手头现有资源甚至那些他人眼中的废弃物来突破资源约束，从而帮助企业实现“无中生有”。

“拼凑”（bricolage）一词起源于结构主义人类学研究[6]，在引入创业研究领域后便备受关注。回顾现有研究，学者大多认为拼凑需要以一定的经验和资源知识为基础，例如，Weick 认为拼凑活动需要大量的实践和先前经验[7]。Baker 和 Nelson 提出企业在巧妙应用拼凑的能力上存在较大差异[3]。Cunha 认为拼凑者大都具有丰富的经验，从而可基于对资源属性的深入了解创造性地重组现有资源以解决资源短缺问题。因此，相比那些没有经验的人，拼凑更可能由经验丰富的个体予以实施[8]。van de Walle 提出拼凑者拥有大量的实践经验与本地化知识，他们在拼凑的过程中利用自身记忆并需要依赖一系列学习过程[9]。由此可知，拼凑者先前所学习、积累的行业经验与相关知识使其能更加有效地识别手头资源的潜在价值，进而提高拼凑的效果。但反过来，却鲜有研究对拼凑是否会影响学习这一问题给出明确的答案。

其实，作为一种不断尝试、实验与试错的过程[10]，拼凑能够创造出大量与资源属性及其组合方式相关的新知识，其本身的即兴创作性[11]使得这些知识大多带有一定的隐性色彩，隐性知识对于企业构建可持续性竞争优势尤为关键[12-13]。研究表明创业成功离不开持续的创业学习[14]。因此，如果新创企业能够通过有效的学习获取、吸收并真正掌握这些知识将对其生存与成长发挥重要作用。然而在现实中，新企业极低的存活率和成长速度却充分说明其学习效率与效果并不理想[15]，加之拼凑本身以“奏效”为目的，上述资源知识作为这一行动中的副产品很可能被创业者所忽视。因此，创业者是否能够学习到拼凑所带来的隐性知识并为其所用仍需进一步验证。此外，由基于不同的学习效率与学习水平所带来的大相径庭的学习效果可推知[16]，并非所有的创业者都能在拼凑中进行同等程度的学习：重视这一学习的创业者往往能充分发挥其主观

---

① 本文原载《科学学与科学技术管理》，2017，38（10）：135-146. 作者：刘人怀，王娅男.

能动性，积极反思并不断总结、分享、交流拼凑经验，并积累相关资源知识；而不重视的创业者则对拼凑采取得过且过的态度，只将其作为一种获取临时解决方案的行为方式，而并不会对其中所包含的智慧、知识和信息进行学习与总结。那么，怎样的情况会更有利于创业者在拼凑活动中的学习呢?

基于上述问题，本文研究了创业拼凑对创业学习的影响；同时，由于具有较强创业导向的企业通常更倾向于追求市场主导地位，重视获取、创造异质性知识[17]，且创业导向的创新性和风险承担性能为学习提供良好的内部环境[18]，因此本文将创业导向的三个维度——创新性、风险承担性与超前行动性作为创业拼凑与创业学习之间关系的调节变量，以探索影响创业者从拼凑活动中学习的情境因素。研究不仅丰富了创业拼凑理论的研究框架，同时也为创业活动提供了更加直接的实践指导。

一、理论基础

1. 创业拼凑

“创业拼凑”这一概念由 Baker 和 Nelson 于 2005 年正式提出，主要指为了解决新问题或应对新挑战，将就着对手头现有资源进行创造性地重新组合[3]。其中，将就（making do）强调其行动偏好（a bias toward action）以及不屈从于现有约束（a refusal to enact limitations）的行为特征；创造性重组通常与企业创新紧密相关；手头资源（the resources at hand）包括创业者所拥有的各种资源和从外部获取的免费或廉价资源，如一些“零碎儿”、工具、想法、知识和技能等[19-20]。在之后的研究中，学者不断丰富着这一概念，如 Duymedjian 和 Ruling 认为拼凑是一种理念/思维，是由实践、潜在世界观和认识论相互作用而形成的[19]；Senyard 等将拼凑定义为一种解决问题的能力[21]；di Domenico 等提出，即兴创作性也是创业拼凑的重要组成部分[11]。本文认为，创业拼凑主要指创业者在资源短缺的情况下，通过重新组合手头资源并立即采取行动以解决新问题或开发新机会的一种行为方式。

通常，拼凑者基于“迟早会有用”的原则收集资源，通过评价资源的性质、替代性、种类，对其进行创造性重组与应用，从而实现对现有资源的实验、修补、操控和重新配置。这一过程在挖掘资源潜在价值的同时深化了拼凑者对资源属性及其相互关系的理解[2]。尽管拼凑是一种不断试错与尝试的过程[10]，但却并非无休止地进行下去。它以“满意”为原则，即当某一方案得以解决当前问题时便结束尝试[19]，并通过后续的不断调整来完善现有方案。

2. 创业学习

创业学习伴随着创业研究从静态特质论向动态行为论的转化而出现，是创业过程的核心[22]。目前，学者主要从机会识别、知识、经验和认知等视角对其进行界定。Rae 和 Carswell 认为创业学习是个体在识别、开发创业机会以及建立和管理新企业过程中构建一系列新方法的过程[23]；Minniti 和 Bygrave 将其定义为创业者不断积累并更新知识的行为过程[14]；Politis 认为创业者通过不同方式将经验转化为创业知识的经验性过程为创业学习[24]。本文将其定义为创业者在创业过程中不断获取并积累内外部经验，积极反思并创造独特新知识的过程，并认为它主要包括经验学习、认知学习和实践学习三种方式[25]。其中，经验学习指通过不断地反思与试错将已有经验转化为创业知识；认知学习也可称为模仿性学习，强调通过观察、模仿他人行为获取并吸收知识的过程；实践学习指根据需要调整以往积累的知识与现实情境之间的差异，并创造新知识的过程。

现有研究对创业学习影响因素的探索包括个体层面的创业者先前经验、自信心、创业认知和创业网络[24-26]，组织层面的领导风格、组织文化、组织结构和战略导向[16,27,28]以及环境动态性、地区文化氛围等外部环境影响因素[29]。尽管学者从多角度、多层面探讨了创业学习的影响因素，但却忽视了拼凑对创业学习所产生的影响。调查表明，创造性拼凑已成为中国新创企业应对资源短缺采取的一种普遍性行为[30]，而这一行为对学习活动所产生的影响却尚不清晰。

二、研究假设

1. 创业拼凑与创业学习的关系

创业拼凑是一个不断试错、尝试和实验的过程[10]。在拼凑的过程中，创业者利用现有资源进行实验、修补创意、重组知识、建构意义，并通过收集、检验、替代、配置或重新配置现有资源来创造解决方案[31]。因此，拼凑并非一蹴而就，而是建立在大量试错与失败的基础之上，这为创业者提供了良好的学习机会，也是创业学习的重要来源[32]。因为每一次实验与试错都会加深拼凑者对现有资源的理解，这种理解不仅是对资源属性及其局限性的理解，更是对资源之间的相互联系与组合方式的理解；而每一次失败也敦促着创业者反思现有知识与能力，并通过积极探索与实践不断创造知识。其实，拼凑不仅实现了资源组合的渐进性调整，同时也在不断改善着创业者的既有知识，它通过挖掘既有知识的新用途、探索其新组合为创业者带来了大量的新知识[33]。此外，拼凑强调不屈从于现有约束，敢于打破常规与惯例，其对于失败较高的容忍度、对于模糊性和不确定性较高的承受能力均有助于培育开放、包容的组织氛围，这既为企业成员的发散思维提供了肥沃的土壤，同时也使其具有更大的意愿与其他组织成员进行沟通与交流，进而通过向他人学习来不断增加自身的知识存量。因此，基于上述分析本文提出以下假设。

H1：创业拼凑有利于创业学习。

2. 创业导向的调节作用

创业导向是企业在创业过程与创业行为中表现出的一种战略态度[34-35]，主要由创新性、风险承担性和超前行动性三个维度表征[34,36]。

创新性是创业导向的重要特征，它反映了企业支持那些可能产生新技术、新产品和新服务的创意、实验和创造性活动的倾向[37]。知识是创新的核心要素[38]，创新性的理念与行动需要大量知识资源的支持才能得以实现[39]。然而，新创企业大多缺乏丰富的知识资源，在较高创新性的作用下，这种短缺所带来的“饥饿感”会使企业对知识产生热切的渴望，致使它们积极探索知识获取路径，并利用包括拼凑在内的一切可能的机会进行充分学习以汲取知识养料。同时，拼凑中不断的实验与试错以及强调打破常规与既有惯例的行动准则，也使拼凑者从中所学到的知识更具新颖性。此外，创业拼凑本身具有即兴创作性[11]，创造性地使用现有资源使得通过拼凑所创造的知识大多带有一定的隐性色彩，隐性知识是促进企业创新的重要因素[40-41]。因此，通过拼凑活动的学习所获取的资源知识在一定程度上满足了由较高创新性所导致的大量（隐性）知识需求。

综上所述，基于拼凑所创造的知识对于创新的重要作用以及由新创企业较高的创新性所产生的强烈的学习动机将有效促进创业者在拼凑活动中的学习。因此，基于上述分析本文提出以下假设。

H2a：创新性正向调节创业拼凑对创业学习的影响，即创新性越高，越有利于创业者从拼凑行为中学习。

风险承担性是指企业采取大胆行动，并对失败和不确定性结果做出最大资源承诺的程度[42]。尽管有学者很早便将风险承担作为区分创业者与非创业者的重要标志，但研究表明成功的创业者并非一味地接受风险，而是设法界定所需承担的风险[43]，并在谨慎评估那些高不确定性项目的基础上对其进行选择、实施，以尽可能降低风险[44]。从决策角度看，创业者需要收集大量的信息与知识并对其进行分析来管理额外的风险，随后做出决策。因此，上述过程产生了大量资源需求，包括物质、人力、技术、市场以及知识资源的需求，这促使新创企业通过各种途径获取所需资源。而拼凑强调对现有资源、知识和创意的重新配置，通过打破常规使新创企业能够识别出现有资源的潜在价值并进一步深化对资源及其相互关系的理解，从而在克服资源功效传统认知的基础上学习到更多新知识[19]，以满足风险承担导向实施过程中的资源与知识需求。此外，较高的风险承担性有助于构建开放、包容的组织氛围以及乐于接受他人观点与经验的行为态度[45]，这为组织成员进行拼凑经验的交流与分享创造了有利条件；包容的组织氛围则使企业更能容忍失败，并在无形中促进了拼凑过程中的不断试错，从而有利于创业者从错误中学习。因此，基于上述分析本文提出以下假设。

H2b：风险承担性正向调节创业拼凑对创业学习的影响，即风险承担性越高，越有利于创业者从拼凑行为中学习。

超前行动性反映了一种机会搜寻的前瞻性观点，指企业预期未来需求变化并率先采取战略行动的倾向[42]。对于具有较高超前行动性的新创企业来说，为有效预期需求变化，须始终对市场、竞争、技术环境保持一定的敏锐性和警觉性，这种状态需要丰富的顾客信息、市场知识和前沿技术予以支撑[46]，但企业本身在资源、知识和能力上的欠缺却使其很难满足这种知识和信息需求。因此，二者之间的供需矛盾推动着企业积极扫描环境、学习知识，而拼凑则成为企业获取相关资源知识的重要途径。同时，超前行动性强调识别并开发那些具有前景的商业机会[13]，这同样离不开创业者丰富的经验、知识和敏锐的市场洞察力，而由此所产生的学习动力将促使拼凑者从试错中不断学习、增加自身知识存量，以有效提高机会开发的速度和效果。此外，基于前瞻性观点，超前行动的新创企业其领先性不仅体现在先于竞争对手推出新产品、新技术的意愿上，更体现在获取知识与信息的速度上[46]。而拼凑行为所带来的资源知识大多烙有“企业专有”的标签，积极学习并获取这些知识是企业知识获取领先性的重要表现。因此，基于上述分析本文提出以下假设。

H2c：超前行动性正向调节创业拼凑对创业学习的影响，即超前行动性越高，越有利于创业者从拼凑行为中学习。

综上，本文的研究模型如图 1 所示。

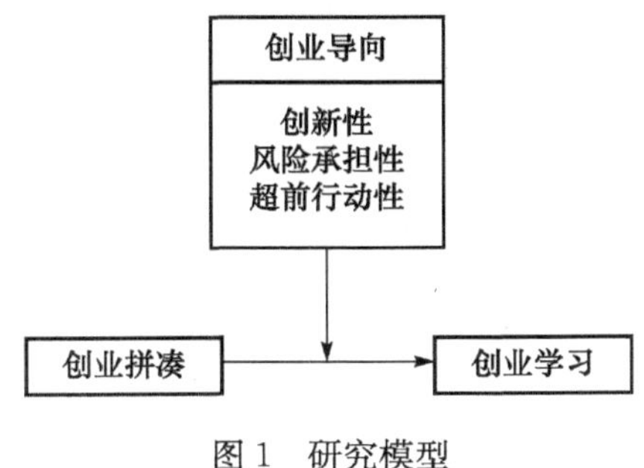

图 1　研究模型

三、研究设计

1. 样本与数据收集

本文采用问卷调查的方式获取研究数据，主要程序如下：①为确保问卷具有良好的信度与效度，通过翻译-回译的方式将英文量表翻译成中文，从而形成初始量表；②对广州、湖南等地的十余位企业家进行实地访谈，在访谈结束后请他们协助填写初始量表，并根据他们的反馈意见对题目的内容和措辞进行调整，从而得到初始问卷；③利用此问卷进行小样本预测试，分析预测试样本数据；④基于预测试的分析结果对问卷做进一步调整并形成正式问卷，进行大样本调查。

本文的研究对象为新创企业，参考 Zahra[47]、McDougall 和 Robinson[48] 的研究，将新创企业界定为成立时间小于八年的企业。按照经济市场的不同发展水平，样本主要分布在我国的东部地区、东北地区、中部地区和西部地区，调查对象主要是新创企业的 CEO、总经理和创业团队的核心人员。问卷采取现场发放和网络发放两种方式：①在相关政府部门、商会的协助下向新创企业发放问卷；②通过微信版问卷星与电子邮件将问卷发放到关键联系人手中，利用他们广泛的社会关系发放问卷。问卷发放时间为 2016 年 6 月至 2016 年 9 月底，历时三个月，共发放问卷 390 份，实际回收 247 份，回收率为 63.33%，剔除其中 38 份填写不完整或明显带有规律性的无效问卷后，最终得到 213 份有效问卷，有效回收率为 86.23%。样本基本情况见表 1。

表 1　样本基本情况汇总

基本特征	具体类别	比例/%	基本特征	具体类别	比例/%
性别	男	62.44	地区分布	东北地区	22.54
	女	37.56		东部地区	31.92
				中部地区	35.68
				西部地区	9.86

续表

基本特征	具体类别	比例/%	基本特征	具体类别	比例/%
教育程度	高中及以下	13.62	企业规模	5 人及以下	20.66
	大专	31.92		6～20 人	46.95
	本科	42.72		21～50 人	16.43
	硕士及以上	11.74		51～100 人	8.92
				100 人以上	7.04
工作年限	3 年以下	12.68	行业	制造业	22.54
	4～6 年	23.94		服务业	16.43
	7～9 年	12.21		信息技术行业	26.76
	10 年以上	51.17		商贸行业	12.21
				生物及医药	2.82
				新能源	3.76
				其他	15.49

注：数据之和不为 100%是数据修约所致

2. 变量测量

本文采用 Likert 5 点打分法测量变量，1～5 分别代表完全不同意、不同意、不确定、同意、完全同意。要求问卷填写者根据题目与自身及样本企业实际情况的相符程度进行打分。

1）创业拼凑

本文对于创业拼凑的测量采用 Senyard 等开发的量表[49]，包含八个测量题项，分别为“当面对新挑战时，通过利用现有资源我们有信心能够找到有效的解决方案（EB1）”“相比其他企业，我们能够利用现有资源去应对更多的挑战（EB2）”“我们善于应用任何现有资源去应对创业中遇到的新问题或机会（EB3）”“我们通过整合现有资源和从外部获得的其他廉价资源来应对新的挑战（EB4）”“当应对新问题或机会的时候，我们假设能够找到可行的解决方案并采取行动（EB5）”“通过整合现有资源，我们能够成功应对任何新的挑战（EB6）”“当面对新挑战时，我们通过组合现有资源获得可行的解决方案（EB7）”“我们通过整合那些原本计划用于其他目的的现有资源来成功应对新挑战（EB8）”。

2）创业导向

本文参考 Covin 和 Slevin 的研究[34]，从创新性、风险承担性和超前行动性三个维度测量创业导向。测量题项分别为“本企业更加倾向于强调研发、技术领先和创新（EO1）”“本企业在最近三年内推出了许多新产品或新服务（EO2）”“本企业自创立后，对产品和服务进行了大幅度的创新（EO3）”“本企业更倾向于选择高风险、高回报的项目（EO4）”“基于外部环境，本企业更倾向于在日常经营中采取大胆、迅速的行动来实现企业目标（EO5）”“当面对不确定情境进行决策时，本企业更倾向于采取大胆、积极的态度去把握潜在机会（EO6）”“在同行竞争中，通常是本企业先采取行动，然后其他竞争对手再跟进或回应（EO7）”“本企业倾向于作为‘领导者’，经常比同行率先推出新产品、抢先进入新市场或率先应用新的管理模式或新技术（EO8）”“本企业先于其他竞争者引入新产品/新服务（EO9）”。

3）创业学习

本文参考单标安等的研究[25]，从经验学习、认知学习和实践学习三个维度对创业学习进行测量。具体的测量题项为“不断反思过去的失败行为（EL1）”“失败行为并不可怕，关键是能从中吸取教训（EL2）”“已有的经验（管理经验、创业经验等）对本企业的创业决策并不重要（反向）（EL3）”“经常与企业成员进行交流（EL4）”“经常参与企业内部或外部各种正式或非正式的讨论会（EL5）”“经常阅读相关书籍和文献以获取有价值的创业信息（EL6）”“创业过程中持续收集有关内外部环境的信息（EL7）”“注重在创业实践中深化已有的创业知识（EL8）”“通过持续的创业实践来反思或纠正已有的经验（EL9）”。

4）控制变量

为了控制其他因素对研究结果的影响，本文根据现有研究选取企业年龄、产业、企业规模，创业者受教育程度、工作年限和创业经验作为控制变量。用企业员工人数近似代表企业规模，使用创业次数代表创业经验。

四、数据分析与结果

1. 同源偏差和无应答偏差分析

由于调查问卷中的所有问题都由同一被试填写，可能出现同源偏差问题。因此，本文使用 Harman 单因子检验法进行检验[50]，即对问卷中的所有测量题项进行探索性因子分析，结果显示在未旋转成分矩阵中，第一主成分对总方差的解释率为 25.847%，进而说明同源性误差对于本研究样本数据的影响并不严重。

为检验问卷的无应答偏差，本文参照 Armstrong 和 Overton 的建议[51]，按照问卷回收的先后顺序选取早期回收和晚期回收两部分样本，并对它们的企业规模、企业年龄和所处行业等客观变量进行独立样本 t 检验。结果发现两组样本不存在显著性差异，从而表明无应答偏差问题并未对本文造成严重影响。

2. 信度和效度检验

1）信度检验

本文使用 Cronbach's α 系数检验问卷信度。通常，Cronbach's α 系数达到 0.8 以上说明信度达到要求，问卷具有较强的可靠性。本文的信度检验结果显示，创业拼凑量表的 Cronbach's α 系数为 0.828，创业导向为 0.860，创业学习为 0.839。因此各量表的题项之间具有较好的内部一致性，信度良好。

2）效度检验

效度检验通常包括内容效度和结构效度。由于本文的测量量表主要借鉴国内外主流期刊上被广泛采用的成熟量表，并基于小样本预测试进行了相应的调整，因此具有较高的内容效度。对于结构效度检验，罗胜强和姜嬿认为检验有理论预期或前人已有的成熟量表时可直接采用验证性因子分析[52]。因此，本文使用结构方程模型对创业拼凑、创新性、风险承担性、超前行动性和创业学习进行验证性因子分析，检验结果如表 2 所示。相比其他竞争模型，五因子模型的拟合效果最好，说明变量之间具有较好的区分效度。

表 2　验证性因子分析结果

模型	x^2	df	x^2/df	RMSEA	RMR	CFI	IFI
EB，IN，RK，PV，EL	438.461	289	1.517	0.049	0.031	0.920	0.922
EB，IN+PV，RK，EL	448.738	293	1.532	0.050	0.031	0.917	0.918
EB，IN+PV+RK，EL	451.115	294	1.534	0.050	0.031	0.916	0.918
EB+IN+RK+PV，EL	886.655	298	2.975	0.097	0.053	0.686	0.691
EB+IN+RK+PV+EL	1199.398	299	4.011	0.119	0.072	0.520	0.527

注：创业拼凑=EB，创新性=IN，风险承担性=RK，超前行动性=PV，创业学习=EL，卡方自由度=x^2/df，渐进残差均方和平方根=RMSEA，残差均方和平方根=RMR，比较适配指数=CFI，修正拟合指数=IFI

3. 相关性分析

从表 3 可看出，创业拼凑与创新性（$r=0.322$，$P<0.01$）、风险承担性（$r=0.187$，$P<0.01$）、超前行动性（$r=0.251$，$P<0.01$）和创业学习（$r=0.434$，$P<0.05$）显著正相关；创业学习与创新性（$r=0.207$，$P<0.01$）、风险承担性（$r=0.192$，$P<0.01$）和超前行动性（$r=0.214$，$P<0.01$）显著正相关。因此，初步验证本文假设，可进一步进行假设检验。

表 3　描述性统计与变量间相关系数

变量	1	2	3	4	5	
1CAG	1					
2IND	0.174*	1				
3CS	0.295**	0.076	1			
4EDU	−0.056	0.014	0.092	1		
5WY	0.080	−0.112	0.026	−0.109	1	
6EEX	−0.056	0.007	−0.015	−0.180**	0.137*	
7EB	−0.012	0.173*	0.078	0.102	−0.096	
8IN	0.018	−0.020	0.114	−0.059	0.046	
9RK	0.010	−0.030	0.189**	−0.111	0.060	
10PV	0.040	−0.048	0.120	−0.013	0.126	
11EL	−0.004	0.182**	0.063	−0.030	0.038	
M	1.99	3.30	2.35	2.54	3.03	
SD	0.841	2.012	1.117	0.893	1.124	
变量	6	7	8	9	10	11
6EEX	1					
7EB	0.024	1				
8IN	0.071	0.322**	1			
9RK	0.098	0.187**	0.575**	1		
10PV	−0.027	0.251**	0.667**	0.563**	1	
11EL	0.208**	0.434*	0.207**	0.192**	0.214**	1
M	0.86	3.86	3.72	3.41	3.52	4.07
SD	1.131	0.429	0.669	0.661	0.637	0.429

注：CAG 表示企业年龄，IND 表示产业，CS 表示企业规模，EDU 表示教育程度，WY 表示工作年限，EEX 表示创业经验，EB 表示创业拼凑，IN 表示创新性，RK 表示风险承担性，PV 表示超前行动性，EL 表示创业学习；$n=213$

** 为 $P<0.01$，* 为 $P<0.05$

4. 假设检验与结果

本文采用阶层回归分析法进行假设检验，结果如表 4 所示。模型 1 为控制变量对创业学习的回归分析，模型 2 为主效应检验，结果表明创业拼凑对创业学习（β=0.470，P<0.001）有显著的正向影响，H1 得到验证。模型 1、3、4 用于检验创新性的调节作用，模型 4 的结果表明乘积项“创业拼凑×创新性”的回归系数为负且显著（β=−0.529，P<0.05），这说明创新性在创业拼凑与创业学习之间起负向调节作用，H2a 没有得到验证。模型 1、5、6 用于检验风险承担性的调节作用，模型 6 的结果表明乘积项“创业拼凑×风险承担性”的回归系数为负但不显著（β=−0.055，P>0.05），H2b 没有得到验证。模型 1、7、8 用于检验超前行动性的调节作用，模型 8 的结果表明乘积项“创业拼凑×超前行动性”的回归系数为负且显著（β=−0.975，P<0.001），这说明超前行动性在创业拼凑与创业学习之间发挥着负向调节作用，H2c 没有得到验证。

表 4　阶层回归分析

变量	模型 1	模型 2	模型 3	模型 4	模型 5	模型 6	模型 7	模型 8
CAG	−0.218	−0.122	−0.122	−0.128	−0.103	−0.110	−0.129	−0.224
IND	0.360**	0.227	0.236	0.236	0.239*	0.240*	0.247*	0.284**
CS	0.224	0.117	0.095	0.095	0.050	0.049	0.074	0.058
EDU	−0.009	−0.173	−0.149	−0.118	−0.117	−0.114	−0.150	−0.072
WY	0.117	0.220	0.209	0.241	0.205	0.210	0.161	0.175
EEX	0.682**	0.614**	0.605**	0.630**	0.593**	0.594**	0.638**	0.546**
EB		0.470***	1.527***	1.517***	1.536***	1.533***	1.479***	1.390***
IN			0.240	0.241				
RK					0.366	0.362		
PV							0.469	0.491**
EB×IN				−0.529*				
EB×RK						−0.055		
EB×PV								−0.975***
R^2	0.081	0.245	0.249	0.272	0.253	0.254	0.258	0.334
ΔR^2	0.081	0.164	0.168	0.024	0.172	0.000	0.177	0.076
F	3.029**	9.519***	8.439***	8.436***	8.654***	7.663***	8.890***	11.336***

注：CAG 表示企业年龄，IND 表示产业，CS 表示企业规模，EDU 表示教育程度，WY 表示工作年限，EEX 表示创业经验，EB 表示创业拼凑，IN 表示创新性，RK 表示风险承担性，PV 表示超前行动性，EB×IN 表示创业拼凑×创新性，EB×RK 表示创业拼凑×风险承担性，EB×PV 表示创业拼凑×超前行动性；n=213

*** 为 P<0.001，** 为 P<0.01，* 为 P<0.05

五、讨论

基于数据分析的结果可明显看出，H1 通过验证，这表明创业拼凑可以促进新创企业的创业学习，创业者能够从拼凑行为中学习到相关知识。而 H2a、H2b 和 H2c 均没有得到验证，结果显示创新性和超前行动性在创业拼凑与创业学习的关系中发挥着负向调节作用，这刚好与前文的理论假设相反，而风险承担性并没有起调节作用。因此，这值得我们进行深入的分析与讨论。通常，创新性倾向较高的企业重视创新能力的提高，会将更多的资源投入到研发活动上[53]，此类企业对于知识的专业性要求较高，从而增加了对专业性技术知识的需求，而对与其他资源相关知识的需求可能会相应地减少。因此，在创新性较高的新创企业中，创业者会减少在拼凑行动中的学习，进而减弱了创业拼凑对创业学习的影响。超前行动性较高的新创企业为了能更好地对未来需求的变化做出预期，通常较为关注市场与环境变化，它们不断进行环境扫描[18,54]，以及时获取市场与环境信息[55]。因此，这类企业往往对与市场和外部环境相关的知识具有较高的需求，从而减弱拼凑行为中的学习。风险承担性较高的企业通常将主要精力投入在最大限度地提高投资回报率上，以尽可能在短期内实现企业成长，因此对于企业内部相关的学习活动往往给予较少关注甚至不关注[56]，进而表现为风险承担性在创业拼凑与创业学习的关系中并没有显著的调节作用，而这也表明高风险行动可能并一定有利于创业学习[18]。

六、研究结论与展望

1. 研究结论

创业拼凑有利于新创企业战胜资源短缺困境，并在其不断的尝试与试错过程中为企业带来资源相关的隐性知识，但是创业者是否能够学到这些知识呢？在怎样的情况下才更有利于其对拼凑活动的学习呢？为了解答上述问题，本文研究了创业拼凑对创业学习的影响，并根据现有研究将创业导向的三个维度——创新性、风险承担性和超前行动性作为创业拼凑对创业学习影响的调节变量。最终的研究结果表明创业拼凑有利于创业学习，创业导向的创新性和超前行动性维度在创业拼凑与创业学习的关系中发挥着负向调节作用，而风险承担性在其中并没有起调节作用。

2. 理论贡献

本文的理论贡献主要体现在以下方面。首先，以往研究大多关注拼凑者先前积累的经验与知识对创业拼凑所发挥的重要作用，忽视了拼凑本身作为一种不断尝试与试错的过程对于学习活动所产生的影响。本文从学习和知识的角度切入，探究创业拼凑对新创企业创业学习的内在影响，从而丰富了创业拼凑的现有研究。其次，尽管以往研究对于创业学习影响因素的探讨日趋丰富，包括创业者的认知与先前经验、创业网络、战略导向、外部环境等，但却忽略了拼凑所带来的知识资源同样具有较高的学习价值。因而，通过本文的研究，明确了创业拼凑与创业学习的关系，完善并丰富了创业学习相关理论。最后，以往研究多将创业导向作为创业学习的驱动因素却较少考虑当深入到具体领域的学习活动时，创业导向所产生的影响。本文将创业导向作为创业拼凑与创业学习关系的调节变量，通过实证研究发现当深入探讨某一领域的学习行为时，创业导向所发挥的作用还应具体情况具体分析。

3. 管理启示

通过本文启示创业者，在采取拼凑行为后应积极反思，获取拼凑经验，学习并真正掌握相关的资源知识。尽管本文的研究结论表明创新性在创业拼凑与创业学习的关系中发挥着负向调节作用，并认为主要原因在于创新性较高的企业对于专业性技术知识的高需求使其减少了对创业拼凑的学习。但其实，专业化知识具有较高的知识深度（knowledge depth），这种专业领域的深层知识可能会产生认知惰性[57]，不利于企业的高程度创新，只会将其局限于当前市场或对现有技术细枝末节的变动中[58]。相反，多元化的知识更可能带来前沿理念和新颖的知识组合[59]，这有助于企业理解新信息和潜在变化，提高企业洞悉前沿技术或市场机会的能力，从而促进创新[60]。因此，偏好创新的创业者应该积极拓展知识的获取方式与途径，不应只关注专业性知识，而应该利用一切可能的机会学习，尤其对于像拼凑这种能够产生隐性知识的创业行为更应该重视，以丰富自身及企业的知识库，从而更好地为企业生存服务。此外，新创企业的生存与成长都需要大量知识资源予以支撑，但本文的结论表明高风险承担性策略并不能对拼凑行为中的学习产生促进作用，从而无法弥补企业的知识缺口[59]，因此新创企业在实施风险承担性战略导向的同时应该重视学习与相关知识的获取。

4. 研究不足与展望

尽管本文具有一定的理论和实践意义，但仍存在着一些不足之处。首先，本文仅考察了创业导向在创业拼凑与创业学习关系中所发挥的调节作用，而未考虑外部环境因素对于这种学习行为所产生的影响，如环境动态性。目前我国正处于经济转型时期，这一背景下的新创企业所面临的环境动荡性较强，因此有必要研究环境因素对于拼凑行为中的学习活动所产生的影响。其次，本文主要关注的是个体拼凑行为，现有研究已将创业拼凑上升到团队和组织层面，因此未来有必要考虑如何在团队与组织层面开展对拼凑行为的学习，如共同拼凑涉及跨组织的拼凑活动，那么采取怎样的学习方式才能使参与拼凑的企业从中所获得的知识最大化将会成为一个有趣的研究问题。最后，本文使用的是横截面样本数据，这可能会导致潜在的内生性问题，比如，变量之间的逆向因果关系。本文发现创业拼凑会对创业学习产生积极影响，但反过来，通过创业学习获取的经验与知识同样能够指导创业者实施更加有效的拼凑。因此，未来可考虑采用纵向研究设计对本文变量之间的因果关系进行更加深入的探讨。

参考文献

[1] Aldrich H E. Organizations Evolving. Los Angeles：Sage，1999.

[2] 方世建，黄明辉. 创业新组拼理论溯源、主要内容探析与未来研究展望. 外国经济与管理，2013，35（10）：2-12.

[3] Baker T，Nelson R E. Creating something from nothing：Resource construction through entrepreneurial bricolage. Administrative Science Quarterly，2005，50（3）：329-366.

[4] Senyard J M，Baker T，Steffens P，et al. Bricolage as a path to innovativeness for resource-constrained new firms. Journal of Product Innovation Management，2014，31（2）：211-230.

[5] 祝振铎，李非．创业拼凑对新企业绩效的动态影响——基于中国转型经济的证据．科学学与科学技术管理，2014，35（10）：124-132.

[6] 列维-斯特劳斯 C．野性的思维．李幼蒸，译．北京：中国人民大学出版社，2006.

[7] Weick K E. Making Sense of the Organization. New Jersey：Wiley-Blackwell，2000.

[8] Cunha M P. Bricolage in organizations. Lisboa：Universidade Nova de Lisboa，2005.

[9] van de Walle S. Building resilience in public organizations：The role of waste and bricolage. Innovation Journal，2014，19（1）：1-18.

[10] Steffens P R，Baker T，Senyard J. Betting on the underdog：Bricolage as an engine of resource advantage. Montreal：Proceeding of Annual Meeting of the Academy of Management，2010.

[11] di Domenico M L，Haugh H M，Tracey P. Social bricolage：Theorizing social value creation in social enterprises. Entrepreneurship Theory & Practice，2010，34（4）：681-703.

[12] Nonaka I. A dynamic theory of organizational knowledge creation. Organization Science，1994，5（1）：14-37.

[13] Wiklund J，Shepherd D. Knowledge-based resources，entrepreneurial orientation，and the performance of small and medium-sized businesses. Strategic Management Journal，2003，24（13）：1307-1314.

[14] Minniti M，Bygrave W. A dynamic model of entrepreneurial learning. Entrepreneurship Theory & Practice，2001，25（3）：5-16.

[15] 朱秀梅，孔祥茜，鲍明旭．学习导向与新企业竞争优势：双元创业学习的中介作用研究．研究与发展管理，2014，26（2）：9-16.

[16] 尹苗苗，刘玉国．新企业战略倾向对创业学习的影响研究．科学学研究，2016，34（8）：1223-1231.

[17] 陈文婷，惠方方．创业导向会强化创业学习吗——不同创业导向下创业学习与创业绩效关系的实证分析．南方经济，2014，（5）：69-81.

[18] Wang C L. Entrepreneurial orientation，learning orientation，and firm performance. Entrepreneurship Theory & Practice，2008，32（4）：635-657.

[19] Duymedjian R，Ruling C C. Towards a foundation of bricolage in organization and management theory. Organization Studies，2010，31（2）：133-151.

[20] Rönkkö M，Peltonen J，Arenius P. Selective or parallel? Toward measuring the domains of entrepreneurial bricolage//Corbett C，Katz J A. Advances in Entrepreneurship，Firm Emergence and Growth. Bingley：Emerald Group Publishing，2013：43-61.

[21] Senyard J M，Davidsson P，Steffens P R. Venture creation and resource processes：Using bricolage sustainability ventures. Queensland：Proceedings of 7th AGSE International Entrepreneurship，2010.

[22] Smilor R W. Entrepreneurship：Reflection on a subversive activity. Journal of Business Venturing，1997，12（5）：341-346.

[23] Rae D，Carswell M. Towards a conceptual understanding of entrepreneurial learning. Journal of Small Business & Enterprise Development，2001，8（2）：150-158.

[24] Politis D. The process of entrepreneurial learning：A conceptual framework. Entrepreneurship Theory & Practice，2005，29（4）：399-424.

[25] 单标安，蔡莉，陈彪，等．中国情境下创业网络对创业学习的影响研究．科学学研究，2015，33（6）：899-906，914.

[26] Holcomb T R, Ireland R D, Holmes R M, et al. Architecture of entrepreneurial learning: Exploring the link among heuristics, knowledge, and action. Entrepreneurship Theory & Practice, 2009, 33 (1): 167-192.

[27] Rowe P A, Christie M J. Civic entrepreneurship in Australia: Opening the "black box" of tacit knowledge in local government top management teams. International Journal of Public Sector Management, 2008, 21 (5): 509-524.

[28] 彭杨. 高技术企业创业学习及其影响因素研究. 杭州：浙江大学，2004.

[29] Chandler G N, Lyon D W. Involvement in knowledge-acquisition activities by venture team members and venture performance. Entrepreneurship Theory & Practice, 2009, 33 (3): 571-592.

[30] 胡望斌，张玉利. 新企业创业导向转化为绩效的新企业能力：理论模型与中国实证研究. 南开管理评论，2011，14 (1)：83-95.

[31] Senyard J M, Davidsson P, Baker T, et al. Resource constraints in innovation: The role of bricolage in new venture creation and firm development. Maritz: Proceedings of the 8th AGSE International Entrepreneurship Research Exchange, 2011.

[32] Petkova A P. A theory of entrepreneurial learning from performance errors. International Entrepreneurship & Management Journal, 2009, 5: 345-367.

[33] Tsai W, Wu C H. Knowledge combination: A cocitation analysis. Academy of Management Journal, 2010, 53 (3): 441-450.

[34] Covin J G, Slevin D P. Strategic management of small firms in hostile and benign environments. Strategic Management Journal, 1989, 10 (1): 75-87.

[35] Lumpkin G T, Dess G G. Clarifying the entrepreneurial orientation construct and linking it to performance. Academy of Management Review, 1996, 21 (1): 135-172.

[36] Miller D. The correlates of entrepreneurship in three types of firms. Management Science, 1983, 29 (7): 770-791.

[37] Wales W J, Patel P C, Parida V, et al. Nonlinear effects of entrepreneurial orientation on small firm performance: The moderating role of resource orchestration capabilities. Strategic Entrepreneurship Journal, 2013, 7 (2): 93-121.

[38] Grant R M. Toward a knowledge-based theory of the firm. Strategic Management Journal, 1996, 17 (S2): 109-122.

[39] Schoonhoven C B, Eisenhardt K M, Lyman K. Speeding products to market: Waiting time to first product introduction in new firms. Administrative Science Quarterly, 1990, 35 (1): 177-207.

[40] Nonaka I, Takeuchi H. The Knowledge-Creating Company. New York: Oxford University Press, 1995.

[41] 何一清，崔连广，张敬伟. 互动导向对创新过程的影响：创新能力的中介作用与资源拼凑的调节作用. 南开管理评论，2015，18 (4)：96-105.

[42] Lumpkin G T, Dess G G. Linking two dimensions of entrepreneurial orientation to firm performance: The moderating role of environment and industry life cycle. Journal of Business Venturing, 2001, 16 (5): 429-451.

[43] Drucker P F. Post-Capitalist Society. London: Routledge, 1993.

[44] 刘预. 创业导向对新企业资源获取的影响：基于中国转型经济背景的研究. 长春：吉林大学，2008.

[45] Slocum J W，Jr，Mcgill M，Lei D T. The new learning strategy：Anytime，anything，anywhere. Organizational Dynamics，1994，23 (2)：33-47.

[46] Rhee J，Park T，Lee D H. Drivers of innovativeness and performance for innovative SMEs in South Korea：Mediation of learning orientation. Technovation，2010，31 (1)：65-75.

[47] Zahra S A. A conceptual model of entrepreneurship as firm behavior：A critique and extension. Entrepreneurship Theory & Practice，1993，17 (4)：5-21.

[48] McDougall P，Robinson R B，Jr. New venture strategies：An empirical identification of eight 'archetypes' of competitive strategies for entry. Strategic Management Journal，1990，11 (6)：447-467.

[49] Senyard J M，Baker T，Davidsson P. Entrepreneurial bricolage：Towards systematic empirical testing. Frontiers of Entrepreneurship Research，2007，29 (5)：1-14.

[50] Podsakoff P M，Organ D W. Self-reports in organizational research：Problems and prospects. Journal of Management，1986，12 (4)：531-544.

[51] Armstrong J S，Overton T S. Estimating nonresponse bias in Mail Surveys. Journal of Marketing Research，1977，14 (3)：396-402.

[52] 罗胜强，姜嬿. 管理学问卷调查研究方法. 重庆：重庆大学出版社，2014：149.

[53] Lee D H，Choi S B，Kwak W J. The effects of four dimensions of strategic orientation on firm innovativeness and performance in emerging market small-and medium-size enterprises. Emerging Market Finance & Trade，2014，50 (5)：78-96.

[54] Daft R L，Weick K E. Toward a model of organizations as interpretation systems. Academy of Management Review，1984，9 (2)：284-295.

[55] Levinthal D A，March J G. The myopia of learning. Strategic Management Journal，1993，14 (S2)：95-112.

[56] 林筠，孙晔，何婕. 吸收能力作用下创业导向与企业成长绩效关系研究. 软科学，2009，23 (7)：135-140.

[57] Tripsas M，Gavetti G. Capabilities，cognition，and inertia：Evidence from digital imaging. Strategic Management Journal，2000，21 (10-11)：1147-1161.

[58] Christensen C M，Bower J L. Customer power，strategic investment，and the failure of leading firms. Strategic Management Journal，1996，17 (3)：197-218.

[59] Taylor A，Greve H R. Superman or the fantastic four? Knowledge combination and experience in innovative teams. Academy of Management Journal，2006，49 (4)：723-740.

[60] Chesbrough H W. Open Innovation：The New Imperative for Creating and Profiting from Technology. Boston：Harvard Business School Press，2003.

澳门会展业的经济效应与发展策略研究①

一、前言

会展业作为现代服务业的重要组成部分，其对本地经济的带动效应十分显著，学术界一般认为会展业的经济效应带动系数在 1∶10～1∶5。Kim 等运用投入产出模型评估了韩国会议业在产出、就业、收入和增加值等方面的经济影响，研究结果显示会议业是韩国国民经济的高产出领域和重要经济带动源泉[1]。之后 Kim 和 Chon 对韩国展览业也进行了同样的经济影响分析，得出的结论与前者类似[2]。国内研究中，胡平和杨杰以上海新国际博览中心为例，通过实证研究得出上海展览业的经济效应带动系数为 1∶8.4[3]。石美玉和王春才则对北京会展旅游的经济效应进行了测算，结果显示会展旅游对相关行业的经济效应带动系数达到了 1∶8.44[4]。

早在 2006 年，澳门特区政府就明确提出以会展业为主业促进澳门经济多元化，2016 年 9 月，澳门特区政府颁布《澳门特别行政区五年发展规划（2016—2020 年）》，继续坚持澳门着力发展会展业以推动澳门经济适度多元化的既定方针。近十年来，澳门会展业得到了全面和长足的发展，“国际基础设施投资与建设高峰论坛”“中国-葡语国家经贸合作论坛部长级会议”“APEC 亚太经合组织旅游部长会议”等多个高规格会议在澳门召开。与此同时与澳门会展业相关的学术研究也蓬勃兴起，但多数都偏重于宏观政策研究，而较少运用统计数据进行定量分析，有关澳门会展业经济效应的学术研究则更加缺乏。其中较有代表性的是陈章喜和王江运用灰色关联分析方法，对澳门会展业（2000～2009 年）进行经济效应的实证研究，结果表明会展业对澳门经济和相关产业的带动作用是十分显著的[5]。

2010 年之后澳门会展业增长加速，至 2015 年统计数据显示“会展活动与会及入场人次”已达到近 252 万，约是 2009 年的 4 倍。此外，在“一带一路”倡议和粤港澳大湾区建设的新形势下，澳门经济及会展业迎来了新的历史发展时期。因此，本文将运用灰色关联分析方法，并根据 2010～2015 年澳门经济及会展业的相关统计数据，对澳门会展业的经济效应进行实证分析，并结合当前新发展形势提出若干对策建议，为澳门经济及会展业的发展政策提供理论支持。

二、澳门会展业经济带动效应的实证分析

（一）研究模型与理论基础

陈章喜和王江认为，产业之间的关联关系是经济带动效应评价的理论基础，即某种产业的运作会波及其他相关产业，直接或者间接导致其他产业的变化，进而对国民经济整体运行产生影响[5]。Kim 等[1]以及 Kim 和 Chon[2]对韩国会展业经济效应的量化

① 本文是刘人怀学术交流促进会 2017 年年会暨澳门工程科技及管理科学学术研讨会专题报告，澳门，2017 年 12 月 16 日．作者：刘人怀，张书莲，冉丽．

分析则涉及零售、餐饮、酒店、娱乐、旅游、广告、交通以及设备租赁等多个行业。因此，会展业与其他产业之间的关联度可以反映出会展业的经济带动效应，同理，会展业与国民经济主要指标之间的关联度也可以体现其对国民经济的影响效应。

本文采用的研究方法是灰色关联分析法，其基本思想是根据序列曲线几何形状的相似程度来判断其联系是否紧密，曲线几何形状越接近则关联度就越大，反之就越小[6]。关联度是因素之间关联性大小的量度，关联度分析事实上是动态过程发展态势的量化比较分析。灰色关联分析法的最大优点是它对数据量没有太高的要求，即使是少量的数据也可以进行分析。其计算过程如下。

步骤一，对原始数列进行无量纲化。一般采用均值化处理，即用各个序列的平均值除以各个序列的原始数据，所得到的商为新的数据列。即令 $X_i(k)=\dfrac{x_i(k)}{\overline{x}_i}$，其中 $\overline{x}_i=\dfrac{1}{n}\sum_{k=1}^{n}x_i(k)$，$k$ 为时刻，i 为因素指标。

步骤二，计算各个时刻会展业序列与其他序列的绝对差，并求出最大差值和最小差值。$\Delta_i(k)=|X_0(k)-X_i(k)|$，进而得出 $M=\max\limits_i\max\limits_k\Delta_i(k)$，$m=\min\limits_i\min\limits_k\Delta_i(k)$。

步骤三，求出会展业与其他相关因素指标的关联系数。即 $r_{0i}(k)=\dfrac{m+\xi M}{\Delta_i(k)+\xi M}$，其中，$\xi$ 为分辨系数，$\xi\in[0,1]$，一般取值 0.5，引入它可减少极值对计算的影响。

步骤四，计算会展业与其他相关因素指标的关联度。由于关联系数的数比较多，不便于比较，因此需要将各个时刻的关联系数集中到一个值，一般是取平均值。公式表达为：$r_{0i}=\dfrac{1}{n}\sum_{k=1}^{n}r_{0i}(k)$。

（二）澳门会展业对经济总体的带动效应分析

1. 指标选取与数据来源

评价澳门会展业对澳门经济总体的带动效应，须计算会展业与经济总体之间的关联度。澳门会展业的发展指标主要有两个："会展活动数目"和"会展活动与会及入场人次"，澳门经济总体则选取"本地生产总值""就业人数"和"固定投资形成总额"等三个指标，分别反映澳门经济总量、就业和投资的发展情况，本次研究的时间段为2010～2015年，所有原始数据均来源于澳门统计暨普查局公布的统计资料（表1）。

表1　澳门会展业与本地生产总值、就业、投资发展情况（2010～2015年）

年份	会展活动数目/个	会展活动与会及入场人次	本地生产总值/百万澳门元	就业人数/千人	固定投资形成总额/百万澳门元
2010	1 512	809 575	226 941	314.8	28 357
2011	1 035	1 277 309	295 046	327.6	36 614
2012	1 015	1 612 596	348 216	343.2	46 518
2013	1 024	2 033 619	411 839	361	54 928
2014	1 050	2 614 473	443 468	388.1	83 227
2015	1 263	2 516 092	368 728	396.5	85 290

资料来源：澳门统计暨普查局

2. 实证检验结果与分析

运用灰色关联分析法对澳门 2010～2015 年会展业和经济总体的相关数据进行关联度测算，所得的检验结果如表 2 所示。从表 2 可以看出，会展业与本地生产总值、就业和投资之间均有较强的关联度，最低为 0.6187，最高可达到 0.8985，这说明会展业的发展可以显著带动澳门经济总量、就业和投资的增长。

表 2 澳门会展业与经济总量、就业以及投资的关联度

会展业	本地生产总值	就业人数	固定投资形成总额
会展活动数目	0.7057	0.8134	0.6187
会展活动与会及入场人次	0.7189	0.6237	0.8985

从会展业的两个发展指标来看，会展活动与会及入场人次对固定投资形成总额的带动效应最为显著（关联度 0.8985），同时其对本地生产总值的带动效应（关联度 0.7189）也略高于会展活动数目（关联度 0.7057）。这是因为会展活动与会及入场人次直接影响消费需求的规模和增长，人次越多则对会展场馆、酒店客房、交通运输以及饮食零售等相关行业的消费需求越大，由此带动相关行业的投资规模和经济增长。另外，会展活动数目对就业人数的带动效应（关联度 0.8134）高于会展活动与会及入场人次（0.6237），也就是说会展活动数目的增加可更有效地提高本地就业率。

（三）澳门会展业对相关产业的带动效应分析

1. 指标选取与数据来源

评价澳门会展业对其他相关产业的带动效应，须计算会展业与相关产业之间的关联度。本文选取的相关产业包括旅游业、酒店业、饮食业、零售业、运输及仓储业、通信业、博彩业和广告业等八个行业。会展业指标则选取“会展活动与会及入场人次”。研究的时间段为 2010～2015 年，所有原始数据均来源于澳门统计暨普查局公布的统计资料（表 3）。

表 3 澳门会展业与相关行业发展情况（2010～2015 年）

年份	会展活动与会及入场人次	旅游业/千人次	酒店业/千澳门元	饮食业/千澳门元	零售业/千澳门元	运输及仓储业/千澳门元	通信业/千澳门元	博彩业/百万澳门元	广告业/千澳门元
2010	809 575	24 965.4	6 554 238	5 505 457	29 753 947	12 999 322	4 162 470	190 621	344 587
2011	1 277 309	28 002.3	8 456 206	6 094 202	40 159 463	14 814 232	5 843 190	270 113	368 997
2012	1 612 596	28 082.3	10 249 228	6 899 498	53 314 680	16 171 904	7 064 416	306 487	399 078
2013	2 033 619	29 324.8	11 895 243	8 402 675	65 908 544	17 425 031	7 531 815	362 745	558 561
2014	2 614 473	31 525.6	13 601 921	9 625 680	68 749 103	19 412 168	8 386 788	353 645	854 029
2015	2 516 092	30 714.6	12 214 636	10 068 461	65 899 000	18 942 000	9 748 000	233 229	917 000

资料来源：澳门统计暨普查局

注：旅游业数据为入境旅客人数，酒店业数据为客房租金

2. 实证检验结果与分析

一项会展活动的开展将能够带动本地多个相关产业的经营活动，尤其是酒店、饮食、零售、旅游、交通物流以及广告等，此外由于澳门博彩业在经济结构中的重要地

位，也需将其包括在内。运用灰色关联分析法对澳门 2010～2015 年会展业和相关行业的经济数据进行关联度测算，所得的检验结果如图 1 所示。由图 1 的关联度排序我们可以看到，会展业对饮食业的带动效应最大，充分体现了“民以食为天”的特点。其次是通信业和零售业，二者相差无几，反映了会展活动对通信业务和物品销售的巨大消费需求。排在第四位的是酒店业，其关联程度与前几位差别也不大。会展业与广告业的关联度也在 0.7 以上，说明前者对后者也具有较强的带动效应。

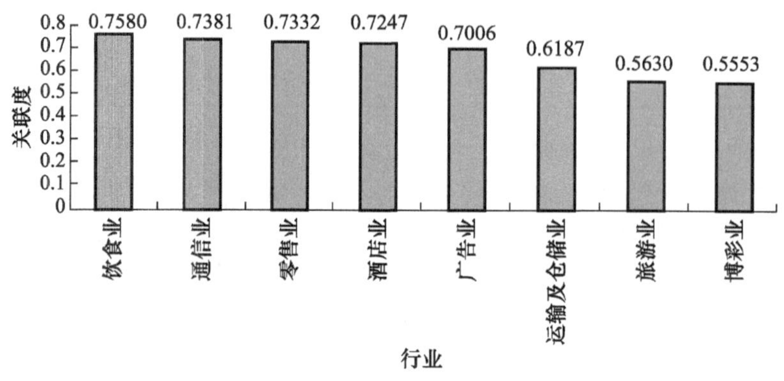

图 1　澳门会展业与相关行业的关联度

在所有测算的八个行业中，会展业与运输及仓储业、旅游业和博彩业的关联度排名最后三位，但最低也在 0.55 以上，从绝对值看仍然具有较强的带动效应。其中，会展业与运输及仓储业的关联度为 0.6187，仍有提升的空间；会展业与旅游业的关联度为 0.5630，说明会展旅游市场尚未完全开发；会展业对博彩业的带动效应最低，这反而有利于改变过去澳门经济博彩业一家独大的局面，提振第三产业内其他现代服务业的发展，从而更快实现澳门适度经济多元化发展的战略目标。值得一提的是，在陈章喜和王江[5]利用 2000～2009 年数据进行测算的实证研究中，澳门会展业和博彩业的关联度是 0.7136，高于会展业和其他产业的关联度，但在本次研究中则降低至 0.5553，排名最后。这是因为 2010 年之后澳门会展业发展迅速，在第三产业中的产值比例不断提升，同时会展业也带动第三产业其他行业同步增长，最终使得博彩业的重要地位逐步降低。

三、澳门会展业经济带动效应的发展策略启示

从上述会展业经济带动效应研究结果我们可以看出，澳门会展业与第三产业内多个行业均有较强的关联性，会展业能够带动其他相关行业的繁荣，而其他行业的经营状况也会反过来影响澳门会展业的发展前景。澳门的城市发展定位是将澳门建设成为“世界休闲旅游中心”和“中国与葡语系国家商贸合作服务平台”，要实现这一目标，不仅需要大力扶持会展业成为新的经济增长点和核心产业之一，同时也要完善第三产业内其他行业的市场环境，从而最大化发挥会展业的经济带动效应。当前正处于“一带一路”和“粤港澳大湾区”建设的历史发展期，澳门应抓住这一机遇，对内完善自身硬件与软件环境，对外加强与粤港以及内地之间的合作，尽快实现自身城市发展目

标。具体来说，有以下几点对策启示。

（一）加强粤港澳大湾区会展合作，开拓“一带一路”沿线国家市场

粤港澳大湾区由“9＋2”城市组成，即广州、深圳、佛山、肇庆、东莞、惠州、珠海、中山、江门等9市和香港、澳门2个特别行政区，其中，广州、深圳、香港是大湾区的经济核心。根据每个城市的发展优势，各个城市的定位与功能也不一样，如香港的定位是“全球金融中心及物流中心”，深圳的定位是“国际创新服务中心”，澳门的发展定位则是“世界休闲旅游中心”，澳门大力发展会展业不仅与自身城市定位十分契合，而且有助于摆脱博彩业一家独大的局面。因此，澳门应抓住大湾区城市群规划契机，发挥本地制度、地理位置以及自然文化资源的优势，根据各个城市的经济发展特点进行资源对接，积极寻求与湾区城市的会展合作机会，争取更多的会展项目在澳门举办。此外，澳门邻近香港和珠海横琴新区，可与之合作举办会展项目。

在“一带一路”倡议下，我国和沿线国家的经贸投资往来快速增长，澳门是海上丝绸之路的重要支点，应当跟随国家战略的发展方向，积极开发“一带一路”沿线国家市场，让澳门成为“一带一路”沿线国家的“世界休闲旅游中心”。在具体策略措施上，澳门应首先调研沿线国家与我国的经贸投资重点领域方向，然后制定有针对性的会展开拓策略，在“中国与葡语系国家商贸合作服务平台”的基础上，将澳门打造成内地和沿线国家企业合作与交流的重要平台之一。在具体会展项目方面，澳门可在已有基础上优先发展几大领域，如基础建设投资、旅游、食品、中医药以及文化交流等，并将之做大做强成为澳门自有展会品牌。

（二）培养会展龙头企业，加强产业联盟合作

在澳门会展业近十年的发展历程中，澳门特区政府作为该行业的领导者，发挥了良好的政策制定与行为引领作用并取得了显著的成效，为澳门会展业的未来成长奠定了坚实的基础。但在市场经济框架下，一个行业的繁荣最终是依靠企业的生命力和竞争力，因此未来政府角色应逐渐向战略导向和市场协调方向转变，减少具体事项的直接参与，扶持培养若干具有国际竞争力和较强实力的大型会展龙头企业。例如，新加坡作为享誉世界的会展之都，拥有新加坡会展服务有限公司以及新加坡会议与展览管理服务有限公司等全球50强会展业巨头，在新加坡、亚太及其他地区成功举办了大量会展项目，拥有丰富的行业经验和顶级的服务水准，在新加坡会展经济发展中发挥了重要的行业标杆作用。

扶持培养会展龙头企业可以考虑中外合资形式，与国际知名会展服务企业合作，在引入资本的同时也提升了澳门本地会展业的服务水平。中外合资形式也有利于开拓国际市场，如在竞投会展项目方面具有品牌优势，因为举办者的综合服务能力也是影响竞投成功的重要因素。此外，澳门本地会展及相关行业企业也可以采用战略联盟的方式合作，以大型企业为核心聚集一批中小企业成立综合性的会展服务集团，化单兵作战为团队协作，通过整合力量提升澳门会展业的综合竞争力。澳门与粤港之间既有竞争又有合作，特别是邻近的珠海和香港，澳门企业可与之合资办企或共同投资成立新的公司。

（三）拓展会展项目类型，适当增加赛事活动和奖励旅游

会展项目类型不仅包括会议与展览，还包括奖励旅游及节事活动。近十年来，在中央政府的大力支持下，澳门特区政府遵循“会议为先”的方针推动会展业发展取得了显著成效，越来越多的国际大型会议在澳门召开，同时以“会”带“展”带动食品、旅游、婚庆、玩具以及汽车等博览会在澳门举办。此外澳门也会因地制宜举办一些节事活动，根据澳门旅游局网站公布的信息，2017 年就有艺术节、荷花节、文化旅游节、美食节等节庆活动，以及龙舟赛、世界女排大奖赛、高尔夫公开赛、国际马拉松等赛事活动。随着近两年澳门会展项目数量及人次增长趋缓，澳门可尝试加大奖励旅游和节事活动的开拓力度，既可以为澳门会展业寻找新的经济增长点，又可以提升城市形象和知名度。

在赛事开拓方面，可以选择一些观众参与度、观赏度和知名度较高的赛事项目，吸引国内外游客来澳门观赛，如足球、乒乓球、网球、帆船等体育赛事以及音乐、舞蹈类文化赛事。在赛事承办方面，如因场地等因素限制不能独立承办，可以和粤港两地合作承办或者作为分会场举办地。近两年澳门加大奖励旅游开发和宣传力度，如 2016 年首次派团参加“亚太区奖励及会议旅游展”。未来不仅要走出去宣传，更要引进来参观，如组织企业代表团体来澳门考察体验，将奖励旅游宣传活动深入到具体企业协会，与当地会展旅游企业加强合作关系等。此外还可以提供奖励旅游个性化定制服务，增强澳门作为奖励旅游目的地的竞争力。

（四）加强交通设施建设，提升交通便利水平

澳门地理面积狭小，交通工具主要依靠公交和的士，此外还有各口岸码头的免费酒店巴士。随着澳门会展旅游业的进一步发展以及与内地的更紧密合作，尤其是“一带一路”和粤港澳大湾区建设所带来的经贸往来，当前的交通运力已明显不足。早在 21 世纪之初，澳门就已经开始计划修建轻轨来完善市内交通条件和舒缓交通拥挤，由于各种原因，工程虽已经开始但何时建成尚未确定。因此，澳门目前可着力先完善公共交通汽车体系，提升交通服务便利水平。

首先是优化当前公交线路，特别是各口岸码头、酒店和旅游景点之间的无缝连接，根据客流量适当增加班次密度，如设立节假日专车、在旅游旺季增加运力等。其次是加强旅游提示服务，在各车站、酒店、景区设立地图和交通指示牌，以便外地来客迅速掌握交通讯息。再次，澳门公交按段收费且不设找零，给外地游客带来不便，可多设立澳门通销售点或零钱兑换点以方便乘车。最后，还可在重要公共区域（如景区、车站及码头）提供免费 Wi-Fi 服务，方便游客查询交通旅游资讯，拥有更周到的商务或旅游服务体验。

（五）完善旅游配套服务，丰富会展旅游产品种类

会展业与旅游业二者关系密切，学术界常常将会展旅游视为一种新兴业态加以研究。梁明珠等在对澳门会展旅游模式的研究中指出，会展旅游发达的城市是因“会”而“游”，而澳门由于博彩业的支柱地位是因“游”而“会”[7]。近十年来，澳门大力发展会展业以降低对博彩业的依赖程度，取得了显著的成效，逐渐从因“游”而“会”向因“会”而“游”转变。但从前述检验结果中我们看到，澳门会展业对旅游业的带

动效应仍然相对较低，二者之间的联动作用尚未充分发挥。因此，在大力发展会展业的同时，还需开发新的旅游产品及服务，最大化宣传和利用本地旅游资源。

首先是旅游景区配套设施的建设问题，目前在澳门的标志性景区，相关的休憩、饮食、零售、游乐和交通设施并非十分充足，大多依赖原有自发的民间经营，政府并未对此进行专项投资建设。这一方面会造成旅游来客的消费需求不能充分满足和挖掘，影响旅游体验和满意度，另一方面澳门旅游业收入和未来发展也受到了限制。其次，澳门可结合本地历史文化统筹各类旅游资源，深入挖掘本地特色开发建设会展旅游新项目或景区，将会展与旅游紧密结合，以增强澳门作为会展举办地的吸引力。再次，可以开发旅游套餐产品，产品涉及行业可以包含会展、饮食、交通、零售、酒店、景点等项目，如参加某一会展项目，可以获得饮食优惠券、赠送礼品、酒店优惠、景点门票等。最后，还需加强对澳门旅游活动项目的宣传力度，特别是互联网移动终端宣传，可以和内地知名平台合作投放植入网络广告，甚至在主要旅游来源地城市投放地面广告等。

（六）注重会展项目质量，提升会展目的地竞争力

自 2007 年开始澳门会展数目大幅增长，年均数量保持在 1000 个以上，比 2006 年增长 3 倍有余。至 2015 年与会及入场人次已达到 250 多万，是 2006 年的 44 倍，2007 年的 8 倍，会展项目平均与会及入场人次也从 2010 年的 535 人次提高到 2015 年的 1992 人次。自 2013 年之后会展数量增长明显放缓，这表明澳门会展业的发展阶段已经从初期的数量快速增长向以质取胜的成熟期转变。因此，未来澳门会展业应当在数量稳步小幅增长的同时，更加注重会展项目的质量水准，提升会展目的地竞争力。

澳门特区政府和相关企业应加强对会展服务质量与绩效的评价，深入开展客户需求和满意度调查研究，及时了解自身不足并制定相应措施予以改进，对具体会展项目的评价也应纳入更多的服务质量与绩效评价指标。例如，王晓敏等以展览企业参展商服务为研究对象，认为展览服务质量评价指标可归纳为物理环境、专业服务及核心利益三个维度[8]；而周杰和何会文则以专业观众认知为研究对象，将会展服务划分为现场、就餐与住宿环境、目的地引导服务、会场交通服务以及公共通信服务等六个因素[9]。由于会展服务涉及第三产业内多个行业，因此需要澳门特区政府统筹规划，制定相应的激励和竞争机制，引导行业提升服务质量与绩效。

参考文献

[1] Kim S S, Chon K, Chung K Y. Convention industry in South Korea: an economic impact analysis. Tourism Management, 2003, 24 (5): 533-541.

[2] Kim S S, Chon K. An economic impact analysis of the Korean exhibition industry. International Journal of Tourism Research, 2009, 11 (3): 311-318.

[3] 胡平，杨杰. 会展业经济拉动效应的实证研究——以上海新国际博览中心为例. 旅游学刊，2006，(11)：81-85.

[4] 石美玉，王春才. 会展旅游带动效应的统计研究——以北京为例. 经济管理，2013，(8)：116-125.

[5]　陈章喜，王江. 澳门会展业的经济效应与粤澳会展业合作. 产经评论，2012，(2)：73-81.
[6]　杜栋，庞庆华，吴炎. 现代综合评价方法与案例精选（第 3 版）. 北京：清华大学出版社，2015.
[7]　梁明珠，钟金凤，廖奇辉. 澳门会展旅游的发展模式及其推进路径. 旅游科学，2012，(2)：77-84.
[8]　王晓敏，胡兵，凌礼. 服务主导逻辑下服务质量评价模型构建与实证研究——以展览企业参展商服务为例. 软科学，2017，31 (3)：111-114.
[9]　周杰，何会文. 会展专业观众的服务认知结构研究——兼论参展动因对服务认知的影响. 旅游学刊，2011，26 (10)：75-81.

创业拼凑、创业学习与新企业突破性创新的关系研究①

1. 问题的提出

新创企业在经济增长中发挥着关键作用[1]，创新是新创企业的核心特点[2]，突破性创新更是对提高新创企业的成功率、激发在位企业采取行动并加快技术变革发挥着重要作用[3-5]。但缺乏经验丰富的人员、社会资本、市场和技术知识、营销技能和资金[6]，以及缺乏合法性、历史绩效记录等一系列“新创劣势”（liability of newness）使得新创企业创新变得困难重重[7]。那么资源有限的新创企业应该如何进行创新呢？创业拼凑理论为解决这一问题带来了启示。

创业拼凑是对手头现有资源的创造性重组[8]，创新是各种要素的重新组合[9]，因此拼凑也是新创企业的一种创新方式[10]。尽管现有研究探讨了创业拼凑对创新的影响[10-11]，但对于拼凑所带来的创新类型学者却有着不同的观点。Ciborra 和 Patriotta[12] 认为拼凑创造的是简单、渐进或不起眼的创新，而非突破性创新。这是因为突破性创新通常需要复杂的资源组合、不同领域的知识与技能，但在很多情况下新创企业很难获取这些资源[13-16]。Senyard 等[17]认为由于拼凑主要是为解决新问题或迎接新挑战创造出新颖的解决方案，因此拼凑所产生的结果就是一种创新，但是这种创新通常是相对平凡的，即使是那些记录在案的最有价值的创新（如丹麦风力涡轮机产业的发展）也并非是突破性创新，相反在很大程度上是渐进性的甚至是临时的应对方案。因此，在资源限制下试图通过拼凑产生突破性创新不太可能成功，反而会削弱由选择性拼凑所带来的一些优势。但越来越多的研究认为“必要性是发明之母”，如何使用资源和拥有何种资源是同样重要的[18]。有学者提出新创企业更适合进行突破性创新[19-20]，这是因为受到资源约束的新创企业具有一定的灵活性，这不仅能提高资源的利用效率，同时也能够为企业“瘦身”，所以资源限制反而会促进企业创新。Bicen 和 Johnson[21] 认为在动荡的市场环境中，资源有限的企业同样能够进行突破性创新，关键在于思考如何创新性地使用这些资源。从而表明，资源约束并非企业突破性创新的阻碍因素，拼凑同样能带来突破性创新。目前，上述观点大多以案例分析为主[22-24]，尚未有学者从实证角度探索创业拼凑与突破性创新的关系及其内部作用机理。

知识观将企业看作知识的集合体，知识是企业核心竞争力的主要源泉，企业通过有效的知识管理，可以不断增强其自身的创新能力，提高应变能力。创业拼凑具有即兴创作性，是一种不断尝试、试错和实验的过程，通过这一过程，创业者可以更加深入地理解现有资源的属性及其组合方式，学习更多的相关资源知识并提高新创企业的创新能力。因此，本文基于知识观的相关内容，探讨创业拼凑对新创企业突破性创新的影响，并将创业学习作为中介变量，探索二者关系的影响路径。

---

① 本文原载《科技管理研究》，2017，37（17）：1-8. 作者：刘人怀，王娅男.

2. 文献回顾

（1）创业拼凑

“拼凑”（bricolage）一词最初由法国人类学家 Lévi-Strauss 在其 1967 年的著作《野性的思维》中提出，强调“利用手头资源将就”。2005 年，Baker 和 Nelson[8] 正式提出了“创业拼凑”的概念，即为了应对新问题和开发新机会，将就着对手头资源进行创造性的重新组合。其中，手头资源（the resources at hand）主要指创业者所拥有的各种资源以及从外部获取的免费或廉价资源，包括一些“零碎”、工具、技能、想法和知识等；将就（making do）强调了一种行动偏好（a bias toward action）以及不屈从于现有约束（a refusal to enact limitations）的行为特点，但将就并不意味着拼凑只能创造出不完美、次优的解决方案，有时它也会带来“意料之外的卓越成果”[25]；为新目的而重组资源是一种创新的驱动机制。此后，学者不断丰富“拼凑”的概念。Duymedjian 和 Rüling[26] 将拼凑定义为一种理念/思维，强调创业者在面临挑战时的行动导向、突破固有局限和对手边资源的重视，并且认为拼凑不仅是一种实践活动，它其实是由实践、认识论和潜在世界观的相互作用而形成的。Senyard 等[27] 认为拼凑是一种解决问题的能力，决定着现有资源的配置、协调、整合和部署，并驱动着企业的创新成果与竞争优势的产生。本文基于 Baker 和 Nelson 提出的概念，将创业拼凑定义为在资源匮乏的情况下，创业者重新组合手头资源并立即采取行动，以解决新问题或开发新机会的一种行为。

（2）创业学习

创业成功离不开持续的创业学习[28]，创业学习是创业过程的核心。现有研究对创业学习的界定主要基于 3 种视角：①机会发现视角，即认为创业学习的目的在于创业机会的识别与开发，如 Rae 和 Carswell[29] 认为创业学习是个体在识别和开发机会以及组建和管理新创企业过程中构建新方法的过程。②知识视角，即认为个体通过创业学习获取、积累并创造出相应的创业知识，如 Minniti 和 Bygrave[28] 认为创业学习是创业者累积并更新知识的行为过程。③经验与认知学习视角。这一视角既将创业学习视为经验在创业背景下的转化，同时也认可认知因素的关键作用，如 Young 和 Sexton[30] 认为创业学习是用于获取、留存并应用创业知识的各种经验性和认知性过程。本文将创业学习定义为：创业者在企业创建中不断获取并积累相关的创业经验，积极反思并创造出独特新知识的过程。

创业学习包括经验学习、认知学习和实践学习[31]，3 种学习方式相互补充[32]。其中，经验学习是创业学习的主要方式，强调通过不断反思、试错，将已有经验转化为创业知识，这些经验包括创业经验、管理经验和行业经验[33]。同时，在创业中仅凭自身经验是不足够的，还需要适当地借鉴他人的间接经验[34]，即认知学习，它强调通过观察和模仿他人的行为来获取并吸收知识的过程。此外，创业者在不同创业阶段所面临的情境存在较大差异，直接经验和间接经验在大多数情况下并不能直接应用到现有的创业情境中，需要创业者通过实践学习来调整过去所积累的创业知识与现实情境之间的差异，从而不断创造新知识[35]。

（3）新企业突破性创新

越来越多的证据表明，相比在位企业，新创企业的创新性更强。这是因为在位企业的官僚主义[10]、年龄和预测破坏性变化方面的劣势抑制了企业创新[4]，而新创企业通常不被既有惯例束缚，组织结构灵活，其创新不会蚕食现有产品且资源“黏滞性”较低[10]，因此更加适合进行突破性创新[20]。

突破性创新强调产品的根本性改变与技术的革命性变化[36-37]。有学者将知识的变化也加入到突破性创新的定义中，如 Benner 和 Tushman[38]认为突破性创新主要指获取新知识并为新顾客或新兴市场开发新产品。然而，对于突破性的程度却很难在概念上界定，因此同一现象可能被称为不连续性技术、新兴技术、架构创新甚至是破坏性技术[39]。例如，Govindarajan 和 Kopalle[40]认为创新的突破性与破坏性不同，前者基于技术维度，后者基于市场维度；而 Garcia 和 Calantone[41]则认为突破性创新会带来宏观上的市场不连续性和微观上的技术不连续性。综上，本文认为突破性创新主要指，对产品和技术的根本性变革以及新知识的创造与获取，它包括市场与技术上的不连续性。

3. 研究假设

（1）创业拼凑与突破性创新的关系

创业拼凑的目的在于解决新问题或应对新挑战，这种问题解决行为包括创意或知识要素的创造与重组，多数有创意的行为和产品都是现有知识的拓展和组合，而这种知识重组机制尤为适合开发突破性创新，特别是在新兴技术领域[42-44]。Ravishankar 和 Gurca[24]通过对新兴市场中电动汽车企业的案例研究发现，拼凑会为企业带来不连续性技术。Mascitelli[45]认为突破性创新是基于意料之外的创造力与洞察力的飞跃，并且来源于利用个体所拥有的隐性知识，因此突破性创新与隐性知识密切相关。创业拼凑强调即兴创作[46]，现有资源可被视为“隐性资产”（hidden assets），这些资源确实存在但可能暂时尚未被识别或创业者不知如何使用它们[47]，而这种“使用”更加强调不同于传统用途的使用方式，因此当拼凑行为发生后，创业者所获取并创造的相关资源知识往往具有一定的隐性特质，这便为新创企业进行突破性创新创造了有利条件。此外，Senyard 等[17]提出，如果企业对“足够好”的解决方案采用较高标准，那么拼凑会成为创造突破性创新的一种机制；Lévi-Strauss 认为拼凑并非只能带来次优方案，有时也会产生“令人意外的卓越成果”[25]。因此本文提出假设如下。

H1：创业拼凑有利于新创企业的突破性创新。

（2）创业学习与突破性创新的关系

创业学习是获取、积累创业相关经验并不断创造新知识的过程，通过这一过程获取的知识包括功能导向型知识，如销售、市场、生产等领域的知识，和战略导向型知识，如战略和竞争分析、成长管理和商业环境评估等方面的知识[48]。知识是创新的核心要素[49]，而突破性创新是高度个人化知识的产物，并源于个体的经验与学习[50]，因此创业学习与新企业突破性创新密切相关。创业者通过对过去经验的反思与学习，能够获取与产品、商业技能、网络关系等相关的隐性知识[33]，隐性知识是企业技术创新的有效促进者[51]，并且由经验转化而来的创业知识是企业创造力的主要来源[52]，进而促进新企业的突破性创新。同时，通过观察、模仿他人行为获取的间接经验，有助于克服经验学习所形成的路径依赖性[28]，使企业成员以更加开放的态度去学习他人的经验与知识，这一态度有助于其在创新活动中打破常规，以别具一格的方式创造突破。此外，创业者自身也能从他人的行为中得到启发，通过认知过程激发个人的灵感与创造力[53]。因此，本文提出假设如下。

H2：创业学习有利于新创企业的突破性创新。

（3）创业拼凑与创业学习的关系

创业拼凑是一个不断试错、修补的过程，通常要经过活跃的试错性学习[54,55]，因此“试错”原则是这种行为的重要组成部分。通过不断反思试错中的成败，能够为创业者积累更多的拼凑经验与创业知识，尤其是试错过程中所沉淀的隐性知识。创业者在试错和尝试的过程中，通过选择性地关注不同的拼凑方案，有意识地在资源环境演变过程中不断试探资源环境约束拼凑行为的边界，在保留有意义的信息、知识和拼凑方案的同时[56]，也学到更多资源及其组合的相关知识。创业拼凑强调打破常规、惯例，创造性地解决现有问题，并能容忍失败，这有助于为新创企业建立起一种开放、包容的组织氛围，使企业成员的思维更加开放，愿意相互沟通与交流，进而有更多机会获取外部的信息、经验和知识，借鉴与模仿他人的行为，并在未来规避类似的失败行为[57]。此外，拼凑者大多是有着广泛自学能力的多面手[58]，拼凑过程中的失败促使其不断反思现有的知识和能力可否应对不利因素，当其意识到所拥有的技能与知识不足以解决当前问题时，便会积极探索、不断实践，以创造出新知识、新能力来解决这一问题[31]。因此本文提出以下假设。

H3：创业拼凑有利于创业学习。

（4）创业学习的中介作用

在拼凑中，企业利用现有资源实验、修补创意、组合知识、建构意义[59]，资源被收集、检验、丢弃、替代、配置和重新配置，每一次实验与修补，都能使创业者更加深入地理解手头资源，这种理解不仅包括资源是什么，同时还包括它们是如何联系的以及如何组合，因此，不断尝试的行为过程既增加了创业者的拼凑经验，也为其在资源的使用和组合上提供了一定的灵活性。创业者则通过学习，将这些经验转化为多元化的创业知识为企业创新服务。创业拼凑具有一定的即兴创作性[46]，强调打破既有惯例与规制，不屈从于现有约束，从而为新创企业创造了许多新知识，这些知识外显为创造性的问题解决方案。从这一角度看，创业者通过对拼凑所带来的独特创意、解决方案甚至全新的技能进行不断积累与学习，增加了企业的隐性知识，隐性知识提高了企业的创造力，并对突破性创新发挥着决定性的作用。此外，从他人处获取间接经验，既培养了企业成员开放的行为态度，同时也有助于激发个体的创造力与灵感，从而有利于企业的突破性创新。因此，本文提出假设如下。

H4：创业拼凑对突破性创新的影响通过创业学习而产生作用。

综上，本文的研究框架如图 1 所示。

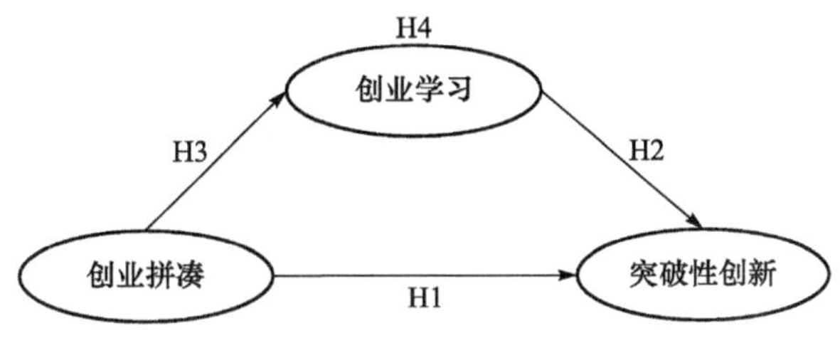

图 1　本文的研究框架

4. 研究设计

(1) 样本与数据收集

本研究采用问卷调查的方式收集收据。为保证问卷的信度与效度，本文对英文量表采用翻译-回译的方式将其翻译成中文，并在此基础上编制初始问卷，然后对广州、湖南等地的企业家进行实地访谈，在访谈结束后请其填写初始问卷，根据他们的反馈意见对题项的内容、措辞进行调整，从而得到文章的初始问卷。对此问卷进行小样本预测试，基于预测试的分析结果进一步调整问卷，并利用最终得到的正式问卷进行大样本正式调查。

根据 Zahra[60]、McDougall 和 Robinson[61]的研究，本文将新创企业界定为成立时间小于 8 年的企业，调查对象主要为新创企业的 CEO、总经理以及创业团队的核心人员。按照经济市场发展水平划分，样本收集范围包括我国东北地区、东部地区、中部地区和西部地区，并主要借助关键联系人广泛的社会关系和相关政府部门发放问卷。问卷发放期为 2016 年 6 月至 2016 年 9 月底，历时 3 个月，累积发放问卷 390 份，实际回收 247 份，回收率为 63.33%；剔除其中填写不完整和明显带有规律性的无效问卷 34 份后，最终得到的有效问卷数量为 213 份，有效回收率为 86.23%。样本基本情况参见表 1 所示。

表 1　样本基本情况汇总　　单位:%

基本特征	具体类别	比例	基本特征	具体类别	比例
性别	男	62.44	地区分布	东北地区	22.54
	女	37.56		东部地区	31.92
				中部地区	35.68
				西部地区	9.86
企业规模	5 人及以下	20.66	行业	制造业	22.54
	6～20 人	46.95		服务业	16.43
	21～50 人	16.43		信息技术行业	26.76
	51～100 人	8.92		商贸行业	12.21
	100 人以上	7.04		生物及医药	2.82
				新能源	3.76
				其他	15.49

注：数据之和不为 100%是数据修约所致

（2）变量测量

本文的变量测量均采用李克特5点量表，1～5分别表示从“完全不同意”到“完全同意”，并要求被试根据题项与自身和企业实际情况的符合程度进行打分。

① 创业拼凑

参考Senyard等[17]的研究，本文使用8个题项测量创业拼凑，具体包括“当面对新挑战时，通过利用现有资源我们有信心能够找到有效的解决方案（EB1）”“相比其他企业，我们能够利用现有资源去应对更多的挑战（EB2）”“我们善于应用任何现有资源去应对创业中遇到的新问题或机会（EB3）”“我们通过整合现有资源和从外部获得的其他廉价资源来应对新的挑战（EB4）”“当应对新问题或机会的时候，我们假设能够找到可行的解决方案并采取行动（EB5）”“通过整合现有资源，我们能够成功应对任何新的挑战（EB6）”“当面对新挑战时，我们通过组合现有资源获得可行的解决方案（EB7）”“我们通过整合那些原本计划用于其他目的的现有资源来成功应对新挑战（EB8）”。

② 创业学习

对于创业学习的测量，本文采用单标安等[35]在中国情境下开发的量表，将创业学习划分为经验学习、认知学习和实践学习3个维度，测量题项分别为“不断反思过去的失败行为（EL1）”“失败行为并不可怕，关键是能从中吸取教训（EL2）”“已有的经验（管理经验、创业经验等）对本企业的创业决策并不重要（反向）（EL3）”“经常与企业成员进行交流（EL4）”“经常参与企业内部或外部各种正式或非正式的讨论会（EL5）”“经常阅读相关书籍和文献以获取有价值的创业信息（EL6）”“创业过程中持续收集有关内、外部环境的信息（EL7）”“注重在创业实践中深化已有的创业知识（EL8）”“通过持续的创业实践来反思或纠正已有的经验（EL9）”。

③ 突破性创新

本文参考He和Wong[62]的研究测量突破性创新，测量题项包括“本企业在产品的研制上经常会引入新的理念/创意（RNO1）”“本企业凭借新产品/服务开辟了全新的市场（RNO2）”“本企业不断创新销售渠道（RNO3）”“本企业在行业中率先开发和引入全新的技术和工艺（RNO4）”“本企业的新产品/服务/技术给产业带来了重大影响（RNO5）”。

④ 控制变量

为避免其他因素的干扰，本文选取行业、企业年龄和企业规模作为控制变量。这是因为：不同行业在竞争强度、动荡程度和资源属性方面的差异会对企业创新产生影响；不同规模的企业往往有着不同的学习、创新水平，小规模企业的组织结构更加灵活，因此开展拼凑、学习和创新等行为的阻力较小。同时，企业年龄和企业规模会通过直接影响企业能力而对新创企业的创新产生影响[63]。通常，企业规模可用销售额、员工人数和资产总额表示[64]，本文使用员工人数代表企业规模。

5. 数据分析与结果

（1）同源偏差分析

由于本研究的调查问卷均由同一主体填写，为避免同源偏差问题对研究结论造成影响，本文采用Harman单因素检测方法检验所回收数据[65]，即对问卷中的所有问项进行探索性因子分析，结果显示在未旋转时第一主成分只解释了变异量的 16.847%，说明本研究不存在严重的同源偏差问题。

（2）信度和效度检验

① 信度检验

本文使用 Cronbach's α 系数检验各量表信度。根据吴明隆[66]的观点，整份量表的 Cronbach's α 系数最低要达到 0.7。本研究信度检验结果显示，创业拼凑量表的 Cronbach's α 系数为 0.828，创业学习为 0.839，突破性创新为 0.848，从而说明量表具有较好的内部一致性，问卷信度良好。

② 效度检验

效度检验主要以内容效度和结构效度为主。本文的量表主要参考国内外主流期刊上广泛使用的成熟量表，并在小样本预测试中进行了调整，因此具有较高的内容效度。在结构效度检验中，当有理论预期或是检验前人已有的成熟量表时，可直接使用验证性因子分析[67]，因此，本文使用结构方程模型对创业拼凑、创业学习和突破性创新进行验证性因子分析。结果如表 2 所示，相比其他 3 个竞争模型，三因子模型的拟合效果比较好，说明文中的 3 个变量具有较好的区分效度。

表 2　验证性因子分析结果

模型	x^2	df	x^2/df	RMSEA	RMR	CFI	IFI
EB，EL，RNO	350.241	206	1.700	0.057	0.029	0.911	0.913
EB+EL，RNO	582.133	208	2.799	0.092	0.038	0.770	0.773
EB，EL+RNO	692.640	208	3.330	0.105	0.064	0.702	0.706
EB+EL+RNO	924.434	209	4.423	0.127	0.069	0.560	0.565

注：创业拼凑=EB，创业学习=EL，突破性创新=RNO

（3）相关性分析

本文在进行假设检验之前，首先分析了变量之间的相关性。结果显示（表 3），创业拼凑与创业学习（$r=0.434$，$P<0.01$）、突破性创新（$r=0.231$，$P<0.01$）显著正相关，创业学习与突破性创新（$r=0.273$，$P<0.01$）显著正相关，因此可进一步做假设检验。

表 3　描述性统计与变量间相关系数

变量	均值	标准差	1	2	3	4	5	6
1. 企业年龄	1.99	0.841	1					
2. 行业	3.30	2.012	0.174*	1				
3. 企业规模	2.35	1.117	0.295**	0.076	1			
4. 创业拼凑	3.86	0.428	−0.012	0.173*	0.078	1		
5. 创业学习	4.07	0.429	−0.004	0.182*	0.063	0.434**	1	
6. 突破性创新	3.69	0.657	−0.046	0.106	0.120	0.231**	0.273**	1

** 为 $P<0.01$，* 为 $P<0.05$

（4）假设检验与结果

本研究采用阶层回归分析的方法进行假设检验，结果如表 4 所示：模型 1 和模型 3

分别展现出控制变量与创业学习、突破性创新的作用关系：模型 2 和模型 4 的结果表明创业拼凑对创业学习（β=0.464，P<0.001）和突破性创新（β=0.198，P<0.001）具有显著的正向影响，假设 H1 和 H3 得到验证；模型 5 的结果显示创业学习与突破性创新（β=0.215，P<0.001）之间存在显著的正相关关系，假设 H2 得到验证；在模型 6 中加人创业学习后可看出，创业拼凑对突破性创新（β=0.118、P>0.05）的作用不再显著。由此可知，创业学习在创业拼凑对突破性创新的影响中发挥着完全中介的作用，假设 H4 得到验证。

表 4　阶层回归分析

变量	创业学习			突破性创新		
	模型 1	模型 2	模型 3	模型 4	模型 5	模型 6
企业年龄	−0.255	−0.125	−0.42	−0.364	−0.365	−0.343
产业	0.358***	0.217	0.186	0.125	0.109	0.088
企业规模	0.225	0.104	0.420**	0.368	0.371	0.350
创业拼凑		0.464***		0.198***		0.118
创业学习					0.215***	0.172**
R^2	0.038	0.201	0.034	0.075	0.095	0.108
ΔR^2	0.038	0.163	0.034	0.041	0.061	0.033
F	2.766**	13.094***	2.488	4.213**	5.482***	4.988***

*** 为 P<0.001，** 为 P<0.01

6. 研究结论与展望

创业拼凑会为新创企业带来突破性创新吗？其中的内在作用机制是什么？基于上述问题，本文从知识观角度研究了创业拼凑对突破性创新的影响，同时选取创业学习作为创业拼凑与新企业突破性创新相互作用的中介变量，进而探索了创业拼凑对突破性创新影响的内部作用路径。最终的研究结果表明，创业拼凑对新创企业的突破性创新有正向促进作用，创业学习在创业拼凑与突破性创新中发挥着完全中介的作用。

在理论上，本文的研究结论从一定程度上解决了学者对于创业拼凑与突破性创新关系所持观点的不一致。尽管学者普遍认为创业拼凑促进企业创新，但对于创新的程度却有着不同的观点：一些学者认为拼凑带来的并非突破性创新，而只是一些渐进性的、微不足道的创新；另外一些学者认为拼凑并非只是产生次优方案，有时它也会带来不连续性技术等一系列“意外的卓越成果”。本文通过实证研究发现，创业拼凑有利于突破性创新，但这种影响需要通过创业学习这一中介变量实现，即创业学习在其中发挥着完全中介的作用。此外，本文的研究结论丰富了创业拼凑理论的研究框架。创业拼凑强调对手头资源的创造性重组，因此其本身就是一种创新。现有研究认为，创业拼凑有利于企业创新[10,68]，但却没有深入探讨其中的作用机制。创业拼凑是一个不断试错、尝试和实验的过程，通过这一过程，拼凑者更加了解现有资源及资源间不同的组合方式，并学到相关的资源知识，而知识对于企业突破性创新发挥着至关重要的作用，尤其是隐性知识。因此，本文从知识视角切入，将创业学习作为创业拼凑与新创企业创新的中介变量，从而探索了二者相互影响的内部作用机理。

在实践上，本文的研究结论表明，创业拼凑对新创企业的突破性创新发挥了积极作用。创新是新创企业打破在位企业垄断并获取竞争优势的关键手段[69]，对企业生存发挥着重要作用。在我国，新创企业的成长空间小、失败率高已成为不争的事实，新而小的新创劣势以及贫乏的资源禀赋均给企业创新尤其是突破性创新带来了困难，并限制了大多数新创企业的创新能力[10]。本研究为受到资源限制的新创企业突破性创新带来了一种新的行为模式——拼凑。因此，企业不应把资源约束作为突破性创新的阻碍因素，相反，应积极探索现有资源的潜在用途与价值，通过资源重组创造更多的创新性成果，进而变被动为主动，将"客观上资源是否够用"的"获取"问题转化为"主观上如何使用资源"的"创造"问题[70]，真正做到"无中生有"。此外，创业学习所发挥的中介作用也启示创业者在拼凑的过程中应不断学习、反思、总结，加强成员间的沟通交流，主动分享拼凑经验，在深化资源知识的同时创造出更多的新知识，进而通过不同的学习方式提高组织成员学习的效率和效果，为企业创新提供更加坚实的知识基础。

尽管本研究具有一定的理论与实践意义，但仍存在着一些不足之处。首先，本文只从知识视角出发，揭示了创业学习在创业拼凑与新企业突破性创新之间的中介作用，可能忽略了发挥中介作用的其他变量，例如，创业拼凑有利于企业动态能力的提升[71]，而动态能力又进一步影响企业创新，因此未来可从能力视角拓展相关中介机制的探究，如资源整合能力、创新能力等。其次，本文在揭示创业拼凑对突破性创新的影响路径时未能考虑一些调节变量的作用，如不同的组织氛围、机会类型和创业环境等。再次，本文只是初步探讨了创业拼凑与新创企业突破性创新的关系，未来可进一步分析中国新创企业利用拼凑获取突破性创新的相关案例，从而为企业提供更加具体化的实践指导。最后，本文的实证研究使用的是企业的横截面数据，因此未来有必要采用纵向研究设计进一步测量变量之间的因果关系。

参考文献

[1] SCHUMPETER J A. The theory of economic development: an inquiry into profits, capital, credit, interest, and the business cycle. Cambridge: Harvard University Press, 1934: 36.

[2] 杨勇，袁卓. 技术创新与新创企业生产率：来自 VC/PE 支持企业的证据. 管理工程学报, 2014, 28 (1): 56-64.

[3] SONG L, SONG M, PARRY M. Economic conditions, entrepreneurship, first: product development, and new venture success. Journal of Product Innovation Management, 2010, 27 (1): 130-135.

[4] SONG M D, BENEDETTO C A. Supplier's involvement and success of radical new product development in new ventures. Journal of Operation Management, 2008, 26 (1): 1-22.

[5] SORESCU A B, CHANDY R K, JPRABHU J C. Sources and financial consequences of radical innovation: insights from pharmaceuticals. Journal of Marketing, 2003, 67 (4): 82-102.

[6] ANTOLÍN-LÓPEZ R, CÉSPEDES-LORENTE J, GARCÍA-DE-FRUTOS N, et al. Fostering product innovation: differences between new ventures and established firms. Technovation, 2015, 41-42: 25-37.

[7] STINCHCOMBE A L. Social structure and organization//MARCH J C, CRUSKY O. Handbook of organizations. Chicago: Rand McNally, 1965: 142-193.

[8] BAKER T, NELSON R E. Creating something from nothing: resource construction through entrepreneurial bricolage. Administrative Science Quarterly, 2005, 50 (3): 329-366.

[9] OLSON E M, WALKER O C, RUEKERT R W. Organizing for effective new product development: the moderating role of product innovativeness. Journal of Marketing, 1995, 59 (1): 48-62.

[10] SENYARD J, BAKER T, STEFFENS P, et al. Bricolage as a path to innovativeness for resource-constrained new firms. Jounal of Product Innovation Management, 2014, 31 (2): 211-230.

[11] CUO H, SU Z, AHLSTROM D. Business model innovation: the effects of exploratory orientation, opportunity recognition, and entrepreneurial bricolage in an emerging economy. Asia Pacific Journal of Management, 2015, 33: 1-17.

[12] CIBORRA C U, PATRIOTTA G. Groupware and teamwork in R&D: limits to learning and innovation. R&D Management, 1998, 28 (1): 43-52.

[13] SCHOONHOVEN C B, EISENHARDT K M, LYMAN K. Speeding products to market: waiting time to first product introduction in new firms. Administrative Science Quarterly, 1990, 35 (1): 177-207.

[14] GREEN S G, WELSH M A, DEHLER G E. Advocacy, performance and threshold influences on the decisions to terminate new product development. The Academy of Management Journal, 2003, 46 (4): 419-434.

[15] ROTHAERMEL F T, DEEDS D L. Alliance type, alliance experience, and alliance management capability in high-technology ventures. Journal of Business Venturing, 2006, 21 (4): 429-460.

[16] SWINK M, SANDVIG J, MABERT V A. Customizing concurrent engineering processes: five case studies. Journal of Product Innovation Management, 1996, 13 (3): 229-244.

[17] SENYARD J M, BAKER T, DAVIDSSON P. Entrepreneurial bricolage: towards systematic empirical testing. Frontiers of Entrepreneurship Research, 2009, 29 (5): 1-14.

[18] HANSEN M H, PERRY L T, REESE C S. A Bayesian operationalization of the resource-based view. Strategic Management Journal, 2004, 25 (13): 1279-1295.

[19] CHRISTENSEN C M, BOWER J L. Customer power, strategic investment, and the failure of leading firms. Strategic Management Journal, 1996, 17 (3): 197-218.

[20] HAMILTON W F, SINGH H. The evolution of corporate capabilities in emerging technologies. Interfaces, 1992, 22 (4): 13-23.

[21] BICEN P, JOHNSON W H A. How do firms innovate with limited resources in turbulent markets. Innovation Oraganization & Management, 2014, 16 (3): 4207-4240.

[22] CIBORRA C U. The platform organization: recombining strategies, structures and surprises. Organization Science, 1996, 7 (2): 103-118.

[23] GARUD R, KARNØE P. Bricolage versus breakthrough: distributed and embedded agency in technology entrepreneurship. Research Policy, 2003, 32 (2): 277-300.

[24] RAVISHANKAR M N, GURCA A. A bricolage perspective on technological innovation in emerging markets. IEEE Transactions on Engineering Management, 2016, 63 (1): 53-66.

[25] 列维-斯特劳斯. 野性的思维. 李幼蒸，译. 北京：中国人民大学出版社，2006.

[26] DUYMEDJIAN R，RÜLING C C. Towards a foundation of bricolage in organization and management theory. Organization Studies，2010，31 (2)：133-151.

[27] SENYARD J，DAVIDSSON P，STEFFENS P. Venture creation and resource processes：using bricolage sustainability ventures// Langan-Fox J. Proceedings of the 7th AGSE International Entrepreneurship Research Exchange. Queensland：University of the Sunshine Coast，2010：637-648.

[28] MINNITI M，BYGRAVE W. A dynamic model of entrepreneurship learning. Entrepreneurship Theory & Practice，2001，25 (3)：5-16.

[29] RAE D，CARSWELL M. Towards a conceptual understanding of entrepreneurial learning. Journal of Small Business and Enterprise Development，2001，8 (2)：150-158.

[30] YOUNG J E，SEXTON D L. What makes entrepreneurs learn and how do they do it?. The Journal of Entrepreneurship，2003，12 (2)：155-182.

[31] 单标安，蔡莉，陈彪，等. 中国情境下创业网络对创业学习的影响研究. 科学学研究，2015，33 (6)：899-906，914.

[32] GREENO J G，COLLINS A M，RESNICK L B. Cognition and learning//BERLINER D C，CALFEE R C. Handbook of educational psychology. New York：Simon & Schuster Macmillan，1996：15-46.

[33] POLITIS D. The process of entrepreneurial learning：a conceptual framework. Entrepreneurship Theory & Practice，2005，29 (4)：399-424.

[34] HOLCOMB T R，IRELAND R D，HOLMES JR R M，et al. Architecture of entrepreneurial learning：exploring the link among heuristics，knowledge，and action. Entrepreneurship Theory & Practice，2009，33 (1)：167-192.

[35] 单标安，蔡莉，鲁喜凤，等. 创业学习的内涵、维度及其测量. 科学学研究，2014，32 (12)：1867-1875.

[36] DEWAR R D，DUTTON J E. The adoption of radical and incremental innovations：an empirical analysis. Management Science，1986，32 (11)：1422-1433.

[37] ETTLIE J E，BRIDGES W P，O' KEEFE R D. Organization strategy and structural differences for radical versus incremental innovation. Management Science，1984，30 (6)：682-695.

[38] BENNER M L，TUSHMAN M L. Exploitation，exploration，and process management：the productivity dilemma revisited. The Academy of Management Review，2003，28 (2)：238-256.

[39] HURMELINNA-LAUKKANEN P，SAINIO L M，JAUHIAINEN T. Appropriability regime for radical and incremental innovations. R&D Management，2008，38 (3)：278-289.

[40] GOVINDARAJAN V，KOPALLE P K. The usefulness of measuring disruptiveness of innovations ex post in making ex ante predictions. Journal of Product Innovation Management，2006，23 (1)：12-18.

[41] GARCIA R，CALANTONE R. A critical look at technological innovation typology and innovativeness terminology：a literature review. Journal of Product Innovation Management，2002，19 (2)：110-132.

[42] KATILA R，AHUJA G. Something old，something new：a longitudinal study of search behavior and new product introduction. Academy of Management Journal，2002，45 (6)：1183-1194.

[43] MERTON R K. The sociology of science: theoretical and empirical investigations. Chicago: University of Chicago Press, 1979.

[44] KEUPP M M, CASSMANN O. Resource constraints as triggers of radical innovation: longitudinal evidence from the manufacturing sector. Research Policy, 2013, 42 (8): 1457-1468.

[45] MASCITELLI R. From experience: harnessing tacit knowledge to achieve breakthrough innovation. Journal of Product Innovation Management, 2000, 17 (3): 179-193.

[46] DI DOMENICO M, HAUGHH, TRACEY P. Social bricolage: theorizing social value creation in social enterprises. Entrepreneurship Theory & Practice, 2010, 34 (4): 681-703.

[47] LINNA P. Bricolage as a means of innovating in a resource-scarce environment: a study of innovator-entrepreneurs at the BOP. Journal of Developmental Entrepreneurship, 2013, 18 (3): 1-23.

[48] ROXAS B G, CAYOCA-PANIZALES R, DE JESUS R M, et al. Entrepreneurial knowledge and its effects on entrepreneurial intentions: development of a conceptual framework. Asia-Pacific Social Science Review, 2008, 8 (2): 61-77.

[49] GRANT R M. Prospering in dynamically-competitive environments: organizational capability as knowledge integration. Organization Science, 1996, 7 (4): 375-387.

[50] LEONARD D, SENSIPER S. The role of tacit knowledge in group innovation. California Management Review, 1998, 40 (3): 112-132.

[51] KOSKINEN K U. Tacit knowledge as a promoter of success in technology firms// Proceedings of the 34th Annual Hawaii International Conference on System Sciences. Hawaii: IEEE Computer Society Washington, 2001.

[52] 朱秀梅，孔祥茜，鲍明旭. 学习导向与新企业竞争优势：双元创业学习的中介作用研究. 研究与发展管理，2014，26 (2)：9-16.

[53] 单标安. 基于中国情境的创业网络对创业学习过程的影响研究. 长春：吉林大学，2013.

[54] STEFF ENS P R, BAKER T, SENYARD J M. Betting on the underdog: bricolage as an engine of resource advantage// Proceedings of Annual Meeting of the Academy of Management. Montreal: Academy of Management, 2010.

[55] SENYARD J M, BAKER T, STEFFENS P R, et al. Entrepreneurial bricolage and firm performance: moderating effects of firm change and innovativeness// Proceedings of Annual Meeting of the Academy of Management. Montreal: Academy of Management, 2010.

[56] 方世建，黄明辉. 创业新组拼理论溯源、主要内容探析与未来研究展望. 外国经济与管理，2013，35 (10)：2-12.

[57] LUMPKIN G T, LICHTENSTEIN B B. The role of organizational learning in the opportunity-recognition process. Entrepreneurship Theory & Practice, 2005, 29 (4): 451-472.

[58] SENYARD J M, DAVIDSSON P, STEFFENS P R. Environmental dynamism as a moderator of the relationship between bricolage and firm performance// Proceedings of Academy of Management Annual Meeting. Vancouver: Academy of Management, 2015.

[59] SENYARD J M, DAVIDSSON P, BAKER T, et al. Resource constraints in innovation: the role of bricolage in new venture creation and firm development// Proceedings of the 8th AGSE International Entrepreneurship Research Exchange. Melbourne: Swinburne University of Technology, 2011: 609-622.

[60] ZAHRA S A. A conceptual model of entrepreneurship as firm behavior: a critique and extension. Entrepreneurship Theory & Practice, 1993, 17 (4): 5-21.

[61] MCDOUGALL P，ROBINSON R B，JR. New venture strategies：an empirical identification of eight 'archetypes' of competitive strategies for entry. Strategic Management Journal，1990，11 (6)：447-467.

[62] HE Z，WONG P，Exploration vs. exploitation：an empirical test of the ambidexterity hypothesis. Organization Science，2004，15 (4)：481-494.

[63] 许晖，李文. 高科技企业组织学习与双元创新关系实证研究. 管理科学，2013，26 (4)：35-45.

[64] 李雪灵，韩自然，董保宝，等. 获得式学习与新企业创业：基于学习导向视角的实证研究. 管理世界，2013，(4)：94-106，134.

[65] PODSAKOFF P M，ORGAN D. Self-reports in organizational leader reward and punishment behavior and research：problems and prospects. Journal of Management，1986，12 (4)：531-544.

[66] 吴明隆. 问卷统计分析实务：SPSS 操作与应用. 重庆：重庆大学出版社，2010：244.

[67] 罗胜强，姜嬿. 管理学问卷调查研究方法. 重庆：重庆大学出版社，2014：149.

[68] SALUNKE S，WEERAWARDENA J，MCCOLL-KENNEDY J R. Competing through service innovation：the role of bricolage and entrepreneurship in project-oriented firms. Journal of Business Research，2013，66 (8)：1085-1097.

[69] 李翔，陈继祥. 新创企业技术创新与商业模式创新的交互作用研究. 现代管理科学，2015，(3)：109-111.

[70] 张建琦，安雯雯，尤成德，等. 基于多案例研究的拼凑理念、模式双元与替代式创新. 管理学报，2015，12 (5)：647-656.

[71] 李非，祝振铎. 基于动态能力中介作用的创业拼凑及其功效实证. 管理学报，2014，11 (4)：562-568.

数字经济　创新未来[①]

欢迎大家莅临杭州电子科技大学创新与发展研究院，就数字经济与创新话题进行研讨。

到今年底，改革开放已经四十年，中国的经济与社会发展突飞猛进，中国已成为世界第二大经济体，对世界经济增长的贡献率超过 30%。浙江在全国率先提出“数字经济”1 号工程，体现了浙江省委、省政府对新时期信息技术发展的前瞻性、战略性部署。

对一个国家而言，科学技术的理解深度关乎其兴衰成败。

“数字经济”并不是今天才有的概念。早在 20 世纪 90 年代中期，美国经济学家就出版了名为《数字经济》的著作。但到了 2001 年，新经济神话突然破灭，纳斯达克股指跌到 1000 多点，许多互联网公司倒闭。当时很多经济学家争论数字经济有没有提高生产率、互联网经济是不是泡沫。

过去几十年，数字经济发展迅猛，在经济发展中的引领和主导作用不断增强。预计在 21 世纪上半叶，数字经济仍将唱主角。目前，我国数字经济正处于从跟跑向并跑，甚至领跑转变的关键时期，应特别注重加强信息技术自主创新，不断增强数字经济对发展的推动作用。

2016 年的二十国集团（G20）领导人杭州峰会首次提出全球性的《二十国集团数字经济发展与合作倡议》，表明发展数字经济已成为全球共识。

当前，大数据、云计算、物联网、人工智能技术的发展将数字经济又推向高峰。2004～2007 年，全球经济年均增长速度高达 5%左右，是近 30 年来增长最快且最为平稳的一段时间。2008 年的国际金融危机使全球经济特别是传统金融业遭受重创，但苹果、脸谱、谷歌、微软、亚马逊等数字公司基本上毫发无损。我国的阿里巴巴、百度、腾讯等数字企业受影响也不大，为我国经济稳定增长做出了贡献。

当前，我国经济发展已进入新常态，需要寻求新的动力。大数据、人工智能、虚拟现实、区块链等技术的兴起为人们带来了希望，世界各国不约而同地将这些新的信息技术作为未来发展的战略重点。

新技术的发展越来越快，我们必须快速响应。石器时代经历了数万年的演进，印刷术的推广耗费了一个世纪的时间，电视机的普及花了几十年，而微信的普及只用了几年的时间。人类社会技术进步以指数速度发展，国际上将这一规律称为技术进化的加速回报定律。

埃森哲公司预测，数字化带来的社会效益可以远远超过其创造的行业价值。到 2025 年，各个行业的数字化转型有望带来 100 万亿美元的社会及商业潜在价值。其中，仅汽车、消费品、电力、物流四大行业的数字化转型就将为社会和行业带来超过 20 万亿美元的潜在累积价值。

---

① 本文是企业数字化转型论坛开幕式讲话，杭州，2018 年 11 月 30 日.

数字化已成为经济发展的主要动力，发展数字经济是使世界经济焕发新活力的良方。虽然信息技术的前景十分光明，但在推动经济发展和人类文明进步方面还有很长的路要走。有专家断言，对整个信息时代而言，人类现在的信息处理能力还只是相当于工业革命的蒸汽机时代。

我们一定要抓住数字经济发展的战略机遇，在新的时代条件下，从分工与合作、供给与需求、经济结构的复杂性、思想文化以至深化改革等多个维度全面理解数字经济。

目前，我国数字经济发展已取得骄人成绩。数字经济为我国更好更快地完成工业化任务提供了新的引擎。但从普及信息技术到在各行业取得实际效果并不是水到渠成的事，需要付出许多艰苦细致的努力。同样，信息技术用于智能制造等行业，也需要一个磨合和熟悉的过程才能发挥实效。目前，我国信息技术主要应用在电子商务和社交等生活类产品上。今后应更加重视将信息技术应用于生产领域，推动提高生产效率，对产业升级和节能环保做出更大贡献。

发展数字经济，既需要科技工作者潜心研究，也需要“政产学研金”各部门紧密合作，努力做出重大发明和创新，推动数字经济发展跃上新的高峰。发展数字经济前景光明、任重道远。必须认真贯彻落实习近平同志有关重要论述和省委、省政府的决策部署，做大做强数字经济，加快传统产业数字化、智能化，努力在新时代的全球经济竞争中抢占先机。

互补性资产对双元创新的影响及平台开放度的调节作用①

一、研究背景

平台经济成为21世纪全新的经济模式，它的繁荣造就了大量互联网平台企业。互联网平台企业成为资源配置和利益协调的重要载体，也是推动技术创新和商业创新的重要组织[1]，由此，越来越多的学者研究平台企业[2]，尤其是创新生态系统中的平台企业。

平台企业的创新对社会经济发展产生重要影响，但当前学术界对平台企业的研究主要集中在价格和竞争方面，对平台企业创新的研究还未充分重视[1]。双元创新是提高企业创新绩效的重要途径，但目前学术界讨论最多的是传统企业的双元创新与企业绩效的关系[3]，对平台企业双元创新机制的研究较为缺乏。平台企业的双元创新建立在开放式创新生态系统基础之上，通过平台包络不断向周边邻近行业渗透或者向全新领域进行颠覆[4]，重构覆盖区域的资源和能力，从而进行利用式创新和探索式创新，这样的创新方式既能防止陷入“创新陷阱”，保持整个平台的均衡发展，又能促进更加优越的新平台代替旧平台，不断完善平台及其创新生态系统[5]。

随着开放式创新成为主流的创新范式，平台及其生态系统理论的研究不断深入，学者逐渐意识到互补性资产的重要性。然而，对生态系统中互补性资产的理论研究还处于探索阶段，尚未对互补性资产的重要作用进行深入探讨。平台生态系统中的互补性资产跨越了企业边界，分布在平台领导、消费者、卖家、供应商、开发者、内容提供商、广告商等众多主体之中[6]。互补性资产是提升平台企业地位和能力的关键因素，是实现平台生态优势的重要基础[7]，也是各个主体之间实现资源互补、跨界创新和共赢共生的重要支撑[8]。

平台开放是平台治理的重要手段，同时关系到平台企业的创新方式及创新绩效。目前，学术界对平台开放度的研究，主要集中在平台开放度对企业绩效的影响方面，且大多数学者认为，平台开放度的设置是一个复杂的问题，需要将平台开放度置入具体的情境，才能解释平台开放度与企业绩效的关系。平台开放度的设置会影响平台参与主体的数量和类型[2]，关系到平台企业创新方式的选择、创新过程及竞争优势的获得[9]，而目前鲜有文献研究平台开放度对互补性资产与双元创新关系的调节效应。

综上分析，本文将在前人研究的基础上，探索互补性资产对双元创新的作用机制，分析平台开放度对互补性资产与双元创新关系的调节作用。

二、理论分析与研究假设

1. 互补性资产与探索式创新

目前，学术界在研究互补性资产与创新的关系时，主要局限于互补性资产对组织

---

① 本文原载《管理学报》，2019，16（7）：949-956. 作者：刘人怀，张镒.

合作绩效的影响，且大多数文献的研究对象是传统企业。例如，Hoskisson 等[10]指出，互补性多过相似性，则更有助于企业进行创新合作；Stieglitz 和 Heine[6]认为，激进式创新离不开新资源和互补性行为的动态调整；Roy 和 Cohen[11]分析了在颠覆性技术变革过程中，企业下游的互补性资产存量对其产品创新的影响，并发现具有更多下游互补性资产的企业很可能成为产品创新的领导者；Kani 和 Motohashi[12]指出，通过整合企业内部和外部理念可以创造新价值，实现开放式创新。可见，关于互补性资产与平台企业双元创新关系的研究有待进一步补充和完善。

开放式创新理论认为，创新生态系统中的互补性资产是平台企业进行探索式创新的重要支撑：①平台企业可以利用生态系统中创新主体具有的先进生产设备或技术，以补充平台企业实现架构创新，发挥平台企业在生态系统中的引领示范作用，进行系统性的变革与颠覆[4]，有效应对外部突破性技术变革带来的冲击；②通过整合具有竞争力的生产能力或服务水平，以改变传统线性企业的层级结构，探索兼具层级和网络双重属性的平台组织架构，对企业的业务流程进行模块化设计[12]，促进创新要素跨越组织边界，实现在平台生态系统中自由流动和配置；③利用外部创新主体提供的新工艺或新方法进行生产，为生产制造提供新技术，对价值创造流程进行重构[13]，帮助平台企业获得更多资源进行设计研发，提供新的产品或服务，促进平台企业进行颠覆式或重大创新；④利用互补性资产在用户中建立良好的品牌形象，在充分摸清用户需求的基础上，不断挖掘用户的潜在消费需求，开拓新销售渠道，拓展新消费市场，开发新产品[14]；⑤利用互补性资产吸引新的平台用户，并与新用户建立良好的合作关系，跨越制造和服务实践社区之间的传统界限，创造新知识[15]。

综上，本文认为，互补性资产不仅可以作为平台企业创新所需要的补充要素，还可以促进平台企业自身资源和能力的整合，产生出对平台企业创新发展有利的新资源和新能力，从而有利于平台企业进行一系列的探索式创新活动。基于此，提出如下假设。

假设 1：互补性资产正向影响探索式创新。

2. 互补性资产与利用式创新

开放式创新理论认为，互补性资产是平台企业进行利用式创新的重要补充。平台领导凭借自身的连接功能，将大量用户聚集在特定的时空网络中，并为需求方和供应方提供交易介质，形成了一个以平台架构为基础、大量平台用户（包括供给端、需求端及第三方的企业及个人）共同参与构建的开放式创新生态系统。生态系统中的大量创新主体在自身领域具有一定的经验或实力，如拥有先进的生产设备或技术、具有竞争力的生产能力或服务水平、采用新工艺或新方法进行生产、在用户中具有良好的品牌形象、与客户建立了良好的合作关系等，这些资源或能力作为平台企业进行交易与创新活动的重要补充，加上平台企业具备的强大连接功能和整合能力，足以将这些外部主体携带的资产整合成为平台企业可利用的资产。

就互补性资产与平台企业利用式创新的关系而言，有如下解释：①外部创新主体的互补性资产，有助于平台企业快速积累发展所需的资源和能力[13]，能够迅速掌握先进的生产设备或技术，提升自身的生产能力或服务水平；②互补性资产的获得有利于平台企业采用新的工艺方法进行生产，不断促进平台企业对产品、服务及技术等方面的优化和升级，提高各个流程的效率，降低生产成本[5]；③外部创新主体补充的生产能力或服务能力，可以弥补平台企业在某些方面的不足，有利于平台企业改善现有的产品及服务，不断提高用户的体验水平，并结合平台企业具有的信息优势提供有针对性的服务[14]；④大量品牌商在消费者中具有良好形象，此时平台用户进入平台生态系统，有利于提升平台企业的声誉[16]，可以稳定平台已有客户基础或已占领的市场；⑤生态系统中的互补性资产，有利于平台建立良好的客户关系及管理机制，帮助平台企业了解现有用户的需求，细分目标市场，改进并完善现有产品和服务，不断丰富平台内容。

综上，本文认为，平台生态系统中的互补性资产，有助于平台企业快速积累创新发展所需要的资源和能力，帮助平台企业快速实现创新收益，实现平台企业的短期目标和绩效。换言之，平台企业可利用的互补资产越多，越有利于开展利用式创新。基于此，提出如下假设。

假设 2：互补性资产正向影响利用式创新。

3. 利用式创新的中介作用

学术界对双元创新的界定已基本形成共识，但是对探索式创新与利用式创新之间的关系还存在争议。有学者认为，探索式创新和利用式创新之间是相互矛盾的，是一种替代关系，二者存在此消彼长的关系[17]；也有学者认为，二者是相互补充的，不存在竞争关系，可以同时进行两种活动[18]。产生以上两种分歧的原因在于：企业是否有能力平衡好这两种类型的创新。基于结构双元观点，本文认为，互联网平台企业可以同时进行探索式创新和利用式创新，且利用式创新有利于推动探索式创新。

互联网平台企业进行利用式创新会推动探索式创新的发展。探索式创新较为注重创造新的知识、开发新产品或新服务，或者为新产品增加全新的功能，具有较高的不确定性和风险性[3]。利用式创新则注重对现有知识、技术等资源的挖掘和利用[3]，关心的是企业的短期目标，考虑如何在短期内快速实现收益目标或财务绩效，能够对创新成果进行有效预见，降低未来的不确定性，风险相对较小。利用式创新作为探索式创新的基础，企业在进行创新选择时，可以先充分了解已有的业务领域和存量知识，通过深入挖掘现有资源和能力，不断优化现有流程[13]，不断提升自身能力和技术水平，通过渐进性创新不断积累，实现量变到质变的突破，再创造新知识或技术、开发新产品或新服务，实现突破式发展[19]。综上所述，平台企业的利用式创新有利于促进探索式创新活动的开展。据此，提出如下假设。

假设 3：利用式创新正向影响探索式创新。

平台生态系统中的互补性资产是由大量创新主体提供的资源和能力，有助于平台企业在研发、生产、销售及售后等环节的优化和创新。互补性资产包括先进的生产技术设备、新工艺或新方法、生产能力、品牌、客户关系等，平台企业可以利用这些互补性资产，对现有的产品和服务进行改善或升级，对现有的技能和业务流程进行优化，提高产品和服务的供应效率，降低交易成本[20]；深耕原有市场，为现存用户提供扩展服务，不断丰富现存产品和服务的功能和类型，扩展产品和服务在相关领域的应用[13]。在此基础上，平台企业可以不断稳定市场用户，维持整个生态系统的更新与升级，不断完善原有平台结构和生态治理机制。通过前期的利用式创新来减少创新风险，不断积累资源，提升自身应对外部环境的综合能力，通过量变实现质变，可以为客户提供超越企业现有供应范围的产品或服务；同时，注重探索和利用新的市场机会，开拓全新的细分市场，形成新的用户基础，采取同类型企业没有采用过的战略决策，以便实现企业的长期战略目标及可持续发展，实现平台企业的探索式创新[18]。综上所述，平台企业可以利用互补性资产进行利用式创新，然后循序渐进，在利用式创新的基础上进行探索式创新。换言之，利用式创新是平台企业利用互补性资产进行探索式创新的重要路径。据此，提出如下假设。

假设 4：利用式创新在互补性资产与探索式创新的关系中起中介作用。

4. 平台开放度的调节作用

学者对平台开放度的研究，主要集中于探讨平台开放度对企业绩效的影响，如 Laursen 和 Salter[21]认为，创新开放度与开放式创新绩效之间呈倒“U”形关系；Parker 和 van Alstyne[2]研究得出，平台主导企业可以通过开放度的最优化选择，从而影响主导企业以及下游企业的创新；Gebregiorgis 和 Altmann[22]研究表明，平台开放有利于顾客减弱对锁定的担忧，开放水平在新兴服务提供商采用 IT 服务平台方面起着重要作用。平台开放度是平台企业对内部及与其他创新主体进行治理的关键作用，是平台企业创新发展过程中重要的调节变量，学者对平台开放度调节效应的研究较为少见[2]，尤其是缺乏关于平台开放度对互补性资产与双元创新之间关系的调节作用的研究。

本文认为，平台开放度的大小，决定了平台企业撬动和利用平台生态系统中创新主体蕴藏的资源、能力、知识等要素的多少和程度，进而关系到平台企业整合这些创新要素进行利用式创新：①平台开放度较高，平台接口连接的创新主体数量就越多[23]，不同种类和不同端口的用户数量也就越多，它们携带的大量互补性资产就会聚集在平台生态系统中，互联网平台企业能够整合和利用的互补性资产就会越多，有利于补充平台企业的现有技术和能力；②平台开放度较高，互补性资产就越多，可以帮助平台提高产品和服务质量，提升供应效率，降低交易成本[14]，且平台生态系统中的创新主体类型越丰富多样，越有利于对平台企业进行补充，促进企业不断新增现有产品和服务的功能；③平台开放度越高，对创新主体之间的限制就会降低，有利于平台企业在相关领域进行产品和服务的拓展应用。综上，本文认为平台开放度较高时，平台企业可以利用的互补性资产就越多，越有利于平台企业进行利用式创新。据此，提出如下假设。

假设 5：平台开放度对互补性资产与利用式创新的关系起正向调节作用。

本文同时认为，在平台开放度较高时，有利于平台企业利用创新生态系统中的互补性资产进行探索式创新：①平台开放度较高时，平台生态系统中的互补性资产就会越多，平台在进行创新时就有充足的创新资源进行补充，这有利于平台企业克服资源束缚，根据客户需求接受超越现有产品或服务生产的需求订单，并尝试开发新产品、提供新服务或开发新功能[2]；②平台开放度较高时，大量用户进入平台生态系统内，会出现新的市场和新的机会，平台企业可以开拓全新的细分市场，吸引新用户，利用新机会；③当平台开放度较低时，平台企业对用户进入平台生态系统的门槛设置相对较高，会排斥一部分外部创新主体进入系统[2]，这样就会导致平台企业能够利用的互补性资产减少，平台企业受限于资源或能力，探索式创新活动的开展就会随之减少[22]。

由此可见，当平台开放度较高时，互联网平台企业可以利用的互补资产就越多，更加有利于平台企业进行探索式创新。据此，提出如下假设。

假设 6：平台开放度对互补性资产与探索式创新的关系起正向调节作用。

综上分析，提出本文的理论模型（图 1）。

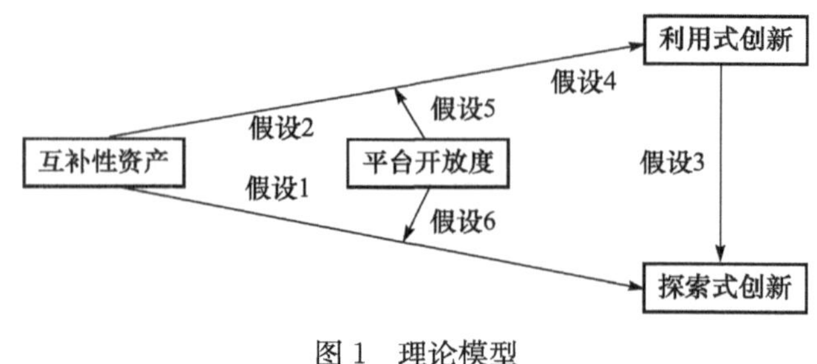

图 1　理论模型

三、研究设计

1. 数据收集

本文选择调研样本企业的标准，主要包括以下四条：①属于互联网平台企业，有别于传统企业，且互联网在企业的经营中起到非常重要的作用；②平台企业设有技术研发部门，对技术创新非常重视；③企业建立的互联网平台对外开放，且有用户在平台上交易，有别于企业内部使用办公平台；④平台企业要有用户基础，以营利为目的，区别于公益类平台组织。调研中较为典型的互联网平台企业，如尚品宅配、中设智控、顺博商城、十记庄园等，这些企业具备互联网平台企业的典型特征，也重视对创新的投入，在业界也具有一定知名度。问卷集中发放时间为 2018 年 8～9 月，共发放问卷 380 份，实际收回 296 份，回收率为 77.89%。为了保证样本是互联网平台企业，确保研究结果的可靠性，一方面，笔者对样本问卷进行一一检查核对，去掉不是互联网平台企业的问卷，去掉基层员工填写的问卷，删除规律作答、填写不完整等无效问卷后，得到 206 家互联网平台企业的问卷，有效问卷回收率为 69.59%；另一方面，通过统计分析，得出的结论符合实际调研过程中的企业创新行为，也与研究假设基本一致，可以反推出收集的问卷具有有效性，能够反映互联网平台企业的特征。样本基本情况见表 1。

表 1　样本基本情况（N=206）

基本特征	具体类别	样本数	比率/%	基本特征	具体类别	样本数	比率/%
职务	中层管理者	155	75.2	员工规模	1～99 人	49	23.8
	高层管理者	51	24.8		100～499 人	107	51.9
成立年限	<5 年	17	8.3		500～1000 人	38	18.4
	5～10 年	85	41.3		>1000 人	12	5.8
	11～15 年	67	32.5	发展阶段	创业初期	12	5.8
	16～20 年	29	14.1		成长阶段	103	50.0
	>20 年	8	3.9		成熟阶段	84	40.8
					转型阶段	7	3.4

注：数据之和不为 100%是数据修约所致

2. 变量测量

为了保证测量工具的信度和效度，研究尽量采用已有的成熟量表，并根据平台互联网企业的具体情况进行微改和调整，以保证测量的准确性。问卷采用 Likert 5 点打分法测量变量，要求问卷填写者根据互联网平台企业自身运营情况进行打分，1～5 分代表“完全不符合”至“完全符合”。调查问卷中主要包括以下量表。

（1）互补性资产。参考 Christmann[24]、Sarkar 等[25]对互联网平台企业互补性资产的测量，主要包括平台企业撬动生态系统中的创新主体“进行销售渠道的建设与维护”等八个测项。

（2）探索式创新。本文根据 Jansen 等[26]、Wei 等[19]的研究，将双元创新划分为探索式创新和利用式创新。用六个测项来测量探索式创新，包括“接受超越企业现有产品或服务的要求”等。

（3）利用式创新。依据 Wei 等[19]、Jansen 等[26]的研究，用六个测项来测量利用式创新，包括“对现有产品或服务进行改良，以适应当前需要”等。

（4）平台开放度。参照 Laffan[27]、Laursen 和 Salter[21]对平台企业的开放度进行测量，包括“与很多外部创新主体建立了合作关系”等六个测项。

（5）控制变量。为了控制其他因素对研究结果的影响，根据已有研究成果和平台企业实际运营情况，本文选取平台企业职务、平台企业营业收入主要模式（赚取差价、生产研发、增值服务、广告）、员工规模、成立年限、发展阶段作为控制变量。

3. 数据分析方法

本文首先采用相关分析进行假设检验的初始测试，然后采用学术界广泛使用的阶层回归，进行中介效应和调节效应的检验。数据处理过程中，为了得到更加简约的模型、保证模型参数估值的稳定性，在进行中介效应假设检验前，将测项进行均值处理；在进行调节效应假设检验之前，对变量进行去中心化处理，目的是尽可能减少多重共线性问题。

四、数据分析与结果

1. 同源性偏差问题检验

为了避免同源偏差对研究结论造成影响，本文采用 Harman 单因素检测方法检验收回的数据，即对问卷中的所有问题项进行探索性因子分析，结果显示：在未旋转时第一主成分只解释了变异量的 34.00%，说明本文不存在同源性偏差问题。

2. 信度和效度检验

变量的信度和效度分析结果见表 2。由表 2 可知，互补性资产、探索式创新、利用式创新和平台开放度的 Cronbach's α 系数分别为 0.831、0.697、0.734、0.805；各变量内部一致性系数均达到 0.7，符合量表值度的可接受标准，表明量表具有较好的内部一致性，问卷信度良好。所有变量的 CR（composite reliability，组合信度）值均大于标准值 0.7，互补性资产与平台开放度的 AVE（average variance extracted，平均变异数萃取量）大于 0.36；探索式创新和利用式创新的平均变异数萃取量 AVE 虽然较低，但是“即使有超过 50%的变异来自测量误，单独以建构信度为基础，研究者可以做出构念的聚合效度是适当的”，由此，探索式创新和利用式创新的量表仍具有可接受的聚合效度，即探索式创新和利用式创新的测量模型之内在质量在可以接受的范围。本文的量表主要参考国内外公开期刊发表和使用的成熟量表，并根据互联网平台企业的具体情况对部分测项内容进行了微调，还进行了小样本预测，因此问卷具有较好的内容效度。

表 2　变量信度和效度分析结果（N=206）

变量	测项数	KMO	Cronbach's α	Bartlett	CR	AVE
互补性资产	8	0.871	0.831	0.000	0.834	0.387
探索式创新	6	0.776	0.697	0.000	0.701	0.286
利用式创新	6	0.775	0.734	0.000	0.740	0.323
平台开放度	6	0.846	0.805	0.000	0.808	0.416

注：KMO 为 Kaiser-Meyer-Olkin，凯泽-迈耶-奥金；Bartlett 为巴特利特

3. 相关性分析

在进行假设性检验之前，对研究中的各变量的均值、标准差和相关系数进行分析，结果见表 3。由表 3 可知，互补性资产与利用式创新（$r=0.698$，$p<0.01$）、探索式创新（$r=0.741$，$p<0.01$）及平台开放度（$r=0.624$，$p<0.01$）显著正相关；探索式创新与利用式创新（$r=0.641$，$p<0.01$）和平台开放度（$r=0.674$，$p<0.01$）显著正相关；利用式创新与平台开放度（$r=0.580$，$p<0.01$）显著相关。初步表明构建的模型、提出的研究假设及选取的控制变量都具有一定的合理性。研究还进行了 VIF（variance inflation factor，方差膨胀因子）检验，均小于 3，远远低于标准值 10，不存在多重共线性问题。

4. 假设检验与结果

（1）主效应及中介效应检验。阶层回归分析结果见表 4。由表 4 可知，模型 1 和模型 2 是以利用式创新为因变量，控制变量、互补性资产对利用式创新的回归分析，结果表明，互补性资产对利用式创新（$\beta=0.652$，$p<0.001$）有显著的正向影响，假设 2 得到验证。模型 3～模型 5 是以互联网平台企业探索式创新作为因变量的回归分析，其中模型 3 是控制变量、模型 4 是互补性资产、模型 5 用于检验利用式创新在互补性资产与探索式创新之间的中介作用。由模型 4 可知，互补性资产对探索式创新产生显著正向影响（$\beta=0.675$，$p<0.001$），假设 1 得到验证；由模型 5 可知，利用式创新对探索式创新产生显著正向影响（$\beta=0.236$，$p<0.001$），假设 3 得到验证；加入利用式创新变量后，互补性资产对探索式创新的影响系数由 0.675 下降为 0.522，且呈显著正相关关系，说明利用式创新在互补性资产与探索式创新之间起着中介作用，验证了假设 4。此外，利用 Amos 软件画出互补性资产、利用式创新与探索式创新的作用路径，描述了利用式创新的中介作用（图 2）。

（2）调节效应检验。平台开放度的调节效应阶层回归分析见表 5。由表 5 可知，模型 1～模型 4 用于检验平台开放度对互补性资产与利用式创新关系的调节作用，模型 4 引入了交互项后，方差解释力 $R^2=56.60\%$依旧没有变化，ΔR^2 从 3.30%降到 0.10%，

表 3　描述性统计与变量间相关系数($N=206$)

变量	均值	标准差	1	2	3	4	5	6	7	8	9	10	11	12
1	2.25	0.43	1											
2	0.48	0.50	0.062	1										
3	0.72	0.45	0.059	−0.311**	1									
4	0.64	0.48	−0.080	−0.148*	0.154*	1								
5	0.09	0.28	0.061	0.153*	−0.036	0.127	1							
6	2.06	0.81	0192**	−0.014	0.263**	0.072	0.061	1						
7	2.64	0.96	0.086	−0.100	0.138*	0.106	0.008	0.490**	1					
8	2.42	0.66	0.064	−0.103	0.235**	0.174*	0.013	0.364**	0.497**	1				
9	3.94	0.55	0.084	−0.023	0.265**	0.285**	−0.031	0.285**	0.098	0.157*	1			
10	3.95	0.51	0.038	−0.033	0.271**	0.220**	−0.003	0.207**	0.093	0.092	0.741**	1		
11	4.07	0.53	0.024	−0.086	0.273**	0.332**	0.023	0.104	0.059	0.164*	0.698**	0.641**	1	
12	3.84	0.60	−0.033	−0.013	0.187**	0.208**	0.029	0.233**	0.175*	0.160*	0.624**	0.674**	0.580**	1

注:1 职务;2 赚取差价;3 生产研发;4 增值服务;5 广告;6 员工规模;7 成立年限;8 发展阶段;9 互补性资产;10 探索式创新;11 利用式创新;12 平台开放度

**、* 分别表示 $p<0.01$、$p<0.05$

表 4　阶层回归分析结果（*N*=206）

类别	利用式创新		探索式创新		
	模型 1	模型 2	模型 3	模型 4	模型 5
职务	0.038	−0.007	0.015	−0.031	−0.030
赚取差价	0.032	−6.028	0.071	0.010	0.016
生产研发	0.258**	0.112	0.261**	0.109	0.083
增值服务	0.330***	0.127*	0.209**	−0.001	−0.031
广告	−0.029	0.085	−0.073	0.045	0.025
员工规模	0.011	−0.099*	0.096	−0.018	0.005
成立年限	−0.029	0.003	−0.004	0.029	0.028
发展阶段	0.067	0.053	−0.032	−0.047	−0.059
互补性资产		0.652***		0.675***	0.522***
利用式创新					0.236***
R^2	0.168	0.532	0.131	0.560	0.588
ΔR^2	0.168	0.364	0.131	0.429	0.029
F	4.990***	24.803***	3.709***	27.676***	27.846***

注：表中系数为非标准化系数

***、**、*分别表示 $p<0.001$、$p<0.01$、$p<0.05$

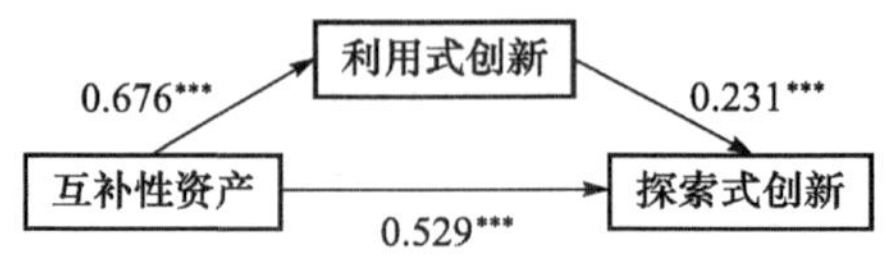

图 2　利用式创新的中介作用

*** 表示 $p<0.001$

交互项“互补性资产×平台开放度”的回归系数不显著($\beta=-0.013$，n.s)，平台开放度对互补性资产与利用式创新关系的调节作用不显著，假设 5 未能通过检验。模型 5～模型 8 用于检验平台开放度对互补性资产与探索式创新关系的调节作用，模型 8 引入了交互项后，方差解释力 R^2 从 63.10%上升到 63.90%，ΔR^2 从 7.20%降到 0.80%，交互项“互补性资产×平台开放度”的回归系数为正且显著($\beta=0.042$，$p<0.05$)，这说明平台开放度对互补性资产与探索式创新关系的调节效应显著，假设 6 得到验证。

表 5　平台开放度的调节效应阶层回归分析（*N*=206）

类别	利用式创新				探索式创新			
	模型 1	模型 2	模型 3	模型 4	模型 5	模型 6	模型 7	模型 8
职务	0.038	−0.007	0.023	0.023	0.015	−0.031	0.011	0.013
赚取差价	0.032	−0.028	−0.032	−0.031	0.071	0.010	0.004	0.000
生产研发	0.258**	0.112	0.107	0.112	0.261**	0.109	0.103	0.088
增值服务	0.330***	0.127*	0.126*	0.121*	0.209**	−0.001	−0.003	0.011
广告	−0.029	0.085	0.062	0.064	−0.073	0.045	0.014	0.008
员工规模	0.011	−0.099*	−0.101*	−0.101*	0.096	−0.018	−0.021	−0.021
成立年限	−0.029	0.003	−0.012	−0.011	−0.004	0.029	0.008	0.005
发展阶段	0.067	0.053	0.052	0.049	−0.032	−0.047	−0.048	−0.041

续表

类别	利用式创新				探索式创新			
	模型 1	模型 2	模型 3	模型 4	模型 5	模型 6	模型 7	模型 8
互补性资产		0.360***	0.282***	0.275***		0.373***	0.263***	0.284***
平台开放度			0.128***	0.125***			0.179***	0.187***
互补性资产×平台开放度				−0.013				0.042*
R^2	0.168	0.532	0.566	0.566	0.131	0.560	0.631	0.639
ΔR^2	0.168	0.364	0.033	0.001	0.131	0.429	0.072	0.008
F	0.499***	24.803***	25.408***	23.045	3.709***	27.676***	33.396***	31.240*

注：表中系数为非标准化系数

***、**、*分别表示 $p<0.001$、$p<0.01$、$p<0.5$

为了确保平台开放度在互补性资产与探索式创新之间起调节效应的可行性，进一步检验调节变量的调节方向，根据学术界的常规做法，选取均值加减一个标准差作为临界值代入回归方程，获得调节变量平台开放度对互补性资产与探索式创新关系的调节效应图（图 3）。由图 3 可知，在平台开放度较高时，互补性资产对双元创新的正向影响比平台开放度低时的效果更加明显，即平台开放度对互补性资产与探索式创新之间的关系起正向调节作用。

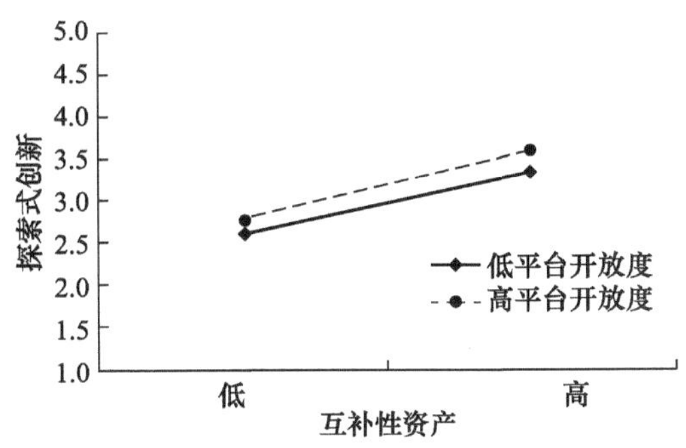

图 3　平台开放度对互补性资产与探索式创新的调节效应图

五、研究结论与讨论

1. 研究结论与理论贡献

通过上述分析与检验，本文有如下发现：①互补性资产正向影响探索式创新和利用式创新。一方面，生态系统中的互补性资产可以帮助平台企业快速积累创新发展所需要的资源和能力，有助于平台企业实现创新收益，达到短期目标；另一方面，互补性资产有助于创造新资源和新能力，从而有利于平台企业在平台架构、产品、工艺、规则等方面进行创新。由此，平台企业可以利用的互补性资产越多，越有利于进行利用式创新和探索式创新。②利用式创新正向影响探索式创新，利用式创新在互补性资产与双元创新之间起部分中介作用。平台企业利用互补性资产进行利用式创新，如对现有产品或服务进行升级、提高供应效率、拓展应用领域、丰富产品种类、扩展服务功能；同时，平台企业通过这些利用式创新，不断积累创新所需资源和能力进行探索式创新，如开发新产品、提供新服务、增加新功能、采用新战略或战术。由此，互补性资产有利于平台企业进行利用式创新，进而促进探索式创新的发展。③平台开放度对互补性资产与探索式创新之间的关系具有正向调节作用。即当平台开放度较高时，互补性资产对探索式创新的正向影响将会得到强化；当平台开放度较低时，互补性资产对探索式创新的正向影响相对减弱。

本文的理论贡献和创新点主要包括：①明晰了平台创新生态系统中的互补性资产对平台企业双元创新的影响机制；②研究平台开放度对互补性资产与探索式创新关系的调节作用，拓展了影响互补性资产与平台创新之间关系的边界条件知识。

2. 研究局限

本文还存在一些局限：①前文提出的假设 5 未能得到验证；②平台开放度还可以细分为平台开放深度和平台开放广度两个维度，这两个维度对互补性资产与双元创新之间关系的影响尚未进行探讨。由此，未来可以收集数据对这些问题进行深入研究。

参考文献

[1] MCINTYRE D P，SRINIVASAN A. Networks，Platforms，and Strategy：Emerging Views and Next Steps. Strategic Management Journal，2017，38 (1)：141-160.

[2] PARKER G，VAN ALSTYNE M. Innovation，Openness，and Platform Control. Management Science，2017，64 (7)：3015-3032.

[3] 吴亮，赵兴庐，张建琦. 以资源拼凑为中介过程的双元创新与企业绩效的关系研究. 管理学报，2016，13 (3)：425-431.

[4] EISENMANN T，PARKER G，VAN ALSTYNE M. Platform Envelopment. Strategic Management Journal，2011，32 (12)：1270-1285.

[5] TIWANA A. Evolutionary Competition in Platform Ecosystems. Information Systems Research，2015，26 (2)：266-281.

[6] STIEGLITZ N，HEINE K. Innovations and the Role of Complementarities in a Strategic Theory of the Firm. Strategic Management Journal，2007，28 (1)：1-15.

[7] 张镒，刘人怀，陈海权. 商业生态圈中平台企业生态优势形成路径——基于京东的纵向案例研究. 经济与管理研究，2018，39 (9)：114-124.

[8] JACOBIDES M G，CENNAMO C，GAWER A. Towards a Theory of Ecosystems. Strategic Management Journal，2018，39 (8)：2255-2276.

[9] ALEXY O，WEST J，KLAPPER H，et al. Surrendering Control to Gain Advantage：Reconciling Openness and the Resource-Based View of the Firm. Strategic Management Journal，2018，39 (6)：1704-1727.

[10] HOSKISSON R E，EDEN L，WRIGHT L M，et al. Strategy in Emerging Economies. Academy of Management Journal，2000，43 (3)：249-267.

[11] ROY R，COHEN S K. Stock of Downstream Complementary Assets as a Catalyst for Product Innovation during Technological Change in the US Machine Tool Industry. Strategic Management Journal，2017，38 (6)：1253-1267.

[12] KANI M, MOTOHASHI K. Determinants of Demand for Technology in Relationships with Complementary Assets among Japanese Firms. China Economic Journal, 2017, 10 (2): 244-262.

[13] FANG C, KIM J H. The Power and Limits of Modularity: A Replication and Reconciliation. Strategic Management Journal, 2018, 39 (9): 2547-2565.

[14] ALVAREZ-GARRIDO E, DUSHNITSK Y G. Are Entrepreneurial Venture's Innovation Rates Sensitive to Investor Complementary Assets? Comparing Biotech Ventures Backed by Corporate and Independent VCs. Strategic Management Journal, 2016, 37 (5): 819-834.

[15] WU B, WAN Z X, LEVINTHAL D A, Complementary Assets as Pipes and Prisms: Innovation Incentives and Trajectory Choices. Strategic Management Journal, 2014, 35 (9): 1257-1278.

[16] 汪旭晖，张其林. 平台型电商声誉的构建：平台企业和平台卖家价值共创视角. 中国工业经济，2017，(11)：174-192.

[17] FLOYD S W, LANE P J. Strategizing throughout the Organization: Managing Role Conflict in Strategic Renewal. Academy of Management Review, 2000, 25 (1): 154-177.

[18] LAVIE D, ROSENKOPF L. Balancing Exploration and Exploitation in Alliance Formation. The Academy of Management Journal, 2006, 49 (4): s111-s132.

[19] WEI Z, YI Y, YUAN C. Bottom-up Learning, organizational Formalization, and Ambidextrous Innovation. Journal of Organizational Change Management, 2011, 24 (3): 314-329.

[20] KOSTOPOULOS K C, BOZIONELOS N, SYRIGOS E. Ambidexterity and Unit Performance: Intellectual Capital Antecedents and Cross-Level Moderating Effects of Human Resource Practices. Human Resource Management, 2015, 54 (S1): 777-780.

[21] LAURSEN K, SALTER A. Open for Innovation: The Role of Openness in Explaining Innovation Performance among UK Manufacturing Firms. Strategic Management Journal, 2006, 27 (2): 131-150.

[22] GEBREGIORGIS S A, ALTMANN J. IT Service Platforms: Their Value Creation Model and the Impact of Their Level of Openness on Their Adoption. Procedia Computer Science, 2015, 68: 173-187.

[23] LEE C, LEE D, HWANG J. Platform Openness and the Productivity of Content Providers: A Meta-Frontier Analysis. Telecommunications Policy, 2015, 39 (7): 553-562.

[24] CHRISTMANN P. Effects of "Best Practices" of Environmental Management on Cost Advantage: The Role of Complementary Assets. The Academy of Management Journal, 2000, 43 (4): 663-680.

[25] SARKAR M B, ECHAMBADI R, CAVUSGIL S T, et al. The Influence of Complementarity, Compatibility, and Relationship Capital on Alliance Performance. Journal of the Academy of Marketing Science, 2001, 29 (4): 358-373.

[26] JANSEN J J P, VAN DEN BOSCH F A J, VOLBERDA H W. Exploratory Innovation, Exploitative Innovation, and Performance: Effects of Organizational Antecedents and Environmental Moderators. Management Science, 2005, 52 (11): 1661-1674.

[27] LAFFAN L. A New Way of Measuring Openness: The Open Governance Index. Technology Innovation Management Review, 2012, 2 (1): 18-24.

系统论视角下旅游学科提升发展的思考①

20 世纪 80 年代，旅游学科在引进国外部分研究成果以及对世界旅游情况介绍的基础上得以创立，中国高等院校出现了最早的旅游管理本科专业，此后，尽管对于旅游学科的科学性存在质疑，但旅游学科建设问题一直受到业内学者的关注与重视。在我国高等院校本科教育以“学科门类”“学科大类”（一级学科）“专业”（二级学科）构成的专业层次设置体系中，旅游管理很长一段时间都是工商管理学科大类（一级学科）下的一个专业（二级学科）。直至 2014 年，在教育部发布的《普通高等学校本科专业目录》中，“旅游管理”升格为一级学科门类，与“工商管理”类平级，实乃旅游学术共同体在旅游学科建设上多年努力的成效见证。但是，目前在国务院学位办的学科专业目录里，旅游管理仍然是工商管理下的二级学科，这样一种本科专业和学位学科级别不同的安排或称为错位学科设置，会给旅游管理学科的进一步发展带来混乱，不利于学科建设和人才培养[1]，因此，提升学科地位有助于实现更加高质量的旅游学科建设与发展。

一、旅游学科提升发展的价值驱动

任何一个学科的发展都建立在其价值基础之上，20 世纪 80 年代之前，中国高校的管理学科几乎没有什么基础，但是社会经济发展的迫切需要凸显了管理学科的价值，可以说，中国管理学在过去 40 年里所取得的学科发展成绩是来自社会所赋予该学科的核心价值体现。而当前的旅游学科正面临着同样的社会需求和价值表现。

1. 旅游产业广泛实践彰显了旅游学科的社会价值

旅游业是全球第一大产业，根据 2017 年世界旅游组织的数据，旅游业已成为世界上增长最快、最重要的经济产业之一，对全球 GDP 的贡献约为 10%，并为国际社会创造了 10%的就业机会，惠及了全球各旅游目的地和所在社区。就中国而言，2018 年全国旅游业对 GDP 的综合贡献初步测算为 9.94 万亿元，占 GDP 总量的 11.04%，旅游就业占全国就业总人口的 10.29%[2]，旅游业已经成为中国的国家战略产业，凸显了其在新时代国家经济社会发展全局中日益重要的功能与作用。同时，旅游业不但有经济效用，而且可以净化人们的思想，给公众带来快乐，是公认的幸福产业，也是最能体现大众对美好生活向往的产业。可以说，旅游已经嵌入现代性社会下人们的思维方式和观念转变中，在中国传统社会中被赋予了全新的时代意义，不但改变着社会经济产业格局和国家的国际影响力，而且正在改变着千千万万普通群众的日常生活和幸福观念。如同早期的管理学科一样，产业发展赋予了旅游学科越来越重要的社会价值，这一价值迫切要求以旅游活动和旅游产业为研究对象的旅游学科为更加广泛的产业实践提供更多、更好、更全面的支撑保障，而进一步提升旅游管理学科的学科地位即是重

① 本文原载《旅游学刊》，2019，34（12）：1-3. 作者：刘人怀，刘小同，文彤.

要途径之一。

2. 旅游研究持续发展奠定了旅游学科的理论价值

旅游学科是一门综合性学科，它是自然科学、技术科学、工程科学、社会科学相结合的学科，管理学、经济学、地理学、社会学、人类学、历史学、生态学等不同学科都参与旅游研究当中，是非常特殊的学科[3]。在旅游科学研究发展历程中，中国旅游学术圈历经了几代人的学术传承，从第一代学者以其他学科背景跨界进行旅游研究到现在不断壮大的旅游管理科班青年学者，旅游领域的研究正在实现从多学科向跨学科的转变。经过多年的探索和讨论，中国旅游研究不仅明确了自身独立的研究对象，构建了清晰稳健的基础理论体系，发育了卓有成效的方法工具，并呈现出家族化、哲学化和知识溢出的明显走向[4]。旅游研究从基础理论、跨学科交叉研究、案例实证与方法论等三方面体现出了独特的研究视角与系统化的研究方法[5]；从旅游研究对象、知识体系及研究方法上看，旅游学科具有极为明显的、区别于工商管理以及其他管理类一级学科的独立体系[6]。旅游学者坚持不懈的深入研究已经促使旅游学科在一定程度上形成了区别于相关学科的创新与差异，奠定了自身学科的理论价值，并且开始向其他传统基础学科形成一定的知识输出[1]。因此，中国旅游学科的提升发展有助于形成更加完整的知识体系，促进旅游领域的自主理论创新，进一步扩大夯实学科的理论价值，构建与社会价值遥相呼应的全面格局。

二、旅游学科提升发展的系统论思考

钱伟长先生曾经专门题词指出，要“以系统工程的科学理论指导旅游事业的运行”。从系统论的视角来看，旅游产业实际上是一个整体的外向系统，不同子系统互相合作产生的原动力，外部关联系统融合支撑产生的辅动力，共同组成旅游发展的动力，而内部不同子系统、外部不同关联系统之间往往又存在局部的关系和问题，这就使得旅游的问题既可能是大系统引发小系统产生的，也可能是小系统反馈大系统形成的[3]。旅游学科的发展是一个系统工程，需要关注加强每一个子系统的独立发展和互动联合，才能在社会价值和理论价值共同驱动下实现进一步的地位提升。

1. 打造内核，反哺外联，凝练提升学科理论贡献

在系统工程学中，某一系统与外部关联系统之间主要通过要素的输入与输出发生关系，形成该系统的输入流与输出流，其中，输出流是系统功能的表征，也是衡量系统的核心指标。虽然旅游学科已经产生一定的知识输出，但过往更多以其他学科对旅游学科的知识导入为主，这也是旅游学科“遭受质疑”的主要原因，未来需要更广泛、更深入的知识溢出来平衡与其他学科的关系地位。实际上，每一个学科的建立与发展，每一个学术问题的提出和解决，都是在质疑之中砥砺前行的，正是这种“质疑”给旅游研究提供了宽广的理论空间。全球经济一体化和社会流动性加速正在引发新的消费变化和产业演变，旅游发展中这个特征尤为突出，不同产业与旅游的融合带来的新业态、技术进步衍生出的旅游消费创新等都可成为旅游研究的新热点[3]。更进一步来讲，系统工程学强调在综合把握系统全局的基础上聚焦重点问题，以关键关系流带动整个系统的发展。新时代的中国旅游研究需要在过去 40 年形成的多元化格局基础上提炼若

干具有引领性的学科研究核心，形成多个权威的旅游研究团队或流派，打造在学术生态圈中能够掷地有声的理论内核，代表并引领整个旅游学术共同体在学术圈中争取更多话语权，强化知识溢出功能，通过凝练和深化理论贡献树立旅游学科的学术威信，才能真正直面“质疑”，提升旅游学科地位。

2. 重视教育，培养人才，构建强化学科社会价值

正如系统工程学的一般系统结构理论中强调基层次对系统功能的重要性一样[7]，所有学科建设的根本目标是提供更好的学科教育、培养人才。回顾20世纪50年代石油勘探开采对地质专业人才的需求以及20世纪80年代改革开放对管理专业人才的需求，我们可以看到，为满足国家经济建设和产业发展需要而大力培养专业人才是地质学科、管理学科在当时得以实现全面大发展的根本原因。因此，旅游学科的提升发展不能仅关注理论研究，更应该强调学科的教育本质，为国家培养旅游产业人才，尤其是实践型人才。由于旅游专业缘起的学科背景不同，我国高等院校中旅游专业归属确实存在明显的多元化错位，这一现象有其相应的历史渊源与当时情境下的现实合理性，但也会造成一种学科专业设置与管理混乱的印象，可能引起大量生源的流失[8]，导致现有旅游专业人才供给不足，直接制约旅游产业的进一步发展。人才是未来旅游学科的“种子”和旅游产业的主力军，人才的数量与质量是任何一级学科的立足根本，旅游学科提升发展必须坚持“两条腿”走路，除了理论体系构建、学术队伍建设之外，还应该加强课程、教材、实践等专业教学体系的建设，通过独立的学科教育培养来增强各层次学生的旅游专业归属感，进而影响其学习生涯与职业选择，促使更多旅游专业人才在旅游行业就业，为旅游产业的强势发展储备高专业黏性的旅游人才资源，强化实现旅游学科的社会价值，才能借助清晰有序的旅游学科教学研究格局促进社会各界对旅游学科的认识与认同，促成旅游学科地位的真正提升。

参考文献

[1] 保继刚．建设旅游管理一级学科，加快旅游人才培养．旅游学刊，2015，30（9）：1-2.

[2] 2018年旅游市场基本情况．https：//www.gov.cn/xinwen/2019-02/13/content_5365248.htm [2019-02-13].

[3] 刘人怀，文彤，闫婷婷．系统论视角下的旅游发展与旅游研究：中国工程院工程管理学部刘人怀院士访谈．社会科学家，2014，(11)：3-6.

[4] 马波．旅游学科升格的理路探析．旅游学刊，2016，30（10）：25-32.

[5] 刘苏衡．关于旅游一级学科建设的几点思考//中国旅游研究院．中国旅游科学年会论文集．北京：中国旅游研究院，2017：233-242.

[6] 左冰，林德荣．交叉与融合：旅游管理一级学科与其他学科之间别的关系．旅游学刊，2016，31（10）：21-23.

[7] 林福永，孙凯．复杂网络关系流与行为关系定理：一般系统结构理论在复杂网络中的应用．系统工程理论与实践，2007，(9)：136-141.

[8] 保继刚，赖坤．旅游管理学科内涵及其升级必要性．旅游学刊，2016，31（10）：14-16.

Guest editorial①

中文摘要

本期特刊社论标题为“新经济时代企业家职业的冲突管理”，就是为学者和管理者对企业家职业的冲突管理提供新的洞察力和视野。

Introduction to a special issue on conflict management in entrepreneurship in the new economy

This special issue of conflict management in entrepreneurship offers new insights and new perspectives on conflict management in entrepreneurship for both scholars and practitioners.

The first paper is "Perspectives on disruptive technology and innovation: Exploring conflicts, characteristics in emerging economies" by Wan Liu, Renhuai Liu, Hui Chen and Jet Mboga. Their study summarizes and compares the characteristics of disruptive technologies/innovation from the dimension of the nature of technology, the institution and the market trajectory in the context of western developed countries and emerging economies through an analytical review of disruptive technologies/innovations. They discuss the value and uniqueness of Christensen's theory applied to the Chinese context and take China's high-speed train as an example to illustrate the main ideas.

The second paper in the special issue is "How harmonious family encourages individuals to enter entrepreneurship: A view from conservation of resource theory" by Weichun Zhu, Jinyi Zhou, Wai Kwan (Elaine) Lau and Steve Welch. With two large longitudinal data sets from different countries, the findings of the paper are twofold. First, personal support from original family and current family can both enhance and increase their entrepreneurial activities. Particularly, not only maintaining a good relationship with parents in adolescence can help individuals gain more support but also a harmonious relationship between parents can provide individuals with more emotional and psychological resources that can help them to start their own businesses. Second, support from the current family can also provide more resources for the individual. Specifically, a more balanced and harmonious relationship with a spouse can reduce the work stress of an individual and motivate he/she to pursue more adventurous career and start his/her own entrepreneurial endeavor.

The third paper is "Entrepreneur's political involvement and inter-organizational

---

① Reprinted from *International Journal of Conflict Management*, 2020, 31 (3): 309-311. Authors: Renhuai Liu, Steven Si, Song Lin, Dean Tjosvold, Richard Posthuma.

conflict resolution in China's transition economy" by Aiqi Wu, Xiaotong Zhong and Di Song. This paper explores the impact of entrepreneurs'political involvement of private-owned enterprise (POEs) on the selection of two modes of interorganizational conflict resolution (private and public) in the transition economies. The paper uses POEs in the transition period of China in 2000 as samples and the results show that the higher the degree of political involvement of entrepreneurs, the more likely they are to rely on the public approach. Furthermore, POEs are generally more satisfied with conflict resolution using the private approach than the public approach. The study provides some interesting findings, which shed light on how the increase of the degree of regional marketization will weaken the positive relationship between POEs' adoption of the private approach and satisfaction of resolution.

The fourth paper is "Team leader's conflict management styles and innovation performance in entrepreneurial teams" by Jielin Yin, Muxiao Jia, Zhenzhong Ma and Ganli Liao. The study arrives at three conclusions. First, team leaders'cooperative conflict management style promotes team passion that can effectively improve team innovation performance. Second, team passion mediates the relationship between team leaders' cooperative conflict management style and team innovation performance to some extent. Third, team emotional intelligence plays a moderating role in the relationship between team leaders' cooperative conflict management style and team passion.

The fifth paper is "Star (tup) Wars: Decoupling task from relationship, conflict" by Kozusznik Malgorzata. The authors conducted three studies of startup members in three countries to explore the moderating effect of conflict behaviors and related coping strategies on task and relationship conflicts in startup teams.

The sixth paper is "Exploring the relationship between entrepreneurial failure and conflict between work and family from the conservation of resources perspective" by Xiaoyu Yu, Xiaotong Meng, Gang Cao and Yingya Jia. Using the data of entrepreneurial enterprises in China, the research examines the relationship between entrepreneurial failure and work-family conflict (WFC)/family-work conflict (FWC) and the moderating role of perceived time control and organizational slack. They found that entrepreneurial failure has a significant positive impact on FWC, whereas entrepreneurial failure has a positive but not significant impact on WFC. In addition, perceived time control and organizational slack significantly weaken the positive impact of entrepreneurial failure on WFC/FWC.

The seventh paper is "The effect of digital transformation strategy on performance: The moderating role of cognitive conflict" by Hecheng Wang, Junzheng Feng, Hui Zhang and Xin Li. The authors investigated how digital transformation strategy (DTS) affects short-term and long-term financial performances and the impact of boundary conditions of TMT cognitive conflict on organizational performance improvement delivered by DTS. Based on the study of 156 Chinese enterprises, the authors found that DTS can promote short-term and long-term financial performances. The results show that the relationship between DTS and short-term financial performance is inverted u-shaped moderated by the cognitive conflict of TMT, and there remains a positive moderating effect on the relationship between DTS and long-term financial performance.

The eighth paper is "Conflicts between business and government in bike sharing system" by Hong Yang, Yimei Hu, Han Qiao, Shouyang Wang and Feng Jiang. There are task and interest conflicts between bike-sharing companies and government officials, and the conflict process is dynamic and interactive because of different perspectives. They analyze the dynamic interaction between the regulations of governments and strategies of bike-sharing companies from the perspective of evolutionary game theory and provide some management insights and policy implications.

The ninth paper is "Effects of CEO humility and relationship conflict on entrepreneurial performance" by Yi Li, Feng Wei, Siyue Chen and Yushan Yan. Based on a questionnaire survey on CEOs and their entrepreneurial team members of 171 start-ups in Shanghai, China, the authors make four main conclusions. First, the humility of the CEO can reduce relationship conflict in entrepreneurial teams. Second, in entrepreneurial teams, CEO political skills moderate the link between CEO humility and relationship conflict. The weaker the CEO political skills, the stronger the influence of CEO humility on relationship conflict. The stronger the political skill of CEO, the weaker the effect of CEO humility on relationship conflict. Third, relationship conflicts in entrepreneurial teams are negatively related to entrepreneurial performance. Finally, CEO political skill moderates the mediating effect of entrepreneurial team relationship conflict on the CEO humility-entrepreneurial performance link.

The last paper is "How social entrepreneurs' attention allocation and ambidextrous behavior enable hybrid organization" by Wenzhi Zheng, James Bronson and Chunpei Lin, This paper, based on the paradox theory, investigates the allocation of social entrepreneurs' attention and their resource actions that lead to hybrid organizations. The authors make an empirical research on social entrepreneurship in China and have several findings. First, the attention focus of social entrepreneurs influences the way in which organizations acquire resources. An external attention focus is conducive to the acquisition of external resources and the improvement of social performance, whereas internal attention focus will lead to ambidextrous behavior and the acquisition of both internal and external resource advantages, to improve the overall social entrepreneurship performance. Second, the internal attention and paradoxical thinking of social entrepreneurs are more conducive to high entrepreneurial performance. At last, normative pressure positively moderates the relationship between social entrepreneurs' internal attention and strategic human resource management.

The rapidly growing research on conflict management covered a variety of research topics in the field of conflict management over the past decade, one of which is the relationship between conflict and entrepreneurship. Despite the fact that there are a number of studies on how conflict management stimulates entrepreneurship research, how to relate conflict management with entrepreneurship and how to deal with the conflict management issues in entrepreneurship management in the current digital economy remain to be better answered and further explored. This special issue has addressed these important but underexplored issues with different perspectives on this research area - Conflict Management in Entrepreneurship and provided new research findings of conflict and entrepreneurship or new ventures for both emerging economies and matured economies.

发挥社会创业的重要作用①

社会创业无处不在，在我国新发展时期，无论是新发展理念还是新发展具体内容，社会创业都是一种不可缺少的推动与发展要素。改革开放 40 多年来，我国产生了华为、腾讯等著名企业，它们通过创新产品与服务填补了市场的需求与大众的需要。但是过去几十年来，我国在推动发展经济创业并取得伟大成就的同时，社会创业没有取得同步的发展。其实，我国的大众创业，应该将经济创业与社会创业有机结合起来，它们是我国创业推动经济发展与美好社会建设的两个方面，只有二者共同发挥作用，才能使我国的经济与社会得到同步的发展，实现两个一百年中第二个一百年的奋斗目标，即到 2049 年中国建国 100 年之际，实现中华民族的强国梦。

社会创业的作用与功能是多方面的。很多社会问题从社会创业的视角来看，就是发现创新点，通过创新与实践社会创业解决社会中存在的种种问题，并可以发现社会创业的机会和潜在的市场所在。社会创业创新推动人类生活更美好，我们的社会更美好，这是社会创业的意义与价值所在。早在 18 世纪，亚当・斯密在《道德情操论》中便提出，市场经济的繁荣并不会伴随着社会公益的增长。近几十年，世人过分追逐经济的发展，给社会的持续发展留下了问题，如环境污染、气候变暖、贫富分化等。然而，越来越多的现象证明，仅依赖于经济创业的成功，会出现诸多的社会问题。无论是政府还是企业是难以有效解决的。因此在推进发展经济创业的同时，相应的社会创业必不可少。尤其在经济欠发达的发展中国家，社会福利体系不完善的情况下，社会问题会显得更加严峻。因此社会迫切需要通过各种创新的模式来填补市场的漏洞，这些为社会创业的发展提供了大量空间与社会创业机会。

基于上面的情景与重大国家社会需求，浙江大学求是讲座教授斯晓夫等中外社会创业领引学者结合中外社会创业前沿研究理论与实践案例，编著了《社会创业：理论与实践》第二版。这本全国性社会创业教材的最大特点是基于中国国情论述中国创业问题，理论联系实际。该书运用了世界上最新的社会创业研究成果，同时采用了大量的中国社会创业案例。我阅读之后，深深感到该书各位专家学者的用心，以及各自对于我们国家美好社会建设的责任。该书的领头人斯晓夫教授在 30 多年前是我担任上海工业大学副校长兼经济管理学院院长时的助理，我们经常在一起讨论管理问题，我们还在 1990 年也是机械工业出版社一起出版了管理专著。

最后，特别感谢该书的组织者斯晓夫教授以及各位作者邀请我为这本社会创业管理教材写序。我在序里谈了一些我对社会创业的看法，供读者参考。

① 本文是《社会创业：理论与实践》序，北京：机械工业出版社，2022.

第十三章　科技管理

谈谈科研中的几个问题①

党的教育事业需要的是德才兼备的教师。作为一名合格的大学教师，政治上，要坚持党的教育方针，坚持四项基本原则；业务上，要既搞教学，也搞科研。教学与科研好比一个人的两条腿，我们必须两条腿走路。仅搞教学而不搞科研，或者仅搞科研而不搞教学，都不能算是一个合格的大学教师。只有把科研搞好了，才能促进教学的发展。今天，我想结合自己在科研中的一些经历与体会，谈几个问题。

任何人搞科研，其目的都是要获得科研项目的成功。而要达到这一目的，能在科研上有所建树和有所贡献，每个人可以根据自己的具体条件，走不同的道路，这是由每个人的素质、性格、基础知识以及所处的环境（包括大、小环境）等各种因素决定的。比如，从上海到北京，可根据每个人的不同情况选择不同的路线，可以坐火车，也可以乘飞机，可乘早班机，也可乘晚班机，可以走济南的航线，也可以走合肥的航线。搞科研也是如此，要达到目的，应根据个人的具体情况与特定条件，走不同的道路。对别人的经验，只能参考、借鉴，而不能完全套用。

党的十一届三中全会以来，搞科研的环境好了，主管部门和学校领导都很支持大家搞科研，为我们开创了一个较好的外部条件。因此，现在的关键就在于个人的内在因素。从个人内在因素的角度来看，要取得科研项目的成功应该具备哪些条件呢？

第一，要有远大的志向。这表现在两个方面：①应该把搞科研看成是我们民族的需要、社会主义祖国的需要。对这一点，在前几次有关的会上，我都反复地强调过。我们中华民族是世界上人口最多的民族，应该在世界上有一定的地位，但目前我们国家在科学技术上还比较落后，国家和人民迫切需要我们拿出更多的科研成果来。每个人都要为中华民族的振兴、祖国的腾飞而奋斗，要在科学研究中贡献自己的力量。搞科研，就要从这样的高度来认识。②要树立在各自的业务领域中有所创新、有所建树的雄心壮志。现在，在每个科技领域中都有许多前沿科学的重大课题，如我国已经提出了高技术范围内七个重大科技领域中的十五个重大科研项目。每个人都应该具有在各自领域的科研中，力求达到国际水平的理想与抱负，力争自己在社会主义四个现代化中作出贡献。就好比每个体育运动员在奥运会上，都要力争在各自参赛的项目中创纪录、拿金牌，为国争光。我们应该有这样的远大目标：把我们学校建成符合社会主

---

① 本文是在上海工业大学经济管理学院科学报告会上的讲话（按录音整理），上海，1987 年 4 月 7 日；高教研究，1987，1（2）：15-20.

义现代化要求的高等学府，并力争建成具有国际水平的第一流大学。每个教师，要争取拿到科研上的金牌。当然，要求每个人的科研成果都达到国际水平，都能拿金牌，是不可能的。但是，每个教师首先必须有搞科研的积极性，要有远大的志向、奋斗的目标，不能只安于现状，只求把书教好就行了。一个安于现状、不求有所作为的教师不能算是一个合格的大学教师。

第二，要有百折不挠、艰苦奋斗的精神。搞科研并不是一件轻松愉快的事情，而是一件十分艰苦的事业，将会遇到许多想象不到的困难和障碍。这一点，老同志都是有体会的。十一届三中全会以来，在党中央的正确指引下，大气候变了，搞科研的环境很好，当然，小障碍还是有一些，如同志之间、学校之间、学派之间还是会有些矛盾的，但总的形势是好的。但是，不管遇到什么障碍，即使是天大的困难，在科研的道路上，也必须以大无畏的精神和排山倒海的意志，冲破急流险滩，一往无前。1964年，我搞了个精密仪器方面的项目。当时，我在兰州大学工作不久，只有二十三四岁，去一个工厂里找课题，厂里告诉我，在60年代初期，我国曾击落了一架侵入我国领空的美国“P2V”型低空侦察机，该机中有一台测高度的精密仪器，它是该机的心脏部件，其核心是一块锯齿形波纹圆板，它的功能是实现飞机的超低空飞行，帮助飞机躲过我国警戒雷达的监视。研制这台精密仪器，关键是解决一个力学上的问题，这个问题被列为国家的攻关项目。我跃跃欲试，回校后向党支部书记做了汇报，但没得到支持，理由是我教学任务重又兼任班主任工作，应该把本职工作搞好，哪有精力去搞什么研究。我只好违心地答应放弃这项研究课题，但我仍利用业余时间进行研究。由于得不到学校的支持，加上课题本身难度又大，搞了一段时间没取得什么进展，又因1965年下乡搞“四清”，就更没有什么时间了。接着“文革”开始，我由一位“红专旗手”突然变成了“修正主义苗子”和“牛鬼蛇神”，被关进了“牛棚”，一待就是两个多月，从“牛棚”出来后，因靠边站而不准工作，我就利用这段“空余”时间，又暗暗地开始了已经中断很久的研究。谁都知道，这要冒很大风险的，当时，我们学校抓阶级斗争特别厉害，已经整了几百人，谁也不敢搞什么研究。作为知识分子，我坚信从事科学研究是国家繁荣富强的需要，搞研究是为了国家而并非为了个人。因此，我躲在家里继续研究这个课题，为了掩人耳目在家门口挂了个竹帘子，这可以挡住来人的视线，而我却可以洞察到室外的一切。终于在1968年我完成了自己的研究课题，并写出了论文，但论文却无处发表，更谈不上推广应用了。直到1969年我又为工厂搞成了几个项目后，当时又正值学校要求我编一本论文专集，我才有机会把自己的文章拿出来。但万万没有料到，系里“造反派”的头头竟说它是什么基础理论的东西，以不宜刊登为借口，又退了回来。直到1972年、1973年有两本全国性的属于我这一行的科技杂志（《数学的实践与认识》和《力学杂志》）复刊后，我才算又有机会投递出自己的文章。很快，我的文章被两份杂志的创刊号录用，但令人费解，却迟迟未予发表。到1976年底，“四人帮”倒台后，新来的党总支书记，一位正直的老干部，才把事实真相告诉我，原来《力学杂志》接连来过5封信，要学校出具我的政治身份证明信，就是那位“造反派”的头头，虽不能在我“解放”之后再说什么，但却扣压信件，用拖延时间的办法，把这件事给无限期地拖压下来。直到那位老干部担任总支书记以后，我的文章，即《波纹圆板的特征关系式》，才在“文革”后恢复的《力学学报》创刊号上登了出来。这篇在“文革”中“偷偷”写成的科研论文得了中国科学院重大科研成果奖。由此可见，在过去，要搞科研不但困难重重，还要冒政治上的风险。现在大气候改变了，但个别的压制情况还是有，特别是个别的基层领导有时压制得很厉害。因为我有过这方面的痛苦经历，所以我作为管理学院的第一任院长，坚决支持大家搞科研，愿意做大家的铺路石，决不允许有人设障碍、搞压制；有受压制的，可以直接向我告状，我给你们撑腰。当然，对于个人，要有百折不挠的奋斗精神，一方面，要经受得起在科研中碰到的种种困难，另一方面，还要承受得住别人打击、压制造成的痛苦。“文革”中，有人批判我是“走资产阶级成名成家的道路”，甚至还攻击说我的论文“就是给人擦屁股人家也不要”。你听了这些话就不干了？那不行，要顶得住，不能因流言蜚语而退缩不前，应该抱着为国家作贡献的坚定信念，勇往直前。这样，科研才能成功，才能出成果。

第三，要具备一定的基础知识。在这方面，每个教师都应当自觉地积极地锻炼自己，不断地提高自身的科研能力，打好基础，这当中包括外语、数理化、计算机、文学和本专业的技术基础和专业知识等。

从外语基础讲，外语应懂得越多越好，越精越好。当然事实上对每个人来说，不可能掌握很多种外语。因此，在学习外语方面应有一个战略，首先，要擅长、精通一门外语，在这个基础上再去搞第二门、第三门。首先要精通哪一门外语？我认为以英语为好。现在全世界的科技文献是每十年增长一个数量级，其中，60%是英文科技文献，主要的发达国家的科技文献大都以英文出版。所以，选择英语作为第一门外语，对搞科研很有好处。当然，如果做翻译又另作别论。比如，你去学一门尼泊尔语，这作为翻译是必要的，但对搞科研则用处不大。如果学第二门外语，我建议学俄语。全世界科技文献中俄语版约占 20%，居第二位。两个超级大国，在经济、科技上都是领先的，首先学好这两门外语，对搞科研是大有帮助的。很难想象一个不懂外语的科技人员，能在 80 年代的今天搞出重大的科研成果。当今世界，时间就是效益，要等翻译家为你翻译出科技文献来，那就晚得多了。全世界 40 多亿人口，那么多的国家，那么多的大学和科研机构，同一个课题就有许多不同肤色的人同时在搞，谁搞得快，胜利就属谁，谁就得金牌。所以，一定要具备扎实的外语基础，要能直接看外国科技文献，缩短收集信息的时间，才能赶在前头。

关于数理化基础。当前的科学可以分成两大类型：一类是以文字描述性为主的科学；一类是以数学工具为主的科学，又叫精密科学。你们基础教研室的许多课程属于精密科学。我院的预测研究所，在预测技术中要搞数学模型等，也是精密科学。我们管理学院的专业也要向这个方面发展，充分运用数学工具。数学王国很大，一个人不可能都精通，但懂得太少也不行。同样一篇科研论文，是否利用先进的数学工具加以描述、论证，对论文的质量影响很大。理化基础也很重要，当然，不同专业有不同的要求，有的专业理化基础要求厚实一些，有的专业则要求低一些，但最好是都适当地学一点、懂一些。学科之间都是相通的，知识面宽一点，对搞科研大有帮助。还有计算机，现在是微机普及的时代，新的科技革命是以计算机的应用为主要标志的，没有计算机的基础知识，就会影响你的科研质量。现在我院还有一部分教师不会使用计算机，所以，我们要办普及班，要求全院教师学会用计算机，要下功夫去掌握它，迅速补上这一课，使我们学院的教学、科研，包括行政管理，都建立在应用计算机的基础之上。

文学基础对搞科研也很重要。一篇科研论文，要表达清楚，有条理，有逻辑性，还要简练，那就非有一定的文学基础不行。有许多问题要靠文字叙述，即使你有很多精辟、独到的见解，但词不达意，表达不清，别人就不能确切理解，就起不到应有的推广效果。现在有的论文语言表达太差，语病很多，文理不通，甚至错别字连篇，连标点符号都不会用。这样的论文，请人家去审阅，往往通不过，很容易就被否定了。所以不管你搞哪一领域的科研，文学基础很重要。

第四，要掌握一定的技巧。每个人都应该有自己的绝招，有擅长的东西。打个比喻，古代的将军对十八般武艺并不是样样都精通的。关公擅长的是大刀，回马刀是他的绝招，如果换成箭、换成枪，就不一定行，而罗成擅长使枪，秦琼善于用锏，各有各的特长。搞科研也是这样，每个人要有自己擅长的东西，有自己的绝招，要扬长避短，才可能有更快的速度和有更多的成功机会。现在全世界有那么多的大学，美国有三千多所，中国也有一千多所，再加上业余性质的，同一领域的科研有很多人同时在搞，你要胜过别人，强人一手，没有绝招、长处不行。要了解自己，扬长避短，发挥自己的长处与特点，而不要拿自己的短处去同别人比，否则，你就会失败。有人对伟大科学家下过一个定义，认为在科学上创造一个伟大概念的人才算是伟大的科学家。一个诺贝尔奖获得者只是创造一个伟大概念的人。爱因斯坦是现代最伟大的科学家，他一共提出了七个伟大概念。不过他也不是样样都行，而是有几个方面的特长，但他充分发挥了自己的特长，所以他成功了。体育比赛有第一名、第二名和第三名，而搞科研，同一个课题，却只能取一名。谁先成功，谁就是第一名，所以你没有绝招，不善于扬长避短，就不可能摘取金牌。

再说查文献的技巧。课题确定以后，首要的就是查文献，以此来论证课题的正确性。题目出得对不对是很关键的，题目出对了，就等于科研成功了一半。只有通过查文献，才能了解在这一方面世界上已经做到了什么程度，以此来验证你的课题是不是超前的。如果百米赛跑的世界纪录已经是 9.9 秒，你不了解，还在那里搞什么 10 秒，就是搞出来也没有意义。通过查文献，如果发现这一课题人家已经做得很好了，你不能超越人家，就不要再做了，把人家最高水平的东西拿来用就是了。如果发现人家做的东西不够，你就加以改进。如果发现这一课题人家根本没有做过，那就是空白，谁最早做出来，谁就能填补空白。现在，每个专业在国际上都有一本专门的文献杂志，并分成两个集团，一个是苏联的，一个是英美的。例如，《应用力学评论》就是一本力学方面的权威文献杂志，把全世界公认的高水平的论文都收进去了，并附有评论，你要翻阅有关力学方面的文献，就可以看这本杂志。我们要查文献，除查出英美的文献杂志和苏联的文献杂志外，还要查国内的文献杂志，只有通过这几个方面的查找，才能论证所选题目的可行性，以便科学地把课题最后定下来。

第五，还要善于拜老师。老师有两个，一个是书本，一个是人。现在着重讲后一个老师。人类科技的进步，都是后人在前人成功的道路上继续前进的。从来没有搞过科学研究的人，会把科研看得神秘莫测、高不可攀。实际上科学研究也是人间的烟火，没有什么了不起。只要攻下了这个堡垒，也是感到平平而已。窗纸捅破了就是那么回事。只要你善于拜老师，把这个老师的绝招学到手的话，就可以跟着老师去做，从查文献起，到写成论文，完成实验等，很快就可以入门。拜老师最好要拜名师，“名师出高徒”很有道理。能多拜几个老师更好，除了个别行业老师很少，一般行业的老师是很多的。我们中国培养人才，常常是从大学生开始，然后读硕士、博士，一直在一个老师身边，要到四五十岁才放出去，这叫近亲培植。发达国家在这方面的经验值得借鉴，他们培养博士生，常常不找自己的学生，而到另外学校去找。毕业后，自己学校不能留他，而是逼他出去到外校任教，到外单位工作。美国的科技人员一般一生要调动五次工作，这群人才流动以后，走了几个地方，就是找老师。多拜几个老师，就可多学到几个绝招，这样才能开创出自己的路子，才能标新立异，去采“星星”、摘“月亮”，拿下大的科研成果。

第六，要有拼搏精神，切忌老换题目。看准方向后，选中一个题目，就要钻进去，沿着选定的方向，始终不渝地前进。不要因难而退，中途改换题目，这不会给你带来成功。当然，如果题目确实选错了，需要改变方向，那是另一回事。要想在科研上有所作为也绝非轻而易举，关键是要知难而进，不能退缩，只要“拼命”去干，就会成功。比如，陈景润连走路都在思考问题，撞到了树上。我也有过这样的经历。1971 年，一机部要搞一套生产高压聚乙烯的设备，这套设备的压力是 2300 个大气压，是我国最高压力的容器要是出问题，一个小城市就完了。当时要我去研究其中一个难度较大的课题，即计算自动保护装置的应力，并规定三个月一定要完成。我这个人也比较好强，就一口承担下来，当时资料室还没开放，我只好到杂乱无章的资料室的书堆中去找资料，接连几天几夜马不停蹄地去干，但一无所获。真是心急如焚、坐立不安，吃不下饭，睡不好觉，日夜想着自己的课题，长期的奔波使我病倒了，尽管医生关照必须安静休息，但我还是坚持搞下去，最后终于找到了办法，如期完成了任务。所以，搞科研常常是苦头吃足的，绝不会一帆风顺的。这正如运动员一样，在历经千辛万苦和付出大量汗水后，才能夺得金牌，搞科研也同样要有这种拼搏精神。

要争时间、抢速度。搞科研要分秒必争，争取以最快的速度拿出科研成果来。最近，国际上出现了超导热，许多国家同时在那儿搞，谁搞得快，早出成果，哪怕是提前一天就是胜利，就是为祖国争得了荣誉。要写好科研论文，论文的文字要简洁，论点要明确，层次要分明，并要及时发表和推广。整个科研过程，有的要几个月，有的要几年，甚至几十年。例如，搞种子的优化，从选出好的种子，到大面积的培育，再到喜获丰收，周期很长。陈景润虽摘取数学王冠，但还有一步未做，这种大题目需要很多年。有的科学家喜欢多方面搞，而有的科学家一生就守着自己的一个领域。但不管怎样，都要争时间、抢速度，要不停顿地搞下去，就好像兔子与乌龟赛跑，只要乌龟不停地走，它最终就可以走在停停走走的兔子的前面。科学在日新月异地发展，搞科研也要有紧迫感，要不停顿地前进。

最后，谈谈管理学院基础课的教师怎么选课题、搞科研。基础课教师的教学工作量大，这是事实。但一定要打破没有时间搞科研的思想，高等学校普遍存在这个思想。作为一个教师本身应该具有教学与科研这两种职能，基础课教师也不能例外。我以前也担任基础课教师，当时，我开材料力学课，教学任务很重，课时排得很满，还要改作业，带学生下厂实习、劳动，另外兼带十个人的毕业论文和班主任工作，但我还是千方百计地抽时间搞科研，时间是挤出来的。拿我们学校的暑期来说，暑期不等于休假，它主要是给你著书立说的时间、搞科研的时间。国外大学里，教授每人要上两门课，如果没有科研成果，没有论文，就没有科研经费，就不能招收研究生。研究生实际上又是助手，粗活都要研究生去做，上计算机、普通实验等要研究生去做。国外的教授也不是轻轻松松地在搞科研。所以，基础课教师一定要打破这一观念，要善于挤时间和利用假期，积极地搞科研。我认为以后中青年基础课教师没有进行科研就不能提职称，老年基础课教师则给予适当照顾，因为这是历史造成的。

现在相当多的课题是集体性的，个人性的很少。丁肇中教授得到诺贝尔奖，实际上也是好多人一起做的，是共同的成果，丁肇中不过是总的负责人。他在联邦德国汉堡市有一个研究中心，就有几百人围着他转。基础教研室可以同其他系、室结合起来搞，也可以与总校、其他学校搭档进行。比如，我们的化学实验室与名牌大学的有较大差距，要搞大的项目是不可能的，怎么办？可以同总校或其他学校联合起来搞，以弥补我们自己的不足。

我们学院的基础课教师搞科研要注意一个问题，即尽量少做纯基础的课题，而多做应用工程性的课题。我们在上海这个经济中心，已经培训了那么多的厂长、总工程师、干部专修班的学生，要充分利用这个有利因素，到工厂去找课题，多搞应用工程性的科研。我们现有的设备与资金，很难为基础性的科研提供条件，因此不宜众人都去搞这方面的课题，当然，国家急需的项目除外，即使我们赔钱也要搞。

最后我还要特别强调两条。一条是搞科研一定要有求实的精神，诚实的态度，不要搞虚假。搞虚假，既害人，又害自己，别人拿了你虚假的科研成果去用，是害了别人，虚假的东西终究要被拆穿，结果身败名裂。许多基础课教师以前没有搞过科研，可以从小事扎扎实实地做起，一点一滴地积累，不断地锻炼和提高自己的能力，这样，既利于己，更利于国家。另一条是要善于同别人合作，相互尊重，取长补短，不能互相拆台，更不能在背后说些不负责任的话，因为合作是为了增强力量，而不是互相削弱，否则就难达到成功的目的。另外，合作中要发扬风格。对于成果的分享既要实事求是，也要发扬风格，即使同事间出现矛盾，也要真诚解决，坦然处理。总之，希望大家要互相支持、互相谅解，团结奋斗，为振兴中华、为2000年的翻两番、为实现社会主义四个现代化，做出经济管理学院应有的贡献。我今天的讲话，如有不妥之处，请大家批评指正。谢谢大家。

青年人的奋斗方向[①]

一、教学与科研

一个“健康”的大学教师，必须具备这样两种“基本功能”：其一是能教学；其二是能搞科研。研究生的毕业，只不过是进入科学王国的第一步。青年人应珍惜自己在青年时代所特有的优势，即才思敏锐、思想闪光。纵观科学大师的成长道路，谁不是在二十岁至三十岁这段黄金时期，具有了与众不同的思想和方法。之后，他们便是无止境地追逐与探索，继而终于达到自己的目标。科学的重大突破是建立在思想观念新基础上的。爱因斯坦伟大，因为他提出了七个伟大的概念；空气动力学家冯·卡门伟大，因为他提出了四个半的伟大概念。

从事科学研究工作，就像进入一个迷宫一样。如果没有线索，思路没有开窍，就很容易让人感觉到枯燥无味，仿佛是走进一条没有阳光的深渊。这就是通常所说的科研工作本身具有的障碍和困难。除此之外，还有一些非正常的干扰。不是常常听说谁搞一个项目，总有讥笑、打击等一类相伴随的现象吗？这是出自部分人自身的秉性：不理解加上嫉妒。因此，要在科研工作中取得成就，就必须具备足够能力来处理工作中出现的正常和非正常的因素导致的问题。要有决心和信心，积累知识从少到多，持之以恒，不断扩大知识面，为今后的突破准备。

当人们形容一个武林高手的功夫时，常常有“十八般武艺都会”等一类的说法。但是他精通的，也就是得心应手的，只可能有一门，这一门他可以达到炉火纯青的程度。也就是说他有一个绝招。做学问也同样如此。通过不断地积累和探索，你就能在科研工作中形成自己独特的方法和体系，并使其日臻完善；相反，要是研究人员没有自己独特的方法和体系，就是再努力，人际关系再好，也只能是甩了西瓜捡芝麻而已。在科研方法上要力求创新，因为这种创新是永恒的。另外，对权威也不必畏惧，要敢于向它挑战。今天被众人认为的荒唐设想，明天就可能是现实。

总结一下，就是思想观念要创新、实际工作要有坚忍的精神、方法论上要有突破。真要是做到了这些，那么，当你漫游在科学的迷宫中时，还能觉得迷惑和彷徨吗？不可能会这样了，继而代之的是得到无穷的乐趣。

二、国家的振兴与青年人的责任

不少人抱怨我们这个社会，向往西方的文明。这是一个不用回避的事实。开放政策以来，我有幸以联邦德国洪堡研究会员的身份在联邦德国工作了两年，并访问和考察了西欧、中欧和南欧的一些国家。对中西方国家的差别和比较，自然比那些仅从一个窗口看外国但又自命不凡的牢骚者更有发言权。我们应该清楚地认识到中国今天相

① 本文是在上海工业大学研究生座谈会上的讲话（按录音整理），上海，1987 年 4 月 8 日.

对于西方发达的资本主义国家的落后，不能归咎于某个集团或是某个政党，而是因为在中国这块土地上有着根深蒂固的封建思想。要冲破这种封建思想的束缚，还得需要几代人坚持不懈的努力。历史上不少有识之士都视封建落后的思想为中国腾飞的“绊脚石”，而进行了艰难的改革。明末时，中国已有了新兴的资本主义萌芽，但后来，落后的清王朝又在中国进行了长达三百年的封建统治，而且每况愈下。这简直是历史的倒退！由于袁世凯的叛变，戊戌变法未能成功，使中国失去了一次改良的机会；孙中山所推行的民主革命，等到在思想理论上和武装力量上都较全面之时，已到了他生命的晚年，革命的成果又落入军阀的手中，我们国家又失去了一次改革的机会。

青年人要有为中华民族振兴和社会主义社会繁荣昌盛而奋斗献身的精神，这不是一句空话。要有时代责任感和远大理想。以前，总是有一些青年对此不屑一顾。其实，这是多么幼稚和天真。一个人如果不想为人类社会的进步做一点什么，那么生活还有什么意义呢？当然，人世间也有一些与自己意愿不相符合的事情，但这并不能成为我们青年人不进取、从而对社会采取不负责任态度的理由。恰恰相反，青年人应该不断积累知识、积极奋斗、耐心等待。众人的奋斗，就能形成推动社会前进的巨大力量，中华民族就一定会腾飞、振兴！战后，西德和日本在经济上的腾飞绝非偶然。那是因为它们重视经济管理、科学和教育。尊重知识、尊重人才不能不说是他们经济上飞速发展的一个重要因素。

三、广泛的兴趣与爱好

我不主张读死书和死读书。读书要博而精。博，即读不同专业的书；精，即读重要的书、本职工作的书。同时，要多参加社会活动，以便得到锻炼。早在兰州大学就读时，我任学生会主席。干得非常轻松愉快。从不因为多参加了社会活动，而觉得对学习有什么大的影响。相反，社会工作于我大有好处，工作能力和知识面都相应提高了。

人是需要交流和理解的，而礼貌、文明、真诚正是交流和理解的纽带。回忆起在西德的许多往事，除了从事正常的科研和学术交流工作之外，还发起组织了鲁尔大学中国科学家和留学生联合会，并任第一任会长。各学科的华裔学者和专家近 80 人参加了联合会。大家为振兴中华、统一祖国在一起讨论交流。特别是在 1983 年春节，我们主办了一次在鲁尔地区很有影响的具有中国特色的春节招待会。会员做出了具有中国各地风味的传统美味佳肴。鲁尔大学的知名教授和该地区的政界要人、工业巨子以及中国台湾来的科学家等应邀参加了招待会。通过这次招待会，不仅增进了东西方人们的友谊和了解，而且也加强了中国大陆与台湾两边人民之间的联系。真是收获不少啊！青年人就是要多参加这样一些有益的活动。要能从书本中、研究工作中解脱出来，加强横向联系，多为社会为人民做一点实事。

“科学技术是第一生产力”的趋势与我国高新技术发展的战略①

面对世界新科技革命的蓬勃发展，经济、科技在世界竞争中的地位日益突出的态势，10 年前邓小平同志高瞻远瞩地提出：“科学技术是第一生产力。”其最重要的含义就是科技是第一位的变革力量，是变化中的主导因素。“科学技术是第一生产力”的思想不仅符合中国国情，也同样具有世界意义。因为，“实现人类的希望，离不开科学；第三世界摆脱贫困离不开科学；维护世界和平也离不开科学。”邓小平同志特别重视“发展高科技，实现产业化”，他 88 岁高龄南巡途中，满腔热情呼吁“搞科技，越高越好，越新越好”，表达了他对科技发展的殷切期望。未来的 21 世纪将是科学技术广泛交叉融合、迅速转化的世纪，是不断印证邓小平“科学技术是第一生产力”思想的新世纪。审时度势及时制定我国高新技术发展战略，是时代之必然。

一、“科学技术是第一生产力”的国际趋势

自第二次世界大战结束以来，在世界范围内兴起并仍在继续进行的新科技革命，是人类历史上规模最大、范围最广、层次最高、影响最深远的一次革命，是科技革命史上全新的时代。这场革命分为五个阶段[1]：1945～1955 年，第一个 10 年，以原子能的释放和利用为标志，人类开始了利用核能的新时期；1955～1965 年，第二个 10 年，以人造地球卫星的成功发射为标志，人类开始摆脱地球引力向外层空间进军；1965～1975 年，第三个 10 年，以重组 DNA（脱氧核糖核酸）实验的成功为标志，人类进入了可以控制遗传和生命过程的新阶段；1975～1985 年，第四个 10 年，以微计算机大量生产和广泛使用为标志，揭开了扩大人脑能力的新篇章；从 1985 年至今，以软件开发和大规模产业化为标志，人类历史迎来了信息革命的新纪元。在信息时代，最重要的资源是信息，人们主要利用脑力劳动创造社会财富，服务业成了信息时代最庞大的经济部门。

21 世纪将是科学技术全面发展的时期。以信息技术、生物技术和材料技术三大前沿技术为代表的高新产业开始占据主导地位，并将和能源、航空航天、海洋开发以及农业、医疗保健、制造业和环境保护等方面的技术一起，对经济、社会以及人们的生活方式产生重大影响。

新科技革命缓解了资源在国家实力中的重要地位，减轻了经济发展和军事对资源的依赖；新科技革命吸引各国重视高精尖技术的运用，变追求数量为立足于质量；新技术革命削弱了军事冲突的作用，提供了武力以外的新的抗衡手段。国家实力内涵的更新，迫使世界各国完成了从生存意识到发展意识的飞跃，把推进高科技的发展确定为国家战略重点，由此引发新一轮激烈的国际竞争。当今国际社会“科学技术是第一

① 本文原载《邓小平科技理论与广东实践》（广东省科学技术协会、广东省科学技术委员会主编），广州：暨南大学出版社，1998，32-42.

生产力”的竞争有如下趋势。

1. 技术产业化

高新技术产品的广阔市场和高额利润，使许多国家特别是发达国家产业界在加强科技发明和创造同时，把重视技术的应用作为突出措施来抓，以加快高新技术产业化进程，因而在科技进步日新月异的今天，科技转化的周期越来越短，高新技术的转化更为明显。20 世纪 50 年代，一项高新技术从实验到产品问世，一般为 10 年，长的要 20～30 年，如地球卫星从研制成功到上天运行，花了 30 多年。如今，发达国家更新产品周期已缩短至 3～5 年，而在微计算机领域，大约每隔半年就有新的机型面世。诚然，加快科技成果产业化并非易事，如技术上的发明需要三分努力，应用开发就需要七分努力，而使它投入批量生产则需加倍的力量。

由于高新技术产品拥有广阔市场和丰厚利润，引致发达国家和地区不遗余力地推动高新技术的产业化进程。其有效的措施是增强开发投资，加快科技优势向商品优势的转化，如美国在 20 世纪 80 年代末，用于研究性开发的投资即达到 1300 亿美元。进入 90 年代，随着冷战结束后形势的变化，又逐步调整有关基础研究和应用研究的传统观念以及政策法规和资金分配计划等，关闭一些无实际意义的国家级重点实验室，以保证国家级高新技术成果的转化。兴建科技园也是发达国家转化高新技术成果的有效途径。高科技园是科研、高校和产业部门联手合作的共同体，这种共同体的最大优势就是能加快高新技术产业化。

2. 投入集约化

高新技术的竞争，除了人才、技术因素外，最重要的是经济实力的竞争，即使人才、技术也与经济基础密切相关。高新技术由于研制的艰巨性、复杂性及其大量人力物力的消耗，使技术的前期投入比较高，从而产业界获得技术专利的支出相当昂贵；高新技术产品的独创性和高精密特点则要求生产设备、检测手段、工作环境和原料选用都达到较高水平，因而高新技术产业化的一个重要条件是资金密集、投入集中。为了在国际高科技竞争中不至于落后，各国竞相增加科研经费的投入。近几年，日本科研经费的增长速度居世界第一位，1995 年的经费总额占国民生产总值的 3.19%，到 2000 年将上升到 3.5%。美国准备把今后 5 年的国家科研基金预算提高 1 倍。

3. 科研前沿化

发达国家积极向科学研究的前沿领域进军。以日本的计算机为例，目前世界上使用的最先进的计算机是第四代大规模集成电路计算机。科学家正在着手研制第五代智能计算机和光子计算机，而日本已在研究生物计算机和第六代仿人脑计算机的理论了。科研前沿化需要大批优秀的人才，现在不少国家到国外用重金聘用人才。日本计划 1/3 的研究人员准备从国外引进。美国、加拿大、澳大利亚和新西兰为了在国外网罗人才，再次修改了移民政策。

4. 科技保密化

对高技术采取严格的保密措施，这在美国表现得最为明显。美国为了恢复和保持在高技术领域里的优势，采取了许多“技术堡垒”措施，如制定《知识产权保护法》、不允许外国人在美国商用数据库检索资料、限制外国在战略领域里的投资等。科技保密化趋势使各国高科技的发展必须建立在吸引国外高新技术与本国自行开发相结合的基础上。

5. 经营全球化

全球性经营战略不仅反映在产品的国际化销售不断扩大，也反映在跨国公司设置的全球化进程日益加快。跨国公司已有几十年历史，但以往主要集中在传统产业的扩张。高新技术产业公司由于技术因素，外扩势头不强。近几年，随着高新技术的日益成熟，生产能力的不断扩大，大公司急需寻求国外合作者。与此相适应，世界经济的蓬勃发展和购买需求的不断增强，以及消费层次的相应提高，为高新技术产业提供了更多更大的国际市场。欧美和日本的许多大公司不失时机到国外布点抢滩，加快实施产销一体化、经营国际化的战略。例如，美国的国际商用机器公司、摩托罗拉通信设备公司等国际知名的高新技术企业，近两年已在中国、南非等10多个国家设立了组装整机、生产配件的专业工厂。继美国企业之后，日本的高新技术企业也纷纷兴办"国际企业"。日本大举蚕食美国的高科技地盘，近10年，日本大规模集成电路的世界占有率，已从28%上升到50%。

6. 银企联合化

金融业的发展特别是金融与企业的关系一定程度上决定着国家及企业经济的发展。高新技术产业由于资金需求量大、市场风险大，更需要银行部门的支持。因此，许多国家发展高新技术产业的又一举措，是加强银行与企业的联合，借助银行的资金优势、财会优势和管理优势，推进企业购买新技术，加快发展速度。

二、我国高新技术发展的基本状况

我国发展高新技术及其产业的基本思路有两条：一是沿着高新技术基础研究→高新技术成果开发→高新技术商品化→高新技术产业化的发展方式，这是由高新技术及其产业自身发展特点所决定的；二是运用高新技术改造面广量大的传统产业，最终使产业结构合理化、高级化，从而增大高新技术对经济增长的贡献率。我国发展高新技术及其产业的主要内容包括直接采用高新技术生产更新换代产品，促使传统产业向高级化发展；利用高新技术改造旧的工艺（流程）；运用新材料技术开发替代传统材料的具有新功能、高附加值的新材料；运用具有较高技术含量的装备替代传统设备以及嫁接式运用高新技术或综合运用高新技术改造传统产业等。

1988年8月经党中央、国务院批准实施的火炬计划，经过10年艰苦努力，已经取得了显著成效，顺利地度过了初创期，开始进入新的发展阶段。全国已认定了1.2万家高新技术企业，组织实施了近2000项国家级火炬计划和近5000项地方级火炬计划项目。52个国家级高新技术产业开发区已初具规模。高新区在近130万人就业规模上，实现了年人均产值近17万元。在市场经济条件下，高生产率意味着高产出、高收益和高竞争力。从国家总体经济来看，在1994～1995年我国工业增加值平均增长5.07%，而同期高技术工业增加值增长43.66%，高技术增加值占全部工业增加值比例已上升到10.55%，成为促进产业结构升级和拉动国民经济持续、快速、健康发展的新生力量。在一些关键领域，一批著名企业已创出了自有知识产权的名牌产品，具备了参与国内外市场竞争的实力。1992年，52个高新区年收入上亿元的企业仅有39家，而1996年已发展到390家，超10亿元以上的有30家。高新技术企业从小到大，显示出强大的生命力。

广东近年经济社会取得的巨大成就，无不得益于科学技术所起的重要作用。1991年广东明确提出今后的发展目标：工业中的科技含量要从20%提高到50%，农业要从20%提高到40%，高科技产品比例要从3%提高到7%～8%。广东已建立县以上科研和技术开发机构1600多家，“七五”至“八五”时期，广东投入127亿美元，先后引进7000多项先进技术项目。130多万台（套）技术设备，对全省工业进行大规模技术改造，使广东在较短时间内缩小了与国内先进省市和发达国家的差距。广东科技进步对经济增长的贡献率由1978年的19%增至目前的39%，广东GDP由1978年的184.7亿元增至6097.42亿元。广东有2/3的新增产（行）业是由近20年来的科技进步带来的，劳动生产率的提高也有过半数得益于此，科技进步对广东新增GDP和新增GDP对新增工业产值的贡献率已分别达61.04%和66.54%。至1996年止，全省已认定高新技术企业569家。1995年，广东高新技术产品进出口总额达100.9亿美元，其中出口额为53.39亿美元，居全国之首。同年，广东大中型工业企业高技术产品销售185亿元，居全国第2位。广东利用高新技术发展农业，全省农业综合商品率达80%，“三高”农业产品已占农产品总量一半以上，农产品及其加工产品出口产值占农业总产值的30%[2]。广东实践无不证明科技进步是推动国民经济增长的首要因素。

由于我国高技术发展起步较晚，整体上仍处于发展的初级阶段，我国工业技术只有20%达到发达国家20世纪80年代的水平，与发达国家相比差距甚远。政府对高新技术发展的支持力度不够，政策不完善，特别是投入不足，资金短缺已成为制约我国高技术快速发展的关键因素之一。发展高科技资金短缺的原因：第一，国家对高科技的投资偏弱。高科技投入的总经费，我国不仅在绝对值上不足美国的5%，全国人均经费仅为第三世界国家人均的1/2；财政科技投入占国家财政支出的比重远远低于西方发达国家、新兴工业化国家和许多发展中国家。发达国家对高技术研究开发经费的投入占国民经济总产值的2.5%，而我国不到1%，20世纪末要达1.5%也不容易。第二，高科技投入的渠道偏少。我国高科技投入缺乏社会化的投资渠道，尤其缺乏大型企业的有效介入和科技风险基金的有效形成与运作，科技投入过分偏倚政府。企业本应是高技术发展的主体，发达国家的企业在高技术投入中一般占其总投入的50%，而我国企业尚未完全成为真正独立的经营主体，缺乏对风险投资的激励机制和承担能力，尽管企业在银行的存款达4000多亿元，但仍不愿向高风险、投资大、见效长的高技术产业投资。第三，高科技投入结构不当。由于投入额不足，条块分割，难以启动重大科研，只能进行低水平的重复。据统计，我国的科研项目，40%与国外重复，管理软件的低水平重复率达70%。第四，银行功能没有充分发挥。我国银行的资金规模已达12 000亿元，是支持高科技产业发展的重要力量。但由于银行体制仍未根本理顺，不愿将其资金投向高风险企业或产业。第五，利用外资不够。虽然我国高技术市场前景广阔，对外商颇具吸引力，却由于我国市场机制不健全和投资效益低等缘故，使一些外商望而却步。

高技术及产业化是两个层次的问题，目前在高技术研究方面我国取得一定进展，而产业化则非常薄弱。我国的科技攻关能力在世界上是第一流的，如两弹一星的成果，而高技术产业化的水平却大大低于科学研究的水平。我们要积极探索一条中国式的技术进步新路子，发展高科技、实现产业化，全面推进“科技是第一生产力”的发展战略。

三、我国高新技术发展的战略选择

世纪之交，各发达国家着眼于21世纪的经济和科技的新一轮竞争，纷纷调整和制定高科技发展战略，为保持领先地位打下坚实基础。我国作为世界大国，任何时候“都必须发展自己的高科技，在世界高科技领域占有一席之地”（《邓小平文选》第三卷，第279页）。我国已制定了《高技术研究发展计划纲要》和《国家中长期科学技术发展纲要》，并启动和正在实施“863”计划、国家科技攻关计划、火炬计划等，组织各方面力量在经济建设主战场、高技术研究及产业化、基础性研究三条战线展开攻坚战。

我国发展高新技术的战略是“跨越发展，开拓创新，系统集成，示范带动”。跨越发展是后起国家经济起飞的成功之路，世界格局从冷战转向经济竞争及我国改革开放20年的快速发展，为我们创造了跨越发展的历史机遇。要实现跨越发展，必须开拓创新。目前技术发展的国际化趋势，为我们技术引进提供了便利。21世纪初，我国将进入技术引进的第三阶段，在技术引进的基础上，要及时转向消化吸收和创新，以自立于世界高技术之林。系统集成是系统技术的需要，也是高新技术与传统产业集成、与社会要素结合的需要。示范带动是应用高技术促进经济发展的成功经验，只有采用示范带动的点源式样板，才可能激励、启示和引导广大企业积极跟进，最终达到以高科技研究和发展带动企业技术进步的目的。

发展我国的高科技，总的原则是“一靠政策，二靠投入，三靠人才”，要突出高新技术发展战略上的创新。

一是把培植高新技术产业作为科技立国的重大举措。高科技发展要以提高我国国际经济竞争力和市场占有率为目标。高科技发展要体现国家的产业政策，有所为和有所不为，多方联合，优势集中，有序分工。要实现我国的产业经济发展战略重心大转移，由重视传统的劳动密集型、机械密集型产业向科技密集型、技术密集型产业为主导的新型产业经济模式转变，使我国在不长时间内成为经济大国、科技强国；要站在世界科技发展的新前沿和能够驾驭未来竞争的制高点来审视、谋划我们的科技战略，通过培育高技术产业，加快传统产业现代化的改造，探寻新的产业经济增长点，使我们的综合国力有较大的提高，使科技进步跃上新台阶。

二是集中优势力量主攻科技难关。我国科技基础薄弱，尤其是高新技术起步晚、力量弱，许多方面难成特色。现有的高新技术企业有相当部分是与国外企业合资合作，借他人的实力和专利发展，有些电子、音像类技术还是从非正常渠道获取。因而时与外国企业，特别是欧美企业发生知识产权的矛盾和纠纷，给企业和国家造成不良的政治影响和经济损失。因此，从积极接受挑战、主动驾驭竞争的需要出发，我们应集中优势力量，主攻高新技术的重点。从技术力量而言，应发挥三方面的作用，即专业科研机构、高校科技力量和大中型企业的科技人员的作用。我国现有科技人员2800万，其中1/5是高级科技人才，有足够的开发创造实力。但由于人员分散、专业限制、政策束缚和条件滞后等因素，导致信息不通、力量分散、各自为政，甚至专业人才无用武之地，直接影响到科技力量的有效发挥。因此，一方面要加快产、学、研一体化进程，协调好国家和地方、科研和企业的合作关系，重大项目由国家或省统一部署，各参与单位分工负责，协作攻关，重点突破；另一方面加快与国际接轨，扩大信息量，及时掌握国际经济技术的新变化，调整我国的主攻方向，避免无效劳动、重复劳动，提高科技成功率。

三是规划跨世纪科技人才的培养。现代科学技术向高精尖、跨学科发展，产业向知识或技术密集型发展，战略资源不再是物质资本，而是人力资本，是知识。为保证我国未来经济健康、持续发展，培养跨学科人才是一项重要工程。培养此类人才，第一，要提倡导师指导和集体培养相结合的方法；第二，实行国内外、校内外联合培养研究生；第三，鼓励跨学科跨专业报考研究生。要改革教学方式，以大学为基地，将教育融入科研，在研究过程中培养学生，使学生能在产、学、研合作的实践中吸收更多有益的经验。

四是普及全民创造教育。创造力是科技和社会发展的一种动力，没有创造性思维的人，是不可开拓进取的；没有创造精神的民族，是难以实现繁荣与发展的；没有创造性的时代，必将是一个黯淡而平庸的时代。创造力和其他技能一样，是可以通过教育、训练而激发出来，在实践锻炼中不断提高的。创造力开发正是通过把关于创造的理论和方法，转化成劳动者的创造素质和技能来推动科技进步和经济发展。截至目前，全世界已有的创造技法，已达数百种，创造力测评方法达 100 多种，制定出创造力训练教学模式达 10 多种，已有 40 多个国家进行创造力开发教育。要使我国的人才真正形成创造优势，成为推动高科技发展的生力军，就必须从战略高度上重视全民创造力开发，为科技和经济的发展注入活力，增强后劲。

五是把有限的科技优势转变为竞争优势。我国的总体实力已接近中等发达国家现有水平，但我国的人均占有率较低，高校、科研单位的自创增值能力偏弱，企业积累能力太小，加之各地基建摊子铺得太大，资金饥渴情况普遍；各级财政十分紧张，预算内的科技经费十分有限；有限的科技资金过于分散，上不了大项目。解决投入问题已是我国高科技产业可持续发展的至关重要任务，应全方位、多层次增加科技投入。不止于国家有限拨款，坚持国家、地方都随经济发展而按比例增加科技投入；不止于财政有限拨款，坚持财政拨款、企业提留双向并进；不止于外部有限的输血，坚持放活高校、科研单位，让其自己增资创收；不止于国家有限投入，坚持对外开放科研，利用国外资本解决科技经费不足的问题。在增加科技投入的基础上，必须实行集中投入、保证重点开发。同时要鼓励有发展前景的高新技术企业兼并没有前途的中小企业，以利于现有的生产要素、扩大经营规模，带动产业产品结构的优化，以新的优势与国外企业竞争国内市场，再反弹到国际市场，扩大我国高新技术企业的生存空间。只要推动一批大型高新技术企业运作，我国的高科技实力和综合国力就能确确实实地进入世界技术强国之列。

参考文献

[1] 宋健. 现代科学技术基础知识. 北京：科学出版社，1994.
[2] 李超，杨颖柔. 邓小平的科技观与广东科技发展的实践. 现代哲学，1997，(3)：20-25.

人才是振兴国家的关键[①]

新春之际，惠风和畅。今天应邀在此华堂出席盛会，会见老相识，结交新朋友，并受到各位首长的接见，非常荣幸。在此，谨感谢中国人才杂志社，给我们提供了这样一个难得的机会。

忝列《中华名流》并作为封面人物，实在心有惴惴。在座德宏才羡、事业大成者比比，可谓名副其实之“中华名流”。置身其中，不由得感受到一股强烈的进取精神涌动。这种精神，对我来说，尤其值得珍视。

如果仅从年龄来看，我们这一代，已过了人们常说的成就事业的“最佳年龄区”，但作为一个在青少年时代就抱定科教兴国理想，近四十年从事科教事业的我来说，“最佳时期”的逝去，并不意味着不能取得“最佳成绩”，关键要看有没有保持一种“最佳精神”，正如眼前大厅里所充盈的那种进取精神。正因为保持了这种精神状态，达尔文古稀之年后，仍完成了《植物的运动本领》等许多重要著作；德国自然科学家洪堡在75岁时才正式动笔撰写《宇宙》这部最重要的著作；我国唐代名医孙思邈的医学巨著《千金翼方》，是100岁时完成的。所以人的精神尤为重要，它可以弥补诸多已失去的优势，激发那些已不具备最佳的条件。换句话也可以说，它可使人生的最佳时期延长，最佳时机增多，从而创造出最佳成果。

近年来，我对此体会尤深。我在学校现在大部分时间花在党政领导岗位上，事务繁杂，教学科研专业工作几乎成了“副业”。忙碌劳顿之时，仍然受到一种强大的力量的鼓舞。一种理想，一种精神，促使我尽力协调好两方面工作，并努力做得完满一些。去年，当学校的各项工作得到社会肯定（《光明日报》在教师节之际，于1998年9月10日对暨南大学曾有专门报道文章）的同时，广东省1997年自然科学奖唯一一等奖证书和广东省教学成果一等奖证书，也双双颁发，最近又获得国务院侨办第二届科技进步奖一等奖，令人格外欣慰。这是进取精神促动的结果。我当以这一精神激励后半生，更积极地进取，努力为国家、为社会作出更大的贡献。

21世纪即将来临，那是知识经济的世纪，人才是振兴国家的关键，大家作为入选《中华名流》的跨世纪人才，理当在新世纪为中华崛起作出应有贡献。请各位珍视自身，多多保重。最后，感谢《中华名流》编辑同志的热情帮助和支持。

祝各位身体永康，万事胜意！并向各位拜个早年，祝各位春节快乐！

---

① 本文是在《中华名流》首发式上的讲话，北京，1999年2月6日.

传播科学思想 提高国民素质[①]

两年前，当暨南大学出版社与清华大学出版社将联袂出版的一套院士科普丛书策划方案报到我手中时，我就深感这一选题所蕴藏的特殊能量与时代特征。两年来，经中国科学院、中国工程院的精心组编，清华大学出版社、暨南大学出版社的精诚合作，以及两院院士的大力支持，中国第一部完全由院士撰写的大型科普精品——“院士科普书系”，今天正式面世了。这是中国科普工作的一件大事，它对于传播科学思想，提高国民素质，促进中国社会的现代化有着重要的意义。正因为如此，江泽民总书记在百忙之中，专门为这部科普书系撰写了题为《提高全民族的科学素质》的序言。这是党和国家领导人对我们这部书系的极大肯定，也是对我们工作的极大鞭策。

近代科学400多年的发展历史，一直是在与形形色色的反科学、伪科学的社会势力的斗争中艰难前行；中国自五四运动以来，科学与民主就一直是横扫一切封建腐朽势力的旗帜；党的十五大把“科教”提到了“兴国”的高度，“科教兴国”是中华民族越千年、跨世纪的战略，而科普工作正是科教兴国的重要组成部分，是知识传播的主要内容。在我们这个全民科学文化素养亟须提高的国度里，社会对科普的需求十分强烈，即使在社会日益文明的今天，“法轮功”之类的反科学、伪科学的邪教余孽，仍在毒害着一些缺乏科学思想的善良民众。崇尚科学、反对迷信，我们还有一段相当长的路要走。

这次由100多名德高望重的两院院士撰写的“院士科普书系”本本皆精彩，篇篇高水准。在院士们的笔下，科学原理严谨准确，技术内容通俗易懂；精练生动的语言，融人文教育于科学教育；重视揭示科学方法，展现科技最新成果和发展前景。读者在领略院士科学家独特思考的同时，更将获得广泛而深层的思想启迪。院士参与科普工作，意义重大，正如江总书记在序言中所言：“科教兴国，全社会都要参与，科学家和教育家更应奋勇当先，在全社会带头弘扬科学精神，传播科学思想，倡导科学方法，普及科学知识。”我相信，“院士科普书系”必将成为我国当今最高水平的系列科普品牌。感谢两院及院士们对本书的支持，感谢各级领导对本书的关心，也感谢两校出版社同仁们的齐心协力！

---

① 本文是在“院士科普书系”首发式上的讲话，北京，2000年6月4日；一个大学校长的探索，北京：高等教育出版社，2011，325.

大力重视非开挖工程技术①

由广东省非开挖技术协会筹划的“2000年广州非开挖技术报告会”现在开会。

省科协钟世伦副主席，中国香港非开挖技术协会理事长蔡群威先生，国内有关单位和研究部门，广东省、广州市的有关政府部门、施工企业、中外厂商、研究单位和高等学校共80余人出席了本会，我谨代表广州非开挖技术协会向各位表示热烈的欢迎，对支持、赞助本会的有关单位表示诚挚的感谢！

现代非开挖地下管线施工技术是20世纪70年代末才起步的，其中经历了初期的徘徊和艰难，至20世纪80年代中，在发达国家才逐渐为人们所认可和接受，终于在1986年于伦敦成立国际非开挖技术协会（International Society for Trenchless Technology，ISTT），包括中国在内，已有25个国家或地区协会参加，ISTT拥有近2000名个人会员和1200名公司会员。

我国的非开挖工程技术发展大致为三个阶段，即前期、专门设备引进期（20世纪80年代中至90年代中）和自主研制创业期（20世纪90年代初）。我国对非开挖技术十分重视，已列入国家863计划（属自动化）。

广东省正是在1997年专门设备引进期，开始进入水平导向钻（horizontal guide driller，HGD）非开挖铺地下管线的快速发展阶段，到今年11月份为止，广东共引进各级水平导向钻机8台，国产2台，自制设备一套，分别在省内（广州、深圳、东莞、惠州、韶关等）各地开展了水平导向钻非开挖铺管工程近250项。稍后，你们还将会听到广州市政维修处广州南洲互通式立交桥的配套工程，3米直径、长888米的顶管技术报告。在广东展开的各级非开挖施工技术，充分显示了它在城市化、城市现代化的优越和巨大的威力，一个巨大的高新技术产业市场轮廓已经出现。

总之，广东的非开挖工程技术亦是处于一个活跃期，有大量的工作等待我们去做。

进行技术交流，提出议题，商讨解决正是本次报告会的目的。

祝大会圆满成功！

① 本文是2000年广州非开挖技术报告会开幕词，广州，2000年12月16日.

认真评选科技奖①

现在，我向大家报告第七届广东省丁颖科技奖评选和第八届中国青年科技奖候选人推荐的情况。

按照《广东省丁颖科技奖条例》和《广东省丁颖科技奖实施细则》的规定，全省级学会、省直有关单位、中央驻粤单位及各市科协共推荐了第七届广东省丁颖科技奖候选人 100 名，其中工科 35 人，理科 22 人，农科 14 人，医科 29 人。

省科协组织联络部拟定了“第七届广东省丁颖科技奖评审工作方案”。省科协常委会学术交流工作委员会提出了第七届丁颖科技奖评审工作委员会、专家评审委员会及其下设的学科专家组成员建议名单。评审工作委员会由 15 人组成，主任由卢钟鹤主席担任；专家评审委员会由 19 人组成，主任由本人担任。专家评审委员会聘请了 29 位专家组成理、工、农、医四个学科专家组，组长分别由许宁生（中山大学理学院院长）、陈克复（华南理工大学造纸与环境工程学院院长）、廖森泰（省农科院副院长）、徐复霖（省中医药管理局副局长）兼任。“第七届广东省丁颖科技奖评审工作方案”及评审机构设置和人员组成建议名单，均经报请卢钟鹤主席审查同意。

7 月 22 日，学科专家组召开了评审会议。各位专家按理、工、农、医四个学科分别对第七届丁颖科技奖的 100 名候选人的推荐材料进行评审。每位候选人的材料经两位专家评审，并由主审专家向学科组说明推荐理由，专家们按照客观、公平、严格的原则，推选出 39 名第七届广东省丁颖科技奖人选。

7 月 24 日，我主持召开了专家评审委员会会议。各学科专家组组长汇报了评审情况，逐个说明了丁颖科技奖人选的推荐理由。经专家评审委员会委员充分酝酿，然后以无记名投票方式，产生了 28 名（按得票数顺序排名）第七届丁颖科技奖初选名单。同时，专家评审委员会根据《中国青年科技奖条例》规定，在第七届丁颖科技奖初选名单中，初选出 8 名第八届中国青年科技奖候选人推荐名单。

8 月 17 日，卢钟鹤主席主持召开了评审工作委员会会议。本人详细报告了第七届丁颖科技奖初选和第八届中国青年科技奖候选人推荐的情况，评审工作委员会委员经过充分酝酿和讨论，一致赞同专家评审委员会关于第七届丁颖科技奖初选名单和第八届中国青年科技奖候选人推荐名单。

关于第七届丁颖科技奖评选和第八届中国青年科技奖推荐情况报告完毕，请各位常委予以审议。

---

① 本文是广东省科学技术协会常委会上的汇报，广州，2003 年 8 月 20 日；教育与科技管理研究，北京：科学出版社，2016，349.

与时俱进　开拓创新[①]

为了进一步发挥科协系统的整体优势，为省级学会和科技工作者建立一个多学科、综合性、开放性的学术交流平台，以满足广大科技工作者参加学术交流的基本需求，繁荣我省的学术交流，促进学科发展、科技创新和人才成长，广东省科协六届二次全委会议决定举办广东省科协首届学术活动周活动。

首届学术活动周活动的主题是：科技创新与全面建设小康社会。活动周内，省科协和省级学会将联合政府有关部门、科研院所、高等院校、企事业单位以及有关国家级学会，围绕广东全面建设小康社会、加快率先基本实现社会主义现代化目标，围绕广东科技创新、经济建设和社会发展的重点、热点问题，组织科技工作者广泛开展各种形式的学术活动，交流科学技术发展的最新成果，研究和探讨广东加快发展、协调发展和全面发展的对策和建议。

为办好本届学术活动周活动，省科协成立了“广东省科协首届学术活动周组委会”。组委会名誉主席由省人大常委会主任、省科协主席卢钟鹤同志担任，主席由刘人怀院士担任，副主席由罗富和、孙玉、陈勇、钟南山、骆世明、梁明、谢明权、颜泽贤等同志担任，并成立了组委会执行委员会。首届学术活动周的工作得到了宋海副省长的高度重视，今天宋海副省长在百忙中出席开幕式并作重要讲话，为我们办好学术活动周的活动带来极大的鼓舞和鞭策。

目前，在各级领导和社会各界的关心支持下，广东省科协首届学术活动周的各项筹备工作已全面就绪，今天正式开幕。学术活动周的时间为 2003 年 11 月 11 日至 17 日。活动周将举办 6 场院士报告会，将由 33 个省级学会举办 35 场各种形式的学术会议、学术论坛和学术成果展示会，涉及理、工、农、医和综合学科等自然科学领域，有近 6000 名科技工作者参加学术交流，国内有关著名科学家、专家学者将出席有关会议并作学术报告。

同志们，广东省科协首届学术活动周的举办，必将对广东的科技进步、经济建设和社会发展产生积极的推动作用。让我们紧密地团结在以胡锦涛同志为总书记的党中央周围，坚持以邓小平理论为指导，深入学习“三个代表”重要思想、党的十六大精神和党的十六届三中全会精神，进一步解放思想，与时俱进，开拓创新，努力奋斗，为广东全面建设小康社会、加快率先基本实现社会主义现代化作出新的贡献。

预祝广东省科协首届学术活动周活动圆满成功！

① 本文是广东省科学技术协会首届学术活动周开幕词，广州，2003 年 11 月 11 日.

以科技创新加快推进全面建设小康社会步伐[①]

当今社会，人类业已进入21世纪和知识经济时代，以发展高技术和加速高新技术产业化为主要标志的科技经济竞争愈演愈烈，经济全球化进程在日益加快。改革发展中的中国面临着新的发展机遇和严峻挑战，中国应主动、快速应对全球知识革命的挑战。正如党的十六大报告所明确指出的："我国进入全面建设小康社会、加快推进社会主义现代化的新的发展阶段。国际局势正在发生深刻变化。世界多极化和经济全球化的趋势在曲折中发展，科技进步日新月异，综合国力竞争日趋激烈。"[②] 改革开放20多年来，中国取得了举世瞩目的成就，在世界舞台上来自中国的声音越来越响亮。但我们也必须清醒地意识到，"我国正处于并将长期处于社会主义初级阶段，现在达到的小康还是低水平的、不全面的、发展很不平衡的小康，人民日益增长的物质文化需要同落后的社会生产之间的矛盾仍然是我国社会的主要矛盾"。[②]如何抓住21世纪头20年这一经济社会发展和科技发展的重要战略机遇期，加快推进社会主义现代化，把"低水平的、不全面的、发展很不平衡的"[②]小康社会，建设成为"经济更加发展、民主更加健全、科教更加进步、文化更加繁荣、社会更加和谐、人民生活更加殷实"[②]的小康社会，科技创新在其中起着关键作用。可以说，科技的发展与创新既是全面建设小康社会的重要目标，也是实现经济社会各项战略目标的根本性措施。下面，我想着重谈一下在我国迈向全面建设小康社会的进程中，我国科技发展与创新所取得主要成就、存在的主要问题，以及我国科技创新的主要内容与思路，旨在为加快推进全面建设小康社会步伐建言献策。

一、科学技术是全面建设小康社会的重要支柱

人类社会发展进程表明，科学技术是第一生产力。工业革命以来的200多年，科技的发展与创新一直推动着世界历史向前发展。英、美等国能成为世界强国，是与其持续领先与强大的科技创新能力不无关系的。18世纪60年代，以英国蒸汽机的发明和应用为主要标志的第一次科技革命，使社会生产力发生了革命性变革，引导人类进入机器时代。19世纪和20世纪之交，以美国等国电机、电动机的发明和应用为主要标志的第二次科技革命，把社会的工业化提高到一个新阶段，使社会生产力进入电力时代；发生于20世纪中期的以原子能、电子计算机和空间技术的发展为主要标志的第三次科技革命，使科学技术对经济社会影响的广度和深度进一步拓展。今天，人类社会已经进入到一个知识不断创新、科技突飞猛进的新时代，科学技术的更新速度日益加快，

① 本文是在广东省科学技术协会首届学术活动周上的主题报告，广州，2003年11月11日；一个大学校长的探索，北京：高等教育出版社，2011，331-336.

② 《全面建设小康社会，开创中国特色社会主义事业新局面》，https://www.gov.cn/test/2008-08/01/content_1061490.htm[2024-08-09].

科技成果商品化、产业化的周期大大缩短，科技创新正在成为经济和社会发展的主导力量。对于像我国这样的发展中国家而言，经济迅速发展与社会进步，本身就是对科技发展的最大挑战，也会产生最大的社会需求，能否紧紧抓住新科技革命的机遇，是发展的关键所在。全面建设小康社会，就离不开我国的科技进步和创新。

1. 我国在科技发展与创新方面的主要成就

新中国成立半个多世纪以来，尤其是改革开放以来，我国科学技术取得了飞速发展，有力地促进了我国社会经济的发展和综合国力的提升，并为全面建设小康社会，加快我国向科技强国迈进奠定了坚实的基础。这些成就主要表现为下述四方面。

(1) 科技创新能力逐步增强。通过实施国家 973 计划、863 计划、国家科技攻关计划等一系列计划，我国科学知识生产数量增长很快。如我国科学论文在科学引文索引 (Science Citation Index，SCI)、工程文献索引 (Engineering Index，EI) 和国际科技会议论文索引 (Index to Scientific & Technical Proceedings，ISTP) 所占总数，20 世纪 90 年代前五年一直在第 15 名左右徘徊，2002 年已跃居第 6 名。又如国内专利申请受理量和授权量，2001 年分别达到 16 万余件和近 10 万件，分别比 1991 年增加了 263% 和 364%。尤其是一批具有重要意义和影响的原始性创新成果相继涌现，譬如水稻基因组精细图绘制成功，13.1 万亿次并行机研制成功，10 兆瓦高温气冷核反应堆并网发电成功，神舟五号载人飞船发射成功等，表明我国在当今若干科学前沿领域取得了重要进展，某些重点和关键领域已接近或达到国际先进水平。

(2) 科技对经济社会发展的贡献不断增大。高新技术产业的蓬勃发展，已经成为拉动国民经济增长的重要力量。1991～2001 年，我国高新技术产业的工业总产值从 3000 亿元左右增加到 18 000 亿元左右，年均增长 20%左右，超过同期全部工业年均增长速度 10 多个百分点。在国民经济构成中，高新技术产业所占比例由 10 年前的 1%提高到目前的 15%。特别是通过持续不断的科技攻关，我国在产业技术研究方面取得了多项重大突破，有效促进了产业结构的优化调整，促进了社会可持续发展。我们先后制定了《农业科技发展纲要 (2001—2010 年)》和《可持续发展科技纲要 (2001—2010 年)》，建立国家工程技术研究中心 200 多家，在全国 27 个省份的近 2000 家企业推动制造业信息化工程，解决三峡工程、西气东输等国家重大工程建设急需的关键技术和设备问题，推动清洁能源汽车、洁净煤技术等的开发应用，开展水资源及其污染治理等方面的研究和科技攻关。

(3) 科技体制改革取得实质性进展，科技与经济脱节的“两张皮”问题得到基本解决。我国企业研究与开发 (research and development，R&D) 投入占全社会研究与开发投入比重超过 60%，已经成为研究与开发活动的主体。应用型科研机构向企业化转制，形成了以市场需求为主要导向的研究与开发新格局。社会公益类科研机构分类改革，一支稳定服务于社会公益性事业的精干科研队伍正在加速形成。

(4) 科技投入显著增加。2002 年，全国研究与开发投入 1161 亿元，比 1990 年的 125 亿元增长了 8 倍多；全国 R&D 投入占 GDP 比例自 2000 年开始超过 1.0%，实现了历史性突破。

2. 我国在科技创新方面存在的主要问题

在看到成绩的同时，我们也要清醒地看到，与欧美日等发达国家相比，我国科技水平整体上仍然相对落后，在相当程度上制约着我国现代化建设和全面建设小康社会进程。这主要表现在下述三方面。

(1) 中国科研产出数量占世界科学知识生产数量的比重仍然较小，尤其是原创性成果较少。从各国占科学引文索引论文数量的比例看，美国基本稳定在30%，英国、日本也都在8%，而我国仅占3%。我国大陆学者在国际上发表科技论文的引用次数与美英德日等国相比差距较大，只与韩国和我国台湾接近。特别需要我们增强危机意识的是，在一系列关系国家现代化建设全局，关系国家经济、国防安全的重大高新科技领域，我们拥有自主知识产权的科技成果还不能满足日益紧迫的需要，甚至在某些领域与发达国家的差距仍在继续扩大。原始创新能力不足，已成为制约中国可持续发展的突出矛盾。

(2) 产业发展对外技术依赖程度过大。这一问题集中表现在具有战略意义的航空设备、精密仪器、医疗设备、工程机械等高技术含量和高附加值产品。近年，我国每年形成固定资产的上万亿元设备投资中，60%以上用于进口。中国光纤制造装备的100%，集成电路芯片制造装备和石油化工装备的80%以上，轿车制造、数控机床、纺织机械等的70%被国外产品占领。这使我国在工业化进程中付出了过高的经济成本，导致我国产业结构调整和升级易于受制于人，如不加快解决，就有可能被长期锁定在国际产业分工的末端。

(3) 研究与开发（R&D）人力资源薄弱，人均经费少于发达国家。我国R&D人力资源在绝对数值的比较上，居于世界前列，但在相对量的比较上，与发达国家相差甚远。1987～1997年，我国每10万人口中R&D科学家和工程师人数为454人，而日本为4090人，美国为3676人，俄罗斯为3587人，韩国为2193人，我国与这些国家相差5～10倍。从R&D科学家和工程师人均占有经费看，中国远远落后于其他国家。按当年汇率折算，2000年中国从事研究与开发人员的年平均经费为1.2万美元，而韩国是8.9万美元，日本是15.8万美元，刚刚达到约韩国的1/7和日本的1/13。由于我们投入过低，从事研究与开发人员的潜力无法得到充分发挥，大大影响了我国研究与开发的效率。

二、加快科技创新的主要内容和思路

在全面建设小康社会的进程中，科学技术的发展与创新，要着重研究经济建设、社会发展、人民健康和国家安全相关的重大战略需求，及时把握物质科学、信息科学、生命科学、数学、认知科学以及高技术的前沿理论与方法，制定中长期发展规划，用先进的科学技术理论与方法解决重大战略需求，登攀科技高峰，改革创新体制，培养、吸引和组织创新队伍，革新科技管理与文化，建设国家创新体系，实现我国创新能力的跨越式发展。

1. 科技创新的主要内容

在国家安全方面，要适应当代军事变革和现代战争的特点，为国防现代化建设提供技术支撑。在现代农业方面，要发展面向全面小康社会需求的生态农业。在信息科技方面必须满足安全、高效、多样化、网络化、智能化服务需求。在材料与先进制造的发展方面，应以提高我国产品国际竞争力，满足我国经济社会发展与国防战略需求为目标。在空天与海洋方面，要把握制空天权，认知海洋、开发海洋、保卫领海权益。在资源、生态、环境方面的重点是：使我国可持续发展能力不断增强，生态环境得到改善，资源利用效率显著提高，促进人与自然的和谐，推动整个社会走上生产发展、生活富裕、生态良好、文明发展之路。另外，还要在人口、健康与生物安全，城镇化与城乡基础设施，战略高技术，公共科学、技术与支撑平台，以及公共科学平台等方面，进行科技创新，从整体上极大地提升我国的国家创新能力。

2. 科技创新的主要思路

在21世纪头20年，我们要在建立和完善适应社会主义市场经济体制的科学技术体制并在形成合理的科学技术布局的基础上，使我国逐步进入“科技大国”“科技强国”行列，培养强大的自主创新能力，在科学和高技术领域占有重要的一席之地，掌握一批重要知识产权，形成支撑我国核心竞争力的知识创新和技术创新基础。具体而言，我认为，我国的科技创新可在以下八方面寻求突破。

(1) 加快高新技术产业化的步伐，大力提升和改造传统产业。走新型工业化道路必须发挥科学技术作为第一生产力的重要作用。推进产业结构优化升级，形成以高新技术产业为先导，基础产业和制造业为支撑，服务业全面发展的产业格局。目前，我国产业技术自主创新能力不足，尽管是“世界工厂”，但大量核心产业技术仍掌握在跨国公司手里，在国际分工中被固化在低技术、低附加值的环节。在世界制造业加速向我国转移的过程中，我们不能一味地强调发挥劳动力的比较优势。而应当通过传统产业的高技术化获得新的竞争优势，通过制造业的信息化，包括农业的信息化，为振兴传统产业提供强有力的技术支撑。

(2) 大力加强农业和农村科技工作。党的十六大报告提出建设现代农业，发展农村经济，增加农民收入，是全面建设小康社会的重大任务。可以说，没有广大农村的小康，就不可能有真正意义上的小康社会。我国的农业和农村经济已经步入了一个新的发展阶段，同时面临着新的农业科技革命正在全球范围兴起的机遇和挑战，农业和农村工作中的中心任务是进行结构的战略性调整和增加农民收入。农业科技工作必须做到四个转变，即从主要注重数量向更加注重质量和效益转变；从生产服务向生产加工与生态协调发展转变；从强调资源开发向资源开发和市场开发相结合转变；从主要面对国内市场向面对国内、国际两个市场转变。我们要努力为改善农产品的品质，增加农民收入提供科技支撑；为保障国家食物安全提供科技支撑；为缓解资源短缺的压力，保护生态环境，发展可持续农业提供科技支撑；为应对加入WTO的挑战，提高我国农产品的比较优势，增加国际竞争力提供科技支撑。此外，还要加强农业科技成果转化环节的工作，继续加大力度实施星火计划；为农业和农村经济结构的调整，为农民增收作出实质性的贡献；强调加强农业科技开展园区的示范和带动作用，增强其辐射能力。深化农业和农村科技体制的改革，加强第一线的农业科技力量，建立和完善农业科技推广和创新体系。制定有效政策，大力增加农业科技的投入，加强农业科技能力的建设，包括人才培养、科研基地建设和一些基础性工作。

(3) 积极推进社会可持续发展领域的工作。社会发展领域的工作，包括环境、资源和人口与健康等方面。我们必须坚持以人为本的思想；突出科技创新为人类发展服务，把提高人民的生活质量作为社会可持续发展工作的根本出发点。重点是要突出保障食物安全、生态环境安全、水资源安全、油气资源安全、战略矿产资源安全。

(4) 加强基础研究和战略高技术研究，提高我国原始创新能力。应当说，我国在一些特色领域，如生命科学、信息科学、材料科学、环境科学、能源、航空航天技术等方面，取得了卓越成就，已经进入了世界科技前沿，甚至取得了突破性进展。但和欧美日等发达国家和地区相比，我们在科技的整体实力和创新水平上，仍存在一定差距。我们必须奋起直追，通过不断的努力来提高原始性创新能力，要弘扬勇于探索的科学精神，要凝聚和培养创新型人才。同时各级政府要创造良好的环境，社会要营造新的文化，包括容忍失败、鼓励创造。

(5) 深化科技体制改革，推进国家创新体系建设。我们必须把科技创新和体制创新结合起来，以“三个代表”重要思想为指导，以加强国家创新体系建设为目标，按照党中央、国务院决定的要求，加大深化改革的工作力度，在大幅度提高各类创新主体创新能力的同时，着力转变体制和机制。我们应坚持体制和机制创新，坚持以人为本，坚持改革与发展相结合，坚持配套进行改革。科技体制改革的根本目的，是要建设国家的创新体系，提高我国战略创新的能力和产业的核心竞争力。在建立国家创新体系时应遵循以下思路：进一步以深化改革为动力，发挥政府宏观调控功能和市场配置资源的基础性作用，优化配置国家创新资源，激发创新主体的内在活力，集成社会创新力量，形成多种所有制并存，国家和社会共同推动的创新格局，大幅度提高国家创新能力，加速科技成果的转化和应用，提高国家竞争力和人民生活质量。

(6) 加强战略研究，制订科学和技术的长远发展规划。在战略研究方面，为了迎接加入 WTO 的挑战，科技部已会同有关职能部门提出了人才、专利、技术标准三大战略，并且形成了几个重大专项实施人才战略，就是要切实贯彻人才资源是第一资源的战略思想，把以人为本和知识有价落到实处。实施专利战略就是要努力提高原始性创新能力，掌握核心科技，增强科技、经济的竞争力。实施技术标准战略就是要尽快研究建立国家技术标准体系，打破别国技术壁垒，争取经济主动权。

(7) 加强科技基础设施建设。我国在科学技术的基础设施方面，要建立与时俱进的、适应当前科技发展水平的、与我国财力增长相适应的科技条件的大平台，包括科研基地、重点实验室、工程技术中心，科技资源的共享和一系列的科技条件，也包括在科技期刊上我们有新的进展。

(8) 加强科学技术的普及工作，弘扬先进文化。代表先进文化的前进方向是我们科技界面临的一项非常重要的任务。一个国家的科技实力，既表现在提高方面，也表现在普及方面；既体现在攻坚和创新的能力上，也体现在科学技术的普及程度和公众的科学素质方面。应该说，后一个方面是我们国家非常薄弱之处，我们必须坚持两个方面，普及科学知识，弘扬科学精神，在全社会形成崇尚科学，鼓励创新，反对迷信和伪科学的良好氛围。

目前，科学技术正以加速度推进人类社会的发展，形成一个鼓励创新、勇于创新的社会氛围，对一个社会的发展起着至为关键的作用。诚如江泽民同志所指出的“创新是一个民族进步的灵魂，是一个国家兴旺发达的不竭动力”[①]，全面增强国家的创新能力，既是实施科教兴国战略的需要，也是为了加速推进全面建设小康社会的步伐。而科技创新反映了现代科技发展的本质规律，在国家的创新体系中占有举足轻重的战略地位。全面建设小康社会的伟大事业，热切期盼着科技的发展与创新为之提供科技支撑和强大动力。我相信，广大科技工作者一定会以自强不息、百折不回、勇攀科学高峰的精神，勤奋工作，开拓创新，为圆满完成党的十六大提出的各项战略任务，为实现中华民族伟大复兴，为我国经济建设、国防安全和社会可持续发展，贡献自己的智慧与汗水。

① 《创新是民族进步的灵魂，是国家兴旺发达的不竭动力：学习江泽民同志关于科技创新的有关论述》，https：//www.dswxyjy.org.cn/n1/2019/0625/c427785-31187716.html[2023-11-02].

加强基础研究 实现科技强省[①]

基础研究是人类文明进步的动力，是科技与经济发展的源泉和后盾，是新技术、新发明的先导，也是培养和造就科技人才的摇篮。在综合国力竞争中，基础研究的发展水平已经成为一个民族的智慧、能力和国家科学技术进步的基本标志之一。以基础研究及其所孕育的高新技术原始性创新为主要标志的科技自主创新能力的竞争，已经成为当今世界科技竞争的制高点，乃至国家竞争成败的分水岭。基础研究作为原始创新的源泉、高新技术及其产业发展的先导，对当代科学技术的整体发展、新兴产业群的崛起以及经济和社会的变革产生了巨大的不可估量的推动作用。

世界上的发达国家无不对基础研究给予了极大的重视。世界头号科技经济强国——美国，凭借其强大的经济竞争力，形成世界上领先的整体科技实力。美国政府和民间机构所作的调查都得出相同的结论：美国政府对基础研究进行的长期、稳定的支持，是维持美国科技经济竞争力的根本。

教育和科学水平是衡量一个国家实力的重要指标。俄罗斯与发展中国家的根本差别不在于其拥有核武器、石油和原材料，而在于其具有非常高的教育水平。尽管最近10年俄罗斯基础研究实力有所下降，但俄罗斯是除美国以外的在所有科学领域都进行科学基础研究的国家，研究领域面大而宽，基础研究根基雄厚。英国政府科技投入中的60%用于基础研究。保持基础性研究的高水平是当前德国科技政策的核心之一。20世纪80年代始，德国的研究与发展经费中约有20%用于基础性研究，这一比例远超过美国的12%、日本的13%，此后持续增长，1992年以后，德国基础研究经费达到并基本保持在25亿欧元（占29%）。日本大力推进战略性基础研究，强调“关键的关键，是创造出自己的新技术”。韩国政府在2003年投资1696亿韩元加强基础科研和人才培养，以促使基础研究获得长足进步。

过去，基础研究的动力主要来自科学发展过程中的内部矛盾。在社会不断发展的今天，国家、社会方面的需求动力越来越大，越来越直接成为推动基础研究不断发展的力量，使基础研究处于一个非常重要的地位。当一个国家基础研究水平远远落后于别的国家，而这个国家又想很快地在经济上崛起，那么，直接引进就是最有效的途径，日本、韩国、新加坡等都是靠直接引进别人的科学成果发展起来的。但现在，我国年GDP总量已超过一万亿美元，经济总量已经排世界第六位，开始向中等发达国家水平的目标前进了，我国的基础研究应该到了要提升的新阶段。激烈的国际竞争，对中华民族的自主创新能力提出了新的、更高的要求。在尖端高科技领域，在最前沿的科学领域，在涉及商业利益的高技术领域，在国防科技领域，没有人会把最先进的技术和成果转让给我们，一个基础研究实力薄弱的发展中大国不可能在科技方面掌握自己的命运。忽视基础研究，将会使我国与发达国家的差距继续拉大，未来世界就没有我们

① 本文原载《科技管理研究》，2004，24（5）：1-3.

的地位。金融风暴中那些缺乏自主创新和知识产权、主要依赖直接引进别人技术而发展起来的经济的虚弱和不堪一击，已经向我们昭示了这一点。

知识经济对人才结构、人才的素质提出了新的要求。科技创新的关键是人才，基础研究是充满活力的创新活动，是培育具有创新精神和创新能力人才的摇篮，激烈的人才竞争使基础研究重要的战略地位更加凸显。首先，考察科学发展史，不难发现，杰出的人才需要培养造就。通过基础研究，不但要出成果，也要培养出高层次的科技人才。科技活动逐渐成为高水平大学培养人才的重要方式。如果一个国家不搞基础研究，就不可能有高的教育水平；凡是教育水平高的高校，教师都在积极从事基础科学研究。其次，基础研究为培养、稳定一支高水平的科研队伍发挥了积极作用。近 10 年来，我国科研队伍建设得到重视和加强，特别是对优秀中青年科技人才的培养达到了前所未有的程度，各种人才计划相继出台，这些人才计划的实施，使大批优秀青年学者获得了较强的经费支持，他们中的许多人正在逐步成长为我国多学科领域的学术骨干和学科带头人。目前我国有一支近 8 万人的基础研究队伍，其人员的年龄结构已发生明显变化，中青年科技骨干的比例正在迅速上升，45 岁以下的国家自然科学基金项目负责人比例从 1986 年的 12%提高到 70%，在国家重点实验室工作的中青年科技骨干的比例已接近 50%，参与 1998 年启动实施的《国家重点基础研究发展规划》项目的 35 岁以下的青年学者和研究生已占 60%以上。

基础科学在整个自然科学体系中有十分重要的地位和作用。基础研究包括对科学本身的基础研究（纯基础研究）和应用科学技术基础研究（应用基础研究）两个部分。前者是以认识自然现象、探索自然规律、增加人类知识为目的的科学研究，后者是围绕重大应用目标或某种应用技术而进行的基础性科学研究，两者既有区别，又有联系。基础研究是社会与科学发展的基础，由基础科学研究产生的大量新思想、新理论、新效应等为应用科学提供了理论基础。

自 20 世纪 90 年代后期以来，随着经济建设的迅速发展和工业生产水平的不断提高，劳动密集型产业结构赖以生存和发展的环境发生了根本的改变，广东原有的工业生产低成本和技术领先的优势开始丧失；从市场环境看，市场结构已由卖方市场逐渐转向买方市场，人民消费水平迅速提高，使得消费市场的热点不断向高层次发展，广东必须推动产业结构从劳动密集型转向技术密集型；从经济增长方式来看，随着经济的发展，广东在过去 20 多年的发展过程中所采取的速度型、粗放型增长方式的弊端日显。此外，国家在财税、金融、外贸、价格等方面的改革，使得粗放型经济增长方式的条件不复存在，宏观环境的变化和工业本身的发展规律，都要求广东经济向以高新技术产业为主的效益型、内涵型发展模式转变。

知识经济建立在知识和信息的生产、分配及使用上，高新技术产业以高新科技为最重要的资源依托。由于知识创新和技术创新的速度加快，劳动者需要不断更新自身的知识与技能，教育与培训有走向终身化的趋势，这些都需要基础研究的支撑。

目前，广东的经济实力和高新技术产业在国内还处于领先地位，但随着长三角、京津等地区的高速发展，开发大西北、振兴东三省步伐的不断加快，广东面临着的知识、技术、人才竞争将越来越激烈。“人无远虑，必有近忧”，如果不大力加强基础研究的力度，从根本上解决知识、技术、人才的来源问题，再过若干年，广东的高新技术产业乃至整个经济的发展将会缺乏后劲。

长期以来，我国的基础研究大多还是属于跟踪性的创新，分散重复，缺乏重大科学发现和技术发明，原始创新不多，年轻后备力量不足。但近年来，我国的基础研究开始取得了一些令人振奋的成果：神舟五号载人航天实验取得圆满成功，成为标志着我国进入世界先进航天大国行列的里程碑；我国国际科技论文数量持续稳定增长，2003年已跃居世界第五位；2003年来自国内的发明专利申请数量八年来首次超过来自外国的申请；产生连续两届国家自然科学奖一等奖。这一切都表明，多年来困扰我国科技发展的原始性创新能力不足的状况正在得到改观。

更重要的是，我国在以下三个方面奠定了基础研究发展的基础。

(1) 认识方面，基础研究工作的重要性得到了共识。国家、政府各级领导极为重视基础研究。江泽民同志指出："基础研究很重要"；全国技术创新大会提出："重大突破性创新要着眼于从基础研究抓起，不断形成新思想、新理论、新工艺，为应用研究和技术开发提供源泉，增强持续创新的能力"；高校在认识、观念方面的转变，使高校充分发挥了在科技特别是基础研究方面的潜能，成为科技创新基础研究的一支主力军；广东省也极为重视基础研究，在全国率先建立了省级自然科学基金，累计投入经费超过2亿元。

(2) 项目到人，观念上有了一个质的改变。"以人为本""人比项目更重要"的思想开始被接受并有所体现。教育部组织实施了"跨世纪人才工程"，近年又实施《面向21世纪教育振兴行动计划》中的"高层次创造性人才工程"；中国科学院组织实施了"百人计划"，近年在知识创新工程推动下，又推出了"引进海外杰出人才计划"；国家基金中杰出青年基金等人才类板块项目的比重加大。广东省基金的类别结构，一开始只有面上项目层。20世纪90年代，广东开始设立青年项目，目前形成了研究团队项目（培养高层次人才队伍）、重点项目、自由申请项目、博士启动项目（培养青年后备力量）等四个层次项目。

(3) 投入大大增加，结构趋向合理。2002年，我国基础研究方面的投入达60多亿元，占R&D经费的比重为5%左右，到2005年将上升到8%～10%，接近中等发达国家的水平。2001年以来，广东省每年投入基础研究的经费超过3000万元。建立了一批科技基础条件平台，现有国家、部门重点实验室16个，省重点实验室72个，以通用的实验仪器为例，实验室已达到发达国家中等水平，对人、财、物等方面的投入与建设均达到了一个较高的水平。

但是，目前广东的基础研究还存在以下几个问题：①在国家层面上的基础研究中，重大创新缺乏，国家级大项目、高级别奖项不多，国家重点实验室、工程中心数量偏少，影响力、辐射力偏弱，对广东省经济社会发展中的关键技术的支撑不够；②投入产出效益不高，以论文发表情况为例，"九五"期间，被三大索引收录的广东省科技论文数量在全国的排位（第11位），落后于广东省各项基金资助（基础研究）产生的论文数量在国内的排位（第6位），更落后于广东省发表的科技论文数的国内排名（第4位），这表明广东省的基础研究投入产出效益尤其是高水平成果的产出率不高；③优秀学术带头人，尤其是中青年学术带头人缺乏（国家杰出青年科学基金 1 年资助的人数为 160 人，而自这项基金设立 10 年以来，广东省获资助的人数总共不到 50 人），人才培养无论是在数量上还是在质量上，都远远不能满足广东经济、社会发展的需要。在这样的形势下，针对如何在科技强省战略中使基础研究促进广东省社会、经济向更高层次发展，我们提出如下一些建议。

（1）打造具有区位特色的科学中心。这是科技强省战略实施过程中，使基础研究发挥作用的关键。纵观世界经济中心转移的历程，无论它是在英国、德国还是美国，无不与科学中心的转移密切相关。广东是经济大省、高技术产业大省，随着竞争的日益激烈，要取得更大的发展，必须加强国家、地方目标对基础研究的“需求牵引力”，打造具有区位特色的科学中心，使之成为知识、人才、技术的聚集地和辐射中心。

（2）通过加强基础研究，建设有自己特色的创新文化，这是建设文化大省的一个重要组成部分。创新文化有三个层次，第一层次是外部形象、规模、队伍、场地、设备等；第二层次是规章制度和行为规范；第三层次，也是最重要的层次，是价值、观念、道德，这些是创新文化的核心。要克服急于求成、急功近利等各种浮躁思想，大力培养有棱有角、敢于创新的品格，提倡养成“十年磨一剑”的耐性，树立甘于寂寞、以质取胜的精神。希望所有的科研人员，特别是青年科研人员，在这些方面有强烈的意识。写 100 篇平平常常文章的人，在学术舞台上可能很快就会消失；而写一篇有分量的文章，能引起大量后续研究，使大家都跟着去做，可能会在学术界树立一个里程碑。著名数学家吴文俊先生在获国家最高科学技术奖后讲了一句话：“什么是创新？创新就是要别人 follow me，让别人跟着我做，如果我 follow 别人，就不是创新了”。这样一些观念，要成为我们创新文化的精髓。

（3）优化创新环境，实现基础研究资源的合理配置，提高科研的效率。资源配置中最重要的是讲求实效，就是绩效优先，这是非常非常重要的。虽然与发达国家相比，我国目前对基础研究的投入水平并未达到理想程度，但政府近年来的投入水平已创新高。正因为如此，科学界和管理部门在讨论基础研究的问题时，不再像过去那样把“投入”当作第一位的问题了，而是讲求怎么把钱花好，把钱用得正确。在很多国家，一些非常贵重的科学仪器，一个大城市只有一套；而我们，现在就有很多套，然而每一套都使用得不足，在每一套上，我们都没有创造出很多新的成果。有时候去参观一些实验室，这些单位要叫人专门去拿钥匙才能把实验室门打开，然后揭开用白布盖着的贵重设备，介绍它的指标有多高，这实在令人忧虑，因为如果再放五年，这些就成为落后的设备。做好规划和布局、增强学科间交叉、注重科学的评估、减少非研究性负担、开放共享的平台，让基础研究产生更好的效益，更好地为科技强省服务，是目前摆在我们面前的紧迫任务。

我国的工业化比欧美晚了约 200 年。100 多年来的痛苦探索，使我们明白了一个真理：一个没有强大科技实力的国家，不可能屹立于世界民族之林。在 21 世纪之初，瞻望我国宏伟的发展目标，我们应该更加重视基础科学研究，以实现中华民族伟大复兴的夙愿。

成立广东省仪器仪表学会是紧迫的历史使命①

现代仪器仪表界定为测量和控制仪器仪表与系统、科学仪器、医疗仪器，以及其相关的传感器、元器件和材料。根据统计数据，我国现有各类仪器仪表企业6000多家，职工总数88万人。我国仪器仪表已经形成门类品种比较齐全，具有一定技术基础和生产规模的工业体系，成为亚洲除日本以外第二大仪器仪表生产国。

应当清醒地看到，虽然我国仪器仪表工业有了一定的发展，但远远不能满足国民经济、科学研究、国防建设以及社会生活等各个方面日益增长的迫切需求。我国仪器仪表产品，绝大部分属于中低档技术水平，而且可靠性、稳定性等关键性指标尚未全部达到要求。高档、大型仪器设备几乎全部依赖进口，中档产品以及许多关键零部件，国外公司同样占有国内市场60%以上的份额。据海关统计，除去随成套工程项目配套引起的仪器仪表不计，每年进口各类仪器仪表总额接近我国仪器仪表工业总产值50%。此外，在6000多家企业中，年销售收入超过1000万元的不足1000家，全行业经济效益低下。

更为严重的是，低水平重复生产异常突出。无论是工业自动化仪表或是科学仪器，几乎每一个主要产品都有几十家甚至几百家企业重复生产，乃至产品的型号规格都完全相同，其产品的技术水平多只相当国际20世纪80年代初期水平，而且产品质量全面过关的微乎其微。低水平重复生产肆意扩展，对我国仪器仪表工业的发展造成了极大的危害：第一，低水平重复生产，耗费了大量人力、物力和财力，导致我国仪器仪表工业的发展不能集中优势力量，捏紧拳头，难以摆脱长期分散落后的局面；第二，大量低水平重复生产，破坏了市场竞争的有效秩序，致使许多企业不能集中力量提高产品质量和发展生产；第三，低水平重复生产的结果，必然出现大量产品质量低劣，给用户造成恶劣的影响，使民族仪器仪表工业的发展受到沉重的打击。

我国仪器仪表工业在科技创新及产业化方面跟国际潮流相比滞缓太多，形势非常严峻。在已经跨入21世纪的今天，我国仪器仪表的普遍水平还停留在20世纪80年代初国际水平上；大型和高档仪器设备几乎全部依赖进口；许多急需的专用仪器还是空白；中低档产品保证质量上还有许多难关需要攻克……科技创新及其产业化进展滞缓，是制约我国仪器仪表工业发展的一个“瓶颈”。

仪器仪表产业其产值虽然在国民经济总产值中所占比例很小，据统计在美国约占4%，在我国不到1.5%，但现代仪器仪表在当今社会的重要作用怎样评估都不为过。现代仪器仪表是信息产业的源头和重要组成部分，又是信息产业带动其他产业发展的桥梁。现代仪器仪表的应用领域，已经涉及“农轻重、海陆空、吃穿用”，无所不在。因此，发达国家都把发展仪器仪表产业提到战略高度，充分予以重视和支持。我国仪器仪表工业的发展还处在脆弱和幼稚的阶段，而且面临着三重压力：一是国际上现代

① 本文是向广东省科学技术协会的报告，广州，2005年9月11日.

仪器仪表正向着微型化、集成化、智能化和总线化迅速发展，必须快速跟上；二是国内市场被外商挤占，特别是我国已经加入 WTO，竞争将更趋激烈，必须争夺生存和发展的空间；三是我国仪器仪表工业的发展还受制于“瓶颈”的制约，必须尽快挣脱“瓶颈”的约束加快发展的步伐。如果我们不能在跨入 21 世纪、加入 WTO 的 5 年到 10 年内振兴我国仪器仪表工业，那么我国仪器仪表工业将在三重压力下被彻底压垮，其后果不堪设想。因此，加快发展仪器仪表产业是紧迫的历史使命。

现代仪器仪表集成了光、机、电、计算机等各种新技术和应用了多种基础学科的研究成果。现代仪器仪表应用在国民经济各种产业、科学研究、国防建设，以及社会生活的各个领域，因此，仪器仪表不归属在某个具体的产业部门。针对仪器仪表产业，建立一个专业学会，在制定发展政策和统一规划协调方面作为政府的参谋和咨询组织是十分必要的，也符合国家改变管理职能，充分发挥中介组织在行业发展中起协调管理的作用的方针。专业学会对全行业的发展，包括发展规划、布局、政策和重大项目的建议，基地建设和重点企业的发展，组织产、学、研相结合等许多方面都可以发挥指导作用。

目前，国内仪器仪表行业所面临的严峻形势在广东省尤其突出。作为一个经济大省，各类仪器仪表企业众多，迫切需要成立一个能为仪器仪表企业服务的学术组织——广东省仪器仪表学会。广东省仪器仪表学会将是广东省的仪器仪表科技工作者组成的以促进仪器仪表及自动化科学技术发展和普及为宗旨的，具有公益性、学术性的社会团体。本会将认真执行党的方针、政策；贯彻“百花齐放，百家争鸣”的方针；坚持民主办会原则，充分发扬学术民主，开展学术上的自由讨论；提倡辩证唯物主义和历史唯物主义，坚持实事求是、开拓创新、与时俱进的科学精神、科学态度和优良学风；团结和组织广大仪器仪表科技工作者，在改革开放中为繁荣我省仪器仪表事业、提高我省仪器仪表学术水平、振兴我省仪器仪表工业、加速实现我省社会主义现代化作出贡献。

1. 振兴仪器仪表产业是国家鼓励发展的政策

为了振兴我国仪器仪表产业，国务院决定由国家计委、国家经贸委[①]、科技部、财政部等有关部门协商制定必要的扶植政策。扶植就是鼓励发展，国家希望尽早制定出鼓励仪器仪表产业发展的有关政策。成立广东省仪器仪表学会是响应国家的号召，振兴仪器仪表行业的重要措施。为此中国仪器仪表学会负责人专程来到暨南大学，希望挂靠暨南大学应用力学研究所成立广东省仪器仪表学会。

2. 广东省科研院所、高校、企业积极支持成立广东省仪器仪表学会

2005 年 7 月 31 日在暨南大学应用力学研究所召开了广东省仪器仪表学会第一次筹备会议。会议由暨南大学校长刘人怀院士主持。参加会议的代表共 20 多人，有省内高校及科研院所和企业的代表，他们分来自暨南大学、华南师范大学、中山大学、华南

① 国家计委是国家计划委员会的简称，1998 年的机构改革中，国家计划委员会更名为国家发展计划委员会，2003 年《国务院机构改革方案》将国家发展计划委员会改组为国家发展和改革委员会，不再保留国家经济贸易委员会。

理工大学、工业和信息化部电子第五研究所、广东省电子学会及广州、深圳、佛山、东莞等地的企业。代表们踊跃发言，纷纷表示要为广东省的仪器仪表事业多做实事，盼望尽快成立广东省仪器仪表学会。目前已经收到广东省仪器仪表学会会员登记表 50 多份。

3. 成立广东省仪器仪表学会可以实现政、产、学、研、金的有效结合

仪器仪表科技创新需要一批既有学识又有经验的边缘科学和应用技术的人才。仪器仪表行业这类人才本来不多，近年来又大量流失，人才匮乏成为仪器仪表行业科技创新及产业化的阻碍。仪器仪表科技创新，尤其是实现创新成果产业化，必须有官方、企业、高校、科研院所和金融界的有效结合。在目前我省仪器表企业面临的处境下，成立广东省仪器仪表学会可以提供行业信息，科技服务企业，协助行业成功实现产业化。

综上所述，为了加快发展仪器仪表产业，成立广东省仪器仪表学会是紧迫的历史使命，成立广东省仪器仪表学会的条件成熟、依据充分。

培养青少年的创新精神[①]

在这生机勃勃、百花争艳的大好春光里，第二十一届广东省青少年科技创新大赛隆重开幕了。首先，我对本届青少年科技创新大赛的举行表示最热烈的祝贺！向参加大赛的老师、同学们和大赛评委会的同行们致以诚挚的问候！向为本次大赛付出辛勤劳动的广东省青少年科技中心、惠州市科协、惠州市科技局、惠州市教育局表示衷心的感谢！

青少年科技创新大赛是广东省科学普及活动的一次盛会，也是广大青少年展现科技才能的一次盛会，能够有幸参加此次大赛的活动，为推动青少年科技创新教育作出贡献，我感到非常高兴。因为青少年强则国家强，青少年兴则民族兴，在同学们朝气蓬勃的精神风貌和创造发明的良好志趣中，我看到了科学的明天和希望，看到了中华民族的未来和希望。

英国科学家培根在几百年前提出了"知识就是力量"的口号，这就像一盏黑夜中的明灯，驱散愚昧和无知的浓雾，引导着无数有志青年追求真理、献身科学。特别是在知识经济时代来临的21世纪，科学技术飞速发展，科技创新日新月异。知识更新的速度越来越快。创造性思维及科技创新在人类社会和生产力发展中的原动力作用日益凸显。有关资料表明，现在每年的技术淘汰率是20%，也就是说一项技术的寿命周期平均只有5年，在一定意义上讲，当前国家之间综合实力的竞争，关键在创新，核心在人才；"科技是第一生产力""人力资源是第一资源"是提升国家核心竞争力和综合实力的一项战略举措。中华民族伟大复兴，不但需要当代人的努力，更决定于未来一代人的素质，青少年的综合素质特别是科学素质如何，将直接关系到我们能否巍然屹立于世界民族之林，能否实现中华民族伟大复兴，培养和提高青少年的科学素质比任何一个时期都显得更为迫切。

也许有些青少年认为，创新是很高深的事情，是科学家和专业人士才能从事的特殊工作，中小学生在创新方面哪能有什么作为？其实不然，创新能力并非少数科学家和专业人士所独有，也不是到了成年后才形成，古时候的曹冲称象、司马光破缸救同伴就是青少年创新能力的良好体现。中小学生在学习中找到解决问题的新方法，写出具有新意的作文，也是创造性的一种表现。而且青少年时期是人的原创力发展的重要阶段，好奇心重、求知欲强，正是培育创新意识和创新能力的大好时机。同学们在学习和生活中应形成尊重科学、崇尚创新的科学思想，向老师学习、向一切有知识的人学习，培养深入探究、勇于实践的科学素质，积极投身科技创新活动，为将来的提高和发展打下良好基础。

第二十一届广东省青少年科技创新大赛的举行，不但为同学们学科学、用科学良

---

① 本文是第二十一届广东省青少年科技创新大赛开幕式上的讲话，惠州，2006年4月7日；教育与科技管理研究，北京：科学出版社，2016，370-371.

好学风的形成提供了社会环境，而且为同学们创新精神和实践能力的培养提供了锻炼机会。我希望广大青少年朋友珍惜科技创新大赛的交流和学习机会，相互切磋，取长补短，提高自身的科技创新素质。我更热切地希望青少年朋友们在今后的时间里，勤奋学习，积极探索，大胆实践，勇于创新，努力使自己成为振兴广东、振兴中华的栋梁之材！

最后，预祝第二十一届广东省青少年科技创新大赛取得圆满成功！

祝参加本次大赛的选手们取得优异成绩！

大力推动科技事业的发展①

第三届“看中国”系列丛书出版座谈会暨《科技创新与品牌》杂志首发式开幕了，我代表“看中国”系列活动组委会、《科技创新与品牌》杂志社，向各位代表表示热烈的欢迎。

“看中国”活动以贯彻党的十六大精神，树立科学发展观，搭建各界学习、交流、对话合作的平台为宗旨；以弘扬科学家与中华优秀人才及改革开放以来各界取得的卓越成就为主旋律，是一项高层次的大型公益活动。该项活动由中国科技报研究会“看中国”系列活动组委会牵头隆重推出，已成功举办了两届，出版了5卷大型彩色系列丛书。

“看中国”活动举办以来，得到各级领导与社会各界的好评，同时得到了布赫、何鲁丽、顾秀莲、路甬祥、周铁农、徐匡迪等党和国家领导人的亲切关怀与支持，并分别为“看中国”系列活动丛书题写书名、题词或出席首发式。

今年出版“看中国”丛书中的两本分别为《中国新农村建设方略》和《人才强国》。《中国新农村建设方略》是一部根据中共中央两个一号文件以及胡锦涛总书记、温家宝总理在省部级主要领导干部建设社会主义新农村专题研讨班上的重要讲话精神，由省部级有关领导直到农村基层干部撰文组成的专辑，其主要内容是：当前新农村建设中的问题和政策解读、规划方案、取得的成绩、农村新面貌以及在新农村建设中涌现出的先进人物和先进事迹。

《人才强国》主要体现我国的“科教兴国”和“人才强国”的战略，书中集中了一大批社会精英，主要是在各行各业做出突出贡献的专家、领导干部等。为了实现强国梦，他们不为名不为利、不辞辛劳、夜以继日地工作在第一线，是许多国家级重大课题的带头人。本书集中反映了他们的工作、生活及其动人事迹。

《科技创新与品牌》杂志是在“看中国”系列活动的基础上构想出来的，于2006年底由新闻出版总署批准正式创刊，是一本由中国科学技术协会主管，中国科技新闻协会主办，面向国内外公开发行的中央级综合性科技月刊，杂志有由袁隆平等20位院士组成的强大编委会及高素质的编辑记者队伍，其办刊宗旨为：介绍科技人才成就、科技创新成果、品牌范例及品牌战略在经济建设中的重要作用，古今中外的创新经验，为社会主义科学技术发展服务。

本刊将以独特的新闻视角提供大容量的科技信息，为各界架起信息传递、沟通合作的桥梁，使创新成果尽快转换成生产力，造就具有自主知识产权的知名品牌，打造中国的国际品牌。

通过对科技工作者、科学家在国家经济建设中取得的卓越成就及奋斗拼搏精神的

---

① 本文是在第三届“看中国”系列丛书出版座谈会暨《科技创新与品牌》杂志首发式上的讲话，北京，2007年7月6日.

报道，提高人们对科技创新成果与品牌战略在经济建设中重要性的深刻认识，为推动科技事业的发展起到积极的促进作用。

《科技创新与品牌》的创刊号已于 7 月 1 日正式出版，就是大家手里的这本，希望大家喜欢她、支持她，把她打造成真正的期刊品牌。

由于时间仓促，书、刊中的差错在所难免。请各位代表多提宝贵意见。

最后，祝各位代表北京之行愉快，身体健康！

祝大会圆满成功！

迎接新挑战①

第九届全国振动理论及应用学术会议暨中国振动工程学会成立 20 周年庆祝大会胜利召开了！金秋十月，丹桂飘香，来自全国各地广大从事振动工程研究的专家学者和工程技术人员欢聚在美丽的西子湖畔，相互交流切磋，共同探讨各个前沿领域的最新科研成果和发展动态，共同迎接 21 世纪对我们广大科技工作者的新挑战。

中国振动工程学会经过 20 年的发展，已成立了 14 个专业委员会、18 个省级地方学会，在学术交流、期刊出版和人才举荐等方面取得了显著成就。学会成功地举办了第 3 届至第 8 届全国振动理论及应用学术会议和专业学术交流活动，举办了多次振动工程领域国际学术会议；学会主办的《振动工程学报》和《振动与冲击》期刊均被确定为国家中文核心期刊，近几年来，其影响因子逐年提高并多次获奖，已成为国内颇具影响力的知名期刊；学会多次推荐两院院士候选人，数次推选中国科协全国代表大会代表和委员候选人，推荐优秀科技工作者，推荐优秀科协工作者等。

振动工程是一门新兴的交叉学科，它的涉及面相当广泛，几乎涵盖了机械、船舶、车辆、航空、航天、建筑、土木、水利、地震、能源、海洋、生物等各个工程技术领域。此次会议收到了来自全国各地各部门科研院所、高等院校和工业企业界的论文 530 余篇，较全面地反映了我国在这些领域的科研和应用现状，体现了我国振动工程学科近年来的学术技术水平。一篇篇论文犹如一朵朵鲜花在国家级的学术殿堂开放，象征着振动工程界又一个繁花似锦、万紫千红的春天。

本次学术会议的论文，经由本次会议特邀的全国各地近 50 位专家学者认真评审，选出了 350 多篇较优秀的论文，以《第九届全国振动理论及应用学术会议论文集(2007)》的方式，由浙江大学出版社正式出版。其中论文摘要印刷成书，而全文汇编在光盘中。论文集共分为 13 个部分，即大会报告、非线性振动、随机振动、结构动力学、机械动力学、转子动力学、振动与噪声控制、结构抗震与振动控制、动态测试与信号分析、模态分析与试验、故障诊断理论与应用、振动利用工程、微纳机电系统动力学。

这次会议得到了相关全国学会、高等院校、科研设计部门和企事业单位的大力支持。浙江大学应用力学研究所、机械电子控制工程研究所和现代制造工程研究所为本次学术会议论文的征集、评审和编辑做了大量的工作，浙江大学出版社的同志为本论文集的出版付出了艰辛的劳动，谨在此一并表示衷心的感谢！

在本论文集的征集、评审、编辑和出版工作中定有欠妥之处，恳请作者、读者予以谅解，并提出宝贵意见。

祝愿本次学术会议取得圆满成功！

---

① 本文是《第九届全国振动理论及应用学术会议论文集（2007）》序，杭州：浙江大学出版社，2007，1-2.

同心同德 开拓前进[①]

正当全国各族人民喜迎党的十七大隆重召开，为加快构建社会主义和谐社会努力拼搏之际，中国振动工程学会第六次全国会员代表大会在美丽的城市杭州隆重召开了！这是我国振动科技界的一件大事和喜事，请允许我代表中国振动工程学会第五届理事会，对来自全国各地的会员代表、对莅临会议指导的浙江省、浙江大学的有关领导和来宾，表示热烈的欢迎和衷心的感谢！

同志们，中国振动工程学会第五届理事会是在2003年选举产生的。在过去的四年时间里，理事会在中国科协的领导下，在加强组织建设，开展国内外学术交流，组织科技期刊出版，推动国家和地方的经济建设，以及促进学科的发展等方面做了许多卓有成效的工作，使学会的影响进一步扩大，知名度进一步提高。在这里，我们要向为此作出积极贡献的第五届理事会的各位领导和全体理事表示诚挚的谢意！根据中国科协的指示精神和学会章程的规定，经中国科协批准，我们在这里举行学会第六次全国会员代表大会，进行理事会的换届改选工作。出席会议的全国会员代表来自我国20多个省（自治区、直辖市）的80多个单位，基本覆盖了我国从事振动工程研究、教学、开发、生产和使用的主要单位，具有较为广泛的代表性。他们绝大多数是具有副高级技术职务以上的科技人员，具有一定的学术造诣，并热心学会工作，为开好这次全国会员代表大会，为我会今后的兴旺发达奠定了坚实的组织基础。

现在，我们全国会员代表济济一堂，共商学会发展大计。这次全国会员代表大会的主要任务是：

（1）审议学会第五届理事会工作报告。

（2）审议学会章程及其修订说明。

（3）选举产生学会第六届理事会。

（4）举行学会成立20周年庆活动。

（5）讨论交流学会工作。

因此这次会议的任务十分繁重，而时间又极为短促。请各位代表团结一致，集中精力，切实完成会议的各项任务，把这次会议开好。

代表们、同志们！党的十七大号召我们深入贯彻落实科学发展观，坚持党的基本理论、基本路线、基本纲领、基本经验，继续解放思想，坚持改革开放，推动科学发展，促进社会和谐，为夺取全面建设小康社会的新胜利而努力奋斗。十七大向我们提出了很高的要求，作为与我国各个工程技术领域有着广泛联系的国家一级学会，我们肩负着光荣的历史使命。让我们在以胡锦涛同志为总书记的党中央领导下，高举邓小平理论的伟大旗帜，全面实践“三个代表”重要思想，更加深入学习贯彻十七大精神，同心同德，开拓前进，为完成我国的第三步战略目标，为实现中华民族伟大复兴，做

① 本文是中国振动工程学会第六次全国会员代表大会开幕词，杭州，2007年10月18日.

出我们积极的贡献!

中国振动工程学会自成立以来，一直受到挂靠单位南京航空航天大学领导在人力、物力、财力等多方面的大力支持。在召开第六次全国会员代表大会之际，我们谨向南京航空航天大学的领导和参加学会工作的同志们表示衷心的感谢! 同时，浙江大学为开好这次全国会员代表大会，举行中国振动工程学会成立 20 周年庆祝活动，以及举办第九届全国振动理论及应用学术会议，作了较长时间的准备，提供了良好的工作条件和生活条件。为此我们特向浙江大学领导的大力支持和承担具体工作同志的辛勤劳动表示深切的感谢!

最后衷心祝愿中国振动工程学会第六次全国会员代表大会圆满成功!

创新路上的感想[①]

创新这个题目，是个很大众化的题目，是人人都知晓的题目，也是个很难讲的题目。今天讲的题目是“创新路上的感想”，是讲自己碰到的方方面面有意义的事情，讲自己参与创新的经验和体会。

创新是一个相当宽泛的概念，创新不仅仅是技术创新，也可以指理论创新，还可以是观念、体制的更新等，创新应该包括人类生活、社会发展取得进步的各个方面，其核心要素是取得新的认识或获得新的成果，是在突破原有认识基础上的一种创造性的智力活动。

纵观人类社会的发展史，实际上就是创新的过程、创新的历史。从石器时代到信息时代、从原始社会到工业社会，在人类数百万年的社会发展过程中，创新起到了极其重要的作用。新的科学理论、新的发明、新的技术、新的工程、新的材料、新的思想、新的文化、新的制度等，层出不穷，渐次递进，推动人类社会向前发展。因此，创新仍然是当代的主旋律，仍然在社会发展和人类进步中发挥着重要作用。

今天主要讲两个问题：第一个问题讲创新是社会进步的不竭动力，第二个问题是创新的一点感想。

一、创新是社会进步的不竭动力

1. 自然科学的创新

社会发展核心的问题是要创新，它是社会发展的动力。实际上人类社会的发展史就是创新的过程、创新的历史。自然科学理论的创新十分重要，对社会的关系甚大。从 15 世纪起，产生了现在的自然科学。遗憾的是，尽管我们中华民族创建了世界最优秀的历史文化，但自然科学却没在我们中国产生。这是我们的制度、管理出了毛病，造成自然科学没在中国产生。除南欧外，欧洲的其他地区历史比较短，在 10 世纪以前都还处于蛮荒时期，10 世纪以后才比较好一些。我们中国在宋朝、明朝时经济总量已经是全世界第一，全世界经济总量一半以上在中国，可惜我们错过了机会。第一枚火箭是中国人发明的，宋朝就有火箭了，但是我们只把它作为娱乐的工具，没有把它作为科学技术。因为我们传统的管理把科学技术作为一种低档的工作，以前的技术人员被称为匠人，这些匠人只能够做低档的工作，连参加科举考试的资格都没有。所以，我们虽然早有科学技术，但不受重视，因而不可能产生自然科学。为什么当时世界最强大的国家是英国、西班牙、葡萄牙这些国家，就是他们先搞自然科学。

自然科学始于意大利文艺复兴时期，开始了数学、力学、物理学、化学等理论研究，核心部分是数学微积分、高等数学的诞生。伟大的文艺复兴名家达·芬奇有一句

① 本文是应上海市中国工程院院士咨询与学术活动中心的邀请所做的报告（按录音整理），上海，2008 年 5 月 27 日；科技创新与品牌，2009，(1)：10-16；科技创新与品牌，2009，(2)：8-11.

名言："数学是科学的皇后，力学是数学的天堂。"直至500年后的今天，这句名言依然意义深远。自然科学包括数学、物理学、化学、天文学、地理学、生物学、力学7门科学。这些学科内容从表面上看似乎对人类关系不太大，但是实际上每一个理论的出现对世界的影响都是巨大的，甚至使社会实现跨越式的发展。库仑提出电的库仑定理，使我们进入今天的电子时代。所以自然科学一个理论的产生，将对人类社会产生翻天覆地的变化。我们要重视自然科学基础理论，这与人类的发展关系很大。近100年来，是人类几百万年历史中发展最快的时期，原因就是自然科学理论有出色表现，特别是20世纪初爱因斯坦相对论的提出。我们今天可以在天空飞行，特别是1969年7月20日美国"阿波罗"飞船载人登上了月球，更显示了自然科学理论的伟大作用。我们中国晚了几十年，现正在赶上来，神舟七号载人飞船的发射就是一个有力的证明。大家都遗憾中国到现在都没有诺贝尔奖获得者，其原因就是我们的基础科学薄弱，加之长期以来又不受重视。

2. 技术和工程科学的创新

我们对技术科学和工程科学的创新要抓紧。人类改变世界就是通过技术科学和工程科学来实现的。技术中首先就是生产工具方面要有创新。我们中国一万年前是石器时代，是把石头打磨成劳动工具；然后进入铜器时代，进行青铜的冶炼、制造。不知道大家去过四川省广汉市三星堆博物馆看过那些铜器展品没有？非常精致，我们中国的铜器制造在世界同一个时代是最强的。铜器时代过了是铁器时代，铁比铜更坚硬。后来是蒸汽时代、电气时代，然后到今天的信息时代，发展就更快了。以前是几千年一个时段，现在是10来年就是一个时段，一个技术创新可以很快地带来时代进步。1946年，美国出现了第一台计算机，把人类社会大大地推进了一步。20世纪60年代互联网的出现，代表了全世界地球村时代的来临。中国出现互联网才10年多的时间。过去大家对另外一个国家感觉很遥远，现在大家随时可以跟全世界的人通话，信息交流非常方便了。所以技术创新使生产力和人类生活进步很快。

科学理论和技术的集成便是工程科学。工程科学的创新更是推动社会迅速向前迈进的动力。著名的登月工程、三峡工程等就是典型的杰作。

3. 创新推动社会制度的变革

第三个方面就是创新推动社会制度的变革。比如说人类早期的原始社会，原始社会过了是奴隶社会，接着是封建社会以及资本主义社会，最后是社会主义社会。这些社会制度的变化，带来了生产力的进一步发展，使人们的生活水平迅速提高。

我出生在抗日战争时期，是四川成都人，亲眼看到日本飞机轰炸我们祖国。那时国家非常落后，尽管日本飞机飞得很低，我们也拿它没办法。我的祖父是1903年清朝政府派到日本东京振武学校留学的第一届学生，同盟会会员，他去的时候把辫子剪掉，回国的时候没法回来，要装一条假辫子。那个时候是中国非常落后的时代，后来辛亥革命推翻了清王朝。再后来中国共产党又把蒋介石推翻，建立了社会主义社会。几十年来社会进步非常快，特别是1978年3月18日，全国科学大会在北京举行，邓小平同志说"科学技术是第一生产力""知识分子是工人阶级的一部分"，这两个著名论断也是制度的创新。老同志都知道，那时说这两句话是相当不容易的。过后，以经济建设为中心的30年，中国经济实现大幅度跨越，解决了温饱问题。与我们年轻时相比，现在食品丰富，居住条件改善，交通便利，各方面都有巨大进步，这是社会主义初级阶段制度的创新带来的。小平同志带着我们中华民族进入了盛世时代，2007年中国的经济总量GDP是世界第四。

4. 文化上的创新

第四方面，文化上的创新也推动着人类社会的进步。我们中华民族在春秋战国时是百花齐放、百家争鸣时期，文化非常灿烂，以至于现在从人们基本的生活方式到思想，还是受那个时代的影响。2001年世界100多位诺贝尔奖获得者聚会发表宣言："21世纪，地球上人类的进步要依赖孔子的思想。"先进的文化一定要传承，要发扬光大。孔子非常有远见，他的许多思想在两千年以后的今天仍然值得中华民族坚持，外国人都看到了这一点。我们国家现在开始重视了，已在世界上建立了200多家孔子学院，把孔子的思想推向世界。

世界上的文化也是一样的。欧洲18世纪启蒙运动的文化创新，揭开了思想解放的序幕，声势大，影响远，其中代表人物是法国思想家孟德斯鸠。他的思想使得三权分立制度在欧洲建立，如法国、英国等，使欧洲成为世界的科技、经济强大地区，使美国成为强国，形成了西方文明。

所以我们不要把创新仅仅看成科技创新，其他方面也很重要，缺一不可，每个方面都要创新才行。但是对于我们每个人来说，要具备什么条件才有可能进行创新呢？这就是我要讲的第二个大问题。

二、创新的一点感想

首先，一个人希望进行创新就必须要有"三个心"：责任心、耐心和好奇心。其次，还要具备三个条件：宽厚的知识积累、勤奋和善于抓住机遇。

1. 要有"三个心"

1）责任心

因为创新是大脑的智慧活动，要实现创新需要许多精力和时间，甚至要经受许多艰难和曲折，所以搞创新的人首先要有责任心，做事要负责任，做事要自始至终坚持，做事要一丝不苟，做事要为国家和民族的事业着想。没有责任心的人，他不可能做出创新的事来为大家服务。

2）耐心

没有耐心的人，今天想到个问题，提出新的设想，明天可能就把它丢掉了，那是做不下去的。一定要坚持，要有恒心。创新的概念在提出的时候往往是大家都不支持，甚至受到挖苦、打击，很多困难使人做不下去，因为你的创新大家都没认识到。如果你提出来一个所谓新的东西大家都支持，那基本算不上创新。很多人都不认识的时候，你提出来，有的人会说你这个人骄傲、异想天开，领导也不一定喜欢。所以，一个人要进行创新，一定要有忍耐的精神，没有忍耐的精神你是做不出创新的。

我自己一辈子用耐心来要求自己，一辈子的很多事之所以能够成功就是靠耐心、靠忍耐。忍字当先，没有忍耐，今天回想起来，许多事是做不成的。20世纪70年代末，那时的中国人才学会访问我“你成功的秘诀是什么?”我马上就回答是“忍耐”。尤其是我们中国人，做事就更要忍耐。

3）好奇心

作为一个要有创新思想的人，好奇心很重要，必须要对事物好奇，要有兴趣。一个人对生活没有兴趣，根本就不可能提出各种问题：生活中的问题、工作中的问题、国家的问题、世界的问题。只要有兴趣，你才能在这种兴趣中找到一些新的思想。

2. 要具备三个条件

1）宽厚的知识积累

创新，是要在前人基础之上得到一个飞跃，智慧的飞跃。创新，是人家没有做过的没有想过的事情。人类已经有几千年的知识积累了，积累了很多优秀的东西，你如果没有什么知识，不知道这些东西，你提出来，而别人早就提过了，早就有人做过了，这不是创新。牛顿说过他是在巨人的肩膀上成长起来的。前人的东西已经非常丰富，所以你想要创新，就得有丰富的知识，不仅要有本专业的知识，还要有其他专业的知识。

2）勤奋

这个条件大家很明白。所有的科学家、成才的人都说他的成才的基本点是勤奋，不勤奋的人在这个世界上是做不出成就的。关于勤奋，有很多名言，比如唐代文学家韩愈说：“业精于勤，荒于嬉；行成于思，毁于随。”著名数学家华罗庚说：“聪明出于勤奋，天才在于积累。”英国文学家、历史学家卡莱尔说：“天才就是无止境刻苦勤奋的能力。”东西方的科学家、成功人士都会讲勤奋，这个是基础。

3）善于抓住机遇

这个条件也是很多人讲的，机遇就是你提出创新的机会。机遇往往是突然来临的，是没有先兆的，突然来到你的面前。所以，机遇总是喜爱那些提前准备好的人。有些人老是羡慕别人比自己强，其实这些人都是抓住了机遇。很多人都遇到过机遇，但是很多人都没有抓住机遇。因为机遇到他面前的时候，他没有具备条件。法国科学家巴斯德说过：“机遇只偏爱那些有准备、有头脑的人。”我们绝不能等到机遇来了才临时抱佛脚，那是不能成功的，要创新必须要善于抓住机遇。

下面就讲讲自己的体会，生活上的、管理上的、科学上的，都是自己的亲身经历，提供给大家参考。

3. 创新的一些故事

1）大学校长的管理创新

今天在场的有很多是大学老师和大学生，大家都知道一个大学的管理是非常难的。在中国目前的大学校长圈里，流行着一句话：“大学的校长不是人当的。”一般人会说：“当大学校长多光荣啊，社会地位多高啊，多受尊敬啊！在国外的大学校长还要厉害，可以去竞选市长、省长甚至国家总统。”实际上，中国大学校长的确不好当。我在高校里面当了 20 年的校领导职务，亲身体验到做好校领导之苦，越是有心把学校办出水平，日子会过得越苦。但是，又觉得大学这个舞台，可以为国家多作些贡献，是一件十分有意义的事情。想通了这一点，我才有勇气承担校长职务。积 20 年办学之经验，要把工作搞好，要把学校办出水平，就必须在管理上要有创新思想。

我原来在上海工业大学任副校长，做著名科学家钱伟长校长的助手，后来上级把我调到广州暨南大学任副校长，接着又任校长，前后长达 20 年之久。新中国成立以前，暨南大学在上海真如办学。当时这个学校在全国排行大概在第五到第九名之间，是中国著名的大学。那时，上海有 4 所著名大学：上海交通大学、同济大学、复旦大学、暨南大学。这个学校在中国是非常有特色的大学，今年（2008 年）已有 102 年的历史，是中国历史最悠久的大学之一，是清朝光绪皇帝亲自签字批准建立的，是中国第一个对海外办学的大学。从清朝开始办学就招海外学生，一直到现在都是华侨大学。但是这个学校又是多灾多难的大学。老一辈人都知道，在“文革”以前，有海外关系对个人来说是很可怕的事。所以一解放，这个学校就关闭了。后来，一些中央的领导意识到华侨仍然很重要，同时侨胞们也要求复办，于是就在广州复办。因为广东是中国最大的侨乡，中国的华侨有一半是广东人。暨南大学在广州复办的第一任校长是陶铸同志，这说明中央很重视。“文革”时期，学校再次停办，直到 1978 年重又恢复。这个学校在 102 年的时间长河里面停办 3 次，停办一次就换一遍老师，只剩个校名了。

尽管办学这样曲折，但是学校仍然培养了很多杰出人才。据不完全统计，有 13 位国家领导人、10 多位院士，还有外国领导人，以及国内外许多的杰出人士。总之，对国家的贡献很大。

华侨对中国的贡献很大。从辛亥革命开始到现在，几乎每一次中国的巨大前进的跨越，华侨华人都出了大力。没有孙中山先生领导的华侨同胞，我们的辛亥革命无法成功。30 年的改革开放又是靠华侨华人、靠港澳台同胞出了力，取得了巨大成功。这 30 年来中国引进的外资，70％是华侨华人、港澳台同胞完成的，不是靠洋人。许多西方人是不喜欢中国上去的，他们不会发善心希望中国上去。我们的华侨华人和港澳台同胞爱国、爱故乡。全世界没有哪个民族像我们中国人，他就是改了名字，拿了外国的身份证，仍然爱这个祖国、爱这个故乡。祖国经过 30 年努力，成功了，成了世界第四经济大国。

20 世纪 90 年代初，上级调我到暨南大学任职，来到了广州。可是，进了这个学校就让人伤心。学校办得太差了，已没有好名声。离开上海的时候，同事们都问我是不是调到山东济南去了。许多人都不知道这个学校，学校的知名度很低。广东和上海不一样，上海的高校办得比较好，而广东的高等学校在老百姓的心目中却办得很不好。都被取了很难听的绰号，这在全国都很罕见。老师和学生都不愿戴校徽，大家不愿在外面说自己是暨南大学的老师、暨南大学的学生，嫌丢人。我在上海工作得好好的，去了个名声不好的学校当领导，你说心里咋能高兴？我一向服从组织安排，只得耐着性子继续待下去。于是，我就团结大家、依赖大家，想办法要把这个学校办成质量好的学校，以“侨校＋名校”为发展目标，以“从严治校、从严治教、从严治学，依法治校，实事求是”（简称为“严、法、实”三字经）为方法，开始一次一次的改革。大概做了 20 多项改革，许多是全国高校第一次进行，共制定了 327 个管理制度。

这个学校的新生进校学习的程度很不一样。它的办学宗旨是“面向海外、面向港澳台”，是全国两所实行校长负责制的学校之一，是个特殊学校。学生来源是全世界的华侨华人的子弟、港澳台的学生和大陆的学生。全世界高中毕业不是同一个水平的，不可能像跳高一样规定一个同样高度，不能统一高考。美国的学生跟德国的学生不一样，跟非洲的学生不一样，跟毛里求斯这些国家的学生又不一样。学风也不好，我1993年秋天分工管教学，检查时发现早上8：00上课，到8：30还有很多学生没到教室，老师也不管。因为一些老师喜欢“揾钱”，就是到外面赚钱，对本职工作不负责，教风不好。我听了几节课碰到几件怪事。一位老师到了教室，上了讲台就问：“同学们，你们是本科生还是大专生?”他走上讲台还不知道听课对象。有一个老师还问：“你们上节课上到哪儿了?”他连上一节课上到哪儿都忘记了。我又检查了第4节课的下课情况，严重到无法想象的程度，本来是12：00结束，一位教师在11：10就下课了。

要想办法把学校办好，首先就要使学风变好。为改变学生平时不努力学习，仅靠一次次补考来过关的现象，对学生就采取了第一项改革措施：取消补考、实行重修。1993年，这个学校就没有补考了。这在全国是第一次，以后就在全国推广了。学生成绩不及格，不需记入档案。非常宽容，允许重修三次。学生如果不想要低分成绩，愿意要高分，也可以重修。因为重修不仅要多花时间，还要多花钱，所以不仅学生重视，而且家长也会帮学校管好学生。这样一来，许多学生就努力读书了。

第二项改革就是从1993年开始实行标准学分制。这个制度规定本科学生拿到160学分就可以本科毕业，并允许提前毕业。70%是必修课，30%是选修课。一年两个学期，每学期20周，一学期20学分。优秀生可以读快一点，差的学生读慢点，还允许学生休学。总之，学生可以按照自己情况读3年、4年、5年，甚至6年、7年都行。我觉得大学生就像产品一样，对社会来说符合质量条件出去就行了。可以有的人慢点，有的人快点。实行这个新制度之前，学生的学习负担很重，有的专业周学时竟高达40学时，以致学生学习疲于奔命，既学不好，又使学生非常厌烦学习。所以就把学分降下来，搞少而精，跟国际一致。学生在校学习一年40周，每学期16周上课，2周复习，2周考试，学生有备考时间。一门课可以同时由几个老师主讲，学生既可以挑上课老师，也可以挑上课时间，就是根据自己的情况读书。传统的中国大学是保姆式大学的培养人才方式，这不是优秀的制度，优秀的制度应该是使学生根据自身情况和爱好来上课。这样，我们实施了中国第一个弹性的学分制。加上取消补考的措施，学生从学期开始到结束都努力用功，而且关怀学生，在考试前给学生2周时间复习功课，以使学生在考试前有充分时间准备。这样一来，学生的学风好转了，学生的学习质量也提高了。

我们的考试制度非常严格，在全国应该是最严的学校了。大家都知道中国过去的学校，包括在我们小时候，都没有作弊的现象。到了20世纪80年代，出现了考试作弊现象，而且越来越严重，这对培养中华民族的接班人很不利。大家作弊的结果就是不讲信用，作假。而且有的老师也跟着作弊，学生给老师送个礼就把分数改过来，这样搞下去我们的民族还有什么前途。诚信是一个民族最重要的基本素质。我们花费了很多精力，设计了一套完整的考试制度，以消除作弊现象。期末考试时，实行大考场制度，全校学生都在体育馆里面考试。840人考试，大考场，学校统一监考，这件事《人民日报》等媒体都报道了。大考场中，考生看不到周围同学的相同内容的试卷。考试桌子也做了专门设计，桌子没法私放东西。进考场的时候，学生所有东西都要存放。进到大考场以后，学生要靠自己的智商和能力来完成考试，在那种情况下作不了弊。自从设立大考场以来，便没有学生作弊。学生无法作弊，考试质量就很好，而且公正、公平。学生要考试好就得平时学习好。平时学习不好就不行，所以整个学习过程都能够管住。我们学校每年要把一些优秀学生送到国外去读书，交中国的学费到国外留学一年，学生如果作弊拿好成绩选拔出国留学，显然有失公平。由于我们的考试管理严格，有一次，我们省就把干部的考核放在我们学校举行，请我们监考，当场抓出了两个作弊的副厅长，后来听说省里免掉了他们的职务。

还有阅卷制度改革，首先考试题出法就改了，用试题库。出题是一群老师出的，每人出一道题，然后组成试题库。对试题库每套试卷题编好号码，考试前两天由主管校长随机选号抽题，此时谁也无法知道考题内容。然后请几个人去印卷子，这几个印卷的人跟外界不能来往。改试卷是集体阅卷，每位老师改一道题，不能拿回家由一个老师来改。所以严格而公平的考试制度会使学生的学风变好。

为改善教风和提升教学质量，从1993年开始，我们便要求教授上本科基础课。同时，对老师的课堂教学管理使用了三重评估制度。我在外国呆过几年，看到国外大学的学生在期末课程结束后要根据教师的授课质量给老师打分，以评估教师的授课质量。我就把这个制度借用过来，让学生给老师打分。对上课的内容、教材、教法等项目，由学生打分。开始时老师不服，说因为教学生很严，学生便给打低分。后来我就搞了三重评估，第一个就是学生评老师；第二个就是校、院、系领导评老师，并具体规定了领导每人一学期听多少课；第三个就是专家评老师，全校请了40位专家，大多是教课教得好的刚退休老师，返聘回来，每周规定他们每人听课8小时。他们在全校任何时候随机听课，不通知任何人，就变成一个随机的抽样检查，这是用数学办法，运筹学的概念。像我去听课，临上课前一两分钟才进入教室，坐在教室后座，不让讲课老师发现，以免老师紧张，又可看到真实情况。听一节课下来就给老师打分了，看他讲得好不好。领导、专家、学生三方面考察，综合评定等级。然后每年全校表扬10位最佳授课老师，大照片挂在学校公告栏里面，树立榜样让大家向他们学习。同时，升工资，给激励。对于每年考核不合格的老师就亮黄牌，警告；再不及格，就下岗。下岗后，必须重新学习才能再上岗或改做其他工作。通过这样的办法管理老师，老师的教风很快就好了。

与此同时，我们在全校推行新的分配制度——暨南大学量化考核制度。中国高校现在普遍实行两种分配制度，其中一个典型是北大、清华的九级岗制度，另一个就是暨南大学的量化考核制度。当年《光明日报》同一天报道了这三个学校的两种分配模式。清华、北大模式是把老师分成一至九等，发放岗位工资。这种制度实施可能比较麻烦。我们是按照公平和激励原则，不用奖金概念，不鼓励院系部处创收，而是由学校直接筹款来增加教职工收入。我们将全校教职工每人的收入分为两部分，一部分是国家规定的，称为国家工资；一部分是学校发的：称为校内工资。校内工资采取量化考核办法，把每一个老师的工作细分成很多项，再按项计分。比如说上课，上本科生、硕士生、博士生、成人教育的课有不同分数。甚至上短期培训班的没有学历、学位证书的都有分数。学生数量多了也有加分，重复班也有分数，搞得比较公平。你教不同的课程，有不同的分数。星期日上课比平时上课要加一点分，在外地上课要加一点分，在海外及港澳台地区上课也多加一点分，晚上上课要加一点分，中午上课也加一点分，分得很清、很细。指导教学实验、批改作业都有分数，当班主任也有分数。还有社会任职的分数，在外面学术界任个什么职务，对学校知名度有提高，也给分数。然后，出版著作、发表论文也有分数。在国内外不同级别杂志发表的文章各有不同分数。做的科研成果被转化为产品，有分数。拿到科研项目也加分。在外面获得奖励、作了贡献也有分数。学年结束时，评为优秀的，也要对原始分数乘以大于 1 的系数。多少分，就多少钱，也不封顶。这样下来，大家都愿意干活了。很快，学校科研高速发展，教学质量稳步上升，教职工收入大大增加。当我 2006 年初从校长岗位退下的时候，以校本部教职工为例，2005 年全年人均收入为 8.89 万元，是 1995 年人均收入 8000 元的约 11 倍。在全国高校校长开会的时候，有人说我们学校工资是全国高校第一了。周边学校领导私下跟我说："我们的压力太大了，请校长手下留情。"发给教职工的工资总额在全校总支出中仅占 30%，增发的余地还很大。

这个分配制度的变化还促进了工作变化，办学水平提高了。我刚到学校工作时，暨南大学居然没有一个省部级的重点学科。现在不仅有 15 个省部级重点学科，而且还有 2 个国家重点学科，博士点由 7 个增加到 54 个，研究生由 615 人增加到 6567 人，海外及港澳台学生由 1982 人增加至 10 270 人，整体上去了，由一般大学成为国家"211 工程"大学，成了中国的名校。以中国网大排行榜为例，在全国 2000 多所高校中，2005 年排在第 42 位。走到海外，你们碰到在中国读过书的，你问他在哪读书？很多回答是读暨南大学的。2000 年，中国大学校长代表团到葡萄牙访问，桑帕约总统给我们安排一个翻译，那个翻译走到我面前说："刘校长，我是你的学生。"她是我们学校外国语学院的毕业生，在葡萄牙工作。世界各国有很多我们暨南大学的学生。据统计，2005 年，暨南大学有 2.2 万名多学生，其中海外及港澳台学生占一半，有 81 个国家和地区的学生在这里留学，海外及港澳台华侨华人和港澳台学生占全国大学同类学生总数的一半以上。

现代化大学的显著特色是老师们既能教好书，又能做好科学研究。但是，我刚去时的暨南大学的老师却不是这样，几乎无高水平的学术论文，科研水平较差，一年里全校仅有几篇 SCI（Science Citation Index，科学引文索引）和 EI（Engineering Index，工程索引）论文。于是从 1996 年开始，我提出用"教学、科研"双中心目标取代了过去单一的"教学中心"目标，并用"不做科研的老师是残疾老师"的讲话来转变大家的观念。接着，又采取了一系列措施，如大力引进博士，引进优秀人才，改革职称评审制度，加强研究生质量管理，对老师的科研成果进行量化考核等，从而使学校的科研项目、科研经费、科研论文和成果推广快速增长。科研经费从 1995 年的 400 万元上升到 2005 年的 1.3 亿元，增长近 32 倍。三大索引（SCI、EI、ISTP①）收录的学术论文从 1995 年的 9 篇上升到 2005 年的 297 篇，增长约 32 倍。

---

① ISTP 英文全称为 Index to Scientific & Technical Proceedings，译为科技会议录索引。

一所大学，如果经费太少，那是很难办好的。1995 年，全校经费短缺，实验设备费仅 50 万元，基建费仅 300 多万元，而且全校从上到下忙于搞创收，严重影响学校的办学质量和发展。为此，从 1996 年开始，实行财务集中管理的措施，把各院系和部处资金集中起来实行一级财务管理，各院系部处不再搞创收，由学校直接发放校内工资，让教职工集中精力安心做好本职工作，这一方法在全国亦具有开创性质。学校实行开源节流原则，以开源为主、节约为辅。学校通过多种方法筹措资金，一方面请求中央和省政府加大投入，另一方面是调整结构，增加收入，同时又强调勤俭节约，不准浪费。对各单位的账户进行了清理，并加大了监管力度，不准做假账，不准做两本账，禁止搞“小金库”，实行“收支两条线”既有助于廉政建设，保护干部，预防腐败，又使学校的办学经费大幅增加。同时，又要求大家严格遵守国家及相关部门的财务制度，不准乱收费。这一系列的改革，使学校财务状况越来越好，保证了学校教育事业健康和高速地发展。2005 年，全校总收入达到 13 亿元，约为 1995 年 2.26 亿元的 5.8 倍。同时，2005 年，学校的固定资产总值达到 16.3 亿元，约为 1995 年 3.26 亿元的 5 倍。学校的校园面积达到 174.43 万平方米，约为 1995 年 112.32 万平方米的 1.5 倍。已经建成和正在建设的校舍面积共 83.81 万平方米，为 1995 年 46.39 万平方米的 1.81 倍。图书藏量增至 270 万册，为 1995 年 135 万册的 2 倍。教学科研设备增至 2.7 亿元，为 1995 年 4980 万元的 5 倍多。我卸任时，在校本部，还给后任校长留下积余的近 3000 万元现金。

暨南大学历史上很有特色，出了很多知名人物。可是，很多人不知道暨南大学，同学们甚至连暨南大学的“暨南”是什么意思、怎么来的都不知道。我做了校长以后，派人到中国历史档案馆去查，查出学校申办时光绪皇帝的批准签名，查出哪天成立的。最后我们便办了校史馆。校史馆可以鼓励学校师生们热爱自己的学校，珍惜前人留下的宝贵遗产，不忘记历史。大家的反映很好，知道了自己的历史，更爱自己的母校。当时全国一个校史馆都没有。1998 年，在暨南大学举行全国 100 所“211 工程”大学校长第一次聚会，大家看了我校的校史馆后说很有特色。现在，很多大学都建了校史馆。

1998 年，我们接受珠海市委和市政府的热情邀请，经上级领导口头批准同意，在珠海唐家湾创建了暨南大学珠海学院，既是在珠海经济特区创办的第一所高等学校，又掀开了新中国异地办学新的一页。建立的时候，有人就说：“刘校长你要犯政治错误了，中国哪能允许异地办学。”我觉得中国大学太少，办大学于国家有利，于人民有利，又不是贪污腐败，应该大胆地去办学，即使犯了罪，坐了牢，未来也会平反的。暨南大学珠海学院在唐家湾办了两年后，又迁到珠海前山。可以说，办学的酸甜苦辣均尝尽，靠勇气坚持，靠办学质量坚持，靠珠海市委、市政府黄龙云书记等领导的大力支持，才得以坚持至今。我们始终坚持珠海学院与广州校本部是同水平招生、同水平办学、一样的毕业文凭。前不久，教育部周济部长来校视察，赞扬说：“暨南大学珠海学院是暨南大学校本部的延伸。”做这些创新的事情，很多时候是冒风险做的，你只有认定目标，坚持下去才能成功。创新肯定跟周围不一样，但是创新的目的是进步。

我们中国为什么在世界上没有有名的大学，其主要原因之一是中国的大学都是关门办学，很不开放，主要招自己的学生，外国的学生愿者上钩，到中国来几个算几个。

这么办国际化大学是不可能成功的。现在地球是一个村，人家先进的东西要学习，你有先进的要输出，才能够提高，大学是为社会培养现代化人才，要培养有世界观念的人才，所以要主动送你的学生出去，要欢迎人家的学生到你这儿来。

我第一次出国，到了国外一看，感觉我们太落后了。1981 年初出国的时候是从中国科技大学出去的，第一批全国仅有 8 个人，由联邦德国直接挑选的洪堡学者，作为世界优秀青年科学家出去。出发前，受到著名科学家钱三强先生的亲切接见和热情鼓励。到了联邦德国，发现西方的管理在变化，大学办得不错。应该把人家好的东西学回来，提高自己，只有这样才能说中华民族是伟大的民族，才能把我们的国家变成伟大的国家。当时我就在反思，怎么才能把中国的大学办好，为中国培养优秀人才，我们中国才有希望。从 1995 年底任暨南大学校长后，便开始在校内狠抓办学质量，并实行开放式办学，走国际化之路，花很大力气在世界五大洲建立姊妹学校，全是有名的大学。经过 10 年来的努力，我校在世界五大洲共建立姊妹学校 46 个。比如说我到埃及以后，想和他们的名校建立合作关系，当时非洲和中东地区最好的大学肯定是埃及的开罗大学，我就去拜访开罗大学校长。他不知道暨南大学在何方，我跟他讲了以后，开始他不愿意，我讲了半小时就把他讲同意了，这样，我们便与埃及的开罗大学建立成姊妹学校。要使世界的主要国家、华侨华人多的国家尊重我们，承认我们的文凭，使我校学生在该国能站住脚、求职顺利、工作发展顺利，就必须让人家知道中国有一个办学质量相当不错的暨南大学，才有可能。与该国著名大学结成姊妹学校是一条好途径。现在，海外及港澳台学生到中国来读书，大多选择暨南大学。一道风景线是，全校 1 万多海外及港澳台学生，来自世界五大洲 81 个国家和地区。

暨南大学众多学生来自海外及港澳台，到广州来读书，不仅要学好中文，同时也应该掌握好英文，成为双语的人才。对于内地生说来，由于国内从小学到大学的英语教育不太成功，时间虽花得多，到大学毕业时，许多学生仍不能开口讲英文，致使国内非常缺乏英语好的专业人才，只好靠留学归国人才来满足需要。考虑到上述情况，我们从 1996 年起，便在全校院系开始双语教育，部分专业课程用英文授课。2001 年，更设立国际学院，学院内的全部专业一律使用英文授课，现已开设 6 个专业：临床医学、药学、食品质量安全、国际经济与贸易、会计学、行政管理。这个学院既为海外及港澳台学生所欢迎，又令内地生喜爱，成为内地首家兴办的不用出国留学就享受留学效果的大学。

暨南大学是新中国第一个设立医学院的大学，但是只有一个附属医院，学生培训、实习的地方严重不足，而且我们又无钱再办医院。经过认真研究，我们选择跟深圳特区最好的一所三甲医院——深圳市人民医院谈判，他们很高兴，就签字了，成为我们的第二附属医院，这是中国第一次大学与医院联合办学，受到时任国家卫生部副部长殷大奎的高度赞扬。后来，我们一共建了 8 所附属医院，都是广东的好医院。他们成为附属医院以后，重视科研，医疗水平也上去了，病人愿意到那里去看病。我们学校的学生到那里实习也不花钱，实习的条件也很好。那些医生还免费给我们培养研究生。现在很多大学都在采用我们这个办法。

还有与广东省人民检察院实行合作预防职务犯罪。几乎每一个校内重要会议我都要讲一次预防腐败。我虽然苦口婆心地讲，但是仍有个别干部不听忠告继续搞腐败进了牢房。为此我就想了一招，加大反腐力度，请求广东省人民检察院张学军院长支持，合作预防腐败，包括暨南大学建房子、采购、招生等，这些比较容易出现腐败的地方，请广东省人民检察院派人来监督。广东省人民检察院将此事上报中央，中央批示表扬我们做得好，这也是全国高校第一次做此事。因为高校不是避风港，也有腐败。有些学校连校长、书记都被抓过，腐败确实严重。检察院进来以后，他们每件事都参与，威慑作用很大。从那以后，暨南大学就比较太平，我也比较放心。这是管理上的创新。

下面，我再讲一个我自己的有趣的故事，急中生智救治垂危病人。

2）生活中的一次创新经历

这个故事发生在甘肃省平凉地区泾川县黑河公社马漕生产队，这是个非常落后的山区。

“文革”时期，我被下放到泾川县插队落户当农民。下放的地方十分落后。黑河公社的地形是东西长 30 公里、南北宽 5 公里，黑河从中间经过，是山区。我所在的是东边第一个生产队，距离公社所在地有 30 里[①]路。我们 10 个老师编成一个班，在一个生产队。刚到的时候，发现当地人基本上都患有大骨节病。男人全都是矮子，手足全部是大关节。很可怜的一个生产队，人口数在新中国成立后都没变化，始终是 107 个人。全村由于人矮炕小，就没有一个炕可以让我睡觉。我近 1.8 米的身高，没法睡觉，每晚只好头睡里面，把腿悬空在外面，十分辛苦。但我是被改造对象，不能提要求。后来，生产队长发现了，关心我，找一个炕让我一个人睡，我就睡对角线，这才解决了我的睡觉问题。这个生产队缺医少药，农民治病困难，考虑到我有一些医学知识，我就大胆地自愿当一名义务赤脚医生，一边劳动一边给他们治病。在那里待了半年，我成了全科医生，什么病都看，仅一个病人没处理，有一个妇女难产，我不敢看，就婉拒了。他们平时不太用药，吃一点药，就很灵光，很快就好了，所以我的威望越来越高。后来我就看出名了，成了“神医”。

1968 年冬天的一个半夜里，很冷，凌晨两点钟，有人叫我快起来，说有一个人昏死过去了。我就赶快去救人，我也是胆大，什么医疗器械都没有，就靠知识。现场一看，一个二十三四岁的小伙子昏死过去了。检查完毕，发现睾丸肿胀很大，判断他是疝气发作，已经快没气了。按理说这个病必须立即做手术，把小肠送回腹腔才能救活。可是这个地方离县城医院有 80 里，没有公路，只能抬担架，要走一整天，现在人已经休克，显然此法不可取。我就另想出路。因为我是学力学出身，就想利用重力原理，让小肠返回腹腔。我就叫几个农民把他头朝下，脚朝上地提起来，他已经没气了，死马权当活马医。5 分钟后，他就有呼吸了，10 多分钟后就活过来了。大家欢呼说：“刘老师，你是神医啊!”这是关键时候没有办法的办法。但是如果你胆小怕事，就救不了人。我觉得在生活中也有机会创新。自那件事以后，方圆几十里的人都来找我看病。

3）高压超高压容器试制的科技创新

现在讲自己在科技上的创新。我 18 岁时在“581”工程中担任中国第一颗人造地

---

① 1 里=0.5 千米。

球卫星一个研制小组的组长，在科研上得到许多锻炼。但是两年以后，由于上级说我的政审不合格，因为我家里既是归侨家庭，又是台属家庭，所以被迫离开了人造地球卫星研制小组。“文革”中的 1969 年夏天，我在兰州汽车修配厂当工人时，机会就来了，得以在兰州石油化工机器厂做科技创新。

这个石油化工机器厂是当时中国石油化工机械的第一号工厂，是苏联援助中国的 156 项工程之一。这个厂二分厂当时的技术科长，就是第十届人大常委会委员、曾任湖北省委书记的贾志杰同志。他是留苏学生，喜欢搞技术革新。他仿照世界领先水平的美国设备，在中国第一次试制了 6 台生产航空煤油的钳重整装置。试制出来以后一检验，发现不合格，没成功。因为这是个高温中压产品，工作的时候是 300 摄氏度高温，要 80 大气压力①，但是他们的试制产品只能达到 60 大气压力。一台设备价值几千万元，这个问题如果不能解决，花费巨大人力物力研制的设备将全部成为废品。在兰州的第一号有名大学是兰州大学；也是当时全国 27 所重点大学之一。他们就到兰州大学求援，找到军宣队队长。兰州大学没有化工机械专业，没有压力容器专业，找来找去就只能是我们力学系。力学系里面能够与这个产品有点关联的就只有从事板壳理论研究的我和我的老师叶开沅先生。叶先生正在蒙冤挨批斗，无法接受研究任务。于是军宣队将任务交给我一个人。我那时 29 岁，没有搞过实际研究，唯一的长处是我有一点理论基础。我是中国除了北大以外第一批学力学的大学生，搞过一点理论研究，发表过 3 篇学术论文。但是，我读书时没有得到机械工程知识的培养，学校办学条件差，没做过一次实验，没学过制图，要承担这个任务，难度可想而知。

我立即去了工厂，向贾志杰科长报到。他们就给我说这个问题，图纸我也看不懂，我只学过板壳力学，没有学过压力容器。我觉得这个设备重要，这是个机会。我不能说我不干，就说我好好努力吧。那时正是知识分子接受改造的时期，因此我就自己决定半天跟工人师傅一起劳动，当铆工、当焊工，半天搞研究。这个装置封头是开孔椭球壳，我没有学过，文献中也找不到相关理论，于是我就自己创造理论，给出了计算公式。经过 3 个月的日夜奋战，我终于把这个问题解决了。我给的结论是：试制产品封头开孔处强度不够，应予补强。随后，贾科长他们又做了试验验证，所得试验值与我的计算值吻合，证实了我所建立理论的正确性。于是接受我的建议，在封头开孔的地方沿着孔焊一个 32 毫米厚的钢板，1 台花了几千元钱。弄上去以后就行了，就能达到 80 大气压力了。他们好高兴，我也是好高兴。

我是本科生，又不是研究生毕业，知识面就是那么一点点，但就是我热爱科学、有创新的思想使我为中国第一台生产航空煤油的装置试制成功贡献了力量，不然试制产品就成废品了。我们那个时候做科研也没什么好处，既没有补助，也没有奖金，也没有表扬。《人民日报》报道的时候，只说是兰州大学力学系一位老师的帮助，名字都没有。但是我觉得只要对国家有好处，就值得。如果一个人一生只想着名利、想着升官发财，那是成不了功的，要抛弃这些名利才能成功。

这位贾科长，也确实了不起，他做了很多创新。当时国家以农业为基础，但化肥

---

① 1 大气压力＝1 千克/厘米^2。

不足，尿素是最好的肥料，他便仿照世界王牌产品——荷兰设备试制中国第一台大型尿素合成塔，22 米高，压力高达 220 大气压力。这种装置一旦出问题，周边地方都要遭殃，很吓人的。我年轻时就是胆大，我跟他们做的时候都没怕过，死了人要负责任，但是当时都没考虑过这些。这个产品就是吸取了前面钳重整装置试制的经验，前一台封头弄得太薄了，这台就弄很厚。可是厚了以后封头孔边就产生很多裂纹，一个高压容器带有很多裂纹就不安全，不能用于生产。他们又请我赶快把这个问题解决。为此，我又创立了世界上第一个实用的厚壁球壳理论，给出了设计公式。计算后就做试验，理论结果居然对上了。于是，他们又接受我的建议：将球形封头减薄，去掉开孔地方的补强圈，使产品试制成功。

接着就研制我国第一台大型换热分离氨组合设备，其中的水冷却器是高压高温设备，它的厚管板属于关键部件。厚管板是圆形的厚板，厚 230 毫米，直径 700 毫米，上面要开 745 个孔洞，每个孔洞的直径约为 10 毫米，孔洞中接上管子，用管子作冷热交换。石化和炼油厂中大约有 40%都是这种热交换器。世界上这一领域的最权威学者是波兰的乌班诺夫斯基。他创立的是薄管板理论，而厚管板理论全世界没有。当时要设计制造 230 毫米厚的管板，全中国没有一个厂能加工。中国最大加工能力只能做到 190 毫米，再厚一点就做不了。他们又找到我，看能不能把管板减薄。做了半年，我自己建立了厚管板理论，给出了解析解，完成了这一课题。“文革”结束后，经仔细查阅文献，发现还是全世界第一个厚管板理论。根据理论计算结果，建议管板按 150 毫米厚度设计制造就可以了。这个产品是联邦德国的专利，厚度是 230 毫米，比较保险。我建议厚度为 150 毫米，工厂不敢做，太薄了出事怎么办呢？他们就采用了中国当时能加工的最大厚度的 190 毫米来加工管板。最后，产品设计制造成功。

后来我又参加研制中国最高压力容器，按世界王牌产品英国设备研制 2300 大气压力的高压聚乙烯反应器。反应器需要在筒体侧边开个孔，压力太高不好处理，要我计算孔边应力。我拿着这个题目做了两个月都没有任何头绪。甚至生了场病，眼睛都红肿了，以至到医院做手术。朝思暮想，一个晚上在似睡非睡的时候终于想出来了一个解决办法，采用复变函数方法来处理这个问题，给出了解析解，从而使中国第一台最高压力容器顺利研制。

由于在兰州石油化工机器厂做了一点创新工作，校军宣队就表扬我，并且在兰州大学为我出了一本科技专刊，这在当时是罕见的。我非常感动，觉得给国家做了点好事。科技创新非常重要，中国的第一批高压超高压容器产品就是这样研制出来的。

4）精密仪器仪表波纹膜片弹性元件设计理论的创新

现在再讲一个科学理论上的创新故事。这是我一生刻骨铭心又十分心酸的一个故事，也是我的成名作，这就是波纹圆板特征关系式。在精密仪器仪表中，波纹圆板被称为波纹膜片。波纹圆板一般是用铍青铜和不锈钢等金属做的，非常薄，一般仅有 0.1 毫米厚度，用到卫星上不到 1 厘米直径大，用到其他仪表中则有几厘米大，是一些精密仪器仪表的心脏元件。1963 年在兰州大学毕业后留校任教，我想做科研，便到兰州的工厂去调研。找到一家国家级的仪表厂，当时他们正好有一个任务，就是研制锯齿形的波纹圆板。

那时，美国有两种飞机经常侵略中国的领空，一种是 U2 飞机，一种是 P2V 飞机，成天在中国的上空飞行。U2 飞机飞得太高了，20 000 多米高。我们打不着。P2V 飞机只飞几百米高，但我们的雷达找不到它。后来我们设法击落了一架 P2V 飞机。击落以后，我们国家对这架飞机的高度表进行仿制研究。高度表的心脏元件就是一个锯齿形的波纹圆板，很薄，只有 0.1 毫米左右的厚度，根据大气压力的不同产生的变形不同来测量高度。这个东西中国过去不能制造，我所到的这家厂就要制造这个东西，以供中国军事飞机使用。

这个厂无法研制，便请我给他们先计算一下。我没学过这个理论，不知道怎么计算，可是我很喜欢这个课题，愿意做。厂里同意我做，我就回去汇报。汇报的结果是“你这么忙还到外面去干什么?”并给我戴了顶“不务正业”的帽子，把我批评了一通。领导不同意，就没办法签订课题合同。但是我觉得这是国家急需的重要课题，我们的军事工业、航空工业要上去，精密仪表是先行官。于是，我就利用业余时间悄悄进行研究。

从文献上看到，苏联科学家巴诺夫院士研究这个问题最早，接着是他的两个学生，他们都只研究很简单的问题，且所获得的理论结果精确度很差，远不能满足设计要求。后来又有几位日本和美国科学家做这方面研究，也没能解决这个问题。美国是理论加经验做出这种产品。这是个世界难题，是一个非线性数学问题，理论难度很大，全世界就几个大人物做。我那时是初生牛犊不怕虎。世界上没人做的才好嘛，国家也需要，我一直做到 1965 年夏天。后因参加农村“社教”和“文革”开始而被迫停止，但是只要有空，我就躲在家里继续做这项研究工作。

当时家里很穷，还要自己掏钱买纸买笔。研究计算工作量很大，连手摇计算器都没有。尽管系计算室有，但不让用。我要用五位数字运算加、减、乘、除、开方等，用手算，打算盘，翻对数表，用了几麻袋计算纸张。那个时候做业务是要被批判斗争的，于是我爱人保护我，她经常在门口帮我看着，有人来，我就马上收起研究资料，在外面放一本《红旗》杂志。没有办法，做科研还要搞秘密活动。终于在 1968 年 4 月算出来了，不仅计算方法、程序和公式简单，而且计算结果的精确度远远超过了苏联的世界领先水平的成果。他们算出来的误差大约是 30%，我的误差只有 5%左右。在大学的时候学了 2 年数学专业，加上 3 年力学专业，所以我是靠自己努力把这个问题攻克了。但是攻克了以后，没有马上发挥作用，工厂也不要你的成果，因为没跟工厂签课题合同。从 1966 年夏天开始，所有的中国学术期刊都停办了，没有地方发表，也没有地方汇报，你说了还要挨批判，我只好把成果锁在家里。以后虽有两次发表的机会，但因各种干扰，始终没有发表成。

直到 10 年后的 1978 年 3 月，我的波纹圆板的学术论文才得以在《力学学报》发表。

这篇论文发表时，我已经 37 岁，正好调到中国科技大学近代力学系飞行器结构力学专业教研室任教。1978 年 12 月，第五届全国精密仪器仪表弹性元件学术会议在上海召开，会议筹备委员会发现了我的这篇论文，便请我在大会上做特邀报告。大会报告后，会议评价这篇学术论文达到国际水平，立即引起轰动，《文汇报》马上报道，中国科技大学表扬，中国科学院发文件通报全院表彰，国务院副总理方毅也签字表扬。以后，我又将自己的波纹圆板的一系列成果在全国许多精密仪器仪表厂和研究所推广，从而改变了我国以往的经验设计的历史。

1978 年 3 月份发表这篇文章，距我开始做这个题目的时候，已经长达 14 年。按道理说，这么多年这个课题应该早有人做过了，可居然世界上还没有人攻克这个理论难题，仍然是世界前沿的作品。“文革”前，我在《科学通报》发表了 3 篇文章，这是当时中国的权威杂志，也是非常好的，但是我的成名作还是这一篇。这个创新故事，我是吃足了苦头才获得成功的。这就是科技的创新，只有经过坚忍不拔的努力，还要有知识的积累，还要抓住机遇才能成功。

5）提出“东水西调”工程创意设想

最后，再讲一个最近的事情。2004 年，通过中国工程院和广东省人民政府参事室，我向国家提出“东水西调”工程建议。以后，又陆续在一些地方作过讲座，报纸也有报道。最近又获得国家科研经费资助，正式开始了这一课题的研究。

21 世纪制约我国经济发展的重要因素之一是水资源缺乏问题。我国人均水资源约为世界平均水平的 1/4，其中华北和西北地区的水资源更是极度贫乏。北方的黄河、淮河和海河三大主要流域的河川径流量不到全国的 6%，而耕地面积却占了全国的 40%。水资源的不足，不但影响工农业生产，影响人民群众的日常生活，而且制约着我国经济社会的可持续发展。一些北方城市已长期对居民实行水的限制供应，有的城市因大量开采地下水，造成了地面大幅下沉，存在很大的隐患。同时，由于生态环境的恶化，我国北方的一些沙漠迅速扩展。为此，国家开始实施南水北调工程，以解决北方缺水的困难。然而，中国南方淡水也不足，故不是长久之计。

海洋占地球面积 70%，海水取之不尽。故我提出“东水西调”工程的建议，把渤海和黄海的海水由东向西用管道输送到首都、华北和西北各省区市，以永解当地之“渴”。在防止土地盐碱化的前提下，在沿线建立若干咸水湖，并在湖边按需建立海水淡化工厂，供当地人民淡水需要。据了解，目前海水淡化的成本是 4 元/吨，若技术更新，当可进一步降低成本。海水还可综合应用，比如建化工厂，提炼盐和稀有元素，综合效益可观，更可造福人类。咸湖水的自然蒸发，亦可改善当地气候，进一步可增加耕地，变沙漠为良田。

“东水西调”工程，可以根据财力，逐步向西推进。当务之急，是首先解决首都北京的缺水问题。按目前我国的工程技术能力与财力来看，启动“东水西调”工程已不是困难之事，希望此创新得到大家的支持，以更快地造福于人类。

我在这里讲一生的创新故事，就是要告诉大家我的感想。我们人类的进步要靠创新，无论是科技、制度、政治、文化方方面面都需要创新，没有创新人类就不能进步，是无数人的创新才使我们的社会进步。但是也不可能人人每天都在创新，应该说多数情况下，为了保证产品和办事的质量，大家都要按标准、按部就班地工作。只是当生活、工作有需要更新的时候，才用创新去解决问题。天天都在创新，那是要搞坏的。现在人们对创新谈得很多，不注意这个问题。我们多数人、多数时候应该按标准化的工作方式进行工作，只是社会应该给创新一个宽容的气氛。我们整个中华民族应该鼓励有创新思想、创新概念的人，要鼓励他们、帮助他们，要给他们减少障碍，要宽容，即使创新失败了还要宽容。

为提高全民族科学素质做出新贡献①

今天，我们召开广东省科普志愿者协会成立大会和首届会员代表大会，在各级领导的亲切关怀和全体代表的共同努力下，会议取得了圆满成功。首先，我要感谢各位代表对我的支持和信任，选举我担任协会首届会长。下面，我讲几点意见。

一、要充分认识科普志愿工作的重要意义，增强使命感和责任感

2008 年 12 月，胡锦涛总书记在纪念中国科协成立 50 周年的重要讲话中提出，希望我国广大科技工作者大力普及科学技术，积极为提高全民族素质作出新贡献。② 广大科技工作者要把普及科学技术、促进广大人民群众深入了解科技知识作为义不容辞的社会责任，把贯彻落实科学技术普及法和全民科学素质行动计划纲要作为科技工作的重要方面，努力成为科学知识的传播者、科学方法的实践者、科学思想的倡导者、科学精神的弘扬者。要积极参与科普活动，开展科普创作，充分发挥自身优势和专长，把科研和科普有机结合起来，通过多种渠道、多种方式积极主动地向公众介绍科研最新发现、展示科技创新成果②。

在科学技术快速发展的今天，一个国家、地区科学技术的普及程度，从根本上决定这个国家、地区生产和文化的发展水平，决定着这个民族的创造能力。科技创新是国家、地区综合竞争力的关键，而科学普及是科技创新的前提和基础。科技创新是在科技前沿不断取得新的突破，科学普及则是科技创新的一个重要基础。而科技成果只有为全社会所掌握、所应用，才能发挥出推动社会发展进步的最大力量和最大效用。普及科学技术，提高全民科学素质，既是激励科技创新、建设创新型国家的内在要求，也是营造创新环境、培育创新人才的基础工程。

因此，从事科普志愿活动，促进科学技术的普及推广，是一项非常崇高的事业。我们成立广东省科普志愿者协会，就是希望通过这个组织的建立，让更多有着志愿精神和社会责任感的科技工作者参与进来，实施跨地区、跨行业、跨学科的协作与联合，实现人力、物力、智力的共享与发展，建立一支阵容大、素质高、相对稳定的科普工作队伍，为提高公民科学素质，为我省全面建设小康社会、建设创新型广东贡献应有的力量。我希望理事会全体成员和全体会员充分认识科普志愿工作的重要意义，进一步增强使命感和责任感，勇挑重担，团结协作，共同努力，把科普志愿者协会建设成为一个充满生机活力，健康发展的科技团体。

我们的协会刚刚成立，工作千头万绪。我们要在省科协的领导下，紧紧围绕我省

---

① 本文是在广东省科普志愿者协会成立大会上的讲话，广州，2009 年 3 月 24 日；教育与科技管理研究，北京：科学出版社，2016，356-358.

② 《胡锦涛在纪念中国科协成立 50 周年大会上的讲话》，https://www.gov.cn/ldhd/2008-12/16/content_1179001.htm[2023-11-03].

科普工作的部署，重点抓好以下几方面的工作。

二、关于今年协会的主要工作意见

一是要认真学习宣传贯彻落实《中华人民共和国科学技术普及法》和《全民科学素质行动计划纲要（2006—2010—2020年）》。

2002年6月，我国颁布实施了《中华人民共和国科学技术普及法》（简称《科普法》），标志着我国的科普事业进入了一个崭新的历史阶段。《科普法》明确了科普工作的目标是提高公民的科学文化素质。2006年2月，国务院颁布了《全民科学素质行动计划纲要（2006—2010—2020年）》（简称《科学素质纲要》），明确了我国公民科学素质建设的指导思想、工作方针和目标任务。目前，深入贯彻落实《科普法》和《科学素质纲要》，是开展科普工作的首要任务。协会的工作要紧紧围绕这一主线，积极主动地配合我省贯彻实施工作的部署和要求，创造性地开展各项工作。

二是要面向公众开展各项科普志愿活动，积极打造协会的科普品牌。

开展科普志愿活动是我们协会生存与发展的基础，只有通过开展科普活动，才能取得更多的社会支持。一方面，我们需要积极承担省科协交办的大型科普活动，策划贴近群众、贴近生活、贴近时事的经常性科普活动，如组织科普专家志愿者进社区开展专题科普讲座，围绕“节能减排、生态文明、安全健康、新农村建设”以及当前发生的热点事件和突发事件，联合各地市、县（市、区）科协，依托科研院所、高等院校专家及大学生科普志愿者的专业知识和力量，举办“科普大讲堂”等活动，为群众关心的问题提供咨询服务。另一方面，我们要充分利用协会团体会员和会员所在单位的力量，积极开拓协会活动空间。同时，利用互联网技术和渠道，组织科普志愿者在线访谈，打造科普活动品牌。

三是要突出会员的主体地位，认真做好为会员的服务工作，大力发展科普志愿者队伍。

会员是协会的根本，是协会生存与发展的重要基础。突出会员的主体地位，做好各项服务工作，是协会的一项根本任务。目前全省注册登记的各类科普志愿者已达10万多人，这批有着志愿精神和社会责任感的科技工作者和参与科普工作的社会活动力量正活跃在科普工作的各条战线，推动科普事业发展。我们要加强与这些科普志愿者的联系，主动开展各项服务工作，加强对会员的培训，维护他们的合法权益。同时，进一步发展会员，动员更多的科技工作者和社会热心人士加入科普志愿者队伍，指导地方发展科普志愿者组织。

四是要加强协会自身建设，坚持民主办会。

协会章程是我们开展科普志愿服务的指引，我们一定要依法依章办会，充分发挥理事会成员的作用，加强横向联系与沟通，广泛发动各成员所在单位开展科普志愿服务，提高公民科学素质。同时，要加强制度建设，不断完善协会的各项规章制度，建立科普志愿者个人服务档案，保持队伍稳定性，加强与外界的公共沟通，树立良好的公众形象，增强协会的公信力、凝聚力和影响力。

同志们，科普是旨在提高公民科学文化素质的一项基础性、长久性、公益性的事业，是实施科教兴国和走可持续发展之路的重要基础。作为一名科普志愿者是光荣的；作为科普志愿者协会，我们的任务是艰巨的。相信科普志愿者协会在今后的工作中一定能做出可喜的成绩，一定会为广东省科普事业的发展做出重要的贡献。

稳步推进科普志愿服务事业①

一、2010 年的主要工作情况

2010 年，在省科协、省民政厅的正确领导下，认真学习贯彻党的十七届三中、四中全会精神和胡锦涛总书记在纪念中国科协成立 50 周年大会上的重要讲话精神及省委十届五次、六次全会精神，按照省科协七届四次全委会议的工作部署和要求，紧扣我省实施《科学素质纲要》的各项工作，充分利用会员力量，积极开拓协会活动空间，面向公众开展各项科普志愿活动，打造协会的科普品牌；加强协会自身建设，坚持民主办会，认真做好为会员的服务工作，大力发展专业的科普志愿者队伍及专业委员会。

1. 加强协会自身建设

（1）稳步推进协会各层次组织的自身建设。据 2009 年协会牵头开展全省科普志愿者登记注册反馈情况，广州、深圳、东莞、中山等市已拥有较完善的科普志愿者服务队伍。2010 年，协会把加强科普志愿者组织建设作为深化科普志愿服务工作的重要抓手，加强引导与服务，进一步提高科普志愿者的凝聚力，如广州市科普志愿者协会举办以“科普志愿者在社会活动中的作用”为主题的培训班，提升科普志愿者的综合素质。

同时，协会本部以重点吸纳各类基金会、企业领导入会为推手，积极筹措协会组织建设及发展所需的人力、物力、财力，推进协会可持续发展，如 2010 年，新发展常务副会长 1 名，常务理事 3 名。

（2）成立光动媒技术专业委员会。2010 年 10 月，光动媒技术专业委员会正式获省民政厅批准成立。“光动媒”科技是秦兆年副教授和刘达莲老师十几年的研究成果，该成果获得三项国家专利，对于促进国家科教发展，特别是贫困地区的科教发展具有重大意义。成立光动媒技术专业委员会，应用“光动媒”科技，能让投影、幻灯片等设备得到充分利用，缩小城乡教育水平差距，促进优质科普教育资源的开发与普及。

（3）筹建低碳生活委员会。哥本哈根会议后，低碳生活成为全球新共识，但在日常生活、工作、学习中还普遍存在认识不到位等问题。为进一步推进节能减排、低碳生活与构建节约型社会，促进资源再生利用，由广州岭南教育集团发展研究中心主任刘丹青所长、中山大学龚隽教授等发起组建，基本架构已经完成，待省民政厅审核批准。

（4）编印《科普志愿者》会刊。会刊主要反映科普志愿服务的情况，包括重大活动、协会动态、项目研究、人物专访、会员园地、它山之石等栏目，会刊已于 2010 年 11 月创刊发行，主要发送给协会会员、全国科协系统、省（市）纲要办成员单位、社区及大学生科普志愿者服务团队。

---

① 本文是广东省科普志愿服务协会第一届理事会工作报告，广州，2011 年 3 月 20 日；教育与科技管理研究，北京：科学出版社，2016，365-370.

(5) 建设科普志愿者服务网站及开通 QQ 群服务。科普志愿者网站及时发布科普志愿服务信息、会员动态信息，宣传会员典型做法和经验，为科普志愿者招募、报名、注册、交流、博客以及寻找高层次科普专家提供一站式服务。同时，开通 QQ 群服务号"13252491"，及时传递为会员服务信息。

(6) 申报 2010 年省科技厅科技攻关项目。协会组织专家申报省科技厅科技项目"一站式科普人才及资源库服务平台建设"，并获省科技厅立项支持。建立一站式科普人才及资源库服务平台，是以科普资源开发为主线，紧密结合传统科普工作，运用互联网络信息技术，整合广大科普志愿者，为公众提供一站式科普人力和资源服务。公众通过该平台，既能获取科普资源，又能与科普志愿者建立互动，并获得个性化的科普服务。

(7) 积极申报省科协开展第三批社团创新试点项目。根据省科协有关工作安排，协会积极组织申报省科协第三批创新试点，并获省科协批准。协会将重点围绕自身组织建设，完善社团法人治理机制，完善以会员为主体的组织体制，制定会员发展规划，强化为会员服务，落实会籍管理，建立会员数据库和信息管理系统，探索科学的会员制度。

2. 贯彻落实《科学素质纲要》

(1) 组织开展"5·12 防灾减灾日"等重大科普活动。受省科协、省全民科学素质纲要实施工作办公室、省地震局、南方电监局等单位委托，组织开展"5·12 防灾减灾日"、科技进步活动月、百人千场安全用电社区行活动、全国科普日等重大科普活动，积极与媒体合作，扩大活动的影响力，多次活动都在广东电视台、广州电视台及《羊城晚报》等省(市)媒体宣传报道，取得较好的社会影响。

(2) 组织编写系列科普图书。围绕"节约资源能源、保护生态环境、保障安全健康、促进创新创造"主题，协会联合省科普信息中心开展科普挂图、科普图书、科普动漫等科普资源的开发、集成工作，宣传与人民群众工作、生活密切相关的科学常识、科学思想和科学方法。截至 12 月底，协会已组织开发"春夏养生食疗法""侵害防范应急自救""意外伤害应急自救""野外逃生自救"等科普挂图 100 多种。

同时，为发挥协会人才智力优势，组织协会专家编写针对领导干部公务员科普图书 2 种、社区居民科普图书 10 种和青少年科普图书 6 种。如根据《中共中央 国务院关于加强青少年体育增强青少年体质的意见》(中发〔2007〕7 号) 及教育部《中小学学生近视眼防控工作方案》，切实加强学生视力保护工作，实现《中共中央 国务院关于加强青少年体育增强青少年体质的意见》提出的通过 5 年左右的时间，使我国青少年近视的发生率明显下降的工作目标。根据省教育厅、省科协工作安排，协会组织专家编印《珍爱眼睛 预防近视》科普图书 30 万册、标准对数视力表 5 万张及《珍爱眼睛 预防近视》科普动漫。

(3) 承办全国社区科普工作会议等活动。受中国科协科普部、省科协委托，承担全国部分省区市社区科普工作座谈会，研讨"十二五"开展社区益民计划的试点工作。同时，为推进我省社区科普"五个一"创建活动，根据省科协工作安排，承担省级科普示范社区创建及评比活动。先后评出 50 个省级科普示范社区，创建了 500 多支社区科普志愿者服务队。

（4）举办科普大讲堂。一是承办神农健康大讲堂，由协会会员单位广州白云山和记黄埔中药有限公司赞助，每月举办一场面向城镇居民的有关健康养生知识科普讲座，深受群众喜爱与欢迎。二是发动地市基层科普组织，联合承办科普大讲堂。

（5）组织承担第 25 届全国青少年科技创新大赛宣传联络工作。第 25 届全国青少年科技创新大赛于 2010 年 8 月 7～13 日在广东省广州市举行，本届创新大赛由中国科协、教育部、科技部、国家发展改革委、环境保护部①、国家体育总局、共青团中央、全国妇联、国家自然科学基金委员会、广东省人民政府共同主办。广东省科协、广东科学中心承办。

按照省科协的工作部署，协会承担大赛筹备工作执委会宣传联络部工作，成立物料组及新闻中心工作组，完成大赛会徽吉祥物的征集及会徽的设计任务。做好赛前、赛中、赛后的新闻宣传工作。建立大赛专题网站，利用广东广播电视台、南方电视台、广州电视台等主要新闻栏目频道播出大赛专题节目，在《南方日报》《广州日报》《羊城晚报》等省内主要报纸媒体的头版及主要版面刊登我省青少年科技创新成就的报道。同时，发动新华网、中新网、网易、新浪等网站对大赛进行了报道和转载。

此外，协会还参与了大赛科普志愿者的培训工作，并组织人员积极参与大赛后勤保障部工作，保障大赛每日 1500 人的用餐，得到与会领导和参赛嘉宾选手的一致好评。

3. 组织协会会员交流与合作

为加强海峡两岸社会志工发展的交流，受台北市服务教育协会的邀请，协会组织部分常务理事组成赴台考察团，由协会副会长郑文丰带队，共 9 人，于 2011 年 1 月 7 日至 14 日应邀前往台湾考察志愿（志工）服务状况和科普工作。

二、2011 年主要工作安排

2011 年是我国深化改革开放、加快转变经济发展方式、实施“十二五”规划的开局之年；是我省加快转型升级、建设幸福广东的关键年；是中国共产党建党 90 周年，也是省科协成立 50 周年。按照省科协七届五次全委会的工作部署和要求，坚持围绕中心、服务大局、深入基层，团结动员广大科普志愿者，积极开展科普志愿服务，为提高我省公民科学素质，加快转型升级、建设幸福广东，促进社会和谐发展作出新的贡献。

1. 进一步加强协会自身建设

（1）加强科普志愿服务信息管理系统开发与推广。以协会申请省科协社团创新试点为契机，开发完善科普志愿服务信息管理，强化协会信息化应用，提高为会员服务能力。

（2）发展高层次科普志愿者。一是协会本部继续以重点吸纳各类基金会、企业领导入会为推手，积极筹措协会组织建设及发展所需的人力、物力、财力，推进协会可持续发展。二是发展高层次科普专家，组建专业科普志愿者服务队，为基层科普组织提供智力支撑。

---

① 2018 年《国务院机构改革方案》不再保留环境保护部，组建生态环境部。

（3）继续筹建低碳生活委员会。

（4）试点联合部分地市科协组建协会办事处（或地方科普志愿者协会）。

（5）推进协会自身服务体系建设。

会刊将在已有内容基础上，增设“健康养生”及“科普漫画”栏目；网站将以完善会员管理及信息内容共建为突破口，开展会员信息共建评优活动，及时发布会员动态信息；QQ群“13252491”将及时传递为会员服务信息。

2. 进一步贯彻落实《科学素质纲要》

（1）继续做好防灾减灾日等品牌主题科普活动。围绕“节约能源资源，保护生态环境，保障安全健康，促进创新创造”主题，整合社会力量，采取大联合大协作方式，深入开展各种主题性科普活动。组织科普专家讲师团、科普志愿者服务队深入参与防灾减灾日、科技进步活动月、科技下乡、全国科普日、低碳生活及全民健康科技行动等活动，推动公民科学素质稳步提高。

（2）继续实施农民和居民等重点人群科学素质工作。按照中国科协和省科协的要求，探索实施“社区科普益民计划”，广泛开展各类有针对性的技能培训，倡导和普及节约资源、健康生活、防灾减灾、低碳环保等观念和知识，提升社区居民科学素质，促进形成科学文明健康生活方式。深入开展科教进社区、全民健康科技行动、社区科普大讲堂等社区科普活动，依托社区公共服务设施，拓展和发挥科普功能。通过社区居民科学素质行动和科普人才建设工程，加强社区科普志愿者服务队、科普宣传员建设，深入开展省级科普示范社区创建评选活动，培养造就一支专兼结合、高水平的科普人才队伍，为全民科学素质工作提供人才支撑。

围绕省委、省政府建设“幸福广东”的要求，推进城乡科技教育服务均等化，开展“绿色电脑扶贫，送科普信息下乡”活动，通过回收城市闲置的电脑设备，经过整修调试、外观处理、消毒等专业的绿色环保处理环节后赠送给贫困落后地区建立电脑室。配备农民生产生活息息相关的科普信息，帮助贫困地区农民掌握生产知识脱贫致富；普及环保、饮水、饮食、疾病防治等科普信息，帮助农村改善生活环境，提高农民身体素质。同时，组织大学生科普志愿者服务队利用节假日及寒暑假时间，深入农村开展科普志愿服务，宣传普及环保、饮水安全及疾病防治等知识。

（3）继续开展科普资源开发工作。配合省科普信息中心继续做好科普图书、科普挂图、科普动漫及岭南科普网等科普资源开发与共享服务工作。充分利用现代技术手段，不断丰富科普资源的表现形式。如开发针对学前儿童的有声科普图书，在欠发达地区普及推广，实现农村与城市学前儿童科普教育的均等化；继续开展“光动媒”科普资源的开发和普及推广工作。

（4）开展低碳、生态科普旅游及健康科普服务。探索借助绿道，构建科普志愿服务驿站。

努力开创仪表元件分会工作的新局面①

今天，我们在这里召开中国仪器仪表学会仪表元件分会第五届理事大会。首先我代表第四届理事会对各位同仁多年来对分会的支持和对仪表元件行业所做的工作表示真诚的谢意！对本次理事大会的召开表示衷心的祝贺！现在，我代表第四届理事会做大会工作报告，请大会审议。

一、第四届理事会工作回顾

仪表元件分会第四届理事会在总会的指导下，在会员的不断努力和挂靠单位沈阳仪表科学研究院的大力支持下，分会主要做了以下几个方面的工作。

1. 学术交流和信息交流工作

从全球经济角度看，主要有两个重要趋势——“全球化”和“技术创新”。“全球化”的主要特点是生产要素在全球的重新配置，以及生产要素在全球配置速度和流动速度的加快。“技术创新”又分为：研发、试用利用、推广、商业化应用等几个阶段。科学技术是第一生产力，是可流动的生产要素，其创造的效益远远超过自然资源。目前，世界上 95%的新专利为发达国家所拥有，我们要缩小这个差距的一条重要途径就是扩大学术交流，并通过交流催生“技术创新”。

学术交流和信息交流是仪表元件学会重要工作内容之一。多年来，分会多次组织会员单位参加座谈会、博览会、展销会。从 2002 年至今，共召开学术交流会 20 多次，参加会议技术人员达 900 多人，组织参加展会 30 多次。

（1）2002 年 9 月 19 日，在第 13 届多国仪器仪表展览会期间，在北京理工大学组织了“MEMS 技术”② 学术报告会，参加会议人数达 100 人。大连理工大学唐祯安教授、清华大学王晓浩博士、北京理工大学李科杰教授、中国科学院上海微系统与信息技术研究所李昕欣研究员和沈阳仪表科学研究院赵志诚副院长分别做了“微传感器中的微尺度热特性研究”“微系统和纳系统”“微小型无人系统的发展及其对测试技术的挑战”“MEMS 传感器静/动态性能提高相关问题研究”“MEMS 技术在硅电容加速度传感器中的应用——兼介绍采用 MEMS 技术的大片硅腐蚀装置”学术报告。报告内容充实、丰富，报告形式生动、有趣，会场气氛积极、热烈，达到了技术交流的目的。同时，对于推进我国 MEMS 技术、微系统/纳系统的研究与发展起到了积极的作用。

（2）为了进一步发挥分会的作用，不断加强分会与企事业单位的交流，经过多方协作，召开了两次新产品推广会。2005 年 6 月 11～13 日，“六氟化硫产品鉴定推广会”在沈阳金剑大厦召开，共有 40 多位行业内的专家、专业人士和分会会员单位参加了这

---

① 本文是中国仪器仪表学会仪表元件分会第四届理事会工作报告，扬州，2011 年 5 月 26 日；教育与科技管理研究，北京：科学出版社，2016，332-343.

② MEMS 英文全称为 micro-electro-mechanical system，译为微机电系统。

次推介会。会上，专家们对沈阳仪表科学研究院成功研制的六氟化硫产品进行了鉴定和推广，并给予了较高的评价。会议取得了预期的效果，在直接用户之间进行了很好的推广。目前，该产品已畅销国内，取得了较好的经济效益和社会效益。

2005 年 11 月 29～30 日，“硅电容/硅压阻复合传感器新技术、新产品推研会”在沈阳金剑大厦召开，共有 150 位行业内的专业人士参加。沈阳仪表科学研究院张治国、唐慧、张春晓分别作报告，讲解了这些产品在科研与生产中遇到的问题和解决办法，以及用户及专家的评价等。这次会议的成功召开标志着我国首台采用国际前沿技术，具有自主知识产权的硅电容/硅压阻复合传感器问世。

国内数十家产品用户、合作厂商以及科技日报社、沈阳日报社、沈阳人民广播电台的代表参加了推研会。会议在热烈的讨论中持续了两天，达到了预期的目的，向仪器仪表制造企业与用户介绍了沈阳仪表科学研究院在硅电容/硅压阻复合传感器及相应仪表产品的技术成果和开发成果，并为成果进一步商品化、产业化提供了交流与合作的平台。

这两次会议还得到了沈阳市人大、辽宁省科技厅、辽宁省经贸委、辽宁省发改委、沈阳市科技局、沈阳市经贸委等部门的关心和大力支持，会议推出的系列产品引起了业内外的广泛关注。

(3) 2006 年 7 月 18 日，举办了组合电器波纹补偿器生产基地剪彩仪式及技术介绍会。为了进一步发挥分会的作用，不断加强分会与企事业单位的交流，更便捷地为企事业单位提供信息技术服务，携手共同发展中国仪器仪表事业，经过多方协作，2006 年 7 月 18 日，举办了“组合电器波纹补偿器生产基地剪彩仪式及技术介绍会”。

沈阳仪表科学研究院和沈阳汇博热能设备有限公司中的骨干力量都是分会的重要成员。2006 年，沈阳仪表科学研究院和沈阳汇博热能设备有限公司共同出资 800 余万元，建成了国内组合电器波纹补偿器最大的生产基地，面积为 2100 平方米，年生产能力 8000 万元。此次剪彩仪式，应邀前来的均为国内在高压电器生产方面的领军单位和有重大影响的制造厂商，到会嘉宾 30 多名。

这次会议为供需双方搭建起互相了解、交流合作的平台，达到了预期的目的。这次活动还得到了沈阳市人大、辽宁省科技厅、辽宁省经贸委、辽宁省发改委、沈阳市科技局、沈阳市经贸委等部门的关心和大力支持，生产基地及技术介绍会引起了业内外的广泛关注。

(4) 为了进一步发挥分会的作用，不断加强分会与政府、企事业单位之间的交流，携手共同发展中国仪器仪表事业，2007 年 7 月 12 日，仪表元件分会承办了沈阳仪器仪表产业发展座谈会。该会议由沈阳市委主办，沈阳铁西区人民政府、沈阳仪表科学研究院、仪表元件分会共同承办。参加会议的人员有：国内仪器仪表行业重点企业代表，沈阳市委、市政府领导，沈阳铁西区人民政府领导，沈阳仪表科学研究院院长，仪表元件分会理事等 50 余人。

会议通过观看电视纪实片《凤凰涅槃》，参观考察企业，参加座谈会等形式，使与会代表充分了解沈阳亟待发展仪器仪表产业的现状。

座谈会由沈阳市副市长王英主持，参会代表与沈阳市委、市政府的领导们共同商议了我国仪器仪表产业的发展趋势，并就如何建设发展沈阳仪器仪表产业、如何参与沈阳仪器仪表产业建设发展、如何与沈阳装备制造企业协作配套等问题进行了专题座谈。最后，辽宁省省长陈政高同志讲了话。

这次会议充分体现出在沈阳工业全面振兴、经济社会和谐发展的大环境下，仪表元件分会与沈阳市委、市政府一起，把发展仪器仪表产业提升到事关沈阳长远发展的战略地位，并为仪器仪表产业的发展作出了努力和贡献。

（5）积极组织参加在沈阳召开的中国国际装备制造业博览会。第五、六、七、九届中国国际装备制造业博览会在沈阳召开，许多分会会员单位参加了各届博览会，该博览会是经国务院批准的国家级展会，由商务部、国家发改委、中国贸促会和辽宁省政府主办。分会常务理事长徐开先作为制博会的专家组成员之一，参加了参展仪器仪表产品的评审工作。在 2008 年 9 月1～5 日召开的第七届制博会上，经过专家组的认真评审，河南汉威电子股份有限公司生产的气体敏感元件及传感器产品被授予银奖。

（6）2008 年 11 月 29 日，在暨南大学举行了由仪表元件分会理事长单位暨南大学主办的广东省仪器仪表学术会议。首先由本人致开幕词，接着又作了《创新路上的感想》的大会报告。报告结束后，会议分为两组进行学术交流，分组报告中大家互相讨论、互相交流，气氛热烈。会议将收到的 26 篇论文编辑成论文集《21 世纪的仪表科学与技术》，其中 16 篇论文做了学术报告。会议还评选了 6 篇优秀论文，并进行了表彰。

（7）引领高端可靠性技术、探索环测行业科技之路。2009 年 5 月 15 日，由仪表元件分会参与的中国仪器仪表学会泛珠三角区域分会联盟、广东省仪器仪表学会、华南国家计量测试中心、广东省计量科学研究院、星球国际资讯（香港）有限公司、《环境技术》杂志社联合主办的“首届 2009 年华南（东莞）环境与可靠性试验技术论坛”在东莞厚街广东现代国际展览中心举办。

本次论坛将中国计量科学研究院、中国工程院张钟华院士、中国仪器仪表学会副理事长兼秘书长吴幼华、华南国家计量测试中心、广东省计量科学研究院副院长潘嘉声等国内科研院所、企业的权威专家以及众多业内精英齐聚一堂，探讨国内最新针对产品环境试验、适应性与可靠性技术的研究和应用，促进行业技术的交流与进步。参会的代表来自中兴、华为、富士康、广汽、比亚迪、中国电器科学研究院、中国赛宝等电气电子、汽车、测试、计量，以及环境与可靠性试验设备和配套商等研究院所和企业，共计 240 余人积极参与了本次技术论坛。本次论坛由仪表元件分会和广东省仪器仪表学会主持。

（8）权威汇聚——直击全球认证测试技术。2010 年 5 月 14 日，为了更好地普及全球认证测试技术、促进珠三角地区认证测试技术交流与应用，由仪表元件分会参与的中国仪器仪表学会泛珠三角区域分会联盟，广东省仪器仪表学会，广东省家电商会携手 UL（Underwrites Laboratories，美国保险人实验室）、SGS（societe generale surveillance，通用公证行）、BV（Bureau Veritas，必维国际检验集团）、Intertek（天祥集团，质量和安全服务公司）、TÜV（Technische Überwachung s-Verein，技术检验协会）、中国赛宝，华南国家计量测试中心、《仪器仪表商情》等联合主办的 2010 全球认证测试技术论坛在东莞厚街广东现代国际展览中心举行。

本次论坛，秉持“为企业质检部、技术部、实验室人员、采购部、外贸部等提供高质量增值平台”的宗旨，政府领导、行业领导、认证测试行业专家，各大专院校及科研院所学员和科技人员出席了论坛。众多业内精英齐聚一堂，探讨全球最新认证测试技术研究和应用，促进行业技术的交流与进步。参会的代表包括来自 UL、SGS、BV、Intertek、TÜV、CVC、格力电器、比亚迪、中国电器科学研究院、中国赛宝等电气电子、汽车、测试、计量，以及认证测试技术试验设备等科研院所和企业，共计 300 余人参与了这次技术论坛。这次论坛由仪表元件分会和广东省仪器仪表学会主持。

(9) 第二届华南（东莞）环境与可靠性试验技术交流论坛。2010 年 5 月 15 日，由仪表元件分会参与的中国仪器仪表学会泛珠三角区域分会联盟、广东省仪器仪表学会、中国赛宝实验室可靠性与环境工程中心、星球国际资讯（香港）有限公司联合主办的“第二届华南（东莞）环境与可靠性试验技术交流论坛”，于东莞厚街广东现代国际展览中心开幕。这次会议旨在搭建一个国内产品环境与可靠性试验技术方面的专家和企业之间的沟通平台，增进国内在产品环境条件、环境试验、适应性和可靠性试验技术的交流与进步。

(10) 第六届国际仪器仪表与测控自动化高峰论坛。2010 年 11 月 18 日，由深圳市科学技术协会、中国仪器仪表学会仪表元件分会、广东省仪器仪表学会、深圳市仪器仪表学会、深圳中国工程院院士活动基地、《仪器仪表商情》等单位联合举办的“第六届国际仪器仪表与测控自动化高峰论坛”暨 2010 移动互联终端检测与手机制造测试研讨会，在深圳国际会展中心举行。这次论坛是让广大客户了解 3G 测试的最新发展趋势，增强用户对移动互联终端检测与手机制造测试技术的认识，为关注 3G 技术发展的厂商和企业提供一个学习、交流的平台。现场有学术报告，媒体、专家、厂商汇聚一堂，是一年一度的行业盛会。

(11) 积极承办第十一届全国敏感元件与传感器学术会议（STC 2009）。由仪表元件分会参与的全国敏感元件与传感器学术团体联合组织委员会主办，沈阳仪表科学研究院及传感器国家工程研究中心承办的第十一届全国敏感元件与传感器学术会议（STC 2009）于 2009 年 11 月 4 日至 7 日在杭州成功召开，来自全国各地的专家、教授、学者、工程技术人员 180 余人参会。

这次大会的宗旨是“举办高水平的学术活动，创造有特色的产品交流，促进传感器业界联合，推动传感器产业发展”，围绕敏感元件与传感器行业现状和发展等方面进行了深入广泛的学术交流和产品展示。

这次会议的主题是“传感器与产业化”。会议期间，除了优秀会议论文交流外，还邀请了有关领导、著名专家、企业家作关于传感器对我国仪器仪表行业的影响，敏感元件与传感器的国内外现状和应用，传感器产业化等特邀报告和专题报告，并组织了敏感元件、传感器产品展示。

会议期间，按照既定日程分别进行了大会开幕式、大会特邀报告、优秀论文大会交流、大会专题报告、产品信息交流、获奖论文颁奖仪式等。沈阳仪表科学研究院院长、第十一届全国敏感元件与传感器学术会议主席庞士信、全国敏感元件与传感器学术团体联合组织委员会主席宋宗炎分别做了讲话，浙江大学党委副书记郑强教授为大会致欢迎辞。大会开幕式由沈阳仪表科学研究院原院长、第十一届全国敏感元件与传感器学术会议顾问、第十一届全国敏感元件与传感器学术会议论文评奖委员会主任徐开先主持。

这次会议的召开，得到了广大与会者的首肯，无论在把握会议的宗旨和方向，特邀报告、专题报告的选题，产品展示平台的安排，还是论文评奖程序等方面，都采取了与以往敏感元件与传感器学术会议不同的举措和尝试，得到了与会者的认可。这次大会达到了预期目的，取得了圆满成功。

（12）加强国际合作，协助承办国际交流、研讨会议。仪表元件分会副理事长单位大连理工大学组织了“IEEE 电子器件协会大连研讨会”①。此次研讨会邀请了来自欧洲、美国、日本、印度等的 8 位 IEEE 电子器件方面的知名学者，分别就各自在微电子领域的最新科研进展及取得的成果做了特邀报告。其中，IEEE 会士（IEEE fellow）、IEEE 比利时分会主席科尔·克拉埃（Cor Claeys）教授做了题为“Trends and Challenges in Micro and Nanoelectronics for the Next Decade”的报告，介绍了下一代微纳电子器件的趋势和挑战；来自日本 CTIF 公司的 Shingo Ohmon 博士做了下一代互联网的模型和设想的报告。该研讨会受到 IEEE 的高度评价。

大连理工大学还举办了“中国—新加坡双边微电子研讨会”，邀请来自新加坡南阳理工大学的四位教授和多位大连理工大学的教师进行了研讨，并初步形成了一个合作研究项目的框架。该项活动还得到了科技部的经费支持。

（13）2010 年 10 月沈阳仪表科学研究院和传感器国家工程研究中心协办了 2010 年首届国际磁电子器件及产业化研讨会，这次会议由中国船舶重工集团公司主办，大会邀请了国内外磁电子领域知名专家杭州电子科技大学/东方微磁科技有限责任公司钱正洪教授、美国约翰霍普金斯大学 ChiMing Chien 教授、美国明尼苏达大学 Jianping Wang 教授、法国 Spintec 中心 Claire Baraduc 博士、中国科学院物理研究所韩秀峰研究员、英国约克大学 Yongbing Xu 教授等出席大会并做大会特邀报告。此外，还有众多国内外知名学者、专家及研究人员以各类学术报告、海报及展台形式发布他们的最新研究成果。这次大会对磁电子材料、磁电子器件设计制造、器件二次开发及磁电子器件产业化前景等议题进行总结、交流和讨论。本次国际会议对于促进磁电子领域的学术交流和合作、推动国内磁电子产业发展及产业技术升级换代，起到积极的推动作用。

（14）组织会员单位积极参加其他学会的活动。沈阳汇博光学技术有限公司是国内领先的滤光片与反光镜的基础研究以及专业设计与制造单位，尤其在生物医学应用滤光片、影像投影应用反光镜方面是国内行业的带头企业。承担国内 70%的生物医学滤光片与影像投影反光镜设计与制造，是光学薄膜协会委员单位，多年来连续参加 SPIE（Society of Photo-Optical Instrumentation Engineer，国际光学工程学会）、中国光学协会、光学薄膜专业委员会组织的各种学术会议，与世界范围内的专家进行技术交流。廖邦俊、费书国担任光学薄膜专业委员会委员。近年参加的学术会议有 2009 年光学薄

① IEEE 英文全称为 Institute of Electrical and Electronics Engineers，译为电气电子工程师学会。

膜前沿国际会议（International Conference on Frontiers of Optical Caotings 2009）、中国光学学会 2010 年光学大会（天津）、2010 年的第七届国际薄膜技术与应用会议（The 7th International Conference on Thin Film Physics and Applications）（上海）。

2. 参与仪器仪表元器件标准制修订工作

多年来，仪表元件分会与沈阳仪表科学研究院一起，共同组织并参与了有关仪表元件标准的编制修订工作。分会参与编制修订的各项标准共 108 项，其中国际标准 5 项，国家标准 16 项，行业标准 87 项。分会各会员单位对仪表元器件标准起草十分重视，取得了较好的效果，提高了分会的行业地位。

3. 做好宣传工作，树立分会形象

（1）组织会员单位，积极参加国际展会。参加了美国西部光电展、德国光电展、美国临床医疗展、德国医疗器械展、德国分析设备展、德国“SENSOR＋TEST”展会、韩国国际自动化展会等，分别针对生物医学光学滤光片、舞台灯产品、传感器、波纹管等产品进行宣传推广，使得众多国内外客户对我国产品从陌生到熟识再到信赖，成为开拓国外市场最行之有效的途径。在德国分析设备展和美国临床医疗展上，主推了新产品高端荧光显微镜用滤光片。

通过德国“SENSOR＋TEST”展会，集中展示了传感器及测试测量技术方面从简单的元器件到成套的系统产品。展会上展出的展品包括工业自动化仪表与控制系统、自动化、IT 解决方案及软件、科学仪器、电子与电工测量仪表、仪表材料、仪表元器件及附件、传感器、仪器仪表工艺装备及加工设备等应用于各个行业、各领域的仪器仪表及测量控制系统。“SENSOR＋TEST”展会，也是中国传感器公司了解并打开欧洲市场的最好平台，更是了解最先进的传感器技术的第一媒介。通过参展，和国外先进传感器公司进行交流，学习国外先进的传感器技术，同时也对外宣传了我国的产品，融入传感器国际市场大舞台，开拓国际市场。

韩国国际自动化展览会是韩国自动化领域最大的展览会，也是在亚洲规模最大且具有影响力的涉及产业自动化所有领域的专业展会之一。该展会以自动化产品为主，传感器产品仅仅是其中的一部分。此次展会汇集了韩国、德国、法国、英国、美国、日本、中国等 20 多个国家 300 多个企业自动化行业的精英，展示了他们在自动化领域的最新技术、产品、材料及设备，提供了一个了解最新动态、寻找最新技术、采购最新产品、交流最新信息的最佳平台。

积极组织会员单位参加国内展会，如中国国际光电博览会、上海传感器展及北京多国仪器仪表展览会。从展会上了解到，智能通信传感器、智能变送器、车用传感器、医疗器械用传感器产品等在展会上备受关注，为企业研发新产品的市场调研打下了基础。

（2）在“仪表技术与传感器”网站及《仪表技术与传感器》杂志上宣传。在秘书处挂靠单位沈阳仪表科学研究院网站和“仪表技术与传感器”网站上分别建立了“行业动态”专栏，扩大分会影响，提高知名度，积极宣传分会。将《仪表技术与传感器》杂志作为发表平台。

4. 科学技术普及活动

2009 年，分会常务副理事长徐开先，编写了《科技论文撰写与科技管理》一书，发放到青年科技工作者手中，鼓励并指导青年科技工作者撰写科技论文及参与科技管理。

5. 协助主管部门开展行业性工作

2002 年，按照总会的要求，按时完成了《中国仪器仪表学会专业分会基本情况调查表》《理事情况调查表》、中国仪器仪表学会六次全委会会员代表及理事候选人的推荐上报，并为第 13 届多国仪器仪表展览会学术会议推荐了报告人王立鼎院士。

在认真完成总会部署的任务的同时，分会秘书处按时参加总会定期召开的秘书长及专职干部工作会议，并能认真对会议精神进行汇报和传达。

由于工作成绩显著，仪表元件分会连续获得 2002 年度、2003 年度优秀学会二等奖，获得 2008～2009 年度中国仪器仪表学会优秀组织工作奖。

6. 咨询服务工作

对会员单位及行业进行技术咨询服务是分会的日常工作。分会秘书处每年都收到很多咨询电话、信件及电子邮件。2002 年至今，共进行 560 多次咨询服务，如产品导购、产品助销、求助人才、难题解答等。对此类工作，分会本着用户至上，有求必应，竭诚服务的原则，尽量满足要求，得到了很好的评价。分会也多次和国外企业联系，为引进技术、合作经营、合资办厂、产品销售、提供信息、引线搭桥，与美国 IC 公司、美国 Omega 光学公司、硅微结构公司、硅国际公司等建立了联系，利用我们的优势，为会员单位和行业尽力做一些工作。

7. 组织建设工作

在总会的支持下，分会每两年就对仪表元件行业进行一次摸底工作。

近年来，由于企业兼并、转制、破产、改名，会员退休、离职、调动等，分会结构变化很大，仪表元件分会这支队伍急需整顿。秘书处共发函 300 多封，征求行业内科技人员和其单位的意见，了解行业情况，把那些愿意参加分会活动的科技人员重新组织起来，强化和壮大分会队伍。

分会秘书处又专门印制了《会员登记表》、《中国仪器仪表学会仪表元件分会工作条例》及《中国仪器仪表学会仪表元件分会第五届理事会理事候选人登记表》等，并多次借参加行业展会的机会，多方面征求科技人员与行业单位的意见，了解行业情况，为本次换届选举做了充分的准备工作。从去年年底至今，共吸收新会员 51 名。同时，秘书处对会员提出的合理化建议进行了汇总，以便今后工作的有效开展。

8. 积极组织培训讲座，为地方经济发展服务

(1) 面向全国高校教师举办了“纳米电子技术课程师资培训班”。在国家外专局的资助下，大连理工大学先后组织派遣 4 个团共 30 多人分别到美国、欧洲进行半导体技术培训和交流，并邀请十几位国外著名半导体器件专家前来讲学。主要有：邀请了美国普渡大学教授马克·伦德斯特伦（Mark Lundstrom）和穆哈马德·艾莱蒙（Muhammad Alam）来访，面向全国高校教师举办了“纳米电子技术课程师资培训班”。Mark Lundstrom 教授就纳米电子器件的模型与理论做了四个专题讲座；Muhammad Alam 教授就纳米电子器件的可靠性相关理论做了四个专题讲座。参加讲座的学员来自国内 20 余所高校相关专业的教师和研究生 50 余人。

（2）为地方政府处级以上干部培训。2007 年 6 月 12 日，分会与沈阳市直机关工委、沈阳市经济委员会共同组织了“先进装备制造业系列讲座”，首讲由沈阳仪表科学研究院庞士信院长主讲，时任辽宁省委常委、市委书记陈政高，市委常委、秘书长顾春明，市委常委、宣传部部长马占春以及来自市直机关 21 个单位的主要领导和 200 多名处级干部参加了讲座。

这次讲座旨在完成振兴老工业基地的历史使命，敏感元件及传感器是先进装备制造业的先行官，发展敏感元件及传感器技术对发展先进装备制造业起着不可低估的重要作用。讲座内容从敏感元件及传感器的基础工艺到仪器仪表，最后引申到先进装备制造业，内容丰富，使市委、市政府重要参谋决策部门和经济部门了解了敏感元件及传感器、仪器仪表及装备制造业的基础知识、国内国际产业发展方向和趋势、沈阳的优势及发展方向，使市直机关广大党员干部做好了承担振兴重任的思想战略储备和知识储备。

（3）创新理论与方法培训工作。承办中国机械工业集团有限公司东北区的“创新理论与方法”培训工作，参加培训人数 40 人。通过 4 个单元系统的“TRIZ 理论和方法”① 培训学习，从学员每期反馈的情况可以看出，随着培训的深入，学员的思路逐渐清晰和条理化，解决具体问题的能力大大提高。学员中最普遍和深刻的感受就是：TRIZ 理论提供了行之有效的解决问题的思路和方法，为今后具体技术问题的解决提供了方法论，为实际工作奠定良好的基础。为进一步加深对创新理论的认识，丰富培训内容，为学员集体购买了《TRIZ 入门及实践》书籍。一系列的课外推进与保障措施，为培训工作奠定了坚实的基础。“自主创新，方法先行”，该项工作的最终目标是：通过三个阶段创新理论和方法的培训与推广，利用先进的技术创新理论和工具，培养一批掌握创新理论与方法的技术骨干，形成一批以专利技术为重点的创新成果，攻克一批关键技术，进一步提升自主创新能力和综合研发能力。

（4）“公司知识产权的保护和管理”培训。2010 年 5 月 6 日，邀请沈阳亚泰专利商标代理有限公司郭元艺总经理，对工程技术人员进行“公司知识产权的保护和管理”培训。从“知识产权的概念及范围、专利的概念及客体分类、知识产权保护与企业核心竞争力的提升途径、企业专利申请及运用的整体策略、专利申请及专利经营对策、专利权利的取得及法律保护、专利生产的途径、专利侵权诉讼”等八个方面进行了详细讲解。通过培训，丰富了知识产权方面的知识，了解了专利在企业发展中的重要作用。

9. 为企业的产学研合作牵线搭桥

（1）为落实国务院《关于加快培育和发展战略性新兴产业的决定》，促进科技创新，加快企业经济发展和社会进步，充分利用高等院校的技术、人力等资源以及先进成熟的技术成果，利用企业的生产条件，提高学校的科研能力，将科研成果尽快地转化为生产力，为加强双方的联系与合作，充分发挥大专院校在高端人才聚集、科研实力强等方面的优势，通过分会的牵线搭桥，2010 年 10 月 13 日，沈阳仪表科学研究院同大连理工大学电子科学与技术学院签署了战略合作框架协议，建立长期战略合作关

① TRIZ 理论指发明问题解决理论。

系，以提升沈阳仪表科学研究院创新能力、促进经济发展。

(2) 积极推动会员单位同国外厂商进行战略合作。沈阳汇博光学技术有限公司2011年4月与美国的国际光学薄膜领先厂商Omega形成战略合作，就3D投影（三维投影）、生物医学、太阳能等领域开展广泛的中美技术与市场开发合作。

(3) 促成了沈阳仪表科学研究院与沈阳市浑南区人民政府、浙江纽顿流体控制有限公司战略合作框架协议的签署。

二、存在的问题

分会工作主要存在以下不足。

(1) 由于理事所在的单位变化很大，不仅仅是转制的问题，专业方向等也发生了变化，这给学术活动、技术交流的开展造成了一定的困难。加之，分会经费缺乏，分会的工作困难较多，做的工作还不够理想。

(2) 近年来，仪表元件行业变化很大，出现了一些新的企业和部门，如合资企业、合作企业、乡镇企业、民营企业，他们是一部分不可忽视的行业力量。他们身居行业学会之外，并愿意向行业学会靠拢，但受到各方面的制约，分会还没有把他们吸收进来、组织起来、没有在行业中发挥作用。

三、下一届理事会工作建议

这次大会的一项重要任务是民主选举产生新一届理事会，完成分会的换届工作。经过近一年的充分酝酿准备，分会发展了一部分新会员，面貌焕然一新。新一届理事会将是一个强有力的领导集体，理事遍及全国各地，有大专院校、科研院所及企事业单位，并考虑到仪表元件的所有技术领域。为了更好地肩负起社会赋予分会的历史使命，我代表第四届理事会向新一届理事会提出以下建议。

1. 分会工作的指导思想

分会将在总会和新一届理事会的领导下，勤奋务实，开拓创新，成为交流主渠道、科普宣传主力军、国际民间交流的主要代表、科技工作者之家。

2. 遵循分会宗旨，主动搞好双向服务

仪表元件工作的系统性强、范围广，因此分会除了为政府部门做好助手外，还要主动加强与其他各有关科研院所、高校与学会、协会等方面的广泛交往，发挥仪表元件分会的桥梁与纽带作用，形成合力，共同把仪表元件分会的技术创新、新产品开发与新技术推广活动搞得更加灵活多样，进而把仪表元件工作和技术服务提高到一个新水平。

3. 促进技术交流与技术合作

分会聚集了国内一流的科研机构、大专院校和企业的专家、学者，要充分发挥学会的媒介、咨询作用，构建技术交流与技术合作平台，营造良好学术氛围，共同促进仪表元件技术的发展。要加强跟踪国际先进水平，加强与国外学者的技术交流和与外企的技术合作，缩短与国际水平的差距。充分发挥专家、学者、企业家对分会的指导作用和对行业发展的促进作用。建议分会促进专家、学者、企业家加强联系，站在行业的角度，把会内、会外组织起来，使分会成为促进仪表元件行业发展和提高行业总

体水平的活动中心。

4. 抓好自身建设，加强组织建设

加强组织建设是分会发展的根本保证。仪表元件分会是科技工作者的组织，是人才荟萃的团体，加强分会的基础建设，首先要扩大会员队伍。随着我国市场经济体制的不断完善，科技工作者的工作环境也发生了深刻的变化，纯“事业”型的科研体制已不复存在，在国有、民营和“三资”企业中也汇集着相当一批优秀的科技人才。因此，会员发展工作应及时对传统的模式进行调整，聚焦新领域，研究新动向，采用新机制，及时满足科技工作者的学术与科技需求，把更多优秀的科技人才吸引、凝聚到分会中来。

分会在组织发展会员的同时，要充分考虑发挥分布于企业各个层面的仪表元件工作人员的作用。

搞好分会理事会的组织建设，不断加强分会自身发展，建立、健全规章制度，改进工作作风，提高工作水平，实现管理的规范化。不断拓展分会的活动领域，进而扩大服务范围和社会影响。

通过组建新一届的理事会，分会能够加强和促进组织建设，增强凝聚力，为今后工作的开展奠定基础；通过技术与信息的交流，活跃思想，开阔思路，能够为推动行业的发展与技术进步起到积极的作用。

在各项工作中，我们必须甘于奉献、主动服务、加强联系、搞好协调，充分发挥分会的纽带作用，使仪表元件分会真正成为仪器仪表行业的参谋和益友。同时，通过上述工作，实行“人本战略”，即以人为本，以会员为本。

仪表元件分会是纯学术性组织，涉及学科、领域较多，分会如何有效运作才能发挥最大效能，推动技术进步和发展？如何在市场经济大潮中提高仪表元器件的地位和作用？受大环境和市场经济影响，分会活动难以开展，需要全体理事的共同努力，开创分会工作的新局面。分会的国际性合作如何开展？要加强与独资和合资企业合作，举荐相关理事补充到分会来，把分会办成开放式的学会。

建议每年一次的理事会由理事长单位、副理事长轮流举办，理事单位也可申请举办，以促进仪表元件技术的交流与融合。提请本次理事会讨论，请各位理事提出建议与意见。

我相信这次会议一定能给仪表元件领域带来新的生机。在新一届理事会的带领下，仪表元件分会一定能够加强仪器仪表新工艺、新技术、新材料、新装备的交流和推广，促进“政、产、学、研、金”结合，提高仪器仪表行业的设计技术、制造技术和检测技术等方面的技术水平，为振兴我国的仪器仪表工业作出努力。

仪表元件分会第四届理事会在改革的道路上走过了不平凡的历程，虽然一路艰辛，但也获得了发展，创下了业绩。

仪表元件工作艰巨而又光荣，仪表元件事业任重而道远，让我们在新的理事会的组织下，努力工作、不断创新，为开创仪表元件工作的新局面，作出更大新的贡献！

关注世界科技创新态势[①]

一、前言

科学技术和工程是人类在认识和改造自然的伟大实践中获得的丰厚知识财富。它们分别对应于认识自然、改造自然的不同环节，因此，彼此之间有区别，更有联系。现代科学技术体系可划分为三个部分。

(1) 科学。科学又分为自然科学和社会科学两大类，它们分别是对应于自然界和人类社会的科学理论。特别是自然科学，包括数学、物理学、化学、天文学、地理学、生物学和力学共七个大类，这是认识自然现象、探索物质运动的客观规律所形成的基本理论、概念或原理。

(2) 技术。技术是以自然科学的理论为基础，针对工程中带有普遍性的问题统一处理而形成的科学分支。即是说，是运用自然科学理论，为提高效率、节约资源和开辟新生产领域而发展的方法和手段。过去工科学院所设专业，大部分是指技术专业，如焊接技术、铸造技术、光电子技术、集成电路工艺与设计技术、计算机技术、生物技术等。

(3) 工程。工程是综合应用自然科学、社会科学和技术的知识，使自然资源最佳地为人类服务而发展起来的一类专门综合技术，即在生产实践中产生的设计、工艺、流程、装备和质量控制等，它的任务是改造客观世界并取得实际的成果。因此，工程避不开客观事物的复杂性，必然要应用各个有关学科的成果。同时，工程又受经济建设和社会发展需求的拉动，直接为经济建设和社会发展作贡献。复杂性和综合性是工程的主要特点。随着当代科学技术和生产实践活动的发展，科学、技术与工程三者之间的联系越来越密切，科学向技术、工程的转化和渗透更加快速、广泛，工程、技术对科学进步的推动作用日益强烈。它们相辅相成、相互促进，共同构成一个不断发展的庞大知识体系，成为现代生产力最活跃的因素。

二、20世纪自然科学的伟大成就

现代自然科学是从15世纪下半叶开始，在摆脱了神学的统治之后，伴随资本主义工业大生产的兴起而发展起来的。意大利的达·芬奇（1452～1519）是代表人物，他用试验确定了金属丝、杆和梁的强度，论述了数学和力学的作用和地位，他有一句名言："数学是科学的皇后，力学是数学的天堂"，至今意义仍然深远。

19世纪和20世纪之交，不能不提到哥廷根学派。以德国哥廷根大学的数学家费利克斯·克莱因（Felix Klein）（1849～1925）和物理学家路德维希·普朗特（Ludwig

① 本文是在贵州工业强省院士专家论坛上的报告，贵阳，2011年8月29日；教育与科技管理研究，北京：科学出版社，2016，312-327.

Prandtl）（1875～1953）创建应用力学为标志。他们明确提出资源贫乏的德国要在实力上称雄世界，就必须将自然科学的研究成果用于生产，由于当时力学最成熟，就选择力学，叫应用力学，这就是新的科学前沿。

哥廷根学派取代了牛津、剑桥的地位，成为世界学术中心，这个学派的杰出传人德国物理学家爱因斯坦（1879～1955），被称为 20 世纪最伟大科学家，另一个杰出传人力学家冯·卡门（1881～1963）被称为 20 世纪第二大科学家。

中国的“三钱”，对中国 20 世纪科学发展起了重要作用。钱学森、钱伟长、钱三强均是哥廷根学派传人。钱学森是冯·卡门的博士生，钱伟长是加拿大应用数学家辛格（J. L. Synge）的博士生，又是冯·卡门的助手，辛格也是哥廷根学派传人。钱三强是法国物理学家约里奥·居里（Joliot-Curie）夫妇的博士生，约里奥·居里夫妇也是哥廷根学派的传人。

人类知识宝库中有 80％的科学发现、技术发明和工程建设是 20 世纪的科学家和工程师们创造的。20 世纪的科学无论在深度，还是在广度上，都远远超过了 19 世纪以前的几千年总和。物理学中的相对论、量子力学，生命科学中的 DNA 双螺旋结构及遗传密码的发现，不仅带动了一系列科学技术和生产领域的发展，而且进一步深刻地改变了我们对世界的认识，向科学真理更进一步逼近。

（1）相对论和量子力学是 20 世纪科学最伟大的发现。1895 年，德国物理学家威廉·伦琴（Wilhelm Röntgen）（1845～1923）在研究阴极射线时，意外地发现了 X 射线。这项发现如此重要，使伦琴在 1901 年成了第一位诺贝尔物理学奖获得者。此后第二年、第三年，相继发现了放射性和电子。

X 射线、放射性、电子三项发现，动摇了经典物理学的大厦，引发了 20 世纪最初 30 年的物理学革命。先是黑体辐射的实验测量促使德国物理学家普朗克（1858～1947）提出了量子力学，接着，光速的测量导致爱因斯坦提出相对论。

没过多少年，放射性和相对论促成了原子弹、氢弹爆炸，震撼世界，而电子和量子理论则促成了微电子、激光、计算机、超导、互联网等的出现，把人类带进了前所未见的信息时代。

（2）DNA 双螺旋和遗传密码的发现打开了理解生命奥秘的大门。1953 年，美国遗传学家和生物物理学家沃森（Watson）和英国生物物理学家克里克（Crick）（1916～2004）揭示了生物大分子 DNA 的双螺旋结构，DNA 分子是由 4 种碱基核苷酸分子所组成的两条长链缠绕而成。生物体遗传基因的密码由 4 种碱基的特定排序所编码，人体中独立的基因数量约为 2 万多个，每个基因含几百到上万个碱基对，共有 31 亿个核苷酸碱基对，分布于 23 对染色体的 DNA 中。DNA 犹如一条磁带，它用遗传信息谱写、记录、传播着一支美妙的生命之歌。科学家在分子水平上建立了生物世界多样性（100 多万种动物、30 多万种植物和很多种微生物）和生命物质一致性的辩证统一观，对生命现象的认识已深入到核心层次。

在揭示生命本质取得实质进展的同时，一门以上述成果为基础的综合应用性学科——生物技术，自 20 世纪 70 年代后获得迅速发展，已在医药、食品、化工、能源、农业、环境保护中广泛应用。

近 100 年来，人类不仅取得了上述三项划时代的科学成就，而且其他各门科学技术都取得了长足发展。可以说，20 世纪是人类历史上辉煌的科技世纪。

三、20 世纪工程技术科学的伟大成就

20 世纪是个奇迹的世纪，其工程技术科学的伟大成就主要显示在以下四个方面。

1. 电气化

19 世纪称为蒸汽时代。

20 世纪则称为电气时代。

工程技术的划时代成就之一就是电的广泛应用，是社会最重要的技术基础，对人们和货物的生产方式和生活方式产生了重大影响，电气化的程度为现代化的最显著标志。电力是最便于输送和分配、最清洁和最便于使用和控制的优质能源。

20 世纪之初，全世界的人们生活在没有电灯、没有电话、没有自来水、没有广播、没有电视、没有煤气的时代。

电力改变了城市的面貌，改变了人生活的方式，更加舒适。改变了生产方式，节省了体力。现在简直无法想象，没有电，我们怎么生活！

1879 年美国发明家托马斯·爱迪生（Thomas Edison）电灯的发明，促成了 1882 年第一座商用电厂的产生，但是用直流电，输送区域很小。经过美国发明家特斯拉（Tesla）和美国电气工程师查尔斯·斯泰因梅茨（Charles Steinmetz）的努力，成功地实现了交流电的商品化，能远距离传输高压电。1903 年，美国发明家查尔斯·柯蒂斯（Charles Curtis）研制出 5000 千瓦的汽轮发电机。1925 年，在美国波士顿建成了世界上第一座高压蒸汽发电厂。1927 年，132 千伏传输线路在美国铺设。

接着电网系统形成，一个国家，甚至国际互联电网，远距离输电成功。发电形式也在多样化，有火力发电、水电、核电。

美国的远距离输电，在 1969 年达到 765 千伏。我国的三峡电站是世界装机容量最大的水电站，装机容量将达到 2250 万千瓦。

2. 汽车和飞机

19 世纪末，世界上一般人一生的旅行距离就在自己出生附近的区域里，主要是采用步行的方式。

到 20 世纪末，汽车已经成为世界上人们出行和运货的主要交通工具，而且成为反映一个国家经济状况好坏的主要标志，很少有像汽车那样的机器为人们广泛使用。现在，全世界汽车大约有 5 亿辆，其中 1/3 在美国。美国汽车的销量已占美国商品批发销售总额的 1/5，零售总额的 1/4 还多。美国汽车工业已跃居制造业之首，促进了钢铁、橡胶、石油化工、石油生产、涂料、平板玻璃制造等工业的发展。

1769 年，法国军事工程师尼古拉·约瑟夫·屈尼奥（Nicolas-Joseph Cugnot）制造了世界上第一部汽车，实际上是一辆蒸汽动力三轮车。1885 年，德国机械工程师卡尔·本茨（Karl Benz）制造了第一辆汽油动力汽车。1900 年，一辆汽车仅 100 个零部件，今天却有 14 000 个左右。

现在公路、高速公路上到处是汽车，把人们和货物带到很远的地方。

飞机将人们快速带到世界上任何地方。飞机制造业也促进了科学、技术、经济的繁荣和发展。

1903 年，美国飞机发明家莱特进行了第一架有动力装置飞机的成功飞行，连续飞行了 12 秒，飞行距离大约 36.6 米。

第一次世界大战（1914～1918 年）和第二次世界大战（1939～1945 年）加速了航空的发展。

航空技术的伟大之处就在于它开阔了人们的视野，促进了科学技术经济的繁荣，以美国为例，它的收入占美国 GDP 的 6%以上。

3. 计算机和互联网工程

计算机和互联网的发明是 20 世纪技术科学的最伟大成就，是 20 世纪的重要象征，它带动了一次世界性的新技术革命，使信息化时代来临。

由于电气化，汽车、飞机，技术创新改进了人类的生活。人们摆脱了繁重劳动，轻松自由地远行。随后，计算机出现了，它令世界震惊，使人们从令人生厌的计算或流水线上的繁重劳动中解放出来。在 20 世纪之末，计算机已成为各行业不可缺少的组成部分，并通过互联网开始打通向新世界的大门，深刻地改变了人们的生活和工作方式，人们能够不分地理位置和政治信仰，共享信息，相互学习。

1946 年，第一台数字电子计算机（Electronic Numerical Integrator and Computer，ENIAC）在美国投入运行，每秒可进行 5000 次加减计算。

1981 年，美国 IBM 公司推出个人计算机（personal computer，PC）。

在 20 世纪 80 年代，美国的超级计算机每秒可进行 100 万亿次计算。如果用过去的计算器计算，则要花 1000 万年才行。

1962 年，美国麻省理工学院（Massachusetts Institute of Technology，MIT）的一名研究生伦纳德·克兰罗克（Leonard Kleinrock）发明互联网技术。1971 年，雷·汤姆林森（Ray Tomlinson）推出了电子邮件，20 世纪 90 年代在世界迅速发展。1989 年，万维网（World Wide Web，WWW）诞生，在互联网上首次进行音频与视频多播图像。1999 年，互联网有 1.5 亿用户，有 8 亿多个网页可以访问。

互联网技术不由任何公司、企业或国家占有或控制。它通过计算机、光纤、卫星和电话线把全世界的人民即时连接起来，正在改变文化模式、商业活动、工业方式以及研究和教育工作。

网上读书、网上查阅资料、网上购物、网上聊天交朋友……正在深刻地改变世界。

4. 宇航工程

宇航工程大大开拓了世界知识的宝库，是人类历史上不朽的成就。

1957 年，苏联发射的世界上第一颗人造地球卫星，穿越大气层，震惊了世界。

1961 年，苏联尤里·加加林（Yuri Gagarin）上天绕地球飞行一周，是世界上第一位宇航员。

1969 年 7 月 20 日，美国阿波罗 11 号飞船登月，人类第一次到达了月球表面着陆行走。阿姆斯特朗成为在月球上行走的第一人。

地球轨道卫星的出现，引起了全球通信的革命，为全球提供即时通话和视频服务，为天气预报提供服务，为飞机、汽车和舰船提供导航服务。

四、目前世界科技的前沿形势

目前，世界科技领域充满生机，成果层出不穷。

1. 宇宙与物质结构（宏微观世界）

目前，在这一领域一直保持着活跃状态。人类的科学研究始终要面对的最大问题就是宇宙起源和进化与物质结构。

宇宙是如何起源和进化的，若干年来逐渐形成了许多不同的宇宙学理论，现有的基本理论认为，大爆炸是宇宙的开端，但对于宇宙的最终结局并无定论。2002 年，美、英科学家提出了循环宇宙模型理论，认为宇宙将永远不会结束，将永远处于从生长到消亡的循环过程中。大爆炸不是宇宙的起点，也不是终点，而只是宇宙不同阶段的“过渡”。2008 年，欧洲核子研究中心大型强子对撞机的启用，无论从所涉及的人力、物力、财力还是从规模上说，都堪称人类历史上最大的物理实验计划，该实验将把高度活跃的质子以超快速度撞击到一起，上演微缩版的“宇宙大爆炸”，目的是揭示宇宙起源。

接着，2003 年，特别大的有里程碑意义的成就是美国科学家获得了宇宙“婴儿期”的照片，这是目前最清晰的宇宙微波背景图，由此精确地测出了宇宙的年龄为 137 亿年，误差不超过 1%。有力地支持了大爆炸产生宇宙的学说。宇宙的成分 4%是由原子构成的普通物质，23%是目前尚不了解的暗物质，73%是暗能量。2010 年 7 月，欧洲空间局公布了由普朗克太空望远镜收集的数据所绘制的首张宇宙全景图。

2005 年，德、英科学家测得月球的准确年龄为 45.27 亿年。

2006 年，俄罗斯科学家测得地球的准确年龄为 45.67 亿年。

而且在 2006 年，国际天文学联合会通过新定义，将位居太阳系九大行星末席 70 多年的冥王星驱逐出了行星家族，降级为“矮行星”。

在黑洞问题研究方面，英国著名科学家霍金推翻自己以往的观点，在 2004 年说，黑洞似乎并非吞噬一切。2007 年，观察到银河系中有 1000 多个超大质量黑洞，从黑洞中逃离的热气体可能是宇宙生命种子的来源，这突破了黑洞是“终极毁灭”的形象。2008 年，科学家根据美国航空航天局观察站的数据发现，黑洞释放能量的过程并非爆发性的，而是温和而有规律的。

1992 年，苏梅克-利维 9 号彗星在第一次接近木星时，被木星强大引力撕碎成 21 块碎片，并排成一列飞行星体，像一列火车在太阳系内奔驰。1994 年 7 月，这列星体再次接近木星，并撞击木星，释放出相当于 20 亿颗原子弹的能量。这次撞击在木星表面上留下了相当于地球数倍体积的庞大坑洞。木星体积较地球大 1500 多倍。若是撞上地球，那么人类的家园将毁灭。这对人类是一次警告。从 2006 年开始，由美国发起并主办了国际小行星搜寻活动，防患于未然。

在寻找有生命的行星的热潮中，澳大利亚科学家在 2004 年发现，银河系内有 1/10 的星体具备适合生命存在的条件，它们较太阳系要早 10 亿年诞生。英国科学家认为，太阳系以外的星系中，有 1/20 拥有类似地球的行星，可能有生命存在。2008 年取得了很多激动人心的进展，如美国阿雷西博天文台利用射电望远镜首次在距离地球 2.5 亿光年的 Arp 220 星系中测到了构成生命的两种最基本物质：甲亚胺和氢氰化物。

2002 年，英国科学家认为太阳表层可能存在大量暗物质。通过对数千个遥远类星体进行观察，发现了宇宙中存在暗能量的新证据。2008 年，美国亚利桑那大学天文学家已估算出了隐藏在太阳系中的暗物质的总重量。美国麻省理工学院科学家研制出一种新型探测器，可鉴别出暗物质粒子。

2002 年，美、德、加研究小组首次成功地对反物质原子的内部结构和物理特性进行了研究，并初步断定反氢原子与氢原子在内部结构上似乎没有什么差别，而且成功地制造出约 5 万个低能量状态的反氢原子，这对比较物质与反物质的差别，解答宇宙构成等问题有重要意义。接着在 2006 年，日本科学家生成稳定的接近绝对零度的反氢原子。俄罗斯科学家直接观察到“活”的反氢原子，为反物质的研究开辟了新道路。

周期表是化学理论的骨架，有 103 个已知的元素。近几年，俄罗斯科学家合成了门捷列夫元素周期表上的第 113 号和第 115 号元素；美、俄合作，合成了第 118 号超重元素；俄罗斯和德国又合成了第 114 号元素。

2007 年，法国科学家发明了一种光子盒，2.7 厘米见方，可以在 1/7 秒内捕捉并监控 1 个光子，并监控光子从产生到消失的全过程。

在粒子物理研究方面，2006 年，俄、比、德科学家合作，观察到中子衰变的新方式，一个自由中子衰变成质子、电子、反中微子和光子。而且俄罗斯科学家在 2005 年对中子的半衰期作了精确测定，为 878.5 秒，较公认的少 7.2 秒。这将对宇宙构成的了解产生重大影响。2008 年，美国 IBM 公司与德国科学家借助原子力显微镜，发现让单个钴原子在光滑的铂平面运动需要 210 皮牛的力，这是人类首次测定推动单个原子所需的力。

2010 年，美国科学家使用高分辨率蝇眼阵列望远镜，首次搞清楚高能的宇宙射线由质子组成。宇宙射线起源于银河系外部，是宇宙中一种具有相当大能量的带电粒子流，对其研究已经成为探索宇宙起源、发展历史、天体演化、空间环境等科学之谜的最重要途径。

2. 生命科学与生命技术

基因的研究是最热门领域。

2001 年 2 月 12 日，中、美、英、日、德、法六国科学家联合小组宣布人类基因组图谱。次年，又绘制完成了人类基因组序列。人类基因数量为 2 万～2.5 万个，较前面预计的少得多。人类基因 99.5%是相同的，人类基因密码的差异为 0.5%。人类基因组有 31.746 亿个碱基对。2007 年，世界首份个人 DNA 图谱面世，DNA 分子结构的双螺旋模型发现者之一，美国生物学家詹姆斯·沃森成为自己研究的受益者——他成为世界上第一份完全破译的“个人版”基因组图谱的拥有者。接着，出现了数种更为简便、快捷、经济的基因测序新技术，使这一领域的工作快速发展。我国科学家完成了第一个中国人基因组序列图谱——“炎黄一号”的绘制和分析。美、英等国科学家完成了最大规模人类遗传多样性调查，其样本涵盖了全球 50 多个不同的地理学族群，大多数人的遗传祖先追溯起来都不限于一个大陆。

科学家又开始进行染色体的破译工作。例如，14 号染色体有约 1000 个基因；7 号染色体包括 1.53 亿个碱基对和 1150 个蛋白质编码基因。几年来，已全部完成染色体破译工作。

接着，对动物、植物、微生物的基因测序工作喜讯频传。例如，大鼠 2.5 万个基因，与人类 90%相同。黑猩猩与人类基因相似程度达到 96%以上。美国科学家成功破译史前庞然大物猛犸象 80%的基因组，使科学家在复活猛犸象的道路上又向前迈进一步。2009 年，加拿大科学家使用已灭绝的猛犸象的一个冷冻样本的 DNA，再造出了他们的血液，与原来的猛犸象的血液完全一样。

基因的研究成果对基因治疗、基因药物研究带来了希望。

干细胞的研究是又一热点。

干细胞是人体内能够分化成多种细胞的基本细胞，即可在人体内转化为任何细胞。由于这一领域蕴藏着广泛的医疗前景和巨大的商机，故国际竞争日趋激烈。干细胞的各种获取渠道以及它们的分化功能研究是当前的热点。过去只能从脐带中获取，现已发现可以从胎盘、骨髓、肌肉、大脑、皮肤、脂肪以及多种胚胎组织等渠道获取干细胞。美国科学家从胎盘中提取大量干细胞的技术，比脐带中获取多 10 倍。以色列科学家诱导人类干细胞转化为胰岛细胞，对 1 型糖尿病治疗迈出关键一步。中国科学家从干细胞分化克隆出每分钟跳动 30～120 次的自律跳动心肌细胞团，向克隆完整人体心脏迈出第一步。

2008 年，美国科学家实现了发育生物学家长久以来的梦想——直接将一种体细胞转变成另一种体细胞，而无须借助胚胎干细胞。

2001 年，美国科学家克隆出含有 6 个细胞的人体胚胎。这又引起克隆人的伦理道德争论。联合国教科文组织公报，生殖性克隆人是违背人类尊严的。但世界医学界认为，克隆干细胞具有很高的医疗价值，可治疗包括糖尿病、帕金森氏症、艾滋病、癌症等绝症。

接着在 2003 年，美国科学家用冷冻 20 多年的爪哇野牛的皮肤细胞成功克隆出两头爪哇野牛，为克隆保护濒危动物展示了新的希望。2006 年，美国科学家在实验室中培养出膀胱，并顺利转移到 7 名患者体内。由于膀胱是自身体内细胞培养产生，所以不会发生免疫排斥，这是国际上首例从实验室培育的完整器官移植。而且，在美国，在 2007 年人造生命诞生，用化学合成了人工染色体，并成功地移植到了另一个没有染色体的细胞中，创造出了有史以来第一个人造生命。2008 年，美国科学家给处理后的动物尸体心脏注入活细胞，成功地使这些心脏恢复了跳动。英国、意大利和西班牙联合小组利用成人干细胞培养的气管成功地为西班牙女患者实施了器官移植。2010 年，德美合作，利用干细胞培育出与人类耳蜗内毛细胞非常相似的细胞，从而向利用再生医疗方式治疗失聪迈出了重要一步。

2010 年 5 月，美国科学家文特尔通过化学合成“丝状支原体丝状亚种”的 DNA，并植入去除了遗传物质的山羊支原体内，创造出世界上首个“人造单细胞生物”。

同时，在 2007 年，还有首只人兽混种羊产生，15%的人类细胞，85%的绵羊细胞，这一进展向动物器官用作人体移植的目标前进了一步。

随着人类基因组织测序的完成，生命科学研究重点转向细胞中的蛋白质。在2007年，第一张人类器官蛋白质组图谱完成。我国科学家实施了“人类蛋白质组计划”，测定出6788个中国成人肝脏蛋白质，有望攻克肝癌诊治难题。同时，我国科学家还发现构成DNA的第六元素，即硫，原来5种是碳、氢、氧、氮、磷。

在脑科学这一国际热门领域中，一直认为脑细胞是不能再生的，但是在美国，2006年科学家首次发现人类大脑细胞拥有与干细胞一样的自我更新能力。

同时癌症的防治一直是研究重点。2006年，英国科学家首次研究证实一种恶性肿瘤能在狗之间传播，而且澳大利亚塔斯马尼亚岛上有袋目动物中可能也存在一种传染性癌症，这是对人类的提醒！癌症可能有传染性。

2006年，在美国，宫颈癌疫苗“加德西”上市，这是人类研制成功的第一种癌症疫苗。

关于地球的生命起源和发展研究也是一个前沿领域。

地球年龄45亿年了，过了10亿年后，在35亿年前，生命在地球上诞生。

35亿年来，地球生物已经经历了5次大灭绝。第5次灭绝是6500万年前恐龙突然灭绝，科学家对灭绝原因有不同说法，尚未有定论。灭绝前，恐龙是最凶猛的动物，统治地球超过1亿年，剑龙、梁龙、暴龙是陆地上的霸主，蛇颈龙是海中王，拥有巨大羽翼的翼龙则成为天空的统治者。可是，它们突然灭绝了。有人说，是流星撞地球，伽马射线爆发，饿死、冻死，气候改变，海平面上升，淹死恐龙等。最糟糕的一次灭绝发生在2.5亿年前的二叠纪末期，被称为大灭绝，地球上95%的物种都灭亡了。一些科学家认为，我们正濒于第6次大灭绝的边缘。几万年以来，人类开始支配地球，由于工业发展和人口增多，到21世纪末人口将达到100亿人，发生了温室效应，地球气候变暖、目前物种灭绝数量加快。根据科学家研究，生物物种灭绝95%原因为气候变暖、污染加大所造成，估计到21世纪末，人口达到100亿人之时，地球的半数物种将灭绝，因此要走低碳建设之路，要拯救地球！拯救人类！防止第6次大灭绝到来！

3. 信息与通信技术

各种新型计算机的研制方兴未艾，计算机的超微型化和超强计算能力一直是各国竞相追逐和比拼的目标。2008年，美国研制出细菌计算机和生物分子计算机。日本也研制了由17个分子组成的分子计算机。新加坡科学家基于热和“声子”来设计新一代计算机，称作声子计算机。

美国在超级计算机领域继续领先，日本、欧洲奋力追赶。美国IBM在超级计算500强排行榜上继续保持绝对的统治地位。2006年，IBM和美国能源部研制的蓝色基因/L以每秒280.6万亿次的运算速度排列榜首，遥遥领先。2008年，IBM公司造出更快的每秒1105万亿次的超级计算机。2010年，美国又造出世界最快的每秒1750万亿次超级计算机。

我国也在奋力追赶，2008年6月，研制出“曙光5000A”型每秒160万亿次超级计算机，排在世界第七位。2010年6月，我国又研究出曙光“星云”型每秒1270万亿次超级计算机，为世界第二位。2010年11月，我国国防科技大学终于超过美国，研制出“天河一号”每秒2570万亿次超级计算机，为世界第一位。

在存储技术方面，有许多进展和突破。美国2007年罗切斯特大学在光学存储研究领域取得突破，利用新开发的单光子技术，能使整张图像进行编码和储存，并使信息完整再现。美国宾夕法尼亚大学开发出一种新型纳米器件，能储存10万年的海量电脑数据，探索速度较现有的存储设备快1000倍，且更省电，存储空间更小。

在晶体管、光刻和芯片技术领域、美国继续领先。

在通信技术方面，日本和美国十分出色，2007年，联合开发出（10千兆赫兹时钟周波数）量子暗号系统，利用单一光子水平光的量子暗号密钥成功在200公里光纤距离传输信号，达到目前世界速度最快、传输距离最长纪录。

2008年，世界最大规模的“网格计算”网络在欧洲正式启动，用于大型强子对撞机数据分析和处理。“网格”是指通过互联网整合成一台巨大的超级计算机。

4. 纳米科学技术

纳米技术在很大的程度上影响人类社会的生产，改变人类的生活方式。实际上，纳米技术的奇妙和潜力目前已经有所显现，人们在市场上几乎每天都可以看到带有纳米字样或具有纳米特性的新产品、新系统、新材料，它们影响着医疗保健、计算机、日常消费品、能源、国防、食品等各个方面。

纳米技术就是在纳米尺度上（1纳米到100纳米）研究物质的特性和相互作用，并利用它来制造具有特定技术的产品。也就是说，要对原子和分子进行操纵，利用在如此小的尺度上物质可能显现出的极不寻常的特性来开发人们所需要的新材料、新产品和新技术，它正在成为全球关注的重点。由于原子和分子行踪不定，又那么小，要将它们捕捉住，作为产品，其难度可想而知。

近20年发展起来的纳米技术是逆向思维创造性地提出来的，来自1955年的理论物理学家理查德·费曼。过去，人类从磨尖石头，到光刻芯片，即从整体切掉许多材料到有用形状，现在是用单一分子和原子来制造零件。

纳米科学技术包括四个方面：①纳米材料；②纳米动力学，制造微型电动机械系统，作传感器、执行器、光纤通信、医疗诊断仪器；③纳米生物学和纳米药物学，使药物粒子全部溶于水，最大限度发挥药效；④纳米电子学，使原来电子器件发展到更小、更快、更冷。

2001年，美国科学家首次研制了一种与骨头非常类似的新材料——骨状纳米纤维，从而打开了合成骨替代的大门。

2002年，美国科学家制造出单个分子大小的原子级纳米晶体管，其电流由1个或2个金属原子输送，利用该技术，可使计算机电路再缩小6万倍。

2004年，中国科学家成功地开发了先进纳米燃料技术。它能把汽油等燃料完全变成纳米燃料，实现充分燃烧。该技术用于汽车，可净化尾气50%～90%，节约燃料20%～30%，提高动力10%～30%，降低发动机噪声，延长发动机寿命，对缓解环境污染和能源危机有重要意义。同年，美国科学家利用DNA碎片制造了两条腿仅10纳米长的微型机器人，这一重大突破可望使纳米制造业的梦想变成现实。

2005年，美国科学家研制出世界上第一个纳米阀门和世界上最小、能像真车一样滚动的纳米汽车，可用来进行原子或分子的运输。

2006年，意大利科学家制成了世界上最快的纳米电动机，目前用来将药物送进细胞的分子引擎，未来可用来建造“化学计算机”。同年，美国科学家利用氧化锌纳米线的弯曲和伸直时的压电效应将机械能转化为电能，该技术可用于体内自供能量的纳米传感器。美国科学家用碳纳米管制成的纳米刀，可用来更精确地切割和研究细胞。美国科学家将分子磁共振成像技术的灵敏度提高了1万倍，有望成为医学诊断的有力工具。

2007年，德国和法国科学家借助将氮化硼颗粒的大小从微米级降低到纳米级，使材料的硬度提高到接近钻石，而且其断裂韧度和抗磨能力都大大高于钻石。同年，英国、俄罗斯和荷兰的科学家，用单原子厚的石墨薄片制成能探索单个有毒气体分子的传感器，其灵敏度比任何迄今演示过的气体探测器高百万倍，可探测隐藏的爆炸物或致命的一氧化碳。美国科学家开发了由多层银和氧化铝组成的超级透镜，它使科学家不仅能看见细胞核，而且能观察到活细胞中个别分子的动作和行为。

2008年，美国科学家制成氧化锰纳米线，其直径只有20纳米，它对石油有极大的亲和力，能吸收相当于其自身重量20倍的石油。这种材料可用来清除海洋中的石油泄漏，也可用于水的净化。在纳米科技的研究工具方面，德国科学家制造了一台可用来测量纳米世界中极小作用力（如纳牛顿或皮牛顿量级）的原型实验装置。

5. 能源与环境

世界经济的迅速发展，加之人口的迅速增加，使能源消耗和环境污染加剧，因此可再生能源开发利用与环境保护就成为21世纪各国政府及科学家和经济界共同关注的焦点问题。

我们居住的地球已诞生了45.67亿年。多少年以来，仅仅是地球科学家、天文学家，将它作为行星关注。20世纪后期，日益成为热门话题。这是由于人类活动对自然系统的影响迅速增大，经济增长，人口膨胀，需求扩大，使人类与其赖以生存的自然环境之间的矛盾日益尖锐。1804年，全球人口为10亿人。1927年，全球人口增加到20亿人。1960年，全球人口为30亿人。1975年，全球人口为40亿人。1987年7月11日，全球人口达到50亿人。1999年，全球人口为60亿人。2007年，全球人口达到67亿人。1910年，我国人口已达4亿人。2007年，我国人口已增至约13.2亿人，约占世界总人口的1/5。

在地球上，我们赖以生存的生物圈有多大？薄薄几米厚的土壤，几公里厚的大气层，几公里深的海洋。目前，无论土壤、大气，还是水资源、能源资源（石油、煤），都在恶化中。

目前，石油、天然气仅有50年用量，煤炭仅有200年用量。

地球表面只有11%土地适于耕种，其中大部分已使用，而且土壤流失严重。世界沙漠化的扩展速度，每年大约在5万～7万平方千米。每年全球土壤损失量高达254亿吨，其中中国43亿吨，约占世界1/6。研究表明，在自然力作用下，每生成1立方厘米原土壤，大约需100～400年的漫长岁月。因此，每年全球的土壤损失量已超过新土壤的形成量。尤其我国10多年城市化和经济发展，引起耕地的减少，我们更应爱惜耕地。

清洁的大气几乎成为城市居民的奢侈品。目前世界约 1/4 人口居住在空气烟尘超标的地区。过量的城市烟尘及其夹带的氮、硫氧化物，造成呼吸疾患、癌症和男女不育。冰箱、空调机中的氯氟烃类制冷剂的大量使用和排放，造成大气臭氧层减薄，特别是出现了南极上空的臭氧空洞，地球的保护罩没有了，紫外线辐射加大，更威胁了人类和生物的生命！工业发展、燃爆增加、森林和植被减少，使大气中二氧化碳浓度增大，造成温室效应，使全球气候变暖。100 年来，已升高 1℃，且升温速度在加剧。全球的降雨量分布发生变化，从而引发水、旱灾情的概率加大。地球水资源总量 15 亿立方千米，其中 97%是海洋，淡水仅有 3%。由于人类淡水用量激增和水资源污染，对工农业发展、生活质量改善造成严重制约。

我国人均淡水占有量仅为世界平均值的 1/4。近 100 年来我国地表平均气温明显增加，升温幅度约为 0.5～0.8℃，比同期全球平均值略高。

滥用资源、破坏资源、扩大生态赤字，无异于自我毁灭。“生态文明”“绿色科技”“绿色制造”“清洁能源”“低碳社会”正为社会接受。

在能源方面、环境方面，世界科技发展很快。

在开发可再生能源方面，当今最大的热点是太阳能。太阳能电池方面，2008 年年初，美国硅谷建成了世界上最大的太阳能电池制造厂，一年内可生产出 43 万千瓦的太阳能电池。

太阳能发电厂方面，2009 年，德国在莱比锡兴建欧洲最大的太阳能发电厂，发电功率 4 万千瓦，总投资 1.3 亿欧元，可满足 1 万户家庭用电要求。以色列和美国合作在美国加利福尼亚沙漠建造世界最大的太阳能发电厂，最大发电能力 55.3 万千瓦，相当于 1/3 个大亚湾核电站发电量，可为 40 万户家庭提供电力。2007 年 9 月，在内蒙古鄂尔多斯市郊，我国建成了 205 千瓦太阳能电站。

风能是目前世界上增长速度最快的能源，这是清洁的新能源。

目前，英国走在世界海上风能的前沿，2008 年成为世界海上风力发电站装机容量最多的国家，总装机容量达到 870 万千瓦。我国 2007 年年底的风电建设已突破了 600 万千瓦，居世界第五位。

海浪发电方面，以色列于 2006 年发明了一种海浪发电装置，效率可达 75%。2008 年，爱尔兰研发和安装了一台海浪能源转换装置，发电能力超过 1000 千瓦，足以满足 1000 户家庭需要。2008 年，英国科研人员研发成功世界首台潮汐能涡轮发电机，将它放入海湾时涨时退的潮汐中，就能产生可供 1140 户居民使用的电力。

核能发电方面，正在进行可控热核聚变发电研究，欧盟、美、日、中、俄、韩、印七方代表从 2006 年开始合作。

全球生物燃料生产呈快速发展趋势，受到各国高度重视。生物燃料技术呈多样化发展趋势。

英国将在威尔士建造世界上最大的生物发电厂，即 35 万千瓦木屑燃料发电厂，造价 4 亿英镑，寿命为 25 年，预计近期建成。2007 年，世界首座第二代生物柴油加工厂在芬兰投产，使用原料为菜籽油、棕榈油、大豆油、动物脂肪等，成本低，其二氧化碳排放量仅为传统柴油的 16%～40%，尾气排放量也降低了 30%左右。2007 年，美国开发出家禽废弃物转化生物柴油装置，生物柴油的产量约为废物总量的30%～50%。美国还开发出用玉米和甘蔗制造丙烷的工艺。以色列利用海藻高效净化环境并制造生物燃料。2008年，巴西科学家发现象草，植株高4米，易于生长，无须施肥，产量高，种一年可以收获20年。干草经简单加工就能制成燃料，1公顷象草产生的能量能替代36桶石油，是植物中首屈一指的替代生物燃料。

燃料电池在2007年也有喜人表现，英国研制出一种锂化合物，可使机载燃料电池用在汽车上，连续行驶483千米。

在节能减排领域，也有一系列新进展。

2007年，加拿大使用节碳器技术，产生的富氢天然气可用于以天然气为燃料的内燃机、燃气炉具和燃气轮机，其燃烧排放的二氧化氮可降低50%～60%，二氧化碳可降低7%。

2007年，美国通用电器公司建造了一台目前最洁净的火车，能回收刹车时的能量，用于爬坡和加速，从而把燃料消耗和温室气体排放都减少50%。

2010年LED成为照明行业的新宠儿。此节能灯的寿命是传统白炽灯的50到100倍，耗电量仅为白炽灯的1/8，荧光灯的1/2。

垃圾和废物处理与再利用工艺不断有创造性成就。如美国生产的Hawk10循环机能用微波将有机废物的分子链撕裂，并合成为油料或天然气，制造出的能量是消耗掉能量的18倍，生产过程完全是零排放，而且处理之后的废物体积减小了65%以上，每小时可把10吨汽车垃圾（轮胎、杯子、塑料袋等）转变成能提供428.5万千卡热量的天然气。

6. 航天技术

航天技术历来是各国激烈竞争的领域，21世纪以来的10年，它对人类社会进步起到更为巨大的推动作用。

探月是一个亮点。1969年7月20日，阿波罗飞船载人登上月球。2004年11月，欧洲“智慧1号”进入绕月轨道。2006年9月，欧洲航天局“智能1号”以打水漂形式撞击月球，分析月球表面物质和月球历史。2007年9月，日本“月亮女神”绕月飞行。2007年10月24日，我国嫦娥一号发射成功，绕月飞行，观测了月球的每一寸土地，于2008年11月12日公布了世界上最完整的一幅影像图。2008年7月，美国宣布，对阿波罗带回的岩石样本分析表明，月球早期有水存在。2008年10月，印度首颗探月卫星升空。

载人航天是又一个亮点。2003年10月15日、2005年10月12日和2008年9月25日，神舟五号、神舟六号和神舟七号先后载人上了太空，使我国成为俄美之后第三个成功实现载人航天和掌握空间出舱活动技术的国家。

彗星考察也十分热门。2004年1月，美国发射“星尘号”，擦过“维尔特二号”彗星，带着上百个彗星尘埃粒子的返回舱，于2006年1月返回地球，这是人类首次把月球以外的星球样本送回地球。欧洲航天局也在2004年3月2日，发射了“罗西塔”彗星探测器，它将经过10年长途跋涉，进入“楚留莫夫—格拉西门克”彗星的轨道，并着落，这在人类航天史上也是前所未有的。2005年，美国又做了一个惊人壮举，发射

“深度撞击号”探测器，击中坦普尔 1 号彗星，了解到彗核很小且是分层的，表面多种地貌，坑洼和平地均有。表面覆盖着 10 多米深的细粉状物质。彗星在靠近太阳时会喷发。彗星内部存在大量含碳、氮的有机分子。

对火星、土星、金星、水星也发射探测器，取得不凡成果。特别是对火星，2004 年，美国发射“勇气号”与“机遇号”两辆火星探测器登陆考察，发现有冰以及火山爆发证据。欧洲航天局也发射了“火星快车”。2008 年，美国的凤凰火星着陆器已发现火星上有水。这是惊人的发现！美国还准备在 2030 年让人登陆火星。

人类还在寻找地球外生命，除关注火星外，也关注太阳系外的星球。2008 年年底，一艘美国的飞船第一次走出太阳系，向太阳系外飞去。

2001 年 3 月 23 日，是世界航空航天史上一个令人难忘的日子。地球上空 350 千米处运行了 15 年之久的俄罗斯“和平号”空间站，在完成其历史使命后，它的全部碎片安全地坠入预定的南太平洋海域，它创造了一系列辉煌业绩。它是 20 世纪质量最重、寿命最长、载人最多、技术最先进的航天器。它绕地球飞行近 8 万圈，行程 35.2 亿千米，先后 109 次与其他飞船对接，共有 11 个国家 104 名宇航员到访过，进行了 3 万项研究、实验。

目前，从 1998 年开始建设的新国际空间站，重达 300 吨，内部空间超过 700 立方米，已成为最大的宇宙飞船。仅在 2007 年，就有三架航天飞机，3 艘货运飞船，2 艘载人飞船到来过。美、俄宇航员共进行 23 次太空行走，创下单年太空行走次数的新纪录。10 年来，已绕地球飞行 573 000 圈，飞行里程 21 亿千米，共有 15 个国家的 167 人造访过。2009 年，乘员从 3 人增加为 6 人。空间站总投资达 1000 亿美元，2010 年已全部建成。

2008 年，美国首次成功预测小行星撞击地球事件，误差只有 35 秒。

迄今为止，全世界已有 130 多个国家和地区参与了航天活动，近 30 个国家和地区形成了自己的航天工业。全球已进行了 124 次探月活动，18 人登上月球。我国共发射了 90 多颗各类人造地球卫星，4 艘无人飞船，3 艘载人飞船，1 个月地球探测器。

推进振动工程学科发展[①]

在丹桂飘香的金秋十月，中国振动工程学会第七次全国会员代表大会暨第十届全国振动理论及应用学术会议在南京隆重召开了！这是我国振动科技界的一件大事和喜事。现在，请允许我代表中国振动工程学会，对莅临会议指导的上级领导、南京航空航天大学的领导和各位来宾，对来自全国各地的代表，表示热烈欢迎和衷心感谢！

中国振动工程学会积极发挥学术社团的作用，为推进我国振动工程学科发展、科技创新、学术交流、科普教育等方面作出了积极的贡献。学会已经建立了 14 个专业委员会，18 个省级地方学会。学会成功举办了第 3 届至第 9 届全国振动理论及应用学术会议和各项专业学术交流活动，举办了多次振动工程国际学术会议。学会主办的《振动工程学报》和《振动与冲击》期刊均被确定为国家中文核心期刊，已成为国内颇具影响力的知名期刊。学会多次推荐两院院士候选人，数次推选中国科协全国代表大会代表和委员候选人，推荐全国优秀科技工作者，评选学会青年科技奖，推荐中国青年科技奖，很好地为会员服务。多年来，学会致力于促进科学技术繁荣和发展，很好地为经济社会发展服务；致力于促进科学技术普及和推广，很好地为提高全民科学素质服务；致力于促进科技人才成长和提高，很好地为科技工作者服务。学会着眼于建设科技工作者之家，当好科技工作者之友，加强自身建设。在这里，我们要向为此作出积极贡献的第六届理事会的各位领导和全体理事表示诚挚的谢意！

根据中国科协《关于同意中国振动工程学会召开第七次全国会员代表大会的批复》（科协函学字〔2011〕111 号）和学会章程的规定，中国振动工程学会召开第七次全国会员代表大会，审议学会第六届理事会工作，进行理事会的换届选举。同时，我们举行第十届全国振动理论及应用学术会议。来自全国各地从事振动工程研究的专家学者和工程技术人员欢聚在一起，相互交流切磋，共同探讨各前沿领域的最新科研成果和发展动态。出席会议的代表来自我国 20 多个省（自治区、直辖市）的 80 多个单位，基本覆盖了我国从事振动工程研究、教学、开发、生产和使用的主要单位，具有较为广泛的代表性。其中，会员代表绝大多数是具有副高级技术职务以上的科技人员，具有一定的学术造诣，并热心学会工作，为开好这次全国会员代表大会，为我学会今后的兴旺发达奠定了坚实的组织基础。现在，我们全国会员代表济济一堂，共商学会发展大计。这次全国会员代表大会的主要任务如下。

（1）审议学会第六届理事会工作报告、财务报告。

（2）为 2010 年度学会青年科技奖获得者颁奖。

（3）选举产生学会第七届理事会。

（4）同期进行第十届全国振动理论及应用学术会议和振动工程技术仪器设备展览。

---

① 本文是中国振动工程学会第七次全国会员代表大会暨第十届全国振动理论及应用学术会议开幕词，南京，2011 年 10 月 27 日；教育与科技管理研究，北京：科学出版社，2016，327-328.

这次会议的任务十分繁重，而时间又极为短促。希望各位代表团结一致，集中精力，切实完成会议的各项任务，把这次会议开好。

代表们、同志们！我国科技事业发展的奋斗目标是，到 2020 年时使我国进入创新型国家行列，到新中国成立 100 年时使我国成为世界科技强国。广大科技工作者责任重大、使命光荣。让我们更加紧密地团结在以胡锦涛同志为总书记的党中央周围，高举中国特色社会主义伟大旗帜，以邓小平理论和“三个代表”重要思想为指导，深入贯彻落实科学发展观，在全面建设小康社会的伟大历史进程中，为加快建设创新型国家作出新的更大的贡献！

中国振动工程学会自成立以来，一直受到挂靠单位南京航空航天大学领导在人力、物力、财力等多方面的大力支持。本次会议又由南京航空航天大学承办，学校科协部门、学会办公室和南航机械结构强度与振动国家重点实验室的同志们为本次会议的筹备做了大量工作，付出了辛勤的劳动。借此机会，我们谨向南京航空航天大学的领导表示衷心的感谢！向承担会议工作的全体同志表示深切的感谢！

最后，衷心祝愿中国振动工程学会第七次全国会员代表大会暨第十届全国振动理论及应用学术会议圆满成功！

积极投身科技创新活动[①]

在这生机勃勃的初夏季节，2012 年理工学院学生学术科技节隆重开幕了。首先，我对本届理工学院学生学术科技节的举行表示最热烈的祝贺！向参加大赛的研究生、本科生同学致以诚挚的问候！

本次理工学院学术科技节的主题是“扬起科技之帆，献礼建院九十年”，包括学院 6 个系的特色品牌活动，主题鲜明，内容丰富，形式多样，既体现了理工科的专业特点，也体现了活动的创新要求。学生学术科技节的举行，不但为同学们学科学、用科学良好学风的形成提供了外部环境，对拓宽学术视野、丰富校园文化起到了积极作用；而且为同学们创新精神和实践能力的培养提供了锻炼机会，为同学们提供了一个增强学术科技素养、提高自身能力的舞台，在研究生、本科生激发科研热情、活跃学术思维等诸多方面，具有积极的现实意义和长远意义。

知识经济时代来临的 21 世纪，科学技术飞速发展，科技创新日新月异，创造性思维和科技创新在人类社会和生产力发展中的原动力作用日益凸显。在一定意义上讲，当前国家之间综合实力的竞争，关键在创新，核心在人才。“科技是第一生产力”“人力资源是第一资源”是提升国家核心竞争力和综合实力的一项战略举措。中华民族伟大复兴，不但需要当代人的努力，更决定于年轻人的素质，当代青年的综合素质特别是科学素质如何，将直接关系到我们能否巍然屹立于世界民族之林，能否实现中华民族伟大复兴，培养和提高青少年的科技素质比任何一个时期都显得更为迫切。

大学学习的青年时期是人的原创力发展的重要阶段，好奇心重、求知欲强，正是培育创新意识和创新能力的大好时机。同学们应该在学习和生活中形成尊重科学、崇尚创新的科学思想，向老师学习、向一切有知识的人学习，培养深入探究、勇于实践的科学素质，积极投身科技创新活动，为将来自身的提高和发展打下良好基础。

同学们，学术科技节的举办时间虽然只有两个月，但是科技创新将需要我们终生为之奋斗，正所谓“仰之弥高，钻之弥坚”。在多年的科研工作中，我的体会是要在科技创新方面取得成就，需要我们耐得住寂寞，要有耐心、好奇心和责任心，要有宽厚的知识积累，要勤奋，要有机遇。在此，我衷心地希望广大的研究生、本科生珍惜学术科技创新大赛中的交流和学习机会，相互切磋，取长补短，在活动中将专业知识与创新实践相结合，努力开展协同创新，真正锻炼、提高自己的创新能力、科研能力和动手能力，提高自身的科技创新素质！同时，我更热切地希望同学们将本次学术科技节作为新的起点，在今后的时间里，勤奋学习，积极探索，激扬学术，积极创新，勇于实践，努力使自己成为振兴广东、振兴中华的栋梁之材！

最后，祝 2012 年理工学院学生学术科技节取得圆满成功！

---

① 本文是 2012 年暨南大学理工学院学生学术科技节开幕词，广州，2012 年 5 月 12 日；教育与科技管理研究，北京：科学出版社，2016，294-295.

认真做好“2011 计划”重大战略部署工作[①]

首先，我代表教育部科学技术委员会管理学部，热烈欢迎大家来到这里，参加由教育部科学技术委员会主办、管理学部以及北京工业大学承办的“2011 计划”管理创新论坛。

“2011 计划”要求建立一批“2011 协同创新中心”，以加快高校机制体制改革、转变高校创新方式、集聚和培养一批拔尖创新人才、产出一批重大标志性成果，充分发挥高等教育作为科技第一生产力和人才第一资源重要结合点的独特作用，在国家创新发展中作出更大的贡献。

“2011 计划”将成立领导机构、专家咨询委员会、第三方评审监督机制，对计划的实施进行管理。这一管理机制如何有效运行，权责如何划分，应该有怎样的协同创新人事制度、怎样的创新拔尖人才培养模式、怎样的衡量其创新质量和贡献的评价制度，建立怎样的机制以监督其资源得到有效配置，这些都是关系到“2011 计划”能否顺利实施的关键问题。

今天，我们特邀各位专家，共同探讨“2011 计划”的评审认定办法、管理机制体制、绩效评价办法、监督机制，为“2011 计划”的顺利实施提供宝贵的意见和建议。在此，我希望各位专家在本次论坛上能够畅所欲言、相互沟通交流，贡献自己的专业知识和专家建议，使本次论坛能得到有效的、丰富的成果。

此外，北京工业大学为我们的论坛举办从资金、会务、场所等各方面提供了大力的支持。我代表管理学部和本次论坛的专家，特别感谢北京工业大学！

最后，预祝论坛取得圆满成功。

---

① 本文是教育部科学技术委员会第一届论坛开幕词，北京，2012 年 6 月 6 日；教育与科技管理研究，北京：科学出版社，2016，281.

组建“政产学研金”合作平台 推动协同创新迈上新台阶[①]

胡锦涛总书记在庆祝清华大学建校 100 周年大会上的讲话中强调：“要积极推动协同创新，通过体制机制创新和政策项目引导，鼓励高校同科研机构、企业开展深度合作，建立协同创新的战略联盟，促进资源共享，联合开展重大科研项目攻关，在关键领域取得实质性成果，努力为建设创新型国家作出积极贡献。”[②] 把协同创新提升到关系国家前途的高度进行阐述，集中体现了党中央对教育事业的高度重视，这是对大力提升高等学校创新能力的一种号召，是对高等学校支撑创新型国家和人力资源强国建设的一种期待，也开启了我国高等教育由大到强的新征程。

在这个伟大的新征程中，教育部从深入贯彻国家教育、科技、人才规划纲要要求的高度出发，大力推进科教兴国战略和人才强国战略实施，制定、颁布了《关于进一步加强高等学校基础研究工作的指导意见》《关于实施高等学校创新能力提升计划的意见》《关于全面提高高等教育质量的若干意见》等相关文件，提出了加强基础研究、提升创新能力、建设一流大学、提高教育质量、支持社会又好又快发展的新思路。

当今世界，创新已成为经济、社会发展的主要驱动力。面对日新月异的科技进步，迫切需要中国高等教育转变创新理念和模式，促进社会各类创新力量的协同创新，以提高国家整体创新能力和竞争实力。

高校拥有天然的多学科优势、丰富的人才资源以及多功能特性，作为科技第一生产力和人才第一资源的重要结合点，在国家创新发展中具有十分重要地位和独特作用，必须肩负起协同创新的时代重任。高校的优势学科群与科研院所、行业企业、地方政府以及国际社会等建立深度合作，形成协同创新的有机整体，可以解决诸多方面的国家重大需求和重大科学问题，这是提升国家创新能力的一个有效途径，也是为国内外诸多协同创新实践所证明的。

但是，高校在积极推进协同创新工作中，需要升华现有的理念和思路，强化“创新”的办学理念，深化“协同”的办学思路，充分发挥高等院校的科研创新优势，大力协同社会各种创新力量和创新资源，构建一批集科技创新、成果转化、服务社会为一体的协同创新平台，将原来的“产、学、研”三方面的两头，分别再加上政府和金融机构，即应改为“政、产、学、研、金”的协同创新合作平台。

具体言之，政府应负责营造协同创新良好环境，强化政策的激励引导作用，为企业、高等院校、科研院（所）营造有利于协同创新的政策环境，引导创新要素向

① 本文是教育部科学技术委员会第一届论坛主题报告，北京，2012 年 6 月 6 日；中国高等教育，2012，(20)：1.

② 《胡锦涛：在庆祝清华大学建校 100 周年大会上的讲话》，http://news.cntv.cn/china/20110424/104095_1.shtml[2023-11-14].

企业汇聚，在科技投入、成果奖励、创新平台建设等方面加大扶持力度，同时还应强化考核的促进保障作用。企业居于协同创新的应用主体地位，在提升自主研发和技术创新能力的同时，逐步使企业成为研发投入主体、技术创新主体、创新成果应用主体，提升企业的核心竞争力，使企业健康发展，进而更好地服务社会。高等院校既培养科技创新人才，又提供科技创新成果，是协同创新的重要源头和坚强后盾。科研院（所）是重要支撑体，应和高等院校合作，为企业提供能应用的科研成果。金融机构可为协同创新提供投入，提供风险管理服务，这是协同创新平台的强大支柱。

如果专就高等院校而言，承担着培养高级专门人才、发展科学技术文化、促进社会主义现代化建设的重大任务。但是，我国的高等教育和科技研究在一定范围内还存在着理论和实际脱离、科技与经济脱节的问题，没有根据国家、民生的需要提出问题、组织有效攻关，有时会游离于企业和市场之外。至今仍存在这样的现象：一方面是高校大量的科研论文和科技成果被束之高阁，另一方面是企业技术匮缺，国家原本有限的科技投入，没有发挥应有的效益。为了改变这种现状，高等院校应深入推进协同创新，主动积极与其他有关的学校、院（所）、企业合作，让成果更贴近企业需求，让成果走出实验室，并下大力气着重解决“中试”的薄弱环节。

在高校传统的管理体制下，科研人员的晋升、待遇一般取决于科研人员的论文数目多少、项目经费如何，科研成果转化则基本列入“忽略不计”的范围，从而导致了高校科研人员重论文重经费、轻推广轻应用的思想倾向。如果说轻推广轻应用的思想倾向还是主观因素的话，改变起来相对容易，那制约成果转化的“中试”薄弱环节，解决起来则有相当的难度。现在，一般的高校实验室远不具备“中试”的场地条件和设备条件，而且大部分的科研项目中也没有“中试”环节的经费支持，对于此类无法直接用于生产的科技成果，以市场经济原则运营的企业一般不会接受。鉴于此，“政、产、学、研、金”协同创新平台亟须解决的核心问题是要完善组织模式和运作方式，让“中试”顺利进行。首先，政府相关部门制定协同创新的相关法规、政策，建立科研成果向企业转化的中介机构，通过体制机制创新和引导合作，建立协同创新的战略联盟，为“政、产、学、研、金”协同创新平台提供大的平台，依靠法治来保障这种长远的合作利益。我们也欣喜地看到，教育部、财政部在这方面已经迈出了一大步。其次，应该明确各方的利益和责任，尤其是明确规定政、产、学、研、金五个方面的利益分配原则，同时兼顾五个方面内部所有参与者的权益，无论五个方面内部参与者的排名先后、贡献大小，均应按照一定的分配原则享有一定利益，以充分调动参与者积极性和能动性。再次，产、学、研之间的协同工作需进一步加强，高校在强调科研成果的同时，更加重视科研成果的转化工作，并逐步完善以加快科研成果转化和产业化为主导的激励机制，促进高校与企业的有效合作，更好地为国家和地方经济建设、为社会发展服务。企业方面应更广泛、更深入地同高等院校合作，将科研成果尽快地应用到生产过程中，不仅可以通过运用高新技术降低生产成本，提高生产效益，而且可以促进企业的转型升级，更好地增强企业的竞争力。最后，如果政府的平台建设、产、学、研的协同工作运转融洽、高效，那金融机构的参与、支持将是水到渠成。

如此，在协同创新工作中，政府起引领作用，金融机构起支柱作用，产、学、研三个方面组成一个平台，各个方面均要受益才算成功。由于五个方面紧密联系，有机衔接，一荣俱荣，一损俱损，所以政、产、学、研、金的每个方面都会协调、都会尽力，推动协同创新工作向着更高的层次发展。

开拓创新 做好科普工作[①]

今天6月29日，是《中华人民共和国科学技术普及法》（以下简称《科普法》）颁布实施10周年的纪念日。《科普法》是我国科普事业发展的一个新的里程碑，对加强科学技术普及，提高公民的科学文化素质，推动经济发展和社会进步，产生了重要而深远的影响。在《科普法》颁布实施10周年之际，我们一同回顾贯彻落实《科普法》的历程并展望科普事业的美好未来。

一、《科普法》实施的回顾与展望

十年前，《科普法》于2002年6月29日由第九届全国人大常委会第二十八次会议通过，中华人民共和国〔2002〕第71号主席令发布自公布日起施行。这标志着我国科普工作进入了法治化、规范化的新阶段。《科普法》分“总则”“组织管理”“社会责任”“保障措施”“法律责任”“附则”共六章三十四条，明确提出了“实施科教兴国战略和可持续发展战略，加强科学技术普及工作，提高公民的科学文化素质，推动经济发展和社会进步”的总体目标，指出“科普是公益事业，是社会主义物质文明和精神文明建设的重要内容”“发展科普事业是国家的长期任务”“国家保护科普组织和科普工作者的合法权益”“国家支持社会力量兴办科普事业”“科普工作应当坚持群众性、社会性和经常性，结合实际，因地制宜，采取多种形式”来普及科学技术知识、倡导科学方法、传播科学思想、弘扬科学精神，并强调“科普工作应当坚持科学精神，反对和抵制伪科学”“任何单位和个人不得以科普为名从事有损社会公共利益的活动”。

应该说，《科普法》是一个创造，一种创新，是针对中国国情所采取的一个具有中国特色的重大举措，其目的就是提高公众的科学文化素质，实现全面建设小康社会、和谐社会和创新型国家的宏伟目标，充分体现了以人为本的科学发展观和社会主义核心价值体系。

胡锦涛总书记早在2004年6月2日于中国科学院、中国工程院两院院士大会上的讲话中就明确指出：“科技创新和科学普及，是科技工作的两个重要方面。”在2006年1月9日全国科学技术大会上的讲话中又进一步指出：“把科技创新与提高人民生活水平和质量紧密结合起来，与提高人民科学文化素质和健康素质紧密结合起来，使科技创新的成果惠及广大人民群众。”“要在全社会广为传播科学知识、科学方法、科学思想、科学精神，使广大人民群众更好地接受科学技术的武装，进一步形成讲科学、爱科学、学科学、用科学的社会风尚。”

① 本文是广州市纪念《中华人民共和国科学技术普及法》颁布十周年高端论坛的主题演讲，广州，2012年6月29日；教育与科技管理研究，北京：科学出版社，2016，359-362.

二、《全民科学素质行动计划纲要（2006—2010—2020年）》颁布实施，全面推进公民科学素质建设

经过国家层面的深入调研论证，2006年2月6日，国务院颁发了《全民科学素质行动计划纲要（2006—2010—2020年）》（以下简称《全民科学素质行动计划纲要》），要求“各省、自治区、直辖市人民政府，国务院各部委、各直属机构”“结合本地区、本部门实际，认真贯彻实施”。

从《科普法》到《全民科学素质行动计划纲要》，从依法科普到科学素质建设，我国的科学教育、传播和普及工作又跨上了一个新的台阶。

近年来，广东省委、省政府一直高度重视全民科学素质工作。“十一五”期间，按照国务院的有关工作要求，我省认真贯彻实施《全民科学素质行动计划纲要》，坚持“政府推动、全民参与、提升素质、促进和谐”的工作方针，采取“大联合、大协作”的工作方式，广泛开展主题科普活动，认真组织实施重点人群科学素质行动，扎实推进科学素质基础工程建设，建立完善纲要实施工作机制，取得了显著成绩。科学素质工作的公共服务能力明显增强，公众学习科学的机会与途径明显增多，公民科学素质明显提高，2010年我省公民基本具备科学素质的比例为3.3%，高于全国平均水平3.27%，比2005年提高了1.7个百分点，较好地实现了我省“十一五”全民科学素质工作的目标任务，为我省加快建设创新型广东、促进经济发展方式转变作出了积极贡献。

在看到成绩的同时，我们也要清醒地认识到：我省公民具备基本科学素质的比例还不高，与发达国家和国内先进地区相比有一定的差距，与我省经济社会发展水平不协调。同时，公民自觉运用科学知识处理日常工作、生活问题的习惯尚未形成，科学应对各种自然灾害和重大突发事件的能力还不强，社会上不科学、不文明的现象和行为时有发生。科普为公众服务的能力还不强，公众接受科学传播、教育、普及的渠道和机会有待增加等。这些都需要引起我们足够的重视，加强公民科学素质建设任重而道远。

三、整合资源，创新思路，为提升公民科学素质服务

我是一名科研工作者，也是一名科普工作者。2009年5月13日，广东省科普志愿者协会正式成立，作为广东省科普志愿者协会会长，我有责任和义务带领我省广大科普志愿者深入贯彻实施《科普法》和《全民科学素质行动计划纲要》，积极动员和组织更多有着志愿精神和社会责任感的科技工作者参与进来，实施跨地区、跨行业、跨学科的协作与联合，实现人力、物力、智力的共享与发展，建立一支阵容大、素质高、相对稳定的科普工作队伍，为提高公民科学素质，为我省全面建设小康社会、建设创新型广东贡献应有的力量。

近年来，广东省科普志愿者协会围绕省委、省政府的中心工作，认真履行职责，开拓创新，整合资源，加强自身建设，提升服务能力，积极开展多项与社会管理、公共服务等有关科普服务，稳步推进科普志愿服务事业和公益科普服务项目发展，着力

打造协会科普品牌项目，初步形成了凝聚科普志愿者的知识和力量、以科普资源开发和科普活动开展为依托、面向重点人群深入社区、学校、基层开展科普志愿服务的模式。

近两年来，协会牵头实施了“垃圾分类，从我做起”公益科普宣讲行动、“万座爱心科普漂流书屋”工程与快乐阅读行动、“创新人才培养与校外实践”公益行动、“关爱青少年视觉健康”行动和“幼儿护苗素质传播”行动等五大行动，得到中国科协、省委和省政府有关领导的高度评价及社会媒体的广泛关注。

为贯彻落实国家、省、市有关城市生活垃圾分类文件精神，促进生活垃圾分类减量及再资源化，在广大科普志愿者开展生活垃圾分类调研的基础上，2011 年 7 月，我专门向时任广州市委书记提交了“以科普资源为抓手，推进‘垃圾分类，从我做起’进校园、进社区科普宣传公益活动的建议”，当时的广州市委书记批示，同意协会有关工作；9 月，我又向时任广东省政府省长黄华华同志提交了“关于支持省科普志愿者协会开展‘垃圾分类，从我做起’科普资源包研发及公益科普宣讲活动的建议”，黄华华同志、朱小丹同志、宋海同志都先后做重要批示，同意拨专款支持协会开展“垃圾分类，从我做起”科普资源包研发及公益科普宣讲活动。目前，协会已研发了“垃圾分类，从我做起”科普挂图、图书、动漫游戏、动漫宣传片、益智玩具等 25 种，先后在广州空军直属机关幼儿园、广州荔湾康有为小学、穗园小区等地开展了 100 多场“垃圾分类，从我做起”公益科普宣讲活动，宣传善待地球的环保意识，着力提高公众对生活垃圾分类的认识与参与度。

2011 年 9 月 17 日，在广东省“全国科普日”启动仪式上，省科普志愿者协会正式启动了“科普漂流书屋”工程。“十二五”期间，拟在全省主要城乡、社区、中小学校建立 10 000 座“科普漂流书屋”，推动社区居民、青少年参与图书漂流和读书活动，促进科学知识传播、交流与分享。2011 年 12 月，省政府办公厅印发《广东省全民科学素质行动计划纲要实施方案（2011—2015 年）》（粤府办〔2011〕88 号），明确把“科普漂流书屋”工程建设，作为科普基础设施建设的重要任务。2012 年 2 月 9 日广东省副省长陈云贤到省科协视察调研时，听取“科普漂流书屋”工程实施情况，并给予高度评价。

2011 年 9 月 27 日，省科普志愿者协会与深圳市慈善总会青少年素质教育公益基金联合，正式启动“青少年国际创新人才培养与校外实践”公益行动项目，通过青少年的国际视野拓展、科技创新、艺术展演、社会公益、未来商业科学体验 5 大方向，培养未成年人综合素质能力。

当然，这些活动、项目的有效开展，成绩、经验的不断积累，与广大科普志愿者能够充分认识《科普法》《全民科学素质行动计划纲要》的重要意义是分不开的。科普是旨在提高公民科学文化素质的一项基础性、长久性、公益性的事业，是实施科教兴国和走可持续发展之路的重要基础。在《科普法》颁布实施 10 周年这个值得纪念的日子里，作为一个科普志愿者是光荣的，但我们的任务是艰巨的。让我们携起手来，勇挑重担，团结协作，共同努力，为我省科普事业的蓬勃发展、为全民科学素质的稳步提高贡献自己的力量，用我们的实际行动向党的十八大献礼。

复合材料创新成果丰硕①

现代高科技的发展离不开复合材料，复合材料对现代科学技术的发展有着十分重要的作用。复合材料的研究深度和应用广度及其生产发展的速度和规模，已成为衡量一个国家科学技术先进水平的重要标志之一。人们预言，21 世纪将是复合材料的世纪；进入 21 世纪以来，全球复合材料市场快速增长，亚洲市场增长较快，目前亚洲复合材料市场占世界复合材料总产值的 36%（2006 年总产值 190 亿欧元），预测未来 10 年亚洲复合材料市场市均以两位数增长，产值将会占到世界复合材料总产值的一半。复合材料应用领域，如航空、运输业、消费品、风能、管道、建筑、电子设备等行业在亚太地区的增长比世界上任何地区的增长都要快。而在亚洲，中国在“十一五”期间，复合材料的发展尤为迅猛。从应用领域的角度，主要表现为以下方面。

（1）汽车领域。新能源汽车与新材料；电池材料壳体、隔膜、储能；轻量化：整车、电池。

（2）轨道交通领域。车辆内饰复合材料需求旺盛；无砟轨道充填式复合材料垫板在高速铁路上得到成功应用。地铁用第三轨保护罩（拉挤与接触成型）及其支座（SMC 模压）、电缆支架（SMC 模压）、逃生平台（拉挤酚醛玻璃钢型材）等得到广泛应用。

（3）风电领域。风电每年以 25%以上的速率递增：单机容量越来越大；叶片越来越长、自重越来越大；刚度、强度要求越来越高；由玻纤转向碳纤。目前可达4～5MW，塔高 70m、叶片可长达 50～60m。叶片长度每增加 6%，发电量可增加 12%。

（4）大飞机领域。中国航空工业集团哈飞建成大型复合材料基地，具备年产 150 架份飞机复合材料构件的能力，可年产各种复合材料构件 1000 多个品种。

（5）环境工程领域。脱硫环保工程，继 2007 年第一个烟塔一体工程之后，有多个烟气脱硫烟塔一体工程相继建设（烟道 8.5m×42m，烟囱 4m×180m）。

（6）沿海油气田领域。先进复合材料在沿海油气田领域应用极为广泛：岸基和沿海、井上结构和井下系统。应用方向包括：管道系统、油箱、油罐、围栏、扶手、通道，特别是采油的竖井和将平台锚固到海底的安全索、功能缆线与绳索等。

（7）基础设施建设领域。基础设施修复、更新、加固已构成复合材料目前极重要的应用领域；全复合材料房屋生产企业落户成都；复合材料第一座跨线公路桥、两座混凝土组合梁复合材料桥在石家庄建成；碳纤维芯复合电缆、复合材料输电塔架已挂线试运行。

（8）电力领域。复合材料杆塔的绝缘等级高（30kV/mm），可以大幅度地缩小输电通道走廊宽度，且应用复合材料的耐腐蚀特性，使沿海地区输电线路的维护要求显著降低。输电线路复合材料杆塔可提高 60%防雷性能．耐受电压提高 2.8 倍，全寿命

① 本文是《复合材料“十一五”创新成果荟萃》前言，北京：中国科学技术出版社，2012，1-2.

周期成本低。目前国内尚没有一条全复合材料杆塔批量投入使用。国家电网公司拟进一步开展复合材料在输电杆塔上的应用研究，解决应用瓶颈问题。

(9) 缠绕管道领域。目前总产量达 70 万吨，增长 150%。管道出口量和境外管道工程承包项目增多。夹砂管道、工艺管道、脱硫管道、高压管道、水处理用管道、船用管、电缆保护管等应用领域不断拓展。吸收塔立式提升缠绕工艺，弥补了原有的立式成型工艺一次成型高度小的缺陷，使一次成型的筒身高度从 6 米提升到 10 米。玻璃钢管道是海水淡化工程中唯一公认的管道产品。目前我国定长缠绕设备代表国际先进水平。管道产品未来发展方向：高刚度低成本新型管材，海水淡化玻璃钢管道，高压管道等。

2012 年年初，工业和信息化部发布了《新材料产业“十二五”发展规划》，可以预见，“十二五”期间，我国产业结构深层次升级调整，必将促使航空、航天、船舶、风电、建筑、交通运输、体育用品等各产业领域对复合材料的需求持续上升，推动复合材料行业更快发展。所有这些成就的背后都凝聚着复合材料人的心血，在“十一五”结束之际，需要对“十一五”期间中国复合材料取得的各项成果做全面的回顾和总结。从而明晰当今中国复合材料科研生产和服务的现状，更好地对接市场需求，在“十二五”工作中稳步发展。为了回顾 2000 年以来，尤其是“十一五”期间中国复合材料在诸方面取得的巨大进展，总结业界同仁为行业的发展做出的巨大贡献，中国复合材料学会组织出版《复合材料“十一五”创新成果荟萃》一书。本书共分为三大部分，第一个部分从政策方面做了简单回望，并就复合材料行业细分专业“十一五”期间的发展做了详细的回顾，提出了未来可能遇到的问题和发展预期；第二部分，从四个方面收录了中国在“十一五”期间复合材料及复合材料交叉学科完成的科研项目、取得的科研成果、专利以及部分省部级以上的奖励；第三部分，附上新材料“十二五”规划以及中国部分复合材料专家的简要信息，供参考交流。

本书由中国复合材料学会理事长杜善义院士领衔，由我和陈祥宝院士、张立同院士、孙晋良院士、黄伯云院士、周国泰院士、张博明教授等专家共同编纂。学会办公室王军文主任负责本书的策划和统筹，苏志老师负责编辑工作：北京航空航天大学材料科学与工程学院朱明明、秦兴等负责部分内容信息采集和校对工作。

在本书编写过程中得到了国内复合材料领域诸多知名专家的关心、支持和帮助。其中，浙江富阳特种纤维应用研究所所长申屠年教授对本书的策划提出建设性建议，航天材料及工艺研究所、温州宏丰电工合金股份有限公司大力支持了本书的印刷。为了使本书能够尽快出版，国内主要复合材料生产厂家及科研单位为本书提供了大力支持，在此表示衷心感谢！特别提出感谢的是，为本书第一章“回顾与展望”撰稿的各位专家，他们的文章当是本书浓墨重彩的一笔。

“十一五”期间，中国复合材料行业取得成就巨大，编纂本书所需信息量巨大，需要对不同地区、不同细分专业、不同类型的信息进行收集、纠错和鉴别整理，难免出现不完整甚至错误之处，还请批评指正！

肩负使命与责任　继续向前[①]

近十年来，随着广东省大炼油、大化工生产基地的建成，中海壳牌、中海炼油及原有的炼油大厂茂名石化、广州石化的扩建装置的相继投产，对压力容器技术的需求日益迫切。广东省压力容器技术发展也需要与经济发展共同进步，因此广东省压力容器学会也肩负着自己使命与责任。自2010年组成新一届理事会，学会主要完成了以下工作。

一、积极参与国家级课题、加强与行业主管部门的联系

广东省压力容器学会积极参与与压力容器相关的国家级课题，参与标准化组织，编制标准规范。暨南大学、华南理工大学、广东省特种设备检测研究院、广州市特种承压设备检测研究院等单位都承担了国家、省、市各级的与压力容器相关的科研项目。

学会积极加强与国家压力容器学会联系。

二、积极组织对重大共性技术及失效案例的分析，提出对策与建议

伴随着压力容器技术的进步，压力容器在材料、设计、管道组成件与支承件制造、容器现场制作与安装以及在用压力容器检测、完整性评估、失效分析、工程风险评价与控制等领域取得了积极的技术进展。但是近年来我国压力容器、压力管道的事故也有所增加，特别是管道事故发生的原因错综复杂，有管件制造质量问题，有管道安装或使用管理等问题，对管道的失效分析将为避免管道重大事故的发生起到积极的作用。广东省压力容器学会发挥其技术人才密集、跨部门跨行业的优势，根据需要组织有关技术力量和人员会同国家技术监督局开展事故失效分析，提出对策与建议。

随着广东省大炼油大化工装置的投产，企业在压力容器压力管道使用中会碰到许多技术难题，学会将组织专家与企业一起共同研究，进行课题攻关，帮助企业解决实际问题，推动省内压力容器技术的进步与发展。

三、跟踪国际压力容器技术领域的研究新动向开展学术研究

世界发达国家对压力容器压力管道的研究较早，研究涉及生产装置用压力容器、工艺管道、化工行业用长输管道、民用工程中的水、煤气管道等，已经取得很多科研成果，领先制订了有关压力容器压力管道缺陷评定标准，广东省压力容器学会积极组织专家跟踪国际压力容器压力管道技术进步和研究前沿，了解国外对有关压力容器压力管道的技术标准和技术法规，以满足省内压力容器压力管道的检验、安全分析与评定的需要。先后有60余人次参加了在用压力容器基于风险的检验关键技术、压力容器检测新技术、新工具、新方法、球罐的不开罐检测、管道完整性管理（pipeline integrity

---

① 本文是在2012年广东省压力容器学会学术交流会上的讲话，广州，2012年11月2日.

management，PIM）的技术研讨。具体内容如下。

（1）球形容器在石油、化工、冶金、城市煤气等工业领域被广泛应用于储存液化石油气、液化天然气、液氨、液氮、液氢、液氧、天然气、城市煤气、压缩空气等物料；在原子能发电站作核安全壳；在造纸厂用作蒸煮球；在化工行业作反应器等。随着广东省工业的发展，球罐的数量不断增加，其技术上的需求越来越迫切。通过交流，对连续性生产的球罐，通过不开罐检测技术降低了停工的成本。

（2）管道完整性管理：是指管道运营公司对管道潜在的风险因素进行识别和评价，并采取相应的风险控制对策，将管道运行的风险水平始终控制在合理的和可接受的范围之内。

管理平台非常适用于长输管道、城镇燃气和热力、厂区内的工业管道、厂区地下管网。随着广东省石化基地的建成投产，此技术的应用很好地解决了管道合理检验的问题。

（3）针对常规检验中存在的问题，如：不清楚容器的失效模式、失效机理，造成检验不足、盲目追求“全面”；不清楚容器失效发生的可能部位，造成检验无效、过度检验；容器的重要度划分考虑因素较少重点不突出，造成检验周期确定依据不足，过频或过长等，积极探讨压力容器检测新技术、新工具、新方法，避免传统检验的某些不足、确保本质安全。目前承压设备检验的先进技术有：声发射检测技术、射线实时成像检测技术、超声端点衍射波检测技术（time of flight diffraction，TOFD）、超声C扫描检测技术、超声相控阵检测技术、磁记忆检测技术、漏磁检测技术、红外热成像检测技术、远场涡流检测技术、电磁涡流裂纹表面检测技术、脉冲涡流检测技术等。新技术的推广应用对改善目前压力容器常规检验存在的问题起到了很好的作用。对广东省石化基地设备的安全起到积极的推动作用。

自2010年换届以来，学会的会员在 *Journal of Pressure Vessel Technology*、《压力容器》、《华南理工大学学报（自然科学版）》、《环境工程》、《中国有色金属学报》、《机械设计与制造》、《焊接学报》、《中国安全生产科学技术》、《中国粉体技术》等杂志上发表一百多篇与压力容器相关的论文。如我和宁志华副教授2011年在 *Journal of Pressure Vessel Technology*（SCI，EI收录）发表了《焦炭塔瞬态温度场的分析》的论文；2012年王小聪、李茂东等在《环境工程》上发表了《生物质颗粒层燃工业锅炉节能减排技术分析与测试研究》的论文；2012年杨景标、郑炯等在《中国安全生产科学技术》上发表了《承压类特种设备系统性风险研究》的论文。

四、做好国家标准的宣贯和培训工作

国家压力容器压力管道技术管理方面近年来投入大量的人力物力进行开发研究，在技术管理的更新上更是日新月异，新技术新标准不断推出。学会正在利用专家云集的优势，及时组织标准的宣贯，新技术的介绍推广。

五、建立了广东省压力容器学会网站

随着网络技术的发展，网站的建立有利于技术的推广，信息的传播，学会于2011

年3月建立自己的网站广东省压力容器学会网站，搭设起了技术交流平台，增强了学会各会员单位的联系，营造了良好的学术气氛。网站分学会介绍、学会新闻、学会通知、学会论文、压力容器法规、学会服务、学会章程等模块。

六、积极筹办第五届全省压力容器学会学术会议

学术交流是学会工作的重点，也是学会工作的基本任务和根本宗旨，经过积极努力由学会主办、广州市特种承压设备检测研究院承办了这次第五届全省压力容器技术学术会议。本次会议的内容很丰富，有几个重要报告，中国特种设备检测研究院贾国栋副院长的《在役承压设备法规标准体系建立及进展》的专题报告；广东省特种设备检测院的《大型石化装置承压设备定期检验》专题报告；暨南大学宁志华副教授、博士的《焦炭塔鼓胀变形与开裂几个问题的研究》学术报告；广州市特种承压设备检测研究院的《承压设备检验检测新技术》专题报告。

让我们共同努力，为广东省压力容器的发展和安全运行做出应有的贡献！

去除浮躁现象　建立公正科技评价机制①

1978年3月18日，在全国科学大会上，邓小平同志[1,2]提出了两个著名论断：知识分子是工人阶级的一部分；科学技术是生产力。随后，他又进一步指出，科学技术是第一生产力。由此开始，中国才真正有了“尊重人才”和“尊重知识”的良好风气。35年来，我国教育和科技受到国家的高度重视，大学生在校人数居世界第一，科学论文数也跃居世界第二。由于科技的促进，国家经济实现了高速发展，一个落后的社会变成几百年来的第一次太平盛世，经济总量跃居世界第二。

随着我国经济快速发展，高等教育在获得发展良机的同时，也出现了急功近利的浮躁现象，这对高等教育和国家科技发展极其不利，对实现中华民族伟大复兴的中国梦不利。

目前，浮躁之气弥漫整个中国的大学校园，几乎所有的大学都在“大干快上”，各种考核、评奖、争项目、夺排名，目不暇接，以致全校上下都在为考核、评估、荣誉疲于奔命，师生们没有了认真读书思考的时间。教师们一年到头，忙于填各种表格，忙于申报项目跑项目，忙于向领导汇报，忙于找“关系”疏通渠道，反而挤压了自己的主要工作——培养学生和进行科学研究，高校的教师已很难安静下来事做学问。现在大学校园里，几乎没有人散步，全都一路小跑，好像赶地铁，这样的氛围，对大学的长期发展不利。以前进了大学校园，会觉得很清静，现在进了校园，觉得和外面没有多少区别。

在这种浮躁的氛围下，不正之风不可避免地出现了。高校教师做学问的时间少了，学问做得不深，而且学术不端行为和学术腐败现象越来越多、越来越严重；一些科研学术人员在名利的诱惑面前心态浮躁，某些制度性的严重缺失又为他们的学术不端行为甚至造假打开了方便之门。现在的高校行政系统和科研项目一荣俱荣，一损俱损，毕竟学校要排名，要综合实力评价，需要科研成果、著名学者，而从造假起步的科研项目一旦做大、作出名声，就更难发现其中的问题，容易导致高校科研出现“造星运动”。同时，高校、科研机构还“运作”出来许多权威专家给“造星”和项目打包票，科研成果就更加真假难辨。这是中国高校的耻辱！长此下去，何谈科技创新，何谈实现伟大的中国梦想！

高等学校、高校教师之所以浮躁，有很多原因，比如社会大环境的影响、高校管理体制存在定位问题、大学行政化等，但其实质原因是目前科技管理和评价出现了一些问题，是我们的“科技指挥棒”出了问题。比如高校的各种评审、科技评估过多，教师职称评审、教师申请科研项目评审、教师承担项目的中期评审和结题评审、学生学位评审、学术期刊等级评审、科技奖励评审、专利评审、学校等级评审、实验室评

① 本节内容是教育部科学技术委员会第二届论坛主题报告，大连，2013年6月14日；教育与科技管理研究，北京：科学出版社，2016，284-285.

审、学科评审、课程评审、论文评审、优秀教师评选、杰出青年评选、长江学者评选、千人计划评选、千百十人才评选等，上述有些评审是需要的，有些评审是不需要的，有些是需要改进评审程序和方法的。总之，要改革这种现状，必须在科技管理上有所创新，制定出更合理的评价体系和奖励制度，废除急功近利的科研评估制度。科技评价，首先应该是公正，其次应该不要牵涉基层领导和教师太多时间，甚至不需要领导和教师参与，这样才能切实保证高校的环境利于培养人才，利于科技创新，使教师能够安静下来做学问，我们民族复兴才有希望!

参考文献

[1] 邓小平. 科学技术是第一生产力. 邓小平文选（第三卷）. 北京：人民出版社，1993：274-276.

[2] 邓小平. 在全国科学大会开幕式上的讲话 . 邓小平文选（第二卷）. 北京：人民出版社，1983：82-97.

切实做好科技评价工作①

经过一天紧张而热烈的发言和讨论，由教育部科学技术委员会主办、管理学部和东北财经大学承办的“科技评价专题论坛”圆满落幕，论坛取得圆满成功。

在今天的论坛上，委员、专家们，各抒己见，就大学科技评价发表了真知灼见。专家们就高校科技评价的现状进行了深刻反省，对所存在的问题及产生的原因进行了深度剖析，对评价机制体系、评价方法及评价成果的合理使用等问题以及今后的改革方向，提出了启发性、方向性的建议和意见。

论坛的成果丰富，专家们畅所欲言。接下来，我们会把今天的研讨内容进一步深化，汇总成会议纪要，并凝练成专家建议，上报教育部领导。希望能对我们的高校科技事业发展有所贡献。

最后，我代表教育部科学技术委员会以及管理学部对参加会议的教育部各位领导、管理学部的各位委员、战略研究基地的专家们以及我们的特邀来宾表示感谢，感谢大家提出了很好的建议。同时，对承办此次论坛的东北财经大学表示衷心的感谢！感谢你们提供了优美的会议环境，优质的会议服务以及资金支持，使得本次论坛得以顺利进行，取得圆满成功，请大家再次以热烈的掌声，感谢东北财经大学！

现在，我宣布，教育部科学技术委员会科技评价专题论坛闭幕！

① 本文是教育部科学技术委员会第二届论坛闭幕词，大连，2013 年 6 月 14 日；教育与科技管理研究，北京：科学出版社，2016，284.

专家学者不妨多点科普和传播意识[①]

程虹教授新著《中国质量怎么了》今年六月甫一面世，七月即和余华等的著作一道荣登畅销书排行榜。《中国质量怎么了》抓住了备受人们瞩目的“质量”这一话题，写得既深刻又通俗，既严谨又俏皮，该书得以畅销，理所当然，程虹教授是质量领域重要的专家，他在质量研究特别是宏观质量研究方面多有创建。把这些创建以喜闻乐见的形式普及给广大民众，并且作为提升公民素质的一部分。我认为，程虹教授做了一件非常有意义的事情，显示了他的独到之处和应有的担当。由此我也产生了几点联想。

联想一：专家学者要搞点科普

经过一代又一代人的辛勤耕耘，可以说，我国已经阔步迈上了科技大国的快车道；但一个不争的事实是，我国离科普大国还有较大的距离。我把它概括为“三不多”——参与科普的专家不多，有影响力的科普著作不多，公众的科普知识不多。

目前我国科普作品就数量而言，还很不够，特别是好的科普作品更是远远不够，而且，这些为数不多的科普作品往往面临着读者不多，甚至没有读者的局面。为什么科普作品会出现如此尴尬的境地？归根结底还是专业、优秀、有影响力的科普作者太少的缘故。作者少—作品少—读者少，科普正是陷入了恶性循环的“三少”窘境；要走出这一窘境，好的科普作者是关键。

好的科普作者必须是专业的，他（她）应该有深厚的专业学养，又谙熟科普规律，精通科普写作。由于专家学者站在研究的前沿，论述高屋建瓴，剖析问题举重若轻，若他们躬身参与科普工作，其深度和影响力非一般作者可比，创作的作品才可能摆脱低端、拼凑的嫌疑，显示出科普作品应有的高端大气。《中国质量怎么了》就很好体现了这一点。

中国质量问题不仅是我国公众关注的焦点，也成为世界关注的焦点。这是一个纷繁复杂的现象，也是一个庞大丰富的系统。要科学解剖中国质量，厘清其内在规律，拿出令人信服的答案，必须是质量领域造诣深厚的专家，才能游刃有余。

程虹教授是我国颇具影响力的宏观质量科学研究和人才培养机构——武汉大学质量发展战略研究院院长，也是我国宏观质量领域唯一的学术期刊《宏观质量研究》的主编。他是宏观质量领域颇为勤奋的开拓者：出版了第一部专著《宏观质量管理》，带领团队建立了众多质量领域的观测基地；基于消费者感知的角度，发布了《2012 年中国质量发展观测报告——面向“转型质量”的共同治理》；领导团队完成了多个重大质量研究课题。可以说，《中国质量怎么了》是程虹多年来重要研究成果的聚合，是科普化的学术专著。专业的高度奠定了科普的厚度，这也是一部科普著作成功的要素。

---

① 本文原载《宏观质量研究》，2013，1（2）：1-3.

科普创作是有一定难度的，弄不好会让人觉得力有不逮，视为畏途（即使是一个卓有成效的专家学者，要写出普及面广、有效性强、受到大众普遍欢迎的科普著作，也不是一件易事。好在程虹教授驾轻就熟，他把笔触深入到复杂而又矛盾的中国质量现象背后，条分缕析地发掘藏而不露的诸多质量影响因素，譬如经济社会发展、法律与政策、公民行为与文化、大数据时代的信息等。由于是专门的质量研究专家，司空见惯的质量现象，在他的笔下常常发人之所未发，所提问题新颖独特，解决问题的方法科学务实，时常让人眼前一亮，受益多多。

从全球化和大数据时代的质量着眼，透过由大国质量、二元质量和转型质量相互叠加而形成的我国独特的质量现象，作者自如地在政府、企业、消费者之间转换角色，从政府的角度探讨质量，从专家的视域解读质量，从公民的角度感知质量，透过纷繁复杂的表象，剖析质量领域深层次的问题。通过层层解剖，隐藏于政府、企业以及消费者背后三方面质量问题的经济命题豁然开朗；而且，通过廓清职责，政府、企业、社会在书中各安其位，各守其责，而又齐心协力致力于中国质量的共同治理。该书特别强调社会、市场、消费者在质量治理中的作用和职责。对于目前我国的质量评价，人们大多倾向于认为：这是一个糟糕透顶的时代；而该书基于实证分析得出的结论却是："质量，这是一个最好的时代，也是一个最坏的时代。"每当出现质量问题时，人们总是抱怨政府管得太少，而该书却出语惊人："质量体制中最大的问题不是政府没有尽责，而是管了太多由市场、社会管的事。"在质量建设中，消费者并不是无所作为的旁观者，也不能满足于仅仅作为一个义愤填膺的批评者，该书特别倡导："人人都是中国质量的建设者"，"中国消费者质量自觉之日，就是中国质量全面振兴之时"。一个个来自于质量研究前线的前卫观点，挑战、瓦解着人们固有的观念，让人们感到既是思想的激荡，又是智慧的享受。这种吸引力自然是无法抗拒的。在现实和网络中，读者好评如潮："发人深省的一本好书！爱不释手！""如醍醐灌顶，几句话便解开心中迷雾……""本来是买来写论文参考的。没想到完全着迷，让我对未来有了全新的思考空间，突然就觉得思路豁然开朗，很有启发。""作者用数据说话，观点令人耳目一新，很有冲击力！"

联想二：专家学者对学术论文和专著可以有畅销意识

我国每年产出的论文和专著汗牛充栋，但不少都传播有限，昙花一现，最终躺在故纸堆和网络的某个角落默默蒙尘，不为人知。论文和专著传播有限，当然和他们的专业性较强有关。但是，从另一方面来说，我们的专家学者是不是也存在着一个观念的盲区，即多少缺乏点传播意识？对于那些蕴含真知灼见、启人心智、极具现实意义的论文和专著，却没有用科普的形式去进一步推广和传播，任其湮灭，是非常可惜的。曾经有一位老科学家为了推广科普，提倡应该用科普文章对应解释每一篇论文；当然，在目前的形势下，这不现实，也无须如此苛求。但是，有关经济社会，有关国计民生，有关公民素质和思想解放的课题，值得好好做做科普大文章，以便最大限度发挥其效益。

古语有云："言而无文，行之不远。"对于学术论文和专著的科普工作，我想可以加上这样一句："文而不俗，行之不远。"这里的"俗"当然不是俗气的"俗"，而是通俗的"俗"，要"俗"得深刻，"俗"得有质量和品位。

科学的难点在于把复杂的问题简单化。科普著作创作自有其固有的规律。我想不外乎这么几点：直面生活，贴近现实；深刻其里，通俗其表；形式活泼，语言晓畅。程虹和他的团队一直倡导直面真实问题的研究。作为学术类科普著作，既要站得高，着眼于从科学上进行规律性的解释；又要立得稳，对现实中的各种现象进行饶有兴味的实证分析。《中国质量怎么了》很好地做到了这一点，把深刻的质量命题蕴含在日常生活生动的现象中，把非常专业性的质量问题，化繁为简，化难为易，化深为浅，让专业之外的读者一看就明白。"一个国家三个质量""为何女性的衣橱永远差一件衣服?""为什么互联网再怎么发达，哈佛大学、耶鲁大学这些常春藤学校永远不可能被替代?""中国的质量究竟谁说了算？"" '价廉'可以'物美'吗?""技术领先的公司为何会倒闭?"看看书中这些片段，其知识性和趣味性跃然纸上。

在《中国质量怎么了》一书中，作者与其说是一个著名的质量专家，不如说更像一个阅历丰富、可亲可敬的导游。他带领读者徜徉于国内外一处处质量"风景"，用通俗易懂、趣味横生的语言娓娓道出其前生后世，引人入胜，启迪人们遐想和沉思。捧读这部书，既有阅读的快乐，又有漫游的趣味；对于读者而言，有什么理由拒绝双重的快乐呢?

联想三：科普工作需要有激励机制和责任意识

在文件中，科普工作的地位已经很高了。1994 年，中共中央、国务院文件《关于加强科学技术普及工作的若干意见》中，把科普工作提到关乎社会主义现代化事业的兴旺，关乎民族强盛的战略高度。2012 年颁布的《中华人民共和国科学技术普及法》，明确了政府及有关部门在科普方面的职责，指出科普工作是全社会的共同任务。

但现实中，科普工作和它应有的地位之间有着巨大的落差。科普不仅陷入了恶性循环的"三少"窘境，不少专家学者对科普工作也并不关心。一个科研人员，揽到了课题，课题最终验收通过了，任务似乎也就完成了；课题越多，就越容易确立其在业界的地位。在科研的过程中，人们很难想起科普。毫无疑问，科普成了被遗忘的角落，被边缘化了。我记得，几年前中国科学院的一项调查显示，近八成科技工作者认为：科普和自己关系不大，科普工作有专人做。

为什么出现这种状况呢? 主要是科普工作缺乏激励机制，制约了专家学者参与科普的积极性。科普工作往往很难有看得见的绩效，既不算成果，职称评定时往往毫无分量，经济上也没有什么效益；因而科技工作者科普动力普遍不足，甚至认为搞科普是上不了台面的低层次的事情，似乎也就不足为怪了。

其实，科技工作应该是一个相辅相成的完整过程，包括科学研究、科学应用和科学普及工作。科技既要创新，又要普及；创新和普及，是科技工作的"一体两翼"，不可分割。科普工作，对公众抛弃成见，引发、点燃其潜藏的创新火花，不可或缺。但现实中，由于种种主客观原因，往往把血肉相连的"一体两翼"活生生割裂开来，重创新，轻普及。

2010 年，第八次中国公民科学素养调查结果表明，我国具备基本科学素养的公民比例为 3.27%。这个数字是不容乐观的。近年来，一些突发公共事件为什么常常引发一个个热点科普话题？例如“非典”与公共卫生、日本核泄漏与“盐荒”、食品质量事故与食品安全等，莫不如是。当科普往往只有在突发事故中成为热点、成为人们关注的中心时，只能昭示这样一个事实：公众科普知识贫乏，我们的科普工作头痛医头，脚痛医脚，严重滞后于现实。最近，一个所谓的“气功大师”牵扯的名人之多，让人震惊。提高公民科学素养与人力资源水平，是我国现代化过程中绕不过去的话题。科普工作确实任重而道远。

大力推广科普工作，亟须国家增加激励机制，据悉，国家有关部门正筹划或者已经行动起来，在培养科普专业人才、增加科普经费、把科普课题列入国家科技计划等方面，都有所动作。科普一方面需要激励，另一方面也需要使命感和责任意识。《中国质量怎么了》的作者意识到：科普不仅是社会文化的一部分，也应该是国家科技创新的一部分；搞好科普工作，对推进现代化进程是不言而喻的。为消费者提供有价值和免费的质量数据，重塑公民的质量观念与行为，倡导全社会共同参与，做中国质量的建设者——这是《中国质量怎么了》的写作初衷和责任。

科普不是简单普及科学知识，更要普及科学精神、科学品质、科学思维和科学方法。《中国质量怎么了》一书说：“质量是科学更是信仰。”对于科普工作，在该书作者看来，不仅是造福公民的公益事业，也是饱含信仰的慈善事业。在搞科普不能获得多少切身利益的现有体制下，程虹教授能够沉下心来，把科普当作一项事业，确实需要有一种信仰，他走遍大江南北，举办了无数场广受欢迎的讲座，只为让科学的质量观念深入人心；可以说，《中国质量怎么了》是他以另一种形式在延续质量科普讲座，撒播科学的质量观念。欣喜的是，他的这一良苦用心，很快获得了读者的回应和认同：“质量，突然让我觉得肩上的一份责任，质量之路，既在共同治理；质量之路，也在每一个人脚下。”

在读《中国质量怎么了》的过程中，我既品尝了喜悦，也读到了责任。对于这样一部优秀的学术科普著作，我向读者大力推荐，希望更多的人能够一睹为快。

推动力学学科发展①

在中国力学学会 2014 年全国会员代表大会暨第九届、第十届理事会扩大会议召开之际，我谨代表中国力学学会第九届理事会向各位嘉宾、各位代表的到来，表示热烈的欢迎和衷心的感谢!

中国力学学会是由钱学森先生、周培源先生、钱伟长先生、郭永怀先生等老一辈力学家发起成立的全国力学工作者的群众组织。钱令希先生、郑哲敏先生、王仁先生、庄逢甘先生、白以龙先生、崔尔杰先生、李家春先生等老一辈力学家曾先后领导中国力学学会工作，团结全国力学工作者，为推进我国力学事业的发展作出了重要贡献!中国力学学会现有 2 万余名会员，31 个分支机构，主办 18 份学术期刊，已经成为在我国学术界享有崇高声誉、在国际上具有重要影响的科技社团。2010 年 10 月，中国力学学会会员代表大会选举产生了以胡海岩院士为理事长的第九届理事会。四年来，在全国广大力学工作者的共同努力下，开拓创新，锐意进取，在推动国家科技创新体系建设和提升我国力学学科的国际影响力方面成效显著。这些成绩的取得离不开各级领导和广大力学工作者一直以来的关心和支持，离不开以钱学森先生为代表的我国老一辈力学家的奠基性、开创性贡献。在这里，请允许我代表中国力学学会第九届理事会，向长期以来给予中国力学学会关心、帮助和支持的中国科学技术协会、国家自然科学基金委员会、中国科学院力学研究所等有关单位和同仁致以最真挚的感谢!

本次会议在上海召开，得到上海大学的大力支持，在这里，请允许我代表第九届理事会，感谢上海大学的支持!感谢会务组全体同志的辛勤工作!

最后，预祝大会圆满成功!

① 本文是中国力学学会 2014 年全国会员代表大会暨第九届、第十届理事会扩大会议开幕词，上海，2014 年 11 月 15 日；教育与科技管理研究，北京：科学出版社，2016，331.

汇聚人才为广东发展效力[①]

首先，我谨代表广东院士联谊会对自广东院士联谊会发起筹备、成立以来给予关心和帮助的中国科学院、中国工程院和广东省委、省政府，及相关部门表示衷心的感谢，对为首届广东院士高峰年会顺利召开给予大力支持的广州市人民政府、广东省教育厅、共青团广东省委员会等单位表示由衷的谢意。特别要感谢路甬祥副委员长、马兴瑞副书记、徐德龙副院长、陈云贤副省长、陈建华市长、李婷局长等领导莅临大会指导。

广东院士联谊会是由在粤工作院士和粤籍院士自主发起，由院士自愿组成的全省性、联合性、学术性、非营利性社团组，也是全国第一个院士自愿组成的社团。

一、广东院士联谊会是应时而生

党的十八大报告中新提出的国家战略就是创新驱动发展战略。在习近平总书记系列重要讲话精神中，创新驱动发展是高频词和关键词。做创新驱动发展的排头兵，是习近平总书记对广东的殷切期望，也是决定广东未来发展前途命运的关键。十八届三中全会之后，广东在全国率先颁布实施了《中共广东省委 广东省人民政府关于全面深化科技体制改革加快创新驱动发展的决定》。前不久，广东省政府又重磅推出粤府〔2015〕1号——《中共广东省委 广东省人民政府关于加快科技创新的若干政策意见》。人才是科技创新最关键的因素，创新驱动实质上是人才驱动。相较于北京、上海，广东的院士较少，但广东籍院士数量并不少，多达101名（在外工作87名）。更重要的是，在外工作的广东籍院士都有一颗心怀家乡发展的赤子之心，都希望并十分乐见有一个专业的平台更好地为家乡的经济社会发展贡献智慧力量。为顺应政府培育和扶持社会组织发展的大趋势及国家科技体制改革对科技社团发挥作用赋予的新期待，广东院士联谊会应运而生，于2014年6月在北京成立。目前，广东院士联谊会登记院士会员已达186名（含在粤双聘）。

二、广东院士联谊会将顺势而为

广东省委十一届四次全会明确提出，要把创新驱动发展作为推动经济结构战略性调整和产业转型升级的总抓手和核心战略抓好落实。为贯彻落实省委加快创新驱动发展的战略部署，广东院士联谊会将致力于团结和凝聚国内外广东籍及在广东工作院士的力量和智慧，并以此平台通过“院士引院士、院士团队”汇聚国内外高端创新要素，更高层次地促进技术与经济结合、更高水平地促进广东科技智库的建设、更高效率地加快本土高端科技人才的培养，为广东创新驱动发展提供有力的科技和

① 本文是首届广东院士高峰年会开幕词，广州，2015年3月29日；教育与科技管理研究，北京：科学出版社，344-345.

人才支撑！这也是广东院士联谊会助力打造广东院士高峰年会这一学术品牌活动的目的和意义所在。

我们相信，在中国科学院、中国工程院和广东省委、省政府的关心指导下，在院士们的共同努力和大力支持下，按照习近平总书记对广东提出的殷切期望，广东要努力成为发展中国特色社会主义的排头兵，深化改革开放的先行地，探索科学发展的实验区，率先全面建成小康社会，率先基本实现社会主义现代化。广东院士联谊会一定会为广东早日实现“三个定位、两个率先”的总目标作出更多更大的贡献。

扎实开展科普志愿服务工作，为全面提高我省公民科学素质而努力①

我受第一届理事会委托，向大会作工作报告，请代表们审议。

一、第一届理事会工作回顾及主要成绩

自第一次会员代表大会以来，在省科协、省民政厅的领导支持下，协会已有25万会员。全体会员认真学习贯彻党的十七大、十八大和各届全会精神以及省委、省政府系列重要文件精神，紧紧围绕中心工作，坚持"整资源、想事干、创新招、务实效"的工作态度，全面贯彻落实《广东省全民科学素质行动计划纲要实施方案（2011—2015年）》的任务和要求，团结和动员全省科普志愿者，面向基层、关注民生，广泛开展科学普及、传播和宣传志愿活动，统筹推进多项与社会管理、公共服务等有关的科普志愿服务项目，深化和拓展公益科普品牌项目建设，提高科普志愿者服务水平，为提高公民科学素质发挥了重要作用。

协会成立以来，主要开展了以下三方面的工作。

1. 抓住热点问题，深入实施"四大"公益科普行动

1）组织开展"美丽城乡，从垃圾分类做起"科普资源包研发及千场公益科普宣讲活动

从2012年起，协会紧紧围绕省政府办公厅印发《关于进一步加强我省城乡生活垃圾处理工作的实施意见的通知》（粤府办〔2012〕2号）的精神，重点依托"四支"科普志愿者队伍（垃圾分类科普资源包研发队、垃圾分类科普志愿者宣讲队、科普艺术团志愿者服务队及绿色家园科普志愿者服务队），一是研制适合五类重点人群的"垃圾分类"科普资源包23种；二是联合广州市等珠三角地区（9市1区）科协、住建局（城管局），积极实施千场"美丽城乡，从垃圾分类做起"进社区、进校园的公益科普宣讲活动。经4年来活动现场随机调查统计，市民对生活垃圾分类的知晓率提高了30%，参与率提高了25%，投放准确率提高到80%，有效促进我省城乡生活垃圾处理工作。

同时，通过院士直通车，协会会长刘人怀院士先后向时任省委书记汪洋同志提交"关于推进'餐厨废弃物变废为宝'的建议"，向朱小丹省长提交"关于支持省科普志愿者协会开展'垃圾分类'公益宣讲与培训服务的建议"，向时任省长黄华华同志提交了"关于支持省科普志愿者协会开展'垃圾分类，从我做起'科普资源包研发及公益科普宣讲活动的建议"，时任省委书记汪洋同志、朱小丹省长、黄华华省长等省领导均做出重要批示。

2）积极实施科学素质教育体验馆及科普书屋公益项目

在省纲要办、省科协及各地科协等有关单位的大力支持下，从2011年起协会与省

① 本文是广东省科普志愿者协会第一届理事会工作报告，广州，2015年11月9日.

科普中心结合国家“社区科普益民计划”“省级科普示范社区”“省级科学教育特色学校”等项目，先后建立了150座科学素质教育体验馆（含科普书屋），登记漂流的科普图书已达20万册，惠及100万市民群众。

项目启动以来，得到社会各界的关注、支持和新闻媒体的广泛报道。省委常委、统战部部长林雄同志在参加广东省社区科普工作经验交流会时对科普书屋项目给予高度评价；中国科普出版社、可口可乐公司提供项目或资金支持我省开展科学素质教育体验馆及科普书屋工程建设，南方日报、广东电视台等300多家媒体给予广泛报道。

3）积极实施“千乡万村科普惠农行动”公益项目

为提高农村农民科学素质，在省纲要办、省科协支持下，从2013年起，协会连续承办了三届“千乡万村科普惠农行动”——大学生下乡科普惠农服务实践活动，先后动员我省高校近3000多名大学师生下乡开展“保护环境、垃圾分类、食品安全、疾病预防、健康生活”等社会公共科普宣传教育与调查活动，为广大农民朋友提供技术咨询与培训服务，引导农民用科学知识指导生活与生产，促进农民增收创效，提高农民科学素质，为我省公民科学素养水平达到6.91%（第九次中国公民科学素质调查结果显示：广东省具有科学素质的公民比例达6.91%，位居全国第六位）做出了积极贡献。

4）积极实施“预防近视，珍爱眼睛”公益项目

针对青少年近视发展率高等问题，2009年12月起，在省教育厅、省科协支持下，协会正式启动“预防近视，珍爱眼睛”公益行动。多年来，围绕教育部关于青少年近视防控等文件精神，先后联合中山大学眼科中心、省视光学会、省野光源视力保健研究院、省视力眼镜商会等单位建立了珍爱眼睛科普志愿者服务队，在广州、深圳、汕头、江门、韶关等6个地市免费配发《珍爱眼睛 预防近视》科普图书30万册、科普挂图10万张、Flash科普动漫光盘5万张，举办预防近视科普报告200多场。

2. 加快转型，积极承接政府职能转移项目

近年来，党委政府不断加强社会组织建设，协会紧跟形势，清晰定位，积极探索承接政府购买服务。2012年起，协会被省民政厅确定为第二批具备承接政府职能转移和购买服务资质的社会团体。

1）积极承担和参与主题性大型科普活动

协会成立以来，与省科普中心组织开展了国家“基层科普行动计划”项目，文化科技卫生“三下乡”，“5•12防灾减灾日”，科技进步活动月，全国科普日，“美丽城市，从垃圾分类做起”千场公益科普活动，“科普书屋工程”，“千乡万村科普惠农行动”，岭南科普大讲堂，“创先争优学雷锋，志愿服务我先行”，“健康养生社区行”，第十七届中国科协年会等系列科普志愿服务活动，发挥科普志愿者的积极作用。

2）组织承担公民科学素质系列读本编印工作

受省纲要办委托，2011年起，协会与省科普中心承担了“公民科学素质系列读本”编印出版工程，先后组织编写出版了系列科普丛书40多部，得到时任副省长宋海同志的高度评价，丛书先后荣获省科普作品创作大赛一等奖、二等奖等多个奖项。

3）组织承办年会科普志愿服务工作

2015年5月，中国科协和省政府联合在广州成功举办了以“创新驱动先行”为主

题的第十七届中国科协年会。协会承办了本届年会的志愿者服务工作，招募了500多名大学生科普志愿者，他们以积极饱满的热情、周到细致的服务，发扬了“奉献、友爱、互助、进步”的志愿服务精神，表现出高度责任心、奉献精神、良好素质和吃苦耐劳的品质，圆满完成了本届年会的志愿服务工作，受到主办方和参会国内外嘉宾的高度赞赏。

4）实施最高海拔地区科普支教，树立科普慈善教育典范

在省纲要办支持下，2012年协会秘书长郑文丰到拉萨向普玛江塘乡完小学捐赠了一批科普教学设备和点读笔，深受当地学校师生欢迎。同年，协会副秘书长刘冰带领19名科普志愿者，骑行爬上5200米的珠穆朗玛峰大本营，在世界上海拔最高的乡镇小学支教，开展支教、心愿征集等一系列微公益活动，树立了山区科普教育慈善的典范项目。

5）实施“同心同行·梦想启航”广东公益行服务

为促进提升公民科学素质和能力，特别是帮助提升普通院校毕业生、复转军人、待业青年等的就业能力和科学素养，2012年协会携手中国光华科技基金会，在广东深入开展“同心同行·梦想启航”公益行活动，并得到基金会捐赠价值两千万的网络学习卡的物资支持，为协会开展公益服务提供重要支持。

6）承接“青少年国际创新人才培养与校外实践”公益项目

在省科协、省教育厅的支持下，2011年协会联合青少年素质教育公益基金，启动了“青少年国际创新人才培养与校外实践”公益项目。该项目先后为1000名教师开展创新人才培养和校外实践课程培养提供系列公益辅导及培训。

3. 加强自身建设，提高服务能力与水平

多年来，协会紧密联系省级科普示范社区科普志愿者服务队、地市科协、大学生科普志愿者服务队等，努力提高广大科普志愿者的服务能力，借助“互联网+”，提升科普服务水平。

1）举办科普志愿者服务培训班

自2011年3月21日起，结合“科普书屋工程”、“广东省科普示范社区创建”等项目，协会举办了六届科普志愿者服务培训班，先后培训省级科普示范社区科普志愿者服务队、大学生科普志愿者服务队等共约3500人。

2）借助“互联网+”，推动科普志愿服务信息化

为实现科普志愿者招募、审批、排班、调度、考勤、志愿时记录，及科普专家查找、科普资源学习与竞答、科普公益项目组织与服务为一体，结合年会科普志愿服务工作需要，协会联合省科普中心建立了科普志愿服务一站式综合信息平台。

3）宣传先进典型，组织开展广东科普志愿服务奖评选活动

为突出科普志愿服务先进组织和个人的典型榜样作用，加强他们之间的交流与学习，协会每两年组织开展“广东科普志愿服务奖”评选活动。在广大会员的积极参与支持下，共有20个会员单位被评为“广东科普志愿服务先进单位”和50位会员被评为“广东科普志愿服务先进个人”。

4）积极筹集资金，发展科普志愿事业

（1）积极参与财政项目申报。多年来，协会先后申报省科技厅、省民政厅、省住

建厅、省科协、广州市科创委等单位有关项目，并得到立项支持项目 8 个，项目资金 170 万元。

(2) 积极承接政府职能转移项目。多年来，协会先后受省科协、省纲要办、省科普中心委托，参与承担系列科普服务、学术论坛、海智服务等项目，筹集项目资金达 800 万元。

(3) 会费交纳情况。自协会成立以来，在广药集团、东莞科技馆、中山科协、海珠科协等会员单位支持下，先后筹集会费 20 多万元。

5) 成绩与荣誉

(1) 协会组织参加第七届、第五届广东省科普作品创作大赛，并分别荣获广东省科协、广东省科技厅颁发一等奖 1 项、二等奖 1 项、三等奖 1 项。组织参加广州地区科普作品大赛，荣获广州市科创委颁发二等奖 1 项。组织参加中国科教电影电视协会、国家新闻出版广电总局①科蕾奖作品创作大赛，荣获三等奖 1 项。

(2) 2011 年，协会被省科协评为省级科技社团创新发展试点单位，并被授予 2012—2013 年度省级学会先进集体。

(3) 2011 年，协会派李之宁等 3 位科普志愿者代表广东参加第四届全国公众科学素质大赛喜获季军。

(4) 2012 年，协会科普艺术团志愿者服务队陈美榈团长、光动媒技术专业委员会主任刘冰、副主任秦兆年、刘达莲等被团省委评为“五星级志愿者”。

二、主要经验与存在的问题

总结第一届理事会成立以来的实践经验，我们深刻体会到以下几点。

(1) 协会的发展离不开党委政府的重视和社会各界的支持。科普志愿服务是公益事业，科普志愿者工作是公益性工作，离不开政府坚实的政策和财政的支持。科普志愿服务又是一项基础性、长期性和系统性的工作，功在当代，利在长远。因此，应努力争取党委政府的重视，以及在政策和资金方面给予的支持。同时，科普志愿者工作是一项工作量大、辐射面广的社会性工作，需要社会各界的理解、支持与配合。本届理事会在这些方面都作了相应的工作，取得了一定成效，但仍然还不够，还要继续努力。

(2) 团结协会各成员单位特别是理事会成员单位共同协作才能做好工作。协会理事会既然是由各相关单位和部门的代表所组成和参与的，理事会就要注意听取他们的意见和建议，公平妥善地处理会务，团结协会各成员单位，相互支持，共同努力，协会才能发展壮大，广东省科普志愿服务事业才能蒸蒸日上。

(3) 改进工作机制，加大会员服务力度。会员是协会的主体，协会理事会是由会员选出来，由全体会员授权处理会务的。因此，必须注意听取会员的意见和建议，反映会员的要求和愿望，从而凝聚人气，团结力量，谋求更大、更远的发展。今天科普

---

① 2018 年《国务院机构改革方案》不再保留国家新闻出版广电总局，在国家新闻出版广电总局广播电视管理职责的基础上组建国家广播电视总局。

志愿者综合平台也将开通运行，该平台集成为会员服务，实现精准推送。

在回顾和肯定成绩的时候，我们也应清晰地意识到目前本会还存在的一些问题。一是协会自身建设力度不足，工作机制还不够完善，活动组织能力还有待进一步加强。二是协会对科普志愿者工作的经验总结不够及时、全面，理论研究不深入、不扎实。三是协会位于广东，毗邻港、澳、台，但与港、澳、台地区的同行们交流、合作与学习很少。这些问题直接影响着广大会员和社会各界对协会的认可，制约着协会事业的发展，需要在今后的工作中采取有力措施，切实加以解决。

三、对今后努力方向的建议

(1) 加强发展规划研究。要进一步解放思想，切实把加强科普志愿服务工作作为推进我省公民科学素质建设的基础性工程，要以提高公民科学素质为突破口，制订发展规划，采取有力措施，推进我省科普志愿服务工作迈上新台阶。

(2) 加强协会自身建设，积极探索社团发展新路子。要按照社会组织改革发展的要求，开拓进取，勇于创新，积极改进和完善协会的组织形式和运行方式，以改革谋出路，以改革求发展，加强协会自身建设。同时，在条件成熟时要建立健全协会党组织，进一步增强协会的服务能力，不断探索具有广东特色的科普志愿者社团发展新路子。

(3) 壮大我省科普志愿者队伍，努力提升服务能力与水平。科普志愿者队伍的建设是一项长期的、艰巨的任务。面对新常态下的全民科学素质的要求，不仅要发展和壮大科普志愿者队伍力量，更要努力提升广大科普志愿者服务能力和水平。要采取多种途径和方法，积极组织开展科普志愿者培训与交流工作，尤其要重视本省社区科普志愿队和农村地区科普志愿者的培训工作，要呼吁和争取政府有关方面给予更多、更实在的资助。此外，充分利用互联网手段，实现科普信息化，特别是用好科普志愿者综合信息平台，整合科普资源，推广科普菜单式服务，实现公民科学素质建设的精准服务。

(4) 创新活动形式，丰富活动内容，积极推进全省科普志愿服务全面发展。要大胆创新活动形式，丰富活动内容，结合广东实际，积极创办和引导地市协会组织开展低成本、普及广的科普活动，让更多的人有机会、有条件参与活动，激发公民的科技兴趣和创新热情，培养他们的创新精神和实践能力，全面提高公民科学素质和人民的生活水平。

各位代表，加强科普志愿服务工作，培养公民的科技创新精神和实践能力，全面提升公民的科学素质，任重道远，使命光荣。让我们团结引导广大科普志愿者，改革创新，为推动创新型国家建设、协调推进“四个全面”战略布局，全面提高我省公民科学素质，实现中华民族伟大复兴而努力奋斗！

应该高度关注青少年的视觉健康问题[①]

党的十九大报告指出，我国要在本世纪中叶建成富强民主文明和谐美丽的社会主义现代化强国。创新是引领发展的第一动力，是建设现代化经济体系的战略支撑。需要培养造就一大批具有国际水平的战略科技人才、科技领军人才、青年科技人才和高水平创新团队。这为我们开展青少年科技教育，培养合格健康人才指明了方向。

《国民视觉健康》白皮书研究显示，2012 年我国 5 岁以上近视的患病人数在 4.5 亿左右，约占全国总人口的 1/3。预期到 2020 年，近视人口将达 7 亿，患病率近 51%。视力缺陷的所带来的隐患非常严重，高度近视高发不仅危害当代人口素质，而且也影响我国未来的人口素质，如在航空航天、精密制造、军事等行业领域，符合视力要求的劳动力可能面临巨大缺口，这将直接威胁到国家安全。

早在 2016 年，中共中央、国务院印发的《“健康中国 2030”规划纲要》中指出要“加强学生近视、肥胖等常见病防治”。

现实中，家长及老师对青少年的视觉障碍问题普遍认识不足，对孩子近视发生无可奈何，对孩子高度近视潜在隐患普遍认识不足，缺乏有效预防近视的方式方法，如学习照明环境欠佳，近距离用眼时间过多，户外运动时间不足，饮食安排不合理，睡眠时间不够等。

回忆我本人在德国学习工作经历，德国学校、工作场所照明环境是非常舒适的，患近视的人很少。我近视也是在国内读书时，与照明环境欠佳有很大关系。

因此，作为一名老科技工作者，借眼调节训练灯二类医疗器械注册证发布会的契机，我呼吁全社会都来关注青少年的视觉健康问题，特别要关心关注学习场所的照明环境，改善照明条件，为我国建成社会主义现代化强国提供健康人才储备做出积极贡献。

祝眼调节训练灯二类医疗器械注册证发布会圆满成功！祝大家新年身体健康，万事如意！

① 本文是在眼调节训练灯二类医疗器械注册证发布会上的讲话，广州，2018 年 1 月 30 日.

为中华民族伟大复兴做出更多更大的贡献[①]

大家下午好。我讲两点。

第一、感谢。

感谢大家对我的信任！我要感谢广东省委、省政府、中国工程院、中国科学院、省委组织部、省民政厅、省科技厅，特别感谢胡春华副总理、李希书记、马兴瑞省长、周济院长、李晓红书记、黄宁生副省长，还有时任副省长陈云贤、袁宝成一直以来对我们工作的关心和帮助，感谢深圳市委、市政府、广州市委、市政府多年来对院士联合会工作的支持。我要感谢第一届理事会领导集体为联合会发展的辛勤付出，特别感谢联合会的主要发起人，还有院士联合会秘书处的同志们。

谢谢大家！

第二、谈谈今后的重点工作。

当前，广东正在推进重大产业化成果转化落地、广东高水平大学和高水平理工科大学建设，出台了《广深科技创新走廊规划》，深入实施重大人才工程、高新技术企业培育等八项举措，扎实推进重大创新平台建设，全面启动实施省实验室建设计划，建设网络空间科学与技术、再生医学与健康、材料科学与技术、先进制造科学与技术等4个省实验室，适时建设第二批省实验室，力争申建国家实验室；大力推进国家大科学装置建设，积极创建综合性国家科学中心……还有粤港澳大湾区战略等，这些都给了我们广阔的舞台，是广大院士到广东创新创业最好的机遇。

一是我们将努力在重大产业化成果转化落地有所作为。我们将重点考虑在条件成熟的地方建设院士团队科技成果转化基地（园），并在政府引导基金框架下，聚焦一两个领域推动发起一两支院士团队创新创业基金，为院士专家重大科技成果转化找空间、找资金、找资源。接下来，我们将进一步梳理有产业化意愿、有明确产业化资源需求的成果，助力院士团队重大科技成果转化。

二是我们将努力在省实验室建设方面有所作为。我们将重点围绕省实验室的建设，主动对接需求，为实验室引入高水平的团队。我们已经在行动，基于我们跟佛山市的战略合作关系，我们正在为季华实验室对接葛昌纯等院士团队。广东省在省重点实验室建设上是花很大本钱的，是干事的大平台。有兴趣的院士我们可以协助开展对接。

三是我们将努力在粤港澳大湾区和广深科技创新走廊建设中有所作为。进一步加强与在香港院士的联络沟通，深化粤港院士的交流，深化港澳院士专家与广东省委、省政府、产业界的沟通，从战略咨询层面献计大湾区建设。

四是我们仍将努力在院士之家建设有更大的作为。进一步建立健全各项规章制度，申报并力争通过5A社团评审，以评促建，深入推动制度化、规范化、可持续发展。

此外，通报一下今年重点工作：①配合中国创交会组委会，办好6月22—24日在

---

① 本文是广东院士联合会第二次会员大会讲话，北京，2018年5月27日.

广州举办的中国创交会系列活动——院士专家创新创业成果展、花城科技论坛暨新材料产业创新发展峰会；②围绕粤港澳大湾区战略实施，立足深港合作，办好 9 月上旬在深圳举办的深圳院士专家峰会；③根据马兴瑞省长指示，集中力量办好 11 月上旬在东莞举办的第四届广东院士高峰年会（按惯例，同期召开第二届第一次理事会会议），请各位院士、各位理事预留时间，请各位院士支持和参与。

各位院士，联合会因我们而诞生，靠大家大力支持才能发展，靠大家积极参与才能发挥更大的作用。我相信，在大家的共同努力下，广东院士联合会一定能够发展得更好，能够为广东经济社会发展、能够为中华民族伟大复兴作出更多更大的贡献！

全省科技工作者应履行科普社会责任[①]

习近平总书记在2016年全国“科技三会”上的重要讲话中指出：“科技创新、科学普及是实现创新发展的两翼，要把科学普及放在与科技创新同等重要的位置。”“希望广大科技工作者以提高全民科学素质为己任，把普及科学知识、弘扬科学精神、传播科学思想、倡导科学方法作为义不容辞的责任，在全社会推动形成讲科学、爱科学、学科学、用科学的良好氛围，使蕴藏在亿万人民中间的创新智慧充分释放、创新力量充分涌现。”[②] 总书记的号召，为广大科技工作者履行科普社会责任指明了方向，提出了更高的要求。今天，省科协、省科技厅等部门联合举办首届广东科普嘉年华活动，这是我省贯彻落实习近平新时代中国特色社会主义思想和党的十九大精神的一项重大举措。值此机会，我们向全省科技工作者发出倡议。

一、勇于担当，做科普工作的主要推动者

科学普及是面向全体公众的一项重要的公益性事业，是科技工作的重要组成部分。科技工作者作为科技创新工作的主体，也是科普工作的主要推动者。我们要切实履行科普的社会责任，自觉地把科普工作融入日常科技创新活动中去，采用公众易于理解接受的方式，宣传科技创新成果，让公众更好地理解科学支持科学，分享科技发展成果，推动社会形成创新发展的良好氛围。

二、严于律己，做科学精神的传承者

科普工作，不仅仅是传播科学知识，更重要的是传播科学研究中蕴涵的科学精神、科学思想和科学方法。我们要继承和发扬求真务实、勇于创新的科学精神，不畏艰险、勇攀高峰的探索精神，团结协作、淡泊名利的团队精神，报效祖国、服务社会的奉献精神。倡导敢于创新、勇于竞争、诚信合作、宽容失败的精神，恪守坚持真理、诚实劳动、亲贤爱才、密切合作的职业道德，维护科技界的良好社会声誉，努力成为良好学术风气的维护者、严谨治学的带头者、优良学术道德的传承者，让科学的殿堂更加圣洁。

三、乐于奉献，做科学知识的传播者

我们要适应新时代人民对美好生活向往的科普需求，积极投身于提高全民科学素质的社会事业，紧跟时代潮流，瞄准科技前沿，聚焦重点领域，面向四大重点人群，深入农村、社区、学校、机关，开展贴近基层、贴近群众的科普活动，大力弘扬科学

---

① 本文是在首届广东科普嘉年华启动仪式上的讲话，广州，2018年9月15日。

② 《习近平：为建设世界科技强国而奋斗》，http://news.cnr.cn/native/gd/20160531/t20160531_522287749.shtml[2024-08-14].

精神，普及科学知识，传播科学思想，倡导科学方法，推动全社会进一步形成讲科学、爱科学、学科学、用科学的良好风尚。

四、坚持真理，做维护科学的守望者

科技乃国之重器。关系到国家安全、生态安全、粮食安全，关系到人民福祉。我们要坚持真理，自觉维护科学技术的正确导向，坚决反对弄虚作假、各种伪科学、反科学的行为，引导公众运用科学技术知识正确处理日常工作生活中的问题，自觉抵制科学谣言及腐朽落后文化的侵蚀，让科学更好地服务经济社会发展和广大人民。

坚守初心 坚持创新[①]

今天，我们在这里举行粤港澳大湾区院士峰会暨第四届广东院士高峰年会。在此，我谨代表广东院士联合会对中国科学院、中国工程院，广东省委、省政府以及港澳相关机构给予此次活动的大力支持表示由衷的谢意！对为院士峰会筹备做了大量工作的东莞市委、市政府领导和相关部门的同志们表示真诚的感谢！衷心感谢各位领导拨冗而来、莅临指导，衷心感谢各位院士专家远道而来、积极参与。

今年两会，习近平总书记在参加广东代表团审议时强调，人才是第一资源[②]。在今年的两院院士大会上习近平总书记发出了新时代建设世界科技强国的“动员令”，又用了很大篇幅来谈人才问题，充分体现了总书记“人才是创新的第一资源”的理念，让我们备受鼓舞！前几天，习总书记又亲临广东视察并到我的所在单位暨南大学视察，再次强调创新和人才的重要性。

广东院士联合会发起成立于2014年两院院士大会，初心在于凝聚和团结国内外院士专家力量，为广东创新驱动发展提供有力的科技和人才支撑。目前，联合会会员院士234人，联络服务高端科技人才超过1300人。其中港澳会员院士超过50人。粤港澳天然的地缘关系和亲缘关系催生了粤港澳院士大湾区院士峰会与第四届广东院士高峰年会的联袂上演。

东莞作为广东重要创新城市和广深科技创新走廊的腹地，出台了《关于打造创新驱动发展升级版的行动计划（2017—2020年）》，提出从服务自身发展为主向支撑国家重大战略需求转变，为广东和全国创新驱动发展提供支撑中作出更大贡献，充分体现了东莞作为、东莞担当!

作为院士，我们将始终牢记习近平总书记的殷殷嘱托，把个人理想自觉融入国家发展伟业，紧紧围绕国家重大战略需求，积极投身粤港澳国际科技创新中心建设，勇于攻坚克难，矢志不渝自主创新，抢占科技竞争和未来发展制高点。

作为院士联合会，我们将更加积极地加强与粤港澳高校、科研院所和企业的互动、交流、合作，将以开放、务实的态度搭建粤港澳大湾区院士专家创新创业联盟，为院士专家创新创业、企事业单位创新发展、区域创新体系建设提供支撑，助力东莞打造成为粤港澳大湾区的创新高地和具有全国影响力的科技产业名城。

同志们，朋友们，广东院士联合会的发展离不开大家的关心和帮助。特别是离不开广东省委、省政府的正确领导，离不开中国科学院、中国工程院的关心指导，离不开广大院士专家以及相关部门、社会各界的鼎力支持。这里，我们还要再次特别感谢自始至终支持我们广东院士联合会的马兴瑞省长、周济院长。

---

① 本文是2018年粤港澳大湾区院士峰会暨第四届广东院士高峰年会开幕词，东莞，2018年11月2日.

② 《习近平：发展是第一要务，人才是第一资源，创新是第一动力》，http://www.moj.gov.cn/pub/sfbgw/gwxw/tpxw/201803/t20180307_163017.html[2023-11-14].

众人拾柴火焰高。我相信，在大家一如既往的支持下，在广大院士的共同努力下，我们一定能够为高水平建设粤港澳大湾区国际科技创新中心，为我省打造国家科技产业创新中心和科技创新强省，早日实现“四个走在全国前列”作出更多更大的贡献！

追求真理　奉献国家①

中华民族历经风雨，经过改革开放四十年，在中国共产党的领导下，进入中国特色社会主义新时代。我们中国工程院作为中国工程科学技术界最高荣誉性、咨询性学术机构，更需要有新气象新作为。

我从1963年开始担任大学教师，一直从事教学科研工作，投身国家工程科技、工程教育事业到现在，经历了改革开放，直到今天改革开放40周年。我常说，这一辈子是伴随着国家兴衰而成长的一辈子。1999年，我被选为中国工程院机械与运载工程学部院士，又于2000年再被遴选为中国工程院工程管理学部首批院士。从第一天作为院士开始，我就认识到，这首先是祖国和人民给我的光荣称号，应该万分珍惜，自己今后的行为只能为院士的称号增光增色，决不能增添丝毫脏丑，应该更严格要求自己，做到以下两点。

（1）坚决遵守《中国工程院章程》，切实履行自己的义务，完成中国工程院交给自己的各项工作和任务。

（2）切实完成党和人民交给的任务，做好自己的本职工作。

从参加我国第一颗东方红人造地球卫星科学研究开始，我从事科学研究已经60年了。我一直严格要求自己，不怕困难，踏踏实实做学问，包括对自己的学生和学术团队成员，都要求他们不畏艰险，不搞虚假，实实在在做学问，为国家为人民多做贡献。

成为工程院院士后，我更加严格要求自己，要珍惜院士称号，要更多地为国家为人民多做贡献。我为自己立了一个座右铭“多做好事，多做善事，不做坏事!”不管是在暨南大学校长、党委书记任上，还是从岗位上退下来，都按这个座右铭要求自己。

特别是在中国工程院每两年一次增选院士的时期，我更加注意。坚持守正扬清，严把院士增选入口关，选出合格的院士，对守护院士称号的荣誉性至关重要。这是为我们工程院增添生力军的重要时刻，使命重大，意义非凡!

关于如何守正扬清，我的体会是，从“严法实”这“三字经”入手，维护中国工程院这块净土。何谓“严”，我认为就是“严把入口关”，尽力坚持高标准、高质量推选院士候选人。何谓“法”，就是指依法办事、照章办事、干净办事。在政策层面，要做到“有法可依”，在执行层面，要做到“违规必究”。每一位院士都要坚决抵制不正之风，决不接受贿赂，防止任何学术不端行为污染院士称号和我们神圣的学术殿堂。何谓“实”，就是坚持实事求是，保证当选院士的道德质量、学术质量双双合格，经得起人民与历史的检验。

作为一名科技工作者，科技报国是我们肩负的历史使命，我们要牢牢记住习总书记的指示，沿着前辈科学家们的道路，记住历史，坚守科学道德，坚持守正扬清，扎扎实实，脚踏实地，用勤劳的双手托起中华民族伟大复兴的中国梦。

---

① 本文是在中国工程院“守正扬清”主题宣讲活动会议上的讲话，武汉，2018年11月23日.

让未来祖国的科技天地群英荟萃①

今天，第 34 届广东省青少年科技创新大赛在汕头市澄海音乐厅隆重开幕了，这是我省青少年展示交流科技创新活动成果的重大盛会。我代表大赛评委会，向大赛隆重开幕表示热烈的祝贺，向全体参赛师生致以诚挚的问候！

我省青少年科技创新大赛走过了 34 个不平凡的春秋岁月，已经成为发现和培养科技创新后备人才的一个重要平台，越来越得到社会的广泛重视和支持。作为一个老科技工作者，我有幸见证了十多年来的省创新大赛。每一年参加这个活动，我都感到非常激动、也非常开心。看到了同学们富有创新的成果，感受到了同学们投身科技创新的热情，也看到了我们国家未来科技发展的希望。我们为同学们的进步和成长感到由衷的高兴，为同学们充满奇思妙想、聪明智慧的成果感到欢欣鼓舞！实践证明，广泛组织开展青少年科技创新活动，对于激发青少年的科学兴趣、培养青少年的创新精神、提升青少年的科学文化素质，培养一代又一代社会主义事业的建设者，具有十分重要的意义！

习近平总书记在中国科学院第十九次院士大会、中国工程院第十四次院士大会上的讲话中指出："当科学家是无数中国孩子的梦想，我们要让科技工作成为富有吸引力的工作、成为孩子们尊崇向往的职业，给孩子们的梦想插上科技的翅膀，让未来祖国的科技天地群英荟萃，让未来科学的浩瀚星空群星闪耀！"② 总书记的讲话，为我们深入开展青少年科技创新活动指明了方向。一年一度的创新大赛，既是展示青少年科技创新活动成果的舞台，也是青少年放飞科学梦想、追求科学真理的实践平台。我们衷心希望同学们珍惜参赛的机会，遵守大赛纪律，尊重评委、尊重老师、尊重同学，把参赛当作一次检验自己、学习先进的重要机遇，成绩好坏并不重要，关键在通过参与体验，学会做人、学会求知、学会创造、学会合作，以天下为己任，从小树立科学精神，把兴趣作为开启科学之门的钥匙，把想象作为科技创新的翅膀，把勤奋作为持续探索的动力，努力使自己成为德智体全面发展的社会主义建设人才！

最后，预祝大赛取得圆满成功！

---

① 本文是在第 34 届广东省青少年科技创新大赛开幕式讲话，澄海，2019 年 3 月 30 日.

② 《习近平：在中国科学院第十九次院士大会、中国工程院第十四次院士大会上的讲话》，https://www.gov.cn/xinwen/2018-05/28/content_5294322.htm[2024-08-14].

众人拾柴火焰高[①]

承蒙大家的信任，广东省院士联合会承担了联盟前期筹备工作，并将在未来一段时期内承担联盟秘书处工作。承蒙大家的厚爱，我非常荣幸担任首任联盟主席。同时，衷心感谢大家对联合会工作的支持，特别感谢中国科学院、中国工程院、省委、省政府、香港中联办、澳门中联办及省直相关部门对联盟筹备工作给予的指导和关心，特别感谢各位领导莅临指导。

今天上午的会议，我们完成了一系列的审议表决事项，充分讨论了联盟工作的方向和重点，各位领导也发表了讲话，为联盟的发展提出了很多宝贵的建设性意见。这里我讲三点意见。

站位要高、格局要大。粤港澳院士专家创新创业联盟的发起成立既是“港澳所需”，响应港澳地区广大院士专家积极投身粤港澳大湾区建设的赤诚心愿，也是“湾区所向”，搭建粤港澳大湾区院士专家重大科技成果转移、转化合作的大平台。我们要站在服务粤港澳大湾区战略的高度，要有“世界湾区格局”，时时、事事对标“中央要求”“省委要求”“港澳所需”“湾区所向”，尽“广东所能”，要尽联盟所能，尽院士联合会所能。

定位要准、工作要实。粤港澳院士专家创新创业联盟是粤港澳高校科研院所合作交流的民间渠道，是非政府组织，是战略联盟，更是工作联盟。联盟是虚，工作可以很实。我们要坚持需求为导向，为联盟成员服务、为粤港澳院士专家服务，特别是要为院士专家创新创业铺路搭桥。

依章治盟、群策群力。联盟是大家的联盟，联盟未来作用的发挥既取决于秘书处工作，更有赖于大家的积极参与，群策群力，齐抓共管，规范管理，加强工作监督，提高工作的透明度，提升联盟的组织力。

同志们、朋友们，众人拾柴火焰高。我们相信，在中国科学院、中国工程院、港澳中联办、港澳特区政府的关心指导下，在广东省委、省政府的正确领导下，在联盟各单位的共同努力下，我们一定能够为院士专家创新创业提供有力支撑，在助力粤港澳大湾区建设成为国际科技创新中心和国际一流湾区的伟大事业中作出更多更大的贡献！

最后，让我们再一次以热烈的掌声感谢各位领导的莅临指导。

---

① 本文是在粤港澳院士专家创新创业联盟第一次主席团会议讲话，广州，2019 年 4 月 21 日.

努力推动压力容器技术进步①

中共中央、国务院印发《粤港澳大湾区发展规划纲要》中指出，要将粤港澳大湾区“建成世界新兴产业、先进制造业和现代服务业基地，建设世界级城市群”，“加快推进珠三角大型石油储备基地建设，统筹推进新建液化天然气（LNG）接收站和扩大已建LNG接收站储转能力”，为大湾区压力容器领域提供了大好的发展空间。国家质监局在“特种设备2025科技发展战略研究报告”中一再强调压力容器安全的重要性。压力容器分会努力发挥自身平台的功能，促进企业、高校和科研机构之间进行友好的合作交流。

本世纪以来，我国压力容器科技工作者解决了过程工业生产规模大型化、介质环境苛刻化及安全长周期运行所面临的诸多技术难题，建立了基于全寿命周期风险可控的设计制造与维护技术方法，使得我国重要压力容器基本不再依赖进口，万台设备失效事故率逐年下降。然而，过程工业装置大型化导致的压力容器重型化问题越发突出，不仅造成耗材、耗能费工、加工制造困难，而且可能存在一定的安全隐患。如何在确保本质安全的前提下实现压力容器轻量化绿色设计制造，如何减少灾害条件下压力容器失效对环境造成的破坏污染，是我国“十二五”期间两个重点研究解决的突出问题。同时，近年来物联网、大数据等现代信息技术与人工智能技术快速发展，给我国压力容器传统技术带来了前所未有的机遇与挑战。如何与现代信息和人工智能技术深度融合，促进我国压力容器技术向数字化、网络化、智能化方向发展，也是我国压力容器技术工作者在“十二五”期间探索解决的一个重要问题。因此，结合中央战略要求，“十三五”期间人工智能发展背景下开展以压力容器材料基因组与增材制造、网络协同制造与智能工厂、智能运行与维护等为压力容器领域的重要研究方向。

近十多年来，随着广东省大炼油、大化工生产基地的建成，中海壳牌、中海炼油及原有的炼油大厂茂名石化、广州石化的扩建装置的相继投产，对压力容器技术的需求日益迫切。广东省压力容器技术发展也需要与经济发展共同进步，因此压力容器学会也肩负着自己使命与责任。自2010年组成第五届理事会，压力容器学会发挥学会跨学科、跨部门和跨地区的特点，结合生产实际的需要，积极开展学术交流，宣传和推广应用科技成果，努力推动我省压力容器技术进步；充分利用我会的人才优势、智力资源和团队力量、搭建联系平台、组织联合申报课题。围绕企业的需求，利用学会优势，积极开展技术服务、技术培训和技术咨询工作，为企业多做实事，认真抓好学会的基本工作，夯实基础，凝聚力量，确保我会稳步发展。

一、积极参与国家、省市课题、标准化建设

压力容器学会积极参与与压力容器相关的国家级课题，参与标准化组织，编制标

① 本文是广东省压力容器学会第五届理事会工作报告，广州，2019年7月4日.

准规范，理事长刘人怀院士成为国家非金属压力管道标准委员会名誉主任，多次参加标准的修订。暨南大学、华南理工大学、广东省特种设备检测研究院、广州市特种承压设备检测研究院等单位都承担了国家、省、市各级的与压力容器相关的科研项目。

学会积极加强与中国机械工程学会压力容器分会联系。

二、跟踪国际压力容器技术领域的研究新动向开展学术研究

压力容器学会积极组织专家跟踪国际压力容器压力管道技术进步和研究前沿，了解国外对有关压力容器压力管道的技术标准和技术法规，以满足省内压力容器压力管道的检验、安全分析与评定的需要。先后有 60 余人次参加了在用压力容器基于风险的检验关键技术、压力容器检测新技术、新工具、新方法、球罐的不开罐检测、管道完整性管理的技术研讨。

学会理事长刘人怀院士出版了压力容器专著《压力容器和压力管道的分析与计算》，由科学出版社于 2014 年出版。

2013 年学会理事长刘人怀院士、学会副理事长王璠教授“典型板壳结构的理论与计算”的研究获广东省科学技术奖励自然科学类二等奖。

积极举办第五届全省压力容器技术学术会议。学术交流是学会工作的重点，也是学会工作的基本任务和根本宗旨，经过积极努力由学会主办、广州市特种承压设备检测研究院承办了第五届全省压力容器技术学术会议。会议于 2012 年 11 月 2 日在广州大厦召开，本次会议的内容丰富，中国特种设备检测研究院贾国栋副院长作了《在役承压设备法规标准体系建立及进展》的专题报告；广东省特种设备检测院李绪丰作了《大型石化装置承压设备定期检验》专题报告；暨南大学宁志华副教授、博士的《焦炭塔鼓胀变形与开裂几个问题的研究》学术报告；广州市特种承压设备检测研究院的《承压设备检验检测新技术》专题报告。学会有 12 人参加了于 2013 年 11 月 9 日至 11 日在安徽省合肥市召开的第八届全国压力容器学术会议，来自国内外有关压力容器科技工作者 600 余人参加会议。学会的 8 位理事在六个分会场分别从容器用钢、材料性能、腐蚀、断裂、蠕变、疲劳、设计、制造、使用管理、标准等方面宣读了论文。

学会的会员在 *The International Journal of Engineering and Science*、《压力容器》、《华南理工大学学报（自然科学版）》、《环境工程》、《中国有色金属学报》、《机械设计与制造》、《焊接学报》、《中国安全生产科学技术》、《中国粉体技术》等杂志上发表 60 多篇与压力容器相关的论文。

学会理事 4 人参加了 2013 年 8 月在西安举行的 2013 年全国力学大会，学会理事长刘人怀院士作为主席团成员参加大会，3 位理事在分会场作了学术报告。

学会理事 4 人参加了 2018 年 5 月在南宁举行的振动工程全国学术大会，并在分会场作了学术报告。学会理事 10 人在广东省特种设备学术年会上作了压力容器相关的学术报告。

三、开展产学研合作研究

暨南大学、广州市特种承压设备检测研究院、中国石油化工股份有限公司广州分

公司、共同开展了“焦炭塔的在线健康监测研究”，于 2013 年 5 月对中国石油化工股份有限公司广州分公司 3 号焦炭塔进行了停产检测，积累了健康检测的数据；2015 年 5—9 月对中国石油化工股份有限公司广州分公司 3 号焦炭塔进行了现场检测。形成了一套在线监测软件，授权了一项发明专利。发表了学术论文“Probabilistic life prediction for crack initiation on coke drums based on Monte Carlo simulation”，《基于外壁监测数据的焦炭塔内壁应力推算方法》；培养了研究生，论文题目《15CrMoR 钢焦炭塔疲劳裂纹萌生与扩展的预测分析》（导师：王璠教授）。“焦炭塔的在线健康监测研究”研究成果获得了中国腐蚀与防护学会科学技术二等奖，中国特种设备检验协会科学技术三等奖，广东省机械工程学会科技进步三等奖。

四、做好国家标准的宣贯和培训工作

国家压力容器压力管道技术管理方面近年来投入大量的人力物力进行开发研究，在技术管理的更新上更是日新月异，新技术新标准不断推出。《中华人民共和国特种设备安全法》于 2014 年 1 月 1 日正式施行，《中华人民共和国压力容器工程师手册》也于 2013 年正式施行。学会与广东省特种设备协会合作，积极进行《中华人民共和国特种设备安全法》《中华人民共和国压力容器工程师手册》培训工作。2013 年开展了“压力容器设计及技术管理”培训。

五、积极推动 RBI 技术在新装置建设前期的应用

在用压力容器及压力管道基于风险的检验（risk-based inspection，RBI）技术在近些年在广东省大型石化企业开展得很好，为企业的安全、稳定、长周期生产起到了很好的作用。学会积极推动 RBI 技术在新装置建设前期的应用工作。通过早日介入，对设计的情况进行分析，在装置投用后对实际检测数据进行对比分析，提出的检验、检修建议更具有针对性。目前在广州石化新建 20 万吨/年高性能聚丙烯装置应用，取得较好的效果。对广东省石化企业设备的安全管理技术起到积极的推动作用。

六、建立了压力容器学会网站

随着网络技术的发展，网站的建立有利用技术的推广，信息的传播，学会于 2011 年 3 月建立自己的网站广东省压力容器学会网站，搭设起了技术交流平台，增强了学会各会员单位的联系，营造了良好的学术气氛。网站分学会介绍、学会新闻、学会通知、学会论文、压力容器法规、学会服务、学会章程等模块。

大力加强高校科技创新[①]

自 1978 年党的十一届三中全会以来，我们国家社会经济迅猛发展，取得了巨大的成就，成为世界第二大经济国，给世界带来了巨大的震撼！中华民族终于再次站起来了！

我们国家的高等学校在这一时期高速发展，为国家培养了大批人才，奉献了许多科研成果。

同时，我们还注意到，我国的制造企业中，百分之九十多无自主知识产权产品。在国际市场出售的商品中，有很多是“中国制造”，但成绩背后的隐忧却不应忽视，因为这些产品所产生的利润中，中国工厂所能获得的仅达 10%，其余 90%都被国外的创业公司或设计公司拿走。

因此，加强我国高校科技管理实在重要。要提高高校科研成果的创新要素，取得世界科技前沿的成果，既要加强基础理论研究，又要更重视将理论成果转换为产品，使我们国家紧跟世界前沿转化为引领世界科技潮流。

“产学研”结合是一个跨行业、跨部门、跨地区合作的系统工程。过去，这结合不太成功，使我国产品的创新性较差。

为此，建议改变我们过去的“产学研”合作模式为“政产学研金”模式，以便实现中华民族的伟大梦想。

政府的地位十分重要，是关键。政府应负责营造科技创新的良好环境，强化政策的激励引导作用，为企业、高校、科研院所营造有利于科技创新的良好环境，引导创新要素向企业集聚。在科技投入、成果奖励、创新平台建设等全面加大扶持力度，同时还应强化考核的促进保障作用。

企业是主体地位，应提升自主研发和技术利用的创新能力。

高校是企业的坚强后盾，要多多提供科技创新成果。要改变过去教师只写论文、只报专利、不重视成果转化的现状。

金融机构要为“产学研”提供经济支持，这是强大支柱。

只有充分发挥“政”和“金”的作用，我国高等学校的科研成果才会真正转化为高科技产品，为我国走向繁荣富强做出更大的贡献。

① 本文是花城论坛讲话，广州，2020 年 7 月 6 日.

创新争先　自立自强①

在深入学习贯彻习近平新时代中国特色社会主义思想之际，在党的二十大即将召开之际，我们迎来了第六个全国科技工作者日。作为一名科技工作者，我很荣幸受到“最美科技工作者”光荣称号表扬，我要万分感谢党和人民的信任！

1978年，邓小平同志在全国科学大会上做出了“科学技术是生产力”② 的重要论断，接着又提出“科学技术是第一生产力”③ 的重要论断，这对推动中国改革开放和社会发展起着决定性的推动作用。

64年前，1958年9月，我到兰州大学读书，正是1956年第一次全国科学大会后提出“向科学进军”的时代，我有幸进入科学研究工作之中，参加“581”工程，成为我国第一颗东方红人造地球卫星兰州大学研制小组的组长。随后，1960年，又参加数学大师华罗庚先生提出的优选法的推广工作。接着，1962年，又开始从事美国宇航之父、力学大师钱学森先生的博士生导师冯·卡门开拓的20世纪板壳非线性力学前沿领域的世界难题：“扁薄球壳的非线性稳定问题”的科学研究工作。由于得到力学大师钱伟长先生的大弟子叶开沅老师的直接指导，我一生的科研工作都与此相关。感谢党和国家的培养和关怀，感谢学校和老师的指导，让我走上了正确的科研道路，为国家和人民做了一点有意义的工作。年轻时，尽管生活条件很艰苦，但对科学前沿的孜孜追求一刻也没有放松。由于家庭和学校的教育，我养成了有好奇心的好习惯。1963年，当大学老师刚开始，我就在完成助教的教学工作和学生政治辅导员工作之外，自己抽时间到工厂访问调查，想寻找有实际工程意义的课题。

那时，美国P2V低空侦察机经常侵略我国领空，我们很难对付。恰好，经过千辛努力，1961年在辽宁庄河市击落一架P2V低空侦察机。这种飞机的关键仪表是高度表，引起关注，国家决定由兰州135厂仿制。高度表的核心元件是波纹圆板，其问题是力学问题。我便大胆地表态，愿意承担任务。他们正在想办法，却无从下手，所以很高兴地同意了我的请求！

经历了千辛万苦，那时又无法用计算机，是靠手算完成了这个任务。14年多的努力，总算成功。国际上波纹圆板非线性弯曲理论研究工作的难题终于得到解决。

几十载科技事业痴心不改，六十多年坎坷岁月矢志不渝，后来我又参加神舟飞船、航空母舰歼15B舰载机、精密仪表、高压容器制造等科研项目，一直做到现在。

我还担任大学校领导多年，特别是担当暨南大学的领导，更深知办侨校的重要性。通过实施“侨校＋名校”发展战略和一系列教学改革创新措施，努力将海外华侨和港

---

① 本文是在广东“最美科技工作者”发布仪式座谈会上的讲话，广州，2022年5月30日.

② 《在全国科学大会开幕式上的讲话》，http：// www. qstheory. cn/books/2019-07/31/c_1119484755_17. htm [2023-11-08].

③ 《【党史声音日历】“科学技术是第一生产力”重要论断提出》，http：// news. cnr. cn/dj/20210905/t20210905_525590964. shtml[2023-11-08].

澳台学生培养成为拥护“一国两制”，为港澳繁荣稳定、祖国和平统一作贡献的专业人才，使华人学生成为知华、爱华、友华的专业人才。

虽已耄耋之年，我仍然听从党和国家的召唤，在科研领域工作。这是使命，更是责任。

现在我正在积极投身餐厨垃圾处理的事业之中。因为，我深知“绿水青山就是金山银山”① 科学论断的深刻内涵。13 年前，省科协创建广东省科普志愿者协会，我有幸被科协领导聘为会长，并与省科协郑文丰处长合作，那时垃圾围城，便开展垃圾分类科普工作。由于好奇心，我发现垃圾处理是一个问题，目前技术未能处理完美，特别是餐厨垃圾治理更是大事。那时，广州的餐厨垃圾每天就有万吨之多，对城市污染，对人民健康，影响巨大。实际上，人类五千多年来，都未找到好方法解决。所以，我立即组建团队，跨界融合，历经十二年的艰苦努力，终于完成了餐厨垃圾的最先进处理技术——“联合生物加工技术”的研发，通过 36 小时发酵，把餐厨垃圾转化为非粮乙醇、燃料油脂、酒糟蛋白粉等系列产品，实现了对餐厨垃圾处理的完美无害化和 100%资源化，既与人民健康大大有关，又能产生巨大经济效益。这是属于国际领先的科技创新成果，第一个示范工厂（每日处理 100 吨餐厨垃圾）已于 2020 年 7 月在四川成都建成，目前正在全国推广。作为广东人，我希望我们的成果能尽快在广东落地，能为广东作更大贡献。

科技兴则民族兴，科技强则国家强。重温中国百年来科技发展的历史，让我们万分感慨！自立自强是中华民族立于世界民族之林的奋斗基点，自主创新是我们攀登世界科技高峰的必由之路，创新争先让我们在通往未来的道路上行稳致远。我们广东是国家改革开放的排头兵，作为一名科技工作者，为能在广东做点贡献而感到万分高兴。让我们一致努力，为实现中华民族伟大复兴而努力奋斗！

---

① 《“绿水青山就是金山银山”理念的科学内涵与深远意义》，http：//www.xinhuanet.com/politics/2020-08/14/c_1126366821.htm[2023-11-08].

刘人怀院士文集

第五卷

刘人怀◎著

科学出版社
北京

内 容 简 介

本文集由著者在力学和管理科学领域60多年中所发表的文章汇编而成，分为上、下篇。上篇是力学部分，结合工程需要，对单层板壳、波纹板壳、双层板壳、夹层板壳、复合材料层合板壳和网格扁壳等六类板壳的非线性弯曲、稳定和振动问题，以及厚、薄板壳弯曲问题进行理论探索。下篇是管理科学部分，联合实际，对工程管理、公共管理、工商管理、科技管理和教育管理的问题进行了研究。

本文集可作为高校力学和管理科学专业，以及相关专业老师、研究生、本科生的参考书，适合政府、高等学校、科研院所、科技社团、企业等领域的领导、管理人员和科研人员参考阅读，又可作为航天、航空、航海、机械、建筑和交通等工程设计师与工程师的设计制造指导书。

GS京（2025）0637号

图书在版编目（CIP）数据

刘人怀院士文集 / 刘人怀著. -- 北京 ：科学出版社，2025. 4. -- ISBN 978-7-03-080897-4

Ⅰ. O3-53；C93-53

中国国家版本馆CIP数据核字第2024JH1723号

责任编辑：陈会迎/ 责任校对：姜丽策

责任印制：张 伟/ 封面设计：有道设计

科 学 出 版 社出版

北京东黄城根北街16号

邮政编码：100717

http://www.sciencep.com

北京中科印刷有限公司印刷

科学出版社发行 各地新华书店经销

*

2025年4月第 一 版 开本：787×1092 1/16

2025年4月第一次印刷 印张：153 3/4

字数：3 646 000

定价：698.00元（全五卷）

（如有印装质量问题，我社负责调换）

目　　录

第十四章　教育管理 …… 1921
欢迎新同学们 …… 1921
谈谈课堂教学中的几个问题 …… 1923
谈谈大学的学习生活 …… 1924
前进中的经济管理学院 …… 1927
情牵母校 …… 1930
（日本）新世纪中文电视学校校长致辞 …… 1931
当代科技发展与大学理念和人才培养 …… 1932
加快我国高等教育进入世界先进行列 …… 1940
关于高考的一点浅见 …… 2019
认真做好高校力学教学指导工作 …… 2021
我国力学专业教育现状与思考 …… 2022
浅谈高等学校科学管理“三”字经 …… 2027
我的语文观 …… 2040
爱低碳生活　创绿色校园 …… 2041
深化力学课程改革 …… 2043
从学生会主席到大学校长之路 …… 2045
试答“钱学森之问” …… 2048
认真开展高等学校教育教学研究 …… 2055
加强重大工程结构安全领域国际合作 …… 2056
知识交融　创新成才 …… 2058
从教感想 …… 2065
珍惜时间　勤奋成才 …… 2068
以我个人经历及家庭经历浅谈家庭教育 …… 2072
教化育人　师法天地 …… 2075
感谢党和国家的培养和教育 …… 2078
澳门民办高校发展现状与前景 …… 2081
八十有感 …… 2087
新都二中——我的母校 …… 2091
推动暨南大学“重大工程灾害与控制”教育部重点实验室跨越式发展 …… 2092
高等教育管理实例：暨南大学“侨校＋名校”之路（1995—2006） …… 2101
1. 暨南大学校长任职仪式讲话 …… 2101
2. 加强基础　从严治校　培养高素质人才 …… 2104

3. 根据侨校特点改进教学工作　采取有效措施提高教学质量 …… 2107
4. 提高认识　办好成人教育 …… 2116
5.《红杏枝头春意闹——暨南大学成人高等教育毕业生业绩选》序 …… 2121
6. 暨南大学“211 工程”部门预审汇报 …… 2122
7. 在暨南大学“211 工程”部门预审总结会上的讲话 …… 2136
8. 庆贺九十华诞　创建一流大学 …… 2137
9. 开拓创新　共建附属医院 …… 2138
10. 办出特色　办出水平 …… 2139
11. 总结经验　深化改革　加快科技发展 …… 2147
12. 积极主动地为侨务工作服务 …… 2150
13. 用现代化管理促进高等教育事业的发展 …… 2156
14. 深化教育改革　提高教学质量 …… 2158
15. 暨南大学兴办高等华侨教育的历史回顾与展望 …… 2160
16. 华文学院越办越好 …… 2164
17. 在高校党的建设中贯彻落实邓小平“从严治党”的思想 …… 2165
18. 脚踏实地　循序渐进 …… 2169
19. 举行全国 100 所“211 工程”学校赠书仪式 …… 2175
20. 转变观念　量化考核　优劳优酬 …… 2176
21. 同质同水平异地办学 …… 2181
22. 暨南大学国际化之路 …… 2183
23. 面向新世纪的创新教育 …… 2186
24. 坚持社会主义办学方向　办好华侨高等教育　为海内外培养高素质人才 …… 2191
25. 辉煌与梦想 …… 2203
26. 狠抓办学质量　走“侨校＋名校”之路 …… 2204
27. 暨南大学的特点和优势 …… 2214
28. 弘扬中华民族文化 …… 2216
29. 珍惜“暨南人”光荣称号 …… 2219
30. 饮水思源 …… 2221
31. 为祖国侨务事业和华侨高等教育做出新的贡献 …… 2223
32. 建设国际化、现代化、综合化的高水平社会主义华侨大学 …… 2227
33. 深深感谢校友深情 …… 2229
34. 办好研究生教育至关重要 …… 2231
35. 进一步提高干部人事档案工作的管理水平 …… 2233
36. 胸怀世界　放眼未来 …… 2237
37. 持之以恒　依法治校 …… 2240
38. 全球化进程与华侨高等学府的重要使命 …… 2241
39. 努力完成高校扩招任务 …… 2247

40. 答谢珠海人民 …… 2254
41. 坚决反对和防止腐败是学校重大的政治任务 …… 2255
42. 寄语中层干部 …… 2257
43. 暨南大学的创新发展之路 …… 2266
44. 弘扬中华文化　发展华文教育　传播华夏文明　促进文化交流 …… 2274
45. 为迎接暨南大学百年庆典增添新的光彩 …… 2276
46. 发挥优势　深化改革　保证重点　改善条件　提高质量 …… 2277
47. “侨校＋名校”的发展定位 …… 2283
48. 教学是学校的生命 …… 2290
49. 大力发展具有侨校特色的研究生教育 …… 2298
50. 贺暨南大学澳门校友会会长就职 …… 2306
51. 忍耐是一个人成功的秘诀——与暨南学子谈成才 …… 2307
52. 标准学分制的研究与实践 …… 2313
53. 积极服务海外华侨华人社会 …… 2318
54. 暨南大学百年校庆公告（第一号） …… 2320
55. 面向海外　面向港澳台　为祖国统一大业服务 …… 2321
56. 坚持大力发展研究生教育 …… 2331
57. 广纳贤才　全球招聘院长 …… 2333
58. 学会普通话　走遍天下都不怕 …… 2336
59. 亚洲　青春　竞技 …… 2337
60. 贺暨南大学香港社会学同学会成立 …… 2338
61. 立足侨校　服务学生　全面推进我校学生德育工作 …… 2340
62. 图书馆是大学的心脏 …… 2343
63. 生逢其时的学院 …… 2344
64. 述职报告 …… 2345
65. 衷心祝愿暨南大学的明天更美好 …… 2361

附录一　刘人怀大事年表 …… 2364
附录二　刘人怀主要论著目录 …… 2384

第十四章　教育管理

欢迎新同学们[①]

在三十年国家大庆佳节的前夕，在中国科学技术大学校庆二十一周年来临之际，你们从祖国的四面八方，从祖国北面的翠绿的兴安岭，从南面的碧波万顷的南海之滨，从西面的边陲新疆到东海之滨，迢迢千里，以优异的成绩来到合肥，来到亿万中国青年万分向往的中国科学技术大学求学读书，请让我代表中国科学技术大学的全体教师，热烈地欢迎中国科技大学的新同学们！

你们在过去的中小学生活中，勤奋刻苦，不仅有为"四化"贡献青春的崇高理想，而且勤于为理想而艰苦奋斗，从而把理想与现实在奋斗中统一起来。祝愿每个有理想的青年，都从眼前的现实起步，以艰苦卓绝的奋斗，作为通往理想境界的阶梯，为祖国实现"四化"进而实现共产主义贡献毕生精力。

今天，我们为实现"四化"而奋斗，这是中华民族历史上从未有过的创举，其任务之艰，难度之大，更需要亿万人民，特别是青年一代，艰苦奋斗，克勤克俭，一往无前！

理想的阶梯，属于珍惜时间的人。富兰克林有句名言："你热爱生命吗？那么别浪费时间，因为时间是组成生命的材料。"鲁迅以"时间就是生命"的格言律己，献身无产阶级文学艺术事业 30 年，始终视时间如生命，笔耕不辍。

理想的阶梯，属于刻苦勤奋的人。

理想的阶梯，属于迎难而上的人。奋斗的必要，恰恰是由于困难的存在。在通往"四化"的征途上，坎坷、曲折、荆棘、浪涛是不会少的。只有以不懈的韧劲，一级级攀登阶梯，才能一步步接近那光辉的理想之巅！

在这举国上下、万众一心渴望早日实现"四化"的时刻，我们对同学们有着特别殷切的期望，这期望将鼓舞着千千万万新中国青年成长。

实现"四化"的重任，历史地落在你们这一代青年肩上。为了适应这个伟大转变，为了使这一代青年的青春，在"四化"的进程中变得绚丽多彩，你们必须成为具有真才实学的专家和各种人才。

青年一代的知识水平和科学素养的高低，直接关系到"四化"事业的成败。科学的成果是美妙的，而探索科学的道路是崎岖的。科学之宫的大门，对于勤奋学习、勇

---

① 本文是中国科学技术大学开学典礼上代表教师的讲话，合肥，1979 年 9 月 1 日；教育与科技管理研究，北京：科学出版社，2016，143-144.

于探索的勇士是敞开的，慷慨的。只有勤奋地学习，刻苦地学习，才能掌握科学，从而掌握在“四化”中大显身手的武器。

祖国“四化”的前景给你们青年同学们开启了无限美好的未来，也带来了艰巨的责任。新同学们，新的生活开始了！祖国在期待你们！九百六十多万平方公里的锦绣河山，正等待你们去描绘新图。投身到这场火热的斗争中去吧，你们奋斗的青春将为祖国，增添光彩！

谈谈课堂教学中的几个问题[①]

学校是培养人才的“工厂”。我们要为四个现代化早出人才，多出人才，就必须大力提高教学质量。在提高教学质量这一问题上，讲究课堂的教学法是一个很重要的方面。下面，就这一问题浅谈几点个人看法。

一、关于课堂上的“因材施教”问题

辩证唯物主义认为，世界万物都存在着差异，差异就是矛盾。同一班级的学生，存在着很大的差异。“因材施教”正是在承认这种差异的基础上，采取的唯物主义的教育原则。在课堂上，如何贯彻“因材施教”的原则呢？笔者认为，应该把教学的着眼点放在大多数学生上面，但同时，又要适当照顾优等生和较差的学生。在这一原则指导之下，要使讲授的内容为多数人掌握，同时，又讲授一点高、难的东西，使优等生也有“偏饭”可吃。而在讲授的重点内容上，又要尽量细讲，以使较差的学生也能吃透。这样，优等生可以深造，较差的学生也会逐渐弥补缺陷，增加信心，赶上全班的队伍。

二、切忌满堂“灌”，注意“少而精”

笔者在刚当老师初讲课时，有这样的教训。觉得交代全面、细致，尽量多讲。才能使学生将所讲授的内容学到手。哪知，适得其反，学生往往越学越糊涂。后来，笔者反复揣摩老教师的讲课方法，才渐渐体会到面面俱到的坏处、满堂“灌”的弊病和“少而精”的妙诀。从一堂课的内容看，总是有主次之分，轻重之别。我们必须把讲授的精力放在教材的精华、关键、重点、难点上，引导学生牢牢把它们抓在手里。相对来说，对次要地方只要略微讲讲就可以了。这样讲课，表面上看来讲得“少”，但是却抓住了教材的重点和精华，即抓住了主要矛盾。学生听课后将会理解透彻、记忆深刻，其余问题也会迎刃而解。

三、注意讲课的语言

讲我们力学方面的课程，一般说来，内容比较枯燥。如若不重视自己的课堂语言，更会使讲课效果大受影响。讲课时，如果语调平板、语病很多、势必使学生厌烦、昏昏欲睡。结果是你出了一身大汗，他觉得是受罪。如果你声调始终很高，语速又快，就像机关枪射击，那也会使学生大脑如遇洪水一般，毫无思维余力，从左耳进来，又很快从右耳跑出。所以，重视课堂语言实在不是一件可忽略的小事。因此，讲课时，首先需要语言简练、准确、生动。这样，学生头脑会始终清醒，能激起思维的兴趣。其次，要留有适当的间隙，使学生有思维的时间和精力，充分发挥学生的主观能动性。总之，课堂上知识的传授主要是靠语言这个工具来完成的，所以大有我们重视的必要。

① 本文原载《中国科学技术大学简报》，第103期，1979年9月27日；教育与科技管理研究，北京：科学出版社，2016，142-143.

谈谈大学的学习生活[①]

一、大学生活的重要性

一个人一生的各个时期，对于读书人来说，没有比大学这个黄金阶段更重要的了。青年人是初升的太阳，是未来的科学家，是未来的明星，而大学阶段是你们今后为人类服务、为祖国四个现代化贡献力量的基础。俗话讲："万丈高楼平地起"，就是这个道理。今后的高楼，也就是你用自己的学识去建造，去为大家使用、居住。这个基础，就是马拉松赛跑的起跑点，是科学研究赛跑的起点，最后，就看胜利者是谁?

胜利者将是一些特殊的青年，而不是赛跑起点的全体选手。而且，重要的是你参加哪种赛跑，是县级、省级或国家级！同学们，你们选择得很好，是我国称之为重点之重点的中国科技大学。这里为你们提供了好的环境，好的设备和资料，好的老师。这将为你们的进步提供最好的动力。中国科学技术大学是培养研究人才的基地，是中国科学院唯一的一所院校。当然，不否认，科学家也可以是自学成才的，也可以来自非名牌大学。

同学们，你们首先就要弄清楚，你们的奋斗方向，并为此而努力。

科学技术有多方面内容，俗话说："七十二行，行行出状元。"当然，有的行当要重要些。而且，在重要的行当上，作最重要的贡献，这是每一个人都喜欢的。

我们近代力学系包括四个专业：高速空气动力学专业、飞行器结构力学专业、工程热物理专业和爆炸力学专业，这四个专业都非常重要，属于力学和物理学科，涉及的面很广，包括航天、航空、航海、建筑、水利、交通、机械、采矿、冶金、化工、石油等领域。它们深刻地改变着工程设计的思想，为工业、农业和国防服务，同时，也为认识自然建立功劳，如宇宙论、天体演化、地球起源、星系结构、天体爆炸、太阳风、行星磁场、大气、洋流、海浪、地壳运动、地幔对流等。力学之所以有这样的广泛性，起因于力学是研究自然界中最基本、最简单的运动形式，即位置移动。力学学科相当重要，正如马克思说，力学是"大工业的真正科学的基础"；恩格斯说，"力学是最基本的自然科学"。当然，不能说"力学可以包打天下"。学习力学，有前途。今后内容甚多，任务很繁重，要向宏观、微观进军，包括岩体力学、地球力学、物理力学、等离子体力学、化学流体力学、爆炸力学、生物力学、理性力学等。

力学所起的作用很大。从19世纪末到20世纪前60年，力学工作者对当时的航空技术和航天技术作出了震撼世界的成果。以飞机为例，人类最初是幻想。在人类的历史长河中，一直都想像鸟那样能飞上天。在1500年前，《述异记》卷中就写了一个美丽的神话："鲁班刻木为鹤，一飞七百里。"通过神话幻想对飞行作了科学预见的描绘。又如"嫦娥奔月"，可以说是达到了神话幻想的顶峰，竟然想到吃了不死药的美丽妇

① 本文在中国科学技术大学近代力学系新生会上的讲话，合肥，1980年11月3日；教育与科技管理研究，北京：科学出版社，2016，144-148.

女，飘飘然飞上了月宫，幻想月亮是可以住人的地方。

接着，人类就想模拟鸟飞行。在古代中国和欧洲，都有人尝试用鸟羽毛做成人的翅膀，绑在人身上，想用扇动翅膀的方法来实现飞行。

两千年前，我国西汉时代的韩信就创造了风筝。风筝比空气重，飞上了天。

1783 年，使用热空气，出现了飞上天的气球。

1882 年，在工业革命后，俄国的莫查伊斯基以风筝为榜样，制成了世界第一架用蒸汽发动机和螺旋桨推进的飞机，飞行时速达 40 千米/小时。

1903 年，美国莱特兄弟又前进一步，制成装有活塞式发动机和螺旋桨的飞机。

1943 年，飞机时速达 620 千米/小时。

1953 年，飞机时速达 1300 千米/小时。

1962 年，液体火箭发动机时速达 6600 千米/小时。

与航空技术有关的两个重要理论：升力理论和附面层理论，都是力学的突出贡献。有了它们，才有航空技术成立的根本条件，才有了研究飞机的基本原理，从表面不光滑、刚度小的结构进展到刚度大的全金属的流线性结构。接着，解决了声障问题。飞机速度接近声速时，产生了激波，飞机的阻力很快增大。为了解决这个问题，花了六七年时间，从 1100 千米/小时到 1300 千米/小时，产生了气动力学，也叫可压缩流体力学，才解决了这个问题。到了 20 世纪 50 年代，洲际导弹需求，促使航天技术发展，产生了再入大气层的加热问题，达到几千度高温，用烧蚀防热的办法解决了这个问题。

在这一阶段中，产生了许多英雄豪杰。近代著名的科学家有两人，第一个是德国的近代物理学家爱因斯坦，提出了广义相对论，奠定了近代引力理论的基础。第二个是近代力学家冯·卡门，1963 年，美国白宫将美国第一枚国家科学勋章授予他。他在工程学、自然科学以及教育事业的领域中作出了无与伦比的贡献。他对我们人类的当今生活的影响，大概胜过当代任何一个科学家、工程师。喷气式飞机每小时飞行上千公里，导弹可以打击上万公里以外的敌人，火箭能探测遥远的行星，都已成为现实，其中许多关键的环节是因他成功的。中国著名的“三钱”，其中两个人与他有关：我们的系主任钱学森是他的第一个中国博士研究生；我的祖师爷钱伟长先生在获得了应用数学博士学位后，就去到他的身边，成为他的助手。

归结起来，在我们近代力学系读书，是很有前途的。我们系是值得大家热爱的！

二、在大学阶段如何学习

在本科阶段，作为一名学生，必须要重视以下几点。

1. 要有科学献身的精神

要在未来有所成就，就必须在学习中，不怕吃苦，要勤奋。读书要有孜孜不倦的精神，着迷的精神，穷追的精神，就像古人那样，“头悬梁，锥刺股”似地读书。

明朝的李时珍，走万里路，访问上千位老农、土医、渔民和猎人，有时几天不下山，饿了吃干粮，天黑了在野地里过夜，用了 27 年时间，才完成巨著《本草纲目》。

居里夫妇用了几年时间，才从三四吨的沥青矿渣中提取出 0.1 克氯化镭。

冯·卡门为了献身科学，终身不婚，平常不修边幅，不烫裤子，衣服口袋常露出皱纸片。为怕上课迟到，雇人专门提醒。

陈景润为了攻克科学难题，时常不修边幅，走路都碰到树上。

牛顿想问题时曾把放在鸡蛋旁边的怀表放在水锅里当鸡蛋煮。

法国大科学家安培去教室途中，路过塞纳河边时，捡起一块鹅卵石放在衣袋中。边走边想问题，隔一会儿，漫不经心地把衣袋里的怀表扔进了河里。到了教室，从口袋里取出来一看，发现是一块鹅卵石。

2. 要善于使用时间

要重视时间的使用，不要浪费时间。俗话说："一寸光阴一寸金，寸金难买寸光阴。"

苏联一位著名的生物学家，叫柳比歇夫（1890～1972），从 26 岁开始，实行"时间统计法"，整整 56 年，直至 1972 年去世。他每天都要把度过的时间结算，每月一大结，年终一总结。阅读《人类的进化》，全书 372 页，花 6 小时 45 分钟。开会，看电影，坐车……都统计。他具有强烈的时间感，往往不看表，误差极小。非完整时间，像"坐车"时，就用来学英语。他一生出版 70 多部学术著作，写了一万二千五百张打字稿的论文和专著。

3. 要有独立的创造精神的培养

在大学读书时，继承是主要的，但要培养思想的火花。我读大学一年级时，参加了中国第一颗东方红人造卫星的研制小组工作，对我培养创造精神影响极大。科学就是创造，不断创新，不断独立思考。即使外界压力很大，也要继续往前，要有批判的精神，要多加问号。

爱因斯坦在 1921 年获得诺贝尔物理学奖。他的获奖，不是由于他的最出色的贡献、提出了相对论，而是因为量子理论。当时，许多人还不承认他提出的相对论。

4. 要尊敬老师，善于继承

我们国家有一句名言，叫"名师出高徒"。杰出的科学家和其领导的研究机构，往往可培养出几代优秀的科学家。如冯·卡门和其学生钱学森、钱伟长。又如美国的 100 多位诺贝尔奖得奖者，半数以上都向名师学艺。1906 年，英国剑桥大学的物理学家汤姆逊，因发现电子而获诺贝尔奖，以后他培养的 9 个人也获得了诺贝尔奖。这 9 个人之一，英国核物理学卢瑟福在 1908 年因元素蜕变的研究获诺贝尔奖，其后他的学生中又有 11 个人获诺贝尔奖。卢瑟福的学生丹麦物理学家玻尔，于 1922 年因原子理论获诺贝尔奖。随后，他的学生中又有 7 个人获诺贝尔奖。

5. 要有坚实基础和广博的知识

在大学学习阶段中，首先要把主要精力和时间用在你所选择的专业方向上，以便打下坚实的知识和能力基础。同时，也要掌握一些其他学科的知识。因为在未来的工作中，往往会碰到较复杂的问题，这时，你就能得心应手去完成。

6. 要成为德智体合格人才

学习中，首先要加强自己的道德修养，学好政治课，确立为中华民族献身科学事业的精神，这将是你未来学好专业功课和奉献给祖国的最大学习和工作动力！

然后，学好专业课，并注意体育锻炼，成为德智体全面发展的人才，这样你才有知识和能力的专长，加上健康的身体，去为祖国实现社会主义现代化奉献力量。

前进中的经济管理学院①

上海工业大学经济管理学院于1985年3月5日经上海市人民政府批准正式成立。它的前身是上海机械学院冶金分院，于1978年由上海机械学院、上海市冶金局、闸北区三方联合创办，属上海市冶金局主管。设有冶金、冶金机械和工业自动化等三个系，属工程技术类院校。1980年起更名为“上海工业大学分校”，改设了企业管理工程系，并开始注重管理专业的建设。

由于上级领导的关心和重视，社会各方的支持，在全院教师、职工共同努力下，学院从建院至今各方面有了很大发展。尤其是近几年来发展较快，学院已初具规模。1985年建院初期学院仅有一个系和一个专业，现已设有三个系、三个专业、一个研究所。

学院现有教职工215人，其中，教师112人，职工103人。

学院现有各类学生786人，其中，研究生9名，本科生448名，专科生37名，劳模学生27名，夜大学生195名，总工程师岗位职务班37名，厂长经理培训班33名。

为适应社会对各类管理人才的需要，学院以培养研究生、本科生为主，还开设了干部专修科和劳模大专班。同时开设了总工程师、厂长、经理培训班，大力开展干部培训和成人教育，多层次地培养管理人才已形成了经济管理学院办学的一大特色。

1980年以来已为上海培养了管理人才3786名。其中，研究生33名，总工程师培训310名，厂长、经理培训2354名，劳模和干部专修科410名，夜大本科171名。

在研究生培养方面，我院近几年来有了很大进展。自1983年起，我院的管理专业就开始招收研究生。这对新办的专业来说确有一定的难度。刚开始时，研究生生源不足，要靠从报考其他院校的研究生生源中调剂。经过我们的努力，1987年起情况有了很大的变化，生源充足，报考数与招生数的比例不断上升，近年来始终在15∶1左右。研究生的论文质量也得到了同济大学等专家教授的肯定与好评。

对本、专科生的培养，学院领导始终把学生思想政治工作放在首位。经常对学生进行时事政治教育，进行理想、道德、情操和纪律教育，进行爱国主义和普法教育。近年来，通过建立班主任工作汇报制度及由部分干部教师参加的督导员制度，并根据我院学生走读，在校时间少，与社会家庭接触多的特点，采取家庭访问和邀请家长来院座谈等形式互通信息，形成一个学校、家庭齐抓共管的局面。同时，学院领导针对学生的思想热点，经常邀请现任厂长、经理的校友来院作“大学生成才之路”“怎样当好一名厂长”“企业家应有的素质”等报告，使学生了解社会对人才的需求。通过这一系列教育措施，学生的政治思想素质有了普遍提高。

为了加强对学生的基础理论教育，使之更好地适应上海经济建设和社会主义发展的需要，在加强基础理论课，拓宽专业知识面和提高实践能力等方面采取了必要的措

① 本文原载《工大三十年》，上海工业大学校刊编辑室主编，上海，1990，129-134.

施。组织教师制定合理的教学计划，并根据发展需要及时修订计划，增设各类选修课。到目前为止，全院已先后开设了三十多门选修课。同时加强了学风、校风的建设，树立了良好的学习风气。特别在计算机和外语教学方面成绩显著，以全国大学生四级统考为例，我院历年都取得较好的成绩。1987 年我院学生及格率为 55.5%，超过了全国重点高校平均 37.2%的及格率。1990 年及格率达 58.9%。同时，学院还重视学生身体素质的提高，积极开展体育活动。学院师生克服场地狭小、体育设施简陋的困难，因地制宜开展小型多样的体育活动，取得了良好的效果。近年来，在全校性体育项目比赛中，我院学生取得了较好的成绩，如长跑、篮球、足球等曾多次获得校冠军，在 1987 年、1988 年全市高校篮球联赛中，我院学生男子篮球队连续荣获高校乙组冠军。

学院根据社会对管理人才的需求，结合自身的专业特色，为上海培训了一批高层次的管理人才，赢得较好的社会声誉。从 1980 年起，我院就开始试办企业的厂长、经理培训班，每期半年左右，使在职的厂长、经理学习了现代化企业管理的知识，提高了企业管理水平。十年来，学院共培训了厂长、经理 2354 名，经我院培训的厂长、经理在全国统考中取得优异成绩，获得优秀成绩人数的比例高于全市平均水平。

近年来，我们还走向社会，分别在普陀区和宝山区开办厂长、经理培训班，受到了高教局、市经委的嘉奖，被评为市成人教育的先进典型，还召开了现场会。1985 年受中央组织部和国家教委委托，我院承担了培训总工程师的任务，现已举办了 10 期，培训了 310 名总工程师，使我院成人教育的办学层次又提高了一步。

为了满足企业不同层次管理人才的需求，我院还开设了两年制的干部专修科，使工矿企业中的一些中层以上的干部通过成人高考来我院脱产学习两年，取得大专学历。实践证明，这些学员具有丰富的实践经验和管理能力，学得好、学得快，掌握了管理理论知识，回企业后马上能用，深受企业的欢迎。从对我院专修 811 班的跟踪调查来看，全班 46 人毕业后分别担任了企业各级领导，挑起了企业管理的重担。

根据市总工会的要求，我们还举办了劳模大专班，已有 130 名劳动模范先后经过两年时间的学习取得大专学历。

对于高等院校来说，教学和科研犹如两条腿，缺一不可。学院在搞好教学的同时，重视科研工作的开展，发挥研究所、研究室等专业机构人多力量强的优势，承接国家和省市的重大科研项目和纵向课题。还鼓励教师利用厂长、经理等校友积极开发横向课题，1985 年来，共承接科研项目 67 项。其中，“动作因素分析法”获 1985 年上海市科技进步奖三等奖；“高技术开发区论证”获 1987 年上海市科技进步奖二等奖；“崇明县 2000 年经济、科技、社会发展规划”获 1987 年上海市科技进步奖二等奖。1985 年以来全院科研经费达 138.1 万元，其中 1987—1989 年合计为 126.4 万元。

随着科研工作的广泛开展，教师的学术水平有了较大提高。1985 年以来在国外和国内一级刊物上发表的论文有 63 篇，出版专著 12 部，译著 7 部 502 万字。为了总结广大教师的科研成果，学院组织了专门力量收集、整理我院教师的论文，组织出版《上海工业大学经济管理学院论文汇刊》。1989 年学院在资金紧张的情况下还抽出 5 万元经费作为学院提供给全院教师的科研基金，目的是使更多的教师有机会参加科研，1989 年度有 73 人次教师提出了 19 项科研项目。

近几年来，学院在抓紧搞好教学、科研工作的同时，还加强了师资队伍的建设。师资队伍是办学的基础，办学历史短、新开专业多，师资不足是我院的突出矛盾。1985 年经济管理学院建院初期，我院仅有教师 75 名，其中副教授 3 名、讲师 46 名。通过 5 年的努力，师资队伍有了很大的发展，现有教师 112 名，其中教授 5 名、副教授（包括副研究员和高级工程师）16 名、讲师、工程师 71 名。几年来，我们依靠上级的支持及自身的努力，发展了师资队伍，同时采取了各种途径来弥补因师资不足而影响的教学质量。如：通过上级领导的支持，从其他院校调进部分骨干教师；选用部分本校和外校的本科生和研究生；聘请兄弟院校、科研单位或社会上学有专长的人才作为我院的兼职教师；通过国际学术交流的途径，先后请美国、加拿大、英国、荷兰、日本、中国香港等国家和地区的专家教授来我院讲学或直接为学生授课，几年来已有 14 批外籍专家和教授来院讲学、上课，加强和开阔了学生的外语水平和知识面；选派教师在国内或去国外有关院校进修。几年来，我院已选派了 3 名教师在国内、13 名教师去国外进修学习，为中、老年教师配备好助手，以老换新，提高师资水平，建立学术梯队，以便后继有人。

几年来，学院在教学、科研、师资队伍建设等各方面的工作得到很大发展，取得一定的成绩，这是与艰苦创业、自强不息，改善办学条件的努力分不开的。经济管理学院的校舍是由一所中学改造的，面积小（占地面积 8.88 亩），校舍差（校舍总面积 7000 多平方米），办学条件非常艰苦。这些年来，在上级领导的关心和社会支持下，依靠全体师生员工的努力，我们的办学条件已有了较大的改善。如：1983 年由市经委支持拨款建造一幢 3200 平方米的教学大楼。1987 年以来，节约开支，利用有限的经费在原有 8 台微机的基础上，陆续增添了 26 台微机，改造了学院计算机房，成立了计算中心。计算中心在教学科研中充分发挥作用，使每个学生上机时间达 50 学时以上，并承担了 17 项科研项目（其中已完成 12 项）。1988 年学院改造了外语语音室，购置了先进的语音设备，使我院的外语教学设施有了改善。学院还因陋就简盖起了 200 平方米的简易室内体操房，添置了体育设备，使体育教学条件得到改善。同时，自己动手完成北楼的加层，增加了 200 多平方米的办公用房，缓解了办公用房紧张的局面。通过三年多的不懈努力，用自筹资金建造的 3376 平方米的七层教学、食堂综合楼目前已破土动工，一年后建成，这将使学院的办学条件得到更大的改善。

通过对经济管理学院近几年工作的回顾，我们看到，在校党委和钱伟长校长的领导下，学院各方面都有较快的发展。但是，由于学院底子薄，建院历史短，各方面条件差，又远离校本部，所以学院在继续发展的道路上还会遇到很多的困难。但我们相信，在校领导的关心支持下，只要全院师生员工团结一致，同心同德，发扬自强不息、艰苦奋斗的精神，管理学院的各项工作定将会做得更好，学院的前景将是十分美好的。

情牵母校①

温江中学是我的母校。

斗转星移，岁月流淌，转眼间，阔别母校已经 38 载。母校，是我永远魂牵梦萦的地方。那里，尽管我没有亲人，可是无数次在梦中都看到了母校。三年的学习生活，使我终生难忘，带给我一生的影响。我珍爱母校的心情经常在胸中荡漾。

今年四月，我有幸作为四川联合大学“211 工程”预审专家回到成都。在会议上认识了四川省教委王可植主任。他在会议结束的第二日，亲自陪同我回到了久别的母校。

再度踏进母校的校门，我的心情格外激动，思绪如潮涌。

星期日的校园，显得静谧、舒适。徒步在校园的路上，沿途的一草一木，一砖一石，一弯一角，不少都是成长时的烙印。这次重游，许多被遗忘的往事如泉涌般闪现在眼前。想起昔日，我们班级是留苏预备班，我曾任班长，同学们团结友爱，学习功课勤奋积极。任教的老师和班主任工作相当认真尽责。学习俄语课时，徐德福老师见我笨嘴拙舌，不会发颤音“p”，就用激将法说我“永远学不会”。为此，我苦练了三天，终于发出正确的颤音。为了达到劳卫制体育锻炼标准，全校跑得最快的黎秉昌同学为我领跑，使我百米跑的成绩迅速提高到 13.7 秒。每一个影像来得那么清晰，令我记忆犹新，使我回忆起年轻时曾拥有的热情。

见到了教我数学的钟石钧老师，教我音乐的庹老师。他俩的课教得真好，令我终生受益。在这里，我要再次说，感谢你们，尊敬的钟老师、庹老师和其他未拜见的老师们。

竹贵荣校长和张诗德副校长等同志详细地向我介绍了母校的变化和进步。现在，我的母校已面貌一新。教室以及教学设施很好，教学质量优良。我为母校享誉四川省而自豪!

我怀念在母校的短暂而难忘的三年学习生活。在母校建校七十周年到来之际，我衷心祝愿拥有悠久历史的母校焕发青春，早日建成为既有时代气息，又有自己特色的蒸蒸日上的中学。祝愿我的母校，发扬优良传统和好的校风，培养出更多更好的学生，为祖国的繁荣富强作出更大的贡献。

---

① 本文是四川省温江中学建校 70 周年的祝词，广州，1996 年 8 月 5 日；教育与科技管理研究，北京：科学出版社，2016，257-258.

（日本）新世纪中文电视学校校长致辞[①]

在跨入21世纪之际，新世纪中文电视学校应运而生，这是华侨华文教育史上具有重要意义的大事，对推动海外华文教育的进一步发展，必将产生积极而深远的影响。

中华语言文化具有五千年的历史，博大精深，对人类社会的进步作出了伟大的贡献。中华民族素有重视教育的传统。开展海外华文教育，是关系到中华语言文化能否在海外三千万华人华侨中传承和延续的问题，对他们的生存、发展具有重要意义，对华侨、华人侨居（入籍）国多元化文化的发展和经济繁荣、社会进步也必将起到积极的促进作用。

最近二十多年来，中国实行改革开放政策，经济建设取得举世瞩目的成就，与各国的友好关系和经济、文化交流不断发展，中华语言文化的国际地位日益提高，海外华文教育重现生机，并有了新的发展，在世界范围内出现了“华文热”。

当今时代，科学技术突飞猛进。新世纪中文电视学校采用现代化手段，进行华文教育，为广大海外华人华侨子女以及其他族裔人士学习中华语言文化创造了更加良好的条件，这对弘扬中华语言文化，对促进中国人民与世界各国人民的友谊和文化交流，必将作出重要贡献。

暨南大学是一所具有93年历史的华侨高等学府，一向以弘扬中华文化、培养华人华侨子女为己任，致力促进海外华文教育的发展。我们希望海外华侨华人子女在中、小学阶段努力打好中华语言文化基础，将来前来暨南大学求学。

祝愿新世纪中文电视学校面向新世纪，不断开拓进取，为海外华文教育多作贡献！

祝同学们好好学习，天天向上！

① 本文原载《中文导报（日本）》，2000年3月16日（新世纪中文电视学校是海外第一所中文电视学校，兼任校长）。

当代科技发展与大学理念和人才培养①

人类进入21世纪，世界各国都给予高度重视，从政府首脑到黎民百姓，从专家学者到仁人志士，无不关注未来世界政治、经济、社会、文化、教育的走向。在这共同的思考之中，人们从不同的视角，以不同的思维和方式来研究与探讨今后国际社会发展的趋势。尽管，在我们眼中的当今世界仍有局部战争、恐怖主义、贫困饥荒、环境污染等困扰的存在，但我们还是应该看到和平与发展仍然是世界的两大主题。建立世界政治经济新秩序，共同对付已经出现或可能出现的各种危机，通过科技与经济、科技与社会、科技与教育结合谋求更大的发展，以满足人们日益增长的物质文化生活需要，这是世界各国人民的共同愿望。作为创造思想、发展科技与培养人才的摇篮——现代大学，既面临着经济全球化和科技经济一体化所带来的发展机遇，又需应对结构调整、社会转型、观念变化所带来的严峻挑战。如何抓住21世纪给我们带来的发展机遇，重塑现代大学的精神与理念，需要我们大家共同面对和共同探讨。

一、现代科学技术发展的特点与趋势

现代大学对莘莘学子进行科技教育，培养他们的科学素养，增强科技意识，提高科技水平和创新能力，除了一般地了解科学技术发展的历史和规律外，还应该懂得和掌握现代科学技术发展的特点与趋势，从而使现代大学的科技教育适应社会的发展，满足21世纪现代科技对人才的需要。

19世纪末开始的物理学革命拉开了现代科技的帷幕。以相对论和量子力学的一系列突破性进展为先导，现代自然科学在广度和深度上、在思想方式和研究方法上、在学科体系结构上、在科学与技术及科学与社会的关系等方面都出现了质的飞跃。现代高新技术不仅以基础科学为先导，还以某种方式同基础科学、技术科学联成一体。科学的整体化、技术的综合化和科技一体化等已成为现代科学技术发展过程中新的特点和主要趋势[1]。

1. 现代科学发展的主要趋势

目前，自然科学发展中呈现的第一个主要趋势为：一方面物质科学继续揭示自然界更深、更广、更久远的层次和各种极限状态下的物质运动规律；另一方面系统科学与生命科学正逐步阐明与人类的关系更密切的各类复杂系统的行为规律。由于物质科学、生命科学和系统科学这三大综合性科学内部和相互之间各个分支学科的相互渗透，使得现代科学发展出现第二个主要趋势：现代科学在高度分化的基础上产生了高度的综合，综合表现为多层次、多维度的学科交叉与渗透，表现为横向学科和综合性的学科群不断涌现。

---

① 本文原载《暨南高教研究》，2002，(1)：1-8.

2. 现代技术发展的主要趋势

第一，以基础自然科学新成果为先导的高新技术成为现代技术体系的带头技术，如信息技术的基础是微电子技术。第二，各门类技术相互渗透，相互促进，并在某些技术领域围绕一个重大问题或重大目标形成庞大的综合性技术群。光通信技术就是激光技术与通信技术相互渗透的产物。第三，综合应用多种门类技术的复杂大系统的研制开发成为技术发展的主要途径之一。美国空间技术发展史上的两个里程碑——阿波罗登月飞船和航天飞机的研制成功，都涉及数千个技术开发项目，其范围囊括了现代技术所有主要领域。第四，以软科学为理论基础的多种社会技术成为现代科技体系门类。管理技术、决策技术、经济运行宏观调控技术等已超越了经验加随机应变的前技术化阶段，初步实现了理论指导下的优化的程式操作。第五，大多数技术创新出现于新产品的研制过程中。

3. 科学技术一体化的趋势

首先，科学的技术化。科学活动中包含着大量的技术科学研究，技术发展研究和技术应用研究作为其辅助部分。这些辅助的技术活动并非用于科学研究成果向相应技术领域的转化，而是服务于科学研究活动自身的需要。其次，技术的科学化。技术上升到技术科学，通过相应基础科学的指导，形成系统的技术知识体系，反过来完善和提高已有的技术。最后，科学技术连续体的形成，一般通过两种途径：一是科学的技术化与技术的科学化两个过程相对展开，衔接后由于实践需要的推动相互渗透与融合而成；二是由于科学实验装置的技术原理符合某种实践需要，科学的技术化连续演变成新技术。

4. 科学技术与人文科学相结合的趋势

当代社会历史的客观进程以及当代任何重大的科学技术问题、经济问题、社会问题和环境问题等所具有的高度的综合性质，不仅要求自然科学、技术科学和社会科学的各主要部门进行多方面的广泛合作，综合运用多学科的知识和方法，而且要求把自然科学、技术和人文社会科学知识结合成为一个创造性的综合体。当代人类面临需要解决的问题的高度综合性质，决定了当代自然科学和技术与人文社会科学结合，这是当今科学技术发展的新趋势和新特点。

二、大学科技教育的改革与发展

现代大学开展科技教育是随着现代科学技术的进程而变化的。当现代科技呈现科学与技术、科技与社会、科技与人文以及理论与应用相互融合、渗透并相互促进等特征的时候，世界各国大学的科技教育也正从组织到机构，内容到形式发生了相应的变化。单科性大学向多科性直至综合性发展，狭窄性向宽口径、厚基础专业演变；单一的人才模式向复合型、创新型转化，文理渗透、理工医交叉，跨学科培养人才，已成为现代科技教育改革与发展的新潮流。

1. 单科性大学向多科性或综合性大学发展

大学是培养高层次人才的摇篮，人才的规格、类型与大学的类型密切相关。有什么样的大学，就可能培养出什么样的人才。纵观世界各国，特别是发达国家高等学校类型的变化，其轨迹与现代科技的变化几乎同出一辙。现代科技从宏观到微观，又从微观到它们之间的融合、交叉形成中观，再逐步发展成为新的宏观。同样从少数大学的产生，随着经济社会的发展，出现大量单科性的高等院校；然后，单科性大学又逐步向多科性或综合性大学发展。我国高校同样经历过这种变化。1949 年，仅有高校 205 所，综合大学 49 所，工业院校 28 所，农业院校 18 所，医药院校 22 所，师范院校 12 所，语文院校 11 所，财经院校 11 所，体艺院校 18 所，其他院校 29 所。[2] 1952 年为重点培养工业建设人才和师资，发展了一些专门院校；1958 年建立了一批以物质资源为代表的新的工科院校，如航空学院、邮电学院、钢铁学院、石油学院、地质学院等。20 世纪 60～80 年代地方办学和行业办学有了很大的发展，这对于满足地方经济和行业人才的需求起了积极的作用。但它带来了单科性院校过多，人才培养模式单一的矛盾。这不仅与科学技术教育的精神相悖，阻碍高校学科发展和科研水平的提高，更由于人才培养模式的单一，发展后劲不足，不能适应现代化建设的需要。进入 20 世纪 90 年代，加快了高校结构、类型的改革步伐。从 1992 年到 2000 年的 8 年间有 450 所高校（其中普通高校 319 所，成人高校 131 所）合并组建为 188 所高校（其中普通高校 181 所，成人高校 7 所）。[3] 北大与北医合并、复旦与上医大、浙大与浙医、浙农、杭大合并等，通过合并、重组建立起了一大批真正意义上的综合性大学。最近两年，随着高等教育管理体制改革的深化，高校的结构、类型有了很大的调整，不少单科性、小规模的院校进行了合并、重组，多科性、综合性的大学在增加，总体的高校数量得到控制，到目前为止，我国共有高校 1042 所。暨南大学在 20 世纪 70 年代末就依据世界高等教育和科学技术发展的趋势，在大学内设立了医学院，是国内改革开放后设立医学院最早的高校。现在暨南大学拥有理、工、医、文、史、经、管、法、教育等 9 大学科门类，16 个学院。学科间文理渗透、理工医结合，为培养厚基础、宽口径、高素质的复合型人才提供了良好的条件。

2. 狭窄性专业向宽广性拓展

随着人类对物质世界的认识和科学技术的发展以及社会经济生活中部门、行业的分工，大学的教育逐步建立在门类、学科、专业之上，专业教育和专门人才的培养成为高等教育的基本特征。而专业的设置大都又依据各国的国情，不同的经济结构、产业结构需要不同的人才结构和专业结构。我国的专业划分主要是根据学科门类，专业的设置是根据国家建设事业的需要。从 20 世纪 50 年代初到 80 年代初本科专业从 215 个增加到 1039 个，其中文科由 19 个增加到 60 个，理科 16 个增加到 158 个，工科 107 个增加到 537 个，农科 16 个增加到 60 个，林科 5 个增加到 22 个，医科 4 个增加到 29 个，师范 21 个增加到 40 个，财经 13 个增加到 40 个，政法 2 个增加到 8 个，体育 1 个增加到 8 个，艺术 11 个增加到 63 个。[4] 这些专业的设立在特定的形势下，适应了当时社会各项事业对应用型人才的急需，形成了以培养实用人才为主的教育模式。但同时暴露出了存在的缺陷，一是专业口径狭窄；二是知识结构单一；三是培养规格统一。为了解决这些不适应科技教育和人才培养的矛盾，我国先后于 1987 年、1993 年对本科专业目录进行了两次大的修订，目录内专业由原来的 1300 多种缩减到 500 余种。但是，仍然存在着专业划分过细过窄、专业宏观结构不尽合理等方面的不足。1998 年再次对本科专业目录进行修订，本次修订遵循“科学、规范、拓宽”的原则，从社会对高等学校人才的需要出发，按照教育规律，结合国情和教改实际，吸收和借鉴国外有益的经验，进一步拓宽专业口径，增强专业的适应性。在不影响增设一些新兴学科、边缘学科和交叉学科专业的前提下，使本科专业由504种减少至249种，调减幅度为50.6%。本次专业主要是按11个学科门类进行划分，其中哲学3种专业，经济学4种，法学12种，教育学9种，文学66种，历史学5种，理学30种，工学70种，农学16种，医学16种，管理学18种。[5]实施4年后新的专业目录逐步显示出它的优越性，它将为培养出基础扎实、知识面宽、能力强、素质高适应社会需要的人才奠定基础。

3. 人文教育与科学教育相融合，培养学生的全面素质

人文教育与科学教育是现代教育不可或缺的重要组成部分。人文教育指的是培养人文精神的教育，它是以人格教育和道德教育为主要内容；而科学教育是指以征服和改造自然，促进物质财富增长和社会发展为目的，向人们传授自然科学技术知识、开发人的智力的教育。从人类教育发展的历史轨迹可以看出，人文教育与科学教育是不可分割的，科学教育注重教育的直接社会功能，开拓了人的智慧与知识，拓展了人们的认识领域，促进了生产力的发展，使人类在开发自然的过程中获得了巨大的物质财富；人文教育重视人性的完善，提升人的道德精神，对促进人们确立正确的世界观、人生观和价值观具有重要的作用。因此，科学教育和人文教育在促进人的全面发展中具有不同的功能和作用。

科学技术本身是一种与人类理想和自由密切相关的高层次文化，它集中体现了人类对知识和真理的追求，是人类文明的重要组成部分。一个人的科技知识素养在很大程度体现了这个人的文化素养和整体素质，而科学技术活动作为一种理性活动，对于推动人的理性思维和智力发展有着巨大而深远的作用。在科学发展过程中形成的科学精神和科学方法不仅缔造了科学本身，推动了技术发展，而且改变了人的认识能力，创造了现代文明。21世纪在现代大学中科技教育与人文教育的界限将会不断淡化，无论理科的学生，还是文科的学生，了解对方的知识，懂得对方的理论将会成为他们学习知识的需要。从20世纪90年代开始，我国的高校高度重视在大学生中同时进行科技教育和人文社科教育，一些理工科院校大多建立了素质教育基地，成立了人文学院或教育学院，为大学生系统地开设人文社会科学课程，社会心理学、公共关系学、社会环境学等成为理工科学生选修的课程，而高等数学、科技发展史、计算机科学与技术、网络技术等已成为文科学生的公共基础课程。大学不仅在学科建设、课程设置上充分考虑科技知识与人文知识的传授，而且在教学过程中广大教师也十分注重从社会背景和科技、经济等因素去提出问题、分析问题、研究问题，使学生能够全方位地思考问题和解决问题。暨南大学是一所综合性大学，在全校学生中同时进行科技和人文教育有着得天独厚的条件，9大学科可任学生选课。除此外，暨南大学的学生还可在广州石牌地区6所高校（如华南理工大学、华南师范大学、华南农业大学等）选课。选课的自由度，体现了学生学习的自主性，充分拓展了他们的知识面。

实践表明，科技教育对大学生的素质发展起着积极的作用。第一，在生理素质方面，科技教育帮助大学生正确认识人体结构及其各器官的生长特性和发展规律，从而

形成正确的生活方式，开展有效的体育锻炼，促进生理素质发展。第二，在心理素质方面，科技教育能够促进大学生的智商发展，提高认知能力和直觉思维能力，培养他们的创造能力；同时科技教育还可以激发学生探索真理的热情和对科学事业的崇敬感，促进优良性格的形成。第三，在社会文化素质方面，科技教育有利于增强学生的社会责任感。科技应用的社会效应具有两面性，既能给人类带来幸福，也能给人类带来灾难。科技功能的两面性要求每一个从事科学活动的人都具有强烈的社会责任感，以便保证科学技术沿着人类服务的方向发展。科技教育对学生审美素质发展也能起着促进作用。科学活动，包括科学学习，导致科学活动主体美感的产生。这种美感来源于科学活动的对象（自然现象和事物）、科学活动的结果（科学理论、概念、定律等）、科学方法本身所具有的审美特征。学生在科学学习、科学研究的过程中必然受到美的熏陶，形成科学美感。第四，科技教育还具有思想教育的功能。科技教育可以为学生科学世界观的形成提供充分的科学依据，因为科学理论课程中的知识内容是在基本世界观思想的概念结构基础上加以组织的，这些世界观原理按照各基本学科系列间与系列内的联系加以了条理化的分类。[6]所以，学习科学知识能够促进科学世界观的形成。

4. 理论与应用相结合，造就具有创新精神和实践能力的人才

大学生对科学技术的认识，主要还是通过自己的学习和教师的传授，掌握书本知识和前人的经验，而这种对事物的认识只是停留在一种理性认识之上，他们通过思维，运用概念、定义、逻辑、推理、演绎来分析、判断事物。在人类对事物的认识中需要这种认识方法，但仅仅限于这种认识方法又是很不够的。人们要获得真知灼见必须经过由感性认识到理性认识，再从理性认识到实践的二次飞跃。因此，实践对于大学生掌握知识极其重要。所谓实践，就是学生们将自己所学到的理论知识应用于解决实际问题之中。目前，学生在大学的实践，主要还是通过科学实验。但是，我们现在的这种实验大多又是验证性的，而且还是教师设计由学生来完成的实验。在开初的时候可以这样做，但到了大学三四年级，如果还是这样，我们培养的学生只会照葫芦画瓢，缺乏独立动手解决实际问题的能力。21 世纪需要的是具有创新精神和实践能力的人才。如果现代大学的教育教学改革最终不能解决这个问题，那就要落伍于现代科学技术和现代社会。

造就大学生的创新精神，我认为就是要培养学生具有创造性的思维、创造性的品格和创造性的能力。创新精神不要当作一般的口号来提，而是要在大学整个的教育教学过程中得到实实在在的落实。首先，解决思维的模式问题。我们现在的思维方式，包括教师和学生大多是形式逻辑，程式性的思维，想问题、做事情都是按部就班，不敢有突破，不敢有超越，非常缺乏发散性的思维。诚然，思维方式不仅仅就是在大学形成，从小学到中学，一直到大学，养成了目前的思维习惯。其次，改革教学方法。目前陈旧、刻板的教学方法是现代大学培养创新人才的最大障碍之一。学生很多的思维习惯、思维方法都是受到课堂教学的影响，如果课堂教学没有创新，不进行深刻的变革，培养创新精神难以奏效。最后，大学生的实践问题。现在普遍认为内地的大学生与国外大学生相比，主要差距在动手能力和创造能力方面。我们回顾一下，大学 4 年，学生真正接触社会实践和科学实验有多少时间？只要检查一下教学计划就可以看

到学生的大部分时间和主要精力是用于理论学习。再看看学生的毕业论文就可知道他们的科研水平和实践能力。因此，培养创新精神和实践能力是现代大学与传统大学的根本区别，是当今科技教育要达到的根本目的。

三、21世纪现代大学的理念

理念是一个精神、意识层面的综合性结构的哲学概念，是人们经过长期的理性思考及实践所形成的思想观念、精神向往、理想追求和哲学信仰的抽象概括。一般是指人们对于某一事物或现象的理性认识、理想追求及所持的思想观念或哲学观点。因此，大学理念应该是人们对大学的理性认识、理想追求及其所持的大学教育思想观念和哲学观念。[7]

1. 大学理念的内涵

大学理念包含了大学理想、大学观念、大学精神、大学使命和大学目标。大学理想既是空灵抽象的，又是现实具体的。它既是牵引大学各项工作的精神力量，又是与具体工作融合在一起的。大学理想的实现有赖于每一位学生和教师的实践。对于大学理想的认识，不同的国度、不同民族的学者，由于其不同的价值观形成了多种分类。有的认为大学理想是人文主义理想、功利主义理想、科学研究理想、教育机会均等理想以及科学民主民族的理想等。大学理念是一种教育思想观念，是人们通过观察和思考而获得的对教育现象的理性认识。它实际上是大学的办学指导思想，是观察、分析、论述和处理大学改革与发展问题时所处的角度或采取的态度，包括大学价值观、质量观和发展观。

大学精神是一所大学整体面貌、水平、特色及凝聚力、感染力和号召力的集中反映，是大学的理想、信念、情操、行为、价值和道德水平的标志，是一所大学的支柱和灵魂。广义的大学精神是指大学所普遍存在的优良校风、相对稳定的群体心理优势和精神状态；狭义的大学精神是一所大学在长期的教育实践中沉淀的特定的人格化和个性化精神最富典型意义的特征。有的学者将“学术自由、文化创新、真理至上”作为大学精神，但更多的高校是结合学校的历史沉淀、办学风格、治学态度和育人标准而形成具有自身特色的大学精神。大学使命，最根本的是以新的思想引导和推动社会，以新的人才和新的知识成果服务于社会。21世纪赋予大学的历史使命更具有强烈的社会性、时代性和国际性。为了不辱使命，现代大学必须立于信息革命和知识经济的前列，不断创造新思想和新知识，以引导社会沿着人类可持续发展的道路前进；应注重高质量人才的培养，使他们具有创新精神和实践能力，能够跟上时代发展步伐，掌握社会需求的知识技能，逐步成为社会发展的领导力量。大学目标，是指大学培养人才所特有的种类、层次、规格和要求。它是国家总体教育目的在高等教育领域的具体化，具有鲜明的时代特征。不少的专家学者提出大学教育要做到：通识教育与专才教育、授业解惑与启思导创、人文教育与科学教育、全面发展与个性培养、学校教育与社会教育相结合。大学应追求的目标是，培养通专结合，既有人文精神又有科学素养，能够适应复杂多变的未来社会的国际性通用型的创新人才。

2. 大学校长的教育理念

讨论现代大学理念的同时，我们不可不研究大学校长的教育理念。大学校长的教育理念对于办学治校有着重要的影响。大学校长的教育理念：一是对现代大学的价值判断，要回答“大学是什么？大学干什么?”，大学的使命与校长的责任是否形成一致？二是对现代大学的地位确立，进入21世纪高等学校的三大功能在社会进步中将发挥更加积极的作用。但是，具体对于自身的学校应该办成为一个什么样的大学，达到什么样的水平，形成什么样的特色，在校长的教育理念中应该有一个清晰的思路和明确的目标。三是对现代大学精神的认识标准，大学的灵魂是什么？大学的精神支柱是什么？大学校长对于自己所领导的大学该用什么思想来统一全体师生的意志和信念，以什么精神来形成全校性的凝聚力、感召力和推动力。四是对现代大学办学目的的客观评价，人才培养、科学研究和社会服务，既是大学的使命，也是办学的目的。但对于某一所大学，由于类型不同、基础不一样，其办学的目的则有所侧重，有的以科研为主体；有的是科研与教学并重；还有的以教学为主。大学校长将依据自己的教育理念作出抉择，确定学校的发展方向。五是对现代大学人才模式的准确设计，“通用型、创新型和国际型”是21世纪人才模式的新标准，随着科学技术一体化和经济全球化对人才培养的类型、规格提出了更高的要求，作为一校之长对自己大学培养什么样的人才应该“心中有数”，从概念设计到教育实践，以及所涉及的培养目标、专业设置、教学内容、教学方法等都要有一个科学的设计和正确的选择。六是对现代大学科学民主管理的实施，现代管理是现代大学的标志之一。大学是一个结构和功能复杂、其工作任务和组织成员充分体现了智力劳动特性的学术教育机构。这造成就大学管理自身而言，不仅是一个实际操作问题，也是一个需要整体思维、宏观把握的管理哲学问题。大学校长若无明确的治校理念，其学校不乱也难以有效率。[8]校长用人的标准、处事的方式、平日的作风、个人的品格等在其办学与治校的过程中都充分体现出来。现代大学的理念需要大学校长予以理解、接受并转化成自己的思想、观念，且在办学与治校的实践中不断地深化和提高，最终形成自己的教育理念和管理行为。

四、科技教育与现代大学理念

今天，科学技术已成为推动社会发展的力量源泉和重要动力，无论是经济增长，还是人民群众物质文化生活的改善都离不开科学技术的贡献。科学技术与人类社会的关系越来越紧密。生活在这个现代社会的青年，是这个时代的接班人和建设者，掌握与应用现代科学技术将成为现代人的基本任务和必要的技能，也是他们未来生存与发展的必备条件。现代大学是向社会培养和输送人才的“工厂”，向青年传授科学知识和生产技术已成为大学的使命，让每一位大学生具有科学素养与实践能力，使他们能更好地服务于社会，造福于人类。科技教育的这种目的是与大学理念中的大学目的是一致的。正因为有了科技教育对人才培养所达到的良好效果，才使大学哲学的思想、观念有着牢固的现实基础。

科学技术是一项事业，热爱并献身于这项伟大的事业，要有一种强烈的事业心和社会责任感以及甘于奉献的思想品质，缺乏这种信念是难以在科学事业中有所作为、有所成就的。当大学生踏入科学殿堂的第一步就要有足够的思想准备，否则的话，那只有另做选择。今天，进入现代大学第一天的青年就要立志，就要坚定不移地树立热爱科学、热爱专业的思想，没有这个思想基础就不能成“大器”。科学技术研究和应用也是一项十分艰巨的工作，要耐得住寂寞，“十年磨一剑”，要有百折不挠的精神和坚忍不拔的毅力，切忌浮躁、急功近利。这种在科学实践中所培养的脚踏实地的作风，既是科技工作者的风范，也是大学生们所要追求的精神境界。这种科学精神与大学理念中的大学精神是完全吻合的，大学精神从某种意义上说是科学精神的升华，一代又一代的师生不断传承与发扬光大，使之铸成了今天的大学精神。

从事科学技术还要有一种不唯书、不唯上、不信邪的信念，只相信被客观事实所证实的真理。追求真理，既是科学技术的目的，也是科技工作者的目标。在年复一年的科技实验中，这种信念就像一盏明灯一样引领科技工作者刻苦钻研、不断拼搏，渐渐地形成了他们应有的品格与道德。在科技教育中向大学生灌输这种意识、思想和观念，并在实践中让他们体会、感受，得到熏陶，逐步成为他们的行为准则。久而久之，一代代的沉淀就形成了大学的精神和理念。一所大学要立于世界大学之林，除了具有一定的物质条件外，最重要的还是这所大学的精神和理念。这是用再多钱也买不来的，它是靠几十年，甚至几百年积累，靠成千上万人的品格、作风、思想、精神、业绩、成就所铸造的。21 世纪的今天，我们要弘扬现代大学的精神，重塑现代大学的理念。

参 考 文 献

[1] 刘大椿，何立松. 现代科技导论. 北京：中国人民大学出版社，1998.

[2] 余立. 中国高等教育史下册. 上海：华东师范大学出版社，1994.

[3] 左春明. 一场深刻的历史性变革. 中国教育报，2000-05-25.

[4] 《中国教育年鉴》编辑部. 中国教育年鉴（1949～1981）. 北京：中国大百科全书出版社，1984.

[5] 中华人民共和国教育部高等教育司. 普通高等学校本科专业目录和专业介绍：1998 年颁布. 北京：高等教育出版社，1998.

[6] 梅德维杰夫. 培养学生世界观的综合方法//彼得罗卡斯基. 苏联德育心理论集. 陈会昌，译. 北京：教育科学出版社，1989.

[7] 潘懋元. 多学科观点的高等教育研究. 上海：上海教育出版社，2001.

[8] 睦依凡. 大学校长的教育理念及其与治校的关系. 教育研究，2000（7）：11-17.

加快我国高等教育进入世界先进行列①

一、形势

1. 我国高等教育的形势分析

改革开放以来，我国高等教育事业发展很快，初步形成了适应国民经济建设和社会发展需要的多种层次、多种形式、学科门类基本齐全的社会主义高等教育体系，为社会主义现代化建设培养了大批高级专门人才，在国家经济建设、科技进步和社会发展中发挥了重要作用。

1）我国高等教育所取得的成就

A. 高校招生规模与在校生规模大幅度增加

改革开放以来，我国高等教育事业规模有了很大的发展，如表1所示（港、澳、台数据尚未计人，下同）。尤其是从1999年开始，普通高等学校扩大招生规模，其增长幅度之大，是前所未有、世所罕见的。1999年，普通高校扩招51.3万人，相当于当年新办了50多所万人大学。1998年招生108.4万人，2004年招生420万人，六年几乎翻两番。2003年，普通高等学校在校大学生1108.6万人，加上成人高等学校在校生，两者之和超过1900万人，毛入学率为17%；2004年，普通高等学校与成人高等学校在校生之和超过2000万人，毛入学率为19%[1,2]。已有资料报道，2005年，普通高等学校将招生475万人，比上一年增加55万人，增长率为13.1%；加上成人高等学校招生数，两者之和超过510万人[3]。每万人口中普通高等学校在校生数2003年为86.3人，约是1998年27.3人的3.16倍，约是1978年8.9人的9.7倍，如果加上成人高等学校人数，增长倍数则更为可观。一些专家认为，我国的高等教育正在由精英教育向大众化教育过渡[4]。

表1 我国高等教育事业基本情况

指标	1978年	1980年	1985年	1990年	1995年	1998年	1999年
学校数/所	598	675	1016	1075	1054	1022	1071
专任教师/万人	20.6	24.7	34.4	39.5	40.1	40.7	42.6
招生数/万人	40.2	28.1	61.9	60.9	92.6	108.4	159.7
在校生数/万人	85.6	114.4	170.3	206.3	290.6	340.9	413.4
毕业生数/万人	16.5	14.7	31.6	61.4	80.5	83.0	84.8
每万人口中在校大学生数/人	8.9	11.6	16.1	18.0	24.0	27.3	32.8

① 本文是全国教育事业“十一五”规划研究课题报告，2005年；教育与科技管理研究，北京：科学出版社，2016，22-113；作者：刘人怀，纪宗安，方丽，孙东川，贾益民，熊卫华，孙红萍，温碧燕，王雄志，李朝晖，宗世海，廖仕湖，钟国胜，李东生.

续表

指标	2000 年	2001 年	2002 年	2003 年	2004 年	2005 年	
学校数/所	1041 * **1813**	1225	1396	1552 * **2110**			
专任教师/万人	46.3	53.2	61.8	72.5			
招生数/万人	220.6	268.3	320.5	382.2	420	475 * **510**$^{+}$	
在校生数/万人	556.1	719.1	903.4	1108.6 * **1900**$^{+}$	* **2000**$^{+}$		
毕业生数/万人	95.0	103.6	133.7	187.7			
每万人口中在校大学生数/人	43.9	56.3	70.3	86.3			

注：1. 表中 2004 年、2005 年数据，是根据本月份教育部网页上的材料，其余数据均来自文献［1］

2. 表中打 * 的粗体字数据包含成人高等教育，其余数据均为普通高等教育；上标＋表示“超过”，例如 * 2000^{+} 表示 2004 年普通高等教育与成人高等教育的在校学生数之和超过 2000 万人

但是，全国各地区高等教育办学规模有明显的差异，东部和华北等地的办学规模要强于西北、西南地区。表 2 列出了 2003 年普通高等学校在校学生数按地区排序前 12 名，同时列出了它们的招生数和毕业生数。它们的三项指标分别都占全国的 63％以上；实际上，前 9 名所占比例都在 50％以上，而前 4 名所占比例在 25％左右。可见，全国高等教育的分布情况是很不均衡的。

表 2　2003 年普通高等学校学生情况

地区	在校学生数/人	比例/％	招生数/人	比例/％	毕业生数/人	比例/％
全国	11 085 642	100.00	3 821 701	100.00	1 877 492	100.00
江苏	(1) 859 674	7.75	256 595	6.71	137 048	7.30
山东	(2) 761 417	6.87	273 894	7.17	117 253	6.25
湖北	(3) 721 513	6.51	250 198	6.55	119 118	6.34
广东	(4) 587 779	5.30	225 837	5.91	105 533	5.62
河北	(5) 575 542	5.19	203 826	5.33	113 442	6.04
河南	(6) 557 240	5.03	190 214	4.98	108 975	5.80
湖南	(7) 537 220	4.85	193 830	5.07	90 035	4.80
辽宁	(8) 514 191	4.64	163 802	4.29	98 908	5.27
四川	(9) 512 663	4.62	180 308	4.72	74 307	3.96
陕西	(10) 499 017	4.50	168 127	4.40	79 785	4.25
浙江	(11) 484 135	4.37	168 167	4.40	78 685	4.19
北京	(12) 454 480	4.10	141 790	3.71	82 828	4.41
12 省份合计	7 064 871	63.73	2 416 588	63.23	1 205 917	64.23

目前，我国高等教育在校生数量居世界第一，教育质量在世界上也是比较先进的，人民群众接受高等教育的机会在短短 6 年内翻了两番。2004 年，我国高等教育毛入学率达到了 19％，2005 年的毛入学率还会有所提高。即便若干年之后毛入学率达到 40％，我国的高等教育还有很大的发展空间。中国内地与港、澳、台地区相比，也有

不小的差距。台湾 2003 年高等教育（18～21 岁）的"粗入学率"为 90.2%，每千人口中高等教育学生数为 58.3，而大陆同年每万人中在校大学生数仅为 86.3[1]。

B. 高等教育体制改革和结构调整紧密结合，层次结构和科类结构逐步趋向合理

高等教育的层次结构和科类结构逐渐趋于合理，才能与产业结构调整和经济社会发展相适应。在高等教育的三大层次中，研究生教育经过两次专业调整与合并，逐渐趋向合理。本科专业则在 1993 年的调整中，由总数 813 种减少为 504 种，1998 年又调整为 249 种，这就拓宽了专业面，增强了社会适应性，更好地满足产业结构对人力资源的要求。专科教育作为高等教育的开端和重要组成部分，在提高国民素质和适应产业结构调整而进行转业、转岗和再就业培训方面发挥了重要作用。尤其是成人高等教育，则逐渐向高层次岗位培训、大学后继续教育方面发展。国家采取了多方面的措施，如大力推进成人高等教育管理体制的改革，强化成人高等学历教育质量控制机制，合理调整成人高等学校的设置和布局，鼓励、支持社会力量办学等。尽管学历教育在目前来说是成人高等教育的重要组成部分，但未来的成人高等教育的重点将更多是进行大学后教育。

C. 高等教育的对外开放与国际交流合作逐渐扩大

1978 年至 2000 年，我国向 100 多个国家和地区派出各类留学人员 27 万人，已有 9 万人学成回国；接收各国来华留学生 21 万人；高等学校积极聘请外籍教师，吸收和借鉴国外的有益经验，促进了教学质量和科研水平的提高。

1996～2004 年 9 月底，国家留学基金管理委员会共派出了留学人员 18 167 人，到期应回国 15 610 人，实际回国 15 092 人，按期回国者占应回国人员的 96%以上。教育部发起的"公派出国留学效益评估"课题研究报告表明，公派留学不仅带来巨大社会效益，且经费投入与直接经济收益比为 1∶18 左右，经济效益显著。

表 3 为 1978 年以来我国留学生情况（以及研究生培养情况）[1]。其中"出国留学人员"包括公派和自费出国留学。

表 3　我国研究生培养和留学生情况　　（单位：人）

年份	研究生数			出国	学成回国
	在学人数	招生数	毕业生数	留学人员	留学人员
1978	10 934	10 708	9	860	248
1980	21 604	3 616	476	2 124	162
1985	87 331	46 871	17 004	4 888	1 424
1986	110 371	41 310	16 950	4 676	1 388
1987	120 191	39 017	27 603	4 703	1 605
1988	112 776	35 645	40 838	3 786	3 000
1989	101 339	28 569	37 232	3 329	1 753
1990	93 018	29 649	35 440	2 950	1 593
1991	88 128	29 679	32 537	2 900	2 069
1992	94 164	33 439	25 692	6 540	3 611

续表

年份	研究生数			出国	学成回国
	在学人数	招生数	毕业生数	留学人员	留学人员
1993	106 771	42 145	28 214	10 742	5 128
1994	127 935	50 864	28 047	19 071	4 230
1995	145 443	51 053	31 877	20 381	5 750
1996	163 322	59 398	39 652	20 905	6 570
1997	176 353	63 749	46 539	22 410	7 130
1998	198 885	72 508	47 077	17 622	7 379
1999	233 513	92 225	54 670	23 749	7 748
2000	301 239	128 484	58 767	38 989	9 121
2001	393 256	165 197	67 809	83 973	12 243
2002	500 980	202 611	80 841	125 179	17 945
2003	651 260	268 925	111 091	117 307	20 152
总计	3 838 813	1 495 662	828 365	537 084	120 249

2004 年，国家留学基金管理委员会共录取各类留学人员 3987 人，涉及 48 个国家，派出人员总体水平高于前两年，在录取的高级研究学者、访问学者、进修人员中具有硕士以上学历的占 80.04%，具有副高级以上专业技术职称的人员占 70.73%。2005 年，国家留学基金管理委员会将进一步扩大选派规模，公派留学生人数将比 2004 年增加近 1 倍，以多种资助方式在全国选拔各类出国留学人员 7245 人，提高层次成为 2005 年的主要努力方向；2005 年公派留学将重点资助七大学科领域：通信与信息技术、农业高新技术、生命科学与人口健康、材料科学与新材料、能源与环境、工程科学、应用社会科学与 WTO 相关学科。七大领域占公派留学经费的 70%左右。七大领域下设 137 个专业方向，都是我国亟待发展的学科。2005 年公派留学人员层次也将提高，高级研究学者以及博士后、博士生比重进一步加大[5]。

D. 民办高等教育开始崛起

我国民办高等教育起步于 20 世纪 70 年代末 80 年代初，1992 年邓小平同志南方谈话发表之后，民办高等教育进入了一个崭新的发展时期，迅速形成燎原之势。据不完全统计，各地民办高校从 1991 年的 450 所增至 1995 年的 1209 所，其中新增民办高校 800 余所（由于有的学校合并或撤销，所以总数不等于简单求和）。民办学校办学条件逐渐改善，办学规模不断扩大。进入 90 年代后，我国民办学校开始把举办重点转向中、高等职业教育和职业培训。我国的办学体制改革迈开了较大的步伐，取得了突破性的进展，民办教育从为公办教育“拾遗补缺”发展成为社会主义教育事业不可缺少的组成部分。2001 年年底，民办高等教育机构为 1202 所，注册学生 113 万人，其中具有颁发学历文凭资格的民办高校 105 所，在校学生 15 万人。

E. 国家对高等教育的投入有所增加

1992～2001 年，全国普通高等学校生均预算内事业费从 4092 元上升到 6816 元，增加了 2724 元，如图 1 所示。2001 年为 6816 元，比上年减少 6.75%，主要原因在于吉林、河北、河南、甘肃、贵州等中西部地区有所减少（北京、浙江、广东等地有不同程度的增加）。

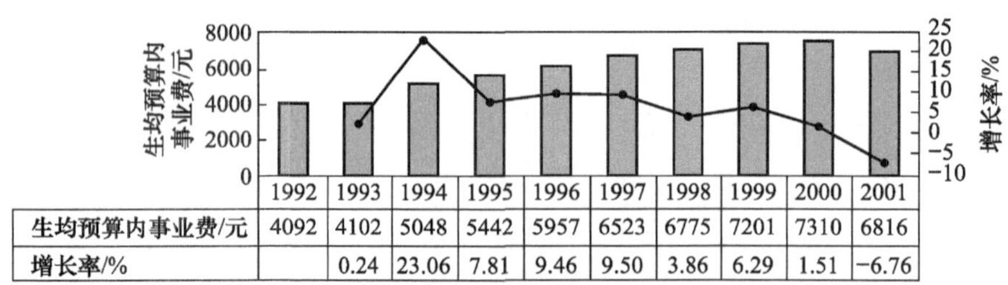

	1992	1993	1994	1995	1996	1997	1998	1999	2000	2001
生均预算内事业费/元	4092	4102	5048	5442	5957	6523	6775	7201	7310	6816
增长率/%		0.24	23.06	7.81	9.46	9.50	3.86	6.29	1.51	−6.76

图 1　1992～2001 年普通高校生均预算内事业费及增长率

表 4 反映了 1991～2002 年各年度我国教育经费情况[1]，并且计算了两个百分比。经费基本都是逐年增长的，“国家财政性教育经费”占 GDP 的百分比在 1991～1995 年是呈下降趋势的，1995～2002 年逐年上升，2002 年为 3.32%。

2）存在的主要问题

尽管我国高等教育经过多年的发展已取得了不少的成绩，但从目前来说，仍然存在着不少问题。例如，高等教育经费仍然不足，没有确保“三个增长”；高等教育体系有待于进一步完善，职业技术教育仍然未能真正在“职业教育”上下功夫；高等教育的专业结构不尽合理，仍未能满足社会主义市场经济建设的需求；地区发展不平衡，东西部差异大；民办高等教育发展仍然困难重重。

这里对于经费问题着重予以说明。

表 5 列出了 2002 年全国高等教育经费情况[1]。可以看到：在“国家财政性教育经费”中，绝大部分是“地方”投入的，“中央”∶“地方”=1∶8.9，就是说，“中央”的投入约占 1/10。我们还计算了“普通高等学校经费/成人高等学校经费”，如表 5 的最下面一行所示，看来，成人高等学校经费太少了。

根据表 5 分析的情况，这里先提两点建议。

（1）建议中央财政加大对于高等教育的投入。

（2）在确保普通高等学校经费继续较快增长的同时，成人高等学校经费应该有更大幅度的增长。

总而言之，我们的教育经费投入还是不足，还要继续有较快较多的增长。台湾地区“公共教育经费占本地居民生产总值”的百分比在 1991 年以来，有 9 年在 6.1%及以上，最高年份 1993 年为 7.0%，最低年份 2000 年为 5.5%，2003 年为 5.9%[1]。

表 6 再次列出表 4 中的数据，并且作了一些计算。由表 6 可知。

（1）“各年度的教育经费合计（A）”1991 年以来是逐年增长的，增长幅度不一。

（2）其中“国家财政性教育经费（B）”占 A 的百分比逐年下降，2002 年比 1991 年下降了约 20 个百分点。

再作两项计算：

2002 年各年度的教育经费合计/1991 年各年度的教育经费合计≈7.49；

表 4　各年度我国教育经费情况

年份	合计	国家财政性教育经费（A)/万元	预算内教育经费/万元	社会团体和公民个人办学经费/万元	社会捐资和集资办学经费/万元	学费和杂费/万元	其他教育经费/万元	A 占合计的比例/%	GDP/亿元	A 占 GDP 的比例/%
1991	7 315 028	6 178 286	4 597 308		628 210	323 476	185 057	84.46	21 617.8	2.86
1992	8 670 491	7 287 506	5 387 382		696 285	439 319	247 380	84.05	26 638.1	2.74
1993	10 599 374	8 677 618	6 443 914	33 323	701 856	871 477	315 100	81.87	34 634.4	2.51
1994	14 887 813	11 747 396	8 839 795	107 795	974 487	1 469 228	588 907	78.91	46 759.4	2.51
1995	18 779 501	14 115 233	10 283 930	203 672	1 628 414	2 012 423	819 760	75.16	58 478.1	2.41
1996	22 623 394	16 717 046	12 119 134	261 999	1 884 190	2 610 391	1 149 798	73.89	67 884.6	2.46
1997	25 317 326	18 625 416	13 577 262	301 746	1 706 588	3 260 792	1 422 783	73.57	74 462.6	2.50
1998	29 490 592	20 324 526	15 655 917	480 314	1 418 537	3 697 474	3 569 741	68.92	78 345.2	2.59
1999	33 490 416	22 871 756	18 157 597	628 957	1 258 694	4 636 108	4 094 901	68.29	82 067.5	2.79
2000	38 490 806	25 626 056	20 856 792	858 537	1 139 557	5 948 304	4 918 352	66.58	89 468.1	2.86
2001	46 376 626	30 570 100	25 823 762	1 280 895	1 128 852	7 456 014	5 940 766	65.92	97 314.8	3.14
2002	54 800 278	34 914 048	31 142 383	1 725 549	1 272 791	9 227 792	7 660 099	63.71	105 172.3	3.32

表 5　2002 年全国高等教育经费情况　（单位：万元）

类别	合计	国家财政性教育经费（A)	预算内教育经费	社会团体和公民个人办学经费	社会捐资和集资办学经费	学费和杂费	其他教育经费	A 占合计的比例/%	A 占 GDP 的比例/%
总计	54 800 278	34 914 048	31 142 383	1 725 549	1 272 791	9 227 792	7 660 099	63.71	3.32
中央	6 581 621	3 531 595	3 042 307		206 891	872 793	1 970 341	53.66	0.34
地方	48 218 657	31 382 452	28 100 077	1 725 549	1 065 900	8 354 998	5 689 758	65.06	2.98
高等学校	15 832 129	7 875 176	7 548 856	417 624	279 514	4 264 517	2 995 298	49.74	0.75
普通高等学校	14 878 590	7 521 463	7 243 459	331 363	278 253	3 906 526	2 840 985	50.55	0.72
成人高等学校	953 539	353 713	305 398	86 261	1 261	357 991	154 314	37.09	0.03
普通/成人	15.60	21.26	23.72	3.84	220.66	10.91	18.41		

表 6　各年度全国教育经费情况

年份	各年度的教育经费合计(A)/万元	国家财政性教育经费（B)/万元	B占A的比例/%	预算内教育经费（C)/万元	C占A的比例/%	学费和杂费(D)/万元	D占A的比例/%	B占GDP的比例/%
1991	7 315 028	6 178 286	84.46	4 597 308	62.85	323 476	4.42	2.86
1992	8 670 491	7 287 506	84.05	5 387 382	62.13	439 319	5.07	2.74
1993	10 599 374	8 677 618	81.87	6 443 914	60.80	871 477	8.22	2.51
1994	14 887 813	11 747 396	78.91	8 839 795	59.38	1 469 228	9.87	2.51
1995	18 779 501	14 115 233	75.16	10 283 930	54.76	2 012 423	10.72	2.41
1996	22 623 394	16 717 046	73.89	12 119 134	53.57	2 610 391	11.54	2.46
1997	25 317 326	18 625 416	73.57	13 577 262	53.63	3 260 792	12.88	2.50
1998	29 490 592	20 324 526	68.92	15 655 917	53.09	3 697 474	12.54	2.59
1999	33 490 416	22 871 756	68.29	18 157 597	54.22	4 636 108	13.84	2.79
2000	38 490 806	25 626 056	66.58	20 856 792	54.17	5 948 304	15.45	2.86
2001	46 376 626	30 570 100	65.92	25 823 762	55.68	7 456 014	16.08	3.14
2002	54 800 278	34 914 048	63.71	31 142 383	56.83	9 227 792	16.84	3.32

2002 年学费和杂费/1991 年学费和杂费＝28.53。

显而易见，第二个倍数比第一个大得多。可能有人认为，2002 年和 1991 年都是当年价格，不可比，但是，这一问题在两个倍数的计算中是同样的，所以，两个倍数还是具有一定的可比性的，那么，是否说明学生及其家庭的负担加重了呢?

下面，我们作一些国际对比。表 7 列出了世界各国或地区“大学生粗入学率”2000 年在 40%以上者。中国只有台湾和澳门的粗入学率能够超过 40%。实际上，中国台湾 2003 年的粗入学率为 90.2%。

表 7　各国或地区“大学生粗入学率”2000 年在 40%以上者

国家或地区	1990 年	2000 年	2000 年排名	国家或地区	1990 年	2000 年	2000 年排名
中国内地	3.0①	7.5①②		波兰	21.7	55.5	11
中国香港	19.1	27.4③		荷兰	39.8	55.0	12
中国澳门	25.4	52.1	15	法国	39.6	53.6	13
中国台湾*	37.9**	68.4	4	以色列	33.5	52.7	14
韩国	38.6	77.6	1	意大利	32.1	49.9	16
美国	75.2	72.6	2	阿根廷	38.8	48.0②	17
新西兰	39.7	69.2	3	日本	29.6	47.7	18
俄罗斯	52.1	64.1	5	德国	33.9	46.3⑧	19
澳大利亚	35.5	63.3	6	新加坡	18.6	43.8	20
加拿大	94.7	60.0②	7	乌克兰	46.6	43.3⑧	21
英国	30.2	59.5	8	保加利亚	31.1	40.8	22
西班牙	36.7	59.4	9	印度	6.1	10.5②	
白俄罗斯	47.6	56.0	10				

注：①中国内地数据来源于世界银行，②1999 年数据，③1997 年数据，④1995 年数据，⑤1991 年数据，⑥1996 年数据，⑦1994 年数据，⑧1998 年数据，⑨1993 年数据，⑩1992 年数据，这些数据均来自文献[6]，*为中国台湾的数据，来源于文献[1]，**为 1991 年数据

表 8 列出了“公共教育经费支出占国内生产总值比重”2000 年在 4.0%及以上者。

中国只有台湾的“公共教育经费占本地居民生产总值”在 4.0%以上（5.5%），台湾这一指标在 1993 年为 7.0%。

由表 7、表 8 可知：我国内地的“大学生毛入学率”和“公共教育经费占 GDP 的比例”还是太低了，而且，两项指标都比印度低。

种种问题都有待于在改革与发展中继续下功夫去解决。加强教育的对外开放与合作交流是其中的一个重要方面，本节旨在对此进行分析，提出若干对策建议。

表 8　各国或地区“公共教育经费支出占国内生产总值比重”2000 年在 4.0%及以上者

国家或地区	1990 年	1995 年	2000 年	2000 年排名	国家或地区	1990 年	1995 年	2000 年	2000 年排名
中国内地	2.3①	2.5①	2.9①		荷兰	5.7	5.0	4.8④	9
中国香港	2.8	2.9			澳大利亚	4.9	5.2	4.7	10
中国澳门	1.7		3.6		巴西		5.0	4.7	10
中国台湾	6.5	6.6	5.5	6	德国		4.7	4.6④	11
以色列	6.3		1.3		西班牙	4.2	4.7	4.5④	12
马来西亚	5.1	4.4	6.2	2	英国	4.8	5.2	4.5④	12
新西兰	6.1	7.0	6.1	3	意大利	3.1	4.6	4.5④	12
白俄罗斯	4.8	5.5	6.0④	4	伊朗	4.1	4.1	4.4	13
法国	5.3	6.0	5.8	5	捷克		5.2	4.4	13
南非	5.9	5.9	5.5	6	俄罗斯	3.0	3.6	4.4④	13
加拿大	6.5		5.5④	6	墨西哥	3.6		4.4④	13
泰国	3.6	4.1	5.4	7	印度	3.7	3.1	4.1④	14
波兰		4.9	5.0④	8	阿根廷	10.0	3.6	4.0	15
美国	5.1		4.8	9					

注：①中国内地数据来源于世界银行，②1994 年数据，③1998 年数据，④1999 年数据，⑤1991 年数据，⑥1996 年数据，⑦1992 年数据，⑧1993 年数据，这些数据均来自文献［6］。中国台湾的数据来源于文献［1］，其指标名称为“公共教育经费占本地居民生产总值的比例”，为 1991 年数据

2. 高等教育国际化与 WTO 对我国高等教育的影响

1）经济全球化与高等教育国际化

经济全球化已成为当今世界经济发展不可逆转的客观趋势，其主要特点是生产的全球化。经济全球化不仅为发达国家所积极倡导，而且也增加了发展中国家外资进入的自由度。科学技术进步是经济全球化的客观依据。改革开放 20 多年来，我国经济有了很大发展，经济全球化离不开我国的发展，同时，经济全球化必然给我国的政治、文化等领域带来重大影响。在经济全球化时代，高等教育国际化趋势明显加快。考察教育活动，不难发现，一方面，各国按照自己的需要和利益，建立了富有本国文化传统和特色的教育制度，形成了多样化的教育体制和教育目标，另一方面也形成了一个越来越强劲的大趋势，出现了教育国际合作与交流越来越频繁的普遍现象。进入 21 世纪以来，高等教育国际化的势头更为迅猛，受到各国政府和教育机构的高度重视，留学生人数快速增长，国际教育大市场开始形成。

目前发达国家高等教育大众化历程已基本完成，在商业利润的驱动下，世界一流大学正在通过其优质的教学资源、卓越的全球声誉，吸引着来自全球的优质生源；一般院校因国内生源不足，为缓和财政紧张的状况，也力图抢占全球生源市场，尤其是通过各种手段招收我国优秀学生，从而使我国生源不断外流。这样使得我国高等教育在国际化过程中面临着更大的挑战。

2）加入 WTO 对我国教育的影响

A. 我国加入 WTO 的教育服务承诺[6]

WTO 将服务贸易分为 12 大类，教育服务是其中一类。据《服务贸易总协定》的有关规定，除由各国政府彻底资助的教育活动以外，凡收取学费、带有商业性的教育活动，均属教育服务贸易范畴。

教育服务贸易有四种方式：①跨境交付，指一个成员方在其境内向任何其他成员方境内的消费者提供的服务，如通过网络教育、函授教育等形式提供教育服务。②境外消费，指服务的提供者在一成员方境内向来自另一成员方的消费者提供的服务，如出国留学和培训。③商业存在，指一成员方的服务提供者在另一成员方境内设立商业机构或专业机构，如在其他成员方境内设立办学机构或合作办学。④自然人流动，指一成员方的服务提供者以自然人身份进入另一成员方的境内提供服务，如外籍教师来华任教、我国教师到国外任教。

我国加入 WTO 的教育服务承诺主要包括以下四个方面：①对于小学、初中教育以及军事、警察、政治和党校教育，我国没有作出开放市场的承诺。②对于出国留学和培训，接受其他成员国来华留学生没有限制。③对于高等教育、成人教育、高中阶段教育、学前教育和其他教育我国作出了有限开放市场的承诺。允许其他成员国来华开办合作办学性质的教育机构或进行其他形式的合作办学，并允许外方在合作办学机构中控股；其他成员国在我国要以商业存在方式开展教育服务，只能以合作办学方式进行，不能独立在我国境内向我国公民提供教育服务；在我国境内的中外合作办学必须遵守《中华人民共和国中外合作办学条例》的规定。④外籍个人教育服务提供者受到我国学校和教育机构的聘用或邀请，可以到我国提供教育服务，但外籍个人教育服务提供者必须具备学士或学士以上的学历，从事本专业工作两年以上，具有相应的资格证书或专业职称。

B. 加入 WTO 提供的机遇

加入 WTO 以后是一个统一与多样、合作与冲突的发展过程。以市场经济为基础、以自由贸易和公平竞争为核心的 WTO，为我国高等教育的发展提供了良好机遇。概括而言，这种机遇主要表现为：提高了我国高等教育的国际化程度，扩大我国教育市场；加快和促进教育体制改革步伐，推进我国现代教育管理制度的建立；增大教育投资力度，拓宽教育投资渠道，社会向教育事业投入的渠道将更加多元化，投资数量会持续增长；加快教育结构的调整和升级，引发人才培养模式的改革；加强产学研贸合作，进一步强化高校的社会服务功能；给我国高等教育提供了强大的机制效能，如为机制转换、结构调整、政策改革和宏观控制等深层矛盾提供了强大的冲击动力。

C. 加入 WTO 对我国高等教育的挑战

以市场经济为基础，以自由贸易和公平竞争为核心的 WTO，对我国高等教育发展也提出了严峻的挑战。

教育市场的竞争将日趋激烈。入世，就意味着认同国际规则和开放市场。虽然我国教育事业没有列入先期开放的承诺表中，但这种保护性的封闭只是短暂和有限的。随着我国经济逐步融入世界经济体系之中，我国教育市场的开放将是必然的，而且由于其巨大的市场潜力，早就为发达国家所觊觎。近年来，许多国外跨国公司和教育机构凭借其经济实力和教育与科技的优势更是“抢滩登陆”。看好我国教育市场的不仅是西方发达国家，就连东南亚地区的部分国家也争先恐后，纷纷进军我国“摆摊设点”，例如，近年来西方发达国家在我各大城市中举办的教育巡回展惊人火爆。美国、英国、德国、澳大利亚等国家纷纷出台境外办学或招聘国外人才的新政策，吸引了世界不少优质教育资源。

2000 年 7 月 5 日，《中华读书报》刊载了一篇题为《“托福”——美国人设置的中国教育成果收割器》的文章。该文认为，美国教育最成功的一项措施、最得意的一笔交易就是在中国设置了一个中国教育成果收割器——托福。美国人不必对中国教育付出什么代价，中国人已经把尖子人才培养好了，并且通过层层考试筛选已经集中到大学，特别是重点大学里来了；美国人只需在中国设置一个“中国教育成果收割器”，就轻轻松松地把中国教育培养出来的尖子收割走了。据报道，目前我国通过各种渠道移居美国的本科以上的各类专业人才已达 45 万人。

由此可见，我国的教育，特别是高等教育在国内教育市场上的竞争将是空前激烈的，竞争的结果将取决于自身的竞争实力和对入世后教育市场的清醒认识和正确反应。

我国高等教育体制将面临调整和创新。入世后我国的高等教育市场将逐步对外开放，按照服务总协定第三款第 76 条“高等教育服务”、第 77 条“成人教育服务”规定，只要两国间订有这方面的协议，国外办学主体将以各种形式参与我国的办学，使办学主体更加多元化，高等教育市场竞争更加激烈，现有的高等教育体制将面临调整和创新。

对高等教育人才培养质量要求将更高。入世后，教育国际化日趋明显，我国高等教育能否在国际人才市场找到自己的立足点，关系到我国科技水平、综合国力以及在世界的总体水平地位。国际高校间如何相互承认学历、承认学分，各种形式的转学教育、升学教育、终身教育、远程教育如何与国际接轨，这些直接影响人才质量的评价与提高。适应 WTO 的公平、公开、公正三大原则，在多元化的教育质量观指导下，培养国际型人才刻不容缓。因此，我国高等教育在发展已有专业的同时，需要考虑市场的需求，调整专业结构，创新培养模式，提高人才培养质量，以培养出能被国际国内市场接受的人才。

对高等教育教学内容和方法提出了更高的标准。入世以后，市场经济活动要遵循 WTO 的基本原则，如市场开放原则、公平贸易原则、透明度原则和非歧视原则等，对于这些原则，我们并不熟悉，更缺乏从事这类全球化经济活动的国际型人才。所以我国高等教育要及时更新教学内容，设置相关课程，加快课程改革和教材建设，改革教育方法和手段。

总之，我国高等教育在国际教育市场上的竞争将是空前激烈的，竞争的结果将取决于自身的竞争实力和对入世后教育市场的清醒认识和正确反应。

二、我国出国留学教育的对策

留学教育是国际教育服务贸易的主要形式之一，也是我国教育，特别是高等教育的重要组成部分。我国多年的留学工作成绩表明，派遣优秀人才赴国外留学是我国高等院校进行国际交流和培养创新人才的重要途径。随着经济全球化进程的加快和我国加入 WTO，我国的高等院校将日益向国际化方向发展。留学工作也将面临着新的机遇和挑战。本节旨在通过分析我国出国留学教育的现状和发展趋势，总结改革开放以来我国留学工作取得的成绩，揭示目前我国出国留学教育存在的问题，为我国出国留学工作提出一些政策性的建议，从而推动我国出国留学教育的发展。

1. 我国出国留学教育的现状和发展趋势

我国政府一直十分重视出国留学工作。重用留学生的传统可以追溯到晚清时期的洋务运动。民国时期，孙中山先生也是十分重视留学生的。在他组建的南京临时政府里，部长和次长当中 80%是留学生。新中国成立之初，我国曾大量派遣留学生到苏联和东欧国家留学。

党的十一届三中全会以来，我国进入改革开放和社会主义现代化建设的新的历史时期。我国出国留学和留学回国工作也进入了一个新的发展阶段。1978 年 6 月，在党的十一届三中全会召开之前，邓小平同志就发表了关于扩大向外派遣留学生的重要讲话，明确提出留学生的数量要增大，他说“要成千成万地派”①，“要千方百计加快步伐”①。1992 年，国家出台了“支持留学，鼓励回国，来去自由”的工作方针。党的十六大又提出了“尊重劳动、尊重知识、尊重人才、尊重创造”的重大方针。

近年来，随着自费留学热潮的兴起，我国出国留学人员数量迅速增加，留学回国的比例有所提高，留学专业分布更为广泛，留学目的地集中在发达国家。在各类留学人员中，国家公派出国留学人员的层次一直保持比较高的水平，自费出国留学人员出国前受教育程度有所下降。与公派留学相比较，自费留学回归率较低。我国出国留学教育表现出“留学教育自费化、专业选择多元化、自费留学低龄化”的特点。

1）出国留学人员总量

据统计，自 1978～2002 年年底，我国内地出国留学人员达 58.3 万人，遍及世界 100 多个国家和地区[7]。1996 年，我国全面实施国家公派留学改革。表 9 和图 2 列出了 1996～2002 年，我国出国留学人员数量及其增长变化情况。如图表所示，20 世纪 90 年代末期以来，我国出国留学人员数量迅速增加。

---

① 《纪念邓小平扩大派遣留学生讲话发表 40 周年国际学术研讨会举办》，http://www.xinhuanet.com/world/2018-05/29/c_129882135.htm[2023-11-09].

表 9 1996～2002 年我国各类出国留学人员数量及增长变化情况[7]

年份	出国留学总人数		国家公派出国留学		单位公派出国留学		自费出国留学	
	人数/人	增长率/%	人数/人	增长率/%	人数/人	增长率/%	人数/人	增长率/%
1996	20 905	—	1 905	—	5 400	—	13 600	—
1997	22 410	7.20	2 110	10.76	5 580	3.33	14 720	8.24
1998	17 622	−21.37	2 639	25.07	3 540	−36.56	11 443	−22.26
1999	23 749	34.77	2 661	0.83	3 204	−9.49	17 884	56.29
2000	38 989	64.17	2 808	5.52	3 888	21.35	32 293	80.57
2001	83 973	115.38	3 495	24.47	4 426	13.84	76 052	135.51
2002	125 000	48.86	3 500	0.14	4 500	1.67	117 000	53.84

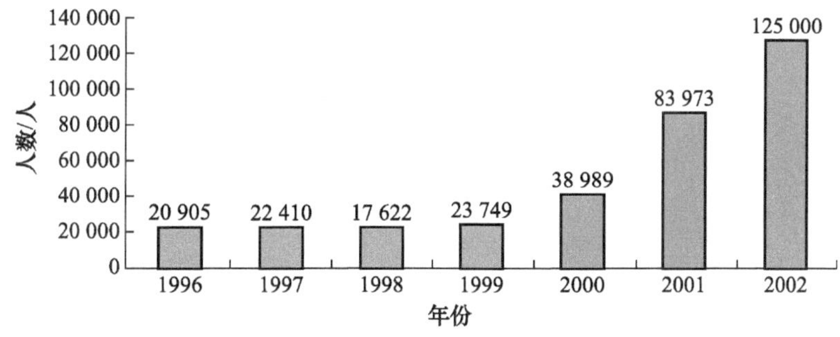

图 2 1996～2002 年我国出国留学人员总数

2）各类出国留学人员的构成

根据出国留学的经费来源，出国留学分为公派出国和自费出国两类，公派出国又分为国家公派和单位公派。20 世纪 90 年代下半期以来，自费出国留学人数在出国留学总人数中所占比例均超过 50%（表 10），从 1999 年起，我国自费出国留学比例逐年上升（表 10），自费留学人员数量以年增长率超过 50%的速度迅猛增长，2001 年自费出国留学人数的增长率更高达 135.51%（表 9）。

表 10 1996～2002 年各类出国留学人员比例构成

年份	出国留学总人数/人	派出比例/%		
		国家公派	单位公派	自费留学
1996	20 905	9.11	25.83	65.06
1997	22 410	9.42	24.90	65.68
1998	17 622	14.98	20.09	64.94
1999	23 749	11.20	13.49	75.30
2000	38 989	7.20	9.97	82.83
2001	83 973	4.16	5.27	90.57
2002	125 000	2.80	3.60	93.60

注：表中数据进行过修约，故存在合计不为 100%的情况

从图 3 和图 4 可以看出，与自费出国留学相比较，公派出国留学（包括国家公派和单位公派）人员数量变化相对比较稳定。1996～2002 年国家公派和单位公派出国留

学人员数量一直在 2000～6000 人的范围内浮动，而每年自费出国留学人数都超过 1 万人。可见，自费出国留学已经成为我国公民出国留学的主要途径。近年来我国出国留学人数迅速增长的主要原因是自费出国留学人员大幅增加。20 世纪 90 年代末期起，我国又兴起了新的一轮自费出国留学热潮。

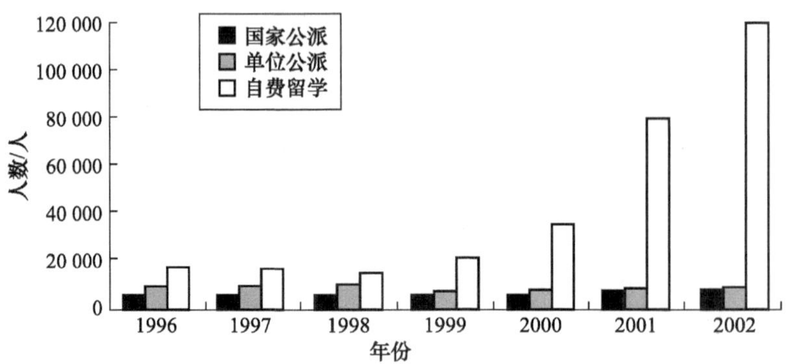

图 3　1996～2002 年我国出国留学人员数量

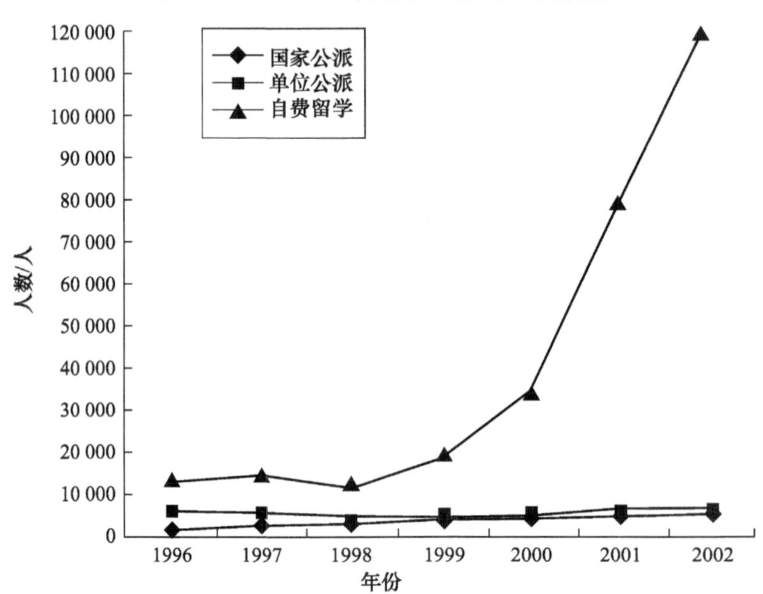

图 4　1996～2002 年我国各类出国留学人员数量变化情况

3）出国留学人员的学科构成

改革开放之初，我国公派出国留学生以学习理工科为主。在 1978～1981 年国家公派出国的 7456 名留学生中，赴海外学习理工科的有 6039 人，占 81.0%；学习语言和其他学科的有 1150 人，占 15.4%；学习人文科学和社会科学的只有 267 人，只占 3.6%[8]。

20 世纪 90 年代下半期以来，这种局面有所改变。我国留学人员出国留学专业向多元化方向发展。近几年，我国留学基金委员会适当调整了国家公派留学录取人员学科构成，相应提高了人文、社会科学的选派比例。2001 年，国家留学基金管理委员会录取人员按学科统计的情况是：理科 267 人，占 12.9%；工科 596 人，占 28.8%；医科

319 人，占 15.4%；农科 199 人，占 9.6%；经管 205 人，占 9.9%；文科 379 人，占 18.3%；非通用语种 105 人，占 5.1%（图 5）[9]。

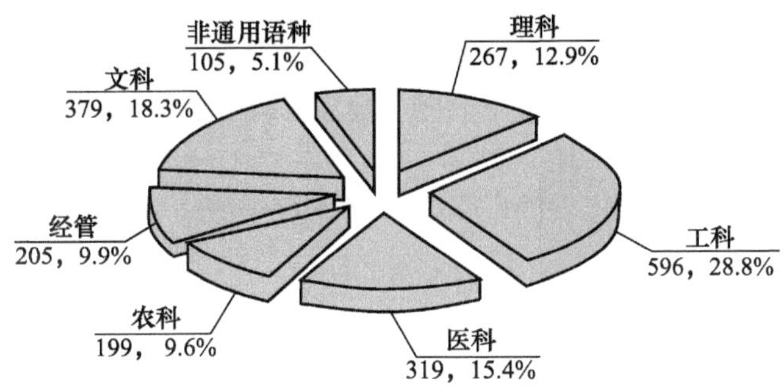

图 5 2001 年出国留学人员专业构成

笔者在与留学服务中介机构工作人员的访谈中了解到，近年来，自费留学生的学科选择也逐渐趋向多样化。据介绍，过去许多自费留学生都是抱着出国镀金或移民的目的留学，在学科选择上带有很大的盲目性。他们往往根据学科的难易程度选择留学专业，因此选择读商科的比较多。近几年，自费留学生在学科选择上更加理性。他们会从自身的兴趣爱好、专业背景、所学专业今后的发展前景、国际和国内劳动力市场的需求状况等多个方面考虑，选择留学专业。

4）出国留学人员的层次

改革开放以来，我国出国留学人员的受教育程度比较高。有关学者在 1998 年 9～10 月对部分归国留学生进行了一次问卷调查。调查结果表明，调查对象出国前受教育程度的比例分别是高中 8%、大中专 7%、学士 38%、硕士 32%、博士 15%[8]。

在各类出国留学人员中，国家公派留学人员的层次一直保持在比较高的水平。我国全面实施国家公派留学改革以来，在历年国家公派留学人员中，大部分都具有硕士以上学历或副高职以上职称，如表 11 所示。

表 11 1997～2000 年国家留学基金管理委员会选派留学人员学历构成

年份	具有硕士以上学历者所占比例/%	具有副高职以上职称者所占比例/%
1997	72.4	59.1
1998	67.1	66.5
1999	71.0	66.0
2000	73.6	66.2

资料来源：《中国教育年鉴》编辑部. 1998～2001 年的中国教育年鉴. 北京：人民教育出版社

此外，近年来，我国自费留学人员受教育程度有所下降。以北京市为例，2001 年北京市申请自费留学人员 5919 人，比 2000 年增加 104 人。具有研究生以上学历 2411 人，占 40.7%，比 2000 年减少 267 人，其中，博士毕业生 738 人，占 12.5%，比 2000 年减少 42 人；硕士毕业生 1621 人，占 27.4%，比 2000 年减少 195 人。具有本专科学历人数 3508 人，比 2000 年增加 371 人，其中，本科毕业生 2281 人，占 38.5%；专科毕业生 728 人，占 12.3%，比 2000 年增加 130 人[10]。

中小学生逐渐成为我国自费留学潮中的重要角色。北京市教育部门于2003年年初对北京市部分中学进行了一次调查。调查结果表明，在被调查的五所中学中，30%以上的学生希望出国留学。1996年，上海办理出国留学的中小学生为30多人，1997年为60多人，1998年为100多人，1999年和2000年增至1000多人，2001年之后每年竟以1500多人的速度快速增长。以上数据在一定程度上表明，我国自费留学教育有低龄化的趋势。

5）出国留学人员留学目的地分布

图6所列是1978～1995年我国留学生在国外的分布[11]。如图6所示，我国留学人员留学目的地集中在发达国家。

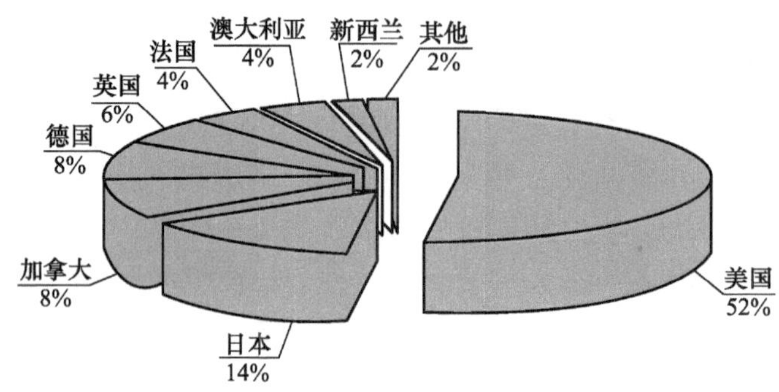

图6　1978～1995年我国留学生在国外的分布

6）留学人员回国比例

自1978～2002年年底，我国出国留学人员总数为58.3万人，回国15.3万人。仍在国外留学的43万人中，有27万人尚在国外高等教育机构学习。出国留学人员学成者中，有近一半人员回国工作，有超过一半人员留在国外。表12列出了1995～2002年，我国留学回国人员数量及其构成。如表12所示，20世纪90年代中期以来，我国留学回国人员数量逐年增长。

表12　1995～2002年我国留学回国人员数量及其构成

年份	留学回国人员总数/人	增长率/%
1995	5 000	—
1996	6 570	31.40
1997	7 130	8.52
1998	7 379	3.49
1999	7 448	0.94
2000	9 121	22.46
2001	12 243	34.23
2002	18 000	47.02

资料来源：1997～2003年的中国教育年鉴

尽管近年来我国留学回国人员数量有所增长，但与出国留学人数的大幅增长相比较，回国人数的增长显得非常缓慢，如图7和图8所示。

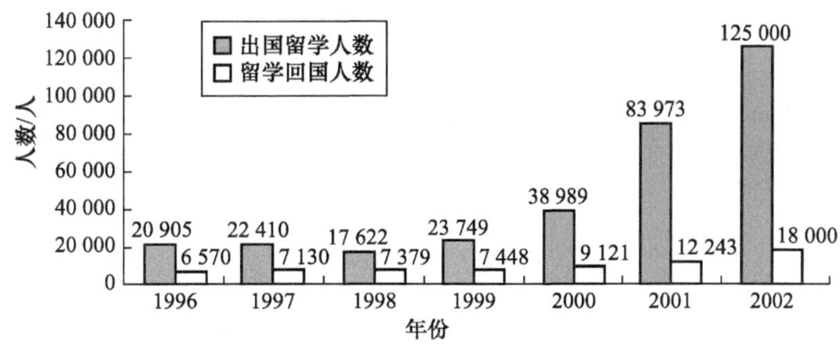

图 7 1996～2002 年出国留学人员和留学回国人员

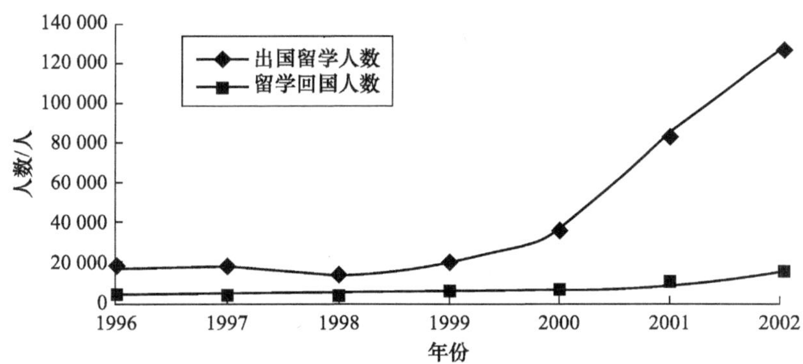

图 8 我国出国留学人数和留学回国人数变化趋势

在公派和自费两类留学人员中，公派留学人员的回归率比较高。国家公派留学人员的回归率在 1996 年以前达 70%以上，1996 年国家实施公派留学改革以后，国家公派留学人员按期回归率占应回国人员的 90%以上[12]，说明这次改革在提高留学人员回归率方面是卓有成效的，见表 13。

表 13 1998～2001 年国家留学基金管理委员会资助公派留学人员回归率

截止日期	应到期回国人数/人	已回国人数/人	按期回归率/%
1998 年 10 月	1178	1089	92.4
1999 年 11 月	3657	3300	90.2
2000 年 10 月	1999	1855	92.8
2001 年 12 月	2514	2427	96.5

资料来源：1999～2002 年的中国教育年鉴

自费留学回国人员近年来虽然明显增加，但其比例依然很低。如图 9 所示，1998～2001 年，我国自费出国留学人数以每年 50%以上的幅度快速增长，2001 年自费出国留学人数达到 7.6 万人。但这几年自费留学回国人员数量增长相对比较平缓，每年回国人数不超过 7000 人。

以上数据表明，自从我国全面实施公派留学改革以来，绝大部分公派留学人员都能按期回国。国家公派留学改革大大地提高了我国公派留学人员回归率。自费留学回归率过低是导致我国留学人员回国率较低的主要原因。

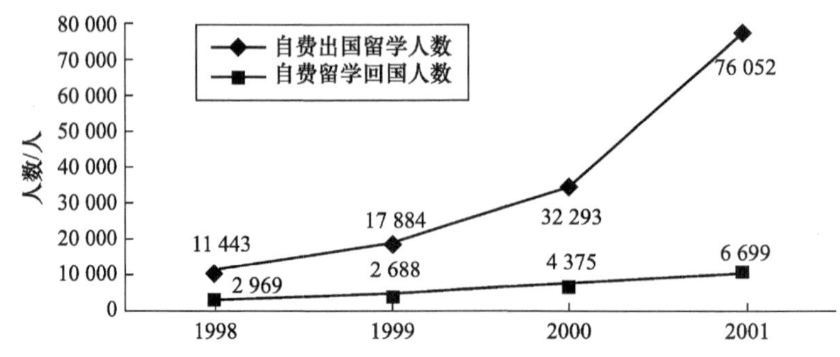

图 9　自费出国留学和留学回国人数变化趋势

资料来源：《中国教育年鉴》编辑部. 1999～2002 年的中国教育年鉴. 北京：人民教育出版社

2. 留学教育的积极作用和存在问题

1）留学教育的积极作用

（1）留学教育为我国公民就学提供了更多的选择，在一定程度上缓解了我国基础教育和高等教育供给不足的问题。目前，我国基础教育和高等教育的整体规模均不能满足社会发展和公民个人的需要，与发达国家还有很大一段差距。下面我们将我国的高等教育状况与美国作一个简单的比较。我国国土面积与美国相近，但我国人口却是美国人口的近五倍。1999 年，我国共有高等院校 1942 所，本专科在校生 718.91 万人，毛入学率约为 10.5%[13]，而 1997～1998 年度美国共有高等院校 4064 所，在校生约 1450 万人，18～19 岁年龄段的入学率约为 62.2%。根据我国第五次全国人口普查提供的数据，我国目前受过高等教育的人口仅占我国适龄劳动人口的 5.2%，而 1995～1997 年，美国的这一比例是 46.5%[14]。

近几年，我国大大加快了教育改革和高等教育大众化的进程。但当前我国教育体制还存在许多问题，教育改革遇到了许多困难：教育投入不足、国内高中和大学数量少，学校教学和科研设备落后；教育发展不平衡，教育资源分配不均，优质教育资源奇缺；应试教育盛行，升学竞争激烈等。据统计，我国每年有将近 10%的小学毕业生、50%的初中毕业生、75%的高中毕业生不能升入高一级的学校学习，而 90%的家长却迫切期望自己的孩子能够接受高等教育。

在国内教育服务严重供不应求的情况下，一些人选择出国留学，到教育资源相对优厚的发达国家学习。出国留学教育为我国公民的就学提供了更多选择，在一定程度上缓解了我国基础教育和高等教育供给的不足。

（2）留学归国人员在科学技术、教学科研和经济贸易等领域为推动我国现代化建设作出了重大贡献，为我国国民经济创造了巨大的财富。我国学子留学海外，是我国追踪世界科技发展潮流，缩短与世界发达国家差距的一个有效途径。在庞大的留学军团中，有相当部分的留学生学成以后选择归国创业，为祖国服务。目前我国留学回国人员已达 15.3 万人。这些归国留学生利用从国外学到的先进科学技术和管理经验，为我国的发展建设事业服务，成为国家建设和科技发展的中流砥柱。

目前，留学归国人员在教学、科研等领域的重要岗位上占有很大的比例，他们已成长为这些领域的骨干力量或尖端科技的带头人。据统计，党中央、国务院和中央军委表彰的“两弹一星”23个功臣中，有21个是留学人员；1999年中国科学院士中留学人员占81%，中国工程院院士中留学人员占54%。国家重大科技攻关项目、“863”计划、人事部①等七部委组织实施的“百千万人才工程”、中国科学院的“知识创新工程”等，入选的人才一半以上都有留学经历。

此外，越来越多的留学人员选择回国创办企业，为国作贡献。目前，全国有70多家留学人员创业园，入园的留学人员企业4000多家，入园留学人员15 000多人，这些创业园的年产值逾100亿元。以上这些事实和数据都说明，留学人员在高素质人才队伍中占有重要的分量，他们直接或间接为我国国民经济创造了巨大的财富。

2001年，北京大学教育学院与中山大学高等教育研究院合作承担了由教育部国际合作与交流司和财务司共同发起并资助的“改革开放以来我国公派留学效益评估”课题的研究。该课题的研究结果表明，出国留学的收益包括个人收益和社会收益、经济收益和非经济收益、显性收益和隐性收益。而且，公派留学的社会收益远高于个人收益，非经济收益远高于经济收益，长远的隐性收益远高于眼前的显性收益。根据该课题组的研究结果，公派留学归国者带来的收益主要体现在以下几个方面：一是为我国教育科技领域培养了能够与国际学术界进行对话的新一代学术领导群体；二是培养了一批具有国际管理经验的高等院校和科研机构的领导骨干；三是使我国几乎所有学科的知识，包括学术思想、理论和研究方法在很大程度上都得到了更新，创设了一大批国内曾经空白的学科，陆续引进了大批新教材及新的教学方法，极大地提高了我国学科建设和高等教育的水平，对高等院校人才培养的质量产生了重大影响；四是我国的科研水平有了显著的提高，大大缩短了我国与国际水平的差距，一些学科已经达到国际领先水平；五是留学归国人员通过承担国际合作和委托项目，通过科研成果转化以及通过决策支持研究为国家创造了巨大的直接经济效益；六是留学归国人员建立了广泛的国际学术交流网络，促进了中外学者的交流与国际合作[12]。

（3）海外优秀留学人员作为外国了解我国的一个窗口，起到了弘扬中国优秀文化，宣传我国经济建设和发展良好局面的作用。文化方面的差异是造成人类分歧和冲突的主导因素之一。目前，在西方国家，许多人不了解中国，只知道中国是一个古老而神秘的东方国家。有些人对中国有一定的了解，但由于中西方文化的差异，或受到西方一些媒体对我国的一些负面报道的影响，对中国内地存在一些偏见，形成了不良的印象。理解是正视并尊重的前提。交流是理解的一个直接途径。留学人员在海外有机会与外国人直接接触和交流。他们可以用自己的言行，向外国人展示中国的优秀文化传统。他们可以向外国人介绍我国对外开放的政策，宣传我国经济建设的良好发展势头和人们安居乐业、国泰民安的良好社会局面，提高我国的声望，树立我国在外国人心目中的良好形象。

事实上，我国的大部分出国留学人员在国内都是优秀学生或各行各业的优秀人才。

---

① 2008年国务院机构改革，将人事部、劳动和社会保障部的职责整合划入人力资源和社会保障部。

他们在海外留学期间都表现不俗。据统计，2002 年，中国留学生获得英美博士学位的数量居世界之首。在英国获得博士学位的外国留学生中，中国有 208 人，名列第一，比第二位的德国（146 人）和第三位的马来西亚（141 人）高出许多。在美国，获得博士学位的外国留学生中人数最多的还是中国，总数有 2187 人，是第二位的印度（888 人）和第三位的韩国（738 人）的约两到三倍。国外著名高校对中国留学生的评价都是比较高的。外国人对中国留学生的评价肯定在一定程度上会影响他们对中国的评价。

2）出国留学的消极影响及我国留学工作存在的问题

对于发展中国家来说，留学教育是一把双刃剑，既有积极有利的一面，同时又有消极不利的一面。当前，我国留学工作仍存在一些问题。

（1）目前我国出国留学人员回国率还比较低。国际研究数据表明，发展中国家在经济起飞阶段，2/3 的留学生归国效力，1/3 留在国外工作学习、沟通信息[8]。使留学人员回归率保持在 2/3 这个比例是比较合理和有利的。我们来看看我国的情况。如前所述，我国有 58.3 万留学人员，目前有 27 万人还在国外学校读书，58.3 万人减去 27 万人是 31.3 万人。在这 31.3 万人中，回国的是 15.3 万人，也就是说只有一半不到的人回国了，离 2/3 的理想比例还有一段差距。

有学者认为，出国留学人员学成不归会导致人才和资金的外流，引发两个恶性循环：一是人才流失的恶性循环，二是教育投资匮乏的恶性循环[15]。首先，学成后留居国外的留学生人数大于回国留学生人数是人才外流的表现。决定世界各国未来发展的是一个国家的人才储备。人才外流就可能会导致国家经济发展中的智力支持不足；智力支持不足，就意味着经济增长速度缓慢，而经济增长缓慢会导致国家和社会为各类人才提供的薪酬待遇和工作条件不足，最终必然引发人才的继续外流。这种恶性循环必将对整个国民经济发展带来巨大的负面影响。其次，出国留学往往需要投入大笔资金。据留学中介服务机构有关人员的介绍，各个留学目的国的自费留学费用不尽相同，一般不少于 8 万～10 万元，有些国家留学费用高达 20 万元以上。如果以一个自费留学生留学费用 10 万元计算，2002 年自费出国的 11.7 万留学生共花费 117 亿元。这笔庞大的资金如果投入到国内的教育领域，无疑将对我国教育事业产生巨大的推动作用。但这笔资金却完全投入到了国外的教育部门，造成国内教育资金投入不足的恶性循环。教育资金投入不足，则教育水平难以快速提高，教育质量难以满足求学者的需要，进而使更多的求学者选择国外教育部门，导致更多的教育资金外流，阻碍国内教育事业的良性发展。可见，留学回国工作的成败对我国经济和教育的发展都有重大影响。

事实上，我国历来都非常重视留学人员回国工作。近年来，党和国家制定了一系列方针政策，采取了许多措施，吸引、鼓励留学人员回国。国家分别就公派和自费出国留学、加强对留学人员的管理、鼓励海外留学人才回国工作、以多种形式为国服务等问题制定和颁布了一系列政策，推出和实施了“新世纪百千万人才工程”“百人计划”“长江学者奖励计划”等与留学回国工作密切相关的项目和政策措施；中央和地方各部门为了进一步改善留学回国人员的工作、生活条件，投入了大量资金，设立专项基金，创办“留学人员创业园”，出台了一些针对留学人员回国创业的优惠政策，切实帮助留学回国人员解决家属就业、子女就学等问题。此外，各地还先后多次组织招聘团到海外招聘优秀留学人员，与留学人员座谈，向他们介绍国内相关政策，吸引、鼓励他们回国或在海外为国服务。应该说，党和国家采取的这一系列措施都取得了一定的成效。近年来，我国留学人员回国率有所上升。自1996年以来，我国留学回国人员数量逐年增长。

但是，我们必须看到，目前我国留学回国工作还存在一些没有解决的问题。近年来，各地吸引留学人才的政策大多重技术引进，重开办科技企业，对带高新技术回国的留学人员奉为上宾，对回国留学人员创办的企业实行税收优惠。这些措施一方面会让那些从事科研、教育工作，或在各种大型企业和外资企业任职的留学回国人员产生不公平感，另一方面各地竞相出台各种优惠政策，也是地区间恶性竞争的表现。此外，各地的优惠政策在执行过程中也存在许多问题。例如，某些地区制定的优惠政策相当吸引人，但在实际执行过程中却由于有关部门协调不够而无法落实，使留学回国人员对政府的诚信产生怀疑。以上这些问题值得国家有关部门重视，进一步制定和完善与留学回国工作有关的政策和法规，并采取有效措施，使政策得以切实执行。

(2) 随着我国自费出国留学人数的急剧增加，留学中介服务行业的管理成为我国留学工作的一个新的重点，也是一个难点。过去，出国留学主要指公派出国留学。改革开放的政策改变了我国只有公费留学的历史。20世纪80年代末，20世纪90年代初，随着我国留学政策的放开和“支持留学，鼓励回国，来去自由”留学工作方针的确定，越来越多人有了自费出国留学的愿望。但由于信息不通，申请手续繁杂、经验不足等问题，很多个人自费留学申请者很难达到目的。他们希望有专门的中介机构为他们提供服务。就是在这种情况下，留学中介服务机构应运而生。开始的时候，这个行业相当混乱。据统计，1994年上半年到1997年上半年，仅北京就有上千家所谓的“留学中介”。有些所谓的中介机构其实只有一部电话，一张办公桌，一把椅子。

1997年下半年，国家开始整顿留学中介行业。1999年，教育部联合公安部、国家工商行政管理局制定和颁布了《自费出国留学中介服务管理规定》及其实施细则，并且先后两次对全国270家留学中介机构进行了资格认定。此外，教育部还对留学中介的市场准入、与境外机构的合作、广告宣传等方面作了明确的规定。

近年来，自费出国留学中介机构通过与国外高等院校、教育机构开展合作，帮助众多的学子实现了出国留学的梦想，使许多人通过接受国外教育，成为社会发展所需要的人才。以2002年为例，全国有12.5万人出国留学，其中自费留学生有11.7万人，占当年出国留学总人数的94%，其中80%的人是通过留学中介办理出国的。应该说，出国留学中介机构较好地满足了人们出国自费留学的需要，在中外学生交流方面起到了举足轻重的作用。

但近年来留学中介机构违规操作，欺骗、坑害消费者的情况也时有发生。一些中介机构模糊国外学校的性质和资质：把国外的专科学校说成是综合性大学，或某大学下属的学院；把与世界名牌大学没有任何关系的语言学校说成是正规预科学校，是进入这些大学的必由之路和捷径。还有一些中介机构巧立名目，多收中介费：通过中介办理出国留学的，出国前一般需交两笔费用，一是中介服务费，二是境外服务费。前者一般是明码标价，而后者“学问”就大了。一些中介事前不说明这些费用作何用途，就算说了，当事人也无法搞清楚自己交的费用是不是真正用到这些方面了。而且，一些当事人缺乏自我保护意识，没有向中介机构索要正式发票，一旦出了问题要打官司也拿不出有效凭据，无法通过法律途径挽回损失。还有一些中介机构用虚假宣传误导当事人。比如说，一些国家（如南非）法律明确规定不允许留学生打工，但不法中介却谎称当地政府允许留学生边打工边学习，以吸引那些家境不太富裕，想走勤工俭学途径的留学当事人。而后，还有一些没有取得资格的中介机构仍在从事留学中介活动，给消费者造成巨大损失。因此，国家教育部门和其他有关部门应进一步制定相关政策，采取有效措施，治理和整顿留学中介服务市场，促进我国自费留学的有序性和有效性。

（3）一般层次的留学回国人员回国后将面临巨大的就业竞争压力。尽管许多留学人员出于专业发展和个人生活方面的考虑选择了留在国外，但近几年，由于美国、日本等国家经济发展水平的滑坡，加上一些西方国家受恐怖主义的威胁，社会出现不稳定，以及我国国内经济的持续发展，越来越多的留学人员学成以后选择回国。近几年留学人员回国比例逐年递增，就是一个证据。据全国青联海外学人工作部与《青年参考》报最近联合主办的“海归搜索行动——海外留学与归国人员现状大调查”的结果显示，近九成（87.7%）留学人员有回国的意愿。

对海外归来的留学人员民间有种称呼，把他们称为“海归”。“海归”派的加入，令我国本来就竞争激烈的就业市场面临更大的压力。过去，“海归”是各个用人单位抢着要的“宝贝”。“海归”常常与优厚的待遇、可观的收入联系在一起。然而，最近一两年，从一些沿海发达城市的人才市场传出的信息表明，“海归”的吸引力开始减弱，许多“海归”回国以后难以找到理想的工作，有的“海归”甚至在求职中故意隐瞒自己的海外求学经历。“海归”似乎不再像从前那么吃香了。这种现象的出现是由多种原因引起的。首先，大批“海归”回国，人们对“海归”的认识逐渐平淡化。所谓物以稀为贵，“海归”人数的逐渐增多，这种资源的供给增加，资源的价格就会下降。其次，并不是所有的“海归”都能满足用人单位的要求。有些人仅仅是为了获得留学经历而留学，以为出去镀镀金，回来以后就会身价百倍。这部分人往往是自费出国，在国外读的是一般的学校，专业选择上也比较随便，哪个要求低，比较好读就读哪个。他们希望凭着手里的镀金文凭，在国内谋得理想的工作。殊不知，国内许多用人单位经过前几年聘用留学人员的经验，对留学人员的聘用已经有了比较理性的认识，已经从原来的盲目引进变成慎重选择。一些没有真才实学的“海归”往往很难进入重要岗位或者通过试用期。最后，随着我国高等教育质量的提高和中外合作办学的逐渐兴起，本土人才的国际化水平大大提高，竞争力大大增强，“海归”派不再是一枝独秀。在以上几点原因的作用下，我们就不难理解为什么最近一些回国留学人员在求职过程中会受到挫折。

目前，我国留学回国工作重点还是在如何吸引留学人员回国上，对回国人员的就业和安置方面的措施也是主要针对那些高层次的回国人员。广大的一般层次的留学回国人员在就业和安置方面将面临较大的竞争压力。尽管这个问题才刚刚出现，尚不严重，但如果我们不能正视，找出应对策略，就可能会挫伤海外留学人员的回国积极性，最终造成人才和资金的外流。

3. 关于对策措施的建议

1）进一步扩大公派留学人员规模，适当调整公派留学人员类别

A. 增加国家公派留学人员数量

从 1978 年至 2004 年，我国国家公派出国留学人员约 7 万人，每年平均数只有约 2600 人。这与邓小平同志当初在关于扩大向外派遣留学生的重要讲话中提出的“要千方百计加快步伐”① “要成千成万地派”① 的设想还有相当大的差距。近几年，国家有意识加大了公派留学人员规模，但由于基数较小，所以总量还是不大。随着自费留学热潮的兴起，自费留学人员的数量迅猛增加，国家公派留学人员在每年出国留学人员总数中所占的比例迅速缩小（表 10），国家公派出国留学在我国出国留学教育中逐渐变为配角。

从留学人员回归率来看，国家公派留学人员回归率远远高于自费留学人员回归率。1996 年全面实施国家公派留学改革后，国家公派留学人员回归率一直保持在 90%以上。在自费留学人员回归率仍然偏低的情况下，公派出国留学是我国培养具有国际知识和经验的高级人才的重要途径。笔者认为，有关部门应当采取措施，进一步扩大公派留学人员的规模，以缓解国家对高级人才的迫切需要。

B. 增大高学历人才派出比例，延长派出时间

改革开放以来，我国公派留学生类别经历了“主要派遣本科生”“逐渐减少以致基本上停派本科生”“逐渐增派研究生和进修人员”“减少派遣研究生”“主要派遣进修人员和访问学者”的几个阶段。国家主要是从回归率、派出效益等方面考虑，实行这种变化和调整的。事实证明，公派进修人员和访问学者的回归率确实很高。相比之下，改革开放初期派出的本科生回归率较低。但是，我们还应看到事情的另一面。进修人员和访问学者的留学时间一般比较短，大多在一年左右，有的甚至只有半年或几个月，而且许多人是初次出国，到了国外还需要一些时间来熟悉环境，克服语言困难，有时候还需要转换导师、调整课程等，等到他们基本上适应了当地的环境，真正进入角色，准备静下心来学习和研究时，离回国的时间已经不远了。因此，许多短期出国进修的回国人员私下里与别人谈起自己的留学经历时都表示，出国一趟，主要的收获是开了眼界，长了见识，至于学术水平和科研能力的提高则是次要的。因此，笔者认为有关部门应适当调整公派留学人员类别，提高本科生和研究生，特别是研究生的派出比例，延长派出人员的留学时间，以便留学人员在国外有足够的时间学习进修和参与科研。

2）鼓励自费出国留学，创造良好的法制和社会环境

A. 对自费出国留学的专业进行适当引导

对于公派留学生的学科比例，国家有关部门是可以根据国家需要进行直接控制和调整的。对于自费留学生出国留学的专业选择，国家则无法进行直接干预。近年来，自费留学生在学科选择上逐渐变得理性起来。他们中的许多人（如到非移民国家留学的留学生）是打算学成以后回国工作的，所学专业今后在国内的发展前景、国家今后

① 《纪念邓小平扩大派遣留学生讲话发表 40 周年国际学术研讨会举办》，http://www.xinhuanet.com/world/2018-05/29/c_129882135.htm[2023-11-09].

对各种专业人才的需求状况都是他们在选择留学专业时考虑的重要方面。笔者认为，国家应该采取措施，对自费留学生的学科选择进行适当的引导。有关部门可以通过媒体、各中学和高等院校、留学服务中介机构，向自费留学人员宣传国家的长远发展规划和未来几年各种专业人才的供需变动情况，让留学人员有计划地选择留学专业、安排学习内容，形成一支适应国家未来发展需要的、专业结构合理的留学人才队伍，为我国今后的发展储备大量人才。

B. 建立公平竞争、高效有序的留学人才市场

近年来，自费留学回国人员数量逐年增加。尽管目前这个数字还很小，每年只有几千人（图 9）。但笔者估计，随着我国国内经济的迅速发展和我国留学回国政策的进一步完善和深化，在若干年以后，自费留学回国人员数量一定会大幅增加，留学生回流潮也会逐渐形成。到那时，广大的留学回国人员，特别是一般层次的自费留学人员将面临巨大的竞争压力。如前所述，这种现象近两年在部分沿海发达城市已经初见端倪。笔者认为，国家应当未雨绸缪，采取有效措施，努力建立一个公平竞争、高效有序的留学人才市场。可以采取的措施包括：建立留学归国人才库；对归国留学生进行资格认证；在自愿参加的原则上，举办一些考试，测试留学生的专业和外语水平。这些措施一方面可以为用人单位挑选聘用优秀留学人才提供参考，促进留学归国人员公平竞争，实现优胜劣汰。另一方面，这些措施还可以让那些希望通过留学出国镀金的留学生和那些因望子成龙、望女成凤而打算把年幼子女早早送到国外的留学生父母认识到，国家需要的是有真才实学、内功扎实的国际型人才，促使他们更加理性地看待自费出国留学，降低自费出国留学的盲目性，减少教育投资的外流。

C. 加强对自费留学中介市场的宏观调控力度，建立留学中介行业协会

如前所述，目前某些留学中介服务机构片面追求经济利益，采取提供虚假信息、发布虚假广告、多收中介费等方式，侵害留学当事人的合法权益。针对这些现象，国家有关部门做了大量的工作：教育部于 2002 年年底成立了教育涉外监管处，专门对我国各类教育涉外活动实施监管；对境外教育机构进行资质认证，公开发布外国教育主管部门认可的国外学校名单，并定期对所公布的名单进行修改和更新；建立教育涉外监管信息网，将自费留学中介相关政策、国外政府和权威机构认证的高等学校、合法中介机构与国外合作项目名单、合法中介机构名单及其业绩、受表彰或受处罚情况等上网公布等。

笔者认为，在市场经济条件下，留学中介服务机构作为自主经营的企业实体，政府主管部门对中介机构的业务经营活动不宜太多地进行直接干预。政府应在加大宏观调控力度的基础上，鼓励留学中介服务机构建立行业协会，推行行业自律。据笔者了解，一些地区（如北京）的留学中介服务机构已经成立了行业协会，另一些地区却因种种原因而未建立协会。如在广东，有关法律规定 30 个企业以上才能建立类似协会的组织，而广东目前通过资格认证的留学中介机构只有 13 家，还没达到建立合法协会的标准。笔者建议，在全国范围内建立留学中介行业协会，并在各地建立分会。通过行业协会制定出国留学中介机构工作流程的规范版本及服务标准，使中介在咨询服务、申请入学服务、境外服务和签证服务等方面都有章可循，服务收费明码标价。以行业

协会的名义组织培训，提高留学中介服务从业人员的专业素质和业务水平。此外，行业协会还可以在处理投诉、打击非法经营和违规操作、评估服务质量等方面开展工作，促进留学中介服务行业的优胜劣汰。

3）采取行之有效的措施，吸引优秀留学人才回国

A. 深化国家公派留学制度改革

留学人员回归率是测量一国留学教育工作成败的关键指标。吸引和鼓励留学人员回国工作一直是我国留学工作的重点。二十多年来，我国采取了大量措施，吸引海外优秀留学人员回国。至今，我国留学学成人员中，有将近一半回国。近年来，留学回国人员数量逐年增加。我国留学回国工作取得了喜人的成绩。

我国留学回国工作成绩最突出的是在公派留学上。1995 年，国家对公派留学制度进行改革，按“个人申请，专家评审，平等竞争，择优录取”的办法选拔留学人员，并实行“签约派出，违约赔偿”，促使留学人员学成回国。所谓“签约派出，违约赔偿”，是要求出国留学人员在出国前必须与国家留学基金管理委员会签订出国留学协议书。协议书的内容包括：出国留学人员出国前必须按规定向留学基金会交付押金；留学期满按期回国者，押金全部退还本人；留学期满不能回国服务者，按规定赔偿其出国留学期间国家为其支付的全部经费。事实表明，这一办法确实行之有效。1996～2002 年，公派留学人员的回国率都达到 90%以上。可以说，这个办法基本上解决了公派出国留学人员按期回国服务的问题。今后应继续执行和进一步深化公派留学改革。

B. 想方设法提高自费留学人员回国率

近年来，我国自费留学人员迅速增加，但自费留学回国人员数量却未见同步增长。据调查，目前尚在美国和日本的我国留学人员在 20 万人以上，至少 2/3 的人已经完成学业，有将近 7 万人已申请了美国的绿卡或在日本长期居留。在拿到美国绿卡和在日本长期居留的人员中，自费留学人员占 90%以上[12]。笔者认为，留学回国工作的下一个重点应该是如何提高自费留学人员回国率。

国家应该继续坚持“支持留学，鼓励回国，来去自由”的 12 字方针，对公派留学和自费留学一视同仁。国家可以通过向优秀自费留学生发放奖学金或提供贷款等方式，帮助他们解决留学资金上的困难，让他们感受到国家对自费留学人员的深切关怀和信任。2003 年，我国向首批 95 名优秀自费留学生颁发了“国家优秀自费留学生奖学金”。奖学金的数额并不是很大，但它的意义和作用却是巨大的。许多获奖人员都表示，要把获奖作为一个新的起点，不负祖国和亲人的众望，刻苦攻读，完成学业后报效祖国。国家今后应继续推行自费留学生奖学金制度，并不断扩大奖学金的影响力。此外，国家还可以参考公派留学的办法，以协议的方式为自费留学生提供抵押贷款，学成以后不能按期回国的按协议赔偿。贷款的对象不限于申请自费留学的学生，还包括那些在国外留学过程中遇到经济困难的留学生。

4）继续鼓励海外留学人员以多种形式为国服务

目前，我国还有大量的留学人员在海外学习或工作。他们当中已有 1/3 的人获得了在当地国的永久居留权。这些人大多是具有较高学历的优秀人才。他们虽然因为各种复杂的历史和现实原因而没有回国，但他们当中的绝大多数人都有相当强烈的爱国

之情和报国之志。许多留学人员都在想方设法，采取各种可能的方式为祖国作贡献。

2001年5月14日，国家人事部、教育部、科技部、公安部、财务部联合公布了《关于鼓励海外留学人员以多种形式为国服务的若干意见》后，海外留学人员掀起了一股为国服务的热潮。有的留学人员带着项目和资金回国创办高新科技产业，为祖国引进外资和人才；有的留学人员开办国际贸易公司，开展跨国贸易，帮助国内提高创汇和增加就业机会；有的留学人员采用"双基地"模式，每年定期做回国讲学、短期访问、合作研究等，提高国内研究和科技水平；还有一些留学人员在海外主动承担起文化使者的工作，开展中外文化交流，进行民间外交，组织民间国际交流，为祖国带来巨大的非经济收益。

事实表明，鼓励海外留学人员以多种形式为国服务的政策是有效的。国家有关部门今后应继续坚持并进一步加大力度贯彻和执行这一政策。当务之急是要建立一个面向全体海外留学人员的、具有权威性的网站，在网站上公开我国政府的有关方针政策，展示我国改革开放进展实况，发布国内外人才供求信息，作为海外留学人员与国内有关政府部门和企事业单位之间、回国留学人员和海外留学人员之间、海外留学人员之间相互沟通、传递信息的桥梁。

5）加快我国基础教育和高等教育的改革和发展，积极推进多种形式、不同层次的中外合作办学

当前，我国基础教育和高等教育整体规模还不能满足社会发展和公民个人的需要。国家应采取有力的改革措施，加大对教育的投入，提高我国教育服务的供给能力。

推行中外合作办学是我国引进国外优质教育资源，缓解国内教育经费不足的一个有效方法。一方面，多种形式、不同层次的中外合作办学将为广大求学者提供更多选择的机会，满足他们对教育服务的需求；另一方面，我国通过引进先进的教学设施、教学手段、教育的方式方法和组织形式，学习外国先进的办学理念、管理体制，改善办学条件、提高教学水平、优化办学模式，促进教育改革的进程。笔者相信，这种成本低、收效大的"不出国门的留学"教育方式，将会受到越来越多学生和家长的青睐。

三、发展来华留学教育的对策

来华留学教育是我国高等教育的一个重要组成部分，也是我国政府的一项重要工作，是实现国家外交政策的重要部分和应尽的国际主义义务，同时也是促进我国改革开放，增进与各国教育、文化交流合作，利于吸收和利用国外智力为我国现代化建设服务，利于促进国际理解并维护国际和平环境的具有战略意义的工作。

1. 发展来华留学教育的意义

到了20世纪90年代，面对近20年国际教育风起云涌的发展态势，中国高校经过深思对发展来华留学教育的意义有了更为理性和深刻的认识。笔者以为，主要有以下四个方面。

1）发展来华留学教育符合国家利益，是国家利益的要求，具有战略意义

接受和培养外国留学生是国家利益的需要。不少国家从全球战略或地区战略的高度将其作为本国对外文化政策的一个重要组成部分，其着眼的长远打算是把外国留学

生作为今后长期促进东道国发展的有利的人员基础和人力资源储备。正如教育部副部长韦钰同志在 1994 年全国来华留学生工作会议上指出："进一步提高对来华留学工作战略意义的认识是做好今后一个时期工作的重要前提。"我国作为世界政治舞台一支重要的和平力量，长期以来为建立公正合理的国际政治经济新秩序而努力。来华留学生工作历来是我国外交工作的一个组成部分，与我国外交工作密切相关。通过大力发展来华留学教育工作，培养一支了解、热爱、支持中国的友好力量，有利于加强中国与世界各国的联系和交往，更利于保障和平的国际环境。在以经济建设为中心的新时期，我们应当将吸引来华留学生作为中国对外科学与经济合作的"面向未来"的投资，当来华留学生学成归国承担起领导职务后，我们这些现时的"投资"无疑会产生巨大的作用，某种程度上来说，不啻找到了开辟中国与来华留学生祖国的政治、经济、文化全面合作的钥匙。

2）发展来华留学教育是高等教育国际化的内在要求和高等教育新生职能的体现

所谓高等教育国际化，就是加强国际高等教育的交流合作，向各国开放国内教育市场，并充分利用国际教育市场，培养有国际意识、国际交往能力、国际竞争能力的人才，在教育内容和教育方法上不断适应国际交往和发展的需要。当中国高校的校园中出现越来越多的来自世界不同国家和地区的来华留学生，无疑对促进中国大学在培养国际化人才中实现东西方跨文化交流、碰撞，加深中国学生对异域文化的理解和包容，开阔双方的思维视野有着毋庸置疑的积极作用。通过彼此的取长补短，达成不同国家、地区、民族的文化精华的融合和共存，并最终形成我国高等院校发展来华留学教育，推进国际化进程的内生原动力。

3）发展来华留学教育是办学层次提高的体现和中国大学进入世界一流大学的关键

世界一流大学必然是拥有较大比例"国际学生"和"国际师资"的开放型、研究型、国际型大学。按目前国际通行的标准是，世界一流大学外国留学生人数至少应占在校学生总数的 15%。目前，国际化程度较高的世界知名大学，如美国哈佛、英国剑桥、日本东京大学，其外国留学生占本国学生比例均达 15%以上。

在高等教育国际化趋势影响下，在我国，接受来华留学生人数已成为衡量一所大学或学院国际化程度的重要因素。例如，"211 工程"建设评估的评价体系中已将接受和培养外国留学生数量列为条件之一，规定进入国家"211 工程"建设立项的高等院校，其外国留学生数应占学生总数的 5%～10%。而来华留学生学科分布和层次结构，还标志着这所高校学科水平高低和其在国际上的地位，标志着学科对外开放程度的高低。

随着我国扩大开放和综合国力逐步增强，高校对外交流的广度和深度日益扩大，高等院校通过实施跟进国际留学生教育发展战略，促进教学质量、科研水平逐年提升，实现高层次研究型大学的办学水平、办学层次快速提升，中国大学成为世界一流大学指日可待。

4）发展来华留学教育是高校增加办学资金来源的渠道和赢得国际人才竞争的途径

在我国，高等教育的政府投入少于年教育经费占当年 GDP 总量 4%的世界平均值。高等教育规模快速扩张和办学经费普遍短缺的矛盾突出。为了解决政府投入较少而带来的困难，各高校不得不积极拓展经费筹集的渠道，而发展留学生教育，通过接收自费来华留学生，获得学费和部分生活费的收益，至少可以部分缓和对外交流方面的资金缺口。这部分收益可以用来邀请海外知名学者来华短期讲学和研究，推进大学自身的国际学术交流。中央电视台报道，来华留学生的增加给中国教育市场带来一定的经济收入，仅北京市 2001 年一年此项收入就有约 9 亿元，折合美元约 1.088 亿元。

另外，大批外国留学生来到中国的大学和学院学习，不仅给高等教育发展急需的资金增加了重要的补充渠道，而且会吸引并可能使更多有国际教育经验的专家留在中国。美国每年获得博士学位的外国留学生中约有 60%留在美国工作，而从事研究与开发（R&D）的科学家和工程师中，外国出生的人所占的比率高达 20%。从 1949～1969 年，美国共引进 40 多万名高级科技人员，其中 37.5 万名来自发展中国家，约占 3/4，主要办法就是“引进留学生，深加工‘半成品’战略”。

2. 来华留学教育的现状与特点

1）来华留学教育的现状

我国培养外国留学生的历史悠久，早在春秋战国时期就有留学生教育。到盛唐时期，由于我国有注重研究社会伦理的历史传统，形成了优秀的传统文化，对世界特别是对亚洲产生了极大的吸引力，因而成为当时的留学大国。当时在长安学习的外国留学生就达 1000 多名，这些留学生主要来自日本、朝鲜和西域各国，受当时大唐高度发达的文化所吸引，主要进入国子监学习。到明朝万历年间来华留学的人数达到 3000 人，除了日本和朝鲜学生外，还有来自欧洲诸国的学生，来华留学的热潮一直持续到清末。

随着新中国的成立，来华留学教育作为我国高等教育的一个重要组成部分掀开了新的篇章，但走过了一段不平凡的历程。

1950 年，我国尚处于经济恢复时期，百废待兴。面对当时东西方两大阵营严重对峙的局面，为加强与社会主义国家之间的交流与合作，我国政府同意与波兰、捷克斯洛伐克、罗马尼亚、匈牙利、保加利亚 5 国互派留学生。这批留学生 1950 年年底和 1951 年年初来华，共 33 人，入清华大学中国语文专修班学习，开始了我国接受和培养外国留学生工作。随着我国与周边国家关系的发展，以及亚洲、非洲、拉丁美洲国家民族独立运动的兴起，越南、朝鲜以及其他亚非拉新独立国家的留学生相继来华，苏联、日本、美国、英国、法国等国也派出留学生来华学习。据 1950～1966 年的统计，我国共接受和培养了 68 个国家的 7239 名留学生。这些留学生主要来自社会主义国家，占 90.8%；日本、欧美等发达国家留学生人数很少，仅占 1.9%。

从 1966 年开始，由于“文化大革命”的原因，我国接受和培养外国留学生的工作一度中断。到 20 世纪 70 年代初，随着世界政治形势的变化和我国国际地位的提高，经国务院批准，我国于 1973 年恢复了接受和培养外国留学生工作。当时来华留学生仍以阿尔巴尼亚、越南、朝鲜、罗马尼亚等社会主义国家的学生为主，也接受了亚非拉国家的部分学生，由我国政府提供奖学金。当年共接收 44 个国家的 383 名来华留学生进入我国高校学习。

从 20 世纪 70 年代开始，世界政治格局发生了巨大变化，东西方两大阵营对峙的格局为三个世界的格局所代替，因而来华留学生的国别构成发生了很大变化。虽然在20世纪70年代仍以亚非国家的留学生为主，但到20世纪80年代，第一世界、第二世界国家的来华留学生人数也迅速增长；同时，我国开始接受自费来华留学生，其人数也在逐年增加。据1973～1989年的统计，我国共接受129个国家享受奖学金的留学生15 978人；此外，这一时期来华自费留学生（包括长期生和短期生）人数迅速增长，达2万余人，超过了享受奖学金的留学生人数。

进入20世纪90年代，随着苏联的解体，世界政治格局由美苏争霸的局面向多极化方向发展，冷战时代结束。我国与世界各国在各个领域的交往大大增加，为发展来华留学生教育创造了良好的外部环境。同时，我国改革开放的深入，也给来华留学生教育带来勃勃生机。例如，政府转变职能，扩大了高校办学自主权，激发了高校搞好留学生教育的主动性；社会主义法治建设的加强，使对外国留学生的管理逐步走上法治轨道；来华留学生教育管理体制也日趋完善合理。这些都有力地促进了来华留学生教育的发展。据1990～2003年的统计，1990年，我国接收留学生8495人，1991年接收130个国家（地区）12 557人，1995年接收150个国（地区）37 205人，1999年已达到164个国家44 000人，2002年达到175个国家的85 829人（表14）。其中，亚洲国家由于日本、韩国及其他周边国家来华学生人数发展较快，占总数的66.8%，非洲国家来华学生由于我国政策调整而人数急剧下降，仅占2.4%；自费来华留学生人数迅速增长，已占留学生总数的88.2%。2003年的来华留学生是1980年的30多倍，是1990年的10多倍。近20年来，我国来华留学教育人数和规模及生源国都有极大发展。

表14 来华留学教育情况（1980～2002年）

年份	1980	1985	1986	1987	1988	1989	1990	1991	1992
人数/人	1 381	3 250	4 343	4 408	6 400	4 993	8 495	12 557	13 993
年份	1994	1995	1996	1997	1998	1999	2000	2001	2002
人数/人	26 000	37 205	41 211	43 712	43 084	44 000	52 150	61 869	85 829

2）当前来华留学教育的特点

A. 留学教育形成市场化运作状态

目前，来华留学生主要以自费生为主，政府派遣生数量很少。面对日益扩大的自费留学群体，产品供需关系在逐渐形成，大学拓展市场的积极性在逐渐增强。

B. 来华留学教育层次不平衡

来华留学教育需求层次和教育产品输出层次偏低，留学教育主要局限在低水平的汉语言教育和普通进修生为主，其中约30%为语言进修生和普通进修生，博士生、研究生和本科生之和仅占全部留学生数量的20%多一点。

C. 来华留学的发展潜力大

虽然来华留学生的数量落后于发达国家的留学生教育水平，但中国的增长速度在国际留学市场中是最快的，近年来的平均增长率约为30%。目前在全球留学生数量排名中处于前十位，学生生源已扩展到170多个国家，而且需求市场从过去以落后国家的学生为主向较发达国家学生为主转变。

D. 具有一定的优势教育产品

来华留学教育的优势主要集中在传统学科领域，分别为历史、中文、考古等社科类学科和中医学专业，其中发达国家和周边国家的学生来我国学习传统优势学科的人数增长最快。

E. 与发达国家留学教育相比，来华留学教育的学科类别失衡

重点在文科和医学类，理工科需求量相对少。见表 15 的对比数据。

表 15　中日两国 1998 年留学学科人数及比例分布

项目		文科	医科	理科
中国	人数/人	36 550	4 730	1 724
	比例/%	85	11	4
日本	人数/人	16 869	10 533	10 564
	比例/%	44.4	27.7	27.8

F. 来华留学教育地区分布不均

接受来华留学生的院校从开始时的单一情况，发展到现在的地区分散化，分布在全国 31 个省（自治区、直辖市）的 353 所高等院校和其他教学机构。但来华留学生主要集中在北京、上海等经济发达地区。

3. 来华留学教育的优势分析

回顾 50 年来我国接受和培养外国留学生的历史，成绩是显著的。我们探索出了一条有中国特色的、比较系统的来华留学生教育模式，一方面为许多国家，尤其是发展中国家培养了大批科技、教育、外交、商贸和管理人才，另一方面为我国在世界各国发展外交和经贸关系，开展文化教育与科学技术交流作出了重要的贡献，其具有的潜在发展优势有以下几点。

1）传统学科的优势

在当前的留学教育中，争取留学生主要靠大学学科优势。我国高等教育有悠久的发展历史，一些具有中国特色的学科，如中文、历史、哲学、建筑、中医等学科力量雄厚，具备一定的教育实力，可以吸引相当数量的学生来华留学。而且受邻近和同属东亚文化圈的影响，周边国家的学生都愿意来我国留学。从学科分布看，该类留学生中 90%是自费，主要集中在中文、中医等传统学科。可以预料，东南亚国家将成为我国今后来华留学的重要组成部分。另外，由于一些落后国家与发达国家在科技水平上有差距，这类国家学生根据自己国家实际需要，可能只需要学习针对解决现实问题的实用技术，我国的重点大学某些学科比发达国家的同类学科更具有针对解决现实问题的实用性，因而对落后国家的留学颇具吸引力。例如，很多中国的农业类学科发展就更适应第三世界国家的需求。

2）高等教育规模和体系的优势

改革开放以后，我国高等教育质量方面有很大的提高，整个高等教育中有一批重点大学和特色学院具备吸引留学生的规模和体系优势。普遍认为，留学生在本国高校

本科生中所占比例能达到10%，效果较好，而且在研究生的留学生中需要比例更高，而我国目前的状况离这个数据相差较远。总体来看，我国留学教育生产能力不差，产出却相对较低，留学生教育生产潜力尚待进一步开发。在发达国家，理工科的留学生比例比文科高，但中国恰恰相反，说明我国重点大学还有扩大留学生规模的空间。

3）费用方面的优势

与西方高成本留学教育相比，我国的留学费用包括学费、生活费比西方国家便宜很多。对于经济落后国家的学生来说，来我国进行工程学科方面的留学不仅实用性相对较强，而且费用相对很低。由于我国的整体消费水平较低，教师劳动力报酬也相对便宜，工程类学科的留学学费标准也比英法国家普遍要低，因而在我国留学，总体费用比发达国家便宜得多，我国高等教育在费用方面具有明显的竞争优势。

4. 来华留学教育发展的问题或制约因素

来华留学人数增长速度很快，潜在需求较大。但我国才刚进入留学教育市场发展阶段，来华留学需求市场中的学生数量、学历层次以及学科类别结构上还远未达到市场化发展所要求的稳定的需求规模。而国内高等教育整体质量又相对偏低，大学的学科优势也不明显，这都导致来华留学教育缺乏强大而持续的竞争力。剔除国家经济和政治等大环境因素，从大学角度来看，制约来华留学教育发展的因素有以下几个方面。

1）缺乏政府指导、政策扶持和更为明确法律法规的保障

由于我国来华留学生事业一直以来与国家对外政策的关系十分紧密，因此，及时、周到的政府指导而非行政命令更符合教育体制的发展方向和来华留学教育的发展需要；另外，来华留学生事业的美好发展前景和现实弱小态势，还需要政府的政策扶持和鼓励，需要更为完善的法律法规的保护。

2）高等教育总体学科水平还不高，制约了来华留学教育发展中高层次学生数量发展

理工类学科水平总体发展不平衡，英语授课和英语教科书运用太少；文科类学科缺少国际性课程；农业和生物类学科除水稻、淡水养殖等少量学科外缺乏领先方向；医药类学科除中医中药外，较少世界领先的技术和学科发展方向。

3）教育开放程度与接收留学生数量居国际前列的国家相比还较低

笔者认为，对教育开放程度的把握必须坚持辩证唯物论，注意开放性的“双刃剑”效应。世界各国对高等教育的开放都持较为谨慎的态度。既要防止没有梯次的“彻底开放观”，又要防止“既放又收”“突放突收”的不当方式。

4）接受来华留学生大专院校的硬件整体水平还有待提高

许多高校的留学生宿舍、教室老旧、短缺，活动场所少，低于国际标准水平线。应该学习日本为发展留学生教育将留学生宿舍写入行动纲要的态度，认真对待来华留学生的生活条件、生活设施。

5）对外汉语教育的教学水平和覆盖范围、布局水平还亟待提高

尽管全世界说汉语的人口最多，平均每5个人中就有一个会说中文，但考虑到发展来华留学生教育的主要对象是西方发达国家的青年，因而在对外汉语教育的全球化布局中必须分清主次、重点；此外，对外汉语教育教师还应尽量多进行双语教学实践

和自身的海外教学实践。

6）来华留学教育的宣传途径单一，招生方式不够灵活

目前，我国大多数院校主要通过少数境外代理和留学中介介绍，以及留学生上门咨询等被动方式来招生，机会损失较多，招生宣传方式也有同样的弊端。

7）放弃了对毕业回国留学生的跟踪回访和感情投入

没有发挥出来华留学生对我国各项事业发展的可能贡献能力，同时也就放弃了进一步发展来华留学生规模的可能。

5. 关于来华留学的对策措施建议

针对上述七个方面的制约来华留学教育发展的因素，笔者提出如下的对策措施建议。

（1）不断完善对教育国际化的有关立法，加强政府对教育的国际化。特别是来华留学教育的宏观指导和政策调控，及时发布相关的动态信息。条件成熟时可制定《来华留学教育管理条例》以规范和保障全国来华留学教育工作可持续发展。

（2）加快我国高校迈向世界一流水平的步伐，促使我国整体学科水平上一个新台阶。少数国际先进学科专业实施赶超战略，实现国际领先。加强各学科英语教材建设和英语授课师资的培养，培养一大批可以按照国际通用教材开展教学活动的教师，推行“双语教学”。

（3）按照 WTO 规则，逐步开放教育服务市场。处理好按计划实行教育开放和国家行使教育方针权、教育行政管理权和教育教学实施权之间的关系，加强教育活动中采用国际通用的德育教育原则，坚持民族特色和文化背景的教育内容，加强国际性能力培养的课程建设。

（4）重视对来华留学生生活条件的硬件建设。采取国际惯例进行分流安排，学制在 1～2 年的留学生集中住宿、集中管理、统一安排住宿和大型业余活动，学制在 2 年以上的留学生由他们自己选择可以在公安部门的指导下分散住宿，以此缓解住房矛盾；同时辅之以经济手段鼓励和约束住宿的选择。在新建留学生住宿楼时注意合理设计，如突出教室的小班化、宿舍和卫生的单人化，以及食堂餐厅的便利、廉价等特色，以此吸引留学生来校学习。留学生的设施既不必奢侈也不要太小气，应当以经济、实用、方便、周到为配置留学生生活设施、生活用品的标准和原则。

（5）发展政府主导力量，借助社会力量，扩大对外汉语教学的覆盖面，途径多样化。由国家组织对外汉语教师的培养和有计划的国际交流；强化对外汉语教师对外国人学习和掌握汉语的教学研究，培植对外汉语的交流基础；多与各国的华人学校、双语学校建立长期的交流计划，开设中国文化和语言课程；进行长期或短期的汉语培训、文化交流等活动；建设对外汉语的基地和设施，有计划地支持、指导、交流双语教学的经验；通过华人团体、国际交流与合作的外方学校帮助在境外设教学点；自主投资或合作办学到境外开展汉语教学学校，甚至可以实行免费授课。这一方面法国政府的做法值得借鉴，法国政府曾派出 3 万多名法语教师到世界各地教授法语，以使学生学会法语后有兴趣到法国留学。宣传和招生的突破在于运用先进的信息通信技术和委托代理的商业运作模式，广泛推介和争取一切可能的渠道。如通过地方政府外经委、外商投资企业协会来招收外商及家属子女学习汉语；设立网站和网上报名；通过国际教

育协作组织发布招生信息；广泛调动各教师和基层单位的积极性参加招生工作；或借助我国驻外使馆的窗口。

(6) 加强与来华外国留学生回国后的联络和交流。来华留学生这座沟通中国与留学生派遣国的桥梁，应该变成中国与世界各国人民在政治、经济、文化各方面开展广泛的国际交流与合作越走越宽的通途。

(7) 丰富留学生课余生活，让留学生真正融入中国的日常生活。在提高我国高校的学术、课堂教学水平同时，广泛开展进入中国家庭的教学实践环节十分必要。这对加深来华留学生对中国更为全面的了解和加强民间联系十分重要。也可以与旅行社共同合作，使留学生业余生活内容更丰富多彩。这也会给相关部门或单位带来一定的经济效益。

(8) 设立奖学金，吸引更多留学生，特别是中等发达国家的留学生。

(9) 保持较低的学习费用，开放对留学生从事劳务的限制，吸引更多的来华留学生。随着中国加入 WTO，教育部门应该对来华留学政策作出新的调整，比如允许来华留学生在一定限制范围内开展合法的勤工助学。合理的勤工助学范围界定，既可以避免对中国劳动力市场的冲击，又可以丰富留学生交流的深度和广度，并减少他们的经济负担。

总之，世界各国教育愈加趋向国际化，各国的人才市场愈加开放。只要我国把自己的教育窗口打开，宣传工作做好，创造好留学生的生活条件，中国的来华留学生教育也会具有更大的吸引力。

四、中外合作办学的对策

随着信息技术与大众传媒的迅猛发展，世界各国之间的经济、政治合作领域日益扩大，不同文化间的交流也日益增多，世界正迈入一个全球化的时代。而所谓全球化，实质上是一个“现代性从西方发达社会向世界扩展的过程”，主要表现为西方发达国家试图以其自身的文化以及价值观念为标准和范型，去同化其他一切“异质的”、非西方的文化，从而寻求一种文化上的普遍与同一趋势。然而，进入新千年以来，震惊世界的“9·11”事件及其以后的发展已经充分显示了世界不同文化之间存在的深刻的矛盾与差异。将西方文化凌驾于一切其他文化之上并对其加以齐一化，已经被历史的发展证明是不可行的。从 20 世纪 50 年代以来后现代主义文化思潮的兴起表明人们已经在思考不同文化之间的差异性问题，并力图通过对西方“现代性”的文化霸权的深刻思考与批判来维护与拯救物种、价值观念以及文化类型的多样性，以此保持世界的无限丰富的可能性与整合性。意大利著名哲学家翁贝托·埃科（Umberto Eco）在博洛尼亚大学成立 900 周年的纪念大会上曾说：“后现代主义的一个显著特征就是，不在于减少差别而是更多地发现差别。人们发现的差别越多，能够承认和尊重的差别越多，就能生活得更好，就能更好地相聚在一种相互理解的氛围之中。”

在这种背景下，教育作为文化选择、传承、交流的一种重要手段，也不可避免地走向国际的联网合作；在这种合作中，我们追求的不是全盘的西化，不是以西方的标准为唯一标准，而是在保持自身民族性与独立性基础上的一种国际的品质。20 世纪 80

年代中期，日本临时教育审议会首先提出了高等教育国际化的概念，提出“要培养世界通用的日本人”。在欧洲，欧洲一体化进程极大地推动了欧洲高等教育的国际化发展，1992年便开始实施《欧洲共同体促进大学生流动计划》，并拨出专款，鼓励本国师生在会员国之间留学，相互承认学历、文凭、学位等，以此推动区域内的人才流动。美国在第二次世界大战后就通过《国际教育法》，致力于促进高等教育的国际化；近年来随着经济一体化、环境生态问题的全球化和多元文化的交融与冲突，其国际化步伐大大加快，在《美国2000年教育目标法》中明确提出了“要通过国际交流，努力提高学生的全球意识、国际化观念”的战略目标。在我国，在改革开放尤其是加入WTO以来，人们习惯已久的生活方式受到了异质文化的强烈影响与冲击，各类高层次人才，尤其是通晓国际“游戏”规则人才的缺乏，对高等教育提出了严峻的挑战，人们越来越意识到，国际的交流与合作将成为新时期高等教育发展的一种新的思路与方向，而在教育对外交流合作的过程中，如何促进不同文化之间的交流而又保持自身文化的民族性与独立性，如何在引进国外优质教育资源的同时又加强对于教育质量的监控，便成为亟待我们思考的问题。

1. 中外合作办学及相关法规的基本情况

所谓中外合作办学，按照《中华人民共和国中外合作办学条例》第三条的界定，即指“外国教育机构同中国教育机构在中国境内合作举办以中国公民为主要招生对象的教育机构的活动”，是指外国法人组织、个人以及有关国际组织同中国具有法人资格的教育机构及其他社会组织，在中国境内合作举办以招收中国公民为主要对象的教育机构，实施教育、教学的活动。

在改革开放之初，我国便已经开始探索与举办各种形式的中外合作办学活动。20世纪80年代中期，中国人民大学、复旦大学等高等院校相继举办了中美经济学、法学培训班。随后，天津财经学院与美国俄克拉何马市大学合作举办MBA班，南京大学与美国约翰斯·霍普金斯大学合作创建中美文化研究中心等。80年代末90年代初，我国境内中外合作办学机构逐渐增多。与此同时，国家教委有关部门就中外合作办学问题进行了大量的调查研究，并在此基础上起草并于1993年6月30日下发了《关于境外机构的个人来华合作办学问题的通知》。通知明确指出，多种形式的教育对外交流和国际合作是我国改革开放政策的一个重要组成部分；要有条件、有选择地引进和利用境外于我有益的管理经验、教育内容和资金，有利于我国教育事业的发展；开展中外合作办学应坚持“积极慎重、以我为主、加强管理、依法办学”的原则，遵守我国的法律，贯彻我国的教育方针，经过教育主管部门批准并接受其监督和管理。文件还对合作办学的范围、类别、主体等作出了相应的规定。

1993年下半年国家教委开始着手拟订《中外合作办学条例》，最后形成了《中外合作办学暂行规定》并于1995年1月26日正式颁布实施。《中外合作办学暂行规定》就中外合作办学的意义和必要性、应遵循的原则、合作办学的范围、主体及审批权限和审批程序、办学机构的领导体制、发放证书及外国文凭、学位授予等问题都作出了明确的规定。中外合作办学进入了一个新的时期。据不完全统计，1998年前后，全国各类中外合作办学机构已达300多个，仅上海市就有60个，北京市、江苏省和山东省也各有 40 个左右。与我国合作办学的有 20 多个国家和地区，列前几位的有美国、澳大利亚、日本、加拿大和我国的香港、台湾。截至 2000 年年底，我国共有中外合作办学机构 562 家，其中与美国合作的有 115 家，与澳大利亚合作的有 105 家。学历教育机构占有较大比例，有 316 家，工商管理、会计等类占有较大比重，为 38%；语言类占 38%；两类合计 60%。我国加入 WTO 后，中外合作办学呈现加速发展的势头，沿海地区申报的数量大大增加，中西部地区的中外合作办学项目开始起步。总体上讲，办学层次有所提高，合作形式也日益多样化。截至 2002 年年底，全国共有中外合作办学机构和项目 712 个，与 1995 年年初相比，增加了 9 倍多，覆盖了 28 个省（自治区、直辖市）。到 2003 年 12 月底，授予国外学位与香港特别行政区学位的高等学校合作办学在办项目达 137 个。新的发展催生了新的问题，也促使了新的法律法规的出台。2003 年 3 月 1 日中华人民共和国国务院令第 372 号发布了《中华人民共和国中外合作办学条例》并于 2003 年 9 月 1 日起施行。2004 年教育部讨论通过《中华人民共和国中外合作办学条例实施办法》自 2004 年 7 月 1 日起施行。

简要回顾历史，我们可以看到，中外合作办学经历了一系列的发展与变化过程，并且呈现以下特点。

（1）办学主体多元化。改革开放之初，投资主体与办学主体是有限的几个经济发达国家；现在，参与中外合作办学的主体有来自世界各大洲的教育机构或非教育机构、政府或个人、跨国集团或政治经济合作组织。

（2）办学形式多样化。从单一的培训班形式发展为各种形式并举的格局。第一类，中外有关院校之间的合作。在教育主管部门的批准下，境外大学在境内高校上课，大学生毕业可获得本校文凭和境外大学文凭，这是最主要的形式。第二类，境内高校与境外教育或非教育机构之间的合作办学，由境内高校具体安排课程设置，教材、教师等，外方提供资金，学生就读与读一般本地课程无异，这主要是引进国外财力办学。第三类，由中外政府之间进行教育合作项目，主要是开展两国在外交、文化、经济、贸易政策等之间的交流。

（3）办学层次多样化。从办学层次分布看，学历教育机构占多数。截至 2002 年年底，中外合作办学机构中，学历教育机构共有 372 个，其中，初中 2 个、高中 40 个、职业学校 69 个、中等专业学校 36 个、大学专科层次 82 个、大学本科层次 69 个、研究生层次 74 个。

（4）办学模式多样化。从合作办学的具体模式来看，目前主要存在三种办学模式，即融合型、嫁接型和松散型模式。

融合型模式就是在人才培养过程中把境内高校的教学模式和国外合作学校的教学模式完全地融合在一起。首先，引进对方的教学计划、教学大纲、教材和相关教学手段；其次，聘请对方教师来境内高校讲课，派遣境内高校教师去对方进修；再次，引入对方的教学方法，如课堂讨论、实践环节、案例教学等；最后，以双语授课。通过全面引进国外的教学模式，做到在国内培养出适应国际市场需要的合格人才，同时也进一步推动境内高校教学内容、教学方法的改革。融合型模式受到了普遍欢迎，首先，满足了人们渴望接受国外高水平教育的需求；其次，避免了在国外学习的巨大经济压力；最后，在知识传授过程中，克服了与国外教育的巨大梯度，能逐步适应国外的教学方法，进而达到国际认可的教学质量。

与融合型模式相比，嫁接型模式具有自身的特点。嫁接型模式主要是充分保留各自的教学模式，通过双方各自对对方学校开设课程的评估，互认对方学校的学分，学生获得双方学校规定的学分，即可获得双方学校颁发的毕业证书和学士学位证书。嫁接型模式主要有 2-2 模式、1-2-1 模式和 3-2 模式。从嫁接型模式看，由于该模式结合了中西方的教育优势，能直接出国接受国外的教育，对国外的人文背景、生活方式有较深层次的了解，除专业知识外，还能在很大程度上提高学生的外语水平，同时相对高中毕业直接去国外学习而言，可节省很多费用，且能较好地开展思想政治教育和我国优秀民族文化传统教育，帮助学生确立正确的世界观、人生观、价值观。

松散型模式就是通过聘请国外教师来境内高校讲学，境内高校教师去国外学习、借鉴国外的教学经验，学生去国外短期学习、实习等手段，实现教学与国际接轨的方法。这是一种渐进的模式，也是目前我国大部分高校所推行的教育国际化的尝试。从某种意义上说，这种模式对我国高等教育的教学改革更具有普遍意义和可操作性，我们可以以这种模式通过不同渠道吸取国外办学的先进经验，利用国际教育资源，提高我们的办学水平和办学质量，培养适应国内外市场需要的复合型人才。

2. 目前中外合作办学过程中存在的问题

1）教育主权与西方发达国家的文化渗透以及不同文化之间的冲突问题

“教育主权是指一国固有的处理其国内教育事务和在国际上保持教育独立自主的最高权力。”在对外体现我国教育主权时，我国《教育法》第六十七条规定：“教育对外交流与合作坚持独立自主、平等互利、相互尊重的原则，不得违反中国法律，不得损害国家主权、安全和社会公共利益。”然而，合作办学在为发展中国家缩小与发达国家的差距提供了机遇的同时，也对发展中国家的主权和民族传统文化造成不容忽视的冲击。在中外合作办学过程中，也不可避免地会有一些国家或政府借合作办学之名，对学生施加文化政治影响，进行文化渗透，而教育行为往往也在一定程度上成为国家（政府）的一种政治策略与行为。同时，如亨廷顿所言，不同的价值文化已经成为民族与民族、国家与国家之间冲突的根本或主要来源。在中外合作办学过程中，不同的文化传统、价值理念必然导致合作各方在教学资源、教育理念，甚至教育策略、教学方法等各个方面的选择上存在着极大的差异。如何弥合不同文化之间的隔阂，切实在引进国外优质教育资源、促进我国教育改革创新的同时又保持自身文化的民族性与独立性，便成为我们不得不面对的问题。

2）有关法规条例的滞后性与办学主体的利益驱动以及利益实现之间的问题

中外合作办学除受到上述法规条例的规范与限制以外，按照 WTO《服务贸易总协定》的精神，其作为一项涉外服务贸易活动，还受到《中华人民共和国宪法》、《中华人民共和国对外贸易法》和《中华人民共和国教育法》的制约。《中华人民共和国宪法》第十九条规定，国家鼓励集体经济组织、国家企业事业组织和其他社会力量依照法律规定举办各种教育事业。《中华人民共和国对外贸易法》第十条规定，“从事国际服务贸易，应当遵守本法和其他有关法律、行政法规的规定”。《中华人民共和国教育法》除了重申宪法第九条的规定外还具体涉及了“中外合作”的概念。《中华人民共和国教育法》第二十五条规定，国家鼓励企业事业组织、社会团体、其他社会组织及公民个人依法举办学校及其他教育机构。任何组织和个人不得以营利为目的举办学校及其他教育机构。而在这里，有关法律对教育的举办的若干限制明显与目前的实际发展状况不相符合。

首先，中外合作办学不得以营利为目的规定与教育作为服务贸易的性质相左。《中华人民共和国教育法》第二十五条规定，任何组织和个人不得以营利为目的举办学校及其他教育机构。《中华人民共和国中外合作办学条例》第三条“中外合作办学属于公益性事业”重申了这一原则。这一限制既同教育服务的实际不符，也同作为国际贸易的教育服务的性质相悖。而《中华人民共和国民办教育促进法》则明确规定，“营利性民办学校的举办者可以取得办学收益”。显然，各种法律法规之间对同一问题的处理存在着不一致甚至相悖谬的地方，在实际操作过程中就不可避免地使人产生政出多门、无所适从的感觉。

从法律的意义来说，“营利”是指为取得超出资本的收益进行经营，并将收益分配给投资人的行为。营利性是区别企业与其他社会组织的重要标志。根据法律规定，学校和其他教育机构作为非营利性组织，不能把教育教学作为可以取得超出资本的收益并将其分配给投资人的业务来经营，也就是说不能把获取经济利益作为追求的首要目标，但并不限制、更没有禁止其通过各种途径或形式获取经济收益。然而在实践中，我国对非学历教育的合作办学机构是要征收营业税的。如果所有的教育机构都不营利，那么也就不应该要求其纳税。对于合作办学机构来说，在国家不能提供资金、不承认学历的情况下，如果不能获得一定的经济收益，就会丧失竞争力和吸引力，进而面临生存危机。这在实践中显然是不利于人们投资教育的积极性的。因此，应将合作办学机构与国民教育序列（或公立学校）区别对待，根据合作办学机构类型上的差异（法人型、非法人型、项目型、学历型、非学历型等），允许合作办学机构把办学目的定位于：在追求社会效益的同时，可以获取一定的经济利益。即以追求社会效益为主，取得经济效益为辅。也就是说，对于在办学过程中取得的经济收益，除绝大部分必须用于对教育教学的再投入外，还应当给予合作者（特别是外方以资金投入为主时）适当的回报或返还。这在性质上不属于营利性行为。反之，如果要求合作办学机构只能把社会效益作为唯一追求的目标，对投资者没有适当的回报或返还，就会在一定程度上影响合作办学各方的利益和积极性。因此，要尽快修改和完善中外合作办学的法规体系，如对“不得以营利为目的”的限制要改为允许取得合法回报，限制牟取暴利，为中外合作办学扫清障碍。

其次，在有关中外合作办学主体的规定方面，下位法与上位法不一致。《中华人民共和国教育法》第二十五条和《中华人民共和国高等教育法》第六条均承认公民作为办学的主体。《中华人民共和国中外合作办学条例》对此却并未涉及，而只是指出“申请设立中外合作办学机构的教育机构应当具有法人资格”。根据法的效力层次的一般原则，等级高的主体制定的法，效力高于等级低的主体制定的法。作为全国人大制定的《中华人民共和国教育法》和全国人大常委会制定的《中华人民共和国高等教育法》在效力上自然高于国务院各部门制定的规章。因此，《中华人民共和国中外合作办学条例》理应明文承认中国公民以及外国公民参与合作办学的主体地位。

最后，《中华人民共和国教育法》关于"汉语言文字为学校及其他教育机构的基本教学语言文字"的规定不利于国外教育资源的引进；《中华人民共和国中外合作办学条例》将中外合作办学定义为以招收中国公民为主在中国境内实施的教育教学活动，缺乏概括性，从而使得新出现的办学形式无规章调整。例如，按现有定义，目前合作办学活动中较为流行的"2＋2"模式，就不完全属于"中外合作办学"的范畴。这些，都应该进行进一步的调整和规范。

3）合作对象的资质评估与办学质量的监管问题

中外合作办学作为我国政府大力提倡和支持的教育活动，也是社会广泛欢迎的一种教育服务。我们应该积极总结经验，根据未来的发展需求，制定出一套严格而合理的游戏规则，以保证参与者的公正和公平；在此前提下，以制度监管和考核为基础，以市场选择为尺度，促使中外合作办学能够在中国健康地发展壮大。然而，现实情况是，一方面合作办学的审批时间太长、手续繁琐，不利于合作办学的正常开展；另一方面，对非法合作办学的整治、打击力度不够，影响了合法的中外合作办学的发展。虽然中外合作办学机构已经有了近二十年的发展历史，大部分已经步人规范化办学轨道，但是，以"合作办学"为幌子，行"圈钱"之实的种种违法的、不规范的行为依然存在，而我国现在在这一方面尚缺乏明确的准人标准与质量控制标准，缺乏必要的质量监控体系与透明有效的社会评价体系，同时，对合作办学教师资格与水平也无法进行有效的认证与评定，致使合作办学机构的设立缺乏可靠的质量标准。因此，对合作对象的资质评估与办学质量的监管，如对中外合作办学过程中出现的违规、违法行为的监管，对中外合作办学机构财务运作情况的监控，对中外合作办学质量的监督等就成为我们需要重点关注的问题。

4）在学校（办学机构）的具体运作的微观层面，教育观、人才观的变化以及相应带来的课程、教学、管理与评估的改革也是我们需要关注的问题

（1）教学计划与课程的衔接。目前我国的教学体制和国外高校有很大的不同，在人才培养方式、规格方面存在较大差异，在课程设置、课程内容、教学方式等方面也不尽相同。因此，在合作过程中，我们必须对目前相关的教学计划进行修订，进行较大程度的教学改革，以适应国际化的要求。同时要重视和保证马克思主义理论教育和我国民族传统文化教育，并坚持在吸取国外高校成熟经验的同时，努力形成自身的特色。

（2）学制改革与学籍管理。我国高等学校目前普遍实行学年学分制度，而国外多为弹性学分制，在与国际教育接轨时有一些困难。教育部已提出在国内尝试推行弹性学分制，即学生可在八年内修完学士要求的学分，获得学士学位。这样，就为国际接轨创造了条件，也为高校的国际合作办学创造了条件，学生可在更长的时间内选择在国内或国外学习，有较长的时间适应东西方文化的差异，提高外语水平。

（3）师资队伍建设。拥有与国际化高等教育接轨的教师队伍，才能培养适应国际市场需要的优秀人才。目前，大多数高校教师仍然缺乏国际化教育的理念和专门的知识，同时外语水平较低，不能适应外语授课的要求，不能完全达到培养国际化人才的

要求，从而使教学质量受到不同程度的影响。

(4) 学生管理工作。在国际合作办学过程中，学生管理非常重要。学生是教育质量的集中体现，无论学生在本国抑或是去国外学习，我们都要加强对学生的思想政治教育，特别是爱国主义、人生观、价值观和法律、法规的教育，使学生能在学习科学文化知识的同时，具备良好的品行。

3. 关于对策措施的建议

为了促进中外合作办学良性发展，下面提出 7 条关于对策措施的建议。

(1) 加强民族传统文化教育，坚持“以我为主，加强管理，依法办学”的原则。

(2) 尊重教育主体多元化，充分融入国际教育市场，打破公立教育垄断局面，给予合作办学以国民待遇。

(3) 建立教育试点单位或地区，以点带面，促进中外合作办学的健康与良性发展。中外合作办学是我们面临的一个新的课题，由于各国国情与发展状况的不一致，我们无法简单地套用其他国家与地区的经验和做法。因而，我们有必要仿照我国经济改革的一些举措，建立一些“教育特区”，将其作为教育改革的“试验田”。可以为这些“教育特区”赋予更大的权限，鼓励其进行教育体制的改革与创新，摸索中外合作办学的新的途径与路子，并在合适的时期将其经验向外推广。

(4) 建立科学、合理而透明的质量监控标准体系以及社会评价机制。教育拨款要依据有关高校的质量评估与社会评价，同时，要规范许可证制度，办学审批权适当下放，简化审批程序。只要有合法的办学宗旨、健全的组织机构、合格的师资、必要的场所、设备等条件、必需的资金保障，都应鼓励合作办学。建立中国自己的教育认证机构，使中国的高等教育在学历、学分等方面与国际接轨，而对那些非法的合作办学要坚决予以整顿，取消其合作办学资格。可以考虑对合作办学设置独立的教育评估体系，并建立常规性的教学质量检查制度。

(5) 推动中外合作，深化教育改革。中外合作办学不仅仅是引进国外的资金、技术与人才，更重要的是要通过中外合作办学引入先进的教育理念和办法，推动中国教育改革有实质性的突破，把过于强调社会价值的教育价值观转变为适应社会价值与尊重人的主体价值相结合的教育价值观，使我们的教育为整个经济、社会的发展服务，为提高全民族的素质服务，为每个受教育者的成长与发展服务。同时，这种改革不仅是高等教育体制本身的改革，也要带动与引导基础教育领域内的改革，如对高考制度的改革等。

(6) 推动教育管理改革，给予高校更大的自主空间。在管理体制方面，要逐步建立与完善中央与地方两级管理、以省级政府统筹管理为主的管理体制。从政府管理的角度来看，重点则要转变职能，扩大高校办学自主权。淡化政府与高校的行政附属或纽带关系，变办学行为的政府集中决策为各高校的分散决策，激发高校个体的创造性。政府重在制定中长期高等教育发展规划，搞好依法治校等宏观调控；高校则成为面向国内外自主办学的独立法人，享有各种依法自主办学的权利。

(7) 加强教育输出，繁荣中华文化。目前，我们所讲的“合作办学”指的仅仅是国外教育资源的输入，而一种真正的合作必然在“输入”的同时进行“输出”。因而，有条件的机构或高校应该积极寻求赴境外办学的途径，以海外课程、专门机构或海外分校甚至网络教育等各种形式出口教育，通过教育“输出”来提高本国教育的竞争力，显示本国高等教育在国际上的潜力和优势，抢占国际市场的份额。在教育“输出”中，不仅输出自己的优势学科与特色项目，同时也通过不同形式的交流活动，宣扬中华文化，在中国了解世界的同时，也让世界更真实地了解中国，了解中国的历史与现状，了解中国的发展理念与愿景。

五、赴境外办学的对策

教育全球化是现代性的根本后果之一。像现代性的其他后果一样，教育全球化既是一种激励人建构的想象，又是一个充满矛盾和对抗的过程，它在赋予人们改变全球教育市场的同时也改变着人自身。在教育全球化过程中，中国教育市场日渐成为“香饽饽”，成为各国教育向外拓展的“必争之地”。在应对这种局面的时候，中国也充满了突出重围、从“边缘”走向“中心”的热情和冲动，其中就包括赴境外办学。

当赴境外办学问题受到理性关注的时候，有两点应当强调：一方面应保持清醒的中国教育现代性的问题意识，另一方面必须确立一个广阔的跨文化视界。这两点构成了我们分析中国赴境外办学困境与出路的基点。

我们将从以下几个基本问题出发，分析中国赴境外办学的意义、面临的问题和应采取的策略。

1. 中国必须赴境外办学

在分析中国赴境外办学问题时，我们首先面临着一个前提性问题，即中国必须赴境外办学吗？它涉及我们讨论的“中国赴境外办学”是不是一个真问题，而其本质在于中国赴境外办学究竟有什么意义。我们将从四个方面回答这一问题。

（1）赴境外办学是中国由大国走向强国的必然路径。自清末以来，由大国走向强国就一直是国人的梦想，时至今日，我们已经取得了举世瞩目的成就，但离世界强国依然有一定的差距。建立强国的目的并不是为了恃强凌弱，而是为了摆脱被欺凌的命运。

从世界发展史看，强国之所以成为强国的途径基本有三种，即依托军事的“力”、依托经济的“利”以及依托文化的“理”，现在，依托文化的“理”成为最强势、最有效的途径。

中国在军事、经济方面取得了巨大的成就（虽然并不是足以称“强”），但文化，尤其是作为文化核心的教育发展明显滞后于经济和军事建设，仍然与发达国家之间存在相当大的差距。因此，在很大程度上讲，中国依然行进在从世界的“边缘”走向世界的“中心”的路上，中国教育市场在繁荣的背后潜藏着被分割、包围的危险。强势的经济贸易并不能直接获得国际“教育市场”，民族文化的昌盛和复兴也不能没有国际教育市场。

作为开拓国际教育市场的最直接途径，赴境外办学不仅是经济发展到一定状态下的必然选择，是在全球化与本土化互动脉络中稳固本土化的必要条件，也是中国由大国走向强国的必然选择。

（2）赴境外办学是中国教育实现第二次突飞猛进的必然选择。1999 年以来，中国教育，尤其是高等教育，实现了历史性的突飞猛进式发展，创造了中国高等教育蓬勃发展、充满生机和活力的大好局面。中国教育如何进一步发展，以取得“第二次”突飞猛进是摆在国人面前的一个现实问题。综观关于谋划改革的新突破，实现发展的新跨越的思路时，我们发现，频率最高的两个词是“质量”和“国际化”。事实上，在众多的思路中，这两个词具有密切的联系，即教育质量应是“具有国际（化）水准的质量”“走向国际化必须依靠质量”。

从发达国家占有国际教育市场的途径看，无论是吸引留学生，还是赴境外办学，依靠的主要是具有国际水准的质量。赴境外办学，在一定程度上讲，比吸引留学生具有更大的难度，因为它依托的不仅仅是教学质量，还包括学习环境建设的质量和管理的质量。赴境外办学本身，就显示出对高质量办学水平的自信。在中国狂热的出国留学潮中，以及带有“迷信”色彩的国际合作中，中国教育事实上处于明显的弱势，而强化这种弱势的主要因素是缺乏对中国教育质量（包括管理质量）的自信。如何建立对中国教育的自信，最直接的莫过于赴境外办学。因此，赴境外办学的最重要的意义，不是占有国际教育市场份额本身，它的意义更多的是在文化心理上建立对中国教育的自信心，而这种自信本身必然激发中国教育进一步提升质量的热情，它无疑是中国教育改革的“助推器”。中国教育的第二次突飞猛进离不开这种自信，因此，赴境外办学将是实现中国教育“第二次突飞猛进”的必然选择。

(3) 缺乏赴境外办学的教育国际化体系是不完整的体系。从教育国际化的发展脉络看，体系完整的教育国际化包括两个层面，即“教育的国际化”和“国际化的教育”，前者指依托国际优质教育资源提升自身水平以达到国际水准，后者指优质的教育资源为国际共享以提高全球教育水平。这即是一些学者所说教育国际化的“现场卷人”和“跨距离互动”两个维度，体现了教育国际化“全球地方化或地方全球化”以及“多边互动性”的特点。

一个国家的教育国际化大体可以分成三个阶段，第一阶段是倡导和鼓励出国留学，第二阶段是在引进国际优秀资源在国内合作办学，第三阶段是本国优质资源直接进入其他国家和地区。虽然这三个阶段并不总是存在先后顺序，但都包括本国优质资源进入其他国家，这样教育国际化才可能在“全球—人类”的单位模式上展开，实现教育输出国和教育输入国的“双赢”，从而塑造出“全球共同认可的教育价值观”。

在当前中国教育国际化的体系中，教育资源输出绝大多数是输出留学生和教师输出，因而整个体系不完整，也很难说是“双赢”，因此，需要赴境外办学，将优质资源输出，促进“全球共同认可的教育价值观”的形成。

(4) 延误赴境外办学的时机将失去更多的国际市场份额。自中国改革开放以来，中国参与国际教育合作的机会越来越多，取得了巨大的成就，但同时，输出留学生和进入中国的国外教育机构也越来越多。尤其是加入 WTO 以来，输出留学生和进入我国的外国教育机构数量迅猛增长，与我国吸引的外国留学生和进入外国的教育机构相比，明显不对等，中国的教育市场面临着极大的冲击，中国占有的国际市场份额也没有达到预期的目标。

在这种形势下，我们不能只将视野局限于国内而过分乐观地看待中国教育改革的成就。教育国际化的本质就是要用全球和全人类的眼光看待本国的教育改革和发展。因此，我们应采用包括赴境外办学在内的多种形式开拓国际教育市场，争取更多的国际教育市场份额。

基于上述分析，我们认为，中国必须赴境外办学。

2. 中国赴境外办学存在的障碍

根据教育部的不完全统计数据，截至2002年年底，中国共有中外合作办学机构和项目712个，与1995年年初相比，增加了9倍多，覆盖了28个省（自治区、直辖市）。中外合作办学呈现加速发展的势头，沿海地区申报的数量大大增加，中西部地区的中外合作办学项目开始起步。总体上讲，办学层次有所提高，合作形式也日益多样化。但到目前为止，中国赴境外办学的机构数量增加相当缓慢，就笔者目力所及的资料看，只有上海交通大学、暨南大学、国家对外汉语教学领导小组办公室等不足10家，并且不是实体性办学，总体规模很小，尚不足500人，这与我国的地位、办学水平等很不相称。因此，我们必须回答一个基本问题：中国赴境外办学的障碍在哪里？或者说，中国赴境外办学缺少哪些必要条件？

事实上，这个问题牵涉面非常广泛，几乎涉及宏观和微观的各个层面。在此，我们从以下5个方面试图回答。

（1）总体上看，在社会文化心理上，缺乏赴境外办学的心理准备。对于中国而言，国际教育合作是新鲜事物，目前形成的“集体意识”主要集中在引进国际优质教育资源、提高国内教育水平和输送留学生的层面，即定位在“请进来”，而对赴境外办学缺乏必要的自信和心理准备，如作为国际教育合作的核心法律《中华人民共和国中外合作办学条例》，最重要的原则和出发点是“扩大开放”“引进外国优质教育资源”，有利于提高办学质量，从而全面提高我国教育的国际竞争力。

这与客观条件不够成熟有关，也与对完整的国际教育合作体系的理解不够有关，同时也与缺乏用全球、全人类的眼光看待中国教育有关，在处理“民族—国家”“国际化—本土化”等矛盾时，更多地从“地域”意义上的国家角度思考和看待国际教育市场，缺乏“国际意识”，因而不能很好地在“全球—人类”的单位模式下构建中国教育体系。

（2）在宏观政策安排上，缺乏必要的制度规范。到目前为止，关于国际教育合作的制度、规范主要有：中华人民共和国国务院令第372号《中华人民共和国中外合作办学条例》，中华人民共和国教育部令第20号《中华人民共和国中外合作办学条例实施办法》，《中外合作举办教育考试暂行管理办法》，教育部、公安部、国家工商行政管理局令第5号《自费出国留学中介服务管理规定》，教育部、公安部、国家工商行政管理局令第6号《自费出国留学中介服务管理规定实施细则（试行）》，教育部令第15号《高等学校境外办学暂行管理办法》，《教育部关于设立和举办实施本科以上高等学历教育的中外合作办学机构和项目申请受理工作有关规定的通知》，《教育部关于做好中外合作办学机构和项目复核工作的通知》以及《教育部关于启用〈中外合作办学机构申请表〉和〈中外合作办学项目申请表〉等事项的通知》。而其中的三个通知及《中外合作举办教育考试暂行管理办法》事实上不过是执行政策的相关规范。因此，从总体上看，关于赴境外办学的“绝对”制度规定缺乏。

另外，关于国际教育合作的相关制度，采用的是《中华人民共和国民办教育促进法》的思路，在很大程度上能够起到扩大对外开放、引进优质教育资源的作用，但它们对赴境外办学的相关制度，如对赴境外办学机构的产权问题，是否可营利以及盈利的处置问题，准入、运行、退出、激励、抗风险等机制缺乏必要的规范和安排，因此，关于赴境外办学的“相对”制度也缺乏。在这种情形下，赴境外办学事实上具有很大的风险。

（3）我国总体办学思想、办学水平、管理体制、办学模式、人才培养模式等与发达国家之间存在较大差别。在一些学者看来，中国教育管理体制和人才培养模式是苏联体制和模式的移植。但我们认为，主导中国教育管理体制和人才培养模式的更多的是在解放区创立、在新中国成立后得到强化的体制和模式，其中比较多地借鉴或移植了苏联的一些相关做法，而且自改革开放以来，大量借鉴和吸收了美国的相关措施，如贯穿高等教育教学管理制度改革全过程的学分制。因此，其内核是中国的而不是国际的，这在客观上与发达国家之间存在很大的差别，其根本差别在于：虽然历经了艰难的改革，但中国教育思想、管理体制和人才培养模式仍然烙着“计划”的印迹，“计划”模式仍然在主导着教育管理体制和人才培养模式，而发达国家则是以市场为主导的管理体制和人才培养模式，虽然任何国家都不完全排斥计划。

毋庸讳言，在借鉴和吸收发达国家的相关做法改进我国教育管理体制和人才培养模式时，我国教育也相应地具有了更多的“后发”和“外铄”性质。因此，当以发达国家的标准衡量时，中国教育管理体制、办学模式、人才培养模式和总体办学水平都存在较大的差距。尤其是中国很难进入国际主流的质量认证体系，因此，学分、学位都难以得到国际承认，在直观上造成中国教育质量差的印象，中国失掉了本来存在的比较优势。

（4）在“民族—国家意识”得到强化的“后现代”时期，各国都存在有意识的教育市场保护体系。“教育是改变人的最有力的武器”，仅仅从市场的角度理解教育国际化是片面的。在“民族—国家意识”得到强化的“后现代”时期，在教育市场开放的背后都隐含着民族和国家利益。虽然，我们并不完全承认和接受有些学者提出的受民族文化、宗教、种族文化主导的民族矛盾是国家与国家之间的根本矛盾，但我们认为，这些观点部分本真地反映了客观事实，在霸权和反霸权并存的世界中，对外来文化尤其是外来教育机构的抵制、排斥心理和或明或暗的制度安排几乎在任何国家都存在。因此，在中国国力还不足以在世界称强、在汉语还不是强势语言的背景下，赴境外办学存在很大的难度。

（5）赴境外办学难以解决师资、场地、设备、生源等问题。师资、场地、设备、生源等“硬件”是制约赴境外办学的瓶颈问题，多数试图赴境外办学的机构因为在这类问题上难以解决而导致赴境外办学的计划搁浅。

3. 加速赴境外办学进程的对策措施建议

通过上述分析，我们不难发现，当前中国赴境外办学的现状是，规模小、进程慢、难度大。然而，赴境外办学如在弦之箭不得不发，否则将贻误机遇。因此，我们面临的问题是：为加速赴境外办学的进程，中国应当采取什么措施？我们将试图从以下八个方面提出对策措施建议。

(1) 激发民族自信心，充分展现民族文化的比较优势。诚如上文分析，自信心不够是制约我国赴境外办学的重要原因，因此，激发民族自信心是加速我国赴境外办学的首要问题，而自信心来源于“比较优势”，因此，充分展现比较优势，是提升民族自信心、更多地占有国际教育市场份额的关键。

我们认为，在当前形势下，中国教育的比较优势集中在民族文化传统上。因此，在参与国际教育合作过程中，我们不应过多地强调“与国际接轨”，因为，越强调“接轨”，中国教育就越具有“外烁”性质，越失去自身的特色和比较优势。在加速赴境外办学方面，我们应强化“越是民族的就越是世界的”观念。

(2) 政府应将国际教育合作作为新的经济增长点，切实加强制度建设，建立以高水平学校为主体的多样化的赴境外办学体系。政府应以“全球—人类”的视野，谋划中国教育改革。在加速赴境外办学的过程中，政府的主要职责是建立制度规范、营造氛围和维护赴境外办学机构的利益。在当前，政府应切实树立“教育是投资而不是消费”的观念，将国际教育合作作为新的经济增长点，加强与其他国家和地区的高层接触，切实推进中国教育加入国际教育质量认证体系的进程，进一步促进中国与其他国家、地区学分互认、学历互认的进程；应进一步完善关于国际教育合作的制度，尤其应建立赴境外办学的激励措施，应明晰赴境外办学机构的产权，明确规定是否可营利以及盈利的处置规则，明确规定赴境外办学机构的准入、运行、退出等规范，建立切实可行的抗风险机制。

在加速赴境外办学的进程中，政府应赋予教育机构应有的自主权，激发教育机构的参与意识，建立以高水平学校为主体，各级各类公立、民办教育机构广泛参与的赴境外办学体系。

(3) 各级各类教育机构应努力建立具有特色和先进性的办学模式与教育模式。在当前的研究材料和政策文本中，特色受到了应有的重视，但在同时，先进性遭到了忽视。事实上，特色和先进性犹如一枚硬币的两面，相辅相成。我们博大精深的民族文化本身既具有特色也具有先进性，但在挖掘时，我们常常侧重于宣扬特色，而忽视了先进性。

在教育领域，民族文化的特色和先进性应体现在办学模式、教育模式、课程内容、师生关系等各个方面，因此，需要我们通过教学各个环节体现出来，从而构建具有特色和先进性的办学模式、教育模式等，否则我们的比较优势仅仅是一个抽象概念。

已经与其他国家、地区建立教育合作关系的教育结构，应充分利用现有平台，吸收先进经验，寻找赴境外办学的契机，积极向外拓展。在办学模式上不应固守“校校合作模式”，在人才培养模式上，也不应固守为适应国内市场的模式，而应探求更多、更可行的模式。

(4) 深入研究包括 WTO 相关条款在内的国际教育市场规则，广泛借鉴发达国家赴境外办学的经验。综观国际领域的经济合作，失败的常常是对国际规则不熟悉、不能熟练运用的机构，这一点应充分引进参与教育合作的机构的重视。无论是政府，还是教育机构，都应深入研究包括 WTO 相关条款在内的国际教育市场规则，并深入研究准备赴办教育机构的国家的文化传统、产业结构及其对人才的需求、教育制度、生源情况等；应当广泛借鉴发达国家赴境外办学的经验，其中尤其应当借鉴英国、美国等国家实行的参与国际教育竞争的制度安排，借鉴斯坦福大学（该校已经在国外办有 8 所分校）等学校赴境外办学的招生、专业设置、课程体系构建等具体措施。

（5）加强中介机构建设，充分利用媒体宣传。从我国参与国际教育合作的经验看，中介机构和媒体宣传功不可没。因此，应充分发挥中介机构和媒体宣传的功能，积极寻找赴境外办学的机会。但到目前为止，中介机构和媒体宣传的功能主要集中在拓展出国留学的市场上，已经举办的教育论坛、教育巡回展等也都集中在“请进来”的层面。为加速我国赴境外办学的进程，政府应有意识地转变媒体宣传和各项活动的重心，将其重点转变到“走出去”的宣传和努力上来。

（6）充分利用网络，积极抢占教育信息化高地。从国外发达国家进入中国市场的经验看，远程教育是吸引中国生源的一柄“杀手锏”，有些国家和学校甚至不惜代价地在网络上展出其优秀课程内容以激发学生的兴趣。建议有关教育机构，尤其是高水平大学利用网络，充分展示其教育实力，通过先发展远程教育的途径，做大做强自己的声誉，进而寻找机会赴境外办学。

（7）树立成功典范，避免贪大求全。上海交通大学、暨南大学、中国国家对外汉语教学领导小组办公室等机构已经开了走出国门，赴境外办学的先河。政府应当采取优惠措施，积极扶持这些在境外办学的机构及其境外教育机构，使其成为赴境外办学的成功典范，而不应过多地追求规模效应，贪大求全。

（8）赴境外办学的进程和重点。在当前形势下，赴境外办学应选择合适的进程和重点。从教育层次看，应优先发展高等教育和幼儿教育；从教育类型看，应侧重发展非学历教育；从学校类别看，应优先发展“侨校”和其他有鲜明特色的学校；从教育内容上看，应优先发展华文教育和汉语教学；从地域上看，应侧重于华侨华人密集或汉语占有重要地位的地区；从合作机构看，应侧重于与企业或培训机构合作，参与职业培训。

六、发展华文教育的对策

1. 华文教育的含义和意义

华文教育指对华侨或华人子弟的中国语言文化教育。广义的华文教育可以包括国内的对外汉语教学，请看表 16。

表 16　华文教育与相关概念

项目	1. 华裔		2. 非华裔	
	（1）非学历	（2）学历	（1）非学历	（2）学历
Ⅰ. 国外	华文教育 （一般为儿童）	汉语教学	华文教育 （一般为儿童）	汉语教学
Ⅱ. 国内	语文教学（国民） 华文教育	语文教学（国民） 华文教育/所有专业	汉语教学（国民） 对外汉语	汉语教学/所有专业（国民） 对外汉语/所有专业

国内学界对华文教育与非华文教育（对外汉语教学等）并没有很清楚的界定和统一的认识，但大体上大家同意把国外对华裔儿童甚至非华裔儿童（也在华文周末班、华校读书的）的中国语言文化教学称为华文教育；同时把国内针对华裔子弟（一般是青年、成人）的中国语言文化教育（包括仅学汉语言文化的和其他专业的）称为华文教育。华文教育与国内的对外汉语教学有差异，因为它不一定是第二语言的教学，有的属于第一语言教学，有的是双语教学的性质；华文教育中中国文化的教育是重要的组成部分；华文教育的对象主要是儿童。本节一般不区分华文教育和对外汉语教学，但以狭义的华文教育为核心。

华文教育是海外华侨和华人传承和弘扬中华文化、保持民族性的有效手段，是凝聚侨心，促进华族良性发展的内在动力；华文教育也是使世界各国人民了解中国的重要途径；华文教育还是促进祖国统一、推进祖国建设的重要手段。总之，华文教育事业是一项有利于华侨华人社会存续和发展，有利于国际友好和世界和平，有利于中华民族振兴的重要事业，是一项长期的、宏大的工程。

2. 国外国内华文教育的发展现状

1）海外华文教育蓬勃发展

A. 发展近况

据报道，目前海外华人有 3000 多万人（2001 年台湾侨委会统计为 3580 万人）。海外华校有 3000 多所（2001 年台湾侨委会统计的是 3032 所）。另有上百个国家的 2300 多所学校开设汉语课。海外学中文的已超过 3000 万人。

下面选择美国、加拿大、法国、澳大利亚、印度尼西亚、马来西亚、新加坡几个有代表性的国家作些说明。

a）美国

根据最近在广州召开的第四届国际华文教育研讨会的有关资料[16]，美国社区中文学校超过 1000 所，学习中文的人数有 15 万人；美国有 300 多所中小学设有中文课，开设中文课程的公立、私立大学有 1000 所，占美国高校的 1/3。华文教育在美国有较长的历史，目前专门的华文学校基本上有三类：第一类是历史最早的中华学校，学生以广东籍后裔为主；第二类是台湾背景的华校，主要是 20 世纪 60 年代从台湾移民美国的华人的后裔；第三类是 20 世纪 70 年代末以后从中国内地移民美国的新移民办的中文学校。从下面几组数字可以看出美国华文教育的发展势头（表 17，图 10，表 18，图 11）。

表 17　华盛顿希望中文学校[17]

年份	学生/人	教师/人	班次
1993（创办时）	26	6	不详
2003	2000 多	200 多	从学前班到高中

b）加拿大

据文献报道，加拿大中文学校协会有 35 所中文学校；不列颠哥伦比亚省有 200 多所私立中文学校，学生约 2 万名，教师 500 多名；多伦多市有 110 所公立学校开设了中文课；仅蒙特利尔的佳华中文学校就有近千名学生，40 多名教师。

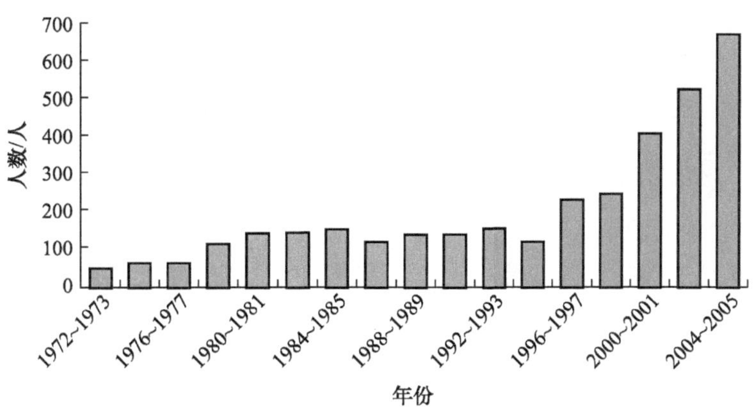

图 10 底特律中文学校历年招生发展情况[18]

根据文献［18］重新制表，个别数据因为原文看不清楚可能不准确

表 18 美国中文学校教学研讨会变化表[17]

年份	学校/所	教师代表/人
1998	10	19
2000	20	35
2002	72	117（含国内去的 20 多人）

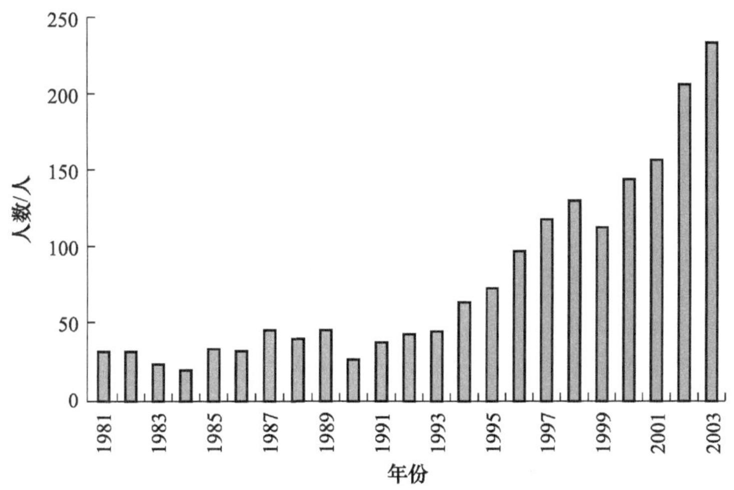

图 11 俄克拉何马大学汉语项目学生注册人数增长情况[19]

根据文献［19］重新制表

在加拿大的魁北克省，汉语正取代法语，成为第二外语。

c）法国

法国华侨华人会中文学校有 696 名学生，19 个班级。据法国媒体报道，中文是目前法国中学教育中发展最快的外语，2000～2002 年全法选中文的学生每年增加 30.28%，仅巴黎就有 6 所小学，141 所中学开设中文课；2004 年开学，有 7600 名学生选修中文。几乎所有高等专业学院和多数综合性大学都把中文列为高等教育课程之一。法国国立东方语言文化学院 2004 年注册学习中文的学生近 1900 人，比上一年增加 20%以上。据法国教育部汉语教学总督学白乐桑讲[20]，有 90%的汉语学习者母语为法语。“汉语热”在法国比起在欧洲其他国家，更多地被视为普遍趋势的代表。

d）澳大利亚

1981 年澳大利亚“社区语言教学项目”在澳大利亚新南威尔士州公立学校启动。目前估计全澳学华文的学生有 5.8 万人，华校 485 所，其中中学 256 所，小学 139 所，其他 90 所，这还未包括其他各类职业培训课程和社区周末华文学校的学生。仅维多利亚州就有 25 000 多名中小学生在课堂学习中文；教华文的中小学有 150 多所；另外 50 多所为周末华文学校。悉尼也有周末华文学校 50 多家，为超过 10 万名来自华裔及非华裔社区的学生提供中文教育[21]。中国侨网 2003 年 12 月 12 日报道，澳大利亚统计局 2002 年 11 月 19 日发表的 2001 年人口普查最后一期报告称，中文已成为澳大利亚第二大语言，全澳讲中文的人口有 40 多万人，占总人口的 2.1%。从而使中文一跃成为澳大利亚最大的外语语种。

e）印度尼西亚

印度尼西亚是海外华人最多的国家。20 世纪五六十年代印度尼西亚的华文教育非常兴盛，曾经拥有 1500 多所华文学校。1967 年以后全部被封闭。20 世纪 90 年代印度尼西亚的华文教育开始复苏[22]。1999 年华文解禁以后各地华文补习班如雨后春笋般涌现。仅泗水市的新中补习班就有 1500 名学生在学华文。经济较为落后的西加里曼丹省山口洋市华人占当地人口的 42%，近两年来创建了 22 所华校，吸收了当地近 7000 名乡村孩子。为了解决师资问题，当地的小教师只有 16 岁就走上了讲台！2001～2002 年广东汉语专家团两次巡回印度尼西亚各地培训华文教师，报名人数是 3000 人，参加培训的有 1400 人，其中，最大的 70 多岁。2001 年 10 月印度尼西亚举行首届汉语水平考试（Hanyu Shuiping Kaoshi，HSK），报名人数约 1200 人，其中雅加达 600 名。根据我们的调查[23]，2003 年印度尼西亚共有超过 3000 位华人在做华文教学工作，其中绝大多数为家教，有 100 多家补习班或补习学校（课后加周末教学），有至少 56 所正规中小学、幼儿园开设中文课，有 23 所大学开设中文课。印度尼西亚教育部最近制定了 2004～2007 年四年计划，将逐步在全国各地 80 个城市的 8000 所国民高中设汉语选修课，积极发展以印度尼西亚文、英文和华文为主的三语学校。2004 年 7 月 24 日中文正式成为印度尼西亚国民高中课程。如果所有高中开设中文，需要 3 万名教师。2004 年雅加达和东爪哇先后有多所三语学校开办。2004 年 8 月我们听说印度尼西亚教育部接到了 103 所大学开设汉语专业的申请，其中绝大多数因为缺乏师资而没有得到批准，他们的标准是每所学校要有自己的华文老师 6 名，其中学士 4 名，硕士 2 名。印度尼西亚是华人大国，也是华文教育需求最大的国家。对印度尼西亚的华文教育应当给予特别的重视。

f）马来西亚

马来西亚的华文教育有很好的基础。2004 年董教总麾下的华文小学有 1200 多所，华文独立中学 60 所，华文大专学院 3 所，在校生 70 多万人，形成完整的华文教育体系，其中在各华小就读的非华裔儿童接近 7 万人。中国侨网 2004 年 8 月 17 日报道，

2000 年全国有 643 所中学（马来文中学）开办华文班，学生 62 680 人；2004 年升为 800 多所，学生猛增至 23 万人。中国侨网 2004 年 10 月 19 日又报道，2004 年 8 月 20 日中国政府准予 60 所独立中学的毕业生赴中国深造时免汉语水平考试。

g）新加坡

新加坡华人比例很大，占总人口的 76.7%。但由于历史原因，很多华人不会讲汉语。1989～1998 年，新加坡华人家庭讲汉语的比例从 69%减少到 56%，跌幅大约每年一个百分点。但是新加坡华文教育的命运在 2004 年出现了重大转机，从政府总理李显龙到内阁资政李光耀都发表了大量重新审视华教政策、提升华教地位的讲话。政府成立了华文教育检讨委员会，并在一个月内完成一份华文教育检讨白皮书，提交国会讨论。到目前为止，已经看到的政策是 2005 年 1 月起开办双文化课程，分别在三所高中和一所学院开设双文化课程，旨在培养了解中国历史和文化的华文精英，即中英双语精英。其中华侨中学、南洋女中、德明政府中学中修读双文化课程的中三学生从 2005 年起将有 2～3 个星期的时间到美国、英国、澳大利亚、新西兰的学府进行学习，在中四及中五两年则有 5～6 个月的时间到中国学府学习。这三校的双文化生可直接入读中国知名大学。所有 2004 年的小一和小四生将率先从 2008 年开始接受新的华文课程。一些小学则在 2005 年开始先进行“先认后写”汉字的教学方式。中国侨网 2004 年 10 月 14 日报道，2004 年的汉语水平考试，新加坡考生比往年多出 11 倍，仅次于韩国和日本。其他国家也大同小异，总之都呈现出持续增长的“华文热”[24]。

B. 海外华文教育持续升温的原因及远景预测

一般都认为华文持续升温是中国经济发展的结果。这样说不错，但不尽完善和详细。这里做一些补充，特别重视引述外国人的看法。

a）海外华裔人口增加

首先是新华人移民的增多。20 世纪 60 年代，中国台湾移民大批移居海外；20 世纪 70 年代末以后中国内地移民大批移居海外。据报道在全球 3000 多万海外华人中新移民占到 1000 万人，与此同时，美国等国家领养中国儿童的比例增加不少，比如美国加利福尼亚州有 2 万多个被领养的中国儿童，这些孩子以及他们的家长都加入了华文学习的群体，出现了“家长班”“爸爸班”的现象。

b）中国综合国力的强盛

移民人口的增长和领养中国儿童的增长并不能解释所有的华文持续升温现象，尤其是不能解释为什么还有越来越多的非华裔开始热衷学习华文的现象。中国政府 1978 年以来实行改革开放政策，国力日渐强盛，创造了和平崛起的辉煌成就，使越来越多的国家、人民重新认识华文的经济价值，重新审视中华文化。

面对不断增加的经济全球化的压力，对美国来说，向外发展已成为其必然的选择。中国这个拥有世界人口 1/4 的发展中国家对许多发达国家来说都是潜力巨大的理想市场，从而，作为与中国经济往来的一个“副产品”的汉语项目，在美国许多高校里被日益看好。以俄克拉何马州为例，俄克拉何马州和中国日益加强的经济交往，提高了对汉语人才的需求。1994 年以来不断有企业和未开设汉语项目的大学向俄克拉何马大学咨询学习汉语的信息[18]。同时，为了增强学生的就业竞争力，商务专业的学生纷纷开始辅修汉语专业，带来了正规大学汉语课程的日益兴旺。美国弗里蒙特地区资深中文教师杜丽玉说[19]：“中国已经成为新兴的世界强国，中文成为第二外国语也成为一个必然趋势，包括美国在内的许多国家都把汉语列为主要外语，目前全球学中文的人口已经超过 3000 万人。”中国侨网 2004 年 9 月 22 日报道，为了使美国公众了解英语以外的多种语言在经济全球化的进程中的重要地位，使语言学习（英语以外）成为每个美国人生活的一部分，美国国会把 2005 年定为语言年，这一举措反映了美国国家语言政策上的变化和美国公众对于英语以外其他语言和文化在认识上的相应改变。

法国教育部汉语教学总督学白乐桑说：“法国学中文的人越来越多，是因为随着全球化的加快，有的小语种在消亡，但汉语的国际化程度却是越来越高，汉语的使用价值也越来越大。当然，中国的全方位开放，使汉语超出了一种语言和一种文化本身的魅力，从此代表了华文世界的经济实力，懂中文成了法国青年就业的一张王牌，也成了法国人眼中‘异国性的象征’”。

在德国，多种汉语培训班颇受欢迎。对不少大学生来说，在德国就业市场不景气的情况下，掌握汉语更成为他们未来获得工作的重要砝码。

在新时期，蒙古国和中国两国的经贸合作迅速扩大。中国目前是蒙古国的第一大贸易国。近年来，每年到中国的蒙古国公民有 35 万～40 万人，正是在这样的背景下，蒙古国的华文教育开始步入“快车道”。

新加坡前总理李光耀在 2004 年 6 月提议新加坡要进入中国这个大市场，必须好好认识中国，了解中国文化和中国语言。（新加坡《联合早报》，转引自中国侨网 2004 年 9 月 7 日）新加坡的张苇[25]指出：“中国的崛起是推动新加坡华文发展的最主要外在因素，华文政策以搭中国经济起飞的顺风车为导向可以说是无可避免的。”

C. 华文需求远景预测

马来西亚董总主席郭全强说：“不但华人，其他国家也开始一起学习华文，相信在 21 世纪末，华文将成为强势语言，在包括教育经济和政治领域扮演举足轻重的角色。中国的崛起，让东南亚各国的教育、经济领域都有一定的改变，20 世纪以英文为主导，21 世纪已有所改变，即英文和华文并重。”

英国语言学家戴维·葛拉多尔（David Graddol）则在 2004 年 2 月《科学》杂志上发表文章说：中文正逐渐跃升为全球仅次于英文的新强势语言，未来十年，新的必学语言可能是中文。他表示，到 2050 年，最普遍的语言前 3 位将分别是中文、印度乌尔都语、阿拉伯语，全球以英语为母语的人将由 20 世纪中叶的 9%降到 5%。

另有网络专家预测[20]，目前约有七至八成的互联网使用英文，一成用中文，至 2050 年，中文应用比率将上升到三至四成。

2）中国的政府、高校、学院所作的努力

A. 中国政府对华文教育非常重视

近年来，国内外华文教育、对外汉语教学工作发展很快，华文教育在世界范围内呈上升趋势。我国政府非常重视华文教育工作。2000 年 9 月 30 日，江泽民主席为北京华侨补校题词：“发展海外华文教育，促进中外文化交流。”2004 年 3 月胡锦涛主席在参加全国政协会议联组讨论时对华文教育工作作出重要指示：“无论从优秀传统文化的传承角度考虑，还是对骨肉同胞的亲情考虑，都应对海外侨胞开展华文教育给予帮助和支持。要加大政府的投入，动员各方面的力量来支持这件事情。”① 为贯彻落实胡锦涛主席的指示精神，由国务院侨务办公室（简称国务院侨办）牵头的“国家海外华文教育工作联席会议”已在北京正式成立，并制定了《2004—2007 年海外华文教育工作规划》，“中国华文教育基金会”也已注册成立并召开了第一届一次理事会议。这充分体现了国家对华文教育的重视和支持。

B. 国务院侨办的工作

国务院侨办是专门从事华侨华人工作的政府机构，华文教育是其工作中一部分重要内容。国务院侨办以 28 个华文教育基地为龙头开展工作；具体工作有立项支持研究、开发，支持国内华文教学，主持召开国际华文教育研讨会，资助开展师资培训、校长培训，开展“寻根之旅”夏令营活动，作文大赛，捐建海外“华星书屋”等。

C. 国家汉语办公室的工作

国家汉语办公室 2004 年启动了汉语桥工程。这是一个经国务院批准，由教育部等中央 10 部委联合建设的对外汉语教学大工程，其目标是让汉语走向世界，使汉语在世界主要国家和地区尽可能广泛而深入地传播，特别是在学校语言教学中，更多成为主要外国语言课程之一，成为 21 世纪新的强势语言。汉语桥工程包括 9 个重点项目：①在世界各地设立孔子学院；②中美网络语言教学；③教材和音像、多媒体制作；④国内外汉语教师队伍建设；⑤对外汉语教学基地建设；⑥汉语水平考试；⑦世界汉语大会和“汉语桥”比赛；⑧“汉语桥”基金和援助国外中文图书馆；⑨组织保障和基本建设。另外设立了 13 个支持周边国家汉语教学重点院校：推行了国际汉语教师志愿者计划等。

D. 国内高校的努力

20 世纪 90 年代初，国内华文教育主要由国务院侨办领导的暨南大学、华侨大学和北京华文学院（华侨补校）、集美华文学院（华侨补校）、昆明华文学校（华侨补校）等院校在做。从事“对外汉语教学”的有北京语言学院及暨南大学、南开大学、南京大学、上海师范大学等。后来国家放开了政策，所有的高校都可以从事华文教育了。据国家留学基金管理委员会统计，2003 年全国来华留学人员共有 77 715 人，分布在全国 31 个省（自治区、直辖市）的 353 所高等学院校和其他教学机构（未包含港澳台的相关数据）。这些留学生分别来自 175 个国家。其中、学历生 24 616 人，非学历生 53 099 人，前者从专科到本科、硕士研究生到博士研究生，他们所修的专业不详，应有不少是专修中文的。

国内高校在华文教育方面所做的工作主要有以下几个方面：华文教育学科理论建设，学术交流；留学生教育和对外汉语教学；承办汉语水平考试；华文教材和课件的编写开发及函授远程教学；华文师资的培训、培养和外派；合作办学和海外办学；海外孔子学院的承建。这些工作主要由国务院办公厅的 28 个基地和国家汉语办公室（以

---

① 《背景资料：海外华文教育概况》，https://www.chinanews.com/hwjy/news/2009/10-20/1919272.shtml [2023-11-11].

下简称国家汉办）的对外汉语教学基地、支援周边国家重点院校做的，但留学生教育遍布全国 34 个省（直辖市、自治区）和特别行政区，有近 400 家教学机构。暨南大学是我国开办的华侨最高学府，已有 99 年校史，从创办以来一直是个外向型、国际型的学校。现在暨南大学是国家“211 工程”重点建设大学，国务院侨办华文教育基地和国家汉办支持周边国家重点院校，其工作具有较高的代表性，加之数据保存完整，所有下文的举例以暨南大学的为主，其他院校的暂时只能就有限的调查举些重要的例子，不能做穷尽性统计。

a）华文教学科学理论建设和学术研究、交流

（1）学科理论建设和学术研究。2000 年以来，暨南大学华文教育基地[26]（以暨南大学华文学院、暨南大学华文教育研究院为主体）教学科研人员共承担了国家社科、国务院侨办、国家汉办、教育部等省部级科研项目 47 项，其中多数是华文教育、对外汉语学科领域的重要课题或基础课程。比如“海外汉语教材编写及研究”“华文教育学”“网上学中文”“初级华语网络课程的研制”“汉语教学法研修教程”“现代远程教育在海外华文教师培训中应用研究”“印度尼西亚汉语使用及汉语教学现状调查”“印度尼西亚华文教师汉语基础教材项目”“面向印度尼西亚的汉语师资培训现状、内容、方式、策略研究”等。4 年来该基地教学科研人员共发表学术论文 368 篇，其中核心期刊以上 134 篇。

在科学研究的基础上，学院还开设了“华文教育概论”“华文教材编写研究”“华文教育心理学”“对外汉语教学法研究”等新课程。

（2）对其他华文教育基地、对外汉语教学基地、支持周边国家重点院校也作了大量研究。据了解，国务院侨办从 2000 年以来已下达科研立项 30 余项；国家汉办从 2000 年以来已经下达科研立项及调研立项 75 项。

（3）专门研究机构。目前国内已有至少 4 家对外汉语研究所（中心）、华文教育研究中心，它们是暨南大学华文教育研究院、厦门大学海外华文教育研究所、华侨大学华文教育研究院、北京语言大学对外汉语研究中心。

（4）学术期刊。为了推动华文教育学与对外汉语研究，凝聚科研力量，暨南大学于 2001 年正式创办了华文教学与研究专业期刊《暨南大学华文学院学报》（华文教学与研究），同时将《广州华苑》改版为学术研究与综合并重的刊物。《暨南大学华文学院学报》有国际、国内刊号，创刊以来以鲜明的特色、明确的定位和较高的编审印校质量赢得了国内外同行的认可，并被列为暨南大学“211 工程”首期建设标志性成果。

到目前为止，国内有正规华文教育专业学术期刊 3 种，包括世界汉语教学学会主办的《世界汉语教学》和北京语言大学主办的《语言教学与研究》，学汉语读物一种（《学汉语》，北京语言大学）。另有非正式的华文教育刊物两种：《海外华文教育》（厦门大学）、《国外汉语教学动态》（北京外国语大学）。《云南师范大学学报》于 2003 年增加了“对外汉语教学与研究版”（双月刊）；北京大学对外汉语教育学院拟于 2005 年 7 月出版《汉语教学学刊》；上海师范大学拟于 2005 年 3 月创办《对外汉语研究》。

（5）华文教育研讨会。国务院侨办近年来先后主办了 4 届国际华文教育研讨会，第四届于 2004 年 12 月 13 日至 17 日在广州召开，由暨南大学承办。会议以“开拓华

文教育新思路，共谋华文教育大发展”为主题，围绕海外华文教育的现状与发展趋势、如何集中海内外力量共谋华文教育大发展、开展海外华裔青少年的工作思路与措施、有效发挥华文教育基地的作用等问题进行专题研讨。

2002年11月召开了首届印度尼西亚华文教育与华文文学国际研讨会。这是国内第一次在中国境内为某一国家的华文教育召开专门研讨会。

为推动华文教育的深入发展，暨南大学与中央民族大学、香港中文大学等合作，于2005年1月主办了“第四届国际双语学研讨会”，由暨南大学华文学院承办。华文教育与双语教育关系密切，这次会议的召开，标志着华文教育研究的又一次深入。

b）留学生教育

暨南大学的留学生教育一直走在全国前列。1953年广州华侨学生补习学校就开始大量招生。1986年暨南大学对外汉语系就开始了“对外汉语教学”方向硕士研究生教育。暨南大学华文学院现有非学历教育（汉语速成班、商贸汉语班、粤语班等）、专科（下辖对外汉语、海外华文教育两个专业）、本科、硕士等层次，并依托汉语言文字学博士点招收“汉语应用”“现代汉语语法”博士生，形成了丰富而比较完整的办学体系。除此以外，还举办了各种短期班和夏令营，满足了各层次人士的学习需要。该学院的本科“对外汉语教学”专业获广东省名牌专业荣誉；“华文教育课程建设”获广东省教学改革成果一等奖。最近本基地正在组织力量申报国务院侨办重点学科“华文教育学”本科专业和“语言学及应用语言学”博士点。

2000～2001年在华文学院就读的外国留学生人数是371人，2001～2002年577人，2002～2003年661人，2003～2004年708人（不含来华培训人数和海外函授学历学员及研究生班）。2004年下半年在校留学生582人，来自40个不同的国家。

国内很多大学、学院的留学生教育近年蓬勃发展。2002年我国除台湾、香港、澳门地区外招收各种留学生85 179名。2003年略有下降，为77 715人。2004年尚无统计数字，估计会超过历史水平，因为2003年中国和亚洲遭受过非典的影响。来华留学生多数是学习汉语的，以2003年统计为例，非学历生53 099人，学历生24 616人；前者可以肯定是专修汉语的，后者中还包括不少专科、本科生是攻读汉语专业的。对这些留学生我们未做种族分类，但估计华裔为主体。特别是学汉语的学生，例如，2003年来华留学生中亚洲学生达42 190（非学历生）＋21 482（学历生）＝63 672人，占总数的82％，而亚洲学生中除韩国、日本外，东南亚留学生以华裔为主，这从暨南大学华文学院、华侨大学集美学院、厦门大学海外教育学院的留学生构成中可以清楚地看出。目前在中国国内有395所高校招收外国留学生，外国留学生的总人数达到了8.5万人，其中约有6万人专门学习汉语，占外国留学生总数的70％以上。来华留学生中有很大一批是华裔，他们中多数人又是专门学习汉语的。对华裔学生的教学具有不同于对非华裔学生教学的规律，上述几所专门大学的学院在这方面作出了持久而深入的探索，摸索出了丰厚的经验。

c）承办汉语水平考试

国内汉语水平考试在31座大中城市设有考点，每年开考3次。参加考试的主要是外国留学生，外国在华公司的随从人员，其中有高级考权的考点有3个，分别在北京、上海和广州。暨南大学华文学院是设在广州的全国有高级考权的考点之一。近年考生不断增加，最多一次参加考试的人数是 850 人。

d）教材、课件的编写开发及函授远程教学

暨南大学长期以来注重海外华文教材和教学课件的编写与开发。面向欧美的基础教材《中文》和面向柬埔寨的基础教材《华文》获得成功。《中文》是针对欧美儿童学习汉语的系列教材，主教材共 12 册，加上学生练习册 A、B 各 12 册和教师手册共计 48 本。它是目前海外公认的最好的华文教材，已在世界各国发行达 450 多万套，并在印度尼西亚获准免费复制。为适应海外某些繁体版学校的需求，2004 年教材编写组又完成了《中文》繁体字版的写作。现又完成了国家汉办针对印度尼西亚的特别项目中两个分支项目《攻克汉字难关》《基础华文》的编写任务，已交出版社出版。

适应海外学习者的要求，暨南大学还开发了华文学习多媒体系列课件。包括《中文》多媒体（光盘）版，该课件 2003 年荣获第七届全国多媒体教育软件大奖赛二等奖；2003 年 9 月《中文》网络版已经全部开通，供世界各地的学习者采用；《汉语》多媒体（光盘）版。2003 年用于成人零起点学习者远程学习的《初级华语》文字教材和网络课程完成，并在美国威斯康星奥克来尔大学试用。该项目是教育部“新世纪网络课程建设工程”项目，即将正式出版。

暨南大学华文学院还组织编写了专门用于师资培训的教材。完成了国家汉办项目《汉语教学法研修教程》，已于 2004 年由人民教育出版社出版。2003 年出版了《华文教育心理学》。针对印度尼西亚特别国情策划编写的“印度尼西亚华文教师培训汉语基础教材项目”一套 6 本教材也已编写完毕，正在试用。

编写院内长期班或短期班使用的教材。2004 年，暨南大学华文学院自编的系列教材“快捷汉语”一套 8 本已经完成，正在出版中。国家汉办项目《对外汉语教学语法释疑 201 例》也已于 2004 年由商务印书馆出版。1999 年《最新实用汉语口语》由北京大学出版社出版，现已被韩国再版使用。此外学院还编写了多本用于院内短期班教学的教材，如《游学在中国》等。

2004 年 7 月“海外华文教材建设”项目获得暨南大学校级教学成果一等奖。

在教材编写方面，北京语言大学作出了突出的成绩，其教材不但在国内使用，有的也被国外所采纳；北京大学、复旦大学、华东师范大学都编写了不少教材。被海外采用较多的有《新实用汉语课本》《汉语会话 301 句》。北京华文学院编写的《汉语》是针对东南亚全日制中文学校的教材，难度大，反响良好。国家汉办立项之一的“乘风”是针对中学生的中美网络语言教学项目，由北京大学与美国方面合作开发；湖南师范大学与美国中文教学研究会合作编写了拼音教学简明教材《学习中文的钥匙》。

国内北京语言大学、华东师范大学、厦门大学开展了网络远程华文教育。

e）华文师资的培训、培养与外派

华文师资的培训和培养。2000 年以来暨南大学采取请进来、走出去以及开办学历教育等方式，在海外华文师资培训、培养方面做了很多开拓性工作，取得了显著成绩。2001 年，该校派出 3 位博士、硕士作为广东汉语专家团团长和主要成员赴印度尼西亚雅加达、泗水、棉兰、万隆 4 座大城市进行了近 3 个月的师资培训，接受培训的学员达1600名。这是印度尼西亚禁止华文长达32年之久后首次公开的师资培训，对于推动印度尼西亚华文教育产生了重大影响。2002年1月、2003年7月又先后派出3位讲师从事第二、第三期培训。此外还派出教授赴印度尼西亚、文莱等从事华文师资培训，发挥了重要作用。2000年以来，来暨南大学接受培训的海外华文师资共有12期，331人，分别来自印度尼西亚、马来西亚、澳大利亚、泰国和老挝。

适应东南亚华文教育的需要，暨南大学还将海外华文师资培训与学历教育结合起来，于2002年开办了面向印度尼西亚华人学习者（主要是华文教师）的汉语言专科函授教育，目前已招收两批学员，共计194人。该函授班采用自学与派教师前去面授辅导相结合的方式，效果良好。2002年起又率先与新加坡华夏管理学院合作培养华文教育方向硕士研究生班，现有3届共23名学员在读，第一批学员即将于2005年暑假毕业。

国内院校为外国培养学历华文师资多是零星进行的。只有印度尼西亚东爪哇华文教育统筹机构与中国福建省侨办、广东省侨办合作，分别由福建师范大学、广东幼儿师范专科学校代为培养华文本科师资或幼儿师资，其学生来自华族或友族，以合同制管理，学生学成回国后至少在华校或政府学校服务5年。他们的经验值得借鉴。与国外合作培养硕士人才的有华东师范大学与美国宾夕法尼亚大学联合举办对外汉语硕士班，招收对象主要是美国的华文教师[27]。厦门大学海外教育学院开办了面向海外的对外汉语教学硕士学位班，已有多名学员就读，自学与回厦门大学接受集中授课相结合。

外派华文师资和志愿者支持海外华文教学。为了解决海外华文师资紧缺困难，暨南大学还向美国、波兰、蒙古国、印度尼西亚、巴基斯坦等国派遣汉语教师，帮助当地大学进行汉语教学。2004年国侨办外派培训团10个。

2004年7月20日，暨南大学首批20名国际汉语教师志愿者前往印度尼西亚，执行中国国家汉办与印度尼西亚教育部中教司签订的合作协议。这20名志愿者被安排到印度尼西亚9个省20所国民中学，在其高一教授汉语课。这标志着印度尼西亚汉语教学正式进入国民教育系统。据调查，我国目前已向印度尼西亚（20名，暨南大学）、泰国（21名，云南师范大学）、菲律宾（18名，福建师范大学）、毛里求斯（5名，国家汉办从网上征集）等国派出了4批共64名志愿者。

f）合作办学和海外办学

除了暨南大学外，北京师范大学、南京师范大学、北京语言文化大学与美国加州州立大学长滩分校合作，内容之一是美国方面派学生到中国留学一两年。云南大学为缅甸举办了华文师资2年制大专函授班。南京大学与马来西亚韩江学院合作开办中文硕士课程，2004年已招生4届[24]。九江学院送中文等专业毕业生赴泰国清迈实习。美国俄克拉何马大学曾于2003年设立一个“中国之旅”项目，以提供奖金的方式派符合条件的汉语学习者到中国北大、复旦、西安外国语学院、云南师大进行为期一个月的语言学习及旅游访问。

中国有两所私立教育机构赴国外独立或合作办学。北京新东方教育科技集团2004年在加拿大多伦多市兴建了第一所中文学校，其宗旨是大力促进普通话中文教学，秉承优质、高效及寓教于乐的教学传统，竭诚为华人及其他社区服务，其校长刘濂曾20年全职任教于加拿大外交部语言学院和美国国务院外交语言学院，并相继在大学和教育局所属中文学校兼职达 18 年之久。中国侨网 2004 年 9 月 14 日报道，北京新亚研修院与印度尼西亚坤甸市 WIDYA DHARMA（维迪亚哈尔马）大学合作，派 4 位教师入住执教。坤甸为西加里曼丹省首府，这里与附近的山口洋市华人集中，而且保留潮州话、客家话，学生有良好的学习汉语基础。

g）海外孔子学院的承建

国家汉语桥工程项目之一的孔子学院建设分别由国内大学承建。孔子学院是一个非营利性的社会公益机构。孔子学院总部设在北京，具有法人地位，境外的孔子学院都是其分支机构，目前主要采用中外合作的形式来开办。它以孔子学院总部提供的教学模式、课程产品等作为主要教学资源，主要开展包括多媒体及网络在内的汉语教学，大、中、小学中文教师培训，实施汉语水平考试，与国内院校相衔接的中文学历教育，以及赴华留学咨询等内容。目前已建成美国的“马里兰大学孔子学院”“韩国汉城孔子学院”、肯尼亚的“内罗毕大学孔子学院”、瑞典的“北欧斯德哥尔摩孔子学院”、乌兹别克斯坦的“塔什干孔子学院”。其中拟建于印度尼西亚首都雅加达的孔子学院拟由暨南大学华文学院承建，越南孔子学院拟由云南师范大学承建。时机成熟时中国要在全球开办 100～500 所孔子学院。

3. 当前国外国内华文教育的特点和问题

特点和问题不是一回事，但海外华文教育的特点和问题密切相关，所以本节把两者放在一起讨论，重点是讲问题。我们的基本看法是海外华文教育因国情不同而呈现出明显的复杂多样性，相应地它所存在的问题也不能一概而论。

1）海外华文教育办学形式不一，学习人员多样化

a）办学形式不一，有走向正规化的趋势

海外华文教育办学形式大体上分为两类。

A 类：家教，周末班或周末学校、正规华校或国际学校，双语学校、三语学校。

B 类：即国民学校，包括幼儿园，小学、初中、高中，大学。

大体上来说，A 类形式以华裔子弟为主体[23]，教师也全部是华人，其中家教数量不好统计。在印度尼西亚，华文家教占有很大的比重，有 4000 多人。周末班在东南亚多称补习班，多数在下午和周末进行；在美国却普遍表现为周末中文学校。正规学校有不少是多年延续下来的华文全日制学校，得到了政府认可，专为华人子弟而开，如马来西亚的 1000 余所华文小学和 60 所独立中学。国际学校如德国有完全按照中国九年制教学体系使用中国普通小学教材进行教学的华文学校。三语学校如印度尼西亚 2004 年新成立的几所以印度尼西亚语、英语、华文进行教学的学校。B 类学校中的华文课基本上属于选修外语，大学则还有专门的华文专业、华文系。泰国正在筹建一所华文学院。美国不少正规中学和大学开设了中文课；东南亚有的国民学校华人子弟集中，从幼儿园到高中都开有华文选修课。印度尼西亚最大的一间国民学校棉兰苏东中学有学生 17 000 多人，绝大多数为没有华文基础的华人子弟，他们的中文教师大约有 120 名。

可见，不同的教学机构，华文教学的对象、任务悬殊。一般而言，全日制的华文学校教学最为容易，因为这样的学校有充分的母语教学环境，有大量的华文学习、使

用的时间；中学、大学自愿选修的学生（比如美国俄克拉何马大学经贸专业学生的辅修汉语生）学习有积极性，但一般课时太少、效果不理想；最令华文教师和华人家长头疼的是那些周末班、课后班的华人孩子，尤其是没有条件在家讲华语的孩子，他们学习既没有动力，又没有足够的时间练习，效果十分差，学生逃课和退学现象严重。

海外华文教育有个重要的问题就是走向正规，与国民教育接轨，以及进入国民教育系统。这方面美国作出了可贵的努力，其方法之一是争取将周末班的中文学习成绩转换为政府认可的学分。二是争取中文进入学业评价测验（Scholastic Assessment Test，SATⅡ），这是美国大学委员会委托教育测试服务社定期举办的世界性检测，作为美国各大学申请入学的参考条件。1994 年起中文能力已列入 SAT Ⅱ外语考试语种。三是 2003 年 12 月美国“大学汉语和中国文化预修课程及考试项目”（简称 AP 中文）正式启动。这标志着中国语言文化教学开始进入美国国民教育主体学校。根据计划，2006 年开始教授 AP 中文课程，2007 年正式举办考试[28]。这几项工作使美国华文教育有了更大的发展空间，使学生的学习积极性大为提高，同时也提高了对老师的资质要求和师资数量。前文说过印度尼西亚从 2004 年 7 月开始将华文教育正式列入国民教育系统，拟在 8000 所国民中学开设汉语。仅这一项，就需要 30 000 名华文教师。显而易见，华文教育融入主流社会是一个正确的方向，它非常有利于华文教育的生存和发展，同时也带来了华文教育师资、教材等的更大需求。

b）华文学习者多样，华文基础和学习动力差异大

一般而言，华裔学习者多为儿童，非华裔学习者多为成人，前者不少是非自愿的学习，后者则全是自愿学习的。由于历史原因，海外华裔华语基础及其家庭语言背景差异很大。

从表 19 可知，第 1 种即家长与孩子都会普通话的情况是最理想的，美国新移民的家庭多有这个优势；第 5 种，家长与孩子都既不会普通话也不会方言，新加坡等东南亚国家多有这种情况，印度尼西亚 1967 年华校关闭后出生的华人和他们的孩子就是这种情况；第 2 种适用于印度尼西亚坤甸、山口洋，马来西亚，缅甸（北部）等。对这类家庭中儿童的华文教育其实仅仅是母语教育或双语教育；第 3、第 4 种情况适用于华人家长没有精力照顾孩子、孩子由原住民保姆带的情况，印度尼西亚部分华人家庭就是如此。明白了这些区别，才能明白海外华文教育的生源差异。

表 19　华文学习者家长和孩子本人的华文基础

项目	家长		孩子	
	普通话	方言	普通话	方言
1	√	—	√	—
2	—	√	—	√
3	√	—	—	—
4	—	√	—	—
5	—	—	—	—

注：√表示会，—表示不会

据记载，英国举行了一种华文考试，华人子弟普遍觉得太容易，从而更放松了学习；同样是《中文》，美国孩子嫌难，泰国学生觉得太浅，印度尼西亚坤甸、山口洋的孩子肯定也会觉得太浅，原因是学生的华文基础和背景不同。澳大利亚在这一点上做得好，他们针对华裔与非华裔学生分别制定了不同的标准，这才是正确的出路。进一步要做的工作是：把华族孩子也分不同情况区别对待，开不同的课，用不同的教材，按不同的标准考试。

2）海外华文师资、教学管理人员普遍水平较低

海外华文师资水平不高、后继无人的报道近十年来屡见报端，根据我们的调查，至今无多大改变，其中欧美的华文教师素质不错，但大多数不是学中文的，更不是学师范的，而是搞理工科的，是家长义务代劳；东南亚的则是因为普遍的历史原因而造成人才断层，而华裔学生数量庞大。加上华文教育纷纷融入主流社会，进入国民教育系统，友族人士纷纷开始学华文（法国有 90%的华文学习者为非华族），华文师资紧缺的问题还会更加突出并持续多年；仅印度尼西亚国民教育系统就需要 3 万名华文教师。与此同时，懂华校管理的人才更是奇缺。这些都是制约海外华文教育发展的瓶颈，必须予以高度重视。

3）资金缺乏，校舍和教学设施落后

这个问题在东南亚比较普遍和严重。由于国家经济基础薄弱，又没有政府支持，不少华校校舍特别简陋。在欧美，无论是华校还是学生家里，计算机、网络等齐备，而在东南亚，电话都不能普及，不少家庭十分贫穷。与此相应的问题是开不出工资，聘不到老师。在东南亚不少国家，比如泰国、印度尼西亚，学华文的华裔青年不少，但愿意当老师的很少。在泰国，华文人才如果能去公司就职，可以挣到每月约 1.5 万泰铢，而如果当华文教师才能拿 8000 多泰铢。在印度尼西亚某些地方，虽然华文教师待遇很低，但华文教师每节课的课酬约等于原住民同节课酬的 10 倍。

在这样的国家，靠收学费支持华校生存并不能解决问题。而且，资金问题也影响到教材费用和图书资料建设费用。

4）华文教学的协调与管理机构有待加强或新建

目前美国有“全美中文学校协会”“全美中文学校联合总会”这样的全国性行业协会，马来西亚有全国性的华文教育领导机构董教总，其他国家多数没有这样的行业协会和管理机构。这不利于当地华文教育行业的内部交流，不利于华校机构与所在国政府的沟通，同时也给中国前去这些国家提供帮助造成不便。

5）国内政府和学校的问题

这些年无论国家还是地方政府、国务院侨办、国家汉办，侨办基地院校、国家汉办基地院校或国家汉办支持周边国家重点院校，都在华文教育方面做了很多工作，并且近两年支持力度明显加大，但我们认为还有诸多不足。

一是对复杂多样、迅猛升温的华文教育和汉语教学调查不够，研究不够。比如对海外华文教学机构、华文教育人物、华文教育资料没有详尽调查。由于没有详尽的调查分析和充分的估计，我们所做的一些工作针对性不强，我们所提供的帮助不够平衡，或者显得杯水车薪（特别是合格教师的支持、教材支持和图书支持方面）。由于资金等

条件的限制，学界所做的研究只能局限于来华学生课堂教学的部分探索，对海外华文教育的研究才刚刚开始摸底，对各种各样教学机构、学校群体缺乏深入研究，于是，我们所能提供的学术支持也十分有限。这一点如果与美国的 TESOL 协会所做的工作相比[29]，就会看得更加清楚。他们以专业、务实的精神开展工作，作出了非常大的成绩。具体来看，国家对华文教育科研投入太少。比如，一个跨国的研究项目教育部只资助 8000 元到两三万元人民币；一个投入 20 多人费时两年多、得了优秀奖的网络课程，教育部只立项不资助，经费自筹。

二是缺乏比较长远的统一规划和部署。目前国家级的规划只有孔子学院比较具体，其他的大多要求比较笼统。

三是国内与国外的信息沟通不畅。由于国外华人分布面十分广，又多数没有统一的组织，所以我们很难把有关信息在短期内传播到海外，加之国内网络化程度不高，很多信息在海外难以得到，比如网上查不到国家汉办认定的对外汉语教学基地名单。信息不畅是最近的第四届国际华文教育研讨会上呼声较高的一个方面。

四是对华文教育的管理还比较粗放，需要加强协调，增强管理和教学的专业化。表现之一是基本信息的统计整理和共享不够。例如，国务院侨办和国家汉办推出了那么多师资培训项目，但国家没有完整的（至少没有公布出来）记录，致使有的海外华文教师先后五次来华，参加不同系统的、不同院校的、中央的或地方的培训。

表现之二是基地评估等管理比较看重表面指标，忽视了国内华文教育的历史传统和特殊性。有些华文教育开展得很一般的内向型院校也跻身对外汉语基地，甚至人为造成为了争取基地一夜之间冒出多个新的海外教育学院、对外汉语学院的奇怪现象。

表现之三是对国内外华文教育课程设置、教学大纲特别是测试评估的忽视，从而导致了华文教学没有分门别类的质量标准和科学化的考试。要支持和发展华文教育，必须重视语言教学、双语教学、第二语言教学规律。目前所推行的汉语水平考试只是汉语水平测试，而不是学习成绩测试，更不是教学评估测试。对国内外华文机构教学水平的科学评价和效益管理并未提上议事日程。目前新东方和新亚学院都加入了国内或海外华文教育，相信他们的不俗作为会对政府支持和正规高校操作的华文教学形成质量效益挑战。

4. 推进华文教育国际化现代化的对策措施建议

华文教育本身就是国际现象，目前要做的工作是使它进一步国际化；华文教育的现代化主要指管理观念的更新，服务能力的专业化和服务手段的现代化。

根据前文的分析，并借鉴美国 TESOL 协会的经验，我们认为推动华文教育国际化现代化应采取如下具体对策措施建议。

1）华文教育国际化现代化的总体目标

更有针对性、更高质量、更高效率地满足海外华文教育和来华留学的要求，并开拓更大的海外市场。

2）海外华文教育的发展及华文师资的培训培养方向

海外华文教育的发展方向应当是正规化及融入所在国教育体系。这是形势发展的客观要求，也是能否保持华文教育生存发展的唯一方向。

海外华文教师培训培养的方向是职业化和本土化。因为不向职业化发展，质量无法保证，也不便于管理，不能保证把享受过国内国外培训机会的人士吸引到大量需要人才的正规教育机构中去，也不利于教师待遇的提高。众所周知，美国正规学校的老师（有执照）比没有进入这个系统的老师待遇和生活保障相差很大。泰国的正规教师享受文官待遇。

3）解决华文教育资金问题的方针

国内要以国家投入，基金会融资，公立、私立学校共同参与等方式推进国内外华文教育工作。我国高等教育不主张向产业化发展，但是华文教育部分可以采取政府投资、融资与产业化发展并举之路。要把钱花在最重要的地方：其一是东南亚、非洲等落后国家、地区，特别是华文教育需求极大而条件非常落后的印度尼西亚等国家。其二是优先投入到华文师资的培训和培养资助上。其三，对于西方发达国家和正规学校的学生教材可以免予赠送，但对东南亚的孩子则应千方百计满足他们基本的教材需求。

鼓励地方政府、民营教育机构共同为海外华文教育发展出钱出力。云南省 2004 年计划每年设 180 万奖学金奖励东南亚华文学习者来云南学习，其做法值得提倡。

华文教育要有成本效益观念，不管是国家、高校还是海外华校，都要核算成本，评估效益。条件较好的地方完全应该收学费，而且优质优价。

4）优先开展的几项具体工作

a）在深入调查研究的基础上制定中近期规划

包括科研攻关规划，国外华文师资培训、培养和支援规划，教材、教辅材料、教具、多媒体课件开发规划，培养或协助各不同国别、不同地区、不同族群、不同目的学习者的数量与质量规划，管理、服务水平发展规划等。

b）加大调查研究投入，优先资助对海外华文教育市场需求、发展方向和教学规律的调研

鼓励国内研究者与国外各级各类华文机构、华文教研人员及有关正规中小幼学校、大学人员合作研究，提高研究、开发的针对性和质量。目前国内研究人员普遍侧重于做对外汉语研究、汉语本题研究，而很少作出有价值的华文教育研究，原因之一是研究者缺乏研究海外华文教育的资金条件。一些重要研究项目如下。

（1）各国语言政策、文化风俗、跨文化交际问题研究。

（2）华文教育在各国融入主流社会的可能性及方法步骤研究。

（3）各海外华文教育机构、任务和华文教育资料搜集整理和数据库建立（既是管理、服务的需要，也能为更多的研究人员提供资料方便）。

（4）针对各级各类学校的华文教学体系、课程设置、教学目标、教学原则研究，制定多种有针对性的课程规划和考试大纲（马来西亚、印度尼西亚西加里曼丹省及缅甸北部的华文教育实为母语教育，但与国内的语文教学应有不同；泰国开设了双语文科班、双语理科班，它们的课程设置、教学原则方法以及考试等都应有不同的方案，法国拟自己编写汉语教学大纲，印度尼西亚华文教学与协调机构制定了自己的教学大纲，我们应主动参与研制和完善）。

(5) 海外华文教师培养、培养方案研究（分门别类的培训、培养方案。美国拟参考我国对外汉语教学资格认定办法编制华文教师认证标准，我们都应主动参与)。

(6) 针对不同族群、不同年龄、不同学习环境（学校、家庭）下学习者的学习特点、难点研究。

(7) 汉字教学及与繁简字有关的教学实践（克服“汉字难学论”，开发汉字学习速成方法和课件；有人说是十个美国汉语学习者六个怕汉字，所以他就不教汉字了；有人主张只认不写，或不写字而打字；菲律宾某华文教育机构负责人考察了上海实验小学集中识字的教学经验，得到了极大鼓舞)。

(8) 海外学生汉语学能测定研究（HSK 只在高等部分考学生的口头作文，而没有书面汉语水平测试，留学生能不能跟上全汉语课堂教学？是否具备记笔记能力？马来西亚独立中学学生入读中国大学免试中文；新加坡双语文化生也将如此；其他学校的学生水平会越来越高，如何处理有关水平认定问题？美国 content-based 教学、学习的实验已有多年，华文教育领域如何在学习其他专业课程内容的同时学习外语，应用这种双语教学方法?)。

(9) 学习、教学和课程、教学机构效率、绩效评估研究及测试软件的开发。

(10) 汉语与学生所在国语言（第一语言）对比研究，为教学和教材编写提供了支持。

(11) 海外华语研究，包括海外汉语普通话与方言的面貌、使用状况与发展变化，华文媒体语言，为提高华文教学质量、改善华文教学大环境服务。

c) 沟通协作

加强与海外华文教学协会、华文教学协会管理机构的联络，及时沟通信息，研究工作、协同行动。

d) 建立大型网站

5) 师资的培训培养和支援是重中之重，应当在这方面投入最大的资金，作最持久的努力

这样做原因有二。一是海外华文教师缺口太大。有的出现严重老龄化现象，青黄不接；普遍水准不够，不具备应有的资质，更难以进入正规教育系统；新增长的华文教师需求强劲。二是教师是华文教育存在和发展的关键，也是提高教学质量的关键。具体做法如下。

(1) 现有师资的培训。走出去的培训要依赖所在国家地区协调机构和华侨和华人社会团体，作极有针对性的培训；不要简单重复，变成扰民；重点培养愿意去正规学校服务的人士，而不是只为自己挣钱的家教；重点是学历教育，像暨南大学、厦门大学、华侨大学等所做的专科、本科、硕士学位教育那样。海外来华受训也要优先接受正规学校的汉语教师和愿意去正规学校服务的人士，不管他们是华人还是非华人，因为他们是华文教育的主力，可为所在国培养更多的学生，为进一步兴办本国师范教育打下良好基础。

(2) 来华接受学历教育和高校教师来华进修。要从愿意从事华文教育且基础较好的华裔和其他族裔的高中毕业生中挑选较大批量的人士来华读幼儿师范、华文教育或

对外汉语中专、大专、本科学校，并以合作管理的方式送他们回国为华文教育服务；学费以中国国家奖学金为主，有能力的海外华侨和华人社会团体也可以支持一部分。同时接受国外高校华文教师（多数是非华裔）来华进修、访学或攻读高一级学位。

（3）赴海外合作办学培养师资。国内华文教育和对外汉语基地单位特别是有华文教育的对外汉语硕士、博士授权点的大学要重点与海外有中文师范专业的大学合作，在当地培养师资，像华东师大与美国宾夕法尼亚大学的合作那样。这种努力主要以解决国外大学、中学华文师范教学水平的提升为目的，为华文教育在海外的继续发展打基础。

（4）大力培养储备教师以援外。国外每所有中文专业的大学至少应有一位较高学历职称和经验的人做教学和顾问工作；每所学生集中的正规中学也应有一名中国合格的中国华文教师执教。华文教育需求大的国家（地区）驻外使领馆应当配备至少一位懂华文教育的专职干部。要储备一批（上千名）合格的志愿者以满足各国持续增长的教师需求。

（5）加快推进海外华文教师资格认证。

（6）以项目管理的方式，鼓励国内教研人员与海外同行合作从事华文教材的修订、升级、新编，华文读物、教具和教学课件的编写开发，拓宽华文教育现代化之路。

目前的重点：一是优秀海外华文教材的升级和修改；二是编写更多的本土化华文教材；三是多样化通俗易懂有趣的华文或双语读物的编写出版；四是系列使用华文教具、挂图的开发；五是制作新的教学课件，推动多种华文课程上网，努力探索海内外合作进行网上教学的新路子，提高教学效率。应鼓励国内外华文教育工作者与书商合作，以产业化的手段生产大量适销对路的教材、读物、教具和多媒体产品，加快华文教育产业化步伐。

（7）大力捐赠华文图书。华文图书是教师教学与学生自学之必需，也是华文学习环境建设之重要部分，且费用不是很大，应在华人和华文教学集中的海外各种学校大力捐赠。

（8）“寻根之旅”夏令营活动需要改进。继续坚持“寻根之旅”夏令营活动，但形式可以改进。不一定搞大型的、统一时间的；不一定住宾馆，而可以探索“家住”模式，像美国俄克拉何马大学与清华大学那样，以及福建省侨办所创办的结对家住方法一样。

（9）加强质量效益监控。在科学研究的基础上对国内各高校的华文教育、汉语教学进行质量效益监控，以便提高质量、提高效益；鼓励先进，鞭策落后。对海外华文教育机构做一定测评，以便分门别类，作不同的支持和指导。

七、发展海外与港澳成人高等教育的对策

暨南大学是国务院侨办主管的重点综合性大学，是海内外知名的华侨最高学府。发展海外与港澳的成人高等教育，为海外与港澳培养人才，是暨南大学的一项重要任务。此项任务已卓有成效，不仅在办学上具备了相当的规模，而且探索和总结了办学经验，为暨南大学进一步发展海外与港澳成人高等教育创造了条件。

1. 办学概况

暨南大学于1985年开始在港澳开办成人高等教育，教学对象是港澳在职人员，为他们提高学历层次，主要是开办高中起点专科和专科起点本科。随着我国改革开放的继续深入，我国经济的腾飞和持续高速发展，世界上不少国家，尤其是东南亚各国正在掀起学习汉语的热潮，出现了“中国经商热”，粤港澳的经济合作和文化交流也不断加强，海外与港澳许多在职人员纷纷要求来暨南大学学习、深造，促进了暨南大学面向海外、面向港澳举办成人高等教育的进一步发展。20年来，暨南大学在海外与港澳相继开设了中国经济与管理、特区经济、外向型经济、口腔医学、中医骨伤、护理学、对外汉语、社会学、英语、中医学、计算机科学与技术、环境科学、会计学、法学、物流管理、应用心理学（犯罪心理学和社会心理学方向）等专业，共招生3482人，已有毕业生1463人，其中大专毕业生1220人，本科毕业生243人。暨南大学根据海外与港澳成人教育的特点开展教学活动，培养的学员们毕业后在工作中更上一层楼，譬如，1996届社会学专业本科毕业生大多成为澳门社团的骨干，有许多毕业生担任澳门重要的社会职务和社团职务。有一部分毕业生已成为港澳社会各界的精英。例如，香港临时立法会议员罗叔清、香港工会联合会会长郑耀棠，还有王如登、李泽添等都是香港特区的第十届全国人大代表。又如，澳门中华教育会理事长李沛霖和澳门街坊联合总会理事长姚鸿明等都是澳门特区的第十届全国政协委员，潘玉兰是澳门工会联合总会会长、澳门特区的第十届全国人大代表，梁庆庭是澳门立法会议员、澳门街坊联合总会副会长。

近几年抓住了机遇，在海外与港澳的办学有了长足的进步，在开设专业和制订培养方案时，都会考虑到海外与港澳的社会实际和发展需求，为满足港澳社会工作人员迫切需要开设社会学、中医学、会计学、法学等专业；为填补澳门护理行业高学历层次空白开设护理学专业；针对澳门市场急需人才开办物流管理、应用心理学等专业；为满足海外华侨华人学习中文的迫切需求，从2002年开始连续三年在印度尼西亚开办对外汉语专业。

目前，印度尼西亚、港澳办学规模首次突破1000人，在印度尼西亚、港澳的协办单位有印度尼西亚万隆福清同乡基金会、香港大学、香港专业进修高等学院、香港工会联合会业余进修中心、澳门业余进修中心、澳门暨育服务中心等；本校教学协办院系有社科部、商学系、环境工程系、法学系、中医系、外语学院和华文学院等；开办对外汉语、社会学、英语、中医学、计算机科学与技术、环境科学、会计、法学、物流管理、应用心理学、行政管理、知识产权等专业；在学学生1161人，其中印度尼西亚学生297人。

暨南大学在海外与港澳举办成人高等教育，坚持“侨校＋名校”的发展战略，在实践中探索、寻找规律，经过20年的努力，已形成一套具有特色、较为完善的办法。

（1）用足国家给予我校单列申报招生计划的优惠政策。由于面向海外、面向港澳办学的政策性很强，因此，我们要吃透上级有关文件精神，在海外与港澳每办一个班，招生计划都要获得国务院侨办和教育部的批准，根据港澳有关高等教育法例，及时准备大量的办学资料通过协办单位向当地政府职能部门申报，获得批准后才能招生，俗话说，万事开头难，招生计划的落实才是良好的开端。

（2）适应市场需求及时开设专业和调整课程。按市场机制运行，加强与海外、港澳协办单位沟通，要求他们关注市场热点，把握发展机遇，认真做好市场调查，利用综合性大学学科门类齐全的优势，打破院、系、专业的界限，瞄准市场需求多开专业，开新专业，增强我校在海外与港澳招收在职人员的吸引力，逐步扩大办学规模。要结合港澳、海外的需求调整课程，适当开设一些符合当地实情的课程，删除一些不适宜面向海外与港澳学生开设的课程，力求突出课程的先进性和适用性。例如，社会学专业本科班，开设社会服务行政管理、社区组织及社会规划等课程，这样，对提高在职从业人员的业务素质和专业技能起到较大的作用，因此深受学生和用人单位的欢迎。

（3）招生办法灵活多样。暨南大学在招生录取工作中，根据国务院侨办的指示精神，遵照教育部有关招生科目、招生对象的条件等规定，单独组织考试。做到既遵照有关规定，又符合海外与港澳的实际情况，有的专业，如香港会计学专业、法学专业，参照香港有些院校的做法，中学会考四门课程及格以上成绩的考生可申请免试入学；有的专业，如澳门经济管理专业，采取笔试的办法，考试科目为高中语文、数学和英语；有的专业，如澳门护理学专业，采取面试与笔试相结合的办法。按考试成绩经资格审查后进行筛选，从中择优录取，报广东省考试中心备案。

（4）在海外与港澳有得力的协办机构。为了照顾港澳在职人员的特点，我们尽量利用晚上、星期六及星期日安排在港澳集中授课，从组织生源、提供教学场地、教学设施，到安排教师食宿和管理学生等一系列工作都需要海外与港澳协办机构完成。因此，在海外与港澳寻找合作伙伴就显得特别重要。暨南大学与海外、港澳联合办学的方式多样化：①与港澳普通高校联合办学，如与香港大学联合举办社会学专业，招收港澳在职的社工人员，双方都承担教学任务；②与港澳成人教育管理机构联合办学，如与澳门业余进修中心、澳门暨育服务中心联合举办护理学、物流管理等专业，这种方式由暨南大学派教师到办学点授课，对方负责提供教学场地和设施，安排教师食宿；③与港澳成人教育院校合作，如与香港专业进修高等学院联合开办法学、会计学等专业；④接受港澳学术团体委托联合办学，如与香港中医骨伤学会联合举办中医学专业；⑤与海外群众团体，如印度尼西亚万隆福清同乡基金会联合开办对外汉语专业。

（5）按照“点”“面”结合的办法选派教师。“点”是指配备一支相对稳定的主干课程教师队伍，这批教师连续多个学期为港澳班上课，熟悉境外的学生特点，了解港澳教学内容和教学方法，建立他们的个人档案，保证他们在港澳授课的连续性，这方面的工作需要办学院系的紧密配合。“面”是指某些课程所选派的教师覆盖面广，让更多的教师，尤其是学术水平高、教学效果佳的年轻教师利用教学空隙进行调研，了解港澳社会现状。由于港澳高校拥有较丰富的教育资源，国际化程度高，有许多东西值得学习和借鉴，从而增长他们的见识。把上述两种情况结合起来，既保证了面向港澳办学的教学质量，又有利于培养、锻炼外向型的教师。

（6）保证任课教师顺利出境。近年来，根据广东省人民政府港澳事务办公室和广东省公安厅联合下发的文件精神，教师赴港澳授课要办理工作签证，同时由原来的因公审批办理证件调整为因公、因私审批办理证件，这是一项政策性和时效性强且很烦琐的工作，要保证教师顺利出境必须认真对待。

(7) 组织教师编写辅导资料。根据教育部规定，面授教学时数只占全日制高等学校同层次同专业授课总学时的30%左右，因此，教师必须专门为函授学生编写自学指导书，指出该门课程的重点、难点，补充新知识和新信息，布置思考题等，帮助学生理解和消化教学内容，对港澳学生来说，更是必不可少的，要把自学指导书连同教材及时发放到学生手中。

(8) 强化实践性教学环节。在切实加强基础理论教学的同时，充分重视加强实验、实习、社会调查等实践性教学环节，注重培养理论与实践相统一的人才。安排理工类如环境科学专业学生回校本部上实验课；安排法学专业学生到广东省高级人民法院进行旁听庭审活动；组织会计专业学生到广州会计师事务所实习；组织中医专业学生回内地中医院实习；组织新闻广告专业学生做社会调查等。

(9) 与协办单位、班委会齐抓共管。充分发挥协办单位管理人员的作用，同时发挥班委自我管理作用，检查学生出勤情况，抓好课堂纪律和考试纪律，反馈教师授课情况，不断改进教学工作，积极、主动地为教学为学生服务。

暨南大学在海外与港澳举办成人高等教育的成效和特色为港澳有关领导和人士所瞩目。中央人民政府驻香港、澳门联络办各级领导一贯重视暨南大学在港澳举办成人高等教育的相关事宜，有关领导经常亲自参加暨南大学在港澳举行的各种成人高等教育活动，港澳有关行业和协办机构也给予大力支持和密切配合，港澳新闻单位经常宣传报道暨南大学在港澳办学的状况及毕业生的业绩，从而扩大暨南大学在港澳的影响。

2. 存在问题

暨南大学在海外与港澳举办成人教育，是前人未做过的事业，无前人经验可借鉴，加上目前出国出境仍有诸多不便，使我们无法经常到那里进行调查研究，从而限制了办学针对性，影响了办学规模和速度。目前，在海外与港澳举办成人高等教育遇到的主要问题如下。

(1) 在海外与港澳开办新专业本科班，上级主管部门审批时间较长并加以限制。

(2) 申请学士学位的学生要参加广东省外语统考，而港澳学生不太适应这种外语应试方法，造成生源流失。

(3) 在香港举办成人教育的国内外院校很多，竞争激烈，退学人数较多，办学规模难以扩大。

(4) 在东南亚国家中只有印度尼西亚开展华文教育，教学函授点较多，每个教学函授点的学生人数偏少，教学成本偏高，向学生收取的学费太低。

3. 关于对策措施的建议

成人高等教育的对象广泛，办学形式多样，而且有针对性，社会需求量大，具有普通高等教育不能替代的作用和功能，因而有广阔的发展前景。海外与港澳成人高等教育也不例外。暨南大学必须面对教育国际化的竞争和挑战，引进境外先进的教育观念、管理经验和办学模式，面向海外、面向港澳培养具有国际竞争意识和竞争能力的专门人才。我们除了必须发挥暨南大学作为侨校的优势，用足一切有利客观条件外，还必须具有开拓创新意识，探索出一条切实可行的办学路子，一步一个脚印地向前走，为海外与港澳成人教育工作迈上新台阶而努力。为此，提出如下建议。

(1) 海外与港澳成人高等教育本科学生不再参加广东省学位外语统考。请求广东省教育厅同意暨南大学根据外招生的特点采用分流方式自行组织海外与港澳学生学位外语考试，其课程学习成绩、学士学位外国语考试成绩和毕业论文均达到普通本科生教学计划及申请学士学位的各项要求，经审核合格者可授予学士学位。

(2) 重视拓展海外成人高等教育。由于我国已加入 WTO，形成了全方位开放的格局，世界上不少国家，尤其是东南亚各国华侨、华人迫切需要学习中国法律、中国对外贸易、中国语言文学、中医学和针灸学等，我国政府可通过外交途径与有关国家尤其是东南亚各国在文化交流方面达成共识，力争从政策上、法律上允许暨南大学与有关国家尤其是东南亚各国高等院校、文化团体联合办学，从而为暨南大学开拓海外成人高等教育提供根本保障。同时，争取尽快在东南亚各国举办成人高等教育。

(3) 扩大暨南大学对外办学的权限范围。暨南大学成人高等教育要办出华侨大学的特色，上级主管部门就要制定配套政策和灵活措施，实行单列的招生计划，允许先招生后备案，允许开设本科新专业。

(4) 简化教师出国出境面授的审批手续，保证教师依时出境。

(5) 加大开展海外华文教育的力度，以满足有志于学习中国汉语言及文化的海外与港澳人士的要求。

(6) 建立并完善海外华文教育基金会，以解决部分教学经费问题。

八、珠海大学园区的发展及对外合作交流的对策

1. 珠海大学园区发展现状

珠海大学园区创办 5 年来，得到了国家、省、市、有关学校和部门的大力支持，先后引进了 9 所大学（其中 7 所已投入使用）和 5 个产学研基地。珠海普通高校在校生由 1998 年暨南大学一所大学的 100 人变为现在的 4.2 万人，大手笔地改写了珠海高等教育的历史，创造了中国高教发展史上的奇迹。作为一个城市，珠海拥有的大学生和高校数量在广东省位居第二，仅次于广州市。珠海已经成为广东省一个重要的高等教育基地。

1）珠海大学园区的特点

与国内其他大学城（园区）相比，珠海大学园区主要有以下特点。

(1) 以省内外优质高等教育资源为主。目前国内的一些大学城，如南京仙林大学城、上海松江大学城、广州大学城、宁波高教园区等，大多是本地高校或省内高校的延伸，而珠海大学园区的高校来自省内外，既有部属名牌大学，如中山大学、北京师范大学、暨南大学等，也有省属高校，如遵义医学院，并以名牌大学为主。

(2) 初步形成了规模适中、结构合理、层次较高的地方高等教育体系。目前广东省内的大学城（园区）各具特色。如深圳大学城以研究生教育为主，佛山大学园区以高等职业教育为主，东莞大学城以理工教育为主等。比较而言，珠海大学园区具有更强的多元性和综合性，学科覆盖面广，专业设置较为齐全。在类型上，普通高等教育

与职业高等教育并存；在层次结构上，重点院校与一般院校同在，以全日制本科教育为主，同时发展研究生教育；在布局上，按照珠海城市建设规划和产业发展规划，相对集中分布于金湾区和香洲区，有利于校地联动发展。从园区各高校的质量、办学规模、层次结构、专业覆盖范围来看，珠海大学园区在全国大学城（大学园区）中，具有特殊的代表性。

（3）突出体制创新，积极探索地方高等教育发展新模式。珠海大学园区从建立伊始，就把体制创新定为自身的鲜明特色，力争把珠海大学园区建设成为广东省重要的高等教育和产学研基地，成为全国引进大学创办大学园区的范例和我国高等教育对外开放和合作交流的窗口，努力探索区域性高等教育发展的新模式，其中包括多渠道、多元化的投资和融资模式，高等学校异地办学的管理模式，普通高校独立学院的发展模式。

珠海大学园区已形成多样化的校区管理格局。其中既有校本部延伸管理的校区型管理模式，如暨南大学珠海学院、中山大学珠海校区、遵义医学院珠海校区；也有主体搬迁到珠海的主体校区模式，如广东科学技术职业学院；也有独立运作、自主办学的独立学院模式，如北京师范大学珠海分校、北京理工大学珠海学院、珠海科技学院；还有以科技开发为主的高校产学研基地，如清华科技园、哈工大新经济资源开发港。珠海大学园区为各高校探索新的办学思想和管理模式提供了极佳的试验场所。

（4）独特的区位优势，成为各高校开展对外合作与交流的窗口。珠海与香港隔海相望，与澳门陆地相连，影响力辐射东南亚，具有独特的区位优势。许多大学选择珠海，正是看中这一点，希望借助这一优势，积极开展对外高等教育和科技的合作与交流。珠海大学园区日益成为我国高等教育对外合作与交流的一个重要窗口。

2）珠海大学园区在提高城市核心竞争力中的作用

围绕珠海经济结构和产业定位，引进国内著名大学在珠海设立校区（学院、分校）或产学研基地，推动科教事业和社会经济发展，是珠海创办大学园区的基本出发点。其作用主要表现是以下几方面。

（1）为经济发展提供强有力的智力支持和人才保障。珠海坚持大力发展高新技术产业，努力做大做强电子信息、电气机械、精密机械制造等现有支柱产业，重视发展石化、能源、装备等重型工业，加快现代服务业的发展。根据珠海社会经济发展的实际要求，每年需要引进各类专业人才 8000～10 000 人。但“十五”期间，实际每年仅引进了 3000 多人，人才的匮乏已成为制约经济发展的瓶颈。大学园区可以为珠海的经济发展提供强有力的智力支持和人才保障。

（2）搭建科技创新平台，推进珠海科技产业发展。大学与科技紧密相连，一流大学拥有一流的科研队伍和科技成果。珠海大学园区各高校正在发挥桥梁作用，把校本部的科研力量、科研机构、科研项目、科技产业引入珠海，逐步形成产学研一体化，为珠海搭建科技创新平台，推动珠海的科技进步和产业发展。目前，园区各高校结合珠海社会经济发展的需要，积极开办新专业、兴办与珠海产业发展密切相关的研究所，参与本地企业的科技攻关和产品研发。2003 年高校在珠海申报的科研项目达 40 项，2004 年达到 100 多项。珠海的科技创新海岸依托大学园区，聚集了近百家以软件研发为主的高新科技企业，正逐渐发展成为一个有较高成果孵化、转化和较强辐射带动能力的高新技术产业带。

(3) 提升珠海城市文化的水准，推动珠海“文化盛市”的建设。较之于人才、科技、经济方面的贡献，大学园区对城市文化发展的贡献，更是有深远影响和重大意义：一是大学园区各高校的建筑聘请国内外著名的设计师设计，融国内外风格于一体，具有国际化气息，体现了一种城市物态文化；二是大学是优秀文化的传播载体，有大学的地方就容易形成追求真理、创造知识的浓厚风气，形成讲道德、讲文明的良好社会规范，大学引领社会风气，改善市民结构，提高市民素质，赋予城市文化以鲜明的时代特色；三是大学所引导的开放性、多元性、创造性城市文化氛围、文化环境，将会为城市的进步与发展提供最重要的人文条件和永不枯竭的动力；四是校园文化是一个城市文化的重要组成部分，大学园区各高校为珠海提供一批高质量的文化场馆，比如体育场、图书馆等。特别是中央音乐学院落户珠海，将直接推动珠海“文化盛市”的建设。

(4) 加快珠海的城市化进程。大学园区的建设打破了珠海城市原有的格局，使珠海的城市布局更合理，城市功能更完整、更完善，大大延伸了城市的骨架，形成了现代化城市的新格局。大学园区已成为城市建设的一个新亮点，形成一个个新型社区。各高校的校园，过去是大片的荒地、洼地或烂尾楼，现在却是一座座建筑风格各异的花园式校区。珠海大学园区各高校的选址和布点，符合珠海城市建设规划和发展要求，完善了珠海城市功能。校园建设与珠海的城镇建设协调推进，与珠海的产业发展布局互相依托，加快了唐家湾镇和金湾区的城镇化进程和经济发展步伐。

(5) 提升珠海的国际化水平。珠海大学园区各高校注重走国际合作办学的道路，与国外及港澳地区开展多层次、多方位的教育和科技合作，这些交流与合作扩大了珠海的国际影响。

总之，与珠海大办科技工业园，大力发展高新技术产业的战略相比，珠海大办大学园区的战略，更侧重于对珠海可持续性发展的影响和对珠海未来的影响。因此，建设大学园区是一项具有巨大现实作用和长远历史意义的重大举措。

3) 珠海大学园区存在的一些问题

(1) 大学园区管理体制亟待理顺。大学园区既有部属高校，也有省属高校，既有延伸管理的校区，也有独立学院。大学园区面临着如何划分和理顺教育部、省教育厅和地方以及校本部对学校的管理责任和范围的问题。

(2) 资源共享方面有待于进一步完善。珠海大学园区在资源共享方面做了一些创新和探索，除了学分互认、教师互聘等资源共享之外，更强调软件的共享，比如设立珠海大学园区资源共享网，在珠海市人力资源中心网站设立大学园区专区等，促进高校与高校之间、高校与社会之间的资源共享。但是由于各高校设立的时间不长，在资源共享方面还有许多工作有待进一步完善。

2. 珠海大学园区对外合作交流情况

1) 珠海大学园区具有的独特优势

(1) 区位优势。珠海是我国最早设立的经济特区之一，毗邻港澳，依托珠三角经济发达地区，自然条件优越，环境优美。改革开放以来，珠海建成了较为完善的基础设施、四通八达的海陆空立体交通网络。港珠澳大桥建成后，珠海与香港的交通往来更加便捷。

（2）名牌大学的优势。珠海大学园区引进的高校以名牌大学为主，依托他们，可以更好地与港澳和国外大学开展合作与交流。

（3）办学机制的优势。珠海大学园区各高校都是新设立的，既具有校本部传统的优势，更具有办学机制的优势，在教师的聘用、学校的管理、人才的培养、专业的设置等方面具有灵活性，容易跟国际接轨，没有历史包袱。

2）取得的成效

（1）合作办学形式多样。比如北京师范大学已经与香港浸会大学签订合作协议书，由香港浸会大学投入资金，在北京师范大学珠海分校内设立国际学院，以本科教育为主，开设研究生教育，除了招收内地学生外，还招收港澳学生。清华科技园与加拿大祥达旅游学院开展非学历教育，学习酒店业、旅游服务业技能培训及外语培训等。

（2）留学生人数逐年增加。目前暨南大学珠海学院和中山大学珠海校区有2000多名的港澳台和东南亚学生。

（3）扩大了学生与教师的交流互访。目前珠海大学园区各高校都聘有国外或港澳的教师，各高校的教师和学生与港澳的教师和学生之间交往频繁。

（4）到国外办学。北京师范大学珠海分校计划到新加坡办学。

（5）产学研合作。比如哈尔滨工业大学新经济资源开发港与香港和国外的大学或公司合作，设立科技孵化基地，利用香港和国外在信息方面的优势承接订单，到孵化基地开发软件。

3. 关于对策措施的建议

（1）将珠海大学园区作为“中国高等教育特区”，与教育部共建，纳入教育部的管理范围。从园区各高校的质量、办学规模、层次结构、专业覆盖范围来看，珠海大学园区在全国大学城（园区）中，具有特殊的代表性，兴办最早，适合进行各类高等教育发展的探索，是设立高等教育特区的理想场所。

（2）将珠海大学园区作为我国高等教育对外合作与交流的试验田。希望赋予珠海大学园区一些政策，大胆进行对外合作与交流的探索。比如香港和澳门作为我国的特别行政区，又纳入泛珠三角的合作范围，港澳的高校与大学园区的高校合作不应适用《中华人民共和国中外合作办学管理条例》，希望给予更为宽松的政策，甚至在政策上视同为国内高校之间的合作。

从园区各高校的质量、办学规模、层次结构、专业覆盖范围来看，珠海大学园区在全国大学城（园区）中，具有代表性，适合进行各类高等教育发展的探索，是设立高等教育试验田的理想场所。

九、关于对策措施的若干建议

下面提出18条关于对策措施的建议。

1. 实行大开放、大交流、大合作，加快我国高等教育的大发展、大提高，使之迅速跻身于世界先进行列

把对外开放与合作交流作为今后一个时期我国教育改革与发展的重点，首先是作为“十一五”计划的重点。改革开放以来，我国国民经济建设蒸蒸日上，发展势头继续看好，教育事业也有了很大的发展，特别是 1998 年以来，高等教育可谓是突飞猛进，在规模位居世界第一而且具有较高水平的情况下，现在有条件、有实力走出国门、走向世界。通过高等教育的大开放、大交流、大合作，可以加快我国高等教育的大发展、大提高，迅速跻身于世界先进行列。

加强教育对外开放与合作交流的目的，在于落实科学发展观，加快推进我国教育事业的现代化与国际化，更好地实施科教兴国战略。

落实科学发展观，是教育改革与发展一次新的历史机遇。抢抓机遇，乘势而上，这是经济领域各行各业、各地区取得成功的一条重要经验。实现教育现代化的目的，是提高教育水平，使得教育更好地促进与国民经济发展和社会全面进步。教育国际化是促进我国教育现代化的重要手段，是实现我国教育现代化的“催化剂”“加速器”。

在推进我国教育事业现代化与国际化的同时，应该注意民族化、多样化、综合化。这里说的“民族化”是指具有中国特色，保存优秀的传统文化，培育与社会主义市场经济相适应的先进文化，体现我国改革开放的辉煌成就等。民族的、先进的、有特色的，才是世界的、一流的，否则，可能是随波逐流，跟在别人后边亦步亦趋，人家走弯路我们也走。所说的“多样化”是指教育思想、教学内容要有包容性，“海纳百川，有容乃大”，多多吸取外国的优秀文化，少一些“禁区”和“壁垒”。所说的“综合化”是指科学教育与人文教育并重，理论与实践并重，学生会动脑与会动手并重；单科性大学要向含有人文学科的多科性大学转变（但是，并不一定都要办成综合性大学）；要培养“完整的人”、全面发展的人，而不是有所偏废的“半个人”、畸形发展的人[30]。

2. 我国高等教育进一步发展的关键在于开放与搞活

只有开放，才能加快高等教育的国际化进程；搞活就是要建立适应国际教育市场的运行机制和行之有效的管理体制。

搞活的关键在于政府自身的合理定位与充分放权。应该积极贯彻“宏观调控，微观搞活”的方针，尽量放开，以“学校行为”为主，“政府行为”为辅。作为政府部门，教育部、教育厅的职能应该主要是：制定法规，提出办学的准入条件，提供信息与服务，组织检查与评估，其他的事情应该放权，放给大学校长。教育行政部门应该发挥“穿针引线”的作用，发挥类似于“招商引资”的作用，组织各种各样的“洽谈会”“信息发布会”“参观考察团”，也可以几所大学采取联合行动，“一致对外”，保障国际交流合作的有序性。

3. 教育的对外开放与合作交流要注意双向性、平等性、互利性

从双向性而言，既要鼓励出国留学，又要积极吸收外国人来华留学；既要重视积极引进优质教育资源，开展中外合作办学，又要重视积极到海外办学。相对而言，在当前，应该把我国公民出国留学和多种形式到海外办学作为重点。

要注意对等原则，坚持平等互利。在引进对方到我国办学的同时，应该向对方提

出我方到对方办学的对等要求，促进我国教育走出国门，走向世界。拒绝“单边主义”“霸权主义”“教育侵略”。

在国际合作的“价值链”中，要注意保护和提升我们自己的价值。要警惕国际“教育贩子”和投机倒把的“奸商”；还要避免我国地区之间、学校之间无序竞争、“削价竞争”、恶性竞争，让外国人坐收渔利。

教育对外开放与合作办学的重要内容之一是“按照国际惯例办事”“与国际接轨”。例如，学位的互相承认与对接，包括同等学位的互相承认与对接，以及我国内地目前所没有的“副学士”学位、“副博士”学位，如何恰当定位，给予承认和衔接，以及学分的互相承认、学制差异的弥合、知识产权问题等。

“与国际接轨”也具有双向性。应该积极宣传我国教育的伟大成绩与大好形势，这有两方面的作用：第一，鼓舞士气，提高民族自信心和自豪感，防止一些人有意无意夸大我国教育的落后面；第二，向世界展示我国教育的面貌，宣传中国的和平崛起，吸引更多的外国留学生到中国来。

我们认为：“与国际接轨”的双向性，外国人是不难接受的，难的倒是我们的某些同胞瞧不起自己的好东西。试看经济领域，外国人制造产品很注意适合中国人的习惯，外资企业很注重在中国的“本土化”，这值得我们学习。

4. 积极引进外国的优质教育资源，引进优秀的外国教育家来中国办教育，包括任教和当校长

对外合作办学是要引进优质教育资源而不是良莠不分。要注意对方学校的档次，必须有利于提高我方的教育质量和教育水平；应该讲究“门当户对”“眼睛向上”“攀高亲”，不能“低就”“下嫁”、损公肥私。

办好一所大学，关键在于校长与教授，而优秀的校长和教授可以在国内外招聘引进（尤其要重视从国外引进）。这是香港办大学的一条成功的经验，也是国际惯例。英国的著名大学已经引进我国学者担任校长（例如，复旦大学杨福家教授 2004 年 1 月已经赴英国担任诺丁汉大学第五任校长），我们应该有更大的气魄。

欧洲和其他西方国家的一些大学把具有国外留学和教学科研经历作为大学教授岗位聘任的重要条件之一。在德国如果一所大学的教授岗位出现空缺，将在全世界范围内公开招聘，实现了大学师资来源国际化。我国的一些大学也开始采用向国外招聘人才的举措，目前招聘到的主要是海归学者，还要下功夫招聘“洋教授”。尤其要积极引进欧洲国家的优质教育资源，促进我国高校与欧洲强校在职业教育、海洋研究等领域的合作与交流。

大学校长要努力成为出色的教育家，而不是企业家或者其他勉为其难的角色。

中外大学校长论坛应该继续积极举办，可以是全球的、多边的，也可以是双边的。

现在，我国很多大学都提出要办成世界一流大学，实际上离世界一流大学还差得比较远，甚至连一流大学是什么样子的、自己要办成什么样子的也不甚了了。建议各校（尤其是“211 工程”的学校）到外国去“找对象”——选择一所世界一流大学结成“姐妹学校”，多多向对方学习。此事需要各校自己采取主动，也需要教育部允许我国各校办学模式多样化，而不是千校一面。

5. 迅速而有计划地分批分期组织高校领导和骨干教师出国访问和进修，让他们开阔视野、增长才干

提这一条建议的理由是显而易见的。如果对外部世界不了解，不具备世界眼光，坐井观天夜郎自大，是很难开展国际化现代化建设的。多年来，教师有计划有步骤地出国访问甚少，这种现象现在应该迅速扭转了。

6. 积极到海外去办学，开办多种学校，大力推进教育输出

我们高兴地看到：在国外开办孔子学院已经取得实质性进展，国务院已经批准汉语桥工程，推广“乘风汉语”和“长城汉语”。

赴境外办学可以由近及远展开，重点先在受汉文化影响较大的国家（东南亚、东北亚）开展，逐步扩展到大洋洲、欧洲、美洲；重点先放在发展中国家，逐步扩大到西方发达国家；重点先放在外国的首都，逐步扩大到其他地方。

随着我国的综合实力和国际影响力的不断增强，国际地位日益提高，我国与世界各国在政治、经济、文化、教育等领域的交流越来越多，汉语的应用价值大大提升，“汉语热”“中国热”正在世界范围内悄然兴起。我国在继续积极向国外推出“孔子学院”的同时，可否考虑推出“孙文学院”（或“逸仙学院”），推出“小平学院”或“改革开放学院”（包括面向海外华人华侨，面向外国人）。

7. 在教育对外开放与合作交流中，应该发挥计划机制的优越性，应该与国家的侨务工作紧密结合

应该发挥计划机制的优越性，增加教育对外开放与合作交流的有序性，在国际价值链中保护和提升我国的利益。

计划机制不等于传统的计划经济体制，后者必须抛弃，而恰当地使计划机制与市场机制相结合，实行“宏观调控，微观搞活”——这是系统管理普遍适用的一条基本原理。

教育对外开放与合作交流应该分地区、针对不同的对象提出不同的侧重点。

教育对外开放与合作交流应该与国家的侨务工作紧密结合，争取海外华人社团的协助与支持，而且把海外华侨作为重要服务对象。应该看到：国民党一直是比较重视侨务工作的，他们的做法可以参考和借鉴，在侨务工作上，可否开展“国共合作”，形成合力?

8. 鼓励公派和自费出国留学，出国留学人员数量在“十一五”期间大幅度增长

改革开放以来，我国向国外派遣了数量较多的留学生和访问学者，其中，2004 年国家留学基金管理委员会共录取各类留学人员 3987 人。据报道，2005 年国家留学基金管理委员会将进一步扩大选派规模，公派留学生人数将比 2004 年增加近 1 倍，以多种资助方式在全国选拔各类出国留学人员 7245 人，提高层次成为 2005 年的主要努力方向[31]。鉴于我国出国留学的基数还较小，与我国的人口基数和经济发展水平不相适应，建议进一步扩大选派规模，在“十一五”期间大幅度增长。那么，到 2010 年，总的增量大约是 25 万人，减去回国的人数，净增 10 多万人。

与此同时，积极鼓励自费留学。如果能够翻两番甚至更多，也是值得高兴的。不要顾忌资金外流，这种外流是值得的。它是居民的教育投资，正当消费，比之于现在

大量存在的“黄色消费”“黑色消费”“愚昧消费”（大办婚丧嫁娶、封建迷信等），不可同日而语。其中有一个小留学生问题很引人关注，我们认为：一是要加以适当引导（而不是阻止），让家长和孩子增加理性思考，对坚持出国者，帮助他们选择好学校；二是对于在国外的小留学生要运用使领馆的力量和当地华人社会的力量加以保护。清朝末年曾经组织过幼童留学，清政府的做法是值得嘉许和借鉴的。

由于目前我们的基数并不大，以上两个翻两番（5 年增长 4 倍）甚至每年翻一番（每年增长 1 倍，5 年增长 5 倍），对于 13 亿人口的大国，恐怕还是嫌少不嫌多。

不要担心他们不回国。钱学森院士曾经现身说法地讲：他们好比是当年的钱学森，我钱学森不是也回来了吗？现在，留学回国人员（海归派）已经越来越多，这种趋势还会加强。即便一些人长期不回来，也没有什么坏处，相反，增加多少万海外华人华侨也是好的，相信他们绝大多数人会具有一颗中国心。

9. 加大吸引海外优秀学生来华留学的力度，在“十一五”期间大幅度增长

吸引海外留学生的总量需要大幅度增加。在“十一五”期间翻两番或者每年翻一番，恐怕也是可能的，而且是应该的。留学生的双向流动，应该是“大出大进”，才能有利于我国高等教育的国际化和现代化。

外国正在加大吸引中国留学生的力度，它们的做法值得我们借鉴。例如，英国、澳大利亚、新西兰等国家在 2005 年的政府奖学金项目中，都开设了专门面向国际留学生的项目，以期吸引更多的国际留学生，其中对中国留学生倾斜。苏格兰国际奖学金项目，全球 22 席中国占 9 席；中澳奖学金项目，亚洲名额的 2/3 以上给中国学生，针对研究生的奖学金项目为“澳大利亚长江奋斗奖学金”，主要面向在澳大利亚和亚洲地区的高等院校从事短期研究的研究生和博士后研究人员，每个课题的研究人员可获得总金额为 2.5 万澳元的奖学金；新西兰国际学生奖学金计划，目标锁定研究生，奖学金主要集中在信息技术、生物技术、语言等重点项目，其中很大一部分将颁发给中国学生。针对博士生的奖学金将达到每年 2 万美元，同时新西兰还鼓励博士留学生在新西兰的学校教授一些课程，获得工资报酬，奖学金申请办法将于不久后公布。

在积极吸收外国留学生的同时，积极推动教育高层互访，与有关国家建立稳定的工作磋商机制；继续做好学历学位互认工作，开拓“强强合作”的领域和范围。

10. 要为教育的改革开放设置必要的“防火墙”，但是又不必过于谨小慎微、缩手缩脚，要有足够的民族自信心

面对西方的教育侵略、腐朽文化，“防火墙”不可不设。但是，对于教育开放的负面影响不必过多顾虑和担忧，其理由如下。

（1）相信我们自己的“抵抗力”。

（2）相信中国的固有文化、传统文化是优秀的，相信孔夫子不会输给耶稣，相信大多数中国人不会被“西化”。事实上，基督教传入中国也已经有 200 多年，尽管有不少人信教，但是，中国仍然是孔夫子的影响最大；佛教传入中国已有 1000 多年，连佛教也实现了“中国化”。改革开放以来，西方思想和文化大量涌入，对我国有很大的冲击，但是并没有从根本上动摇我们的固有文化，相反，有利于发展我国的先进文化。中国留学生到美国、日本、欧洲，并没有被完全“西化”，他们学成回国或者仍然留在

外国，仍然保持一颗“中国心”“魂系祖国”。

（3）相信大多数来华留学的外国学生是友好的、善良的，对于中国不会怀有恶意，而且他们很多人会受到中国文化的影响，甚至被中国文化所同化。不妨对比一下满族入关之后被汉族同化的情形。犹太人在全世界都没有被同化，唯独在中国是例外。所以，如果说“和平演变”的话，谁演变谁？很可能是中国人演变别人。

章开沅教授说[32]：“西方传教士来华兴教办学，目的当然是‘（教）化中国’，亦即使中国‘基督（教）化’，但结果更为明显的却是自身的‘中国化’。”原来企图“化中国”，结果反而是“中国化”，充分说明了中国文化强大的生命力、同化力。我们要有足够的民族自信心。

11. 大力加强华文教育和对外汉语教育，并且注意对外汉语教育与华文教育的统筹安排

胡锦涛同志 2004 年 3 月在全国政协致公党、中国侨联组政协委员联组会议上，就进一步推动新时期海外华文教育工作做出了重要指示，表示在加大政府投入的同时建立华文教育基金会，动员社会力量支持华文教育事业。2004 年 9 月，中国华文教育基金会正式注册成立。中国华文教育基金会以“弘扬中华文化，促进华文教育事业发展，加强中外文化交流”为宗旨，业务主管单位是国务院侨办，理事单位包括中央统战部、中央对外宣传办公室、全国人大华侨委员会、外交部、国家发展改革委、教育部、财政部、文化部①、国家广播电视总局、国家新闻出版总署②、国务院侨办、全国政协港澳台侨委、中华全国归国华侨联合会、中国致公党中央委员会、国家语言文字工作委员会、暨南大学、华侨大学。大力开展海外华文教育，弘扬中华文化，有利于海外侨胞保持民族特性、增进与祖（籍）国的联系和感情，有利于在经济全球化进程中增强海外侨胞自身竞争力，有利于中国走向世界、世界了解中国，有利于促进祖国和平统一和世界的和平与发展。

根据实际情况，对外合作办学常常是“对外汉语教学”和“华文教育”打头阵，两者都应该积极发展，并且做好统筹安排。要编出好的教材，“寓教于文”即注重教学内容的知识性、故事性、趣味性，而不是标语口号、政治标签式的说教。

在华侨华人中开展的华文教育，不仅仅是一种语言教育，它还负有民族文化薪火相传的使命。华文教育的实质就是借助中华民族语言的推广、传承、弘扬中华文化，保持华侨华人的民族特性。开展华文教育，不能只限于语言功能的传授，要将语言文字的学习与文化的传承有机地结合起来，使受教育者在学习语言的过程中，了解、继承和发扬中华优秀传统文化。

20 世纪 60 年代台湾、20 世纪 90 年代大陆都有大批移居海外的“新移民”，不要让他们成为“化外之民”。我国到外国举办的学历学位教育，应该允许“新移民”及其

---

① 2018 年《国务院机构改革方案》提出不再保留文化部，将文化部、国家旅游局的职责整合，组建文化和旅游部，作为国务院组成部门。

② 2013 年《国务院机构改革和职能转变方案》将国家新闻出版总署、国家广播电影电视总局的职责整合，组建国家新闻出版广电总局。2018 年《国务院机构改革方案》不再保留国家新闻出版广电总局。

子弟上学，发扬“宏教泽而系侨情”（1906 年暨南学堂创办的初衷）的历史传统。

目前，华文教育和对外汉语教学分别属于两个部门主管，两项工作虽然有所区别，更多的是共同点，应该加强联系与合作，统筹有关事宜，形成强大的合力。

12. 充分发挥教育在“一国两制”和祖国统一大业中的作用，建议实施“港澳台万千百工程”和“港澳台千百十工程”，并且按照一个中国原则来调整宣传口径

人才决定未来，青年决定未来，教育决定未来，“百年大计，教育第一”。应该充分发挥教育在“一国两制”和祖国统一大业中的作用。

港澳已经回归多年，时至今日，不宜再把港澳事务作为“外事”，把与港澳的合作办学“视同于中外合作办学”。与港澳的合作办学应该去除许多“框框”，大力推进，借助于港澳的优势，推进内地学校的现代化与国际化。与港澳的合作办学是在“一国两制”架构下的国内不同地区的合作办学，可以更加开放与搞活。

台湾是中国的领土，与台湾的合作办学也不宜“视同于中外合作办学”，应该视同于与港澳的合作办学。教育在“一国两制”和祖国统一大业中可以发挥巨大作用，母校对于校友具有长久的、巨大的影响力和感召力，“同学情”“师生谊”可以化为凝聚力、“向心力”。现在，大陆与台湾有好几所同名学校，如清华大学、交通大学、中山大学、暨南大学（台湾有“国际暨南大学”，在南投，1992 年创办）、苏州大学/东吴大学，可以积极开展多种多样的对口交流，做统战工作。

可否开放教育工作者“港澳台自由行”？现在手续繁多，很费时间，往往耽误事情。中国人在中国的土地上行走还不大容易，层层关卡，何苦呢？

香港回归以来，还存在不少问题有待于妥善解决。台湾的情况更是令人担忧。2004 年 3 月台湾的“大选”尽管扑朔迷离，有一点是很清楚的：“台独”的势力在增长，尤其是在年轻人中很有市场。年轻人，尤其是受过高等教育的年轻人，在很大程度上决定台湾民意，必须重视对年轻人的工作，这是高等教育的十分重大的历史使命。

为了中华民族的统一大业，建议对港澳台实行大幅度的优惠政策：实施“港澳台万千百工程”和“港澳台千百十工程”。“港澳台万千百工程”是说：每年奖励一万名台湾来大陆学习的优秀本科大学生，一千名香港来内地学习的优秀本科大学生，一百名澳门来内地学习的优秀本科大学生，他们的学习与食宿费用全免——“万百千”与台港澳的大学生人数相比，百分比其实很小。“港澳台千百十工程”是说：每年奖励一千名台湾来大陆学习的优秀博士研究生，一百名香港来内地学习的优秀博士研究生，十名澳门来内地学习的优秀博士研究生，他们的学习与食宿费用全部减免。

根据参考文献[33]，2002 年的数据：全国普通高校生均教育经费支出 15 119.56 元，其中个人部分 5979.69 元；中央部门普通高校生均教育经费支出 23 884.75 元，其中个人部分 8956.33 元；博士生 1 个折 3 个，硕士生 1 个折 2 个，函授夜大学生 3 个折 1 个，来华留学生 1 个折 2.5 个（四舍五入）。根据这些数据进行简单计算可知：两个工程的总经费大约是 9 亿元，以今天的国家财力，这笔经费应该是可以筹措的。

如果吸引力还不够大，可以加大奖学金的资助范围和资助强度。

此外，可否设想：目前台湾有 100 多所大学，而大陆有 1500 多所普通大学[1]，大陆的大学与台湾的大学结成“姐妹学校”，可以是“多对一”，也可以“一对一”为主，

辅以“多对一”。内地的“211 工程大学”就有大约 100 所，它们应该发挥较大的作用。台湾的大学有些水平不是很高，尤其是历史比较短，大陆的水平较高、历史较长的大学与台湾的大学结成友好对子，一般而言，会受到后者的欢迎。

为了做好对港澳台的工作，建议在宣传口径上加以调整：凡属中国的统计资料，应该包括内地和港澳台。例如，报道 GDP 时，建议说“全中国的 GDP 是多少，其中内地多少，台湾多少，香港多少，澳门多少”；报道奥运会奖牌数时，建议说“全中国获得的奖牌是多少，其中大陆是多少，台湾是多少……”，总之，充分体现“世界上只有一个中国”的基本原则。

在华文教育和对外汉语教学方面积极开展海峡两岸的交流与合作，加强两岸华文教育工作者的合作，共同推动中华文化发扬光大。

海峡两岸都是中国人，都肩负着弘扬中华优秀文化的历史使命。长期以来，台湾地区的华文教育工作者，为华文教育作出了很大努力。他们长期为海外华校编印和赠送教材及图书资料，帮助他们改善办学条件，还招收优秀华裔学生到台湾地区学习、深造，为海外华校培养了大量师资。虽然海峡两岸在许多方面有分歧，但在帮助华侨华人在海外生存、发展，推动和促进中华文化在世界传播的问题上，海峡两岸还是有较多的共识和较强的互补性。

作为中华民族大家庭的组成部分，海峡两岸既然可以在经济、科技、文化、体育方面进行交流与合作，也完全可以在发展华文教育这个问题上加强合作与交流。两岸可以合作编写教材，合作培训师资，交流在华文教育方面取得的经验和心得，在语言、文字等学术领域内共同研讨，相互借鉴，求同存异，共同推动华文教育事业发扬光大。

13. 积极借鉴经济领域改革开放的成功经验，重点大学应该走在现代化国际化的前列

我国的经济改革取得了伟大的成功，经济建设取得了伟大的成就，教育改革比较滞后，现在，教育改革可以参考经济改革和经济发展的经验（当然不是照搬照抄）。在一定的意义上，政府办的学校不妨比作“国有企业（国营企业）”，民办学校不妨比作“乡镇企业”“民营企业”，合作办学不妨比作“三资企业”，那么，这些企业在改革与发展中的经验与教训都是可供学校参考的。

类似于制定产业政策，可以对于教育的不同领域、不同学科专业、不同地区的教育，制定不同的准入条件、审批条件，五年左右修订一次。

重点大学应该走在现代化国际化的前列，而不是故步自封。不要重走某些国企大厂的老路，不要重走某些名牌产品的老路，如中华牙膏、上海手表、永久/凤凰自行车等，现在“风光”如何？恐怕是“今非昔比”吧。

建议 4 和建议 12 为重点大学提了两个“一对一”、两种“姐妹学校”，合起来是 100 多对“三姐妹”，既是为重点大学增加了压力，也是为它们提供了机遇。

教育主管部门要解放思想，慎重而有序地开放国内高等教育市场，积极引进外资，兴办中外合资的以职业技术培训为主的各级教育。在引进外资办学的问题上，必须坚持以下原则：一是解放思想，大胆实践；二是先行试点，循序渐进，合理布局；三是坚持中外合资办学；四是加快与此相关的法律法规建设，确保高等教育自主权的完整性和中国文化传统的继承性，增强与世界文化教育多样性的交流与融合。

14. 继续坚持高等教育切实按照教育规律办事不动摇

这一条建议与建议 2、建议 14 以及后面的建议 16 并不矛盾，而是相辅相成的两个方面。如果只要一个方面，那是不完整的，难免会有偏颇。

坚持教育事业按照教育规律办事，这是教育部和各级教育行政部门必须恪尽职守的重大责任。我们高兴地看到，教育部领导同志郑重申明：从来没有赞成过“教育市场化”之类的提法。面对“教育市场化”的阵阵声浪，作为国家政府部门的教育部应该成为中流砥柱。

教育事业是公益性的，非营利性的。教育领域需要引进的是市场经济中的竞争机制，而不能简单化为“市场机制”或者“教育市场化”“教育商品化”。必须警惕某些人在这些貌似有理的口号下的“敛钱行为”，警惕他们借教育开放与合作办学之名“发教育财”，借洋人之手赚同胞的钱。必须坚持竞争机制与计划机制相结合。

“教育市场”一词与“教育市场化”有所不同，但是，也应该界定在一定的范围之内论及，而不宜大肆渲染，到处乱用。据报道，一些国家（如新加坡、新西兰、澳大利亚、加拿大）已经制定了“雄心勃勃的”“进军中国教育市场”的计划，这是我们要加以警惕的。而不是像有一些人那样盲目跟随，把“教育市场化”作为先进的世界潮流，把中国的教育市场交给西方国家操纵。

15. 类似改革开放之初搞经济特区一样，开辟“教育特区”——深化教育改革的试验区：可以是某些高校，如暨南大学；也可以是某些地区，如珠海大学园区

暨南大学的传统和特点适合于这样做。暨南大学的前身是 1906 年创办于南京的暨南学堂，“宏教泽而系侨情”。改革开放以来，坚持“面向海外、面向港澳台”的办学方针，暨南大学有了很大的发展与提高，在海内外享有很高的声誉。“华侨最高学府”暨南大学是中国拥有海外以及港澳台学生最多的大学。在校 22 000 名全日制学生中，共有来自世界五大洲 57 个国家和港澳台 3 个地区的学生 8966 人，数量居全国高校第一，其中，台湾学生 524 人，约占全国在大陆的台湾学生总数的 1/8，而来自海外及港澳台地区的研究生 741 人，约占全国总数的 1/4。每年报考暨南大学并被录取的海外及港澳台学生，均大于全国其他所有高校的总和。

暨南大学积极推进各项改革，努力实施“侨校＋名校”发展战略并且取得了显著的成绩，学校在全国高校中的声誉和地位迅速上升，目前是在前 50 名之内。暨南大学不但在广州有校本部，而且，在经济特区有深圳旅游学院和珠海学院，这是“本部办学的延伸”（周济部长 2004 年夏天来广东视察时对我校办学模式的充分肯定）。暨南大学作为我国高等教育改革开放的“教育特区”，作为加强对外开放与合作交流的重点试验学校，应该是很合适的。

珠海大学园区“异军突起”。在全国目前众多的“大学城/大学园区”之中具有特色，可谓独树一帜。目前已经有来自全国各地的十多所大学，大部分是重点大学。其中，有综合性大学也有单科性大学，有理工科大学也有艺术类大学，有普通高校也有职业技术学校，有的学校以办学为主也有的学校以创办科技产业为主，争奇斗艳，各有特色，生机勃勃。珠海的高等教育从零开始，仅仅五六年时间就跃居广东省第二位，这是很不容易的。除了几个直辖市以外，在一个市域之内有如此丰富多样的高等教育，

在全国恐怕是数一数二的。建议如同改革开放初期创办经济特区一样，把珠海市办成教育特区。

珠海的地位与形势，珠海大学园区的结构与特点，适合于这样做。例如，在北京不好办的事情，不妨到珠海来办，在本部不好办的事情，不妨到珠海校区（分校）来办；办好了，探出一条路子，万一办不好，影响也不会太大。

在教育特区里，可以进行多种试验，如民办教育，独立学院和其他公办新机制新模式，引进外资办学，股份制办学，以及探索“异地办学”的经验与教训等。

16. 继续大幅度增加对教育的投入

首先是政府加大投入，同时，鼓励民间资金投入，其次，吸收外资投入。

最近几年来，教育经费增长较多，但是仍然不能满足教育事业提高质量和持续发展两个方面的需求。当前，我国财政性教育经费占 GDP 的比例仍然偏低，极大地制约了科教兴国战略和人才强国战略的实施。据统计，2003 年国家财政性教育经费占 GDP 的比例为 3.28%，比 2002 年的 3.32%下降了 0.04 个百分点，这种下降是 1995 年以来的第一次，必须引起我们高度关注[3]。

台湾地区 2003 年“公共教育经费”占“本地居民生产总值”的比例为 5.9%。[1]

现在我国民间资金很多，民营企业家集聚了大量的资本要寻找出路，建议制定相应的政策，鼓励民间资金投入（捐赠或投资），对他们给予一定的回报。经济领域改革开放的成功因素之一是积极引入了大量的外资，教育领域也可以考虑引进外资。

世界银行下属的国际金融公司在发展中国家已向 11 个民办教育机构提供 4400 万美元的贷款，但是在中国却因为限制太多，无法办理。

改革开放以来，我们已引进了这么多外资来振兴经济，现在，应该积极考虑引进外资来振兴教育。

17. 积极开展高教研究，包括国际对比研究、历史对比研究和对于教会大学的研究

各校都应该有高教研究所（或者高教研究中心），切实开展研究工作，为本校的发展出谋划策，也为国家的高等教育提供建议。

高教研究要重视国际对比研究，研究国际一流大学的办学经验。

高教研究要重视对我国高教的历史对比研究，总结经验和教训，特别是北大、清华、西南联大、哈军工等校。建议开展教育史研究、校史研究、教育家和著名校长研究。

教会大学的出现是中国历史上的一种重要现象，也是目前世界高等教育的一种现实。从技术角度，教会大学有没有可取之处？能不能从中得到一些警戒和借鉴？章开沅教授说：“我曾经参观过香港、台湾地区多所基督教大学，也曾经参观过欧美、东亚许多国家素有基督教背景的著名大学，觉得这些学校至今仍然可以作为我们高教工作，特别是作为学生人格教育的重要参考。”[32]此外，毋庸讳言，改革开放以来，一些以前具有宗教背景的学校实际上已经部分地得到了恢复。例如，中山大学获准成立了“岭南（大学）学院”，北京大学设立了“燕京学堂”，南京师范大学设立了“金陵女子学院”等；而且，随着香港的回归，香港岭南大学、香港浸会大学更是不容回避的现实。所以，在警惕教育侵略的同时，对于昔日的教会大学，可否从历史的角度，以开放的心态重新作一些研究？

18. 澄清并克服“专升本”误区，大力发展职业技术教育

这一条建议似乎与教育对外开放有一点距离，其实，也可以从国际对比中得到借鉴。因为西方发达国家大多很重视职业技术教育，尤其是高等职业技术教育。

我们认为，“专升本”的提法本身在理论上就是一个误区。大专与本科都是高等教育，是同一个层次上的不同培养模式，培养不同类型的高级专门人才，不存在什么高低之分和升不升的问题。这个误区不利于发展我国的职业技术教育，尤其是高等职业技术教育，不利于培养高技能人才。

现在国内大学都是一个模式，通通向着哈佛、剑桥的通才教育、精英教育看齐，从教育模式到专业目录都是“千校一面”，造成了人才的严重同构化，很难满足就业市场各种层次、类型的需要。

高技能人才紧缺已成全国性问题，据国家劳动部门统计，在我国 7000 多万技术工人中，高技能人才仅占 4%，远低于发达国家 35%的水平。专家认为，高技能人才短缺已成为“严重影响我国经济持续健康发展”的瓶颈。近年来，一些地方的“民工潮”已经开始变成了“民工荒”，其实是“技能型人才荒”，应该引起教育部门的高度重视。

有识之士已经指出，要重视培养高技能人才，高校也要出能工巧匠。现在，在普通高校大学毕业生就业难的同时，有一技之长的高职毕业生却很紧俏，很多企业以 6000～8000 元的月薪都招不到人。据介绍，深圳职业技术学院的毕业生就业率连年在 97%以上。深圳每年需增加高技能人才 7400 人，而培养和引进的高技能人才仅有 1100 人，缺口相当大。

为了克服这一误区，应该对专科教育作政策倾斜，加大技能型人才培养力度。具体建议有两点。

(1) 把义务教育的范围扩大到职业技术教育。针对多年来我国教育界不重视职业技术教育以及当前我国职业技术人才奇缺的局面，把义务教育（或者“半义务教育”，即部分地减免他们的学费，并且补贴部分生活费）扩大到中技、中专和大专教育范围，严格控制“中专学校升大专”“大专学校升本科”的一窝蜂现象。

(2) 把专科教育与研究生培养直接挂钩。现在，专科毕业一般是不能报考研究生的，唯独 MBA 有所例外。可否类似于 MBA 的做法（也是有种种约束条件的，其中有些是必要的，有些则要适当放宽），让专科毕业生也能报考工程硕士和其他若干专业学位的研究生？这样，让专科毕业生的前途“上不封顶”，才能够对他们及其家长起到激励作用。

参考文献

[1] 中华人民共和国国家统计局. 中国统计年鉴—2004. 北京：中国统计出版社，2004.

[2] 教育部. 2004 年中国教育改革与发展取得新进展. 2004-12-19.

[3] 周济. 用科学发展观统领教育工作全局. 2004-12-19.

[4] 教育部. 教育部 2005 年工作要点. 2005-01-04.

[5] 教育部. 我国 2005 年公派留学重点资助七大学科. 2004-12-21.

[6] 吴岩. 从教育承诺看 WTO 的影响. 北京市教育科学研究院报告，2000.

[7] 中国教育年鉴编辑部. 中国教育年鉴 2003. 北京：人民教育出版社，2003.

[8] 刘权，董英华. 中国赴美留学生专业分布及其留美倾向分析. 暨南高教研究，2003，(1)：112-115.

[9] 中国教育年鉴编辑部. 中国教育年鉴 2002. 北京：人民教育出版社，2002.

[10] 北京市教育委员会. 2002 年北京教育年鉴. 北京：开明出版社，2002.

[11] 演杨. 当代自费留学潮成因中的经济因素分析. 三门峡职业技术学院学报，2002，1 (2)：9-12.

[12] 陈学飞. 留学教育的成本与收益：我国改革开放以来公派留学效益研究. 北京：教育科学出版社，2003.

[13] 中国教育年鉴编辑部. 中国教育年鉴 2000. 北京：人民教育出版社，2000.

[14] 曾庆红. 充分发挥广大留学人才在全面建设小康社会中的独特历史作用：在全国留学回国人员先进个人和先进工作单位表彰大会上的讲话. 神州学人，2003，(12)：3-8.

[15] 李红波. "出国留学潮"的经济学分析. 辽宁教育研究，2004，(2)：41-43.

[16] 孙清忠. 近年来国外华文教育的发展状况概述. 第四届国际华文教育研讨会，广州，2004-08-17.

[17] 石慧敏. 关于加强美国中文教师培训的一点思考：从美国的中文教学现状谈起. 国外汉语教学动态，2004，(3)：38-41.

[18] 李功赋. 校际中文竞赛对促进中文教育的作用. 第四届国际华文教育研讨会，广州，2004-08-17.

[19] 桂明超，符红云. 美国俄克拉何马大学汉语项目成功因素分析. 国外汉语教学动态，2003，(4)：39-44.

[20] 曹纬. 法国南方华人协会. 第四届国际华文教育研讨会，广州，2004-08-17.

[21] 黄磊. 澳大利亚华文教育概况. 华大国际华文教育研讨会，泉州，2003.

[22] 陈汉龙. 印尼西加地区华文教育现状调查报告. 第四届国际华文教育研讨会，广州，2004-08-17.

[23] 宗世海，李静. 印尼华文教育的现状、问题及对策. 暨南大学华文学院学报，2004，(3)：1-13，15.

[24] 李祖清. 由缅甸的现状看世界华文教育的趋向. 第四届国际华文教育研讨会，广州，2004-08-17.

[25] 张苇. "华文精英"计划延伸的文化思考. 华声报，2004-08-17.

[26] 贾益民. 华文教育学学科建设刍议：再论华文教育学是一门科学. 暨南学报（哲学社会科学），1998，20 (4)：46-53.

[27] 华霄颖. 海外华文师资的地区特点及培训模式初探. 第四届国际华文教育研讨会，广州，2004-08-17.

[28] 李竞芬. "中文学校"成为"社区资源中心"的理想. 第四届国际华文教育研讨会，广州，2004-08-17.

[29] 宗世海，刘晓露. 他山之石，可以攻玉：美国的第 38 届 TESOL 年会综述. 暨南大学华文学院学报，2004，(4)：71-76.

[30] 杨叔子. 科学与人文的融合. 科学报告厅　科学之美，北京：中国青年出版社，2002：259-288.

[31] 国家留学基金委员会. 国家留学基金委第九次全委会提出明年公派留学扩大规模提高层次. 中国教育报，2004-12-21.

[32] 章开沅. 基督宗教与中国大学教育. 序言. 北京：中国社会科学出版社，2003.

关于高考的一点浅见[①]

目前我国实行一年一度的普通高等学校招生全国统一考试制度，其对中国大众的影响之巨已是人所共知。在一定时期内，高考仍然不失为选拔人才的一种较好方式。作为一种客观存在，急需我们研究解决的是如何使高考制度进一步完善，从而实现内容科学、形式灵活、录取公平的高考，将学生引向身心健康发展的正确方向上，努力发挥出高考的积极导向作用。为此，我提出以下两点建议。

一、采取由国家统一命题、由教育部和各省市协商招生数量和划定录取分数线的做法

目前，已经进行的高考内容改革主要体现在命题方面，诸多省市可以自行命题。由此产生的多套命题班子的运作，多种高考试卷的印制，不但耗费了大量的人力物力，造成了成本的增加和监控难度的加大，而且高考试卷难度系数的不同，使高考分数难以反映省市间的教育水平差异，从而导致国家宏观管理的困难。教育发达的地方，如北京、上海等地，院校多，名额多，考分低；教育不发达的地方，如西藏、新疆、青海、内蒙古等地区，教育水平较低，考分低；其他省份的考分则普遍较高。考核标准的不统一，导致国家教育部门对各地的基础教育状况难以有准确的了解和把握，不利于对基础教育的宏观把握和指导。

实行由国家统一命题、由教育部和各省市协商招生数量和划定录取分数线的做法，不但可以保证试卷对考生的原始公正性，保证试卷考核内容与教学大纲的一致性，而且可以兼顾不同省市的教育水平，保证本地域有一定数量的学生被高校录取。由于试卷一致，对各省市基础教育水平状况的衡量便有了一个客观标尺，从而使作为教学成效检测手段的成绩一目了然，更加具有客观性与可比性。一方面，国家教育管理部门可以借助高考这种形式了解国家各地域的教育水平和教育质量，从而更好地发挥高考的导向作用；另一方面，也可以根据考试成绩找出教育基础薄弱的地区，有针对性地加大指导和扶植力度，从而逐步缩小地域间的教育水平差距，提高我国基础教育的整体水平。

二、采取全国统一考试科目，只考语文、数学、外语、综合四门课程的做法

1999 年，教育部印发《关于进一步深化普通高等学校招生考试制度改革的意见》，启动了新一轮的高考改革，也就是“3＋X”科目设置方案。“3”指语文、数学、外语，“X”指由高等学校从高中科目中（包括综合）自行确定一科或几科考试科目。1999 年，广东省率先实行了“3＋X”科目设置方案，2000 年推广到 5 个省市，2001 年扩大到 13 个省市，2002 年全国各省市都实行了“3＋X”科目设置方案，全面进入了新一轮高考科目设置改革。

---

① 本文原载《科学中国人》，2005，(9)：13.

在各省市实行的“3＋X”高考科目设置方案归纳起来主要有三种：“3＋文科综合/理科综合”“3＋不同专业的考试科目要求”“3＋大综合＋学生自选科目”。上述种种“3＋X”方案的设置，要么是提早进行文理分科，不利于学生形成全面的知识结构，在一定程度上影响了学生的素质教育；要么是选考的学科数目太多，变相地增加了学生的学习负担，不利于创新能力的培养，与呼声渐高的学生减负相背离。

有鉴于此，我建议采取只考语文、数学、外语、综合（历史、地理、物理、化学、生物）四科的做法，在考试科目数量和考试内容方面做一理性的规定。因为从知识结构方面讲，考试科目数量只有设置适当，才能反映学生的知识水平。同时，应该利用较少的考试内容来达到高等学校在选拔新生方面对相关能力的要求。考试科目的改革和考试内容的改革是紧密关联的，只有将两者全盘考虑、整体进行，才是完整意义上的改革，才有可能比较全面地做到有助于中学推进素质教育，有助于我国基础教育健康发展。

认真做好高校力学教学指导工作[①]

首先，我代表新一届力学教学指导委员会，向出席 2006—2010 年教育部高等学校力学教学指导委员会第一次全体会议的各位委员和领导表示热烈的欢迎和衷心的感谢！

2006—2010 年教育部高等学校力学教学指导委员会于今年 3 月正式成立，共有委员 64 人，保留了上届委员 22 人。力学教学指导委员会下设“力学类专业教学指导分委员会”，有委员 26 人，主任委员是北京大学苏先樾教授；“力学基础课程教学指导分委员会”，有委员 38 人，主任委员是上海交通大学洪嘉振教授。

本次大会的议程和任务主要有四个方面：①向新一届力学教学指导委员会委员颁发聘书；②听取和审议伍小平院士、何世平教授、洪嘉振教授代表上一届力学教学指导委员会所做的工作报告，听取相关工作经验介绍；③听取和审议由我代表新一届力学教学指导委员会所做的本届工作设想；④研讨两个分委会今后五年的工作安排，并交流各分委会工作经验和教改经验。

在今年 3 月教育部高等学校有关科类教学指导委员会成立大会上，周济部长指出，“十一五”是我国社会主义现代化建设承前启后的重要时期，要站在科学发展观的战略高度，准确把握新时期高等教育发展的历史任务。一是必须把高等教育工作重心放在更加注重提高质量上来；二是进一步提升高等学校的科学研究和社会服务水平；三是扎实推进高等学校的各项建设，以加强建设增强实力，促进发展。

周济部长更期望新一届教学指导委员会要努力成为教育部宏观管理教学工作的依靠力量、指导高校教学工作和教学改革的骨干力量、实现高校规范教学管理的推动力量、加强师资队伍建设的引领力量，希望新一届教学指导委员会委员自觉成为致力于教学工作、投身教学改革、加强自身学习、倡导优良学风的带头人。

我们有充分的理由相信，通过本次大会和各位委员的共同努力，力学教学指导委员会一定能在“十一五”期间，在力学类专业发展战略研究、力学专业规范制定、国内外力学教育状况系列调研、力学教师培训、教材建设以及力学类专业教学改革经验交流等方面做出应有的贡献，推动全国高等学校力学学科的全面发展。

① 本文是 2006—2010 年教育部高等学校力学教学指导委员会第一次会议开幕词，广州，2006 年 7 月 25 日；教育与科技管理研究，北京：科学出版社，2016，134-135.

我国力学专业教育现状与思考①

“力学课程报告论坛”在全国高等学校教学研究中心、全国高等学校教学研究会、教育部高等学校力学学科教学指导委员会、中国力学学会教育工作委员会和高等教育出版社的发起和组织下，在大连理工大学的支持下，顺利召开。我们有理由相信，“力学课程报告论坛”将对提高全国力学课程教学质量起到积极的推动作用，我代表教育部高等学校力学学科教学指导委员会就我国力学专业教育现状与思考谈一些看法。

一、目前高等教育面临的任务与挑战

目前，我国高等教育在学总人数超过了2300万人，规模位居世界首位，毛入学率达到21%，在一个较短的时间内实现了历史性跨越，进入了国际公认的大众化发展阶段。“十五”期间，高等教育教学改革不断深化，人才培养质量稳步提高，科学研究水平全面提升，社会服务能力显著提高，国际合作交流日益广泛，国际地位明显提高，各项改革取得突破性进展，为各行各业输送毕业生1397万人，高等教育迎来了生机勃勃的崭新局面。但是，高校人才培养面临不少困难，存在许多薄弱环节，深化改革的任务相当艰巨。

在2006年4月2006—2010年教育部高等学校教学指导委员会成立大会上，周济部长强调“十一五”是我国社会主义现代化建设承前启后的重要时期，要站在科学发展观的战略高度，准确把握新时期高等育发展的历史任务。

根据教育部的总体要求和目前高等学校面临的历史机遇与挑战，我们认为要以科学发展观统领高校教学工作，必须紧紧抓住高等教育质量这一生命线。育人是高等学校的根本任务。培养德智体美全面发展的一代新人，必须要充分发挥教学的主渠道作用，切实提高教学质量。必须加大教学投入，强化教学管理。要加强学风建设，营造良好育人环境；要加强教学评估，完善质量保障体系，这是保证教学质量行之有效的手段，今后必须坚定不移地开展下去；要加强教师队伍建设，深化教学改革。要以培养学生的创新精神和实践能力为重点，不断深化人才培养模式、课程体系、教学内容和教学方法的改革，推进教学改革向纵深发展。

在北京召开的教育部各类教学指导委员会主任委员会上，与会者认为，对于高等学校教学工作应该在六个方面“进一步重视和加强”：一是进一步重视和加强高等学校育人根本任务的实施；二是进一步重视和加强本科教学在学校工作中的地位；三是进一步重视和加强素质教育：四是进一步重视和加强学生思想道德和人文修养的教育；五是进一步重视和加强学生实践能力和创新能力的培养；六是进一步重视和加强国家优秀教学成果、精品课程以及各种教学改革成果的推广和应用。

---

① 本文原载《中国大学教学》，2007，(1)：32-34.

二、我国力学教育的现状

力学学科是历史悠久而又充满活力不断发展着的学科。力学发展的活水源头一共有三个：生产与工业的需求，同其他基础学科的渗透，以及力学内在发展的矛盾提出的新课题。时代不同了，力学的研究内容、手段也在变化。从近 20 年的趋势来看，两个特点必须认识到，一是计算机科学和力学点结合，二是非线性力学提到突出的地位。

力学人才，来自高等学校力学专业。2003 年，据高等学校理工科教学指导委员会统计，我国理学类理论与应用力学专业点 17 个，工学类的工程力学专业点 64 个，工程结构分析专业点 2 个，力学专业总数达 8 个。我国高等学校力学专业曾经历过辉煌，也面临过困境，20 世纪 80 年代中后期至 90 年代渐渐被冷落了，这种冷落是全社会对力学淡忘的反映。它反映在优秀学生不报考力学专业；反映在一部分力学专业纷纷报名换招牌；反映在力学学生专业分配不吃香；反映在力学家中部分人认为力学不需要单独办专业；等等。它是整个理科教育衰落的一个侧面。

中国力学教育的特点是，许多大学都办力学系，但数理基础教育的质量近年有所下滑。我们的学生数学基础比较薄弱，在其他课程如物理学基础和能力培养上也存在很大差距。我国高等学校在校学生超过 500 万人，其中需要每年将力学课作为基础课的理工学生近 50 万人。

关于我国力学专业的教学质量的评估，就扩招前已设置力学专业的 39 所高校而言，已经建立了一个基本的质量保障体系；但是 1999 年扩招以后，大部分新建专业还很难说能保证力学人才的教学质量。20 世纪 80 年代以来，力学专业历届指导委员会建立了一整套力学人才的培养目标、教学计划，数理基础与力学主干课程的设置、培养学生实践和创新能力的教学环节等逐渐规范并完善，全国力学教育界各主流学校形成共识，还制定了要求明确、简便易行的专业评估方法，并在二十多所学校的工程力学专业中进行了三次评估，在课程设置方面，确定 7 门基本的力学课程为主干课程，制定了课程的基本要求和大纲，组织编写、出版与推荐了一批好教材，组织力学教师暑期培训班，并对各校弹性力学、流体力学等课程进行了课程评估。由于国家的投入和各校的努力，近十年以来上述 39 所高校力学专业的办学条件有了很大改善，建立了 5 个力学教学实验基地，学生应用计算机的条件大大改善，师资队伍得到了更新与发展，目前 45 岁以下的青年教师已占 55%以上，其比例远高于其他专业。

与本专业的过去相比，近十年来力学专业所培养的人才质量总体来说有所提高，特别是计算能力、外语能力有所提升，知识面拓宽，但由于各种因素的制约，理论分析能力有所下降。总体说来，由于本专业对人才培养坚持了基础扎实与重视实践的指导思想，力学人才在数理基础、综合素质方面比国内一般工科专业强，但与欧美高校，特别是一流大学相比，仍有差距。欧美顶尖的大学非常注意大学生数理基础培养，相比之下，我国目前大多数高校的力学系，大学数学课一般只安排 4 学期（两年），比 20 世纪五六十年代与 80 年代减少许多，使学生的数理基础与分析能力受到了较大的削弱。

力学专业学生的优势在于，首先目前力学专业本科生招生人数比扩招前增加了约一倍，远低于其他工科专业本科生、研究生扩招人数的倍数，而由于中国高校总规模的扩展，仅一般工科院校力学师资一项就有很大需求，力学及各种工程专业对于研究生生源也有很大需求，力学本科生的培养有利于提高工科研究生与基础力学师资的质量。其次，鉴于当今科学技术发展迅猛，而今后高校本科生培养着重于通式教育，力学专业学生基础好、计算机能力强、适应面宽，与我国目前高校所培养的单一工程领域的工科学生相比，较容易转换服务领域。

针对上述情况，上届力学教学指导委员会对我国高等学校力学专业发展提出以下几点建议。

(1) 在稳定招生总人数的前提下，设置力学类专业的学校数目应当做到稳定规模、提高质量，进一步调研新办力学专业的办学质量，加强督导。

(2) 国内各高校针对力学专业本科生培养模式提出了适应社会需求，多层次、多模式、多渠道培养力学人才的改革方案。人才市场是波动的，专业人才培养却是相对稳定的，所以必须从宏观和微观两个方面来考虑问题，即使是毕业生供小于求也要进行改革。

(3) 不论是工科力学专业还是理科力学专业，均有培养模式呈多元化、课程设置模块化的趋势。目前并没有一种统一的做法，但总体而言，仍认为数学和力学的基础要宽一点、厚一点。

(4) 要加强对以下问题的研究：首先是复合型力学人才培养；其次是 21 世纪的力学教育体系；最后是研究型力学人才创新能力培养基地。

三、创新力学专业教育的思考

力学专业改革与发展的总体思路是进一步拓宽力学人才的知识面，培养交叉型、复合型人才，以满足新世纪对力学人才的需求。其主要着眼点在于：

(1) 在现有理科力学专业的基础上，发展新的交叉学科方向，如力学与生命科学的交叉、力学与材料科学的交叉等，以培养新的交叉型力学人才。

(2) 在现有工科力学专业的基础上，以我国的大规模工程建设、大科学工程为背景，发展复合型的力学专业，扩大力学的领域，推动力学的发展，培养大工程需要的复合型力学人才。

(3) 研究新形势下的力学人才培养模式、课程体系及内容，研究在新形势下如何提高人才培养质量等问题。

根据目前新形势下力学学科专业发展需要，我认为应当在以下几个方面开展进一步的工作。

1. 按照教育部的要求，充分发挥力学教学指导委员会在力学专业教学指导中的作用，推动力学专业教学改革

教学指导委员会的具体工作应该包括以下五个方面的内容。

1) 理论指导

教学指导委员会要进一步组织并加强教育教学理论研究、本学科的发展战略研究、本学科专业的质量保障研究等，用研究成果来指导大学本科及高职高专教育。

2) 政策指导

把教育部有关教育教学方面的政策及时转化为教学规范，对高校的教学工作起到

指导的作用。

3）质量指导

“十一五”规划明确提出把高等教育发展的重点放在提高质量和优化结构上。这要求教学指导委员会要进一步强化质量意识，加强教学质量保障措施的研究与制定工作。

4）经验指导

积极推广教学改革的成功经验，推广优秀教学成果，促进本学科领域先进教育理念、教育方法、质量保障措施的推广运用。

5）信息指导

采取各种形式，及时收集本学科领域教学、科研、招生和就业等方面的信息，加强各科类教学指导委员会的经验交流，构筑信息交流的平台，为高校提供信息服务。

2. 根据新的人才培养形势和要求，组织全国高校力学专业合作，进一步完善与充实我国“力学专业发展战略研究”报告和其他三个专业评估规范

2006 年教育部向力学教指委下达了 4 项“高等理工教育教学改革与实践项目”。我们准备在广泛调查研究基础上高质量完成“力学专业发展战略研究”报告以及“力学类专业指导性专业规范研制”、“力学基础课教学基本要求研制”和“力学类专业专业评估研究与实践”三个项目，为教育部提供准确、客观、可靠的咨询意见、建议和决策依据。

为体现分类指导的原则，调查研究应考虑不同地域、不同层次、不同类型的高校，尤其还要考虑没有教指委委员的省份的情况，要加强与他们的联系。要召开针对地方院校的力学专业办学的研讨会和力学专业人才培养的研讨会。同时讨论地方院校应如何进行专业划分，以利于学生的就业。

3. 加强教材研讨建设和国家精品课程建设，将创新人才培养提高到一个科学的水平上

教材建设与研讨是提高力学专业教学质量的一个重要因素，国家精品课程在力学专业教学中的示范作用已经为大家所广泛接受。今后我们将在各门力学基础课程内容之间的衔接与融合、力学基础课程教学与创新人才培养的关系、提高力学基础课程作为技术基础课的地位，以及名优教材建设方面开展研究。从各专业创新人才培养的角度，组织基础课教师、专业课教师、专业第一线资深的学者与工程技术人员，对现有的教材进行深入的研讨，真正将教材建设推向一个新高度。

在国家精品课程建设上，坚持“宁缺毋滥”的原则，充分发挥力学各专业教学指导分委会的作用，制定关于精品课程推荐的程序，将此项工作规范化、制度化。

4. 高度重视力学类专业学生的实验能力培养问题，重视实训基地建设与实验室建设问题

实验是基础力学教学中不可或缺的一个重要环节，是学生素质教育与能力培养的重要环节，目前不少学校的力学实验教学现状与本科生培养目标是相矛盾和不协调的，影响了力学专业的教学质量，应该引起我们的高度重视。

加强基础力学实验室建设，要结合本科教学评估要求，呼吁学校加大对基础力学实验室的设备经费的投入，要改革实验教学内容，提高实验教学的质量，将现代化教

学手段引入实验教学，实行开放教学，提高实验室有限资源的有效利用率，力学教指委将组织专门的研讨会，就目前我国力学类专业中在实验教学和实验室建设问题展开专题研讨，就该问题提出专门的调研报告。

5. 进一步加强中青年教师的培养

我们应当清醒地看到：目前青年教师虽然学历普遍提高，但教学经验欠缺，对课程的体系与教学内容了解得不深，在教学的严谨性和教学方法方面也有待于进一步提高和改进。为了将已经取得的力学教学改革成果应用到教学中，在提高教学质量中发挥作用，加强中青年教师的培训成为当务之急。

我们要重视先进教学手段使用与技术开发的交流与培训，定期举办中青年力学教师的专题培训和研讨、全国力学青年教师讲课比赛等得到大家认同的活动，使这项工作制度化。

6. 建立力学教学指导的信息门户网站，实现优质教育资源共享

经过教育部工作部署和前几届教指委积极响应及贯彻，已经在力学网络课程、国家级精品课程、立体化教材、教学素材库和题库等方面形成了一系列优质教育资源。力学教学指导委员会将建立信息门户网站，并与中国力学学会网站紧密合作，向全国的力学工作者和学生提供一个强大的信息共享平台和交流平台，充分发挥力学基础课程教学改革取得的重要成果，特别是国家级精品课程以及获国家级教学名师奖的优秀教师的示范、辐射作用，充分地发挥优质力学教育资源的作用，实现资源共享，从而促进力学类课程教学质量的大幅度提高。

7. 加强力学专业评估与考察，开展相关研究工作和质量监控工作

今后5年我们将接受教育部委托开展大量的力学专业评估与考察工作，教指委认真组织实施，将评估工作与相关研究工作结合起来，与质量监控结合起来，真正发挥教学指导委员会在教学质量评估与监控方面的主导作用。

浅谈高等学校科学管理“三”字经[①]

如何做一个管理者，如何把一所大学的管理工作做好，笔者想根据个人的经历谈谈这个问题，而不是谈书本上那些理论。大学管理有它的特殊性。大学有不同的类型，有研究型大学、教学型大学、本科大学、职业学院，每个学校的定位不一样，管理也不一样。但是，没有一本专门的书讲这些问题。特别是在中国，管理更是很长时期没有得到重视。在20世纪80年代以前，中国就不强调这个问题，大学不培养管理人才，学管理的地方都没有。对于任何人，比如让你在管理岗位上工作，你上来就做领导，马上就处理事情，那时的管理基本上是凭脑袋凭经验。

实际上，真正好的管理者是需要理论指导的。管理科学本身又随不同民族、不同国家、不同时代、不同体制而有所不同。管理是科学，但是管理相当多时候又是艺术，七分科学，三分艺术。每个人的管理方法都不一样。比如说今天是刘书记执政，学校可能是这样管理，明天换了一个书记，可能又是另外一种管理，人世间找不到完全同类型的管理。怎么能把事做好，这要看管理方面的本事。管理科学理论用得好，这所学校必定发展得快，品牌好、知名度高，培养的学生质量高、受社会欢迎。

笔者年轻时候喜欢读书，古今中外，都有涉猎，以后又读了一些管理方面的书。实践经验是从学生时代开始得到的。当老师后，又先后担任室、系、院和校各级领导职务，并在国内外六所大学工作过。

担任这些职务时有这些经历，便知道怎样能把高校搞好。当然这些想法也不是一开始就形成的，形成最后的理念时，已到了50多岁。因为我们这代人，比较简单，就是唯上，领导讲的就绝对是正确的，从来没去怀疑过哪一点不对。过去是唯上，现在是既要唯上，同时又要唯真理。归纳起来，笔者拙见，作为一个高校的管理人员，要把学校办好，就必须具备以下四方面的条件。

一、学校管理人员必须具有三个“心”

（一）要有爱心

一个人当领导，无论是当学校里哪级领导，首先要有爱心，这是必须具备的，没有爱心就做不好工作。什么叫爱心？这很简单，也就是说要爱祖国，最起码对自己的祖国要热爱，对自己的民族要热爱，爱中华民族，爱学生。你不爱祖国，不爱民族，不爱学生，在这个岗位就待不下去。具备了这个条件，才具备了上岗的基本品质条件。我们想想，一个不爱国的人，他怎么能够培养无产阶级事业的接班人，怎么能够培养我们国家优秀的人才，让我们国家能够得以传承，这是不可能的。当领导就一定要有一个广阔的胸怀，要爱自己的祖国，要爱自己的民族，要爱学生，要忠诚于我们国家的教育事业。

---

① 本文原载《科技创新与品牌》，2008，(10)：14-17；(11)：10-13.

（二）要有责任心

要有责任心，这一点是非常重要的。不管在什么岗位，都要把这个岗位的事情做好。不要今天刚到这个岗位，马上就想要升职升官，要调什么岗位，这个岗位不好，等等。既然到了这个岗位，就要把工作做好，千万不要见异思迁。这件事其实很简单，但是要做到却是非常难的。我们观察周围的人，很多人是不负责任的。到他手上的事情，他不负责，他不种自己的地，却常常要去耕别人的田。不管是什么岗位，哪怕是不喜欢的岗位，也要认真，要尽力把工作做好。责任心非常重要。今天高校内非常浮躁，浮躁的原因就是一些人责任心不够，老是成天想着要升职、要升教授等。其实只要认真做自然有机会升职。如果不好好做事，老出纰漏，当然就升不了职，升了职也得下来。责任心对每一个领导干部都非常重要，必须把岗位工作真正当自己的事情干，把它干好。

（三）要有耐心

做领导干部更要有耐心，为什么呢？因为做很多事情，可能别人非常急，国家也非常急，环境要求你要很快做决定，要把它做完做成。但是实际上，有些事情急于求成是做不成、做不好的。比如说学校要变成一所名校，这不是三五天的事，是要慢慢做的，这是急不得的事情。很多事情都急不得。即使别人很急，火烧眉毛，你也要非常沉着，必须有耐心。对很多困难的事情，半途而废，不行！必须要忍耐才能过这个关，要有耐心，这个耐心还包括忍耐，否则很多事情做不成。回想起来，笔者这一生很多事情都是靠忍耐过了关。

首先举个学术例子，笔者成名作的题目是“波纹圆板的研究”。1963 年大学毕业以后，笔者很想在科研方面为国家做贡献，所以就去兰州市的一些工厂搞调查研究，寻找科研课题。后来找到一家航空仪表厂，该厂正好有一个仿制美国飞机高度表的任务，其核心问题是要研制高度表的核心元件，即一个锯齿形波纹圆板，属于板壳非线性力学研究领域。可是，该厂找不到人研究这个元件。笔者是初生牛犊不怕虎，勇敢地答应试试，他们很高兴。

笔者非常高兴地拿着这个科研课题回到学校，立即向教研室支部书记请示汇报。支部书记猛批评了一顿，不准去做这一研究课题，还给笔者戴了一顶“不务正业”的帽子。但是笔者觉得这个课题太好了，对我们国家太重要了，所以便不顾领导意见，在业余时间偷偷地做。从 1964 年做到 1965 年的夏天，差不多快做好了，突然接到上级通知参加“农村社教运动”，不到一年，开始了“文化大革命”。笔者受迫害当了“牛鬼蛇神”，课题一直做不成，直到 1966 年 8 月。当时，笔者除参加学校活动以外，还冒着风险悄悄做这个课题。这就是忍耐。组织不让做，做了也没什么好处，而且做这个课题所用的经费还是自己掏的。同时，做课题计算所用的电动和手摇计算机也无法从大学里借出来，只好借助算盘和对数表，用手算完成巨大的计算任务，草稿用了几麻袋。“文化大革命”期间，笔者受爱人保护，偷偷在家里做业务。外文也是不能学的，后来就买了《毛主席语录》英文版、俄文版，巩固自己的外语。那个时期没有做业务的条件，国内所有的学术刊物全部停止出版。做业务是没有好处的，也没有想到要升职称，什么荣誉，什么奖，都没有的。只是觉得，这些研究未来对国家、对民族有用，完全是靠忍耐才做下去。这个研究课题完成于 1968 年，直到 1978 年才在学术刊物上发表，立即引起学术界的轰动。这一课题从开始到完成再到公开整整经历了 14 年的时间。

其次，笔者到暨大工作后，也是很需要有耐心的。1991 年 11 月 27 日，上级调笔者到暨南大学任副校长。20 世纪 90 年代初，在改革开放前沿的广东，老百姓认为暨南大学和其他广东高校都办得差，并取了难听的绰号。为此我十分伤心，真是不高兴，一百个不高兴！但是要服从组织安排，只得继续待下去。于是暗暗下决心要把这个学校办好。因为笔者排在领导班子第五位，笔者就从第五位的工作做起。

1993 年秋天开始分管教学，第一天管教学就到教学大楼检查上课情况。8 点钟上课，7 点 50 分就在大楼门口站着，到了 8 点开始数有多少迟到的学生。那个时候全校有 3000 多名本科生和 2000 多名专科生，居然有 1500 多名学生迟到，最严重的到 8 点 40 分才来。随后又检查了第四节课的下课情况，本来是 12 点结束，可一位教师在 11 点 15 分就下课了。同时，笔者还随机去教室听老师讲课。一个老师到了教室，上了讲台就问："同学们，你们是本科生还是大专生?"他走上讲台还不知道听课对象。有一个老师还问："你们上节课上到哪儿了?"他连上一节课上到哪儿了都不知道。他是如何备课的，真是天知道！教学的管理非常混乱。要改变这种现象，阻力不小。这可急不得，得一件一件地做，一定要把教学管好，要用很多办法去解决，很需要耐心。差不多花了 10 年的工夫，才把暨大办学名声不好的帽子摘掉了。

从 1993 年开始，笔者搞了一系列改革，通过改革找出路。一个改革、两个改革、三个改革，逐步改革。因为自己担任的是管教学的副校长，开始时便做学分制的改革[1-3]。做自己权力范围内的事，改革成功率高些。做事情只有非常有耐心，才能够真正把事情做好。做每一件事情，不管是大事还是小事都要有耐心。

二、学校管理人员必须树立三个"第一"的观念

（一）学生第一的观念

做学校领导一定要树立学生第一的观念，没有学生就没有学校，学校是学生成长的摇篮。现在中国的很多学校还不是很明白这个道理。一些学校好像以教师为主，或者以领导为主，部分人认为我是校长，我是书记，应该以我为主，其实学校里面应以学生为主。校长、院长、处长、老师，都是为学生服务的。没有学生就没有学校，因学生存在才有老师。假如说全是我们老师在一块儿，那是研究院，不是学校。学生是学校的主体部分，我们是为学生服务的。所以，在心目中应该认为学生第一，要牢固地树立这个观念。

要围绕学生开展工作，首先就要变我们中国传统的保姆式教育为自主式的教育。以学生为第一，就是学生在学校里要受到充分的尊重，要让学生自己决定学什么专业，需要学些什么东西，要怎么成才，要快还是慢，怎么选择老师，等等。就是说，要让学生来选，让其有挑选的余地。按照学生的愿望、学生的兴趣、学生自己的目标来确定他的发展方向，不要由我们老师、领导强迫他去做。我们中国传统教育是强迫式的，是一种家长管理式的教育。这种教育培养不出创新人才。我们现在提倡创新，提得很

好，但是我们的方法不对，我们学校的管理体制不对，是保姆式教育，保姆式教育是容不下这种创新的，因为创新的人才往往是要异想天开的。聪明的孩子他要做贡献，他要想一些人家都不敢想的事情，这才能创新。

其次，要为学生创造成才的氛围和最好的条件。以暨南大学为例，笔者根据学校的办学任务和实际情况，提出了“侨校＋名校”的发展战略[4,5]。笔者的本意是用最少量的字告诉全校师生，暨南大学是侨校，这是我校的性质，将暨南大学办成侨校是我们的使命，同时我校的目标是成为名校，只有成为名校，侨校的任务才能完成得更好。这便决定了我们暨大应该招什么样的学生，应该培养什么水平的学生，应该怎样培养学生，应该为学生提供什么样的学习环境和条件。华侨华人对我们国家帮助很大，从辛亥革命开始到现在，几乎每一个中国前进的跨越，往往都是华侨华人做了巨大贡献。没有孙中山先生领导的华侨华人，辛亥革命成功不了。今年，也就是 2008 年，是改革开放 30 年，这 30 年来中国引进的外资，70％是华侨华人完成的，不是靠洋人，他们不会发善心让中国上去。中国 2007 年的 GDP 成了世界第四，核心是我们的改革开放引进的外资，这些外资主要是我们华侨华人来完成的，所以华侨华人做了重大贡献，那我们暨南大学就要为他们服务好，满足他们的愿望，多培养他们的子弟，并使之成才。

为此，笔者一再在学校强调要多招海外及港澳台学生。可是一些老师、一些领导不喜欢，跟笔者诉苦。在暨南大学当老师、当干部确实辛苦，苦在哪里？大家都知道中国内地的高中和我们国家的香港、澳门、台湾的高中不一样，跟欧洲、美国的不一样，跟其他发展中国家的也不一样，由此造成我们学校的新生受教育程度相差甚大，习惯也不同，面对这种复杂的教学对象，教书相当难。所以，有的老师只愿意教内地的学生，不愿意教海外及港澳台学生。笔者觉得暨南大学既然是侨校，就应该累一点，人家华侨华人给我们做了这么大的贡献，当然应该多招海外及港澳台学生。为了有利于华侨华人和港澳台学生成长，经再三研究，我们将内外学生比例定为 1∶1，50％海外及港澳台学生，50％内地学生。刚当校长时全校只有来自 10 多个国家和地区的 1000 多名海外及港澳台学生，现在已发展到 1 万多名（国内高校第一），来自世界五大洲 80 多个国家和地区。全世界主要国家和地区都有学生来了。学校牌子太差，校园环境太差，办学质量太差，谁愿意来读这个学校？学生家长和学生本人都希望读好学校，要好文凭。让几万里远的人知道这个学校，知道学校的牌子好，这谈何容易！令人欣慰的是，经过艰苦的努力，校园焕然一新，办学质量大大提升，暨南大学开始名扬四海，成了学子向往的大学。达到了 1∶1 的数量目标，而且在全国 2000 多所高校里，暨南大学从一般学校一跃成为一所国内排名前 50 位的名校。

没有名气，不是名牌，一个学校便招不到质量好的学生。如果办得差，每年招生的时候就十分辛苦，没人来读。一个学校要办成名校很不容易，在一般的意义上，名校是指世界的一流学校，或者全国的一流学校。实际上，每个领域、每个层次学校中的好学校也是名校。人们接受的教育是不同层次的教育，社会需要不同层次的人才。每个学校都要找好自己的位置，做好定位。有了定位，才能做好“学生第一”的有关工作。

（二）质量第一的观念

质量是生命。任何单位、产品都要讲质量，办学中要抓的问题有很多，但首先要讲质量第一[6]。可是质量第一的观念却很难树立起来。在每个人的工作里面，在每个部门的工作里，质量的含义都是不一样的。学校的整体工作，很难做好。我们现在有一种很不好的社会风气，那就是做事不认真，往往差不多就行了。“差不多”先生太多。处事呢，大而化之，马马虎虎。“好像”“几乎”“大约”“大概”这种词汇很多，工作上全是这种处理方法，数字也是如此，所以无法保证工作任务高质量地完成。

中国在质量方面不注意的问题太多了，我们工程上的问题很多，而且有时候还掺杂着腐败在里面。我们学校发生过一件案子，教训很深刻。有一年，学校修建家属住房，负责建造的工程队不顾工程质量，偷工减料。有三栋房子的基础地桩按设计要 22 米，有人举报这个桩弄短了，我便要主管副校长调查。基建处一些负责人，不仅对抗、抵制调查，而且威胁我们。笔者坚持要现场调查，马上那几个负责人的脸色都发白了。挖了几个小时后，发现 22 米的桩只有 14 米。既腐败，又没有质量概念。特别是发现那根 1 米长的皮尺，被人为地缩短了，实际只有 80 多厘米。

每一件事，包括考试，质量的管理问题都很多。当时学校的学风不好，学生对考试都无所谓，学生中 40％～50％都有不及格课程，那就补考，过了 30％，再补考，一补、二补、三补、四补，老师出题都出烦了，给你 60 分算了。管理人员也不注意，无所谓，报表一看过关就行了。学生成长过程的每一个关口，很多时候对质量的监控都是不到位的，产生这个问题的原因是多方面的。

要有好的质量，就要遵守质量管理的四个原则。

第一个原则，就是质量要符合要求。应该让全校的老师、干部明白要求是什么。不只是院长、书记、处长知道是什么原则，你下面的人都要知道是什么要求，什么是优秀，什么是良好。要大家都明白这个才能保证质量。从前，对于我们的质量标准，领导知道应该怎么做，下面的干部不知道，老师不知道，那你就做不好。保证质量，就是要让管理人员知道要求是什么，什么叫符合要求，这是第一原则。

第二个原则，质量系统的关键在于预防。不要等事后来检验、来评估，靠检验和评估来保证质量是保证不了的。首先应该预防，根据这个事情的程序，找出哪些东西可以预防，事先采取措施。

第三个原则，工作的标准必须是零缺陷。零缺陷就是没有缺陷，而且是第一次就把事情做好，不是要第二次、第三次来做。一次就把事情做对做好，要提倡零缺陷，这是我们质量管理的核心。

第四个原则，质量是用不符合要求的代价来衡量的。在工作里面，不符合要求的那部分工作量的花费，就是你的代价。

如果我们在全体教职员工中都树立质量第一的概念，那我们的教学工作、科研工作和行政管理工作就一定做得很好。

（三）管理第一的观念

一个国家、一个省、一个城市、一个单位，甚至一个家庭，怎么算是做好？笔者认为做到四个字就行了。对国家来说，是强国富民。国家强大，老百姓富裕那就很好。

强国是我们宏观条件好，富民是老百姓生活好，包括住房、收入、环境、精神上都很好。对学校来说，是强校富民。学校很强，学校品牌好，大家生活幸福，师生才会爱这个学校，愿意为这个学校奋斗，愿意为学校工作，愿意为学校做贡献。怎么做到这四个字呢？核心的问题是领导者要树立管理第一的观念，懂管理，会管理。管理实在太重要了，过去我们不太强调管理。

管理的核心是什么？是发展战略。决定好发展方向以后，如果因为环境和人的关系，今后发展慢了也没有太大关系，慢了但也没有走错。把发展战略搞错了，不符合环境，不符合时代，不符合国家要求，那你越努力越坏，学校就办得越不好。一定要首先确定好发展战略，其次再做科学的决策，第一件做什么，第二件做什么，第三件做什么。还有用人，要用优秀的老师、优秀的干部，等等。

改革开放以来的 30 年，是我们中国几百年来第一次出现的盛世！按照邓小平理论，国家高度重视管理，才有了今天这么现代化的环境，国家强大，老百姓也生活得很好。回想 20 世纪的“大跃进”、三年困难时期以及后来的“文化大革命”，都是管理出了问题，弄得几乎经济崩溃，教训深刻。

管理十分重要，管理不能随心所欲，一定要科学管理。管理的内容很多，包括发展战略、科学决策、科学用人、科学机制，等等，还有很多，这里面的内容非常丰富。学校领导要管好学校，就一定要科学管理，就一定要首先树立三个“第一”的观念：学生第一的观念、质量第一的观念和管理第一的观念。

三、管理学校的三条原则

（一）因材施教的原则

在学校里一定要因材施教。这里包括两个方面，一方面，要根据学生个体的差异因材施教；另一方面，每个学校由于性质不一样，层次不一样，专业不一样，所以要按照学校自己的定位去培养人才。例如，对于一流大学而言，要培养的是精英人才，那就要按照精英人才的要求安排教学和实验。对于职业学院，定位是培养应用型的科技人才，毕业的学生不是去从事基础理论创新，那就要按照应用型人才的要求去培养。这样一来，教学的安排，教材的安排，课时的安排，实习实验的安排，德育的安排，等等，都要根据学校的性质来决定。不要都用一流大学的方式，而是要根据学生的情况用自己的方式，这才是因材施教。

（二）有教无类的原则

有教无类，这是孔夫子的思想，就是学生即使差一点，甚至有一些瑕疵，老师也仍然要教育他。不要歧视学生，不歧视落后的学生，不歧视失败的学生，不歧视有瑕疵的学生，要宽厚。这一点，一些老师、一些干部没有做到。要考虑学校不是惩罚人的地方，是培养人才的地方，应把不同的人都能够培养成人才。不要去埋怨学生，有差错的时候也不要歧视，一定要关爱他们。

（三）奖惩分明的原则

奖惩分明，就是要以奖为主、惩罚为辅。一定要搞清楚奖惩的关系，不要以惩为第一，应该以奖为第一。例如，处长对科室里面做得好的要及时奖励、激励；对做得不好的，首先要扶一扶。在惩罚的时候，要留点余地，还要宽厚一点，对事要严肃处理，对人的处理要宽厚一点。提倡善良，就是说在惩罚的时候，还希望他改正错误，给他关爱，留有余地，可重可轻的时候要从轻处理。应该实行疑罪从无的方法。事情未搞清楚的时候千万不要去处理人家。前些年，只要认为你犯了错误或犯了罪，你就一定是犯了错或犯了罪，所以在政治运动中搞了很多的冤假错案。这些冤假错案产生的原因都是把怀疑变成了人家的罪行、人家的错误。要激励同志做好事，奖励优秀的同志。激励为主，表扬为主。

四、管理学校的三个方法

（一）从严治校的方法

在家里培养孩子，需要严格才能使子女成才。在系统比较乱的情况下就要用严格的办法。在暨南大学，为使校风、教风、学风好转，笔者[7]提出了“三从严”的原则：从严治校，从严治教，从严治学。坚持了多年，取得了成功。

大学里面，核心是两件事：老师是教课，学生是听课。老师的教学工作主要是用上课质量来检查的，学生的学习质量主要是靠考试来检查的。所以领导要抓两个重要方面：教和学。学生方面把考试抓住，当然平常也抓，平常有很多程序，核心是抓考试。中国在作弊方面自古以来都管得很严格，我从小学到大学就没见过作弊，哪怕两个人考试时座位很近，都不敢去看人家的考卷。西方也是这样，一个大学生作弊后，一辈子都找不到好工作，所以在西方诚信很重要。我们为什么现在有这个毛病呢？很大的问题就是诚信差了，就是不严格造成的。一个社会失去了诚信，在管理上的代价就很大，当领导就很苦，说一不是一，说二不是二，很麻烦。

为了杜绝考试作弊，笔者就想出大考场的办法，这在中国是第一次。全校学生期末考试在一个考场里。我们学校最大的房间是体育馆，学生在体育馆里考试，这件事《人民日报》都报道了。在大考场中，每一行或每一列都不是一个专业的，排梅花形的座位，每位考生见不着周围同卷子的人。考试桌子设计得很特别，私存夹带完全没有可能。进考场的时候，学生无关考试的东西要存放。进到大考场以后，学生要靠自己的智商和能力来完成考试，在那种情况下作不了弊。自从设立大考场以来，便没有学生作弊。学生无法作弊，考试质量就很好。学生要考试好就得平时学习好，所以整个学习过程都能够管住。

另外还有考题问题，学生巴结老师送礼，有的老师就会漏题，为此，就搞试题库。全校每门课程都搞试题库。一个试题库不是一个老师做，是几个老师分开做，而且一套题是几个老师的试题混在一起的，最后收集起来一门课程起码有十几套考题。主管教学的副校长到考前两天才选考题。我们把试卷编成号码 1、2、3、4……他随机抽号码，但不知考题的内容。每个专业都抽好后，由另外的人去印刷，参与印刷的只有几个人，这几个人不带手机，跟外界不联系，就在学校里面或者郊区去做这个事情。在两天之内把考卷全部都印好封存。包括硕士生、博士生的考试全部都这样做，所以暨南大学考试非常严格、公平。题目漏出去，作弊，这是最大的对人才选拔的不公正，选不出优秀人才。

关于阅卷工作，我们也想了办法。教师集体阅卷，阅卷后试卷不能由老师带出阅卷室。阅卷要管好，整个流程都注意，每一步都科学管理。

考试严格以后，省里对我们比较相信，就把几次干部考试放在我们学校举行。有两位副厅长作弊，当场被我们抓住了，副厅长的官都丢了。不诚信的人就绝对不能做领导，作假的人肯定做不了好领导。今天我们社会存在部分贪污腐败、产品质量差的问题，很多都是因为不严格、不注重质量，祸害太大了。

在学校这样一个培养人才的地方一定要严格管理，大家首先就要要求学生不作弊，要讲诚信。在学校的时候，如果允许学生作弊及格，那么他到社会上就更作弊了。所以我希望我们的领导，在管理学校的时候一定要严格，符合质量。符合质量就不要怕得罪人。我觉得人活一辈子一定要有人格。一辈子多做善事，多做好事，一定不要做坏事。全中国高校都应该严格管理，严格以后，我们培养的人才质量就高了。这些学生到社会后就会体会到严格带来的价值、严格带来的质量，严格带着他们最后走向成功，严格带来的是民族的诚信和兴旺。

（二）依法治校的方法

我们国家最近一直在强调法治，因为中国过去是人治的国家，现在是由人治走向法治。这一点说起来容易，其实是很难很难的事情。20 世纪 80 年代初，我在德国待了两年多，发现德国管得比较好，社会治安比较好，经济搞得也比较好。经济上通货膨胀率低，基本是在 1%左右，物价稳定。每年每个人的工资增长率是超过通货膨胀率的。人们很热爱自己的工作，都愿意好好地干活，找到工作就很好。他们说的好日子就是可以旅游，可以去玩。有钱人开奔驰，开好车，穷人也有自己的车，就是差一点的车。这是差别。社会管理得比较好，很大的原因是法治，就没人敢犯罪。大家都不犯错误，都循规守法，不偷税漏税，该交多少税就交多少税，而且办事不求人，每一个人都做好自己的工作。我要办什么事情都按照制度，找张三办、李四办都一样。不像我们学校的一些部门，要先研究研究，你得求我，你得巴结我，你得送礼，才能办事。人家的高校使用法治这样的方法，我们差得比较远。现在老百姓办个事很困难，该办的事情，一些部门有时都不给你办。所以我们应大力提倡，老百姓要办的事情，只要是制度上允许办的事情，人家来找你办，你就要立即去办，不能有拖拉，不要刁难人家，哪怕这个人平常跟你关系不好。因此，学校方方面面都应首先建立健全制度，包括学校整体的管理制度，教学的管理制度，科研的管理制度，后勤的管理制度，住房制度，教师管理的制度，卫生、保卫制度，等等。

笔者在学校一任职，就开始搞制度建设。每个部门都建制度，甚至每个小方面都有制度。最后搞了 300 多个制度，编成两本文件集：《行政管理卷》[8]和《教学科研卷》[9]。另外还有党务方面的制度。然后发到学校各个单位，每一个科室都有，任何人办事都看这个制度，照着程序做。就不要临时考虑是找张三还是找李四。哪个事情属于哪个部门负责，都要告诉老百姓。电话要公开，大家好办事。中国传统求人的制度要改变，传统的拉关系习惯要改变，绝不能谁有权力，就拿在手里面，别人就得求他。所以一定要改变我们的管理制度，要真正为老百姓服务，要按制度来办事。

制度不要经常变，尽量少修改，一次就搞好一点。制订制度首先是科室先搞个草案，其次征求多方面意见，最后领导班子集体讨论定稿。制度首先要符合宪法，其次要符合我们的教育法。如果一个学校搞几百项制度，包括干部的选拔等都搞成制度的话，我想最后制度就不会因为领导的改变而改变。过去，老百姓认为只有优秀的校长、优秀的书记来领导，学校才有好日子过。现在，张三校长换了没关系，李四来当还是按照这个制度，这个学校就会健康发展。我们国家如果制度化、法治化了，就会科学地向前发展。

从中国传统的人治走向法治，希望我们的制度建设加快一点。希望每个高校建立制度，每方面都搞制度，小制度，大制度，从整体到局部都搞，包括干部选拔。比如说，选举制的干部怎么选，任命制的干部怎么任命，考评怎么考，选举制的干部考评和任命制的干部考评，应该是不同的。过去，都是一锅儿煮。任命制的干部要对上负责。校长任命处长，处长自然要对校长负责，考评处长的分数应该以直接领导他的分管副校长和校长的评分为主。现在任命制的干部往往是让群众来投票决定他的去留，那么这个干部就不敢管群众了。如果他管严了一点，许多人就会投他的反对票。优秀的干部往往很严格，下面有的人就给他打不及格。我们应以任务完成得好坏来评价一个人。选举制的干部应该以群众的分数为基准，因为你是选出来的，你要对群众负责，这个不能由学校领导来决定，应该由群众的分数来决定。现在的情况是考评干部都一样，任命制干部、选举制干部、业务干部、党的干部都进行一样的考核。不应该一样。还有教师的考核，等等，要有不同的办法，不要一个办法对付所有的教职员工。

制度设计非常重要。去暨南大学的时候，笔者希望把它办成名校。因为这个学校有百年历史，今年是建校 102 年，是中国最早的 7 所大学之一，肩负重要的办学任务，理所当然它应该办成名校。但是，笔者去的时候，这所学校办得比较差，4000 名教职员中，连笔者在内仅有 8 位博士生导师，老师里面只有 8 个人有博士学位，被三大索引（SCI、EI、ISTP）收录的学术论文一年仅有几篇，1992 年只有 3 篇，而北京大学当时有数百篇。这 3 篇中，我一个人要占 2 篇。

为了提升学校学术水平，笔者想了三个主意。第一，让原有教师人人都搞科研，提出“不搞科研的教师是残疾的老师”这个口号，施加压力，逼迫老师都要搞科研。第二，大力引进人才，招名牌学校的教授以及博士进来。第三，制订制度，特别是制订学校的分配制度，以激励老师搞科研。

当时的暨大很穷，校机关到 1995 年的时候连一个季度 50 元奖金都发不出来。所以笔者就要想办法挣钱。首先，亲自负责全校财务工作，宣布院系部处不搞创收，挣钱是校长的事，校长有责任搞来钱。院长应该做院长的工作，处长应该做处长的工作，系主任应该做系主任的工作，每个人做好自己的本职工作。其次，改变分配，原来是各系发奖金，现在是学校统一发奖金，而且改名为校内工资，由此搞了个新的分配制度——暨南大学量化考核制度[10]。

改革以后，在暨南大学，每人的收入由两部分组成：国家工资和校内工资。国家部分我们改不了，教授拿多少钱，处长拿多少钱，那是固定的。学校部分在改革后是这样分配的，比如说老师，你给学校做多少贡献，学校就给你多少奖金。我们不能说你是多少就是多少，而是把工作进行量化。比如说科研方面有多少论文、多少科研项目，以及成果的推广，等等，我们把每件事情都量化。把国家的项目、地方的项目等不同的项目分级；还有就是你拿的是100万元的项目，还是10万元的项目，当然不一样；而且各学科也不一样，搞理工的项目经费要多一些，搞文科的搞个大钱不容易，文科的分数跟理科有区别，项目也有区别。对于发表论文，在世界、在中国不同的杂志发表，哪些是著名杂志，哪一类杂志多少分，把这个划分好，不同专业都不同，划得很细。对于教学，你上本科生、硕士生、博士生、成人教育的课都有不同分数，甚至上不颁发学历、学位证书的课，上短期培训班的课，都有分数；学生人数多了也有加分，重读班也有分数；教不同的课程有不同的分数，礼拜天上课比平时上课多加一点分，在外地上课要多加一点分，晚上上课要多加一点分，中午上课也多加一点分，分得很清、很细；开展教学实验、批改作业都有分数；当班主任、做学生工作的都有分数；还有社会任职的分数，你给学校带来了名气，在外面任个什么职务，对学校知名度有提高作用，也给你分数。

把这些方方面面的分数加到一块，一年里，你做了些什么事情，填个表格，就是你的校内工资。我们开始设1分1元钱，大家很高兴，以后又提升分值，最后升到1分1.6元。大家看到好处，有的人一年可以拿到几十万元。笔者任校长十年来，全校教职员工人均年收入由1995年的8000多元上升到2005年的88 900元，收入增长了10倍。总之是鼓励大家多做事。这下，我校的科研论文数量大大增加，特别是被三大索引（SCI、EI、ISTP）收录的论文逐年增加，现在已经达到四五百篇，在全国名校里面排得上号了。

学校要鼓励大家好好工作，就要制订激励的制度、严格管理的制度、出成果的制度。所以笔者就先抓住这个关键的分配制度。以前人人都不愿意多上课，特别难的课都不愿意上，愿意上简单课。现在是人人都在抢着上课。过去系主任求张三上什么课，还要拜托拜托才能上课。现在没这个现象了，大家愿意上课，而且都愿意做科研。现在暨大大部分老师都搞科研，做不了科研的老师则赶紧读书，读硕士、读博士。经过训练他就能做科研，能够既有科研本事又有教学本事，那名牌学校就办出来了。

大学是个学术机构，有三个功能：培养人才，出科学成果，为社会服务。你的制度方方面面都可以搞，每件小事都可以搞，但是要抓关键，对学生就是抓考试，对老师就是抓教学。前面我已讲过有些老师原来上课的时候不认真备课，笔者想办法对此情况进行改变。考虑到老教授的示范性功能，所以我在1993年就制订制度，要求教授上本科基础课。学生进校以后，一年级是培养他品质最好的时间，学风最好是在一年级培养。那时，他爱学习，他要上进，他要打好基础，为此安排教授级老师上课最好。好几年后，我才看到教育部在全国要求执行这样的制度。

然后，要防止老师上课不认真备课。上课跟演戏一样，剧本有了，就是表演，是在课堂上表演。老师在课堂上表演的时候，只有学生能监督他。但是，因为学生要考试，老师要管着他，学生的监督作用也有限。所以笔者就提出三重评估制度便把教师教学管住了。西方的学校是每一学期期末都要学生给老师评分。暨大在20世纪90年代初就搞了，结果老师不服，说因为教学生很严，学生便给打低分。对此，笔者提出再请专家评估。全校请了40位专家，大多是教课教得好的退休老师，将其返聘回来，每周规定他们每人听课 8 小时。他们在全校任何时候随机听课，不通知任何人，这就变成一个随机的抽样检查，这里用了数学中的运筹学方法。全部听完所有的课程是不可能的，那样量太大。40 位专家对老师给一个评价，主要针对这个老师的备课情况、教材情况、讲课情况、跟学生的互动情况等。这个专家不一定是本行专家，也不可能做到哪一行都懂。因为我们学校现在是 60 多个专业，而且一个专业的课程那么多，应该说每个人不可能懂很多课程，但是基本道理懂得，所以对于这 40 个专家不要求他听本专业的课，就是让其在全校广泛地听，听了以后给一个评分。结果老师还是不服，说专家不懂他的课，给的评分不正确。笔者后来想了想，再搞公平一点，让领导听课，从教研室主任、系主任、院长至校长、书记，每人都规定听课任务。比如说校长、书记，一学期很忙，听 4 节课；副校长听 8 节课；系主任多听一点；教研室主任再多听一点，这些人对听过的课程都打分。他们也是随机的，不通知任何人，这样会发现很多问题。

听课的时候，一般 8 点的课程，笔者到 7 点 59 分才悄悄进入教室，而且坐在最后一排。很多时候上课老师都没发现笔者，听完课以后老师才发现。这个办法很灵，见到了很多真实情况。校长去听课，不干扰老师，老师未发现也不会紧张，他很紧张反倒不好。

这样一来，我们就设立了三重评估制度：领导的评估、专家的评估和学生的评估。开始时大家还是不服，我们便连续做了两年统计，结果是三个评估分数近 80%是一致的，还不错。教师所得评估的分数如果不及格，我们就亮黄牌，告诉他上课质量太差了，请他赶快改进。如果连续两年被亮黄牌，就不能授课，就要下岗了，那就到人才交流中心去等着，或者调走，或者去进修，或者改做非教师的工作。对优秀的上课教师，则每年在全校隆重表彰 10 位优秀授课教师，既给荣誉又给奖金。于是，学校的授课质量大大提高，教风迅速好转。

做事就要找到每一样事情的关键，对于教师来说上课就是关键，对于学生来说考试就是关键。在学校里面，不应该有作假作弊的事情发生，应该打造一个诚信的校园，讲质量的校园。但这些管理光靠人是做不好的，几个人忙不过来，所以制度是关键。这些制度不是专对老王的，也不是专对老张的，是对着所有人的。这个制度管着质量，要订得细一点，不要订得太粗，每个部门都根据自己的情况认真做。每所高校应根据自己办学多年存在的教学问题、后勤问题和干部问题等，认真治理，要讨论好，要依法治校，笔者觉得这个太重要了。中国的落后，落后在法治上。绝对不能搞成今天张三当校长，张三一套；明天李四当校长，李四一套。形成了一个好的制度以后，这个学校就能够健康地往前发展。

（三）实事求是的方法[11]

我们在工作中要讲究真实，千万不要搞假的东西。一个学校如果不实事求是，搞假的东西，工作是做不好的，出发点就不对。所以，笔者在暨大期间，一直强调干部、教师给我作报告、汇报工作要说真话，要说真的数字，不能搞虚假。

如果你的工作、你的数据是假的，你做的决定就不可能正确。如果不是真的财务情况，你管财也管不好。在财务方面，笔者主张开源节流。开源为主，节流为辅。千万不要以节流为主，应该以开源为主，以创造财富为主，以节流为辅，不要浪费，这是第一个原则。第二个原则就是不能做假账。绝对不能做假账，不能去骗领导骗群众，去逃税。只搞一本账，不搞小金库。实际上，现在有些单位搞了一些假的东西，对上面一套，对下面一套。只有数据真实，不搞虚假，才有真实的前提，才能够做出好的决定。应该要求所有的干部，特别是领导干部实事求是，再不要去吃这个假的亏了，不要去吹牛。我们是什么水平就是什么水平，我们是什么状况就是什么状况。

可是，在今天仍然看到有些地方还有这种作假现象，并且是个严重的现象。有些人看到领导需要什么数字就给什么数字，领导喜欢什么就说什么；领导希望说大，他就说大，领导希望说小，他就说小。这样子会害党、害国家、害民族、害学校、害自己。笔者 18 岁入党，已有 50 年的党龄了，50 年来，看到很多人就是喜欢作假，所以便悟出了这个道理，一定要实事求是，只有以真实的事情为基准，才能有科学的管理。

但是，真正做到实事求是很难。因为有时候领导不喜欢你说真话，有时你周围的人不喜欢你说真话，但是你应该坚持，这是十分重要的。这是基础，这是原则，这是方法，这至关重要。不实事求是，既造成人们不团结，又造成社会落后、经济落后。传统的“逢人只说三分话”，就是我们中国人不喜欢讲真话的写照。要建成和谐社会，关键是要每个人真心待人，真实地把你的思想表露给别人。这样的话，关系也好处，工作也好做，管理也好做。实事求是，对我们做学问的人尤其重要，尤其是在培养人才方面，来不得半点虚假。

五、小结

上述四个方面，高校领导如果做到，就一定能团结全校教职员工，把学校工作搞好。

一个国家需要不同层次的人才，需要办多样的高校。中国方方面面的事业，不仅需要最优秀的人才，还需要各种岗位、各种类型的人才。为了完成这一培养人才的任务，还需要注意以下几点。

首先需要发挥每所学校原有的优势，避开自身的劣势。其次，就是要有重点思想，要保证重点。学校不可能方方面面平均花钱，要重点突出，要把财力用在重点上。最后，要改善条件。办学条件要改善，老师和学生的生活条件要改善。当学校管理者，就是要强校富民，这样才能够激励老师都爱学校，都愿意为学校付出，都愿意为学校变得更加美好做出自己的贡献。

做管理是个很辛苦的事情，而且是个非常难的事情，要做到前面几点不容易。高校是基层单位，做负责人是很苦的工作，这些工作不是好干的工作，但又是非常光荣值得自豪的工作。中华民族要成为世界伟大的民族，中国要成为现代化的国家，首先就要办好教育。先是基础教育，后是高等教育。高等教育的工作里包括了德育、智育和体育。如果我们培养出的都是道德很好、业务很好的人才，那我们国家方方面面的重要岗位都是这样的人在工作，我们社会的很多弊端就会去掉了，我们的社会就一定会健康地向前发展。

参考文献

[1] 杨德广，王锡林. 中国学分制. 上海：上海科学技术文献出版社，1996.

[2] 刘人怀. 暨南大学积极推行学分制管理. 高等工程教育信息，2002，(11)：1-2.

[3] 刘人怀. 标准学分制的研究与实践. 中国大学教学，2004，(3)：41-43.

[4] 刘人怀. 暨南大学的发展战略. 暨南教学，2001-07-08.

[5] 刘人怀. 狠抓办学质量 走“侨校＋名校”之路. 暨南高教研究，2001，(2)：10-19.

[6] 刘人怀. 质量：命中靶心，一次就把事情做对. 深圳：中国青年科技企业家管理论坛，2002.

[7] 刘人怀. 发挥优势 深化改革 保证重点 改善条件 提高质量：暨南大学校长任职仪式讲话. 暨南大学校报，1996-06-15.

[8] 暨南大学校长办公室. 暨南大学文件汇编（行政管理卷）. 广州，2002.

[9] 暨南大学校长办公室. 暨南大学文件汇编（教学科研卷）. 广州，2003.

[10] 刘人怀. 转变观念 量化考核 优劳优酬：暨南大学教学科研人员考核与分配体制的改革. 高教探索，2000，(1)：5-8.

[11] 刘人怀. 用现代化管理促进高等教育事业的发展. 暨南教育，1998，(1)：80-88.

我的语文观[①]

“什么叫语文？平常说的话叫口头语言，写到纸面上的叫书面语言。语就是口头语言，文就是书面语言。把口头语言和书面语言连在一起说，就叫语文。”这是我国著名作家、教育家叶圣陶先生说的话。由此可见，一个人从孩提时期的字词咿呀，到学习时期的知识汲取、工作时期的文字应用，语文岂可须臾与我们离开！在我们一生奔腾不息的生活之河中，语文是最活跃的元素，它滋养着生活之河，也为生活之流所滋养。

身为炎黄子孙，学习语文自我们呱呱坠地就开始了，而且伴随着整个人生历程，所以语文不是可学不可学的问题，而是非学不可的问题，是必须要学好的问题。我们的母语是世界上最美的语言之一，它积淀着深厚的人文底蕴和文化价值。学习语文，可以使我们获取蕴藏在母语中的知识精华，热爱沉淀在母语中的民族情感、民族文化和民族精神，提高自身修养。同时，博大精深的中华文化承载着华夏民族五千年的古老文明，而我们了解中华文化的重要渠道之一就是通过文字。在学好语文的过程中，我们对中华文化的了解也将登堂入室，步入一个更加绚丽夺目、繁花似锦的世界。

在学习过程中，我们要注意培养对语文的热爱之情。古人云：“知之者不如好之者，好之者不如乐之者。”兴趣对学习有着神奇的内驱动作用，能变无效为有效，化低效为高效。我出生在一个知识分子家庭，曾祖父 24 岁中举；祖父 12 岁考取清末最后一代秀才；父亲语文功底深厚，并具较好的英语、日语、德语基础。我出生后看到的第一件东西就是书，自小便和语文结下了不解之缘，三岁便由母亲教我背诵《三字经》等童蒙读物。因为喜欢语文，至今我还记得一年级语文第一课的内容：“来来来，来上学，大家来上学”。如果培养起对语文的兴趣，学习语文将不再是学习简单的字词、文章，而是一种阅读的享受，一种心灵的洗礼，一种精神的渗透。

学习语文的同时，要更好地应用语文。语文在人们的口语交际、知识获取、阅读写作、思想交流、科学研究等方面，起着非常重要的桥梁和工具作用。语文是一门基础学科，是工具学科。理解能力不好，不会分析问题，便极大地影响学习其他科目。我虽然是理工科出身，但一直深爱语文，每天坚持阅读，并从中受益良多。在中学时代，我曾获学校演讲冠军。走上学校领导岗位后，语文在我的科研、工作、演讲、交流等方方面面发挥了重要作用。正因为如此，当我读出高妙玄奥的道理、讲出妙语连珠的言辞、做好语惊四座的演讲、写出梦笔生花的文章时，更是深深体会到语文学习与应用的相得益彰。

语文学习无止境，素质提高恒久远。最后，我想以伟大的浪漫主义诗人屈原所作《离骚》中的名句作结语，并与大家共勉：“路漫漫其修远兮，吾将上下而求索。”

---

① 本文原载《语文月刊》，2009，(4)：1.

爱低碳生活　创绿色校园①

2009年12月，哥本哈根气候变化大会向全世界70亿人宣布：要保护地球！全球随即进入低碳经济时代。

冰冻三尺，非一日之寒！

在茫茫宇宙中，地球是人类赖以生存的唯一的得天独厚的乐园。一万年来，人类与自然一直和谐相处。

但是，18世纪末，人类进入工业时代以后，就进入了一个新时期。人类从大自然的手中夺得了权力，成为地球的主人，要征服自然！我们国家还出现“与天斗，其乐无穷”的誓言和行动。

200年后的今天，全球能源告急！资源告急！环境告急！全球70亿人的需求几乎要把自然拖向崩溃的边缘。

再看看我们国家，从1978年底改革开放以来，获得了举世瞩目的成就，经济持续高速发展，经济总量跃升为世界第三，即将登上世界第二的宝座，中华民族终于再次站起来了。

但是，我国经济的高速发展，主要靠物质投入的传统发展方式，在一定程度上，是以环境资源的巨大牺牲以及经济发展的扭曲与不合理为代价的。以推动国家经济增长的消费、投资、出口这三驾马车为例，长期以来，我国经济越来越严重地依赖出口这一驾马车的拉动。外贸依存度高达60%，珠三角尤为严重。特别是我国的外贸出口，主要依赖大量廉价劳动力，大量资源、能源，随之而来的是环境污染的巨大代价。农民是廉价劳动力，目前用度已到顶。我国的资源十分有限：我国的石油储量仅占世界的1.8%，天然气占0.7%，铁矿石占9%，铜矿石占5%，铝土矿占2%。在人均资源方面，我国人均工业化所需要的45种主要矿产资源为世界平均水平的1/2，人均耕地、草地资源为1/3，人均淡水资源为1/4，人均森林资源为1/5，人均石油资源仅为1/10。以污染造成全球变暖为例，我国面临减排二氧化碳等温室气体的严峻形势。从2003年至2006年，我国四年能耗增量超过了以前25年能耗增量的总和。2003年以前，我国发电能力不到2万亿千瓦，现在已经增加到7万亿千瓦。2003年以前钢产量为2亿吨，现在高达7亿吨。这些都要大量耗费碳资源。随之，我国二氧化碳排放量激增，2006年超过美国成为世界第一排放国，排放量为64.37亿吨，占全球排放量的22%，与1990年相比，几乎翻了三番。我国的排放量激增，最大的推动力是发电、交通运输、制造业以及现在人们的生活方式。

大量的研究表明，大气中的二氧化碳是地球升温的祸首。世界著名气候科学家、美国宇航局戈达德空间研究所詹姆斯·汉森认为：“我们的排放水平早在20年前就早

① 本文是在绿色澳门建设研讨会上的发言，澳门，2010年6月5日；一个大学校长的探索，北京：高等教育出版社，2011，246-248.

已超标。”

超过临界点并不意味着马上发生重大的灾难，但意味着一种从“可能”到“必然”的转变。

如果全球气温上升 5℃，暴风和干旱等极端天气出现的概率会急剧增大，海平面将上升 10 米，海中的一些岛屿与大陆沿海地带，特别是一些大城市会被淹没，全球一半物种可能面临灭绝。

如果气温上升 6℃，地球上的生命将会遭到毁灭性打击。

我国是升温最为严重的国家之一，近 100 年来气温平均升高 1.1℃，略高于同期全球平均升温幅度，近 50 年来变暖尤其明显。特别是广州地区，已经是连续第 20 个暖冬，近 50 年来，平均温度升高了 0.6℃，是全球平均增温速率的 2 倍。发生极端天气与气候事件的频率和强度出现了明显变化。因此我们应尽快行动起来，热爱低碳生活，做好节能减排。

要建设低碳社会，我们的首要任务是要建设以低碳排放为特征的产业体系和生活模式。要改变高能耗、高污染的粗放型增长方式以及生活方式，以形成有利于我国可持续发展的经济和社会发展模式。

在这期间，建设绿色校园十分重要。

高等教育是国家重中之重的事业，既要为国家培养人才，又要为国家提供科技成果。要使这些人才和成果符合低碳社会的要求，学校就必须改变培养方式，引入先进的科学技术知识，设置新的专业和新的实验室，编写新的教材，充实原有教材内容。

低碳生活，对于学校师生来说，关键是态度。

如果整个社会是大海，每个人就是一滴水；大海遭受污染，影响着每一滴水；同时，也需要每一滴水的努力，帮助净化污染。

低碳生活需要人人参与！

学校应通过课堂主渠道，把“节粮、节电、节水、节材、节油”等知识纳入教学内容，使学生对什么是低碳生活、为何要选择低碳生活、怎样做到低碳生活等有更深入的理解，从而培育其“低碳生活是每一个人应有的生活态度”的意识，获得实现低碳生活的能力。

老子说：“合抱之木，生于毫末；九层之台，起于累土；千里之行，始于足下。”创建绿色校园，让我们从课堂开始，从餐饮开始，从穿衣开始，从起居开始，从出行开始。让我们每个人从我做起，从身边小事做起。倡导“节粮、节电、节水、节材、节油”，成为低碳生活的倡导者，成为低碳理念的传播者，成为实现低碳生活的技能的拥有者！

熄灭的是灯光，点亮的是意识！

哥本哈根离我们很远，校园就在我们身边！

让我们为缓解全球变暖做出自己的贡献！

爱低碳生活，创绿色校园！

深化力学课程改革①

今天，来自全国各地200多所高校的近300位老师，相聚在美丽的成都市。我谨代表论坛组委会，对各位老师出席此次论坛表示诚挚的感谢和热烈的欢迎。

2007年以来，涉及力学、数学、物理、化学化工、计算机、电子电气、机械、环境、地球科学、生命科学等共10个学科的全国课程报告论坛已经举办了46次，参加报告论坛的老师人数超过2万人，涉及高校近千所。今天，新一届的力学、数学、物理、计算机等4个学科的课程报告论坛同时在4个城市举办，不得不说这是我们高校教学领域的一件盛事。课程是大学培养人才的最基础的工作，大家这样重视，说明我们国家的教育事业有望，国家走向现代化有望。

大家知道，中国的“三钱”中的两钱：钱学森和钱伟长都是我们国家力学学科的开创人，他们曾分别在中国科技大学和清华大学主持力学专业的建设和发展，而且亲自授课，能引人入胜，几十年来，一直让学生津津乐道。他们是我们授课的榜样。

力学课程报告论坛从2006年开始，已在大连、广州、南京和西安先后成功举办了四届。论坛既有专门聘请的力学教学和研究方面的知名专家做关于学科前沿和教学领域最新进展的报告，又有针对当前高校力学课程建设的热点问题而经组委会组织评选的有代表性的研讨报告，全国许多高校的力学课程的一线教师参加了论坛。论坛所讨论的问题逐步向纵深发展，它的规模和影响在不断地扩大，这充分说明我们这个论坛是有生命力的。

今年三四月间，论坛组委会分别在北京和上海召开了论坛的主题凝练会，专门就今年论坛的主题进行了研讨。四五月，又分别在江苏、湖南、山东举办了3个省的分论坛，近200位教师参加了会议，并提出了自己对力学课程报告论坛的相关需求、意见和建议。

在上述工作的基础上，本届论坛确定的主题为“深化力学课程改革，适应多样化人才培养模式需要”。为使广大教师能够更多地参与交流和讨论，本届论坛沿袭上届论坛的形式，延长分组研讨的时间，内容具体到课程，按课程组织专题深入研讨。为使研讨落在实处和有效，我们围绕理论力学、材料力学、工程力学、结构力学、弹性力学、流体力学六门课程和力学专业建设主题，根据不同专题分别进行论文征集，并从投稿中选取有代表性的成果在这次分会场组织交流和研讨。

一年一届的论坛周而复始，得到了广大力学课程教师的积极响应与大力配合。我们知道，仅仅依靠两天的会议来全面深入探讨力学课程教学中遇到的所有问题，是远远不够的，但我们希望借助这两天，选取有代表性的一个或几个问题，或者集中的一个或几个热点，来自不同地区、不同类型的具有不同特色的学校的教师能够展开充分

① 本文是第五届全国力学课程报告论坛开幕词，成都，2010年11月6日；教育与科技管理研究，北京：科学出版社，2016，141-142.

的交流与研讨，相互启发，由此延伸开来，以点带面，共同促进力学课程教学质量的持续提高。

在这里我要代表论坛的组委会向各位赴会的专家和全体教师表示衷心的感谢和崇高的敬意，向承办这次论坛的西南交通大学的领导和师生表示衷心的感谢。通过大家的努力，这届论坛一定会取得圆满成功。

从学生会主席到大学校长之路①

我奉上级命令调来暨南大学已经整整 20 年了，今天应邀来做一个命题式的报告，既高兴又有一点别扭。高兴的是能与同学们见面，别扭的是我当校学生会主席时并未有当大学校长的目标，我过去所任的学校各层级的领导职务，不是群众推举的就是领导直接任命的，我从未有当大学校长的雄心壮志。今天，只好勉为其难来讲一讲这一个问题。

大学时的学生会干部是学生群体的精英和核心，在不同时代承担不同的职责，但归根结底，应当是学习和工作双丰收，成为学生群体的表率。

下面，谈谈如何做好学生会干部。首先，学生会干部应认清自己的职责，既然是学生的表率，就必须培养自己具备以下条件。

一、应有三颗心

1. 爱心

要热爱祖国，热爱中华民族，热爱自己读书的大学，热爱自己所学的专业，热爱自己的社会工作，尊老爱幼，爱憎分明，痛恨人间一切丑恶之事。具有爱心，才能热爱生活、热爱事业，才能为祖国的美好未来奉献自己的所有力量。

2. 好奇心

对自己周边的事要爱问为什么？有了问题，才会有兴趣去寻找解决问题的方法，得到问题的正确答案，才可能做创新的人。必须虚心向他人请教，从而学到知识、学到本领。

3. 耐心

世上本无平坦的路，要做好事、善事，一定会常遇到障碍，会让别人不理解，甚至遭到反对。因此，一定要咬着牙，挺过去，有了忍耐心，就会渡过困难，追求到真理！

二、应有三个性格

1. 诚实

对祖国忠诚，对人民忠诚。作为当代大学生，一定要有诚实的品格。要说真话，不讲假话。做事要实事求是，要清清白白做人。做人要与人为善，多做好事和善事。一个具有诚实品格的人，一定会一生一世受人尊敬！

2. 不怕苦

一个人应该有不怕苦的性格。遇事要吃苦在前，享乐在后。要先天下之忧而忧，后天下之乐而乐！

---

① 本文是在暨南大学学生会干部会上的讲话，广州，2011 年 4 月 10 日；教育与科技管理研究，北京：科学出版社，2016，165-167.

只有具有不怕苦的性格，才能在遇到困难时勇于面对，顺利渡过难关，从而获得成功。

3. 乐观

人们常说“人生不如意事常八九”和“家家有本难念的经”。因此，人遇到困难时，常会愁眉苦脸，会被困难折磨、打倒。因此，一个人应具有积极向上的乐观性格，勇敢面对困难，从而顺利渡过难关。凡事要看远点，在冬天时要想到春天，在雨天时要想到晴天，在夜晚时要想到晴朗的白天！要笑口常开，要一生开朗。那样一来，你一定是一个幸福的人，做事一定容易成功！

三、应有自己的人生目标

进大学后，应有自己的人生目标，今后无论做何种工作，都应为祖国的繁荣富强作出奉献，要多做好事，多做善事，不做坏事。现阶段作为学生会干部，一定要在同学中事事起表率作用。

四、自己成长的故事

在小学、初中、高中阶段，我就担任学生干部：小学时，担任过班长、校学生会副主席；初中时，担任少先队小队长、中队旗手、大队长；高中时，担任留苏预备班的班委、班长。经过这些职务的锻炼，我学会了如何团结同学，如何组织同学完成老师交给的任务。更认识到，要当好学生干部，就必须处理好学习与工作的矛盾，自己必须事事做表率。

进入大学后，任过以下职务：副班长、班长、团支部委员、团总支宣教委员、党小组长、党支部委员、政治辅导员、校学生会主席、校务委员会委员等。对于每个岗位，我都热爱，认真工作、办事公正，同时保持全班学习成绩第一的位置。特别是在学生会主席岗位上，花去的时间特别多，对我的锻炼特别大，培养了我的战略眼光和管理能力。

我在大学一年级时，于 1959 年 3 月 13 日光荣加入了中国共产党，1959 年 10 月 1 日又被学校授予“红专旗手”称号，学校把我树为全校学生的标兵。党对我的教育和培养影响了我一生，“党的需要就是我的志愿”成为我的座右铭。在读大学时，正碰上“大跃进”时期和三年困难时期，加之又在艰苦的大西北，更是苦中又苦。我在这一时期，既保持了全班学习成绩第一的名次，又做好了繁重的社会工作，事事都起带头作用，我未叫过一声苦，事事走在前面，饿着肚子还要每月节约 2 斤粮食支援同学。带好全班同学，挺过困难时期，终于在大学毕了业。大学时任学生会主席等社会工作的锻炼为我的未来发展奠定了一个坚实的基础。

毕业后留校任助教，还下乡参加农村社会主义教育运动，任过工作队领导小组成员、组长、工作队团委书记等职。1978 年秋天，在中国科技大学近代力学系飞行器结构力学专业教研室被选为副主任，当时这一个教研室有 42 位教师，是中国科技大学最大的一个教研室。此时，钱学森先生兼任我们的系主任。接着，我的《波纹圆板的特征关系式》论文被学术界誉为“达到国际水平”而受到中国科学院通报表彰，中国科技大学师生誉称我为“中国科技大学的一颗明星”。后来，我又脱颖而出，成为中国赴西德留学的第一批洪堡学者之一，并被推选为鲁尔大学中国科学家和留学生联合会首任主席。1983年4月回国后，便任中国科技大学近代力学系副系主任，接着，在1986年被调往上海工业大学任副校长，成为钱伟长先生的助手。6年后，又来到暨南大学任副校长和校长，前后共任副校长和校长20年。在1999年11月和2000年9月，又先后被遴选为中国工程院机械与运载工程学部和工程管理学部的院士。

实际上，在任大学学生会主席时，我就坚持要在学生群体中做表率，要吃苦在前，享乐在后，决不谋私利。因此党组织喜欢我，群众拥护我。到了教师岗位后，我仍然坚持这样做，既把业务做好，又把社会工作完成好。而且在“文化大革命”时期，尽管受难，我也不做坏事、不随波逐流。在兰州大学、中国科技大学、上海工业大学、暨南大学四个不同类型的大学中，从学生会主席走上大学校长岗位，我都未改变性格，坚持讲真话，坚持公正廉洁办事，坚持只做好事、善事和不做坏事，坚持为中华民族伟大复兴的事业奋斗终身。让我特别感到欣慰的是，2006年初，在我和全校师生的共同努力下，暨南大学实现了“强校富民”，学校已由一般的大学上升为一所研究型大学，成为一所“211工程”国家重点大学，成为一所有国际影响的大学。在全国2000多所高校中，排名由几百位上升到第36位。

试答“钱学森之问”[①]

“钱学森之问”，就是著名科学家钱学森院士晚年在各种场合不止一次提出的问题：为什么我们的学校总是培养不出杰出人才?

这个问题，钱老自己其实是有答案的。2005 年 7 月 30 日，钱学森曾向温家宝总理进言：“现在中国没有完全发展起来，一个重要原因是没有一所大学能够按照培养科学技术发明创造人才的模式去办学，没有自己独特的创新的东西，老是‘冒’不出杰出人才。这是很大的问题。”

钱学森之问和钱老自己的回答，振聋发聩，实际上指出了中国的教育所存在的问题。本文试图对目前教育界的一些不良现象进行分析，深入探究其产生的背景和原因，并进而提出若干化解之策和可操作的计划。

一、教育问题的现状

新中国成立 60 多年来，国家进入了盛世时代，教育事业也取得了辉煌的成就。特别是改革开放 32 年来，教育事业更是突飞猛进，中国正处于由教育大国向教育强国发展的进程之中。但是，我国教育仍存在许多问题。

1. 教育体制和管理有待改善

在高等教育方面，目前的病态表现为：从新中国成立初期至今，教育革命和教育改革始终不断，但一直未形成科学的、稳定的人才培养体系，难以拥有杰出人才成长环境；学术浮躁；学校缺乏办学自主权，办学千篇一律，许多大学无特色；上级管理名目繁多，对学校、教学和科研的评估、考核、检查太过频繁，基层穷于应付，甚至弄虚作假；学校管理过分行政化，行政干涉学术过多；学校存在产业化问题，错误强调学校科研要产业化，使基础科学研究萎缩；学校存在关系化问题，社会和学校人治大于法治，师生办事常常要找关系才能办成，使杰出人才成长受限；“官员”型校长多，优秀校长少；教师学术不端行为和学术腐败现象愈演愈烈，甚至涉及部分党委书记和校长以及“知名”学者，但处理惩治既慢又不严；学生考试作弊现象未得到有效管治；教学和科研奖励以及科研项目申请中，拉关系现象时有发生，造成公正性缺失，制约了杰出人才出现；学校从上到下搞创收，教师无法专心致志做学问；许多教授喜欢做官，不喜欢做学问，不喜欢承担教课任务……以上这些问题都亟待教育体制和管理的尽快改善。

2. 教育经费投入不足

中国的教育投入水平不但与发达国家有着较大的差距，也低于印度、印度尼西亚等国家。例如，中国教育经费投入占 GDP 的百分比，1999～2008 年的多年平均数字为

① 本文是教育部科学技术委员会管理学部专家报告，原载《中国高校科技》，2011，(10)：4-7，14. 作者：刘人怀，郭广生，徐明稚，陈劲，陈德敏.

2.91%，相当于沙特阿拉伯的44%、美国的52%、印度的78%、日本的81%、俄罗斯的82%，这个比例与我国多年GDP高速增长的状况不相适应。

联合国教科文组织统计网站所列的216个国家或地区（其中37个无数据），1999～2008年教育经费投入占GDP的比重的多年平均数字为4.71%，中国在其中名列第143位，相当于世界平均数字的61.78%。即使在2012年底如期完成4%的目标，也仍低于平均数字，与中国的综合国力很不相称，与中国GDP总量居世界第二的位置何啻天壤之别。

3. 整个社会崇尚教育的氛围有待提高

现在，国家和老百姓都比30多年前富裕得多了，但是青少年的人生价值取向也改变了许多，许多人只追求享乐、时髦，不大讲高尚的道德、理想、情操。目前社会上存在着“向钱看、向官看”的不良倾向，整个社会对教育的态度存在着一定的功利性。

例如，学生进入大学读书，愿意学软科学专业，不愿意学硬科学专业；部分学生大学毕业后，以做大官、挣大钱为奋斗目标，不愿从事学术研究工作；大学生厌学的多了，刻苦读书的少了。

大学生是全社会最敏感的群体。社会上有多少病态现象，大学生中就有多少病态表现。

教育问题不光是大学的问题，也是中小学乃至幼儿园的问题，同时也是学生家长和社会的问题。尽管推行了多年的素质教育，但目前从小学到大学的整个教育体系仍摆脱不了应试教育的尴尬。中小学生过于沉重的学业负担和教育功利化倾向引起全社会的担忧。许多学校只把高考升学率、优秀率高低作为衡量学校整体教育质量的标准。道德教育跟不上，年轻人缺乏热爱生活、热爱祖国、热爱民族的激情，不愿意做学问，不愿搞学术。例如，有的学校和家长“唯升学率马首是瞻”，有的学校为激励高中学生，甚至提“三年地狱，一生天堂”的荒谬口号；幼儿教育小学化，甚至发展到“胎儿教育”。

二、若干化解之策

1. 建立可持续发展的教育体系

要建立可持续发展的教育体系，需要从社会、家庭、学校等多个方面进行。

应该提高各级领导的整体素质。领导的思想品德和素质将会影响教育和科技的发展。应该加大打击贪官污吏的力度，保障社会公平与正义。这将净化社会环境，对青少年树立正确的人生目标和远大理想起到积极的作用。

家庭教育是建立可持续发展的教育体系的重要环节，应鼓励父母对孩子进行正确的引导，注重家庭对孩子的影响，不要让孩子负担太重，上各种各样的“课外班”，让孩子为自己“圆梦”。同时，社会要营造鼓励、支持优秀青少年成长的环境。

基础教育要鼓励青少年“好好学习，天天向上”，树立热爱祖国、服务人民、献身科学的理想。幼儿教育小学化、中小学生负担太重的现象要尽快解决，让中小学生在快乐中成长。基础教育要鼓励青少年德智体美劳全面发展，不要盲目追求考试分数和升学率。

社会应强调公平竞争，降低人为因素在选拔人才过程中的比重。完善奖学金、助学金及勤工助学制度，让不富裕家庭的孩子也有深造的机会，使全民族的素质得以提高。

2. 教育去行政化，摆脱行政体制对教育的束缚

《国家中长期教育改革和发展规划纲要（2010—2020年）》提出“推进政校分开、管办分离”，逐步取消高校行政级别，确定了教育去行政化的目标。教育去行政化有两方面含义，一是减少行政对教育的干预，二是减少“官本位”对教育的误导。高等教育应充分体现“百花齐放、百家争鸣”的原则，创造自由、开放的学术氛围，让优秀人才脱颖而出。

减少行政对教育的干预，不要让行政事务干扰做学问，不要让功利性诱惑干扰做学问。要建立与完善中国特色的大学制度，必须造就一大批杰出的教育家，应适时引入校长职业化制度，让校长尽心尽力管理好学校，一心一意搞好教育。

子夏说“仕而优则学，学而优则仕”，这两句话构成一个良性循环，是具有正面意义的。但是，现在只突出前一句话而忽视后一句话，就变成负面意义了，以至于“官本位”气息弥漫教育界和学术界。很多教授热衷于当官，十几个、几十个教授“竞聘”一个处长岗位。因为部分人认为当了官可以搞“权钱交易”，可以获得不当官的教授难以获得的“红利”，如利用职权获得项目和经费。“官本位”现象加剧了浮躁。

应该在全社会提高教授的地位，全社会尊重知识，尊重人才。教授也要自重、自爱、自律，不要趋炎附势、随波逐流。鼓励教授“慎独”，“富贵不能淫，贫贱不能移，威武不能屈”。鼓励教授移风易俗，做精神文明建设的带头人。

在全社会提高教授的地位，首先要在高校提高教授的地位。机关人员一定要树立服务意识，要有谦逊精神，尊重知识，尊重人才，不要对教授吆喝过来、吆喝过去，弄得他们团团转，消磨他们的时间和精力。

3. 教育去产业化，摆脱市场化对教育的干扰

现代社会是专业化分工的时代，一个组织应注重提高核心竞争力，尽量少从事或不从事自己不擅长的业务。企业是如此，大学更是如此。教书育人、选拔人才是学校的首要和根本任务。即使在发达国家，科研成果的转化工作也是由企业的研究机构而不是直接由大学来完成的。大学就是教育与科研中心，不应该成为资本运作中心和利润中心。

应该说，中国的大学教师尤其是教授，现在的收入不低了，生活比较安定和舒适，超过了小康水平，不少人达到了富裕水平，比“文化大革命”期间陈景润躲在小房间里搞“1＋2”研究的条件好了不知多少倍，比马克思写《资本论》、曹雪芹写《红楼梦》的条件好了不知多少倍，但是现在的教授出了多少优秀成果呢？近几十年来，出了几位大师呢？现在，有些教授的钱袋子鼓鼓的，但是他们还是不满足，除了挣钱还是挣钱，跑来跑去挣钱，飞来飞去挣钱；哗众取宠的“大报告”多得很，“出场费”高得很；写论文、写书也成了市场炒作，粗制滥造的很多，精品很少。

现在的教育界和学术界的最大弊端可以归结为两个字：浮躁！除了浮躁还是浮躁。今天的中国，还有没有陈景润式的人物？还有没有曹雪芹式的人物？关键是有没有他们耐得清贫、甘于寂寞、潜心做学问的精神。如同“劣币驱逐良币”，“铜臭气”驱逐了书香气，“铜臭气”弥漫于教育界和学术界。

古人云：淡泊以明志，宁静以致远。做学问不能急功近利，不能到处凑热闹、赶

时髦。为了把学问做好，不妨鼓励教授和博士“好高骛远”，鼓励他们“两耳不闻窗外事”，鼓励他们“不食人间烟火”，远离名利场，远离市井尘嚣，洗净市侩习气。

鼓励学习陈景润，当“书呆子”；鼓励学习马克思、曹雪芹，耐得清贫，甘于寂寞，潜心做学问；鼓励学习陈寅恪，特立独行，把学问做深做大。

建议不要过分炒作“大学生创业”。大学生的首要任务是学习，不是创业。能够创业的大学生，少而又少。不能因为一个比尔·盖茨获得了成功，就认为千千万万大学生都能成功创业，都能成为比尔·盖茨。比尔·盖茨只是凤毛麟角，是可遇不可求的稀有人物。如果大学生能够普遍创业，还要研究生干什么？工程师、博士和教授也都要下岗才是。大学生的第一要务是静下心来好好完成学业，为以后创业打下基础，而不是马上创业，立竿见影！

“教育产业化”的提法混淆了是非，偷换了概念。教育本来就是产业——第三产业，还需要什么“产业化”？“教育产业化”的一些做法其实是错误地在搞“教育市场化”和“教育商品化”，这就改变教育是基础产业和公益事业的基本属性了。

4. 教育去关系化，让创新人才脱颖而出

大学应创造宽松的学术氛围。鼓励学术探索和学术争论。在爱国家、爱中华民族的前提下，追求不同的治学模式。大师之所以成为大师，其重要因素就是在成长为大师的过程中，比其他人花更多的时间潜心做学问，而不是熟谙人情世故处理复杂的人际关系，这也就是人们往往认为大师不善交往、“脾气古怪”，甚至难以接近的原因。应该创造宽松、公正的氛围，避免学者为上下左右的关系所困；不仅要尊重每一个学者的研究成果，还要善待每一个处于研究过程中的人。

大学应创造浓郁的文化氛围，让学生受到优秀文化的熏陶。学生不仅应学习理工，还应学习人文；不仅应了解现代，还应了解古代；不仅应知道外国，更应知道中国。避免大学生和研究生成为“有知识，没文化”的人。

大学应该宽容对待失败。科学探索与创新的过程中成功与失败并存，不能以成败论英雄。对于前沿性的探索，应该允许出现差错、允许失败。失败的教训也是有益的，可以避免在同一个地方重复跌倒。

5. 建立科学的、公正的、多样化的考评机制，确立长远的学术目标

一所大学之所以能够成为名校，是与它长期的文化沉淀、学术积累和人才济济分不开的。大学的水平是不可能仅通过短期、片面的指标考核就能够提升的。现在的一些考核指标只重数量、不重质量，发表论文只讲篇数、不讲水平。不要以数量代替质量，搞片面性的“数量化考核”，让填写各种各样的考核表格消耗教师的时间和精力。

要重视基础科学的研究工作，不要仅以研究成果的使用价值和经济效益作为考评的主要依据。不能过分强调科研成果的转化尤其是快速转化。一方面，要重视科研成果转化，促进产业发展和行业进步；另一方面，如果在研究过程中过分强调成果转化，研究的深入性和探索性就无法得到保证，必然导致短期行为和急功近利行为。陈景润的研究成果“1＋2”至今还没有看到什么转化的迹象，但是，这项研究成果受到了全世界数学界的重视，为中国赢得了声誉。如果中国多一些陈景润式的学者，多一些“1＋2”式的研究成果，无疑是一件大好事。

当务之急，对评奖和科研项目申请，要建立严格的评审制度和程序，保证公平，决不能让搞关系者得利，为杰出人才涌现创造条件。

三、让一部分人先静下心来做学问

“让一部分人先静下心来做学问!”这是仿照30多年前邓小平同志提出的“让一部分人先富起来”的口号。

回答钱学森之问，出杰出人才、出大师，必须有一批人认真搞研究，做学问；这就必须戒除浮躁、静下心来。中国的学术界、教育界现在太浮躁了，从南到北，从东到西，从一般高校到“211工程”“985工程”高校，鲜有不浮躁的；从助教到教授，鲜有不浮躁的。心态浮躁就坐不下来，就不能认真做学问，就不能出大成果、成大器。一部分人先静下心来做学问，将会带动越来越多的人静下心来做学问。戒除浮躁，静心做学问，潜心做学问，必须抵住市场化、商品化的诱惑，要挡住“挡不住的诱惑”。

要努力将教师从“以利驱动”的思维模式中解脱出来。大环境“静”不下来，学校的小环境一定要设法“静”下来，如果小环境也“静”不下来，教师个体是很难“静”下来的。

建议实施两个计划：宁静致远计划、老骥伏枥计划。

1. 宁静致远计划

宁静致远计划就是让一批人宁静以致远：在若干年后成为杰出人才和学术大师。宁静致远计划也可以称为“答问计划”，即回答钱学森之问。如果实行该计划，相信少则3年，多则5年，全国可以出现一批高水平的研究成果。如果全国有1万名学者纳入宁静致远计划，设想3年之后有10%的学者出了水平较高的成果，那就是1000项；5年之后如果有30%的学者出了水平较高的成果，那就是3000项。

2. 老骥伏枥计划

“老骥伏枥，志在千里”，这是赞扬和勉励老年人的美好的语言。老骥伏枥计划的对象是55～70岁的老教师，他们有的已经退休，有的即将退休。他们在教学和研究的岗位上工作了几十年，积累了丰富的经验和学识。他们是学校的宝贵财富而不是可以甩掉的“包袱”。他们多数希望“老有所为”，愿意“发挥余热”，继续服务于学生、服务于学校和社会。

他们可以再工作3～5年或者更长一段时间，主要做以下工作：指导青年教师，包括指导教学、指导科研；为学科建设、学校发展献计献策；编写教材。

对老骥伏枥计划的老教授，要给予工作津贴，而且适当从优，使得其“工作津贴＋退休工资”高于退休之前的收入。这样容易调动积极性，而且，学校实际增加的开支并不多。

四、大学管理模式优化，试办教育特区

1. 探索大学管理模式优化

(1) 全国高等学校，无论公立与民办，一律改由省市管理，其中一流大学的经费继续由中央财政支持，以利于教育部腾出手来，从宏观角度来改善教育环境，同时又

有利于竞争，利于杰出人才涌现。

（2）建立学费分类制度，按不同学校档次及专业类型区别收费，提高重点学校的学费，降低普通学校及基础学科的学费。对师范、农业、矿业、国防等专业全部免除学费，并适当补贴学生在学期间的生活费用，以利于来自经济困难家庭的学生上学。

（3）建立更好的奖学金制度，对有志于学的优秀学生加大资助力度；完善社会诚信体系，加强在校学生的诚信教育；建立有效的勤工助学和助学贷款制度以利于家庭贫困学生上学。

2. 试办教育特区

（1）建立政府评议与社会第三方评议相结合的民办高校评级制度，对高质量的民办高校给予招收本科层次学生的权力，条件特别好时，也应允许招收硕士、博士研究生。这些学校将与公办学校竞争，使我国高等教育事业更易向高水平发展，更利于涌现杰出的创新人才。

（2）在经济发达地区的部分高校中试行校长负责制，如在环渤海、长三角和珠三角这三个经济较发达的地区分别从“985 工程”、“211 工程”、教学型以及职业教育高校中，选择 3～5 所试行校长负责制。为确保贯彻党的方针政策，学校设党组，校长兼任党组书记。学校领导班子要少而精，只设副校长 1～3 人即可。

大学校长应该是教育家（起码应该是想要成为教育家的热心人士），且职业化，全心全意办大学。如果要求大学校长“既是教育家，又是政治家和企业家”，恐怕是要求过高，求全责备。试想，可否要求企业家同时也是教育家和政治家？可否要求政治家同时也是企业家和教育家？恐怕不能，所以，对大学校长也不能要求过高。

大学校长如果不成其为教育家（或者不想成为教育家），恐怕其他方面再好也不能算称职。现在的大学校长之中，教育家不多，“官员型”的校长、“商人型”的校长较多，他们能够办好大学吗？

校长负责制应该与教授治校相结合。教授治校是西方国家成功的办学模式，对于我国高校建设具有重要的借鉴意义。在我国实行教授治校大概需要有个过程，建议先实行教授治学。

教授治学，可以成立教授治学委员会，其职责如下：凡是学术问题均须经过教授治学委员会讨论；教授治学委员会成员可以列席校长办公会议或校务委员会会议，参与决定学校发展和管理中的重大事宜；监督教风学风，防止学术不端行为，对于学术造假行为给予鉴别和提出处罚意见，提交学校领导执行。

教授治学不但可以在教育特区的高校实行，也可以在党委领导下的校长负责制的高校中实行，就是说，普遍实行，无一例外。一所高校，如果不是教授治学，还有谁能治学呢？现在许多高校是机关人员“治学”，教授的地位并不高，有些机关人员常常对教授吆喝来、吆喝去，把学术研究变成行政事务，这是违背学术研究规律的。

五、让中国的伯乐更多一点

大学乃至整个教育体制既要建立良好的人才培养机制，也要建立有效的人才选拔机制。教育界不仅应该人才济济，同时也应该伯乐济济。伯乐济济的环境，易使优秀人才脱颖而出，从而会有大师型人才出现。

1. 选拔伯乐型官员

政府的各级教育及科研主管部门，学校的行政及教学单位，都应注重选拔伯乐型的人才担任领导职务。他们敢于选拔任用优秀人才，让其担当重要的学术研究和科研攻关任务。

各级领导、各级教学与科研管理人员应该甘当绿叶、扶持红花，以选拔、推动人才的成长为己任。

2. 注重学术团队、学术梯队的建设

在当今的信息时代，科学研究不是个人单枪匹马可以完成的。个人的知识面和眼界毕竟有限，没有团队的集体探讨、“协奏”和交流，很难把握最新的研究动向和研究趋势。因而，大学应积极鼓励和倡导、大力支持学术团队的建设。学术团队可以是自组织的，依据共同的兴趣爱好和相同的或互补的学术背景而自发形成，学校对于此种情况应该鼓励，创造与之相适应的氛围和物质条件；如果需要，人才可以在校内自由流动。

对于优势学科，应重视由不同职称和不同年龄层次组成的学术梯队的建设。学术带头人应该有博大的胸怀和敏锐的洞察力，善于听取不同意见，允许学术争论，宽容年轻学者的失败和错误，不以自己的喜好和取向限制其他人的具体研究方向。

大学应注重培养专才（包括“偏才”与“怪才”）。要培养专才，提拔优才，就要拓宽吸纳人才的途径，让有特殊才能、有志于学的学子顺利进入大学的殿堂。

要不拘一格，让优秀人才脱颖而出。要通过非常规途径选拔人才，同时，要有相应的措施防止可能产生的腐败。

“百年大计，教育为本。”现今，中国的很多高校已经是“百年老校”，更多的高校已经办学一个花甲子（60年），时间不短了。我们应该清醒地认识到，中国高等教育百年之路并不是平坦的，其中经历了长期战乱，政权变革；新中国成立以后政治运动频繁，全国院系大调整、1957年“反右”、十年“文化大革命”，以及改革开放以来的市场化冲击等，都对中国的高等教育造成了严重的干扰或破坏。目前，中国的教育同时受到两方面的批评：一是“太保守了”，求稳怕乱，是“计划经济体制最后的顽固堡垒”，希望把高等教育放开；二是搞得太乱了，面对的“乱办学”“乱收费”现象，希望“管一管”。

现在，认识到大学存在问题、认识到创新人才重要性的有识之士越来越多，钱学森之问引起大家思索，这正是中国高等教育转型的征兆，是高等教育振兴的希望。与世界名校相比，中国高校与其差距是很大的，需要奋起直追。希望尽快缩小差距，提高办学水平，培养出更多杰出的创新人才，不辜负党和国家的期望，不辜负时代赋予的使命。

认真开展高等学校教育教学研究①

受顾秉林理事长委托，我主持今天下午的第五次常务理事会。大家都很辛苦，有的是专门来参加这个会的，如黑龙江教育厅的魏厅长、江苏省教育厅的丁厅长（他还要连夜赶回去）；还有的是明天要参加教学论坛。所以开个短会，一小时左右，主要内容是请杨祥秘书长将近年，特别是研究会一年来的工作介绍一下，因为研究会明年届满，打算换届，秘书处做了一些调研，有一些想法，请大家讨论。

教学研究会第一届理事会成立于1998年底，理事长为曲钦岳院士；第二届理事会成立于2007年，现任理事长是顾秉林校长。近年来，教学研究会根据我国高等教育教学改革实际，作为中介组织、专家组织，在组织开展群众性教育教学研究与实践方面做了许多探索，发挥了应有作用，得到了广大高校的响应和支持，在高校中影响很大。特别是研究会秘书处创造性地探索研究会、高教研究中心与《中国大学教学》杂志“三位一体”的工作机制，创设大学基础课程系列报告论坛（至今有50多场）和中国大学教学论坛。这些论坛场场爆满，很受欢迎，已成为品牌论坛。研究会的这些探索也得到了教育部领导的肯定，得到了高教司的支持。为此，高教司拟进一步做实高教研究中心的工作，进一步发挥研究会的作用。

最后，谢谢大家多年来对研究会工作的支持，也希望大家今后继续积极参与研究会的有关工作，为推动和促进我国高等教育的教育教学改革，为提高人才培养质量做出更大贡献！

① 本文是第二届全国高等学校教学研究会第五次常务理事会开幕词，合肥，2011年11月18日.

加强重大工程结构安全领域国际合作[①]

首先，我代表暨南大学“重大工程灾害与控制”教育部重点实验室对各位来参加“城市生命线工程结构安全”国际联合实验室和“广东省城市生命线工程结构力学应急技术研究中心”揭牌仪式表示最衷心的感谢！

暨南大学“重大工程灾害与控制”教育部重点实验室与美国加州大学圣地亚哥分校著名的鲍威尔结构实验室于 2009 年 12 月签订了合作协议，在重大工程结构安全领域展开合作。两年来双方已经成功实现了人员互访和项目合作，在广东的科技企业中成功设立了由广东省科协授牌的企业院士工作站，引进了美方的超级展示系统等世界领先的科技产品，并积极参与中国高速铁路建设。美国工程院院士、中国工程院外籍院士、美国加州大学圣地亚哥分校鲍威尔结构实验室主任、暨南大学荣誉教授 Frieder Seible 先生数次访问暨南大学。在这些合作基础上，报请暨南大学同意，双方决定成立国际联合实验室。

国际联合实验室的建设目标是：针对城市生命线工程面临的安全风险与隐患，充分利用和发挥美国加州大学圣地亚哥分校鲍威尔结构实验室和暨南大学“重大工程灾害与控制”教育部重点实验室各自的研究基础和优势，在城市生命线工程安全风险评估、结构安全设计、结构安全检测、结构安全修复与加固等方面开展高水平的理论、计算与实验研究，设计制造城市生命线工程结构安全预警系统与产品，为减少和防范城市生命线工程的重大安全事故提供理论支持和技术保障，将实验室建设成为城市生命线工程安全领域具有国际水准的高水平联合实验室。

美国加州大学圣地亚哥分校鲍威尔结构实验室和暨南大学“重大工程灾害与控制”教育部重点实验室在各自的研究领域已取得一些成绩，Seible 院士将在稍后的大会报告中对鲍威尔结构实验室的成绩给大家做汇报。“重大工程灾害与控制”教育部重点实验室近年在先进复合材料结构非线性力学问题与工程应用、振动与冲击下结构的非线性动力响应、重大工程结构健康监测理论与应用、包装力学等研究方向取得了较好成绩并获多项奖励。下面我简单介绍其中两个项目的情况。

“复合材料基本力学问题的理论研究”项目是我近二十年来带领助手和学生在复合材料结构方面开展的应用基础理论研究。主要特色在于从增强相、基体及其界面，铺层与层间界面，以及整体结构等不同层次对复合材料结构的变形、振动、稳定性、损伤等的线性和非线性行为进行系统的多尺度研究。本项目的一些成果已经和正在应用于纤维缠绕固体火箭发动机壳体的结构设计、含裂纹桥梁的复合材料修复加固技术以及新型歼击机复合材料的缺陷和损伤分析，为我国自主创新设计制造最新型歼击机提供了复合材料设计依据，为飞行器、机械、交通等工程设计提供了重要指导。项目成果获 2005 年度广东省科学技术奖一等奖。本项目发表学术著作 7 本，学术论文 225 篇，

① 本文是暨南大学“城市生命线工程结构安全”国际联合实验室揭牌仪式开幕词，广州，2012 年 1 月 6 日.

主要发表在《国际非线性力学杂志》（*International Journal of Non-Linear Mechanics*），《国际固体与结构杂志》（*International Journal of Solids and Structures*），《复合材料科学与技术》（*Composites Science and Technology*）等国际权威刊物上。SCI 收录 125 篇，EI 收录 123 篇。被世界五大洲 1000 多位其他学者 SCI 正面引用 1192 次，在复合材料基础理论研究方面产生重要国际影响，被国际上一些著名学者称为“原创性工作”和“先驱性工作”。

“典型工程结构损伤检测方法与加固技术”项目围绕桥梁在使用过程中的移动载荷识别、结构的损伤评估与加固技术等三个方面展开，涉及力学、结构工程、材料工程、传感器技术、信息技术以及计算机技术等多门学科，复合度高。课题组成员在 10 多年的时间里，共获批 24 项科研项目，其中，国家级项目 4 项，省部级项目 11 项，横向课题 9 项，总经费近 400 万元。发表论文 200 余篇，其中，37 篇被 SCI 收录，78 篇被 EI 收录，32 篇被 ISTP 收录。发表专著 1 部。单篇他人引用次数最高达 215 次，单篇 SCI 他人引用次数最高达 85 次，累计 SCI 他人引用次数达 412 次，所检索的论文累计他引 1034 次。研究成果获得学界好评，多项研究成果应用于建筑监测、桥梁加固和移动载荷识别，测试效果良好，为保障结构安全提供了强有力的技术支持。项目成果获 2011 年度广东省科学技术奖二等奖。

各位来宾，老师们、同学们，相信在中美双方科研人员努力下，在各级领导部门的大力支持下，双方的合作一定能够在城市生命线工程结构安全领域取得突出的成绩，我们一定不辜负大家的期望，早出标志性成果，报答各位的厚爱。

最后，再次感谢各位领导、嘉宾在年初的百忙之中抽时间参加揭牌仪式和论坛。祝大家新的一年身体健康，万事如意！

知识交融　创新成才①

进入大学读书后，如何学习？这是一个很有意义的问题。我细细思考后，归结于要打好深厚知识的基础，要在知识交融中成长。下面，结合自己的实践，讲讲这个问题，供大家借鉴。

一、大学的求学阶段

1. 专业知识的学习

我进入兰州大学时，学习的第一个专业是数学专业。从 1958 年 9 月至 1960 年 7 月，整整学了两学年数学专业课程。接着，学校将我转入新成立的力学专业，做固体力学方向第一班的学生，读固体力学的课程，一共是三年，直至 1963 年 7 月毕业。显然，我读的专业涉及自然科学的两个学科：数学和力学。同时，我自己苦读文学书籍，常看小说、散文和诗歌，也喜读历史、地理、考古等期刊书籍，有空常进图书馆，因此知识面较宽。

2. 科研锻炼

在大学五年的学习生活中，我获得了三次科学研究的锻炼。

1）参加“581 工程”

1958 年，国家确定兰州为中国的科学城，要在西北研制中国第一颗东方红人造地球卫星，被称为“581 工程”。我被批准加入这一科研项目，并担任研制小组长，组员是同班的两个成绩特别优秀的同学：余庆余和周永良。这是一个秘密工程项目，又无老师指导，全是我们自己动手研制。做了两年后，学校欲将我送到西北工业大学进修深造，可惜，因我的“海外关系”，未能成功。但是这一段刻骨铭心的经历，令我受益匪浅。这段科研经历培养了我的研究兴趣，拓宽了我的知识视野，锻炼了我的独立从事科学研究的能力。

2）推广优选法

从 1958 年开始，著名数学家华罗庚在数学界推动了“理论联系实际”和“数学直接为国民经济服务”的活动，他注意到国际上刚出现的最优化方法，提出优选法。我受系领导委派，于 1960 年夏天，带领全年级 60 多位师生赴甘肃省陇西县文峰人民公社推广优选法，将全年级师生分成 10 多个组，对全公社农田进行规划。这一项科研活动又让我经受了锻炼，既使我提升了科学研究的能力，又进一步锻炼了我的科技管理能力。

3）撰写本科毕业论文

读五年级时，著名力学家钱伟长先生的大弟子叶开沅先生教我学习《板壳力学》

① 本文是百年暨南文化素质教育讲堂报告，第 145 期，广州，2012 年 9 月 27 日；教育与科技管理研究，北京：科学出版社，2016，157-164.

课程并指导我做毕业论文，叶先生出的题目是“球面扁薄圆壳的稳定性问题”。这一题目由两个人做，另一位同学叫刘法炎。这个题目源自著名力学家钱学森先生（师从世界杰出科学家冯·卡门）于1939年发表的划时代论文。由这时开始，我便触接到了力学研究的前沿领域，走上了力学科研的正确道路。所以，我特别感谢叶开沅先生，感恩叶开沅先生的谆谆教诲！

3. 品德锻炼

一个人的一生，良好品德的具备是首要的，是最重要的。

1）家庭教育

我出生在一个经济窘迫的书香之家，是教师世家。曾祖父刘声远24岁就中了举人，但中举后不久就去世了。祖父刘良幼时被称为神童，12岁考取秀才，两年后因为成绩优秀补为廪生。在20岁时被清政府送到日本振武学校留学，后加入同盟会。回国后在四川法政学堂等校任教，并参加辛亥革命，于32岁时病逝。此时，我的父亲身为长子才12岁。他毕业于著名的成都石室中学，后任中小学老师。抗日战争时投笔从戎，上前线打日本侵略军。外祖父扬升之也是清末秀才，是私塾老师。母亲扬晴岚读私塾三年，粗通文墨，持家勤俭，特别善良。父母在向我传授知识的同时，首先教育我要清清白白做人，认认真真做事，要宽以待人，严于律己。加上我的三哥刘人慰于新中国成立前参加中国共产党的地下工作，更使我受到革命教育。感恩我的祖父母、外祖父母、父母和兄弟，感恩我的老师，让我从小就受到了以下教育：“热爱祖国”“热爱中华民族”“尊敬老师”“尊老爱幼”“多做善事，多做好事”“己所不欲，勿施于人”“助人为乐”“身教重于言教”“勤劳”“诚实”等。加上我幼时正遇上日本侵略我国，目睹国家的贫弱，所以我从小就严格要求自己，希望能为祖国的繁荣富强做出贡献。

2）艰苦锻炼

九岁以前，生活在旧社会，自小便跟着父母过着困苦的生活，没有穿过一件新衣服，没有照过一张照片，没有吃过一顿好饭。从1959年夏天开始，又碰到国家经济困难。1960年，在大学食堂用餐，整整一年未见过一片肉。每月粮食定量才26斤，一半是粗粮，细粮的面粉还是全麦面，带着麸皮。而且，我作为学生干部，还带头每月节约两斤送给班上饭量大的同学。这时，年级有十多位同学饿得扛不住了，便弃学回家了。我在班上带头不叫苦，同时还把学习和社会工作做好。

3）勇气锻炼

尽管我从小体弱多病，但我自己却尽量锻炼自己的勇气。我18岁刚进兰州大学读书的第3天，1958年9月6日下午，全校师生坐火车去定西县参加“大跃进”中的“引洮上山工程”建设。全年级60人住在一个山间小庙里，附近见不着农民。半夜时，一位叫高怀一的同学癫痫病突然发作。我当时被系领导任命为副班长和工地安全委员，考虑到自己的职责，便独自上路去20里外请医生来看病。哪知在半路，在野地里突遇两条狼拦住了去路，我遇到了一生中最严酷的考验，立即镇定下来，左手开着手电筒直对着狼，右手立即解开裤腰带提在手中抖动，对峙半小时后终于吓走了狼。7年以后，1965年冬天，也是在定西县，参加农村社会主义教育活动，又一次在半夜遇到狼，这次是一条狼，狭路相逢，我又用同样办法吓走了狼。一个人遇事镇静，有勇气，就一定能迎难而上，就会不怕困难，就会战胜困难，迎来胜利。

4）助人为乐性格的培养

活在世界上，要一生活得值得，就必须认识人生的价值。我们生活在群体中，作为中华民族一分子，就必须首先做到凡事都要“吃苦在前，享乐在后”，要从身边做起。从读小学开始，一直到大学阶段，我都是班上成绩第一名，但我不骄傲，而且主动帮助成绩差的同学，耐心地回答他们的问题。由于新中国成立后我父母经济好转，我便从高中起，几乎每月都用现金帮助家庭困难的同学，而且不要他们归还，不要他们回报。在1960年经济困难时期，我自己也饿惨了，但我仍每月节约2斤粮食给同学。逢下乡艰苦时节，我都选最差的住处给自己，而做事时，又把最重最苦的活留给自己。

我活到现在，越活越快乐，笑容天天留在面孔上，留在心中。

5）学习上刻苦的锻炼

读书时，首先要热爱自己所学的专业，只有这样才有学习的热情，才有学习的动力。先要学好专业的基础课，然后又要学好专门化课程。学习中，要认真。要多做作业，要做好实验。注意学习方法，该记熟的东西一定要记熟，熟能生巧。

6）社会工作的锻炼

从小学开始，一直到大学阶段，老师和同学都选我做学生干部。我乐于担任，任劳任怨，不仅做了奉献，而且培养了我的好性格，培养了我的管理能力和服务能力。进大学后，我先担任副班长和系团总支宣教委员。为了做好宣教委员的工作，完成主办系板报的任务，我还自费订阅《新闻战线》月刊（20世纪50年代国内唯一一份新闻刊物）。一年级结束时，恰逢祖国十周年大庆，我被学校授予“红专旗手”称号。二年级时，就担任党小组长。那时学校内无专职学生工作的干部，我要负责两个年级120名学生的管理工作。从三年级起直到大学毕业，我担任校学生会主席和兰州大学第一届校务委员会委员工作，任务更加繁重，但还是圆满完成了任务，同时，学习成绩仍保持第一。

二、知识交融的创新故事

1. 旅游学方面的创新

（1）1984年8月7日，安徽省计划委员会召开全省旅游发展规划座谈会。我当时被杨纪珂副省长聘为顾问，受他的委托在会上致辞。在会上，受到安徽省旅游事业管理局局长的热情邀请，临时作了一个关于欧洲旅游见闻的报告，并建议安徽省首先开展黄山旅游区的规划工作。随后，又提出建立安徽航空公司和建设黄山机场的建议。这些建议受到省政府重视，并开始了黄山旅游区规划工作。

（2）1986年5月，由钱伟长先生推荐，我担任上海工业大学经济管理学院首任院长。上任后，经过调研，发现我国高校本科教育缺乏旅游专业。在上海，仅有上海旅游专科学校，于是创办旅游本科专业。

（3）1986年下半年，鉴于旅游学方面的研究状况，我提出用系统工程理论研究旅游学，创建“旅游工程学”学科。同时，建立上海旅游工程学会，任理事长之职。以后，我又被推选担任上海旅游协会、上海旅游学会和上海旅游教育研究会副理事长等职务，从理论和应用上研究旅游问题。

(4) 1988 年，我承担了上海市旅游事业管理局委托的课题“上海旅游交通研究”，并写出论文《上海旅游交通的症结与对策研究》，于 1990 年在著名旅游期刊《旅游学刊》上发表。

(5) 1989 年，我承担了时任上海市市长的朱镕基同志交办的上海对外经济贸易委员会课题，撰写了《上海华亭集团旅游宾馆摆脱困境的对策研究》报告。

(6) 1989 年，主持了关于“旅游工程学”的学术会议，出版了文集《旅游工程原理与实践》，著名科学家钱伟长先生为此书题写了书名和赠言，给了我极大的支持与鼓励。

(7) 1993 年，接受国务院侨务办公室主任廖晖同志的委托，在深圳市创办我国第一个旅游学院——暨南大学中旅学院（后改名为暨南大学深圳旅游学院）。

(8) 1994 年，我应邀前往日本神户商科大学①作学术报告“开展旅游工程学研究，促进旅游事业发展”。

(9) 至今，我已经培养旅游管理博士 15 人，出版著作 1 本，发表文章 20 余篇。

2. 管理科学方面的创新

(1) 1986 年 5 月，受著名科学家钱伟长先生委托，创办上海工业大学经济管理学院，担任首任院长和预测咨询研究所所长。此时，我才发现，在我国最大的城市，会计学科仅有上海立信会计专科学校在培养专科学生，为此，我筹建了会计学本科专业。

随后，我被推选为上海高校经管学院院长联谊会首任会长以及上海技术经济和管理现代化研究会（后改名为上海管理学会）第一副会长。

从此，我在管理科学的理论研究和应用上做了一些工作。

(2) 1986 年，担任上海市科学技术委员会课题“崇明岛经济、科技、社会发展战略研究”的顾问。

(3) 1987 年，担任上海市科学技术委员会重大项目“浦东新区建设工程”的课题组长，于 1989 年完成了课题报告。

(4) 1990 年，出版了专著《工业企业管理岗位要素设计》。

(5) 2003 年，受中共中央政治局委员、广东省委书记张德江同志委托，担任中共广东省委和广东省政府的重大课题“广东省信息化调研”副组长。

(6) 2004 年 6 月 2 日于澳门，在张德江同志主持的泛珠三角区域合作与发展论坛上作主题演讲——“泛珠三角：推进科技、教育和文化的区域合作”。

(7) 2009 年，担任中国工程院中国工程科技中长期（2010～2030）发展战略研究项目子课题“公共安全相关工程科技发展战略研究”的组长。

(8) 至今，我已经培养企业管理博士 18 人和管理硕士 99 人，出版著作 5 本，发表文章 170 余篇。

---

① 2004 年与其他学校重组为兵库县立大学。

3. 华文教育创新

1993 年 3 月 9 日，在广州华侨学生补习学校会议室，国务院侨务办公室廖晖主任亲自主持会议，讨论该校并入暨南大学的发展规划。会上，廖晖主任点名我提出意见，我即席发言，提出三条建议：将广州华侨学生补习学校改名为暨南大学华文学院；负责华文学科的本科和研究生教育；将校本部的对外汉语系和预科部整体并入华文学院以加强办学力量。廖主任当场表态同意我的意见，我国第一个华文学院随即按此三个建议兴办。

4. 高校异地办学的尝试

1998 年 2 月，珠海市政府在筹建珠海大学的申请未被国家批准后请我帮忙在珠海办大学，我接受邀请，冲破当时高校不能异地办学的禁令，于当年 8 月在珠海市唐家湾建立了珠海特区第一所大学——暨南大学珠海学院（后迁自珠海市区前山），开了我国高校异地办学的先河。

5. 兴办大学城的建议

2000 年春节期间，我向珠海市委黄龙云书记（现任广东省人大常委会主任）建议，为把珠海特区建设得更好、更有特色，可引进几个著名大学到珠海办学，形成大学城特色，如美国的波士顿、英国的牛津。黄书记愉快地接受了我的建议，珠海市兴办了我国第一个大学城。

6. 暨南大学体育学科的崛起

1993 年夏天，我开始分管学校的体育部工作。经过了解，我才知道，暨南大学的学生在广东办学时期从未参加过全国的比赛。于是，我采用鼓励和加强管理的办法，促使学生积极参加国内大学生的体育比赛。负责此项工作的第二个月，就让学生参加了在杭州举行的第二届全国大学生田径锦标赛，第一次参赛就获得了一枚银牌。第二年，1994 年夏天，我亲自领队参加了在山东师范大学举行的第三届全国大学生田径锦标赛，仅在第一天上午就获得 5 枚金牌，令全国高校体育界震惊。1995 年暨南大学受到教育部重视成为国家试办高水平运动队的 53 所学校之一，并在新加坡高校乒乓球友谊赛上赢得第一。2005 年 11 月 7 日至 11 日，接受亚洲大学生体育联合会和中国教育部的任务，暨南大学承办了亚洲第一届大学生田径锦标赛，我任总指挥。这是国际大学生体育运动在我国举行的首场正式比赛。19 个国家和地区共 40 所高校参赛，运动会开得圆满成功。暨南大学运动员共获 14 枚金牌、9 枚银牌、7 枚铜牌，是亚洲第一的高校。主管体育的教育部章新胜副部长当场表扬了我的工作。

随后，暨南大学设立体育专业。到我离开校长岗位之时（2006 年 1 月），暨南大学的学生共获 290 余枚金牌，其中国际比赛金牌 56 枚。

7. 开校医联合办学先河

1995 年底担任暨南大学校长后，我了解到医学院学生临床实习的艰难。学校仅有一所附属医院，远远不能满足本校医学专业学生实习的需要。为此，我找到深圳市卫生局周俊安局长和深圳市人民医院杨建国院长，他们立即同意，帮助我校解决教学实习困难。双方签了协议，深圳市人民医院成为我校第二附属医院，接收我校实习的医学专业学生；其医生又成为我校教师，还可指导研究生；该医院加入高校序列，加强了医学科学研究工作。这是双赢的事情，皆大欢喜。在签署协议之时，国家卫生部殷大奎副部长表扬我们在全国开了先例。此后，我校又先后将 6 所地方医院变成了我校的附属医院。从此，我校医学专业的学生得到了更好的培养条件。

8. 暨南大学国际化

1995 年底，在我刚任暨南大学校长时，学校仅有 16 个国家和地区的境外学生 1952 人。

这与国家对我校的“面向海外、面向港澳台”办学的要求差距较大。为此，我提出要在华侨华人定居的主要国家与该国著名大学建立姊妹关系。到我卸任校长之时，已建立姊妹学校的国家有以下 24 个。

亚洲：越南、泰国、马来西亚、新加坡、柬埔寨、缅甸、印度尼西亚、文莱、菲律宾、印度、日本、韩国。

欧洲：法国、英国、德国、俄罗斯、丹麦。

非洲：埃及、南非、毛里求斯。

美洲：美国、秘鲁、巴西。

大洋洲：澳大利亚。

与此同时，还加强与我国驻外使领馆和华侨华人社团以及海外校友的联系，争取得到他们的支持，同时加强招生宣传和组织工作。在 2005 年，暨南大学中来自境外的五大洲 77 个国家和港澳台 3 个地区的学生已增加到 10 609 人，占在校生的一半，超过全国其他学校的总和。

同时，在 2001 年，创办了中国第一所全英语教学的国际学院，使学生在国内就能接受同留学生一样的培养。

9. 信息化建设

1992 年，鉴于互联网的发展情况，我向学校提出立即接入互联网的建议，经批准后，由我负责，在罗伟其教授的协助下，于 1994 年，使暨南大学成为广东省第一所接入世界互联网的高校。

2004 年，接受中国工程院委托，我任坛主，主办了世界工程师大会的中国网上论坛。

10. 大学校史馆的创建

为激励师生员工的爱国爱校之情，在迎接暨南大学九十周年校庆的 1996 年，我提出建立暨南大学校史馆的建议，在暨南大学校友捐款后得以建成。

1998 年，在我的倡议下，全国高校的“211 工程”学校的校长来到我校聚会，教育部韦钰副部长和上述各校校长参观了我校校史馆，大家赞不绝口。此后，许多学校也纷纷建立校史馆。

11. 反腐倡廉工作的创新

鉴于学校办学任务增加和基建工程浩大，反腐形势严峻，我主动与广东省检察院检察长张学军同志联系，得到他的热情支持，签署了《共同预防职务犯罪协议书》，使暨南大学成为我国第一个与检察院合作共同预防职务犯罪的高校，受到上级领导好评。此后，学校遵纪守法形势好转。

12. 暨南大学由普通高校提升为名校

担任暨南大学校长后，根据学校的情况，我提出“侨校＋名校”的发展战略，提出中国第一个弹性学分制——标准学分制和第一个绩效考核制，实行“严”（即从严治党、从严治校、从严治教、从严治学）、“法”（即依法治校）、“实”（即实事求是）三字治校方针，采用“学生第一”“管理第一”“质量第一”的原则，实施一系列改革措施，使暨南大学从一般高校进入名校之列，成为国家“211 工程”重点大学，名次达到全国第 36 位，在全国被称为异军突起。教职工年均收入由 1995 年的 8254 元增至 2005 年的 88 900 元，增长 10 倍；教职工人均住房面积由 13.5 平方米增至 23.74 平方米，增长近 1 倍，每位教职工均享受了福利分房待遇。这使我感到欣慰，实现了“强校富民”的理想。

13. 精密仪器仪表领域的创新

1964 年夏天，我到社会上调研以寻找新的科研项目。在兰州万里机电厂，得知他们正在仿制研究美国 P2V 低空侦察飞机的测高计，其核心元件是一个锯齿形波纹圆板。我经过 14 年的艰难曲折的努力，终于创建了世界第一个精确的设计公式，以后，又做了一系列这一领域的研究。

14. 压力容器领域的创新

1969 年秋天，应兰州石油化工机器厂二分厂技术科长贾志杰同志（后任过湖北省委书记）的邀请，受兰州大学军宣队的派遣，我参加了仿制世界领先水平的美国设备在中国第一次试制的生产航空煤油的铂重整装置的研制工作，提出了一种新的实用的厚壁圆柱壳理论，提出了椭球封头开孔厚壁接管根部的应力计算公式，提出了建议，使产品试制成功。此后，又做了一系列高压和超高压容器及压力管道的研究工作。

15. 铁路桥梁工程的创新

1974 年，铁道部①第一设计院邀请我对陕西白水河铁路大桥的高桥墩进行设计计算。该桥是当时全国最高的桥，桥墩高 69.4 米，我提出一种新的实用的变厚度截头锥壳理论，提出了强度和刚度的计算公式，供设计院使用。

16. 东水西调工程的建议

从 2004 年开始，为改善我国首都北京和华北、西北地区水资源极其贫乏的状况，在国家实施南水北调工程之外，我提出建议，再采用东水西调工程，即从长江口取淡水、从渤海和黄海取海水，用管道将淡水和海水或海水淡化的水输向北京、华北和西北，以彻底解决我国北方缺水的状况。

---

① 2013 年实行铁路政企分开，不再保留铁道部。

从教感想①

我先后在五所大学任大学教师 50 年，担任大学本科生毕业论文指导老师也是 50 年，任研究生指导教师 30 年，任高校领导 22 年，培养了 53 名博士（固体力学、工程力学、企业管理、旅游管理等专业）以及 124 名硕士（固体力学、工程力学、管理工程、工商管理、高级工商管理等专业），我深深感到任教授光荣，任教师令人自豪。

特别地，我更认识到研究生导师岗位的重要性。其任务是为国家培养科技人才，出科技成果。这一任务，十分光荣，十分伟大。这一任务，对于国家说来，是核心事情。

下面，我来谈谈任研究生指导教师的几点体会，供大家参考。

一、担任研究生指导教师的要求

做一名研究生指导教师，首先必须要做到为人师表。即是说，无论是做人，还是做事，都要是学生的表率。

做人，是首要的条件，主要包括：①爱国精神；②忠厚、真诚；③只做好事，不做坏事。

做事，主要包括：①责任心；②勤劳；③创新。

实例如下。

（1）1982 年，在西德鲁尔大学，台湾“科技规划委员会顾问”×××对我进行策反，我立即进行坚决斗争，并赶到波恩向我国驻德大使馆报告，受到大使馆表扬。

（2）1981 年 3 月，在西德哥廷根歌德学院学习德语时，授课的德语老师讲到台湾时将其说成国家，我立即举手发言，指出老师的错误，声明：“台湾仅是中国的一个省！”

1982 年，在瑞士日内瓦，参加一个国际会议时，又遇到此情况，再次发言申明。

（3）我这一生，不管在何场合，都讲真话，干实事，我甚至在任校长时，公开在教师大会上讲这要求。

当然，在特殊情况下，如“文化大革命”非正常时期，我则采取沉默等“消极”方式。例如，当我的校长江隆基同志、老师叶开沅先生先后挨批斗时，我认为他们是好人，所以我不写大字报、不发言、不造反。由此自己被戴上“牛鬼蛇神”“大走资派江隆基的孝子贤孙”“大右派钱伟长的徒子徒孙”的帽子，挨批斗，几乎被整死，我也未后悔。

甚至造反派要求公安局枪决叶先生，我还组织群众联名反对。叶先生坐牢 8 年出狱后回校，造反派不收留，我还向学校写保证书，让老师回校工作，终于得到平反。

---

① 本文是在国防科技大学为参加全军研究生导师培训的新增导师所作的报告，长沙，2013 年 7 月 4 日；教育与科技管理研究，北京：科学出版社，2016，167-170.

(4) 我一生中，教过上千名本科生，培养了 53 名博士生（属于固体力学、工程力学、企业管理和旅游管理等专业），124 名硕士生（属于固体力学、工程力学、管理工程、工商管理、高级工商管理等专业）。

我对学生，无论是上课还是指导学位论文，都很认真，只有符合质量才允许毕业。

讲一个本科教学管理提升质量的故事。1993 年 10 月，在暨南大学我提出并实行“取消补考、实行重修”的改革措施，全国首创，以保证质量。不怕大家反对，不怕学生闹事，就是为了侨校的办学质量和声誉。这样一来，学风迅速好转，加上其他改革措施，暨南大学跨出一大步，由一般高校成为全国重点大学，由全国排几百名提升到第 36 名。

(5) 1970 年夏天，承担了我国最高压力容器“高压聚乙烯反应器”研制工作，设计压力 $2300kg/m^2$，要对筒体爆破帽孔洞处应力集中进行理论分析计算，整整一个月，想尽一切办法，我都无法下手，急得眼睛红肿，住进医院，进行眼睛手术。我夜以继日地工作，终于在一次似睡非睡的梦境中找到了方法，用复变函数映射法解决了问题，那是我当“牛鬼蛇神”挨批斗之后的事情。造反派曾当众污蔑我说：“你的论文还不如擦屁股的卫生纸!”而我仍然硬着头皮继续做科研、做创新。

(6) 我现在仍然在科研第一线工作，除了原有方向外，目前在做三个题目，即东水西调工程、地效飞行器研制、垃圾分类与处理，都很有意义。我一定要坚持创新，活到老，学到老，做到老!

二、因材施教培养学生

每个学生有每个学生的优势和劣势，要善于发挥学生的优势，弥补其劣势，决不能歧视暂时落后的学生。

实例如下。

我带学生时，不仅了解学生，也了解其家庭，争取全面了解学生，按其特长培养学生。对理论功底好的学生，就导向其去做基础科学前沿难题。对动手能力强的学生，就导向其去做偏工程、偏应用的课题。培养的学生可早毕业，也可晚毕业。总之，要达到质量标准才能毕业。我带的博士生，有的不到 2 年就毕业了，有的 8 年才毕业。

三、指导学生走上正确的科研道路

培养研究生，就是培养研究生如何做人，如何做科研。

这里主要讲如何指导学生走上正确的科研道路。对于硕士生，是培养其进入科研的大门，使其懂得科技创新如何实现，为何有价值。对于博士生，则是培养其独立从事科研的能力，让其完成一项具有系统创新性质的课题。

科学研究包括三类题目：①新问题、新方法；②新问题、老方法；③老问题、新方法。

总之，要在前人工作基础上往前进一步就行。当然，要选对人类、对国家有意义的题目来做，要选科学前沿问题做，要选有实际价值的问题做，即选“顶天立地”题目做。

实例如下。

以我自己为例。1962年，我的本科毕业论文题目“球面扁薄圆壳的稳定性问题”是叶开沅先生拟定的。我查询文献后，才知道题目出得好，其来自钱学森先生（师从冯·卡门）所写的于1939年发表的划时代论文。20世纪，国际力学界顶尖的领域，就是板壳非线性力学，相当难！因此，我在本科阶段就被著名力学家钱伟长先生的大徒弟叶开沅先生领入科学的前沿。我真诚感谢钱伟长先生、叶开沅先生，这使我终生在这一有意义的领域里工作，既做了基础科学课题，又做了工程实践课题。我的许多论文都在国际上开了先河。我也用这样的方法指导学生，很多学生成才。

珍惜时间　勤奋成才①

大家刚进入暨南大学大门，我作为一名老师，向大家表示热烈的欢迎！

从现在开始，大家告别了中学的学习生活，进入大学读书。你们的生活方式将发生大变化，没有人再叫你起床，没有人再催你去学校读书，没有人再督促你完成作业。从这一刻起，你既要面对丰富多彩的生活，也要面对捉摸不定的未来，一切将由你自己做主。从这一刻起，你将度过1400多天的大学生活（个别五年制专业学生是1900多天）。大学生活，有人说它丰富多彩，有人说它枯燥、艰辛。我认为，它是人生中最亮丽多彩的时段，是你未来成功的基础。同学们，你们应该珍惜大学生活的每一天！

我们的祖先关于时间留下了许多名言：一刻千金；争分夺秒；分秒必争；日月如梭；一寸光阴一寸金，寸金难买寸光阴；一日之计在于晨，一年之计在于春；时间是生命；等等。归结起来，就是时间重要！大学时间如何管理，确实是一个重要问题。

一、立志

首先，应该弄清楚，进入大学是为什么？

大学是大学生个人实现人生价值的平台，是个人知识积累、能力培养和素质塑造的关键过程，即是说，要立德修身，成为治国之才！读了大学，就要立志为国家服务，为中华民族奉献，以国家的利益、人民的利益为最高利益！

1. 钱伟长考进大学时的故事

著名科学家，我们暨南大学的董事长、名誉校长钱伟长先生被誉为中国的“三钱”之一，他是我的老师。现在讲一讲他进入清华大学读书的故事。

1931年9月16日，钱先生[1,2]进入清华大学读中文系。恰好进校之后第三天，碰上“九一八事变”，日本帝国主义侵略我国东三省。他义愤填膺，决心弃文学理，走“科学救国”之路。他便马上去找当时的物理系主任、著名科学家吴有训先生，要求转专业到物理系读书。吴主任见他物理考得太差，仅15分，化学也是20多分，这怎么能读物理专业呢？便一口拒绝了他。钱先生是文科成绩好，中文和历史都是100分，是偏才生。但他不死心，整整缠了吴有训先生一个星期。吴先生早上一离开家，钱先生就在那里等他，一路说到系里；上课时在教室外等他，一下课就缠住他。弄得吴先生一点办法也没有，只好屈服，同意钱先生在物理系试读一年。附加条件是：只有一年后每门课达到70分，才能转为正式学生，否则回中文系读书。从此，钱先生走上了“科学救国”之路。

钱先生说：“这是我一辈子中一个重要的抉择。”

---

① 本文是在暨南大学新生入学教育会上的讲话，广州，2014年9月23日；教育与科技管理研究，北京：科学出版社，2016，153-157.

2. 我入学的故事

我出生在旧中国的一个经济窘迫的书香之家，当时生活十分困苦，但受到了较好的家庭教育，知道祖父加入同盟会参加辛亥革命以及父亲到抗日前线参战的报国之举，又跟着参加共产党地下工作的三哥颠前跑后，朦胧地知道神圣的革命事业，在幼小的心灵里种上了爱国主义种子。1955 年考上温江中学的四川省留苏预备班，成绩一直是全班第一。毕业前夕，由于我有“海外关系”属于“政审不合格”学生，因而留苏未成，临时转为参加高考。因为家学渊源的关系，我从小喜欢文学。但 1956 年，我们国家发出“科学大进军”号召，要走现代化之路。学校张光汉校长找我谈话，要求我学理科。我从小听党的话，党叫干啥就干啥。所以我在高考入学志愿处，就填写了数学等离政治远一点的理工专业。尽管我的高考平均成绩达到 89.7 的高分，但仅被第七志愿的兰州大学数学专业录取。入学时，我的成绩是兰州大学 600 名新生中的第一名。

如果没有政审问题，我就走上了青年人最喜欢最向往的留学生道路。加之，又未能进入北京大学、清华大学这些名校和物理、无线电等专业读书，因此，进入兰州大学时，尽管兰州大学也是当时全国 27 所重点大学之一，我的心情仍然相当不好，但是，听党的话是我的做人原则，我坚强地忍耐下来，开始了我的大学学习生活，走上了“科学救国”之路。

二、科学管理时间

确定了读大学的志向后，就要对时间科学管理，进行战略设计。要对照自己的身体情况、学习基础、经济状况，选好自己喜欢的专业，确定是慢慢读还是快快读。要用好时间。千万不要虚度光阴，不要浪费时间，要做一名品学兼优的学生。

首先，要抓好宏观的时间管理，就是四年（或五年）的时间管理。这里要注意以下事情：专业的选定；学分制课程的选定；无论是提前毕业、推迟毕业还是按时毕业，一定要符合质量才能毕业。

其次，要抓好微观的时间管理，就是每一学期、每一月、每一星期和每一天的时间管理。要按自己的情况，决定选读几门课，是多选还是少选，是什么时间上课，是什么时间参加体育运动，是什么时间参加课外活动。总之，要把时间用好，要认真读书，认真上课，认真做实验，认真做作业，认真锻炼身体。国家需要时，就挺身而出。

1. 钱伟长先生的故事

钱伟长先生[1,2]入学转专业成功后，早晨 6 点就起床开始读书，天天如此，终于在一年后继续留在物理系读书。全班同学共 11 人，毕业时只剩下 8 人。

钱先生说：“我在大学本科四年中，得到了终生难忘的良好教育。”他的老师有：吴有训、叶企孙、萨本栋、赵忠尧、周培源、任之恭等。另外，还遇到短期来做讲座的世界知名教授：波尔、冯·卡门等。

钱先生还说：“在这四年中，我不仅在数学、物理、化学方面建立了较广泛的基础，而且找到了一整套自学的科学方法，并树立了严肃的科学学风，为我一辈子科研教学工作打下了一个坚实的基础。”

1935 年毕业时，他与顾汉章同学完成了毕业论文《北京大气电的测定》，并在当年青岛举行的全国物理学年会上宣读。这是钱先生的第一篇论文。

毕业时，钱先生考上清华大学物理系研究生，以全国考生第一名的成绩获得商务印书馆高梦旦总经理授予的研究生奖学金。

1935 年冬天，在日本入侵华北的压力下，北京以及全国的学生掀起了著名的“一二·九”运动。钱先生有强烈的爱国热情，参加了南下自行车宣传队，行程几千里，积极进行抗日救亡运动。

同时，钱先生因家贫，身体十分瘦弱，进大学时身高才 1.49 米。在清华大学著名体育老师马约翰的指导下，积极参加体育运动，成为清华大学田径队、足球队等项目代表队队员，还参加过全国田径运动会。毕业时，他的身高已升到 1.65 米。身体健康，为他的科学研究成功打下了坚实基础。

2. 钱学森先生的故事

著名科学家钱学森先生被誉为中国“三钱”之一，他是我在中国科技大学近代力学系任教时的系主任。现在介绍一下他读大学时的故事。

1929 年，钱先生[3]以优异成绩从北京师范大学附中毕业。他做出了自己人生的第一次选择，要像詹天佑一样，为落后的中国修铁路，以总分第三名的成绩考入在上海的国立交通大学机械工程学院铁道机械工程专业，走“实业救国”之路。

入学后，钱先生认真用好时间，认真读书，成绩均在 90 分以上，位居全班第一。

在交通大学一直流传着钱先生“两个 100 分”的故事。

第一个 100 分，是 1933 年的一次水力学考试，老师给他 100 分的成绩。

试卷发下来后，他自己发现了一个不起眼的笔误，把“N_s”写成了“N”，连老师都未注意到。他立即退回考卷，要求老师扣分。这让老师甚为感动，于是改为 96 分。这份试卷被老师精心保存 47 年，到 1980 年被送到上海交通大学档案馆。1996 年，上海交通大学百年庆典上，该试卷作为珍贵资料首次得以公开展示。

由这一故事可见，钱先生读大学时是多么严格要求自己！

第二个 100 分，是一份热工实验报告。钱先生特别重视实验课，十分认真和仔细，这份报告长达 100 多页，完整、详尽地记录了他在实验中观察到的现象和细节，并有不少创见，且书写整洁，作图清晰。实验老师看了这份报告后大为惊叹。这份实验报告也就成为交大机械工程系历史上最佳的学生实验报告。

1934 年，钱先生以全专业第一名的成绩，从交通大学毕业。他在大学本科学习阶段通过用好时间、认真读书，打下了未来发展的坚实基础。

3. 我自己的故事

1958 年 9 月 3 日，宝成铁路被大雨冲坏，使得我晚了两天才到达兰州大学报到。从进校第一天开始，一直到五年后毕业，我都用好时间，积极参加学校安排的学习活动。

从 1958 年 9 月至 1960 年 7 月，我在数学专业读书。1960 年 9 月，转入力学专业读书，直到 1963 年 7 月毕业。

我读大学的这五年时间，即 1958～1963 年，是新中国的一个特殊时期，是开展反右派、“反右倾”、“拔白旗”、“抓阶级斗争到底”等政治运动的特别时期。现在看来，许多事情搞错了。但当时的实况，常常是惊心动魄的，日子难熬。要读好书，是十分困难的。

五年中，我们入学的 60 名同班同学中，先后有近 20 人因为饥饿离开了学校，但我仍坚持认真读书，认真听课，认真完成作业，认真做实验，认真参加实习。同时，我还要担负繁重的学生工作，先后任副班长、团支部委员、团总支宣教委员、党小组长、政治辅导员、校学生会主席、校务委员会委员等职，花掉了许多时间。尽管如此，除了我一年级第一学期刚从南方来到寒冷的西北，身体素质差使得体育课不及格外，我年年是班上学习成绩第一名。

1959 年 3 月 13 日，我与陈广才同学一起，光荣地加入中国共产党，是班上第一批入党的党员。到毕业时，全班还有一个同学入党。

1959 年 10 月 1 日，我被学校树立成标兵，被授予“红专旗手”称号。

1960 年和 1962 年，我又先后被学校授予“先进工作者”和“优秀生”称号。

虽然兰州大学环境艰苦，又碰上国家特殊的困难时期，但我遇到了好的校长江隆基同志，遇到了好的老师钱伟长先生的大弟子叶开沅先生和陈文源、陈庆益等，还有许多好领导，如林迪生副校长、校党委宣传部崔乃夫部长（后任国务院民政部部长）、高炳兰（系党支部书记）、薛玉庸（系党支部委员）等，是他们直接培养和指导了我的学习和工作，使我在大学生阶段能顺利成长。同时，我还参加了我国第一颗东方红人造地球卫星的研制工作，参加了著名数学家华罗庚“优选法”的推广工作，在做毕业论文时得到了叶先生指导，开始了国际力学前沿——板壳非线性力学的研究工作，并与刘法炎同学一起完成了第一篇学术论文《球面扁薄圆壳的稳定性问题》，在 1963 年 10 月学校学术讨论会上宣读。

大学阶段的学习生活为我的一生打下了坚实的基础，使我有了好的知识基础和强大的解决困难的能力，我要感恩我的母校，感恩我的好领导和好老师。

参考文献

[1] 钱伟长. 八十自述. 深圳：海天出版社，1998.

[2] 钱伟长. 教育和教学问题的思考. 上海：上海大学出版社，2000.

[3] 奚启新. 钱学森传. 北京：人民出版社，2014.

以我个人经历及家庭经历浅谈家庭教育[①]

泱泱中华民族，传承着五千多年的历史文明文化。在悠长的历史岁月长河中，出现了许多伟大的、优秀的著名人物：有政治领袖、军事策略家、科学发明家、教育家、作家、诗人、名医以及民族英雄等，这些名垂史册的杰出人物的成就，至今仍闪耀着灿烂的光辉。这些名人的成名、成功，应归功于教育。古时候，没有现在的大型的分段式的学校集体教育，家族式的家庭教育是教育的主要成分。因此历史上的著名人物的成长、成功、成名，主要是家庭教育的作用。孟母三迁、岳母刺字就是优秀的家庭教育典范。

我自己的成长和成功，也能进一步证明家庭教育的重要性。我出生在新繁镇一个普通家庭，小时候家里很穷，父母生下我们兄弟姐妹七人，我是最小的一个，生活艰辛，我们却都适龄入校读书，这在当时贫穷落后的旧中国是少见的。我们刘氏家族的宗谱，明确列出家训十六条，要求子孙后代必须遵守。这十六条是：①孝父母；②友兄弟；③勤耕织；④崇节俭；⑤课读书；⑥肃闺训；⑦敦忠厚；⑧重庭谊；⑨谱牒（未经合谱、嫁娶不得相混）；⑩防止异性乱宗；⑪阴阳两宅水、树保护；⑫子嗣必须是同宗之人；⑬祭扫祖坟；⑭春秋祭拜祠堂，子孙均要参加；⑮对不种田而读书者奖励50两，对乡试中试者奖励进京差旅费白银200两，会试中试者奖白银300两；殿试中试者奖励500两；⑯族中有人做官后，学正在每次考试时应捐给祠堂白银50两，知县每年应捐给祠堂白银200两。

刘家这份家族家庭要求，体现了家庭的教育理念，重视读书，重视品行修养，重视传承优良传统。因此，从明朝末年到现阶段，刘氏家族中出现了进士、文魁、武魁、举人、贡生、廪生、秀才，以及学士、硕士、博士、教育部新世纪优秀人才、院士等诸多人物。

我的祖父刘乾初在这种家庭理念的教育下，在12岁时考中秀才，被誉为“成都神童”。于清末的乱世中，青年有志，为寻求救国之道，1903年，20岁即奔赴日本振武学校留学，是四川最早的留学日本的学生，在日本他参加了孙中山先生的同盟会，回国后在学校任教，在报纸社编辑报纸，在革命军政府工作，惜32岁英年早逝。但他的青年壮志，爱国爱民、报效国家的精神，至今仍鼓舞刘家子孙后代。

我的父亲刘伯言秉承家族遗训，酷爱读书学习，从小到老，读书不辍，他只有中学文凭，通过自学进修，成为一名优秀的中学语文、英语老师。抗日战争时，为保卫祖国弃笔从戎，加入川军到抗日前线参加战斗。退休以后还自学日语、世界语及中医学，做到可阅读外语报纸，给别人开处方治病。他为子女树立读书学习的榜样，而且为家庭购买了许多书籍，为子女创造读书的条件。

我的母亲杨晴岚是我们兄弟姐妹一致认为的优秀母亲，她身上含蕴着中国传统妇

① 本文是成都市新都区纪委教育材料，成都，2017年10月20日.

女的诸多美德，孝顺、勤劳、坚强而又温和，博爱、乐于助人。旧时代，父亲工作无定到处飘荡，有时音讯全无，七个子女的生活和教育、家庭重担都压在母亲一个人肩上。每天，天刚亮，她第一个起床，首先到祖母卧室问候安好，然后烧火煮饭，照料子女吃早餐、上学，然后做家庭清洁，并想法找活赚钱，如纺线，织布，编粽叶书包，到乡下种田，到市场卖旧货、米、油等。晚上，照顾祖母和子女睡下后，她还在灯下缝补衣裳，纳做鞋底。如此繁重的家务劳动，如此沉重的家庭负担，母亲从不叫苦叫累，不怨天尤人，最难能可贵的是她不打骂子女出气。

虽然贫穷，母亲却乐于助人，每逢赶场天，乡下的亲戚都爱到我家休歇，母亲都毫无怨言热情接待。记得因抗日战争，逃难到新繁的廖姆姆带着三个年幼子女，举目无亲，我母亲却伸出友情之手帮助他们、接待他们。新中国成立后，我到北京遇到廖姆姆在海政文工团工作的女儿，谈起往事感动不已。

母亲还有着良好的生活习惯，爱清洁、讲卫生，家中环境整齐清洁，大人孩子的衣服虽然破旧，却常常换洗，不因贫穷而褴褛。

我的父母亲为我们子女树立了爱读书、爱学习、坚强、勤劳、友爱、不怕苦累的榜样，又为我们创造了清洁卫生的读书环境，以及安静和谐的氛围，因此七个子女都会读书学习，个人成绩优良，参加工作后，在不同的工作岗位上表现优秀。

我个人从新繁小学、新繁初中、温江中学一直到兰州大学毕业，读的都不是名牌学校，但从小学到大学我的学习成绩都名列前茅，工作后不断取得进步和好成绩，获得现在的成就。我认为这是良好的家庭教育的效果。

父母性格善良温和，家中无争吵，无打骂。家中安静和谐，使我读书复习时能够心无旁骛，专心思考，迅速记忆。父母从不干预孩子的学习，也不在别人面前炫耀孩子，使我安心读书，把学习作为责任和快乐之源。

在人生的道路上，我也曾遇到许多困难和烦恼，但母亲敢于面对困难、不怕苦累的坚强意志品质，潜移默化地影响了我，使我闯过一个个难关，克服一个个困难，一步步走向成功。记得在高中毕业时，我在留苏预备班，因我二哥在台湾，政审后被取消资格，又说保送直升大学，也未成功。我因此没有参加高考复习。在这样的突然情况下，我没有抱怨，没有颓废，没有放弃，参加高考仍然取得了好成绩。接着第二个考验又来到我面前，我的高考成绩可上重点大学，可政审后又被几所重点大学拒收，最后被分配到偏远的兰州大学。那年，我的高考成绩是全校 600 名新生中的第一名。进校后，在新中国成立十周年的大喜日子，我被学校授予“红专旗手”称号。五年的学习过程中，我依然认真努力学习，完美完成大学阶段的学习任务，迎接工作后的各种挑战。1978 年底，由于科技创新上的成功，我被中国科技大学誉为学校的“一颗明星”。1981 年初，成为第一批由西德挑选的洪堡学者，学术成果被誉为“国际板壳力学理论的最高水平”。1999 年当选为中国工程院机械与运载工程学部院士，2000 年再当选为中国工程院工程管理学部首批院士。

以上我的祖父、父亲、母亲以及我个人的经历，证明了家庭教育的作用和重要性。目前，许多家庭教育有着误区，家长望子成才，想方设法让孩子进好学校、重点学校，要求孩子考试必须得高分，单纯追求学习成绩增加孩子学习内容，让孩子的一切休息时间都用在补习功课上，孩子压力大，读书成为精神负担，这种偏向的学习，不可能造就社会国家需要的人才。

我认为好的家庭教育应该具备下列几点要素。

(1) 正确的家庭教育理念：重点培养做人素质，培养子女优良品德行为以及良好的意志品质，培养子女良好的学习、生活习惯，根据孩子的个性特点及水平，合理对孩子提出人生追求目标。

(2) 良好的安静温馨的家庭气氛，整齐清洁的家庭环境，有文化阅读的空间，是孩子读书学习、健康成长、心理良好发展并成长成才的重要条件。

(3) 父母要在子女面前树立良好的学习榜样。要求子女读书好，父母也要爱读书。要求子女优秀，父母的品德行为就必须高尚。如果父母是一个做事认真、工作有成绩、博学广闻、对人热情诚信的人，子女必然耳濡目染，成为优秀的人、有成就的人。

关注家庭教育，重视家庭教育，让我们的下一代健康成长！

教化育人 师法天地①

教育与人才选拔制度真正产生密切的关系，是从隋文帝开始的。中国的教育，源于生产生活的传习，初衷是对人的培养。《孟子·尽心上》：得天下英才而教育之，三乐也。教育产生初期，读书与做官的关系并不密切。商周时期，读不读书无关紧要。只要你是担任王室或诸侯国官职的官员的嫡长子，你就不用发愁，将来你老爹的位置就是你的。这种"世卿世禄"制，公门有公，卿门有卿，农门有民，所以，读不读书，努不努力，基本上是无用的。战国时期要打仗，军功制成为底层草根进入上流社会的狭窄通道，所以从军打仗成了有抱负年轻人的首选。这个时候读书人不多。秦统一后虽然有保举、吏道、通法、征士等措施，但都没有形成与读书有关的气候，但教育参与人才选拔的作用，较之世袭制已有了进步。比如通法一途，即指凡通晓法令者，即有可能入仕。要想当公务员，就得读法律条文而且能解会用，读书有用了。汉代制度主要是察举和征辟。这样的制度，主要看负责举荐的官员知不知道，欣不欣赏，考察的是"孝廉"。应该说这是进步。看你是否符合标准，一般有四个指标，一是看德行，二是看学识，三是看决狱，四是看才干。这四种指标中有两个可以考察，即学识和决狱。学识是指对经书的知晓程度，决狱是看你对法令的理解程度。有了这两条，只要德行不亏，才能出众，就可以当官了。历史发展到这个时候，大家应该意识到读书的重要性了。东汉中期以后，地方豪强地主势力膨胀，国家行政能力日趋瘫痪，选人用人制度由官僚家族支配，参与者则侧重名声，沽名钓誉。民谣曰："举秀才，不知书。察孝廉，父别居。寒素清白浊如泥，高第良将怯如鸡。"好的制度一旦腐朽，则必然走向反面。所以，220 年曹丕称帝后，接受颍川世族陈群的建议，将人才划分为上上，上中，上下，中上，中中，中下，下上，下中，下下九个等级，是为"九品中正制"，这种制度很快由士族把持。到西晋时，门第和家世成了选才的唯一标准。"上品无寒门，下品无士族。"百姓子弟再努力，再读书，也都在下上，下中，下下之间沉浮。士族子弟，当然无论怎样也都位居上上，上中，最次也不过上下。谁还读书？隋文帝杨坚即位后，废除了"九品中正制"，选官不问门第，规定各州每年向中央选送三人，参加秀才、明经等科的考试，通过后录取为官。

隋炀帝杨广史评不好，祖君彦恨之入骨，"罄南山之竹，书罪未穷；决东海之波，流恶难尽。"就是这么一个坏蛋皇帝，在即位第一年即下诏云："君民建国，教学为先，移风易俗，必自兹始"。很快恢复了被杨坚废除的国子监、太学以及州县学，并在大业二年（606 年）增设进士科，肇始了行有 1300 年之久的、人类历史上影响最大的考试活动：科举。将教育与人才选拔紧密结合起来，不但推动了波澜壮阔的时代发展，也书写了悲欢离合的万千人生。读书做官，千百年来成为标配。当然，这个制度在从成长、壮大到没落的过程中，在从真老虎到纸老虎的演变中，出现了许多失去公允、泯

---

① 本文是《中国当代教育名家》序，北京：社会科学文献出版社，2018，1-5.

灭良知甚至是黑暗腐朽的历史污点，但将读书与选人制度紧密配合起来，其时间长度和范围广度，当属世界之最。可见即使是杨广这样的千夫所指的皇帝，也会做点有意义的事。科举制度开了读书做官之门。读书做官经常被人诟病。我们不能一味强调读书做官，但也不能全面否定读书做官，我们要十分重视人的直接经验和实践活动，也要特别重视人的学识品格。最好是让正直、清廉、信仰坚定、能力超群、理论功底扎实、实践经验丰富的人来担任各级领导职务。读书人不从政，难道让文盲来当官？只是当了官的读书人，不要失了本真。

起于隋，盛于唐，行经宋元明清，至清光绪年间，科举考试因程式僵化而陷于末途。1905 年清廷下诏："著即自丙午科为始，所有乡、会试一律停止，各省岁科考试亦即停止。"科举制"寿终正寝"。这个时候，近代中国积贫积弱，外敌入侵，军阀内战，偌大华夏放置不下一张安稳的书桌。但在这个过程中，在科举制结束后兴起的新学也在一点点成长，至新中国建立时，已基本形成了"六三三四"的学制。

新中国成立后，我国的教育事业发生了翻天覆地的变化。中国人民在站起来的同时，深深感觉到文化教育对民族发展、社会进步、个人提升的重要意义，人民群众学习文化的积极性空前高涨。1959 年 10 月召开的全国扫除文盲、业余教育工作会议资料显示，1949—1958 年 10 年间，全国扫除青壮年文盲 7502.9 万人。其间，千百万教师田头看书，挑灯备课，风雨中奔波讲授。在普通教育领域，各级各类学校如雨后春笋般兴起，呈现出生机勃勃的发展局面。虽然知识分子在各种运动中受到一些误解和委屈，但是他们以天下为己任，从未放弃传道受业的历史使命。

改革开放以来，国家富起来的潮流成就了一个个财富人生。从改革开放之初的"万元户"到后来的"千万富翁""亿万富豪"，财富的气场越来越强大。这些富翁、富豪都有自己的老师。富人富了，千百万教师还在讲台上，青灯黄卷，耕耘不辍。他们眼中真正的财富，是经济和社会发展的人才；他们追求的理想，是培养中华民族伟大复兴的脊梁。

党的十八大以来，以习近平同志为核心的党中央吹响了强国的号角，中国特色社会主义进入新时代。教育的使命和着国家民族事业的兴旺使得教育焕发了新的生机，教育工作者秉承自己的使命，不忘初心，砥砺奋进，坚守教育的基本规律，积极探索新时代中国教育面临的新形势、出现的新问题，使自己的工作紧贴时代的脉搏。党和政府也高度关心广大教师的工作和生活。《中共中央　国务院关于全面深化新时代教师队伍建设改革的意见》指出，坚持兴国必先强师，深刻认识教师队伍建设的重要意义和总体要求。指出教师承担着传播知识、传播思想、传播真理的历史使命，肩负着塑造灵魂、塑造生命、塑造人的时代重任，是教育发展的第一资源，是国家富强、民族振兴、人民幸福的重要基石。要全面贯彻落实党的十九大精神，以习近平新时代中国特色社会主义思想为指导，紧紧围绕统筹推进"五位一体"总体布局和协调推进"四个全面"战略布局，坚持和加强党的全面领导，坚持以人民为中心的发展思想，坚持全面深化改革，牢固树立新发展理念，全面贯彻党的教育方针，坚持社会主义办学方向，落实立德树人根本任务，遵循教育规律和教师成长发展规律，加强师德师风建设，培养高素质教师队伍，倡导全社会尊师重教，形成优秀人才争相从教、教师人人尽展其才、好教师不断涌现的良好局面。习近平总书记赋予人民教师崇高的地位，他指出，“一个人遇到好老师是人生的幸运，一个学校拥有好老师是学校的光荣，一个民族源源不断涌现出一批又一批好老师则是民族的希望”。广大教育工作者不辱使命，为新时代中国教育的健康发展担当尽责，默默奉献。

当代教育和着国家和民族的命运在兴衰起伏，但总的是在发展。1949 年，全国各级各类学校在校生数为 2554.70 万人，仅占总人口的 4.72%。到 2017 年，全国各级各类学校在校生达到 3.60 亿人，占到全国人口总数的 22.01%。68 年间，在校生增加了 3.34 亿人，总数是当年的 14.09 倍，占比是当年的 4.66 倍。这么快的增长速度，这么大的体量，反映了当代中国教育发展的规模成就。

教师队伍是取得这样成就的重要的职业群体。2017 年，全国各级各类学校专任教师达到 1627.26 万人。在党中央、国务院的领导下，广大教师和教育工作者在平凡的岗位上敬业勤奋，教书育人，默默耕耘，无私奉献，培养了一批又一批人才，为教育发展、民族振兴、国家富强作出了突出贡献。

2017 年 11 月，中国教育学会、中国高等教育学会、中国职业技术教育学会、中国教育电视台、中国教育报刊社、人民教育出版社等 6 家单位第一次联合开展了评选中国当代教育名家活动。经过广泛发动推荐、严格审核遴选、开展专家初评、专家委员会终评四个环节，通过严格审核材料、集体讨论审议，以认真负责的态度，以公平、公正的立场，经过热烈的讨论和反复权衡、精心斟酌，从 400 名入围候选人中选出当代教育名家终评会议的候选人和递补候选人名单，最终推选了 90 位当代教育名家。

教育正在回归本初，关于人的全面发展的理念已经突破应试教育的强大磁场，发出越来越强大的能量。正是为了实现教育在社会进步、人类发展当中的真正作用，各级各类学校教师中的优秀人才在不断思考，不断实践，不断改革，他们以锲而不舍的奉献精神在平凡的岗位上探求实现学生全面发展的途径。心系国家和民族命运，饱念爱国和教育情怀，生发强烈的使命感和责任感，忠诚于党和人民的教育事业，全面贯彻党和国家的教育方针，坚持社会主义办学方向，不断探索，求正归真。他们是学生学会做人、学习知识、创新思维、成长成才的引路人，也在这样的过程中，将自己锻炼成教育名家。他们信念坚定，情操高尚，品德高洁，学识扎实，既是人师，又是楷模；他们教育理念清晰，教育思想丰富，致力于在教育教学实践中构建中国当代教育思想理论体系；他们创新教育模式和教育方法，形成了自己鲜明的育人风格；在师生员工中乃至社会上享有崇高声誉。人格魅力和学识魅力产生的影响力，获得广泛认可。

由黄河科技学院组织专门机构，通过对 90 位当代教育名家的访问联系，在获得 25 位具有广泛代表性的当代教育名家的同意和支持后，将他们的事迹择要撰写，编著成书。其间文稿已和名家本人反复交流并获得认可。在编排体例上突破了以往此类书籍单调僵化的框框，对每个名家的事迹介绍，附有一篇普通教师或学生的文章。虽然这些教师和学生不是名家名人，但他们对这些名家的崇敬感佩之真情，呼之欲出，在一定意义上反映了名家的人格魅力和学识魅力，反映了他们无私奉献所产生的巨大影响力。

本书出版之际，受黄河科技学院董事长胡大白先生和本书主编杨雪梅校长之邀，我很高兴为之作序。

感谢党和国家的培养和教育[①]

今天，在我从教 56 年的时刻，校党委和学校按照中共中央统战部 2019 年 4 月 11 日通知为我召开退休座谈会，感激万分，谢谢林书记和宋校长！

今年，恰逢我入党 60 年，入党时受到党的教导，确立了“党的需要就是我的志愿”“党叫干啥就干啥”“党指向哪里我奔向哪里”“吃苦在前，享受在后”“多做好事，多做善事，不做坏事”的人生格言。

也是 60 年前，兰州大学授予我“红专旗手”的光荣称号。回顾 60 年的历史，感慨万千，我没有白过。按照党和国家的培养和教育，在 60 年红与专的道路上，勤勉一生，做了一点奉献。

深深感谢党的教导和培养，没有党的培养，就没有我自己！

我来到暨南大学已经快 28 年了，在领导和同志们的关怀和支持下，为国家为人民为党尽心尽力做事，完成了党交给的任务。

一、第一阶段（1991. 11—1995. 12）

这段时期，我任副校长，按周耀明校长分给我的工作认真做事。第一年，负责学校行政、档案、外事、图书馆工作，联系经济学院。半年后，因罗国民副校长调走，又将他分管的财务处、总务处、基建处、建筑设计事务所、校办产业办公室、校园管理、爱卫会和全校群众性体育工作交给我代管。再过了半年，领导班子又重新分工，让我分管本科、专科、成人教育、预科、研究生理工医科、政治理论课和德育课的教学管理以及教务处、体育部、外事处、电教中心、现代管理中心、综合档案室，筹办旅游学院和华文学院，联系经济学院。

来到暨大，可以说，举目无亲，不认识一个人。我按照入党后，党对我的教育，吃苦在前，享乐在后，尽心尽力，将党交给我的任务做好。同时，要干干净净做事。

那时，暨大尚是一所二本招生的普通本科院校，被人视为二三流学校，贬称为“花花公子”大学。

我提出取消补考、实行重修，教学三重评估和弹性学分制等改革措施，得到领导和教工、同学们的支持，使教风和学风有了好转。同时还完成了筹建深圳中旅学院和华文学院等任务。

二、第二阶段（1995. 12—2006. 1）

这段时期，我任校长，中间又兼任三年多校党委书记。负责全面工作。

我提出“侨校＋名校”发展战略，提出“从严治校、从严治教、从严治学，依法治校和实事求是”的“严、法、实”三字办学原则，在领导的关心和大家的支持下，

---

① 本文是在退休座谈会上的讲话，广州，2019 年 5 月 10 日.

克服困难，忍辱负重，改革向前。1996 年 6 月，学校由一般院校提升为“211 工程”国家重点大学，走向国际化、现代化、综合化的高水平大学之路。在全国不同机构的综合实力排行榜中的位置不断上升，以中国网大的中国大学综合实力排行榜为例，从 1998 年至 2005 年的排名，依次为 87 位、72 位、60 位、40 位、37 位、36 位、51 位、42 位。我校在不同机构的综合实力排行榜中的位置不断上升，2001—2005 年，已连续 4 年在全国 1577 所高校中位居前 50 所高校之列。创建了珠海学院，使全校学院由 7 个增加到 20 个。

“侨”校性质鲜明，海外和港澳台学生来源地由 1995 年 19 个国家和地区增长为 2005 年五大洲 74 个国家和地区，海外和港澳台学生由 1982 人增加到 10 609 人，2005 年境外生占全校学生接近一半。不花一分钱，学校校园面积由 727 466 平方米增加到 1 744 302 平方米，增长一倍。

学校建筑物面积由 46 万平方米增长到 107 万平方米，翻了一番，而且新造面积达到 83.81 万平方米。

师生员工生活工作环境、学习条件得到大大改善。教职工都享受了福利分房待遇，教职工收入增加了 11 倍。

在这一阶段，1999 年 11 月，我当选为中国工程院机械与运载工程学部院士；2000 年 9 月，再当选为中国工程院工程管理学部首批院士，为学校和国家侨务系统实现院士零的突破。

三、第三阶段（2006.1—2019.5）

在组织关怀下，2006 年 1 月，我从校长岗位退下，再次全身心投入到业务工作中。此前，白天基本未搞业务，仅仅晚上和休息日搞一些业务，加之暨南大学又无力学专业，虽然在 1993 年 1 月，建立了力学研究所，但开始除我以外，仅有一位力学老师（王璠），由于我是校领导，我不能利用权力发展自己的学科，所以力学学科未得到发展，直到 2004 年建立力学本科专业后，力学学科才稍微大了一些，这时，我已快离开领导岗位。退下来后，力学专业由小到大，现在终于建设成为一个拥有 3 个一级学科专业的学院，只是，力学专业教师队伍仍可怜！仅有 13 个人。

我在暨南大学从事教学和科研工作已经 28 年了，许多业务工作是在第三阶段做的。

在教学方面，从事“板壳非线性力学”“旅游工程学”教学。培养研究生方面，至今，已培养工程力学博士 6 人，企业管理博士 24 人，旅游管理博士 21 人，固体力学硕士 1 人，工程力学硕士 8 人，EMBA 硕士 25 人，管理科学与工程硕士 2 人，MBA 硕士 82 人。

在科研方面，先后承担和完成国家自然科学基金重点和一般项目、国务院侨务办公室科技基金项目、教育部科学技术研究项目、国家“211 工程”重点建设项目、广东省科技厅项目、中国工程科技发展战略广东研究院重大项目以及来自其他研究院所、企业的项目。

力学：出版专著 7 本，主编著作 1 本。在国内外发表学术论文 176 篇，其中英文论文 65 篇。

管理：出版专著 6 本，主编著作 5 本。在国内外发表文章 378 篇，其中英文文章 10 篇。

与此同时，我还兼任中国工程院工程管理学部副主任和首席咨询专家、中国工程院机械与运载工程学部常委、中国工程院院士广州咨询活动中心主任、广东院士联合会会长、粤港澳院士专家创新创业联盟主席、教育部科技委员会委员、战略指导委员会委员、学风建设委员会委员、管理学部主任、教育部高等学校力学教学指导委员会主任委员、中国科协中国梦讲师团成员、国家自然科学基金委员会科学部专家咨询委员会委员、国家质量监督检验检疫总局①和国家标准化管理委员会委员、中国振动工程学会理事长、中国力学学会副理事长、中国复合材料学会副理事长、全国高校教学研究会副理事长、中国仪器仪表学会常务理事和仪表元件学会理事长、广东省政协历届委员联谊会名誉会长、广东省人民政府参事、广东省科协副主席等职，为国家和社会服务。

今天借此退休时刻，我要感谢中共中央统战部和广东省委、省政府的领导的关心，感谢林书记和宋校长的关怀，感谢全校师生多年的支持，特别是要感谢力学与建筑工程学院的同志长期以来的帮助和支持，这使我在暨大 28 年中为国家为人民为党做了一些有意义的事情。

由于本人能力有限，有一些事情尚未办好，还请大家谅解。

祝愿我们暨南大学在中共中央统战部的直接领导下，更加灿烂辉煌！

① 2018 年组建国家市场监督管理总局，不再保留国家质量监督检验检疫总局。

澳门民办高校发展现状与前景①

一、引言

澳门的高等教育发展仅有 30 多年，但自 1999 年澳门回归以来，政府大力推行改革，开辟了高等教育发展的新纪元（冯增俊等，2010）。经历了固本延续、调整拓展、特色提升三个阶段后（张红峰，2014），澳门的高等教育无论是在规模还是在质量方面，都取得了巨大的进步。美国总人口数约 3.2 亿人，2017 年高等教育入学人数约 200.4 万人，仅占总人口数的 0.63%②，而澳门总人口数约 60 万人（劳动人口 39.7 万人、就业人口约 20.6 万人）③，2017 年的高等教育本地生的入学人数占比达到 3%，由此可见，澳门高等教育发展其实卓有成效。

澳门目前的高等教育体系中公办高校有 4 所，包括澳门大学、澳门理工学院④、澳门旅游学院⑤、澳门保安部队高等学校；民办高校有 6 所，包括澳门科技大学、澳门城市大学、澳门镜湖护理学院、澳门管理学院、中西创新学院、圣若瑟大学。截止到 2016 年 10 月底，澳门民办高校学生总人数是 17 964 人，占澳门所有高校学生人数（32 750 人）的 54.85%。其中创建于 2000 年的澳门科技大学，如今是澳门规模最大的综合性大学，连续 6 年入围上海交通大学高等教育研究院发布的“中国两岸四地大学排名”百强大学⑥。相对而言，内地的民办高校虽然经过 20 年的发展，已具备一定规模，但在校生约为 610.9 万人，仅占高校在校生总人数（约 2625.3 万人）的 23.27%⑦，更没有能进入百强的高校。综上所述，澳门民办高校在其独特的历史与文化背景下，对当地人力资本建构和经济发展起到了关键的作用，有着不同于内地民办高校的角色与定位（吴江秋和晏阳，2014）。

随着 2016 年美国经济放缓、英国脱欧公投等全球不确定性因素的增加，加上中国经济进入新的高速转轨格局，2016 年人均地区生产总值位列中国第一的澳门目前也正面对着产业与经济转型的巨大挑战。2016 年 9 月 8 日，澳门特区政府公布《澳门特别行政区五年发展规划（2016—2020 年）》，明确指出建设“一个中心，一个平台”（世界旅游休闲中心，中国与葡语国家商贸合作服务平台）的战略思路，重点培育会展、中医药、文化创意三大新兴产业，加快当地经济适度多元化与可持续发展。鉴于产业多

① 本文原载《中国管理科学的研究与实践——第四届中国管理科学论坛文集》，北京：科学出版社，2019，315-323. 作者：刘人怀，尹涛，呼玲妍，覃大嘉，江峰.

② https://nces.ed.gov/fastfacts/display.asp?id=372.

③ 时间序列资料库. http://www.dsec.gov.mo/PredefinedReport.aspx?ReportID=6&Lang=zh-CN.

④ 2022 年 3 月 1 日更名为澳门理工大学。

⑤ 2024 年 4 月 1 日更名为澳门旅游大学。

⑥ 科大概况. http://www.must.edu.mo/about-must-tw/intro.

⑦ 国家统计局. http://data.stats.gov.cn/easyquery.htm?cn=C01&zb=A0M0404&sj=2015.

元化与转型需要关键人力资本的支撑，本文聚焦于探讨作为当地重要人才培养摇篮的澳门民办高校，应如何把握新时代的机遇，建构可持续发展的战略规划。

二、澳门民办高等教育发展现状

1. 澳门民办高校发展历史

如前所述，澳门自回归以来，经济飞速增长，根据世界银行数据，2016 年澳门人均地区生产总值位列中国第一。其经济发展之迅猛，引起了愈来愈多中外学者的关注（吴江秋和晏阳，2014）。人才是我国经济社会发展的第一资源［见中共中央、国务院印发的《国家中长期人才发展规划纲要（2010—2020 年）》］，而人才的培养离不开高等教育（刘林等，2013）。于是，诸多中外学者在感叹澳门经济发展之快的同时，也对其高等教育展开了深入研究，其中不乏对民办高等教育的探索。澳门民办高等教育历史悠久，1594 年，罗马传教士在澳门建立圣保禄学院，这所大学是澳门私立大学的起源，它促进了中国先进知识成果的推广，为加强东西方文化的交流做出了贡献（张红峰，2014）。

自 20 世纪 90 年代以来，随着特区政府对民办高等教育的扶持，一批民办高校得到创办或升级，在数量和规模上都进一步扩大。1988 年，从澳门东亚大学分离出的澳门城市大学诞生，壮大了民办高校队伍；1996 年，澳门高等校际学院成立，2009 年更名为圣若瑟大学，这所天主教大学，充分体现了澳门文化的包容性；还有一所民办高校是澳门镜湖护理学院，它的前身为 1923 年创办的镜湖高级护士学校，1999 年被认可为私立高等教育机构并升格为学院。随着澳门回归，民办高校呈现出欣欣向荣的态势，澳门科技大学于 2000 年建校，澳门管理学院于 1988 年创建，中西创新学院于 2001 年创办。至此，六所民办高校队伍正式形成，凭借各自的特色为民办高校的壮大做出了贡献。

总体而言，20 世纪 90 年代之后，澳门民办高等教育一片繁荣，这离不开中央政府及特区政府的扶持。正是由于中央政府通过划拨土地、鼓励澳门高校到内地招生、创造澳门与内地合作办学的机会等方式，澳门民办高校在短期内获得较大发展，并为澳门经济贡献力量（高飞，2016）。

2. 澳门民办高等教育发展特色与战略定位

与公办院校相比，澳门民办高校已经具有一定的优势。截止到 2016/2017 学年，澳门民办高等教育学生注册数为 17 739 人，占澳门所有高校学生注册数的 54.2%；近 5 年来，澳门民办高等教育学生注册数呈现出蓬勃向上的趋势（图 1）。在研究生培养上，澳门民办高校也不逊色于公办院校（表 1）。澳门 6 所民办高校相关情况，见表 2。

澳门科技大学、澳门城市大学都拥有博士、硕士和学士三级学位授予权；澳门科技大学崇尚学术自由，支持教师和研究人员开展多元化的学术研究，鼓励团队协作，促进学术多元发展。

总体而言，澳门民办高校的发展特色与战略定位可以总结为如下几点。

1）有丰富、多元的跨文化课程

澳门有其独特的文化背景，一方面，澳门作为中国的领土，深受华夏文明的影响，

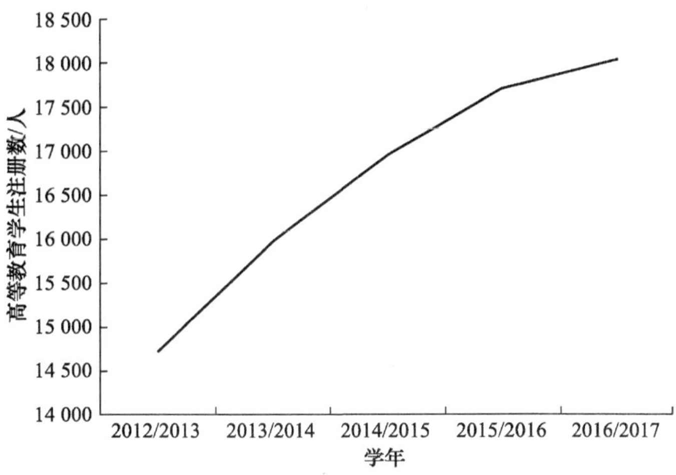

图 1　澳门民办高等教育学生注册数

资料来源：高等教育辅助办公室 2016 年高教统计数据汇编

表 1　2016/2017 学年澳门高等院校研究生人数及比率

高等院校	研究生人数/人	总学生人数/人	研究生人数比率
澳门大学	3 359	10 029	33.49%
澳门城市大学	1 373	5 834	23.53%
圣若瑟大学	202	1 063	19.00%
澳门科技大学	2 772	10 373	26.72%

资料来源：高等教育辅助办公室 2016 年高教统计数据汇编

表 2　澳门 6 所民办高校相关情况

项目	澳门科技大学	澳门城市大学	圣若瑟大学	澳门镜湖护理学院	澳门管理学院	中西创新学院
成立年份	2000	1988	1996	1923	1988	2001
热门专业（特色学科建设）	注重特色优势研究领域，如中药质量与创新药物、澳门媒体研究等	国际款待与旅游业管理博士、硕士学位课程是澳门唯一获联合国旅游组织 TedQual 优质教育素质认证的特色课程		护理学	采用融合式教学，灵活富有弹性；多重互动渠道，促进师生互动与交流	
教职工人数	有讲座教授、教授、特聘教授、副教授、助理教授等数百人，大部分教研人员拥有博士学位	355 人，聘请海内外知名专家及学者进行授课与指导	201 人	65 人，9 个客座教授、7 个荣誉教授、2 个访问教授	34 人	62 人，其中荣誉教授 18 人
注册学生数	全日制学生 10 373 人	全日制学生 5 524 人，非全日制学生 310 人	全日制学生 1 063 人	全日制学生 287 人，非全日制学生 10 人	非全日制学生 215 人	全日制学生 182 人

资料来源：各高校官网

珠三角南端的地理位置，使其具有了岭南特色的海洋文化，明清后不少福建人的迁入为澳门注入了闽台文化，如现在的“妈祖文化”。另一方面，自明代开始澳门就成了中外交流的门户，不少西方传教士通过澳门进入中国并为澳门带来了全新的西洋文化，葡萄牙人侵后更为澳门注入了属于拉丁文化系统的葡萄牙文化。幸运的是，中华文化与拉丁文化都具有极强的包容性，因此我们今天才有幸看到在澳门多种文化共荣共生、多个宗教和谐并存和中西合璧的建筑随处可见的奇特景象。“博物馆”式的澳门历史城区，也因此在 2005 年成功被联合国教科文组织列为世界文化遗产（吴江秋和晏阳，2014）。

澳门兼容并蓄的文化背景成为其高等教育产业成长与发展的重要支撑。较公办高校而言，民办高校的政策更为宽松，因此文化包容性更高。圣若瑟大学设有基督宗教研究中心和少数族裔研究中心也说明了这一点。换言之，澳门的民办高校具有较大的自主权，可以设立更具文化特色的课程。例如，自 2008 年开始，澳门旅游学院新增了“澳门文化遗产概要”和“澳门文化导游课程”两门新课程，这使澳门高等教育在面向世界的同时也反映了本土文化特征（吴立保和胡雅菁，2014）。

2）善于培养多语种国际人才

受历史原因的影响，澳门从中小学校开始，就呈现出“三文四语”的现象，即汉语学校、葡语学校、中葡语学校和英语学校。在主流的汉语学校中使用汉语，中文是书面语，广东话是教学用语；在葡语学校里，葡文既是书面语又是教学用语；在中葡语学校，同时教授中文和葡文；英语学校使用英语作为媒介，英语充当书面语和教学用语（马早明，2010）。

同样地，澳门高等教育系统不仅在教学用语上，而且在学制及其课程标准上，都具有多样化的特征（马早明，2009）。澳门高等教育具有以多语教学、学制迥异和学校自治为特征的多元化特点，使得澳门在中葡英等多语人才方面优势明显（van Schalkwyk and Hoi，2016；张红峰，2014）。鉴于此，民办高校在师资聘用、教学与课程设计等相关方面的弹性较大，也更能不遗余力地培养多语种国际人才，这些优秀的人才毕业后，多半能在需要国际交流与合作的商业、外交等方面有杰出表现，这有助于扩大中国与葡语、英语等国家在各领域的合作范围。

3）具有多元、国际化的财政资助来源

回归后澳门特区政府坚定贯彻“一国两制”的基本方针，积极施政。在此期间，我们发现澳门特区政府在整个经济社会的发展中扮演的是“服务者”而非“管理者”的角色。特区政府充分尊重市场经济的发展规律，尽量地减少干预，发挥澳门经济自由灵活的优势，维系经济的活力，特区政府更多的是作为一个“服务者”为经济的整体运行创造一个良好的环境（吴江秋和晏阳，2014）。

毫无疑问，前述的特区政府角色与体制，在很大程度上促进了民办高校的发展。简单地说，澳门特区政府秉承谁投资谁管理谁受益的原则，准许一批行业机构或专业协会积极投资办学。例如，澳门镜湖护理学院是澳门镜湖医院慈善会的下属机构，而中西创新学院是由澳门博彩娱乐有限公司投资创办的。民办高校是民间需求的产物（刘芳，2016），它们的存在使澳门高校“公私并举，投资多样”，为打造多元文化教育之都做出了巨大贡献。

三、澳门民办高等教育未来发展前景

综上所述，本文在考虑澳门本地高等教育发展现状与战略定位的前提下，对澳门民办高校的可持续发展战略规划，提出对策与建议如下。

1. 突出特色学科的同时，适度多元化

2016 年 9 月 8 日，澳门特区政府公布《澳门特别行政区五年发展规划（2016—2020 年）》，明确指出建设“一个中心，一个平台”（世界旅游休闲中心，中国与葡语国家商贸合作服务平台）的战略思路，重点培育会展、中医药、文化创意三大新兴产业，加快当地经济适度多元化与可持续发展。

由于澳门人口和地域空间小，不宜开办太多的课程，否则容易导致各高等院校的资源浪费（黄发来，2015；刘祖云和孙秀兰，2012）。因此，为响应“一个中心，一个平台”战略，澳门高等教育的学科布局围绕博彩业、旅游业的同时，依据适度多元化的发展战略，发展会展、中医药、文化创意等新兴学科领域（吴立保和胡雅菁，2014）。

2. 借助粤港澳大湾区平台，开展学术合作

《中华人民共和国国民经济和社会发展第十三个五年规划纲要》的出台，预示着打造粤港澳大湾区国际都市圈的时机已成熟。这一平台无疑将为粤港澳的学术交流创造便捷条件。港珠澳大桥的通车，使珠海与港澳站在一个新的历史起点上，珠港澳三地的合作与互动会更深层次地展开，三地高等教育资源的流动成本将进一步降低，提高了珠海高校与港澳高校合作的可能性。

粤港澳高等教育合作的核心是人才培养。人才培养的合作模式，可以在三地高等院校中展开：找出广东企业急需哪些方面的人才，据此为三地各合作高校的学生配备相关专业的教师。这样的模式对于港澳学生来讲，可以充分利用广东高校的优质资源，可以更好地提前了解内地企业的专业知识和实际情况，便于毕业后在内地工作，在一定程度上也能缓解澳门地区的就业压力。对于内地的学生来讲，也能充分利用港澳的教育资源，获取多方学位。对于合作的院校而言，有利于促进学术交流，最终通过优势专业的合作，推动高水平大学品牌的形成。

3. 加强与葡语国家的学术交流

众所周知，英语是当前世界上最流行的语言。当前，澳门各高校普遍重视英语在教学中的使用，英语的地位明显得到提升。同时，独特的历史背景使澳门具有葡语的优势。《中华人民共和国国民经济和社会发展第十二个五年规划纲要》提出，加快建设中国与葡语国家商贸合作服务平台。这一政策促进了中葡的商业合作和学术交流。澳门有许多葡语相关培训中心，民办高校可以加强与中葡论坛培训中心的合作，通过开办中葡语言研修班等方式，促进与葡语国家高等院校的交流与合作，使其高等教育质量以及国际化水平提升，成为葡语国家留学、进修、交流的目的地（佚名，2012）。

四、结语

随着澳门经济进入平稳发展阶段，这个只有61万居民的微型经济体正站在新的历史起点上，面临着机遇与挑战。澳门特区行政长官崔世安强调，特区政府将会继续坚持“教育兴澳、人才兴澳”的施政理念，将会持续提升澳门高教水平，构建终身学习型社会，提高澳门人文素养和整体竞争力。此外，鉴于目前应试教育的创新性不足会给学生较大压力，他也提到了注入创新教育模式的可能性。证据显示，2015—2017年，特区政府不断地增加对教育尤其是高等教育资源的投放，2017年教育预算已达到115.6亿澳门元。总而言之，中西包容的文化氛围加上“一带一路”“粤港澳大湾区”等政策利好，已为当地民办大学的未来发展与前景创造了独特的战略条件，澳门民办高等教育应紧紧抓住这一难得的历史机遇，加快发展，提高竞争力。

参考文献

冯增俊，江健，郭华邦，等. 2010，澳门回归十周年教育发展战略与未来走向. 教育研究，31（1）：69-74.

高飞. 2016. 澳门私立高等教育的历史进程、发展现状与治理特征分析. 浙江树人大学学报（人文社会科学），16（4）：5-9.

黄发来. 2015. 澳门高等教育：发展与质量. 大学（研究版），（11）：47-55.

刘芳. 2016. 论民办高校的核心价值取向及办学理念. 西部素质教育，2（19）：33，41.

刘林，郭莉，李建波，等. 2013. 高等教育和人才集聚投入对区域经济增长的共轭驱动研究：以江苏、浙江两省为例. 经济地理，33（11）：15-20.

刘祖云，孙秀兰. 2012. 澳门经济结构与教育结构失衡研究. 亚太经济，（5）：144-148.

马早明. 2009. 澳门高等教育的历史沿革//纪念《教育史研究》创刊二十周年论文集（14）：中国地方教育史研究（含民族教育等）. 北京：《教育史研究》编辑部：1046-1050.

马早明. 2010. 文化视野下的澳门高等教育变迁. 高教探索，（2）：31-35.

吴江秋，晏阳. 2014. 回归后“澳门模式”的经济特征及其可持续发展. 经济地理，34（8）：1-7.

吴立保，胡雅菁. 2014. 小微地区高等教育发展的路径选择：回归后澳门的经验与借鉴. 中国高教研究，（12）：33-37.

佚名. 2012. 澳门，积极搭建中国与葡语国家商贸合作服务平台. 时代经贸，（255）：60-64.

张红峰，2014. 回归十五年澳门高等教育的回顾与展望. 广东社会科学，（6）：101-108.

van Schalkwyk G J，Hoi K K M. 2016. Youths'reasoning about higher education in Macao//King R B，Bernardo A B I. The Psychology of Asian Learners. Berlin Springer：71-84.

八十有感[①]

光阴似箭，日月如梭，我已年满八十岁了。借此机会，谈谈自己的一些感想。

一、简况

我在国家最危难的1940年出生，经历了艰苦卓绝的抗日战争和异常困苦的旧中国，终于来到欣欣向荣的新中国。1958年9月，考入兰州大学数学专业。1960年9月，转入力学专业学习。1963年7月毕业。先后在兰州大学（1963—1978）、中国科技大学（1978—1986）、德国鲁尔大学（1981—1983；1993—1994）、上海工业大学（1986—1991）、暨南大学（1991—至今）、澳门科技大学（2008—2010）任教。

从1958年秋天开始，参加国家“581工程”，担任兰州大学关于我国第一颗东方红人造地球卫星的研制小组组长，开始我人生第一次科研课题的研究工作。至此，一直参加科研，已经有61年。

1960年7月，学校让我参加招生工作，随即又让我担任数学力学系1960级学生的政治辅导员，开始负责政工干部工作。

1963年8月，留校任教，至今已有56年。先后为本科生、硕士生和博士生授课，指导本科生毕业论文、硕士生和博士生的学位论文，直接培养上千名本科生、138名硕士和67名博士。同时，还担任教研室、系、院、校各级领导职务，任校长、党委书记、副校长总共22年。1999年11月，当选为中国工程院机械与运载工程学部院士。2000年9月，再当选为中国工程院工程管理学部首批院士。

我出生在一个教师之家，我的曾祖父、祖父、父亲都是教师。通过培养人才和从事科学研究，我深深感到做教师光荣，令人自豪！

二、科教兴国的重要性

回顾一下一百多年来的国家状况。

1840年，第一次鸦片战争爆发。远隔万里，来了4000名英国兵，打败了拥有4亿人口的清朝政府，签下了历史上第一个屈辱的不平等条约《南京条约》，赔款2100万银圆，我国沦为半殖民地半封建社会。

接着，1856年的第二次鸦片战争，1894年的中日甲午战争，1900年的八国联军侵华战争，1931年至1945年的抗日战争，使我国大片国土被侵占，差点亡国。

从1840年至1949年，旧中国与外国共签署了1182个不平等条约，平均一个月签一个不平等条约，割让土地数百万平方公里，仅仅在清朝就赔给外国16亿两白银。旧中国被称为“东亚病夫”。

更为可恶的是，日本还想灭掉我们中国！

---

① 本文是在广东院士联合会和“重大工程灾害与控制”教育部重点实验室（暨南大学）主办的院士人生励志报告会上的讲话，广州，2019年7月18日.

早在 1584 年，日本的丰臣秀吉就计划在 1593 年占领北京，然后就在 1594 年迁都北京。只是由于他病死，日本的迁都计划才未成功。这样的思路为日本后来的历代统治者所继承。

第二次世界大战于 1945 年 8 月结束后，美军缴获了日本的绝密作战计划，由于敏感，一直被封存。直至 2006 年才在小范围内公布。

1935 年，日本内阁会议记录上记载，在完全征服中国后，就将日本首都从东京迁到北平。在迁都的同时，从日本移民 100 万。

国弱遭人欺凌，日本统治者何其毒也！

我们国家，那时科教落后，几乎亡国啊！

从元朝开始，知识分子被贬为“臭老九”，一直到 1949 年。不尊重知识、不尊重人才，所以国家落后挨打！

我国的第一个大学，直到中日甲午战争后，1895 年 10 月 2 日，北洋大学（现为天津大学）才在天津成立。旧中国的教育、科技太落后了，生活中人们见到的东西，都带一个“洋”字，铁钉叫“洋钉”，伞叫“洋伞”，布叫“洋布”，火柴叫“洋火”，人力车叫“洋车”，蜡烛叫“洋蜡”。我们的科技太弱了。科技弱，则国弱。我小时候，亲身尝到了痛苦，目睹抗日战争时的国弱挨打的现实！

1949 年 10 月 1 日，新中国成立了！在中国共产党的领导下，我们国家才开始重视科技和教育，国家一步一步走向富强！特别是改革开放 40 年来，重视科技和教育，经济高速发展，成为 GDP 排名世界第二的国家，让全世界震惊！

三、热爱科学、热爱教育

我于 1958 年 9 月进入大学读书，正是苏联发射世界上第一颗人造地球卫星的第二年，毛主席下令，“我们也要搞人造地球卫星”。为此，中央决定搞“581 工程”，研制我国东方红第一颗人造地球卫星。兰州被命名为科学城，兰州大学成为参加单位。“两弹一星”工程被评为我国 20 世纪最伟大的第一个工程。我有幸参加了这一伟大的科研项目，使我经受了锻炼，认识到科研的艰难和重要性，这引导我走上热爱科学的道路！

1963 年 8 月担任老师后，更感到“科教兴国”的重要性。

做一位大学老师，我首先要求自己做到“为人师表”，无论是做人还是做事，都要做学生的表率。

早在大学一年级，我 18 岁时，就加入了中国共产党，到现在党龄已有 60 年。1959 年新中国成立十周年国庆节时，我被学校选为全校的“红专旗手”。我一直要求自己，做人要：①有爱国精神；②忠厚、真诚；③只做好事，不做坏事。做事要：①有责任心；②勤劳；③创新。

实例如下。

(1) 1982 年，在西德鲁尔大学，台湾“科技规划委员会顾问”孟某某受台湾“外交部”沈副部长派遣对我进行策反，我立即进行坚决斗争，并赶到波恩向我国驻德大使馆报告，受到大使馆表扬。

(2) 1981 年，在西德哥廷根市歌德学院学习德语时，授课的德语老师讲到台湾时说成国家，我立即举手发言，指出老师的错误，“台湾仅是中国的一个省”！

（3）1982年，在瑞士日内瓦参观联合国日内瓦总部，西德裁军代表团瑞尔先生向我们介绍裁军方面的国际形势，讲话中提到台湾时，将台湾与中国大陆并列，成为两个国家。我立即举手发言，指出他的错误，申明台湾是中国的一个省，不是一个国家。他当场承认错误，向我致歉。

（4）1983年，鲁尔大学校长欢迎我留在该校工作。我谢绝了高薪聘请，提前返回祖国，要为祖国的教育和科研发展做贡献。

（5）我一生从小到大，都讲真话、干实事。即使在“文化大革命”非正常时期，我也做实事，但采用沉默等“消极”方式对待。例如，我们的校长江隆基同志、老师叶开沅先生挨批斗时，我认为他们是好人，所以我不写大字报、不发言、不造反。由此自己被戴上“牛鬼蛇神”“大走资派江隆基的孝子贤孙”“大右派钱伟长的徒子徒孙”的帽子，挨批斗，几乎被整死，我也不后悔。

甚至造反派要求公安局枪决叶开沅先生，我还组织群众联名反对。叶先生坐牢八年出狱后，造反派拒绝他回校。我还向学校写保证书，让老师回校工作，终于得到平反。

（6）我对学生，无论是上课还是指导学位论文，都很认真，只有符合质量要求才允许学生毕业。

（7）1991年底到暨南大学工作后，于1993年开始负责教学管理，发现学风太差，便于当年10月提出“取消补考、实行重修”改革措施，属全国首创。不怕大家反对，不怕学生闹事，就是为了提升侨校的办学质量和声誉。这样一来，学风迅速好转。加上采取其他一系列改革措施，暨南大学声誉逐渐变好，促使暨大由一般高校成为全国“211工程”重点大学。

（8）再讲一个我的成名作故事，刻骨铭心又十分心酸。

1963年8月，留校任教后，我想联系实际，找一个国家需要的科研问题进行研究，便到兰州的工厂去调研。自己花时间，自己花钱买车票，组织上也不表扬。经过几个月努力，终于在135厂（航空仪表厂）发现一个好课题，研制飞机高度表的心脏元件：锯齿形的波纹圆板。

那时，美国的P2V低空侦察机，经常侵略我国领空，我们只能抗议。经过解放军的努力，采用密集的群海战术，终于在1961年11月6日于辽宁省庄河县击落一架。这架飞机上关键的精密仪表就是高度表，135厂接到仿制任务，但他们无从下手。我是一个喜欢搞科研的初出茅庐的年轻人，尽管是第一次见到波纹圆板，但马上感到这是一个好课题，只要国家需要，我就应去努力研究。于是，我马上答应下来，待我回校请示批准后再正式接这个任务。

回校后，我立即向教研室党支部书记汇报此事，不幸的是，立即受到他的严厉批评：刚任老师，任务很重，怎么去干外单位的事，不务正业！无法得到领导批准，只好作罢。过后，我想不通领导的意见，但又觉得这个课题事关我国航空工业，意义重大，于是决定在完成校内任务后悄悄做这个课题。

当年10月，利用去北京出差的机会，专门到首都图书馆查文献资料。由此了解到，波纹圆板的理论研究问题至今仍未解决，造成这种状况的主要原因是波纹圆板本身形状复杂，参数很多，特别是大挠度非线性微分方程在数学上求解极其困难。

第一位研究者是苏联科学院巴诺夫院士，他于 1941 年发表论文，但未得到满意的结果。接着，1945 年，苏联的费奥多谢夫院士又进行了尝试，其解答与实验值误差高达 39%。1955 年，苏联的安德列娃院士继续研究，但结果仍不理想。同年，日本的赤坂隆教授改用能量法研究，仅获得类似结果。全世界仅他们四人进行了理论研究，而我国尚无人研究。于是，我断定，这个问题从理论和实际角度上都值得花力气探索。

这个问题的关键之处是如何求解非线性微分方程组，困难极大。我先改变之前的解决方法，得到的数值结果仍不行。此时，已是 1965 年 9 月，我下乡参加农村社教工作，无法继续进行研究。接着，又碰上“文化大革命”，挨批斗，但仍未磨灭我搞科研的兴趣，我深信科研对国家的重要！所以当年摘掉“牛鬼蛇神”帽子后，我便躲在家里继续偷偷进行这项研究。大量的数值计算都靠手算，终于在 1968 年 4 月，使用自己提出的修正迭代法，给出了解析解的数值结果，公式既简单，精确度又高，与实验值十分吻合，误差仅为 0.90%，完全可以供精密仪器仪表波纹圆板元件设计使用。这是国际上第一次对波纹圆板的特征关系式给出圆满的解答。但当时，全国无学术期刊可以公开发表。等到 1972 年兰州大学《科技专刊》和国家《力学》杂志给予发表机会时，又遭造反派压制扼杀。

1976 年 10 月，一举粉碎了“四人帮”，党和国家获得了新生。我的好运也来了。我们系的工宣传长换成兰州石油化工机械厂二分厂的周师傅，他认识我，十分了解我的品格。他主动找我谈话，说《力学》杂志编辑部已经多次给系里来信，要求出具我的政治身份证明，以便发表我的波纹圆板论文。但系里造反派领导采用压制战术，始终不给回答，致使论文不能发表。周师傅说，你是优秀老师，应该让你发表论文。这时，正好学校又派来一位领导干部，张映怀同志。他原在学校团委工作，在我任校学生会主席时，已同他熟悉。他也支持周师傅的意见。于是，系的证明寄给《力学》杂志编辑部。《力学》杂志在“文化大革命”以后，于 1978 年 1 月又正式改回原来的名称《力学学报》，我的论文被放在该刊恢复名称后的第一期发表。此时，距这项工作的开始时间已整整 14 年。看到这本刊物时，真是喜极而泣！

这篇文章发表以后，迅速引起关注，我收到国家第一机械工业部仪器仪表局举办的第五届全国仪器仪表弹性元件学术会议的邀请，请我到会做特邀学术报告。1978 年 12 月 20 日，我到上海在这个会议上做了报告，立即引起很大反响。《文汇报》1978 年 12 月 30 日第一版报道我的论文达到国际水平。消息传到中国科技大学校园，正好是党的十一届三中全会公告发布之时，也许这是“文化大革命”后第一次报道该校老师论文达到国际水平的消息，引起全校瞩目，大家称我是“中科大的一颗明星”。

1979 年 8 月 29 日，《中国科学院简报》第 64 期颁发中国科学院文件：“中国科技大学讲师刘人怀在基础理论研究方面取得新成果”，通报全国表彰。国务院副总理兼中国科学院院长方毅亲笔在文件上批示“应予表扬”。

关于这个创新故事，我是吃足了苦头才获得成功。

我现在已经是年过 80 岁了，仍然在教育和科研第一线工作。为了实现中国梦，我一定要同大家一起，继续坚持创新，活到老，学到老，做到老！

新都二中——我的母校①

四川省成都市新都区第二中学（简称新都二中），是我的初中母校，令我永远为之骄傲。

新都二中的历史可以追溯到宋代创办的新繁县学，清代乾隆年间更名为繁江书院，清末改制为县立高等小学，民国时期称为新繁县初级中学。在新都二中千年以上的办学历史长河中，培养的学生可谓群星璀璨，人才辈出，例如宋代的廉吏梅挚、元代的宰相张惠、明末清初的学者费密、清末民初的反封建斗士吴虞、民国时期的文豪艾芜、新中国成立后的神话学大师袁珂，均是优秀学子的杰出代表。在办学过程中，新都二中注重培养学生"修身齐家治国平天下"的家国情怀，不仅带动了当地的教育文化事业，更引领了尊师重教的社会风尚。

我于1940年出生于新都二中的所在地——四川省新繁县繁江镇，1952年考入了新都二中，当时叫新繁县初级中学。

我家是书香门第，教师世家，世代都重视教育，并积极投身教育事业。明朝时，始祖言传身教，勉励读书；高祖更是深信"诗书继世长"的古训，考上监生；曾祖24岁中了举人。祖父刘良12岁考中秀才，后来成为廪生，20岁时被派往日本公费留学，在那里参加了孙中山先生创立的同盟会，1906年回国后从事教育，并任《天民报》主笔。父亲刘伯言读书刻苦，学识渊博，长期任教于新繁县高等小学、县初级中学，担任中学语文教研室主任，培养出神话学大师袁珂这样的得意门生，以自己的努力和成绩为新都二中赢得了荣誉。

我和二哥、四哥、三姐在新都二中都受到了良好的教育，成绩一直优秀，这得益于母校千年的文化积淀和良好的教风学风。我曾两次被学校选为省长任团长的四川省慰问荣誉军人代表团十名成员之一，初二时被选为少先队大队长，获得高中免试保送资格。1954年生日那天，我光荣地成为一名共青团员。在新都二中求学期间，学校的全面培养，帮我打下坚实知识基础；师长的谆谆教诲，使我树立未来人生志向，以致离开母校多年后，我仍能深切感受到学校的温暖和帮助。

扬帆致远逐未来，千年二中愈芳华。往昔成就令人感奋，今日发展使人鼓舞，学校的师资力量雄厚，教学设施先进，新都二中已经发展成为一所现代化的高级中学，成为广受赞誉的川西名校，昂首步入辉煌奋进的明天。

值此母校《简史》编成之际，谨以自己的以上回忆为文，并代为序。

① 本文是《新都二中简史》序，成都，2021年1月23日.

推动暨南大学“重大工程灾害与控制”教育部重点实验室跨越式发展[①]

现在，我代表“重大工程灾害与控制”教育部重点实验室，向各位专家做2020—2021年度工作报告，请予审议。

一、实验室基本情况

“重大工程灾害与控制”教育部重点实验室（暨南大学）于2007年2月12日经教育部发文批准同意立项建设。2015年1月通过教育部组织的重点实验室建设评估，被教育部批准正式成立。2015年10月22日教育部正式下文通过验收。2016年6月召开了2014—2015年度会议，2018年1月召开了2016—2017年度会议，2019年3月召开了2018年度会议，2020年9月召开了2019年度会议。

“重大工程灾害与控制”教育部重点实验室主要以重大工程结构（包括板壳和大跨度网壳结构、管道结构、高层建筑、大型桥梁、城市生命线工程等）和重大工程装备（高温发动机、航空发动机、核电装备等）为研究对象，应用先进的非线性力学理论，结合材料科学和工程技术的最新成果，对工程结构和装备中的灾害进行机理分析、状态监测、诊断和智能控制，研究并解决重大工程结构和重大装备的灾变机理及其控制的关键理论和技术问题。

重点实验室通过严格科学的管理，营造浓厚的学术氛围，凝聚一支高水平的学术团队，培养造就多名科研业务熟练、学术水平高超、在国内外有一定影响力的学术带头人。

实验室目前有固定工作人员79人，包括中国科学院院士1名，中国工程院院士1名，教育部“长江学者奖励计划”特聘教授1名，国家杰出青年科学基金获得者、中国科学院“百人计划”入选者1名，国家高层次人才1名，教育部新世纪优秀人才1名。博士生导师31人，正高级人员（含教授、研究员和教授级高级工程师）32人，副高级人员（含副教授、副研究员、高级工程师和高级实验师）38人；博士学位获得者占72%。

二、实验室的定位和研究方向

1. 依托学科

“重大工程灾害与控制”教育部重点实验室涵盖暨南大学力学、光学工程、信息工程、包装工程、材料学和土木工程等学科。其中力学学科下的工程力学为广东省攀峰重点学科、国务院侨办重点学科，2015年入选广东省高水平大学重点建设学科，工程力学在“板壳结构非线性力学”领域处于国际领先水平；土木工程在“城市生

① 本文是“重大工程灾害与控制”教育部重点实验室2020—2021年度工作报告，广州，2022年9月5日.

命线工程结构检测、监测和预警”方面卓有成效，在重大工程结构的灾变机理及控制关键技术、结构加固技术、绿色建筑材料等方面的研究在国内处于先进水平；光学工程是广东省优势重点学科，在“光纤传感技术”和“高速光通信”领域处于国内领先水平；包装工程在“运输包装动力学理论与设计、包装材料中有害物迁移动力学研究”方面处于国际先进水平。

实验室作为高层次科研平台，有力推动了依托学科的建设与发展。2020 年度土木工程专业成为广东省一流本科专业建设点，2021 年度土木工程和工程力学专业成为国家级一流本科专业建设点，2021 年度结构工程硕士点升级为土木工程一级学科硕士点。

2. 实验室定位

“重大工程灾害与控制”教育部重点实验室是华南地区依托力学学科建设的唯一的一个教育部重点实验室，不仅是华南地区重大工程灾害控制的重要研究平台和人才培养基地，对华南地区的有关大型工程结构的行业协会及学会起到了良好的支撑作用；而且还承担或协助了全国五十多家单位的科研任务，为国家科技需求和地方社会经济发展做出了重要贡献，成为国家重大工程灾害控制研究的主要平台之一。

3. 实验室研究方向

实验室的主要研究方向包括以下几个。

（1）重大工程结构的非线性分析及安全风险评估。主要研究大型工程结构的非线性稳定性研究、安全评估和应急处置，带缺陷板壳结构的损伤分析和疲劳寿命预测，含损伤复合材料结构的多尺度力学破坏机理分析，含损伤工程结构剩余强度预测，工程结构非线性动力学响应分析等。

（2）重大工程装备的事故预警与控制。主要研究装备关键部件材料强度性能的微纳观机理、装备关键材料在多场耦合复杂环境中的性能演化及材料性能测试与表征技术、材料建模与结构强度及安全评估的统一直接途径、复杂环境中重大工程装备安全状况的预警控制集成化技术等。

（3）重大工程结构的损伤检测与健康监测。主要研究重大工程结构荷载识别、结构损伤检测、监测及评估，含损伤工程结构的加固理论与修复技术、用于结构健康监测的光纤传感和信号传输技术、基于大数据分析的重大工程结构灾害预警等。

（4）岩土力学与重大工程地质灾害。主要研究现代岩土力学理论及其工程应用、重大工程地质灾害快速识别与防控，深基坑支护变形、地基承载力和沉降计算、软土地基沉降与边坡渐进破坏等。

三、科研活动和成果情况

1. 承担任务概况

2020—2021 年度实验室获批纵向项目共 64 项，获批经费 5437 万元；横向项目 120 项，立项经费 8553 万元。其中获批国家级项目 17 项，总获批经费 4947 万元；获批 100 万元以上的重大科技项目共 11 项，总经费 6249 万元。

目前实验室承担的项目科研经费共计 2.21 亿元，其中纵向科研经费 1.37 亿元，横向科研经费约 8400 万元。

2. 研究成果概况

2020—2021 年度重点实验室的研究成果涵盖了四个主要研究方向，不仅包含多个前沿领域的理论研究成果，还涉及不同领域的工程应用及社会服务成果。共发表论文 130 篇，其中 SCI 收录论文 84 篇，包括中国科学院 1 区或 TOP 期刊论文 11 篇；EI 收录论文 25 篇。出版专著 3 部。获发明专利授权 12 项，申请公开发明专利 43 项。获得行业学会科学技术一等奖 2 项、二等奖 1 项。

3. 代表性科研成果简介

1）研究方向一：重大工程结构的非线性分析及安全风险评估

A. 大型工程结构的非线性稳定性研究、安全评估和应急处置

发展了压力管道结构失效的核心理论与关键测试技术。该成果针对一系列的单壁钢管、保温夹层管、弹性补偿波纹管受不同荷载作用下的失效问题，考虑几何非线性和材料非线性，建立适用于管道失效分析的非线性理论体系。还研制了一套适用于压力管道的测试设备，可以监测管道所处的外压环境，测试高静水外压或其他复杂工况载荷下管道结构的屈曲、屈曲扩展、破坏以及缺陷检测等力学性能。该项成果获得 2021 年度广东省仪器仪表学会科技进步奖一等奖。

B. 含损伤复合材料结构的多尺度力学破坏机理分析

（1）发展和丰富了含脱层复合材料结构的屈曲分析方法。

提出了含多个贯穿脱层的纤维增强复合材料层合板的屈曲性能分析方法。该方法不仅简化了传统方法烦琐的推导过程，而且揭示了深层次的力学机制，可推广到对含多分层损伤层合板的非线性力学性能的评估，为航空航天先进复合材料的结构设计和力学分析提供有力的技术支持。

基于一阶剪切变形理论，提出了湿热环境下含脱层复合材料中厚板的屈曲和后屈曲响应的理论分析方法，并建立了基于偶应力理论的剪切变形微纳米板的尺度效应分析模型。同时，还开展了考虑纤维桥联效应的复合材料层合板屈曲及分层扩展行为的研究。

（2）建立了多个纤维增强复合材料的损伤本构模型。

建立了多个用于纤维增强复合材料的损伤演化及强度预测的损伤本构模型，包括考虑基体弹塑性效应的三维损伤模型，考虑剪切非线性效应和损伤累积导致材料属性退化的三维损伤模型，以及双轴受压的高延性纤维增强水泥基复合材料本构模型。这些损伤模型不仅能表征复合材料的非线性行为，而且在损伤演化及强度预测方面均具有优异的表现。

上述研究成果发表在包括复合材料领域的国际权威期刊《复合结构》（*Composite Structures*）和“中国科技期刊卓越行动计划”期刊《复合材料学报》上。

C. 工程结构非线性动力响应分析

在工程结构非线性动力响应分析方面取得了丰富的成果，主要如下。

（1）提出了两种新的结构动力学分析方法。

第一种是具有可控数值耗散的线性两步时间积分方法，该方法同时具有稳定性以及位移、速度和加速度能同时达到二阶精度等优点。该项成果发表在《国际工程数值方法杂志》（*International Journal for Numerical Methods in Engineering*）上，并成

为该期刊2020—2021年度的高被引论文。

此外，还提出了谱一致起步线性两步法及其等效单步法。该新算法避免了传统线性两步法在起步计算中谱性能不可控的问题，使位移、速度和加速度能同时达到二阶精度，适用于求解含物理阻尼、外载荷、非线性等各类特征的结构动力学问题。

(2) 开展了基于精准数值模拟方法的三体问题三体碰撞解研究，不仅获得了1646个全新的三体碰撞解，还获得了三体碰撞解渐近特性的数值证据。

(3) 开展微电子机械系统（micro electro mechanical systems，MEMS）陀螺真空高低温振动特性测量方法与振动激励方式研究，所搭建的MEMS振动测量平台形成了应用示范，可为MEMS陀螺的可靠性优化设计提供有益的指导。

(4) 提出一种新的基于可靠度的结构优化性能度量法——BFGS（Broyden Fletcher Goldfarb Shanno）平均值方法。与已有的典型能度量方法相比，该方法收敛快、精度高，数值计算稳定。

2) 研究方向二：重大工程装备的事故预警与控制

A. 材料建模与结构强度及安全评估的统一直接途径

在发展有效的新本构模型来表征金属强度演化和预测金属疲劳失效行为及其高效数值算法等方面取得多项创新成果。

(1) 解决金属材料软化表征疑难，首次提出从强化至软化的完整准确表征。

(2) 提出反例首次直接阐明屈服概念无法表征循环失效响应的根本局限性，摒弃屈服概念，提出全程耗散新弹塑性模型，实现在普遍意义上自动统一地表征单调和循环破坏效应，无须涉及额外的损伤变量和裂纹演化以及特设判据。

(3) 提出高效数值算法全程模拟金属疲劳破坏效应，突破通常算法由于超高耗时无法处理高周及超高周疲劳问题的计算瓶颈。

(4) 基于创新型弹塑性模型，实现记忆合金从伪弹性到塑性的转变的直接统一模拟，无须涉及相变表征及其判据。

关于上述研究成果在国际权威期刊上共发表SCI论文10篇，其中中国科学院1区论文及TOP期刊论文2篇，并获得一项优秀论文奖励。

B. 复杂环境中重大工程装备安全状况的预警控制集成化技术

基于时变理论，开展对垂直悬挂提升结构振动的监测数据分析。该研究成果解决了曳引绳端部振动特性模型简化的边界条件处理问题，为曳引系统振动特性分析提供参考，为电梯运行的安全监测技术研发提供理论指导。

3) 研究方向三：重大工程结构的损伤检测与健康监测

A. 结构损伤检测、监测及评估

在结构损伤检测及健康监测方面发展了多种新方法，取得了丰硕的成果，主要包括以下几项。

(1) 提出基于迁移学习的对结构损伤敏感且对环境因素鲁棒的结构特征提取方法。该方法通过引入迁移学习技术，设计双分支结构深层域适应神经网络，提取对损伤敏感但对建模误差与环境因素不敏感的结构特征，为结构损伤检测的工程实践中遇到的环境不确定性问题提出了一种全新的思路，具有广泛的工程前景。该成果发表在土木工程领域顶级期刊《计算机辅助土木和基础设施工程》（*Computer-Aided Civil and Infrastructure Engineering*）上。

（2）提出基于主成分分析的损伤识别方法，并在此基础上开发了基于少量传感器信息的桥梁结构安全监测方法。这一成果提出利用少量传感器信息的桥梁高分辨率全桥完备模态识别方法，突破了传统模态识别方法由于传感器数量的限制只能得到极其稀疏的模态振型的瓶颈。关于该项成果发表了 5 篇 SCI 论文，其中 3 篇 TOP 期刊论文，并获得 4 项发明专利授权。

（3）在结构健康监测领域提出了一种新的结构状态评估指标——加权传递率保证准则（weighted transmissibility assurance criterion，WTAC），制定并研究了结构连续和不连续损伤的正则化策略，以及基于加权传递函数灵敏度的结构损伤多状态检测策略。关于该项成果发表了 3 篇 SCI 论文及 1 项发明专利授权。

（4）提出一系列新的结构损伤识别方法，包括：基于变分自动编码器无监督特征提取的方法、基于改进加权迹最小绝对收缩与选择算子（least absolute shrinkage and selection operator，Lasso）正则化的结构损伤检测混合蚁狮优化与改进 NM（Nelder Mead）方法、基于加权模态数据与柔度置信准则的鲸鱼优化算法（whale optimization algorithm，WOA），以及基于包含改进 Nelder-Mead 算法的蚁狮优化器（ant lion optimizer with Improved Nelder Mead algorithm，ALO-INM）与加权迹范数的方法。这些方法均具有较高的识别精度和较好的噪声鲁棒性，为结构损伤识别提供了丰富的成果。

B. 结构荷载识别及其评估

（1）基于压缩感知与冗余字典的移动荷载识别。

利用移动荷载先验信息构造用于稀疏表达移动荷载的冗余字典；根据移动车载与车致响应关系定义构造字典的间接方式；基于压缩感知理论，直接利用响应压缩系数推导荷载识别方程，采用稀疏正则化求解方程。此方法能明显缩短测量响应数据长度，可有效缓解响应采集与能耗之间的矛盾；能直接利用响应压缩系数识别移动荷载，具有较强鲁棒性。

该项成果发表在中国科学院 1 区期刊《机械系统与信号处理》（*Mechanical Systems and Signal Processing*）上。

（2）定义了一种移动荷载识别的半凸函数，提出了一种预条件范围限制的广义最小残余（generalized minimum residue，GMRES）移动荷载识别新算法，确定了移动荷载识别病态性的病态性指标（ill-posedness index，IPI）指标与正则化方法参数选择准则，并基于字典扩展和稀疏正则化方法对武汉某多跨连续桥梁进行了移动车载现场识别，采用理论发展、数值仿真和现场实验相结合的方式，研究和验证了所提多种移动荷载识别理论方法的正确性、鲁棒性和现场适用性，取得移动荷载识别的新突破。

C. 含损伤工程结构的加固理论与修复技术

（1）动态载荷作用下加固结构的界面性能分析。

对广泛应用于工程结构加固的单搭接头开展动态载荷下的界面力学行为分析。通过使用拉普拉斯变换法与有限傅里叶余弦级数方法相结合解决了单搭接头受三角脉冲加载时的界面力学行为。研究结果可启发合理设计趋于平衡的单搭接头（single-lap adhesive joints，SLJ），从而提高接头承载能力。

对 FRP① 混凝土界面性能开展研究，发现往复变幅疲劳影响对构件疲劳寿命的影响较大，并建立了修正的 Miner 破坏准则预测构件疲劳寿命和裂缝增长模型。相关研究成果为工程应用中 FRP 黏结界面性能分析评价和提升，提供了科学解释和理论方法。

（2）玻纤复材固废粗纤维增强混凝土的研发。

基于玻纤固废高值化再生利用理念，提出将玻纤复材固废处理成粗纤维，并将所得粗纤维替代钢纤维制备纤维混凝土，大幅提升了混凝土的抗折强度和韧性、抗压强度及抗劈裂性能。所研发的混凝土尤其适用于海洋环境、化冰盐使用环境等特殊场合。研究成果发表在中国科学院 1 区期刊《建筑和建筑材料》（*Construction and Building Materials*）上。

4）研究方向四：岩土力学与重大工程地质灾害

在现代岩土力学理论及其工程应用方面，开展了软硬复合地层条件下盘形滚刀破岩机理研究。

为解决复合地层条件下盾构机刀盘刀具异常磨损等设计与施工问题，采用理论分析、有限元数值模拟和试验相结合的方法，对软硬复合地层条件下的盘形滚刀破岩力学机理进行深入研究。基于实验室的千吨立柱式压力机，设计研制了盘形滚刀线性切割试验装置。通过该盘形滚刀线性切割试验装置，进行了模拟岩样及灰岩试样的滚刀破岩试验，得到了复合地层岩层界面处的盘形滚刀破岩特性。

该研究成果获得两项发明专利授权。

四、队伍建设与人才培养

1. 队伍建设

肖衡教授在 2020 年和 2021 年先后入选爱思唯尔（Elsevier）发布的“材料力学”和“力学”学术领域的“中国高被引学者”榜单。这是肖衡教授连续 7 年被该榜单收录。袁鸿教授荣获 2021 年度广东省仪器仪表学会杰出科技人才奖。多名教师获得国家级、省级和校级优秀教学成果和教学奖励。

2020—2021 年度，实验室引进优秀青年人才 5 名，包括第四层次正高 1 名，第五层次副高 3 名，第六层次讲师 1 名。

2. 人才培养

2020—2021 年度共有 3 名博士后进站、5 名博士后顺利出站。目前共有在站博士后 5 名。博士后人员共获批国家自然科学基金青年科学基金项目 4 项。

2020—2021 年度招收力学博士研究生 9 人，毕业 9 人；招收学术硕士研究生 50 人，毕业 43 人。2020 年起专业硕士学位点调整为“土木水利”，2020—2021 年度共招生 87 人，年均招生规模比 2019 年扩大近一倍。

2020—2021 年度，研究生为主发表的（研究生第一作者或导师第一作者、研究生第二作者）科研论文共 120 多篇，其中 SCI 论文 70 多篇。研究生参与完成的科研成果

① FRP 即 fiber reinforced polymer，纤维增强聚合物。

获得广东省仪器仪表学会科技进步奖一等奖、广东省力学学会自然科学奖一等奖。博士生吴嘉瑜的学位论文获得2020年度“广东省力学学会优秀博士学位论文”；博士生王思宇获得2021年暨南大学研究生创新论坛“优秀创新奖”；硕士生谢勇康、刘超获得国际会议论文奖。

2020—2021年实验室继续加强创新型工程科技人才培养，加大力度推进工程力学“钱伟长创新班”、“卓越未来”结构工程师创新计划和国际CDIO① “同心并力班”人才培养项目三大举措。本科生在各类国家级、省级科技活动及学科竞赛中斩获近40项大奖。

五、开放交流与运行管理

1. 对外交流与合作情况

实验室积极开展学术合作与交流，与国外知名高校：美国加州大学圣地亚哥分校鲍威尔结构实验室、亚利桑那州立大学、英国贝尔法斯特女王大学、澳大利亚墨尔本大学、格里菲斯大学、西澳大学、法国巴黎第六大学、日本茨城大学、西班牙拉科鲁尼亚大学等高等院校建立了长期稳定的科研合作关系。

与国内多所科研院所，如兰州大学西部灾害与环境力学教育部重点实验室、太原理工大学、重庆大学、中国科技大学、香港理工大学、同济大学、青海大学、工业和信息化部电子第五研究所等保持稳定的合作关系。

2020—2021年度，实验室主办及承办学术会议10次；参加国内国际学术会议并受邀作学术报告60余人次，邀请国内外著名学者来访并作学术报告近30场。

同时，实验室继续加强与东莞理工学院、广东省地震局以及深圳防灾减灾技术研究院的联合共建，在共同申报重大科研项目、开展科学研究、承接重大工程技术服务及产学研攻关方面取得新的突破。

实验室在促进学科交叉和新兴学科发展方面也开展了多项工作。将大数据、人工智能、3D打印等新兴学科引入实验室“城市生命线工程结构检测、监测和预警”“工程结构及其关键部件的力学与工艺性能”等特色研究领域，开展学科交叉与融合。2020年，实验室联合应急管理学院、光子技术研究院，共同申报应急管理学部重点实验室。2021年，实验室参与“苏炳添速度研究与训练中心”建设。

2020—2021年度，实验室承担或协助了越秀地产开发有限公司、广州汽车集团股份有限公司、贵州大自然科技股份有限公司、北京临近空间飞行器系统工程研究所、中广核研究院有限公司、广东省地震局等五十多家单位的科研任务，为国家科技需求和地方社会经济发展做出了重要贡献。

同时，实验室对华南地区的有关大型工程结构的行业协会及学会起到了良好的支撑作用，协调并很好地组织了中国仪器仪表学会仪表元件分会、广东省仪器仪表学会、广东省非开挖技术协会、广东省机械工程学会压力容器分会等年度会议及学术活动。实验室还承担了广州市科协科普项目“超材料设计与3D打印制备科普宣传活动”，开展了40余场科普活动。

---

① CDIO，即conceive（构思）、design（设计）、implement（实施）、operate（运行）。

2. 公共平台情况

2020—2021 年度购置新设备 3 台，软件 2 套，共计 83 万元。主要包括静态数据采集仪（27 万元）1 台、分析仪器全息声发射仪（17.78 万元）1 台、微机控制电液伺服压力试验机（16 万元）1 台、深基坑支护结构设计软件（12 万元）等。

为迎接教育部评估，目前实验室正在对土木实验楼进行升级改造和修缮建设，增加实验室使用面积，以确保达到评估对实验场地安全性和场地面积的要求。

随着人员规模的扩大和科研水平的提高，实验室所承担的科研任务日益增多，实验室的大型实验设备的使用率普遍较高，美特斯工业系统公司（Material Test System，MTS）材料试验机、电液伺服疲劳试验机、惠普（HP）计算服务器长期满负荷运行；落锤冲击试验机、霍普金森杆冲击动力系统等已完成多个重要课题的冲击动力测试任务；高速拉伸试验机、电动加载式蠕变试验机、非接触式应变测量系统、复合材料热压机、红外热像仪等在材料力学性能测试、结构关键部件应力分析、复合材料结构损伤检测等方面发挥着重要作用。

3. 运行管理情况

"重大工程灾害与控制"教育部重点实验室实行主任负责和学术委员会评审制。设主任 1 名、常务副主任 1 名、副主任 3 名，建立了由 5 名成员组成的分工合作管理班子；同时配备 3 名专职管理人员，包括重点实验室秘书 1 名、科研助理 1 名和设备管理人员 1 名。形成完善的实验室运行支撑和服务体系，充分保障实验室日常工作的顺利开展。

重点实验室学术委员会，作为实验室的学术领导机构，主要负责决定实验室的科研方向、审定研究课题、监督经费使用、制定开放基金项目申请指南、审批开放基金申请及组织项目结题答辩等。

重点实验室实行开放、流动、联合、竞争的运行机制，实验室接受国内外科研人员以客座研究人员身份来实验室工作。积极开展多种形式的国际合作和学术交流，如合作研究、科研人员互访和讲学、联合培养研究生等。

4. 依托单位支持情况

依托单位暨南大学在多方面给予了"重大工程灾害与控制"教育部重点实验室大力支持，在人才引进方面给予政策倾斜。重点实验室是人财物相对独立的科研实体，实验室空间相对集中。学校在广东省高水平大学建设中，给予了重点实验室一定的经费支持。

"重大工程灾害与控制"教育部重点实验室的进一步发展，还需要学校继续给予支持和帮助。相信在学校的大力支持和实验室全体成员的共同努力下，今后一定能取得更加辉煌的成绩，为国家和地方经济发展做出更大的贡献。

六、目前存在的问题和下一步的发展思路

目前存在的主要问题如下。

（1）博士研究生招生人数依然偏少。2020—2021 年度，力学一级学科博士点共招收了 9 名博士研究生。目前博士点共有博士导师 14 名，博士生招生指标仅为 4 人。博士研究生是实验室科研工作的主力军，是高水平科研成果产出的主要贡献者之一。博

士研究生的不足，极大影响了实验室的发展。实验室一方面充分利用招生指标，加大力度吸纳优质生源；另一方面也希望学校在研究生招生指标方面进一步给予政策倾斜和大力支持。

(2) 引进人才指标过少。2020—2021 年度仅引进人才 5 名，2022 年实验室引进人才指标 0 或 1 名。实验室依托力学与建筑工程学院，为引进优秀人才做了大量的工作，但由于指标的限制，引进的人才难以满足实验室发展的需求。今后将进一步加强与学校相关部门的沟通，积极争取学校支持。

下一步实验室将保持并稳步提升几个主要研究方向的水平，进一步突出在“重大工程灾害与控制”方面已形成的鲜明特色和优势；以粤港澳大湾区建设为契机，积极开展三地重大科研项目合作、联合共建优势学科和实验室、进行科研成果分享转化等；结合实验室的主要研究方向，瞄准前沿科学问题和重大科技需求，加快科学研究和技术研发；参照国内“双一流”高校相同学科培养要求，提升人才培养质量。组织科研团队申报重大科研项目，培育重大科研成果，推动实验室跨越式发展。

高等教育管理实例：暨南大学“侨校＋名校”之路（1995—2006）

1. 暨南大学校长任职仪式讲话[①]

今天，国务院侨务办公室刘泽彭副主任在此宣读了任命决定，又听了赵阳司长语重心长的讲话，对此，我深深地感谢领导和同志们对我的信任、关怀和支持！

我来暨南大学已经4年多了，在领导和同志们的支持和帮助下，做了一些工作。现在又要承担更重的担子，深感自己能力不够，担子太重，真是诚惶诚恐！

办好一所大学，实在太不容易了。面对这一重担，我只有坚持务实的精神，尽心尽力，鞠躬尽瘁，团结好班子内的全体同事，团结好全校师生员工，将党和人民交给的任务完成好！

我们学校是国内1080所高校中历史最悠久的学校之一，建校90年来，特别是1978年复办以来，在国务院侨办的直接领导下，我校各方面工作有了许多可喜的进步，取得了许多成就，为国家包括港澳台地区以及各国华侨、华人社会培养了数万人才，在国内外已成为一所有一定影响和知名度的华侨最高学府。

按照“面向海外、面向港澳”的办学方针，我校担负着培养华侨、华人社会以及国家包括港澳台地区的高级专门人才的重要任务。面对这一任务，根据国内外一流大学和我校多年办学过程中积累的办学经验及形成的办学特色，我们应该积极探索新时期社会主义华侨大学的办学模式，突出教学中心和科研中心，争取在建校100周年（2006年）或稍长一段时间内，将我校建设成中国以及东南亚的著名大学，以至国际有影响的一所大学。

为了实现这个目标，我们需要客观地认识自身的优势和不足。我校90年发展过程中所形成的特色和优势，概括起来讲有以下五个方面。

（1）有一支较强的教师队伍。教师的质量决定学校的水平。现在，我校1036名教师中有国家级突出贡献专家1人，博士生导师13人，教授134人，副教授325人。

（2）学科比较齐全，是我国第一所拥有医科的综合大学。全校有31个本科专业，53个硕士专业，加上两个博士生培养点，共9个博士专业。我校研究生专业数排在全国50名左右。有一个国家重点建设的文科基地，是国家试办高水平运动队的学校。省部级重点学科11个，省重点课程4门。

（3）拥有一定水平的教学和科研设施。有各类实验室47个，省级重点实验室1个。特别在1995年全省实验室评估中，我校是第一所合格的高校。在1995年全省电化教学评估中，也列为全省第一名。计算机校园网络已经建成，广东省高等教育厅即

① 本文是在暨南大学校长任职仪式上的讲话，广州，1996年1月4日；暨南大学校报，1996年1月15日.

将在本月对我校校园网络进行验收，我校也是第一所验收学校。获得国家自然科学基金多少是衡量一个学校学术水平高低的重要标准，1995 年申报国家自然科学基金项目获准数是 1994 年的 3 倍多，命中率在全国高校中排第 3 位。同时，我们还拥有一座藏书丰富，达 130 万册的图书馆，一座华侨医院，以及一座现代化的邵逸夫体育馆。

(4) 我校的国际性质。我校校友遍布世界五大洲。同时，在港澳台地区有众多校友。这是全国其他高校所没有的自己的特色。仅本学年度，就有 27 个国家和地区的学生来我校求学，学生来源于五大洲主要国家和地区。

同时，我校与美国、英国、德国、日本以及东南亚许多大学有紧密学术交往，从而使我校的学术水平跻身于国际行列。

(5) 90 年办学中，形成了侨校的办学特色和传统，积累了较丰富的办学经验，特别是校董事会的设置更具特色。我校在国内外已有一定的声誉和影响。

上述优势和特色，正是我们暨大以后进入“211 工程”，成为著名大学的基础和条件。

与此同时，也应看到，与国内外一流大学相比，与国家对我们的要求相比，我校还存在很大差距和许多困难。主要如下。

(1) 我校的教育经费偏少，不仅远少于国外知名大学，就连广东省几所大学，如中山大学、华南理工大学我们也比不上。

(2) 我校的教风、学风还不够好，有待根本好转。

(3) 我校的教育质量、科学研究和管理水平也不够理想，有待进一步提高。

这些差距和困难说明，暨南大学今年要完成“211 工程”预审，用较短时间来成为一流大学，任务是十分艰巨的，需要我们作艰苦的努力。

按照上述情况，我们应该实行 20 字方针，即“发挥优势、深化改革、保证重点、改善条件、提高质量”。

发挥优势——就是要发扬暨大 90 年积累的优良传统，从总体上以较少的经费，获得最好的办学效益。

深化改革——就是继续深化教学、科研、后勤以及管理体制的改革，使学校的发展更符合客观规律的要求。当前，要研究学校基金分配制度的改革，要把学分制搞得更好，抓专业改革，向应用型专业发展，并采取措施使科研管理有利于学校学术水平的提高，同时，理顺后勤保障系统。

保证重点——就是把有限的人力、物力和财力，用在重点建设和发展关系全局方面，确保重点学科、重点实验室、重点课程、博士点建设、重点师资，特别是营造使中青年优秀人才脱颖而出的环境和条件，以带动全校各方面的建设和发展。同时，抓好基础课建设，特别是“三语”(汉语、英语、计算机语言) 课程，其中首要的是大学英语课程。

改善条件——就是尽最大努力，改善师生员工的工作条件、学习条件和生活条件。首先，搞好校园规划，使校园卫生、文明、美丽，减轻校园商业气氛。其次，使教室、实验室、图书馆条件改善，并抓好学生宿舍和教师住宅的建设。关心师生员工生活，以便最大限度地调动教职员工的办学积极性。

提高质量——办学质量是学校的生命。因此，我们必须坚持“三严”的办学方针，即从严治校、从严治教、从严治学。努力提高人才质量、科研质量和学校管理水平，坚持法治，不搞人治。

以上是个人的一点初步想法，我上任之时，不烧三把火，只扎扎实实地工作。我们的工作目标就是多方筹措经费，提高学校管理水平，与党委一起，抓住机遇，千方百计，为把暨南大学办成广大华侨、华人、港澳台地区以及国内青年求学的好地方而努力奋斗！

同志们，我衷心希望得到何校长、周校长以及全体老同志的支持，虚心向他们学习宝贵的管理经验。在国务院侨办的领导下，团结全体同志，把工作搞好。我衷心希望全校师生员工加强团结，振奋精神，从我做起，做好本职工作，采用务实精神，一步一个脚印地前进，使暨南大学在20世纪最后的几年中，为祖国的统一大业，为祖国的现代化和繁荣富强，为培养更多更优秀的德智体全面发展的人才，作出我们应做的最大的贡献。

2. 加强基础　从严治校　培养高素质人才①

1996 年是暨南大学建校 90 周年。经过几代暨南人的辛勤耕耘，学校已为海内外培养了 5 万余名暨南学子。在世纪之交的今天，如何继往开来，进一步办好华侨高等教育，提高教学质量，使暨大的事业在“九五”再上新台阶，以崭新的姿态迎接 21 世纪的挑战，是海内外暨南校友与关心侨教事业的热心人士关注的话题。

高等教育的本质特征是培养高级专门人才。现在在校的大学生作为跨世纪的一代，他们的素质如何，不仅反映和代表了一所高校的办学水平，而且在很大程度上影响中国的综合国力能否大幅度提高。教学工作是学校经常性的中心工作，是学校工作的主旋律和永恒的主题。重视教学质量，大力提高办学水平，是高校最重要的工作之一。学校就是要搞教育，全校教师与各级领导，时刻不能忘记学校最基本的活动是教学和科研，而不是其他。暨大作为一所“面向海外、面向港澳”的华侨学府，不仅历史悠久，积累了丰富的办学经验，而且在海内外享有盛誉。但今日之暨大与学校辉煌的过去相比，离国家“211 工程”的具体要求尚有差距。为了重铸暨大的辉煌，创一流大学，必须从严治校，团结全校师生员工，同心同德，脚踏实地，真正把学校的事情办好。华侨高等教育不仅是我国整个高等教育事业的重要组成部分，而且也是贯彻国家侨务政策的重要方面。立足广东，面向海内外，适应国家侨务工作的需要培育英才，是暨大义不容辞的职责。下面，就如何抓住机遇，深化改革，迎接挑战，提高学校的办学水平与教育质量谈几点思路。

一、拓宽口径，重视基础课教学

作为一所开放型、外向型的综合大学，必须十分重视基础课教学。大学本科教育是通才教育，它所培养的不是高级专家，而是复合型的通用人才。因此，我们应该在本科教学中给学生以比较广博的知识，使学生毕业后能够适应社会变化的需要，能够适应当今科学变化与发展的需求。

（1）调整专业结构。专业是构成高校教学的基本单元，专业要适应社会主义市场经济和未来科学技术发展的需要，根据我校特殊的办学任务与具体情况，应坚持“应用性和涉外性”的原则，努力拓宽专业口径。对于纯理科和冷门专业，要逐步缩减，理科要往应用方向转移，要理工结合，以工为主。对一些专业面过窄或招生分配困难的老专业，应分情况采取减少招生、停止招生等措施，并积极创造条件新设一些社会急需的专业。在现今这种情况下，专业调整的任务非常紧迫，也非常艰巨，既要保证学校的长远发展，也要考虑社会的需要与学生的意向。

（2）突出“三语”“两课”教学。“三语”课程指大学英语、大学语文、计算机应用基础，“两课”指政治理论课、德育课。这 5 门课是学校覆盖面最广的课程，是每一个暨大学生都应该学好的课。抓好这 5 门课的教学，对于提高人才培养质量，树立暨

① 本文原载《暨南教育》，1996，(1)：1-3.

大在海外的声誉至关重要。“两课”的目的是培养学生良好的品德。如果我们培养的学生熟练掌握了英文和中文，加之象征 20 世纪文明的计算机语言，又具有良好的品德，暨大在海外肯定会赢得良好的声誉。

(3) 坚持教授上基础课。教授上基础课对提高教学质量、激发学生的学习兴趣有积极作用，要长期坚持下去，使之成为制度。同时要进一步完善学分制，部分专业课逐步采用英语教学。

二、办学质量是学校的生命

要办好学校，离不开严格、科学化的管理。严师出高徒，只有从严才能建设优良的校风、学风和教风。办学质量是学校的生命。提高高校人才培养质量，是世界各国面向 21 世纪共同思考的主题。没有质量就没有效益，在激烈的竞争中，学校就会无立足之地。我在各种场合多次强调，要从严治校，从严治教，从严治学。坚持用“三严”来管理学校，而强调“三严”关键是要处理好教与学的关系，要有一支结构合理、素质较高的教师队伍。有了一支好的教师队伍，很多问题都会迎刃而解。所以，从严治教，抓好教风，这个环节是第一位的，是最关键的，是最值得学校重视的问题。只有在抓好教风的前提下，才能抓好学风。作为教师要为人师表，要有敬业、乐业、精业的奉献精神。教师不得提前下课，不得对教学工作应付了事。对学生要实行严格的考勤制度，对考试作弊者要严肃处理。要使考试作弊这一可耻的现象能够在向一流目标前进的大学里减少直至被杜绝。对在工作、学习方面表现突出的师生要予以重奖。

要办好学校，不能靠“人治”，要靠“法治”，学校制订的规章制度任何人都得自觉遵守。管理出效益。对于管理学校，我主张遵循“分口分级管理”的原则，即校长主持全面工作，根据校领导的具体分工，日常工作由主管校领导分口负责，各级都有相应的责、权、利，用规章制度规范管理，从而避免“人治”。

三、一切为了学生

学生是学校的主人。我们要树立“学生第一”的思想，应该明确在大学里学生是第一位的，因为办大学是为大学生服务的，是为了培养、造就高素质的人才。1995 年 10 月我到美国访问时，在约翰逊威尔士大学，我发现一个特别的现象，这所学校从校长到每个办公室的秘书直至学校的所有教师，每个人胸前都佩戴一枚徽章，章上写着“student first”(学生第一)。对此我感触良多，也深受教育。

暨大是一所华侨学府，学生来自世界五大洲和祖国各地。他们来到暨大求学，是为了接受中华文化的熏陶，是为了学好专业知识，学有所成。无论生源、成绩，我们都应该一视同仁，并因材施教。我们绝对不能厌弃成绩差的学生，在 20 世纪 90 年代我们要继续发扬“有教无类”的传统教育思想。

“给学生以机会”“给学生以关怀”“给学生创造环境”“一切为了学生”“使学生全面发展”，是我们必须遵循的原则。尽管学校的工作千头万绪，但是以教学和科研为中心，一切为了学生的原则必须坚持。只有我们培养的学生成才了，暨大的声誉才会提高，暨大才会成为海内外莘莘学子向往的求学胜地。

四、以科研促教学

要进一步提高教学质量，必须大力提高科研水平。教学与科研是互相联系、互相促进的。全校教学及科研人员，要树立热爱科学、献身科学的精神，以科研促进教学水平的提高。一流大学的关键是科学研究。一流大学要出一流人才，但一流人才的培养必须在一流的科研氛围中进行。

在科学技术飞速发展的今天，如果一位教师不从事科研工作，不进行知识的更新，不深入到学科的最前沿，要想提高教学水平，开阔学生的视野，激发学生的思维，是不可想象的事情。只有从事科研工作，才能转变教学思想，更新教育观念，吸收最新的教学内容，改革课程体系，从而使课程结构和课程内容整体优化，把学生培养成为有较强的自学、适应能力，知识结构合理的高素质人才。

要提高学校的办学水平，办法和措施有很多，在此我不再一一赘述。作为一校之长，我深感责任重大。暨大作为“华侨最高学府”和成立较早的少有的著名的大学之一，在我国高等教育史上占有不可或缺的重要一席。我们应该站在这样一个定位上，即立志把暨大办成国内一流，并在海外有良好声誉的著名学府。尽管由于屡遭播迁与停办，加上某些人为因素，学校曾失去了好几次发展的契机，但是令人欣喜的是，现今正值国家实施“211 工程”，学校 90 华诞，面临世纪之交，1996 年是“九五”的开局之年，加之国家即将对港澳恢复行使主权，学校又一次面临着重大的发展机遇。

学府巍巍，历史悠悠。暨大已走过了 90 年的沧桑历程，暨大的事业是美好的、永恒的。在今天办好华侨高等教育，挑战与机遇、困难与希望同在。我们要振奋精神，励精图治，抓住“天时、地利、人和”的难得机遇，为海内外培养高素质的人才，把一所声誉卓著、办学一流的暨大带入 21 世纪，为我国华侨高等教育事业在 21 世纪初叶的飞速发展奠定坚实的基础。

3. 根据侨校特点改进教学工作 采取有效措施提高教学质量①

我的报告讲三个大问题：①教学和教学改革工作的回顾；②根据侨校特点改进教学工作，从多方面采取有效措施提高教学质量是我们面临的紧迫任务；③下一阶段教学和教学改革工作的思路。

一、教学和教学改革工作的回顾

自去年5月召开全校教学工作会议至今已过去了一年半时间。在过去的一年半里，我们以《中国教育改革和发展纲要》（以下简称《纲要》）的精神为指导，认真贯彻德、智、体全面发展的教育方针和“两个面向”的办学方针，从严治校，从严治教，从严治学，深入进行教学改革，加强教学管理，进一步改善教学条件，使教学工作发生了许多新的变化。上次教学工作会议提出的加强基础课教学等多项任务得到了较好的落实。这一年半我们所做的主要工作和取得的成效有如下几个方面。

（一）深入学习《纲要》和《教师法》，进一步增强了我们搞好教学工作的信心

学习中共中央和国务院印发的《纲要》是我们上一次教学工作会议所定的三项任务之一。《纲要》进一步明确要把教育摆在优先发展的战略地位，这对我们教育工作者是一个巨大的鼓舞和鞭策。华侨高教事业是我国整个教育事业的一个组成部分，然而又是一个比较特殊的部分。搞好华侨教育工作，对国家的现代化建设将起到特殊的作用，具有特殊的意义。《纲要》对我们更好地贯彻德、智、体全面发展的教育方针具有实际的指导作用。

《中华人民共和国教师法》（以下简称《教师法》）的制定和颁布，不但有重大的现实意义，而且有深远的历史意义，是体现党和国家以尊重知识、尊重人才为核心的知识分子政策的重大举措。《教师法》的颁布，使保护教师的合法权益和对教师的管理等走上了法制化的轨道。

总之，通过学习《纲要》和《教师法》，我们看到了教育事业发展更美好的前景。我们搞好教学和教改工作的方向更明确了，信心更足了。

（二）进一步完善学分制，为更好适应“两个面向”办学创造了更为有利的条件

我校是全国较早实行学分制的高校之一，但由于原计划经济体制对办学的制约和其他某些条件不具备，学分制不够完善。学分制的主要特点是将学分作为计算学生学习分量的单位。学分制与学年制的主要区别有两条：一是突破学制年限规定，学生何时毕业，在学业上仅看是否修够了专业规定的必修学分和总学分，而不问修学年限；二是在一定范围内学生可以自由选课。正是学分制的这两大优点，能够较好克服学年制条件下统得过死，本科教育模式单一，学时偏重，专业面过窄，不利于学生创造能

① 本文是在暨南大学1994年教学工作会议上的讲话，广州，1994年11月24日；暨南教育，1994，(2)：1-8.

力的培养等弊端。采用学分制，可以更多地照顾到学生个体的志趣、性向、智力水平、学业起点不同等因素，从而更好地贯彻因材施教原则。尤其像我们这样的侨校，学生来源广，特殊情况多，更应该采用学分制这种比较灵活的教学管理制度。这样做，也便于我校的教育与国际教育接轨。基于这些考虑，我们采用了逐步完善学分制的做法，使学分制一步一步地与学年制脱钩。从 1993 年开始，我们采用了标准学分制度，规定每个学期本科学生须修读的标准学分数为 23 个学分。采用标准学分制是进一步完善学分制，规范教学管理和深化教学改革的一个重大步骤。这也是国外实行学分制学校的普遍做法。在标准学分制条件下运作，学年的概念将淡化，学分的概念将强化。我们做了“新生新办法、老生老办法”的规定，把重修、副修、修读第二学位等的学分纳入标准学分数内。重修不另收费。根据标准学分制的规定，各专业修订了教学计划，均衡了各学期的教学任务，压缩了部分课内学时，为今后按学分收费上学打下了基础。特别值得一提的是，因为标准学分制的出台，重修不另收费，曾经矛盾比较突出的重修问题得到了比较迅速和比较顺利的解决。

（三）加强教学质量监控，使教风和学风有了一定好转

我们在多种场合，包括在上次教学工作会议的报告中都提到，教学质量是我们暨大的生命。本科是暨大办学的主体，如何提高本科层次的教学质量，尤为全校上下所关注。要保证教学质量就必须对教学的每一环节进行严格把关，其中课堂教学和考试是两个非常关键的环节。课堂教学目前仍是理论课教学的主要形式，因此，我们采取了评估、听课和检查相结合的措施来加强对课堂教学质量的监控。

（1）评估、听课和检查。为在评估上采用现代手段，去年学校教务处购置了光标阅读器，经各方面征求意见选取了教学态度、教学内容、教学方法和教学效果等 4 项内容作为课堂教学质量评估的一级指标，选取“倾注爱心、教书育人”等 12 项内容作为二级指标，并印刷了适合用光标阅读器阅读的评估表，在全校的本科课堂进行了两个学期的试评。每次有 3000 多名学生参加评估，被评估教师 400 多人。凡认真按照要求组织学生评估的系，评估的结果可信度是比较高的，反映也比较好，对提高课堂教学质量起了较大作用。教师中不认真备课，课堂讲授比较随意的情况未再发现了。个别教师（占被评估教师的 3%左右）讲课被学生评价为不及格，对他们震动较大，能积极与学生沟通，更新教学内容，改进教学方法，提高授课水平。

为避免学生单一评估可能出现的偏差，学校组织了文、理、经、医几个专家听课组，以便掌握教学第一线情况，同时对某些教学质量有争议的课堂教学给出较权威的结论，对某些疑难问题进行验证。专家组 20 位专家已听课 167 门、510 节，被听课教师 211 人。专家通过听课获得了大量的教学信息，反映了不少宝贵意见。与此同时，学校还规定了校院系和教研室负责人等听课的制度；教务处的同志开展了定期和不定期的日、夜间检查和巡视课堂教学情况的活动。

（2）严肃考试纪律。与早些年比较，这两年学生的考风有了较大的好转，上个学期全校只发现 5 起考试作弊行为。学校对命题、考场管理、试卷交印和评阅、成绩登记及作弊处理办法等都做了具体规定，并严格执行。特别是从去年开始，对考试问题做了两项更为严格的规定：一是取消补考，不及格课程必须重修；二是对作弊学生一

律不授予学士学位，对再次作弊的学生给予开除学籍的处分。考试制度严了，促进了学风好转，用学生自己的话来说就是“再也不敢掉以轻心了”。现在学生白天抓紧学习和自觉晚修的增多了。

（四）突出抓好重点课，特别是全校公共基础课的教学和建设，取得了不少喜人的成果

大学英语、计算机应用基础和大学语文是全校最主要的几门公共基础课，学校统称“三语”教学，始终将这“三语”教学作为课程教学和建设的重点来抓。从德、智、体全面发展考虑，有关的政治理论课（包括对华侨和港澳台学生开的爱国主义教育课）和体育课，也是学校关注的重点课。另外，还有为部分学院开的公共基础课（如高等数学），或同一学院内各专业的共同基础课等，多数都已列为校一级的重点建设课。校级重点课第一批有 68 门、第二批有 40 门。学校对重点课实行政策倾斜，从人、财、物等方面加大投入。包括 1988 年的投入，学校已投入重点课建设经费 150 万元。目前，学校正组织专家组对到今年建设期满的第一批校级重点课进行评估验收。待验收完毕连同上个学期验收的一道公布结果，并研究下一步的做法，下达第三批重点课建设经费。

（1）加强基础课教学。为抓好基础课教学，学校采取了几项具体措施。第一，安排教授上本科基础课。教授在学术上造诣深厚，教学经验丰富，由他们开设基础课对本科低年级学生思维空间的开拓，基本知识、基础理论和基本技能的掌握，学习兴趣的激发，乃至蓄积足够的学习后劲，提高这一层次的整体教学质量，都是十分重要的。根据统计，教授上本科课程的百分比已由原来的 18%上升到目前的 68.7%。第二，推广经济学院教师挂牌上课的经验。经济信息管理和会计等系部分课程实施挂牌上课，在校内外反响热烈。全院上个学期有 8 个课堂 600 多名学生自由选课；本学期还有 3 个课堂继续挂牌上课。挂牌上课是完善学分制必然要采取的做法，是实行平等竞争，打破论资排辈，保证良好教学效果的一项措施。它与教授上本科基础课相辅相成。经济学院在这方面先行一步，值得肯定。学校对挂牌上课积极鼓励，给予支持。第三，扩大大学英语强化教学试点。大学英语强化教学原只在计算机系试行，从去年开始扩大到医学、国际金融、国际新闻等 7 个专业。根据外语中心提供的资料，强化教学总的效果是好的，强化班比非强化班的四级统考通过率一般可高十多个百分点。

（2）抓好体育课教学。我校是全国 106 所体育课程优秀高校之一，并是中国电化教育协会普通高等学校体育电教专业委员会理事长学校，也是广东 7 所高水平运动队学校之一。前些时候，广东省高教局还向国家教委①申报我校为全国 30 所高水平运动队学校之一。我校如能跻身于全国高校体育 30 强之列，这将对提高学校的声誉和推动各项工作产生重大影响。在今年暑假举行的全国大学生运动会中，我校成果甚丰，计有：①田径方面，我校获金牌 5 枚，银牌 1 枚，铜牌 6 枚，这是我校在全国大学生运动会上金牌“零”的突破（另在今年夏天日本东京国际田径明星赛上我校马丽雯同学获得女子 100 米银牌）；②乒乓球男子团体和男、女混合双打第 3 名；③网球女子团体第

① 1998 年更名为教育部。

3 名。我校不但体育参赛成果较喜人，而且体育部的教学改革也很有特色，体育电教等成果获省级以上 3 次奖励。根据侨校的特点，体育部规定在学生学完一学期基础课以后可自行选修学习。这种选项教学法与经济学院试行的挂牌上课有些类似，学生学得主动活泼，教学效果明显增强。

（3）医学院的“人体解剖”和中文系“现代汉语”两门课建设成果突出，已列为省级重点课。企管系的“管理学原理”、中文系的“文学概论”和计算中心的“计算机应用基础”3 门课已获得省级重点课评选资格。经济信息系的“统计学原理”、医学院的“生理学”和社科部的“马克思主义原理”等课程建设成果也比较突出。

（五）加强电化教育和教学设备建设，教学条件有所改观

由于学校的争取和国务院侨办的支持，从 1992 年开始港中旅将连续 5 年每年资助我校 680 万港元用于添置和更新教学和教学管理的设备等。从去年全校教学工作会议至现在，学校教学设备固定资产纯增值 960 万元，等于教学设备固定资产总值增加了 20%，同时，部分教学设备的档次也有所提高。由于教学设备条件改善，现在基本上能按教学大纲要求开出实验。去年以来，全校新增实验项目 163 个，同时实验质量也明显提高。例如，医学院病理解剖课等，学生原使用已老化磨损的显微镜观察组织细胞形态，有时无法分清是病态的细胞还是正常的细胞。现在显微镜更新了，学生反映视野清晰了，根据观察的内容也可以作出正确判断了。特别是由于投入约 435 万元购置了 100 多台微机和添置了四套进口 WE-7900 型语音设备，计算机和大学英语的教学条件有了很大改善。计算机课目前可以达到理论/上机 1∶1 的教学要求；大学英语课增加了 192 个座位的语音训练设备，可供每人每周在语音室上课增加 2 节。

我校的电化教育近年来在普及和提高两个方面都有了较大发展。全校现有功能齐全的多媒体组合教学课室两间；有普通电教课室 80 间 3120 个座位；有语言实验室 19 套 760 个座位；有各类电视教材 3841 部 9683 小时。全校有 273 门课程有配套音像教材，有 70%的教师利用电教手段上课，仅去年就有约 8 万人次（包括师生）接受电化教育。发展电化教育意义重大深远，在传统教学中引入现代视听结合的手段进行教学，大大缩短了教学时间，拓宽、深化和丰富了教学内容，提高了教学效果，是教育的一场革命和飞跃。今年暑假，学校投入二十几万元在几个主要的教学楼的 80 间课室安装了投影仪，并在这些教学楼备有幻灯和无线扩音系统，进一步改善了电教条件，受到师生的欢迎。教师利用现代媒体教学的积极性和热情越来越高，要求制作配套电教教材的学科越来越普及。我校在“遗传学”课程获省多媒体组合教学试验奖励后，今年又有“空间解析几何”课程得奖。还有“中国古代史”、“生理”和“无机化学”等 3 门课参加省的多媒体组合教学试验立项。

（六）招生与考试、培训工作

学校顺利完成了 1994 届毕业生毕业工作和 1994 级新生入学工作；顺利组织了去年中、去年底和今年中的大学英语四、六级统考工作，以及今年 10 月全省高校的计算机应用基础课统考工作。此外，学校还组织办了 4 期共 300 多名教师参加的使用电教设备培训班。

(七) 教改收到效果

全校有不少单位和个人积极进行教学和教学改革探索，并在不同程度上收到了效果。

(1) 经济学院和文学院决心落实李岚清副总理今年 5 月 3 日来我校视察时的讲话精神，从抓打通部分基础课开始实行通才与专才结合的教育。经济学院各系已基本商定从 1995 级开始打通 21 门经济类课程，其中包括国家教委高教司规定的财经类学生必修的 11 门核心课程。文学院也已初步确定在全院开通三至五门公共必修课，以后逐渐增开。该院还从上学期开始了分步实施多层次课程内容检查工作，检查教学计划、教材选定、讲授内容等各个教学环节。上个学期已完成本科层次的检查，本学期进行大专层次检查，下个学期拟检查研究生层次及重点课建设情况。通过检查，促进了教风和学风好转。

(2) 新闻系学生无论到全省哪个单位实习，对方都不再收实习费，走通这条路子是很有价值的，现在全国高校中能做到这一点的极少。计算机、化学等系结合教学设法吸收学生参与可能的科研的做法大受学生欢迎。社科部部分哲学课教师将一贯的灌输式教学法改为专题讨论式教学，收到了比较满意的教学效果。该部针对爱国主义教育课“中国古代文化”出版的新教材《智圆行方的世界——中国传统文化新论》各方面反映较好。中文系坚持多年的导师制，即一名导师指导几名学生写作，包干数年，颇有成效。

(3) 在教学、教改工作中，许多教师的实绩突出。今年教师节我校表彰的教师中有魏中林等 39 名国务院侨办授奖的优秀教师；有宋献中等 4 名广东省授奖的“南粤教坛新秀”。教师队伍中新进了大批博士，派出海外及港澳台进修和回校的教师数增多，教师队伍素质进一步提高。

(4) 从学生方面看，品学兼优和单项成绩突出的比率都有所增大。今年有 30 名学生办理修读双学位手续，有 27 名学生申请副修专业，有 28 名学生可免试升读硕士学位。在近年两项影响较大的全国高校新闻奖活动中，“可口可乐杯”赛 37 所高校各交 3 篇作品参赛，我校新闻系学生夺得了唯一的特等奖和 5 个一等奖中的 2 个；在“韬奋新闻奖”中，我校学生也夺得一、二等奖各一项。要求较高的“计算机世界奖学金”，每届奖六七十名，我校每届有 2 名学生获奖。1992 级内地（大陆）学生四级统考通过率的前 3 名为：金融系、经济信息管理系、商学系；海外及港澳台学生通过率的前 3 名为：医学院、经济系、新闻系。

(八) 新成立的华文学院和中旅学院克服困难开展了各具特色的教学工作

第一届中旅学院学生已进校学习，该院还对 38 名新生进行了为期两周的军事训练。对学生进行军训，在我校尚属首次。

华文学院已基本理顺各方面的关系，使教学工作走上正轨。今年对外华文教育招生有较大发展，有来自 19 个国家和地区的 197 名学生来求学，是我校这类招生历年来最多的一年。今年还召开了海外华文教育交流会，有十多个国家和地区的代表到会，扩大了对外影响，加强了对外交流。华文学院办学正朝着上档次、上层次的方向发展。

另外全校的专科教学也稳步发展。

二、根据侨校特点改进教学工作，从多方面采取有效措施提高教学质量是我们面临的紧迫任务

我校从1978年复办以来已招收和培养各类各层次学生26 635人，其中海外及港澳台学生4468人。目前在校的有（包括成人教育）12 506人。我们办学16年，能够为社会输送这么多高级专门人才，是值得骄傲的事。但我们也必须看到，我们的教学质量距离应有的要求还有不小的差距；同时，我们面临的办学压力却越来越大，任务越来越繁重。办学压力包括港澳回归的日期越来越近，台湾已创办新的暨大，港澳也增办新大学；原来对外招生我们与华大两争天下的局面已被打破，而且会进一步打破；由于“211工程”的实施，众多高校争上质量、争上水平，等等。在这些压力面前，我们如果缺乏有力的抗争措施，我们办学的前景就不容乐观。现在除大学英语、计算机统考外，还有多门课也已列入统考范围。另外，上级组织的各种各样的评估也在逐渐展开。从学、教和管几方面看，还存在不少问题。我们已长期习惯于“满堂灌”“抱着走”的教学方法。教学内容有的已陈旧、有的则过多重复，教学手段也较落后，远远跟不上形势发展的要求，适应不了不同学生的学习需要。

谈到教学质量问题，许多人埋怨我校学生来源复杂、参差不齐。这的确是影响我们的教学质量的关键因素。但我们又不得不正视现实。这个现实就是：首先，暨大作为一所华侨大学，不可能不招海外及港澳台学生，不但要招而且要尽可能多招；其次，暨大目前在海外及港澳台一些国家和地区还只能招到学业水平较一般的学生。在上次教学工作会议报告中，我谈了倾注爱心，促进教学相长的问题。今天我想再讲一讲这个问题。我们招收的海外及港澳台学生由于来自不同的国家和地区，学业水平参差不齐，确实给教学工作带来很多困难。但我们也要看到，所有来暨大上学的学生，都还是经过了考核挑选的，不是基础最差的学生，有的学生甚至很优秀。而且这些学生绝大多数都是满怀对祖国的感情和对暨大的信赖来上学的，因此，我们不能有嫌弃情绪，不能使他们失望，要满腔热情关心他们的进步，千方百计促使他们成才。凡古今中外的教育家和优秀的教育工作者，都是主张教育的平等性和慈爱性的。我们进行的是普通高等教育，不仅仅是培育出类拔萃的英才。一名真正优秀的教师，他的才华不仅要反映在能教好基础好的学生，更重要的是要反映在能发现学生个体的特长，因材施教，教好基础差的学生。海外及港澳台部分学生开始基础差一些，学得吃力一些，但他们的动手能力和思考问题的能力不一定比内地（大陆）的学生差，这是大家都公认的。我们能否针对海外及港澳台学生的特点在教学以及教学管理上更灵活一些，更切实际一些呢？这个问题大家可以展开讨论。但无论采取何种方式，我们都必须在海外及港澳台学生的学习上花费更多的精力。暨大办学的宗旨就是培养海外及港澳台学生，如果不招他们，不培养他们，暨大也就没有存在的必要了。李岚清副总理讲暨大这个牌子很响亮，中国就这么一所很有特色的华侨大学。特色也好，响亮也好，都离不开一个“侨”字，离开了就没有特色，也不响亮了。对外招生是党和政府交给我们暨大的任务。这对祖国的统一、富强和振兴有重大的战略性意义。因此，无论从教育工作者的一般责任来讲，还是从完成政治任务的高度来讲，我们都必须把招收来的海外及港

澳台学生培养好。可以肯定地说，暨大绝不可能靠多招内地（大陆）的学生再造辉煌，而只能靠经过不懈地努力把教学质量搞上去，靠质量，靠办学水平吸引更多的海外及港澳台学生再造辉煌。

三、下一阶段教学和教学改革工作的思路

下一阶段或今后一年教学和教学改革工作总的指导思想和总的任务是要继续认真贯彻《纲要》的精神和德智体全面发展的教育方针；要认真贯彻今年全国和全省教育工作会议精神；要强调根据侨校的特点进行教学改革，改出成效；要坚持从严治校、从严治教、从严治学的方针，使校风、教风和学风有明显好转，使教学质量上一个新台阶（“三严”“三转”“一上”）。

具体的任务和要求如下。

（1）继续提倡尊师爱生，使教风和学风进一步好转。转变教风是转变学风进而形成良好校风的关键。转变教风要从爱生入手，从备课、教材选择、课堂讲授、实验（实习）指导、教学方法的运用等各个教学的具体环节入手，看教学态度，看教学水平，看教学效果。要务实，不务虚。

转变学风要看到上课、完成作业、参加实验（实习）、讨论发言和自觉早、晚修等情况，要看学习态度，看是否遵守学习纪律，看学习效果。

在教与学这对矛盾中，教是主导，学是主体。教师是人类灵魂的工程师，是办好学校的主力军，理应受到全校的尊重；反之，教师应当关心和爱护学生，为人师表，教书育人。我们应当通过各种宣传方式使尊师爱生的口号深入人心，使尊师爱生蔚然成风。爱生与从严要求学生是对立统一的，两方面缺一不可。

（2）要以课堂教学为核心，进一步强化教学质量管理，措施有四条。

（a）进一步完善和扩大学生评估工作。各单位要把学生评估课堂教学质量当作一项重要的工作来抓，尤其要训练新生，使之开始就抱积极的态度，当作正常的教学活动来完成任务。评估工作每学期都在第 15 周至第 16 周进行。由于采用光标阅读器阅读评估表，评估的要求必须严格遵守。这项工作要派责任心强的同志经办。本人有任课任务的要回避。老师应加强与学生的交流，让学生熟悉自己的姓名，以便准确填写评估表，避免张冠李戴。评估的结果必须送达每位老师，及时纠正评估中可能出现的差错。

学生评估本学期拟从本科课堂扩展到专科课堂，今后还要创造条件评估实验、实习课。

（b）坚持专家组听课和评估。专家组要通过听课和评估沟通更多教学信息，验证疑难，纠正学生评估中可能出现的某些偏差。

（c）落实各级领导的听课制度。领导听课要恰当安排，听课节数要用适当方式予以公布。

（d）奖优罚劣。明年教师节前要评选课堂教学优秀的教师。同时，要挑选教学效果特别好的课堂举行观摩教学活动；课堂教学质量要与职评等挂钩，评估不及格者黄牌警告，不予晋级。对特差者要出示红牌，调整其工作。

(3) 进一步健全学分制。总的构想是要达到“四按”“二制”“三自”的目标，并分步到位。

(a)“四按”指按标准学分制规定注册，按标准学分制规定交费，按标准学分制规定选课，按总学分规定毕业。

(b)“二制”指导师、导生制和电脑选课制。导师的主要职责是指导学生选课，关心学生平时的学习和思想表现，引导学生德、智、体全面发展。导生从研究生或高年级的优秀学生中选拔，主要职责是协助管理学生宿舍生活和社团活动，并帮助低年级学生顺利选课。学分制逐步完善，学生打乱自然班上课的情况会越来越普遍，学生人手有一张自己的课表，只有用电脑选课，并在网络条件下运作，才能在较短的时间内理顺大量的选课关系。为此，要编排和出版全校课程手册，要制作和试验电脑选课系统，任务很艰巨，需要较长时间准备。

(c)“三自”指学生可以自主确定专业方向，自主选择上课教师，自主安排学习进程。自主确定专业方向，指同学在低年级时不分专业，修完学科基础课以后，可在一定条件下根据个人志趣与学习状况在本学科范围内确定专业方向。如同一课程有多位教师开课时，学生可选择听自己满意的一位教师的课。只要符合标准学分制的规定，学生可以多修、少修、缓修、免修各类课程。

李岚清副总理说搞学分制是已定下来的方向，暨大可以走得快一点。因此，我们要在已有的基础上积极推进，但是要避免盲目性和急于求成。

(4) 要根据侨校特点积极开展教学研究和教学改革试验。这是本次教学工作会议的主要议题。1997 年和 1999 年香港和澳门将先后回归祖国。我们不但要研究过渡期港澳台学生的教育问题，还要研究港澳回归后的教育问题。

(a) 希望每个系一级教学单位，尤其是海外及港澳台学生较多的单位明年能拿出一两项比较有特色的教学改革成果。

(b) 明年要为 1996 年全国第三届优秀教学成果评奖做准备（最近广东省高教局已就此下发通知）；我们要及早做好报奖工作，要重奖教学工作成绩突出的单位和个人。

(c) 今后优秀的教学研究论文将被视为教育科研成果，在评职称晋级时给予肯定。我们准备拿出一笔经费支持教师们进行教学研究。

(5) 要进一步做好新教师上岗培训和尚未掌握电教技术的教师的培训。要研究电教设备和课室如何方便教师上课使用的管理问题，切实提高各类教学仪器设备的利用率。由于目前课室数量不足，今后晚上要适当排课，请各单位和老师要理解和支持。

(6) 要进一步抓好课程建设工作。各系一级教学单位都要把重点课程建设纳入自己的教学管理任务范围，有些单位把此视为课程承担人个人的责任是不对的。课程建设要重点与一般相结合。学校拟实行校、院、系分级管理分级建设课程的办法。重点课程要突出抓好教师梯队建设、教材建设、教学内容更新和采用现代化教学方法、手段等环节，使课程教学质量有明显提高。

(7) 要进一步改善教学条件，加强实践教学环节管理。实践课比重较大的单位要严格规范实践环节管理，保证实践教学的效果。明年要就此进行一次专门调查研究，总结经验，提出改进措施。

(8) 要加强艺术教育，艺术中心要开设更多的艺术课程。要按照广东省高教局和国家教委有关通知的要求，对学生进行健康的高雅的艺术教育，培养学生的审美情趣，提高他们对艺术的鉴赏力。现在在校的学生是下一世纪的人才，他们需要有较全面的知识结构，要有艺术修养。我们要充分认识艺术教育在普通高等教育中培养德、智、体全面发展人才的重要地位和作用，搞好有关的管理和建设。

(9) 要进一步加强考试管理，从严从快处理各种考试舞弊行为。从 1994 级开始，学校将用光标阅读器阅读学生成绩登记表，请老师们积极配合。

(10) 这次大会除印发了《关于进一步健全学分制的若干意见》外，还印发了《暨南大学教学工作规范》《暨南大学预订教材管理办法》《暨南大学教研室工作条例》和加强电化教育方面的教学工作文件，经过大家讨论修改后，要加以贯彻实施。教学工作是学校经常性的中心工作，我们必须进一步增强教学意识，花时间来研究解决教学上的一些重大问题。我们要进一步树立信心，创造更多的教学工作业绩向建校 90 周年献礼，为把我们暨南大学办得更有特色、更有水平而努力奋斗！

4. 提高认识　办好成人教育[①]

在普通高校中开展成人教育，是国家规定的普通高校基本任务之一。我校是担负此项任务的一所综合性大学。成人教育的教学工作是我校教学工作不可分割的一个重要组成部分。因此，我们将成人教育工作列入这次全校教学工作会议的重要议题，用半天时间专门讨论、研究成人教育工作。现在，我就这个问题谈几点意见。

一、进一步提高对成人教育在经济发展和社会进步中的地位与作用的认识，重视成人教育

今年六月，党中央和国务院主持召开了改革开放以来的第二次全国教育工作会议。在实现现代化建设第二步战略目标和建立社会主义市场经济体制的关键时期召开的这次重要会议，研究教育工作如何进一步适应加快现代化建设步伐和建立市场经济的需要，全面部署和动员全党全社会实施《中国教育改革和发展纲要》，实现 20 世纪 90 年代的教育改革和发展目标，研究和解决教育改革和发展中的重大问题。在这次会议的主要任务与议题中，职业教育和成人教育不仅被列入而且被提到前所未有的高度去加以重视。会议认为，职业教育和成人教育是现代化教育的重要组成部分；大力发展职业教育和成人教育，是加快提高劳动者素质、振兴经济的必由之路。因此，会议将这两种教育的发展明确规定为我国教育发展的三大任务之一，将其放到调整宏观教育结构的突出位置之上，并将其确定为推行教育改革的重点。

在全国教育工作会议上对职业教育和成人教育所取得的这种共识，是对现代教育潮流把握的结果，是从我国国情出发借鉴发达国家发展教育经验的结果，也是对我国近十几年来发展成人教育的实践进行总结的结果。

第二次世界大战以后，世界性的教育潮流发生了很大的变化。由于科学技术的迅猛发展、信息量的增加和知识更新速度的加快，传统的“一次性教育”思想，已越来越不适应传播新知识、学习新知识和新技能、不断提高人的自身素质和竞争能力的需求，因而逐渐为那种将教育过程贯穿于人的一生的“终身教育”思想所取代。由于一次性教育思想向终身教育思想的转变，也由于科技进步所带来的教育手段、教育方法的创新，传统的、比较单一和比较刻板的办学形式也向着现代的、多元的和开放化的方向发展。成人教育在这种变化中可以说是起了一个带头的作用。

纵观世界，成人教育的发展程度大体上与一个国家科技、经济发展的程度相一致。几乎所有的发达国家不仅有比较完善的普通教育制度，而且有比较成熟的成人教育制度。德国在这方面的情况就更为突出些。我国正在向现代化的目标迈进，发达国家在教育方面的举措与经验，是人类的共同财富，当然值得我们去借鉴和学习。更何况成人教育比普通教育花钱要少些，发展成人教育适合我国的国情，也符合我们穷国办大

---

① 本文是在暨南大学 1994 年教学工作会议上的讲话，广州，1994 年 11 月 24 日；暨南教育，1994，(2)：9-12.

教育的需要。

党的十一届三中全会以来，我国成人教育发展很快，取得了可喜的成绩。全国现有成人高校 1300 多所，此外，还有近 700 所普通高校开展了成人教育。十多年来，全国共培养了本、专科毕业生 364 万名，占全国本、专科毕业生的 46%。按照《中国教育改革和发展纲要》的要求，到 2000 年，这个比例还要提升到 48%，通过成人教育所培养的本、专科毕业生人数也要大量增加。这就预示着今后数年我国成人教育的规模将要进一步扩大，发展速度也将进一步加快。

我校的成人教育，自 1982 年始办以来也有很大的发展。历年来从我校各类学历班毕业的成人教育学生已达 12 000 多名。这些毕业生，在各自的工作岗位上大都为经济建设和社会进步作出了可喜的贡献。目前，我校在学的成人教育学生人数已达 4600 多名，占了我校学生总数的三分之一。我校的成人教育，在办学层次上，已由专科发展到本科和双学位；在招生区域方面，已由省内发展到省外，乃至境外。正在香港开办的有中医骨伤大专班，社会学专科起点本科班；在澳门开办的有护理大专班，社会学专科起点本科班；正在新加坡招生办班的有中文本科班；正在马来西亚招生办班的有中文本科班和商学本科班。目前我校已初步形成了多层次、多形式、多渠道和灵活的成人教育办学体系。

总的来说，我校的成人教育是有成绩的。我校成人教育的教学质量有一定的保证，成人教育的毕业生大都得到用人单位的好评，如香港社会学专业大专班毕业生郑耀棠（班长）是香港工联会主席、人大代表、香港议员，毕业后在社会工作中起到很大作用；又如澳门工联会正、副理事长均是我校大专班学生，毕业后被选为议员的有好几个。这些学员的良好表现，使我校成人教育在社会上具有较高的声誉，并为我校的成人教育带来较为充足的生源。所有这一切都为我校成人教育今后的发展提供了良好的基础和创造了良好的条件。

在党中央、国务院高度重视下，我国的成人教育必将有个大的发展。在全国的大好形势下，我校的成人教育也将发展到一个新规模和攀升到一个新的台阶。然而，我校成人教育今后的发展，还有待于我们对成人教育认识的进一步深化，也有待于目前我校成人教育中存在问题的克服。

应该说，我校各院、系、所对开展成人教育基本上是重视的。有些院系对于此项工作更是当作一种事业来看待，认识明确、领导重视、精心策划、一丝不苟，任务完成得相当好。在上次教学工作会议上和这次会议上介绍经验的单位，就是在开展成人教育工作中做得比较好的单位。然而，也应该指出的是，还有一些同志对成人教育存有偏见，认为成人教育是一种低水平的非正规的教育，更有的人将成人教育仅仅看作是一种创收的手法。这些认识和看法，都是偏颇的、错误的，如不加以纠正和提高，必将成为我校成人教育发展的障碍。

目前，我校成人教育的运转可以说是基本正常，但在其中也还存在着不少问题。

(1) 我校成人教育的办学条件，从硬件来说，还不够好。特别是在课室、宿舍和教学设备等方面跟兄弟院校相比较，我们的条件较差。今后，学校将争取尽快改善成人教育的办学条件。

（2）成人教育的教学质量有待提高。教育学院的同志曾对前后六届的成人教育毕业生质量进行了跟踪调查，发现前几届的毕业生质量要比后几届的毕业生质量高。

（3）成人教育管理体制还没有完全理顺。1991 年学校颁发了《暨南大学成人教育管理暂行规定》，明确规定全校各类成人学历教育和成人非学历教育归由成人教育学院统一管理。但该规定并未得到彻底执行，各行其是的问题时有发生。

以上问题的存在，不利于我校成人教育质量的提高，不利于我校成人教育的健康发展。这些问题的解决，正是我们这次会议要讨论研究的重要内容之一。

二、认真学习中央指示，进一步明确我校成人教育的任务和方向

中央在 1983 年给予我校指示："学制要灵活多样，除招收本科生和研究生外，还可办预科和函授教育，实行学分制。"国务院侨务办公室在 1985 年又对我校指示："同意你校设立函授部，对国外华侨、港澳同胞、外籍华人和归侨、侨眷开办函授教育。"

上述两个指示，非常明确地指出了暨南大学成人教育的任务和方向，也就是说，暨南大学"面向海外、面向港澳"的办学方针，不只是对全日制而言，而且包括成人教育。所以努力开拓港澳台、海外成人教育，是我校成人教育的基本任务，也是我校成人教育发展的方向。

就当前港澳台和海外的现实情况来看，也迫切需要我们加快开展港澳台、海外的成人教育。首先，港澳台同胞、海外侨胞迫切要求能在不离开现职工作岗位的情况下，通过函授加面授，学习祖国的文化，学习汉语，学习中医、针灸，学习中国的对外贸易政策，学习中国的法律，等等；其次，当今世界上掀起的"中国热"和"汉语热"，必将推动汉语（中文）函授教育的发展；最后，这是港澳向"一国两制"过渡的需要。鉴于香港 1997 年、澳门 1999 年回归祖国日益临近，为能顺利过渡以及过渡后能实现港人治港、澳人治澳的方针，人才是关键。为此，培训在港澳工作的干部，提高他们的文化素质和学历层次的任务要求，已迫在眉睫。担负此项任务，对我们华侨大学来说，是责无旁贷的。

综上所述，我校积极开拓港澳台、海外成人教育，不但必要，而且非常迫切。今后要集中优势兵力向外发展，立足港澳，着眼海外。

为此，全校各单位对这个问题要有一共同的认识，要同心协力，充分发挥我校综合性大学科类齐全、师资雄厚、设备先进、信息灵通的优势，打破院、系、专业的界限，都把强项拿出来，形成"拳头"，切实办出一些具有我国、我校特色的专业，以增强对外招生的吸引力。

在"走出去"的同时，还要"接进来"，即接收海外和港澳在职成人或华侨子女来校参加各种长短不等的专科班、进修班、经济考察班、文化旅游班等。

三、加强管理，理顺关系，进一步提高我校成人教育的质量

教学质量是成人教育的生命线，是我校成人教育能否生存和发展的关键。我校成人教育的质量，从总体上来说，是有保证的，用人单位的反映是好的，并基本上得到社会的认可。但是，我们要看到当代科技的发展非常迅猛，改革开放不断深化，高科

技产业愈来愈多，因而对人才的要求也愈来愈高。在这种新形势下，我们所培养的人才与社会的需要还有一定距离，并且根据追踪调查资料，发现我校近几年来的教学质量有下降的趋势。这个问题不能不引起我们高度的重视。

从今年开始，广东省高教局成人教育办公室加强了对教育质量的调控管理，今后对全省成人学历教育的办学条件、教学质量要进行评估，对部分基础课和专业课进行全省统考、抽考，并在一定范围内公布成绩和名次。

为了切实保证我校成人教育教学质量，为了我校在省成教办统考、抽考中取得较好成绩，从而进一步提高我校成人教育在社会上的声誉，我们必须认认真真地做好如下几项工作。

1. 各级领导要重视成人教育工作

我校成人教育招生计划是国务院侨办、国家教委正式批准下达的任务，是全校招生计划的一个重要组成部分，并非计划外招生。各级领导要把成人教育任务纳入本单位教学任务之内，要指定一位领导兼任此项工作，并成立一个成人教育小组，经常研究、检查本单位承担的专业班的课程教学情况，对教学上存在的问题，要及时给予解决，对本单位开出的课程的教学质量全面负责。

2. 要重视成人教育教师队伍的配备和管理

提高教学质量的关键在教师，建立一支具有良好政治业务素质、结构合理、相对稳定的教师队伍，是提高我校成人教育质量、改革和发展我校成人教育的关键所在。对这个问题，我们必须取得共识。

首先，各单位在选派教师教授成人教育课程时，一定要注意质量，具有讲师以上职称的教师，不能少于70%，凡未上过讲台的新教师（含在学高年级研究生），一定要通过试讲合格后，在老教师的指导下上课。对于干部培训班的课程，要选派有实践经验、业务水平较高的教师。

其次，为了有利于积累和总结成人教学经验，各单位选派教授成人教学班课程的教师，要稳定三届不变，如因特殊原因需要换教师时，需经教育学院审核后送主管校长批准。

3. 加强基础课建设，逐步建立公共课试题库

（1）抓好基础课建设。确定第一批重点建设课程为“公文写作”“英语”“计算机应用”“经济数学”“政治经济学”等五门。按学校对重点课程建设的要求进行建设，每门课程由教育学院拨给一定的建设费用。

（2）抓好部分专业基础课建设。在基础课建设取得经验的基础上，再重点建设一批专业基础课，如市场学、会计学原理、统计学原理、管理学基础、公共关系等课程。

（3）抓好试题库建设，对重点课程逐步实行全校统考，以促进学风、教风的好转。

（4）要进一步理顺管理体制，加强管理队伍建设，逐步实现管理科学化、现代化。

开拓海外成人教育，任务重、难度大，必须由学校统一组织和安排，各单位分工合作。为此，要进一步理顺我校成人教育管理体制，要坚决贯彻执行《暨南大学成人教育管理规定》，做到应统则统，尤其是涉外的成人学历教育和非学历教育，均应由教育学院代表学校统一管理。

要加强对管理队伍的培训和考核，提高管理人员的素质，不断提高工作效率，改善服务态度，主动与各教学单位加强联系，搞好协作；要健全和完善各项规章制度(本次会议印发了五种规章制度，请大家认真讨论和修改补充)，逐步试行学分制；进一步完善计算机管理学籍的软件，充实计算机等硬件，尽快实现管理电脑化。

我相信，只要全校上下通力合作，共同努力，我校成人教育质量将会迈上一个新的台阶，我校向外开拓成人教育的步伐将会加快。

5.《红杏枝头春意闹——暨南大学成人高等教育毕业生业绩选》序[①]

近几年来，成人高等教育发展很快，为我国的社会主义建设培养了一大批人才，受到了社会各阶层的欢迎和肯定。但是对成人高等教育所造就的人才，我们的舆论还没有给予足够的宣传，因此从总体来说，社会上对这方面的了解是比较有限的。为了取得社会对成人高等教育的理解和支持，国家教委和省高教局都正在组织编写、出版《成人高等教育优秀毕业生业绩选》。这充分体现了国家和我省的教育部门对成人高等教育的重视。我相信它必将对促进我国的成人高等教育的发展起到重大作用。

为了配合国家教委和高教局的这一举措，为了取得社会各界对我校成人高等教育的理解和支持，并对我校成人高等教育的成就有进一步的了解，我校教育学院决定出版《红杏枝头春意闹——暨南大学成人高等教育毕业生业绩选》一书。

我校自 1978 年复办以来，已经培养了成人高等教育学员 1 万多人。他们毕业后大都回到原单位工作，成了单位的业务骨干，并做出了很大的成绩，有的甚至做出了突出的贡献，得到了社会上的充分肯定。由于时间、人力、篇幅所限，这本书不可能把所有的毕业生都加以介绍，在这里他们根据各部门和一些单位的推荐，组织人员采写了几十位毕业生，从中选出 51 篇，编成了这本书。也许会有一些表现突出的毕业生这次没有列入，我想今后可在教育学院办的刊物《暨南成人教育》上加以介绍。

本选集的文章大部分写成于 1992 年，所介绍的毕业生有的可能已经调动工作，有的可能已经再创佳绩。由于种种原因，已经来不及改写或增补材料。虽然如此，我认为就目前介绍的业绩，还是十分感人，十分有启迪意义的。

本选集所写的事迹生动感人，文字也不错，但也有个别文章，文字略嫌粗疏，不过，瑕不掩瑜，因此它不失为一本有意义的、值得一读的好书。是为序。

---

① 本文是《红杏枝头春意闹——暨南大学成人高等教育毕业生业绩选》序. 广州：华南理工大学出版社，1994，1-2.

6. 暨南大学“211工程”部门预审汇报①

暨南大学申请“211工程”部门预审得到国务院侨务办公室的批准、广东省人民政府的支持和国家教委的同意，全校师生员工受到巨大鼓舞。这是学校十分难得的发展机遇。我谨代表全校一万七千名师生员工向各位专家和各位领导不辞辛劳前来我校进行预审工作，表示热烈的欢迎和衷心感谢！

下面，我从暨大的历史、现状和发展规划三个方面，向各位专家和领导进行汇报。

一、暨南大学是一所具有辉煌历史和独特地位的华侨最高学府

（一）暨南大学是我国历史最悠久的华侨高等学府

暨南大学的前身是暨南学堂，1906年（清光绪三十二年）由清政府两江总督端方创建于南京，是我国第一所由国家创办的华侨学府。“暨南”二字出自《尚书·禹贡》“朔南暨，声教讫于四海”，其含义是要把声教——声威教化，即中华文化，远远传播到海外。

1918年，暨南学堂改名为国立暨南学校；1927年，国立暨南学校改名为国立暨南大学。

由于暨南大学建校九十年来从未改变华侨学校的性质，因而在海内外有着非常广泛的影响。

（二）暨南大学在中国高等教育发展史上具有独特的地位

暨大作为华侨学校，自创办以来，始终受国家重视，有着辉煌的过去。1921年，国立暨南学校与国立东南大学在上海合办了中国历史上的第一所商科大学——上海商科大学。这是当时全国仅有的5所国立大学之一。

暨大曾吸引大批知名学者和政府要员从事管理和教学工作。比如，第一任校长郑洪年，为当时全国最高学术教育行政机构大学委员会排名第3的委员；最早的校董会董事有黄炎培、张謇、陈立夫、林森、孙科、宋子文、孔祥熙等。曾应聘到暨大任职任教的有郑振铎、马寅初、曹聚仁、叶公超、夏丏尊、潘光旦、顾颉刚、黄宾虹、潘天寿、王统照、邓初民、李达、梁实秋、夏衍、钱钟书、周建人、严济慈、楚图南、许德珩、潘序伦、洪深、胡愈之、俞振飞、孙大雨、刘大杰、周谷城、周宪文、王亚南、刘佛年等一大批国内著名学者；其中黄炎培、许德珩、周谷城、周建人、严济慈、楚图南、胡愈之7位在新中国成立后担任了全国人大常委会副委员长等要职。

在暨大的历史上，不仅教职员队伍中名流荟萃，而且学子中涌现出一大批非常杰出的人才。国内的有李岚清副总理，吴学谦（政协）副主席；全国人大常委会委员、长期担任宋庆龄秘书的罗叔章，曾任天津市委第一书记、中央政法委员会副书记的陈伟达，曾任湖北省副省长的邓垦，中共中央候补委员、海南省副省长王学萍，广东省人大常委会副主任侣志广，中国科学院院士、历史地理学家谭其骧等；还有一些学子为

① 本文是在暨南大学“211工程”部门预审会上的汇报，广州，1996年6月12日.

中国人民的解放事业献出了宝贵的生命，如革命烈士江上青就是其中之一。境外的有曾任泰国国会主席的许敦茂，国际奥委会委员、现任台湾红十字组织负责人徐亨，新加坡国立大学首任校长李光前，印尼侨领司徒赞等。

暨大在旧中国高教界有着比较突出的地位：在1931年12月南京政府教育部公布的17所国立大学排序表上，暨大居第10位；在1948年《第三次中国教育年鉴》公布的32所国立大学中，暨大居第9位。

新中国成立后，党和政府非常重视华侨高等教育事业。1958年国务院批准在广州重建暨南大学。国家领导人廖承志、荣毅仁先后担任暨大董事长；现任董事长钱伟长是著名科学家、全国政协副主席；中共中央中南局书记陶铸、广东省副省长杨康华和省长梁灵光先后兼任过校长。

在改革开放新的历史条件下，党中央、国务院将办好暨南大学的意义提到了战略的高度，1983年6月20日批复了中央宣传部、教育部、国务院侨务办公室《关于进一步办好暨南大学和华侨大学的意见》（中共中央〔1983〕24号）。24号文件深刻地指出：进一步办好暨南大学和华侨大学“对于做好台湾回归祖国、收回港澳主权，完成祖国统一大业……对于开展拥护祖国统一的爱国者的广泛统一战线，都具有重大意义”；文件将暨南大学列为“国家重点扶植的大学”。

最近，中央和省的主要领导同志纷纷题词祝贺暨大建校90周年。江泽民总书记的题词是“爱国爱校，团结奋进”；李鹏总理的题词是“坚持面向海外、面向港澳办学”；乔石委员长的题词是“努力办好有特色的华侨最高学府，为统一祖国，振兴中华作贡献”；李瑞环（政协）主席的题词是“发扬爱国传统，致力振兴中华”。

党中央、国务院的文件和中央、省的领导同志的题词，给暨南人以巨大的鼓舞。

暨大在“文化大革命”前是高教部直属院校，现在是国务院侨务办公室所属重点大学；国务院侨办与广东省人民政府签订协议共建暨大。

二、暨南大学贯彻“两个面向”的办学方针，各项事业蓬勃发展，是一所基础较好的综合性华侨大学

暨南大学的各项事业目前正处于前所未有的大好发展时期。

（一）华侨华人和港澳台学生数居国内高校首位，人才培养质量享誉海内外

暨大九十年来培养华侨华人和港澳台学生18 000人，其中1978年以来培养6814人；目前在学的境外学生有2849人。本科学生中，境外学生约占1/3。改革开放以来，境外学生来自五大洲61个国家和中国港澳台地区，而且境内学生大多数也是华侨华人和港澳台眷属子女。暨大学生来源之广，华侨华人和港澳台学生人数之多，在国内高校中一直居于首位，成为我国对外办学的一大基地。

1. 各类毕业生分赴海内外，普遍受到欢迎与好评

暨大不仅招收的华侨华人和港澳台学生最多，而且培养质量有保证，毕业生受到社会好评。

1）境外

根据对香港、澳门毕业生的跟踪调查，用人单位普遍反映：暨大毕业生热爱祖国，

基础知识扎实，工作积极肯干。

例如，在香港的暨大历届毕业生逾 5000 人，其中 65%是 1982 年以后毕业的。毕业生中许多人在当地已有相当地位和影响，如社会学专业毕业生王国兴，1995 年当选为香港市政局议员，并担任筹委会预委会委员、港事顾问；工业经济 1994 届博士毕业生黎鸿基担任香港沙宣证券公司高级分析员，并被国务院技术经济研究中心聘为顾问。

又如在澳门，暨大的历届毕业生近 2000 人，其中 80%以上是 1982 年以后毕业的。校友中有 1 人担任了澳门基本法起草委员会委员兼秘书，并任澳门基本法咨询委员会委员；有 3 人当选为澳门立法会议员及澳门市政议会议员；澳门政府医院——仁伯爵综合医院的医生一半以上是我校医学院毕业生（150 人）。

我校的港澳校友，还有多人当选为广东省人大代表和广东省政协委员。

暨大毕业的港澳学生在不同的岗位上为港澳地区的稳定繁荣和平稳过渡做出了积极的贡献；华侨华人毕业生回原居住地均表现良好，为促进当地各项事业和实业的发展做出了自己的努力。

2）境内

1982 年以来，我校培养的国家计划境内学生 80%以上留在广东工作，社会反映良好，如 1982 届企管系毕业生谢树文任广州造纸厂厂长，成为广东十大杰出企业家之一，评为全国优秀大学毕业生。有一定社会知名度的省内校友还有全国十佳青年优秀乡镇企业家潘子怡、广东省粮食企业集团公司总裁董富胜、香港商报副总经理邝景廉等；省外的有 1991 年获“做出突出贡献的中国博士学位获得者”荣誉称号的丘进、当代著名青年诗人汪国真等。

2. 侨教改革和建设取得可喜成绩

1）加强德育工作，创造性地建立了侨校学生思想政治教育体系

近年，我校加强学生德育工作的针对性，采取了分类别、分层次、分年级开展德育工作的办法，产生了良好效果。对境内学生，按照国家教委规定，进行系统的马克思主义理论和建设有中国特色社会主义理论教育，培养他们成为社会主义事业的建设者和接班人；对华侨、港澳台学生，按照中央 24 号文件精神，主要进行爱国主义教育；对外国留学生（多数是华裔学生），主要是使他们了解当代中国，增进对我国的友好感情。学校还开设哲学等政治理论课供境外学生选修，组织他们到外地参观、考察，了解改革开放的大好形势，结合社会实践进行思想教育。

2）深化教学改革，加强教学建设

（1）健全和完善学分制。我校是全国实行学分制较早的高校之一。从 1994 级开始，我校采用了国外高校通行的标准学分制度。四年制本科总学分数 170 学分；每个学期标准学分数 23 学分；选修课的学分占 30%以上。学习能力稍差的同学允许少修学分，但不能低于 18 学分；学习能力较强的根据上一个学期的平均学分绩点情况，可分别奖励多修 5—10 学分。从而更好地体现了因材施教原则，推进了我校教育与国际教育进一步接轨。

（2）对课堂教学质量进行全面评估。由学生、听课专家组与校、院、系和教研室负责人三方评估教师课堂教学效果，采用光标阅读器和电脑处理评估资料。评估结果与教

师的奖惩等挂钩，1995 年学校从中评出“十佳授课教师”予以重奖；若连续两次评估不及格则取消任教资格。评估制度化，有力地促进了课堂教学质量的提高和教风的好转。

(3) 严格考试制度，形成良好学风。我校从 1993 年起就取消了补考制度，学生学习成绩不及格必须重修。同时加强思想教育，严格执行考试制度，对考试作弊者当场给予严厉处分，因而考试纪律已根本好转。学生已养成勤奋学习习惯，学风良好。

(4) 重视外向性专业建设。我校先后增设了国际新闻、国际商务、信息工程等十多个外向性与应用性较强的专业，较好增强了对外招生吸引力。以外向性为目标的专业改造亦取得明显成效，如医学专业的课程结构已基本符合英联邦医学教育的要求；会计学专业的课程结构已达到国际会计专业要求的标准。1994 年 12 月，国家教委组织专家组对我校“应用化学”专业进行评估，评估结果为优良；还对该专业针对不同来源的学生采用不同方法施教和重视青年教师培养工作特别给予加分。

(5) 重视课程建设。“八五”以来，学校先后投入 195 万元重点建设 100 门课程，部分课程已取得明显的效果，有 7 门课程已被评为省级重点课程，其他多数课程已达到校级优秀课程标准。

(6) 重视教材建设。全校先后自编、出版教材 203 种，引进外文原版教材 57 种。在全国高校优秀教材评奖中，我校有 4 本教材获国家级奖，4 本获部委级奖。

(7) 加强基础和实践环节教学。我校强调教授上本科基础课。目前，上本科基础课程的教授人数已达教授总人数的 70%。

全校现有固定的境内、外教学实习基地 67 个，境内、外教学实习、见习医院 22 所；能按教学大纲开出实验项目，保证学生的基本技能训练。

(8) 全面规划旅游、华文和成人教育，拓宽对外办学渠道。中旅、华文是近几年新办的学院，发展势头良好。学校重视这两个学院的全面建设，以形成我校对外办学的新特色。今年 5 月，广东省将我校作为试点单位，对我校成人教育进行评估，评估结果优良。

3) 教学质量提高，获得大批教学奖项

(1) 英语统考成绩提高。近年我校各层次学生品学兼优和单项成绩突出的比率有较大幅度提高。学生参加全国大学英语四、六级统考通过率逐年上升，如四级统考通过率 1991 级 60%，1992 级 69%，1993 级 72%。

(2) 研究生教育质量提高。我校是全国首批获得研究生学位授予权的学校，形成了较为完整的研究生质量管理制度。从 1978 年至 1995 年共培养了境内外研究生 1264 名，成为我国内地为港、澳、台和东南亚地区培养研究生最多的高校。我校切实保证研究生培养质量，如 1994 年全国 26 个招收工商管理硕士（MBA）的高校，统一进行入学后英语考试，我校名列第二；同年，国务院学位办对 1991—1993 年在职授予学位的人员进行检查，我校 15 名人员质量全部过关，受到国务院学位办好评。我校还是国家教委指定的对外招收研究生的唯一试点单位。

(3) 多项参赛成绩优异。我校学生参加亚洲田径锦标赛等国内外体育大赛成绩优异，仅 1995 年，在国外比赛中就获金牌 12 枚，在全国和全省的比赛中获金牌 70 枚。1992 年，获国家教委授予的“全国普通高校体育课程评估优秀学校”称号；1995 年被定为全国 53 所试办高水平运动队高校之一。体育课程质量进入全国高校先进行列，为国内外所注目。1995 年，在广东省大学生运动会上，我校勇夺团体总分、金牌总数和奖牌总数三项第一，并获得体育道德风尚奖，被广东省高教厅授予“广东省普通高等院校体育先进单位”称号。我校学生参加国内外其他业务竞赛也取得不少突出成绩。例如，在去年日本国驻广州领事馆和广东省高教厅共同举办的第 11 届日本语演讲大赛中，我校夺得团体冠军；在国家教委和团中央举办的第四届“挑战杯”全国大学生课外学术科技作品竞赛中，我校获团体总分第 8 名，一、二、三等奖各一项；在全国大学生数学建模竞赛中，我校代表队获广东赛区一等奖；最近又在美国大学生数学建模竞赛中获两项三等奖。

(4) 一批教学成果获国家级和省、部级奖励。在全国高校优秀教学成果评奖中，我校的“结合使用录像、语言实验室与电脑辅助英语教学”等两项成果获国家级奖；有 13 项成果获省级一、二等奖。

(5) 电化教育名列广东高校第一。我校 96%的教室装配了电教设备，是广东高校中第一所有组织有计划地开展计算机辅助教学（computer assisted instruction，CAI）研究与开发工作的学校，70%的教师已使用电教手段上课。

我校一直是“中国电化教育协会普通高等学校体育电教专业委员会”① 理事长学校，在全国有较大影响。

1995 年，在广东省高校电化教育评估中，我校被评为优秀学校；且总分在全省 10 所规模最大的高校中排名第一。

（二）科学研究和学科建设“侨”字特色鲜明，整体水平不断提高，具有较大优势

1. 科学研究成果丰硕，多项成果水平居国内外领先地位

(1) 人文社会科学研究独具特色，成果显著。我校人文社会科学研究具有鲜明的“侨”字特色，设有东南亚研究所、华侨华人研究所、特区港澳经济研究所等一批研究境内外问题的专门机构。这些单位为国务院和省有关部门制定对外政策，开展外事活动，提供了不少有价值的研究报告；出版的大批华侨华人等问题研究专著广受社会好评；华侨教育研究居国内领先水平。

“八五”期间，全校文科类出版著作 313 部，发表论文 3700 篇；获省、部级以上奖励 77 项；承担各类科研课题 310 项，其中国家社会科学基金和自然科学基金项目 16 项，国家教委及广东省项目 52 项；广东省高教厅及广州市项目 63 项。

(2) 理、医科科研水平不断提高，取得多项在国内外居领先水平的成果。我校理、医科科研水平近年迅速攀升，“八五”期间获得的科研经费比“七五”期间增加近 3 倍，获省、部级以上科技成果奖数亦有增加。

理、医科科研取得一批重要成果，其中在国内外居领先水平的有：由齐雨藻教授主持的“中国东南沿海赤潮发生机理研究”，1994 年通过了国家自然科学基金委专家组的验收，有多项指标达到国际领先水平；“中国沿海典型增养殖区有害赤潮发生动力学及防治机理研究”项目最近又通过了国家自然科学基金委专家组的论证，为我国“九

① 中国电化教育协会于 2002 年 11 月改名为中国教育技术协会。

五”首批立项的13个重大项目之一。刘人怀教授创造性地研究了6类板壳非线性理论，其中多项研究在国际领先；特别是他的精密仪器仪表弹性元件设计理论，结束了几类弹性元件（如波纹膜片等）依靠国外公式设计的历史，刘人怀教授成为我国该领域的开拓者和带头人。由李辰教授主持的异种角膜移植居国内领先水平；异种角膜镜片术的研究跻身于国际先进行列。由林剑教授主持的“重组碱性成纤维细胞生长因子（bFGF[①]）的研究”，为国家“八五”攻关项目，提前一年通过了国家技术鉴定，获得了国家卫生部颁发的国家一类新药证书；这是广东省第一个基因工程一类新药，也是我国第二个基因工程一类新药，目前已设厂投产。刘学高教授主持的“HCG避孕疫苗研究”，已研制成功国产第一代避孕疫苗，通过了国家验收，被称为我国避孕疫苗研究的一个里程碑，获得国家计生科技攻关二等奖；他本人也被评为全国6名“计划生育科技功臣”之一，所在单位生殖免疫中心被评为全国计划生育科研先进集体。

2. 学科整体水平提高，形成文、经、理、医四大学科门类各自综合优势

我校现有7个博士点，53个硕士点，33个本科专业，覆盖文、史、经、管、法、理、工、医8个学科。博士点数在全国高校中列第72位，硕士点数列第46位。全校现有国务院侨办批准的办属重点学科9个，广东省重点学科2个。

我校几大学科门类综合优势具体如下。

（1）经济学科整体实力居华南各省高校前列。暨大不仅是全国设立经济学科最早的高校，也是改革开放以来最先设立经济学院的高校，已形成以工业经济博士点为龙头、以10个硕士点为主干、以13个本科专业为基础、外向特色和综合优势明显、结构也比较合理的经济与管理学科群。

经济学科对境内外学生都有较大吸引力，已培养博士生7名，硕士生388名；目前在读博士生11人，在读硕士生180人。

（2）文、史学科基础雄厚，部分领域研究居全国领先地位。我校文、史学科有一个国家文科基地，有文艺学、现代汉语、专门史（中外关系史）等3个博士点，10个硕士点，8个本科专业，7个研究所（室），具有基础雄厚、传统学科与新兴外向学科齐头发展等特色。例如，著名汉语方言学家詹伯慧教授和李如龙教授的方言研究已延伸到港澳和东南亚地区，在全国处领先地位。比较文艺学专家饶芃子教授在“中西文艺学比较”等多个新的研究领域有较高造诣，在我国和东南亚各国都有较大影响。

（3）医学学科形成独特的办学优势。暨大是改革开放以来全国第一所在综合性大学里设立医学院的高校，也是全国少数几个在世界卫生组织注册的医学院校之一。医学学科本科学制六年，与国际教育接轨。该学科有内科学（血液病）、眼科学和妇产科学3个博士点，有药理学等19个硕士点和2个本科专业；共有高级职称人员230人，具有专家密集、整体水平高等特点和优势。

十多年来，医学学科已毕业和在读的境外学生937名，是我校招收培养境外生最多的学科。

（4）理学学科改革取得明显成效，初步形成多个新兴学科发展方向。我校进行

① bFGF，即basic fibroblast growth factor，碱性成纤维细胞生长因子。

“纯理”专业改革，取得显著成效。

第一，走出了一条产、学、研结合和理、工结合的成功之路。

第二，促进跨学科合作，初步形成了优势和特色明显的生物化学、生物医学和信息技术三大发展方向。

第三，已拥有电子工程等 5 个工科专业，而且“工”的比重在不断增大，应用学科实力呈上升态势。

（三）师资队伍素质良好，拥有一批学术声望较高的教授

我校现有专职教师和专职科研人员 1125 人，其中教授和研究员 152 人（内有 1 名国家有突出贡献中青年专家，19 名博士导师）；副教授和副研究员 437 人。

师资队伍具有两个特点：一是整体较年轻，平均年龄为 44 岁；二是高学历教师所占比例较大，博士、硕士占 49.1%，其中博士占 10.2%。

我校先后培养了三批重点教师共 162 人；1995 年，为落实广东省的“千百十工程”，又选拔 58 名中青年学术带头人和重点教师进行重点培养。

学校拥有一批学术造诣较高的教授、专家，如有获国家有突出贡献中青年专家称号的固体力学专家刘人怀教授；眼科学专家李辰教授、徐锦堂教授；血液病学专家郁知非教授；工业经济学专家黄德鸿教授、云冠平教授；汉语方言学专家詹伯慧教授、李如龙教授；文艺学专家饶芃子教授；妇产科学专家王自能教授；藻类学专家齐雨藻教授；生殖免疫学专家刘学高教授；生物遗传学专家林剑教授；金融学专家张元元教授；病理生理学专家李楚杰教授；生物材料学专家邹翰教授等。

我校大批青年骨干教师迅速成长，一批青年学者崭露头角，如近 6 年来，先后有 7 人获得霍英东教育基金会基金类和教学类奖励；有 6 名年轻教授被遴选为博士生导师；由青年教师主持的“功能性内窥镜鼻窦外科”的研究、计算机图像处理研究、证券投资组合研究和肿瘤病理研究等多项研究处于国内同行研究的前列，各学科的发展充满了希望。

（四）暨大充分发挥自身办学优势，挖掘潜力，为广东侨乡服务

我校通过三大途径积极为广东侨乡服务。

(1) 挖掘办学潜力，为广东培养急需人才。15 年来，我校为广东培养了近 10 万名人才，其中学历教育学生超过 3 万人，非学历教育学生超过 6 万人，为提高广东全省人口中大学生的比率做出了贡献。

(2) 利用智力资源优势，为广东改革开放献计献策。我校充分利用经济学科门类较齐全，并拥有研究境外和特区问题的大批人才资源优势，为广东改革开放贡献聪明才智。例如，特区港澳经济研究所从 1979 年开始就参与对经济特区的理论和重大方针、政策的研究，为广东、也为全国经济特区的建设与发展做出了重要的理论贡献。

我校是以群体学科综合优势参与广东省和广州市发展战略研究的主要高校之一，如在 20 世纪 80 年代初，承担并完成了广东省政府首批的 9 项科技发展战略研究，占该批研究项目的一半。近几年，我校一批专家又参与广东省委、省政府组织的珠江三角洲经济区发展规划工作，并完成了乐昌[①]和连南瑶族自治县等一批山区贫困县的发展战

① 1994 年撤县设市。

略研究任务。

(3) 救死扶伤，为广东城乡人民提供医疗保健服务。我校医学院附属医院（广州华侨医院）多次受到广东省卫生厅、市政府卫生局的表扬和嘉奖。医院每年接诊病人30多万人次，急诊3万多人次，收治住院病人1万多人次，进行手术5000多例，为广东城乡人民的身体健康提供了良好服务。

（五）全面建设，为学校未来发展打下坚实基础

暨大目前设有文、理工、经济、医、中旅、华文和教育7个学院，21个系，34个科研机构；在校教职工3601人；在校学生13 012人，其中研究生615人（内含博士研究生52人），本、专科生7849人，其余为成人和华文教育等学生。

我校是全国唯一的一所分别在港澳两地设立MBA研究生教学点的高校。

最近，经国家教委批准，我校成为全国37所有接收本科保送生资格的高校之一。

暨大占地总面积121.8万平方米，建筑总面积49.1万平方米；图书馆藏书134万册，已开通了计算机管理集成系统和光盘检索系统；全校现有教学设备总价值8298万元；有一个世界银行贷款建立的“组织移植与免疫”实验室和47个普通教学实验室；有一座多功能现代化大型体育馆；学校已于去年底建成校园计算机网，实现与中国教育和科研计算机网（China Education and Research Network，CERNET）及Internet联网。

我校医学院拥有一所有650张床位、已通过卫生部评审的三级甲等附属医院（广州华侨医院）。学校还有一家初具规模的出版社。

今年内，教职工住房可全部解困，达到广东省规定的住房标准。

暨大地处改革开放较早的侨乡广州，又毗邻港澳；有一个规格高、成员分布于近10个国家和地区、关心和支持学校建设的董事会；已与美国、日本、英国、德国等十多个国家及中国港澳地区的29所高校和学术机构签订了学术交流合作协议；学校还与加拿大和英国注册会计师协会分别合作办学，已走出一条与境外合作办学的成功之路；十多年来，学校收到海内外各界热心人士支持办学的各类捐赠，总值逾1亿元；广东省近7年给我校的财政补贴和拨款4745万元；1993年，在国务院侨办支持策划下，筹资7000万元改善了华文学院办学条件；在国务院侨办领导下，多家中旅和深圳华侨城指挥部共同筹资近1亿元，与我校合作在深圳华侨城兴办了中旅学院；近年，在国务院侨办帮助指导下，香港中旅集团资助我校3500万港元购置教学仪器设备。这些说明，暨大具有对外办学难得的人缘、地缘优势，为今后的发展打下了良好的基础。

我校领导班子注意抓大事、议大事，坚持党的基本路线，全面贯彻党的教育方针，全心全意为办好学校服务；在认真抓物质文明建设的同时，抓好精神文明建设；加强党的建设和教职工思想政治工作，坚持正确的办学方向。学校还开展了多方面的校内管理体制改革尝试，提高管理水平和办学效益。全校师生奋发向上，恪守“忠信笃敬”校训，具有良好的精神风貌。

我校在系统回顾办学的成功经验和为社会所做贡献的同时，非常清醒地认识到：与国内外先进高校相比，与国家和地方发展的要求相比，还有一定的差距；要使学校进一步办出特色、办出水平，也存在某些方面的困难。

例如，教育经费偏少，办学条件与侨校要求还不相适应。教育投入不足，学校的基本建设、教学设备、图书资料、生活设施等从总体上看，尚未达到中央 24 号文件提出的“标准应该适当高于国内其他大学”的水平。

又如，与 20 世纪五六十年代相比，今天的华侨华人和港澳台青年就学取向已发生很大变化，如何适应境外这方面的变化，进一步拓宽对外办学渠道，更好地完成国家赋予的“两个面向”的办学任务，还要做艰苦的努力。

三、面向 21 世纪，全面提高办学水平，再创暨大辉煌

各位专家、各位领导、同志们，再过两天，暨南大学就要庆祝九十周年华诞了。暨大九十年走过的光辉历程说明，华侨教育是一项伟大的事业。办好暨南大学，是全球华侨华人和港澳台同胞的热切希望。通过侨教联系侨情，扩大国际交往，对于我国的改革开放事业，具有十分重要的意义，是一项战略性的任务。为此，我校做出了从现在起至 2010 年的整体建设与发展规划，决心发扬暨南优良办学传统，再创暨大新的辉煌。

（一）发展目标

我们的战略目标如下。

经过 15 年左右的努力奋斗，到 2010 年前后，使暨南大学的教育质量、科研水平和办学效益等方面进入国内高校先进行列，成为我国“面向海外、面向港澳台”办学和传播中华文化的重要基地，成为在香港、澳门、台湾地区和华侨华人社会具有重要影响的社会主义华侨大学。

上述目标拟分两个阶段实现。

第一阶段（1996—2000 年），通过“211 工程”部门预审，并以此为动力，加快推进各项改革和建设，使我校的教育质量、学科建设、科学研究、管理和社会服务水平上一个新的台阶，使学校整体实力显著增强；对外办学层次明显提高，外招学生数有大幅度增加，成为我国“面向海外、面向港澳台”办学、培养高层次人才的重要基地，为总体目标的实现打下坚实基础。

第二阶段（2001—2010 年），适应 21 世纪科学技术发展和对外办学任务的需要，加速教育现代化进程，使学校的办学水平与整体实力进入国内高校先进行列，成为教育和科研两个中心，全面实现中央 24 号文件提出的要求和学校“211 工程”规划的总目标。

（二）实现目标的任务和改革措施

1. 人才培养与教育改革

我校人才培养坚持“面向海外、面向港澳台”的方针，贯彻以培养“应用型”与“外向型”人才为主的原则，全面提高人才培养的质量。到 21 世纪初，使我校发展成为立足广东、面向海外和港澳台培养高级专门人才的重要基地。

主要措施如下。

1）妥善处理稳定教育规模与发展的关系

在稳定办学总规模的基础上调整在校学生的层次结构，做到：大力发展研究生教育，适度发展本科教育，控制和压缩专科教育，努力发展华文教育，积极发展成人教育。

到 2000 年，在校生规模由目前的 13 012 人发展到 14 600 人；到 2010 年，发展到

16 400 人。

2）深化各层次教育改革，切实提高人才培养质量

（1）研究生层次。大力发展研究生，重点是博士研究生教育，积极创造条件建立研究生院。到 2000 年，在校研究生数达到 1200 人。除一部分学科专业继续培养理论研究型硕士生外，大部分硕士点将转向培养应用型硕士生。先在国家文科基地试行“硕、博连读”的培养方式，然后在其他博士点逐步推广。

（2）本科层次。以提高学生全面素质为改革重点，强化基础教学，深化专业改革，加强专业的外向性和应用性，拓宽专业知识面，增设工科专业和体现侨校特色的专业。

（3）办好中旅学院。依托校本部的师资力量和学科优势，发挥中旅系统经济实力雄厚和实习场所条件优越等综合优势，借鉴特区的改革经验，将中旅学院办成全国一流的旅游管理人才培养基地，为我国旅游事业输送各层次专门人才。

（4）办好华文学院。造就能熟练运用汉语的、对我国友好的海外人才，把华文学院办成我国南方进行对外汉语教学、传播中华文化的重要基地。

（5）成人教育。在积极为广东地方培养人才的同时，通过函授等多种成人教育形式，开拓境外办学渠道，更好完成我校“两个面向”办学任务。

2. 学科建设与改革

“九五”期间直至 21 世纪初，我校将尽力加强重点学科建设，以提高学校整体办学水平。

1）学科建设的基本思路

（1）重点建设领先学科。例如，文艺学和工业经济学两个学科是国务院侨办和广东省所属的重点学科，我校力争在 2000 年前将这两个学科建设成为国家级重点学科；并争取在这两个学科中建立博士后流动站。

（2）大力扶植优势学科。现代汉语、眼科学为国务院侨办所属重点学科，“九五”期间首先将这两个学科建设成为广东省的重点学科，然后，争取尽快建设成为国家级重点学科。同时，加强内科学（血液病）、专门史（中外关系史）和妇产科学 3 个博士点的建设工作，拓宽这些学科的研究方向，积极创造条件，使它们发展成为广东省的重点学科。

（3）扩大重点学科规模。国际金融、投资经济、新闻学、水生生物学、遗传学、病理学、病理生理学等学科，有的是国民经济建设急需的，有的是历史悠久、在内地（大陆）和港澳台地区有较大的影响。“九五”期间，我校将积极扶植这批学科，以发挥其优势，提高水平，争取把其中的 2—3 个建设成为博士学位授予点。

（4）促进形成新兴、交叉学科。例如，继续促进计算机软件、电子学与信息系统、管理信息系统和通信工程等学科间的交叉渗透，使之形成信息工程学科。

（5）积极组建学科群。以重点学科为龙头，相关学科为支撑，组建经济与管理、中国语言与文化和生物医学学科群。

（6）保持基础学科优势。我校将继续抓好数学、力学、物理学、化学、生物学等基础理论学科建设，以保持综合性大学应有的理论优势。

（7）合理调整学科布局。稳步发展工学、管理学和法学学科，使我校学科门类更

为齐全，结构更为合理。

2）学科建设的目标

力争在“九五”期间实现以下目标。

建立2个国家级重点学科、2个博士后流动站。

增建2—3个省级重点学科、2—3个博士学位授予点。

组建3—5个学科群。

使我校2—3个学科达到全国领先水平，在国际上有一定的影响；部分学科进入国内高校先进行列，学科整体水平和实力位于华南地区高校的前列。

培养2—3名国家级学科带头人和10名左右具有较高学术造诣、在国内外有一定影响的中青年学科带头人。

3）学科建设的主要措施

(1) 加强学科队伍建设。近期，我校遴选与增补了文艺学、工业经济、内科学（血液病）、专门史（中外关系史）等4个博士点6名年轻博士生导师，成为这些学科的带头人。我校还将进一步配合实施广东省的“千百十工程”，加快学科梯队建设的步伐。

(2) 加大学科建设的投入。一方面积极争取国家和广东省人民政府对我校重点学科建设增加投入，使其成为学科建设投入的主渠道；另一方面学校将重点学科作为建设的重中之重，计划“九五”期间投入建设经费2500万元；同时，要求每个学科自身也要多渠道筹措学科建设经费，并做好人员、物资等方面的配套工作。

(3) 搞好学科环境建设。主要是认真做好政治思想工作，搞好内部团结，调动各方面的积极性，创造良好的治学环境。

(4) 加强学科建设的管理。一是加强对学科建设的领导，实行统一规划、分级目标管理；二是分层建设，滚动发展。

3. 科学研究的发展与改革

1）自然科学研究的发展目标

立足世界科技发展前沿，面向经济建设主战场，紧密结合“两个面向”办学方针，建立起比较完善的科技工作体系，力争全校的科学技术工作到2010年前后进入国内高校先进行列。

主要指标如下。

预算外科研经费以每年15％的速度递增，到2000年达到1000万元，2010年超过2000万元。

到2010年，科学研究项目和经费有1/5来源于境外。

在核心期刊发表的论文数每年递增10％。

获省部级以上奖励的成果数逐年增加，至2010年，获奖成果数比现在翻一番。

2）人文社会科学研究的发展目标

以邓小平同志建设有中国特色的社会主义理论为指导，针对港澳台和华侨华人社会的发展变化及广东侨乡经济建设的需要，加强对华侨华人社会、港澳台和珠江三角洲的政治、经济、文化、教育等问题的科学研究，争取出一批高水平的科研成果，为国家制定侨务政策和广东建设侨乡起到“思想库”“智囊团”作用；把我校建设成为华侨华人社会、港澳台及广东侨乡重大问题的研究中心。

4. 师资队伍建设与改革

1）师资队伍建设的目标

建立起一支数量适当、结构合理、素质较高、既能认真搞好教学又能积极从事科研工作的师资队伍。

2）师资队伍分期建设任务

(1) 近期（“九五”前2年），使校本部教职工人数与学生人数比例从现在的1∶3.7达到1∶4.5；专任教师人数与学生人数的比例由1∶9达到1∶10；使教师中有研究生学历的比例从现在的49.1%达到55%，其中有博士学位的由10.2%达到15%。

(2) 中期（“九五”后3年），基本完成学术骨干队伍的更替工作，专任教师人数与学生人数的比例达到1∶12。具有研究生学历的教师达到教师总数的75%，其中有博士学位的教师达到30%。每个学科、专业均应有1—2名学术造诣较高的学科带头人。

(3) 长期（2001—2010年），教师队伍的整体素质达到国内高校先进水平；一批学科带头人的学术水平居国内领先地位，并在国际上有一定的影响。

3）师资队伍建设的主要措施

一是控制队伍的发展规模。在20世纪末，专任教师、专职科研人员编制严格控制在1300人以内，同时减少固定编制，适当增加流动编制。

二是拓宽师资队伍人才补充渠道，优化队伍结构，争取每年至少引进50名博士。

三是加强对中青年教师的培养，重点做好58名中青年学术骨干的培养工作。

四是转换管理模式，增强队伍建设的活力。实行“非升即走”“专兼结合”的聘任管理办法。

5. 校内管理体制的改革

我校管理体制改革的目标是：完善领导科学决策程序；创造人才竞争环境；体现公平分配原则；建立高效运行机制；提高自我发展能力。

1）充分发挥董事会的作用

要在坚持办学方向、制定学校发展规划和筹措办学资金等方面，进一步充分发挥董事会的积极作用。

(1) 认真加强学校与校董的联系，强化董事会的职能。

(2) 面向华侨华人社会，遴选董事，优化董事会成员结构。

(3) 发动董事资助办学，为学校办学现代化贡献力量。

2）健全校、院、系三级管理体制

(1) 进一步明确校、院、系之间的职责。

(2) 调整校部机关机构设置，精减非教学、科研人员，建立高素质、高效率专职管理的干部队伍。与此同时，进一步搞好干部人事制度、校内分配制度和财务管理制度等改革，加快校办产业发展，努力提高办学效益。

6. 贯彻《共建协议》，增强为广东服务的能力

在未来15年中，进一步贯彻落实《共建协议》（即《国务院侨办和广东省人民政

府共同建设暨南大学协议》)，认真完成《共建协议》中赋予我校的人才培养、科学研究、社会服务等方面为广东服务的任务。

1）加快教育发展步伐，为广东建设教育强省做出贡献

“九五”期内，研究生教育每年将以15%的速度增长，从1995年的615人发展到2000年的1200人，在校研究生数比1995年翻一番；本科教育从1995年的5377人发展到2000年的8100人，增长51%；成人高等教育从1995年的3500人发展到2000年的4600人，增长31%。同时，继续办好新闻学等4个专业的高等自学考试班，扩大办学规模，为实现广东省计划到2000年每万人中有56名大学生的目标做出贡献。

2）优化专业结构，为广东，特别是广州市培养应用型人才

1996年至2000年，我校将根据广东省、广州市经济建设的目标与任务，着力办好应用性专业，培养各层次，特别是高层次应用型人才。

3）大力发展科学技术，为广东省、广州市的经济建设和社会发展服务

“九五”期间广东的经济增长要求科技进步的贡献率达到50%，我校在邮电通信、信息基础工程、医药工业、第三产业、卫生保健、计划生育、环境保护等方面有较强大的技术力量，能为广东省和广州市的发展继续做出贡献。“九五”期间，我校将为广东开发一批生物医学材料，并充分利用医学院附属医院的科研与医疗条件，为广东建立窥镜中心、移植中心、康复中心及肿瘤基因诊治中心等，并完善小儿麻痹后遗症矫治中心建设工作。

7. 对外学术交流与合作

(1) 2000年前，我校在巩固已有境外交流与合作关系的同时，力争再与美国、英国、德国、日本等发达国家各1—2所有声望的大学或研究机构建立学术交流和合作关系。

(2) 我校已与泰国、越南、马来西亚、新加坡等国部分高校建立了学术交流与合作关系，力争与其他东南亚国家的高校尽快建立学术交流与合作关系。

(3) 进一步丰富与境外的联合办学成果，开辟国际合作办学的新路。

8. 党建工作、思想政治教育与校园精神文明建设

第一，要充分发挥党组织的政治核心与保证监督作用，要进一步加强党组织的思想、组织、作风和制度建设；第二，要进一步做好党政干部的培养、选拔、管理工作，按干部队伍“四化”要求加强干部队伍建设；第三，从实际出发，抓好师生员工的思想政治教育，维护学校的安定团结；第四，坚持抓好校园精神文明建设，营造良好的育人和治学环境，形成爱国爱校的一代新风；第五，要继续搞好校园美化、绿化、净化工作，保证校园良好的治安秩序，使我校校园跻身全国文明校园行列。

（三）重点项目建设经费概算与筹资渠道

初步测算，学校完成“211工程”项目的建设需15年左右的时间，投资4.5亿—6亿元；而“九五”期间需投入2.6亿元。重点建设的项目除重点学科外，还有公共服务体系如图书馆、重点实验室、教育和科研计算机网等建设项目及基本建设项目。

上述项目的主要筹资渠道包括以下几个。

(1) 国务院侨务办公室在正常拨款的基础上，每年增加1750万元作为“211工程”建设经费(其中含香港中旅集团资助750万港元)。

（2）根据《共建协议》，广东省人民政府每年对暨南大学投入 1500 万元建设经费。

（3）争取校董、校友捐资每年 1000 万元。

（4）学校自筹每年 1000 万元。

从上述渠道，预计每年可筹措 5250 万元作为“211 工程”建设的专项经费。只要各方努力，可以实现筹资的目标。

各位专家、各位领导、同志们，暨南大学历经沧桑，在“文化大革命”中又遭浩劫，能发展到今天的规模和水平，实属不容易，是几代暨南人不懈努力的结果，是国务院侨办、国家教委、广东省和广州市人民政府以及海内外各界热心人士大力支持的结果，在世纪和千年交替的重要变革时期，暨大有幸接受“211 工程”部门预审，并承蒙各位专家和领导亲临指导，我们相信，这定会成为暨大实现腾飞的新起点。

面向 21 世纪，我们全体暨南人充满信心，决心为实现“211 工程”的宏伟目标奋力拼搏，把暨南大学建设成为国内一流水平的华侨高等学府，向全市、全省、全国人民、海外侨胞和港澳台同胞交一份满意的答卷。

诚恳希望各位专家、各位领导和同志们提出宝贵意见。

7. 在暨南大学“211 工程”部门预审总结会上的讲话①

经过专家们两天半紧张而有序的工作，按照国家教委预审规定的要求，专家们已经圆满地完成了对我校“211 工程”部门预审的工作。刚才，专家组长、中国科学院院士王梓坤教授宣布了专家组通过我校“211 工程”部门预审。对这一大的喜事，我感到十分的兴奋和由衷的高兴。对全体暨南人而言，6 月 14 日是我们学校大喜的日子，是暨南大学历史上具有里程碑意义的日子，将载入暨南大学的史册！借此机会，我代表暨南大学的党政领导和全校的师生员工，对全体专家和各位领导，表示衷心的感谢！

两天半以来，专家们以认真负责的态度，科学严谨的作风，实事求是地评价了我校总体建设规划和各项工作。专家们不辞辛苦地连续工作，一丝不苟的精神，令我们非常敬佩，值得我们大家学习。专家们对我校的评审意见和刚才在会上的发言都是十分中肯的，贴切的，语重心长的。专家们的真知灼见，对暨南大学今后的改革与发展，有着重要的指导作用，我们将很好地领会、消化专家的意见，虚心向兄弟院校老大哥们学习，认真修改、不断完善我们所做的规划。按照从严治校、从严治教、从严治学的办学思想，进一步落实各项整改措施。

接受“211 工程”部门预审，既是受教育的过程，也是增强全校凝聚力的过程。全校上下都在认真地思索，在本世纪末和下个世纪初，如何把我们暨南大学办得更有特色，办出更高的水平，更好完成党中央、国务院交给我们学校的任务。

我们现在面临着暨南大学发展得最好的时期，我们一定要抓住这次机遇，迎接挑战。我们校领导班子全体成员和全校师生员工，满怀信心，坚信在国务院侨办、国家教委和广东省人民政府的领导下，在各位专家、兄弟院校和海内外社会各界人士的关心支持下，暨南大学一定会以崭新的面貌迈向 21 世纪。明天，将是我们学校 90 周年大庆的日子，今天是我们“211 工程”部门预审通过的日子，两件大的喜事将给予我们最大的激励。

最后，让我们再一次感谢各位专家和各位领导。祝专家们、领导同志们身体健康，万事如意。

---

① 本文是在暨南大学“211 工程”部门预审总结会上的讲话，广州，1996 年 6 月 14 日；暨南教育，1996，(2)：12.

8. 庆贺九十华诞 创建一流大学①

在暨南大学昨日顺利通过“211工程”部门预审之际，海内外校友与全校师生员工迎来了建校90周年的大喜日子。今天，我们在这里举行隆重的校庆庆典活动。这是一次历史性的盛会，对暨南大学今后的建设和发展必将产生重要的作用和深远的影响。首先让我代表全校17 000余名师生员工，谨向光临庆典的中央、国务院侨办和省、市领导，向来自海内外的各位嘉宾、校董和校友表示最热烈的欢迎。向所有关心、支持我校建设和发展的上级领导、校董、校友、海外侨胞、港澳台同胞和国内外的朋友们表示衷心的感谢和崇高的敬意。

暨南大学作为“华侨最高学府”和成立较早的著名国立大学之一，在我国华侨教育史和近代高等教育史上都占有独特的地位[1]。暨南大学一向致力于弘扬中华文化，为海外、港澳台地区培养人才。建校90年来，已为海内外培养各层次毕业生50 000多人。在海内外校友中，不少人在事业上卓有成就，贡献突出，享誉中外。可谓人才辈出，誉传五洲。

党和国家对暨南大学历来都高度重视、无限关怀。为庆贺暨南大学建校90周年，党和国家领导人江泽民、李鹏、乔石、李瑞环、荣毅仁、李岚清、钱其琛、李铁映、谢非、吴学谦、钱伟长、霍英东、马万祺等在百忙中为我校题词，充分体现党和国家对广大海外侨胞、港澳台同胞的无限关怀和对华侨高等教育事业的高度重视，我校全体师生员工和海内外校友深受教育和鼓舞。

经过90年的风风雨雨，艰苦创业，学校现已成为一所实力雄厚、学科齐全，涵盖文、史、经、管、法、理、工、医等学科的新型综合性大学。学校声誉远播，在海内外的影响不断扩大。可以说，暨南今日之成就，凝聚了几代暨南人的辛勤耕耘。今天，当我们举行建校90周年庆典之际，缅怀暨南先辈们为暨南的建设与发展艰苦创业的业绩，校友们感到十分亲切与鼓舞。

暨南的事业是美好的、永恒的。在世纪之交的今天，学校面临着千载难逢的重大发展机遇。在这里，我谨代表学校衷心希望海内外的暨南校友、校董和朋友们，紧密团结起来，一如既往地关心、支持母校的发展。全校师生员工将在国务院侨办的直接领导下，团结拼搏，分步实施“211工程”总体建设规划，争取在2006年母校百年校庆之际，将暨大建设成为一所名副其实的全国一流大学，并成为在海外与港澳台地区有重要影响的社会主义综合性华侨大学。

参考文献

[1] 暨南大学校史编写组. 暨南校史（1906—1996）. 广州：暨南大学出版社，1996.

① 本文是在暨南大学建校九十周年庆祝大会上的校长致辞，广州，1996年6月15日；暨南大学校报，1996年7月15日.

9. 开拓创新　共建附属医院[①]

今天在深圳市人民医院举行深圳市卫生局和暨南大学共建暨南大学深圳附属医院签约仪式，令我十分高兴。

深圳市是我国第一个经济特区，经济建设成就卓著，享誉中外。深圳市人民医院是深圳特区最大最好的医院，是一所拥有 800 张病床的现代化的大型综合医院，在许多学科方面已形成自己的特色，已经拥有一支医疗技术过硬、教学经验丰富的医疗队伍，历年来，承担了多家高等医学院临床教学任务，这对今后进一步加强共建双方在医疗、科研和教学等全方位的合作奠定了一个良好的基础。

暨南大学是中国第一所由国家创立的华侨高等学府，已有 90 年历史，一直以“忠信笃敬”作为校训，坚持“面向海外、面向港澳台”的办学方针，仅对 1978 年以来进行统计，共培养来自世界五大洲 64 个国家和港澳台地区的学生 5000 余人。暨南大学现设有医学院、文学院、理工学院、经济学院、华文学院、中旅学院和教育学院，共 7 个学院，33 个本科专业，7 个博士授权学科，53 个硕士授权学科。现有各层次学生 12 557 人，其中海外及港澳台学生达 2577 人，为中国高校第一。1996 年 6 月，暨南大学通过“211 工程”部门预审，成为全国面向 21 世纪 100 所重点建设的大学之一。

暨南大学医学院成立于 1978 年 4 月，是全国第一所综合性大学里的 6 年制高等医学院校，设有临床医学专业和口腔医学专业，拥有 3 个博士授权学科，同时还建立了一所具有现代医疗设施并得到联合国世界卫生组织认可的医院，即华侨医院。

学校实行从严治校、依法治校方针，学校声誉日渐上升。

暨南大学与深圳市政府达成共建协议，开创了大学和地方联合办医院的先河，将使深圳市人民医院成为我校医学院的第二附属医院，从此开始，它将担负起临床教学任务，不仅承担本科实习生的教学工作，而且承担招收和指导博士、硕士生，以及开展科学研究的任务。这会对提升深圳市人民医院的医疗质量起到积极作用，又会促进暨南大学的办学质量和水平的提高，双方定会双赢！最后，我衷心希望深圳市人民医院成为我校医学院第二附属医院后，共建双方本着互利互惠、共同促进、共同提高的原则，进一步加强在各个领域的合作，共同把共建工作抓好，使共建工作结出丰硕成果。

---

① 本文是在暨南大学和深圳市卫生局共建暨南大学医学院第二附属医院签约仪式上的讲话，深圳，1997 年 1 月 25 日；一个大学校长的探索，北京：高等教育出版社，2011，169.

10. 办出特色　办出水平①

一、我校“211 工程”立项的思路、目标和任务

1996 年 6 月 14 日，暨南大学顺利通过国家“211 工程”部门预审。随后，学校即结合预审工作总结进行立项准备。经过一年左右的准备，立项工作取得了三方面的成果：

（1）明确了思路。

（2）选定了项目。

（3）设计了方案。

（一）立项工作的基本思路和项目遴选

根据国家“211 工程”立项的有关文件精神及国务院侨办和广东省人民政府的要求，结合我校实际，经过反复讨论，我们确定了立项工作基本思路：面向 21 世纪，以学科建设为核心，保证重点，突出特色，立足服务，提高水平，促进合作，打好基础。

我校现有 1 个国家级文科基地，9 个部级（国务院侨办）重点学科，2 个省级重点学科，1 个世界银行贷款项目实验室，1 个省级重点实验室，7 个博士点，53 个硕士点，34 个本科专业，覆盖文、史、经、管、法、理、工、医八大学科门类。显然，“211 工程”立项不可能将这么多学科专业全部包容进去。我们遴选学科建设项目所持的指导思想和原则是：一要保证上水平，二要体现侨校特色，三要增强为地方和行业服务的能力，四要有很好的发展前景。

具体讲，“九五”期间，我校抓学科建设的指导思想是：

（1）把在广东处于首位、在全国同行处于先进、在某些研究方向处于国内领先地位的学科摆在学科重点建设的首位，尽最大努力争取把这些学科发展成为国家重点学科，以提高学校的学术地位和学科水平，扩大我校在海内外的影响。

（2）大力扶植具有良好发展潜质又独具自身特色的优势学科，争取将它们建设成为本行业系统内或区域性的先进学科，然后，逐步发展成为国家重点学科，增强为行业和地方服务的能力。

（3）积极组建以重点学科为龙头、相关学科为支撑、能够联合开展重大科研攻关和合作培养复合型高级专门人才的学科群体，形成我校新的综合性的学科体系。

（4）积极扶植和促进边缘学科和新兴学科的发展，适应 21 世纪科学技术发展和现代化建设需要。与此同时，进一步合理调整现有学科的结构，稳步发展工学、管理学和法学，使我校学科门类更为齐全，结构更为合理，充分发挥综合性大学的优势。

根据这样的指导思想和原则，我们通过学科自我申报、学院初选推荐和学校论证遴选三个程序，并经广东省高教厅和国务院侨办文宣司领导的多次具体指导，最后确定“文艺学与汉语文学”、“产业经济与工商管理”、“汉语言文字学与海外华文教育”、

① 本文是在暨南大学“211 工程”建设项目立项可行性论证会上的汇报（摘要）；暨南大学校报，第 223 期，1997 年 7 月 30 日.

“生物技术与生物医学工程”、“计算机信息与通信技术”、“中外关系史与华侨华人”及“生殖科学与计划生育”等7个学科项目作为“211工程”立项项目。对这7个学科建设项目，学校的意见是：

（1）“文艺学与汉语文学”、“产业经济与工商管理”以及“汉语言文字学与海外华文教育”3个学科项目列为学科建设项目的“重中之重”项目。

第一，比较而言，在上述7个学科项目中这3个项目立项的基本条件较充足，较符合学校“211工程”立项原则的要求。

第二，“文艺学与汉语文学”和“产业经济与工商管理”两个学科项目的核心学科都属省级重点学科，经建设后有可能跻身国家重点学科行列，并企望分别建成博士后流动站。另外，“产业经济与工商管理”学科项目的核心学科“产业经济”博士点及该项目的MBA硕士点在全国居于先进水平；同时，该项目群体优势突出，理论和应用支撑学科较齐全。加强该项目的建设，有利于提高学校为地方经济建设与社会发展服务的能力。“文艺学与汉语文学”项目的“比较文艺学”研究方向居于全国领先水平，中国语言文学专业是国家文科基础学科人才培养和科学研究基地，为该项目有力的依托。“汉语言文字学与海外华文教育”则是我校的特色学科项目，加强这一项目的建设有助于更好体现侨校办学宗旨，提高华文教育水平，配合国家侨务工作布局需要，增强我校为侨务系统服务的能力，为弘扬中华文化、推动海外华文教育发展作出更大的贡献。

（2）“生物技术与生物医学工程”和“计算机信息与通信技术”项目欠缺的是仍未有博士点作核心学科，但在我校理工学院，它们是具有较强发展潜质的学科。前者承担国家级项目的研究能力强，取得的成果水平高。具有较强的学科交叉性，能较好开展理、工、医多学科合作研究，反映我校设有医学学科的综合性大学优势。后者能适应现代社会对计算机和通信技术人才需求飞速增长的形势，紧密结合广东通信事业迅猛发展的需要和侨务系统管理工作现代化的迫切需要，并结合学校的公共服务体系建设和为其他项目提供必要的技术服务。该项目侧重应用技术的研究和开发，有产学研结合的良好基础，具有较强的自我筹资能力。“中外关系史与华侨华人”项目培养高层次人才的能力和综合研究华侨华人等问题的能力较强，成果积累丰硕。建设该项目将增强我校为国家外事工作特别是为侨务工作决策服务的能力。“生殖科学与计划生育”项目的学科带头人在国内外避孕疫苗研究领域有较高的学术地位和取得了较有影响的研究成果。加强这一项目建设将充分发挥我校设有医学院的作用，实现理、工、医优势互补，为我国特别是为广东计划生育工作作出新的贡献。

（3）强调各个项目要把队伍建设摆在首位，采取培养、引进等方法配备好年青学术带头人和学术骨干，以增强学科发展的后劲。

（4）经费统筹安排，不搞平均分配。

当然，作为“211工程”的学校，除了学科建设外，还应考虑到学校的整体建设。因此，学校在公共服务体系、基础教学设施和基本建设等方面也遴选了项目，列入“九五”立项建设规划。

（二）建设总目标

我校“211 工程”建设的总体目标是：经过 15 年左右的努力奋斗，到 2010 年前后，使暨南大学的教育质量、科研水平和办学效益等方面进入国内高校先进行列，成为我国面向海外、面向港澳台办学和传播中华文化的重要基地，在香港、澳门、台湾地区和华侨华人社会具有重要影响的社会主义华侨大学。

（三）“九五”期间改革建设的目标和任务

我校“211 工程”“九五”建设目标是：以通过“211 工程”部门预审为动力，加快推进各项改革和建设，学校的人才培养、学科建设、科学研究和管理水平上一个新的台阶；加大教育投入，加快重点项目和基础设施建设步伐，大力改善办学条件，全面提高办学效益，使我校整体实力有显著增强；加强对外宣传，扩大学校影响，外招学生有大幅度的增加，对外办学的层次和水平有明显的提高，成为我国“两个面向”办学、培养高层次人才的重要基地。

“九五”期间，我校改革和建设的主要任务包括以下 8 个方面。

（1）人才培养。我校人才培养坚持“面向海外、面向港澳台”的方针，主动适应 21 世纪对人才的需求，全面提高各层次教育质量和学生的综合素质。在稳定办学总规模的基础上重点调整学生的层次结构，积极发展研究生教育，适度发展本科教育和成人教育，压缩专科教育，大力发展华文教育。“九五”期间发展规模（在校生人数）规划如下：

年度	研究生/人	本科生/人	专科生/人	华文学院（含预科）/人	成人教育学生/人	合计/人
1996 年	795	5 972	1 685	321	3 517	12 290
2000 年	1 200	8 200	0	800	4 400	14 600

注：护校中专学生未列入

“九五”期间，研究生人数以每年 12%的速度增长，到 2000 年博士生、硕士生之比由目前的 8.5∶100 提高到 14.3∶100，海外及港澳台研究生与内地研究生之比由 18∶100 提高到 25∶100。

（2）科学研究。我校科学研究工作要紧密结合“两个面向”的办学任务，为实现学校总体办学目标作贡献；为侨务系统解决教育、科技、文化等重大问题作贡献；为国家特别是为广东的经济建设和社会发展作贡献。要使我校逐步成为教学和科研中心。到 2000 年，人文社会科学研究总体水平要进入全国综合性大学先进行列，部分研究领域达到国内领先水平，少数研究项目在国际上有一定影响。自然科学和工程技术科学研究工作要贯彻重视基础研究、突出应用研究的方针。主要指标是：预算外科研经费以年度 15%的速度递增；发表于核心期刊的论文每年递增 10%；获省部级以上成果奖数逐年提高，到 2000 年，获成果奖数比现在增加 20%以上。

（3）学科建设。学科建设的目标是：力争在“九五”后期建成 2 个国家重点学科，2 个博士后流动站，再建 2～3 个省级重点学科。争取新增 2～3 个博士学位授予点，使现有博士点研究方向增加一倍以上，并完善硕士、博士培养体系。组建 6～8 个学科群体，使我校有 1～2 个学科建设项目整体达到全国先进水平；同时，培养 2～3 名相对较年青的国家级学科带头人和 10 名左右具有较高造诣、在国内外有一定影响的中青年学科带头人。

（4）师资队伍建设。师资队伍建设要从注重数量发展转向重视质量提高。到 20 世纪末，队伍的综合素质得到显著提高。在研究生规模有较大幅度增长、学校总规模适当扩大的情况下，专任教师人数与学生人数的比例达到 1∶10 左右。具有研究生学历的教师由目前占教师总数的 45％提高到 70％左右，其中具有博士研究生学历的教师达到教师总数的 20％。

（5）办学物质条件建设。“九五”期间要进一步开辟筹资渠道，经费的年增长率应超过 12.5％。要加强各项基础设施建设，使我校的办学物质条件逐步达到中央 1983 年提出的“标准应该适当高于国内其他大学”的水平。学校要加大科技产业建设力度，提高科技成果开发和转化能力。到 2000 年，校办产业上交学校的利润要有较大幅度的增长。要进一步改善学生住宿条件；教职工的住房面积达到广东省规定的标准。

（6）对外学术教育交流与合作。在 2000 年前，在与美国、英国、德国、日本等主要发达国家的高校和研究机构广泛进行学术教育交流合作的基础上，再争取 10 所左右高校及研究机构和我校建立学术教育交流和合作关系。并在与泰国、越南、马来西亚、新加坡等国高校合作的基础上，再争取与其他东南亚国家高校建立学术教育交流关系。

“九五”期间，每年聘请海外及港澳台长期专家 5 名以上，短期专家 10 名以上；派赴海外及港澳台留学、进修和进行学术教育交流的教师数要有较大增加。

（7）校内管理体制改革。要通过深化改革完善领导的科学决策程序，引进竞争机制，为人才成长创造良好的环境，体现按劳分配原则，形成自我发展、自我完善和自我约束的机制。

改革的主要任务是：①进一步加强董事会工作；②进一步完善校长负责制；③健全校、院、系三级管理体制和搞好校部机关的改革；④深化干部人事制度和分配制度改革；⑤加快后勤管理和财务管理体制改革。

（8）党建、思想政治教育与校园精神文明建设。我校党的建设要把加强党的思想、组织、作风和制度建设放在突出位置，充分发挥党委的政治核心作用、党支部的战斗堡垒作用和党员的先锋模范作用，为实现我校办学的总目标和总任务而奋斗。

根据我校地处改革开放前沿和侨校教职工及学生结构的特点，按照《暨南大学教工思想政治工作条例》、《暨南大学师德建设工作规范》和《暨南大学学生德育实施办法》等文件的规定，坚持长期地、有步骤地对教职工和内地学生进行建设有中国特色的社会主义理论和党的基本路线教育，世界观和人生观教育，社会主义法治和道德教育；对港澳台和华侨学生加强爱国主义教育；对华人学生和外国留学生加强了解中华文化和对我友好的教育。全面提高师生的政治思想素质。

抓好校史和“忠信笃敬”校训教育，培养尊师爱生、互敬互爱、讲文明、守纪律等良好习惯，形成“爱国爱校、团结奋进”的一代新风，营造校园内浓厚的学术气氛，使我校校园跻身全国先进高校文明校园行列。

二、“九五”期间重点建设项目

1. 重点学科建设

1）文艺学与汉语言文学

整体目标：力争建立以“比较文艺学”为核心和特色的“文艺学”国家重点学科；在目前培养硕士生和博士生的基础上，联合“现代汉语”博士点及有关学科，争取建立中国语言文学博士后流动站；建设一支适应本学科发展要求的、高水平的教学、科研队伍。

主要指标：计划培养内地博士生 20 名，培养海外及港澳台博士生 4 名。在建成博士后流动站以后，计划每年进站 1 人。争取国家社会科学项目 4 项，国家教委项目 4 项。

2）产业经济与工商管理

整体目标：形成以产业经济（企业管理）为龙头，理论经济和工商管理为支撑，有金融、投资、财务、营销管理、粤港澳产业经济等相关学科相配套，整体实力显著增强的应用经济和工商管理学科群。争取工业经济学科跻身于国家重点学科行列，为建立经济学博士后流动站创造条件；学科整体达到国内先进水平，多个研究方向达到国内领先水平。

主要指标：在学博士生达到 20 人以上，硕士生（含 MBA）达到 480 人。“九五”末期发表文章数量要比 1996 年增长 20%，其中在核心期刊发表的增长 10%。各类国家科研项目尤其是国家级科研项目要有所增加，同时积极争取地方政府和企业委托的横向科研课题。

3）汉语言文字学与海外华文教育

整体目标：组织一支高水平的教学、科研队伍，在汉语言文字学与海外华文教育的结合研究上达到国内领先水平，其成果能够在国际上产生重大影响；把我校建成为在海内外具有较大影响的对外汉语言文化教学的重要基地，国内外华文教育研究和资料中心；为海外培养各个层次的汉语言文化研究和教学的专门人才，为弘扬中华语言文化作出贡献。

主要指标：计划培养博士生 15 人，硕士生 45 人；在校学习华文的留学生达 400 人以上；培养对外汉语及华文教育师资 100 人以上。争取省部级以上社会科学研究项目不少于 10 项。

4）生物技术与生物医学工程

整体目标：争取把现有的中试基地办成培养国内基因工程药物研究和产业化高层次人才基地。为建成国家生物材料及制品的重点开发单位及建立广东省介入疗法导管生产基地做好基础性工作。遗传学与生物医学工程两个硕士点将联合申报博士学位授予权。

主要指标：承担省部级以上项目 15 项，其中国家级重点项目 5 项，培养硕士生 30 名。

5）计算机信息与通信技术

整体目标：建立居国内一流水平的计算机信息和通信技术人才培养基地，争取计算机软件硕士点达到博士点水平；学科整体达到国内先进水平。

具体指标：承担省部级以上科研项目 8 项，国内外核心期刊发表论文约 100 篇，SCI 收录论文 15 篇左右，培养硕士生 40 名。

6）中外关系史与华侨华人

整体目标：争取本学科初步成为培养高层次人才的一个基地和华侨华人问题的科学研究中心；有多个研究方向达到国内领先水平，使国际政治或国际关系硕士点达到申报博士点水平。

具体指标：争取获得 2 项以上的国家级科研项目和 8 项省部级项目，获得一批有分量的科研成果；培养 2～3 名在国内同行中取得学术权威地位的学者；拟出版 3 套本领域的学术丛书、5～6 本学术专著。

7）生殖科学与计划生育

整体目标：争取把本学科群建设成为我国南方生殖科学与计划生育科学研究及人才培养基地。巩固现有博士点和硕士点，增加新的博士点研究方向。在生殖免疫调控机理研究、避孕疫苗的开发研究以及与优生优育相关的研究领域，整体上要达到国际先进水平；通过对生殖科学的基本理论研究为生殖调控及优生探索新途径；在生殖免疫调控基础研究方面作出有国际影响的工作。

具体指标：培养 8 名博士和 10 名硕士；争取获得省部级以上科研项目 8 项。

2. 公共服务体系建设

1）图书馆建设

到 2000 年，建成比较现代化的、与国内外教育和科技信息网接轨的文献信息服务中心，整体水平进入国内高校先进行列，其中华侨华人教育和科研的信息资源系统应达到国内领先水平。

2）校园计算机网络建设

要在已经接通中国教育和科研计算机网（China Education and Research Network，CERNET）及国际交互网 Internet 的基础上开发校园教育和科研管理信息系统，组织技术力量研究和跟踪计算机信息网络的新技术、新方法，使该网络不断完善。我校在 1995 年底已建成校园 FDDI（fiber distributed data interface，光纤分布式数据接口）主干网，目前已将全校用于管理、教学、科研的 12 栋大楼用光纤联网。今后的建设目标是：到 1997 年底继续完善校园主干网建设，增置主要设备，开发网络应用软件等；到 2000 年，将 FDDI 升级为 ATM 主干网，完成校园网全方位应用软件的研制、开发工作。

3）公共实验服务中心建设

要通过重点建设使我校测试服务水平进入国内高校先进行列，成为我校重点学科建设项目的重要支撑条件，为国家侨务工作和地方经济建设提供高水平的分析测试服务。

3. 基础教学设施建设

（1）计算机基础教学设施建设。

（2）本科外语基础教学设施建设。

（3）公共基础教学实验室更新与改造。

(4) 专业基础教学实验室建设。

4. 基础设施建设

(1) 水、电、通信、广播设施建设。

(2) 消防安全设施建设。

(3) 基建(包括新建第二教学大楼、科学馆、学生宿舍和扩建医学院大楼等)。

三、"九五"建设项目资金筹措和分配方案

1. "九五"重点建设项目资金概算及来源

"九五"期间,我校"211 工程"建设需经费 1.98 亿元。经费来源如下:

(1) 国务院侨办专项资金 1 亿元(其中学科建设经费和基建经费各 5000 万元)。

(2) 广东省专项资金和"共建"经费 5000 万元。

(3) 学校自筹 5000 万元。

2. 各项资金分配方案

"九五"期间 1.98 亿元建设经费的安排是:重点学科建设费 8300 万元(占总建设经费的 41.9%);基础教学设施建设费 1650 万元(占 8.3%);公共服务体系建设费 1750 万元(占 8.8%);基础设施建设费 7950 万元(占 40.2%)。

四、预期效益分析

我校"211 工程""九五"重点建设项目完成后,将取得显著的建设成效,学校各个方面的面貌将发生重大变化。

1. 学校总体水平

通过几年的高投人、大建设,为学校在 21 世纪腾飞打下良好基础,学校总体办学水平和办学效益有显著提高,侨校特色将更加鲜明,教育质量、学科建设、科学研究和各项管理工作均步上一个新台阶;良好校风、教风和学风进一步形成,学校在国内外的声誉大大提高,对内对外招生吸引力进一步增强,华侨、华人、港澳台学生比例显著提高。学校充满蓬勃发展的生机和活力。

2. 学科建设水平

通过分层分步建设,学科的各个方面都将发生根本性变化,为促进祖国统一和改革开放,为侨务工作和为地方现代化建设服务的能力将显著增强。实力最强的两个学科有可能进入国家重点学科行列并建成博士后流动站;省级重点学科数量将有新的增加;基础学科将得到进一步加强,应用学科、新兴交叉学科和作为侨校的特色学科将有较大发展;现有博士点研究方向将拓展一倍。学科结构将更合理,门类更齐全。各级重点学科的学科梯队更加完善和加强。综合性大学的优势将更好得以发挥和体现。

3. 人才培养效益

到 2000 年,在校生规模达 14 600 人。学生层次结构将发生重大变化,研究生尤其是博士研究生数将有较大幅度增加,专科层次将基本停办,人才培养质量将有明显提高。

4. 科研工作贡献

"九五"后期,我校科研工作在总体上将接近国内高校先进水平,与海外及港澳台

地区的学术教育交流与合作将在更高层面、更大范围内展开，大大提高为国家统一大业、现代化建设和为侨务工作、地方社会经济发展服务的能力。

5. 队伍建设

教师队伍和管理干部队伍建设进一步趋于年轻化、高学历化，整体政治、业务素质明显提高。若干名在国内外有较高学术声望的中青年专家将脱颖而出，成为学校跨世纪的学术中坚骨干，其中的最突出者将成为国家和省重点培养的学术骨干，有的可能成为中国科学院院士。

6. 办学条件改善

（1）公共服务体系建设项目的完成将极大地改善全校的教学、科研条件，提高现代化水平。图书馆现代化文献信息中心及校园网的全面建成，将加速实现与国际学术环境接轨，为我校开展华侨教育提供较优越的条件。

（2）基础设施建设全面改善，为教学、科研和师生生活提供更为可靠的后勤保障。有关设施的建设朝中央 1983 年提出的“标准应该适当高于国内其他大学”的水平跨进了一大步。校园实现文明、安全、有序管理，育人环境进一步优化。

11. 总结经验　深化改革　加快科技发展①

参加今天会议的有各院、系（所）分管科研工作的领导、科研项目的主持人和一些教研室的负责人，部署1998年度理医科研工作，就某一年度的科研工作由学校进行统一部署在我校是第一次，这主要是我校进入“211工程”，科研工作被列入学校的中心工作后，面临着如何按照211工程的总体建设目标，强化改革力度、加快发展的新问题。下面，我就理医科研工作的改革和1998年度的工作安排谈几点意见。

一、1997年科研工作的基本总结

1997年，是“九五”科技规划的第二年，也是暨南大学发展史上不平凡的一年。一年来，我校科技工作以“211工程”建设为机遇，积极贯彻国家的科技工作方针，努力探索创新，加快发展，经过广大科技人员和管理人员的共同努力，到目前为止，科技工作取得了一定的成绩，这些成绩主要体现为：

（1）一年来，着力构建暨南大学科研激励机制，首次兑现了科研的各种配套奖励，取得良好效果。1997年，我校评选了19位在“八五”时期做出显著成绩的科研人员，包括文科科研人员，授予暨南大学“八五”期间杰出科研工作者称号，并举行了隆重的颁奖表彰仪式，产生了良好效果。1997年，我校还颁布了暨南大学科学基金配套奖励办法、暨南大学科研成果配套奖励办法和暨南大学四大索引收录论文配套奖励办法等，并完成了首次兑现工作，兑现金额达到11万元多，这是我校历史上第一次拨出这么多款项对科研人员进行奖励，在广大科研人员中产生了积极的影响，将对暨南大学科研工作的长期、稳定发展起到巨大的推动作用。

（2）科研成果工作继续取得优秀成绩。1997年的科技成果工作，除继续做好面上项目成果外，对一些重大成果申报高层次科技奖励进行了一些探索，力图有新的突破，经过努力，我校历史上第一次有科技成果获得广东省自然科学奖一等奖的奖励，填补了我校在这方面的空白。我校在申报国家自然科学奖中也进行了一次尝试，尽管由于各种原因未取得突破，但也为将来的申报提供了经验。

（3）申报科研项目和科研经费取得一定成绩。1997年，我校积极向国家计委、科委、自然科学基金委和广东省科委、高教厅、卫生厅、广州市科委等部门申报科研课题，到目前为止，我校共获得国家、省、市等各类科研项目59项，共获得约454万元的科研经费。

（4）对科研项目的管理进行改革和探索。创立了科研项目结题验收指标体系，完成了国务院侨办科研项目的结题验收工作。积极探索科研项目的跨部门联合管理新模式，与产业办共同对国家计委重点科技项目进行联合管理，保证了该项目超额完成预定指标，取得显著成绩，为今后有产业前景的重大科研项目开辟了新的管理模式。

---

① 本文是在1998年暨南大学理医科研工作部署会议上的讲话，广州，1997年12月9日；一个大学校长的探索，北京：高等教育出版社，2011，141-143.

二、1998 年理医科研工作的若干设想和安排

我校的科研工作，尽管取得了一些成绩，但对比国内一些重点大学，对比国家对“211 工程”高校的建设要求和目标，还有相当大的差距，我们要有非常清醒的认识。我们这届学校领导，首次将科研工作列入学校的中心工作，提出建设暨南大学“教学、科研两个中心”的设想，就是希望通过科学技术研究，全面提高我校的学术水平，扩大我校在国际、国内的影响，完成侨校的办学任务。1998 年及以后的两年，是决定我校科技工作以什么姿态进入 21 世纪的关键三年，我们科研工作在未来三年的总体思路是：营造良好的外部关系，进一步深化改革，构建我校比较完善、科学的科技运行机制，促使我校科技工作健康、稳定和高速发展。为此，对 1997 年底和 1998 年的科技工作做如下设想和安排。

(1) 各院、系（所)，一定要加强对科研工作的领导。科研工作既是学校的中心工作之一，又是系统工程，学校和科研处要加强领导，这是毫无疑义的，但单单依靠学校和科研处的努力还远远不够，大家一定要转变观念，不要以为科研工作只是科研处的事，而应该将科研工作列入系（所）日常工作的范围，经常过问和研究本系（所）的科研工作，并采取措施，促进本系（所）科研工作的开展。我们将按照国家教委的规定要求，在“九五”计划完成后，对专职的科研机构进行考核排序。科研处准备建立科研工作年报制度，将各系（所）在每一年度科研工作的基本情况如实通报给全校，并折算成相对值和平均数进行排序，希望通过这项工作，增加一点工作的压力，使科研工作真正受到重视。

(2) 建立最低科研工作量制度、科研工作个人业绩评估制度和科研工作个人业绩档案。在高等学校不搞教学就不能当教师这是天经地义的真理，在“教学、科研两个中心”的高校，不搞科研的教师不会是高水平的老师，也同样不是一个合格的教师。去年我还在学校大会上讲过，不搞科研的教师就是“残疾”老师。科研处的统计发现，“八五”期间，全校理工医科只有 156 人主持过各类科研项目，而理工医科具有副高职称以上的人员接近 400 人，除有部分老师参与到一些合作科研项目中之外，还有一些具有副高以上职称的人员在“八五”期间没有主持过科研项目或参与过科学研究，甚至有的同志在“八五”期间没有发表过论文。像我校这种进入了“211 工程”的学校，如果继续维持现状，肯定与我校“211 工程”建设的步调不一致，影响我校“211 工程”建设目标的实现。我们已经初步构建了科研激励机制，并将继续推出一些科研奖励办法，切实使科研做出成绩的科技人员能从中得到精神和物质的奖励，甚至重奖，同时我们将在“九五”期间着手建立人人都搞科研的管理机制。科研处正与学校人事部门一起，联合制定和实施一个最低科研工作量制度，具体要求正高、副高、中级和初级职称人员的最低科研工作量，“九五”过后，我们就以这一标准对每一个教师进行衡量和公布结果。科研处还将从今年起对每一个教职工试行科研业绩评估，建立科研业绩档案，试行方案成熟后，以后有关科研的考核、职称工资晋升、先进评选、特殊津贴和福利享受等，都将与科研业绩评估结果挂钩。

(3) 以申报各级科研项目和科研成果为明年科研工作的突破口。1998 年度的科研

工作安排，将以申报和承担各级科研项目（包括横向项目）、申报和获得更多科研成果奖、争取更多的科研经费资助为中心开展。科研处安排了几个人，就明年申报各级科研项目和科研成果进行咨询指导，请各单位积极主动做好安排。对拟申报的各个科研项目和成果，科研处还将组织人员进行审查、修改或重组，增加申报成功的机会，并将就某些重点领域，组织集团力量联合申报重大课题。为充分发挥综合性大学的优势，鼓励跨学科门类的科研交叉合作，跨学科门类的科研合作课题的各方分课题负责人经科研处审核同意，都可以享受总课题负责人的一切待遇，科研配套奖励中也增加了这类课题的配套比例。

（4）做好科研课题的年终检查、结题验收工作。我校目前承担的国家级重点或重大项目，是我校科研工作的支柱，完成情况的好坏直接关系到我校的声誉和未来发展，我们已经开始到这些单位去现场办公，听取意见，加强支持的力度，予以重点保证。此外，我们将在年底前对在研的其他项目进行检查，对于今年内到期的项目进行结题验收，请各类课题负责人提前做好准备。

各位领导、专家、同志们，1997 年底和 1998 年的科研工作任务相当繁重，只要我们端正认识、加强领导、团结努力、共同奋斗，是完全可以完成这些任务的。

12. 积极主动地为侨务工作服务①

高等华侨教育不仅是我国高等教育事业的重要组成部分，而且也是贯彻国家侨务政策的重要方面。暨南大学是我国最早创办的一所华侨最高学府，建校 92 年来，一直以向海外传播中华文化，培养华侨华人子女为己任。特别是 1978 年复办以来，暨大在中央〔1983〕24 号文精神指导下，在国务院侨务办公室的正确领导下，认真贯彻“面向海外、面向港澳台”的办学方针，通过侨教为国家的侨务工作服务，为促进祖国的现代化建设和统一大业，以及中外经济文化交流作出了重要贡献。

在世纪之交的今天，侨务工作面临着新的形势、新的任务与新的挑战。如何以邓小平理论为指导，认真贯彻落实党的十五大精神，抓住契机，深化高等华侨教育改革，更好地为侨务工作服务，是摆在海内外暨南人面前的一项崭新课题。下面，对我校为侨务工作服务的做法、取得的成绩及其新的思路，作简要叙述。

一、暨大为侨务工作服务的主要做法

暨大主要是通过侨教，为侨务工作服务，为海外华侨华人社会和港澳台地区培养高素质人才服务，主要措施有下述三点。

1. 始终把招收华侨华人和港澳台青年作为招生工作的首要任务

1）抓好本科招生工作

我校和华侨大学 1982 年即享有独立对外招生权。在国务院侨办的统一领导下，成立了“暨南大学华侨大学联合招生委员会”及其办公室，统筹两校的对外招生事宜，负责两校联合招生考试工作的具体实施。自 1983 年以来，已形成了一套类似内地高考的严格的招生考试管理办法。1978 年以来，学校先后共招收海外和港澳台地区的本专科学生 7000 多人。1997 年学校在对港澳台及海外招生中，仅澳门 1800 多名考生就有半数以上报考我校。

此外，学校还以香港会考成绩录取香港学生；在澳门招收推荐保送生；以马来西亚华文独中统考成绩录取新生。既保证新生质量，又大力拓展对外招生渠道。

2）招收境外研究生

学校于 1984 年开始向海外和港澳台招收研究生，把向外拓展研究生教育当成一项战略性任务，予以高度重视。1988 年，在全国高校中率先推出“兼读研究生制”，受到学员的普遍欢迎。现共有 185 名港澳台和海外生在校攻读博士、硕士学位。到 1997 年 7 月，已有 100 余名获得博士、硕士学位。

3）大力拓展港澳成人教育

学校于 1985 年开始举办港澳成人高等教育，先后在港澳开设了特区经济、中国经济与管理、社会学、外向型经济、口腔医学、中医骨伤和护理学等专业。为了理顺管

---

① 本文是在全国侨务办公室主任座谈会上的发言，北京，1998 年 1 月 19 日；教育与科技管理研究，北京：科学出版社，2016，185-191.

理体制，学校在教育学院还专门设立了海外教育部。

2. 探索适合外招生特点的教育管理模式

学校针对港澳台和海外学生的特点，积极进行教学研究和教学改革试验，努力探索新时期高等华侨教育规律。由于外招生来自不同的国家和地区，教育制度不同、社会文化背景不同，其政治态度、思维方式、生活习惯、学习目的和知识水平都与内地生不尽一致。但他们热爱祖国，热爱故乡，热爱中华文化，内心深处潜藏着很深的民族意识和爱国爱乡情感；思维活跃，动手、社交与自学能力较强；希望了解、认识中国。

1）针对学生特点，积极进行教学改革

（1）加强基础和实践环节教学。针对港澳台和海外学生的特点，学校始终把抓好“三语”（中文、英文、计算机语言）教学，作为提高教学质量的关键。如开设大学语文、写作课程，还单独开设普通话实习训练班，加强普通话训练，提高其应用中文的能力。在大学英语教学方面，1992 年建立了广东省第一家高校外语教学无线电发射台，我校语言实验室的数量亦居广东省高校前茅。针对这些学生英语水平参差不齐的情况，入学测试后分层次进行教学，并开设英语重修班和强化试点班。全校无论是理、工、医还是文、经的学生，一律要上计算机课程，而且特别注意实践训练。学校还坚持教授上基础课的制度，努力帮助这些学生打好扎实的基础。

在加强基础教学的同时，学校重视抓好学生的实践教学，提高学生的动手能力，保证学生的基本技能训练。学校现有固定教学实习基地 67 个，教学实习、见习医院 24 间。

（2）针对学生兴趣，改革体育教学。学校重视学生德、智、体全面发展，重视体育教学。我校体育课程在全国高校中率先实施选课教学，使学生学得积极主动。学校针对港澳台和海外生好动、活跃这一特点，开展丰富多彩的体育活动，以利其身心健康发展。

（3）改革专业课程设置，开设涉外课程。学校每年制订（修订）一次教学计划，使各专业的课程结构适应对外办学和形势发展的需要。近 9 年内，学校淘汰旧课程 600 门，增设新课程 700 门。大部分课程具有较强的涉外性与应用性，同时又兼顾了学科发展的需要。

（4）改革教学方法，加强教学的外向性。从 1996 年开始，学校提倡对港澳台和海外学生采用英文教材，用英语授课。学校规定，本科非英语专业学制年限内，至少有一门专业课程采用最新英文版教材，用英语授课；对于学生较多的国际金融、国际新闻、计算机软件以及医学等专业，要求每个专业每学年都有一门课程采用英文教材授课，而且逐渐增多这样的课程。此外，还对外招生、分班教学，课后予以个别辅导。

（5）为体现侨校特点，从 1996 年下半年开始，我校对境内外学生的大学英语课程实行分流教学。根据港澳台和海外生的特点，制定教学大纲、编写系列教材，也按照一至四级分级教学，每段考试均为校内统一命题。境内学生仍实行国家大学英语四级统考。

2）适应“两个面向”办学需要，加强教学管理和教学建设

（1）建立适外的教学管理制度，不断完善学分制。学校在全国高校中较早实施学分制，以体现因材施教的原则，使教学管理与国际高等教育接轨。目前，实行标准学分制，规定四年制本科总学分为 160 学分（六年制医学为 280 学分），其中选修课占

30%以上。

(2) 重视外向型专业建设。学校原有专业都是在计划经济体制下设立的文理基础学科专业。学校根据中央24号文件要求，通过调查论证，相继增设了国际新闻、国际金融、国际商务、国际经济、投资经济、经济信息管理、广告学、税务、信息工程、电子工程、计算机软件、保险学、汉语言等外向型与应用型专业。同时拓宽专业面，培养复合型、应用型、涉外型的人才。

(3) 严格考试制度，既有利于良好学风的形成，又有利于港澳台和海外学生的要求。学校相继制定了《暨南大学考场纪律》、《暨南大学处理考试作弊的规定》和《暨南大学学生学业成绩考核管理细则》等文件。从1993年起取消补考，实行重修制。对考试作弊者当场给予严肃处分。

(4) 严格的课堂质量评估制度。近年来，我校逐步建立起较完善的学生、专家与领导课堂教学评估体系。教学评估形成制度，评估结果与教师的奖惩、晋级等挂钩，并据此每年评出“十佳授课教师”予以表彰。评估制度化，有力地促进了课堂教学质量的提高。

(5) 改进教学手段，推进教学手段的现代化，提高教学质量。学校电化教育水平居广东省高校第一名，96%的教室装设了电教设备，70%的教师使用了电教手段上课。

3. 对不同类别的学生进行不同的德育课教学

暨大根据学生生源的不同性质，制定了德育培养的分类目标，对境内生进行马列主义、邓小平理论教育，使他们成长为德、智、体全面发展的社会主义事业的建设者和接班人；对港澳台侨生进行爱国主义教育，先后编写出版了《智圆行方的世界——中国传统文化新论》《当代中国研究》教材，并对港澳生开设了香港、澳门“基本法”课程，旨在将他们培养成为热爱祖国，能为当地社会进步和繁荣稳定作出贡献的人才；对外籍学生（主要是华裔），主要进行中华文化教育，旨在使其认识、了解中国，热爱中华文化，将其培养成为对我国友好，能为当地各项事业作出良好服务的人才。

二、暨大为侨务工作服务的主要成绩

1. 为海外和港澳台地区培养了近2万名优秀人才

暨大建校92年来，共培养了65 000多名毕业生。其中1949年前培养8000名，1958～1970年培养5630名，1978年以来培养50 000余名。在这中间约有20 000名海外及港澳台学子，暨大校友遍布世界各地和港澳台地区。学校成为海外华侨华人和港澳台子弟报考境内高校的首选学校。在海外的著名校友有：泰国国会前主席、副总理许敦茂，新加坡国立大学首任校长李光前，印度尼西亚侨领司徒赞，泰国知名企业家颜开臣。在港澳台的著名校友有：国际奥委会委员、台湾红十字组织负责人徐亨，港澳知名企业家李秀恒、马有恒等。这些暨南学子恪守“忠信笃敬”的暨南校训，为居住地的社会发展、繁荣稳定，为促进中外经济文化交流和祖国的早日统一，作出了很大贡献，在海外和港澳台地区为暨大赢得了良好的社会声誉。

2. 为港澳台的顺利回归和繁荣稳定作出了重大贡献

面向港澳台办学，是暨大办学方针的具体实施。根据对港澳毕业生的跟踪调查，用人单位普遍反映：暨大毕业生热爱祖国，基础知识扎实，工作积极肯干。在香港的暨大历届毕业生逾5000人，其中65%是1982年以后毕业的。其中不乏佼佼者，譬如社会学专业毕业生王国兴，1995年当选为香港市政局议员，并担任预委会委员、港事顾问；工业经济1994届博士生黎鸿基担任香港沙宣证券公司高级分析员，并被国务院经济技术研究中心聘为顾问。在澳门历届毕业生达2000人，其中80%是1982年后毕业的。仅政府公务员就达400人。校友中有1人任澳门基本法起草委员会委员兼秘书，并任澳门基本法咨询委员会委员；有3人当选为澳门立法会及市政议会成员。港澳地区校友还有多人当选为全国和广东省人大代表，并担任许多重要团体（如工联等）的领导人。以香港预委会为例，400名委员中，与暨大有关的委员就达19人。同时很多暨大学生担任编辑、记者。从暨大毕业的这批学子在不同的岗位上，为港澳地区的稳定繁荣和平稳过渡作出了积极贡献。而正是从这些校友的言行中，反映了暨大对祖国统一工作所作的贡献。

三、暨大为侨务工作服务的新思路

华侨教育是一项伟大的事业，办好暨大，是海外华侨华人和港澳台同胞的热切希望。通过侨教联系侨情，扩大国际交往，促进我国的改革开放事业，具有十分重要的意义，是党和国家赋予我校的一项战略任务。值此世纪之交，全校上下要进一步在学校各项工作中强化为侨务工作服务的意识，高举“侨”字这面大旗，打好“侨牌”，以开拓、务实的精神，开创高等华侨教育工作的新局面。暨大拟在下述四方面采取行之有效的措施更好地为侨务工作服务。

1. 努力拓展外招生源，进一步深化教学改革

1）采取得力措施，进一步提高外招生的比例

暨南大学的办学方针是“面向海外、面向港澳台”，旨在为华侨华人社会和港澳台地区培养大批高素质人才。1997年，我校港澳台和海外学生人数已增加到2701人，占全校学生数的30.6%，但距国务院侨办提出的内外招生数要达到5∶5的比例要求，尚有差距。为了早日达到这一目标，学校决定实行学制改革，对外实行一年春秋两次的招生制度。学校要抓住世界“华文热”与中国国际地位大幅度提高，以及香港顺利回归，澳门将于1999年回归这一难得的历史机遇，积极主动地做好对外招生宣传工作，采取灵活多样的措施，在保证质量的基础上，吸引更多的港澳台及海外学生来校学习。目前，学校要重点拓展东南亚生源，尤其要做好华文教育基础较好的马来西亚的招生工作。在港澳台地区要巩固原有成果，同时要考虑到暨大地处广东侨乡，要兼顾到归侨、侨眷子女等的入学要求。

2）深化教学改革，把暨大办出特色与水平

暨大要在海内外享有盛誉，关键是要创名牌，凭借特色与实力吸引学生。为此，学校采取了一系列措施，深化学校教学改革。如增开午间选修课，进一步完善学分制；大力强化“三语”教学、基础课教学和分流教学；陆续对非英语专业的部分课程采用英文教学；加强专业改造与课程内容的更新以及教学手段的现代化；从严治校，依法治校，提高学校的管理水平，从而增强学校的涉外性、应用性，并逐步与国际高等教育接轨。

2. 大力发展华文教育，向海外传播中华文化

几千万华侨华人遍布于世界各地，在思想观念上已从“叶落归根”转变为“落地生根”。但作为中华儿女，他们对中国仍怀有深厚的感情。随着我国改革开放事业的深入发展，国际性的“华文热”日益升温，华侨华人迫切希望学习华文，接受中华文化的熏陶。其他外籍人士亦渴望通过学习华文，以加深了解、认识中国，加强与中国的经济文化联系，这就为我校开展华文教育提出了新的要求和机遇。暨大华文学院经过几年来的建设和发展，已成为我国开展华文教育的重要基地，并成为我国具有对外汉语水平考试资格的三所高校之一。从 1997 年开始，学校专门设立了供外国人学习汉语的汉语言本科专业，现已正式招收第一届本科生。同时还设立了各种学习中华文化的短期学习班，每年有来自世界 20 多个国家和地区的学生前来学习。受国务院侨办委托，我校华文学院先后主持编写了柬埔寨与北美版华文教材。特别是北美版华文教材，甫一运抵美国，即很快为各中文学校推广使用，受到广泛的欢迎和好评。它的出版发行为北美地区中文学校提供了一套系统、科学、适合北美地区华侨华人儿童生活特点的全新的中文教材，大大推动了海外华文教育事业的发展，扩大了暨大在海外的知名度。

3. 抓好“211 工程”建设，提高为侨务工作服务的水平

暨大已相继通过国家“211 工程”部门预审和专家立项论证，进入面向 21 世纪全国重点建设的 100 所高校行列，这也是全国侨务系统唯一一所进入国家“211 工程”行列的高校。以“211 工程”建设为中心，推动学校其他工作的开展，是学校工作的重中之重，也是在新形势下侨务工作发展的客观需要。已获准立项重点建设的学科有 7 个，其中“汉语言文字学与海外华文教育”“中外关系史与华侨华人”学科，主要是直接为侨务工作服务。在“九五”期间，上述两个学科将分别投入 400 万元人民币的重点建设经费。华侨华人学科是体现暨大办学特色的优势学科。暨大在华侨华人研究方面有着优良的传统，取得了令人瞩目的成绩。通过深入研究侨史、侨情，掌握侨务动态，既可以为国家制定侨务政策提供咨询，又可以通过学术交流，加强海内外中华儿女之间的联系和中外交往。其他 5 个学科，通过重点建设之后，将会出成果、上水平，在整体上提高暨大的办学质量、水平和效益，增强综合实力，从而保证学校具有雄厚的实力，更大的知名度，吸引更多的港澳台和海外华侨华人青年来我校求学，以便更好地为侨务工作服务。

4. 做好校友、校董工作，积极主动地为侨务工作服务

校友、校董是办好高校的巨大力量。暨大较早设有董事会、校友会。作为一所外向型的华侨大学，校友、校董遍布海外华侨华人社会和港澳台地区。作为一笔丰富的人才资源、信息资源、财力资源和社会资源，做好校友、校董工作，不仅可以为暨大带来人、财、物诸方面的支持，而且还可以他们为桥梁，进一步促进中外经济文化交流，推进祖国的统一大业。

再过 700 多天，人类将步入 21 世纪和下一个千年。为侨务工作提供高效、优质的服务，是暨大义不容辞的职责。暨大将不负重托，坚持以邓小平理论为指导，并将这一理论与暨大工作的实践紧密结合起来。我们有信心、有能力，在国务院侨办的正确

指导下，通过海内外暨南人的共同努力与团结奋斗，全面实施“211工程”总体建设规划，在21世纪初叶把暨大办成名副其实的一流大学，把一个结构优化、办学一流、质量更高的高等华侨教育全面推向21世纪，从而为祖国的侨务工作以及实施“科教兴国”战略作出更大的贡献。

13. 用现代化管理促进高等教育事业的发展①

“有朋自远方来，不亦乐乎!”这是出自中国的儒家经典《论语》中的一句话。今天，我有幸能与来自万里之遥的英国以及省内的同行，欢聚一堂，共同切磋高等教育管理方面的问题，的确是一件“不亦乐乎”的事情。

英国是世界工业革命的摇篮。英国的高等教育对世界各国都产生了重要影响，牛津大学、剑桥大学是享誉世界的名牌学府。去年，我曾访问过英国的几所大学，尽管来去匆匆，但英国高等教育科学、有序的管理方式，仍给我留下了深刻的印象。成书于中国春秋时代的《诗经》中，有“它山之石，可以攻玉”一词，我们在这里借用这个典故来说明，要以英国山上之石，用作琢磨我国玉器的砺石，意即中国要借鉴、学习英国高等教育的成功经验。

再过600多天，人类即将步入下一个世纪和下一个千年。传授、创新知识与作育人才的高等教育，对一个国家的兴盛发展与综合国力的提高，起着越来越重要的作用。随社会经济的发展和高等教育功能的扩展，高校已走出昔日的“象牙之塔”，与社会经济、政治、科技、文化的联系越来越密切。高等教育正由社会的边缘进入社会活动中心，日益成为以知识为基础的“知识经济”时代社会发展和进步的重要推动力。

借此机会，我想结合暨南大学的实际，谈一谈我对高等教育管理方面的认识，目的是抛砖引玉。

首先，请允许我简要介绍一下暨南大学的情况。

暨南大学创办于1906年，先后在南京、上海、福建建阳和广州办学。暨南校名源自《尚书•禹贡》：“朔南暨，声教讫于四海。”意思是说要向海外尤其是南洋传播中华文化。作为“华侨最高学府”，暨南大学有着辉煌的过去，在中国现代高等教育史上占有不可或缺的独特地位。1923年，暨南学校与东南大学合办的上海商科大学，为当时中国仅有的5所国立大学之一。1927年，升格为国立大学，是当时少有的成立较早的国立大学之一，在当时的国立大学排序中，暨大位居10名之内。在抗日战争的艰难岁月里，学校先是在上海租界坚持办学，1941年太平洋战争爆发后又迁至福建建阳办学，弦歌不辍，为东南一带大学教育守最后之壁垒。特别值得一提的是，英国著名的中国科技史专家李约瑟博士曾到建阳暨大访问。1958年，暨大在广州重建。经过几代暨南人的辛勤耕耘，今日之暨大已发展成为一所涵盖文、史、法、教育、经、管、理、工、医诸学科的综合性华侨大学。暨南大学凭借其侨校特色和综合实力，已先后通过国家“211工程”部门预审和立项论证，现正进人建设阶段。所谓“211工程”，是指面向21世纪中国重点建设100所高校和一批重点学科。

在中国1020所普通高校中，暨大有着自己的特色与优势。综括起来主要有以下三点：其一，“侨”字是最大的特色。暨大贯彻“面向海外、面向港澳台”的办学方针，生源来自海外、港澳台地区和内地。自1978年以来，学生来源于五大洲67个国家和

① 本文是在中英高等教育管理研讨会上的发言，广州，1998年4月8日；暨南教育，1998，(1)：80-82.

地区。本学年度，来自31个国家和港澳台地区的学生达到近3000人，约占全日制在校生的1/3。这种特色即是暨大的优势所在。

其二，暨大办学历史悠久，学科门类比较齐全，是中国最早建立的国立大学之一。建校92年来，已为海内外培养了6万余名高素质人才。暨大声誉远播，桃李遍五洲。

其三，暨大是一所开放型的国际性大学，这主要表现在以下五个方面：一是生源来自世界各地，从她成立的第一天开始，即招收华侨学生；二是暨大与欧美、东南亚等地的十多所高校签订了学术交流协议；三是课程及专业设置具有涉外性；四是实施标准学分制，提倡英语授课；五是实行春秋两季招生制度。

我认为，要办好一所高校，除了具备一流的师资、拥有大师级的学者、充裕的办学经费外，一流的管理是至关重要的。而在管理的实施过程中，校长处于枢纽地位。一所高校要高效、有序地运作，既要有行之有效的规章制度，又有赖于校长的果断、学术威望与人格魅力。我在管理暨大时，实行"分口分级"的管理原则，做到权力的逐级分解与条块有机结合，避免办事互相推诿。在具体实施过程中，可以用"严""法""实"三个字来概括我的管理高校的理念。"严"指"从严治校、从严治教、从严治学"，用严来保证教学质量的提高。"法"指要"依法治校"，搞法治，不搞人治，既要遵守国家的法规，又要完善校内规章制度，规范和约束学校的办学行为。"实"指要实事求是，要务实，踏实，不要搞形式主义，不要搞浮夸风。

成天下之才者在教化，教化之所本者在学校。我认为，大学生是一所高校的主人。在教育过程中，要提倡"有教无类"的思想，使每一个学生拥有平等受教育的机会。同时，对大学生除了实施专业素质与人文、科学以及心理素质教育外，还要帮助学生学会如何做人，如何与人相处，从而具备儒家所推崇的"修身齐家治国平天下"的本领。

教学工作是高校工作永恒的主旋律。重视教学工作，一切以教学工作为中心，是一个成功的高校管理者所应具备的素质。同时，要大力开展科研工作，合理配置在基础研究、开发研究与应用研究方面的资源，使教师进入学科最前沿，具备广阔的学术视野和雄厚的学术实力，从而促进教学水平的提高。要管理好高校，还要善于弹钢琴，要抓住主要矛盾，要抓住重点。以"211工程"建设为中心，搞好7个重点学科建设就是暨大工作的重中之重，一切工作要围绕并服务于这一全局性的工作。

"知识经济"时代已经悄然来临，高等教育可谓任重道远。一个国家要实现现代化，离不开教育与科技的现代化。21世纪是教育的世纪，高等教育将肩负起更加重要的社会职责，社会亦将更加重视高校的特殊作用，社会经济的发展在更大的程度上依赖于高校积极主动的参与及其卓越贡献。

当然，高校在走出"象牙塔"之后，在积极主动适应社会经济发展之际，还具有引导、促进社会经济发展以及人类自身日臻完善的职责。大学是思想最活跃、最富创造力和活力的学术殿堂。作为人类神圣的精神家园，大学有着更崇高的理想与追求，这就是《诗经》中所描述的"高山景行"的美好境界，"虽不能至，心向往之。"让我们以此共勉吧。

14. 深化教育改革　提高教学质量[①]

今天，暨南大学教学工作会议隆重召开了。这次学校教学工作会议意义特别，第一，党的十五大发出了“高举邓小平理论伟大旗帜，把建设有中国特色社会主义事业全面推向21世纪”的号召，这是我国高等教育事业改革和发展的又一次重大历史机遇。我校的这次教学工作会议，就是要从时代和社会改革的高度出发，高举邓小平理论的伟大旗帜，把握难得的好时机，全力推动学校教学改革，努力使之有新的突破性进展。第二，两周前，由教育部召开的第一次全国普通高等学校教学工作会议（这是新中国成立以来，首次由国家教育行政部门召开的全科类教学工作会议），明确了今后一段时间高校深化教学改革的基本思路，即：贯彻教育方针，更新思想观念，拓宽专业口径，改革内容方法，加强素质教育，提高教育质量。我校的这次教学工作会议的议题正是“转变教育思想，深化教育改革，提高教学质量”，与全国高校教学改革总思路相一致。第三，我们暨南大学自1978年以来，通过全校教职员工和几届领导班子的共同努力，学校的教学工作取得了较大成绩，教学改革积累了丰富的经验，近年我校教育的整体发展可以说是历史上最好的时期之一。这为我校面向21世纪的教学改革打下了比较扎实的基础。因此，我特别寄希望于这次教学工作会议，希望它开得热烈，开得卓有成效。在此，我谨代表学校党政领导，对暨南大学教学工作会议的召开，表示热烈祝贺。

在此世纪之交，我校的人才培养工作正处于一个历史性的关键时期，我们的教学工作也面临着新的挑战和困难。切实深入的教学改革，是我们迎接挑战、战胜困难的唯一出路。对于高校来说，培养人才是根本任务，教学工作是主旋律，提高教学质量是永恒的主题，教学改革是各项改革的核心，本科教育是其基础。我校在深化教学改革，提高教学质量的同时，一定要明确“教学是主旋律”与“211工程建设是中心工作”的并行不悖、密不可分的关系。学校“211工程”建设得好，将直接带动学校整个教学工作的发展，诸如现代教育技术的发展，实验条件的改善，各类实践基地建设的加强等。最近，从国家“211工程”主管部门获悉，从“十五”计划起，“211”项目的遴选先定学科、后定学校，届时评不上国家重点学科的“九五”的“211”学校，将被滚动出列。我们面临的压力是巨大的，进入“211”来之不易，保住“211”更需要全校教职工的继续齐心协力。在“211”高校行列中建设我们的学校，它的“主旋律”将更加优美嘹亮。

教学改革，一定要紧紧抓住培养什么样的人和怎样培养人这两个根本问题。长期以来，我国高等学校的本科专业设置过窄，实行的是一种狭隘的专业教育。随着科技的发展及社会主义市场经济体制的建立，这类“专才”毕业生越来越与社会的需要不相适应。因此，我们必须拓宽本科教育的口径，减少专业种类，扩展学科基础。今年上半年，教育部将颁布新的本科专业目录，全国本科专业总数将调减一半。学校各院

---

① 本文是在暨南大学教学工作会议上的讲话，广州，1998年4月9日；暨南教育，1998，(1)：1-3.

系及教务部门一定要以此专业调整、改造和重组为契机，调整人才知识、能力和素质结构，拓宽基础，整合课程，构建新的专业平台，增强专业方向设置的柔性。不断深化教学改革，努力培养出基础厚、口径宽、能力强、素质高的适应21世纪的新人才。

我校的学分制在全国实行较早，也有一定的影响。但目前也存在学分偏多、课时偏多的问题。学生负担较重，不利于学生综合素质的培养。因而，我们的教育思想必须随学科发展和社会需求变化而转变，要制订合理的培养方案和教学计划，大力压缩课时总量、减少总学分，减轻学生负担，为学生的全面成长留出必要的时间和空间。同时，图书馆、实验室等要充分向学生敞开，要开展多种形式的课外科技、文化和社会实践活动，开辟“第二课堂”。

我们学校是一所“面向海外、面向港澳台”的华侨大学，“侨”字就是我们的特色。目前，我校外招生人数已近3000人，占学校学生数的30%以上。不久，将努力达到国务院侨办提出的内外招生数各占50%的比例要求。为此，我校已进行了系列改革，如两季招生、午选课、分流教学等。特别是尝试对非英语专业的部分课程采用英语教学。这项尝试应着力抓下去，并要加快步伐。作为侨校，理应在这一方面走在全国前列，我们的方向，应是“双语教学”的学校。因此，我们每个系、每个专业可否增加一门英语授课课程，并努力做到逐年递增。这是一项艰巨而现实的工作，也是我们这所海外及港澳台学生即将占50%的华侨学府的必然而迫切的要求。

教学的改革，要有与之相适应的教学管理体制，教学管理是一门科学，兼有学术管理与行政管理双重职能，学校各级领导干部都应学习教育理论，研究教育思想，懂得教学规律，熟悉教学管理。我们在深化教学改革的同时，也要进一步深化校内管理体制的改革，管理跟不上，教学改革也将落不到实处。因此我们要继续坚持“发扬优势，深化改革，保证重点，改善条件，提高质量”的办学原则，并把它落实到“严”“法”“实”三个字上。“严”，就是要严格管理、严格要求；“法”，就是要依法治教、依法治校；“实”，就是抓大事、干实事。这样，才能使我校的教学改革有保障，教学质量得以提高。各位老师，各位同志：21世纪将成为一个教育世纪，21世纪的经济将是知识经济，我们今天在此召开的暨南大学教学工作会议，正是面向这一伟大世纪而进行的一次教育思想改革的探讨，我相信这次教学工作会议，将会对我校今后教学的发展起到极大的推动作用。希望全校教职员工心往一处想、劲往一处使，努力使我校的数学质量、教学水平再上一个新台阶。

预祝本次教学工作会议取得圆满成功。

15. 暨南大学兴办高等华侨教育的历史回顾与展望①

发轫于20世纪初叶的中国高等华侨教育，已经走过了一条坎坷与辉煌的历程。暨南大学素称“华侨最高学府”，她的前身暨南学堂创办于清光绪三十二年（1906年），这是由中国政府创办的第一所华侨学校。经过几代暨南人的辛勤耕耘，建校92年来，暨大已发展成为中国对外办学和向海外传播中华文化的重要基地。当人类即将步入下一个世纪和下一个千年之际，回首暨大兴办高等华侨教育的历程，并且未雨绸缪，大力拓展高等华侨教育，以崭新的姿态迎接21世纪的挑战，是一件有重要意义的事情。

暨大具有鲜明的侨校特色，蕴含中国传统文化底蕴。暨南大学立校的宗旨是“宏教泽而系侨情”。校名“暨南”一词，源自《尚书・禹贡》：“东渐于海，西被于流沙，朔南暨，声教讫于四海。”意即：中华民族优良的道德风范和文化教育，以中国为中心辐射、传播到四面八方，其影响遍及四海。考虑到侨生主要来自南洋，学校的创办人、清两江总督端方以“暨南”作为校名。暨南校训“忠信笃敬”，亦出自《论语・卫灵公》，它既是儒家思想的重要组成部分，也是中国知识分子修身养性、砥砺品行所追求的一种思想境界。自此，暨南成为向海外华侨华人传播中华文化，维系海内外中华儿女交往以及中外交流的一条重要的文化纽带。

如果我们追溯一下暨南校史，不难发现，暨南大学尽管饱经沧桑，屡遭停办和播迁，但她始终抱着弘扬中华文化、培养华侨华人子女之使命，以顽强的生命力，不断发展壮大，开拓前进。暨南大学以其侨校特色和学术声誉，在中国现代高等教育史尤其是高等华侨教育史上，占有不可或缺的独特地位。

1921年，暨南学校与东南大学合办的上海商科大学，为当时中国仅有的5所国立大学之一。1927年，暨南学校升格为国立大学，系当时成立较早的少有的国立大学之一。在民国政府教育部的国立大学排序中，暨大位居10名之内。在抗日战争的艰难岁月里，学校先是在上海租界坚持办学，1941年底太平洋战争爆发后又迁至福建建阳，弦歌不辍，独力维系我国的高等华侨教育，为东南一带大学教育守最后之壁垒。1958年，暨大在广州重建，进入了一个新的发展时期。今日之暨大已发展成为一所涵盖文、史、法、教育、经、管、理、工、医诸学科的综合性华侨大学。学校现设有文学院、理工学院、经济学院、医学院、管理学院、中旅学院、华文学院和教育学院，设有12个博士点，58个硕士点。暨南大学凭借侨校特色和综合实力，已先后通过国家“211工程”部门预审和立项论证，进入面向21世纪中国大陆重点建设的100所高校行列。

在中国大陆1020所普通高校中，暨大有着自己的特色与优势，综括起来主要有以下三点：

其一，“侨”字是最大的特色。暨大贯彻“面向海外、面向港澳台”的办学方针，生源来自海外、港澳台地区和内地。自1978年以来，学生来源于五大洲64个国家和

① 本文是在台湾暨南国际大学“华侨教育学术研讨会”上的特邀报告，南投，1998年6月2日；暨南学报（哲学社会科学），1998，20（4）：1-4.

港澳台地区：1997～1998学年度，来自31个国家和港澳台地区的学生达到近3000人，约占全日制在校生的1/3。这种特色即是暨大的优势所在。

其二，暨大办学历史悠久，学科门类比较齐全，是中国最早建立的国立大学之一。建校以来，已为海内外培养了6.5万名高素质人才。暨大声誉远播，桃李遍五洲。

其三，暨大是一所开放型的国际性大学，这主要表现在以下五方面。一是生源来自世界各地，从她成立的第一天开始，即招收海外学生；二是暨大与欧美、东南亚等地高校有学术交流协议；三是课程及专业设置具有涉外性；四是实施标准学分制，提倡用英语授课；五是对海外实行春秋两次招生制度。

据统计，从暨大毕业的6.5万名学子中，有近2万名来自海外和港澳台地区。暨大校友遍布世界五大洲和港澳台地区，学校成为海外华侨华人和港澳台学生报考内地高校的首选高校。学校在海外的著名校友有：泰国国会前主席、副总理许敦茂，著名爱国侨领、新加坡国立大学首任校长李光前，印尼侨领司徒赞，泰国企业家颜开臣；在港澳台的著名校友有：国际奥委会委员、台湾红十字组织负责人徐亨，港澳企业家马有恒等。这些暨南学子恪守“忠信笃敬”的暨南校训，为居住地的社会发展、繁荣稳定，为促进中外经济文化交流和港澳的顺利回归，做出了很大贡献，在海外和港澳台地区为暨大赢得了良好的社会声誉。

暨大在兴办高等华侨教育的过程中，主要是通过侨教为侨务工作服务，为海外华侨华人社会和港澳台地区培养高素质人才。

(1) 始终把招收华侨华人和港澳台青年作为招生工作的首要任务。

自从1907年3月，首批21名印尼侨生回暨南学堂求学以来，已有近2万名华侨华人子弟和港澳台青年在暨南完成学业。在不同的历史时期，不论时局如何变迁，暨大始终秉承这一理念，大力拓展外招生源，积极创造条件，招收、培养华侨华人子弟和港澳台青年。近20年来，伴随着中国改革开放政策的实施，暨南的事业进入了一个全新的发展时期。学校在海外和港澳台地区，不仅大量招收全日制本科生和成人教育学生，而且还招收博士生、硕士生。

(2) 探索适合外招生特点的教育管理模式。

海外学子负笈大陆，远离父母。由于长期生活在海外和港澳台地区，他们在语言、学习、宗教信仰和风俗习惯诸方面仍有不少短期内难以适应之处，暨大针对外招生的上述特点，有针对性地采取了一些措施。

其一，在学习上因材施教。针对外招生的实际情况及学业水平之差异，延聘教师为其补习功课，并对学业程度不同的学生依次分班授课。考虑到他们学成后要回居住地发展，学校特为其开设南洋概论、华侨华人史、当代中国研究等课程。在专业设置与课程体系安排方面，也充分考虑到涉外性与应用性，并鼓励用英语授课。学校还实行标准学分制、主辅修制，以培养宽口径、厚基础、具有综合性素质的人才。在体育方面，暨南富有优良的体育传统。考虑到外招生爱好运动，且家境富裕，学校开展了丰富多彩的体育活动，以利其身心全面发展。

其二，在生活上关怀备至。学校对外招生根据特点，适当照顾。如早在南京时期，学校规定学生除自备衣着和零用钱外，学费和食宿费一概豁免，每年另由学校发给冬夏两季制服。在上海、建阳时期，郑洪年、何炳松两任校长，对学生嘘寒问暖，在生活方面予以无微不至的关怀，很多暨南学子在几十年后，每当回忆起在暨南的求学生涯，都十分感念两位老校长。随着在新时期学校物质条件的全面改善，学校大力改善外招生的生活条件，如专门为他们提供侨生公寓，在吃、行及通信诸方面提供便利。

其三，在德育教育方面实行严格管理，注重对学生进行中华文化与品行修养方面的教育。如在南京时期，侨生所修功课中就有修身、经学、历史等 3 门教会侨生为人处世知识的课程。1994 年，学校还恢复了"忠信笃敬"的校训，学校对外招生开设中国传统文化概论课程，旨在帮助学生认识、了解中国，热爱中华文化，同时学校建章立制，实施"三严"（从严治校、从严治教、从严治学）方针，狠抓"三风"（校风、教风、学风）建设，以便学生养成良好的习惯与严谨的作风。

华侨教育是一项伟大的事业。办好暨大，是海外华侨华人和港澳台同胞的热切希望。通过侨教联系侨情，扩大国际交往，促进改革开放事业和现代化建设，具有十分重要的意义。值此世纪之交，暨大要继续高举"侨"字这面大旗，打好"侨牌"，以开拓、务实的精神，开创高等华侨教育工作的新局面。暨大拟在下述四方面采取行之有效的措施办好高等华侨教育。

（1）继续贯彻"两个面向"的办学方针，努力拓展外招生源，进一步深化教学改革。

一是采取得力措施，进一步提高外招生比例

1997～1998 学年，我校港澳台和海外学生人数已占全校学生数的 30.6%，但距国家提出的外招生数要达到 50%的比例要求，尚有差距。为了早日达到这一目标，学校决定实行学制改革，对海外实行一年春秋两次的招生制度。学校要抓住世界"华文热"与中国国际地位大幅度提高，以及香港顺利回归，澳门将于 1999 年回归这一难得的历史机遇，积极主动地做好对外招生宣传工作，采取灵活多样的措施，在保证质量的基础上，吸引更多的港澳台及海外学子来校学习。目前，学校要重点拓展东南亚生源，尤其要做好华文教育基础较好的马来西亚的招生工作。在港澳台地区要巩固原有成果，同时要考虑到暨大地处广东侨乡，要兼顾到归侨、侨眷子女等的入学要求。

二是深化教学改革，把暨大办出特色与水平

暨大要在海内外享有盛誉，关键是要创名牌，凭借特色与实力吸引学生。为此，学校采取了一系列措施，深化学校教学改革。如增开午间选修课，进一步完善学分制；大力强化"三语"（中文、英文、计算机语言）教学、基础课教学和分流教学；陆续对非英语专业的部分课程采用英文教学；加强专业改革与课程内容的更新以及教学手段的现代化；从严治校，依法治校，提高学校的管理水平，从而增强学校的涉外性、应用性，并逐步与国际高等教育接轨。

（2）大力发展华文教育，向海外传播中华文化。

几千万华侨华人遍布于世界各地，在思想观念上已从"叶落归根"转变为"落地生根"。但作为中华儿女，他们对中国仍怀有深厚的感情。随着中国改革开放事业的深入发展，国际性的"华文热"日益升温，华侨华人迫切希望学习华文，接受中华文化的熏陶。其他外籍人士亦渴望通过学习华文，以加深了解、认识中国，加强与中国的经济文化联系，这就为学校开展华文教育提出了新的要求，带来了新的机遇。暨大华文学院经过几年来的建设和发展，已成为大陆开展华文教育的重要基地，并成为具有对外汉语水平考试资格的3所高校之一。从1997年开始，学校专门设立了供外国人学习汉语的汉语言本科专业，现已正式招收第一届本科生。同时还设立对外汉语本科专业以及各种学习中华文化的短期学习班，每年有来自世界20多个国家和地区的学生前来学习。暨大华文学院先后主持编写了柬埔寨与北美版华文教材。特别是北美版华文教材，甫一运抵美国，即很快为各中文学校推广使用，受到广泛的欢迎和好评。它的出版发行为北美地区中文学校提供了一套系统、科学且适合北美地区华侨华人儿童生活特点的全新的中文教材，大大推动了海外华文教育事业的发展，扩大了暨大在海外的知名度。

(3) 抓好“211工程”建设，提高学校的教学、科研水平。

暨大已相继通过国家“211工程”部门预审和立项论证，她是侨务系统唯一一所进入国家“211工程”行列的高校。以“211工程”建设为中心，推动学校其他工作的开展，是学校工作的重中之重，也是在新形势下侨务工作发展的客观需要。已获准立项重点建设的学科有7个，其中“文艺学与汉语文学”“中外关系史与华侨华人”学科，主要是直接为侨务工作服务。华侨华人学科是体现暨大办学特色的优势学科。暨大在华侨华人研究方面有着优良传统，取得了令人瞩目的成绩。通过深入研究侨史、侨情，掌握侨务动态，既可以为国家制定侨务政策提供咨询，又可以通过学术交流，加强海内外中华儿女之间的联系和中外交往。其他5个学科，通过重点建设之后，将出成果、上水平，在整体上提高暨大的办学质量、水平和效益，增强综合实力，从而保证学校具有雄厚的实力、更大的知名度，吸引更多的港澳台和海外华侨华人青年来校求学，以便更好地为侨务工作服务。

(4) 做好校友、校董工作，积极主动地为侨务工作服务。

校友、校董是办好高校的巨大资源。暨大较早设有董事会、校友总会。作为一所外向型的华侨大学，校友、校董遍布海外华侨华人社会和港澳台地区。暨南校友关心、热爱母校，对母校心怀眷念。作为一笔丰富的人才资源、信息资源、财力资源和社会资源，做好校友、校董工作，不仅可以为暨大带来人、财、物诸方面的支持，而且还可能凭借他们作为桥梁，进一步促进中外经济文化交流，大力推进大陆的现代化建设。

历史的风风雨雨，社会的革故鼎新，观念的进退守舍，浓缩在暨大92年的校史中，又汇聚在新世纪面前。暨南人在新的机遇与挑战面前，既要承接悠久传统，又要着力开拓未来。暨南人将秉承“朔南暨，声教讫于四海”的办学使命，把一个结构优化、办学一流、质量更高的高等华侨教育学府推向21世纪。

16. 华文学院越办越好[①]

今天我们欢聚在这里，共同庆祝华文学院建院5周年、广州华侨学生补习学校建校45周年。我谨代表暨南大学党政领导班子，表示最热烈的祝贺！并对各位领导和远道而来的校友们，表示诚挚的欢迎和衷心的感谢！表示最崇高的敬意！

广州华侨学生补习学校有辉煌的过去，更有优良的传统。她在华侨教育史上立下过汗马功劳，不愧是一所优秀的华侨学校。

1993年，华侨补校并入暨南大学，成立华文学院。新生的华文学院不仅继承了华侨补校面向海外华侨、华人及港澳台学生开展中华语言文化教育的优良传统，而且以发展对外汉语教学、弘扬中华文化为己任，继往开来，继续为国家的侨务工作、华文教育和对外汉语教学事业作贡献。

华文学院建院5年来，大力抓好校园建设，改善教学环境与条件，完善教学体系，提高教学质量，加强学科建设，取得了显著成绩。学院不仅为海外华侨华人及其他外国学生开展非学历教育，而且开设了外国留学生汉语言专业及面向海内外培养对外汉语教师的对外汉语专业，同时培养对外汉语教学研究方向的硕士研究生。此外，华文学院还为海外编写华文教材，为推动海外华文教育事业的发展，作出了积极的贡献。这些，都离不开上级领导的关心和支持，也离不开许许多多关心、热爱华侨补校，关心华文学院的校友们的热情帮助，更离不开华文学院全体教职员工的辛勤劳动。现在，华文学院已初具规模，正在努力办成中国华南地区规模最大、条件最好、办学体系最完善的对外汉语教学基地和华文教育基地。我想，这是值得每一个华侨补校、华文学院以至暨南大学的新老校友们高兴的事情。暨南大学将积极支持华文学院的各项工作，把华文教育搞得更好。

最后，让我们共同祝愿：华文学院越办越好！

---

① 本文是在暨南大学华文学院建院五周年庆贺大会上的讲话，广州，1998年8月16日；广州华苑，1998，12：4.

17. 在高校党的建设中贯彻落实邓小平“从严治党”的思想[①]

高校是培养社会主义现代化宏伟事业建设者的摇篮，也是社会主义精神文明建设的重要阵地，它在21世纪科教兴国的战略中将起至关重要的作用。所以，高校的改革与发展，是时代的呼唤，也是中国社会发展的需要。而高校为实现其历史使命，加强党的建设，为高校的改革提供强有力的思想保证、政治保证、组织保证，这是必然的要求。对广东的高校来说，在回顾20年的改革成就及展望未来的发展时，深刻领会与贯彻邓小平的党建思想，对办好我省的高等教育具有重要的指导意义。

对一所具有92年历史的暨南大学来说，曾几次因时局变化而迁徙与停办，可谓历经风雨。1978年，暨南大学乘改革的春风在广州复办。20年来，学校在党的正确领导下，已由复办时的10个系、12个专业发展到今天的23个系、33个本科专业、71个博士和硕士学位授权点的、综合性的国家“211工程”重点建设大学。这一成果的取得，无疑是广东高等教育20年改革与发展的写照，也是高校党建工作的成果。对认真贯彻邓小平的教育思想和党建思想，学校党委有深切的体会与感受。特别是近年来，以学习和贯彻邓小平“从严治党”的思想作为党建工作的切入点，大力改善学校党的领导，为学校的改革带来了焕然一新的面貌。回顾起来，把邓小平的有关理论与学校改革实践紧密结合，是学校弥足珍贵的经验，也是学校今后要坚持的党建方向。

一、深入学习邓小平的党建理论，把握邓小平“从严治党”思想的特征

党的十一届三中全会以来，邓小平同志始终坚持从严治党的主张，他多次强调这种迫切性与必要性。对如何从严治党，他主要有三个方面的论述。

首先，坚持从严治党，要按照党章办事，严格遵守党的纪律。邓小平同志指出：“国要有国法，党要有党规党法。党章是最根本的党规党法。没有党规党法，国法就很难保障。”[②] 他要求共产党员无论功劳、职位如何，都要严格按党章办事，严格维护党的纪律，在纪律面前人人平等。

其次，坚持从严治党，要开展积极的思想斗争，坚决克服软弱涣散现象。邓小平同志指出：“党内不论什么人，不论职务高低，都要能接受批评和进行自我批评。”[③] 邓小平同志认为要拿起批评和自我批评的武器在党内开展思想斗争，但要避免“左”，不能一讲思想斗争就是“搞围攻、搞运动”，要坚持实事求是，以提高领导者的威信。

最后，坚持从严治党，要建立健全党内监督制度，对领导干部实行严格监督。邓小平同志曾指出：我们需要实行党的内部监督，也需要来自人民群众和党外人士对于我们党的组织和党员的监督。他还提出要从制度上作出适当的规定，以便于对党的组织和党员实行严格的监督；并强调党内监督的重点是各级领导机关和领导干部，同时，

① 本文原载《新的伟大工程》，广东经济出版社，广州，1999，436-442.

② 《邓小平文选》第2卷，第147页.

③ 《邓小平文选》第3卷，第38页.

为加强监督，必须提供有效的法规和制度依据。

学校党委深入学习邓小平从严治党思想，并牢牢把握住其思想特征。学校党委认识到：要加强高校党的建设就要把从严治党体现在党的建设的各方面，即体现在党的思想建设、政治建设、组织建设、作风建设和制度建设上。同时，从严治党，最终必须落实在正确的党建目标上。在高校的党建工作中，我们正是按这一思想，促进侨校各项改革并取得了显著成效。

二、在学校党建的各方面落实邓小平“从严治党”的思想

近年来，在《光明日报》、《人民日报·海外版》、中国教育电视台等多家媒体对学校的报道中，一直肯定暨南大学“四从严”的治校方略，即把“从严治校，从严治教，从严治学”和“从严治党”作为学校上水平、上质量的重要保证。其中，学校党委特别注意把邓小平“从严治党”思想作为办好社会主义华侨大学的指导原则，将其贯彻于党建工作的方方面面。

（一）在学校党的思想建设中，严格要求，严密筹划，使从严治党的思想深入人心

落实思想建设的目标，学校党委在每年的教职工、学生思想滚动调查基础上，采取“三个结合”的方式，以达至全方位的思想渗透。

一是把党员的政治教育与师德建设结合起来，使“严”之有道。经过1989年春夏之交的政治风波后，学校党委开始反思，高校教师，特别是党员教师的人生观、世界观对年青一代的影响是至为深刻的。学校对上讲台的教师严格要求、严格把关；对新上岗的教师进行培训和严格考核；对新党员，党校组织进行人生观、世界观方面的辅导；对各级党组织，宣传部在每期制订的学习计划中，落实这一方面的教育。学校党委特别注意把这一方面的教育与师德建设融合在一起。从1996年开始，深入开展了“讲师德、练师能、树师表、铸师魂”的建设工程。通过党校、党课、组织生活、民主生活等形式，采取建立宣传栏，举行报告会、讲座、党的知识竞赛、演讲比赛、征文比赛、影视观摩，广泛开展“双学”活动和师德建设活动。在师德建设中树立正确人生取向；在政治学习中升华师德水平，使广大党员自觉履行“爱国爱校、为人师表、严谨治学、敬业奉献”的师德规定，使从严治党“严”之有道。

二是把学习邓小平的理论与学校精神文明建设结合起来，使“严”之有效。从严治党，搞好党风，是实现党对精神文明建设的领导前提，也是切实开展精神文明建设的有效保证。为此，学校党委把从严治党的思想放到邓小平的整个理论体系中加以把握，学校成立了精神文明研究中心和邓小平理论研究中心，不定期出版精神文明建设通讯和编写邓小平思想研究专著，并于1998年10月召开了全国性的邓小平理论研讨会。根据中央和省委的有关精神，学校把“师德建设工程”“培育文明大学生工程”“文明校园建设工程”作为一体化工程统一筹划，把邓小平理论作为精神文明建设健康发展的保证，在“四从严”的要求下，努力把学校建设成为社会主义精神文明的重要基地和示范区，努力实现全体党员以邓小平理论为主要内容的政治素养的显著提高。

三是校、院、系党的领导干部带头学习与全体党员全面学习结合起来，使“严”之有方。党要管党，一要管好干部，二要管好党员，这对办好社会主义的华侨大学，

坚持“面向海外、面向港澳台”的办学方针尤为重要，否则，三级领导之间及领导和普通党员之间难以沟通，难以统一认识，从严治党便会无从落实。为此，校领导每学期进行2～3次的中心组学习，党委常委定期召开民主生活会；每学期定期召开中层干部会和不定期党员大会，在会上传达和学习邓小平的教育思想和从严治党思想，再加上党校的干部培训和基层组织生活，全校形成了遵守党规党纪的舆论氛围和良好风气。

（二）围绕党的政治建设，严格贯彻党的教育方针，在办学方向上体现从严治党的方针

高校党的政治建设，主要应落实到贯彻党的教育方针、保证社会主义大学正确的政治方向上。为培养“德、智、体”全面发展的人才，学校将德育列为首位，并针对境内外学生的不同特点，因材施教。我校境外生占学生总数的40%，对他们则大力宣传邓小平爱国主义思想和“一国两制”思想，以祖国的优秀文化、壮丽河山和改革开放的非凡成就，激发侨生和港澳台学生的爱国主义情怀和报效祖国的责任感，对境内生则通过“两课”使邓小平理论进教材、进课堂、进头脑，以培养社会主义事业的建设者和接班人。

在办学方向中，强调“四从严”，既保证了人才的又红又专，又弘扬了正气，改善了校风，端正了党风。由于“严”字当头，学校在迈向“211工程”的进程中，政治建设发挥了极大的凝聚力和战斗力，全校上下，万众一心，为学校“211工程”的预审、立项打下了坚实基础，同时，也为学校贯彻党的基本路线、贯彻党的教育方针作出了政治上的保证。

（三）在党的组织建设中，严格管理，为从严治党提供有力保障

学校从抓好党组织的日常工作入手，不懈地进行组织建设。注意建立健全民主集中制，实行了党支部建设目标管理评估制，并重视党员发展工作，特别是在改革学校的人事制度上，不断完善用人机制，使干部能上能下。学校在坚持党管干部的原则上，引入竞争、激励机制，使能者上、庸者下。为了在学校疏通能上能下的渠道，从1994年起，就正式推行干部三年一聘的聘任制，全校党政干部按平等竞争、按需择优的原则聘任上岗。1997年，完成了270多名中层干部的聘任，免去了一些群众基础差、能力弱的干部的职务，并按公开、公平、公正原则招聘了7名副处级干部。为使学校与个人免除能上能下的后顾之忧，学校原则上不再作非领导职务提升，更多地将选拔“双肩挑”干部到各级领导岗位上任职，在逐步提高干部个人素质的同时，着重扭转干部队伍官多兵少的“倒三角”现象。另外，把科学的考核措施和严格的岗位责任制结合起来，作为严格评价、奖惩的依据。在考核中采取个人总结同群众评定、组织考察相结合，定性评价与定量打分相结合的办法，以期对干部作出公正、客观、全面的评价。评定结果与奖惩、任免、升降挂钩，并与校内奖金分配制挂钩。这一系列改革措施的推行，提高了干部队伍的素质，增强了干部队伍的活力，达到严格管理的目的。

（四）在党的作风建设中，严格监督，严肃执行纪律，在反腐保廉上体现从严治党的力度

有人认为高校是反腐斗争中的“净土”和“绿洲”。我校党委牢记邓小平同志从严治党的要求，批判了这种忽视高校存在腐败现象的实际情况，认真贯彻中央的有关精

神，既突出重点又整体推进学校的党风廉政建设。近两年来，我校重点抓系处以上中层领导干部廉洁自律的工作；进一步加大查处处以上干部的违纪案件的力度；进一步强调从“自重、自省、自警、自励”的“四自”标准教育干部自觉廉政的有关规定；进一步清理通信工具；认真抓好校院两级领导干部廉洁自律专题的民主生活会。在突出重点、狠抓落实的基础上，学校党委注意以点带面，不断推进其他工作的开展。我校两年来，坚持严肃执行纪律，有 14 人受到党纪政纪处分，查出违纪案例 14 宗，清理出违纪金额 882 144.39 元，通过办案为学校挽回经济损失 901 189.30 元。

（五）在党的制度建设中，严格遵照党规党纪，为从严治党做好建章立制工作

学校在依法治校、依法行政方面，建立健全了一系列规章制度，先后制定了《暨南大学党委职责范围和工作制度》《暨南大学党委常委议事规则》《暨南大学学校领导班子成员民主生活会制度》《暨南大学系总支职责和工作制度》《暨南大学关于建立健全党风廉政制度建设的意见》等一系列规章，从制度上保证党的纯洁性，防止违法乱纪现象的发生。

三、从严治党，为实现我校的“五个一”党建目标而努力

根据邓小平从严治党的思想，从严治党的落脚点是为了实现正确的党建目标。对于高校来说，这一目标是围绕学校中心工作、围绕学校办学方向而设定的，暨南大学在贯彻党的十五大精神、第七次全国高校党建工作会议和我省第八次党代会精神的基础上，结合华侨大学的办学特色，以努力达到如下“五个一”的党建目标。

建立一个坚决贯彻执行党的路线与方针政策、符合江总书记提出的“努力成为社会主义的政治家、教育家”要求的领导班子。

建立一条符合《中华人民共和国高等教育法》、贯彻“两个面向”、能主动服务于地方经济发展的办学路子。

建立一支爱国爱校、团结奋进、精于业务、严谨守纪的以党员为骨干的教职工队伍。

建立一种适应学校对外办学、保证学校党组织充分发挥作用的良好机制。

建立一套加强党员教育管理、及时解决自身所存在的矛盾与问题、不断增强凝聚力和战斗力的工作制度。

围绕这“五个一”目标建设，学校党委将继续贯彻邓小平从严治党思想，把暨南大学带进一个辉煌的新世纪。

18. 脚踏实地　循序渐进[①]

新年伊始，暨南大学董事会隆重举行第四届第一次会议，共商新形势下，在国务院侨务办公室和广东省人民政府的领导下发挥董事会在暨大办学中的作用，推进学校“211 工程”建设和发展，努力把暨南大学办成名副其实的一流大学。

我就两年半来学校工作的主要情况和今后工作的初步设想向各位作汇报。

一、两年半来学校主要工作回顾

过去的两年半，学校深入贯彻“面向海外、面向港澳台”的办学方针，坚持“从严治校”和“依法治校”的原则，以“211 工程”建设为中心，带动各项工作开展，在各位董事的大力支持和上级主管部门的领导下，经过全校上下的共同努力，我校的办学质量和水平大大提高，教学、科研力量不断增强，学校的综合实力更上一层楼。

（一）努力工作，加快步伐，抓好“211 工程”建设

1994 年 1 月召开的暨南大学董事会三届一次会议通过了学校提出的争上国家“211 工程”的初步设想后，学校即将此列为奋斗目标。1996 年 6 月 14 日，学校顺利通过了国家“211 工程”部门预审，进入了全国面向 21 世纪重点建设的 100 所大学的行列。1997 年 9 月 12 日，我校“文艺学与汉语文学”等 7 个学科项目获准立项。10 月 23 日，我校“211 工程”拟购仪器设备清单获通过。同年 12 月，提前启动“文艺学与汉语文学”学科。

1998 年是我校“211 工程”建设步入实施的重要阶段。主要完成了以下八个方面工作。

（1）学校“211 工程”重点学科项目全部启动建设。

（2）调整了“211 工程”建设的决策机构和管理机构，确定各机构的组织形式、成员组成及主要功能，对我校“211 工程”建设实行分层、分块、分项目领导。

（3）制定了《暨南大学“211 工程”建设项目管理办法》等 4 个管理办法。

（4）配合国务院侨办和广东省政府有关部门，进一步落实建设资金。三年来，共计 7530 万元人民币已投入到位。

（5）积极准备并上报有关材料，学校已于 10 月 28 日正式得到了国务院侨办和广东省政府《关于暨南大学“211 工程”建设项目可行性研究报告的批复》。

（6）及时上报了《关于审批“211 工程”拟建重点学科建设项目和拟购大型仪器设备的请示》，争取我校“211 工程”建设早日正式由国家批准立项。

（7）布置了我校“211 工程”公共服务体系各子项目制订规划和任务书的工作。

（8）布置了侨办和省重点学科以及校内重点学科进行学科自评估的工作。

---

① 在暨南大学董事会第四届第一次会议的报告，原载《暨南大学校报》，第 251 期，1999 年 3 月 8 日.

（二）贯彻“两个面向”，突出“侨”字特色

1. 针对学生特点，深化教学改革

港澳台和海外学生有其自身特点，因此，学校始终把抓好“三语”教学作为提高教学质量的关键。1996 年，学校首先对“大学英语”课程实行境内外生分流教学；1997 年又对“大学语文”实行分流教学；去年“高等数学”也进行了分流。分流教学既按境内外生源分流，又按学生程度和学科分流，真正做到了因材施教。

1996 年开始，学校提倡对港澳台和海外学生采用英文教材，用英语授课。学校规定，本科非英语专业学制年限内，至少有一门专业课程采用最新英文版教材，用英语授课。为鼓励教师用英语授课，学校还特别制定了具体奖励办法。此外，学校还根据华侨、港澳台学生和华人学生的不同情况，制定了德育培养的分类目标，专门为外招生编写适合其特点的教材。

2. 发展华文教育，传播中华文化

我校华文学院是我国开展华文教育的重要基地，并成为我国具有高级程度汉语水平考试资格的三所高校之一。从 1997 年始，学校专门设立了供外国人学习的汉语言本科专业，同时还设立了各种学习中华文化的短期学习班，每年有来自 20 多个国家和地区的学生前来学习。我校华文学院编写的北美华文教材，甫抵美国，即为 60 所中文学校争相使用，反响热烈。

3. 积极主动宣传，扩大对外招生

招收华侨华人和港澳台青年是我校招生工作的首要任务。近两年多来，学校加强对外宣传，对外招生工作取得了显著成绩。1997 年境外学生报考我校达 1578 人，录取 974 人（含预科），为历年录取境外生人数之最。

1998 年，外招形势更加喜人。本科外招生报名达 1701 人，特别是澳门考生，占澳门高中毕业生人数的 60%以上。报考学生分布世界 30 个国家和地区。共录取 1174 人（含预科）。

外招研究生成效突出。1997 年外招研究生达 102 人，为 1996 年的 2.5 倍。1998 年外招研究生又增加到 112 人，比 1997 年增加 10%；另招博士生 5 名。

此外，我校还于去年春季率先在全国实行对外春季招生。

4. 发挥侨校优势，服务侨务工作

教学科研为侨务工作服务，是学校的重要任务之一。我校“211 工程”获准立项的 7 个项目中，就有 3 个项目直接服务于侨务工作。

另外，学校的成人教育也积极拓展港澳成人教育工作。

（三）深化教育改革，提高教学质量

为适应社会需求及对外办学需要，学校及时调整专业，1996 年以来共增设了“投资经济”“保险学”“护理学”“对外汉语”“行政管理学”等 5 个新专业。

我校的学分制改革在原有成果基础上，大大向前推进了一步，在校内外引起较大反响。

（1）减轻学生学习负担，培养创新人才。学校决定调低总学分，4 年制本科总学分由 160 学分压缩为 150 学分。同时压缩专业必修课学分，增加选修课学分。

（2）改按学年收费为按学分收费。

（3）实行导师制。

（4）加强基础课教学，打通院系基础课、专业基础课，开出晚选课，实行教师挂牌上课等一系列与学分制配套的措施。

调整办学结构，优化办学层次。学校本着稳定发展本科教育，积极发展研究生教育，逐步压缩专科规模的思路，在校专科生的规模从1995年的2472人减少到1998年的623人；本科生的规模逐年扩大，1995年在校本科生5377人，1998年增加到7218人。招生质量也不断提高，学校去年录取考生中，高分段学生比前一年成倍增加，文理平均分比前一年提高了10分。

研究生教育长足发展。1997年报考我校研究生的考生超过2000人。1998年报考我校研究生的考生达2516人，录取499人，其中博士生32人，比前一年增加33.3%。去年，我校MBA无论是报名人数，还是第一志愿上线人数，均在全国56所试办MBA高校中排列第4位，影响较大。我校研究生总人数于1997年突破千人大关，1998年达1309人。

去年，我校新增列了金融学等5个博士点，国际贸易等10个硕士点。同年6月，我校的国家文科基地顺利通过教育部的中期检查评估，评审结果为优秀。

成人教育成绩显著。学校继1996年省函授教育评估第一名，夜大学教育评估第二名后，1997年该两项评估获全省总分第一，并被国家教委授予“全国成人高等教育评估优秀学校”。

在办学条件方面，1996年，学校校园网顺利通过广东省验收，成为全省最早完成的高校之一；在当年全省高校实验室评估第一的基础上，1997年学校新建了作为省重点的现代电子技术实验室；我校积极发展电化教育，1996年学校电化教育在全省高校评估中名列重点院校总分第一。1998年，我校被评为“全国电化教育工作先进单位”，是全国十所高校之一。学校还积极向外延伸，改善办学条件。1996～1997年，先后接纳深圳市人民医院和珠海市人民医院为我校医学院第二、三附属医院。1月初，经国务院侨办批准，广州市红十字会医院和清远市人民医学院将分别成为我校医学院第四和第五附属医院。四家三甲医院的加盟，大大增强了我校的临床教学以及科研工作的力量。去年，我校还与珠海市人民政府合作，在该市共建暨南大学珠海学院（正式批文下达前，称珠海教学点）。

体育竞赛成绩突出。去年7月，在第三届全国大学生羽毛球锦标赛上，我校共获得4金7银1铜和4项第四名。我校3名运动员被教育部决定选派组成中国大学生羽毛球男队参加当年9月在土耳其举行的世界大学生运动会，获得2金1银1铜。8月，在第六届全国大学生田径锦标赛上，我校共夺得5金5银2铜，刷新一项大会纪录，并获女子团体总分第七、男子团体第八、男女团体总分第八的佳绩。在去年底结束的第13届亚运会上，我校3位学生代表中国作为主力参加女子足球赛，获得金牌，另2位同学分获男子标枪和女子三级跳远银牌。

（四）加强科研管理，开展科学研究与学术交流

我校的科研工作是紧密结合“211工程”建设，结合人才培养等工作一起进行的。

学校重视师资队伍建设，注意引进培养学术骨干和学科带头人。1996 年至 1998 年，学校共引进和在职培养博士 65 名，另外引进知名专家 5 人，目前全校共有博士 147 人，硕士 564 人。学校鼓励教师攻读博士、硕士学位，现共有 86 人在境内外攻读博士、硕士学位。我校有 6 人被遴选为“千百十工程”省级培养对象，52 人为校级培养对象。科研队伍加强了，学校及时结合学科优势，增设科研机构。1997 年，学校先后成立了金融研究所等 6 个研究所（中心），并把数学力学研究所更名为应用力学研究所。

近两年半来，我校申报科研项目及项目成果成绩突出。1997 年，我校自然科学共获国家、省、市各类科研项目 59 项，获资助经费 450 多万元。人文社会科学获国家社会科学基金项目和国家教委专项课题 13 次，获准数远高于中大和华师大等单位。在广东省“九五”社会科学规划立项中，我校有 25 项课题被批准立项，无论是被批准数还是获资助的金额数，均居全省首位。当年我校共有 19 项科技成果获奖，国家级 1 项，省部级 5 项，厅局级 13 项。在最近公布的全国普通高等学校第二届人文社会科学研究成果奖名单中，我校获二等奖两项，三等奖一项。去年，我校自然、人文社会科学共获各级科研项目 218 项，经费 712 万元。特别是在国家基金委管理科学部“防范金融风险快速反应”项目招标中，我校凭雄厚的实力一举夺标，成为全国 4 家立项单位之一。随后，我校又成功地争取到教育部的重点科技项目的重点实验室向我校开放，这是教育部首次破例在部属外高校以我校为唯一试点单位受理申报，我校申报的 3 个重点科技项目，均获得批准和资助。在国家基金委的 6 个科学部中，我校在生命科学部和管理科学部中已有其特色和优势，去年我校在生命科学部获得资助科研项目数在全国综合性大学中排列第 5；管理科学部获资助科研项目数一直居广东省高校榜首。成果奖励方面，截至去年 12 月已揭晓的统计，学校共获省部级奖 8 项，厅局级奖 29 项。

学校还着力构建了科研激励机制，颁布了学校科研基金项目、科研成果及四大索引收录论文等配套奖励办法。学校还先后制定了 27 个科研管理办法。目前正在制订学校科研中长期发展规划等。

近两年多来，学校有关人员对欧、美、日、东南亚等多个国家和地区的十余所大学进行了访问，同时接待了十余个国家的专家学者数十人次的来访，签订了一系列学术交流协议。特别是与越南胡志明市经济大学和印度尼西亚雅加达达尔玛帕沙达大学结为姊妹学校，成为我国第一个与这两个国家大学建立学术交流关系的大学。

（五）抓紧配套改革，做好各项工作

1. 财务管理及分配制度

我校 1996 年开始进行财务制度改革，实行新的学校基金管理制度，统一全校财务结算，全校财务管理实现了良性循环。去年，学校结算中心成立，使学校的财务运作及管理更加规范化。与新基金制度的实行相适应，学校对原来由各单位小范围的奖金分配制度也进行改革，实行校内工资制度。

2. 干部人事制度方面

1997 年，学校对全校处科级干部进行了考核，并对 234 名处级干部，367 名科级干部进行重新聘任或任命，有 23 人落聘。1998 年初，学校对部分中层领导干部职位实行全校公开招聘选拔，使一批优秀人才脱颖而出。学校对全体干部严格实行任期制、责任制和轮岗制。

学校从1996年开始，陆续对各单位进行人事定岗定编，同时对全校教职工实行聘任制，以调动积极性。学校定编工作的总原则是：提高办学效益，优化队伍结构，保证学科建设。

核编与岗位设置协调，适当向教学、科研以及重点学科倾斜。学校定编工作的目标是教学科研编制占50%以上，教辅人员占30%，行政后勤不超过20%。

3. 积极推动后勤改革

我校后勤改革的目标是实行事企分开，两权分离，创造企业化、产业化的后勤保障体制和服务体系，使学校后勤工作逐步社会化。1997年，学校首先将电话管理社会化，提高了通信效率和办学效益，为学校节省了资金，给教职工带来了实惠。去年，学校遵照国家的房改政策，圆满完成了全校1283户的售房任务。学校的教职工住宅区将逐步实行住户自我管理的模式。

4. 校园环境、教职工生活方面

两年多来，校园卫生、绿化、治安等各方面有了明显的好转，学校一跃而成为广州市花园单位、卫生模范单位、先进单位和天河区综合治安先进单位。学校还邀请广州市规划局，对校园进行了整体规划。

自1996年11月至1998年10月，学校共进行了三次新旧房的分配，近1700户教职工喜迁新居，基本解决了教职工住房困难问题。学校还斥资千万元进行了全校性的电力扩容和自来水加压，缓和了学校用电用水的紧张。

另外，曾宪梓科学馆也于去年封顶并进入最后装修。学生宿舍楼工程进展良好。

二、今后工作的设想

“211工程”建设，在今后10年内，都将是我校的一项中心工作。我校“211工程”建设的总体目标是：“经过10年左右的努力奋斗，使暨南大学的教育质量、科研水平和办学效益等方面进入国内高校先进行列，成为我们面向海外、面向港澳台办学和传播中华文化的重要基地，在香港、澳门、台湾地区和华侨华人社会具有重要影响的社会主义华侨大学。”因此，我校将在2000年以前，主要做好以下八个方面的工作。

1. 人才培养

“九五”期间，研究生人数以每年12%的速度增长，到2000年博士、硕士之比由目前的8.5∶100提高到14.3∶100；境外研究生与境内研究生之比由18∶100提高到25∶100；研究生与本科生之比，由目前的18.1∶100提高到20.5∶100。

2. 科学研究

到2000年，人文社会科学研究总体水平要进入全国综合性大学先进行列，部分研究领域达到国内领先水平，少数研究项目在国际上有一定影响。自然科学和工程技术科研工作要贯彻重视基础研究、突出应用研究的方针。主要指标是：预算外科研经费以年度15%的速度递增；发表于核心期刊的论文每年增加10%；获省部级以上成果奖数逐年提高，到2000年，获成果奖数比现在增加20%以上。

3. 学科建设

力争在“九五”后期建成 2 个国家重点学科，2 个博士后流动站，再建 2～3 个省级重点学科。争取新增 2～3 个博士学位授予点，使现有博士点研究方向增加一倍以上，组建 6～8 个学科群体，更充分发挥我校的优势和特色，使我校有一至二个学科建设项目整体达到全国先进水平；同时，培养 2～3 名相对年青的国家级学科带头人和 10 名左右具有较高造诣、在国内外有一定影响的中青年学科带头人。

4. 师资队伍建设

到本世纪末，专任教师人数与学生人数的比例达到 1∶12 左右。具有研究生学历的教师由目前的 51.5%提高到占教师总数的 60%左右，其中具有博士研究生学历的教师达到教师总数的 15%。

5. 办学物质条件建设

“九五”期间经费的年增长率应超过 12.5%。加强各项基础设施建设，使我校的办学物质条件逐步达到中央〔1983〕24 号文件提出的“适当高于国内其他大学”的标准。提高科技成果开发和转化能力，到 2000 年，校办产业上交学校的利润要有较大幅度的增长。

6. 对外学术教育交流合作

再争取 10 所左右高校及研究机构和我校建立学术教育交流和合作关系。

7. 校内管理体制改革

改革的主要任务是：①进一步加强董事会工作；②进一步完善校长负责制；③健全校、院、系三级管理体制和搞好校部机关的改革；④深化干部人事制度和分配制度改革；⑤加快后勤管理和财务管理体制的改革。

8. 思想品质教育与校园精神文明建设

根据我校的侨校特点，学校将长期地、有步骤地对教职工和境内学生进行邓小平理论和党的基本路线教育，世界观和人生观教育，社会主义法治和道德教育。对港澳台和华侨学生加强爱国主义教育。对华人学生和外国留学生加强了解中华文化和对我友好的教育。

我们相信，在国务院侨办和广东省委、省政府的直接领导和各位董事、各位朋友的热心关怀、指导下，经过全校师生员工的共同努力，一定可以克服困难，缩小差距，把暨南大学办得更有特色，更有水平，以全新的姿态跨入 21 世纪，全面完成“211 工程”总体建设规划的任务。

19. 举行全国100所“211工程”学校赠书仪式①

今天，暨南大学董事、香港石汉基先生向全国100所“211工程”学校赠书仪式在我们这所有93年历史的华侨高等学府隆重举行，使我们全国100所“211工程”大学校长第一次有机会欢聚一堂，接受赠书，也使我校有机会向各兄弟院校学习，我们感到非常高兴。在此，我谨代表暨南大学领导班子以及全校3万名师生员工向石汉基先生表示由衷的感谢，对各位领导和各位同仁的到来表示热烈的欢迎。

石汉基先生是石景宜先生的长子。石景宜先生也是暨南大学董事，热爱图书文化事业。我国实行改革开放政策以来，他向祖国内地许多学校、公共图书馆和文化单位捐赠了大量图书，支持祖国教育、文化事业的发展。他还向台湾的一些学校和文化单位捐赠图书。在大陆和台湾举办图书展览，促进海峡两岸的文化交流。他的事迹，已在海内外传为美谈。

石汉基先生继承父业，立志走父亲的道路，赠书报国。从1986年以来，他已先后向广东、福建、山东、西藏等省区的学校、公共图书馆、文化单位和全国政协赠书50多万册。石汉基先生为了支持国家“211工程”建设，这次他又向全国100所“211工程”学校捐赠港台版图书7万多册。石汉基先生热爱祖国，热爱教育事业，他慷慨解囊，积极支持祖国教育事业的发展。对他所作的重要贡献，我们将永远感念在心，并把它化为振兴和发展我国教育事业的动力。

这次赠书仪式，得到教育部和国务院侨办的高度重视和关怀。全国政协叶选平副主席、陈俊生副主席和孙孚凌副主席，教育部韦钰副部长和教育部“211”办等部门的负责同志，国务院侨办刘泽彭副主任及有关部门负责同志，广东省任仲夷老书记、梁灵光老省长、政协郭荣昌主席和广州市的各级领导在百忙中出席这次赠书仪式，在此我谨向各级领导致以衷心的感谢。

这次赠书仪式，我校能作承办单位，深感荣幸。我们希望各兄弟院校对我校的工作多给予指导和帮助，我们也力争把接待和其他会务工作安排好。

① 本文是在香港石汉基先生向全国100所“211工程”学校赠书仪式上的讲话，广州，1999年11月8日；教育与科技管理研究，北京：科学出版社，2016，225-226.

20. 转变观念　量化考核　优劳优酬①

面对新世纪的来临，在科学技术突飞猛进、知识经济已见端倪、国力竞争日趋激烈的形势下，国家提出了科教兴国的伟大战略决策，对我国高等教育的发展，对高校人才培养的质量提出了更高的要求。但我们的教育观念、教育体制、教育结构、人才培养模式、教育内容和教学方法相对滞后，高校教师队伍的整体水平和质量仍有很大的差距，而且，传统的分配体制难以适应建设高质量高校教师队伍的要求。建立一支政治业务素质良好、结构合理、相对稳定的教师队伍，是高教改革和发展的根本大计。

一、建立优劳优酬分配体制的必要性

传统的高校人事管理模式不能适应新时期建设高质量教师队伍的要求。这种带有计划经济色彩的管理模式有两大弊端。

一是通过国家各种人事制度确立了教师职务终身制。现行的户籍制度、人事档案制度、职务评聘制度、退休养老制度、失业与社会保险制度等，实际上都是为人才的单位所有制和用人的终身制保驾护航的。

二是国家高度集中的指令性工资制度。这种工资制度是以整体的低待遇和平均主义为其基本特征的。这种制度难以体现按劳取酬、效率优先、优劳优酬的激励原则。

自改革开放以来，国门打开，人们有机会与国外同行交流比较，他们发现自己的劳动付出、业绩贡献与所获得的报酬是不相称的。

教师待遇问题一直是困扰高校发展和人才成长以及学科建设的一个尖锐而迫切的问题。一方面，国家直接投入不足和学校自筹资金能力有限，教师特别是青年教师工资待遇偏低。另一方面，在分配体制中平均主义、吃“大锅饭”现象比较突出。平均主义是严重挫伤优秀人才积极性、造成人才流失的重要原因，从根本上制约了高校高质量教师队伍的建设与发展。这些年来，我们仍然还是在传统的高校人事管理模式的框架内进行修修补补，还没有革命性的突破。

我们必须针对传统的高校人事管理模式的两个基本弱点进行大胆改革，争取有大的突破。新型的高校人事管理模式应该是“按需设岗、公开招聘、竞争择优、按岗聘任、优劳优酬、聘约管理、开放流动”，实行这种管理模式的前提和基础是“按岗聘任、优劳优酬”。这种模式实现了教师“身份管理”向“岗位管理”的转变，破除了“终身制”，形成具有激励竞争机制的用人制度。这是具有生机和活力的高校留人用人机制。这是国家人事部门正在推行的专业技术人员和事业单位人事制度改革的核心内容。

在教师待遇相对优厚，各种后顾之忧可以通过较高收入从社会得到解决的情况下，高校的人才市场就会变成“用方市场”，到那时，要想在高校谋一个职位，必须凭借个人的水平和能力，通过激烈的竞争才能得到。也只有到那时，学校才不必再为教师的住房、配偶、子女等琐事操心，而是把主要精力用在招收优秀的学生、招聘优秀的教

① 本文原载《高教探索》，2000，(1)：5-8.

师、制订好的人才培养方案、抓教学科研管理、向社会争取更多的办学资金、开展更多的国际合作与交流等。这样，高校才能从根本上走出今天的困境，才谈得上“与国际接轨”。

二、建立教学科研人员业绩量化考核指标体系

为把暨南大学建成教学、科研型大学，需要建立对工作业绩进行评估的指标体系。暨南大学参考了国内多所学校的考核方法并受到国外高校管理的启发，在考核小组、院系干部及教学科研人员反复讨论的基础上，制定出台一套全新的量化考核指标和管理方法——《暨南大学教职工业绩考核暂行办法》、《教学科研系列考核计分标准（试行）》和《1999—2000年度校内工资发放方法》作为新分配体制试行。从而比较客观、全面地对业绩进行量化，模糊了教学科研人员单一的职能界限，充分发挥每个人的潜能，优化学科队伍，合理利用人力资源，调动职工的积极性。

评优指标体系由以下几方面构成。

1. 教学工作

以本科生理论课每节2分为基准，范围在1.0～3.0分，以区分学生数和课程类别，控制班学生规模；本科生毕业论文参照学分，按不同学科制定计分标准；研究生按年级、学科、培养类型，适当考虑导师组成员计分。凡已享受授课酬金的课程，如MBA、成人教育、研究生课程班、海外及港澳台授课等，按30%计分。

为了保证教学效果，原则上以专家组、系级教学指导小组和学生对该课程的评估成绩总和的平均值，以75分为系数1.0，进行乘积计算。同时对讲授基础理论课的教授、博导、院士另按1.1～1.3系数计算。对教学中出现事故者，按轻重程度扣分。教学成果参照科研成果的奖项下调一级计分。

对实验室建设，按经费来源相应计成绩。

2. 科研工作

按课题、学术论文、成果奖和专利、创收等内容进行量化。

科研课题、经费：按科研课题来源和经费到位金额，核算成总分，由课题负责人按贡献大小分给课题组成员，凡立项无经费者，只计相当于常规立项课题的1/10立项分。

学术论文：按四大索引收集、核心刊物、公开发表论文及其他四种类别分别计分。最高300分/篇，最低2分/篇，对第二作者等，按相当于第一作者的50%～20%计分。

学术著作、教材：共分3类，除主编、副主编占一定的分外，按0.5～2分/千字计分，获奖者另加分。

科技成果：国家最高奖第一获奖人1500分～厅级三等奖20分，第二完成人及以后占第一获奖人的30%～5%不等计分。

专利成果：按发明专利、实用新型专利、外观专利分别计分。

科研管理：对不能如期完成或弄虚作假，甚至科研道德败坏者，扣100～500分或重罚。

3. 加分

为活跃校园学术风气，对校、院、系学术报告会的主讲人和听众均适量计分。

对获党、政部门集体奖或个人奖的个人均予加分。

对兼任各级党、政、民主党派，工、青、妇，省级以上各学术团体负责人，校学术委员，教研室主任，党支部书记，系主任助理，本科生、研究生工作秘书适当加分。

对“双肩挑”干部按其业务技术职称和行政岗位级别给予标准工作量补贴 1/2 以上分。

对院士、教学科研中突出贡献者可免考核，按良好等级计分。

三、建立考核成绩与校内工资挂钩的新分配体制

教学科研人员实行定性考核，即按德、能、勤、绩等综合表现，由基层考核小组用定性方式评出优、良、中、及格和不及格，其系数是 1.1、1.0、0.85、0.75 和 0，与业绩量化考核分乘积为实得分；行政干部考核等级系数按优、良、称职、基本称职和不称职分五等，其等级系数为 1.1、1.0、0.85、0.75 和 0。

根据各方面构成业务量化考核成绩和定性考核等级，按不同职称，从高分中评优。

(1) 考核工作量能较全面反映每位教学科研人员上一学年的工作成绩，用此成绩与下一学年的校内工资挂钩。在价值上高、中、初级职称略有差距，并以各种职称人员的全校平均工作量的 90%左右为标准工作量，如正高 1000 分，副高 850 分，中级 700 分，初级 600 分。

由于是第一年试行，对于成绩不理想者，给予一学年的保护期，最低分段大约 10%的人员保证能获得不低于前一学年的校内工资额。

(2) 对其他系列的人员，制定相应的职称、级别等级的校内工资标准，并与综合考试成绩系数挂钩，其乘积即为本人可获得的校内工资。

(3) 对于“双肩挑”处级以上干部，可以领取专职干部的校内工资额，或者以业务考核成绩加岗位补贴分与考核等级系数乘积领取工资。

(4) 行政管理人员根据其参加考核岗位类别，按与考核等级系数乘积领取工资；新参加工作未定级人员靠近相应学历定级人员领取相应校内工资。

(5) 以学院为单位，可按各自工作量计算的金额领取工资，也可以按各系列标准值领取工资。以标准值领取工资的单位原则上要进行二次分配。

四、教学科研人员业绩全面量化考核与分配体制挂钩的积极作用

最近，教育部下发的《关于当前深化高等学校人事分配制度改革的若干意见》文件，力度之大前所未有。高校在 2 至 3 年的时间内，将全面推行教师聘任制和全员聘任合同制。这就意味着教授、副教授等职称头衔将不再是一朝加冕荣耀终身。

目前，舆论的作用已把人们的视线引向高校教师待遇的热点问题上。清华大学率先大幅度提高教师的待遇，拉开了高校分配制度改革的序幕。“大学教授升值了，滥竽充数不行了”，“大学薪金大革命”，等等，一场分配制度的“改革风暴”在大学校园刮起来。

新一轮人事分配制度改革，在“按需设岗，按岗聘任，择优上岗”的前提下，实行“优劳优酬，多劳多得”的制度，力图把过去那种强调身份，以职务级别为主的分配方式，转化为强化岗位，以岗位职责、业绩、贡献为主的分配方式。

暨南大学在实施第一轮校内分配制度改革的初期，阻力是很大的。为此，学校主要领导面临很大压力并做了大量的工作，力求以事实证实其改革的方向是正确的、意义是长远的。应该说明，学校现今的分配体制改革之所以能够跨前一大步，得益于1996年开始的校内基金管理和分配体制改革。

暨南大学实行的教学科研人员业绩全面量化考核与分配体制挂钩的新一轮改革已产生了积极的影响，大多数教职工的收入水平比过去有了较大提高，个别教授的校内工资一年可达5万元，加上津贴、基本工资和课酬，年薪可超过10万元，达到全国重点学校教师收入水平。

暨南大学分配制度的改革特别向教学、科研一线的队伍倾斜，学校的校内分配制度基本上解决了他们的后顾之忧，使他们专心教学、科研工作。学校分配制度的改革取得了明显的效果，概括起来，这种激励机制具有如下几方面的积极作用。

1. 激励作用

根据考核成绩，校内工资在同系同职称中可以拉大差距达3～8倍，实现优胜劣汰，合理指导优先上岗，改变1/3的人干、1/3的人看、1/3的人不干的局面，起到奖勤罚懒的作用；可指导评优争先，选拔骨干教师和学科带头人；有利于发现教学、科研优秀人才，为优秀人才脱颖而出提供条件。

2. 平等竞争

是在不讲学历、资历，没有歧视背景情况下的平等竞争，尊重每个人的劳动，极大地调动了广大教职工的工作积极性，有人说，从中找回了做人的尊严，做事公平。

3. 减员增效

改变过去重复进人，多进人，上大课或有课无人上，出勤不出力的局面。学科的优化整合，改变教学科研人员功能专一化的现象，让那些教学科研能力强和爱岗敬业者脱颖而出，为把学校办成教学科研双中心的目标做出贡献。同时控制了校内单位盲目的进人，合理布局教学、科研力量。校内分配制度改革为公共教学、跨院系选课、自选课提供了条件，使学校的人力资源得到充分利用，也为学校实施专业目录调整、教学科研力量的重组、系所合一等工作提供了条件。

4. 促进管理

首先是对干部队伍的管理。将考核等级系数与校内工资或考核成绩挂钩，突出干部管理工作成绩。其次，为学校定编定岗提供了重要依据，为实施真正意义上的聘任制及聘后的管理提供可靠的保证，对教职工延聘、返聘、退休均有指导作用。通过对工作量的综合测评，还可掌握各单位工作总量的情况。促进学校人事、教学、科研、研究生等部门管理工作的规范化，提高科学管理水平。

5. 优胜劣汰

现已出现找课上，寻事做，学术讲座蔚然成风的新局面。对不胜任现职，无心向教、向研的人员有明确的控制指标，有利于转岗分流、岗位竞争，促进人员合理流动，合理使用。

6. 吸引人才、留住人才

相信新分配制度的实施，必将对人才具有相当大的吸引力，吸引校外优秀人才来校工作，有利于人才引进和校内人才队伍的稳定。

在计划经济加速向市场经济过渡的今天，我们必须逐步树立“成本核算、量化管理”的观念。我校率先进行的业绩全面量化考核，不单是“算工分”和“分奖金（校内工资)”，正如大多数教职员工所说的那样，尽管还有不够完善的地方，但大方向是正确的。

当然，我们也看到了改革方案存在一些不足。第一，学科之间的权重不平衡以致难以掌握准确的评估尺度，量化考核标准还不够完善，难以体现综合性大学多学科、多类型的特点；第二，考核过多考虑量的计算且易造成斤斤计较，如何与质量很好结合起来，也是一个问题；第三，量化考核工作量庞大，由于涉及教职工的切身利益，务必防止弄虚作假。所以，学校花费了有关职能部门大量的精力，并将数据用计算机处理。

21. 同质同水平异地办学[①]

这次驱车来珠海，心情格外舒畅，可谓过春风十里，尽草木青青；美丽的珠海更是花枝排比，嘤嘤贺喜；因为今天，一所拥有近百年历史的高等学府将与一座崭新的现代化滨海城市再度联袂，共同谱写珠海经济特区教育发展史的新篇章。在此，在这个春意融融的时刻，我代表暨南大学师生员工，对出席今天暨南大学与珠海市人民政府合作建设珠海学院签字仪式的各位领导、各位嘉宾和各界朋友表示衷心的感谢，对珠海市政府的周到安排和热情接待表示诚挚的谢意。

暨南大学是一所具有 94 年悠久历史的华侨高等学府，在中国高等教育发展史中，有着不可替代的地位，她曾创造了中国高校的三个“第一”：中国第一所由国家创办的华侨大学，中国第一所面向世界办学并招收留学生的大学，也是新中国第一所设有医学院的综合性大学。肩负着历史的重托，不辜负党和人民的期望，我们暨南人，致力于“两个面向”的办学方针，努力提高学校的办学层次和办学水平，使学校跨入了国家面向 21 世纪重点建设的 100 所“211 工程”大学的行列。学校依法治校，“严”字当头，注重教学质量，始终把提高教学质量、提高学生综合素质作为办学之首要，使暨南大学在海内外声誉日隆。今日之暨南大学已拥有 10 个学院（其中医学院在广州、深圳、珠海、清远四个城市设有 5 家附属医院，均为国家级“三甲”医院），26 个系，36 个本科专业；在研究生教育方面，有 12 个博士点，59 个硕士点，1 个博士后流动站；目前在校全日制学生 11 000 多人，其中来自世界五大洲 30 个国家和地区的海外及港澳台学生 3600 人，博士、硕士研究生 1600 人，本科学生 8700 人。暨南大学在学校发展的蓝图中，早早地相中了中国最早的两个经济特区，继 1993 年在深圳开中国校企联合办学的先河，成立了暨南大学中旅学院后，1998 年我校应珠海市人民政府要求，在珠海共建暨南大学珠海学院（初期对外简称教学点），并于当年开始招生。这是珠海市第一次拥有自己的高等学校，而在此之前的 1997 年，我校已与珠海市人民政府共建珠海市人民医院，为暨南大学医学院第三附属医院，随后招收了 15 名临床医学专业的硕士研究生，这也是珠海经济特区第一次拥有自己培养的研究生。现在，暨南大学与珠海市人民政府共瞻前景，在过去携手办学的基础上，进一步上规模、上层次，在更高的层面上，再度合作建设暨南大学珠海学院，这无论对暨南大学还是珠海特区未来的发展都有重要的意义和深远的影响。

新的暨南大学珠海学院，依山傍水，草木葱茏，是莘莘学子求学的好地方。这有赖于珠海市政府对新珠海学院关怀有加，将位于市中心 40 余万平方米的土地，连同价值 7000 万元人民币的两幢建筑物以及诸多教学、生活设备永久无偿提供给暨南大学珠海学院，并每年投入一定的教育经费。暨南大学也将加大珠海学院的投入，在 5 年内把暨南大学珠海学院建成可以容纳 3000 名全日制本科学生的综合性学院，同时发展硕

① 本文是在珠海市政府和暨南大学合作建设暨南大学珠海学院签字仪式上的讲话，珠海，2000 年 4 月 29 日；暨南大学校报，第 283 期，2000 年 5 月 15 日.

士、博士研究生的高层次教育。我相信，在珠海市人民政府的大力支持下，暨南大学珠海学院一定会办成一所具有鲜明特色的现代化新型高等学校，她不仅对珠海特区乃至广东省在新世纪增创新优势、更上一层楼发挥重要作用，也必将对海内外的经济文化发展作出应有的贡献；同时，她还将有利于暨南大学更好地贯彻“面向海外、面向港澳台”的办学方针，成为沟通港澳台地区及海外的桥梁之一，为暨南大学进一步走向世界作出贡献。

寄语珠海风日道，明年春色倍还人。我诚恳地邀请在场的各位领导、嘉宾、朋友们明年在暨南大学珠海学院主体工程竣工剪彩之日，再度光临珠海，光临暨南大学珠海学院。

22. 暨南大学国际化之路①

今天有幸在这里演讲，我感到非常高兴，谢谢贵校校长的周到安排和热情接待，谢谢大家出席这次演讲会。今天我所讲的题目是“暨南大学国际化之路”。

阿尔弗雷德·诺斯·怀特黑德（英国哲学家、数学家）在其《大学和它的职责》一文中指出，“大学的任务是将想象力和经验融合在一起”。那么我想，从一个大学国际化的角度出发，我们同样可以说“大学的任务是将不同的文化融合在一起”。同时，提供一个地方，让来自不同文化背景、有着不同思维方式、经历和生活习惯的人们可以相互了解、相互尊重，并且相互欣赏。

现在越来越多的人士，无论是教育界还是工商界，都逐渐认识到让他的毕业生或雇员有一个“国际化视野”的重要性，因为只有通过这一点，他们才能从保守、封闭的惯性思维中解放出来，而代之以一种灵活、开放的思维，也只有通过这样的方式，他才能够跟上今天这个更加开放的国际社会的潮流，跟上当今网络时代的脚步。

所以，现在我们面临的问题就是：如何实现国际化？我认为，一个大学的国际化就像一部交响乐，大概可以分为下面几个乐章。第一步，应该打开校门，在意识和作风上实现开放；第二步，在自己与外部的大学和其他机构之间搭建起交流的桥梁，把自己同外部世界联系起来，用今天的话说，也就是“上网”。任何一个大学都应该认识到没有谁可以独立发展。当这些外部的条件成熟后，下一步要做的工作就是创造。它应该通过吸收来自不同地方的文化，创造自己独特的校园文化，“海纳百川，有容乃大”。只有通过国际化，一个大学才能够不断地前进和发展。

在这里，我想以“暨南大学外招生的教育和管理”为侧重点，简单介绍一下暨南大学国际化的成功经验，并希望通过此次的交流与更多的学校建立起广泛的联系。

暨南大学成立于 1906 年，是中国历史最悠久的几所大学之一。自成立第一天起，她就是以招收海外华人子弟和传播中华文化为已任。“面向海外、面向港澳台”是暨南大学的办学方针。在中国的教育史上，暨南大学是一所招收海外留学生的高等学府，也是中国第一所设立医学院的综合性大学。现在全校设 10 个学院：文学院、理工学院、生命科技学院、医学院、经济学院、管理学院、华文学院、教育学院、中旅学院、珠海学院；有 5 所大学附属医院。学生 20 000 多名，其中全日制学生 11 000 名（包括 1680 名博士和硕士研究生），3600 名来自港澳台地区和世界五大洲 30 多个国家（包括秘鲁）的海外学生，约占全日制学生总数的 1/3。迄今为止，已经先后有来自世界上 72 个国家的学生在暨南大学求学。他们中有很多人在毕业后成为他们所在国家的栋梁之才，其中包括泰国国会前主席、曼谷银行董事长许敦茂先生，新加坡国立大学首任校长李光前先生，以及其他许多政府、工商界和文教界的著名人士。仅在过去的 20 年中，暨南大学就已经培养海外及港澳台等地区各类层次的学生一万多人，堪称门生遍

① 本文是在暨南大学与圣马丁大学合作签约仪式上的讲话，利马，秘鲁，2000 年 5 月 25 日；一个大学校长的探索，北京：高等教育出版社，2011，18-20.

五洲，桃李满天下。每年暨大招收的港澳台及海外学生人数比国内其他大学外招生的总人数都要多。暨南大学，作为中国国际化程度最高的高等院校之一，正在海内外，特别是东南亚地区赢得越来越高的声誉。

暨南大学对外开放的办学传统，临近港澳近 100 多公里的地理位置为她的国际化提供了得天独厚的条件。国际化不仅是她过去和现在吸引海外学子的原因，而且也是她自身的发展目标。为了实现这个目的，暨大采取了一系列行之有效的措施。在外招生方面，暨大实行春秋两季招生，并在香港、澳门等地区设立了办事处。学校还有组织的派招生工作组前往东南亚国家了解生源情况，扩大对外宣传，这使得暨南大学在学生工作的第一个环节上实现了时空双重意义上的开放。为了同国际惯例接轨，暨大在国内率先实行了学分制，并取消了补考制度，使优秀的同学可以提前毕业。语言是交流的基础，语言教学在暨南大学受到了高度的重视，大学内开设有英、法、德、日、葡等语种的课程。学校加强了对国内学生的英语教学，规定每个专业至少有两门课程要用英语授课。外语不仅是学生顺利毕业获得学位的要求，同时也是授课教师晋升职称的标准。学校还聘用了多名外籍教师教授外国语言及文化，向中国学生介绍国际学术、科技界的最新动态。

学校下设华文学院，专门对来中国学习的留学生教授中国语言文化，课程内容包括普通话、粤语、书法、美术、中国哲学、中国历史、中国社会、妇女问题等诸多方面。学校还不定期地邀请有经验的教师进行讲座，使外国学生能充分了解中国文化的精髓。

在做好校内工作的同时，暨南大学还积极在海外传播中华文化，真正实现暨南大学“声教迄于四海”的创校理念。华文学院为北美地区和柬埔寨国家的华文学校编纂了汉语教材，仅在美国就有 170 多间学校采用，并在当地受到极大的欢迎。因此可以说，暨南大学已成为一扇窗口，中国大学生从这里看见一个无限广阔的世界，海外大学生也透过这里看到了中国。

因为港澳台及海外生在暨南大学学生人数中占了很大比重，因此暨大的教学管理呈现出了自己独特的风格。简而言之，就是一种“求同存异”的校园文化。根据学生不同来源和不同要求，我们设置了相应的课程，并根据学生入学时的不同层次，在授课时采取分流教学，使学生根据不同的起点，可以选听不同程度的课程，使得他们能够在不同的起点起飞，却飞翔在同一片天空下。

在学习中，内地生和港澳台及海外生奋飞在同一片蓝天里；在生活中，他们也可能居住在同一宿舍里。根据外来学生的要求，他们可以选择同国内生住在一起。这种同住的宽松氛围，可以使彼此之间不再陌生，更加充分地交流。

不同产生了文化的多元，而相同则是和谐的来源。来自西方国家的学生，来自中西方文化交融共生的港澳地区的学生，与在中华传统文化背景下成长起来的内地学生在一起，创造出暨南大学独特的校园文化风景。海外学生举办的“迷你马来西亚风土人情”“泰国文化风情展”等文化活动使国内学生不出国门就感受到异国的风土人情，而同时国外学生也被博大精深的中华文化文明深深吸引着。这种双向流动的交流在促进国际凝聚力、增强国际一致性方面所起的作用是不可低估的。

除了上面所说的这些软环境之外，暨大同时注重改善硬件设施，使学生可以通过现代科技实现真正意义上的国际化，所以暨大再次率先在国内行动，正着手将国际互联网接入学校宿舍，使学生能随时随地置身于国际社会日新月异的发展变化之中。现代科技在学生管理和教育方面的应用，再一次缩短了学校与学校之间、国家与国家之间的距离。

在国际化的进程中，暨大还努力开拓与国外学术界的合作。截至目前，它已与国际上超过30所规格较高的文化科研机构建立了交流关系。交流形式包括科研项目合作、教授互访活动及学生交换项目等。在学生交换项目中，学校注入了很多精力，提供了很大的便利，使得更多的中国学生有机会走出国门，走向世界。

“路漫漫其修远兮，吾将上下而求索。”迈向国际化的道路是漫长的，充满了竞争与挑战，但它也是一条充满机遇和激情的道路。暨大有着悠久的开放传统，并且一直在这条实现学校现代化的必由之路上坚定地前进着，我坚信，暨南大学一定会有更加美好的未来。我衷心希望这次对贵校的访问，与贵校结成姐妹学校，将会促进暨南大学更进一步国际化，将会促进两校学术的繁荣和进步！

23. 面向新世纪的创新教育①

暨南大学是中国第一所由国家创办的华侨学府，“暨南”二字出自《尚书·禹贡》“朔南暨，声教讫于四海”，意即将中华文化远远传播到海外。学校的前身是1906年（清光绪三十二年）由清政府创立于南京的暨南学堂。1921年与东南大学在上海合办中国首所商科大学——上海商科大学。1927年在上海建成国立暨南大学。暨南大学素有“华侨最高学府”之称。早在建校初期，学校即制定校训：“忠信笃敬”，注重以中华民族优秀的传统道德文化培养造就人才。多年来，学校贯彻“面向海外、面向港澳台”的办学方针，从1978年以来共培养各层次毕业生5万余人，其中培养海外、港澳台地区各层次的学生1万余人，他们来自世界五大洲72个国家和港澳台地区，堪称桃李满天下。

今天，面向新世纪办好国际化、现代化的华侨大学，我们要适应世界教育发展的新趋势，转变教育观念，不仅要重视高等教育的规模扩大，更要强调教育质量的同步变革，着力开展创新教育。

一、新的观念：全面推进创新教育

21世纪是以知识创新和应用为重要特征的新经济时代，创新精神和实践能力已成为能否在国际竞争中赢得主动权的关键因素。教育在培育民族创新精神和培养创造性人才方面，肩负着特殊使命。

在中国建设一流的华侨大学，必须面向世界、面向未来、面向现代化，适应世界教育发展的趋势，树立先进的创新教育思路。为此，我们一定要适应社会、经济、科技发展提出的新的要求，迎接新的机遇和挑战。20世纪80年代以来，不少国家以“国情咨文”“白皮书”“蓝皮书”等形式表达了对21世纪高等教育发展趋势的关注，国际关于科技教育的学术交流十分频繁。一系列教育创新的文件及活动，向我们展示了高等教育的发展新趋势，新的教育思路和教育观念。例如，在人才培养上更加注重能力、素质的培养，特别是创新能力的培养；在课程设置上注重人文及广博知识的教育，推崇“通识教育”；在教学过程中，倡导科学精神与人文精神的融通；在教学方法上，注重灵活性和启发性，倡导“和谐教育”；在学习的时间与空间跨度上，延伸教育的功能，发展远程教育，倡导终身教育等。为了适应世界教育的新趋势，形成先进的教育理念，暨南大学从1998年至1999年，在全校开展了“面向新世纪的暨南创新教育”大讨论，力求从传统的教育模式束缚下解脱出来，认识新世纪大学的功能、使命和作用，积极培养适应世界经济、社会、科学技术发展要求的创造性人才。

我们要在全校形成一种教育的新观念：一是把人才培养作为系统工程，真正把提高人才培养质量当作学校生存和发展的价值目标。二是树立尊重个性、鼓励个性发展

① 本文是在海峡两岸面向21世纪科技教育创新研讨会上的发言，武汉，2000年10月22日；海峡两岸面向21世纪科技教育创新研讨会论文集（华中科技大学和台湾大学主编），2000，15-18.

的观念，把实施素质教育、培养拔尖人才作为衡量学校教育质量、教学水平的重要指标。三是树立大教育观，将人才培养融入整个社会背景之中，课内和课外结合、校内和校外结合、当前和未来结合，为人才成长提供广阔的舞台。

我们借鉴世界著名大学的经验，根据高等教育发展规律和暨南大学的实际情况，提出了“综合性、教学科研型、国际性”的办学模式。

综合性是大学的基础。世界一流大学多是综合性大学，现在中国重点大学亦逐步走向综合。就世界趋势而言，综合化已成为当今世界科学技术发展的主要特征。大学学科门类的设置与科学技术文化的发展趋势应该是一致的。从教育规律来讲，学生在各种学科互相渗透、各种文化互相耦合的环境下才能较好地发展全面素质，深厚的文化底蕴和强大的学科综合是建设一流华侨大学必不可少的条件。暨南大学一直是学科门类比较齐全的综合性大学，且是全国第一所有医学院的大学。近年，暨南大学按照综合性的要求进行学科调整和建设，提出要“发展文经管已有优势，突出生命学科和医学科，加强理科特别是工科的建设”的方针，进一步优化学校的综合性学科设置。

教学科研型是主导。为实现培养海内外杰出人才的目标，我们把教学科研并重放在突出地位，强调科研与教学相结合，在不断做出科技创新成果的同时，培养出众多有创造性的学生。我们认为只有从事创造性科学和技术探索工作的教师，才能教出好的学生。我们明确提出：不从事科学研究的教师是“残疾”的教师。我们还从管理制度及分配制度上引导教师积极从事科学研究，实施了优劳优酬的校内工资分配制度。我们还在提高研究生特别是博士生培养质量的同时，进一步增大研究生和本科生的比例，在校博士生、硕士生与本科生之比由 1995 年的 1/6 增加到目前的 2/9，研究生总数增加了 3 倍多。同时，我们还积极探索本科生早期参与科学研究的途径，推行大学生科研活动计划。

国际性办学是学校的特色所在。一方面是向海外开放，向港澳台开放，吸引更多海外学子来校学习，成为东西方文化和科学技术交流的桥梁；另一方面是向内地开放，把学校的发展与国家和地区的经济、社会发展紧密结合，使学校不仅成为培养人才的基地，而且成为新思想、新技术的源泉和信息汇集的交流中心。近几年，学校坚持办学的国际性，来自五大洲 30 个国家和地区的海外及港澳台学生已达 4000 人，占学生总数的 4 成，且频繁开展国际学术交流与合作，与 39 所世界知名大学建立姐妹学校关系，互派访问学者及相互交换学生。我们还在深圳、珠海两个对外开放的窗口设立学院，更便捷地为港澳台地区培养各类人才，使学校的功能在时间和空间上都大为拓展。

二、新的突破：教学科研互动发展

暨南大学有悠久的历史，其建设和发展得到国家的高度重视和亲切关怀。1996 年 90 周年校庆，江泽民同志为我校亲笔题词：“爱国爱校，团结奋进。”目前，学校全日制在校学生已达 11 000 人，其中研究生超过 2000 人。学校重视师资队伍建设，注意培养学术骨干和学科带头人，师资结构质量已名列全国师资最优的 25 所高校行列，排名第 13 位。我们完全有基础、有条件、有能力、有信心，将学校办成具有先进水平和侨校特色的一流大学。

构筑面向21世纪一流的学科专业体系、建设一流的学科专业体系是一流大学的基础和龙头。目前，我校拥有“211工程”建设的7个重点项目，在文、史、法、经、管、教育、理、工、医等9个学科门类中覆盖了1个博士后流动站，12个博士学科专业，59个硕士学科专业。这些学科专业具有鲜明的前沿学科和华侨教育特色，但与一流大学所要求的综合协调发展、整体水平高的标准还有一定差距。我们将根据21世纪科学技术发展的趋势和国家现代化建设对人才的需要，在现有的基础上精心设计、科学规划，加强基础学科，培植新兴学科，扶持重点学科，大力推进学科的综合发展和整体水平的不断提高，逐步构建和形成特色鲜明、层次清晰、融会贯通、发展空间大的一流大学的学科专业体系。

实施“高层次创造性人才工程”，努力将学校办成培养海内外高层次、高水平、高素质人才的基地。我国现代化建设和华侨在国际上的地位提高需要大量各类优秀人才，尤其是高层次的创造人才。我校的办学思路定位于：积极拓展研究生教育，稳步发展本科生教育。我们要通过认真实施“高层次创造性人才工程”，进一步解放办学思想，挖掘办学潜力，拓宽办学渠道，提高办学质量。要在坚持办好本科基础教育的同时，大力发展研究生教育，特别是要积极发展博士生教育，办好博士后科研流动站，为国家和海外华侨、华人社会培养出大批能够适应将来社会发展的高级人才和高科技前沿的科技帅才。

积极参与国家科技创新体系的建设，努力将学校办成高科技研究和社会科学研究基地。我校确立了“重视基础，突出应用”科研指导原则，科学研究面向世界科技前沿，面向国家与地区的国民经济主战场，紧密结合“211工程”建设和人才培养等工作，无论在广度和深度上都得到了长足的发展。我校认真贯彻“科教兴国”的战略，坚持以教学科研为中心，建立健全科技进步机制，形成教育科研相互促进的良性循环。我校要进一步立足前沿，面向海外，面向港澳台，突出重点，在高科技和社会综合发展的交叉学科领域开展攻坚，并将研究成果尽快转化为生产力，为国家的科技进步、知识创新做出应有的贡献。

优化教师队伍，培育一流教师。争创一流大学必须具有一流的教师队伍。我校制定和实施吸引和稳定高级人才的优惠政策，采取送学深造、在岗锻炼、重点培养等措施，造就出一批教育家和学术学科带头人。江泽民同志在教育工作会议上要求，“教育者必先受教育，不但要学专业知识、科学文化知识，还要学政治知识、实践知识，以不断丰富和提高教师的教书育人的水平”。21世纪所需的教师素质，既包括高水平的思想政治素质、敬业爱岗的职业道德，也要求教师通过不断地学习和接受培育，充实自己和更新知识，提高水平和能力，成为教书育人的专家和从事教育教学研究的复合型人才。在教学过程中，教师既要发挥主导性作用，又要大力弘扬“教学相长”的优良传统，建立相互学习、相互切磋、相互启发、相互激励的和谐的师生关系，这是全面实施素质教育至关重要的环节。

三、新的姿态：开拓进取培育人才

一所大学能否在社会中产生重大影响，关键看它培养出的人才能否具有很强的创新精神和创新能力，在知识创新、科技进步和社会发展中有所作为，成为开创国家未来的杰出人才。正如江泽民同志所指出的“综观国内外大学的历史情况，各国的著名大学为数不多，其所以著名，固然有各方面的条件。如经费、师资、实验室、校舍等，但归根到底，要靠其培养出来的学生在社会上建功立业，日久天长，形成社会公认的信誉”。我们争创一流大学，重中之重的工作，就是培养和造就出大批高素质的创造性人才。

强化素质教育。素质教育是培养创造性人才的基础。素质教育不是着眼某一方面或某几方面的素质，而是全面发展的综合素质，是思想道德素质、文化素质、业务素质、身体和心理素质的整体提高。综合素质的培养要面向全体学生，而不是部分有兴趣、有愿望的学生或少数尖子学生；同时，我校要造就浓厚的氛围和良好的环境，以促进每一个学生的综合素质的提高。综合素质的培养要注重个性的发展，要鼓励学生充分发展自己的个性，结合自身的特点来发展自己的素质强项，有所侧重地进行综合素质的培养和提高。我校要因材施教，积极培育学生兴趣，调动学习积极性；我校要不断加大课程改革力度，尽可能多传授新知识、新技术、新信息，让学生不断接受高新技术前沿的熏陶；我校要进一步改革教学方法和手段，大胆探索启发式教学、讨论式教学和直观形象教学的有效方式。课堂讲授要突出重点，少而精，给学生留下充足的思维空间，引导学生独立思考；我校要加强人文科学和美育等方面的教育，激发学生的灵感，培养学生的洞察力、想象力，以开阔视野，活跃思维，触类旁通。我校要积极在大学生中开展学术活动，如学术讲座、科技影展、出版学生学术刊物等，引导学生热爱科学。设立大学生创新活动专项基金、鼓励科研实践。继续开展高品位、高层次的人文社会科学系列讲座、高雅艺术活动、读书活动等，营造浓郁的校园文化氛围。坚持开展社会实践活动，加强科技下乡、技术推广活动，为大学生提供创新的舞台。

突出办学特色。“侨”字是我校的特色，为侨服务是我校的主要任务，学校各方面的工作都应立足“侨”字。一要针对学生特点，深化教学改革。要特别重视教学手段和教学方法的改革，压缩学时要通过利用网络技术和多媒体技术来解决。我校要始终把抓好“三语”（中文、英文、计算机语言）教学作为提高教学质量的关键。要根据华侨、港澳台学生和华人学生的不同情况，进一步完善分层教学。二要多层次发展华文教育，使我校成为国家开展华文教育的重要基地。三要积极扩大对外招生，坚持稳定香港和澳门招生，拓展海外招生的工作思路，逐年增加外招生数量。四要发挥侨校优势，服务侨务工作。学校“211 工程”获准立项的 7 个项目中，已有 3 个项目直接服务于侨务工作，学校的教学科研要尽力为侨务工作服务，更多地面向海外及港澳台办学，更多地将科研成果运用于侨务工作。

加快“211 工程”建设。我校要坚定不移地将“211 工程”建设作为全校的中心工作，并将重点学科作为龙头带动其他方面的建设。在人才培养方面，更加注重培养学生适应时代要求的整体素质和综合能力，建立新型的人才培养机制；在学科建设方面，优先发展领先学科，重点建设优势学科，积极组建学科群体，进一步强化文、史、经、管、法、理、工、医相互渗透，探索在提高综合实力的基础上建立边缘学科和新兴学科，重组我校学科的体系；在科学研究方面，强化研究工作在学校建设与发展中的地位，充分发挥科学研究在提高教学质量、推动学科建设、增强办学实力中的主导作用；在队伍建设方面重点培养和造就一支高素质、高水平，以中青年教师为骨干的师资队伍，同时，建立一支精干、优良、高效的管理干部队伍，使两支队伍成为办好暨南大学的中坚力量；在管理体制改革方面，积极探索与国外高校管理模式接轨，充分发挥董事会的作用，进一步完善校长负责制，健全校、院、系三级管理体制，推进干部人事制度、校内分配制度及后勤社会化的改革，充分体现华侨大学的管理特色。

提高学校的办学品位。学校的校风、学风对于学生的整体素质提高，特别是文化素质的提高非常重要。学校的文化氛围、人文环境会对学生产生极大的潜移默化的影响。我校提出从严治校、从严治学、从严治教就是为了优化学校的文化氛围，提高学校的办学品位。一个学校的办学品位是长期积累、逐步优化的结果。我们既要注意发挥课堂主渠道对加强大学生文化素质教育的作用，也要注意发挥校园文化的教育功能，把校园文化氛围和校园环境建设作为文化素质教育重要而潜在的课堂，将文化素质教育渗透到学生丰富多彩的校园文化生活中。

再经过 6 年时间，学校将迎来 100 周年校庆。学校力争用 6 年左右的时间，在教育观念、教育模式、教学内容、课程体系、管理体制、教育方法和技术等方面在改革上有新的突破，初步形成暨南大学培养高素质创新人才的新体制、新机制。

24. 坚持社会主义办学方向　办好华侨高等教育为海内外培养高素质人才[①]

百年大计，教育为本，教育是一个民族兴旺发达的根本大计。作为教育龙头的中国高等教育担负着为社会主义现代化建设培养高素质专门人才的重任，在迎接21世纪的挑战、实施科教兴国战略和国家创新体系工程中，发挥着日益重要的作用。21世纪综合国力的竞争，归根到底是人才与科技的竞争。在高校深入开展“三讲”教育，是新形势下加强高校领导班子建设，提高高校领导干部素质的一项重大举措。高校“三讲”教育解决的问题可以说很多，但核心是办学方向和培养什么人的问题。通过“三讲”教育，可以帮助我们全面正确地理解和贯彻党的教育方针，坚持社会主义办学方向，深化教育改革，重视素质教育，培养德智体美等方面全面发展的社会主义建设者和接班人。目前，我校的“三讲”教育，在国务院侨办党组、广东省委的正确领导下，在巡视组的指导、把关下，正在深入展开。总的看来进展顺利，发展健康，成效比较显著。我们在“三讲”教育过程中，要以江泽民同志“三个代表”重要思想为指导，集中精力，抓住重点，通过“三讲”教育着力解决好学校的办学方向和培养什么人这一根本性问题。

下面，我想结合10多年来从事高校管理工作的实践，以及在“三讲”教育期间的学习心得，围绕如何坚持社会主义办学方向，办好暨南大学这一问题，谈几点学习体会。我的讲话主要分两大部分。

一、历史回顾，特别是新中国成立51年来，始终以坚持社会主义办学方向作为教育方针的重要内容

办学方向是由教育方针决定着的，有什么样的教育方针就有什么样的办学方向。而教育方针是国家根据政治、经济的要求，为实现教育目的所规定的教育工作总方向，它是教育政策的总概括。教育方针的内容，一般包括教育性质与任务、教育目的，以及实现教育目的的基本途径。教育目的是培养人的总目标，它所指明的是在一定社会中，要把受教育者培养成为什么样的人的根本问题。

我国在不同的历史时期，根据社会发展的要求、政治形势任务、指导思想，制定了不同的教育方针。20世纪上半叶，我国尚处于旧民主主义革命和新民主主义革命时期。1906年，清政府学部所颁布的教育宗旨（方针）是忠君、尊孔、尚公、尚武、尚实。它既保留了封建的忠君和尊孔的教育思想，又从德国、日本搬来了公民教育、军国民教育和科学技术教育思想。30年代，中央苏区规定的苏维埃文化教育总方针为：“在于以共产主义的精神来教育广大的劳苦民众，在于使文化教育为革命战争与阶级斗争服务，在于使教育与劳动联系起来，在于使广大中国民众都成为享受文明幸福的人。”新中国成立后，进入了社会主义建设时期，教育方针不断随着形

① 本文是在暨南大学干部会上的讲话，广州，2000年12月20日.

势的变化被赋予新的内涵。为了说明新中国成立 51 年来，始终坚持社会主义办学方向这一问题，我想在这里对党的教育方针的历史演变作一个简要的回顾。在新中国成立后 51 年的发展进程中，教育方针或具有方针意义的指示、规定主要有以下 15 种表述形式。

(1) 1949 年 9 月 21 日至 30 日，中国人民政治协商会议第一届全体会议在北京举行。会议一致通过了《中国人民政治协商会议共同纲领》。《共同纲领》第五章规定："中华人民共和国的文化教育为新民主主义的，即民族的、科学的、大众的文化教育。人民政府的文化教育工作，应以提高人民文化水平、培养国家建设人才、肃清封建的、买办的、法西斯主义的思想、发展为人民服务的思想为主要任务。"并规定"提倡爱祖国、爱人民、爱劳动、爱科学、爱护公共财物为中华人民共和国全体国民的公德"。还规定"中华人民共和国的教育方法为理论与实际一致。人民政府应有计划有步骤地改革旧的教育制度、教育内容和教学法"。《共同纲领》所规定的新中国的教育方针，在 1949 年 12 月下旬，教育部在北京召开的第一次全国教育工作会议上得以贯彻和确认，会议将上述方针确定为教育工作的总方针。

(2) 1951 年 3 月下旬，教育部在北京召开了第一次全国中等教育会议。会议就普通中学的宗旨和教育目标作出明确规定：使青年一代在智育、德育、体育、美育各方面获得全面发展，使之成为新民主主义社会自觉的积极的成员。这次会议的意义，在于首次提出了儿童、少年、青年要在德、智、体、美诸方面全面发展的目标。

(3) 1953 年 6 月 30 日毛泽东在接见青年团第二次全国代表大会主席团时，提出要使青年"身体好、学习好、工作好"。毛泽东"三好"的指示，后来在 1955 年 8 月中华全国学生联合会第十六次代表大会上形成决议。大会号召："贯彻毛主席'身体好、学习好、工作好'的指示，把自己培养成为具有高度的社会主义觉悟，能够掌握现代科学知识，身体健康的全面发展的社会主义建设者。"毛泽东"三好"的号召，不是以教育方针的形式提出，但实际上成为教育工作培养目标和广大青少年学习成长的行为指南。

(4) 1957 年 2 月 27 日，毛泽东在最高国务会议第 11 次会议上的讲话中，提出"我们的教育方针，应该使受教育者在德育、智育、体育几方面都得到发展，成为有社会主义觉悟的有文化的劳动者。要提倡勤俭建国。要使全体青年们懂得，我们的国家现在还是一个很穷的国家，并且不可能在短时间内根本改变这种状态，全靠青年和全体人民在几十年时间内，团结奋斗，用自己的双手创造出一个富强的国家"。1958 年 9 月 19 日中共中央、国务院发布《关于教育工作的指示》。《指示》提出："党的教育工作方针，是教育为无产阶级政治服务，教育与生产劳动相结合；为了实现这个方针，教育工作必须由党来领导。"这一方针，在 1961 年中共中央批准试行的《教育部直属高等学校暂行工作条例（草案）》中，进一步得到贯彻执行。《条例》规定："高等学校的基本任务，是贯彻执行教育为无产阶级的政治服务、教育与生产劳动相结合的方针，培养为社会主义建设所需要的各种专业人才。"1958 年中央指出的方针，成为全国各级各类学校和全国人民公认的教育方针，沿用了 20 多年。

(5) 1966 年 5 月 7 日，毛泽东在审阅了人民解放军总后勤部《关于进一步搞好部队农副业生产的报告》后，写信给林彪。信中在讲到人民解放军应该是一个大学校后，提出“学生也是这样，以学为主，兼学别样，即不但学文，也要学工、学农、学军，也要批判资产阶级。学制要缩短，教育要革命，资产阶级知识分子统治我们学校的现象再也不能继续下去了”。后来这封信简称“五七指示”，尽管不归教育方针之列，但在长达10多年的时间内，起着指导学校发展与改革指标的作用。“文化大革命”中，教育遇到挫折，均与此指示有关。

（6）1978年3月5日，五届人大第一次会议通过的《中华人民共和国宪法》第十三条规定：“国家大力发展教育事业，提高全国人民的文化科学水平。教育必须为无产阶级政治服务，同生产劳动相结合，使受教育者在德育、智育、体育几方面都得到发展，成为有社会主义觉悟的有文化的劳动者。”

（7）1981年6月27日，在中共中央十一届六中全会通过的《中国共产党中央委员会关于建国以来党的若干历史问题的决议》中，提出“坚持德智体全面发展、又红又专、知识分子与工人农民相结合、脑力劳动与体力劳动相结合的教育方针”。这是在“真理标准”讨论与党的十一届三中全会后，在纠正了“文化大革命”中及以前的“左”倾错误，确立党的马克思主义思想路线、政治路线和组织路线后，新提出的教育方针。

（8）1981年11月30日，在五届人大第四次会议上，《政府工作报告》提出了今后经济建设十条方针，其中第九条中明确指出：“我们教育的基本方针是明确的，这就是使受教育者在德育、智育、体育几方面都得到发展，成为有社会主义觉悟的有文化的劳动者和又红又专的人才，坚持脑力劳动与体力劳动相结合，知识分子与工人农民相结合。现在的任务是要根据现代化建设中的实际情况来进一步贯彻执行这个方针。”

（9）1982年12月4日，五届人大第五次会议通过并公布施行的《中华人民共和国宪法》中，有关教育的条款有：第二十四条　国家通过普及理想教育、道德教育、文化教育、纪律和法制教育，通过在城乡不同范围的群众中制定和执行各种守则、公约，加强社会主义精神文明的建设。国家提倡爱祖国、爱人民、爱劳动、爱科学、爱社会主义的公德，在人民中进行爱国主义、集体主义和国际主义、共产主义的教育，进行辩证唯物主义和历史唯物主义的教育，反对资本主义的、封建主义的和其他的腐朽思想。第四十六条　中华人民共和国公民有受教育的权利和义务。国家培养青年、少年、儿童在品德、智力、体质等方面全面发展。这些条文并没有明确规定教育方针，但可以说具有方针的意义。

（10）1980年5月26日，邓小平为《中国少年报》和《辅导员》杂志题词：“希望全国的小朋友，立志做有理想、有道德、有知识、有体力的人，立志为人民作贡献，为祖国作贡献，为人类作贡献。”后来，1983年4月29日，邓小平在会见印度共产党员（马克思主义）中央代表团时的谈话中说道：“过去很长一段时间，我们忽视了发展生产力，所以现在我们要特别注意建设物质文明。与此同时，还要建设社会主义的精神文明，最根本的是要使广大人民有共产主义的理想，有道德，有文化，守纪律。”这就是全国人民共知的“四有”人才标准，被教育界公认的“四有”教育目标。

（11）1983年10月邓小平为北京景山学校题词：“教育要面向现代化，面向世界，面向未来。”自题词后，在有关教育的文件中，在国家领导人有关教育的讲话中，都把“三个面向”当作教育的方针，教育事业的发展也沿着“三个面向”前进。

(12) 1985 年 5 月 27 日，中共中央在《关于教育体制改革的决定》中提出：“教育必须为社会主义建设服务，社会主义建设必须依靠教育。”《决定》认为这是教育体制改革的根本指导思想，这一提法是党和国家在总结社会主义建设的历史经验的基础上，对教育本质与社会功能认识的飞跃，具有划时代的意义。

(13) 1986 年 4 月 12 日，第六届全国人民代表大会第四次会议通过并颁布施行的《中华人民共和国义务教育法》中规定：“第三条　义务教育必须贯彻国家的教育方针，努力提高教育质量，使儿童、少年在品德、智力、体质等方面全面发展，为提高全民族的素质，培养有理想、有道德、有文化、有纪律的社会主义建设人才奠定基础。”

(14) 1993 年 2 月 13 日，中共中央、国务院正式印发的《中国教育改革和发展纲要》中，总结了 40 多年教育曲折发展的历程所得出的宝贵经验，提出了建设具有中国特色社会主义教育体系的八项基本原则，其中第一、第二、第三项是直接关于教育方针的规定：“第一，教育是社会主义现代化建设的基础，必须坚持把教育摆在优先发展的战略地位。第二，必须坚持党对教育工作的领导，坚持教育的社会主义方向，培养德智体全面发展的建设者和接班人。第三，必须坚持教育为社会主义现代化建设服务，与生产劳动相结合，自觉地服从和服务于经济建设这个中心，促进社会的全面进步。”它说明了教育在社会主义现代化建设中的基础地位，说明教育必须为社会主义现代化建设服务，说明了社会主义教育的培养目标和办学方向，以教育总则的形式规范了教育的发展。

(15) 1995 年 3 月 18 日八届人大第三次会议通过了《中华人民共和国教育法》。在《教育法》总则中明确规定了教育方针：“第五条　教育必须为社会主义现代化建设服务，必须与生产劳动相结合，培养德、智、体等方面全面发展的社会主义事业的建设者和接班人。”1998 年 8 月 29 日九届人大四次会议通过了《中华人民共和国高等教育法》，再次重申了这一教育方针。在原国家教委政策法规司所编《〈中华人民共和国教育法〉释义》一书中，明确说明：“本条是关于国家教育方针的规定。”应该说，这是以教育基本法的形式以最准确文字对国家教育方针最完整的表述。这是全国人民经过长期求索得出的科学成果。它将指导中国社会主义教育事业的健康发展。

以上我们简要回顾了新中国教育方针的演变进程，发现不同的表述是与时代紧密相连的。当今教育方针最明显的变化，就是从教育为无产阶级政治服务到教育为社会主义现代化建设服务的转变。这个转变是党的伟大历史转折的反映，是 20 世纪第三次思想解放运动在教育上的最大成果。为社会主义现代化建设服务，培养社会主义事业的建设者和接班人，是教育方针的最新的科学概括，它表明了新中国的教育应该坚持社会主义办学方向，服务于社会主义建设，以及为社会主义事业培养“四有”新人的问题。

二、坚持正确的办学方向，办好华侨高等教育

确立什么样的办学思想和办学方向，是关系到培养什么样的接班人的大问题。对

此中共中央政治局委员、广东省委书记李长春同志多次指出，高校“三讲”教育要解决培养什么人和怎样培养人的问题。高校的办学方向，当前很重要的是应该把握好以下几个方面：第一，是否对马克思主义理论教育，对新形势下做好思想政治工作有足够的重视？第二，是否全面理解和贯彻党的教育方针，全面推进素质教育，培养真正合格的社会主义接班人？第三，是否明确学校定位，深化改革，加快发展，切实提高办学水平、办学质量，在“科教兴国”“科教兴粤”战略中发挥应有的作用？第四，是否树立起阵地意识，自觉抵制各种错误思想和腐朽文化的侵蚀？是否理直气壮、旗帜鲜明、毫不含糊地对违反经济建设为中心、违反改革开放政策、违反四项基本原则的错误思想、政治观点，进行批评和澄清？

暨南大学是国家最早建立的一所华侨学校，经过 94 年来几代暨南人的辛勤耕耘，迄今已发展成为一所独具特色、声誉远播的“华侨最高学府”。新中国成立后，党和国家对华侨高等教育事业极为重视。1958 年国务院批准在广州重建暨南大学，由中共中央中南局书记陶铸兼任校长。1963 年学校重建董事会，由中央侨委主任廖承志担任董事长。当年，学校还被定为国务院高教部的直属院校。改革开放以来，中央专门下发文件，对暨南大学的特殊性和办好暨大的战略意义十分重视。1983 年 6 月 20 日中共中央、国务院批复了中央宣传部、教育部、国务院侨办《关于进一步办好暨南大学和华侨大学的意见》(〔1983〕24 号)。文件明确指出：进一步办好暨南大学“对于做好台湾回归祖国，收回港澳主权，完成祖国统一大业”，“对于开展拥护祖国统一的爱国的广泛统一战线，都具有重大意义”，并“将暨南大学和华侨大学列为国家重点扶植的大学”。为贯彻、落实中央指示，国务院办公厅于 1984 年 10 月 4 日转发了国务院侨务办公室《关于办好暨南大学、华侨大学的报告》(国办发〔1984〕91 号)，决定对我校实行相应的特殊政策和灵活措施，赋予更多的办学自主权，以利于办出特色，办出水平。下面，我想就暨南大学在跨世纪、越千年的新形势下，如何坚持社会主义办学方向，办好华侨高等教育谈几点看法。

1. 迎接新世纪侨校教育的机遇与挑战

放眼新世纪国际形势，全球经济一体化的趋势正无可阻挡，扑面而来，中国也即将加入世界贸易组织。中国正融入世界，世界也正走向中国。国与国之间的壁垒日益削弱，相互交流日益频繁，对于我们侨校拓展华侨教育事业，是一个难得的机遇。实践表明，越是坚持改革开放，扩大世界交往，走国际化、现代化的道路，就越能吸引更多的海外华侨华人及港澳台学生前来学校就读。以近 5 年境外青年报读我校情况为例（仅本科及研究生）：1996 年报考人数 1350 人；1997 年 1580 人；1998 年 1830 人；1999 年 2020 人；今年 2150 人。报考人数的逐年递增，也预示着新世纪我校侨教事业的光明前景。

但机遇与挑战并存。在看到光明前景的同时，我们也看到新世纪的困难。首先，随着中国教育改革的深入，我们“侨校”原有的某些特殊政策受到了其他高校的挑战，在外招生方面，侨校将不再是“特保儿”。目前全国已有 140 多所高校，特别是数十家著名高校直接对外招生，与我们“侨校”形成了强有力的竞争。其次，海峡对岸的台湾也出于政治目的，大开绿灯，倾力与我们争夺生源，甚至于数年前成立了与我校同名的“暨南国际大学”。再次，随着学校生源的日益广泛，学生成分更加复杂。学生多元文化背景的不同学习要求和新时代对高校教育质量与水平的更高要求，将使我们侨校教育难度愈来愈大，无论在教育观念、办学模式、管理体制、招生体制、专业设置、运作机制和教育内容、教学方式等方面，都将对我们提出新的挑战。

因此，新世纪的新形势，已把我们“逼”上了办国际化、现代化的名牌华侨高等学府的快车道。我们只能迎接这一挑战，才能完成党和国家交给我们暨南大学的特殊办学使命，再创21世纪暨南新辉煌。

2. 建立具有侨校特色的思想政治工作机制

党和国家赋予暨大的“面向海外、面向港澳台”办学任务，本质上，就是培养人才，争取人心的工作。为落实这项任务，学校要认真贯彻党的基本路线和党的教育方针，站在传播中华文化、促进祖国统一、扩大中外交流、推进我国现代化建设的高度，努力按照侨校的特殊要求办学，并认真贯彻落实今年3月学校党委制定的《中共暨南大学委员会关于加强和改进思想政治工作的若干实施意见》，将境内学生培养成为德、智、体全面发展的社会主义事业的建设者和接班人；将港澳台侨学生培养成为热爱祖国，能为当地社会进步和稳定繁荣作出贡献的人才；将外国留学生培养成为对我国友好，能为当地各项事业作出良好服务的人才。

为保证办学任务的顺利完成和各类学生培养目标的全面实现，我校要积极抓好党的建设，认真做好师生员工的思想政治工作，使学校沿着正确的方向前进。

1）练好内功，发挥党的政治核心作用

我校是一所试行校长负责制的社会主义华侨大学，党委要始终把坚持社会主义办学方向和党的政治领导放在学校一切工作的首位。党委要加强党的思想、组织和作风建设，充分发挥导向、保证和监督作用，发挥基层党组织的战斗堡垒作用和广大党员在各项工作中的先锋模范作用。全校党员要通过学习邓小平理论、江泽民“三个代表”的论述和党的方针、政策，提高对改革开放等重大问题的认识。特别是通过反腐倡廉教育，有效地促进学校的党风廉政建设。建立和健全两周一次党组织生活制度、民主评议党员制度、评选表彰党内先进的制度、支部建设目标评估制度以及培养、发展党员等规章制度。各级党的组织和广大党员要在完成学校各项重大任务中充分发挥先锋作用。

2）抓好师生员工的思想政治工作，形成具有侨校特色的思想政治工作体系

我校师生员工思想政治教育工作要采取的主要措施有以下四点：一是对教职工和境内学生坚持开展党的基本路线和建设有中国特色的社会主义理论教育。学校要采用集中学习讨论、组织观看录像、举行知识竞赛、进行专题辅导等多种多样的形式开展党的基本路线和建设有中国特色的社会主义理论教育。二是将爱国主义作为对教职工、境内学生进行思想政治教育的主要内容。大批归国华侨、港澳台同胞以及他们的眷属汇集暨南，共同为华侨高等教育事业作出贡献，爱国主义是激励他们做好工作、搞好学习的内在动力；数千名港澳台侨学生能和境内学生同在暨南园求学，是因为有着共同的目标——学习掌握中华民族的传统文化，学习掌握现代化知识。因此，学校要始终将爱国主义教育作为思想政治教育的重要内容。通过多种方式的教育，使广大师生员工进一步了解改革开放与现代化建设给国家带来的巨大变化，增强爱国情感和民族

自豪感，激发努力工作和勤奋学习的热情。三是将“忠信笃敬”校训作为对全体师生员工进行思想品德教育、规范言行的重要内容。“忠信笃敬”校训出自《论语·卫灵公》，校训一直是暨南人恪守的信念和遵守的道德规范。在新的历史条件下，学校要赋予它新的内涵，用以教育和团结师生员工，发扬暨南人的优良传统，为学校的腾飞作出新的贡献。四是分类别、分层次、分年级对学生开展思想政治教育。根据我校的特殊情况，学校要对学生的思想政治教育继续采取分类、分层次和分年级进行的办法，创造性地建立侨校学生思想政治教育体系。马克思主义理论和思想品德课（简称“两课”）是对学生进行思想政治教育的两门主课，我校要针对境内生、港澳台侨生和外国留学生三类学生的不同培养目标，按照博士研究生、硕士研究生和本科生等不同层次的不同要求，分年级安排有关教育内容。对境内学生，按照原国家教委规定的课程、教材和学时授课，开展教学活动，系统进行马克思主义理论和邓小平理论以及爱国主义、集体主义等方面的教育。对港澳台侨学生，按照中央24号文精神，不要求他们上系统的马列政治理论课，而是根据其特点，确定思想教育的要求：一是热爱祖国，热爱母校，关心、支持对港澳恢复行使主权和祖国统一大业，拥护“一国两制”，执行《香港基本法》和《澳门基本法》；二是了解祖国的历史和现状，关心和支持改革开放事业；三是增强社会责任感，能为居住地区的稳定、繁荣作出贡献；四是具有良好的公民意识、职业道德。对外国留学生（主要是华裔学生），按照有关政策规定，主要是使他们了解当代中国，增进对我国的友好感情，提高个人的品德修养。

3. 用正确的理想信念构筑国内学生的精神支柱

国内学生占全校学生人数的2/3，要在暨大牢牢坚持正确的办学方向，就要对他们进行理想信念的教育。而理想信念是人们对未来的向往和追求，是人们的政治立场和世界观在奋斗目标上的集中体现。崇高的理想信念是一种强大的精神力量，它在激发人们的主动性和创造性、鼓舞斗志、振奋精神等方面能产生巨大的能动性。社会主义和共产主义的理想信念是中国共产党人的精神支柱、力量源泉。然而，在改革开放不断深入，发展社会主义市场经济的条件下，特别是国际风云变幻，共产主义运动处于低潮的形势下，一些人对社会主义的理想信念产生了动摇，对社会主义的命运表示忧虑和担心。存在着“培养建设者的意识强，培养接班人的意识弱；办学意识强，阵地意识弱；教书意识强，育人意识弱”和忽视意识形态问题的现象。面对各种严重的挑战，如何坚定我们的理想信念，树立正确的世界观、人生观、价值观成为一个严峻的问题。我校担负着培养德、智、体、美全面发展的社会主义建设者和接班人的战略任务，在学校坚持正确的理想信念，以崇高的理想信念构筑广大青年学生的精神支柱，使他们肩负起承前启后、继往开来的历史使命，对于回应各种挑战，全面推进有中国特色的社会主义伟大事业具有深远的意义。

坚持正确的理想信念，就要把握社会主义的办学方向，增强阵地意识。学校承担着培养人才、知识创新与传播、社会服务三大重任。在经济全球化、政治多极化的国际局势下，学校又是国内外敌对势力同我们争夺青年、争夺群众的前沿阵地，是意识形态领域马克思主义和各种非马克思主义激烈交锋的汇聚地。因此，在学校坚持正确信念，用社会主义和共产主义崇高理想信念牢牢地占领学校这一重要阵地，是学校最大的政治和最高的党性。正如邓小平同志指出的："毫无疑问，学校应该永远把坚定正确的政治方向放在第一位。"所以，在我校的体制改革与发展过程中，必须坚持正确的办学方向，理想信念是管方向、管根本的。这就必须用社会主义的崇高理想信念为学校的改革和发展提供政治保证和价值导向，努力把学校办成先进科学技术和知识创新的重要阵地，传播马克思主义和社会主义精神文明建设的重要阵地，培养成为社会主义现代化建设者和接班人的重要阵地，对国家经济建设和社会发展具有相当影响力的思想库、智慧库。

坚持正确的理想信仰，就要全面贯彻党的教育方针，以理想信念教育为核心，全面抓好学校的素质教育，培养社会主义现代化建设的合格人才。全面推进素质教育，是学校培养社会主义现代化建设需要的各级各类高层次人才的内在需求。这里讲的"素质"，包括思想品德素质和科学文化素质两个主要的方面，其中"思想政治素质是最重要的素质"。因为，教育的根本是对人的心灵、智慧的开发，对人的性情的陶冶，独立自由精神的养成。对此，我们要全面深刻领会党的教育方针，把教育的使命感落实到素质教育的探索中，落实于教育观念的更新中，转变育人观念，树立以理想信念价值为核心的综合素质培养的育人思想，重德能、促智能，着力于学生身心和谐发展的教育质量观。在意识形态和价值观冲突日趋激烈的复杂背景下，更加注重广大青年的理想信念修炼，在帮助广大青年学生树立崇高理想和升华思想境界上下功夫，努力构筑当代大学生的精神支柱，树立正确的世界观、人生观、价值观，全面推进素质教育，促进学生全面发展、持续发展；努力培养和造就一批政治坚定、思想敏锐、知识渊博，有创新精神和实践能力，又善于与人合作的全面发展的社会主义现代化建设者和接班人，使广大青年学生在建设有中国特色的伟大事业中发挥主力军作用。

坚持正确的理想信念，关键在于加强学校党的建设。发展社会主义华侨高等教育事业，关键在人，关键在党。要保证我国的华侨高等教育事业能坚定社会主义理想信念，沿着社会主义方向健康地发展，关键在于加强学校党的建设，尤其是学校领导班子和领导干部队伍建设。首先，要从政治大局的战略高度加强学校党的思想建设。通过生动活泼、卓有成效的理论教育活动加强广大党员、干部的理论学习，提高理论修养和政治素质，提倡勤于学习、勤于思考，自觉刻苦学习马列主义、毛泽东思想，尤其是学习邓小平理论，理论联系实际，善于运用马克思主义基本立场、观点、方法观察和分析形势，分析改革开放重大现实问题，引导广大党员干部从理论思考中坚定马克思主义信仰和追求。其次，以"三个代表"的要求加强学校领导班子和领导干部的政治建设、组织建设和作风建设，树立政治意识、大局意识、责任意识，不断提高政治素质、领导水平、业务能力、管理艺术和服务水平，从而使社会主义的理想信念根植于党的建设的丰富实践中，落脚于学校的改革发展进程中，使广大师生员工从切身的生活实践中更加坚定社会主义理想信念，坚定马克思主义不动摇。

4. 树立"侨校十名校"的办学观念，突出侨校特色，深化教育改革，培养高素质人才

暨大是一所侨校，侨字是我们最大的特色。我们要坚决贯彻实施国家的侨务政策，将侨校特色作为办学的重要发展目标，将侨校特色视为学校生存和发展的基础。我们

深知，发展特色，就是发展优势，这也是所有暨南人的永远追求。我们所说的“侨校”与“名校”，两者不能截然分开，它是一个事物的两个方面，相互补充、互相促进。“侨”是基础，是母体；“名”离不开“侨”。所谓“皮之不存，毛将焉附”。我们的名校意识，是建立在侨校特色之上的，因此，不能简单化地拟之于北大、清华等。“侨”是我们办学的任务和目标，“名”是为了我们更好地完成任务和达到目标。前面提到，新世纪我校将面临艰巨的挑战。除真正意义上的教科实力竞争外，在某些方面，实际上正潜在地进行着一场政治角力，我们的“侨”字，包含着这一政治任务。这里还涉及一个“点”与“面”的问题。一方面，我们要在数量上尽量多地招收海外华侨华人及港澳台学生，努力普及热爱中华民族、中华文化、对我友好的基本面；另一方面，也应树立“精英”意识，努力培养造就一批未来的侨界精英、领袖式人物，从而在世界华人华侨社区发挥重大影响力。造就精英，更需要名校效应。可以预见，一所国际化、现代化、响当当的名牌华侨高等学府，会对党和国家的侨务事业作出多大贡献！因此，创名校正是为了进一步为办好华侨高等教育事业服务。实施“侨校＋名校”的办学思路，即是“坚持社会主义办学方向”在暨大的具体体现。

坚持正确的办学方向，需要不断深化学校的改革。办好暨大，是党和国家赋予学校的一项战略任务，也是海外华侨华人和港澳台同胞的热切希望。通过侨教联系侨情，扩大国际交往，促进我国改革开放事业和现代化建设，具有十分重要的意义。暨大将继续高举“侨”字大旗，打好“侨”牌，走国际化、现代化的办学道路，认真贯彻《中共中央、国务院关于深化教育改革全面推进素质教育的决定》，查找学校在办学方向和培养人方面还存在哪些问题与不足，并通过“三讲”边整边改，切实解决存在的主要问题，以努力开创华侨高等教育工作的新局面。学校拟在下述五方面采取行之有效的措施深化华侨高等教育改革。

1）继续贯彻两个面向的办学方针，努力拓展外招生源，进一步深化教学改革

一是采取得力措施，进一步提高外招生比例。目前，我校港澳台和海外学生人数已占全校学生数的1/3，但距国务院侨办提出的外招生与国内生数要达到1∶1的比例要求，尚有差距。为了早日达到这一目标，学校将继续进行学制改革，加大春季招生力度。学校要抓住中国国际地位大幅度提升，香港、澳门顺利回归，中国即将加入WTO所带动的世界“华文热”这一难得的历史机遇，积极主动地做好外招生宣传工作，采取灵活多样的措施，在保证质量的基础上，吸引更多的港澳台及海外学子来校学习。学校将在确保港澳台地区生源数量的基础上，重点拓展东南亚及世界各国华人华侨生源，目前外交部有关部门已开始通过我国驻外使领馆帮助我校作好宣传，这将有力地促进我校的外招生工作。同时，我校也要加强对内的“三侨子女”的招生工作，努力争取各省区市侨办继续对学校进行更多、更深入的宣传与介绍。在这方面，同时也要大力提高研究生外招生的数量与质量。

二是深化教育改革，把暨大办出特色与水平。暨大要在海内外保持并进一步提高声誉，关键是要创名牌，凭借特色与实力吸引莘莘学子。学校要积极倡导先进的教育思想和观念：首先是把人才培养作为系统工程，真正把提高人才培养质量当作学校生存和发展的生命。其次是树立尊重个性、鼓励个性发展的观念。把实施素质教育、培

养拔尖人才作为衡量学校教育质量、教学水平的重要标志。再次是树立大教育观，将人才培养融入整个社会背景之中，课内和课外结合、校内和校外结合、当前和未来结合，为海内外学生的成才提供广阔的舞台。为此，学校将深化已进行的系列改革措施，如增开午间晚间选修课，进一步完善标准学分制；大力强化“三语”（中文、英文、计算机语言）教学、基础课教学和分流教学；加大对非英语专业的课程采用英语授课。作为一所外向型的大学，必须大力提高英文教学的比例，尽快地把暨大办成双语制教学学校；要加强专业改革与课程内容的更新以及教学手段的现代化；要开展丰富多彩的境内外学生互动交流的课外活动；从严治校，依法治校，提高学校的教育管理水平，从而增强学校的涉外性、应用性，并逐步与国际高等教育接轨。

2）大力发展华文教育，向海外传播中华文化

几千万华侨华人遍布于世界各地，作为中华儿女，他们对中国普遍怀有深厚的感情。随着我国改革开放事业的深入发展，国际性的“华文热”日益升温，华侨华人迫切希望学习中文，接受中华文化的熏陶。其他外籍人士亦渴望通过学习中文，以加深了解、认识中国，加强与中国的经济文化联系，这就为学校开展华文教育提出了新的要求和机遇。学校华文学院经过几年来的建设和发展，已成为我国开展华文教育的重要基地，并成为我国具有对外汉语水平考试三个等级（基础、中级、高级）考试资格齐全的四所高校之一。今年9月又被国务院侨办授予国家华文教育基地称号。学校将充分利用这一特色与优势，进一步加强语言学与应用语言学硕士点、对外汉语本科专业以及各种学习中华文化的短期学习班和夏令营的招生与教学，努力吸引更多的海外华侨华人及其子女来我校学习、进修。另外，还要主动延伸与扩大华文教育，华文学院在为海外编写华文教材方面成效显著，特别是1999年全部完成的12册北美版华文教材，为北美地区140余所中文学校争相使用。学校将继续加强海外华文教材的研究与编写，争取为世界各地的华侨华人及其子女提供系统、科学且适合当地生活特点的全新的中文教材，以推动海外华文教育事业的发展。

3）抓好“211工程”建设，继续提高学校的教学、科研水平

暨大作为历史悠久、学科门类比较齐全的“211工程”大学，应该在国家振兴教育与科技创新的发展战略中有所作为，因此，我校正以“211工程”7个重点学科建设项目为中心，构筑面向21世纪一流的学科专业体系，并将以此为基础和龙头，推动学校教学、科研工作的开展。这也是新形势下侨务工作发展的客观需要。在重点建设的7个学科项目中，就有“汉语言文字学与海外华文教育”“中外关系史与华侨华人”“计算机信息与通信技术”等3个学科，直接为侨务工作服务。如华侨华人学科是体现暨大办学特色的优势学科，已取得了令人瞩目的成绩。最近，华侨华人研究所申报的国家人文社会科学重点研究基地已获教育部专家组评审通过。通过深入研究侨史、侨情，掌握侨务动态，既为国家制定侨务政策提供咨询，又可通过学术交流，加强海内外中华儿女之间的联系和中外交往。其他学科，通过重点建设之后，将出成果、上水平，既能将有关成果运用于侨务事业，运用于国家及广东省的国民经济主战场，又能在整体上提升暨大的办学质量、水平和效益，从而保证学校具有雄厚的实力，更大的知名度，吸引更多的海外华侨华人和港澳台青年来校求学，更好地为侨务工作服务。

4）进一步优化办学结构，提高办学重心，大力发展研究生教育

近几年来，我校的办学结构得到优化，办学重心得到上移。学校将继续遵循“规模、结构、质量、效益”的原则，大力发展研究生教育。目前有在校研究生 2007 人。事实证明，研究生教育是强校之路，我们要抓住机遇，扩大研究生在学校整个学生中的比例。同时，采取得力措施，申报新的博士点、硕士点和博士后流动站，以大幅度提升学校的综合实力，并力争早日申报研究生院成功。

5）做好校友、校董工作，积极主动地为侨务工作服务

校友、校董是办好学校的巨大资源。暨大较早设有董事会、校友总会。作为一所外向型的华侨大学，校友、校董遍布海外华侨华人社会和港澳台地区。作为一笔丰富的人才资源、信息资源、财务资源和社会资源，做好校友、校董工作，不仅可以为暨大带来人、财、物诸方面的支持，而且还可能凭借他们作为桥梁，进一步促进中外经济文化交流，推进学校的建设与发展。

5. 学校在 21 世纪初叶的发展目标

1996 年学校在“211 工程”预审过程中曾制订了学校 15 年的发展规划，经过 4 年的运作，前 5 年计划的主要指标已提前完成，因此，进行中段调整修订十分必要。目前学校将在 1996 年整体规划基础上修订、调整，制订学校新的 5 年发展计划，并提交即将于明年初召开的学校教代会讨论。

“两个面向”是学校的大旗。“侨”字特色，要贯穿办学始终，学校在这方面要下大力气，华文教育及华侨华人研究应放到相当高度。

在目前修订的 5 年计划中，学校拟初步作如下定位：坚持“面向海外、面向港澳台”的办学方针，大力发展研究生教育，适度发展本科教育，积极发展华文教育，稳定成人教育规模，着力提高教育质量。学校综合实力力争达到全国普通高校前 50 名行列。

学校拟重点发展如下学科：

（1）继续保持和加强我校经济、管理学科综合实力强的优势，重点发展产业经济学、工商管理、会计学、金融学、管理科学与工程等学科。

（2）进一步突出与强化我校文史学科的特色，重点发展文艺学、华侨华人、华文教育、专门史、国际关系、汉语言文字学、新闻学等学科。

（3）大力发展我校有一定优势，且与国民经济建设紧密相关的学科，重点发展生物医学工程、水生生物学、计算机与信息工程学、工程力学、区域经济学、旅游管理等学科。充分发挥我校综合性大学的优势，加强理工医交叉学科的建设，重点发展眼科学、妇产科学、内科学、病理学与病理生理学、药理学等。

学生规模：本科生将控制在 11 000～12 000 人，研究生突破 3000 人。

学科建设：一级学科博士授权点 3 个，博士点 18 个，硕士点 70 个，国家级重点学科 2 个，省级重点学科 14 个。

另外，要继续提高本科教学质量，加强教学管理队伍建设，实行研究生担任助教的制度。明年，既是 21 世纪的开端，也是国家“十五”规划的开局之年。我们希望全校师生为修订好学校“十五”规划献计献策，为学校在 21 世纪再铸辉煌奠定坚定的基础。

千年更替，世纪之交。创办于 20 世纪初叶的暨南大学面临着新的发展机遇与挑战。我们要紧紧抓住当前正在我校开展的“三讲”教育的契机，联系学校面临的新形势、新任务，联系学校面临的改革、发展、稳定的实际，切实加强我校的党建和思想政治工作，牢牢坚持社会主义办学方向，使我校更好地担负起时代赋予的重任，更好地为国家的侨务政策和国家的现代化建设服务。

以上我提出的一些尚不成熟的看法，希望能引起大家的进一步深入探讨，错漏之处，也恳请大家批评指正。

25. 辉煌与梦想[①]

世纪约会，千年相交，人生逢此境，可谓真正的“千年等一回”。今年的冬季特别温暖，今天的校园格外温馨，今夜的气氛如此火热，我们暨南大学师生欢聚一堂，用高昂的歌声礼赞祖国的辉煌，用绚丽的舞姿抒发新世纪梦想，用优美的乐符描绘神州大地人民心中的如意吉祥。

弹指一百年。回想世纪之初，中华民族跌跌撞撞，在八国联军的声声铁蹄中进入本世纪门槛。世纪的开门礼，竟又是一项辱国丧权的“庚子赔款”条约，然后是半个世纪的纷飞战乱，启蒙救亡。新中国的崛起，改写了中华民族屈辱的历史，东方的睡狮开始觉醒。世纪之末短短 20 年的改革开放，中华民族创造了前所未有、世人震惊的经济腾飞奇迹，携着龙年的龙威，我们又站在了新世纪的起跑线上，前面虽然荆棘丛生，但我们看到的仍是一片灿烂的阳光。越千年，跨世纪，我们暨南大学也充满新的希望，学校已确定“侨校＋名校”的办学目标，挑战新世纪。挑战本身就蕴藏着制胜的机遇。我们无须豪言壮语，只是抱定一个信念：为圆一个多世纪中华民族的强国之梦，我们勤奋工作，努力向前，踏踏实实地为 21 世纪的教育事业作出暨南人应有的贡献。

今天的晚会主题是“辉煌与梦想”。今夜是满天星光，辉煌已定格在今天，美好的梦想将随着我们晚会悠扬的欢歌曼舞，迎着新世纪的钟声，信心百倍地走进新时代。

最后，让我们一起欢度今夜良宵，健康快乐地走进新世纪。

① 本文是在暨南大学迎接新世纪文艺晚会上的贺词，广州，2000 年 12 月 29 日；暨南大学校报，第 299 期，2001 年 1 月 5 日.

26. 狠抓办学质量　走“侨校＋名校”之路[①]

今天，是教学工作会议结束的大会。我已在前面的大会上讲过两次话，今天是第三次讲话。今天的讲话，主要从整体高度谈谈我校的发展，以便把教学工作搞好。

前面已讲过，教学是高校的主旋律，是我们的生命线，在学校处重要地位，是基础。特别本科教育是基础。从宏观讲，关于我们暨大如何发展，教学工作如何做，还有目前工作的方向，这里谈谈自己和校领导班子的看法。

一、科技与教育发展的重大作用

小平同志曾多次反复说过：实现社会主义现代化，科技是关键，教育是基础。这讲到了科技与教育在我国现代化建设中的重要意义。中央已确定，我国到2050年基本实现现代化，同时已把全国各省区市达到现代化的时间做了预测。上海第一个，广东大概是第四位。我们暨大只有按中央领导同志高瞻远瞩的决策来确定学校的发展，才能很好完成中央给我校的办学任务。

世纪之交，江泽民同志说：当今世界，以信息技术为主要标志的科技进步日新月异，科技成果向现实生产力的转化越来越快，初见端倪的知识经济预示人类的经济社会生活将越来越取决于教育的发展、科学技术和知识创新的水平。教育将始终处于优先发展的战略地位。可见科技和教育在社会经济发展中的重大作用。

目前，全国的一千多所高校，拉响了改革号角。顺应21世纪发展战略，创建学科更加齐全、结构更加优化、综合实力更强、办学效益更高的大学，摆在了我们每一所大学面前。各校都在努力。从中央电视台播的消息和看各大报纸关于高校的消息就知道。特别是《人民日报》最近一个一个介绍名牌大学情况。只要注意就可知道，全国高校处于白热化竞争之中。特别以清华、北大为首，两校要建成世界一流大学，跟随的还有多所高校。广东省委、省政府也确定要建两所全国一流、世界著名的大学。在这种情况下，我们暨大的路如何走？值得我们所有干部、教职工深思。不能只看到自己的一份工作，要看到国家和世界形势的发展。

21世纪已过了几个月，大家是否注意到新世纪前沿科学是什么？作为大学老师，我们要清楚，前沿科学是什么，前沿研究是什么。科学的每一项发现，使社会发生翻天覆地的变化。过去的20世纪，归纳起来有三项伟大科学成就，促进20世纪的整个变化。一是相对论的发现，二是量子力学的发现，三是DNA螺旋结构生命学的发现。这三大发现带来整个20世纪人类文明。大家看古代史，漫长的人类社会，五六千年人类文明史，到20世纪才有翻天覆地的变化。简单举一例，看北京故宫，可知今天有些普通百姓的生活比当时皇帝的生活还好，是因为科学发明创造带动了整个社会进步。其中高校起了核心作用。

---

① 本文是在2001年度教学工作会议闭幕式上的讲话（根据录音整理），广州，2001年3月15日；暨南高教研究，2001，(2)：10-19.

20世纪文明从哪开始？过去中国人突出的成就，使中国人在世界上有独特的地位。在过去的20个世纪中，中国人领先了14个世纪。到了15世纪下半叶，世界科技中心从中国转到西方。明朝时期，我国闭关自守，科技进步太慢，科技中心转到英国。英国大约领先三四百年历史，有三四百年的辉煌。工业化使小小的英国号称为“日不落”国家。英国很小，物产资源也不丰富，但它在世界上占了那么多土地，这是因为有实力。这是因为科技从达·芬奇、文艺复兴时开始，接着有牛顿力学，英国科技、工业上去了。英国大学有四五百年历史。今天到剑桥、牛津这些大学看，也会叹为观止，感受到英国的实力。她一个学校就能出三四十个诺贝尔奖获得者，全世界才一百多个啊。英国是世界第二个科技中心。19世纪后半叶到20世纪初中心转到德国。科技核心以哥廷根学派为首，其中著名人物、物理学家马克斯·普朗克（Max Planck，1858～1947年）带动一个学派，集中了现代科技的精英，有爱因斯坦这些大科学家，从而使整个世界发生变化。应该说，20世纪德国的哥廷根学派带动了整个世界的变化。世界两次大战是德国发动的，德国是相当于我们广东省地盘大小的国家，它发动两次世界大战，是有经济实力、工业实力，是由于科技发达。20世纪20年代后，科技中心转到美国，现在科技中心还在美国。世界三大发现推动了美国科技发展，产生了计算机、人造卫星、宇宙飞船、原子弹、氢弹高精尖技术。20世纪发生天翻地覆的变化，科技与教育立下了汗马功劳。

那么今天的科技前沿是什么？目前归纳起来有信息科学、生命科学、环境科学、材料科学。高校如抓不住前沿，永远在传统学科里做工作，是没有前途的。世界在干什么，人家在干什么，如果不清楚，每个老师、科技工作者只顾在自己原来学的领域里工作，那我们暨大在这场竞争中就要失败。名牌就要变成非名牌，就会淘汰出局。所以，我们学校一定要盯住科技前沿发展。21世纪信息科学发展，将使世界10年一大变。如果谁不跟上发展步伐，他掌握的许多东西未来都不会有用。我们读大学时，在上世纪50年代，还是用手摇计算机。有些大学甚至还在用算盘。尽管40年代第一台计算机已出现了，但未普及。当时中国大概只几台M3型计算机，一台就有一层楼面大，但功能不如今天一个微机。到20世纪末，一年半年一个变化。目前美国计算机最快的已达每秒12.7亿次运算速度，但我国最快的只有7千多万次，差距很大。我们中国科技还落后。如何赶上去，要抓科技和教育。不然，我国无法现代化。

二、暨南大学的闪光点

在世界这种局势面前，我们暨大走什么样的路？教学工作是基础。我们如何走？这要看暨大今天的位置，知己知彼，百战不殆。我们不能盲目骄傲，也不能盲目悲观。要了解自己，有何可自豪的地方。不要一味只认为别的学校好，自己学校不好，一味埋怨，看不到暨南人、暨大闪光的东西。老埋怨怎么进步？要看到我校在全国一千多所普通高校中至少有10个闪光点，也就是10个中国第一。一所大学，有一样东西能在全国数第一，就了不起，而我校就有10个第一，因此，不要看不起自己学校。我们师生、特别今天在座的干部是学校的核心层，应该了解。

（1）中国政府设立的第一所侨校，是first。1906年清政府建暨大，到今年95年了。95 年的大学，在中国来讲是值得骄傲的。因为大学要较长历史才能办好，一流大学短时间办成是不可能的，因为名大学要有沉淀。在我国教育史上，暨大排在全国第七位。中国历史悠久的大学较少。在本世纪初，中国还有私塾。欧洲四五百年历史的大学很多，培养了精英，所以成为近代文明中心。

（2）全国第一所开放型大学，是第一所招收留学生的学校。这也不简单，不是封闭办学。现在有些学校说要办成一流名校，我看是空谈。一所一流名校，必须是国际化的学校，要全世界承认才成。如果只招中国学生，哪怕都是拔尖学生，也不能成世界名校。世界名校要有世界各地的学生，在世界上产生影响。从国际化讲，我校在中国最强。当然，这是说一个方面，不是说全部。我校是第一个招留学生的大学，1907 年印度尼西亚的侨生乘船来求学。华侨是特殊群体，中国人到了外国，还非常向往自己的故乡，希望祖国开办一所为他们服务的学校。这样暨大应运而生，招海外学生。那时中国大学也刚开始办，我校一开办就招了国外的学生。

（3）我校是目前中国境外生最多的大学。目前在校境外生 4401 人，这在全国是第一，包括 34 个国家和港澳台地区，五大洲都有，面很广。最近，学校准备在教学大楼旁造一堵万国墙作纪念，把 1978 年以来在暨大读书的同学，是哪个国家来的，加以记载，成为暨大一景。近 20 多年来，学生来自五大洲 77 个国家和港澳台地区，这在全国大大超前，在全世界也是很好的，是国际化的显示。学生千里、万里来此求学，是学校有特色、有闪光的地方才来。

（4）出了许多名学生，培养了社会精英。精英标志是培养了社会领袖、政治家和科学家。如哈佛大学培养了多少总统，我国清华大学培养了多少部长。暨大培养了 3 位中外副总理，这也是我校特色，有中国的、也有外国的领导人。其他学校没有。如我国的吴学谦，先后在学校两次读书共 8 年，奉命做地下工作，后来成为外交部部长、副总理；还有李岚清，当时因暨大经济会计有名，投考我校。后因暨大停办，合并到复旦，他从复旦毕业，但还是算暨大校友，现在是中央政治局常委、副总理；泰国的副总理、议长巴实•干扎那越先生（中文名许敦茂），在泰国当了领袖，也是不容易的事情。这也非偶然，是学校办得好的结果。我们学校的老师，后来也出了不少名人，当了领导。不说董事会的，就有 6 位校友新中国成立后成为副委员长，1 位成为副总理。一位副总理是黄炎培先生。6 位副委员长是严济慈、周谷城、周建人、楚图南、许德珩、胡愈之。一般学校出不了这么多领导人。我们的许多学生、老师在社会成为名人。

（5）第一所设立董事会的学校。我校 1922 年设董事会，历史悠久，面广。我们现在董事不仅内地有，港澳有，还有日本、美国、秘鲁、欧洲有，这是董事会特色。

（6）学科齐全。我校是新中国第一所设立医学院的综合性大学。学科齐全，培养人才氛围好，单科性大学难以培养杰出人才，人才素质高低与学校专业多少有关系。多学科大学对培养高素质人才有很好的作用。好多学校近年合并才有医学院，如北医大合到北大。有医学院的高校多数只有一两年历史，我们已二十多年了。

（7）行政架构现代化。校院系一体化，与国际上大学体制一样。国内高校原来基本上无学院，近年才有，我们管理体制国际化走在前面。

(8) 第一所春秋两季招生的学校，招生改革走在前面。现在其他学校才开始两季招生，我校 1997 年已开始。

(9) 全国第一所取消补考的学校。我校从 1993 年开始取消补考，是非常大的转变，是创造既严且宽的体制，是贯彻学分制中实行的，是现代化办法。现在有的学校才开始这样做。

(10) 第一所有选手参加国际奥运会体育竞赛项目的高校。1936 年柏林第十一届国际奥运会我校有 14 个选手参加，这是全国高校绝无仅有的。我校体育有特色，新中国成立前、新中国成立后到现在体育都好。

总之，要在一千多所高校中排一个第一不容易，这说明暨大有自己的基础，有值得自豪的地方，我们要在此基础上往前发展，找准自己的位置。

三、暨南大学的发展近况

这几年在全校教职工的努力下，学校有了较大发展。我们大力发展研究生教育，适度发展本科教育，稳定成人教育规模，用这一政策管理学校，使之向高层次发展。从 1996 年起校本部不招专科学生，办学重心上移。办专科不可能成为名校。当时专科生 2472 人，现校本部没有了，非校本部在校专科生 643 人。再过两年，全校就不会有专科生了（包括华文、中旅、珠海学院），因为从 2001 年起全校不再招专科生。本科规模在扩大，1995 年全校本科生 5377 人，现在 9498 人，增加了将近一倍。硕士生增加幅度更大，从 1995 年的 563 人发展到现在的 1839 人，增加了 2 倍多。博士生从 1995 年的 52 人发展到现在的 188 人，也增加了 2 倍多。现在研究生共 2027 人，在全国排第 32 位。这不容易，原排在全国 100 多位。但总体实力排在 60 位，还不行。

博、硕点排在全国高校第 42 位。这 5 年有长足发展，博士点由 7 个发展到 14 个，翻了一番。其中还增加了一个一级学科博士点，包含多个二级学科博士点，增加的 7 个大于原来 7 个的覆盖面。硕士点由 50 个发展到 66 个。以 1995 年的 50 个算，数量上看只增加了 16 个。但实际上增加了 20 个硕士点，是由于医学的三级学科硕士点取消、合并了。

本科专业由 30 个发展到 39 个，增加了应用性、科学前沿性的专业。学校专业面在增大，体现侨校特色的外招生数有了很大发展，外招生 1995 年为 1982 人，现为 4401 人，翻了一倍多。现有境外研究生 413 人，占全国同类生的 1/4，有举足轻重的作用。但研究生结构欠佳，博士生数偏少，只 188 人，去年申报研究生院，这一项把我校拉下来了，要 300 个博士生才能批。实际去年新批的 22 个研究生院有四五个学校的研究生总数还比我校少。我们有些点博导少，招生少，造成博士生偏少。从这可看出，学校专业的发展、学校的发展应在什么地方。

学院也在调整。学院由 1995 年的 7 个发展到今天的 16 个。这是适应 21 世纪科学发展变化的需要。学科分界更加清楚，哪个是热门领域，哪个是发展领域，哪个是关键领域。由于《中华人民共和国高等教育法》的实施，今年我校取得院系审批权。以前上级管，不能调整学院。为什么先成立生命科技学院，因它是领先学科。生命科技学院组建两年，生龙活虎发展，在我校很有特色。有领导、外宾来，首先让他们看生命科技学院，就知道我校的水平。他们取得了许多研究经费，为学校增了许多光。这里表扬生命科技学院的领导和老师们。

还有管理学院，也非常有特色。该院研究生占全校的近一半，本科、研究生几乎已达 1∶1 的比例，研究性学院会慢慢办成。管理学科在今天是非常重要的学科。我们中国在过去 50 年中发生的许多重大失误都是管理上的。一个领袖如有一个错误思维、错误决策，影响就很大，就把全国带入深渊。如当年的“大跃进”，毛主席想快，想几年赶上日本，几年赶上美国，提这样的口号，全国热情高涨，好像很快可以实现共产主义。但目标的设定基础怎么样，没有科学分析和预测，使得 1960 年以后整个国家经济困难，生活困难。1958 年全国还把麻雀当害虫驱打。有人当时出主意，中国人多，每人同时拿一根竹竿、拿一个盆子敲，赶麻雀，使麻雀不能休息。全国从南到北、从东到西赶，麻雀累了就会掉下来。《刘少奇回忆录》中记载，当时北京中南海也赶麻雀，只有刘少奇没有出来赶，躲在书房看书。今天看这些是大笑话。从环保看，一个地方如果没有了鸟叫声，会是很糟糕的事，说明环境很差。这说明管理科学重要。改革开放前，全国几乎没有管理专业，我们成立管理学院，在社会上声誉很好。特别是工商管理，大家不要嫉妒他们研究生多，这是暨大的光荣和特色。管理学院成立促进了学校发展，提高了学校声誉。

新组建的还有信息科技学院、药学院、外语学院、新闻与传播学院、法学院、珠海学院。特别要说明，我们还有国际学院也已成立。我们是国际性大学，要与世界潮流接轨。以往学生来，只能全部听中文讲课。尽管中文是世界第一大语言，可学生回当地不可能马上使用。中文没有世界性，世界语言是英语。有的境外生来暨大报到，一听说是用中文授课，马上提出不读了。我们姊妹学校的交换学生来暨大全部学中文也不愿意。所以，我们要在世界上立足，成为好的侨校，必须要用英语授课，提倡双语教学。这几年每个专业都有用英语授课的课程，有了进步，但步子太慢。所以，我们改换思维，成立国际学院，在这个学院全部用英语教学，中文作外语。希望全校各学院支持，办成我校的招牌。这是难度很大的事。目前如哪个学院组建一个用英语教学的专业，就放在国际学院，老师仍属原学院，由各学院包专业，由各学院办。现已有医学院、管理学院和经济学院报名。希望今年 9 月份能开学，成为现代化学院。

信息科学是 21 世纪的核心学科，国家和广东省都非常重视。我们组建信息科技学院，要使之成为我校龙头。

新闻、外语、法学、药学院，还有珠海学院，都是新办的。新学院要老学院支持。珠海学院是我校最大的学院，将办到 5000 人的规模。很想办环境工程学院，这是世界前沿学科；还有材料工程学院，也组建不起来，目前没有力量。希望这些学科的教师努力。今后一定要发展新兴学科，不要仅固守在传统学科努力，仅搞传统学科是没有大出息的。要在国际前沿学科发展，我们暨大才会有成绩。

校区也有发展。目前是四个校区，校本部、华文、深圳、珠海校区。校本部今后以发展研究生教育为主；华文学院发展华文教育为主；中旅学院办旅游学科；珠海学院紧邻港澳，能更近地为港澳服务，“面向海外、面向港澳”。以广州一个大本营，加毗邻港澳前沿的深圳、珠海来发展学校。新区磨蝶沙，打算今年动工，办成低年级学生区。校本部教学区四五百亩地，已有一万多名学生，太拥挤，考虑把本科往外移，减轻本部人口密度太大的压力。也有建议搬几个学院到将来的新区。学校已规划好，国务院侨办也通过了，广州市原来也答应，但现在广州市规划局要求征我校这块地，要求换地，目前正在谈判，不在磨蝶沙就可能在东边一块离校区不远处换几百亩地，就会是第五个校区。

附属医院也增加发展了，有了 5 个附属医院。分布在广州、深圳、珠海、清远，覆盖面比较大。医学教学条件有了改善。

我校学科的水平也在发展。刚获批医药生物技术研究开发中心，是国家工程中心；还有一个国家级华侨华人人文社会科学基地；原有一个教育部文艺学文科基地；有两个博士后流动站，一个教育部重点实验室。这些都是“零”的突破，使学校地位上升。

学校土地变多了。经过努力，珠海市政府无偿赠送我校市区土地。1995 年我校为 110 万平方米土地，现增至 161 万平方米土地。还有蓝色图保护区 10 多万平方米。学校土地增加了，这是笔巨大财产。我校在广州市的土地在减少，修华南大道等占用了我校部分土地。但整体在扩大，没花多少钱，是靠学校的努力。

学校资产也增加了，5 年来学校固定资产有很大增长。1995 年全校固定资产是 2.7 亿元，现发展到 13.3229 亿元。仪器设备由 4985 万元发展到 1.0815 亿元。

省重点学科由 2 个发展到 10 个。目前正在冲刺国家重点学科。我们还缺两个“大东西”，这两年在奋斗。一个是建研究生院，一个是建国家重点学科。要争取“零”的突破，其他已突破。我们最近申报 8 个国家重点学科，正在通讯评审。从上个世纪 80 年代后就没有搞过国家重点学科。我们从 1996 年进“211 工程”，成为全国 100 所重点大学之一，才有这些“零”的突破。还有“零”没突破，需要全校努力。

学生情况有了好转，去年从二表招生变成全国一表招生，从一般大学招生变成重点大学招生。我校在广东早几年已是重点招，在全国重点招去年才成功。

科研经费也由原 300 多万元增加到去年的 5356 万元，增长了 10 多倍。

除了吸收了几所附属医院外，去年还吸收了交通部信息技术研究所，成为我校的研究所。去年 8 月中央已批准，侨办上个月批准，正在办移交。

学校结构、专业、系在调整，学校土地在扩大，学院在扩大，许多“零”的突破，学生素质在提高，办学重心在上移，学校经费也在增加，通过这些努力，才总体上使暨大在全国的排位从 1998 年的第 87 位到 1999 年的第 72 位，2000 年的第 60 位。我们跑得很靠前，人家称我们暨大是目前中国高校的一匹黑马，跑的速度很快。这些数字说明暨大在快速前进。大家看报纸，可以看到暨大在国内的形象，几乎所有重要报道，都要拉暨大说话。5 月 31 日《光明日报》上登了一篇《声誉：渐成高校竞争焦点》的文章，第二个学校就讲我们暨大。最近我去北京，中央电视台要播 70 所名校，马上找到我要播暨大。现在高校许多大事都要报道我校，这是全校上下、全校党政共同努力的结果。

四、暨南大学“侨校十名校”的发展战略

我们已到了这个阶段，今后是进还是退？我想大家一致意见是进，继续前进。我们有 10 个中国第一的特色，已有了这么好的基础，理所当然要继续往前走。那最重要

的、核心的问题是质量，办学的质量是关键。

我们要成为名校，有什么战略？说得简单一点是“侨校＋名校”。首先，我校性质是侨校，95年来三个政府都定我校为侨校，我们不能把“侨”字去掉。我校“侨”的特色要鲜明。没有“侨”，学校就将没有活力。侨生、境外生要多。还有培养模式、管理工作等，各方面都要体现侨校特色。我们是校长负责制，体现侨校特色。但具体到每个方面，每一位领导都要时刻想到侨校，哪怕搞错一点，都有可能成为大事故。1995年我们出了马来西亚学生一个事件，使我校至今在东南亚都没有翻身，使一件小事情变成了大事情。同时，是名校才有吸引力。不然，学生家长干吗要送子女不远万里来暨大读书？差的学校人家会愿意来上吗？名校才能吸引优秀学生。办成名校是我们义不容辞的责任。现在有些领导、教师困惑，感到侨生学业差。但随着办成名校，学生质量就能上去，就可招到更多优秀学生。但目前想全部招优秀海外学生不可能，十多年二十年都可能办不到。如果有北大、清华的声誉，那就好办多了。但我们还没有这样的基础和实力。所以，要艰苦卓绝努力才能把暨大办成名校。只有名校办好了，才能办好侨校的事情，这是相辅相成的。

我校对港澳办学任务并不因港澳回归而减轻，而会更重。我们当年培养港澳人才，是为了港澳回归。港澳回归后，是社会主义祖国的特区，但实行“一国两制”。港澳仍实行资本主义，因此港澳的大学不可能培养社会主义人才，还要靠我们。

今天还要面对祖国最后一块土地台湾回归统一问题。中华民族统一大业是中国的头号政治任务，我们暨大做什么？要义不容辞承担台湾回归有关任务，要培养台湾学生成为爱国者，反对“台独”的人，培养有能力的人。还要使更多华侨华人精英学生来我校读书，使世界华侨华人心向祖国、祖籍国，反对“台独分子”，支持中国现代化，那我国现代化就更快了。所以按中央24号文件，我们责任很大，要走“侨校＋名校”的道路才能完成任务。

何谓名校、一流大学？有三条：①要有许多一流的学科。在世界有名是世界一流，在中国有名是中国一流。如我们经济管理学科在华南最好，就在华南有名；如果在全国数得上，就会是全国一流；到国际上有名，就成国际一流。所以学科有名是关键。就如医院，有几个科有名，医院才会有名一样。②要有名师，要几位著名大师、名教授。③要有高素质、高质量的学生。有这三条才能成名校。暨大还有距离。我们名学科、名师、名学生都还不足，需要扎实努力才能实现。

现全国企业都在打造品牌。计划经济下不需要品牌。在市场经济下，企业的产品靠品牌生产。现在《中华人民共和国高等教育法》颁布，把高校推向了市场。高校的品牌战略使高校在白热化竞争，成为全国的态势，都注意声誉。是北大、清华、复旦的还是暨大的，学生入校后一辈子也离不开这个学校的名称。过去，我们听说有学生入校后不愿意戴暨大的校徽在街上走，就是因牌子还不够好。名声特别重要，一所学校追求名牌战略成功的话，就能站住脚，那时，我们的学生毕业就好找工作。品牌好的学校，社会抢它的毕业生；品牌差的学校，毕业生找工作都很难。所以，现在把毕业生分配受不受欢迎作为学校品牌的一个标志。我校近年发展是上升的，所以毕业生就业比较好。受欢迎情况，我校在网上的排名是第18名。我校学生一次就业率高，学生不需费什么气力，凭学校品牌用人单位愿意接收。早些年就不是这样，深圳人事局曾有通知，要哪些学校的毕业生，其中没有我们暨大，不要我校毕业生。我们今天是在推学校前进，创学校品牌、名牌。只有名校才有生命力，才能完成侨校的任务。

要注意哪些问题？近年提出若干办学措施、原则，都是为了学校品牌，为了办好暨大。我们提出了“严、法、实”三个字。治校“四从严”：从严治党，从严治校，从严治教，从严治学。“严”字的核心就是使质量有可靠保证。我们的毕业生是合格的，科研成果是高质量的，“放水”不对。从入学、中间过程直到毕业，每一关都要严格。无论考试、各项制度、校风、教风、学风、管理等都要严格，不能只是招生关严格。不允许学校学术腐败。但学术要民主，要有学术氛围。

要依法治校。学校所有业务领域，教学，科研，研究生、本科生、成人教育，以至党的系统，这些年都在制定各方面制度。这次教学工作会议搞了11个制度文件，已原则通过，就是要依法治教学，不搞人治，搞法治。同时要实事求是，虚的、假的东西最后要穿帮的，是没有价值的。要扎扎实实做事，才能把学校搞上去。所以，要继续坚持“严、法、实”三个字，同时要走“三化”道路：现代化、国际化、综合化。

无论办学条件、办学思想、管理，都要现代化。大家看到校园在绿化、美化，就是搞现代化，住房、生活条件、教室、运动场也要现代化。还要修医学院大楼，行政大楼，游泳池，等等。校本部、珠海学院都在搞一系列基建，是搞硬件现代化。但软件、管理也要现代化。

我们还要国际化，五大洲学生在这里读书，不走国际化道路不行。我们要进行国际学术交流，教学内容要国际化，三类学生的教学要清楚。本次教学工作会的核心是把华侨、港澳台生，内地生和外国留学生三类学生的教学问题搞清楚。多年来，这方面做得不太成功，老师教课困难。成绩好的和成绩差的学生差距太大，没有找到好办法教。这次教学工作会来集中研究，面对现实，必须国际化。要把各种学生培养成才，要因材施教，有教无类，德智体都要教好。特别德育要强调，做人教育是第一的。大学教知识，更重要的要教会学生做人。三类生的德育要分类教，不要混为一谈。持外国护照的学生不能用马列、四项基本原则教育他们。对他们要进行爱中华教育，让他们了解中华文化博大精深，与中国友好，遵纪守法。不要看到都是黄皮肤、黑眼睛、黑头发，但政策要求不一样。侨生、港澳生虽持中国护照，但他们生活在资本主义制度下，也不能按内地生教法教四项基本原则。三类生教育必须科学化。国际化内容很多，方方面面，教务工作、行政工作，都要注意这个特色。内地生要培养成社会主义事业接班人，强调政治和思想教育。国际化还有学科知识，教材要先进，要外文原版教材。还有专业设置等要国际化。我们是综合性大学，学科要发展热门的、前沿的、国际化的，不要抱着冷门传统学科不放。要适应时代要求，不断调整。

要通过这些努力，使我校成为校风、教风、学风好的学校，要靠在座的各级领导团结一致，克服一切困难。大家要以暨大为荣，首先做好自己的本职工作。有个别人利用各种机会骂学校，把小事当大事，这是不利办名校的。谁有见北大的老师出来骂北大的吗？暨大有个别人就骂暨大。自己在这个学校，实际也骂了自己。要像维护生命一样维护校誉。学校在发展中非常困难，我们不像北大、清华等名校，办得很顺利。

如人家申请国家重点学科很容易，我们就很难。我们经费上面投入偏少，但有人花钱像流水。要知道我校没有特殊经费。我们搞“211 工程”主管部门没有投入，比其他校困难。我们得到的投入少，又要办好学校，如何办？只能艰苦奋斗。

我们采取了若干措施，学校收益较好。政府投入只占学校经费的 1/3。我们要从管理要效益。钱多不一定能办好事情，但钱少了办不成名校。越有名才能越有钱。我们进到“211 工程”，有了名，进到前 100 名内，日子才好过一点。下一个 5 年，我们要争取使学校进到前 50 名。所以，要爱护学校。有的人造谣说学校没钱了，我告诉大家，学校还没借过国家一分钱。学校与银行签了 10 亿元授信合同，就是贷 10 亿元不要担保。这是考虑学校发展，预支未来经费，学现代办法。有一则报道，有一幅漫画，说东西方两个老太太谈天。东方老太太说现在很高兴，人老了，终于住上了新房子。几十年节衣缩食，七八十岁了，快死了，终于有了一套自己的新房子。西方老太太说，我从二三十岁就住上了新房子，一直住到现在，是分期付款、贷款方式买的新房子。一个二三十岁就住了新房子，一个到七八十岁才住上新房子，你说谁的日子过得好？这是观念问题。学校持新观念，把未来的钱提前用，使学校尽快现代化，尽快改善师生生活条件。住宿、教室、实验室条件在改善。但有的人挖苦，实际是落后观念，用落后骂先进，用不正确的观点攻击正确的。

现在回过头看，学校一些改革成功了，是在骂声中成功的。所以，要支持新生事物，全校师生员工要互相理解、团结，增强凝聚力，学校才能发展得更快。如果有人给暨大抹黑，要去掉污点，就要花很大精力。要提倡爱国爱校，团结奋进。要将自己的生命与学校联系起来，因为你是暨南人，想与暨大分开也分不了。你在暨南大学毕业或工作过，即使离校，也与这所学校分不开了。暨大强了，你走到哪里名声也好，要维护学校声誉。还要搞好工作，提高学校声誉。这样，我们就能更好实现学校“十五”计划。

我们“侨校+名校”有两个指标比例。首先，关于侨校，国家要求我校侨生达 50%，任务非常艰巨。侨生一般是 2∶1 或 3∶1 录取，成绩太差的不能录，录了学校声誉马上下降。所以，既要增加侨生，又要保证学校声誉，要实际些。如柬埔寨每年有 5000 名华文初中毕业生，华文教育只到初中，没有高中，我们能否把预科办低些，扩大到初中，学生可来预科读高中，再上大学？预科和华文学院要赶紧做这件事。办预科初、中、高级班。初级班相当于高一的学生，从高一培养。要多思维办学，多方面办侨校，作为一件大事来抓。

其次是名校问题，名校就要走研究型大学之路。目前我们强调教学中心、科研中心，双中心，把教学、科研紧密结合。科研要上去，教学要上去。这次是教学工作会议，就是要全校把教学质量搞上去，要很多措施。这次教学工作会讨论 11 个制度，核心是标准学分制，要执行好。再不能发生一个学生 17、18 学分的课程不及格，还不重修的事。一定要及时重修，成绩差的不能超学分选修。计算机选课要拦住那些成绩差的，不能让他们多选课。学分制是既严又宽的制度，特别符合我校学生实际，要执行好，有利我校提高教学质量。同时，辅以其他制度，如考试制度，还有名师讲课制度。正教授上基础课，我校 1993 年已开始。强调教授要上本科基础课，只上研究生课是不行的。这样才能使我校学生得到素质最高的培养。最近教育部根据李岚清副总理的指示，作大工程，在抓教授上基础课，我们已抓了 8 年了。各院系要再强调这一点。教授一定要上本科基础课，这是提高质量的关键。

教学上要严格把关，不允许考试作弊，才能保证质量。还有教学事故处理，办法、条文不适合的可修改，但要坚持教学事故的处理。有人提出要允许教师坐着讲课，这荒唐。全世界老师讲课都是站着的，要不就不当老师。这是当老师的规矩。如果生病了，站不了，可以坐着，那是例外。但正常的都要站着讲课。如允许坐着讲课，我们暨大的声誉肯定马上一落千丈。我们教学上的基本原则一定要把握住。这次教学工作会我反复强调教学质量，讲学分制，讲教学思想。请大家注意，教学这个基础必须通过若干严格化过程管理才能成功，光学分制还不行。要注意学生是第一，从我校长开始，所有老师，所有领导，都是在为学生服务，学生是主体。一切学校工作都要为学生服务，没有学生成不了大学。老师起主导作用，老师是关键。一所学校办得好不好，要看老师。教学工作会任务要落实到老师身上。老师要选好教材，备好课，上好课。

这个体系如何保证？首先改善办学条件、教学工作条件，同时搞好评估。要坚持“三重评估”制度，不要丢掉这一特色。学生评估老师一定要坚持。这次教学工作会还听到有人不同意学生评老师。大家看报纸，最近有一篇报道清华大学特色的，说清华启动学生评老师。清华这一点比我们做得慢，我们 1993 年就全面实施开了。暨大这几年为何声誉上去？是我们这几年采取了若干超前的、别人没采取的措施。清华才开始学生评老师，说这个制度好。这次会上还有人说学生评老师不公。我看好老师绝对不会怕学生评，只有教得不好的老师才会怕学生评。一个严格的老师，高质量教学的老师，学生会终生崇拜你，感谢你一辈子。我们想一想，自己今天当了老师、当了领导，怀念什么样的老师？怀念小学、中学、大学的好老师。好老师是严格的。如果是“放水”的老师，考试时学生要多少分给多少分，试题也给，我想这样的老师在学生心中会是一钱不值的。只有严格把关、好的老师才会得到学生尊重。所以，一定要坚持严格评估制度。对那些荒唐的理论大家一定要批驳，要群起而攻之，不能让坏的思想、谣言、理论满天飞而不加制止。包括网上不健康的东西要鞭打。我们全体领导要坚持原则，才能把学校办好。

教学是主旋律，是基础。只有通过制度建设，领导、管理讲科学化，才能做好。这是个系统工程，非常难做，还须奋斗若干年。“211 工程”第一期即将结束验收，还要迈向“211 工程”第二阶段；我们还要接受教学评估。这些关口不过，那我们学校又要走向低潮。大家都希望学校名声变好一些，在高校激烈竞争中我们能脱颖而出往前走，走到前 50 名再往前，到 40 名、30 名都是可能的，一定能成功的，成为研究型大学。我们设计近期达到 3000 个研究生，进而再达到 4000 个，然后发展到 6000 个。本科生稳定在 12 000 人多一点。研究生和本科生比例达到 1∶2，以这个指标为导向。一个是前面讲的侨生指标，一个是刚才讲的研究生和本科生指标，两个发展目标。希望大家围绕我校的发展目标，每人做好自己的本职工作。从校领导班子，到各院系领导班子，各部处领导班子，大家团结一致，努力把自己管理的工作做好，无论教学、科研、管理工作都做好，我想，我们暨南大学一定能成功，一定能胜利。

27. 暨南大学的特点和优势[①]

暨南大学是一所国家“211 工程”重点建设的综合性大学，是在海内外享有盛誉的国际型华侨高等学府。简而言之，是一所“侨校＋名校”。她有三大特点和优势。

一、国际化

暨南大学创办于上个世纪初叶的 1906 年，今年将迎来九十五华诞。这所历史悠久的大学虽然经历了清朝、中华民国和中华人民共和国三个不同的历史时期，但是“面向海外、面向港澳台”的办学宗旨从来没有改变。

暨南大学的校名有着丰富的内涵和文化底蕴。“暨南”二字典出《尚书·禹贡》：“朔南暨，声教讫于四海。”意在将中华文化远播海外。第一批暨南学子，全部来自印度尼西亚。95 年来，暨大已为海内外培养了 8 万多名各类专才，56 个校友会分布于世界各地。1978 年改革开放以来，就有来自五大洲 77 个国家以及我国港澳台地区的学生先后在暨大就读。目前，每年在校的境外、海外生都保持在 4000 多人。

暨南大学不仅是中国历史上第一个华侨高等学府，第一个向海外招收留学生的大学，也是目前中国大陆高校中招收境外、海外学生规模最大、人数最多的高校。境外、海外本科生数占全国第一，境外、海外研究生数占全国 1/4。

暨大每年都通过公开、公平的竞争选拔部分学生到美国、英国、加拿大、日本、澳大利亚、瑞典、德国、新加坡等国交流学习。

二、综合化

暨南大学在 95 年的发展中，虽历经坎坷，但坚持办综合性大学的方针，执意改革发展的方针不变。

暨南大学从初期的 5 个学院：商学院、文学院、理学院、法学院、教育学院发展为今天的 15 个学院：理工学院、信息科技学院、生命科技学院、医学院、药学院、经济学院、管理学院、法学院、文学院、新闻与传播学院、外国语学院、华文学院、中旅学院、珠海学院和教育学院。暨南大学是新中国最早拥有医学院的综合性大学，它为暨大发展生物医学交叉新型学科奠定了坚实的基础和广阔的前景。目前，暨大生物工程、生物医学工程中的部分研究项目不仅列入国家“九五”规划之中的“重中之重”的项目，有些科研项目的研究水平已处世界先进水平。如获国家一类新药证书的碱性成纤维细胞生长因子（bFGF）等。综合性大学带来的优势就是多学科并驾齐驱发展竞争和相互渗透，相互交融孵化繁衍新型交叉学科。同时也给学生提供了拓展知识面的大好机会。目前暨南大学已有 39 个专业，涵盖文、史、理、工、医、经、管、法、教育等多个学科。校园面积 151 万平方米，坐落在南方美丽的广州文化区。除广州天河校本部校区外，暨南大学还在深圳、珠海两个特区分别建立了中旅学院和珠海学院，

---

① 本文是在中央电视台《高考咨询——名校面对面》中的演讲，北京，2001 年 5 月 13 日.

以方便更多的海外学子，尤其是港澳台同胞子女升读暨南大学。

三、现代化

今日暨南，已跻身国家“211 工程”行列，有一级学科博士学位授予权专业 1 个，二级学科博士学位授予权专业 13 个；有硕士学位授予权专业 63 个，专业硕士学位授予权专业 3 个；博士后流动站 2 个。是有 MBA、临床医学硕士和口腔医学硕士授予权的高校之一，还有教育部、财政部定点的基地专业 2 个。暨大是新中国第一所设立医学院的综合性大学，医学院设有 5 家三级甲等附属医院和 1 个直属医院。

暨南大学师资力量雄厚，专任教师 994 人，其中 60%以上的教师拥有博士、硕士学位。暨南大学能够吸引如此众多的海内外学子云集暨南园，不仅仅是因为它拥有悠久的历史、深厚的文化积淀，也不仅仅是它“面向海外、面向港澳台”的办学方针所带来的各种机遇，更重要的是“暨南精神”，是“从严治校、从严治教、从严治学”的校风、教风、学风。

暨南大学紧跟世界科技的发展，从 1984 年起就开始在本科实行学分制，并不断加以改进和完善，成为与国际教育制度接轨的标准学分制。为适应海内外学生的需要，为了在 21 世纪培养高素质的专门人才，暨南大学从招生、教学、管理、分配等一系列问题上进行大胆地探索与改革。如暨南大学率先在中国实行春秋两季招生制度；率先实行一周七天，每天从早到晚排课的“全天候”教学试点；率先试行教师挂牌上课，学生跨校、院、系选课的制度；率先实行优秀学生选读双专业、双学位以及提前毕业的做法，有困难的学生也可以推迟毕业。

暨南大学还建立了与国际互联网连接的先进的校园网络，实行多媒体教学、网上选课、批注作业等。学生在宿舍里也能享受网络教学带来的丰富信息资源。学生在宿舍里还可以打世界各地的电话，紧紧把握新时代的脉搏，及时了解各种信息。

新世纪是知识的世纪，新世纪是人才的世纪。暨南大学积极倡导先进的教学思想和理念，真正把提高教育质量、培养高素质人才当作学校生存和发展的生命线。不同国籍、不同肤色、不同习俗的青年在多元文化的背景下，在生机盎然的校园中，融洽相处、健康成长。暨大毕业生分布于祖国各地和世界 77 个国家以及港澳台地区。他们以较扎实的理论知识和富于创新、勇于开拓的精神面貌，得到用人单位的好评，学生受欢迎程度在全国排第 18 位，其中不乏优秀人才，如泰国国会前主席许敦茂、我国在任的中央政治局常委李岚清副总理以及国务院原副总理吴学谦。许多优秀毕业生成为香港、澳门特区政府公务员等。内地毕业生一次性就业率在 90%以上，每年有 300 多家单位到暨南大学来要人，如普华永道、德勤、安达信等世界著名五大会计师事务所，中国联通、金山软件、爱立信有限公司、宏碁电脑公司、汇丰电子资料广州公司、广州海关、美的集团、康佳集团等大公司，中联办、港中旅、香港南洋商业银行、广东省四大银行、亚洲电视、广东新闻系统的所有报刊、电台、电视台等单位，英国驻广州领事馆等。我校选送到广东省、地、市各级政府中的优秀毕业生人数在广东高校名列前茅。

站在一个新的起跑线上，暨南人豪情满怀，将继续遵循“侨校＋名校”的办学思路，为新世纪培养高素质人才谱写新的篇章。

28. 弘扬中华民族文化①

首先，请允许我在此，真诚感谢世界各地华侨华人社团及在座各位，长期以来对我们侨校事业的热情帮助和支持。

受国务院侨办委托，我在此代表国务院侨务办公室所属三所院校——北京华文学院、华侨大学和暨南大学，向各位领导、各位朋友简要介绍我们三所兄弟院校办学的基本情况。

三校虽然校与校之间远隔千里，但彼此都有着共同的办学特色，即“侨”字特色。“面向海外、面向港澳台”是我们的办学方针。弘扬中华民族文化，传播现代科技知识，凝聚侨心，维系侨情，是我们的办学宗旨。下面我分两个部分作简要介绍。

一、学校概况

(一) 北京华文学院

北京华文学院坐落于北京阜外大街，其前身是成立于1950年的北京华侨学生补习学校，由著名侨界领袖何香凝、廖承志亲自发起创办。她是中国最早开展华文教育的学校之一。2000年底，经教育部、国务院侨务办公室批准改建为华文学院。学院通过开办多种长短期培训、研修班、夏（冬）令营，外派教师和编写教材等形式进行华文教育，拥有一支具有丰富教学经验和海外教学经历的教师队伍。学院教学与生活设施较先进。目前，一座能容纳560名学生，建筑面积近2万平方米的现代化学生宿舍楼正在加紧施工，预计明年6月交付使用。建校50年来，共有4万余名华侨华人学生在该校接受各种形式的培训。学院成为首都地区面向海外，全方位为广大华侨华人提供中国语言和文化培训的华文教育基地。2000年9月，国家主席江泽民为学院建校50周年题词：“发展海外华文教育，促进中外文化交流。”

(二) 华侨大学

华侨大学是在周恩来总理亲切关怀下于1960年由国家创办的一所华侨高等学府。学校现有理、工、文、经、管、法、哲、艺术等30个本科专业，9个省部级重点学科。学校专任教师540人。100多位教师具有或正在攻读博士学位。

学校校园占地面积69万平方米，建筑面积40万平方米。学校本部坐落于泉州市郊，另一校区设在厦门市集美学村。学校建有现代教育技术中心、信息网络中心及25个实验室和32个研究所（室）；图书馆藏书近百万册，中外期刊万余种，电子图书馆即将启用。海外侨胞、港澳同胞捐建的陈嘉庚纪念堂等一批建筑设施，为师生提供了学习和生活的优美环境。建校40年以来共培养海内外各类人才5.6万余人，其中境外毕业生3.2万多名。学校现有各类在校生近12 000人，其中境外生1500余人，分别来自世界30多个国家和地区。

① 本文是在海外侨团联谊大会上的讲话，北京，2001年6月22日.

（三）暨南大学

暨南大学创办于1906年。“暨南”二字典出《尚书·禹贡》：“朔南暨，声教讫于四海。”意即面向南方，将中华文化远播海外。暨南大学不仅是中国历史上第一个华侨高等学府，第一个向海外招收留学生的大学，也是目前中国大陆高校中招收境外学生规模最大、人数最多的一所综合性大学。境外生数占全国第一，境外研究生数占全国1/4。95年来，暨大已为海内外培养了6万多名各类专才。

学校为国家“211工程”即面向21世纪国家重点建设的100所大学之一，学科涵盖理、工、医、文、史、经、管、法、教育九大门类，共设16个学院、35个系、39个本科专业、66个硕士学位授权学科、14个博士学位授权学科、2个博士后流动站、1个国家工程中心、2个国家和部级重点基地、3个国家和省级重点实验室、10个省级重点学科。拥有46个研究机构，60个实验室。图书馆藏书151万册，设有华侨华人信息资料中心。暨大是新中国第一所设立医学院的综合性大学，医学院附设有现代化的5间国家级三级甲等附属医院和一个直属医院，拥有病床3500张。

学校有专任教师994人，其中中国工程院院士1人，国家“长江学者”特聘教授2人，广东省“珠江学者”特聘教授2人，教授171人，副教授416人。专任教师中60%以上拥有博士、硕士学位。师资力量居全国高校第13位。

学校占地面积156万平方米，建筑面积73万平方米。学校本部坐落于广东省广州市市区，另有三个校区分布在深圳市、珠海市及广州市广园东路。学校现有各类学生21 365人，其中全日制学生12 407人，包括全日制本科学生9498人，博士和硕士研究生2027人。全日制学生中，华侨华人和港澳台学生4401人，约占全日制在校学生的1/3。仅1978年改革开放以来，就有来自五大洲77个国家以及港澳台地区的学生先后在暨大就读。

二、华文教育

世界上现有3000多万华侨华人，他们人在海外，情系祖国、情系故乡，因此，华文教育既是一项语言教育事业，更是一项弘扬中华民族文化的社会事业，是关系到中华文化在海外华侨、华人中承传的事业，也是关系到定居海外的中华儿女基本权益的事业。三校作为侨校，义不容辞地担负起这一历史使命。学校根据各自优势，采取多种形式，大力开展华文教育。除专门从事华文教育的北京华文学院外，暨南大学、华侨大学也把华文教育作为办学的一个重要组成部分，两校重视华文教育和对外汉语教学，分别设有独立于校本部之外的华文学院。两校华文学院除了开展各类培训、研修、夏（冬）令营，外派汉语教师及编写华文教材外，还进行多层次的办学，形成了初中、高中、预科、专科、本科多层次的办学格局。还分别是国家汉语水平考试在华南地区和福建省的考点。去年，两校华文学院与北京华文学院一同成为国务院侨务办公室首批华文教育基地。

华侨大学集美华文学院，于1997年2月，由创办于1953年的集美华侨学生补习学校与华侨大学对外汉语教学部合并而成。校址在厦门集美。学院现有专任教师50人。专职对外汉语教师23人，其中副教授8人，讲师14人，助教1人；硕士7人，在读硕士 5 人；持有对外汉语教师资格证书的 19 人。

暨南大学华文学院，于 1993 年 6 月，由创办于 1953 年的广州华侨学生补习学校与暨大对外汉语教学系、预科部合并而成。校址在广州市广园东路。学院设应用语言学系、对外汉语系、预科部、函授部和华文教育研究中心共 5 个系级建制的教学科研单位，是国家对外汉语教师培训点和对外汉语教师资格证书考试点。现有大学预科、先修班学生近 400 人。学院拥有教师 97 人，其中教授 5 人，副教授 10 人，高级教师 8 人，讲师 41 人，助教 27 人，平均年龄 45 岁，中青年教师大部分有博士、硕士学位。

三、普通高等教育

21 世纪是科学技术迅猛发展的世纪。未来世界的竞争，归根到底就是人才的竞争。暨南大学、华侨大学作为华侨高等学校，十分强调培养适应国际竞争力的应用型、外向型人才，注意调整适合海外学生的专业设置和学科方向，实行学分制，并重视英语、汉语和计算机语言“三语”教学。同时，也注重开展国际学术交流与科技合作，分别与世界各地数十所高校和科研机构建立了合作关系。两校根据自身特点，在办学方面各有侧重。

华侨大学在学科、专业方面以工科为优，在人才培养上，强调“加强基础、拓宽专业、重视实践、培养能力”。通过深化教学改革，建立更加完善的教学科研与管理机制，以更好地培养外向型应用人才。

暨南大学学科较为齐全，注重多学科的相互渗透，在重视应用型的同时，更强调培养复合型人才。学校确立了“大力发展研究生教育，适度发展本科教育，积极发展华文教育”和教学、科研“两个中心”的办学思路。根据学校国际化、综合化、现代化的三大特点和优势，暨南大学提出了“侨校＋名校”的发展战略。为完成这一战略目标，学校持续强化教风、学风、校风建设，以“严、法、实”三个字来抓落实。“严”就是从严治教、从严治学、从严治校；“法”就是依法治校；“实”就是实事求是。有此保障，暨南大学一系列卓有成效的改革，必将进一步提升学校在海内外的地位，为海内外学子提供一流的教育。

以上是三所侨校的办学情况。希望各位朋友多多给予支持、指导，以利我们为祖国的侨务事业做出更大的贡献。

29. 珍惜“暨南人”光荣称号[①]

九月清秋，暨南大学珠海校园山林葱茏，丰草争茂。不久前，我们在此刚刚送走珠海学院第一届毕业生，今天又在学院首期工程如期竣工之际，迎来了 2001 年秋季入学的新同学。借用唐代刘希夷的诗句：“年年岁岁花相似，岁岁年年人不同。”你们这一批风华正茂的新生，的确有着与往年不同之处，那就是，你们是伴随新世纪钟声与 21 世纪同行的大学生。经过刻苦的学习、激烈的竞争，你们以优异的成绩考上了暨南大学，成为光荣的暨南人！在此，我谨代表学校领导和全校 3 万名师生员工，向来自全国各地、世界五大洲 26 个国家及港澳台地区的 4000 余名新生中，就读于珠海校区的 750 多名新同学表示衷心的祝贺和热烈的欢迎！借此机会，也向一直以来关心和支持珠海学院建设的珠海市委、市政府及香洲区前山街道办事处表示衷心的感谢！

暨南大学成立于 1906 年，是中国第一所由国家创办的华侨学府，是中国第一所面向世界招收留学生的学校，也是我国历史最悠久的几所大学之一。“暨南”二字源自《尚书·禹贡》“东渐于海，西被于流沙，朔南暨，声教讫于四海”，意即将中华文化远远传播到海外。早在建校初期，即制定校训：忠、信、笃、敬，注重以中华民族优秀的传统道德文化培养造就人才，要求学子们做到语言忠诚老实，行为敦厚严肃。学校坚定不移地贯彻“面向海外、面向港澳台”的办学方针，以弘扬中华文化，培养华侨华人子女为己任。创办以来，已为海内外输送了毕业生 6 万余人，其中不乏在事业上出类拔萃、声名远播的佼佼者，如中共中央政治局常委、国务院副总理李岚清，全国政协副主席、国务院原副总理吴学谦，泰国国会前主席、副总理许敦茂，新加坡著名侨领李光前等，均曾在暨南大学就读。

1996 年 6 月，暨南大学跨入中国大学百强行列，成为全国面向 21 世纪重点建设的 100 所大学之一。根据 1998 年开始的全国高校综合实力排名，我校当年名列全国高校第 87 位，1999 年位居第 72 位，2000 年上升至第 60 位，今年，我校已跃居全国第 40 位。今天的暨南大学，学科门类齐全，综合优势突出，师资力量雄厚。学校设有 16 个学院，即文学院、新闻与传播学院、外国语学院、理工学院、信息科学技术学院、经济学院、管理学院、法学院、医学院、药学院、生命科学技术学院、国际学院、中旅学院、珠海学院、华文学院和教育学院。我校是新中国第一所设有医学院的综合性大学。医学院现有 5 所附属医院：广州华侨医院、深圳市人民医院、珠海市人民医院、广州市红十字会医院、清远市人民医院，均是国家级三甲医院，另有一所直属医院，即深圳华侨城医院。目前，全校拥有 2 个博士后科研流动站、14 个博士学科专业、66 个硕士学科专业、39 个本科专业。全校现有全日制学生 13 000 余人，其中博士、硕士研究生 2500 人、本科生 10 000 多人，来自世界五大洲 30 多个国家和港澳台地区的学生 4500 多人（这在国内高校中是首屈一指的）。此外，学校还有继续教育学生近 10 000 人。学校现有专任教师 979 人，其中教授 171 人、副教授 416 人。学校在广州、

① 本文是在暨南大学珠海学院 2001 年秋季新生开学典礼上的讲话，珠海，2001 年 9 月 14 日.

深圳、珠海有 4 个校园，总面积 160 余万平方米。

林深鸟众，花香蝶来，名校自有名校效应。近年来，我校招生状况持续形势大好。今年，我校继续成为海内外学子报考的热门学校，生源质量优异，数量充足，共录取各层次全日制学生 4000 多人，其中研究生 914 人，本科生 3222 人，预科等其他学生 609 人。在内招生方面，继去年四川、安徽等省逾千人第一志愿争报我校十余个名额的情形之后，今年安徽省和重庆市又出现近千人踊跃以第一志愿报考我校的盛况。在外招方面，我校是港澳台和海外华侨华人社区青年学生在中国大陆求学的首选学校，今年共录取各类外招生近 2000 人。

同学们，在你们成为暨南大学新成员的同时，我想向你们提几点希望：

首先，希望同学们珍惜“暨南人”这一光荣称号，发扬“爱国爱校、团结奋进”的暨南精神，以暨南大学为荣，为暨南大学争光，以校训“忠信笃敬”作为立身处世的道德规范，树立正确的世界观、人生观、价值观，努力学习文化科学知识，学习做人的道理，成为一个对社会有用的人才。

其次，大学阶段的学习要求同学们更具主动性和创造性，要在学习中注意培养自己分析问题、解决问题的能力，尤其要注意培养自己的创造能力。光阴似箭，希望同学们珍惜在大学里学习的日子，珍惜青春年华。

同学们，你们来自海内外，在暨南大学新的环境里，希望你们尊敬师长，遵守校纪，互助互爱，共同进步。同时，也希望老师们给予新同学以更多的关心和照顾，让同学们充分感受到“暨南”家庭般的温暖，让我们一起来继续建设好这所校风、教风、学风优良的大学。

21 世纪是知识的世纪，是人才的世纪，和平与发展仍是时代的主题。作为新世纪第一批大学生，更应把握时代潮流，为世界和平与发展贡献自己的光和热。世界是属于你们的，未来也是属于你们的。

最后，我想套用一句流行语并转赠给同学们——“今天我以学校为荣，明天学校以我为荣”——我相信，同学们一定有这种精神与勇气。

30. 饮水思源[①]

在暨南大学建校95周年前夕，我们在这长江之滨，六朝古都，美丽的南京隆重集会，举行"暨南学堂纪念碑"揭碑仪式。在此，我谨代表位于祖国南大门广州的暨南大学3万余名师生向莅临揭碑仪式的各位领导、嘉宾和校友表示热烈的欢迎，向长期以来关心、支持暨南大学发展的海内外热心人士致以诚挚的谢意！

上一个世纪之交的1906年，暨南大学的前身暨南学堂由清朝两江总督兼南洋通商大臣端方在南京创办。1907年年初，从南洋开往上海的一艘普通邮轮上载有21名特殊的乘客，他们风华正茂，踌躇满志，带着家人乃至所有海外侨胞的殷殷希望，踏上了负笈暨南之路。他们就是学校的首批侨生。同年3月23日，暨南学堂在南京鼓楼薛家巷正式开学，端方亲临学校，并与师生合影以示重视，从而开启了我国华侨教育之帷幕。可以说，南京是暨南大学办学的起点，是我国华侨高等教育的摇篮，也是我国第一个招收留学生的学校。

"暨南"一词源自《尚书·禹贡》："东渐于海，西被于流沙，朔南暨，声教讫于四海"，意即将中华文化传播四海，将华夏精神名扬世界。对传统文化造诣极深且有考古癖的端方，对江宁提学使陈伯陶倡议用既深且僻的"朔南暨"典故，作为华侨学校的校名十分欣赏，即刻拍板允诺，暨南这个名字由此叫响。自此，学校一直秉承"朔南暨，声教讫于四海"的办学使命，以"宏教泽而系侨情"为办学宗旨，为海内外培育了万千英才。曾担任过新加坡国立大学首任校长的李光前先生，泰国国会前副总理许敦茂，我国国务院前副总理吴学谦，现任中共中央政治局常委、国务院副总理李岚清，著名革命烈士江上青等，早年曾求学于暨南大学。

悠悠岁月，弹指之间，近一个世纪之后的南京暨南学堂旧址已经找寻不到多少痕迹。回顾历史，当时只有几十人的暨南学堂，尽管迭遭播迁与停办，但它仍以顽强的毅力不断发展壮大，经过近一个世纪的历程，暨南学堂这棵幼苗已长成一棵参天大树，业已根深叶茂，植根于中华沃壤与海外华侨的心田。拨开历史的尘烟，回顾95年来的风风雨雨，我们从中领略到了暨南创办时的阵痛与磨难，成长时的蹒跚脚步，以及办学的艰辛与顿挫。在21世纪的第一个秋天，我们迎来了学校建校95周年这一喜庆时刻。近一个世纪以来，暨南与世纪伴行，与中华民族盛衰的历史相交织，与新中国奋斗崛起的历史相辉映。历经了晚清、民国和新中国三个不同发展时期的暨南大学，在中国高等教育史上烙下了深深的印记。

今天，我们从广州踏上寻根之源，在学校的发祥地南京举行"暨南学堂纪念碑"揭碑仪式，就是为了饮水思源，温故知新。暨南的发展凝聚了几代暨南人的心血与汗水。他们筚路蓝缕，以启山林之功，将永载暨南史册。暨南之所以名闻遐迩，是与其历史文化的积淀且拥有一流的师资不无关系的。暨南的骨架就是由一大批学术素养丰厚的杰出学者构成的，正如梅贻琦所言："所谓大学者，非谓有大楼之谓也，有大师之

① 本文是在暨南大学原址"暨南学堂纪念碑"揭碑仪式上的讲话，南京，2001年11月8日.

谓也。”黄炎培、严济慈、郑振铎、曹聚仁、钱钟书以及从暨大校友中走出的5位副总理，7位全国人大常委会副委员长，早已浓缩于暨南校史上，放射出耀眼的光芒。

阅尽95年沧海桑田，巍巍暨南，再次焕发勃勃青春。今日之暨南，已是国家面向21世纪“211工程”重点建设的100所大学之一，现拥有16个学院，35个系，39个本科专业，66个硕士学位授权学科，1个一级学科博士学位授权学科，13个二级博士学位授权学科，2个博士后流动站。学科门类齐全，暨南大学是新中国第一所设立医学院的综合性大学。学校师资力量雄厚。在广州、珠海、深圳分别设有4个校区，校园面积1.74平方公里。根据网大《中国大学排行榜》2001年综合指标排名，暨大位居全国高校综合实力第40名。国际化、现代化、综合化，是当代暨南大学作为华侨最高学府的三大特色，它吸引着海内外学子纷至沓来。改革开放20多年来，所培养的学生来自世界五大洲78个国家和港澳台地区。目前，在校各类学生共有23 809人，其中全日制本科生10 537人，研究生2529人，华侨华人、港澳台学生4893人，高居全国第一，无愧为“华侨学子的摇篮”。

从旧中国成立较早的著名国立大学，已成为国家重点建设的100所高校之一。暨南大学近百年的征程，虽历经劫难，却仍然雄风重振。这源于千千万万暨南人的不懈追求。这是薪火相传凝聚而成的“暨南精神”。何为“暨南精神”，我认为，校训“忠信笃敬”就是对它的深邃而简洁的阐释。而江泽民主席为我校校庆的题词“爱国爱校，团结奋进”则是对这一精神通俗而精辟的概括。“暨南精神”是所有暨南人的共同财富，拥有这一财富，暨南大学的未来将更加灿烂辉煌。

再过8天，11月16日，即是暨南大学建校95周年的喜庆日子。今天，我们在南京寻根溯源，举行“暨南学堂纪念碑”揭碑仪式，正式拉开了95周年校庆活动的帷幕。在此，我再一次代表暨南大学并以我个人的名义，感谢江苏省、南京市领导对暨南大学的关心与帮助，感谢暨南大学南京校友会为建立“暨南学堂纪念碑”所作的努力。最后，祝各位领导、校友、嘉宾身体健康、万事如意。

31. 为祖国侨务事业和华侨高等教育做出新的贡献①

领导同志能在百忙中与大家一起座谈，当面听取我们的汇报，我感到非常荣幸，先向各位领导表示由衷的感谢。下面，我就暨南大学近年来的发展情况及存在的主要问题作简要的汇报。

暨南大学是一所“面向海外、面向港澳台”办学的高等华侨学府。改革开放以来，尤其是“九五”期间，在党和国家及广东省委、省政府的关怀和重视下，学校在许多方面都取得了令人振奋的进步。自 1996 年成为国家“211 工程”重点建设大学后，学校在网大的全国高校综合实力排名持续上升：1998 年为第 87 位，1999 年为第 72 位，2000 年第 60 位，今年已跃居第 40 位。

目前，学校拥有 4 个校区、16 个学院、36 个系，涵盖了理、工、医、文、经、史、管、法和教育等九大学科门类。有 2 个博士后流动站，1 个一级博士学位授权学科，13 个二级博士学位授权学科，66 个硕士学位授权学科，39 个本科专业。5 所国家级三甲附属医院，1 所直属医院。共有各类在校学生 23 809 人，其中全日制本科生 10 537 人，研究生 2529 人；来自 38 个国家和地区的境外学生 4893 人，居全国第一位。

“九五”期间，国务院侨办与广东省政府就共建暨南大学达成了共识，使学校得到了更加有力的支持。学校依照“从严治校、从严治教、从严治学”的原则，坚持依法治校。在实践“侨校＋名校”的发展战略中，循着国际化、现代化、综合化的办学思路，以研究型、高水平社会主义华侨大学为目标，我们进行了一系列的调整和改革。

为突出“侨”字特色，学校针对三类不同来源的学生，制定了不同的培养目标。为满足境外学生的学习需要，学校 1996 年就开始实行双语教学，要求每个专业至少有一门课程用英语教学，目前为止，已有 56 门课程实现了英语教学，并率先在全国创设了全英语教学的国际学院，现已有两个专业（临床医学和国际经济与贸易）。为了照顾内外学生水平的差别，学校采用了弹性学分制，对不同的学生进行调节。

5 年来，学校共修建教职工住宅 160 000 平方米，修建学生宿舍 60 000 平方米，教学科研用房 50 000 多平方米，另外，大量图书、设备的充实和更新，以及电化教学手段的运用，使基础设施和教学科研环境有了很大改善。

与 1996 年“九五”初期相比，我校博士学位授权学科、在校学生总数和全日制本科生都分别增长了一倍，研究生增长了 4 倍多，境外学生增长了 2.5 倍，科研经费增长了 10 倍。同时，学校还填补了院士、博士后科研流动站、一级博士学位授权学科、国家级重点实验室、重点基地、重大科研项目等多项空白，在许多方面实现了零的突破。

我们知道，学校的每一项发展和进步都凝聚着党和国家的关心，广东省委、省政府的支持，同时，我们也得到了珠海特区、深圳特区、深圳华侨城、中国中旅、香港中旅、福建中旅等单位的大力帮助。我校“211 工程”即将转入以学科建设为重点的二期建设，我们热切希望教育部、广东省人民政府能与国务院侨办一起对我校“211”二

① 本文是在广东省部分高校校长座谈会上的发言，广州，2001 年 11 月 11 日．

期工程进行共建。

接下来，我就学校发展过程中遇到的问题提几点建议，并期望得到解决。

一、关于建立研究生院问题

近年来，随着学校办学重心的不断上移，办学结构不断优化，我校的研究生教育得到了长足的发展，现正处于全面上升的态势。学校现有中国工程院院士 1 人，2 个博士后流动站，80 个博士、硕士学位授权学科，已形成了含博士后、博士、硕士和本科教育在内的完整的人才培养体系。目前，有在校研究生 2529 人，其中外招生 461 人。在校研究生规模居广东第 3 位，全国前 30 位。1999 年 11 月，我校被教育部和国务院学位委员会评为“全国学位与研究生教育管理先进单位”。

如今，我校的办学水平、整体实力和研究生教育水平在全国同类高校中居于前列，在海内外尤其是在港澳台与东南亚地区具有良好的声誉，已基本具备试办研究生院的条件。同时，我校是侨校，是国家对外窗口学校，应与国际接轨，以便为国家的统一大业和现代化事业做出更大贡献。我们希望教育部能提供这个机会。

二、关于办学经费问题

1. 希望提高外招生拨款定额标准

暨南大学有特殊的办学使命，担负着培养海外华侨华人和港澳台青年的重任。随着我国综合国力的不断增强，特别是在加入 WTO 以后，港澳台和海外学生到我校求学的人将越来越多。但长期以来，由于国家给我校外招生拨款定额太低，仅按国内学生标准拨款，与所需投入相差太大，导致学校相应的基础设施建设滞后，至今没有留学生宿舍。对此，留学生意见很大，以致影响了我校的海外生源，同时也削弱了学校与台湾暨南国际大学争取海外生源的竞争能力。鉴于此，根据我校境外学生人数已占全日制学生人数 1/3 的实际，恳请国家能提高我校外招生经费人均定额标准，达到国家标准。

为此，我校建议从明年开始，参照教育部所属高校外招生的现行拨款标准，提高暨南大学外招生经费拨款人均定额，以后随教育部所属高校拨款标准的提高而提高。以今年我校外招生数量计算，两种标准的收费总差额高达 3500 余万元。

2. 关于解决海外及港澳台地区研究生收费标准问题

暨南大学现有外招研究生 461 人，对这部分外招生，学校给予了许多照顾与关怀，特别是学费标准多年来都低于教育部公布的收费标准。近年来，因办学成本越来越高，因此，学校根据教育部教学〔2000〕15 号文有关招收港澳台学生收费标准的规定，拟于 2002 年初适当提高收费标准。但今年 10 月份我校收到教育部教学〔2001〕16 号文，要求“自费全日制研究生的学费标准由招生学校参照本校招收同类专业委托培养研究生的标准收取”。

我校非常理解和支持国家的政策和规定，但我校外招生若参照委托培养国内生按年人均 10 000 元的标准收费，则比 2001 年之前的标准还低。在上级拨款没有对外招部分相应增加，而办学成本一直上升的情况下，我们认为本校办学方向与教育部所属院校有不同之处，外招研究生数量占全国总数的1/4，而且，我校是唯一一所经教育部批准单独对外招收研究生的大学，收费标准如果按以上规定收取，不仅会影响各专业点的发展，同时，在无相应拨款的情况下，将给我校研究生的培养带来相当大的经费困难。为此，我校希望国家能考虑到我校特殊的办学性质，酌情解决，或增加拨款，或按原办法执行。

3. 关于华侨医院的行政归属问题

广州华侨医院于1978年10月经国家计委批准建立，1982年经教育部批准成为暨南大学医学院附属医院。一直以来，党和国家非常重视华侨医院的建设和发展，中共中央、国务院在〔1983年〕24号文中明确指出，由暨南大学代管的广州华侨医院是为了解决海外华侨回国看病和医学院教学需要而建立的，并特别提到“对暨南大学代管的广州华侨医院的基本建设投资及卫生事业费，要给予保证”，但因隶属关系，医院的资金来源一直没有得到落实。医院每年的财政补贴只有床位费150万元，远远低于广州市同级医院的补贴标准。由于缺少专项经费，医院建设速度很慢，远远不能满足对海外学生教学和临床的需要。

随着医改的实施和深入，旧的问题尚未解决，医院又面临新的困难：首先，医院未被纳入到广东省和广州市的区域卫生规划，得不到省市的支持；其次，第一附属医院属国家办的非营利性医院，价格要受当地政府的控制，只能进行保本经营或亏损经营；再次，医院以前赖以生存和发展的药品收入将从医院分割出去，各类检查收费标准也要下调，医院得不到后续发展资金；最后，医院建院初期建设的房屋和购买的设备逐渐陈旧，大批人员都要离退休。

为此，我校在国务院侨办的支持下曾多次前往国家计委、卫生部和财政部要求解决医院的财政补助户头问题，但都因归口管理问题而未能落实。现建议如下，以期解决。

（1）学校通过与广东省卫生厅共建形式，解决第一附属医院在广东省区域卫生规划地位问题。

（2）根据医院的性质和使命，广州华侨医院应定为享受政府补贴的非营利性医院。按照《关于卫生事业补助政策的意见》精神，政府对非营利性医院的经费支持应包括财政补助和基本建设投资。

4. 关于启动货币分房资金问题

根据省市住房制度改革的要求，我校已于1999年底全部完成了房改工作。现为让未参加过房改的教职工早日享受国家规定的优惠政策，使教职工安居乐业，根据广东省政府《关于加快住房制度改革实行住房货币分配的通知》，我校拟启动住房货币分配。依照上文的规定，住房货币分配的资金来源应是“各级财政纳入预算，原用于住房建设、维修和房租补贴的资金”。经测算，我校住房货币分配一次性启动资金为2900万元，恳请国家财政予以划拨。

一直以来，暨南大学坚持“面向海外、面向港澳台”的办学方针，致力为海外华侨华人和港澳台同胞培养人才，是祖国大陆向海外传播中华文化、维系海外中华儿女亲情和开展中外文化交流的一条纽带，为港澳回归及海外华侨华人的教育事业做出了重要贡献。教育体制改革后，我校在国务院侨办仍保留着特殊的政治意义。1983年，中共中央和国务院专门就如何办好暨南大学和华侨大学发了第 24 号文，确定两校为“国家重点扶持的大学”，“对这两所华侨大学，应当作为教育战线的重点项目进行投资”。但多年来，教育经费投入不足已经严重影响学校的发展，尤其是教学实验基础设施和侨生的住宿条件，已影响了我校在海外的招生数量。因此，我们期望国家能加大对我校的投入，给学校以大力支持。我校决不辱使命，为祖国侨务事业和华侨高等教育做出新的贡献。

32. 建设国际化、现代化、综合化的高水平社会主义华侨大学①

在21世纪的第一年这激动人心的时刻，来自海内外的各位嘉宾、校友和社会各界热心人士聚首暨南园，共贺暨南大学95周年校庆。在此，我谨代表暨南大学3万名师生员工向出席今天庆典的各级领导、校董、校友和来宾表示热烈的欢迎！向长期以来关心和支持暨南大学建设和发展的海内外朋友致以最诚挚的谢意！同时，我们仍不忘缅怀和感谢曾在暨南大学创业和工作过的先辈们，并对他们表示深深的敬意。

从1906年创办的暨南学堂，到今天跻身于中国高校百强、享誉海内外的高等华侨学府，近百年来，暨南大学可谓是历经沧桑，风雨兼程。她与岁月同歌，与中华民族的命运同沉浮。虽屡经变迁曲折，但总是顺应时代的潮流而不断前进。学校始终以“宏教泽而系侨情”为办学宗旨，以“朔南暨，声教讫于四海”为办学使命。建校95年来，共为祖国大陆和海外、港澳台地区培养各类高素质人才7万余人。仅改革开放以来，就有来自世界五大洲79个国家以及港澳台地区的学生先后在暨南就读过。

暨南大学是祖国大陆向海外传播中华文化，维系海内外中华儿女亲情和开展中外文化交流的一条重要纽带，在我国高等教育史上有着特殊的地位。她不仅是中国历史上第一个华侨高等学府，是中国历史上第一个招收留学生的大学，是1952年全国院系调整以后第一所设立医学院的综合性大学，也是目前中国大陆高校中招收海外及港澳台学生人数最多的一所综合性大学。改革开放以来，暨南大学的建设和发展得到了党和国家的高度重视，同时也赢得了校董、校友、海外华侨华人、港澳台同胞及社会的广泛支持。暨南大学的发展一日千里，在许多方面都取得了令人振奋的进步，综合实力日渐增强，办学规模不断壮大，办学层次稳步提高，在海内外的影响日益扩大。学校1996年成为国家“211工程”重点建设的大学后，在全国高校的综合实力排名不断上升：1998年为第87位，1999年为第72位，2000年第60位，今年已跃居第40位。今日之暨南大学已是一所学科门类齐全，师资力量雄厚，科研实力强劲，誉满海内外的综合性华侨大学。

目前，学校拥有4个校区、16个学院、36个系，涵盖了理、工、医、文、经、史、管、法、教育等九大学科门类。有2个博士后流动站，1个一级博士学位授权学科，13个二级博士学位授权学科，66个硕士学位授权学科，39个本科专业，5所附属医院，1所直属医院。共有各类在校学生23 809人，其中全日制学生13 789人，包括研究生2529人；来自38个国家和地区的海外及港澳台学生4893人，居全国第一位。与1996年建校90周年时相比，我校博士学位授权学科、在校学生总数和全日制本科生都分别增长了1倍，研究生增长了4倍多，海外及港澳台学生增长了2.5倍，科研经费增长了10倍。同时，学校还填补了院士、博士后科研流动站、一级博士学位授权学科、国家级重点实验室、重点基地、重大科研项目等多项空白，在许多方面实现了零

① 本文是在暨南大学建校九十五周年庆祝大会上的校长致辞，广州，2001年11月16日；暨南大学校报，2001年12月3日.

的突破。

系侨情，任重道远；宏教泽，只争朝夕。我们在回顾总结暨南大学 95 年风雨沧桑与辉煌成就的同时，更对新世纪暨南大学的光辉前景充满着期盼与渴望。“侨校＋名校”是暨南大学的发展战略，走研究型大学的道路是暨南大学的发展方向，我们正循着国际化、现代化、综合化的高水平社会主义华侨大学的目标奋进。面对新的机遇与挑战，我们信心百倍。我们坚信：在国务院侨办和广东省人民政府的直接领导下，在海内外校友、校董和所有热心华侨、华文教育事业的朋友们大力帮助和支持下，暨南大学全体师生员工恪守“忠信笃敬”的校训，弘扬“爱国爱校，团结奋进”的精神，团结一致，开拓进取，暨南大学的明天一定会更美好，暨南大学的前景一定会更灿烂。让我们一起努力，再创辉煌，迎接暨南大学的百年华诞！

33. 深深感谢校友深情[①]

大家不辞辛劳，从国内外各地前来参加母校 95 周年校庆活动和校友总会理事会会议，体现了校友们对母校的一片深情。我谨代表母校师生员工向大家表示热烈的欢迎，向会议表示热烈祝贺，并对你们以及通过你们向全世界校友 95 年来对母校的大力支持表示衷心感谢！

在历史刚刚跨进 21 世纪之际，我们举行暨南大学建校 95 周年的庆祝活动，既是为了总结过去，更是为了开创未来。暨南大学建校以来，始终以弘扬中华文化、培养华侨子女为办学宗旨。暨南的兴衰荣辱、变化发展，都与海外侨胞的命运紧密相连。暨南大学为国内和海外华侨教育事业作出了重大贡献，在海外侨胞中享有很高声誉。改革开放以来，暨南大学进入了一个新的蓬勃发展的时期，学校坚定不移地贯彻“面向海外、面向港澳台”的办学方针，不断深化教育改革，扩大对外开放，致力为海外和港澳台地区培养人才。自从 1996 年暨大进入国家“211 工程”行列以来，学校各方面工作有了很大发展。学校不仅办学规模扩大了，而且优化了办学结构，提高了办学层次，改善了办学条件，教学科研工作上了一个新台阶。学校现设有 16 个学院，36 个系，39 个本科专业，66 个硕士点，14 个博士点（其中一级学科博士点 1 个），博士后科研流动站 2 个。全校博士、硕士研究生达到 2529 人，本科生突破万人大关。学校办学实力和对外竞争力、吸引力大大增强，华侨、华人、港澳台学生报考暨大人数逐年增多，现全校在读的华侨、华人、港澳台学生共有 4893 人，是全国大学中华侨、华人、港澳台学生最集中、人数最多的大学。95 年来我校已培养各层次毕业生 7 万多人，暨南校友分布于世界五大洲、祖国各省市。现国内外共有暨南校友会 56 个。

广大暨南校友对母校怀有深厚的感情。校友总会成立以来，大大加强了国内外校友与母校的联系，促进了各地校友会之间联谊活动的开展，互相沟通信息，增进了相互了解和友谊。各地校友会致力弘扬暨南精神，团结校友，互助共进。校友总会对团结全球各地校友会和广大暨南校友，发挥了枢纽、桥梁、纽带作用，大大增强了广大暨南校友对母校的向心力和凝聚力。各地校友会和广大暨南校友心系母校，多方关心和帮助母校工作，积极参加母校的重要活动，为母校的建设和发展作出了重大贡献。母校所取得的成就，与各地校友会和广大暨南校友的支持是分不开的。在此，我谨向各地校友会和广大暨南校友致以衷心的感谢。

校友总会成立以来，进行了卓有成效的工作，取得很大成绩。黄旭辉理事长已作了全面总结。今天将选举产生新一届理事会。我相信，在第二届理事会领导下，校友总会将继承和发扬暨南的优良传统，高举爱国爱校的旗帜，进一步推动全球暨南校友的大团结。校友楼已经建成，它将是校友会活动的重要基地之一，有利于加强总会与各地校友会的联系，也便于开展校友的联谊活动。无论校友总会，还是各地校友会，

① 本文是在暨南大学校友总会第二届理事会成立大会上的讲话，广州，2001-11-16；暨南校友，2001，4：6-7.

都应成为广大校友温暖的大家庭，大大互相关心，互相帮助，共同进步，为国家、为社会多作贡献，为母校争光。

21世纪，学校实施“侨校＋名校”发展战略，要把暨南大学办成水平一流、享誉国内外的大学。要实现这一目标，既要依靠全校师生员工的团结奋斗，也有赖于校友们的帮助和支持。母校的声誉与校友们的利益息息相关，母校的声誉提高了，校友们在社会上就会得到人们的尊重和重视。我相信，暨南大学的校友会，对推动社会进步和学校的发展，必将能发挥积极的作用。

希望各地校友会今后进一步加强与母校和校友总会的联系，欢迎校友们方便的时候多到母校看看，对母校的工作多给予帮助、指导。

祝大家身体健康、家庭幸福、事业成功！

谢谢大家。

34. 办好研究生教育至关重要①

21世纪的第一个初冬已经来临，但美丽的南国花城依旧鲜花盛开，气候宜人。刚刚度过95周年华诞的暨南大学，如期迎来了“海峡两岸和港澳地区学位与研究生教育研讨会”的胜利召开。在此，我谨代表暨南大学对各位领导以及来自内地和港澳台地区的各位专家、教授的光临表示热烈的欢迎！

暨南大学是中国第一所由国家创办的华侨高等学府，第一所向海外招收留学生的大学，是新中国第一所设立医学院的综合性大学，在中国高等教育史上享有重要地位。学校始终以“宏教泽而系侨情”为办学宗旨，以“朔南暨，声教讫于四海”为办学使命。建校95年来，共为祖国大陆和海外、港澳台地区培养各类高素质人才7万余人。仅改革开放以来，就有来自世界五大洲79个国家和港澳台地区的学生先后在暨南大学就读过。

近年来，尤其是“九五”期间，学校得到了党和国家的高度重视，同时也赢得了社会各界的广泛支持，在许多方面取得了令人振奋的进步。1996年，暨南大学成为国家“211工程”100所重点大学之一，正式跻身于中国高校百强。自1998年开始，学校在全国高校的综合实力排名不断上升：1998年为第87位，1999年为第72位，2000年第60位，今年已跃居第40位。今日之暨南已是一所学科门类齐全，师资力量雄厚，科研实力强劲，誉满海内外的综合性华侨大学。

目前，学校拥有4个校区、16个学院、36个系、39个本科专业，涵盖了理、工、医、文、经、史、管、法、教育等九大学科门类，拥有3500张病床的5所国家级三甲附属医院和1所直属医院。学校共有各类在校学生23 809人，其中全日制学生13 789人，包括来自世界38个国家和地区的境外学生4893人，居全国第一位。

“九五”期间，我校高度重视学位与研究生教育，办学重心不断上移，取得了令人瞩目的成绩。与“九五”初期相比，学校的博士学位授权学科增长了1倍，研究生规模扩大了4倍多。填补了学校过去无院士、博士后科研流动站、一级博士学科等多项空白。学校现有2个博士后流动站，1个一级博士学位授权学科，13个二级博士学位授权学科，66个硕士学位授权学科。有在校研究生2529人，另外还有研究生课程进修班学生2500人；研究生中有来自10多个国家和港澳台地区的学生461人。

在新的世纪里，我校将继续努力构建面向新世纪、体现学校办学优势与特色的专业学科群。通过建设一个高水平、高素质的师资队伍，若干个一流的学科，全面提高人才培养质量、科研水平和办学层次。按照“侨校＋名校”的发展战略，将暨南大学建成一个新知识、新思想、新理论的摇篮，建成一所国际化、现代化、综合化的高水平社会主义华侨大学。

本次会议在暨南大学召开，我们深受鼓舞，但更感鞭策。出席今天会议的都是在

① 本文是在海峡两岸和港澳地区学位与研究生教育研讨会开幕式上的讲话，广州，2001年11月30日.

学位与研究生教育方面取得突出成就的高校，是海峡两岸暨香港、澳门研究生教育的主体力量，我们希望得到大家的支持和帮助，对我们的工作多提宝贵意见。

最后，预祝大会圆满成功！祝各位身体健康，生活愉快！

35. 进一步提高干部人事档案工作的管理水平[①]

今天，国务院侨务办公室在我校举办侨办系统干部人事档案培训班，这不仅为我们提供一次相互学习、促进交流、共同提高的机会，也为侨办系统最终实现干部人事档案的制度化、规范化、科学化管理奠定了基础。在此，我谨代表暨南大学向前来参加培训班的领导和同志们表示热烈的欢迎！

暨南大学与世纪同行，已走过了95年的风雨历程，不仅为海外培养和输送了大批人才，而且为海外华文教育的发展做出了重要贡献。在国务院侨办的领导和关怀下，在广大海外侨胞、港澳台同胞和校董、校友的大力支持下，经全校师生员工的努力奋斗，暨南大学在改革开放的道路上阔步前进，成为全国面向21世纪重点建设的100所大学之一，学校在教学、科研等方面都有了长足发展，干部人事档案目标管理工作也取得了可喜的成绩。

1996年，中央组织部制定了在全国实行干部人事档案工作目标管理的方针，提出“努力实现2000年干部人事档案工作目标的任务要求”，我校按照广东省委组织部的部署，在国务院侨办的直接领导下，认真、细致、积极有效地开展了达标升级的各项工作。经过三年的努力，我校人事档案管理工作有了实质性的进展。2001年1月8日，顺利通过了国务院侨办、广东省委组织部6人专家考评小组的验收，并经中央组织部审核，国务院侨务办公室以人发〔2001〕117号文正式批准我校为“干部人事档案目标管理一级单位”。这也是广东省高校第一个干部人事档案“目标管理一级”单位。

下面我将从管理体制、规章制度、队伍建设、计算机管理以及达标后如何对档案的动态实行规范化管理等方面，谈谈我校实施干部人事档案目标管理的具体做法。

一、领导重视是实现干部人事档案工作目标管理的前提

学校在开展干部人事档案工作目标管理的过程中，得到各级领导的高度重视和大力支持。学校专门成立了档案工作领导小组，建立了以系、部、处为单位的档案工作联系网络，把实现干部人事档案工作的目标管理提上学校的重要工作日程，制订出5年工作规划和具体实施方案，并逐项逐年检查落实情况。

(1) 校长亲自负责档案工作领导小组的领导工作，人事处长兼任档案工作领导小组副组长。针对人力不足、设备短缺、库房拥挤等问题，学校及时拨资金、添设备、配备人员、扩建档案用房，改善办公条件。此外，校领导还亲自察看档案室办公条件的改善情况和档案整理情况，对档案管理工作给予大力支持。副组长除定期听取汇报外，还积极参与研究部署档案管理和干部档案信息库的建立工作，亲自督促、检查达标落实情况，使人事档案管理逐步走上现代化管理的轨道。

(2) 加强档案管理的“硬件”建设，为实现目标管理创造条件。由于领导重视，资金到位，工作条件在以下几个方面得到彻底改善。

---

① 本文是在全国侨务系统干部人事档案培训班上的讲话，广州，2001年11月30日.

库房达标。学校对档案库房进行了全面的装修，扩建了阅档室、储藏室。库房环境良好，布局合理，“三室”分开，总面积达230平方米，完全符合规定标准。

设备齐全，措施到位。学校对档案铁柜、打孔机、切纸机等档案设备全部进行了更新，配置了气体消毒柜、复印机、碎纸机、空调、抽湿机、温湿度计及灭菌杀虫药物，落实了防火、防盗、防光设施。此外，学校还将在职干部档案卷夹全部换成了中央组织部规定的新型档案卷夹。

各级领导的重视和鼓励大大激发了管档干部的工作热情，他们勇于创新，积极进取，不仅自行改编创建了“汉字对角号码”用于档案的科学编排，而且还不断健全管理制度，完善工作网络，立足收集整理，狠抓基础建设，充分利用计算机管理干部档案，积极主动地开展工作。

二、健全和遵守各项规章制度，是实现目标管理运作制度化的根本保证

干部人事档案管理的八项工作制度，是开展干部档案工作的保障，但如果这些制度仅限于写在纸上、挂在墙上，那只能是纸上谈兵，只有在工作中认真付之于行动，才能确保档案管理质量的提高。对此，我们的具体做法是：

(1) 在建立健全各项规章制度的基础上，学校成立了“暨南大学干部人事档案工作领导小组”，下设组长1名、副组长2名、组员13名、档案联络员83名，档案联系网络遍布各院、系、部、处。我们积极利用工作网络开展工作，及时收集所缺材料，以确保档案的完整性。

(2) 认真履行工作职责，严格执行查借阅制度。除组织、人事、保卫和纪监审部门外，其他部门查档一律凭查档介绍信。针对个别单位领导派出一般工作人员前来查阅干部考核成绩的现象，我们要求各单位认真组织中层干部学习《干部人事档案查借阅制度》，严格按章办事。

(3) 针对学校某些部门材料收集中存在的不足，我们有效利用联系网络，加强沟通与联系。当发现材料不齐的档案时，我们采取分类登记造册，及时分发到形成档案材料的相应部门共同追索。对确实索取不到的材料，要求单位档案联络员通知本人提供证书原件，尽可能保证档案材料的齐全。

(4) 为把好送交归档材料质量关，我们下发了“送交归档材料工作制度”，并对归档材料提出了具体的要求。对不合格的材料，直接向档案联络员提出要求，合格后补充归档。同时，在收集材料的过程中，我们还认真做好真伪材料的鉴别，把好档案材料归档关。所有材料须经有关部门审核，复印件盖章后才可归档。拒收个人或非组织渠道报送的材料，对有疑问的材料，及时请示汇报，交由组织、纪检部门协助调查处理，以确保材料的真实性。由于掌握了收集材料的主动权，改变了下面送什么我们收什么的被动局面，使档案材料的质量从根本上得到了提高。

三、建设高素质的档案队伍，增强改革创新意识，是实现干部人事档案目标管理的根本途径

(1) 组织档案管理人员参加广东省委组织部举办的“干部档案管理培训班”和本

校的计算机培训学习，使大家了解中央关于开展干部人事档案目标管理的有关精神，熟知人事档案工作的有关规定和政策，掌握档案材料的收集、鉴别、分类、编排、归档、保护等各个环节的技术操作，运用现代化的手段努力做好干部档案的管理工作。

（2）认真组织协助档案整理的各院系总支秘书、支委及退休党员干部进行培训学习。学校先后共举办了干部档案整理培训班六期，组织参训人员观看广州市委组织部录制的《干部人事档案管理》录像，自行编写和印发了“整理人事干部档案操作程序”“档案目录书写格式范例”“剪切装订技术加工”等材料，作为培训班主要授课内容。培训工作对档案整理起到了重要的指导作用，使全校在职干部人事档案的质量和外观得到了根本改观。

（3）为提高工作效率，我们在四角号码的基础上自行改编创建了“汉字对角号码”。此法的特点是：简便、快捷，看名知号。结合全校人员姓氏的分布，经过几十次的拆分组合，增设了撇、折、框和三点水，使各类档案数目相对平均。去年，我们根据这种方法将工资单、干部履历表和年度考核表等 20 000 多份材料进行编码分类，仅用了两周时间就全部归入个人档案，工作效率大大提高。“汉字对角号码”的采用，增强了利用档案号进行检索、查询、统计的功能。

（4）认真研讨和处理工作中出现的新情况、新问题。针对年度考核表逐年增加而且利用率高的现象，我们本着既保证利用又保护人事档案的原则，经学校及省委组织部档案部门批准，我们建立了“业务考核专卷”，单独装袋，独立装柜，使这个问题得到妥善解决。

（5）为及时掌握干部的变动情况，我们于 1998 年组织全校统一填写了一次《干部履历表》，由各单位档案联络员负责把好填写第一关，档案室复查合格后，交组织部和人事科审查盖章后归档。填写合格率达 99%以上。

（6）注意收集新信息。通过学校简报、校刊、会议等多种渠道，档案室及时收集新的材料。譬如，我们就曾利用学校福利分房的机会，主动收集到一些出国定居的教职工名单和某些退休回乡的老教工去世的证明材料，保证了档案的完整性。

四、应用计算机进行数据化管理，为开发高校人事档案信息资源做好基础工作

为使干部人事档案逐步实现现代化管理，我校人事处建立了“人事管理信息系统”，该系统具有档案信息查询、检索、统计等功能，系统内有较完整的干部个人资料。人事处内部实行系统联网，各科室及时输入本部门产生的最新数据，做到资源共享，优势互补，以更好地发挥群体优势。

（1）人事管理信息系统给查询工作带来了极大的方便。我们把职工的基本数据和信息输入计算机，人事处各科室从系统上便可直接了解职工的基本情况。根据规定，我校每年的技术职称晋升和两年越级晋升工资工作都需要出具 2～5 年的考核成绩，以往是由各单位领导逐个查档，既费时费力，档案损害程度又高。现在只需进入系统调出便可查看，既保护了档案，又提高了工作效率。

（2）解决了档案中姓名用字不规范的问题。在信息输入的过程中，发现现用名和曾用名混用，使用同音字、形近字的现象，我们立即通知单位联络员与本人联系，由

其提供本人的身份证和户口簿，复印后审核盖章归档，使姓名用字统一规范。

(3) 用于人事档案数据的统计。根据需要输入相关条件进行检索、筛选、排序、统计，不仅方便快捷，还能及时掌握职工单位变动情况，方便了档案的分类管理。

五、重视达标后的档案管理工作，促进高层次目标管理的实现

我校的干部人事档案管理工作，努力做到了制度健全、操作规范，已初步实现干部人事档案的目标管理，并通过了一级考核标准的检查验收。但是，人事档案的管理是动态的，即使已经达标，管理工作也不能只停留在原来的水平上，应根据情况的变化而不断增加新内容。为此，我校现以充实档案内容、提高卷宗质量、加强人事信息管理为重点，制定了新的归档操作步骤：

(1) 接收档案登记。

(2) 陈旧材料的消毒处理。

(3) 变动信息及时输入人事信息管理系统。

(4) 编码分类归入个人档案。

(5) 整理装订及时入库。

人事档案工作是党和国家人事工作的重要组成部分，它为组织上全面考察了解和选拔使用人才提供依据。因此，我们将进一步加强和提高干部人事档案工作的制度化、规范化、科学化管理水平，更好地为学校的组织、人事、教学、科研和管理工作服务。

最后，希望大家对我们的工作提出宝贵意见。

36. 胸怀世界 放眼未来①

金秋九月，风清气爽，硕果飘香。为迎接来自内地31个省区市、香港和澳门特别行政区、台湾，以及海外32个国家的攻读博士、硕士、学士学位的莘莘学子，今天，我们在这修葺一新的现代化校园里隆重举行暨南大学2002年秋季新生开学典礼。在此，我谨代表学校3万名师生员工对新同学的到来表示热烈的欢迎。同时，也向关心暨南大学建设和发展的广东省委、省政府和大力支持珠海学院建设的珠海市党政领导和珠海人民致以由衷的谢意和崇高的敬意，向为完成珠海学院二期工程付出大量心血的所有教职员工表示衷心感谢和诚挚问候。

“暨南”二字源自《尚书·禹贡》“东渐于海，西被于流沙，朔南暨，声教讫于四海”，意即朝向南方（东南亚）办学，这就是说要将中华文化远远传播到海外。成立于1906年中国第一所由国家创办的华侨学府由此而得名。暨南大学是中国第一所面向世界招收留学生的学校，也是我国历史最悠久的几所大学之一，她的创办为中国教育史涂上了浓墨重彩的一笔。早在建校初期，学校就制定校训：忠、信、笃、敬，注重以中华民族优秀的传统道德文化培养造就人才，要求学子们做到语言忠诚老实，行为敦厚严肃。长期以来，学校坚定不移地贯彻“面向海外、面向港澳台”的办学方针，“宏教泽而系侨情”。96年来，暨南大学名师荟萃，巨匠云集，为海内外输送了大批人才，如中共中央政治局常委、国务院副总理李岚清，全国政协副主席、国务院原副总理吴学谦，泰国国会前主席、副总理许敦茂，新加坡著名侨领李光前等，就是其中的代表。

暨南大学是祖国大陆向海外传播中华文化，维系海内外中华儿女亲情和开展中外文化交流的一条重要纽带，在我国高等教育史上有着特殊的地位。改革开放以来，学校的建设和发展得到了党和国家的高度重视。近日，我校向国家提出的加大投入、改善现状的请示在党中央和国务院领导的直接关怀下，已得到批准。国家决定在“十五”期间对我校给予5.38亿元人民币的投入，为我校实现“十五”总体目标提供了资金保障。

1996年6月，暨南大学跻身中国大学百强，成为全国面向21世纪重点建设的大学之一。以此为契机，学校抓住机遇，深化改革，成果显著。今年上半年，我校被教育部《中国高等教育评估》杂志列为中国77所研究型大学之一，名列第53位；在网大《中国大学排行榜》的高校综合实力排名中，我校已从1998年第87位跃居至第37位。学校现已圆满完成“211工程”建设一期任务，从9月16日开始正式进入“十五”“211工程”第二期建设时期。今天的暨南大学，学科门类齐全，综合优势突出，师资力量雄厚。学校设有16个学院，即文学院、新闻与传播学院、外国语学院、理工学院、信息科学技术学院、经济学院、管理学院、法学院、医学院、药学院、生命科学技术学院、国际学院、中旅学院、珠海学院、华文学院和教育学院。其中中旅学院是

① 本文是在暨南大学2002年秋季新生开学典礼上的讲话，珠海，2002年9月19日.

内地首家通过世界旅游管理专业教育质量认证的高等旅游学院。我校是新中国第一所设有医学院的综合性大学。医学院现有 6 所国家级三甲附属医院，一所直属医院，一所专科医院，共有病床 3945 张。目前，全校拥有 2 个国家重点学科、2 个博士后科研流动站、14 个博士学位授权学科、67 个硕士学位授权学科、43 个本科专业。最近，我校还成为国务院学位办批准的全国 30 所开展高级管理人员工商管理硕士（EMBA）的学校之一。全校现有各类学生 25 000 余人，其中全日制博士、硕士研究生 3245 人，本科生 11 000 多人，来自世界五大洲 44 个国家和港澳台 3 个地区的学生 6000 人。学校现有院士 2 人，专任教师 982 人，其中教授 155 人、副教授 407 人。学校的 4 个校区分别坐落在广州、深圳、珠海，总占地面积 170 余万平方米。

暨南大学既是一所学科齐全的综合性大学，同时也是一所传统体育强校。在 96 年的办学历程中，我校在体育方面荣获过不少殊荣。早在 1936 年，暨南大学学生就代表中国出征在德国柏林举行的第十一届奥运会。1995 年，我校被教育部批准为试办高水平运动队的大学。此后多年来，我校同学分别在各类国际国内比赛中夺金摘银，名扬海内外。将于 9 月 29 日至 10 月 14 日在韩国釜山举行的第十四届亚运会上，我校有包括“中国飞人”之称的徐自宙等 8 名同学将分别参加四个项目的角逐。

林深鸟众，花香蝶来，暨大上下同心戮力创名校的结果是，吸引了众多优秀学子纷纷前来求学。今年，我校继续成为海内外学子报考的热门学校，生源不仅数量充足，而且质量稳步提高，在录取的 7069 人中，有研究生 1153 人，本科生 4261 人。内招生方面，各地生源继续出现火爆局面。在安徽省，仅第一志愿报考我校的就有 6100 多人，重庆也出现逾千人争报我校几十个名额的情形。外招方面，我校已成为港澳台和海外华侨华人社区青年学生赴中国大陆求学的首选学校，共录取各类外招生 3600 多人。外招生的踊跃加入使得我校的外招生录取人数自 1978 年以来首次超过内招生录取人数。顺时代之需，应运而生的我校国际学院，在全国率先采用全英语教学，有“国际经济与贸易”、“会计学”和“临床医学”三个专业，尽管去年才成立，但今年已经显示出它迅猛的发展势头，招生人数比去年翻了一番。

同学们，当置身于这美丽的现代化园林式校园中时，你们可曾想到，两年前，这里还只是一片未曾开垦过的山地。也就是说，学校仅用了两年多的时间，就完成了这所集教学、运动、饮食、住宿等功能于一身的全新校园的建设。这是团结的力量，是智慧的结晶，更是珠海市委、市政府和珠海人民大力支持的结果。暨南大学珠海学院是一所综合性学科的学院，创建于 1998 年，是中国第一所已在珠海培养出本科生和专科生的大学。目前学院有博士、硕士和大学一至四年级的本科学生 4200 人，已形成了一个完整的人才培养体系。

同学们，今天的开学典礼隆重而又特别。为改善本部的校园面貌和学生住宿条件，本学期学校在校园建设方面做出了重大调整。学校要充分利用珠海学院优越的硬件设施，把新生全部安排到这里学习，以便对校本部那些破旧的学生宿舍进行彻底改造。这是对学院的信任，同时也是对学院的考验。学校将会对学院的工作予以大力支持。在学校的支持和珠海学院的努力下，我相信同学们在这里必将有一个崭新而又充满希望的开始。借此机会，我对同学们的学习和生活提几点希望。

首先，在同学们成为光荣的暨南人以后，我希望大家恪守“忠信笃敬”的校训，以其作为立身之本和行事准则，做到言忠信，行笃敬。因为只有先学会做人，而后才能做好学问。要遵纪守法，修身自爱，积极向上，弘扬暨南文化，光大暨南精神，做一个名副其实的暨南人。

其次，大学阶段是一个特殊的学习阶段，它不仅对同学们的自觉性提出高要求，而且还需要同学们在学习、科研中发挥独立自主、勇于探索、勇于创新的精神。在学校为同学们提供的相对自由的空间里，希望同学们好好珍惜，本着对自己、对学校、对社会负责的态度努力学习，使自己在德、智、体方面得到全面提高。

21 世纪是知识的世纪、信息的世纪，和平与发展仍是时代的主题。作为新世纪的大学生，应该胸怀世界，放眼未来，力争做一个对社会、对时代有用的人。

同学们，你们来自世界各地各个不同的地方，大家相聚在暨南园非常难得。希望你们遵守校纪，互助互爱，共同营造一个美好的学习、生活氛围。同时，也希望老师们给予新同学以更多的关爱和照顾，让同学们充分感受到暨南园家庭般的温暖，大家一起携手建设好这所校风、教风、学风优良的大学。

最后，祝同学们身体健康，学习进步。

37. 持之以恒　依法治校[①]

我校自1996年6月成为全国“211工程”重点大学以来，不断深化各项改革，并根据国家有关法律法规、政策和上级有关规定以及学校实际情况制定了一系列管理文件，现汇编出版，这是经过长时间实践、探索和反复酝酿的结果。在高等教育竞争日趋激烈和高校管理工作日益规范的今天，加强建章立制、提高管理水平已显得更为迫切和必要。本书的出版，是我校对高校行政管理工作的有益探索，也是我校加强行政管理工作的重要措施。

“九五”以来，我校在“211工程”建设、教学与科研等方面都取得了显著成效，办学规模进一步扩大，办学层次与水平、综合实力不断提高，软硬件设施得到了大幅改善。当前我校呈现良好的发展态势，但也感受到了外部激烈竞争所带来的压力。要使学校在现有基础上继续向前发展，我们除了在“211工程”建设、教学与科研、学科建设、师资队伍建设、人才培养模式等方面下功夫以外，还必须进一步加强学校的行政管理工作，在学校的建设和发展中更好地发挥组织、协调、促进作用。

多年来，我校一直在倡导和坚持“从严治校、从严治教、从严治学”的办学原则，大力推进“校风、教风、学风”建设；致力于建立“依法治校”的管理机制和“实事求是”的工作作风，力求实现行政管理工作的制度化、现代化和科学化。应该说，这项工作是有成效的，它所带来的积极影响也是有目共睹的。但是，这并不说明我们的管理工作没有偏差或漏洞，我们的规章仍需进一步健全，制度还要不断完善，干部的管理水平需要不断提高。

今年是中央提出的“转变作风年”，这项工作非常及时和必要。作风建设是一项长期的系统工程，不可能一蹴而就。“三讲”教育以后，学校针对全校师生所提的合理意见和建议，就管理工作特别是机关工作作风进行了大范围、大力度的整改，有成效，但仍不尽如人意。在人人都在追求工作质量、办事效率和品牌效应的今天，我们的管理工作不应成为制约学校教学科研发展的瓶颈，我们应该也必须从根本上改变“门难进、脸难看、事难办”的官僚作风，改变人浮于事的懒散作风，改变以人治代法治的不正之风，以管理促质量，以管理促效益。

当前，依法治校已成为高校行政管理工作的一项必然选择。这是一个双向互动的管理过程，它既要求管理者依法办事，也要求管理对象遵纪守法；既是一个管理的过程，同时也是一个监督的过程。依法治校的精髓就是要求所有管理工作人员都必须依照法治精神和规章制度从事各种管理活动，同时也要求从事教学科研的教职员工依法实施自己的行为。《暨南大学文件汇编》是学校各单位和师生员工监督、管理学校的基本依据。出版《暨南大学文件汇编》的目的，是使我们每一位教职员工和学生在日常的管理和被管理活动中，都能够有章可循，有据可查。因为只有这样，才有法治可言，学校才能在科学、合理、公正、有序的管理机制保证下健康发展。我们期待这种现代化管理机制的建立，并为之不懈努力。

---

① 本文是《暨南大学文件汇编》序，广州，2002年9月25日.

38. 全球化进程与华侨高等学府的重要使命[①]

当今中国高等教育的发展面临诸多因素的影响，有来自国内的：一是中国经济社会的发展向高等教育提出了更高的智力要求和人才保障，希望能提供更多优质的人力资源，推动经济社会的快速发展；二是教育发展的规律则要求高等教育必须坚持规模、结构、质量、效益的综合、协调、可持续的发展。同时，也有来自国际方面的，经济全球化引起高等教育的国际化，地区之间和国家之间高等教育的相互交流、合作与渗透日趋活跃。尤其随着中国加入世界贸易组织后，对外开放的广度与深度可谓前所未有。这些来自国际方面的因素，在计划经济时代我们高校也许可以无须多虑，照章办学。但是，今天不行了，无论是办学的理念，还是治校的韬略；无论是人才培养的模式，还是教学的内容与方法，都必须适应国内与国际两个人才市场的需要。开拓国际视野，将高等教育的发展和高校教学的改革放入全球化的背景之下来思考、谋划是新世纪对大学校长们提出的新要求、新课题，也是新的挑战。因此，需要我们共同探讨与交流，寻求对策，从必然王国到自由王国。

一、全球化与高等教育国际化对中国高教所形成的影响

迄今为止，人们对于全球化的认识可谓众说纷纭，很难达成共识。而本次研讨会的主题也并不是对“全球化”进行专门的研讨，因此，我们只需了解全球化的一般概念，掌握其特征，提高对高等教育国际化的认识。综观各种形式与内容的全球化，它们的共性大多表现在时空的变化方面。所以，全球化的根本特征是各国经济社会联系的普遍化与密切融合。全球化就是人类不断跨越空间障碍与社会障碍，在全球范围内（物质的和信息的）充分沟通，是达到更多共识与共同行动的过程。全球化是当代人类社会生活的活动空间日益超越民族主权版图界限，在世界范围内展现出全方位的沟通、联系、交流与互动的客观历史进程及趋势。

全球化绝非单一的经济全球化，而是全面的全球化。由于经济全球化追求生产要素的全球配置和经济收益的全球获取，所以它不仅会冲击各国的经济活动，要求其在统一世界市场的框架中调整原有的种种行为，而且会造成广泛而深刻的政治、文化影响。这种政治与文化的影响，在很大程度上就是政治与文化的全球化。同理，经济全球化对生产要素的全球配置包括了高级人才在世界范围内的组合与聘任，从而构成人力资源的激烈竞争；加上高等教育市场成为今天发达国家获取利润的重要渠道之一，与跨国集团一样，跨国的教育蓬勃而起。由此可见，高等教育国际化是全球化在高等教育领域中的表现。

高等教育国际化是“跨世界、跨民族、跨文化的高等教育交流与合作，即一个国家向世界发展本国高等教育的思想理论、国际化活动以及与他国开展的相互交流与合作”。高等教育国际化的核心是现代大学正在逐步走向世界。当今的世界是一个开放的

---

① 本文原载《中国高教研究》，2002年，(9)：33-36.

世界，其本质是文化的开放。高等教育国际化的主要要求是加强国际交流与合作，开放教育市场，培养具有全球意识和开放意识、具有较高文化品位、能够参与国际事务和具有国际竞争能力的创造型人才。其中，基础是普遍增强所有现代大学的全球意识和参与国际竞争的能力，重点是着力办好一批世界一流水平的现代大学，关键是建立以“学校自治”为根本特征的现代大学制度。如果以往我们对全球化和高等教育国际化的认识，只是停留在对一种教育现象观望和感悟的基础上，那么，中国加入世界贸易组织后，参与全球化和高等教育国际化就成了今天高校实实在在的事情。下面，我们将分析全球化和高等教育国际化对我国高等教育和大学教学所带来的主要影响。

其一，加剧高等教育和高素质人才的竞争。发达国家凭借他们优质的教育资源和良好的环境条件，分割发展中国家的教育市场。出国留学的普遍化、低龄化和多样化充分反映了国家之间高等教育发展不平衡所带来的冲击。人才的竞争不仅表现在数量上，更表现在质量上。这种竞争迫使我们提高办学水平，使得一批高校和一些学科专业加快建设，力争在较短的时间内接近或基本达到国际公认的水平；同时，提高大学的教学质量，改革人才培养模式，尤其要把培养学生具有进取精神和创新能力放在突出的位置，造就更多宽口径、厚基础的复合型人才，以及具有全球视野和国际交流能力，能够把握国际经济、政治或科技发展趋势的高素质、高层次人才。

其二，促进高等学校贴近人才市场与经济建设。随着新的竞争机制引入，大学生的就业将越来越受到就业市场供求关系的影响。人才的价值与其就业层次、岗位、薪酬挂钩越来越紧，价值也会较多地通过价格来体现，人才的薪酬水准和质量越来越向国际市场水准靠近。高校毕业生在就业市场的表现，将逐步成为评估大学水准和质量的重要标准。经济全球化带动了全球性经济结构、产业结构、技术结构大规模的调整，随之而来的却是高等教育结构、人才培养结构，甚至学科专业结构大范围的变化。高等教育必须适应经济建设，高等学校的专业设置和学科结构调整应当满足高新技术发展和加入世界贸易组织对新型人才的需求。

其三，推动高等学校教学内容与教学方法的改革。毋庸置疑，教育具有鲜明的民族性和本土性，为民族的振兴与繁荣是教育的出发点和归宿点。但是，随着时代的进步和发展，特别是全球化进程，现代教育不仅表现出其民族性，而且还反映出所具有的国际性。向学生灌输的教学内容，展现的知识视野，培育的思想观念，训练的实践能力都必须具有面向全球（包括本土在内）的理念。当前全球性问题日益严重，如和平与安全、环境保护、南北关系、控制国际犯罪、打击恐怖活动、对付艾滋病等。对这些全球性问题的认识与解决，不仅是整个人类社会的责任，也是各种教育不可回避的内容。因此，应该对现行的教学内容进行调整和充实，增加具有国际意义的课程，如国际政治、经济、文化、历史、地理等。通过这些课程的学习，让学生了解世界各国地理位置、经济优势、历史演变、著名人物、风土人情、文化艺术等知识，理解不同社会制度国家之间民族文化和价值观念的差异，具有宽阔的胸襟和包容性，能够与不同国家、不同民族的人民一道从事共同的事业。

二、华侨高校的办学特性顺应高等教育国际化的潮流

暨南大学 1906 年创建于南京，是我国政府最早兴办的一所面向世界的华侨高校。1927 年，暨南大学首任校长郑洪年指出："鉴于侨胞处于殖民政府铁蹄之下，受尽帝国主义之蹂躏，暨南教育非提高程度，扩充为完善大学，不足以增进侨胞之地位，不足以谋适应其特殊环境，不足以使华侨父老咸达自由平等之目的。"暨南大学的办学，所招收的学生来自五洲四海，而培养的学子学成毕业后又分赴世界各地，谋生图发展。这种跨地域、跨国界的教育，就是国际高等教育的形式之一。从某种意义说，暨南大学的教育具有国际化的特征。这种教育将中国的文化向世界传播，即通过学子们将所学的中华文化带到他们的居住地，与当地的民族文化进行交流、融合，产生新的文化，适应于他们生活的地区和一道劳作的民众，逐步演变成为世界文化的一部分。

由于是为全球华侨华人服务的教育，无论教育的形式，还是教学的内容，均体现"中西合璧"的教育精神。中国传统文化，如中国的历史地理、语言文字、伦理道德、民俗风情等无疑为侨校教学的主要内容。因为，学子们需要掌握民族文化的真谛，在海外继承、弘扬与光大中华民族的优良传统。但是，学校并未固守这一教学内容，而是有一种全球视野，从国际经济发展的形势和工业革命对产业调整的要求，以及海外学子在异国他乡的生存与发展等诸多因素考虑，除学习中华文化外，还需掌握与现代科技相关的专业知识和专门技能，如工商、会计、机械、建筑、化工、生物、中医中药等方面的知识。在教育的形式上既有专业教育，也有职业教育；既有大学教育，又有预科教育。

华侨高等教育的形成与发展与其他类型高等教育的产生具有明显的不同。华侨高等教育是在国家出于对本国侨民教育考虑的基础上，由基础教育逐步发展起来的高等教育。这种教育具有浓烈的民族传统教育的色彩，她肩负着向海外传播中华文化的重任，又有鲜明的西方教育的"风味"，教育教学中的民主性、科学性和开放性，充分体现近代高等教育的时代特点。由于这种教育的特点，引发了在教育对象、培养目标、教学内容和教学方法上众多的不同。因此，在兴办华侨高等教育的过程中，华侨高校不仅要依据国内的教育方针与政策，而且还要遵循高等教育的发展规律，符合国际惯例，如学校最早于 1922 年在领导体制方面实行董事会制度；在教学管理方面也是最早实行学分制的高校之一；在内部管理结构方面也是最早实行校院系三级；在学科建设方面也是综合性大学中较早设立医学学科的高校，等等。这一切表明，暨南大学由于特殊的使命，使得其办学从一开始就与国外著名大学办学的思路、方法相"对接"。

三、华侨高校对外办学成效显示应对高等教育国际化的实力

改革开放以来，在党和国家的亲切关怀下，在海外侨胞、港澳台同胞的积极支持下，国内华侨高等教育得到了迅速的恢复和快速的发展，办学规模、层次、结构、质量和水平均取得了显著的成绩和长足的进步。仅从 1978 年开始，暨南大学就培养了来自海外的毕业学生 6280 名。目前，来自五大洲 44 个国家和地区的 5061 名海外及港澳台学生在校就读，占学校全日制学生人数的 36%，是国内拥有海外及港澳台学生最多的高校。20 多年的发展，不仅使"面向海外、面向港澳台"的办学方针得到全面的贯彻、落实，为港

澳回归、祖国统一，为增进中国人民和世界各国人民的友谊作出了贡献；而且在教育教学的理论与实践中形成了一整套既适应中国国情，又符合国际惯例的对外办学思想、观念以及教学体系和管理办法，为暨南大学今后更大范围、更宽领域、更高层次的对外办学奠定了坚实的基础。暨南大学与国内其他高校相比具有不同的使命，因此在招生、教学、管理和对外交流等方面有着一定特殊性，从而也就形成了自身的特色。

（1）灵活多样的对外招生。党和国家要求华侨高校“招生的主要对象是华侨、港澳、台湾和外籍华人青年”。为了广开渠道，扩大学生来源，学校采取灵活的形式，广泛招收海外及港澳台的学生。国家批准华侨高校在港澳设立办事处，由学校单独命题，自主在港澳地区招生。同时，学校还通过我国驻外使领馆、华侨华人社团以及校董招收海外学生。对于在港澳地区中学就读的成绩优良的学生，有的可以通过校长推荐直接入读；通过中七统考的可以免试入学。学校率先实行对外春秋两季招生，方便海外及港澳台春季毕业的学生随时进入学校就读。通过多年的努力，暨南大学已实行全方位、多层次的对外招生，本科教育、研究生教育、成人教育和预科教育全部面向海外及港澳台，形成了各类、各层次联合对外办学格局。

（2）对接国际的办学模式。暨南大学从创办到恢复与发展，一直是按国际高等学校的架构来设立和运行。学校是一所人文、社会、自然以及工程技术科学相融合的综合性大学，文学、历史学、新闻学、经济学、管理学和法学学科颇具特色，生命科学、医学和药学融为一体，数学、物理、化学、计算机、环境、土木、食品等自然科学和工程技术学科相互依托又相互交叉融合。华文学院是国务院侨务办公室华文教育的基地，肩负对外办学的重任；也是国家汉办唯一批准在广东设立的汉语水平考试举办单位。多年来，学院通过多种形式的教育，既向海外广泛地普及汉语言文字，传播与弘扬中华文化；又培养大批对外汉语方面的高级专门人才，向海外的华文教育输送高水平的师资。学校的预科教育在对外办学方面也发挥了重要作用。为了适应中国加入世界贸易组织的需要，2001 年学校新建国际学院，实行全英语教学，全力培养国际型人才。华侨高校实行有董事会的校长负责制，充分发挥海内外知名人士、工商首领、著名专家学者的聪明才智和优质资源共同办好侨校。学校实行校院系三级管理，同时，引入竞争和激励机制，始终保持旺盛的生机与活力。

（3）适应海外及港澳台的教学改革。华侨高校培养的学生，学成毕业后绝大多数返回他们各自不同的居住地区就业。毕业生就业的这些国家和地区之间存在许多不同的方面，如经济发展水平不同、产业结构不同、人才需求不同和就业程度不同。因此，学校除了一般的教育外，还要考虑他们的充分就业，根据本校海外及港澳台学生就业国家和地区的社会经济发展情况、就业形势和人才结构设置专业。海外及港澳台学生可以根据自己的基础、能力和兴趣自主选择专业，也可转换专业。

由于海外及港澳台学生与内地学生在社会制度、文化背景、基础知识等方面存在较大的差异，学校采取因材施教、分类教学。对于公共基础课程的设置和教学从以下三个方面考虑。

一是从人生观和价值观的角度，开设一些有关如何认识社会、认识人生的课程，帮助他们树立正确的人生观和价值观。无论来自何地的青年学生，也不管他们生活在什么样的社会制度之下，他们大学毕业后都要走上社会，面对人生道路的选择。学校主渠道的教学工作应该起到正确引导和积极帮助的作用。通过课程的学习使他们认识到成为合格的公民、做一个有利于社会的人是大学生的基本准则。

二是从形成爱国思想和建立民族情感这个角度，开设一些有关中国语文、历史和文化等方面的课程。让他们了解中华民族的悠久历史、灿烂文化与优良传统以及中华民族对人类所做的贡献，使他们有一种民族自豪感和自信心。

三是从谋生与创业的角度，设置一些适应于他们回到居住地能顺利就业的工具性或技能性的课程，如外语、计算机。过去，相当多的华侨华人学生继承父业，从事商务。近几年来，发展中国家都在进行产业结构的调整，知识密集型产业的比重在不断地提高，加上科技革命的影响，新的技术、新的行业迫切要求他们掌握更多的现代科技知识。因此，不少的外招学生不像以往那样主要选择商学，而是选读电子计算机、电子工程、生物工程等与高新技术相关的专业。

(4) 内外有别与适当照顾的管理方法。由于海外及港澳台学生和内地学生在生活方式、学习规律、作息习惯等方面多有不同，如果对他们的管理完全用一个模式，难以奏效。因此，在管理上不同的事务采用不同的方式。对于执行学校的规章制度两类学生一视同仁，不搞特殊，以体现从严治校；对于学校组织的活动两类学生一道参加，比如社会实践、毕业实习、科技文化体育等活动，以增强同学之间的感情交流；对于生活管理方面尽量考虑外招学生的习惯，如住得宽松一点，等等。另外，海外及港澳台学生可组织自己的社团，开展活动；可以在校园过自己民族的节日；可保留自己的宗教信仰，但不能搞宗教活动。

四、华侨高校对外办学战略体现融入高等教育国际化中的作为

加入世界贸易组织有利于促进我国高等学校的双向开放，在 WTO 规则的保护下积极实施“走出去”战略。高等教育国际化的进程将鼓励我国有条件的高校，充分发挥比较优势和后发效应，主动走出去办学。另外，在一些地区，与某些发展中国家的高等教育相比，我国的高等学校也有相对的优势，在尊重这些国家主权和民族习俗等方面也有较好的基础，完全可以吸引这些国家来华的留学生。

暨南大学经过近百年的办学，有着悠久的对外办学历史，在海内外享有盛誉。学校长期积淀的办学经验、形成的办学体系和建立的良好办学信誉，完全能够在我国实施的“走出去”战略中有所作为。进入 21 世纪，华侨高校对外办学的思路更加清晰，发展的战略更加明确，即在对外办学中“立足港澳台，面向东南亚，走向全世界”。

(1) 突出重点，充分利用华侨高校的优势，巩固海外及港澳台办学成果。改革开放以来，学校贯彻“面向海外、面向港澳台”的办学方针，重点为港澳回归与祖国统一大业培养人才。因此，港澳台地区学生在我校外招学生中一直占有较大的比重。随着香港、澳门的回归，除暨大和华大外，内地在港澳地区招生的高校已有一百多所；加上，香港自身有较宽裕的高教资源以及国外高校的插足，一场无硝烟的高校生源争夺大战早已在港澳打响。尽管如此，我校凭借多年的经验和众多校董与校友的帮助，在港澳地区招生仍然保持一定优势，2002 年共招收录取了新生 1450 名。在巩固海外及

港澳台办学的基础上，我们逐步把对外办学的目光投向更加开阔的地区。

（2）抓住机遇，拓展东南亚地区的华侨高等教育和华文教育。东南亚地区是我国的相邻地区，自古以来这个地区的国家和人民与我国保持友好交往和传统友谊。这个地区也是华侨华人居住最多的地区之一。我国政府历来重视与这个地区国家之间在政治、经济、文化、教育等领域的交流与合作。1997 年亚洲金融危机后，由于中国在这场亚洲金融危机中表现出负责任的态度，赢得了这些国家的高度赞誉和信任。加上，我国经济保持着持续增长的态势，与这些国家的经贸往来日益活跃，如泰国即使在金融危机之时，与我国的贸易都没有下降。经济活动推动着这些国家和政府对华语地位与作用的重新认识，加上华侨华人的强烈要求，使华文教育的开展出现了新的局面。一些国家解除对华文的限制，容许华文教育的普及，还有的要求政府官员学习和掌握汉语；汉语基本上与英语、当地语具有同样的地位，在一些学校普遍开设；有的国家采取积极、灵活的措施开放教育市场，欢迎外国大学人境办学。可以说，这是改革开放后华文教育在东南亚地区迎来的又一个春天。

暨南大学以其敏锐的办学直觉，迅速地抓住机遇，大力开展面向东南亚地区的办学。近几年，来自东南亚地区的学生逐年增多，人数在我校外招学生中排列第二。学校还不失时机地利用我国与周边国家良好的关系，通过校董、校友以及华侨华人社团开展对外办学。一方面，编写华文教材，帮助这些地区解决华文教育中所遇到的困难；另一方面，分别在马来西亚、泰国、缅甸、越南、老挝等国家设立对外办学的招生点，便于学校对外招生宣传和当地学生的报考。

（3）积极筹划，努力开辟北美地区的华文教育。改革开放以来，中国的海外移民大幅增加，经济发达的北美地区成为新移民的首选地，美国华人今已超过 230 万，其中来自中国大陆者逾 50 万；加拿大的 100 万华侨华人中大陆新移民也占 1/5。大陆新移民不仅数量可观，而且其文化素质之高也是有目共睹的。发达国家普遍地通过调整新移民政策以吸引本国所需要的职业和技能移民。以留学人员、科技人员、技术移民为主的大陆新移民，其科学文化水平不仅远高于祖籍国而且高于所在国同龄人口的平均水平，如在新移民较为集中的美国新泽西州，华人平均 9 人中就有一个是博士。这些众多新移民子女的华文教育，一直成为他们日益关注的问题。加上这一地区有比较宽松的文化、教育政策，近年华校发展之快，被业内人士以“雨后春笋”所形容。这个地区一种新型中文学校非常活跃，以其教育教学的师资、形式、内容和质量受到当地华侨华人的欢迎。这种华文教育的氛围和环境，为华侨高校对外办学提供了良好条件。目前，我校正在积极研究进入这一地区办学的途径和策略。通过提供教材和举办夏令营的形式保持与中文学校的友好交往，逐步发展培养师资，输送人才，打好基础；再通过侨团、侨社、校董、校友合作办学，开辟新的对外办学渠道。

总之，华侨高校就其办学渊源体现出国际性，就其办学特性也与全球化和高等教育国际化的潮流相一致，就其办学成效也充分显示参与高等教育国际化的竞争能力，就其办学方向也可在实施“走出去”战略中有所作为。因此，我们殷切期望在全球化和高等教育国际化的进程中，党和国家能够把暨南大学建设成为面向 21 世纪具有国际影响和竞争能力的高水平大学。

39. 努力完成高校扩招任务①

江泽民同志在党的十六大报告中指出："教育是发展科学技术和培养人才的基础，在现代化建设中具有先导性全局性作用，必须摆在优先发展的战略地位。"在全面建设小康社会的过程中，要"坚持教育创新，深化教育改革，优化教育结构，合理配置教育资源，提高教育质量和管理水平，全面推进素质教育，造就数以亿计的高素质劳动者、数以千万计的专门人才和一大批拔尖创新人才"，为未来高等教育的发展提出了更高的要求，指明了新的方向。中共中央政治局委员、广东省委书记张德江同志刚到广东，就深切关心广东高等教育的发展状况，全校教职员工深受张书记讲话精神的鼓舞，纷纷表示一定要积极贯彻十六大精神及实践张书记的重要指示，积极为广东高等教育的发展，为广东实现"争创新优势，率先实现基本现代化"的远大目标贡献力量。现在，我将学校现状和发展情况进行汇报。

一、暨南大学简要发展历程和现状

暨南大学是我国第一所国家创办的华侨大学，其前身是1906年创建于南京的暨南学堂。先后在上海、福建建阳等地办学。学校是应当时侨居海外的华侨之需而设，旨在"宏教泽而系侨情"。这一办学宗旨从未改变。新中国成立后，为实施国家侨务政策，弘扬中华优秀传统文化，满足海外华侨华人和港澳台地区青年的求学需要，党和国家十分重视华侨高等教育的发展。1958年，暨南大学被列为国务院高教部的直属院校。1983年，中共中央、国务院专文批复中宣部、教育部、国务院侨务办公室，将暨南大学列为国家重点扶植的大学。1994年，国务院侨办与广东省人民政府签署协议——合作共建暨南大学。1996年暨南大学成为国家面向21世纪重点建设大学。今年10月，学校顺利通过"九五""211工程"验收，成功进入"十五""211工程"建设新时期。

"九五"期间，在国务院侨办及广东省委、省政府的领导下，我校坚持从严治校和依法治校的原则，全校教职员工认清形势，正视差距，扬长避短，开拓创新，以"211工程"建设为龙头，大力推进学校各项事业不断进步，在许多方面发生了从无到有、从小到大、从弱到强的可喜变化，成功实现跨越式发展。

（一）办学规模变大

学生人数显著增加。与"九五"之初相比，学校各类学生由13 012人增加到27 383人，增长110.4%。学校全日制本科生由5377人增加到12 263人，增长128.1%。研究生由615人增加到3245人，增长427.6%。

（二）校园和建筑面积扩大

校区由原来的3个（广州石牌校本部和广园西路校区，深圳华侨城校区）增加到4

① 本文是在广东高等教育2003年度工作会议暨发展咨询会议上的发言，肇庆，2002年12月28日；一个大学校长的探索，北京：高等教育出版社，2011，78-85.

个（加上珠海校区），现还有 1 个新校区正待开发。校园占地面积由 112 公顷增加到 174 公顷，增长 55.4%，校园建筑面积由 506 991 平方米增加到 961 088 平方米，增长 89.6%。

（三）办学层次提高

学校研究生与本科生之比由 1995 年的 1∶8.74 上升到 1∶3.78，专科生由 2472 人减到 109 人，明年这个数字将变为零，而且校本部从 1997 年开始即没有专科生。

（四）办学特色更加鲜明

海外及港澳台学生由 1982 人增加到 6894 人，增长 247.8%，是中国海外及港澳台学生最多的大学。1995 年，只有 16 个国家的学生来校学习，而今天已上升为世界五大洲的 53 个国家。建立了姊妹关系的大学遍及世界各地，是中国第一所在世界五大洲建有姊妹大学的学校。另外，学校作为国务院侨办华文教育基地和国家汉办支持周边国家汉语教学的重点单位，编写全套中文教材共 48 册，现已被多个国家使用，共发行 300 多万套，深受华侨华人和外国人的欢迎和好评。

（五）科研实力增强

学校科研经费由 400 万元增加到 8000 万元，增长近 19 倍。获得的科研项目在“973”“863”等国家重点项目方面实现了零的突破。获得的专利从无到有，现在获专利授权 20 项。学术论文数增加了 1.3 倍，其中被三大索引（SCI、EI 和 ISTP）收录的论文数增加了 10.7 倍。获省部级科技奖励增加了 5 倍多。科研成果的应用也有较大的进步。

另外，学校在国家重点学科、国家研究基地、教育部重点实验室、工程研究中心等方面都实现了零的突破。新增 2 个国家重点学科，1 个国家人文社会科学重点基地、1 个教育部重点实验室、1 个教育部工程研究中心、8 个广东省重点学科、1 个广东省教育厅重点实验室。

（六）学科结构优化

本科专业由 1995 年的 30 个增加到 43 个，硕士学位授权学科由 1995 年的 50 个增加到 67 个，博士学位授权学科已由 7 个增加到 14 个，并且在一级学位授权学科方面实现了零的突破。博士后站实现零的突破，达到 3 个。教学系由 21 个增至 37 个，学院数由 7 个增至 16 个，涵盖的学科门类更加广泛。

（七）师资队伍结构改善

1093 名专任教师中，有研究生学位者 852 人，占 78%，其中博士 281 人，博士学位获得者占专任教师的总数由 1995 年的 5.8%增加到 25.7%。博士生导师 63 人，教授 183 人，副教授 572 人。新增院士 2 人，填补了学校无院士任教的空白。获教育部设置的“长江学者奖励计划”和广东省设置的“珠江学者”岗位计划的特聘教授岗位各 2 个。

（八）为祖国统一大业和广东经济发展服务的能力增强

自 1978 年至今，我校已接收过来自世界五大洲 91 个国家和港澳台 3 个地区的学生前来学习，为港澳的顺利回归和广泛团结世界华侨华人做出了积极贡献。同时，作为广东高等教育的重要组成部分，自 1995 年以来，我校共为广东省培养各类人才 27 549 人，其中全日制本、专科生 12 022 人，研究生 4895 人，继续教育学生 10 632 人。这期间，我校还为广东地区招收研究生课程进修班学生 6281 人，短期培训各类学员 3 万余人（表 1～表 3）。同时，所培养的华侨华人和港澳台学生也大多原籍广东。

表 1　全日制本、专科生招生情况

时间	外招人数	内招人数		广东生源人数			备注（全为广东学生）
		总数	本科	总数	本科	专科	
1995 年	356	1 772	997	1 652	847	775	预科 30
1996 年	432	1 683	1 318	1 519	1 168	321	预科 30
1997 年	520	1 625	1 384	1 436	1 214	192	预科 30
1998 年	619	1 660	1 450	1 367	1 187	150	预科 30
1999 年	703	2 561	2 503	2 084（含 600 走读生）	1731	295	预科 58
2000 年	771	1 872	1 862	1 291	1 161	120	预科 10
2001 年	866	2 003	2 003	1 276	1 276	0	开始停招专科生
2002 年	1 256	2 500	2 500	1 397	1 397	0	0
合计	5 523	15 676	14 017	12 022	9 981	1 853	188

表 2　研究生招生情况

时间	招生总数	广东生源人数		
		博士	硕士	总数
1995 年	228	7	61	68
1996 年	299	8	114	122
1997 年	421	18	139	157
1998 年	583	16	242	258
1999 年	596	38	263	301
2000 年	701	58	264	322
2001 年	914	84	327	411
2002 年	1 153	91	211	302
合计	4 895	320	1 621	1 941

表 3　继续教育在校生情况

时间	成人教育（含函授、夜大、成人脱产）	非学历教育（含研究生进修班、自考班、进修培训）
1995 年	3 500	702
1996 年	3 517	187
1997 年	3 798	6 410
1998 年	4 409	5 031
1999 年	5 036	3 634
2000 年	5 716	3 416
2001 年	6 582	3 429
2002 年	7 117	4 065

（九）综合实力增强

学校的固定资产总值由 2.7 亿元增至 14 亿元，增加了 4 倍多，图书藏量由 135 万册增至 170.7 万册，教学科研仪器设备由 4985 万元增至 1.5 亿多元，增加了 2 倍多。

学校在不同机构的综合实力排行榜中的位置不断上升，今年被教育部《中国高等教育评估》杂志评为77所研究型大学之一，名列第53位。按中国网大的中国大学综合实力排名，我校排名逐年上升。1998年第87位，1999年第72位，2000年第60位，2001年第40位，2002年第37位。

（十）学校的办学效益提高

全校的教职工和专业教师人数基本没有变化，1995年分别为3601人和1036人，目前为3649人和1093人，且学校所获上级经费投入并未大幅增加，但学校完成的任务却成倍增加，显然办学效益已更加优良。在2002年广东管理科学研究院“中国‘211工程’大学教师人均效率排名”中，我校排在第41位。

我校上述进步与国内一流大学相比，仍有很大差距，我们正在努力缩短距离。“九五”期间的进步，为我校“十五”期间的进一步发展，为顺利完成国家及地方赋予的各项任务奠定了坚实的基础。

二、暨南大学“十五”规划发展目标

为保持良好的发展势头，顺利实现“侨校＋名校”的发展战略，为国家和地方多做贡献，我校委托中国国际咨询公司制定了“十五”《暨南大学总体发展规划》（简称《规划》），该《规划》根据实际情况对我校“十五”期间的发展方向、目标以及具体建设项目进行了详尽的可行性论证，得到国家计委的肯定。国家计委批准在“十五”期间向我校投入5.1亿元人民币专项资金实现这一规划。

此外，国家“211工程”协调办公室也对我校“十五”“211工程”项目投入2800万元。

下面，我主要介绍一下“十五”期间我校《规划》中有关人才培养和基础设施建设的内容。

（一）人才培养

按照“大力发展研究生教育，适度发展本科教育，积极发展华文教育，稳定成人教育规模”的发展思路，计划到2005年，学校本科生规模达到13 000人，硕士生达到4500人，博士生达到600人，海外及港澳台学生与内地学生的比例达到1∶1。

（二）基础设施建设

根据《规划》及国家计委的批复，我校“十五”期间利用5.1亿元专项资金应完成的项目见表4。

表4 “十五”期间专项资金应完成的项目

序号	项目	“十五”建筑面积/米^2	“十五”投资/万元
一	广州本部		
1	主要工程		
1.1	教室、实验室实习场所	28 000	5 740
1.2	图书馆	22 844	7 424
1.3	校、系行政用房	9 000	2 304
1.4	游泳馆	1 800	569

续表

序号	项目	“十五”建筑面积/m²	“十五”投资/万元
1.5	曾宪梓科学馆	13 500	4 806
1.6	医学院	25 000	7 550
1.7	学生食堂	5 000	775
1.8	教工宿舍	4 000	740
1.9	教工食堂	2 800	462
1.10	生活及其他附属用房	9 100	1 957
1.11	运动场改造	20 000	320
1.12	旧楼改造	60 000	3 000
2	公用设施配套改造工程		
2.1	拆除工程		197
2.2	附属设施		713
2.3	室外道路及广场		882
2.4	校园绿化		260
2.5	室外管网改造		1 248
2.6	土方工程		200
3	其他费用		7 360
	小计	201 044	46 507
二	磨碟沙校区		
1	主要工程		
1.1	教室、实验室实习场所	26 522	5 437
1.2	图书馆		
1.3	校、系行政用房	5 000	1 280
1.4	会堂	1 120	392
1.5	风雨操场	1 613	526
1.6	学生食堂	3 500	543
1.7	教工宿舍	1 044	193
1.8	教工食堂	1 030	170
1.9	生活及其他附属用房	4 000	860
2	公用设施		752
3	其他费用		2 355
	小计	43 829	12 508
	总计	244 873	59 015

三、完成扩招任务的措施及实际问题

（一）措施

发展高等教育，提高入学率，将广东建成教育强省，这是省委、省政府做出的重大决策，我们坚决拥护。为完成省教育厅下达给我校2005年在校本科生达到20 000人的任务，我们将采取如下主要措施。

（1）积极引进师资，扩大师资队伍规模。努力创造良好的生活、工作环境，提供力所能及的优惠政策。根据招生计划、专业设置及现有师资队伍结构和数量，面向海内外，一方面吸收优秀的硕士、博士毕业生，一方面引进一些学术带头人，以解决扩招后的师资不足问题。

（2）抓紧修建教学大楼、实验楼、图书馆、运动场、教工住宅、学生宿舍、学生食堂等基础设施。因我校现有 4 个校区的建筑物和学生容量已趋饱和，现有设施无法满足扩招所带来的更大需求，因此我校当务之急是紧急修建所需的基础设施。

（3）筹措资金，解决扩招所带来的一系列投入问题。目前，我校已没有财力支持师资引进和基础设施所需的资金投入，唯有通过向银行贷款或社会捐赠以及引资共建来解决，我们将以此为突破口，寻找合适、有效的途径，解决资金问题。

（二）实际问题

（1）师资问题。师资是制约我校扩大招生规模的瓶颈。这一方面表现在现有师资的数量不能满足未来扩招的需要。根据扩招人数及教育部规定的师生比计算，我校必须引进专任教师 500 人，加上管理和后勤人员，总数近 1000 人。另一方面在于教师住宅楼的缺乏限制了学校引进师资的数量和速度。根据广州市的政策，凡教师住宅楼一概不予报建，我校拟建的引进师资的教师周转房因此而搁浅。目前，我校急需引进的师资都因住房问题无法解决而迟迟不能到位。同时，学校因缺乏资金，目前无法启动货币分房工作。

（2）校园用地和基础设施问题。要完成扩招任务，就必须增加约 217 380 平方米房屋建筑面积，但我校现有的 4 个校区已无法容纳更多的建筑物。因此，我们不仅没有扩招所需的建筑用地，就是已经纳入《规划》的计划安排 4000 名学生的磨碟沙校区至今也不能开发。因为该地块已被广州市政府征用，而市政府置换给我校的土地目前已基本落实，即等面积置换到广州氮肥厂区域。据悉，该地块至少要到 2004 年 3 月才能全部交付我校使用。如按照广州市正常的报建速度，我校必须到 2005 年初才能完成报建工作。这样一来，不仅扩招的学生无处安置，就是原计划安排的 4000 名学生也无处容纳。

（3）资金问题。“十五”期间，国家给予我校的专项资金都必须按规划项目投入，因此我校那些没有纳入国家《规划》而又必须建设的项目以及扩招所需增加的基础设施项目都必须另筹资金解决。为完成国家《规划》外的项目，包括珠海校园的所有工程、校本部的 10 栋学生宿舍等，需经费 6.5 亿元，学校希望省委、省政府帮助解决，不足部分由我校自筹和贷款解决。而为完成扩招任务所需投入的各类资金仅仅基建部分估计还需 3.5 亿元左右，这已远远超出了学校的承受能力。

（三）建议

（1）中央已给我校“十五”期间专项投入 5.38 亿元，故再请广东省委、省政府给予相应配套经费。同时，请为将要扩招的 7000 名大学生也给予相关的经费投入。

（2）由于校园分散，影响办学质量和效益，根据张书记关于“校园土地可以置换”的指示，我校请求省委、省政府将华文学院（225 亩）和磨碟沙（354 亩）共约 580 亩的土地置换到学校本部南大门对面的广州跑马场区域。这样既可以解决校园分散，难于规划、难于管理的问题，又可以加速基础设施建设，为提高办学质量和完成扩招任务提供坚实的保障。

（3）发展我省的高等教育，完成扩招任务需要全省上下的通力支持，我们渴望省委、省政府能与广州市政府进行协调，对一些不利于完成扩招任务的政策予以变通，应允许特事特办。如对教师住宅楼的限制报建问题，基建工程报建手续及进度问题，等等。

因为我校隶属于国务院侨办，对进入广州大学城参加扩招一事，经请示，侨办领导认为我校并不具备扩招的能力，因此，恳请省委、省政府能酌情解决我们的困难，以便我校能够完成国家批准的“十五”建设规划以及省里下达的扩招任务。

40. 答谢珠海人民[①]

今晚，灯火辉煌的香洲文化广场显得格外美丽。到处洋溢的新年气氛和即将开始的精彩演出都在预示，这将是一个美好的夜晚。在此，我谨代表暨南大学 3 万余名师生，感谢珠海市党政领导和珠海人民几年来对珠海学院的支持、帮助，并向所有对珠海学院付出心血和劳动的教职员工表示感谢，祝大家在新的一年里工作顺利，万事如意！

在即将过去的一年里，暨南大学在全体教职员工的共同努力下，各项事业蒸蒸日上。今年上半年，我校被教育部《中国高等教育评估》杂志列为中国 77 所研究型大学之一，名列第 53 位；在网大《中国大学排行榜》的高校综合实力排名中，我校已从 1998 年的第 87 位跃居至今年的第 37 位。广东管理科学研究院今年的“中国‘211 工程’大学教师人均效率排名”中，我校排在第 41 位。目前，学校已圆满完成“九五”“211 工程”建设任务，顺利进入“十五”“211 工程”建设全新发展时期。

与此同时，珠海学院在各方面也有了长足的发展。学院已开设汉语言文学、英语、新闻学、广告学、计算机科学与技术、信息管理与信息系统、金融学、国际经济与贸易、法学、行政管理、财务管理、会计学、工商管理和市场营销等 14 个本科专业，现有博士、硕士和大学一至四年级的本科学生 4200 人，已形成一个完整的人才培养结构。学院的各项设施设备先进、齐全：多功能餐饮中心、活动中心，美观舒适的学生公寓，新颖的运动场地，等等，这些，都为珠海学院的学子们提供了良好的学习、生活环境。短短的几年里，珠海学院已成为一个充满生机的现代化的综合性学院。

作为有 96 年历史的中国历史最悠久的大学之一，我们感到十分荣幸，在 1998 年 8 月，暨南大学与珠海市人民政府合作开办珠海学院，开创了珠海特区历史上开办全日制高等教育的先河。2000 年 4 月，暨南大学再次与珠海市人民政府达成共识，将暨南大学珠海学院迁址珠海市中心，并由珠海市人民政府永久无偿提供 866 亩土地作为暨南大学珠海学院新校园的建设用地。值得骄傲和欣慰的是，我们没有辜负珠海人民的厚爱和期望。除 1997 年我校与珠海市人民医院即暨南大学第三附属医院为珠海合作培养第一批研究生外，2001 年 7 月，珠海土地上培养的第一批大专生从我校珠海学院毕业；2002 年 7 月，珠海土地上培养的第一批本科生依然毕业于我校珠海学院。

在珠海学院茁壮成长的背后，是珠海市党政领导的热心关怀，是珠海市人民的鼎力支持。珠海学院有如此美好的今天，是珠海市委、市政府和珠海市人民扶持的结果，是所有关心珠海学院发展的中央、地方各级党政领导、干部以及社会各界人士支持和帮助的结果。为此，我们在这除夕之夜，由我校师生呈上一场特别的广场式露天文艺演出，向珠海市政府和人民表达我们深深的谢意！

---

① 本文是在答谢珠海市委、市政府和珠海人民汇报演出暨新年文艺晚会上的讲话，珠海，2002 年 12 月 30 日；一个大学校长的探索，北京：高等教育出版社，2011，204-205.

41. 坚决反对和防止腐败是学校重大的政治任务[①]

今天，我们在这里隆重举行广东省人民检察院与暨南大学共同开展预防职务犯罪工作签约仪式。此事受到中央领导、教育部领导和国务院侨办领导的高度重视，各级领导都做出了重要批示。此举将开全国检察院与高校进行同步预防之先河，具有十分重要的现实意义和示范作用。

在这里，首先请允许我向张学军检察长表示衷心的感谢，感谢张检察长对“同步预防”工作的直接领导，并感谢张检察长今天在百忙之中将为我们作的精彩报告。也衷心感谢为广东广大百姓做出突出贡献、辛勤劳动的省检察院各位检察官对我校的关心与支持。

经过改革开放20多年的实践，我国已初步建立起了社会主义市场经济体制。加入WTO后，我国的经济运行和人们的社会生活正在全方位和国际接轨。在这个过程中，由于受封建主义、资本主义腐朽思想的侵蚀，一些工作人员计较个人得失的风气滋长，世界观、人生观、价值观扭曲，甚至有些领导干部也道德沦丧，滥用权力，以权谋私，违法乱纪，走上职务犯罪的道路。

随着市场经济的发展，高校已不再是一片净土，从以往的科学研究、成果鉴定评奖、课程考试、招生培养等方面出现的学术违规现象，到当前暴露的基建、后勤、物资采购等领域的职务犯罪，表明高校已成为“新的腐败灾区”。

就我校而言，作为有97年历史中央部属国家“211工程”重点建设高校和国务院侨办与广东省政府共建学校，近些年的确取得了令人瞩目的成绩，提前并以优秀评价完成“九五”“211工程”建设任务，学校博士点、重点学科和名牌专业增多，科研实力大大提高，学校形象与品牌得以提升，暨南大学已经发展成为一所在海内外享有盛誉的名校。按照中国网大中国大学综合实力排行榜，我校于2002年已跃居全国高校第37名。在学校蓬勃发展的大背景下，有着发展的喜悦和激励，同时也存在着严峻的考验。尽管近几年来我校加强了党风廉政建设，从严治校，依法治校，采取了一系列有力措施，加强了对教职工的教育，但仍有个别人员没能坚守住拒腐防变的思想防线，毁了自己，也损害了学校的利益和声誉。

党的十六大报告指出：坚决反对和防止腐败，是全党一项重大的政治任务。并强调要从源头上预防和解决腐败问题。权力的滥用就是职务犯罪的本质特征，开展职务犯罪的预防工作重在治本，重在促进检察职能活动的政治效果、社会效果和法律效果的统一。去年，最高人民检察院职务犯罪预防厅正式成立，中国检察机关有了第一个专门的预防国家工作人员职务犯罪的机构。由此可见，国家非常重视职务预防犯罪工作。我省今年初开始对全省“十五”规划部分重点工程项目实行同步预防，目前已初见成效。随着国家教育体制改革的深入，我校“十五”“211工程”建设工作的进展顺

① 本文是在广东省人民检察院与暨南大学开展同步预防职务犯罪工作协议签约仪式上的讲话，广州：2003年4月16日；暨南大学校报，第354期，2003年4月30日。

利，教学规模及招生人数将不断扩大。国家特别为我校的发展投入 5.38 亿元专项资金。学校基本建设、实验设备物资采购等项目繁多，工程数量多、投资大，管理任务重。在这种情况下，党风廉政建设和反腐败工作就愈加要坚定不移地开展。因此，为了从源头上预防职务犯罪的发生，增强广大干部、教职员工廉洁自律的自觉性，营造勤政廉政、遵纪守法的工作氛围，我校主动请求省人民检察院帮助我校搞好廉政建设，实施预防职务犯罪工作。显而易见，这一件事对我校的发展建设是非常有必要的，是非常有意义的。

省检察院将发挥检察机关职能作用，为我校依法从事管理工作提供法律保障。重点对基建工程建设、物资采购和招生等重要环节是否严格遵守国家法律和有关规定制度，参与监督和检查，帮助我校建立健全有关制度和廉政监督制约机制，确保预防职务犯罪措施落实到位。

我校将大力支持检察机关依法独立行使检察权，如发现有贪污、受贿等职务犯罪线索及时举报，并主动配合检察机关对职务犯罪案件的查处工作。我校将进一步制订廉政和法治教育方案，不断在干部和教职员工中开展廉政教育和法治教育，切实增强干部和教职员工的法治意识，及时防止和纠正苗头性、倾向性问题，防微杜渐，构筑起牢固的思想道德防线。同时还要落实检察机关提出的检察建议，积极发现和堵塞机制、制度上的漏洞，不断完善内部管理规章制度和监督制约机制，减少犯罪的机会，建立机制防线，促进公职人员依法履行职责，遏制和减少职务犯罪的发生。

校园环境干净十分重要，而学校校务教务的干净就更加重要！我们希望，在花园般的暨南园里，海外和港澳台以及内地学生能健康成长！

42. 寄语中层干部①

今天，我们在这里召开新干部任命大会，这是一次意义不同寻常的会议。作为新一届中层干部，你们到了新的岗位，将要承担新的任务。这是组织的信任、群众的信任，也是你们自己的光荣。在这里，我代表学校党政领导向同志们表示衷心的祝贺！

这里我讲几个问题：第一，我们学校面临的发展形势；第二，我们学校能够实现跨越式发展的原因；第三，学校今后的发展目标；第四，对大家的希望。

一、学校面临的发展形势

讲形势的目的是要在座的干部珍惜今天的大好局面。我国目前处在一个非常好的阶段，我们走上了一条繁荣富强的道路，走上了一条强国富民的道路。

我们国家去年已成为世界上第六大经济强国。我国的GDP居世界第六位，再用一两年的时间，可能达到第五位；五年内可能达到第四位。现在，我们已步入了全面奔小康的阶段，全国政通人和，中国人几百年来一直期望的日子变成了现实。就广东省而言形势也特别好。广东省是中国的经济大省，GDP总量占全国的1/10。我们暨南大学处在这样一个盛世，处在一个中国经济最发达的省份，这是我们的机遇。一个学校要发展，如果处在一个非常贫穷的地方，那是非常困难的。天时地利人和决定了我们暨南大学能够发展，我们处在好的时代、好的地点。我校自1996年以来，一直保持着蓬勃的发展态势，改革取得了胜利。

今年上半年面临“非典”疫情的考验，全校师生员工都表现得很出色，团结一致抵抗这场灾难，使学校得以顺利度过那段非常时期。昨天上午，省高教系统召开表彰大会，我们学校有一个先进集体——校医院、7位个人受到了表彰。前段时间，我们学校还受到了广东省、广州市有关部门的表彰。这说明在发展中遇到曲折的时候，我们也能挺过。大家还记得在前一个月，学校发现了一个可能是“非典”的人，学校迅速将南门的一栋楼腾空，作为隔离区，单位和个人都按要求立即搬出来，被隔离的83个人连夜住进去了，大家都很支持，没有怨言。我想这也是学校每件事都能顺利完成的原因。省教育厅上星期五晚上电话通知我，说我们学校应该列为全省抗击“非典”的先进集体，根据要求，我们连夜总结经验组织材料上报。但昨天大会宣布，由于高校系统评先进集体困难，很多学校都表现得很好，所以后来教育厅临时决定表彰全省高校。我校非常受省里关注，因为我们外招生多，影响大。

到目前为止，我们学校的标志性成果是什么呢？第一，是我们“211工程”建设的胜利，这是我们暨南大学最核心的工作。去年我们的“211工程”顺利通过验收，7个子项目全部评为优秀，以总体优秀的成绩通过，获得专家肯定，而且顺利进入第二期。我们进入第一期应该说比较艰难，当时，教育部的领导视我们学校是差学校，但当我

---

① 本文是在中层干部目标责任签字仪式上的讲话（按录音整理），广州，2003年7月11日；暨南高教研究，2003，(2)：1-8.

们邀请教育部的领导——韦钰副部长来学校视察以后，她对我校刮目相看，肯定暨南大学是一所比较好的学校。现在我们顺利进入了第二期。我们在第一期获得的经费支持很少，只有省政府给了 5000 万元。就凭这 5000 万元，学校实现了跨越式发展。而第二期到目前为止我们已获得了 5.68 亿元，是一期工程的 11 倍多。我想，这是因为我们第一期做得好，才有现在的第二期，这是全校上下特别是诸位领导干部——校级领导和中层干部共同努力的结果。

第二，我们学校“211 工程”建设的一个重要措施是办学重心上移。大家记得，在 1996 年以前，我们各院系每年的焦点是向学校申报多办专科，而现在争办专科的时代已过去。从 1996 年我们进入“211 工程”开始，校本部首先不办专科，现在全校各校区都没有专科，是一个本科教育层次以上的学校。我们遵循大力发展研究生教育，适度发展本科教育这样一个指导思想来发展。我们研究生和本科生人数比例从 1996 年的 1∶8.74 上升到今年的 1∶3.7，现在正在向 1∶2 的目标发展，不断增加研究生的比重，真正朝研究型大学发展。办学重心上移，开始很多人还想不通。大家看到，我们学校往前发展是非常困难的，虽然今天看起来很容易。在实现跨越式发展的过程中，我们遇到了非常大的阻力，很多人反对，很多人想不通，认为学校会走偏，学校会变穷。其实这条道路走对了，现在学校的收入更多了，教职工的收入更多了，学校品牌更好了。

第三，我们的学科建设得到了大发展。学科建设是学校办学的核心。据最新消息，我们学校在这次博士点评审中，获得了 2 个一级学科博士点，6 个二级学科博士点，加起来 14 个二级学科博士点。而我们今年上半年还只有 1 个一级学科博士点，13 个二级学科博士点，现在总数翻了一番。这说明我们这次申报是成功的。回想在进入“211 工程”以前，我们只有 7 个博士点，我们在头 4 年翻了一番，为 14 个博士点，再过了三年——今年又翻了一番。也就是说，我们的博士点经过“211 工程”近 7 年的建设翻了两番，有了 28 个博士点，为原来的 4 倍。这样一个数量站在全国高校中间，我们在面子上都好过一点。我们的硕士点建设也很不错，从 50 个硕士点变到现在的 88 个硕士点，增加了 38 个硕士点，增长了 76%，覆盖更加全面。我想再过两年以后，我们学校硕士点就能跨过 100 个大关。这期间我们还实现了很多零的突破。我们过去没有国家重点学科、国家重点基地、教育部重点实验室，现在我们都有了。所以说，我们的学科建设取得了很大的胜利。

第四，我们的科研实力增强了。学校科研经费的变化反映了我们的科研发展情况。我们科研经费到去年为止增长了 19 倍，过去一年仅 400 万元，现在是 8000 万元，我们现在向 1 亿元大关进军。我们的学术论文数增加了 1.3 倍，入选三大索引的论文增加了 10.7 倍，跨过了 100 篇大关，达 105 篇。还记得 10 年前，我刚到暨南大学的时候，全校一年入选三大索引的论文只有几篇，现在是 105 篇。学校高水平科研成果增多了，国家重点重大项目如“973”、“863”、国家自然科学基金等项目都能拿到，项目水平显著提高。横向项目水平也不断提高，经费的数量也在增加，这是非常了不起的。我校已从一个教学型的学校发展成为一个教学科研双中心的学校，实现了质的转变。

第五是学校的整体实力增强了。首先我们的校园面积增加了 55%。学校要发展，校园是关键。因为珠海市政府免费给我们提供办学用地，因此虽然广州市的校园面积一直在缩小，但现在校园总面积还是扩大了。校园的建筑面积也翻了一番，从1996年的50万平方米发展到今年已超过100万平方米。校园的建筑，无论是教室、实验室，还是老师和学生的住宅都发生了显著的变化。

几年来，我们学校的收入增多了，与1995年相比，现在已翻了两番。1995年全校包括附属第一医院总收入是1.96亿元，去年我们的总收入是8.28亿元，增长了3倍多，这是非常不易的。1996年我们实行财务集中管理，取消了二级财务，当时许多院系都有意见。顶着这个压力，我们坚持走过了这条改革之路，事实证明，我们的改革是对的。去年，我们的财政拨款总计2.9亿元，但我们的总收入是8.28亿元，这就是说我们自己挣了5亿多元。

我们的固定资产也增加得很快，从1995年的2.7亿元增加到目前的14亿元，增长了4倍多，我们的家底变厚了。我们的学生也增加了很多，本科生从5000人增加到12 000多人，研究生从615人增加到3445人。研究生的数量在全国高校排在20来位。

第六是我们学校的办学特色更加鲜明了。我校现在的海外及港澳台学生6854人，来自于世界五大洲56个国家和港澳台3个地区，这个面是非常宽的。而在“211工程”以前，学校只有16个国家和港澳台地区的学生。我们的学校也走向了世界，参加了世界大学校长年会，而且在世界五大洲主要国家建立了姊妹学校，这是目前中国唯一一所学校做到的。学校的“侨”字特色更加鲜明了，海外及港澳台学生占总学生数的40%。

第七是教职工生活水平不断得到提高。几年来，教职工人均住房面积翻了一番，人均收入增加了7倍。从9月份开始，教职工的待遇将进一步得到改善，考核分值将从去年的1.2元/分提升到1.5元/分。

第八是学校的品牌更好了。一个学校办得好坏，在于别人客观上怎么看。我们不能在家里谈自己变好了，要让别人来评价。无论是政府的评价，还是社会团体的客观评价，对于我们都是很好的。广东省政府这几年来一直把我们排在全省高校的第三位，在进入“211工程”以前，我们是七八位甚至十位。教育部搞的“211工程”我们进了100强。从1998年开始，全国团体评估中国大学排行榜，仁者见仁，智者见智，社会各界对此有不同的声音。有的喜欢评，有的怕评。美国每年由社会团体公布大学排行榜，我认为是好事，因为可以通过评估让公众了解学校。中国最早是网大在评估，比较权威。在其每年的评估中，我们学校都在进步，现已由1998年的87位上升到今年的36位。暨大由一个不知名的大学，由被人家误以为是山东的“济南大学”到全国高校的36位，是非常令人欣慰的。我们既不是教育部也不是广东省直属的学校，我们是国务院侨办直属的学校。侨办只有两所学校，不太引人注意，能够有今天这个局面，大家一定要珍惜！

第二个评估单位是广东管理科学研究院，每年评选中国大学100强。它从2000年开始，第一次评我们是第81位，我们逐渐进步，今年已是第49位。不管怎样，这些民间团体对我们的评价都在全国前50位。这是相当不容易的成果。一个学校的品牌，决定着它的发展，决定着它的生命和健康状况。之所以学校目前在海内外受到欢迎，是因为我们的品牌好了。近日，广东省和全国的招生工作正在进行，我们在广东省招收1350名学生，但有2640名过重点线的学生报读我校，为招生人数的近2倍。这是我们近年来招生形势最好的一年，我校也因此成为全省高校生源形势最好的学校。这说明广东省学生愿意读暨大，暨大是热门学校。这不是在报上发表几篇文章就能做到的！这么多的学生考暨大，这是从来没有过的。学校进入“211工程”前，每年只有10%～20%第一志愿的考生。这种情况从1997年才真正得到改变，1996年第一志愿学生也只达到80%。生源能够反映一所学校的地位。现在日子好过了，每个院系招生都很容易，都是高分，今年广东考生670分以上才能进我校。对本校教职工子女我们给予照顾，达到643分的全部录取，这是学校给大家的一个福利，如不给予照顾，很多教职工的子女进不来，这是学校经过大力争取得到的。我们在港澳地区的招生形势也非常好，比去年超出200多人。今年上半年因为“非典”的问题我们还担心今年的招生情况差，因为没有出去宣传。但是现在暨南大学的地位已经稳固了，信誉在外，所以海外生才会蜂拥而来，这是我们学校品牌好的标志。不仅本科生这样，我们的海外博士、硕士学生的情况也不错。我们现在有700位海外及港澳台研究生，占全国海外及港澳台学生的1/4，我们已成为最受海外及港澳台学生欢迎的大学，这就是学校的品牌和质量引起的变化。今天这个形势，这7年多来实现跨越式发展是来之不易的，不是说随随便便就可以过来的。大家想一想，这7年多来，大家吃了很多苦，我们一定要珍惜这大好形势。我希望有朝一日我们跨到二十几位、十几位。这不是没有可能的，只要大家努力！

二、学校实现跨越式发展的原因

回顾过去，我们发现学校的变化来自改革，是改革改变了我校的状况，改革改变了我们落后的地位，改变了我们暨南大学不为人知的历史。现在很少有人说我们是山东的“济南大学”了。邓小平同志的改革开放，发展是硬道理的论断在我们学校得到了证实。我们采取了很多改革措施，从学分制到分配制度、财务制度、干部考核等一系列改革，使得我们学校有了这样大的变化。而我们改革的核心是什么呢？是我们师资队伍的变化。一个大学最关键的是师资队伍，显然我们现在的师资队伍比过去好得多。我刚到暨南大学时的两个数据至今让我记忆犹新，一个是全校教师里只有8个人是博士，跟名校相比差得远；另一个是博士生导师的数量，全校的博导加上我也是8个。全校只有8个博导！今天我们有70多位博导，300多位博士，我们现在已经有1/4以上的教师有博士学位。师资队伍质量的提高，决定了学校的发展变化。办大学主要是靠师资、人才，尤其是高层次人才的聚集；靠着我们坚持“从严治校、从严治教、从严治学”，加上依法治校，实事求是，我将其归纳为“严、法、实”三字原则。我们要用严格的管理办法，用法治代替人治，要坚持实事求是的原则，不要搞假大空的东西。我们就凭着这些一步一步向上攀登，才有了今天这个局面。这归根结底，是由于我们高举中央24号文的旗帜，在毛泽东思想、邓小平理论和“三个代表”重要思想的指引下，在国务院侨办和广东省委、省政府的直接领导和关心下，经过全体校级领导干部和中层干部的努力，取得了今天这些成绩。今天我们对学校新一届中层干部进行任命。回顾过去我感到很高兴，也借此机会向大家表示崇高的敬意和衷心的感谢！

三、未来发展目标

由于我们在座的都是新一届中层干部，因此要了解我们有什么目标。待会儿我还要跟大家签订目标责任制的协议书。因为没有目标，我们就成了“瞎子”，这是不行的。我们要坚决实施“侨校＋名校”的战略，使学校成为一所名副其实的研究型大学。尽管《中国高等教育评估》去年评我们为研究型大学，但我们离真正的研究型大学还有距离。首先学生指标还差一点，研究生与本科生的比例要达到 1∶2。因此学校要按照国家计委制定的计划，到 2010 年全校达到 8000 名研究生、16 000 名本科生、400 名预科生、600 名华文教育学生、3000 名继续教育学生，一共是 28 000 人的规模。我们校本部到 2010 年有 7000 名研究生、3800 名本科生、3000 名继续教育学生；珠海学院的规划是 1000 名研究生、4000 名本科生；旅游学院是 1500 名本科生，华文学院是 2700 名本科生、400 名预科生、600 名语言生；磨碟沙江南校区 4000 名本科生，这是学生的分布情况。但是从目前的形势来看，这种分布还有变数，变数在什么地方？是因为我们江南校区被广州市征用。去年底我提出了换跑马场的计划，上半年原定换广氮。换跑马场是很难的事情，但我一直是抱 1%的希望，用 100%的努力去做。到今天为止，仍然是 1%的希望。我们做了很多努力，得到很多支持，但关键的广州市政府不点头，所以我们还得继续努力。今天告诉大家这些，是要让你们清楚这个情况。

2010 年设计的目标是把暨南大学建成具有优势学科，达到或接近世界先进水平，在海外特别是在东南亚地区知名的、一流的综合性大学。这是中央给我们的定位。

我们现在的口号是什么？是成为一所国际性的大学，成为一所现代化的大学，成为一所综合性大学。要达到这个目标，难度很大，当然我们经过这些年的发展，形势还是不错的。我们现在拥有 28 个博士点，尤其是经济和管理学科，在广东省乃至华南都是首屈一指的，这是暨南大学的传统优势学科，是从建校以来的强项，要继续发展下去。我们也希望文学、物理、临床医学、生命科学等学科在下一次硕士、博士点申报的时候有更大的突破，首先是在博士点、在一级学科博士点上有突破，使我们学校能迅速站到新的高度。要达到这个目标，我们校级领导班子、中层干部队伍责任非常重大，任务非常艰巨，为此，我向大家提几点希望。

四、几点希望

第一是希望大家努力完成上述目标，使暨南大学继续向前发展，品牌更好。大家是新一届干部，要做好这些工作，首先是要求大家加强学习，学习好毛泽东思想、邓小平理论和“三个代表”重要思想。加强学习是做好工作的前提和基础，对一个领导干部来说这是工作的需要。要加强学习就必须端正学风，坚持理论联系实际。我们只有学好理论，才能够辩证地看问题，才能把问题看得深、看得透，遇到问题才能果断处置，才能够防止形而上学，才不会人云亦云。

第二是希望我们全校的干部都要忠于祖国，忠于人民，忠于神圣的教育事业。一个人没有祖国的概念，没有人民的概念，我们教育岗位上的领导干部，如果没有神圣教育的概念，就不可能做好自己的工作，不可能襟怀坦白，他一定会斤斤计较，他不会无私奉献，不会努力为人民服务，他只会想到自己。如果一个人心里能够总是想着祖国，不管在任何情况下，把祖国的强大、人民的幸福装在心中的话，我们的教育事业一定能够做好。

第三是希望大家要求真务实，注重实效，也就是我经常所说的实事求是。我一直要求我们的统计员要把数据搞准确，不能搞假数字，别搞歪风邪气。做事情立足于什么样的基础，我们在什么情况下发展要清楚。做什么事情都要实事求是。我们的干部要说实话，办实事，求实效。搞虚假第一次可能会成功，但第二次可能就会失败。说穿了那是害党、害人民、害学校。在今天的会上，我可能是年纪最大的人，我经历了很多。假大空非常害人，我在工作中一直强调这一点，这是肺腑之言，希望大家记住，不要搞假的东西，我们一定要求真务实，讲实效。

第四是要继续坚持“从严”原则，即“从严治校，从严治教，从严治学”，要把它贯彻到我们每一个人的工作中去。“从严”谈得容易，做起来非常难，特别是在这个讲关系的年代，要严格谈何容易。我们中国人有时有一个差不多的思想，什么事做得差不多就行了，不要求完善。实际上只有工作完善，才能确保工作的质量。我们今天很多改革措施非常好，但是我们有一些改革还不够成功，就是因为某个环节不严格。我举两个例子。比如我们的学分制。我们绝对是中国最早实行弹性学分制的学校，但这里面有些核心东西一些院系到现在还搞不明白，我们的职能部门也有不明白的。学生不及格必须重修，这件事情从 1993 年取消了补考以后开始做，要继续坚持。学分制的核心是必须坚持这一点，就是某门课程不及格以后，必须及时重修，及格以后再继续往前修。这是最基本的东西，但是我们的教务员，我们的老师，我们的管理部门都不愿管。学分制与我们的每一个老师、每一个领导干部都相关，我们每一个领导都要去熟悉它。学生要根据自己的经济状况、身体状况、智商情况来安排自己的学习。但这个已讲了好多年的问题，仍然没有得到很好的解决。我们一定要管住这一点，学生不能带着很多不及格的课程前行，这是违背学分制的本意的。我们现在要实行导师制，就是希望导师能严格管理，今后不再出现上述现象。再如我们的考试。这些年来我们暨南大学在考试方面做了很多工作，制定了很多办法，应该说是成功的，学生的作弊现象大为减少。我们暨南大学的特点是从考卷、试题库，到考场管理，一直到考试结果抽查，都有一套办法，考虑得非常周到，但是这里面任何一项小的改革都冲击着很多人。今年上半年博士生入学考试启用试题库，很多博导、很多院长都在反对，我听了感到很滑稽。公平是考试的第一原则，但我们的老师们希望自己掌握考试的权利。那怎么行呢？他们想要谁就是谁，那不就走向腐败了吗？这些事有些院长就不敢管。不管就不严，不严就会产生不公平。学校应该是一个公平的地方，要以分数来衡量。只有公平，学校的知名度才会高。如果学校能够做到公平，就会有更多的好学生进来。好多人因为设试题库的事威胁我说，这样搞以后没有学生来考我们学校。事实怎么样呢？更多人考暨南大学！学校有今天的知名度，绝对是从严的结果。只有从严才能保证质量，才能给社会提供高水平的人才，暨南大学就会凭这个特点立足。英国的伊顿公学大家知道吧，培养英国领导人的贵族学校，它是非常严格的。在剑桥、哈佛等世界名牌大学读书是混不过关的。如果一所大学能混到文凭的话，这所学校就是下等的学校。所以，我再次提出“严格”的口号，这是为了暨大的前途。只有把暨大“花花公子”的帽子摘掉，学校才能够往前走。今年春节前，原主管教育的副省长、现省人大常委会主任卢钟鹤同志到我家里慰问我时说的第一句话就是：刘校长，我感谢你把暨大“花花公子”的帽子摘掉了。十多年前，我刚到暨南大学的时候，听说暨大是个“花花公子”学校，我非常伤心，这是耻辱。只有从严治校，学校才有前途，因此大家在每一个程序、每一个方面都要严格。严格可能会得罪人，但得罪的只是少数人，可这样能够保证我们学校的办学质量。

第五是希望大家要坚持依法治校，敢于负责，恪尽职守。所谓依法治校，就是把中国传统的人治社会变成法治社会，使我们学校方方面面的管理走向制度化，这样才能科学地管理学校。从1996年以来，由于各职能部门和各院系的努力，我们把大部分制度搞出来了，不管怎么样，应该说这些制度对我们学校的发展起到了重要作用。这个星期我们印发了《暨南大学文件汇编·教学科研卷》，今年初我们印发了《行政管理卷》。

这两本书就是学校的“法典”。现在发到每一个院系的教研室、图书馆及学校的每一个单位，全校师生应凭这两本书来管理学校，只要人人都按制度办，按原则办，学校的管理就能搞好。科学的管理一定要依法办事。今天我把书带到会场，就是要让大家记住这个。前几天有人问我一个问题，我说你去看看这两本书，就知道怎么做，书里面从办事的程序到怎么办都告诉了大家，每件事你都会在这两本书里找到答案。有了这两本书，我们要考虑的就是怎么管理，怎么按制度执行。如果有什么不合理的地方我们再修改，但是不能经常变。只有把我们的“法治”搞好了，学校这台机器才可以顺利地往前走。希望各位干部在有了这两本“法典”以后要敢于负责。现在有的干部，就是做老好人，什么事情都是批“同意”报上来，他（她）明知那个同意是不对的。前几年有一个院长告诉我：我的同意就是不同意。我问他，这件事情明显不对你还批同意到我这儿干什么，他说，我也不敢签不同意，我的同意就是不同意。把责任全部往学校推。

现在我们全校大概有290多位中层干部，每个人都要对所管的领域负责。如果你们对不应同意的事情批同意，最后报到校领导那儿，校领导不清楚情况一旦批错了，那就贻误大事。学校是个大集体，是个系统工程组成的大集体，一个零件不对，这个系统就走慢了，甚至停止运转。我们许多小事情最后都造成了大错误。有些人他不知道自己在做什么，随意处理，不负责任，不加考虑。学校每一个岗位都是非常重要的，一定要在处理事情的时候想远一点，想宽一点。把握不准的事情要请示，对重大事情要请示！再如我校外语学院最近出了计划生育的事情——一位职工多生了个孩子。就院领导来说我觉得也情有可原，因为这位职工以前得过癌症，得癌症后经常请假，并且随便开假条，最后假条也没人看，也没人去家里看她。领导说去看她，她说不要过去看。因为她怕人去看她。就这样她就把孩子生下来了。这是暨南大学的违纪，计划生育的违纪。计划生育是国策。如果我们的领导、院领导、系领导工作细一点，是能够看得见的。生活是复杂的，你三番五次要去看她，她为什么不让你去看她，你就得打个问号。如果去她家里看，问题就能及时解决。外语学院这是第二次违反计划生育。

原来的一位职工是外语中心的，也多生了一个，最后被开除了。学院没有吸取经验教训。本以为我们大学校园里面，大家文化层次高，计划生育都能自觉遵守，但是还是有爆冷门的地方。这是因为我们的工作不严格、不细致、不负责任。计划生育是国策，这次我们也处理了一些干部。我也很同情大家，虽然大家都很忙，但大家一定要重视，这是国策。尽管我们是办学，但对国家政策要注意，不要认为大家都会自觉遵守，就有个别人不愿遵守。因为出了这件事，今年就没有计划生育奖金，而且学校也会为此挨批。大家要注意这些问题，要恪尽职守。每个人自己的工作一定要做好，想周到一些。现在有些干部连制度都弄不清楚。这次在竞争上岗的时候，我参加了几场，当李兴昌主任问大家纪检方面的问题，有的人根本就答不出来。学校的制度要弄清楚，你的职务、岗位、要求是什么，你应该做些什么。各单位对老百姓应该热情接待，人家老在投诉，说我们有些部门不热情。

第六是希望大家廉洁自律，艰苦奋斗，团结协作。作为一名领导干部，不论是校领导干部，还是我们中层干部，你的一言一行群众都在关注，只有廉洁奉公，干净做事，洁身自好，你的形象和权威才能在群众中树立起来。如果你自己言行不正，你的做法就得不到群众的支持。我们过去几年几乎每次会议都讲廉政建设，反腐败，但还是出现了个人违纪的事情，出现了李××、周××等人坐牢的事情，这是我们不愿见到的。李××这样一位名教授因为 22 万元触犯刑律。希望我们的领导干部一定要小心使用手中的权力，这是人民给你的权利，一定要使用好，不能把权利变成金钱。我们学校比较大，钱比较多。违法人员中还有一名工人。总务处的一名工人都贪污了 30 万元。我们的中层干部，甚至一名工人，如果管得不严，就会产生腐败现象。我们过去不光是讲，也采取了措施，但是还是出现了腐败，这是一件遗憾的事情。我们与省检察院共同预防职务犯罪，这是保护我们的干部不犯错误，使我们暨南大学变得更加干净，有利于我们培养人才。廉洁自律靠自己，组织给你讲了你不听，不论多大的干部，只要违法，照样要受到处罚。艰苦奋斗是我们党的优良传统，但是现在很多人忘记了这个传统，我们学校也存在这种情况。铺张浪费、吃喝玩乐、奢侈腐化、讲排场，一个小小的会议都要拉到外面去开。今后凡是到外面开会，一定要得到校领导的批准，要省点钱。尽管学校从国家拿到将近 6 个亿，尽管学校挣了很多钱，但是我们需要钱的地方还很多，差距还很大，大家要继续保持艰苦奋斗的作风。西方一些国家很有钱，但人家请客不像我们这样花钱，我们有些人动不动到外面花几千块钱请客，其实用不着。不花钱也能办好事。我提醒大家一下，请大家注意这个问题，包括房子的装修。新房子不断建成，有的装修太豪华了。我们是学校，不是宾馆！

最后我讲一下团结协作，这一点很重要。每一个院系、每一部门以一把手为中心组成为一个领导班子，这个领导班子一定要团结，一定要有协作精神，没有团结，你这个班子就没有战斗力。首先是班子里面要互相理解，互相谅解；要补台，不要拆台，什么事情张三想不到，李四就给他补一补。一个人的大脑不是计算机，想了这面就忘了那面，这是难免的。不要一个部处一个工作做得不好，就批评指责，就把它搞垮。不要轻信谣言，传谣更不行；拆台，更不行。我们学校个别人成天制造谣言，把同志之间的信任和团结搞得荡然无存。班子内首先是坚持原则。这方面一把手是核心，一把手起着关键的作用，负全面的责任。一个单位的好坏，归根结底很大程度上在一把手。因为我们是一个实行校长负责制的大学，学院是院长负责制，系里是系主任负责制。院长、系主任是各单位的核心，要团结院系党的一把手共同把工作搞好，要用民主集中制的原则，实现你的目标。那就是要少数服从多数，个人服从集体，下级服从上级，民主集中制的核心内容要记得，不能只抓住其中一点而不顾其他，要三点统一，要用民主集中制原则保证我们班子的战斗力。同时，我们要走群众路线，按照我们今天签订的责任制把工作做好。团结是胜利的保证，今天我要拜托各位一把手，各层次的一把手，要把你这个班子带好。副院长、副主任是一把手的助手，要清楚自己的位置。现在有些班子不团结，是正手和副手的职责不清楚，正手做不好正手，副手没做好副手。副手是助手，副手与正手争权是错的。今天是新一届的中层干部上任的开始，希望从今天开始，大家要认真履行自己的工作岗位职责，同时，要以个人品德上的表率带动本单位全体群众前进。这两天要搞好新旧交接，赶快把工作交接好。新班子刚刚上任，从今天起责任就交给你们，出了事就找新班子。我们实行的是校院系的体制，是校、部（处）、科的体制，分层分级管理体制，要层层落实。大家责任在身，要把工作做好。老班子安排的是正确的就不要调整，要前后相延续。这次中层干部换届，是在学校发展非常关键的时期。另一项是我们学校已进入“十五”“211 工程”建设阶段，这是本届班子的核心工作，要承担起这一非常光荣的任务。同时，你们这届班子也是我们暨南大学建校 100 周年校庆那年交班的一届。迎接建校 100 周年，这在我们校史上是件大事。建校 100 周年时，你们正处在关键的工作岗位上，也是一个难得的机遇。大家要以出色的工作迎接 100 周年校庆的到来，使得我们暨南大学在 100 岁诞辰的时候，有新的成绩展示在全国人民面前，展现在全世界华侨华人面前，使我们学校真正实现“侨校＋名校”的战略目标，完成党和国家赋予我们暨南大学特别的办学任务。

最后，再一次地拜托诸位。要带领全校 3 万名师生更上一层楼，到一个新的境界，就靠我们这些人。拜托大家把工作做好！

43. 暨南大学的创新发展之路①

教育是培养人才和提高民族创新能力的基础，必须放在现代化建设的全局性战略性重要位置。作为培养高层次创新人才和增强民族创新能力重要阵地的高等院校，在我国经济建设和社会发展中扮演着越来越重要的角色。面对复杂多变的国际形势，我国高等教育要为国家在激烈的国际竞争中保持优势而发挥其应有的作用，要完成为国家现代化建设培养优质人才的使命，要培养具有全球意识和开放意识、具有较高文化品位、能够参与国际事务和具有国际竞争能力的创造型人才，就必须积极采取应对措施，积极开拓国内国际教育市场，大力推进创新教育，不断提升办学层次和办学水平，提高高等教育的国际竞争力以及为经济发展和社会进步服务的能力。

华侨高等教育是中国高等教育的重要组成部分，肩负着为海外华侨华人社会和港澳台地区培养人才的特殊使命，其作用和地位不可替代。面对高等教育领域竞争日趋激烈的紧迫形势，华侨高等教育着力突出“侨”字特色，提高办学水平，更好地为侨服务，为祖国统一大业服务。作为华侨高等教育重要代表的暨南大学为此提出了“侨校＋名校”的发展战略。

一、竞争激烈的外部环境表明唯有创新才有发展

“九五”以来，党和国家对高等教育给予了前所未有的重视和关注，除加大投入外，同时还采取了两项重大举措，这在中国高等教育史上可以说是空前的。一是面向21世纪重点建设一批大学和学科，二是进行结构调整，整合办学力量，集中办学资源，有重点地合并一批院校。这两项措施都是有针对性、有重点地提升一批基础好、实力强的院校的办学层次和办学质量，为建设一流大学和高水平大学奠定基础。暨南大学作为华侨高等教育的重要代表，是我国高等教育的一个组成部分，其在全国高等院校隶属关系调整后仍隶属于国务院侨务办公室，这进一步明确了学校为侨服务的责任和使命。然而此时，国内外的办学形势和环境已发生巨大的变化，一直在国家政策保护下而具有生源优势的华侨高等教育已面临着来自国内外的冲击和挑战。一方面，国家对高等学校办学自主权逐步放开，香港、澳门回归以后，除暨南大学和华侨大学以外，内地已有许多高校开始在香港、澳门招生。另一方面，港澳地区及一些华侨华人数量较多的国家和地区看到华文教育的广阔前景，纷纷开办华文教育，与我国华侨高校争夺生源。另外，台湾的一些高校也纷纷采取减免学费等相关优惠政策或其他方式吸引华侨华人学生和港澳地区学生前往学习。

对暨南大学来说，生源及办学的特殊性决定了我们的竞争对手是来自多方面的，它们既有本系统（华侨院校）的高校，也有系统外的高校；既有国内的高校，同样还有来自其他国家和地区的高校。如果我们的办学水平和办学质量止步不前的话，其结

① 本文是在广东省高校领导干部暑期读书班上的报告，珠海，2003年7月28日；一个大学校长的探索，北京：高等教育出版社，2011，174-182.

果可想而知。因此，我们不仅要坚持特色，而且还要实施名牌战略，大力提高学校的办学层次和办学质量，提高学校的竞争力和吸引力，才能在国内外教育市场的激烈竞争中立于不败之地。

二、曲折而又辉煌的历史赋予我们创新的动力

即将迎来97年诞辰的暨南大学，在其悠久的历史上，既有时代留下的沧桑，也有智慧释放的光芒。作为中国第一所由国家创办的华侨学府，其前身是1906年清政府创立于南京的暨南学堂。在随后的近一个世纪里，学校经历了4次易址、3次停办。1958年，暨南大学在广州重建，直属教育部，由时任广东省委书记陶铸兼任校长。1970年，学校因"文化大革命"而停办。1978年，暨南大学在广州原校址复办。1983年，中共中央、国务院24号文件批复中央宣传部、教育部、国务院侨务办公室，将暨南大学列为国家重点扶植的大学，隶属于国务院侨务办公室。屡次变迁或停办，阻碍了学校的持续发展，影响了学术风格、学术环境的形成。直到1978年以后，暨南大学才真正有了一个和平、安宁的发展环境。然而，历史的动荡曲折并没有影响大批教育界仁人志士对教育的忠诚和追求。在暨南大学不平凡的历史上，曾有如黄炎培、马寅初、郑振铎、梁实秋、周谷城、钱钟书、周建人、夏衍、许德珩、胡愈之、严济慈、楚图南、黄宾虹、潘天寿等一大批著名学者，在暨南大学任教，他们培养了诸如国务院前副总理吴学谦，国务院前副总理李岚清，知名人士江上青，著名侨领、新加坡大学首任校长李光前，泰国国会前主席、副总理许敦茂，新加坡中华总商会前会长陈共存，院士谭其骧、曾毅、邓锡铭、侯芙生，以及近年来内地和港澳台地区的王学萍、佀志广、颜开臣、马有恒等许多官员，工商及文教界著名人士。

一所百年老校应该具有与其历史相称的实力和品牌，因为其在办学经验、人文环境、文化底蕴以及历史积淀等方面都具有那些新办大学所无法比拟的优势，后继者有责任和义务将这些优势进行传承和弘扬。曲折的经历和深厚的历史积淀表明了暨南大学在各个历史时期的重要意义和举足轻重的地位，这既让全体暨南人自豪，也给了我们重铸暨南辉煌的信心，给了我们改革创新、发展暨南的动力。我们深切地感受到，要继承未竟的事业，要延续暨南精神，要开创暨南大学的美好明天，就要审时度势、开拓创新。

三、正视现状，着眼未来，全面实施改革创新

"九五"以来，我校按照"发挥优势、深化改革、保证重点、改善条件、提高质量"的发展思路，坚持"严、法、实"的办学思想（即"从严治校、从严治教、从严治学""依法治校""实事求是"），转变观念，认清形势，正视差距，扬长避短，开拓创新，以"211工程"为龙头，以教学科研为中心，大力推进学校各项事业不断进步，在实施"侨校＋名校"的发展战略中，学校在许多方面发生了从无到有、从小到大、从弱到强的可喜变化，成功实现跨越式发展。

（一）从学分制入手，推动教学改革

1. 以学分制改革为核心，不断提高教学质量

我校实施学分制改革的指导思想是"以人为本"，将学生的利益放在第一位。在这

种制度下，学生可以自主选择课程，自主选择教师。学分制最大的优点在于能给学生提供更主动的学习条件，最大限度地发挥学生自主学习、向上发展的能力，它适应我校生源世界性的特征，有利于教育目标的实现。这是由传统的保姆式教育向现代自主教育转变的重要体现。我校在 1978 年复办之初即开始试行学分制，但那时是初级阶段，还处于学年学分制状态。1993 年起在全国率先取消补考，施行标准学分制。标准学分制可以说是一种完全从学生利益出发的教学制度，它既可以使学生学到真正的知识，又可以使学生不必为补考及将补考记录载入档案而担心；既能保证学生的学习质量，又能因重修交费、重花时间读书而给学生施加一定的压力。我校不仅是广东省最早实行学分制的高校，而且是全国最早实行弹性学分制的高校。经过 20 余年的实践，学分制的各种管理办法已日臻完善，对贯彻我校“两个面向”的办学方针，以及培养学生素质均有着积极的促进作用。

为配合学分制的实施，我校根据内外两类学生的不同特点，对学生的培养目标进行定位，合理设置专业，突出创新精神与实践能力的培养，使受教育者得到全面发展。我校对海外及港澳台学生的教学要求是“面向世界、应用为主”，对内地学生的教学要求是“加强基础、目标上移”。

为突出基础课教学，保证基础课教学质量，使新生在低年级即获得坚实的基础知识，我校自 1993 年就开始实行教授必须上本科基础课的制度，延续至今，效果良好。

为提高课堂教学质量，我校自 1985 年开始实行课堂教学评估制度，1993 年开始实行课堂教学三重评估制度，即每学期分别由学生、校院系领导和听课专家组对课堂教学质量进行评估。这一评估制度的完整、公平、公正性受到师生的欢迎。学校根据评估结果采取了一系列奖惩结合、以奖为主的措施，激励先进，鞭策落后，对不合格教师进行淘汰，优秀教师则成为学习标兵。这在一定程度上加强了师生间教学信息的交流和教师对课堂教学质量的重视，有效地提高了课堂教学质量。三重评估制度在国内由我校首创，且效果良好，现已被国内许多高校参照使用。

2. 加强教风、学风建设，大力推进考试改革

为进一步加强校风、教风、学风建设，提高教学质量，严肃考试纪律，我校采用建试题库，分 A、B 卷考试，加强考场管理，抽查各专业试卷内容等措施，促使教师认真对待考务、学生认真对待学习和考试。从 2001～2002 学年上学期开始，我校还设立了可一次容纳 840 人的大型考场，杜绝了作弊现象，保证了考试的公正性。这一全国首创性举措受到了许多高校及新闻媒体的广泛关注。在今年的博士生入学考试中，我校也采用大型考场考试，学生及社会反响良好。

3. 突出国际性特征，鼓励和扶持双语教学

为贯彻“两个面向”的教育方针，培养国际性人才，我校从 1996 年开始提倡和鼓励教师用英语进行本科专业课教学，经过几年实践，此项工作逐步得到推广。目前我校非英语专业用英语教学的课程已达 28 门，使用全英语教材的课程达 34 门、中英对照教材的达 40 门。以此为基础，2001 年 6 月，我校在全国高校中第一个成立了采用全英语教学的国际学院。目前该学院已开办了临床医学、国际经济与贸易、会计学等 3 个专业，最近又新增设了食品质量与安全专业。

为办好以上全英专业，学校采取措施大力鼓励教师开设全英课程，极大激发了教师开设全英课程的积极性，对办好国际学院，进一步扩大国际学院规模起到了重要的促进作用。

4. 调整办学重心，优化办学结构

为进一步提高办学层次，优化办学结构，按照“大力发展研究生教育，适度发展本科教育，稳定成人教育规模”的办学思路，我校校本部从1996年开始停止招收专科生，到今年7月，全校已没有全日制专科生。同时，学校根据社会发展和学生求学、就业的需要，不断调整专业结构，及时增设适应国内外学生要求的热门专业，对优化我校专业结构，培养社会急需人才，起到了积极的推动作用。

目前，学校研究生与本科生之比由1995年的1∶8.74上升到1∶3.73。

与此同时，学科结构进一步得到优化。本科专业由1995年的30个增加到48个，硕士学位授权学科由1995年的50个增加到86个，博士学位授权学科已由7个增加到28个，并且在一级学位授权学科方面实现了零的突破，现有3个一级学科。博士后流动站实现了零的突破，达到3个。教学系由21个增至37个，学院数由7个增至16个，涵盖的学科门类更加广泛。

（二）以学科建设为中心，大力鼓励教师从事科学研究

科研水平是衡量一所大学办学水平的重要标志。为鼓励教师在做好教学工作的同时，积极投身科学研究，学校以“211工程”重点学科建设为中心，大力推进科研工作。从1996年开始，学校用“教学、科研”双中心目标取代了过去单一的“教学中心”目标。同时，学校在分配体制和财务管理方面实施了一系列有助于科研工作开展的举措。学校将教师获得的课题级别、数量，科研论文数量及其发表刊物的档次等科研成果直接与校内工资挂钩，对其进行量化考核。如教师在《自然》（*Nature*）或《科学》（*Science*）上发表1篇文章，可获得12万元的奖励。这些举措充分调动了教职工教学科研积极性，激发了教师的科研潜能，使教师的科研论文和科研项目数量大幅上升，使学校的科研经费得到快速增长，同时也促进了教学质量的提高。

另外，实行集中管理财务政策，由学校统一颁发校内奖金，减轻了教师通过创收谋取福利的压力，使教师将主要精力投入到教学科研中，成效显著。2002年，我校科研经费已由1995年的400万元增加到8000万元，增长20倍；在“973”“863”等国家重点项目方面实现了零的突破；专利从无到有，现在获专利17项；学术论文数增加了1.3倍，其中被三大索引（SCI、EI、ISTP）收录的论文数增加了10.7倍，去年首次突破100篇大关，达105篇；获得的省部级科技奖励增加了5倍多；科研成果的应用也有较大的进步。

另外，学校在国家重点学科、国家研究基地、教育部重点实验室、工程研究中心等方面都实现了零的突破。新增2个国家重点学科、1个国家人文社会科学重点基地、1个教育部重点实验室、1个教育部工程研究中心、8个广东省重点学科、1个广东省教育厅重点实验室。

（三）确立“严、法、实”办学思想，推动管理体制创新

1. 合理调整人员结构，稳步推进机构改革

本着“精减、效能、统一”的原则，“九五”至今，我校先后进行了两次大规模机

构改革，行政处级机构已由1999年以前的29个减为18个，与此同时，学校对科级机构也进行了整合，优化了结构，提高了效率。为保证教学和科研这两个重点，几年来，学校注重引进专业教师，控制行政人员数量，原则上不进行政干部。虽然学校的办学规模不断扩大，但人员结构得到优化，尽管机构数量比改革前少，可效率却更高了。

1995年至今，全校的教职工和专业教师人数基本没有变化，1995年分别为3601人和1036人，目前为3649人和1093人，且学校所获上级经费投入并未大幅增加，但学校完成的任务数量却成倍增加，显然办学效益已更加优良。在2002年广东管理科学研究院公布的“中国89所‘211工程’大学教师人均效率排名”中，我校排在第41位。从收入方面看，几年来，学校的经济效益有了大幅提高。1995年，全校的总收入为1.96亿元，其中国家财政拨款0.37亿元；2002年，全校的总收入达8.28亿元，其中国家财政拨款2.93亿元。总收入与1995年相比，增长3.22倍。

2. 整顿机关作风，加强廉政建设

为配合“三讲”教育的开展，切实改进机关作风，学校各机关部处针对“门难进、脸难看、事难办”的现象，结合群众的意见和建议，制定有效措施，开展了以“内强素质，外树形象”为主旨的边整边改工作。机关各单位进一步加强建章立制，实行岗位职责、办事程序公示制；各级领导干部切实改变议事的方式方法，削减“文山会海”，简化办事程序，增强服务意识，提高管理水平和工作效率；制定了新的公文运转办法，对公文写作要求、呈递程序和批复时限做出相应规范，同时强化督办职能。通过上述一系列措施，机关作风从根本上得到好转。

为搞好廉政建设，净化干部队伍，学校一直把反腐倡廉作为一项重要工作来抓。自1995年以来，学校积极查处违纪案件及涉案人员，有效惩治了腐败。为加强对各级财务、基建工程和设备采购的审计监督工作，今年上半年，我校已与广东省检察院签订了《共同预防职务犯罪协议书》，这一举措在全国高校也是首创。同时，学校还在学术领域大力开展反对学术不端行为的斗争，净化我校的学术环境。

为加强依法治校，7年来，我校根据党和国家及上级管理部门的有关规定，先后制定了近200个制度性文件，对教学、科研及行政管理工作进行规范。为便于师生学习、了解和遵守文件精神，我校现已将那些仍在执行的文件编纂成《暨南大学文件汇编》(分为行政管理卷和教学科研卷)，作为学校日常管理的主要依据。

3. 加强横向联合，开展合作办学

1993年，我校在深圳开办旅游学院（原中旅学院），开创了校企联合办学的先河。1998年，为了有利于在港澳地区办学，我校与珠海市政府在珠海合作办学，我校成为中国第一个在珠海开办全日制高等教育的大学，珠海本地培养的首批专科生和本科生分别于2001年和2002年毕业于我校珠海学院。另外，为加强医学类学生的实践教学，我校在全国首创了校医联合办学的共建道路，从1997年开始，分别在广州、深圳、珠海、清远、江门等地共建了7所附属医院，成功走出了一条学校和社会联合办学的路子。合作办学的不断发展，扩大了办学规模，提升了办学实力，有效树立了学校在港澳地区的影响。

4. 实行春秋两季招生，方便学生入读

为在更大程度上方便海外及港澳台学生报读暨南大学，我校于 1998 年率先在全国实行春秋两季招生。另外，我校还是全国第一个在海外如越南、泰国、老挝、美国等国家设立招生报名点的大学，方便了学生咨询和报名。

5. 改善师资结构，提高师资质量

近几年来，学校尽力引进博士及学科带头人，大力培养中青年骨干教师。为创建名师工程，建设一支高水平的师资队伍，学校于去年开始实施特聘教授岗位责任制。该制度的实施一方面可为优秀拔尖人才提供更好的学术环境，另一方面可以激发广大教师的上进心，有利于在全校形成健康向上的学术氛围。

目前，全校 1093 名专职教师中，有研究生学位者 852 人，占 78％，其中博士 281 人，博士学位获得者占专任教师的比例由 1995 年的 5.8％增加到 25.7％。博士生导师 70 人，教授 183 人，副教授 572 人。新增院士 2 人，填补了学校无院士任教的空白。获教育部设置的“长江学者奖励计划”和广东省设置的“珠江学者”岗位计划的特聘教授岗位各 2 个，已引进珠江学者 1 人。

6. 实行人事分配制度改革，调动教师科研积极性

1998 年，针对传统的高校人事管理模式中存在的教师职务终身制和国家高度集中的指令性工资制度，我校进行了大胆改革，制定了一套全新的量化考核指标和管理方法，开始实行新的分配体制。新的分配体制充分发挥了个人潜能，优化了学科队伍，调动了教职工教学科研积极性。这一分配模式因属全国高校首创，故被媒体称为“暨大模式”。

7. 推进财务制度改革，集中财力办学

改革以前，由于资金分散，可供学校支配的资金有限，削弱了学校的投入能力；同时，各院系由于将精力过多地投入创收，不同程度地影响了办学质量，延缓了学校提高整体办学水平的速度，而且容易滋生腐败。1996 年，学校把各独立核算单位资金账户集中起来统一管理，集中了学校财力，加大了监管力度。通过一系列配套改革措施的实施，学校的办学结构得到调整，办学水平和办学质量不断提高。

四、创新赢得了发展机遇，学校综合实力不断提高

“九五”期间，学校以“211 工程”建设为核心，采取了改革举措，并不断改善基础设施，提高教师福利待遇，从而保证了教学质量，扩大了办学规模，提高了办学层次和办学水平，成效显著。

1. 办学规模变大

学生人数显著增加，办学规模扩大了 1 倍，相当于在原有基础上新办了一所暨南大学。与“九五”之初相比，学校各类学生由 13 012 人增加到 27 211 人，增长 109.1％，其中海外及港澳台学生由 1982 人增加到 6854 人，增长 245.8％。全日制学生由 8824 人增加到 17 450 人，增长 97.8％，其中本科生由 5377 人增加到 12 532 人，增长 133.1％；研究生由 615 人增加到 3359 人，增长 446.2％；海外及港澳台学生由 1766 人增加到 6080 人，增长 244.3％。

2. 校园和建筑面积扩大

校区由原来的3个（广州石牌校本部，广园西路校区，深圳华侨城校区）增加到4个（新增的珠海校区开办于1998年），现还有1个新校区（磨碟沙校区）正待开发。校园占地面积由112公顷增加到174公顷，增长55.4%，校园建筑面积由506 991平方米增加到961 088平方米，增长89.6%。

3. “侨”校特色更加鲜明

海外及港澳台全日制学生由1766人增加到6080人，增长244.3%，占全校学生比由20.5%增长到37.4%，是中国海外及港澳台学生最多的大学。特别是在我校攻读博士和硕士学位的海外及港澳台学生目前已达612人，约占全国总数的1/4，较1996年的124人增长393.5%。1995年，只有16个国家的学生来校学习，可今天来自世界五大洲53个国家的学生在我校求学。学校建立的姊妹大学遍及世界各地，是中国第一所在世界五大洲建有姊妹大学的学校。另外，学校作为国务院侨办华文教育基地和国家汉办支持周边国家汉语教学的重点单位，编写全套中文教材共48册，现已发行300多万套，被40多个国家使用，深受华侨华人和外国人的欢迎和好评。

4. 为祖国统一大业和广东经济发展服务的能力增强

自1978年至今，我校已接收来自世界五大洲90个国家和港澳台3个地区的学生前来学习，为香港、澳门的顺利回归和广泛团结世界华侨华人做出了积极贡献。同时，作为广东高等教育的重要组成部分，自1995年以来，我校共为广东省培养各类人才6万多人，其中全日制本科学生9981人，专科生1853人，研究生1941人，研究生课程进修班学生6281人，继续教育本、专科学生10 632人，短期培训各类学员3万余人。另外，我校通过管理咨询、科技成果转化等方式为广东省的经济发展和进步也做出了积极贡献。

5. 综合实力和办学质量更高

学校的固定资产总值由2.7亿元增至14亿元，增加了4倍多，图书藏量由135万册增至170.7万册，教学科研仪器设备由4985万元增至1.5亿多元，增加了2倍多。

学校在不同机构的综合实力排行榜中的位置不断上升，去年被《中国高等教育评估》杂志评为77所研究型大学之一，名列第53位。在中国网大的中国大学综合实力排行榜中，我校已由1998年的第87位上升到第36位。在广东管理科学研究院每年的“中国大学100强”中，我校的排名由2000年的第81位上升到第49位。在广东省71所高校中，我校办学实力和水平已处于第三位。

6. 教职工生活水平改善

1995年，我校本部教职工家庭住房总面积为169 775平方米，人均住房面积为13.5平方米；2002年，我校本部教职工家庭住房总面积为320 343平方米，人均住房面积为23.74平方米。与1995年相比，上述两项面积分别增长89%和76%。“九五”至今，我校教职工的工资待遇也不断提高，人均工资已由1995年的8254.12元提高到2002年的65 694.57元，增长近7倍。

五、持续改革创新，逐步开拓国际教育市场

暨南大学的根本任务是通过高等教育为侨服务，为国家服务，根据这一指导思想，我们将不断提高海外及港澳台学生比例，预计在“十五”末期，学校全日制学生中海外及港澳台学生与内地学生数之比达到1∶1。为实现这一目标，我校将继续加强内涵建设，努力适应国内国际教育市场对人才的需求，进一步提高办学层次和办学水平，以吸引更多的海外及港澳台学生来校学习。

暨南大学的办学渊源已体现了其国际性特征，其办学特性符合高等教育国际化的潮流，其办学成效充分显示了其参与高等教育国际化竞争的实力，我们有理由相信，暨南大学能够在实施“走出去”战略，开辟新的对外办学渠道，开拓国际高等教育市场方面有所作为。

改革创新取得的成果固然令人振奋，然而回首这些年所走过的路，可谓举步维艰。几乎每一次改革都伴随着非议、质疑、讥讽和责难。10年来，我尝够了改革的艰辛，但我凭着对党和国家的忠诚，对事业的热爱，对暨南大学的历史、未来以及学校全体教职员工负责的信念，忍辱负重，勇往直前，以无可辩驳的事实证明了暨南大学实施改革创新的紧迫性和必要性。改革固然艰辛，唯有把这种艰辛视为自己的职责和使命，唯有把党和国家及学校、师生的利益放在第一位，我们对此才会义无反顾。温家宝总理上任之初曾提过林则徐的两句诗“苟利国家生死以，岂因祸福避趋之”，并以此作为他今后的工作准则。我想这不仅是总理，也是我们所有领导干部应有的工作态度和工作准则。

44. 弘扬中华文化　发展华文教育　传播华夏文明　促进文化交流①

初冬的羊城，虽已渐透丝丝寒意，但此刻我们的心情却是格外温暖。值“暨南大学华文学院建院50周年庆典”隆重召开之际，我谨代表暨南大学对华文学院表示衷心的祝贺，对国务院侨办领导、各位嘉宾、校友的光临表示热烈的欢迎和诚挚的谢意！

暨南大学创办于1906年，是一所历史悠久、享誉海内外的国立高等学府，也是国家面向21世纪重点建设的100所大学之一。

长期以来，学校坚定不移地贯彻“面向海外、面向港澳台”的办学方针，“宏教泽而系侨情”。1993年6月，在国务院侨办的大力支持和领导下，已经走过40年历程的广州华侨学生补习学校和暨南大学对外汉语教学系以及预科部合并，组建成立了暨南大学华文学院，继续弘扬中华文化，发展华文教育，传播华夏文明，促进文化交流。

从华侨补校到暨南大学华文学院，经历了整整半个世纪。在这半个世纪里，不管中间经历了多少历史变故，补校和暨南大学总保持着某种渊源和感情。

华文学院作为暨南大学面向海外开展汉语和中华文化教育的专门学院，每年有来自世界几十个国家和地区的学生在这里学习汉语，研修中华文化，华文学院成为海外学生学习汉语和中华文化的重要基地。

除了对外汉语教学，预科部也是华文学院办学的重点之一。

暨南大学是中国向海外传播中华文化，维系海内外中华儿女亲情和开展中外文化交流的一条重要纽带，在我国高等教育史上有着特殊的地位。华文学院则是暨南大学向世界各地传播中华文化的一个极为重要的窗口。无论是当年的补校，还是现在的华文学院，她的成长都一如既往地得到了国务院侨办的直接领导、关心和支持。暨南大学也将进一步加大投入力度，继续支持华文学院的建设和发展。为充分发挥华文学院的窗口作用，我们已建立了一系列行之有效的制度，如教学商榷制度、外出调研制度、学术交流制度等。以此为基础，我们还将加强对外汉语教师队伍的建设，加大人才引进力度，吸引更多的优秀人才加入我们的队伍，实现教学师资高学历化、专业化，梯队结构合理化，使华文学院成为能承担各级各类对外汉语教学和重点、重大科研任务的教学研究中心，在学科理论和与对外汉语教学相关的汉语本体研究中形成一支较强的科研队伍，并进一步帮助学院扩大对外招生和合作办学，通过各种渠道加强对外宣传，加强与国外著名大学、华文教育机构、驻外使馆等的联系与合作。同时，支持华文学院发展远程华文教育。

暨南大学华文学院作为华南地区对外汉语教学窗口，我校已将其作为国家对外汉

① 本文是在暨南大学华文学院建院50周年庆典上的讲话，广州，2003年11月15日；暨南大学校报，第369期，2003年11月30日.

语教学基地建设目标列入学校“211 工程”建设规划及年度工作计划，以发挥自身优势，推动国家对外汉语教学事业的发展。

今日校友汇聚暨华园，抚今追昔，但见桃李争妍，硕果流丹！让我们一起祝福华文学院，共同创造华文教育更加美好的前景！

45. 为迎接暨南大学百年庆典增添新的光彩[①]

暨南大学董事会第五届第一次会议今天开幕了。在此我谨代表暨南大学董事会，向出席会议的全体董事和嘉宾表示最热烈的欢迎和衷心的感谢。

目 1999 年 1 月暨南大学第四届董事会第一次会议召开以来，在四年多的时间里，暨南大学取得了令人瞩目的成绩，办学水平、办学质量有较大提高，学校的综合实力、整体竞争力迈上了一个新的台阶，实现了跨越式的发展。学校在全国大学的排名逐年提高，进入全国前四十名的一流行列。去年，暨南大学通过国家“211 工程”“九五”建设项目验收，顺利进入“十五”“211 工程”建设阶段。暨南大学能取得如此喜人的成绩，与国务院侨办、教育部和广东省委、省政府的正确领导密不可分，与全体董事的支持和帮助密不可分，同时与广大海外侨胞、港澳同胞和其他热心华侨教育事业的仁人志士的关心和努力紧密相连。在此，我再一次感谢各级领导、全体董事、各位嘉宾和广大热心华侨教育的人士，感谢你们长期以来给予暨南大学的热心帮助和大力支持。同时，也希望大家一如既往地关心和支持暨南大学的建设和发展。

暨南大学第四届董事会由 77 位董事组成，四年来，已有 6 位董事先后逝世，有 15 位在政府工作的董事退休或工作调动。为了加强暨南大学董事会的力量和健全董事会的组织机构建设，根据国务院侨办关于选聘董事的有关规定，第五届董事会将增聘一些热心华侨教育的各界人士加入我们董事会队伍，充实董事会的力量，使董事会在暨大的发展中发挥更大的作用。

今天的会议，除了请各位董事审议《暨南大学工作报告》和《暨南大学第四届董事会工作报告》外，还将着重讨论如何加强董事会自身建设，增聘董事，以及如何进一步发挥董事会在暨大建设和发展中的作用等问题，希望大家踊跃发言，共同为暨南大学今后的发展出谋献策。

根据暨南大学董事会章程，暨南大学第四届董事会已圆满完成了历史使命，在此，我再一次向第四届董事会全体董事表示崇高的敬意和诚挚的谢意！相信新一届董事会在各位董事的共同努力下，一定能够积极发挥各自的优势和能力，协助暨南大学早日实现建成海内外知名的高水平社会主义华侨大学的目标，为迎接暨南大学百年庆典增添新的光彩！

最后，我预祝会议圆满成功！并祝大家身体健康！

谢谢！

① 本文是暨南大学董事会第五届第一次会议开幕词，广州，2003-11-16；第五届董事会特刊，2004，22.

46. 发挥优势 深化改革 保证重点 改善条件 提高质量[①]

一、四年来工作回顾

过去的四年时间里，学校按照“发挥优势、深化改革、保证重点、改善条件、提高质量”的发展思路，深入贯彻“面向海外、面向港澳台”的办学方针，坚持“严、法、实”（即“从严治校、从严治教、从严治学”“依法治校”“实事求是”）的办学思想，以“211工程”建设为龙头，以教学科研为中心，大力推进学校各项事业不断进步，成功实现了跨越式发展。

（一）主要措施

（1）解放思想，转变观念。根据国内外高等教育的发展形势及学校的处境和地位，学校坚持改革创新，并以此为基础，确立了两个重要原则：学生第一原则，“严、法、实”原则；一个基本观念：不搞科研的教师是“残疾”的教师。

（2）坚持以人为本，不断推进学分制改革。我校在全国率先实施学分制改革的指导思想是以人为本，将学生的利益放在第一位，给学生提供更主动的学习条件，最大限度地发挥学生自主学习、向上发展的能力。学分制是由传统的保姆式教育向现代自主教育转变的重要体现。经过不断实践，我校学分制各种管理办法已日臻完善。

（3）合理制定培养目标，科学采用因材施教原则。根据生源的不同特点，学校制定了不同的培养目标：将内地学生培养成为德、智、体等方面全面发展的社会主义事业建设者和接班人；将香港学生培养成为热爱祖国，拥护“一国两制”，拥护香港基本法的专业人才；将澳门学生培养成为热爱祖国，拥护“一国两制”，拥护澳门基本法的专业人才；将台湾学生培养成为热爱祖国，拥护“一国两制”，反对“台独”的专业人才；将华侨学生培养成为热爱祖国，维护祖国和平统一大业的专业人才；将华人学生培养成为热爱中华文化，热爱故乡的专业人才。为配合学分制的实施，学校根据境内外两类学生的不同特点，对学生的培养目标进行定位，合理设置专业。我校对海外及港澳台学生的教学要求是“面向世界、应用为主”，对内地学生的教学要求是“加强基础、目标上移”。

（4）保证课堂教学质量，继续推行我校创新的“三重评估”制度。学校每学期先分别组织学生、校院系领导和听课专家组对课堂教学质量进行评估，然后再根据评估结果采取相应的奖惩措施，激励先进，鞭策落后。这在一定程度上加强了师生间教学信息的交流和教师对课堂教学质量的重视，有效地提高了课堂教学质量。

（5）突出基础课教学，坚持教授上基础课制度。为突出基础课教学，使新生在低年级即获得坚实的基础知识，形成优良的学风，我校坚持实行教授必须上本科基础课的制度，效果良好。

---

① 本文是暨南大学董事会第五届第一次会议报告，广州，2003年11月16日；暨南大学校报，第370期，2003年12月2日.

(6) 适应市场需求，抓好“三语”教学。为主动适应市场需求，体现我校外向型办学特征，学校将汉语、英语和计算机语言列为学生的必修课，同时还采取了一系列有力措施，要求学生学好“三语”。

(7) 加强学风建设，推进考试改革。为严肃考试纪律，培养学生诚信品质，我校采用建试题库等措施，促使教师认真对待考务、学生认真对待学习和考试。从2001～2002学年上学期开始，我校设立了可一次容纳840人的大型考场，为杜绝作弊、保证考试的公正性起到了重要的作用。在今年的博士生入学考试中，我校也采用大型考场考试，学生及社会反响良好。

(8) 突出国际性特征，鼓励和扶持“双语”教学。为贯彻“两个面向”的教育方针，培养国际性人才，学校提倡和鼓励教师用英语进行本科专业课教学。目前我校非英语专业用英语教学的课程已达28门，使用全英语教材的课程达34门，使用中英对照教材的课程达40门。不仅如此，2001年6月，学校还在全国高校中第一个成立了采用全英语教学的国际学院。目前该学院已开办了临床医学、国际经济与贸易、会计学、食品质量与安全等4个专业。

(9) 调整办学重心，优化专业结构。学校按照“大力发展研究生教育，适度发展本科教育，积极发展华文教育，稳定成人教育规模”的办学思路，以及社会发展和学生求学、就业的需要，不断调整专业结构，及时增设满足国内外学生要求的热门专业。

(10) 实行春秋两季招生，方便学生入学。我校1998年率先在全国实行春秋两季招生。另外，我校还是全国第一个在海外，如越南、泰国、老挝、美国、秘鲁、厄瓜多尔等国家设立招生报名点的大学，方便了学生咨询和报名。

(11) 加强横向合作，开展联合办学。为了有利于面向港澳办学，我校在1998年与珠海市政府进行了合作，成为中国第一所在珠海开办全日制高等教育的大学，并在珠海培养了首批专科生、本科生和硕士研究生。由珠海市政府无偿划拨近900亩土地的暨南大学珠海学院于2000年6月正式迁入现址，现有全日制学士、硕士、博士学生5000余人。目前，珠海学院正逐步发展成为学校的一个新亮点。

(12) 积极合作共建，首创校、医联合办学之路。为加强医学类学生的实践教学，我校在全国首创了校医联合办学的共建道路。几年来已分别在广州、深圳、珠海、清远、江门等地先后共建了7所附属医院，成功走出了一条学校和社会联合办学的路子。

(13) 发挥对外办学优势，开拓国际教育市场。为提高海外及港澳台学生的比例，学校除加强内涵建设，努力适应国内国际教育市场对人才的需求外，在颜开臣董事的倡议和大力支持下，还积极着手在泰国首都曼谷开办暨南大学曼谷学院，这也是我国大学走向世界的第一次尝试。

(14) 以学科建设为中心，大力鼓励教师从事科学研究。为鼓励教师在做好教学工作的同时，积极投身科学研究，学校以“211工程”重点学科建设为中心，大力推进科研工作。学校用“教学、科研”双中心目标取代了过去单一的“教学中心”目标，并将教师的科研成果直接与校内工资挂钩，对其进行量化考核。

(15) 改进财务管理模式，集中财力办学。改革以前，由于资金分散，可供学校支配的资金有限，削弱了学校的投入能力；同时，各院系将精力过多地投入创收，不同程度地影响了办学质量，延缓了学校提高整体办学水平的速度，而且容易滋生腐败。通过一系列配套改革措施的实施，学校的办学经费大幅增加，办学结构得到调整，办学水平和办学质量不断提高。

（16）改革人事分配制度，调动教师工作积极性。针对传统的高校人事管理模式中存在的教师职务终身制和国家高度集中的指令性工资制度，我校进行了大胆改革。新的分配体制充分发挥了个人潜能，优化了学科队伍，调动了教职工教学科研积极性。这一分配模式因属全国高校首创，被媒体称为“暨大模式”。

（17）改善师资结构，提高教师质量。近几年来，学校积极采取措施，尽力引进有博士学位的教师和学科带头人，大力提升教师水平。同时，学校还于去年启动了名师工程，实施特聘教授岗位制。

（18）合理调整人员结构，稳步推进机构改革。本着“精减、效能、统一”的原则，通过机构改革，学校党政处级机构由 1999 年以前的 29 个减为 18 个。为保证教学和科研这两个重点的发展，几年来，学校注重引进专业教师，控制党政人员数量，原则上不进党政干部。虽然学校的办学规模不断扩大，但人员结构得到优化，尽管机构数量比改革前少，可效率却更高了。

（19）加强建章立制，推进依法治校。为加强依法治校，我校将近几年制定的仍在执行的近 200 个制度性文件，编纂成《暨南大学文件汇编》（分为行政管理卷和教学科研卷），作为学校日常管理的主要依据。

（20）加强廉政建设，净化干部队伍。今年 4 月学校与广东省检察院签订了《共同预防职务犯罪协议书》。我校是全国第一所为加强廉政建设而采取与司法部门联合预防腐败措施的高校。同时，学校还在学术领域大力开展反对学术不端行为的斗争，净化校内学术环境。

（二）主要成绩

（1）实现了“九五”“211 工程”建设目标，顺利进入“十五”“211 工程”建设阶段。

1999 年底，学校顺利完成了“211 工程”中期检查工作。2002 年 7 月，学校“九五”“211 工程”子项目以全优的成绩通过验收。10 月，我校“九五”“211 工程”建设项目验收总体评价为优秀。11 月，顺利通过《“十五”“211 工程”建设项目可行性研究报告》的专家论证，成功进入“十五”“211 工程”建设阶段。

（2）办学规模变大了。

与 1999 年相比，学校各类学生已由 19 988 人增加到 26 881 人，增长 34.5%，其中海外及港澳台学生由 3568 人增加到 7484 人，增长 110%。全日制学生由 11 568 人增加到 19 659 人，增长 69.9%，其中研究生由 1583 人增加到 4236 人，增长 167.6%；海外及港澳台学生由 3113 人增加到 6566 人，增长 110.9%。

（3）校园和建筑面积扩大了。

校区由原来的 3 个增加到 4 个，现还有 1 个新校区正待开发。校园占地面积由 112 公顷增加到 174 公顷，增长 55.4%，校园建筑面积由 506 991 平方米增加到 1 074 732 平方米，增长 112%。

(4) 办学层次提高了。

学校研究生与本科生之比由 1999 年的 1∶5.47 上升到 1∶3.31，专科生由 693 人变为零，校本部从 1996 年开始即不再招收专科生。

(5) 国际性特色更加鲜明了。

全日制海外及港澳台学生占全校学生比例由 26.9%增长到 34%，是中国海外及港澳台学生最多的大学。在校海外及港澳台博士和硕士学生约占全国总数的 1/4。1999 年，只有 26 个国家的学生来校学习，今天已有来自世界五大洲 52 个国家的学生在我校求学。学校是中国第一所在世界五大洲建有姊妹大学的学校。学校编写的全套中文教材现已向 40 多个国家和地区发行 300 多万套，被 40 多个国家使用，深受华侨华人和外国人的欢迎和好评。

(6) 科研实力增强了。

四年来，我校获得各级各类项目 881 项，其中国家级项目 123 项、省部级项目 397 项；获国家、省部级奖励 55 项。科研经费由 1999 年的 3230 万元增至 2002 年的 8375 万元，增长了 1.6 倍。学术论文，特别是被三大索引（SCI、EI、ISTP）收录的论文数较 1999 年增加了 4 倍多，达 105 篇。1999 年，我校首次主持了国家自然科学基金管理学部重点项目、“863”高新科技项目及“973”子项目。获得的专利从无到有，现在获专利 17 项。

新增 2 个国家重点学科、1 个国家人文社会科学重点基地、1 个教育部重点实验室、1 个教育部工程研究中心、1 个广东省教育厅重点实验室。获批 8 个侨办重点学科，7 个广东省重点学科。

(7) 学科结构优化了。

2000～2002 年，学校先后成立了生命科学技术学院、珠海学院等 8 个学院。其中，国际学院是全国第一所采用全英语教学的综合性学院。新增了生物医学工程、土木工程等 9 个本科专业。我校电子信息工程等 5 个专业成为广东省高校名牌专业。目前，本科专业有 48 个；硕士学位授权学科有 88 个；博士学位授权学科有 28 个，在一级博士学位授权学科方面实现了零的突破；博士后站达到 6 个。另外，学校获得了医学专业学位（硕士）、工程硕士学位和全国 30 所首批高级管理人员工商管理硕士（EMBA）专业学位的授予资格。

(8) 师资队伍结构改善了。

现有的 1363 名专任教师中，有研究生学历者 909 人，占 66.7%，其中博士 305 人，博士学位获得者占专任教师的比例由 1995 年的 5.8%增加到 22.4%。博士生导师 71 人，教授 225 人，副教授 560 人。新增院士 2 人，填补了学校无院士任教的空白。获教育部设置的“长江学者奖励计划”和广东省设置的“珠江学者”岗位计划特聘教授各 2 人。

(9) 办学效益提高了。

四年来，全校的教职工和专业教师人数基本没有变化，且学校获得的经费支持并未大幅增长，但学校完成的任务数量却成倍增加。在 2002 年广东管理科学研究院“中国‘211 工程’大学教师人均效率排名”中，我校排在第 41 位。

（10）教职工生活水平改善了。

1999年，我校本部教职工家庭住房总面积为235 803.35平方米，人均住房面积为11.56平方米；2003年，我校本部教职工家庭住房总面积为320 343.15平方米，人均住房面积为23.24平方米。与1999年相比，上述两项面积分别增长35.9%和101%。我校教职工的收入水平不断提高，2002年较1999年增加了146%，达65 694.57元。

（11）综合实力变强了。

学校的固定资产总值由6.2亿元增至15.5亿元，图书藏量由148.04万册增至184.26万册，教学科研仪器设备由7828万元增至17 141万元。

学校去年被教育部《中国高等教育评估》杂志评为77所研究型大学之一，名列第53位。在中国网大的中国大学综合实力排行榜中，我校的排名已由1999年第72位跃居至2003年的第36位。在今年的研究生教育综合实力排名中，我校名列第46位。

（12）在海内外的影响扩大了。

从去年开始，学校录取本、预科海外及港澳台人数已连续两年超过内地招生录取人数，这是我校自1978年以来首次出现这种良好局面。今年报考的海外及港澳台学生人数已达3631人，比1999年增长87.8%；实际入学的本、预科海外及港澳台学生人数1860人，较1999年增长97.9%。四年来，共有来自38个国家的1965名留学生来我校学习汉语。在内地招生方面，我校在大部分省份的出档线都高出国家重点线30分左右，有的专业甚至高出70分以上。今年报考我校研究生的人数达5200人，比1999年增长123%；入学人数为1495人，较1999年增长151%。至此，我校全日制在校研究生规模已达4236人，较1999年增长167.7%。

（13）为华侨华人和港澳台服务的能力提高了。

2000年9月，“国务院侨办华文教育基地”在我校正式挂牌，使我校成为国家首批、广东省唯一一个华文教育基地。在过去的四年里，学校先后对来自泰国、美国等6个国家的300余名华文教师进行了培训；同时，学校还派出教师分赴美国、瑞典等11个国家开展华文教育。

四年来，学校在开展海外合作办学方面也取得了实质性成果。2002年6月，我校与新加坡华夏管理学院合作招收“华文教育”硕士研究生，开我校在海外办学之先河。学校在印度尼西亚万隆设立的首个函授点，目前已有144名注册学员。经国务院侨办和教育部批准，我校将在颜开臣董事的支持下筹办泰国曼谷学院，开办中国语言文学、工商管理等专业，直接为华侨华人服务。曼谷学院将成为我国高校在国外设立的第一个全日制本科学院。

（14）基础设施和公共服务体系完善了。

四年来，校本部完成了曾宪梓科学馆、校友楼、24栋教工住宅、13栋学生宿舍楼、标准田径场、标准网球场、标准游泳池、学生食堂扩建、医学院大楼、附属第一医院门诊楼加层等总计23.6万平方米的建筑工程。珠海学院完成了包括学生宿舍、餐饮中心、运动中心、教学大楼、图书馆等建筑面积为23.1万平方米的工程。

为完善公共服务体系建设，学校投资90多万元用于图书馆网络服务器升级。完成实验室环境改造70项，维修面积8000多平方米。

二、今后的工作设想

（一）人才培养

按照“大力发展研究生教育，适度发展本科教育，积极发展华文教育，稳定成人教育规模”的发展思路，计划到 2005 年，学校各类学生规模达到 25 000 人，其中本科生 13 000 人，硕士生 4500 人，博士生 600 人。到 2010 年，学校各类学生规模达到 26 000～28 000 人，其中本科生 16 000 人，硕士生 6500 人，博士生 1500 人，研究生与本科生的比例达到 1∶2。

（二）师资队伍

到 2005 年，教师队伍由现在的 1093 人增加到 1400 人，编内在岗教师中具有研究生学历者达 80％以上，其中具有博士学位者达 35％以上；具有国内一流或领先水平的学科带头人 20 人以上，拥有两院院士 3 人以上，培养出 30 名左右在国内有一定影响的，具有较高学术水平、较大发展前途和潜力的学术骨干。

（三）科学研究

争取建成 1 个国家级重点实验室和 2 个省级重点实验室，使学校的科学研究力量得到进一步增强。到 2005 年，年度科研经费达到 8000 万元；年发表科技论文（国家四大检索系统收录）数 1500 篇左右，其中被三大索引收录的论文数为 100 篇左右；获国家级、省部级重大科技成果奖 3～4 项。目前，在重点实验室建设、科研经费的获取和被三大索引收录的论文数方面，我们已经完成了规划要求，比原计划提前了 3 年。

（四）基础设施建设

根据《暨南大学总体发展规划》及国家发展改革委的批复，我校“十五”期间要建设的项目约 30 项，总建筑面积约 164 873 平方米，总建设资金达 5.9 亿元。由于国家给予我校的 5.1 亿元专项资金都必须按规划项目投入，因此，那些没有纳入《规划》而又必须建设的项目以及为完成国家扩招任务所需增加的基础设施项目，都必须另筹资金解决。为完成《规划》外的项目，包括珠海校区的所有工程、校本部的 10 栋学生宿舍等，约需经费 6.5 亿元。另外，为完成扩招任务所进行的基建部分还需经费 3.5 亿元，这些都需要通过贷款自筹经费和捐款来解决。

47. "侨校＋名校"的发展定位①

华侨高等教育是中国高等教育不可或缺的重要组成部分，肩负着为海外华侨华人社会和港澳台地区培养人才，发展国家高等教育，促进国家及地区经济发展和社会进步的特殊使命，其作用和地位不可替代。在新的世纪里，面对高等教育领域竞争日趋激烈，越来越多的非华侨教育的大学竞相开拓海外华侨教育市场的紧迫形势，华侨高等教育如何找准自身的正确定位，切实贯彻国家的侨务政策，着力突出"侨"字特色，不断提高办学水平，更好地为侨服务，为祖国统一大业服务，作为华侨高等教育重要代表的暨南大学为此进行了积极探索和尝试，创造性地提出了"侨校＋名校"的发展战略。

一、正确理解侨校与名校的关系

要准确理解并切实贯彻"侨校＋名校"的发展战略，首先必须认识侨校与名校之间的关系，只有把二者之间的关系弄清楚了，才能正确运用和实践这一战略，我们的工作才不会迷失方向。事实上，侨校与名校之间是一种个性与共性的关系，即侨校是暨南大学固有的属性，同时也是学校办学的立足点，这是学校自开始办学的第一天就具备了的属性和特点；而名校则是高等院校的共同追求和目标，这不论在国内还是国外均无例外。以暨南大学为代表的华侨高等教育作为中国高等教育的特殊组成部分，也必须顺应国内国际高等教育的发展潮流，不断上层次、上水平。学校坚持发挥"侨"字特色，立足于其个性，充分利用其特有而其他高校所不具备的优势大力发展华侨高等教育，是为了不断提高综合实力和品牌影响力，以便更好地为侨服务。建设名校，则要求学校的办学质量、办学层次和办学水平不断提高，并且在其提高的过程中，使特色和优势得到更好的发挥。换句话说，只有实现了共性，个性才能得到张扬。因为只有不断提高办学质量、办学层次和办学水平，吸引更多的学生前来学习，学校的"侨"字特色和优势才能更加突出和鲜明。同时，也只有在保证个性不断发展的基础上，共性才有可能实现。因为只有学校的"侨"字特色和优势得到充分展示和发挥，学校的实力和品牌影响力才能不断提升，名校的目标才能实现。

在理解和实施"侨校＋名校"发展战略的时候，我们应该知道，"侨"和"名"都是暨南大学未来发展中必不可少的，二者是统一的、相辅相成的。侨校和名校作为同一个载体上的两个主要内容，二者是兼容、相辅相成、相互促进的，并非矛盾的。对暨南大学来说，为侨服务是本源属性和使命，无论学校如何发展，或者说无论发展到何种程度，只要国家赋予的使命和任务不变，学校就必须立足侨校办学，否则，学校的办学就成了无源之水，无本之木，正可谓皮之不存，毛将焉附，而名校一说就更无从谈起。但如果面临日趋激烈的竞争和来自国内外的双重冲击与挑战，单纯强调一

---

① 本文是全国高等学校教学研究会报告，广州，2003年12月20日；一个大学校长的探索，北京：高等教育出版社，2011，105-112.

个“侨”字，不遵循高等教育的发展规律和改革潮流办学，不根据社会发展和就业市场的需要办学，不改革，不发展，不升级，学校面对竞争就会没有竞争力，面对市场就没有吸引力，其办学特色的发挥和办学宗旨的实现就会受到影响。

因此，唯有将二者紧密结合起来，实践“侨校＋名校”的发展战略，既有特色，又有品牌，才能相互促进，共同提高，学校才能在保证特色的同时不断提高竞争和持续发展的能力，才能在日趋激烈的国内国际高等教育市场中站稳脚跟，吸引更多的华侨华人和港澳台地区的青年来校学习，更好地完成国家赋予暨南大学的历史重任，学校成为海内外知名的社会主义高等华侨大学的目标才能实现。

二、实施“侨校＋名校”发展战略的必要性

（一）立足侨校，突出特色的需要

97 年前，清政府应当时侨居海外（主要是东南亚地区）的华侨之需，在南京创办暨南学堂，旨在“宏教泽而系侨情”。97 年来，学校的办学宗旨从未改变。新中国成立以后，为弘扬中华优秀传统文化，满足海外华侨华人和港澳台地区青年的求学需要，党和国家十分重视华侨高等教育的发展。1958 年，暨南大学被列为国务院高教部的直属高校。1983 年，中共中央、国务院将暨南大学列为国家重点扶植的大学。根据党和国家赋予的特殊办学使命和办学任务，暨南大学以为侨服务为宗旨，以满足华侨华人和港澳台学生的需要为第一要务，坚持“面向海外、面向港澳台”的办学方针，致力为海外和港澳台地区培养人才。仅改革开放至今，学校已为海外 94 个国家和地区的华侨华人培养各类优秀人才 10 000 余人，为贯彻国家侨务政策，弘扬中华优秀传统文化，广泛联系和团结海外华侨华人，促进香港、澳门的顺利回归做出了积极贡献。

人们常说，世界上只要有海水的地方，就有华侨居住。随着新生代华侨队伍的逐渐扩大，需要接受华文教育的人也将越来越多。据统计，全世界共有华侨华人 3000 多万人，因此，在新的世纪里，作为联系全世界中华民族子孙，传播中华优秀传统文化和现代科学知识的重要桥梁，暨南大学必须立足侨校特色，积极贯彻国家的侨务政策，充分发挥国家华文教育基地的辐射作用，针对海外及港澳台学生的特点，进一步深化改革，创造更加有利的学习环境和学习条件，努力提高学校的综合实力和办学品牌，巩固其作为全球华侨华人和港澳台同胞子女来内地求学首选高校的地位，加倍做好对世界华侨华人的教育服务工作，团结更广泛的力量，为海外和港澳台地区培养和造就具有爱国、爱乡意识与民族自豪感，能为当地社会发展和稳定繁荣做出贡献的人才，从而为实现祖国的统一大业和中华民族伟大复兴做出新的贡献。

（二）谋求发展，保持特色的需要

中国或世界一些高等院校的发展经验告诉我们，学校要具有强劲的竞争力和吸引力，最重要的是它必须在某些学科具有优势，而且这种优势是大众公认的优势，通俗地说，就是一个知名的品牌。其实这是一个非常浅显易懂的道理。因为在这一点上，学校与企业没有什么两样，它们的生存和发展都需要一个知名的品牌。企业追求名牌效应，为的是占领市场，创造更大的经济利润；而大学追求卓越，为的是吸引更多的教育对象，为社会培养更优质的人才，创造更高的社会效益。

“九五”以来，党和国家对高等教育进行了巨额投入，同时还采取了两项重大举措，这在中国高等教育史上可以说是空前的。一是面向 21 世纪重点建设一批大学和学科；二是进行结构调整，整合办学力量，集中办学资源，有重点地合并一批院校。这两项措施都是有针对性、有重点地提升一批基础好、实力强的院校的办学层次和办学质量，为建设一流大学和高水平大学奠定基础。暨南大学作为华侨高等教育的重要代表，是我国高等教育的一个组成部分，是国家重点建设的大学，必须顺应建设一流大学的总趋势、总要求。

早在 1927 年，时任国立暨南大学校长的郑洪年先生曾说：“鉴于侨胞处于殖民政府铁蹄之下，受尽帝国主义之蹂躏，暨南教育非提高教育程度，扩充为完善大学，不足以增进侨胞之地位，不足以谋适应其特殊环境，不足以使华侨父老咸达自由平等之目的”。如果说郑洪年欲“提高教育程度，扩充为完善大学”的观点，是基于他对当时国内国际形势及侨胞所处社会地位和生存状况的深刻认识，是为了“使华侨父老咸达自由平等之目的”，那么我们今天实施“侨校＋名校”的发展战略，一方面是为了更好地贯彻国家的侨务政策，更好地为侨服务；另一方面，这也是学校自身发展的需要。

在过去很长的一段时间里，暨南大学在立足为侨服务的同时，忽视了学校的办学质量和办学水平，致使学校的实力和水平与一所近百年历史的大学所应具有的实力和水平相去甚远。一所百年老校应该具有与其历史相称的实力和品牌，因为其在办学经验、人文环境、文化底蕴以及历史积淀等方面都具有那些新办大学所无法比拟的优势，后继者有责任和义务将这些优势进行传承和弘扬。经近年来的不懈努力，学校的办学水平、办学层次以及综合实力得到迅速提升，但离当前高等教育发展及学校自身发展的要求仍有很大的差距。同时，多年来，一直在国家政策保护下而具有生源优势的华侨高等教育正面临着来自国内外的双重冲击和挑战。一方面，随着社会主义市场经济体制的建立和完善，国家对高等学校办学的自主权逐步放开，香港、澳门回归以后，除暨南大学以外，国内已有 100 多所高校在香港、澳门招生，其中还包括一些在国内乃至在世界上都颇有影响的大学，它们利用各种优势和特点吸引海外华侨华人前往学习，已对高等华侨学校的外招生源构成了威胁。另一方面，港澳及一些华侨华人数量较多的国家和地区看到华文教育的广阔前景，利用自身的教育资源或以与内地高校合作办学的形式开办华文教育，与我国华侨高校争夺生源。在中国加入世界贸易组织以后，这种争夺变得更加激烈。另外，台湾地区的一些高校出于政治上的需要，也纷纷采取减免学费等有关优惠政策或其他方式吸引华侨华人学生和港澳地区学生前往学习。

面对上述严峻的形势，我们如果不适时调整办学理念，加大创新教育的力度，尽快与世界高等教育接轨，提高实力，创造优势，从某种程度上说，学校的生存和发展将会受到威胁，特色也将难以保持。这是因为，在社会主义市场经济条件下，在中国加入世界贸易组织以后对教育所做承诺的前提下，中国的所有大学（军校、警校、党校等除外），不管其性质如何，它都必须遵守教育市场的竞争规则。对暨南大学来说，生源及办学的特殊性决定了我们的竞争对手是来自多方面的，它们既有本系统（华侨院校）的高校，也有系统外的高校；既有国内的高校，又有来自其他国家和地区的高校。如果说大家都在努力上台阶、上水平，而我们的办学水平和办学质量却止步不前的话，其结果可想而知。因为海外及港澳台学生没有理由放弃居住国或地区的优质教育，而不远千里来暨南大学求学。因此，我们不仅要坚持特色，而且还要实施名牌战略，大力提高学校的办学层次和办学质量，提高学校的竞争力和吸引力，这既是保证特色的需要，同时也是学校自身发展的需要。学校只有底子厚了，实力强了，名气大了，才能保证在华侨乃至中国及世界高等教育领域应有的地位，才能在国内外教育市场的激烈竞争中立于不败之地，吸引更多的海外及港澳台青年前来学习，更好地为贯彻国家的侨务政策及为侨服务多做贡献。

三、实施“侨校＋名校”战略的成效和可行性

“九五”期间，暨南大学在国务院侨务办公室及广东省政府的正确领导下，审时度势，开拓创新，根据中国高等教育的发展趋势及国际高等教育的发展经验，结合自身的实际情况，调整办学理念，开创性地提出了“侨校＋名校”的发展战略，力争在“十五”末期使学校成为一所海内外知名的社会主义高等华侨大学。根据这一战略思想，暨南大学以“211 工程”建设为龙头，积极贯彻“面向海外、面向港澳台”的办学方针，重点突出“侨”字特色，从严治校，从严治教，从严治学，优化专业结构，调整办学重心，围绕学校的建设和发展，在学分制、学生培养、教学质量评估、基础课教学、用英语讲授专业课、学生考试、专业设置、招生时间、合作办学、机构改革、人事分配制度、师资队伍建设、教师科研、财务管理、后勤社会化、机关作风建设、廉政建设等方面采取了一系列改革措施，取得了显著成效。

在暨南大学“九五”“211 工程”验收总结大会上，专家在感叹暨南大学在短时期内所取得的成绩的同时，深刻地指出，暨南大学之所以能够发生如此大的变化，保持这一可观的发展速度，其根本原因在于办学理念的转变，在于确立了“侨校＋名校”的发展战略。也正是在这一科学、合理、切合实际的办学理念的指导下，近几年来，暨南大学凭借一日千里的发展速度，整体实力得到大幅提高，在教学、科研、人才培养等方面取得了令人瞩目的成绩。

（一）办学规模变大

2003 年与 1995 年相比，各类学生总数已由 13 012 人增加到 26 881 人，增加 1 倍多。研究生已由 615 人增加到 4236 人，增加了近 6 倍。其中，硕士研究生由 563 人增加到 3709 人，博士研究生由 52 人增加到 527 人，博士生人数增长了 9 倍多。本科生由 5377 人增加到 14 025 人，增加了 1.6 倍。海外和港澳台学生由 1982 人增加到 7484 人（包括研究生 747 人），增加了 2.8 倍。综合以上数据，学校当前的规模相当于在 1995 年的基础上多办了一所暨南大学。校园面积进一步扩大，校园占地面积由 112 万平方米增至 174 万平方米，增加了 55.4%。建筑面积由 46 万平方米增至 107 万平方米，增加了 1.3 倍。另外，学校在合作办学方面也有了长足发展，现分别在深圳、珠海两地设有校区，与当地政府合作办学。尤其是珠海学院，现有全日制博士、硕士和本科学生 5200 多人。学校医学院以共建形式新增 8 所附属医院，其中国家三级甲等医院 6 所，8 所附属医院共有病床 4085 张，增加了 4 倍。

（二）专业结构优化

随着市场经济体制的建立，社会发展对人才的要求不单是数量上的增加、质量上的提高，更重要的是对专业领域的要求更为广泛。国外一些发达国家由于科技水平和教育水平较高，对就业人员的专业知识和基本技能的要求与中国相比更为严格，因此，为适应国内国外学生学习和就业的需要，学校对专业设置进行了调整和优化。到目前为止，暨南大学的本科专业已由1995年的30个增加到48个，其中电子信息工程、会计学、新闻学、汉语言文学、生物技术等5个专业被评为广东省名牌专业。硕士学位授权学科已由1995年的50个增加到88个，博士学位授权学科已由1995年的7个增加到28个。教学系由1995年的21个增加到38个。自1995年至今，学校在文学院、理工学院、医学院、经济学院、华文学院、深圳旅游学院（原中旅学院）、教育学院的基础上，本着优化结构，促进发展的原则进行了专业调整，先后新增了管理学院、生命科学技术学院、珠海学院、信息科学技术学院、新闻学院、外国语学院、法学院、药学院、国际学院。其中国际学院是我国第一所采用全英语教学的综合性学院，深圳旅游学院是内地首家通过世界旅游管理专业教育质量认证的高等旅游院校。现有的16个学院38个系的48个专业涵盖了文、史、经、管、法、理、工、医、教育等九大学科门类。专业结构的调整和优化，使一些新兴或热门学科得到及时、自由、充分的发展，适应了市场、学生及学校自身发展的需要。

（三）科研水平提高

近年来，学校一直以“211工程”建设为龙头，以教学科研为中心，努力加强学科建设，科研实力不断增强。学校的科研经费已从“八五”末期的400余万元增加到2002年的8000万元，增长了19倍。学校的科技论文和专利申报数量都有大幅增长。教职工发表的学术论文数从1995年的502篇增长到2001年的1311篇。2002年，学校教职工发表的被三大索引收录的论文数达105篇，与“八五”末期相比增加了近11倍。“八五”期间，学校的专利申请数几乎为零，但自1996年以来，共申请专利86项，获得专利授权20项。获得的省部级奖励由“八五”期间的10项增加到“九五”末期的61项。

7年来，学校新增了2个国家重点学科、7个广东省重点学科，1个国家人文社会科学重点研究基地，1个教育部重点实验室，1个教育部工程研究中心，1个广东省重点实验室和1个教育厅重点实验室，设有国务院侨务办公室华文教育基地。除广东省重点学科和教育厅重点实验室以外，其他方面均属于实现零的突破。结束了无博士后科研流动站的历史，现有6个博士后工作站。

（四）师资队伍结构改善

师资队伍建设是学校发展中的重要内容，师资队伍的水平是体现一所大学办学水平的重要标志。经过几年的努力，学校的师资队伍结构有了很大改善，师资队伍质量有了大幅提高。学校现有专任教师1363人。教师中有研究生学位者909人，占专任教师总数的66.7%。有博士生导师71人，教授225人，副教授560人。学校新增了两院院士2人，实现了零的突破。新增博士生导师63人，获教育部设置的“长江学者奖励计划”特聘教授岗位2个，广东省设置的“珠江学者”岗位计划特聘教授岗位2个。

有 11 人增选为广东省“千百十”省级培养对象。师资队伍的梯队建设已见成效，一支老中青相结合，年龄、学历结构平衡的队伍已初步建成。同时，学校加大投入，利用一系列工作、生活上的优惠政策吸引海内外的专家学者及一批有朝气、有知识、有作为的创业者来校工作，为学校的未来发展做好了人才准备。

（五）“侨”字特色更加鲜明

随着实力的增强，学校在外界的影响和名气也更大了，虽然竞争日益加剧，但生源形势却是越来越好。到目前为止，学校有来自世界 52 个国家和港澳台 3 个地区的海外及港澳台学生 7484 人，占在校生总数的 40%多，而 1995 年，学校还只有来自 16 个国家和港澳台 3 个地区的外招学生 1982 人。正是由于办学层次和水平的提高，暨南大学已成为海外华侨华人和港澳台青年来内地求学的首选学校，海外及港澳台学生人数一直稳居全国高校榜首。2002～2003 年，暨南大学的海外及港澳台学生录取人数连续两年超过了在内地的招生人数，总数达到 3500 多人，这也是学校自 1978 年以来外招生的录取人数首次超过在国内的招生人数。这同时也证明了一个事实，即学校的层次和水平提高了，知名度扩大了，侨校的特色不是削弱了，而是增强了。在实力不断壮大，知名度日益提高的同时，学校积极开展对外交流与合作，成为中国第一所在世界五大洲都建有姊妹大学的大学。

（六）办学层次和办学效益更高

按照“大力发展研究生教育，适度发展本科教育，积极发展华文教育，稳定成人教育规模”的发展思路，学校合理调整办学结构，取得了显著成效。目前，学校研究生与本科生之比由 1995 年的 1∶8.74 上升到 1∶3.31，专科生由 2472 人减为零。为向高层次办学，校本部从 1996 年开始即不再招收、培养专科生。在提高办学层次的同时，学校的办学效益也持续得到提高。与“九五”初期相比，全校的教职工和专业教师人数基本没有变化，1995 年分别为 3601 人和 1036 人，目前为 3649 人和 1363 人，且学校所获上级经费投入并未大幅增加，但学校完成的任务数却成倍增长，这必须有优良的管理质量和办学效益做保证。在 2002 年广东管理科学研究院“中国‘211 工程’大学教师人均效率排名”中，暨南大学排在第 41 位。

（七）综合实力和办学质量更高

在实施“侨校＋名校”发展战略的过程中，暨南大学的综合实力不断增强。1996 年，暨南大学被教育部列为国家面向 21 世纪重点建设的大学，2002 年，暨南大学“九五”“211 工程”建设的七个重点学科及公共服务体系全部以优秀的成绩通过专家组的验收。同时，学校的“十五”“211 工程”顺利通过立项论证，并获国家 2800 万元的专项投入（“九五”期间没有获得投入）。网大（http://www.netbig.com）的中国大学综合实力排行榜调查数据显示，暨南大学在中国高校的综合实力排名现已由 1998 年的 87 位上升到 2003 年的 36 位。另外，人民网（http://www.people.com.cn）近几年也对中国的大学做了排名，即“中国大学 100 强”，暨南大学在其中的排名自 2000 年至今依次为 81、68、55 和 49 位。尽管上述两个评选机构由于评估体系、评价指标方面的差别以及其他原因，最后得出的结论有一定差别，但不管最后的排名结果如何，有一点是毋庸置疑的，那就是正如这两组数据所显示的，暨南大学在这段时间内，确确实实是在上升、在进步、在发展，而且这一发展是在高等教育内外竞争日趋激烈的大环境下，在各高校都在努力上层次、上台阶的情况下取得的，这对一所以华侨高等教育为特色的侨校来说能在上述条件下保持这样一个发展速度并不容易。有鉴于此，学校于2002年被《中国高等教育评估》杂志列为研究型大学，在入选的77所大学中名列第53位，今年名列第46位。另外，学校的固定资产总值由1995年的2.7亿元增加到2002年的15.5亿元，增加了近5倍。图书馆藏书量由1995年135万册增至2002年的184.26万册。教学科研仪器设备值由1995年4985万元增至2002年1.7亿多元，增加了2倍多。

以上成绩得益于“211工程”建设的带动，得益于全体教职员工的不断努力，但更重要的原因是有了一个全新的办学理念，有了一个正确的指导思想，那就是“侨校＋名校”的发展战略。实施“侨校＋名校”发展战略既保证了学校的“侨”字特色，也促进了学校自身的发展。侨校与名校是一个统一的共同体，相辅相成，相得益彰。这一战略思想是建立在对教育规律和时代特征以及学校实际情况的深刻理解和认识基础之上的，而且事实已经并将继续证明，暨南大学实施“侨校＋名校”发展战略是有成效的，是切实可行的。“九五”以来，国家并没有为学校的建设和发展加大人力、财力和物力投入，也没有为学校提供更多政策上的优惠和支持，但学校取得了以上一系列看得见、摸得着的成绩，这是思想解放、实事求是、改革及时、措施得力、管理得体的结果，同时也是对“侨校＋名校”发展战略可行性的实践和有力论证。在未来的发展道路上，学校要乘着党的十六大的东风，根据《十六大报告》提出的开展创新教育的要求，继续按照“侨校＋名校”的发展战略，大力加强软硬件设施建设，把承载着党和国家以及全世界无数华侨华人希望的暨南大学真正建成一所海内外知名的社会主义高等华侨大学。

48. 教学是学校的生命①

教学是学校的基础工作，是学校的生命，是学校的主旋律。我们招进了学生，第一位的任务就是教育他们。教学工作至关重要，无论何时都应该重视。不要因为我们现在提了“侨校＋名校”，我校成了名牌学校，成了研究型大学，就忽视教学。成了名校之后，不注意教学，就会把学校的名声搞坏。所以我们学校这些年一直坚持每年召开一次教学工作会议。每年的教学工作会议有一个主题，今年的主题是实践教学问题，强调抓实践教学。在讲实践教学问题之前，我想先强调一下几个问题。我们学校确定了“侨校＋名校”的发展战略，这是我们学校的定位。这四个字，我们的上级领导同意，其他的高等院校听了我们的介绍之后也非常赞赏。办一所大学首先是定位问题。在昨天刚刚结束的全国高校教学研究会议上我也作了这个发言，会议也非常欣赏我们这个定位。因为现在很多高校定位问题没解决，所以这次全国高校教学研究会议就研究高校的定位问题。现在我们学校确定了“侨校＋名校”的发展战略，就应该按照这个定位搞好各项工作。全校所有的教师、所有的职工、所有的部门都应该围绕这四个字做好工作，实现这个发展战略。

第一个词是“侨校”，这个很明白，这是我们的特色，是从 1906 年建校以来就确定的特色。暨南是中国政府办的第一所侨校，如果摆脱了“侨”字，就不能成为侨校。侨校的特色是什么呢？就是要侨生多。所以我们应该注意侨生的比例，而且要高比例才行。但是应该说，经过若干年的努力，到现在侨生的数量仍然偏少。我们现在有 7400 个境外生，其中 3500 个香港学生，2000 多个澳门学生，500 多个台湾学生。我们的华侨华人学生偏少，只有 1000 多人。这个数目，与学校学生总的数量相比也是偏少的。但我们学校的性质，新中国成立后给我们的定位是“面向海外、面向港澳台”，因此，港澳台学生也是我们重要的服务范围。那我们加强港澳台的招生工作、培养工作也是非常重要，甚至是特别重要的。所以我们追求的首先是港澳侨台学生的数量，其次是质量，培养他们成才。我校自从进入“211 工程”学校以来，在“侨”字上做了大量工作，全校的海外及港澳台学生由 1900 人发展到今天的 7400 人，8 年里的发展速度还是很快的，但离我们的目标还有一段距离。我们希望全校内地与海外及港澳台学生的比例达到 1∶1。目前全校是 14 000 名本科生，4300 多个研究生，还有一些预科生和华文教育学生，总共是 19 000 个全日制学生。但只有 7400 个海外及港澳台学生，还不到 40％，还有点距离。侨生比例偏小，无论是学校的招生办，研究生部的招生办，还是成人教育的招生办，都应该把大量的力气用到海外、港澳台去，这是我们的任务。最近中央派出了调查组到我们学校来调查研究，为了香港和澳门的稳定、繁荣、发展，希望暨南大学更多地做出贡献。这只有把学校办好才有可能，所以仅仅提“侨校”还不行，为此，这几年我们提出了“名校”。既然要完成为港澳台培养人才，为心系祖

① 本文是在暨南大学 2003 年教学工作会议上的讲话（按录音整理），广州，2003 年 12 月 22 日；暨南高教研究，2004，(1)：1-6.

国、心向故乡的世界华人华侨服务的任务，如果不是一流的名校，我们就绝不可能吸引全世界的华侨华人把孩子送到我们这里来读书。试问，哪个家长愿意把孩子送到一个办得很差的学校去读书？每一个家长，每一个青年人，都愿意去读一所好的学校，获得一个好的文凭。因此，暨南大学不办成名校，侨校的任务就无法完成。祖国的统一大业，祖国的现代化事业这个伟大任务，要求我们暨南大学成为一所名校，这是客观的需要。我们前几年没提，那是因为我们还不太强，这几年学校有了巨大的发展，得到了国内外的好评，我们才提出这个“侨校＋名校”的战略。所以今天这个战略应该让每一个师生都知道，为这四个字来做贡献，凝聚我们所有的力量，发挥我们所有的优势，创造一切的条件，来完成这个伟大的任务，为我们祖国的统一大业，为我们祖国的现代化事业做更多贡献。这四个字我们应很好地理解，很好地执行，这是时代的需要。那么，由这四个字又引申出另外几个字的要求，我们要走国际化、现代化和综合化的道路。

第一，国际化。国际化的道路是名校所需要的。一所名校必须与国际进行紧密的学术交流，文凭在国际上得到认可。所以我们一直在打国际牌，并完成了几个方面的工作。

一是招生。现在一共有 52 个国家的学生在我们学校求学，世界五大洲主要的发达国家及发展中国家都有学生在我们这里读书。我们教学大楼前面的万国墙，每年都在增加国家数。暨南大学更多地为世界所了解，远在千里万里都愿意到这里来读书，说明有吸引力。国际化要求生源要国际化，我们的学术工作要国际化，人才要国际化，所以我们教师队伍也开始从海外招聘，甚至有外国人到学校来任教。我们的学术工作一定要以世界的前沿学科为准，以前沿的水平为准，来发展我们的科学研究工作。应该说我们现在已经在世界上找到了朋友，我们加入了世界大学校长联盟，我校是世界大学校长联盟的成员。

我们在世界五大洲建立了姊妹学校，同时我们又靠着侨校的性质，在华侨华人最多的东南亚每一个国家都找到姊妹学校，明年可以实现这个目标。这样的结果，就是我们的学生可以到对方那里去学习，对方也可以到我们这里来学习。交往的结果就是互相承认学分，承认学校。我们找的又是好大学、一流的大学做我们的姊妹学校，那么我们的文凭就会在那个国家得到承认。“文化大革命”前的外交战略，影响了我们与东南亚国家的关系，工作是非常困难的。在越南我们大概花了三四年的工夫才得到批准得以进入。当然，即使晚进去，我们也是国内第一个进去的学校。比如，印度尼西亚、马来西亚、柬埔寨、老挝，我们都是国内第一个在那里找到地盘的学校。国际交流合作处这些年来做了大量的工作，各院系配合，我校与越南的大学，与老挝唯一的一所大学老挝大学结盟，柬埔寨金边皇家大学也与我校结盟。金边皇家大学是柬埔寨的第一号大学。印度尼西亚的第二号大学等几所大学也已与我校结盟。这次我们去了菲律宾，菲律宾教育比较普及。菲律宾的第五号大学，即将与我们签署姊妹学校的合作协议。文莱唯一的一所大学也将与我们结盟。我们在东南亚各个国家或地区基本上都找到了建立合作关系的姊妹学校，现在就剩缅甸和独立不久的东帝汶了。东帝汶现在还有维和部队在那里，包括中国派驻的维和部队，高等教育现在还不知道怎么样。

缅甸的仰光大学长期关门，目前我校只能招生。现在东南亚 10 国包括新加坡，都有好的学校、最好的学校与我们结盟。所以，东南亚华人华侨最多的地方我校已基本覆盖了，这给我们学校为华人华侨服务带来了好的条件。

二是学术研究。这些年来，科研工作发展非常快，科研经费是我们在进入“211 工程”前的 20 倍，科研论文数也成倍增加，国家许多重大重点项目我们都能拿到，但我们和名校相比还有较大差距。北大、清华的科研经费都在 10 亿元左右，而我们还不到一亿元。首先我们要奋斗过一亿元大关。我们那么大一个学校过去科研经费才 400 万元，那是说不出口的事情。我们科研的题目也不错，科研的项目、论文、专利、应用、产业化都走出了一大步，甚至我们科研的推介会还去泰国开过，这在国内其他高校从来没有过。哪个大学科研的推介会跑到国外去开？说明我们的科研走出了国门，我们的科研迅猛地飞跃前进，为我校国际化增加了光彩。但我们在国际化的征途中还有很多工作要做。

要提高我们的学术水平。你学术水平不高，发达国家的大学不承认你，如我们学校就无法与美国一流的前几所大学结盟，美国前 10 名的大学根本就不认可你。只是有一次机会，我们错过了。我们和美国一流大学的关系仅仅在管理学院做 EMBA 教育方面，邀请斯坦福大学的教师为我们的学生上课，现在我们的学生正在斯坦福大学上课。一个大学如果能够和世界的名大学在教学方面进行交往，就会提高这个大学的品牌影响力。我们一定要和一流的科学家、一流的学者进行交往。今年诺贝尔奖获得者能够在学校的讲台上做学术讲座，这就是一个标志。如果你这所学校很差，人家不会到你学校来。所以我们要请名家、名师，国际的、国内的来我校讲学，切磋学术，要进行学术的合作，这样我们学校才能走向世界的前沿。国际化包含了非常多的方面，这些我只是从高度上说，具体的还要制度的改变。方法、策略、管理都要国际化。我们的学分制管理走向了国际化，但我们还有差距，最大的毛病是我们的选修课太少，必修课太多，不利于培养具有创新素质的人才，所以我们的学分制还要继续前进。尽管在国内我们是不错，每个学校都很赞赏，但在国际上我们还有差距，还要改革。所以在国际化上我们要大做文章，首先是要加强招生工作。几个招生办都做得很不错，无论是研究生层次、本科生层次还是教育学院都在努力做这个工作。这个工作还要继续加强，我们在国外设了很多招生办，要采取若干措施，重视发挥招生办的作用，不仅是职能部门，学校各院系都要配合。

第二，现代化。首先是学校硬件的现代化，校园的现代化。现代化的管理、现代化的思维、现代化的措施和现代化的硬件设施都要搞好，都要符合标准。我们的校园还不够好，还有很多很差的房子，规划得不好，现在还来不及做。现代化的管理水平也不够高，人治的问题还很严重。前几天我接到一个台湾学生给我的来信，批评我们两个部门的管理人员的工作。这个学生写得语重心长，他毕业了，在离校的时候给我留了几句话。我们有些部门同志的态度从来就没有好过，无论是科长还是科员没有一个好面孔对学生，这根本就不是现代化管理。现代化的管理从企业的角度来说就是顾客第一的思想，学生是学校的第一，有了学生才有我们老师，才有我们校长，才有我们干部。你不好好为学生服务，谈何把他培养成为高素质的人才？所以我们机关干部的服务工作，尽管经过了“三讲”教育，好多部门仍然做得不够，有极个别的老师也非常糟糕。最近我们在处理一个教授，他向自己的研究生勒索，要研究生出钱帮忙买东西，干这干那。为人师表到哪去了？这哪里是现代化的思想。所以我们的服务工作要更进一步地改善，除了我们现有的硬件以外，实验室、图书馆也要搞好。我们希望在不久的将来，每一个老师都有工作间，为成为一个名校创造基本条件。要把实验室搞好。要建图书馆大楼，让我们所有的图书资料都能尽快地与读者见面，让读者可以自由地借阅。我们现在只有 30 万册的图书在开架，很多图书没有开架，没有地方，所以图书馆要新建，要现代化，实验室的设备要改进。校园也偏小，我们要争取把校园扩大。最关键的是现代化的观念，没有现代化的观念，就没有现代化的管理，没有现代化的办学，没有现代化的环境，所以全校特别是今天在座的各位领导、教授，要用现代化的观念来发展我们的学校，做好工作，特别是教学工作。学校也投了大量的钱来做这件事情，我们希望在“211 工程”第二期结束的时候，现代化的局面更好一些。现代化的观念还要强调依法治校，今年校办出版了两本文件集，希望大家按照制度办事，按制度办事，学校才能科学地发展。我们把制度订好，不好的制度我们就改，把它改成现代化的制度，我想暨南大学若干年后就会成为一所很好的名校。

第三，综合化。这是因为我们是综合性大学。世界上的高等院校发展的历史证明，综合性的大学在培养人才方面能够形成最好的氛围，单科性大学的学生，其思维、观念、学识都比较单一和狭隘，所以要强调学校综合性的重要性。我们学校以前只有 30 个本科专业，现在已发展到 52 个本科专业，专业面扩大了，我们还要有更多的专业。如果要与硕士点、博士点相平衡的话，我估计要有 60 个专业。我们目前缺乏的是工科型专业，工科太弱，文经管历来比较强，理工医差一些。理工医学院要大力加强自己的建设，成为名牌学院，与名牌学校相适应。理工医为什么吸引不了海外生？就是因为知名度差，硬件差，教师素质差，那就要找出弱点，奋起直追。这几年来，我们的生命科技学院有了大踏步的飞跃，给学校争了很多的光彩，但也要更进一步地发展，其他学院要跟进。没有博士点的学院赶紧争取，想办法下次能申请到博士点。博士点太少的学院，要增加博士点。整个学校二级学科的博士点是 28 个，跟我们学校的名声还不相称。尽管比过去好了很多，但是数量偏少，国家重点学科偏少，这两个问题需要每个学院努力去奋斗。要把综合性的内涵做好。对于比较冷门的专业，要想办法办成热门，热门专业要维持热门的地位。只有这样，国内外的学生才愿意报考我们学校。这几年招生办的日子觉得比较好过，每年招生的时候我们不用去求人，不是等着哪个学校录取完了以后，才开始我们的录取。我们现在第一批录取第一志愿的就足够了，所以日子好过。对海外及港澳台学生而言，尽管有 137 所高校在港澳台竞争，许多高校条件都比我们好，但大部分考生还是考我们学校，所以我们能招到好学生。在莫斯科举办国际高等教育展的时候，我们暨南大学的摊位，最多人来询问。我们在海外的知名度高，这就是我们学校办学质量高的结果。否则到海外举目无亲，怎么办？就要靠你的知名度和你在海外的校友。

“侨校＋名校”，国际化、综合化、现代化的办学道路要坚持。这几年来学校实行了一系列改革措施，这些改革措施是我刚才讲的，要能够有所进步。没有改革是不可能变好的，但在改革中我们花了很多的精力，很多人不理解，很多人反对。学校班子忍受着这些痛苦，才使学校有所进步。大家可能还记得我校在取消补考的时候，多少人不理解，上千人反对。当我们实行财务集中管理制度的时候，当我们砍掉专科的时候，当我们实行量化分配制度改革的时候，非常的困难，一步一个脚印才走到今天这个地步，所以每一个干部，特别是今天在座的干部要珍惜我们今天改革的成果，要坚持我们改革的措施。有些措施还没有实行好，比如学分制，本来应该是最亮的亮点。但是对个别学生重修的门数，很多院系管教务的人不管。今年我们有一个出现状况的学生，读了10年书，11门课重修。有这种学生但从来没汇报到学校来。他怎么读得下去呢？读了十年书，本科毕不了业，无路可走嘛！怎么能让他背那么重的包袱呢？这是有些人在学分制的管理上出了毛病。从教学开始，哪门课不及格就得及时重修，不能积累。从1993年开始实行标准学分制时就强调过这点，但很多单位怕得罪人，又允许学生去选课，他11门课长期不及格，他吃得消吗？他心理压力大，分管学生工作和管教学的系主任、班主任没来汇报过，让一个好好的青年人出现问题，即使他读不完大学，他还是可以去好好工作嘛。所以我们各层次的领导干部、教师，要把学生当作自己的孩子来爱护。不是每一个人都可以成大才，他读不完大学，这辈子还可以好好做人嘛，对不对？我们的思想工作不到家，学分制度没实行好。所以我希望大家能理解我们的改革，严格去贯彻。如果我们把学分制真正实行好了，就不会拖到那么晚。又比如我们的量化考核，校内工资的量化，鼓励教师的积极性，能者多劳，多劳多酬，优劳优酬，这是现代化的管理制度。但现在还有很多人想不通。我们调查了一下，都是很多工作量完成不了，做不了科学研究的人。我最近打听了一下，我们学校有些人做了十几年的讲师，还在那个岗位上，你说他能教好课吗？就是这些人在说话。所以我们一定要加强我们的量化改革工作，淘汰不合格的人。我们在元月3日、4日开教代会，要把定编定岗的方案交给全校教工来讨论，讨论以后我们就执行。我们除了干部定岗，科级干部、处级干部定岗以外，员工要定岗，教师也要定岗，不合格的我们就要淘汰，到人才中心或调出去做其他工作。要坚持我们的一些制度，实在差的人，系主任就要赶快告诉学校，这个人已不适合在这个工作岗位。现在有些领导怕得罪人。怕得罪人，你这个系的工作就搞不好。要使学校的工作做好，就必须使所有的员工都能很好地在自己的工作岗位上工作，完成他岗位上的任务，这样的干部，这样的教师才能继续聘用。我们实行的聘用制必须要坚持。还有我们强调的教授上基础课。我们从1993年就开始了，教育部2001年才下文件，我们超前了。我们在许多领域超前，才换得学校的进步。我们有些老师当了教授就不上基础课。教授不上基础课，本科生的质量就很成问题，学生进校以后得不到很好的打基础的训练。因为教授知识渊博，积累了很多好的经验，他可以把最先进的经验很浅显地告诉学生，以他为师表对学生进行培养，将会使我们的新生得到很好的培养，使其养成读书求学、做学问的好习惯，或者学会怎么做人，学到一些本事。所以我们坚持让教授上基础课，希望这次会议以后去查一查，各个院系你的教授上基础课没有？我希望大部分教授都能承担本科的基础课程，保证我们暨南大学基础课的质量。我们的“三语”（汉语、英语、计算机语言）教学改革是非常不错的一个改革措施。但是我现在发现我们在英语方面还有差距，还需要加强。当然我们比过去好多了，我看到我们的英语四级统考通过率达到90%多了，我刚来暨南大学时及格率不到50%。但是我们英语的口语方面不行，看可以，答卷可以，说话不行，这是我们整个中国英语教学的失败，从名牌大学到我们学校都有这个问题。现在幼儿园都在教英语了，读到大学毕业都开不了口，怎么教成这样？“哑巴”语言。我们是侨校，我们跟国外打交道最多，但是我们的人都开不了口，那怎么行。所以英语的教学，应加强再加强。比如我们可实行强化训练和课堂上教学方法的改革。许多高校在英语方面都非常努力，但是在这方面我们走得比较前，办国际学院和全英语教学的专业，这是我们暨大的亮点。我们因此才能跟人家对口，他的学生才会到我们这来读书。最近我刚从菲律宾、文莱回来，包括我去访问菲律宾军事学院，这是菲律宾最高等的军事学院，已有一百年历史，军事人才全是这个学院培养的，将军院长在跟我们见面的时候，知道我们学校的情况，他要派学生到我们学校来读书。全中文教学怎么能吸引人？我们的国际学院是英语教学，所以我们的国际学院要扩大，每个学院要去支持它，使得暨南大学成为双语的学校，既要让人们到了暨南大学以后受到五千年的中华文化的熏陶，同时又能使他走向世界，使他容易寻求职业，使他的工作能做得更好。基础要靠英语，英语的教学，无论是公共课的教学，还是专业的教学都要进一步加强。我们尽管做了改革，但做得还不够。我们的管理要跟上，刚才讲的思想工作等问题还差得很远。我举个简单的例子。这两天全国高等学校教学研究会议在我校召开，周远清副部长，还有高教司副司长刘凤泰，许多名牌大学的校长、院士来我们学校开会，这是很好地让人家了解我们的机会，我们有意把校史馆打开，让人家看看。一看就知道我们暨南大学了不起，因为很多人第一次来暨南大学，对我校不了解。周部长讲，他到了很多学校，有些高校真的不像高校，走进去到处摆摊卖东西。他说我们学校不错，真像好的高校。我们学校漂亮在哪里？我们有几条道路比较好，绿化很好；再就是我们的湖泊很好，如日月湖和南湖。但是听说这几年我们管南湖的离退休处，学期没结束就把水放干了。我看了以后，觉得真是不会管理。每一个部门，即使是小部门都要想到学校的荣誉、学校的面孔。从进大门开始，每一个角落，食堂的每一张桌子都要弄好，你这个学校才是好学校。我去国外，有些大学给我留下了终生难忘的印象。上次我去英国看剑桥大学，一看那真是了不起，你一走进那个学校就感觉到那真是世界名校。就看它的面孔不去看教学，看环境，看它的校园，看它著名的剑桥河（学生比赛的那条河），看它的草地。那里游客很多，我跑到草地上想去坐一下，刚坐下去，马上就有管理人员跑过来干涉我，告诉我不能坐。我们学校的一些人，没有这个意识，把湖水放干了，不知道要干什么，那样的事情在假期里做嘛。学习期间要让学校非常漂亮，让师生员工生活在漂亮的环境里面。有些人不知道围绕着学校的名声去服务。我希望全校每个人、每一个部门的管理工作都要想到学校。我举这个例子，不是说这个管理部门犯了什么大错误，但对学校的荣誉确实不太好。我们的改革要加强，做了就把它做到底。

还有考试的严格管理。今年8月份全国各大报纸新闻出现“作弊学生告倒暨大”这样的大标题，全国各大报纸、媒体、电视台播这个事情。我们有的人惊慌失措以为这下麻烦了，给学校的声誉带来麻烦。我正在开院士大会，我看了，我不怕。中央电视台叫我去做节目回答这个问题，各个媒体找我谈话，要我发表校长声明。我说，如果作弊的人都能把一个名牌学校、好学校告倒的话，那还有什么天理！一个星期以后，我们教务处收集的信息 80%都是支持暨大的，只有少数学生说同情那个作弊的学生。显然公理会战胜谬误，进一步人家会说你暨大了不起。所以只要我们坚持改革，坚持从严治校、从严治教、从严治学，坚持依法治教，坚持实事求是，我们学校就一定能办好。所以，我们的改革，切记要把它做得更好更完善。

今天的会议是讨论实践教学，这是很重要的问题，因为长期以来在我们中国的高校普遍存在着重理论、轻实践的现象，各大学几乎都这样。教实验课的教师地位低，教理论课的地位高，这成为一个常规。实验课受到轻视，实验课的效果就会受到影响，再加上这些年来学校经费不足，实验课的投入仍然不够，设备不够先进，设备的数量不符合我们扩大招生的要求，怎么把实验教学搞好是个关键。其实人人都明白，理论一定要和实践结合才能取得成功。培养高素质的人才，必须实验、实习、实践都加强，而且要特别重视。我们学校的定位是“加强基础、突出应用”，我们主要培养应用型的学生。根据暨大生源的情况，我们这个学校主要不是培养科学家，更多的是培养应用型人才。应用型人才就是实践重要，实验重要。就像我们医学院一样，医学院学生走到医院里面，有一番理论，治不了病，看不好病行吗？病人就是检验他学习好坏的标准，能不能看好很重要。所以实验、实习非常的关键。这次会议强调这个事情，为我们学校培养学生“加强基础、突出应用”这八字努力，所以今天开完会后，希望大家重视我们的实践教学，重视实习，把这些工作搞好。尽管我们现在钱不够，设备不够，但大家要想办法把学生培养好，培养学生成才，成为有素质的人才。这次会议时间不长，刚好在期末，各方面的工作都比较繁重，又加上今年上半年“非典”的危害，使得我们学校上半年的许多工作压在了下半年，社会上的任务又压到学校里面来，所以我们学校的工作就比较紧，会比较多，希望大家全心全意把这次会议开好。教学是生命，教学是基础，教学是主旋律，而且是我们这所大学的第一需要。我们现在被国内各大媒体的排行榜排为研究型大学，按道理“侨校＋名校”就是要做一个研究型大学的侨校。什么是研究型大学？就是我们研究生层次的数量要与本科生层次的数量大体相当。现在国内通行的说法，什么是研究型大学？第一个指标就是学生比例，研究生与本科生的比例必须达到 1∶1，即一个研究生，一个本科生，或者最少是 1∶2，就是一个研究生，两个本科生，我们离这个比例还差一段。1995 年是 1∶8.7，现在是 1∶3。根据国家给我们的发展规划，至 2010 年要达到 1∶2。我们现在努力实现这个目标，实现研究型大学的目标。我们的本科生从 4000 多人发展到 16 000 人，而且今年已达到这个标准。我们已录取 4000 个本科生，目前我们不再扩大，维持每年招收 4000 个本科生的规模。研究生现在有 4300 多人，要翻一番到 8000 人，还差 3700 人，还需要大家努力。我们现在有 88 个硕士点，只有 28 个博士点，博士点偏少，博士生数量偏少，就硕士生与博士生的比例来说，硕士生偏大，博士生偏小。所以学校各院系都要努力增加博士点，使博士生的数量增加，使研究生整体的数量增加，尽快争取达到 8000 个，使研究生与本科生的比例达到 1∶2。

第二个指标是国际型学生，海外生要占 15%以上，这个可以算是达到了。我们比其他学校要好。全国很少有学校达到这个数字，我们这个达到了，但我们海外及港澳台地区生源的素质还不够高。要吸引优秀的学生，就像美国把我国优秀的学生吸引到美国去读书，这才是一流大学要做的。你把差的学生吸引进去，不算有本事，要把海外及港澳台的优秀学生吸引到你这里来读书，这个还有一段距离，我们还需要努力。还有我们的科学研究要再上台阶，我们名师太少，院士太少。一个学校没有名师，没有院士，就谈不上是研究型大学。尽管你其他指标都达到了，但缺乏名师，缺乏名学科，这是不行的，学校核心的支撑是名师、名学科。要我们全体领导、所有的教授都为总体的目标而努力，把我们学校办成一个研究型的大学，一个侨校性质的研究型大学，这是我们未来的任务。希望大家进一步努力，发挥优势，继续进行改革，完善各项措施，加强管理，提高质量，把品牌打造好，为我们国家的统一大业，为我们国家的现代化，做出我们暨南大学的贡献。

49. 大力发展具有侨校特色的研究生教育①

春回大地，万象更新。新学期伊始，我们会聚一堂，在这里召开2004年暨南大学研究生教育工作会议，共商研究生教育和学校发展之大计。这是本学期也是2004年学校工作的头一件大事。时隔数载，当我们的研究生教育规模和水平发展到一个新阶段，在面临今后如何进一步发展的关键时刻，召开这次研究生教育工作会议，其意义非同一般。在此，我代表学校对这次会议的召开表示热烈的祝贺！

几年来，我校研究生教育的发展取得了令人瞩目的成绩。大家出于对今后研究生教育发展的极度关切，也感到研究生教育中存在一些矛盾和问题，如研究生的培养质量问题、扩招带来的教育资源短缺问题、导师队伍问题、学科专业发展的不平衡问题，等等。存在矛盾和问题并不奇怪。用辩证法的观点看，矛盾是普遍存在着的，矛盾既存在于事物发展的一切过程中，又贯穿于一切过程的始终。研究生教育也是如此。研究生教育在快速发展的道路上，会不断出现新情况、新问题、新矛盾、新困难，解决这些矛盾和问题的过程，实际上就是发展和提高的过程。问题是如何抓住主要矛盾和矛盾的主要方面，解决最紧迫的问题，使其他问题迎刃而解。我今天讲话的主题是，大力发展研究生教育，努力建设具有侨校特色的高水平研究型大学。

一、回顾过去，研究生教育的快速发展使我们深切认识到“发展是硬道理”

我校的研究生教育发展到现在的规模和水平来之不易。今天，我们应该回过头来审视我们的发展历程，从中总结出经验和需要思考的问题，为今后的发展提供依据和借鉴。总体说来，暨南大学研究生教育的发展大体可分为两个阶段。第一阶段为暨南大学复办到20世纪90年代中期，可视为起步阶段。暨南大学复办后，我校成为全国首批具有研究生学位授予权的高校之一。1981年，我校首次获得的全国第一批硕士学位授权的学科、专业达9个（分别是政治经济学、文艺学、比较文学与世界文学、英语语言文学、基础数学、分析化学、眼科学、药理学）。1984年，我校首次获得的全国第二批博士学位授予权的学科、专业有2个［专门史、内科学（血液病）］。其后，我校博士学位授权学科点的建设发展速度缓慢：第三批获得博士学位授权学科点2个（1986年），第四批2个（1990年），第五批1个（1993年），第六批为0（1995年）。此时，全校仅有9位博士生导师，7个博士学位授权学科点，50个硕士学位授权学科点（含三级学科），相当于二级学科点43个，其中包括MBA（全国首批26所试点单位之一）。在校研究生数量为615人，其中博士生仅52人。

20世纪90年代中期即我校成为“211工程”重点大学至今为第二个发展阶段，可视为快速发展阶段。为了抓住机遇，谋划暨南大学未来的发展，新的校领导班子对学

---

① 本文是在暨南大学研究生教育工作会议上的讲话，广州，2004年3月10日；一个大学校长的探索，北京：高等教育出版社，2011，227-235.

校的办学思路进行了重新审定，1996 年提出以教学和科研双中心代替以往以教学为中心的低目标的发展思路。同时，对办学结构也进行了调整，果断砍掉了专科教育，大力发展研究生教育，积极加强学科建设和学科点建设，使我校研究生教育驶上了发展的快车道。由于确立了正确的发展战略，当 1999 年全国高等教育扩招时，我校已经具备了很好的发展基础。我校也因此能够抓住新机遇，再上新台阶。经过第七次、第八次申报工作的努力，到 2000 年，我校博士学位授权点的数量达到了 14 个，与 1996 年相比翻了一番。“生物医学工程”博士点的批准，实现了我校一级学科博士学位授权学科点零的突破。

去年，第九批申报博士学位授权学科、专业结果公布，我校获得“应用经济学”“工商管理”2 个一级学科博士点；“工程力学”“病理学与病理生理学”“中国古代文学”“中西医结合临床”4 个二级学科博士点，使我校博士学位授权学科点的数量升至 28 个。与 2000 年相比，我校博士学位授权学科点的数量又翻了一番，与 1996 年以前相比，则翻两番。硕士学位授权学科点的数量也升至 89 个，与 1996 年相比，数量也翻了一番。目前，又传来喜讯，国务院学位办公布了 2003 年在一级学科范围内自主设置学科、专业的名单。我校在“生物医学工程”一级学科内自主设置的“生物材料与纳米技术”“生物医学信息技术”“生物医药工程”“生物与医学物理”“细胞与组织工程”5 个二级学科，在“工商管理”一级学科内自主设置的“财务管理”二级学科，共 6 个博士学位授权学科获得批准。这样，使我校可以开展招生的博士学位授权学科点的数量升至 34 个。

作为研究型大学标志之一的博士后流动站的数量去年也翻了一番。学校原有“应用经济学”（1999 年）和“临床医学”（2000 年）两个博士后流动站和一个“暨南生物医药开发基地”博士后工作站。去年我校“工商管理”“中国语言文学”“生物学”三个博士后流动站获得人事部批准，使我校博士后站的总数达到 6 个。

在校研究生的规模也呈现出跨越式发展。2000 年，在校研究生的数量为 2007 人，其后三年的年增长速度在 30%左右，2003 年在校研究生的数量达到 4236 人，其中硕士生为 3709 人，博士生为 527 人。与 1995 年相比，研究生总数增长了近 6 倍，其中博士生数量增长了 9 倍多。在研究生中，港澳台、华侨、华人研究生的人数达到历史新高，为 747 人。

学校师资力量也明显加强。现有专职教师 1363 人，其中中国工程院院士 2 人，中国科学院院士 1 人，教授 225 人，副教授 560 人。近期经过研究生导师遴选，现有博士生导师 117 人，为 1995 年的 13 倍，硕士生导师 636 人。研究生导师队伍也开创了暨南大学有史以来的新高。

所有这些数据都生动表明，暨南大学的研究生教育在过去几年中呈现跨越式的发展态势，办学规模不断扩大，办学层次不断提高，学校的整体实力在明显提升。

回顾研究生教育的发展历程，使我们深切认识到发展是解决学校所有问题的关键，发展是硬道理，是第一要务。我校研究生教育取得了重大的成就，一条重要经验就是在全国扩招前学校已做好了从思想到办学条件方面的充分准备，因而当扩招机遇来临的时候我们能够抓住发展机遇，扩大研究生办学规模，促进学校整体实力提升。事实表明，抓住了发展这个主题，我们就可以高屋建瓴、统领全局，学校就会呈现出朝气蓬勃、欣欣向荣的大好局面。

回顾研究生教育的发展历程，我们清楚地看到，关键时刻办学思路的确立决定着学校具有重大意义的发展举措。办学的思想、宏观的战略，决定着前进的方向，决定着学校发展的前景和未来的地位。只有加强宏观思考和战略分析，才能总揽全局，把握方向，谋划好学校改革与发展的未来。

回顾研究生教育的发展历程，我们看到发展过程中存在的问题。例如，①我校至今尚未被批准试办研究生院；②我校博士学位授权学科点和博士后流动站的数量整体上仍然偏少；③我校最早获得硕士学位授予权的若干学科、专业，至今未能获得博士学位授予权；④在研究生的数量中，代表一个学校研究生教育水平的博士生的数量偏少；⑤我校至今还没有论文入选全国百篇优秀博士论文；⑥在发展的同时，有些点上的培养质量管理工作还存在问题；等等。所有这些问题都值得大家认真地思考。

二、与时俱进，大力发展研究生教育是实施“侨校＋名校”战略的基石

我们应该清醒地看到，暨南大学在前进，广东高校乃至全国高校也都在前进。近年来，国内高校都在抢抓历史机遇期，集中智慧、精心谋划，实现学校跨越式的发展。高校之间的竞争空前激烈，其势头如千帆竞发。为了提升自己的实力，一时间许多学校合校，高校阵营发生变化。据统计，到目前为止，全国有 500 多所高校进行了整合，最终形成了 1533 所新的高校。合并结果使得强者更强，羽翼更丰，实力更加雄厚。在这种形势下，我们要问：暨大怎么办？

为了自立于全国高校的强手之林，暨南大学的应对战略是：始终如一地贯彻执行“面向海外、面向港澳台”的办学方针，实施“侨校＋名校”的发展战略，走国际化、现代化、综合化的道路，大力发展研究生教育，朝着建设具有侨校特色的高水平研究型大学的目标迈进。

“侨校”是我们的特色，是我们的生存之本。任何时候我们都必须固本强基，牢记为侨服务的宗旨。不仅如此，我们还要充分发挥侨校的特色，特色就是竞争力，特色就是战斗力，在“侨”字上大做文章，发挥优势，以特色取胜。暨南大学的学位与研究生教育在发展中初步形成了自己的特点，总结起来有这样几点。

（1）暨南大学是最早获得教育部批准（1984 年），面向港澳台地区和海外自主招收研究生的高校。

（2）暨南大学是目前招收港澳台研究生数量最多的高校（在校研究生 747 人），并为香港、澳门的顺利回归做出贡献。

（3）暨南大学是首创招收兼读制研究生的高校（1989 年）。

（4）暨南大学是在海外及港澳台地区多处设立兼读制硕士面授点的高校（在中国香港、澳门、台湾地区和新加坡等地设立了 8 个面授点）。

（5）暨南大学是采取一年两次研究生招生，招生方式灵活的高校。

我们今后不仅要发扬光大这些特点，而且还要继续开创新思路，使我们的特点更鲜明、更突出、更全面。

“名校”是什么？名校就是高水平的研究型大学。什么是研究型大学？国内外对此有不同的评价标准。归纳起来，研究型大学至少应有以下四个特征：拥有相当规模的学科点和学位授权点；能够获得相当可观的科研经费；拥有相当数量的高水平学术成果；拥有一支高水平的学术队伍。《中国大学评价》杂志在2002年提出了中国研究型大学的三条分类标准。其一为：将全国所有大学的科研成果得分按降序排列，并从大到小依次相加至得分超过全国大学科研成果总得分的70%为止，各被加大学就是研究型大学。

应当承认，将暨南大学建成研究型大学，而且是高水平大学，是对我们的挑战。要实现这一目标，学校必须与时俱进，大力发展学位与研究生教育，在现有研究生教育发展水平的基础上，采用多种形式，继续扩大招生规模，进一步提高培养质量，使研究生教育水平处于全国前列。通过深化改革，优化学科结构，加强导师队伍建设，构筑研究生培养的质量保障体系，改善办学条件，全面开创研究生教育的新局面。从现在起到2010年，是我校研究生教育发展的第三阶段，可称为“更上一层楼”阶段。为此，我们要在大力发展研究生教育上下功夫。

(1) 大力发展研究生教育，必须继续扩大研究生教育的规模。

现在我校研究生的规模是4236人，到今年秋季新生入校预计可超过5000人。这离学校提出的到2010年发展到8000人的规模，实现本科生和研究生的比例为1∶2的要求尚有一定的差距。要实现上述目标，我们应当对以下几方面的影响因素给予充分考虑。第一，目前全国研究生数量发展处于稳定期，我校扩招的空间在缩小。第二，扩招要有一定数量学科、专业点支撑，目前我校博士学位授权学科点偏少，这会在一定程度上影响我们的扩招工作。这就涉及第十次博士点的申报工作。全面启动第十次博士点申报工作是我们这次会议的重要内容。总结经验，表彰先进，未雨绸缪，争取再创佳绩是我们的目的。希望在座的各位对此要高度重视，积极出谋划策，及早进行准备。第三，国家针对过去十年博士研究生招生年平均增长21.3%的发展情况，对研究生教育的层次结构进行调整，提出今后将稳步发展博士研究生规模，大力发展硕士研究生规模，使授予博士学位与硕士学位之比接近1∶10。这与我校博士生教育前期发展较慢，希望今后予以重点发展的愿望相左，加大了博士生扩招的工作难度。今年下拨的博士生招生指标（155人）与我们上报的数目相差甚远就是例证。如何处理好这一矛盾需要认真思考。第四，国家对学位研究生的培养在进行类型调整，提出专业学位研究生培养的规模必须相应有较大的发展。这应该说是我们扩大招生规模的一个重要途径和机遇。目前，“会计”硕士专业学位获得批准，但总体上我校具有的专业学位种类较少，招生规模较小，这方面的工作有待进一步加强。第五，我们不受在外招收研究生数量指标的限制，对于如何充分利用这一优势进一步扩大海外及港澳台生源，如何调动院系领导和全体研究生导师招收海外及港澳台学生的积极性等，需要认真思考和拿出对策。

(2) 大力发展研究生教育，就必须抓好学科建设。

我们知道发展是硬道理，但发展要靠实力，实力要靠建设。建设不仅包括基础设施建设，更重要的是学科建设。学科建设是高校发展的龙头。高校的人才培养、科学研究和为社会服务三大任务，都是以学科为基础进行的。在学校的整体建设中，要坚持以学科建设为主线，以重点学科建设为核心，学科建设的根本任务就是凝练学科方向、汇集学科队伍、构筑学科基地。但要注意，在规划学科建设时一定要与研究生教育的发展紧密相连，要统筹考虑，统一规划，不能脱离研究生教育独立设计学科建设。

关于学科建设与申报学位授权点的关系，可以说，两者是相辅相成的。学科建设是获得学位授权点必要的前提条件；而获得学位授予权是学科建设“瓜熟蒂落”的结果。博士点、硕士点的获得是学科建设的必然结果，是学科建设获得成就的具体表现形式。一方面，我们在学科点申报过程中要积极地付出努力，千方百计地去争取；另一方面，我们更应该注重平时对学科建设的不懈努力和长期积累。

（3）大力发展研究生教育，就必须加强导师队伍建设。

导师队伍对研究生的培养质量和学校发展的水平至关重要。我们和世界一流大学的差距主要体现在导师队伍的水平方面，新世纪高校之间的根本竞争还是导师队伍的竞争。导师不仅要在学术上严格要求学生，更要在思想上严格要求，具有实事求是、严谨的治学学风，科学的思维方法和良好的道德品质。优秀的导师必须具有三个特点：第一，要有创造性的工作方法、创新精神，做一些别人做不出来的成果；第二，必须是严师，对学生要求严格，鼓励学生大胆创新，勇于冒风险；第三，导师要做到教书育人，将自己的好作风、好思想、严谨的治学态度潜移默化地传授给学生，把肩负的政治历史使命融入教育当中。这三个特点，可以概括说成是“严师出高徒”。因此，我们必须不断建立严谨治学的研究生导师队伍，把培养高水平的导师队伍作为一项重要的工作来抓。

（4）大力发展研究生教育，就必须成功申办研究生院。

试办研究生院是发展我国研究生教育的一项重要举措。我国于 1984 年首批试办研究生院，加上 1989 年和 2000 年的 2 批，以及特批的 3 所，共批准 56 所高等学校成立了研究生院（其中合并 2 所），占全国高等院校的 5%。试办研究生院应具备一定的条件：学科、专业比较齐全，科学研究基础好；有较多能够指导博士生和硕士生的教授、副教授和学科、专业授权点；有多年培养研究生的工作经验，管理制度比较健全；有供博士生、硕士生使用的专业实验室，并配有必要的实验设备和测试手段，图书资料比较齐全等。

研究生院的实质是一种研究生的培养制度，确立了一种符合高层次人才教育规律的培养机制。研究生院的功能主要表现在提高大学的学术地位和层次，争取高水平的研究人员和师资，争取各种科研项目和经费，制定研究生教育管理政策，保证研究生的培养质量，促使研究生教育服务于国家。

研究生院的设立和建设对研究型大学的形成和发展具有不可替代的作用。为研究型大学学科间的高度融合、交叉、渗透和整合提供广阔的操作平台。我校要办成研究型大学，首先就要成立研究生院。因此，我殷切地希望我们全校同志，团结一致，上下努力，力争尽早成功申办研究生院。

三、从严治校、从严治教、从严治学，努力提高研究生的培养质量

研究生培养质量是高等学校教育和科学研究综合实力的体现，代表了教师群体的科学认知水平和创造性，是衡量大学水平及学校在国内外地位的综合指标。我们要大力发展研究生教育，就要从严治校、从严治教、从严治学，努力提高研究生的培养质量，尤其提高博士生的培养质量。博士生教育是高等教育的最高层次，博士生的培养质量是一个学校乃至一个国家高等教育水平的重要标志。我们一定要高度重视博士生的培养工作，强化质量意识，在资源配置上向博士生教育倾斜。始终坚持“育人为本、质量第一”的观点，正确处理发展、创新、质量的关系，弘扬创新主旋律，牢记质量生命线，确保培养质量的不断提高。

强化过程管理，做好研究生培养过程每一环节的工作，是提高研究生培养质量的关键。过程管理是比较微观具体的管理，根据不同学科、不同专业的特点，提出一些具体的实施措施和方式。要把工作重点放在包括修订培养方案、重视课程教学和学位论文的研究写作等阶段的管理上来，通过对各培养环节的强化管理和考核，以利于高质量创新人才培养工作的顺利进行。

（1）建设一支高水平的导师队伍，是提高研究生的培养质量的根本保证。研究生教育的特点决定了导师在研究生培养过程中的中心地位和主导作用。“师者，所以传道受业解惑也。”研究生导师素质直接关系到研究生的培养质量；导师的学术水平、思维方法、治学态度、思想作风等直接熏陶着研究生。这些都涉及导师队伍的建设。

研究生导师是培养研究生的重要工作岗位，而不是一个固定的职称、职务和荣誉称号。上学期，学校根据学科、专业的发展和研究生招生工作的需要，遴选出一批博士生导师与硕士生导师，同时出台了《暨南大学审核研究生导师招生资格的暂行规定》，除要求导师承担有一定科研项目、经费充足、有稳定的研究方向、有连续的科研成果和强大的科研能力外，还对导师每年的招生人数进行了限制：“原则上每人每年招生的人数不超过 3 人；同时兼招博士生和硕士生的导师招生人数一般不超过 5 人”。这将为保证研究生的培养质量起到积极的作用。

导师的知识构成和学术水平直接影响着研究生创新能力的培养。高水平的导师能够站在学科前沿，预见学科发展的未来，具有创新性的思维方法、工作经验，治学严谨，对研究生要求严格。研究生对这样的导师也十分敬仰、爱戴，把导师看成自己心中的楷模。这表明导师对研究生的影响在其培养与成长过程中起着关键性的作用。为了能使新遴选的导师尽快进入角色，这学期我们要试行新导师上岗培训制度。一方面，通过培训使新导师了解、熟悉研究生管理的规章制度；另一方面，邀请学术造诣高，培养、指导研究生富有成功经验的导师现身说法，使新导师尽快进入指导研究生工作的角色，使他们身为人师，发挥传道、受业、解惑的作用。对个别不负责任的导师，除了用规章制度来约束外，还要建立对导师的评估机制，做到优胜劣汰。

（2）严把招生关，确保生源质量。优秀的生源是提高研究生培养质量的重要保证。近年来，学校的扩招也使报考研究生的人数急剧增加，在这种背景之下，如何保证生源质量，客观、科学地测评生源水平，如何把研究生的入学考试成绩与优秀人才的选

拔结合起来，将素质好、有创造潜能的学生选拔进来，是亟待解决的问题。为了适应社会对高层次人才的需求，教育部调整了全国硕士研究生入学考试科目。从2003年开始，将硕士研究生入学考试中的初试科目由5门改为4门（政治理论课、外国语、基础课和专业基础课），将专业课调整到复试中去进行，专业课的考试形式和内容自定。这种做法体现了复试是进一步考查考生的综合素质和能力的重要功能。我们要认真研究和加强复试工作，积极探索和完善重在考查学生的素质和综合能力的复试形式和办法，逐步提高复试比例，加大复试权重，制定和完善复试标准，形成一套在研究生录取过程中从初试到复试的科学、规范的质量把关体系，特别是严格执行复试程序，进一步规范管理，以提高复试的公平性、公正性和有效性，严防招生中的失范行为。

（3）科学的研究生培养方案的制订，是提高研究生培养质量的前提保证。培养方案是研究生培养过程的指导性文件。由此确定研究生的培养目标和研究方向，明确研究生培养的过程和环节，其科学与否直接关系到高层次人才培养质量。研究生培养方案的修订和实施是一项复杂的系统工程，需要调动培养单位的领导、导师、研究生秘书等的积极性，使其齐心协力、集思广益，根据不同学科类别和专业的特点，跟踪学科前沿，拓宽研究生知识面，在充分讨论的基础上制订出切合本学科专业实际的培养方案。

新修订的培养方案，要力争从研究生教育规律出发，坚持科学性、先进性、可行性与创造性相结合的原则，进一步明确培养目标，认真确定研究方向，拓宽培养口径，注意研究生能力的培养，在学制、学习量、学分要求、必修课课程结构等方面进行修订和调整，使研究生的培养立足于较高的起点和学科发展的前沿。在课程设置上，强调知识的系统性、前沿性和研究方法的科学性、创新性，进一步创造条件为研究生开设一系列拓宽知识面和应用性的选修课程，并加大对课程检查评估的力度。

对于研究生在学期间发表一定数量的学术论文问题，我们应给予客观的评价和思考。一方面，这是研究生的研究成果得到社会公开检验的体现，也是提高研究生培养质量的一个重要环节，并可为提升学校的实力和排名做出一定贡献。另一方面，这一规定应制定得合理可行，使研究生通过努力既能达到目标，又不至于难以实现而出现学术失范行为。因此，要本着实事求是的精神，制定出切实可行的规定，对不同层次、不同类型、不同专业的研究生要分别考虑，不搞一刀切。对超过规定在权威刊物上发表论文，或被三大索引收录的，要给予适当的奖励。

（4）重视课程建设，改进教学方法。课程教学对研究生基础理论和专业理论水平的提高具有重要的意义，因此，为保证和不断提高研究生培养质量，对课程教学这一重要环节必须加强。不仅需要在教学秩序等方面进行严格管理，更应强调课程内容的拓展和更新。鼓励教师根据本学科的特点，紧密结合国际学术前沿和科研实践，推进研究生的课程建设，在充分调研和论证的基础上提出一个更合理的设置方案。学科组应对课程设置方案的针对性、先进性、课时等严格审查把关。同时鼓励高水平的教师编写研究生课程教材，力争入选全国研究生优秀教材。教学过程着重围绕培养研究生的科研能力、创新能力和解决问题的能力来进行，教师采取形式多样的授课方式，如讨论式、问题式、参与式等来组织教学活动，并注意加强对现代化教学手段的引进。采用这些有效的教学方式和手段不仅要让研究生理解和掌握现有的理论知识，更要引导他们懂得这些知识是如何获得的，使他们努力发现新知识。在教学中不仅要让研究生学会分析问题，更要让他们对所研究的问题进行综合归纳和比较，在学会发现新问题、寻找解决问题的突破口的同时，演绎出新的知识和结论。

（5）严把学位授予质量关。学位授予质量是衡量一个学校研究生培养质量的重要标志。研究生撰写学位论文的主要目的是通过科研工作实践，提高自身独立从事科研工作的能力，这一过程包括在导师（导师组）的指导下，查阅文献资料，开展调查研究，了解课题的历史现状及前沿动态，进行深入思考，精心设计实验（试验），反复讨论交流，对主要矛盾进行分析解决，破解关键问题，最后总结成文，并通过答辩。显然，这是一个相当全面的训练，综合体现了对研究生培养能力的要求。学位论文应该是研究生科研能力和学术水平的集中体现。要完善对学位论文的监控措施，做好论文的中期检查、评阅和答辩工作，逐步加大双盲评审的力度。在学位授予过程中，充分发挥导师、学位点、学科组、学位评定分委员会及校学位评定委员会的把关作用，从各个环节上加强管理和监控，把好研究生培养质量的最后一关，确保学位授予的质量。

同志们，我校这次研究生教育工作会议的召开，为大家提供了一次总结审视过去、设计规划未来的交流平台，这对于我校研究生教育今后的发展意义重大。我希望在座的各位珍惜机会，开阔视野、启迪思维、畅所欲言、认真讨论。全校上下要抓住机遇、为大力发展我校的研究生教育，将我校建成高水平研究型大学而共同努力！我相信，在大家的共同努力下，2004 年暨南大学研究生教育工作会议一定能够取得圆满成功！

50. 贺暨南大学澳门校友会会长就职[①]

值此老友新朋相聚的美好时刻，我谨代表暨南大学 3 万余名师生员工向澳门校友会 2004～2006 年度新任会长、理事和监事表示诚挚的祝贺，向暨南大学澳门校友会的各位校友致以亲切的问候和美好的祝愿。

创办于 1906 年的暨南大学，在近一个世纪的风雨历程中，逐步建成具有“侨”字特色和自身优势的一流大学。如今，学校已桃李芬芳天下，暨大的校友遍及祖国的大江南北和世界各地。尤其是澳门的校友，活跃在澳门的各行各业，已成为澳门社会的重要力量，为澳门的回归和繁荣稳定作出了积极贡献。学校为你们的业绩深感骄傲和自豪。校友们身在澳门，但心系母校，多年来你们通过不同的方式对母校给予捐助、支持，体现了大家对母校的深情。在此，我代表学校对你们表示诚挚的谢意。

暨南大学作为国家“211 工程”重点建设的大学，正向新的征途奋进。学校确立了把暨南大学建设成高水平的研究型大学的目标，任重而道远。这更需要我们广大校友的鼎力支持和社会各界的关心。希望校友们更多关心母校，为母校发展献计献策。

让我们所有“暨南人”更加紧密地团结起来，心向母校，迎着新世纪的曙光，共建更加美好的锦绣校园，共铸暨南大学新的辉煌。让我们用实际行动和丰硕的成果迎接 2006 年百年校庆的到来。祝澳门校友会在新任会长及理事、监事的带领下更有作为，继续为澳门的繁荣发展作出新的贡献！

① 本文是在暨南大学澳门校友会 2004—2006 年度会长、理监事就职典礼上的讲话，澳门，2004 年 4 月 2 日；一个大学校长的探索，北京：高等教育出版社，2011，211.

51. 忍耐是一个人成功的秘诀——与暨南学子谈成才[①]

今天，校团委邀请我来谈如何成才这个问题，这是一个老生常谈而又历久弥新的话题。在这里，我想结合自身的成长经历谈一点切身体会，希望对同学们能有所教益。

人类在地球上已经存在了几万年，但真正成为有思想的人还是近几千年的事。我们中华民族具有5000年的悠久历史和灿烂文化，经历了农业经济、工业经济和知识经济时代。时代的划分与科技发展紧密相连，工业革命时期蒸汽机的出现，加快了人类发展的步伐。20世纪是电气时代，21世纪是信息时代，由于邓小平理论为我们指明了一条光明大道，从1978年党的十一届三中全会开始，中华民族进入了一个伟大的时代。同学们，你们就正处于这个伟大的民族复兴时代，具备一个人成功所需的天时、地利、人和的大好环境。“天时”，指我国已进入全面建设小康社会阶段，到2050年要达到中等发达国家水平；“地利”，指我们地处中国最发达的三大地区之一的珠三角中心城市——广州；同时我们有极好的“人和”因素，暨南大学的各项事业处于全面上升态势。我出生在抗战烽火四起的成都郊区，那是一个物质十分匮乏的年代，到我上大学的时候条件还非常艰苦，与当今的大学生活是无法同日而语的。每个人都梦想着成功，都渴望实现自身的人生价值。古往今来，无数杰出人士为我们留下了许多成功的范例。在我看来，一个人要成才是要具备诸多条件的。

一、志存高远：树立远大的理想

具体说来，成才的首要条件就是要树立远大的理想。一个人如果没有理想，绝对不会成为有用之才。理想是一种激励我们前行的动力，一个人在人生的每个阶段都应有对理想的正确定位，而且要适度，定位太高就有点好高骛远，所以必须要确立经过努力可以达到的理想目标。我生活在艰苦的抗战时期，当时大半个中国已经沦陷，我的家乡虽未被日军占领，但也被日军的轰炸机不停轰炸，当时做教师的父亲要上前线抗战，家里没有经济来源，只能靠变卖东西维生，所以当时幼小的我就深刻体会到了生活的艰辛和落后就要挨打的严酷现实。由于我出生在书香门第，家里的长辈都是世代教书，所以家庭的环境从小就灌输给我要有“精忠报国”的思想，这在我幼小的心灵里埋下了长大后立志报国的种子。在新中国的红旗下长大的我更加深切体会到只有科技才能救国的真理，立志要做一名科学工作者。正是树立了这个理想，支撑我不畏艰辛，走上攀登科学高峰的道路。我从18岁开始做科研工作，大一时参加当时中国第一颗卫星——东方红一号卫星的试制工作，这对我的一生都有着十分重大的影响。我认为只要有理想，通过努力，自然会找到合适的道路。我如今依然在搞科研，最新的论文是有关复合材料壳体振动的问题，我做了几十年的科研工作，仍乐在其中，觉得这是对国家、对民族都有贡献的事业。现在国家提倡“科教兴国”，更加证明了我所做的工作、当初所确立的理想是正确的。

---

① 本文是在百年暨南讲堂首讲上的讲话，广州，2004年4月15日；暨南青年，第47期，2004年7月8日.

二、读万卷书，行万里路

要成才就要有浓厚的兴趣，有爱好，要勤于读书。我从小的兴趣就是看书，渴望从书本中吸收丰富的营养。从三岁就开始识字，七八岁开始读小说。我十分喜爱读小说，中学前就读完了所有的中国名著，之后又阅读了许多外国名著，包括法、美、英、俄乃至像拉脱维亚这样的小国家出版的小说。通过阅读小说，我了解世界各国的生活习俗、先进的科学和做人做事的方法等。再大一点时我就读名言警句。我 18 岁加入中国共产党，开始看关于党史的资料，也去图书馆看诸如《新观察》《旅行》类的新闻、地理杂志。同时，还读经济杂志。现在做了校长事务繁杂，仍坚持博览群书，获取多方面的信息，比如我最近在看《康熙大帝》，看现在你们年轻人都喜欢看的《往事并不如烟》，还有一些传记和回忆录。同时，每天还要看十多份报纸。“读万卷书”，就是要在年轻的时候读大量的书，从中获取大量信息，日后能在自己所从事的领域里助自己一臂之力，获取更多有益的知识；“行万里路”，就是要在不断实践的过程中，做到多思善谋，见多识广，视野开阔。读书让我有了比一般人更多的处世智慧，我会用不同的方法来解决一个问题，同时可以利用看到的知识旁征博引，举一反三。所以，我认为爱读书，有个人的兴趣爱好，并勤于思索，是成才的必经之路。

三、勤学好问

孔子云：“学而不思则罔，思而不学则殆。”科学工作就是不断发现问题和解决问题，自古以来，只有不断发问并勤于耕耘的人才能攀上科学高峰。一个大家很熟悉的例子就是牛顿通过苹果坠地发现了万有引力定律，他善于对身边每个人都习以为常的事情质疑，这样才有了为自然科学奠基的定律。我本人就是一个好问问题的人，经常会打破砂锅问到底，问到别人觉得“理屈词穷”，到了各地都喜欢发问，所以很容易掌握一个地方的情况。一个有所成就的人，绝对是一个爱问问题、爱好思考的人，只有具有打破砂锅问到底的探索精神，才能找到开启科学大门的钥匙。

四、要惜时如金

“时间就是生命”，这个道理大家都明白，但其实大多数人面对宝贵的时间是熟视无睹的。光阴似箭，“逝者如斯夫！不舍昼夜。”等我们白了少年头，才后悔当初没有好好珍惜时间，然后就教导自己的后代要珍惜时间，这样一代又一代下去，却没有什么成就。拿我所处的那个时代来说，十年的“文化大革命”，视知识如粪土，认为“知识越多越反动”，不准看书，否则会被当成反革命，那时候只有《毛泽东选集》可供阅读，于是我就买英文、德文、俄文版本来学习外语。当时我所在的兰州大学的一些知识分子只能靠做家具来打发时间，我周围差不多 99％的人都没做学问，而我就关起门来搞科研，做了好几个有意义的课题，使自己的科研能力有了很大提高，我成为兰州大学凤毛麟角的搞科研的人，校军宣队还破例为我出版了个人的科研论文专集。所以我要奉劝同学们，一定要好好珍惜青年时代的大好时光，做有意义的事，争分夺秒，只争朝夕，为日后成功打下坚实基础。

五、忍耐是成功的秘诀

我的个人格言就是“忍耐是成功的秘诀”。我之所以能成功，除了前面说的几个因素外，还有一个很核心的问题就是我很会忍耐。在我所处的那个时代，我感到了个人力量的渺小，因此这也磨砺出了我能忍耐、甘于寂寞的品质。还有就是做事要执着和富有韧性，这是我的特点，我会坚持自己认为对的观点，哪怕是和“强权”人物抗衡我也从不畏惧。同时还要勤奋，天下没有任何事是可以轻轻松松就有所成就的。所以，忍耐、执着、勤奋都是成功必备的品质。许多科学家都认为，一个人的成功，99%靠勤奋，1%靠天赋，我不是一个非常聪明的人，所以我做事靠的就是勤奋二字。我可以和大家分享一则关于我成名的故事。

我当初读高中时，是四川留苏预备班的班长，但因为政治背景的原因，无法去苏联深造。学校要求我只能参加高考，当时距离高考仅剩几天的时间，我靠着平时的勤奋积累，以全校第一名的成绩考入了兰州大学数学系。毕业之后，我很想到科研单位工作，便向组织申请，但作为学生会主席的我，深得老师的信任，组织上让我不要有个人主义，要求我留校任教，于是我又服从组织安排，认真从事教学工作。但是，我仍然十分向往科研工作。我找到一个当时只有三个国家在做的课题，这也就是我的成名作，关于波纹圆板特征关系式的问题。但当时向党支部汇报这个想法时，却遭到了强烈反对。于是我只能私底下偷偷做，但这个研究的过程是非常崎岖、坎坷和艰难的。

从 1964 年到 1968 年，我用了整整四年的时间研究波纹圆板这个课题，这四年的探索过程是相当艰苦的。我要偷偷看业务书，要做高达 5 位数的繁冗计算，要反复审核结果，不过当成果最后终于出来时，我十分高兴，我的成果要比当时美国、苏联研究的还要好。我帮我国工厂改变了依赖经验和外国公式设计产品的历史，创造性地提出了精密仪器仪表的心脏——弹性元件设计公式，被工程应用，这是我的第一个科研成果。之后我又从事尿素合成塔之类的化工机械产品的开发研制，提出了厚板壳弯曲理论用于高压换热器超高压容器试制和大型储油罐新型网格顶盖、大型减压塔、铁路高桥墩、新型钻头的设计依据，被工程应用。关于波纹圆板的研究的文章却是我的得意之作，我很想尽快公之于世，但又受到“造反派”的压制，说该论文专搞理论，轻实践，不准发表。英雄无用武之地，我的理想就此搁浅。直到 1972 年，全国学术刊物复办，并向各高校征集稿件，我的关于高压管板的文章才在《数学的认识与实践》上发表，然而原本答应我发表关于波纹圆板的文章的《力学》却未能如期发表。事后才知道在四年间，该杂志先后发出六封信，要求我所在的系出一份关于政治审查的证明都被“造反派”拒绝了。直到 1978 年，《力学学报》终于发表了我的论文，随之在国内引起了轰动。在当年上海召开的学术会议特邀我去做学术报告，我是这个研究领域里全世界撰写论文最多的作者，我的论文也被世界著名力学家称为“代表当代国际板壳理论领域科学工作现状的最高水平。”

从 1964 年到 1978 年，经历了长达十五年的时间，我的理想才实现。所以通过这个故事，我想让大家明白“功夫不负有心人”的道理，这是我的切身体会。同学们，今天你们所处的时代与过去已无法相比，国家鼓励科学研究，没有人再压制你。我们那个时代，是谁搞科学研究谁就倒霉的时代，我为了科研工作当了“牛鬼蛇神”，被关进了“牛棚”，那是非常可怜的事啊！所以我希望同学们要珍惜时间，在最好的青春年华，在创新能力最强的时候，自强不息，奋力拼搏，任何时候都不要气馁，要勇往直前，越挫越勇。

六、做好准备，抓住机遇

今天我还想奉献给大家的，就是一个人要想取得成就，要想成才，就要在机会还没到来前做好准备，不要等机会到来时，才临时去抱佛脚，那样机会也往往错过了。一个普通的人，一生的机会是不多的、短暂的，当机会来临时，如果你没有能力和条件抓住它，机遇就会擦肩而过，转瞬即逝；而条件已准备好，你如果能抓住机遇，就成功了。机会永远只属于有准备的人。就像现在毕业求职一样，你们准备好了吗？你只有具备了足够的思想与业务条件、身体条件、心理条件，做好了各种应试准备，才能找到理想的工作。

我的一生，总是提前做好准备，才抓住了很多机遇。如果我的基础没打好，初中毕业那年就垮了。初中毕业，学校要保送我读高中，结果没有保送成，于是临时叫我去考试，我以优秀的成绩考入高中，还考进了全省留苏班。高中毕业，又没保送成，又要临时考大学。如果没有良好的基础、足够的准备，临时考大学那是多么困难的事情！进入大学，我又抓住机会，参加了人造卫星研制小组，尽管后来被迫离开了，但是我能靠坚实的基础抓住每一次机会。比如学习外语，我是学俄文出身的，没学过英语，可我觉得做科学研究仅仅掌握俄文是不行的，还要学好英文。但那个时候不允许学英文，不允许用英语发表文章，于是我就在大四时，自己选修英语，正好碰到吴青（冰心女儿）老师任教，她现在是北京外国语大学教授。其实我一生的英语基础就学了七十二个小时，但好老师为我打下了坚实的基础。找不到好的英文书籍，我就看英文版的《毛泽东选集》来学英语。毕业留校任教后，我又夯实我的外语基础，利用业余时间上夜校学德语，学了四十个小时。1978 年，我被调到中国科技大学，那是我一生中的黄金时期，我被称为当时中科大的四颗科技明星之一。当时中国要走向世界，邓小平同志决定把中国的科学家派到国外去学习深造，正好世界上最著名的几个基金会要吸收中国的科学家去工作，中科大当时也在挑选优秀教师赴德国。我刚到中科大，也不敢报名，我知道我从来都是在政治上吃亏的，政审总不合格。可系主任鼓励我，因德国非常重视科研业务，在业务上我是全系最好的，就把我报上去了。当时学校反复讨论，从四十多人中选拔出六个人，把我们六个人的材料送到德国，同时还要上交英文论文。我仅有一点英语基础，论文翻译起来很辛苦。材料送到德国，中科大只有我一个人通过。中国当时选派了八名学者，是第一批经德国审查的洪堡学者，那是非常光荣的事情。从此我走上了世界的学术舞台。在德国做研究工作的第一个月，我就把我的文章写成了英文，在国际排名第一的《国际非线性力学杂志》上发表。风华正茂的同学们，从这些例子来看，我所抓住的每一次机遇都有着充足的准备。你们一定要有远大的理想和抱负，要提前做好准备，如果没条件，没准备好，即使机会摆在你眼前，也是没用的。

七、名师出高徒

同学们，读书，要读名著；找老师，要找好老师，找名师。名师是成才关键。特别在今天这个时代，科学技术非常发达，分工非常精细，如果你的老师没有什么名气，你很难有大成就。大家都知道我国三位著名的科学家——“中国三钱”。钱学森先生追随的导师就是冯·卡门，冯·卡门是20世纪最伟大的科学家之一。钱学森的博士论文，被称为划时代的论文，他的研究成果是在老师指导下完成的；钱伟长先生出国后先师从世界应用数学大师，博士毕业后又到冯·卡门身边担任助手。他们都得到冯·卡门的学术真传。所以中国的力学和应用数学，到今天仍走在世界的最前沿。钱三强先生，他的导师是居里夫人的女儿和女婿。他们带着世界领先的科学技术回到祖国，极大地繁荣了中国自然科学。可见，真是名师出高徒！同学们，没有老师就没有我们自己，我们就很难到达科学的高峰。每一个科学家都是站在前人的肩膀上，攀上的学术高峰。所以我们一定要尊敬老师，按照老师的教导去做，特别是研究生同学，你要找到名师指引，才能学习到当前的热门领域，才能在前沿的科学领域里遨游。

我非常有幸，遇到了几位好老师。其中，我的老师——叶开沅教授，他师承钱伟长先生。我的第一个研究课题为什么能同钱学森、钱伟长的课题一样？为什么能走到世界力学的前沿呢？我就是在叶老师的带领下，创立了求解非线性微分方程的修正迭代法，才系统创造性地研究了波纹板壳、夹层板壳、复合材料板壳、网格扁壳、单层板壳、双金属扁壳六类非线性弯曲、稳定和振动问题，后来我又在钱伟长先生身边工作了六年，跟他学业务，做科学研究，对我的帮助非常大。在德国做洪堡学者时，我又找到了名师，那就是策纳（Zerna），他是欧洲第一号力学权威，德国的核电站全部都是他最后签字审查通过的。他欢迎我的到来，跟随着这位名师，我见了很多世面，进入了世界力学最前沿的研究领域，还第一次受到了德国总统的接见。所以说，同学们，有机会一定要找到好老师，要找到名师带你上路，这是成才的捷径。

八、强健体魄和过硬心理素质

最后一个条件，要具备强健的身体和过硬的心理素质。做科学研究，是非常艰苦的，如果身体不好又没有好的心理素质，那肯定完不成研究任务。所以要成才，就要锻炼好身体，培养好心理素质。从我年轻时到我成名，没有沾到科学研究的一点经济效益。如果科学家是为了追求金钱，那他是不会成功的。“文化大革命”时我搞科研主要是因为兴趣，而不是为了多挣一点钱，反而还要自己掏钱出来，倒贴腰包买纸张计算！那时候我们经济很困难，家里很穷，我十五年没升过工资，还养两个孩子，每个月微薄的工资都入不敷出，经常靠借钱过日子。当时老师也没有公文纸、笔墨什么的可以报销，都要自己准备。为了计算，光草稿纸我就用了几麻袋，但就是自己贴钱搞科研我还要挨批判，被当成“牛鬼蛇神”打入“牛棚”。一个“左派”批判我，说我的那些用尽心血写出来的科研论文一文不值，连卫生纸都不如。他用最恶毒的话攻击我，侮辱我，我听了非常寒心。其实我图的是什么？我做科学研究图的是什么？我没有低头，不怕别人的侮辱和攻击，我不为成名，我不为金钱，父辈教导我要精忠报国，是国家和人民需要我这样去做，就是这种精神支撑我继续坚持研究下去，就是凭着对科研工作的热爱和坚强的毅力，我才走向成功。同学们，要笑对人生，要把黑夜当成白天，要把雨天当成晴天，要把冬天当成春天！

过去我们国家积贫积弱，太落后了，中国科技不进步将会永远挨打。同学们，21世纪是中华民族伟大复兴的时代，也是中华民族扬眉吐气的时代，你们有这么好的条件，更应该及早做好准备，学习好，锻炼好，努力创造成才条件，为国家、为人民、为我们伟大的民族作出更多有意义的事情，无愧于时代的重托。

祝同学们早日成才！

52. 标准学分制的研究与实践[①]

暨南大学将“侨校＋名校”作为发展定位。根据这一定位，形成了一整套适应自身办学特点的教学体系和管理制度，标准学分制就是其中一个主导性的教学管理制度。下面就其做一些介绍。

一、标准学分制的含义、缘起及实施意义

1. 标准学分制的含义

学分制有多种模式，“标准学分制”是暨南大学推行的学分制模式。这种模式把不同学制学生取得学位必须修满的总学分按学期均分（注：我校有四年、五年、六年三种本科学制）。以四年制本科为例，即把学生取得学位必须修满的 160 学分按 8 个学期均分，每个学期的标准学分为 20 学分，每个学年为 40 学分。在保证每学年 40 标准学分不变的情况下，同时规定学生每学期必须修习的学分的最低标准；如学习成绩优秀，则有相应的免费奖励修读规定。这样既防止了学生过分少修学分，保证了教学资源的有效利用，又能控制学生过分多修学分，保证学习质量，并可使学生能根据自己的情况提前或延迟毕业，从而实现真正意义上的弹性学分制。

2. 标准学分制的缘起及实施意义

作为“面向海外、面向港澳台”办学的暨南大学，外招学生已达在校学生的一半。由于外招的学生来源广（来自 50 多个国家和地区），受教育的背景不一，要满足他们的学习要求，需要采用灵活的教学管理制度。暨南大学从 1978 年开始试点学分制，1983 年全面试行学分制；1985 年试行主、副修制和双学位（双专业）制。从 1993 年开始对学分制实行较大的改革，采用标准学分制，并取消补考，实行重修制等。2000 年 8 月，我校“高等学校学分制管理制度的改革与实践”项目被批准为世界银行贷款资助项目，根据项目的研究目标，我们对学分制的一些核心问题做了进一步的深入调查研究，出台了多项新的措施，使标准学分制得以进一步完善。

二、完善标准学分制的主要措施

1. 进一步完善学籍管理规定，推出校长免费学分奖励金评选办法

我校推出的校长免费学分奖励金评选办法规定，学习成绩好的学生不仅每个学期可超过标准学分多修课程，而且实行免费多修，暂定：一个学年结束时修够 38 个总学分，同时平均学分绩点在 3.0 以上的学生就有资格进入奖励选拔范围。学校将在各个专业中按学分绩点排序，选取前 15％的学生（国家基地班和名牌专业为 18％）享受这个奖励金。其中成绩在本专业列第一名的学生奖励 10 个免费学分并享受特等奖励金；列本专业前 3％的学生奖励 8 个免费学分并享受一等奖励金；列 4％～10％的学生奖励 4 个免费学分并享受二等奖励金；列 11％～18％的学生奖励 2 个免费学分并享受三等奖励金。

---

① 本文原载《中国大学教学》，2004，(3)：41-43.

2. 根据学分制的内在要求，抓好选课工作，打造多功能网络技术平台

1）抓好选课改革建设工作，提供充足的高质量选修课

学分制以选课制为基础，学生选课自由度的大小是衡量学分制完善程度的关键性指标，因此，我校近年进一步加大了选课工作的改革建设力度，要求不仅要做到有足够数量的高质量选修课可供学生选择；同时要使选课的过程没有技术障碍。目前，全校同年级选修课如按学制年限教学计划计，可提供的选修课程逾 1600 门。

为提高选修课（包括晚选课）质量，学校制定了《公共选修课暂行管理办法》，建立了公共选修课专家审查制度。同时，对全校的公共选修课实行优存劣汰制，每个学期评估成绩排在最后的 3～5 门课程将被淘汰；同时全校所有公选课每 3 年重新评估一次，不合格的予以取消。

2）完善网上选课系统，提供现代选课手段

为顺利实现个性化的学分制教学计划，我校在 2000 年 12 月研制开通了网上选课系统，为完善学分制提供了不可缺少的技术支持条件。经过多个学期的选课实践和不断完善，目前系统运行稳定，速度快，选课效率高，学生可在我校联网的任何一台计算机上选课，不受时空的限制。

3）开发网上多功能系统，构建先进的学分制管理运行技术平台

为改进学分制管理手段，构建更为优越的教学管理技术平台，我校进一步改善了排课系统，以方便学生选课：一是实行全天候排课（中午、晚上及星期六、星期日都开课）；二是尽量使同一学期开设的相同课程在白天和晚上分别开出；三是对实验室和语音室实行全天候开放，并改变原来上午多安排理论课，下午多安排实验课的做法，将理论课和实验课错开安排，使教室资源得到充分利用，学生选课自由度进一步加大。此外，对一些覆盖面较大的课程安排进行调整，凡有两个专业以上的学生选修的课程，上、下学期都开课；在假期继续开设公共必修和专业必修课，给学习成绩突出的学生和重修学生更多选课机会。

在开发网上选课、排课系统的同时，我校还完成了网上排考及查询系统、学籍管理系统、考试成绩录入系统等。网上管理系统的多功能开发，进一步破除了教学管理技术层面的障碍，摆脱了人工管理的落后局面，使我校学分制管理上了一个新台阶。

3. 出台系列教学改革措施，构建学分制条件下人才培养质量保障体系

1）实施“大平台”招生培养和分流教学新模式

从 2003 级开始，我校在经济、管理、外语、华文、新闻与传播等 5 个学院首次实行了按学院招生统一培养的“大平台”新模式。这一措施的实施，使学生“自主选择专业、自主选择教师、自主选择课程、自主安排学习进程”的学分制要求得以进一步全面落实。

（1）实施“大平台”教学的学院公共课学分统一，前 3 个学期或 4 个学期安排全校公共基础必修课和各学院学科基础必修课。

（2）学生一年半或两年后根据已修完的学分总数和学分绩点等选择专业。

（3）学科基础必修课的学分数最低不得少于 30 学分，最高值由各学院决定。

（4）对相关学科、专业、课程体系等进行整合。例如，计算机、大学语文、大学

英语等公共必修课程改变了以往“齐步走”的教学模式，根据学生已有程度，采用必修模块与选修模块相结合的课程体系，实施分级教学新模式。

(5) 内、外招两类学生在“大平台”教学期间采用两套教学计划，实行分流教学，待其选择专业后统一培养。

2) 适应侨校办学和培养国际化人才要求，成立国际学院，采用全英语教学

2001 年 6 月，我校率先在全国高校中成立了用英文原版教材、用全英语教学的国际学院。国际学院已设 5 个专业，全英语授课教师在全校选拔，部分聘请外籍和校外专家。几年来国际学院招生形势很好，教师的教学积极性和学生的学习积极性都非常高，国际学院学生的英语能力明显高于其他专业。

3) 为加强对学生选课和学习方法的指导，推出本科生导师制

我校从 2003 级秋季新生开始全面实施导师制。导师的职责是对学生“导向”“导学”“导心”，以“导学”为主。要求导师帮助学生了解我校学分制教学管理制度的具体规定以及相关专业教学计划的要求，根据内、外招两类学生的不同特点、基础、特长和兴趣，对学生进行选课和学习方法的指导，审批学生的选课单，引导学生处理好学习的质和量的关系，以避免学生随意选课和凑学分等不利于学习的行为发生。导师制规定导师要通过与学生的接触交流了解学生的所思所想，排解学生的心理障碍，培养学生乐观向上的心理品质和自立、自强的人格。

导师工作要纳入年度考核，考核结果折算成教学工作量计入校内工资，并记入本人业务档案，作为晋升职称和有关奖励的条件之一。

4) 组织大型考场，严格考试管理

2002 年 7 月，我校对期末考试的组织工作进行了重大改革，在学校体育馆设立大型考场。考场可同时容纳 840 名学生考试。不同专业、不同课程的学生混编，前后左右保证是不同试卷和课程，监考教师也实现统一安排，有效地杜绝了考试作弊现象。自实行大考场考试制度后，学生作弊现象基本杜绝。大型考场的设立开了全国高校考试管理改革的先河，对整肃考试纪律，提供公正、公平的考试环境，保证试卷质量、考试质量，促进学风的转变等都起到重要作用，并得到全校师生一致肯定。

5) 完善课堂教学质量“三重评估”制度

我校从 1993 年开始抓课堂教学质量“三重评估”工作（学生、专家、领导评估）。近年，在修订评估指标体系的同时，改进了专家和领导听课办法。例如，改专家自由、独立听课为小组有针对性地听课，然后进行集体评议，使总的评估结果更公正合理；同时，改革过去反馈听课意见不及时的做法，做到专家和领导每听完一堂课，即时把书面或口头意见转达给讲课的老师，有效地促进了课堂教学质量的提高。

6) 落实教授上本科基础课规定

我校从 1993 年开始抓教授上本科基础课，近年进一步在考核等方面强化这项工作。目前，教授上本科课程的比例已达 90%以上。

7) 增开系列学术讲座

为活跃本科学术气氛，近年我校在校院两级都设立了名师讲座，开设名师系列课程。现在，一个学期的大小学术讲座全校近 200 场。

4. 深化人事、分配等制度改革，进一步完善标准学分制相关配套措施

1）人事、分配制度改革

近3年来，我校教师和科研人员工作量考核管理办法不断得到改进。例如，2001年，对全部用英语授课的教师，按普通授课的3倍计算工作量，极大地调动了教师用英语授课和选用全英语教材的积极性。

对各系列职称评定条例进行了修改，新的评审条件更加注重对教师教学业绩和教学研究的探讨。例如，晋升教学系列的高级职称必须有教研论文，课堂教学评估不及格者一票否决等。

修订《暨南大学教师进修管理条例》，加强了教师，特别是青年教师的岗前培训。对教学科研骨干，则加强了到国内外其他高校的进修培养。同时，进一步完善了有关激励机制。

2）学生管理工作改革

我校根据标准学分制下学生工作出现的新特点，逐渐打破以前学生按系、专业安排宿舍的做法，按照标准学分制要求，对不同专业的学生混合安排住宿，并对学生实行社区化管理，学生根据本社区情况积极开展课外活动。

加强学生德育工作，积极开展第二课堂教学。2000年8月，成立了暨南大学学生工作指导委员会，采取班主任与学生工作秘书相配合的办法，采用多种措施，切实加强对全校学生德育工作的指导。

建立健全学生奖学金和助学贷款制度。目前我校共有优异学生奖学金、优秀学生奖学金等10余种奖学金，特别是2003年推出的《暨南大学校长奖励金管理办法》，其规定的受奖面广、奖金额度高，设奖较以前更合理，极大地提高了学生学习积极性。为帮助我校经济困难的学生完成学业，根据财政部、教育部有关文件精神，制定了《暨南大学国家助学贷款实施办法》。2003级新生入学报到时，特别开辟“绿色通道”，让那些经济困难的学生可暂缓交费，及时办理入学手续。

3）后勤社会化和图书馆管理等改革

为配合标准学分制运作，以学生为中心，提高服务质量，学校积极推进后勤社会化改革。2000年，我校成立了“暨南大学后勤集团”，设饮食服务中心、生活服务中心、学生宿舍管理中心等。后勤服务进一步改善，餐厅基本上做到全天供应。学生能随时吃上热饭热菜，图书馆自早上8时至晚上10时开放，自修室自早上7时30分至晚上12时开放。

4）收费制度改革

在收费方面，拟改革原来按学年收费的办法，目前已制定出按学分收费的管理条例。

三、实施标准学分制的几点体会

1. 学分制要与时俱进，不断完善和创新

从实践过程看，像我校这样在生源复杂的情况下实施学分制在全国高校是不多见的。我校学生包括中国港澳台和内地学生、华侨、华人以及外国留学生几大类，要保证各类学生顺利完成学业，达到较高的质量，更好地适应新世纪的要求，除需要按照

学分制的基本要求制订方案、政策外，更重要的是还要根据本校的特点和情况变化对学分制不断加以创新完善，这样，学分制才能充分发挥其应有的作用。自 1993 年推行标准学分制以来，特别是通过近几年的努力，我校已建立起了比较完整、具有侨校特色的学分制管理体系。

我校外招学生数量较多，因此，学分制的每一项改革措施都会引起海外及港澳台较广泛的关注。例如，2003 年我校“大平台”招生措施出台后，报考我校 5 个“大平台”招生学院的港澳学生就激增了近 3 倍，创外招生数历年的最高纪录，而且外招生数自学校 1978 年复办以来首次超过了内招生数。

2. 实施标准学分制必须严格执行“标准”

标准学分制规定了每个学期和每个学年学生应当修习的学分数，这绝不是像有人说的在搞“计划经济”，而是一种符合学生学习实际和学校办学的客观实际的制度规定。实践证明，推行学分制如果不规定每个学期和每个学年修读的学分标准数，或者规定了而不严格执行，对学生各学期和各学年修读的学分不加任何限制，就必然会造成教学管理上的混乱。最突出的问题是大量学生积累大量的重修课程学分，无法按正常教学要求修读后续的课程学分。

3. 实施学分制必须完善相关的配套措施

实施学分制是一项系统工程，不仅与上级的政策和社会对人才的需求情况等相关，而且对牵涉的教学设施和教学资源管理服务部门的保障要求也要相应地提高。例如，需要校内图书馆、实验室、微机房、食堂等各个部门和单位的配合，以延长开放时间，更灵活周到地服务师生。

总之，2003 年“大平台”招生措施的推出，标志着我校的标准学分制发展进入了一个新阶段。在此基础上，将推出“菜单式”学位课程“模块”教学计划，届时，标准学分制定会推进到一个更高的层面。

53. 积极服务海外华侨华人社会①

久闻欧华联会在促进欧洲华侨华人社团的团结协作、促进欧洲各国与中国的经贸合作和文化交流、反对“台独”、推动祖国和平统一大业方面始终进行着积极的努力，并取得了显著的成绩，我感到由衷的敬佩！今天，能够受邀参加第十二届欧华联会，我感到非常的高兴！

21 世纪是知识经济的时代，华侨华人作为各居住国的一个有机组成部分，在经济全球化的浪潮中，同样无法规避知识经济带来的巨大挑战。华侨华人只有培养优良的综合素质，提高自身的竞争力，很好地融入知识经济发展的潮流，才能谋求更好的发展空间。培养高素质人才，已成为华侨华人在 21 世纪成功进步的关键。为此，暨南大学进行了长期不懈的努力。

暨南大学是中国第一所由国家创办的华侨学府，素有“华侨最高学府”之称，由国务院侨务办公室直接领导。暨南大学是中国第一所面向海外及港澳台招收留学生的大学，是目前中国拥有海外及港澳台学生最多的大学，也是学科最齐全的大学，是包括理、工、医、文、史、经、管、法、教育专业的综合性大学。“暨南”二字出自《尚书·禹贡》篇：“东渐于海，西被于流沙，朔南暨，声教讫于四海。”意即向中国南方（华侨华人分布最广的东南亚地区）传播中华文化。学校的前身是 1906 年清政府创立于南京的暨南学堂。1923 年迁往上海。1927 年更名为国立暨南大学。抗日战争期间，迁址福建建阳。1946 年迁回上海。1949 年 8 月合并于复旦大学、上海交通大学等学校。1958 年在广州重建，直属教育部，由时任广东省委书记的陶铸兼任校长。1970 年至 1978 年初，学校因“文化大革命”的影响而被迫停办。1978 年，暨南大学在广州原校址复办。

历经百年风雨洗礼的暨南大学，已是国家面向 21 世纪重点建设的大学，在广东省 90 多所高校中位居第 3 位，在全国 1500 多所高校的综合实力排行榜中已跃居前 50 名左右。学校设有 17 个学院，38 个系，63 个科研机构，52 个本科专业，89 个硕士学位授权学科，34 个博士学位授权学科，6 个博士后站；现有专职教师 1373 人，其中中国工程院院士 2 人，中国科学院院士 2 人，博士生导师 115 人，教授 255 人，副教授 559 人；在校各类学生 28 968 人，其中全日制学生 19 996 人，包括研究生 4372 人，本科生 14 112 人，海外及港澳台学生 7772 人，生源来自世界五大洲的 55 个国家和港澳台 3 个地区。

目前，学校在广州、深圳、珠海三地共有 4 个校区，校园占地总面积 174 万平方米。校舍建筑面积 107 万平方米，图书馆藏书 180 余万册。学校设有 6 所国家级三甲附属医院，即广州华侨医院、深圳市人民医院、珠海市人民医院、广州市红十字会医院、清远市人民医院和江门市五邑中医院，1 所专科医院即深圳市眼科医院，1 所直属医院

① 本文是在欧洲华侨华人社团联合会第十二届年会上的讲话，伯明翰，英国，2004 年 8 月 16 日；一个大学校长的探索，北京：高等教育出版社，2011，375-376.

即深圳华侨城医院。附属医院共有职工 6399 人，病床 4319 张。

本着“宏教泽而系侨情”的办学宗旨，暨南大学始终积极贯彻“面向海外、面向港澳台”的办学方针，恪守“忠信笃敬”之校训，将为“侨”服务作为学校的神圣使命，注重以中华民族优秀的传统文化培养造就人才。自建校至今，暨南大学共培养了来自世界五大洲 95 个国家和港澳台 3 个地区的各类人才 10 余万人。他们从暨南大学毕业后奔赴五湖四海，走向世界各地，在促进海外华侨华人团结、谋求更好的发展空间、促进祖籍国的经济建设和社会进步方面，取得了巨大的成绩；他们在大力弘扬中华民族优秀传统文化、积极推动中西文化的合作与交流、促进人类文明的发展方面，进行了不懈的努力；他们在反对任何形式的分裂祖国的行为和言论、为早日实现国家统一方面，做出了积极的贡献。

同时，学校面向科技发达以及华侨华人分布较多的国家积极开展对外学术和教育交流，与世界五大洲 22 个国家和地区的 60 多所高等院校和文化机构签订了双边协议或建立了学术交流关系，是中国第一所在世界五大洲都建有姊妹学校的大学。暨南大学充分利用中华文化这一促进海外华侨华人团结发展、增强凝聚力的有效动力，在开展对外交流的过程中，大力传播中华文化，促进世界华侨华人社区团之间的了解与协作，促进华侨华人的发展。

当此民族振兴、盛世再现之际，海外华侨华人的进步和发展，不但有利于居住国的经济繁荣和社会进步，而且有利于推进中国的现代化建设，有利于促进中国的统一和民族复兴。我们深信，在世界华侨华人的努力下，在社会各界的大力支持下，暨南大学一定能够为提升海外华侨华人的综合竞争力做出更大的贡献！

今天参加欧华联会，向大家介绍暨南大学，这是你们的学校，希望你们关心这个学校的成长和发展，多多在侨胞中宣传这所学校，并将自己的子女送到我们的学校来读书，我们一定用最好的方式将他们培养成才。

54. 暨南大学百年校庆公告（第一号）[①]

巍巍暨南，百年学府。2006 年 11 月 16 日，素有“华侨最高学府”之称的暨南大学，将迎来百年华诞。在此，我们谨向长期以来关心和支持暨南事业发展的各级领导，社会各界人士，海内外华侨华人、校董、校友致以诚挚的谢意！

声教讫四海，桃李遍五洲。1906 年，暨南学堂在南京创立，承担起了将中华文化传播四海，将华夏精神弘扬世界的伟大使命，开创了我国华侨教育历史之先河。自此，一代代暨南人薪火相传，奋斗不息。从沧海中的一条船，到今天蜚声中外的华侨最高学府，暨南大学的发展历程与祖国的前途和民族的命运休戚相关，荣辱与共。在国务院侨务办公室的直接领导下，1996 年，暨南大学成为全国面向 21 世纪国家重点大学以后，成功实现跨越式发展。近百年的文化传承，暨南大学已是誉传内外，桃李芬芳。迄今为止，学校已接纳过来自世界五大洲 97 个国家和港澳台 3 个地区的莘莘学子。目前，学校共有全日制学生 22 000 人，其中博士和硕士研究生 5008 人，来自世界 57 个国家和港澳台 3 个地区的学生 8966 人。

2006 年的百年校庆，既是暨南大学发展史上的一个重要里程碑，也是学校继往开来、再创辉煌的新机遇。我们将以此为契机，总结经验，发挥优势，为实现“侨校＋名校”发展战略，为创建海内外知名的高水平研究型大学再续新篇！

暨南大学百年校庆的各项筹备工作已正式启动，庆典活动将于 2006 年 11 月 16～18 日举行，我们盛情邀请并热忱欢迎各级领导、海内外校董、校友和各界朋友聚首暨南，共襄盛典。

特此公告

校长　刘人怀　党委书记　蒋述卓

2004 年 11 月 11 日

---

① 本文原载《暨南校友》，2004，4：1.

55. 面向海外　面向港澳台　为祖国统一大业服务①

暨南大学是中国第一所由国家创办的华侨学府，是中国第一所面向海外招收留学生的学校，是中国最早设立医学院的综合性大学，现为国家“211 工程”重点大学。其前身是 1906 年创立于南京的暨南学堂。学校曾先后建址于上海、福建建阳等地，1958 年在广州重建。

近百年来，暨南大学作为国家联系中华民族海内外华侨华人亲情的文化桥梁和纽带，一直致力为海内外华侨华人和港澳台地区培养造就人才。新中国成立以后，尤其是进入改革开放时期，暨南大学在国家侨务工作、建立和巩固爱国统一战线工作中发挥了重要作用。1983 年，中共中央专门下发关于办好暨南大学的文件（中共中央〔1983〕24 号），确定了两个“面向”（面向海外、面向港澳台）的办学方针，要为祖国统一大业服务，并将暨南大学列为“国家重点扶植的大学”，隶属于国务院侨务办公室。

一、改革和发展

为贯彻落实“24 号文件”精神，忠实履行党和国家赋予的特殊办学使命，切实为港澳台同胞和海外华侨华人服务，努力加强对港澳台学生的培养，给港澳台学生创造更好的学习、生活条件，学校按照“发挥优势、深化改革、保证重点、改善条件、提高质量”的办学思想，坚持“严、法、实”（即从严治校、从严治教、从严治学，依法治校和实事求是）的办学原则，以“211 工程”建设为龙头，以教学科研为中心，采取了一系列改革措施。通过改革，学校在许多方面发生了从无到有、从小到大、从弱到强的可喜变化，成功实现了跨越式发展。现遵照教育部的要求，将我校在办学实践中摸索出的一些经验和体会予以总结，以期通过与兄弟院校的交流进一步做好新时期的工作。

（一）改革创新的措施

（1）解放思想，转变观念。根据国内外高等教育的发展形势、学校生源的特点及学校的实际情况，认识到，唯有改革才能发展，唯有创新才能进步。为顺利完成党和国家交给我校的光荣任务，办出特色，我校制定了一个战略目标：侨校＋名校。确立了两个重要原则：学生第一原则，“严、法、实”原则。树立了一个基本观念：不搞科研的教师是“残疾”的教师。办学要走“三化”道路：国际化、现代化和综合化。

（2）坚持以人为本，坚决推行学分制改革。学分制最大的优点在于能给学生提供更主动的学习条件，最大限度地发挥学生自主学习、向上发展的能力，它适应我校生源的世界性特征，即不同水准和不同国家、地区背景，有利于教育目标的实现。在这种制度下，学生可以自主选择教师、自主选择课程、自主选择学习进度、自主选择学习时间。这是由传统的保姆式教育向现代自主教育转变的重要体现。我校在 1978 年复办之初即开始试行学分制，但那时是初级阶段，还处于学年学分制状态。自 1993 年起

① 本文是在教育部港澳台工作座谈会上的发言，北京，2004 年 12 月 11 日；教育与科技管理研究，北京：科学出版社，2016，191-203.

在全国率先施行标准学分制，即学生修满规定的学分数，就可毕业，这样学生可以提前毕业，也可推迟毕业。标准学分制可以说是一种完全从学生利益出发的教学管理制度。我校不仅是广东省最早实行学分制的高校，而且是全国最早实行弹性学分制的高校。经过20余年的实践，学分制的各种管理办法已日臻完善，对贯彻我校两个“面向”的办学方针，改善学风以及培养学生素质均有着积极的促进作用。

（3）合理制定培养目标，科学实践因材施教原则。为配合学分制的实施，我校根据国内外两类学生的不同特点，对学生的培养目标进行定位。对内地学生的培养目标是：将学生培养成为德、智、体等方面全面发展的社会主义事业的建设者和接班人。对香港学生的培养目标是：将学生培养成为热爱祖国，拥护“一国两制”，拥护香港基本法的专业人才。对澳门学生的培养目标是：将学生培养成为热爱祖国，拥护“一国两制”，拥护澳门基本法的专业人才。对台湾学生的培养目标是：将学生培养成为热爱祖国，拥护“一国两制”，反对“台独”的专业人才。对华侨学生的培养目标是：将学生培养成为热爱祖国，维护祖国和平统一大业的专业人才。对华人学生的培养目标是：将学生培养成为热爱中华文化，热爱故乡的专业人才。同时，合理设置专业，突出创新精神与实践能力的培养，使受教育者得到全面发展。我校按学生特点，对港澳台和海外学生的教学要求是“面向世界、应用为主”，对内地学生的教学要求是“加强基础、目标上移”。

（4）保证课堂教学质量，推行“三重评估”制度。为提高课堂教学质量，我校自1985年开始实行课堂教学评估，1993年开始在全国率先实行课堂教学“三重评估”制度，即每学期分别由学生、校院系领导和听课专家组对课堂教学质量进行评估。这一评估制度因具有完整、公平、公正性而受到师生欢迎。学校根据评估结果采取了一系列奖惩结合、以奖为主的措施，激励先进，鞭策落后，对不合格教师进行淘汰，优秀教师则成为学习标兵。这一措施有效促进了师生间教学信息的交流，提高了教师对课堂教学质量的重视程度，课堂教学质量明显提高，教风也因此得到进一步改善。“三重评估”制度在国内由我校首创，且效果良好，现已被国内许多高校参照使用。

（5）保证基础课教学质量，坚持教授上基础课制度。为保证基础课教学质量，使新生在低年级即获得坚实的基础知识，我校自1993年就开始在全国率先实行教授必须上本科基础课的制度，延续至今，效果良好。

（6）加强基础课教学，抓好“三语”课程。为使学生扎实掌握基础知识，体现我校外向型办学特征，学校将英语、计算机语言和汉语列为学生的必修课，同时还采取了一些有力措施，推动学生学好“三语”。

（7）加强学风建设，推进考试改革。为进一步加强校风、教风、学风建设，提高教学质量，严肃考试纪律，我校从1993年开始，率先在全国取消补考，实行重修制度。这项改革措施已在全国推广。同时，强化考试过程的管理，建试题库，分A、B试卷，加强考场管理、抽查各专业试卷等措施，促使教师认真对待考务，学生认真对待学习和考试。从2001～2002学年上学期开始，我校还设立了可一次容纳840人的大型考场，杜绝了作弊现象，保证了考试的公正性。这一全国首创性举措受到了许多高校及新闻媒体的广泛关注。在最近两年的博士生入学考试中，我校也采用了大型考场安排考试，学生及社会反响良好。

（8）突出国际性特征，实行“双语”教学。为贯彻两个“面向”的教育方针，培养国际性人才，我校从1996年开始提倡和鼓励教师用英语进行本科专业课教学。经过几年实践，此项工作逐步得到推广，特别受到港澳台和海外学生欢迎。目前我校每学年实行全英语教学的课程达53门，双语教学的课程达73门。在推广全英和双语教学的基础上，2001年6月，我校在全国高校中第一个成立了采用全英语教学的国际学院。目前该学院已开办了临床医学、国际经济与贸易、会计学、食品质量与安全和药学等5个专业。

（9）调整办学重心，优化办学结构。为了适应港澳台和海外学生的求学需要，我校通过采取大力发展研究生教育、积极发展华文教育、稳定本科教育、不办专科教育等措施适时调整办学重心，优化办学结构，以增强我校为海内外特别是为港澳台培养出更多优质骨干人才的能力。

（10）实行春秋两季招生和毕业制度，尽力为学生提供方便。为在更大程度上方便港澳台和海外学生报读暨南大学，我校于1998年率先在全国实行春秋两季招生制度。另外，我校还在境外，如中国港澳台地区、越南、泰国、老挝、马来西亚、菲律宾、美国等国家和地区设立招生报名点，方便了学生咨询和报名。同时，也实行春秋两季毕业制度，以利于学生及时就业。

（11）改革预科教育，增加港澳台和海外学生生源。为了更好地集中精力完成学校的办学任务，我校从2001年开始，停止招收预科内招生，而且按港澳台和海外学生的实际情况，从单一的一年制预科，改为三种学制：半年制、一年制和三年制。这样一来，初中毕业的境外学生就可以来校学习，因而增强了预科教育对境外学生的吸引力。

（12）加强横向合作，开展联合办学，率先在全国进行异地办学，以便更直接地服务港澳。1993年，我校在深圳开办旅游学院，开了校企联合办学的先河。1998年，为了进一步方便港澳青年入学，我校与珠海市政府合作办学成立珠海学院，成为中国第一个在珠海开办全日制高等教育的大学，已为珠海特区培养了首批专科生、本科生和硕士研究生。同时，也为深圳特区培养了首批硕士研究生。2004年10月，我校又与广东省知识产权局签订合作协议，共同创办了广东省乃至华南地区首所集教学和科研为一体的知识产权学院。

（13）积极合作共建，首创校医联合办学之路。我校是全国最早设立医学院的综合性大学，由于我校经费缺乏，办一所附属医院已十分困难，无法满足医学类学生实习的要求，而医学专业又是港澳台和海外学生最喜爱就读的专业，因此，为加强医学类学生的实践教学，我校在全国首创了校医联合办学的共建道路。几年来已在广州、深圳、珠海、清远、江门等地先后共建了7所附属医院，其中6所是国家级三甲医院，成功走出一条同社会合作的经济实惠的办学道路。

（14）发挥对外办学优势，开拓国际教育市场。根据规划，学校到“十五”末期，全日制学生中港澳台和海外学生与内地学生数要达到1∶1。为实现这一目标，我校不断加强内涵建设，提高办学质量，努力适应国内外教育市场对人才的需求。同时，采取具体措施，减少全日制本科和继续教育的内招生数量，鼓励多招港澳台和海外学生，多到港澳台和海外办学，因而外招学生人数逐年大幅增加。2004 年，我校外招生入学人数已超过内地学生入学人数。

(15) 只有学校品牌好了，才能吸引更多更好的港澳台和海外学生前来我校读书。故此，我们以学科建设为中心，大力鼓励教师从事科学研究，走现代化大学之路。为鼓励教师在做好教学工作的同时，积极投身科学研究，学校以“211 工程”重点学科建设为中心，大力推进科研工作。从 1996 年开始，学校用“教学、科研”双中心目标取代了过去单一的“教学”中心目标，并对教师的科研成果进行量化考核，将考核结果直接与校内工资挂钩。这一举措收效显著，我校教师的科研论文、科研项目和成果推广应用数量大幅上升，学校的科研经费快速增长，同时也促进了教学质量的提高。

(16) 改革人事分配制度，调动教师工作积极性。1998 年，针对传统的高校人事管理模式中存在的教师职务终身制和国家高度集中的指令性工资制度，我校进行了大胆改革，制定了一套全新的量化考核指标和管理方法，开始实行新的分配体制，即校内工资制度。新的分配体制充分发挥了个人潜能，优化了学科队伍，调动了教职工教学科研积极性。这一分配模式因属全国高校首创，被媒体称为“暨大模式”。

(17) 改善师资结构，提高教师质量。近年来，学校积极采取措施，尽力引进有博士学位的老师和学科带头人，大力提升教师水平。目前，学校教师中有两院院士 5 人，博士生导师 118 人，硕士生导师 589 人，其中有博士学位者 405 人。填补了无院士任教的空白，改变了教师学位偏低的局面。同时，学校还启动了名师工程，于 2002 年实施特聘教授岗位制。该制度的实施一方面可为优秀拔尖人才提供更好的学术环境，另一方面可以激发广大教师的上进心，有利于在全校形成健康向上的学术氛围。

(18) 改进财务管理，集中财力办学。改革前，由于资金分散，可供学校支配的资金有限，削弱了学校的财力；同时，由于院系将精力过多地投入创收，不同程度地影响了办学质量，而且容易滋生腐败。1996 年，学校把各院系和各部处资金集中起来统一管理，集中了学校财力，有利于加大对教学和科研的投入，而且加大了监管力度，使在第一线工作的领导和教工将精力集中用于本职工作，使学校办学经费大幅增加，从而保证了办学水平和办学质量不断提高。

(19) 合理调整人员结构，稳步推进机构改革。本着“精减、效能、统一”的原则，“九五”至今，我校先后进行了两次大规模机构改革，党政处级机构由 1999 年以前的 29 个减为 18 个。与此同时，学校对科级机构也进行了整合，优化了结构，提高了效率。为保证教学和科研这两个重点，几年来，学校注重引进专业教师，控制党政人员数量，原则上不进党政干部。虽然学校的办学规模不断扩大，但人员结构得到优化，教师数量比重增大，而教职工总数仍保持不变。尽管机构数量比改革前少，可效率却更高了。

(20) 加强建章立制，推进依法治校，以便使学校走上科学管理轨道。为加强依法治校，8 年多来，我校根据党和国家及上级管理部门的有关规定，先后制定了 200 多个制度性文件，对教学、科研及行政管理工作进行规范。为便于师生学习、了解文件精神，我校已将这些文件编纂成《暨南大学文件汇编》(分为行政管理卷和教学科研卷)，作为学校日常管理的主要依据。

(21) 加强廉政建设，净化环境和队伍。自1996年以来，学校一直把反腐倡廉作为一项重要工作来抓，先后查处了30多起违纪案件及一批涉案人员，有效惩治了腐败。为进一步加强对财务、招生、基建和设备采购的审计监督工作，学校与广东省检察院签订了《共同预防职务犯罪协议书》。这是全国第一所高校为加强廉政建设所采用的措施。同时，学校还在学术领域大力开展反对学术不端行为的斗争，净化校内学术环境。因此，校风、教风、学风有了明显好转。

(二) 改革创新的成效

(1) 实现了“九五”“211工程”建设目标，顺利进入“十五”“211工程”建设阶段。

1999年底，学校顺利完成了“211工程”中期检查工作。2002年7月，学校“九五”“211工程”子项目以全优的成绩通过验收。10月，我校“九五”“211工程”建设项目验收总体评价为优秀。11月，学校顺利通过了《“十五”“211工程”建设项目可行性研究报告》的专家论证，成功进入“十五”“211工程”建设阶段。鉴于我校“九五”期间的建设成果，“十五”期间，国务院决定投入5.1亿元专项资金用于我校的基础建设。国家“211工程”部际协调小组办公室决定向我校投入2800万元专项资金，改变了“九五”期间国家对我校“211工程”建设分文未投的状况。同时，广东省政府也进一步加大了对我校“211工程”建设的投入，由“九五”的5000万元增加到“十五”的8000万元。

(2) 办学规模变大了，港澳台学生更多了。

2004年与1995年相比，学校学生人数显著增加，办学规模扩大了一倍，相当于在原有基础上新办了一所暨南大学。学校各类学生已由13 012人增加到28 427人，增长118%，其中港澳台和海外学生由1982人增加到8966人，增长352%。香港研究生由18人增至254人，增长13倍；香港本科生由425人增至2753人，增长5.5倍。澳门研究生由55人增至172人，增长2倍；澳门本科生由691人增至1967人，增长1.8倍。台湾研究生由5人增至238人，增长46.6倍；台湾本科生由128人增至251人，增长近1倍。全日制学生由8824人增加到21 473人，增长143%，其中本科生由5377人增加到15 335人，增长185%；研究生由615人增加到5008人，增长714%。

(3) 校园面积和建筑面积扩大了。

自1995年以来，校区由原来的3个（广州石牌校本部，广州广园西路的华文学院，深圳华侨城的旅游学院）增加到4个（新增了珠海学院）。在靠近港澳的两个特区建立校区，更有利于直接为港澳台服务。另外，还有1个新校区（广州磨碟沙校区）正待开发。校园占地面积由112公顷增加到174公顷，增长55.4%，校园建筑面积由46万平方米增加到107万平方米，增长133%。

(4) 办学层次提高了。

学校研究生与本科生之比由1995年的1∶8.74上升到今年的1∶3.06，专科生由2472人减为零。

(5) 境外特色更加鲜明了。

作为中国港澳台和海外学生最多的大学，我校港澳台和海外学生已由1982人增加到 8966 人，占全校全日制学生比由 20％增长到 41.8％。特别是在我校攻读博士和硕士学位的港澳台和海外学生目前已达 741 人，约占全国总数的 1/4，较 1996 年的 124 人增长 498％。尤为值得可喜的是，在校台湾学生已达 524 人，占全国总数的 1/8。1995 年，只有 16 个国家和地区的学生来校学习，而今天的在校学生则分别来自世界五大洲 57 个国家和地区。同时，学校在面向科技发达以及华侨华人分布较多的国家开展对外学术和教育交流方面也取得了突出成效。迄今为止，已同世界五大洲 23 个国家和地区的 60 多所高等院校和文化机构签订了双边协议或建立了学术交流关系，是中国第一所在世界五大洲均建有姊妹大学的学校。

（6）学科结构优化了。

本科专业由 1995 年的 30 个增加到当前的 52 个，其中电子信息工程、会计学、新闻学、汉语言文学、生物技术等 5 个专业被评为广东高校名牌专业，“有机化学”“货币银行学”被评为广东省精品课程。硕士学位授权学科由 1995 年的 50 个增加到 89 个，博士学位授权学科已由 7 个增加到 34 个，并且在一级学位授权学科方面实现了零的突破，已达到 3 个。博士后站实现零的突破，达到 6 个。教学系由 21 个增至 39 个，学院数由 7 个增至 19 个。其中深圳旅游学院是内地首家通过世界旅游管理专业教育质量认证的高等旅游院校。知识产权学院是广东省乃至华南地区首所集教学和科研为一体的知识产权学院。学校现有的学院涵盖了文、史、经、管、法、理、工、医、教育等九大学科门类。

（7）师资队伍结构改善了。

学校现有专任教师 1367 人。教师中有研究生学位者 1054 人，占专任教师总数的 77.1％，较 1995 年增长 132.7％，其中，具有博士学位的教师 405 人，占专任教师总数的 29.6％，较 1995 年增长 393.9％，有博士生导师 118 人，较 1995 年增长 807.7％，正高职称 238 人，副高职称 502 人。学校新增了两院院士 5 人，实现了零的突破。获教育部设置的“长江学者奖励计划”特聘教授岗位 2 个和广东省设置的“珠江学者”岗位计划特聘教授岗位 2 个。有 11 人增选为广东省“千百十”工程省级培养对象。

（8）学生素质提高了。

暨南大学毕业生素以工作适应能力强，思想活跃，视野开阔而深受社会各界欢迎。连续几年来，我校本科学生就业率均达 90％以上，2003 年达 93.62％。2000 年，在全国最受欢迎的大学评比中，我校名列第 18 位。同年，在由教育部等单位联合主办，全国 45 所重点高校参与的大学生电脑节活动中，我校的《自然保护概论》荣获“教学方案设计竞赛”全国二等奖。在 2000 年举行的第五届和 2001 年举行的第六届全国大学生（包括研究生）科技作品“挑战杯”决赛中，我校报送的作品全部获奖，名列全国第六，广东第一。2001 年，在团中央、中国青年志愿者协会举办的第四届中国青年志愿者行动中，我校 Warm Touch 青年志愿者服务队被评为“中国百个优秀青年志愿服务集体”。在 2004 年“泰豪杯”全国大专辩论会中，我校学生辩论队经与 15 所全国著名院校激烈角逐，获得亚军。

作为一所传统的体育强校，我校运动员在许多国际（包括奥运会）国内重大体育赛事上都有杰出表现。自 1996 年以来，我校运动员在各类国际国内比赛中共荣获金牌

214 枚，其中在国际比赛中荣获金牌 42 枚。在第 19～21 届世界大学生运动会上，我校运动员都获得了金牌和银牌；在 2000 年举行的第六届全国大学生运动会上，我校荣获全国高校总分第二名，仅次于上海交通大学；在前国家女子足球队中，我校学生高红、赵利红、邱海燕均为主力队员；在第九届全运会上，我校健儿共夺得 27 枚奖牌，其中金牌 14 枚；在第十四届亚运会上，我校运动员获得 2 枚金牌、2 枚银牌。在香港举办的首届国际武术邀请赛上，我校运动员摘取了 14 枚金牌，成为荣获金牌最多的代表团。在 2003 年 10 月结束的第八届世界大学生国际象棋锦标赛上，我校运动员获男子冠军、女子季军。

（9）科研实力增强了。

学校科研经费由 1995 年的 400 万元增加到 2003 年的 1.092 亿元，增长 26 倍。获得的科研项目在“973”“863”等国家重点项目方面实现了零的突破。教职工发表的学术论文数较 1995 年增长了近 2.3 倍，总数达 1300 多篇。其中被三大索引（SCI、EI、ISTP）收录的论文数达 105 篇，与“九五”初期相比增加了近 12 倍。获省部级科技奖励的数目增加了 6 倍多。科研成果的应用也有较大的进步。

8 年来，学校新增了 2 个国家级重点学科，先后有 25 个学科被评为省部级重点学科。同时，学校还增加了 1 个国家人文社会科学重点研究基地，5 个省部级重点实验室，1 个教育部工程研究中心。设有国务院侨务办公室华文教育基地。除广东省重点学科和省教育厅重点实验室以外，其他方面均属于实现零的突破。同时，也结束了学校无博士后科研站的历史，现有 6 个博士后站。

（10）为祖国统一大业服务的能力增强了。

自 1978 年至今，我校已接收过来自世界五大洲 97 个国家和港澳台 3 个地区的学生前来学习，为港澳的顺利回归和广泛团结港澳台同胞与世界华侨华人作出了积极贡献。尤其是“九五”以来，我校港澳台和海外学生数量不断增长，现已从 1995 年的 1982 人增加到 8966 人。同时，对港澳台和海外研究生的培养工作正在日益加强。

（11）综合实力和办学质量更高了。

学校固定资产总值由 1995 年的 2.7 亿元增至去年的 15.7 亿元，增加了 4.8 倍多；图书藏量由 135 万册增至 215 万册，增长了 59%；教学科研仪器设备由 4985 万元增至 2.08 亿多元，增加了 3.17 倍多。

学校在不同机构的综合实力排行榜中的位置不断上升，2002 年被教育部《中国高等教育评估》杂志评为 77 所研究型大学之一，名列第 53 位；今年，名列第 46 位。在中国网大的中国大学综合实力排行榜中，我校排名自 1998～2003 年依次为 87 位、72 位、60 位、40 位、37 位、36 位、51 位。在广东管理科学院每年的“中国大学 100 强”中，我校的排名已由 2000 年的第 81 位上升至 2004 年的第 51 位。

（12）办学效益提高了。

全校的教职工和专任教师人数基本没有变化，1995 年分别为 3601 人和 1036 人，目前为 3784 人和 1367 人，且学校所获上级经费投入并未大幅增长，但学校完成的任务数却成倍增加，显然办学效益已更加优良。在 2002 年广东管理科学研究院“中国‘211 工程’大学教师人均效率排名”中，我校在全国高校中排名第 41 位。

(13) 教职工生活水平及学生住宿条件改善了。

1995 年，我校本部教职工家庭住房总面积为 169 775 平方米，人均住房面积为 13.5 平方米；2004 年，我校本部教职工家庭住房总面积为 320 343 平方米，人均住房面积为 23.74 平方米。与 1995 年相比，上述两项面积分别增长 89%和 76%。“九五”至今，我校教职工的工资待遇也不断提高，自 1995 以来，人均年收入分别为：1995 年 8254.12 元，1996 年 12 545.69 元，1997 年 20 210.31 元，1998 年 22 354.60 元，1999 年 26 710.35 元，2000 年 35 157.82 元，2001 年 43 734.80 元，2002 年 65 694.57 元，2003 年 78 000 元，2003 年与 1995 年相比增长近 8.45 倍。

与此同时，我校学生的住宿条件也有了根本改善。1995 年，学校学生宿舍面积为 81 686 平方米。“九五”至今，学校新建学生宿舍 31 栋共计 200 878 平方米，增长 145.9%。目前我校博士生 1 人住 1 间房，港澳台和海外学生 2 人住 1 间房，已达国家标准。

目前，学校共有教职工 3784 人，各类学生 28 472 人，其中全日制本科生 15 335 人，博士研究生 648 人，硕士研究生 4360 人。我校作为中国境外生最多的大学，有来自世界五大洲 57 个国家和港澳台 3 个地区的学生 8966 人，其中研究生 741 人。学校现设有 19 个学院，39 个系，65 个科研机构，71 个实验室，52 个本科专业，89 个硕士学位授权学科，34 个博士学位授权学科，其中 3 个博士学位授权一级学科。有 5 个博士后流动站（应用经济学、临床医学、中国语言文学、生物学、工商管理），1 个博士后科研工作站，2 个国家级重点学科（产业经济学、水生生物学），15 个省部级重点学科。拥有国务院侨办华文教育基地、国家人文社会科学华侨华人重点研究基地、教育部中国语言文学人才培养和科学研究基地、教育部国家大学生文化素质教育基地。有教育部工程研究中心 1 个、教育部重点实验室 1 个、国家中医药管理局重点实验室 1 个，广东省重点实验室 3 个。同时，我校还是招收和培养高级管理人员工商管理硕士（EMBA）、工商管理硕士（MBA）、会计学硕士（master of proffessional accounting，MPACC）、临床医学硕士、口腔医学硕士、工程硕士试点学校以及教育部试办高水平运动队的学校。

学校在广州、深圳、珠海三地共有四个校区，校园占地总面积 174 万平方米，其中广州校本部 727 466 平方米。校舍建筑面积 107 万平方米，图书馆藏书 215 万册。学校设有 6 所国家级三甲附属医院，即广州华侨医院、深圳市人民医院、珠海市人民医院、广州市红十字会医院、清远市人民医院和江门市五邑中医院，另有 1 所专科医院（深圳市眼科医院），1 所直属医院（深圳华侨城医院）。附属医院共有职工 6116 人，病床 4320 张。

二、对港澳顺利回归祖国和祖国统一大业的贡献

作为香港、澳门地区爱国爱港澳力量及治港治澳人才的重要培养阵地，暨南大学曾为港澳的顺利回归和社会稳定作出了积极贡献。自 1978 年复办至今，暨南大学共在港澳台地区招收各类学生 22 010 人，其中研究生 1314 人，本科生 12 917 人。长期以来，从暨南大学毕业的许多学生曾经或正在港澳社会的主要部门担任要职，是港澳地

区爱国爱港澳的中坚力量，港澳地区左派爱国群众团体的领袖绝大多数是我校校友。尤其是在澳门，我校有1000多名校友在特区政府任公务员，其中处级以上领导300多人，副局级以上的高层领导有30多人；澳门工联总会和街坊总会的主要领袖为我校毕业生；在澳门的医疗卫生系统中，约70%的医护人员曾就读过暨南大学。澳门立法会有5人、行政会有2人是暨南大学的校友；澳门地区的全国人大、政协代表中，有5位毕业于暨南大学；在澳门首届特区政府推选委员会中，有17位委员是暨南大学的校友；在澳门特区第二届行政长官选举委员会中，我校有7位校董、18位校友当选。我校董事会的主要董事都集中在港澳地区，他们绝大部分都任职港澳乃至国家领导部门，是我校联络的另一支爱国爱港澳的重要力量。

港澳回归以来，我校通过人才培养的方式积极配合党和国家在港澳地区贯彻和维护“一国两制”，为港澳地区的繁荣稳定作出了积极贡献。一方面，根据国家要求为港澳培养爱国人才，另一方面为港澳需要开展工作。今年九月，为支持香港特区第三届立法会选举，我校通过多种形式向3500多名在校香港学生宣传贯彻基本法和坚持“一国两制”的重大意义，引导学生为维护香港稳定贡献力量。在学校的精心组织下，香港学生纷纷自觉回港投票选举立法会爱国爱港议员，为巩固和扩大爱国力量在本届立法会中的席位做了大量工作。

暨南大学招收和培养港澳台和海外学生的规模一直位居全国高校之榜首。目前，学校共有港澳台和海外学生8966人，其中香港学生4324人（含博士生37人，硕士生217人，本科生2753人，预科生997人，继续教育学生320人）；澳门学生2735人（含博士生20人，硕士生152人，本科生1967人，预科生54人，继续教育学生542人）。我校目前港澳学生情况表明，我们实际上是为香港和澳门办了一所培养爱国爱港澳人才的大学。

近年来，为台湾地区培养高层次人才，为台湾顺利回归祖国准备爱国力量，已成为我校对台办学的方向和工作重点。2004年，我校在港澳台和海外地区录取的学生中，高层次人才结构比重的增大以台湾地区的学生最为突出。目前，我校共有台湾学生524人，其中本科层次以上的学生489人（包括博士生83人，硕士生155人，本科生251人），占台湾学生总数的93.3%。

自2001年以来，由我校华文学院与香港警察员佐级协会、本地督察协会合办的香港警务人员普通话培训班已连续开班50期，培训学员1175人，包括总警司1名，高级警司1名，警司5名，总督察20名，高级督察45名，督察8名，探长315名。为了更好地团结香港警察队伍，我还亲自出任暨南大学香港警察同学会的名誉会长。

与此同时，我校人文社会科学研究在为国家有关部门制定侨务、港澳台政策，了解侨情方面提供了许多有重要价值的咨询材料，仅提供华侨华人问题咨询材料就有30余份。

三、今后工作的重点

服务于国家统一大业和侨务事业，是我校的中心工作。为此，我校将在巩固以往发展成果的基础上，重点做好以下几项工作。

1. 坚定不移地贯彻中共中央〔1983〕24号文件精神

坚持“面向海外、面向港澳台”的办学方针，把对港澳台和海外地区学生的招收和培养作为学校的头等大事来抓，切实为港澳台服务，为“侨”服务。

2. 努力提高办学质量和办学水平

“九五”以来，正是由于办学质量和综合实力的稳步提升，暨南大学才能在与全国众多高校争取港澳台和海外生源的竞争中保持绝对优势，才能持续保持港澳台和海外学生北上求学首选学校的地位，不断吸引众多港澳台和海外学生前来求学。

总结以往经验，结合当前国际国内高等教育的发展形势，我们深刻认识到，只有创造优质的办学水平和办学质量，学校才会为越来越多的家庭和学生所认同，学校的生源数量和质量才能得到保障。这样，才能为港澳台提供优质人才，并使这些人才在将来成为该地区的栋梁之才。

3. 继续扩大港澳台和海外学生招生规模

今后，暨南大学将进一步总结经验，加大招生宣传和人才培养力度，积极遵照党和国家对我校的要求，面向海外、面向港澳台办学，为港澳地区的繁荣稳定，为和平统一台湾作出新的贡献。我校力争在近几年内，使全日制港澳台和海外学生与内地学生总数的比例达到1∶1。

4. 大力为港澳台地区培养高层次人才

扩大在港澳台和海外地区的研究生招生规模，提高外招学生的培养层次，为海外和港澳台培养更多优质骨干人才、领袖式人才已成为我校近年外招工作的重点和目标。因为相对而言，高层次人才更容易融入社会高层乃至决策层，其对社会的影响和作用更为关键。

21世纪是中华民族实现伟大复兴的世纪，肩负着为海外华侨华人和港澳台地区培养人才光荣使命的暨南大学更加任重道远。在21世纪，暨南大学将继续按照邓小平理论和“三个代表”重要思想的要求，在国务院侨办、教育部和广东省委、省政府的直接领导下，弘扬“爱国爱校、团结奋进”的暨南精神，树立“国际化、现代化、综合化”的办学理念，与时俱进，不断创新，为实施“侨校＋名校”发展战略，为建设海内外知名的高水平研究型大学而奋斗。

要完成上述任务，学校还存在许多困难，如国家经费投入不足，广州校区分散在三地、不利于办学等，还请上级帮助解决。

56. 坚持大力发展研究生教育[①]

跨入21世纪，世界各国更加重视学位与研究生教育和拔尖创新人才的培养，均从国际视野和战略高度谋划与推进本国高层次人才战略计划。2001年，美国《加强21世纪美国竞争力法》提出，三年内每年吸引19.5万名优秀高新科技人才。欧盟各国积极应对，每年吸引8000名至20 000名高新科技人才。日本、韩国分别实施“人类新领域计划”和“21世纪头脑开发计划”，引进短缺的高新科技人才。我国基于“科学技术是第一生产力”和“人才资源是第一资源”的科学论断，及时制定与实施“科教兴国”战略和“人才强国”战略。

今日之世界，竞争呈现出新的特点：全球战略资源争夺由物质资源转向人力资源、由自然资源转向科技资源，拔尖创新人才已成为国家之间竞争的核心和焦点。处于和平崛起和发展中的我国，始终将学位与研究生教育作为国家发展和事业进步的人才库与智力库，通过科学研究与人才培养相结合的方式，自主地培养经济建设、社会发展、科技进步和文化繁荣所需要的拔尖创新人才和高层次、复合型专门人才，为国家的发展和民族振兴作出贡献。

同样，作为肩负国家“面向海外、面向港澳台”办学重任的暨南大学，在新的世纪里为实现党和国家的“三大任务”，将责无旁贷地通过学位与研究生教育为海外华人华侨社会和香港、澳门、台湾地区培养和造就各类高层次人才，为促进中国人民和世界各国人民的友好交往，为香港、澳门地区的稳定、繁荣，为实现祖国的统一作出我们应有的贡献。

暨南大学和我国其他著名高校一样，是改革开放以来最早开展学位与研究生教育的高校之一，也是国家实施“面向海外、面向港澳台”办学方针，培养高层次人才，对海内外形成最具影响的华侨高等学府。学校历来非常重视学位与研究生教育和高层次专门人才的培养。尤其是1996年学校通过“211工程”预审，成为国家重点建设的大学后，学校的办学类型发生了变化。这种变化意味着学校将从一所“以教学为中心的高校”向“教学与科研并重的高校”转变。我和学校领导班子清醒地认识到这种深刻变化带来的艰巨任务和对学校未来发展产生的深远影响。站在时代的高度，审时度势，果断决策，适时地提出了“退出专科教育，稳定本科教育，大力发展研究生教育”的战略构想和具体措施。

10年过去了，“实践是检验真理的唯一标准”。我们可以欣喜地看到以下变化。

坚持大力发展研究生教育，使学校的办学层次、办学水平得到了迅速提升，实现了办学类型的转变。从1999年开始，学校再也没有全日制专科毕业生。在校研究生与全日制本科生的比例，由1995年的1∶8.7发展到2004年的1∶3。目前，在校研究生达5300多人，其中博士生648人，海外和港澳台研究生741人。

① 本文是《引路者论道：研究生指导教师学位与研究生教育研究论文选》序，暨南大学研究生部编，广州：广东人民出版社，2005，1-3.

坚持大力发展研究生教育，使我校师资队伍的素质和结构发生了重大变化。1995年教师队伍中具有博士学位的仅有82人，现在专任教师中具有博士学位的有521人，占教师人数的1/3；而具有研究生学历的达到76.98%。目前，学校有院士5人，博士生导师124人，硕士生导师478人。

坚持大力发展研究生教育，促使我校科学研究和科学技术的水平得到了空前提高。目前，学校有65个科研机构、2个省（部）级工程（研究）中心、5个省（部）级重点实验室、71个实验室、250名专职科研人员，科研经费已达数亿元。

坚持大力发展研究生教育，提高了学校的整体实力和竞争能力。1995年，学校仅有7个学院、21个系、7个博士点、50个硕士点、30个本科专业。现在，学校拥有20个学院、39个系、5个博士后科研流动站、1个博士后科研工作站、3个博士学位授权一级学科、28个博士学位授权点、36个博士招生专业、90个硕士点、52个本科专业、2个国家级重点学科、15个省（部）级重点学科。

坚持大力发展研究生教育，为我校实施“侨校＋名校”战略奠定了坚实的基础和拓展了更大的发展空间。暨南大学素称华侨最高学府，已有百年的办学历史，在海内外享有盛誉。但是，暨南大学要办成高水平的著名大学，如果仅仅维持原有的办学层次，就难以吸引海内外高素质的生源和师资，而大力发展学位与研究生教育则是实施“侨校＋名校”战略的重要举措。

从以上几个方面，我们可以认识到，学位与研究生教育和学校的地位、声誉紧密相关。研究生教育作为科学研究创新的基地、高层次人才的培育中心和推动社会变革的动力站，不仅与国家的前途命运紧密相连，而且也与暨南大学的荣辱兴衰紧密地联系在一起。无论过去、现在，还是将来，我们仍需坚持大力发展研究生教育。

我们肩负着历史的重任，任重道远。暨南大学作为唯一一所代表国家“面向海外、面向港澳台”办学的“211工程”高校，在一个新的百年里，将以更高的水平、更好的质量和更强的实力立于世界高等学校之林，参与高等教育的国际化，参与拔尖创新人才的培养。我们正在绘制新的蓝图，制定新的目标，设想分为两步，第一步，争取成功申办研究生院，实现学位与研究生教育新的跨越；第二步，能够在不久的将来，把暨南大学办成具有侨校特色、享誉海内外的高水平研究型大学。

57. 广纳贤才 全球招聘院长①

今天，美丽的暨南园到处洋溢着欢庆的气氛，我们怀着无比喜悦的心情，在这里隆重举行仪式，聘任来自世界各地、在各自领域具有高深造诣的10位院长。首先，我代表学校党政领导对各位来宾踊跃参加今天的聘任仪式表示真诚的感谢！对10位院长的竞聘成功表示衷心的祝贺！对10位院长的到来表示热烈的欢迎！

21世纪鲜明的时代特征，显示了中国高等教育的改革发展在很大程度上将直接决定中国先进生产力、先进文化的发展进程和最广大人民根本利益的实现程度，“三个代表”重要思想把高等教育的地位和作用升华到新的更高境界。以“面向海外、面向港澳台”为办学方针的暨南大学，为了在中华民族伟大复兴中，在推动海外华侨华人和港澳台地区发展先进生产力和先进文化的进程中，发挥更加积极而重要的作用，就必须进行改革，全面提升为社会服务的实力。本次面向海内外招聘外国语学院、华文学院、经济学院、管理学院、法学院、理工学院、信息科学技术学院、药学院、珠海学院、艺术学院10个学院的院长，就是暨南大学为此而采取的一项大力度改革措施。

总体看来，本次活动取得了圆满成功。经过激烈的竞争、认真的考察、严格的筛选，最终，李从东教授、符启林教授、马宏伟教授、龚建民教授、王玉强研究员、王志伟教授、张铁林先生、卢植教授、班弨教授、冯邦彦教授脱颖而出，分别获聘为10个学院的院长。他们的能力强、水平高、业务精，在演讲、答辩中，发挥了很好的水平，充分展示了自己的精神风貌和卓越才华。他们的到来，为阅沧桑而奋进的暨南大学输入了新鲜血液，为“宏教泽而系侨情”的华侨最高学府注入了新的活力。

暨南大学是中国第一所由国家创办的华侨学府，素有“华侨最高学府”之称，是中国历史上最悠久的大学之一，是中国第一所招收留学生的大学，也是目前内地拥有海外及港澳台学生最多的大学。学校的前身是1906年清政府创立于南京的暨南学堂，1927年更名为国立暨南大学，1949年合并于复旦大学、上海交通大学等学校，1958年在广州重建，由时任广东省委书记陶铸兼任校长，1970年因“文化大革命”的影响而被迫再次停办，1978年复办。在98年的办学历史中，学校坚定不移地贯彻“面向海外、面向港澳台”的办学方针，已为海内外输送了各级各类人才18万人。据不完全统计，我校校友中，已有2位中共中央政治局常委，5位副总理（包括外国），1位外国议长，6位全国人大常委会副委员长，以及一些院士、著名学者、著名企业家等。

历近百年风雨洗礼而弥新的暨南大学为了谋得更好发展，在20世纪末就确立了“侨校+名校”的发展战略，努力将暨南大学的办学水平提高到一个新的层次。为此，学校确定了“严、法、实”三字的办学原则（即从严治校、从严治教、从严治学，依法治校和实事求是），按照“发挥优势、深化改革、保证重点、改善条件、提高质量”的发展思路，与时俱进，开拓创新，采取了许多敢为天下先的改革措施，如率先施行

① 本文是在暨南大学面向世界公开招聘10位学院院长仪式上的讲话，暨南大学校报，第403期，2005年1月21日.

与国际接轨的弹性学分制（即标准学分制）；在全国首先实行课堂教学三重评估制度；率先取消补考，实行重修制度；首先实行教授必须上本科基础课制度；率先实行春秋两季招生、春秋两季毕业制度；首先改革预科教育，将一年制改为半年制、一年制和三年制三种；狠抓英语、汉语、计算机语言“三语”教学；根据内地、海外及港澳台学生的不同特点，制定不同的培养目标和培养方案；实行全新的量化考核指标和分配体制，即校内工资制度，被媒体称为“暨大模式”；首创一次可容纳 800 余人的大型考场，最大限度杜绝作弊；作为第一个与省检察院签订《共同预防职务犯罪协议书》的高校，加强廉政建设；第一个成立全英语教学的国际学院；第一个实行校医联合办学（如深圳市人民医院、珠海市人民医院、清远市人民医院、广州市红十字会医院、江门市五邑中医院、深圳市眼科医院等附属医院）；第一个在世界五大洲均建有姊妹学校的大学，并已向国外的姊妹大学交换了 128 名学生。

上述改革措施的实行，使暨南大学在许多方面都取得了令人振奋的进步：综合实力日渐增强，办学规模不断壮大，办学层次稳步提高，在海内外的影响日益扩大。学校现已是国家“211 工程”重点大学，已连续 4 年在全国 1577 所高校中位居前 50 所名校之列；设有的 20 个学院 39 个系 52 个本科专业，涵盖了文、史、经、管、法、理、工、医、教育等九大学科门类。在 20 个学院中，深圳旅游学院是内地首家通过世界旅游管理专业教育质量认证的高等旅游学院，国际学院是我国第一所采用全英语授课的学院。我校也是新中国第一所设有医学院的综合性大学。医学院现有 6 所国家级三甲附属医院，1 所专科医院，1 所直属医院，附属医院共有职工 6399 人，病床 4320 张。学校拥有 89 个硕士学位授权学科，34 个博士学位授权学科，6 个博士后站。我校还是招收和培养高级管理人员工商管理硕士（EMBA）、工商管理硕士（MBA）、会计学硕士（MPACC）、临床医学硕士、口腔医学硕士、工程硕士试点学校以及教育部试办高水平运动队的学校。

学校现有专职教师 1477 人，其中中国工程院院士 2 人，中国科学院院士 3 人，博士学位拥有者 438 人，占专职教师的 30%。学校现有全日制学生 22 000 人，其中本科生 15 335 人，研究生 5008 人；来自世界五大洲 57 个国家和我国港澳台 3 个地区的学生 8966 人，数量居全国高校第一；在我校就读的海外及港澳台地区的研究生 741 人，约占全国总数的 1/4；台湾学生 524 人，约占全国同类学生总数的 1/8。学校办学国际化特色明显。特别是近两年报考我校并被录取的海外及港澳台学生，均大于内地其他所有高校的总和。

在科技发展突飞猛进的 21 世纪，党的十六大报告明确指出当前教育的主要任务是“提高教育质量和管理水平，全面推进素质教育，造就数以亿计的高素质劳动者、数以千万计的专门人才和一大批拔尖创新人才”。这是时代赋予高校的神圣使命！当前高校间的竞争又异常激烈，百尺竿头，更进一步，尤其困难。学校要激流勇进，取得更大进展，为海外及港澳台地区培养更多的优秀人才，更好地向世界传播中华文化，就必须实行人才强校战略。因为高校间的竞争归根结底是人才的竞争，人才问题是战略问题，只有一流的人才加一流的管理，方能成为一流的学校。为了提高干部的管理水平，学校于去年 9、10 月间，派出了由党政机关一把手 25 人组成的暨南大学赴美培训与考察团，进行了为期 25 天的培训与考察活动，取得了显著效果。今年，学校还计划将各学院院长派往英国学习，以期在提高院长管理水平的基础上，实现提高学校管理水平的目的。

为了进一步引进海内外人才，并提高学院的教学科研和管理水平，学校进行了这次大力度的改革——面向世界招聘 10 个学院的院长。因为，这不仅可以使学校的教学、管理更好地适应在校学生的世界性特征，将学生培养成为国际型人才；而且可以建立更广阔的办学平台，招纳贤才，使学校更富现代化、国际化；更可以充分发挥海内外著名专家学者的学术特长和影响力，提高学校教学科研水准和管理水平，进而大幅提升学校在海内外的影响及竞争力。

自去年 9 月 30 日学校陆续在《人民日报》海外版、《光明日报》、《南方日报》、香港《大公报》、《澳门日报》、《欧洲时报》、美国《侨报》等著名报纸及相关媒体上发布招聘信息后，海内外人士报名踊跃，共有来自英国、德国、芬兰、哈萨克斯坦、美国、加拿大、新加坡、日本、中国（包括香港地区）9 个国家和地区的 92 位学者前来应聘，其中海外学者 32 人、香港特区 2 人、内地学者 40 人、校内 18 人。应聘者有来自哈佛大学、剑桥大学等世界一流大学的专家，也有来自北京大学、北京师范大学、中国人民大学、中国政法大学、南京大学、复旦大学、上海交通大学、兰州大学、西安电子科技大学、天津大学、吉林大学、中山大学、华南理工大学、香港理工大学等 20 余所全国知名大学的学者。此次应聘人员呈现出学位层次高、年纪轻（均在 50 岁以下，最年轻的 27 岁）、学术造诣深厚、同行专家认可度高等特点。

学校专门成立了由校长和书记亲自负责的招聘专家委员会，对应聘人员进行严格的资格审查和筛选，向 46 人发出了面试通知。在去年 12 月 1 日～9 日连续 9 天时间里，对 10 个学院的院长候选人进行了面试，每位应聘人的陈述和答辩时间为 1 个小时。通过竞聘演说和答辩，并经学校党委常委会讨论后，确定了 15 名考察人选，由组织部和人事处组成考察组前往应聘者的学习或工作单位进行考察，同时聘请同行专家对他们的科研学术成果进行鉴定，最后确定了 10 个学院的院长人选，其中海外 3 人、内地 4 人、校内 3 人。

10 位院长的到任，不仅是学校当前的一件大事，而且对暨南大学的将来也具有十分重要的意义和深远的影响。对 10 位新院长来说，这意味着有了一个新的起点；对全校来说，也有了一个新的起点。因为我们今天聘任的 10 位院长均是在各自领域颇有建树的专家，具有很强的业务能力和工作能力。因此，我希望大家能够确立适应时代发展要求的办学思想，积极探讨大学的规律、功能和使命，树立以学生为中心、以学术为主导的管理理念；同时要严于律己，廉洁奉公，勤俭办学，力争在今后的工作中展示新风貌、开拓新局面、创立新业绩。

同时，我也希望诸位院长紧紧围绕“侨校＋名校”的发展目标，针对当前高等教育的发展趋势和学校本身的特点，根据学院的具体情况，积极探索并认真实施具有侨校特色和显著效果的人才培养模式和管理方式，促进学校整体教学科研水平和管理水平的不断提高，从而在加快建设海内外知名高水平研究型大学的步伐中，树立新的典范！在实现“侨校＋名校”战略目标的过程中，做出自己的卓越贡献！

58. 学会普通话　走遍天下都不怕[①]

普通话是13亿中国人共同的语言，也就是说，是世界上最多人使用的语言，“学会普通话，走遍天下都不怕”。学会它有特别重要的意义，它将会使香港的明天更美好。长期以来，暨南大学在普通话培训方面与香港的有关单位进行了多次积极有效的合作，已经与香港警务督察协会合作成功举办了56期“香港警务人员普通话培训班”，在香港公务员队伍中引起了强烈反响。经过香港民政事务总署和暨南大学的共同努力，今天，我们又迎来了第一期“香港民政事务总署公务员普通话研习班”开学典礼！各位新学员的到来，不仅为美丽的暨南园增添了诸多喜庆气氛，也为具有百年历史的暨南大学输入了新鲜血液。

华文学院是暨南大学面向海外开展汉语、中华文化及预科教育的专门学院，其前身是成立于1953年的广州华侨学生补习学校。学院现为国务院侨务办公室华文教育基地，被中国国家对外汉语教学领导小组办公室确定为“支持周边国家汉语教学重点院校”，也是中国华南地区办学规模最大、办学实力较强的对外汉语教学和华文教育的专门学院。学院现设有留学生非学历教育、学历教育、海外汉语教师培训、对外汉语本科和硕士研究生教育、预科教育，目前在院学生近2000名，分别来自40多个国家和地区。在这里我要向各位宣布一个好消息，我校申报的“华文教育”本科专业已经获得教育部批准，成为目前国内高校开设的首个华文教育本科专业！她的设立符合当今海外华文教育和侨务工作形势发展的迫切需要，符合我校学科建设和发展的需要，充分体现了我校的办学特色和优势。

各位新学员，你们现在来到了暨南大学学习，虽然学习时间短，但仍然是暨南大学历史中的一名学生，因此，你们从今天开始就成为一名光荣的暨南人。我真诚地希望大家珍惜在暨南大学的这段学习时光，充分利用大学的优越条件，坚持学习的自主性、学习的开放性和学习的可持续性，在努力学习普通话的同时，注意改善自己的知识结构，提高自己的综合素质，做到学有所成、学有所用，为将来在工作岗位上做出更大成绩打好基础。学校从事华文教育的教学、科研和管理人员，也一定以高度的责任感和使命感，重视这次普通话研习班，以现代的教育理念和在过去普通话培训中取得的成绩指导我们的教学和管理工作，从而在这一次的普通话培训中取得更好效果。同时，我也希望各位新学员珍惜“暨南人”这一光荣称谓，恪守“忠信笃敬”的校训，并将其作为立身之本和行事准则，大力弘扬中华文化，积极发扬光大暨南精神，做一个名副其实的暨南人。

最后，祝第一期“香港民政事务总署公务员普通话研习班”取得圆满成功。

---

① 本文是在香港民政事务总署公务员普通话研习班开学典礼上的讲话，广州，2005年3月21日；一个大学校长的探索，北京：高等教育出版社，2011，383.

59. 亚洲　青春　竞技[①]

今天，整洁美丽的暨南园彩旗飘扬，英姿勃发的亚洲青年笑语喧阗，在这喜庆的气氛中，第一届亚洲大学生田径锦标赛隆重开幕了。在此，我代表暨南大学 3 万余名师生员工向参加此次运动会的全体运动员、裁判员、教练员和各位来宾、各位朋友表示热烈的欢迎！向一直热情支持第一届亚洲大学生田径锦标赛的各级领导和社会各界表示衷心的感谢！

作为承办方的暨南大学，能够迎来亚洲的一批新朋友，我们感到由衷的高兴！你们的到来，更为学校明年的百年华诞增添了新的荣耀！因为素有“华侨最高学府”之称的暨南大学，本身就是一个国际大家庭，她是中国第一所由国家创办的华侨学府，是中国历史最悠久的大学之一，是中国第一所招收留学生的大学，是中国第一所在五大洲建有姊妹大学的学校，更是目前中国拥有海外及港澳台学生最多的大学，现有来自世界 71 个国家和港澳台地区的 10 892 名学生在校学习。自创办以来，暨南大学一直为世界的和平与发展积极做着力所能及的贡献。在 99 年的办学历程中，已为世界五大洲 106 个国家和港澳台地区输送了各类人才 20 余万人。暨南大学的海外学子学成回国后，积极致力于经济发展，始终勤勉于社会进步，不少人成为祖籍国的名人贤达和社会政要，为居住地的经济发展和社会进步做出了突出成绩，如著名侨领、新加坡大学首任校长李光前，泰国国会前主席、副总理许敦茂，新加坡中华总商会前会长陈共存等，便是其中的杰出代表。他们在展现卓越才华和成就骄人的事业的同时，也使学校赢得了社会各界的广泛好评和高度赞誉。

现在，第一届亚洲大学生田径锦标赛在暨南大学隆重举办，又为暨南大学做出新的贡献提供了有利契机。这不仅可以推动亚洲各国大学生体育运动的更快更好发展，而且可以进一步促进亚洲各国青年之间的了解，增进亚洲各国青年之间的友谊。此次运动会又是亚洲大学生的首届田径锦标赛，暨南大学全体师生员工怀着无比喜悦的心情，将承办此项赛事作为学校百年校庆的重要活动之一，对此给予了高度关注和大力支持，力争使本项赛事成为出色的一届大学生田径锦标赛，以隆重、热烈、圆满、精彩的效果载入史册。我也相信，在本届田径锦标赛上，来自亚洲 19 个国家和地区的 40 所高校的大学生运动员们一定能够秉承“亚洲、青春、竞技”的宗旨，大力发扬艰苦奋斗、不怕困难的拼搏精神，全面弘扬团结友爱、互帮互助的合作精神，奋力拼搏，共同提高，赛出水平，赛出风格，全面展示当代大学生积极向上、朝气蓬勃的无限活力和时代风采！

最后，祝第一届亚洲大学生田径锦标赛圆满成功！

祝所有选手在今后的比赛中取得佳绩！

---

① 本文是在暨南大学举行的第一届亚洲大学生田径锦标赛开幕词，广州，2005 年 11 月 7 日；一个大学校长的探索，北京：高等教育出版社，2011，240-241.

60. 贺暨南大学香港社会学同学会成立①

在这天朗气清、秋风和畅的美好日子里，暨南大学的新老校友欢聚一堂，隆重举行暨南大学教育学院香港社会学同学会成立暨第一届委员会就职典礼，我由衷地感到高兴。

多年来，诸位校友在香港社会努力营造和谐环境，积极关注国计民生，致力于经济发展和社会进步，为香港的发展做出了突出的贡献。特别是在香港回归以后，社会学的广大校友在维护香港的稳定方面，在反对任何形式的分裂祖国的行为和言论方面，扮演了积极的角色，并且在促进香港与祖国内地的交流合作方面进行了长期不懈的努力，为香港的稳定繁荣、为两地的经济发展和社会进步做出了新的贡献。校友们在展现卓越才华和成就骄人的事业的同时，也使母校赢得了社会各界的广泛好评和高度赞誉！

诸位校友尽管已经离校，但仍然时刻关注母校的发展，积极为学校的发展出谋献策、捐款捐物，借此机会，我也代表学校对所有给予学校捐助、支持的校友表示诚挚的谢意！正是在广大校友和海内外各界人士的长期帮助和大力支持下，勇于开拓、敢于创新的暨南大学在 1996 年成为国家“211 工程”重点大学后，为了取得更好发展，采取了许多敢为天下先的改革措施。

全面推行学分制改革，率先实行与国际接轨的弹性学分制（即标准学分制）；对内地、海外及港澳台两类学生制定不同的培养目标、培养方案和教学要求；在全国高校中第一个建立校史馆；实行教授必须上本科基础课制度；实行课堂教学三重评估制度；率先取消补考，实行重修制度；狠抓英语、汉语、计算机语言“三语”教学；率先实行春秋两季招生、春秋两季毕业制度；改革预科教育，将一年制改为半年制、一年制和三年制三种；率先在世界五大洲设立报名点；实行全新的量化考核指标和分配体制，即校内工资制度，被媒体称为“暨大模式”；较早地实现了一级财务管理；全面推进校园信息化建设，是广东省第一所接入世界互联网的高校；重视提高干部的管理水平，2004 年、2005 年分别把机关部、处、直属单位一把手和各学院院长送到美国威斯康星大学、澳大利亚格里菲斯大学等进行管理培训和经验交流；面向海内外招聘 10 位院长；第一个成立全英语教学的国际学院；2004 年创办了华南地区第一家集教学和科研为一体的知识产权学院；第一个实行学校医院联合办学；国内第一个与省检察院签订《共同预防职务犯罪协议书》的高校，加强廉政建设；第一个在世界五大洲均建有姊妹学校的大学，并已向海外的姊妹大学交换了 161 名学生。

上述改革措施的实行，使学校在近几年的工作中取得了显著成绩，成功实现了跨越式发展：学校已在全国 1700 多所高校中跃居前 50 名左右，科研经费已连续两年过亿元。目前，学校设有 20 个学院 44 个系 56 个本科专业，涵盖了文、史、经、管、法、理、工、医、教育等九大学科门类；拥有 131 个硕士学科专业，54 个博士学科专业。

① 本文是在暨南大学教育学院香港社会学同学会成立会上的讲话，香港，2005 年 11 月 27 日；一个大学校长的探索，北京：高等教育出版社，2011，209-210.

在1484名专职教师中，有中国工程院和中国科学院院士7人，博士学位拥有者502人，占专职教师的34%，高于全国设有研究生院大学的平均水平。学校现有全日制学生23 752人；其中，本科生16 336人，研究生6074人，研究生与本科生之比为1∶2.7，目前，共有来自世界五大洲71个国家和港澳台3个地区的学生近11 000人在校学习，数量为全国高校之冠。暨南大学正逐渐成为海外及港澳台学生到大陆求学的首选高校，报考我校并被录取的海外及港澳台学生数，连续三年均大于全国其他所有高校的总和。

明年，暨南大学将迎来百年华诞，我衷心地希望校友们踊跃地以实际行动来为母校的百年华诞庆典献礼。同时，我也真诚地邀请校友们在百年校庆时回母校看看，届时，我们又可以一起共道人生历程，展望美好未来。

最后，衷心希望暨南大学教育学院香港社会学同学会在第一届委员会的带领下不断发展，大有作为，为香港的繁荣发展做出新的贡献!

61. 立足侨校　服务学生　全面推进我校学生德育工作①

2005 年暨南大学学生工作会议今天隆重开幕了。这是我校在新世纪召开的第一次学生工作会议。开好这次大会，对于学校更好地贯彻落实中共中央、国务院中发〔2004〕16 号文和广东省委、省政府粤发〔2005〕12 号文精神，全面推进我校的大学生德育工作，积极营造大学生健康成长的良好环境，为实施“侨校＋名校”的发展战略提供强大的思想保证，具有十分重要的现实意义。在此，我谨代表学校党政领导并以我个人的名义，对大会的胜利召开表示热烈的祝贺！向长期以来在教书育人、管理育人、服务育人岗位上辛勤工作，为我校研究生、大学生的健康成长付出心血与汗水的同志们，表示亲切的问候！

大学生是社会宝贵的人才资源，是民族的希望、社会的未来。青年大学生历来是社会上最富有朝气、最富有创造性、最富有生命力的群体，我们党总是把关注的目光投向青年大学生。党和国家的主要领导人，历来十分重视青年的成长。毛泽东同志把青年看作早晨七八点钟的太阳，指出青年是推动国家发展和社会进步的生机勃勃的力量。邓小平同志在青年身上寄寓了我们事业兴旺发达的光辉前景。江泽民同志指出，青年兴则国家兴；青年强则国家强；青年有希望，未来发展就有希望。胡锦涛同志对青年则提出了勤于学习、善于创造和甘于奉献的要求。这些精辟论述，集中反映了党对青年的高度重视、热情关怀和殷切希望。这对于学校从全局和战略高度，认真贯彻落实中发〔2004〕16 号文和粤发〔2005〕12 号文精神，以胡锦涛总书记、省委张德江书记在全国、全省加强和改进大学生思想政治教育工作会议上的重要讲话精神为指导，做好新形势下侨校的大学生德育工作，在暨南园形成爱护青年、关心青年、鼓励青年成才，支持青年干事业的浓郁氛围，具有十分重要的现实意义。

高校是培养青年大学生的重要阵地，其办学水平和办学层次的不断提升是培养高素质人才的重要保证。为创造一个优良的育人环境，学校自“九五”开始，坚持“从严治校，从严治教，从严治学”的办学原则，优化专业结构，调整办学重心，在学分制、学生培养、教学质量评估、基础课教学、英语授课、学生考试、专业设置、招生时间、合作办学、机构改革、人事分配制度、师资队伍建设、教师科研、财务管理、后勤社会化、机关作风建设、廉政建设等方面采取了一系列改革措施，取得了显著成效。学校的跨越式发展，为青年大学生的成长成才提供了更好的成长环境和发展空间。为做好学校的大学生德育工作，下面，我讲几点意见。

一、恪守“学生第一”的办学理念，为学生提供最优质的教育资源

“学生第一”的办学理念反映了一所学校办学的价值取向，从总体上规范着学校的各种办学行为。在学校的教学、科研、管理和服务工作中，我极力倡导恪守“学生第

① 本文是在暨南大学学生工作会议上的讲话，广州，2005 年 12 月 25 日；暨南大学校报，特刊，2005 年 12 月 31 日.

一”的理念。没有学生，就没有大学，学校的一切工作都是为了学生。作为一所传承、创造知识的场所，大学是为学生而设的。所以，我们要把学生看成学校的生存之本，人类社会的发展之本，把促进学生的全面发展看成学校的发展之本，把一切为了学生作为推动学校各项工作改革的动力之本。

“学生第一”的办学理念要求我们为学生提供最优质的教育资源，追求质量第一，教育质量是学校的生命，学校的一切工作只有在追求质量第一的基础上才能不断前进。“学生第一”就是要求育人第一，学校的本质是育人，无论是教学、科研、管理服务活动，还是环境建设、文化氛围都是为了作育人才，只有坚持育人第一，才能在学校中形成人人都是育人工作者、处处都是育人环境的局面，才能最大限度发挥学校的育人功能；“学生第一”就是要求责任心第一，将造就人才作为第一位的任务，以培养创新型人才、拔尖人才为己任。

“学生第一”的理念，其基本立足点是充分尊重和关心师生，核心是致力于培养和造就人才，既要培养高质量的学生，也要造就高水平的教师。学校工作应当围绕并服务于培养人才，服从于培养人才这一大局，全心全意致力于促进人才的全面发展。“学生第一”的理念，体现了现代教育的本质要求。它不仅是对我校百年来办学经验的历史总结，是我校认真学习贯彻马列主义、毛泽东思想、邓小平理论和“三个代表”重要思想的生动体现，同时也是实现学校跨越式发展的根本需要。在建设高水平研究型大学进程中，全校上下必须牢固树立这一理念，并在实践中不断予以丰富和发展。

二、关爱学生，有教无类

善待学生，爱生如子，是每一个教育工作者的基本职责，也是建立新型师生关系的前提。卢梭在《爱弥儿》一书中说：“教育就是爱。”苏联教育家赞科夫也说过：“当老师必不可少的，甚至几乎是最主要的品质就是热爱儿童。”现代教育社会学的研究表明，师生间的人际关系是整个学校教学过程中全部人际关系的最主要、最基本的部分。健康向上的教育教学氛围与和谐的师生关系是紧密联系的。和谐的师生关系能启迪智慧、激发创造，从而使师生在愉悦的氛围中完成教学和学习的任务。热爱学生是建立民主、平等、和谐的师生关系的基础。教师真挚地爱学生就能对学生尊重、宽容、理解和信任，师生之间才能思想相通，情感交融。

著名教育家、清华大学老校长梅贻琦曾说：“所谓大学者，非谓有大楼之谓也，有大师之谓也。”他提出，一所名校，大师比大楼重要。在现代大学校园，我们还要崇尚“大爱”的理念，营造一个以人为本的和谐校园氛围，真诚地爱护每一个教师、学生，使人才辈出而服务于社会，推动国家、民族的进步，这就是大爱。大师的言传身教就是在传播对全社会、全人类的大爱。大爱更源于大学的管理者——办教育的人少一分急功近利，大学师生方能多一分钻研创新。

今年，我校的生源来自世界五大洲 71 个国家和港澳台 3 个地区，不论来自何地，不论是干部、富商大贾的子女，还是平民百姓的孩子，抑或智力存在差异的学生，我们都要平等对待，一视同仁。这与孔子在《论语·卫灵公》中所推崇的“有教无类”的教育理念是一脉相承的。孔子“有教无类”教育理念指：在教育对象内容方面，不分种类，即不分贵贱、庶鄙，不分善恶，不分阶级、阶层、年龄和地域，也不分个性差异，凡是愿意来学习的，统统收为弟子施予教育。我希望全校的教育工作者能弘扬孔子“有教无类”的思想，对学生多一份关爱与尊重，因为被尊重与平等对待是学生的一大愿望，能大大激发学生的学习潜能，给他们以自信和力量。

三、立足侨校，更新教育管理与德育工作的观念

暨大最大的特色是“侨”字，这也是学校的立身之本。办好华侨高等教育，为国家的侨务工作、统一大业和现代化建设培养大批高素质人才，是党和国家赋予学校的特殊办学使命。侨校学生的教育管理和德育工作，既遵循着高等教育的普遍规律，又彰显着独特的个性。下面，我想简要谈几个观点。

一是要立足侨校，始终不渝地坚持用中华文化对莘莘学子进行教育，使其能肩负起传承优秀的中华文明的使命，实现“朔南暨，声教讫于四海”的办学宗旨。二是全校教育工作者要肩负“传道、受业、解惑”的重任，受业是传授知识，解惑是启迪思维，传道是升华精神，三者不可或缺且依次递进，构成了育人的一个完整体系。三是要教会学生如何做人，这是立身之本。胡锦涛总书记在年初全国加强和改进大学生思想政治教育工作会议上明确提出：“培养什么人，如何培养人，是我国社会主义教育事业发展中必须解决好的根本问题。”可见这一问题的重要性。从人的成长发展来看，要成就小事，主要靠业务本领，而要成就大事，得靠思想品德和综合素质。教育的根本问题是培养人的健康人格，而这应从培养良好的行为习惯着手，正如亚里士多德所说的：“播种一种行为，收获一种习惯；播种一种习惯，收获一种品格；播种一种品格，收获一种命运。”四是要教会学生如何做事，由传统保姆式的管理变为自主式的管理。中国传统的教育过多强调、要求学生听话、顺从，大小事情全由家长与教师一手包办，这种教育管理方式严重束缚了学生个性的发展与积极性、主动性的发挥以及创新能力的培养。我们要创造条件，积极引导学生进行自我教育、自我管理，使其在尝试、实践及与他人相处、合作、共事的过程中，锻炼能力、增长才干、经受磨砺，培养其自信心、进取心、意志力、自制力，以及自觉的社会责任心、强烈的历史使命感和自强不息的人生追求。

不断加强和改进大学生德育工作，深入贯彻实施“育人为本、德育为先”的工作理念，是新形势下党和国家对高校思想政治教育工作提出的新要求。学校即将迎来百年华诞，跨入第二个百年征程。暨南大学面临着千载难逢的发展机遇。“逆水行舟，不进则退”，学校正处于一个在实施“侨校＋名校”发展战略进程中的爬陡坡阶段。“上下同心，其利断金”；“人心齐，泰山移”。我希望全校师生万众一心，以主人翁的姿态为学校的发展贡献才智，以创一流的佳绩向百年华诞献礼。让我们以本次学生工作会议的胜利召开为契机，着力探讨、积极构建侨校学生教育管理与德育工作的新机制，圆满完成党和国家赋予暨大的特殊办学使命，为推进国家的侨务工作、统一大业和现代化建设而努力奋斗！

62. 图书馆是大学的心脏①

今天，我们在施工现场隆重举行暨南大学新图书馆封顶仪式，我感到非常的高兴和振奋。因为有人曾经说过“教授是大学的灵魂，图书馆是大学的心脏”。在一定程度上讲，图书馆的发展水平就是大学发展水平的一个间接体现，暨南大学新图书馆的兴建正可以说是学校发展历史的一个新起点，也是学校实现跨越式发展的一个新标志。

自进入“211 工程”建设以来，学校一直在多方筹集办学资金，积极进行基础设施建设，努力改善办学条件，在学校进行可能是有史以来最大规模基建的同时，学校采取的各项创新性改革措施仍然取得了可喜的效果，使暨南大学成功实现了跨越式发展。现在，暨南大学共有各类学生 30 499 人，其中全日制学生 23 752 人，博士、硕士研究生跨过 6000 大关，达到 6074 人，较去年增长 20%，是 10 年前的 10 倍，办学层次有了进一步提高。在校的海外及港澳台学生突破万人大关，共有来自世界五大洲 71 个国家和港澳台地区的各类学生 10 609 人，数量为全国高校之冠，较去年增长 20%，较 10 年前增长 4.4 倍。拥有 131 个硕士学位授权学科，较去年增长 47%，较 10 年前增长 1.6 倍；新增了 3 个一级学科博士点、12 个二级学科博士点，一级学科博士点不但较 10 年前实现了零的突破，数目也达到了 6 个；二级学科博士点达到 40 个，较 10 年前增长 5 倍；加上自主设置的二级博士学科专业，学校共有 54 个博士学科专业，较去年增长 59%，较 10 年前增长 6.7 倍。高水平科研论文数量持续上升，被三大索引收录 297 篇，较上年增长 30%，较 10 年前增长 31 倍，科研能力有了进一步提高；据 2005 年中国网大的中国高等学校综合实力排名，暨南大学居 42 位，继续稳定在中国 2000 多所高校的名校之列。

图书馆建设一直是我校“211 工程”基础设施建设项目之一，学校始终对图书馆的硬件软件建设倾注大量心血，并给予了大力支持。现在，一座新的更大规模的图书馆即将落成，这对于更好地为教学科研提供支撑，丰富学生第二课堂，拓展学生的知识面，必将起到重要作用；同时，对于进一步促进学校“211 工程”建设，也为我校迎接即将到来的教育部本科教学评估打下了坚实的基础。希望图书馆能抓住发展的良好机遇，进一步树立现代图书馆的新理念，努力开拓创新，以一流的环境、一流的管理创造出一流的服务、一流的业绩，为学校的发展做出新的贡献。

在竞争日益激烈的 21 世纪，我们的面前会有不少的机遇和挑战，这就需要每一位暨南人担负起历史赋予的责任，以纵观学校总体、服从战略部署的眼光，从学校事业的高度来考虑各项工作。我诚恳地希望全校教职员工能以新图书馆的建设为契机，团结一致，努力拼搏，以昂扬的姿态、崭新的面貌投入到工作中去，为开创我校工作的新局面，为把暨南大学早日建设成为一所国际知名的高水平研究型大学而努力奋斗。

---

① 本文是在暨南大学新图书馆封顶仪式上的讲话，广州，2006 年 1 月 14 日；一个大学校长的探索，北京：高等教育出版社，2011，297-298.

63. 生逢其时的学院①

我在 1991 年 11 月调任暨南大学副校长前，担任上海工业大学副校长兼经济管理学院首任院长，又兼任上海高校经济管理学院联谊会首任主席。及至 1995 年 12 月担任暨南大学校长后，一直关注经济管理学科的发展。

适逢此时教育部对学科分类及专业进行了调整，管理学作为独立的学科门类从经济学科划分出来，为该学科的发展提供了更为广阔的空间。学校适时就管理学院的成立向各方面征求意见。由于暨南大学经济管理学科是学校的强势学科，历史悠久，学科关系较为复杂，一些领导和教师担心管理学院的分设会削弱经济管理学科的整体力量，所以持有不同的意见。学校领导班子经过几年努力并多方听取意见及进行充分论证后，认为成立管理学院将利大于弊，于是报请国务院侨办批准，于 1998 年 9 月暨南大学管理学院正式挂牌成立。

虽然管理学院成立的时间较晚，但由于国家颁布了学科调整的新规，对管理学科的发展极为有利，加上历史的积淀和一支优秀的师资队伍，学院成立不久即获得了企业管理专业和会计学专业的博士学位授予权。至今学院已拥有 2 个博士授权一级学科，各项事业也蒸蒸日上，我甚感欣慰。

① 本文原载《声教四海 商科一脉》，暨南大学管理学院编，2006，70.

64. 述职报告①

根据国务院侨办调令，我于1991年11月27日来到暨南大学任副校长，1995年12月28日又任校长；其间，在1996年12月～2000年2月又兼任了校党委书记。转眼之间，校长任期10年，业已结束。整整10年时间里，在邓小平理论指引下，按照“面向海外、面向港澳台”的办学方针，在国务院侨办和广东省委、省政府的直接领导与关怀下，我和学校领导班子的同志们一道，带领、团结全校教职员工积极努力、开拓进取、勤俭办学，采取了许多在全国高校中具有首创性的改革措施，使学校的综合实力不断增强，办学水平不断提高，为国家侨务和港澳台工作服务的能力不断增强，在海内外的影响不断增大，成功实现了跨越式发展。暨南大学由一般高校上升为国家重点大学，在近5年内连续稳居全国1700余所高校的前50所名校之列，在2005年的网大排名中居第42位。

概括来说，在过去的10年时间里，我在用分口分级负责制注意抓好日常工作的同时，提出了多项改革措施，其主要目的在努力做好两件事：一是竭力提升学校的品牌，丢掉“花花公子”大学的坏名声，以便使暨南大学校风、教风和学风变好，成为国际化、现代化、综合化的高水平研究型大学，为中华文化的更好传播、祖国统一大业的更快完成和现代化建设的早日实现做出更大贡献；二是大力改善教职工的工作、生活条件。今天，我可以欣慰地向党和人民报告，这两项工作均取得了令人满意的结果。

一、主要改革性措施

对于一所高校而言，能否取得更快更好地发展，关键在于是否制定了正确的发展战略和思路。对此，我有着深刻的认识。所以，在1996年1月4日的任职仪式上，我明确提出了“发挥优势、深化改革、保证重点、改善条件、提高质量”的发展思路，在工作中注重特色，强调改革，在发展中追求质量。自带领学校于1996年6月成为国家“211工程”大学以来，我在学校的教学、科研、招生、人事、组织、财务、后勤等方面倡导进行了多项改革。

（一）明确学校的战略目标和办学原则

在1996年开始的第一任期内，我提出暨南大学应由教学为中心的学校向教学和科研双中心的学校发展，在暨南大学这样有90年历史的高校，第一次提出重视科学研究的思想。2000年开始的第二任期内，我为学校确立了“侨校＋名校”的战略目标，因为暨南大学是有着95年历史的侨校，负有特殊使命，特色是为侨服务，没有理由不把它办成培养优质人才的学校，“侨校”和“名校”是不可分割的一个整体。在工作中，我强调“严、法、实”（即从严治校、从严治教、从严治学，依法治校和实事求是）的办学原则，坚持法治、而不是人治；坚持按照客观规律办事，而不是根据主观臆断办事；使学校更好地得到健康有序地发展，以便改变过去校风、教风、学风差的面貌，

① 本文是送交国务院侨务办公室的工作汇报，广州，2006-07-10.

甩掉在广州被老百姓称为是“花花公子”大学的坏名声。

（二）坚决推行学分制改革

为改进学风，更有利于人才培养、适应学校生源的世界性特征，我在学校本科原有学年学分制的基础上，在全国率先提出并施行与国际接轨的弹性学分制，即标准学分制，学生可以自主选择教师、自主选择课程、自主选择学习时间、自主选择学习进度，学生修满规定的学分数（目前是 160 学分），就可毕业，这样学生可以提前毕业，也可推迟毕业，有力促进了由保姆式教育向现代化教育的转变，这项制度已在全国推广。

（三）对境内外两类学生制定不同的培养目标、培养方案和教学要求

为改变过去内外学生一样培养的缺陷模式，更好地进行因材施教，我提出对海外及港澳台学生采取“面向世界、应用为主”，对内地学生采取“加强基础、目标上移”的教育目标培养人才；将内地学生培养成为德、智、体全面发展的社会主义事业建设者和接班人；将香港学生培养成为热爱祖国，拥护“一国两制”，拥护香港基本法的专业人才；将澳门学生培养成为热爱祖国，拥护“一国两制”，拥护澳门基本法的专业人才；将台湾学生培养成为热爱祖国，拥护“一国两制”，反对“台独”的专业人才；将华侨学生培养成为热爱祖国，维护祖国和平统一的专业人才；将华人学生培养成为热爱中华文化，热爱故乡的专业人才。

（四）实行规范学期制

为了更有利于教师安排教学、学生安排学习时间，也为了配合学分制的实施，为了和世界发达国家教育接轨，我在国内两学期的学制中率先提出将每学期的时间固定为 20 周，其中 16 周授课时间、2 周复习时间、2 周考试时间。

（五）率先实行教授上基础课制度

为保证基础课教学质量，使新生在低年级即打下坚实的知识基础，我提出教授必须上基础课的制度，效果良好。

（六）强调英语、计算机语言和汉语的“三语”教学

为在教学中更好贯彻“两个面向”的方针，使学生有更好的基础知识，我提出将英语、计算机语言和汉语（简称“三语”）列为全校学生的强化课程，同时还采取了一些有力措施，推动学生学好“三语”。

（七）在全国首先实行课堂教学三重评估制度

教学质量的高低与否，是衡量“名校”的重要标准之一。为了改善过去不好的教风、提高教学质量，我在全国率先提出每学期组织学生、校院系领导和听课专家组对教师的课堂教学质量进行评估，根据评估结果采取了一系列奖惩结合、以奖为主的措施，使教风好转、课堂教学质量明显提高，现已被国内许多高校参照使用。

（八）大力加强校风、学风、考风建设

我为改善过去不好的学风，于 1993 年率先在全国提出取消补考，实行重修制度，这项改革措施已在全国推广。自 2001～2002 学年上学期开始，我提出建立一次可容纳 800 余人的大型考场，将不同专业、不同年级的学生混排考试，最大限度杜绝作弊，保持诚信美德，这一全国首创性举措受到了许多高校及新闻媒体的广泛关注。同时，还

通过强化考试过程的管理，采取建试题库、分A、B试卷、加强考场管理、抽查各专业试卷等措施，促使教师认真对待考务。

（九）大力调整办学重心

我在学校提出“大力发展研究生教育、适度发展本科教育、积极发展华文教育、稳定成人教育规模”的人才培养思路，不办专科教育，不断提升办学重心，优化办学结构。

（十）更新观念，积极探索新的办学方式

从2003级开始，我与分管校长一起，在经济学院、管理学院、外国语学院、华文学院、新闻与传播学院实行按学院招生统一培养的“大平台”新模式。这一措施的实施，使学生“自主选择专业、自主选择教师、自主选择课程、自主安排学习进程”的学分制内涵得以更好地体现，也为学校内外招学生的分流教学创造了有利条件。同时，还全面开始推行导师制，对学生进行“导向”“导学”“导心”，以“导学”为主。导师制的实施，不但有利于对学生进行因材施教，更有利于学生的全面发展。

（十一）第一个成立实行全英语教学的国际学院

为适应国家经济的高速发展，适应海内外学生学习需要，我于2001年在我国率先提出并建立了全英语教学的国际学院。目前，该学院已成为我校在社会上很有显示度的学院，已开办了临床医学、国际经济与贸易、会计学、食品质量与安全、药学、行政管理等6个专业。

（十二）改革预科教育

从2001年开始，我根据招生实际情况，为预防腐败，提出停止招收预科内招生；并根据海外和港澳台学生的实际情况，将单一的一年制预科，改为三种学制：半年制、一年制和三年制，增强了预科教育对境外学生的吸引力。

（十三）实行春秋两季招生、春秋两季毕业制度

我于1998年率先在全国提出并实行春秋两季招生、春秋两季毕业的制度，以适应境外学生需要，利于学生及时入学、就业。

（十四）开拓多种渠道，扩大海外及港澳台生源

为在更大程度上方便海外及港澳台学生报读暨南大学，我提出在境外多设立报名点，进一步加强港澳台招生工作，在台湾、海外其他国家（如越南、印尼、柬埔寨、老挝、缅甸等）设立了招生办事处。同时，学校加入世界大学校长联盟（IAUP），扩大学校的国际影响。另外，加强了董事会工作，提出了增加港澳董事和海外董事以及增设卫生部、外交部、港澳办、台办领导任副董事长的措施，这对境外学生的增加都起了积极作用。

（十五）实行交换生制度

为了更好地培养国际化、现代化人才，培养精英人才，我极力主张并实施了交换生制度。我校学生只需交纳国内大学学费，便可到国外姊妹大学接受一年的教育。迄今为止，我校已向国外的姊妹大学（美国、英国、法国、日本、韩国、菲律宾等国大学）派出了200余名学生。

（十六）大力鼓励教师从事科学研究

调来暨大工作后，看到全校几乎无高水平学术论文（SCI、EI、ISTP）、科研水平较差的情况，于是从1996年开始，我提出用“教学、科研”双中心目标取代了过去单一的“教学”中心目标，并对教师的科研成果进行量化考核，将考核结果直接与校内工资挂钩，使学校教师的科研项目、科研经费、科研论文、成果推广快速增长，同时也促进了教学质量的提高。

（十七）努力提高师资质量

调来暨大工作后，我看到全校1000多名教师中，仅有7位博士生导师，8位获博士学位的老师，师资力量很差，便提出并采取多种措施，大力引进有博士学位的老师和学科带头人，积极提升教师水平。

（十八）有计划地对教师进行培训

为提高教师水平，我提出培训教师计划，学校每年划拨一定经费选派有培养潜力的教师到国外进修；同时，学校还准许每年有15%的教师攻读更高一级学位。

（十九）注重学科建设

为适应国家现代化建设发展需要，适于21世纪的世界科技发展态势，使校内一些学科得到更快发展，我提议新成立了13个学院。在“211工程”和国家、省部级重点学科建设中，将优势力量集中、组合，并投入大量的资金和人力，采取许多有力措施，注意了管理、经济、生命科技、信息工程、医学、文学、历史、华文教育、华侨华人等学科，终于形成了暨南大学的学科优势。

（二十）深入改革人事分配制度

看到全国高校80年代以来实行的创收制度，影响了高校的办学质量，滋长了腐败，故于1996年在全国率先提出量化考核的公正合理又有激励精神的校内工资分配制度。制定了全新的量化考核指标和管理方法，实行新的分配体制，即校内工资制度，以便优劳优酬、多劳多酬，充分发挥个人潜能，调动教职工教学科研积极性，被《光明日报》称为“暨大模式”，成为全国两种分配模式之一（另一种是北大、清华的九级岗位模式）。

（二十一）稳妥地调整教职工比例

为实现工作效益最大化，我提出分流政策，增加教师数量，减少党政干部数量。经过多年调整，学校在现有教职工4000人的规模下，除附属医院职工和直属后勤职工外，使专职教师、教辅人员、党政干部的比例已经达到6∶2∶2的优良结构比例。

（二十二）面向海内外招聘10个学院的院长

为改善学校管理，引进杰出人才，我于2004年提出公开招聘院长的建议。最后聘任了10位院长，其中海外3人、国内4人、校内3人，充分表明了学校容纳优秀人才的信心和胸襟。

（二十三）稳步推进机构改革

本着“精减、效能、统一”的原则，我先后提出并进行了两次大规模机构改革，裁减了11个部处和一些科级机构；原则上不调进党政干部，而只引进专业教师，使教师数量比重增大，而教职工总数仍保持不变，承担的任务却大大增加了。

（二十四）实行干部轮岗制度

从1996年开始，在中层干部中，我提议并开始实行轮岗制度。此举不仅大大激活了干部的活力、创造力，有利于干部的全面发展和工作的开拓创新；而且有利于让年轻有为的干部脱颖而出，发挥才干，为学校的发展培养了高素质的干部队伍。

（二十五）重视提高干部的管理水平

为使我校管理干部适应学校国际化发展的形势，我提出干部出国培训计划。2004年，学校组织部处和直属单位一把手赴美国威斯康星大学等大学进行管理培训和经验交流，成效显著。去年10月，学校又组织各学院院长到澳大利亚格里菲斯大学进行管理培训。今年，学校又组织各学院党委书记到澳大利亚格里菲斯大学进行管理培训。

（二十六）大力推进依法治校

学校原来制度缺乏，科学管理很差、对此，我提出“依法治校”口号，根据国家政策，制定我校各方面的制度100余种。最近又把近年来制定的有关教学、科研及行政管理的200多个制度性文件，编纂成《暨南大学文件汇编·行政管理卷》《暨南大学文件汇编·教学科研卷》，作为学校日常管理的主要依据。

（二十七）集中进行财务管理

看到学校经费严重短缺，国家教育经费太少，实验设备费每年仅50万元，基建费每年仅300万～400万元，而且校财务管理十分混乱，我提出集中进行财务管理的措施。从1996年开始，学校实行把各院系和各部处资金集中起来实行一级财务管理、奖金全由学校统一发放以有利于教职工安心做好本职工作的改革措施。实行开源节流原则，以开源为主，通过多种方法筹集资金，首先是请求中央和省政府加大投入；其次是调整结构，增加收入，同时又强调勤俭节约，不准浪费。对各单位的账户进行了清理，并加大了监管力度，不准做假账，不准做两本账，禁止搞“小金库”，实行“收支两条线”，既有助于廉政建设，保护干部，预防腐败，又使学校的办学经费大幅增加。同时，要求学校严格遵守国家及相关领导部门的财务制度，不准乱收费。

（二十八）全面推进校园信息化建设

调来暨大后，看到学校信息化建设很差，便提出接入互联网和加强电化教学的意见。现在，学校的网络接入了所有的教室、教工和学生宿舍、办公场所，每间教室均装有多媒体教学设备，各校区之间也实现了联网，并且是广东第一所接入世界互联网的高校；逐步提升校园的网络化管理程度，积极倡导无纸化办公；同时，大力加强电化教学和多媒体资源库建设，近年来一直走在全国高校前列。学校在信息化建设方面取得的突出成绩也引起了社会的广泛关注，自2004年10月校园网站改版后，日平均访问量和浏览量分别达到1.5万人次左右；而且，还成功承担了2004年世界工程师大会的中国网上论坛。

（二十九）在全国高校中第一个建立校史馆

根据我校历史悠久、特色鲜明的特点，我提出建立校史馆，在展现博大精深的中华文化和学校辉煌成就的同时，也在潜移默化中培养了境内学生以及华侨和港澳台学生的爱国爱校精神，增强了海外华人学生对中华文化的认同感，这种方式已在全国得到推广。

（三十）加强图书馆、实验室、出版社和档案建设

调来暨大后，看到学校图书馆书刊少、管理落后；实验室设备差，连本科一些基本实验也做不出来；出版社管理更混乱；档案工作也做得不好。为此，我对这些工作在增加经费投入和调入专业人员的同时，提出了许多改革措施，实现了图书馆的现代化管理，合并了分散的实验室，建立了全校的开放实验室，加强了出版社的财务和业务管理，注意了档案的科学管理。采取这些措施后，相关方面的工作得到很大好转。

（三十一）实行政企分开，抓好校园建设工作

来暨大工作后，发现校园内商业店铺林立，房屋陈旧，为此，我采取了系列措施，对后勤和产业采取政企分开方法，成立了后勤集团和科技产业集团；关闭了全校众多商店，建立了集中的商业服务中心；赶走赖在学校的许多外面的工程队；对校园进行规划，筹集资金，兴建教学楼、行政楼、图书馆楼、学生和教师宿舍、学生食堂、中小学和幼儿园楼；改变了过去校园用土地换房屋的思路。在我任职期间，既未丢失一寸土地，又无花费地增加了校园土地。同时，也抓好了校园绿化工作，成为广州市绿化先进单位。

（三十二）稳步加大基建筹资力度，加强基建工作

10 年来，学校共计划筹集基建资金 16.01 亿元，其中国家基建拨款 6.14 亿元，学校自筹基建资金 9.87 亿元。到位的基建拨款 10.39 亿元，其中国家基建拨款 3.38 亿元，学校自筹基建资金 7.01 亿元。学校实际支出基建资金 10.35 亿元，其中国家基建拨款 3.39 亿元，学校自筹基建资金 6.96 亿元，有力地支撑了学校大规模基础建设工作的进行。

由于“211 工程”建设成绩优异，国家发展改革委向我校投入 5.1 亿元建设经费，目前已到校 1.3 亿元；仍有 3.8 亿元建设经费可供使用。

（三十三）率先开展异地联合办学

由珠海市领导邀请，我积极筹划，1998 年与珠海市政府合作办学，得到珠海市无偿赠予土地 57 万多平方米，以及相当多的资金和固定资产（已超过 1 亿元），建立了珠海学院，成为我国第一个在珠海开办全日制高等教育的大学。办这所学院采用了勤俭节约的方针，仅贷款 2.85 亿元建立了这所学院，并且由珠海学院自筹资金还贷。珠海学院的办学质量与校本部一样，现有博士、硕士、学士学生约 6000 人。教育部周济部长赞誉它“不是异地办学，而是校本部的延伸”。正由于有了珠海学院，不仅增加了对境外学生的吸引力，而且也为校本部基建提供了可能。学校已为珠海特区培养了首批的当地专科生、本科生和硕士研究生。鉴于知识产权学科的重要性，2004 年 10 月，学校又与广东省知识产权局签订合作协议，共同创办了华南地区首所集教学和科研为一体的知识产权学院。

（三十四）全国第一个实行校医联合办学

由我提议并操作，已在广州、深圳、珠海、清远、江门等地先后共建了 7 所附属医院，其中 5 所是国家级三甲医院（深圳市人民医院、珠海市人民医院、广州市红十字会医院、清远市人民医院、江门市五邑中医院），不花钱，为医学院改善了办学条件，既提供了优质的医学学生实习基地，又使这些医院为我校增加了培养研究生的力

量。这种模式已在国内推广。

（三十五）中国第一个在世界五大洲均建有姊妹学校的大学

为使大学走向国际化、现代化，更能吸引世界五大洲华侨华人学生来暨大读书，我提出了与世界更多交往，并提高交流层次，学校现已同世界五大洲26个国家（包括美、英、法、日等发达国家和华人华侨最多的东南亚10个国家）和港澳地区的72所高等院校和文化机构建立了学术交流关系，使暨南大学的文凭在世界名牌大学中普遍得到承认。

（三十六）采取多种措施全面进行反腐倡廉

在1996年，由我提议设立了信访办公室，并实行每周校长接待日制度，接受群众监督，解决实际问题。为进一步加强对财务、招生、基建和设备采购等方面的审计监督，学校制定了《暨南大学党风廉政建设责任制实施办法》等一系列文件，从制度上防止腐败现象的滋生。为切实做好反腐倡廉工作，由我提议，学校在2002年与广东省检察院签订了《共同预防职务犯罪协议书》，成为全国第一个与省检察院合作共同预防职务犯罪的高校，受到上级领导和多方面好评。同时，学校还在学术领域大力开展反对学术不端行为的斗争，净化校内学术环境。为了带头做好反腐倡廉工作，10年来，我没有与基建工程队、物资供应商直接签订合同，没有参加招投标，没有与他们吃喝过；在招生、基建、采购等工作中，没有接受过任何贿赂。连我的领导班子同志们，我也没有单独请他们吃过饭。

二、强化日常工作和深化改革措施对提升学校品牌所起的作用

在日常工作得到强化的同时，上述的许多改革措施不仅具有针对性，适应了学校自身的侨校特色；而且具有普遍性，符合了当前高等教育发展规律；加上领导的支持和全体师生员工的共同努力，使学校在近几年的工作中取得了显著成绩。与我就任前的1995年相比，暨南大学在许多方面都取得了令人振奋的进步，成功实现了跨越式发展。

（一）“211工程”建设明显进步了

1996年6月，学校成功地通过了国家“211工程”部门预审，第一次成为国家100所重点大学之一。2002年7月，学校“九五”“211工程”子项目以全优的成绩通过验收，11月成功进入“十五”“211工程”建设阶段。鉴于“九五”建设的优异成绩，“十五”期间，国务院投入了5.1亿元专项基建资金，国家“211工程”协调办公室投入2800万元专项资金，改变了“九五”期间国家对我校“211工程”建设分文未投的状况。同时，省政府的投入也由“九五”的5000万元增加到“十五”的8000万元。2006年5月，学校“十五”“211工程”项目圆满通过验收。

（二）对海外及港澳台学生的吸引力增大了

由于工作努力，海外及港澳台学生数持续增长。1995～2005年秋季开学后的海外及港澳台学生数分别为1982人、2577人、2667人、3380人、3568人、4323人、4893人、6894人、7484人、8966人、10 609人，目前为10 270人，较1995年增长4.2倍；特别是在校攻读博士、硕士学位的海外及港澳台研究生达868人，约占全国总数

的 1/4。1995 年，只有 16 个国家的学生在校学习，今天的在校学生分别来自世界五大洲 77 个国家，增长 3.8 倍。而且，据教育部统计，在大前年、前年和去年连续三年的对外招生工作中，暨南大学均取得骄人成绩，报考并被学校录取的海外及港澳台学生数，均大于全国其他高校的总和。

（三）为港澳的稳定繁荣和祖国统一大业的服务能力增强了

伴随学校办学能力和办学水平的提升，暨南大学完成国家赋予的特殊历史使命的能力也有了突出表现，在加强国家侨务工作、建立和巩固爱国统一战线方面，在促进港澳回归及繁荣、台湾的早日回归、祖国统一大业的早日完成方面，已经并正在发挥着其他高校无法比拟的重要作用。

学校的诸多港澳校友曾经或正在港澳地区的主要部门担任要职，是港澳地区爱国爱港澳的中坚力量，港澳地区左派爱国群众团体的主要领袖是我校校友。据不完全统计，在澳门特区政府的公务员中，有 1000 多人毕业于暨南大学，担任中高层职务公务员（含司、局、厅、处及相应级别以上）的暨南校友就有 300 多人，其中担任局长、副局长职务的有 30 多人。澳门特首何厚铧、全国政协原副主席马万祺均担任我校董事会副董事长。澳门行政会、立法会中，各有 5 位暨南大学校友，分别占行政会、立法会成员的 1/2 和 1/5。拥有 8 万会员的澳门左派群众组织工联会主席以及街坊联合总会理事长都是我校校友。在澳门卫生医疗系统中，70％的医护人员曾就读暨南大学。在金融、教育、旅游等行业中，暨南大学的澳门校友也占有相当比例。因此，可以毫不夸张地说，对于澳门的回归和稳定，暨南大学做了相当大的贡献。

在台湾，由于地方政府不承认大陆学历，导致我校毕业生在政府任职的很少。但是，香港地区最大的左派爱国群众团体（香港工联）的领袖大多是我校校友，成为贯彻“一国两制”，促进香港繁荣稳定的中坚力量。另外，我校董事会的多数校董集中在港澳地区，他们绝大部分是社会知名人士，如香港的霍英东、曾宪梓、霍震寰、余国春等先生，任职于香港特区重要机构乃至国家领导部门，是我校另一支爱国爱港的重要力量。在台湾，我校有校友会，创立会长是前奥委会委员、台湾红十字组织负责人徐亨先生，有较大影响。在我校学习的台湾学生学成回台后，在反对“台独”、拥护祖国统一、宣传“一国两制”方面发挥了重要作用，产生了广泛的社会影响，为祖国统一大业的早日完成准备了骨干力量。

目前，暨南大学共有海外及港澳台学生 10 270 人，其中，香港学生 5001 人，澳门学生 2828 人，台湾学生 642 人。在校就读的台湾学生数占全国高校同类学生总数的 1/7。

（四）校园面积和建筑面积扩大了

校区由原来的 3 个增加到 4 个，新增了珠海学院，校园占地面积由 112.32 万平方米增至 174.43 万平方米，增长 0.6 倍；校园的总建筑面积由 46 万平方米增加到 107 万平方米，增长 1.3 倍。1996 年至今，已经建成和正在建设的校舍面积共计 83.81 万平方米，总造价为 14.68 亿元人民币；截至 2005 年底，实际支出 10.35 亿元，其中上级拨款 3.39 亿元，学校自筹 6.96 亿元。所建校舍中，已经竣工的校舍面积 60.59 万平方米，包括进入“211 工程”建设以前动工、于 1996 年竣工的校舍建筑面积 13.73 万平方米，进入“211 工程”建设以后动工并完成的校舍建筑面积 46.86 万平方米；正在施工的校舍建设面积 23.22 万平方米。

校本部完成科学馆、行政办公大楼、土木工程实验楼、医学院大楼、理工学院大楼、校友楼、华景新城暨大科技产业大厦、成教楼、附中楼、暨大附小、幼儿园等教学行政用房；金陵苑 3 栋、金陵苑 4 栋、新真如苑 23 栋、新真如苑 26 栋、新真如苑 A 区 B 区、新建阳苑 1～7 栋等学生宿舍；羊城苑 31～40 栋，新明湖苑 1～12 栋，苏州苑 11、12、13、28、29 栋，华景新城 11、13、15 栋，陶育路 131、133、135、137、139、141、143、145、147、160、162、164 栋，虹口街 8、10、12、16、18、23、25、27、29 栋，明湖 22 栋扩建，苏州苑 1 栋扩建等教工宿舍；学生食堂改建扩建、教工食堂扩建、第二游泳池、田径运动场、网球场、田径运动场辅助用房等，总面积为 40.27 万平方米的建筑工程。珠海学院完成了包括学生宿舍、餐馆活动中心、运动场和游泳池、体育馆、教学综合大楼（含现代教育技术大楼、图书馆）、行政办公大楼以及校园道路、环境、地下网管等配套工程，总建设面积为 17.92 万平方米的建筑工程。华文学院完成了办公大楼、教学大楼以及教工周转房建设等总计 2.22 万平方米的建筑工程。

当前，校本部正在进行教学大楼、管理学院教学楼、第二理工教学楼、第二文科楼、图书馆、出版社印刷厂大楼、外聘教师周转公寓楼、大礼堂、博物馆等工程建设，珠海学院正在进行行政楼等工程建设，总建设面积共计 23.22 万平方米。

上述基建工程情况见附表 1。

附表 1　暨南大学新建校舍情况（1996～2005 年）

单位	学校产权建筑面积					
	建筑物名称	建筑面积/米^2	动工时间	竣工时间	造价/元	备注
校本部	科学馆大楼	12 722.40	1997.11	2000	44 795 481.53	其中：国际会议厅 777.60 平方米
	土木工程楼	1 441.00	2003.02	2003	5 118 821.10	
	运动场辅助用房	4 840.00	2001.09	2003		
	新医学院大楼	24 359.00	2002.04	2004	58 401 622.47	
	新理工学院大楼	16 500.00	2003.03	2004		
	新成教楼	16 794.00	2004.01	2004	26 122 716.32	
	新附中楼	5 626.00	2004.03	2004	7 529 015.81	
	华景新城暨大科技产业大厦	9 996.30	1995	1996		尚未办理产权，因学校原来地换房、房产公司向银行抵押的历史原因造成
	新行政办公大楼	24 294.00	2003.03	2005	42 448 305.93	
	金陵苑 3 栋　学宿 14	11 530.00	1998.05	1999	12 177 913.25	其中：加装电梯 177 913.25
	金陵苑 4 栋　学宿 11	16 500.00	2000.07	2001	22 637 913.25	其中：加装电梯 177 913.25
	新真如苑 23 栋　学宿 6	8 025.00	2001.09	2004	7 281 993.97	

续表

单位	学校产权建筑面积					
	建筑物名称	建筑面积/米²	动工时间	竣工时间	造价/元	备注
校本部	真如苑 26 栋　学宿 8	4 519.00	2001.07	2002		
	新建阳苑 1-7 栋　学宿 17-21	23 411.00	2002.12	2003	33 155 731.73	
	真如苑 A 区　学宿 1	15 527.00	2002.11	2003	60 102 243.66	
	B 区 1　北座学宿 2	6 462.00	2002.11	2003	135 355 795.86	含 B1、B2、B3 区
	B 区 2　南座学宿 5	7 359.00	2002.11	2003		
	B 区 3	11 500.00	2002.11	2003		
	学生食堂 1 期	600.00	1995.11	1996		
	学生食堂 2 期	496.00	2002.04	2002	320 084.77	
	教工食堂扩建给学生用	2 365.10	2005	2005	2 632 659.61	
	电话楼加建	280.00		2001		
	暨大附小	4 230.00	1995	1996		
	幼儿园	3 300.00	2003.04	2004	4 906 165.81	
	游泳馆机房	120.00	2001.08	2002		
	校友楼	2 430.00	1999.10	2002	6 600 000.00	
	明湖 22 栋扩建	193.00	1999.10	1999		
	苏州苑 1 栋扩建	1 100.00	1999.07	1999		
	校本部小计	236 519.80			469 586 465.07	
华文学院	办公用房（综合大楼）	9 609.20	1995.06	1996.08	14 935 518.09	
	办公用房（教学大楼）	9 609.20	1995.06	1996.08	14 935 518.09	
	供应设施用房（配电房）	43.52	1994.09	1996.09	1 759 664.21	
	居住用房（周转房）	2 973.00	2000.01	2000.12	6 037 094.94	
	华文学院小计	22 234.92			37 703 795.23	
珠海学院	教学楼	52 793.00	2001 03	2001.08	111 427 082.13	
	体育馆	7 863.00	2003.02	2003.11	26 000 000.00	
	餐饮中心	12 078.00	2001.03	2001.08	31 429 175.97	
	后勤楼	1 533.00	2001.08	2002.02	1 993 000.00	
	一期学生宿舍	38 803.00	2001.03	2001.09	61 973 174.64	
	二期学生宿舍	24 218.00	2002.04	2002.08	28 143 421.15	
	侨生宿舍	16 210.00	2002.10	2003.08	22 668 133.95	
	交配电	8 630.00	2001.06	2003.10	9 852 951.66	
	教工住宅（一期）	14 841.47	2002.11	2003.06	14 000 000.00	
	珠海学院小计	176 969.47			307 486 939.50	
第一临床医学院	门诊五、六层加层	4 030.00	2002.05	2003.04	8 400 000.00	
	第一临床医学院小计	4 030.00			8 400 000.00	

已售出的公房建筑面积

单位	建筑物名称	建筑面积/米²	动工时间	竣工时间	造价/元	备注
校本部	羊城苑 35 栋	3 030.47	1995.05	1996.03	6 650 000.00	含 35、36、37 栋
		羊城苑 36 栋	2 766.18	1995.05	1996.03	

续表

单位	建筑物名称	建筑面积/米²	动工时间	竣工时间	造价/元	备注
校本部	羊城苑 37 栋	3 914.49	1995.05	1996.03		
	华景新城 11、13、15 栋	10 237.70	1995.05	1996.07	8 030 000.00	
	陶育路 131、133、135、137 栋	16 959.62	1994.11	1996.02	8 130 000.00	
	陶育路 160、162、164 栋	10 902.82	1995.11	1996.09	10 180 000.00	
	陶育路 139、141 栋	7 745.82	1995.11	1996.09	7 230 000.00	
	陶育路 143、145、147 栋	11 403.41	1995.11	1996.09	10 690 000.00	
	虹口街 8、10、12 栋	14 311.92	1995.11	1996.09	13 420 000.00	
	虹口街 16、18 栋	7 659.58	1995.05	1996	7 160 000.00	
	虹口街 23、25、27、29 栋	14 843.91	1995.11	1996	13 840 000.00	
	羊城苑 31	2 597.00	1999.08	2000	13 330 000.00	含 31、32、33、34 栋
	羊城苑 32	2 597.64	1999.08	2000		
	羊城苑 33	2 598.88	1999.08	2000		
	羊城苑 34	2 596.72	1999.08	2000		
	羊城苑 38	3 009.19	1998.06	2000	9 480 000.00	含 38、39、40 栋
	羊城苑 39	2 600.54	1999.08	2000		
	羊城苑 40	2 600.54	1999.08	2000		
	苏州苑 11	1 941.62	1999.08	2000	9 160 000.00	含 11、12.13 栋
	苏州苑 12	2 917.69	1999.08	2000		
	苏州苑 13	2 916.50	1999.08	2000		
	苏州苑 28	2 745.32	1999.08	2000	6 260 000.00	含 28、29 栋
	苏州苑 29	2 544.84	1999.08	2000		
	新明湖苑 1	2 643.25	1999	2001	14 760 000.00	含 1、3、5、7 栋
	新明湖苑 2	3 552.57	2000.05	2001	15 110 000.00	含 2、4、6、8 栋
	新明湖苑 3	2 268.17	1999	2001		
	新明湖苑 4	2 457.77	2000.05	2001		
	新明湖苑 5	2 369.38	1999	2001		
	新明湖苑 6	2 457.77	2000.05	2001		
	新明湖苑 7	2 406.90	1999	2001		
	新明湖苑 8	2 457.00	2000.05	2001		
	新明湖苑 9	2 457.77	2000.08	2001	14 150 000.00	含 9、10、11、12 栋
	新明湖苑 10	2 457.77	2000.08	2001		
	新明湖苑 11	2 605.35	2000.08	2001		
	新明湖苑 12	2 605.35	2000.08	2001		
	合计	166 182.09			167 580 000.00	

续表

正在施工面积						
单位	建筑物名称	建筑面积/米^2	动工时间	竣工时间	造价/元 （施工中标价）	计划投资
一、校本部	管理学院教学楼	25 217.00	2005.05	在建	54 136 672.52	70 560 000.00
	图书馆	38 180.00	2005.06	在建	90 770 546.16	118 860 000.00
	外聘教师周转公寓楼	53 000.00	2005.10	在建	90 946 829.65	90 000 000.00
	第二理工教学楼	30 000.00	2005.11	在建	38 909 188.00	90 000 000.00
	第二文科楼	12 500.00	2005，11	在建	18 872 588.38	45 000 000.00
	出版社印刷厂大楼	1 245.00	2005.11	在建	1 632 331.34	3 000 000.00
	教学大楼	31 463.00	2005.12	在建	77 466 018.17	106 560 000.00
	礼堂	11 217.00	2005.12	在建	27 969 236.06	50 000 000.00
	校本部小计	202 822.00			400 703 410.28	573 980 000.00
二、珠海学院	行政楼	29 332.00	2004.09	在建		54 620 000.00
	一、二项合计	232 154.00				628 600 000.00

注：总共新建校舍 83.81 万平方米，总造价 14.08 亿元

（五）硬件设施改善了

学校固定资产总值由 1995 年的 2.7 亿元增至 17.6 亿元，增加 5 倍多；图书藏量由 135 万册增至 270 万册，增长 1 倍；教学科研仪器设备由 4985 万元增至 2.7 亿元，增长 5 倍多。

（六）办学规模变大了

各类学生数由 13 012 人增加到 32 284 人，增长 1.5 倍；全日制学生由 8824 人增加到 23 892 人，增长 1.7 倍，相当于在原有基础上新办了一所暨南大学。其中，来自世界五大洲 71 个国家和港澳台 3 个地区的各类学生 10 270 人，数量为全国高校之冠，较 1995 年增长 4 倍多。

（七）办学层次提高了

学校博士、硕士研究生跨过 6000 大关，达到 6567 人，较 10 年前 1995 年的 615 人增长 9 倍多；研究生与本科生之比由 1995 年的 1∶8.74 上升到今年的 1∶2.4；专科生由 2472 人减为 0。在《中国高等教育评估》的 1999 年中国大学在校研究生规模排序（前 80 位）中，我校已经名列 32 位。

（八）学科建设水平提升了

本科专业由 30 个增加到 61 个，较 1995 年增长 1 倍多；硕士学位授权学科由 50 个增加到 133 个，较 1995 年增长 1.7 倍；博士学位授权学科由 7 个增加到 54 个，较 1995 年增长 6.7 倍；博士一级学位授权学科实现零的突破，达到 6 个；博士后站实现零的突破，达到 6 个；而且实现了国家重点学科、国家工程中心、国家重点基地零的突破，教学系由 21 个增至 44 个；学院数由 7 个增至 20 个，涵盖了文、史、经、管、法、理、工、医、教育等九大学科门类，学校还成为招收和培养高级管理人员工商管理硕士（EMBA）、工商管理硕士（MBA）、会计学硕士（MPACC）、公共管理硕士（MPA）临床医学硕士、口腔医学硕士、工程硕士试点学校。

（九）师资队伍结构改善、水平提高了

现在学校工作的（包括双聘的）中国科学院和中国工程院院士 7 人，实现了无院士的突破；具有博士学位的教师 530 人，较 1995 年增长 4 倍，在专职教师中的比例达 1/3，超过全国设有研究生院大学 1/4 的平均水平。已获 2 项国家教学成果奖，实现了国家教学成果奖零的突破。

（十）学生素质提高了

连续几年，学校本科学生就业率均达 90%以上。2000 年，在全国最受欢迎的大学评比中，学校名列第 18 位．学生的科技创新能力有了较大提高，在 2004 中国-宁波科技创业计划大赛中，共有来自美国、加拿大、英国、法国留学生和国内 20 个省市的 869 个项目参赛，我校报送的 5 件作品全部进入了复赛，其中《暨鹰生物股份有限责任公司创业计划》获得国家新秀创业计划奖，这是我校学生创业计划作品首次在科技部主办的创业大赛中获奖，实现了历史性突破。在历届“挑战杯”科技创新大赛中，学校均取得优异成绩，如在第八届“挑战杯”全国大学生课外学术科技作品竞赛中，我校报送的 6 件作品获得 3 个二等奖、3 个三等奖，总成绩位居广东高校第二名、全国高校并列第十五名，并成为下一届“挑战杯”竞赛的发起高校，学生的信念、责任和服务意识也有了显著增强，2001 年，我校 Warm Touch 青年志愿者服务队被共青团中央、中国青年志愿者协会授予“全国百个优秀青年志愿服务先进集体”荣誉称号。在中央电视台举办的“2004 泰豪杯全国大专辩论会”中，我校辩论队获得亚军。近 10 年来，学校一跃成为一所体育强校，学校运动员在国际国内比赛中共荣获金牌 290 余枚、其中国际比赛金牌 56 枚，拥有多名世界冠军，特别是在近三届全国大学生运动会中均名列全国高校前八名。

（十一）科研实力增强了

学校科研经费已连续三年过亿元，去年更达到 1.4 亿元，较 1995 年的 400 万元增长 34 倍。学校教职工发表的各类论文及撰写的科研报告总数有 3000 多篇，较 1995 年的 500 余篇增长近 5 倍；其中被三大索引（SCI、EI、ISTP）收录的高水平论文 297 篇，较上年增长 1.3 倍，较 1995 的 9 篇增长 32 倍。据不完全统计，10 年来，学校共有 275 项科研成果获省部级以上政府奖励，包括一等奖 40 项、二等奖 67 项、三等奖 168 项。

（十二）办学效益更加优良了

全校的教职工人数变化小，1995 年和当前的人数分别为 3601 人和 3925 人，人数仅增加了 324 人，为 9%，未超过编制数；且学校所获上级经费投入并未大幅增长，但学校完成的任务却成倍增加，显然办学效益更加优良。在 2002 年广东管理科学研究院“中国‘211 工程’大学教师人均效率排名”中，我校在全国高校中排名第 41 位。

（十三）学校的总体收入增加了

2005 年，学校四个校区加上附属第一医院的总收入达到 13 亿元，较 1995 年的 2.26 亿元增长 4.8 倍；学校的教育事业收入达到 7.38 亿元，较 1995 年 0.76 亿元增长 8.7 倍。随着学校财务状况的好转，我在 2001 年带领相关部门清理了 1978 年复办以来至 1995 年期间所有完工但都未结算的基建工程，共计 15 项，总共支付款项 2702 万元。

10 年的学校财务收支情况见附表 2。

附表 2　1995～2005 年暨南大学收支情况统计表

单位：万元

年度	总收入									总支出	收支差额
	教育事业收入	教育事业收入中的部分款项			基建拨款	华侨医院收入	产业、后勤等收入	香港暨南大学基金会收入（港币）	总计		
		中央拨款	地方拨款	捐赠收入							
1995 年	7 643	4 020	1 232	12	500	9 416	4 663	390	22 613	20 209	2 404
1996 年	12 739	4 975		168	1 900	10 292	5 887	992	31 809	26 984	4 825
1997 年	17 288	6 299	2 024	249	1 100	11 941	6 155	1 727	38 211	32 138	6 073
1998 年	18 496	7 396	2 962		2 930	14 255	5 094	1 020	41 796	31 999	9 797
1999 年	26 279	7 730	2 818	570	1 148	15 444	4 818	1 045	48 734	45 354	3 380
2000 年	30 977	10 560	4 298	650	965	18 707	7 777	1 078	59 504	52 304	7 200
2001 年	46 115	12 579	5 345	1 220	1 144	21 326	8 934	1 087	78 606	75 473	3 133
2002 年	50 617	14 504	8 679	178	3 429	23 610	8 160	1 675	87 491	79 488	8 003
2003 年	57 019	16 608	9 950	360	5 329	25 300	8 052	1 658	97 358	92 033	5 326
2004 年	68 470	19 656	13 069	109	1 429	27 983	8 596	1 099	107 578	111 219	−3 641
2005 年	73 785	20 444	14 986	1 212	14 378	31 288	9 494	1 066	130 011	127 580	2 431

注：在 2004 年出现了超支 3641 万元的情况，这是因为国家发展改革委支付的基建资金还未到位，学校又进行大规模基础建设，预先垫付了基建款项 6936 万元造成的。截至 2005 年 12 月 31 日，学校基建工程共向银行贷款 32 500 万元（包括珠海学院贷款 28 500 万元），按现行会计制度核算要求，并没有列入学校支出

(十四) 海内外声誉进一步提高了

学校在不同机构的综合实力排行榜中的位置不断上升,已连续4年在全国1700多所高校中位居前50所名校之列。2002年,学校被教育部《中国高等教育评估》杂志评为77所研究型大学之一,名列第53位;2004年名列第46位。在中国网大的中国大学综合实力排行榜中,学校在1998～2005年的排名依次为87位、72位、60位、40位、37位、36位、51位、42位。

伴随学校综合实力和海内外声誉的提高,暨南大学在去年11月7日～11日承办了第一届亚洲大学生田径锦标赛,这是国际大学生在我国举行的首次比赛,更是第一次由一所高校承办国际性体育赛事。在大家的共同努力下,整个比赛以高水平、无作弊、无投诉、无意外、无损伤的圆满效果赢得了与会人员的高度评价。而且,我校运动队以14枚金牌9枚银牌7枚铜牌的优异成绩位居参赛的19个国家和地区40所高校之首,此次暨南大学承办的亚洲大学生田径锦标赛得到了各方的一致好评,大赛竞赛指导委员会副主任刘锦钊(澳门)在点评此次大赛时,用了"奇迹"这个词来表达他的观点。2006年3月23日～26日在阿联酋首都阿布扎比召开的亚洲大学生体育联合会执委会会议中,"与会者多次就学校承办上述赛事表达了个人的欣喜和体会,赞美之词不绝。"教育部副部长、亚洲大学生体育联合会主席章新胜认为:"这次委派大学承办锦标赛事的实践,为今后亚大体联举办其他项目赛事奠下了一个良好的模式,同时,认同暨南大学在承办赛事的组织工作上做出了卓越的成果,为亚洲树立了好的榜样"世界大学生体育联合会会长佐治·桥霖(George E. Killian)先生高度赞扬:"赛事能在高规格、公平公正裁判队伍和合理的奖励制度下完成。"对于暨南大学成功承办"亚大体联历年来最盛大的锦标赛事,深深被感动"。

三、师生员工生活工作环境、条件的改善

经过10年努力,随着我所倡导的改革工作的深入进行和"侨校+名校"战略的逐步实施,学校的整体水平有了较大提升,办学效益有了大幅提高,也拥有了可以进一步改善教职员工生活的经济实力。而且,在号召广大教工爱岗敬业、乐于奉献的同时,我也注意从工作、生活方面关心广大教职员工,使师生员工的生活工作环境、学习条件得到明显改善。

第一,我校教职工的工资收入不断提高。1995～2005年,以校本部教职工为例,人均年收入分别为8254.12元、12 545.69元、20 210.31元、22 354.60元、26 710.35元、35 157.82元、43 734.80元、65 694.57元、78 000元、88 000元、88 900元,2005年与1995年相比增长近10倍。

第二,教职工的住房条件有了大幅改善。自1996年进入"211工程"建设以来,学校兴建了总面积为23.8万平方米的教工住宅。1995年,校本部教职工家庭住房总面积为169 775平方米,人均住房面积为13.5平方米;2005年,校本部教职工家庭住房总面积为320 343平方米,人均住房面积为23.74平方米。与1995年相比,上述两项面积分别增长近0.9倍和0.8倍。

第三,我校附中、附小、幼儿园的新校舍均已建成,教学质量也有了较大提高,

加上对教职工子女上大学读书的照顾，进一步解决了教职工的子女教育问题，解除大家的后顾之忧。同时，对全校教职工的医疗、福利，也给予了较好的保障。

第四，学校还努力为教职员工创造一个安静、安全、舒适的生活环境，积极进行住宅小区物业管理工作，并拨出专款补贴教职员工的物业管理费用，以使广大教职员工享受到专业的物业管理服务。

第五，为了改善教师的工作条件，正在全校实施教授有单独工作室、副教授有半间工作室、讲师助教有工作小间的计划。

此外，对全校 2000 多名离退休教职工的生活、医疗、福利条件，也逐年予以改善，他们的收入也稳步提高。

而且，学生的生活条件也有了较大改善。1995 年，学校学生宿舍面积为 8.17 万平方米，现在的学生宿舍面积为 28.26 万平方米，较 1995 年增长 2.5 倍。学生食堂建筑面积由 1.17 万余平方米增至 2.65 万平方米，增长 1.3 倍，就餐环境也有了大幅改善。

四、结语

在大家的共同努力下，暨南大学已由一所一般大学成为国家“211 工程”重点大学，成为一所名校，为侨务工作做出了贡献。虽然工作中留下了一些遗憾，如原想为暨大的未来提供更好的发展基础与机会，却没有办成珠海市和深圳市先后附带优厚条件赠予我校的 5800 亩土地和 4.79 平方公里土地；但是，当我看到学校的进步和教工生活的改善时，便忘记了这些遗憾和工作中的诸多艰辛。

作为一名暨南人，虽然离开了暨南大学校长的职位，我仍然会永远支持学校的建设和发展。在此，我衷心地期望，在国务院侨办的领导下，在新一届领导班子的带领下，把暨南大学这所侨校早日建设成为一所海内外知名的高水平研究型大学，为中华文化的传播、为国家统一大业的完成、为祖国现代化建设事业的实现做出更大的贡献。

65. 衷心祝愿暨南大学的明天更美好[①]

今天宣布新任校长的任命，在此，我衷心地向胡军同志表示热烈的祝贺！

我来到暨南大学工作，转眼已经14年多了！抚今追昔，感慨万千！借此机会，向领导和同志们表达我的真诚感谢，同时汇报一下自己的感想。

当年离开上海时，许多朋友以为我调往山东济南工作。当时，我想，暨南大学这样悠久的历史已被人遗忘，真是太可惜了。来校后，又了解到这所学校被人称为“花花公子”大学，校园面貌也不好，让人感到特别伤心、特别可惜。

为此，我感到肩上担子沉重，责任大。我暗下决心，要办事公正、公平、公开，要认真、务实，要多为国家、为人民、为全校师生员工多办好事，力争尽快把学校形象改变过来。同时，要在任期内干干净净做事。

在国务院侨务办公室和广东省委、省政府的领导下，在前任校长所打的基础之上，在领导班子的合作共事之下，得到全校师生员工的大力支持，我担任了4年副校长、10年校长，并兼3年多党委书记，现在我很愉快地离开校长岗位。回顾这14年，可以说，事做得十分辛苦、十分艰难，在改革时甚至还要冒着“枪林弹雨”。归结起来，做了让人感到欣慰的两件事。

一、把学校办成名校，为侨务工作、为祖国统一大业工作作更多贡献

按照“面向海外、面向港澳台”的办学方针，根据学校的实际情况，在教学、科研、行政管理、干部工作、党风廉政等方面提出了一系列改革措施，提出了“侨校＋名校”的发展战略，提出了“从严治校、从严治教、从严治学，依法治校和实事求是”的“严、法、实”三字的办学原则，在领导的关心和大家的支持下，应该说，办名校之事有了成效，令我感到欣慰。

学校已由一般学校提升为国家重点大学。由教学型大学提升成研究型大学，学校在国内外的声誉大大提升，在全国1577所高校的综合实力排行榜中，已连续5年排在50位以前，2005年网大排名第42位。在广东省104所高校中升至第3位。

学校在保持教职工4000人的前提下，提高效率，做大做强。在校各类学生已达30 499人，较10年前的13 012人翻了一番多；本科生由5377人增至16 336人，增长2倍；研究生由615人增至6074人，增长近9倍；为祖国现代化和侨务工作、港澳台工作做了更多实事，培养了更多人才。

学校的侨校特色更加鲜明，为侨务工作、为香港与澳门回归和巩固发展工作、为台湾回归工作、为祖国统一大业以及中华文化传播做了贡献。今日的海外及港澳台学生跨过1万人大关，达到10 609人；生源国家和地区遍布全球五大洲，达到71个；其数量是10年前的5倍和4.5倍。特别是近3年海外及港澳台的新生数，均大于全国其

---

① 本文是在离任暨南大学校长会上的讲话，广州，2006年1月14日；一个大学校长的探索，北京：高等教育出版社，2011，396-398.

他高校的总和。国际化性质更加突出。

学校办学层次上升。早在 1996 年就停办了专科。研究生数量得到快速增长，研究生与本科生数量之比由 1∶8.7 上升为 1∶2.7。一级学科博士点实现了零的突破，达到 6 个；二级学科博士点达到 40 个，为 10 年前的近 6 倍；硕士点达到 131 个，为 10 年前的 2.6 倍。国家重点学科、国家重点基地、国家工程中心等实现了零的突破。科研经费达到 1.5 亿元，为 10 年前的 38 倍。三大索引（SCI、EI、ISTP）的学术论文达到近 300 篇，为 10 年前的 32 倍。

学校的师资质量水平上升，实现了零院士的突破，有博士学位老师达 634 人，为 10 年前的近 8 倍。

学校的财力有了较大好转，去年全校总收入达到 13 亿元，较 10 年前的 2.26 亿元增长近 5 倍。而且学校未欠债，这在全国高校中罕见，仅学校下属二级学院，即珠海学院借款 2.85 亿元。学校的固定资产达到 17.2 亿元，较 10 年前的 2.7 亿元增长 5 倍多。学校的净资产达到 18 亿元，较 10 年前的 3.6 亿元增长 4 倍。学校的科研教学设备资产达到 2.33 亿元，较 10 年前的 0.49 亿元增长 3.8 倍。学校的图书达到 278.53 万册，为 10 年前 134.7 万册的 2 倍。校园土地达到 174.4 万平方米，为 10 年前 112.3 万平方米的 1.6 倍，增加的土地不仅分文未花，而且新办的珠海学院增强了学校的办学实力，使广州校本部校园重建改造成为可能。记得当时有人给我提意见，说珠海学院是“拖垮暨南大学的陷阱”。有的甚至说，异地办学要“犯政治错误”。今天，事实胜于雄辩。正如去年教育部周济部长视察我校珠海学院时，当场表扬了我校：“这不是异地办校，这是暨南大学校本部的延伸。”十年来，学校新建房屋面积达 83.8 万平方米（包括上届领导班子开始修建且未完工的 13.7 万平方米和正在修建尚未完工的 23.2 万平方米），校园建筑面积达到 107 万平方米，为 10 年前 46.4 万平方米的 2.3 倍。新建房屋总投资达 16 亿元。

二、师生员工生活条件得到了改善

经过十年的努力，全校教职员工的平均年收入由 10 年前的 8254.12 元增长为去年的 88 900 元，增长 10 倍。

全校教职工家庭人均住房面积由 10 年前的 13.5 平方米增长到现在的 23.74 平方米，增长近 1 倍。

为附属的中学、小学和幼儿园新修了校园，实施了照顾教职员工子女的上学优惠政策，解除了教职员工的后顾之忧。

大学生的生活和学习条件也得到了大大的改善。

在此，我要再次感谢侨办和省委、省政府领导的关心和支持，感谢领导班子同志的合作以及全校师生员工的共同努力和支持！

除了上述两点欣慰以外，我还有两点遗憾。

（1）2001 年，珠海市人民政府赠送我校 5800 亩唐家湾土地以及 1 亿元建校经费开办新的校园。

（2）最近，深圳市人民政府赠送我校 7200 亩大梅沙东面土地以及校园建筑物免费修建，以开办新的校园。

上述两件事未获领导批准，失去了未来发展机遇，对此深表遗憾。

记得5年前，某市还看不起我校，曾发文件宣布不招收我校的毕业生，而现在盛情邀请我们去办学，令人感慨。

由于自己能力有限，还有许多事情未做好，请大家谅解。

最后，我衷心希望新的领导班子把工作做得更好，衷心祝愿暨南大学的明天更美更好，为国家现代化和侨务工作作出更大贡献！

深深感谢各位领导的关心和鼓励，深深感谢领导班子同志的倾力协作，深深感谢全校师生员工以及校友的热情支持！

附录一　刘人怀大事年表

1940 年

7 月 20 日，出生于四川省新繁县繁江镇（现为四川省成都市新都区新繁街道）。

1944 年

9 月，入新繁县繁江镇第二小学读书。

1945 年

3 月，转至新繁县繁江镇第一小学读书。因患眼疾，休学半年。

1946 年

秋天，因患病，再休学近一年。

1947 年

获四川省新繁县小学竞考优胜奖。

1948 年

再次获四川省新繁县小学竞考优胜奖。

1951 年

被选为新繁县繁江镇第一小学学生会副主席。

1952 年

7 月，小学毕业，考入新繁第一初级中学读书。

12 月 9 日，加入中国少年儿童队（1953 年改名为中国少年先锋队）。

1953 年

学校授予优秀生称号。

被选为四川省八一建军节慰问荣誉军人代表团成员，在省长带领下慰问荣誉军人。

1954 年

被选为学校少先队大队长。

再次被选为四川省八一建军节慰问荣誉军人代表团成员，在省长带领下慰问荣誉军人。

1955 年

初中毕业，考入温江中学留苏预备班。

1956 年

11 月 13 日，加入中国新民主主义青年团（后改名为中国共产主义青年团）。

1957 年

被选为班长。

1958 年

高中毕业，考入兰州大学数学系数学专业。

参加兰州大学有关国家“581 工程”（即中国第一颗东方红人造地球卫星项目）的

研制小组，任组长。

1959 年

3 月 13 日，加入中国共产党。

国庆 10 周年，获学校共青团委员会“兰州大学红专旗手”称号。

1960 年

获学校党委和校务委员会“兰州大学一九六〇年先进工作者”称号。

任党小组长，负责数学系 58 级和 59 级学生工作。

带领全班 60 名同学到陇西县文峰人民公社推广华罗庚先生的优选法。

9 月，转入固体力学专业第一班读书。

任政治辅导员，负责数学力学系 60 级学生工作。

1961 年

被选为兰州大学学生会主席，并任兰州大学校务委员会委员。

1962 年

进行毕业论文“球面扁薄圆壳的稳定性问题”研究工作，指导老师为叶开沅先生。

任兰州地区大学生纪念“一二・九”学生运动大会主席。

获“兰州大学优秀学生”称号。

1963 年

在北京中国建筑科学研究院结构研究室实习，导师是著名建筑工程大师何广乾先生。

大学毕业，留校在数学力学系力学教研室任教，任叶开沅先生助手。

负责“板壳理论”课程辅导工作。

担任 63 级力学专业的政治辅导员。

到清华大学照澜院 16 号拜访钱伟长先生，聆听教诲。

到兰州 135 厂调研，接受研制被我国击落的美国 P2V 低空侦察机高度表的核心元件——波纹膜片的任务。

1964 年

编写《扁壳理论》讲义，为 60 级同学讲课。

1965 年

在国内顶级学术期刊《科学通报》第 2 期和第 3 期连续发表论文 3 篇：1. 在对称线布载荷作用下的圆底扁薄球壳的非线性稳定问题，2. 圆底扁薄球壳在边缘力矩作用下的非线性稳定问题，3. 在内边缘均布力矩作用下中心开孔圆底扁球壳的非线性稳定问题。

任定西县葛家岔公社葛家岔大队农村社教工作队领导小组成员和岘口梁生产队农村社教工作组组长。

1966 年

任榆中县马坡公社农村社教工作队办公室机要秘书、工作队共青团团委书记和工作队机关共青团支部书记。

6 月，“文化大革命”开始，被批斗为“反革命修正主义苗子”，关入“牛鬼蛇神棚”长达两个月之久，随后获平反。

1967 年

与甘肃省中医院诸凤鸣医生结婚，后有两个男孩，刘昊和刘泽寰。

1968 年

编写《弹性力学》讲义，为 63 级力学专业学生授课。

到泾川县黑河公社马槽生产队当农民。

1969 年

到黑河公社兰州大学五七干部营任秘书。

在兰州汽车修配厂当工人。

接受校军宣队命令，前往兰州石油化工机器厂二分厂，为试制产品进行科研。

1970 年

完成“‘9091’产品封头内接管间焊缝的应力计算”、“尿素合成塔底部球形封头开孔的应力计算”、“高压聚乙烯反应器径向开孔的应力计算”、“换热分离氨组合设备中水冷却气管板的应力计算”和“换热分离氨组合设备三角垫密封圈的近似计算”等课题，分别为我国第一台生产航空煤油和润滑油的铂重整反应器、第一台大型尿素合成塔、第一台最高压力（2300 大气压力）高压聚乙烯反应器和第一台换热分离氨组合设备试制成功提供了依据。

1971 年

到兰州大学一条山农场劳动。

患病住院治疗，前后一年之久。

1972 年

参加在兰州举行的全国换热器技术经验交流会，经过申辩，保护了“高压固定式热交换器厚管板弯曲理论”成果。

1973 年

论文“高压固定式热交换器管板的应力计算——复变元圆柱函数的应用”发表在《数学的实践和认识》第 1 期。

兰州大学出版《科技专刊》，专门发表在兰州石油化工机器厂完成的 3 篇论文。

1974 年

编写《材料力学》讲义，为 72 级力学专业学生授课。

继续为兰州石油化工机器厂服务，完成“加氢反应器双锥面密封中螺栓的受力分析”课题，为我国第一台大型加氢反应器试制成功提供了依据。

1975 年

完成“加氢反应器顶部厚壁球形封头大孔边缘的应力分析”课题，为我国第一台大型加氢反应器试制成功提供了依据。

完成“500 万吨/年常减压装置减压塔的截头圆锥壳的稳定性和下端部分壳体的应力分析”课题，为我国最大的洛阳炼油厂的减压塔试制提供了依据。

1976 年

编写《板壳力学》讲义，为 73 级力学专业学生授课。

1977 年

完成铁道部第一设计院（兰州）的“中国最高铁路大桥——陕西白水河铁路大桥近百米高桥墩的强度和刚度计算”课题，为白水河大桥的建成提供了依据。

完成“双层套箍式厚壁压力容器环沟部位的应力状态”课题，为我国第一台大型加氢反应器试制成功提供了依据。

1978 年

3 月，受邀调入中国科学技术大学近代力学系飞行器结构力学教研室任教，校长是郭沫若先生，系主任是钱学森先生。

到校后几天，临时奉命要求讲授“数理方程”课程，助学校渡过难关，受到《中国科学院简报》通报表扬。

经过 14 年多的压制后，论文“波纹圆板的特征关系式”终于在“文化大革命”后恢复的《力学学报》第 1 期发表。接着，受邀到上海参加第五届全国仪器仪表弹性元件学术会议作特邀报告，论文被评价“达到了国际水平”。

升任讲师，并被选为教研室副主任。

1979 年

《中国科学院简报》第 64 期发文“中国科技大学讲师刘人怀在基础理论研究方面取得新成果”，通报全院表彰。国务院副总理方毅亲笔批示：“予以表扬。”

任《应用数学和力学》（中、英文版）编委。

1980 年

升任副教授。

1981 年

作为西德洪堡基金会挑选的中国首批 8 名洪堡学者之一，前往西德留学。行前，中国科学院钱三强副院长专门接见，给予鼓励。

3 月，先到西德哥廷根市歌德学院分校学习德语。8 月，到波鸿鲁尔大学结构工程研究所任客座教授，所长是著名力学家策纳（Zerna）教授。

1982 年

专程到特里尔市和伍珀塔尔市拜谒伟大导师马克思和恩格斯的故居。

在波恩市西德总统官邸受到卡斯滕斯 Carstens 总统接见，次日西德《总报》刊载总统接见的大幅照片。

在中国驻西德大使馆受到中国科学院卢嘉锡院长（后任全国人大常委会副委员长）和著名数学家吴文俊先生的亲切接见。

波鸿鲁尔大学中国同学联合会成立，被选为首任会长。

受邀前往希腊亚里士多德大学讲学。著名力学家帕纳格奥托朴洛斯（Panagiotopoulos）教授高度赞扬学术报告：“这些工作能够体现国际板壳理论领域科学现状的最高水平。”

波鸿鲁尔大学伊普森校长专门宴请，表扬为波鸿鲁尔大学做了有益的事。学校负责外事工作的热赫女士代表学校正式邀请留在波鸿鲁尔大学任教，予以婉谢。

台湾派人邀请赴台工作，当场拒绝，并立即向中国驻西德大使馆汇报。

“弹性元件的理论研究”成果获中国科学院重大科技成果奖二等奖。

1983 年

以波鸿鲁尔大学中国同学联合会名义，主持春节庆祝活动，邀请校长、教授等近 100 人出席，表达中国留学生的谢意。次日，《鲁尔消息报》专门报道、赞扬。

在波鸿鲁尔大学工作期间，共完成 5 篇英文论文，随后陆续发表。

英文论文“双层金属旋转扁壳的非线性热稳定问题”发表在国际非线性力学顶级期刊 *International Journal of Non-Linear Mechanics*（《国际非线性力学学报》）1983 年 18 卷 5 期。美国科学院院士、《国际非线性力学学报》主编纳什（Nash）教授评价：“是原创新成果”，使用的“修正迭代法”是“优美的数学方法”。

4 月，告别西德，返回祖国。

1984 年

英文论文“中心载荷作用下波纹圆板的大挠度问题”发表在英文国际期刊《国际非线性力学学报》1984 年 19 卷 5 期。

英文论文“具有平面边缘区域的波纹圆板的大挠度问题”发表在国际固体力学权威期刊 *Solid Mechanics Achieves*（《固体力学文汇》）1984 年 9 卷 4 期。此论文被《固体力学文汇》列为该刊十年来发表的最优秀论文。

论文“波纹环形板的非线性弯曲”发表在我国顶级学术期刊《中国科学》A 辑（中英文版）1984 年 3 期和 1984 年 6 期。

论文“在复合载荷作用下波纹环形板的非线性弯曲”发表在《中国科学》A 辑（中英文版）1985 年 6 期和 1985 年 9 期。

任中国科学技术大学学术委员会委员。

受安徽省杨纪珂副省长的邀请，受聘为省长顾问，并兼任安徽省政府新办的远东工程咨询公司的副总经理（后兼任总经理）。向省政府提出黄山旅游规划等建议，并参与规划研讨。

1985 年

升任教授，任近代力学副系主任和《中国科学技术大学学报》副主编。被选为安徽省力学学会首届副理事长，接着任代理事长。任《应用数学和力学》中英文版常务编委，珠海市人民政府咨询委员。

到上海参加国际非线性力学大会，做大会报告“在均布和中心集中载荷联合作用下的波纹圆板的非线性弯曲”。

学校授予“中国科学技术大学先进工作者”称号。

1986 年

2 月，受钱伟长先生邀请，前往上海工业大学和上海市应用数学和力学研究所工作。继续兼任中国科学技术大学教授。在研究所内，任钱伟长先生助手，指导力学博士生，承担国家“七五”科技攻关第 51 项子课题“波纹管与膜片设计”。

5 月，在市中心恒丰路创办上海工业大学经济管理学院，任首任院长，并兼任预测咨询研究所所长。

6 月，任上海工业大学副校长，分管研究生教育工作，一年后又增加后勤和基建管理工作。

在安徽九华山举行第三届华东固体力学学术讨论会，任大会主席。

完成中国石化总公司北京设计院课题“新型储油罐网格顶盖的设计计算”。

“波纹圆板和双金属扁壳的非线性弯曲理论”成果获中国科学院科技进步奖三等奖。

1987 年

中华人民共和国李先念主席在上海视察，受到接见。

被选为上海市力学学会常务理事。创办上海旅游工程学会，任首任会长。

全国近代数学和力学讨论会在上海举行，任执行主席。

承担上海市科委重大科研项目“浦东新区建设工程”。作为顾问，参与完成上海市科委项目“崇明县经济、科技、社会发展战略规划研究”。

英文论文“对称线布载荷作用下夹层圆板的非线性弯曲”发表在英文国际著作 *Progress in Applied Mechanics*（《应用力学进展》），荷兰马提勒斯·尼霍夫出版社，1987。

文章“有效利用外资　扩大对外开放”发表在《国际商务研究》第 6 期。

1988 年

上海市朱镕基市长（后任国务院总理）来校看望钱伟长先生，陪同接待。

被选为上海市高校管理学院院长联谊会首任会长，上海总工程师联谊会首任会长，上海市技术经济和管理现代化研究会副理事长，中国行为科学学会创办的中国公共关系专业委员会首届副主任和中国仪器仪表行业协会传感器分会首届常务理事等，任《上海力学》副主编。

承担上海旅游事业管理局科研项目“上海旅游交通研究”。

9 月，受著名复合材料力学家贾春元邀请，前往加拿大卡尔加里大学土木工程系，任访问研究科学家，承担加拿大国家科学基金课题“复合材料层合扁球壳在均布载荷作用下的非线性稳定分析”，完成论文“考虑横向剪切的对称层合圆柱正交异性扁球壳的非线性稳定问题”，发表在《中国科学》A 辑（中英文版），1991 年第 7 期和 1992 年第 6 期。11 月底，提前完成任务，返回祖国。

1989 年

任《仪表技术与传感器》编委会副主任，被选为上海市旅游教育研究会副会长。承担上海市对外经济贸易委员会科研项目“上海华亭集团旅游宾馆摆脱当前困境的对策研究”。

英文论文“波纹圆板的非线性弯曲和振动”发表在英文国际期刊《国际非线性力学学报》1989 年第 3 期，英文论文“轴对称分布载荷作用下开顶扁球壳的非线性稳定问题”发表在英文国际期刊《固体力学文汇》1989 年第 2 期。

“板壳非线性问题的理论与计算”成果获国家教育委员会科技进步奖二等奖。

获国务院侨务办公室和中华全国归国华侨联合会的“全国优秀归侨、侨眷知识分子”称号以及上海市精神文明建设委员会、中共上海市委宣传部、上海市妇女联合会的“上海市‘五好’家庭”表彰［上海市闸北区（现归属静安区）仅有一户］。

1990 年

任国务院第四批博士生导师，被选为中国力学学会理事、上海市旅游协会（学会）（现更名为上海市旅游行业协会）首届副会长。

专著《板壳力学》和《工业企业管理岗位要素设计》由机械工业出版社出版。

文章“上海旅游交通的症结与对策研究”发表在《旅游学刊》第 2 期。

获人事部“中青年有突出贡献专家”称号，英国国际人物传记中心国际荣誉奖和国际学者名人奖。

1991 年

到加拿大多伦多出席国际第 18 届名人大会。

受邀到京接受国务院侨务办公室廖晖主任（后任全国政协副主席）召见，接受邀请前往暨南大学工作。8 月 27 日，任暨南大学副校长。因病住院，迟至 11 月 27 日才到任，分管行政、档案、外事和图书馆，联系经济学院。年底，又被选为学校工会主席。

被选为中国仪器仪表学会仪表元件学会理事长，任南京航空航天大学兼职教授，承担国家自然科学基金项目“复合材料板壳的非线性理论计算”。

英文论文“计及高阶影响的复合材料层合矩形板非线性弯曲的简化理论”和“方形网格扁球壳的非线性稳定理论”发表在英文国际期刊《国际非线性力学学报》1991 年第 5 期。论文“夹层矩形板的非线性振动”发表在《中国科学》A 辑（中英文版）1991 年第 10 期和 1992 年第 4 期。

主编著作《旅游工程原理与实践》由上海百家出版社出版。

获国务院政府特殊津贴，美国传记协会的国际终身杰出成就金像奖和杰出导师奖。

1992 年

帮助暨南大学企业管理系申报工商管理硕士（MBA）学位点，得到批准。

增加分管工作：后勤和基建，并负责深圳旅游学院的筹建。

被选为广东省力学学会副理事长、广东省复合材料学会副理事长、广东省高校价值工程研究会会长，任同济大学兼职教授。

获上海市总工会、新民晚报社、上海电视台二台、《现代家庭》杂志社授予的“上海‘十佳’现代好丈夫”称号。专著《板壳力学》获机械工业出版社优秀图书一等奖。

1993 年

鉴于群众意见，校领导重新分工，分管本科、专科、成人教育、预科、研究生理工医科、政治理论课和德育课的教学管理工作及教务处、体育部、外事处、电教中心、现代管理中心、综合档案室。参加筹办旅游学院和华文学院，联系经济学院。

任兰州大学兼职教授，被选为兰州大学校友总会副会长。

经国务院侨办批准，成立暨南大学数学力学研究所，任所长。

年底，接受德国洪堡基金会盛情邀请，以高级洪堡学者身份再赴德国波鸿鲁尔大学静动力学研究所从事科研工作，承担洪堡基金会课题“复合材料层合扁壳的非线性稳定与振动问题”。

英文论文“复合载荷作用下具有刚性中心和光滑边缘的波纹环形板的非线性弯曲”发表在英文国际刊物《国际非线性力学学报》1993 年第 3 期。专著《夹层壳非线性理论》由机械工业出版社出版。

1994 年

在德国期间，完成两篇英文论文：1. 考虑剪切影响的对称层合圆柱正交异性扁锥壳的非线性屈曲；2. 考虑剪切影响的对称层合圆柱正交异性开顶扁球壳在均布压力作用下的非线性屈曲，都发表在英文国际期刊《国际非线性力学学报》1996 年第 1 期。

3 月底，完成任务，提前返回祖国。

8 月，带领校学生运动队前往济南，参加第三届全国大学生田径锦标赛。仅开幕式后半天比赛便获得了 5 枚金牌。经受磨难，受到国家教委关怀，次年 5 月批准暨南大学为全国 53 所试办高水平运动队的大学。

受饶芃子副校长委托，前往国务院侨办和国家教委申报暨南大学中文系成为全国第一批文科基础学科人才培养和科学研究基地。虽有波折，但最终申请获得通过。

上述两件事情的成功，是暨南大学国家教委重点学科零的突破。

被选为中国力学学会常务理事和中国复合材料学会理事，任西南交通大学兼职教授。

“夹层板壳的非线性理论与计算”成果获国务院侨务办公室第一届科技进步奖一等奖。

1995 年

12 月 28 日，任暨南大学校长。

英文论文“边缘均布力矩作用下圆底夹层扁球壳的非线性屈曲”和“夹层扁锥壳的非线性振动”先后发表在英文国际期刊《国际非线性力学学报》1995 年第 1 期和第 2 期。

专著《夹层壳非线性理论》获机械工业出版社优秀图书一等奖。

广东省教育工会授予“优秀职工之友”称号。

1996 年

2 月，兼任暨南大学董事会副董事长、秘书长。

前往国家教委拜会领导，正式提出暨南大学成为“211 工程”重点大学的申请。经过艰辛努力，终于同意申请。回校后立即召开全校教职工大会，要求大家全力以赴，迎接“211”工程预审。6 月 14 日，国家“211 工程”专家组予以通过，学校成为“211 工程”重点大学。

6 月 15 日，暨南大学 90 周年校庆，双喜临门！

党和国家领导人江泽民总书记、李鹏总理、乔石委员长、李瑞环主席等领导纷纷亲赐墨宝祝贺，钱伟长副主席、廖晖主任、卢瑞华省长还亲自到校庆贺。

年底，被选为校党委书记，又任广东省科学技术协会副主席。

“全面深化教学改革，严格教学管理，促进校风、教风、学风建设”成果获广东省教学成果奖一等奖。

1997 年

向国务院侨办和广东省教育厅的暨南大学“211 工程”立项可行性论证专家组汇报 7 个建设项目立项事宜，获得批准。学校投入 1.96 亿元，正式启动“九五”“211 工程”建设。

为解决学校医学院办学困难，得到深圳市政府批准，使深圳市人民医院成为我校第二附属医院，开创大学和地方联合办医的先河。接着，又将珠海市人民医院、广州市红十字会医院、清远市人民医院、江门市五邑中医院、深圳华侨城医院和深圳市眼科医院建成学校的附属医院。

应邀参加广州市委、市政府召开的“穗港澳地区十二所高校校长与广州市领导恳谈会”，发言“实施城市管理系统工程建设　开创广州可持续发展新格局”，被林树森市长（后任贵州省省长）评价为“会议最好的发言，非常有启发，非常有针对性。”

年底，参加中国赴葡萄牙大学校长代表团去里斯本访问，受到桑帕约总统的热情接待，友好交谈。

被选为中国海外交流协会常务理事和广东省海外交流协会副会长。

承担学校“211 工程”重点学科子课题“传感器的理论研究与应用”和辽河石油勘探局课题“大直径扩孔钻头受力分析及计算机模拟研究”。

3 篇英文论文：1. 夹层壳的非线性理论：（1）中厚度壳的精确运动学；2. 夹层壳的非线性理论：（2）近似理论；3. 具有边缘大波纹的波纹环形板的非线性弯曲，发表在英文国际期刊 *Applied Mechanics and Engineering*（《应用力学和工程》）1997 年第 2 和第 3 期。英文论文“考虑横向剪切的对称层合直线型正交异性椭圆板的大挠度弯曲”发表在英文国际期刊 *Archive of Applied Mechanics*（《应用力学汇刊》）1997 年第 7 期。

“板壳非线性理论与计算”获广东省自然科学奖一等奖。

1998 年

创办暨南大学邓小平理论研究中心，任主任。任全国邓小平思想研究会顾问，承办由全国邓小平思想研究会主办的邓小平理论与改革新发展理论研讨会，乌杰会长（国家经济体制改革委员会原副主任）亲自主持会议。

在广东省委组织部和广东省党的建设学会主办的“纪念党的十一届三中全会二十年，广东党的建设理论”研讨会上，作报告“在高校党的建设中贯彻落实邓小平‘从严治党’的思想”。

应邀参加在台湾暨南国际大学举行的“华侨教育学术研讨会”，作主题报告“暨南大学兴办高等华侨教育的历史回顾与展望。”

8 月，与珠海市政府签署“共建暨南大学珠海学院协议”。接着，在珠海市唐家湾校区举行隆重的开学典礼，开创国内异地办学的先河。2 年后，平白无故遭难，被迫迁往珠海前山办学。

经过艰辛努力，在暨南大学创办管理学院。

被选为中国力学学会副理事长，中国振动工程学会常务理事和全国高等学校教学研究会理事。

承担国务院侨办科技基金项目“精密仪器仪表弹性元件的非线性理论与计算”。

英文专著 *Study on Nonlinear of Plates and Shells*（《板壳非线性力学研究》）由科学出版社和暨南大学出版社在美国纽约发行。英文论文“对称层合圆柱正交异性扁球壳的非线性动态屈曲”发表在英文国际期刊《应用力学汇刊》1998 年第 6 期。

“复合材料结构的分析与计算”成果获国务院侨务办公室第二届科技进步奖一等奖。获国务院侨务办公室“优秀教师”称号。因对海外中文教育事业和全美中文学校协会工作的大力支持和帮助，获全美中文学校协会感谢奖。

1999 年

专程到京拜访中华人民共和国副主席、暨南大学董事会名誉董事长荣毅仁先生，汇报学校工作。

创办生命科学技术学院。

应邀到京参加国庆五十周年大庆活动。

发起组织国家“211 工程”100 所大学校长第一次聚会，在暨南大学进行经验交流。

暨南大学加入世界大学校长联盟，到比利时布鲁塞尔出席世界大学校长联盟第 12 届大会，与世界各国大学校长进行交流。

应邀到香港参加“1999 年香港研究生教育会议”，作主题报告“面向 21 世纪的研究生教育”。

年底，广东省高教厅和国务院侨办组织专家组对学校“211 工程”进行中期检查，认为学校办学水平显著提高，上了一个台阶。

任日本新世纪中文电视学校校长，教育部科学技术委员会数理学部委员。

承担辽河石油勘探局课题“增程抽油井受力分析计算”。

获广东省教育厅“广东省南粤教书育人优秀教师”称号，英文专著《板壳非线性力学研究》获中国出版协会优秀著作一等奖。

11 月 20 日，收到中国工程院宋健院长来函，获知当选为中国工程院机械与运载工程学部院士。

12 月 31 日晚，在广东省肇庆市举行“全球大学生跨世纪大联欢”文艺晚会，迎接新世纪来临。来自 30 个国家和地区的 2000 多名大学生和嘉宾到会，任会议主席。

2000 年

2 月 15 日，国务院侨办宣布，继续任暨南大学校长，不再兼任党委书记。

年中，参加中国工程院第五次院士大会，党和国家领导人江泽民总书记等接见、合影，并作指示。在机械与运载工程学部会议上，作学术报告“板壳分析与应用”，汇报 37 年来的学术成果。

中国工程院第五次院士大会批准成立工程管理学部，经过评审，9 月 28 日，被中国工程院主席团选为工程管理学部首批院士。

组织 100 多位两院院士编写科普作品，汇成《院士科普书系》，第一辑 25 本由暨南大学出版社联合清华大学出版社发行。

承担国家自然科学基金项目“双层网格扁壳的非线性理论及稳定性分析”，中国工程院咨询项目“工程科学技术在社会生产力发展中的作用和地位研究”和教育部 21 世纪初高等教育教学改革项目“高等学校学分制管理制度的改革与实践”。

英文论文“夹层扁锥壳的非线性稳定性”发表在英文国际期刊《应用力学和工程》2000 年第 2 期，英文论文“变厚度 U 型波纹管的非线性变形分析”发表在英文国际期刊《应用力学汇刊》2000 年第 5 期。主编文集《应用力学研究与实践》由暨南大学出版社出版。文章“转变观念　量化考核　优劳优酬”发表在《高教探索》第 1 期。

中共广东省委、省政府授予“广东省劳动模范”称号。澳大利亚格里菲斯大学授予名誉教授。

被选为广东省机械工程学会压力容器分会理事长。

2001 年

创办全英语教学的国际学院。创建药学院、信息科技学院、新闻与传播学院、外国语学院和法学院。

鉴于原有校庆存在不足之处，进行改革。谱写新校歌，并将校庆时间改为 11 月 16 日。为增强凝聚力，设立万国墙，记录 1978 年以来，学生来源国家；塑立伟大思想家孔子和伟大科学家爱因斯坦塑像；在暨南大学办学原址南京鼓楼公园芳草地塑立暨南大学纪念碑。

党和国家领导人荣毅仁副主席，钱其琛副总理，钱伟长、霍英东、马万祺副主席等为暨南大学 95 周年校庆题词致贺。钱伟长、马万祺副主席等领导出席校庆大会。

任教育部高等学校力学教学指导委员会副主任委员，中国科学技术协会全国委员会委员，广东省政府教育咨询小组成员，广东省政协常委。

承担广州白云国际机场公司课题“广州新白云国际机场主航站楼钢结构强度、稳定性校核计算”和珠海市香洲区前山镇政府课题“珠海市前山镇（街）转型与社区建设研究”。

工程力学硕士点开始招收硕士生。

“暨南大学学分制改革的实践与研究”成果获广东省教学成果奖一等奖。获日本创价大学最高荣誉奖。

2002 年

10 月，国务院侨办和广东省政府组织专家组对学校“九五”“211 工程”建设项目进行整体验收，给予优秀评价。

接着，国务院侨办和广东省政府又组织专家组对学校“十五”“211 工程”建设项目进行立项论证，给予批准。

被选为全国高等学校教学研究会副理事长，广东省非开挖技术协会首届理事长和广东省机械工程学会副理事长。

承担广州市荔湾区政府科技局项目“以价值工程方法全面提升荔湾区商旅核心竞争力”。

著作《20 世纪我国重大工程技术成就》由暨南大学出版社出版。

为暨南大学达到科技事业单位“国家二级”档案管理标准做出突出贡献，获国家档案局荣誉证书。

2003 年

为加强反腐倡廉，与广东省检察院张学军院长签约，共同开展预防职务犯罪。

任广东省政协常委和教科文卫体委员会副主任，广东省政府参事室参事和经济组名誉组长，中国工程院院士广州咨询活动中心主任。被选为中国振动工程学会理事长，中国复合材料学会副理事长，中国仪器仪表学会常务理事。任《振动工程学报》主编，《复合材料学报》副主编。

承担教育部科研重点项目“大型建筑网壳结构在强台风作用下的非线性稳定分析

与设计应用”，国务院侨办重点学科建设项目“工程力学重点学科建设”，中国石化茂名炼油化工公司课题“螺纹锁紧环损伤分析”，中共广东省委、省政府重点项目“广东省信息化调研”。

工程力学博士点开始招收博士生。

文章“广东省发展高新技术的若干意见和建议”发表在《政协广东省第八届委员会委员建言选编》（广州）。

2004 年

面向世界公开招聘 10 位学院院长，共有 9 个国家和地区的 93 位学者应聘。

创办知识产权学院。

国家邮政局发行《侨乡新貌·暨南大学》特种邮票，是全国第二所上榜邮票的大学。

出席 9 省 2 市泛珠三角区域合作与发展论坛，演讲题目“泛珠三角：推进科技、教育和文化的区域合作”。

出席在英国伯明翰举行的欧洲华侨华人社团联合会第十二届年会，演讲题目“积极服务海外华侨华人社会”。

出席在上海举行的世界工程师大会，向大会汇报“发展中国家的工业化道路”论坛讨论情况。

“工程结构故障诊断”广东省高校科研型重点实验室获批成立，任主任。

工程力学专业开始招收本科生。

任教育部科学技术委员会管理科学部主任，广东省政府科学技术咨询委员会委员，西安交通大学兼职教授。被选为中国工程院工程管理学部常委，广东省行政管理学会副会长。

承担国家自然科学基金项目“焦炭塔变形机理分析及剩余寿命的研究”，教育部全国教育事业“十一五”规划研究项目“加强教育对外开放与合作交流的对策研究”，广东省委、省政府专题战略研究项目“现代服务业科技发展问题研究”。

文章“标准学分制的研究与实践”发表在《中国大学教学》第 3 期，“中国制造业的生存哲学”发表在《科技中国》创刊号。

因在中国学位与研究生教育学会建设中做出显著贡献获建设贡献奖。

2005 年

创办暨南大学艺术学院。

受教育部推荐，亚洲大学生体育联合会将第一届亚洲大学生田径锦标赛交给暨南大学承办。任大赛总指挥。11 月 7 日晚，开幕式隆重举行，19 个国家和地区的 300 多名运动员出席。全国政协副主席钱伟长，亚洲大学生体育联合会主席、教育部副部长章新胜，国务院侨办副主任刘泽彭，世界大学体育联合会会长佐治·桥霖等领导和嘉宾到会。经过三天比赛，进行了男子 22 项、女子 20 项角逐，完美结束。暨南大学获金牌 14 枚，亚洲第一。章新胜和佐治·桥霖对比赛给予高度评价：“组织工作卓越，为亚洲树立了榜样！”

成立暨南大学 100 周年校庆筹委会。宣布百年华诞将在 2006 年 11 月 16 日举行。

任教育部高等学校力学教学指导委员会主任委员，国家科技奖（国家自然科学奖）

评审专家和广州市政府决策咨询专家。

为国家神舟飞船研制，承担中国科学院长春光学精密机械与物理研究所课题“高级航天光学遥感器自由振动的模态分析”。另外，还承担教育部高等学校博士学科点专项科研基金项目“深海采矿软管集矿机系统动力特性和空间形态研究”和广东省教育厅项目“工程结构故障诊断重点实验室建设”。

英文论文“波纹扁球壳的非线性稳定性”发表在英文国际期刊 *International Journal of Applied Mechanics and Engineering*（《国际应用力学和工程学报》）2005 年第 2 期。主编文集《中国制造业企业国际化战略》由暨南大学出版社出版。

“深化教学改革，优化培养模式，造就高质量海外和港澳台人才的探索与实践”成果获广东省教学成果奖一等奖和国家级教学成果奖二等奖。获香港理工大学“杰出中国访问学人”称号。

2006 年

组织专家组进行学校“十五”“211 工程”6 个重点学科建设项目和 2 个公共服务体系建设项目的验收工作，均以优秀的成绩通过验收。

1 月 14 日，不再担任暨南大学校长，回到应用力学研究所和管理学院任教。在全校中层干部大会上致谢词，感谢领导和全校师生员工大力支持，才做了令人欣慰的成绩：使学校由一般学校成为国家“211 工程”重点大学，在全国高校网大排名升为 42 位。学生由 13 012 人增至 32 284 人，研究生由 615 人增至 6074 人，海外和港澳台学生由 16 个国家和港澳台 3 个地区的 1982 人增至 71 个国家和港澳台 3 个地区的 10 609 人。停办了专科，博士点由 7 个增至 54 个。实现国家重点学科零的突破。学校的年收入由 2.26 亿元增至 13 亿元，校园土地由 112.3 万平方米增至 174.4 万平方米，建筑面积由 46.4 万平方米增至 107 万平方米。教职工年均收入由 8000 元增至 88 900 元，教职工人均住房由 13.5 平方米增至 23.74 平方米。建立了高中，为附中、附小和幼儿园建设新校园，解除大家的后顾之忧。学生的学习和生活条件得到极大改善。

被选为全国科技大会代表，出席大会，听取胡锦涛总书记讲话。

10 月 16 日，应邀到浙江省绍兴市，参加 2006 年中国越剧艺术节开幕式。会前，受到中共浙江省委书记习近平（后任中共中央总书记）单独亲切接见。

被选为中国工程院工程管理学部副主任，广东省政协历届委员联谊会名誉会长，中国力学学会副理事长，全国力学课程报告论坛首届组委会主任。

专著《精密仪器仪表弹性元件的设计原理》和《复合材料层合板壳理论探索》由暨南大学出版社出版。文章“关于尽快制定国家统一法的建议”和“关于改善我国北方水资源缺乏的一个建议”刊登在《参事建言（2004—2005 年）》，中国评论学术出版社出版。

“复合材料基本力学问题的理论研究”成果获广东省科学技术奖一等奖，“关于允许市民在重大节日中有条件燃放烟花爆竹的建议”获广东省政府参事室优秀议政奖。为中国教育技术协会普通高校体育专业委员会做出重大贡献，获感谢状。学校授予“暨南大学终身贡献奖”。

2007 年

获教育部批准，“重大工程灾害与控制”教育部重点实验室（暨南大学）立项建设，任主任。

成立暨南大学战略管理研究中心，任主任。

任澳门大学校长遴选委员会委员，《科技创新与品牌》总编辑，广东省仪器仪表学会理事长。

专著《夹层板壳非线性理论分析》由暨南大学出版社出版。文章“工程管理是管理对国民经济的深度介入”发表在《中国工程管理环顾与展望——首届工程管理论坛论文集锦》，中国建筑工业出版社出版。文章“我国力学专业教育现状与思考”发表在《中国大学教学》第 1 期。

“在推进和谐社会建设中，切实解决‘农民工’身份问题”获广东省政府参事室优秀议政奖。

2008 年

国家质量监督检验检疫总局和国家标准化管理委员会授予“中国标准化科学家”称号，并任中国标准化专家委员会委员。

任沈阳仪表科学研究院顾问，同济大学顾问教授，华中科技大学兼职教授，合肥工业大学兼职教授。

受澳门科技大学校监和董事长廖泽云的热情邀请，先任澳门科技大学校监顾问，从 7 月 1 日起任澳门科技大学常务副校长。

承担中国特种设备检测研究院课题“应对技术性贸易措施检测标准研制”和广州市政府决策研究专家研究项目“关于城市基础设施建设投融资体制改革研究。”

文章“浅谈高等教育学校科学管理‘三’字经”发表在《科技创新与品牌》第 10 期和第 11 期，“谈谈创建现代管理科学中国学派的若干问题”发表在《管理学报》第 3 期，“再谈创建现代管理科学中国学派的若干问题”发表在《中国工程科学》第 12 期。

“关于将清明节设为国家法定假日的建议”获广东省政府参事室优秀议政奖。

2009 年

任教育部第六届科技委员会委员、管理科学部主任、战略指导委员会委员、学风建设委员会委员，中国工程院工程管理学部首席咨询专家，国家质量监督检验检疫总局特种设备安全技术委员会副主任委员，澳门特区政府科技委员会委员，澳门科技大学董事，广东省科普志愿者协会首任会长。

为我国第一艘航空母舰“辽宁号”舰载机歼 15B 的研制，承担中国航空工业集团 601 研究所课题“带缺陷碳纤维复合材料层合矩形板力学行为的理论研究与数值模拟技术”。

另外，还承担中国工程院工程管理学部咨询项目“东水西调工程研究”。

英文论文“夹层环形板的大挠度问题”发表在英文国际期刊 *Journal of Mechanics and MEMS*（《力学和微机电系统学报》）第 2 期，“考虑横向剪切的对称圆柱正交异性层合截头扁锥壳的非线性稳定”发表在英文国际期刊《国际应用力学和工程学报》第 3 期。论文“膜盒基体的理论与设计”发表在《澳门科技大学学报》第 1 期。“三谈创建现代管理科学中国学派的若干问题：四条定义和三点建议”发表在《中国工程科学》

第8期。

“新形势下金融风险及防范对策的思考”获广东省政府参事室优秀议政奖。

2010年

2月8日，广东省委、省政府领导：中央政治局委员、省委书记汪洋（后任中央政治局常委，全国政协主席），省长黄华华，省人大常委会主任欧广源，省政协主席黄龙云等接见院士专家，做首席发言“大力推动‘政产学研金’合作创新　为广东省经济社会发展作贡献”，受到汪洋书记赞赏。

任《澳门商报》顾问委员会名誉主席。

6月底，卸任澳门科技大学常务副校长职务。

承担“211工程”三期建设国家级重点建设项目“重大工程结构的非线性力学问题”。另外，还参加中国工程院和国家自然科学基金委员会联合项目“中国工程科技中长期（2010—2030）发展战略研究”，并负责子课题“公共安全相关工程科技中长期（2010—2030）发展战略研究”。为国家高铁工程服务，承担株洲南车时代电气公司课题“高速列车表面风载研究及牵引变流器动态响应特性分析”。

文章“工程管理信息化的内涵与外延探讨”发表在《科技进步与对策》第19期。

“大规模引进和培训人才为广东产业结构优化升级服务”和“关于将香港、澳门特别行政区的所有统计数据纳入全国性统计数据的建议”均获广东省政府参事室优秀议政奖。

2011年

任中国（澳门）综合发展研究中心首任执行会长，第二届全国科学技术名词审定委员会委员，广州市突发事件应急管理专家。

承担国家自然科学基金重点项目“纤维增强先进复合材料及其结构失效机理的多尺度力学研究”，教育部高等学校博士学科点专项科研基金项目“表面形态对生物材料与细胞相互作用影响机制的力学研究”和广东省科技厅科技项目“广东省城市生命线工程应急技术研究中心建设”。

专著《网壳结构的非线性弯曲、稳定和振动》由科学出版社出版，《一个大学校长的探索》由高等教育出版社出版。主编文集《国际化视野与本土化关注——MBA战略管理案例精选集》由科学出版社出版。

文章“试答‘钱学森之问’”发表在《中国高校科技》第10期，论文“弹性元件国内外理论发展概况”发表在《仪表技术与传感器》第9期。

2012年

被中国科学技术协会聘为“科学道德与学风建设宣讲团专家”，先后在呼和浩特、包头、广州、西安、宁波等地为高校学生做“献身科学　追求真理”报告。

为推动中国管理科学的创建与发展，于12月12日，在澳门发起和组织由教育部科学技术委员会管理科学部主办的第一届中国管理科学论坛，海峡两岸暨香港、澳门近百位专家学者出席，任论坛主席，做主题报告“现代管理科学中国学派研究综述”。

中国人民解放军总政治部和中国工程院组织院士军营行，受到中央军委委员、空军司令员许其亮上将（后任中央军委副主席）和空军政委邓昌友上将的热情接待。

经国务院侨办批准成立“城市生命线工程结构安全”国际联合实验室，任主任。

任武汉大学质量发展战略研究院学术委员会主席、东北大学兼职教授、工业和信息化部电子第五研究所兼职专家。

文章“组建‘政产学研金’合作平台 推动协同创新迈上新台阶”发表在《中国高等教育》第20期。

“关于实行九年一贯制办校的建议”获广东省政府参事室优秀议政奖。

2013年

2月，在海南三亚参加中国科学技术协会党组书记陈希（后为中央政治局委员，中央组织部部长）主持的全国科学道德和学风建设宣讲教育专家研讨会，接受任务，在“中国科学技术协会2013年度弘扬科学道德 践行‘三个倡导’奋力实现‘中国梦’”报告会上宣讲。这一年，先后在贵阳、沈阳、济南做“百年追梦 科技兴国”报告。同时，在贵阳为贵州省党政领导建言：“以创新推动贵州绿色崛起。”

任重庆大学航空航天学院名誉院长，西南交通大学特聘专家，中国仪器仪表学会名誉理事。

著作《20世纪中国知名科学家学术成就概览（管理学卷）》第一分册和第二分册由科学出版社出版，文章“传统文化基因与中国本土管理研究的对接：现有研究策略与未来探索思路”在《管理学报》第2期发表。

上海大学授予杰出贡献奖，“典型板壳结构的理论与计算”成果获广东省科学技术奖二等奖。

2014年

继续执行中国科学技术协会任务，先后在西宁、乌鲁木齐、石河子、海口和呼和浩特做“百年追梦 科技兴国”报告。

参加云南党政领导与院士专家座谈会，做首席发言“充分发挥桥头堡作用 推动孟中印缅经济走廊建设”。

为促进广东科技发展，发起组织广东院士联合会。第一次院士会员大会于6月在北京会议中心举行，81位院士参加，被选为执行会长。

10月，以“面向实际的中国管理科学”为主题，在成都西南交通大学举行第二届中国管理科学论坛，任论坛主席。

被选为中国工程院机械与运载工程学部常委。任国家发展改革委国家信息中心旅游研究规划中心学术委员会主席，天津市院士专家工作发展促进会顾问，国家自然科学奖评审委员会委员。

专著《压力容器和压力管道的分析与计算》和《旅游工程管理研究》由科学出版社出版。文章“‘技工荒’困扰中国高尔夫球具代工产业”发表在《香港信报财经月刊》第4期。

2015年

教育部批准，“重大工程灾害与控制”教育部重点实验室（暨南大学）正式成立，任主任。

广东院士联合会和广州市政府联合主办第一届广东院士高峰年会，70位院士出席，任会议主席。院士会员已有 186 名。接着，又在东莞举行广东院士团队科技创新成果展示暨院士专家东莞行。

任上海大学特聘专家，杭州电子科技大学创新与发展研究院院长，兰州大学管理学院战略委员会主席。

承担广东省高水平大学重点学科建设项目“非线性力学及其工程应用”。

专著《工程管理研究》和《工商管理研究》由科学出版社出版。

广州市城市管理委员会授予“广州垃圾分类公益形象大使”称号。

2016 年

5 月，经过艰辛努力，暨南大学力学与建筑工程学院终于获批成立，形成“本科—硕士—博士—博士后”完整人才培养体系。

第三届中国管理科学论坛在杭州电子科技大学举行，会议的主题为“‘互联网＋’生态下的中国管理科学”，海峡两岸暨香港、澳门 100 多位专家学者到会，任论坛主席。同时，为主编英文国际期刊 *Chinese Management Studies*（《中国管理研究》）的特刊，使中国管理科学的声音走出国门，特在论坛中加设英文论文论坛。

由广东院士联合会和深圳市政府联合主办的第二届广东院士高峰年会在深圳举行，会议的主题是“聚焦未来，创新驱动，绿色发展”，任会议主席，共组织了 12 大项 42 场专项活动。

任南京航空航天大学直升机旋翼动力学国家级重点实验室学术委员会主席，中船海洋与防务装备公司专家顾问。

专著《现代管理的中国实践》和《教育与科技管理研究》由科学出版社出版。

广东省科学技术协会、广东省教育厅、广东省科学技术厅和广东省知识产权局授予广东省青少年科技教育荣誉奖。

英文论文“用中国的阴阳协调观点认识劳动冲突”发表在英文国际期刊 *International Journal of Conflict Management*（《国际冲突管理学报》）2015 年 26 卷第 3 期，获该刊杰出论文奖。

2017 年

由广东院士联合会和佛山市政府联合主办的第三届广东院士高峰年会在佛山举行，会议主题为“开放引领，创新驱动，智造未来”，53 位院士出席，任会议主席。中央政治局委员、广东省委书记胡春华出席，高度赞扬广东院士联合会：“年会主题充分体现了习近平主席的要求，完全符合广东实际。”

第四届中国管理科学论坛在杭州电子科技大学举行。论坛主题为“创新驱动竞争力：中国转型背景下的宏观微观管理”，海峡两岸暨香港、澳门 100 多位专家学者参加，任论坛主席。为纪念和总结论坛的成功举行，特将论坛的优秀论文汇成文集出版。

被选为广东省老科学技术工作者协会会长，广东省智能制造专家委员会委员。

主编英文国际期刊《中国管理研究》2017 年 11 卷 1 期特刊，正式发行。英文文章“现代管理科学中国学派的研究”和“中国社会经济转型中极其重要的管理问题——对战略和创新的多种科学展望”在本期发表。论文“复合材料层合板壳非线性力学的研究进展”在《力学学报》第 3 期发表。

2018 年

5 月，广东院士联合会第二次会员大会在北京会议中心举行，120 位院士会员出席，选举了第二届领导，被选为会长。广东院士联合会院士会员已达 230 人。

第五届中国管理科学论坛在长春吉林大学举行，论坛主题为“新技术革命与产业变革下的中国管理”。海峡两岸暨香港、澳门 150 多位专家学者出席，任论坛主席，做主题报告“关于改善我国北方水资源缺乏的建议”。

广东院士联合会和东莞市政府联合主办的第四届广东院士高峰年会在东莞举行，会议主题是“拥抱新时代，抢抓新机遇，激发新动能”，61 位院士出席，任会议主席。

发起和组织中国工程院工程管理学部论坛“城市矿产工程前沿技术论坛”在重庆举行，100 多位专家学者出席，任论坛主席，做大会主题报告“保护人类健康 实现美丽中国”。

任中国工程科技发展战略广东研究院学术委员会主席。

承担中国工程科技发展战略广东研究院重大咨询研究项目“广东智能制造发展战略与实施路径研究”。

广州市委宣传部、广州市科技创新委员会和广州日报报业集团颁授“广州创新英雄”称号。

英文论文“中国社会经济转型中极其重要的管理问题——对战略和创新的多种科学展望”获英文国际期刊《中国管理研究》高度赞美。

2019 年

为推进粤港澳大湾区建设，发起组织粤港澳院士专家创新创业联盟，成立大会于 4 月 21 日在广州南沙举行，被选为联盟主席。

由广东院士联合会、粤港澳院士专家创新创业联盟和东莞市政府联合主办的“2019 粤港澳大湾区院士峰会暨第五届广东院士高峰年会”在东莞举行，58 位院士出席，任会议主席。会议主题为“智领新时代，智汇大湾区”，共举行了 14 大项 50 场专项活动。

第六届中国管理科学论坛在上海大学举行。论坛主题为“数据赋能，创新管理”，近 200 名专家学者参加，任论坛主席。为主编英文国际期刊《国际冲突管理学报》，增加英文论文论坛。

承担中国工程科技发展战略广东研究院重大咨询研究项目“广东省‘一核一带一区’产业布局研究”，中国工程院工程管理学部咨询研究项目“中国餐厨垃圾无害化、资源化、减量化治理战略研究”和“以人工智能技术提升粮食流通的治理能力和粮食安全水平的战略研究”。

主编著作《中国管理科学的研究与实践——第四届中国管理科学论坛论文集》由科学出版社出版。文章“绿色再制造的探索”和“珠三角城乡生活垃圾统筹治理战略研究”载入书中。

经过十年努力，就餐厨垃圾治理难题进行摸索和攻克，在成都市新都区实现科技成果转化，在“减量化、无害化、资源化”三个方面实现了对传统处置方式的颠覆性技术超越。

香港国际智慧城市研究院、广州城市矿产协会和澳门国际绿色环保产业联盟授予“2018 年度粤港澳大湾区城市矿产领域领军人物”称号。

2020 年

7 月，四川利兴龙环保科技有限公司以“联合生物加工技术”为核心的餐厨垃圾处理项目在四川省成都市新都区工业东区兴能路正式投产运行，每日处理 100 吨餐厨垃圾，能使餐厨垃圾在 36 小时内转化成非粮乙醇、工业油脂、酒糟酵母蛋白粉等高附加值市场产品，零污染、零排放，不再影响人类健康。作为问题发现者，十年前组织工程团队。核心技术是世界领先创新技术，由教育部新世纪优秀人才支持计划获得者、暨南大学生命与健康工程研究院副院长刘泽寰教授培育出“噬污酵母”，拥有近 70 项专利，包括美国、欧盟和日本专利，研究论文发表在国际生物权威刊物 *Bioresource Technology*（《生物资源技术》）。父子同心，经过 10 年艰苦卓绝的努力，得到新都区委区政府和兴城建设投资有限公司的大力支持，才使创新技术成果转化成功。目前，正在推广中。最近，刘泽寰又研发了新一代“噬污酵母”，效益提升了一倍。

年底，由广东院士联合会、粤港澳院士专家创新创业联盟和东莞市政府联合主办的 2020 粤港澳院士峰会暨第六届广东院士高峰年会在东莞举行，大会主题“科学引领，跨界创新，融合发展”，54 位院士出席，2000 多人参加，任会议主席。联合会拥有院士会员 278 位。

第七届中国管理科学论坛在广西南宁广西大学举行。论坛主题“全球疫情下经济转型与管理创新”，海峡两岸暨香港、澳门近 200 名专家学者出席，任论坛主席。

承担中国工程院工程管理学部咨询研究项目“强制分类背景下城市生活垃圾治理和评价机制研究”。

主编英文国际期刊《国际冲突管理学报》2000 年 31 卷 3 期特刊，正式发行。英文论文“关于杂乱的技术和创新的展望——探索经济中出现的冲突特征”在本期发表。

2021 年

代表广东院士联合会和广东省人民医院签署合作协议，使广州市九所医院都能直接为院士会员和配偶就诊看病带来方便。

7 月，出席在贵阳举行的“2021 年生态文明贵阳国际论坛”，做报告“保护人类健康　实现美丽中国”。

接着，在成都参加“学习习近平总书记‘七一’重要讲话精神”报告会暨 2021 年中国工程院院士四川行活动，介绍利兴龙项目，受到大家关注。

年底，2021 粤港澳院士峰会暨第七届广东院士高峰年会仍在东莞举行，任会议主席，因病未能出席。

任广东制造强省建设专家咨询委员会副主任和国家市场监管总局特种设备局安全委员会顾问。

2022 年

5 月 30 日，广东省授予“广东最美科技工作者”称号。

截至 2022 年，共出版学术著作 15 本（力学 8 本，管理科学 7 本），主编著作 8 本（力学 1 本，管理科学 7 本），发表学术文章 702 篇（力学 238 篇，管理科学 464 篇）。

培养博士 71 人（力学 19 人，管理科学 52 人），其中澳大利亚 3 人，香港 1 人，澳门 3 人，台湾 9 人；培养硕士 143 人（力学 22 人，管理科学 121 人）。

《刘人怀自传》在 2023 年 8 月出版。

2024 年 8 月，因成都绿色循环科技产业园餐厨废弃物资源化利用项目获 2023 年度中国碳达峰碳中和十大科技创新奖。

附录二　刘人怀主要论著目录

一、力学

著作：

1. 刘人怀．板壳力学．北京：机械工业出版社，1990.
2. 刘人怀，朱金福．夹层壳非线性理论．北京：机械工业出版社，1993.
3. Liu Renhuai. Study on Nonlinear Mechanics of Plates and Shells. New York, Beijing, Guangzhou: Science Press and Jinan University Press, 1998.
4. 刘人怀．精密仪器仪表弹性元件的设计原理．广州：暨南大学出版社，2006.
5. 刘人怀．复合材料层合板壳理论探索．广州：暨南大学出版社，2006.
6. 刘人怀．夹层板壳非线性理论分析．广州：暨南大学出版社，2007.
7. 刘人怀．网壳结构的非线性弯曲、稳定和振动．北京：科学出版社，2011.
8. 刘人怀．压力容器和压力管道的分析与计算．北京：科学出版社，2014.

主编文集：

1. 刘人怀，傅衣铭，黄小清，扶名福，王璠．应用力学研究与实践．广州：暨南大学出版社，2000.

论文：

1965—1969

1. 刘人怀．在内边缘均布力矩作用下中心开孔圆底扁球壳的非线性稳定问题．科学通报，1965，(3)：253-255.

1970—1979

2. 刘人怀．高压聚乙烯反应器厚壁筒体径向开孔的应力计算．国家第一机械工业部和化工部联合设计小组科研项目报告，1970 年 7 月 27 日；压力容器和压力管道的分析与计算，北京：科学出版社，2014，80-94.
3. 刘人怀，程昌钧，陈庆益，陈文崛，张建国．高压固定式热交换器管板的应力计算——复变元圆柱函数的应用．数学的实践与认识，1973，(1)：52-64.
4. 刘人怀，陈山林．尿素合成塔底部球形封头开孔的应力计算．科技专刊（兰州大学），1973，(1)：1-13.
5. 刘人怀，陈山林．铂重整装置反应器椭球封头中心开孔接管的强度问题．科技专刊（兰州大学），1973，(1)：14-28.
6. 刘人怀．有限元法及其在薄板弯曲问题中的应用．化工炼油机械通讯，1973，(6)：22-34.
7. 刘人怀．在轴向压力与均匀外压力共同作用下薄壁截头圆锥形壳的稳定性．兰州大学学报，1975，(2)：16-25.

8. 刘人怀. 加氢反应器顶部厚壁壳体的应力分析. 化工炼油机械通讯，1975，(6)：40-55.
9. 刘人怀. 双锥密封中的内力分析. 化工炼油机械通讯，1975，(6)：55-57.
10. 刘人怀. 厚壁圆柱壳轴对称变形近似理论的应力公式. 化工炼油机械通讯，1975，(6)：58-59.
11. 刘人怀，王凯. 500万吨/年常减压装置减压塔下端部分壳体的应力分析. 压力容器，1975，(3)：1-28.
12. 刘人怀. 在边缘载荷作用下中心开孔圆底扁薄球壳的轴对称稳定性. 力学学报，1977，(3)：206-212.
13. 刘人怀. 双层套箍式厚壁压力容器环沟部位的应力状态. 兰州大学学报，1977，(4)：9-25.
14. 刘人怀. 波纹圆板的特征关系式. 力学学报，1978，(1)：47-52.
15. 刘人怀. 具有光滑中心的波纹圆板的特征关系式. 中国科学技术大学学报，1979，9 (2)：75-86.

1980—1989

16. 刘人怀. 厚壁球壳的弯曲理论及其在高压容器上的应用. 化工炼油机械通讯，1980，(3)：1-10.
17. 刘人怀，魏俊. 厚管板的设计. 化工炼油机械通讯，1980，(4)：1-12.
18. 刘人怀. 在边缘力矩作用下夹层圆板的非线性轴对称弯曲问题. 中国科学技术大学学报，1980，10 (2)：56-67.
19. 刘人怀. 双层金属截头扁锥壳的热稳定性. 力学学报，1981，特刊：172-180.
20. 刘人怀. 双层金属中心开孔扁球壳的非线性热稳定问题. 中国科学技术大学学报，1981，11 (1)：84-99.
21. 刘人怀. 夹层圆板的非线性弯曲. 应用数学和力学，1981，2 (2)：173-190.
22. Liu Renhuai. Nonlinear bending of circular sandwich plates. Applied Mathematics and Mechanics：English Edition，1981，2 (2)：189-208.
23. 刘人怀，施云方. 夹层圆板大挠度问题的精确解. 应用数学和力学，1982，3 (1)：11-23.
24. Liu Renhuai and Shi Yunfang. Exact solution for circular sandwich plate with large deflection. Applied Mathematics and Mechanics：English Edition，1982，3 (1)：11-24.
25. 刘人怀. 固定式厚管板的弯曲问题. 力学学报，1982，(2)：166-179.
26. Liu Renhuai. Non-linear thermal stability of bimetallic shallow shells of revolution. International Journal of Non-Linear Mechanics，1983，18 (5)：409-429.
27. 刘人怀. 波纹环形板的非线性弯曲. 中国科学，A辑，1984，(3)：247-253.
28. Liu Renhuai. Nonlinear bending of corrugated annular plates. Scientia Sinica：English Edition，Series A，1984，27 (6)：640-647.

29. Liu Renhuai. Large deflection of corrugated circular plate with a plane central region under the action of concentrated loads at the center. International Journal of Non-Linear Mechanics，1984，19 (5)：409-419.

30. Liu Renhuai. Large deflection of corrugated circular plate with plane boundary region. Solid Mechanics Archives，1984，9 (4)：383-406.

31. 刘人怀. 在复合载荷作用下波纹环形板的非线性弯曲. 中国科学，A 辑，1985，(6)：537-545.

32. Liu Renhuai. Nonlinear analysis of corrugated annular plates under compound load. Scientia Sinica：English Edition，Series A，1985，28 (9)：959-969.

33. Liu Renhuai. Nonlinear bending of corrugated circular plate under the combined action of uniformly distributed load and concentrated load at the center. Proceedings of The International Conference on Nonlinear Mechanics，Beijing：Science Press，1985，271-278.

34. Liu Renhuai. The study on nonlinear bending problems of corrugated circular plates. The Symposium of Alexander von Humboldt Foundation，Shanghai，1987-10-17.

35. Liu Renhuai. On large deflection of corrugated annular plates under uniform pressure. The Advances of Applied Mathematics and Mechanics in China，1987，1：138-152.

36. Liu Renhuai. Nonlinear bending of circular sandwich plates under the action of axisymmetric uniformly distributed line loads. Progress in Applied Mechanics，Dordrecht：Martinus Nijihoff Publishers，1987，293-321.

37. 刘人怀，成振强. 集中载荷作用下开顶扁球壳的非线性稳定问题. 应用数学和力学，1988，9 (2)：95-106.

38. Liu Renhuai and Cheng Zhenqiang. On the nonlinear stability of a truncated shallow spherical shell under a concentrated load. Applied Mathematics and Mechanics：English Edition，1988，9 (2)：101-112.

39. 刘人怀，李东. 均布载荷作用下开顶扁球壳的非线性稳定问题. 应用数学和力学，1988，9 (3)：205-217.

40. Liu Renhuai and Li Dong. On the nonlinear stability of a truncated shallow spherical shell under a uniformly distributed load. Applied Mathematics and Mechanics：English Edition，1988，9 (3)：227-240.

41. 刘人怀. 复合载荷下波纹圆板的非线性分析. 应用数学和力学，1988，9 (8)：661-674.

42. Liu Renhuai. Nonlinear analysis of a corrugated circular plate under combined lateral loading. Applied Mathematics and Mechanics：English Edition，1988，9 (8)：711-726.

43. 刘人怀，朱高秋. 夹层圆板大挠度问题的进一步研究. 应用数学和力学，

1989，10 (12)：1041-1047.

44. Liu Renhuai and Zhu Gaoqiu. Further study on large deflection of circular sandwich plates. Applied Mathematics and Mechanics：English Edition，1989，10 (12)：1099-1106.

45. Liu Renhuai and He Linghui. On the nonlinear stability of a truncated shallow spherical shell under axisymmetrically distributed load. Solid Mechanics Archives，1989，14 (2)：81-102.

46. Liu Renhuai and Li Dong. On the non-linear bending and vibration of corrugated circular plates. International Journal of Non-Linear Mechanics，1989，24 (3)：165-176.

47. 刘人怀，聂国华. 板壳非线性力学研究的最新进展. 上海力学，1989，10 (3)：19-32.

48. 刘人怀，宗赴传. 集中载荷作用下具有平面边缘的波纹圆板的非线性弯曲问题. 首届全国敏感元件与传感器学术会议论文集，上海：哈尔滨市科委编辑，1989，547-550.

1990—1999

49. 刘人怀，何陵辉. 轴对称分布载荷作用下开顶扁球壳的非线性稳定问题. 上海工业大学学报，1990，11 (5)：407-419.

50. 刘人怀. 板壳非线性力学. 自然科学年鉴 (1989)，纪念创刊十周年专辑，上海：上海翻译出版公司，1990，3. 71.

51. 刘人怀，何陵辉. 四边简支对称正交层合矩形板的非线性弯曲问题. 应用数学和力学，1990，11 (9)：753-759.

52. Liu Renhuai and He Linghui. Nonlinear bending of simply supported symmetric laminated cross-ply rectangular plates. Applied Mathematics and Mechanics：English Edition，1990，11 (9)：801-807.

53. 刘人怀，李东. 波纹圆板的非线性弯曲和振动. 上海工业大学学报，1990，(4)：283-294.

54. 刘人怀. 对称圆柱正交异性层合扁球壳的非线性稳定问题. 应用数学和力学，1991，12 (3)：251-258.

55. Liu Renhuai. On the nonlinear stability of symmetrically laminated cylindrically orthotropic shallow spherical shells. Applied Mathematics and Mechanics：English Edition，1991，12 (3)：271-279.

56. 刘人怀. 考虑横向剪切的对称层合圆柱正交异性扁球壳的非线性稳定问题. 中国科学，A 辑，1991，(7)：742-751.

57. Liu Renhuai. Nonlinear stability of symmetrically laminated cylindrically orthotropic shallow spherical shells including transverse shear. Science in China：English Edition，Series A，1992，35 (6)：734-746.

58. 刘人怀，吴建成. 夹层矩形板的非线性振动. 中国科学，A 辑，1991，(10)：

1075-1086.

59. Liu Renhuai and Wu Jiancheng. Nonlinear vibration of rectangular sandwich plates. Science in China: English Edition, Series A, 1992, 35 (4): 472-486.
60. Liu Renhuai and He Linghui. A simple theory for non-linear bending of laminated composite rectangular plates including higher-order effects. International Journal of Non-Linear Mechanics, 1991, 26 (5): 537-545.
61. Liu Renhuai, Li Dong, Nie Guohua and Cheng Zhengqiang. Non-linear buckling of squarely-latticed shallow spherical shells. International Journal of Non-Linear Mechanics, 1991, 26 (5): 547-565.
62. 刘人怀. 新型跳跃薄膜的研究. 仪表技术与传感器，1991，(3)：10-11.
63. 刘人怀. 波纹膜片理论的研究. 仪表技术与传感器，1991，(5)：9-10；1991，(6)：9-11.
64. 刘人怀，邹人朴. 复合载荷作用下具有刚性中心和光滑边缘的波纹环形板的非线性弯曲. 江西工业大学学报，1991，13 (2)：163-171.
65. 刘人怀，聂国华. 网格扁壳的非线性弯曲理论. 江西工业大学学报，1991，13 (2)：186-192.
66. 刘人怀，聂国华. 网格扁壳的非线性自由振动分析. 江西工业大学学报，1991，13 (2)：193-198.
67. 刘人怀，何陵辉. 层合圆薄板的轴对称弯曲问题. 江西工业大学学报，1991，13 (2)：199-205.
68. 刘人怀，胡俍. U型波纹管非线性变形的刚度和应力分析理论. 全国第九届弹性元件学术会议论文，广州，1992-03-20；压力容器和压力管道的分析与计算，北京：科学出版社，2014，272-287.
69. 刘人怀，成振强. 简支夹层矩形板的非线性弯曲. 应用数学和力学，1993，14 (3)：203-218.
70. Liu Renhuai and Cheng Zhenqiang. Non-linear bending of simply supported rectangular sandwich plates. Applied Mathematics and Mechanics: English Edition, 1993, 14 (3): 217-234.
71. Liu Renhuai and Zou Renpo. Non-linear bending of a corrugated annular plate with a plane boundary region and a non-deformable rigid body at the center under compound load. International Journal of Non-Linear Mechanics, 1993, 28 (3): 353-364.
72. 刘人怀. 对称层合圆柱正交异性复合材料扁球壳的大挠度方程. 应用数学和力学：钱伟长八十诞辰祝寿文集，北京和重庆：科学出版社和重庆出版社，1993，279-284.
73. Liu Renhuai. Large deflection equations of symmetrically laminated composite cylindrically orthotropic shallow spherical shells. Applied Mathematics and

Mechanics：English Edition，Wei-zang Chien Eightieth Anniversary Volume，Beijing：Science Press and Chongqing Publishing House，1993，355-361.

74. 刘人怀，钟诚．考虑横向剪切的对称层合圆柱正交异性中心开孔扁球壳的非线性屈曲．暨南大学学报，1994，15（1）：1-12.

75. 刘人怀，翟赏中．均布载荷作用下波纹环形板的非线性弯曲．华南理工大学学报，1994，22（增刊）：1-10.

76. 刘人怀．波纹管的制造与理论研究概况．全国首届管道技术与设备学术会议的大会特邀报告，南京，1994-10-14；压力容器和压力管道的分析与计算，北京：科学出版社，2014，267-272.

77. Liu Renhuai and Cheng Zhenqiang. On the non-linear buckling of circular shallow spherical sandwich shells under the action of uniform edge moments. International Journal of Non-Linear Mechanics，1995，30（1）：33-43.

78. Liu Renhuai and Li Jun. Non-linear vibration of shallow conical sandwich shells. International Journal of Non-Linear Mechanics，1995，30（2）：97-109.

79. 刘人怀，徐加初．中心受载下具有平面边缘区域的固支波纹环形板的非线性分析．暨南大学学报，1995，16（1）：1-13.

80. 刘人怀，张小果．均布载荷作用下具有硬中心的开顶扁球壳的非线性屈曲．工程力学，1995，3（增刊）：1839-1844.

81. Liu Renhuai. Non-linear buckling of symmetrically laminated，cylindrically orthotropic，shallow，conical shells considering shear. International Journal of Non-Linear Mechanics，1996，31（1）：89-99.

82. Liu Renhuai. On non-linear buckling of symmetrically laminated，cylindrically orthotropic，truncated，shallow，spherical shells under uniform pressure including shear effects. International Journal of Non-Linear Mechanics，1996，31（1）：101-115.

83. 刘人怀，王志伟．变厚度 U 型波纹管非线性变形分析．管道技术与设备，1996，（1）：1-4.

84. 刘人怀．复合材料层合扁锥壳的非线性稳定问题．第九届全国复合材料学术会议论文集，北京：世界图书出版公司，1996，249-255.

85. 刘人怀，王璠．复合材料层合扁球壳的非线性振动屈曲．第九届全国复合材料学术会议论文集，北京：世界图书出版公司，1996，256-261.

86. 刘人怀，梅魁银．中心开孔扁球壳在均布载荷作用下的非线性屈曲．暨南大学学报，1996，17（5）：1-7.

87. 刘人怀，王志伟．复合材料面层夹层板中转动一致有效理论．上海力学，1996，17（3）：222-228.

88. 刘人怀，朱金福，张小果．夹层环形板的非线性弯曲．暨南大学学报，1997，18（1）：1-11.

89. 刘人怀，肖潭. 双层正交正放网格扁壳结构的非线性弯曲理论. 现代力学与科技进步（庆祝中国力学学会成立 40 周年学术会议论文集），第 3 卷，北京：清华大学出版社，1997，1212-1215.

90. 刘人怀，王璠. 复合材料层合扁球壳的非线性强迫振动. 力学学报，1997，29 (2)：236-241.

91. Liu Renhuai and Zhu Jinfu. Nonlinear theory of sandwich shells, Part Ⅰ-Exact kinematics of moderately thick shells. Applied Mechanics and Engineering, 1997, 2 (2): 213-240.

92. Liu Renhuai and Zhu Jinfu. Nonlinear theory of sandwich shells, Part Ⅱ-Approximate theories. Applied Mechanics and Engineering, 1997, 2 (2): 241-269.

93. Liu Renhuai, Xu Jiachu and Zhai Shangzhong. Large-deflection bending of symmetrically laminated rectilinearly orthotropic elliptical plates including transverse shear. Archive of Applied Mechanics, 1997, 67 (7): 507-520.

94. Liu Renhuai and Yuan Hong. Nonlinear bending of corrugated annular plate with large boundary corrugation. Applied Mechanics and Engineering, 1997, 2 (3): 353-367.

95. Liu Renhuai. The effect of the Göttingen school on investigation of flexible plates and shells in China. The Symposium on International Scientific Cooperation for Developing Countries, Bonn, Germany, 1997-04-15.

96. Liu Renhuai and Li Dong. A new approach to nonlinear vibration of orthotropic thin circular plates. Proceedings of the International Conference on Vibration Engineering, Shenyang: Northeastern University Press, 1998, 248-252.

97. Liu Renhuai and Wang Fan. Nonlinear dynamic buckling of symmetrically laminated cylindrically orthotropic shallow spherical shells. Archive of Applied Mechanics, 1998, 68: 375-384.

98. 刘人怀，肖潭. 矩形底双层网格扁壳的非线性弯曲. 暨南大学学报，1998，19 (3)：1-5.

99. 刘人怀，肖潭. 矩形底双层网格扁壳的非线性屈曲. 暨南大学学报，1999，20 (1)：1-6.

2000—2009

100. 刘人怀. 板壳分析与应用. 中国工程科学，2000，2 (11)：60-67.

101. Liu Renhuai and Li Jun. On nonlinear stability of shallow conical sandwich shells. Applied Mechanics and Engineering, 2000, 5 (2): 367-387.

102. 刘人怀，王志伟. 考虑横向剪切应力连续的复合材料面层夹层壳非线性一致有效理论. 应用力学研究与实践，广州：暨南大学出版社，2000，1-12.

103. Liu Renhuai and Wang Zhiwei. Nonlinear deformation analysis of a U-shaped bellows with varying thickness. Archive of Applied Mechanics, 2000, 70:

366-376.

104. 刘人怀，王璠. 浅谈我国生物力学的研究近况. 暨南大学学报，2001，22 (1)：59-63，69.

105. 刘人怀. 多姿多彩的板壳结构. 高校招生，2002，(7)：1.

106. 刘人怀，王璠. 波纹扁球壳的大挠度方程. 钱伟长教授九十华诞祝寿文集，上海：上海大学出版社，2003，15-22.

107. 刘人怀，张卫，王璠，徐加初，郭葆锋. 广州新白云国际机场主航站楼结构的动态分析. 华南理工大学学报，2003，31 (增刊)：7-9.

108. Liu Renhuai and Wang Fan. Nonlinear stability of corrugated shallow spherical shell. International Journal of Applied Mechanics and Engineering，2005，10 (2)：295-309.

109. 刘人怀，李东，袁鸿. 正交异性扁薄球壳的非线性轴对称振动. 振动工程学报，2005，18 (4)：395-405.

110. 刘人怀，宁志华. 焦炭塔鼓胀与开裂变形机理及疲劳断裂寿命预测的研究进展. 压力容器，2007，24 (2)：1-8.

111. 刘人怀. 膜盒基体的理论与设计. 澳门科技大学学报，2009，3 (1)：111-116.

112. Liu Renhuai. Large deflection of annular sandwich plates. Journal of Mechanics and MEMS，2009，1 (2)：145-156.

113. Liu Renhuai and Su Wei. Nonlinear Stability of symmetrically laminated cylindrically orthotropic truncated shallow conical shells including transverse shear. International Journal of Applied Mechanics and Engineering，2009，14 (3)：769-790.

2010—2019

114. 刘人怀，袁鸿. 弹性元件国内外理论发展概况. 仪表技术与传感器，2011，(9)：1-8，29.

115. 刘人怀，薛江红. 复合材料层合板壳非线性力学的研究进展. 力学学报，2017，49 (3)：487-506.

二、管理科学

著作：

1. 刘人怀，斯晓夫. 工业企业管理岗位要素设计. 北京：机械工业出版社，1990.
2. 刘人怀. 一个大学校长的探索. 北京：高等教育出版社，2011.
3. 刘人怀. 旅游工程管理研究. 北京：科学出版社，2014.
4. 刘人怀. 工程管理研究. 北京：科学出版社，2015.
5. 刘人怀. 工商管理研究. 北京：科学出版社，2015.
6. 刘人怀. 现代管理的中国实践. 北京：科学出版社，2016.
7. 刘人怀. 教育与科技管理研究. 北京：科学出版社，2016.

8. 刘人怀. 刘人怀自传. 北京：科学出版社，2023.

主编文集：

1. 刘人怀. 旅游工程原理与实践. 上海：百家出版社，1991.
2. 常平，刘人怀，林玉树. 20世纪我国重大工程技术成就. 广州：暨南大学出版社，2002.
3. 郭重庆，刘人怀. 中国制造业企业国际化战略——第五届中国青年科技企业家管理论坛文集. 广州：暨南大学出版社，2005.
4. 刘人怀，杨东进，朱锋. 国际化视野与本土化关注——MBA战略管理案例精选集. 北京：科学出版社，2011.
5. 王礼恒，刘人怀，郭重庆，王基铭. 20世纪中国知名科学家学术成就概览・管理学卷・第一分册. 北京：科学出版社，2013.
6. 王礼恒，刘人怀，郭重庆，王基铭. 20世纪中国知名科学家学术成就概览・管理学卷・第二分册. 北京：科学出版社，2013.
7. 刘人怀，丁烈云. 中国管理科学的研究与实践——第四届中国管理科学论坛论文集. 北京：科学出版社，2019.

文章：

1979

1. 刘人怀. 欢迎新同学们. 中国科学技术大学开学典礼上代表教师的讲话，合肥，1979-09-01；教育与科技管理研究，北京：科学出版社，2016，143-144.
2. 刘人怀. 谈谈课堂教学中的几个问题. 中国科学技术大学简报，第103期，1979-09-27；教育与科技管理研究，北京：科学出版社，2016，142-143.

1980

3. 刘人怀. 谈谈大学的学习生活. 中国科学技术大学近代力学系新生会讲话，合肥，1980-11-03；教育与科技管理研究，北京：科学出版社，2016，144-148.

1987

4. 刘人怀. 谈谈科研中的几个问题. 在上海工业大学经济管理学院科学报告会上的讲话（按录音整理），上海，1987-04-07；高教研究，1987，1（2）：15-20.
5. 刘人怀. 青年人的奋斗方向. 在上海工业大学研究生座谈会上的讲话（按录音整理），上海，1987-04-08.
6. 刘人怀，史乐毅. 有效利用外资 扩大对外开放. 国际商务研究，1987，(6)：1-5.
7. 刘人怀. 征求意见 完善报告. 在上海市科技委员会“崇明县2000年经济、科技、社会长远发展规划战略研究”重大课题评审会上的讲话，上海，1987-07-29.
8. 刘人怀. 领导用人标准. 在上海市委党校学习时的读书报告，上海，1987-11-13；现代管理的中国实践，北京：科学出版社，2016，100-104.

1989

9. 刘人怀，斯晓夫. 如何防止公共关系庸俗化. 公共关系，1989，(4)：10.
10. 刘人怀，于英川等. 上海浦东新区建设工程. 上海市科学技术委员会重大科研项目报告，上海，1989；工程管理研究，北京：科学出版社，2015，73-108.

1990

11. 刘人怀，于英川，王怡然，王荣璠，汪正元. 上海旅游交通的症结与对策研究. 旅游学刊，1990，5（2）：10-14，47.
12. 刘人怀. 城市政府工作目标管理与治理整顿、深化改革. 学习与实践，1990，（11）：35-36.
13. 刘人怀. 领导科学与领导艺术. 上海工业大学经济管理学院讲稿，上海，1990-05-28；现代管理的中国实践，北京：科学出版社，2016，85-100.
14. 刘人怀. 前进中的经济管理学院. 工大三十年，上海：上海工业大学校刊编辑室，1990，129-134.
15. 刘人怀，厉无畏，范家驹，钱幼森，姜文豹，郑琦. 上海华亭集团旅游宾馆摆脱当前困境的对策研究. 上海市对外经济贸易委员会重大科研项目报告，上海，1990；旅游工程管理研究，北京：科学出版社，2014，110-119.

1991

16. 刘人怀. 人才开发是搞好企业管理的关键. 上海管理科学，1991，（1）：6.
17. 刘人怀. 上海——旅游业的春光与希望. 旅游工程原理与实践，上海：百家出版社，1991，1-6.
18. 刘人怀，钱幼森. 旅游工程理论及其在浦东开发中的应用. 旅游工程原理与实践，上海：百家出版社，1991，43-54.
19. 刘人怀，王荣璠，汪正元，于英川，王怡然. 上海旅游交通中的若干问题. 旅游工程原理与实践，上海：百家出版社，1991，96-126.

1993

20. 刘人怀. 旅游工程学的提出. 日本神户商科大学演讲，神户，日本，1993-04-20；旅游工程管理研究，北京：科学出版社，2014，1-3.

1994

21. 刘人怀.《红杏枝头春意闹——暨南大学成人高等教育毕业生业绩选》序. 广州：华南理工大学出版社，1994，1-2.
22. 刘人怀. 根据侨校特点改进教学工作 采取有效措施提高教学质量. 在暨南大学1994年教学工作会议上的讲话，广州，1994-11-24；暨南教育，1994，（2）：1-8.
23. 刘人怀. 提高认识 办好成人教育. 在暨南大学1994年教学工作会议上的讲话，广州，1994-11-24；暨南教育，1994，（2）：9-12.

1995

24. 刘人怀. 认真抓好基础教学 努力提高教学质量. 暨南教学，1995，（42）：12-31.

1996

25. 刘人怀. 暨南大学校长任职仪式讲话，广州，1996-01-04；暨南大学校报，1996-01-15.
26. 刘人怀. 加强基础 从严治校 培养高素质人才. 暨南教育，1996，（1）：1-3.

27. 刘人怀. 暨南大学“211 工程”部门预审汇报，广州，1996-06-12.
28. 刘人怀. 在暨南大学“211 工程”部门预审总结会上的讲话. 广州，1996-06-14；暨南教育，1996，(2)：12.
29. 刘人怀. 庆贺九十华诞 创建一流大学. 暨南大学建校九十周年庆祝大会校长致辞. 广州，1996-06-15；暨南大学校报，1996-07-15.
30. 刘人怀. 庆九十华诞 建一流大学. 暨南大学校报，1996-06-15.
31. 刘人怀. 弘扬暨南传统 再铸辉煌未来. 暨南大学校报，1996-06-15.
32. 刘人怀. 弘扬暨南精神 创办一流大学. 中国高教研究，1996，(6)：28-30.
33. 刘人怀. 情牵母校. 四川省温江中学建校 70 周年祝词. 1996-08-05；教育与科技管理研究，北京：科学出版社，2016，257-258.

1997

34. 刘人怀. 开拓创新 共建附属医院. 暨南大学和深圳市卫生局共建暨南大学医学院第二附属医院签约仪式讲话，深圳，1997-01-25；一个大学校长的探索，北京：高等教育出版社，2011，169.
35. 刘人怀. 办出特色 办出水平. 暨南大学“211 工程”建设项目立项可行性认证会上的汇报；暨南大学校报，第 223 期，1997-07-30.
36. 刘人怀. 贯彻十五大精神 为 21 世纪培养高素质人才. 高教探索，1997，(4)：10-11.
37. 刘人怀. 实施城市管理系统工程建设 开创广州可持续发展新格局. 穗港澳地区十二所高校校长与广州市领导恳谈会主题报告，广州，1997-12-03；羊城科技报，广州，1998-02-08.
38. 刘人怀. 总结经验 深化改革 加快科技发展. 1998 年暨南大学理医科研工作部署会议讲话，广州，1997-12-09；一个大学校长的探索，北京：高等教育出版社，2011，141-143.

1998

39. 刘人怀. 教学是主旋律. 暨南教育，1998，(1)：1-2.
40. 刘人怀. 积极主动地为侨务工作服务. 全国侨务办公室主任座谈会发言. 北京，1998-01-19；教育与科技管理研究，北京：科学出版社，2016，185-191.
41. 刘人怀. 发挥侨校优势 培养高素质人才. 学位与研究生教育，1998，(1)：10-11.
42. 刘人怀. 用现代化管理促进高等教育事业的发展. 中英高等教育管理研讨会发言，广州，1998-04-08；暨南教育，1998，(1)：80-82.
43. 刘人怀. 深化教育改革 提高教学质量. 暨南大学教学工作会议讲话，广州，1998-04-09；暨南教育，1998，(1)：1-3.
44. 刘人怀. 高举邓小平理论伟大旗帜 加强我校党建工作和党建研究. 暨南大学 1998 年党建工作研究会论文集，广州，1998-04-11.
45. 刘人怀. 华文学院越办越好. 暨南大学华文学院建院五周年庆贺大会讲话，广州，1998-08-16；广州华苑，1998，12：4.
46. 刘人怀. 暨南大学兴办高等华侨教育的历史回顾与展望. 台湾暨南国际大学

"华侨教育学术研讨会"特邀报告，南投，1998-06-02；暨南学报（哲学社会科学），1998，20（4）：1-4.
47. 刘人怀. "科学技术是第一生产力"的趋势与我国高新技术发展的战略. 邓小平科技理论与广东实践（广东省科学技术协会、广东省科学技术委员会主编），广州：暨南大学出版社，1998，32-42.
48. 刘人怀. 努力办出学校特色 为"科教兴国"作贡献. 高教探索，1998，（4）：6-7.

1999

49. 刘人怀. 迎接世纪之光 创建特色大学. 中国经济快讯，1999，（2）：16.
50. 刘人怀. 面向21世纪的研究生教育. 一九九九香港研究生教育会议论文集，香港，1999，51-53.
51. 刘人怀. 在高校党的建设中贯彻落实邓小平"从严治党"思想. 新的伟大工程，广州：广东经济出版社，1999，436-442.
52. 刘人怀. 人才是振兴国家的关键. 在《中华名流》首发式上的讲话，北京，1999-02-06.
53. 刘人怀. 脚踏实地 循序渐进. 暨南大学董事会第四届第一次会议报告；暨南大学校报，第251期，1999-03-08.
54. Lin Renhuai. Succesful internationalization in Jinan University. Proceedings of XIIth Triennial Conference for The International Association of University Presidents, Brussels, 1999-07-11, 100.
55. 刘人怀. 在授予马万祺先生名誉博士学位仪式上的讲话. 广州，1999-10-23；一个大学校长的探索，北京：高等教育出版社，2011，206.
56. 刘人怀. 在曾实梓科学馆落成典礼上的讲话. 广州，1999-10-23；一个大学校长的探索，北京：高等教育出版社，2011，207-208.
57. 刘人怀. 举行全国100所"211工程"学校赠书仪式. 广州，1999-11-08；教育与科技管理研究，北京：科学出版社，2016，225-226.

2000

58. 刘人怀. 新年贺词. 暨南大学校报，2000-01-10.
59. 刘人怀. （日本）新世纪中文电视学校校长致辞. 中文导报（日本），2000-03-16.
60. 刘人怀. 同质同水平异地办学. 珠海市政府和暨南大学合作建设暨南大学珠海学院签字仪式讲话，珠海，2000-04-29；暨南大学校报，第283期，2000-05-15.
61. 刘人怀. 暨南大学国际化之路. 暨南大学与圣马丁大学合作签约仪式讲话，利马，秘鲁，2000-05-25；一个大学校长的探索，北京：高等教育出版社，2011，18-20.
62. 刘人怀. 转变观念 量化考核 优劳优酬. 高教探索，2000，（1）：5-8.
63. 刘人怀. 面向新世纪的创新教育. 海峡两岸面向21世纪科技教育创新研讨会论文集，武汉，2000，15-18.
64. 刘人怀. 开启思想的眼睛.《MBA案例》序，广州：暨南大学出版社，2000，1-2.

65. 刘人怀. 面向新世纪的暨南创新教育. 面向新世纪的暨南创新教育，广州：暨南大学出版社，2000，1-8.
66. 刘人怀. 传播科学思想 提高国民素质. “院士科普书系”首发式讲话，北京，2000-06-04；一个大学校长的探索，北京：高等教育出版社，2011，325.
67. 刘人怀. 执政党建设理论的新发展. 暨南大学校报，2000-07-10.
68. 刘人怀. 建议重视我省仪器仪表工业的发展，以迎接 21 世纪挑战. 在广东省政府主办的 2000 年广东经济发展国际咨询会上的讲话，广州，2000-11-16.
69. 刘人怀. 大力重视非开挖工程技术. 2000 年广州非开挖技术报告会开幕词，广州，2000-12-16.
70. 刘人怀. 坚持社会主义办学方向 办好华侨高等教育 为海内外培养高素质人才. 在暨南大学干部会上的讲话，广州，2000-12-20.
71. 刘人怀. 辉煌与梦想. 暨南大学迎接新世纪文艺晚会贺词，广州，2000-12-29；暨南大学校报，第 299 期，2001-01-05.

2001

72. 刘人怀. 本科教学是基础. 暨南教学，2001-04-09.
73. 刘人怀. 暨南大学的特点和优势. 在中央电视台《高考咨询——名校面对面》中的演讲，北京，2001-05-13.
74. 刘人怀. 弘扬中华民族文化. 在海外侨团联谊大会上的讲话，北京，2001-06-22.
75. 刘人怀. 暨南大学的发展战略. 暨南教学，2001-07-08.
76. 刘人怀. 在敦聘池田大竹先生为名誉教授仪式上的致辞. 东京，日本，2001-07-14.
77. 刘人怀. 珍惜“暨南人”光荣称号. 在暨南大学珠海学院 2001 级秋季新生开学典礼上的讲话. 珠海，2001-09-14.
78. 刘人怀. 饮水思源. 在暨南大学原址“暨南学堂纪念碑”揭碑仪式上的讲话，南京，2001-11-08.
79. 刘人怀. 为祖国侨务事业和华侨高等教育做出新的贡献. 在广东省部分高校校长座谈会上的发言. 广州，2001-11-11.
80. 刘人怀. 巍巍暨南 焕发青春. 暨南大学校报，2001-11-16.
81. 刘人怀. 建设国际化、现代化、综合化的高水平社会主义华侨大学. 暨南大学建校九十五周年庆祝大会校长致辞，广州，2001-11-16；暨南大学校报，2001-12-03.
82. 刘人怀. 狠抓办学质量 走“侨校＋名校”之路. 2001 年度教学工作会议闭幕式讲话，广州，2001-03-15；暨南高教研究，2001，(2)：10-19.
83. 刘人怀. 突出侨校特色 推进创新教育. 中国高等教育，2001，(19)：38-39.
84. 刘人怀. 深深感谢校友深情. 暨南大学校友总会第二届理事会成立大会讲话，广州，2001-11-16；暨南校友，2001，4：6-7.
85. 刘人怀. 办好研究生教育至关重要. 在海峡两岸和港澳地区学位与研究生教育研讨会开幕式上的讲话，广州，2001-11-30.

86. 刘人怀．进一步提高干部人事档案工作的管理水平．在全国侨务系统干部人事档案培训班上的讲话，广州，2001-11-30.

2002

87. 刘人怀．广东省发展高新技术的若干意见和建议．广东省政协八届五次会议大会发言，广州，2002-01-31；政协广东省第八届委员会建言选编，广州：政协广东省委员会办公厅，2003，289-297.
88. 刘人怀．当代科技发展与大学理念和人才培养．暨南高教研究，2002，(1)：1-8.
89. 刘人怀，王学工．对某跨国公司绩效考评系统的评价．暨南学报（哲学社会科学），2002，24（1）：29-33.
90. 刘人怀．生日贺信，钱伟长教授九十华诞纪念文集．上海：上海大学出版社，2003，49.
91. 刘人怀．质量是企业生存的根本．中国工程院和深圳市人民政府主办的中国青年科技企业家管理论坛主题报告，深圳，2002-06-18；中国青年科技企业家管理论坛论文集，中国工程院和深圳市人民政府主办，深圳，2002，30-35.
92. 刘人怀．持之以恒 依法治校．《暨南大学文件汇编》序，广州，2002-09-25.
93. 刘人怀，张永安，傅汉章，林福永，杨英，梁明珠，刘治江．珠海前山镇（街）转型与社区建设研究．珠海市香洲区前山街道党工委和街道办事处科研项目报告，珠海，2002-04-05；现代管理的中国实践，北京：科学出版社，2016，139-166.
94. 刘人怀．全球化进程与华侨高等学府的重要使命．中国高教研究，2002，(9)：33-36.
95. 刘人怀．胸怀世界 放眼未来．在暨南大学 2002 年秋季新生开学典礼上的讲话，珠海，2002-09-19.
96. 刘人怀．暨南大学积极推行学分制管理．高等工程教育信息，2002，(11)：1-2.
97. 刘人怀．努力完成高校扩招任务．广东高等教育 2003 年度工作会议暨发展咨询会议发言，肇庆，2002-12-28；一个大学校长的探索，北京：高等教育出版社，2011，78-85.
98. 刘人怀．答谢珠海人民．答谢珠海市委、市政府和珠海人民汇报演出暨新年文艺晚会讲话，珠海，2002-12-30；一个大学校长的探索，北京：高等教育出版社，2011，204-205.

2003

99. 刘人怀．广东的治安状况与投资环境，广东省政协各界别委员代表座谈会发言．广州：2003-01-12；现代管理的中国实践．北京：科学出版社，2016，137-139.
100. 刘人怀．教育彩票作用惊人．南方日报，广州，2003-01-24.
101. 刘人怀．暨南大学校长刘人怀贺信．钱伟长教授九十华诞纪念文集，上海：上海大学出版社，2003，49.

102. 刘人怀. 统一思想 提高认识 为建设海内外知名的社会主义华侨大学而努力奋斗. 暨南大学校报，2003-03-10.
103. 刘人怀. 严格执行教学制度 进一步规范教学管理. 暨南大学校报，2003-03-25.
104. 刘人怀. 建设广州石牌大学城. 广东省做大做强高等教育座谈会发言，东莞，2003-04-02；教育与科技管理研究，北京：科学出版社，2016，121-123.
105. 刘人怀. 坚决反对和防止腐败是学校重大的政治任务. 广东省人民检察院与暨南大学开展同步预防职务犯罪工作协议签约仪式讲话，广州，2003-04-16；暨南大学校报，第 354 期，2003-04-30.
106. 刘人怀，张永安，傅汉章，谭浩邦，梁明珠，诸凤鸣，王桂林，管宇，雷方. 以价值工程方法全面提升荔湾商旅核心竞争力，广州市荔湾区人民政府科研项目报告，广州，2003；旅游工程管理研究，北京：科学出版社，2014，37-96.
107. 刘人怀. 办好院士之家. 中国工程院院士广州咨询活动中心亮牌仪式讲话，广州，2003-06-18；教育与科技管理研究，北京：科学出版社，2016，286-287.
108. 刘人怀. 寄语中层干部. 在中层干部目标责任签字仪式上的讲话（按录音整理），广州，2003-07-11；暨南高教研究，2003，(2)：1-8.
109. 刘人怀. 暨南大学的创新发展之路. 在广东省高校领导干部暑期读书班上的报告，珠海，2003-07-28；一个大学校长的探索，北京：高等教育出版社，2011，174-182.
110. 刘人怀. 认真评选科技奖. 在广东省科学技术协会常委会上的汇报，广州，2003-08-20；教育与科技管理研究，北京：科学出版社，2016，349.
111. 刘人怀. 求真务实 从严治校 团结协作 再创辉煌. 暨南大学校报，2003-09-15.
112. 刘人怀，叶向阳. 公司治理：理论演进与实践发展的分析框架. 经济体制改革，2003，(4)：5-8.
113. 刘人怀. 与时俱进 开拓创新. 广东省科学技术协会首届学术活动周开幕词，广州，2003-11-11.
114. 刘人怀. 以科技创新加快推进全面建设小康社会步伐. 广东省科学技术协会首届学术活动周主题报告，广州，2003-11-11；一个大学校长的探索，北京：高等教育出版社，2011，331-336.
115. 刘人怀. 弘扬中华文化 发展华文教育 传播华夏文明 促进文化交流. 在暨南大学华文学院建院 50 周年庆典上的讲话，广州，2003-11-15；暨南大学校报，第 369 期，2003-11-30.
116. 刘人怀. 为迎接暨南大学百年庆典增添新的光彩. 暨南大学董事会第五届第一次会议开幕词，广州，2003-11-16；第五届董事会特刊，2004，12.
117. 刘人怀. 发挥优势 深化改革 保证重点 改善条件 提高质量. 暨南大学董事会第五届第一次会议报告，广州，2003-11-16；暨南大学校报，第 370 期，2003-12-2.
118. 刘人怀. 绿色化是中国制造业的必由之路. 2003 年中国机械工程学会年会特邀主题报告，深圳，2003-11-29；工商管理研究，北京：科学出版社，2015，

177-181.

119. 刘人怀. 进一步深化改革 提高质量. 暨南大学校报，第 369 期，2003-12-02.

120. 刘人怀. 培养更多更优质人才. 全国高等学校教育研究会第四次常务理事会暨学术研究会开幕式致辞，广州，2003-12-20.

121. 刘人怀. “侨校＋名校”的发展定位. 全国高等学校教学研究会报告，广州，2003-12-20；一个大学校长的探索，北京：高等教育出版社，2011，105-112.

2004

122. 刘人怀. 立足侨校 服务青年 让青春绽放出绚丽的光彩. 暨南大学校报，2004-01-05.

123. 刘人怀. 为实施“侨校＋名校”发展战略贡献力量. 暨南大学校报，2004-01-08.

124. 刘人怀. 坦诚建言. 广东省人民政府参事、馆员迎春茶话会讲话，广州，2004-01-14.

125. 刘人怀. 旅游教材为旅游教学之本.《现代饭店管理》序，广州：暨南大学出版社，2004，1-2.

126. 刘人怀. 光大华侨文化 建设文化大省. 广东省政协“弘扬岭南文化，建设文化大省”专题座谈会发言，广州，2004-02-11.

127. 刘人怀. 开创科协工作新局面. 广东省科学技术协会第六届委员会第三次会议总结讲话，广州，2004-02-20；教育与科技管理研究，北京：科学出版社，2016，345-346.

128. 刘人怀. 大力发展具有侨校特色的研究生教育. 暨南大学研究生教育工作会议讲话，广州，2004-03-10；一个大学校长的探索，北京：高等教育出版社，2011，227-235.

129. 刘人怀. 中国制造业的生存哲学. 科技中国，2004，创刊号：56-57.

130. 刘人怀. 教学是学校的生命. 在暨南大学 2003 年教学工作会议上的讲话（按录音整理），广州，2003-12-22；暨南高教研究，2004，(1)：1-6.

131. 刘人怀. 贺暨南大学澳门校友会会长就职. 在暨南大学澳门校友会 2004—2006 年度会长、理监事就职典礼上的讲话，澳门，2004-04-02；一个大学校长的探索，北京：高等教育出版社，2011，211.

132. 刘人怀. 为广州科技发展提出建议. 中国工程院院士广州院士咨询活动中心第一期院士沙龙活动开幕词，广州，2004-04-23；教育与科技管理研究，北京：科学出版社，2016，287.

133. 刘人怀. 忍耐是一个人成功的秘诀——与暨南学子谈成才. 在百年暨南讲堂首讲上的讲话，广州，2004-04-15；暨南青年，第 47 期，2004-07-08.

134. 刘人怀. 坚持人类与环境和谐共生 走可持续发展的工业化道路. 2004 年世界工程师大会“发展中国家的工业化道路”专题论坛主席的开坛词，广州，2004-05-08；工程管理研究，北京：科学出版社，2015，4-5.

135. 刘人怀. 泛珠三角：推进科技、教育和文化的区域合作. 泛珠三角区域合作与发展论坛演讲录，澳门，2004-06-02，32-37；现代管理的中国实践，北京，科学出版社，2016，106-109.

136. 刘人怀．推进科教文化交流．人民日报，2004-06-02.

137. 刘人怀．传承岭南文化 服务文化大省．岭南文史，2004，(2)：扉页.

138. 刘人怀．标准学分制的研究与实践．中国大学教学，2004，(3)：41-43.

139. 刘人怀．《系统工程与管理科学研究》序．广州：暨南大学出版社，2004，1-2.

140. 刘人怀．学为人师 行为世范．2004 年暨南大学纪律教育活动警示教育报告，暨南大学校报，2004-07-08.

141. 刘人怀．基础研究是科技进步的标志．广州科技报，2004-07-30.

142. 刘人怀．推进珠三角经济区的工作．光明日报，2004-08-06.

143. 刘人怀．加强振动理论研究与应用．中国振动工程学会 2004 年全国振动工程及应用学术会议开幕词，成都，2004-08-12；教育与科技管理研究，北京：科学出版社，2016，328-329.

144. 刘人怀．积极服务海外华侨华人社会．在欧洲华侨华人社团联合会第十二届年会上的讲话，伯明翰，英国：2004-08-16；一个大学校长的探索，北京：高等教育出版社，2011，375-376.

145. 刘人怀．校长寄语．暨南大学校报，2004-09-16.

146. 刘人怀，韩大建等．促进广东省职业教育发展．广东省第九届政协教科文卫体委员会和民盟广东省委员会向政协提交的调研报告，2004 年 9 月；教育与科技管理研究，北京：科学出版社，2016，124-131.

147. 刘人怀．加强基础研究 实现科技强省．科技管理研究，2004，24 (5)：1-3.

148. 刘人怀．为疾病控制工作添砖加瓦．第三届全国伤害预防控制学术会议欢迎词，广州，2004-10-09.

149. 刘人怀．关于“发展中国家的工业化道路”论坛的讨论．“2004 年世界工程师大会”分组会上的主题演讲，上海，2004-11-05；一个大学校长的探索，北京：高等教育出版社，2011，341-348.

150. 刘人怀．建立粤港澳综合协调机构．政协广东省第九届委员会提案，广州，2004，731；现代管理的中国实践．北京：科学出版社，2016，112.

151. 刘人怀，蒋述卓．暨南大学百年校庆公告（第一号），2004-11-11；暨南校友，2004，4：1.

152. 刘人怀．关于尽快制定国家统一法的建议．广东省政府参事室建议，广州，2004-12-02；参事建言（2004—2005 年），香港：中国评论学术出版社，2006，538.

153. 刘人怀．面向海外 面向港澳台 为祖国统一大业服务．教育部港澳台工作座谈会发言，北京，2004-12-11；教育与科技管理研究，北京：科学出版社，2016，191-203.

2005

154. 刘人怀．广纳贤才 全球招聘院长．在暨南大学面向世界公开招聘 10 位学院院长仪式上的讲话；暨南大学校报，第 403 期，2005-01-21.

155. 刘人怀. 培养理论与实务并重人才. 在暨南大学首届会计硕士专业学位研究生开学典礼上的讲话，广州，2005-02-10.

156. 刘人怀. 务实开拓 追求卓越. 暨南大学校报，2005-03-10.

157. 刘人怀. 学会普通话 走遍天下都不怕. 香港民政事务总署公务员普通话研习班开学典礼讲话，广州，2005-03-21；一个大学校长的探索，北京：高等教育出版社，2011，383.

158. 刘人怀，姚作为. 关系质量研究述评. 外国经济与管理. 2005，27（1）：27-33.

159. 刘人怀. 改革创新 增强竞争力. 暨南大学与香港亚洲电视台全面合作签字仪式讲话，香港，2005-04-25；一个大学校长的探索，北京：科学出版社，2011，170-173.

160. 刘人怀. 抓好教学质量. 暨南大学校报，2005-05-20.

161. 刘人怀. 团结努力 稳步前进 为“侨校＋名校”战略目标的实现共同奋斗. 暨南高教研究，2005，(2)：1-6.

162. 刘人怀. 坚持大力发展研究生教育.《引路者论道：研究生指导教师学位与研究生教育研究论文选》序，暨南大学研究生部编，广州：广东人民出版社，2005，1-3.

163. 刘人怀. 激励民办专科学校升为本科学校. 政协广东省第九届委员会提案，广州，2005 年 6 月；一个大学校长的探索，北京：高等教育出版社，2011，382.

164. 刘人怀. 关于高考的一点浅见. 科学中国人，2005，(9)：13.

165. 刘人怀，纪宗安等. 加快我国高等教育进入世界先进行列. 全国教育事业“十一五”规划研究课题报告，2005；教育与科技管理研究，北京：科学出版社，2016，22-113.

166. 刘人怀. 成立广东省仪器仪表学会是紧迫的历史使命. 向广东省科学技术协会的报告，广州，2005-09-11.

167. 刘人怀. 让象牙塔成为顶梁柱. 中国中部科技创新与风险投资国际论坛主题报告，合肥，2005-09-21.

168. 刘人怀. 市校联合 共同进步. 暨南大学与佛山市人民政府全面合作协议签字仪式讲话，佛山，2005-10-18；一个大学校长的探索，北京：高等教育出版社，2011，377-379.

169. 刘人怀，陈万鹏等. 大力发展我省高中阶段教育. 广东省第九届政协教科文卫体委员会向政协常委会提交的专题议政材料，广州，2005 年 10 月；教育与科技管理研究，北京：科学出版社，2016，247-254.

170. 刘人怀. 亚洲 青春 竞技. 在暨南大学举行的第一届亚洲大学生田径锦标赛开幕词，广州，2005-11-07；一个大学校长的探索，北京：高等教育出版社，2011，240-241.

171. 刘人怀. 贺暨南大学香港社会学同学会成立. 在暨南大学教育学院香港社会

学同学会成立会上的讲话，香港，2005-11-27；一个大学校长的探索，北京：高等教育出版社，2011，209-210.

172. 刘人怀. 加强海峡两岸产业合作. 经济全球化格局下的两岸产业合作研讨会开幕词，广州，2005-12-07.

173. 刘人怀. 承前启后 务真求实 科学编制“十一五”发展规划. 第 16 届四校工作交流联谊会交流材料汇编，广州，2005，54-64.

174. 刘人怀. 欢迎印度、尼泊尔、巴基斯坦新同学. 印度、尼泊尔、巴基斯坦新生开学典礼讲话，广州，2005-12-21；一个大学校长的探索，北京：高等教育出版社，2011，198-199.

175. 刘人怀. 新年贺词. 暨南校报，2005-12-30.

176. 刘人怀. 立足侨校 服务学生 全面推进我校学生德育工作. 在暨南大学学生工作会议上的讲话，广州，2005-12-25；暨南大学校报，特刊，2005-12-31.

2006

177. 刘人怀. 喜获丰硕成果 笑迎百年华诞. 暨南大学第六届教代会暨第十届工代会第二次会议讲话，广州，2006-01-13；一个大学校长的探索，北京：高等教育出版社，2011，387-395.

178. 刘人怀. 生逢其时的学院. 声教四海 商科一脉，暨南大学管理学院编，2006，70.

179. 刘人怀. 图书馆是大学的心脏. 暨南大学新图书馆封顶仪式讲话，广州，2006-01-14；一个大学校长的探索，北京：高等教育出版社，2011，297-298.

180. 刘人怀. 衷心祝愿暨南大学的明天更美好. 离任暨南大学校长会上的讲话，广州，2006-01-14；一个大学校长的探索，北京：高等教育出版社，2011，396-398.

181. 刘人怀. 更好地发挥排头兵的作用. 广东省科学技术协会第六届委员会第五次会议总结讲话，广州，2006-02-28.

182. 刘人怀. 关于改善我国北方水资源缺乏的建议. 参事建言（2004－2005 年），广东省人民政府参事室编，香港：中国评论学术出版社，2006，229.

183. 刘人怀. 关于允许市民在节假日有条件燃放烟花的建议. 参事建言（2004－2005 年），香港：中国评论学术出版社，2006，518-519.

184. 刘人怀. 绿色制造与学科会聚. 学科会聚与创新平台——高新技术高峰论坛，杭州：浙江大学出版社，2006，15-18.

185. 刘人怀. 培养青少年的创新精神. 第二十一届广东省青少年科技创新大赛开幕式讲话. 惠州，2006-04-07；教育与科技管理研究，北京：科学出版社，2016，370-371.

186. 刘人怀. 述职报告. 送交国务院侨务办公室的工作汇报，广州，2006-07-10.

187. 刘人怀. 认真做好高校力学教学指导工作. 2006－2010 年教育部高等学校力学教学指导委员会第一次会议开幕词，广州，2006-07-25；教育与科技管理研究，北京：科学出版社，2016，134-135.

188. 刘人怀. 为促进全国高等学校力学专业发展做出贡献. 2006－2010 年教育部高等学校力学教学指导委员会第一次会议工作报告，广州，2006-07-25.
189. 刘人怀. 推动振动理论和应用技术的发展. 中国振动工程学会 2006 年全国振动工程及应用学术会议开幕词，昆明，2006-08-15；教育与科技管理研究，北京：科学出版社，2016，329-330.
190. 刘人怀. 为力学课程教学质量和水平的提高献计献策. 首届全国力学课程报告论坛开幕词，大连，2006-11-03.
191. 刘人怀. 我国力学专业教育现状与思考. 首届全国力学课程报告论坛主题报告，大连，2006-11-03；中国大学教育，2007，(1)：30-32.

2007

192. 刘人怀. 岁月留声. 岁月留声——院士活动剪影（二），中国工程院院士广州咨询活动中心编，2007，1.
193. 刘人怀. 工程管理是管理对国民经济的深度介入. 中国工程管理环顾与展望——首届工程管理论坛论文集锦，北京：中国建筑工业出版社，2007，260-262.
194. 刘人怀，周裕新，何问陶，吴厚德，毛蕴诗，于正林，陈婉玲，彭赟，刘凤英，黄莹莹. 关于改善财政宏观调控深化分税制财政体制改革的调研报告. 参事建言（2006 年），香港：中国评论学术出版社，2007，75-92.
195. 刘人怀. 在推进和谐社会建设中切实解决“农民工”身份问题. 参事建言（2006 年），香港：中国评论学术出版社，2007，312-315.
196. 刘人怀. 消费者行为研究是营销理论的基石.《服务消费决策行为研究——基于品牌关系的角度》序，北京：中国标准出版社，2007，3-5.
197. 刘人怀. 我国力学专业教育现状与思考. 中国大学教学，2007，(1)：32-34.
198. 刘人怀. 关于将清明节设为国家法定假日的建议. 广东政协，2007，(6)：10-11；参事建言（2007 年），香港：中国评论学术出版社，2008，327-329.
199. 刘人怀. 大力推动科技事业的发展. 在第三届“看中国”系列丛书出版座谈会暨《科技创新与品牌》杂志首发式上的讲话，北京，2007-07-06.
200. 刘人怀. 迎接新挑战.《第九届全国振动理论及应用学术会议论文集（2007）》序，杭州：浙江大学出版社，2007，1-2.
201. 刘人怀. 同心同德 开拓前进. 中国振动工程学会第六次全国会员代表大会开幕词，杭州，2007-10-18.
202. 刘人怀，袁国宏. 旅游中间产品转移价格的确定. 经济问题探索，2007，(11)：107-111.
203. 刘人怀，袁国宏. 我国旅游价值链管理探讨. 生态经济，2007，(12)：102-104.
204. 刘人怀，袁国宏. 旅游业零负团费的运行机制及危害性探析. 商业时代，2007，(25)：87-88.
205. 刘人怀. 搞好力学课程教学. 第二届全国力学课程报告论坛开幕词. 广州，

2007-12-15；教育与科技管理研究，北京：科学出版社，2016，140-141.

206. 刘人怀. 黄石应该在建设特色城市中凸出优势. 同舟行，2007，专刊：5.

207. 刘人怀，袁国宏. 从 CSSCI 旅游研究文献看旅游学学科发展. 人文地理，2007，22（4）：77-81.

2008

208. 刘人怀. 做科技创新的鼓手. 科技创新与品牌，2008，(1)：1.

209. 刘人怀，何问陶. 发展“乡村旅游”促进广东新农村建设. 广东政协，2008，(3)：21-22；参事建言（2008 年），香港：中国评论学术出版社，2008，197-200.

210. 刘人怀. 爱心和匠心. 珠海市卫生系统医学人文精神论坛报告，珠海，2008-02-26；现代管理的中国实践，北京：科学出版社，2016，169-172.

211. 刘人怀，孙东川. 谈谈创建现代管理科学中国学派的若干问题. 管理学报，2008，5（3）：323-329.

212. 刘人怀. 探索的脚步. 科技创新与品牌，2008，(7)：1.

213. 刘人怀，龙先东. 文化生产力：管理的视角. 生产力研究，2008，(7)：54-56.

214. 刘人怀. 创新路上的感想. 应上海市中国工程院院士咨询与学术活动中心邀请所做的报告（按录音整理），上海，2008-05-27；科技创新与品牌，2009，(1)：10-16；(2)：8-11.

215. 刘人怀. 关于治理垃圾的建议. 广东产学研结合高端论坛暨院士专家云浮行活动开幕式专题报告摘要，云浮，2008-08-19；工程管理研究，北京：科学出版社，2015，243-244.

216. 刘人怀. 浅谈高等学校科学管理“三”字经. 科技创新与品牌，2008，(10)：14-17；(11)：10-13.

217. 刘人怀，孙东川. 再谈创建现代管理科学中国学派的若干问题. 中国工程科学，2008，10（12）：24-31.

218. 刘人怀. 科技创新是企业的生命. 广电运通，2008，(31)：31-32.

2009

219. 刘人怀. 为提高全民族科学素质做出新贡献. 广东省科普志愿者协会成立大会讲话，广州，2009-03-24；教育与科技管理研究，北京，科学出版社，2016，356-358.

220. 刘人怀. 关于城市基础设施建设投融资体制改革研究. 决策与咨询，广州：广州市人民政府研究室，2009，(5)：1-6.

221. 刘人怀. 我的语文观. 语文月刊，2009，(4)：1.

222. 刘人怀，孙东川，孙凯. 三谈创建现代管理科学中国学派的若干问题：四条定义与三点建议. 中国工程科学，2009，11（8）：18-23，63.

223. 刘人怀，孙凯，孙东川. 大平台、聚义厅及其他——四谈创建现代管理科学中国学派的若干问题. 管理学报，2009，6（9）：1137-1142.

224. 刘人怀. 旅游交通与航空运输规划.《航空运输规划》序，西安：西北工业大学出版社，2009，1-2.

225. 刘人怀，孙凯，孙东川. 大型工程项目管理的中国特色及与美苏的比较. 科技进步与对策，2009，26（21）：5-12.

226. 刘人怀. 关于将香港、澳门特别行政区的所有统计数据纳入全国性统计数据的建议. 广东省政府参事建议，广州：广东省人民政府参事室，2009，（82）：1-3.

227. 刘人怀. 办好旅游教育至关重要. 第三届亚太地区旅游会展教育与产业发展国际研讨会暨穗港澳会展业对接长三角论坛开幕词，澳门，2009-10-22.

2010

228. 刘人怀. 大规模引进和培训人才 为广东产业结构优化升级服务. 民主与决策，香港：中国评论学术出版社，2010，126-140.

229. 刘人怀. 大力推动“政产学研金”合作创新 为广东省经济社会发展做贡献. 广东省委、省政府领导接见院士专家会发言，广州，2010-02-08；一个大学校长的探索，北京：高等教育出版社，2011，358-359.

230. 刘人怀. 认真做好青少年科技创新作品的评审工作. 第25届广东省青少年科技创新大赛开幕式讲话，广州，2010-04-10；教育与科技管理研究，北京：科学出版社，2016，371-372.

231. 刘人怀. 建设低碳社会关键在制造创新. 第二届深港澳节能减排论坛开幕式主题演讲，深圳，2010-04-24；工程管理研究，北京：科学出版社，2015，16-19.

232. 刘人怀. 爱低碳生活 创绿色校园. 绿色澳门建设研讨会发言，澳门，2010-06-05；一个大学校长的探索，北京：高等教育出版社，2011，246-248.

233. 刘人怀. 旅游标志景区研究有意义.《从极核到集群：旅游目的地标志景区发展研究》序，北京：经济科学出版社，2010，1-2.

234. 刘人怀.《系统工程基本教程》序，广州：暨南大学出版社，2010.

235. 刘人怀，孙东川.《学科目录》第12学科门类与管理科学话语体系——五谈创建现代管理科学中国学派的若干问题. 学位与研究生教育，2010，（8）：67-73.

236. 刘人怀. 深深缅怀孜孜践行. 科技创新与品牌，2010，（8）：封2.

237. 刘人怀. 积极推进知识产权事业发展. 兰州大学管理学院讲稿，兰州，2010-08-14.

238. 刘人怀，孙凯. 工程管理信息化的内涵与外延探讨. 科技进步与对策，2010，27（19）：1-4.

239. 刘人怀. 转变经济发展方式关键在制造业创新. 第七届沈阳科学学术年会报告，沈阳，2010-10-20；现代管理的中国实践，北京：科学出版社，2016，114-116.

240. 刘人怀. 深化力学课程改革. 第五届全国力学课程报告论坛开幕词，成都，2010-11-06；教育与科技管理研究，北京：科学出版社，2016，141-142.

241. 刘人怀. 做好力学学会教育工作. 中国力学学会第九届第一次理事长与秘书

长会议工作报告，北京，2010-12-13；教育与科技管理研究，北京：科学出版社，2016，331-332.

242. 刘人怀. 关于完善我省应对台风灾害预防措施的建议. 广东省政府参事建议，第 81 期，2010-12-20.

243. 刘人怀等. 关于我国公共安全工程的研究. 中国工程院科研项目报告，北京，2010；工程管理研究，北京：科学出版社，2015，109-202.

2011

244. 刘人怀. 积极办好院士专家企业工作站. 广东省“院士专家企业工作站”经验交流会讲话，广州，2011-01-21；工商管理研究，北京：科学出版社，2015，241-242.

245. 刘人怀. 稳步推进科普志愿服务事业. 广东省科普志愿者协会第一届理事会工作报告，广州，2011-03-20；教育与科技管理研究，北京：科学出版社，2016，365-370.

246. 刘人怀. 坚持重点 保持特色. 肇庆市端州区人民政府工作简报，2011，(1)：6-8.

247. 刘人怀. 从学生会主席到大学校长之路. 暨南大学学生会干部会讲话，广州，2011-04-10；教育与科技管理研究，北京：科学出版社，2016，165-167.

248. 刘人怀. 中国的过去和现在. 为德国访华团“中国业务开发专题讲座”作的报告，北京，2011-04-24.

249. 刘人怀. 努力开创仪表元件分会工作的新局面. 中国仪器仪表学会仪表元件分会第四届理事会工作报告，扬州，2011-05-26；教育与科技管理研究，北京：科学出版社，2016，332-343.

250. 刘人怀. 绿色旅游是 21 世纪旅游业发展的永恒主题. 第五届绿色旅游管理国际论坛开幕词，广州，2011-06-25；旅游工程管理研究，北京：科学出版社，2014，168-169.

251. Liu Renhuai. Preface. Proceedings of the Fifth International Symposium on Green Hospitality and Tourism Management, Marietta: The American Scholar Press, 2011, 56.

252. 刘人怀. 推进“垃圾分类，从我做起”科普宣传活动. 呈送广东省委常委、广州市委书记张广宁的报告，广州，2011-07-01；工程管理研究，北京：科学出版社，2015，249-251.

253. 刘人怀. 加强机械动力学的研究. 中国振动工程学会机械动力学分会成立三十周年暨 2011 国际功能制造与机械动力学会议开幕词，杭州，2011-07-30；教育与科技管理研究，北京：科学出版社，2016，330-331.

254. 刘人怀. 关注世界科技创新态势. 贵州工业强省院士专家论坛报告，贵阳，2011-08-29；教育与科技管理研究，北京：科学出版社，2016，312-327.

255. 刘人怀. 开展“垃圾分类，从我做起”科普资源包研发及宣讲活动. 呈送广东省黄华华省长的报告，广州，2011-08-31；工程管理研究，北京：科学出版社，2015，251-253.

256. 刘人怀，孙凯. 工程管理信息化架构研究. 中国工程科学，2011，13（8）：4-9.

257. 刘人怀，孙凯，孙东川. 当前管理科学研究中的若干问题——几个疑点的澄清和两种研究方法的评析. 管理学报，2011，8（9）：1263-1268，1352.

258. 刘人怀，郭广生，徐明稚，陈劲，陈德敏. 试答“钱学森之问”. 教育部科学技术委员会管理学部专家报告；中国高校科技，2011（10）：4-7，14.

259. 刘人怀. 献身科学 追求真理. 暨南大学学风建设报告，广州，2011-10-09.

260. 刘人怀. 切实推进科技进步与创新. 第九届广东省科学技术协会学术活动周开幕词，广州，2011-10-16；教育与科技管理研究，北京：科学出版社，2016，346-347.

261. 刘人怀. 践行低碳 乘胜前进. 给广东省肇庆市人民政府的建议，肇庆，2011-10-15；现代管理的中国实践，北京：科学出版社，2016，166-167.

262. 刘人怀. 推进振动工程学科发展. 中国振动工程学会第七次全国会员代表大会暨第十届全国振动理论及应用学术会议开幕词，南京，2011-10-27；教育与科技管理研究，北京：科学出版社，2016，327-328.

263. 刘人怀. 精益求精 好上加好. 在徐芝纶院士百年诞辰纪念大会上的讲话，南京，2011-10-27.

264. 刘人怀. 认真开展高等学校教育教学研究. 第二届全国高等学校教学研究会第五次常务理事会议开幕词，合肥，2011-11-18.

265. 刘人怀. 助澳门繁荣. 中国（澳门）综合发展研究中心启动仪式开幕词，澳门，2011-12-30；现代管理的中国实践，北京：科学出版社，2016，169.

266. 刘人怀. 广州院士活动中心 2011 年度工作总结. 岁月留声——院士活动剪影，广州，2011-12-31，2-3.

2012

267. 刘人怀. 系统工程与领导科学.《系统工程干部读本》序，广州：华南理工大学出版社，2012，1-2.

268. 刘人怀. 加强重大工程结构安全领域国际合作. 暨南大学“城市生命线工程结构安全”国际联合实验室揭牌仪式开幕词. 广州，2012-01-06.

269. 刘人怀. 推荐科普漂流书屋进校园. 广东省 2012 年科普漂流书屋进校园活动启动仪式致辞. 广州，2012-02-24；教育与科技管理研究，北京：科学出版社，2016，363-364.

270. 刘人怀. 积极投身科技创新活动. 2012 年暨南大学理工学院学生学术科技节开幕词，广州，2012-05-12；教育与科技管理研究，北京：科学出版社，2016，294-295.

271. 刘人怀. 认真做好“2011 计划”重大战略部署工作. 教育部科学技术委员会第一届论坛开幕词. 北京，2012-06-06；教育与科技管理研究，北京：科学出版社，2016，281.

272. 刘人怀. 组建“政产学研金”合作平台 推动协同创新迈上新台阶. 教育部科学技术委员会第一届论坛主题报告. 北京，2012-06-06；中国高等教育，2012，(20)：1.
273. 刘人怀. 为地方高水平大学发展献计献策. 北京工业大学地方高水平大学发展战略研究中心揭牌仪式致辞，北京，2012-06-06.
274. 刘人怀. 开展“垃圾分类”宣讲与培训服务. 呈送广东省朱小丹省长的报告，广州，2012-06-28；工程管理研究，北京：科学出版社，2015，253-257.
275. 刘人怀. 开拓创新 做好科普工作. 广州市纪念《中华人民共和国科学技术普及法》颁布十周年高端论坛主题演讲，广州，2012-06-29；教育与科技管理研究，北京：科学出版社，2016，359-362.
276. 刘人怀. 复合材料创新成果丰硕. 《复合材料“十一五”创新成果荟萃》前言，北京：中国科学技术出版社，2012，1-2.
277. 刘人怀. 关于推进“餐厨废弃物变废为宝”的建议. 呈送中央政治局委员、广东省委书记汪洋的报告，广州，2012-09-01；工程管理研究，北京：科学出版社，2015，258-260.
278. 刘人怀. 友好合作 加强多学科交叉. 第二届海峡两岸暨南大学工程学科跨学科研讨会开幕词，广州，2012-09-04.
279. 刘人怀. 激发企业创新的热情. 2012 年自主品牌大会暨《科技创新与品牌》杂志创刊五周年庆典开幕词，北京，2012-09-06；工商管理研究，北京：科学出版社，2015，242-243.
280. 刘人怀. 开展健康教育. 第 51 期广东科学技术协会论坛开幕词，广州，2012-09-17；教育与科技管理研究，北京：科学出版社，2016，362.
281. 刘人怀. 知识交融 创新成才. 百年暨南文化素质教育讲堂报告，第 145 期，广州，2012-09-27；教育与科技管理研究，北京：科学出版社，2016，157-164.
282. 刘人怀. 关于实行九年一贯制办校的建议. 广东参事馆员建议，2012，(52)：1-3；教育与科技管理研究，北京：科学出版社，2016，254-256.
283. 刘人怀. 肩负使命与责任 继续向前. 2012 年广东压力容器学会学术交流会讲话，广州，2012-11-02.
284. 刘人怀. 建设幸福广东. 第十届广东省科学技术协会学术活动周开幕词，广州，2012-11-12；教育与科技管理研究，北京：科学出版社，2016，347-348.
285. 刘人怀. 促进仪器仪表事业发展. 中国传感器应用与发展技术大会暨 IMCA2012 届第八届国际（深圳）仪器仪表与测控自动化高峰论坛开幕词，深圳，2012-12-04；教育与科技管理研究，北京：科学出版社，2016，343-344.
286. 刘人怀. 以创新推动宁波绿色崛起. 宁波市第七届学术大会，宁波，2012-12-07.
287. 刘人怀. 中国管理科学的现状和走向. 首届中国管理科学论坛开幕词，澳门，2012-12-12；现代管理的中国实践，北京：科学出版社，2016，79.

288. 刘人怀. 现代管理科学中国学派研究综述. 首届中国管理科学论坛主题报告，澳门，2012-12-12.
289. 刘人怀，孙东川，孙凯，朱丽，刘泽寰. 关于实施“东水西调”工程的建议. 中国工程院课题报告（2009—2012 年），北京，2012；工程管理研究，北京：科学出版社，2015，203-242.
290. 刘人怀. 2012 年中国工程院院士广州咨询活动中心工作总结. 岁月留声——院士活动剪影（八），中国工程院院士广州咨询活动中心编，2012-12-31，1-2.

2013

291. 刘人怀. 从教五十年. 刘人怀院士从教五十周年庆贺文集，广州：暨南大学出版社，2013，1-14；
292. 刘人怀. 开展美丽广东科普教育示范户工作. 呈送广东省常务副省长徐少华的报告，广州，2013-02-04；工程管理研究，北京：科学出版社，2015，260-261.
293. 刘人怀，姚作为. 传统文化基因与中国本土管理研究的对接：现有研究策略与未来探索思路. 管理学报，2013，10（2）：157-167.
294. 刘人怀. 专家学者不妨多点科普和传播意识. 宏观质量研究，2013，1（2）：1-3.
295. 刘人怀. 早日实现“美丽广东梦”. 广东省“美丽城市，从垃圾分类做起”公益科普活动启动仪式上的讲话，惠州，2013-05-18；工程管理研究，北京：科学出版社，2015，261-262.
296. 刘人怀. 百年追梦 科技兴国. 中国科学技术协会 2013 年度弘扬科学道德 践行“三个倡导” 奋力实现“中国梦”报告会主题报告，贵阳，2013-05-24；现代管理的中国实践，北京：科学出版社，2016，116-123.
297. 刘人怀. 去除浮躁现象 建立公正科技评价机制. 教育部科学技术委员会第二届论坛主题报告，大连，2013-06-14；教育与科技管理研究，北京：科学出版社，2016，284-285.
298. 刘人怀. 切实做好科技评价工作. 教育部科学技术委员会第二届论坛闭幕词，大连，2013-06-14；教育与科技管理研究，北京：科学出版社，2016，284.
299. 刘人怀. 从教感想. 在国防科技大学为参加全军研究生导师培训的新增导师所作的报告，长沙，2013-07-04；教育与科技管理研究，北京：科学出版社，2016，167-170.
300. 刘人怀. 开展千乡万村科普惠农行动. 广东省“千乡万村科普惠农行动”启动仪式暨科普惠农服务培训班致辞，广州，2013-07-10；教育与科技管理研究，北京：科学出版社，2016，364.
301. 刘人怀. 提高公民科学素质 促进经济社会发展. 河源市委党校报告，河源，2013-09-11.
302. 刘人怀，覃大嘉，梁育民，左晓安. 产业工人的中国梦：从低技能劳工到专业技术工人的人资转型升级战略. 战略决策研究，2013，4（5）：37-50.
303. 刘人怀，覃大嘉，左晓安，梁育民. 中国鞋业代工现状与人资转型发展战略.

新经济，2013，(26)：3-4.

304. 刘人怀. 必须抓好力学教育工作. 中国力学学会教育工作委员会2013年工作报告，汕头，2013-12-01.

2014

305. 刘人怀. 保障用药安全. 广东省医药集团公司“家庭过期药品回收”10周年暨吉尼斯世界纪录认证仪式讲话，广州，2014-03-13；教育与科技管理研究，北京：科学出版社，2016，362-363.

306. 刘人怀，覃大嘉，梁育民. “技工荒”困扰中国高尔夫球具代工产业. 信报财经月刊（香港），2014，(4)：124-127.

307. 刘人怀. 充分发挥桥头堡作用，推进孟中印缅经济走廊建设. 云南省党政领导与院士专家座谈会院士专家发言提纲汇编，昆明，2014-05-25，1-3.

308. 刘人怀. 企业战略管理概要，西南交通大学经济管理学院博士生课程讲稿，成都，2014-06-23.

309. 刘人怀，覃大嘉，左晓安. 广东服装代工提升国际竞争力的人资发展战略研究. 新经济，2014，(25)：32-37.

310. 刘人怀. 珍惜时间 勤奋成才. 在暨南大学新生入学教育会上的讲话，广州，2014-09-23；教育与科技管理研究，北京：科学出版社，2016，153-157.

311. 刘人怀. 结合中国的实际创建中国现代管理科学. 第二届中国管理科学论坛开幕词，成都，2014-11-01；现代管理的中国实践，北京：科学出版社，2016，79-80.

312. 刘人怀. 推动力学学科发展. 中国力学学会2014年全国会员代表大会暨第九届、第十届理事会扩大会议开幕词，上海，2014-11-15；教育与科技管理研究，北京：科学出版社，2016，331.

313. 刘人怀. 建设低碳社会 托起美丽中国梦. 第三十四次中国科技论坛——绿色建设美丽中国论坛致辞，广州，2014-11-19.

314. 刘人怀，文彤，闫婷婷. 系统论视角下的旅游发展与旅游研究——中国工程院工程管理学部刘人怀院士访谈. 社会科学家，2014，(11)：3-6.

315. 刘人怀. 提升农民科学素质. 2013—2014年广东省“千乡万村科普惠民行动”总结与表彰工作会议讲话，广州，2014-12-27；教育与科技管理研究，北京：科学出版社，2016，365.

2015

316. 刘人怀. 让优秀青少年脱颖而出. 第30届广东省青少年科技创新大赛开幕式讲话，河源，2015-03-29，1-4；教育与科技管理研究，北京：科学出版社，2016，372.

317. 刘人怀. 汇聚人才为广东发展效力. 首届广东院士高峰年会开幕词，广州，2015-03-29；教育与科技管理研究，北京：科学出版社，2016，344-345.

318. 刘人怀. 加强互动 促进合作. 在完美（中国）公司座谈会上的讲话，中山，2015-05-04.

319. 刘人怀．中国生产力创新品牌产业联盟成立大会开幕词．科技创新与品牌，2015，(11)：17.
320. 刘人怀．扎实开展科普志愿服务工作，为全面提高我省公民科学素质而努力．广东省科普志愿者协会第一届理事会工作报告，广州，2015-11-09.

2016

321. 刘人怀．旅游城市品牌的塑造．《中国旅游城市品牌个性感知研究——基于广东入境游客视角》序，广州：暨南大学出版社，2006，1-2.
322. 刘人怀．研究传染病突发事件的危机管理十分重要．《澳门传染病突发事件的危机管理能力研究》序，北京：中国文史出版社，2016，1-2.
323. 刘人怀，覃大嘉，杨东进，梁育民．国际化过程中培育的核心动态能力在“反向国际化品牌战略”中的作用——基于中国天生国际化 OEM 企业的实证研究．第三届中国管理科学论坛论文集，杭州，2016-05-07，197-205.
324. 刘人怀．创新不止勇担当、敢为人先宏达人．《广东宏达爆破股份有限公司企业文化理念故事集》序，广州：广东宏达爆破股份有限公司，2016-06-12.
325. 刘人怀．以智能制造促进产业转型升级．应广东省经济和信息化委员会邀请在广东省委党校所作的报告，广州，2016-06-21.
326. 刘人怀．两化深度融合与质量管理．中华人民共和国人力资源和社会保障部高级研修班报告，杭州，2016-11-29.
327. 刘人怀．网络强国战略与实践．《网络强国战略与浙江实践》序，北京：科学出版社，2016，3-4.

2017

328. Liu Renhuai，Sun Kai，Sun Dongchuan．Research on Chinese school of modern GUANLI science．Chinese Management Studies，2017，11 (1)：1-11.
329. 刘人怀．城市遗产旅游景观的研究．《消费主义视角下城市遗产旅游景观的空间生产：成都宽窄巷子个案研究》序，北京：科学出版社，2017，1.
330. 刘人怀．保护遗产至关重要．《公众参与遗产保护的激励机制研究》序，广州：暨南大学出版社，2017，1-2.
331. 刘人怀．应用旅游工程理论探讨阳江旅游发展规划．广东省阳江市党政干部大会报告，阳江，2017-08-09.
332. 刘人怀．大力促进科技与经济融合．深圳先进制造业院士指导委员会揭牌仪式讲话，深圳，2017-09-24.
333. 刘人怀，王娅男．创业拼凑对创业学习的影响研究——基于创业导向的调节作用．科学学与科学技术管理，2017，38 (10)：135-146.
334. 刘人怀．以我个人经历及家庭经历浅谈家庭教育．成都市新都区纪委教育材料，成都，2017-10-20.
335. 刘人怀，王娅男．创业拼凑、创业学习与新企业突破性创新的关系研究．科技管理研究，2017，37 (17)：1-8.
336. 刘人怀．真诚感谢 继续为祖国为人民效力．刘人怀学术交流促进会 2017 年年会暨澳门工程科技及管理科学学术研讨会讲话，澳门，2017-12-16.

337. 刘人怀. 保护人类健康 实现美丽中国. 刘人怀学术交流促进会 2017 年年会暨澳门工程科技及管理科学学术研讨会专题报告，澳门，2017-12-16.

338. 刘人怀，张书莲，冉丽. 澳门会展业的经济效应与发展策略研究. 刘人怀学术交流促进会 2017 年年会暨澳门工程科技及管理科学学术研讨会专题报告，澳门，2017-12-17.

339. 刘人怀. 推动广东非开挖产业快速发展. 广东省非开挖技术协会第二届理事会工作报告，广州，2017-12-15.

2018

340. 刘人怀. 应该高度关注青少年的视觉健康问题. 眼调节训练灯二类医疗器械注册证发布会讲话，广州，2018-01-30.

341. 刘人怀. 大力加快优质高等职业学校建设. 首届全国高职高专“优质校”建设与评价论坛致辞，杭州，2018-03-24.

342. 刘人怀. 为中华民族伟大复兴做出更多更大的贡献. 广东院士联合会第二次会员大会讲话，北京，2018-05-27.

343. 刘人怀. 教化育人 师法天地. 《中国当代教育名家》序，北京：社会科学文献出版社，2018，1-5.

344. 刘人怀. 全省科技工作者应履行科普社会责任. 首届广东科普嘉年华启动仪式讲话，广州，2018-09-15.

345. 刘人怀. 谈谈标准化工作对当前的重要意义. 广东省 2018 年世界标准日宣传庆祝活动主旨演讲，广州，2018-10-14.

346. 刘人怀.《制造业运营管理决策优化问题研究》序. 北京：科学出版社，2018.

347. 刘人怀. 坚守初心 坚持创新. 2018 年粤港澳大湾区院士峰会暨第四届广东院士高峰年会开幕词，东莞，2018-11-02.

348. 刘人怀. 促进城市矿产资源化利用，共建美丽中国. 中国工程院工程管理学部城市矿产工程前沿技术论坛开幕词，重庆，2008-11-10.

349. 刘人怀. 追求真理 奉献国家. 中国工程院“守正扬清”主题宣讲活动会议讲话，武汉，2018-11-23.

350. 刘人怀. 数字经济 创新未来. 企业数字化转型论坛开幕式讲话，杭州，2018-11-30.

351. 刘人怀. 勇担重任 敢攀高峰. 基于 BIM 技术的装配式建造技术应用交流会讲话，郑州，2018-12-24.

352. 刘人怀. 敢为人先谋发展. 我的兰大——人物访谈录 2，兰州：兰州大学出版社，2018，155-168.

2019

353. 刘人怀. 有效推动轨道交通结构健康监测和整治修复技术的发展提升. 第二届粤港澳大湾区轨道交通结构健康监测与整治修复技术学术研讨会讲话，东莞，2019-01-05.

354. 刘人怀. 让人民过上好日子. 采用“联合生物加工技术”治理餐厨垃圾示范性工程动工典礼开幕词，成都，2019-02-17.
355. 刘人怀. 为推动仪器仪表产业发展做出新贡献. 广东省仪器仪表学会第二届理事会工作报告. 广州，2019-03-22.
356. 刘人怀. 让未来祖国的科技天地群英荟萃. 第 34 届广东省青少年科技创新大赛开幕式讲话，澄海，2019-03-30.
357. 刘人怀. 汇聚人才 科技兴国. 2019 年世界工厂青年科学家大会报告. 杭州，2019-04-02.
358. 刘人怀. 热爱航空 科技兴国. 广东省航空科普资源交流会主题报告，珠海，2019-04-20.
359. 刘人怀. 众人拾柴火焰高. 粤港澳院士专家创新创业联盟第一次主席团会议讲话. 广州，2019-04-21.
360. 刘人怀. 感谢党和国家的培养和教育. 退休座谈会讲话. 广州，2019-05-10.
361. 刘人怀. 努力推动压力容器技术进步. 广东省压力容器学会第五届理事会工作报告，广州，2019-07-04.
362. 刘人怀. 从教从研六十年的感想. 院士人生励志报告会. 广州，2019-07-18.
363. 刘人怀. 绿色再制造的探索. 中国管理科学的研究与实践——第四届中国管理科学论坛论文集，北京：科学出版社，2019，1-10.
364. 刘人怀，江峰. 珠三角城乡生活垃圾统筹治理战略研究. 中国管理科学的研究与实践——第四届中国管理科学论坛论文集，北京：科学出版社，2019，11-29.
365. 刘人怀，杨静，段显明，程翠云. 城市餐厨垃圾回收逆向物流系统构建的研究. 中国管理科学的研究与实践——第四届中国管理科学论坛论文集，北京：科学出版社，2019，30-39.
366. 刘人怀，杨颖，覃大嘉，崔鼎昌，江梓力. OEM 员工的人力资源管理实践和离职意向：以衡阳鞋厂为例. 中国管理科学的研究与实践——第四届中国管理科学论坛论文集，北京：科学出版社，2019，96-110.
367. 刘人怀，尹涛，呼玲研，覃大嘉，江峰. 澳门民办高校发展现状与前景. 中国管理科学的研究与实践——第四届中国管理科学论坛论文集，北京：科学出版社，2019，315-323.
368. 刘人怀. 八十有感. 在广东院士联合会和“重大工程灾害与控制”教育部重点实验室（暨南大学）主办的院士人生励志报告会上的讲话，广州，2019-07-18.
369. 刘人怀，张镒. 互补性资产对双元创新的影响及平台开放度的调节作用. 管理学报，2019，16（7）：949-956.
370. 刘人怀，刘小同，文彤. 系统论视角下旅游学科提升发展的思考. 旅游学刊，2019，34（12）：1-3.
371. 刘人怀. 责任重大使命光荣. 2019 粤港澳大湾区院士峰会暨第五届广东院士

高峰年会开幕词，东莞，2019-11-28.

2020

372. Liu Renhuai, Si Steven. Lin Song, Dean Tjosvold, Richard Posthuma. Guest editorial. International Journal of Conflict Management , 2020, 31 (3): 309-311.

373. 刘人怀. 大力加强高校科技创新. 花城论坛讲话，广州，2020-07-06.

374. 刘人怀. “重大工程灾害与控制”教育部重点实验室2019年工作报告. 广州，2020-09-04.

375. 刘人怀. 高端引领 服务地方. 2020粤港澳院士峰会暨第六届广东院士高峰年会新闻发布会讲话，广州，2020-10-27.

376. 刘人怀. 坚守初心 坚持创新. 2020粤港澳院士峰会暨第六届广东院士高峰年会开幕词，东莞，2020-11-03.

377. 刘人怀. 全球疫情下经济转型与管理创新. 第七届中国管理科学论坛开幕词，南宁，2020-12-05.

378. 刘人怀. 加强数字产业化和产业数字化双轮驱动. 2020数字化转型与高质量发展论坛暨浙江省首席信息官峰会开幕词，海宁，2020-12-19.

379. 刘人怀. 关于推进餐厨垃圾治理装备高技术制造业发展的建议. 呈送广东省人民政府建议，广州，2020-12-29.

2021

380. 刘人怀. 新都二中——我的母校.《新都二中简史》序，成都，2021-01-23.

381. 刘人怀. 做好餐厨垃圾处理 促进绿色低碳发展. 2021年生态文明贵阳国际论坛“城乡建设绿色低碳发展”主题论坛演讲，贵阳，2021-07-12.

382. 刘人怀，霍孟军. 高管职业生涯关注对企业创新产出的影响. 科技进步与对策，2021，38（15）：135-142.

2022

383. 刘人怀. 创新争先 自立自强. 广东“最美科技工作者”发布仪式座谈会讲话，广州，2022-05-30.

384. 刘人怀. 发挥社会创业的重要作用.《社会创业：理论与实践》序，北京：机械工业出版社，2022.

385. 刘人怀. 推动暨南大学“重大工程灾害与控制”教育部重点实验室跨越式发展.“重大工程灾害与控制”教育部重点实验室2020－2021年度工作报告，广州，2022-09-05.

386. 刘人怀，谢骞. 平台企业社会责任对消费者购买意愿的影响研究——平台领导力的中介作用. 评价与管理，2022，(3)：21-28.